PERMAFROST

Fourth International Conference

PROCEEDINGS

July 17-22, 1983

Organized by
University of Alaska and
National Academy of Sciences

Organizing Committee

T. L. Péwé
G. E. Weller
A. J. Alter
J. Barton
J. Brown
O. J. Ferrians, Jr.
H. O. Jahns
J. R. Kiely
A. H. Lachenbruch
R. D. Miller
R. D. Reger
A. L. Washburn
J. H. Zumberge
L. DeGoes

NATIONAL ACADEMY PRESS
Washington, D.C. 1983

NATIONAL ACADEMY PRESS 2101 Constitution Avenue, NW Washington, DC 20418

The National Research Council was established by the National Academy of Sciences in 1916 to associate the broad community of science and technology with the Academy's purposes of furthering knowledge and of advising the federal government. The Council operates in accordance with general policies determined by the Academy under the authority of its congressional charter of 1863, which establishes the Academy as a private, nonprofit, self-governing membership corporation. The Council has become the principal operating agency of both the National Academy of Sciences and the National Academy of Engineering in the conduct of their services to the government, the public, and the scientific and engineering communities. It is administered jointly by both Academies and the Institute of Medicine. The National Academy of Engineering and the Institute of Medicine were established in 1964 and 1970, respectively, under the charter of the National Academy of Sciences.

Library of Congress Catalog Card Number 83-63114

International Standard Book Number 0-309-03435-3
Permafrost: Fourth International Conference, Proceedings

Printed in the United States of America

The National Academy Press was created by the National Academy of Sciences to publish the reports issued by the Academy and by the National Academy of Engineering, the Institute of Medicine, and the National Research Council, all operating under the charter granted to the National Academy of Sciences by the Congress of the United States.

Preface

Perennially frozen ground, or permafrost, underlies an estimated 20 percent of the land surface of the earth. It affects many human activities, causing unique problems in agriculture, mining, water supply, sewage disposal, and construction of airfields, roads, railroads, urban areas, and oil and gas pipelines. Therefore, understanding of its distribution and behavior is essential.

Although the existence of permafrost has been known to the inhabitants of Siberia for centuries, not until 1836 did scientists of the Western world take seriously the accounts of thick frozen ground existing under the forests and tundra of northern Eurasia. In that year, Alexander Theodor von Middendorf measured temperatures to a depth of approximately 107 m in permafrost in the Shargin Shaft, an unsuccessful well dug in Yakutsk for the governor of the Russian-Alaskan Trading Company. It was estimated that the permafrost there was 215 m thick. For over a century since then, scientists and engineers in Siberia have been actively studying permafrost and applying the results of their research in the development of the region. Similarly, prospectors and explorers have been aware of permafrost in northern North America for many years, but not until World War II were systematic studies undertaken by scientists and engineers in the United States and Canada.

As a result of the explosive increase in research into the scientific and engineering aspects of frozen ground since the late 1940s in Canada, the United States, the USSR, and, more recently, the People's Republic of China, and Japan, among other countries, it became apparent that scientists and engineers working in the field needed to exchange information on an international level. The First International Conference on Permafrost was therefore held in the United States at Purdue University in 1963. This relatively small conference was extremely successful and yielded a publication that is still used throughout the world. In 1973 approximately 400 participants attended the Second International Conference on Permafrost in Yakutsk, Siberia, USSR. By that time it had become apparent that a conference should be held every 5 years or so, to bring together scientists and engineers to hear and discuss the latest developments in their fields. Thus in 1978 Canada hosted the Third International Conference on Permafrost in Edmonton, Alberta, including field trips to northern Canada. Approximately 450 participants from 14 nations attended, and Chinese scientists were present for the first time. The published proceedings of all three of these conferences are available (see p. 1524).

In Edmonton it was decided that the United States would host the fourth conference, and a formal offer was made by the University of Alaska. Subsequently, the Fourth International Conference on Permafrost was held at the University of Alaska at Fairbanks, July 17-22, 1983. It was organized and cosponsored by committees of the Polar Research Board of the National Academy of Sciences and the State of Alaska. Local and extended field trips to various parts of the state and northwestern Canada, to examine permafrost features, were made an integral part of the conference.

Approximately 900 people participated in the numerous activities, and 350 papers and poster displays from 25 countries were presented at the conference. Many engineering and scientific disciplines were represented, including civil and mechanical engineering, soil mechanics, glacial and periglacial geology, geophysics, marine science, climatology, soil science, hydrology, and ecology. The formal program consisted of panel discussions followed by paper and poster presentations. The panels considered the following themes: pipelines, climatic change and geothermal regime, deep foundations and embankments, permafrost terrain and environmental protection, frost heave and ice segregation, and subsea permafrost.

The U.S. Organizing Committee is indebted to the many sponsors for their financial support, to the technical and professional organizations which participated in the program, and to the local Fairbanks organizers for their efforts, which resulted in a highly successful meeting.

Troy L. Péwé, Chairman
U.S. Organizing Committee
Fourth International Conference on Permafrost

Jerry Brown, Chairman
U.S. Committee on Permafrost
Polar Research Board

Contributing Sponsors

The following sources contributed financial support to the Fourth International Conference on Permafrost. Funds to prepare and publish the volume were derived from contributions made to the University of Alaska Foundation.

Federal Agencies

Army Research Office
Bureau of Land Management
Cold Regions Research and Engineering Laboratory
Forest Service
Minerals Management Service (DOI)
National Aeronautics and Space Administration
National Science Foundation
Office of the Federal Inspector (ANGTS)
Office of Naval Research
U.S. Department of Energy
U.S. Geological Survey

State of Alaska

Alaska Council on Science and Technology
Department of Transportation and Public Facilities
State Legislature

Industry and Private

SOHIO Alaska Petroleum Company
Atlantic Richfield Foundation
EXXON Production Research and EXXON Company USA
John Kiely

Bechtel Petroleum, Inc., and Bechtel Power Corporation
Chevron USA, Inc.
Alyeska Pipeline Service Company
Mapco/North Pole Refining
Alaska Construction Industry Advancement Program
Dresser Foundation
British Petroleum, Ltd., Inc.
Morrison-Knudsen Company, Inc.
Kodiak-Nabors (Anglo-Energy)
Marathon Oil Foundation
Northwest Pipeline Service Company
Shell Oil Company
Union Oil of California Foundation
Cameron Iron Works, Inc.
Alaska Industrial Resources, Inc.
Phillips Petroleum
3-M Company
Alfred Jurzykowski Foundation, Inc.
National Geographic Society
Law Engineering Testing
R & M Consultants
Anaconda Copper Company
Brown and Root (Halliburton)
CONOCO, Inc.
Dames and Moore
Ebasco Services, Inc.
Getty Oil Company
Harza Engineering
Nerco Minerals (Resource Associates of Alaska)
Rocky Mountain Energy
Schlumberger Well Services
Tesoro Petroleum Corporation
Conrad Topp Engineering
Tryck, Nyman, Hayes
Alaska International Constructors
Doyon Ltd.
Shannon and Wilson
R. A. Kreig & Associates, Inc.
DOWL Engineers
Fryer, Pressley, Elliott
Peter Kiewit and Sons
William J. King
Stutzman Engineering Associates
Unwen Scheben Korynta Huettl

Conference Publications

The official proceedings of the conference consist of: (1) the Abstract and Program volume, which was published prior to the conference and is available from the University of Alaska; (2) the present volume of 276 contributed papers by authors from 22 countries; and (3) a final volume that will contain the panel and plenary presentations, additional contributed papers, and a list of participants.

A review group, acting under the U.S. Organizing Committee for the Fourth International Conference on Permafrost, was responsible for reviewing the abstracts submitted in response to the initial call for papers, and later for reviewing the papers themselves. To accomplish these tasks within the span of the year preceding the conference, strict deadlines were imposed and enforced. The review group met twice at the National Academy of Sciences, first in the fall of 1982 and again in April 1983.

The objective of the group was to ensure the preparation of scientifically and technically sound camera-ready copy on a schedule consistent with the desire to have the contributed papers published by late 1983. The task was greatly facilitated by the ability of the authors to prepare their own camera-ready copy within the deadlines.

More than 200 reviewers were called on to ensure an adequate and timely critique of each manuscript. These reviews were carried out with generous cooperation from Canadian specialists. The papers from the USSR were accepted on the basis of Soviet review prior to submission to the U.S. Organizing Committee. The Soviet camera-ready copy was prepared in the United States following revision of their English-language manuscripts. An additional 30 Soviet papers are being prepared for publication in the final proceedings volume. The authors from the People's Republic of China were provided review and editorial comments, and their papers were then returned to them for final preparation.

The members of the review group were:

Robert D. Miller, Chairman, Cornell University
Duwayne M. Anderson, State University of New York, Buffalo (Remote Sensing and Planetary Geophysics)
Robert F. Black, University of Connecticut (Periglacial and Permafrost)
Joseph M. Boyce, National Atmospheric and Space Administration (Lead Reviewer, Remote Sensing and Planetary Geophysics)
Robert F. Carlson, University of Alaska, Fairbanks (Hydrology)
John Dennis, U.S. Park Service (Ecology)
Hugh M. French, Ottawa University (Lead Reviewer, Periglacial and Permafrost)
Laurel E. Goodrich, National Research Council of Canada (Thermal and Mechanical Engineering)
William D. Harrison, University of Alaska, Fairbanks (Geothermal and Subsea)
Hans O. Jahns, Exxon Production Research Co. (Lead Reviewer, Geothermal and Subsea)
C. William Lovell, Purdue University (Lead Reviewer, Civil Engineering)
Virgil Lunardini, Cold Regions Research and Engineering Laboratory (Lead Reviewer, Thermal and Mechanical Engineering)
Howard Thomas, Woodward-Clyde Consultants (Civil Engineering)
Leslie Viereck, U.S. Forest Service (Lead Reviewer, Ecology)
Ted S. Vinson, Oregon State University (Civil Engineering)
John R. Williams, U.S. Geological Survey (Lead Reviewer, Hydrology)
Louis DeGoes, Secretary

The generous editorial assistance of the Cold Regions Research and Engineering Laboratory and the National Research Council during various stages of manuscript review and preparation is gratefully acknowledged. A special note of appreciation is due to Louis DeGoes for his dedicated efforts throughout the planning and review stages of the preparation of the proceedings.

In addition to the proceedings, a series of field trip guidebooks was published by the Alaska Division of Geological and Geophysical Surveys and is available from that agency in Fairbanks. A special permafrost bibliography of over 4,000 citations was published by World Data Center A for Glaciology in Boulder, Colorado, as Glaciological Data Report GD-14. The bibliography covers much of the world literature on permafrost published during the last five years.

R.D. Miller, Chairman
Paper Review and Publications

U.S. Organizing Committee

Troy L. Péwé (Chairman), Arizona State University
Gunter E. Weller (Vice-Chairman), University of Alaska
Amos J. Alter, Technical Council on Cold Regions Engineering, ASCE
Jay Barton, University of Alaska
Jerry Brown, Cold Regions Research and Engineering Laboratory
Oscar J. Ferrians, Jr., U.S. Geological Survey
Hans O. Jahns, Exxon Production Research Company
John R. Kiely, Bechtel Corporation
Arthur H. Lachenbruch, U.S. Geological Survey
Robert D. Miller, Cornell University
Richard D. Reger, Alaska Division of Geological and Geophysical Surveys
A. Lincoln Washburn, University of Washington
James H. Zumberge, University of Southern California
Louis DeGoes, Fourth International Conference on Permafrost

Program Chairman: William D. Harrison, University of Alaska
Paper Review and Publication Chairman: Robert D. Miller, Cornell University
Field Trips Chairman: Oscar J. Ferrians, Jr., U.S. Geological Survey
Local Arrangements Chairman: William M. Sackinger, University of Alaska

Participating Technical Organizations

Alaska Academy of Engineering and Science
Alaska Council on Science and Technology
Alaska Oil and Gas Association
American Association for the Advancement of Science (Arctic Division)
American Geophysical Union (Committee on Space Hydrology)
American Society of Civil Engineers (Technical Council on Cold Regions Engineering)
American Society of Mechanical Engineers (Low Temperature Heat Transfer Committee)
American Meteorological Society (Committee on Polar Meteorology)
Cold Regions Research and Engineering Laboratory
Comité Arctique International
International Geographical Union (Periglacial Commission)
U.S. Man and the Biosphere Program (MAB, Project 6B)

Contents

---

*Papers are presented alphabetically by senior author; see general subject and author indexes for additional information.

PERMAFROST

Fourth International Conference

PROCEEDINGS

ON THE RELATIONSHIP BETWEEN THE STRUCTURE AND CHARACTERISTICS OF BEDROCK MASSES AND THEIR PERMAFROST HISTORY

V. E. Afanasenko, V. N. Zaitsev, V. E. Romanovskii, and N. N. Romanovskii

Geology Department,
Moscow State University, Moscow, USSR

Comprehensive investigations in the areas of permafrost hydrogeology and engineering geology within the Stanovoy and Mongol-Okhotsk folded systems revealed the presence of horizons and zones of severe jointing in the bedrock. They occurred both where there was no permafrost and at the top and bottom of the permafrost. The thickness of the zones involved reaches 100 m and more, and the ice content where the rock is frozen reaches 6-10% by volume. The formation of these horizons of severely-jointed rock is related to the dynamics of permafrost development during climatic changes in the Late Pleistocene and Holocene, specifically to intense frost shattering caused by repeated freeze-thaw. These horizons and zones of severe jointing exert a major influence on the formation of the existing subsurface drainage and of talik zones which operate both as recharge and discharge zones in terms of groundwater and also as lateral transfer zones beneath the permafrost.

The blocks of mountain bedrock in the zone of sporadic permafrost distribution have specific structural characteristics, i.e. the highly jointed horizons of exogenous origin. In the hydrogeological aspect these are hydrogeological blocks of insular and sporadic freezing. Their highly jointed horizons are most commonly the highly saturated zones, the water of which is of great economic value for industrial and municipal water supply.

The structure and characteristics of bedrock blocks were studied in the central part of the Baikal Amur Railway zone, an area subjected to multiple and substantial changes of the frozen state in Pleistocene and Holocene time. Those are Stanovaya and Mongol-Okhotsk mountain regions composed of rocks of the Early Proterozoic age, as well as Paleozoic and Mesozoic folded belts. Similar to other regions of the Baikal region, the Transbaikal region, and the Verkhoyano-Kolymskaya mountain-folded regions, subjected to substantial transformation of the total frozen state in the Quaternary period, it is suggested that the state and characteristics of the hard and semihard rock blocks of the territory in question have been changed by the dynamics of permafrost.

To identify regional characteristics of cryogenic disintegration near the upper and lower boundaries of permafrost, special investigations based on electrical prospecting methods were carried out. The distribution of the most typical structural characteristics of the zone of exogenic jointing of bedrock blocks within the study area, the permafrost thickness, and the density of the component rocks, were primarily established. The techniques suggested by Piotrovskaya and Romanovsky (1982) were used. The most probable values of the thickness of the exogenic density zone and specific electric resistance of the unfrozen jointed bedrock, characterized by extensive areal occurrence, are given in Table 1.

Usually, a normal field of thickness distribution of the exogenic jointed zone of the study area has apparent anomalous areas of two types, within which the values are higher by two or more times than the background ones.

The first anomalous type includes linear zones having faults with a break in continuity. The anomalous areas of the second type are confined to the valleys of large rivers such as the Gilyui, Ilikan, etc., formed within a relatively long period of Early-Mid Pleistocene entrenchment (Piotrovskaya and Romanovsky 1982). Such areas commonly extend one or more km across the whole width of these valleys.

The formation and further development of high thickness of dispersion zones, in the aforementioned valleys, were caused not only by initial jointing along the linear tectonic disturbances, but also by subsequent processes of cryogenic disintegration causing formation of a jointed weakened zone in the upper part of the bedrock blocks. The anomalously great thickness of such jointed zones was observed by us only in ancient-stage valleys and did not occur in younger valleys formed as a result of Holocene entrenchment. Thus, the upper and oldest part of the Gilyui River valley is characterized by a great thickness of jointed zones. The downstream part of the valley was shaped later. The zones of exogenic jointing were either insignificant or absent.

Regional analysis of ice-filling of the jointed zones in the frozen part of the bedrock blocks, using the electrical prospecting techniques and prototype observations, made it possible to identify zones of complete ice saturation, up to their "bulged" state, in the section of frozen bedrock. The latter is indicative of intensive processes of cryogenic disintegration in these parts of bedrock blocks. The disintegration processes are confined to specific relief forms having different terrain expression.

Upon studying regional variability of rock ice-

TABLE 1 Thickness parameters for the dispersion zone and specific electric resistance of the thawed jointed bedrock.

Elementary geological body	Dispersion zone thicknesses (m)			Specific electric resistance of rock		
	Total number of measurements n	Medium θ, m	Standard S $\hat{\theta}$, m	Total number of measurements n	Medium θ, k .m	Standard S $\hat{\theta}$, k .m
1) Dispersion zone of metamorphic rock blocks of Archean age. The surface corresponds to the Pliocene/Mid-Pleistocene relief stage (upper horizon)	26	53.6	1.2	166	1.1	0.02
2) The same (lower horizon)	-	>100	-	85	4.6	0.07
3) Dispersion zone of similar blocks, but the surface formed as a result of the Upper-Pleistocene-Holocene entrenchment cycle (lower horizon)	8	53.3	36	4	4.3	0.25
4) Dispersion zone of Proterozoic granitoid blocks (PR) within the Pliocene/Mid-Pleistocene surfaces	38	51.2	1.0	93	2.3	0.04
5) Dispersion zone of Mesozoic granite blocks (Mz), whose surface corresponds to Pliocene/Mid-Pleistocene stage	7	44.0	2.8	7	4.0	0.38

filling, we used the values of specific electric resistance (SER) of frozen jointed bedrock, obtained in geophysical studies and confirmed by prototype observations. In this case, it seems convenient to use the index of rock ice-filling rate, introduced by us as a function:

$$P_{ric} = \frac{P_j^{fr}}{P_{mon}}$$

where P_j^{fr} = SER of frozen jointed rocks;

P_{mon}^j = SER of the unfrozen "monolith" part of the same rock block.

According to the data, the value of the P_{ric} parameter is different for divides and valley areas, thus confirming the substantial differences in conditions of permafrost formation and its cryogenic structure within the limits of these relief forms.

The values P_{ric} = 0.35 - 0.95 are most characteristic of the lowest part of the section of frozen jointed bedrock on slopes and low (to an altitude of 1100-1300 m) on interfluves. P_{ric} rarely exceeds 1.0 here. It means that the frozen rocks of such areas are, as a rule, characterized by inherited cryogenic structures and low ice content. The joints of such rocks, in the belt of weathering, are only partially filled with ice. This cryogenic structure of jointed rocks points to the weakening of cryogenic disintegration here at present.

The frozen jointed bedrock, at the lower permafrost boundary in the ancient-stage valleys, is most commonly characterized by the values of P_{ric} = 1.2 to 2.3. The values of P_{ric} below 1.0 are extremely rare. Most of the joints are entirely filled with ice, and, in many cases, the rocks occur in the "bulged" state. The high ice content seems to give unequivocal evidence of active cryogenic disintegration both recently and at present.

Studies of the intensity of cryogenic disintegration at the base of permafrost within the so-called bald mountain belt are of particular interest. Due to the effects of altitudinal zonality, such areas were distinguished by the most severe temperature conditions during the whole period of perennial frozen ground existence and development. It resulted in deep freezing of bedrock blocks. It is supported by the results of our geophysical studies on the bald mountain divide surfaces above the altitude of 1400 m in the Soktakhan range, as well as within the Stanovoi range at an altitude of 1300-2000 m. Permafrost thickness in these areas reaches between 200 and 500 m. The presence of highly jointed bedrock immediately under the permafrost layer is characteristic of the section. Besides, in many areas the perennially frozen rock series has a more complex structure. Two intervals of high ice content, upper and lower, interbedded with ice-poor frozen rocks, are common in the section. The structure of this section suggests that cryogenic disintegration of the rock near the base of bedrock blocks of the bald mountain belt, as well as in the upper part of the perennially frozen rock section was most intensive during the Pleistocene and Holocene.

To understand fully the scope of our findings on the intensity and total development of dispersion zones in the cryogenic series, it is necessary to approach the problem from a historical aspect and to relate the principal structural characteristics and properties of bedrock blocks to "permafrost" history. As mentioned above, cryogenic disintegration of bedrock blocks in the "permafrost" zone was caused by multiple processes of partial or complete freezing or thawing of rocks at the base or in the upper part of the frozen section. These processes result in a substantial increase in rock jointing and porosity.

It is common knowledge that one of the basic indicators of continuous existence of frozen series during the Pleistocene was the presence of syngenetic multiple-veined ice in sections in Quaternary deposits of different ages. In our studies and also in those of Alekseyev (1968), Maksimova (1971), Kaplina et al. (1975), Vtyurin (1976), the occurrence of syngenetic multiple-veined ice in sediments of the first, second and third terraces above the floodplain on the floors of the Olekma, Ust-Nyukzha, Zeya, Gilyui, G. and M. Olda and other river valleys, i.e. approximately up to 54° N has been established. The upper surface of the ice veins is, as a rule, "detached" from the present base of the active layer by 0.8-1.5 m. The space over the vein is filled with peaty sandy loam, sand, peat or ice of thaw-lake origin. This structure of the upper parts of syngenetic ice wedges is indicative of the fact that in the warmest periods of Upper Pleistocene and Holocene some parts of valley floors and intermountain areas were characterized by a considerable increase in the mean annual rock temperature within the range of their values below 0°C and in the depth of seasonal thawing (2-3 fold as compared to the present levels). The frozen series did not thaw everywhere or completely. It was largely controlled by high ice content (up to 50-70%) of the mantle of unconsolidated deposits, substantial heat-insulating effects of surface moss-peat covers, the effect of the negative temperature shift at the base of the seasonally-thawed layer and the present inversion in distribution of the mean annual air temperatures from divides to valley and depression floors. The long-term (secular and millenial) increases and decreases of mean annual rock temperature should result in substantial shifts (up to 100 m) in the lower frozen ground boundary. Besides the valley floors, the most favorable conditions for long "preservation" of perennial frozen rock in the periods from the Mid-Pleistocene to the present have occurred within the limits of the cooling belt, most persistent in time, coinciding with the bald mountain part of the Stanovoi, Dzhugdyr, Tukuringra, Soktakhan and Dzhagdy ranges, confined to altitudes of over 1300-1500 m. Multiple and substantial displacements of the lower permafrost boundary occurred here, as in the valley floors. The displacements are indicated by new electric prospecting data on the frozen zone structure of the mountain structures, mentioned earlier, as well as by the data presented by Kaplina et al. (1975). Most high-temerature frozen rock series are formed within the belt, transitional from the inversion to the altitudinal-zonal one. We place in this category the low, smoothed interfluves and gentle slope areas adjacent to the mountains and bordering the Upper-Zeya and Uda depressions, as well as the Amur-Zeya plain. In the Late Pleistocene and Holocene, these areas were evidently noted for multiple neogenesis of frozen series and their thawing. The thawing is indicated by the absence of syngenetic multiple-veined ice on these relief forms, as well as by the present wide occurrence of sporadic frozen series and areal zones with high bedrock jointing traced to a depth of over 50 m, as already mentioned. The syngenetic multiple ice wedges did not occur in either relief form of Amur-Zeya plain, i.e. south of Tukuringra-Soktakhan-Dzhagdy mountain range, although cases of pseudomorphism occurrence have been described for them in the literature. The thawing is most likely indicative of a complete areal degradation of perennial frozen rocks on the Amur-Zeya plain during the Holocene climatic optimum. The thawing is also supported by wide occurrence of large relict thermokarst forms there (up to a few km across), on the interfluves of the upper reaches of the Onon, Ulmin, Olga and Tu rivers, and also by the presence of hummock-and-sink hole microrelief in the Selemdzha and Byssy river basins. According to Mordvinov (1940), the formation of such forms was caused by complete thawing of multiple ice wedges during the Holocene climatic optimum. Our calculations of perennial freezing support the idea that the maximum thickness of frozen ground observed at present on the Amur-Zeya plain (about 50-80 m) could have formed between 3000 and 4000 years ago in the period of Late-Holocene cooling that followed the climatic optimum.

The principal changes in the physical state of hard rock (thawed and frozen) had a substantial effect on rock jointing and porosity in those intervals where the changes were most substantial and recurrent.

Rather extensive weathering crusts were recorded by us and a number of other researchers in the upper part of the MFR section within a wide strip of the submontane area bordering the Upper Zeya and Uda depressions, and Amur-Zeya plain. This weathering crust type, with a thickness reaching 40-50 m, has attracted attention from experts in different fields. Thus, it was reported by Alekseyeva (1978) that vermiculite weathering crusts were found on intermediate and basic igneous rocks that are enriched with biotite and phlogopite. The formation of such weathering crusts is related by the researcher to rock disintegration and leaching. The occurrence of the weathering crusts formed on the account of cryogenic disintegration is supported by the recent findings of "Mosgiprotans" Institute. Thus, weathering crusts were discovered in the sections of a number of mountain hollows and quarries along the foothills bordering the Upper-Zeya depression, commonly with high-pressure ground water that causes active ice formation in winter.

The development of intensive jointing and cryogenic disintegration zones affects the saturation of bedrock, its permeability and the capacity for subsurface drainage. Areas with well washed joints and high rock porosity are developed in areas of intensive subsurface drainage with infiltration and pressure filtration. Additional ground water pressure is developed in the same areas with an attenuated subsurface drainage and joints filled with a fine-dispersed filler, because these rocks constitute poorly conducting media.

In areas adjoining the linear persistent large

faults with a break in continuity, the thickness of dispersion zones and those with highly jointed rocks may reach 100-150 m.

The thickness of the upper horizon of high jointing and rock weathering beyond tectonic disturbances changes within 10-30 m, reaching sometimes over 50 m. The rocks of this horizon are irregularly saturated in their thawed state, because they commonly contain a great amount of finely-dispersed material and serve as confining beds in this case. In cases where the horizon's rocks constitute highly broken hard rocks, but with a small amount of finely-dispersed filler ("collapsible" rock), they have good filtration characteristics and are, as a rule, saturated.

Our electrical prospecting suggests that the zones of cryogenic disintegration are confined to the lower permafrost interval of the rock blocks. The dispersion zones below the base of the permafrost, discovered within the valleys of large and medium rivers of "pre-Holocene" occurrence, as well as within the bald mountain belt (at over 1300 m altitude), are 40-100 m thick. In the frozen part of the section, the rocks of these zones are characterized by relatively high ice content, but in the unfrozen state they are characterized by high saturation.

The principal movement of artesian water takes place at the base of permafrost in highly jointed zones of cryogenic origin within the limits of all hydrogeological blocks of the investigated region, both of sporadic and continuous shallow freezing. They are fed through infiltration taliks in interfluves and from the upper courses of small river valleys, although the discharge occurs through the pressure-filtration taliks in river valleys. The latter results in intensive icing formed by ground water discharge in winter.

REFERENCES

Alekseyev, V. R., 1968, Formation and occurrence conditions of the present polygonal-veined ice in South Yakutia, Transbaikalia and Amur river region, in Voprosy geologii Pribaikalya i Zabaikalya, no. 3(5), Chita, p. 232-234.

Alekseyeva, Z. I., 1978, Manifestations of the weathering crusts in the eastern part of the Baikal Amur Railway (Dzheltulaksky region), in Kora vyvetrivaniya, no. 16, Moscow: "Nauka," p. 206-212.

Kaplina, T. N., Pavlova, O. P., Chernyadyev, V. P., and Kuznetsova, I. L., 1975, The most recent tectonics and formation of perennial frozen rocks and subsurface waters, Moscow: "Nauka," p. 124.

Maksimova, L. N., 1971, On the genesis of cryogenic structures in alluvium of the tributaries of the Amur river middle course, Trudy PNIIIS, v. XI, Moscow, p. 191-198.

Mordvinov, A. I., 1940, Relief and permafrost of the left bank of the Byssy River middle course and the adjacent foothills of the western slope of Turansky range, Trudy komissii po vechnoi merzlote, v. 9, Moscow-Leningrad: USSR Academy of Sciences, p. 57-134.

Piotrovskaya, T. Yu., and Romanovsky, V. E., 1982, Studies of the dispersion zone of rocks in regional investigations, in Inzhenernaya geologiya, no. 6, p. 32-46

Vtyurin, B. I., 1976, Subsurface ice in the valley of the Zeya River, in Sezonniye i mnogoletnemyorzliye porody, Trudy Tikhookeanskogo instituta geografii DVNTs AN SSSR, Vladivostok, p. 42-44.

MEASUREMENT OF PERMEABILITIES OF FROZEN SOILS

Jaime Aguirre-Puente and Jacques Gruson

Laboratoire d'Aérothermique, Centre National de la Recherche Scientifique
4 ter, route des Gardes, F92190 Meudon, France

A method for the measurement of the permeability of frozen soils is proposed. It consists of applying a temperature below 0°C only to the central portion of a cylindrical specimen of a porous medium and a positive temperature to the other two portions. To fix and know the position of the 0°C isotherm it is necessary to impose the same absolute value to the two temperatures. It is possible to use this method without limitation of the explored domain of temperatures. The first apparatus that was constructed is described, and the results of permeability measurements on a silt at -0.3°C are given and discussed. The experience acquired has led to the construction of a second, improved permeameter and to foresee new supplementary protections against small variations of temperature in the experimental room, which were found to have an important influence on the stabilization of the thermal and flowing regimes and reproductibility of experiments.

INTRODUCTION

In a porous medium of fine texture the water contained in the pores undergoes a phase transformation process that is progressive with the evolution temperature below zero degree Celsius. For a given temperature in an isothermal system the ratio of the quantity of liquid water to the mass of the solid matrix adopts a value which corresponds to thermodynamic equilibrium. This equilibrium is assumed to exist in each pore of the system; it depends on the interface phenomena and on the sense of the thermal variation which the medium has undergone to attain an isothermal condition. Water in the liquid state is normally assumed to be present as adsorbed water in a film between the solid matrix and the ice in the pores. A particular characteristic of this water film is its apparent high stress-bearing capacity in a direction perpendicular to the film (Vignes-Adler 1975) despite of the high mobility of the molecules of water.

On the other hand, when a moist porous medium is subject to freezing, a cryogenic suction appears near the ice-water interface existing in the pores. On a macroscopic scale this suction results in a depression of the interstitial water and causes a water flow from the warmer to the colder portions of the system; this moving water solidifies on the ice interfaces encountered during the flow. The water flow takes place through the unfrozen porous medium and also through the films of adsorbed water (Aguirre-Puente et al. 1977). Distillation across the ice crystals formed in the pores also allows the transfer of mass (Miller 1970).

During the freezing of a porous medium a distribution of temperature is established, and a phase-transformation layer appears the thickness of which depends on the texture and the nature of the matrix. Complex interactions between the thermal, physico-chemical, and mechanical parameters determine the intensity of the cryosuction, the rate of water movement, and the position and rate of increase of the ice lenses (Aguirre-Puente et al. 1977). In general, it may be stated that these phenomena are much more intense if the porous medium has a fine texture, if the temperature gradients are small, and if the duration of the phenomena is extended in time.

To acquire a better understanding of the zones of phase transformation and the regimes of water migration induced by thermal gradients, it is necessary to determine the laws governing the cryosuction and also the characteristics of permeability of soils at temperatures below 0°C (Aguirre-Puente et al. 1977).

The purpose of this paper is to propose a method for the measurement of the permeability of frozen soils, and to describe the permeameters fabricated in the Laboratoire d'Aérothermique. The preliminary results obtained for a silt soil, and the difficulties experienced during our research are discussed.

PROBLEMS ENCOUNTERED IN THE MEASUREMENT OF PERMEABILITY

For the measurement of the permeability of frozen soils, the use of a classical permeameter, main tained below 0°C, is excluded because the free water at the inlet and outlet reservoirs of the apparatus will freeze. The ice will then block the flow despite the persistence of capillary and adsorbed water around the particles of the medium.

However, the principle of a classical permeameter may be used under isothermal conditions if temperatures only slightly below 0°C are employed in order to maintain the free water of the reservoirs in a supercooled state. In this case, to obtain the freezing of the soil it is necessary to produce nucleation of ice in the interstitial water by introduction of a nucleating agent that initiates the phase transformation. Membranes permeable to water but impermeable to ice must be placed at the extremities of the test sample to prevent freezing of the supercooled water in the reservoirs. In this method, used by Miller et al. (1975), the temperature range of utilization is limited by the degree of supercooling achieved.

A more elaborate system for eliminating the solidification of water in the reservoirs at the extremities was conceived by Burt and Williams (1976). The system consists of immersion of the permeameter

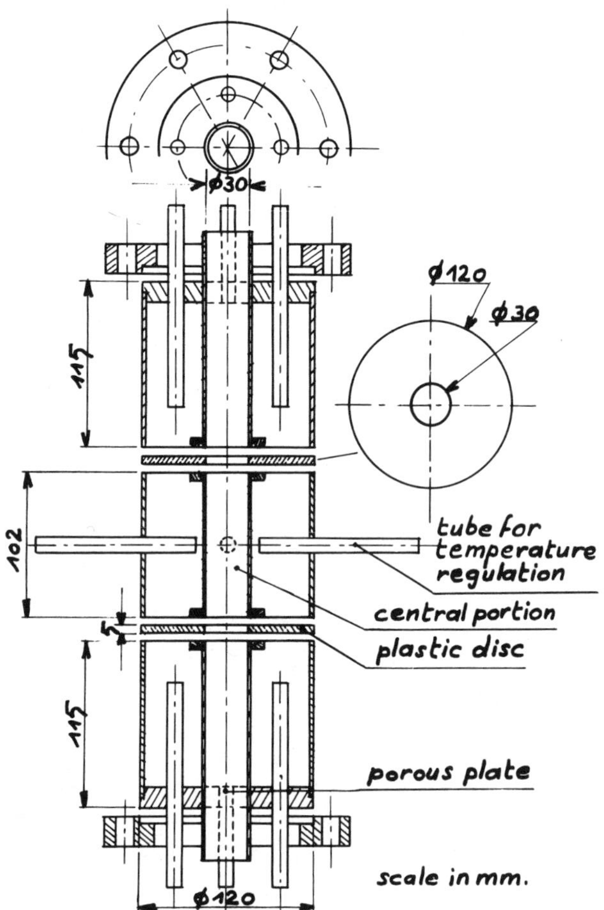

FIGURE 1 First permeameter constructed

in a thermostatic bath. The central portion of the permeameter contains the soil saturated with pure water, and two semipermeable membranes are provided at the extremities to separate the central portion from the end reservoirs. The supercooling of the liquid bath and of the end reservoirs is obtained by using a solution of lactose. The semipermeable membranes permit water to pass through but restrict the passage of dissolved molecules, thereby allowing the water of the sample to remain pure. Freezing of the porous medium is accomplished by a nucleating agent introduced from the exterior. The temperature domain that may be explored with this system is limited by the freezing point of the lactose solution. Despite this limitation however, the lactose system may also be used for the study of nonisothermal systems and to draw conclusions concerning the mechanism of freezing (Williams and Wood 1982).

DESCRIPTION OF THE METHOD AND APPARATUS EMPLOYED

At the Laboratoire d'Aérothermique du CNRS an apparatus for the measurement of permeability of frozen porous media based on the principle of the classical permeameter was constructed. Pure water is used to saturate the porous medium and to supply the reservoirs and channels of the permeameter during the experiment. To eliminate freezing of the free water present at the extremities of the apparatus, only the central portion of the test specimen is kept at the desired negative temperature, while the two extreme portions and the end reservoirs are maintained at a positive temperature. The effective length of the frozen portion to be considered for the calculation of permeability is defined by the two freezing fronts formed in the porous medium. To impose the position and to prevent any major displacement of these freezing fronts, the two extreme portions of the specimen are maintained at a positive temperature having the same absolute value as the negative temperature of the frozen portion. If the precise temperature of the freezing front in the porous medium is known, the symmetry must be accomplished in relation to this temperature.

The system consists of a copper cylinder containing the specimen, which is enclosed by a coaxial cylindrical sleeve. Two plastic plates in the form of a hollow disc having low thermal conductivity (celeron) are placed perpendicular to the axis in order to divide the annular space into three compartments. Cryostats and thermostats are used to maintain the two temperatures symmetrical with respect to 0°C (or the characteristic temperature of the freezing front) in the compartments of the annular space by means of the circulation of liquids.

Figure 1 shows the first apparatus constructed. It was used for the experiments discussed in this paper. The diameter of the test specimen and the effective length of the frozen portion are approximately 3 cm and 10 cm, respectively.

The test material was compacted layer by layer in the cell, and its moisture content and the degree of compaction were rigorously controlled. Pistons provided with porous plates were tightly fixed at the extremities of the specimen after the compaction.

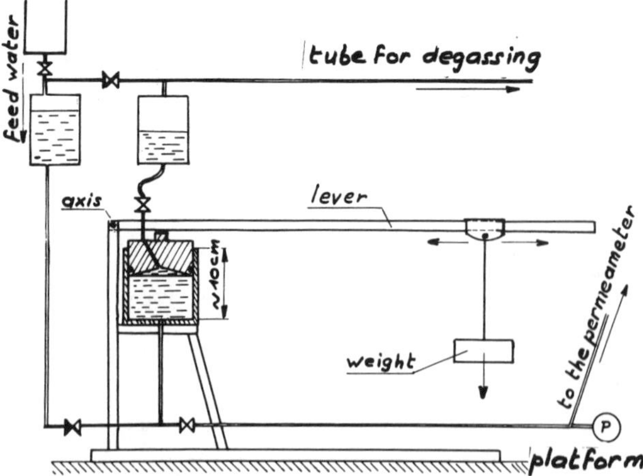

FIGURE 2 Water-supply system

In the preliminary experiment a rapid freezing was used as a first step to avoid the phenomenon of supercooling in the central portion of the sample and the resulting modification of the soil structure. For this purpose temperatures of about -40°C and +40°C were imposed on the central and end portions, respectively. Once thermal stabilization had been obtained, the temperature were increased and decreased progressively in stages up to the desired values always keeping them symmetrical with respect to 0°C.

Temperature control was accomplished systematically in order to verify the position of the 0°C isotherm. For this purpose, thermocouples were placed on the internal wall of the specimen holder, primarily near the insulating plates defining the three compartiments and midway along the length of each portion.

The use of water free of dissolved gases is necessary in such experiments. The supplying of water at the desired pressure to inlet of the permeameter is done with the aid of an airtight system (see Fig.2). This system consists of a cylinder and piston in which the pressure is applied by a lever arm. The length of the lever arm and the weight suspended at its extremity determine the pressure exerted. Finally, observation with a cathetometer of the meniscus formed across a capillary of small diameter by the water leaving at atmospheric pressure allows the measurement of water discharge from the specimen.

EXPERIMENTAL EXAMPLE

Preliminary measurements were conducted on a silt. It was the same soil used at the freezing station of Caen (France) to simulate roads studied during the 1970's by the Laboratoire Central des Ponts et Chaussées and the Centre National de la Recherche Scientifique (Philippe et al. 1970).

Due to the high sensitivity of the freezing phenomena of the porous media of fine texture, long durations are required for the stabilization of the thermal and hydraulic regimes. Despite the poor control of ambient conditions in our experimental room, the experiments performed presented several periods of stabilization during which the rate of water flow was reproductible.

To illustrate the determination of the permeability, the case corresponding to a frozen portion of soil at -0.3°C is presented here.

The characteristics of the soil after compaction in the cell were as follows : density ρ = 2120kgm^{-3}, dry density γ_d = 1820kgm^{-3}, water content ω=0.166 (16.6%), porosity ε = 0.32, total length of the sample L = 0.336m, length of the portion at -0.3°C and ℓ = 0.107 m (supposing that the isotherm 0°C corresponds to the frozen line).

The pressure imposed on the water in the inlet reservoir during this experiment was 0.55 bar. Figure 3 represents five periods of evolution of the meniscus in the fine capillary of the outlet. Regularity and reproductibility are quite satisfactory. The velocity of displacement of the meniscus equal to the slope of the straight line drawn on the experimental points. Taking into account the 2mm-diameter of the capillary, it is possible to determine the rate of water flow going through the

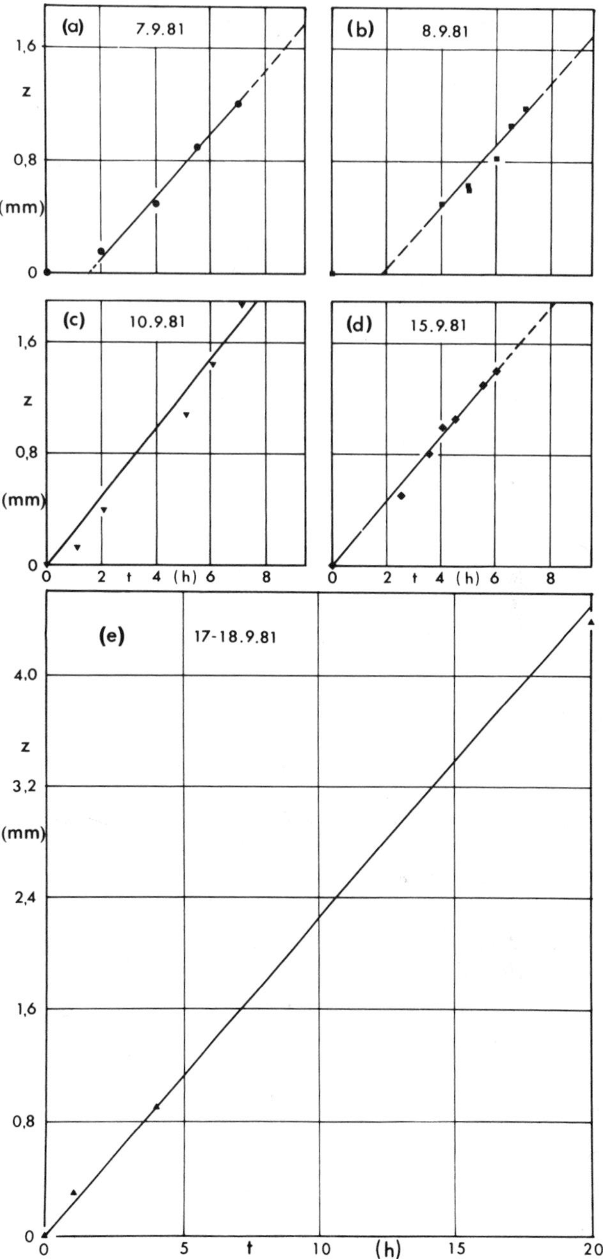

FIGURE 3 Relative position of the meniscus in the capillary vs. time, for five periods of good reproducibility.

specimen. This information is presented in Table 1.

The expression of the generalized Darcy law used in fluid dynamics is

$$\vec{v} = -\frac{K}{\mu}\,\text{grad}\,\hat{P}, \qquad (1)$$

where $\vec{v}$ is the filter-velocity, K the permeability, μ the dynamic viscosity, and $\hat{P} = p + \rho gh$, the driving pressure of the water in the pores (with p, interstitial pressure and h, the level of the considered point with reference to an arbitrarily fixed horizontal plane). In the case of our experiment, we can neglect the gravitational influence.

Table 1

T = -0.3°C		ℓ = 0.107 m		$\Delta p = 0.55 \times 10^5$ Pa	
Date	Duration of the measurement during stable conditions	Velocity of the meniscus		Volume rate of flow	Filter-velocity
	h	mmh^{-1}	ms^{-1}	m^3s^{-1}	ms^{-1}
7.9.1981	7	0.221	6.139×10^{-8}	1.9286×10^{-13}	2.728×10^{-10}
8.9.1981	7	0.222	6.167×10^{-8}	1.9374×10^{-13}	2.741×10^{-10}
10.9.1981	7	0.250	6.944×10^{-8}	2.1815×10^{-13}	3.086×10^{-10}
15.9.1981	6	0.235	6.528×10^{-8}	2.0508×10^{-13}	2.901×10^{-10}
17.9.1981 18	20	0.225	6.250×10^{-8}	1.9635×10^{-13}	2.778×10^{-10}
	Mean rate of water flow			2.0124×10^{-13}	

One point of our research program consists of analyzing the different thermodynamic aspects governing the behavior of the water during its passing through the fine porous media. Indeed, it is probable that the Darcy law is not absolutely valid in the case of media having very fine interstices and when the influence of interstitial pressure could have effects in the phase transition processes concerning, for example, the equilibrium temperature ice-water and the distillation through the ice crystal. Despite the absence of study, we can use the Darcy law and determine an effective of apparent permeability:

$$K = \frac{v\,\mu}{\text{grad } p} = \frac{v\mu\ell}{\Delta p}. \tag{2}$$

We can also neglect the influence of the unfrozen portions that have a very much higher permeability than the frozen portion. The value of this permeability was obtained with the same specimen under isothermal conditions corresponding approximately 20°C and was

$$K_{\text{unfrozen}} = 1.14 \times 10^{-16} m^2. \tag{3}$$

The filter-velocity (v) is the rate of flow (q) divided by the surface of the circular section (A) of the specimen, and we can therefore calculate the permeability in the frozen portion by the formula:

$$K = q\mu\ell/A\Delta p, \tag{4}$$

with μ= 1.8 centipoise = $1.8 \times 10^{-13} kg\ m^{-1} s^{-1}$ for T = 0°C, ℓ = 0.107 m, $\Delta p = 0.55 \times 10^5 Pa$, $q = 2.0124 \times 10^{-13}\ m^3 s^{-1}$, and $A = \pi(0.03)^2/4 = 7.0686 \times 10^{-4} m^2$:

$$K = 10^{-18} m^2 \text{ for } T = -0,3°C. \tag{5}$$

For comparison the values obtained by Burt and William (1976) for Slims Valley silt and Leda clay were, respectively, $K = 10^{-16} m^2$ and $K = 2 \times 10^{-18} m^2$ approximately.

DISCUSSION

The principal problems encountered during the measurements were mainly those coming from the small variations in temperature occurring frequently during the night. Indeed, even small thermal perturbations may cause considerable variations in the quantity of unfrozen water, resulting in a sensible additional flow due to the change of density in the water-ice or ice-water transformations. On the other hand, a small change in temperature will produce a change of temperature of the frozen portion and the unfrozen portions that are in the same sense; consequently, the thermal symmetrical conditions will be disturbed, producing a supplementary displacement of the frozen line, which will take a long time to arrive at a new stabilized condition.

Some practical problèms were inherent to the long periods required for stabilization of temperatures: for instance, there were breakdowns in the electric current and in the temperature-regulating apparatus (cryostats), which disturbed the progress of the experiment.

Finally, we emphasize the need for a more elaborate and complete study of the flow of water in a porous medium when ice is present in the pores and when the interstices are very fine.

CONCLUSION

The apparatus that was constructed and used to determine the permeability of frozen soils confirms the validity of the method proposed, which consists of applying symmetrical temperatures in relation to 0°C (or to the characteristic temperature of the frozen line) to prevent the solidification of the free water present in a permeameter. With this method there is no limitation to the domain of temperature that can be studied. Due to the necessity of reducing the duration of thermal and mechanical stabilization, we have recently constructed the permeameter shown in figure 4 which is an apparatus similar to the one used for the measurements presented in this paper, but which has smaller dimensions. We also foresee the use of a supplementary thermostatic guard at 0°C to protect the system against the thermal variations of the experimental room.

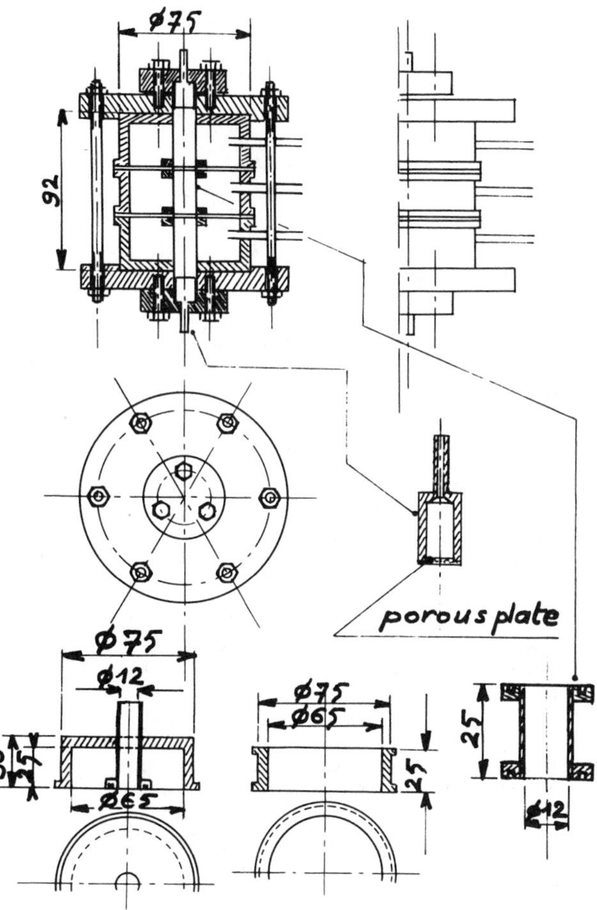

FIGURE 4 Second permeameter constructed.

ACKNOWLEDGMENTS

We appreciate very much the participation of Mr R.N. Sukhwal who helpd us in the translation of this text from the French.

REFERENCES

Aguirre-Puente, J., Vignes, M., and Viaud, P., 1973, Study of the structural changes in soils during freezing: Permafrost--The USSR Contribution to the Second International Conference, Yakutsk, July 13-28: Washington, D.C., National Academy of Sciences, p.315-323.

Aguirre-Puente, J., Frémond, M., and Menot, J.M., 1977, Gel dans les milieux poreux. Perméabilité variable et mouvements d'eau dans la partie à température négative, in Frost Action in Soils, International Symposium, University of Lulea, Proceedings, v.1, p.5-28.

Burt, T.P., and Williams, P.J., 1976, Hydraulic conductivity in frozen soils, in Earth Surface Processes, v1, p. 349-360.

Miller, R.D., 1970, Ice-sandwich: functional semipermeable membrane: Science, v.169, p.584-585.

Miller, R.D., Loch, J.P.G., and Bresler, E., Transport of water and heat in a frozen permeameter, in Soil Science Society Am. Proc., v.39, p.1029 1036.

Philippe, A., Aguirre-Puente, J., and Bertouille, H. 1970, La Station de gel de Caen. Présentation-Première expérimentation: Rapport commun Laboratoire d'Aérothermique, Laboratoire Central des Ponts et Chaussées et Centre de Géomorphologie du C.N.R.S.

Vignes-Adler, M., 1975, Etude fondamentale de la congélation des milieux dispersés à l'échelle du pore: Thèse de Doctorat ès Sciences, Paris, France.

Williams, P.J., and Wood, J.A., 1982, Investigation of moisture movements and stresses in frozen soils: Final Report to the Department of Energy, Mines and Resources Earth Physics Branch.

NOTES ON CHEMICAL WEATHERING, KAPP LINNÉ, SPITSBERGEN

Jonas H. Åkerman

Department of Physical Geography, University of Lund, Sölvegatan 13, S-22362 Lund, Sweden

Surface karst features upon outcrops of dolomitic limestone are described from an area in the western part of Spitsbergen (78°04'N, 13°38'E). The morphometric characteristics of karrenforms are compared with the height above sea level, position in the terrain and upon the outcrops, exposure to prevailing winds, the snow cover. The best developed forms are found between 50 and 150 m a.s.l upon gently sloping rock surfaces where there is a thick snow cover which can provide water during the melt season. Signs of differential chemical weathering are common in the area in the form of elevated quartz-filled joints forming complex patterns on the surfaces of the outcrops. A good correlation is found between the height of these quartz veins and the height above sea level. As the isostatic uplift history of the area is well known, it is possible to estimate the denudation rate of the dolomitic limestone to be 2.48 mm/1000 yr.

INTRODUCTION

Chemical weathering under periglacial conditions has received relatively little attention. Though several authors have remarked that chemical processes have been neglected, very few studies have been devoted specially to the subject (Cailleux 1962 1968, Corbel 1959, Helldén 1974, Rapp 1960b, Troll 1944, Washburn 1969). The opinion that chemical weathering in the periglacial environment is unimportant is, however, today gradually dispelled. It should be observed that Rapp (1960a) in his studies of slope processes in Spitsbergen and, especially in Kärkevagge in northern Sweden has showed chemical denudation processes to be quantitatively dominant (Rapp 1960b p.184-185).

Observations of surface karst features, exfoliation forms, alveoles and taffoni, oxide rinds, carbonate coatings, desert varnish and case hardening as well as active grus weathering are today commonly reported from various periglacial environments, indicating the importance of the chemical processes. In the genesis of the abovementioned features, the chemical action is only a part of a complex mechanism. The karst processes are better known and less complex. The number of studies and observations from the Arctic is limited and the intensity of the karst processes in the arctic environment compared with other climatic regions is still under discussion (Corbel 1952, 1957, Bögli 1956, 1960, Helldén 1974, Sweeting 1964, Jennings 1971).

Corbel initially estimated very high denudation rates for the arctic regions, 40 mm/ 1 000 yr in Spitsbergen (Corbel 1957, 1959), but later, after more field studies, he changed the rate to 15.75 mm/ 1 000 yr. Helldén (1974) studied a part of the strandflat south of Kapp Linné (the same area dealt with in this study), and found values which are very close to Corbel's estimates (11.4 - 15.5 mm/1 000 yr. Studies from northern Canada indicate considerably lower denudation rates - 5.0 mm/1 000 yr (Corbel 1959) and 2.05 mm/1 000 yr (Smith 1965). Great differences in the precipitation climate - Spitsbergen is comparatively wet (200-400 mm) and the Canadian area studied is comparatively dry (130 mm) - make a comparison between the areas difficult. The observations by Helldén (1974) in Spitsbergen are the most reliable figures from this area despite the fact that the investigation area was small and the observation period short.

INVESTIGATION AREA

In the area south of Kapp Linné (Figure 1) Helldén observed that the area was very poor in karst features (Helldén 1974 p.124). Other investigators have also commented on the lack of karst features in this area despite the presence of a suitable bedrock. During the course of studies on periglacial geomorphology and processes the author noted several examples of karst features (Åkerman 1980 p.70-75) and some time was devoted to the subject, although it was not a part of the investigation program.

A major part of the investigation area lies upon the Hecla Hoek series, which here consists of Precambrian-Ordovician rocks (Flood et al. 1971). Like elsewhere along the west coast, the Hecla Hoek series strike more or less parallell to the coast, and the rocks consist of vertical to nearvertical thin strata of schists, quartzites, tillites, dolomites, limestones and conglomerates. The eastern part of the area lies upon younger sedimentary rocks of which Upper Carboniferous limestones are the most important.

The area has an arctic, humid climate, with a mean annual air temperature of -4.8°C (1912-1975) and a mean annual precipitation of 400.8 mm (1934-1975) (cf. Åkerman 1980 p. 14-46). The annual course of air temperature and precipitation is given in Table 1, and from these figures we find that the summer climate is cold and that the period during which the chemical processes may act is short.

OBSERVATIONS

During the geomorphological inventory and mapping of the investigation area, surface karst features were observed in many places. The forms found were not large and conspicuous but common and characteristic of the limestone outcrops. The distribution

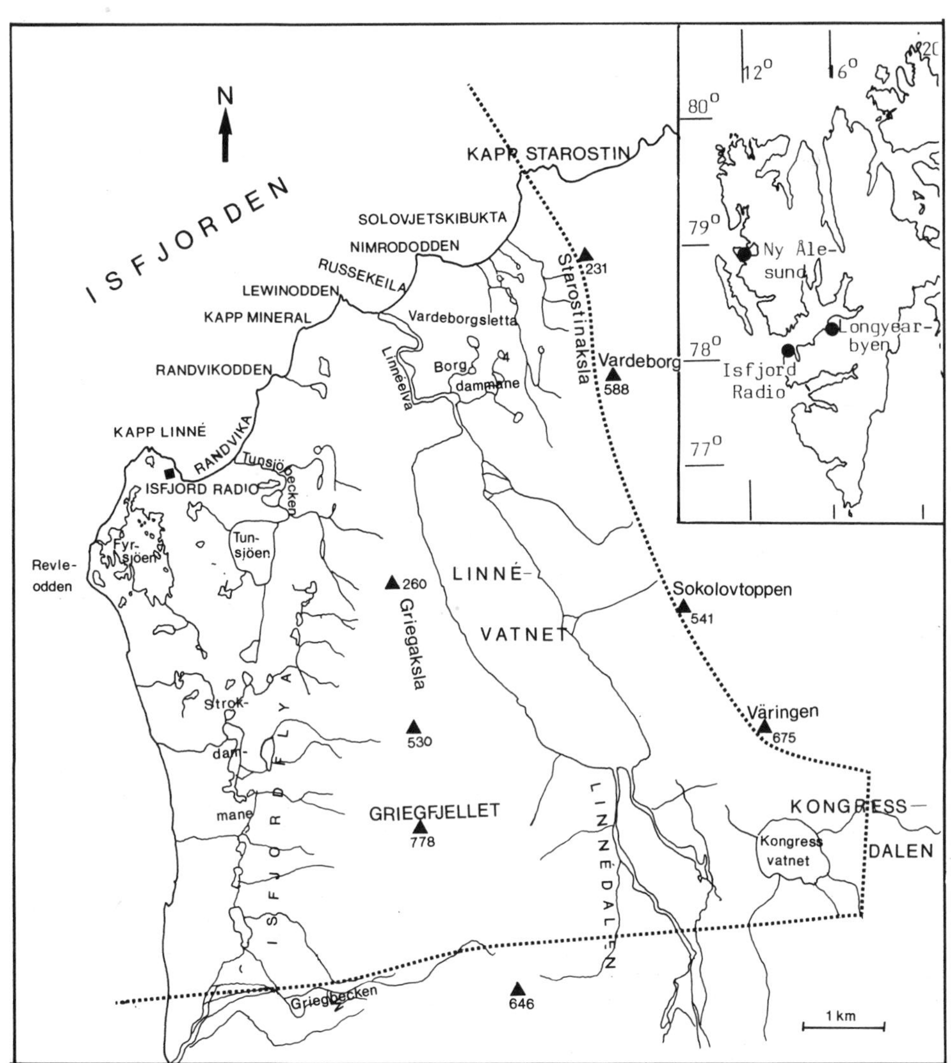

FIGURE 1 Orientation map of the investigation area near Kapp Linné, Spitsbergen. Place names and heights (m) according to Norsk Polar-institutt Map no. B9, Blad ISFJORDEN, original scale 1:100 000.

of the observations, the more important ones of which were included in the geomorphological maps, shows that two separate areas can be distinguished: (1) the central part of the strandflat west of the Linnévatnet Lake and (2) the west facing slopes of the Wardeborg-Väringen mountain ridge and the area around the Kongressvatnet Lake (Figure 2). The observations on the strandflat are connected with the outcrops of dolomitic limestones and the surface karst features east of the Linnévatnet Lake are found in soft Cyatophyllum limestones.

The Kongressvatnet area

The surface forms observed in this area are mainly karren forms, but also weakly developed doline forms were observed around the Kongressvatnet Lake. The karren forms are generally found on fairly steep to vertical surfaces on the mountain walls. They consist of furrows, running in sets straight down the steepest inclination with sometimes sharp, but in most cases rounded, ribs in between. The width ranges between 1 and 10 cm, and the length rarely exceeds 40 cm (Table 2). In figure 4 an example of this type of karren forms from the area around the Kongressvatnet Lake is shown. Also common are small

TABLE 1. Average Monthly Air Temperature at Standard Meteorological Height and Precipitation at Isfjord Radio Station, Kapp Linné, Spitsbergen 1946-1975

	Temperature (°C)	Precipitation (mm)
January	-11.2	31.5
February	-11.5	30.5
March	-12.2	30.9
April	-9.2	22.3
May	-3.5	23.6
June	+1.6	24.8
July	+4.7	35.8
August	+4.2	45.0
September	+1.1	40.6
October	-3.4	41.3
November	-7.1	38.8
December	-9.5	35.6

grikes (solution slots), which are solution-widened joints in the limestone.

To obtain a picture about the surface karst features and their relations with some environmental factors, some simple observations regarding their

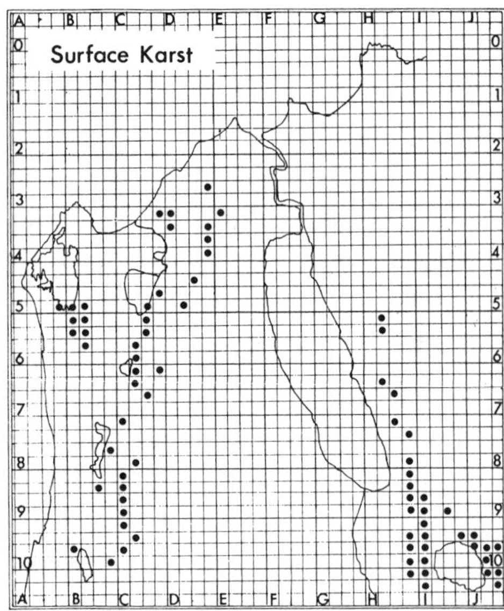

FIGURE 2 The occurrence of surface karst features within the investigation area south of Kapp Linné, Spitsbergen.

position in the terrain were performed. The factors concerned were height above sea level, association with flowing water, rock walls and their exposure, loose blocks, rock outcrops (level surfaces, sloping surfaces, summits), and snow depth.

In figure 3 the relation between the width of the observed karren forms and their height above sea level is shown. There is a tendency for the widest karren forms to occur between 100 and 200 m a.s.l. Below and above this height the number of wide karren forms is decreasing. This may be explained by a shorter time for formation at the low levels (the valley was earlier filled with a glacier), and less favorable formation conditions at the higher levels (the physical weathering dominating?). The exposure and orientation of the rock wall may be an important factor in the chemical weathering processes, sun, wind and precipitation exposure (cf. Dunkerley 1979 p.333). However, the majority of the rock walls of the area are exposed to the west and southwest, and only a small percentage of the rock walls face other directions. Therefore it is difficult to obtain a picture about whether in this area the occurrence of karren forms may be related to orientation and exposure or not.

The karst forms found on the more level sites have been related to their occurrence in association with flowing water, their position upon rock outcrops, the occurrence upon loose blocks, and snow depth (Table 3).

TABLE 2 Morphometric Characteristics of Karren Forms in the Kongressvatnet-Vardeborg Area East of the Linnévatnet Lake, Spitsbergen

	0-2	2-5	5-10	10-25	25-40	> 40 cm
Width	38.2%	49.5%	12.0%	0.3%	0%	0%
Length	0.3%	3.1%	11.2%	47.2%	33.0%	5.2%

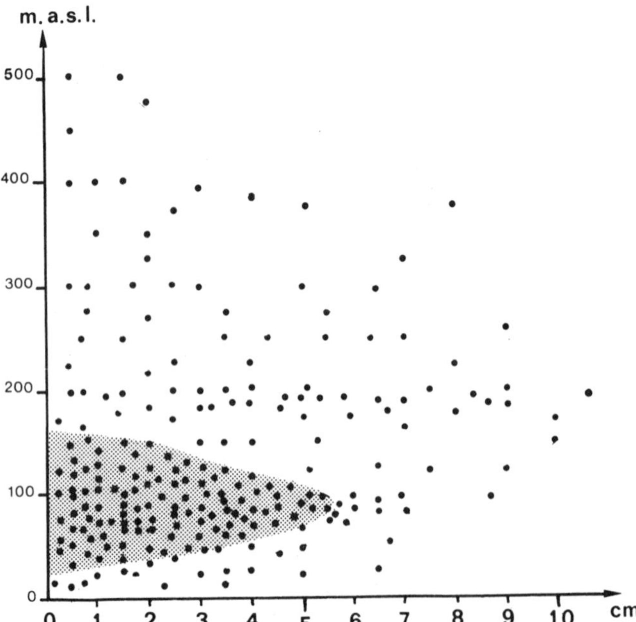

FIGURE 3 The relation between the width of the observed karren and their height above sea level in the Vardeborg-Kongressvatnet area east of the Linné vatnet Lake, Spitsbergen.

TABLE 3 Occurrence of Surface Karst Features on the More Level Sites in the Area East of the Linnévatnet Lake, Spitsbergen

Flowing water	Rock outcrops			Loose blocks
	level surfaces	sloping surfaces	summits	
7.2%	16.6%	51.3%	32.1%	9.7%

The surface karst features found in association with the small streams of the area are few and mainly restricted to the Kongresselva River ravine. The observations upon loose blocks are also few but here the figure is probably an underestimate as only larger blocks were incorporated in the observations. Rock outcrops dominate the observation material, and sloping surfaces or sides of the outcrops are the places where karst features are most frequently found. There are many factors determining this distribution, but in this area the snow depth is of great importance. Level surfaces at the tops of outcrops and more rugged outcrops or summit surfaces are generally blown free of snow. They will therefore be comparatively dry, and are furthermore subject to a more intensive frost shattering which will destroy forms created by chemical processes.

The sites which have a medium thick snow cover (0.5-1 m) are those which have the best developed surface karst features. Here there is a fairly large amount of water released during the snowmelt season. Snowmelt occurs early in the summer, after which the surfaces are exposed to precipitation. Snow insolation affords protection from frost shattering during late spring and late autumn, the periods with the highest frequency of freeze-thaw cycles.

FIGURE 4 Karren forms on a dolomitic limestone outcrop southeast of Kapp Linné, Spitsbergen.

FIGURE 5 Complex net pattern of quartz filled veins in a dolomitic limestone outcrop at 35 m a.s.l. 4 km east of Kapp Linné, Spitsbergen.

The Strandflat Area

Surface karst features are found on the dolomite outcrops, but here the forms are generally smaller and less well developed than was the case in the softer limestones east of the Linnévatnet Lake. In general the karren forms are weakly developed solution flutes, small solution facets or a chaotic, rugged ridge system a few centimetres high and preferably on the flat or gently sloping summits of the outcrops. Welldeveloped Rillenkarren are rare, but some examples are found in the eastern and highest part of the strandflat (Figure 4).

Differential Weathering

More interesting than the weakly developed karren forms is the differential weathering of the dolomitic limestones with quartz veins. The dolomitic limestones of the strandflat have in many cases complex joint systems which are filled with a white or yellowish white quartz. As the weathering rate in the limestone is considerably higher than that in the quartz, the quartz filled joints are left as a complex pattern of elevated veins on the surface (Figure 5). Since it was observed that these veins seemed to be higher and better developed on higher levels, it was concluded that the veins might be used to estimate the limestone denudation of the area (cf. Dahl 1967, Birkeland 1982).

There is a progressive increase in vein height with elevation (Figure 6). The differences in height of the veins at each site are great, but as both the mean and the maximum show a good correlation, it is reasonable to conclude that the correlation with the height above sea level as indicated in Figure 6 is relevant. In the discussion below, the maximum values are used as probably being the most significant in the discussion about denudation rates.

There is a fair amount of data available concerning the isostatic uplift of the area and thus the number of years each outcrop has been above the sea level (Büdel 1968, Schytt et al. 1967, Stäblein 1978 and the authorś 14C datings). In Figure 7 the

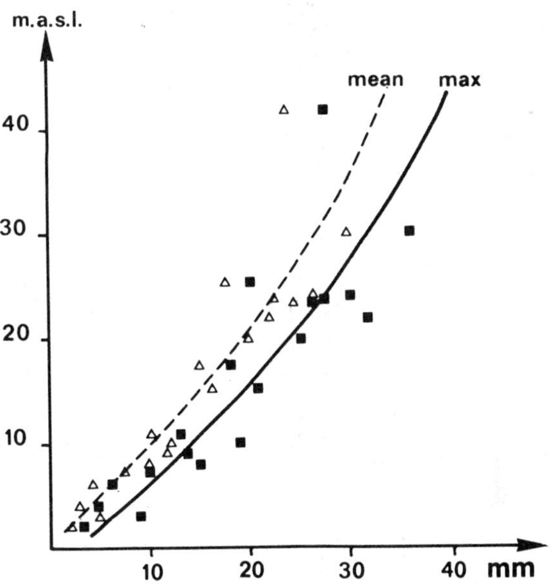

FIGURE 6 The relation between the mean and maximum height of the quartz veins upon dolomitic limestone outcrops and the height above mean sea level for the strandflat area east of the Linnévatnet Lake, Spitsbergen.

correlation between the maximum height of the quartz veins and the estimated number of years the outcrops in question have been above the sea level is shown. If we assume that the height of the quartz veins is a measure of the limestone denudation and use the results above and estimate the denudation rate, we obtain a mean of 2.48 mm/1 000 yr. The maximum value is 3.78 mm/ 1 000 yr and the minimum is 1.5 mm/1 000 yr. These values are considerably lower than those obtained by Corbel (1960) and Hellden (1974), who calculated values between 11.4 and 15.74 mm/1 000 yr. The difference is due to the fact that Corbel and Helldén based their estimates upon hydrochemical data -the amount of $CaCO_3$+$MgCO_3$

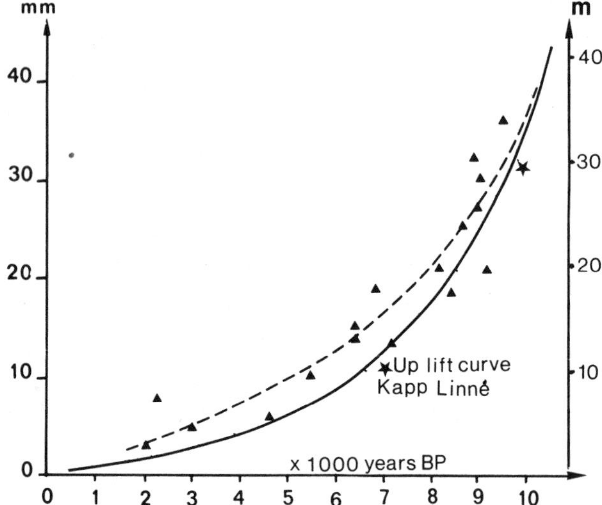

FIGURE 7 The relation between the height of the quartz veins and years each site has been above sea level. The isostatic uplift curve according to Schytt et al. (1967). Two of the authors 14C datings (Lu-1722 and Lu-2020, Dep. of Quaternary Geology, Univ. of Lund) have been added (✶).

transported in solution in the streams. The water in the streams comes partly from rainwater and partly from snow and has been in contact with the bedrock and limestone particles in the soil for a longer time. Therefore the solution/denudation rates based upon this kind of measurements must be considerably higher than that on exposed rock. The rock outcrops are exposed to running water during snowmelt and during rainfall, dew deposition and fog wetting only so solution/denudation is less. Taking this into account, the result obtained is probably fairly representative.

Another result which ought to be considered is that the denudation rate seems to have decreased during the 10 000 yr period the processes have acted in the area (Figure 8). The lowest denudation rates are found for the sites which have been above the sea level during the last 2 000 to 4 000 years. During this period the denudation rate is about 1.5 mm/1 000 yr while those sites which have been above the sea level during 4 000 to 9 000 years have denudation rates of up to 3.5 mm/1 000 yr. This may be explained by (1) the change in surface roughness over time or (2) the climatic optimum during the post glacial warm period.

The first possibility must always be a part of the explanation but it will probably be of great importance only during the earliest stages. The well documented climatic changes during the post glacial thermal optimum provided warmer summerclimates with more and longer periods with precipitation in the form of rain, drizzle and fog wetting each year. The vegetation was also more ample which made the surface water more agressive as a result of a higher concentration of humus acids. This will increase the limestone solution considerably (cf. Helldén 1974).

CONCLUSIONS

The study area, which is representative of the strandflat areas of central west Spitsbergen as regards the geological, geomorphological, climatological and hydrological factors, is rich in forms and features created directly or indirectly by chemical weathering and associated processes. The majority of the forms are small and thereby easily overlooked. Still, the processes and the forms resulting from them are an important part of the periglacial environment.

The most common and widespread process in this respect is the limestone solution and the surface karst features associated with it. There is a clear relation between the karst features observed and their localization in the terrain as well as with the height above sea level. This latter is related to age of exposure.

The correlation found between the height of quartz veins upon dolomitic limestones and the time available for their formation allows of an estimation of the limestone denudation during the last 10 000 years. The figure obtained is 2.5 mm/1 000yr and is considerably lower than earlier studies have shown for the area (11-15 mm/1 000 yr), however, these latter studies were based upon hydrochemical measurements, result from processes active 4-5 months per year. The rock outcrops here studied are exposed to running water only a few days during the snowmelt season and during those days with rain precipitation, surface condensation and fog. Taking this into account (15 days during snowmelt and 46 days of rain -the average number of rainy days June to September), the period during which the process can be active will be restricted to some 1.7 months only. Further, the water acting on the rock outcrops is less agressive than the soil/ground water acting upon the level rock surfaces and on the soil particles. Regarding these factors, the obtained values

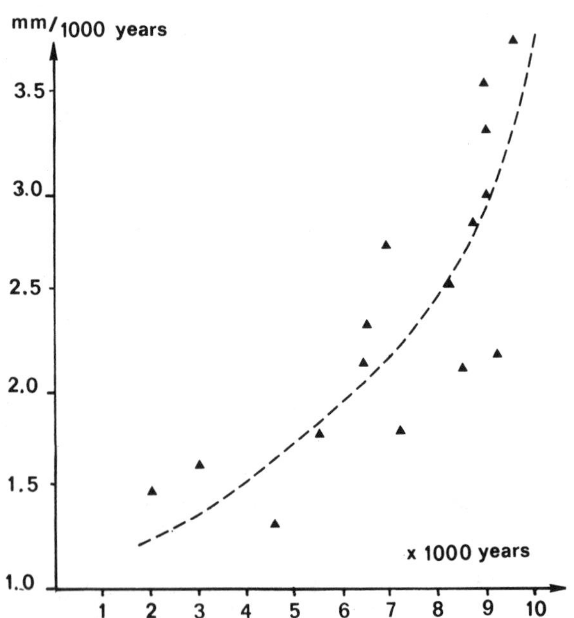

FIGURE 8 The relation between the estimated denudation rates and the years each site has been above sea level.

are probably fairly representative, indicating that the denudation rates here obtained reasonably well correspond to those obtained by Helldén (1974). This means that observations of differential weathering of this type might be used as an indirect tool for estimates of limestone denudation. The result also indicate that the denudation rates seem to have decreased during the 10 000 year period the processes have been acting in the area. Despite the fact that other factors to some extent are involved there are reasons to assume that this may reflect the warmer and wetter climate during the post glacial temperate period and a colder and drier period during the last 2 000-4 000 years.

ACKNOWLEDGEMENTS

This work was supported by grants from the Swedish Soc. for Anthropology and Geography, the P. Westling Fond and the K. A. Wallenberg Fond. I am grateful to Professors A. Rapp, K. Everett, P. Birkeland, G. Miller, H. French, Dr. U. Helldén and Dr. S. Forman for valuable comments on the manuscript. R. Lazlo made the photographic laboratory work.

REFERENCES

Åkerman, H. J., 1980, Studies on periglacial geomorphology in West Spitsbergen. Meddelanden från Lunds Univ. Geografiska Inst. Ser. Avh. LXXXIX.

Birkeland, P. W. 1982, Subdivision of Holocene glacial deposits, Ben Ohau Range, New Zealand, using relative-dating methods. Geological Soc. of America Bulletin, v.93, p. 433-449.

Büdel, J., 1968, Die Junge Landhebung Spitsbergens im Umkreis des Freeman-Sundes und der Olgastrase. Würzburger Geogr. Abh. 22(I):1-21.

Bögli, A., 1956, Der Chemismus der Losungsprozesse und der Einfluss der Gesteinsbeschaffenheit auf die Entwicklung des Karstes. In Lehman ed.: Report of the commission on Karst phenomena. IGU 18th. Int. Geog. Congr. Rio de Janeiro. p. 7-17.

Bögli, A., 1960, Kalklösung und Karrenbildung. Zeitschrift für Geomorphologie, Suppl. 2 p. 4-21.

Cailleux, A., 1962. Etudes de geologie au detroit de McMurdo (Antarctic). CNFRA (Commité national francais recherches Antarctique), no 1, 41 p.

Cailleux, A., 1968, Periglacial of McMurdo Strait. Biul. Peryglac. 17, p. 57-90.

Corbel, J., 1952, Les phenomen karstiques en Suède. Geogr. Annaler 34. Häfte 3-4, p. 204-237.

Corbel, J., 1957. Les karst du nord-ouest de l'Europe Mem. Inst. Etudes Rhodanlennes de l'Universite de Lyon, no. 12, 544 p.

Corbel, J., 1959, Vitesse de l'erosion. Zeitschrift für Geomorphologie N.F. 3, p. 1-28.

Corbel, J., 1960, Nouvelles recherches sur les karst arcriques Scandinaves. Zeitschrift für Geomorphologie, Suppl. 2, p. 74-80.

Dahl, R., 1967, Post-glacial micro-weathering of bedrock surfaces in the Narvik district of Norway. Geografiska Annaler 49A, 2-4, p. 155-166.

Dunkerley, C. D. L., 1979, The morphology and development of Rillenkarren. Zeitschrift für Geomorphologie, N.F. 23, 3, p. 332-348.

Flood, B. et al., 1971, Geological map of Svalbard. Norsk Polarinstitutt Skrifter, no. 154A.

Helldén, U., 1974, Karst. En studie av Artfjällets karstområde samt jämförande korrosionsanalyser från Västspetsbergen och Tjeckoslovakien. Meddl. från Lunds Univ. Geografiska Inst. Ser. Avh. LXXII, 192p.

Jennings, J. N., 1971, Karst. An introduction to systematic geomorphology, vol. 7, The M.I.T Press, 252p.

Rapp, A., 1960a, Talus slopes and mountain walls at Tempelfjorden, Spitsbergen. Norsk Polarinstitutt Skrifter no 119, 96p.

Rapp, A., 1960b, Recent development of mountain slopes in Kärkevagge and surroundings northern Scandinavia. Geografiska Annaler v.XLII, no 2-3, p. 73-200

Schytt, V. et al., 1967, The extent of the Würm glaciation in the European Arctic, a preliminary report about the Stockholm Univ. Svalbard Exp. 1966. International Union of Geodesy and Geophysics, General Assembly of Bern 1967, Publ. 79, p. 207-216.

Smith, D. I., 1965, The solution erosion of limestone in an Artic morphogenetic region. P. 99-110. In "Problems of Karst Denudation", O. Strekl ed., Brno 1965.

Stäblein, G., 1978, Extent and regional differentiation of glacioisostatic shoreline variations in Spitsbergen. Polarforschung 48 (1/2), p. 170-180.

Sweeting, M. M., 1964, Some factors in the absolute denudation of limestone terrains. Erdkunde 18, p. 92-95.

Troll, C., 1944, Strukturboden, Solifluction und Frostklimate der Erde. Geologische Rundschau 34, p. 345-694.

Washburn, A. L., 1969, Weathering, frost action and patterned ground in the Mesters Vig district, North East Greenland. Meddel. om Grönland. Bd. 176, no. 4.

LUG BEHAVIOR FOR MODEL STEEL PILES IN FROZEN SAND

O. B. Andersland and M. R. M. Alwahhab

Department of Civil Engineering, Michigan State University,
East Lansing, Michigan 48824 USA

The adfreeze bond strength for piles embedded in frozen sand includes ice adhesion to the pile, friction between the pile surface and sand particles, and mechanical interaction between frozen sand and the pile surface roughness. Experimental work has shown that the presence of lugs, small protrusions, increased the volume of ice matrix and soil particles involved in the initial rupture, thereby increasing bond due to ice adhesion. Additional pile movement and soil interaction with lugs mobilized sand dilatancy and particle reorientation effects along with soil bearing forces on the lugs. The model pile load was significantly increased. The net result was that creep displacement behavior will differ considerably depending on pile movement. Very small displacements characterize the start of tertiary creep (failure) for applied shaft stresses below those required for initial rupture of ice adhesion. Larger displacements and higher shaft stresses are associated with mobilization of mechanical interaction forces. Experimental relations presented include load displacement curves for constant displacement rate tests; constant load creep curves for a model pile with multiple lugs; and the dependence of creep displacement rates and/or adfreeze bond on lug size, temperature, and sand volume fractions (ice content).

INTRODUCTION

Pile installation in frozen ground often involves dry augered or bored holes with a mixture of sand and water used to fill the annulus around the pile. Load supporting capacity develops when the sand-water slurry is solidly frozen in place. Pile capacity is primarily dependent on the adfreeze bond strength of the frozen soil in the zone adjacent to the pile. This adfreeze strength is dependent on physical, thermal, and rheological properties of the frozen slurry and the pile surface roughness and type. Published load tests (Crory, 1963) have shown that adfreeze bond failure is sudden and abrupt and that strength can only be partially regained after motion stops. Increased pile surface roughness in the form of lugs or corrugations has been used to increase the load capacity. Lack of information on the influence of lug size and spacing, soil ice content, and temperature on increased long-term adfreeze bond strength has limited the use of lugs on field projects.

Load transfer at the pile/frozen soil interface includes ice adhesion, friction between the pile surface and soil particles, and mechanical interaction between frozen sand and pile surface roughness. The presence of a lug increases the volume of ice matrix and soil particles involved in rupture, thereby increasing the bond due to ice adhesion. Development of bearing forces is a function of lug size and requires larger displacements to mobilize the friction and dilatancy components of soil shear strength. Density of the soil particles relates to the soil friction and dilatancy and in turn mechanical interaction forces. Experimental relations are presented showing load-displacement curves for four lug sizes and a standard deformed No. 3 bar (about 9.52 mm, 3/8 in.); typical constant load creep curves; and the dependence of creep displacement rates and/or adfreeze bond strengths on lug size, temperature, and sand volume fractions (dry density).

EXPERIMENTAL PROCEDURES

Model Pile Set-up

The frozen sand sample shown in Figure 1, with a plain steel 9.52 mm bar and one lug, served as a model steel pile section suitable for pull-out tests. Lug heights included h equal to 1.59, 3.17, and 4.76 mm with a length L of 12.7 mm. Plain bars were included to show the limiting condition as lug size goes to zero. The frozen sand sample shown in Figure 1 was representative of a slurry which might be used to fill the annulus around a pile in the field. All sand specimens were prepared in split molds 152 mm (6 in.) in diameter and frozen at close to -20 ^{o}C for about 12 hours. Sample heights were limited to 152 mm or less so that pull-out loads would not exceed the equipment capacity. The sand consisted of subangular quartz particles with a uniform gradation (size range of 0.105 mm to 0.595 mm) and a coefficient of uniformity equal to 1.50. Relatively dense samples were formed by placement of sand in water and tamping until the desired density was obtained. Low sand volume fractions with reasonably uniform particle distribution were obtained by mixing precooled sand with snow and adding precooled water to increase the degree of saturation. Actual sand volume fractions were based on the sample height, dimensions, and weight of sand remaining after thawing at completion of the test.

After removal from the mold, rubber membranes (double thickness) were placed as shown in Figure 1 to protect the frozen sample during immersion in the refrigerated anti-freeze/water coolant mixture. A thermistor embedded in the frozen sample permitted

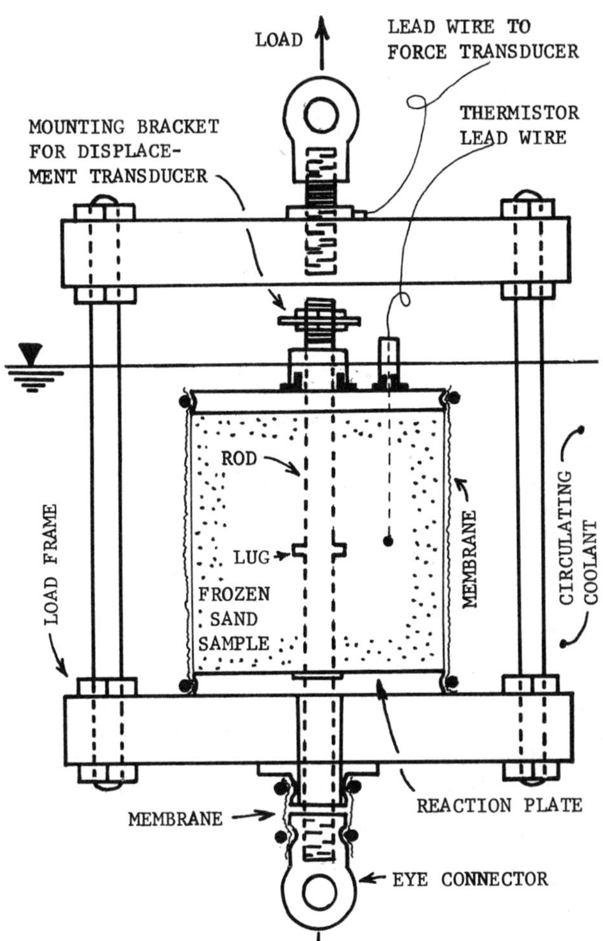

FIGURE 1 Equipment for pull-out tests including sample, steel bar with lug, and loading frame immersed in the circulating coolant.

temperature monitoring before and during the test. A threaded portion at both bar ends provided for easy hook-up to an eye connector and reaction hook at the bottom of the coolant tank and mounting of a displacement transducer at the upper bar end (Figure 1). Clearance between the bar and reaction plate was maintained at close to 1.6 mm for all bar diameters by use of a removable steel washer in the reaction plate.

Test Procedure

Both constant load (creep) and constant displacement rate pull-out tests were conducted using the loading frame shown in Figure 1. The initial bar stress prior to loading was limited to a very small value represented by the weight of the bar, lug, and eye connector. Constant loads were applied using a lever system and dead weights. Constant displacement rates were produced using a Graham variable speed gear box mounted on a 44.5 kN (10 kip) capacity Soiltest load frame. Movement of the bar relative to the lower reaction plate and soil sample was monitored using a displacement transducer mounted on the bar above the coolant liquid with the core supported by a rod extending down to the reaction plate. Loads were monitored by the force transducer attached to the top of the loading frame shown in Figure 1. Sample temperatures were controlled by circulating an antifreeze/water coolant mixture from an external refrigerated bath around the protected sample and loading frame.

EXPERIMENTAL RESULTS

Constant Displacement Rate Tests

Load displacement curves for a plain steel bar with a single lug are given in Figure 2 for an average displacement rate of 15.2 um/min and a temperature close to -10 °C. Very little displacement (less than 50 um) was observed up to the first break in the displacement curves (initial rupture of the ice matrix). Transducer limitations prevented more accurate measurements for the small displacements. After rupture the load quickly dropped to a small residual value for the plain bar (h = 0). Introduction of a lug increased the ice matrix volume and soil particles involved in the initial rupture, thereby increasing the adfreeze bond due to adhesion. Additional bar displacement permitted bearing forces to develop in front of the lug as shear atrength was mobilized in the sand. The drop in load after initial rupture decreased as lug size was increased. The ultimate load, which was mobilized at around 5 mm displacement, greatly exceeded the load for no lug and is a function of lug size.

A series of lugs on a standard deformed No. 3 bar in frozen sand at the same sand fraction (64 %) and temperature (-10 °C) gave the load displacement curves shown in Figure 3. The sample height H was reduced so that the equipment load capacity would

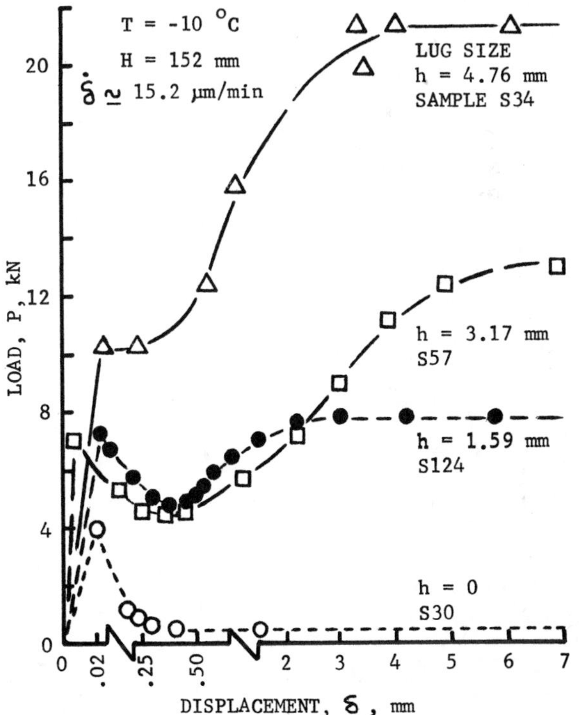

FIGURE 2 Load-displacement curves for a 9.52 mm steel bar and four lug sizes in frozen sand at -10 °C and a sand volume fraction of 64 % (data from Alwahhab, 1983).

not be exceeded. Initial rupture in the ice matrix appears to occur at larger displacements. The load continued to increase up to an ultimate value after which it decreased. This decrease in load, as compared to no decrease for a single lug, appears to be due to an overlap of pressure bulbs in front of the lugs and to formation of a void space behind the displaced lug. Preliminary work has used Boussinesq's equations to estimate pressure bulb overlap from consecutive lugs. Examination of the failure zone in front of single lugs after a test showed crushed sand particles extending forward a distance of 2 to 3 times the lug height.

Constant Load (Creep) Tests

Displacement-time curves for a plain 9.52 mm steel bar with multiple lugs (h = 0.397 mm) are given in Figure 4 for a temperature close to -10 °C. The sample height was limited to 76 mm (10 lugs) so as not to exceed equipment load capacity. Preliminary creep pull-out tests on plain bars showed that failure (tertiary creep) associated with ice adhesion was initiated at very small displacements, values close to 0.1 mm or less. For the standard No. 3 bar with multiple lugs the failure displacement of about 3 mm (Figure 4) may correspond to mechanical interaction forces. With a larger single lug experimental data indicated that secondary creep would continue with no failure for large displacements (80+ mm) as long as boundary conditions did not interfere. It appears that overlap of pressure bulbs and formation of a void space behind the lugs may be responsible for development of tertiary creep at close to 3 mm bar displacement. The conventional creep curves in Figure 4 show a small initial elastic displacement followed by a plastic (irreversible) creep up to steady-state creep, represented by the straight line portion of the curve. To reduce the number of samples required, a step loading procedure was used as shown by the

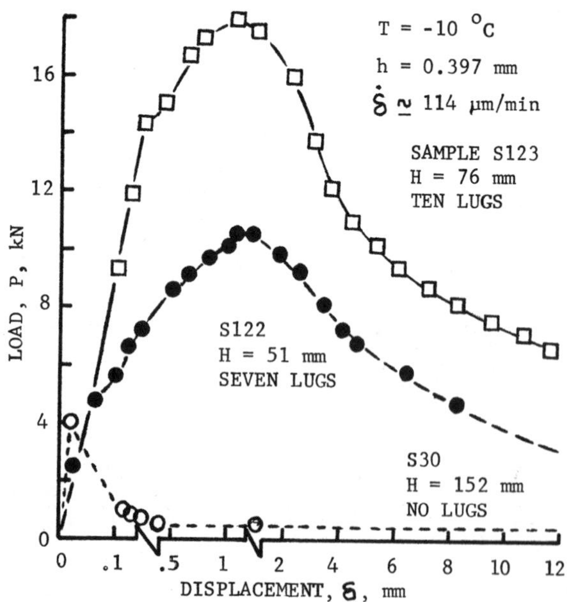

FIGURE 3 Load-displacement curves for a standard deformed No. 3 bar in frozen sand at -10 °C and a sand volume fraction of 64 % (data from Alwahhab, 1983).

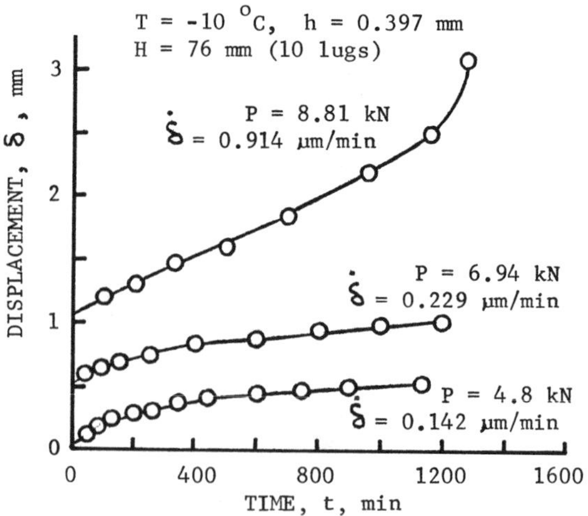

FIGURE 4 Creep curves for step loading a standard No. 3 deformed bar with ten lugs in frozen sand at -10 °C and a sand volume fraction of 64 % (data from Alwahhab, 1983).

additional two curves in Figure 4. A small elastic and plastic displacement developed as each load increment was applied. Using duplicate samples and a single lug, it was shown that creep rates based on incremental loads were in agreement with creep rates for the first load. To insure development of secondary creep it was important that the load be greater than the long-term bond strength corresponding to the current displacement.

DISCUSSION

Adfreeze Bond Strength

Load transfer at the interface between the model pile and frozen sand involved three bond components: (1) ice adhesion to the pile, (2) friction between sand particles and the pile surface, and (3) mechanical interaction between frozen sand and the pile surface roughness. Different load transfer mechanisms are involved. Ice adhesion is the result of increased attraction between water molecules and the pile surface at freezing temperatures. Surface type (steel, concrete, etc.) and contaminants (salts, minerals, etc.) in the ice significantly influence adhesive forces at the interface and cohesive forces in the ice. Load displacement curves in Figure 2 showed that very little movement mobilized ice adhesion forces and that rupture occurred for displacements of 0.1 mm or less. During creep, ice adhesion may continue for larger displacements. The second component, friction, is dependent on the coefficient of friction (or friction angle) between the soil particles and pile surface and is proportional to the normal stress pushing particles against the pile. An upper limit would be the long-term strength, $\tau_{\ell t}$, of frozen soil which is comprised of both cohesive, $c_{\ell t}$, and frictional, $\phi_{\ell t}$, components and can be expressed by the Mohr-Coulomb relation,

$$\tau_{\ell t} = c_{\ell t} + \sigma \tan \phi_{\ell t} \qquad (1)$$

where σ is the normal stress on the shear plane. Weaver and Morgenstern (1981) indicated that the normal stress on a pile is typically less than 100 kPa so that the frictional component of adfreeze bond would generally be very small. Surface roughness of the pile introduces mechanical interaction forces when pile movement occurs relative to the sand particles. Roughness is dependent on surface geometry and actual heights of aspirities on the pile surface. Lugs on the pile surface were used to increase surface roughness and to permit mobilization of bearing forces acting on the frozen sand. Bearing forces on the single lug in Figure 2 prevented the load from decreasing at larger displacements to the smaller residual load shown for the plain bar (h = 0). Use of multiple lugs (Figure 3) altered the load-displacement curves raising questions as to lug spacing for a given lug height.

Lug Size Effect

The use of lugs provided an increase in surface roughness of the model piles and increased both ice adhesion and mechanical interaction during movement in the frozen sand. The effect of increasing lug size on load displacement curves is shown in Figure 2. The lug increased surface area for ice adhesion but more importantly the lug increased the volume of ice matrix and soil particles involved in the initial bond rupture. This rupture involved both ice adhesion at the interface and cohesion in the ice matrix resulting in an increased adfreeze bond strength. After initial rupture the pull-out load decreased to a small residual value for plain bars. The presence of lugs and additional bar movement permitted mobilization of mechanical interaction forces. For a model pile with a single lug the ultimate load developed at from 2 mm to 5 mm displacement (Figure 2). The ultimate load for multiple lugs (Figure 3) developed at about 1.4 mm displacement and then decreased, probably due to pressure bulb overlap and formation of a void space behind the lug.

Data for constant load (creep) pull-out tests are summarized in Figure 5 with load and displacement (creep) rates both plotted on logarithmic scales. The data suggest a straight line relationship for each lug size for which the adfreeze load P can be expressed as

$$P = P_{c\theta}\left(\frac{\dot{\delta}}{\dot{\delta}_c}\right)^{1/n} \qquad (2)$$

where $\dot{\delta}$ is the creep displacement rate, n is the creep parameter, and $P_{c\theta}$ is the adfreeze proof load for an arbitrarily selected displacement rate $\dot{\delta}_c$. The adfreeze strength can be readily calculated based on the 9.52 mm bar diameter and 152 mm embedment length. Some data points at the slower creep rates appear to fall to the left of their respective lines in Figure 5. It appeared that loads for these points were below the long-term strength for the corresponding bar displacement rate. Secondary creep would not have been reached for the time given to that load increment and the recorded creep rate would be larger than the true creep rate.

Temperature Effect

Temperature, thru its influence on cohesive forces in the ice matrix and adhesive forces at the pile/frozen sand interface, has a significant effect on adfreeze bond strength. This influence was shown by the pull-out load versus creep displacement rate data summarized in Figure 6 for tests on 9.52 mm diameter model piles with a single 3.17 mm lug. Temperatures of -2, -6, -10, and -15 °C are represented by the four lines. Again some observed creep rates are larger than their true values for the reason explained in the previous section. The slope and hence the creep parameter n are the same as in Figure 5. Selection of one creep displacement rate, plotting these loads, $P_{c\theta}$, against $(1 + \theta\,\theta_c^{-1})$, both on logarithmic scales, permits evaluation of the exponent ω (Figure 7) in the equation

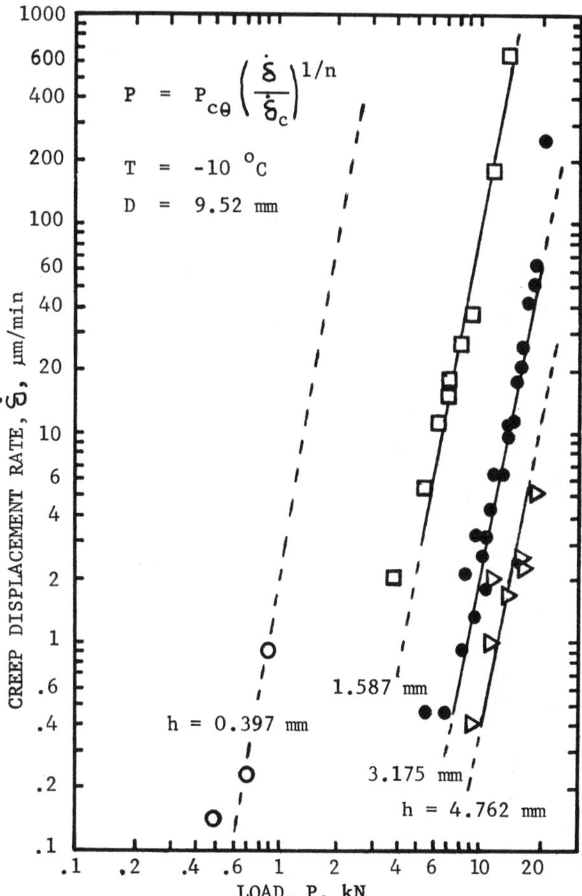

FIGURE 5 Lug height effect on load for creep of plain 9.52 mm bars in frozen sand at -10 °C with a sand volume fraction of 64 % (data from Alwahhab, 1983).

$$P = P_{c\theta}\left(\frac{\dot{\delta}}{\dot{\delta}_c}\right)^{1/n} = P_{co}\left(1 + \frac{\theta}{\theta_c}\right)^{\omega}\left(\frac{\dot{\delta}}{\dot{\delta}_c}\right)^{1/n} \qquad (3)$$

where θ is the absolute value of the negative temperature in degrees Celsius, θ_c is an arbitrary temperature, usually 1 °C, and P_{co} is $P_{c\theta}$ for a freezing temperature close to 0 °C. Other symbols are the same as in equation 2.

Bond Dependence on Ice Content

Adfreeze bond strength at the pile/frozen sand interface will vary with ice content. The one extreme will include only ice adhesion at the

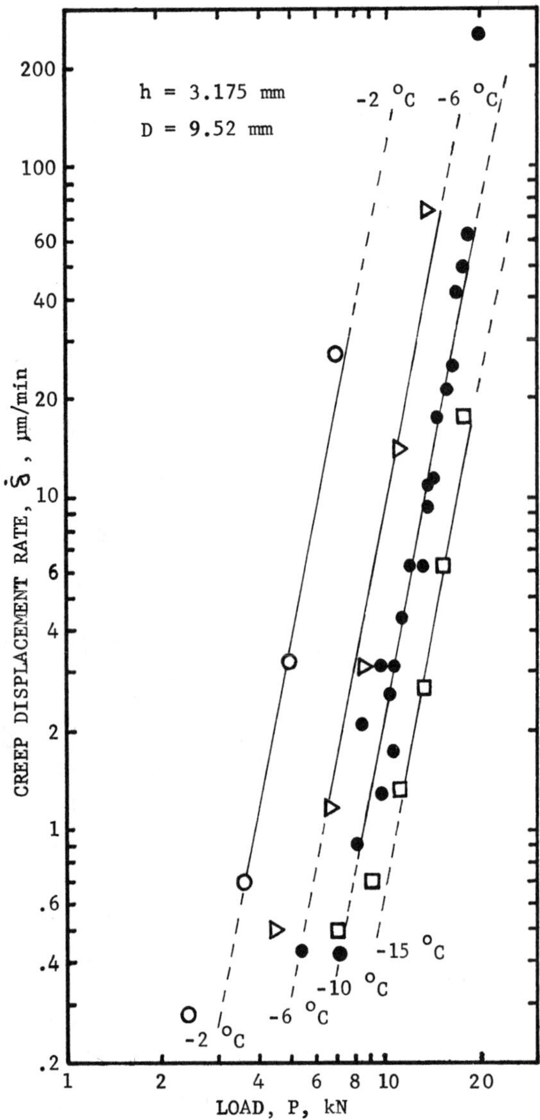

FIGURE 6 Temperature effect on load for creep of plain 9.52 mm steel bars with a 3.175 mm lug in frozen sand with a sand volume fraction of 64 % (data from Alwahhab, 1983).

interface and cohesion within the ice. The other extreme, with very dense sand, will include large contributions from mechanical interaction between sand particles and the pile surface roughness. A series of constant load (creep) pull-out tests on frozen sand with different ice contents (or sand volume fractions) provided the load versus creep displacement rate data summarized in Figure 8. The 9.52 mm model steel pile included a single 3.17 mm lug. The line representing a 64 % sand volume fraction was also included in Figures 5 and 6 for h = 3.175 mm and T = -10 °C, respectively. It appears that the creep parameter n (based on slope of the lines) is independent of both temperature and ice content for the range of creep rates shown.

The influence of ice content on creep pull-out loads is shown in Figure 9 for two constant creep displacement rates. The zero sand volume fraction, represented by polycrystalline snow-ice, gave the

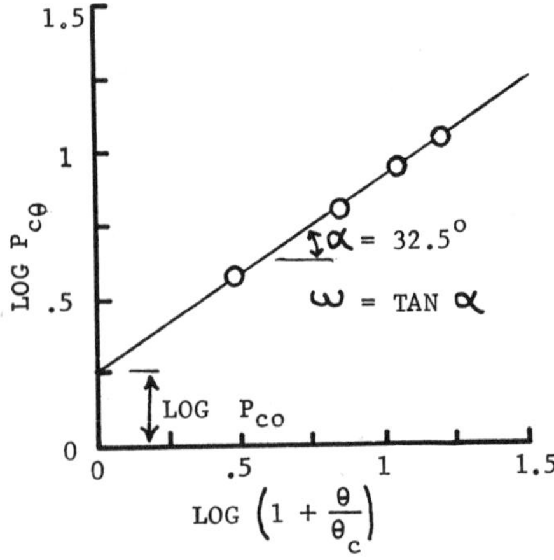

FIGURE 7 Temperature dependence of $P_{c\theta}$ for a creep displacement rate $\dot{\delta}$ equal to 1 µm/min (data from Figure 6).

smallest adfreeze strength. Most ice found in soil pores is of a polycrystalline type with random crystal orientation. The addition of dispersed sand particles to this ice increased the adfreeze strength, perhaps due to the greater stiffness of the particles which involves more ice volume at the pile/frozen sand interface in the failure process.

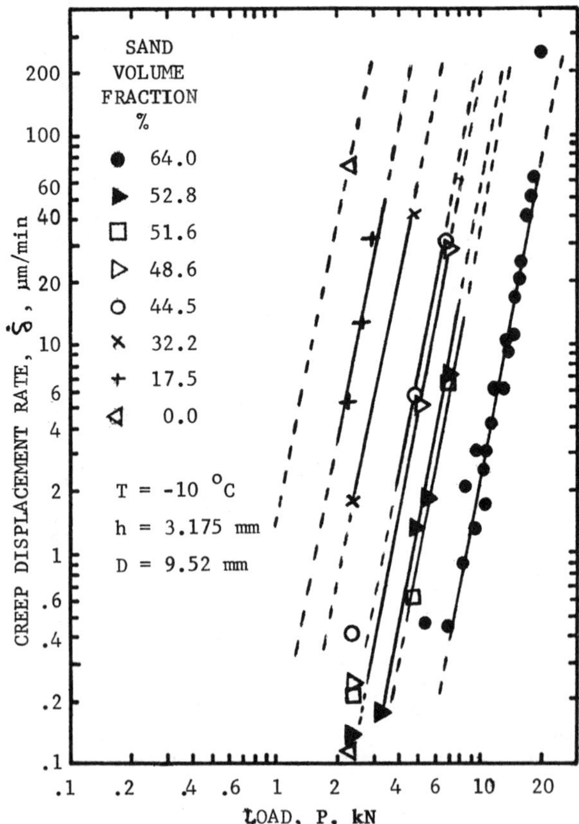

FIGURE 8 Effect of sand volume fraction on load for creep of plain 9.52 mm steel bars with a 3.175 mm lug in frozen sand at -10 °C (data from Alwahhab, 1983).

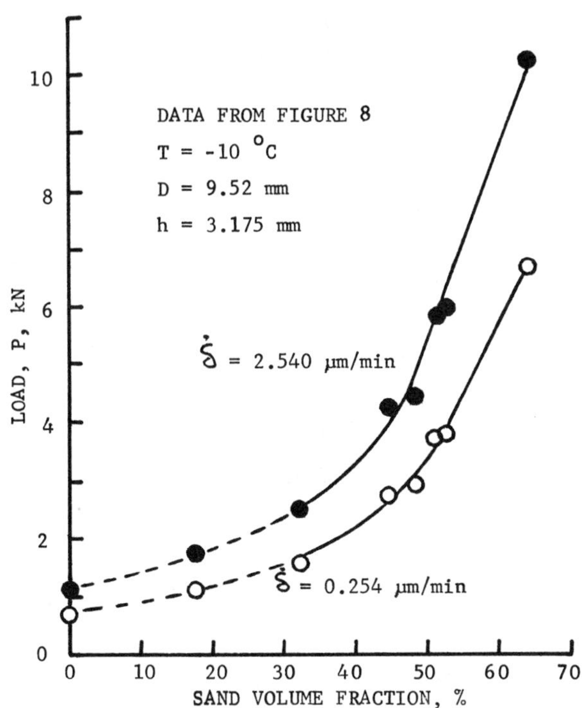

FIGURE 9 Load dependence on sand volume fraction during creep for plain steel 9.52 mm bars with a 3.175 mm lug at -10 ^{o}C.

Beginning at a sand fraction close to 42 %, sand particle interaction with pile surface roughness rapidly increased the pull-out load and adfreeze strength. The curves in Figure 9 clearly show the benefits of using a high density sand slurry mixture during field placement of piles.

CONCLUSIONS

1. Ice adhesion was mobilized at very small displacements with rupture at less than 50 µm for constant displacement rate tests. The presence of sand particles and more surface roughness (lugs) increased the ice matrix volume involved in rupture and thereby significantly increased the adfreeze strength.

2. With an increase in surface roughness (addition of a single lug) load-displacement curves showed the initial rupture of ice adhesion followed by mobilization of mechanical interaction forces. Use of multiple lugs showed an increase in the ultimate pull-out load in a direct ratio to the number of lugs. Suitable lug spacing appears to be dependent on pressure bulb overlap for consecutive lugs and formation of a void space behind lugs.

3. Creep displacement behavior will differ considerably, depending on pile movement. Very small displacements characterize the start of tertiary creep for applied shaft stresses below those required for initial rupture of ice adhesion. Larger displacements and higher shaft stresses are associated with mobilization of mechanical interaction forces.

4. Sand volume fraction (or ice content) significantly influenced the adfreeze strength for a given creep rate, bar and lug size. For ice-rich samples (sand volume fraction < 42 %) the adfreeze strength depends primarily on ice adhesion and behavior of the ice matrix. Mechanical interaction of pile surface roughness with sand particles was important primarily for ice-poor samples (sand volume fraction > 42 %).

5. Methods used to account for temperature effects on the adfreeze strength (or pull-out loads) agreed very well with the experimental data. The creep parameter n was independent of temperature and ice content over the range of creep rates observed.

ACKNOWLEDGMENT

This study was supported by National Science Foundation Grant No. CME-7915784 and by the Division of Engineering Research, Michigan State University. Their support is gratefully acknowledged.

REFERENCES

Alwahhab, M. R. M., 1983, Bond and slip of steel bars in frozen sand, unpublished Ph.D. thesis, Michigan State University, East Lansing, Michigan.

Crory, F. E., 1963, Pile foundations in permafrost, Proceedings--Permafrost International Conference, NRC Publ. 1287: Washington, D.C., National Academy of Sciences.

Weaver, J. S., and Morgenstern, N. R., 1981, Pile design in permafrost: Canadian Geotechnical Journal, v. 18, p. 357-370.

INCREASE WITH DEPTH OF FREEZE-BACK PRESSURES ON THE SIDES OF DRILLHOLES ASSOCIATED WITH FREEZING OF A FLUID IN A CAVITY

V. I. Antipov and V. B. Nagayev

Research Institute of Petrochemical and Gas Industry
Moscow, USSR

Recently, there has been a considerable growth of interest in the problem of secondary freezing of thawed materials around a drillhole. This is associated with the fact that the process of secondary freezing leads to the development of considerable freeze-back pressures on the walls of drillholes, which may lead to their collapse. Analysis of studies of secondary freezing has shown that no satisfactory theory has been advanced thus far which would permit reliable forecasting of a stressed condition in the freezing materials around a drillhole. The present paper suggests an approximate theory for calculating maximum freeze-back pressure σ_{max} on the walls of a drillhole as the materials freeze around it and as drilling mud filtrates freeze in a cavity. Identification of the dependence of σ_{max} on depth is based on the solution of the problem of the axially-symmetrical plane on the development of the freeze-back pressure $\sigma_{(t)}$.

SOLUTION FOR FREEZE-BACK PRESSURE

In this paper an approximate theory is proposed on the basis of permafrost soil and ice equations derived from experiments. Following the work of Dubina (1977) it is easy to obtain the freeze-back pressure on well casings (Figure 1), if the equations of ice state are based on the Mellor and Smith (1966) isotherm and Royen (1922) hypothesis which considers dependence of ice properties upon temperature τ:

$$\sigma_{(t)} = -n(KSv)^{\frac{1}{n}} A^{-\frac{1}{n}} [S_{(t)}^{-\frac{2}{n}} - S_o^{-\frac{2}{n}}] - \frac{\xi K^m \operatorname{Sin}[\pi(1-\lambda)]}{3^{\frac{m+1}{2}} \pi\lambda m S_o^{2m}} \times \int_0^t \frac{d}{d\tau}(S_o^2 - S_{(t)}^2)^m (t-\tau)^{-\lambda} d\tau , \quad (1)$$

where t = time;
d = radius of the casing;
S_o = radius of a cavity;
$S_{(t)}$ = inner radius of the ice-water interface in a cavern;
v = velocity of the ice-water interface;
n, A = the parameters of ice state equation;
k = coefficient of volumetric expansion when ice forms from water;
m, ξ, λ = the rheological parameters of permafrost soil.

To determine the maximum pressure on a casing σ_{max}, which develops if there is complete freezing of mud filtrate in a cavity, it is necessary to assume $S_{(t)} = a$, the radius of the casing and $t = t_a$ in equation 1, where t_a is time of complete freezing of fluid in a cavity.

$$\sigma_{max} = -n(Kav)^{\frac{1}{n}} A^{-\frac{1}{n}} (a^{-\frac{2}{n}} - S_o^{-\frac{2}{n}}) - \frac{\xi K^m \operatorname{Sin}[\pi(1-\lambda)]}{3^{\frac{m+1}{2}} \pi\lambda m S_o^{2m}} \times \int_0^{t_a} \frac{d}{d\tau}(S_o^2 - a^2)(t-\tau)^{-\lambda} d\tau . \quad (2)$$

The calculation of σ_{max} at various depths according to equation 2 gives the values which decrease with the depth, but do not agree with observations. In field experiments Ruedrich and Perkins (1974) observed that the value σ_{max} increases as the depth Z increases. According to the plane axially symmetrical problem just solved, this is an unlikely result.

However, it is taken into account that during freezing of a fluid in a cavity a frozen girt (Figure 1) is formed in the top of the cavity, then a relationship σ_{maxZ} which agrees with the field data may be obtained (Figure 1). This girt does not allow the fluid remaining to flow upwards under pressure and thus pressure builds up in the system. Eventually new and thicker ice layers form up to the well casing, thus increasing the volume of the frozen fluid. This results in continuous compression of the fluid remaining in a cavity, which accumulates stresses transformed to the well wall. So to determine value σ_{max} at a depth Z it is necessary to add value σ_{max} derived from the solution of the axial-symmetrical problem at the same depth to the pressure that has accumulated in the fluid adjacent to the casing in the frozen overlying layers.

Thus, to determine σ_{maxZ} at a depth Z (Figure 2) the pressure $\Delta\sigma_{maxZ}$ adjacent to the well casing where an ice layer is built up and the hydrostatic pressure $\Delta\sigma_r$ corresponding to the difference in the elevation of Z and the elevation of the previous calculation I, should be added to the value σ_{max1}:

$$\sigma_{maxZ} = \sigma_{max1} + \Delta\sigma_{maxZ} + \Delta\sigma_r \qquad (3)$$

A computer-aided numerical solution of the first two quantities of this expression was developed according to the equations of Bulatov (1979) for cavities in the upper part of the profile of the well DS4-6 described by Ruedrich and Perkins (1974). Figure 3 illustrates the resulting theoretical relationship σ_{maxZ}, obtained with equation 3 and the experimental values obtained from transducers on the well casing.

CONCLUSION

In the framework of the model under consideration it is quite natural to expect freeze-back pressure build-up with depth. The comparison of computed freeze-back pressure values calculated with equation 3 with the field experimental data convinces us that the theoretical model chosen is appropriate.

When choosing the optimum technological production conditions of wells under permafrost one should consider build-up of freeze-back pressure according to equation 3 rather than equation 2, as the σ_{max} increases with depth.

REFERENCES

Bulatov, A. I., Pustilnik, Ya., and Vidovski, A. L., 1979, Calculation of pressure on casings at freezing of fluid in a cavity: Neftyanoye khozyaistvo, no. 3, p. 25-27.

Dubina, M. M. et al., 1977, Thermal and mechanical interaction of construction with permafrost soils: Novosibirsk, "Nauka," p. 138.

Mellor, M., and Smith, J. H., 1966, Creep of snow and ice: U.S. Army CRREL Research Report, no. 220.

Royen, N., 1922, Istryck vid temperatur högningar. Hyllning shrift tilägned F. Vilh. Hausen pa sextioarsdagen: Stockholm.

Ruedrich, R. A., and Perkins, T. K., 1974, A study of factors influencing the mechanical properties of deep permafrost: J. Petrol Technology, v. 26, p. 1167-1177.

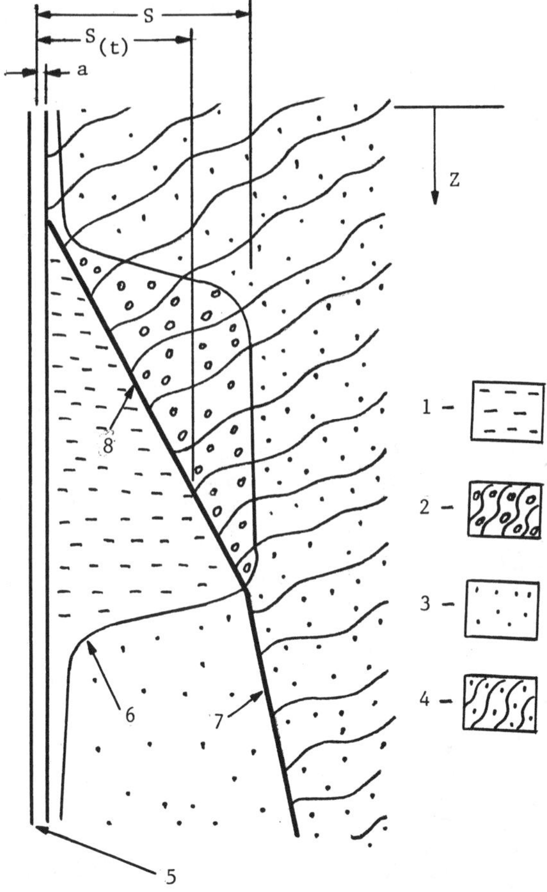

FIGURE 1 Freezing of fluid in a cavity. (1) filtrate, (2) frozen filtrate, (3) thawed soil, (4) frozen soil, (5) a well, (6) boundary of cavity, (7) ice-water interface, (8) girt (ceiling).

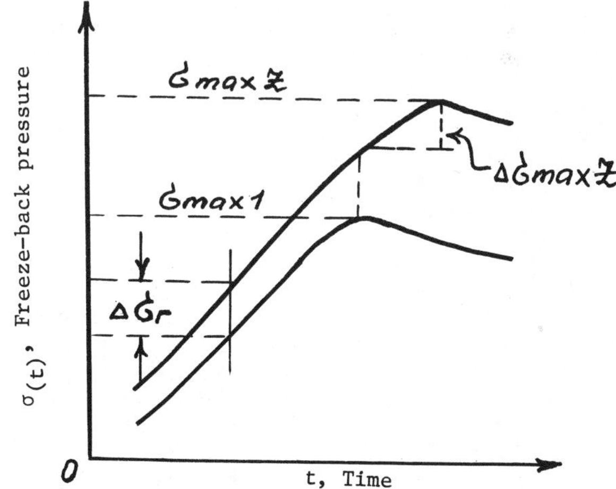

FIGURE 2 Calculation of freeze-back pressure at various depths.

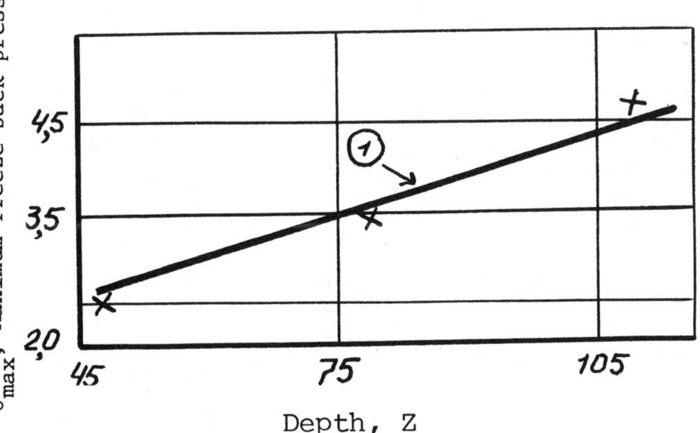

FIGURE 3 Relationship between the maximum freeze-back pressure of the well wall and the depth. 1 - theoretical relationship; XX - experimental values σ_{max} measured by sensors on the well casing.

THERMAL ABRASION OF COASTS

F. E. Are

Permafrost Institute, Siberian Branch
Academy of Sciences, Yakutsk, USSR

Thermal abrasion of the shores of any water body or stream is controlled by the same basic principles and is distinguished by peculiarities dictated by differing hydrological regimes. The main influence of the frozen state of the materials on the development of thermal abrasion is determined by their ice content and reveals itself when the materials slump on thawing. The greater the potential for settlement, the greater is the rate of thermal abrasion and the limiting value of shore retreat. The frozen condition of the materials does not prevent destruction of the shores. In the case of large water bodies the thermal abrasion rate is dictated by the removal of the thawed materials from the shore zone, while in small water bodies and in shores developed in massive ice bodies the limiting factors are the thermal processes. During recent decades climatic cooling has slowed the rate of thermal abrasion by the sea. One of the possible controls on the orientation of thaw lakes is the irregular development of thermal abrasion due to spatial variations in the distribution of sediments prone to thermal abrasion collapse in their shores. Rates of erosion in river banks developed in similar materials are identical whether or not permafrost is present.

Coastal thermal abrasion of water bodies (seas, lakes, storage reservoirs, rivers) is an integrated process of destruction of coasts, composed of frozen materials that follows the same general principles. Destruction is due to the combined hydromechanical and thermal impact of water masses on coasts. The hydromechanical effect is exerted by waves and currents which develop in various combinations on all water bodies. When considering the erosion of coasts of seas and other large water bodies, the destructive effect of waves has commonly been invoked. But choppiness is always accompanied by currents, the velocity of which is of the same order of magnitude as that of river currents. It is these currents that perform major work on wash-out of coasts and longshore drifts. On the other hand, large rivers commonly produce severe roughness which plays an important role in the destruction of their shores. For example, in the area near the mouth of the Lena River such strong gales occur that even lake-class vessels are compelled to take cover. Thus, there is no difference in the principles of erosion among coasts of seas, lakes, reservoirs and rivers. The same situation applies to the thermal effect of water masses on coasts. Just the same is the development of littoral benches, i.e. part of the littoral zone above the water on reservoirs and rivers. Processes of destruction of coasts of different types of water bodies differ only in some features produced by the different hydrological regime. This makes it possible to develop a unified theory of thermal erosion process for water bodies of all types.

Observations made by a number of researchers indicate that even under the most severe climatic conditions, in the spring thermal erosion starts at above-zero temperatures of water within water bodies. The published data on the depth and character of seasonal thawing on the littoral shelf, measurements by this author on the coasts of the Lena, the Aldan, the Vilyui reservoir, and the Laptev Sea, as well as calculations of seasonal thawing depth on the littoral shelf lead us to conclude that from the earliest days of the ice-free period the underwater slopes of coasts attacked by thermal erosion are covered by a layer of unfrozen drift. Therefore, for a wide range of natural conditions thermal erosion results from the wash-out of an unfrozen surface layer. Direct wash-out of frozen ground is observable only in the upper part of the littoral shelf at the time of storms, high water and flash floods. Exceptions are coasts bordered by ice and those of some thermal-karst lakes in the tundra.

We should distinguish between the two kinds of thermal effect of water bodies on permafrost: a direct effect and an effect transmitted through a layer of unfrozen ground. The direct effect leads to a maximum impact. The theory of heat transfer lacks any solution for its calculation. To infer the order of magnitude, one can resort to the formula for computing convective heat transfer at longitudinal flow of a turbulent current along a plate. For the case of minor roughness (ripple 0.3 m high), these formulas give a value of the coefficient of convective heat transfer of the order of 1100 $W/(m^2\ °C)$. With such a coefficient of convective heat transfer and the temperature of water +1°C, a sheer ice precipice would recede about 30 m during three months of an ice-free period (Are 1979a).

O. N. Vinogradov and A. N. Krenke (1964) have determined the rate of ice coast retreat due to thermal erosion on Franz Josef Land (Sedov glacier) to be the difference in observed velocities of the glacier movement and the advance of a vertical ice cliff into the sea in the period between the subsequent separations of icebergs. They have obtained a value of 20 m per year. In the area under consideration, the sea is predominantly ice-free from 2 to 4.5 months in the year, but throughout the summer it abounds in drift ice. The water temperature does not exceed 3°C. The Sedov glacier is descending into a relatively closed bay. Under these conditions strong choppiness is seldom observed (Sukhodrovsky 1967).

Based on laboratory and in situ measurements, Pekhovich and Shatalina (1970), and Khodakov et al. (1978) have derived identical equations of convective heat transfer for the case of a water flow moving along ice plates. The coefficient of convective heat transfer, inferred from these equations for conditions of minor agitation, is 2000-3000 W/(m² °C).

The calculations of W. Morgan and W. Budd (Schwerdtfeger 1979), based on a generalization of ground-based and spacecraft observations of near-Antarctic water temperatures drift velocity and decrease of icebergs in size as these move away from the Antarctic coast, have shown that in the course of three months at a water temperature of +1°C, between 10 and 20 m of ice melt off the underwater surface of icebergs.

The above mentioned results of calculations, experiments and in situ observations indicate that the thawing rate of ice and primarily of exposed frozen ground in contact with water exceeds by a few or many times the observed rates of wash-out of coasts attacked by thermal erosion which in the Arctic seas normally do not exceed 10 m per year (Are 1980). Thus, the potential for ground thawing on a littoral shelf is very great, but because of the presence of a surface layer of unfrozen debris that slows down the thawing it is generally not fully implemented.

Under the severe climatic conditions of the Arctic coast at the subzero annual mean temperature of water, the bottom seasonal thawing near thermal abrasive coasts, composed of thaw-collapsible permafrost, causes the bottom surface to lower substantially. Every year this occurs until the permafrost surface accumulates a layer of thawed and settled sediments, whose thickness is equal to the depth of seasonal thawing. Mathematical simulation, on a computer, of the process in question has been carried out. In selecting the input data, it was assumed that thawing applies to a sea bottom composed of the ice complex (silty sediments of high ice content and a high content of polygonal ice wedges), whose properties are listed in Table 1. The surface temperature of the bottom is specified in two variants (Table 2). Variant A corresponds to warmer sea conditions in which the mean annual temperature of the bottom surface tends to be zero. Variant B roughly corresponds to more severe conditions near the coast of Kotelny Island (Table 3). Under normal conditions, except for extreme variant 1-A, the process under consideration declines rapidly. For the first two years, more than a half of the possible subsidence was completed, and that frequently reached 2 m. The time required for it to reach an equilibrium state seldom exceeds 10 years.

The permafrost temperature effect in the development of thermal abrasion is small because, as soils are thawing, the amount of heat required

TABLE 1 Properties of Ice Complex.

Variant	Ice volume content	Relative thaw subsidence	Heat conductivity coefficient W/(m² °C)		Heat capacity kJ/(m³ °C)	
			State of ground			
			frozen	unfrozen	frozen	unfrozen
I	0.90	0.85				
II	0.78	0.66	2.3	1.6	2050	2950
III	0.57	0.32				

TABLE 2 Temperature of Surface of Bottom and of Permafrost, °C

Variant	Month X-VI	VII	VIII	IX	Temperature of ground
A	-1.4	2.0	6.3	3.1	-11
B	-1.6	1.0	4.4	0.4	-13

for their temperature to rise is generally much smaller than that required for ground ice to melt.

The main influence of a frozen state of soils upon the development of thermal erosion is determined by their ice content, and it manifests itself when ice content exceeds the porosity in an unfrozen state, that is, when soils involve thaw settlement. This effect implies two aspects.

On one hand, the volume of sediments coming into the water body when the coast composed by thaw collapsible ground is being destroyed, is smaller than that resulting from destruction. This diminishes the work of water bodies, required for them to wash out and transport soils, in comparison with coasts composed of similar thaw-uncollapsible ground.

On the other hand, thawing of collapsible soils forming the bottom is attended by a deepening of the water body and, therefore, by an increase of wave energy. In the absence of agitation, thermal settlement of the bottom itself may cause coastal recession.

For that reason, other conditions being equal, with increasingly excess ice content, there is an increase of the rate of thermal erosion and of the limiting value of coastline recession.

If ice content of soils composing the coast exceeds a critical value, the coast then recedes continuously. Critical ice content is that ice content in soils at which their complete thawing produces a settlement of the ground surface down to the water level in the water body (G_b). It is matched by the critical level of the water (G_k), which coincides with the surface level of soils constituting the coast, following their complete thawing and settlement. The possibility of

TABLE 3 Results of Calculations of Perennial Seasonal Thawing Below Sea Bottom (Limiting State)

Variant	Thawing depth, m	Thickness of unfrozen layer after subsidence, m	Bottom subsidence, m	Time required for thawing to reach limiting state, years
I-A	4.9		4.2	17
II-A	2.2	0.7	1.5	9
III-A	1.1		0.4	9
II-B	1.5	0.5	1.0	4
III-B	0.75		0.25	4

unlimited destruction of the coast is determined by the relation $G_b \gtrless G_k$. If its left-hand side is less than the right-hand, coastal recession has a limit, and, if vice versa, recession becomes continuous, and one of the main principles of the theory of development of sea shores is violated, that at constant sea level erosion is restricted by the development of a limiting profile of equilibrium.

Perennially frozen unconsolidated deposits differ by having greater content of silt particles than their unfrozen counterparts. Particularly high silt content occurs in the ice complex of the Yana-Indigirka depression and the Kolyma-Alazeya shoreland where the predominant silt content amounts to 60-80%, and the maximum is around 95%. Silt content exerts the same influence upon development of thermal erosion as the ice content of thaw-collapsible soils because fine particles are easily suspended and transported by water (Are 1980). As for large shallow water bodies, the shores of which are formed by permafrost of high silt content, wash-out and transport of bottom sediments develop throughout the water area, and the concept of a shore slope below the water is virtually invalid here.

Perennially frozen unconsolidated deposits are able to withstand the destructing impact of water masses only in a frozen state. Once unfrozen, its wash-out capacity is equal to or greater than that of unfrozen soils of similar composition.

Calculations of the shore wash-out coefficient, according to Kachugin (1959) for several areas of sea and lake shores attacked by thermal erosion are formed of an ice complex, have shown that the values of the wash-out coefficient, inferred by neglecting thaw settlement, are in excess of $300 \cdot 10^{-9} m^3/yr$. Thus, the ice complex, or permafrost with the highest ice content, is, according to Kachugin's classification, a soil readily yielding to wash-out. Its wash-out rate is equal to or larger than that of unfrozen soils of similar composition: fine and silty sands, loess-like sandy loam and loam (Are 1979a). Permafrost wash-out rate depends on its cryogenic texture. For instance, in the case of net-like texture it is one order of magnitude higher than for a massive one (Ershov et al. 1979).

The results of the investigations discussed so far indicate that, in the case of destruction of shores attacked by thermal erosion, ground thawing occurs before wash-out. Consequently, the frozen state of soils, in all practicality, does not hinder shore destruction. For large water bodies, the rate of thermal erosion is limited by wash-out of unfrozen products of littoral zone destruction, rather than by thermal processes. Therefore, agitation on large water bodies plays the same role in the development of shores as outside the permafrost zone. Thermal-erosional shores of small bodies of water, with a water level exceeding a critical one, low energy of agitation, and weak currents recede mainly due to thermal destruction. This is the least intensive type of thermal abrasion.

The influence of the ice cover of water bodies upon thermal-erosional shores leads, in the main, to preventing their destruction. In this case, the most important role is played by the duration of the ice cover, and in seas also by the amount of drift ice in summer. The destructive effect of ice on shores is not great and does not substantially influence the shoreline position.

Thermal erosion is defined by its irregular discontinuous development. Its maximum rate is observed for extremely high water levels. In the seas, particularly strong storms, accompanied by a precipitously high level storm surge, produce shore destruction that would require many years to accomplish at normal rates of erosion.

The rate of retreat of thermal-erosional sea shores varies greatly in space and time. Even the values averaged over 20-25 years are most irregularly distributed along the littoral. There are no regular differences in retreat rate of thermal-erosional shores formed by yedoma and thaw-lake sediments.

The exceptionally rapid retreat of thermal-erosional sea shores, with a rate of 100 m or more per year, is generally recognized. In the last 150 years, the maximum rate of thermal abrasion was observed in 1944-1946 on Semenovsky Island in the Laptev Sea to be at least 55 m per year. The average long-term rates are far lower. For shores formed by fine-grained soils of high ice content (ice complex, alas sediments, etc.) the maximum mean rate from many years of observations is roughly 10 m per year, while the predominant mean values for shores of mainlands and large islands vary from 2 to 6 m per year. The mean estimated thermal abrasion rates along the large lengths of the coast (tens of kilometers) usually do not exceed 2 m per year.

The rate of thermal erosion of sea shores formed

by ice-rich, fine-grained soils, under comparable conditions seems to exceed by 3-4 times that of shores formed by unfrozen ground of similar lithologic composition. But annual rates of shore retreat are close to each other because during most of the year the Arctic Seas are ice-covered.

The last decades have witnessed a tendency of the thermal erosion rate to decrease. A likely cause of this is the increase of drift ice amount in the Arctic Seas, as a consequence of the fall of temperature since 1940.

One of the main features of thermal erosion of lake shores is its dependence on dimensions of water area. This dependence is different for shores formed by soils of different ice content. If ice content is much lower than a critical one (thaw uncollapsible ground), shore wash-out can then develop for a geologically long time only for very large lakes, the hydrological regime of which is close to that of seas. Therefore, most of the lakes of this type have stable shores. If ice content of shore-forming soils exceeds a critical one, then thermal erosion proceeds irrespective of the water area dimensions. These factors have the implication that thermal erosion is presently observed mainly for thermokarst lakes. This has, in particular, been pointed out by Tomirdiaro and Ryabchun (1973).

Thermal-erosion induced destruction of the shores of thermokarst lakes can develop irregularly due to the different force and duration of differently directed winds. This non-uniformity is believed to be one of the possible reasons for the formation of oriented lakes. But its influence depends on some other factors, discarded so far when considering the problem of oriented lakes. These include wash-out and thaw-collapsible rates of soils, the thickness of thaw-collapsible ground and its hypsometric position with respect to the water level. These factors can determine a given orientation of thermokarst lakes if they obey regular spatial variations and conversely, they can be responsible for the diversity in shape of lakes if their spatial variations have a random character.

The mean rates of many years of wash-out of the banks of the Lena River have been determined for the last 18-26 years (Are 1979). In the middle flow on a floodplain the total average rate of bank retreat is 6.5 m per year. The rate of retreat of steep banks of principal river branches amounts to 19-24 m per year. The front parts of islands are washed out with a rate of up to 40 m per year. The rate of wash-out of a sand terrace above floodplain, 25-30 m high, reaches 11 m per year in the area of the steep bank of one of the principal river branches.

In the Olenek channel of the Lena River delta the banks composed of muskegs up to 9 m thick, split by wedge ice and supported by a sand base, retreat at a mean rate of 1.7 m per year. In the sharpest curves the maximum rate is as high as 6 m per year.

Comparison of wash-out observations of the banks of the Ob River (Zemtsov and Burakov 1972, Trepetsov 1973) and the Lena River has shown that under quite similar hydrological conditions the floodplain and low terraces above the floodplain, formed of sand and sandy-loamy unfrozen and perennially frozen thaw-uncollapsible soils, are washed out approximately with the same intensity.

Under comparable conditions, the annual values of retreat of river banks of the permafrost zone can be smaller than those outside it because of the short duration of active wash-out due to the peculiarities of the hydrogeological regime of rivers in the permafrost zone. These peculiarities are most pronounced in small plainland rivers, fed by atmospheric precipitation.

One of the possible reasons for the abnormally broad river valleys in the permafrost zone is the thaw-collapsibility of soils of the upper part of the geological cross-section. The influence of this effect decreases with the deepening of the valley because the ice content of soils drops off with depth.

REFERENCES

Are, F. E., 1979, The influence of a frozen state of soils upon wash-out of the river banks of the permafrost zone, in Regional and cryolithological investigations in Siberia, Yakutsk: IM SO AN SSSR, pp. 107-117.

Are, F. E., 1979a, Water bodies under conditions of permafrost. Abstracts of papers in All-Union Scien.-Tech. Meeting on Dynamics of Shores of Water Bodies, their Protection and Rational Utilization, in Book 6, Introductory Papers, Cherkassy: Ukrain. Branch of CNIIKIVR, pp. 69-82.

Are, F. E., 1980, Thermal erosion of sea shores, Moscow: Nauka, 159 p.

Ershov, E. D., Kuchukov, E. Z., and Malinovsky, D. V., 1979, Wash-out rate of frozen ground and criteria for evaluating thermal erosion hazards of an area, "Vestnik Mosk, un-ta," ser. geol., no. 3, pp. 67-76.

Kachugin, E. G., 1959, Engineering and geological explorations and predictions of the development of shores of water bodies, in Recommendations on the study of the development of water body shores, Moscow: Gosgeoltekhizdat, pp. 3-89.

Khodakov, V. G., Gordeichik, A. V., and Moiseeva, G. P., 1978, On the rate of ice thawing in the water, "Materials of glaciological research," v. 33, Moscow: Inter-Departmental Geophysical Committee of the USSR Academy of Sciences, pp. 196-200.

Pekhovich, A. I., and Shatalina, I. N., 1970, Thawing of ice under conditions of forced convection. Proceedings of Coordination Meetings on hydraulic engineering, v. 56, Leningrad.

Tomirdiano, S. V. and Ryabchun, V. K., 1973, Lacustrine thermokarst on the Lower-Anadyr lowland, in Internat. Conf. on Permafrost. Reports and Communications, v. 2, Yakutsk: Kn. izd-vo, pp. 58-67.

Trepetsov, E. V., 1973, Deformation of the banks of the Ob River in the Tyumen Region, in Soil erosion and river bed processes, v. 3, Moscow: Izd-vo Mosk. un-ta, pp. 276-284.

Schwerdtfeger, P., 1979, Review on icebergs and their uses, "Cold Regions Science and Technology," v. 1, no. 1, pp. 59-79.

Sukhodrovsky, V. L., 1967, Relief-formation in

periglacial conditions, Moscow: Nauka, 120 p.
Vinogradov, O. N., and Krenke, A. N., 1964, Morphology and evolution of ice shores (from exploration on Franz Josef Land), "DAN," v. 155, no. 4, pp. 795-798.
Zemtsov, A. A., and Burakov, D. A., 1972, Lateral erosion of the Ob River and the potential for its prediction, "Geomorfologiya," no. 4, pp. 61-64.

PREDICTING FISH PASSAGE DESIGN DISCHARGES FOR ALASKA

William S. Ashton and Robert F. Carlson
Institute of Water Resources, University of Alaska,
Fairbanks, Alaska 99701 USA

Improper placement of highway culverts may selectively or totally block fish migration, thereby decreasing available spawning and rearing habitat. Blockage will occur with a combination of excessive culvert water velocities, lack of fish resting areas upstream and downstream of the culvert, and scour at the culvert outlet, creating perched conditions. Previous studies of fish passage culverts have determined fish swimming abilities and profiles of culvert water velocity. There are limited studies of the hydrologic relationship among frequency, duration, season, and magnitude of discharge for the design of fish passage through culverts in northern regions. We analyzed streamflow records from 14 gaging stations in south central, interior, and arctic Alaska (drainage area <260 km^2) to determine the highest consecutive mean discharge with 1, 3, 7, and 15 day durations. Streamflow during three periods were analyzed: spring, April 1 to June 30; summer, July 1 to August 31; and fall, September 1 to November 30. The Lognormal distribution, using the Blom plotting position, predicted flood frequency values. Regionalization, with the index-flood method, of single station values provides a method for predicting discharges from ungaged drainage basins. Regressions developed to predict the 2-year return period discharge found that the significant basin characteristics were basin area and forest cover in the spring period, and basin area, forest cover, and mean annual precipitation in the summer and fall periods.

INTRODUCTION

Four criteria must be considered for effective and practical design of highway culverts to pass fish: the flow regime of the stream, the hydraulic properties of the culvert (i.e., shape, roughness, or length), the swimming abilities of the fish species and age classes present, and the time of year of fish movement and migration in the stream. Understanding the flow regime is important for determining the relationship among the frequency, duration, season, and magnitude of flow. The frequency is important to understanding the risk or probability that a given magnitude of flow will occur. The duration of time for which a given magnitude of flow is exceeded provides the time a fish species might be delayed in its normal migration. The season of the flow tells whether a given magnitude flow will occur during a critical period in the life cycle of a fish species. This three-dimensional representation provides a more detailed description of the flow regime than the highest annual instantaneous discharge.

Development of arctic oil and gas resources in the United States and Canada has provided an impetus to study the effects of highway culverts on fish passage in northern latitudes (Dryden and Jessop 1974, MacPhee and Watts 1976, Katopodis et al. 1978, and Elliott 1982). Prevention of extended delay of fish in their upstream migration is an important design criterion. Delay of spawnable fish can cause them to spawn at less suitable spawning sites, affecting spawning success and causing stress and physical damage. This makes them more vulnerable to disease and predation (Dryden and Stein 1975). Dryden and Stein estimate a maximum delay of 3 days during the mean annual flood may be tolerated by a spawning run without serious biological consequences. During the design flood, with a return period of 50 years, they recommend a maximum delay period for spawners of 7 days. These maximum delay periods have been recommended in culvert designs for the MacKenzie Highway in the Northwest Territories (Dryden and Stein 1975). The State of Alaska Office of the Pipeline Coordinator (State Pipeline Coordinator's Office (SPCO) 1982) design requirements for the proposed Alaskan Northwest Natural Gas Transportation System specify use of the instantaneous mean annual flood. Allowable delay time is not mentioned. During 3 years of study on Popular Grove Creek, Alaska, spring upstream migration of arctic grayling (*Thymallus arcticus*), the primary design fish in interior Alaska, preceded the spring peak discharge in 1973, followed it in 1974, and occurred about the same time in 1975. Consideration must be given to spawning migrations of other species occurring during the summer and fall (Watts 1974). Morrow (1980) describes the periods of spawning migration generally as: arctic grayling in the spring, chinook and sockeye salmon during the summer, and Dolly Varden and coho salmon during the fall. To predict the effect of high flows on fish passage through culverts, knowledge of the flow regime during these fish passage periods is required.

Regionalization of single station discharge data is a technique for predicting discharge from ungaged drainage basins. Of the methods available, empirical approaches such as the index-flood method are practical in regions with few long-term stations (Benson 1962). One of the problems with the index-flood method is the effect of drainage basin size on the slope of the frequency curve. To reduce this effect, only basins smaller than 260 km^2 are considered. To account for the differently timed spawning migrations of different species, three periods of analysis are used: spring, summer, and fall. Since most of the concern for fish passage is at low return period flows, only flows with return periods of 20 years or less are used.

METHODS

We used streamflow data from continuously recording U.S. Geological Survey gaging stations in the hydrologically similar area (Area II) defined by Lamke (1979; Figure 1). Stations

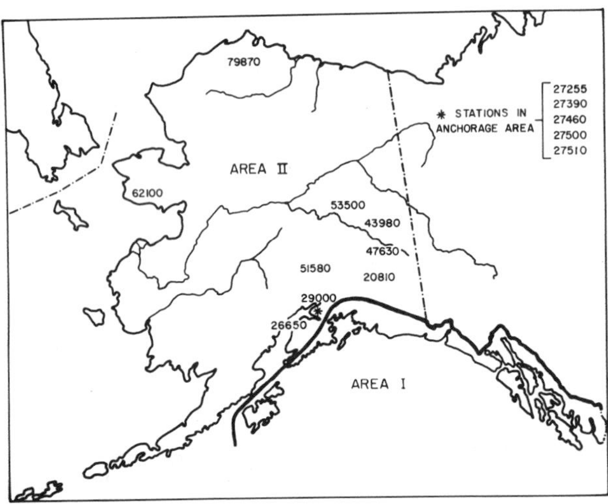

FIGURE 1 Areas of Alaska used in the flood-frequency analysis. Station numbers show approximate location of the gaging stations used in this paper (from Lamke, 1979).

within this region were deleted from further consideration if the basin area was greater than 260 km^2, 15% or more of the basin area was covered by glaciers, the streamflow was regulated, or there were less than 5 years of record as of November 1981. Aleutian Island stations, although within Lamke's region definition, were deleted from consideration. Outliers and stations with zero flow periods are treated as described in Kite (1977). Three periods of the year were selected for streamflow analysis: spring, April 1 to June 30; summer, July 1 to August 31; and fall, September 1 to November 30. The highest consecutive mean discharge with durations of 1, 3, 7, and 15 days were computed for each period. Predicted discharge values were computed using a Canadian flood frequency program (Condie et al. 1976). Four frequency distributions are available with this program: Gumbel (Extreme Value Type I), Lognormal, Three-Parameter Lognormal, and the Log-Pearson Type III. Each of the distributions uses the maximum likelihood method of fitting. In the event a true solution is not found by this method, a moment fit is used. Condie et al. revised the program in 1981 to replace the Weibull plotting position formula with a generalized plotting position formula developed by Adamowski (1981). The Lognormal distribution was selected for further analysis of the streamflow data. The program was modified to provide the 95% confidence limits for each predicted value, and to use the Blom plotting position formula

$$F_i = (i - 0.375)/(N + 0.25), \qquad (1)$$

where F is the plotting position of ith smallest as a probability value of the flow being less than the associated flow, i is the serial number of ith smallest in a sample size N, and N is the sample size (Cunnane 1978).

We regionalized the single station data using the index-flood method as modified for use with computers, the Lognormal distribution and computation of confidence limits (Coller 1963, as cited in Kite 1977). The mean annual flood was defined by the predicted flood with a probability of occurrence of 0.50. Only stations with 10 years or more of record were used. If a station had 10 years of record during at least two of the three periods, the station was included in the analysis, otherwise it was dropped from further consideration. Dimensionless flood ratios were computed by dividing predicted floods with return periods of 2, 5, 10, and 20 years by the predicted 2-year discharge. Floods with return periods higher than 20 years were not considered. Dimensionless ratios were computed for the four durations within the three periods, and median ratios were selected for dimensionless plots. Confidence limits for each ratio were determined by dividing the upper and lower 95% confidence flood value at each return period by the predicted 2-year return period discharge. Median values were selected from these ratios and plotted as confidence limits on the flood ratio plots. Multiple regression equations were developed using basin characteristics and climatic parameters to predict the 2-year discharge on ungaged drainage basins. Regression coefficients were computed using an HP 9845 stepwise regression program. Due to limited data, a maximum of three independent variables were considered for each equation. A variable was included in the model if it was significant at the 5% level and increased the R^2 by at least 5%. Variables considered in regression analysis were drainage area, mean annual precipitation, basin storage area, forest cover, glacier area, mean minimum January temperature, mean basin elevation, and mean annual snowfall.

RESULTS

Nineteen gaging stations met the criteria of basin size, percent of area as glaciers, and length of record. A base period of 1965-1981 was selected, reducing the number of stations to 14 (Table 1). Dimensionless flood ratios were computed and plotted with 95% confidence limits (Figure 2). The 2-year return period flood for ungaged basins is predicted using a regression

TABLE 1. Gaging Stations.

Station No.	Station Name	Permafrost Zone[1]	Drainage Area (km^2)	Period of Record
15208100	Squirrel Cr. at Tonsina	Discontinuous	182.60	7/65-9/75
15266500	Beaver Cr. near Kenai	Discontinuous	132.00	10/67-9/78
15272550	Glacier Cr. at Girdwood	Discontinuous	160.60	8/65-9/78
15273900	SF Campell Cr. at canyon mouth near Anchorage	Discontinuous	65.30	10/66-9/81
15274600	Campell Cr. near Spenard	Discontinuous	180.50	6/66-11/81
15275000	Chester Cr. at Anchorage	Discontinuous	51.80	4/65-9/76
15275100	Chester Cr. at Arctic Blvd. at Anchorage	Discontinuous	70.40	6/66-9/81
15290000	Little Susitna River near Palmer	Discontinuous	160.30	4/65-9/81
15439800	Boulder Cr. Central	Discontinuous	81.10	5/66-9/81
15476300	Berry Cr. near Dot Lake	Discontinuous	168.60	7/71-9/81
15515800	Seattle Cr. near Cantwell	Discontinuous	93.80	10/65-9/75
15535000	Caribou Cr. near Chatanika	Discontinuous	23.80	10/69-9/81
15621000	Snake River near Nome	Discontinuous	222.00	9/65-9/81
15798700	Nunavak Cr. near Barrow	Continuous	7.23	10/71-9/81

[1] after Brown and Péwé.

equation with the form

$$Q_{mn} = aA^bP^c(F + 1)^d, \qquad (2)$$

where

Q_{mn} = dependent variable, the peak mean discharge for the mth period, where s is spring, su is summer, and f is fall, and the nth duration where 1 is 1 day, 7 is 7 days, and so forth, m^3/s;
a = regression constant;
b, c, and d = regression coefficients for the independent variables (basin characteristics);
A = drainage area, km^2;
P = mean annual precipitation, mm;

(F + 1) = percent forest cover, expressed as a whole number.

The regression coefficients, with their associated percent standard error, are given in Table 2. Basin characteristics used in the analysis are from Lamke (1979) and SPCO (1981). Basin characteristics not included in these publications are taken from the sources cited by Lamke.

DISCUSSION

Regionalization of single station data illustrates the hydrologic complexity of small basins and regions with sparse data networks. Hydrologic similarity among drainage basins is required for regionalizations. In practice, however, the degree of dissimilarity permitted within a homogenous region is not well defined (Molsey 1981). The regionalization by Lamke (1979) determined significant basin characteristics in the flood hydrology of Alaskan streams. Among these are the mean minimum January temperature and percent forest cover. These characteristics are believed to be indirect indicators of the presence or absence of permafrost. As more data become available, refinements are made of these initial assumptions. With five additional years of streamflow data, Janowicz and Kane (1983) and SPCO (1981) have determined, for instantaneous annual peak flows, that Lamke's Area II can be subdivided. Because of insufficient data and the generalized nature of the data we do have, we cannot determine the effect of permafrost on

TABLE 2. Regression Coefficients for Predicting 2-Year Return Period Discharge.

Dependent Variable Q_{mn}	Regression Constant $a \times 10^{-5}$	Regression Coefficients			Percent Standard Error
		b	c	$d \times 10^{-3}$	
Q_{s1}	9,400	1.097	--	-9.1	31
Q_{s3}	8,000	1.107	--	-9.4	29
Q_{s7}	6,200	1.128	--	-9.5	28
Q_{s15}	260	0.884	0.643	-8.8	22
Q_{su1}	4.88	0.764	1.310	-6.9	23
Q_{su3}	4.64	0.770	1.294	-7.2	21
Q_{su7}	3.94	0.758	1.303	-7.3	19
Q_{su15}	2.46	0.776	1.333	-7.1	18
Q_{f1}	1.22	0.627	1.559	-5.8	14
Q_{f3}	1.30	0.667	1.495	-5.7	13
Q_{f7}	1.40	0.696	1.437	-5.6	12
Q_{f15}	1.64	0.722	1.370	-5.5	13

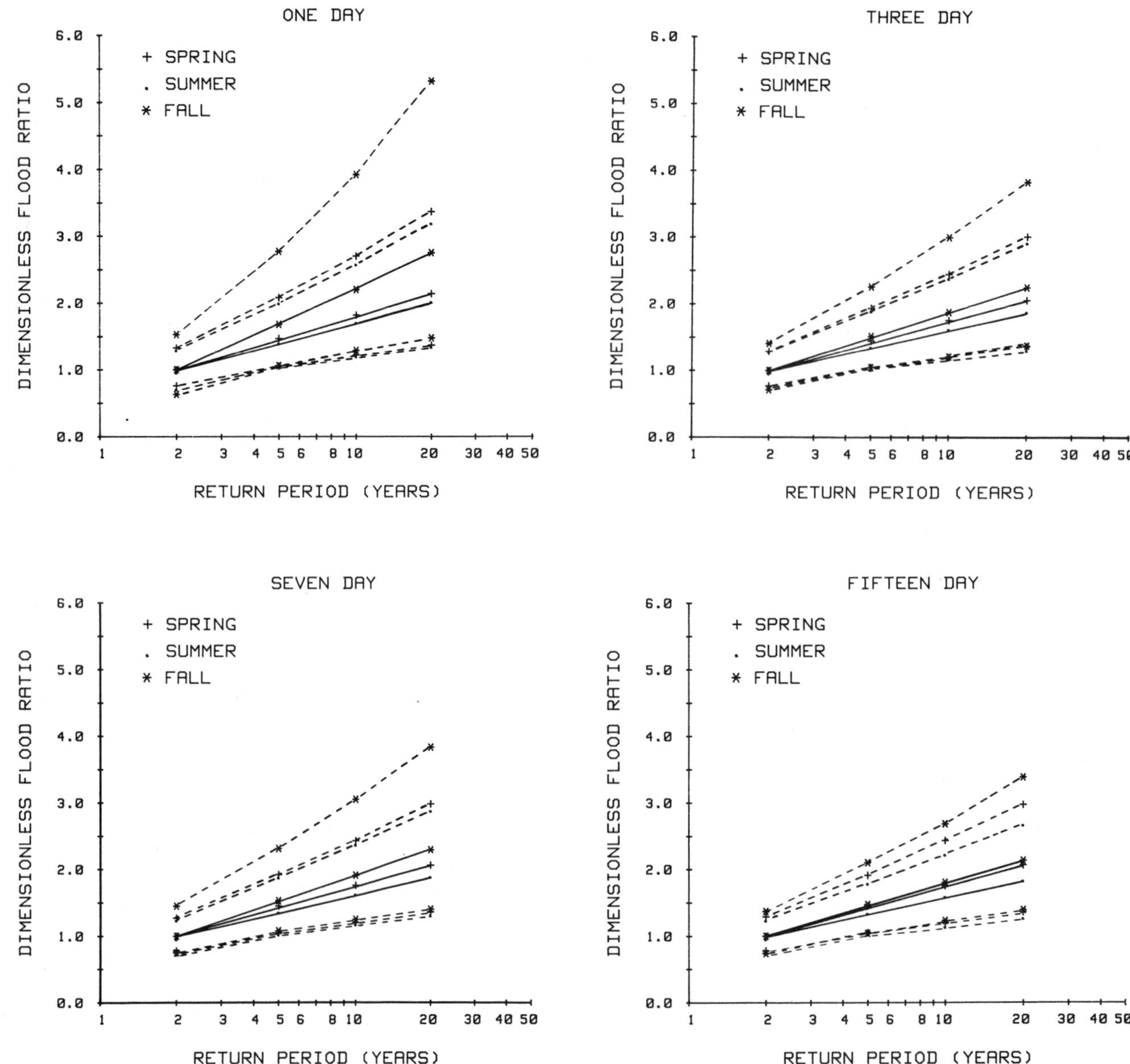

FIGURE 2 Regional fregional curves for the four durations. The predicted line (*———*) for the three periods is shown with the respective 95% confidence limits (*-----*).

arctic and subarctic statistical flood hydrology.

The regionalization presented in this paper predicts flood magnitudes for small drainage basins. These flood magnitudes can be predicted for the season of the year, flow duration, and frequency of occurrence of interest. This method provides reasonable estimates for watersheds with the following limitations: the basin area should be <260 km^2 and glacier cover should be <15% of basin area; predicted spring 1-day flows can exceed Lamke's predictions for some combinations of basin characteristics; and since spring flow predictions for streams in the Anchorage bowl are overestimated, multiply estimate by 0.4 (Lamke recommends 0.5 for instantaneous peak flows regardless of season).

For the design of culverts for fish passage, Watts (1974) recommends use of the probable discharge at the time of fish migration. Timings of fish migrations vary with species and geographic location. Of the fish passage culvert design methods (Watts 1974, and Dryden and Stein 1975) that recommend a fish passage discharge less than the instantaneous annual peak, none provide a method for predicting discharges on ungaged watersheds. The regionalization described in this paper provides culvert designers with a method for predicting the fish passage discharge on ungaged watersheds in Alaska.

ACKNOWLEDGMENTS

This paper presents results from research funded by the Federal Highway Administration and administered by the Alaska Department of Transportation and Public Facilities, project number F36202. The authors are grateful to D. Thomas for answering our statistics questions.

REFERENCES

Adamowski, K., 1981, Plotting formula for flood frequency. Water Resources Bulletin, v. 17, p. 197-202.

Benson, M. A., 1962, Evolution of methods for evaluating the occurrence of floods: Washington, D.C., U.S. Geological Survey, Water Supply Paper 1580-A, A1-A29.

Brown, R. J. E., and Pewe, T. L., 1973, Distribution of permafrost in North America and its relationship to the environment: A review, 1963-1973: In Permafrost-Second International Conference, Yakutsk: Washington, D.C. National Academy of Sciences.

Condie, R., Nix, G. A., and Boone, L. G., 1976, Flood damage reduction program flood frequency analysis: Ottawa, Engineering Hydrology Section, Inland Waters Directorate.

Cunnane, C., 1978, Unbiased plotting positions -- A review: Journal of Hydrology, v. 37, p. 205-222.

Dryden, R. L., and Jessop, C. S., 1974, Impact analysis of the Dempster Highway culvert on the physical environment and fish resources of Frog Creek: Winnipeg, Resource Management Branch, Fisheries and Marine Service.

Dryden, R. L., and Stein, J. N., 1975, Guidelines for the protection of the fish resources of the Northwest Territories during highway construction and operation: Winnipeg, Resource Management Branch, Fisheries and Marine Service.

Elliott, G. V., 1982, Final report on the evaluation of stream crossings and effects of channel modifications on fishery resources along the route of the trans-Alaska pipeline: Anchorage, Special studies, U.S. Fish and Wildlife Service.

Janowicz, J. R., and Kane, D., 1983, Peak runoff comparisons for watersheds with varying amounts of permafrost. In: Proceedings of the Fourth International Conference on Permafrost (in press): Washington, D.C., National Academy of Sciences.

Katopodis, C., 1977, Design of culverts for fish passage. In: Proceedings of the Third National Hydrotechnical Conference: Quebec, Canadian Society for Civil Engineering.

Katopodis, C., Robinson, P. R., and Sutherland, B. G., 1978, A study of model and prototype culvert baffling for fish passage: Winnipeg, Western Region, Fisheries and Marine Service.

Kite, G. W., 1977, Frequency and risk analyses in hydrology: Fort Collins, Water Resources Publications.

Lamke, R. D., 1979, Flood characteristics of Alaskan streams: Anchorage, U.S. Geological Survey, Water Resources Investigations, 78-129.

MacPhee, C., and Watts, F. J., 1976, Swimming performances of arctic grayling in highway culverts: Anchorage, Final Report to U.S. Fish and Wildlife Service.

Molsey, M. P., 1981, Delimitation of New Zealand hydrologic regions: Journal of Hydrology, v. 49, p. 173-192.

Morrow, J. E., 1980, The freshwater fishes of Alaska: Anchorage, Northwest Publishing Company.

State Pipeline Coordinator's Office, 1981, A regional flood frequency analysis for that portion of the proposed Alaska Natural Gas Transportation System Route in Alaska: Fairbanks, Alaska Department of Natural Resources.

State Pipeline Coordinator's Office, 1982, Analysis of acceptable grayling passage velocities for use in culvert design along the Alaska Northwest Natural Gas Transportation Service Company route: Fairbanks, Alaska Department of Natural Resources.

Watts, F. J., 1974, Design of culvert fishways: Moscow, Idaho, Water Resources Research Institute, University of Idaho.

PERIGLACIAL GEOMORPHOLOGY IN THE KOKRINE-HODZANA HIGHLANDS OF ALASKA

Palmer K. Bailey

Department of Geography and Computer Science
United States Military Academy, West Point, New York 10996 USA

The portion of the Kokrine-Hodzana Highlands near the Arctic Circle and adjacent to the Dalton Highway is an easily accessible area containing a diverse assortment of periglacial features. The region is within the discontinuous permafrost zone and probably has never been glaciated. Bedrock is principally metamorphic with major intrusions of porphyritic granites and some mafic and ultramafic rock. The nature of the bedrock profoundly affects the type and character of the local landforms. Finely jointed and foliated bedrock tends to promote development of cryoplanation terraces. Terraces are most fully and distinctly developed on metamorphic bedrock, especially schist and phyllite, at elevations above 600 m. Terraces are usually obscure or absent on igneous bedrock. Widely jointed massive bedrock tends to form the best tors. Of the 128 tors physically inspected in the field, 88 are composed of highly weathered porphyritic granite distributed over a wide range of elevations. Both the diameter and the particle size of sorted polygons are likewise determined in part by bedrock type. Other periglacial landforms include felsenmeer, sorted circles, ice-wedge polygons, thermokarst depressions,open-system pingos, extensive gelifluction slopes, and anthropogenic mounds.

INTRODUCTION

The portion of the Kokrine-Hodzana Highlands adjacent to the Dalton Highway is an easily accessible area containing a diverse assortment of periglacial features. During the summer of 1982 an area approximately 6 km wide from a point 39 km north of the Yukon River extending 85 km northward along the highway to Gobbler's Knob was investigated (Figure 1). The area is astride of the Arctic Circle and is within the discontinuous permafrost zone. Permafrost, where present, is generally found between 0.5 and 3.0 m from the surface (Kachadoorian 1971). Drill holes associated with construction in the area often bottomed in permafrost at depths up to 20 m. The maximum depth of the permafrost is unknown. The area is beyond the maximum extent of the last major glaciation and has probably never been glaciated (Hamilton 1982). Bedrock of the region is principally Precambian and lower Paleozoic metamorphic rocks with large intrusions of Cretaceous porphyritic granite. There are also smaller intrusions of Mesozoic mafic and ultramafic rocks (Kachadoorian 1971, Brosgé and Patton 1982). The mean annual temperature at the tree line on Gobbler's Knob is approximately -3°C (Haugen 1982). Mean temperatures at lower sites in the region such as the Bonanza Creek, Fish Creek, and Kanuti River valleys are probably significantly colder because of persisting winter temperature inversions. The coldest official temperature in Alaska, -62°C, was recorded in 1971 at Prospect Creek, only 5 km north of Gobbler's Knob.

CRYOPLANATION LANDFORMS AND TORS

Cryoplanation has been an important process in the development of landforms in this region. Excellent examples of ridgecrest, hillside, and summit cryoplanation terraces (Reger 1975) are visible from the Dalton Highway. Terraces are most fully and distinctly developed on metamorphic bedrock, especially schist and phyllite. Terraces are usually obscure or entirely absent on igneous bedrock. The majority of the residual rock tors, however, consist of igneous rocks. There is an average of 0.13 tors per km^2 in metamorphic bedrock and 0.46 tors per km^2 in granite. The distribution of tors shown in Table 1 tends to indicate that their occurrence is much more closely related to bedrock type than to elevation. The bimodal distribution of granite tors is simply a reflection of the mean elevation of two large outcrop areas, Finger Mountain and Bonanza Creek.

TABLE 1 Elevation and Composition of Tors Near the Dalton Highway in the Kokrine-Hodzana Highlands, Alaska. MN = Metamorphic Nonfoliated, MF = Metamorphic Foliated, IG = Igneous Granite, IM = Igneous Mafic or Ultramafic.

	Bedrock Type			
	MN	MF	IG	IM
Percent of Area	<1	62	32	5
Number of Tors				
800-900 m	-	5	-	-
700-800 m	-	4	10	-
600-700 m	4	4	26	-
500-600 m	5	16	16	-
400-500 m	2	-	10	-
300-400 m	-	-	23	-
200-300 m	-	-	3	-
Totals	11	29	88	0

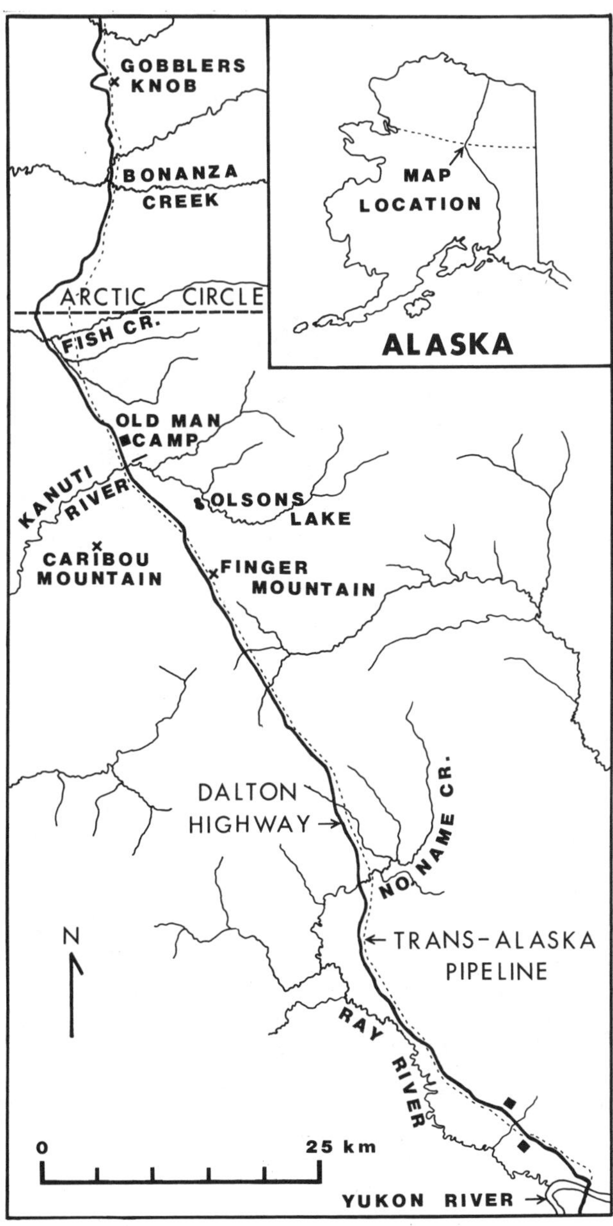

FIGURE 1 Location map.

Features on Igneous Bedrock

Of the 128 tors physically inspected in the field, 88 are composed of a highly weathered porphyritic granite containing large phenocrysts of feldspar (Cook and Bailey 1983). At many of the sites the weathering has been so intensive that rock specimens may be crushed in the hand. The deepest sample obtained during this study was from the base of a 7 m high quarry face near Bonanza Creek. This sample crumbled when tapped with a hammer. Drilling in the area has revealed weathered bedrock to a depth of up to 16 m. There is no apparent correlation between depth of weathered bedrock and permafrost conditions. The weathered rock and the abundant grus in the region may be the products of either frost weathering or chemical disintegration by hydration of biotite (Reger, personal communication). Deep weathering must have occurred under conditions different than the present environment. The permafrost existing in most of the area inhibits both mechanical and chemical weathering below the active layer. Linton (1955) proposed a two-stage hypothesis of tor formation requiring deep chemical weathering during a warm climatic period followed by removal of weathered products by gelifluction under periglacial conditions. The great extent of weathering in the area supports this hypothesis but does not exclude other explanations.

Granite crops out at Finger Mountain, near Old Man Camp, and in the Bonanza Creek basin. At Finger Mountain, the bedrock is the least intensely weathered. Widely jointed resistant parts of the rock stand as large prominent tors (Figures 2 and 3). The tors occur at and above the local

FIGURE 2 Granite tor on the north side of Finger Mountain.

FIGURE 3 Finger Rock, a granite tor on Finger Mountain.

tree line. The area between tors is generally a coarse block field, often covered by low shrubs and tundra. The typical boulder diameter is nearly 1 m in places.

North of the Kanuti River near Old Man Camp the granite is more intensely weathered. The surface material is a thick residual mantle of weathered granite. Between Kanuti River and Old Man Camp there are no tors, but numerous small low mounds of disintegrating granite are apparent. These vegetation-free areas may be the residual products of tor weathering. If Linton's two-stage hypothesis applies here, the areas may, however, mark the site of future tor emergence. There are numerous tors north of Old Man Camp. These are much smaller, however, and lack the vertical development of tors at Finger Mountain. Block fields and sorted nets are rare and poorly developed.

In the Bonanza Creek basin the bedrock is similarly deeply weathered, but very large tors are present. They are generally massive hill-top outcrops displaying widely spaced vertical joints and surrounded by decomposed granite (Figure 4). Downslope from the tors the loose surface material is covered by a mat of Cladonia lichens with scattered spruce and birch trees. A dense mixed broadleaf and needle-leaf forest occupies most of the basin. The entire basin is at or below tree line and is predominantly free of permafrost.

FIGURE 4 Granite tor near Bonanza Creek.

Mafic igneous rocks underlie much of the area near the east end of Caribou Mountain and a narrow band on the south side of the South Fork of Fish Creek. Sorted stripes and polygons are excellently developed in both of those areas. Mafic tors were not found in the area studied. Just to the west, however, the summit of Caribou Mountain (Figure 5) displays a prominent ridgecrest of mafic igneous rock (Kachadoorian 1971).

FIGURE 5 Cryoplanation terraces and summit crest on Caribou Mountain.

FIGURE 6 Cryoplanation terraces on metamorphic bedrock near Fish Creek.

Features on Metamorphic Bedrock

Nearly every hill or mountain of metamorphic rock in the area has several levels of cryoplanation surface (Figure 6). In general, the most angular terrace profiles occur at the highest elevations. The more subdued rounded forms tend to occur at lower elevations. This agrees with observation at other Alaskan locations (Reger 1975, Reger and Péwé 1976). All terraces in the area are above 600 m elevation. The normal distribution of cryoplanatin terraces with regard to elevation reported by Nelson (1982) was not observed. The limited maximum elevation in the area perhaps restricts the local distribution to the lower half of a normal curve, thus producing a simple increasing function.

Of the 40 tors of metamorphic rock, 11 consist of a dense greenstone. Some of the greenstone is probably metamorphosed basalt dikes, based on the mineralogy and structure of the band of greenstone east of the highway at the Arctic Circle. Foliated igneous rocks grade into greenstone in several

FIGURE 7 Both cryoplanation terraces and tors on metamorphic bedrock 55 km north of the Yukon River.

FIGURE 9 Large central tor on the mountain shown in Figure 7.

FIGURE 8 Cryoplanation scarp near the summit of the mountain shown in Figure 7.

FIGURE 10 Large tors on the east end of the mountain shown in Figure 7.

tors. About 4 km south of Bonanza Creek several tors of metamorphic rock contain many small quartz veins that tend to increase their resistance to weathering. A few well-developed tors are composed entirely of schist and phyllite. In these tors the foliation is usually horizontal, the least favorable orientation for cryofraction, and joints are widely spaced. The most spectacular examples of these tors are found on a mountain 830 m high located east of the Dalton Highway about 55 km north of the Yukon River. On the summit are several well-developed cryoplanation surfaces with distinct scarps in association with several tors (Figures 7 - 10).

OTHER PERIGLACIAL FEATURES

The greatest variety of permafrost-related features is found on the flats of the Kanuti River near Olsons Lake. The many small lakes on the flats are probably thaw lakes of thermokarst origin. Ice-wedge polygons are present throughout the flats. Several low mounds on the south end of the flats have the form of low pingos (Kreig and Reger 1982). Shallow excavations in these mounds, however, revealed that they consist of clean, well-sorted sand. They are probably relict sand dunes stabilized by vegetation (Figure 11). The mounds have not yet been drilled to confirm the absence of an ice core.

Two large hills are located in a small basin 14 km south of Finger Mountain, 2 km west of the Dalton Highway. Those hills are composed of non-sorted sediment and are probably open-system pingos (Kreig and Reger 1982).

Seasonal frost mounds were observed near the north side of Old Man Camp airfield, near an access road crossing of the South Fork of Fish Creek, and adjacent to the Dalton Highway 1.3 km north of

The North Fork of Bonanza Creek. All three sites indicate an anthropogenic origin similar to those observed by Nelson and Outcalt (1982). Road and airfield construction has disrupted the local drainage, creating areas with excess water available for development of ice cores.

Sorted circles, nets, polygons, and stripes occur commonly in the region, usually above 550 m elevation. Sorted polygons often occur at the base of the concave slopes of cryoplanation scarps. Polygons were observed in all of the bedrock types of the study area. Both polygon size and particle size, however, are affected by lithology. The polygonal granite felsenmeer on Finger Mountain consists of coarse particles over 1 m in size with polygon diameters up to 10 m. Several kilometers south, the sorted polygons of metamorphic rock are no more than 5 m in diameter with a coarse particle size of 0.2 m (Figures 12 and 13).

FIGURE 11 Flats of the Kanuti River near Olsons Lake. The mound in the center of the photo about 100 m long is probably a vegetation-stabilized sand dune.

FIGURE 12 Sorted nets of metamorphic rock 15 km south of Finger Mountain. Photo by Richard Cook.

Large gelifluction slopes are found throughout the region, especially on the north-facing sides of ridges. Some of these are long, uniform slopes covering many square kilometers. Many of the more gently sloping gelifluction surfaces show a remarkably regular spacing of alder shrubs (Figure 14), perhaps due to changes related to frost action (Everett 1981).

CONCLUSIONS

The geomorphic features in the Kokine-Hodzana Highlands have developed during long exposure to a cold, rigorous climate without interuption by glaciation. The type and character of features are determined in part by the nature of the bedrock. In general, widely jointed massive bedrock tends to produce the best tors. There are over three times more tors per unit area in granite bedrock than in metamorphic bedrock. Tors occur at all elevations in the area. Finely jointed foliated bedrock tends to promote cryoplanation. Cryoplanation terraces are most common and most distinct at higher elevations. None were identified below 600 m. Well-developed tors and distinct terraces seldom occur together. Sorted patterned ground occurs on all types of bedrock at elevations above 550 m. The sizes of individual features and their constituent particles, however, are strongly affected by rock type. A wide variety of other periglacial landforms indicates that a diversity of geomorphic processes have been or are active in the region.

ACKNOWLEDGMENT

This study was supported by a project of the U.S. Army Corps of Engineers through the U.S. Army Cold Regions Research and Engineering Laboratory.

FIGURE 14 Stereogram illustrating regular spacing of alder shrubs on a gelifluction slope near Fish Creek.

REFERENCES

Brosgé, W. P., and Patton, W. W., Jr., 1982, Regional bedrock geology maps along the Dalton Highway, Yukon crossing to Toolik, Alaska: U.S. Geological Survey Open-File Report 82-1071, scale 1:500,000, 2 sheets.

Cook, R. P., and Bailey, P. K., 1983, Tors along the trans-Alaska pipeline: Geological Society of America Abstracts with Programs, v. 15, n. 3 p. 114.

Everett, K. R., 1981, Soil-landscape relations at selected sites along environmental gradients in northern Alaska, Contract Report No. DAAG 29-79-C-0160: Research Triangle Park, N.C., U.S. Army Research Office, 357 p.

Hamilton, T. D., 1982, A late Pleistocene glacial chronology for the southern Brooks Range: stratigraphic record and regional significance: Geological Society of America Bulletin, v. 93, p. 700-716.

Haugen, R. K., 1982, Climate of remote areas in north-central Alaska: 1975-1979 summary, CRREL Report No. 82-35: Hanover, N.H., U.S. Army Cold Regions Research and Engineering Laboratory, 110 p.

Kachadoorian, R., 1971, Preliminary engineering geological map of the proposed trans-Alaska pipeline route, Bettles and Beaver quadrangles: U.S. Geological Survey Open-File Report 487, scale 1:125,000, 2 sheets.

Kreig, R. A., and Reger, R. D., 1982, Airphoto analysis and summary of landform soil properties along the trans-Alaska pipeline route: Alaska Division of Geological and Geophysical Surveys, Geologic Report 66, 149 p.

Linton, D. L., 1955, The problem of tors: Geographical Journal, v. 121, p. 478-487.

Nelson, F. E., 1982, Spatial properties of cryoplanation terraces and associated deposits in northwestern North America: University of Michigan, Ph.D. dissertation, 297 p.

Nelson, F., and Outcalt, S. I., 1982, Anthropogenic geomorphology in northern Alaska: Physical Geography, v. 3, p. 17-48.

Reger, R. D., 1975, Cryoplanation terraces of interior and western Alaska: Arizona State University, Ph.D. dissertation, 326 p.

Reger, R. D., and Péwé, T. L. 1976, Cryoplanation terraces: indicators of a permafrost environment: Quaternary Research, v. 6, p. 99-109.

UNCONFINED COMPRESSION TESTS ON ANISOTROPIC FROZEN SOILS FROM THOMPSON, MANITOBA

T. H. W. Baker and G. H. Johnston

Division of Building Research, National Research Council of Canada, Ottawa, Ontario, K1A 0R6

Laminated frozen soils, consisting of alternate horizontal layers of dark brown clay and light brown silt, with ice lenses between the layers, are found in the northern area of ancient glacial Lake Agassiz. Laboratory tests confirmed the results of in situ ground anchor, penetrometer, and pressuremeter tests performed at Thompson, Manitoba, which showed these soils to be stronger in the horizontal direction, parallel to the layers, than in the vertical direction. Undisturbed block samples of these laminated (varved) frozen soils were obtained and test specimens were prepared at various orientations to the direction of layering. Unconfined compression tests, using both naturally frozen and remolded frozen specimens, have shown the extent of strength anisotropy due to the natural anisotropic soil structure.

INTRODUCTION

A series of in situ creep loading tests was undertaken in the frozen varved clay soils at Thompson, Manitoba, between 1967 and 1974. These field tests included circular plate anchors (Ladanyi and Johnston 1974), power-installed screw anchors (Johnston and Ladanyi 1974), pressuremeter (Ladanyi and Johnston 1973) and electric penetrometer tests (Ladanyi 1976, 1982). Creep equations were developed using the theory of an expanding spherical cavity. It was found that the creep exponent deduced from the static penetration tests was much higher than that found in the pressuremeter tests. Still higher values were obtained from creep tests on the ground anchors, performed earlier at the same site. Since the anchor and penetrometer tests loaded the soil in a vertical direction and the pressuremeter loaded the soil in a horizontal direction, this indicated that the frozen varved clay was more resistant to deformation (i.e., stronger) in the horizontal direction than in the vertical direction. This was attributed to the natural anisotropic structure of the frozen varved soils (Ladanyi 1976). A systematic study was undertaken to determine the influence of the anisotropic structure of these soils on the response to the various loading conditions described above. This paper describes a series of unconfined compression laboratory tests on field samples of frozen varved clay from the Thompson area.

INVESTIGATING STRENGTH ANISOTROPY

The term strength anisotropy is used exclusively in this paper to describe the variation of stress-strain behavior in the soil with direction of loading. Most soils exhibit stress-strain anisotropy, due mainly to the nature of the depositional process. This anisotropy becomes marked in layered or laminated (varved) soils formed by cyclic sedimentation and subsequent one-dimensional consolidation. The testing of unfrozen varved soils to determine their strength and deformation properties has been reported by several authors, including Tschebotarioff and Bayliss (1948), Eden (1955), and Metcalf and Townsend (1960). These investigators performed unconfined compression tests at various orientations to the varves and found a distinct anisotropy in shear strength. In testing Lake Agassiz varved clays from Steep Rock, Ontario, Eden (1955) found that when clay layers predominated, the shear strength was higher when the direction of loading was parallel to the layers. When silt layers predominated, the shear strength was higher when the direction of loading was perpendicular to the layers.

Frozen fine-grained soils often have an anisotropic structure, due to the existence of ice lenses formed perpendicular to the direction of freezing. Livingston (1956) studied the strength anisotropy by performing unconfined compression tests on frozen Keweenaw silt at various orientations to the ice lenses. There was a large scatter among the observed stress and strain values, due to specimen variability and to the extreme sensitivity of the soil-ice mixture to slight variations in temperature at the time of the test, particularly in the temperature range (-1 to -2°C) at which the tests were conducted. These tests indicated that the shear strength was generally higher for specimens with ice lenses parallel to the direction of loading.

The strength anisotropy of soils is usually expressed as the ratio of the shear strength with the major principal stress oriented at various angles to the vertical (S_θ) to the shear strength in the vertical direction (S_0). The obvious angles of significance for horizontally layered soils are $\theta = 45°$, where the maximum shear stress is parallel to the layers, and $\theta = 90°$, where the major stress is parallel to the layers.

The shear strength ratios for the Steep Rock varved clay and frozen Keweenaw silt are shown in Table 1. These ratios indicate a definite strength anisotropy which is dependent upon the properties of the individual layers. It was thought that the frozen varved clay from the Thompson area would show an even greater strength anisotropy than the soils mentioned above, due to the combined presence of horizontal varves and ice lenses.

TABLE 1 Strength Anisotropy of Some Soils From the Literature

Soil	$\frac{S_{90}}{S_0}$	$\frac{S_{45}}{S_0}$	Reference
Lake Agassiz varved clay (Steep Rock, Ontario)			Eden (1955)
predominantly clay layers	1.18	0.51	
predominantly silt layers	0.57	0.92	
Frozen Keweenaw silt	1.23	0.63	Livingston (1956)

SITE LOCATION AND DESCRIPTION

The city of Thompson (55°45'N, 97°50'W) is located in northern Manitoba about 640 km north of Winnipeg, within the discontinuous permafrost zone. Perennially frozen ground occurs in scattered small patches or islands varying in size from a few square metres to several hectares (Johnston et al 1963).

Thompson is also located within the northern extremity of ancient Lake Agassiz, which existed during the Wisconsin period of glaciation (Figure 1). Characteristic of the depositional processes associated with glacial Lake Agassiz was the formation of extensive deposits of laminated (varved) silt and clay. The varved soils are fully saturated and often contain ice lenses from hairline to 5-10 cm in thickness, normally oriented parallel to the predominantly horizontal laminations. The alternating horizontal layers of light brown silt and dark brown clay combine with the horizontal ice lenses to produce an anisotropic soil structure.

FIELD PROGRAM

Sampling

A backhoe was used to excavate a test pit to a depth of about 3 m at a site approximately 5.5 km north of the city of Thompson, near the southwest corner of the Thompson airstrip, in March 1978. The pit was located in a large spruce island close to a ground temperature cable installed by the Geological Survey of Canada (GSC) in 1973. The site (GSC 2A) is described in detail by Klassen (1976). The walls of the pit were visually logged, and samples were obtained for water content, organic content and grain size analyses. A log of the test pit is shown in Figure 2.

A layer of distinctly laminated (varved) frozen clay occurred between the depths of 1.25 and 2.25 m. Small (0.3 m^3) blocks of the frozen varved material were wrapped in polyethylene sheeting, placed in insulated boxes packed with ice, and shipped to Ottawa by air. These samples were maintained at -6°C in a cold room until specimens were prepared for testing.

Ground Temperatures

Ground temperatures were measured weekly by NRC staff on the thermistor cable installed at the site by the Geological Survey of Canada. Figure 3 shows ground temperatures obtained in early March 1978 and the maximum, minimum, and mean temperatures for that year. The maximum and minimum temperatures for 1978 were near the extreme values observed for the years 1975–1978. The mean annual ground temperature, at the depth at which the block samples were obtained, was between -1 and -2°C. At the time of sampling, the ground temperature, in the sampling zone, was between -2 and -3°C.

LABORATORY TESTING

Test Specimen Preparation and Description

Cylindrical test specimens 76 mm in diameter and 152 mm in height were machined from the block samples using a band saw and lathe in a cold room. The procedures for sawing, machining and preparing the specimens are described by Baker (1976). Specimens were machined at various angles to the plane of the horizontal varves. "Vertical" specimens contained varves oriented perpendicular to the vertical axis of the specimen. "Horizontal" specimens contained varves oriented parallel to the vertical axis of the specimen. Another set of specimens was machined with varves oriented at 45° to the vertical axis of the specimen. Each specimen contained from 15 to 20 varves, consisting of dark brown clay layers from 8 to 12 mm in thickness and light brown silt layers from 2 to 5 mm in thickness. Ice lenses visible on the sides of the specimens ranged in thickness from hairline to a few millimetres.

Bulk densities were determined prior to testing by measuring the volume and weight of the test specimens. After testing they were weighed, oven dried, and weighed again to determine their total water contents. The densities and water contents are presented in Table 2.

Hydrometer tests on "combined" samples (containing both light and dark layers) indicated that the varved soil consisted of 80-90% clay size (<0.002 mm) material. Ignition-loss tests (ASTM D 2974) indicated an average organic content of 12%. Eight Atterberg Limits tests showed a liquid limit of 52.0%, a plastic limit of 32.7%, and a plasticity index of 19.3%. The average water content was 35.0%, about 2-3% lower than that determined at the field site (Figure 2).

Several remolded specimens were specially prepared in the laboratory to try and remove the anisotropic structure formed by the varves and ice lenses. Some specimens were completely thawed, and the soil was mixed and remolded to densities and water contents similar to the natural samples. These remolded cylindrical specimens were placed in the cold room at -6°C and allowed to freeze from all sides. They were removed from the molds, and their ends were machined prior to testing. Small hairline radial ice lenses could be seen at the ends of the frozen remolded specimens. These remolded specimens were tested for comparison with results obtained on the natural (undisturbed) frozen soils. Frozen bulk

TABLE 2 Properties of Test Specimens

Specimen	Bulk density (kg/m^3)	Dry density (kg/m^3)	Water content (% dry wt.)	Loading direction to plane of varves
Undisturbed				
CS-1	1,799	1,332	35.1	Parallel
CS-2	1,791	1,353	32.4	Parallel
CS-3	1,780	1,334	33.5	90°
CS-4	1,766	1,316	34.2	90°
CS-5	1,761	1,284	37.2	90°
CS-6	1,751	1,298	34.9	Parallel
CS-7	1,769	1,288	37.3	45°
CS-8	1,772	1,318	34.4	45°
CS-9	1,759	1,303	35.0	45°
CS-10	1,828	1,374	33.1	Parallel
Mean	1,778	1,320	34.7	
Stand. Dev.	23	29	1.6	
Remolded				
RS-1	1,776	1,340	32.6	
RS-2	1,852	1,440	28.6	
RS-3	1,761	1,303	35.1	
RS-5	1,768	1,303	35.7	
Mean	1,789	1,347	33.0	
Stand. Dev.	42	65	3.2	

densities and total water contents were determined before and after testing; these are presented in Table 2.

Unconfined Compression Tests

Unconfined compression tests were performed at a cold room temperature of -6°C, using a 250 kN capacity screw-driven universal testing machine. These tests were carried out at nominal strain rates of 0.01 and 1.0%/min. Some tests were performed at an intermediate strain rate of 0.1%/min, but were incomplete due to the limited amount of sample material available. Data from this incomplete test series will not be presented. An extensometer consisting of three displacement transducers located on the specimen, measured axial deformation and tilting (Baker et al 1982). Compliant platens (Baker 1978) were used to reduce end effects and to ensure a uniform normal application of pressure to the specimen ends.

Vertically oriented specimens had the maximum normal stress direction perpendicular to the plane of the varves to simulate the anchor and penetrometer tests performed in the field. Horizontally oriented specimens had the maximum normal stress direction parallel to the plane of the varves, in a manner similar to the field pressuremeter tests. Specimens oriented at 45° had the maximum shearing stress directed parallel to the plane of the varves.

Stress-strain curves for the two nominal strain rates are shown in Figures 4 and 5. The loading direction relative to the plane of the varves for each specimen is indicated. Some tests were duplicated to show the variation in stress-strain behavior between specimens. The maximum axial strain imposed on specimens was between 10 and 15%, due to the limitations imposed by the 83 mm diameter extensometer ring. As a result, none of the tests reached peak stress.

DISCUSSION AND CONCLUSION

All of the specimens exhibited an initial yield at about 0.15% axial strain; the stress then continued to increase until the test was stopped. Although none of the specimens reached failure (peak stress), the value of the stresses at and beyond yield is an indicator of the relative strength of specimens and can be used to determine strength anisotropy. All specimens deformed as right cylinders without central bulging or expansion of individual layers. This indicated that the strains within the specimen were uniformly distributed.

In both ranges of strain rate the horizontally oriented specimens were consistently the strongest, followed by the specimens oriented at 45°. The vertically oriented specimens were the weakest. This relationship was also observed in the limited tests performed at the intermediate strain rate. Strength increase with strain rate seemed to be similar for all specimen orientations. Tests that were duplicated showed very similar stress-strain behavior. This probably reflects the narrow range in the natural physical properties (densities and total water content). The similarities in physical properties between specimens enhanced the reliability of comparing the test results. The remolded laboratory-frozen specimens were as much as 80% stronger than the naturally-frozen undisturbed specimens. Although there was some anisotropy in the structure of the remolded specimens, these results indicated that the natural anisotropic structure acts to reduce the shear strength.

The strength anisotropy is shown in Table 3 for the two strain rates and for axial strains corresponding to yield (0.15%) and 10%. The amount of anisotropy is not pronounced at yield or at small strain rates, but it becomes quite significant at 10% strain. High horizontal strengths are similar

TABLE 3 Strength Anisotropy

Axial strain (%)	Nominal strain rate (%/min)	$\frac{S_{90}}{S_0}$	$\frac{S_{45}}{S_0}$	$\frac{S_0}{S_R}$
0.15 (yield)	0.01	0.93	0.99	0.55
	1.0	1.14	1.18	0.67
10.0	0.01	1.13	1.03	0.72
	1.0	1.35	1.06	0.83

to those observed by Eden (1955) in unfrozen varved clays from Steep Rock when the clay layers predominated. Specimens in this study had a high clay content (80-90%). High shear strengths at 45° orientations have not been seen in any of the data on varved soils in the literature. Ice cementation may resist the shearing stresses acting between the layers. The higher strain rates used in this study are comparable to those used by Eden (1955) and Livingston (1956) in their test programs. Frozen Thompson varved clays appear to exhibit a greater strength anisotropy than either the unfrozen varved clay from Steep Rock (Eden 1955) or the frozen Keweenaw silt (Livingston 1956).

An estimate of the average unfrozen water content of the combined soil at the in situ temperature of -1°C, based on the liquid limits, is 15% (Tice et al. 1976). The unfrozen water content would probably be different in the individual clay and silt layers, due to the surface area of the soil grains, and would influence the relative stress-strain behavior of these layers. It is recognized that the change from the in situ temperature to the laboratory temperature of -6°C, coupled with sublimation of the samples during 2 years of cold storage, would lower the unfrozen water content. The strength anisotropy in these specimens is probably due to the combined effect of the varves and their different unfrozen water contents, and other physical properties, as well as to the ice lenses. No attempt was made to estimate the relative contribution of these various components.

The series of in situ tests performed by Johnston and Ladanyi in an attempt to determine the bearing capacity of the frozen varved soils at Thompson were analyzed using cavity expansion theory to estimate the stress-strain behavior of the soil. Cavity expansion theory assumes that the soil behaves as a homogeneous, isotropic, ideally plastic material (Ladanyi 1963). Horizontal varves and ice lenses make these soils heterogeneous. This laboratory study has shown that these soils are not isotropic but anisotropic in their stress-strain behavior. The stress-strain curves indicate that these soils do not deform in an ideally plastic manner. A modification could be made to the cavity expansion theory to take into consideration strength anisotropy. Davis and Christian (1971) have proposed a modification, taking into consideration strength anisotropy, to the yield criterion which Scott (1963) recommended for predicting the bearing capacity of undrained soils. They show strength anisotropy to be present in most cohesive soils.

A more detailed laboratory study similar to the one presented in this paper would be needed to determine the complete range and magnitude of strength anisotropy associated with a particular soil deposit. One would have to consider the influences of various components of the soil deposit (number and physical properties of the varves, ice lenses, etc.) that could contribute to the anisotropy. The results of field testing using the pressuremeter and penetrometer would provide a useful check on the strength relationships obtained and a qualitative appraisal of the influences of sampling and testing procedures.

ACKNOWLEDGMENTS

The authors wish to thank A. Chevrier and C. Hubbs for their help in preparing the specimens and carrying out the tests. This paper is a contribution from the Division of Building Research, National Research Council of Canada, and is published with the approval of the Director of the Division.

REFERENCES

Baker, T. H. W., 1976, Transportation, preparation and storage of frozen soil samples for laboratory testing, in Soil specimen preparation for laboratory testing: ASTM Special Technical Publication 599, p. 88-112.

Baker, T. H. W., 1978, Effect of end conditions on the uniaxial compressive strength of frozen sand, in Proceedings of the Third International Permafrost Conference, Edmonton: Ottawa, National Research Council of Canada, p. 608-614.

Baker, T. H. W., Jones, S. J. and Parameswaran, V. R., 1982, Confined and unconfined compression tests on frozen soils, in Roger J. E. Brown Memorial Volume, Proceedings of the Fourth Canadian Permafrost Conference, Calgary: Ottawa, National Research Council of Canada, p. 387-393.

Davis, E. H. and Christian, J. T., 1971, Bearing capacity of anisotropic cohesive soil: American Society of Civil Engineering Proceedings, v. 97, No. SM5, p. 753-769.

Eden, W. J., 1955, A laboratory study of varved clay from Steep Rock Lake, Ontario: American Journal of Science, v. 253, p. 659-674.

Johnston, G. H., Brown, R. J. E. and Pickersgill, D. N., 1963, Permafrost investigations at Thompson, Manitoba: Terrain studies: Division of Building Research, National Research Council of Canada, Technical Paper No. 158, 96 p.

Johnston, G. H. and Ladanyi, B., 1974, Field tests of deep power-installed screw anchors in permafrost: Canadian Geotechnical Journal, v. 11, No. 3, p. 348-358.

Klassen R. W., 1976, Landforms and subsurface materials at selected sites in a part of the shield--north-central Manitoba: Geological Survey of Canada, Ottawa, Paper 75-19, 41 p.

Ladanyi, B., 1963, Expansion of a cavity in a saturated clay medium: American Society of Civil Engineering Proceedings, v. 89, No. SM4, p. 127-161.

Ladanyi, B., 1976, Use of the static penetration test in frozen soils: Canadian Geotechnical Journal, v. 13, No. 2, p. 95-110.

Ladanyi, B., 1982, Determination of geotechnical parameters of frozen soils by means of the cone penetration test, in Proceedings of the Second European Symposium on Penetration Testing: Rotterdam, A. A. Balkema, p. 671-678.

Ladanyi, B., and Johnston, G. H., 1973, Evaluation of in situ creep properties of frozen soils with the pressuremeter, in The North American Contribution to the Second International Permafrost Conference, Yakutsk: Washington, D.C., National Academy of Sciences, p. 310-318.

Ladanyi, B., and Johnston, G. H., 1974, Behaviour of circular footings and plate anchors in permafrost: Canadian Geotechnical Journal, v. 11, No. 4, p. 531-553.

Livingston, C. W., 1956, Excavations in frozen ground, Part I, Explosion tests in Keweenaw silt: Boston, U.S. Army Snow, Ice, Permafrost Research Establishment, Rep. 30, 97 p.

Metcalf, J. B. and Townsend, D. L., 1960, A preliminary study of the geotechnical properties of varved clays as reported in Canadian Engineering Case Records, in Proceedings of the Fourteenth Canadian Soil Mechanics Conference, Niagara Falls, Ontario: National Research Council Canada, Technical Memo, No. 69, p. 203-225.

Scott, R. S., 1963, Principles of soil mechanics: Reading, Mass., Addison-Wesley, 440 p.

Tice, A. R., Anderson, D. M., and Banin, A., 1976, Prediction of unfrozen water contents in frozen soils from liquid limit determinations: Hanover, N. H., Cold Regions Research and Engineering Laboratory, Rep. 76-8, 17 p.

Tschebotarioff, G. P., and Bayliss, J. R., 1948, The determination of the shearing strength of varved clays and their sensitivity to remoulding, in Proceedings of the Second International Conference on Soil Mechanics and Foundation Engineering, v. 2: Rotterdam, Haarlem, p. 203-207.

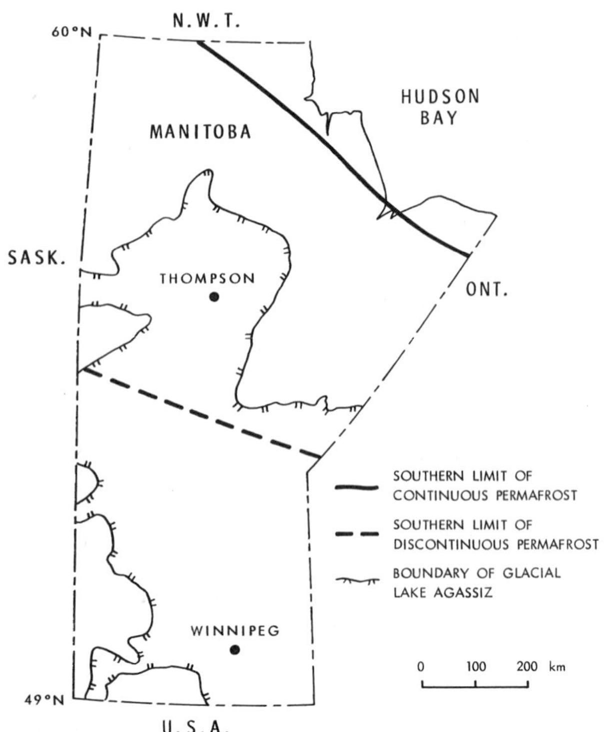

FIGURE 1 Location map.

SOIL DESCRIPTION	SOIL PHASE	ICE PHASE
Living moss, Peat	Pt	None
Dark brown frozen peat	Pt	Nbe
Light brown frozen clay silt (not laminated)	CL	Vr
Frozen varved clay light brown silt layers (1/2 - 2 cm) dark brown clay layers (2 - 6 cm), ice lenses (2 - 4 cm)	CL	Vs
Light brown frozen clay silt, plastic frozen (not laminated)	CL	Nbe

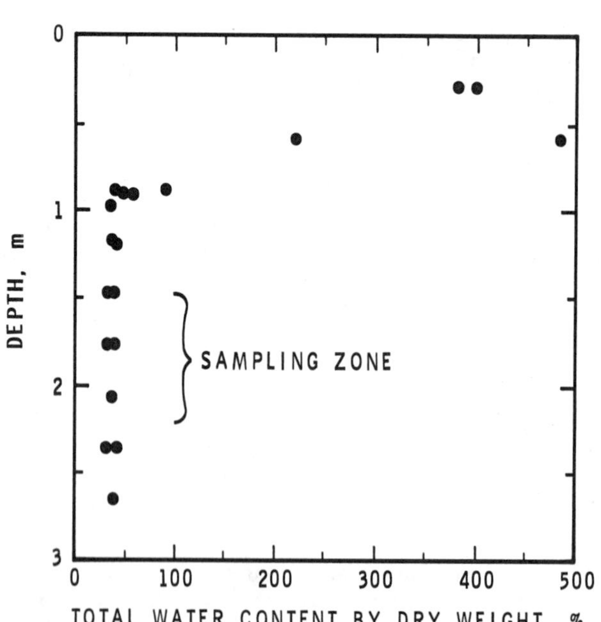

FIGURE 2 Log of test pit.

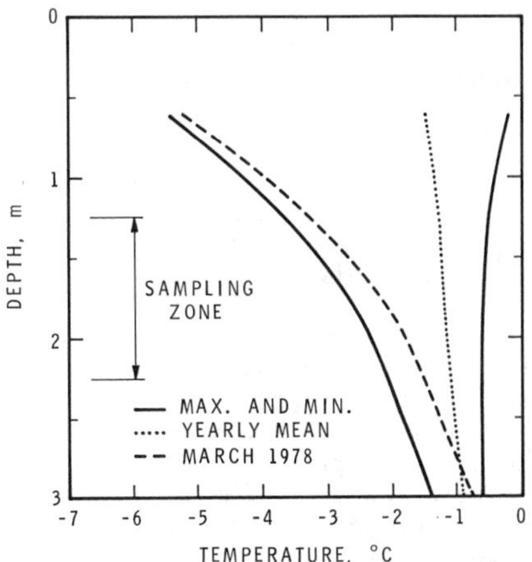

FIGURE 3 Ground temperatures, from temperature cable. GSC site 2A, Thompson, Manitoba.

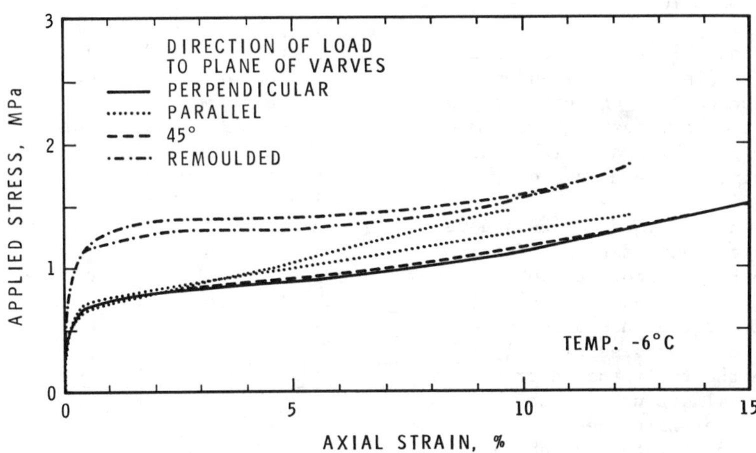

FIGURE 4 Stress-strain curves at 0.01%/min.

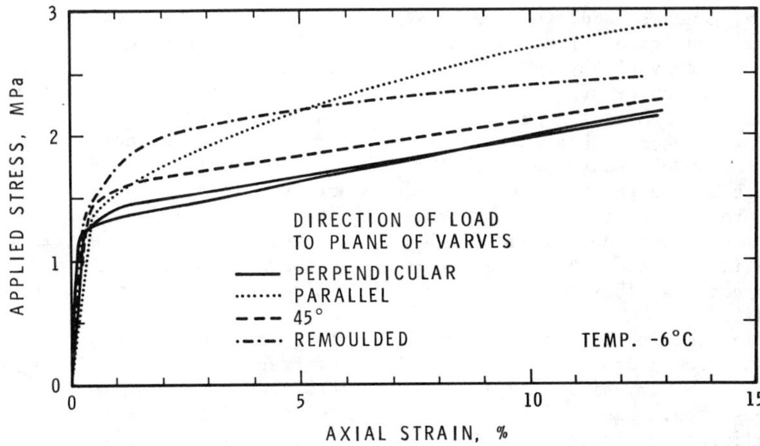

FIGURE 5 Stress-strain curves at 1.0%/min.

MECHANISM AND REGULARITIES OF CHANGES IN HEAT CONDUCTIVITY OF SOILS DURING THE FREEZING-THAWING PROCESS

Ye. N. Barkovskaya, E. D. Yershov, I. A. Komarov, and V. G. Cheveriov

Geology Department,
Moscow State University, Moscow, USSR

The paper discusses the results of comprehensive experimental studies on the mechanism and regularities of heat conductive properties of a variety of soils of different genesis and age. These soils have different granulometric and chemical-mineralogical composition and are in various degrees of salinization and peating; they also have different cryogenic texture. The studies were carried out by the stationary regime method applied during the process of freezing and subsequent thawing in the temperature interval from +25° to -25°C, which encompassed the temperature interval of intensive phase transformation of water in soils.

INTRODUCTION

The study of heat conductivity of rocks and soils is important both scientifically and practically. Prediction and control of such cryogenic processes as freezing-thawing, heaving-settlement, thermokarst, and thermal erosion are impossible without the understanding of thermophysical properties.

During the last decades many researchers have studied thermophysical characteristics. Theoretical principles have been elaborated to describe heat transfer in porous soils, and a considerable amount of experimental material has been collected. The experimental data, however, was normally obtained for ground in the fully thawed or solidly frozen state. There is almost no data on heat conductivity of soils under large negative temperatures, i.e., in the range of intensive phase transition of water. The mechanism and regularities of heat conductivity of soils during the cyclic freezing-thawing process are still unknown. And, finally, a generalizing work has yet to be compiled on the mechanism and regularities of heat conductivity in cryogenic soils of various composition, structure, genesis, and age under different thermodynamic conditions of formation.

The staff of the Laboratory of Physics, Physical Chemistry, and Mechanics of Frozen Grounds, under the Chair of Geocryology of the Geological Faculty of the Moscow State University, has carried out comprehensive research to determine the mechanism and to establish the regularities in the changes of the thermal conductivity of frozen, thawed, and freezing and thawing soils, and further to analyze and generalize the available experimental material. This paper describes several basic results of these studies.

The materials studied were soils (derived from solid to fine-dispersion rock) of different genesis, age, granulometric composition, and chemical-mineral composition; these soils had various degrees of salinization and peating, humidity, density, and cryogenic texture. Samples of disaggregated soils with disturbed structure were specially prepared for the purpose of finding regularities in the changes of their heat conductive properties. The method of comprehensive research was used because it combines the determination of heat conductivity in the temperature interval from +25° to -25°C with estimation of the macro- and microstructure of the solids and of the phase content of included water. This research was based on the method of stationary thermal regime, which permits the measurement of heat conductivity of rocks not only at high negative temperatures up to -0.5°C but also during the cycle of freezing and subsequent thawing; moreover, the method provides an estimation of the anisotropy of thermal properties.

The Mechanism of Changes in Heat-Conductive Properties of Rocks during the Freezing-Thawing Process

The temperature effect on heat conductive properties of rocks has various manifestations. As shown by both laboratory and published data, a decrease of temperature increases heat conductivity in crystalline rocks and reduces it in amorphic rocks. These effects are due to a difference in the mechanism of heat transfer; in crystalline rocks heat transfer occurs by elastic oscillations of the crystalline grid, whereas, in amorphic rocks it is the result of non-elastic dispersion of thermal energy.

Unlike solid rocks, moist soils have non-linear and extremal types of dependence of heat conductivity (λ) on temperature (Fig. 1). Typical parts of the curve may be selected for analysis. The change of λ coefficient in the region of positive temperatures ($t_1 \div 0°C$ and $0°C \div t_8$) is linear; this feature is related to the analogous dependence of heat conductivity of water on temperature. The parts from 0°C to the temperature of the beginning of freezing ($t_2 = t_{43}$) and from the temperature of the end of thawing (t_7) to 0°C correspond to relation $\lambda(t)$ in the region of positive temperatures, a fact attributed to the unchanging phase content of the system. The length of these parts depends on the composition of the soil.

In the course of further temperature decrease in

the interval (t_2 to t_3), corresponding to the region of intensive phase transformation, a considerable increase in λ value is observed. Starting with the temperature of the beginning of freezing (t_2), the value of heat conductivity coefficient at t_3 can be twice its value in the thawed state. This fact is a change in soil composition and to the first appearance of ice crystals, the heat conductivity of which is four times greater than that of water. A study of the microstructure of soils during freezing shows a complicated process of formation of ice-cementing structural ties. The particles and aggregates become larger, thus reducing the contact heat resistivity in soils.

The part of the curve corresponding to the temperature interval (t_3 to t_4), where the intensity of phase transition of water is reduced, the coefficient of heat conductivity somewhat increases to a certain constant value owing to the accretion of pore ice. Structural studies in this temperature range did not reveal any notable changes in the structure of frozen soils.

The decrease of temperature in the part of the curve (t_4 to t_5) causes a reduction of the heat conductive potential of the soil. The study of microstructural transformations in frozen ground has demonstrated that this effect is connected with the appearance of microcracks in ice and with its fragmentation; the near-contact ice-soil zones become more friable owing to thermo-mechanical stresses. The decrease of λ value in this interval of temperature can reach 25-30%. In soils the percentage increases from sands and sandy loams to clays and peats.

In the interval of negative temperatures (t_5 to t_7), the values of λ coefficient of soil during freezing do not coincide with the values for the same soil during thawing. This hysteresis in the $\lambda(t)$ dependence is observed in practically all studied grounds. It is attributed to the development of irreversible structural transformations and to peculiarities in the formation of the phase content of water. The area of the hysteresis loop depends on the content and structure of the solid particles and on the intensity of structural changes; it reduces from peats and clays to sandy loams and sands. The maximal difference in the values of heat conductivity of soils in the freezing-thawing cycle is not more than 30%.

Under conditions of phase transformation of water, the λ value of thawing ground in the range (t_6 to t_7) is more than that of freezing ground in the same temperature interval; this fact is related to a difference in phase content and structure under the same temperatures.

The expected hysteresis of heat conductivity of moist soils is practically absent in the range of positive temperatures, because, notwithstanding the changes in structure during thawing, heat transfer in the liquid phase is the determining factor of heat conductive properties in this temperature range.

Regularities in the Changes of Heat Conductivity during Freezing-Thawing of Soils with Different Composition and Structure

The effect of the chemical-mineralogical composition of soils on heat conductivity is associated with the peculiarity of crystalline-chemical structure of minerals, with the contents of the pores, and with organic admixtures and their quantitative ratios. Heat conductivity of solid rocks mainly depends on the heat conductive properties of minerals composing them; the most common mineral is quartz, whose heat conductivity is much higher than that of many pore-forming minerals. Thus, the increase of quartz content in magmatic rocks from basic to acid is responsible for the observed tendency of λ to increase in that series. Heat transfer in sedimentary-cemented rudaceous, sandy, and sandy loam rocks is also to a large extent determined by heat conductivity of rock-forming minerals. Their role, however, is reduced owing to micro- and macrocracks and porosity.

These effects are confirmed by the results of experiments carried out on sandy loams with different mineral content: quartz, and vermiculite (Fig. 2, curves 1 and 2). Heat conductivity of quartz sandy loam is higher than that of vermiculite sandy loam owing to the larger λ value of quartz. Vermiculite sandy loam is characterized by a stronger hysteresis of $\lambda(t)$, caused by microstructural transformations occurring with greater intensity as compared to quartz sandy loam.

The effect of the heat conductivity of minerals on the general heat conductivity of rocks is reduced with an increase in dispersion, because the role of contact heat conductivity is greater. Thus, the influence of mineral composition on the λ of clay soils is manifested not directly, but indirectly through hydrophilicity. Clays with sharply differentiated crystalline-chemical structures were chosen as the subjects of research: montmorillonite and kaolinite (with similar density values and pores almost completely filled with water) (Fig. 2, curves 3 and 4). In the region of positive temperatures the values for these clays are practically identical (0.5 to 0.65 W/m °C), and the shape of the $\lambda(t)$ dependence is close to linear. The slope of these straight lines is similar to the slope of the straight line depicting the temperature dependence of the λ coefficient of water. Consequently, heat conductivity of thawed clay soils does not depend strongly on the nature of the minerals, but is primarily determined by the heat resistivity of water molecules composing a single water layer. On the contrary, under negative temperatures the difference in the structure of minerals is fairly well manifested by affecting the proportions of ice and unfrozen water. The value of λ and the area of the hysteresis loop of $\lambda(t)$ for kaolinite clay are less than those for montmorillonite clay owing to larger irreversible transformations in the structure of the latter.

The presence of organic impurities in soils is an important factor influencing the development of their heat conductive properties. A special series of experiments on peats and peated grounds was performed to study the effect of the degree of peating and organic decomposition on the heat conductivity of ground. In the temperature range from +15 to -15°C the change in heat conductivity coefficient is in general the same for sands, peated sands, and peats (Fig. 3). When the pores are, for practical purposes, completely filled with water (or ice) the value of λ reduces with an increase of peating, which correlates with data obtained by A. A. Konovalov and L. T. Rotman (1970). Also the λ value in the region of negative temperatures close to 0°C decreases with the degree of peating. This effect is attributed to the difference in the quantitative water-ice ratio and to the nature of thermal contacts formed between ice, sand particles, organic inclusions, and water.

The reduction effect of the λ value for peat and peated sand, in a solidly frozen state, is associated with the development of microcracks in pore ice during cooling of the soil. Results obtained lead to the conclusion that the presence of organic impurities reduces heat conductivity in the +15 to -15°C range because of low heat conductivity of peat, a larger amount of liquid phase, and structural defects appearing as the result of cooling. The hysteresis in the λ(t) dependence for peats under freezing-thawing conditions is caused by intensive physical-chemical and structural transformations. A tendency for an increase in the λ coefficient is noted in the frozen peat series from largely to slightly decomposed; heat conductivity of thawed peats is practically the same.

Salinization of rocks strongly influences heat conductive properties. The effect of NaCl salinization on heat conductive properties of fine quartz sand and medium loam was studied (salinization of frozen soils was Z = 0.5, 1.0, 1.5%). The increase of Z to 1% reduces by half the heat conductivity of water-saturated sands. The obtained experimental data imply a complicated λ(t) dependence (Fig. 4) including the presence of maximums, because under freezing the change in salinization affects not only the phase composition of water and the heat conductivity of porous solids, but causes structural transformations as well. For soils with different salinization the λ(t) dependence is analogous, but the values of their heat conductivity differ by 50% and more. When the temperature decreases to -25°C the heat conductivity value of unsalinized sands (curve 1) monotonically decreases. For salinized sands the decrease of heat conductivity at temperatures below -15°C changes to a slow increase of λ values, owing to better thermal contacts resulting from the freezing-out of solution and the crystallization of salts.

The effect of grain size on heat conductivity was studied by many researchers, among them R. I. Gavriliev, V. T. Balobaev (1978), N. S. Ivanov (1962) and A. F. Chudnovsky (1962). These studies demonstrated a reduction of heat conductivity of soils with an increase of dispersion in the series (rudaceous-sandy-sandy loam-loamy-clay grounds) over the entire studied temperature range, which included the temperatures of the water phase transitions. The character of the λ dependence in the range from 0 to -5°C (Fig. 5) is associated with changes in the area of thermal contacts and in the amount of unfrozen water, both of which increase in the order shown for the dispersion series. In the course of study, data on heat conductivity of loess soils was obtained for the first time. The value of λ in loess soils at similar values of water saturation and density is intermediate between the heat conductivity of loamy sands and of clays. Among loess soils, the λ coefficient increases with an increase in the size of their aggregates. Thus, all other conditions being equal, heat conductivity of coarse-dust loesses is 1.3 to 1.5 times greater than that of fully loesslike loams of the same genesis and age.

As shown by the results of experimental research, the nature of heat conductivity dependence on moisture of loess soils is the same as that of sands and sandy loams, but different from that of loams and clays, a result contrary to expectations based on the high content of dust and clay particles in loesses. This phenomenon can be explained by the model of loess soil structure; the model indicates that with an increase of water saturation up to values of 0.5 to 0.7 (water content reaching 20%) deformation of the surfaces of globules and extension of areas of thermal contacts occurs owing to the swelling of clay shrouds. Subsequent moistening up to 0.7 (water content >20%) does not cause notable changes in heat conductivity of the soil, because heat flow is then almost completely within the water system. The freezing of water-saturated loess soils is accompanied by a considerable growth of heat conductivity as the result of formation of ice-cementing bonds. Consequently, λ of frozen loesslike loam of natural composition at a water content of 30% is three times as large as λ at a water content of 3%.

The problem of the influence of cryogenic texture on soil heat-conductive properties is as yet hardly approached. The study of frozen sandy loam of natural composition shows that the value of λ for soils with massive cryogenic texture (1.8 to 2.2 W/m-degC) exceeds λ for soils with layered and subreticulate texture (1.4 to 1.6 W/m-degC). Numerous defects of ice structure (cracks, air bubbles, organic admixtures, etc.) have been noted in sandy loam soils of layered texture; these defects create additional resistivity to heat transfer.

Varying the conditions of freezing of clay has enabled us to obtain specimens of massive and of layered cryogenic texture with almost identical ice contents. Heat conductivity of clay with massive texture normally exceeds heat conductivity of clay with layered texture by about 25%.

In summary, results derived from experimental research have revealed a complicated mechanism of development of heat-conductive properties of soils and have indicated basic trends in the variations of heat conductivity, which depend on the composition and structure of soils and the temperature conditions.

REFERENCES

Chudnovsky, A. F., 1962, Thermophysical properties of characteristic soils. Leningrad: Izdatel'stvo Fiziko-Matematicheskoy Literatury.

Gavriliev, R. I., Balobaev, V. T., 1978, Experimental studies of thermal properties of rudaceous soils, in Heat exchange in frozen landscapes, Yakutsk.

Ivanov, N. S., 1962, Heat exchange in cryolithozone.

Konovalov, A. A., Roman, L. T., 1970, Dependence of thermophysical characteristics of peats and peated grounds on their physical properties.

Papers of the 6th Meeting on Exchange of Experience in Construction under Formidable Climatic Conditions, Krasnoyarsk.

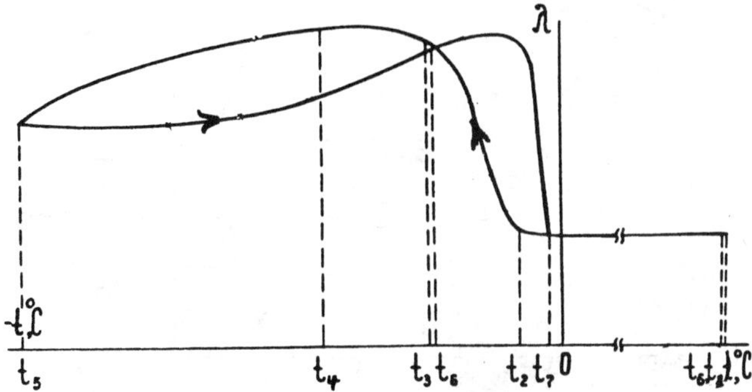

FIGURE 1 General type of curve of heat conductivity coefficient of soils with temperature in the process of freezing (t_2 - t_5) and subsequent thawing (t_5 - t_8).

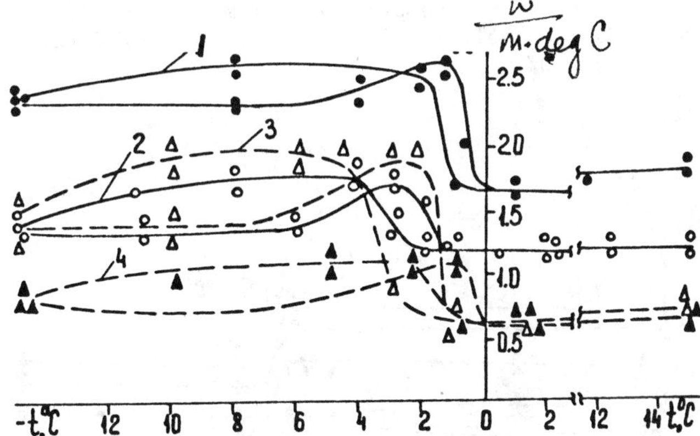

FIGURE 2 Dependence of heat conductivity coefficient on temperature in water-saturated soils of different mineral composition: 1 - heavy quartz sandy loam; 2 - heavy vermiculite sandy loam; 3 - montmorillonite clay; 4 - kaolinite clay.

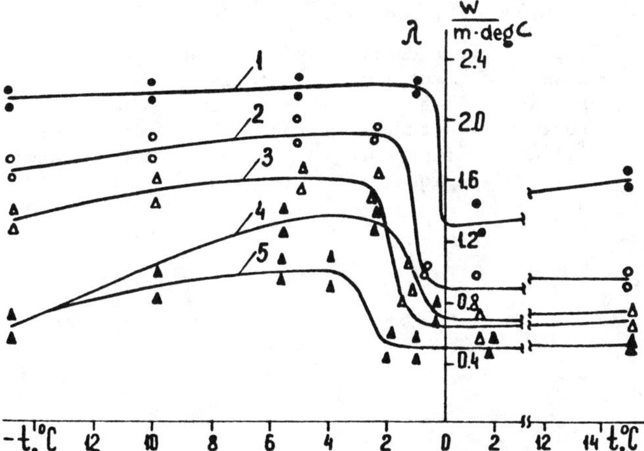

FIGURE 3 Dependence of heat conductivity coefficient on temperature for fine sand with different amounts of peat under freezing (1,2,3,5) and subsequent thawing (4).

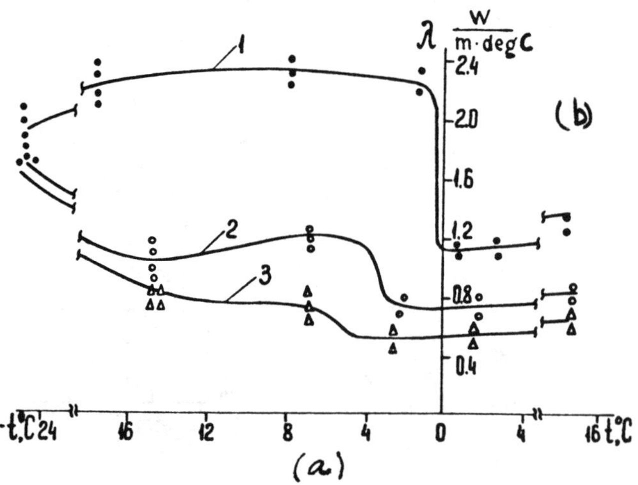

FIGURE 4 Dependence of the amount of unfrozen water (a), and the value of heat conductivity coefficient (b) on temperature for moist fine sand with different salinization: 1 - Z=0.0%; 2 - Z=0.5%; 3 - Z=1.0%.

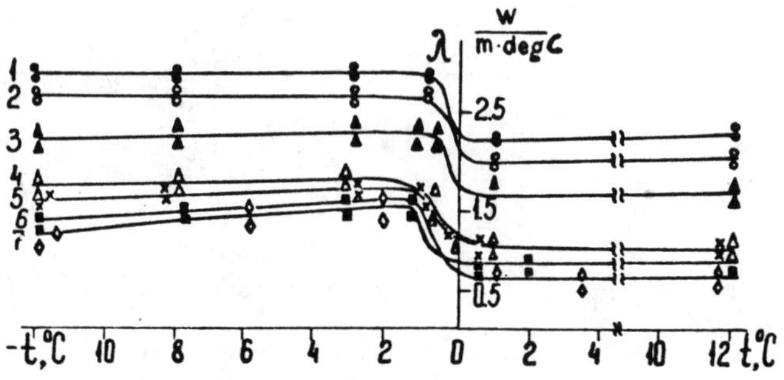

FIGURE 5 Dependence of heat conductivity coefficient of water-saturated soils having different grain size with temperature in the process of freezing. 1 - rudaceous rock debris with sandy loam filling; 2 - fine sand; 3 - light fine sandy loam; 4 - loesslike loam; 5 - medium loam; 6 - clay; 7 - peat.

PROPERTIES OF GEOTEXTILES IN COLD REGIONS APPLICATIONS

J.R. Bell, T. Allen and T.S. Vinson

Dept. of Civil Engineering, Oregon State University, Corvallis, OR 97331 USA

To investigate properties of geotextiles in cold regions applications, a laboratory program was conducted with five geotextiles. Freeze-thaw durability in fresh and saline water was considered. Geotextile strength before and after 300 freeze-thaw cycles showed no serious degradation. Geotextile load-deformation-time relationships were determined at +22°C and -12°C by wide-strip tensile and static creep tests. Temperature had little effect on strength, but creep was significantly affected. Lower temperatures resulted in reduced creep. Both geotextile structure and polymer type were significant to creep. Polypropylene geotextiles were affected to a greater degree by temperature than polyester geotextiles.

A preliminary investigation was performed to determine the potential of geotextiles as capillary breaks in a highly frost susceptible silt. Soil columns were frozen from the top at a constant rate. Free water was available at the bottom for two columns--one contained a geotextile layer and one did not. A third control column was frozen without free water or a geotextile. Heave and water content increases during freezing were determined for each soil column. The results indicate some geotextiles have the potential to significantly reduce frost heave. The effectiveness is different for different geotextiles.

INTRODUCTION

Geotextiles have been widely used in the more temperate areas of the world for the past two decades (Bell and Hicks 1980). More recently, engineers concerned with arctic and subarctic construction have incorporated geotextiles into their designs (Tart and Luscher 1981). To date, geotextiles have been used in their traditional roles to provide filtration, separation, and reinforcement in drains and roadways (Bell and Hicks 1980). They are also being used in erosion control structures in the Arctic (Leidersdorf et al. 1981).

While successful applications of geotextiles in cold regions have been made, questions remain unanswered relative to the influence of the cold environment on their performance. For example, the influence of cold temperatures on geotextile load-strain and creep relationships is largely unknown. Degradation due to freeze-thaw cycles also has had limited study.

In addition to the usual problems, engineering/construction practices for building foundations, roadways, and embankments in cold regions of the world are intimately associated with freezing related phenomena of initially unfrozen ground and thawing related phenomena of initially frozen ground. Freezing of initially unfrozen frost-susceptible soil can result in heave at the ground surface and disruptions of embankments or foundations placed on or wholly within the zone of freezing. Thawing of ice-rich soils results in settlement and loss of bearing strength (termed thaw instability).

The published information relative to geotextile uses in construction to limit or ameliorate problems associated with frost heaving and/or thaw weakening is limited. Kinney (1981) has suggested that a geotextile layer could be used to bridge thermokarsts, but no installations are known. Creep is a very important consideration in this use. Brantman et al. (1977) reported the use of a geotextile as a capillary break in a roadway test section in the USSR but did not include an evaluation of the performance. Sale et al. (1973) and Bell and Yoder (1957) describe encapsulation of roadbed soils in waterproof membranes to limit frost effects.

Roth (1977) performed laboratory soil column freezing tests. He found a 10 cm (3.9 in.) thick layer of gravel sandwiched between two geotextile layers worked as a capillary cutoff and limited frost heave. He also concluded the cutoff layer should be above the free water but below the freezing front for maximum effectiveness. Hollingsworth (1976) and Hoover et al. (1981) performed laboratory capillary cutoff tests using only a fabric as the capillary break. They tested only one fabric (Mirafi 140). Both conclude the geotextile reduced water migration, and Hoover observed that two geotextile layers worked better than one layer. The geotextile properties which influence the performance of a geotextile layer as a capillary break are permeability, pore size, thickness, and wetting characteristics.

Hoover et al. (1981) installed Mirafi 140 as a combined reinforcement layer and capillary break in county roads in Linn County, Iowa, USA. The geotextile was placed on top of a frost-susceptible subgrade soil. They observed limited improvement in roadway stability.

The existing literature shows numerous individuals believe geotextiles to have great potential for use in arctic and subarctic regions. The results to date support this belief but are limited and often inconclusive.

In recognition of the need to determine geotextile mechanical properties and the effectiveness of geotextiles to reduce frost action in cold regions, a research program was conducted at Oregon State University to investigate (1) the influence of freeze-thaw cycles in a freshwater and saline water environment on the load-strain-strength of geotextiles; (2) evaluate the load-strain-strength and creep characteristics of geotextiles at subfreezing temperatures; and (3) investigate the ability of geotextiles to act as capillary breaks and thereby limit frost heave. The results from the research program are summarized herein with emphasis on the frost heave tests. Additional results from the load-strain-strength and creep studies at subfreezing temperatures have been published elsewhere (Allen et al. 1982, Allen 1983).

GEOTEXTILES TESTED

Five geotextiles, each with different structure and/or material characteristics, were employed in the research program. These geotextiles are described in Table 1. The geotextile types were selected to ensure that the load-strain-strength characteristics associated with several geotextile structures and fiber polymer compositions could be compared in a meaningful way.

The geotextiles tested represented woven and nonwoven fabrics. For the nonwovens, needle-punched and bonded materials were tested. Both polyester and polypropylene fibers were included. The geotextiles represented a range of permeability, strength, modulus, and susceptibility to creep.

TEST PROCEDURES AND EQUIPMENT

Tensile Tests

A wide strip tensile test was used to characterize the load-strain-strength behavior of the geotextiles. A 200 mm (8.0 in.) specimen width and a 100 mm (4.0 in.) specimen length between test grips were employed to ensure the results obtained would simulate, to as great a degree as practical, the plane-strain loading conditions which exist in the field (Shrestha and Bell 1982). Five specimens of each geotextile type were tested. Both MTS and Instron test systems were employed in the program. The Instron was used for specimens which exhibited very large strains at failure.

Each geotextile type was tested under the following conditions:

- room temperature [22°C (71°F)] in a dry state
- subfreezing temperature [-12°C (10°F)] in a dry state
- room temperature in a saturated surface dry (wet) state
- room temperature [22°C (71°F)] in a saturated surface dry (wet) state following 300 freeze-thaw cycles in a dry state

TABLE 1 Geotextiles Tested.

Geotextile	Filament	Geotextile Construction	Nominal Weight gm/m^2
Bidim C-34	Polyester Continuous	Nonwoven Needlepunched	272
Stabilenka T-100	Polyester Continuous	Nonwoven Resin Bonded	100
Typar 3401	Polypropylene Continuous	Nonwoven Heat Bonded	136
Fibretex 300	Polypropylene Continuous	Nonwoven Needlepunched	300
Propex 2002	Polypropylene • Slit Film	Woven	150

- room temperature [22°C (71°F)] in a saturated surface dry (wet) state following 300 freeze-thaw cycles in distilled water
- room temperature [22°C (71°F)] in a saturated surface dry (wet) state following 300 freeze-thaw cycles in saline water.

The freezing temperature condition was achieved by placing the geotextiles in a walk-in cold room for 24 hours, and testing in the cold room. The saturated surface dry (wet) condition was obtained by soaking the geotextiles in water for 24 hours and blotting the specimens just prior to testing. Freeze-thaw cycling in either distilled or saline water was accomplished by placing a specimen in a sealed plastic bag filled with either distilled or saline water and placing the bag on a rack in a freeze-thaw chamber.

The tensile strength of each specimen was normalized to a nominal mass per unit area to account for specimen variability. The normalized strengths of five specimens of a given geotextile type for a given test condition were averaged. The significance of the average values when compared with the average normalized strength of other test cases was determined using a student's t-distribution for a 90% confidence level.

Creep Tests

The creep characteristics of the geotextiles were evaluated at room temperature [22°C (71°F)] and subfreezing temperature [-12°C (10°F)] when loaded to various percentages of ultimate wide strip tensile strength. Three test specimens of a given geotextile type for a given test condition were trimmed to a width of 152 mm (6.0 in.) and secured in grips with a spacing of 76 mm (3.0 in.). The three test specimens were connected in series for dead weight loading.

Average normalized strength was used to determine the load required for each set of creep specimens. Each geotextile type was tested at load levels of 20, 35, 50, and 65% of ultimate wide strip tensile strength. The loads associated with a given test condition were applied as rapidly as possible without impact, and deformation readings were taken at 1, 2, 5, 10, 30, 60, 120, 240, 1,440, and 2,880 min. Thereafter, readings were taken every week up to approximately 15 weeks or until rupture occurred. A test was also terminated when no measurable deformation occurred during a period of one week. The subfreezing temperature condition was achieved by conducting the creep tests in a walk-in cold room.

Frost Heave Tests

Frost heave tests were performed in the frost heave chamber shown schematically in Figure 1. The chamber, operated in a walk-in cold room, contained three tapered lucite cups 171 mm (6.75 in.) high with 116 mm (4.17 in.) and 105 mm (4.13 in.) top and bottom diameters, respectively. The soil to be tested was contained in the cups. Two cups had small holes drilled through the bottom and a 6.5 mm (0.25 in.) thick porous stone on the bottom. The third cup had no holes, thereby excluding the entry of water. The three specimen cups were held in a 0.394 m (15.5 in.) diameter disk of high-density styrofoam (brand) insulation to assure uniaxial freezing. The base of the cups extended below the bottom of the styrofoam so water in the bath could be raised above the level of the porous stones. Thermistors were installed through the side of one of the cups with holes in the bottom. The thermistors were spaced vertically at 14 mm (0.56 in.) intervals. Thermistors were placed only at the top and bottom of the other two cups.

The water bath was insulated and the water temperature was maintained at 1°C (34°F). An environmental chamber was used to control the air temperature above the soil specimens at -5°C (23°F).

Three dial gauges were used to measure the heave for each specimen. Thin [0.3 mm (0.012 in.)] rubber membranes were placed around each soil specimen. The inside of each cup was coated with silicone grease to reduce friction. Plastic wrap was placed on the top of each specimen to reduce sublimation during freezing.

The test soil was a nonplastic silt from Fairbanks, Alaska, known to be highly frost susceptible. The test specimens were carefully compacted to 90% of standard maximum dry density (AASHTO T-99) at a water content 2% above optimum.

After the soil specimens were compacted in the cups and the cups were placed in the styrofoam insulation container, the soil was protected against evaporation and the assembly was precooled for 12 hours at 1°C (34°F). After chilling, the assembly was placed on the water bath, the dial gauges were mounted and zeroed, and the environmental chamber was placed over the assembly. The system was allowed to cool until the 0°C (32°F) isotherm had moved a maximum of 6 mm (0.25 in.) below the surface of the soil.

When the freezing front reached a depth of about 6 mm (0.25 in.), initial readings of all thermistors and dial gauges were taken. As freezing progressed, dial gauge and thermistor

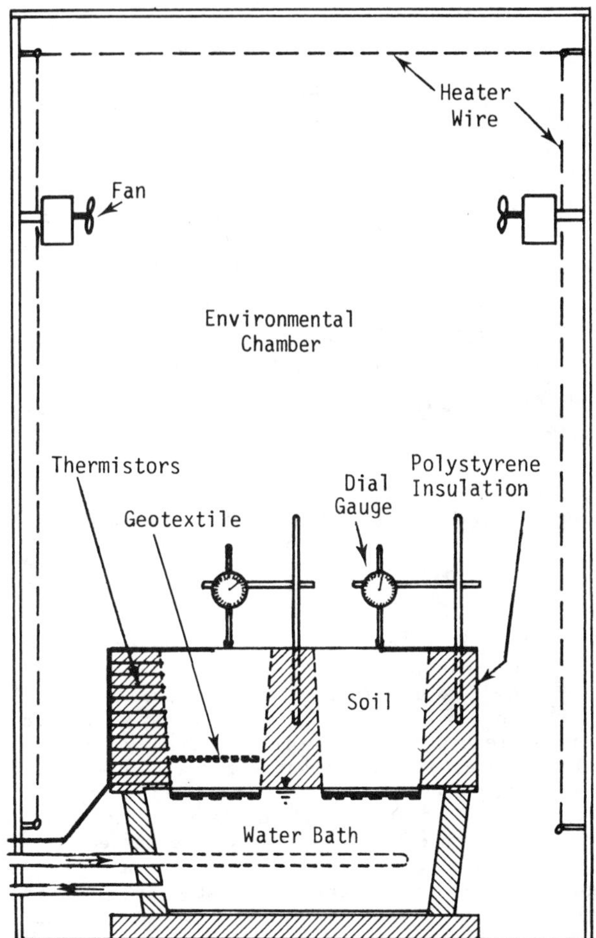

FIGURE 1 Frost heave test cell.

readings were taken at 0.5, 1, 2, 3, 5, 8, 12, 18, and 24 hours. Readings were continued after 24 hours or until the freezing front reached the level of the geotextile layer, whichever was less. At the end of the test the specimens were removed from the cells and moisture content samples taken. Water content was determined immediately above and below the location of the freezing front at the end of the test and at approximately 19 mm (0.75 in.) intervals throughout the length of each specimen.

Three specimens were tested simultaneously with each test. The specimens were prepared as identically as possible except one had a geotextile layer 38 mm (1.5 in.) above the bottom of the specimen. The specimen with the geotextile and one other specimen were open to free water at the bottom. The third specimen was for control and had neither geotextile nor access to free water.

TEST RESULTS

Tensile Load-Strain-Strength

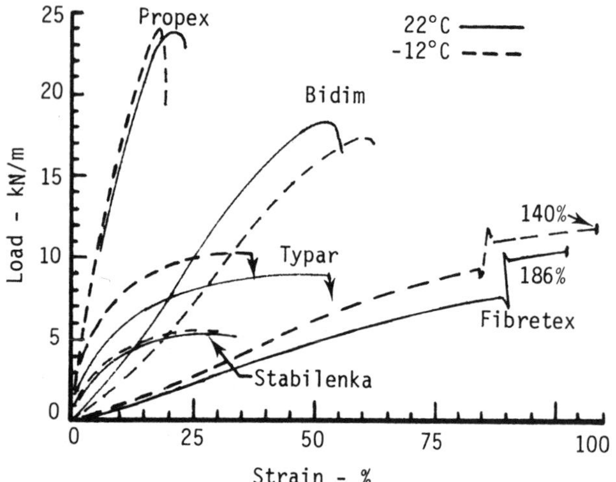

FIGURE 2 Axial load versus strain (dry condition).

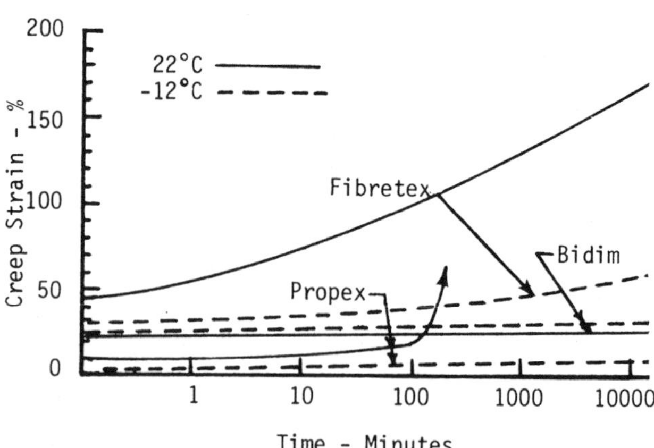

FIGURE 3 Creep strain versus time at 50% of tensile strength.

Tensile axial load versus strain relationships for the geotextiles tested at temperatures of 22° and -12°C (71° and 10°F) are shown in Figure 2. The modulus and strength of the Typar increased with decreasing temperature. The strain at failure decreased significantly for all polypropylene geotextiles. No other statistically significant (90% probability) deviations in load-strain-strength characteristics were observed.

A summary of all the tensile load-strain characteristics of the geotextiles tested is given in Table 2. The strain at failure for the resin-bonded polyester geotextile increased upon wetting. Other results indicate that normalized strength, percent strain at peak strength, and normalized secant modulus at 10% strain do not change appreciably from the initial dry values when the geotextiles considered were tested in a wet condition at room temperature or following 300 freeze-thaw cycles in a dry, distilled water, or saline water environment. An exception was Propex 2002 which exhibited a significant decrease in strength (approximately 12%) after 300 freeze-thaw cycles in distilled and in saline water.

Creep

Typical creep strain versus time relationships for the geotextiles tested at 22° and -12°C (71° and 10°F) at a load level of 50% of the wide strip tensile strength are shown in Figure 3. A reduction in temperature to -12°C (10°F) resulted in a decrease in creep strains for geotextiles with polypropylene fibers. At this load level creep strains for geotextiles with polyester fibers were not significantly influenced by this temperature change.

This difference in the effect of temperature on creep for polypropylene and polyester geotextiles may be explained in terms of the glass transition temperatures of these two polymers. The glass transition temperature for polypropylene is about 0°C (32°F), whereas for polyester it is about 110°C. When the temperature of a polymeric material falls below its glass transition temperature, the ability of the material to deform is severely restricted. Since the glass transition temperature of the polyester is well above the range of temperatures tested in this investigation, and since for the polypropylene it is within this range, the deformation characteristics of polyester geotextiles change much less with temperature than those of the polypropylene geotextiles. At load levels of 20% ultimate strength, the decrease in creep strain with temperature was not statistically significant for any geotextile. The geotextile structure and polymer appear to dominate short-term creep strains, and the polymer and temperature control the long-term creep rates.

The polypropylene materials experienced tertiary creep and failed at load levels of 50% or 65% of wide strip tensile strength at 22°C (72°F). The geotextiles with polyester fibers did not fail at this temperature until the load level was 80% of wide strip tensile strength. Significantly, at -12°C (10°F) none of the polypropylene materials failed at a 50% load level, however, and only one geotextile (Typar) failed at a 65% load level.

Frost Heave

Frost heave rates for all tests are presented in Table 3, and water contents after freezing for two tests are shown in Figure 4. The data in Table 3 show all of the nonwoven geotextiles except the Stabilenka caused a marked reduction in the heave rate. The only woven fabric, Propex, was intermediate in its effect.

Of the three most effective geotextiles, Fibretex and Bidim are both thick and have high permeabilities. Typar, in contrast, is thin and has a low permeability. Typar is also the most hydrophobic and requires a head of about 75 mm (3 in.) of water to initiate flow. Of the fabrics

TABLE 2 Load-Strain Characteristics of Geotextiles.

Geotextile	Property	I	II	III	IV	V	VI
		Test Condition (see legend below)					
Bidim C-34	Normalized Strength (kN/m)	18.6	17.5	16.0	16.7	16.0	15.6
	% Strain at Peak Strength	53.2	60.0	59.9	64.6	58.2	59.0
	Normalized Secant Modulus at 10% strain (kN/m)	33.5	20.3	25.2	28.5	31.7	30.7
Stabilenka T-100	Normalized Strength (kN/m)	5.9	5.6	5.3	5.4	5.4	5.3
	% Strain at Peak Strength	32.6	27.9	44.0	44.1	41.6	43.9
	Normalized Secant Modulus at 10% strain (kN/m)	37.7	41.0	30.8	33.5	33.6	32.6
Typar 3401	Normalized Strength (kN/m)	8.9	10.3	8.8	9.0	9.0	8.8
	% Strain at Peak Strength	53.3	31.3	43.8	40.4	43.4	42.3
	Normalized Secant Modulus at 10% strain (kN/m)	56.9	75.2	61.9	63.8	61.5	63.6
Fibretex 300	Normalized Strength (kN/m)	10.3	-	9.9	10.3	10.6	10.9
	% Strain at Peak Strength	186	-	166	188	192	213
	Normalized Secant Modulus at 10% strain (kN/m)	6.1	7.5	7.5	4.7	5.5	5.4
Propex 2002	Normalized Strength (kN/m)	24.2	23.7	24.4	23.3	21.3	22.9
	% Strain at Peak Strength	21.2	18.0	19.8	22.7	20.7	22.0
	Normalized Secant Modulus at 10% strain (kN/m)	151	163	162	146	144	151

I = Control, Dry Condition (DC), 22°C (71°F) II = Control, DC, -12°C (10°F)
III = Control, Wet Condition (WC), 22°C (71°F) IV = 300 Cycles, Freeze-Thaw (FT), Dry Environ., WC
V = 300 Cycles, FT, Distilled Water Environ., WC VI = 300 Cycles, FT, Saline Water Environ., WC
Note: Specimens subjected to freezing and thawing were cycled between -15° and +15°C (5° and 58°F)

TABLE 3 Freeze Front Penetration and Heave Rates.

Geotextile	Freeze Rate mm/hr	Heave Rate - mm/hr: No Free Water or Geotextile	With Free Water: Without Geotextile	With Free Water: With Geotextile
Bidim C-34	0.91	0.08	0.46	0.18
Stabilenka T-100	0.58	0.01	0.46	0.56
Typar 3401	2.49	0.03	0.54	0.12
Fibretex 300	2.11	0.07	0.43	0.13
Propex 2002	1.75	0.08	0.58	0.32

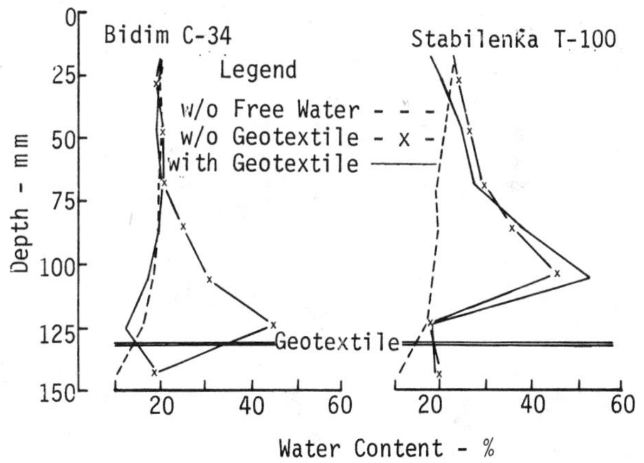

FIGURE 4 Soil water contents after freezing.

tested, only the Stabilenka was hydrophilic and readily adsorbed water.

Figure 4 illustrates the water contents in the soil specimens for the poorest and one of the best results. It is clear that for the conditions of these tests, Stabilenka had virtually no influence on the water migration while Bidim essentially stopped heave completely. These tests are qualitative and cannot be directly related to field conditions; however, it does appear that some geotextiles show potential as capillary breaks to limit frost heave. Different geotextiles have different effects and the factors affecting their effectiveness are not well defined. Additional research into this use of geotextiles is both needed and deserved.

CONCLUSIONS

Based upon the test results presented for the five geotextiles considered in this research program, the following conclusions have been reached: (1) the mechanical properties of geotextiles, in terms of load-strain-strength and creep characteristics, are not adversely affected by subfreezing temperatures in a temperature range associated with many cold regions engineering applications; (2) freeze-thaw cycling in dry, distilled water, or saline water environments has little influence on the load-strain-strength characteristics of most geotextiles; (3) geotextile structure and polymer tend to control short-term creep strains; (4) the fiber polymer and temperature control long-term creep rates; (5) for temperatures warmer than 0°C (32°F), polyester fabrics have much lower creep rates and higher thresholds of tertiary creep than polypropylene; however, for colder temperatures there is much less difference between the two polymers; (6) many geotextiles have considerable potential as capillary breaks to limit frost heave; (7) other factors equal, thick fabrics tend to be better capillary breaks than thin fabrics; (8) geotextiles which are hydrophobic are more effective capillary cutoffs than those which absorb water readily; and (9) geotextiles deserve further testing of their abilities as capillary breaks to limit frost action.

ACKNOWLEDGMENTS

The authors gratefully acknowledge the financial support provided by the Crown Zellerbach Corporation. The contents of this paper reflect the views of the authors, who are responsible for the facts and the accuracy of the data presented herein, and do not necessarily reflect the views or policies of the Crown Zellerbach Corporation. The efforts of Laurie Campbell and Julie Womack, who assisted in the preparation of this paper, are greatly appreciated.

REFERENCES

Allen, T., 1983, Properties of geotextiles in cold regions applications, M.S. Thesis, Department of Civil Engineering, Oregon State University, Corvallis, OR.

Allen, T., Vinson, T.S., and Bell, J.R., 1982, Tensile strength and creep behavior of geotextiles in cold regions applications, in Proceedings, Second International Conference on Geotextiles, V. III, pp. 775-780: St. Paul, MN, Industrial Fabrics Association International.

Bell, J.R. and Hicks, R.G., 1980, Evaluation of test methods and use criteria for geotechnical fabrics in highway applications, FHWA/RD-80/021: Washington, D.C., Federal Highway Administration.

Bell, J.R. and Yoder, E.J., 1957, Plastic moisture barriers for highway subgrade protection, in Proceedings, V. 36, pp. 713-735: Washington, D.C., Highway Research Board.

Brantman, B.P., Kazarnovsky, V.D., Polunovsky, A.G., and Ruvinsky, V.I., 1977, Experiments on the use of synthetic nonwoven materials for road structures, in Proceedings, International Conference on the Use of Fabrics in Geotechnics, V. 1, pp. 35-40: Paris, Ecole National des Ponts et Chaussees.

Hollingsworth, H., 1976, Evaluation of Mirafi 140 fabric as a capillary water barrier, Unpublished report: Casper, WY, Chen and Associates, Inc.

Hoover, J.M., Pitt, J.M., Handfelt, L.D., and Stanley, R.L., 1981, Performance of soil-aggregate-fabric systems in frost-susceptible roads, Linn Co., IA, in Transportation Research Record No. 827, pp. 6-14: Washington, D.C., Transportation Research Board.

Kinney, T.C., 1981, Expanded report on the use of geotextiles to bridge thermokarsts, Unpublished report to Alaska DOT: Fairbanks, AK, Shannon and Wilson, Inc.

Leidersdorf, C.B., Potter, R.E., and Goff, R.D., 1981, Slope protection for artificial exploration islands off Prudhoe Bay, in Proceedings, 13th Annual Conference, V. VIII, pp. 437-447: Houston, TX, Offshore Technology Conference.

Roth, W.H., 1977, Fabric filter for improving frost susceptible soils, in Proceedings, International Conference on the Use of Fabrics in Geotechnics, V. 1, pp. 23-28: Paris, Ecole National des Ponts et Chaussees.

Sale, J.P., Parker, F., and Barker, W.R., 1973, Membrane encapsulated soil layers, in Journal of the Soil Mechanics and Foundations Division, V. 99, SM 12: New York, American Society of Civil Engineers.

Shrestha, S.C. and Bell, J.R., 1982, A wide strip tensile test of geotextiles, in Proceedings, Second International Conference on Geotextiles, V. III, pp. 739-744: St. Paul, MN, Industrial Fabrics Association International.

Tart, R.G. and Luscher, U., 1981, Construction and performance of frozen gravel fills, in Proceedings, Specialty Conference on the Northern Community - A Search for a Quality Environment, pp. 693-704: New York, American Society of Civil Engineers.

EFFECT OF COLOR AND TEXTURE ON THE SURFACE TEMPERATURE OF ASPHALT CONCRETE PAVEMENTS

R. L. Berg[1] and D. C. Esch[2]

[1]U.S. Army Cold Regions Research and Engineering Laboratory
Hanover, New Hampshire 03755 USA
[2]Alaska Department of Transportation and Public Facilities
Fairbanks, Alaska 99701 USA

During the fall of 1981 and the spring of 1982, eight test items were established on an asphalt pavement in Fairbanks, Alaska. The test items were: two sections of untreated pavement, yellow-painted pavement, white-painted pavement, "standard" chip seal, fine-grained "standard" chip seal, chip seal with dark brown aggregate, and chip seal with white marble aggregate. The test items were located on a main road. Surface temperatures were monitored hourly by thermocouples attached to an automatic data collection system. The ambient air temperature, wind speed and direction, amount of precipitation, and radiation balance were continuously recorded at an untrafficked pavement approximately 100 m from the test items. Incident and reflected shortwave radiation measurements were made nearly every weekday over each test item using a hand-held radiometer. N-factors, ratios of surface thawing indexes to air thawing indexes varied from about 1.2-1.3 for the white- and yellow-painted surfaces, respectively, to about 1.4-1.5 for the other surfaces. Daily and monthly n-factors for a particular surface varied depending on wind speed, the durability of the surface treatment, and the amount of incident solar radiation. Approximately 2200 vehicles per day crossed the test sections; turbulence induced by the traffic caused n-factors to be lower than reported by other authors. Abrasion by traffic reduced the effect of the surface treatments.

In the discontinuous permafrost regions of Alaska, extremely thick granular embankments (15-20 ft deep) may be necessary to completely contain seasonal thawing. Due to inadequate quantities of materials, costs of material acquisition and placement, or excessively elevated surfaces, the necessary thick embankments are seldom constructed, and considerable differential settlement frequently results beneath roads and airfields. Three passive methods have been used to minimize seasonal thaw depths and thereby minimize embankment thickness requirements: (1) thermal insulating layers (Berg 1976, Esch 1973); (2) high water content layers (Esch and Livingston 1978); and (3) materials that reduce the surface temperature of the embankment (Fulwider and Aitken 1963, Berg and Aitken 1973).

In this paper we discuss the effects of surface color and texture on the thawing season surface temperatures of an asphaltic concrete roadway pavement in Fairbanks, Alaska. The roadway carries a substantial volume of traffic each day, so this study differs from previous surface temperature studies of pavements that were not subjected to significant traffic-generated air movements and to the effects of tire wear on surface coatings.

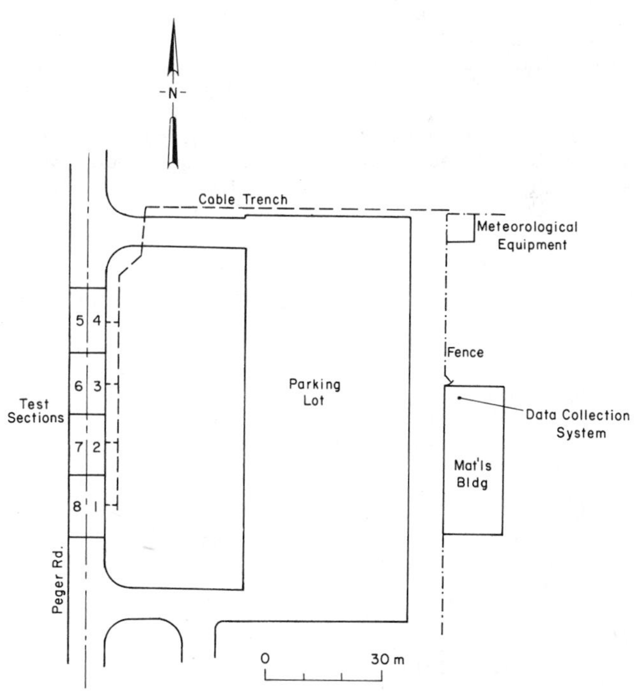

FIGURE 1 Layout of test sections and instrumentation at Peger Road test site.

SITE PREPARATION AND INSTRUMENTATION

The test sections are located on Peger Road between the entrances to the Interior District offices of the State of Alaska Department of Transportation and Public Facilities (DOT/PF) (Figure

1). Peger Road in Fairbanks, Alaska, was reconstructed in the summer of 1981, and in the fall of 1981 and spring of 1982 eight test sections were installed. The test items were: two sections of untreated pavement, yellow-painted pavement, white-painted pavement, "standard" chip seal, fine-grained "standard" chip seal, chip seal with a dark brown aggregate, and a chip seal with white marble aggregate. Prior to placing the 5-cm (2-in.) thick asphaltic concrete pavement, DOT/PF engineers installed four 2.5-cm (1-in.) diameter PVC pipes horizontally near the top of the base course. The pipes extended beyond the pavement on each side by about 0.6 m (2 ft). In the late summer, 15-cm (6-in.) diameter cores were removed from the pavement over the PVC pipes, and two 6-mm (1/4-in.) diameter holes were drilled through the cores from bottom to top. Two copper and two constantan thermocouple wires were fed through the PVC pipe from each core hole to the edge of the road. Two copper constantan thermocouples were then fabricated at each core hole and a thermocouple was epoxied into each hole, with leads extending from the bottom of the cores. The sensing tip of each thermocouple was approximately 3 mm (1/8 in.) from the top of the core. Warm liquid asphalt cement was poured on the exposed base course, and the core containing the two thermocouples was pushed into the asphalt cement. Additional liquid asphalt cement was poured into the annulus between the core and the undisturbed pavement.

Thermocouple wires were buried beneath the ground or laid on the ground surface from the edge of the pavement to the DOT/PF office building (Figure 1), where a data collection system was located. Meteorological equipment consisting of radiometers to measure incident and reflected shortwave radiation and incoming and emitted longwave radiation, equipment to measure wind speed and wind direction, and hardware to measure precipitation and air temperature were located within a fenced area to prevent vehicles from damaging the equipment. The meteorological equipment is located approximately 100 m (300 ft) from the test sections. Output from the meteorological equipment was transmitted to the data collection system and other recorders through extension wires mounted on the fence. Measurements from all of the equipment commenced in March or April 1982 and will continue through early summer 1983. The data collection system recorded the surface temperatures and radiation totals hourly during summer 1982. Average wind speed and direction were also recorded hourly. Between 1100 and 1300 hr each day from April through September 1982, a hand-held radiometer was used to make instantaneous measurements of incident and reflected shortwave radiation over the test sections. The albedo of each surface was computed from these measurements.

The layout of the test sections is shown in Figure 1. Each test section was approximately 15.2 m (50 ft) long by 5.9 m (16 ft) wide. The average daily traffic on Peger Road during the 1982 summer was approximately 2200 vehicles per day in each direction. The speed limit on Peger Road is approximately 22 m/sec (50 mph). Gradations and asphalt application rates for each of the chip seals are shown in Table 1. Different liquid asphalt application rates were necessary to retain the various gradations of chips properly.

DATA ANALYSIS

The mean and design air freezing and thawing indexes from observations at the Fairbanks International Airport (period of record 1950 through 1979) are:

Mean air freezing index	3200°C-days	(5760°F-days)
Design air freezing index	3724°C-days	(6704°F-days)
Mean air thawing index	1844°C-days	(3320°F-days)
Design air thawing index	2104°C-days	(3787°F-days)

The design values were calculated as the average of the three extremes in the 30 years of record.

Thawing indexes from the Peger Road site and the Fairbanks International Airport for the 1982 thawing season are shown in Table 2. Summer 1982 was considerably warmer than normal and the air thawing index measured at the Peger Road site was

TABLE 1 Gradations, colors, and asphalt application rates for Peger Road chip seals.

	Fairbanks S&G		Browns Hill	White marble
	E-chips	C-chips	B-chips	C-chips
Test section	1*	2*	3	4
sieve size	% passing	% passing	% passing	% passing
0.75"	100	100	100	100
0.50"	100	100	53	100
0.375"	93	51	16	86
#4	25	2	4	2
#8	7	1	4	1
#200	1	0	1	0
Color	light gray		dark brown	white
Asphalt				
(gal/yd^2)	0.11*	0.24*	0.43	0.31
(L/m^2)	0.50	1.09	1.95	1.40

* Resealed on 26 August 1982 with asphalt application rate of 1.58 L/m^2 (0.35 gal/yd^2) and C-chips.

TABLE 2 1982 thawing indexes and n-factors, Peger Road and Fairbanks International Airport.

Month	Air index (°F-days)	Fairbanks S&G E-chips Surface index (°F-days)	n	Fairbanks S&G C-chips Surface index (°F-days)	n	Browns Hill B-chips Surface index (°F-days)	n	White marble C-chips Surface index (°F-days)	n
Apr	59.9	198.1	3.31	165.4	2.76	151.4	2.53	133.6	2.23
May	491.7	850.4	1.73	780.5	1.58	748.8	1.52	755.9	1.54
Jun	851.0	1243.3	1.46	1191.7	1.40	1147.2	1.35	1182.8	1.39
Jul	947.7	1369.3	1.44	1345.7	1.42	1309.4	1.38	1360.5	1.44
Aug	758.9	1062.2	1.40	1048.3	1.38	1033.8	1.36	1077.8	1.42
Sep	464.8	607.5	1.31	599.7	1.29	618.4	1.33	638.3	1.37
Oct	1.4	14.6	10.43	12.6	9.0	15.9	11.36	20.5	14.64
Total	3575.4	5345.4	1.50	5143.9	1.44	5024.9	1.41	5169.4	1.45
°C-days	1986.3	2969.7		2857.7		2791.6		2791.6	
Start	23 Apr	12 Apr		12 Apr		13 Apr		13 Apr	
End	2 Oct	7 Oct		6 Oct		7 Oct		8 Oct	
Length	162	178		177		177		178	

n = surface index ÷ air index

Month	White paint Index (°F-days)	n	Yellow paint Index (°F-days)	n	Bare pavement Index (°F-days)	n	Air index, Fairbanks Int'l Airport (°F-days)
Apr	143.4	2.39	172.6	2.88	219.0	3.66	60.0
May	764.8	1.56	796.5	1.62	859.6	1.75	458.8
Jun	967.6	1.14	1047.5	1.23	1254.0	1.47	795.0
Jul	1080.8	1.14	1152.1	1.22	1378.9	1.46	964.1
Aug	875.5	1.15	914.9	1.21	1092.8	1.44	762.6
Sep	533.3	1.15	547.4	1.18	640.5	1.38	519.0
Oct	9.2	6.57	9.4	6.71	18.9	13.50	9.0
Total	4374.6	1.22	4640.4	1.30	5463.7	1.53	3568.5
°C-days	2430.3		2578.0		3035.4		1982.5
Start	15 Apr		12 Apr		12 Apr		23 Apr
End	3 Oct		3 Oct		7 Oct		3 Oct
Length	171		174		178		163

slightly warmer than at the airport, which is about 3 km (2 mi) west of the Peger Road test site.

Monthly and seasonal n-factors, determined from:

n = surface thawing index ÷ air thawing index

are also shown in Table 2.

The n-factor for the bare pavement was 1.53; for the chip seals, n-factors varied from 1.41 to 1.50. N-factors for the white-painted and yellow-painted pavements were 1.22 and 1.30, respectively. On the basis of data from airfield pavement studies in Fairbanks and at Thule, Greenland, the U.S. Departments of the Army and the Air Force (1966) relate summer n-factors to the pavement type and average summer wind speed. The average wind speed measured beyond the zone of traffic-induced turbulence near Peger Road was 0.7 m/sec (1.6 mph) for the 1982 summer. For an asphaltic concrete pavement, an n-factor of about 2.8 is obtained from the above reference.

Air and pavement surface temperatures have recently been measured at several arctic and subarctic locations. Table 3 contains n-factors that were determined from data gathered from asphaltic concrete pavements. In general, airfield pavements and test sections exhibit the highest n-factors and roadway pavements the lowest. The n-factors computed for Peger Road were significantly lower than those calculated for other roadway surfaces.

Airfields and test sections do not receive as much traffic to induce air turbulence as do roadway surfaces. At the location where surface temperatures were measured on the Richardson Highway, approximately 700 vehicles per day use the road. Approximately 4400 vehicles per day use Peger Road. It is unlikely that either the Kotzebue or Inuvik airports have more than 100 aircraft movements per day and probably fewer than 10 vehicles per day use the airfield or roadway test sections. These data imply that turbulence induced by vehicular traffic is an important influence on the n-factor of an asphaltic concrete pavement. In addition, vehicular

TABLE 3 Summer n-factors for asphalt concrete pavements in arctic and subarctic environments.

Location	Reference	Period of record (yrs)	Type of facility	n-factors Range	Average
Fairbanks	Berg & Aitken (1973)	1	AT[1]	2.11-2.28	2.19
Fairbanks	Berg & Aitken (1973)	1	RT[2]	1.72-1.96	1.84
Fairbanks	Berg & Aitken (1973)	1	RT	---	0.98[3]
Fairbanks	Lundardini (1978)	1	AT	1.40-2.13	1.92
Inuvik	Johnston (1981)	5	A[4]	1.70-1.89	1.79
Kotzebue	Esch & Rhode (1976)	2	A	1.49-1.72	1.60[5]
Kotzebue	Esch & Rhode (1976)	2	A	1.66-2.00	1.84[6]
Farmers Loop	DOT/PF data	4	R[7]	1.52-1.64	1.56
Parks Hwy	DOT/PF data	5	R	1.59-1.86	1.70
Richardson Hwy	DOT/PF data	2	R	1.58-1.66	1.62

[1] Airfield test sections
[2] Roadway test sections
[3] White painted pavement
[4] Airfield
[5] 30 cm (12 in.) below pavement surface
[6] Insulated pavement
[7] Roadway

TABLE 4 Average monthly measurements of albedo of test sections (albedo values expressed as a percentage of incident shortwave radiation).

Month	Fairbanks E-chips	S&G C-chips	Browns Hill B-chips	White marble C-chips	White paint	Yellow paint	Bare pavement
Apr*	16	17	17	24	24	28	--
May	11	13	13	17	20	19	--
Jun	10	12	14	15	45[1]	41[2]	15
Jul	8	9	9	11	45	38	13
Aug	12[3]	11[4]	12	14	46	44	13
Sep	17	18	19	19	45	42	11[5]
Avg	12	13	14	17	38[6]	35[7]	13

* Days 12-30 only
[1] Repainted afternoon 8 June (23 on 1-8 June)
[2] Repainted afternoon 8 June (22 on 1-8 June)
[3] Additional liquid asphalt and C-chips added on 26 August (13 on 1-26 August)
[4] Additional liquid asphalt and C-chips added on 26 August (11 on 1-26 August)
[5] Radiometers removed on 24 September
[6] 45% after 8 June
[7] 41% after 8 June

traffic may indirectly cause the pavement to absorb less solar radiation by wearing the black asphaltic concrete coating off a larger portion of the generally lighter-colored aggregate.

Berg and Aitken (1973) reported that the albedo of the asphaltic concrete test sections was 16%. At Peger Road the albedo averaged 13% (Table 4) on the bare pavement; therefore we conclude that the surface albedo did not cause the lower temperatures observed on Peger Road. Turbulence caused by passing vehicles seems to be the most logical explanation. Since increased turbulence causes increased convective heat loss, lower surface temperatures result. Lunardini (1981) also indicates that increased average summer wind speeds cause reduced surface thawing indexes.

Surface temperatures on the white-painted test section on Peger Road resulted in an n-factor of 1.22 for the season. Berg and Aitken (1973) reported an n-factor of 0.98 and an albedo of 66% for the white-painted test section in their study. The albedo for the white-painted test section on Peger

Road averaged 38% for the entire thawing season and 45% after the test section was repainted on 8 June (Table 4). The most probable cause of the reduced albedo on Peger Road was degradation of the paint coating due to traffic.

The yellow-painted test section on Peger Road reflected nearly as much radiation as the white paint. The albedo averaged 35% for the entire summer and 41% after it was repainted on 8 June. The n-factor for the yellow-painted test section was 1.30, only slightly larger than that for the white-painted section.

None of the chip seals reduced the surface temperatures as much as the white or yellow paint. The Browns Hill B-chips were most effective. They were also the largest and were applied with the highest application rate of asphalt. The Fairbanks S&G and the white marble C-chips resulted in about the same n-factor. The white marble chips fractured due to traffic loadings. Had they remained intact they probably would have been more effective than the dark or colored chips. The Fairbanks S&G E-chips were the least effective chip seal. The application rate of asphalt in 1981 was inadequate to properly bond the Fairbanks S&G E-chips to the pavement; approximately one-half of the chips were lost, exposing the bonding layer of asphalt and lowering the albedo. White and yellow paints were applied twice to the roadway; each coat was approximately 0.2 mm (0.01 in.) thick. Annual repainting would be necessary to provide consistent coverage of roadways. Painting should be completed early in the thawing season to avoid damage by studded tires.

CONCLUSIONS

In this study, white or yellow paint applied to the pavement surface reduced surface temperatures and n-factors more than any of the chip seals.

Comparing these results with those reported by Berg and Aitken (1973) indicates that painted treatments may be more effective on airfield pavements where the volume of traffic is inadequate to cause rapid degradation of the paint. Airfield pavements may require repainting once every 5 years or more, but roadways will probably require repainting every year, due primarily to removal of paint by studded tires. Asphaltic concrete airfield pavements probably remain "blacker" due to less tire abrasion removing the asphalt coating from the aggregate.

The data obtained at Peger Road and at other arctic and subarctic sites indicate that traffic-induced air turbulence appears to reduce thawing season n-factors, i.e. the numerical value of the n-factor decreases with increased amount of traffic.

ACKNOWLEDGMENTS

This study was sponsored by the State of Alaska Department of Transportation and Public Facilities. The meteorological equipment and observers were furnished by the U.S. Army Atmospheric Science Laboratory detachment at Ft. Wainwright. Jerry Bower, DOT/PF, changed magnetic tapes and reported equipment malfunctions to the investigator from the U.S. Army Cold Regions Research and Engineering Laboratory (USACRREL). Richard Guyer, USACRREL, progammed and installed the data collection system. Gregor Fellors, Gary DeCoff, and Linda Gee, USACRREL, prepared and executed the computer programs for converting, totaling, and averaging the data.

REFERENCES

Arctic Construction and Frost Effects Laboratory, 1950, Comprehensive report, investigation of military construction in arctic subarctic regions, 1945-1948: ACFEL Technical Report 28: Hanover, N.H, U. S. Army Cold Regions Research and Engineering Laboratory.

Berg, R.L., 1976, Thermoinsulating media within embankments on perennially frozen soil: Hanover, N.H., U.S. Army Cold Regions Research and Engineering Laboratory, Special Report 76-3.

Berg, R.L., and Aitken, G.W., 1973, Some passive methods of controlling geocryological conditions in roadway construction, in North American Contribution, Second International Conference on Permafrost: Washington, D.C., National Academy of Sciences, National Research Council.

Esch, D.C., 1973, Control of permafrost degradation beneath a roadway by subgrade insulation, in North American Contribution, Second International Conference on Permafrost: Washington, D.C., National Academy of Sciences, National Research Council.

Esch, D.C., and Livingston, H., 1978, Performance of a roadway with a peat underlay over permafrost, interim report for period 1973-1977: State of Alaska Department of Transportation and Public Facilities.

Esch, D.C., and Rhode, J.J., 1976, Kotzebue airport, runway insulation over permafrost, in Proceedings of the Second International Symposium on Cold Regions Engineering: Fairbanks, Alaska: Cold Regions Engineers Professional Association, p. 44-61.

Fulwider, C.W., and Aitken, G.W., 1963, Effect of surface color on thaw penetration beneath a pavement in the arctic, in Proceedings, First International Conference on the Structural Design of Asphalt Pavements: Ann Arbor, Michigan, Braun-Brunfield, Inc.

Johnston, G.H., 1981, Design and performance of the Inuvik, N.W.T. airstrip, in Proceedings, Fourth Canadian Permafrost Conference: National Research Council of Canada.

Lunardini, V.J., 1978, Theory of n-factors and correlation of data, in Proceedings, Third International Conference on Permafrost: National Research Council of Canada.

Lunardini, V.J., 1981, Heat transfer in cold climates: New York, Van Nostrand Reinhold Company.

U.S. Departments of the Army and the Air Force, 1966, Arctic and subarctic construction - calculation methods for determination for depths of freeze and thaw in soil: TM 5-852-6.

WEICHSELIAN PINGO REMNANTS (?) IN THE EASTERN PART OF THE NETHERLANDS

S. Bijlsma*

Soil Survey Institute and State Geological Survey,
P.O. Box 98, 6700 AB Wageningen, Netherlands

Three depressions with depths ranging from 4 to 19 m were investigated. It was determined that they are most probably pingo remnants. The pingos formed in fluvial sediments, perhaps between circa 29,000 and 23,000 years BP. Before circa 23,000 years BP, no surface expression of the pingos was present. It is assumed that a buried ice core remained and was covered by eolian and fluvial deposits. Circa 12,000 years BP the buried ice core melted and the overlying material subsided to form water-filled depressions, that subsequently filled with organic matter. The pingo remnants have no ramparts.

*Deceased

INTRODUCTION

Closed topographic depressions interpreted as pingo remnants are common in the till covered area of the northern part of the Netherlands (Figure 1) and were first described by Maarleveld and Van den Toorn (1955). A recent summary of the investigations of these pingo remnants is given by De Gans and Sohl (1981) and by De Gans (1982).

A few pingo remnants have been described from other parts of the Netherlands (Maarleveld 1976, Bisschops 1973).

The purpose of this paper is to describe depressions in the eastern part of the Netherlands that can be interpreted as pingo remnants.

GEOLOGICAL FRAMEWORK

The eastern part of the Netherlands was covered by ice during the penultimate glacial, the Saalian. The ice left a landscape of ice-pushed ridges, basins, and almost level plains with till and glaciofluvial deposits. In the last glacial, the Weichselian (Table 1), the glaciers did not reach the Netherlands: this time periglacial conditions existed.

In the Early and Middle Pleniglacial (Table 1), small rivers laid down deposits of sand intercalated with loam and peat. Layers of fine sand and loam are particularly common in the upper part of the deposits. These fluvial deposits are called (niveo-)fluviatile by Van der Hammen et al. (1967), and fluvioperiglacial deposits (which include fluvial and lake deposits) by Zagwijn and Paepe (1968). Fluvioperiglacial deposits occur in the fieldwork area (Figure 1) in all the lower areas between the ice-pushed ridges. Palynological studies (Van der Hammen et al. 1967, Zagwijn 1974, De Gans and Cleveringa 1981) indicate a severe climate, with some warmer interstadial intervals (Table 1) during the deposition of the fluvioperiglacial deposits. It seems probable that during the Early and Middle Pleniglacial there was permafrost in areas not occupied by rivers.

In the Late Pleniglacial eolian sands were deposited over nearly the whole fieldwork area. These fine grained sands with thin layers of loam are called Older Coversand I.

Overlying the Older Coversand I is the Beuningen Complex. The basal part of the Beuningen Complex is an arctic soil developed in cryoturbated Older Coversand I (Van der Hammen et al. 1967). This soil is often overlain with a layer of coarser (fluvial ?) deposits with large ice-wedge casts (Maarleveld 1976, Kolstrup 1980). The uppermost part of the Beuningen Complex is a thin eolized pebble band called the Beuningen Gravel Bed.

TABLE 1 Stratigraphy of the Weichselian.

Chronostratigraphy				C-14 age	POLLEN ZONE
HOLOCENE					H
				10,000	
WEICHSELIAN	Late		Late Dryas Stad.	11,000	III
			Allerød Interst.	11,800	II
			Earlier Dryas St.	12,000	I c
			Bølling Interst.	13,000	I A/B
	Middle or Pleniglacial	Late		14,000	
				23,000?	
				27,000?	
		Middle		29,000	
			Denekamp Int.	32,000	
			Hengelo Int.		
			Moershoofd Int.		
				50,000?	
		Early			
	Early				
EEMIAN					
SAALIAN					

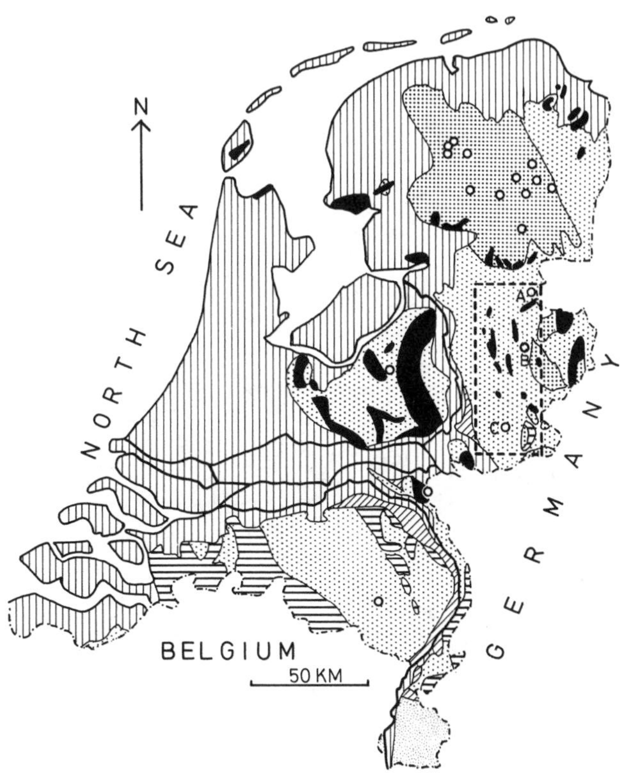

PRE-WEICHSELIAN FLUVIAL DEPOSITS

ICE-PUSHED RIDGES

TILL, OFTEN COVERED BY EOLIAN DEPOSITS

WEICHSELIAN FLUVIAL DEPOSITS

WEICHSELIAN FLUVIOPERIGLACIAL AND EOLIAN DEPOSITS

LOESS

o PINGO REMNANT

STUDY AREA

HOLOCENE DEPOSITS

FIGURE 1 Generalized geological map of the Netherlands showing the location of the investigated pingo remnants and other pingo remnants described in the literature.

Overlying the Beuningen Complex, eolian deposits of Older Coversand II (with the same lithology as the Older Coversand I) are found.

In the Late Weichselian the so-called Younger Coversands were deposited. They are coarser than the Older Coversands and often contain some fine gravel. Loam layers are absent. In contrast to the Older Coversands, accumulation of the Younger Coversands often took place in the form of dunes.

FIELD WORK

Closed topographic depressions are indicated on the Geomorphological map of the Netherlands, scale 1:50,000. This map also gives information on their surface form and present depth. From this map depressions were selected for preliminary field study. As the purpose of the survey was to identify pingo remnants, depressions had to be circular or oval and more than 2 m deep (De Gans 1982). The accessibility of the depression and the thickness of the organic infill determined whether it was selected for a more detailed investigation.

The described depressions are in almost flat plains that were originally covered with Holocene peat. The peat was dug out for fuel in the last centuries and after it had been removed, organic sediments only remained in the lowest parts of the terrain, usually former lakes.

Detailed cross sections were made of the depressions and their surroundings. The lithostratigraphy was established by means of hand drilling equipment. Stratigraphic units were identified and correlated using lithological properties such as grain size, loam content and the presence or absence of gravel. The soil of the Beuningen Complex, an important marker horizon, is not identifiable but the overlying coarse sands and the gravel bed are usually recognizable in borings.

Depression A (Nieuwe Veen, 6°40'00"E-52°33'49"N, see Figure 1) is nearly circular with a diameter of 350 m and an infill of organic material of 19 m thick (Figure 2). In all 45 borings were made to study the depression and the surrounding area. The depression formed in fluvioperiglacial deposits consisting of fine sands, alternating with layers of silt, probably lake deposits. These deposits are more than 4 m thick and the base was not reached by the borings. No rampart is present. Because of its depth the inorganic infill in the deeper parts could not be studied. An infill with Older Coversand I, deposits of the Beuningen Complex, and Older Coversand II were found on the sides of the depression. The lithology and thickness of these deposits seem to be the same as those in the surrounding area.

Some Younger Coversand is intercalated with the organic infill in the depression. Outside the depression the Younger Coversand is well developed. The beginning of organic sedimentation is dated as Earlier Dryas Stadial (pollen zone Ic, Table 1, Figure 5).

Depression B (Mokkelengoor, 6°36'00"E-52°19'20"N, Figure 1) has an oval shape with a short axis of 320 m and a long axis of 430 m. Its depth, including the inorganic infill, is 4 m (Figure 3). 61 Borings were made to study this depression. The depression cuts through fluvioperiglacial deposits but not through the underlying till.

The fluvioperiglacial deposits are coarse grained at the base. The uppermost part of these deposits in the area surrounding the depression are very fine sands with thin silt layers that are sometimes peaty. They are interpreted as lake deposits.

No rampart was found. Older Coversand I is present in the depression as well as gravelly sands interpreted as a part of the Beuningen Complex. Older Coversand II is well developed in the surrounding area but is very thin in the depression. The beginning of organic sedimentation is dated as

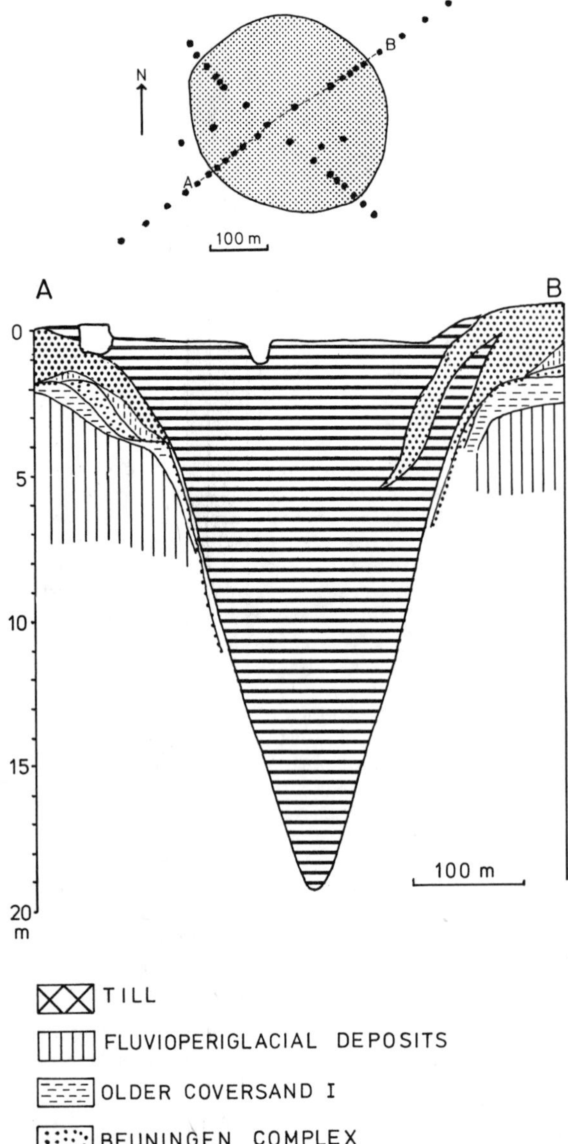

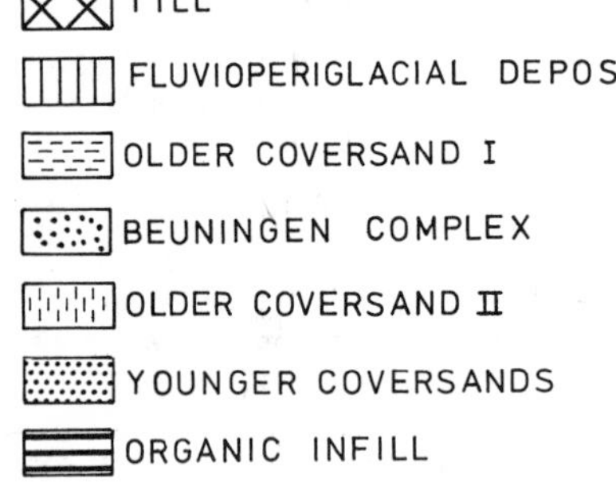

FIGURE 2 Depression A. Upper part: shape of the depression, location of the borings and the cross section. Lower part: cross section.

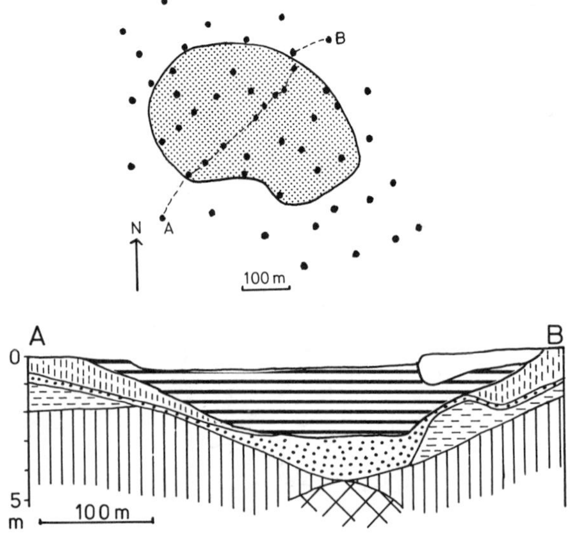

FIGURE 3 Depression B. Upper part: shape of the depression, location of the borings and the cross section. Lower part: cross section. For legend see Figure 2.

Earlier Dryas Stadial (pollen zone Ic, Table 1, Figure 5).

Depression C (Mariënvelde, 6°28'00"E-52°01'00"N, Figure 1) is nearly circular with a diameter of 275 m and a depth, including the inorganic infill, of 8 m (Figure 4). 39 Borings were made to study this depression and the surrounding area. The depression is formed in fluvioperiglacial deposits consisting of fine grained sands. In most of the borings outside the depression, the uppermost layer is composed of very fine sand or silt, that can probably

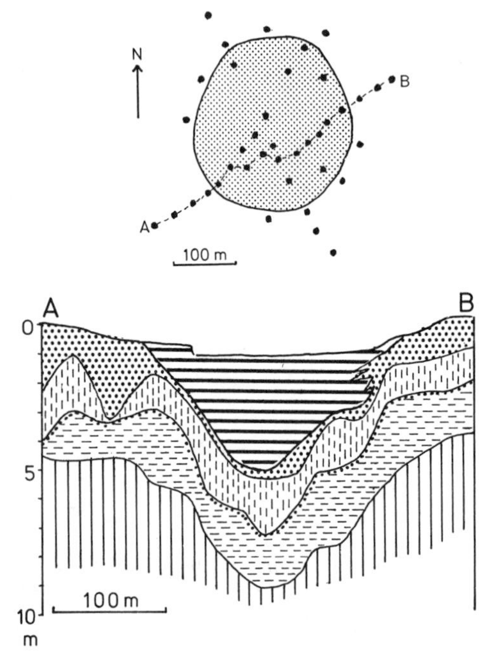

FIGURE 4 Depression C. Upper part: shape of the depression, location of the borings and the cross section. Lower part: cross section. For legend see Figure 2.

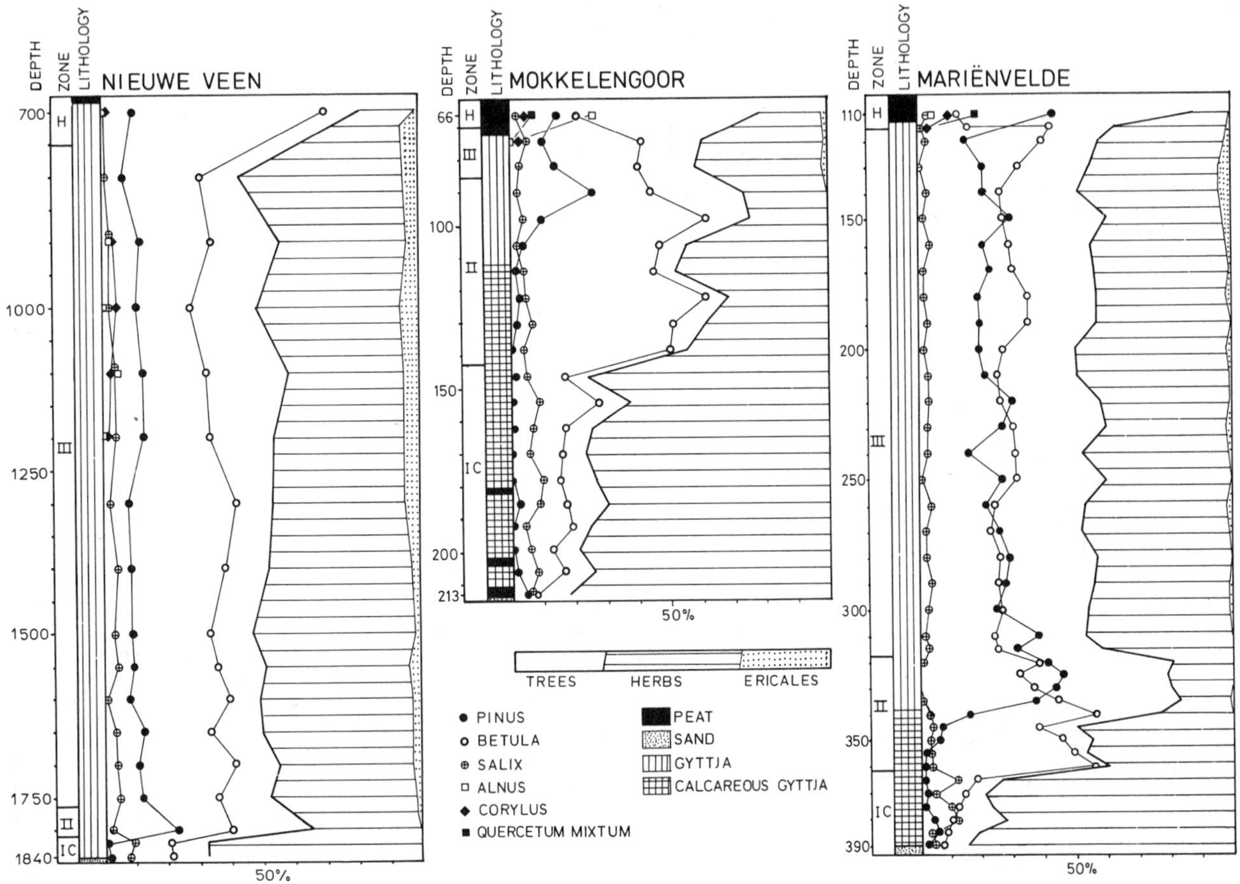

FIGURE 5 Pollen diagrams.

be interpreted as lake deposits.
The depression has no rampart. In the depression, Older Coversand I, a thin gravel layer interpreted as a part of the Beuningen Complex, and Older Coversand II are found. The lithology and thickness of these layers do not differ from those of the corresponding layers in the surrounding area. The Younger Coversands are very thin in the depression but are well developed in the surrounding area. The beginning of the organic sedimentation can be dated as Earlier Dryas Stadial (pollen zone Ic, Table I, Figure 5).

DISCUSSION

Origin

The three depressions have so much in common that their origins do not need to be discussed separately. Although they differ in diameter and depth, they all formed as nearly circular depressions in fluvial deposits. Late Pleniglacial deposits are present in all the depressions but they do not fill them. The beginning of organic sedimentation is roughly synchronous and starts circa 12,000 years BP.

It is often very difficult to identify pingo remnants with certainty (Flemal 1976). According to Flemal, pingo remnants are circular to oval ramparted or non-ramparted forms, with diameters ranging from a few tens to a few hundreds of meters. For areas without permafrost it must be proved that the depression formed in a period in which permafrost might have existed, i.e. in a glacial period.

Their circular form and the pollen data from the basal sections of the organic infill suggest that the depressions described above may be pingo remnants.

Flemal (1976) lists five possible mechanisms that can produce forms resembling pingo remnants: glacial, thermokarst, karst, eolian and anthropogenic processes.

Although the investigated area was covered by glacier ice in the Saalian, a glacial origin can be dismissed. Such depressions would have been filled with Eemian and Early Weichselian deposits, and not solely with Middle and Late Weichselian deposits.

As no soluble material underlies the depressions a karst origin is impossible. Anthropogenic processes are very unlikely, as pollen data from the organic infill indicate that deposition began during the Late Weichselian. Eolian processes can produce nearly circular ramparted or non-ramparted depressions that can reach the same dimensions as the described depressions, but are usually less than 2 m deep (De Gans and Sohl 1981). An eolian origin is therefore unlikely.

Thermokarst processes include the differential

melting of massive segregated ice (pingos, palsas, ice wedges or massive tabular sheet ice) or dispersed ice. It is difficult to envisage large bowl-shaped depressions being formed by differential melting of even large complex ice wedges, massive tabular ice sheets or dispersed ice, as for this the bodies of ice would have to have been nearly circular. Palsas are unlikely to have fossilized, as their structure is such that the former surface will be restored after melting (Lundqvist 1969). Although some other modes of origin cannot be ignored, the most likely explanation of the origin of the depressions is the presence of a large circular ice body that uplifted the overlying deposits. Part of these deposits was subsequently removed by erosion, and after the ice melted, a depression remained. This process very closely resembles the formation of pingos and pingo remnants as described by numerous authors (e.g. Mackay 1979), so an origin as a pingo remnant is assumed.

If the depressions are interpreted as being pingo remnants, the absence of ramparts must be explained. Ramparts are formed during pingo growth and the early stage of decay by creep, mass movement, and surface wash from the uplifted overburden (Mackay 1979). The overburden consisted of fine sands with some silt layers. Because it is elevated, the surface of a pingo is better drained than the surroundings and it seems quite plausible that part of the overburden could be removed by deflation. The remaining part of the overburden is probably present in and around the depression but because no lithological differences exist it cannot be identified in borings.

Age of Formation and Decay

As the supposed pingo remnants form depressions in fluvioperiglacial deposits it can be inferred that the pingos formed either in a late stage of, or after the deposition of these deposits. From the radiocarbon data available in the Netherlands and Belgium, Kolstrup (1980) indicates an age between 28,000 and 27,000 years BP for the uppermost fluvioperiglacial deposits. Most of the fluvial activity then ceased, probably because the climate changed to drier and colder conditions (Wijmstra and Van der Hammen 1971) after the Denekamp Interstadial, about 29,000 years BP. It may be imagined that after the fluvial activity ceased, permafrost developed in the alluvial plains (De Gans 1982). Pingos probably formed on the sites of residual ponds (French and Dutkiewicz 1976, Mackay 1979), or where ponds formed in thaw depressions (Czudek and Demek 1970). In the areas surrounding the pingo remnants lake deposits were found in the uppermost part of the fluvial deposits. Thus it is thought that the pingos formed after 29,000 years BP, but there is no good evidence that they did not form earlier.

Both the age of the organic fill and its intercalation with Younger Coversand show that the ice cores melted completely after the onset of the Late Weichselian, at circa 12,000 years BP. Water-filled depressions were formed. Over time these became infilled with organic matter.

As, especially in the depressions A and C, the development of the Older Coversand I and II and of the deposits of the Beuningen Complex found in the depressions is simular to that of the corresponding deposits in the surrounding areas, it seems that neither a hill nor a depression was present during their deposition. A high hill would have prevented deposition, a depression would probably have been filled completely, resulting in the deposits being thicker than in the surrounding areas. If this interpretation is correct it would mean that during at least the later part of the sedimentation of Older Coversand I, the pingos had already been partly destroyed. The Older Coversand I was deposited between 27,000 and circa 23,000 years BP (Kolstrup 1980). Pingo formation may thus have occurred between circa 29,000 and 23,000 years BP. Partial melting of the ice core so that the pingo became almost level with the surface must then tentatively be assumed to explain the development of the inorganic fill.

Van der Hammen et al. (1967) assume that polar desert conditions existed in the Netherlands during the period from circa 27,000 to 13,000 years BP. Organic deposits are extremely scarce in sediments from this period. Together with the widespread occurrence of eolian deposits, this points to the absence of a closed vegetation cover. Under these conditions deflation would have been important in any dry, finegrained sediment. It can be expected that because of their elevation the pingo skins were well drained, so if they consisted of fine grained sand, deflation would have been easy.

Partial removal of the overburden e.g. by deflation of the upper part of the active layer, may cause the active layer to reach the ice core. The upper part of the ice core then melts. The meltwater can drain through the active layer or over the surface. The excess of moisture in the active layer may slow down or even stop deflation. This process may continue until an almost level surface is reached. Older Coversand I could then be deposited over the remaining overburden and the increasing thickness of the covering material could prevent the active layer from reaching the remaining ice core.

CONCLUSIONS

The depressions described may be pingo remnants, although they have no ramparts. They are all nearly circular crater-like depressions formed in fluvial deposits. They seem to have been formed in alluvial plains at a time of reduced discharge, when permafrost developed in the alluvial plains. It is not possible to date the pingo growth precisely: it occurred probably between 29,000 and circa 23,000 years BP. From the lithology and thickness of the inorganic fill, compared with those of the same sediments in the areas surrounding the pingo remnants it is tentatively concluded that before circa 23,000 years BP no surface expression of the pingos existed any more, but buried ice cores remained. These were covered with eolian and fluvial sediments. From the dating of the lowest parts of the organic fill the final melting of the ice core can be put at circa 12,000 years BP. Subsequently, the deposits covering the ice core subsided and water-filled depressions were formed. These depressions were filled with lake deposits and peat in the Late Weichselian and the Holocene.

REFERENCES

Bisschops, J.H., 1973, Toelichtingen bij de Geologische Kaart van Nederland 1:50,000 Blad Eindhoven Oost (51-O): Rijks Geologische Dienst, Haarlem.

Czudek, T. and J. Demek, 1970, Thermokarst in Siberia and its influence on the development of lowland relief: Quaternary Research v. 1, p. 103-120.

De Gans, W., 1982, Location age and origin of pingo remnants in the Drentsche Aa valley area (The Netherlands): Geologie en Mijnbouw v. 61, p. 147-158.

De Gans, W. and P. Cleveringa, 1981, Stratigraphy, palynology and radiocarbon dating of Middle and Late Weichselian deposits in the Drentsche Aa valley system (Drente, The Netherlands): Geologie en Mijnbouw v. 60, p. 373-384.

De Gans, W. and H. Sohl, 1981, Weichselian pingo remnants and permafrost on the Drente plateau (The Netherlands): Geologie en Mijnbouw v. 60, p. 447-452.

Flemal, R.C., 1976, Pingos and pingo scars: their characteristics, distribution and utility in reconstructing former permafrost environments: Quaternary Research v. 6, p. 37-53.

French, H.M. and L. Dutkiewicz, 1976, Pingos and pingo-like forms, Banks Island, Western Canadian Arctic: Biuletyn Periglacjalny v. 26, p. 211-222.

Kolstrup, E., 1980, Climate and stratigraphy in northwestern Europe between 30,000 B.P. and 13,000 B.P., with special reference to The Netherlands: Mededelingen Rijks Geologische Dienst, Haarlem v. 32, p. 181-253.

Lundqvist, J., 1969, Earth and ice mounds: A terminological discussion, in Péwé, T.L., ed., The Periglacial Environment Past and Present: Montreal, McGill-Queens University Press, p. 203-215.

Maarleveld, G.C., 1976, Periglacial phenomena and the mean annual temperature during the last glacial time in the Netherlands: Biuletyn Peryglacjalny v. 26, p. 57-78.

Maarleveld, G.C. and J.E. van den Toorn, 1955, Pseudo-sölle in Noord-Nederland: Koninklijk Aardrijkskundig Genootschap 2e Reeks LXXII, p. 343-360.

Mackay, J.R., 1979, Pingos of the Tuktoyaktuk Peninsula area, Northwest Territories: Geographie Physique et Quaternaire v. 33, p. 3-61.

Van der Hammen, T., G.C. Maarleveld, J.C. Vogel and W.H. Zagwijn, 1967, Stratigraphy, climatic succession and radiocarbon dating of the last glacial in the Netherlands: Geologie en Mijnbouw v. 46, p. 79-95.

Wijmstra, T.A. and T. van der Hammen, 1971, Outlines of the Upper Quaternary history of the Dinkel valley, in Van der Hammen, T. and T.A. Wijmstra, eds., The Upper Quaternary of the Dinkel valley (Twente, Eastern Overijssel, The Netherlands): Mededelingen Rijks Geologische Dienst, Haarlem v. 22, p. 201-205.

Zagwijn, W.H., 1974, Vegetation, climate and radiocarbon datings in the Late Pleistocene of The Netherlands Part II: Middle Weichselian: Mededelingen Rijks Geologische Dienst, Haarlem v. 25, p. 101-111.

Zagwijn, W.H. and R. Paepe, 1968, Die Stratigraphie der weichselzeitlichen Ablagerungen der Niederlande und Belgiens: Eiszeitalter und Gegenwart v. 19, p. 129-146.

THREE SUPERPOSED SYSTEMS OF ICE WEDGES AT MCLEOD POINT, NORTHERN ALASKA, MAY SPAN MOST OF THE WISCONSINAN STAGE AND HOLOCENE

Robert F. Black
Department of Geology and Geophysics, University of Connecticut, Storrs, Connecticut 06268 USA

Actively growing, surface ice wedges up to 9-m wide in primary, secondary, and tertiary polygons began growing by about 12,000 years B.P. at McLeod Point, 120 km southeast of Barrow, northern Alaska. Below the surface wedges two superposed, buried systems of inactive ice wedges in primary, secondary, and tertiary polygons were both truncated by separate episodes of deep thaw. The younger episode clearly resulted from a thaw-lake cycle. The older is attributed to a thaw-lake cycle, but available evidence is less definitive. The two buried systems of truncated ice wedges are in deposits that are separated by an organic horizon that is >40,000 radiocarbon years B.P. By analogy and correlation with the major events in the Quaternary history of northern Alaska, it is inferred that the three systems of ice wedges and associated sediments may span most of the Wisconsinan Stage and Holocene.

INTRODUCTION

McLeod Point, 120 km southeast of Barrow, northern Alaska, is a 7.2-m bluff of fine-grained sediments and ground ice fronting on the Beaufort Sea. In 1981 the bluff revealed a system of actively-growing, massive, surface ice wedges (in unit III) overlying the truncated roots of two buried, superposed systems of inactive ice wedges in deposits of units I and II (Figure 1). An unconformity locally rich in finely comminuted organic matter separated units I and II.

At least one thaw-lake cycle (Britton 1967, Black 1969, Sellmann et al. 1975) of late Wisconsinan age has modified the sediments, ice content, and chemistry of the younger deposits and truncated the upper system of buried ice wedges. Another thaw-lake cycle, or less likely a marine transgression, truncated the ice wedges of the lowest unit.

Measurements on airphotos taken in 1948 show that the bluff retreated 230 m on the west side and 420 m on the east side during the interval 1948/1981, or 10-18 m per year. Hence, the ice wedges studied in 1949/1950 (Black 1974) were destroyed. However, the small geomorphic entity--an old upland--in which the wedges were exposed is still represented today.

The main purpose of the field study in 1981 was to collect material for radiocarbon dating in order to determine when the large, surface ice wedges in unit III began to grow, and also, if possible, the ages of the older buried ice wedges. The radiocarbon dating indicates that the large, surface ice wedges of unit III began to grow about 12,000 B.P. The lower system of buried wedges is beyond the age of radiocarbon dating. By analogy and inference from the Quaternary history of northern Alaska the three systems of ice wedges may span most of the Wisconsinan Stage and Holocene.

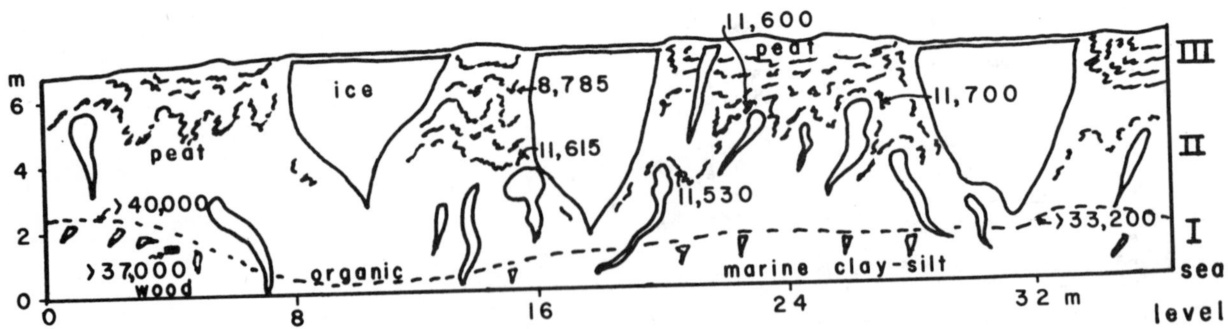

FIGURE 1 Composite diagrammatic sketch of part of the bluff at McLeod Point, showing the three units in which separate systems of ice wedges appear. The boundary between units I and II is a dashed line representing an unconformity with finely comminuted organic matter. The boundary between units II and III is the lowermost broken irregular line representing peat above the ice wedges in unit II.

STRATIGRAPHY

Black (1964, p. 68-69) characterized the lowermost sediments in the general area as marine silty clays and clayey silts with less than 10% very fine sand and correlated them with the Skull Cliff unit of the Gubik Formation of Pleistocene age. Organic matter makes up 6-8% of the total weight and increases in the upper part of the bank. The sediments are generally dark gray to black, greenish gray, or dark brown, but locally have dark yellow-brown, oxidized blotches, and irregular streaks. Ice in grains, granules, veins, dikes (including wedges), and small irregular bodies makes up 25-60% of the sediment by weight and increases upward. The marine sediments are cut into by thaw lakes, some of which extend to sea level. Most penetrate only 1-2 m. They leave characteristic deposits readily distinguishable from the underlying marine deposits by their color, bedding, and sorting. The lake sediments have much greater organic content and roots of water-loving plants in growth position in addition to large undecayed woody fragments and finely comminuted tissues.

The lower hiatus separating the two buried systems of ice wedges was considered in the field to be the bed of a former thaw lake rather than the result of a marine transgression on the basis of: 1. the lack of a lag of coarser sediments at the hiatus, 2. the presence of abundant finely comminuted organic matter, 3. the presence of local oxidized zones below the hiatus, as from groundwater movement, 4. the presence locally of truncated roots of ice wedges directly below the hiatus, and 5. differences in pH, specific conductance, and resistivity above and below the hiatus.

The measurements of pH, specific conductance, and resistivity of materials from the three units (Table 1) were made of the mineral zones rather than of the organic-rich zones. All are from the section at McLeod Point. Measurements from other nearby stratigraphic sections are not included, but showed similar results. Brown (1969) made more comprehensive studies of specific conductance of permafrost with marine and freshwater deposits at Barrow with similar results.

Arithmetic means and ranges of the small populations indicate good separation of values. The lowermost horizon (unit I) is considered to be marine on the basis of its uniform, thinly laminated stratification, greasy feel (typical of that unit), and widespread distribution. The uppermost unit (III) is quite clearly reworked material deposited in a former thaw lake. The drop in specific conductance and rise in resistance from unit I to unit II suggest that electrolytes in the marine deposits were lost during the first presumed thaw-lake cycle.

No diatoms or other fossils of marine or fresh water were seen in samples of either units I or II. They were examined by both binocular and petrographic microscopes before and after treatment with hydrogen peroxide. However, the presence of fossils in similar sediments along the coast is not universal (David M. Hopkins, oral communication, 17 March 1983).

Preliminary laboratory examination of samples from the sediments above and below the hiatus reveals marked differences. Unit I is comprised mostly of well rounded, fine silt and clay with abundant fibrous plant tissues and coal fragments. Heavy minerals are sparse. Unit II is mostly well rounded, coarse silt and clay with a few percent of very angular, fine to medium sand, sparse fibrous plant tissues and coal fragments, and common heavy minerals of considerable variety. However, its fine silt is very similar in shape and composition to that in unit I. The bimodal textural distribution and distinctly different shape and composition of the sand compared with the fine silt preclude either a common origin or much distance of travel by water for the larger grains. A glacial deposit nearby is indicated. Thus, these findings support the interpretation of the local presence of glacial detritus of the Flaxman Formation of early Wisconsinan age (MacCarthy 1958, Hopkins 1979) in the stratigraphy. Presumably unit II with its glacial suite was the lowermost unit exposed in this area in 1949/1950 (Black 1964, p. 68).

A glacial origin for some fine-grained material in the bluff at McLeod Point is suggested also by the presence of occasional large erratics of igneous origin along the shore and in the bluffs. They are found from the Canadian border to Point Barrow and farther southwest (MacCarthy 1958). Leffingwell (1919) correlated them with the Flaxman Formation, a thin deposit of till on Flaxman Island, about 180 km southeast of McLeod Point. A granitic boulder, for example, just east of McLeod Point in 1950 rose above the shallow water about 1 m. It is possible, but not probable, that those erratics in shallow water were ice-rafted in modern times to their present position. However, this is not possible for those erratics occurring in the sediments of the bluffs. Even though none was seen in the bluff at McLeod Point, it is more logical to believe that they

TABLE 1 pH, specific conductance, and resistance of strata at McLeod Point.

UNIT	pH	SPECIFIC CONDUCTANCE	RESISTANCE(000)	NUMBER
III	(5.8) 4.5-7.1	(113) 70-176	(31) 24-40	8
II	(7.1) 6.8-7.7	(513) 255-720	(22.5) 18-26	7
I	(7.2) 7.1-7.4	(970) 800-1110	(19.5) 16.5-22	3

NOTE: Numbers in ( ) are arithmetic means of the ranges. pH was measured on thawed sediment diluted 1:1 with distilled water. Specific conductance in parts per million was measured on thawed sediment diluted 5:1 with distilled water. Resistance in ohms was measured with an electrode spacing of 4 cm on thawed sediment, but with no added water.

were eroded from nearby bluffs, as at Drew Point or Cape Halkett (next west and east of McLeod Point, respectively) where they have been seen (Hopkins, written communication, 18 February 1983).

ICE WEDGES

Truncated remnants of ice wedges in the lowermost system (unit I, Figure I) observed in 1981 were not recognized seaward in 1949/1950 (Figure 2). The ice wedges of unit I occurred only locally along the bluff in 1981. Where present they were less than 1-2 m above sea level and always with a discontinuous, finely comminuted organic zone directly above them. They were not seen where the organic zone approached sea level. The ice wedges were less than 1 m high and 1 dm wide (all widths cited herein are normal to the strike of the wedges). The wedges were spaced generally at intervals of 4-6 m, but only a few would be present in a particular spot. The ice-wedge remnants megascopically displayed vertical fabrics characteristic of other buried wedges (Black 1974). They were not exposed more than 1-2 dm into the bank, but their distribution and strike suggested that primary, secondary, and tertiary polygons were present (Black 1974).

The ice wedges in unit II above the lower organic layer were larger and better displayed than those in unit I, in the bluff as well as in a thermal niche carved by the sea beneath the face (Figures 1 and 2). (Black 1964, p. 68, and 1974, p. 250 reproduce other photographs of the bluff showing ice wedges in unit II). Some wedges were seen in three dimensions to join in polygons with diameters of 3-7 m. Primary, secondary, and tertiary wedges were present. Black (1974) described their fabrics. Many of these wedges penetrated the lower organic zone and upturned that stratum on both sides of the ice wedges (Figure 3). They clearly are younger than the organic zone they penetrated. Some wedges are as much as 1 m wide and several meters high. The larger wedges have flowed upward according to the exterior shape of the wedges, the disruption and distortion of foliation planes generated by inclusions in former contraction cracks, and the recrystallization and realignment of ice crystals. They have flowed upward 1-3 m, like diapirs, to elevate another unconformity with scattered bits of coarse fibrous peat in its upper part. These wedges also seem to have been truncated by deep thaw, according to their distribution, size, shape, and stratigraphic setting.

The upper organic zone, which is more disjunct and distorted than the lower, marks the inception of the low-centered polygon ponds associated with the uppermost, very large, surface wedges of unit III (Figure 1). The surface wedges were 5-7 m wide in the bank exposed in 1981 (Figure 1). One 9-m wide wedge was seen in 1949/1950 seaward of this locality. Open contraction cracks and megascopic fabrics within the ice wedges indicated current activity (Black 1974). Fibrous sedge peat, in part in growth position, is in sedimentary layers aggregating as much as 3.3 m in the former low-centered polygon ponds between the surface wedges. The organic material increased in amount upward.

DATING

Four radiocarbon samples from the base of the upper peat (unit III, Figure 1) between the large surface wedges date the inception of vegetation in the low-centered polygon ponds. The dates range from 11,530 to 11,700 B.P. (Table 2). The vegetation is assumed to have begun growing within some centuries of the initiation of those wedges, when the double-raised rims adjacent to them generated low-centered polygon ponds. Hence, it can be assumed that the largest and oldest surface ice wedges of unit III began to grow about 12,000 radiocarbon years ago or somewhat earlier. Only a brief time would have been necessary to regelate the permafrost after draining of the thaw lake exposed the surface. The ice wedges of unit II were truncated by that thaw lake. How much time was involved in the thaw-lake cycle is not known.

Two dates of >33,200 and >40,000 B.P. (Table 2) on the lower organic zone that separates deposits of units I and II indicate the great antiquity of the oldest system of ice wedges. A spruce log (Picea) in the marine deposits below the ice wedges in unit I dated at >37,000 B.P. (Table 2). However, the same sample gave very low amino acid ratios as did a fragment of willow (Salix) from the same horizon (N. W. Rutter, written communication, 4 October 1982). This suggests a late Pleistocene age. All wood seen is erratic.

The pollen and spores in the two dated samples of the lower organic zone and in 13 other samples in the stratigraphic succession in the general area suggest the presence of herbaceous, wet tundra (Tom Ager and Effie Shaw, U. S. Geological Survey, written communication, 7 October 1982). "No clear evidence of dramatically different past vegetation types" was apparent between the older and younger materials. Preservation of pollen is generally poor. Many are recycled from Cretaceous, Tertiary, and older Quaternary deposits. Hence, none of the pollen provides much evidence of the age of the buried ice-wedge systems or of their host sediments.

The only previously dated, buried ice wedge in northern Alaska is 14,000 B.P. (Brown 1965, p. 39-41). It was at Barrow at a depth of 3 m in "a ′primary surface′, or one that has not been recently reworked by the thaw-lake cycle". In central Alaska several buried ice wedges are dated 24,000-32,000 B.P. (Péwé 1975, p. 53).

At present the earlier ice wedges and strata at McLeod Point can only be dated by analogy with other events and sequences in the region. Previous studies have suggested that the lowermost sediments at McLeod Point range from mid-Wisconsinan to Illinoian. Black (1964, p. 67-68) noted that the lowermost sediments exposed in 1949/1950 in the vicinity of McLeod Point had a glacial component of sand and suggested (p. 89) that they might be Illinoian in age. The upper sediments (p. 69) were considered to be deposits in thaw-lake basins.

Sellmann and Brown (1973, p. 177-178) correlated the Skull Cliff unit of the Gubik

FIGURE 2 Photograph of the bluff at McLeod Point, showing the massive ice wedges of unit III at the surface and the much smaller diapiric wedges in unit II. No ice wedges of unit I are visible. 31 August 1950.

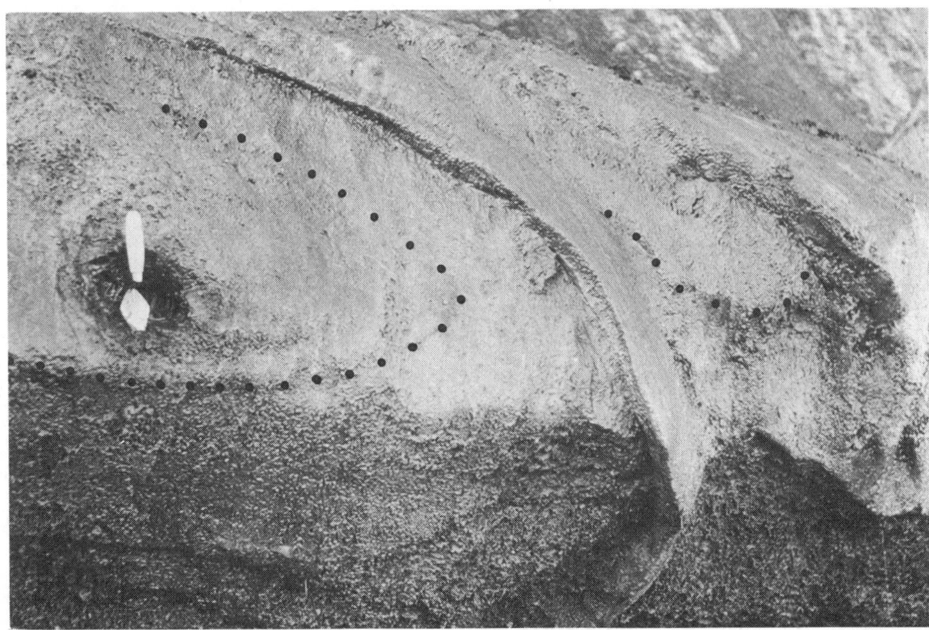

FIGURE 3 Root of a narrow ice wedge in unit II that cuts the organic layer with trowel between units I and II. Much of the surface is obscured by rapidly melting ice and clay-silt. The upturned organic layer is shown by inked dots. 2 August 1981.

Formation at Barrow with the Pelukian transgression (Sangamonian) and the lower marine phase of the Barrow unit of the Gubik Formation with the mid-Wisconsinan transgression. They considered the latter to be represented by the 7.5-8 m elevation attained by large ice-rafted boulders. This includes McLeod Point.

McLeod Point lies north of an extensive, elevated beach dated as more than 51,000 years B.P. (Carter and Robinson 1980). They suggested that the beach deposits and the marine mud to the north (including that at McLeod Point) may be synchronous and may constitute the youngest marine deposits on this part of the coastal plain.

David M. Hopkins (written communication, 18 February 1983) considered the beach deposits to be Pelukian (Sangamonian) in age and that the Flaxman Formation (early Wisconsinan) with its ice-rafted glacial component overlies those deposits at McLeod Point.

All these conclusions provide no clear-cut age assignment for the timing of growth of the ice wedges in units I and II. However, further constraints have materialized recently.

Carter (1981, 1982) describes a sand sea on the Arctic Coastal Plain immediately south of McLeod Point and associated, but older, fossil sand wedges immediately north and east of that sand sea (this volume). The primary sand wedges occur in sediments older than about 12,000 B.P. and have a maximum limiting age in sediments correlated with the Pelukian transgression. He concludes that arid conditions existed during late and middle Wisconsinan time and perhaps throughout the Wisconsinan Stage.

A frequency distribution of 78 radiocarbon dates from northern valleys of the central Brooks Range (roughly 300 km to the south) reveals no finite dates lie between 13,200-30,000 B.P. (Hamilton 1982). The suggestion is that conditions were unfavorable (too cold and dry) for growth of woody shrubs and peat-forming plants.

These interpretations place restrictions on the dating of the older buried ice wedges at McLeod Point as ice wedges require humid conditions and primary sand wedges require arid conditions for growth. It would seem that the sand sea is too close to McLeod Point to have had a distinctly different climate simultaneously. Moreover, the pollen from the interval between units I and II indicate herbaceous wet tundra (Tom Ager and Effie Shaw, written communication, 7 October 1982)--not the arid environment required for primary sand-wedge growth. Therefore, the logical conclusion is that the ice wedges of units I and II must be older than the sand wedges as they cannot be younger nor the same age. If the correlation of the sand wedges with late and middle Wisconsinan time is correct, the two older systems of ice wedges at McLeod Point must be early Wisconsinan or older.

The buried ice wedges are presumed to have grown as large as the present surface ice wedges. This seems necessary in order to provide sufficient ice in the ground to permit thaw lakes to incise to the roots of the wedges. Each system would require at least 12,000 years to grow to such size if the present surface wedges are a guide. The two thaw-lake cycles also require some thousands of years. Thus, many tens of thousands of years are required for the development and destruction of the three systems of ice wedges.

CHRONOLOGY AT MCLEOD POINT

Two almost equally viable scenarios seem applicable to the chronology of the deposits and ice wedges at McLeod Point. If the lowermost sediments in the bank are Pelukian in age, then the ice wedges of unit I logically are early Wisconsinan. This correlation almost demands that the truncation of those wedges occurred during the Flaxman transgression which left the glacial sediments of Unit II. The ice wedges of unit II should have grown in the latter part of early Wisconsinan time in order to be older than the arid middle and late Wisconsinan interval of sand-wedge growth. The thaw lake that truncated the ice wedges of unit II could have started at any time during that long interval. It was drained sometime prior to 12,000 years B.P. to permit the growth of the wedges of unit III.

If, however, the lowermost sediments are Flaxman, then the ice wedges of unit I grew immediately thereafter in early Wisconsinan time. The hiatus that truncated them could be assigned to a thaw-lake cycle, which is preferred over a marine transgression. However, the ice wedges of unit II would still have to grow prior to middle Wisconsinan time in order to preceed the growth of the primary sand wedges to the southeast. The record of middle and late Wisconsinan time is again presumed to be lost in the subsequent thaw-lake cycle that truncated the ice wedges of unit II. The growth of ice wedges in unit III followed about 12,000 years B.P. On the basis of knowledge to date the latter scenario is preferred, but the first cannot be ruled out with data in hand. At least it seems that most, if not all, the Wisconsinan Stage and Holocene are involved in the stratigraphy and ice wedges at McLeod Point and the time span may be longer.

ACKNOWLEDGEMENTS

Grateful acknowledgement is made to the National Science Foundation for funding my field studies in 1981. The logistic support provided by the U.S. Geological Survey in 1981 facilitated immensely the studies at McLeod Point and permitted other areas to be covered as well. Andrew M. Knowlton was a capable field assistant. The project would not have been possible without the background of knowledge I had gained while employed by the U.S. Geological Survey and without the former support of the Arctic Research Laboratory at Barrow and of the Office of Naval Research. Critical review of a draft of this paper by David M. Hopkins, Jerry Brown, and L. David Carter improved it considerably and is acknowledged gratefully.

TABLE 2 Radiocarbon dates, McLeod Point (see Figure 1).

DATE	SAMPLE NO.
8,785 +/- 230	GX-8130
11,530 +/- 170	I-12,129
11,600 +/- 170	I-12,130
11,615 +/- 305	GX-8131
11,700 +/- 170	I-12,131
>33,200	I-12,132
>37,000	I-12,127
>40,000	I-12,128

NOTE: Dates are based on the Libby half life of 5568 years for C14. The age is referenced to the year A.D 1950.

REFERENCES

Black, R. F., 1964, Gubik Formation of Quaternary age in northern Alaska: U.S. Geological Survey Prof. Paper 302-C, p. 59-91.

Black, R. F., 1969, Thaw depressions and thaw lakes--A review: Biuiletyn Peryglacjalny No. 19, p. 131-150.

Black, R. F., 1974, Ice-wedge polygons of northern Alaska: in Coates, D. R., ed., Glacial Geomorphology, Publications in Geomorphology, State University of New York, Binghamton, N. Y., p. 247-275.

Britton, M. E., 1967, Vegetation of the arctic tundra: in Hansen, H. P. ed., Arctic Biology, 8th Annual Biological Colloquium, Oregon State University, 2nd ed., p. 67-130.

Brown, J., 1965, Radiocarbon dating, Barrow, Alaska: Arctic, v. 18, p.36-48.

Brown, J., 1969, Ionic concentration gradients in permafrost, Barrow,Alaska: Cold Regions Research and Engineering Laboratory Research Report 272, 25 p.

Carter, L. D., 1981, A Pleistocene sand sea on the Alaskan Arctic Coastal Plain: Science, v. 211, p. 381-383.

Carter, L. D., 1982, Late Wisconsin desertification in northern Alaska: Geological Society of America Abstracts with Programs, v. 14, no. 7, p. 461.

Carter, L. D., and Robinson, S. W., 1980, Minimum age of beach deposits north of Teshekpuk Lake, Alaskan Arctic Coastal Plain: U. S. Geological Survey Circular 823-B, p. B8-B9.

Hamilton, T. D., 1982, Quaternary stratigraphic sections with radiocarbon dates, Killik River Quadrangle, Alaska: U.S. Geological Survey Open File Report 82-606, 31 p.

Hopkins, D. M., 1973, Sea level history in Beringia during the past 250,000 years: Quaternary Research, v. 3, p. 520-540.

Hopkins, D. M., 1979, The Flaxman Formation of northern Alaska: Record of early Wisconsinan shelf glaciation in the high arctic?: Pacific Science Congress, XIV (Khabarovsk), Abstracts, Additional Volume, p. 15-16.

Leffingwell, E. de K., 1919, The Canning River region, northern Alaska: U.S. Geological Survey Prof. Paper 109, 251 p.

MacCarthy, G. R., 1958, Glacial boulders on the Arctic Coast of Alaska: Arctic, v. 11, p. 71-85.

Péwé, T. L., 1975, Quaternary geology of Alaska: U.S. Geological Survey Prof. Paper 835, 145 p.

Sellmann, P. V., and Brown, J., 1973, Stratigraphy and diagenesis of perennially frozen sediments in the Barrow, Alaska, region: in Permafrost--The North American contribution to the Second International Conference, Yakutsk: Washington, D. C., National Academy of Sciences, p. 171-181.

Sellmann, P. V., Brown, J., Lewellen, R. I., McKim, H., and Merry, C., 1975, The classification and geomorphic implications of thaw lakes on the Arctic Coastal Plain, Alaska: Cold Regions Research and Engineering Laboratory Research Report 344, 21 p.

THE SAGAVANIRKTOK AND ADJACENT RIVER SYSTEMS, EASTERN NORTH SLOPE, ALASKA: AN ANALOG FOR ANCIENT FLUVIAL TERRAIN ON MARS

J.C. Boothroyd[1] and B.S. Timson[2]

[1]Department of Geology, University of Rhode Island, Kingston, Rhode Island 02881 USA
[2]Earth Surface Research, Inc., 114 State Street, Augusta, Maine 03240 USA

The large-scale geomorphic features of the Eastern North Slope, Alaska were formed by the Sagavanirktok and adjacent rivers during Quaternary time. The largest landforms are low hills, here called erosional remnants, situated between active and abandoned river courses. The erosional remnants are flanked by Pleistocene age terraces and abandoned floodplains lower than the remnant surfaces. Formation of remnants and terraces proceeded more rapidly during times of deglaciation when discharge was greatest, augmented by catastrophic drainage of moraine-dammed lakes. There has been extensive modification of valley floors and of remnant and terrace slopes by eolian, cryogenic, and debris-flow activity. Geomorphic processes on the Eastern North Slope have produced a suite of morphologic features characteristic of those seen, or inferred, on Viking Orbiter images of Mars in areas containing smaller outflow channels and depositional basins. The formation of these channels and basins--Ladon Valles (24°S, 29.5°W) is a good example--predates the development of most major outflow channels and may have occurred at a time when the Martian climate was "wetter" and thus more akin to present conditions on the North Slope. The similarity of the two suites of morphologic features suggests that similar processes, in type and magnitude to Quaternary processes that have shaped the Eastern North Slope, may have operated on Mars.

INTRODUCTION

The search for appropriate terrestrial analogs of the channels and valleys of Mars has resulted in the identification of several possible examples for the large outflow channels on Mars; specifically (1) the Channeled Scabland (Baker 1978, Baker and Nummedal 1978), (2) submarine canyons and valleys (Nummedal 1982), and (3) terrain modified by continental glaciation (Lucchitta 1982). It is also clear that the large outflow channels and the smaller channels and valleys have undergone extensive slope and floor modification by various mass-wasting and eolian processes after the original formative events (Mars Channel Working Group, in press). The purpose of this paper is to offer a terrestrial analog, the Sagavanirktok and associated drainage Systems of the Eastern North Slope, Alaska (Figure 1), for the smaller outflow channels and larger valleys on the Martian surface, and to illustrate some mass-wasting and eolian modifications to the fluvial valleys.

We examined LANDSAT images and U-2 color infrared photographs, particularly LANDSAT winter images obtained during times of snow cover and low sun angle (Figure 2). We then visited the Eastern North Slope in August 1979 and August 1981 and chose to work along the Sagavanirktok River because it is the most active river in the area that is easily accessible via the pipeline haul road. We did however, conduct a low-level aerial reconnaisance survey of the area shown in the large box on Figure 1.

GEOMORPHOLOGY OF THE CENTRAL ARCTIC SLOPE

General

The general study area, outlines in Figure 1, lies mostly in the Arctic Coastal Plain and Arctic Foothills physiographic provinces of northern Alaska (Wahrhaftig 1965). The coastal plain is underlain by semiconsolidated sandstone, siltstone, and claystone (continental and marine) of Tertiary age (Beikman 1980), which is in turn overlain by Quaternary age fluvial gravels and marine silt (Péwé 1975). The foothills are underlain by Cretaceous partially to totally consolidated terrigenous rocks that are moderately folded in contrast to the low northward dip of the Tertiary coastal plain sediments (Beikman 1980).

Rivers draining northward from the Brooks Range have created a complex fluvial drainage system on the Tertiary and Cretaceous rocks. The principal rivers, from west to east, are the Colville, Itkillik, Kuparuk, Toolik, Sagavanirktok, and Kadleroshilik (Figure 1). The Colville and Sagavanirktok have large tributaries associated with their stream systems. The present rivers, with the exception of the Colville and the Sagavanirktok, are underfit with respect to valley and floodplain width. Pleistocene-age, alpine glaciers of the Brooks Range advanced northward down the river valleys in the foothills to deposit prominent end moraines at the southern edge of the study area (Coulter et al. 1965, Hamilton 1978) (Figure 1).

Quaternary Landscape Development

The study area is underlain by up to 650 m of permafrost (Péwé 1975). Formation of ground ice has resulted in the development of ice-wedge polygons, other patterned-ground features, and pingos.

A summer active layer is present to a depth of 1-2 m (Péwé 1975).

The northward draining river systems have cut down into the underlying regolith and bedrock during Quaternary time to produce the following suite of landforms. The largest-scale geomorphic features are long, low hills, here called erosional remnants, situated between active and abandoned river courses. The erosional remnant hills are up to 100 km long and rise up to 200 m above presently active fluvial channels (Figure 2). The hills within the Coastal Plain physiographic province (Figure 1) have large, flat, relatively undissected upper surfaces that dip northward with an inclination slightly greater than the active river channels. Franklin Bluffs (Figure 2) is a good example.

The erosional remnants are flanked by Pleistocene-age fluvial terraces and abandoned floodplain segments that are lower in elevation than the remnant surfaces, but are 20-60 m higher than the present river beds. They are labeled (T) on Figure 2. The only structural control of remnant location appears to be a major southwest-northeast trending lineament that bounds the southeast or northwest side of each major remnant hill and is defined by the Colville River floodplain in the southwest (Albert 1978) (Figure 2). Contacts between the various Tertiary and Cretaceous units (Beikman 1980) cut across the remnants and play little role in remnant shape (Figure 2).

The southern older terraces exhibit etched topography on LANDSAT winter imagery and consist in part of drained thaw lakes that were active during late Pleistocene and early Holocene time (Gatto and Anderson 1975, Pewe 1975) (Figure 2). The etched terrain is bounded on the south by the foothills, folded belt of the Brooks Range. The folds trend generally east-northeast in this area. Tongues of morainal disintegration topography, labeled (M) on Figure 2, extended out of the valleys of the foothills and have been dissected by the active rivers.

The Colville and Sagavanirktok are robust braided rivers that are actively downcutting and laterally migrating to modify terrace and erosional remnant shape. However, most of the present rivers, with the exception of the Sagavanirktok and the Colville, are underfit with respect to valley and floodplain width, particularly the Itkillik and the Toolik. It is apparent that much of the remnant and terrace development took place during periods of deglaciation after Pleistocene ice advances. Hamilton (1978) indicates the past existence of glacial lakes dammed behind some of the moraines shown in the lower right on Figure 2. Increased runoff during deglaciation was augmented by catastrophic flooding during drainage of the moraine-dammed lakes.

THE SAGAVANIRKTOK DRAINAGE SYSTEM

Sedimentary Processes

The Sagavanirktok (Sag) is the second-largest river, after the Colville, on the North Slope of Alaska, with a length of 267 km. Maximum discharge is during spring breakup; flow declines to essentially zero during the winter freeze. The Sag is a coarse-gravel, braided river that is degradational

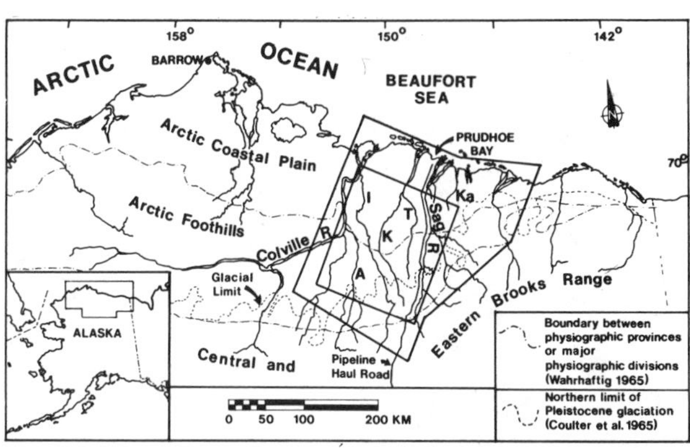

FIGURE 1 Location map of the Eastern North Slope, Alaska. General study area is outlined by large box centered on the Sagavanirktok (Sag) River; Figure 2 (LANDSAT image) is outlined by smaller inset. Rivers indicated by letters are (A) Anaktuvuk, (I) Itkillik, (K) Kuparuk, (T) Toolik, (Ka) Kadleroshilik.

through most of its length, becoming aggradational only on the last 20 km of delta plain.

Active bars are a combination of longitudinal bar complexes and large transverse bars. Many transverse bars are formed at the downstream end of chutes incised into low terraces of the inactive fluvial plain. Formation of chutes is due to blockage of the river by ice derived from icings (aufeis). Icings form by repeated overflows of water on river ice cover that is frozen to the river bed, due to intermittent thawing or, more importantly, a groundwater source (Hall and Roswell 1980, Sloan et al. 1976). Spring breakup results in ice drives that jam up and direct the river laterally to an unoccupied part of the fluvial plain. Thus icings play an important role in controling downcutting and lateral migration of the active channel system.

Valley Modification

The Sag river is still actively downcutting adjacent to the Franklin Bluffs erosional remnant (Figure 2), but other portions of the Sag channel system together with the channel systems of adjacent rivers, have been abandoned and mass-wasting and eolian processes are modifying the older surface. Figure 3, a geomorphic map using U-2 imagery as a base, delineates some of the landforms that result from these modifying processes.

The Sag active bar and channel system is shown as (Fa). Other portions of the Sag channel system are inactive (Fi) or have been recently abandoned (Fb). The inactive channel segments exhibit fresh, unvegetated bar and bedform morphology but were not subject to flow during the latest meltwater seasons. The abandoned channel system is at the same elevation or slightly higher than the active bars and channels but is partially vegetated and modified by cryogenic processes.

Relict bar and channel systems exist as terraces elevated 10-30 m above the active river. Immediately adjacent to the terrace scarp is an eolian levee (El) composed of silt and fine sand derived from the active and inactive bars and channels. Bar

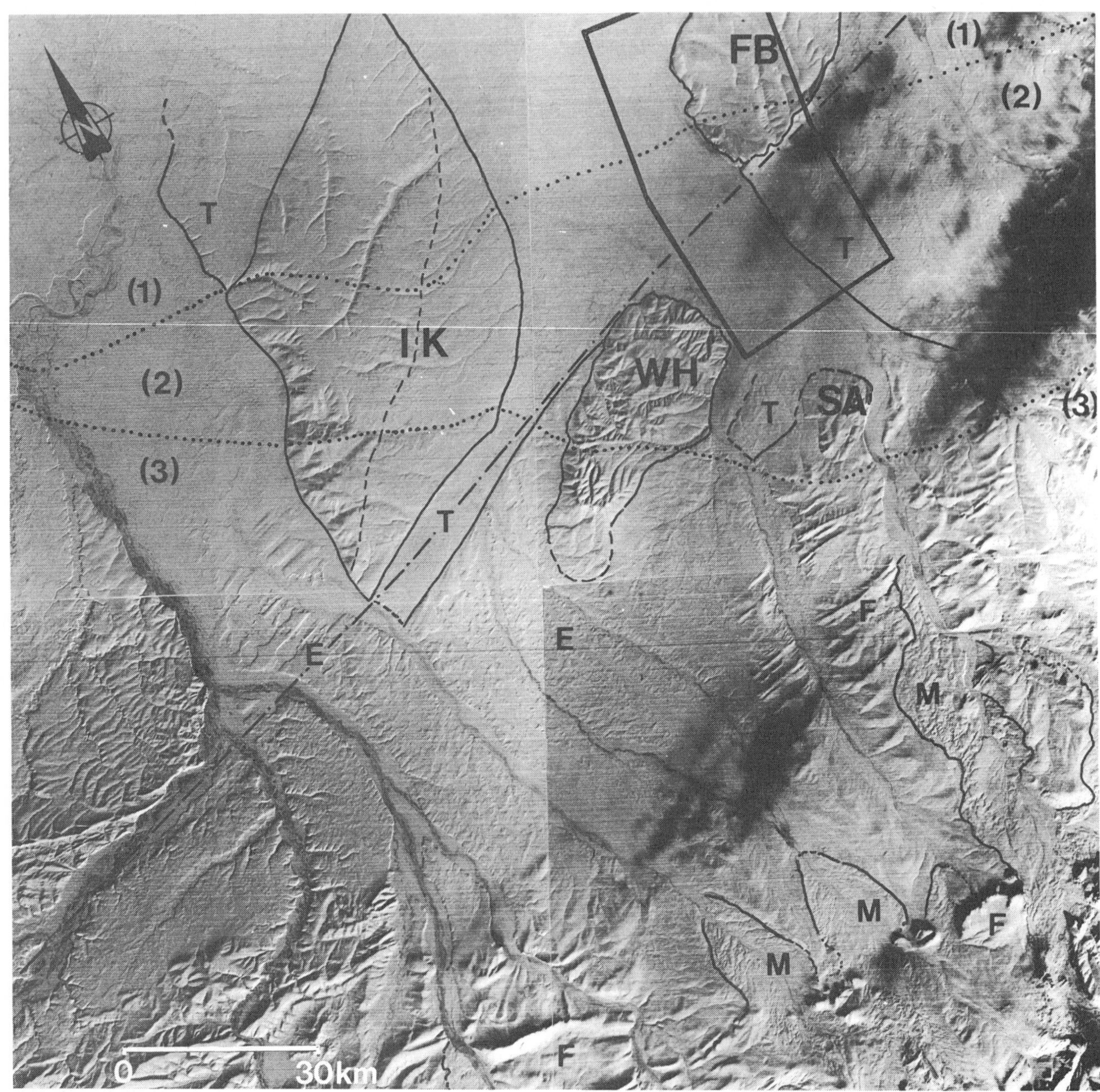

FIGURE 2 LANDSAT band 5 image of Eastern North Slope, Alaska. Erosional remnants identified are: (IK) Itkillik-Kuparuk, (WH) White Hills, (SA) Sagwon, (FB) Franklin Bluffs. Other features are (T) terraces, (E) etched terrain, (F) foothills, (M) moraines. The dashed line on the (IK) remnant delineates a higher surface to the west from a lower surface to the east. Dash-dot line indicates the lineament of Albert 1978. Dotted lines delineate rock-unit boundaries of Beikman (1980); (1) Upper Tertiary marine rocks, (2) Lower Tertiary continental rocks, and (3) Cretaceous rocks. Box encloses map area of Figure 3. Scene E-1217-21235-5 01, February 25, 1973.

and channel topography is still identifiable in places (Er), but much of this terrain contains a lake/pond plain (Ep) of wind-aligned lakes developed in the active layer of the continuous permafrost. The eolian levee has longitudinal dunes oriented in an east-west direction, parallel to the dominant winds. The dunes are 20-30 m long and 1-4 m high. Similar dunes also are present on parts of the abandoned bar and channel system. In some locations, the dunes have been eroded into streamlined, yardang-like landforms.

Still older surfaces, such as the upper surface of the Franklin Bluffs remnant, are low-relief, gently-sloping plains now being modified by solifluction (Csl). Younger, moderate-relief slopes have been formed mainly by solifluction processes (Csm). The slopes of older remnants with greater relief are dominated by mass-wasting processes rang-

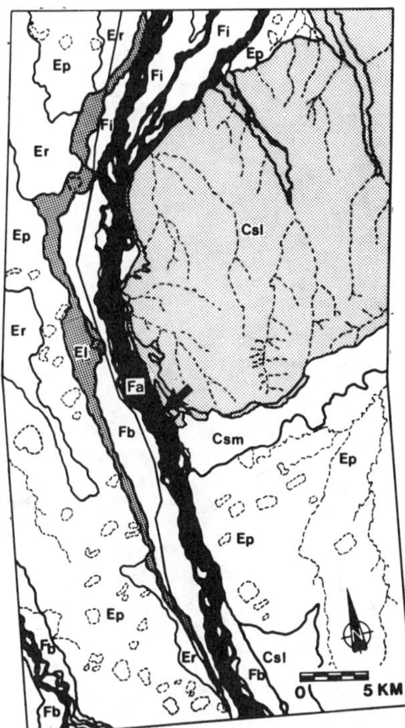

FIGURE 3 Geomorphic map of Franklin Bluffs area, Sag drainage system. Area is shown in Figure 2. Franklin Bluffs erosional remnant is shown by light-shaded pattern. Explanation: Fluvial Channels: (Fa) active (black), (Fi) inactive, (Fb) abandoned; Eolian/Cryogenic: (El) eolian levee, (Er) relict terrace, (Ep) lake/pond plain; Colluvial/Cryogenic: Solifluction Slopes: (Csl) low relief, (Csm) moderate relief. Thaw lakes and streams shown by dashed lines, pipeline haul road by solid straight line segments. Location of alluvial fans (Figure 4) shown by arrow.

ing from solifluction to debris flow that are fed by ground-ice melting and collapse, resulting in the development of thermocirques, debris lobes, and alluvial fans. Small alluvial fans are particularly well-displayed on the flanks of the Franklin Bluffs erosional remnant, where distal fan aprons are encroaching on the inactive fluvial plain of the Sag (Figure 4).

MARTIAN FLUVIAL TERRAIN

Introduction

The gemorphic processes operating in the Sagavanirktok River drainage system combine to produce a suite of morphologic features characteristic of those seen, or inferred, in Viking Orbiter images of Mars of areas containing smaller outflow channels and depositional basins. The Ladon Valles area of the Margaritifer Sinus Quadrangle (MC-19) (U.S. G.S. 1979) is a good example (Figure 5). The formation of these channels and basins predates the development of most of the major outflow channels and may have occurred at a time when the Martian climate was "wetter" (Carr 1980), and thus more akin

FIGURE 4 Small alluvial fans at the southwest corner of the Franklin Bluffs erosional remnant. The fans are growing by debris-flow deposition onto an inactive floodplain segment of the Sag River. Relief of the Bluffs is about 160 m, distance from fan apex to toe is about 300 m.

to present or Pleistocene conditions on the North Slope. The similarity of the two suites of morphologic features suggests that similar processes, in type and magnitude to Quaternary processes that have shaped the Eastern North Slope, may have operated in the Martian fluvial terrain.

Ladon Valles Drainage System

The Ladon Valles drainage system, centered on 24°S, 29.5°W, consists of (1) an array of small outflow channels separated by erosional remnants, (2) a late source area of chaotic terrain adjacent to and northeast of the crater Holden, and (3) a probable depositional basin to the northeast (Figure 5). The presumed depositional plain is inside an older, highly degraded, multiringed basin, the boundaries of which were mapped by Saunders (1979), discussed by Schultz and Glicken (1979) and named the Ladon Basin by Schultz et al. (1982).

The outflow channels and remnants, when viewed on stereo images (611A13, 32, 650A15, 16, 18) show several levels of channels and terraces (Figure 5), indicating a complex degradational history with presumed multiple flow events. The horizontal scale of features in Ladon Valles is comparable to channel and remnant features on LANDSAT imagery of the Sagavanirktok drainage system. The nature of the regolith on bedrock in the Ladon area is not known for certain, but because it is in the older cratered plains province of Mars (Saunders 1979), there could be a deep blanket of crater ejecta that, when combined with presumed permafrost and ground ice, is not unlike the semiconsolidated Tertiary rocks of the North Slope.

Water-release mechanisms are also unknown, but a past wetter climate, combined with the confined aquifer release mechanism of Carr (1979), could have resulted in multiple flows originating in the area northeast of Holden crater from water stored in presumed permafrost prior to the formation of chaotic terrain. This is a situation analogous to the release of water from the Brooks Range during deglaciations. The depositional basin to the north of the Ladon outflow channels functioned as a sedi-

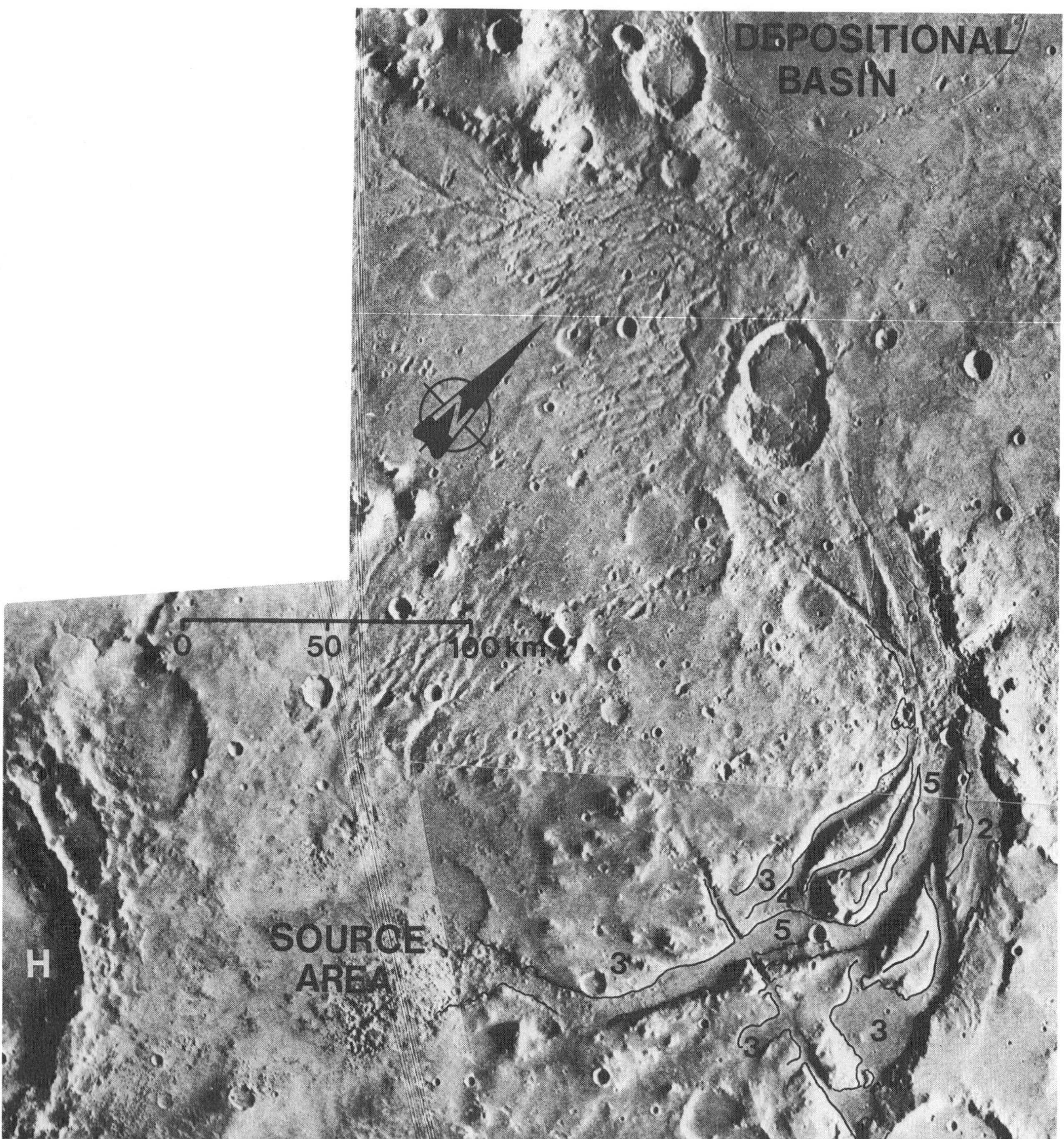

FIGURE 5 Ladon Valles and adjacent area. Drainage flowed from late source area of chaotic terrain near crater Holden (H) to a depositional basin inside an ancient multiringed impact basin. Numbers indicate relative elevation and age of channel floors and terraces (1 is highest or oldest). (611A30, 650A18, 19).

ment sink much like the Sagavanirktok, Colville, and other delta plains and the Beaufort Sea have done on the North Slope.

Modification to Channels and Valleys

The Ladon Valles images are medium-scale resolution (190-220 m per pixel) and do not show fine details of secondary modification to valley walls such as mass-wasting and debris flow although chaotic terrain (Figure 5), on a scale consistent with the morainal disintegration topography (Figure 2) of the North Slope, can be mapped. Thus high-resolution images (30-50 m per pixel) of Martian channels

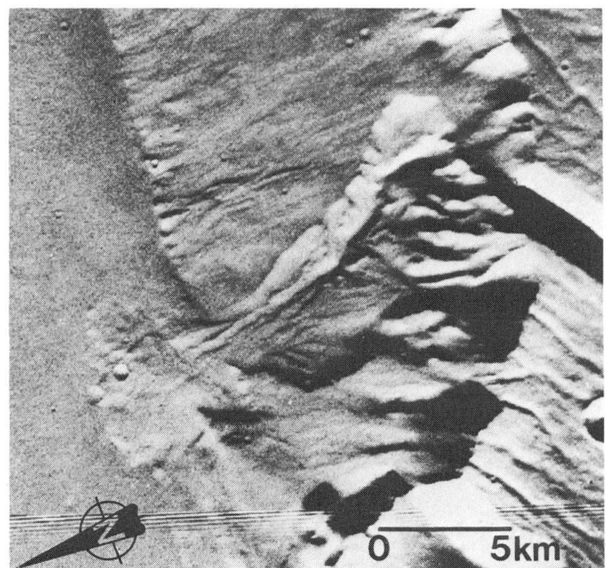

FIGURE 6 Debris fan in the northern channel of Kasei Vallis (27.2°N, 67°W) with source area on the canyon wall, and perhaps from the graben on the upper plateau. Note scale (665A24).

on a scale comparable to U-2 images on the Arctic Slope are used to illustrate valley modification.

Figure 6 shows an alluvial fan that appears to have formed by debris-flow processes in the northern channel of Kasei Vallis (27.2°N, 67°W). The fan extends about 5 km out onto the floor of the channel from a source on the valley wall.

The wall height of the Kasei channel is much greater than even those of the Brooks Range Valleys, but fan size is not inconsistent with the larger debris fans of the Sag drainage basin.

ACKNOWLEDGMENTS

The project was supported by the NASA Planetary Geology Program under grant NSG-7414. Access to study sites by Arco and Sohio oil companies and by the Alyeska Pipeline Service Company is gratefully acknowledged. This work has benefited greatly from discussions with members of the Mars Channel Working Group, particularly V. Baker, B. Lucchitta, H. Masursky, D. Nummedal, P. Patton and D. Pieri. L. Dunne assisted in the field; P. Ladd and J. Grant assisted in the laboratory; S. Ponte typed the manuscript.

REFERENCES

Albert, N. R. D., 1978, Landsat mosaics of Eastern North Slope Petroleum Province, Alaska, with preliminary interpretation of observed features: U.S. Geol. Survey Map MF-928V, scale 1:500,000

Baker, V. R., 1978, The Spokane flood controversy and the martian outflow channels: Science, v. 202, p. 1249-1256.

Baker, V. R., and Nummedal, D., 1978, The Channeled Scabland: NASA Office of Space Science, Planetary Geology Program, Washington, D.C., 186 p.

Beikman, H. M., 1980, Geologic map of Alaska: U.S. Geol. Survey Map, scale 1:2,500,000.

Carr, M. H., 1979, Formation of martian flood features by rlease of water from confined aquifers: Journal of Geophysical Research, v. 84, p. 2995-3007.

Carr, M. H., 1980, The morphology of the martian surface: Space Science Review, v. 25, p. 231-284.

Coulter, H. W., Hopkins, D. M., Karlstrom, T. N. V., Pewe, T. L., Wahrhaftig, C., and Williams, J. R., 1962, Map showing extent of glaciaions in Alaska: U.S. Geol. Survey Map I-415, scale 1:2,500,000.

Gatto, L. W., and Anderson, D. M., 1975, Alaskan thermokarst terrain and possible martian analog: Science, v. 188, p. 255-257.

Hall, D. K., and Roswell, C., 1980, Analysis of the origin of aufeis feed-water on the Arctic Slope of Alaska: NASA Tech. Memo. 81992, Reports of Planetary Geology Program, 1979-1980, 41 p.

Hamilton, T. D., 1978, Surficial geologic map of the Philip Smith Mountains quadrangle, Alaska: U.S. Geol. Survey Map MF-879A, scale 1:250,000.

Lucchitta, B. K., 1982, Ice sculpture in the martian outflow channels: Journal of Geophysical Research, v. 87, p. 9951-9973.

Mars Channel Working Group, in press, Channels and valleys on Mars: Geological Society of America Bulletin.

Nummedal, D., 1982, Continental margin sedimentation: its relevance to the morphology on Mars: NASA Tech. Memo. 85127, Reports of Planetary Geology Program, 1982, p. 256-257.

Péwé, T. L., 1975, Quaternary geology of Alaska: U.S. Geol. Survey Professional Paper 835, 145 p.

Saunders, R. S., 1979, Geologic map of the Margaritifer Sinus quadrangle of Mars: U.S. Geol. Survey Map I-1144 (MC-19), scale 1:5,000,000.

Schultz, P. H., and Glicken, H., 1979, Impact crater and basin control of igneous processes on Mars: Journal of Geophysical Research, v. 84, No. B14, p. 8033-8047.

Schultz, P. H., Schultz, R.A. and Rogers, J., 1982, The structure and evolution of ancient impact basins on Mars: Journal of Geophysical Research, v. 87, p. 9803-9820.

Sloan, C. E., Zenone, C., and Mayo, L. R., 1976, Icings along the trans-Alaska pipeline route: U.S. Geol. Survey Professional Paper 979, 31 p.

U.S. Geological Survey, 1979, Controlled photomosaic of the Margaritifer Sinus South West quadrangle of Mars: U.S. Geol. Survey Map I-1209 (MC-19SW), scale 1:2,000,000.

Wahrhaftig, C., 1965, Physiographic divisions of Alaska: U.S. Geol. Survey Professional Paper 482, 52 p.

PALEOTEMPERATURE ESTIMATES OF THE ALASKAN ARCTIC COASTAL PLAIN DURING THE LAST 125,000 YEARS

Julie K. Brigham and Gifford H. Miller

Institute of Arctic and Alpine Research and
Department of Geological Sciences
University of Colorado, Boulder, Colorado 80309, USA

Beach, nearshore, and shallow-marine deposits of the Gubik Formation in the western portion of the Alaskan Arctic Coastal Plain have been subdivided by stratigraphy, sedimentology, and amino acid geochronology into transgressive/regressive sequences representing six high stands of sea level of Pliocene and Pleistocene age. Combined with paleoclimatic proxy data and assumptions concerning permafrost conditions, amino acid results on mollusks from the last interglacial high sea stand, the Walakpa member (≈ 125 ka B.P.), provide a means of quantifying paleotemperatures in permafrost throughout Wisconsin time and places limits on the possible magnitude of temperature change throughout this period. The results indicate that glacial-age mean permafrost temperatures on the coastal plain averaged -18°C or lower and were more than 8°C colder than today. This implies that mean annual air temperatures probably averaged between -19°C and -24°C and arctic summers were probably cooler than at present. Areas such as Ellesmere Island, N.W.T., or the Dry Valleys of Antarctica may serve as partial modern analogs for the glacial climates of the North Slope.

INTRODUCTION

The Arctic Coastal Plain and adjacent continental shelves of northern Alaska (Figure 1) are mantled by a thin veneer of unconsolidated sediments known as the Gubik Formation (Black 1964). In addition to extensive fluvial, eolian and thaw lake constituent facies, beach, nearshore, and shallow-marine sediments that were deposited as a result of eustatic sea level changes during Pliocene and Pleistocene time compose a significant part of this complex stratigraphic sequence. This paper deals specifically with marine sediments of the last interglacial high sea stand (125 ka B.P.), now recognized as the Walakpa member of the Gubik Formation (Brigham 1983). From the extent of isoleucine epimerization in fossil Mya truncata and Mya arenaria in these sediments the data are used to constrain a series of possible paleotemperature models for the North Slope during the last 125,000 years that attempt to accomodate all of the presently known paleoclimatic proxy data.

METHODS

Amino acid geochronology is a technique that relates the extent of diagenetic changes of indigenous proteins in the carbonate matrix of fossil mollusks to variables of temperature (thermal history since deposition) and time (since death of the organism). Most amino acids may occur in one of two optically-active forms, or enantiomers, which are mirror images of each other. Those forms that rotate polarized light to

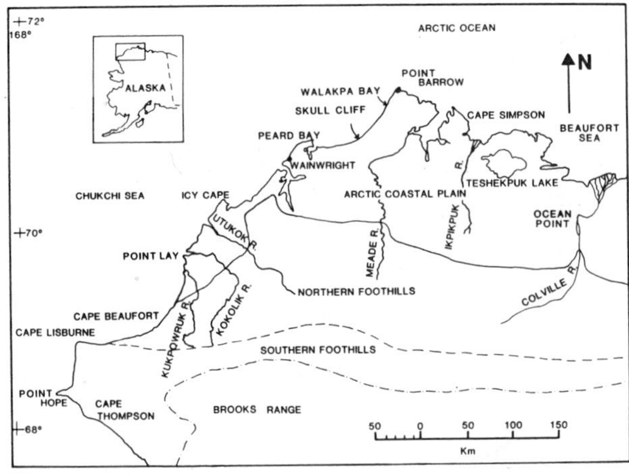

FIGURE 1 Arctic Coastal Plain of northern Alaska, showing the location of study area between Pt. Barrow and Cape Beaufort.

the right are D-amino acids (dextrorotatory) and to the left are L-amino acids (levorotatory). Constituent amino acids that comprise the protein within the carbonate matrix of living mollusks are exclusively of the L-stereoisomer configuration. After death of the organism the L-amino acids reversibly interconvert to the D-configuration until an equilibrium mixture is reached. Thus, fossil shells of increasing age contain a greater porportion of D-amino acids. In this study, the epimerization of L-isoleucine to D-alloisoleucine (hereafter aIle/Ile) in the Total (Free plus peptide-bound) acid hydrolysate was measured in fossil mollusks to determine the relative are of marine deposits exposed on the Arctic Coastal Plain, and to make Late Pleistocene paleotemperature estimates.

Temperature is the most critical rate-controlling factor governing protein diagenesis. In general, a 4°C temperature rise will approximately double the epimerization rate of L-isoleucine. However, for relatively young samples (e.g., 125 ka B.P.) at temperatures lower than about -10°C, the rate of epimerization is so slow that the change in the absolute value of the aIle/Ile ratio, created by fluctuations in temperature of a few degrees, is very small and often less than the resolving power of the technique. Hence, the method cannot detect temperature differences of $< \pm 2$°C in this time/temperature range.

Because rates of amino acid diagenesis vary between different genera, analyses were restricted to Hiatella arctica (Linné), Mya arenaria (Linné), and Mya truncata (Linné). When possible, at least three individuals of each genus were analyzed from each depositional facies in order to detect mixed populations of aIle/Ile ratios indicative of reworked fossils. Samples were prepared for analysis according to Miller (1982 p. 52). Analyses were completed using an ion-exchange liquid-chromatographic amino acid analyzer utilizing O-Phthaladehyde fluorescence detection.

STRATIGRAPHY OF THE WALAKPA MEMBER

The absolute age and extent of amino acid diagenesis of fossil mollusks from the Walakpa member (Table 1) are of critical importance to the discussion of Late Quaternary paleotemperatures. This member, which represents one of the last transgressions on the western Coastal Plain, forms a prominent shoreline of gravel beach deposits up to 10-12 m a.s.l. that are continuously exposed above older transgressive marine units for almost 2.5 km on the southside of Walakpa Bay. Correlative gravel and sand facies on the north side of the bay grade northward into lagoonal sediments containing Macoma balthica. Near Barrow, the Walakpa member is composed of a barrier bar system that is exposed in the bluffs below the village and traceable in coastal bluffs for ≈ 3 km southwest until it pinches out over the Karmuk member. Beach sediments of the Walakpa member also partition Wainwright Inlet from the Chukchi Sea. Due to the low thermal history of the region (current mean annual air temperature at Barrrow, -12.8°C), isoleucine epimerization ratios in shells from the Walakpa member are only slightly above those in modern shells (modern sample = 0.011 ± .002). A radiocarbon analysis on drift wood from the Walakpa member near Walakpa Bay yielded an age of greater than 36,000 years B.P. (Beta-1766) and additional infinite dates on woody materials from this unit at Barrow have been reported by Sellman and Brown (1973, >39,900 years B.P., I-3628) and Coulter et al. (1960, >38,000 years B.P., W-380). A finite date on marine shells collected near Barrow (31,200 ± 810/900 years B.P., DIC-2569) is considered to be a minimum estimate. Correlative beach deposits of the barrier bar system extending from Barrow to east of Teshekpuk Lake have been identified by Carter (Carter and Robinson 1981) and radiocarbon dated to >51,000 years B.P.(USGS-676). The Walakpa member is considered correlative with the Pelukian transgression of Hopkins (1967) and to date from the last open-ocean $\delta^{18}O$ minimum (5e) ≈ 125,000 years B.P. (Shackleton and Opdyke 1973).

Table 1 Extent of Isoleucine Epimerization in the Walakpa member of the Gubik Formation, Western Arctic Coastal Plain, Alaska[1].

MEMBER	Hiatella Total	Hiatella C.V.%	Mya Total	Mya C.V.%
MODERN	0.011 ± .002	(18%)	--	
WALAKPA	0.014 ± .002	(14%)	0.018 ± .002	(11%)

[1]Values represent peak height measurements, mean ± 1δ; F = Free, T = Total amino acid fraction
[2]C.V.% = coefficient of variation = standard deviation/mean X 100%
[3]ND = not detectable, dashed lines indicate no data

PALEOTEMPERATURE ESTIMATES

Equations which express the relationship between the extent of epimerization, absolute time, and diagenetic temperature are reviewed by Schroeder and Bada (1976) and may be expressed as:

$$\ln \left[\frac{1 + D/L}{1 - K' \, D/L} \right] = (1 + K')\, K_1 \, t + C \qquad (1)$$

where D/L = aIle/Ile ratio of the sample; K' = inverse of the equilibrium constant (0.77); k_1= forward rate constant of isoleucine epimerization; t = time since death of the organism; and C = analytical constant that represents the value of the left side of the equation at death of the organism and accounts for laboratory induced epimerization during preparation (C = 0.0195 in samples discussed here).

The relationship between k_1 and temperature is described by the Arrhenius equation

$$k_1 = Ae^{\left(\frac{-Ea}{RT}\right)} \qquad (2)$$

where A = constant for each species; Ea = energy of activation; R = gas constant (1.987); and T = absolute temperature (°K). Radiocarbon dated control samples and pyrolysis data further define the relationship between the isoleucine epimerization rate constant (k_1) and temperature (T) for Mya truncata (Miller et al. 1983. The relationship between k_1 and T for Mya truncata is expressed as:

$$\log k_1 = 16.33 - 6131 / T \qquad (3)$$

and the energy of activation is calculated to be 28.06 kcal $mole^{-1}$. By replacing equation (3) for k_1 in (1) and solving for T, the equation can be written with T, t, and D/L as the only variables.

$$T = \frac{6131}{16.33 - \log\left\{\frac{\ln\left[\frac{1 + D/L}{1 - .77\ D/L}\right] - 0.0195}{1.77t}\right\}} \quad (4)$$

For a sample with a complex thermal history, T can be considered the effective diagenetic temperature (EDT) that is the temperature associated with an effective rate constant estimated by integration through a probable temperature history for a sample. Because of the exponential relationship between the rate constant (k_1) and temperature (T) this thermal integration is more sensitive to periods of high temperatures that accelerate the reactions more than the deceleration effected by an equivalent temperature decrease.

The effects of a variable thermal history on the EDT can be important on both an annual cycle and long-term climatic shifts. The arithmetic mean annual temperature of the ground will deviate from the effective diagenetic temperature in proportion to the amplitude of temperature about the mean. Although the mean annual air temperature at Barrow is currently -12.8°C, for example, integrating the monthly temperature cycle over the year calculates into an EDT of -3.6°C, reflecting the influence of warm summer temperatures. In permafrost, the value of the EDT is influenced by the amplitude of the annual temperature wave at a given depth and approximates the mean annual ground temperature at the depth of zero amplitude. Figure 2a shows the decrease in the EDT at discrete intervals calculated from the current mean monthly temperatures measured at depth in permafrost near Barrow (Lachenbruch et al. 1962). Near the ground surface, the EDT of the permafrost is quite high (≈ -3.0°C at -60 cm) reflecting the warm summer temperatures and large amplitude of the annual temperature wave. With depth, however, the amplitude of the annual temperature wave is increasingly attenuated to a point where the EDT in the permafrost (EDT -9.3°C at -18.5 m) is similar to the mean annual ground-surface temperature (≈ -10.0°C).

We can evaluate the effect of the annual temperature wave and hence, the EDT, on the rates of isoleucine epimerization under the present climate and during colder glacial climates. The effect of the present climate (using the EDTs at depth in Figure 2a) on rates of epimerization is illustrated for a hypothetical case in Figure 2b in which the present climate is maintained for 100,000 years. Using equation (4) and solving for D/L in successively deeper samples, it can be shown that below a depth of approximately -3 m (dashed line in Figure 2b), the variation in the aIle/Ile ratios created by changes in the EDT with depth is 11.5%; this is equivalent to the precision (expressed as coefficient of variation) attainable in isoleucine epimerization (compare with Table 1). For comparison, a series of Holocene samples subjected to the modern climate for 8,000 years would theoretically yield a depth related variation in aIle/Ile ratios of only 2.5% (Figure 2b). From this exercise, we conclude that (1) over relatively short time intervals subtle changes at the EDT's below about -6°C effect only minor variations in the observed aIle/Ile ratios, (2) consequently, paleotemperature estimates for

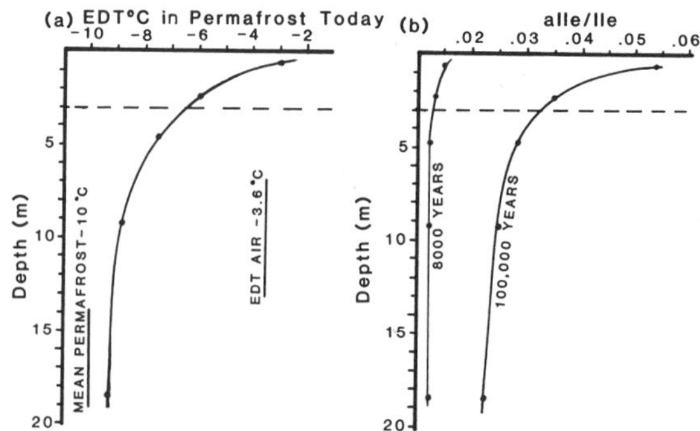

FIGURE 2 Effective diagenetic temperatures (EDT) and aIle/Ile ratios in permafrost. (a) Calculated EDT versus depth in permafrost under the present climate. (b) Effects of the modern EDT profile on the aIle/Ile ratio in samples 8.0 ka and 100 ka years B.P. The variation in ratios observed below -3 m (dashed line) is less than analytical variation in the amino acid method. For 100 ka, the mean aIle/Ile below -3 m = 0.026 ± .003, c.v. = 11.5%; for 8 ka, the mean is 0.12 ± .0003, c.v. = 2.5%.

low-temperature sites are only precise to within -2°to -3°C, (3) that samples collected for amino acid studies in permafrost areas should come from a depth of at least 2.5 m to minimize the effectof the annual temperature wave, and most importantly (4) these data, in turn, allow us to conclude that the mean permafrost temperature at depth is a good first approximation for the long-term EDT experienced by buried shells during each climatic regime over the last 125,000 yrs. In this study, the EDT expressed in equation (4), represents the integration of the thermal history of the upper few tens of meters of permafrost over an entire time period (t), and places constraints on the magnitude of ground and air temperature changes that can have occurred.

PALEOTEMPERATURE RECONSTRUCTIONS

Paleoclimatic proxy data and geomorphic evidence demonstrate that Alaska has experienced a number of broad climatic oscillations during the last 125,000 years, including warm and moist interglacials similar to today, cold and dry glacials, and intermediate interstadial conditions (Hopkins et al. [1982] for reviews). Most of the paleoclimatic reconstructions for this time period are qualitative in approach due to the lack of a means to quantify directly temperature and precipitation parameters. Based upon amino acid paleothermometry, we present here a series of paleotemperature reconstructions for the Alaskan Arctic Coastal Plain during the last 125,000 years that attempt to accomodate all of available paleoclimatic proxy data from the region and to quantify and define limits for the magnitude of temperature change during this time period.

Paleotemperature estimates derived from equation (4) are most accurate when the absolute age of a sample is known. Hence, the most important assumption in this reconstruction is that the Walkpa member is 125,000 years old; the rationale for its age outlined earlier. Younger age estimates can be tested by solving equation (4) for EDT's using younger assumed ages with the observed aIle/Ile ratios. Age estimates for the Walakpa member at 80 ka and 35 ka B.P. predict integrated diagenetic temperatures in the permafrost of -10.9°C and -6.8°C respectively, values similar to or significantly warmer than modern values. In light of paleoclimatic data that suggests the glacial periods in northern Alaska where colder and drier than modern interglacial conditions, we believe that such estimates are unrealistic and that an age of 125,000 is probably most reasonable for the Walakpa member.

AIle/Ile ratios in Mya truncata from last interglacial deposits (Walakpa member, t = 125,000 years B.P.) require an effective diagenetic temperature for the entire time period of ≈ -13.8°C. Using the paleoclimatic record as a guide for interglacial periods of "known" temperature, models can be developed that derive appropriate EDT's for periods lacking proxy data, especially full-glacial episodes. The models employ a computer program that allocates rate constants for intervals of specified climate, thus integrating an effective diagenetic temperature over each interval, and converts the computed rate constant back into a temperature (from equation 3) for an interval of unknown climate. The equation used for these reconstructions is modified from Miller et al. (1983) and written as:

$$k_x = \sum_{i=1}^{n} \frac{k_a t_a - (k_1 t_1 + k_2 t_2 + \ldots k_n t_n)}{t_2 - (t_1 + t_2 + \ldots t_n)} \quad (5)$$

(see Figure 3)

where, k_x = rate constant for a specific time interval of unknown temperature; k_a = rate constant for the entire assumed 125,000 year period; k_1, $k_2 \ldots k_n$ = rate constant for intervals of known temperature; t_a = length of entire time period = assumed 125,000 years; t_1, $t_2 \ldots t_n$ = length of each climatic interval. The effective diagenetic temperature for the unknown interval is calculated using the calculated k_x in place of k_1 in equation 3.

Paleoecological data from numerous Alaskan sources across Beringia suggest that (1) the last interglacial was slightly warmer than today, (2) an early Wisconsin glacial episode (Happy) occurred prior to 65 ka ago, (3) during the mid-Wisconsin (Boutellier Interval, about 65-30 ka B.P.) climate was intermediate between colder, drier glacials and warmer, more mesic interglacial conditions, (4) the late Wisconsin glacial stade (Duvanny Yar) occurred between 30 and 14 ka ago, and (5) during the late Pleistocene/early Holocene (Birch Interval, 14-8.5 ka years B.P.) summer temperatures, were somewhat warmer than the remainder of the Holocene (Hopkins 1982).

To model these climatic intervals, assumptions are necessary concerning the conditions in permafrost, as they may have effected propagation in the annual temperature wave. The most

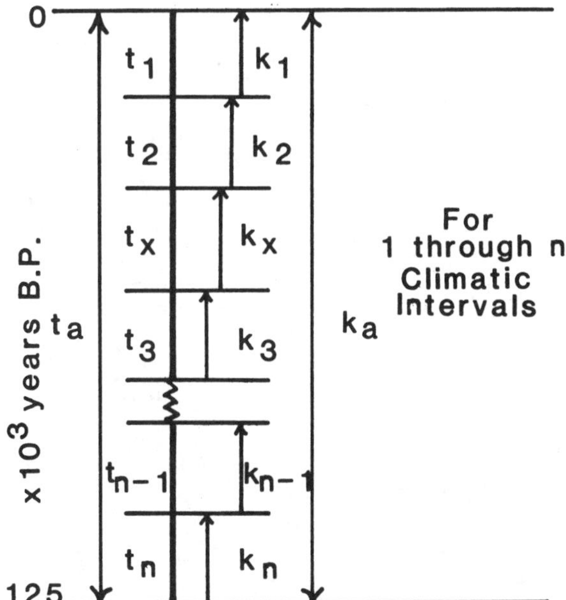

FIGURE 3 Schematic diagram illustrating the parameters used in equation 5 for evaluating past temperature histories. The EDT for the unknown interval x is predicted using k_x in place of k_1 in equation 3. See text for the definition of terms shown here. Any number of known and unknown intervals can be included, but the calculated unknown DTS's will be the average of all the unknown time intervals.

important factors influencing the effective diagenetic temperatures in permafrost are (1) the thermal conductivity of the sediment (which has probably been continuously frozen since the mid-Pleistocene), (2) the depth and insulating value of snow cover, and (3) the amplitude of the annual surface temperature wave. The lack of winter snow cover during dry periods probably allowed greater winter ground cooling. Given the more continental situation of the coastal plain, summer temperatures might be thought to have been warmer, but this may not be true. Sparse vegetation and the high albedo of eolian sand would have inhibited the penetration of summer radiation. The dry soils however, may have had a lower heat capacity and warmed more quickly at given temperatures. It is clear that thaw lakes did not develop during the Duvanny Yar glacial period (30 ka to 14 ka B.P.). Because summer temperatures probably did rise above 0°C in the summer months, the lack of these features must indicate that the upper few meters of permafrost were severely dessicated.

Paleoclimatic models representing mean permafrost temperatures at depth (or EDT's) for known intervals and calculated for glacial intervals are depicted graphically in Figure 4. In each case, equation (5) is used to predict the temperature of the unknown (dashed) interval. In all cases, the temperatures assigned to the intervals are constrained by the fact that an integration of all intervals must yield an average EDT of -13.8°C for the last 125 ka, as calculated earlier.

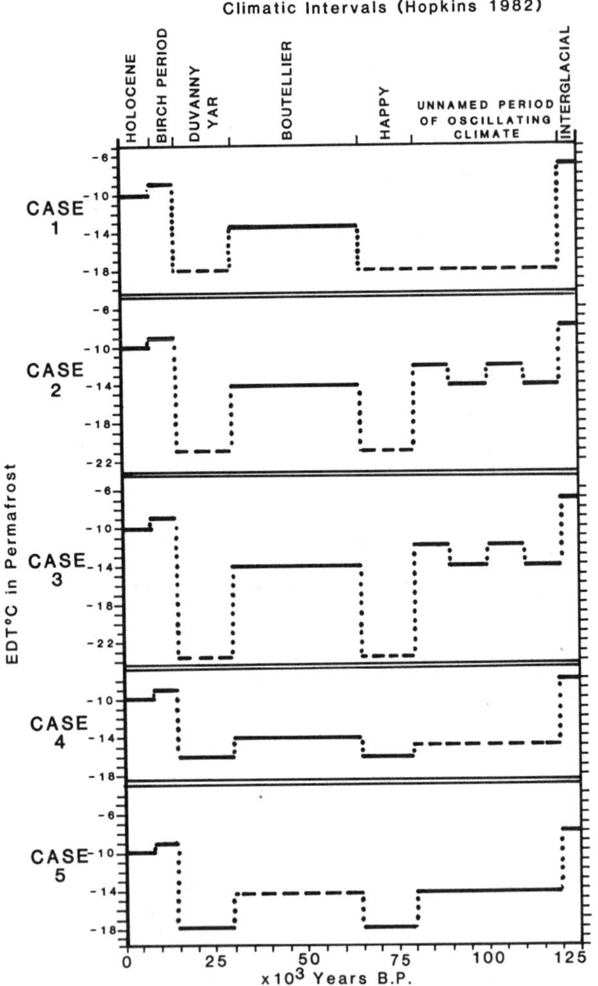

FIGURE 4 Schematic diagram illustrating five possible temperature models calculated from equation 5. Solid lines represent intervals of "known" climate, while dashed lines represent temperatures calculated for intervals of "unknown" climate. All temperatures indicated represent EDT°C at depth in permafrost.

Case 1 is a simple model which partitions the last 125 ka years into only 6 broad climatic intervals. The modern permafrost temperature of -10°C is used to represent the last 8.5 ka years while a temperature of -9°C was chosen for the birch zone (8.5-14ka years) representing a slight warming of climate, accompanied by inferred thicker, more insulating snowcover. The Boutellier interval (30-65 ka years B.P.) is depicted as a period of intermediate climate, and assigned a temperature lower than at present yet above full glacial conditions. Finally, the last interglacial episode is assigned a above full glacial conditions. Finally, the last interglacial episode is assigned a temperature slightly warmer than today as indicated by the extralimital distribution of fossil marine mollusks in deposits of this age across the northern coast. The calculation for case 1 suggests that glacial-age permafrost temperatures (for the Duvanny Yar and Happy (intervals) averaged roughly -18°C, or 8°C colder than today. Because the difference between mean annual ground temperature and mean annual air temperature can differ by as much as 1-6°C, depending on snow-cover (Gold and Lachenbruch 1973), this permafrost temperature suggests past mean annual air temperatures between -19 and -24°C. Because winter snowcover is interpreted to have been thin during the glacial times, the predicted permafrost temperature and mean annual air temperatures were probably similar.

Interpretations of global paleoclimates suggest that the period between 80 and 120 ka years B.P. was marked by oscillating, though not extreme, climatic conditions (Shackleton and Opdyke 1973); hence case 1 is probably oversimplified. In case 2, we have attempted to accomodate these fluctuations in climate with a more complex model by assigning intermediate warm and cold values for the oscillations of $\delta^{18}O$ stages 5a-d. This model effectively provides additional warmth to equation 5 and results in a predicted glacial permafrost temperature (for the Duvanny Yar and Happy glacial intervals) of -21°C. This value is colder than in case 1, and approaches some of the coldest values measured recently in permafrost from the Dry Valleys of Antarctica where the current mean annual airtemperature ranges between -18°C and -25°C (Decker and Bucher 1982). The sensitivity of equation 5 to slight increases in temperature is illustrated in case 3, which is identical to case 2 except the EDT for the last interglacial interval (lasting 5,000 years) is increased by 1°C. This case results in a predicted temperature decrease during the Happy and Duvanny Yar periods (duration totaling 31 ka years) of almost 3°C, reducing permafrost temperatures to -23.8°C. An attempt to model $\delta^{18}O$ stages 5a and 5c with modern temperature values, generated more warmth than could be compensated by any temperature depression during glacial periods.

In case 4, we have attempted to predict an average integrated temperature for the entire time period of $\delta^{18}O$ stages 5a-d (80 ka-120 ka B.P.) assuming glacial permafrost temperatures similar to modern values measured at Alert, N.W.T. (-16.1°C at -15 to -18 m. This reconstruction predicts an average temperature of -14.6°C for this period, or almost as cold as glacial temperatures in that model. Finally, in Case 5 we've used the temperature predicted for $\delta^{18}O$ 5a-d in case 4, and assigned glacial temperatures of -18°C, lower than case 4, to predict temperatures for the Boutellier interval of -15°C. This case also illustrates the sensitivity of the equations and associated rate constants.

CONCLUSIONS

The temperature reconstructions outlined above are intended, as models, to be refined and tuned as additional paleoclimatic proxy data and information concerning permafrost temperatures under variable climates become available. The most important concept that arises from this exercise is that the climte of the North Slope during the glacial episodes was at least 8°C colder than today. Very little warmth can be introduced into the calculations and still accomodate the available data. Any additional warmth (for example, a warmer birch zone) imposed on the illustrated models must be compensated by more intense or more numerous cold periods. Conditions during the latter portion of $\delta^{18}O$ stage 5(a-d) were definitely cooler than at present. Finally, we believe that models for glacial temperatures for the Duvanny Yar and Happy intervals lower than -25° to -28°C become unreasonable.

From these models we conclude that glacial permafrost temperatures on the North Slope probably averaged lower than about -18°C implying glacial mean annual air temperatures in the range of -19° to -24°C. Moreover, it is doubtful that summer temperatures were as warm as today.

The conditions depicted here for the Wisconsin glacial periods are reminiscent of the harsh polar desert climates of northern Ellesmere Island, N.W.T., northeast Greenland, or the Dry Valleys of Antarctica. Possibly these areas could serve as useful modern analogs for reconstructing the glacial age environment of the Alaskan Arctic Coastal Plain.

ACKNOWLEDGEMENTS

The research in this paper represents a portion of J. K. Brigham's dissertation completed at the University of Colorado, and was supported by the U. S. Geological Survey, the Outer Continental Shelf Environmental Assessment Program (BLM/NOAA) and an NSF Doctoral Dissertation grant DPP-81-02967 to Miller and Brigham. Additional support was provided by the Arctic Institute of North America, Geological Society of America, the Explorers' Club and Sigma Xi. Amino acid analyses were performed at the Amino Acid Geochronology Laboratory, INSTAAR, University of Colorado. Samples were prepared by J. Rohde and the manuscript typed by F. Al-Rahim and R. Chavez. We'd like to thank D. M. Hopkins, J. R. Williams, and L. D. Carter, U. S. Geological Survey, and Peter Lea, INSTAAR, for their thorough reviews and useful comments on early drafts of the manuscript.

REFERENCES

Black, R. F., 1964, Gubik Formation of Quaternary age in northern Alaska: U.S. Geological Survey Professional Paper 302-C, p. 59-91.

Brigham, J. K., 1983b, Marine stratigraphy, amino acid geochronology and sea level history of the Gubik Formation, Arctic Coastal Plain, northwestern Alaska: University of Colorado, Ph.D. dissertation, available Dec. 1983.

Carter, L. D., and Robinson, S. W., 1981, Minimum age of beach deposits north of Teshekpuk Lake, Alaskan Arctic Coastal Plain: U.S. Geological Survey Circular 823-B. p. B8-B9.

Coulter, H. W., Hussey, K. M., and O'Sullivan, J. B., 1960, Radiocarbon dates relting to the Gubik Formation, northern Alaska: U.S. Geological Survey Professional Paper 400-B, p. B350-B351.

Decker, E. R., and Bucher, G. J., 1982, Geothermal studies in the Ross Island - Dry Valley region, in Craddock, C., ed., Antarctic Geoscience: Madison, Wisconsin, University of Wisconsin Press, p. 887-894.

Gold, L. W., and Lachenbruch, A. H., 1973, Thermal conditions in permafrost--A review of North American literature, in Permafrost--The North American contribution to the Second International Conference, Yakutsk: Washington, D.C., National Academy of Sciences.

Hopkins, D. M., 1967, Quaternary marine transgressions in Alaska, in Hopkins, D. M., ed., The Bering Land Bridge: Stanford University Press, p. 47-86.

Hopkins, D. M., 1982, Aspects of the paleogeography of Beringia during the late Pleistocene, in Hopkins, D. M., Matthews, Jr., J. V., Schweger, C. E., and Young, S. B., eds., Paleoecology of Beringia: New York, Academic Press, p. 3-28.

Hopkins, D. M., Matthews, Jr., J. V., Schweger, C. E., and Young, S. B., 1982, Paleoecology of Beringia: New York, Academic Press, 489 p.

Lachenbruch, A. H., Brewer, M. C., Greene, G. W., and Marshall, B. V., 1962, Temperature in permafrost: Temperature--Its measurement and control in Science and Industry, v. 3, pt. 1, p. 791-803.

Miller, G. H., 1982, Amino acid geochronology Laboratory 1982 report on current activities: Institute of Arctic and Alpine Research, University of Colorado, 56 p.

Miller, G. H., Sejrup, H.-P., Mangerud, J., Anderson, B., 1983, Amino acid ratios in Quaternary mollusks and foraminifera from western Norway: Correlation, geochronology and paleotemperature estimates: Boreas, in press.

Schroeder, R. A., and Bada, J. L., 1976, A review of the geochemical applications of the amino acid racemization reaction: Earth-Science Reviews, v. 12, p. 347-391.

Sellman, P. V., and Brown, J., 1973, Stratigraphy and diagenesis of perennially frozen sediment in the Barrow, Alaska, region, in Permafrost--The North American Contribution to the Second International Conference, Yakutsk: Washington, D.C., National Academy of Sciences, p. 171-181.

Shackleton, N.J., and Opdyke, N.D., 1973, Oxygen isotope and paleomagnetic stratigraphy of equatorial Pacific core V28-238: Oxygen isotope temperatures and ice volumes on a 10^5 and 10^6 year scale: Quaternary Research, v. 3, p. 39-55.

HYDROLOGY OF THE NAHANNI, A HIGHLY KARSTED CARBONATE TERRAIN WITH DISCONTINUOUS PERMAFROST

George A. Brook

Department of Geography, University of Georgia, Athens, Georgia 30602 USA

Groundwater recharge rates are usually low where permafrost is present. However, in carbonate regions where a karst drainage system predates permafrost development, groundwater recharge and circulation may be vigorous. This is the case in the Nahanni north karst of northwestern Canada. Not only is the volume of groundwater flow in this permafrost region considerable--102×10^6 m^3 of water flows annually to each of two springs which drain the area--but hydrologic activity is spectacular. Between July 19 and 31, 1972, an extreme summer storm deposited 224 mm of rain on the area. First, Second, and Third poljes flooded; maximum water depths were 8.5, 25, and 8 m, respectively; and Third Polje overflowed. The level of Raven Lake (0.25-0.50 km long) rose 49 m at an average 2.9 m/day. Field observations and evidence from LANDSAT images have provided a hydrologic record for the karst for the years 1972-1978. These data indicate that immediately prior to spring snowmelt depressions are dry, and that when snow and ice melts in May, several depressions may flood temporarily because their drainage routes are blocked by ice. The major surface and groundwater activity in Nahanni is not induced by snowmelt but results from unusually high rainfall in the months of June-August.

INTRODUCTION

The presence of permafrost usually retards infiltration into and percolation through soil to recharge groundwater flow systems because the hydraulic conductivity of soil is often several orders of magnitude lower when frozen. Therefore, under permafrost conditions, infiltration through joints in exposed bedrock accounts for a major proportion of groundwater recharge.

Large areas of northern Canada are underlain by limestones and dolomites in areas of both continuous and discontinuous permafrost. Bird (1966), for example, reports that limestone is the most common sedimentary rock in arctic Canada south of Parry Channel, where it covers about 300,000 km^2, or more than 12% of the land surface. Because limestones are soluble, any water penetrating joints will rapidly increase the secondary permeability of the rock, encouraging development of a subsurface drainage system. However, it appears that in some karst regions an efficient karst drainage system may have developed before the onset of the present conditions of permafrozen ground. In these areas, groundwater recharge and circulation is more vigorous than in other high-latitude regions of carbonate bedrock (Brook and Ford 1980, van Everdingen 1981). This is the case in the Nahanni karst of northern Canada, discovered in August 1971 (Figure 1).

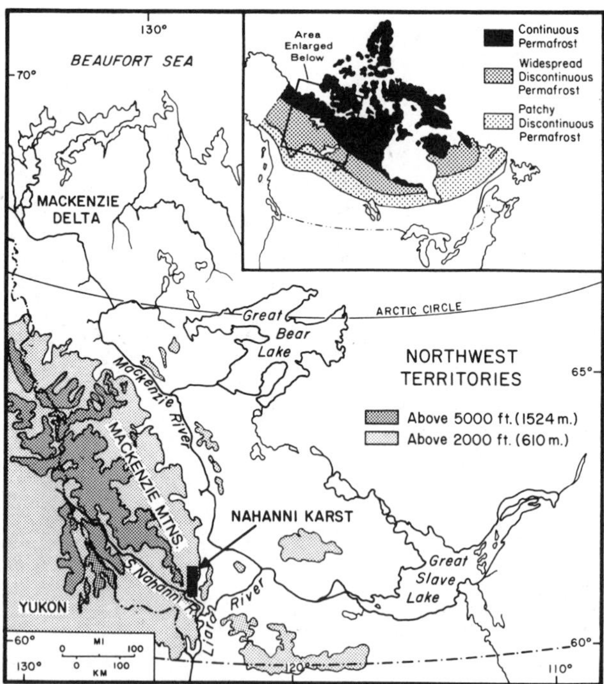

Figure 1 Topography and permafrost in the District of Mackenzie, N.W.T., Canada, and the location of the Nahanni karst. Permafrost boundaries are after Brown (1978).

THE NAHANNI KARST

In the Nahanni karst the predominant landforms are vertical-walled solution dolines, and remarkable networks of solution streets, platea, residual limestone towers, and poljes that are successive stages in the evolution of labyrinth karst (Brook and Ford 1978). The most spectacular labyrinths are located on a narrow structural col (the "north karst"), which connects the upwarped domes of Nahanni and Ram plateaus (Figure 2). With depths exceeding 200 m these are among the most accentuated solutional labyrinths known. In addition, a number of closed fluvial canyon networks up to 1,000 m deep dissect the eastern flank of the south Nahanni Plateau and the southwest flank of the Ram Plateau. All of these canyon systems are blocked by glacial deposits, and today their drainage is through underground caves in limestone and/or dolomite.

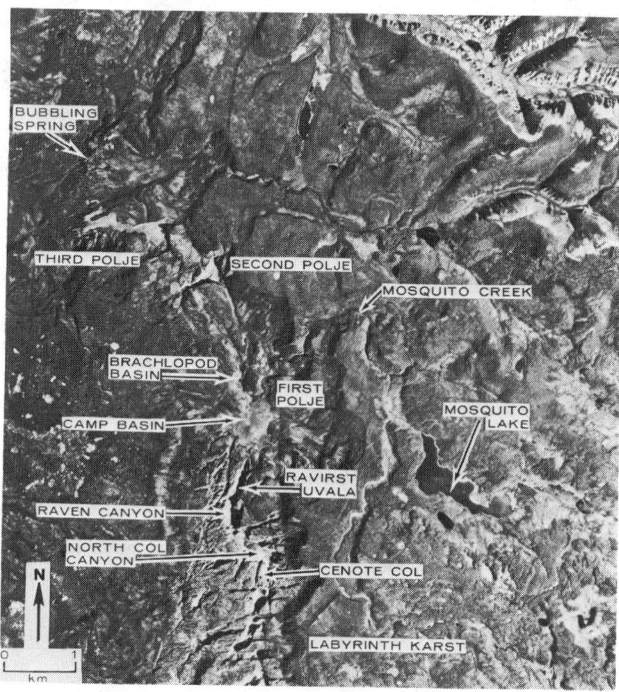

Figure 2 The closed depressions and springs of the Nahanni north karst.

Karst in Nahanni is found at elevations of 840-1,400 m a.s.l.; the entire region falls within the zone of discontinuous permafrost. Calculated mean annual temperature is -4.5°C; precipitation, 566 mm/yr. Vegetation below 1,200 m is northern boreal forest with tundra at higher elevations. Solutional landforms are developed in limestones of the Middle Devonian Nahanni Formation and in argillaceous limestones and calcareous shales of the underlying Headless Formation. Together these are 150-200 m thick. Beneath them are fine-grained dolomites up to 815 m thick, which transmit groundwater. Shales of the Fort Simpson Formation overlie Nahanni limestones in some areas and support surface streams at spring snowmelt or after rainfall. In many cases, water collecting on shales drains on to limestones, where it sinks underground.

Using the Turc (1954) equation, effective precipitation in Nahanni is estimated at 394 mm; evapotranspiration, 172 mm. Effective precipitation migrates underground as both diffuse and conduit recharge. Only two major springs appear to discharge water from the karst terrain. White Spray in the north wall of First Canyon, South Nahanni River, emerges from dolomites, and Bubbling Spring, 30 km to the north, emerges close to the top of the Nahanni Formation limestones (Figure 3). The catchments of Bubbling Spring and White Spray have been delimited on the basis of dye traces, and on geological and topographical evidence. Each is about 260 km^2. The northern boundary of Canal Creek drainage is the approximate divide between the two recharge areas. Canal Canyon cuts through the crest of the Nahanni Plateau structural dome. To the north and south, groundwaters are believed to follow the stratal dip (Figure 3). Runoff into Canal Canyon, the mouth of which is blocked by a glacial end moraine, sinks at the base of the south wall. Water drains into dolomites at 823 m elevation and is thought to remain in cave systems until it reaches White Spray at 274 m a.s.l.

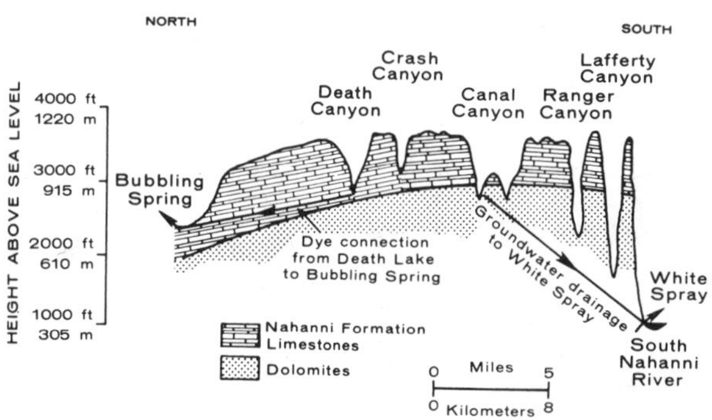

Figure 3 Topographic and geologic section through the Nahanni karst from Bubbling Spring to White Spray.

With 394 mm effective precipitation, approximately 102×10^6 m^3 of water must drain to each spring annually at an average discharge of 3.2 m^3/s. Because both springs were discharging >5.0 m^3/s in summer 1973, discharge must vary seasonally. Assuming that flow averages 5.0 m^3/s during the 6 months from April to September, mean discharge during the rest of the year must be 1.5 m^3/s. White Spray is known to flow in winter, and its discharge has been estimated at 1.0 m^3/s, confirming that when there is no recharge to the aquifer, discharge is much less than during the spring and summer months.

EVIDENCE OF PERMAFROST

The Nahanni karst falls close to the boundary between widespread and patchy discontinuous permafrost (Figure 1). There are various indications of permafrost conditions in the Nahanni. For example, short, relict cave conduits exposed in fluvial canyons and in street walls are often sealed by ice or permafrozen silt at, or close to, the limit of sunlight penetration. In addition, many of the large relict caves display a distinct climatic zonation in the summer months, which gives some indication of temperature variations with depth in the carbonate bedrock. The entrance zone of Grotte Valerie, high in the north wall of First Canyon, South Nahanni River, is characterized by temperatures of 0°-2°C. Water can penetrate the rock above the cave, and small speleothems are being deposited. Further from the entrance is a second zone where rock temperatures are close to 0°C. This zone is characterized by a layer of warmer air near the ceiling, which overlies colder air at the floor. The boundary between the warm and cold air advances into the cave as the summer progresses. The warm air near the ceiling allows water to penetrate the overlying rock, but once it reaches the layer of cold air it freezes, forming ice sheets and ice stalagmites on the cave floor. The third climatic zone is found deeper in the cave where passages descend below the level of the cave entrance and temperatures are thought to remain constant all year. Air temperatures in this zone are -3°C. Passages are dry and dusty because no water can penetrate the roof of the cave (Ford and Brook 1976). Igloo and Ice Curtain caves in the walls of Death Canyon display a basically similar climatic zonation to Grotte Valerie. The coldest temperatures in Ice Curtain Cave were -1.5°C, and in Igloo Cave -4.0°C when measurements were made in July 1973.

The numerous vertical-walled dolines in Nahanni, some of which contain ponds up to 17 m deep, also suggest permafrozen conditions at depth. Ponds in these depressions do not reflect the upper level of a body of groundwater, for pond levels between closely spaced individuals may differ by as much as 33 m. The ponds are thought to be perched above subsurface ice bodies blocking the doline drainage routes. In late June, 1973 Surprise and Hidden dolines (later found to be connected by a short cave) on Cenote Col (Figure 2) contained ponds 17-20 m deep. In early August these dolines emptied abruptly presumably because the ice seal had melted. Perched Basin, another doline on Cenote Col, drained in mid-July 1972 after 3.5 cm of rain had caused melting of the subsurface ice barrier.

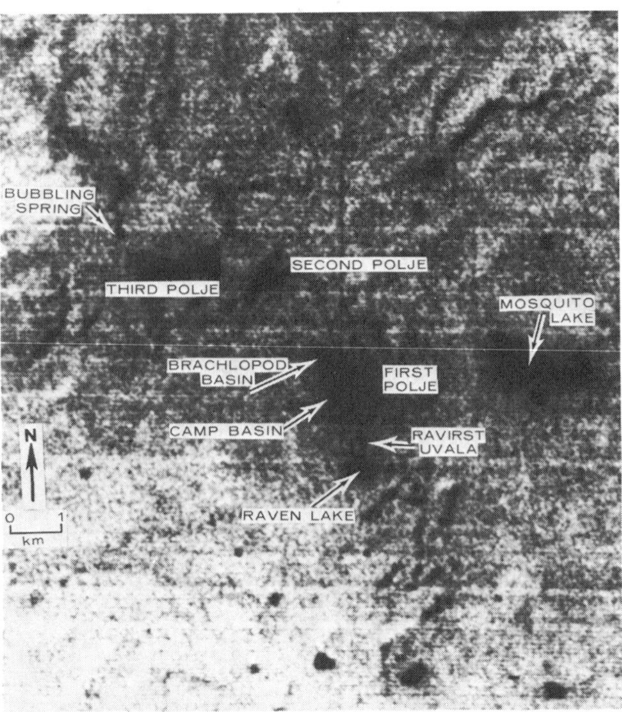

Figure 4 Heavily flooded north karst depressions revealed on an August 22, 1972, LANDSAT image. All three poljes, Ravirst Uvala, Raven Canyon, and North Col Canyon were flooded on this date. Third Polje was overflowing, and there was considerable outflow from Bubbling Spring. See Figure 2 for comparison.

HYDROLOGY OF THE NORTH KARST FROM FIELD AND REMOTELY SENSED DATA

Not only is the volume of groundwater flow in the Nahanni Karst region considerable, amounting to 70% of the annual precipitation, but hydrologic activity is spectacular. This is particularly so in the north karst, which extends for 7 km south of Bubbling Spring. The area includes First, Second, and Third poljes, Ravirst Uvala, Raven Canyon, and North Col Canyon (Figure 2). These depressions are 1.4, 0.7, 1.4, 0.3, 0.5, and 1.0 km long, respectively. Between July 19 and 31, 1972, an extreme summer storm deposited 224 mm of rain on the area. First, Second, and Third poljes flooded; maximum water depths were 8.5, 25, and 8 m, respectively; and Third Polje overflowed. The level of Raven Lake (which varies from 0.25 to 0.50 km long), occupying the floor of Raven Canyon, a deep karst street, rose 49 m at an average 2.9 m/day (Figure 4).

Field data on the hydrology of the Nahanni north karst collected in the summers of 1971-1974 were supplemented using data in LANDSAT 1972-1978 multispectral scanner (MSS) and return-beam Vidicon (RBV) images of the region. Near-infrared band-6 (700-800 nm) and band-7 (800-1,100 nm) images were used to monitor the presence or absence of floodwaters in north karst depressions. The low spectral reflectance of water to infrared radiation makes water bodies easy to identify, as these appear black (Figure 4).

Field observations and evidence from LANDSAT images have provided a reasonably detailed hydrologic record for the north karst for the years 1972-1978 (Table 1). These data indicate that immediately prior to spring snowmelt (surface water bodies become free of ice in May), karst depressions are dry. On April 23, 1977, for example, Raven Lake (70 m deep in July 1972) was a frozen pond 8 m in length. This suggests that surface flow into depressions is arrested as the winter freeze sets in (surface water bodies freeze over in October) and that any water remaining in depressions in October drains underground before the following spring.

LANDSAT images have also revealed that when snow and ice melts in May, several depressions, including First and Third poljes and Raven Canyon, may flood because their drainage routes are blocked by ice. A May 27, 1975, image shows that lakes in the north karst were already free of ice by this

TABLE 1 Hydrologic Conditions in the Nahanni Poljes Based on Field, LANDSAT, and Aerial Photograph Data

YEAR	MONTH	LANDSAT (L), FIELD (F), or AERIAL PHOTOGRAPH (A) OBSERVATION	HYDROLOGIC CONDITION†		
			FIRST POLJE*	SECOND POLJE	THIRD POLJE*
1949	May				
	Jun				
	Jul	A	D	D	PF
	Aug	A	D	F	F
1952	May				
	Jun	A	D	D	D
1961	May				
	Jun				
	Jul				
	Aug	A	D	D	D
1971	May				
	Jun				
	Jul				
	Aug	F	D	D	D
1972	May				
	Jun	F	D	D	D
	Jul	F	F	F	F
	Aug	F & L	F	F	F
1973	May				
	Jun	F	PF	F	F
	Jul	F & L	D	F	F
	Aug	F & L	D	F	F
	Sept	L	D	PF	PF
1974	May				
	Jun	F	D	PF	PF
	Jul				
	Aug	F & L	F	F	F
	Sept	L	D	F	F
1975	May	L	D	D	PF
	Jun	L	D	D	D
	Jul	F & L	D	D	D
	Aug	L	D	F	F
	Sept	L	D	F	F
1976	May	L	PF	D	PF
	Jun				
	Jul	F	D	PF	PF
	Aug	L	D	F	F
	Sept	L	D	F	F
	Oct	L	D	PF	PF
1977	Apr	F	D	D	D
	May				
	Jun	L	D	D	PF
	Jul				
	Aug	L	D	F	F
	Sept	L	D	PF	PF
1978	May				
	Jun				
	Jul				
	Aug	L	D	PF	PF
	Sept	L	D	D	D

† D = dry; PF = partly flooded; F = flooded

* ponds are present at the eastern end of Third Polje and in Brachiopod Basin, First Polje, even under very dry conditions

date and that ponds 250, 100, and 100 m long existed at the eastern end of Third Polje, in Ravirst Uvala, and in Raven Canyon, respectively. A May 12, 1976, image shows that many lakes were still frozen over and much of the north karst was mantled by snow at this time, ponds 500 m long, 150 x 300 m, and 150 m long were present at the eastern end of Third Polje, at the southern end of First Polje, and in Raven Canyon, respectively.

In the 7 years from 1972 to 1978, First Polje is known to have flooded in the period June to September once in every 2-3 years, and Second and Third poljes flooded every year. It is apparent that in most years, north karst depressions are first inundated in June or July and less frequently in August. Flood waters in First Polje drain rapidly, so this depression is rarely flooded in the month of September. Second and Third poljes drain more slowly; once flooded, they almost always remain flooded through August and into September.

In many arctic and subarctic terrains, spring snowmelt is the major hydrologic event of the year. This is the case, for instance, in a karst area to the northwest of Smith Arm, Great Bear Lake, Canada. The northern part of this area lies within the zone of continuous permafrost. van Everdingen (1981) reports that in a given year the maximum flood level in a large depression in this karst is mainly a function of the amount of snow remaining on the ground at the start of snowmelt, and of the rate at which melting proceeds. The duration of flooding is a function of the maximum level reached during snowmelt and the amount and distribution of rainfall during the summer months.

Hydrologic data for the Nahanni indicate that the major surface and groundwater activity is not induced by snowmelt but results from heavy rainfall in the months June-August (Table 1). Unusually high monthly rainfall caused flooding in June 1973, August 1949, and August 1974, when rainfall at Fort Simpson was 307%, 171%, and 232% of normal, respectively; in May 1977, when rainfall at Little Doctor Lake was 245% of normal; in July 1972, when 267 mm of rainfall was measured in the Nahanni karst; and in August 1975 when rainfall at Fort Simpson and Little Doctor Lake was 190% and 168% of normal, respectively. In years with no unusually high monthly precipitation, depressions appear to remain dry or suffer minor flooding.

Flooding of the depressions in the Nahanni karst is believed to occur because of random perching of water above and below ground where drainage conduits are heavily alluviated or where flow is obstructed by subsurface ice bodies.

DISCUSSION

The flow of surface water into ponors in karst depressions in the Nahanni karst is clearly an effective means of recharging groundwater in an area of discontinuous permafrost. However, the recharge system may not have developed under the environmental conditions that exist in this region today.

Fossil caves in Nahanni contain speleothem deposits that have been uranium-series dated to >350,000 years B.P. (Harmon et al. 1977). These caves must, therefore, have been largely abandoned by the drainage water that produced them as much as 400,000 years ago. Furthermore, many speleothem deposits are located in dry cave passages where present temperatures are below freezing. For the speleothems to have been deposited, mean annual temperatures in Nahanni must have been much higher than today. The cave evidence indicates that there were at least four periods during which speleothem growth was enhanced by climatic conditions. These periods at 90-150, 185-235, 275-320, and $\geq$350 ka were almost certainly interglacial phases (Brook 1976; Harmon et al. 1977). It is extremely likely that the present karst drainage system of Nahanni was developed during one or more of these warm phases.

According to Lachenbruch (1968), heat flow from the earth's interior normally results in a temperature increase of approximately 1°C per 30-60 m increase in depth. The mean annual temperature in Nahanni is -4.5°C, suggesting that the base of permafrost could lie between 270 and 135 m--the lower of the two figures being the most likely, based on cave temperature evidence. This suggests that the karst conduits which funnel water to Bubbling Spring and White Spray are largely located within the subpermafrost zone where rock temperatures may be one or more degrees above freezing.

A second important process which helps to keep the karst groundwater system operative is the warming of the surface rocks by recharge waters. In summer 1973 the temperatures of water at White Spray and Bubbling Spring were 4.8° and 4.6°C, respectively. These groundwater temperatures are believed to reflect a mixing in the karst aquifer of snowmelt waters (at, or close to, 0°C), with water derived from summer rainfall. The average temperature of surface waters in the Nahanni in the summers of 1972 and 1973 was 9.5°C. As snowfall accounts for 40% of annual precipitation, groundwater temperatures somewhat below 6°C might be expected. Temperatures of 4.5°C indicate that groundwaters lose heat to the bedrock, thus helping to keep the karst drainage system open.

REFERENCES

Bird, J. B., 1966, Limestone terrains in southern arctic Canada, Proceedings: Permafrost International Conference, NAS-NRC, Publication 1286, 115-121.

Brook, G. A., 1976, Geomorphology of the north karst, South Nahanni River region, Northwest Territories, Canada, unpublished Ph.D. dissertation, McMaster University, Canada, 627 pp.

Brook, G. A., and Ford, D. C., 1978, The origin of labyrinth and tower karst and the climatic conditions necessary for their development: Nature, London, 275(5680), 493-496.

Brook, G. A., and Ford, D. C., 1980, Hydrology of the Nahanni karst and the importance of extreme summer storms: Journal of Hydrology, 46, 103-121.

Brown, R. J. E., 1978, 'Permafrost', Plate 32 in Hydrological atlas of Canada: Ottawa, Fisheries and Environment Canada.

Ford, D. C., and Brook, G. A., 1976, The Nahanni north karst, Northwest Territories of Canada, Proceedings of the 6th International Congress of Speleology, Olomouc, Czechoslavakia, September 1973, II: Academia/Praha, 157-168.

Harmon, R. S., Ford, D. C., and Schwarcz, H. P., 1977, Interglacial chronology of the Rocky and Mackenzie Mountains based upon ^{230}Th-^{234}U dating of calcite speleothems: Canadian Journal of Earth Sciences, 14, 2543-2552.

Lachenbruch, A., 1968, Permafrost, Fairbridge, R. W., ed., Encyclopedia of geomorphology: New York, Reinhold, 833-838.

Turc, L., 1954, Le bilan d'eau des sols: relations entre les précipitations, l'évaporation et l'écoulement: Ann. Agron., Sér. A, 5, 491-595.

van Everdingen, R. O., 1981, Morphology, hydrology, and hydrochemistry of karst in permafrost terrain near Great Bear Lake, Northwest Territories, National Hydrology Research Institute Paper 11, 53 pp., Environment Canada.

OBSERVATIONS ON ICE-CORED MOUNDS AT SUKAKPAK MOUNTAIN, SOUTH CENTRAL BROOKS RANGE, ALASKA

J. Brown[1], F. Nelson[2], B. Brockett[1], S. I. Outcalt[2], and K. R. Everett[3]

[1]Cold Regions Research and Engineering Laboratory
Hanover, New Hampshire 03755 USA
[2]Department of Geological Sciences, University of Michigan
Ann Arbor, Michigan 48109 USA
[3]Institute of Polar Studies, Ohio State University
Columbus, Ohio 43210 USA

Several hundred mounds occur on the lower slope of Sukakpak Mountain. The mean mound height is approximately 1 m and most are elliptical or circular in plan. Clear, massive ice can be found within, below, and adjacent to some mounds. Within and adjacent to one mound, free water under low pressure was observed in late winter. Frozen sediments were found below the water lens. Trees with smooth trunk curvature on top of the mounds suggest long period of stability. Most mounds are found in active drainage channels that develop thick surface icings each winter. As a tentative hypothesis, we suggest that the mounds form by closed-system freezing at sites with higher moisture contents than their surroundings. The causes and frequency of occurrence and annual magnitude of this upheaving are under investigation.

Palsas and ice-cored mounds have been observed in many permafrost regions of the world (Washburn 1983). Only a few have been reported in Alaska, such as those described by Péwé (1975) in the McLaren River valley and by Nelson and Outcalt (1982) on the eastern Arctic Slope, but the dearth of reported Alaskan occurrences of palsas may only reflect a lack of opportunities to make appropriate ground observations. Most palsas reported in the international literature are in regions with terrain and environmental conditions comparable to central Alaska, including discontinuous permafrost, abundance of wet terrain in bogs and on slopes, and slightly negative mean annual air temperatures.

Mackay (1978) suggested that palsas may be distinguished from pingos with the term palsa applying to mounds in which the top of the ice core lies at the base of the active layer. Most published reports suggest that segregation ice, often observed as thin lenticular ice inclusions, is responsible for palsa formation (for example, Ahman 1977, Friedman et al. 1971). Palsas have also been reported that have massive ice (> 50 cm) in their centers (Brown 1973, Sollid and Sorbel 1974). Van Everdingen (1982) reported on seasonal frost blisters, which form when outlets of springs are blocked by freezing of the active layer and formation of surface icings. Progressive freezing of injected water in the active layer causes the overlying surface ice and frozen soil to dome. These mounds usually collapse in summer as the surface icings melt and the ground thaws. The Sukakpak mounds have some of the characteristics of both palsas and frost blisters.

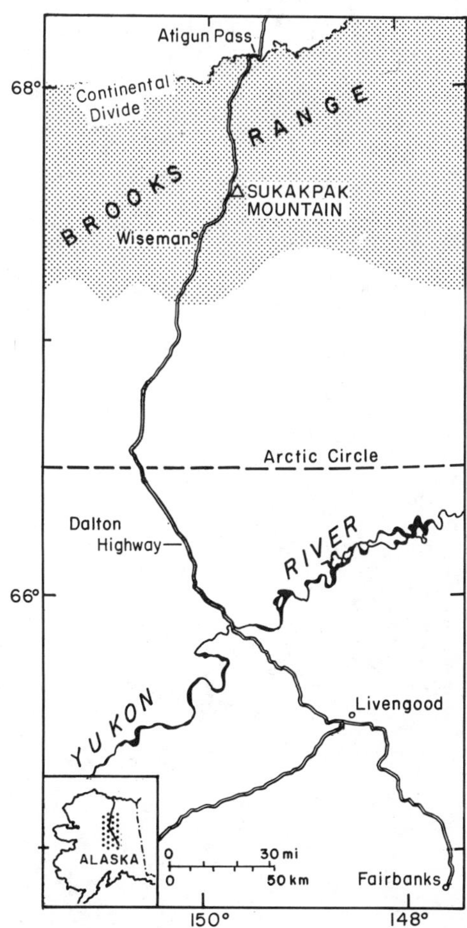

FIGURE 1 Location map for the Sukakpak Mountain study site.

SETTING

The study site is at approximately mile 205 on the Dalton Highway (67°37'N, 149°42'W), some 300 km

Figure 2 (a) Aerial oblique photograph of locality A on October 21, 1982, showing build-up of surface icings and several tilted trees.

(b) Location of study site and the three main localities of mounds (A, B, and C). The Dalton Highway cuts across the lower slope. The elevated oil pipeline is on the terraces of the Middle Fork Koyukuk River.

(c) Ground view of mounds, surface overflows, and icings on October 21, 1982.

(d) Aerial view of icings on October 21, 1982.

(e) A J-shaped spruce tree on a characteristic mound.

(f) A ruptured mound exposing a thick ice layer. Photographed July 21, 1982.

north of Fairbanks (Figure 1). The mounds occur on the mid- and lower west-facing slopes of Sukakpak Mountain at a mean elevation of 400 m; they were noted earlier by Hamilton (1979) and Brown and Berg (1980). The trans-Alaska pipeline and the Middle Fork Koyukuk River form the western boundary of the site (Figure 2b). The site lies within the northernmost zone of discontinuous permafrost (Kreig and Reger 1982). Open-system pingos are found in the south central Brooks Range (Hamilton and Obi 1982).

Sukakpak Mountain is composed of marble and dolomite overlying graphitic and calcareous quartz schist and intruded metabasite dikes (Dillon 1982). Coarse rubble forms large talus aprons at the break in slope between the exposed bedrock and the more gentle lower colluvial slopes (2° to 6°). The lower slopes are composed of water-saturated, medium to coarse gravel and calcareous sand and silt. Boreholes indicate that the depth to bedrock ranges between 5 and 15 m near the road.

The mean annual air temperature is estimated to be -7°C, with a total of 1275 thawing degree-days (C) and 3800 freezing degree-days (C). Annual precipitation averages 400 mm (Haugen 1982 and unpublished data). Ground temperature near the road is approximately -0.5°C at a depth of 12 m (Brown and Berg 1980, and authors' unpublished data). The active layer can exceed 3 m; thaw zones 3 to 5 m thick are found beneath the frozen active layer.

Although the mounds occur throughout much of the site, a large number are confined to surface drainage lines or channels in two areas above the road (Figure 2b, A and B) and below the road (C). These wetter areas generally lack the black spruce (Picea mariana) tree cover of the surrounding areas, and are instead composed of calcareous fens that support a rich flora distinct from those on the mounds and the adjacent spruce-covered upper slopes. Spruce of widely varying ages and sizes are found on most mounds, although some have only a cover of medium-height birch and willow shrubs.

The spruce growing on the mounds display degrees of tilting and bending at various heights along their trunks. On some, the J-shaped curvature begins at the base of the trunk (Figure 2e), while others are curved at the top. Presumably these differences reflect different periods of upheaval and readjustment of growth. The larger trees may be more than 100 years old, which implies considerable stability for the mounds. Some mounds, particularly near the road on the upslope side, have ruptured and are in various stages of collapse (Figure 2f). Such ruptures are probably due to renewed growth induced by the wetter conditions associated with impeded drainage near the road.

MORPHOLOGY AND STRATIGRAPHY

Based on a sample of 102 mounds, heights of the mounds range from 0.35 m to 2.3 m, with a mean of 0.95 m and a standard deviation of 0.4 m (Figure 3). Axial (width/length) ratios indicate that most of the mounds approximate the classic elliptical to circular domed shape (Figure 3). Widths range from 1.8 to 15.1 m, and lengths from 2.6 to 40.2 m. Median width and length are 4.6 and 7.4 m, respectively. Elongated mounds are typically found oriented parallel to the slope in the drainage

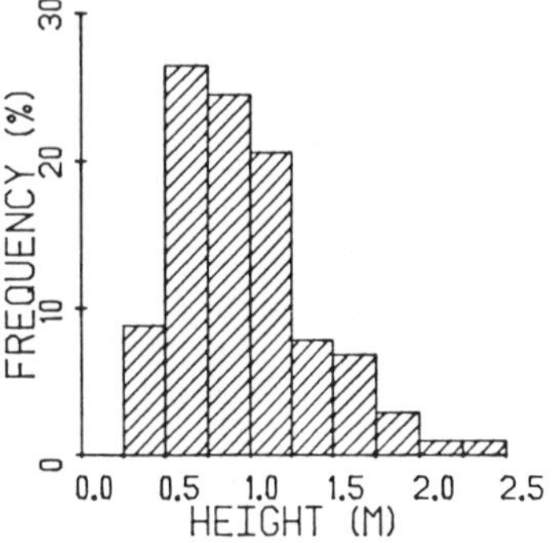

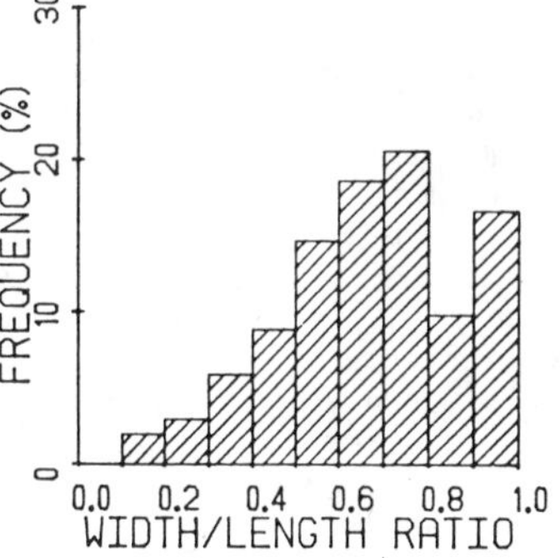

FIGURE 3 Frequency distributions of heights (upper) and ratio of axial lengths (lower) of 102 measured mounds.

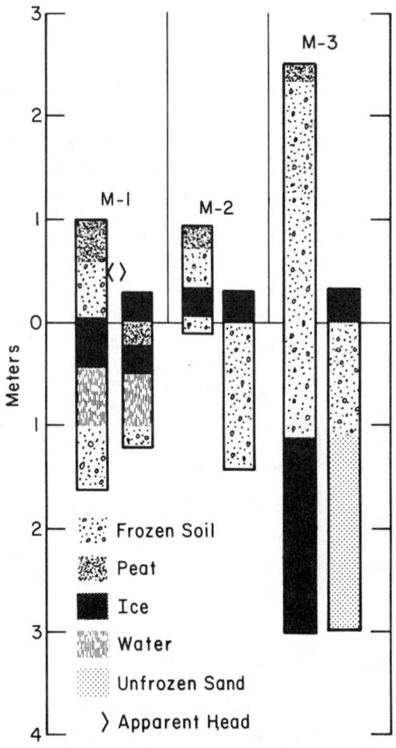

FIGURE 4 Stratigraphy of three mounds and adjacent areas based on April 1982 augering and coring.

TABLE 1 Grain size analysis for mounds in areas A and C.

Sample	Organic (%)	Gravel (%)	Sand (%)	Silt and clay (%)
1	1.0	0.0	57.7	42.3
2	11.0	0.1	46.8	53.1
3	6.0	0.5	35.7	63.9
4	1.0	16.1	47.1	36.8

channels on the upper slope and oriented perpendicular to the slope below the road.

In April 1982, three mounds and adjacent areas were augered and cored. Two of the mounds, situated about 30 m from the road, were approximately 1 m high. The third mound, about 60 m upslope from the road, was 2.5 m high and represented one of the largest mounds found on this site. All three mounds were surrounded by 30 cm of surface icing beneath the snow pack. Figure 4 shows the idealized stratigraphy of these mounds.

One of the two smaller mounds (M-1), had a 50-cm thick ice core underlain by 50 cm of water under pressure. Water rose to within 50 cm of the surface in the hole. A hole cored adjacent to this mound through 30 cm of surface icing encountered artesian water, which rose 15 cm above the ice-covered ground surface. Massive ground ice was encountered above the subsurface water layer. Although some thawed sediments were encountered beneath the water, the deeper underlying sediments were frozen. A second hole 6 m upslope revealed 1.5 m of frozen peat, sand, and gravel over 20 cm of buried clear ice that contained vertically oriented, clean plant roots. There was no water or thawed zone.

A second mound (M-2) contained 20 cm of clear ice at a depth close to the surrounding ground surface. Drilling adjacent to this mound indicated no buried ice or thawed ground. The third mound (M-3) was completely frozen and was composed of relatively dry organic material, sand, silt, and fine calcareous gravel. Massive ice, more than 2 m thick, was encountered below ground level in the mound. This ice changed in character downward from dense to bubbly and porous ice near the bottom of the hole. The lower limit of the ice was not determined. The surface of this mound consisted of large irregular blocks of broken sod and soil and appeared to have undergone forceful growth or subsequent degradation. A hole adjacent to the mound yielded a thawed slurry of fine black sand that extended to a depth of at least 3 m. There was 1.6 m of frozen ground over this thawed zone.

Several additional mounds were sampled in October and the grain size was determined (Table 1).

SLOPE HYDROLOGY

During late spring, summer, and early fall, surface water flows down the slopes along fen-covered drainage lines or channels. Some ponded areas also exist well up to the base of the mountain. These waters are presumably derived from surface runoff provided by melting snow and groundwater emerging from the porous talus slopes and fractured bedrock. Sloan et al. (1976) reported that icings occurred at the base of Sukakpak Mountain before the road was built.

During early fall a thin snow cover develops, but water continues to flow in all drainage channels even as air temperatures drop to -10 and even -20°C. As the active layer refreezes, water was observed seeping from the base of some mounds. Low negative temperatures, overflows, and freezing are observed mainly in the evening. The drainage channels become progressively covered with surface icings; this process continues into the winter and spring. Water was also observed flowing beneath the road in October, apparently between the gravel subbase and the buried peat layer. The area surrounding the mounds in locality C was also covered by a shallow icing. The culverts, which are steamed open in the spring, had already frozen closed in mid-October.

By early spring, icings exceeding 30 cm surround most mounds under a snow cover of 70 cm or more. At that time, free water observed at the base of the snow pack was apparently still flowing from unfrozen zones or from contraction cracks, which criss-cross the slope.

The specific conductance of the surface water during October ranged from 970 to 1400 microsiemens. Melted ice from the newly formed icings ranged from 150 to 460 microsiemens. Water obtained from under the ice layer in M-1 (Figure 4) in April 1982 had a value of 870 microsiemens. The cation composition of that sample in mg/L was Ca = 174, Mg = 72, and Na = 2.7. The overlying ground ice, as expected, is very pure with values of Ca = 0.0022, Mg = 0.0022, and a relatively high value of Na = 0.0078 mg/L. Krothe (1981) analyzed spring water from a location just north of Snowden Creek in the Dietrich River valley. The chemical composition was very similar to M-1 with a conductance of 1399 microsiemens (Ca = 198, Mg = 88, Na = 3; sample S70). This similarity in cation concentrations suggests that the water feeding the mounds may be of deep origin.

DISCUSSION

Most hypotheses on the formation of palsas and ice-cored mounds are quite vague. An explanation representative of much of the earlier literature was given by Brown (1968), who suggested that palsas form where shallow (< 1 m) pools of water freeze to the bottom and underlying water-saturated peat is domed up at "random" locations by "intensive frost action and ice lens growth." Most published reports suggest that ice segregation is responsible for palsa formation, possibly because many palsas display numerous, thin lenticular ice inclusions that may be interpreted as annual growth increments.

A recent experiment performed by Seppälä (1982) in Finland clearly showed the importance of thin snow cover for palsa initiation. After keeping a 5-m^2 area free of snow over a winter, a frozen layer persisted throughout the following summer. After two years, a 0.35-m high palsa had grown, which Seppälä attributed to increased frost penetration depth and segregation-ice formation. From Seppälä's description, it seems likely that soil water is drawn from unfrozen soil toward the freezing region, adding an annual increment of segregation ice to the base of the newly formed permafrost. This palsa clearly gained height in annual increments, supporting the intuitive remarks of many earlier investigators. This type of palsa may be the most common in classic peat-bog palsa sites.

Palsas or mounds with massive ice cores, such as the Sukakpak mounds, have also been reported. The possibility seems not to have been widely considered that such features represent single-year growth with subsequent stability. Under such a hypothesis, the mode of origin of both types might be similar, except that the latter would not necessarily be characterized by repeated growth episodes.

Unlike van Everdingen's (1982) seasonal frost mounds at Bear Rock, N.W.T., the Sukakpak mounds appear to be quasi-stable perennial features, although a few ruptured and collapsed ones were observed on the upslope side of the road. On the stable forms, some trees have a smooth J shape that begins at the base of the trunk, suggesting a single uplift event and subsequent stability (Shroder 1980). The Sukakpak mounds are much smaller than those at Bear Rock, which have axial dimensions ranging from 20 to 65 m and heights that often exceed several meters. The small size of the Sukakpak mounds suggests that a limited water supply is available to individual mounds.

A closed-system freezing situation can be envisioned in fine-grained soils when backfreezing of the active layer is retarded in a super-saturated pocket of soil above an impervious substrate — in this case ice-bonded permafrost. Heaving occurs in response to formation of segregation ice near the base of the active layer and at the bottom of the descending freezing plane. These lenses are fed from the saturated sediments within the confined unfrozen pocket. Such a growth history would be confirmed by observing a desiccated layer of mineral soil separating two layers of nearly pure ice.

In the case of coarser soils, backfreezing of the active layer would again be slowed at the site of pools or supersaturated sediments. Pore water would be expelled from the freezing planes, resulting in an enclosed reservoir under increasing pressure. Complete freezing of the reservoir would result in a massive ice core; in some features, rupture of the overlying frozen soil would release still-unfrozen water. Examples of both these phenomena have been observed at Sukakpak Mountain.

If the water-bearing pocket is completely frozen in a single winter, no further vertical growth is experienced, and the mound may be preserved over a long period of time. Conversely, if some water within the pocket remains unfrozen, vertical growth can be reinitiated in subsequent years. This is particularly likely to occur if the unfrozen pocket is reconnected with surrounding thawed zones in summer. This could also result in the voids or "vaulted chambers" sometimes found in summer beneath the massive ice cores of some frost mounds, including a few observed at Sukakpak Mountain.

CONCLUSIONS

Despite the lack of detailed subsurface and process data, certain tentative conclusions can be drawn.

1. The apparent stability of many of the mounds at Sukakpak is evidenced by smoothly J-shaped trees, which may indicate a single upheaval event followed by a long period of stability. Other mounds at the site may experience multiple-year growth.

2. Mound initiation at this site appears to be heavily dependent on local variations in soil moisture. Subsurface ice-rich strata are present both within and adjacent to the mounds themselves; the mounds probably grow only where especially water-rich sediments are trapped between the seasonal freezing layer and ice-bonded permafrost. Limited prethaw drilling has demonstrated the presence of unfrozen water under pressure within one mound; in concert with the frozen sediments found beneath this feature, this fact is indicative of closed-system freezing.

3. The mounds are concentrated into separate fields occupying two unconnected drainage areas on the lower reaches of Sukakpak Mountain. This, of course, reflects the wetter conditions found at these sites. The chemical composition of the ice cores in the mounds is different from the surface water that forms icings at these sites. Ice within the mounds may be derived from groundwater, or the strong mineralization could reflect a concentration of impurities that accompanies freezing. The presence of surface icings may also be responsible for deeper frost penetration than in surrounding areas that are free of ice cover, since icings have a much higher thermal conductivity than does low-density snow cover (van Everdingen 1982). However, latent-heat effects in the water-rich substrate may offset such effects.

4. The Sukakpak mounds have some characteristics similar to classic peat bog palsas, such as their perennial character, size, and external morphology. However, some are associated with massive ice, occur in different topographic position, and apparently result from closed-system freezing.

ACKNOWLEDGMENT

This study has received partial support from the Department of Energy, The National Science Foundation, and the Cold Regions Research and Engineering Laboratory. Mr. Louis DeGoes provided valuable assistance during the initial spring 1982 observations.

REFERENCES

Ahman, R., 1977, Palsar i Nordnorge. En studie av palsars morfologi, utbredning och klimatiska förutsättningar i Finnmark och Troms fylke: Lunds Universitets Geografiska Institution Meddelanden Avhandlingar, v. 78.

Brown, J., and Berg, R.L. eds., 1980, Environmental engineering and ecological baseline investigations along the Yukon River-Prudhoe Bay Haul Road, CRREL Report 80-19: Hanover, N.H., U.S. Army Cold Regions Research and Engineering Laboratory.

Brown, R.J.E., 1968, Occurrence of permafrost in Canadian peatlands, in Proceedings, Third International Peat Congress: Quebec City, p. 174-181.

Brown, R.J.E., 1973, Ground ice as an indicator of landforms in permafrost regions, in Fahey, B.D., and Thompson, R.D., eds., Research in Polar and Alpine Geomorphology: Norwich, England, p. 25-42.

Dillon, J.T., 1982, Source of lode- and placer-gold deposits of the Chandalar and Upper Koyukuk Districts, Alaska: Alaska Division of Geological and Geophysical Surveys Open File Report 158.

Friedman, J.D., Johansson, C.E., Oskarsson, N., Svensson, H., Thorarinsson, S., and Williams, R.S., 1971, Observations on Icelandic polygon surfaces and palsa areas, photo interpretation and field studies: Geografiska Annaler, v. 53A, p. 115-145.

Hamilton, T.D., 1979, Geologic road log, Alyeska Haul Road, Alaska, June-August 1975: U.S. Geological Survey Open File Report 79-227.

Hamilton, T.D., and Obi, C.M., 1982, Pingos in the Brooks Range, Northern Alaska, U.S.A.: Arctic and Alpine Research, v. 14, p. 13-20.

Haugen, R.K., 1982, Climate of remote areas in north-central Alaska: 1975-1979 summary, CRREL Report 82-35: Hanover, N.H., U.S. Army Cold Regions Research and Engineering Laboratory.

Kreig, R.A., and Reger, R.D., 1982, Airphoto analysis and summary of landform soil properties along the route of the trans-Alaska pipeline system: Alaska Division of Geological and Geophysical Surveys, Geologic Report 66.

Krothe, N.C., 1981, Water chemistry in a permafrost environment, Alaska, in Vinson, T.S., ed., Proceedings, Specialty Conference on the Northern Community: A Search for a Quality Environment, April 8-10, Seattle, Washington: New York, American Society of Civil Engineers, p. 570-590.

Mackay, J.R., 1978, Contemporary pingos: a discussion: Biuletyn Peryglacjalny, v. 27, p. 133-154.

Nelson, F., and Outcalt, S.I., 1982, Anthropogenic geomorphology in northern Alaska: Physical Geography, v. 3, p. 17-48.

Péwé, T.L., 1975, Quaternary geology of Alaska: U.S. Geological Survey Professional Paper 835.

Seppälä, M., 1982, An experimental study of the formation of palsas, in French, H.M., ed., Proceedings of the Fourth Canadian Permafrost Conference: Ottawa, National Research Council of Canada, p. 36-42.

Shroder, J.F., Jr., 1980, Dendrogeomorphology: review and new techniques of tree-ring dating: Progress in Physical Geography, v. 4, p. 161-188.

Sloan, C.E., Zenone, C., and Mayo, L.R., 1976, Icings along the trans-Alaska pipeline route: U.S. Geological Survey Professional Paper 979.

Sollid, J.L., and Sorbel, L., 1974, Palsa bogs at Haugtjornin, Dovrefjell, South Norway: Norsk Geografisk Tidsskrift, v. 28, p. 53-60.

van Everdingen, R.O., 1982, Frost blisters of the Bear Rock Spring area near Fort Norman, N.W.T.: Arctic, v. 35, p. 243-265.

Washburn, A.L., 1983, What is a palsa? in Poser, H., ed., Mesoformen des heutigen Periglazialraumes: Gottingen, Akademie der Wissenschaften (in press).

TESTING PIPELINING TECHNIQUES IN WARM PERMAFROST

L. E. Carlson and D. E. Butterwick

Foothills Pipe Lines (Yukon) Ltd.
1600, 205 - 5th Avenue S.W., Calgary, Alberta, Canada T2P 2V7

A thaw settlement research facility was built by Foothills Pipe Lines (Yukon) Ltd. as part of the permafrost engineering design for the Alaska Highway Gas Pipeline Project. The main purposes of the test facility were (1) to study construction methods for the installation of large diameter pipelines in permafrost, (2) to observe the effectiveness of mitigative designs in minimizing thaw settlement of the pipe and right-of-way surface, (3) to study the behavior of cuts in ice rich hills, and (4) to provide data on the thermal behavior of design modes for comparison with thermal model predictions. This paper discusses the observations of the pipeline design performance and the comparison of that performance with initial predictions.

INTRODUCTION

As a part of the Alaska Highway Gas Pipeline Project, Foothills Pipe Lines (Yukon) Ltd. constucted the Quill Creek Test Facility. The test site is located in the Yukon territory of Canada on the alignment of the proposed gas pipeline approximately 165 km southeast of the Alaska border (Figure 1). The test facility encompassed various programs to study construction techniques and their effects on permafrost terrain and also to test some pipeline design concepts for operating warm gas pipelines across permafrost terrain.

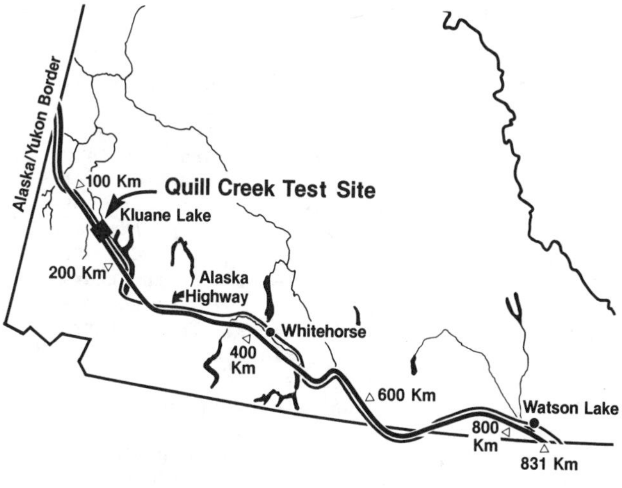

FIGURE 1 Quill Creek test site.

Overall, the test site is approximately 5 km in length and includes a number of separate areas where specific test programs were performed. The site layout as shown on Figure 2 includes (1) the sidehill grading area where various methods of protecting and stabilizing cuts in ice rich hills are studied, (2) the disposal area where the ice rich spoil from the sidehill cuts was stored, (3) the buoyancy control area where select granular fill was tested for its adequacy as a pipeline buoyancy control technique and also excavation procedures could be tested in permafrost terrain, (4) the permafrost mitigative design construction area where problems associated with construction and maintenance of a reasonable length of pipeline with special designs could be evaluated, (5) the mitigative design operational area where shorter test sections are heated to operational temperatures and their performance monitored, and finally, (6) the site power generation, control and data gathering facilities area.

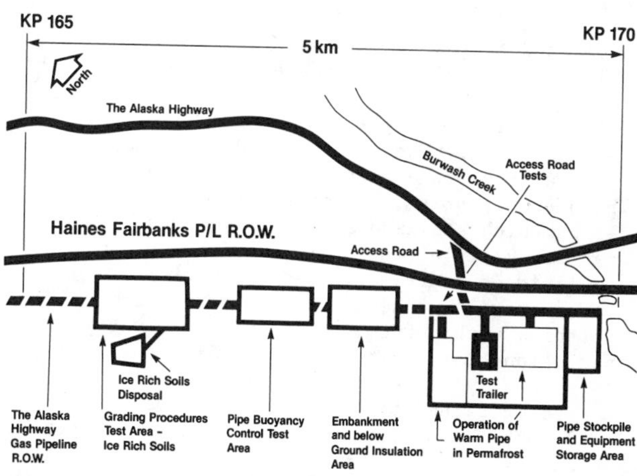

FIGURE 2 General layout, Quill Creek test site.

This paper will deal specifically with observations of the operating design area and the sidehill cut area.

The soil in the operating design area is typically an interbedded sequence of peat, silt, and clay overlaying a thaw stable gravel or gravelly till at a depth of about 5 m. Visual ice content in the top 5 m is 30-50%. The soil in the sidehill cut area consists of a surface blanket of organics of 20-50% ice content, varying in thickness from 1 to 3 m. Beneath this is a clayey till layer with a high percentage of gravel sizes to depths of 4-7 m below ground surface. This material is very rich in ice, having from 30 to 60% visible excess ice. The mean ground temperature across the site is - 1.5°C. The freeze index is 2900°C days and the thaw index is 1300°C days.

PIPELINE DESIGNS IN PERMAFROST

Foothills Pipe Lines is proposing to use either a chilled or warm gas operating mode (gas temperature below or above 0°C, respectively) depending on the relative occurence of frozen and thawed terrain in the discontinuous permafrost region through which the pipeline alignment passes. If the presence of permafrost is relatively infrequent, as occurs along the southernmost three quarters of the alignment in the Yukon, the pipeline will operate warm. Regions of unstable permafrost are still encountered, however, and pipeline design measures are required to mitigate potential thaw settlement of the pipeline foundation soils. To study the behavior of such a mitigative design is one of the purposes of the Quill Creek Test Facility.

The basis of the thaw settlement design is to minimize thawing of the permafrost soils and thus maintain their strength and support capability. Geothermal analysis indicated that no reasonable amount of pipeline insulation could prevent significant thaw of soil supporting a buried pipeline and therefore the choice for this test is an abovegrade design. Geothermal analysis indicated that a gravel pad of an adequate thickness and width can cause the permafrost table to rise up into the granular material, thus stabilizing the subgrade. The amount of gravel, however, can be excessive depending on permafrost temperatures and ambient temperature conditions. The design problem, therefore, was to develop a design which utilized a reasonable amount of granular material and still provided adequate foundation support for the pipeline.

The gravel pad used in this test is nominally 1 m thick with 100 mm (or 50 mm) of polystryrene insulation board. The plastic foam insulation can significantly reduce the quantity of gravel necessary to meet the above thermal performance (Berg 1980). The insulation was laid on a graded gravel pad approximately 300 mm thick and covered with about 600 mm of gravel. The pipeline was then laid on this gravel pad and covered to protect the pipe from environmental and other externally caused damage and from low ambient temperature and to restrain the pipeline from movements due to pressure and temperature loads. Two methods of covering the pipe are used at the test site, one a gravel berm covering the pipe and the other a segmented concrete type of cover. Three test sections are covered with the gravel berm. The test sections differ in the amount of pipeline insulation applied. One section has no insulation, another section has 100 mm of urethane insulation and the other section has 200 mm of urethane insulation. Two additional test sections are covered with the segmented concrete design. These have no pipe insulation as such but the pipe lies on a "saddle" of urethane insulation and the inside of the concrete weights is also insulated prior to placement over the pipe. This latter method of insulating the pipeline was studied as an alternative to continuously insulated pipe. The layout of these five test sections is illustrated in Figure 3.

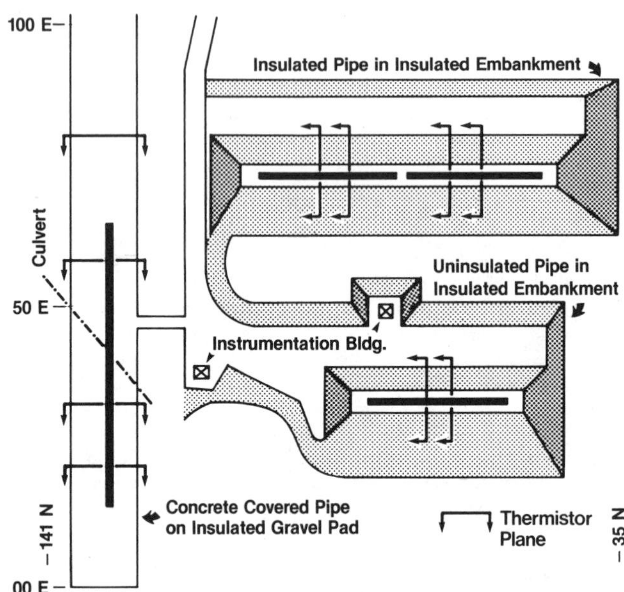

FIGURE 3 Layout of aboveground test sections.

The design variations constructed and monitored in this section of the test site are used to provide verification of the thermal modeling of the designs as well as experimental observation of the design performance. The pipe sections are maintained at a constant temperature of 15°C by an air heating system. Verification of the thermal modeling is the purpose of the varying amounts of insulation on the test sections. The differences in thaw below the test sections will also provide an opportunity to observe the relationship between depth of thaw and amount of settlement and compare it with predicted thaw strains.

Construction of these test sections was performed during two different seasons of the year. The three gravel berm designs were constructed during the winter (Feb.-Mar.), while the concrete covered test sections were constructed during the late summer (Sept.). This affords the opportunity to verify the thermal model with a different set of initial conditions as well as to observe the thaw settlement behavior during construction and operation of gravel pads built on different soil thermal states.

Figures 4, 5, and 6 illustrate the thermal history of the designs through one year of operation. The thermal state of the soil is determined from a matrix of thermistors located below and within the gravel pad. The gravel embankment design without insulation around the pipe (Figure 4) shows a growth of the thaw bulb directly under the pipe. The presence of the warm pipe prevents the seasonal freezing of the material in and under the gravel pad. Our thermal calculations indicate that the thaw bulb will continue to grow with time. The gravel embankment designs with either 100 mm or 200 mm thick insulation around the pipe behaved almost identically. Figure 5 shows complete thawing of the gravel and slight thawing of the peat surface layer below the edge of the embankment during the summer months, but complete freezeback during the winter, for these insulated pipe sections. The insulation around the pipeline effectively negates the thermal influence of the warm pipeline, and the thermal effect is similar to an insulated gravel pad. The concrete covered design (Figure 6) was constructed in the late summer over a thawed active layer. The construction activity and warm gravel placement deepened the thaw under the gravel pad. Before construction the active layer was approximately 1/2 m thick and this was increased to approximately 2 m due to the construction. In the first year of operation the thaw depth has remained relatively constant with seasonal cold ambient temperatures not completely freezing back the active layer.

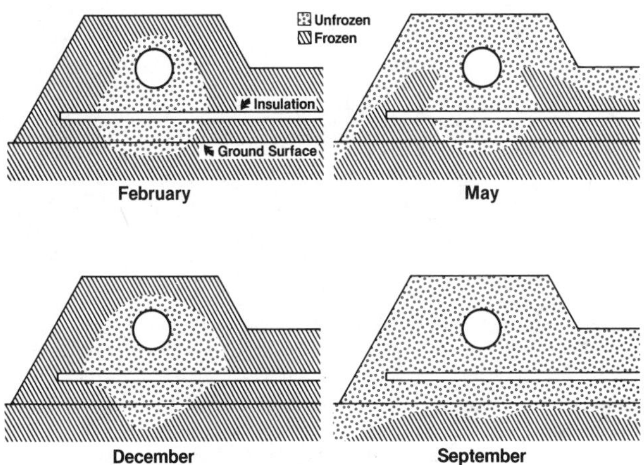

FIGURE 4 Thermal History. Uninsulated pipe in insulated embankment.

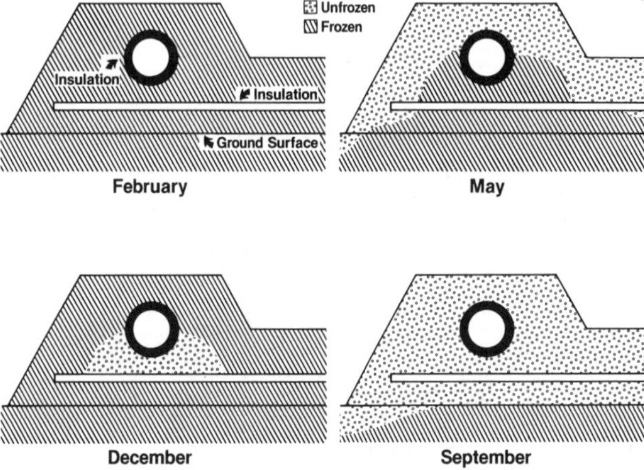

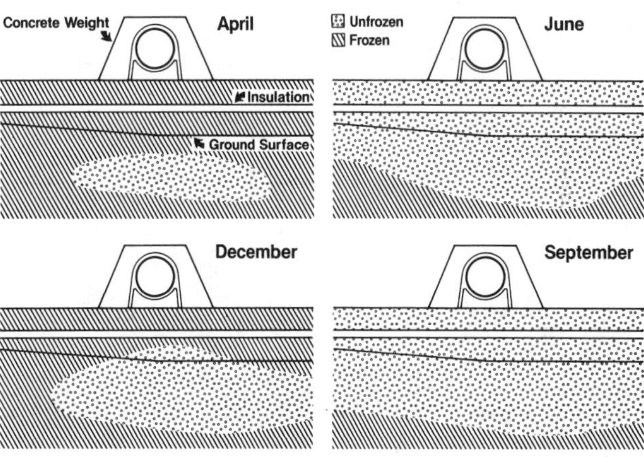

FIGURE 5 Thermal History - Insulated pipe in insulated embankment.

FIGURE 6 Thermal History. Concrete covered pipe on insulated gravel pad.

Computed predictions of the thermal behavior of the various designs have been in good agreement with the measured results. A comparison of these predicted and measured values is illustrated on Figure 7 and 8. These figures show a comparison of the calculated $0^{o}C$ isotherm with inerpolations of measured thermistor temperature eadings. The thermal calculations are performed ith a finite element thermal program that onsiders soil/water latent heat effects and takes nto account changing meteorological conditions Hwang 1976). Thermal parameters used in these imulations are given in Table 1.

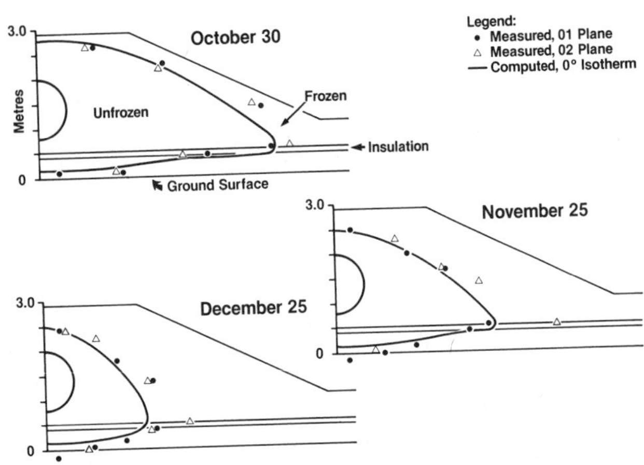

FIGURE 7 Comparison of measured and computed locations of 0° isotherm - uninsulated pipe in gravel embankment.

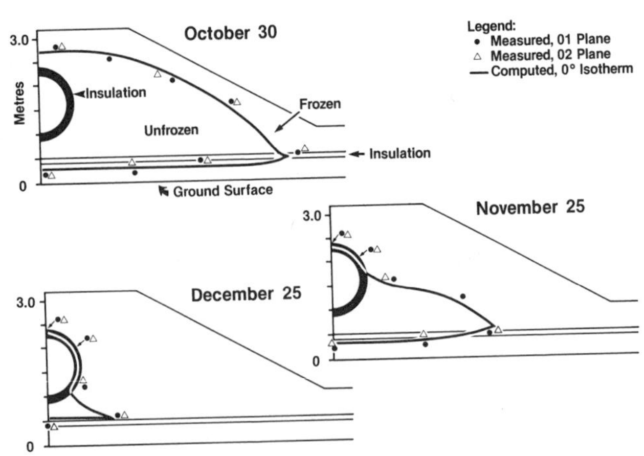

FIGURE 8 Comparison of measured and computed locations of 0° isotherm - insulated (200 mm) pipe in gravel embankment.

Soil Profile		Moisture Content		Bulk	Dry Density		Excess	Thermal Conductivity		Specific Heat		Latent
Depth (m)	U.S.C.	Frozen %	Thawed %	Density $Mg\ m^{-3}$	Frozen $Mg\ m^{-3}$	Thawed $Mg\ m^{-3}$	Ice %	Frozen $Wm^{-1}\,°C^{-1}$	Thawed $Wm^{-1}\,°C^{-1}$	Frozen $kJ\ kg^{-1}\,°C^{-1}$	Thawed $kJ\ kg^{-1}\,°C^{-1}$	Heat $MJ\ m^{-3}$
0.00-0.30	Pt	240	240	1.20	0.353	0.353	0	1.50	0.55	2.01	3.52	283.98
0.30-1.10	ML	45	20	1.56	1.076	1.537	30	1.90	1.27	1.22	1.31	140.66
1.10-1.30	VA	65	27	1.20	0.727	1.039	30	1.80	0.70	1.35	1.47	136.47
1.30-1.50	Pt	115	115	1.45	0.674	0.674	0	1.32	0.51	2.00	3.14	259.81
1.50-2.35	ML	150	59	1.25	0.500	1.000	50	1.55	0.84	1.56	2.02	241.34
2.35-2.60	Pt	120	120	1.20	0.545	0.545	0	1.35	0.50	2.00	3.16	219.22
2.60-2.80	ML	220	103	1.24	0.388	0.775	50	1.55	0.71	1.67	2.49	278.32
2.80-3.15	Pt	400	400	1.20	0.240	0.240	0	1.55	0.56	2.03	3.74	321.79
3.15-3.50	ML	200	91	1.25	0.417	0.833	50	1.55	0.74	1.65	2.38	271.17
3.50-4.15	Pt	400	400	1.20	0.240	0.240	0	1.55	0.56	2.03	3.74	321.79
4.15-4.75	ML	135	92	1.50	0.638	0.912	30	1.57	0.84	1.53	2.39	275.88
4.75-	GC	20	20	2.20	1.833	1.833	0	4.70	2.62	0.93	1.31	122.88
Gravel		4	4	1.95	1.875	1.875	0	1.32	1.75	0.76	0.87	25.14
Styrofoam		0	0	0.04	0.040	0.040	0	0.03	0.03	1.13	1.13	0.00
Urethane		0	0	0.064	0.064	0.064	0	0.024	0.024	0.92	0.92	0.00

TABLE 1 Soil stratigraphy and thermal properties, Quill Creek test site.

The pipe settlement history for the first 18 months of test site operation is shown on Figure 9. The two variations of the gravel embankment design (insulated pipe and uninsulated pipe) have both settled similar amounts with the uninsulated pipe settling slightly more. The presence of the warm (15°C) pipeline has yet to exhibit a significant influence on the settling or the effectiveness of the concept as a thaw settlement mitigative design. A longer embankment section in the construction procedures area of the test facility has settled a similar amount. That test section has the same design cross section as the heated insulated pipe sections but this longer pipeline was not heated. The majority of the settlement of the embankment design to-date can therefore be attributed to consolidation of the surface material under the embankment following the small amounts of thawing occuring seasonally under the embankment due to ambient temperatures.

This is not to say that the design variations will not perform differently in the future. The observation is that the uninsulated pipe causes thaw bulb growth which is sustained through the winter. Increases in thaw depth would likely cause an acceleration of pipe settlement relative to the insulated pipe design.

The concrete covered test sections exhibited significantly less settlement to date than the gravel embankment. This is attributed more to the season of construction than differences in pipe covering. Consolidation of the underlying material occurs more rapidly during construction when the material is thawed and reaches a measure of stability more quickly after the test site is put into operation.

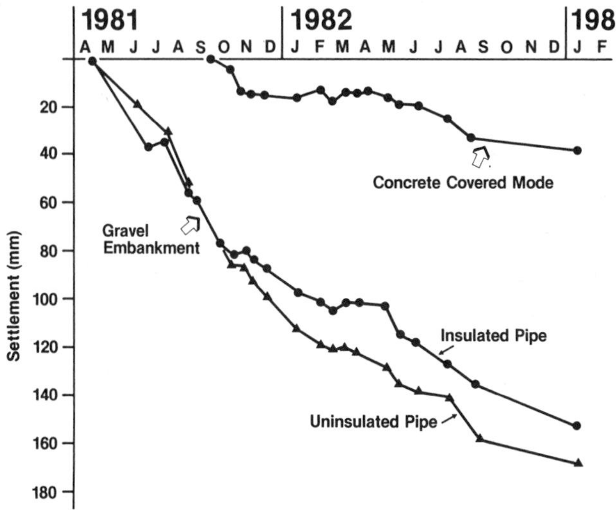

FIGURE 9 Pipe settlement.

SLOPE STABILIZATION

The northernmost area of the Quill Creek Test Facility is dedicated to the evaluation of cut slope stabilization techniques for ice rich permafrost. A side cut facing northeast was made in a hill for approximately 400 m with the cut height varying up to 8 m. As described previously, the soil on this hill has a 20-60% visible ice content.

This test area is divided into six test section where different stabilization methods are used. The first section consists of a vertical cut face approximately 2 m high. The north half of the section was left unprotected and the south half has a wire mesh pegged to the organic cover above the top of the cut. Significant sloughing in the area of the slope face was observed. The organic cover on top of the cut face became undercut during the summer and fell onto the slope along the entire section. No appreciable movement occurred during the winter and little additional sloughing or movement occurred during the subsequent summer. The second test section consists of a 1:1 cut slope varying from approximately 2 to 4 m in height. The north half is protected by wire mesh pegged to the organic cover above the cut and the south half is unprotected. The organic cover developed a small undercut near the top of the cut face and fell onto the slope in late summer of the first year. No significant change occured during the winter. The next summer showed greatly reduced thawing and effective stabilization of the slope face. The third test section consists of a 2:1 slope approximately 8 m high covered by a gravel blanket nominally 2 m thick. There is no visible change to this test section, although approximately 20 cm of settlement has been measured. The fourth test section consists of a 2:1 cut slope approximately 8 m high covered by an organic blanket reinforced by a jute netting. The organic material is nominally 1.5 m thick. There is no visible change to this test section, although settlement in the order of 10 cm has been measured in the cleared area on top of the slope and 15 cm on the peat covered slope area. The fifth test section consists of a 1/2:1 cut slope approximately 4 m high protected by a gravel buttress with an outside slope of 2:1. The general shape of the buttress remains unchanged but settlements ranging up to 30 cm have occurred at the top of the buttress where the gravel abuts the cut face. The sixth test section consists of an unprotected vertical cut approximately 6 m high. Significant changes have occurred all along this section. The organic mat atop the cut face was significantly undercut during the first summer, and this overhang collapsed onto the cut face during September. No changes occurred during the winter months, and minimal changes were recorded to the slope profile during the second summer, however, the organic mat was torn off in some places along the slope. The changing cross-section profile of this test section is illustrated in Figure 10.

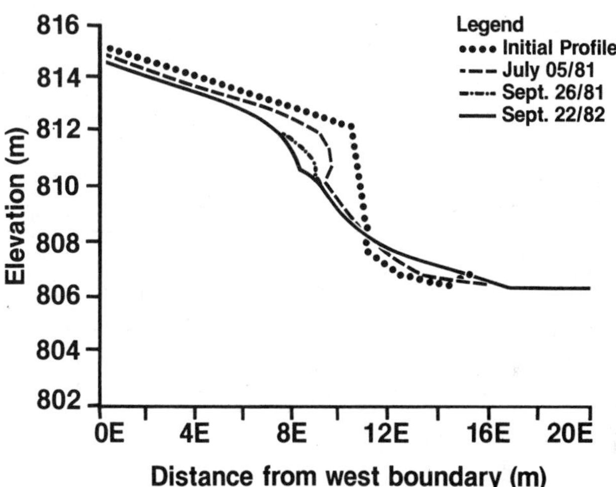

FIGURE 10 Six meter cut profile at Quill Creek test site.

SUMMARY

Two years of operation of the Quill Creek Test Facility has provided valuable data and observations regarding the construction and operation of pipelines in permafrost. The operating test sections have confirmed the pipeline design concepts for permafrost and have confirmed the application of computer thermal modeling in permafrost design work. The observation of the sidehill cuts in ice rich material have confirmed that adequate stabilization can be achieved.

REFERENCES

Berg, R. L., 1980, Road performance and associated investigations, *in* Environmental engineering and ecological baseline investigations along the Yukon River - Prudhoe Bay haul road: CRREL Report 80-19, Chapter 2.

Hwang, C. T., 1976, Predictions and observations on the behavior of a warm gas pipeline on permafrost, Canadian Geotechnical Journal, Vol. 13, No. 4, p. 452-480.

THE GEOLOGY OF MARS

Michael H. Carr

U.S. Geological Survey, Menlo Park, California 94025 USA

Mars has had a long and varied geologic history. Large sparsely cratered volcanoes and vast lava plains attest to sustained volcanism. An extensive system of fractures appears caused by a bulge of continental dimensions in the Tharsis region. Well-integrated valley networks imply slow erosion by running water, while large channels appear formed by catastrophic floods. Collapsed ground, moraine-like ridges, and pitted terrains suggest the effects of ice, landslides and rock glaciers indicate large-scale mass wasting, and dunes and yardangs are evidence of wind action. Despite the similarity in the range of processes that have affected Mars and Earth, differences between the two planets remain large. Under present climatic conditions liquid water is unstable everywhere on the surface and ice is unstable at latitudes below 40°. A permafrost zone is 1 km thick at the equator and 4 to 5 km thick at the poles. Climate changes caused by perturbations in the orbital and rotational motions are small although they may cause alternating cycles of erosion and deposition at high latitudes. Climatic variations during the current epoch appear representative of climates throughout most of Mars' history. Branching valley networks in old terrain suggest different conditions over 3.5 billion years ago. Persistence of climatic conditions that hinder flow of water has prevented cycling of surface materials by erosion, sedimentation, burial, and metamorphism.

INTRODUCTION

The last two decades have witnessed an explosive growth in our knowledge of the Solar System. Spacecraft have made observations on all the planets as far out as Saturn, returning an enormous amount of data. Mars is of special interest to the geologist since most of the geologic processes that operate on Earth have occurred there also, although under quite different environmental conditions. We also know more about Mars than any planetary body other than the Earth and the Moon. Most of its surface was photographed by the Viking orbiters at a resolution of 200 m and large fractions at much better resolutions, ranging down to 8 m. Ground temperatures, atmospheric temperatures and the water content of the atmosphere were monitored planet-wide for over two martian years. In addition, the Viking landers analyzed materials at the surface and recorded meteorological conditions at two locations for several years. This paper examines current climatic conditions, then contrasts them with those that might be inferred with the geologic record in order to determine the extent to which present conditions are representative of those that have occurred in the past. Emphasis is on processes at the surface that involve water or affect its stability. The intent is to provide a context within which the potential for development of permafrost features on Mars can be assessed. For comprehensive summaries of martian geology, see Carr (1981) and Mutch et al. (1976).

PHYSIOGRAPHY AND TOPOGRAPHY

Mars, the fourth planet from the sun, has a diameter of 6788 km, intermediate between that of the Earth (12,756 km) and the Moon (3,476 km). The planet is markedly asymmetrical. Most of the southern hemisphere is heavily cratered like the lunar highlands, whereas much of the northern hemisphere is only sparsely cratered. The heavily cratered terrain stands 1 to 3 km above the reference elevation datum. As on the Moon, several large impact basins are present, the largest being Hellas (1,800 km in diameter) and Argyre (800 km in diameter). Between many of the large craters are what appear to be volcanic plains. Both the craters and the intervening plains are dissected by branching valley networks that superficially resemble terrestrial river valleys and suggest slow erosion by running water. In contrast, much of the northern hemisphere is covered by plains at elevation mostly 1 km below the reference level. The plains have far fewer superposed craters than the southern highlands, and almost no branching valleys. Most of the plains at low latitudes are clearly volcanic, having flows visible on their surface, but in many places, especially at high latitudes, the plains have been considerably modified by the action of wind to produce yardangs and dunes, by water to carve valleys and possibly by ice to form collapse pits, patterned ground and moraine-like features.

The general north-south asymmetry is broken by anomalously high elevations in the Tharsis region, where a broad dome, 5,000 to 6,000 km across and 10 km high, is centered at 115°W on the the equator. Arrayed around the bulge and affecting about a third of the surface area of the planet are numerous radial fractures. At the summit and on the northwest flank of the bulge are several huge volcanoes. Olympus Mons, on the northwest edge, is over 600 km across and 27 km high, and Alba Patera just to the north of Tharsis is 1,600 km across but only a few kilometers high. A smaller bulge 2,500 km across and 5 km high, centered in Elysium at 25°N, 210°W, also has superimposed volcanoes.

To the east of Tharsis is a vast system of interconnected canyons. They extend from close to the summit of the Tharsis dome, eastward for almost 4,000 km. Most of the canyons are over 3 km deep and over 100 km across. In the central section, three parallel canyons merge to form a depression over 7 km deep and 600 km wide. The canyons are mostly flat floored with steep gullied walls. Many contain thick, partly eroded, layered sequences of sediments.

The canyons merge eastward with "chaotic terrain," in which the ground has seemingly collapsed over thousands of square kilometers to form a jumbled array of blocks at a lower elevation than the surroundings. Several large channels emerge from the chaotic terrain east of the canyons and extend northward down the regional slope for over 2,000 km, causing extensive erosion of the volcanic plains of Chryse Planitia. Other channels start north of the canyons and they also converge on, and ultimately die out in, Chryse Planitia. The channels are enormous by terrestrial standards. Many are several tens of kilometers wide; Kasei Vallis, the largest, has eroded a broad swath of terrain in places over 200 km across. All start full-size and have few if any tributaries. These channels have been widely attributed to floods of enormous magnitude although erosion by wind and ice have also been suggested.

The physiography of the poles is distinctively different from the rest of the planet. Extending outward from both poles for a little over 10° in latitude are layered deposits. The layers range in thickness from several tens of meters down to the limiting resolution of the available photographs. Incised into the deposit are numerous valleys and low escarpments that curl outward from the pole in a counterclockwise (eastward) direction in the north and in a predominantly clockwise (westward) direction in the south. Preferential removal of frost from the sunward facing slopes of these valleys gives the poles a characteristic swirl texture in summer. The polar layering and the incised valleys are of considerable interest because they may result from periodic climate change.

CURRENT SURFACE CONDITIONS

Mars currently has a thin atmosphere composed mostly of carbon dioxide. The composition at the surface is 95.3 % CO_2, 2.7 % N_2, 1.6 % Ar, and only small amounts of other constituents (Owen et al., 1977). The average pressure is around 8 mbars, but ranges widely because of the large surface relief. At the summit of Olympus Mons the pressure is only 0.3 mbars; in the deeper parts of the canyons the pressure is around 15 mbars. Surface pressure also varies with season, since about one-third of the atmosphere condenses out to form the winter CO_2 caps. Pressure at the Viking 1 landing site, for example, ranged from 6.9 mbars in late southern winter to 9 mbars in southern summer.

Because the atmosphere is thin it has a low heat capacity. It cools and heats rapidly and has less effect on ground temperatures than on Earth. Ground temperatures have a large diurnal range. Those in southern summer range from around 200°K just before dawn to 280-290°K at midday. Peak northern summer temperatures are 20-30° lower. Assuming a heat flow of 35 erg $cm^{-2} sec^{-1}$ (Toksoz and Hsui, 1978), mean annual temperatures are such that the 273°C isotherm is at a depth of 1 km at the equator and 4-5 km at the poles (Rossbacher and Judson, 1981), so that the planet everywhere has a thick permafrost zone. The temperature profile within the atmosphere is sensitive to the amount of dust it contains. When the atmosphere is clear, heating is largely from the ground below, and diurnal temperature fluctuations are restricted to a thin near surface layer. However, when dusty, absorption and emission of radiation by the atmosphere is enhanced and temperature fluctuations occur throughout the vertical profile (Pollack et al., 1979). One result is increased tidal winds, which blow from the cold nightside to the warm sunlit side of the planet.

A meteorological phenomenon of considerable geologic interest is the annual recurrence of near global dust storms. During spring and summer in the southern hemisphere, local dust storms are common, particularly around the edge of the retreating seasonal cap and in areas with high slopes. In most years storm activity grows to engulf almost the entire planet, obscuring the surface from view. After the storms cease the dust slowly settles out and the atmospheric opacity returns to its pre-storm level in 3-4 months. A number of theories have been proposed to explain the storms. The most plausible is that a feedback mechanism develops between the dust storm and tidal winds (Leovy and Zuruk, 1979). The local dust storms raise dust that increases the contrast between the day and night atmospheric temperatures. This in turn increases the tidal winds that raise more dust, thereby causing a runaway effect. The storms subside when the atmosphere is so opaque that the near surface temperature fluctuations and hence the winds are so reduced that dust can no longer be raised. The storms now occur only during southern summer. At this time diurnal temperature variations are enhanced because Mars is close to perihelion.

The atmosphere contains only minute amounts of water. For most of the year the distribution is fairly even with latitude and the atmosphere typically contains 5-10 precipitable micrometers (Farmer and Doms, 1979). However, the atmosphere is always close to saturation for night temperatures so it holds almost all it can. The only major deviation from the generally uniform latitudinal distribution is during northern summer. At this time the atmosphere at high northern latitudes may contain up to 100 precipitable micrometers, a result of loss of water from a residual northern water-ice cap and a constant illumination that enhances the carrying capacity of the atmosphere.

Pressure and temperature conditions are such that liquid water and ice are unstable almost everywhere on the surface. Liquid water will rapidly freeze or evaporate, depending on local conditions. Ice is stable at the surface all year round within 10° of the poles and for parts of the year down to latitudes as low as 30°. It is never stable at lower latitudes. At latitudes greater than 30°, ground ice is stable all year round at depths of 1 m to 1 km, depending on the specific latitude (Figure 1); at latitudes less than 30° ground ice is unstable with respect to the present atmosphere at all depths (Farmer and Doms, 1979). If ice were introduced into the ground in these equatorial latutudes it would slowly sublime, the rate being controlled by the ability of the water vapor to diffuse to the surface. Despite this disequilibrium, Smoluchowski (1968) suggested that ground ice could survive in the equatorial regions for billions of years if effectively isolated from the atmosphere, such as by a thick, impermeable, fine-grained regolith.

The surface conditions just outlined could alter as a result of changes in the orbital and rotational motions. Mars currently has an orbital eccentricity of 0.093, as compared with 0.017 for the earth. The rotation axis is inclined 25° to the pole of the orbit plane with the southern hemisphere tilted toward the sun near perihelion, thereby causing southern summers to be shorter and hotter than those in the north. However, precession of the rotation axis and the orbit plane causes climatic conditions to alternate between the two hemispheres on a 51,000 year cycle (Leighton and Murray, 1966). In 25,000 years it will be the northern hemisphere that has short, hot summers that trigger global dust storms.

Eccentricity and obliquity also undergo periodic changes. The eccentricity oscillates between 0.01 and 0.4 with characteristic periods of 95,000 and 2 x 10^6 years (Murray et al., 1973). Clearly, when the eccentricity is at a minimum, climatic differences between the two hemispheres are suppressed; when eccentricity is at its maximum, climatic differences are accentuated. Currently, with an eccentricity of 0.093, Mars is in the middle of its eccentricity range. The obliquity (the angle between the equatorial plane and the orbit plane) varies between 11 and 38° with characteristic periods of 1.2 x 10^5 and 1.2 x 10^6 years (Ward, 1973). As obliquity increases polar temperatures rise and equatorial temperatures fall. Because large amounts of volatiles are stored at high latitudes either as ice or adsorbed in the regolith, obliquity variations can significantly affect the volatile inventory of the atmosphere. Mars is at present also in the middle of its obliquity cycle.

These perturbations can affect climatic and geologic processes in a variety of ways. Currently, the north pole may be a region of dust deposition since its seasonal cap forms when the atmosphere is dusty from the southern summer dust storms. In contrast, when the southern cap forms the atmosphere is clear. Thus the northern cap is dirty and dissipates under a clear atmosphere, whereas the southern cap is clean and dissipates under a dusty atmosphere (Kieffer et al., 1976; Pollack et al., 1979). One consequence is that the northern seasonal CO_2 cap dissipates completely in summer to leave a remnant water-ice cap (Farmer et al., 1976), whereas the CO_2 cap in the south does not completely dissipate.

Accumulation of dust scavenged from the atmosphere by formation of the seasonal cap may be the cause of the polar layered deposits (Pollack et al., 1979). Deposition is now taking place in the north but the site of deposition will alternate with the precessional cycle, so the layering may be due in part to precession. The coupling of depositional rates with astronomical perturbations are likely, however, to be complicated and depend in part on the threshold conditions for initiating dust storms (Cutts et al., 1982). At present such conditions are achieved only during southern summer. If eccentricity decreases, the triggering thresholds may not be achieved in either hemisphere and little deposition would occur at the poles. Furthermore, retention of the dust storm debris may depend on the stability of water-ice, which probably acts as a cement. Under conditions of high obliquity ice will be unstable during summer, thereby allowing the dust to dissipate. Thus, while changes in the stability of water and in the rate of dust deposition induced by perturbations of the planet's motions are likely causes of the polar layered deposits, precise correlation of the layers with specific periodic changes is not possible.

Astronomical perturbations may also affect the atmospheric pressure. Carbon dioxide resides in three major sinks: the atmosphere, the polar caps, and the regolith. Several lines of evidence indicate that the present atmosphere represents only a fraction of that outgassed from the planet (McElroy et al., 1977; Pollack and Black, 1979). Particularly significant is the lack of fractionation of oxygen isotopes in the atmosphere despite substantial enrichment in ^{15}N and several mechanisms for exospheric loss of oxygen, which should cause enrichment in ^{18}O. The simplest explanation is that substantial amounts of CO_2 and H_2O exist in the ground close to the surface where they can exchange with the atmosphere and so dilute any fractionation effects. The regolith has a substantial storage capacity for CO_2, most being adsorbed on mineral grains (Fanale and Cannon, 1974, 1979; Fanale et al., 1982). Increase in regolith temperatures will tend to drive off the gas and so increase atmospheric pressures; a decrease in regolith temperature will have the reverse effect. The penetration depth of the annual and diurnal temperature changes is small so that they have little effect on atmospheric pressure. However, temperature changes caused by perturbations of the orbital and rotational motions, which have periods of 10^5 to 10^6 years, penetrate to depths of a kilometer and could significantly affect the atmosphere. Of special importance are obliquity changes that cause large variations in temperatures at high latitudes. According to the modeling of Fanale et al. (1982), during periods of low obliquity the atmospheric pressure may fall as low as 0.1 mbar mainly as a result of growth of a large permanent CO_2 cap; at high obliquities the pressure may rise to 15 mbars. This range is small despite substantial exchange between regolith and atmosphere because of buffering by the polar cap.

The current epoch is thus characterized by a permafrost zone 1 km thick at the equator and thickening to 4-5 km at the poles, instability of liquid water planet-wide, instability of ground-ice at low latitudes, deposition of layered deposits at the poles, intermittent recurrence of global dust storms, and low atmospheric pressures that vary by about a factor of 100. In the next section we will consider the extent to which these conditions are typical of the planet's geologic history.

CHANNELS AND VALLEYS

Lingenfelter et al. (1968) showed that water could flow for considerable distances even on the Moon if discharge rates were high enough. Estimates of the discharge rates of the postulated floods that cut the large martian channels range from 10^6 to 10^9 $m^3 sec^{-1}$ (Masursky et al., 1977; Carr, 1979). Loss by evaporation and freezing under any possible Mars climatic conditions would be negligible in comparison with these discharges. The flood features are thus poor indicators of past climates and could form under present conditions. The branching valley networks are however quite different. They lack bedforms common in the large channels, have a regular hierarchy of tributaries, and start small and increase in size downstream. The networks are thus not channels like the flood features but more akin to terrestrial river valleys. Their presence suggests that streams of modest discharge have at times flowed across the surface.

Over 99 % of the valley networks occur in the densely cratered terrain, which dates from over 3.5 billion years ago (Carr and Clow, 1981). In contrast, the large flood features are found on surfaces of a wide range of ages. The simplest explanation for the restriction of valley networks to old terrains is that conditions required for valley formation existed only very early in the planet's history. Alternative explanations are that the old terrain is more erodible or that water is more available (by seepage) in the old older terrains as compared to the younger. The drainage pattern of the valleys is very open. The old terrain is partly dissected everywhere, but wide undissected interfluves occur between branches in any one network and the drainage basins themselves are widely spaced. Most drainage basins are also small, the distance from the end of the most distal tributary to the mouth of the main valley being generally less than 300 km. These characteristics suggest an immature drainage system in which erosion has not been sufficiently sustained to achieve efficient coverage of the entire landscape or for single drainage basins to dominate large areas at the expense of neighboring networks. Many characteristics of the networks also suggest that groundwater sapping is a prominent process in their development (Pieri, 1980).

A plausible explanation of the valley networks is that early in its history Mars had a thicker atmosphere than at present, and as a consequence, higher surfaces temperatures. The conditions were such as to occasionally permit flow of water at small discharges. At this time impact rates were high and competed with erosion by wind and water in reshaping of the landscape. Around 3.8 billion years ago the impact rates declined, the landscape stabilized and many of the valleys we now see were formed. However, the rate of valley formation appears to have declined rapidly thereafter for valleys are rarely found on younger terrain and mature drainage systems did not develop even on the oldest terrain. Carbon dioxide may have been lost irreversibly from the atmosphere by formation of carbonates and other minerals. Conditions then approached those that presently prevail with a thick permafrost over all of the planet's surface. The large floods that occurred subsequently were possibly the result of break-out of water confined under high pressures beneath the thick permafrost (Carr, 1979). Alternatively, the floods were caused by geothermal melting of ice (Masursky et al., 1977), liquefaction of near surface materials (Nummedal and Prior, 1981) or geothermally induced decomposition of hydrated minerals (Clark, 1978).

Several other surface features support the sequence of events just outlined. An ancient cratered terrain, remarkably little affected by eolian erosion, has survived from over 3.5 billion years ago; the floors of many large craters are almost indistinguishable from those of lunar craters being almost saturated with small (< 1 km in diameter) craters; fine primary depositional features are preserved on impact ejecta and volcanic flows that are billions of years old. Erosion rates have thus been low throughout most of the planet's history. Arvidson et al. (1979) estimated from the preservation craters in Chryse Planitia that erosion rates can have been no more than 10^{-9} m $year^{-1}$ over the last 3.5 billion years. For comparison, Menard (1961) estimated that erosion rates in North America are 3×10^{-5} m $year^{-1}$. While not conclusive, such low erosion rates suggest that the martian atmosphere has been thin, and hence the surface temperatures low, throughout most of Mars' history.

Indirect evidence also suggests that the stability of ice on the surface has not changed greatly throughout Mars' history. If true, then climatic conditions since termination of the very early era of active valley formation have been fairly stable and subject only to the relatively mild changes caused by the perturbations in the orbital and rotational motions. We have seen that at latitudes less than 30-40° ice is currently unstable with respect to the atmosphere for the entire year, whereas at higher latitudes ice is stable for parts of the year. The 30-40° latitude band is also one of transition in erosional style. In the low latitudes primary volcanic and impact features are well preserved. At high latitudes such features are poorly preserved; most of the landscape is modified as if blankets of debris had been repeatedly deposited and removed (Soderblom et al., 1973). Parallel moraine-like features suggest that rates of removal of the debris blankets changed cyclically, possibly indicating again the effects of variations in the orbital and rotational motions (Carr, 1980). The transition in erosional style along the stability boundary of ice is unlikely to be coincidental. Freezing and sublimation of ice may have played a significant role in the stabilization and destabilization of the debris blankets.

The location of the "fretted terrain" appears similarly related to the ice stability boundary. In these terrains most high standing massifs are surrounded by extensive debris aprons, which suggests that rates of mass wasting are unusually high. The terrain is also characterized by large, flat-floored, steep-walled valleys in which debris flows from opposing walls converge on the valley center. Debris flows appear to be particularly mobile in these areas, resulting in enhanced erosion along steep scarps and widening of river valleys into broad flat-floored features. Squyres (1979) showed that the fretted terrain occurs mainly in two latitude belts 25° wide centered on 40°N and 45°S. He noted that these are regions where the maximum amount of water-ice is expected to be precipitated out of the atmosphere in winter. The amounts are small ($<$ 1mm); nevertheless, he suggested that mass wasting is enhanced in these regions because ice mixes with talus debris. The flat-floored valleys appear to have a wide range of ages. If current stability relations of ice are responsible for their formation then their scarcity outside the two belts identified indicates that the stability conditions of ice have not changed dramatically during much of the planet's history, a conclusion which is also consistent with the contrast in erosional styles between high and low latitude. By implication, surface pressures and temperatures are unlikely to have been significantly greater than at present for extended periods after the early period when the valley networks formed, and liquid water is unlikely to have played a major role in the subsequent evolution of the surface.

GLOBAL IMPLICATIONS

A major cause of the differences between Mars and Earth is their contrasting tectonics. The Mars surface is fixed and all those features, such as linear mountain chains, ocean deeps, transcurrent fault zones, and spreading centers that characterize plate junctions are absent. Crustal stability also results in preservation of an ancient geologic record in almost all areas of the planet. The Earth's surface is thus a complicated system in which material is being recycled both through the mantle by subduction and ocean floor spreading, and within the lithosphere and hydrosphere by weathering, metamorphism, erosion and deposition. A second major cause of differences between the Earth and Mars are the climatic conditions which on Mars hinder flow of water across the surface. On earth, water plays an essential role in two major processes: weathering, the chemical breakdown of rock forming minerals into assemblages more in equilibrium with conditions at the surface, and gradation, the steady reduction of surface relief by erosion and transport of materials from high to low areas. Both processes are complemented by opposing processes such as metamorphism and tectonic activity.

The dynamics of the martian crust are totally different. Although the planet has been volcanically and tectonically active throughout much of its history, the lithosphere appears not to have been recycled through the underlying lithosphere. Huge volcanoes have formed, as well as vast fracture systems, but the activity is not concentrated in linear zones as on earth, but rather affects areas of broad regional extent. Furthermore, although water has probably flowed across the surface at times in the past, erosion has been trivial. Where channels are present, they mostly wind between the craters or down the crater rims; rarely has erosion been sufficiently sustained to wear away the craters themselves. Erosion of young features such as the large volcanoes is imperceptible.

The lack of significant sediment transport implies that weathering products are not recycled. Chemical analyses of the surface materials indicate that weathered products, such as clays, are present. How they formed is not known, but liquid water may be abundant below the permafrost. Such water, being highly charged with CO_2, is likely to be an efficient weathering agent. However, without erosion most of the weathered products must remain where they form so are not actively recycled as on Earth. Thus recycling of surface materials and their interaction with the hydrosphere and atmosphere are hindered by climatic conditions that prevent flow of water. The result is that despite sustained internal activity, the planet's surface is relatively stable and features with considerable relief and a wide range of ages are almost perfectly preserved.

REFERENCES

Arvidson, R. E., Guiness, E. A., and Lee, S. W., 1979, Differential eolian redistribution rates on Mars: Nature, v. 278, p. 533-535

Carr, M. H., 1979, Formation of martian flood features by release of water from confined aquifers: Journal of Geophysical Research, v. 84, p. 2995-3007.

Carr, M. H., 1980, The morphology of the martian surface: Space Science Reviews, v. 25, p. 231-284.

Carr. M. H., 1981, The surface of Mars: New Haven, Conn. Yale University Press.

Carr, M. H., and Clow, G. D., 1981, Martian channels and valleys: their characteristics, distribution and age: Icarus, v. 48, p. 91-117.

Clark, B. C., 1978, Implications of abundant hygroscopic minerals in the martian regolith: Icarus, v. 34, p. 645-665.

Cutts, J. A., and Lewis, B. H., 1982, Models of climate cycles recorded in martian polar layered deposits: Icarus, v. 50, p. 216-244.

Fanale, F. P., and Cannon, W. A., 1979, Mars: CO_2 adsorption and capillary condensation on clays - significance for volatile storage and atmospheric history: Journal of Geophysical Research, v. 84, p. 8404-8414.

Fanale, F. P., Salvail, J. R., Banderdt, W. B., and Saunders, R. S., 1982, Mars: The regolith-atmosphere-cap system and climate change: Icarus, v. 50, p. 381-409.

Farmer, C. B., and Doms, P. E., 1979, Global seasonal variation of water vapor on Mars and the implications for permafrost: Journal of Geophysical Research, v. 84, 2881-2888.
Kieffer, H. H., Chase, S. C., Martin, R. Z., Miner, E. D., and Palluconi, F. D., 1976, Martian north pole summer temperature: Dirty water ice. Science, v. 194, p. 1341-1344.
Leighton, R. B., and Murray, B. C., 1966, Behavior of carbon dioxide and other volatiles on Mars: Science, v. 153, p. 136-144.
Leovy, C. B., and Zuruk, R. W., 1979, Thermal tides and martian dust storms: direct evidence for coupling: Journal of Geophysical Research, v. 84, 2956-2968.
Lingenfelter, R. E., Peale, S. J., and Schubert, G., 1968, Lunar rivers: Science, v. 161, p. 266-269.
Masursky, H., Boyce, J. M., Dial, A. M., Schaber, G. G., and Strobell, M. E., 1977, Formation of martian channels: Journal of Geophysical Research, v. 82, p. 4037-4047.
McElroy, M. B., Kong, T. Y., and Yung, Y. L., 1977, Photochemistry and evolution of Mars' atmosphere: A Viking perspective: Journal of Geophysical Research, v. 82, p. 4379-4388.
Menard, H. W., 1961, Some rates of regional erosion: Journal of Geology, v. 69, p. 154-161.
Murray, B. C., Ward, W. R., and Young, S. C., 1973, Periodic insolation variations on Mars: Science, v. 180, p. 638-640.
Mutch, T. A., Arvidson, R. E., Head, J. W., Jones, K. L., and Saunders, R. S., 1976, The geology of Mars: Princeton, N.J., Princeton University Press.
Nummedal, D., and Prior, D. B., 1981, Generation of martian chaos and channels by debris flows: Icarus, v. 45, p. 77-86.
Owen, T., Biemann, K., Rushnek, D. R.,, Biller, J. E., Howarth, D. W., and LaFleur, A. L., 1977, The composition of the atmosphere at the surface of Mars: Journal of Geophysical Research, v. 82, p. 4635-4639.
Pieri, D., 1980, Martian valley morphology, distribution age and origin: Science, v. 210, p. 895-897.
Pollack, J. B., and Black, D. C., 1979, Implications of the gas compositional measurements of Pioneer Venus for the origin of planetary atmospheres: Science, v. 205, p. 56-59.
Pollack, J. B., Colburn, D. S., Flaser, M., Kahn, R., Carlston, C. E., and Pidek, D., 1979, Properties and effects of dust particles suspended in the martian atmosphere: Journal of Geophysical Research, v. 84, p. 4479-4496.
Rossbacher, L. A., and Judson, S., 1981, Ground ice on Mars: Inventory, distribution and resulting landforms: Icarus, v. 45, p. 25-38.
Smoluchowski, R., 1968, Mars: Retention of ice: Science, v. 159, p. 1348-1350.
Soderblom, L. A., Kriedler, T. J., and Masursky, H., 1973, Latitudinal distribution of debris mantles on the martian surface: Journal of Geophysical Research, v. 78, p. 4117-4122.
Squyres, S. W., 1979, The distribution of lobate debris aprons and similar flows on Mars: Journal of Geophysical Research, v. 84, p. 8087-8096.
Toksoz, M. N., and Hsui, A. T., 1978, Thermal history and evolution of Mars: Icarus, v. 34, p. 537-547.
Ward, W. R., 1973, Large scale variations in the obliquity of Mars: Science, v. 181, p. 260-262.

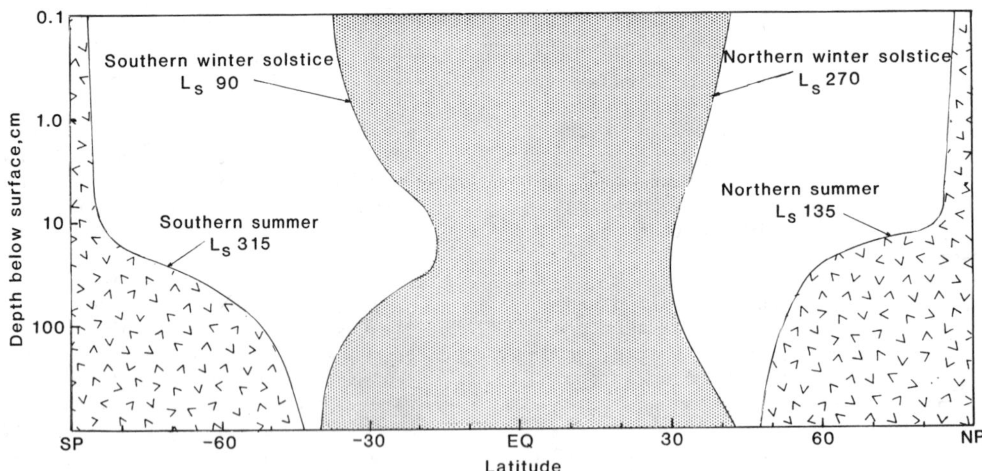

FIGURE 1. Stability relation of ice near the martian surface as a function of latitude. In the equatorial regions that are shaded on the diagram ice is always unstable with respect to the atmosphere under present climatic conditions. In the polar regions that are shaded ice is always stable. In the unshaded areas ice is stable for parts of the year. L_s is the aerocentric longitude of the Sun and is a measure of the time of year (adapted from Farmer and Doms, 1979).

FOSSIL SAND WEDGES ON THE ALASKAN ARCTIC COASTAL PLAIN AND THEIR PALEOENVIRONMENTAL SIGNIFICANCE

L. David Carter

U.S. Geological Survey, 4200 University Drive, Anchorage, Alaska 99508-4667 USA

Fossil sand wedges occur over a broad area of the Arctic Coastal Plain of Alaska upwind of a large field of stabilized linear dunes. The sand wedges are as much as 3 m wide and 7 m deep and form polygonal systems. They developed when wind-driven sand moving across the coastal plain toward the dunes dropped into open thermal-contraction cracks. Radiocarbon dating of organic remains from deposits overlying the sand wedges and from eolian sand in the dune field, together with the age of the deposits in which the sand wedges developed, show that the sand wedges and dunes are Wisconsinan. Growth of the sand wedges could have occurred continuously or episodically throughout the time the dunes were active, but probably was most rapid during the coldest periods, and particularly during late Wisconsinan time. The sand wedges and associated dunes record desert conditions over a significant part of the Arctic Coastal Plain during middle and late Wisconsinan time and perhaps throughout the Wisconsinan Stage.

INTRODUCTION

Sand wedges are wedge-shaped bodies of sand oriented with their apexes pointed downward, that underlie the borders of unvegetated, nonsorted polygons in the drier parts of Victoria Land, Antarctica (Péwé 1959, Black and Berg 1963, Berg and Black 1966) and in some parts of the Sverdrup Islands of arctic Canada (Hodgson 1982). They form in a manner analogous to ice wedges, which commonly underlie the borders of nonsorted polygons in tundra-covered arctic areas and in the more humid parts of Victoria Land. Both sand and ice wedges grow as a result of repeated formation and filling of thermal-contraction cracks in permafrost; ice wedges form when the cracks fill with snow, hoar frost, or meltwater (Leffingwell 1915, Lachenbruch 1962), whereas sand wedges form when the cracks fill with sand that trickles down from the surface (Péwé 1959, Black and Berg 1963, Berg and Black 1966).

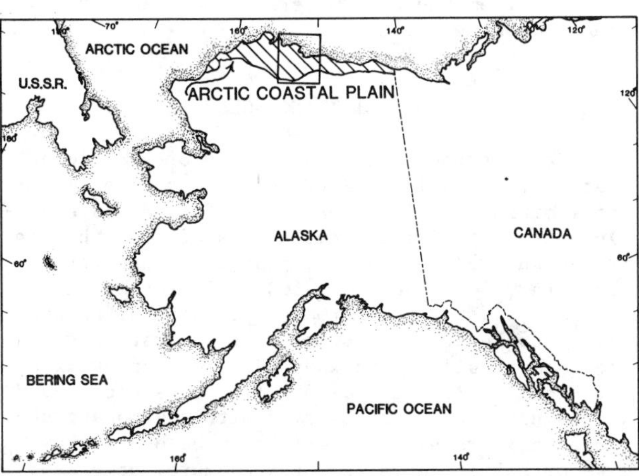

FIGURE 1 Location of Arctic Coastal Plain. Rectangle outlines area discussed in report.

Fossil features similar to modern sand wedges have been reported in Pleistocene deposits at several localities in North America (Berg 1969, Foscolos et al. 1977, Mears 1981), and northern Europe and Asia (Jahn 1975), and in Precambrian deposits in Africa (Deynoux 1982). Recently, fossil sand wedges have been identified in Pleistocene deposits over a broad area of the Arctic Coastal Plain of Alaska (Figures 1 and 2). The Alaskan sand wedges and associated stabilized eolian dunes suggest a dry, barren, and windswept envir-

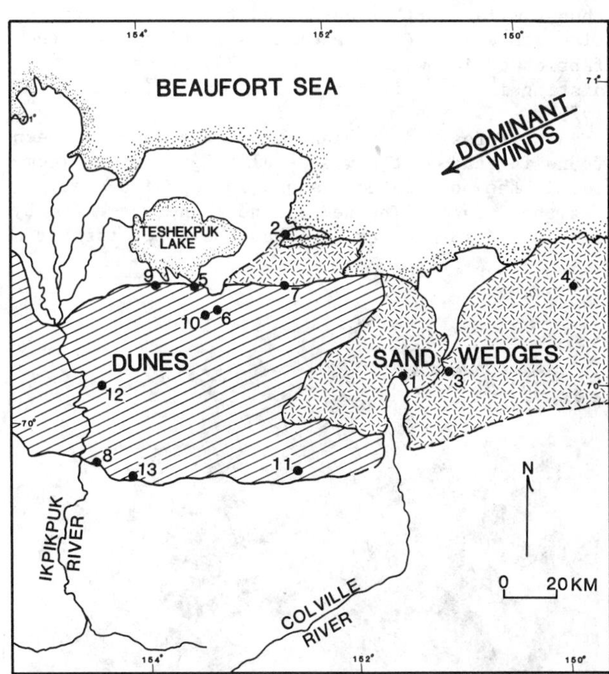

FIGURE 2 Distribution of sand wedges and dunes and location of sites mentioned in text and in Table 1.

onment that is in sharp contrast to the modern tundra-covered, lake-dotted surface beneath which ice wedges are forming.

DISTRIBUTION AND DESCRIPTION

The fossil sand wedges in northern Alaska occur in sandy Pleistocene marine and alluvial deposits north and east of a large area of stabilized eolian dunes (Figure 2). The eastward limit of the sand wedges is undetermined, but they have been observed in the Prudhoe Bay area about 40 km east of Figure 2 (S. E. Rawlinson, personal communication, 1982) and on the north side of Barter Island about 180 km east of Prudhoe Bay (D. M. Hopkins, personal communication, 1983).

Individual sand wedges range up to 3 m wide and 7 m deep (Figure 3), and the distance between the largest wedges ranges up to 15 m. Small secondary sand wedges are locally present between larger wedges, and some large sand wedges are cross-cut by smaller wedges. Wherever exposures are sufficiently extensive, the wedges can be seen to form polygonal systems. Bedding in the host material is invariably upturned at wedge edges.

The wedges possess a distinctive foliated fabric (Figure 4) of near-vertical to vertical laminae in wedge centers grading outward to inclinations that parallel the wedge edges. In many places the laminae are mutually cross-cutting. Wedge apophyses penetrate the host material, and pieces of the host material have been incorporated into the wedges and are cut by laminae that most likely represent individual thermal-contraction cracks (Figure 4). The maximum width of individual laminae is about 5 mm and most are about 1 or 2 mm wide. The fabric of these sand wedges is thus similar to that described for modern Antarctic sand wedges (Berg and Black 1966) and for the fabrics of ice wedges (Black 1974) and serves to distinguish them from ice wedge casts (Black 1976).

Microfossils or plant fragments have not been found in sand of the wedges although they are common in the host materials and in deposits overlying the wedges. The wedge sand is not cemented by ice, and deformed laminae that could be attributed

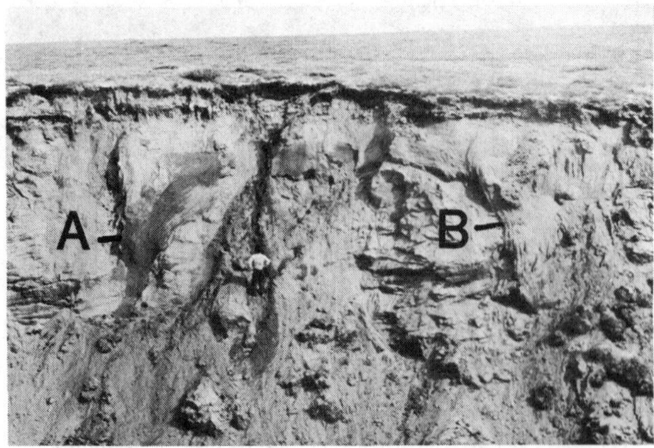

FIGURE 3 Large sand wedges (A and B) at site 1. Man is about 1.7 m tall.

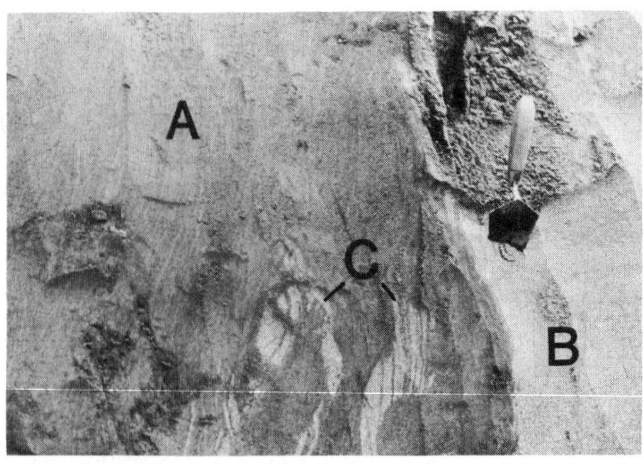

FIGURE 4 Sand wedge fabric (A) and host materials (B) at site 1. Pieces of the host material (C) occur within the sand wedge and are cut by laminae of the sand wedge.

to slumping due to melting of snow or ice have not been observed.

Sand composing the wedges is always texturally distinct from the host materials. This textural difference is illustrated for site 1 (Figure 2) by size analyses for five samples of the host material and samples from five individual wedges (Figure 5). The wedge sand is fine to very fine sand with mean grain sizes ranging from 2.74 to 3.09 phi. The host material, which at this site is predominantly fluvial sand, is distinctly coarser than the wedge sand and has mean grain sizes that range from 1.48 to 2.41 phi.

Although the sand wedges have formed in materials that range from gravelly fluvial sand to silty marine sand, the texture of the wedge sand is remarkably uniform over the broad area of their occurrence. Mean grain sizes for samples from 11 wedges at sites 1 through 4 (Figure 2) range from 2.60 to 3.10 phi (Figure 6). Significantly, the textural analyses for the sand wedges are essentially encompassed by the field of cumulative curves for 20 samples of sand from the dune field (Galloway 1982). Wedge sand is thus texturally distinct from the host materials but similar to sand composing the dunes.

LIMITING AGES

Radiocarbon dating of remains of herbaceous plants at sites 1 and 2 (Figure 2) provides minimum limiting ages for cessation of sand wedge development. At site 1, involutions of peat that intrude sand wedges and thus postdate wedge formation are 9,300 $\pm$ 90 and 9,600 $\pm$ 100 years old (Table 1). At site 2, two samples of herbaceous plants collected in growth position just above the base of silty sand that overlies sand wedges are 9,689 $\pm$ 110 and 10,200 $\pm$ 110 years old. Sand wedges have not been observed within Holocene deposits and are always overlain by Holocene thaw lake or eolian deposits. Furthermore, some sand wedges are cut by large Holocene ice wedges and by Holocene ice wedge casts. The sand wedges are evidently pre-Holocene.

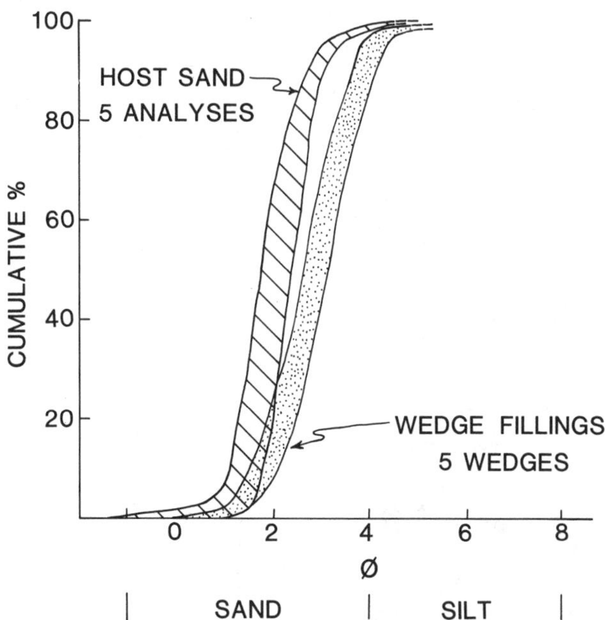

FIGURE 5 Grain-size analyses for five sand wedges and five samples of the host sand at site 1. Stippled pattern indicates the field occupied by cumulative curves for the sand wedge samples, and the diagonal lines indicate the field occupied by cumulative curves for samples of the host sand.

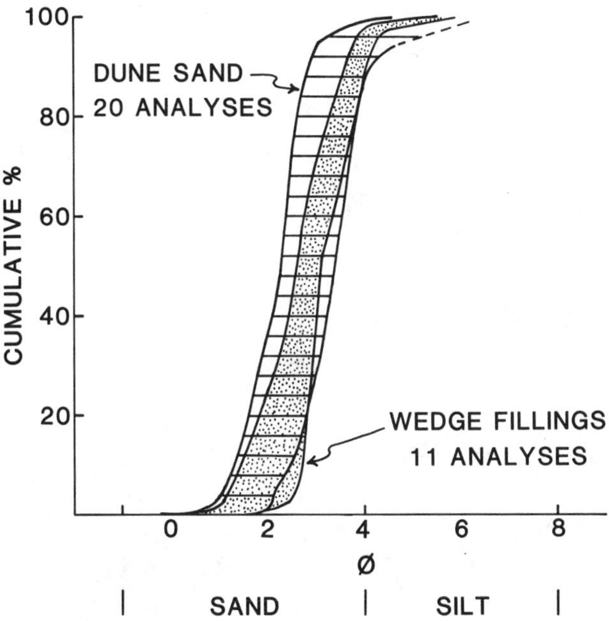

FIGURE 6 Grain-size analyses for 11 sand wedges at sites 1 through 4 and 20 samples of eolian sand from the dune field. Stippled pattern indicates the field occupied by cumulative curves for the sand wedge samples, and the horizontal lines indicate the field occupied by cumulative curves for the dune sand samples.

Stratigraphic relations at site 2 suggest a maximum limiting age for the sand wedges there. At this site, the sand wedges are developed in fossiliferous marine beach deposits composed of pebbly sand. Marine ostracodes in the beach deposits include Cythere lutea O.F. Muller, 1785 and Hemicythere borealis of Hazel, 1967 which occur today in temperate to subpolar waters of the North Atlantic but not in the Beaufort Sea (E. L. Brouwers, written communication, 1980). The mollusk fauna includes Natica janthostoma which today is limited to waters adjoining Japan, Kamchatka, and the Commander Islands (D. M. Hopkins, written communication, 1980). Spruce (Picea) driftwood is locally common in the beach deposits. These facts indicate more open water and warmer climatic conditions than presently prevail, and the deposits are interpreted to represent an interglacial marine transgression. These are the youngest interglacial beach deposits on the Arctic Coastal Plain, and although they are more than 51,000 years old (Carter and Robinson 1981), they are young enough that amino acid ratios (Alle/Ile) determined on valves of Hiatella arctica collected from these deposits are barely distinguishable from those determined on modern specimens of Hiatella arctica (J. K. Brigham, written communication, 1983). This transgression is therefore correlated with the Pelukian (last interglacial) transgression of Hopkins' (1967) Alaskan sequence. If this is correct, then the sand wedges at site 2 developed entirely during the Wisconsinan Stage.

Maximum limiting ages for sand wedges at other sites cannot be determined with current data. However, their relation to the dune field suggests that all the sand wedges reported here are Wisconsinan, although multiple episodes of formation during Wisconsinan time cannot be ruled out.

RELATION TO DUNES

Bedding attitudes in eolian sand and the orientation of dune ridges indicate that the dominant winds during formation of the dune field were easterly to northeasterly (Figure 2), similar to the modern prevailing winds (Carter 1981). The sand wedges thus formed upwind of the dunes. This fact, together with the similarity in grain size between the wedge fillings and the dune sand, suggest that the loose sand that entered the thermal-contraction cracks to form the sand wedges was derived from wind-driven sand moving across the coastal plain toward the dunes.

The dune field consists of large, linear dunes as much as 20 km long, 1 km wide, and 30 m high that are overlain by a sheet of eolian sand a few meters thick (Carter 1981). Small parabolic and longitudinal dunes a meter or so thick and up to 1 km long (Black 1951) formed in the sand sheet sometime after it was stabilized by soil development. Radiocarbon dates of Salix (willow) in growth position within the eolian sand sheet range from 9,640 ± 150 to 11,400 ± 160 years B.P. (Carter 1981), and a paleosol dat- dating from 11,000 to 12,000 years B.P. is locally present at its base (Table 1). Because the sand wedges are overlain and intruded by deposits

TABLE 1 Radiocarbon dates relating to the time of sand wedge development and the activity of linear dunes.

Site	Latitude N & Longitude W	Material	Occurrence	Age and Laboratory Number	Significance
1	70° 04.35' 151° 22.9'	Peat	Involution intruding sand wedge	9,330 ± 90 (Beta-5384)	Minimum limiting age of cessation of sand wedge development
1	"	"	"	9,600 ± 100 (Beta-5383)	"
2	70° 34.72' 152° 31.38'	Herbaceous plants	Growth position in sand overlying sand wedge	9,680 ± 110 (USGS-1377)	"
2	"	"	"	10,200 ± 90 (USGS-1378)	"
5	70° 26.7' 153° 28.9'	Peat	Paleosol separating eolian sand sheet from linear dunes	11,230 ± 170 (I-10,814)	Minimum limiting age for cessation of activity of linear dunes
6	70° 23.3' 153° 12.2'	Wood	Rooted in paleosol separating eolian sand sheet from linear dunes	11,430 ± 170 (I-11,675)	"
7	70° 25.7' 152° 34.23'	Peat	In ice-wedge cast developed in linear dunes	11,700 ± 180 (I-12,177)	"
8	69° 49.37' 154° 24.85'	"	Paleosol separating eolian sand sheet from linear dunes	12,070 ± 100 (USGS-823)	"
9	70° 27.5' 153° 53.1'	Herbaceous plants	Growth position in dune sand	13,140 ± 60 (USGS-1154)	Dates activity of linear dunes
10	70° 18.4' 153° 26.4'	"	Growth position in interdunal pond or lake	13,610 ± 130 (USGS-624)	"
11	69° 48.8' 152° 37.35'	"	Growth position in dune sand	16,490 ± 130 (USGS-1376)	"
8	69° 49.37' 154° 24.51'	"	Growth position in interdunal pond or lake	18,600 ± 210 (USGS-824)	"
12	70° 08.1' 154° 28.6'	"	"	25,200 ± 180 (USGS-1029)	"
11	69° 48.8' 152° 37.35'	"	"	35,900 ± 1,200 (USGS-825)	"
13	69° 48.62 154° 12.5	wood	Detrital wood in alluvium beneath linear dunes	49,400 + 3,000 - 2,200 (USGS-826)	If not redeposited, is most likely minimum limiting age for alluvium beneath linear dunes

the same age as the sand sheet, they are older than the sand sheet. Furthermore, the period from 11,000 to 12,000 years B.P. was a time of dune stabilization, soil development, and ice wedge growth; conditions which are generally unfavorable for the growth of sand wedges. Therefore, the most probable time of development of the sand wedges was while the large linear dunes were active.

Organic remains are extremely scarce within sand of the linear dunes, but radiocarbon dating of the few herbaceous plants that have been found shows that the dunes were active during middle and late Wisconsinan time and perhaps earlier (Table 1). Deposits of an interdunal lake at site 11 that are underlain by eolian sand contain herbaceous plants that yield a radiocarbon age of 35,900 ± 1,200 years B.P. Wood collected from alluvial deposits that underlie the linear dunes at site 13 has been radiocarbon dated as about 49,400 years old, but this may be merely a minimum limiting age for the alluvial deposits. Formation of the dunes thus either began between 36,000 and 50,000 years B.P. or perhaps beyhond the range of radiocarbon dating. However, deposits such as peat beds that would indicate relatively warm, mesic intervals are absent within the dune sand, suggesting that the linear dunes formed entirely since the last interglacial episode, and are Wisconsinan.

TIME REQUIRED FOR GROWTH

Rates of sand and ice wedge growth have been measured in Antarctica (Black 1973, 1982) and rates of ice wedge growth have been measured in Alaska (Black 1974). The measurements from Antarctica seem most appropriate for use here because they include sand wedges. The measurements were made during two intervals that cover a total of about 20 years and they indicate that growth rates are highly variable and are affected by short-term variations in climate and ground conditions (Black 1982). During the first interval, net growth rates averaged 0.79 mm per year, and during the second interval average net growth was only 0.04 mm per year. Black proposed that the reduction in growth rates was due to climatic warming.

Assuming the most rapid growth rate above, the largest of the northern Alaskan sand wedges could have grown in only 3,800 years, whereas the second growth rate would require about 75,000 years. The first growth rate corresponds to the addition of a 2-mm-wide sand lamina every other year. On the basis of observations that the probability of cracking in any one year is less for large wedges than for small ones (Berg and Black 1966), and the assumption that the rate of growth of sand wedges should slow with increasing age (Berg and Black, 1966), this is probably a reasonable short-term upper limit for the growth rate of the Alaskan sand wedges during the coldest periods of their development. The second growth rate corresponds to the addition of a 2-mm-wide sand lamina once every 50 years. Such slow growth rates are unlikely over the long term but, as in Antarctica, could have occurred in response to episodes of climatic warming during the period of wedge development. The system of sand wedges on the Arctic Coastal Plain therefore probably took more than 4,000 years to develop and may have grown continuously or episodically throughout the period that the linear dunes were active.

PALEOENVIRONMENTAL SIGNIFICANCE

Sand wedges indicate dry conditions in which locally there may be a net transfer of moisture from permafrost to the air (Black and Berg 1963), and thus progressive desiccation of the ground. Furthermore, they document an absence of surface water from the time the thermal-contraction cracks form in middle and late winter until they fill with eolian sand (Foscolos et al. 1977); otherwise, ice or composite wedges would form. Therefore, snow cover must have been extremely patchy or absent during middle and late winter or was removed prior to spring thaw. Modern snow cover on the coastal plain is essentially continuous throughout the winter, and represents about half of the 195-252 mm (water equivalent) that falls (Benson 1982). The other half is removed by the wind and deposited in drifts, with an unknown part of this lost by sublimation during transportation (Benson 1982). The area containing the sand wedges could have been kept bare of snow by intense and relatively constant winds, but if this were a primary factor, then snowdrifts should have formed within the dune field and subsequently would have been buried by windblown sand, as is common in cold climate dunes (Ahlbrandt and Andrews 1978). However, neither buried snow nor deformational structures indicative of melted snow or ice have been observed in sand of the linear dunes. Lighter snowfall than now occurs coupled with sublimation induced by desiccating winds seem required to produce the necessary bare ground.

Vegetation must have been extremely sparse or absent in the area containing the sand wedges, because significant plant cover would have inhibited the movement of eolian sand and prevented sand wedge development. The most favorable sites for plant growth would have been in the relatively protected troughs above the wedges. However, had plants grown there, one would expect plant fragments to be incorporated with eolian sand in the wedges. The absence of plant fragments in the sand wedges suggests that the area was completely devoid of vegetation, as in Antarctica where sand wedges are forming today. Vegetation within the dunes downwind of the sand wedges was extremely sparse at best and consisted of scattered clumps of herbaceous plants on the dunes and aquatic bryophytes in interdunal ponds. Scanty snow cover may have been an important factor limiting plant growth both in the dunes and in the area of the sand wedges.

The sand wedges and associated linear dunes thus record dry, barren, and windswept conditions across a significant part of the Arctic Coastal Plain during middle and late Wisconsinan time and perhaps throughout the Wisconsinan Stage. Even so, radiocarbon dating of fossil bones indicates that south of the barren belt defined by the sand wedges and dunes, vegetation was sufficient to support large herbivores such as mammoth, horse and bison (Carter 1982). None of the dates of these mammals, however, is less than 28,000 years B.P. Furthermore, few organic remains have been found in coastal plain deposits between 14,500 and 28,000 years old compared to older and younger deposits, and pollen spectra of this age from the northern Yukon indicate sparse, discontinuous vegetation in the arctic foothills (Cwynar and Ritchie 1980). Desert conditions evidently expanded south of the barren belt during late Wisconsinan time coincident with the expansion of glaciers in the Brooks Range (Hamilton 1982). This arid Wisconsinan cold phase may have coincided with the period of most rapid growth of the sand wedges.

ACKNOWLEDGMENTS

Discussions in the field with R. F. Black and D. M. Hopkins contributed significantly to this study. The manuscript benefited from reviews by O. J. Ferrians, Jr., and D. M. Hopkins. Assistance in the field was provided by J. P. Galloway, who also performed the grain size analyses.

REFERENCES

Ahlbrandt, T. S., and Andrews, S., 1978, Distinctive sedimentary features of cold-climate eolian deposits, North Park, Colorado: Palaeogeography, Palaeoclimatology, Palaeoecology, v. 25, p. 327-351.

Benson, C. S., 1982, Reassessment of winter precipitation on Alaska's arctic slope and measurements on the flux of windblown snow: Geophysical Institute, University of Alaska Report UAG R-288, 26 p.

Berg, T. E., 1969, Fossil sand wedges at Edmonton, Alberta, Canada: Biuletyn Peryglacjalny, v. 19, p. 325-333.

Berg, T. E., and Black, R. F., 1966, Preliminary measurements of growth of non-sorted polygons, Victoria Land, Antarctica, in Tedrow, J.F.C., ed., Antarctic soils and soil-forming processes, American Geophysical Union Antarctic Research Series, v. 8, p. 61-108.

Black, R. F., 1951, Eolian deposits of Alaska: Arctic, v. 4, p. 89-111.

Black, R. F., 1973, Growth of patterned ground in Victoria Land, Antarctica, in Permafrost--The North American contribution to the Second International Conference, Yakutsk: Washington, D.C., National Academy of Sciences, p. 193-203.

Black, R. F., 1974, Ice-wedge polygons of northern Alaska, in Coates, D. R., ed., Glacial geomorphology: Annual Geomorphology Series, 1974, 5th, Proceedings: Binghamton, New York, State University of New York, p. 247-275.

Black, R. F., 1976, Periglacial features indicative of permafrost: Ice and soil wedges: Quaternary Research, v. 6, p. 3-26.

Black, R. F., 1982, Rate of growth of patterned ground in Victoria Land, Antarctica has diminished: Geological Society of America Abstracts with Programs, v. 14, no. 7, p. 444.

Black, R. F., and Berg, T. E., 1963, Hydrothermal regimen of patterned ground, Victoria Land, Antarctica: International Association of Scientific Hydrology, Commission of Snow and Ice, Publication no. 61, p. 121-127.

Carter, L. D., 1981, A Pleistocene sand sea on the Alaskan Arctic Coastal Plain: Science, v. 211, p. 381-383.

Carter, L. D., 1982, Late Wisconsin desertification in northern Alaska: Geological Society of America Abstracts with Programs, v. 14, no. 7, p. 461.

Carter, L. D., and Robinson, S. W., 1981, Minimum age of beach deposits near Teshekpuk Lake, Alaskan Arctic Coastal Plain, in Albert, N. R. D., and Hudson, Travis, eds., The United States Geological Survey in Alaska: Accomplishments during 1979: U.S. Geological Survey Circular 823-B, p. B8-B9.

Cwynar, L. C., and Ritchie, J. C., 1980, Arctic Steppe-Tundra: A Yukon Perspective: Science, v. 208, p. 1375-1377.

Deynoux, Max, 1982, Periglacial polygonal structures and sand wedges in the late Precambrian glacial formations of the Taoudeni Basin in Adrar of Mauretania (West Africa): Palaeogeography, Palaeoclimatology, Palaeoecology, v. 39, p. 55-70.

Foscolos, A. E., Rutter, N. W., and Hughes. O. L., 1977, The use of pedological studies in interpreting the Quaternary history of central Yukon Territory: Geological Survey of Canada Bulletin 271.

Galloway, J. P., 1982, Grain-size analyses of 20 eolian sand samples from northern Alaska, in Coonrad, W. L., ed., The United States Geological Survey in Alaska: Accomplishments during 1980: U.S. Geological Survey Circular 844, p. 51-53.

Hamilton, T. D., 1982, A late Pleistocene glacial chronology for the southern Brooks Range: Stratigraphic record and regional significance: Bulletin of the Geological Society of America, v. 93, p. 700-716.

Hodgson, D. A., 1982, Surficial materials and geomorphological processes, western Sverdrup and adjacent islands, District of Franklin: Geological Survey of Canada Paper 81-9, 44p.

Hopkins, D. M., 1967, Quaternary marine transgressions in Alaska, in Hopkins, D. M., ed., The Bering Land Bridge: Stanford, California, Stanford University Press, p. 47-90.

Jahn, A., 1975, Problems of the periglacial zone (Zagadnienia strefy peryglacjalnej): Warsaw, Państwowe Wydawnictwo Naukowe, 223 p.

Lachenbruch, A. H., 1962, Mechanics of thermal contraction cracks and ice-wedge polygons in permafrost: Geological Society of America Special Paper 70, 69 p.

Leffingwell, E. De K., 1915, Ground-ice wedges; the dominant form of ground ice on the north coast of Alaska: Journal of Geology, v. 23, p. 635-654.

Mears, B., Jr., 1981, Periglacial wedges and the late Pleistocene environment of Wyoming's intermontane basins: Quaternary Research, v. 15, p. 171-198.

Péwé, T. L., 1959, Sand-wedge polygons (tesselations) in the McMurdo Sound Region, Antarctica--A progress report: American Journal of Science, v. 257, p. 545- 552.

RUNOFF FROM A SMALL SUBARCTIC WATERSHED, ALASKA

E. F. Chacho, Jr.[1] and S. Bredthauer[2]

[1]U.S. Army Cold Regions Research and Engineering Laboratory
Fairbanks, Alaska 99703 USA
[2]R&M Consultants, Anchorage, Alaska 99503 USA

Precipitation-runoff ratios were measured on Glenn Creek, a small, second-order, subarctic stream located near Fairbanks, Alaska, in the Yukon Tanana Upland physiographic province. Glenn Creek drains a watershed of 2.25 km^2, of which 70% is underlain by permafrost. A Parshall flume was used to measure streamflow, and a pair of 1.22 m by 2.44 m lysimeters were used to measure precipitation and runoff from the moss-covered permafrost slope. The data from one summer season (1979) and one snowmelt season (1980) indicate the sloping surfaces of the watershed have a very fast response time, long recession, and subsurface runoff prior to complete saturation of the overlying organic material. Glenn Creek streamflow is comparable to the lysimeter runoff with regard to response time and runoff recession, however the watershed precipitation-runoff ratio is much lower. This is attributed to longer travel distances in the watershed, which result in greater evapotranspiration losses, little contribution from the non-permafrost areas, and only partial areas of the watershed contributing to the streamflow.

Watershed studies in the discontinuous permafrost zone of the North American taiga have a relatively short history. The first detailed studies of hydrologic processes in subarctic Alaska were performed by Dingman (1966a,b; 1971; 1973) in the Glenn Creek watershed near Fairbanks, Alaska. Further studies of permafrost-dominated watersheds were initiated in 1969 with the establishment of the Caribou-Poker Creeks Research Watershed 48 km north of Fairbanks, Alaska (Slaughter 1971). Analyses of streamflow from permafrost and nonpermafrost-dominated watersheds of interior Alaska have shown that the presence of permafrost and its associated soil and vegetation types greatly influences runoff generation (Dingman 1971, Ford 1973, Slaughter and Kane 1979, Kane et al. 1981, Haugen et al. 1982). Examination of precipitation and streamflow data by Dingman (1971) for Glenn Creek and Ford (1973) and Haugen et al. (1982) for Caribou Creek resulted in similar observations of (1) prolonged streamflow recessions in permafrost basins, attributable to delayed flow through organic soils associated with permafrost, (2) rapid streamflow response to precipitation, within two hours on the average, and (3) streamflow response time not related to antecedent moisture conditions. Dingman (1971) had reported that the ratio of runoff to precipitation was related to antecedent moisture conditions. This relationship was not found in the Caribou Creek data, but may be due to a lack of precision in the data (i.e. use of mean daily flows) rather than the absence of any relationship (Ford 1973, Haugen et al. 1982).

Previous studies have addressed the infiltration characteristics of both the permafrost and non-permafrost soils typical of Alaska's taiga (Kane et al. 1978, Slaughter and Kane 1979, Kane 1980). In general, it was found that the thermal and hydraulic conditions in the nonpermafrost soils result in infiltration rates high enough to accept all incident moisture (snowmelt or precipitation). No surface or near-surface flows were observed in the field. In the permafrost areas, however, the soils beneath the active layer (depth of seasonal thaw) are essentially impermeable, due to high moisture contents in the frozen soils. The result is saturated conditions in the overlying organic soils and observations of large quantities of water moving downslope through these soils. Dingman (1971) had formed the same conclusions following observations of storm runoff on Glenn Creek. Santeford (1979) studied the moisture-retention properties of the permafrost-associated organic soils by measuring vertical percolation in lysimeters. He concluded that the water-holding capacity of the organic soil had to be exceeded before any drainage occurred.

This paper addresses the processes of runoff production on moss-covered slopes underlain by permafrost through the use of lysimeters installed on a slope in the Glenn Creek watershed. Comparison of the lysimeter data to Glenn Creek streamflow indicates that the overall watershed response to precipitation is lower than that of the lysimeter.

WATERSHED DESCRIPTION

The Glenn Creek watershed is located 14 km north of Fairbanks, Alaska, at latitude 64°57'N, longitude 147°35'W, at the southern edge of the Yukon-Tanana Uplands (Wahrhaftig 1965). The watershed was selected by Dingman (1971) in 1964 for detailed hydrologic studies based on its size, accessibility, lack of human disturbance, and apparent representativeness of the area. The original study basin had an area of 1.8 km^2 (Figure 1). When hydrologic studies were reactivated on Glenn Creek in 1978, the streamflow gauging site was relocated downstream, increasing the study basin area to 2.25 km^2. The drainage area is uniformly distributed between the peak elevation of 493 m at the eastern perimeter and 250 m at the basin outlet. Slope tangents (Figure 2) range from 0 to 0.6, but 80% of the basin ranges between 0.05 to 0.25, with an average basin slope of 0.184. Glenn Creek is a second-order stream for 88% of its total length; the head waters extend up to nearly 400 m. The channel gradient is fairly uniform throughout the lower 1300 m of its course, averaging 0.049 (Dingman 1971). A complete description of the geologic setting of the Glenn Creek watershed is given by Dingman (1971).

The distribution of vegetation types is closely associated with the distribution of permafrost (Figure 1, Table 1). Birch-aspen-white spruce stands, primarily on south-facing slopes, cover 30% of the basin. This vegetation type generally occurs on moderately well-drained silt loams covered with up to 15 cm of thin organic soil. Permafrost is not normally present. The remainder of the watershed (70%) consists of black spruce stands on the north-facing slopes and valley floors underlain by permafrost. These areas generally consist of poorly drained mineral soils overlain by a thick (30-45 cm) organic mat composed of three primary sublayers, each 10-15 cm thick. The living surface layer is predominately sphagnum moss intertwined with various surface plants such as blueberries, cranberries, labrador tea, sedges, and lichens. The middle layer is denser than the surface layer and is composed primarily of dead sphagnum moss. The bottom layer is much denser and is composed of peat and decomposing vegetation (Santeford 1979). The location of the permafrost table occurs at depths from the organic mat/mineral soil interface (Santeford 1979) to 1 m below the surface (Dingman 1971).

The climate of the area is continental, characterized by large diurnal and annual temperature variations, low annual precipitation, low cloudiness, and low humidity. The Fairbanks records show January as the coldest month with a mean temperature

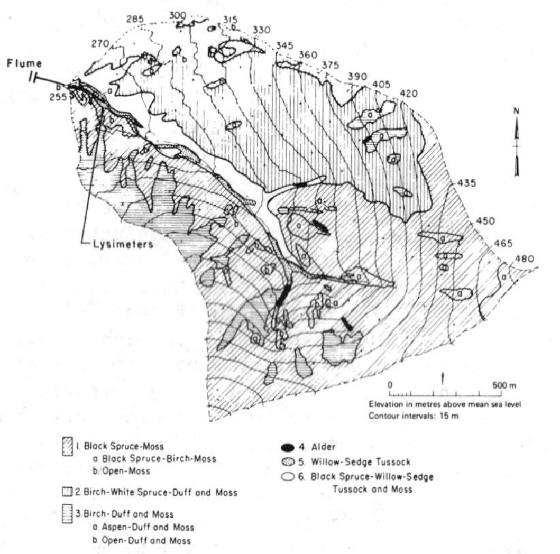

FIGURE 1 Vegetation map of Glenn Creek watershed showing location of flume and lysimeters (map from Dingman [1973]).

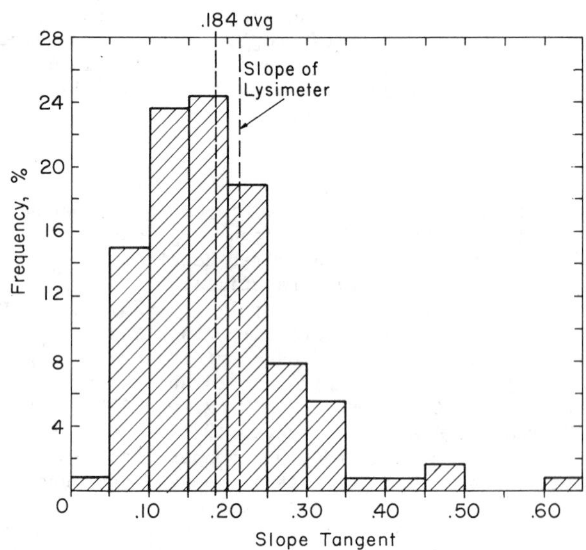

FIGURE 2 Distribution of ground slopes in Glenn Creek watershed showing slope of lysimeters (after Dingman [1971]).

TABLE 1 Areal Distribution of Vegetation Types and Permafrost in Glenn Creek Watershed (after Dingman [1973]).

Vegetation type	Permafrost	Area, km^2	Area, %
1. black spruce-moss	yes	0.855	47.1
1a. black spruce-birch-moss	yes	0.047	2.5
1b. open-moss	yes	0.029	1.6
2. birch-white spruce-duff and moss	no	0.544	30.0
3. birch-duff and moss	yes	0.171	8.4
3a. aspen-duff and moss	no	0.031	1.2
3b. open-duff and moss	no	0.005	0.3
4. alder-bare	yes	0.003	0.2
5. willow-sedge-bare	yes	0.021	1.1
6. black spruce-willow=sedge-bare	yes	0.119	6.6
TOTAL		1.815	100.0

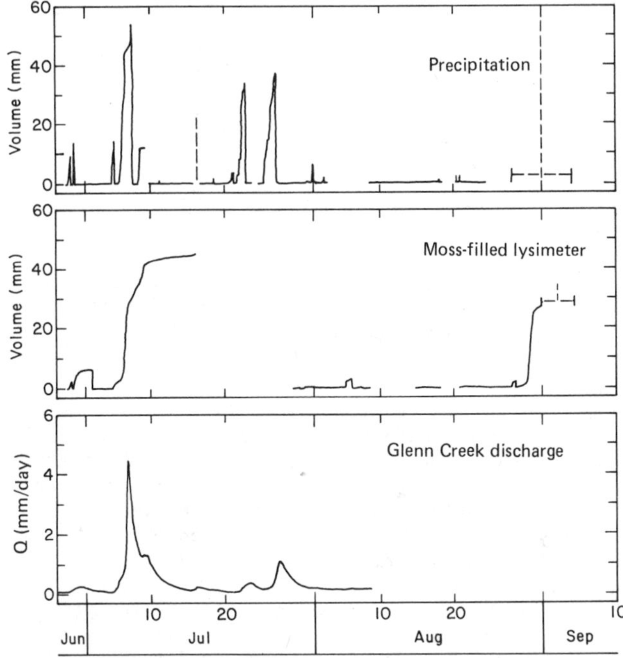

FIGURE 3 Precipitation at lysimeter (top), moss-filled lysimeter runoff (middle), and Glenn Creek discharge for summer 1979. The precipitation and lysimeter plots are of accumulated volumes for individual events. Runoff for Glenn Creek is shown as discharge computed over the entire 2.25 km^2 watershed area. A break in the continuous line indicates periods of missing data where timing of runoff is unavailable; dashed lines show total volumes over the indicated time interval.

of -24.4°C, and July as the warmest month with a mean of 17.1°C. The annual precipitation at Fairbanks averages 285 mm, half of which occurs during the months of June, July, and August (U.S. Department of Commerce 1980). The 1979 summer (May-September) precipitation at Fairbanks was slightly below the mean summer precipitation of 175.5 mm (Haugen et al. 1982).

MEASUREMENTS

Beginning in 1978, stream flow has been measured at the outlet of the basin with a 22.9 cm Parshall flume and F-1-type water level recorder. In the summer of 1979, two lysimeters, each 1.22 m by 2.44 m, bounded on both the bottom and sides, were monitored in the permafrost area with a slope tangent of 0.22, slightly higher than the average ground slope of the basin (Figures 1 and 2). The lysimeters were constructed in September 1978, near the end of the thaw season when the thaw zone was approximately at the organic mat/mineral soil interface. One lysimeter was filled with the in-situ organic soil (moss). The adjacent lysimeter was left empty and functioned as a precipitation gauge. A drain system ran from each lysimeter to a barrel equipped with an F-1-type water-level recorder that measured runoff as a volume.

The available data for the 1979 summer season are plotted in Figure 3. The runoff from the empty lysimeter (precipitation) and the moss-filled lysimeter are shown as accumulated runoff for each event. Runoff at Glenn Creek is shown as discharge in mm/day computed over the entire 2.25 km^2 watershed. The results of selected moss-filled lysimeter measurements and the Glenn Creek data are shown by event in Table 2. The hydrologic response (HR)

TABLE 2 Precipitation, runoff, and hydrologic response to the Glenn Creek watershed, 1979.

	Precipitation at lysimeter		Moss-filled lysimeter runoff				Glenn Creek		
Storm event	Storm period	Total (mm)	Runoff period	Total (mm)	HR	MD (mm)	Runoff period	Total (mm)	HR
1	28 June 1500–29 June 0200	9.6	28 June 1600–29 June 0300	2.4	0.25	7.2			
2	29 June 0800–29 June 1300	14.0	29 June 0800– 1 July 2000	6.6	0.47	7.4	28 June 1900–4 July 1600	0.5	0.02(0.03)[3]
3a	4 July 1200– 4 July 2200	14.6	4 July 1300– 5 July 1300	2.8[2]	>0.19	<11.8			
3b	5 July 1300– 7 July 1200	53.9	5 July 1300– 8 July 0300	31.6[2]	>0.59	<22.2			
3c	8 July 0200– 8 July 1300	11.7	8 July 0300– 8 July 2200	6.7[2]	>0.57	< 5.0			
3d	11 July 0000–11 July 0500	0.8	8 July 2200–15 July 1900	3.6[4]					
3	4 July 1200–11 July 0500	>80[1]	4 July 1600–15 July 1900	44.8	≤0.55	≥36.2	4 July 1600–17 July 0200[5]	8.8	≤0.11(0.16)
4	>15 July 1600–<18 July 1300	21.5	No data				15 July 1900–20 July 0100	0.8	0.04(0.06)
5	20 July 1500–22 July 1800	37.8	No data				22 July 0100–26 July 0900[5]	0.5	0.13(0.19)
6	25 July 0200–26 July 2200	37.3	No data				25 July 0500–2 Aug 0800	2.5	0.07(0.10)
7	17 Aug 1900–18 Aug 0000	1.1	No runoff	0.0	0.0	≥1.1	No data		
8	20 Aug 1900–20 Aug 1900	2.5	20 Aug 1900–21 Aug 0200	0.4	0.16	2.1	No data		
9	27 Aug 1700–<1 Sept 0000	57.0	27 Aug 1800–<5 Sept 1500	34.5	0.61	22.5	No data		

[1] Minimum total--some missing data
[2] Partial runoff volumes by storm event
[3] Values in parentheses based on permafrost area only
[4] Period of long recession, which includes small precipitation event 3d
[5] End of period is extrapolated recession time when runoff events overlap

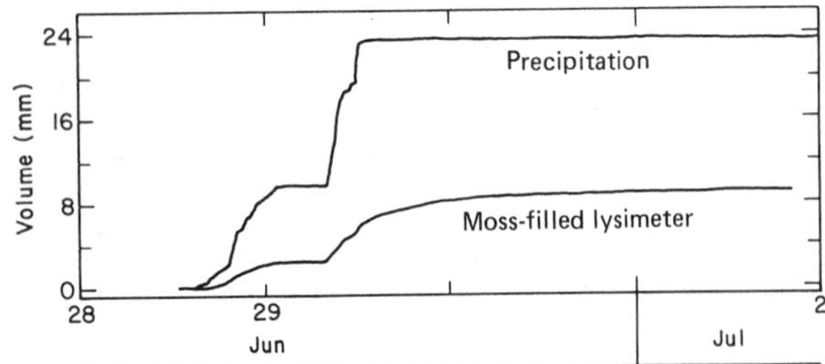

FIGURE 4 Precipitation and runoff from moss-filled lysimeter between June 28 and July 2, 1978 (precipitation and runoff read on same scale).

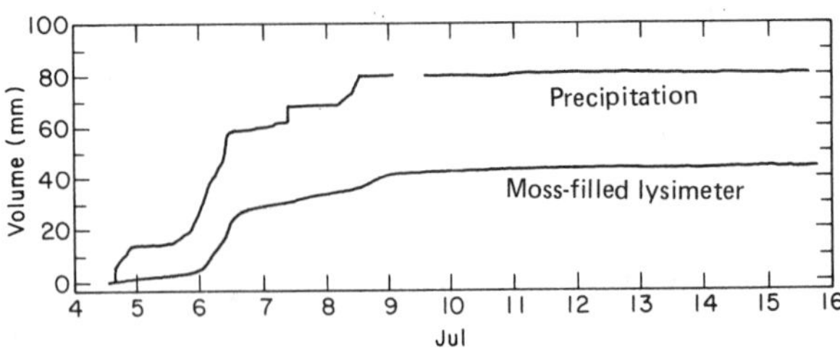

FIGURE 5 Precipitation and runoff from moss-filled lysimeters between July 4 and July 16, 1979 (precipitation and runoff read on same scale).

ratio was computed as the ratio of total runoff volume to total precipitation. The moisture deficit (MD) on the moss-filled lysimeter was computed indirectly as the difference between precipitation received and runoff measured. The runoff volumes of Glenn Creek for individual storm events were separated by plotting the recession discharges vs. time on semi-logarithmic paper and extending the recession curve to an assumed base flow by straight-line extrapolation (i.e. recessions were assumed to follow a simple exponential decay [Dingman 1971]). The beginning of a runoff event was measured from the initial apparent streamflow response; that is, the time at which the water level first began to rise. The two largest events recorded on the lysimeters are replotted in Figures 4 and 5 as total accumulation of runoff over the time periods indicated.

DISCUSSION OF LYSIMETER DATA

The summer 1979 measurements indicate that the response time of the moss-filled lysimeter was very fast, as shown in Table 2 and Figures 4 and 5. Runoff from the moss-filled lysimeter began within one hour of the start of the precipitation in all measured events. It appears that, based on timing of runoff alone, water percolates through the organic mat and moves downslope well before the total moisture deficit of the mat is satisfied. This is also confirmed by inspection of the HR ratios and MD values shown in Table 2. In storm events 1 and 2 (Figure 4), 9.6 mm of precipitation produced 2.4 mm of runoff, resulting in an HR ratio of 0.25 and indicating 7.2 mm of water was retained to satisfy the moisture deficit. In addition, little drainage occurred after the precipitation ended. Event 2 occurred 6 hours later, and had an HR ratio of 0.47 and an MD of 7.4 mm. Since the second event resulted in an additional moisture retention of 7.4 mm, it is apparent the first event did not fully saturate the moss layer, although the higher HR ratio does indicate that the antecedent moisture conditions resulted in less retention of the percolating water.

Events 3a through 3c (Table 2, Figure 5) had greater intensities, larger precipitation volumes, and occurred in such close succession (14 hrs) that the moss-filled lysimeter did not completely drain between storms. The runoff volumes of events 3a, 3b, and 3c have been separated based solely on the timing of precipitation, as shown in Table 2. Separation of runoff events in this way neglects the long-term drainage that did occur, but further separation was impossible without using subjective techniques of hydrograph separation. As a result, the HR ratios are slightly lower and MD values slightly higher than would be computed using the total runoff, and should be considered limits to these values.

Event 3a followed a pattern similar to Event 1 with a high percentage of moisture retained by the moss, a response time of one hour, and runoff apparently occurring before complete saturation of the moss layer. Event 3b was a long intense rainstorm resulting in more than 59% of the precipitation running off and moisture retention approaching 22 mm. Exceedance of the water-holding capacity of the moss-filled lysimeter is indicated, since the rate of runoff equals the rate of precipitation on July 6, as shown by the nearly equal slopes of the runoff accumulation curve and the precipitation accumulation curve (Figure 5). Event 3c, occurring 14 hours later, also resulted in a high HR ratio (>0.57) and a low MD value (<5 mm). Drainage of the moss-filled lysimeter continued for nearly 6 days with an additional runoff of 3.6 mm. Event 3d added 0.8 mm precipitation about midway through this drainage period, but little detectable change in the runoff accumulation curve was observed (Figure 5). Parameters for the overall period, event 3 in Table 2, are shown as limiting values due to an incomplete precipitation record after July 11. Since more precipitation occurred than was recorded, the HR ratio of 0.55 is a maximum and the MD value of 36.2 is a minimum for the total period shown for event 3.

Since the moisture deficit (MD) is computed as the difference between precipitation input and drainage from the moss-filled lysimeter, losses to evapotranspiration would also be included in the moisture deficit value. During long periods of drainage the moisture deficit may, therefore, be overestimated. Pan evaporation at the University of Alaska (14 km south) was 22.1 mm for 4-11 July 1979 (U.S. Department of Commerce 1979). The actual rate of evaporation from the moss is unknown.

Event 7 was of low volume, 1.1 mm, following a dry period in excess of 10 days, and produced no runoff from the moss-filled lysimeter. Event 8 was also of low volume (2.5 mm) but produced 0.4 mm of runoff for an HR ratio of 0.16 and an MD of 2.1 mm. Drainage continued for nearly 7 hrs after the end of the precipitation event. For Event 9, only the beginning time and the precipitation and runoff volumes from the lysimeters were recorded. The HR ratio of 0.61 was the highest recorded for the summer.

The same lysimeters were monitored during the 1980 snowmelt runoff season, as has been previously reported by Kane et al. (1981). The results from that study indicated that of an initial snowpack of 69.9 mm, 21.1 mm were lost to sublimation and evaporation, 24.9 mm were measured as runoff from the moss, and 23.9 mm were retained by the moss. Snowmelt runoff of 5.9 mm had occurred on the empty lysimeter over a 4-day period before runoff first appeared on the moss-filled lysimeter. Assuming that the rate of snowmelt was equal on both lysimeters, the delay in runoff from the moss-filled lysimeter can be attributed to several factors. A portion of the initial snowpack meltwater supplied heat to the organic mat, raising the temperature to the melting point, somwhat analogous to satisfying the cold content of the snowpack. No quantitative data are available for the amount of heat required from the meltwater. The time delay can also be attributed to dessication of the organic mat over the long, cold winter. Santeford (1978, 1979) has reported a redistribution of moisture from the organic mat to the overlying snowpack ranging from 19 to 23 mm over the winter season. Examination of the data from Kane et al. (1981) indicates that only a portion (5.9 mm) of the initial snowmelt was reabsorbed before percolation through the moss layer resulted in measurable runoff. The noteworthy point is that runoff from the moss-filled lysimeter began after only one-fourth of the total moisture deficit of the organic mat had been satisfied.

In previous lysimeter studies on permafrost areas, reported by Santeford (1979), a similar installation was used, but the lysimeters were installed without any slope. Santeford's data show a delayed response in runoff from the moss-filled lysimeter. A model was proposed based on the assumption that no drainage would occur until the water-holding capacity of the moss had been satisfied, at which point all incoming precipitation would percolate through the moss. The data reported here do not support such a model for the sloping portions of the watershed. In fact, the data indicate that the organic mat has a high unsaturated hydraulic conductivity, which results in rapid percolation of precipitation even when a large moisture deficit exists. The concept of a threshold water-holding capacity does not seem to apply. The data suggests that the moisture retention of the moss is a function of precipitation intensity and duration, but the limited data set prohibits any quantitative analysis.

Comparison with Glenn Creek

The HR ratios for Glenn Creek are also shown in Table 2 for the available record. These values fall on the low side of the range reported by Dingman (1971) for Glenn Creek for the summers of 1964-1967 and Haugen et al. (1982) for Caribou and Poker Creeks for the summer of 1977-1979. The fast response times agree with observations made in previous studies (Dingman 1971, Haugen et al. 1982).

The apparent discrepancy of the HR ratios between the lysimeter data and the watershed data can be attributed in part to the short travel distances in the lysimeter compared to travel distances in the watershed. The low HR ratios for the watershed can also be partially explained by the findings of Kane et al. (1978, 1981) and Slaughter and Kane (1979) that nonpermafrost areas contribute virtually no near-surface runoff during the snowmelt or summer seasons because the rate of infiltration exceeds the rate of moisture input. Luthin and Guymon (1974) indicated that the potential evapotranspiration from the birch-aspen forests of interior Alaska (nonpermafrost areas) exceeds the normal summer rainfall and that contribution to baseflow from these areas should be negligible. In the case of Glenn Creek, where 30% of the watershed is nonpermafrost, the runoff-contributing area can be reduced from 2.25 km^2 to 1.6 km^2, which raises the computed HR ratios slightly (Table 2). The bounded moss-filled lysimeter, which allows no subsurface inflow, may be considered representative of the driest parts of the watershed; that is, those parts of the watershed that are fully drained. Assuming uniform precipitation (Dingman 1971) and evapotranspiration throughout the watershed, the lysimeter should, therefore, be representative of the areas with the highest moisture deficit and the lowest hydrologic response ratio. The measurements at Glenn Creek show that the overall watershed hydrologic response is much lower than that in the moss-filled lysimeter. This lends support to the suggestions by Dingman (1971) and Kane et al. (1981) that only portions of the watershed area contribute to overland flow (shallow subsurface flow in the organic mat) and that the partial- and variable-source-area concepts of streamflow generation are applicable to interior Alaska.

SUMMARY AND CONCLUSIONS

An analysis of data on lysimeter runoff from permafrost-associated organic soils indicates that response time is short, drainage of saturated soils is prolonged, and runoff from slopes occurs before the total moisture deficit of the soil is satisfied. Although the slope of the lysimeters was near the average slope of the basin and they were isolated from subsurface inflow, the hydrologic response to storm events was much higher than that computed from the total watershed or from the permafrost areas only. The indication is that large portions of the watershed, including permafrost areas, do not contribute to streamflow.

The implications that runoff is produced from moisture-deficit soils and that partial watershed areas contribute to streamflow are critical in watershed modeling. Clearly, further studies are required to understand the relationship between the flow of water through organic soils and the moisture content of the soil. Lysimeter studies combined with observations of where in the soil profile downslope movement of water takes place may provide some insight to the processes of streamflow generation in permafrost areas.

ACKNOWLEDGMENT

This research was supported by the U.S. Army Corps of Engineers Cold Regions Research and Engineering Laboratory. The authors thank Stephen

Perkins of CRREL for computer programming and help in reduction of streamflow and lysimeter runoff data.

REFERENCES

Dingman, S.L., 1966a, Hydrologic studies of the Glenn Creek watershed, near Fairbanks, Alaska, CRREL Special Report 86: Hanover, N.H., U.S. Army Cold Regions Research and Engineering Laboratory.

Dingman, S.L., 1966b, Characteristics of summer runoff from a small watershed in central Alaska: Water Resources Research, v. 2, p. 751-754.

Dingman, S.L., 1971, Hydrology of the Glenn Creek watershed, Tanana River Basin, central Alaska, CRREL Research Report 297: Hanover, N.H., U.S. Army Cold Regions Research and Engineering Laboratory.

Dingman, S.L., 1973, Effects of permafrost on streamflow characteristics in the discontinuous permafrost zone of central Alaska, in Permafrost: the North American contribution to the Second International Conference: Washington, D.C., National Academy of Sciences, p. 447-453.

Ford, T.R., 1973, Precipitation runoff characteristics of the Caribou Creek research watershed near Fairbanks, Alaska, unpublished M.S. thesis: College, Alaska, University of Alaska.

Haugen, R.K., Slaughter, C.W., Howe, K.E., and Dingman, S.L., 1982, Hydrology and climatology of the Caribou-Poker Creeks Research Watershed, Alaska, CRREL Report 82-26: Hanover, N.H., U.S. Army Cold Regions Research and Engineering Laboratory.

Kane, D.L., Bredthauer, S.R., and Stein, J., 1981, Subarctic snowmelt runoff generation, in Proceedings of the Specialty Conference on the Northern Community, ASCE: Seattle, Wash., p. 591-601.

Kane, D.L., 1980, Snowmelt infiltration into seasonally frozen soil: Cold Regions Science and Technology, v. 3, p. 153-161.

Kane, D.L., Seifert, R.D., Fox, J.D., and Taylor, G.S., 1978, Snowmelt frozen soil characteristics for a subarctic setting: Institute of Water Resources, University of Alaska, Report No. IWR-84.

Luthin, J.N., and Guymon G.L., 1974, Soil moisture-vegetation-temperature relationships in central Alaska: Journal of Hydrology, v. 23, p. 233-246.

Santeford, H.S., 1978, Snow-soil interaction in interior Alaska, in Proceedings, Modeling of Snow Cover Runoff (S.C. Colbeck and M. Ray, eds.): Hanover, N.H., U.S. Army Cold Regions Research and Engineering Laboratory, p. 311-318.

Santeford, H.S., 1979, Toward hydrologic modelling of the black spruce/permafrost ecosystems of interior Alaska, presented at Third Northern Research Basin Symposium Workshop, The State-of-the-Art in Transposable Watershed Models: Quebec, Canada, LaVal University.

Slaughter, C.W., 1971, Caribou-Poker Creeks Research Watershed: Background and current status, CRREL Special Report 157 (AD 726373): Hanover, N.H., U.S. Army Cold Regions Research and Engineering Laboratory.

Slaughter, C. and Kane, D.L., 1979, Hydrologic role of shallow organic soils in cold climates, in Canadian Hydrology Symposium 79 - Cold Climate Hydrology Proceedings: Ottawa, Ont., National Research Council of Canada, p. 380-389.

U.S. Department of Commerce, 1979, Local climatological data, Alaska: Asheville, N.C., National Oceanic and Atmospheric Administration, v. 65, no. 7.

U.S. Department of Commerce, 1980, Local climatological data annual summary with comparative data, Fairbanks, Alaska: Asheville, N.C., National Oceanic and Atmospheric Administration.

Wahrhaftig, C., 1965, Physiographic divisions of Alaska, U.S. Geological Survey Professional Paper 482.

FROST HEAVE OF SALINE SOILS

E. J. Chamberlain

U.S. Army Cold Regions Research and Engineering Laboratory
Hanover, New Hampshire 03755 USA

Theories of ice segregation and frost heave processes in saline soils are briefly examined and modified to explain observations made on clay and sand soils frozen under laboratory conditions. Seawater was observed to reduce the rate of frost heave by more than 50% for both soil types and to dramatically reduce the size of ice lenses. The effect of seawater is to cause the formation of a thick active freezing zone with many ice lens growth sites, each with its own brine concentration. Unbonded brine-rich soil zones between ice lenses are identified as potential zones of low shear strength.

An understanding of the frost-heave and ice-segregation behavior of soils that contain saline pore water is important to the development of offshore petroleum resources in the Beaufort Sea. Understanding the freezing behavior of saline soils is also important to the artificial ground freezing industry. Unfortunately, little is known of ice segregation processes in saline soils that would allow design for frost heave and for changes in physical and mechanical properties.

Mahar et al. (1982) reported a modified form of the Berggren equation to predict frost penetration in saline soils where little or no frost heave occurs. This is important for determining where potential failure planes may occur in artificial islands. If significant amounts of ice segregation and frost heave occur, however, this method may overpredict the depth of freezing. In addition, the potential failure plane may not be forced below the region of freezing as commonly assumed. Partially frozen brine-rich zones within frozen layers may occur and they must also be considered as potential failure zones. A good example of this type of problem was recently observed at a ground freezing site (Maishman, personal communication) where, after excavation, a brine-rich clay layer was observed to slough back to the freezing pipes. Inspection of the site showed that the soil between ice lenses had little ice bonding and thus provided little strength for supporting the excavation.

Inspection of borehole logs obtained by Osterkamp and Harrison (1979) on Reindeer Island in the Beaufort Sea reveals that the occurrence of ice and ice bonding is sporadic and unpredictable, even though temperatures were below the freezing point of seawater.

This paper presents the results of a series of laboratory freezing tests on two soil types saturated with seawater. The freezing behavior is compared with freezing tests on the same soil types saturated with distilled water. Finally, an explanation for the unique freezing behavior of saline soils is offered.

FREEZING PROCESSES IN SALINE SOILS

The freezing of soils that contain saline pore water solutions is a complex process, due to the soluble salts in the pore water fluid. The effects of salts on freezing behavior extend well beyond simply lowering the freezing point. Salts are excluded from growing ice crystals and are concentrated in the adjacent pore fluids, so that ice segregation temperatures are lowered and additional sites for ice nucleation form at or near the original ice segregation temperature.

Hallet (1978) suggested that ice tentacles reach out from a morphologically complicated interface. Domains of solute-rich solutions can become isolated from the unfrozen pore-water solutions and eventually become trapped in solidly ice-bonded material. He suggested that significant solute partitioning will accompany frost penetration in frozen ground, with lower bulk concentrations occurring in ice-bonded layers, and increased concentrations of salt occurring in unfrozen soils beneath a growing ice lens. He further stated that constitutional supercooling in soils will lead to a situation where ice will nucleate and grow in a zone ahead of and separate from the freezing front.

Sheeran and Yong (1975) suggested that ice growth in pores that contain salt solutions requires progressively reduced temperatures because the increased salt concentration due to brine exclusion lowers the freezing temperature of the remaining adjacent pore water. Only partial freezing occurs at the freezing front. Substantial phase change may occur up to a meter behind the frost front as cooling continues, depending on the magnitude of the thermal gradient.

Mahar et al. (1982) concluded that the freezing front progressing through saline saturated soil is characterized by a transition zone of partially frozen soil grading from isolated ice crystals to ice-bonded soil. Continued ice growth requires progressively reduced temperatures. Because of the irregular shape of soil grain boundaries and the complex heat transfer pattern in a pore space, isolated brine pockets develop that may not freeze.

The situation in freezing saline soils is somewhat analogous to the frozen fringe concepts of Miller (1978) and Konrad and Morgenstern (1981). The hydraulic conductivity of the frozen fringe controls the availability of water to growing ice lenses. In the case of saline soils, however, the frozen fringe can be very thick, and ice accumulation probably occurs throughout the zone.

TABLE 1 Properties of the test materials.

Material	Percent finer than						Uniformity coefficient	Liquid limit %	Plastic limit %	Unified soil class
	2.0 mm	0.42 mm	0.074 mm	0.02 mm	0.005 mm	0.001 mm				
Morin clay	100	100	99	84	52	26	16	30	21	CL
Dartmouth sand	96	68	40	18	6	1	29	25	25	SM

TABLE 2 Specimen properties.

Material	Compaction properties			Saturated water content %
	Water content %	Dry density Mg/m^3	Void ratio	
Morin clay	9	1.51	0.84	30
Dartmouth sand	10	1.72	0.56	21

EXPERIMENTAL STUDIES

Two frost-susceptible soils, Dartmouth sand and Morin clay, were selected for this study. Dartmouth sand is a well-graded granular material containing numerous fines. The Morin clay material is classified as a clay of low plasticity. The properties of each of these materials are given in Table 1.

Replicate specimens of both of these materials were prepared using distilled water and seawater solutions. The samples were compacted in layers in 150-mm-diameter multiring Plexiglas cylinders, 150 mm high, lined with rubber membranes. The specimen properties are shown in Table 2. After compaction, the test specimens were saturated with the appropriate water solution under vacuum.

Calibrated thermocouples were inserted at 25-mm intervals through the sides of the rings to monitor the progress of freezing. The appropriate distilled water or seawater solution was made freely available through a porous stone at the base of the sample. Cooling plates connected to refrigerated circulating baths were placed at the top and bottom of each sample, and the entire assembly was placed in an insulated box in a cold room with a 0°C ambient temperature. A small surcharge of 24 kPa was placed on top of each sample.

Freezing was accomplished in three stages: (I) constantly decreasing boundary temperatures; (II) fixed boundary temperatures; and (III) rapidly decreasing boundary temperatures.

During Stage I, a temperature gradient of approximately 0.025°C/mm was propagated downward through the test specimens at a frost penetration rate of approximately 1.5 mm/hr until the zone of freezing was within the middle 50 mm of the test specimen. During Stage II, the boundary temperatures were held fixed, imposing the same temperature gradient for at least 100 hr. Finally, in Stage III, the samples were frozen rapidly to the bottom. The purpose of this freezing procedure was to establish frost heave and brine exclusion characteristics under a constant rate of frost penetration, then to force the growth of ice lenses and the exclusion of salts, and finally to trap the excluded brine for further analysis.

PRESENTATION AND DISCUSSION OF RESULTS

Four freezing tests were conducted on each of the clay and sand materials, two each for the distilled water and seawater solutions. Typical results for each of the soils and pore water solutions are illustrated in Figures 1 through 4. Individual test results are discussed below.

Morin Clay

The Morin clay samples saturated with distilled water heaved rapidly at nearly 12.5 mm/day during Stage I (Figure 1). The frost-heave rate during Stage II of freezing gradually slowed to approximately 2 mm/day. Many small ice lenses developed during Stage I (Figure 1), and a very large ice lens, nearly 50 mm thick, developed during Stage II. The growth of the ice lens during Stage II was shut off by the depletion of the water content in the unfrozen zone below to the plastic limit w_p.

The frost-heave rate for the Morin clay saturated with seawater was 60% lower than that for the distilled water (Figure 2). During Stage I of freezing, the heave rate for the seawater was approximately 4.5 mm/day; it fell below 1 mm/day during Stage II. No large ice lenses were observed to form during either of these stages. Water contents were limited to 50% in the ice-lensed zone (vs. nearly 400% for the distilled water case), and to 35% in the zone frozen rapidly during Stage III. Salinities were considerably partitioned by the freezing process, being reduced to as little as two-thirds of the original value (34.6%) in the ice-lensed zone and increased by nearly 25% in the zone beneath the active ice lens growth. It should be noted that the salinities were determined for 15-mm-thick slices and represent bulk salinities of the soil, ice, and unfrozen water system. Salinity partitioning within smaller elements of each slice was not determined.

Dartmouth Sand

The Dartmouth sand froze with a 50-mm zone of fine- to medium-size ice lenses (Figure 3), most of which formed during Stage I of freezing. The fixed boundary conditions of Stage II could not induce the growth of large ice lenses. The frost-heave rate during Stage I was approximately 8 mm/day, but it fell to nearly 2 mm/day during Stage II.

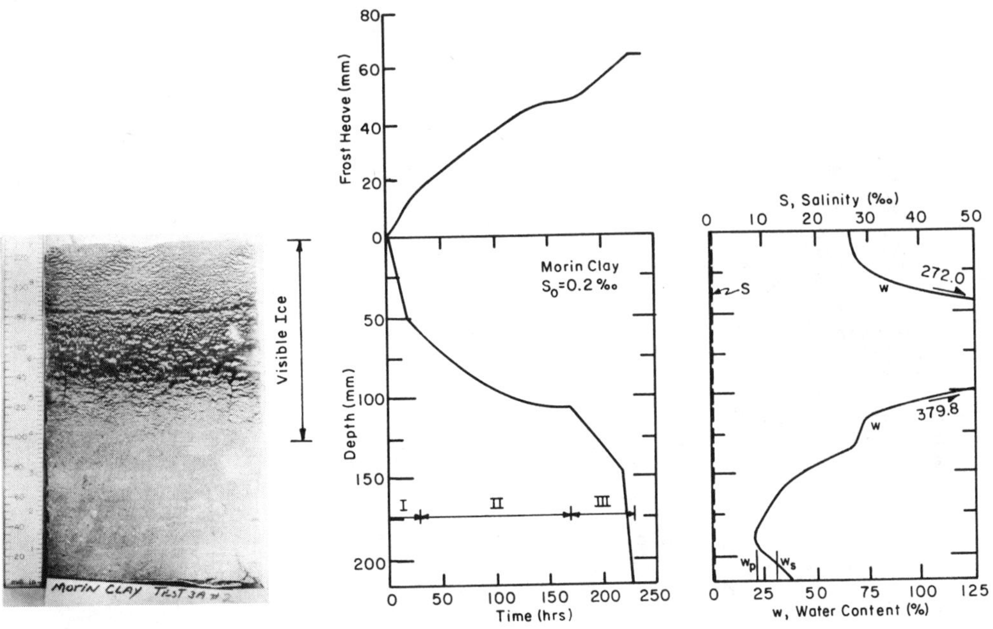

FIGURE 1 Freezing test results for Morin clay saturated with distilled water. Natural salinity S_o is 0.2%. Plastic limit water content w_p and saturation water content w_s are shown for comparison.

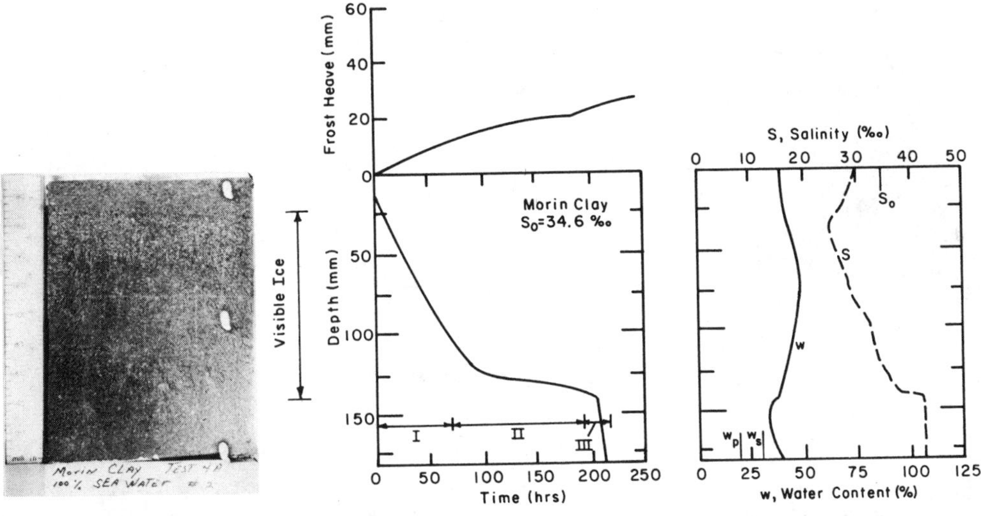

FIGURE 2 Freezing test results for Morin clay saturated with seawater. Initial salinity S_o is 34.6%.

Water contents in the visible ice region were as much as 60%, whereas in the zone rapidly frozen during Stage III they were reduced to less than the original saturated water content of 21%.

Addition of the seawater solution to the Dartmouth sand markedly reduced the frost-heave rate and changed the ice segregation characteristics as it did with the Morin clay. The heave rate during Stage I was less than 3.5 mm/day and it fell to less than 0.5 mm/day during Stage I (Figure 4). Little visible ice was evident. The maximum water content was 30%, which indicates that some ice segregation occurred (the initial saturation water content was 21%). The water content in the rapidly frozen zone formed during Stage III was only slightly above the initial water content, which indicates that little ice segregation occurred there.

The salinity profile (Figure 4) is partitioned to approximately 85% of the original salinity (S_o = 35.7%) in the upper portions frozen during Stage I and the early part of Stage II, and increases to 120% of S_o in the region frozen during Stage III.

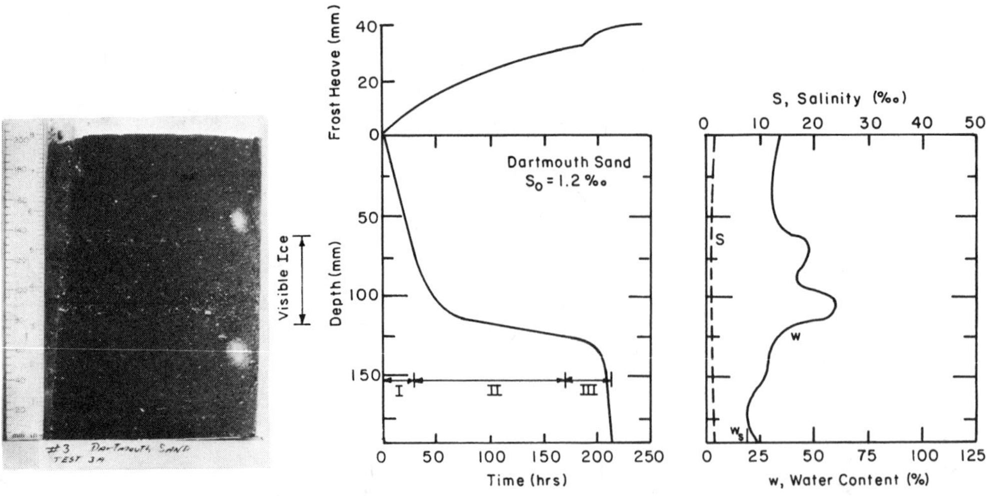

FIGURE 3 Freezing test results for Dartmouth sand prepared with distilled water. Natural salinity S_o is 1.2%.

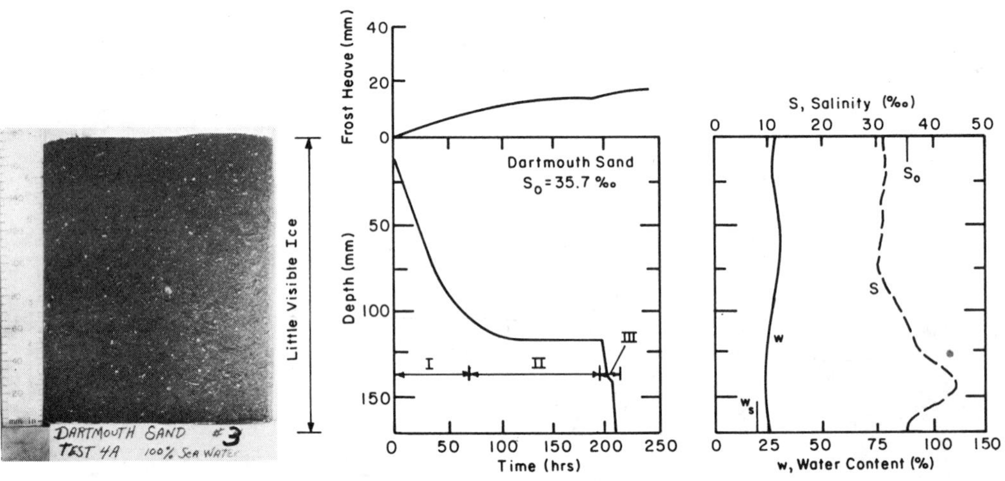

FIGURE 4 Freezing test results for Dartmouth sand prepared with seawater. Initial salinity S_o is 35.7%.

CONCEPTUAL CHARACTERIZATION OF THE EFFECTS OF SALTS ON FROST HEAVE

A few explanations for the effects of salts on frost heave in soil materials were briefly reviewed earlier in this paper. A refinement of these explanations is suggested to account for the reduced ice segregation potential of saline soils and to explain the apparent simultaneous formation of several ice lenses witnessed by Mahar et al. (1982) and shown in the recent freezing tests.

As Hallet (1978) suggested, the solute concentration in saline soils is partitioned by freezing, with solutes being rejected to the surface of the growing ice lens. A schematic diagram of the freezing process is shown in Figure 5. Ice first nucleates when the temperature of the saline pore water T_w falls below the initial equilibrium freezing temperature T^o_{eq} (Figure 5a). As T_w is lowered with time, the ice lens grows, excluding salt into the unfrozen zone below while trapping some brine within the ice.

A second ice lens formation site develops (Figure 5b) when T_w falls below the equilibrium freezing temperature beneath the brine concentration. Ice will continue to grow at the first site at progressively reduced temperatures, concentrating salts in the unfrozen soil layer between the two ice growth sites until the hydraulic conductivity of this layer is insufficient to meet the demand. As T_w continues to be lowered, a second brine concentration forms and a third ice growth site develops below it (Figure 5c). Continuation of this process will result in the growth of many ice lenses concurrently. Growth of the uppermost active ice lens will stop when water is no longer available.

The average equilibrium freezing temperature for all the equivalent moisture within the frozen

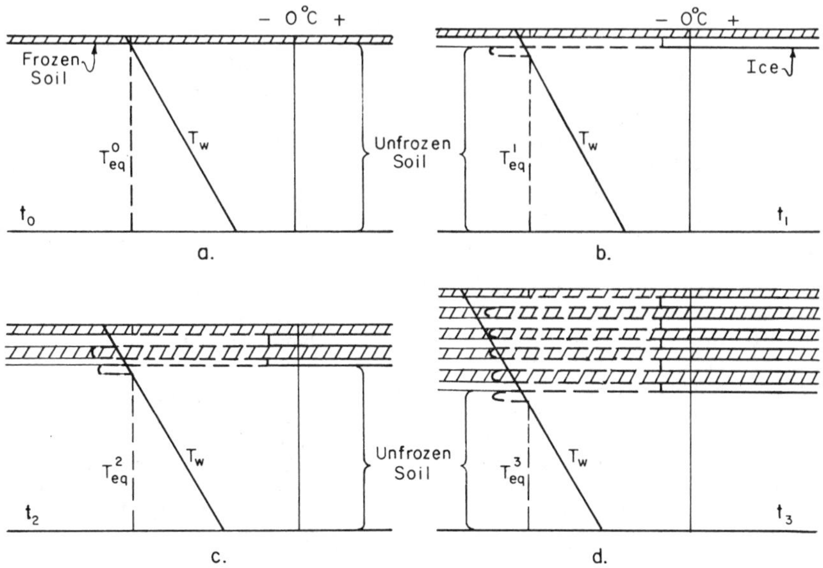

FIGURE 5 Schematic diagram of ice segregation process in saline soils. (a) At time zero (t_o), the pore water temperature profile is T_w and a thin soil layer has frozen; (b) at time t_1, an ice lens is growing and a freezing point depression bulb has formed beneath it due to the exclusion of salts. The segregation freezing temperature at the first ice lens growth site decreases because of the brine concentration, and a second ice nucleation site forms; (c) at time t_2, a third ice nucleation site has formed below the freezing point depression bulb formed by the second ice lens; (d) the process continues at time t_3, with many ice lens growth sites. Ice lens growth at the original nucleation site is shut off by reduced hydraulic conductivity.

zone is higher than the initial condition T^o_{eq} because of the net exclusion of brine. The equilibrium freezing temperature within the soil elements is lower than T^o_{eq} because of brine concentrations in the unfrozen water.

The growth of large ice lenses is restricted by this process because the many small ice lenses and entrapped soil layers lower the hydraulic conductivity within the active freezing zone. Furthermore, the increasing concentration of salts near each ice lens growth site requires progressively lower temperatures to continue growth, whereas progressively lower temperatures reduce the hydraulic conductivity. The flow of water to the uppermost growing ice lens is thus shut off before the lens can develop significant thickness.

FURTHER DISCUSSION OF RESULTS

The water content profiles in Figures 1 through 4 give some evidence of the reduced hydraulic conductivity of the freezing zone in saline soils. For Morin clay prepared with distilled water (Figure 1), the water content in the soil below the zone of high ice segregation has been reduced to the plastic limit water content, which is the practical limit to which a soil can be desiccated by the freezing process (Chamberlain 1981). For the case of the saline Morin clay (Figure 2), the water content in the soil beneath the active ice segregation zone actually increased over the initial saturated (w_s) value during the passing of the freezing front. This indicates that there is a tendency to desiccate the soil elements within the high suction zone of freezing in saline clay soils and to reduce the hydraulic conductivity there, whereas in the clay soil prepared with distilled water the highly desiccated zone lies below the active freezing front.

IMPLICATIONS

Under natural freezing conditions with very small temperature changes with depth, such as within islands in the Beaufort Sea, thick zones of active freezing can occur with alternating zones of ice-bonded and salt-enriched unbonded soils.

From an engineering design point of view, the reduced frost-heave susceptibility of saline soils is a positive factor. However, if the design of an artificial island is predicated on the development of the ice-bonded strength of the fill material, unfrozen zones with brine concentrations may cause weak zones to form above the freezing front.

Under artificial freezing conditions for construction purposes, unfrozen zones of brine-enriched soil can also occur, particularly in clay soils. These unfrozen zones can cause failure of a freeze-wall, even if they are only a few millimeters thick.

CONCLUSIONS

The frost-heave susceptibility of sand and clay soils is significantly reduced by saline pore water. Salts also dramatically reduce the ice lens size and decrease the segregation water content.

The effect of saline seawater on frost heave and ice segregation can be explained in terms of a thick active freezing zone with many ice lens growth sites, each with its own brine concentration. The thick active freezing zone reduces the hydraulic conductivity, and thus the rate of frost heave.

The brine concentrations can cause zones of unbonded soil to occur within an ice-bonded material. These zones of brine concentration can cause planes of low shear strength to occur in frozen structures. These potential failure planes must be considered by engineers designing artificial islands to resist ice forces or ground-freezing projects to stabilize excavation sites.

ACKNOWLEDGMENTS

This work was supported by the Mineral Management Service of the U. S. Department of the Interior as part of their Outer Continental Shelf Oil and Gas Operations Research and Development Program. The author wishes to acknowledge the assistance of Richard Roberts with the laboratory studies.

REFERENCES

Chamberlain, E.J., 1981, Overconsolidation effects of ground freezing: Journal of Engineering Geology, v. 18, p. 97-110.

Hallet, B., 1978, Solute redistribution in freezing ground, in Proceedings of the Third International Conference on Permafrost, v. 1: Ottawa, National Research Council of Canada, p. 86-91.

Konrad, J.-M., and Morgenstern, N.R., 1981, The segregation potential of freezing soil: Canadian Geotechnical Journal, v. 18, p. 482-491.

Mahar, L.J., Vinson, T.S., and Wilson, R., 1982, Effects of salinity on freezing of granular soils, preprint prepared for the Third International Symposium on Ground Freezing: Hanover, N.H., U.S. Army Cold Regions Research and Engineering Laboratory.

Maishman, D., 1983, Personal communication: Rockaway, N.J., FreezeWALL Inc.

Miller, R.D., 1978, Frost heaving in non-colloidal soils, in Proceedings of the Third International Conference on Permafrost: Ottawa, National Research Council of Canada, v. 1, p. 708-713.

Osterkamp, T.E., and Harrison, W.D., 1979, Subsea permafrost; probing, thermal regime and data analysis, in Environmental Assessment of the Alaska Continental Shelf, Annual Reports of Principal Investigators; Washington, D.C., National Oceanic and Atmospheric Administration, U.S. Department of Commerce, v. IX, p. 493-580.

Sheeran, D.E., and Yong, R.N., 1975, Water and salt redistribution in freezing soils, in Proceedings, First Conference on Soil-Water Problems in Cold Regions, May 6-7: Calgary, Alberta, Canada, p. 58-69.

HOLOCENE TEMPERATURES IN THE UPPER MACKENZIE VALLEY DETERMINED BY OXYGEN ISOTOPE ANALYSIS OF PEAT CELLULOSE

Stephen C. Chatwin

Agriculture Canada, Tanzania-Canada Wheat Project,
P.O. Box 6160, Arusha, Tanzania.

Oxygen isotope ratios in nonexchangeable plant cellulose extracted from samples of Sphagnum are positively correlated with the Mean Annual Temperature at their respective growth sites. Furthermore, the isotope ratio of aquatic plant cellulose is independent of the plant species. These results were used to extract a paleotemperature record, spanning the last 10,300 years from an Upper Mackenzie Valley peat core. Temperature does not appear to have been an impediment to permafrost aggradation in this area for the period of record. A comparatively cool climate between 10,000 and 7,000 years B.P. was followed by a warming trend between 6,000 and 5,000 years B.P. Cooling followed, reaching minimum Holocene temperatures approximately 3,500 years B.P. Preliminary results suggest this temperature was 3°-4° cooler than present day Mean Annual Temperatures. Steady warming since that date has resulted in widespread permafrost degradation. Present day temperatures are the warmest that have occurred during the Holocene for the area.

INTRODUCTION

Paleoclimatic information is vital in deciphering the history of permafrost aggradation and degradation in Sub-Arctic peat lands. Recent studies suggest paleotemperature data may be inferred from isotope analysis of plant cellulose. The natural deuterium and heavy oxygen contained in plant cellulose reflects the isotopic ratio of the water incorporated by the plant during synthesis of cellulose (Epstein et al 1976, Thompson and Gray 1977). Furthermore, variation in the ratios of heavy to light isotopes in meteoric water is largely a response to temperature changes at the site of precipitation, decreasing with mean annual temperature (Dansgaard 1964). It follows that the isotopic variations in plant cellulose may reflect the temperature of precipitation at the growth site.

The climatic significance of D/H ratios (Epstein and Yapp 1976) and $^{18}O/^{16}O$ ratios (Gray and Thompson 1976) have subsequently been demonstrated for tree rings grown in different climates. The technique was further applied to buried wood samples in an effort to examine late Wisconsin climates (Yapp and Epstein 1977).

This paper examines the isotopic relationship between climate and peat forming plants and then uses the technique, through analysis of a palsa core, to estimate paleotemperatures in the Upper Mackenzie Valley over the past 10,000 years.

METHODS OF ANALYSIS

Oxygen isotope analysis was conducted on 15 samples taken from a 3.5 m core of frozen peat that was collected from a palsa complex near Ft. Simpson, N.W.T. (latitude $61^{o}28'$N, longitude $120^{o}55'$E). The upper 1.5 m of the core was slightly decomposed, the lower 1.3 m was moderately decomposed, and the bottom 0.2 m was well decomposed. Seven isotopic analyses were conducted on the aquatic plant species presently growing at the palsa site. Additional Sphagnum samples from various North and South American localities were supplied by D. Vitt, University of Alberta, and were also subjected to isotopic analysis.

Cellulose was extracted from the plant samples using a modification of a method described by Theander (1954). The method yields a greyish-white to light grey substance. An X-ray diffraction pattern of the substance reveals strong peaks in the range characteristic of alpha-cellulose. Following extraction, approximately 25 mg of vacuum dried sample is weighed into a nickel boat, placed under vacuum, then combusted in a reaction vessel at 1150°C for 1 hour (see Thompson and Grey 1977). The reaction produces a mixture of CO, CO_2, and H_2 gases. The H_2 gas diffuses through the walls of the reaction vessels, forcing the reaction to completion. After combustion, the gases are trapped in a silica gel maintained at liquid nitrogen temperature. The CO fraction is converted to CO_2 and C by sparking between platinum electrodes. The CO_2 fractions are combined and the yields measured manometrically. The yields averaged 90-95% of the theoretical predictions, indicating trace amounts of contaminant. All peat samples were run as replicates; if replicate results did not agree, a third sample was run. The precision of the replicate analysis was better than $\pm$ 0.4%. Isotope measurements were made on a 90° sector, 25 cm double collecting isotope ratio mass spectrometer. All isotope measurements are expressed as a value (δ/oo) with respect to Standard Mean Ocean Water (SMOW).

Radiocarbon determinations were made by the isotope laboratory of Waterloo University. The samples received standard treatment, including washing with distilled water, removal of organic acids with NaOH, and removal of carbonates with HCl. The ages are corrected and given as years B.P.

RESULTS

Effect of Different Plant Species

Peat is usually composed of a variety of plant species. To compare the oxygen isotope ratios of individual peat forming species, cellulose from samples of all the major aquatic species growing at the palsa site was analyzed. The results (Table 1) show that within the limits of analytical error, all the species have very similar oxygen isotope values. This is a very convenient result, as it means cellulose from bulk peat samples can be analysed irrespective of species composition.

TABLE 1 Isotopic Composition of Modern Peat Forming Plants

Plant Species	$^{18}O_{SMOW}$ $\pm 0.2^o/oo$
Sphagnum fuscum	18.65‰
Sphagnum recurvum	18.51‰
Drepanocladus	18.37‰
Carex aquatilis	18.40‰
Carex (various spp.)	18.47‰
Duckweed	18.39‰
Mixture	18.48‰

Analysis of the Palsa Core

Results from replicate analysis of 15 samples from various depths along the frozen peat core are listed in Table 2. The isotopic results are plotted with respect to depth, together with a radiocarbon age-depth profile and an illustration of the peat stratigraphy (Figure 1).

TABLE 2 Peat Oxygen Isotope Results

Sample	Depth (m)	Replicates	$^{18}O_{SMOW}$	Standard deviation
E 3-1	.05	3	16.80	.40
E 3-2	.20	2	17.02	.14
E 3-3	.51	2	16.58	.39
E 3-4	.77	2	16.17	.10
E 3-5	1.00	2	16.66	.33
E 3-6	1.25	2	16.44	.32
E 3-7	1.42	2	16.05	.19
E 3-8	1.63	2	15.38	.40
E 3-9	1.82	3	15.85	.15
E 3-10	2.12	2	16.36	.35
E 3-11	2.30	3	16.84	.20
E 3-12	2.50	2	16.45	.31
E 3-13	2.75	4	15.55	.27
E 3-14	3.15	2	15.70	.15
E 3-15	3.45	3	15.50	.41

The forest peat is a mixture of Ledum groenlandica leaves, Picea mariana needles, Vaccinium spp. and Cladonia spp. Forest peat is presently forming subaerially on the frozen palsa surfaces. No isotopic measurements were made of the upper forest peat layer. The Sphagnum layer is very homogenous with only occasional twigs of Ledum. Present day Sphagnum bogs are only found in "thaw bogs" entirely surrounded by the palsa complex. Brown moss and sedge peat are typically found in the unfrozen fens and string fens surrounding the palses. Aquatic peat includes well decomposed aquatic mosses, duckweed and algae as is found in open shallow ponds within both the palsa complex and the fens.

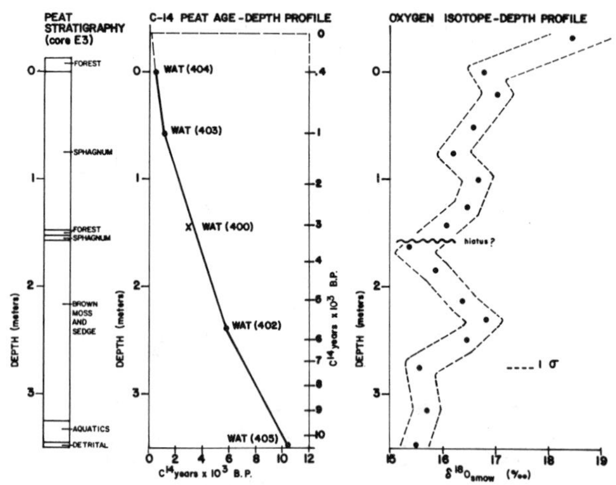

FIGURE 1 Change in oxygen isotope ratios of peat cellulose with depth of sample in the palsa. Five C-14 measurements provide time-depth control. Peat stratigraphic names are explained in the text.

In constructing the age-depth profile, rates of peat accumulation were assumed linear between dated samples. Peat accumulation began 10,380 years B.P., terminating approximately 400 years B.P. Results from analysis of the aquatic plants growing at the palsa site were used to extrapolate the isotope curve to the present day.

The oxygen isotope curve is characterized by relatively little variation throughout the profile. Isotopic values were approximately 3 $^o/oo$ lower between 10,000 and 7,000 years B.P. This was followed by a warming trend culminating between 6,000 and 5,000 years B.P. Minimum isotopic values are recorded between approximately 4,000 and 3,000 years B.P. Interestingly, this corresponds with a change in peat type that suggests temporary freezing of the bog. A gradual increase in the oxygen isotope ratios has occurred in the upper 1.5 m of the core, suggesting steady climatic warming. Plants growing at the palsa site have higher isotope ratios than any of the peat samples.

Calibration of the Isotope Curve

Before paleo-temperature interpretations can be made of the oxygen isotope fluctuations in the peat column, a temperature coefficient, linking temperature change to isotope change, must be found for peat cellulose. A temperature coefficient of 1.3°/oo ^{18}O/°C for Edmonton spruce (Gray and Thompson 1976) was found by comparing isotopic results from individual tree rings with modern temperature records. Unfortunately, peat cannot be calibrated in the same way, as it is impossible to separate yearly accumulations.

A method of deriving a temperature coefficient is to plot the $^{18}O/^{16}O$ ratio of plants from various locations versus the mean annual temperature at those sites.

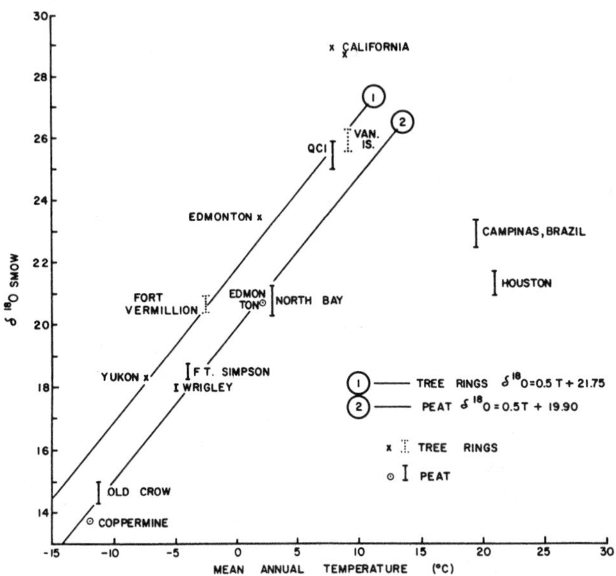

FIGURE 2 The oxygen isotope ratio of Sphagnum and tree rings as functions of the mean annual temperature of their growth sites.

Sphagnum samples from a wide variety of climatic zones were analysed; the results are plotted against Mean Annual Temperature (average of last 5 years) in Figure 2. For some locations, Sphagnum samples were available from more than one bog; the range of results at these locations is plotted as a bar. On the same graph, oxygen isotope values of modern tree ring cellulose from various sites are plotted (P. Thompson, unpublished results). This data set includes the previously mentioned Edmonton spruce analyzed by Gray and Thompson (1976). The isotopic values from northern continental North America (Edmonton, Alberta: North Bay, Ontario; Fort Simpson, N.W.T.; Old Crow, Yukon; and Coppermine, N.W.T.) are reasonably defined by the linear relationship

$$\delta^{18}O = 0.5T + 19.90 \qquad (1)$$

The values for subtropical and tropical localities (Houston, Texas; Campinos, Brazil) fall significantly below this line, while values from a Pacific maritime area (Queen Charlette Islands, B.C.) fall above the line.

DISCUSSION

The results summarized in Figure 2 establish that oxygen isotope ratios for peat cellulose increase as Mean Annual Temperatures increase. That is, the temperature coefficient is positive. The regression equation (1) cannot be used to directly infer paleo-temperatures, however. While Mean Annual Temperature is the prime determinant of the isotopic ratios of precipitation at a particular site, other factors such as thermal history of the air mass (Dansgaard 1964) the seasonality of the annual precipitation, and the evaporation and evapotranspiration rates (Epstein et al 1976) will locally alter the general temperature coefficient. This becomes evident with a comparison of Gray and Thompson's (1976) site specific temperature coefficient of 1.3°/oo/°C for Edmonton tree rings and the regression line of 0.5°/oo/°C connecting the various sites (Figure 2).

The parallel regression lines for tree rings and Sphagnum suggest that their temperature coefficients may also be similar. If 1.3°/oo/°C is a roughly valid coefficient for peat forming plants, then a minimum Holocene temperature of approximately 3-4°C colder than present day is indicated by the paleo-isotope curve (Figure 1) for the Ft. Simpson area. Considerably more work in controlled environment growth chambers is required before reliable temperature coefficients can be established.

Any large change in the humidity or evapotranspiration rate at the Ft. Simpson bog during the Holocene will also affect the paleo-isotope curve. In peatlands the humidity is always high, however, and calculated changes in evapotranspiration over a 3°-4°C temperature range are also very small. These effects therefore should not markedly change the shape of the curve (Figure 1). If enrichment due to increased evapotranspiration rates has occurred, the effect will be to overestimate Mean Annual Temperature.

Climate changes during the Holocene may have included changes in the seasonality of the annual precipitation. Such changes will alter the δ values in peat cellulose, which may or may not be related to the Mean Annual Temperature.

There is also a possibility that oxygen exchange occurred between the C=O bonds in the peat cellulose and percolating bog waters or with bacteria during decomposition. If such exchange did occur, the result would be to homogenize the isotope values in the peat. Experimental evidence (Gray and Thompson 1976, Epstein et al 1976,

Chatwin 1981) suggests that such exchange is very unlikely, as the C=O molecular bond is covalent and very strong. Furthermore, Yapp and Epstein (1977) found no evidence for isotope exchange having occurred in wood that was buried in wet soils for 11,800 years.

Holocene temperature trends indicated by the oxygen isotope curve are somewhat at variance with some paleoclimatic studies completed in the Mackenzie Delta area, approximately 1,500 km to the north. Here, there is reasonably strong evidence for mean summer temperatures having been higher than present day (Ritchie and Hare 1971, Mackay 1978). Others (Delorme et al 1977) have suggested very warm conditions in the early Holocene. This study indicates climatic warming beginning in the early Holocene, up to 5,500 years B.P.; however, the predicted mean annual temperatures are less than present day values. Part of this discrepancy may be due to previous investigators having examined changes in mean summer temperatures, whereas this study is concerned with mean annual temperatures.

Another possibility for this discrepancy is that in the early Holocene glacial meltwaters may still be contributing to the hydrology of the peatland. This reservoir of light δ value water would result in peat isotope ratios lower than expected from the mean annual temperature.

CONCLUDING STATEMENTS

Oxygen isotope analysis of peat cellulose may be a powerful tool in deciphering the history of permafrost aggradation and degradation. Analysis of a palsa core from Ft. Simpson in the Upper Mackenzie Valley suggests that temperature has not been an impediment to permafrost formation during most of the Holocene. There is no evidence for Mean Annual Temperatures warmer than present day during the past 10,000 years. Other factors, in particular the local hydrology or vegetative cover, may be responsible for the limited evidence for permafrost in this area in the early Holocene. Maximum permafrost development probably occurred approximately 3,500 years B.P. when minimum Holocene temperatures were experienced in this area. Preliminary estimates are that Mean Annual Temperatures were approximately 3°-4°C colder than present day. The recent widespread degradation of permafrost is the result of steady climatic warming, in particular over the past 400 years.

ACKNOWLEDGMENTS

This project was funded by the Boreal Institute for Northern Studies and the Geological Survey of Canada. John Gray and Peter Thompson, University of Alberta, provided laboratory facilities and valuable discussion.

REFERENCES

Chatwin, S.C., 1981, Permafrost Aggradation and Degradation in a Sub-Arctic Peatland. MSc Thesis, University of Alberta, Edmonton.

Dansgaard, W., 1964, Stable isotopes in precipitation. Tellus, v. 16, p. 436-468.

Delorme, L.D., Zoltai, S.C. and Kalas, L.L., 1977, Freshwater shelled invertebrate indicators of paleoclimate in northwestern Canada during late glacial times. Canadian Journal of Earth Sciences, v. 14, p. 2029-2046.

Epstein, S. and Yapp C.J., 1976, Climatic implications of the D/H ratio of hydrogen in C-H groups in tree cellulose. Earth and Planet Science Letters, v. 30, p. 252-266.

Epstein, S., Yapp, C.J. and Hall, J.H., 1976, The determination of the D/H ratio of non-exchangeable hydrogen in cellulose extracted from aquatic and land plants. Earth and Planet Science Letters, v. 30, p. 241-250.

Gray, J. and Thompson, P., 1976, Climatic information from $^{18}O/^{16}O$ ratios of cellulose in tree rings. Nature, v. 262, 481-482.

Mackay, J.R., 1978, Freshwater shelled invertebrate indicators of paleoclimate in northwestern Canada during late glacial times (discussion). Canadian Journal of Earth Sciences, v. 15 p. 461-463.

Ritchie, J.C., and Hare, F.K., 1971, Late Quaternary vegetation and climate near the arctic tree line of northwestern North America. Quaternary Research, v.1, p. 331-342.

Theander, O., 1954. Studies on sphagnum peat III. A quantitative study on the carbohydrate constituents of sphagnum moss and sphagnum peat. Acta Chemica Scandinavica Series B, v. 8, p. 989-1000.

Thompson, P. and Gray J., 1977, Determination of $^{18}O/^{16}O$ ratios in compounds containing C, H and O. International Journal Applied Radiation Isotopes, v. 28, p. 411-415.

Yapp, C.J., and Epstein, S., 1977, Climatic implications of D/H ratios of meteoric water over North America (9500-22,000 B.P.) as inferred from ancient wool cellulose C-H hydrogen. Earth Planet Science Letters, v. 34, p. 333-350.

INFLUENCE OF PENETRATION RATE, SURCHARGE STRESS, AND GROUND WATER TABLE ON FROST HEAVE

Chen Xiaobai, Wang Yaqing, and Jiang Ping

Lanzhou Institute of Glaciology and Cryopedology, Academica Sinica
People's Republic of China

To protect structures in permafrost and seasonal frost regions from damage, the authors studied the factors that influence the frost-heave process by conducting experiments on various types of soils, such as loess, clayey loam, loam, sandy loam, and sand, both in the laboratory and in situ. The results showed that the frost penetration rate, surcharge stress, and groundwater table are important factors, in addition to soil particle size. Two critical penetration rates divide the extent of the frost-heave ratio into intensive, slow, and no changing stages, and indicate whether ice will be segregated in the soil. Under surcharge stress, because the freezing point of the soil water is lowered and the specific suction water ratio in the soil at the freezing front decreases, the heave ratio descends, as shown by Equation 7. The influence of the ground water table on frost heave of soils is obvious and must be considered in practice.

Protecting structures from frost damage is one of the important problems in cold regions, promoting the study of the factors influencing frost heave processes, such as particle size, frost penetration rate, surcharge stress and groundwater depth. Many investigators are engaged in these subjects (see references). In this paper all of the above factors, except particle size, are discussed.

TEST CONDITIONS

The experiments were conducted both in laboratory and in situ, including permafrost and seasonal frost regions.

In the laboratory, the samples were made of loess (Lanzhou), red clayey loam (Qingzang Plateau), loam (Zhangyi) listed in Table 1, and medium (for building) sand. The sample cell for the frost susceptibility test is made of plexiglass with a height of 13.0 cm, a top diameter of 11.6 cm and a bottom diameter of 11.0 cm. A permeable plate was put under each sample for water supply in open system. The cell was surrounded with a plastic foam insulator to ensure one-dimensional freezing. Under surcharge stress, besides above condition, the samples were tested in consolidation apparatus using a plexiglass cell with a height of 8.9 cm, a top diameter of 8.5 cm and a bottom diameter of 8.3

TABLE 1 Physical Properties of the Soil Samples

No	Soil Type	Location	Percent by weight for the following particle sizes (mm)							Liquid limit	Plastic limit
			0.5-0.25	0.25-0.10	0.10-0.05	0.05-0.01	0.01-0.005	0.005-0.002	0.002	%	%
1	Loess	Lanzhou		15.83	23.94	25.45	14.50	6.08	14.20	23.0	15.5
2	Red Clayey loam	Qingzang Plateau		13.11	10.90	21.14	11.15	15.15	28.55	25.1	16.0
3	Loam	Zhangyi		2.7	17	38	12.3	29.7		33.6	23.0
4	Sandy loam	Zhangyi		34	30	21	4	11		27.0	19.5
5	Fine sand	Zhangyi		37.3	59.4	2.8				23.7	19.0

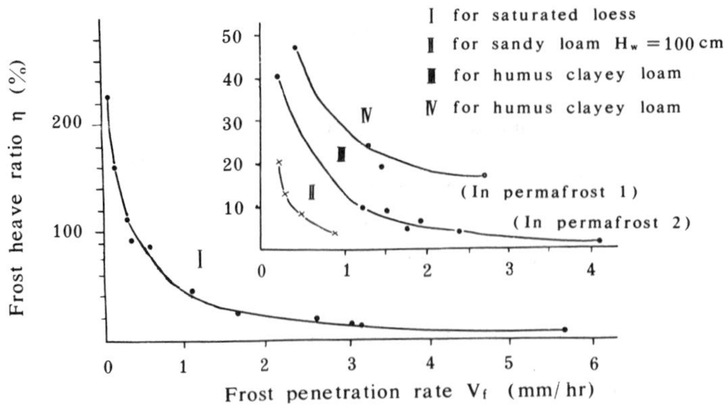

Figure 1. Frost heave ratio vs frost penetration rate.

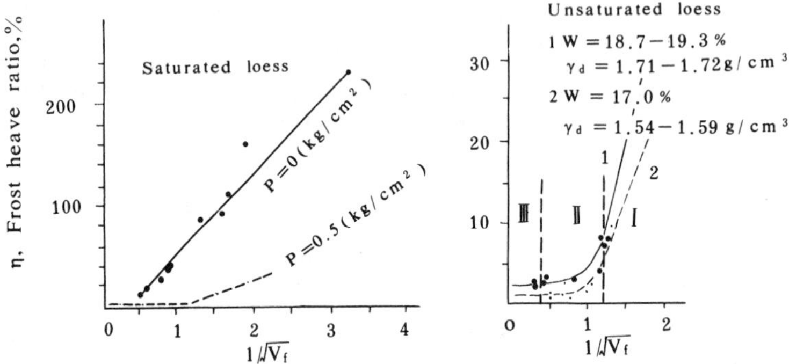

Figure 2. The relationship between frost heave ratio and frost penetration rate.

cm. A plexiglass tube with a diameter of 11 cm and a height of 120 cm was used to measure the redistribution of moisture along depth under various water tables during freezing.

The field work was conducted mainly in the Qilian Mts. (a permafrost region), and in the Zhangyi district (a seasonal frost region) with humus clayey loam, clay, clayay loam, loam, sandy loam and fine sand etc. (some of which are listed in Table 1). The process of frost penetration, air and ground temperature, frost heave (by layer), ground water table, and moisture content along depth, etc were observed.

RESULTS AND ANALYSES

Frost Heave Ratio η and Frost Penetration Rate Vf

The results of our experimental research show that whether for saturated or un-saturated soil, and in laboratory or in situ including permafrost and seasonal frost regions, there is a good relationship between the frost heave ratio η and frost penetration rate V_f (see Fig. 1, where curve I is for saturated loess in an open system). The frost heave ratio decreases with the increase of the penetration rate, which can be expressed as follows,

$$\eta = K / V_f \qquad (1)$$

where K=const. For unsaturated soil in the laboratory or in situ, because of water migration when frost penetrates, the moisture condition along the depth will vary slightly. So even though the curve shapes are similar, there are still some differences from that of the saturated case (see Fig. 1, curves I to IV).

If the relation between η and V_f is as shown in Figure 2, we may divide the development of frost heave ratio into three stages:

I. when $V_f \leq V_{f1}$, i.e. $(1/\sqrt{V_f}) \geq (1/\sqrt{V_{f1}})$,

then $\eta = \eta_0 + \Delta\eta + c\left[1/\sqrt{V_f} - 1/\sqrt{V_{f1}}\right]$ (2)

II. when $V_{f2} \geq V_f \geq V_{f1}$, i.e. $(1/\sqrt{V_{f1}}) \geq (1/\sqrt{V_f}) \geq (1/\sqrt{V_{f2}})$,

then $\eta = \eta' = \eta_0 + \Delta\eta \left[(1/\sqrt{V_f} - 1/\sqrt{V_{f2}}) / (1/\sqrt{V_{f1}} - 1/\sqrt{V_{f2}})\right]^3$ (3)

III. when $V_f \geq V_{f2}$, i.e. $(1/\sqrt{V_f}) \leq (1/\sqrt{V_{f2}})$,

then $\eta = \eta_0$ (4)

For Lanzhou loess and red clayey loam, the relation between their frost heave rate

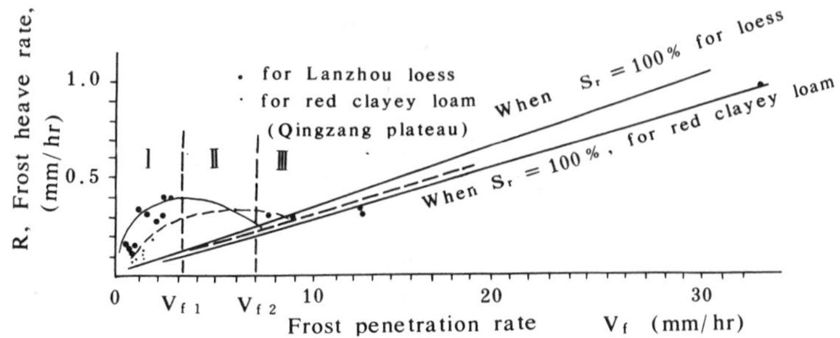

Figure 3. Frost heave rate vs frost penetration rate.

R and penetration rate V_f is shown in Figure 3. Comparing Figure 3 with Figure 2, we see that: Firstly, when $V_f < V_{f1}$, R consists of two parts, one is produced by freezing of pore water, another is produced by ice segregation. It rises rapidly with the increase of V_f. Secondly, when $V_{f2} \geq V_f \geq V_{f1}$, R also includes above two parts, but it decreases with the increase of V_f. Thirdly, when $V_f > V_{f2}$, R is only produced by freezing of pore water in situ and is proportional to V_f. Thus, there is no water migration in this section.

Frost Heave Ratio η and Effective Surcharge Stress P

It is known that according to the thermodynamic equilibrium principle the freezing point of soil water will be depressed when external pressure P and water potential P_w are raised, which is expressed by

$$\Delta T = T_o (V_w P_w - V_i \Delta P_i)/L \qquad (5)$$

where ΔT=the freezing point depression, ^oC
T_o=freezing point of pure water, ^oK
V_w=specific volume of water
ΔP_i=ice stress
V_i=specific volume of ice
L=latent heat of fusion for water

In an open system, $\Delta P_w=0$ and ΔP_i equals the effective surcharge stress P. Based on the above equation, the relationship between ΔT and P for saturated loess was measured and can be expressed approximately by a linear equation,

$$\Delta T \simeq 0.07P \qquad (6)$$

In addition, when the effective surcharge stress rises, the density of soil will increase and the permeability will decrease. As a result, the specific volumetric suction of water ξ_w during freezing decreases (Figure 4). For saturated loess ξ_w can be expressed as follows,

$$\xi_w = \xi_{wo} \exp(-aP) \qquad (7)$$

where ξ_{wo} is specific volumetric suction water when P=0, a=const.
Because of mentioned above, the higher the surcharge stress in soil, the lower the water migration rate to the frost front will be. As a result, the heave will be lower. When the surcharge stress

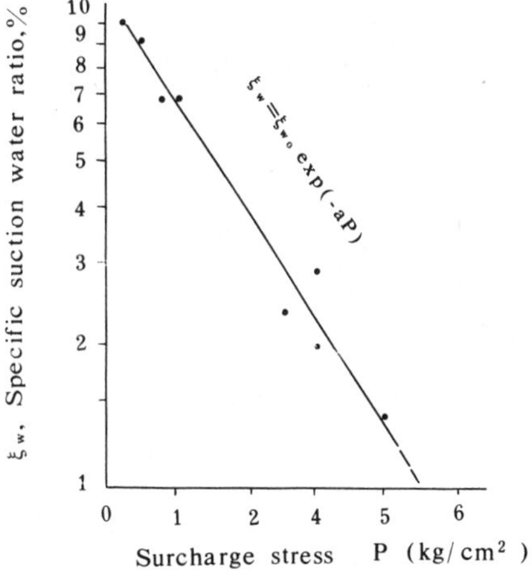

Figure 4. ξ_w vs P of Lanzhou loess.

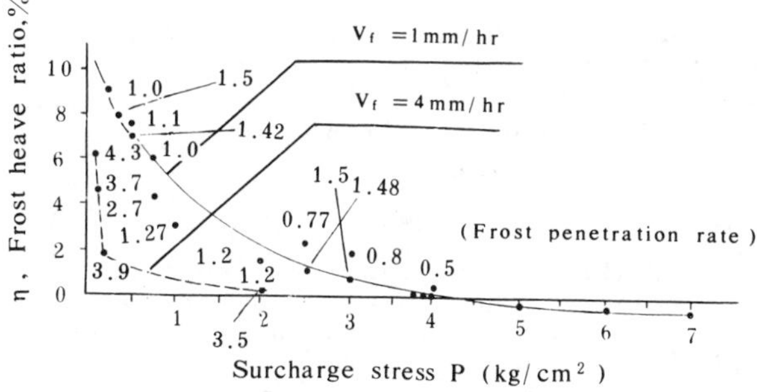

Figure 5. Frost heave ratio of saturated loess vs its surcharge.

reaches the shut-off pressure P_o, there will be no water flow to the frost front, and thus no frost heave will be measured. When the stress P is greater than P_o, the pore water at the front will be expelled during freezing. As shown in Figure 5, the shut-off pressure for saturated loess frozen in an open system is about 4.6 kg/cm^2 (for $V_f \approx$ 1 mm/hr).

The semi-exponential relation between the frost heave ratio η or heave rate R and the penetration rate Vf is shown in Figure 6.

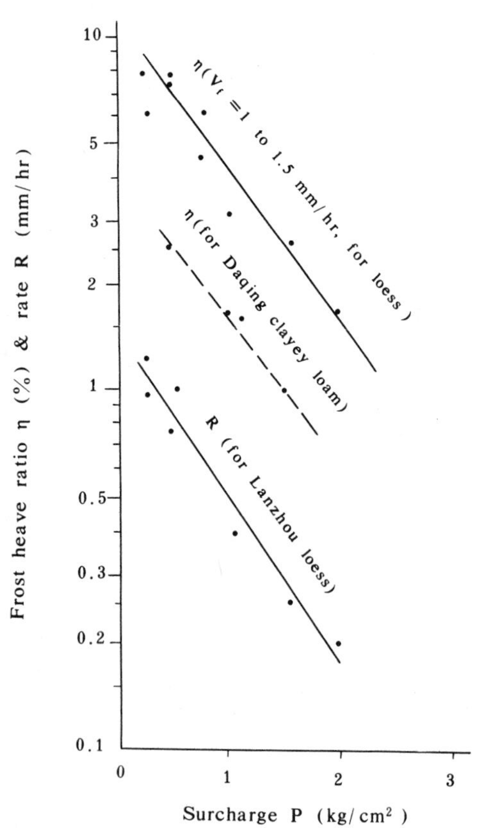

Figure 6. Frost heave ratio and rate vs surcharge.

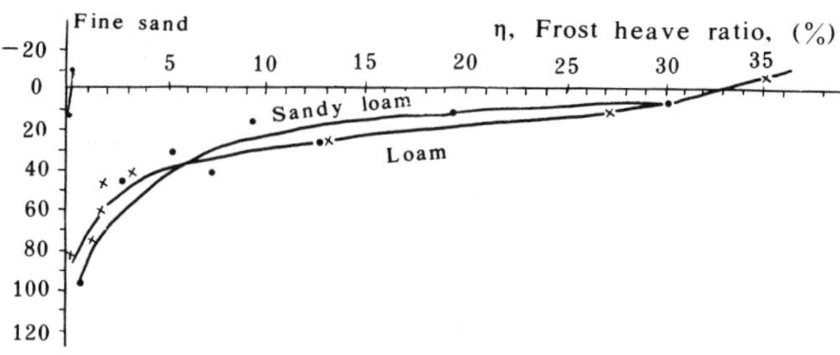

Figure 7. Frost heave ratio η vs distance from ground water table to frost front H_w (Gansu).

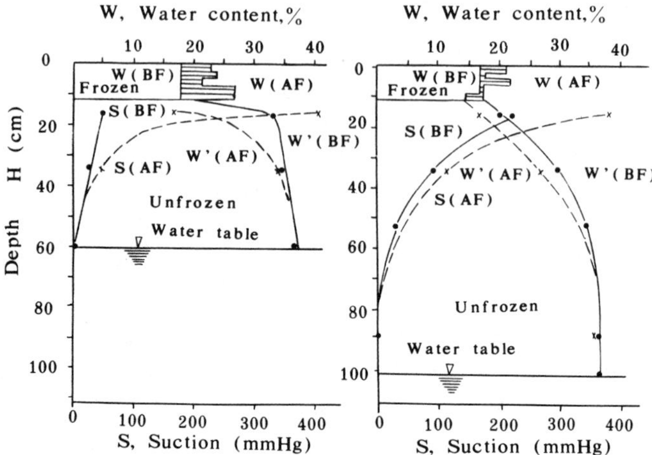

Figure 8. Variation of water content and suction during freezing of a clayey loam (soil no. 3, Table 1). AF = after freezing. BF = before freezing.

It is obvious that the relationship for not only a saturated loess in the laboratory but also for a clayey loam in situ can be described as follows,

$$\eta = \eta_0 \exp(-bP) \quad (8)$$

$$R = R_0 \exp(-DP) \quad (9)$$

Above equations show that the frost heave of a foundation soil may be controlled by surcharge. Conversely, the normal frost heaving force acting on foundation can also be calculated if the given frost heave is known.

Frost Heave and Ground Water Table Hw

In addition to the water content of soil, the groundwater table is an important factor for evaluating the relative frost susceptibility of soil in open system. Some observations obtained at Zhangyi Station on the effect of ground water table are presented in Figure 7. It is shown that the heave ratio for a clayey soil increases rapidly as the groundwater table rises. With a rather deep water table, because the capillary rise in clayey loam is higher than that of sandy loam, its heave ratio is larger. However, when the water table is rather shallow, the heave ratio of sandy loam will be larger because of its higher permeability. Zhu Qiang (1981) derived the following linear equation for evaluating the frost heave ratio of a silty loam with varying groundwater table H_w at a test canal in the Zhangyi district as follows,

$$\eta = 64.4 - 29.6\ H_w \quad (10)$$

The redistribution of moisture with depth under various water tables during freezing was obtained in the laboratory and in situ. Figure 8 gives two examples for clayey loam (see Table 1, No.3) tested in the laboratory. It is shown that there is an intensive water migration zone beneath the frost front with a range of about 30 cm. Before freezing, the soil water suction at the same depth for different water table depths is quite different. However, after freezing, their suction forces near the frost front are nearly the same. As a result, the suction gradient for a shallow water table is much greater than for a deeper water table. This results in more water migration to the frost front and greater ice segregation for the shallow water table condition.

In addition, for the deeper water table, because the water can not be sufficiently supplied to the freezing front, the moisture content at the base of the frozen portion is almost always less than before freezing (see Figure 8).

CONCLUSIONS

1. Two critical frost penetration rates divide the development of the frost heave ratio into intensive, slow, and un-changing stages, and determine whether ice segregation occurs.
2. Under surcharge stress, the freezing point is depressed and specific volumetric suction water during freezing decreases. As a result, the frost heave ratio decreases exponentially with an increase of the surcharge stress.
3. Before freezing, the soil water tension at a given depth is much larger for a shallow water table than for a deeper water table. However, after freezing, the moisture tensions near the frost front are almost the same, thus causing much more water migrate to the freezing front under shallow water table condition.
4. Therefore, besides the particle size, the effect of frost penetration rate, surcharge stress, and groundwater table should be considered for evaluating the frost susceptibility of soil during freezing.

REFERENCES

Aitken, G.W. 1974, Reduction of Frost Heave by Surcharge stress, CRREL TR 184.

Anderson, D.M. and Morgenstern, N.R. 1973, Physics, Chemistry and Mechanics of Frozen Ground: A Review, Proc. of 2nd Int. Conf. on Permafrost.

Arvidson, W.D. and Morgenstern, N.R., Water Flow Induced by Soil Freezing, Can. Geotech. J. 14(2), 1977.

Berg, R. and Ingersoll, L. 1979, Frost Heave in An Instrumented Soil Column, Cold Regions and Technology 3(1980).

Chamberlain, E.J. 1981, Frost Susceptibility Criteria, Presented at National Transportation Research Board Meeting, Jan, 1981.

Corte, A.E. 1969, Geocryology and Engineering---Part II Experimental and Applied Research, "Reviews in Engineering Geology" Vol.II, pp.119-186.

Kaplar, C.W. 1974, Freezing Test for Evaluating Relative Susceptibility of Various Soils. CRREL Technical Report 250.

Kinosita, S., Suzuki, Y., Horiguchi, K. and Fukuda, M. 1978, Observations of Frost Heaving Action in the Experimental Site, Tomakomai, Japan, Proc. of 3rd Int. Conf. on Permafrost.

Loch, J.P.G. and Kay, B.D. 1978, Water Redistribution in partially frozen, saturated silt under several temperature gradient and overburden loads, Soil Sci. Soc. Am Proc.,

Penner,E. 1972, Influence of Freezing Rate on Frost Heaving, Highway Res. Board Rec. No 393.

Penner, E. and Ueda, T. 1977, The Dependence of Frost Heaving on Load Application, Proc. Int. Symp. on Frost Action in Siols.

Penner, E. and Ueda, T. 1978, A Soil Frost-susceptibility Test and A Basis for Interpreting Heaving Rates, Proc. of 3rd Int. Conf. on Permafrost.

Tsutomu Takashi et al, 1978, Effect of Penetration Rate of Freezing and Confining Stress on the Frost Heave of Soil, Proc. of 3rd Int. Conf. of permafrost.

Wang Xiyan, 1982, Frost Heave and Its Distribution in Different Layers Influenced by Shallow Groundwater, J. of Glaciology and Cryopedology, Vol. 4, No. 2.

Zhu Qing, 1981, Property of Frost Heaving in Irrigation Ditch Lined With Concrete and the Measure of Replacing Subsoil with Sand Gravel in Gansu, Proc. of 2nd Nat. Conf. on Permafrost.

VERTICAL AND HORIZONTAL ZONATION OF HIGH-ALTITUDE PERMAFROST

Cheng Guodong

Lanzhou Institute of Glaciology and Cryopedology, Academia Sinica
People's Republic of China

Alpine permafrost cannot be classified into continuous, discontinuous, and island types, since no alpine permafrost is actually continuous. According to its stability, indicated by its thickness and the mean annual ground temperature, the permafrost on the Qinghai-Xizang Plateau should be divided into three zones: upper zone (extremly stable), middle zone (stable, substable, and transitional), and the lower zone (unstable and extremely unstable). The distribution of alpine permafrost shows latitudinal variation. The lower limit rises northwards to its extreme value at latitude 25°22'N, then descends as the latitude continues to increase. Data show that the elevation of the permafrost lower limit is related to the aridity of the area. A region with higher precipitation usually has more cloud cover, which reflects both outgoing and incoming radiation. North of about 40°N, the incoming radiation is less than the outgoing so that the main effect of the cloud cover is to heat the air, raising the alpine permafrost's lower limit; south of 40°N, however, the incoming radiation is greater than the outgoing, so the main effect of the cloud cover is to cool the air, lowering the permafrost's lower limit.

The permafrost above a certain elevation to the south of the permafrost southern boundary is called high-altitude permafrost, but the permafrost to the north of permafrost southern boundary is called high-latitude permafrost. There are many mountains and plateaus in China. The area of high-altitude permafrost in China is up to 1,732,000 km^2 which occupies 80.6% of the total permafrost area of China, and 74.5% of the total permafrost area of Northern Hemisphere.

Summarizing the recent data of high-altitude permafrost distribution reported from the whole world with special reference to the data of China, it has been found that there is an obvious three-dimensional zonation in permafrost distribution, i.e. vertical zonation, aridity or longitude zonation, and latitude zonation.

VERTICAL ZONATION

High-altitude and high-latitude permafrost are quite different in distributional pattern. There is an obvious vertical zonation in the distribution of high-altitude permafrost, therefore the projection of so-called "continuous permafrost" on plane is usually not continuous but isolated. Thus, according to the projections on plane, there are islands of "continuous permafrost," "discontinuous permafrost," and isolated permafrost (Figure 1). The ground temperatures and thicknesses of these permafrost islands are quite different. Under these circumstances, the division of permafrost into zones by using continuous coefficient and the names such as continuous, discontinuous, and isolated are not convenient to high-altitude permafrost. On the other hand, the high-altitude and the high-latitude permafrost have common characteristics in origin. They all are the products of heat and mass exchange between the earth's crust and the atmosphere, and exist in a zone with a lower level of energy. Based on these features, and with special reference to the data of Qinghai-Xizang Plateau, a scheme for classifying high-altitude permafrost into zones has been suggested by using mean annual ground temperature as the main index as shown in Table 1.

The vertical permafrost spectra in various climatic regions are different. In continental regions where the mean annual air temperature near snow line is about -12°C (Daxue Shan, Qilian Shan), there are three zones. In subcontinental regions where the mean annual air temperature near snow line is about -8.5°C (Lenglong Ling, Qilian Shan), there are two zones, and in the maritime regions, the mean annual air temperature near the snow line is about -4°C (Guxiang, Tibet), as a result there is only one zone (Figure 2). On the other hand, where the base zone is the upper zone we have only one zone; where the base zone is the middle zone we have two zones; and where the base is the lower zone, we shall have three zones. Therefore, various combinations of elevation, latitude, and distance from ocean render different permafrost vertical spectra, so that the regional distribution of high-altitude permafrost could be better described than before by using the concept of permafrost vertical spectra.

TABLE 1 Classification of Permafrost of Qinghai-Xizang Plateau.

Name of the zones		Mean annual ground temperature (°C)	Thickness of permafrost (m)	Relationship between elevation of zone boundary (Y) and latitude (x)	Mean annual air temperature at lower limit (°C)
Upper zone	Extremely stable	-5.0	>170		
				Y_5=111.1-1.72x	-8.5
Middle zone	Stable	-3.0 ~ -5.0	110 ~ 170		
				Y_4=108.6-1.72x	-6.5
	Substable	-1.5 ~ -3.0	60 ~ 110		
				Y_3=106.2-1.72x	-5.0
	Transition	-0.5 ~ -1.5	30 ~ 60		
				Y_2=104.2-1.72x	-4.0
Lower zone	Unstable	+0.5 ~ -0.5	0 ~ 30		
				Y_1=101-1.72x	-2.0 ~ -3.0
	Extremely unstable				

TABLE 2 Altitude of Lower Limit of High-Altitude Permafrost in the Northern Hemisphere.

Area	Locality	Latitude	Lower limit (m)	References
China	Himalaya	28°10'	5200	Zhou Youwu and Guo Dongxin (1982)
	Gulu	30°40'	4800	Cheng Guodong and Wang Shaoling (1982)
	Sidaoliang	31°08'	4720	Wang Jiacheng et al. (1979)
	Maintenance Squad 125	31°41'	4640	Wang Jiacheng et al. (1979)
	Maintenance Squad 121	32°00'	4610	Wang Jiacheng et al. (1979)
	Xidatan	35°40'	4150	Wang Jiacheng et al. (1979)
	Qilian Shan	38°20'	3500	Guo Pengfei (1980)
	Tian Shan	43°00'	2700	Qiu Guoqing et al. (1975)
	Altay Shan	48°00'	2200	Tong Boliang and Li Shude (1981)
Nepal	Khumbu Himalaya	27°55'	4950	Fujii and Higuchi (1976)
	Mukut Himalaya	28°45'	4950	Fujii and Higuchi (1976)
USSR	Tien Shan	42°00'	2700	Gorbunov (1976)
	Pamir Plateau	38°00'	3700	Gorbunov (1978)
	Dzhungarskiy Alatau	45°00'	2500	Gorbunov (1978)
Alps	Berner Oberland	46°30'	2350	Barsch (1978)
	Wallis/Graubünden	46°30'	2550	Barsch (1978)
	Flüelapass	46°45'	2300	Haeberli (1975)
Mexico	Citlatepetl	19°00'	4600	Lorenzo (1969)
America	Tesuque Peak	35°47'	3720	Retzer (1965)
	Niwot Ridge	40°00'	3500	Ives and Fahey (1971)
	Plateau Mt.	50°16'	2224	Harris and Brown (1978)
	Garibaldi Part	49°58'	1830	Mathews (1955)
	Near Cassiar	59°18'	1370	Brown (1969)

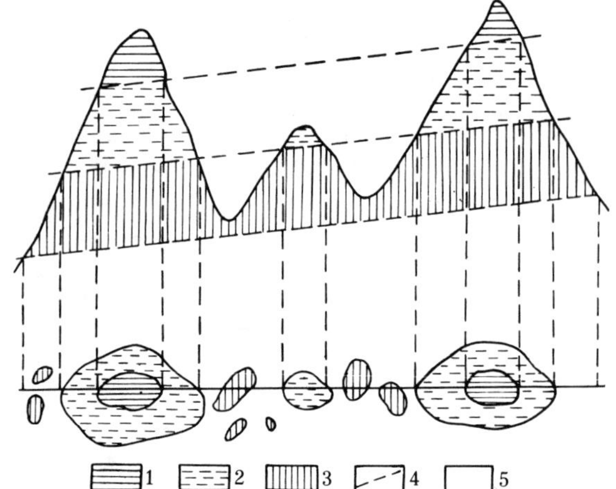

FIGURE 1 Sketch map showing various permafrost islands
1. continuous permafrost; 2. discontinuous permafrost; 3. isolated permafrost; 4. boundary of zones; and 5. seasonal frozen ground.

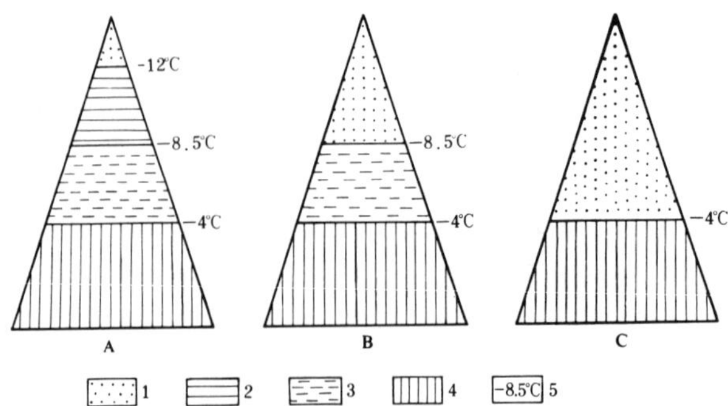

FIGURE 2 Vertical spectra of permafrost in various climatic regions
1. glacier; 2. upper zone; 3. middle zone; 4. lower zone; 5. mean annual ground temperature.
A. continental climatic region;
B. sub-continental climatic region;
C. maritime climatic region.

LATITUDINAL ZONATION

The changes of timberline, snowline, boundary of mountain cold desert soil, and permafrost lower limit with latitude are quite similar (Niu Wenquan 1980; Cheng Guodong and Wang Shaoling 1982; Jiang Zhongxin 1982). By fitting the Gauss curve to the data of high-altitude permafrost lower limit in Northern Himisphere (Table 2), the empirical correlation between lower limit of high-altitude permafrost (H) and latitude (ϕ) has been obtained as follows:

$$H=3650 \exp[-0.003 (\phi-25.37)^2]+1428 \qquad (1)$$

Tentative application gives satisfactory results (Figure 3).

Different from those linear mathematical models obtained before, this function has an extreme value and a point of inflection. As indicated by it, starting from the equator, the permafrost lower limit rises with increasing latitude, and reaches its extreme value of 5078 m at the latitude of 25°22'N. Then it descends with increasing latitude. At the initial stage, the descending slope is steeper, and reaches its maximum at latitude 38°N, then becomes gentle with increasing latitude. These features of the function are closely related to the features of mean latitudinal distribution of earth's radiation budget. Starting from the equator, the incoming radiation also intensifies with increasing latitude, and reaches its maximum at about 25°N of latitude, then reduces with increasing latitude. The difference between incoming and outgoing

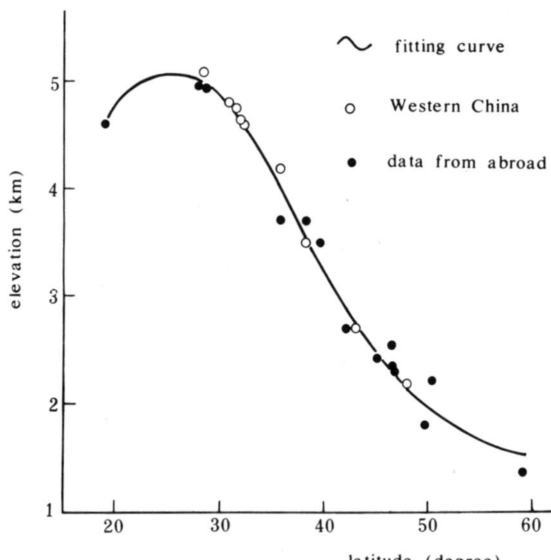

FIGURE 3 Comparison of permafrost lower lower limit in Northern Hemisphere between observed and calculated value.

radiation also reaches its maximum at about 25°N (Critchfield 1974) (Figure 4).

The mean annual ground temperature of permafrost measured on Qinghai-Xizang Plateau has been divided into four groups with reference to the practical conditions on the plateau. Then these four groups of value have been correlated with elevation and latitude, which gives a fairly good result (Figure 5). Applying the linear

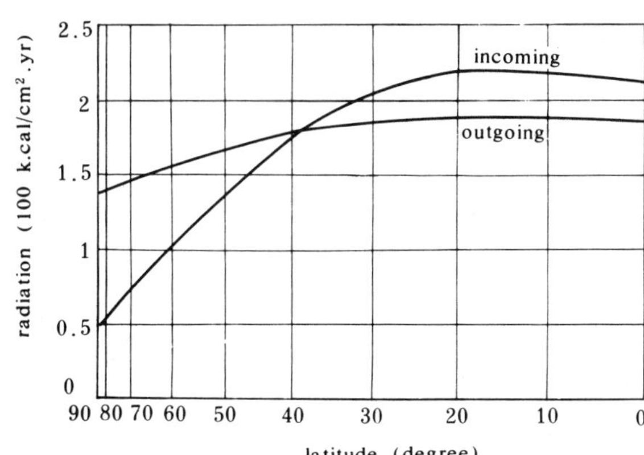

FIGURE 4 Mean latitudinal distribution of earth's radiation budget. (after Houghton)

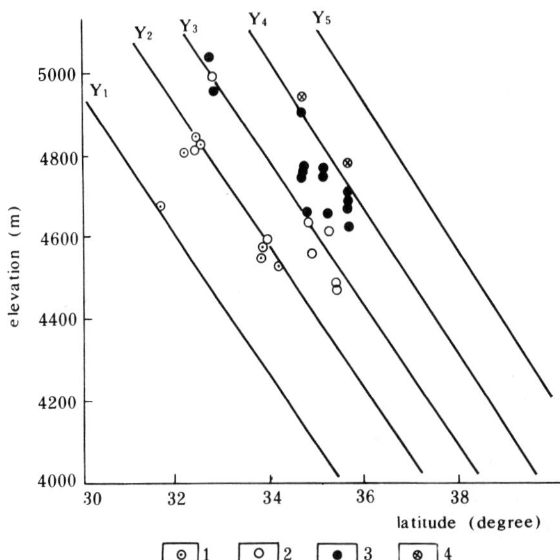

FIGURE 5 Mean annual ground temperature of permafrost on Qinghai-Xizang Plateau vs. elevation and latitude
mean annual ground temperature:
1. +0.5~-0.5; 2. -0.5~-1.5;
3. -1.5~-3.0; 4. -3.0~-5.0.

equations of Y_1 to Y_5 in Figure 5, we can roughly divide the high-altitude permafrost into zones by using elevation and latitude (Table 1). Making a comparison between the practical permafrost zones on Qinghai-Xizang Plateau and tentatively divided zones by using five equations of Y_1 to Y_5, the results are satisfactory (Cheng Guodong and Wang Shaoling 1982).

ARIDITY ZONATION

Former research works have determined that the permafrost lower limit descends with increasing aridity (Haeberli 1978; Harris and Brown 1978; Qiu Guoqing et al., 1983). But the works on Qinghai-Xizang Plateau have arrived at a contrary conclusion: the permafrost lower limit descends with increasing precipitation. As an example, the following binary regression equation has been obtained for Qilian Shan Mountains.

$$H=7789.5-121.3x+8.3\omega \qquad (2)$$

where: H is the elevation of permafrost lower limit (m);
x is the latitude (degree);
ω is the angle of precipitation continentality (degree),

$$ctg\omega=\frac{\text{annual precipitation (mm)}X10}{\text{elevation (m)}}=\alpha$$

$$\omega=arcctg\alpha$$

The correlation coefficient of equation 2 is R=0.95, and residual standard error s=53.7 (m). The results of checking computations are in Table 3, its accuracy is enough for use.

Equation 2 also indicates that the permafrost lower limit descends with increasing precipitation, which is contrary to the former conclusion. To this contrast, we try to explain as follows.

Based on the statistical analysis of meteorological data in Western China, the correlation between mean annual air temperature (T) and elevation (X_1), latitude (X_2), as well as precipitation (X_3) has been obtained.

$$T=b_0+b_1X_1+b_2X_2+b_3X_3 \qquad (3)$$

where: b — coefficient

The values of coefficient b in different latitude sections and corresponding correlation coefficients R and residual standard errors S are listed in Table 4.

Table 4 shows that the values of coefficient b_3 related to precipitation are positive to the north from 40°N, but negative to the south from 40°N. It means that to the north from 40°N the increase of precipitation makes the mean annual air temperature rise, but quite the contrary to the south from 40°N. The reason for this phenomenon is as follows. The region with more precipitation usually has more cloud cover which can reflect both incom-

TABLE 3 A Comparison of Permafrost Lower Limit between Values Measured and Calculated with Equation 2.

Location	Latitude, X (degree)	Angle of precipitation continentality ω (degree)	Elevation of permafrost lower limit, H (m a.s.l.) Measured	Calculated	Difference (m)
Baihuazhang of Lenglong Ling	37°43'	34.9	3450	3504	+54
Baishekou of Lenglong Ling	37°58'	34.9	3500	3473	-27
Garang of Lajishan	36°14'	41.8	3700	3742	+42
Xiangpi Shan Pass of Qinghai Nanshan	36°40'	40.8	3678	3680	+ 2
Dangjin Shankou	39°21'	77.6	3650	3660	+10
Datouyang coal mine in Daqaidam Shan	37°47'	74.4	3850	3861	+11
Reshui coal mine of Datong Shan	37°40'	34.0	3480	3502	+22
Yijia Sheepfold of Zoulang Nanshan	39°14'	54.0	3520	3479	-41
Heka Nanshan	35°49'	41.2	3820	3768	-33
Northern slope of Ela Shan	35°40'	41.2	3850	3805	-45

TABLE 4 Values of Coefficient b in Different Latitude Sections

Latitude	Number of station	b_o	b_1	b_2	b_3	R	S
≥44°	20	58.41	-0.42	-63.13	3.25	0.82	1.53
≥42°	38	66.70	-0.58	-71.46	1.41	0.91	1.55
≥40°	54	65.88	-0.57	-70.44	1.20	0.92	1.52
40° ~ 32°	55	59.60	-0.57	-62.02	-3.84	0.98	1.03
38° ~ 32°	39	67.32	-0.58	-73.87	-4.15	0.99	0.81
36° ~ 32°	19	70.04	-0.59	-84.75	-4.22	0.97	0.80
34° ~ 32°	11	82.88	-0.66	-94.23	-5.36	0.96	0.88

ing radiation to cool the air and outgoing radiation to heat the air. As shown in Figure 4, to the north from about 40°N, the incoming radiation is less than the outgoing, so that the main effect of the cloud cover is to heat the air, therefore the mean annual air temperature rises with increasing cloud cover or precipitation, correspondingly the permafrost lower limit rises. But south of 40°N, the incoming radiation is greater than the outgoing one, so that the main effect of cloud cover is to cool the air, thereby the mean annual air temperature descends with increasing cloud cover, correspondingly the permafrost lower limit descends.

Because the Qinghai-Xizang Plateau is situated south of 40°N, the permafrost lower limit rises with increasing aridity. But the majority of other high-altitude permafrost are situated to the north of 40°N, therefore in these regions the permafrost lower limit descends with increasing aridity.

As mentioned above, the permafrost lower limit does not simply monotonously rise from continental to maritime regions, but changes as shown in Figure 6. South of 40°N, both the snow line and permafrost lower limit descend from continental to subcontinental climate, but the descending magnitude of the lower limit is less than that of the snow line; the snow line continuously descends from subcontinental to maritime climate, but the lower limit turns to rise with the increasing effect

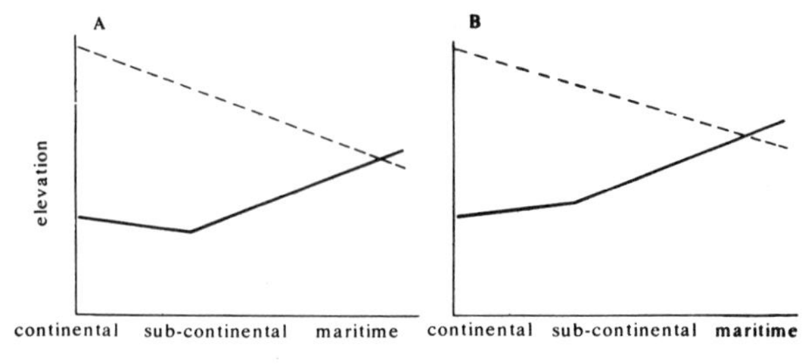

FIGURE 6 Change of permafrost lower limit with aridity
a. to the south of 40°N;
b. to the north of 40°N.
1. snow line; 2. permafrost lower limit.

of snow cover. North of 40°N, the lower limit rises from continental to subcontinental climate, which is against the changing trend of snow line; and the rising magnitude of lower limit increases with the increasing effect of snow cover from subcontinental to maritime climate.

REFERENCES

Cheng Guodong and Wang Shaoling, 1982, On the zonation of high-altitude permafrost in China: Journal of Glaciology and Cryopedology, v. 4, No. 2, p. 1-17.

Critchfield, H. J., 1974, General climatology: Prentice-Hall Inc.

Haeberli, W., 1978, Special aspects of high mountain permafrost methodology and zonation in the Alps, in Proceedings of the Third International Conference on Permafrost, v. 1: Ottawa, National Research Council of Canada, p. 378-384.

Harris, S. A., and Brown, R. J. E., 1978, Plateau mountain: a case study of alpine permafrost in the Canadian Rocky Mountains, in Proceedings of the Third International Conference on Permafrost, v. 1: Ottawa, National Research Council of Canada, p. 385-392.

Jiang Zhongxin, 1982, A discussion on the mathematical model of physico-geographical zonation: Acta Geographica Sinica, v. 37, No. 1, p. 98-103.

Niu Wenyuan, 1980, Theoretical analysis of physico-geographical zonation: Acta Geographica Sinica, v. 35, No. 4, p. 288-298.

Qiu Guoqing, Huang Yizhi, and Li Zuofu, 1983, Basic characteristics of permafrost in Tian Shan, China, in Proceedings of Second National Conference on Permafrost: Lanzhou, Gansu People's Press, p. 21-29.

DISTRIBUTIVE REGULARITIES OF HIGH ICE-CONTENT PERMAFROST ALONG QINGHAI-XIZANG HIGHWAY

Cheng Guodong and Wang Shaoling

(Lanzhou Institute of Glaciology and Cryopedology, Academia Sinica)

Investigations along the Qinghai-Xizang Highway show that in similar geomorphologic units, the lower the mean annual ground temperature, the better the development of high ice-content permafrost. In the same permafrost-temperature zone, the development of high ice-content permafrost has such an order in degree: the most in low mountain and hill regions, second in middle and high mountain regions, and the least in valley and plain regions. In the first region, the development of high ice-content permafrost on north-facing, gentle slopes and low parts of slopes is better than on south-facing, steep slopes and middle and upper parts of slopes; In the valley and plain regions, except the high plain with widespread lacustrine deposits, commonly the high ice-content permafrost does not usually develop well and only exists in swampy sections consisting of fine-grained soil and the backside of high terraces that have an adequate content of fine-grained soil and supply of suprapermafrost water. In the depression between fens, there is commonly high ice-content permafrost.

INTRODUCTION

High ice-content permafrost is a group term including rich-ice permafrost, saturated-ice permafrost and ice layer containing soils (Wu Ziwang 1982). Discussion in this article is restricted to depths of 0 to 1 m under the permafrost table.

The ice content of the permafrost relates to two groups of factors: first, geological-hydrogeological factors which mainly include grain size and suprapermafrost water, and second, thermophysical factors which may be expressed by mean annual ground temperature approximately (Vtyurin 1975). Generally speaking, when other conditions are equal, the lower the mean annual ground temperature, the more favorable the formation of high ice-content permafrost (Cheng Guodong 1982). Because the mean annual ground temperature obeys certain zonation laws, therefore the distribution of high ice-content permafrost also follows certain zonation laws. The high ice-content permafrost may exist in fine-grained soil having appropriate water content within the continuous permafrost zone, but it is only encountered in swampy fine-grained soil having humus within the sporadic permafrost zone.

The first group of factors is not zonal, but the second is. The combination of these two groups of factors in concrete geographical environments controls the regional distribution of high ice-content permafrost. This distribution is reflected in regular change of ice content of permafrost in various geomorphologic units and topographic positions.

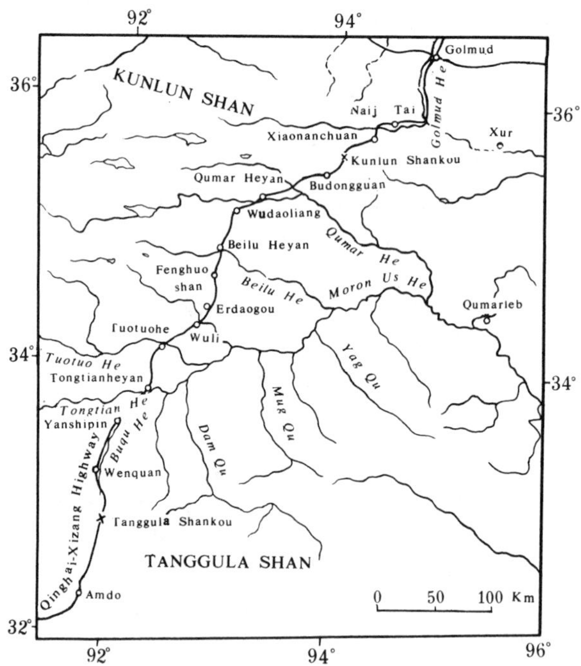

FIGURE 1 Location of the study area

DISTRIBUTION IN SIMILAR GEOMORPHOLOGIC UNITS BUT DIFFERENT PERMAFROST-TEMPERATURE ZONES

In the permafrost region along the Qinghai-Xizang Highway, three kinds of macro-geomorphologic units can be discerned, i.e. middle and high mountain regions, low mountain and hill regions, and valley

TABLE 1 Development of Ground Ice in Different Mean Annual Ground Temperature Zones*

Regions			Mean annual ground temperature (°C)	Thickness of permafrost (m)	Relative development of ground ice
I Middle and high mountain permafrost regions	I_1	Near southern boundary of continuous permafrost	0 ~ -0.3	< 20	A little granular ice
	I_2	Between 114 and 115 maintenance squads	-0.7 ~ -1.0	40 - 50	A little granular ice on south-facing slope; layered ground ice on north-facing slope
	I_3	In Taorjiu mountain	-1.0 ~ -1.2	50 - 60	Thickness of ground ice up to 10 cm near permafrost table
II Low mountain and hill permafrost regions	II_1	Near southern boundary of continuous permafrost	0 ~ -0.3	< 20	A small amount of ice
	II_2	Around 114 maintenance squad	-0.6 ~ -1.0	30 - 50	Layered ground ice just below permafrost table, ice content 30-50%
	II_3	Between 112 and 113 maintenance squads	-1.0 ~ -1.2	50 - 60	Layered ground ice

* After Wang Jiacheng et al., 1979.

and plain regions. When these three kinds of geomorphologic units belong to different permafrost-temperature zones, the development of ground ice in each zone is different in degree.

The development of ground ice in valley and plain regions can be illustrated by examples of Beilu He and Tuotuo He. The first terraces of Beilu He and Tuotuo He consist mainly of mudstone, marle, and siltstone of the Tertiary System. The first terrace of Beilu He lies in the zone with mean annual ground temperature at -0.5° to -1.5°C, while the first terrace of Tuotuo He lies in the zone above -0.5°C. Drill hole CK 5 in the first terrace of Beilu He reveals that the thickness of ground ice layers below the permafrost table is 3-4 cm, and the ice content is 70-80%. Drill hole No. 6 in the first terrace of Tuotuo He (mean annual ground temperature at -0.2°C) reveals that the permafrost is of net cryogenic structure at 2.5 to 3.4 m below permafrost table, and the volumetric ice content is only 10-20%. Obviously, when other conditions are equal, the ground ice in the valley and plain regions lying in the lower mean annual ground temperature zones forms better than in the higher mean annual ground temperature zones.

In both the middle and high mountains, and the low mountain and hill regions, this can be illustrated by the example of the continuous permafrost zone on the southern slope of Tanggula Shan (Table 1).

As shown in Table 1, the mean annual ground temperature decreases northwards, and the ice content of permafrost correspondingly increases.

TABLE 2 Distribution of Various Kinds of Permafrost in Different Geomorphologic Units from K139+00 to K693+00 along Qinghai-Xizang Highway*

Geomorpho-logic units	Milepost	Distance		Distribution of various kinds of permafrost (km)					
				Rare-ice	Some-ice	Rich-ice	Saturated ice	Ice layer containing soil	Talik
Kunlun Shan	139+000—191+000	52.0		4.0	10.0	9.0	3.5	20.7	4.8
			%	7.7	19.2	17.3	6.8	39.8	9.2
Qumar He	200+000—253+700	53.7		6.5	5.6	7.2	12.4	21.4	0.6
			%	12.2	10.4	13.4	23.0	39.9	1.1
Hoh Xil Shan	253+700—306+600	52.9		3.74	7.1	6.0	14.8	21.4	0
			%	7.1	13.4	11.3	27.9	40.3	0
Beilu He	306+600—317+650	11.1		0.8	2.7	0.7	5.1	1.6	0.2
			%	7.2	24.3	6.3	46.4	14.0	1.8
Fenghuo Shan	317+650—371+050	53.4		7.3	9.6	3.7	19.1	13.8	0
			%	13.7	17.9	6.9	35.7	25.8	0
Tuotuo He	371+050—438+000	67.0		16.5	4.4	10.0	13.2	0	22.9
			%	24.6	6.6	14.9	19.7	0	34.2
Kaixin Ling	438+000—468+500	30.5		0	7.4	4.7	4.1	11.3	3.0
			%	0	24.3	15.4	13.4	37.0	9.9
Tongtian He	468+500—490+000	21.5		6.8	0	1.1	3.3	2.0	8.3
			%	31.6	0	5.1	15.4	9.3	38.6
Buqu He	490+000—587+560	97.6		16.8	2.1	3.8	12.3	0	62.6
			%	17.2	2.2	3.8	12.6	0	64.2
Tanggula Shan	587+560—643+000	55.4		9.5	9.9	13.7	17.9	3.4	1.0
			%	17.2	17.8	24.7	32.4	6.1	1.8
Taorjiu Shan	643+000—693+000	50.0		11.0	4.0	24.0	3.0	8.0	0
			%	22.0	8.0	48.0	6.0	16.0	0
Total	139+000—191+000 200+000—693+000	545.0		83.0	62.7	83.8	108.6	103.5	103.4
			%	15.2	11.5	15.4	19.9	19.0	19.0

* The kind of permafrost is determined by the total water content of 1 m permafrost just below the permafrost table.

DISTRIBUTION IN SIMILAR ZONES BUT DIFFERENT GEOMORPHOLOGIC UNITS

The ice content just below the permafrost table in the continuous permafrost zone is higher than in the sporadic permafrost zone. The distribution of various kinds of permafrost in different geomorphological units from K139+00 to K693+00 along the Qinghai-Xizang Highway listed in Table 2.

From Table 2 we can see that, in the continuous permafrost region, the development of high ice-content permafrost is highest in the low mountain and hill regions, second in the middle and high mountain regions, and lowest in the valley and plain regions. This is because there is an obvious vertical zonation of permafrost distribution on the Qinghai-Xizang Plateau and precipitation increases with rising elevation. This makes the mean annual ground temperature lower and precipitation greater in the low mountain and hill regions. Moreover the solifluction deposits are well developed in this region and provide suitable fine-grained soil, and moisture conditions for syngenetic ice. Therefore, the low mountain and hill regions have high proportions of high ice-content permafrost. Although middle and high mountain regions offer the lowest mean annual ground temperature, which is favorable to the growth of ground ice, and more precipitation, the proportion of high ice-content permafrost is less than in the low mountain and hill region because of thinner surficial deposits and coarser soils, which are unfavorable for the growth of ground ice. The valley and plain regions possess the lowest proportions of high ice-content permafrost due to lower elevation, higher mean annual ground temperature, the thermal effect of rivers, and wide-spread coarse-grained surficial deposits. All are unfavorable for the growth of ground ice. Only in the high plains and in some basins consisting of lacustrine deposits, which possess thick deposits with much fine-grained soil, less thermal effect of rivers, and lower ground temperature, is the proportion of high ice-content permafrost high.

DISTRIBUTION IN SIMILAR GEOMORPHOLOGIC UNITS BUT DIFFERENT TOPOGRAPHICAL POSITIONS

Low Mountain and Hill Region

Orientation of slope. In the same low mountain and hill region, where other conditions are equal, the south-facing slope has more radiation, higher ground temperature and evaporation. Thus the ground ice on the north-facing slopes forms better than that on the south-facing slopes. E.g. the ground ice on the north-facing slopes of Dongdagon in Fenghuo Shan possesses larger thickness and extent than on the south-facing slopes. But in margins of the continuous permafrost and the sporadic permafrost zones, it usually shows a difference in quantity and distribution of ground ice, i.e. there is high ice-content permafrost on the north-facing slopes but not on the south-facing slopes. For example, drill holes CK 116-5 and CK 116-6, situated to the north of Amdo, have the same elevation and reveal similar types of soils. CK 116-6, situated on a north-facing slope with more moisture and vegetation, reveals a soil profile as follows: 2.6 m, permafrost table; 2.6-2.8 m permafrost of net cryogenic structure; 2.8-4.4 m, permafrost of moderate to thick layered cryogenic structure with thickness of ice layer up to 15-20 cm. But CK 116-5, on a south-facing slope with rare moisture and vegetation, only reveals permafrost of massive cryogenic structure with no high ice content. Thus, because of different slope orientation, there is a difference in the quantity of ground ice.

Slope. Steep slopes, with thin soil cover and poor drainage are unfavorable to the growth of ground ice. Gentle slopes, with adequate moisture and good vegetation cover are favorable for the growth of ground ice. According to data along the Qinghai-Xizang Highway on slopes of gradients less than 10°, ground ice is usually formed. Slope gradients of 4°-8° are especially favorable to the growth of ground ice. On slopes with gradients of 10°-16° the conditions for growth of ground ice become bad. On slopes with gradients over 16° usually no thick layered ground ice is found. When slope gradients are over 25°, denudation is dominant and there is only cracking ice.

Position. In different positions on slopes with similar orientations, the distribution of ground ice shows a regular change. On the top where bare bedrock or thin soil cover exists, because of coarse soils and good drainage, there is only permafrost of massive and crack cryogenic structure, mostly rare-ice and some-ice permafrost. From the upper to lower positions on slopes, the loose deposits become thicker and water accumulates easily. Thus, the ground ice becomes thicker or the types of permafrost change downslope from rare-ice permafrost to ice layers containing soil.

Basins with fine-grained soils and adequate moisture are favorable for the growth of ground ice, e.g. in the broad area of alluvial-pluvial plain and pluvial fen edges between Maintenance Squads 121 and 122 with smooth terrain, many bogs and 1 m thick humus, the development of ground ice is fairly good, although the mean annual ground temperature is higher. In basins consisting of lacustrine deposits, ground ice amounts are high, e.g. to the south of Maintenance Squad 113, a surface layer 1-2 m thick is composed of contemporary alluvial-pluvial deposits. Below is clay and silty clay containing gravel of lacustrine origin of Early Quaternary age. Near the permafrost table there is thicker layered ground ice with ice thicknessess of 4-10 cm and volumetric ice-contents of more than 70-80%.

The Valley and Plain Region

The taliks are through-going under large river beds, but not under moderate and small river beds. Owing to relative lower elevation, the thermal effects of river water, higher mean annual ground temperatures and coarse-grained soils, usually sand and gravel, there is only rare-ice and some-ice permafrost in these places. In the high flood plains and first terraces there are also only rare-ice and some-ice permafrost. Even in sections consisting of fine-grained soils, only rare-ice and some-ice permafrost develop. Only in swampy sections with well developed vegetation and humus, is there high ice-content permafrost. For example, drill hole CK 114-3 situated on the high flood plain on the eastern bank of Jiebu Qu as far as 1.5 km south of Maintenance Squad 114 reveals mainly rare-ice permafrost, because of spare vegetation and thin permafrost in spite of fine grained material. But drill hole CK 114-6 situated on the eastern bank with 50% vegetation cover and swampy conditions, reveals the following profile: 3.0 m, permafrost table; 3.0-3.4 m, layered ice with ice thicknesses of 4-5 cm and volumetric ice-content of 50%; 3.4-6.6 m, layered ice with lower volumetric ice content than upper part. On the second and third terraces there are rare-ice and some-ice permafrost. Sometimes, one can find high ice-content permafrost at the back of high terraces on the gentle slope of pediments, if there is supply of supra-permafrost water and fine-grained soils are present. For example, drill hole D-14 on the third terrace on southern bank of Tuotuo He reveals that the permafrost table lies at a depth of 2.3 m and the volumetric ice-content is 50-70% from a depth of 2.3-3.5 m.

In the middle and upper parts of tilted plains consisting of pluvial and outwash fens, there is mainly rare-ice and some-ice permafrost due to coarse sediments, e.g. in the broad outwash-pluvial fen between Maintenance Squads 104-107 with gravel sediments and spare vegetation, there is permafrost of massive cryogenic structure, with occasional thin, layered cryogenic structure. The ice content is usually less than 10-15%. But high ice-content permafrost may occur, if fine-grained soils and swampy conditions exist, e.g. in the tilted plains of the outwash fen in Nanshan of Xidatan, there is high ice-content permafrost. In the low part of pluvial and outwash fens and depressions between fens, ground ice amounts are higher than in the middle and upper parts, because it lies in the overflow zone of ground water and has adequate water and a high percentage of fine-grained soils. For example, in the depression in front of the outwash-pluvial fen between Maintenance Squads, 101-104, there is permafrost of thin layered cryogenic structure with 1 cm thick ice lenses in clay silt and silty clay.

REFERENCES

Cheng Guodong, 1982, The forming process of thick layered ground ice: Scientia Sinica (Series B), v. XXV, No. 7, p.777-788.

Vtyurin, B. I., 1975, Underground ice in USSR, Moscow, Science Press.

Wang Jiacheng, Wang Shaoling, and Qiu Guoqing, 1979, Permafrost along the Qinghai-Xizang Highway: Acta Geographic Sinica, v. 34, No. 1, p. 18-32.

Wu Ziwang, 1982, Classification of frozen soils in engineering construction: Journal of Glaciology and Cryopedology, v. 4, No. 4, p. 43-48.

TRITIUM IN PERMAFROST AND IN GROUND ICE

A. B. Chizhov[1] and N. I. Chizhova[1]
I. K. Morkovkina[2] and V. V. Romanov[2]

[1]Geology Department, Moscow State University
[2]Institute of Water Problems, Academy of Sciences
Moscow, USSR

The use of tritium analysis made it possible to establish the existence of considerable amounts of water in permafrost rocks and in ground ice in the zone of annual temperature fluctuations. The moisture was found to migrate from the active layer into the permafrost, resulting in the formation of ice lenses. High tritium concentrations caused by the inflow of meteoric water under 30 years of age are inherent in the growth of wedge ice and golets ice. However, tritium is absent from buried wedge ice and in lenses of primary ice. Tritium analysis offers new possibilities for studying laws governing the formation and age of ground ice in permafrost.

INTRODUCTION

Being a component of a water molecule, tritium is a splendid natural indicator of moisture transportation processes. Its radioactivity makes it possible to determine the age of natural waters. Because of that, tritium analysis has found wide application in meteorology, hydrology and hydrogeology. Since 1976 we have been studying the content of tritium in permafrost and ground ice in Southern Yakutia and the north-western part of the Amur region of the Aldan, Olekma and Giliui river basins. In total, over fifty analyses have been performed. Estimates of tritium concentrations in the samples of permafrost materials from two wells in the Mackenzie river-valley were made in Canada (Michel & Fritz, 1978). As a result, new information was obtained about the insufficiently known processes of water-transportation in the permafrost ground water system, about the formation of the cryogenic structure of the upper horizons of permafrost sediments and the age of ice inclusions.

TESTING PROCEDURE

Selection of samples for tritium analysis was performed as part of a study of complex permafrost and hydrogeological conditions. Samples of frozen material and ice from bore holes, excavations and exposures were thawed out on site, and the water was collected in polyethylene jars. The volume of samples was no less than 0.5 ℓ. All the samples were collected from permafrost of contemporary and Late Quaternary deposits. The maximum sampling depth was 10 to 11 m. In addition, samples of atmospheric precipitation, surface and underground water, and of icings were collected. General chemical analysis of the water and the ice was also carried out.

Concentrations of tritium were determined in the isotope research department at the Institute of Water Problems of the USSR Academy of Sciences (IWP) by using a liquid-scintillation spectrometer after preliminary electrolytic enrichment of water samples. The measurement error in most cases did not exceed ±10%. Where this was not the case, the value of the error is indicated. Results of measurements are presented in tritium units (one tritium unit—T.U. equals a concentration of one atom of tritium per 10^{18} atoms of protium).

RESULTS AND DISCUSSION

The results of the analyses revealed that the concentration of tritium in frozen materials and underground ice of various geneses ranges from 0 to 350 T.U. In liquid atmospheric precipitation in the area it amounted to 170-390 T.U., in snow 80-320 T.U., in surface and suprapermafrost water 150-290 T.U., and in icings 250-400 T.U. Concentration of tritium in subpermafrost water ranges from 426-402 to 8 T.U., depending on its age (Afanasenko et al., 1981). The lowest values are inherent in the sources of deep mineralized water and subpermafrost water in the zone of restricted water exchange. The intake of tritium into the permafrost hydrological system is performed through atmospheric precipitation. Data on average annual tritium concentration in the atmospheric precipitation of the area are presented by I. K. Morkovkina, V. V. Romanov, A. I. Tiurin, and A. B. Chizhov (1981). The considerable increase in tritium concentration in the atmospheric precipitation, which started in 1953, reached its maximum (5200 T.U.) in 1963, then declined to a level of 180-150 T.U.

When interpreting the results it is necessary to take into account the relatively short half-life of tritium (about 12.5 years). Therefore, the absence of tritium points to the isolation of the tested object from access to atmospheric moisture at least for the past 50 years. Concentrations of tritium in excess of 5 T.U. point to the presence of moisture younger than 1953-54 when thermonuclear tritium started to accumulate. And by relating the tritium concentration in the samples to its average concentration in precipitation over the period from 1953 to the date of testing it is possible in some cases to consider in the samples the approximate content of the contemporary water enriched by thermonuclear tritium. The average concentration of tritium in the total amount of precipitation by the time our research began (with due regard for radioactive disintegration) was about 320 T.U.

By 1982 this amount decreased to 270 T.U. Below we shall consider the data obtained on the concentration of tritium in various types of frozen materials and underground ice.

Frozen Unconsolidated Deposits and Ice in the Layer of Seasonal Thawing and Freezing

The uppermost part of the permafrost is bounded by the layer of seasonal thawing. The thickness of this layer at the time of maximum thawing (early October) changes from 0.5-0.7 m in peat, to 2.5-3.0 m in sand. It is to the layer of seasonal thawing above permafrost that the suprapermafrost water is timed. In winter the layer usually freezes completely. There with schlieren of segregation ice as thick as 5 cm are formed in dispersion rocks. Congelation ices are abundant in the layer of seasonal thawing in block deposits of rock glaciers. Tritium concentrations in frozen dispersion deposits and peat varied from 64 to 350 T.U. (10 samples). This is connected both with perennial and annual variability of tritium content in atmospheric precipitation. The lowest values (64 T.U.) were in schlieren ice samples in sandy loam from a depth of 0.55-0.8 m on August 9, 1982 (the depth of thawing at the moment of research was 0.5). These values are consistent with the precipitation in June, 1981. But in the summer of 1981 and 1980 tritium concentrations at the same depths reached 258 and 326 T.U. Tritium in rock glaciers in 1977-78 was between the limits of 80-167 T.U., which approximately corresponded to its content in snow. This is well in accord with the current notion about formation of this type of ice as a result of refreezing of snow melt water.

Strata and lenses of injection and injection-segregation types of ice are widespread in the layer of seasonal freezing above floodplain taliks. Tritium concentrations in the taliks correspond to its content in river water or are somewhat higher. The higher values are caused by the influx of subpermafrost water into alluvium.

Permafrost and Enclosed Ground Ice

Permafrost in the studied area is predominantly discontinuously distributed. Its annual mean temperature varies from 0°C to -4°C; the thickness reaches 200-300 m. The stratigraphic column of perennially frozen deposits is composed of an upper unit of ice-bearing Quaternary deposits underlain by Mesozoic and older bedrock. The thickness of the Quaternary deposits rarely exceeds 20-30 m. They include deposits of bedded and recurrent wedge ice.

High tritium concentrations (80 T.U. and over) are found at depths of 0.2-0.3 m and below the upper boundary of permafrost. Tritium concentrations are related to the existence of an upper transitional layer of permafrost which melted over the last 25-30 years only in the warmest seasons. An idea about the character of tritium distribution at great depths is given by core samples from bore-hole 9 drilled in a 7.5 m high frost mound in the Gorbyliakh river-valley (Fig. 1A). From the surface it is composed of peat which is frozen at a depth of 0.3 m. At a depth of 2.9 m the boring penetrated a meter-thick layer of ice which had a dissolved solids content of 150 mg/ℓ. The ice was underlain by sandy loam with ataxitic cryogenic texture (volumetric moisture content equals 40-70%). The revealed distribution of tritium concentrations may be explained by moisture migration (diffusion) from the layer of seasonal thawing into perennially frozen material in the direction of lower temperatures. To test the hypothesis about the existence of moisture migration from the base of the seasonally thawed layer deep into the permafrost zone, calculations of changing tritium concentrations with depth were made on the basis of solving the equation of diffusional transportation

$$C_{(z)} = C_{(z=0)} \exp - \sqrt{\frac{\lambda}{k}}\, z \qquad (1)$$

in which $C_{(z)}$ is tritium concentration at a depth of z, k is a coefficient of diffusion, λ is a constant representing radioactive disintegration of tritium (0.056 $year^{-1}$).

The theoretical curve of the diffusional distribution of tritium concentrations coincided with the actual one, on which tritium concentrations were placed in the middle of the tested intervals using $k = 6.10^{-4}$ cm^2/s (Morkovkina et al., 1982). This quantity does not exceed the limits of certain well-known values of k in dispersion rocks by laboratory estimates. Approximate estimate of the rate of migrational flow gives a value of 10^{-2} kg/m^2 per day. Time of its existence is determined by the period from the beginning of seasonal thawing to the refreezing of seasonally thawed layer.

Perhaps only by the moisture migration from the layer of seasonal thawing of high tritium concentration in the strata of underground ice in the peatland of the Niukzhi river valley can be explained (bore hole 63, Fig. 1A). In the nearby bore hole 65, tritium concentration in sandy loam with ataxitic cryogenic texture at a depth of 3.2-4.2 m was 21 ± 3 T.U. Essentially the same value (20 ± 3 T.U.) was obtained for a sample of frozen sandy loam, underlying peat bogs at a distance of 0.5 km from bore hole 65 (exploratory shaft 73, Fig. 1B). Attention should be drawn to the fact that the results of the analyses of the samples from exploratory shaft 73 do not reveal regular depth variations of tritium concentrations. Similar findings are also observed for peat bogs in the Derput brook-valley (exploratory shaft 1, Fig. 1B). Low tritium concentrations in its exposed section are evidently conditioned by the character of the temperature regime. The layer of winter freezing does not extend to the upper surface of the perennial frozen peat. This results in its very insignificant temperature gradients and the thawed material at the contact with perennial frozen ones is dehydrated. All this deteriorates the conditions for moisture migration.

Both the areas of bore hole 65 and shaft 73 are composed of slightly to moderately decayed peat with rare interlayers of ice as thick as 5-7 cm. Volumetric moisture capacity of the peat is 70-90%. The age of the lower peat horizon in the area of the Derput brook is estimated by the radiocarbon method at 9,500 years; the age of the upper horizon is 1,800 years. The freezing of peat occurred 2,000 to 1,500 years ago.

Tritium was not discovered in the layers of segregational ice 15-16 cm thick, deposited at a depth of from 2 to 3.5 m in powder-fine sands (Upper Pleistocene alluvium). These are primary types of ice which were formed simultaneously with enclosing perennial frozen sediments, which include extensive wedge ice. Ice wedges of this type are characterized by a concentration of dissolved solids that is usually less than 50 mg/ℓ.

The obtained data and the results of investigations in the Mackenzie river-valley (Mickel & Fritz, 1978) are indicative of intensive penetration of contemporary atmospheric moisture from the layer of seasonal thawing into upper horizons of permafrost. Data on tritium concentration make it possible to approximately determine that in the ground ice there is about 20-30% of moisture content younger than 25-30 years of age and up to 7-10% in perennially frozen unconsolidated deposits. The conclusion about the existence of moisture migration from a seasonally thawing layer into the permafrost zone with formation of icy schlieren agrees well with the results of laboratory experiments (Ershov et al., 1976). This process is hindered by the existence of primary strata of slightly mineralized ice in perennially frozen unconsolidated deposits which have practically no liquid phase. The latter proposition is confirmed by the results of the tritium analysis of ancient recurrent wedge ice.

Recurrent wedge ice (RWI). Underground wedge ice is widespread in alluvium, peat bogs, and proluvium deposits in the north of the Amur region and more seldom in Southern Yakutia. One distinguishes contemporary growing RWI and ancient buried RWI, which are deposited at a depth of more than 0.5-1 m below the layer of seasonal thawing. The depth of penetration of contemporary RWI commonly does not exceed 2-2.5 m, although the depth of the ancient ones is up to 5-7 m.

In contemporary growing ice wedges, tritium concentrations vary within the limits of 20-230 T.U. (9 samples). High tritium concentrations (over 70-80 T.U.), to all appearances, indicate the formation of the great bulk of ice in the last 25-30 years. Lower values are typical of wedges which began forming long before the outset of the thermonuclear tritium intake. The greater part of the ice in this case contains tritium in concentrations less than 5 T.U., which is within the limits of error of the analysis. This is confirmed by the change of the tritium concentrations in one of the veins from 98 T.U. in the middle part to 21 ± 3 T.U. at the lateral boundary. Decrease in tritium concentrations in ice veins from the upper part to the lower one should also be noted. The distribution of tritium inside the veins is evidently determined by conditions of their formation.

In contrast to the contemporary growing ice wedges, tritium in the ancient buried wedges is actually non-existent (the error of observation in most cases did not exceed ±5 T.U.). This is the case even at the upper boundary of the buried ice wedge at a depth of 1-1.5 m. In August, 1982 studies were made of a section of an ice wedge complex in the Lake Tchikun depression, where a modern wedge 10 cm wide at a depth of 1.1 m penetrates into buried wedge ice and is distinctly traced within it to a depth of 2.3 m. Tritium concentration in the younger ice wedge at its point of entrance into the buried one is 64 T.U. In the same contemporary wedge, within the body of the buried one at a depth of 1.5-1.6 m, the tritium concentration equals 20 ± 9 T.U. No tritium has been found in the sample selected at this depth from a buried vein at its point of contact with a contemporary one. It confirms that there is no moisture transportation in the ground ice, in which dissolved solids do not exceed 50-70 mg/ℓ).

Golets ice. This type of ice extends under the layer of seasonal thawing in rock glaciers. It is mainly formed in the process of freezing snow melt water. Fig. 1C shows the changes in tritium concentrations in a golets ice cross-section through one of the rock glaciers in Southern Yakutia (Morkovkina et al., 1981). Ice to a depth of 2.4 m with tritium concentrations of 140-167 T.U. forms a part of the seasonal thawing layer. The presence of tritium further down is connected with the moisture inflow during the period 1953-77. This might be brought about by the movement of rock fragments and ice under the influence of gravity force and of ground heaving. Calculation performed from the data on tritium content in atmospheric precipitation in the layer of perennial golets ice gave the amount of its feeding about 6 mm/year. About 20% of ice was formed during the last 25 years.

CONCLUSIONS

1. Comparatively high tritium concentrations were established in frozen dispersion deposits in the zone of annual temperature fluctuations. Tritium content in frozen rocks of the layer of seasonal thawing and in the transitional layer is near the concentration of atmospheric precipitation. Downward, as a rule, the concentration is less where connected with presence of moisture older than 25-30 years of age.
2. Data obtained from tritium analysis show the existence of a considerable amount of moisture migration from the layer of seasonal thawing into permafrost. This process to a great extent influences the formation of the cryogenic structure and ice content in dispersion permafrost within the zone of annual temperature fluctuations. As a result, the age of segregation ice may be considerably less than the time that has passed since formation of permafrost. The intensive growth of ice layers and formation of cryogenic textures may also be modern.
3. Tritium is an indispensable component of contemporary growing ice wedges. Yet, it does not exist in ancient buried wedge ice and in primary ice strata, which are not influenced by the process of moisture migration.
4. About 20% of golets ice was formed in the process of freezing contemporary atmospheric water enriched by thermonuclear tritium. It is indicative of intensive water exchange in rock glaciers and of comparatively young age of golets ice.
5. Employment of tritium analysis in permafrost studies opens new possibilities in the study of moisture transfer processes in the system perma-

frost-natural water, for determining the nature of moisture exchange in permafrost rocks and for age assessment of ice inclusions.

REFERENCES

Afanasenko, V. E., Chizhova, N. I., Morkovkina, I. K., Romanov, V. V., and Buldovich, S. N., 1976, Ispol'zovanie tritievogo metoda dlia opredelenia vozrasta podzemnykh vod zony intensivnogo vodoobmena Chul'manskoy vpadiny [Use of tritium method for age assessment of underground waters in the zone of intensive water exchange in the Chulman depression]. "Vestnik Moskovskogo universiteta", seria geologicheskaya, no. 2, p. 82-86.

Ershov, E. D., Cheverev, V. G., and Lebedenko, Yu. P., 1976, Eksperimental'noe issledovanie migratsii vlagi i l'dovydelenia v merzloi zone ottaivaiuschikh gruntov [Experimental research of moisture migration and formation of ice in the permafrost zone of thawing ground]. "Vestnik Moskovskogo universiteta", seria geologicheskaya, no. 1.

Morkovkina, I. K., Romanov, V. V., Tiurin, A. I., and Chizov, A. B., 1981. Issledovanie kurumoobrazovania izotopnymi metodami [Investigation of rock glacier formation by isotope methods], *in* Issledovanie prirodnykh vod izotopnymi metodami, Moskva: "Nauka", p. 90-96.

Morkovkina, I. K., Romanov, V. V., Chizov, A. B., and Chizhova, N. I., 1982. Issledovanie soderzhania tritia v merzlykh porodakh i podzemnykh l'dakh [Investigation of tritium content in permafrost and ground ice]. "Vodnye resursy", no. 5, Moskva: "Nauka", p. 164-168.

Michel, F. A. and Fritz, P., 1978, Environmental isotopes in permafrost related waters along the Mackenzie valley corridor. Proceedings of the Third International Conference on Permafrost, vol. 1, Ottawa, p. 207-212.

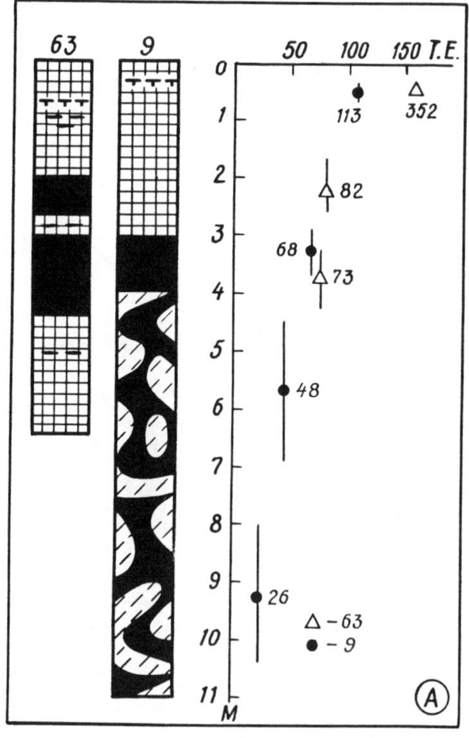

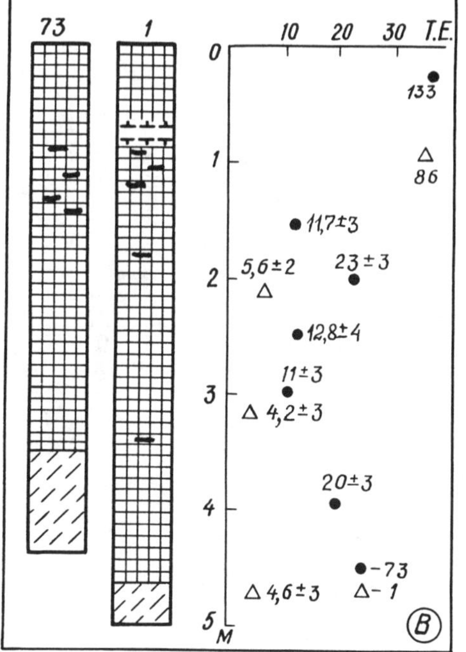

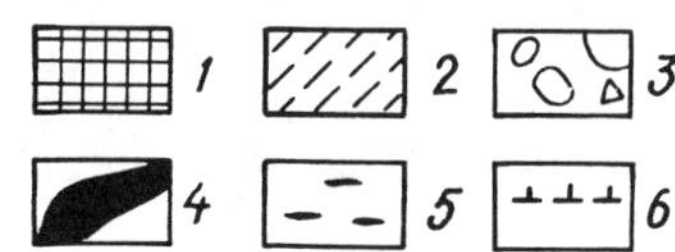

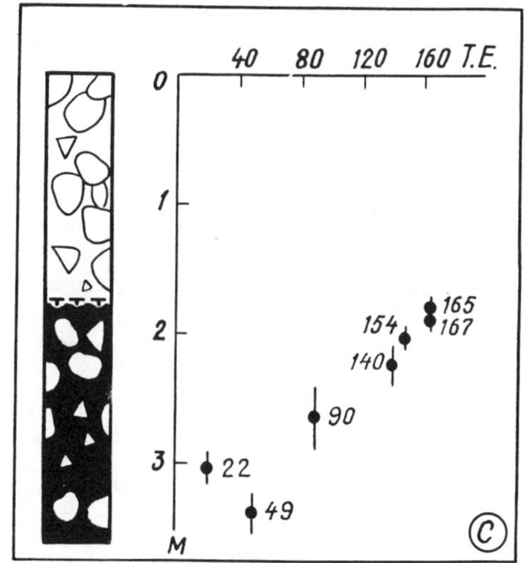

FIGURE 1 Depth variation of tritium content
1 - peat
2 - sandy loam
3 - blocks and rock fragments
4 - ice
5 - ice streaks 1-10 cm thick
6 - boundaries of frozen rocks

THE RECOGNITION AND INTERPRETATION OF *IN SITU* AND REMOULDED TILL DEPOSITS, WESTERN JYLLAND, DENMARK

Leif Christensen

Department of Geology, Aarhus University, DK-8000 Aarhus C, Denmark

Major engineering projects in western Jylland, Denmark are greatly influenced by the extent and thickness of remoulded till deposits. A remoulded till study was carried out on the southern slopes of the Saalian landscape of Skovbjerg Bakkeö north of Skjern using airphoto interpretation and test hole drilling. Remoulded deposits can be recognized on airphotos revealing deformed large-scale crop-polygonal patterns. Apparent shear strength parameters determined by vane-tests for clayey and silty deposits indicate lower strengths in remoulded deposits and higher strengths in *in situ* deposits. Thickness of remoulded deposits varies between 0,5 to 4 meters in the study area. Approximately 40% of the study area is underlain by remoulded deposits.

INTRODUCTION

Intensification of engineering activities in Denmark during the 1970's has resulted in increasing concern about the distribution of remoulded and *in situ* Saalian till deposits in western Jylland. The presence of compressible layers in remoulded deposits and of compressible organic Eemian deposits below remoulded deposits influence major engineering projects such as heating plants and sewage plants. Since such construction projects encounter numerous transitions from compressible to incompressible deposits it is essentiel that the distribution of remoulded deposits can be mapped as accurately as possible. This concern has prompted an investigation to accurately determine the extent of *in situ* Saalian till deposits and remoulded deposits of till origin in a selected area built up of clayey till deposits north of Skjern, western Jylland, Denmark (Figure 1).

Preliminary terrain studies including the interpretation of airphotos and test hole drillings indicated, that remoulded deposits were extremely widespread and variable and that more detailed investigations were necessary. These studies included detailed large-scale airphoto interpretation of crop-marks and test hole drillings.

DESCRIPTION OF STUDY AREA

The study area is located on the southern slopes of Skovbjerg Bakkeö. It is commonly accepted that "Bakkeöerne" - the landscape of Saalian till and meltwater deposits in western Jylland - have experienced active periglacial modifications throughout the Weichselian Glaciation, when the inland ice reached central Jylland (Figure 1). Solifluction has reduced the lower slopes in angle and extended the slopes in length, while the upper parts of "Bakkeöerne" suffered erosion and retreat and near stable slopes were ultimately attained.

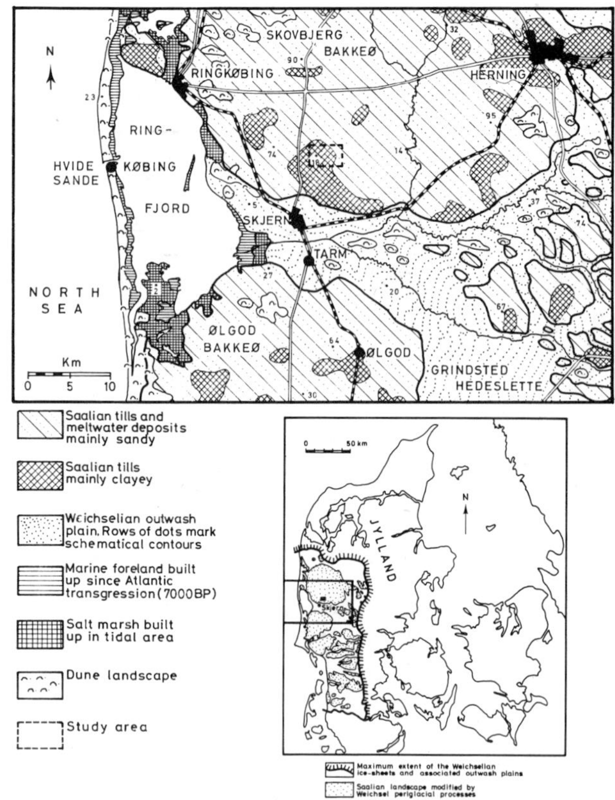

FIGURE 1 Location of the study area.

DELINEATION OF REMOULDED AREAS USING AIRPHOTO MAPPING

Airphoto mapping and interpretation of crop-marks is aided by dry spells. During these periods water-stress conditions reveal differences in plant growth and development which appear as crop-marks in northwest European fields (Shotton 1960, Williams 1964, Morgan 1971, Svensson 1964a, 1964b,

FIGURE 2 Random orthogonal polygonal crop-patterns in a field of spring barley south of Saedding. Photo 9.7.1975.

FIGURE 3 Stripes and elongated orthogonal crop-patterns oriented downslope in a field of rape east of Saedding. Photo 9.7.1975.

FIGURE 4 Pseudomorph of ice-wedge fractured after transition into a fossil state. Central part on Figure 3.

FIGURE 5a. Pseudomorph of ice-wedge deformed after transition into a fossil state.

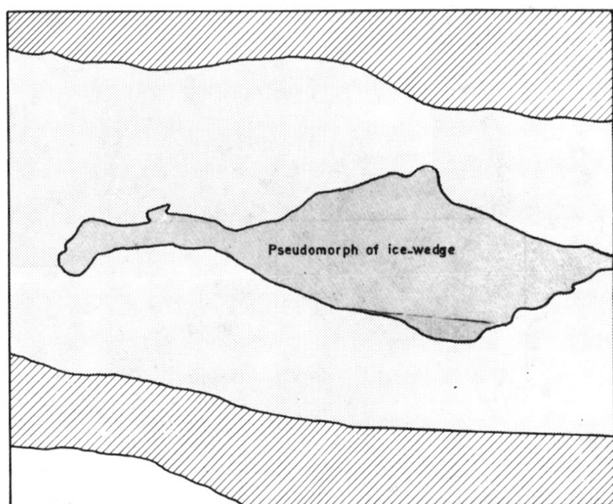

FIGURE 5b Sketch of Figure 5a.

1972, 1973 and 1976, Christensen 1974 and 1978, Jakob & Lamp 1980). In spring and early summer when evaporation exceeds precipitation, positive and negative crop-marks are especially well developed in cereals such as spring barley, wheat and oats (Christensen 1978). Crop-marks can be observed too in root crops and maize during dry spells in late summer and early autumn. Continous mapping of an area therefore demands crop rotation and airphotos from several different years with different periods of water deficiency during the season of growth. The increased use of irrigation by Danish farmers since the mid 1970's has reduced the value of more recent airphotos.

Crop-marks are better primary indicators of *in situ* and remoulded deposits than terrain features because remoulded fossil solifluction deposits of till origin have lost their topographic forms (Christensen 1982).

Three categories of polygonal crop-patterns are commonly found. Random orthogonal and hexagonal patterns are found on flat ground (Figures 2 and 6). On gradients of 1^0 the patterns change to orthogonal, elongated and are oriented downslope. On gradients of 2^0 or more only stripes are found (Figu-

FIGURE 6 Vertical airphoto showing different categories of polygonal crop-patterns around Saedding. Photo 11.5.1954. Copyright Tactical Air Command, Karup.

res 3 and 6).

In situ deposits are commonly found in areas revealing hexagonal and random orthogonal patterns, while remoulded deposits are commonly found in areas revealing elongated patterns and stripes.

The soil structures revealed in excavations through these crop-marks have been interpreted as pseudomorphs of ice-wedge polygonal patterns (Christensen 1974,1978 and 1982, Svensson 1973 and 1978).

In areas with elongated patterns and stripes excavations have revealed fractured, deformed and overturned pseudomorphs of ice-wedges below crop-marks (Figures 4 and 5).

DELINEATION OF REMOULDED DEPOSITS USING DRILLINGS

In delineating clayey remoulded deposits apparent shear-strength parameters as measured by vane tests in drillings can be used (Christensen 1982). The apparent shear-strength values - c_v intact and c'_v remoulded - are markedly lower in remoulded clayey deposits. Figure 7 illustrates that the apparent shear -strength values - c_v and c'_v - as measured by vane responds to changes in the thickness of the remoulded deposits and changes in soil type.

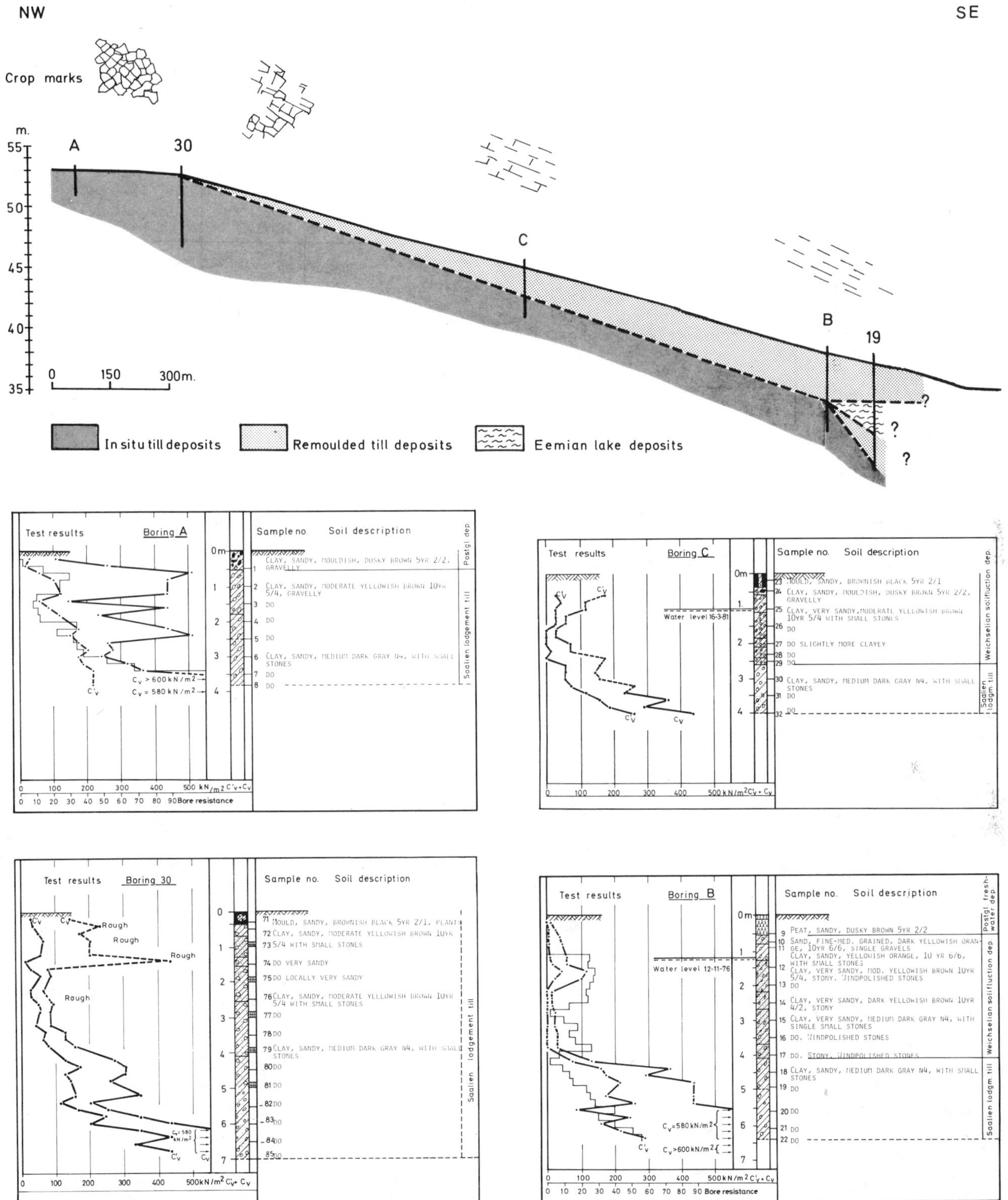

FIGURE 7 Some typical relationships of *in situ* and remoulded deposits to terrain and crop-marks with the corresponding apparent shear-strength values for each drill hole except drill hole 19. Vertical exaggeration x 30.

Downslope the remoulded deposits of till origin become more sandy than the *in situ* deposits. The resultant sandy deposit can often appear quite different than that from which it has derived.

LABORATORY STUDIES

Undisturbed samples from drillings have been investigated for unit weight of soil γ, water content w, grain size composition and consolidation tests.

The unit weight of soil γ is higher and water content w lower in *in situ* tills than in the remoulded deposits of till origin (Table 1).

TABLE 1
Unit weight of soil γ and water content w of *in situ* till deposits and remoulded deposits of till origin in the Saedding area.

	γ t/m^2	w weight %
in situ till deposits	2,07-2,24	11,5-19,5
remoulded deposits of till origin	1,89-2,12	14,8-24,7

Grain-size distributions of *in situ* tills and remoulded deposits of till origin, show that the latter material is more silty and sandy, the clay content being lower (Figure 8).

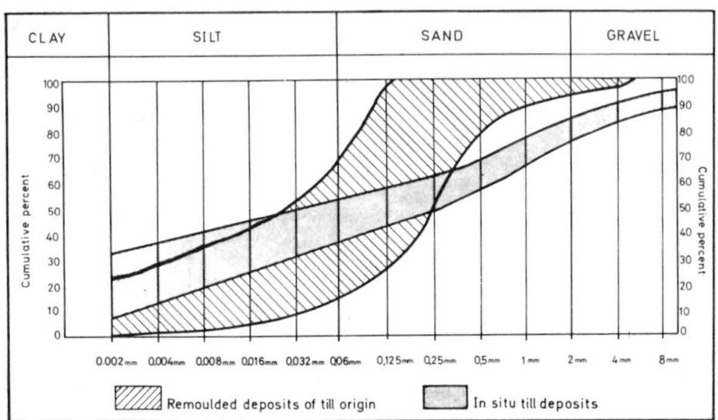

FIGURE 8 Grain-size distribution ranges for *in situ* till deposits (38 samples) and remoulded deposits of till origin (83 samples) in the Saedding area.

Consolidation test-curves have been made on selected samples. A marked difference between *in situ* clayey till materials and clay materials from remoulded deposits is found (Figures 9 and 10). The clays of the example test curve in Figure 9 is considered to be preconsolidated, while the clays of the example test curve in Figure 10 are probably normally consolidated.

It is generally considered that glacial conditions favour consolidation of sediments.

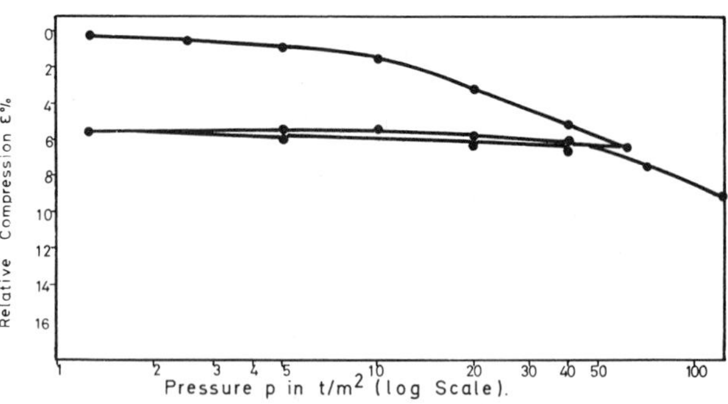

FIGURE 9 Working curve from consolidation test of preconsolidated *in situ* till.

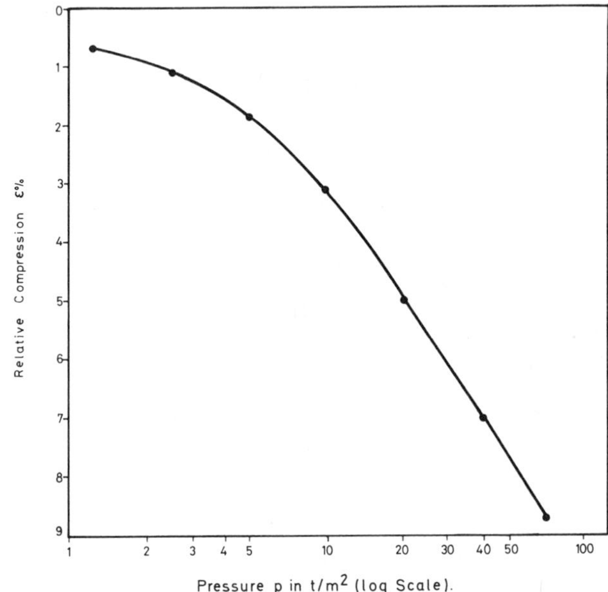

FIGURE 10 Working curve from consolidation test of clay layer in remoulded deposit of till origin.

DISTRIBUTION OF REMOULDED DEPOSITS

Remoulded deposits were present over 40% of the study area and ranged from a low of approximately 5% areal extent in flat areas with gradients less than 1^0 to around 60%-70% on sloping terrain.

An example of a survey map (Figure 11) shows the distribution of *in situ* and remoulded deposits in a section of the study area east of Saedding.

CONCLUSIONS

Remoulded deposits of till origin in the study area are confined to areas with a slope of 1^0 or more. These areas occupy approximately 40% of the study area. Remoulded and *in situ* till deposits can be recognized on airphotos revealing deformed large-scale crop-patterns. Only drillings can estimate the thickness of remoulded deposits. Drillings therefore are useful for providing data to confirm airphoto interpretations.

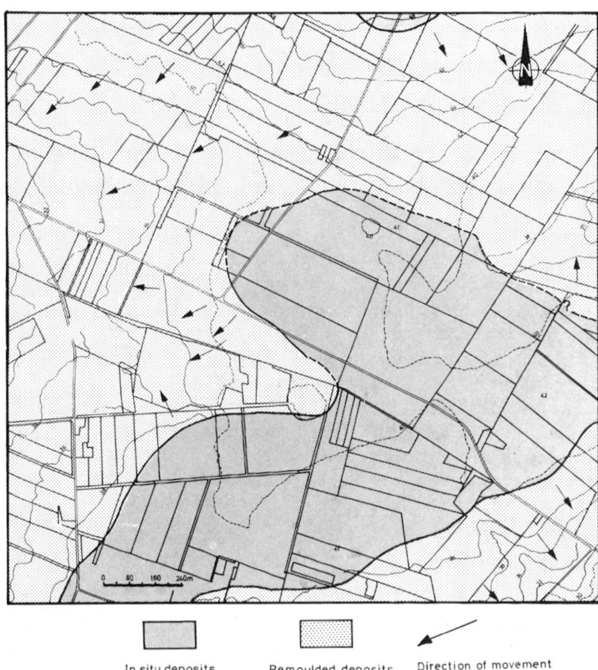

FIGURE 11 Survey map of *in situ* and remoulded deposits east of Saedding. Contour interval 2 meter. See also Figure 6.

ACKNOWLEDGEMENTS

This work was financed by the National Research Council of Denmark. Major assistance was given by Dansk Geotaknik A/S, Aarhus and Svend Melgaard of the Department of Geology, Aarhus University has assisted with this project. David Penney kindly improved the English of the text.

REFERENCES

Christensen, L., 1974, Crop-marks revealing large-scale patterned ground structures in cultivated areas, southwestern Jutland, Denmark. Boreas 3, 153-180.

Christensen, L., 1978, Waterstress conditions in cereals used in recognizing fossil ice-wedge polygonal patterns in Denmark and northern Germany. Third Internat. Conf. on Permafrost, proceedings vol. 1, 255-261.

Christensen, L., 1982, Faststaaende og omlejrede Saale-moraenelersaflejringer. Et eksempel fra Saedding-omraadet, Vestjylland. Geogr. Tidsskr. 82, 91-94.

Jakob, J.A. und Lamp, J., 1980, Fossile Frostpolygonmuster auf Luftbildern Mittelholsteins und ihre bodenkundliche Bedeutung. Meyniana, 32, 129-134.

Morgan, A.V., 1971, Polygonal patterned ground of Late Weichselian age in the area north and west of Wolverhampton, England, Gergr. Ann. 53 A, 146-156.

Shotton, F.W., 1960, Large-scale patterned ground in the valley of the Worcestershire Avon. Geol. Mag. 47, 404-408.

Svensson, H., 1964a, Aerial photographs for tracing and investigating fossil tundra ground in Scandinavia. Biul. Peryglac. 14, 321-325.

Svensson, H., 1964b, Fossil tundramark paa Laholmsslätten. Sver. geol. Undersökn. C598, 1-29.

Svensson, H., 1972,The use of stress situations in vegetation for detecting ground conditions on aerial photographs. Photogrammetria 28, 75-87.

Svensson, H., 1973, Distribution and chronology of relict polygon patterns on the Laholm plain, the Swedish west coast, Geogr. Ann. A 54, 159-175.

Svensson, H., 1976, Relict ice-wedge polygons revealed on aerial photographs from kaltenirchen, Northern Germany. Gergr. Tidsskr. 75, 8-12.

Svensson, H., 1978, Ice-wedges as geomorphological indicator of climatic changes. Dansk Met. Inst. Klim. Medd 4.

Williams, R.B.G., 1964, Fossil patterned ground in eastern England. Biul. Peryglac. 14, 337-349.

ICE-SOIL MIXTURES: VISUAL AND NEAR-INFRARED REMOTE SENSING TECHNIQUES

Roger N. Clark

Planetary Geosciences Division, Hawaii Institute of Geophysics,
University of Hawaii, Honolulu, Hawaii 96822 USA

The spectral properties of ice-soil mixtures are reviewed. Spectra of intimate mixtures of soil and ice are highly complex, nonlinear functions of the optical properties of ice and soil. Water ice has an absorption coefficient that varies by several orders of magnitude in the visual and near infrared (0.4–3μm) and has several prominent overtone absorptions (at 2.0, 1.5, 1.25, and 1.04 μm). Thus, different wavelengths can be used to probe to different depths in the surface as well as for different mineral impurity concentrations. Empirical and/or theoretical models might be used for deriving characteristic grain size of ice or impurity minerals and for deriving abundance of the ice and rock or soil components. Quantitative analysis of remotely obtained reflectance spectra can only be performed by using the absorption features in the spectra of ice and soil and not by broadband response. This might be done by special selection of several narrow band filters in the near infrared that will adequately define the ice absorptions.

INTRODUCTION

Understanding the optical properties of water ice and mixtures of ice and soil are essential for remote sensing studies. The amount of sunlight reflected from a surface composed of ice and soil depends on the abundances and grain sizes of each mineral component, the absorption coefficients of the minerals and the physical state of the ice-soil association. The spectral properties of an ice-soil intimate mixture (e.g., like frozen mud) are a highly complex (nonlinear) combination of the end member spectral properties. Intimate mixtures are one end of the range of possibilities that may be encountered. The other end of the range is called an areal mixture, where the materials are optically isolated patches of the pure end members located on the surface. The spectrum of the latter is a simple linear addition of the spectra of the optically isolated patches of minerals weighted by the fractional areal coverage of each patch in the field of view of the detector. There is a continuous range of mixture possibilities in between the intimate and areal mixtures, as well as vertically stratified cases such as a frost layer on a soil. Inversion of a remotely obtained spectrum of a surface to determine mineral abundance is a formidable problem, if a quantitative and unique answer is desired. Fortunately, water ice has optical properties that are very different from those of other derived from remotely obtained spectra. This paper reviews the current understanding of the spectral properties of ice and mixtures of ice and soil for the purpose of remote sensing.

ICE

H_2O ice contains absorptions in the infrared owing to O-H stretches (3 μm), H-O-H bend (6 μm), and many translation and rotational modes at longer wavelengths. In the spectral region of reflected solar radiation, ~0.4 to ~2.5 μm, ice displays overtones of the above modes at 2.02, 1.52, 1.25, 1.04, 0.90, and 0.81 μm. In reflectance, scattering controls the light returned from the surface to the detector, and scattering can occur from ice-air interfaces (grain boundaries or crystal imperfections) or from impurities mixed in the surface such as a soil. Thus, even in a pure ice or snow, the effective grain size greatly controls the appearance of the reflectance spectrum (Figure 1). The words frost and snow are used interchangeably in this paper since there are no distinguishable spectral differences.

The mean path length that photons travel in a mineral is determined by scattering and by the absorption coefficient of the material. For instance, at 3.075 μm the absorption coefficient of ice is 14008 cm^{-1} (Irvine and Pollack, 1968), and a path length of 1 μm results in absorption of 25% of the light, whereas at the center of the 1.04-μm ice absorption, the absorption coefficient is 0.34 cm^{-1} and a path length of 4.1 cm is required for the same 25% absorption. Thus, different wavelengths probe to different depths in an icy surface. This large variation in optical path length is an advantage for remote sensing studies since each absorption is sensitive to different amounts of impurities and to scattering due to different grain sizes.

SPECTRAL PROPERTIES OF ICE-SOIL MIXTURES

There has been considerable advancement in the understanding of the optical properties of ice and snow and the spectral properties of ice-soil mixtures over the last several years. Warren (1982) reviewed the optical properties of snow, concentrating on how the visual reflectance is affected by small amounts of impurities. Clark (1982) discussed the implications of using both visual and near infrared broadband detectors for remote sensing studies. The visual reflectance of a pure frozen water can vary from 1.0 for a fine grained frost to ~0.4 for an ice with little internal scattering that is characteristic of some terrestrial ice in the arctic and antarctic. With small fractions of a weight percent soot or dust, the visual reflectance can vary from unity to less than 0.1 (Warren, 1982; Clark, 1982). Thus the visual reflectance of ice varies so much with scattering conditions due to grain size and small amounts of impurities that it is unreliable to use the visual albedo alone for any quantitative remote sensing study.

Clark (1981a, b) studied the spectral properties of ice blocks, frosts of various grain sizes, and ice-mineral mixtures. Impurities in an ice or frost layer can radically change the appearance of the reflectance spectrum in the near infrared, as well as in the visual part of the spectrum. The addition of mineral grains dusted on the surface of a frost layer decreases the 1.5- and 2.0-μm absorptions roughly proportional to the fractional areal coverage. However, since the surface of the frost is not a flat layer but a "fairy castle" structure, the mineral grains are somewhat mixed in the uppermost part of the surface, which results in a higher probability of a photon encountering a soil grain. This results in reduced scattering and path length in ice. Also, the albedo of the surface may be drastically reduced many times more than expected for a fractional areal coverage by using an additive model for areal mixtures. Thus, mineral grains on a frost layer are detectable in very small quantities. For example, a fractional areal coverage <0.005 is detectable if the mineral has suitable absorptions outside the major water ice features and the data are of sufficiently high quality, 0.5% precision (Clark, 1981a).

In the case of a frost layer on a soil, a thick frost layer (>1 mm) is required to mask the soil at wavelengths shortward of 1.4 μm owing to the low absorption coefficient of ice but also depends on the mineral reflectance features and the frost grain size. Frost on a very dark surface (e.g., reflectance 6%) is easily seen even when only a few micrometer layer of frost is present.

Spectral studies of minerals with adsorbed water show that the physically and chemically bound water absorptions, which occur at 1.4, 1.9, and 2.2 μm, are distinguishable from the broader water ice absorptions at 1.5 and 2.0 μm and do not shift appreciably in wavelength (<0.01 μm) over the temperature range from 273K to 150K (Clark, 1981b).

In the case of intimate mixtures, a dark material intimately mixed with water ice can completely mask the water absorptions shortward of 2.5 μm (Clark, 1981a). The higher overtones (0.8 μm to 1.25 μm) are more readily masked than the stronger 1.5- and 2.0-μm absorptions; in fact, these weaker absorptions become suppressed with minerals dusted on frost. Even relatively high reflectivity (e.g., 0.8) grains, intimately mixed in a frost layer, can greatly reduce the scattering, thus limiting the path length and suppressing the higher overtone absorptions. These effects cause a decrease in the ratio of higher to lower overtone absorption band depths (e.g., the depth of the 1.25-μm band/1.5-μm band or 0.8-μm band/1.5-μm band). This is in contrast to a thin frost layer on an ice surface whose spectra show an increase in the corresponding apparent band depth ratios (Clark, 1981b).

More recently, Clark and Lucey (1983) have studied the spectral properties of mineral particulates intimately mixed in ice blocks in more detail, and one of their series of spectra are shown in Figure 2. Note how the ice bands at 2.0, 1.5, 1.25, 1.04, and 0.9 μm change with the particulate content. The relative depth of an absorption is defined with respect to a continuum (a smooth line such as a cubic spline polynomial) fit to the reflectance peaks between absorption bands (Clark, 1981b; Clark and Roush, 1983). If R_b is the reflectance at the absorption band center and R_c is the reflectance of the fitted continuum at the same wavelength as R_b, then the band depth D is (Clark and Roush, 1983)

$$D = (R_c - R_b)/R_c \qquad (1)$$

The band depths for the ice absorptions increase as the mean optical path length increases until significant absorption occurs in the wings of the absorption bands, then the bands become saturated and the depth decreases. A saturated band may almost completely disappear as seen with the 1.5- and 2.0-μm bands in the pure ice spectrum in Figure 2.

In order to help provide a calibration of ice purity from reflectance spectra, Clark and Lucey (1983) derived the curves of growth for several ice bands as a function of the photon mean optical path length in ice by using the methods of Clark and Roush (1983). A curve of growth is the relationship between an absorption band depth and increasing optical path length through the material. This is shown in Figure 3 and is similar to curves of growth produced from laboratory spectra in Figure 2. The albedo of the particulate impurity greatly affects the spectrum and the curves of growth. The darker the particulate, the less scattering takes place in the surface, which results in a lower optical path length in ice and a smaller band depth. Clark and Lucey (1983) found that the band depth D, divided by the continuum reflectance R_c, removed the effects of particulate albedo (Figure 4). Polynomials fit to such data for the 1.04-, 1.25-, 1.52-, and 2.02-μm bands provide calibration curves to abundance.

REMOTELY SENSED SPECTRA

The previous discussion implies that some quantitative information can be obtained from remotely sensed spectra of an icy surface, given adequate spectral coverage and data precision. In remote sensing studies of the earth's surface, the atmosphere provides an additional difficulty, owing to many absorptions, mainly from water vapor, oxygen, and carbon dioxide (the latter two are relatively weak). Nevertheless, McCord and Clark (1979) have shown that the extinction due to atmospheric absorptions in this region can be removed in at least some cases. To do this, the same physical area must be measured for different path lengths through the atmosphere at each wavelength. From a satellite or airplane this might be accomplished by changing the look direction while flying over the spot. However, this adds another complication in that the viewing geometry with respect to the surface normal and the solar phase angle will also change. Thus, the scattering within the surface will also change and the spectral properties will change slightly. This is a second order effect, however, and will not significantly affect the crude abundance calculations outlined in the last section.

Recently, Hapke (1981) presented a theory for computing bidirectional reflectance of intimately-mixed, multimineralic surfaces from the complex indices of refraction, grain sizes, weight fractions, and densities of the mineral components. Most presentations of reflectance theories are in terms of deriving the absorption coefficient from laboratory spectra of a powdered sample, given the grain size of the material. However, Clark and Roush (1983) presented a method for deriving abundance from reflectance spectra. In a remote sensing study, if the materials present can be identified from absorption features in the reflectance spectrum, then the absorption coefficients of the materials are known (or can be measured). The reflectance theories can be used to derive the average particle single scattering albedo from observations of the surface at many phase angles. Using the equations to compute the single scattering albedo, when the absorption coefficient is unknown can only determine the product of grain size and absorption coefficient. If the absorption coefficients are known, however, then a non-linear least squares algorithm can be employed to solve for the grain size and the weight fraction of each component.

The ice absorptions may be adequately defined with as few as 9 wavelengths centered near 0.95, 1.04, 1.12, 1.25, 1.40, 1.52, 1.83, 2.02, and 2.24 μm. With these wavelengths, either the previously discussed theoretical or empirical methods might be used to derive abundance from a reflectance spectrum. If a detector system were to obtain images of a planetary surface at wavelengths such as those listed above, in principle it is possible to map the abundance of ice and other minerals.

In practice, for a terrestrial remote sensing project of this sort, an ideal data set would involve ~5 parameters: intensity at many wavelengths (~100 spectral channels from 0.5 to 2.5 μm), position (2 spatial), equivalent atmospheric thickness (air mass), and solar phase angle. The air mass data are required to correct for absorptions in the atmosphere, and solar phase angle data are required to determine the scattering properties (and thus microstructure) within the optical surface. Many spectral channels are required for mineral identification from observed absorption bands. Current technology is adequate for obtaining such a data set. A NASA project will acquire such a data set on a planetary scale beginning in 1989/90 with the Galileo mission to Jupiter where three satellites (ranging in size from Mercury to our moon), whose surfaces contain abundant ice, will be mapped at 204 wavelengths and at a spatial resolution as high as 1 km.

CONCLUSIONS

Reflectance spectra of ice-soil mixtures are very complex, nonlinear functions of the optical and physical properties of the components comprising the surface. Fortunately, these spectra can be understood, and it appears that some quantitative information regarding mineral abundance can be derived. The mineral abundance from remotely obtained spectra can be crudely determined from empirically derived relationships of the optical properties of ice-soil mixtures. A potentially more powerful technique involves a nonlinear least squares analysis of spectra obtained at several different viewing geometrics.

ACKNOWLEDGMENTS

This study was supported by NASA grant NAGW 115.

REFERENCES

Clark, R. N., 1981a, The spectral reflectance of water-mineral mixtures at low temperatures: Journal of Geophysical Research, v. 86, p. 3074-3086.

Clark, R. N., 1981b, Water frost and ice: The near-infrared spectral reflectance 0.65-2.5 microns: Journal of Geophysical Research, v. 86, p. 3087-3096.

Clark, R. N., 1982, Implications of using broadband photometry for compositional remote sensing of icy objects: Icarus, v. 49, p. 244-257.

Clark, R. N., and P. G. Lucey, 1983, Spectral properties of ice-particulate mixtures: Implications for remote sensing I: Intimate mixtures: Journal of Geophysical Research, submitted.

Clark, R. N., and T. Roush, 1983, Reflectance spectroscopy: Quantitative analysis techniques for remote sensing applications: Journal of Geophysical Research, submitted.

Hapke, B., 1981, Bidirectional reflectance spectroscopy 1. Theory: Journal of Geophysical Research, v. 86, p. 3039-3054.

Irvine, W. M., and J. B. Pollack, 1968, Infrared optical properties of water and ice spheres: Icarus, v. 8, p. 324-360.

McCord, T. B., and R. N. Clark, 1979, Atmospheric extinction 0.65-2.50 microns above Mauna Kea: Publication of the Astronomical Society of the Pacific, v. 91, p. 571-576.

Warren, G. S., 1982, Optical properties of snow: Review of Geophysics and Space Physics, v. 20, p. 67-89.

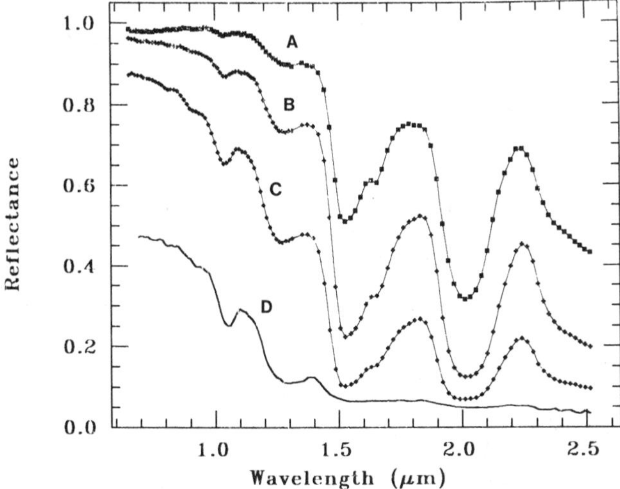

FIGURE 1 The near-infrared spectral reflectance of (a) a fine grained (~50 μm) water frost, (b) medium grained (~200 μm) frost, (c) coarse grained (400–2000 μm) frost and (d) a ice block containing abundant microbubbles are shown. The larger the effective grain size, the greater the mean path that photons travel in the ice, and the deeper the absorptions become. Frost data from Clark (1981b). Ice data from Clark and Lucey (1983).

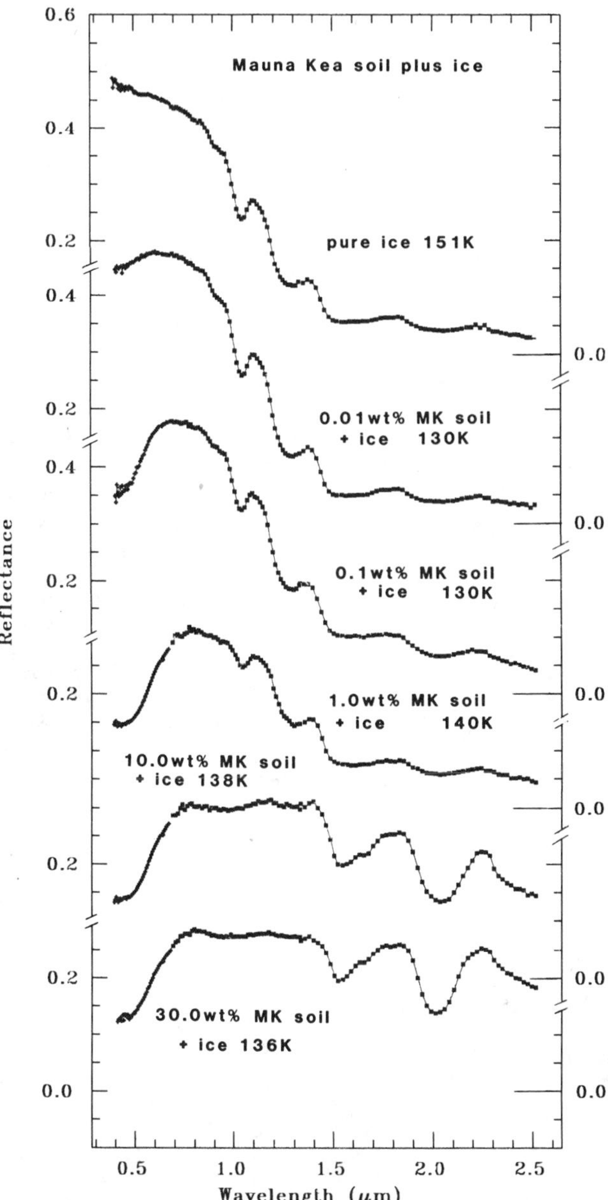

FIGURE 2 The reflectance spectra of ice blocks containing Mauna Kea soil of 15 μm mean grain size are shown (from Clark and Lucey, 1983). Note how the water ice band depths at 2.0, 1.5, 1.25, 1.04, and 0.90 μm change as a function of the weight fraction of the particulates in the ice. The particulate soil limits the photons from penetrating into the ice, thus a greater particulate weight fraction results in less photon path length in the ice and smaller absorption bands. The particulate reflectance also limits the photon path with darker grains reducing the multiple scattering and also the absorption band depths.

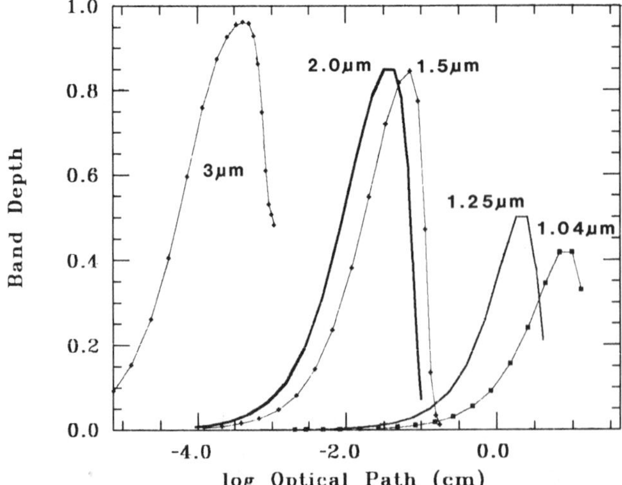

FIGURE 3 Theoretically, derived curves of growth (from Clark and Lucey, 1983) for five ice absorptions are shown. The optical path is the photon mean path length in ice at the absorption band center. Since ice is less absorbing at wavelengths outside the absorption bands, the mean optical path lengths are as much as an order of magnitude greater than at the band centers.

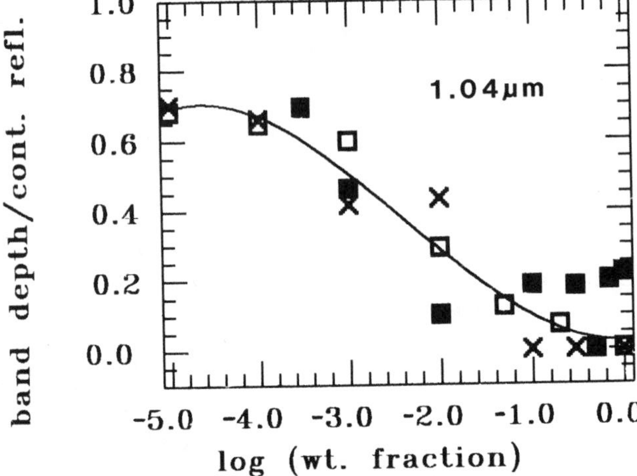

FIGURE 4 The data from Figure 2 (crosses) plus reflectance data for ice-charcoal mixtures (solid boxes) and ice-kaolinite mixtures (open boxes) were used to derive the 1.04-μm ice absorption band depth divided by the continuum reflectance (from Clark and Lucey, 1983) as a function of the log weight fraction of particulates in the ice. Although there is a factor of about 20 between the reflectance of the bright kaolinite grains and the dark charcoal grains, division of the band depth by the continuum reflectance removes the effects of particulate contaminant reflectance. Thus the curve can be used for crude abundance determinations from remotely obtained reflectance data. Other ice absorption bands are sensitive to different amounts of impurities. Some scatter in the charcoal data is due to near zero band depths and very low continuum reflectance, thus approaching zero divided by zero.

GROUND ICE IN THE EQUATORIAL REGION OF MARS: A FOSSIL REMNANT OF AN ANCIENT CLIMATE OR A REPLENISHED STEADY-STATE INVENTORY?

Stephen M. Clifford

Department of Physics and Astronomy, University of Massachusetts, Amherst, Massachusetts 01003 USA

Various lines of morphologic evidence suggest that substantial quantities of ground ice have existed in the equatorial regolith of Mars throughout its geologic history. However recent calculations on the stability of ground ice in this region suggest that any ground ice, emplaced earlier than 3.5 billion years ago, may have long since been lost by sublimation to the atmosphere. It is proposed that one possible explanation for the continued existence of ground ice in the equatorial regolith is that it may be replenished by subsurface sources of H_2O. The existence of a geothermal gradient in the Martian crust could provide the means necessary to thermally cycle H_2O between a deeply buried reservoir of groundwater, similar to those found in cold regions on Earth, and the base of a near surface layer of ground ice. Calculations indicate that a geothermal gradient of 25 K/km would be sufficient to replenish a layer of ground ice 1 km thick over the course of Martian geologic history.

INTRODUCTION

Consideration of the partial pressure of H_2O in the Martian atmosphere and the range of mean annual temperatures at the Martian surface suggests that the occurrence of ground ice in equilibrium with the water vapor content of the atmosphere is restricted to the colder latitudes poleward of $\pm 40^{\circ}$ (Fanale 1976, Farmer and Doms 1979). However, photographs taken during recent spacecraft missions to Mars have provided considerable geomorphologic evidence that the distribution of ground ice on Mars may be global. Much of this evidence is based on the identification of Martian analogs to cold-climate features found on Earth, such as patterned ground, debris flows, and thermokarst (e.g., Carr and Schaber 1977, Rossbacher and Judson 1981, Luchitta 1981). Additional evidence for the presence of regolith H_2O is based on the occurrence of an unusual type of crater morphology called "rampart" craters which is peculiar to Mars (Figure 1). The distinctive lobate nature of the ejecta blankets which surround these craters is thought to originate from the fluidization of ejecta material during an impact into a water or ice-rich regolith (Carr et al. 1977, Johansen 1978).

The existence of substantial quantities of ground ice in the equatorial region of Mars presents a major problem, for it is difficult to reconcile both the initial origin and the continued survival of ground ice in an area where present mean annual surface temperatures exceed the 198 K frost point temperature of the atmosphere by more than 20 K. The problem is further compounded by a recent study of the climatic history of Mars (Toon et al. 1980) which indicates that, throughout most of its history, conditions on the planet have closely resembled those we observe today. These conditions imply that equatorial ground ice should experience a net annual depletion, as H_2O is transfered from the "hot" equatorial regolith to the colder poles (Flasar and Goody 1976).

How, then, is it possible to account for the existence of ground ice within the equatorial regolith? Currently, the most widely accepted explanation is that it was emplaced very early in Martian geologic history and under substantially different climatic conditions. Evidence for such a period comes from the Martian valley networks (Figure 2)--features which bear a close resemblance to terrestrial runoff channels and which are found almost exclusively in the ancient ($\sim$4 billion year old), heavily cratered terrain of Mars (Pieri 1980, Carr and Clow 1981). This evidence suggests that conditions on early Mars were suitable for the global distribution of water within the near-surface regolith; therefore, significant reservoirs of ground ice may have formed as the climate gradually cooled.

If equatorial ground ice on Mars is a relic of some ancient climate, then we must assume it has survived in disequilibrium with the Martian atmosphere for nearly 4 billion years. But is such an assumption reasonable? Those who have addressed this question often suggest that the sublimation of a fossil ground ice layer could be substantially retarded by the diffusion-limiting properties of a relatively shallow layer of fine-grained regolith (Soderblom and Wenner 1978, Arvidson et al. 1980, Luchitta 1981, Rossbacher and Judson 1981). This explanation is based on the work of Smoluchowski (1968), who showed that under certain limited conditions of porosity, pore size, temperature, and depth of burial, a 10 m thick layer of ground ice in disequilibrium with the Martian atmosphere might survive for as long as a billion years. However, based on data obtained from recent spacecraft missions, a new study suggests that the physical properties of the Martian regolith may be inadequate to preserve equatorial ground ice for the required period of time (Clifford and Hillel 1983). In the following discussion the results of this study, and its implications for the existence of equatorial ground ice on Mars, will be reviewed.

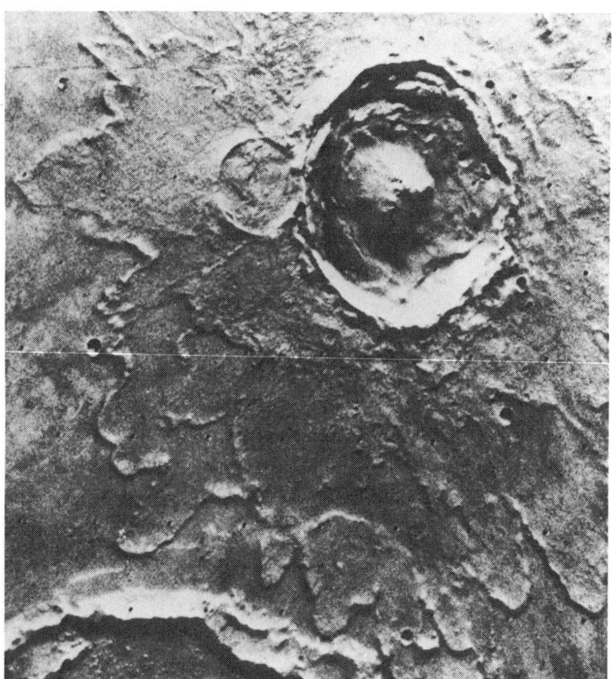

FIGURE 1 A Martian rampart crater.

FIGURE 2 The Martian valley networks.

THE STABILITY OF EQUATORIAL GROUND ICE

The stability of ground ice in the equatorial region of Mars is governed by the rate at which H_2O molecules can diffuse through the soil and into the Martian atmosphere. On Mars this process is complicated by the fact that the atmospheric mean free path of an H_2O molecule is approximately 8 microns (Farmer 1976). Therefore, when molecular transport occurs in very small pores ($r/\lambda < 0.1$), collisions between the diffusing H_2O molecules and the pore walls will far out-number the collisions which occur with other gas molecules; this is the region of Knudsen diffusion. Only in soil pores where the ratio of the pore radius, r, to the mean free path, λ, is large ($r/\lambda > 10$), will ordinary molecular diffusion will be the dominant mode of transport. Since the boundary between ordinary molecular diffusion and Knudsen diffusion is not distinct, the contributions of both processes must be considered when molecular transport occurs in pores of intermediate size ($0.1 < r/\lambda < 10$) (Youngquist, 1970).

To illustrate this dependency more clearly, the effective diffusion coefficient of H_2O on Mars has been plotted as a function of pore size in Figure 3. For pore radii less than about 10 microns, the effective diffusion coefficient of H_2O falls rapidly with pore size--this is the region of Knudsen diffusion. For radii greater than 100 microns, the domain of ordinary molecular diffusion, the effective diffusion coefficient is independent of pore size. Since very small soil pores can significantly retard the diffusive transport of H_2O on Mars, gaseous transport will occur preferentially within the larger pores of any given soil (Smoluchowski 1968, Clifford and Hillel 1983). This effect is clearly seen in Figure 4, where the diffusive flux of H_2O has been plotted as a function of pore size for three different pore size distributions.

To investigate the stability of ground ice in the equatorial region of Mars it was necessary to make certain assumptions regarding the physical properties of the regolith as well as the quantity and vertical distribution of ice within it (Clifford and Hillel 1983). Observations made by orbiting Mariner and Viking spacecraft and the results of the limited physical investigations of the Martian surface carried out by the Viking Landers have given us at least a rudimentary understanding of the structure and physical properties of the Martian regolith. By comparing these properties with soils on Earth that possess similar characteristics, a probable range of pore size distributions for Martian soil was inferred. Based on this type of analysis, high-porosity terrestrial silt and clay-type soils appear to provide the best match to the known Martian surface properties. Theoretical arguments (Pollack and Black 1979) suggest that the global inventory of H_2O on Mars is equivalent to a 100 m layer of ice averaged over the surface of the planet. In our analysis, it was assumed that this ice occupied the available soil pore space in a single massive layer which initially extended from a depth of 100-300 m. To maximize the lifetime of this buried ice layer, the top 300 m of the Martian regolith were considered isothermal. Finally, the temperature range which was considered in our analysis (215-220 K) corresponds to the range of mean annual surface temperatures for the latitudes lying between $\pm 30^\circ$.

Fine-grained soils, such as those thought to exist on Mars, typically exhibit broad pore size distributions. For this reason, the evaporative loss of ground ice was calculated by means of the parallel-pore model of gaseous diffusion (Younquist 1970), where a porous solid is viewed as a collec-

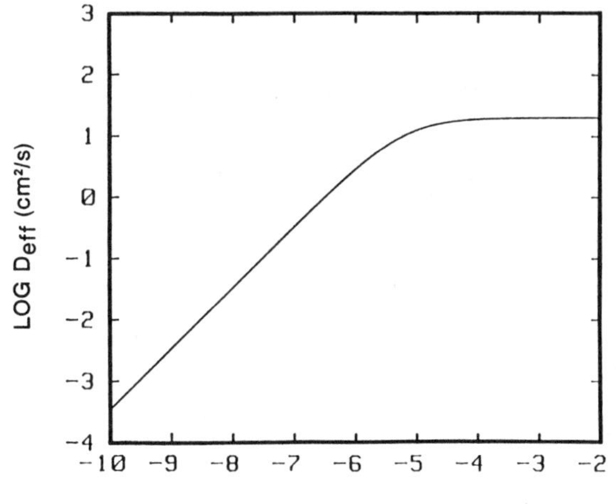

FIGURE 3 The effective diffusion coefficient of H_2O on Mars as a function of soil pore size (Clifford and Hillel 1983).

tion of parallel cylindrical pores of constant radii. The recession of the evaporating upper surface of the ground ice from an initial depth of 100 m to a final depth of 300 m was taken into account by recalculating the flux of H_2O at 1 m intervals. The total lifetime of the buried ice layer was then determined from the sum of these individual contributions.

Calculations based on the above model suggest that, on a global scale, it is unlikely that an unreplenished layer of ground ice could have survived in disequilibrium with the Martian atmosphere for as long as 3.5 billion years. However, it should also be noted that in view of the likely range of geologic environments which may be found on Mars, the necessary conditions for the long-term survival of equatorial ground ice may exist in isolated regions.

Therefore, the fundamental problem that originally motivated this study still remains: how does one account for the seemingly abundant morphologic evidence that substantial quantities of ground ice

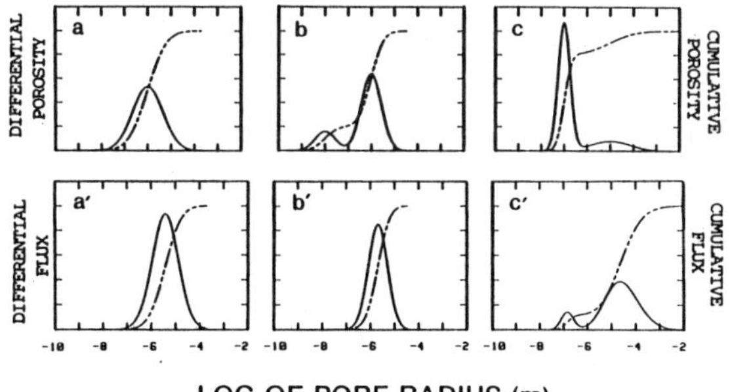

FIGURE 4 Figures 4a, 4b, and 4c, represent three hypothetical pore size distributions whose differential porosity is given by the solid line and whose cumulative porosity is given by the dashed line. Figures 4a', 4b', and 4c' depict the corresponding flux curves of H_2O through these distributions.

presently exist within the equatorial regolith? The possible answers appear to be: (1) the actual physical characteristics of the Martian regolith, and/or the quantity of H_2O stored in the regolith as ground ice, may differ substantially from the model parameters chosen for this study; (2) the accepted interpretation of certain Martian surface features as morphologic indicators of ground ice may be in error; or, (3) equatorial ground ice may have been replenished by subsurface sources of H_2O. While none of these three alternatives can be safely ruled out, the possible subsurface replenishment of equatorial ground ice appears to have some important points in its favor, for it is consistent with our present understanding of Martian geomorphology, surface properties, and climatic history.

POSSIBLE EVIDENCE OF GROUND ICE REPLENISHMENT

As noted earlier, under present climatic conditions, the atmospheric loss of ground ice from the equatorial regolith appears irreversible--once ice has evaporated, or been removed by some other process, it is difficult to see how that ice could be replenished by any atmospheric means. Yet, within the equatorial region of Mars there appears to be morphologic evidence that, at least in some areas, ground ice has in fact been replenished.

First, consider the results of a major impact in the near equatorial regolith. The estimated thickness of permafrost in the equatorial region of Mars is on the order of 1 km (e.g., see Fanale 1976, Rossbacher and Judson 1981); therefore, as first noted by Allen (1979), the production of a crater many tens of kilometers in diameter should result in the excavation of any ground ice which existed, in the region interior to the crater walls, prior to the impact (Figure 5a). While backfilling and melting of nearby ground ice may partially replenish some of the lost H_2O near the crater periphery, it appears unlikely that its lifetime would be very long given the high temperatures and porosity of the postimpact environment. Therefore, it is interesting to note that within a number of large impact craters in the equatorial region of Mars there appear clearly defined rampart craters (Allen 1979). As discussed earlier, the distinctive morphology of this type of crater is thought to result from an impact into a water or ice-rich regolith (Johansen 1978). If one accepts this interpretation, then the presence of rampart craters within the interiors of numerous older craters in the equatorial region of Mars appears to present a problem in that it is difficult to conceive of a scenario by which any ground ice that existed prior to the original cratering event could have managed to survive to produce the distinctive fluidized ejecta pattern often seen as the result of a second and sometimes third (Figure 5b) consecutive impact.

As before, several explanations are possible: (1) contrary to popular belief, rampart craters may not be specific indicators of ground ice (Schultz and Gault 1981), (2) the original emplacement of ground ice in the near-equatorial region may have occurred sometime after many of the earlier impacts but prior to a subsequent period of rampart cratering (although such an explanation runs counter to the findings of both Johansen (1978) and Allen

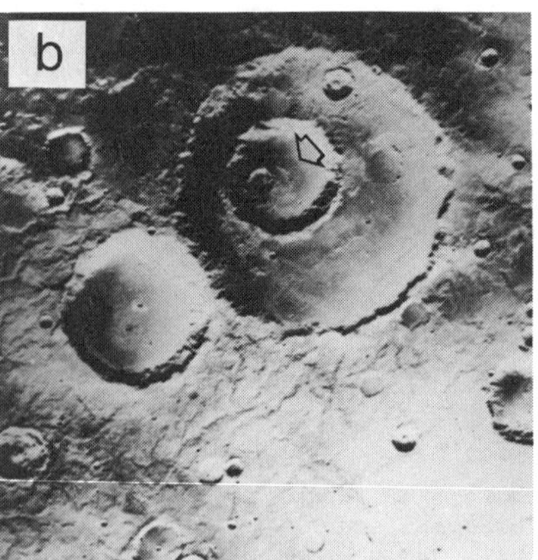

FIGURE 5 (a) The probable ground ice distribution resulting from a major impact in the equatorial region of Mars. (b) A rampart crater within two earlier and concentric impact features (20°S, 203°W) (Clifford and Johansen 1982).

(1979)), or (3) ground ice which was removed by the early impacts was thereafter replenished by some nonatmospheric means (Clifford 1980, 1982, Clifford and Johansen 1982).

PROCESSES OF REPLENISHMENT

It has been proposed that one possible explanation for the continued existence of ground ice in the equatorial regolith is that it may be replenished by subsurface sources of H_2O. Based on current estimates of the inventory of H_2O on Mars (Pollack and Black 1979, J. S. Lewis, personal communication, 1980), it has been suggested that Mars may possess an extensive network of deeply buried aquifers similar to those found in Siberia, the Antartic, and other cold regions on Earth (Carr 1979, Clifford and Huguenin 1980, Clifford, 1980). Alternatively, the occurrence of large amounts of subsurface water may be restricted solely to regions of past major volcanic activity such as Tharsis (10°N, 120°W) (Carr 1979, Fanale, personal communication, 1983). In either case, the vertical transport of H_2O, from such subsurface reservoirs, appears to be a viable mechanism for the replenishment of equatorial ground ice on Mars, even when the vertical distances separating the groundwater from the base of the ground ice layer are measured in kilometers.

Of the various mechanisms which may result in the vertical transport of H_2O on Mars, perhaps the most important has been the process of thermal moisture movement (Clifford 1980, 1982). Like Earth, Mars is thought to be radiating internal heat which will give rise to a geothermal gradient within the planet's crust (Fanale 1976). Clearly, when a temperature gradient is present in a moist porous medium, it will give rise to a corresponding vapor pressure gradient. As a result of this pressure difference, water vapor will diffuse in the Martain crust from the higher temperature (high vapor pressure) depths to the colder (lower vapor pressure) near-surface regolith.

Philip and deVries (1957) and Cary (1963) have formulated two different but widely accepted models for calculating the magnitude of this type of thermally driven moisture transfer. The Philip and deVries approach is based on a mechanistic description of the transport process. They suggest that the exchange of water vapor in an unsaturated soil occurs between numerous small "islands" of liquid which exist at the contact points of neighboring soil particles. The existence of a temperature gradient in the soil causes water from the warmer liquid islands to evaporate, diffuse across the intervening pore space, and condense on the cooler liquid islands at the opposite ends of the pore. In this fashion both moisture and latent heat are transfered through the soil (Figure 6a).

It should be noted that the microscopic temperature gradient across an individual pore is not the same as the measured macroscopic temperature gradient. The macroscopic gradient is an average of the individual gradients which exist across the volumes of air, water, and solids, present in the soil. Of these gradients, the gradient through the air in the pores will be the largest. Since it is the air temperature gradient which governs the magnitude of vapor diffusion, the total amount of H_2O transport through the soil can be several times greater than that predicted on the basis of the macroscopic temperature gradient alone (Philip and deVries 1957).

The Cary model of moisture transport differs from the Philip and deVries approach in that it makes no assumptions regarding the actual mechanism of vapor transport but merely attempts to provide a phenomenological description of the process based on the thermodynamics of irreversible process (Cary, 1963). Despite the differences in approach, it has been shown that the final form of the flux equations for both models are identical (Taylor and Cary 1964, Jury and Letey 1979). After Cary (1966), we can describe the thermally driven vapor flux by the equation

$$J_V = - \frac{\beta\, D_{AB}\, P_S\, H}{R^2 T^3} \frac{dT}{dz} \qquad (1)$$

where D_{AB} is the binary diffusion coefficient of H_2O in CO_2, P_s is the saturated vapor pressure of H_2O at a temperature T, H is latent heat of vaporization, R is the universal gas constant, and β is a dimensionless factor whose value (typi-

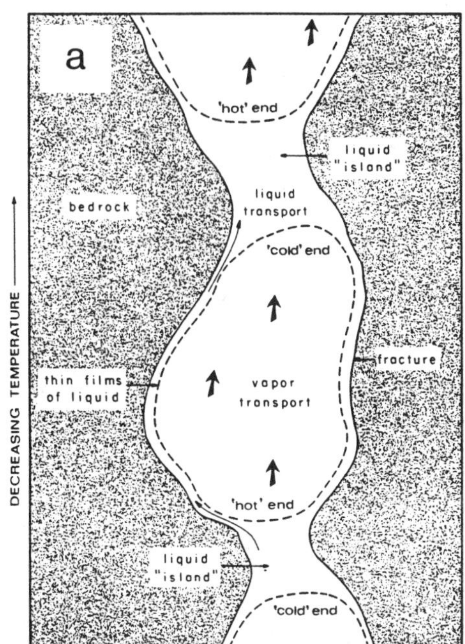

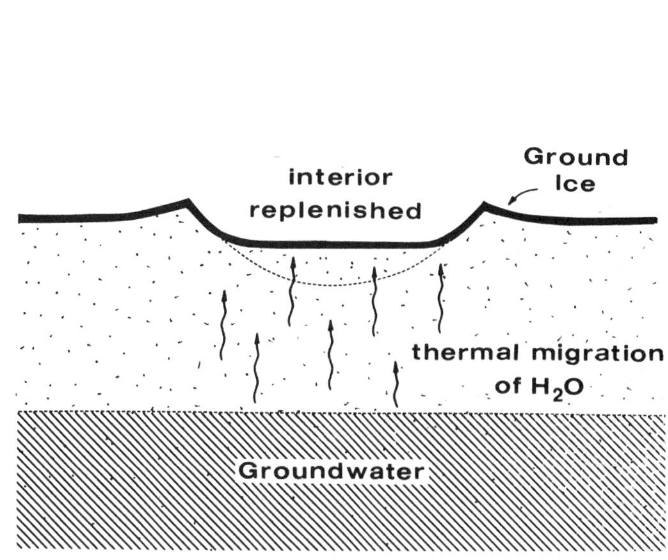

FIGURE 6 (a) The process of thermal moisture transfer proposed by Philip and deVries (1957). (b) The thermally driven transport of H_2O in the Martian regolith may provide the mechanism whereby the interiors of equatorial craters are recharged with ground ice, thus explaining the occurrence of features such as depicted in Figure 5b (Clifford 1980, Clifford and Johansen 1982).

cally around 1.83 (Jury and Letey 1979)) depends on regolith temperature, porosity, and water content. Making use of this equation, we find that for a Martian geothermal gradient of 25 K/km, the calculated flux of H_2O to the freezing front at the base of the ground ice layer is approximately 4 x 10^{-5} g/cm^2 per Martian year. At this rate, over the course of Martian geologic history, the geothermal gradient could supply enough H_2O from a subpermafrost groundwater system to replenish a layer of ground ice over 1 km thick. It should be noted that this transport will occur on a continuous and global basis for as long as there exists a geothermal gradient and subsurface reservoir of H_2O.

Inspection of equation 1 reveals that for the higher temperatures expected at depth the flux of H_2O leaving the groundwater surface greatly exceeds that which finally reaches the freezing front at the base of the ground ice layer. This difference is due to the decrease in saturated vapor pressure which occurs with decreasing temperature. As shown by Jackson et al. (1965), once a closed system has been established (i.e., the ground ice capacity of the near-surface regolith has been saturated under conditions where the ice exists in stable equilibrium with the atmosphere), a dynamic balance of opposing fluxes is achieved. As water vapor rises from the warmer depths to the colder regolith above, it will condense, creating a circulation system of rising vapor and descending liquid condensate (Figure 7).

CONCLUSION

Identifying the processes responsible for the existence of equatorial ground ice on Mars may be a key factor in understanding the nature of the Martian hydrological balance. Theoretical calculations on the stability of equatorial ground ice seem to indicate that, with the possible exception of isolated regions possessing unusual conditions of low porosity and small pore size, any ground ice present in the equatorial regolith should have long since been lost by sublimation to the atmosphere. Yet various lines of morphologic evidence suggest that substantial quantities of ground ice may still reside in the equatorial regolith. One possible explanation for this evidence is that equatorial ground ice has been replenished from sources within

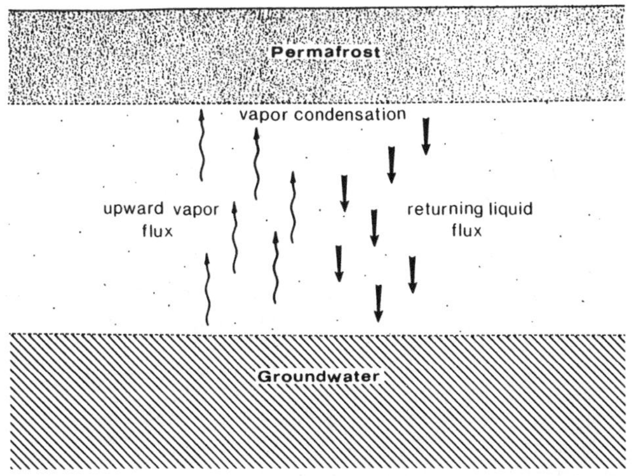

FIGURE 7 The presence of a geothermal gradient in the Martian crust may give rise to a convective cycling of H_2O between a subsurface reservoir of groundwater and near-surface ground ice.

the planet's crust. The proposed thermal migration of H_2O from these subsurface reservoirs is consistent with both our current knowledge of the physical and thermal properties of the Martian crust and with our understanding of the physics of soil water movement in the presence of a temperature gradient.

ACKNOWLEDGMENT

This research was supported under a grant from NASA's Planetary Geology Program (NSG 7405). Additional funding, arranged by J. Boyce (NASA) and J. King (SUNY, Buffalo), is also most gratefully acknowledged. The author is also indebted to F. Fanale, R. Huguenin, D. Hillel, R. D. Miller, and an anonymous reviewer, who have provided many thoughtful and constructive suggestions during the course of this research.

REFERENCES

Allen, C. C., 1979, Areal distribution of Martian rampart craters: Icarus, v. 39, p. 111-123.

Arvidson, R. E., Goettel, K. A., and Hohenberg, C. M., 1980, A post-Viking view of Martian geologic evolution: Reviews of Geophysics Space Physics, v. 18, p. 565-603.

Carr, M. H., 1979, Formation of Martian flood features by release of water from confined aquifers: Journal of Geophysical Research, v. 84, p. 2995-3007.

Carr, M. H., and Clow, G. D., 1981, Martian channels and valleys: Their characteristics, distribution, and age: Icarus, v. 48, p. 91-117.

Carr, M. H., Crumpler, L. S., Cutts, J. A., Greeley, R., Guest, J. E., and Masursky, H., 1977, Martian impact craters and emplacement of ejecta by surface flow: Journal of Geophysical Research, v. 82, p. 4055-44065.

Carr, M. H., and Schaber, G. G., 1977, Martian permafrost features: Journal of Geophysical Research, v. 82, p. 4039-4055.

Cary, J. W., 1963, Onsager's relation and the non-isothermal diffusion of water vapor: Journal of Physical Chemistry, v. 67, p. 126-129.

Cary, J. W., 1966, Soil moisture transport due to thermal gradients: Practical aspects: Soil Science Society of America Proceedings, v. 30, p. 428-433.

Clifford, S. M., 1980, Mars: Ground ice replenishment from a subpermafrost groundwater system, in Proceedings of the Third Colloquium on Planetary Water, State University of New York, Buffalo, p. 68-75.

Clifford, S. M., 1982, Mechanisms for the vertical transport of H_2O within the Martian regolith, in Papers presented to the Conference on Planetary Volatiles, LPI Contribution 488, Houston, p. 23-24.

Clifford, S. M., and Hillel, D, 1983, The stability of ground ice in the equatorial region of Mars: Journal of Geophysical Research, v. 88, p. 2456-2474.

Clifford, S. M., and R. L. Huguenin, 1980, The H_2O mass balance on Mars: Implications for a global subpermafrost groundwater flow system, NASA TM 81776, p. 144-146.

Clifford, S. M., and L. A. Johansen, 1982, Splosh craters: Evidence for the replenishment of ground ice in the equatorial region of Mars, Lunar and Planetary Science Conference XIII, p. 121-122.

Fanale, F. P., 1976, Martian volatiles: Their degassing history and geochemical fate: Icarus, v. 28, p. 179-202.

Farmer, C. B., 1976, Liquid water on Mars: Icarus, v. 28, p. 279-289.

Farmer, C. B., and Doms, P. E., 1979, Global and seasonal variation of water vapor on Mars and the implications for permafrost: Journal of Geophysical Research, v. 84, p. 2881-2888.

Flasar, M. F., and Goody, R. M., 1976, Diurnal behaviour of water on Mars: Planetary and Space Science, v. 24, p. 161-181.

Jackson, R. D., Rose, D. A., and Penman, H. L., 1965, Circulation of water in soil under a temperature gradient: Nature, v. 205, p. 314-316.

Johansen, L. A., 1978, Martian splosh cratering and its relation to water, in Proceedings of the Second Colloquium on Planetary Water and Polar Processes, U. S. Army Cold Regions Research and Engineering Laboratory, Hanover, p. 109-110.

Jury, W. A., and Letey, J. Jr., 1979, Water vapor movement in soil: Reconciliation of theory and experiment: Soil Science Society of America Journal, v. 43, p. 823-827.

Luchitta, B. K., 1981, Mars and Earth: Comparisons of cold-climate features: Icarus, v. 45, p. 264-303.

Philip, J. R. and deVries, D. A., 1957, Moisture movement in porous materials under temperature gradients: Transactions of the American Geophysical Union, v. 38, p. 222-228.

Pieri, D., 1980, Martian valley morphology, distribution, age and origin: Science, v. 210, p. 895-897.

Pollack, J. B., and Black, D. C., 1979, Implications of the gas compositional measurements of Pioneer Venus for the origin of planetary atmospheres: Science, v. 205, p. 56-59.

Rossbacher, L. A., and Judson, S., 1981, Ground ice on Mars: Inventory, distribution, and resulting landforms: Icarus, v. 45, p. 39-59.

Schultz, P. H., and Gault, D. E., Ejecta emplacement and atmospheric pressure: Labboratory experiments, in Papers presented to the Third International Colloquium on Mars, LPI Contribution 441, Houston, p. 226-228.

Smoluchowski, R., 1968, Mars: Retention of ice: Science: v. 159, p. 1348-1350.

Taylor, S. A. and J. W. Cary, 1964, Linear equations for the simultaneous flow of matter and energy in a continuous soil system: Soil Science Society Proceedings, v. 28, p. 167-172.

Toon, O. B., Pollack, J. B., Ward, W., Burns, J. A., and Bilski, K., 1980, The astronomical theory of climatic change on Mars: Icarus, v. 44, p. 552-607.

Youngquist, G. R., 1970, Diffusion and flow of gases in porous solids, in Flow Through Porous Media, American Chemical Society Publication, Washington, D.C., p. 58-69.

DETECTION AND EVALUATION OF NATURAL GAS HYDRATES FROM WELL LOGS, PRUDHOE BAY, ALASKA

Timothy S. Collett

Department of Geology/Geophysics, University of Alaska, Fairbanks, Alaska 99701 USA

The purpose of this study is to develop techniques for the detection and evaluation of in-situ gas hydrates from well log data and to determine possible geologic controls on the occurrence of hydrates in the North Slope region of Alaska. Several new methods of evaluation for subsurface gas hydrate were developed and incorporated with existing techniques. For each of 125 wells examined as part of this study the geothermal gradient was determined and the theoretical stability zone for methane hydrate was calculated. Among these, there was 102 apparent hydrate occurrences in 32 wells. A subsurface structural-stratigraphic framework was established to a depth of 1,000 meters. This sediment package is characterized by three deltaic depositional sequences. The high frequency of hydrate occurrences in the structurally up-dip region of the Kuparuk Oil Field suggests that upward migration of free gas preceded hydrate development in the zone of hydrate stability.

INTRODUCTION

Significant quantities of gas hydrates have been detected in many permafrost regions of the world; in western Siberia, in the Mackenzie Delta of Canada, and on the North Slope of Alaska (Kvenvolden et al. 1980). In 1970 Makogon reported that the Messoyakha field in western Siberia had reserves in the billions of cubic meters of methane gas frozen as gas hydrates (Figure 1). Although gas hydrates have been identified in many regions, their significance and geographical extent have seldom been studied. Work done in the Messoyakha field showed that injections of methanol into the hydrate zones could increase gas reserves by 54% above what would be expected in an equal volume of reservoir rocks filled with free gas (Makogon 1981).

In-situ hydrates can occur in permafrost and can also occur below the base of the permafrost at temperatures above the freezing point of water (Figure 2). Various schemes for hydrate development have been postulated. One theory suggests that a gas hydrate could be part of a preexisting gas reservoir, which was frozen in place. It has been suggested that a hydrate body could form by a flux of free gas into a zone of methane hydrate stability and be frozen in place. Another theory states the possibility that free gas could be trapped at the base of the permafrost and be frozen into hydrate. Hydrates can also be found in association with decaying biomatter, such as coal, which would serve as a source for the methane needed for hydrate development (Pratt 1979).

The overall study of in-situ gas hydrates has been limited, with only several preliminary studies completed (Pratt 1971, Bily and Dick 1974, Kvenvolden et al. 1980). The work of Bily and Dick (1974) provided the most conclusive study to date on the occurrence and detection of in-situ natural gas hydrates. Bily and Dick incorporated well logs including dual induction, sonic, and mud logs as potential hydrate detection devices. Published research to date into the gas hydrates of the Prudhoe Bay region is limited to the work of Osterkamp and Payne (1981), and also Pratt (1979).

CONSIDERATIONS FOR HYDRATE OCCURRENCE

The depth and thickness of the zone of potential hydrate stability can be calculated if the mean annual temperature, geothermal gradient, lithostatic pressure gradient, and gas density are known. The past mean surface temperature at Prudhoe Bay was -10.9°C and the lithostatic pressure gradient is 9.84 kPa/m (Lachenbruch et al. 1982). The mean value of -10.9°C can be viewed as the average surface temperature with which the deep permafrost is presently in equilibrium.

The high number of methane gas shows on the mud logs and several sets of detailed gas analysis from drill stem tests suggest that the dominant gas type in the upper units in Prudhoe Bay is methane.

The geothermal gradients needed to predict the thickness of the hydrate are not generally available. Lachenbruch et al. (1982) observed three geothermal gradients from different bore holes along the Alaskan arctic coast. The lowest geothermal gradient was calculated at Cape Thompson, which was 2.0°C/100 m, at Pt. Barrow the gradient was 2.3°C/100 m and the geothermal gradient at Cape Simpson was 3.3°C/100 m (Lachenbruch et al. 1982). The variance in the geothermal gradient is due to differences in thermal conductivity.

The variance in the geothermal gradient indicates that no one regional gradient should be used to calculate the thickness of the zone of potential hydrate occurrence. The geothermal gradient was calculated separately for each well, the base of the permafrost was assumed to be at -1°C (Lachenbruch et al. 1982), and a temperature gradient was calculated to the base of the permafrost.

Formerly, the base of the permafrost was assumed to be in equilibrium at 0°C, but due to freezing point depression the interface is now believed to be at equilibrium at -1°C with an error of ±.5°C (Lachenbruch et al. 1982). Freezing point depression is related to several factors that may affect the thermal stability of the phase boundary. These factors include the presence of salt ions in solution, the existence of freeze-back pressure, and variations in the types of solids and fluid saturation levels.

Obviously it is impossible to make a reliable temperature estimate at the base of the permafrost in a well bore. However, the error in a regional geothermal gradient would be more significant than using a calculated gradient for each well bore, in which the base of the permafrost is assumed to be at -1°C. The values for the depth to the base of the permafrost were taken directly from the work of Osterkamp on permafrost thickness evaluation from well log data on the North Slope.

Figure 3 illustrates how the depth to the base of the permafrost and a methane hydrate stability curve were used to determine the depth and thickness of the zone of potential methane hydrate stability. In this example the depth to the base of the permafrost is 532 m with a mean annual ground temperature of -10.9°C. Assuming that the temperature at the base of the permafrost is -1°C, the geothermal gradient within the permafrost for this bore hole would be 1.9°C/100 m. In Figure 3 a 1.0°C/100 m geothermal gradient within the permafrost has been plotted along with a calculated 3.2°C/100 m gradient below the base of the permafrost. The geothermal gradient changes abruptly at the base of the permafrost due to a change in thermal conductivity. Therefore, the geothermal gradient was modified below the base of the permafrost in the calculation of the thickness of the zone of potential hydrate stability. The ratio used to manipulate the gradient below the base of the permafrost was given by Lachenbruch et al. (1982). This ratio indicated that the gradient increase by a factor of 1.73 from within the permafrost to the unfrozen strata below the base of the permafrost. A methane hydrate stability curve has also been plotted using a hydrostatic pressure gradient of 9.84 kPa/m (Lachenbruch et al. 1982). The lower boundary of the zone of methane hydrate stability is marked by the lower intersection of the geothermal gradient with the methane stability curve. The upper boundary of the zone of methane hydrate stability is defined by the upper intersection of the methane stability curve and the geothermal gradient. In Figure 3 the upper boundary of the zone is marked by H1 at 177 m. While the lower hydrate boundary is marked by H2 at 957 m, delineating a zone of potential hydrate occurrence 780 m thick.

The methane hydrate stability curve indicates that the geothermal gradient must be less than 3.7°C/100 m for methane gas hydrate to form at Prudhoe Bay. For the geothermal gradient to intersect the methane hydrate stability curve the gradient must be equal to or less than 3.6°C/100 m, which would correspond with a minimum permafrost depth of 290 m. In other words methane hydrate should not exist at Prudhoe Bay if the depth to the base of the permafrost is less than 290 m. The dashed line in Figure 2 represents the 290 m depth contour on the base of the permafrost. If the above line of reasoning is correct, methane hydrate occurrences should be limited to areas north of this contour. The thickness of the zone of potential hydrate occurrence ranged from nothing at the 290 m permafrost contour to more than 1,000 m near Mikkelsen Bay, where the methane hydrate was found to be potentially stable to a depth of 1,119 m.

LOG EVALUATION

The recognition of gas hydrate in well log data is not straightforward, and often the zones of potential hydrate occurrence are not logged, or the quality of the logs may be poor. Another problem in the evaluation of hydrates from well log data is the lack of prior quantitative work.

The work by Bily and Dick (1974) on the evaluation of natural occurring gas hydrates in the Mackenzie Delta is one of only a few papers dealing with the detection of in-situ natural gas hydrates using wire line logs. Bily and Dick discovered that when a hydrate zone was penetrated during drilling, there was a marked increase in the amount of gas in the drilling mud. The hydrate units recorded a relatively high resistivity on the dual induction log and a slight spontaneous potential deflection in comparison to a free gas. Sonic logs also indicated an increase in acoustic velocity.

The first confirmation of the existence of in-situ natural gas hydrate was not until 1972, when ARCO/EXXON were successful in recovering the first natural gas hydrate in a frozen state. The sample was recovered from a depth of 666 m in the Northwest Eileen State #2 well in Prudhoe Bay. The Northwest Eileen well was drilled with cool drilling muds in an attempt to reduce thawing of the permafrost and hydrate. The methane hydrate saturated sample was recovered in a pressurized core barrel, and a simple test was devised to check for the presence of hydrate. The pressure within the core barrel was allowed to equilibrate with the surface pressure, and the core barrel was resealed and warmed above in-situ temperatures. The pressure within the barrel began to rise, indicating the presence of thawing hydrate. This process was repeated several times with similar results. The hydrate sample had a gas composition of 99.17% methane (P. Barker, personal communication, ARCO Alaska Inc., Anchorage, Alaska).

The confirmed hydrate occurrence in the Northwest Eileen well presents itself as an ideal starting point for the development of log evaluation techniques in a hydrate zone. Log responses for the hydrate zone in the Eileen well are graphically represented in Figure 4.

The following list summarizes various log responses, incorporating the methods developed in this study with the evaluation techniques developed by Bily and Dick (1971) in the Mackenzie Delta. The ability of each log to distinguish hydrates from free gas and ice bearing permafrost is also indicated.

1. Mud Log On a mud log there is a pronounced gas kick associated with a hydrate, due to thawing during drilling. The mud log serves as the best tool available for the differentiation of a hydrate saturated unit from gas-free ice-bearing permafrost.
2. Dual Induction Log There is a relatively high resistivity deflection on the dual induction log in a gas hydrate zone, in comparison to that in a free gas zone. The long normal is separated from the short normal due to thawing next to the bore hole. If a unit were hydrate saturated within ice-bearing permafrost, the resistivity response on the dual induction log for the hydrate unit would not be significantly different than the log responses for the surrounding ice-bearing permafrost. Hence, it is impossible without the usage of the mud log to distinguish between hydrate and permafrost. Below the base of the permafrost the high resistivity deflection associated with the hydrate is distinct from the surrounding non ice-bearing zones, but may be similar to that of a free gas.
3. Spontaneous Potential (SP) There is a relatively lower (less negative) spontaneous potential deflection in a hydrate zone when compared to that associated with free gas. The frozen hydrate limits the penetration of mud filtrate thus reducing the negative spontaneous potential.
4. Caliper Log The caliper log in a hydrate interval usually indicates an oversized well bore due to spalling associated with the decomposition of a hydrate. Because the caliper log also indicates an enlarged bore hole in ice-bearing permafrost, it is only useful in detecting hydrates below the base of the ice-bearing permafrost.
5. Sonic Log Acoustic velocities in hydrate are relatively high ranging from 3.1 km/s to 4.4 km/s. Because the sonic velocity of ice-bearing permafrost is very similar to that of gas hydrate, the sonic log cannot be used to detect hydrates within the upper ice-bearing permafrost zone, but it is helpful below the base of the ice-bearing permafrost.
6. Neutron Porosity In a hydrate zone there is an increase in the neutron porosity; this contrasts with the apparent reduction in neutron porosity in a free gas zone. If a unit is hydrate saturated and occurs within the ice-bearing permafrost zone the neutron porosity log would theoretically indicate an increased or reduced neutron porosity, depending on the amount of free gas associated with the hydrate in comparison to that of the surrounding ice-bearing permafrost. Below the base of the permafrost a hydrate unit exhibits a relatively higher neutron porosity compared to water saturated or free gas saturated zones. However, thawing near the well bore complicates the neutron log interpretation.
7. Drilling Rate In a hydrate zone the relative drilling rate decreases, due to the cemented nature of the hydrate. There is a very similar drilling rate response within ice-bearing permafrost, and therefore drilling rate change is not useful as a hydrate detector within the permafrost.
8. Cross Plots In a cross plot of the resistivity and transit time for a series of stratigraphic units saturated with either hydrate or free gas and below the base of the ice-bearing permafrost, there is a grouping of units with similar constituents. Hydrate saturated units fall in a region of relatively higher resistivity and faster transit times while free gas saturated units fall in an area of lower resistivity and slow transit times. Differences are relative and not absolute; the cross plots show a simple clustering of similar properties. A resistivity/transit time cross plot of units that are above the base of the permafrost is not useful as a hydrate indicator, due to the similarity in resistivity and transit time velocities in hydrates and in permafrost.

In the Prudhoe Bay wells the dual induction and mud log are the most valuable tools available for the detection of gas hydrates; caliper and sonic logs are helpful but less definitive. The neutron porosity log showed great promise, but the lack of neutron surveys did not allow adequate assessment of it as a hydrate detection device. Many problems still exist in the evaluation of in-situ hydrates from well log data, and the addition of new evaluation techniques such as the use of cross plots and the addition of the neutron logs has only slightly improved the subjective nature of hydrate detection.

HYDRATE OCCURRENCE IN PRUDHOE BAY

In this study, a structural-stratigraphic framework of 32 key markers within the Tertiary and Upper Cretaceous strata was picked from the gamma ray logs and was established to a depth of 1,000 m. Thirty-three distinct units were defined and described from direct interpretation of the gamma ray logs. The gamma ray surveys were correlated with three complete sets of drill core chips and four petrographic strip logs.

The upper 1,000 m of strata in Prudhoe Bay is characterized by a gentle dip to the northeast, ranging from 20 to 28 m/km and is dominated by three distinct coarsening-upwards deltaic sequences. Howitt (1971) suggested that deposition of the upper units in Prudhoe Bay was more or less continuous in an aqueous environment.

One hundred twenty-five wells were examined for potential hydrate occurrence, with 102 definite occurrences in 32 different wells. Hydrates occurred in relative porous discrete units. Many of the wells had multiple zones of hydrate occurrence, with each hydrate unit ranging from 2 to 28 m thick.

Hydrate occurrences appeared to be regionally isolated to the Kuparuk oil field to the west of Prudhoe Bay, indicated in Figure 1. In the Kuparuk region there are four laterally continuous hydrate saturated sands and two less extensive units. The lateral extent of each hydrate saturated unit has been graphically represented in a three-dimensional block diagram in Figure 5. An east-west cross section through the Kuparuk oil field has been plotted in Figure 6, along with associated hydrate accumulations and inferred environments of deposition.

The presence of a structural control on the occurrence of hydrate is apparent upon close examination of all hydrate zones. With several minor exceptions, all hydrate occurrences are below marker 12, which marks the base of a nonporous prodelta shale. Hydrates were found exclusively between markers 12 and 19.

The sediment package between marker 12 and 19 is described as a deposit of shaly sand with thick interbeds of clean sand and shale, which were deposited in a delta front foreshore environment. There are several notable impermeable shale breaks which act as caps for relatively porous sand units which are hydrate saturated. Due to the interbedded nature of the sediments the hydrate occurs in multiple discrete units, within one well there may be as many as eight different hydrate saturated units.

A subjective A, B or C value has been assigned to each hydrate occurrence in an attempt to quantify the degree of hydrate saturation. The magnitude of the resistivity kick and the gas show associated with each hydrate occurrence was used to calculate the relative saturation of each hydrate. The letter A was assigned to a unit if it appeared to be highly saturated with methane hydrate, B and C indicate a relative decrease in hydrate saturation.

In the cross section in Figure 6 the hydrate appears to be concentrated in the southwest up-dip direction, with a decrease in hydrate saturation down-dip to the northeast. The anomalous occurrence of hydrate in the Kuparuk region, along with the greater saturation of hydrate up-dip, suggests that the free gas necessary for hydrate development may have migrated into place. The source for the free gas may be from either local biological decay in the upper units, or the gas could have migrated from a deeper mature gas zone.

As noted in the cross section of Figure 6, there are a number of hydrate occurrences within the permafrost. The occurrence of hydrate within the permafrost represents a time restraint on the formation of hydrate. Since permafrost is impermeable to gas migration, the hydrate must have developed in the upper intervals before the formation of the permafrost to the present depth.

A possible scenario for the formation of hydrate in the North Slope would begin with free gas migration either from local diagenesis or from depth through a relatively permeable sand unit along the base of an impermeable prodelta shale. The overlying shale unit would act as a cap to vertical gas migration.

The migrating free gas could be trapped in the up-dip direction by a series of different trapping mechanisms. The two most probable trapping mechanisms would be a self-forming hydrate trap and an impermeable ice-bearing trap. The existence of a porosity/permeability trap or a fault trap is unlikely.

The rate of free gas migration and the existence of possible porosity/permeability traps in the hydrate saturated sands are not easily delineated due to the lack of data. The only data available on the porosity/permeability characteristics of the upper units are from stratigraphic logs prepared by the American Stratigraphic Company. The porosity within the hydrate saturated sands varies little, from 38 to 46% (American Stratigraphic Company). The existence of a porosity/permeability trap for the up-dip free gas migration is not likely due to the lack of variation in the lateral porosity in the same units. The unconsolidated nature of the upper units would not lend itself to the formation of a porosity/permeability trap.

The evidence that there is little faulting past the Lower Cretaceous reduces the likelihood of fault traps.

The free gas could have formed a trap when the gas entered the hydrate stability field. As the migrating gas moved into the zone of hydrate stability, the gas would be frozen in place. The frozen hydrate would be impermeable to free gas migration and would continue to thicken as free gas is trapped and frozen in place.

The impermeable permafrost could also form an up-dip trap to free gas migration. As the free gas migrated up-dip along the bedding plane, the gas would be trapped at the base of the permafrost and be frozen in place.

The actual occurrence of hydrate does not favor either the self-forming trapping model or the permafrost trapping model. However, the laterally continuous nature of the hydrate occurrences suggests possible reorganization of the hydrate by multiple periods of freezing and thawing.

CONCLUSIONS

The major findings of this study are:

1. Several well logs have been found to be indicative of the presence of hydrate. Although no single log is definite by itself, used in combination they permit at least a subjective evaluation of hydrate occurrences. For example, the development of new evaluation techniques such as the use of cross plots and the addition of the neutron porosity log as a hydrate detector has reduced the subjective nature of the hydrate evaluation. Hydrate occurrences were identified by a variety of different well logs. The internal consistency of their determination, and their application in a large number of wells, indicate the validity of the method.
2. The recognition of the primarily structural-stratigraphic control on the up-dip gas migrational model for the hydrate accumulations was a significant contribution. The correlation of the actual hydrate occurrences identified in well log data with the structural-stratigraphic framework has allowed the development of a conceptual model for hydrate formation in the North Slope region.

3. The method for determining the zone of hydrate stability was refined to take into account both the difference in thermal conductivity between ice bearing and water bearing strata and the shallow depth limit of the stability zone. The method used for estimating the local geothermal gradients allows estimation of temperature profiles during warmer periods in the Earth's history which, in turn, has provided insight concerning the original formation and accumulation of the actual gas hydrate occurrences.

REFERENCES

Bily, C., and Dick, J. W. L., 1974, Naturally occurring gas hydrates in Mackenzie Delta, N.W.T.: Bulletin of Canadian Petroleum Geology, v. 22, no. 3, p. 340-352.

Howitt, F., 1971, Permafrost geology in Prudhoe Bay, Alaska: World Petroleum, September, p. 28-34.

Kvenvolden, D. A., and McMenamin, M. A., 1980, Hydrates of natural gas: A review of their geologic occurrence: U.S. Geological Survey Circular 825, 11 p.

Lachenbruch, A. H., Sass, J. H., Marshall, B. V., and Moses, T. H., 1982, Permafrost, heat flow, and the geothermal regime at Prudhoe Bay, Alaska: Journal of Geophysical Research, v. 87, no. B11, p. 9301-9316.

Makogon, Y. F., 1981, Hydrates of natural gas: Tulsa, Okla., Penn Well Publishing Company, 237 p.

Pratt, R. M., 1979, Gas hydrate evaluation and recommendations Natural Petroleum Reserve, Alaska: U.S. Geological Survey Special Report TC-7916, 27 p.

Osterkamp, T. E., and Payne, M. W., 1981, Estimates of permafrost thickness from well logs in northern Alaska: Cold Regions Science and Technology, v. 5, p. 13-27.

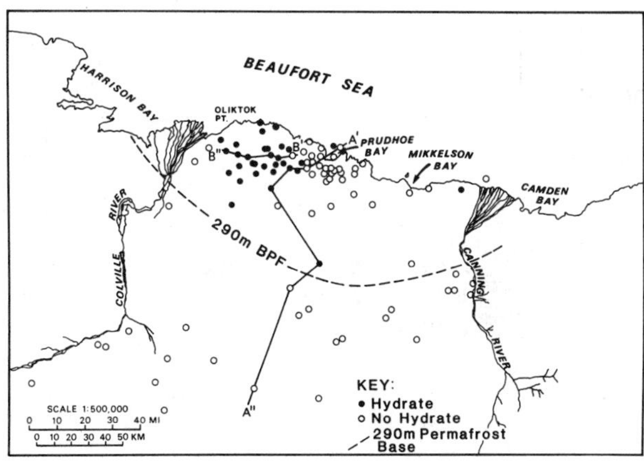

FIGURE 1 Base map of study area Hydrate occurrences have been denoted.

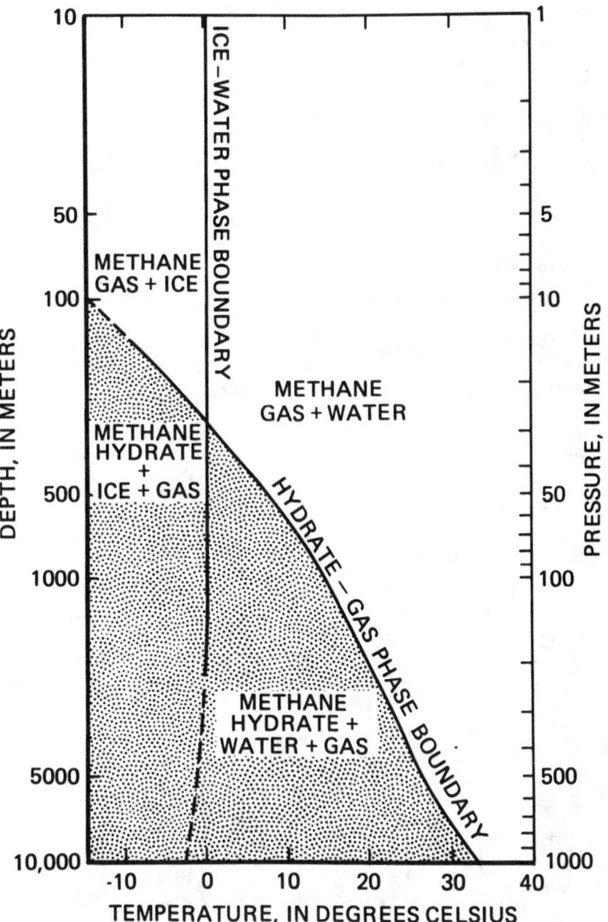

FIGURE 2 Phase boundary diagram showing free methane gas and methane hydrate pattern for a freshwater pure methane system (Kvenvolden et al. 1980).

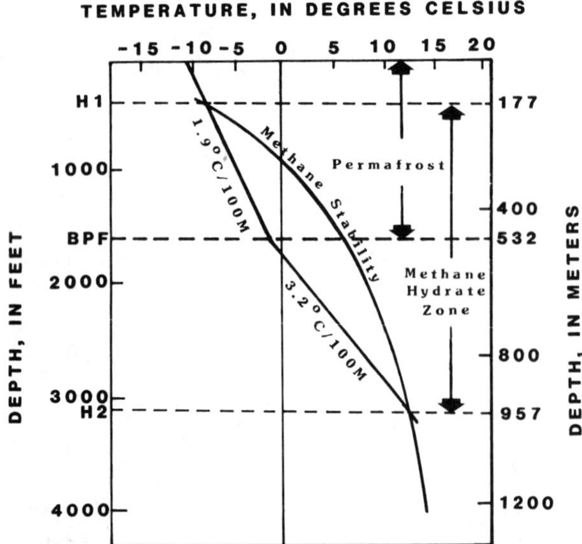

FIGURE 3 Thickness calculation of the methane hydrate stability zone in Northwest Eileen State No. 2 well.

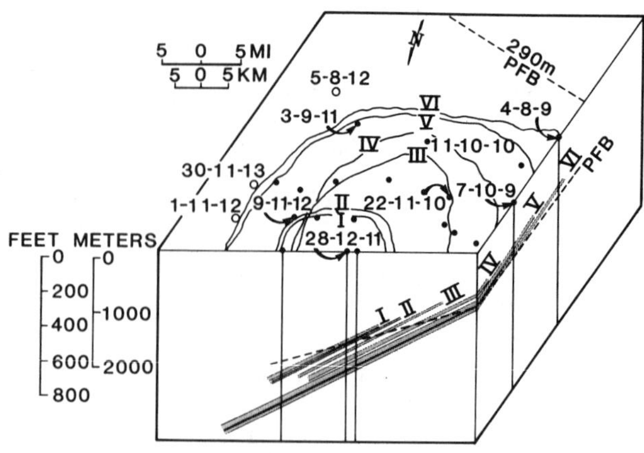

FIGURE 5 Block diagram representation of the hydrate occurrence in the Kuparuk Oil Field.

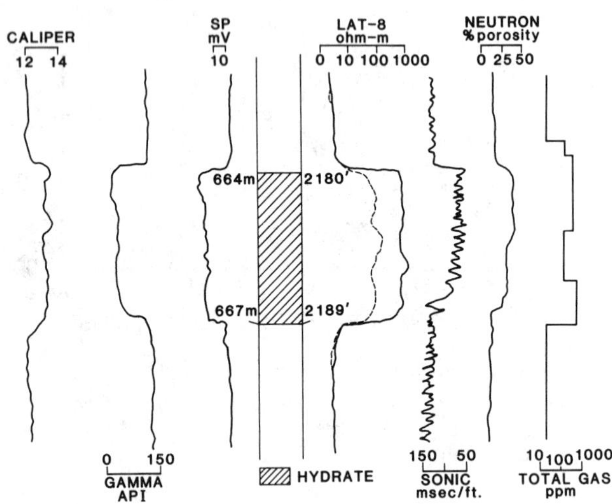

FIGURE 4 Hydrate characteristics in well log data from Northwest Eileen State No. 2 well.

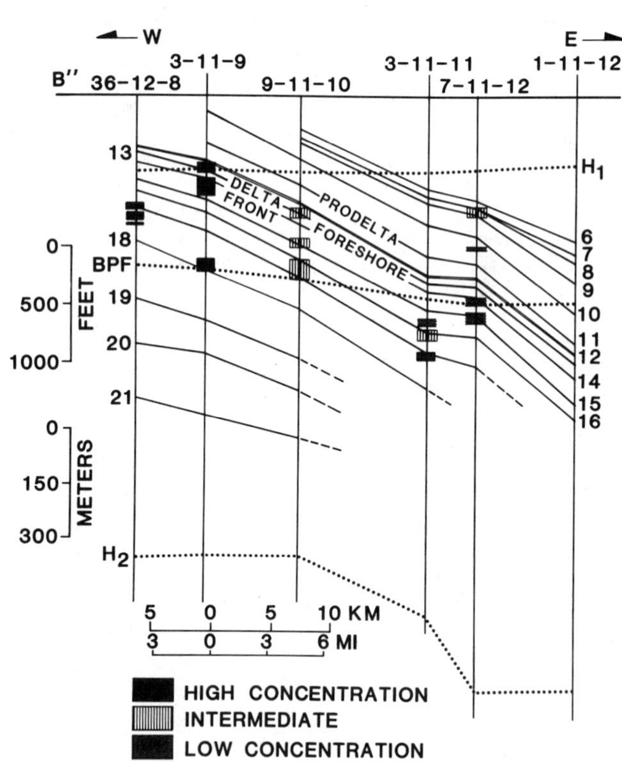

FIGURE 6 East-west cross section through the Kuparuk Oil Field Hydrate occurrences are plotted with degree of saturation indicated.

LONG-TERM ACTIVE LAYER EFFECTS OF CRUDE OIL SPILLED IN INTERIOR ALASKA

C. M. Collins

U.S. Army Cold Regions Research and Engineering Laboratory
Fairbanks, Alaska

Two experimental oil spills of 7570 liters each were conducted at a black-spruce-forested site in February and July of 1976. The long-term effects of the spills on the active layer were directly related to the method of oil movement. The winter spill moved beneath the snow, within the surface moss layer, and the summer spill moved primarily below the moss, in the organic soil. The summer spill affected an area nearly one and one-half times that of the winter spill. Only 10% of the 303-m^2 summer spill area had oil visible on the surface, while 40% of the 188-m^2 winter spill had visible oil. Thaw depths in the summer spill area increased from 1977 to 1980--average thaw depth was 72 cm vs. 48 cm in the control--and remained essentially the same in 1981 and 1982. Thaw depths in the winter spill area continued to increase until 1982 to an average of 92 cm. Summer temperatures 5 cm under the blackened moss are consistently higher than under the undisturbed surface. Presumably the change in albedo due to the surface oil accounts for the increased thaw in the winter spill area.

Over the past decade a number of studies have been conducted on the effects of oil spilled in the arctic and subarctic environment. This is a direct result of the development of the Prudhoe Bay oil field and the construction of the Trans-Alaska Pipeline.

Much of the work on the impact of oil spills has considered the effects of refined petroleum products and crude oil spilled on arctic and subarctic vegetation (Hunt et al. 1973, Deneke et al. 1975, Hutchinson and Freedman 1975, Walker et al. 1978), and several theoretical and laboratory studies have investigated crude oil spills on snow and ice (Raisbeck and Mohtadi 1974, McMinn and Golden 1973). A number of studies examined the biological and chemical effects of small experimental spills on tundra soils and vegetation, especially in the Barrow and Prudhoe Bay areas (Everett 1978, Sextone et al. 1978, Linkins et al. 1978).

In contrast to these mostly small-scale studies, a series of studies conducted near Norman Wells, Canada, attempted to determine experimentally the effects of larger oil spills (Mackay et al. 1974, Hutchinson et al. 1974, Hutchinson and Freedman 1975). A study of similar scope was conducted to determine the physical, chemical, and biological effects of crude oil spilled on a forested permafrost terrain near Fairbanks, simulating as nearly as possible an actual oil spill (Jenkins et al. 1978, Johnson et al. 1980). This paper is based on a continuation of that study and examines the long-term thermal effects of oil spills on the active layer.

Studies other than those on oil spills have investigated the thermal impact of natural and man-made disturbances on the active layer (Brown and Grave 1979). Most deal with the disturbances caused by vehicular traffic, oil development, and related construction activity (Mackay 1970, Lawson et al. 1978). Fire also plays a major role in the natural disturbances of permafrost (Heginbottom 1971; Viereck 1973, 1982; Hall et al. 1978; Racine 1981), where increased thaw of the active layer follows the destruction of the insulating organic layer by fire or the construction of fire lanes. Makihara (1982) studied the modification of surface albedo by coal dust and its effect on the active layer at Prudhoe Bay.

SITE DESCRIPTION

The winter and summer oil spill study was conducted in the Caribou-Poker Creek Research Watershed, within the Yukon Tanana Uplands, 48 km northeast of Fairbanks, Alaska. The study site is located at an elevation of 300 m, on an 8° west facing slope. Two plots, each 10 m by 50 m, were established for oil application. A control plot of the same size was established just upslope (Johnson et al. 1980).

The soil of the study site is a Histic Pereglic Cryaquept with silt loam texture and is typical of the Saulich series found in the lower slopes of the watershed (Rieger et al. 1972). The 01 and 02 layers (moss and peat) average 30 cm thick and overlie a 6-cm A horizon beneath which is a massive, grey silt C horizon that is water saturated above the permafrost. The active layer depth ranges from 20 to 60 cm according to the thickness of the organic (01 and 02) layers.

The vegetation of the study site (Figure 1) is an open *Picea mariana*/*Vaccinium* spp./feathermoss community described by Viereck and Dyrness (1980), a common community type in interior Alaska. Morphologically, it is an open black spruce (*Picea mariana*) stand with a shrub understory of *Ledum decumbens*, *Ledum groenlandicum*, *Betula glandulosa*, and *Vaccinium uliginosum*. Mosses and lichens cover 50% or more of the ground surface (Johnson et al. 1980).

FIGURE 1 Vegetation at study site before experimental spill (August 1975). Note elevated walkways installed for access to plot.

METHODS

Crude oil, heated to 57°C in a closed 7600-liter tank, was applied to the first spill plot on 26 February, 1976, at an ambient air temperature of -5°C. Heated oil was applied to the second plot on 14 July, 1976, at an ambient air temperature of 25°C. In each spill 7570 liters of oil was applied at the top of the plot through a 5-m-long perforated pipe at a rate of about 170 liters/min.

The rate and extent of oil flow downslope following each of the spills was determined by probing with wooden dowels in a 1-m grid pattern (Johnson et al. 1980). Thirty-six thermocouple arrays were installed in each plot, with five thermocouples per array to collect ground temperature data. Maximum depth of each array was 1 m.

The active layer depth was measured by probing with a graduated steel rod at 1-m intervals along six cross-sections located at 1, 3, 6, 9, 14, and 20 m downslope from the spill point in each plot. Thaw depth measurements have been made every September since 1976. Air temperature data are available from several nearby climatic stations maintained in the research watershed (Haugen et al. 1982).

RESULTS

Oil Movement

The timing of the two spills affected the mechanism of oil movement and the area covered by each spill. In addition, the amount of surface impact, which later affected the active layer, was different for the two spills.

During the winter spill, the hot oil melted holes in the 45-cm deep snowpack and then moved downslope under the snow without disturbing the snow surface. Most of the oil movement occurred just above and within the moss (01) layer, above the frozen peat and organic soil (02 and A1) layers. The oil moved 18 m downslope at a decreasing rate for 24 hr before becoming immobile. Following spring snowmelt, the oil became mobile again and moved an additional 17 m downslope. This latter movement was beneath the moss layer and was visually undetectable on the surface.

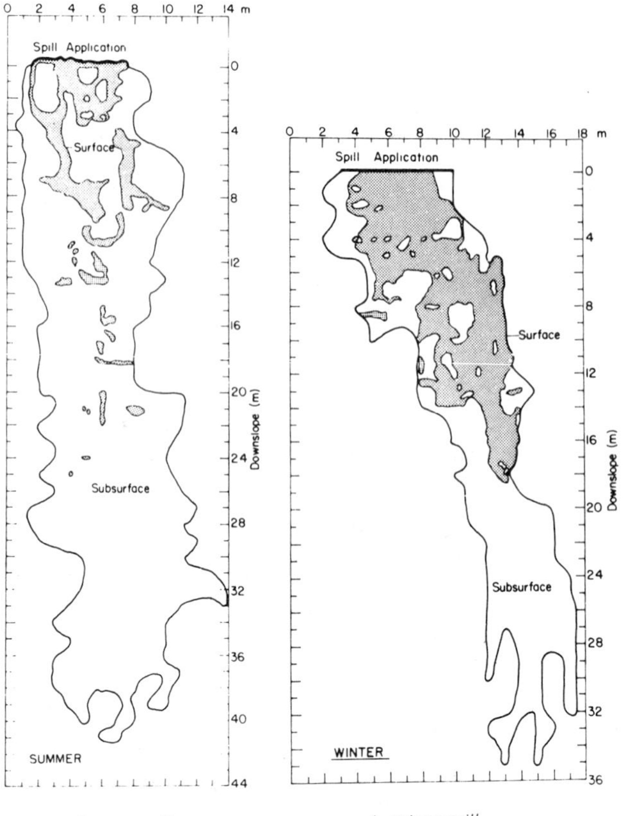

a. Summer spill. b. Winter spill.

FIGURE 2 Plan views of summer and winter spills showing surface and subsurface oiled areas.

The oil of the summer spill rapidly penetrated to the peat (02) layer and moved downslope beneath the moss. The oil was only visible within the first few meters of the plot and in surface depressions downslope. The downslope slope movement of oil was 32 m downslope in 48 hr, with an additional 7 m movement by October 1976.

The winter spill affected an area of 188 m^2 with about 40% of the area having visible surface oil (Figure 2). The summer spill affected 303 m^2 with only 10% having oil visible on the surface.

The average concentration of oil in the impacted area of the winter spill was 41 liters/m^2, considerably more than the 25 liters/m^2 for the summer spill. When viewed in terms of m^2/m^3 (area affected per volume of oil applied), both areas are within the range of 20-100 m^2/m^3 of oil predicted by Mackay et al. (1974) as reasonable for large-scale oil spills. The average concentrations of oil were much greater than many of the small-scale spray spills used to determine the effects of crude oil on vegetation, such as Walker et al. (1978) and Everett (1978).

Thermal Effects on the Active Layer

Effects on the underlying permafrost from the initial spill of the hot oil were not apparent (Johnson et al. 1980). The thermal mass of the warm oil was negligible compared to the thermal mass of

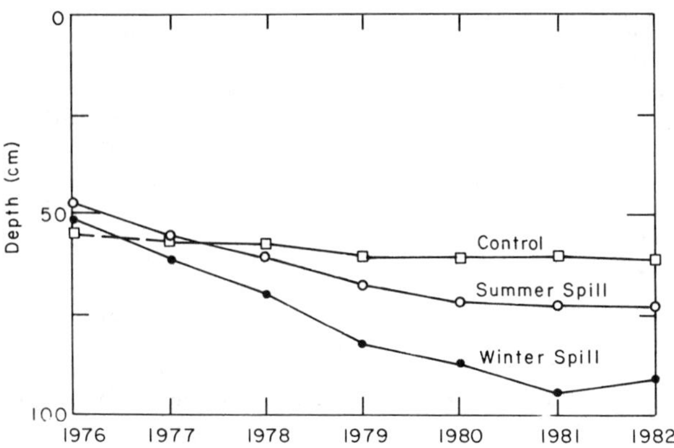

FIGURE 3 Thaw depths for the winter and summer spills and control, averaged from cross-sections at 1, 3, 6, 9, 14, and 20 m downslope from the top of the spill. Thaw depths were measured in September each year, 1976-1982.

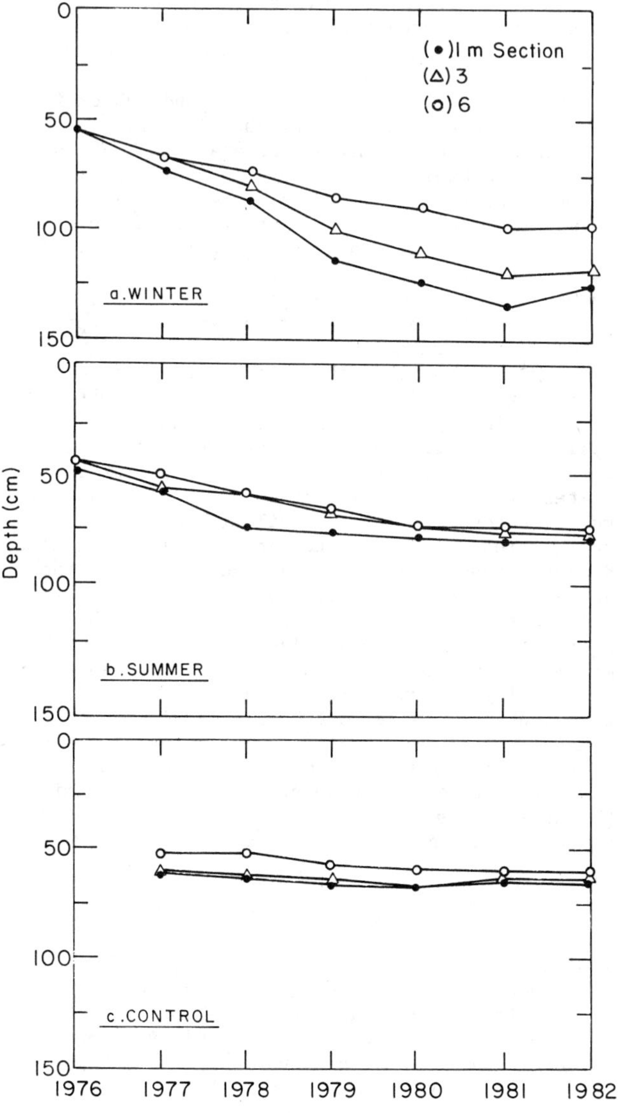

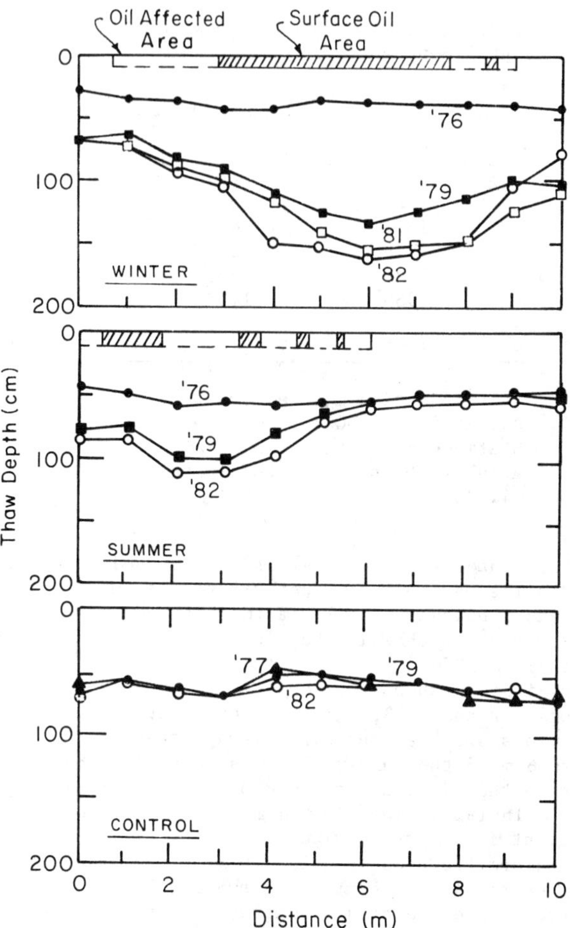

FIGURE 5 Thaw depths of the cross-section 3 m downslope from the top of the spill, in winter and summer spill areas and control, shown for September of different years from 1976 to 1982.

the frozen ground, and during the winter spill, much of the heat of the warm oil went into melting the snow.

Thaw depths for the winter and summer spill areas, as well as the control plot, are shown in Figure 3. The thaw depths are calculated as the average from means of six cross-sections at 1, 3, 6, 9, 14, and 20 m downslope from the top of the spill plot. It should be noted that, 20 m downslope from the top of the spill, thaw depths did not differ significantly for the three plots. Thaw depths in the summer spill area increased each year from 1977 to 1980, with the average thaw depth reaching 72 cm. The average thaw depths did not increase in 1981 and 1982. Maximum thaw in the summer spill reached 111 cm in 1981 and 1982.

Average thaw depths in the winter spill area have increased every year from 1977 to 1981, with a slight decrease noted in 1982. Average thaw depths

FIGURE 4 Average thaw depths for the winter and summer spills and control for cross-sections at 1, 3, and 6 m downslope from the top of the spill. Thaw depths were measured in September each year, 1976-1982.

TABLE 1 Degree days of thawing (°C), April-September.

Year	No. of days
1976	1583
1977	1453
1978	1223
1979	1356
1980	1153
1981	1249
1982	1163

NOTE: Station located 800 m west of spill site at 50 m lower elevation. Station is called Caribou Main, Alaska. Mean annual temperature is -4.9°C.

in the winter spill were 94 cm in 1981 and 92 cm in 1982. The maximum thaw depths have continued to increase, however, to 170 cm in 1981 and 175 cm in 1982. Maximum thaw in the winter plot has increased 165% since 1976.

Figure 4 shows the long-term average thaw depths for the 1, 3, and 6 m cross sections of the two plots and the control. Average thaw in the first 6 m of the summer plot was 77 cm in 1982. Average thaw in the first 6 m of the winter plot was 115 cm in 1982. The first 6 m are the area of greatest impact and maximum thaw.

Along the cross-section at 3 m, the greatest thaw depths in the winter and summer plots generally correspond to the area of surface oiling (Figure 5). This is presumably due to the decreased albedo of the blackened moss in the surface impacted area.

Degree days of thawing for the study area for the period 1977-1982 range from 1200 to 1500 (Table 1). The consistency of the average thaw depths of the control plot indicate relatively consistent natural depths of thaw each year.

Soil temperatures 5 cm below the surface were consistently warmer under the black, surface oil areas compared to the nonsurface-impacted areas within the spills. Maximum temperatures of 31°C were noted under the oil-blackened moss, as compared to a maximum of 18°C under nonblackened moss areas nearby.

DISCUSSION AND CONCLUSIONS

There are conflicting reports on the effects of oil spills on the thickness of the permafrost active layer. For small experimental spill plots at Barrow, Everett (1978) found thaw depth to increase significantly for 2 years after application of oil with a smaller effect in the third year. He also (Lawson et al. 1978) found that after 30 years at the Fish Creek site, in northern Alaska, thaw under diesel spills dating from the late 1940's was almost double that of undisturbed sites, although it should be noted that there was almost no vegetation cover on the diesel spills to act as insulation.

Mackay et al. (1974) observed little thermal effect of oil at a forested spill site, but their observations were based on only one season. Hutchinson and Freedman (1978) also observed that spray spill treatments had only a limited effect on the depth of the active layer, which was not maintained after the first postspill growing season. This may have been due to the small amount of oil applied.

Other types of disturbances have resulted in similar increases in the thickness of the active layer. Following the modification of the organic forest floor by fire and mechanical stripping, Dyrness (1982) observed increases in the thickness of the active layer for 4 years with no indications of a decrease in the rate of thaw. There, the vegetation was a black spruce community similar to the oil spill plots in Caribou-Poker Creek. Heginbottom (1973) reports increases in thaw depths due to forest fires and bulldozed firelines at the 1968 Inuvik fire, with a median thaw of 48 cm for undisturbed areas and 61 and 120 cm for burned and bulldozed areas, respectively. Mackay (1977) reports a slight increase was still occurring in the active layer 8 years after the fire.

Viereck (1982) reports somewhat deeper thaw depths for a burned black spruce community near Fairbanks. Maximum thaw in an unburned area was 40-50 cm; in the burned area, 187 cm in 1980; and in a cleared fireline, 227 cm in 1979 and 200 cm in 1980, as vegetation reestablished itself.

These observations are relevant in light of proposed mechanical clean-up procedures (including burning) following an oil spill, which may cause substantial thawing of the permafrost due to disturbance or destruction of the vegetative cover. The thawing due to clean-up procedures may equal or exceed that caused by the spill. This has to be considered when planning a response to oil spills on permafrost terrain.

The maximum thermal impact on the active layer resulting from the two spills at Caribou-Poker Creek occurred in the first few meters of the winter spill, where oil flowed over the moss and beneath the snow cover. Although the total area of the winter spill was smaller than the summer spill, the much greater percentage of oil-blackened moss on the surface resulting from the winter spill has caused a deeper average thaw. Average thaw depths in the summer plot increased from 1976 to 1980 and have since stabilized. Average thaw depths in the winter plot appear to have stabilized by 1982, although the maximum thaw depth has continued to increase under the heavily impacted areas. These trends contrast with those of smaller oil spills in Alaska, generally of lower application rates and in colder areas, where thaw depths did not increase after the first several years; however, the trend is similar to burned areas in interior Alaska and at Inuvik, Canada, that have similar climates.

ACKNOWLEDGMENTS

This project was originally funded by the Environmental Protection Agency. Subsequent observations were funded by several projects through the U.S. Army Corps of Engineers Cold Regions Research and Engineering Laboratory.

REFERENCES

Brown, J. and Grave, N.A., 1979, Physical and thermal disturbance and protection of permafrost: Special Report 79-5, Hanover, N.H., U.S. Army Cold Regions Research and Engineering Laboratory.

Deneke, F.J., McCown, B.H., Coyne, P.I., Rickard, W., and Brown, J., 1975, Biological aspects of terrestrial oil spills - USA CRREL oil research in Alaska, 1970-1974, Research Report 346: Hanover, N.H., U.S. Army Cold Regions Research and Engineering Laboratory.

Dyrness, C.T., 1982, Control of depth to permafrost and soil temperature by the forest floor in black spruce/feathermoss communities: U.S. Forest Service Research Note PNW-396.

Everett, K.R., 1978, Some effects of oil on the physical and chemical characteristics of wet tundra soils: Arctic, v. 31, p. 260-276.

Hall, D.K., Brown, J., and Johnson, L., 1978, The 1977 tundra fire in the Kokolik River area, Alaska: Arctic, v. 31, p. 54-58.

Haugen, R.K., Slaughter, C.W., Howe, K.E., and Dingman, S.L., 1982, Hydrology and climatology of the Caribou-Poker Creek Research Watershed, Alaska, CRREL Report 82-26: Hanover, N.H., U.S. Army Cold Regions Research and Engineering Laboratory.

Heginbottom, J.A., 1971, Some effects of a forest fire on the permafrost active layer at Inuvik, N.W.T., in R.J.E. Brown, ed., Proceedings of Joint Conference on Permafrost Active Layer, 4-5 May: NRC/ACGR Technical Memorandum Number 113:31-36.

Heginbottom, J.A., 1973, Some effects of surface disturbance on permafrost active layer at Inuvik, N.W.T.: Environmental-Social Committee on Northern Pipelines, Task Force on Northern Oil Development, Report No. 73-16.

Hunt, P.G., Rickard, W.E., Deneke, F.J., Koutz, F.R., and Murrmann, R.P., 1973, Biological effects--terrestrial oil spills in Alaska: Environmental effects and recovery, in Proceedings of Joint Conference on Prevention and Control of Oil Spills, 13-15 March: Washington, D.C.

Hutchinson, T.C., and Freedman, W., 1975, The impact of crude oil spills on Arctic and sub-arctic vegetation, in Proceedings of the Circumpolar Conference on Northern Ecology, 15-18 Sept, v. 1: Ottawa, Ontario.

Hutchinson, T.C., and Freedman, W., 1978, Effects of experimental crude oil spills on sub-arctic boreal forest vegetation near Norman Wells, N.W.T., Canada: Canadian Journal of Botany, v. 56, p. 2424-2433.

Hutchinson, T.C., Hellebust, J., and Telford, M., 1974, Oil spill effects on vegetation and soil microfauna at Norman Wells and Tuktoyaktuk N.W.T.: Environmental-Social Committee on Northern Pipelines, Task Force on Northern Oil Development, Report No. 74-14.

Jenkins, T.F., Johnson, L.A., Collins, C.M., and McFadden, T.T., 1978, The physical, chemical, and biological effects of crude oil spills on black spruce forest, interior Alaska: Arctic, v. 31, p. 305-323.

Rieger, S., Furbush, C.E., Schoephorster, D.B., Summerfield, H., Jr., and Geiger, L.C., 1972, Soils of the Caribou-Poker Creek Research Watershed, interior Alaska, Technical Report 236: Hanover, N.H., U.S. Army Cold Regions Research and Engineering Laboratory.

Sextone, A., Everett, K., Jenkins, T., and Atlas, R.M., 1978, Fate of crude and refined oils in North Slope soils: Arctic, v. 31, p. 339-347.

Viereck, L.A., 1973, Ecological effects of river flooding and forest fires on permafrost in the taiga of Alaska, in Permafrost: North America Contribution Second International Conference: Washington, D.C., National Academy of Sciences, p. 60-67.

Viereck, L.A., 1982, Effects of fire and firelines on the active layer thickness and soil temperatures in interior Alaska, in Proceedings Fourth Canadian Permafrost Conference: Ottawa, Ontario, National Research Council of Canada, p. 123-135.

Viereck, L.A., and Dyrness, C.T., 1980, A preliminary classification system for vegetation of Alaska, General Technical Report PNW-106: USDA Forest Service, Pacific Northwest Forest and Range Experiment Station.

Walker, D.A., Webber, P.J., Everett, K.R., and Brown, J., 1978, Effects of crude and diesel oil spills on plant communities at Prudhoe Bay, Alaska and the derivation of oil spill sensitivity maps: Arctic, v. 31, p. 242-259.

A MATHEMATICAL MODEL FOR THE PERMAFROST THAW CONSOLIDATION

M. Yavuz Corapcioglu

Department of Civil Engineering, University of Delaware, Newark, Delaware 19711 USA

A mathematical model is developed for temporal and spatial distribution of pore water pressure, temperature, ice content and displacements in an unsaturated thawing deformable soil. The water saturation is related to pore pressure and temperature by retention and phase composition curves respectively. The model is simplified for a one-dimensional vertical permafrost thaw consolidation problem. The use of temperature-dependent rheological stress-strain relations is also discussed.

INTRODUCTION

Heat transfer in soils plays an important part in varied fields. Expanding interest in arctic and near-arctic regions is continuing to provide motivation for research on transport processes in frozen and partially frozen soils. Reducing the transfer of pipeline or building heat to the underlying soil and preventing thawing and associated settlement to utilize the high strength of the frozen soil are the main considerations in permafrost regions. Where thawing of the frozen ground cannot be prevented, the rate and extent of degradation are factors that must be taken into account. Such a problem requires a coupled analysis, i.e., thaw, heat transfer, flow and consolidation. This kind of study involves applying coupled differential equations to describe the field conditions.

The process of transfer of heat in soils is unique, since such a medium presents a combination of components, e.g., ice, water and air. The freezing and thawing of soils are complex phenomena, which may take place over a wide range of temperatures, depending upon the soil type and the water content. Such processes require the extraction or addition of the latent heat of fusion that accompany phase changes. The special features of the geotechnical properties of frozen and thawing soil are the dominant creep characteristics under sustained stress and the marked rate dependence of the strength. These characteristics are attributed to the presence of unfrozen water and the various responses of ice to loading. The rate effects in frozen and thawing ground are temperature dependent and vary with the soil ice content.

Several researchers have spent much effort on the mathematical simulation of permafrost thaw consolidation and developed several models. Works by Tsytovich and co-workers appear to be the first studies on this subject (Tsytovich, 1960; Tsytovich et al., 1966). These studies conclude that the thaw consolidation caused by the abrupt change in void ratio that occurs during thawing of the soil ice and results in settlement is approximately equal to the thickness of ice inclusions, is independent of external load, is proportional to depth of thawing, and occurs at a rate proportional to the square root of time. For a considerable period after complete thawing has occurred, settlement proceeds at a reduced rate and is proportional to the logarithm of the time. Later, Tsytovich et al. (1965) have shown that the consolidation of thawing soils occurs with practically constant pore pressure and obeys the theory of consolidation.

One-dimensional study of permafrost thaw and settlement has been investigated by various researchers. Some of them approached the problem from a simplistic and practical point of view (e.g., Zaretskii, 1968; Brown and Johnston, 1970), while others preferred a more rigorous theoretical approach. Morgenstern and Nixon (1971) formulated thawing settlement in terms of the theory of consolidation and provided a solution to a moving boundary thaw consolidation problem where the thaw line is assumed to move proportionally to the square root of time as governed by a parameter α. One-dimensional closed form solutions have been obtained for several cases of practical interest. It is shown that the excess pore pressures and the degree of consolidation in thawing soils depend primarily on the thaw consolidation ratio, $R = \alpha/(2\sqrt{C_v})$, where C_v is the coefficient of consolidation. Later Nixon and Morgenstern (1973) extended their study to incorporate arbitrary movements of the thaw line. This approach employs a finite difference technique to obtain a numerical solution. They also introduced the nonlinear compressibility relation into the theory. McRoberts et al. (1978) considered certain aspects of thaw consolidation effects in degrading permafrost by examining two cases.

There are also various studies to simulate the coupled flow of mass and heat in a freezing, deformable soil system in addition to ones for rigid porous media. Studies on deformable media cover the works of Sheppard et al. (1978) and Del Giudice (1978). Sheppard et al. (1978) developed a model that describes heat and water flow in a freezing soil and has provision for incorporating the effect of overburden pressure and deformation of the matrix. Heave is allowed to occur through the use of deforming coordinates. Del Giudice et al. (1978) present the finite element model, formulated for nonlinear heat conduction, to study ground freezing problem by solving the energy equation only. In view of the difficulty of obtaining closed form solutions for two-dimensional problems Sykes et al. (1974a,b) solved two-dimensional problems by the finite element method to determine the effects of settlement and the significance of variable geometry and forced convection rather than heat conduction only as done previously. The applications of the model include the analysis of a heated oil pipeline buried in permafrost and the analysis of foundations on permafrost. Later, Sykes and Lennox (1976) considered a nonlinear stress-strain relation for thawing permafrost in their model.

Recently, Bear and Corapcioglu (1981) developed a mathematical model for fluid pressure, temperature, and land subsidence owing to temperature and pressure changes in a porous medium. Conservation of mass, energy, and equilibrium equations were developed and the effects of viscous dissipation and compressible work have been included in the formulation. The effects of heat transfer by forced convection as well as conduction were included in the model. Cross-transport phenomena resulting from reciprocal relations had not been considered in their formulation. This paper presents a mathematical model which includes a coupled transport equation in the derivations for a thawing unsaturated porous medium. It is an extension of the theory presented by Bear and Corapcioglu (1981). The mathematical model will be capable of simulating pressure, saturation, temperature, ice content and displacements in a permafrost thaw consolidation problem.

THEORY

The macroscopic unsteady water flow in a partially saturated heterogeneous thawing porous medium can be written as

$$\nabla \cdot \rho_f \underset{\sim}{q} + \frac{\partial(\rho_f \Theta_w)}{\partial t} - \frac{\partial(\rho_i \Theta_i)}{\partial t} = 0 \tag{1}$$

where $\underset{\sim}{q}$ is the specific discharge of water with respect to the fixed coordinates. In (1), ρ_f and ρ_i are the densities of water and ice, respectively. Θ_w is the volumetric liquid water fraction, and t is time. Θ_w is a product of the porosity n and the degree of saturation S_w; that is, $\Theta_w = nS_w$. Θ_i denotes the volumetric ice fraction. Equation (1) has also been used in a similar form by Taylor and Luthin (1978), Sheppard et al. (1978), and Jame and Norum (1980). The frozen component of water, i.e., ice, is assumed to be incompressible and impermeable. In the case of a deforming porous medium where the soil particles move at a velocity $\underset{\sim}{V}_s$, it is the specific discharge of the water relative to the moving solid $\underset{\sim}{q}_r$ that is expressed by Darcy's Law (Bear, 1972, p. 205)

$$\underset{\sim}{q}_r = \underset{\sim}{q} - \Theta_w \underset{\sim}{V}_s \;;\quad \underset{\sim}{q} = \Theta_w \underset{\sim}{V}_f = nS_w \underset{\sim}{V}_f \tag{2}$$

Inserting this relative flux expression in (1) and assuming that under the pressure conditions during thawing in a frozen soil the water is practically incompressible, ρ_f = constant, and the soil particles and ice are incompressible, i.e., ρ_i = constant, ρ_s = constant (however, the soil matrix as a whole is deformable, with changing porosity), we obtain

$$\nabla \cdot \underset{\sim}{q}_r + S_w \frac{d_s n}{dt} + n \frac{d_s S_w}{dt} + nS_w \nabla \cdot \underset{\sim}{V}_s - \frac{\rho_i}{\rho_f} \frac{\partial \Theta_i}{\partial t} = 0 \tag{3}$$

where $d_s(\ )/dt = \partial(\ )/\partial t + \underset{\sim}{V}_s \cdot \nabla(\ )$ denotes the total derivative with respect to moving solid. The fourth term can be obtained from the equation of mass conservation of solid which is given by

$$\nabla \cdot [(1 - n)\underset{\sim}{V}_s] + \frac{\partial(1 - n)}{\partial t} = 0 \tag{4}$$

The change in porosity is related to the volume strain (dilatation) ε of the porous medium by

$$\frac{d_s \varepsilon}{dt} = \frac{1}{1 - n} \frac{d_s n}{dt} = \nabla \cdot \underset{\sim}{V}_s \tag{5}$$

By combining (3) and (5), and an approximation, replacing the total derivatives by partial ones, i.e., assuming $\underset{\sim}{V}_s \cdot \nabla S_w \ll \partial S_w/\partial t$ and $\underset{\sim}{V}_s \cdot \nabla\varepsilon \ll \partial\varepsilon/\partial t$, (3) reduces to

$$\nabla \cdot \underset{\sim}{q}_r + S_w \frac{\partial \varepsilon}{\partial t} + n \frac{\partial S_w}{\partial t} - \frac{\rho_i}{\rho_f} \frac{\partial \Theta_i}{\partial t} = 0 \tag{6}$$

The melting term $\partial\Theta_i/\partial t$ can be expressed in terms of S_w and ε by writing down the conservation of mass equation for ice. If we assume ice moves with the velocity of solid particles $\underset{\sim}{V}_s$

$$\nabla \cdot \Theta_i \underset{\sim}{V}_s + \frac{\partial \Theta_i}{\partial t} - \frac{\rho_f}{\rho_i} \frac{\partial \Theta_w}{\partial t} = 0 \tag{7}$$

By rearranging and making use of (5) and previous assumptions

$$\Theta_i \frac{\partial \varepsilon}{\partial t} + \frac{\partial \Theta_i}{\partial t} - \frac{\rho_f}{\rho_i} \frac{\partial \Theta_w}{\partial t} = 0 \tag{8}$$

The specific discharge relative to the deforming matrix, $\underset{\sim}{q}_r$ is expressed by Darcy's law which states that $\underset{\sim}{q}_r$ is proportional to the total potential gradient. Assuming that linear interactions between the flows are induced by mechanical and thermal effects, the mass flux is not only related to the potential gradient but also to the gradient of temperature. Then

$$\underset{\sim}{q}_r = - \frac{\underset{\approx}{k}}{\mu} (\nabla p + \rho_f g \nabla z) - \underset{\approx}{D}_{MT} \nabla T \tag{9}$$

where p is the pore pressure (suction), $\underset{\approx}{k}$ is the medium's permeability tensor, which depends on n and the absolute temperature T, and μ is the water's dynamic viscosity. The vertical direction is represented by z, and g denotes the gravitational acceleration. $\underset{\approx}{D}_{MT}$ is the thermal liquid diffusion tensor, which is an Onsager's phenomenological parameter. This interference tensor can also be called as the tensor of thermo-osmosis.

The degree of saturation S_w is a function of p and T. Under isothermal conditions, the relation between S_w and p is given by the retention curve. Curves showing the phase composition of the water in the soil for temperatures below 0°C are given by Jame and Norum (1980). Typical curves for S_w are given in Figure 1. Hence

$$\frac{d_s S_w}{dt} = \left.\frac{\partial S_w}{\partial p}\right|_T \frac{d_s p}{dt} + \left.\frac{\partial S_w}{\partial T}\right|_p \frac{d_s T}{dt} \tag{10}$$

The term $\partial S_w/\partial p|_T$ can be determined from the retention curve of the soil. $\partial S_w/\partial T|_p$ can be obtained from the experimental relationship between unfrozen water content and below-freezing temperature. In addition to these two curves we need expressions for the medium's pressure and temperature dependent permeability and viscosity. Usually the dependence of μ on p is much smaller than on T and may be neglected.

The macroscopic energy conservation equation was developed by Bear and Corapcioglu (1981), by starting from microscopic considerations and deriving the macroscopic one by averaging the former over a representative elementary volume of the porous medium. It is assumed that the thermal resistance between water and soil particles is small, hence, local water and matrix temperatures are equal. In what will be given below are the terms due to phase changes that have been added for a partially saturated thawing deformable porous medium.

$$\frac{\partial[(\rho C)_m T]}{\partial t} + L\rho_i \frac{\partial \Theta_i}{\partial t} + \nabla \cdot [(\Theta_w \rho_f C_{\forall} \underset{\sim}{V}_f + \Theta_i \rho_i C_i \underset{\sim}{V}_s + (1-n)\rho_s C_\varepsilon \underset{\sim}{V}_s)T] - \nabla \cdot [\underset{\approx}{\Lambda}_m \nabla T] - p\nabla \cdot \Theta_w \underset{\sim}{V}_f + (1-n)T\gamma \frac{\partial \varepsilon}{\partial t} = 0 \tag{11}$$

where $(\rho C)_m$ is the heat capacity per unit volume of the porous medium, i.e., $(\rho C)_m = \Theta_w C_{\forall} \rho_f + \Theta_i C_i \rho_i + (1-n)\rho_s C_\varepsilon$. $C_{\forall}$ is the water's heat capacity (specific heat) at constant volume. C_i is the heat capacity of ice at constant volume. C_ε is the solid's heat capacity at constant strain. L is the latent heat of fusion at a reference temperature, T^o. $\underset{\approx}{\Lambda}_m = \Theta_w \lambda_f + \Theta_i \lambda_i + (1-n)\lambda_s$ is the coefficient of thermal conductivity of the porous medium as a whole. λ_f, λ_i, and λ_s are the heat conductivities of water, ice, and solid, respectively. Note that $(\rho C)_m$ and $\underset{\approx}{\Lambda}_m$ depend on n, S_w, and Θ_i, which in turn depend on pressure and temperature. Also, $\gamma = \partial \underset{\approx}{\sigma}/\partial T|_\varepsilon$, where $\underset{\approx}{\sigma}$ is the total stress tensor on the solid particles of the soil. In writing (11) we assumed that the vapor phase in the soil has no effect (Fuchs et al., 1978). This assumption enables us to neglect the pore vapor pressure, vapor flux, and the transfer of latent heat by vapor movement.

The first term of (11) represents the change in the total heat content. The second term represents the rate of heat production due to melting. The third term represents the transfer of sensible heat by convection and deformation. Again we assume that ice moves with the velocity of the solid. The fourth term represents the contribution of pure heat conduction.

If we combine (11) with the conservation of mass equations (1), (4) and (7), we obtain

$$(\rho C)_m \frac{\partial T}{\partial t} + L\rho_i \frac{\partial \Theta_i}{\partial t} - \nabla \cdot [\underset{\approx}{\Lambda}_m \nabla T] + [(\rho C)_m \underset{\sim}{V}_s + \rho_f C_{\forall} \underset{\sim}{q}_r] \cdot \nabla T + C_{\forall}\rho_i T \frac{\partial \Theta_i}{\partial t} + C_i \rho_f T \frac{\partial \Theta_w}{\partial t} - p\nabla \cdot \Theta_w \underset{\sim}{V}_f - (1-n)T\gamma \frac{\partial \varepsilon}{\partial T} = 0 \tag{12}$$

Equation (12) may further be simplified by noting that

$$(\rho C)_m \left[\frac{\partial T}{\partial t} + \underset{\sim}{V}_s \cdot \nabla T\right] \equiv (\rho C)_m \frac{d_s T}{dt} \simeq (\rho C)_m \frac{\partial T}{\partial t}$$

where we have $\underset{\sim}{V}_s \cdot \nabla T << \partial T/\partial t$ and assuming $n\nabla \cdot \underset{\sim}{V}_s >> \underset{\sim}{V}_s \cdot \nabla n$ and therefore

$$\nabla \cdot \Theta_w \underset{\sim}{V}_f = \nabla \cdot (\underset{\sim}{q}_r + \Theta_w \underset{\sim}{V}_s) \simeq \nabla \cdot \underset{\sim}{q}_r + \Theta_w \nabla \cdot \underset{\sim}{V}_s \simeq \nabla \cdot \underset{\sim}{q}_r + \Theta_w \frac{\partial \varepsilon}{\partial t} \tag{13}$$

where we introduce the assumption $\underset{\sim}{V}_s \cdot \nabla\varepsilon << \partial\varepsilon/\partial t$. We then obtain

$$(\rho C)_m \frac{\partial T}{\partial t} + L\rho_i \frac{\partial \Theta_i}{\partial t} - \nabla \cdot [\underset{\approx}{\Lambda}_m \nabla T] + C_{\forall}\rho_f \underset{\sim}{q}_r \cdot \nabla T + C_{\forall}\rho_i T \frac{\partial \Theta_i}{\partial t} + C_i \rho_f \frac{\partial \Theta_w}{\partial t} - p\nabla \cdot \underset{\sim}{q}_r + [(1-n)T\gamma - \Theta_w p] \frac{\partial \varepsilon}{\partial t} = 0 \tag{14}$$

The last two terms in (14) express the source of heat due to the internal energy increase per unit volume of porous medium by viscous dissipation and by compression. They may be relatively small in many cases of practical interest.

For a three-dimensional analysis of the problem, we need macroscopic equilibrium equations. Bear and Pinder (1978) develop the macroscopic equilibrium equations by volume averaging the microscopic ones. Neglecting the inertial terms, the total stress tensor $\underset{\approx}{\sigma}$ at a point within the soil satisfies the equilibrium equations

$$\frac{\partial \sigma_{ij}}{x_j} + f_i = 0 \qquad i = 1,2,3 \tag{15}$$

where f_i represents the body force and the summation convention is employed.

Using $\sigma_{ij} = \sigma'_{ij} - S_w p \delta_{ij}$ to express the total stress in terms of the effective stress σ'_{ij} and the pressure p (positive for compression), and separating both σ'_{ij} and p into initial steady values σ'^{o}_{ij} and p^o and consolidation-producing incremental values σ'^{e}_{ij} and p^e, we replace (15) by

$$\frac{\partial \sigma'^{o}_{ij}}{\partial x_j} + f^o_i - \frac{\partial S_w p^o}{\partial x_j} = 0 \qquad (16)$$

$$\frac{\partial \sigma'^{e}_{ij}}{\partial x_j} - \frac{\partial S_w p^e}{\partial x_i} = 0 \qquad (17)$$

where $f^o_i \equiv f_i$. δ_{ij} is the Kronecker delta.

For a thermoelastic porous medium the stress-strain relationships are given by the Duhamel-Neumann relations

$$\sigma'^{e}_{ij} = C_{ijkl}\varepsilon^e_{kl} - \beta_{ij}T^e \qquad (18)$$

where ε^e_{kl} are components of the incremental strain, $T^e = T - T^o$ is the incremental temperature, and the material coefficients

$$C_{ijkl} = (\partial\sigma'^{e}_{ij}/\partial\varepsilon^e_{kl})\Big|_{T=\text{const}}$$

$$\beta_{ij} = -(\partial\sigma'^{e}_{ij}/\partial T)\Big|_{\varepsilon=\text{const}} \qquad (19)$$

as calculated. For the sake of simplicity we shall henceforth limit the discussion to an elastically isotropic porous medium. For an isotropic body, (18) reduces to

$$\sigma'^{e}_{ij} = 2G\varepsilon_{ij} + (\lambda\varepsilon - \gamma T^e)\delta_{ij} \;;\quad \varepsilon \equiv \varepsilon_{kk} \qquad (20)$$

where G and λ are the Lamé constants of the porous matrix and $\gamma\delta_{ij} = (\partial\sigma_{ij}/\partial T)|_{\varepsilon=\text{const}}$. The coefficient γ is also related to the coefficient of volumetric thermal expansion α_T by

$$\alpha_p = \frac{1}{\lambda + \frac{2}{3}G} \;;\; \gamma = (\lambda + \frac{2}{3}G)\alpha_T \;;\; \gamma = \alpha_T/\alpha_p \qquad (21)$$

α_T includes the thermoelastic coupling term, strain in the porous material due to a unit change in fluid temperature, since both have the same local temperatures. Then, by assuming $\varepsilon = \varepsilon(p,T)$

$$\frac{d_s\varepsilon_{ij}}{dt} = \frac{\partial\varepsilon_{ij}}{\partial p}\Big|_T \frac{d_s p}{dt} + \frac{\partial\varepsilon_{ij}}{\partial T}\Big|_p \frac{d_s T}{dt}$$

$$= \alpha_p \frac{d_s p}{dt} + \alpha_T \frac{d_s T}{dt} \qquad (22)$$

Using the usual relationships between ε_{ij} and the components of displacement $U_i (i = 1,2,3)$,

$$\varepsilon_{ij} = \frac{1}{2}\left(\frac{\partial U_i}{\partial x_j} + \frac{\partial U_j}{\partial x_i}\right) \;;\quad \varepsilon = \nabla \cdot \underset{\sim}{U} \qquad (23)$$

we may rewrite (20) in the form

$$\sigma'^{e}_{ij} = G\left(\frac{\partial U_i}{\partial x_j} + \frac{\partial U_j}{\partial x_i}\right) + \left[\lambda \frac{\partial U_k}{\partial x_k} - \gamma T^e\right]\delta_{ij} \qquad (24)$$

Equation (22), together with (17), into which we insert (24), provides four equations for eight variables, U_i, p^e, T^e, S_w, Θ_i, and ε. The mass conservation equations for water (equation (6)) and ice (equation (8)), and the thermal energy conservation equation (14) and expression (10) provide four additional equations in terms of p, S_w, Θ_i, T and ε. Altogether we have eight equations for the eight variables. We have not mentioned $\underset{\sim}{q}_r$ as a variable, as it is easily related to p and T by (9). In addition, we need information on $\underset{\approx}{k} = k(n,S_w)$ with n related to ε, which is given by (5), and $\mu = \mu(p,T) \approx \mu(T)$.

ONE DIMENSIONAL MODEL

For a one dimensional (vertical) elastic consolidation of thawing soils, governing equations will reduce to simpler forms. The conservation of mass equation for water (6) would reduce to the following form with the use of (10) and (22)

$$\frac{\partial \underset{\sim}{q}_r}{\partial z} + (S_w\alpha_p + n\zeta_p)\frac{\partial p}{\partial t} + (S_w\alpha_T + n\zeta_T)\frac{\partial T}{\partial t}$$

$$- \frac{\rho_i}{\rho_f}\frac{\partial\Theta_i}{\partial t} = 0 \qquad (25)$$

where $\zeta_p = \partial S_w/\partial p|_T$ and $\zeta_T = \partial S_w/\partial T|_p$, which are determined from experimentally obtained curves (see Figure 1). Similarly, the conservation of thermal energy (equation (14)) would reduce to

$$\Big\{(\rho C)_m + \alpha_T[(1-n)(\gamma T + C_i\rho_f S_w) - S_w np]$$

$$+ nC_i\rho_f\zeta_T\Big\}\frac{\partial T}{\partial t} + \Big\{\alpha_p[(1-n)(\gamma T + C_i\rho_f S_w)$$

$$- S_w np] + nC_i\rho_f\zeta_p\Big\}\frac{\partial p}{\partial t} + L\rho_i\frac{\partial\Theta_i}{\partial t} - \frac{\partial}{\partial z}[\Lambda\frac{\partial T}{\partial z}]$$

$$+ C_\psi\rho_f \underset{\sim}{q}_r\frac{\partial T}{\partial z} + C_\psi\rho_i T\frac{\partial\Theta_i}{\partial t} - p\frac{\partial\underset{\sim}{q}_r}{\partial z} = 0 \qquad (26)$$

with the assumption of $\partial n/\partial t \cong (1 - n)\partial\varepsilon/\partial t$. The conservation of mass equation (8) for ice would be

$$\Big\{\alpha_p[\Theta_i - (1-n)S_w\frac{\rho_i}{\rho_f}] - n\frac{\rho_i}{\rho_f}\zeta_p\Big\}\frac{\partial p}{\partial t} + \Big\{\alpha_T[\Theta_i$$

$$- (1-n)S_w\frac{\rho_i}{\rho_f}] - n\frac{\rho_i}{\rho_f}\zeta_T\Big\}\frac{\partial T}{\partial t} + \frac{\partial\Theta_i}{\partial t} = 0 \qquad (27)$$

Also, the stress-strain equation would be

$$S_w p^e = \sigma'^{e}_{zz} = (2G + \lambda)\frac{\partial U_z}{\partial z} - \gamma T^e \qquad (28)$$

if the vertical total stress σ_{zz} remains constant, i.e., $\sigma^e_{zz} = 0 = \sigma'^{e}_{zz} - S_w p^e$.

In summary, the one dimensional elastic consolidation of thawing soils can be simulated by the governing equations (25), (26), (27), and (28) to obtain p, T, Θ_i, and U_z. The porosity, n is related to ε by (5), and S_w is related to p and T by (10).

RHEOLOGY OF THAWING SOILS

In the formulation given above we assumed a time-independent elastic stress-strain relation for the medium. This approach is, of course, the most simple one from a mathematical point of view. The phenomena that control the mechanical behavior of frozen and thawing ground are complex. As a result of ice content Θ_i and unfrozen water content Θ_w, thawing soils are characterized by rheological stress-strain properties. Various researchers (e.g., Vyalov, 1963; Stevens, 1973; Aziz and Laba, 1976) have studied the viscoelastic behavior of frozen ground. They have introduced several mechanical models to simulate the behavior of a frozen soil under a load. When frozen ground is subjected to a load, it will respond with an instantaneous deformation and a time-dependent creep behavior. It is therefore convenient to distinguish between the elastic and creep properties, Both will depend on temperature and material properties.

Aziz and Laba (1976) have assumed that the solid soil particles may be considered as an elastic material. Ice, separately, as well as in combination with soil particles, will behave as a viscoelastic solid. Unfrozen water, although incompressible, will flow under pressure gradually with time so that the system will display a certain amount of viscous flow continuing for a long period of time. Accordingly, the proposed model of Aziz and Laba consisted of a spring (to represent elastic deformation), a damper (to simulate viscous flow), and a parallel combination of spring and damper (to represent viscoelastic behavior), with all three elements connected in series. One big disadvantage of their model is that it is not temperature dependent.

To represent a temperature dependent thawing isotropic soil mechanically, we can visualize the behavior of the ground as being equivalent to that of a spring placed in series with a parallel combination of spring and damper (a Kelvin body). Then for a one-dimensional deformation

$$\sigma'^{e}_{zz} = E_1 \varepsilon_s \tag{29}$$

$$\sigma'^{e}_{zz} = E_2 \varepsilon_K + \phi_2 \frac{\partial \varepsilon_K}{\partial t} + \phi_T \frac{\partial T^e}{\partial t} \tag{30}$$

where ε_s and ε_K denote the strains of the spring and the Kelvin body, respectively. The values of E_1 and E_2 are the elastic moduli of the two spring components and ϕ_2 is the viscosity of the Kelvin body. ϕ_T is a thermoviscoelastic coefficient. These parameters have to be determined experimentally. For the series arrangement of the present model the total strain is

$$\varepsilon_{zz} = \frac{\partial U_z}{\partial z} = \varepsilon_s + \varepsilon_K \tag{31}$$

A more convenient stress-strain relationship can be obtained by solving (31) for any known function σ'^{e}_{zz} and T^e of time.

$$\varepsilon_{zz} = \frac{1}{E_1} \sigma'^{e}_{zz} + \frac{1}{\phi_2} \int_0^t \sigma'^{e}_{zz}(\tau) \exp\left[\frac{(t-\tau)}{\phi_2/E_2}\right] d\tau - \frac{\phi_T}{\phi_2} \frac{\partial}{\partial t} \left\{ \int_0^t T^e(\tau) \exp\left[\frac{(t-\tau)}{\phi_2/E_2}\right] d\tau \right\} \tag{32}$$

Note that to obtain (32) we took $\sigma'^{e}_{zz}(t=0) = \varepsilon_{zz}(t=0) = T^e(t=0) = 0$. For a rheological soil the term $\partial\varepsilon/\partial t$ as given by (22) in the governing equations should be evaluated by employing (32) instead of (28). When $d\sigma^e_{zz} = 0$, σ'^{e}_{zz} in (32) is replaced by $S_w p^e$. τ is a dummy integration variable.

Another alternative to (32) would be the use of thermoplastic stress-strain relations. Interested readers can refer to Corapcioglu (1983) for the use of this type of constitutive equations.

INITIAL AND BOUNDARY CONDITIONS

The solution of one dimensional equations (25)-(28) are subject to initial and boundary conditions. Initially, an unsaturated frozen soil column may have certain ice and water contents, a known pore pressure distribution and a uniform or certain temperature distribution, i.e., $S_w(z,0) = f_1(z)$, $\Theta_i(z,0) = f_2(z)$, $p(z,0) = f_3(z)$, and $T(z,0) = T°$ or $T(z,o) = f_4(z)$ where f_1, f_2, f_3, and f_4 are known functions. The undisturbed soil may initially be free of any deformation, i.e., $U_z(z,0) = 0$. Boundary conditions are defined by the particular physical conditions imposed on the soil. On the soil surface, we may have a constant temperature $T(0,t) = T_1 > 0°C$, complete thawing $\Theta_i(0,t) = 0$, atmospheric pressure $p(0,t) = 0$ or a constant surface loading $p(0,t) = p_1$, and saturated soil $S_w(0,t) = 1$. The surface temperature might possibly be also a known function of time $T(0,t) = f_5(t)$. Theoretically at infinity ($z = \infty$), practically at a certain distance from the soil surface ($z = z_a$) the surface effects will diminish. Hence, $S_w(\infty,t) = S_1$, $\Theta_i(\infty,t) = 0$, $T(\infty,t) = T_2$ and $\partial p(\infty,t)/\partial t = 0$, where S_1 is a constant value of water saturation and T_2 is a constant temperature.

CONCLUSIONS

A mathematical model is developed to predict the values of pore pressure, temperature, water saturation, ice content, and displacements for known boundary conditions. The model is simplified for a one dimensional vertical consolidation problem. We should note that the development of the model is subject to various assumptions, such as constant water and ice densities, negligence of pore vapor pressure, identical velocities for ice and solid particles, and others. Most of these assumptions are valid for practical problems.

REFERENCES

Aziz, K. A., and J. T. Laba, 1976, Rheological model of laterally stressed frozen soil. American Society of Civil Engineers, Journal of the Geotechnical Engineering Division, v. 102, p. 825-839.

Bear, J., 1972, Dynamics of fluids in porous media. New York, Elsevier.

Bear, J. and M. Y. Corapcioglu, 1981, A Mathematical model for consolidation in a thermoelastic aquifer due to hot water injection or pumping. Water Resources Research, v. 17, p. 723-736.

Bear, J., and G. F. Pinder, 1978, Porous media deformation in multiphase flow. American Society of Civil Engineers, Journal of the Engineering Mechanics Division, v. 104, p. 881-894.

Brown, W. G., and G. H. Johnston, 1970, Dikes on permafrost: Predicting thaw and settlement. Canadian Geotechnical Journal, v. 7, p. 365-371.

Corapcioglu, M. Y., 1983, Thermo elasto plastic deformation of porous materials. Journal of Thermal Stresses, v. 6, p. 99-113.

Del Giudice, S., G. Comini and R. W. Lewis, 1978, Finite element simulation of freezing processes in soils. International Journal for Numerical and Analytical Methods in Geomechanics, v. 2, p. 223-235.

Fuchs, M., G. S. Campbell and R. I. Papendick, 1978, An Analysis of sensible and latent flow in a partially frozen unsaturated soil. Soil Science of America Journal, v. 42, p. 379-385.

Jame, Y. W., and D. I. Norum, 1980, Heat and mass transfer in a freezing unsaturated porous medium. Water Resources Research, v. 16, p. 811-819.

McRoberts, E. G., E. B. Fletcher and J. F. Nixon, 1978, Thaw consolidation effects in degrading permafrost, in Proceedings of the Third International Conference on Permafrost, v. 1: Ottawa, National Research Council of Canada, p. 694-699.

Morgenstern, N. R., and J. F. Nixon, 1971, One-dimensional consolidation of thawing soils. Canadian Geotechnical Journal, v. 8, p. 558-565.

Nixon, J. F., and N. R. Morgenstern, 1973, Practical extensions to a theory of consolidation for thawing soils, in Permafrost - The North American Contribution to the Second International Conference, Yakutsk: Washington, D.C., National Academy of Sciences, p. 369-377.

Sheppard, M. I., B. D. Kay and J. P. G. Loch, 1978, Development and testing of a computer model for heat and mass flow in freezing soils, in Proceedings of the Third International Conference on Permafrost, v. 1: Ottawa, National Research Council of Canada, p. 76-81.

Stevens, H. W., 1973, Viscoelastic properties of frozen soil under vibratory loads, in Permafrost - The North American Contribution to the Second International Conference, Yakutsk: Washington, D.C., National Academy of Sciences, p. 400-409.

Sykes, J. F., W. C. Lennox and R. G. Charlwood, 1974a, Finite element permafrost thaw settlement model. American Society of Civil Engineers, Journal of the Geotechnical Engineering Division, v. 100, p. 1185-1201.

Sykes, J. F., W. C. Lennox and T. E. Unny, 1974b, Two-dimensional heated pipeline in permafrost. American Society of Civil Engineers, Journal of the Geotechnical Engineering Division, v. 100, p. 1203-1214.

Sykes, J. F., and W. C. Lennox, 1976, Thaw and seepage in nonlinear elastic porous media, in Proceedings of the First International Conference on Finite Elements in Water Resources, Princeton: London, Pentech Press, p. 3.47-3.67.

Taylor, G. S., and J. N. Luthin, 1978, A Model for coupled heat and moisture transfer during soil freezing. Canadian Geotechnical Journal, v. 15, p. 548-555.

Tsytovich, N. A., 1960, Bases and foundations on frozen soils. Highway Research Board, Special Report 58, Washington, D.C., National Academy of Sciences.

Tsytovich, N. A., Yu-K. Zaretsky, V. G. Grigoryeva and Z. G. Ter-Martirosyan, 1965, Consolidation of thawing soils, in Proceedings of the Sixth International Conference on Soil Mechanics and Foundation Engineering, v. 1: Toronto, University of Toronto Press, p. 390-394.

Tsytovich, et al., 1966, Basic mechanics of freezing, frozen, and thawing soils, in Technical Translation 1239, Ottawa, National Research Council of Canada.

Vyalov, S. S., 1963, Rheology of frozen soils, in Proceedings - Permafrost International Conference, NRC Publ. 1287: Washington, D.C., National Academy of Sciences, p. 332-337.

Zaretskii, Yu. K., 1968, Calculation of the settlement of thawing soil. Soil Mechanics and Foundation Engineering, v. 3, p. 151-155.

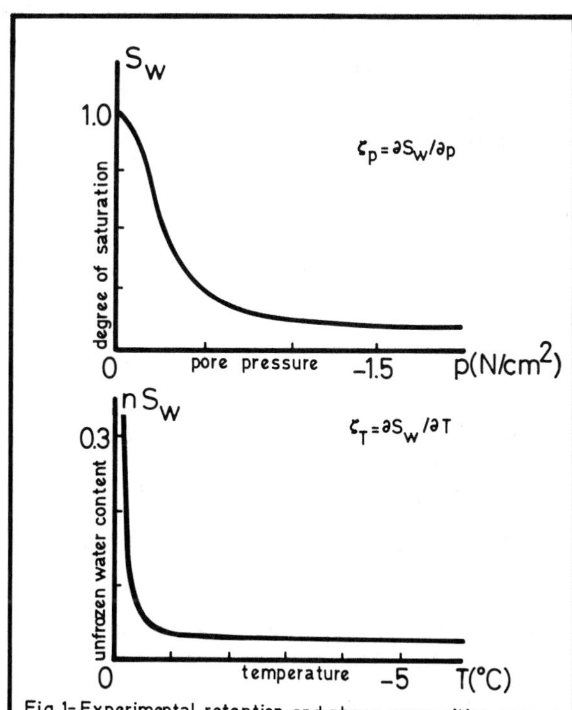

Fig. 1- Experimental retention and phase composition curves to obtain ζ_p and ζ_T respectively for an unsaturated thawing soil.

THERMAL REGIME OF A SMALL ALASKAN STREAM IN PERMAFROST TERRAIN

Samuel W. Corbin[1,2] and Carl S. Benson[1]

[1]Geophysical Institute, University of Alaska, Fairbanks, Alaska 99701 USA
[2]Now at Department of Geology, University of Oregon, Eugene, Oregon 97403 USA

Goldstream Creek near Fairbanks, Alaska is about 8 m wide, 0.25 m deep, and flows 0.3 to 1.3 $m^3\ s^{-1}$. From 1963 to 1973, hydrological and temperature observations were made to determine thermal effects of the stream on the relict permafrost that underlies much of the valley. Soil freezing beginning in October causes build-up of ground-water pressure, and the stream overflows repeatedly until about January, forming aufeis deposits 1 to 2 m thick, depending on depth of soil freezing, which is primarily controlled by the amount of snow cover. The data permitted calculations of the thermal diffusivity of unfrozen soil, using Fourier series, numerical step models and differential analyses; average values ranged from $4 \times 10^{-7}\ m^2\ s^{-1}$ in areas away from the stream to over 15×10^{-7} beneath the stream. Overflows facilitate heat loss from the ice by removing the insulating snow layer. However, they also add heat as the water percolates through fractures in the ice and releases latent heat as it freezes. The mechanism of adding heat is more effective than conduction which is the only mechanism for transporting heat to the upper surface. Although the surface temperature during summer is lower in the stream than in the surrounding ground (11°C compared with > 22°C) the summer heat pulse penetrates deeper under the stream; this is partly due to the topography of the stream channel. The net effect on the permafrost is to lower the relict permafrost table, from about 4 m away from the stream, to nearly 6 m beneath the stream.

INTRODUCTION

Numerous references are devoted to the thermal effects of large bodies of water on permafrost (Brewer, 1958; Gold and Lachenbruch, 1973; Lachenbruch, 1957) the observations presented here show thermal effects produced by a small stream flowing over relict permafrost.

Goldstream Valley near Fairbanks, Alaska has a fairly flat bottom, about 3 km wide, which is underlain by permafrost more than 50 m thick, with an estimated maximum thickness of 75 m. The active layer is generally about 1 m deep. In localized areas, away from Goldstream Creek, the top of the permafrost lies below this depth (it ranges from 1 to 4 m) and during winter pressure builds up in ground water between the seasonally and perennially frozen ground. Under Goldstream Creek, perennially frozen ground is encountered at depths of nearly 6 m. The creek is typically 8 m wide, with well-defined banks 1 to 2 m high, its average water depth is about 25 cm, its gradient is 2.2 m km^{-1}, and its summer discharge varies between 0.3 and 1.3 $m^3\ s^{-1}$. During winter the creek overflows repeatedly to form extensive aufeis deposits. Thickness of the aufeis nearly always exceeds 1.5 m and is sometimes greater than 2 m. The maximum measured cross-sections of aufeis at the end of winter ranged between 11 and 21 m^2, thus they are an order of magnitude greater than the typical pre-freezeup stream cross-section of 2 m^2 (Figure 1).

Goldstream Creek, a tributary of the Tanana River drainage system, is fairly typical of creeks throughout central Alaska and was selected for research due to its easy access from the University of Alaska. The research site, established in 1963, was located 9 km northwest of Fairbanks, Alaska where Ballaine Road crosses the creek, 183 m above sea level. Mean annual air temperature averages about 3°C lower than the -3.4°C observed at Fairbanks International Airport. The mean annual snowcover contains only about 10 cm of water equivalent, but it remains on the ground from October until April. Flat portions of the valley are marshy with numerous icewedge polygons. Soil is mostly silt of low organic content (10 percent or less), vegetation consists mainly of willows and sparse stands of spruce. Hydrological observations, with special attention to winter icing problems, were carried out during the ten years 1963-1973; they lead to three M.S. theses (Kreitner, 1969; Gilfilian, 1973; Corbin, 1977) and one Ph.D. dissertation (Kane, 1975).

This paper deals with the thermal regime of the stream and a cross-section of the land extending 10 to 15 m on each side of the stream center (Figure 2). A network of thermocouples which, at its maximum, contained 98 probes and over 1600 m of wire was arranged in two parts. (1) The first part was in the aufeis itself; before freezing began, a string of thermocouples was placed across the stream bottom. After freeze-up, a string of thermocouples was frozen onto the ice surface. When the original ice surface was buried by a layer of overflow ice a string of thermocouples was frozen onto this ice surface. Five strings of thermocouples were emplaced in this way during four consecutive winters ending in 1971. (2) The second part was a network of subsurface probes arranged vertically in three pits beginning in 1967 and expanded into ten vertical strings in the fall of 1970 (Figure 2). Measurements were made weekly with a Leeds and Northrup, Model 8646, potentiometer using an ice-water bath as reference.

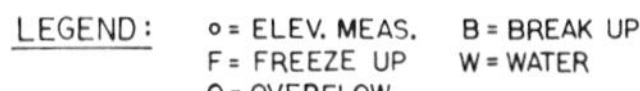

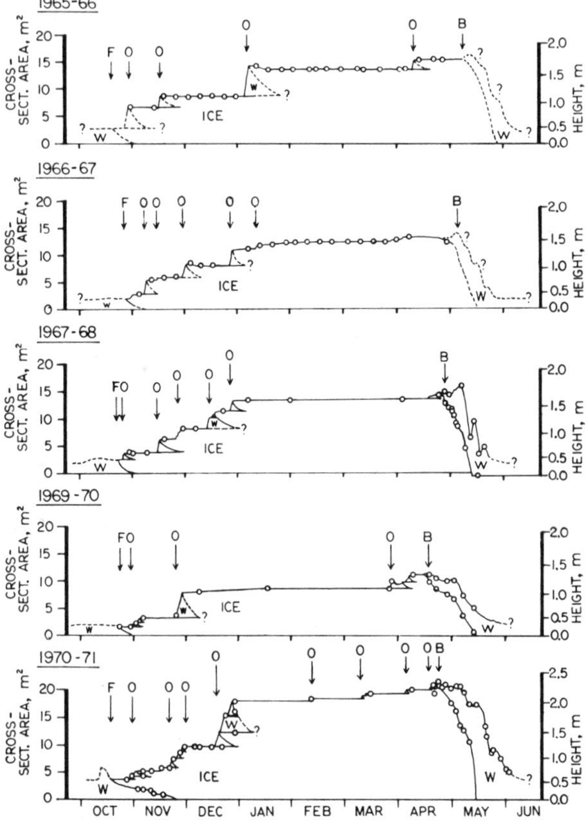

Figure 1. Aufeis formation and break-up (cross-sectional area and height versus time) during five winter seasons -- between 1965 and 1971, no data were taken during the 1968-1969 winter.

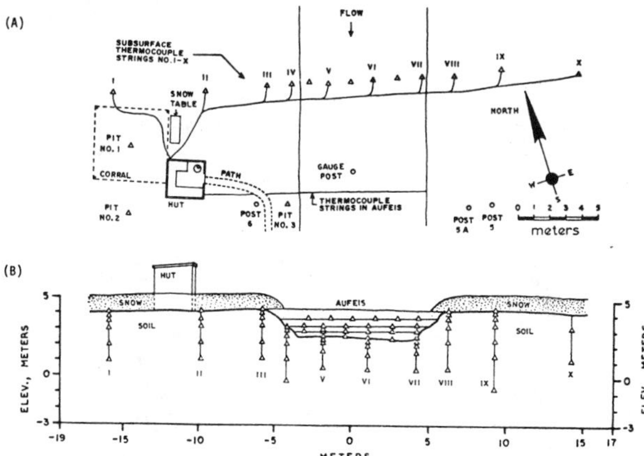

Figure 2. (A) Planar and (B) cross-sectional views of thermocouple network, consisting of pits no. 1, 2, and 3 (installed in fall of 1967) and vertical strings no. I through X (installed in fall of 1970). Elevations refer to an arbitrary datum. The thermocouples in the stream were on a separate cross-section to avoid disturbance of the subsurface thermocouples.

THE ANNUAL CYCLE

A full annual cycle of temperature changes in the aufeis and surrounding soil was measured from October 1970 to December 1971; four representative cross-sections are plotted in Figure 3. A summary of the annual data at the stream center (column VI) and 16 m west of the stream center (column I) is shown, together with meteorological data, in Figure 4.

The freezing process begins in early October at Goldstream Creek with supercooling of the entire stream to temperatures as low as 0.036°C below the freezing point depression of 0.012°C (Gilfilian, 1973). Supercooling followed by underwater ice formation (frazil and anchor ice) was measured during freeze up nearly every winter between 1963 and 1973; the concentration of underwater ice ranged between 0.9 and 4.7 percent of the stream's volume (Benson, 1967, 1973; Osterkamp et al. 1972; Gilfilian et al., 1972; Corbin, 1977). A permanent ice cover was usually in place by 27 October. The date of its formation was consistent with Michel's (1971) soil-freezing model to within 2 days (Corbin, 1977). Overflow of water on top of the ice cover followed by the build-up of aufeis layers began as early as 24 October and maximum thicknesses were usually reached by the end of December. The petrofabric structure of aufeis from the winters of 1965-66, 1966-67 and 1967-68 was studied by Kreitner (1969). Spring break-up, between mid-April and mid-May, generally began with the overflow of organically stained water followed by snow melt-water. The potential for spring flooding is high because the channels are filled with aufeis which begin to melt and undergo mechanical fragmentation. Within three weeks, the entire body of aufeis is removed; the layers calve away from the top down.

Once an ice cover has formed, the cross-sectional area available for stream flow decreases causing water pressure to increase as the ice continues to grow. When freezing is rapid, pressure bulges may form within a week after freeze-up, as in 1966. Water pressure is relieved when the stream breaks through cracks in the ice (cracks due to thermal stress are common in the upper layers) or at the stream margins to overflow the ice surface and produce layers of aufeis (Benson 1967, 1973). Kane measured ground water pressure adjacent to the stream during the 1971-72 winter; he found the hydrostatic head to vary by several meters and observed its control on aufeis formation (Kane et al. 1973; Kane, 1975).

The aufeis develops rapidly to its maximum (Figure 1); the ice surface then remains essentially constant until overflows occur just before or during spring break-up. The lack of overflows between January and April results from a combination of causes. The rate of channel constriction by freezing decreases as the insulating effect of aufeis layers (1 to 2 m thick) increases. Also, the source of water decreases as soil freezing progresses. The flow apparently can be accomodated by the restricted channels and the ice overburden adequately confines the ground water even though its pressure head varies through a range of about 1.5 m (Kane et al., 1973).

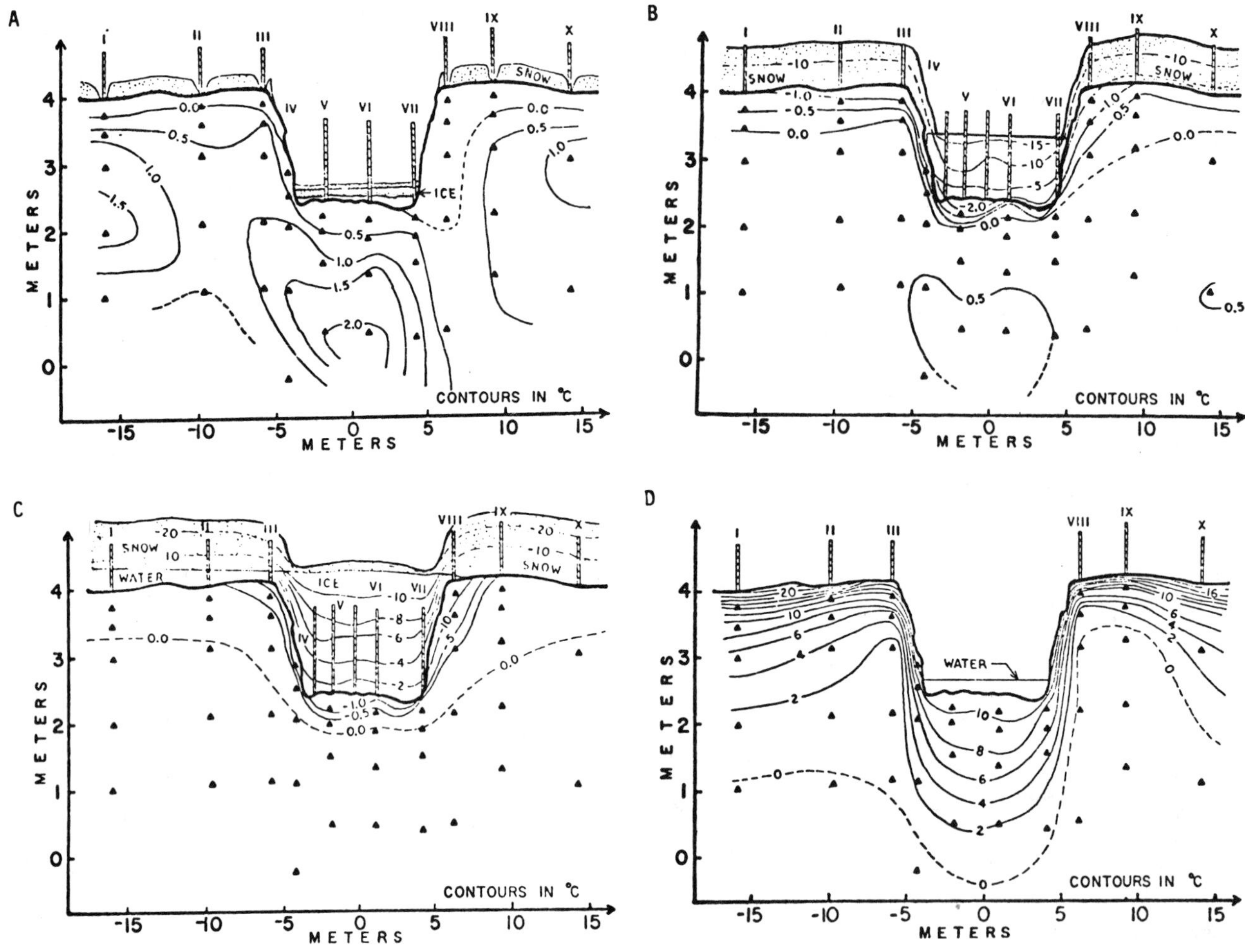

Figure 3. Cross-sectional views of the temperature distributions of Goldstream Creek and surrounding soil during hydrological year, 1970-1971. Parts of Figure 3: (A) 22 October 1970; (B) 15 December 1970; (C) 24 February 1971; (D) 22 June 1971.

THERMAL EFFECTS

The thermal effects of overflows and aufeis formation are complex. If the stream ice is snow covered, the overflow water soaks into the snow, freezes to form an ice layer, and eliminates the insulation which the snow had provided. This contributes to several meters deeper penetration of low temperature isotherms under the stream than under the adjacent snow-covered ground. Conditions varied significantly from one year to the next. The minimum temperature measured on the stream bottom was -10.4°C on 20 January 1967 (between stakes VI and VII, Figure 3); isotherms were inclined in the ice with a gradient of 2.7°C m^{-1} along the stream bed and running water at the bottom on the left side, i.e., near stake IV. In this case, when the aufeis surface reached the 4 m level, overflow water oozed out on the left side only, built a dam which prevented it from moving across the stream and spread into the snow by capillary action on the left side of the stream.

The four winters during which temperature observations were made in the aufeis included extremes of snow cover. The winter of 1969-1970 was one of the mildest on record and total snow depth never exceeded 0.30 m. In contrast, the winter of 1970-1971 was the most severe since 1956-1957. Air temperatures 5.2 C below normal were accompanied by a record-breaking snow depth of 1.10 m on the ground. The insulation provided by the snow was such that the snow-soil interface temperature rarely went below -2°C. But, in spite of the milder conditions during 1969-70, the thin snow cover allowed temperatures as low as -8°C at the snow-soil interface. The depth of freezing in the soil may be a contributing factor in determining the total amount of aufeis formed; deeper freezing over the entire drainage area can be expected to reduce the source of winter flow which is available to be stored as aufeis. This is consistent with the observation that the minimum snow depth (0.30 m) in 1969-70 is associated with the minimum aufeis cross-section of 11 m^2, and the maximum snow

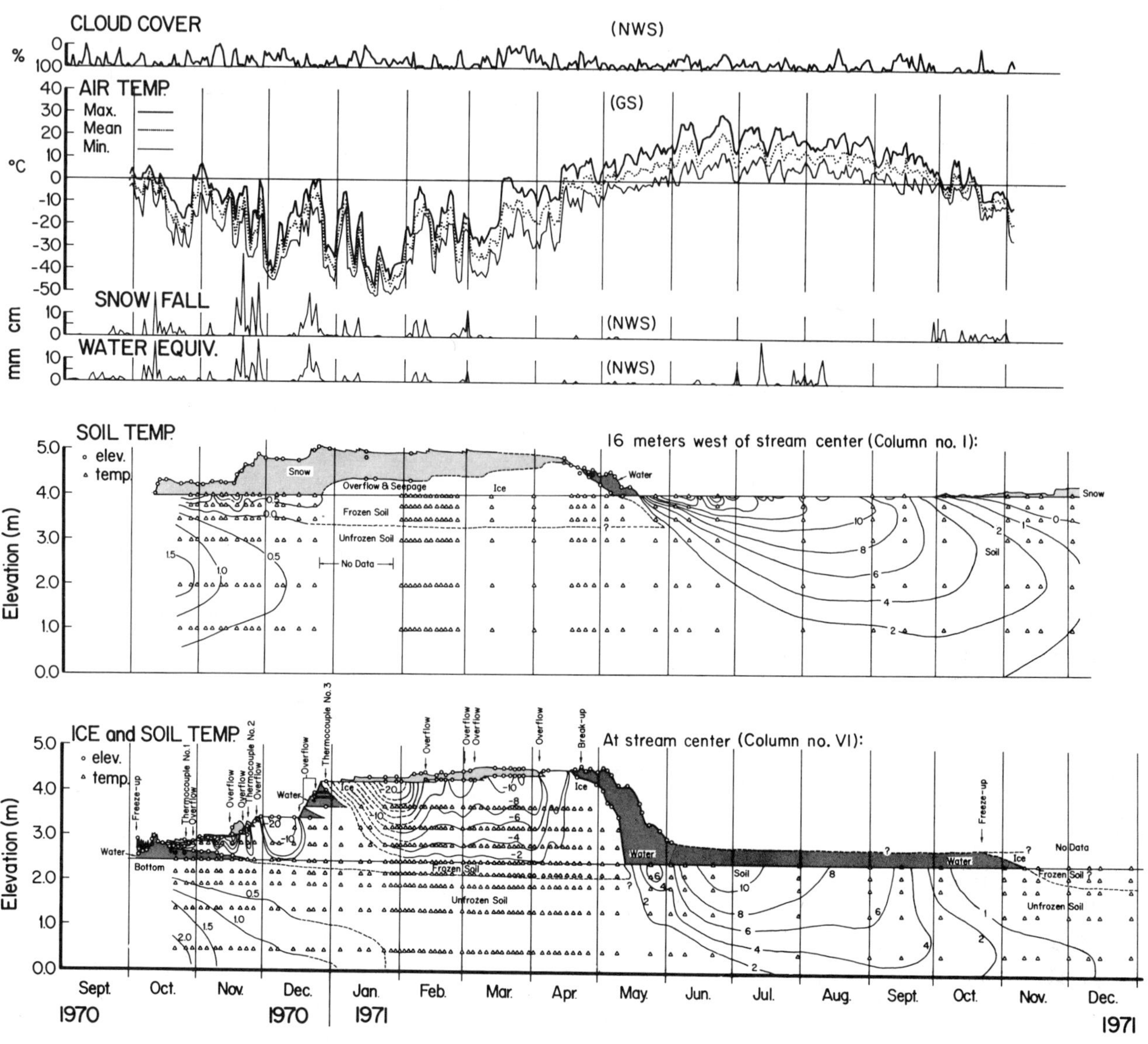

Figure 4. Comparison between temperatures in aufeis and soil at stream center (string no. VI) and temperatures in upper 3 m of soil, 16 m west of stream center (string no. I) from September 1970 through December 1971. Supporting meteorological data are from both the Goldstream Site (GS) and National Weather Service, Fairbanks (NWS). Similar diagrams for the winters of 1967-1968 and 1969-1970 also show the thorough warming of aufeis to 0°C following each overflow.

depth (1.10 m) the following year is associated with the maximum aufeis cross-section of 21 m^2.

Although overflows enhance heat removal by removing the insulating snow cover, their net thermal effect is to add heat to the system. This is primarily by the release of latent heat while the overflow layers freeze. The heat available from freezing a 20 cm thick overflow was more than twice the amount required to bring even the thick-accumulation of aufeis to its melting point. The depth-time diagrams prepared from three years of data (Figure 4 is typical) showed rapid warming to 0°C of the underlying ice following overflows. The rate and thoroughness of the warming was too fast to be accounted for by conduction alone. Apparently water from mid-winter overflows was able to penetrate from above and release latent heat as it froze within cracks or spaces which were sometimes observed between ice layers. Thus the bottom of the growing aufeis is continuous enough to contain the flowing water under pressure but the upper ice surface can be penetrated by percolating overflow waters. This introduces an extremely efficient mechanism of heat transfer into the ice; but heat transfer out of the ice takes place only by conduction.

During several winters, overflows spread over the west bank as well. In January 1971 when this happened, the temperature of the soil below the overflow increased to 0°C and did not freeze again during the winter. Only the lower half of the snow became wetted and turned to ice, the upper half remaind snow and served its role of insulation (Figure 4). Overflows preceding break-up in April rapidly brought ice temperature to 0°C. This shows up best in data from 1968 and 1970, but it is also clear in 1971 (Figure 4); by 10 May 1971 all probes in the soil were isothermal at 0°C.

THERMAL DIFFUSIVITY

The detailed temperature data permitted calculation of the summer thermal diffusivity values in unfrozen soil adjacent to and under the stream, with an assumed zero degree boundary at the top of the permafrost. One-dimensional models were used with thicknesses of infinity, 3.5 and 4.0 m. These one-dimensional models were assumed valid where the isotherms were approximately horizontal, directly under the center of the stream (column VI), and well away from the stream (column I), see Figures 2 and 4. Three methods were used.

First, following procedures outlined by Carslaw and Jaeger (1959), Fourier integration at one month intervals was done for various depths. The thermal diffusivity of a layer between two given depths was calculated using 6-month half-wave pulses. In the top two meters the models were in reasonable agreement with each other and with the data. Below two meters the infinitely thick model deviated significantly from measured values; but the 4m thick slab model continued to agree with the data to the lowest depths of measurement (3 m). Diffusivity values from the 4 m slab ranged from 3×10^{-7} to 7×10^{-7} m^2s^{-1} away from the stream and from 3×10^{-7} to 11×10^{-7} m^2s^{-1} beneath the stream.

Second, using a simple finite difference approximation of the Fourier conduction equation in one dimension, diffusivity was adjusted to give a best fit to the data using 40 cm depth intervals and time intervals of 4 days. The best fits were made by using the 4 m thick slab away from the stream, with diffusivity values of 4×10^{-7} to 6×10^{-7} m^2s^{-1}; beneath the stream center the infinitely thick model worked best with diffusivities of 10×10^{-7} to 12×10^{-7} m^2s^{-1}.

Third, diffusivity was also estimated directly from the data by using the ratio of the derivative of temperature with time to second derivative of temperature with depth. Taking smoothed temperatures during June and July 1971 over intervals of 20 cm and 5 days, diffusivities away from the stream ranged from 2×10^{-7} at the surface to 6×10^{-7} m^2s^{-1} at 2.60 m depth. Beneath the stream, values ranged from 8×10^{-8} to 6×10^{-7} m^2s^{-1} below one meter, but closer to the surface, they were infinite.

The infinite value is a warning that significant, and undetermined, errors exist in these determinations. The uncertainties in the system do not permit detailed calculation. However, if we disregard the infinite value, working averages from the three methods were 4×10^{-7} m^2s^{-1} away from the stream and 11×10^{-7} m^2s^{-1} beneath the stream.

The three-fold difference in diffusivity values may result from several factors. The assumption of a one-dimensional model is questionable under the center of the stream; it is certainly not valid near the edges (Figure 3). However, it seems reasonable away from the stream where the isotherms are essentially horizontal and diffusivities compare favorably with accepted values (Kersten, 1952). Another possibility is latent heat exchange in the soil which would be difficult to detect at both ends of the summer period. Finally, there may be vertical movement of ground water under the stream. Order of magnitude calculations indicate that such flow, at rates of only a few meters per year, would introduce heat transfer rates comparable and added to the conductive rates. This could allow the summer heat pulse to penetrate more rapidly under the stream even though the upper-surface temperatures are lower beneath the stream than beneath the adjacent ground surface (Figure 3D). Similar effects on cooling rates would be expected (Figure 3A, B). This effect would diminish rapidly away from the stream; 16 m away, at column I (Figures 2 and 4) the 4 m-thick slab model is consistent with our temperature data, and with observations by us and others (Kane, et al., 1973; Slater, 1974) that the top of the relict permafrost is about 4 m deep.

SUMMARY

The maximum water temperature in Goldstream Creek during summer is significantly less than the maximum ground surface temperatures; but the topography of the stream channel is such that the ground beneath the stream is warmer than ground adjacent to the stream at the same elevation (Figure 3D). During winter, heat removal from ground beneath the stream is less than from ground adjacent to the stream. This is true even though the adjacent ground is continuously snow covered while overflow water periodically destroy the insulating snow cover on the stream ice. The most significant influence is the large heat source provided by latent heat released as overflow water freezes to form layers of aufeis, and the associated non-conductive heat transfer mechanism. Movement of the freezing water downwards into the ice along fracture patterns provides a more efficient means of heat transfer than conduction which is the only mechanism for transporting heat to the ice surface. In spite of the relatively high thermal conductivity of ice, this provides a heat source as well as a thermal barrier against heat loss from the underlying sediments. Movement of groundwater across isotherms may introduce another mechanism for non-conductive heat transfer; its effectiveness would probably decrease rapidly away from the stream.

The net effect appears to be a lowering of the relict permafrost table from about 4.0 m away from the stream to 6.0 m beneath the stream bottom.

ACKNOWLEDGEMENTS

This work was supported by funds from the Office of Water Resources Research, U.S. Department of Interior (Project A-012-ALAS) by the Earth Sciences Section, National Science Foundation NSF Grant No. 30748, and by the State of Alaska funds. We express special thanks to our colleagues in the field: R. E. Gilfilian, J. D. Kreitner, T. E. Osterkamp and D. C. Trabant and for helpful discussions with W. D. Harrison.

REFERENCES

Benson, C. S., 1966, 1967, 1968, 1973, A study of the freezing cycle in an Alaskan stream, in: University of Alaska, Fairbanks, Institute of Water Resources: Annual Reports, 1966, p. 100-117; 1967, p. 19-22; 1968 p. 13-21; Completion Report, IWR-36, 31 pp.

Brewer, M. C., 1959, the thermal regime of an arctic lake: Transactions of the American Geophysical Union, v. 39, p. 278-284.

Carslaw, H. S. and Jaeger, J. C., 1959, Conductivity of heat in solids 2nd edition: London, England, Oxford University Press, 386 p.

Corbin, S. W., 1977, The thermal regime of a stream in central Alaska, M.S. thesis: University of Alaska, Fairbanks, Alaska 144 p.

Gilfilian, R. E., 1973, Winter history of a small sub-arctic Alaskan stream, M.S. thesis: University of Alaska, Fairbanks, Alaska, 119 p.

Gilfilian, R. E., Kline, W. L., Osterkamp, T. E., and Benson, C. S., 1972, Ice formation in a small Alaskan stream, in Proceedings of the International Symposium on the Role of Snow and Ice in Hydrology: Banff, Canada, World Meteorological Organization, p. 505-513.

Gold, L. W. and Lachenbruch, A. H., 1973, Thermal conditions in permafrost - A review of North American literature. North American Contribution, Permafrost 2nd Int'l Conference 13-28 July 1973 Yakutsk, USSR, U.S. National Academy of Sciences, Washington, D.C.

Kane, D. L., 1975, Hydraulic mechanism of aufeis growth, Ph.D. dissertation: University of Minnesota, Minneapolis, Minnesota, 118 p.

Kane, D. L., Carlson, R. F., and Bowers, C. E., 1973, Groundwater pore pressures adjacent to sub-arctic streams, in Permafrost - The North American Contribution to the 2nd Int'l Conference, Yakutsk: Washington, D. C., National Academy of Sciences, P. 453-458.

Kersten, M. S., 1952, Thermal properties of soils, in Frost Action in Soils, a Symposium: Washington, D. C., National Academy of Sciences, Highway Research Board, Special Report No. 2.

Kreitner, J. D., 1969, The petrofabrics of aufeis in a turbulent Alaskan stream, in Proceedings of the 20th Alaskan Science Conference: College, Alaska, American Association for the Advancement of Science, Alaska Division, p. 332-333.

Lachenbruch, A. H., 1957, Thermal effects of the ocean on permafrost, Bull. Geol. Soc. Am., v. 68, pp. 1515-1530.

Lachenbruch, A. H., 1957, Three-dimensional heat conduction in permafrost beneath heated buldings, U.S. Geological Survey Bulletin 1052-B, p. 51-69.

Michel, B., 1971, Winter regime of rivers and lakes, USA CRREL Monograph III-Bla: Hanover, New Hampshire, U.S. Army Corps of Engineers, 139 p.

Osterkamp, T. E. Gilfilian, R. E., and Benson, C. S. 1972, Observations of stage, discharge, pH, and electrical conductivity during periods of ice formation in a small sub-arctic stream: Transactions of the American Geophysical Union, v. 11, no. 2, p. 268-272

Slater, W. H., 1974, Alaska Department of Highways, Foundation Report, Goldstream Creek at Ballaine Road, Bridge No. 1009, Project No. S-0646(3), Interior District: Fairbanks, Alaska, 6 p.

GEOCRYOGENIC MORPHOLOGY AT SEYMOUR ISLAND (MARAMBIO) ANTARCTICA

A Progress Report

Arturo E.Corte
Instituto Argentino de Nivologia y Glaciologia, Casilla de Correo 330
5500 Mendoza, Argentina

At Seymour Island, 64°14' S.L. and 56°38', Antarctica under a mean annual temperature lower than -10°C, various geocryogenic forms are observed including frost-contraction cracks and polygons, frost domes and pingos, gelifluction features, etc. These first forms require cold permafrost with mean annual temperatures below -5°C. The geocryogenic processes at both sides of the Antarctica Peninsula are compared. In the west part, geocryogenic processes of Deception are compared including sorting and gelifluction features, which formed at -2.8°C. On the east side forms requiring lower temperatures are observed, including thermal-contraction cracks and polygons, pingos and frost domes; they are formed under mean annual temperature of -10°C or lower. The mountain chain of the Peninsula works as a barrier for the cold winds of the Weddell Sea resulting in frigid conditions all along the east sector of the Peninsula. On the west side of the chain, with more prevalent anticyclonic conditions, warmer air is brought from the northeast and west producing higher temperatures. Temperature differences about 8-9°C are observed at the same latitude on the east and west sides. The strong southwest winds with a snow and debris load produces a remarkable phenomenon: snow drift-dammed lakes. The lakes behind such dams can be of several hectares. The growth of mounds displaces drainage lines, this process and the water ponded by the snow drifts also cause continual change in the drainage systems. The fluvial system of the island is too dense for the apparently low precipitation.

INTRODUCTION

This is a progress report on a continuing study of the geocryology of Seymour Island. This report deals with an inventory of the geocryogenic forms based on an aerophoto survey and ground inspections. The next stage of this work will be dedicated to excavations, sampling of soils, and ground ice studies.

Geocryogenic conditions on Seymour Island, shown on Argentina topographic maps as "Vice Com.Marambio", which is located (see map incl.) on the east side of the Antarctic Peninsula, differ greatly in comparison with places at the same latitude on the other side of the Peninsula.

Seymour, like other sectors of the east cost of the Peninsula, has a cold climate, with a mean annual temperature below -10°C (Schwerdtfeger,1975, 1976). By contrast, places situated on the west side of the Peninsula have shown mean annual temperatures 8° to 9°C higher. The low-temperature geocryogenic forms on the east side, i.e. thermal-contraction cracks, pingos, etc. are not observed in the west sector (Corte and Zomoza,1957). Sorted features produced by freezing, thawing and gelifluction occur in places on the west side of the Peninsula where the mean annual temperature is -2.8°C. (Corte and Zomoza,1957).

The geocryogenic features described by Péwé(1959) and Black and Berg (1963) in the area peripheral to the Ross Sea of Antarctica include large thermal-contraction polygons (sand and/or ice wedges), but the variety of forms is very small in comparison with Seymour Island.

In the geocryogenic map (Fig.1) it was considered appropriate to honor the great efforts made by the members of the first Antarctic expedition led by Otto Nordenskjöld to this area in 1901. Besides the names of Bodman, Sobral and Larsen, other topographic features are named after Akerlund, Nordjenskjöld, Ekelof, Johansen.

ENVIRONMENTAL CONDITIONS

Meteorology and climate

The available meteorological records for the east sector of the Antarctic Peninsula are restricted to the following places: Snow Hill (Cerro Nevado), (in Schwerdtfeger,1975) with mean annual air temperatures of -11.5°C; Seymour Island (Marambio, Servicio Meteorológico Nacional) with -12°C (Schwerdtfeger 1975) (Fig.2).

The meteorological conditions on both sides of the Antarctic Peninsula were studied by Schwerdtfeger (1975-1979), who indicated that the mountain chain along the Peninsula acted as a barrier for the cold winds of the Weddell Sea (Fig.2) and that cyclonic conditions were the main cause of the refrigeration of the east sector (Servicio Meteorológico Nacional, 1981) (Fig.3). The west side of the Peninsula is generally under the effects of anticyclonic conditions and receives warmer air from the northwest or west. These conditions account for the difference of about 8-9°C at annual mean temperatures at the same latitude (Fig.3).

Precipitation on Seymour Island has not been measured because of the strong wind; however, the following types of precipitation were observed: snow, graupel (soft hail), rime, and rain. The most significant for this is snow, both directly precipitated and wind blown carried. Though there is no information on their relative importance.

Geology and Soils

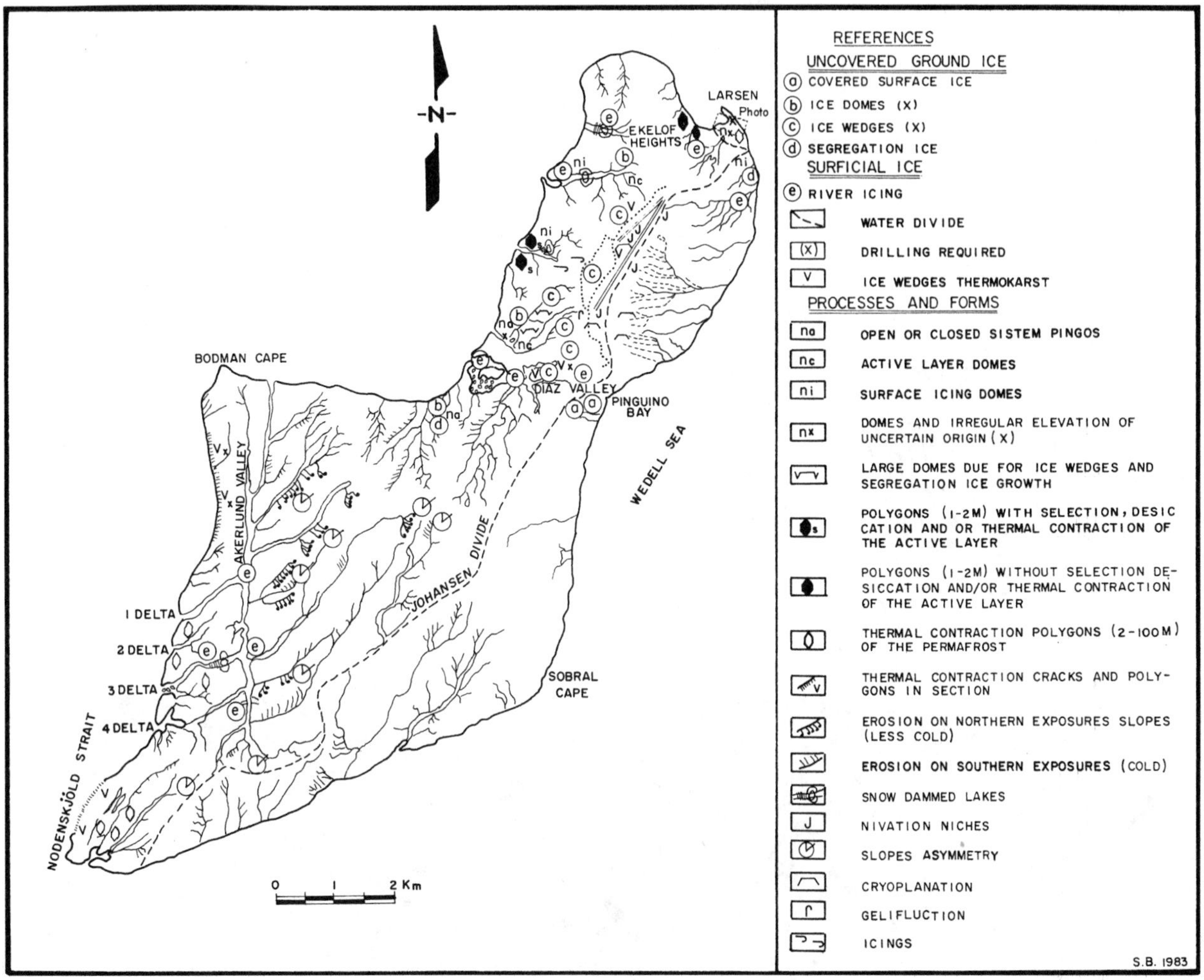

REFERENCES
UNCOVERED GROUND ICE
a COVERED SURFACE ICE
b ICE DOMES (X)
c ICE WEDGES (X)
d SEGREGATION ICE
SURFICIAL ICE
e RIVER ICING
WATER DIVIDE
(X) DRILLING REQUIRED
V ICE WEDGES THERMOKARST
PROCESSES AND FORMS
na OPEN OR CLOSED SISTEM PINGOS
nc ACTIVE LAYER DOMES
ni SURFACE ICING DOMES
nx DOMES AND IRREGULAR ELEVATION OF UNCERTAIN ORIGIN (X)
LARGE DOMES DUE FOR ICE WEDGES AND SEGREGATION ICE GROWTH
POLYGONS (1-2M) WITH SELECTION, DESICCATION AND OR THERMAL CONTRACTION OF THE ACTIVE LAYER
POLYGONS (1-2M) WITHOUT SELECTION DESICCATION AND/OR THERMAL CONTRACTION OF THE ACTIVE LAYER
THERMAL CONTRACTION POLYGONS (2-100M) OF THE PERMAFROST
THERMAL CONTRACTION CRACKS AND POLYGONS IN SECTION
EROSION ON NORTHERN EXPOSURES SLOPES (LESS COLD)
EROSION ON SOUTHERN EXPOSURES (COLD)
SNOW DAMMED LAKES
NIVATION NICHES
SLOPES ASYMMETRY
CRYOPLANATION
GELIFLUCTION
ICINGS
S.B. 1983
-N-
LARSEN
Photo
EKELOF HEIGHTS
BODMAN CAPE
DIAZ VALLEY
PINGUINO BAY
WEDELL SEA
AKERLUND VALLEY
JOHANSEN DIVIDE
I DELTA
2 DELTA
3 DELTA
4 DELTA
NODENSKJÖLD STRAIT
SOBRAL CAPE
0 1 2 Km

Figure 1 ANTARTIC GEOCRYOLOGY-SEYMOUR ISLAND
MARAMBIO 64°14' Lat.South, 56°38' Long.West

I. GEORGIAS DEL SUR
I. SANDWICH
I. CANDLEMAS
DEL SUR
I. SHETLAND DEL SUR
I. ELEFANTE
I. ORCADAS DEL SUR
I. Tte. CAMARA (-2.9°C)
I. BELLINGSHAUSEN (-2.8°C)
DECEP.
(-2.8°C)
MARAMBIO (-10°C)
Co. NEVADO (-11.5°C)(Snow Hill)
I. MELCHIOR (-3.7°C)
I. Tte. MATIENZO (-12.1°C)
I. ARGENTINAS (-5.2°C)
I. ADELAIDA
ALEJANDRO I
MELCHIOR (M.E)
I. ARGENTINAS (ARG.)
ADELAIDA
WEST COST
SNOW HILL (S.H.)
MATIENZO (MAT.)
EAST COST
Tem. °C
5
0
-5
-10
-15
-20
-25
E F M A M J J A S O N D E
S.B. 1983

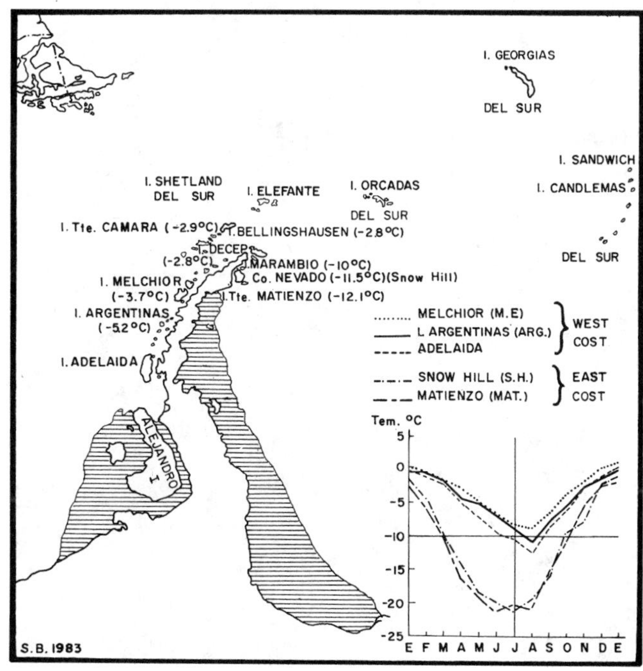

Figure 2 MEAN ANNUAL TEMPERATURES ON THE EAST AND
WEST COASTS OF THE ANTARCTIC PENINSULA
(Schwerdtfeger 1975)

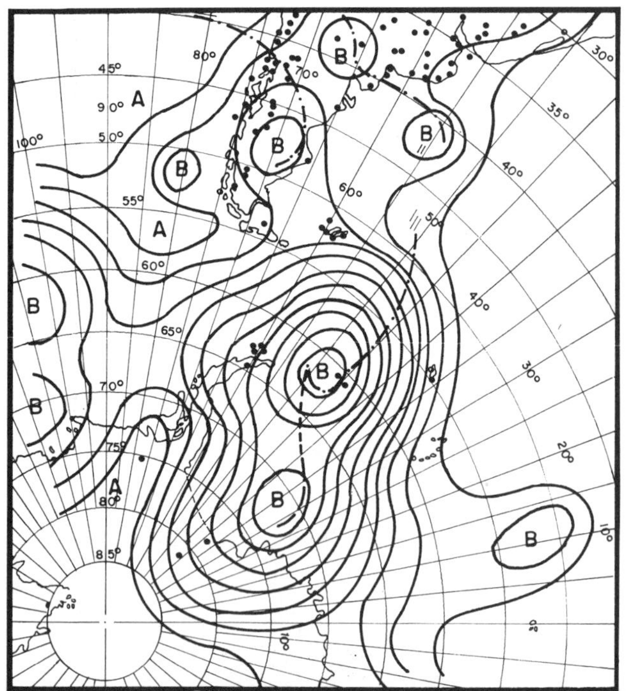

Figure 3 SYNOPTIC SITUATION BETWEEN THE SOUTH POLE AND TEMPERATURE SOUTH AMERICA-11-4-81 A. P.T. Ser.Met.Nac.Buenos Aires

Figure 4 LARSEN DELTA,NORTHERN SEYMOUR ISLAND (MA-RAMBIO) (Servicio de Hidrología Naval)

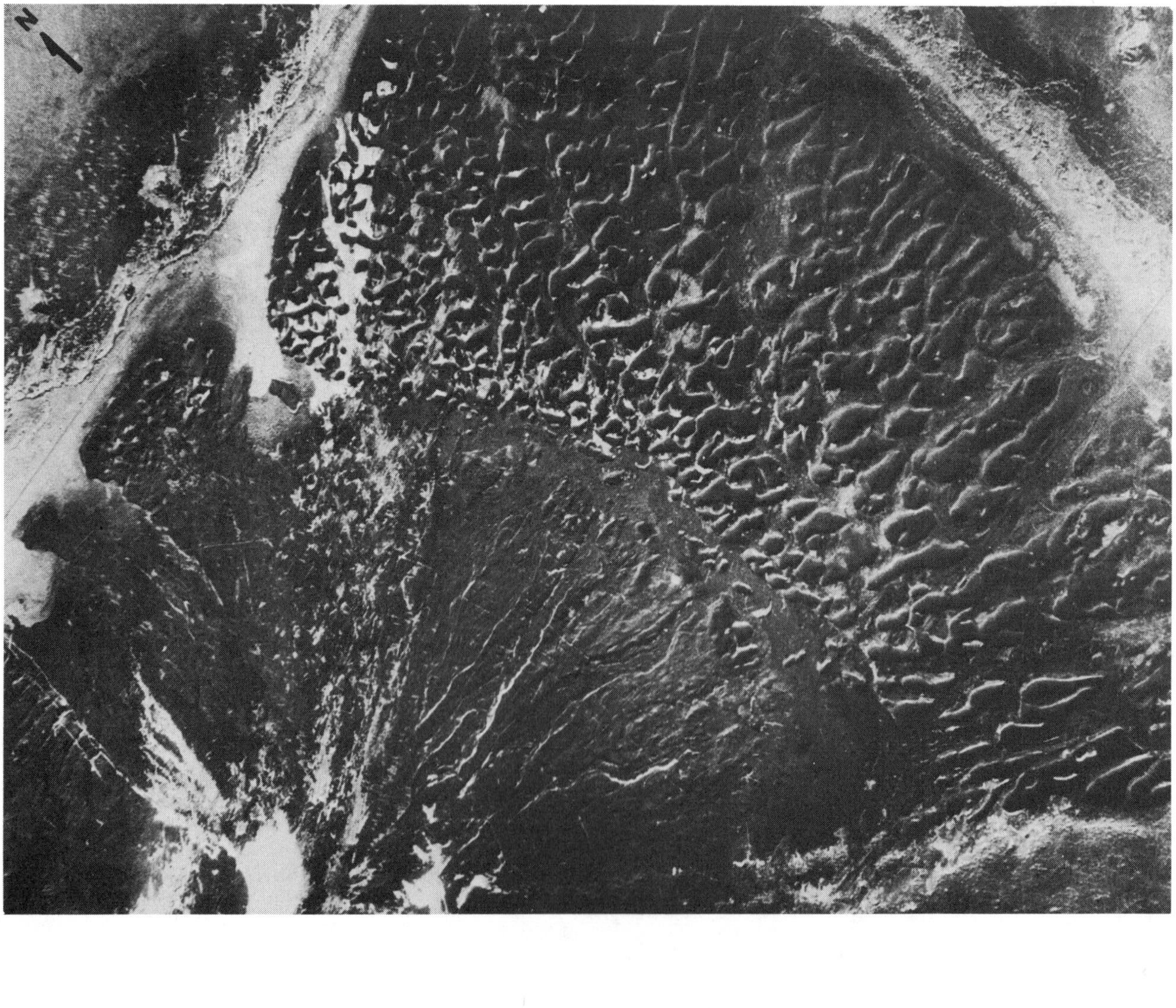

The island is made of Tertiary and Cretaceous sediments with the greatest elevation at 200 m in the Meseta which occupies about 1/20 of the Island's area. Small areas of till occur on the meseta.

Many of the geocryogenic forms are produced in consolidated and non-consolidated Tertiary sediments, most of which are frost-susceptible. Thus frost heaving due to formation of ice lenses must be an important process in these sediments. The wide distribution of troughs left by the melting of ice wedges indicated that the Tertiary sediments also are subject to frost cracking and development of ice wedges.

Air photos of the meseta show that in 1967 thermal-contraction cracking and the formation of large domes and ice-wedge polygons appear to be the main geocryogenic processes. These polygons form nets 100-200 m in diameter at the border of the meseta; the troughs have been emphasized by erosion.

Active Layer and Permafrost

According to meteorological observers (Servicio Meteorológico Nacional), the depth of the active layer at Seymour is greatest at the end of February and is about 40-50 cm. However annual and regional year to year local variations in the depth of the active layer were observed, under varying meteorological and terrain conditions, at the same latitude as Seymour (Brown, 1978).

Permafrost thinkness has not been measured on Seymour Island. However using Lachenbruck's formula (1968) in which permafrost depth can be approximated by multiplying mean annual air temperature (-10°C) by the general teothermal gradient (33 m per Celcius degree), the permafrost depth at Seymour is calculated to be approximately 330 m. This ignores local variations in the geothermal gradient and also the past thermal history.

THE GEOCRYOGENIC FORMS: THE MAJOR FORMS

Frost Domes and Pingos

Frost domes are present in several places but are most abundant and distinctive in Larsen Delta at the north end of the Island. No subsurface sampling has been undertaken yet. The domes of Larsen Delta (Fig.4), are not readily classified as either open systems or closed systems pingos or mounds. They range in diameter from 2 to 10 m. It is intriguing that mounds near the shore contain a circular depression in their center, whereas the mounds near the mountain side do not (Fig.4). The depression is a thermokarst feature (Fig.4).

The domes that have the largest diameter and smallest in height and are closer to the sea (Fig. 4) show the largest thermokarst depressions and appear to be the oldest. On the other hand, the domes that are smaller in area, rounder and more symetrical, and closer to the mountain side show no thermokarst depressions and appear to be the youngest.

The domes are not seasonal mounds; they have existed at least since the first photo flight was made in 1967, (Fig.4). It is possible that these mounds are produced by the freezing of taliks in abandoned channels. (Washburn, personal communication, 1983).

Sections through the domes and depressions and sampling of permafrost ground ice, which is presumed to be present (Corte,1963), would clarify the origin of the domes.

Larger features of six meters or more in high which may be true pingos with a perfect conical shape, were observed in two places on the island, near Quebrada de Díaz and in Jato river (Fig.1). The domes occur in areas in which refreezing of taliks causes local heaving and shifts of river channels.

Thermal Contraction Cracks and Polygons

The mean annual temperature necessary for the formation of thermal contraction cracks and polygons with ice and sand wedges is below -5°C. (Péwé, 1962, Washburn, 1979-1981). Thermal-contraction produces cracks and polygons of various sizes; on the Larsen Delta there are cracks up to 120 m long (Fig. 4) and the polygons have sides of 10 to 30 m (Fig. 4). In the deltas close to the Nordenskjold Strait there are polygons as large as 30 x 100 m (Fig.1).

Thermal-contraction cracks were observed to form after several days with temperatures about -26°C and winds up to 180 km/hr. The observed cracks were 12 mm wide and 15 m long, located in the polygons troughs.

Since these depressions were not trenched, it is unknown if either sand or ice wedges are present. Because snow accumulates in north-south troughs and is blown from the east-west ones, there is a chance that sand and ice wedges may coexist in the same polygons (Romanovskij,1973).

Thermokarst depressions left by thawing ice wedges are present in many cliffs and ravines and are especially well developed in north-facing cliffs between Bodman and Nordenskjöld Strait, and on north-facing cliffs of the meseta. Few such forms are present on south-facing slopes.

Snow Dammed Lakes

Lakes dammed by snowdrifts up to several hectares in size, are a characteristic and very dominant feature of the island (Fig.1). Lakes of such magnitude have not been described previously in the peryglacial literature.

During storms with winds of 100-200 km/hr, the southwest winds carry snow as an almost horizontal curtain. Material from the soft Tertiary and Cretaceous sediments is also picked up by the wind. When the Weddell Sea is frozen the snow and dirt are carried north and accumulate on the lee side of the island as long drifts, sometimes crossing the rivers and damming the drainage. In some places there are snowdrifts 200 m long and as much as 10 m high. Several dams were observed on the north-facing slope of the meseta. These lakes appear to be a more than one year process.

Nivation Niches

Nivation niches were observed on the east slope of Seymour Island. Below the niche a transport slope leads to an accumulation slope characterized by gelifluction lobes.

Icings and Icing Domes

During the fall and early winter, growth of icings on Seymour Island is a dominant characteristic. The largest icings are those related to freezing of surface water. The freezing of the snow-dammed lakes and other lakes produces icing mounds up to half a meter high and several meters across the base. These icing domes are formed by the expulsion of water confined between the permafrost and the overlying thickening ice as it becomes irregularly anchored to the ground.

During this process of formation of the ice-dammed lakes, the icings act as a cementing agent. The water that is ponded up river from the snowdrift moves across the snow dam and, since this barrier is very cold, the water freezes and cements the snow debris and icings into a solid wall.

Gelifluction

Gelifluction (Washburn,1979) is not well developed on the island. Only on the northwest slope of the meseta are there gelifluction forms. These comprise debris from till and Tertiary siltstones and sandstones. Gelifluction is clearly observed in places where a large rock is anchored in the permafrost; the debris flows over the boulder almost as a debris fall.

Slope Assymmetry

Differential erosion as between north-facing and south-facing slopes produces a marked slope assymmetry. The south-facing and southeast-facing slopes (cold) are steep and the north-facing and northwest-facing slopes are more gentle. In these regions there are two factors responsible for the assymmetry; (1) the incoming radiation and (2) the accumulation of snow and wind blown debris in the lee of the mountain.

The drainage divide separates the island into two sectors: a northern one that comprises over two thirds of the island, and a southern sector comprising the rest (Fig.1). If the fluvial system has not been stabilized by a vertical diabase dike (Rinaldi et al, 1978), most of the island would have developed a north-flowing fluvial system.

On the north-facing slopes the accumulated snow promotes dendritic erosion, but on the south-facing slopes the erosion is more linear and drainage lines are therefore less numerous.

If reservoirs are build on Seymour Island, they must be located on the north-facing slope with their greater snow accumulation and larger meltwater flow.

Cryoplanation Surfaces?

Cryoplanation surfaces such as those described by Demek (1969), are low temperature geocryogenic forms, requiring mean annual temperature of -10°C or less and low precipitations according to Reger and Péwé (1976). As indicated by Washburn (1980) though, the temperature significance of cryoplanation terraces is not well established.

The most remarkable feature of Seymour Island is its flat top of "meseta". The meseta has a slight slope to the southeast, and an air strip was prepared with little work by removing some large blocks from the surface. This meseta is sculptured in almost horizontal bedded Tertiary sandstones and siltstones. Glaciers eroded the bedrock and deposited till, which was later subject to cryoturbation. The thickness of the till varies generally from one to three meters, or more in a few places

It is unknown if the glaciers originally left this surface as flat as it is now. For this reason, it is not possible to say that the "meseta" is a true cryoplanation surface. It is interesting to note that at the northwest of Seymour Island, Cockburn Island also has a flat surface with the same slope inclination to the southeast. This surface is also cryoturbated with pingos, sorted nets, circles and gelifluction forms, etc. Nivation is producing a slope with similar orientation to that of Seymour Island. A Pleistocene Pecten conglomerate on the flat Cockburn Island surface indicates that marine processes have been active. (Henning 1924).

The meseta of Cockburn cannot be classified as a cryoplanation surface because it parallels a basalt layer having a slope to the southeast.

Coverd Ice (Burried icings)

Surficial ice, covered with debris, is important in some geocryogenic regions, such as the Central Cordillera of the Andes (Corte,1976); on Seymour Island it was found to be significant only in two places (Fig.1): Penguin Bay and on the northeast slope at the south end of the runway. In these places there are layers of ice one or more meters thick, covered with debris. It appears that these layers are icings covered with slushflows. The thickness of the debris covering the ice is the same as the present active layer about 50 cm.

THE GEOCRYOGENIC FORMS: THE MINOR FORMS

In several places on Seymour Island there are small sorted polygons, 1 m in diameter, with smaller polygons inside. These polygons are developed in weathered Tertiary rock and gravels whose coarser elements are concentrated in the polygonal depressions. Inverse forms are also present: polygons in coarse materials with their troughs underlain by finer sediments. These patterns result from desiccation cracking (Corte and Higashi,1964).

The fluvial network of Seymour Island requires a special treatment. The drainage as represented in the aerophoto survey on the scale 1:10.000 (Servicio de Hidrografía Naval) is notable for the many drainage channels (Fig.1).

During April and May the island shows icings on the small and medium river, and by the end of the fall (end of May) the whole fluvial system of the island, including large rivers is outlined by a mantle of icings. In many cases small creeks appear to flow along ice-wedge troughs. This and other observations suggest that melting of ground ice may be significant in the water balance in the island. Pressumably there are two sources of runoff: (1) external sources graupel, rain and rime and snow and (2) internal sources due to the melting of various kinds of ground ice.

CONCLUSIONS

1. The Antarctic Peninsula is a barrier that separates two different climatic environments: on the east coast, where Seymour Island (Marambio) is located, the mean annual temperatures is -10°C or less; at the same latitude, on the west coast (Melchior), the mean annual temperature is not lower than -3.7°C.

2. The geocryogenic structures on the east coast are cold permafrost forms: thermal-contraction cracks and polygons, and closed systems pingos. These form at -5°C and -6°C mean annual temperature or less. The cryogenic structures on the west coast can be formed at freezing temperatures above -2.8°C.

3. The present climate of the Antarctic Peninsula where different cyrogenic environments exist at the same latitude and elevation on opposite sides of an orographic barrier should be considered when making paleoclimatic reconstructions. It would be interesting to know how the geocryogenic environment differed on opposite sides of the Andes during the cryogenic times of the Pleistocene.

ACKNOWLEDGMENTS

The author would like to thank very sincerely the valuable help given by Dr.Werner Schwedtfeger, Emeritus Professor, University of Wisconsin, Department of Meteorology, Madison, Wisconsin, for information regarding meteorological conditions on the Antarctic Peninsula. The author would also like to thank Dr.Link Washburn for his careful corrections and suggestions to the original manuscript. The final review by William Wayne is also sincerely appreciated. To the reviewers G.Hogdson, and D.A.Hogdson best acknowledgments for thier suggested good style.

REFERENCES

Black,R.F. and Berg T.E.,1963, Patterned ground in Antarctica: Washington Natl.Acad.Sci.Natl. research Council, Proc.Permafrost, Internat.Conf., Pub.no.1287, p.121-128.

Brown,R.J.E.,1978, Influence of climate and terrain on ground temperatures in the continuous permafrost zone of northern Manitoba and Kewatin District: Proc.Third Internat.Permafrost Conf., v.1, p. 16-21.

Corte, A.E.,1963, Relationship between four ground patterns, structure of the active layer and type and distribution of ice in the permafrost: Biul.Peryglacjalny no.12, p.7-90. Also SIPRE Res.Rept.no. 88.

Corte, A.E.,1976, Rock glaciers: Biul. Peryglacjalny no.26, p. 175-197.

Corte,A.E.and Higashi,A.,1964, Experimental research on desiccation cracks in soil: U.S.A. CRREL Res. Rept. no.66, 72 p.

Corte,A.E. and Zomoza, A.L., 1957, Observaciones cripedológicas y glaciológicas en las islas Decepción, Media Luna y melchior: Pub.no.4, Inst. Antártico Argentino, Buenos Aires. p.65-131.

Demek, J.,1969, Cryoplanation terraces, their geographical distribution, genesis and development: Acad.Naklad.Praha, vo.79, no.4, 80 p.

Henning,A.,1924, Le conglomerat a Pecten de l'ille Cockburn: Swedische Sud Polar Expedid.1901-1903, 72 p.

Lachenbruch, A., 1968, Permafrost: Encyclopaedia of Geomorphology, Edit. by Fairbridge - Reinhold Book Co., p. 833-839

Péwé, T.L., 1959, Sand wedge polygons (tesselations) in the McMurdo Sound region, Antarctica: American Journal of Science, v.257, p.550-552.

---1962, Ice wedges in permafrost, lower Yukon river area near Galena, Alaska: Biul.Peryglac. no.11, p. 65-76.

Reger, R.D. and Péwé, T.L., 1976, Cryoplanation terraces, indicators of past permafrost environments: Quat.Res.,v.6, no.1, p.99-109.

Rinaldi, C.A., Massabie,A., Morelli, J., Rosenman, H.L., y Del Valle, R., 1978. Geología de la Isla Vice-Comodoro Marambio: Inst.Antártico Argentino, Contrib. no. 217, 43 p.

Romanovskij,N.N.,1973, Regularities on formation of frost fissures and development of frost fissure polygons: Biul.Peryglac. no. 23, p.237-277.

Schwerdtfeger,W., 1975, The effect of the Antarctic Peninsula on the temperature regime of the Weddell Sea; Monthly Weather Rev.v.103,no.1,p. 45-51.

---1979, Meteorological aspects of the drifts of ice from the Weddel Sea towards middle-latitude westerlies: Journ.Geoph.Res., v.84,no.C10, p. 6321-6328.

Servicio Meteorológico Nacional, 1981, Datos sobre el congelamiento y descongelamiento en Marambio (inedited).

Washburn, A.L.,1979, Geocryology. A survey of peryglacial processes and environments: London, Edward Arnold Publishers, 406 p.

---1980, Permafrost features as evidence of climatic change: Earth Science Rev., v.15,1979-1989, p. 327-402.

DESIGN AND PERFORMANCE OF A LIQUID NATURAL CONVECTION SUBGRADE COOLING SYSTEM FOR CONSTRUCTION ON ICE-RICH PERMAFROST

John E. Cronin

Shannon & Wilson, Inc., Fairbanks, Alaska 99707

In 1976, personnel from the U.S. Navy's Civil Engineering Laboratory erected a building incorporating a subgrade cooling system on ice-rich permafrost near Barrow, Alaska. The cooling system consisted of 15 horizontal, loop-configured, liquid natural convection heat exchangers. The installation at Barrow was heavily instrumented, and data obtained over 3 years allowed calculation of the performance and efficiency of the natural convection heat transfer devices. Experimentation with forced convection (pumping) of the liquid resulted in only a minor improvement of performance. Because seasonal freezing and thawing occurred in frost susceptible material in the subgrade, cyclical settlement and heave were recorded. Since the experiment was completed, foundations incorporating subgrade cooling systems employing air ducts or inclined two-phase heat transfer devices have become increasingly common in the Arctic and sub-Arctic. The liquid system tested compares favorably to two-phase or air duct systems currently in use.

INTRODUCTION

Construction on ice-rich permafrost has relied on pile support and an air space or thick gravel embankments to preserve the integrity of subgrade and structure, but both methods have drawbacks. Elevated support on piles can interfere with the use of the structure, is expensive for heavy floor loads, and can be aesthetically impractical. Prevention of thaw settlement merely by use of a gravel embankment is not feasible. The purpose of this report is to present the investigation of a subgrade cooling system as an alternative means of construction for such structures.

From 1969 through 1976, the U.S. Navy's Civil Engineering Laboratory (CEL) investigated natural convection heat exchange devices for the subsurface thickening of sea ice (Barthelemy 1974). In 1975 CEL began investigating this technology for construction of buildings on permafrost. The primary purpose of the study was to reduce the need for gravel in construction. Laboratory studies (Cronin 1977) resulted in the design of a horizontal, loop-configured, liquid convection device for use in building subgrades. The patent for loop-configured liquid convection devices is held by J. C. Balch. In September 1976 these devices were installed beneath a building erected by CEL at the Naval Arctic Research Laboratory in Barrow, Alaska. The performance of these devices was monitored until September 1979, when research funding cutbacks required termination of the experiment.

DESCRIPTION OF EXPERIMENT

Site Conditions

The site chosen for the building was underlain by ice-rich organic tundra soils covered by a gravel pad on the order of 0.3 m thick. Subsurface investigations indicated generalized conditions as shown in Figure 1. Numerous ice wedges were noted below 0.5 m, the depth of the active layer in Barrow.

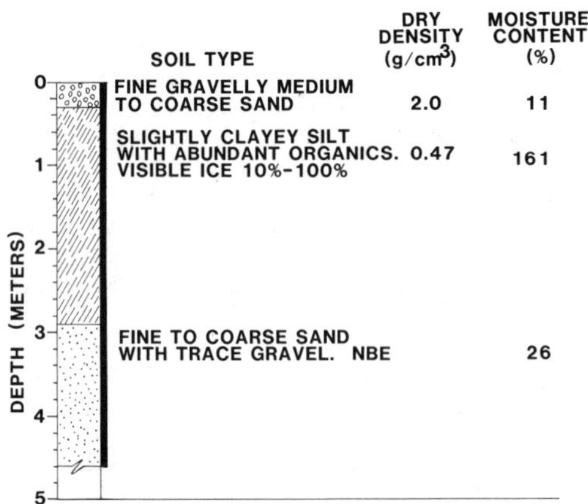

FIGURE 1 Generalized subsurface conditions beneath experimental building in Barrow.

Subgrade Cooling System

The building erected at Barrow was a 12.2 m by 16.5 m metal structure. Fifteen natural convection devices were installed beneath the structure at a 1.2 m spacing, as illustrated in plan view in Figure 2. The below-ground heat "collector" portion of the convection device intercepts heat flow to the ground from the building. Convective flow of an ethylene glycol/water mix in the pipe loop transfers the heat to the above-ground "radiator" portion of the device, where heat is transferred to the cold

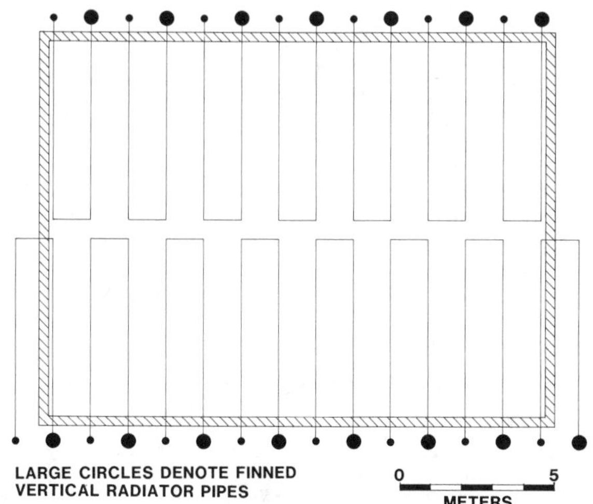

FIGURE 2 Plan view of experimental building in Barrow showing layout of buried heat collector pipes of convection devices.

FIGURE 3 View of completed installation in Barrow showing finned pipe radiators of convection devices.

ambient air. Convective heat transfer continues as long as the collector temperature is warmer than the radiator temperature.

The collector of each device consists of two 6.4 m lengths of nominal 2 inch pipe, connected near the center of the building by a 1.2 m length of the same pipe. The radiator portion of each device, as pictured in Figure 3, consists of a vertical piece of 2 inch pipe on one side of the half-loop, two 2.1 m lengths of nominal 4 inch pipe with welded steel fins on the other side of the half-loop, and a 1.2 m length of 4 inch pipe with welded fins joining the two at the top. Total effective surface area of the radiator (taking into account that fin area is less than 100% efficient for heat transfer) is on the order of 6.2 m^2.

The heat collector portions of the convection devices were buried at a depth of about 0.5 m below the tundra surface, and were sloped at an average of 1.5° downward toward the center of the building to prevent air locking when filled with fluid. The pipes were installed by excavating 20 cm wide trenches into the tundra with a trenching machine and backfilling the trenches with the native ice-rich soils.

Building and Experimental Simulation

The building erected as a test bed for the cooling system was uninsulated and did not require heating for its primary purpose of equipment cold storage. A system of thermostatically controlled electric heating mats was used to simulate heat input to the ground beneath the building. This heating system had a total output when energized of 28,400 watts.

Figure 4 illustrates both the experimental and simulated cross sections of the installation. The heating mats were installed at simulated floor grade, on a 0.3 m thick pad of gravel above the tundra. The thermostat sensors were buried midway between the heating mats and the tundra surface. The mats were covered by gravel and insulation to reduce upward loss of heat.

The experiment was to simulate a building heated to 20°C, incorporating 5 - 10 cm of insulation in the floor, constructed on a 0.3 m gravel pad. Heat input to the ground for purposes of the experiment was controlled by maintaining a constant temperature at the depth of the thermostat sensors. It was realized that this imposed an unreal boundary condition on the experiment, since the temperature at this depth should vary as a function of thaw depth beneath the building. Although the heating mats could produce about 172 W/m^2 when energized, the average heat input (as controlled by the thermostats) was only about 33 W/m^2.

Instrumentation

The building and subgrade cooling system were heavily instrumented to monitor performance. Temperatures were measured by about 150 thermocouples connected to a data logger. Vertical thermocouple strings measured temperature profiles, and a network of 57 thermocouples measured the temperature field in a vertical plane

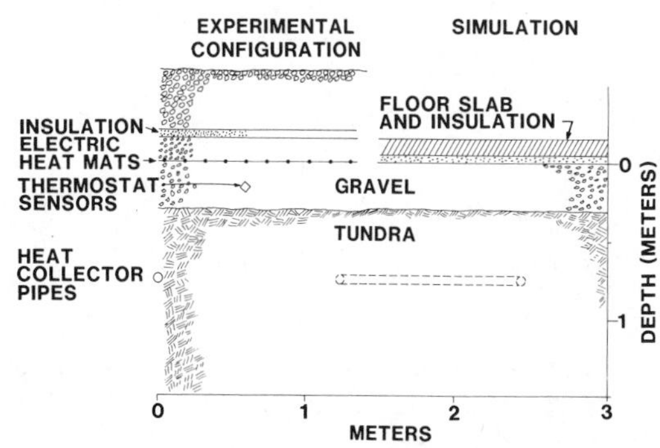

FIGURE 4 Cross section through floor of building, showing experimental configuration and design being simulated.

perpendicular to the collector of a convection device. Additional thermocouples measured collector pipe, heating mat, and air temperatures.

The electric heating mats were divided into eight separately controlled circuits, generally concentric with the perimeter of the building, allowing more uniform response to variable ground temperature conditions. Electric hour meters allowed calculation of heat input to the ground for modeling purposes.

Settlement or heaving was monitored by elevation measurements of points on the foundation and floor of the building referenced to a stable benchmark.

Chronology and Methodology of Experiment

Erection of the experimental building was completed by the end of September 1976. At that time, the depth of thaw beneath the building was about 0.3 m below the tundra surface. With the onset of below freezing air temperatures the following week, the convection devices began freezing the ground beneath the building. The ground was allowed to entirely freeze back beneath the unheated building for 2 months before heat input was initiated.

In early December 1976 the heating mats were energized, cycling on and off to maintain a nearly constant temperature at thermostat sensor depth. This caused the thaw front to recede to a stable location about 0.25 m below the tundra surface (see Figure 5). Analysis revealed that the heat input to the ground was significantly greater than had been planned, simulating an uninsulated floor. This was attributed to thermostat errors caused by exposure to cold ambient air. To prevent excessive summer thawing, it was necessary to limit heat input to a simulation of a building with at least 5 cm of insulation in the floor. Accordingly, the thermostat setting was lowered in

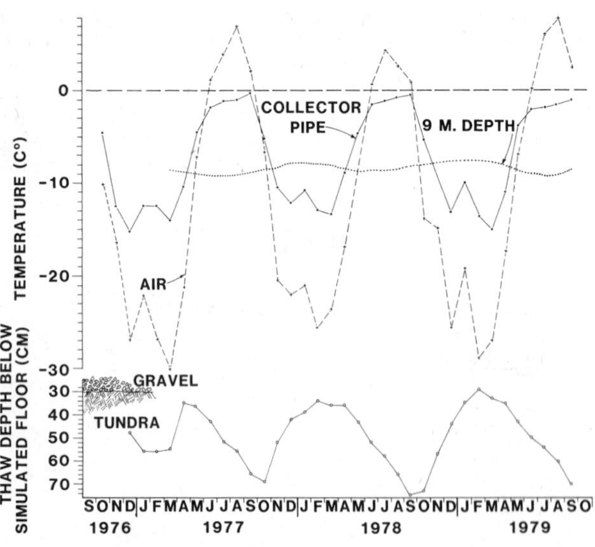

FIGURE 5 Plot showing average monthly temperatures of air, collector pipes, and soil at 9 meter depth beneath building, as well as average depth of thaw during course of experiment.

late March 1977. This resulted in refreezing of the subgrade to near the depth of the tundra surface by the end of April.

Warming air temperatures during May 1977 resulted in thawing, and by late May convective cooling ceased (see Figure 5). The thaw front deepened through the summer until late September, when cooling resumed. Maximum thaw penetration was 0.39 m below the tundra surface, only about 8 cm above the elevation of the collector pipes.

Convection device operation and thaw front location followed a similar pattern during the next 2 years of the experiment.

PERFORMANCE OF SUBGRADE COOLING SYSTEM

Thermal Performance of Convection Devices

Major problems in evaluating the thermal performance of the convection devices were (1) the unrealistic constraint of constant temperature at thermostat sensor depth, (2) the variations in thermostat temperature due to error, (3) the variable ambient air temperature, and (4) the thermal lag in soil temperature variations. Thermal performance is therefore best analyzed by considering average performance over a period of time. All calculations, as well as the derivations in the later section on design factors, are simplified steady-state approximations based on a semi-infinite horizontal plane.

Heat removal rate was calculated for several situations at different times during the year, with results tabulated below:

TABLE 1

Basis for Calculations	Calculated Heat Removal (W/m^2)	Average Air Temp. (°C)
November 1977 freezeback (includes heat mat input, latent heat, and specific heat, averaged for month).	50	-20.5
Total heat mat heat input for one year (Oct. 1977 - Sept. 1978).	52	-18.1 (winter only)
Equilibrium with stable thaw front in March 1978 (heat mat input only).	62	-27
Equilibrium with stable thaw front at higher heat input in March 1977 (heat mat input only).	70	-30

The subgrade cooling system was capable of transferring between 50 and 70 W/m^2 at air temperatures of -18° to -30°C. These values reflect performance at the average wind speed in Barrow, 20 km/hr. Examination of Figure 2 shows that the heat input is intercepted by the equivalent of 13.5 convection devices. Thus the heat intercepted by each device can be calculated, and given the total length of the collector pipe of each device inside the building (12.8 m), the heat rejection by each convection device can be described as 58 - 82 W/m of embedded collector pipe.

Ground Temperature Data

Conclusions regarding ground temperature data are difficult to reach because of the relatively short duration of the experiment. Comparison of ground temperatures at the center of the building and in the natural ground allow the following generalizations: (1) from about June through December natural ground and building center temperatures are almost identical from a depth of 1.5 - 9 m, and (2) from January through May the natural ground temperatures average about 2°C colder through these depths than under the building, although from December through April the two temperatures converge at a 9 m depth. The temperature at a depth of 9 m beneath the center of the building did not show any continuous warming trend during the course of the experiment (Figure 5).

The thaw front beneath the building was in almost the identical equilibrium position during all 3 years of the experiment. Likewise, depth of thaw at the end of the summer was comparable during all 3 years (Figure 5).

Level Data

The results of periodic level measurements on the floor and foundation of the building are plotted in Figure 6. As expected, heaving occurs during the winter as the subgrade freezes, and settlement occurs during the summer as the subgrade soils thaw. Over the limited 2 years measured, both the floor and foundation returned to nearly their original elevations at the end of each year.

The maximum magnitude of this cyclical movement was about 27 mm for the foundation and 60 mm for the floor. The lesser movement of the foundation is probably a result of shallower thawing from less heat input under the foundation, and also the greater load on the foundation. The relatively low magnitude of both is probably due to restriction of freezing and thawing to the original active layer.

While cyclical movements of this magnitude are not ideal, they are minor when compared to the settlement which would have occurred upon thawing of the ice-rich subgrade soils without the convection devices intercepting heat from the building.

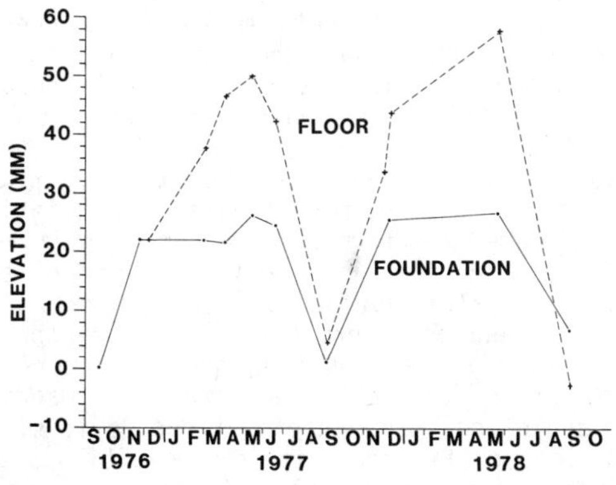

FIGURE 6 Average elevations of floor and foundation monitoring points during course of experiment.

DESIGN FACTORS FOR SUBGRADE COOLING SYSTEMS

Heat Removal

Calculations of the thermal performance of liquid natural convection devices are derived and explained by Barthelemy (1974, 1976) for freezing of sea ice and for soil cooling (Barthelemy 1979). A useful relationship from this latter reference is

$$\theta = \frac{Q_d R_d}{T_a} + 1, \qquad (1)$$

where θ is the pipe-to-air temperature ratio, Q_d is the heat transferred by one convection device, T_a is the ambient air temperature, and R_d is the overall convection device resistance to heat transfer. An example of the calculation of R_d for a convection device is given by Barthelemy (1976).

The temperature ratio θ is defined by Barthelemy as collector pipe temperature divided by ambient air temperature (both in °C). Calculations from the data in Figure 5 show that θ ranges from about 0.4 to 0.8. However, instantaneous temperature readings are not valid for calculating θ because thermal lag in the frozen soils does not allow pipe temperature to respond instantaneously to a change in air temperature. Barthelemy (1979) examines θ both from empirical data during periods of relatively steady air temperature and by calculations based on heat flow. The calculations yield a θ of 0.6, while the empirical θ averages closer to 0.5.

Barthelemy (1979) used a calculated value of R_d = 0.012°C/W for evaluating the observed performance of the convection devices at Barrow, and arrived at a value of θ = 0.6. If one uses the empirical value of θ = 0.5, and an effective number of convection devices inside the building area of 13.5, an empirical value of R_d = 0.015°C/W is derived.

Solving equation (1) for Q_d and incorporating the 14.9 m^2 floor area served by each convection device at Barrow, Q_d can be calculated from the equation:

$$Q_d(W/m^2) = \frac{T_a\ (\theta-1)}{14.9\ R_d}, \qquad (2)$$

where empirical values of θ and R_d are known.

Using this equation and the appropriate air temperatures to check the calculated heat removal capacities from the previous section on thermal performance, values are obtained which are only a few percent lower than those empirically observed.

It should be kept in mind that the values of θ and R_d given above are based on an average wind velocity of 20 km/hr typical of Barrow, and apply only to the convection devices used in the experiment. Both factors will also change with radiator design and with the ratio of effective radiator area to collector area.

Equation (2) is oversimplified in treating θ as a constant for a given convection device. Barthelemy (1976) shows that the value of θ increases with time (as the resistance to heat transfer of the frozen shell around the collector pipe increases). An interesting observation from equation (2) above is that as θ increases, the amount of heat transferred decreases. Thus θ is an inverse measure of the "efficiency" of the convection device. By using calculated rates of heat transfer during the year to solve equation (2) for θ, values for θ as low as 0.36 are obtained (for the 8 month average). Since higher values of θ predict a lower heat transfer rate than observed, the empirical value of 0.5 might be more conservative for design purposes.

The equations above consider only the heat transferred through the convection device from the collector pipe to the ambient air. Considering the frozen/thawed soil interface as the heat source:

$$Q_s = \frac{T_s - T_a}{R_d + R_{sf}} , \qquad (3)$$

where Q_s is the heat transferred from the frozen/thawed interface, T_s is the temperature at the frozen/thawed interface (i.e., 0°C), and R_{sf} is the resistance to heat transfer of the frozen soil. Holding all other values constant, as R_{sf} increases, Q_s decreases. Thus the rate of heat transfer will decrease as the soil around the collector pipe is frozen back, since the frozen soil thermal conductivity is lower than the thawed soil conductivity. Once a stable freezing front position is achieved, the rate of heat transfer will vary only with air temperature.

If the source of heat is viewed as the floor of the building, then

$$Q_f = \frac{T_f - T_a}{R_d + R_{df}} , \qquad (4)$$

where Q_f is the heat transferred from the floor of the building through the devices, T_f is the floor temperature, and R_{df} is the sum of all resistance to heat transfer from the floor to the convection device (floor, insulation, thawed soil, and frozen soil). If one acknowledges that the experimentally observed heat transfer rates presented earlier are actually Q_f, the value of R_{df} in the equilibrium situation in March 1978 can be calculated, and is more than twice the value of R_d. Thus any rigorous design calculations must consider R_{df}. As the depth and spacing of the collector pipes is increased, R_{df} will increase, and the rate of heat transfer will decrease. This becomes a tradeoff which must be considered in the design process.

Natural convection flow rate in the devices was measured to be on the order of 3 - 4 liters/min. by observing the flow rate of particles in the fluid through a clear sight glass incorporated into the radiator. For part of the experiment, three of the convection devices were fitted with pumps which forced convective flow to about 50 liters/min. This resulted in a lowering of collector pipe temperature of only about 1°C, and a slight increase in heat transfer rate. This supports the argument that the convection device resistance, R_d, is primarily a function of the liquid-to-air heat exchange resistance at the radiator, and not of the internal natural convection resistance.

General Considerations

The configuration of the subgrade cooling system tested at Barrow should not necessarily be considered as a pattern for other designs. Gravel pad thickness was reduced to a minimum, with resultant settlement and heaving of the floor and foundation of the building. This resulted in the use of a very close spacing of the convection devices because of the disastrous consequences should the massive ice in the subgrade thaw. In addition, the nature of the experimental simulation was such that (because of thermostat error) a floor with no insulation was modeled during the winter, while a floor with 5 cm of insulation was modeled during the summer. During the winter, the convection devices removed 5-7 times the amount of heat which would have been transmitted through 5 cm of floor insulation.

It remains a design problem to balance a thickness of gravel pad (with a tradeoff between latent heat and frost susceptibility) with a thickness of insulation and with a configuration and spacing of convection devices to arrive at a stable design on a project-specific basis.

OTHER SUBGRADE COOLING SYSTEMS

At the time of the experimental work reported herein, only one other subgrade cooling system using convection devices was known to the author. These were two-phase devices, McDonnell Douglas "Cryo-Anchors," installed beneath a school building at Ross River in the Yukon Territory in 1975 (Haley 1982). In addition, a few air duct cooling systems were in use. A number of recent installations use either two-phase convection devices or air ducts, but there are still no other liquid natural convection horizontal subgrade cooling systems in use which are known to this author.

Comparison of the performance of different types of heat transfer devices is difficult because thermal performance is very sensitive to such factors as the areas and ratio of collector pipe and radiator surfaces, air temperature and wind speed, and the thermal resistance of the soil or other material surrounding the collector pipe. The following data allow an order of magnitude comparison of performance of various devices.

A loop-configured liquid convection device described herein is capable of heat rejection of 58 W/m of embedded length at an average air temperature of about -20°C, wind speed of 20 km/hr, and collector pipe temperature of -10°C (based on empirical data during November 1977). Performance data for Model 500 "Cryo-Anchors" presented by Glover (1975) are for still air, and indicate a power dissipation of about 48 W/m of embedded length for that device under otherwise similar conditions. Data from Davison and Lo (1982) indicate that air ducts at Inuvik had a heat removal capacity of about 9 W/m of embedded length when averaged through the winter.

Thermal performance is not the only criterion for selection of one system over another. Cost, ability to field fabricate, installation requirements, construction and supply logistics, maintenance requirements, and other factors should enter into the selection process. Although a complete comparison of these factors is beyond the scope of this paper, some advantages of liquid convection devices should be considered. Compared to two-phase convection devices, liquid devices have the advantages of (1) installation of the collector pipes horizontally rather than at a slant as most two-phase devices require (this requirement is particularly critical when there is a possibility that frost heaving will cause a two-phase device to slope upwards away from the radiator, which would halt the return of the condensate to the heat collector pipe), (2) the ability to easily supplement natural convective cooling with mechanical refrigeration and forced circulation, and (3) the ability to fabricate the devices in the field without the need for welds capable of sustaining high internal pressures. Compared to air duct systems, liquid devices (1) do not require that the gravel pad be elevated above ground surface to provide air flow or avoid problems with groundwater, (2) do not clog with snow, dust, or water, and (3) do not have to be manually disabled during the summer.

Liquid devices have some apparent disadvantages when compared to two-phase devices. Johnson (1971) found that the heat transfer rate of a liquid loop-configured device fell to zero with a temperature potential (between the air and the collector pipe) of several degrees. Results of the experimental work reported herein suggests that while this effect exists, for the devices tested in Barrow the critical temperature potential for the onset of convection may not be as great as for the devices tested by Johnson. Another apparent disadvantage of liquid devices is the potential for loss of an antifreeze type fluid into the subgrade. However, the 15 devices in Barrow were assembled by inexperienced personnel and no measurable fluid leaks occurred.

CONCLUSION

The experimental work performed by the U.S. Navy's Civil Engineering Laboratory on loop-configured liquid natural convection devices has shown that they are capable of preventing progressive thawing of permafrost beneath heated buildings constructed at grade in the Arctic, and that their thermal performance is comparable to that of other types of heat transfer devices. Incorporation of the information gained in this work with the present state-of-the-art knowledge of insulated pad construction should provide engineers with a useful alternative for Arctic construction.

ACKNOWLEDGEMENTS

The experimental work reported herein was performed with funding provided by the Naval Facilities Engineering Command. Field logistics support was provided by the Office of Naval Research. Mr. J. Barthelemy, formerly of CEL, was instrumental in the design and testing of the convection devices, and in monitoring and interpreting their performance. The staff of the Fairbanks office of Shannon & Wilson, Inc. assisted materially in the preparation and review of this paper. CEL is c/o Naval Construction Battalion Center, Port Hueneme, California 93043. The author was employed as a geologist with CEL 1974-1977.

REFERENCES

Barthelemy, J. L., 1974, Ice engineering-quantification of subsurface ice thickening techniques, CEL Technical Report R-811: Port Hueneme, CA, Civil Engineering Laboratory.

Barthelemy, J. L., 1976, Ice engineering-a heat sink method for subsurface ice thickening, CEL Technical Report R-845: Port Hueneme, CA, Civil Engineering Laboratory.

Barthelemy, J. L., 1979, CEL building and experimental subgrade cooling system, Barrow, Alaska-construction history and performance characteristics, CEL Technical Report R-870: Port Hueneme, CA, Civil Engineering Laboratory.

Cronin, J. E., 1977, A liquid natural convection concept for building subgrade cooling, in Proceedings of the Second International Symposium on Cold Regions Engineering: Fairbanks, University of Alaska.

Davison, D. M., and Lo, R. C., 1982, Preservation of permafrost for a fuel storage tank, in French, H. M., ed., Proceedings of the Fourth Canadian Permafrost Conference: Ottawa, National Research Council of Canada.

Glover, L. W., 1975, Cryo-Anchor-a design approach for arctic foundations, Oil and Gas Pipeline Equipment and Technology Seminar, Moscow, MDAC Paper WD 2593: McDonnell Douglas Astronautics Company.

Johnson, P. R., 1971, Empirical heat transfer rates of small Long and Balch thermal piles and thermal convection loops, IAEE Report 7102: Fairbanks, Institute of Arctic Environmental Engineering, University of Alaska.

Hayley, D. W., 1982, Application of heat pipes to design of shallow foundations on permafrost, in French, H. M., ed., Proceedings of the Fourth Canadian Permafrost Conference: Ottawa, National Research Council of Canada.

AN EXPERIMENTAL STUDY OF THE FROST-HEAVE REACTION

Cui Chenghan and Zhou Kaijiong

The Third Survey and Design Institute, Ministry of Railways
Tianjing, People's Republic of China

The existence of the frost-heave reaction has been recognized by many researchers. To evaluate frost-heave quantitatively for the design of foundations in cold regions, in situ frost-heave tests on model foundations were carried out at Jintao Field Research Station in the Da-Xinganling Mountains of northeast China from 1978 through 1980. Test results showed that the frost-heave reaction around pile foundations exhibits a triangular stress distribution. Its peak value occurs away from the foundation center at one-half of the freezing depth. Based on a simplified triangular stress distribution diagram, the authors present equations for calculating frost-heave reaction in both plane- and space-stress cases. Equations for examining frost-heaving stability and tensile strength of foundations, where the frost-heave reaction is taken into consideration, are also presented for both two- and three-dimensional problems.

Heilungjiang Province of China is located in the southern margin of the Eurasian permafrost area, and the freeze-thaw action there is extremely active. Especially unacceptable frost heave deformation have often occurred on the structures built on frost heave susceptible foundations. As a result some of the structures have been damaged. For example, a pier shaft (the part in earth) of 2-10 m deck plate girder bridge in that area failed because of frost heave. According to our investigation there are several other similar examples. To search the cause of this frost damage, we first of all checked our traditional design method, according to which the pull-out fracture ability of the pier shaft is calculated by subtracting the dead load above the calculated section from the total tangential frost heave force, but the calculated result did not show that the shaft should be broken by the pull-out force. Thus we were engaged in research work on the frost heave mechanism of the structures. Through experimental investigation it was found that the basic cause of the freeze damage mentioned above was the frost heave reaction, which had been excluded from consideration during the calculation in the past. So the question how to count frost heave reaction is the focal point discussed in this paper.

STATE OF ART OF FROST HEAVE REACTION STUDY

Although great achievements have been made on the literature of frost soil mechanics, many important problems have not yet been solved properly in practice; for instance, frost heave reaction has not been taken into account by now in calculating frost heave stability. This is because tests and research work on frost heave reaction have seldom been done. Dalmatov (1957) is the only one who is able to give a quantitative formula. Through modelling tests in laboratory, he confirmed the existence of frost heave reaction and gave the quantitative formula with the flexible length caused by frost heave on the ground surface adjacent to the foundation in the field as frost heave distribution zone and on the assumption that frost heave reaction is distributed triangularly and the principle that action and reaction are certain to be in equilibrium, but this formula has not been seen applied currently for lack of verified information data. The existence of frost heave reaction has also been stated in several other documents but no quantitative methods have been given. To find the correct quantitative methods, we have conducted the following experiments.

EXPERIMENTS ON FROST HEAVE REACTION

Results

For the sake of exploring the distribution rule of frost heave reaction so as to determine on quantity, we have conducted several field tests at Jin Tao Frozen Soil Experimental Station on Da Hinggan Ling Mountains of Northeast China in 1978, 1979 and 1980. The dimension of test piles, action-measuring apparatus and the experimental results are shown in Figure 1. It must be pointed out that the frost heave reaction curve of the first year in Figure 1 is under the influence of the back-fill that has not yet fully settled.

Analysis

Before foundation is embedded into the ground, the base soil at every layer will heave freely upwards when it is frozen, as a result the vertical pressure stress on the frost front will not change round about the freezing period (omitting the influence of water migration).

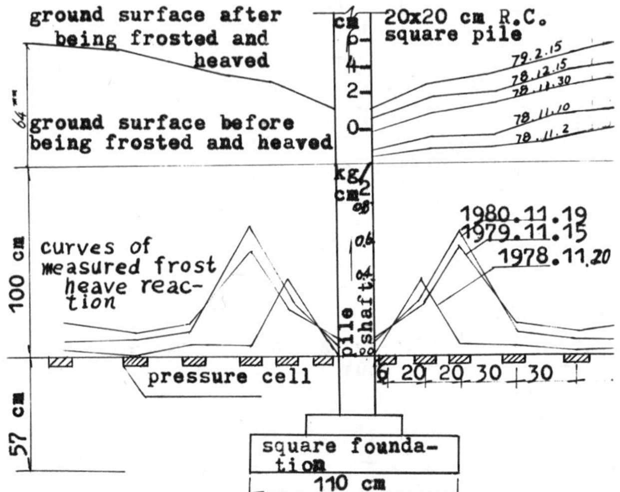

FIGURE 1 Experimental results and apparatus

After foundation is embedded into the ground, the heave of the soil to be cemented with the foundation sides will be bound when the base soil near the foundation is frozen. Thus the interaction force between base soil and the foundation will emerge. The principle of the interaction during frost heave will be dealt with below.

The foundation soil in seasonal frost areas is frozen downwards layer by layer as illustrated in Figure 2. We assume that the first layer soil $\Delta H1$ near the ground surface and the structure sides have already been frozen together and that the soil has achieved its primary volume-heaving quantity in this layer. When the second layer soil $\Delta H2$ gets frozen, the heaving of its volume would be bound by $\Delta H1$, the frozen soil at the first layer above and the thawing soil layer underneath. But the heaving force emerging from earth freezing is so strong that obviously the heaving force between the first layer soil $\Delta H1$ and the masonry is unable to restrict it, thus the first layer soil is sheared and an up-sliding friction force emerges between the first layer soil and the masonry, which is called tangential frost heave force. At the same time the heaving force of the soil at the second layer not only

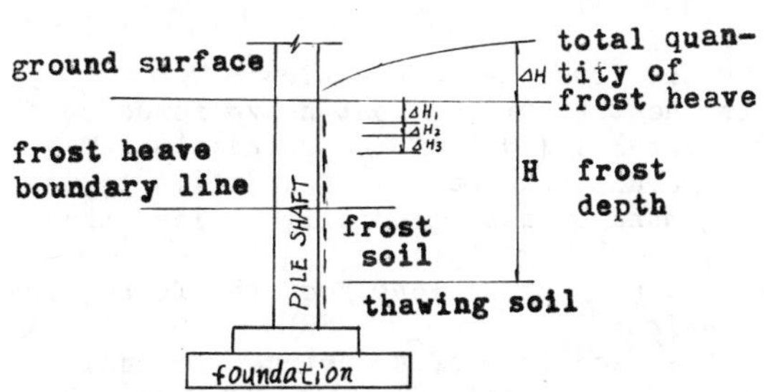

FIGURE 2 The interaction between frost heave soil and foundation.

forces the first layer soil to shift upwards but also compresses the thawing soil layer below. The downward force is called frost heave reaction. And the frost heave force of the third layer soil $\Delta H3$ also pushes the soil at the first and second layers to move upwards. It should be noticed that the frost heave reaction resulting from the tangential frost heave action at the first layer will spread downwards through the frozen soils at the second and third layers (the spreading angle is nearly 45° according to some previous tests). No doubt $\Delta H4$, $\Delta H5$ ... will continuously repeat the above mentioned process --- freezing and heaving. Along with the growth of the frozen layer numbers the tangential frost heave force between the frozen soil and masonry gradually increases. The structure will start to heave when the resultant force exceeds the structure weight, the weight of the soil above the foundation margins and the frost heave reaction force dropping on the margins. Based on the principle of the equilibrium of forces, the tangential frost heave action and reaction always exist simultaneously, equal in quantity but opposite in direction. And the force results from the frost heave action of the soil, so some soil layers will move upwards while some other layers move downwards (compression displacement). Such a boundary line always exists from the beginning to the end of the whole frost heave process, which is called frost heave boundary line as shown in Figure 2. It will move downward gradually along with the deepening of the frost from the beginning of the frost. The pier shaft of the structure bears parts of the pull-out force originating from the frost heave action during the frost heave. The peak value of the tangential frost heave force will appear when the boundary line lowers to a certain position which is the most unfavourable condition for the pier shaft being pulled. This boundary line is observed at the position above 2/3 frost depth in the frost area according to the broken position of a bridge pier in the field.

With regard to the distribution rule of the frost heave reaction, it can be seen in Figure 1 that the ground surface near the structure becomes curved, the length of which is approximate to frost depth. According to the result of the actual measurements for three years in succession, the distribution zone of frost heave reaction is approximate to frost depth too, and its distributing graph appears to be a hump spreading on both sides of the structure or around it, which can be sketched as Figure 3.

From Figure 3 we know that the actually measured frost heave reaction graph is the bdc part, while Dalmatov's (1957) assumed graph is the abc part. As compared with each other, they are different distinctly. This is because Dalmatov (1957) did not take into account the part of the frost heave reaction near the structure that acts on the structure sides through friction between soil at the structure sides

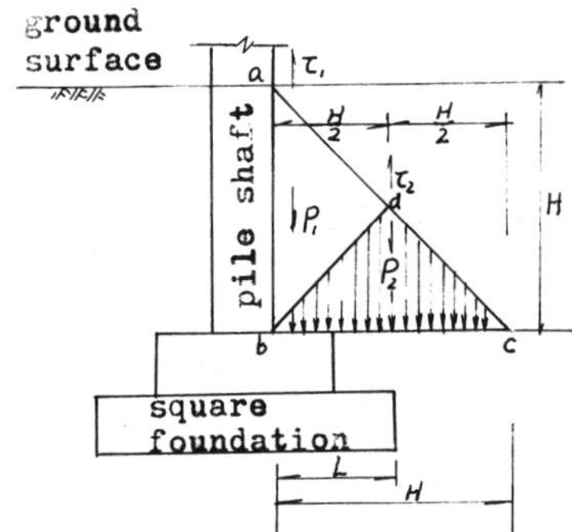

FIGURE 3 A generalized graph of frost heave reaction.

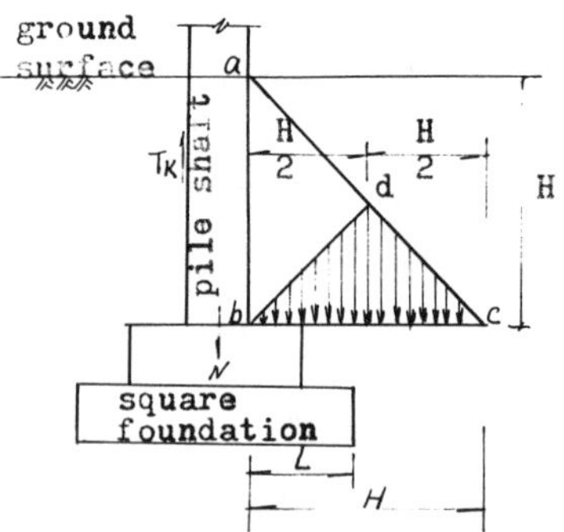

FIGURE 4 Distribution graph of frost heave reaction in two-dimensional problems.

and the soil as well as that between the soil and masonry. If the frictional factor of the structure sides is zero, the stress distribution graph of the frost heave reaction can be generalized in accordance with Dalmatov's (1957) assumption. But the fact is not so. Experiments have proved that the abd part of the frost heave reaction graph acts on the structure sides, and therefore counteracts the tangential frost heave force τ_1 of the same quantity. In fact, the total frost heave reaction should be the sum of the two parts of abd and bdc, i.e. ($p_1 + p_2$) in Figure 3, and the total tangential frost heave force should be ($\tau_1 + \tau_2$), and $\tau_1 = p_1$, $\tau_2 = p_2$, where the two forces, p_1 and τ_1, act on the same structure side equal in quantity and opposite in direction, so they have no influence on the frost heave stability of the structure but take the decisive action on the pull-out failure of the pier shaft. From this we find out the real cause of the pull-out failures of the pier shafts in the fields. And p_2 and τ_2 are the parts that have an effect on the stability of the frost heave action.

Summarizing what is stated above, the frost heave reaction acting on the foundation should be the volume enclosed by bdc stress graph in Figure 3. It should be pointed out that in the actual measurement for the tangential frost heave force index, τ_2 is measurable, but not τ_1. Nevertheless, $\tau_1 = \tau_2$ in two dimensional problems while τ_1 and τ_2 should be calculated proportionally in three dimensional problems. The above analysis also indicates a question to be paid attention to, i.e. when measuring the tangential frost heave force index, shorter piles would produce obviously larger tangential frost heave force. Therefore when measuring in the field, the tested piles must be made close in length to the depth of the actual structure, otherwise the measured data are invalid.

Calculation Formulas of Frost Heave Reaction

Experiments have proved that the stress distribution graph of frost heave reaction appears as a hump distributed on both sides of a structure in two-dimensional problems or around it

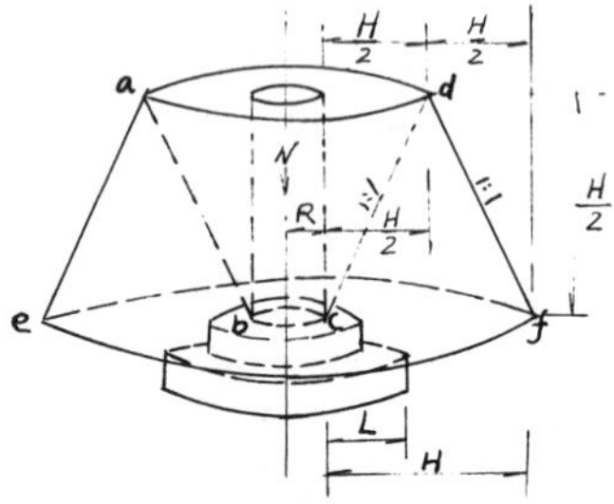

FIGURE 5 A distribution graph of frost heave reaction in three-dimensional problems.

in three-dimensional problems. For the sake of a convenient application, calculation formulas should be as simplified as possible. So we generalize the stress graph measured in the field as Figures 4 and 5.

1. Formulas in two-dimensional problems

a. If $L < \frac{H}{2}$,

both sides:

$$Pn = 4Tk \left(\frac{L}{H}\right)^2 \quad \ldots\ldots\ldots\ldots\ldots\ldots T/M \qquad (1)$$

b. If $L > \frac{H}{2}$,

both sides:

$$Pn = 2Tk \left[\frac{2L}{H}\left(2 - \frac{L}{H}\right) - 1\right] \quad \ldots\ldots T/M \qquad (2)$$

where

Pn is the frost heave reaction; T/M

Tk is the tangential frost heave force on the lateral sides of the foundation; T/M

H is the distance between foundation top and ground surface generally no less than 2/3 of freezing depth; M

L is the projecting length of the foundation skirt; M

N is the dead load of structure (including the weight of soil on the foundation skirt). T

2. Formulas in three-dimensional problems

a. If $L < \frac{H}{2}$,

$$Pn = 8\pi RL^2 Tk \frac{(R + \frac{2}{3}L)}{H^2(2R+H)} \quad \ldots\ldots\ldots\ldots\ldots\ldots \; T \qquad (3)$$

b. If $L > \frac{H}{2}$,

$$Pn = \frac{4\pi RTk}{H(2R+H)} \left[2RL(2 - \frac{L}{H}) - H(\frac{3R}{2} + \frac{H}{3}) + \right.$$

$$\left. + \frac{2L^2}{3}(3 - \frac{2L}{H}) + \frac{H}{2}(R + \frac{H}{3}) \right] \quad \ldots\ldots\ldots T \qquad (4)$$

where R is the radius of column shaft.

The foundation is round. The other symbols are the same as above.

CHECKING COMPUTATION OF FROST-HEAVE STABILITY AND STRENGTH

Checking Computation of Frost Heave Stability

$$n.U.Tk \leq N + Pn + Qm + Qt$$

where

n is the factor of safety 1.1—1.2;

U is the girth of foundation within the effective frost heaving zone above foundation top; M

Pn is the frost heave reaction falling onto the foundation margin on both sides of the foundation or around it; T

Qm is the freezing force between foundation and frozen soil; T

Qt is the friction between foundation and thawing soil;

Other symbols are the same as above.

Calculation of Pull-out Resistance for Structure

1. When calculating in two-dimensional problems

$$P_1 = 2T_kB. \qquad T$$

2. When calculating in three-dimensional problems

$$P_1' = \frac{4\pi RTk}{(2R+H)} (R + \frac{H}{6}). \qquad T$$

3. Calculation for Pull Stress of Pier Shaft Section (Referring to the Bridge Piers with Round Ends)

$$= \frac{Tk \cdot U - N_c + \frac{4\pi RTk}{(2R+H)}(R + \frac{H}{6}) + 2TkB}{F}$$

$$\leq \sigma a \cdot \frac{1}{m}$$

where

σ is the stress on the dangerous section of structure; T/M^2

N_c is the dead load above dangerous section of structure; T

F is the sectional area of the dangerous section of structure; M^2

σa is the limit pull-out strength of concrete T/M^2

B is the length of straight line segment of pier shaft or wall shaft; M

m is the factor of safety (3 is taken).

Other symbols are the same as above.

REFERENCE

Dalmatov, B. I., 1957, Effects of frost heave of soil on foundation of structure

AN INVESTIGATION OF ROCK GLACIERS IN THE KUNLUN SHAN, CHINA

Cui Zhijiu

Department of Geography, Peking University
People's Republic of China

A special form of rock glacier in the continuous permafrost zone is found at high elevations in the Jing Xian Valley of the Kunlun Shan of western China. These rock glaciers generally have a smooth longitudinal profile and lack longitudinal and transverse ridges and furrows. Their gradients are steeper than rock glaciers elsewhere, but their frontal slopes are only slightly steeper than their down-valley gradients. They contain permafrost below a depth of 1.5-2 m, and have interstitial ice contents of up to 57%. These rock glaciers are on north-facing slopes and flow roughly 0.2-3 cm yr^{-1}, primarily because of slope-induced creep facilitated by subsurface water derived from spings above the glaciers. The constituent debris does not originate from rock fragments produced at mountain walls, but instead comes from till and alluvium covering a smooth, rounded ridge just under 5000 m in elevation.

A rock glacier is a tongue-shaped accumulation of rock fragments with a till-like character moving downslope usually in a high mountainous region. Its head may be relatively depressed and its front may be convex, very much like a small ice glacier. Rock glaciers are distributed mainly in the middle and high latitude mountains. The slow downhill creep of rock glaciers may lead to the relatively steep slope of their fronts, generally reaching 20° to 40°, and a few meters to tens of meters in height.

An extensive body of literature exists on rock glaciers throughout the world. Rock glaciers have been reported in the Yukon area of Canada; Alaska, the Sierra Nevada, and the Rocky Mountains of the United States; the Swiss Alps; the mountain areas of northern Scandinavia; as well as in Tian Shan, Qilian Shan, Longitudinal Mountains and Kunlun Shan in China. Probably most rock glaciers can be classified as tongue-shaped or lobate in form, and contain either interstitial ice or a core of glacial ice (White 1976). The rock glaciers in eastern Kunlun (35°49'N, 91°00'E) appear to be a special type and are the subject of this paper.

CHARACTERISTICS OF THE KUNLUN SHAN ROCK GLACIERS

The rock glaciers studied in this investigation are in the Jing-Xian Valley in the eastern Kunlun Shan. This region has both the special Kunlun-Shan-type rock glacier and those formed by the evolution of talus. Of the 18 Kunlun-Shan-type rock glaciers in this valley, 16 are concentrated on its western flank at about 4850 m elevation (Figure 1), and two are on the eastern flank at about 5050 m elevation. The depressions in their upper reaches average about 4800 m elevation, and the average terminus elevation approximates 4680 m. Table 1 presents some dimensions and other information concerning these rock glaciers.

FIGURE 1 One of 16 rock glaciers on the northern slopes of a highland of Kunlun-Shan, central China.

Morphology

All but two of the rock glaciers are tongue-shaped and have a ratio of length to width greater than unity (Table 1); their average length and width is 298 and 99 m, respectively. Although some have a broad tail, others are widest in their middle parts. In some cases, two join together, while the ends of others are branched (Figure 1). In longitudinal profile, some have an upward convex shape with several steps, which suggests that they are relatively active; others are rather smooth and imply relative inactivity. These rock glaciers tend to lack longitudinal and transverse ridges and furrows, and generally have a smooth, gentle longitudinal profile (Figure 2). Their longitudinal gradients of 13° to 30° (Table 1) tend to be steeper than those of rock glaciers in other regions. Although the slope angles of their fronts range from 8° to 40°, most frontal slopes vary between 17° and 24° (Table 1); it can be seen that frontal slopes are only slightly steeper than

TABLE 1 Aspects of the Kunlun-Shan-type rock glaciers.

Location number	Length (L) (m)	Width (w)	L/W	Elevation terminus (m)	Slope orientation	Gradient (°)	
						Longitudinal	Frontal slope
1	350	84	4.17	4670	E	18	20
2	240	24	10.00	4740	NE		
3	396	78	5.08	4700	N45E	18	
4	408	120	3.40	4700	NE	30	40
5	438	120	3.65	4680	N	18	18
6	420	204	2.06	4700	N	19	18
7	408	50	8.16	4670	N30E	21	22
8	276	20	13.80	4710	N30E	15	17
9	396	120	3.30	4650	NE	17	19
10	156	25	6.24	4750	N40E	19	24
11	204	48	4.25	4730	NE	13	8
12	216	96	2.25	4660	NE		
13	108	120	0.90	4760	NE		
14	240	120	2.00	4610	E		
15	324	150	2.16	4570	N20E		
16	180	204	0.88	4610	E		
Average	297.5	98.9	4.519	4681.9		18.7	20.7
dev.	107.8	57.7	3.520	53.4		4.2	8.5
	16	16	16	16		10	9

FIGURE 2 Rock glacier No. 6 illustrating smooth, gentle longitudinal profile.

respective longitudinal gradients. The lower frontal slope angles relative to other rock glaciers may suggest lesser amounts of ice, less cohesive force, and/or slower movement.

Volume estimates were made for two representative rock glaciers—the relatively active No. 4 and the relatively inactive No. 1—using aerial photos for area and electrical surveys for thickness. Data collected from 13 points along two transects on No. 4 reveal that the PS values drop abruptly from 1000 Ω to 200 Ω at about 40 m depth; this depth is considered to represent its thickness. With an area of 4.3×10^4 m^2 and an average depth of 40 m, the volume of No. 4 is approximately 1.9×10^6 m^3. Using the same procedure, the relatively inactive rock glacier No. 1 has a thickness of 25 m which, with an area of 2.4×10^4 m^2, suggests a volume of 0.6×10^6 m^3. These volumes are roughly equivalent or a little smaller than those of the tongue-shaped rock glaciers in the Front Range of Colorado (White 1976).

The electrical survey also suggests the depth to permafrost and the ice content in these rock glaciers. Electrical resistance in the depth range of 1.5-2 m increases from 600 Ω to 1000 Ω, suggesting that the permafrost table is at this depth; this is corroborated by the observation of ice in June and July below that depth. A 1.5-2 m thick active layer on the rock glaciers is in accord with data from nearby. At a depth of 10 m, electrical resistance increases to 4126 Ω and implies a high water content. Bore hole data from rock glacier No. 6 show that the ice content in the active layer is greater than 30%. The ice content at 25 m depth is 29%, and is 57% at 39 m depth, where the temperature is -2°C. At the upper reaches of the rock glaciers, the ice content at depths from 20-30 m is probably less than 30% because of greater slope, coarser grain size, and lower water content. The ice within these rock glaciers is in the form of interstitial frozen rain and snow melt. Thus, and because glacial ice was not observed, they may be categorized as ice-cemented rock glaciers.

Grain Size Distribution

Large boulders seem to be predominant on the Kunlun-Shan-type rock glaciers. Large surface boulders are mostly concentrated on the flanks or inner parts of the lower reaches, thereby presenting a crude striped pattern. Some boulders stand erect due to compression or because of frost heave. In a few locations, the surficial debris is locally sorted into circles or overlapping debris islands.

Rock glacier No. 1 seems to be typical in terms of debris size distribution. On this rock glacier, boulders with a diameter of more than 2 m account for 10% of the total, while boulders with diameters of from 0.5 to 2 m account for 45% of the surficial debris; 30% and 15% of the boulders have diameters of from 0.1 to 0.5 m and less than 0.1 m, respectively. Fine-grained materials are found primarily at depths below 1.5-2 m. In the size fraction less than 2 mm, sand (0.1-2 mm) predominates and accounts for 53-56%; silt (0.005-0.1 mm) constitutes 41-47%, and clay (<0.005 mm) is only 0.07-0.1%. The median granular value (MD) is 0-20 mm, the coefficient of overlap (SO) is 2.6665, and the coefficient of asymmetry (SK) equals 17.095, indicating its poor overlap and multitude of sources.

Source of Debris and Rock Glacier Origin

The debris that makes up these rock glaciers does not originate through the disintegration of bedrock on mountain walls. Instead, it comes predominantly from the middle Pleistocene till deposits on the rounded ridge above them (Figures 2 and 3), and secondarily from Pliocene alluvial gravels underlying the till. The till and alluvium contain such rock types as granite, granodiorite, and pyroxenolite, all of which were derived from the intrusive body of the mountain. The age of the till source shows that these rock glaciers began forming in the late Pleistocene.

The Kunlun Shan rock glaciers developed near the snow line in association with perennial or long-lasting seasonal snow banks. The upper reaches of the rock glaciers are generally at 4800 m elevation, which was also the height of the late Pleistocene snow line. At that time, the ridge crest was only 100 m higher than the snow line. Nivation associated with these snowbanks caused the formation of several wide, shallow depressions that may be considered nivation hollows. Snow melt carries away some finer material and at the same time becomes incorporated in the unconsolidated sediment below the hollows. It is believed that frequent freezing and melting caused the debris on the lower end of the nivation hollows to creep and evolve into either rock glaciers or perhaps large block streams. This origin for the Kunlun-Shan-type glacier is different from that of any other type of rock glacier.

Two periods of rock glacier formation are implied by deposits downvalley from the fronts of the active rock glaciers. The deposit below No. 4 is about 100 m longer than the modern rock glacier; this is also generally true of the other rock glaciers. The history of rock glacier formation agrees well with the local glacial chronology.

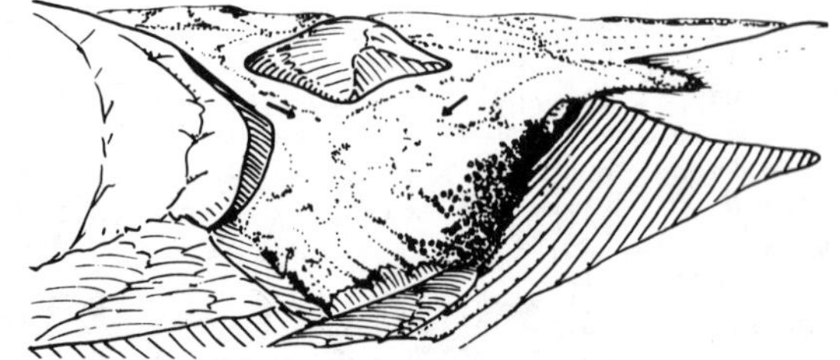

FIGURE 3 Diagram of rock glacier No. 5 originating from rounded ridge.

Magnitude of Activity

The activity of the Kunlun-Shan-type rock glacier is weaker and its velocity is slower than rock glaciers elsewhere. These rock glaciers flow principally by creep initiated by their relatively steep gradients. The widening of the middle reaches of four rock glaciers (Nos. 1, 4, 9, and 10), in conjunction with their dome-like appearance, suggests that the surface flows more rapidly than the basal portions. The available evidence implies that they do not move as a unit. Crevasses in these rock glaciers owe their existence first to rainfall, and second to underground water, and are not necessarily indicators of activity. The Kunlun Shan rock glaciers can be classified into three kinds: active, transitional, and relatively inactive, based on the movement of debris at their fronts. The most active ones are Nos. 4, 5, 6, and 10.

The Kunlun Shan rock glacier typically moves slowly and in various ways. The character and form of movement were clearly indicated by triangulation of painted stones on two representative samples: the active No. 4 and the relatively inactive No. 1. The surface of the middle and lower reaches of the eastward flowing No. 1 became higher on the north side and lower on the south side, and all painted stones moved conspicuously southeastward. The mean annual velocity along the direction of flow of No. 1 was calculated to be 0.225-0.3 cm y^{-1}, but the speed of the middle reach was somewhat faster than the lower reaches. The velocity of the very active No. 4 is estimated to be about 10 times that of No. 1.

In general, the Kunlun Shan rock glaciers are believed to move because of creep, but movement may be facilitated by the flow of subsurface water. For example, in summer, springs flow from the headward depressions (nivation hollows) at the contact between till and alluvium, and the water infiltrates the rock glacier below them. In addition, local freezing and thawing may cause some rolling of surface fragments. Frost heave of stones is apparent, especially on rock glaciers Nos. 4 and 5, and some painted stones were observed to have been raised during the period of triangulation. Some of the painted stones were found to have rolled several meters.

These rock glaciers lack the flow mechanisms of ice-cored or typical ice-cemented rock glaciers and, as a consequence, their flow is rather slow. The greatest speed of the Kunlun Shan rock glacier is considerably slower than those in the Alps and the Colorado Rocky Mountains. The annual average discharge of the Kunlun Shan rock glaciers is estimated to be in the range of hundreds of cubic meters, and is comparable to tongue-shaped rock glaciers in Colorado.

ENVIRONMENT AND FACTORS INFLUENCING DEVELOPMENT

In general, an active rock glacier is a good indicator of the environment, showing not only the lower elevation of permafrost, but also suggesting the snow-line elevation. They form below the snow line where temperatures cause permafrost but where a net accumulation of snow is lacking (Brown and Péwé 1973). Ice-cored rock glaciers in the Alps of Europe extend down to 2700 m elevation, but do not extend more than 400 m below the snow line. In like manner, the lobate rock glaciers on the Kunlun Shan or Tianshan of China flow down to elevations not more than 400-500 m below the snow line; this is also true of the Kunlun-Shan-type rock glacier in the Jing-Xian Valley, where the north-facing snow line is at 5150-5200 m elevation. The average elevation (4682 m) at the ends of the Kunlun Shan rock glacier (Table 1) is about 2000 m higher than the average terminus elevation in the Alps, but the Kunlun Shan snow line is also about 2000 m higher than the snow line of the Alps. The average elevation of the termini of the Kunlun Shan rock glaciers is more than 100 m below the terminus (4800 m) of the glacier on a northern slope 20 km away from the eastern flank of the Jing-Xian Valley.

The Kunlun Shan rock glacier forms in an environment that is colder and more arid than other areas where rock glaciers are common; the mean annual temperature in the Kunlun Shan rock glacier area is -4°C, which is as much as 2°C colder than rock glacier areas of the Alps. The altitudes of the snow line and rock glacier formation are higher in the Kunlun Shan than in the Alps because of the greater continentality of that part of China.

Active rock glaciers are indicators of permafrost because they contain interstitial ice or an ice core. In addition, they may be indicative of the type of permafrost distribution of an area. The elevation of the lower limit of discontinuous permafrost in the western Alps is 2350 m on the northern side of the range and 2550 m in the central areas, where its temperature ranges from 0° to -2°C. Rock glaciers in the Alps have their termini at 2400-2600 m elevation, where their temperatures are lower than -2°C. Because their termini average 50 m higher than the lower limit of discontinuous permafrost, Barsch (1978) held that the presence of rock glaciers is an index of the discontinuous permafrost zone. However, the Kunlun Shan rock glaciers have formed where the lower limit of continuous and discontinuous permafrost is 4550-4600 m and 4200-4300 m, respectively. The ends of the Kunlun Shan rock glaciers are 80-130 m higher than the lower limit of continuous permafrost; therefore, these rock glaciers may be an index of continuous permafrost.

Some of the factors influencing the development of rock glaciers include regional climate, orientation and steepness of slope, lithological character, and geologic structure. Within a favorable area, the slope-related microclimate may play a decisive role, as rock glaciers tend to develop largely on shady slopes. Topoclimatic suitability is corroborated in general by the observed distribution of rock glaciers, and experimentally by Morris (1981) who found that a multiple regression model incorporating radiation shading and an interaction term (representing the multiple effects of elevation, radiation shading, and bedrock jointing) accounts for 72% of the variance in size of rock glaciers in an area of the Sangre de Cristo Mountains of Colorado.

Regarding the formation of the generally north- and northeast-facing Kunlun-Shan-type rock glaciers, the gentle, low-relief hilltops underlain with thick-bedded, unconsolidated sediments also play a significant role. The subsurface water flow favored by the stratigraphy and lithology at the rock glacier source is also a factor. The Kunlun-Shan-type rock glacier is a special mass movement form affected most by these local factors, and does not resemble typical tongue-shaped and lobate rock glaciers elsewhere. They may constitute a unique form, which needs to be more thoroughly investigated.

ACKNOWLEDGMENTS

I would like to thank S. E. White of Ohio State University for his review of an earlier draft of this report. I would also like to thank A. D. Hyers of the University of North Carolina at Wilmington for his efforts at condensing, revising, and rewriting the manuscript. However, I alone, of course, am responsible for the content of this paper.

REFERENCES

Barsch, D., 1978, Active rock glaciers as indicators for discontinuous alpine permafrost. An example from the Swiss Alps, in Third International Conference on Permafrost, Proc., v. 1: Ottawa, National Research Council of Canada, p. 348-353.

Brown, R. J. E., and Péwé, T. L., 1973, Distribution of permafrost in North America and its relationship to the environment: a review, in Permafrost, the North American contribution to the Second International Conference: Washington, D.C., National Academy of Sciences, p. 71-100.

Morris, S. E., 1981, Topoclimatic factors and the development of rock glacier facies, Sangre de Cristo Mountains, Southern Colorado: Arctic and Alpine Research, v. 13, p. 329-338.

White, S. E., 1976, Rock glaciers and block fields, review and new data: Quaternary Research, v. 6, p. 77-97.

VARIATION REGULARITY OF PERMAFROST TABLE BENEATH EMBANKMENTS IN NORTHEAST CHINA

Dai Jingbo

The Third Survey and Design Institute, Ministry of Railways
People's Republic of China

Based on results from an experimental embankment and the experience of two operating railways, the author has analyzed the variation regularity of the permafrost table beneath an embankment and its influencing factors, and proposes measures to protect embankments from damage. When construction is undertaken in the thawing season, the permafrost table beneath the embankment declines to the maximum thawed depth in the same year, begins to rise again in the center of the embankment and on the shaded slope in the second year, rises overall in the third year, and reaches a stable state in 5-6 years. Variations of the permafrost table are closely related to latitudinal zonality, but they also depend on the height and orientation of the embankment, the thermophysical properties of the filling materials, the heat insulation measures adopted, the surface formation after construction, and the hydrological conditions and characteristics of the frozen soil. To protect the embankment from damage, the following measures are available: improving surface drainage, protecting the natural permafrost table near the foot of the embankment slope, and constructing a heat insulation berm.

VARIATION PROCESS OF PERMAFROST TABLE BENEATH EMBANKMENTS IN PERMAFROST REGIONS

An important principle in selecting a railway line in a permafrost region is to use embankment for passage as much as possible. The stability of embankments on permafrost soil depends on the variations in the permafrost table of the foundation during and after construction of the embankment. In the permafrost region in Northeast China, construction is usually carried on in the thawing season, and the cause and process of variation of permafrost table are due to compression of the vegetation peat layer by the filling soil and weakening of the thermo-isolation property. The filling soil carries a considerable amount of accumulated heat, which is dissipated into the foundation. This, together with the heat-absorbing effect of the foundation raises the temperature of permafrost and lowers the permafrost table. There is a general depression of the permafrost table in the year of construction, and a small rise in the center and shaded slope of the embankment in the next year, with a slight further depression in the sunny slope. In the third year after construction, there is an overall rise of the permafrost table. In the sixth year, the permafrost table rises into the embankment filling soil, as the temperature in the embankment approaches the ground temperature of the foundation has reached a new state of thermal equilibrium. After 5 or 6 years, the permafrost table has become relatively stable, as shown in Figure 1.

FACTORS AFFECTING VARIATION OF PERMAFROST TABLE BELOW EMBANKMENT

Effect of Latitudinal Zonality

Latitudinal zonality reflects the combined effect of air temperature and geological and geographical conditions. In the discontinuous permafrost zone at the northern-most latitudes of this region, the soil itself is relatively stable. After construction of the embankment,

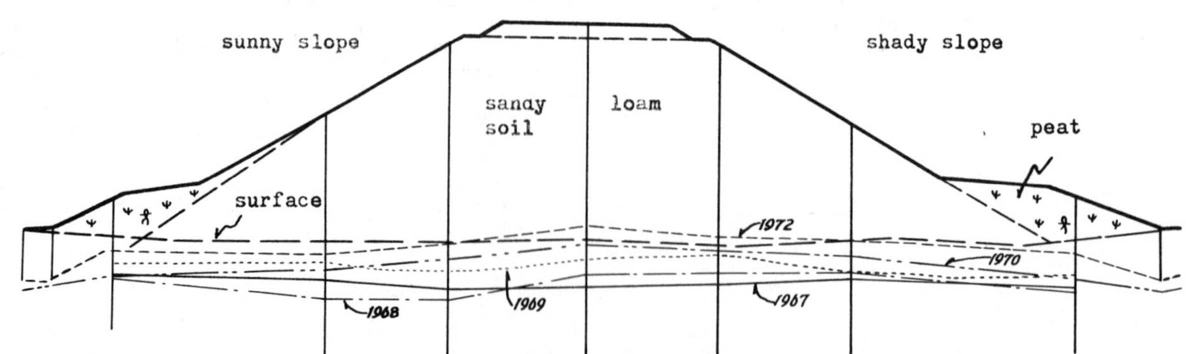

Figure 1. The fluctuations of permafrost table beneath the embankment

the permafrost table rises easily. In the center of the foundation it rises into the embankment filling soil. In permafrost zone in small scattered islands, therefore, permafrost soil is obviously in a state of degeneration and very unstable. After an embankment is built near the southern boundary, it is often difficult to preserve the permafrost soil, and the permafrost table in the embankment foundation and even beyond the slope foot on both sides is depressed considerably. After heat isolation measures are improved in the northern part of this zone, it has been possible to maintain permafrost table in the proximity of the natural table, sometimes with a slight rise above it. In the massive insular permafrost zone, since it is located between the afore-mentioned two zones, once appropriate measures are taken for heat isolation, permafrost table of the foundation will show a slight rise.

Effect of Surface Water

Embankment construction changes the natural condition of surface water flow. When drainage facilities are poor and water accumulates at the foot of the slopes, horizontal seepage often occurs in the foundation. Precipitation in this zone is often concentrated in the comparatively hot months of June through September, and with it a large amount of thermal energy is accumulated. When water seeps through the foundation, there occurs heat exchange between heat released by the water and heat absorption by the frozen soil of the foundation, the result being a drop in the permafrost table. As shown in Figure 2, at the foot of the slope on the left side of the embankment is a big water pit as water seeps through the foundation, considerable lowering of the permafrost table results.

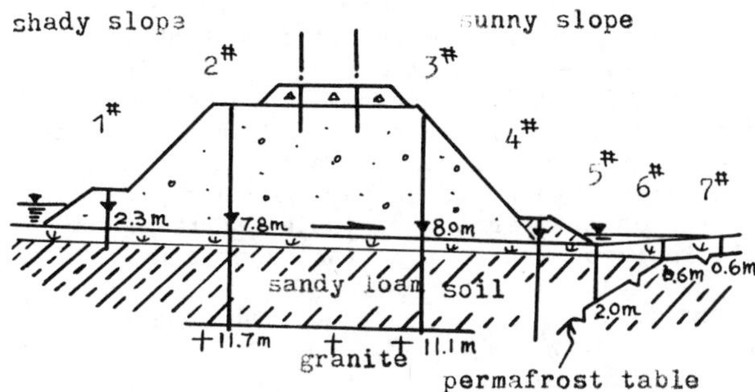

Figure 2. Permafrost table goes downward when surface water passes through the lower part of the embankment

Effect of Orientation, Snowcover and Wind

The direction or run of the line determines the orientation and wind-facing side of the slopes on both sides of the embankment. Therefore, each side is subjected differently to the effect of sunshine, snowcover and wind. The radiation heat of the sun is the principal heat source of permafrost soil. With increasing latitude toward the north, the angle of sunshine is limited; thus there is considerable difference in sunshine time between the sunny and shaded sides of the embankment.

After erection of the embankment, there is a great difference in temperature between the wind-facing and lee sides. When the direction of the line runs NE, cold air carried over by northwest wind in winter is stopped on the shaded side, thus hastening the dissipation of heat of that side. The wind blowing on snow causes great changes in the thickness of accumulated snow at the foot of the slope. Under the combined effect of the above factors, corresponding changes occur in the permafrost table below the embankment. On the line running NS, the difference between east and west slopes is not obvious, so permafrost table of embankment foundation remains a generally stable straight line with little fluctuations. However, when the run is close to EW, the difference between the two sides is quite obvious, as seen from the fact that ground temperature in both the embankment and the foundation is higher on the sunny side than that on the shaded side. Thus, permafrost table variation is higher on the shaded side than on the sunny side. Generally, on the shaded side it either rises or remains constant, while on the sunny side it drops, as shown in Figure 3. Apart from orientation and snowcover, the direction of wind also has a certain effect. In winter, northwesterly winds are quite frequent, and the windward side of the embankment has a temperature which is correspondingly lower

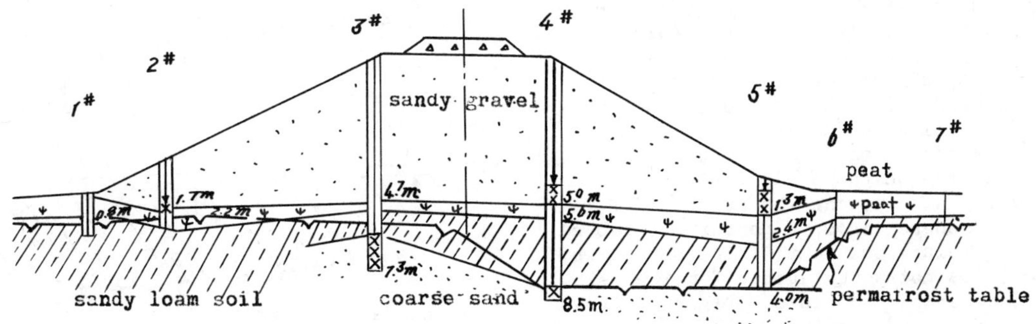

Figure 3. The depths of permafrost table vary with the orientation of the embankments

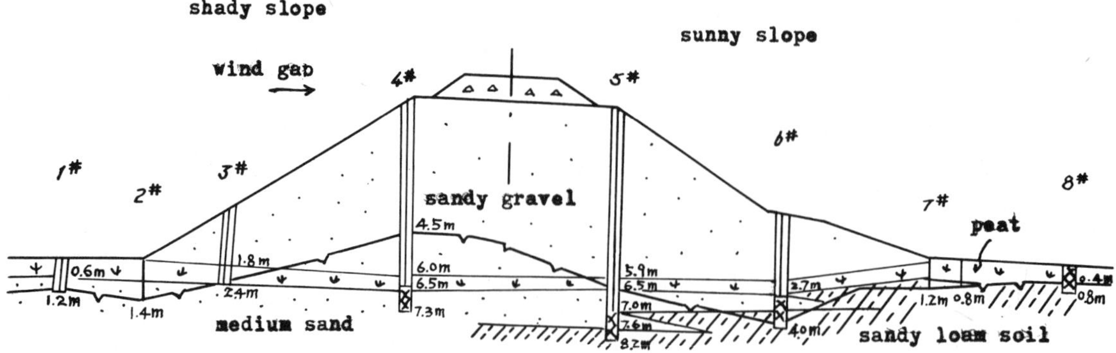

Figure 4. The permafrost table goes upward beneath the shady side of the embankment where cold wind always blows seriously

than that on the other side of the embankment. The Luenlin line at a certain point is located where northwesterly winds are strongest and most intense. The shaded windward slope has no heat isolation berm at the foot of the slope. Here the permafrost table rise into the embankment filling soil is as high as 1.5 meters. It is obvious that northwesterly wind is stopped on the shaded side of the slope, playing an important part in hastening the dissipation of heat, as shown in Figure 4.

Effect of the Height of Embankment

The permafrost table varies widely below embankments of different heights. In the case of embankments of ordinary height, once protection is taken, it is possible to stabilize the permafrost table in line with the natural table, sometimes showing a slight rise. This level is generally close to the original ground, or rises slightly into the embankment filling soil (with the exception of the zone adjacent to the southern boundary). In case of failure to take heat isolation measures on the sunny side, the permafrost table on this side often drops to a certain extent.

Low-fill embankment (usually heights of less than 1.0-1.5 m). Due to compression of surface vegetation and peat layer, there is a reduction in the heat conductivity. Both the radiation heat of the sun and the kinetic energy of the moving train are easily transmitted downward, resulting in a stronger heat exchange than is the case with embankments of more height, causing the permafrost table to drop. In a survey of the Yalin line, conducted in 1965 on the lowering of permafrost table at 16 points of the foundation (the height of fill at 11 places is $\leqslant 1.5$ m, while at 5 places the height of fill is $\leqslant 1.0$ m), a general drop in permafrost table was observed. This means that if the embankment is too low, it is not advantageous for the maintaining of the permafrost table. Figure 5 shows the sections of two adjacent low fill embankments somewhere on the Yalin line (located in a discontinuous permafrost zone). In both cases, there is a drop in the permafrost table. In a low fill with a small area exposed to heat in the entire embankment, the difference between the amount of solar radiation absorbed on each side is not very great, resulting in little variation in permafrost table on both sides of the foundation. The general observation is: the lower the fill, the more likely the permafrost table will drop. Among the different permafrost zones, the drop is most likely to occur in the permafrost zone of small scattered islands.

Embankments with a high fill (usually heights of over 10 m). In such cases, there is an obvious drop of the permafrost table in the foundation. Where the height of embankment does not exceed 18 m, the range of decline is about 1.0-2.5 m in permafrost zones of small scattered islands. When the difference in the level of the permafrost table in the foundation is obvious between the sunny and shaded slopes, decline of the permafrost table on the shaded side is less than that on the sunny side. When

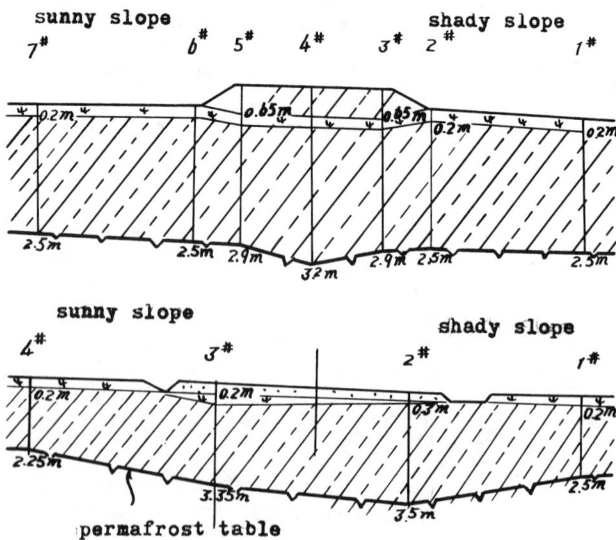

Figure 5. Low-filling embankments cause permafrost table of the foundation to decline

the difference between the two sides is not significant, decline of the permafrost table on both sides in approximately the same. The decline of the permafrost table below high embankments is due mainly to the construction usually being carried out after the spring thaw, when embankment filling carries with it a large amount of heat which is continuously emitted toward the foundation. In winter, though the embankment surface is frozen, its interior remains as loose soil which maintains a temperature always above the freezing point throughout the year. Since the frozen soil of the foundation continually absorbs heat released by the embankment soil, the permafrost table gradually declines under the effect of heat exchange, and then stabilizes at a certain depth after it has reached a new state of thermal equilibrium. In addition, the pressure of a high embankment also has a certain effect on thawing of frozen soil. Once the upper critical height has been exceeded, the higher the fill, the greater the decline of the permafrost table. Among the different permafrost zones, a zone of small scattered islands shows the greatest decline of the permafrost table. The permafrost table will decline more when the embankment is constructed during a warm season than during a cold season.

Effect of the Type of Filling Soil

There are different types of filling soil for embankments, each having a different heat conductivity. This difference depends on the inconsistency of the thermal physical condition of the soil. For example, when clayey loam soil is used, the fill is usually more compact and has a small air volume. Heat conduction is essentially carried on by grains of the soil, and the rate of heat conduction is slow. On the other hand, when coarse-grained soil is used for fill, the porosity is usually higher. Air in the crevices easily produces convection and thus enhances thermal conductivity, resulting in the transmission of thermal energy to a greater depth. Tests of embankments show that, before August in the second year of soil filling, thawing occurs in the sandy soil each month to the extent of 1.0-1.4 m, while thawing in the sandy loam soil is 0.4-1.3 m. The monthly rate of freezing in winter is 1.0-1.5 m for sandy soil, and 0.5-1.1 m for sandy loam soil. Temperature is also lower in loam soil than in sandy soil. The permafrost table will rise again more obviously in a loam soil fill than in a sandy soil fill. Where poor drainage leads to seepage through the foundation, the use of soil for fill may prevent or reduce seepage of surface water into the foundation, thus protecting permafrost soil from the effect of water seepage. However, when drainage is good and under the combined effect of other factors, the permafrost table below different types of soil for embankments of ordinary height may be stabilized in close proximity of the natural table, or with a slight rise above it.

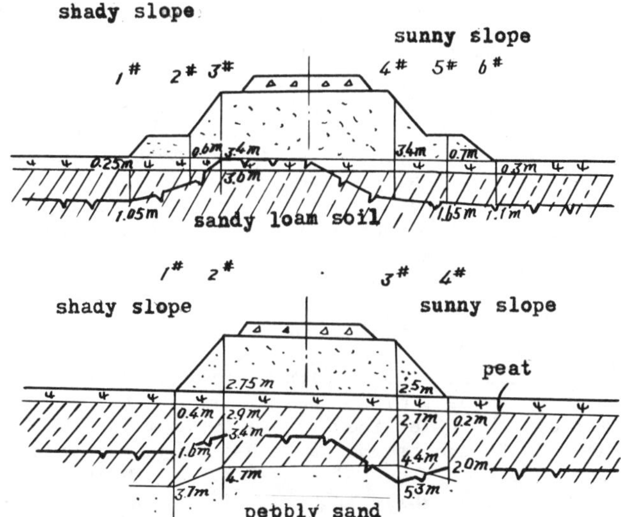

Figure 6. The difference of permafrost table depth between the embankments equipped with and without isolation berm

Effect of Heat Isolation Berm

The function of heat isolation berm is to reduce the effect of temperature produced by thermal conduction to permafrost soil, prevent decline of permafrost table on the sunny side, moderate the rising of permafrost table, and prevent damage to the surface at the foot of the embankment slope by human activity.

Sections of two permafrost soil embankments at a certain point on the Luenlin line, one with heat isolation berm and the other without, are compared in Figure 6. They are located in the permafrost zone of small scattered islands, separated by a distance of 96 meters, essentially similar in geological condition, filling soil, and running direction of the line. The difference in permafrost table of the foundation on the shaded slope side of the two sections is: the one with berm rises 1.5 m and the one without berm rises 0.7 m. On the sunny side, the one with berm rises 0.2 m and the one without berm drops 1.3 m. This shows that heat isolation berm has the effect of preventing the decline of permafrost table in the foundation.

Effect of Vegetation Peat Layer

The vegetation and peat layer is a good heat isolation material. Under natural conditions it has an obvious effect on the soil freezing and thawing depth. It diminishes the effect of air on soil surface, reduces the amount of heat entering into the soil layer, thus giving protection to the permafrost. Therefore, damage to the vegetation and peat layer near the foot of slope and foundation, or compression of its thickness by the embankment, in addition to the peat layer for heat isolation in the foundation, will cause variations in the permafrost table. When the surface vegetation and peat layer is damaged or removed, permafrost table declines. After construction of the embankment, the vegetation and peat layer is subjected to compression and thus becomes thinner and more compact with lower porosity. This condition is not favorable for preservation of the permafrost, since the peat layer heat isolation properties are greatly reduced.

CONCLUSIONS

1. When embankment construction is carried on during the thawing season, the natural permafrost table declines to the maximum thawing depth. In the second year, at the center and on the shaded slope side, permafrost table begins to rise, while there is still some slight decline on the sunny side of the slope. In the third year, permafrost table shows an overall rise. After a period of 5-6 years, the permafrost table has become relatively stable.

2. Variations in the permafrost table of the embankment foundation are closely related to latitudinal zonality and are dependent on the height, orientation, thermal physical properties of the fill material, heat isolation measures adopted, surface formation after construction, hydrologic conditions, characteristics of frost soil medium, etc.

3. Water accumulated at the foot of embankment slopes and intense human activities are the main causes of considerable decline of permafrost table in the embankment foundation and the continuous sinking of the embankment. Therefore, the principles of design for preservation of permafrost soil is to make surface drainage more effective and to protect the natural permafrost table near the foot of embankment slopes.

4. The permafrost table can rise into the embankment filling soil only a small distance. While it rises on the shaded side of the slope, it drops on the sunny side of the slope. Therefore, special attention should be paid to heat isolation on the sunny slope side, for which the primary approach is to build a heat isolation berm.

5. Both low and high embankments cause permafrost table in the foundation to decline. The higher the fill, the greater the decline. To prevent the permafrost table from degradation, the lower critical height of the embankment is 1.5-2.0 m. The upper critical height of the embankment should be determined based on local temperature, fill material, mean annual ground temperature, and construction technology.

MECHANISM AND LAWS GOVERNING THE TRANSFORMATION OF UNCONSOLIDATED ROCKS UNDER THE INFLUENCE OF REPEATED FREEZE-THAW

P. S. Datsko, E. Z. Kouchukov, and Yu. P. Akimov

Geology Department
Moscow State University, USSR

A comprehensive procedure was used to study the mechanisms of the transformation of unconsolidated rocks as the result of repeated freeze-thaw (up to 1,000 cycles) under various external thermodynamic conditions. The latter ensured freezing of samples from both one side and from all sides and also the formation of various types of cryogenic textures. It was possible to establish trends with regard to changes in dispersivity, chemical and mineral composition of the materials, cryogenic structure, plasticity and mechanical properties, depending on their original composition and structure. It was demonstrated that a reduction in the dispersivity of clay soils occurs with an increase in freeze-thaw cycles due to an increase in the amount of secondary silt and secondary sand particles. The most pronounced structural changes are associated with the differentiation of the frozen material into ice and mineral material, thus greatly enhancing the material's heterogeneity. This leads to the formation of a specific structure in terms of porosity and encourages changes in the content of liquid H_2O, lowers the strength and conductivity of the material and increases its susceptibility to saturation and erosion.

The problem of cryogenic transformation of rocks has long been of profound interest to researchers. This is understandable when one considers that solution of the problem will provide the key to the long-term forecasting of cryogenic processes and phenomena. Transformation of dispersed rock/soil under conditions of multiple freeze-melt cycles occurs when interrelated physical, physico-chemical, and chemical processes take place simultaneously, accompanied by changes in the dispersivity, in the chemical and mineral composition of rocks, and in their structure and properties. Conflicting experimental evidence shows that the problem is complex, and that the physico-chemical and chemical aspects of cryogenic weathering have not been adequately investigated.

Results of an all-around cyclical freezing and melting of dispersed rocks and soils have revealed specific transformation features of sand and clay soils. Water-saturated sands are transformed by physical disintegration, which increases the dispersivity, the specific active surface, and the stability of the system. It has been established that repeated freezing and melting tends to reduce the heterogeneity of bi- and poly-dispersed sands, while the non-homogeneity of mono-dispersed sands increases. For example, clean fine-to-medium grained quartz water-saturated sand (sampled at Lyubertzy, Moscow Region) after 50 freeze-melt cycles becomes medium to-fine grained and does not change again, even after 1,000 cycles. The average size of sand particle ($\bar{\alpha}$) becomes smaller, from 400 to 250 μ, whereas the root-mean-square variation in the size distribution of particles (σ) increases from 0.23 to 0.32, indicating the greater heterogeneity of the material. The intensity of dispersion is much higher at the early stages of the test, demonstrating the subsiding nature of the process of disintegration and points to the formation of a near-uniform size of quartz grains under the given thermodynamic conditions. It has taken artifically prepared clean mono-dispersed water-saturated fine-grained quartz sand almost as many cycles (887) to become coarse dust (Minervin and Komissarova, 1979). Many researchers claim that this particle size is critical for quartz, being resistant to cryogenic exposure.

A critical size of cryogenically disintegrated minerals is determined by their crystallo-chemical structure, and, according to Konischev (1981), these sizes can be arranged in a single series: minimum size—amphiboles, pyroxenes and quartz; maximum size—biotite and muscovite; medium size—feldspars; this series indicates the changing degree of dispersion of rocks and soils due to freeze-melt cycles.

This series is consistent with the predominance of quartz particles (0.05-0.01 mm) in cryogenic eluvium. Quartz particles that are smaller display relative resistance to physical disintegration and function as centres of cryogenic aggregation in clay soils, producing secondary silty particles (Poltev, 1972).

Cryogenic aggregation of silty, clay, and colloidal particles, resulting in an accumulation of secondary particles of a size smaller than sand, plays a dominant role in changing the degree of dispersion of water-saturated clay soils as a result of freeze-melt cycles (Figure 1). A good illustration of this effect is the change in the degree of dispersion of a polymineralic clay (m P_2 kv), which, in the course of repeated freezing and melting, showed an increase in the quantity of particles the size of fine sand to coarse dust. Also, medium sized particles increased from 6μ to 13μ while the r.m.s. variations decrease. These results indicate that the particle size becomes more homogeneous. The dispersion transformation, as a function of the frequency of freeze-melt cycles, is also of a subsiding nature. Thus, rock (soil) as a system reaches a state of equilibrium with the thermodynamic conditions of the environment; i.e. a composition of rock resistant to further cyclical cryogenic effects develops.

We have applied methods of mathematic statistics to process the results of our research and the findings of some foreign authors. Figure 1 shows the direction of change in the dispersion of rock (soil) resulting from repeated freezing and melting. The graph clearly indicates that the particles of most sandy and clay soils tend to develop a mean diameter of about 10-30 μ and an r.m.s. variation close to 0.7, which is typical of loess soils. This is further evidence that cryogenic weathering of dispersed material leads to the formation of loess.

Heavy monomineralic clays such as bentonite and kaolin occupy a special position. For example, the degree of kaolin dispersion (eN) increases in response to an increase in homogeneity of granulometric compositions, characterized by a reduction of the r.m.s. variation from 0.72 to 0.64. The change in granulometric composition between 0 and 150 cycles consists of a substantial decline of the coarse dust fraction (up to 40%) owing to its reduction to the size of a clay fraction. On the whole, kaolin dispersion may be described as a characteristic transformation, i.e. the growth of the clay fraction.

Easily-soluble compounds that form part of rock (soil) are subject to fundamental changes as a result of multiple freeze-melt cycles. When judged by the degree of salinity, evaluated on the basis of the dry residue in the water extract, even the classification of the material may change. For example, non-saline kaolin subjected to six freeze-melt cycles becomes slightly saline; a similar treatment turns heavily saline bentonite to excessively saline, while sea clay registers a transition from heavily- to moderately saline.

In general, the quantitative changes regardless of their direction or sign can be said to subside in proportion to the content of unfrozen water in the material, together with a decline in salinity and a reduction in the specific active surface area, in accordance with the mineral composition of the material: montmorillonite > hydromica > kaolinite. The dynamics of these changes corresponds to that of chemical reactions occurring in the material, and is determined by the tendency of rocks (soils) to enhance their resistance to freeze-melt exposure. As with most chemical reactions and solutions, the rate decreases as the steady state is approached. These effects were clearly manifested in the materials studies, in which the rate of the water-extract dry residue transformation during the first six freeze-melt cycles was found to be 10 or more times the rate recorded in subsequent cycles.

When considering the direction of changes in the content of easily-soluble compounds, one should note the growth of such compounds in monomineralic heavy clay soils increasing the salinity, whereas in other materials there is a steady decrease from cycle to cycle. This decrease in the majority of soils is caused by several factors. One factor is the formation of a sizable quantity of crystal hydrates that accompanies the freezing of moisture. For example, the formation of gypsum from anhydrous calcium sulfate, as in the case of polymineralic sea clay (m P_2 kv), leads to a further reduction of sulfate ions and calcium cations in the water extract. Another factor is the aggregation of colloidal particles at the expense of a reduced thickness of unfrozen water film. Because the solubility of such aggregates is diminished, they are not present to a large degree in the water extract.

In heavy monomineralic clays aggregation may result in increased salinity because of the very small thicknesses of films of unfrozen water, which facilitates active water flow. Positively-charged salt colloids, present in an aggregated state, are only slightly soluble. However, as pH goes down and the temperature in the negative region decreases, the colloids increase their charge and disperse, becoming more soluble. Thus, differences in the thickness of unfrozen water films owing to different particle sizes appear to lead to qualitative changes in the chemical and mineral composition of rocks and soils which are subjected to cyclical freezing and melting.

Substantial changes may occur in the mineral composition of the clay fraction. X-ray diffraction analysis and thermoanalysis indicate that after 5 and 50 freeze-melt cycles montmorillonite, hydromica, and kaolinite undergo a fundamental structural transformation in which mixed-layers and minerals with a mobile crystal lattice are produced. With montmorillonite, interlayer spaces d(001) become larger. Montmorillonite that forms part of the clay fraction of polymineralic sea clay (m P_2 kv), shows an increase of the inter-layer space from 1.227 to 1.550 nm as cations of Na^+ are substituted by Ca^+ cations.

The structure of soils (rocks) also undergoes considerable transformations. For example, following a repeated freezing and melting of polymineralic sea clay, the pore space of the frozen material is differentiated owing to the significant growth of the volume of micropores whose radius measures below 0.2 and over 0.5 μ. The number of pores within this band increases, with the mean size of the intraaggregate pores diminishing from 0.20 to 0.15 μ and the homogeneity increasing. The contraction of the intraaggregate pores is due to the "compaction" of clay aggregates and to stronger structural bonds between the particles that constitute the aggregates. The size of large pores, that belong to interaggregate porosity, is a function of the size of the microaggregates themselves. An observed growth of the volume of the pores with a radius over 0.5 μ, and the increase of their percentage in total porosity, are accounted for by the growth of the number of medium-size particles, by more even distribution of microaggregates, and by weak bonds existing between them.

Before freeze-melt cycle treatment, the microstructure of polymineralic sea clay (m P_2 kv) is characterized by the presence of aggregates consisting of dispersed particles measuring from 2 to 5 μ that are uniformly scattered throughout the volume. The degree of aggregation is not high, the aggregates being separated from one another by ice; yet there are no clear inclusions of ice, which would be represented by fine dispersed particles less than 1 μ in size. It takes only 5 freeze-melt cycles for the mineral particle aggregates to grow to 30 μ. A process is taking place wherein the ice rids itself of the finely-dispersed mineral particles diffused in it, i.e. the ice undergoes purification. A similar transformation of microstructure occurs with other kinds of polymineralic and monomineralic frozen rock (soil). Whatever the material, a process of matter differentiation is always involved which is due, on the one hand, to increasing aggregation of the soils and the elimination of ice inclusions from the aggregates as they undergo compaction. On the other hand, the process involves ice concentration, with the number and size of ice sections and crystals increasing. Because of the differentiation into ground and ice, the heterogeneity of frozen rock (soil) as a system increases.

The above-described transformations of material composition and of the porous space structure lead to observable changes in the content of unfrozen water in dispersed rocks (soil) (Vrachev and Datsko, 1980). The amount of unfrozen water in the temperature range from -1 to -10°C increases, while below -10°C it decreases. These changes are caused by the contraction of intraaggregate porous space which determines, along with the specific active surface, the content of the liquid phase of moisture in frozen material. The water contained within interaggregate pores (those over 0.5 μ in radius) freezes at negative temperatures above -0.3°C. It follows, therefore, that an expansion of the interaggregate pore volume by multiple freezing and melting of the material results in the freezing-out of water contained in these pores even at -0.3°C, resulting in a smaller quantity of unfrozen water at temperatures ranging from -0.3 to -1°C.

Transformation of clay soil dispersivity and chemico-mineral composition affects the plasticity properties of clays. Thus, it was typical of all clay soils analyzed (Figure 2) that moisture content in the upper plasticity range was reduced by 3 and 7% for sand loam and loam, respectively, and by 12% for clay. At the same time, the plasticity index for loam decreased from 17 to 10%; in the case of sand loam this index was approximately halved.

Transformation of rock (soil) composition and structure impairs the strength characteristics. The strength of melted samples of polymineralic sea clay (m P_2 kv) decreases from 0.14 MPa to 0.09 MPa when the number of freeze-melt cycles reaches 50. In this case, the coefficient of aggregation of particles smaller than 0.001 mm in size has been found to increase from 1.4 to 1.7, and for particles smaller than 0.005 mm it has increased from 3.1 to 3.8. Such changes of natural dispersivity determine the minimum strength of melted ground after it has been repeatedly subjected to freezing and melting.

The nature of contact interaction influences strength also. The intraaggregate contacts determine only the strength of particular aggregates and their water-resistance characteristics. The strength of the disperse system as a whole depends on the stability of molecular and electrostatic bonding between the aggregates. Therefore, the main type of contact in this case is by coagulation. As the frequency of freeze-melt cycles increases, the number of such contacts reduces because of the larger size of aggregates and eventually leads to a decreased strength of the rock (soil) in a melted state. There is also a lowering of the strength of frozen material (Vrachev and Datsko, 1980) due both to the increased content of unfrozen water and to the specific transformation of the frozen microstructure.

The intensity of rock (soil) transformations is determined by the inherent and external thermodynamic conditions. For example, the intensity of transformation of dispersivity of polymineral sea clay (m P_2 kv) with different initial moisture contents increases progressively from W_r to $1.5W_L$. The most intensive dispersion occurs in samples having maximum moisture content because changes in the thicknesses of adsorbed and unfrozen moisture films in the course of freezing and melting are the greatest. On the other hand, in water-saturated rocks (soils) most of the water turns to ice, whereby maximum crystallization pressure develops which encourages irreversible coagulation.

Clay soils were tested in the temperature range from -5 to +20°C for 48 hours, and in the temperature range from -50 to +50°C for 12 hours. Observed changes of dispersivity indicate that the nature of granulometric composition transformation is similar in clays and in rocks (soils).

The data showed that a repeated alternation of freezing and melting processes results in the formation of lens-type textures in clays and solid cryogenic textures in rocks (soils), while qualitative trends governing the transformation of the material composition and structure are basically the same. The intensity of transformations is more pronounced in the formation of lens-type textures as in this case a high rate of moisture- and mass-transfer is involved.

LITERATURE

Konischev, V. N., 1981. Formation of dispersed rock composition in cryolithosphere. Novosibirsk: Siberian Branch of the USSR Academy of Sciences, Nauka Publ.

Kouchukov, E. Z., and Datsko, P. S., 1979. Transformation of plasticity of dispersed rocks of different geologo-genetic complexes in the process of repeated freezing and melting, in Engineering and Construction Surveys, No. 3(55), Moscow, p. 57-65.

Mazurov, G. P., and Tikhonova, E. S., 1964. Transformation of rock/soil composition and properties in repeated freezing. Journal of LGU, No. 18, Leningrad, p. 64-72.

Minervin, A. V., and Komissarova, N. N., 1979. Formation of structure and texture of subsiding loess soils in Minusinsk intermontane trough. Engineering Geology, No. 1, Moscow, p. 70-83.

Morozov, S. S. et al., 1973. Transformation of loose rock composition and properties under protracted alternating freezing and melting, in Engineering Geology and Soil Science [Voprosy inzhenernoi geologii i gruntovedeniya], Moscow, MGU Publ., p. 140-156.

Poltev, N. F., 1977. Granulometric and microaggregate composition of dispersed rocks (soils) of the North, in Cryogenic research [Merzlotnye issledovaniya]: Moscow, MGU Publ., v. XVI, p. 85-88.

Vrachev, V. V. and Datsko, P. S., 1980. Differentiated evaluation of the strength of frozen rock. Journal of Moscow University, Ser. "Geology," no. 25, p. 73-80.

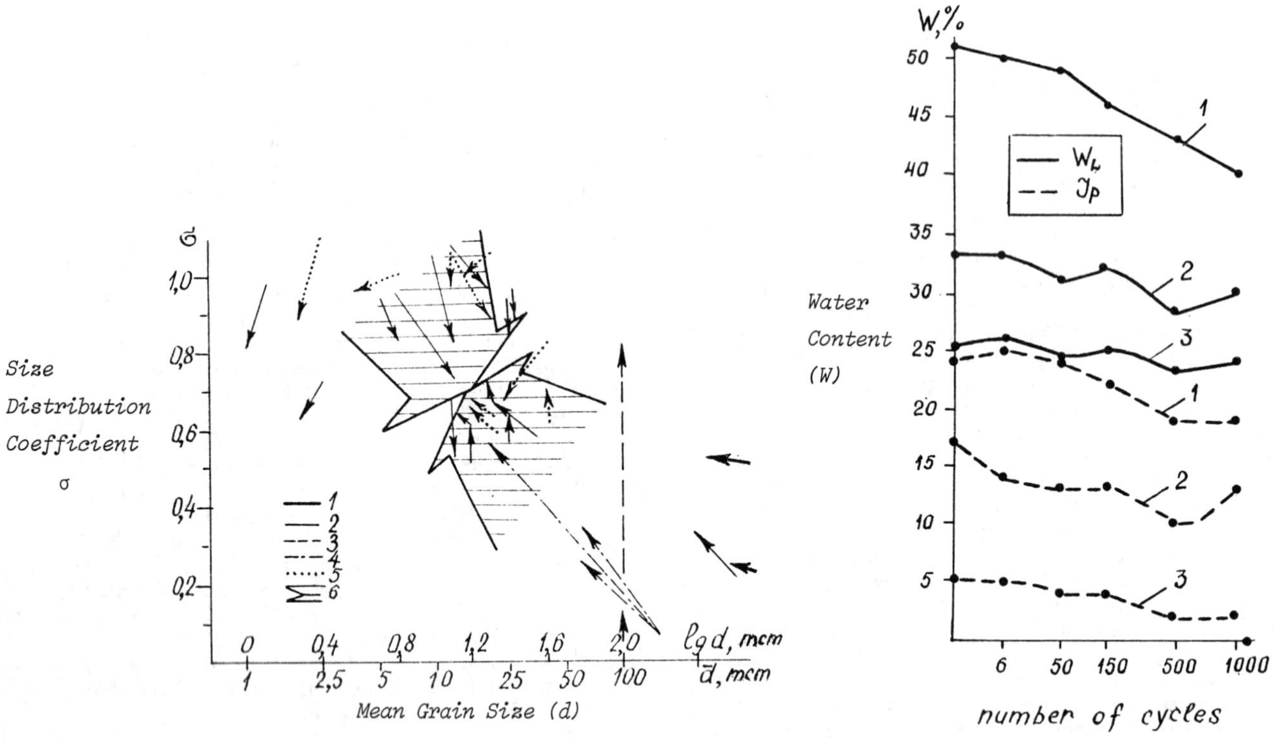

FIGURE 1 Transformation of dispersion or rocks (soils) subjected to multiple (from 50 to 1,000 cycles) freezing and melting, using the following sources: 1 - Mazurov, G. P., Tikhonova, E. S. (1964); 2 - our research; 3 - Konischev, V. N. (1981); 4 - Minervin, A. V., Komissarova, N. N. (1979); 5 - Morozov. S. S. et al. (1973).

FIGURE 2 Dynamics of transformation of plasticity indices in multiple freezing and melting of dispersed soils: clays (m P_2 kv) - 1, loam (Ag II) - 2, sand loam (gm III^1).

PHYSICAL NATURE OF FROST PROCESSES AND A RESEARCH METHOD

Ding Dewen

Lanzhou Institute of Glaciology and Cryopedology, Academia Sinica
People's Republic of China

The frost process, which includes changes in temperature and moisture fields, the formation of ice layers, and frost heave, develops under certain natural conditions, forming geological and geomorphological phenomena in frozen ground regions. The quantitative study of frost processes involves three aspects: the determination of state parameters, the establishment of a mathematical model by model test or system analysis, and the resolution of the numerical or approximate solutions to the model. Based on a group of combination equations related to heat and moisture flow, the frost-heave process, and the heat-water balance at the thaw interface, and on the principles of thermodynamics and the mathematical model, the author presents approximate solutions for calculating the depth of frost penetration, the amount of migrating water, unfrozen water content, and the amount of frost heave.

The prerequisite for the development of further studying the frost process quantitatively and experimentally is to reveal its physical nature. The frost process is composed of some comprehensive physical processes interrelated to each other closely, so that it must be described by a group of combination equations. Modelling test must follow similar conditions of the physical processes, so it is necessary for us to set up similitude criteria completely.

At present, the approximate analysis solution still has its life in the study of frost processes, especially for the open system.

It is worth while to give more attention to the application of thermodynamics to the study of frost processes because of its advantages in the calculation of the subject with phase change and frost heave.

PHYSICAL NATURE IN FROST PROCESSES

The exchange of energy and mass is actively undertaken between the surface layer of the earth's crust and the atmosphere, and between that and the underlying lithosphere, governing the thermodynamic condition of the surface layer itself. As the exchange reaches a critical state, the surface layer goes into a frozen state, and consequently forms a frozen surface layer and a series of phenomena caused by freezing. Therefore the frozen surface layer is a geological and geographical system.

Frozen soils are mainly composed of mineral particles, unfrozen water, ice (with plasticity) and gas. All the components are interrelated and interact with each other closely, resulting in a nonhomogeneous capillary-porous texture. Especially, as there are fine grains, free ions, polarized water molecules, and ice crystals in frozen soils, so we can consider the frozen soils as a physical-chemical system.

Frost action is a comprehensively physical and chemical process. Considering the chemical transformation in soils during freezing being insignificant, the author considers the frost process as only a comprehensively physical process, which includes the transformation of heat, the phase change of water, water migration, as well as various mechanical processes under certain conditions of loading, temperature, and moisture. Among which, the exchange of heat and the phase transition are primary. Each process is independent, and the law of its change depends upon its intrinsic properties. The important thing, however, is that these processes are interrelated to each other, resulting in a complicated feedback system, and governing the characteristics of the whole soil layer. Generally speaking, interactions within a process tend to keep its balance; on the contrary, the actions outside the process tend to destroy the balance. As a result, frost processes irreversibly evolve to an advanced stage.

The study of frost processes includes three main aspects, i.e., the power or mechanism, the state parameters and determination of their values, and the main laws or natures of the process. Above all, frost action is a natural-historical process, i.e., the frost processes proceed under a certain natural-historical condition, resulting in forming the geological and geomorphological phenomena in a frozen ground region. For a seasonally frozen layer, the frost processes include the changes of temperature and moisture field, the formation of the ice layer, and the frost heave process.

One of the most important advances of

current studies on cryopedology is to study frost processes quantitatively and experimentally. The study methods are as follows: Firstly, the state parameters of the process should be determined. For example, according to the thermodynamics view, the parameters of P, V, N, and T should be determined if we do not take into consideration external electrical (or magnetic) field action. Secondly, the mathematical model has to be established, i.e., the relationship between the time- and space-dependent parameters and their influenced factors must be established by means of the physical-mechanical laws so as to found a set of combining equations. And finally, one has to work out the solutions by experiments and analytical methods.

For the frost processes of a seasonally frozen layer, we give the following equations. Among which, the equations referring to the heat and moisture changes, respectively, are

$$\frac{\partial t}{\partial \tau} = a \frac{\partial^2 t}{\partial x^2} \tag{1}$$

$$\frac{\partial w}{\partial \tau} = k \frac{\partial^2 w}{\partial x^2} \tag{2}$$

the equation to the phase transition of water, and to ice or to the change of unfrozen water content is

$$dg_1 = dg_2 \tag{3}$$

and the equation to the frost heave process or the amount of frost heave is

$$dQ = du + pdv \tag{4}$$

where g, Q, and u are all functions of t and w, and the relations between t and w can be expressed by the heat-water balance equations at the frost-thaw interface.

$$\Sigma\lambda \frac{\partial t}{\partial x} = \left[\ell r_0 (w_h + k^+ \frac{\partial w^+}{\partial x}) + Q_c \right] \frac{dh}{d\tau} \tag{5}$$

$$w(h + \Delta h,\ \tau + \Delta\tau) = w_h + \Sigma k \frac{\partial w}{\partial x} \tag{6}$$

The comprehensive processes discussed above cannot be dealt with by an analytic method, but the model test is available. In the light of the nature of frost process, the theory of system analysis is perhaps the most prospective. At present, among the analytic ways with which the various processes can be dealt with independently, both the approximate solution and numerical solution are of great importance. The application of approximate solution will be discussed in the next section.

From eqs 1 and 2, we obtain $\partial\tau/\ell^2$ = idem; $k\tau/\ell^2$ = idem, and from eqs 5 and 4, we have $\lambda t\tau/Lrw\ell^2$ = idem. p/Lrw = idem.

According to the "π" theorem on dimensional analysis, there are nine physical quantities and five dimensions in the above equations altogether, and, therefore, the four criteria above are complete and independent. From the criteria, we can obtain the following equations of similitude multiple:

$$C_\ell^2 = C_a C_\tau, \quad C_\ell^2 = C_k C_\tau,$$

$$C_\ell^2 = C_\lambda C_t C_\tau / C_\ell C_r C_w, \quad C_p = C_\ell C_r C_w.$$

Because of the above four restraining conditions, the similitude multiples have only five variables, and the other three are determined by the above-mentioned conditions.

If the model experiments are performed on the undisturbed soil samples, then we have

$$C_\lambda = C_a = C_k = C_r = C_\ell = 1 ,$$

and from the above, we obtain

$$C_t = 1, \quad C_w = 1, \quad C_\tau = C_\ell^2 .$$

The other similitude multiples can be determined by the dimensional combination method. Here, we only give the similar multiple for stress C_p=1, and that for water flow C_i=$1/C_\ell$.

THE APPROXIMATE CALCULATION OF FROST HEAVE

Determination of Frost Penetration Depth and Amount of Migrating Water

If we take a section of soil layers from the ground surface (X=0) to the top of ground water capillary zone (X=H) as a system, then, according to the nature of transferring heat and moisture in the system, it can be divided into the three interrelated subsystems, i.e., the frozen soil section ($X=X_k$), the freezing soil section (X_k-h) and the thawed soil section (h-H).

To make the subject simple, the author proposes the premises as follows

1. The whole system is homogeneous, the heat transmission takes conduction as the main way, the thermal effect during water migration is reflected only in the latent heat of phase transition of water, and, at the same time, the heat exchange within the frozen soils is taken into consideration.

2. The temperatures at the upper and lower boundary of the system, t_0 and t_H, are taken their average values during the freezing process; the temperature at the freezing front, t_h, is 0°C, and the change of temperature in both the frozen and thawed regions (t^- and t^+) follow the linear laws:

$$t^- = -t_0/h.\ (x-h);$$

$$t^+ = t_H/\ (H-h).\ (x-h). \tag{7}$$

3. Moisture migration takes place in both the freezing and thawed soil regions. The upper boundary of the freezing region, X_k, is determined by the level of the temperature t_k, which corresponds to the critical value (w_k) of unfrozen water in the temperature field

$$X_k = (t_o - t_k)/t_o \cdot h.$$

The water content (w_H) at the lower boundary is a constant, and the water content at freezing front is $w_h = w_o - w'$, where w_o is the initial water content, and w' the amount of water flowing out of the system during thawing. The bonding water content (w_{kp}) of the soils is subtracted from the total water content. The water contents in the two regions, w^- and w^+, change linearly as follows

$$w^- = w_k + (w_h - w_k)/(h - X_k)\cdot(X - X_k);$$

$$w^+ = w_H - (w_H - w_h)/(H - h)\cdot(H - X). \qquad (8)$$

Under the above premises, the balance of heat and moisture at the freezing front can be described by the equations

$$\bar{\lambda}\frac{\partial t^-}{\partial x} - \lambda^+\frac{\partial t^+}{\partial x} = \left(Lr_o(w_h + \Delta w) + 1/2C^-\left|t_o\right|\right)\frac{dh}{d\tau}$$

$$K^+\frac{\partial w^+}{\partial x} - K^-\frac{\partial w^-}{\partial x} = \Delta w\frac{dh}{d\tau} \qquad (10)$$

considering eqs 1, and 2, and letting

$$a = \lambda^-\left|t_o\right| ; \qquad b = \lambda^+ t_H + Lr_oF;$$

$$C = r_o w_h + C^-/2\cdot\left|t_o\right| ;$$

$$E = K^-\left|t_o\right|/t\cdot(w_h - w_k);$$

$$F = K^+(w_H - w_h).$$

we obtain the equation describing the rate of frost penetration from eqs 9 and 10

$$dh/d\tau = \left(aH - (a+b)h\right)/h(H-h)C \qquad (11)$$

Integrating eq 11, we obtain the equation for determining the depth of frost penetration

$$h = -C\left(\frac{bH\cdot h}{(a+b)^2} - \frac{h^2}{2(a+b)} + \frac{abH^2}{(a+b)^3}Ln\left(1-\frac{a+b}{aH}\right)\right) \qquad (12)$$

and that for determining the amount of migrating water

$$Q = \Delta w r_o = r_o C\frac{(E+F)h - EH}{aH-(a+b)h} \qquad (13)$$

From eqs 9 and 10, it can be seen that the applicable range of eqs 12 and 13 is

$$E/(E+F)\cdot H \leq h \leq \lambda^-\left|t_o\right|/(\lambda^-\left|t_o\right| + \lambda^+ t_H)\cdot H \qquad (14)$$

If H is very great, and the moisture migration in the freezing region is not taken into consideration, we have the revised Stephen formula:

$$h = \left(2\lambda^-\left|t_o\right|\tau/(Lr_o w_o + 1/2C\left|t_o\right|)\right)^{1/2} \qquad (15)$$

The condition forming the ice layer can be analysed by eqs 12 and 13. From the full condition forming the ice layer, $r_o w_h + Q \geq r_o w_b$, we can obtain the location where the ice layer is formed.

$$h^1 \geq \left\{a(w_b - w_h) + CE\right\}/\left\{C(E+F) + (a+b)(w_b - w_h)\right\}\cdot H \qquad (16)$$

and, furthermore, obtain the thickness of the formed ice layer

$$L = r_o/\bar{r}_w\ 1.09Cr_o/r_w\int_{h'}^{h}\frac{(E+F)h - EH}{aH-(a+b)h}dh + \int_{h'}^{h}(0.09w_h - w_B)\,dh = r_o/\bar{r}_w\Big((w_B - 0.09w_h - 1.09C\cdot\frac{E+F}{a+b})(h'-h) + 1.09C\left(\frac{EH}{a+b} - \frac{(E+F)a^H}{(a+b)^2}\right)\cdot\left(\frac{aH-(a+b)h}{aH-(a+b)h}\right)\Big) \qquad (17)$$

<u>Determination of Unfrozen Water Content and Amount of Frost Heave</u>

As for the problem of phase change of water in frost process, from the point of view of cryopedology, it is considered as the problem of the formation and distribution of ice, and from that of physics, it is considered as the problem of the balance and transformation of bonding water, unfrozen water and ice in the soil system. This problem is discussed under following premises: Soils system is homogeneous, ice is transformed from pure water (including the migrating water), and the change of gasses in this process is negligible. So the ice in pores of this system possesses the same properties when it is in the free state. The nature of unfrozen water is governed by the features of particle surface, and is characterized by magnetic (thermodynamic) parameters. Therefore, the discussed problem is turned into the problem of phase transformation of water-to-ice in this case without external electrical or magnetic field.

According to eq 3, we have the following relation,

$$(\partial g_1/\partial T)_{PN}dT + (\partial g_1/\partial P)_{TN}dP + (\partial g_1/\partial N)_{TP}dN = (\partial g_2/\partial T)_{PN}dT + (\partial g_2/\partial P)_{TN}dP + (\partial g_2/\partial N)_{TP}dN \qquad (18)$$

According to the well-known thermodynamic relation, eq 18 can be written as

$$(S_2 - S_1)\,dT + (V_1 + V_2)\,dp + (\mu_1 - \mu_2)\,dN = 0 \qquad (19)$$

For the temperature and applied pressure being constant, we have

$$(\partial N/\partial P)_T = (V_1-V_2)/(\mu_2-\mu_1) = -\Delta V/\Delta\mu; \quad (20)$$

$$(\partial N/\partial T)_P = (S_2-S_1)/(\mu_2-\mu_1) = \Delta S/\Delta\mu \quad (21)$$

Since $\Delta S = \Delta H/T$, and $\Delta\mu = RT\, L_n(P_2/P_1)$, we obtain

$$(\partial N/\partial T)_P = \Delta H/RT^2 L_n(P_2/P_1); \quad (22)$$

$$(\partial N/\partial P)_T = -\Delta V/RT\, L_n(P_2/P_1) \quad (23)$$

In eqs 22 and 23, because of $\Delta H<0$, $P_2<P_1$, and $\Delta V>0$, so $(\partial N/\partial T)_P>0$, $(\partial N/\partial P)_T>0$. Thus, unfrozen water content increases with the temperature or pressure increasing, i.e., in this case ice is transformed into unfrozen water.

Integrating eqs 22 and 23, we can obtain the relations for determining the unfrozen water content (or ice content) in soils at a certain temperature and pressure. At present, however, there are some difficulties in the experimental determinations of a few necessary parameters.

If a soil layer suffered to freezing, because of the moisture redistribution (migration) and ice formation, the volume and density of the soil layer must have changed. It is called the frost heave of soils, and the corresponding frost heave increment in the vertical direction is so-called the amount of frost heave.

The increase of ice volume is the necessary condition for producing frost heave, and the full condition is that the increase of ice volume must be greater than the sum of the decrease of pore volume in the soil and the compressive amount of the underlying thawed soils during freezing.

For a semi-infinite ground body suffered an external load of P, if we take the ice within the soil layer with the thickness of dh, which is taking place the phase change as a system, and the upper and lower soil layers as surroundings, the first thermodynamic law must take the form of eq 4.

where $$du = \Delta u r_o\,(w_h + \Delta w)\,dh \quad (24)$$

in which, Δu is the intrinsic energy increment of the unit weight of water during phase change;

and $$dQ = \Delta H r_o\,(w_h + \Delta w)\,dh \quad (25)$$

in which, ΔH is the entropy increment of unit weight of water during phase change.

ΣpdV is the sum of the volume expansion work, and it includes the following two parts:

p_1dV_1 is the expansion work of the system done by overcoming the soil weigh the applied, load, including the compressive deformation work of the underlying soil layer. It takes the form of

$$p_1dV_1 = \left(p + r_o\,(1 + w_h + \Delta w)\right)\,d\ell \quad (26)$$

in which, ℓ is the deformation of the soil layer, i.e., the amount of frost heave of the soil layer.

P_2dV_2 is the expansion work of the system done by expanding to the pore of soil. It states

$$P_2dV_2 = \left(P + r_o\,(Hw_h + \Delta w)\right)\Delta n dh \quad (27)$$

in which, Δn is the change of pore volume of the soil during freezing, and we have

$$\Delta n = -\left(\ell/h - (1.09\Delta w + 0.09w_h)\right)$$

Substituting eqs 24 - 27 in eq (1-4), we obtain the relation for determining the frost-heaving rate.

$$d\ell/dh = \frac{(\Delta U+\Delta H)\ r_o\ \ (w_h+\Delta w)}{p+r_o\ \ (1+w_h+\Delta w)\ h} + \ell/h - (1.09\Delta w + 0.09w_h) \quad (28)$$

Integrating the above relation, we obtain the equation for determining the amount of frost heave

$$\ell = \int_0^h \frac{d\ell}{dh}\,dh \quad (29)$$

If it is known how w varies with depth, the above equation can be solved.

For saturated soils (h=0), the amount of migrating water is a constant, and then the amount of frost heave can be calculated by

$$\ell = \frac{\Delta U-\Delta H}{+w_h+\Delta w}\,L_n\left(\frac{P+(1+w_h+\ w)\ r_oh}{P}\right) \quad (30)$$

CONCLUSIONS

In this paper a mathematical model describing the frost processes in principle has been presented after the analysis of physical nature in frost processes. This model has been founded on the basis of a group of equations. The relationship of similitude multiple, which is used for undisturbed soils in modelling tests, has also been given.

For the open system with the constant underground water level, the obtained approximative solution, which is used for calculating the freezing penetration depth and the quantity of moisture migration, is clear in physical conception. The formula for calculating unfrozen water content, obtained by thermodynamic method, can reflect the effects of temperature, pressure and soil properties, and the formula for calculating the amount of frost heave by the same method is founded under the consideration of factors such as porosity change, solidification of underlying soils, etc.

NOMENCLATURE

X: coordinate of a given soil layer, which original point is at the ground surface (m);
τ: time (h_r);
r_0: dry density of soils (kg/m^3);
r: bulk density of soils (kg/m^3);
t: temperature in soils (^oC);
T: absolute temperature (^oC);
α: thermal diffusivity of soils ($M^2/h_r.$);
λ: thermal conductivity of soils ($k_{cal}./m \cdot h_r. {}^oC$);
c: voluminal specific heat ($k_{cal}./m^3. {}^oC$);
L: latent heat of phase change of water ($k_{cal}./kg$);
w: water content (%);
Δw: amount of migration water of unit volume of soil (%);
k: moisture conductivity ($m^2/h_r.$);
g: isobaric potential;
μ: chemical potency;
Q: quantity of heat;
H: total heat;
U: intrinsic energy;
S: entropy;
P: pressure;
P_i: steam pressure;
V: volume;
N: composition of mass;
R: general fitting constant of gas;
C_l: geometric proportion coefficient;
l: quantity of finalizing geometric type;
-,+: subscripts referring to the frozen and thawed state of soils;
1,2: subscripts referring to the phase of unfrozen water and ice.

ACKNOWLEDGEMENTS

Here, the author heartily wishes to thank his colleagues, Xu Xiaozu, Zhu Yuan-lin, and Peng Wanhual, for their aid in preparation of this paper.

REFERENCES

Ding Dewen and Lo Xuepo, 1979, Theoretical basis of thermotechnical modelling test in ground freezing, 2, Science Notification.

Gao Min and Ding Dewen, 1980, Thermodynamic method used for the calculation of quantity of frost heave in undisturbed ground, Proceedings of the 2nd International Symposium on Ground Freezing.

Lo Xuepo and Ding Dewen, 1978, Study of the thawed calculation of the Building subground in permafrost region, 6, Journal of Lanzhou University.

A STUDY OF HORIZONTAL FROST-HEAVING FORCES

Ding Jingkang

Northwest Institute of the Chinese Academy of Railway Sciences
Lanzhou, People's Republic of China

To investigate the horizontal frost-heaving forces that act on structures, an experimental and a model reinforced concrete retaining wall were constructed at the Fenghou-Shan Field Research Station on the Qinghai-Xizang Plateau in 1976. The experimental retaining wall was 15 m long and 4-5 m high, and the model wall was 3 m long and 1.2 m high. Investigations conducted during three winters (1976-78) indicated that the magnitude and distribution of horizontal frost-heaving forces are substantially governed by the material properties of the backfill and the displacement of the walls. For clayey soil backfill, the maximum horizontal frost-heaving force was 1.4 kg/cm^2 and acted upon the middle or lower middle part of the wall, while for gravel backfill it was 0.76 kg/cm^2 and acted upon the lower part of the wall. Based on the distribution of frost-heaving forces along retaining walls, the author presents three basic models for calculating the horizontal frost-heaving forces acting on structures according to the different working conditions of the structures.

Horizontal frost heaving pressure is one of the frost heaving pressures which act upon foundation of structures during the freezing of soils. According to the condition of emerging and features of functioning, it can be classified into two types---symmetrical horizontal frost heaving pressure and nonsymmetrical horizontal frost heaving pressure (see Figure 1).

The action of symmetrical frost heaving pressures on foundation is the same as that of hydrostatic water pressure so as not to yield direct effect on the stability of foundations. However, because of the symmetrical frost heave pressure the shearing-resistant strength of the interface between permafrost and foundation will be increased, i.e. the tangential frost heaving pressure exerting on foundation will increase and bring about an indirect influence over the stability of foundation.

Nonsymmetrical horizontal frost heaving pressure is the main pressure system acting upon retaining structures in permafrost areas. It is several times, and even tens of times as high as that of Coulomb's ground pressure[1] (see Figure 2). It is, therefore, of great significance to study the magnitude of nonsymmetrical horizontal frost heaving pressure, the law of distribution as well as the condition for emergence while designing and calculating retaining structures in permafrost areas.

To study the problems mentioned above, tests were made on experimental and model retaining walls at site of Qinghai-Xizang Plateau. The experimental retaining walls investigated were reinforced concrete L-shaped ones with a height of 4 m and 5 m respectively and a length of 15m. The soil placed behind the 4 m retaining wall was fine granular soil and that behind the 5 m one was coarse granular soil. The model retaining wall was a multi-layered one made of reinforced concrete with a length of 3 m and a total height

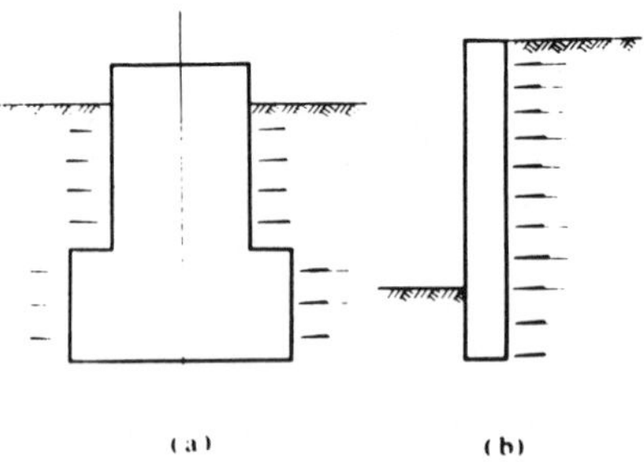

FIGURE 1 Symmetrical and nonsymmetrical frost heaving pressure.
(a) Symmetrical frost heaving pressure;
(b) Nonsymmetrical frost heaving pressure.

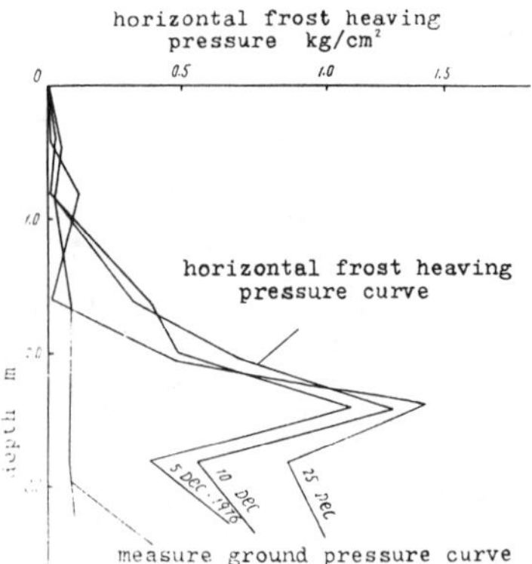

FIGURE 2 Comparison of ground pressure and horizontal frost heaving pressure.

of 1.2 m, the height of each layer being 0.2 m. Fine granular soil was placed behind the wall.

INITIATION OF HORIZONTAL FROST HEAVING PRESSURE

It is well-known that during freezing, soil is accompanied by an expansion of volume, i.e. frost heaving. Frost heaving of soil and the resistance of retaining structures to horizontal loads are necessary for horizontal frost heaving pressures to occur. Horizontal frost heaving pressure will not develop unless frost heaving occurs during freezing of soil behind the retaining structure, or retaining structure can deform freely without preventing frost heaving. The two conditions above, however can be hardly satisfied under normal situations, and as a result, horizontal frost heaving pressures can not be avoided during freezing of the soil behind a retaining wall.

DEVELOPMENT OF HORIZONTAL FROST HEAVING PRESSURE

How does horizontal frost heaving pressure emerge and develop in the course of soil freezing? Figure 3 gives the displacement curve of top of the L-shaped retaining wall during freezing, and Figure 4 presents the developing curve of horizontal frost heaving pressure at different depths on the back of the model retaining wall.

It is seen from Figure 3 that soil body contracts and the ground pressure falls at the beginning of freezing because of decreasing of the soil temperature. The wall deflects towards the fill behind the L-shaped retaining wall (negative displacement) and reaches a maximum (Figure 3 point a) when the ground pressure is reduced to a minimum and the horizontal frost heaving pressure has not yet developed. During this period, the ground surface tends to be stably frozen through exchange of freezing and thawing. After the emergence of a stably frozen state, the horizontal frost heaving pressure develops and increases with depth of freezing. Under the action of the horizontal frost heaving pressure, the retaining wall deflects away from the fill (positive displacement). The curve rises to point b when the frozen depth reaches the seasonally thawed depth. As freezing progresses, the horizontal frost heaving pressure steadily increases, the curve has an increasing slope from point b to c, which means that the horizontal frost heaving pressure rises rapidly along with decrease of temperature of frozen layer. The curve runs smoothly from c to d, which implies that a basic equilibrium exists between the increase and the relaxation of the horizontal frost heaving pressure; the horizontal frost heaving pressure reaching its maximum at this time.

The course of deflection of the L-shaped retaining wall during freezing reflects the process of the emergence and development of the horizontal frost heaving pressure. Through successive observations for three years an analogous curve was obtained.

The above analysis is illustrated by the developing curves of horizontal frost heaving pressure shown in Figure 4. In Figure 4, the horizontal frost heaving pressure developed along with appearance of stably frozen state. With the seasonally thawed layer freezing gradually from the beginning of October to mid-November, the horizontal frost heaving pressure gradually rises (seasonally thawed layer is completely frozen in mid-November). From mid-November to the end of December, horizontal frost heaving pressure rises rapidly to a maximum along with the decrease of temperature of the frozen ground. From the end of December to mid-January, despite the continuous decrease of frozen ground temperature, the increasing and relaxing of horizontal frost heaving pressures basically reaches an equilibrium, the curve fluctuating within a slight range.

It can be seen from the above analysis that

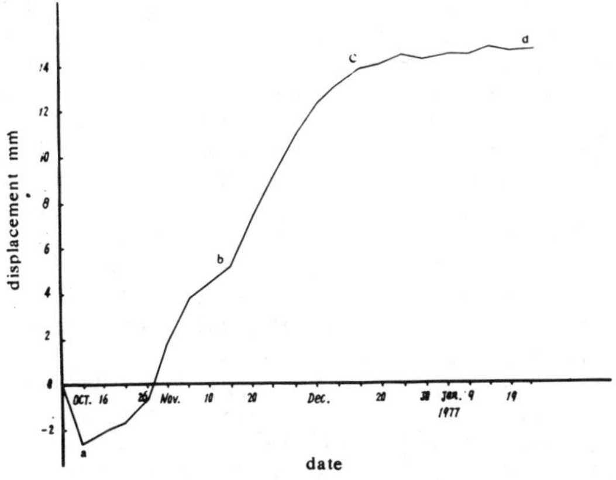

FIGURE 3 The displacement curve of top of L-shaped retaining wall in the freezing.

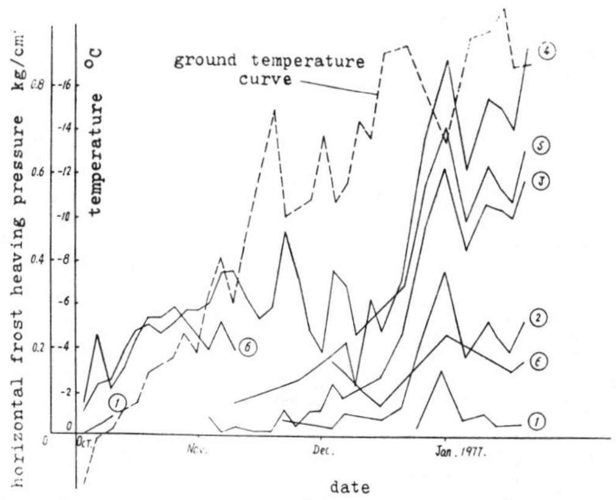

FIGURE 4 The developing curves of horizontal frost heaving pressure on the back of model retaining wall. Six layers were tested; (1) is the top layer.

the action of horizontal frost heaving pressure is nôt synchronous with that of ground pressure. Ground pressure can be considered as zero as horizontal frost heaving pressure develops.

DISTRIBUTION OF HORIZONTAL FROST HEAVING PRESSURE

The distribution of horizontal frost heaving pressure along the back of retaining wall is uneven. Figure 5 gives a distribution curve of horizontal frost heaving pressure along back of L-shaped retaining wall, and Figure 6 that along back of a model retaining wall. It is seen from Figure 5 and 6 that the horizontal frost heaving pressure mainly distributes along the middle and lower parts of the wall. The magnitude of horizontal frost heaving pressure is less along the upper part, and its maximum appears at different places depending upon the fill material used, fine or coarse granular soil. The maximum horizontal frost heaving pressure occurs at the middle or lower middle part of the wall with fine granular soil as fill material, and at the lower part of the wall for coarse granular soil.

The distribution of the horizontal frost heaving pressure is determined by the following factors.

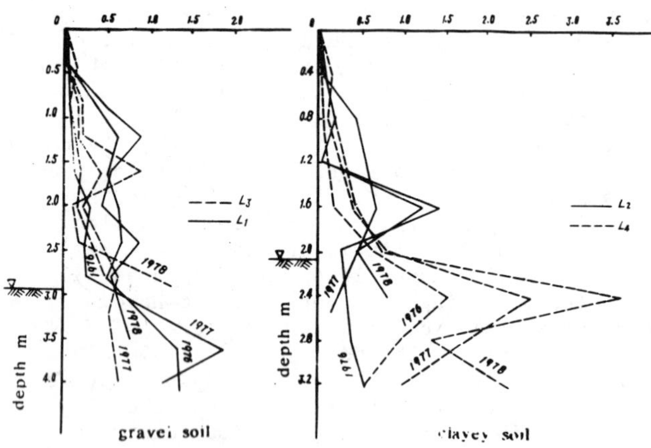

Figure 5 Distribution of horizontal frost heaving pressure along back of a L-shaped retaining wall. L_1, L_2, L_3, L_4 are the number of testing walls.

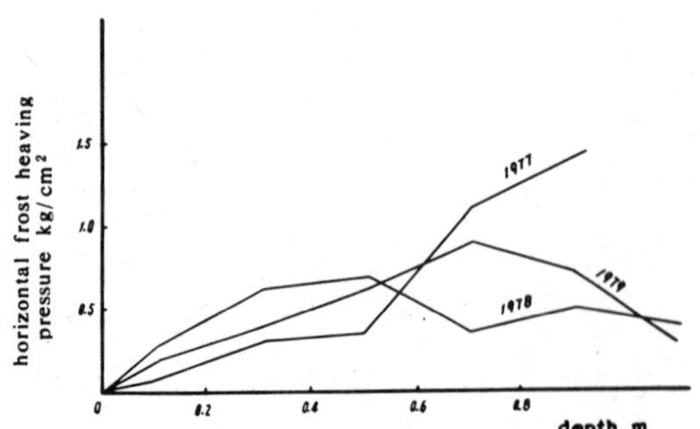

Figure 6 Distribution of horizontal frost heaving pressure along back of a model retaining wall.

Heat Flow Variations

After the retaining wall is established, the condition for heat radiation during freezing of the soil mass changes to a two-dimensional problem. Figure 7 shows the distribution curve of temperature field along the back of the model retaining wall during seasonal freezing.

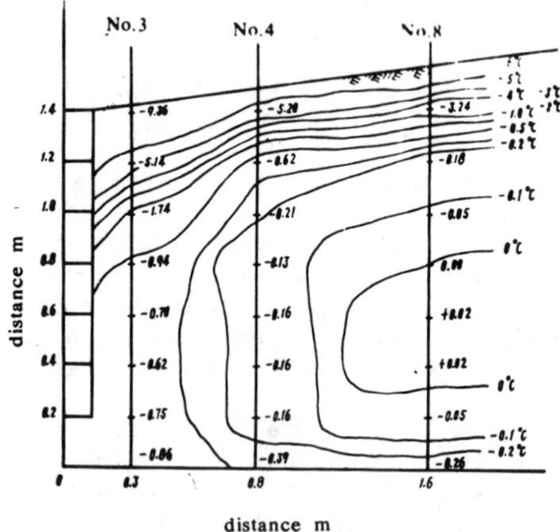

Figure 7 The distribution curve of temperature field along back of the model retaining wall. (October 28, 1978).

It is seen from Figure 7 that during the early stage of freezing the isotherms slope upward from upper part of the wall. This shows that soil body freezes in two dimensions, attaining a maximum rate of freezing. Below the middle of the wall the isotherms are likely to be parallel to wall surface, which indicates that the middle part of the soil mass approximates one-dimensional freezing, its freezing rate being similar to that of the natural ground surface. During freezing the thermal radiation conditions for the soil mass located behind the lower part of the retaining wall (corresponding to the soil mass beneath ground surface in front of the wall) are poor, thus it freezes at a low rate. Such freezing factors of the soil mass behind a retaining wall determine the conditions of moisture flow and ice segregation.

Hydrologic and Geologic Variations

After a retaining wall is built, a new permafrost table inclining toward the wall forms. As

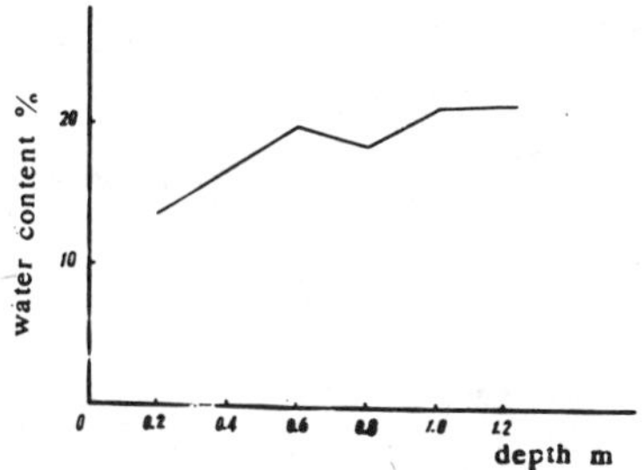

Figure 8 Distribution of water content in soil filled up behind model retaining wall versus depth.

a result, suprapermafrost water converges toward back of the retaining wall above the permafrost table, causing an increase of water content in the soil behind the wall.

On the other hand, the water in the soil behind the wall moves downward under the influence of gravity, which leads to a moisture distribution where the upper part of the fill contains less water, while the middle and lower parts contain more (Figure 8).

Moisture Migration During Freezing

While fine granular soil is freezing, water migrates toward the freezing front. This leads to the increase of water content in soil mass near freezing front. However, during the freezing of coarse granular soil, water flows away from the freezing front, and as a result, a decrease in the water content in the soil near the freezing front tends to occur.

The above three factors determine the distribution of the horizontal frost heaving pressure along back of a retaining wall. With fine granular soil as fill material, the water content behind the upper part of wall is the lowest and the freezing rate is the highest, the water migration rate is slight, and the direction of ice crystal separation slope upward toward retaining wall. This results in the lowest horizontal frost heaving pressures acting upon upper part of retaining wall. At mid depth behind a retaining wall, the freezing rate is similar to that of natural ground surface, because of moderate water content. This increases the flow of water toward the freezing front. Finally, because of unfavorable conditions for thermal radiation from the base of a retaining wall, freezing lags behind that of the soil at mid depth so that there is sufficient time for the water in lower soil mass to migrate to the mid depth freezing front. Therefore, the freezing process at mid depth is analogous to that for an open system having an external water supply. Furthermore, the ice lenses are almost parallel to wall so that the horizontal frost heaving pressure yielded tends to be the highest. As for lower soil mass, the water content decreases because of moisture loss during freezing. In addition, the thickness of lower frozen soil mass decreases so that a smaller horizontal frost heaving pressure occurs.

When coarse granular soil is used as fill material, the development of horizontal frost heaving pressure is dependent upon moisture flow from freezing front downward during the course of freezing, and therefore the maximum horizontal frost heaving pressure occurs at the lower part of the wall.

Main Factors Affecting Horizontal Frost Heaving Pressure

It is evident from the analysis of horizontal frost heaving pressure that the main factors that affect it can be classified into two sorts: first, those determining the heaving properties of fill materials, and second, those deciding the deformation property of the retaining wall. The first is mainly related to different types of soil, water content, temperature, and freezing rate, etc., while the second is primarily related to the type of retaining wall.

Different types of soil have different heaving properties. Under the same boundary conditions, fine granular soils heave more than coarse soils, and therefore higher horizontal frost heaving pressures occur in the former. It can be seen from Figure 5 that both the maximum and mean values of horizontal frost heaving pressure of fine granular soil are larger than those in the coarse granular soil.

The distribution curve of horizontal frost heaving pressure in Figure 5 also reflects the effects of water content. Water content is low behind the upper part of the wall and high in the middle and lower. Correspondingly, the horizontal frost heaving pressures are lower behind the upper part of the wall.

Horizontal frost heaving pressures rise with decrease of ground temperature. From Figure 4, we can see the increasing trend of the frost heave pressure is nearly to the negative increasing trend of the temperature.

The magnitude of the retaining wall deformability is manifested in terms of its resisting effect on heaving. The higher the deformability, the lower the resisting effect, namely, the less the horizontal frost heaving pressure produced.

Distribution of Horizontal Frost Heaving Pressure

It is seen from the observation of the distribution of horizontal frost heaving pressure behind retaining wall that under similar conditions, the frost heave pressures are always low in the upper part and high in the lower, approximating a triangular distribution. Because the locations of the maximum horizontal frost heaving pressure in fine and coarse granular soil are different, the distribution of horizontal frost heaving pressure can be simplified using different triangles (Figure 9). For coarse granular soil, the triangular distribution pattern (Figure 9-3) is adequate for use regardless of buried depth of retaining wall, while for fine granular soil, different distribution patterns---Figure 9-1 or Figure 9-2 are

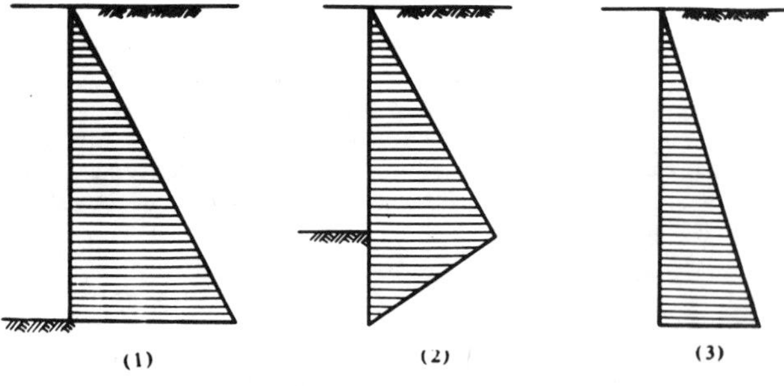

Figure 9 Calculating patterns of horizontal frost heaving pressure distributed along retaining wall back.
(1), (2) are the fine granular soil.
(3) is the coarse granular soil.

adopted in accordance with the practical burying conditions of retaining wall.

Option of Horizontal Frost Heaving Pressure

The horizontal frost heaving pressure obtained by means of multi-layered model retaining wall and the maximum horizontal frost heaving pressures calculated in accordance with the patterns of wall deformation in Figure 9 are listed in Table 1 and 2 respectively.

From the tables, it is seen, the maximum horizontal frost heaving pressure for saturated gravel soil is 0.76 kg/cm^2, while that of fine granular soil whose water content is close to the lower limit of plasticity is 1.4 kg/cm^2.

The maximum horizontal frost heaving pres-

TABLE 1 The Measure Value of Horizontal Frost Heaving Pressure by Model Retaining Wall.

Types of soil	Mean water content before freezing,%	Test year	Maximum horizontal frost heaving pressure						Horizontal heaving pressure (kg/cm^2)	Remarks
			0.0–0.2	0.2–0.4	0.4–0.6	0.6–0.8	0.8–1.0	1.0–1.2		
Clayey soil		1977	0.09	0.30	0.35	1.10	1.40		0.65	Retaining wall built directly on foundation of ground soil
Clayey soil	16.2	1978	0.27	0.62	0.68	0.37	0.50	0.40	0.47	Retaining wall built on 0.3 m foundation of ground soil
Clayey soil	18.73	1979	0.17	0.39	0.62	0.81	0.71	0.30	0.51	

TABLE 2 The Calculative Value of Horizontal Frost Heaving Pressure.

Cross-section under calculation	Types of soil filled	Time	Maximum displacement of wall top (mm)	Maximum horizontal frost heaving pressure (kg/cm^2)	Average horizontal frost heaving pressure (kg/cm^2)
L_4	Clayey soil	1976	13.6	0.57	0.29
		1977	21.5	0.90	0.45
		1978	9.0	0.38	0.19
L_2	Clayey soil	1976	18.9	0.80	0.40
		1977	21.5	0.90	0.45
L_1	Gravel soil	1976	42.1	0.76	0.38
		1977	24.1	0.43	0.22
L_3	Gravel soil	1976	22.8	0.41	0.21
		1977	19.2	0.35	0.18
		1978	28.1	0.50	0.25

sures of fine granular soil were obtained by means of a model retaining wall. The deformation of the model retaining wall was as small as about 0.4 mm, while, in general, the deformations of real retaining wall are larger than this value. Under identical conditions, the maximum horizontal frost heaving pressure in a fine granular soil acting upon a retaining wall will be generally less than 1.4 kg/cm^2. It is, therefore, proper to choose 1.5 kg/cm^2 as the upper limit of maximum horizontal frost heaving pressure.

The maximum horizontal frost heaving pressures of saturated gravel soil were obtained from experimental retaining wall, in which the fine clayey particle content was 5.9%. Therefore, a maximum horizontal frost heaving pressure of 1 kg/cm^2 is adequate for coarse granular soil on account of the possible maximum content of fine and clayey particles.

Based on the above results, it is suggested that under normal conditions the maximum horizontal frost heaving preswures of 1-1.5 kg/cm^2 and 0.5-1.0 kg/cm^2 be adopted for fine and coarse granular soil respectively provided that the calculation are made with triangular distribution patterns for frost heave pressures.

CONCLUSION

Based on the above discussion, we can conclude the followings:

1. There exists a horizontal frost heaving pressure during freezing of soil behind the retaining wall. It is the main pressure system acting upon retaining wall in permafrost areas. It is several times, and even tens of times as high as that of Coulomb's ground pressure.
2. The size of horizontal frost heaving pressure is only related to the heaving properties of soil, and the deformation properties of the retaining wall are unchanged, irrelevant to the height of retaining wall.
3. The action of horizontal frost heaving heaving pressure is not synchronous with that of ground pressure. Ground pressure can be considered as zero as horizontal frost heaving pressure develops.

REFERENCE

Ding Jingkang, 1979, A study of horizontal and tangent frost heaving pressure during freezing of soil in permafrost area of Qinghai-Xizang Plateau.

DESIGN AND CONSTRUCTION OF PILE FOUNDATIONS
IN THE
YUKON-KUSKOKWIM DELTA, ALASKA

Leroy DiPasquale[1], Stephen Gerlek[2], and Arvind Phukan[3]

[1]Design Engineer, Alaska Area Native Health Service, Anchorage, Alaska, 99513 U.S.A.
[2]Design Engineer, Formerly with Alaska Area Native Health Service, Anchorage, Alaska 99513 U.S.A.
[3]Professor of Civil Engineering, University of Alaska, Anchorage, Alaska 99508 U.S.A.

A "reaction beam" type pile load test was conducted in ice-rich silt at Kipnuk, Alaska to determine the load capacity of driven H-piles at the edge of the continuous permafrost region in the Yukon-Kuskokwim Delta. Field test results are compared to existing theories on the design of pile foundations in frozen ground. Additionally, information regarding site conditions, pile installation methods, equipment and construction problems are discussed for the pile foundations of water tanks and auxiliary structures designed and installed in this region at the Eskimo Villages of Kotlik, Hooper Bay, Chevak, Newtok, Tuntutuliak, and Kipnuk. These water tanks ranged from 45,000 to 910,000 liters with accompanying loads of up to 115 kN/pile. Subsoil conditions varied from ice-rich organic silt, and silty sands to poorly graded frozen fine sand with low moisture content.

INTRODUCTION

The Alaska Area Native Health Service designed and constructed water storage facilities, an essential component of most public water supply systems, in the Eskimo villages of Kotlik, Hooper Bay, Chevak, Newtok, Tuntutuliak, and Kipnuk. The design was based on the adfreeze strength method and test results of a pile load test conducted at Kipnuk on driven steel H-piles in ice-rich silt. This paper presents the pile load test results along with theories regarding pile design in permafrost and provides additional information regarding site conditions, methods of pile installation, construction problems, and design parameters at several locations in the Yukon-Kuskokwim Delta of Alaska.

Required water tank capacity was determined on village population, water availability, and quality. Driven or slurried pile foundations were installed to support water storage tanks varying in size from 45,000-910,000 liters with accompanying loads of up to 115 kN per pile.

This paper addresses site conditions and construction consideration that influenced the design and construction of tanks in the region and then will address the pile load test results obtained at Kipnuk.

SITE CONDITIONS

The Yukon-Kuskokwim Delta is located near the southern boundary of the continuous permafrost region of Alaska. The mean annual air temperature is about -1.7°C. The depth of frozen ground, determined from water well logs, was 30-75 m in the six coastal Native villages where pile foundations were installed. These villages span the vast almost flat, fan-shaped delta as shown on the key map (Figure 1). Many meandering streams, sloughs, and shallow lakes cover the area, which is a complex of river, marine, and windblown deposits. The ground surface is generally covered with a wet organic mat, coincidental with the active layer, 0.3-0.6 m thick, composed of a wide variety of low-growing shrubs, grasses, and sedges

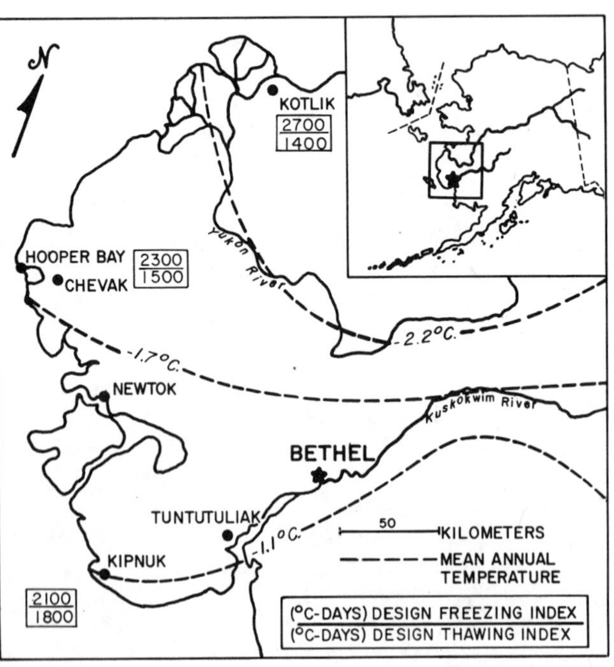

FIGURE 1 Key Map.

rooted in mosses and lichens. Soils typically consist of fine-grained ice-rich silts and sands to the depths investigated (8 m). Soil layers are well defined, ranging from a few centimeters to many meters thick. Typical soil and temperature profiles found at the sites are presented in Figure 2. Soils ranged from well to poorly bonded. Nonvisible and visible ice, both stratified and nonstratified was found. Multi-layer ice lensing with discrete ice lenses of up to 7.5 cm thick was encountered; more massive formations, however, are known to exist.

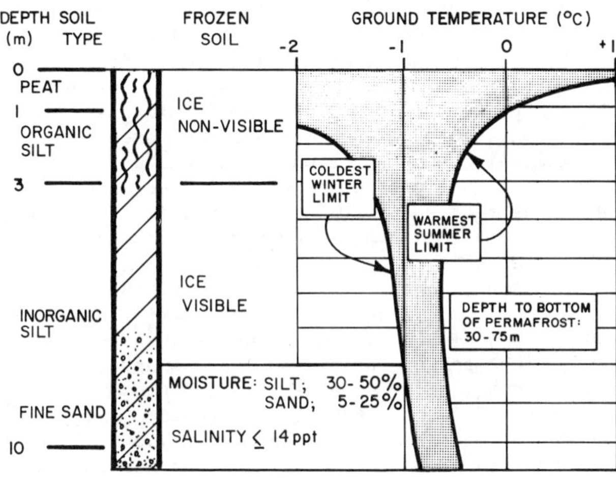

FIGURE 2 Composite Soil Profile and Typical Thermal Regime.

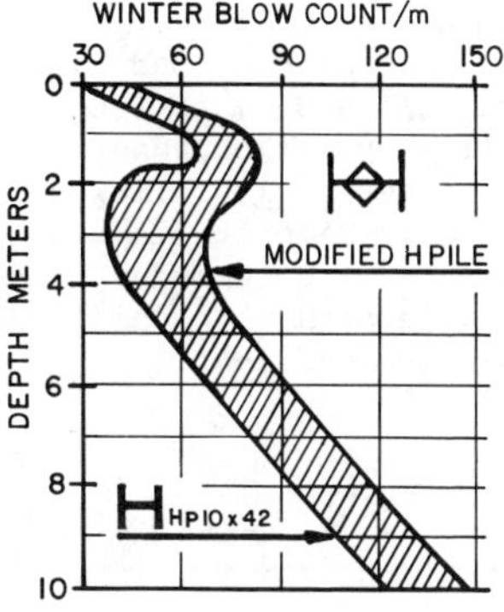

FIGURE 3 Kipnuk Blow Count Record (Drive Energy-2,000 kg-m).

CONSTRUCTION CONSIDERATIONS

Factors such as local transportation, construction season, pile type, and available equipment influence pile foundation installations in this region. Common modes of commercial transportation are barge and airfreight. Overland hauling across frozen trails and rivers can also be used, however, thin ice and severe storms make this method potentially hazardous.

Piling materials are shipped from the Continental U.S. West Coast by barge to village beaches or river bank sites. Pile installation equipment is shipped in a simlar manner from Bethel, the transportation center for the region.

Most village gravel runways cannot accommadate large cargo aircraft. Because the thawed active layer is too soft and fragile to support conventional construction equipment during the summer, pile installation normally begins after freezeup in November and can continue until breakup in May. Shipping and construction periods are well defined and, if missed, work can be delayed as much as a year.

Two general methods of pile installation were used: (1) driving and (2) slurry back. The piles were either driven direct or an undersized predrilled holes. Slurry back piles were installed in an oversized hole and backfilled with a sand-water slurry. Installed costs varied from $2,000 to $4,000 (U.S.A. 1982) per pile. Production rates varied from 2 piles/hour (driven H-piles) to 1 pile/5 hours (slurry back piles).

Compared with the other installation methods used, driving HP 10X42 piles at Kipnuk required heavier equipment. An open-end diesel hammer rated at 2,000 kg-m for 1.5 m stroke height was used. Difficulties were encountered in starting and running the hammer at cold air temperatures.

The plot of blow counts versus depth for two pile configurations at Kipnuk reflects the winter soil temperature whiplash curve (Figure 3). Blow counts increased linearly with depth in the constant temperature region below 4.5 m. Driving resistance was low, and no significant pile deformation was noticed on any aboveground portions of the piles. However, a bent flange was observed after excavating the colder surface soils around the test pile.

The use of cable tool well drilling equipment was used to drill and drive pipe piles in the same manner as well casing and was capable of delivering 520 m-kg of force. A hole smaller than the diameter of the pipe was drilled 1-3 m deep and the pile then driven to the same depth. The drilling and driving process was repeated continuously until the desired depth was reached. The use of a pilot hole allows driving by smaller and more mobile equipment. Water and a bailer was used to remove cuttings from the hole. Time between driving cycles must be short to avoid higher driving resistance and excessive heat addition. The use of a pilot hole allows driving by smaller and more mobile equipment.

The slurry back method of pile installation requires more attention during construction. The hole diameter must be large enough to accept the backfill and allow compaction of the slurry with a concrete type vibrator. A fine sand either imported or mined locally and saturated with +4°C water was used as the slurry materal. Considerable difficulty was experienced when using a solid flight auger to drill to depths below 6 m. Air rotary drilling with compressed air to remove the cuttings proved to be more successful, however, the need for an air compressor added to the logistics problem. Early pile loading was made possible by freezing the slurry with refrigeration equipment.

PILE LOAD TEST

Few pile load tests in North America have been conducted in frozen ground, especially ice-rich "warm" permafrost. The paucity of data can be attributed to testing costs, instrumentation problems, and suitable procedures. In order to obtain better data for design purposes, an axial compression pile load test on driven steel H-piles was performed at Kipnuk during the winter of 1981 at the site where the water tank was planned. Other data on the effects of tensile and lateral loading on driven H-piles were also collected and investigated but are not discussed in this paper.

Load Test Equipment and Instrumentation

A "reaction beam" type load test setup was designed and built of HP 10X42 piles as shown in Figure 4. Two stiffened pile sections welded together and pin connected to the outside support piles formed the reaction beam. The compression test pile was instrumented with 34 thermisters, 5 vibrating strain gauges , and 4 telltales. Protection and access to vibrating strain gages, thermisters and an inclinometer tube on the test piles were provided by welding angles parallel to the H-pile web. Two reference beams for deflection measurement datums were supported by three additional piles. The test facility was covered with a plywood frame building with insulated floor and oil stove.

Vibrating strain gages were accurate to ± 1 micrometer. Immediately after driving, measured residual strains indicated a probable 3.8 cm misalignment in the compression test pile at a depth of 6 m. Dial gages supported by adjustable pipe and accurate to ± .03 mm, were used to measure deflection. Clip angles welded to the test pile provided a deflection measurement bench. Errors caused by surface roughness under the gages were reduced after small mirrors were set and plumbed on each measurement bench. Shelter air temperature was held constant to limit expansion and contraction of gage pipe supports. Pile cap settlement for the compression test was measured by four gages placed symmetrically around the pile.

Loads were applied by a 900 kN hydraulic jack with a 5.7 cm cylinder stroke and calibrated to an accuracy of ±2000 N. The stroke length proved sufficient for the axial load test, but it was too short to apply the necessary load during both the tension and lateral tests. Compressive axial loads became execessively eccentric above 800 kN and the reaction beam twisted beyond the stroke length.

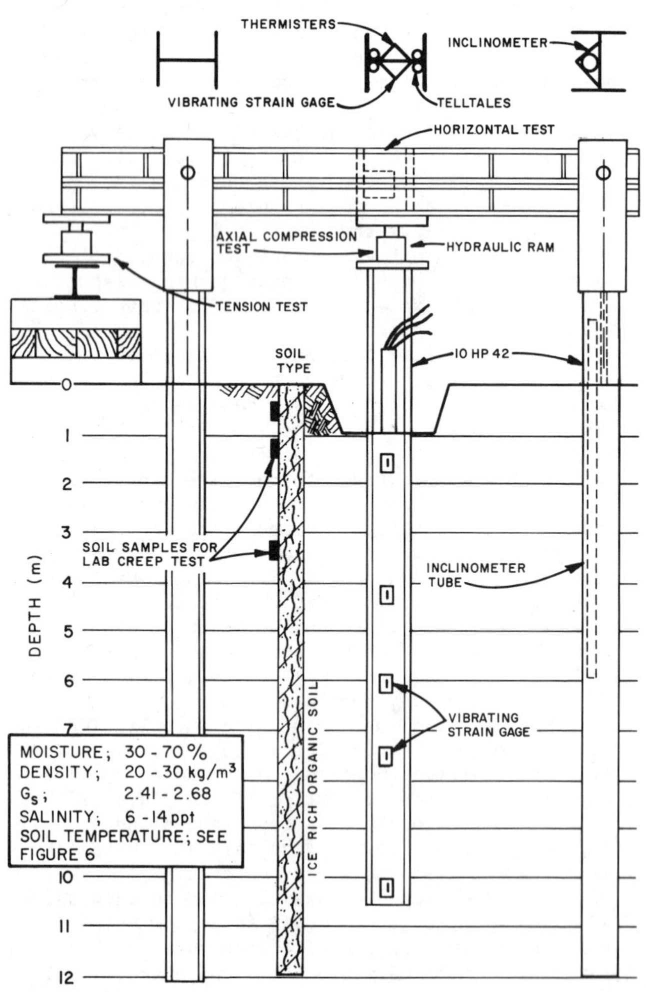

FIGURE 4 Kipnuk Pile Load Test Setup.

Test Procedure and Results

Axial compressive loads were applied to the instrumented test pile in 45 kN increments and relieved in 90 kN increments. Six cycles of loading and unloading were made. Load cycles I and II were to 130 kN, III and IV to 260 kN, and V and VI to 800 kN. Pile cap deflections were read at frequent intervals immediately following application of each load. Pile strain and temperature readings were also recorded. Loads were increased after pile strain and pile cap settlement rates remained constant.

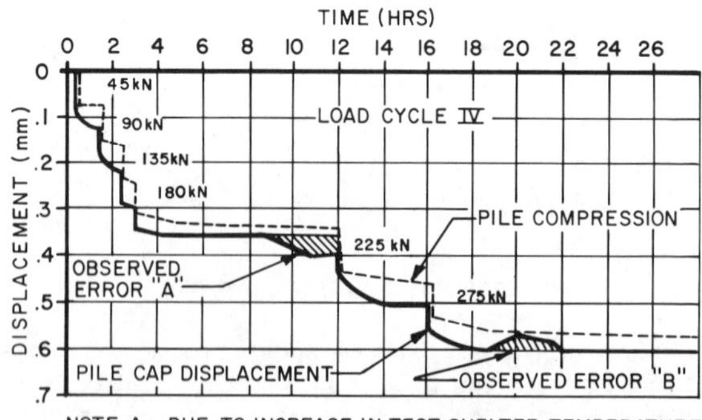

FIGURE 5 Pile Cap Displacement and Shaft Compression for Kipnuk Load Cycle IV.

Measured pile displacements and computed pile shaft compression during load cycle IV are presented in Figure 5. Measured pile shaft strains found for the same load cycle are illustrated in Figure 6. Shaft compression and cap displacement behavior under higher loads (cycles V and IV) are shown in Figure 7. As expected, the magnitude of shaft compression is generally lower than pile cap displacement. It is interesting to note that the pile cap continues to settle while shaft compression is complete.

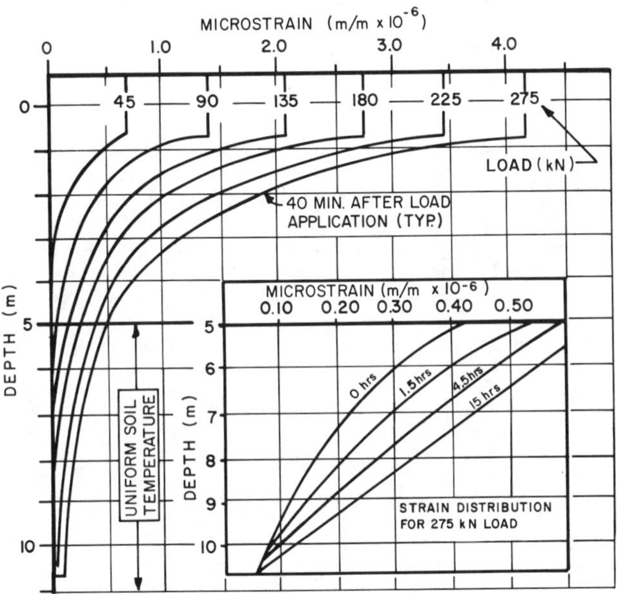

FIGRUE 6 Measured Pile Shaft Strain (Kipnuk Load Cycle IV).

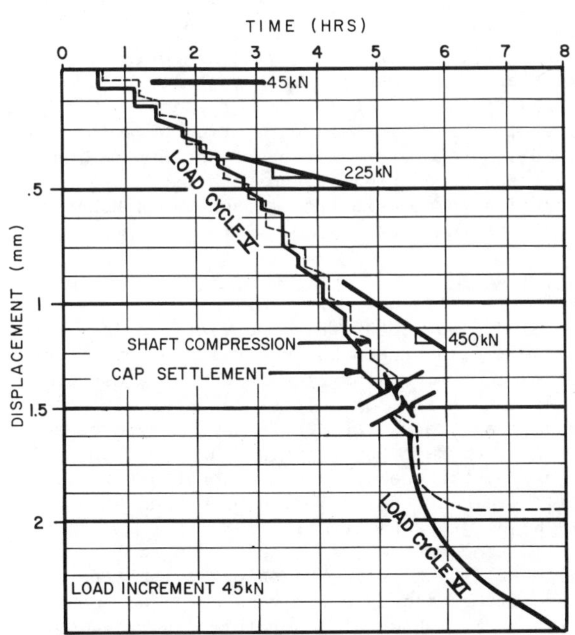

FIGURE 7 Short Term Pile Load Test Results, Kipnuk.

PILE FOUNDATION DESIGN

The design of pile foundations in ice-rich "warm" permafrost has grown from an empirical static analysis approach based on adfreeze strength to a more rational consideration of load-pile displacement behavior based on theoretical creep law models. Attempts to quantify driven pile performance and load capacity in these soils based on dynamic considerations (such as wave equation analysis) lack sufficient data on appropriate soil strength properties and their creep deformation relationships.

Adfreeze Strength Approach

This method is analogous to that used for unfrozen cohesive soils; it is typified by a "floating" pile analysis, makes broad use of engineering judgement and experience, and employs larger factors of safety. Piles are assumed non end bearing unless site conditions reveal otherwise; group effects are also largely ignored. Besides applied load, additional loading due to active layer downdrag and frost heaving are considered.

Known adfreeze strengths based on load tests elsewhere (for example Linell et. al. 1980), are applied to thermally defined increments of depth, and summed to determine capacity. Figure 8 displays this approach for the load test. The circumscribed area of the H-pile was used in this computation to arrive at a short term capacity of 680kN.

Pile capacity may also be assessed directly from short term incremental load test results (which is conventionally the basis for determining unit adfreeze strength values). Using the method of intersecting tangents, the estimated short term capacity found from load cycle V is 650 kN, as shown in Figure 9.

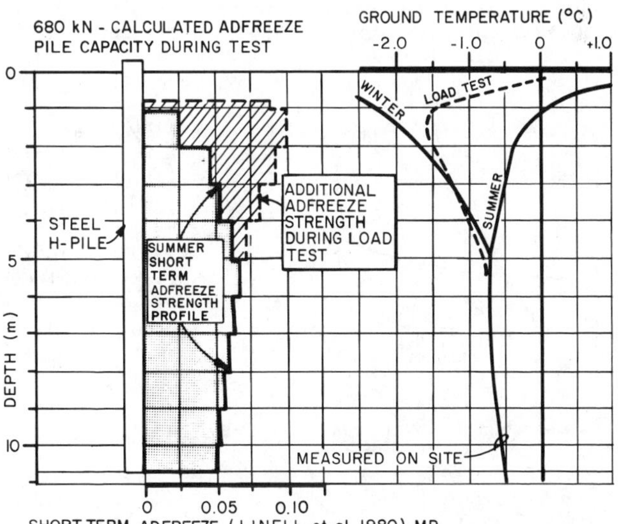

FIGURE 8 Kipnuk Load Test Short Term Pile Capacity Estimated by the Adfreeze Strength Approach Method.

There has been little definitive work on the factors of safety necessary for pile designs based on adfreeze strength in ice-rich permafrost, especially when addressing long term creep. There is a lack of consensus on, and understanding of the transferability of information interpreted from a few load tests applied to sites elsewhere, and to differing pile materials and geometries. For instance, the greater surface area of H-piles does not result in proportionally greater pile load capacity. Between the flanges there is an interaction of shear and adfreeze strength that is not clearly understood.

However, it is known that pile capacities may be increased by mobilizing additional soil shearing stress through the use of protruberances (rings, plates, fins, etc.) or through active or passive ground chilling schemes (Long, 1973).

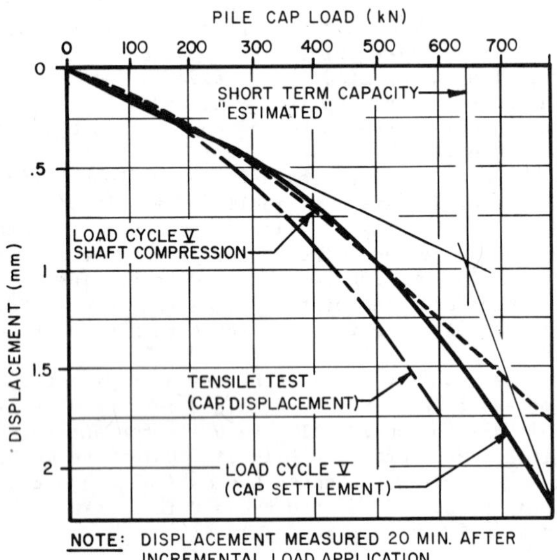

FIGURE 9 Interpretation of Kipnuk Short Term Pile Capacity.

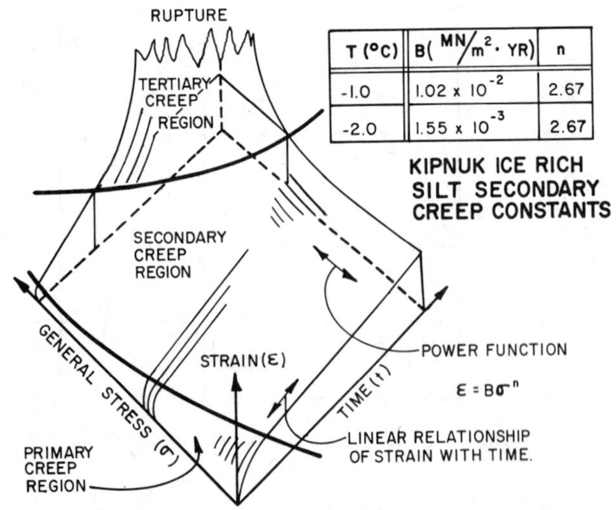

T (°C)	B($MN/m^2 \cdot YR$)	n
-1.0	1.02×10^{-2}	2.67
-2.0	1.55×10^{-3}	2.67

FIGURE 10 Generalized Creep Behavior of Ice-Rich Frozen Soil.

Pile Velocity Approach

Based on observations, the elastic-plastic response of frozen soils to stress can be schematically represented by Figure 10. Full scale field tests on piles have shown similar behavior (Markin et. al., 1973). In "warm" ice-rich silts, practical designs will operate on the dominating secondary creep surface.

Power flow laws have been proposed to describe steady state creep in incompressible materials at constant temperature as a function of multiaxial stress. Applied to permafrost, this relationship has been more simply stated (Vyalov, 1965).

$$\dot{\varepsilon} = \beta\sigma^{n} \qquad (1)$$

where $\dot{\varepsilon}$ is secondary creep rate, σ is the level of a general state of stress, and β and n are temperature dependent constants unique to the frozen soil. The secondary creep constants for Kipnuk ice-rich silt as determined from laboratory uniaxial compression tests are shown in Figure 11.

Equation 1 has been transformed and applied to the idealized shearing of concentric cylinders to arrive at an equation describing pile cap velocity (Ladanyi, 1972).

$$\dot{U}_a = \frac{a3^{(n+1)/2}\beta\tau_a^{n}}{n-1} \qquad (2)$$

Where $\dot{U}_a$ is steady state pile velocity, a is the pile radius, τ_a is shear stress, and β and n are temperature dependent constants determined from uniaxial creep data for the frozen soil in question. This equation is insensitive to changes in normal stress on the lateral surfaces of the pile (i.e., overburden pressure) and to pile compressibility. It assumes that soil properties are homogeneous and isotropic with depth and that there is no slip at the pile-soil interface (i.e., tangential stress is less than adfreeze strength). Equation 2 may be algebraically rearranged and applied incrementally to account for variations in soil creep properties with depth. This allows investigation of velocity-load relationships under different thermal regimes. Equation 2 has been further modified to account for pile compressibility (Nixon and McRoberts, 1976) as shown below:

$$\dot{U}_a = D(z,t)\left(\frac{aE}{2}\right)^{n}\left(\frac{\partial^2 U_a}{\partial z^2}\right)^{n} \qquad (3)$$

With equation 3, the fundamental relationship between pile strain and shaft stress over time may be explored in terms of pile cap settlement. This information is most helpful in interpreting short term load test data. It is generally assumed that when creep data for a particular ice-rich soil are lacking, the use of pure ice constants instead will yield a safe design. (Nixon and McRoberts, 1976; Morgenstern et. al., 1982).
Calculated theoretical creep rates for a 10.5 m pile at Kipnuk (using Equation 2 with site specific secondary creep constants) are shown in Figure 11. However, the ultimate pile capacity must be determined from field testing. Characteristically, pile surfaces have little effect on displacement response at low stress levels. Rough piles will mobilize more tangential

adfreeze shearing strength before shaft slip begins to dominate settlement behavior and the pile plunges.

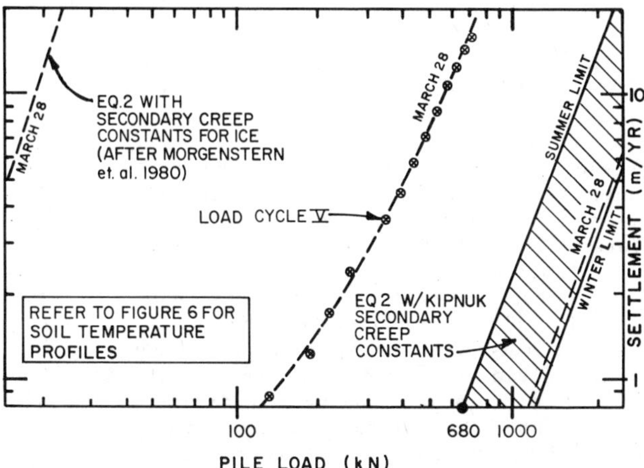

Figure 11 Comparison of Theoretical and Measured Pile Cap Velocities for Kipnuk.

Pile creep may be addressed empirically by long-term static pile load tests. These are usually conducted at higher than anticipated design stress levels to force measurable settlement over practical time periods. A method of extrapolating the results to lower stress levels is by extending the linear relationship of log long-term creep rate versus log pile-load. The load cycle V data (Figure 7) is plotted in this way in Figure 11. Because the test was of short duration, it is not known whether the pile response is in secondary creep at all loads. This may be indicated by the slight curve, thought to represent a primary to secondary creep transition during the early light loads. Further deviation can be expected when the pile eventually plunges at loads higher than reached during the test. An additional problem with interpetation of any pile load test in permafrost is the assessment of creep seasonality. Short term tests yield explicit behavior for a single soil temperature profile. Long term tests tend to meld the pile settlement-load relationship with changing soil thermal conditions at the site. Creep seasonality is more significant at the fringes of permafrost regions where soil temperature variations are larger and extend to greater depths.

CONCLUSIONS

1. The early spring short term capacity for a 10.5 m pile using published empirical adfreeze strengths and Kipnuk soil temperatures is 680 kN. Theoretically, for the same load, the calculated creep rate limits using Kipnuk soil secondary creep constants and equation (2) are 0.17 m/year at early spring and 0.80 m/year at late summer.
2. The prediction of settlement using pile velocity theories with secondary creep constants for ice is conservative in comparison to the pile load test for Kipnuk.
3. The prediction of settlement using pile velocity theories with secondary creep constants for ice-rich silts (Kipnuk) appears to have a similar trend as the plotted LOAD TEST results; however, the magnitude of settlement was found to be less.
4. H-piles may be driven successfully in "warm" ice-rich silts and sands. The principal advantages are speed of installation, increased practical embedment length, minimal disturbance to the environment, and immediate loading.
5. Kipnuk test results show that more field pile load tests are necessary to better understand the behavior of piles in ice-rich frozen soils.

ACKNOWLEDGEMENTS

The authors would like to acknowledge the Alaska Area Native Health Service for permission to publish this work.

REFERENCES

Ladanyi, B., 1972, An engineering theory of creep in frozen soils; Canadian Geotechnical Journal, vol. 9, p. 63-80.

Linell, K. A., and Lobacx, E. F., 1980, Design and construction of foundations in areas of deep seasonal frost and permafrost, SR 80-34: Hanover, NH., U.S. Army CRREL.

Long, E. L., 1973, Designing friction piles for increasing stability at lower installed cost in permafrost. Permafrost: The North American contribution to the second international conference Yakutsk. National Academy of Science, Washington, D.C.

Markin, K. F., and Targulyan, Yu. O., 1973, Bearing capacity of piles in permafrost. Permafrost: The USSR contribution to the second international conference, Yakutsk. National academay of Science, Washington, D.C.

Morgenstern, N. R. Roggensack, W. D., and Weaver, J. S., 1980, The behaviors of friction piles in ice and ice-rich soils, Canadian Geotechnical Journal, vol. 17, p. 405-415.

Nixon, J. F., and McRoberts, E. C., 1976, A design approach for pile foundations in permafrost: Canadian Geotechnical Journal, vol. 13, p. 40-57.

Vyalov, S. S., Ed., 1965, The strength and creep of frozen soils and calculations for ice soil retaining structures U.S. Army CRREL, Translation 76: Hanover, NH., p. 321.

CREEP BEHAVIOR OF FROZEN SALINE SILT UNDER ISOTROPIC COMPRESSION

L. Domaschuk, C.-S. Man, D.H. Shields, and E. Yong

Department of Civil Engineering, The University of Manitoba
Winnipeg, Manitoba, Canada R3T 2N2

Isotropic compression creep tests were conducted on a prepared frozen sandy silt with a pore-fluid salinity of about 60 parts per thousand before freezing. Single-step and multi-step loadings of pressure from 70 to 700 kPa were applied to soil samples at three different temperatures (T = -15°C, -10°C, -5°C). From our results we observed the following: (1) Samples originally in equilibrium underwent attenuating creep upon compression by a superimposed hydrostatic stress; over 70% of the ultimate volume change occurred within the first few hours; a substantial portion of the volume change was not recovered upon unloading. The level of ultimate volume change indicated that volume change during creep could be a significant factor in some problems. (2) A history of loading and recovery had a strain-hardening effect on the frozen soil; so did a history of incremental loading. (3) At -5°C the saline soil was non-bonded but its response to isotropic compression did not seem to be qualitatively different from the way it behaved when it was ice-bonded. This study is part of a broader investigation into the long-term multiaxial creep behavior of frozen saline soils that simulate those beneath the Beaufort Sea.

INTRODUCTION

Frozen soils under conditions of creep are often idealized as incompressible at low temperatures such as in many artificial ground-freezing situations (Vyalov, 1962; Klein and Jessberger, 1979; Klein, 1981). For an incompressible material, a superimposed hydrostatic pressure should have no effect on deformation and flow. For some frozen soils at higher temperatures, experiments indicate that the mean normal stress does have a pronounced effect on creep; yet it is still not uncommon to assume that creep deformation is approximately isochoric, and volume change under creep is neglected. (See, for example, Sayles, 1973; Vyalov et al., 1979.) Perhaps because of this (sometimes implicit) assumption, which we shall hereinafter call the assumption of isochoric creep, many a "general creep equation" in the literature delivers zero strain when the stress is hydrostatic. (See, for example, Ladanyi 1981, Equation (5): Diekmann and Jessberger 1982, Equation (3).) The recent study of Bragg and Andersland (1982), however, shows clearly that the assumption of isochoric creep may have a rather restricted domain of validity. Bragg and Andersland conducted constant-stress uniaxial creep tests on a frozen sand at -6°C, and they measured both the axial and the volumetric strain developed in the samples. Their results showed that while the volumetric strain was quite small when the axial strain was under 2%, it could approach half the value of the axial strain when that value was above 5%.

Our team at the University of Manitoba is currently engaged in a research program to study the long-term multiaxial creep behavior of relatively warm saline frozen soils that simulate those beneath the Beaufort Sea. In our program we will study creep behavior at relatively large deformations (stretch ratio above 5%). For such deformations we question the validity of the following two <u>a priori</u> hypotheses common in works on the multiaxial creep of frozen soils, namely (1) "the assumption of small deformations" under which the infinitesimal strain tensor could be used exclusively as the measure of "strain" and the stretching tensor could be replaced by the time rate of the infinitesimal strain tensor and (2) the assumption of isochoric creep.

In our program we have completed several isotropic compression tests. Data from isotropic compression tests are necessary both for our viscoelastic modeling and for use in Domaschuk's method of estimating settlement of artificial islands.[1] Our results show that frozen soils could experience moderately large volume change even for creep under pure hydrostatic pressure. Thus many "general creep equations" in the literature are in fact severely restricted in scope, and the asumption of isochoric creep once again exhibits its limitations. Because of all these implications, and because, as far as we are aware, isotropic compression tests on saline frozen soils have never been conducted before, we are communicating these results before the completion of our program.

---

[1] This method is an extension of that which Domaschuk and Valliappan (1975) developed to provide a nonlinear approximate solution to the ultimate settlement of structures resting on clay. That solution required an isotropic stress - volumetric strain relationship.

EXPERIMENTAL INVESTIGATION

Sample Preparation

A non-plastic, highly frost-susceptible silt consisting of 34% sand, 56% silt, and 10% clay was used in the study. Sodium chloride was added to the soil to produce a pore-fluid salinity of about 60 parts per thousand.

Samples for testing were formed in a cylindrical, split plexiglass mold, 76 mm in diameter and 200 mm high. The silt was placed in 25 mm layers at Standard Proctor optimum moisture content and was compacted by tamping. An attempt was made to saturate the samples under vacuum.

The samples were frozen unidirectionally at an approximate rate of 25 mm per day by using the procedure adopted by Baker (1976). After the samples were frozen, the portion of the top that had dried out by sublimation was trimmed off.

The water contents, dry unit weights, void ratios, and degrees of saturation of the samples after compaction, after freezing, and at the end of the tests are given in Table 1.

Equipment and Testing Procedure

Isotropic compression tests were carried out in a double-walled triaxial cell similar to that developed by Mitchell and Burn (1971) and later modified by Baker et al. (1982). In such a cell, the change in volume of the sample is measured by a corresponding change in the volume of fluid in the inner cell.

Two single-step constant-stress isotropic-compression creep tests were carried out at -15°C on sample A07 and sample A08 at a stress level of 350 kPa and 700 kPa, respectively. One multi-step creep test was conducted on another sample (A02) at -15°C, using ten stress increments of magnitude 70 kPa each. Before the application of each additional pressure increment, the sample was allowed to compress until no further volume change (within an accuracy of ±0.05 c.c.) occurred within a period of at least 8 hours. After the multi-step test on sample A02 at -15°C was completed, the sample was unloaded, allowed to recover and brought to -10°C. Then another multi-step creep test which followed the procedure of the -15°C test was conducted on sample A02 at -10°C, after which another multi-step test was repeated on sample A02 at -5°C, using the same procedure. In what follows we shall refer to each test by a five-digit code: the first three digits identify the sample, the last two digits the temperature of the test. For example, A0210 refers to the multi-step test on sample A02 at -10°C.

TEST RESULTS AND INTERPRETATION

In each test volume changes were recorded at regular time intervals for each stress increment. A glance at Table 1 reveals that some water was squeezed out from the samples during compression. Thus a sample during creep, strictly speaking, could not be taken as a "body" in the sense of ordinary continuum mechanics. As a simple and convenient way to represent our data, we define for each stress increment the incremental volumetric strain

$$\varepsilon_v^i(t) \equiv (V(0^-) - V(t))/V(0^-) \qquad (1)$$

here t denotes the time elapsed since the increment of stress in question, V(t) is the volume of the sample at time t, and $V(0^-)$ is the volume of the sample just before the increment of stress. Here we want to emphasize that we define the quantity ε_v^i only as a convenient device to represent our data; although we call ε_v^i the "incremental volumetric strain", it carries no further meaning other than that which is defined by (1). In particular, the reader should not confuse it with the "dilatation" or the trace of the infinitesimal strain tensor. We make no hypothesis on the magnitude of the incremental volume change $V(0^-) - V(t)$.

The "incrmental volumetric strain" time graphs of the two single-step tests (isotropic stress increment $\Delta\sigma_m$ = 350, 700 kPa, respectively; temperature T = -15°C) and the first step of the multi-step test at T = -15°C ($\Delta\sigma_m$ = 70 kPa) are plotted in Figure 1. Although the 70 kPa test appears to be of extremely short duration, when the results of this test are plotted under an expanded scale (Figure 3), it is clear that a very low rate of "incremental volumetric strain" had been achieved before the next stress increment was applied.

Similar "incremental volumetric strain" time graphs were obtained for each step of the **multi**-step tests. All the "incremental volumetric strain" time curves are characterized by an "instantaneous" component, a relatively short period of time during which there is a rapid decrease in the rate of "incremental volumetric strain," followed by an attenuating creep at a very low rate. Except for the first few hours after increment of pressure, all the curves are fairly well fitted by an appropriate hyperbolic function of the form

$$\varepsilon_v^i(t) = t/(a + bt) \qquad (2)$$

here a and b are constants to be determined empirically for each curve; b is the reciprocal of the ultimate value of ε_v^i. As examples, the "incremental volumetric strain" time graphs of the two single-step tests are exhibited with their hyperbolic function in Figure 2; the corresponding curve-fitting for the first step of the three multi-step tests on sample A02 ($\Delta\sigma_m$ = 70 kPa; temperature at -15°C, -10°C, and -5°C, respectively) are shown in Figure 3. Examples from later steps of the multi-step tests are shown in Figure 4. "Being the kth step of a test" is identified in the figures by the caption "n = k."

A comparison of the three curves in Figure 3 indicates that sample A02 developed much larger incremental volumetric strains at the test temperature of -15°C than at the other two warmer temperatures. This is contrary to the intuitive expectation that warmer soil would be more compressible. This apparent anomaly could be attributed to the fact that sample A02 had been subjected to a history of loading and recovery before it was tested at the two warmer temperatures. One effect of that history of loading and recovery was that some water and solute were squeezed out during compression; this amount of water and solute was never recovered. Another manifestation of the loading-recovery history was the strain-hardening effect, which gave rise to the apparent anomaly above. In light of this strain-hardening effect, no conclusion could be drawn from our results regarding the influence of temperature on the compressibility of frozen soils undergoing isotropic compression.

An examination of the "incremental volumetric strain" time curves indicates that a major portion of the volume change that would result from an increment of hydrostatic stress occurred within the first few hours after the load increment. If we are interested only in the behavior beyond this brief period instead of the function $\varepsilon_v^i(t)$, a knowledge of the ultimate value of "incremental volumetric strain" should suffice. Since the "incremental volumetric strain" is defined only for a specific stress increment, another quantity should be defined for an arbitrary loading history. The ultimate volumetric strain ε_v^∞ of a sample is defined here for any loading history from time $t = 0$ to time $t = \tau$ as

$$\varepsilon_v^\infty \equiv (V_o - V_\infty)/V_o \qquad (3)$$

where V_o is the original volume of the sample before loading at $t = 0$, and V_∞ is the ultimate volume should the load at time τ be maintained constant from $t = \tau$ to $t = \infty$. We present our data of ε_v^∞ for the three multi-step tests in Figure 5. Here we treat A0215, A0210, and A0205 as if they were tests on three different samples. In other words, sample A0210 at time zero was simply sample A02 after it had been subjected to the multi-step test at -15°C, allowed to recover and brought to -10°C. In Figure 5, for instance, the value of ε_v^∞ at σ_m = 350 kPa for sample A0215 is the ultimate volumetric strain of the sample in question after five increments of pressure, each of which is of magnitude 70 kPa. The data presented in Figure 5 further corroborates the strain-hardening effect incurred on sample A02 by a previous history of loading and recovery.

In the definition of ε_v^∞, the sample in question is supposed to have been subjected to a history of loading that has the following property: for the entire interval $[0, \infty]$ of time under load the isotropic stress takes some constant value σ_m, except perhaps for some finite interval of time $[0, \tau]$ just after loading. For such a loading history we define the ultimate (secant) bulk modulus K^∞ of the sample in question by the expression

$$K^\infty \equiv \sigma_m/\varepsilon_v^\infty \qquad (4)$$

Values of K^∞ for our tests A0215, A0715, and A0815 are shown in Figure 6. In general, K^∞ as defined by (4) depends not only on the ultimate pressure σ_m but also on that part of the loading history before the isotropic stress is kept at the constant value σ_m. In Figure 6, however, K^∞ has more or less a constant value. This suggests that as far as the value of K^∞ is concerned a single-step loading and a multi-step monotonic-increasing loading with the same ultimate value of σ_m might have an almost identical effect; likewise, slightly different samples do not seem to make much difference. Yet our data are too scanty to support any definite conclusion. We hope we can clear up this point with further work.

When sample A02 was allowed to recover after the completion of a multi-step test, only about 50% to 70% of the volume change was recovered. We present the data regarding volume recovery in Table 2. That a significant portion of the volume change was permanent suggests that sample A02 could <u>not</u> be treated as a body of material with fading memory. If this fact could be generalized, it would have important implications for any attempt at a viscoelastic modeling of saline frozen soils. For example, a simple nonlinear hereditary creep constitutive equation (e.g., Vyalov, 1962, Equation (1-47')) that ties up attenuating creep under constant stress with ultimate complete recovery after unloading would be inadequate.

When the temperature of sample A02 was increased to -5°C, it was noted that the sample was no longer ice-bonded. Nonetheless, the behavior of the sample under isotropic compression was not substantially different from its behavior at the lower temperatures. The implication is that under isotropic compression the response of warm non-bonded permafrost, which we often encounter in the Beaufort Sea sediments, might be similar to that of its ice-bonded counterpart. From our results, of course, we cannot draw any conclusion on the response of non-bonded and ice-bonded permafrost to stresses other than hydrostatic.

CONCLUSIONS

1. Each of our frozen saline silt samples underwent attenuating creep upon compression by a constant hydrostatic stress. Except for the first few hours of loading, the creep curve could be well approximated by an appropriate hyperbolic function of the form given in (2) above. Over 70% of the ultimate volume change occurred within the first few hours of loading. The level of ultimate volume change (which came near to 4% of the original volume for our sample A08 at 700 kPa and -15°C) indicated that volume change during creep could be a significant factor in problems involving thick deposits.

2. When an additional hydrostatic stress was superimposed on a sample that was approaching equilibrium after creep under an isotropic stress, the soil underwent a new attenuating creep with characteristics substantially the same as those described in the preceding paragraph.

3. When the frozen saline silt was allowed to recover after a history of isotropic-stress loading, it could ultimately recover only part of the volume change occurred during the previous

loading. Our test results showed that up to 50% of the volume change could remain after recovery. This fact has important implications for any attempt at a viscoelastic modeling. A simple nonlinear hereditary creep constitutive equation that ties up attenuating creep at constant stress with ultimate complete recovery after unloading would be inadequate.

4. A history of loading and recovery had a strain-hardening effect on the frozen saline soil; so did a history of incremental loading.

5. At -5°C the saline soil was non-bonded, but the nature of its volumetric strain-time response to isotropic compression did not seem to be qualitatively different from the way it behaved when it was ice-bonded.

6. Even at a temperature as low as -15°C, water was squeezed out when the soil underwent creep under isotropic compression. This suggests that in modeling the creep behavior of frozen saline soils a quasi one-phase approach might be altogether inadequate; mixture theories should be considered.

ACKNOWLEDGMENT

This research is part of a broader investigation to study, by means of triaxial and pressuremeter tests, the multiaxial creep behavior of frozen saline soils that simulate those beneath the Beaufort Sea. The study has been supported by a grant from the Natural Sciences and Engineering Research Council of Canada.

REFERENCES

Baker, T.H.W., 1976, Preparation of artificially frozen sand specimens, DBR Paper No. 682 (NRCC 15349): Ottawa, National Research Council of Canada, Division of Building Research.

Baker, T.H.W., Jones, S.J., and Parameswaran, V.R., 1982, Confined and unconfined compression tests on frozen sands, in French, H.M., ed., Proceedings of the Fourth Canadian Permafrost Conference (The Roger J.E. Brown Memorial Volume): Ottawa, National Research Council of Canada.

Bragg, R.A., and Andersland, O.B., 1982, Strain dependence of Poisson's ratio for frozen sand, in French, H.M., ed., Proceedings of the Fourth Canadian Permafrost Conference (The Roger J.E. Brown Memorial Volume): Ottawa, National Research Council of Canada.

Diekmann, N., and Jessberger, H.L., 1982, Creep behavior and strength of an artificially frozen silt under triaxial stress state, Paper presented at the Third International Symposium on Ground Freezing, 22-24 June 1982, U.S. Army, CRREL, Hanover, New Hampshire, U.S.A.

Domaschuk, L., and Valliappan, P., 1975, Nonlinear settlement analysis by finite element: Journal of the Geotechnical Engineering Division, ASCE, v. 101, p. 601-614.

Klein, J., 1981, Finite element method for time-dependent problems of frozen soils: International Journal for Numerical and Analytical Methods in Geomechanics, v. 5, p. 263-283.

Klein, J., and Jessberger, H.L., 1979, Creep stress analysis of frozen soils under multiaxial states of stress: Engineering Geology, v. 13, p. 353-365.

Ladanyi, B., 1981, Mechanical behavior of frozen soils, in Selvadurai, A.P.S., ed., Mechanics of structured media, Part B: Amsterdam etc., Elsevier Scientific Publishing Company.

Mitchell, R.J., and Burn, K.N., 1971, Electronic measurement of changes in the volume of pore water during testing of soil samples: Canadian Geotechnical Journal, v. 8, p. 341-345.

Sayles, F.H., 1973, Triaxial and creep tests on frozen Ottawa sand, in Permafrost -- The North American Contribution to the Second International Conference, Yakutsk: Washington, D.C., National Academy of Sciences.

Vyalov, S.S., ed., 1962, The strength and creep of frozen soils and calculations for ice-soil retaining structures: Hanover, New Hampshire, U.S. Army, CRREL, Transl. 76, 1965.

Vyalov, S.S., Zaretsky, Yu.K., and Gorodetsky, S.E., 1979, Stability of mine workings in frozen soils: Engineering Geology, v. 13, p. 339-351.

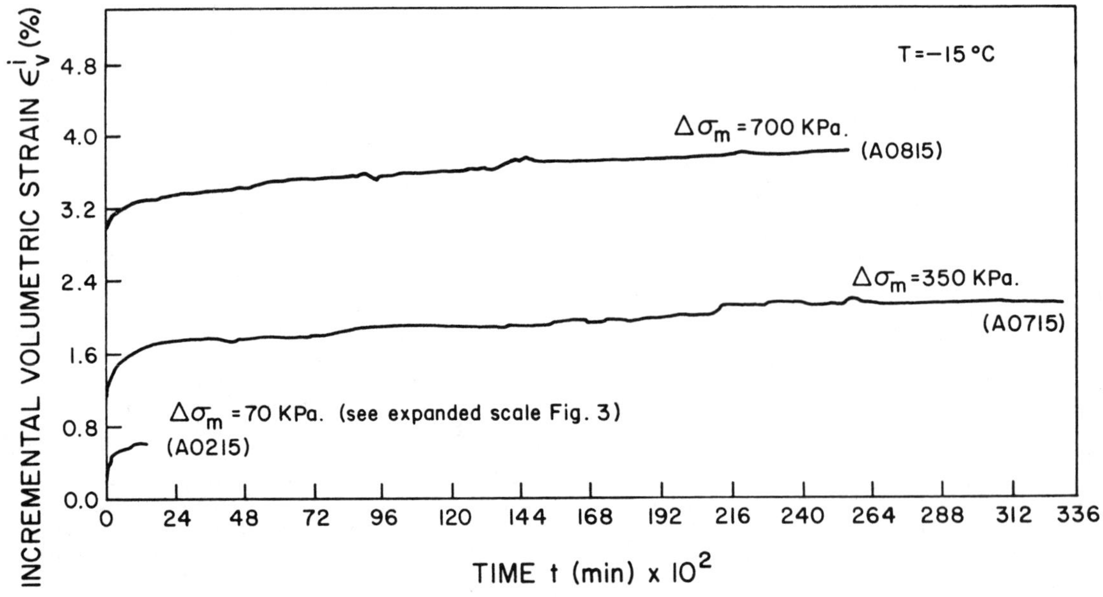

FIGURE 1 Incremental volumetric strain versus time for single-step loading.

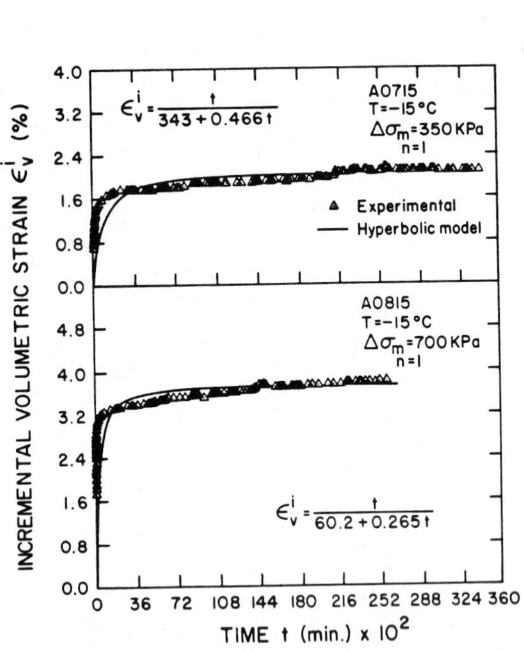

FIGURE 2 Hyperbolic curve-fitting of single-step loading test data.

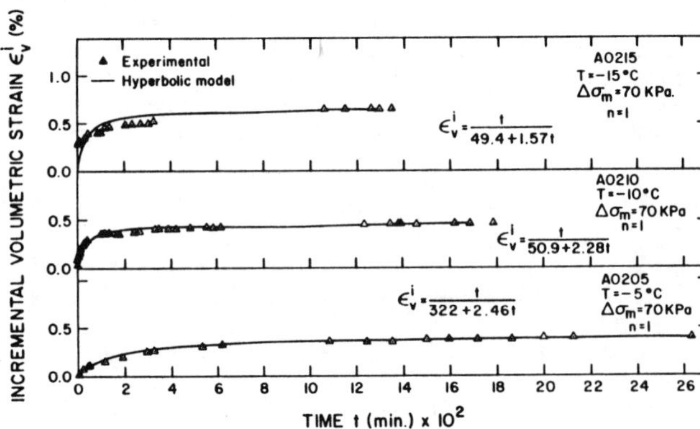

FIGURE 3 Hyperbolic curve-fitting; the first step of multi-step loading test data.

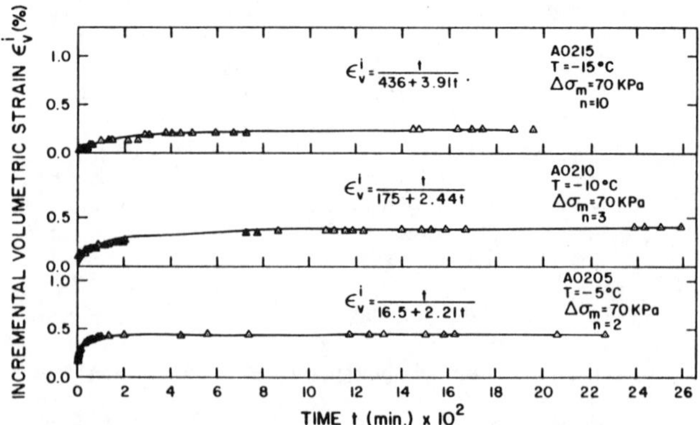

FIGURE 4 Hyperbolic curve-fitting; the nth step of multi-step loading test data.

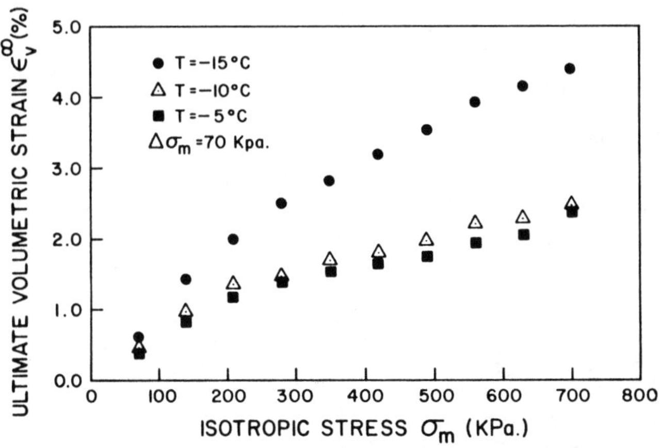

FIGURE 5 Ultimate volumetric strain versus isotropic stress for the multi-step tests.

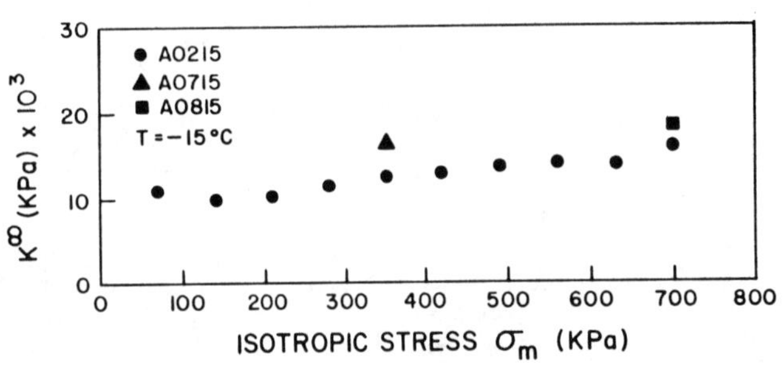

FIGURE 6 Ultimate (secant) bulk modulus versus isotropic stress for T = -15°C.

TABLE 1 Sample Details

	Compacted				Start of Test				End of Test			
Sample No.	W (%)	γ_D (KN/m^3)	e	S (%)	W (%)	γ_D (KN/m^3)	e	S (%)	W (%)	γ_D (KN/m^3)	e	S (%)
A0215	13.9	16.9	0.59	64	22.4	14.8	0.82	81				
A0210						14.9	0.79					
A0205						15.0	0.77		21.7	15.6	0.72	82
A0715	15.1	16.4	0.65	63	20.8	15.7	0.71	87	20.3	16.1	0.66	83
A0815	14.9	16.6	0.62	65	17.7	16.2	0.66	79	17.6	16.5	0.63	76

TABLE 2 Ultimate Volume Change

Sample No.	Vo (CC.)	ΔV_{wt} (CC.)	V_{NR} (CC.)	ϵ_v^∞ (%)	ϵ_p (%)
A0215	644.16	28.27	9.39	4.39	1.46
A0210	634.77	15.66	7.64	2.47	1.20
A0205	627.13	15.35		2.45	

V_{NR} = Non-recoverable volume change; ϵ_p = Non-recoverable part of the ultimate volumetric strain.

ROCK WEATHERING BY FROST SHATTERING PROCESSES

G.R. Douglas[1], J.P. McGreevy[2] and W. Brian Whalley[2]

[1]Menntaskolinn vid Hamrahlid, Reykjavik, Iceland,
[2]Department of Geography, The Queen's University of Belfast,
Belfast BT7 1NN, United Kingdom

We discuss two main types of control on the weathering of rocks by frost shattering. External controls are those concerned with temperature cycling which the rocks experience together with associated moisture variations. With data from Iceland, we show conditions under which air temperatures are a poor indicators of rock temperatures. Rock surfaces can be warmer than the air, even if the latter is below 0°C; or colder than the air. If snow or ice cover the rock surface freezing intensity can be damped. A knowledge of rock properties (internal controls) is important because water may freeze at a temperature other than at 0 ° C. This temperature may be different for different rocks. In the pores of an Icelandic hyaloclastite water froze at about -4°C, water in rock cracks may freeze at rather higher temperatures than this; tests in 'cracks' in plastic blocks suggest about -2 °C.

INTRODUCTION

Frost shattering is central to a consideration of many periglacial phenomena (Washburn,1979) and has traditionally been considered to be the main process responsible for rock breakdown in cold (high altitude and latitude) regions. There is increasing interest in the geomorphic effects of frost shattering and some geomorphologists are beginning to challenge conventional beliefs (e.g. White 1976, Thorn 1979) Here we comment on current knowledge of controls on the operation of frost shattering. We do not examine actual mechanisms of rock breakdown since these are considered elsewhere (McGreevy 1981). Rather, we discuss controls on the mechanisms which we have divided into two categories; external (temperature and moisture regimes at a local level) and internal (properties of the rocks). To illustrate specific points we draw on work conducted on rocks from Iceland. Attention has been focused upon a rock which is common on Icelandic cliffs but which, unlike basalt, is porous rather than jointed.

GENERAL CONSIDERATIONS

Geomorphologists have adopted two approaches in attempting to assess the geomorphic significance of frost shattering. First, an essentially geographic methodology has aimed at establishing global and continental zonations of frost shattering intensity on the basis of gross climatic parameters such as mean annual or monthly temperatures or precipitation (e.g. Peltier 1950, Brochu 1964). An extension of this approach is the use of standard meteorological data to explain patterns of rock breakdown e.g., small scale rockfalls (Church et al 1979), larger scale, perhaps seasonal events (Bjerrum and Jørstad 1957), or even longer term climatic effects (Grove 1972). In such cases, freeze-thaw frequency (as evaluated by recourse to air temperature data) has been directly related to intensity of rock breakdown, presumably by frost shattering. However, although rockfalls may be associated with freeze-thaw cycles this does not necessarily indicate that such cycles are actually causing rock breakdown; it may be that pre-loosened blocks are being released by the ice formation. The distinction between weathering of a rock (its separation from the rock mass) and the actual removal of a weathered rock is rarely made.

The second area concerns experimental determinations of micro-climatic controls from laboratory tests (see McGreevy (1981) for review). In its simplest form, this involves placing rocks in water, alternating the temperature above and below 0°C and observing the results (e.g. Martini 1967). Such investigations tell us something about the temperature and moisture conditions most conducive to frost shattering. This approach too has its limitations. There is no general agreement on testing procedures and various workers or laboratories employ different experimental conditions making comparison of results difficult. In addition, test conditions, although replicable may be unrepresentative of those experienced by rocks in the field. For instance, should the rock under test be of a certain size, be cut in a particular way? Should it be saturated with water under pressure? There is a pressing need to link field observations from various environments to laboratory tests. Thorn (1979) has recently indicated that application of 'Icelandic' or 'Siberian' temperature cycles (Tricart 1956) in experimental studies may be grossly misleading. However, further work is needed to define the extent of discrepancies between field and laboratory freezing patterns.

Both external (rate, intensity, duration and frequency of freezing) and internal (rock properties) controls on rock breakdown by frost shattering mechanisms are only now being fully investigated (Lautridou and Ozouf 1982). We consider this to be an important distinction as it helps to separate the complex factors involved in rock breakdown.

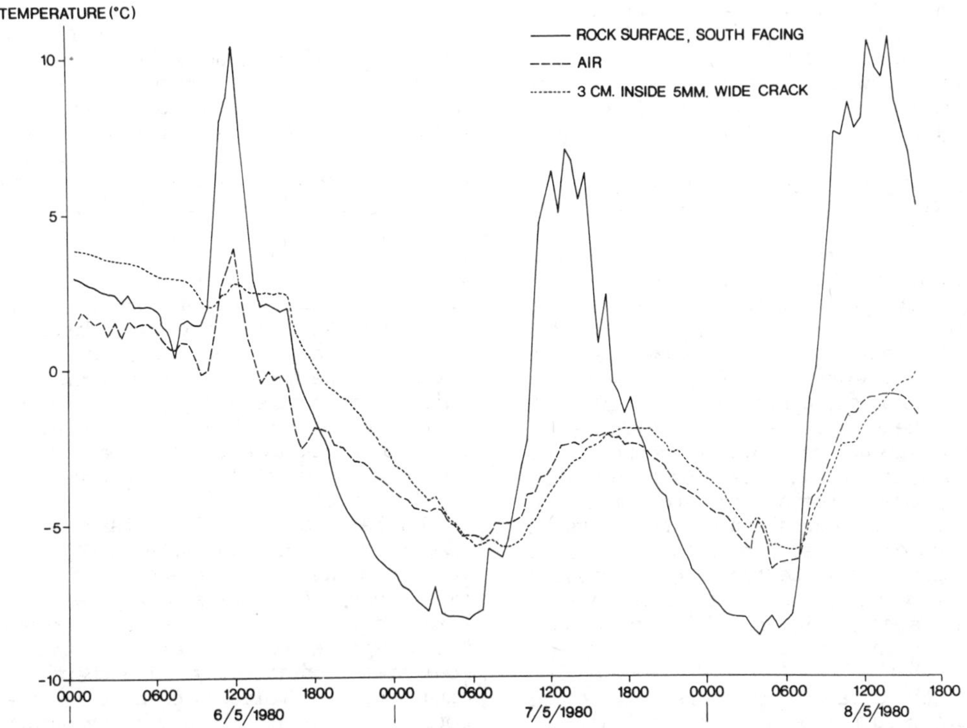

FIGURE 1 Rock surface, crack and air temperatures from Esja, Iceland.

EXTERNAL (METEOROLOGICAL AND CLIMATIC) CONTROLS

It has been realized for some time that air temperature data derived from standard meteorological station observations do not provide good indications of local climatic conditions (Washburn 1979). Several workers have identified difficulties in relating air temperature cycles to ground temperatures (e.g. Fahey 1973, Fraser 1959). Nevertheless, some authors still use temperature data from meteorological stations to infer processes some distance away (e.g. Church et al 1979). Data from Iceland (obtained over a two year period and including continuous records over a four month winter period) illustrate some of these difficulties, even with air tempertures being measured adjacent to the rock surface. Figures 1 and 2 show relationships between rock surface, crack and air temperature variations over selected periods, and illustrate ways in which air temperature data can provide misleading information about rock temperatures. Figure 1 shows a south-facing rock surface at a site in Iceland (Esja, 20km N-E of Reykjavik, 650 m a.s.l.) undergoing temperature fluctuations above and below 0°C in the presence of negative air temperatures. The crack temperature regime reflects changes experienced at the rock surface and does not appear to be closely related to air temperature. Here the use of air temperature to indicate the rock 'frost climate' would underestimate the frequency of rock freezing and thawing. Figure 2 shows the converse: air temperature fluctuates around 0°C quite frequently over a short time whilst the rock surface remains below 0 ° C. Air temperature data therefore overestimates freeze-thaw frequency in such instances. Clearly, use of air temperature data alone would invite erroneous inferences on the nature of geomorphic activity at the study sites.

There are three basic conditions where substantial disparities between rock and air temperatures would present difficulties in the interpretation of air temperatures and rock freezing characteristics viz:

1. Snow covering the rocks where air temperature fluctuations may be damped to give no or few 'frost changes' (ie., oscillations around 0°C).

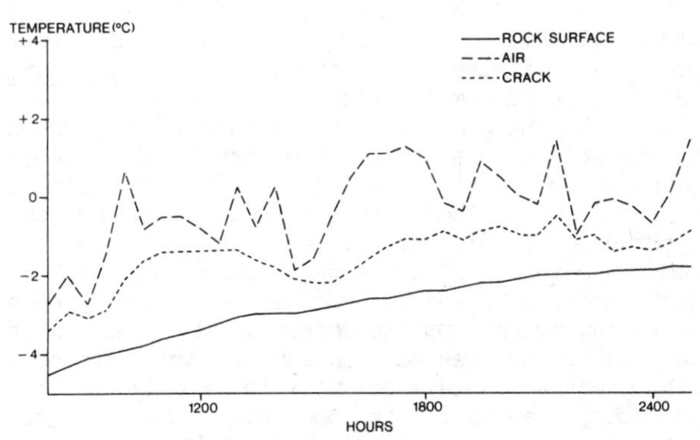

FIGURE 2 Rock surface, crack and air temperatures from Esja, Iceland, 14 December 1981.

2. Ice (verglas) covering the rock surface in certain conditions even where the air temperature is several degrees above 0°C.

3. Conditions where ice or snow does not cover the rock and there is relatively free heat exchange between rock, air and available liquid water.

Rock temperatures are damped with respect to air temperatures in the first two cases and exaggerated in the last.

It should be noted that existing data on air and rock temperature variations relate to short time periods. At present, there is no assessment of their variability over several years. Furthermore, we still require knowledge of antecedent conditions which allow freezing waves to penetrate into the rocks and not just fluctuate at or just below the surface (e.g. in Figure 2). Little is known about the moisture content of rocks. White (1976) has drawn attention to possible relationships between weathering and moisture content. He ventured to question whether rocks in natural situations could ever become >50% saturated with water and then experience freezing. At present we have no data from the sites we are examining in Iceland. However, some preliminary data from a site at 850 m altitude in Central Scotland (McGreevy 1982) suggests that a high (up to 92%) degree of saturation can be attained by rocks which are likely to be subjected to freezing conditions. It is certainly significant that most talus production from cliffs exposed to frost-shattering agencies comes from gully or couloir systems. It is under these conditions that water penetration into rocks is likely to be highest. Once a cliff has a coating of ice (verglas) then this is likely to reduce the amount of thermal change as well as water saturation it can receive. In a couloir however, maximum quantities of free water are available and, hence, frost shattering is also likely to be greatest for a given rock type.

INTERNAL (ROCK PROPERTY) CONTROLS

It has been pointed out (McGreevy 1981, McGreevy and Whalley 1982) that most laboratory experiments use rather small pieces of rock which may not be at all representative of field conditions. The samples used (the 'intact rock' of Attewell and Farmer (1976)) are likely to be quite different from the 'massive rock' (the rock in its <u>in situ</u> state). Tested samples are rarely > 10cm cubes yet this may be the size of material which, when seen in the field, is that which has already been weathered and transported from the <u>in situ</u> position on a cliff. Therefore, such tests ignore the importance of discontinuities such as joints, bedding planes, cleavage and schistosity, all of which provide planar weaknesses on which frost shattering mechanisms can operate. Thus the use of small laboratory-tested rock samples may indicate a much greater resistance to freeze-thaw in the laboratory than if it were possible to test a whole section of cliff. Testing small blocks of rock may yield inappropriate data on which to demonstrate the environmental controls of frost shattering. The testing of large blocks is very difficult and may not be possible for massive rocks.

It should be appreciated that the frost shattering process depends upon water in a rock being frozen (excluding here the hypotheses of hydration shattering or ordered water (White 1976) although these still require water to be in the rock). In this context it has become evident that there is a difficulty in accepting 0 °C as a critical temperature which must be crossed before water will freeze within rocks. Several workers have shown that water may not freeze until several degrees below 0°C in rocks and other materials. Various threshold freezing temperatures have been suggested between -2 and -10 °C (see McGreevy and Whalley (1982) for details and discussion). Our own investigations have provided some further data. In particular we have examined a rock called 'moberg'; this is an hyaloclastite, a subglacially-erupted tuff (Saemundsson 1979, Allen et al 1981). This type of rock is often of great thickness in the geological succession and is thus well exposed on cliff sections in Iceland.

Samples of moberg, taken from near our temperature recording site at Esja, were tested for freezing point depression characteristics. A small core of saturated moberg was contained in a brass capsule and placed in a freezing cabinet. The temperatures of both rock and capsule were monitored continuously using thermocouples and a data logger. Evolution of heat as water freezes in pores is indicated by a rock temperature increase (c.f. Davison and Sereda 1978, McGreevy and Whalley 1982). Figure 3 shows an instance of water freezing at -3.8 °C. Over five tests, the results showed a freezing threshold of -3.8 °C ± 0.5 °C.

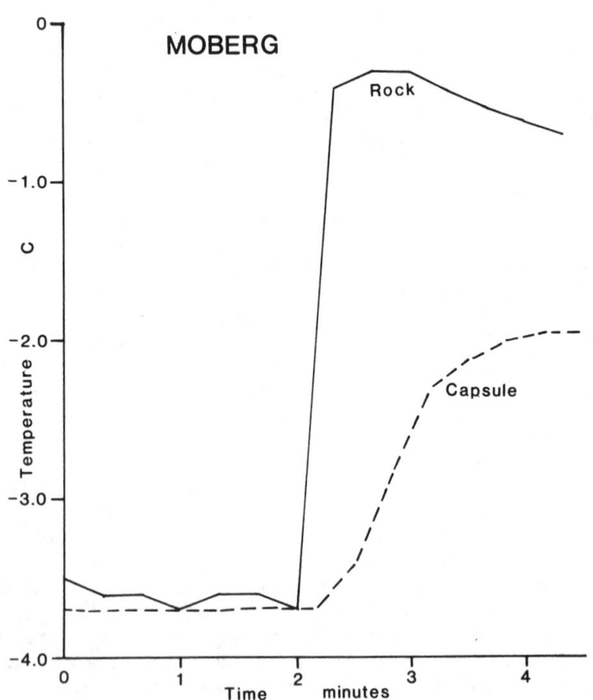

FIGURE 3 Graph of temperature of a saturated sample of moberg during freezing versus time. (The time axis has an arbitrary starting point.)

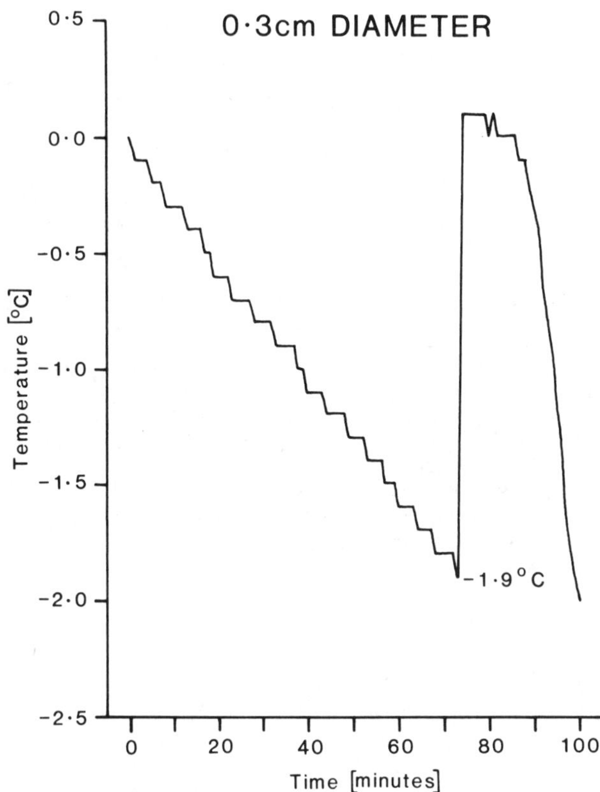

FIGURE 4 Graph of water freezing in a hole in a plastic block, 0.3cm diameter hole, versus time. (The time axis has an arbitrary starting point.)

The inference is that the rock temperature, at a given depth, must fall to at least -3 °C before the water can freeze. Such tests however say nothing about how the presence of cracks which may influence, not only the rock water content, but also the possibly increased susceptibility to damage by virtue of the crack itself. Indeed we do not know how the presence of cracks at various scales affects the breakdown of rocks (again, this is a size problem).

The cliff face at Esja also contains many basalt lava flows as well as moberg. The two rocks differ in terms of the discontinuities within them. The moberg generally lacks all but the regional joint system and mechanical breakdown occurs through the release of small granules of material. The basalt breaks down through the release of blocks of rock, the sizes of which are determined by the heirarchy of discontinuities (faults, joints and microcracks) within the rock mass (Whalley et al 1982). Thus in moberg, water is held and freezes in pores whilst in basalt, cracks of a variety of sizes are more relevant. Water freezing within the moberg can therefore be studied on small samples. As argued elsewhere (McGreevy and Whalley 1982), water freezing to release blocks of basalt is most likely to be associated with larger cracks and fractures rather than fine microcracks. Handling large samples is difficult so we have used a simple model to obtain some information on the freezing of water in cracks.

Instead of using basalt samples a machined perspex (lucite) block (5.5 x 8 x 12 cm) embedded in a 5cm thick jacket of expanded polystyrene was employed. Four holes (0.2, 0.3, 0.4 and 0.5cm diameter and 5cm length) were drilled into one face the face containing the holes remaining exposed. The holes were filled with water and the block placed in a freezing cabinet. Thermocouples inserted into the water-filled holes provided continuous temperature readings and, as with the moberg (Figure 3), showed the temperature increase due to water freezing.

The temperature curves obtained (Figure 4 is one example) show that freezing is initiated between -0.6 °C and -2.5 °C. The freezing of water in rock joints would therefore appear to occur at higher temperatures than in microcracks and pores. Clearly, the use of notches and holes in perspex as a model of rock oversimplifies the situation. For instance, the presence of clay minerals along crack walls could significantly reduce the freezing point of water (Anderson et al 1974). In addition, SEM studies (e.g. Sprunt and Brace 1974) show that cracks rarely, if ever, occur in isolation, and usually comprise one element in an intricate network of cracks of varying shapes and sizes. However, continued experimentation with simple models may throw some light on the effects of freezing in rocks so long as such possible complications are borne in mind.

CONCLUDING REMARKS

Field and laboratory studies of frost shattering must be viewed as complementary if its geomorphic significance is to be properly assessed. The degree to which conclusions based on laboratory studies can be extended to comment on rock weathering behaviour under natural conditions depends upon the degree to which simulated conditions approximate those experienced in nature. This must apply to both external and internal conditions as far as possible. Only by monitoring rock temperature and moisture regimes in situations of geomorphic interest (around snow patches, for example) and subjecting rocks to them can their geomorphic significance be evaluated. Ultimately, the weathering effects of temperature and moisture variations depend upon the properties of the rocks under consideration. For this reason it is important to acquire some knowledge of rock properties which are critical in determining the efficacy of frost shattering.

The relevant literature dealing with cold climate rock weathering leaves undefined the relationships between the variables of climate and rock properties. It is certain that much needs to be done to clarify the exact nature of such relationships.

ACKNOWLEDGMENTS

We thank NATO for a travel grant to assist our observations in Iceland, the Northern Ireland Department of Education for a Research Studentship for J.P.McGreevy and Dr B.J.Smith for constructive comments during our work. Ms J.Orr kindly abstracted data from our chart records, Gill Alexander drew the diagrams and John Allen did the photography.

REFERENCES

Allen, C. C., Gooding, J. L., Jercinovic, M., and Keil,.K, 1981 Altered basaltic glass: a terrestrial analog to the soil of Mars: Icarus, v. 45, p. 347-369.

Anderson, D. M., Tice, A. R., and Banin, A., 1974, The water-ice phase composition of clay/water systems: US Army Cold Regions Research and Engineering Laboratory, Research Report 322.

Attewell, P. B., and Farmer, I. W., 1976, Principles of engineering geology: London, Chapman and Hall.

Bjerrum, L., and Jørstad, F. A., 1957, Rockfalls in Norway: Norwegian Geotechnical Institute, Report F230.

Brochu, M., 1964, Essais de définition des grandes zones périglaciaires de globe: Zeitschrift für Geomorphologie, v. 8, p. 32-39.

Church, M., Stock, R. F., and Ryder, J. M., 1979, Contemporary sedimentary environments on Baffin Island, N. W. T., Canada: debris slope accumulations: Arctic and Alpine Research, v. 11, p. 371-402.

Davison, J. I., and Sereda, P. J., 1978, Measurement of linear expansion in bricks due to freezing: Journal of Testing and Evaluation, v. 6, p. 144-147.

Fahey, B. D., 1973, An analysis of diurnal freeze-thaw and frost heave cycles in the Indian Peaks region of the Colorado Front Range: Arctic and Alpine Research, v. 5, p. 269-281.

Fraser, J. K., 1959, Freeze-thaw frequencies and mechanical weathering in Canada: Arctic, v. 12, p. 40-53,

Grove, J. M., 1972, The incidence of landslides, avalanches and floods in western Norway during the Little Ice Age: Arctic and Alpine Research, v. 4,p. 131-138.

Lautridou, J. P., and Ozouf, J. C., 1982, Experimental frost shattering: 15 years of research at the 'Centre de Géomorphologie du CNRS': Progress in Physical Geography, v. 6, p.215-232.

McGreevy, J. P., 1981, Some perspectives on frost shattering: Progress in Physical Geography, v. 5, p. 56-75.

McGreevy, J. P., 1982, Some field and laboratory investigations of rock weathering, with particular reference to frost shattering and salt weathering: PhD Thesis, Queen's University of Belfast, Belfast, UK.

McGreevy, J. P. and Whalley, W. B., 1982, The geomorphic significance of rock temperature variations in cold environments: a discussion: Arctic and Alpine Research, v. 14, p. 157-162.

Martini, A., 1967, Preliminary experimental studies on frost weathering of certain rock types from the West Sudetes: Biuletyn Peryglacjalny, v. 16, p. 147-194.

Peltier, L. C., 1950. The geographic cycle in periglacial regions as it is related to climatic geomorphology: Association of American Geographers Annals, v. 40, p. 214-236.

Saemundsson, K., 1979, Outline geology of Iceland: Jokull, v. 29, p. 7-28.

Sprunt, E. S., and Brace, W. F., 1974, Direct observation of microcavities in crystalline rocks: International Journal of Rock Mechanics and Mining Science and Geomechanical Abstracts, v. 11, p. 139-150.

Thorn, C. E., 1979, Bedrock freeze-thaw weathering regime in an alpine environment, Colorado Front Range: Earth Surface Processes, v. 4, p. 211-228.

Tricart, J., 1956, Etude expérimentale du problème de la gélivation: Biuletyn Peryglacjalny, v. 4, p. 285-318.

Washburn, A. L., 1979, Geocryology, London,Arnold.

Whalley, W. B., Douglas, G. R., and McGreevy, J.P., 1982, Crack propagation and associated weathering in igneous rocks: Zeitschrift fur Geomorphologie, v. 26, p. 33-54.

White, S. E., 1976, Is frost action really only hydration shattering? A review: Arctic and Alpine Research, v. 8, p. 1-6.

HYDROLOGY OF NORTH SLOPE COASTAL PLAIN STREAMS

Brent Drage, Jeffrey F. Gilman, David Hoch, Leslie Griffiths

Peratrovich, Nottingham & Drage, Inc., Anchorage, Alaska 99503, USA

This paper presents general hydrologic characteristics of coastal plain streams located between the Sagavanirktok River and the Colville River on Alaska's North Slope. Data presented were gathered over the last five field seasons by the authors and others while working under contract with ARCO Alaska, Inc. Following a general discussion of the geographic and climatic setting for these streams, there is a discussion of several important hydrologic parameters which define the uniqueness of streams flowing in the permafrost environment. Permafrost is a key factor in defining drainage basins. From discharge measurements, values for typical slopes, Manning's roughness factors, and Coriolis energy coefficients are presented. Finally, a flood frequency analysis for these streams has been developed using a Log Pearson Type III distribution in conjunction with a regional skew weighting analysis. The regional skew weighting analysis is based on station skews for all gaged streams in Alaska north of 64° N latitude, a region approximately coterminous with the region underlain by continuous permafrost.

CLIMATE AND GEOGRAPHY

The arctic climate and low relief of Alaska's North Slope are two of the dominant independent variables controlling the evolution of the land surface and its hydrology. These strongly influence runoff and sediment yield which in turn affect the drainage network and channel shape. The larger drainages originate in the northern foothills of the Brooks Range. The drainage divides are easily definable and the streams form a dendritic pattern. Moving from the foothills region to the low relief coastal plain, drainage basins and stream patterns become poorly defined. Continuous permafrost and ground ice play a significant role in determining the hydrologic regime of the coastal plain. Surface features and materials are in a continuous state of change, being incorporated into the cycle of thaw-lake development and ice wedge polygon formation, resulting from changes in ground ice volume.

During the short summer season, the land surface is dominated by lakes and drained or partially drained lake basins. These flooded depressions and shallow lakes rimmed with sedge provide approximately one half of the wetland area suitable for aquatic habitat.

There is continuous daylight at this latitude from mid-April through the end of August. Direct solar radiation thus increases dramatically in the months prior to breakup.

Mean annual temperature ranges from -16° to -9°C. The monthly mean temperatures range from -34° to 10°C. The pronounced temperature gradient with latitude found on the North Slope can influence the timing of breakup. Streams with headwaters in the Brooks Range foothills begin to breakup several weeks before streams originating in the coastal plain.

Precipitation is extremely light along the coast, generally between 10 - 15 cm, but also shows a distinct gradient with altitude to higher precipitation in the Brooks Range. Approximately half of the annual amount falls as snow from September through May. Snowmelt runoff is concentrated in the period of late May and early June. Surface winds are strong along the coast but decrease inland. During the ice-free months, the predominant wind direction is from the east. Walker et al (1980) is a standard reference for climatic parameters of the region.

GENERAL HYDROLOGY

Eight small streams between the Kuparuk River and Colville River have been studied. These are shown in Figure 1 and listed in Table 1, with selected hydraulic parameters.

The difficulty of defining hydrologic parameters for North Slope streams is illustrated by the uncertainties attached to delineating drainage areas. The general terrain has little relief and it is extremely difficult to delineate drainage basins on USGS topographic maps. This is particularly true for the streams in the eastern part of the study area. The cycle of thaw-lake development and polygon formation strongly influence the drainage network shape; and due to the dynamic nature of these processes, there is evidence that the drainage patterns intermittently change. Because of this inherent problem, such characteristics as unit runoff lose some of their accuracy.

Characteristics of Larger Rivers

Three major drainages fall within the geographic area of this study; the Sagavanirktok River, the Kuparuk River, and the Colville River. These rivers differ from the smaller coastal plain streams in that their headwaters are in the northern Brooks Range. Approximately 50% of the Sagavanirktok drainage and less than 10% of the Kuparuk drainage lie above 600 m elevation. This characteristic often gives these drainages a hydrological behavior very different from the adjacent coastal plain streams.

Of these three drainages, the Colville is by far the largest with an area of 60,000 km^2. The Sagavanirktok and the Kuparuk follow with 14,898 km^2 and 9802 km^2, respectively. Spring breakup on these rivers generally occurs before the breakup of the coastal plain streams. This characteristic is due to the location of their headwaters in the Brooks Range foothills. The breakup front begins in this relatively warm and sunny region and proceeds northward at approximately 0.5 - 1 km/hour. Like the smaller coastal plain streams, the hydrographs of these rivers peak sharply during breakup. The majority of the year's discharge occurs at this time, with flows falling off dramatically for the remainder of the summer. It has been estimated that 78% of the Kuparuk's annual discharge occurs during the month of June, with flows decreasing dramatically for the remainder of the summer; 34% of the Sagavanirktok's discharge (at the Sagwon gaging station) occurs during June (Peterson, 1983). The decrease of the summer flow in the Sagavanirktok is more gradual than in the Kuparuk.

Freezeup occurs in late September or early October with little to no flow being recorded by January. Although channel flow is essentially halted by freezing, it is believed that some flow continues through the unfrozen riverbed gravels during the winter.

Characteristics of Coastal Streams

Morphology of the coastal plain streams is controlled by permafrost and ground ice. In tundra regions with high ice content and ice wedge formation, a beaded stream pattern is common which is typically composed of deep thaw pools connected by shallow channels (Figure 2). Pools form in localized areas where the ground has a very high ice content or disruption of the tundra has allowed greater depth of thaw. If the pool is large enough, permafrost beneath and along the sides thaws, pools deepen and expand. Water depth up to 6 meters has been measured in some of these pools. In the connecting channels depths typically range from 0.2 - 0.3 meter. During freezeup of the streams, the deep, low velocity pools form an ice cover first, and it can reach thicknesses of 2 meters by the end of winter. It is within the beaded pools that relatively large and thick ice floes can be generated during the winter and flushed down the streams during breakup.

The coastal streams are frozen for all but 3 or 4 months of the year. The flowing water freezes in September or October, and drifting snow makes the drainages virtually indistiguishable from the surrounding country until the end of May or early June when "breakup" occurs.

Coastal plain streams generally break up after the recession of the Kuparuk, Sagavanirktok, and Colville river hydrographs. Due to the lack of a long-term data base, it is difficult to formulate general statements about the timing of this process. This aspect of the breakup phenomenon, along with the concomitant potential for flood frequency prediction, is a facet of arctic hydrology requiring additional data acquisition and analysis.

The breakup of coastal plain streams can be characterized as a sudden event of short duration (Figure 3). Observations made over the past 4 years indicate the average length of a breakup event to be 10-12 days. Peak discharge is generally reached within 4-5 days after initial flow, giving the typical hydrograph an asymmetrical, right-skewed shape. The increase in discharge from virtually zero up to 100m^3/s within a 2- to 5-day period can give the left limb of this hydrograph an extremely steep slope. While bimodal-shaped curves have been observed on the coastal plain streams, this multi-peaked behavior appears to be relatively unusual; breakup, once it occurs, generally advances to a single peak discharge and concludes with a smooth recessional phase of slightly longer duration than the rise.

Breakup proceeds in a south-to-north direction starting with the upper reaches of the streams and proceeding downstream. Pushing slush ice before it, the breakup front has been observed to advance at a rate of up to 1 - 2 m/s. The drainage basins of these streams are choked with deep snow at the time of breakup and the presence of this snow greatly influences the breakup behavior (Figure 4). In its south-to-north migration, the breakup can best be described as a series of surging events: periods of backwater and ponding as water is impounded behind snowdrifts, ice jams, or man-made obstructions (e.g., culverts) alternating with events of relatively rapid advance as flood waters incise and undercut these constrictions (Figure 5). Channel geometry is continuously changing as both the snow and the bottomfast ice is eroded and/or melted by the flood (Figures 6 and 7). Water surface elevation, channel cross sectional area, bed roughness, water velocity, ice load, and backwater effects can change continuously during the dynamic breakup period. The recession of the hydrograph brings the streams back into their banks, and hydrologic parameters attain a stability for the remainder of the ice-free season (Figure 8 and 9).

All coastal plain streams carry an ice load during breakup, although the quantity of this load can vary greatly from drainage to drainage. Likewise, the year-to-year change within a single stream is highly variable. However, in general, the only ice observed of significant strength and thickness originates from the deep pools. It was observed during the 1981 study that the four major western drainages (East Kalubik Creek, Kalubik Creek, Miluveach River, and Kachemach River) maintained a very significant snow cover (Figure 7). The periodic calving of these formations can add greatly to the ice load of these streams, although it should be noted that such slush floes are considerably softer than bottom ice/surface ice floes. Bottom ice has been observed to be present in the beds of all coastal plain streams throughout the breakup period, although it is observed to diminish significantly throughout the 10- to 12-day breakup period.

An extremely valuable parameter for future study and engineering design is the Mannings "n" value, or roughness factor. Based on measurements taken during the 1981 and 1982 field season, "n" values were calculated for the streams observed during breakup. These are presented in Table 1. An interesting phenomenon is the relatively low values calculated for the peak flows, and the higher values for the rising and falling limbs of the hydrograph. One possible explanation for this might be the roughness induced by snowdrifts, channel ice, and tussock floodplain, so that the net effect would be a higher "n" value when the stream is within its banks or covering the floodplain with depths small relative to the tussock heights. The "n" value is continuously changing due to snowpack melt along stream boundaries, calving from snowbanks, and the melting and release of bottom-fast ice.

The consistency of the computed "n" values (Table 1) and the computed energy coefficients for each stream below demonstrate a reasonable level of reliability for the data gathered thus far. The energy coefficient is a measure of velocity distribution over a channel section. According to Chow (1959), experimental data indicate that the value of the energy coefficient varies from about 1.03 to 1.36 for uniform natural channels. The value is generally higher for small channels such as those characteristic of coastal streams and lower for larger, deeper streams.

ENERGY COEFFICIENTS FOR COASTAL PLAIN STREAMS

Stream	Stream Average
Kup Trib 1	1.73
Sakonowyak	1.25
East Creek	1.48
Ugnuravik	1.86

FLOOD FREQUENCY ANALYSIS

As additional data and observations of coastal stream behavior are acquired, it is appropriate to review previous flood frequency analyses and revise them if the new data warrant.

Previous analyses utilized a regional approach based on historical data gathered by the United States Geological Survey at the Putuligayuk River near Deadhorse (drainage area = 456 km^2), Nunavak Creek near Barrow (drainage area = 7.23 km^2), and Happy Creek at Happy Valley Camp near Sagwon (drainage area = 89.4 km^2). Utilizing a modified Log Pearson Type III Flood Frequency methodology, unit runoff values for each recurrence interval were determined and multiplied times the drainage area of each stream to compute the flood magnitudes at each recurrence interval.

It was decided to compute new frequency curves using data from Putuligayuk, Kuparuk Trib. 1, East and Ugnuravik. Nunavak Creek and Happy Creek were eliminated from the analysis because of their distance from the study area and different hydrological regime.

The latest Water Resources Council method as described in their Bulletin 17B "Guidelines for Determining Flood Flow Frequency" (Sept. 1981) was used for the analysis. This technique has been recommended by the Council for use in all Federal projects involving water and related land use and has also been recommended for State and private organizations. Even though one technique may not be applicable to all the various geographical regions within the United States, uniformity of utilizing this technique for comparison with other regions, coupled with our relative sparse data base which makes it difficult to test the technique, it is reasonable to use the Water Resources Council method at this time.

Station skews were computed based on each stream's historical record. These station skews were weighted with a regional skew of 0.0117 based on a mean of station skews for 60 USGS gaged streams north of 64° latitude. The weighted station skews were used in a frequency analysis for the four area streams.

Because the focus of the monitoring program has been on streams with drainage areas between 75 km^2 and 500 km^2, unit runoff values have been defined only for this range. Because streams with smaller drainage areas have different basin characteristics, values presented here should not be extrapolated for computing coastal plain stream flood magnitudes with a drainage area less than 75 km^2.

The upper and lower confidence limits for levels of significance of 5 and 95% are computed by the procedures outlined in Appendix 9 of Bulletin 17B. The confidence limits are expressed in terms of error ranges in Table 2. The 95% confidence limits are excessive due to the small data base.

CLOSURE

The problems in defining the hydrologic characteristics of the coastal streams discussed in this paper only typify the general lack of detailed arctic hydrological data. However, with the existing data in-hand, one can gain an understanding of the general hydrological characteristics of the coastal plain streams. Expanded monitoring programs carried out over a period of years may answer some of the more perplexing problems, and further refinements of the flood recurrence intervals will be possible. It is hoped that the data presented in this paper will serve as a baseline for future hydrologic studies in permafrost dominated regions such as Alaska's North Slope.

REFERENCES

Chow, V.T., 1959, Open-Channel Hydraulics: New York, McGraw-Hill

Lewellen, R.I., 1972, Studies on the Fluvial Environment, Arctic Coastal Plain Province, Northern Alaska: Published by the Author, Littleton, Colorado, 2v., 282 p.

Peterson, L.A., in press, North Slope Water Problems

Scott, K.M., 1978, Effects of Permafrost on Stream Channel Behavior in Arctic Alaska, Geological Survey Professional Paper 1068, 19 p.

Walker, D.A., Everett, K.R., Webber, P.J., and Brown,J., 1980, Geobotanical Atlas of the Prudhoe Bay Region, Alaska: U.S. Army Corps of Engineers Cold Regions Research and Engineering Laboratory, Hanover, New Hampshire

TABLE 1 1981 Hydraulic Parameters: Coastal Plain Streams

STREAM	DATE	DISCHARGE Q (cm/s)	SLOPE** S_o	CALCULATED MANNINGS "n"
Kup Trib 1	6/07/81	13	5.2×10^{-4}	0.040
	6/09/81	88	5.4×10^{-4}	0.028
	6/11/81	39	6.5×10^{-4}	0.029
	6/10/82	94	6.6×10^{-4}	0.021
	6/13/82	17	"	0.054
	6/16/82	-	9.3×10^{-4}	-
Sakonowyak	6/09/81	33	6.2×10^{-4}	0.023
	6/11/81	16	9.3×10^{-4}	0.034
	6/11/82	12	6.7×10^{-4}	0.027
	6/14/82	8.8	"	0.029
	6/18/82	-	4.7×10^{-4}	-
East Creek	6/05/81	3.9	1.1×10^{-3}	0.051
	6/10/81	23	6.0×10^{-4}	0.029
	6/12/81	12	7.8×10^{-4}	0.033
	6/11/82	5.3	7.2×10^{-4}	0.046
	6/13/82	6.0	"	0.030
	6/18/82	-	7.8×10^{-4}	-
Ugnuravik River	6/04/81	5.4	Not Available	-----
	6/09/81	18	1.3×10^{-3}	0.045
	6/12/81	5.7	1.5×10^{-3}	0.051
	6/11/82	3.8	1.4×10^{-3}	0.062
	6/15/82	7.5	"	0.042
	6/16/82	-	1.5×10^{-3}	-
East Kalubik	6/10/81	11	2.3×10^{-3}	*
	6/12/81	8.3	1.9×10^{-3}	*
	6/12/82	4.4	1.8×10^{-3}	0.038
	6/14/82	4.6	"	0.029
	6/18/82	-	1.1×10^{-3}	-
Kalubik	6/08/81	8.8	2.6×10^{-4}	*
	6/11/81	5.1	2.4×10^{-3}	*
	6/12/81	3.4	1.0×10^{-3}	0.030
	6/12/82	2.9	-	*
	6/14/82	3.0	-	*
	6/18/82	-	2.3×10^{-5}	-
Miluveach	6/08/81	38	7.5×10^{-4}	0.026
	6/11/81	45	7.7×10^{-4}	0.020
	6/12/81	32	7.2×10^{-4}	0.019
Kachemach	6/08/81	45	5.3×10^{-4}	0.019
	6/11/81	32	5.6×10^{-4}	0.023
	6/12/81	19	5.2×10^{-4}	0.029

*Due to backwater from snow drifts, Mannings 'n' values are not representative.
**Slopes were not measured in 1982 on same dates discharge measurements were taken.
Slopes used to calculate roughness factor are averages of 1981 and 1982 measured slopes.

TABLE 2 Kuparuk Development Area Frequency Analyses

	Putuligayuk River near Deadhorse		Kuparuk Tributary 1 near Spine Road		East Creek near Spine Road		Ugnuravik River near Spine Road	
Return Intervals	cm/s	cm/km^2	cm/s	cm/km^2	cm/s	cm/km^2	cm/s	cm/km^2
10	167	0.37	166	0.70	62	0.51	72	0.86
25	205	0.46	212	0.89	94	0.77	108	1.28
Drainage Area	456 km^2		241 km^2		124 km^2		85 km^2	
Years of Usable Record	8		5		6		6	
Station Skew	-0.2485		0.0697		-0.2609		0.0666	
Weighted Station Skew	-0.2		0.1		-0.2		0.1	
Confidence Limits								
10 yr.	-27 to 82%		-32 to 182%		-51 to 414%		-45 to 291%	
25 yr.	-29 to 108%		-36 to 281%		-55 to 659%		-49 to 470%	

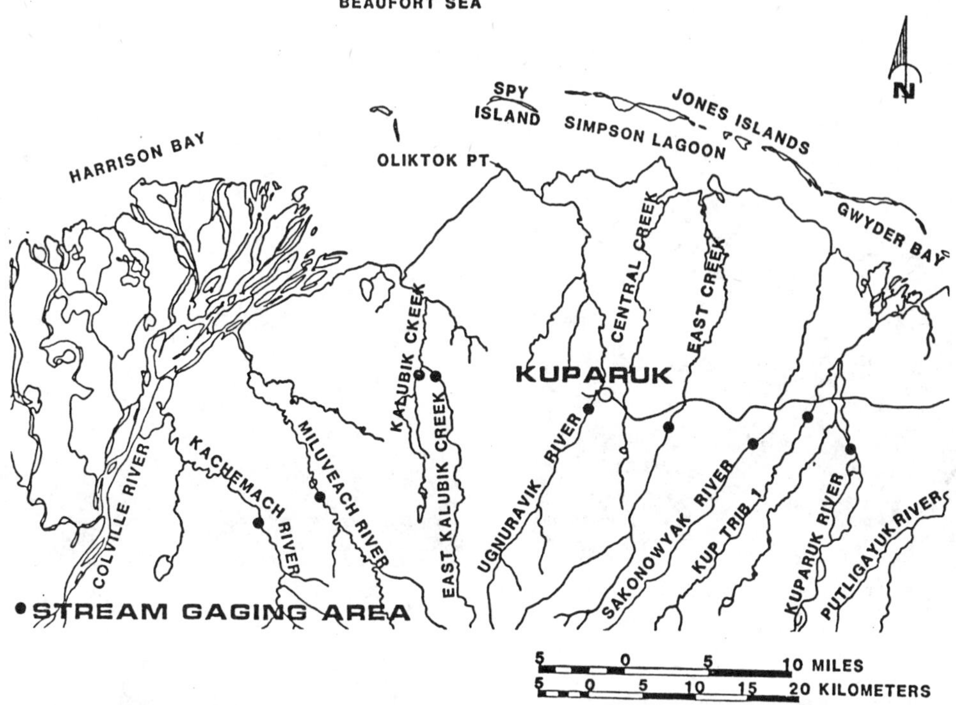

FIGURE 1 Study Area

FIGURE 2 Kalubik Creek June 7, 1982

FIGURE 4 Miluveach River before Breakup

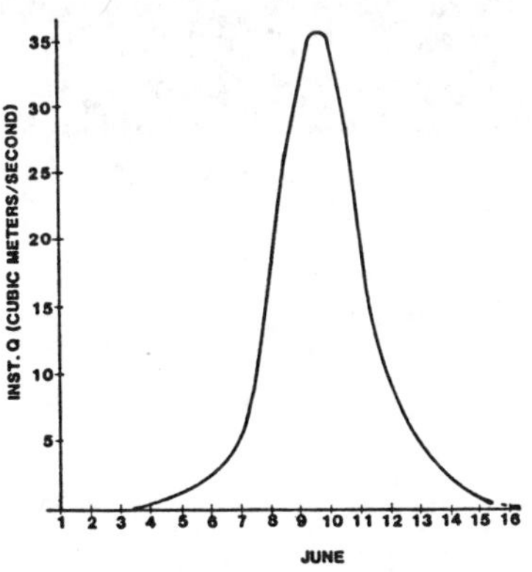

FIGURE 3 Sakonowyak River 1981 Hydrograph

FIGURE 5 Ugnuravik River at Peak Flow 1980

FIGURE 6 Streamgaging on East Kalubik
June 10, 1981

FIGURE 8 Ugnuravik River June 4, 1981

FIGURE 7 Snowbank on east bank of Miluveach River

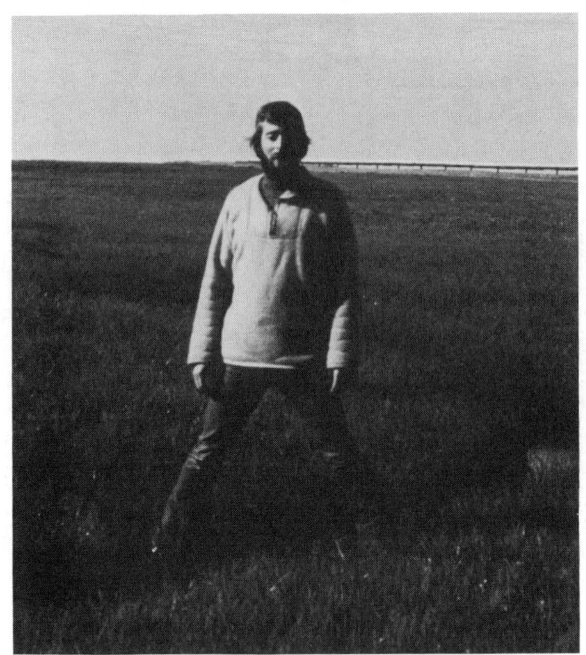

FIGURE 9 D.Hoch astride Ugnuravik River
July 1982

VIBRATORY MODEL PILE DRIVING IN FROZEN SAND

[1]Dufour, S., [2]Sego, D.C. and [2]Morgenstern, N.R.

[1]Geocon Inc.,1311 9th Avenue S.W. Calgary, Alberta, T3C 0H9 Canada
[2]Department of Civil Engineering, University of Alberta
Edmonton, Alberta, T6G 2G7 Canada

Laboratory modeling of vibratory pile driving in permafrost has shown that the strength of the material at the tip has to be overcome by plastic yielding for penetration to occur. This is contrary to the view that melting of the soil at the pile tip dominates the process. Various factors influencing pile penetration have been studied, and the energy balance of the pile penetration mechanism has been developed.

INTRODUCTION

Vibratory driving in permafrost has been applied in the Soviet Union for several years. In North America, vibratory driving was studied in the late 1960s as an alternate piling method on a pipeline project (Huck and Hull, 1971; Hull, 1977). In both these studies and Soviet work (Vyalov and Targulyan, 1968; Vyalov et al., 1969) melting of permafrost was observed around the pile when driving, and it was concluded that penetration was a consequence of soil melting at the tip. Resonant vibratory driving, as used in North America, was singled out as the most efficient method of transforming mechanical energy into thermal energy at pile tips to allow melting and hence pile penetration.

The mechanism that allows transformation of energy at the pile tips is at best unclear. Hence, an experimental program was developed to model vibratory driving of piles in permafrost. The objectives of the program were to find the penetration mechanism of piles driven by vibrations into permafrost and to study the influence of the following variables on the penetration rate (rp): (1) bias surcharge (m_s) (i.e., the dead weight, not acting dynamically and sitting on top of the driver); (2) driving frequency (f); (3) soil temperature (θ); (4) soil density (ρ_d).

This paper first describes the experimental procedures and then presents experimental results and conclusions.

EXPERIMENTAL PROCEDURES

Experimental Facilities

The tests were conducted in a 3.35 x 4.0 m insulated cold room that was kept cool by two fan-operated refrigeration units capable of maintaining the air temperature within 0.8°C of the required temperature. The room could be kept as low as -20°C. A 24-hour electrical alarm system prevented accidental defrosting of the facility. Access to the cold room was through a double door air lock, which reduces inflow of warm air from outside in order to minimize the temperature variation within the cold room.

The electronic equipment controls were kept outside the cold room, from which visual access to the test set-up was gained through a double thermal-pane window.

A buttress support system for the vibratory exciter and a sample pedestal were supported directly on the concrete floor rather than on the insulated false floor (see Figure 1). A 12.5 mm thick aluminum plate was placed on the concrete sample pedestal to ensure a smooth surface. Finally, a cooling coil was embedded into the concrete pedestal near the surface. The coil is intended to circulate a refrigerant to form a barrier to the heat flow emanating from the concrete floor below.

The model pile driver was an electromagnetic vibration exciter type 4812 mounted on body 4801S as made by Bruel & Kjaer. The exciter was bolted upside down to a piston passing through a guiding box coated with teflon to reduce friction. A 76 mm wide webb belt was attached to the top of the piston on one side and, after travelling over the two rollers, to a counterweight support. The details of the system are shown in Figure 2.

In order to minimize intolerable loads on the exciter core of the exciter, a 9.5 mm thick rubber pad was placed between the exciter table and the pile attachment plate. An electric load cell is used to attach the model pile to the attachment plate, which is bolted to the exciter table. Because of the small diameter of the model pile (25.4 mm), the accelerometer used to measure pile acceleration was placed near the edge of the aluminum plate.

A four-channel FM recorder, Bruel & Kjaer type 7003, was used to record the test signals for subsequent analysis. The channels were used as follows: (1) to monitor the pile acceleration; (2) to monitor the pile load; (3) to monitor the body waves through the sample from a triaxial accelerometer embedded into the sample.

Thermocouples were used to monitor the temperature within the sample. Platinum thermistors monitored pile tip temperatures during driving. A data logger was used to record the vertical pile displacement as measured by using a linear voltage displacement transducer.

Finally, a two-channel oscilloscope displayed the load and acceleration traces of the pile during the experiment. A Polaroid Land camera provided photographs of the waves directly from the oscilloscope screen at appropriate time intervals during an experiment.

Experimental Material

Ottawa sand was used as the experimental material because of its known properties. The material was placed in a loose and dense state simply by changing the water content and the placement technique. A typical grain size curve is given in Figure 3. The samples, 500 mm in diameter and 400 mm in height, were frozen uni-dimensionally from the bottom upward under a high temperature gradient by placing dry ice beneath the sample (see Dufour, 1981 for a complete account of the technique and apparatus). This technique ensures uniformity of the ice distribution and physical character of the sample.

The pile used in the experiments was 300 mm long at the start of the experimental program. The pile was tubular and made of steel with a wall thickness of 1.3 mm and an outside diameter of 25.4 mm. The end of the tube was trimmed flat and was not reinforced.

Driving

The resonant frequency of piles is dependent upon the compressional wave velocity in the pile and the length of the pile (Bernhard, 1967; Kovacs and Michitti, 1970). Hence a model pile would have a resonance frequency in excess of 3000 Hz. The best driving was achieved in the laboratory at the pile-soil resonance. This is called rigid body type and has been described by Schmid (1969). Smart (1969) also found that piles essentially behave as rigid bodies at frequencies less than 25% of resonance frequency.

The four parameters that can be controlled (i.e., density, temperature, bias surcharge (m), and driver frequency) were changed to study the influence of each on the pile penetration rate (rp). Thirty-two model pile driving tests were carried out for this experimental program.

EXPERIMENTAL RESULTS

Observations

On the average, loose samples had a dry density and water content of 1.66 tonnes/m^3 and 19.7%, while dense samples had a dry density and water content of 1.80 tonnes/m^3 and 11.7%. The total density of all samples was 2.00 tonnes/m^3.

After driving, it was noticed that the outside of the pile was worn so as to form a sharp edge. There was no evidence of excessive wear on the inside of the pile tip. It is estimated that the pile abraded in length by 2 to 3 mm during about 25 driving tests.

In all tests, a thawed zone formed around both the outside and the inside of the pile. This film was contaminated by steel dust produced by the abrasive action of sand on the oscillating model pile. The dust gave the thawed zone a grey color that allowed easy measurement of the extent of the melted zone. This is illustrated on Figure 4. The melted zone varies in thickness from 1.5 to 2.5 mm. The maximum temperature recorded at the film-pile interface was +4°C. The average temperature was around +2°C. The film temperature was constant throughout the length of the pile. It is interesting to note on Figure 4 that there was no apparent disturbance of the soil ahead of the pile tip. It was also noted that in addition to very fine steel dust in the thawed film, there was a considerable amount of crushed sand particles. Grain size analyses performed on representative samples of thawed film material, presented in Figure 3, show that 12% of the sand grains were crushed.

Pile Acceleration and Soil Response Load

Since the model pile behaved as a rigid body, as previously explained, the load signal recorded at the top of the pile corresponds to the soil response at the pile tip. Typical soil response signals on the oscilloscope are basically composed of sharp impact peaks separated by periods of negligible force applied to the soil, see Figure 5. The pile acceleration signals are also composed of sharp impact peaks occurring at the same time as the load peaks. Between acceleration impact peaks, there is a slightly distorted half sine wave. This distorted half sine wave corresponds to the upward motion of the pile, during which the pile loses contact with the soil. Those two wave forms are characteristic of the Instability Domain of driving force in which the maximum tip resistance is reached (see Schmid, 1969). Hence driving is due to an elasto-plastic deformation of the soil at the pile tip rather than to melting, which would then occur after penetration. The same conclusions developed by Littlejohn and Rodger (1980) for unfrozen soils apply to frozen soils: penetration will occur as long as the total dynamic and surcharge forces are larger than the soil resistance. Previous observations regarding crushed sand particles and pile wear suggest a bearing-capacity type of failure in the frozen soil at the pile tip. No evidence of melting or soil disturbance was apparent ahead of the pile tip. This evidence supports the yield theory as the cause of penetration.

Penetration

Figures 6 and 7 show typical penetration versus time curves. The curves exhibit in general a constant penetration rate (rp). Tests were stopped either because the pile had penetrated its full length or because soil reactions developed at the pile tip were sufficient to prevent penetration. The development of resistance is due mainly to an increase in side friction caused by either a change in pile alignment or by damping of vibrations. Details of all tests are given in Dufour (1981).

Effect of Bias Surcharge

The effect of bias surcharge (m_S) on penetration rate (rp) is summarized in Figure 8. There is a minimum bias surcharge (m_S) below which no penetration can occur. Since impact peaks do not develop on the soil response curve, insufficient energy is delivered to the soil to initiate yielding. As the bias surcharge (m_S) increases, more energy is transferred to the soil and the plastic work done on the soil increases as impact peaks grow. After an optimum bias surcharge (m_S) for a given combination of soil and frequency is reached, a decrease in penetration rate (rp) is observed. This decrease is associated with a decrease in driver efficiency. The vibration amplitude dampens as the pile is held in contact with the soil.

Effect of Soil Temperature

The effect of soil temperature on penetration rate (rp) is summarized in Figure 9. The penetration rate (rp) generally decreases as the soil temperature is lowered because strength increases with decreasing temperature. Depending upon the combination of parameters, the soil can inhibit plastic deformation and hence penetration rate (rp) converges toward zero.

Effect of Driving Frequency

The effect of driving frequency on penetration is summarized in Figure 10. There is a frequency below which no penetration will occur because the bias surcharge (m) becomes too high for the displacement amplitude. As the frequency decreases, the displacement increases but the acceleration decreases. Hence, the load carrying ability of the driver is reduced. This is also reflected in the fact that the force threshold described by Ghahramani (1967) varies with frequency for a given soil and pile. If at the lower frequency threshold the bias surcharge (m) is reduced to allow vibrations to occur, impact spikes in the load versus time plot do not develop and no penetration occurs. If all variables are held constant and the frequency is increased, the penetration rate increases to a maximum and then decreases as the frequency increases. Displacements at frequencies higher than this upper threshold are too small to allow for transfer of sufficient energy to the soil and, as a result, the pile will not penetrate. Also, impact peaks do not develop at higher frequencies.

Effect of Soil Density

Because of small density variation no influence of density on penetration rate (rp) was detected. When one analyzes the soil response signals, as will be discussed later, it turns out that penetration in dense frozen soil required more energy than penetration in loose frozen soil.

ENERGY REQUIRED TO CREATE A PERMANENT DEFORMATION

It has been shown previously that penetration of the pile was due to an elasto-plastic deformation of the soil beneath the pile tip. If a force deformation relationship is known or assumed, plastic work can be estimated.

The amount of energy spent to penetrate the pile 1 cm can be calculated as a function of the soil response load. From an analysis of the wave form data Dufour (1981) has shown that

$E_1 = J/100$

where

J = soil response load (N)

E_1 = energy spent to penetrate the pile 1 cm (Joule/cm)

100 = constant (cm/m). (1)

Hence, the curves on Figure 11 demonstrate that dense frozen soil requires more energy to be penetrated than loose frozen soil. According to the "melting theory," dense frozen sand should have required less energy since its ice content is half that of loose frozen sand.

Analysis reveals that the distribution of energy in the soil-pile system is not simple. Input from the driver is used for strain energy, plastic work, grain crushing, thermal losses from grain to grain and grain to steel friction, and elastic excitation of the soil mass. Some of these features are interactive. Nevertheless, it is apparent that plastic deformation at the pile tip is essential to achieve practical pile-driving in permafrost.

CONCLUSION

This study reveals the dominant mechanism for piles vibrated into permafrost. Several factors affecting the elasto-plastic reaction at the tip were also examined. Only one requirement for penetration transpires from all the studies: the strength of the material has to be overcome by yielding of frozen ground.

The purpose of further research is ultimately to provide good engineering prediction methods for vibratory pile driving in permafrost. A theoretical model that would facilitate prediction of whether or not there will be penetration would allow design of equipment for large scale piling in offshore conditions for example. Understanding of the mechanics of penetration is now sufficient to allow further work on a theoretical basis. The importance of gathering more experimental data should not be overlooked, and every possible field case must be studied to provide a wider base to the present understanding of vibratory driving of piles in permafrost.

ACKNOWLDGMENTS

The authors would like to acknowledge funding from the Canadian Department of National Defense.

REFERENCES

Bernhard, R.K., (1967), "Resonant Curve Analysis," United States Army Cold Region Research Engineering Laboratory, SR 97, Hanover, N.H.

Dufour, S., (1981), "Vibratory Pile Driving in Frozen Sand," M.Sc. thesis, University of Alberta, Edmonton, Alberta, Canada, 211p.

Ghahramani, A., (1967), "Vibratory Pile Driving , Ultimate Penetration and Bearing Capacity," Ph.D. Thesis, Princeton University, Princeton, N.J., 106p.

Huck, R.W., and Hull, J.R., (1971), "Resonant Driving in Permafrost," Foundation Facts, Volume 7, Number 1, pp. 11-15.

Hull, J.R., (1977), "Placing Piles in Permafrost," World Construction, December, pp. 64-67.

Kovacs, A., and Michitti, F., (1970), "Pile Driving by Means of Longitudinal and Torsional Vibrations," United States Army Cold Region Research Engineering Laboratory, SR 141, Hanover, N.H., 17p.

Littlejohn, G.S., and Rodger, A.A., (1980), "A Study of Vibratory Pile Driving in Granular Soils," Geotechnique, Volume 30, Number 3, pp. 269-293.

Schmid, W.E., (1969), "Driving Resistance and Bearing Capacity of Vibro-Driven Model Piles," in Performance of Deep Foundations, American Society for Testing Materials Special Technical Publications 444, American Society for Testing Materials, pp. 362-375.

Smart, J.D., (1969), "Vibratory Pile Driving," Ph.D. Thesis, University of Illinois, Urbana, Ill., 184p.

Vyalov, S.S., and Targulyan, Yu.O., (1968), "Boring and Pile Driving into Permafrost," Osnovaniya, Fundamenty i Mekhanika Gruntov, Number 2, pp. 24-26, English Translation in Soil Mechanics and Foundation Engineering, pp. 115-118.

Vyalov, S.S., Targulyan, Yu.O., and Vsorskiy, D.P., (1969), "Interplay of Frozen Ground with Piles during Vibratory Driving," Technical Translation FSTC-HT-23-944-68, U.S. Army Foreign Science and Technology Center, 13p.

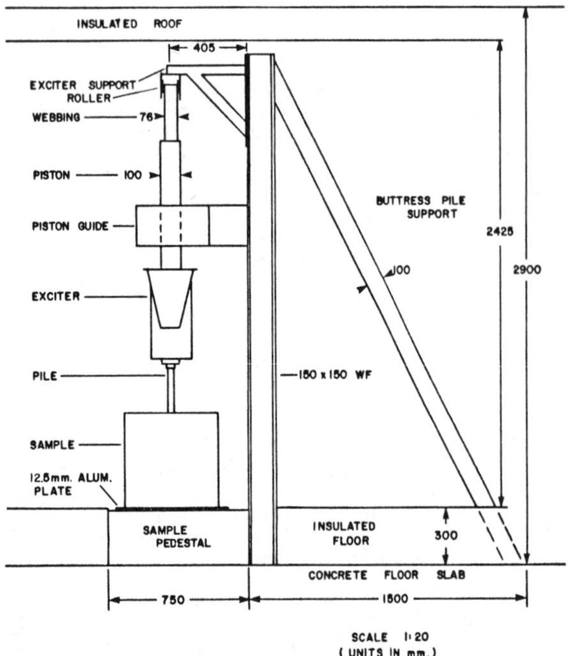

FIGURE 1 Side view of the testing apparatus.

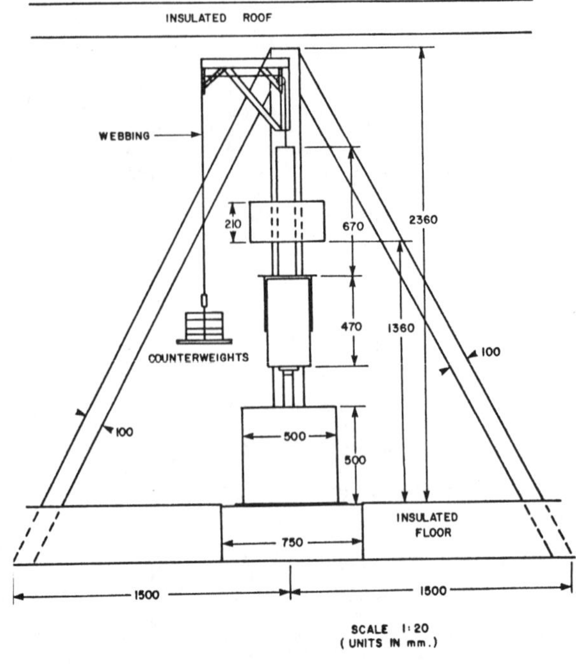

FIGURE 2 Front view of the testing apparatus.

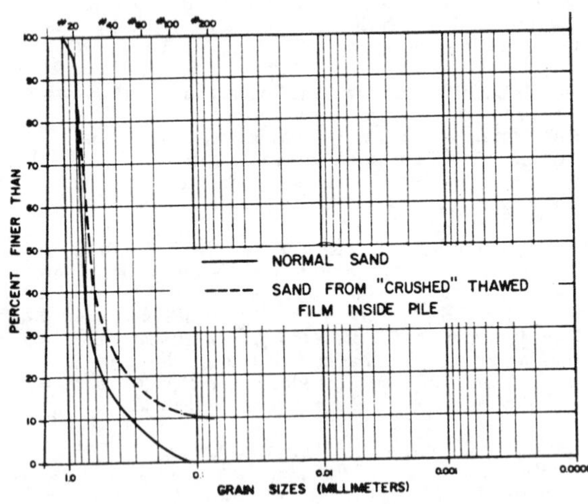

FIGURE 3 Grain size of test material.

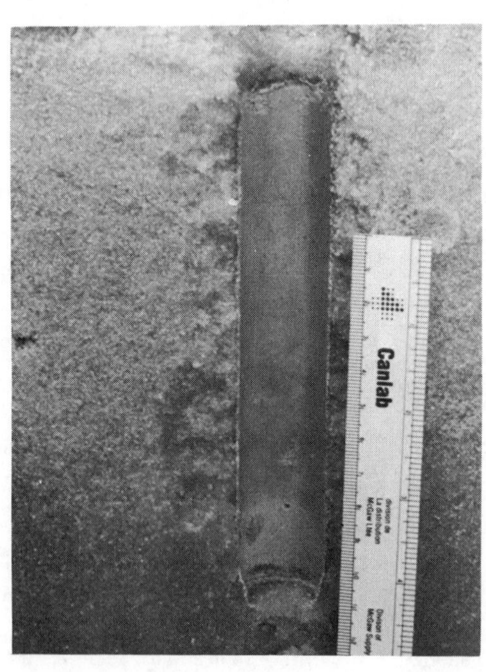

FIGURE 4 Thawed film after driving.

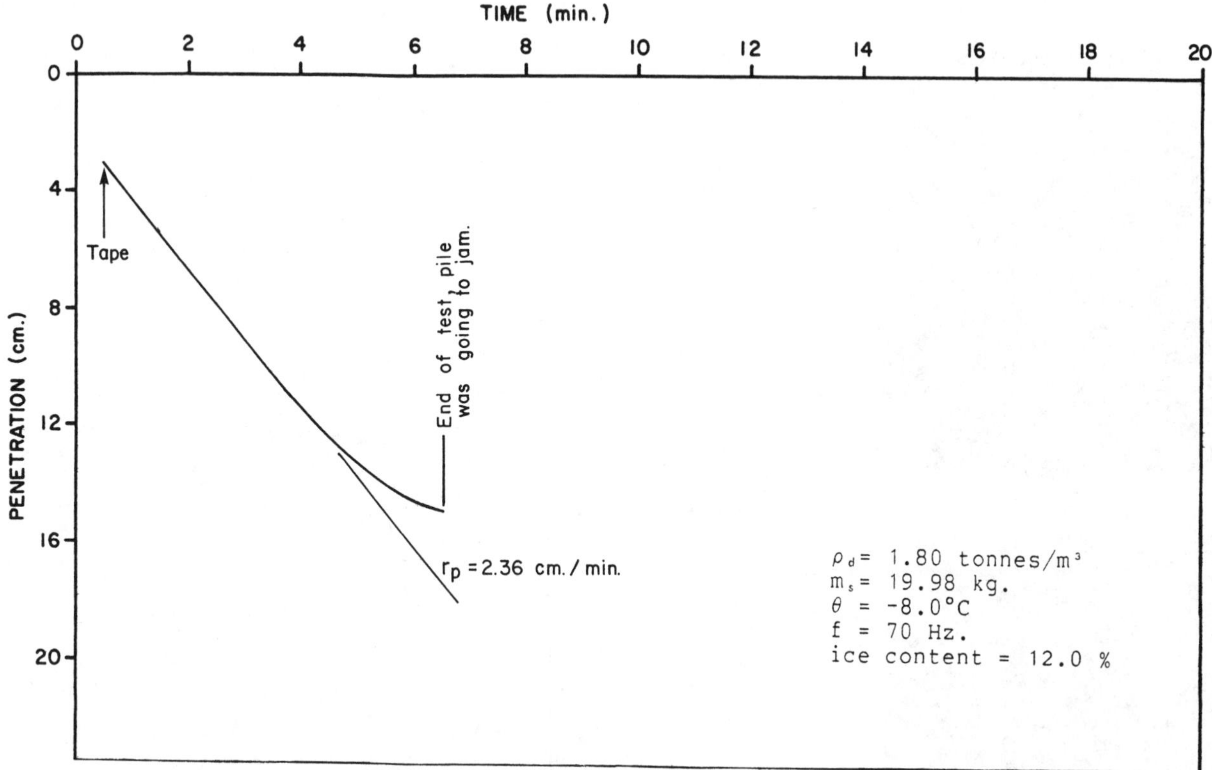

FIGURE 6 Penetration versus time curve, Test DS1-3.

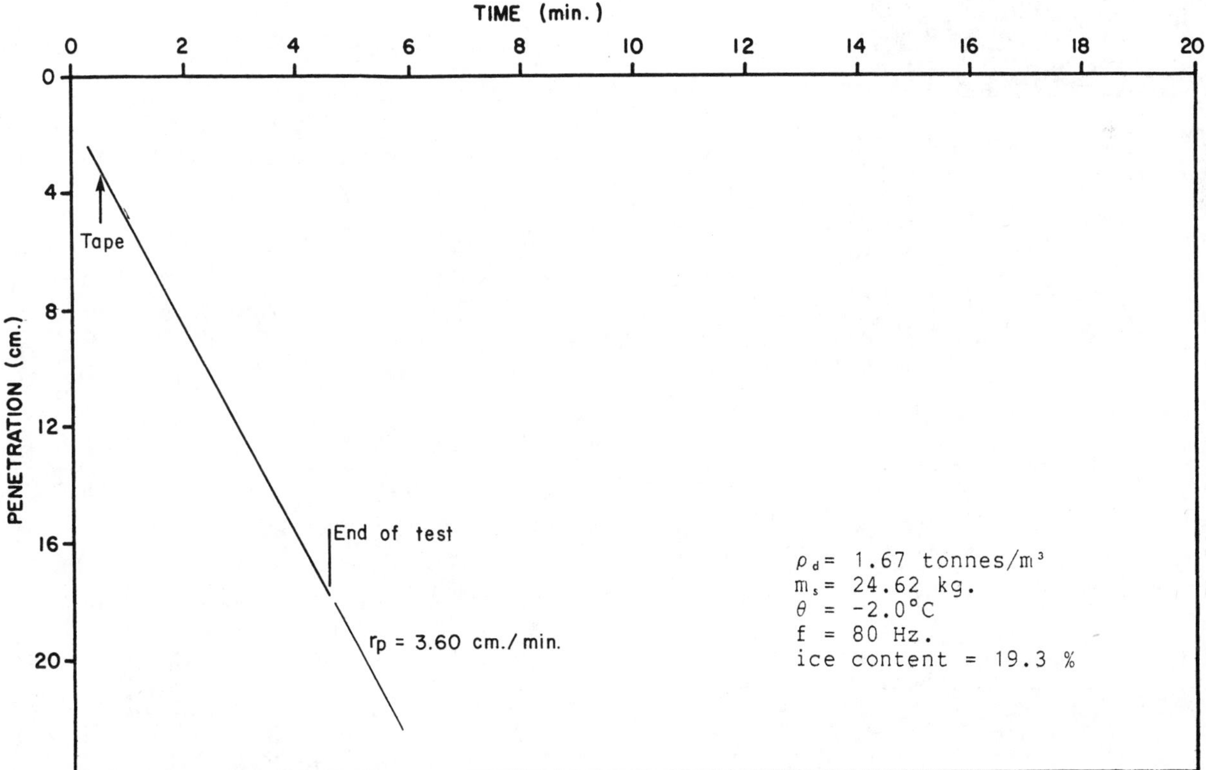

FIGURE 7 Penetration versus time curve, Test LS2-1.

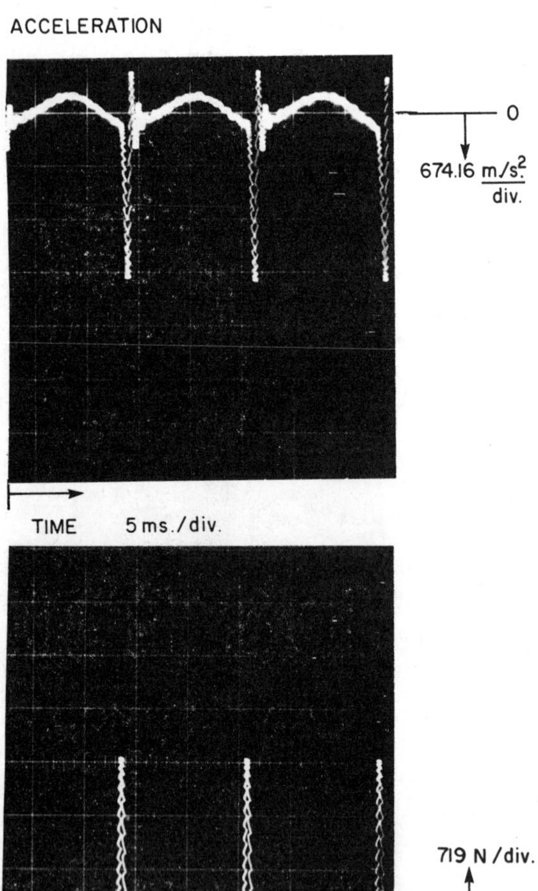

FIGURE 5 Typical recorded data.

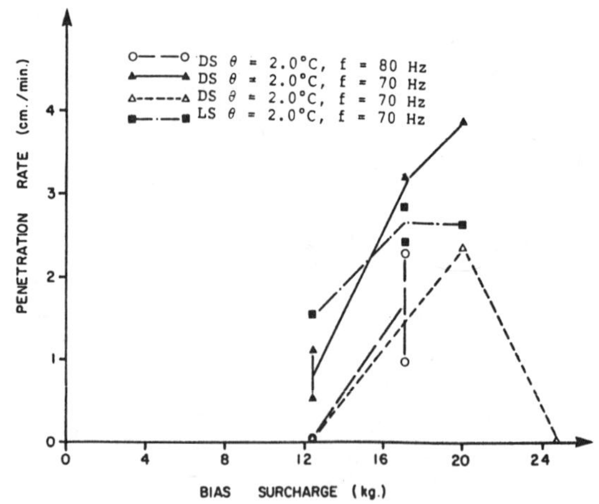

FIGURE 8 Effect of bias surcharge (m_s) on penetration rate (rp).

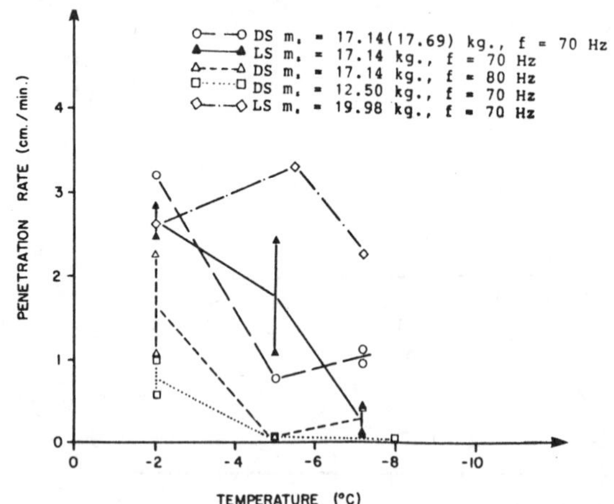

FIGURE 9 Effect of soil temperature on penetration rate (rp).

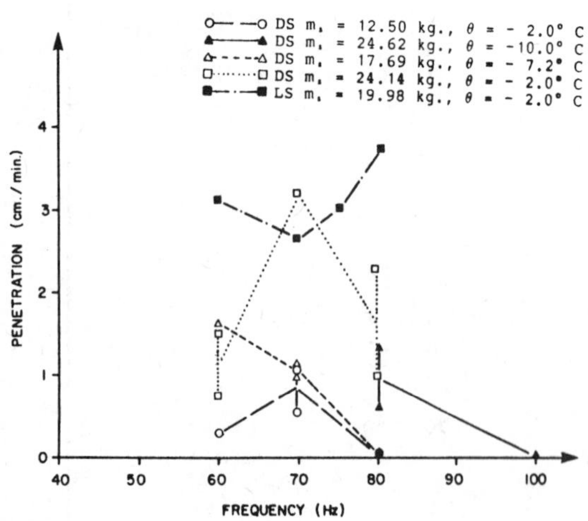

FIGURE 10 Effect of driving frequency on penetration rate (rp).

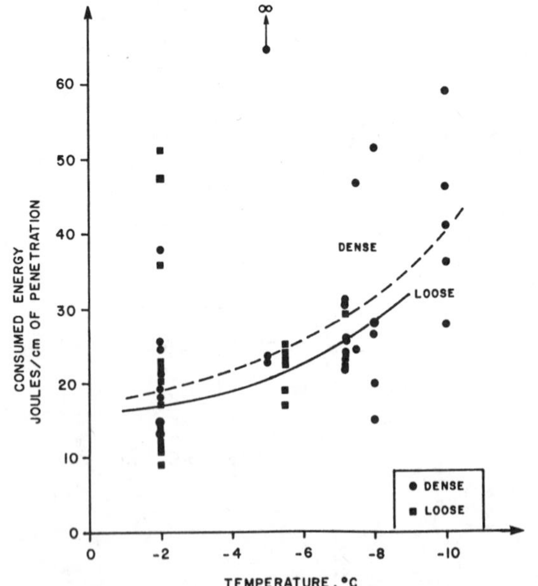

FIGURE 11 Plastic work done on soil to achieve 1 cm of penetration.

PRINCIPLES OF COMPILING MAPS FOR THE CONSERVATION OF THE GEOLOGICAL ENVIRONMENT IN THE PERMAFROST REGIONS

Ye. N. Dunayeva and A. V. Gavrilov

Geology Department
Moscow State University, USSR

The paper describes the principles of compilation and contents of a geological environment conservation map of the cryolithosphere, based on studies of the regime of the permafrost and of its predicted changes. The most important components of the map are: the stability of natural complexes and permafrost-engineering-geological conditions and their reaction to technogenic disturbances; allowable engineering loads; and recommendations for the conservation and optimization of the geological environment in terms of its future use. The evaluation of stability is preceded by a classification of technogenic disturbances of the geological environment, this evaluation is based on forecasts of the permafrost-geological processes induced by these disturbances. The results obtained allow classification of permafrost-engineering-geological sites in terms of response of the geological environment to various technogenic disturbances. The resulting classification is used to subdivide these sites in terms of allowable engineering loads. This, in turn, allows compilation of the map, which reflects recommendations for conservation and optimal use of the geological environment and identifies zones and sites which require specific protection. The model of the geological environment conservation map of the crylithosphere so produced is presented as an example.

Problems of geological environment conservation and the principles of compiling relevant maps are fast becoming very important, as engineering-economic activities in areas underlain by permafrost increase. The principles of compilation and the contents of a small-scale map for the conservation of the geological environment, using the example of the first-priority development of the South-Yakutian territorial industrial complex, were the basis for development of a methodology for the compilation of permafrost-engineering geological maps used to select sites for large industrial complexes and for development of the detailed follow up research programs required for the construction of territorial-industrial complexes along the Baikal-Amur Railway Corridor. This map completed a series of maps which include permafrost-hydrogeological and permafrost-engineering-geological maps, a set of permafrost prediction maps and a map of permafrost-engineering-geological appraisal of an area for land construction. The conditions leading to the formation of permafrost and its predicted changes, which are needed to solve the problems of geological environment conservation in the cryolithic zone, were included in the map. Although the map of the geological environment conservation may apply only to closely related fields, investigations have indicated that the basic elements of this map should be the reaction of geosystems (natural complexes, landscape) and of permafrost to technogenic disturbances; allowable engineering loads on the geological environment and recommendations for its conservation in relation to man's economic activity. This map should also show changes in the geological environment induced by economic development, permafrost-geological processes triggered by these changes and the protected zones and objects.

The authors view the stability of geosystems and of permafrost as a relatively permanent natural complex that includes rock condition, composition and properties, and the dynamics of permafrost-geological processes acting during the natural and historical development of the area. The stability of geosystems and permafrost-engineering-geological conditions (Garagulya and Parmuzin, 1978; Maximova, 1978) depend on their self-regulatory capabilities. The effects of technogenic disturbances are considered negative when they cause changes in a landscape contrary to its natural development. The typical example is the degradation of thermokarst as a result of a snow cover removal in areas composed of ice-rich rocks which then undergo thermokarst-related changes of their properties and composition. When the changes of geosystems and permafrost-engineering geological conditions caused by technogenic and natural and historical factors act in accord, their effects are considered. This positive reaction to technogenic disturbances determine the progress of cryogenic processes, since the natural evolutionary tendency of geosystems becomes enhanced by these disturbances. Thus, vegetation destruction in the areas composed of ice-rich rocks will identify the thermokarst degradation. Such landscapes may become unstable due to vegetation destruction and other changes causing increased summer heat exchange and, on the contrary, remain stable after technogenic disturbances resulting in the increased winter heat exchange in rocks have been inflicted.

In order to evaluate the reaction of geosystems and permafrost-engineering geological conditions to technogenic disturbances, it is necessary to classify these disturbances, to predict the related geocryological changes and to classify the predicted processes using the character and degree of changes introduced into the landscape and permafrost.

CLASSIFICATION OF TECHNOGENIC DISTURBANCES

Two factors govern the reaction of geosystems and permafrost-engineering geological conditions to technogenic disturbances: the inter-component and inter-system relations and the type of a technogenic disturbances responsible for their changes. Thus, changes of inter-component relations in a geosystem caused by the destruction of vegetation on ice-rich rocks or on slopes may bring about irreversible changes in geosystems, e.g. thermokarst development followed by the gravitational processes. Similar reactions are likely to occur when the inter-system relations are affected, e.g. changes in a natural complex on the upper part of a slope can cause changes in its lower part. These examples demonstrate the need for classification of technogenic disturbances to the natural environment based on the changes in inter-component and inter-system relations. The classification of disturbances, obtained for the Aldan-Stanovoi Region, include the following (Table 1): changes in the rock thermal conditions caused by the increased winter or summer heat exchange (by e.g.: snow packing or removal or by vegetation removal); changes caused by the groundwater rise and flooding (such as embankment construction, filling operations); changes in the groundwater regime (excavation of cuts and foundations, installation of drainage systems); changes in the surface water flow (land levelling); changes in the state of stress of rocks on slopes (slope cuts, excavations and slope angle changes).

After the types of technogenic disturbances and the natural and historical conditions were identified, the permafrost prediction was done using the methodology developed at the Permafrost Studies Department, Moscow University (Kudryavtsev, 1974). Forecasts of a temperature field and the depths of seasonal freezing and thawing are essential for predicting permafrost-geological processes induced by technogenic disturbances. In turn, these predictions are helpful in assessing the stability of the geosystems. The identification of the assessment criteria is especially important for small-scale surveys when large areas are mapped and predictions for every natural complex require a large effort. It is therefore necessary to subdivide the predicted processes by their potential effect on the landscape. When the vegetation is removed or other disturbances are inflicted on flat river terraces or water divides, the summer heat exchange of rocks increases; as a result the increased depth of seasonal multi-year thaw, seasonal frost heave or thermokarst development may begin (or intensify). The former has a seasonal rhythm and it effect on a landscape is limited. In terms of predicted changes, the seasonal frost heave should be classified as a process not leading to landscape changes. However, when the permafrost conditions are most conducive to the development of thermokarst, other effects may be expected. Then, if the thermokarst progresses as predicted, the process is classified as landscape-changing. The conditions favouring the initiation and development of progressive thermokarst may be considered as criteria for evaluating the reaction of geosystems to disturbances which result in the increased summer heat exchange of rocks. These conditions are (Table 1): presence of large monomineral ice inclusions or deposits with a high excess ice content; predicted seasonal thaw deeper than the depth of subsurface ice or ice-rich horizons; and the possibility of undrained depressions replacing the thawed ice. Therefore, data on the ice content in rocks and subsurface ice, the depth of their occurrence and their thickness, the predicted depth of seasonal and multi-year thaw and the surface morphology are the most important factors for assessing the reaction of the geosystems to disturbances which involve high summer heat exchange of rocks.

Table 1 presents the assessment of the geosystems and permafrost stability. To classify the natural complexes with varying potentials for change, the extent of the predicted processes was considered (local or areal). The degree or extent of the natural conditions which were potential process-forming factors within the limits of the identified geosystem were used to clarify this classification.

The results of several studies have indicated that while some geosystems and permafrost are stable after the ground regime changes and show instability when the winter heat exchange of rocks increases, others react just the opposite way. Consequently, before a map of a geological environment conservation is compiled, natural complexes should be classified according to their reaction to various technogenic disturbances. The map prepared for the first-priority development South-Yakutian territorial-industrial complex (the model is given in Figure 1) distinguishes five groups of landscapes differing in their responses to various disturbances. These groups are presented in the map legend (Table 1).

The second part of the map legend contains recommendations for the optimum conservation of the geological environment during its economic use. It is divided into three sections. The first section characterizes the allowable engineering loads on the geological environment (Table 2), thus determining the degree of suitability of the geosystems and permafrost for development (Golodkovskaya, Demidyuk, et al., 1975). The map shows groups of geosystems with minimum, average and maximum recommended loads. The first group is highly unstable when subjected to the technogenic disturbances and may be used only in case of extreme need, preferably only for point or linear objects with a localized development zone; the second group has a limited suitability for development; the third group is suitable for any type of development. The construction principles recommended are given for each group, based on the Construction Norms and Standards.

The second section includes nature conservation measures in relation to the economic development of the area.

The third section describes conservation zones and objects. They are divided, based on reasons for their protection, into natural complexes whose disturbance can initiate or suddenly activate permafrost-geological processes of a disastrous nature; natural complexes whose development may cause groundwater pollution; springs of balneological significance; and natural monuments.

The map also shows contemporary changes in the geological environment caused by development activity and the related permafrost-engineering

TABLE 1 Stability of natural complexes with respect to technogenic disturbance of the geological environment in terms of economic development (industrial, civil and linear construction).

Stability categories of natural complexes	Factors initiating processes caused by technogenic disturbances (criteria of potential landscape changeability)	Stability of natural complexes subjected to technogenic disturbances				
		Changes in the heat balance of rocks		Changes in Surface Flow		Decreased slope stability at shoulders of embankments and excavations, bank scouring of reservoirs, etc.
		Due to increased winter heat exchange following snow removal or packing of snow during preconstruction stage, etc.	Due to increased summer heat exchange following vegetation destruction during preconstruction stage, etc.	Due to groundwater rise and flooding during construction of embankments, fills, reservoirs, etc.	Due to vegetation removal during surface grading and other activities.	
1	Absent	+	+	+	+	+
2	Eluvial/diluvial sands and loamy sands on erodable weathered sandstones	+	+	+	Linear erosion (scour channels, ravines, etc.)	+
3	Saturated, thawed biogenic loamy sands and sandy loams	Multi-year freezing, heaving (mounds, etc.) and frost-induced cracking	+	+	+	+
4a 4b	(a) Disaggregated diluvial-colluvial sandy loams and loamy sands, on slopes >5°, diluvial blocks, rock debris of rock streams and structural slopes >15°; (b) landslide and talus blocks, rock debris and grus on slopes >25°	+	+	+	+	Intensified slope processes: (a) solofluction, block sliding, rock streams. (b) avalanches, talus, slides
5a 5b	Diluvial, alluvial-diluvial and alluvial biogenic ice-rich, peaty, sands, loamy sands and sandy loams, a) <3-5m thick b) <3-5m thick	Without stratification and ice wedges +	Multi-year thawing, thermokarst (water logging, depressions, lakes).		+	+
6a 6b		With stratification and ice wedges +			Thermal erosion (scour channels, ravines)	+

Note: + indicates natural complex is stable when subjected to technogenic distrubances.

TABLE 2 Allowable engineering loads on the geological environment.

Groups from Table 1	Category	Permafrost-engineering geological stability of natural complexes and their suitability for development	Appropriate principal uses of frozen ground for building and construction (according to the Construction Norms and Standards)
3, 4b, 5b, 6b	minimum	Highly unstable; suitable for point or linear objects with a localized development zone	I
4a, 5a, 6a	average	Unstable; suitable for limited development	mainly I, sometimes II
1, 2	maximum	Stable; suitable for overall development	II

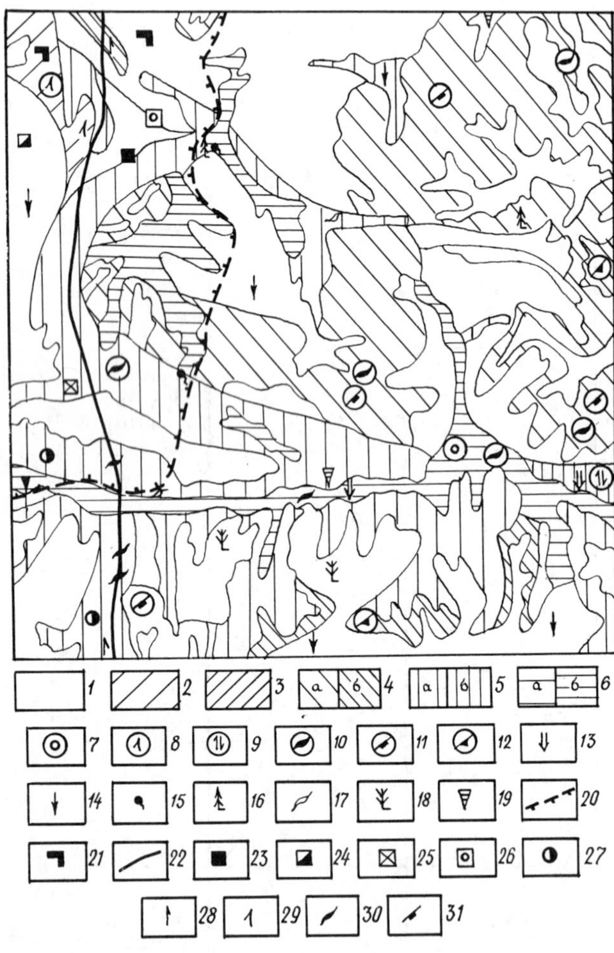

FIGURE 1 Model of the Geological Environment Conservation Map. Description of stability of natural complexes subjected to technogenic disturbances and allowable engineering loads on the geological environment (points 1-6) are given in Tables 1 and 2. Measures for nature conservation in terms of the area of development: 7 - control or prevention of thermokarst or thermoerosional: maintaining rocks in a frozen state; surface drainage; full or partial replacement of icy rocks; biological and technical recultivation; 8 - erosion control: minimizing area vegetation degradation; surface drainage; land stabilization; biological and technical recultivation; 9 - frost heave control: maintaining rocks in a frozen state; regulation of the rate of freezing; surface drainage; grout curtains; full or partial replacement of heave susceptible rocks; 10 - icing control: groundwater lowering and land drainage, groundwater diversion; explosion works; artificial thawing of ice; freezing of groundwater aquifers; 11 - landslide and rock stream control: maintaining rocks in a frozen state; surface drainage; mechanical landslide-control measures; slope transformation; 12 - slide and talus control: slope stabilization; slope transformation; induced caving; trapping and diversion structures. Zones and objects under protection: 4b, 5b, 6b - natural complexes which, if affected, trigger intensified permafrost engineering geological processes of a disastrous nature; areas which, when developed may cause: 13 - pollution of groundwater-seepage taliks; 14 - pollution or freezing of groundwater recharge areas or influent taliks; 15 - springs of balneological importance. Objects of aesthetic value (natural monuments): 16 - landscape of groundwater discharge areas (springs) and rare communities and species of plants; 17 - icings; 18 - suitable for recreation; 19 - picturesque geological outcrops. Geological environment changes caused by economic activity: 20 - border dividing conditionally unchangeable and slightly-changeable natural complexes from highly-changeable (the mark is directed to the latter). Major development projects: 21 - towns and settlements; 22 - roads; mineral deposits in the stage of: 23 - industrial exploration; 24 - industrial development. Building material deposits: 25 - gravel; 26 - stone. Engineering geological processes: 27 - thermokarst and water logging; 28 - heaving and uneven subsidence; 29 - erosion; 30 - icing; 31 - increased mobility of rock streams.

geological processes (Figure 1). The subdivisions of geosystems, based on the major economic projects and related engineering-geological processes used to characterize changes in the natural environment, are described.

In conclusion, it should be stressed that mapping for conservation of the geological environment is fast becoming a very important issue. The authors consider the work done to date to be only the first step in this field. Discussion of the results obtained in this study will contribute to further improvement of the content and methodology of compilation of the natural environment conservation maps of the cryolithosphere.

REFERENCES

Garagulya, L.S. and Parmuzin, S.Yu., 1978, Problems and methods of predicting changes in natural conditions during various stages of geocryological and engineering-geological investigations; Merzlotnye issledovaniya (Permafrost Studies), no. 17. Moscow, Moscow University Press.

Golodkovskaya, G.A., Demidyuk, et al., 1975, Engineering-geological studies for the development of mineral deposits. Moscow, Moscow University Press.

Isachenko, A.G., 1976, Applied Landscape Studies. Leningrad, Leningrad University Press.

Kudryavtsev, E.A., (editor), 1974, Fundamentals of Permafrost Predictions for Engineering-Geological Investigations. Moscow, Moscow University Press.

Maximova, L.N., 1978, On the psssibility of evaluating the stability of geocryological conditions and natural landscapes by study of the types of seasonal thaw (and freezing) of soils and rocks; Merzlotnye issledovaniya (Permafrost Studies), no. 17. Moscow, Moscow University Press.

BIOLOGICAL DECOMPOSITION AND PLANT SUCCESSION FOLLOWING DISTURBANCE ON THE ARCTIC COASTAL PLAIN, ALASKA

J. J. Ebersole and P. J. Webber

Institute of Arctic and Alpine Research
and Department of Environmental, Population, and Organismic Biology
University of Colorado, Boulder, Colorado 80309 USA

Thirty years after abandonment of Oumalik Test Well, vigorous stands of grasses (*Arctagrostis latifolia*, *Poa arctica*) and erect willows (*Salix glauca*, *S. planifolia*, *S. lanata*, *S. alaxensis*) dominate the mesic disturbed areas. We hypothesized that the existence of these communities was due to a higher nutrient availability created by greater decomposition rates in these warm, well-drained sites. Weight loss from three types of materials placed 10 cm deep in disturbed and undisturbed sites for periods of 12 and 13 months showed that decomposition rates in disturbed plots were significantly greater ($p < 0.02$) than in undisturbed plots. Stepwise multiple regression showed that the most important measured environmental factors in explaining the variation in decomposition rates were soil temperature (positively correlated with decomposition rate) and soil moisture (negatively correlated). The greater rates of decomposition presumably indicate increased availability of nutrients which is known to favor species with high rates of turnover such as the grasses and willows. As long as the abiotic conditions favoring high decomposition rates persist, nutrient availability will stay high. We expect that the present communities will not be replaced with communities similar to those of the surrounding undisturbed tundra for several hundred years.

INTRODUCTION

Organic matter decomposes slowly in arctic tundra environments. Low temperature is the primary limiting factor, and moisture is secondarily limiting with decomposition rates being greatest in mesic sites and becoming less in both dry and wet sites (Heal and French 1974). The low rates of decomposition make nutrient cycling slow and produce a system in which nutrient availability limits plant productivity and controls species composition (Haag 1974, Challinor and Gersper 1975, McKendrick et al. 1975, Ulrich and Gersper, 1978).

Disturbance to the tundra surface and vegetation would be expected to affect decomposition rates since disturbed areas have altered temperature and moisture regimes. Damage or destruction of the vegetation layer and peat reduces their insulating value and decreases the albedo of the surface (Brown and Grave 1979). This raises the soil temperature and presumably increases the rate of decomposition (Chapin and Van Cleve 1978). In some cases disturbance improves the drainage of formerly water-logged soils which also would be expected to increase decomposition rates. The work presented here is part of a larger study on the recovery of vegetation following a 30-year-old disturbance. The increased abundance and vigor of erect willows (*Salix glauca*, *S. planifolia* ssp. *pulchra*, *S. lanata* ssp. *richardsonii*, and *S. alaxensis* ssp. *alaxensis*) and the greater abundance of nutrient-loving grasses (*Arctagrostis latifolia* var. *latifolia*, *A. latifolia* var. *arundinacea*, and *Poa arctica* ssp. *arctica*) on mesic areas of the disturbance relative to mesic undisturbed areas prompted us to hypothesize that nutrient availablity in these areas is increased due to enhanced decomposition caused by disturbance. This paper presents the results of the study to test this hypothesis and interprets the development and future of the plant communities on the mesic disturbed areas in light of these results.

SITE DESCRIPTION

The study was done at Oumalik Test Well No. 1, which is located about 125 km from the Arctic Ocean at 69°50'N, 155°59'W on the Arctic Coastal Plain of Alaska (Wahrhaftig 1965). It lies in the subarctic tundra subzone of Aleksandrova (1980) and in the low arctic zone of Polunin (1951). The mean July temperatures in 1979 (apparently an unusually warm summer) and 1980 at Oumalik were 14.4°C and 9.4°C, respectively (Haugen 1982 and unpublished data).

Permafrost is continuous with the upper 3-4 m containing about 50-90% ice by volume (Lawson 1983, this volume). The flatness of the terrain and the presence of permafrost create poor drainage, and most of the area is sedge marsh with water-logged soils. The few upland areas are covered with tussock-dwarf shrub tundra. Even these areas are quite wet, and strong gleying of the soil indicates anaerobic conditions below the upper layers.

The Oumalik well was drilled in 1949-1950 as

part of the oil exploration in the Naval Petroleum Reserve No. 4 (now the National Petroleum Reserve-Alaska). The site was abandoned in 1950 and not disturbed again until work crews removed solid debris in 1980. Lawson et al. (1978) have described a similar site at Fish Creek. All damage to the surface or vegetation tended to cause melting of subsurface ice. Vehicles were driven extensively on the tundra around the well site. Their tracks broke the vegetation mat and damaged the peat layer. Species diversity on these areas was decreased due to the elimination of many species. The new cover is composed of original species which responded favorably to the new environment and species, especially those listed below, which were not on the site before. Bulldozing of the tundra surface was common in the immediate area around the well and caused the most severe disturbance. Areas lowered by bulldozing now frequently have standing water and have been recolonized primarily by the sedges *Carex aquatilis* and *Eriophorum angustifolium* ssp. *subarcticum*. Mounds of soil created by the bulldozing and slopes reworked by bulldozing were colonized primarily by the willows *Salix glauca*, *S. planifolia* ssp. *pulchra*, *S. lanata* ssp. *richardsonii*, and *S. alaxensis* ssp. *alaxensis* and the grasses *Arctagrostis latifolia* var. *latifolia*, *A. latifolia* var. *arundinacea*, and *Poa arctica* ssp. *arctica*. The dominance of the grasses and willows and the extreme vigor of the willows are unusual relative to the surrounding undisturbed tundra.

METHODS

To test the hypothesis that disturbance increased decomposition rates in these sites, we used weight loss of introduced substrates as a simple, integrating measure. Three types of decomposition samples were placed into 9 disturbed and 16 undisturbed plots selected to represent the range of decomposition environments at Oumalik (Table 1). The three types were (1) 2.00 g of *Agropyron cristatum* stems from wasteland in Boulder, Colorado, in a 10 x 10 cm nylon net bag whose largest holes were about 0.7 x 0.3 mm, (2) one cellulose filter paper weighing about 0.56 g in a similar bag, and (3) birch wood tongue depressors weighing 2.4-3.1 g. On July 6, 1980, samples were placed into each of the plots at a depth of 10 cm. The samples were placed at depth because in this study the decomposition of interest occurs in the organic matter mixed into the soil by disturbance rather than in litter on the surface. Ten cm was chosen as the depth of most abundant roots. Each plot contained four replicates of each of the three types of decomposition samples. Two replicates of each type were removed 1 year after emplacement, and two were retrieved at the end of the maximum time available for the experiment on August 3, 1981, which was equivalent to approximately 1.5 growing seasons.

After removal the samples were dried in the field. In the laboratory the tongue depressors and filter papers were rinsed with cold running water to remove adhering mineral soil. The grass was not washed because of the significant water-soluble fraction (15-19%) it contains. The samples were oven dried and reweighed to determine weight loss. Representative used and unused samples of each type of sample were ashed (450°C for 2 hr). The differences in the weight of ash between used and unused samples were used to correct weight loss of all samples for adhering soil. After correction, the percent weight loss was calculated and used as the basis for the numerical analysis.

A soil sample from 10 cm in each plot was analyzed for pH, organic matter, carbonates, water absorption, hygroscopic water, wilting point, field capacity, cation exchange capacity, and total available ammonium, nitrate, potassium, phosphorus, calcium, and magnesium. Soil temperature, depth of thaw, and date of snow meltout were determined for each plot (Table 2).

RESULTS AND DISCUSSION

Despite the limitation of the small sample size, differences between disturbed and undisturbed sites are sharp. The hypothesis that decomposition is greater on bulldozed slopes and in mounds of bladed material than in undisturbed areas is supported by the ANOVA

TABLE 1. Plots Classified by Microrelief/Vegetation Type. Plots in numbered types do not sum to 25 since not all plots fit into the categories listed.

Code	No. of Plots	Description
D	9	Disturbed
U	16	Undisturbed
1	3	mounds of bladed material; dominated by *Arctagrostis latifolia*
2	3	mounds of bladed material; dominated by *Salix* spp. and *Arctagrostis latifolia*
3	2	undisturbed palsas and high-centered polygon centers; dominated by *Betula nana* and *Ledum palustre*
4	2	undisturbed reticulate-patterned ground; dominated by *Dryas integrifolia* and *Carex bigelowii*
5	3	undisturbed marshes; dominated by *Carex* spp. and *Eriophorum* spp.
6	2	undisturbed tussok tundra; dominated by E. vaginatum

TABLE 2. Means and Standard Errors of Selected Abiotic Variables.

Type[1]	Soil temp, °C[2]	Soil mois.[4]	Soil mois.[5]	Thaw, cm[6]	pH	Meltout[7]	NH_4, ppm	NO_3, ppm	P, ppm	Ca, ppm	OM,%
D	8.5 ± 0.5	139.6 ± 36.0	5.7 ± 0.2	58.3 ± 3.6	7.31 ± 0.20	8.3 ± 1.6	18.2 ± 2.9	16.2 ± 4.2	9.7 ± 3.3	4360 ± 250	14.1 ± 4.4
U	5.4 ± 0.6	145.8 ± 39.6	7.2 ± 0.3	41.3 ± 1.9	6.37 ± 0.17	5.6 ± 1.3	34.4 ± 7.9	14.4 ± 2.7	4.0 ± 0.9	6600 ± 880	46.1 ± 7.4
1	9.4 ± 0.8	35.4 ± 0.7	4.7 ± 0.3	62.0 ± 12.2	7.72 ± 0.28	1.0 ± 0.0	9.7 ± 2.7	5.2 ± 0.9	13.3 ± 12.4	3750 ± 680	5.9 ± 0.9
2	9.5 ± 0.7	173.1 ± 76.3	6.0 ± 0.0	59.3 ± 11.4	7.39 ± 0.39	14.0 ± 2.1	217.3 ± 200.4	96.5 ± 72.4	7.9 ± 4.4	4460 ± 720	17.6 ± 7.2
3	3.0 --[3]	370.2 --	5.5 ± 0.5	29.5 ± 1.5	5.42 ± 0.04	1.0 ± 0.0	28.4 ± 6.8	13.9 ± 1.1	5.0 ± 3.0	6880 ± 60	65.2 ± 2.2
4	5.2 --	104.4 --	7.0 ± 0.0	57.0 ± 2.0	7.44 ± 0.18	1.0 ± 0.0	7.8 ± 0.4	4.8 ± 1.4	0.8 ± 0.2	3790 ± 810	8.4 ± 4.1
5	5.6 ± 1.6	106.3 --	9.0 ± 1.0	45.7 ± 2.9	5.84 ± 0.55	4.3 ± 3.3	30.3 ± 4.6	16.5 ± 1.0	2.8 ± 0.2	7080 ± 3380	62.0 ± 4.7
6	2.4 ± 0.4	42.0 ± 4.2	7.0 ± 0.0	35.5 ± 5.5	5.70 ± 0.08	9.0 ± 7.0	15.3 ± 0.6	5.8 ± 1.4	1.2 ± 0.4	2520 ± 460	11.6 ± 1.9

[1] see Table 1
[2] at 10 cm on June 30, 1980
[3] only one measurement so S.E. undefined
[4] % of dry weight, June 30, 1980
[5] scalar with 1 = very dry and 9 = very wet
[6] August 12-14, 1979
[7] days after May 31, 1980, that plot was free of snow

which shows that decomposition rates in disturbed plots were significantly greater ($p < 0.02$) than in undisturbed plots for all three types of decomposition samples over both time periods (Figure 1). Differences between microrelief/vegetation types were not tested for significance because of the small sample size.

The plots with the highest decomposition rates were bulldozed slopes and mounds of bladed material dominated by *Salix* spp. and/or *Arctagrostis latifolia* (Figure 1). These sites have relatively warm soil temperatures (Table 2), and due to their good drainage, the soils of these sites are also well aerated, as indicated by the soil color. The sites with the slowest rates of decomposition were undisturbed tussock tundra, which has cold soil temperatures and anaerobic soil as evidenced by the strong gleying. The greater rates of decomposition in marshes, which also have highly anaerobic soils, were probably due to their higher soil temperatures. The reticulate-patterned ground and the areas dominated by *Betula nana* and *Ledum palustre* were intermediate in decomposition rates, aerobicity as judged from soil color and moisture, and soil temperatures.

The results of the stepwise linear multiple regressions show that temperature is the most important environmental factor in explaining the variation in decomposition rates (Table 3). Other investigations in the Arctic have also shown that temperature is the major limiting factor (Heal and French 1974). Soil moisture is the second most important limiting factor at Oumalik. No samples were in the rare dry sites in which the lack of moisture is limiting so decomposition shows a negative, essentially linear relationship with soil moisture. In studies covering the complete moisture gradient, the response curve of decomposition as a function of moisture reaches a maximum in the mesic sites and declines in both dry and wet sites (Heal and French 1974).

Other factors show weaker relationships to decomposition rate. Phosphorus explains a significant part of the variation and correlates positively with decomposition rate. Heal and French (1974) also found a positive relationship, and Rosswall (1974) reported an increase in decomposition when phosphorus was added to a temperate bog. However, at Oumalik the increased phosphorus may be an effect, rather than a cause, of increased decomposition rates. Depth of thaw at the end of the growing season also correlates positively with decomposition rates. This may be due primarily to the significant positive correlation of thaw and soil temperature and the significant negative correlation of thaw and soil

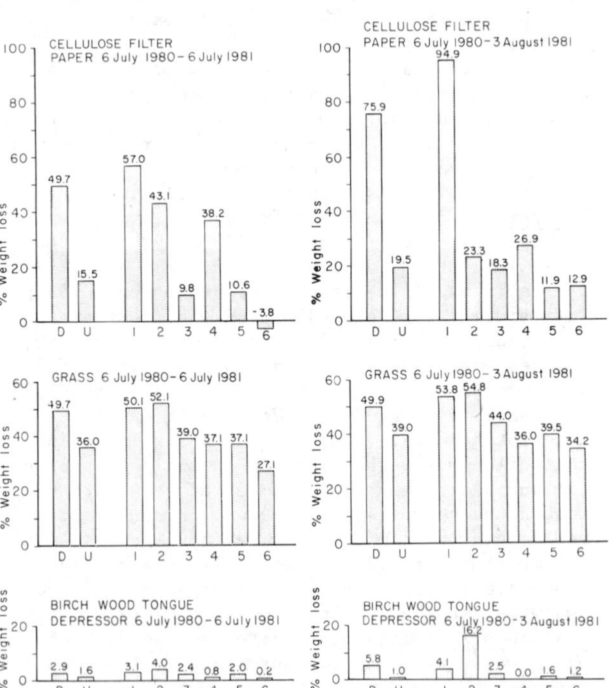

FIGURE 1 Percent Weight Loss of Decomposition Substrates. Plot types are listed in Table 1.

moisture. Plots that become free of snow later have a weak tendency to have greater decomposition rates. One would expect the oposite; that is, plots with an earlier meltout would have greater decomposition rates. This apparent incongruity may be due to the later meltout of high-decomposition disturbed plots which accumulate more snow due to increased surface roughness and greater willow height and density.

The magnitude of the r^2 values for the regressions indicate that 45-78% of the variation in decomposition rates is explained with the variation in the environmental factors measured. The r^2 values tended to be greater for the longer period of time, perhaps indicating that the longer period of time gave a more accurate measure of decomposition potentials of the plots. Possible sources of the variation unaccounted for are nonlinear responses of decomposition rate to variation in environmental factors, interactions among environmental factors in their effect on decomposition, and random variation in decomposition potential or measured environmental factors on a microscale within a plot or habitat type.

TABLE 3. Results of Stepwise Multiple Regressions.

Grass 6 July 1980 - 6 July 1981		r^2
step	variable entered	
1	soil temperature	0.43
2	P	0.46
3	pH	0.50
4	time of meltout	0.56
Grass 6 July 1980 - 3 August 1981		r^2
step	variable entered	
1	soil moisture	0.24
2	soil temperature	0.36
3	thaw	0.53
4	Ca	0.59
Filter paper 6 July 1980 - 6 July 1981		r^2
step	variable entered	
1	thaw	0.30
2	soil temperature	0.36
3	P	0.39
4	soil moisture	0.45
Filter paper 6 July 1980 - 3 August 1981		r^2
step	variable entered	
1	soil moisture	0.33
2	soil temperature	0.43
3	P	0.50
4	time of meltout	0.58
Tongue depressor 6 July 1980 - 6 July 1981		r^2
step	variable entered	
1	soil temperature	0.16
2	thaw	0.40
3	time of meltout	0.45
Tongue depressor 6 July 1980 - 3 August 1981		r^2
step	variable entered	
1	soil temperature	0.26
2	soil moisture	0.45
3	thaw	0.63
4	Ca	0.78

Each step represents a separate equation which includes the variable listed for that step and the previous steps. The r^2 value is for the equation through that step.

Differences between the 12- and 13-month losses varied with the type of decomposition sample (Figure 1). The grass showed only slightly greater losses over the longer time period, apparently because the easily leached or decomposed fraction was gone within the first year, leaving only the more resistant components. Filter paper and birch wood losses tended to increase slightly to substantially in the additional month. Colonization by decomposers may have taken longer in these substrates compared to the grass in which decomposers were probably already resident.

The greater decomposition rates in disturbed areas presumably create a higher availability of nutrients in a system in which production is limited by nutrients, especially nitrogen and phosphorus (Haag 1974, McKendrick et al. 1975, Ulrich and Gersper 1978, Chapin and Van Cleve 1978, Shaver and Chapin 1980). Other studies in the Alaskan Arctic have shown that enhancing the nutrient regime also changes species composition by differentially favoring the growth of high-turnover species (McKendrick and Mitchell 1978, McKendrick et al. 1975, 1980, Shaver and Chapin 1980, Chapin and Shaver 1981). Thus, the enhanced nutrient regime at Oumalik seems to provide a partial explanation for the presence and vigor of the grass and willow communities on the disturbance. The hypothesis to explain the increased decomposition and the presence and abundance of the willows and grasses is presented in Figure 2.

The grasses and willows are preadapted to natural disturbances and originated on the anthropogenic disturbance from their abundant, wind-dispersed seeds. Once established, these high-turnover species probably quickly formed dense stands in response to the favorable nutrient regimes. The dense communities tend to inhibit colonization by other species, especially the slower-growing dominants of the undisturbed tundra. Thirty years after disturbance there is little, if any, evidence that species from the next successional stage are invading these grass and willow communities. As long as the nutrient regime remains enhanced by the rapid decomposition created by good drainage and high soil temperatures, the present communities will persist. Since there is no indication that the factors creating high decomposition rates will change, replacing the grasses and willows with communities resembling those in the undisturbed tundra is predicted to take several hundred years.

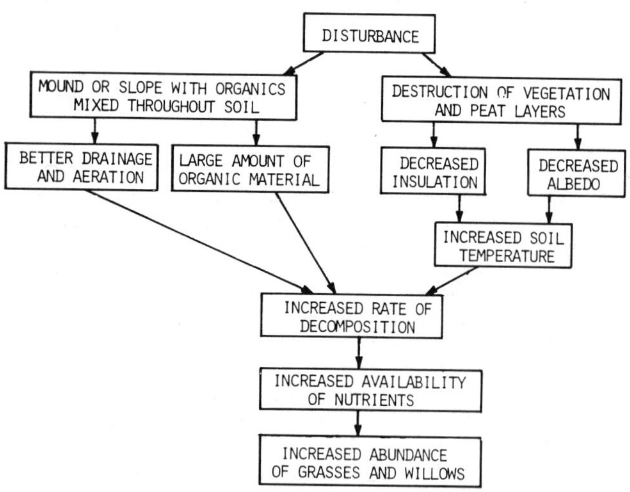

FIGURE 2 Hypothesis to Explain the Increased Decomposition Rates and Increased Abundance of Grasses and Willows of Mesic Disturbed Areas.

SUMMARY AND CONCLUSIONS

Decomposition rates in mesic disturbed areas at Oumalik are significantly greater than in undisturbed areas. Increased soil temperature and decreased soil moisture are the most important factors accounting for the increase. The increased nutrient availability caused by the greater decomposition rates provides a plausible explanation for the persistence of vigorous stands of grasses and willows on the disturbed areas. We predict that the elevated decomposition rates will continue for an extended time and allow the persistence of these communities for many years. In a system such as wet low arctic tundra, in which nutrients are accumulated in organic matter in the soil and in which the slow rate of cycling of these essential elements limits primary productivity and controls species composition, disturbance can enhance the nutrient regime and increase the abundance and vigor of certain plants at the expense of others.

ACKNOWLEDGEMENTS

This work was sponsored by the Cold Regions Research and Engineering Laboratory (CRREL) with funding from the U.S. Geological Survey (USGS).

REFERENCES

Aleksandrova, V., 1980, The Arctic and Antarctic: their division into geobotanical areas, translated by D. Löve: Cambridge University Press. Cambridge, 247 p.

Brown, J. and N. A. Grave, 1979, Physical and thermal disturbance and protection of permafrost, CRREL report SR79-5: U. S. Army Cold Regions Research and Engineering Laboratory, Hanover, New Hampshire, 43 p.

Challinor, J. L. and P. L. Gersper, 1975, Vehicle perturbation effects upon a tundra soil-plant system. II. Effects on the chemical regime: Soil Science Society of America Proceedings, v. 39, p. 689-695.

Chapin, F. S. III and G. R. Shaver, 1981, Changes in soil properties and vegetation following disturbance of Alaskan Arctic tundra: Journal of Applied Ecology, v. 18, p. 605-617.

Chapin, F. S. III and K. Van Cleve, 1978, Nitrogen and phosphorous distribution in an Alaskan tussock tundra ecosystem: natural patterns and implications for development, _in_ D. C. Adriano and I. L. Brisbin, eds., Environmental chemistry and cycling processes: U. S. Dept. of Energy Symposium Series, Washington, D.C., p. 738-753.

Haag, R. W., 1974, Nutrient limitations to plant production in two tundra communities: Canadian Journal of Botany, v. 52, p. 103-116.

Haugen, R.K., 1982, Climate of remote areas in north-central Alaska: 1975-1979 summary, CRREL report 82-35: U. S. Army Cold Regions Research and Engineering Laboratory, Hanover, New Hampshire, 114 pp.

Heal, O. W., and D. D. French, 1974, Decomposition of organic matter in tundra, _in_ A. V. Holding, D. W. Heal, S. F. Maclean, Jr., and P. W. Flanagan, eds., Soil organisms and decomposition in tundra: Tundra Biome Steering Committee, Stockholm, Sweden, p. 279-310.

Lawson, D. E., 1983, Ground ice in perenially frozen sediments, northern Alaska, Proceedings of the Fourth International Conference on Permafrost: this volume.

Lawson, D. E., J. Brown, K. R. Everett, A. W. Johnson, V. Komarkova, B. M. Murray, D. F. Murray, and P. J. Webber, 1978, Tundra disturbances and recovery following the 1949 exploratory drilling, Fish Creek, northern Alaska. CRREL report 78-28: U. S. Army Cold Regions Research and Engineering Laboratory, Hanover, New Hampshire, 91 p.

McKendrick, J. D., G. O. Batzli, K. R. Everett, and J. C. Swanson, 1980, Some effects of mammalian herbivores and fertilization on tundra soils and vegetation: Arctic and Alpine Research, v. 12, p. 565-578.

McKendrick, J. D. and W. W. Mitchell, 1978, Fertilizing and seeding oil damaged arctic tundra to effect vegetation recovery at Prudhoe Bay, Alaska: Arctic, v. 31, p. 296-304.

McKendrick, J. D., V. J. Ott, and G. A. Mitchell, 1975, Effects of nitrogen and phosphorous fertilization on carbohydrate and nutrient levels in _Dupontia fisheri_ and _Arctagrostis latifolia_, _in_ L. L. Tieszen, ed, Vegetation and production ecology of an Alaskan arctic tundra: Springer-Verlag, New York, p. 509-537.

Polunin, N., 1951, The real Arctic: suggestions for its delineation, subdivisions and characterizations: Journal of Ecology, v. 39, p. 308-315.

Rosswall, T., 1974, Cellulose decomposition studies on the tundra, _in_ A. J. Holding _et al._, eds., Soil organisms and decomposition in tundra, Tundra Biome Steering Committee: Stockholm, p. 325-340.

Shaver, G. R and F. S. Chapin III, 1980,

Response to fertilization by various plant growth forms in an Alaskan tundra: nutrient accumulation and growth: Ecology, v. 61, p. 662-675.

Ulrich, A. and P. L. Gersper, 1978, Plant nutrient limitations of tundra plant growth, *in* L. L. Tieszen, ed., Vegetation and production ecology of an Alaskan Arctic tundra: Springer-Verlag, New York, p. 457-482.

Wahrhaftig, C., 1965, Physiographic divisions of Alaska. U. S. Geological Survey Professional Paper 482: U.S. Government Printing Office, Washington, D.C. 52 p.

TRANSIENT ELECTROMAGNETIC SOUNDINGS FOR PERMAFROST MAPPING

R. L. EHRENBARD[1], PIETER HOEKSTRA[2], G. ROZENBERG[3]

[1]SOHIO PETROLEUM COMPANY, DALLAS, TEXAS 75240
[2]GEO-PHYSI-CON, LAKEWOOD, COLORADO 80228
[3]GEO-PHYSI-CON, CALGARY, ALBERTA T2H OH2

Over the years electrical and electromagnetic methods have been successful in mapping the lateral and vertical boundaries of permafrost over the land. Offshore most electrical and electromagnetic methods cannot map top and bottom of permafrost, because of the low resistivity of the brine saturated, unfrozen sediments overlaying offshore permafrost. Transient electromagnetic exploration is a method of electrical resistivity mapping that can detect resistive layers under highly conductive sediments.

In the winter of 1983 a large survey was conducted on lines onshore and offshore in areas west of Prudhoe Bay, Alaska. Over land the bottom of permafrost which generally occurs at a depth less than 650m (2000 ft) was mapped with good accuracy. Offshore top and bottom of permafrost was mapped under as much as 250m of unfrozen, saturated sediments.

For many geoelectric sections transient EM has a higher lateral and vertical resolution than other electrical methods. For example, for a permafrost section over the land, where a resistive layer (permafrost) overlies a conductive layer (unfrozen ground) a 10 percent change in thickness of permafrost causes a 30 percent change in measured signal. Similar sensitivities are not available in direct current, magnetotelluric, or harmonic frequency methods.

INTRODUCTION

Electrical methods to map the distribution of permafrost have been used for a long time. They rely on the fact that frozen and unfrozen ground have large differences in electrical resistivity. In Canada and the United States the two major contributions for deep permafrost mapping are by Daniels, et al. (1), who used harmonic frequency soundings on the North Slope of Alaska, and by Koziar and Strangway (2), who employed magnetotelluric methods in the MacKenzie Delta in Canada.

In this paper the application of transient electromagnetic soundings, with small transmitter-receiver separations (near-zone), to permafrost mapping is discussed. This method to a large extent has been developed in the USSR (3,4). Although several case histories of the use of this method are published (5,6), there appear to be no published reports of its application to permafrost mapping in the USSR. U.S. treatises on the theory of transient electromagnetic exploration are found in the papers by Nabighian (7), and Keller (8).

Since the main purpose of this paper is to illustrate the application of transient soundings to permafrost mapping, the theory of the method is discussed in an illustrative manner. The results shown in this paper are selected typical examples of a large survey over lines onshore and offshore west of Prudhoe Bay, Alaska. The survey was performed in the winter of 1982.

PHYSICAL PRINCIPLES

First, the physical basis of the method is explained. A transient system consists of a transmitter and a receiver. The transmitter configuration can be a non-grounded loop or a grounded line. The configuration used in most of our work has been the non-grounded loop. The sensors, in the Geonics EM37 used in our surveys, are multi-turn coils with effective areas varying from 32 to 1,000 m^2. Figure 1 shows the transmitter-receiver arrangements used for transient soundings.

The current generated in the transmitter is shown in Figure 2. There are two periods of time, time-on and time-off. Measurements are made only during time-off. In accordance with Faraday's law, an electromagnetic induction appears when the current in the transmitter varies with time. When the turn-off ramp is linear, the induced electromotive force is a rectangular pulse.

The electromagnetic induction creates eddy currents in the ground. These induced currents are time-variant and cause a time-varying secondary magnetic field, which is measured as an electromotive force in the receiver coil. It has been shown that the induced currents are horizontal closed rings in the absence of lateral inhomogeneities. There is no vertical component of current flow.

Figure 3 schematically illustrates the

distribution of eddy currents as a function of depth at different times. This figure shows that current maxima move down with increasing time. It must be understood that the currents not only move down, but also out. The current expansion can be described by a diffusion type equation.

The electromotive force measured by the receiver coil is the result of the change of current flow with time. A measurement at time t_o will mainly be sensitive to the resistivity of near-surface layers. With increasing time, the current maxima diffuse down, and the electromotive force will progressively become more sensitive to the properties of deeper layers. Therefore, by making measurements as a function of time, information about the geoelectric section is obtained. In transient soundings, effective exploration depth is dominantly a function of time.

BEHAVIOR OF THE FIELD

In transient soundings, generally only the time derivative of the vertical component of the magnetic field, B_z, is measured. To understand the procedures of data processing and interpretation, the general behavior of the measured field in the center of a transmitter loop over two-layered ground will be briefly discussed. In Figure 4 the measured signal is plotted versus the dimensionless parameter, $\frac{\tau_1}{h_1}$.
The parameter , is given by:

$$\tau_1 = \sqrt{2\pi \rho_1 t \, 10^7}$$

and ρ_1 is the resistivity of the first and t is time after turn-off. For a certain geologic setting the parameter, τ_1 is, thus, proportional to the root of time.

Several stages of time can be distinguished in the behavior of the electromotive force (emf). At early time the emf is independent of time; then there is an intermediate range of time in which the emf rapidly decreases with time, and a late range of time in which the emf falls off as $t^{-5/2}$.

Figure 4 shows the behavior of the field for two-layered sections with large differences in the resistivity of the second layer. Yet is it evident that it is very difficult to visualize the geoelectric section from these data. From the different curves it is difficult to tell which curves have a resistive basement and which ones a conductive basement.

APPARENT RESISTIVITY

In all geophysical uses of electrical and electromagnetic methods it has been found convenient to transform the field measured into apparent resistivities, so as to better visualize how the behavior of the field measured differs from the behavior of the field over uniform half-space. Transient soundings are no exception.

To compare the measured response with the response of uniform half-space it is necessary to briefly consider the field behavior of uniform half-space. At all times the behavior of the field, E, over uniform half-space is described by equation (1).

$$E = \frac{3M\rho}{2\pi r^4}\left[\phi(u) - \sqrt{\frac{2}{\pi}}\, u\left(1+\frac{u^2}{3}\right)e^{-u^2/2}\right] \qquad (1)$$

$$u = \frac{2\pi}{\tau/r}\ ; \qquad \tau = \sqrt{2\pi \cdot 10^7 \rho t}$$

WHERE
r = transmitter - receiver separation
ρ = ground resistivity
t = time after turn-off
M = transmitter moment
ϕ = error function

At "early stage" and "late stage" the field can be closely approximated by the asymptotic expressions of equations (2) and (3).

$$E = \frac{3M\rho}{2\pi r^4}\ ; \qquad \frac{\tau}{r} < 2 \qquad (2)$$

$$E = \frac{\mu^{5/2}\, rM}{40\pi^{3/2} \cdot t^{5/2}\rho^{3/2}}\ ; \qquad \frac{\tau}{r} > 10 \qquad (3)$$

It is evident that the field at late stage is more sensitive to the geoelectric section; the field is proportional to $1/\rho^{3/2}$.

The asymptotic expressions have been found useful in defining apparent resistivity. For practical purposes, two definitions have been found convenient in our work:

1. A definition of apparent resistivity based on the early stage asymptotic expression.
2. A definition of apparent resistivity based on the late stage asymptotic expression.

In equations (4) and (5) apparent resistivities based on early stage and late stage are defined based on asymptotic expressions for the field given by equations (2) and (3), respectively:

$$\rho_a = \frac{2\pi r^4}{3M}E \qquad (4)$$

$$\rho_a = \frac{\mu_o^{\ 5/3}}{4\pi t^{5/3}}\left(\frac{rM}{5E}\right)^{2/3} \qquad (5)$$

When the measured emf corresponds to late stage behavior of the field in the first layer over the critical range of time, it is best to use the definition based on late stage. In cases where measurements correspond to early stage behavior in the first layer, it may be more diagnostic to use the definition based on early stage. This situation, for example, arose when measurements were made over a section with a thick, very conductive surface layer. It could also occur when measurements are made at large transmitter-receiver separations.

Figure 5 shows the apparent resistivity curves based on the late stage behavior for the same sections as used in Figure 4. The apparent resistivity curves are plotted in terms of the dimensionless parameters ρ_a/ρ_1 and τ_1/h_1. It is evident that from these curves the general geoelectric section can be visualized. The figure shows that, at later times, the apparent resistivity approaches the true resistivity of the basement. At intermediate values of time, the values of apparent resistivity increase when $\rho_2/\rho_1 > 1$ and decrease when $\rho_2/\rho_1 < 1$, so that from the behavior of the apparent resistivity curves, the general geoelectric section can be immediately visualized. This is the purpose of introducing apparent resistivities in electrical prospecting.

At small values of τ/h_1, all curves merge into one corresponding to the behavior of the field of uniform half-space of resistivity ρ_1. The behavior of this part of the curve is theoretically fully defined by equation (1), and the value of ρ_1 can be obtained from this part of the curve.

Figure 5 has an illustration of two layer apparent resistivity curves. Several albums of 3, 4, and 5 layered apparent resistivity curves are available from the Soviet literature and we have the ability to forward model any n-layered, horizontal sections.

Interpretation of transient soundings from apparent resistivity curves based on early or late stage proceeds in a manner similar to the interpretation of apparent resistivity curves of magnetotelluric or direct current soundings. Interpretation can be done by matching experimental data with master curves, or by inverse solutions. For transient soundings a detailed interpretation manual has been published by Rabinovich (9).

FIELD SURVEYS AND RESULTS

A large number of soundings were conducted on the North Slope of Alaska west of Prudhoe Bay, Alaska. Measurements were made both onshore over the tundra, and offshore over the sea ice. In all these measurements the transmitter consisted of 500 m by 500 m non-grounded loop of 2 gauge wire. The peak current driven through the loop was about 20 amperes, so that the dipole moment was about 5 x 10^6 ampères - m^2.

The equipment used was a modified Geonics EM37, which allowed measurements over a range of time from about 0.1 to 800 ms in 40 time gates equally spaced on a logarithmic scale of time. In the Geonics EM37 the receivers are multi-turn air coils and the effective areas of the air coils used varied from 32 to 1,000 m^2. It was found that with this system in the Arctic, noise was about 0.2 x 10^{-9} v/m^2.

A very large number of soundings were made in the winter of 1982. From the large quantity of survey data a section has been selected for illustration and discussion of the data collected. Figure 6 shows a geoelectric section over a survey perpendicular to the coast line at Prudhoe Bay, Alaska. Above the geoelectric section, three typical apparent resistivity curves are shown. These curves will be discussed in more detail below. From about station 3900S the survey line runs over the sea ice. The water depth below the sea ice gradually increases to the north, but never exceeded 10 m along the profile. Over land the geoelectric section can be described by two layers consisting of frozen ground with resistivities exceeding 200 ohm-m underlain by unfrozen ground of a resistivity of 2.5 ohm-m. Over the bay the geoelectric section consists of unfrozen, brine saturated sediments with resistivities generally less than 2 ohm-m, frozen ground of high resistivities, underlain again by unfrozen ground of a resistivity of 3.0 ohm-m. The profile shows the top to the permafrost to rapidly decrease with distance from the shore line.

The two transient apparent resistivity curves shown above the section have different behaviors, and the behavior of two of these curves will be discussed next. The experimental apparent resistivity data are superimposed on a best fit master curve. For station A the best fit is for a first layer with a resistivity of 400 ohm-m, and a thickness of 600 m, the resistivity of the underlying unfrozen ground is 3 ohm-m. The Manual of Interpretation by Rabinovich (9) shows that for sections, such as shown for station A, the depth to the conductor in kilometers, H_{km}, can be derived from the left descending branch of the curve by equation (6).

$$H_{km} = \frac{\rho_a^{4/9} \cdot \rho_b^{1/9} (\sqrt{2\pi t})^{10/9}}{3.36\, r^{1/9}} \qquad (6)$$

WHERE ρ_a = apparent resistivity

ρ_b = basement resistivity

t = time

r = transmitter-receiver separation

Equation (6) illustrates several important advantages of transient soundings with small transmitter-receiver separation. These advantages are:

1. Equation (5) shows that ρ_a is proportional to the field measured, E, as $E^{-2/3}$. Combining equations (6) and (4), it is evident that $\rho_a \doteq E^{-8/27}$, where k is a constant. Therefore, a 10% change in depth to the conductor will cause about a 30% change in measured field. That is a sensitivity about 3 three times higher than can be obtained in any other electrical method.

2. The depth to the conductor (H_{km}) is proportional to the basement resistivity to the power $\rho_b^{1/9}$, or a 100% error in basement resistivity will cause only about a 10% error in the value of H_{km} . That is important because in all geophysical methods it is difficult to obtain reliable measurements of basement resistivity. Measurements must be made to very large spacing (direct current), or to very low frequencies (magnetotelluric), or to late time (transient) to obtain a good value of basement resistivity.

Station B is a curve for about station 500S along the profile of Figure 6. Again the experimental data are superimposed on a best fit master curve. The data now must be fitted to a three layer section with a first layer (unfrozen ground) with a resistivity of 2 ohm-m and a thickness of 200 m; the second layer has a high resistivity (frozen ground) and a thickness of 350 m; the third layer (unfrozen ground) has a resistivity of 2.50 ohm-m. The matching of the experimental data of station B has a larger range of equivalence than the section of station A. The solution is not sensitive to the value of ρ_2 but very sensitive to values of ρ_1 and ρ_3.

The Gull Island well off Prudhoe Bay is located near the sounding of station B. The E-log from the Gull Island well showed the top of permafrost at 204 m and the bottom at 530 m, in good agreement with the interpretation. The interpretations along all lines could be verified at three other well locations. The agreement at three wells was good.

DISCUSSION

Transient electromagnetic exploration was found to be an effective tool in mapping top and bottom of permafrost onshore and offshore. Features found particularly attractive of the method are:

Lateral Resolution

The measurements were made with a receiver in the center of a 500 m by 500 m transmitter loop. Depth of exploration is mainly a function of time rather than transmitter-receiver separation. A good lateral resolution is maintained with deep exploration.

Vertical Resolution

It was shown that for a two layer section of frozen ground underlain by unfrozen ground, a 10% change in depth to the conductive basement causes about a 30% change in measured field. This high sensitivity translates into a higher vertical resolution.

ACKNOWLEDGMENTS

The authors wish to thank the Directors of the Standard Oil Company (Ohio) for permission to publish this paper.

REFERENCES

Daniels, J.J., Keller, G.V. and Jacobson, J.J., 1976, Computer-assisted interpretation of electromagnetic soundings over a permafrost section; Geophysics, V.41, No. 41, 752-765.

Kaufman, A.A. and G.M. Morozova, 1970. The Theoretical Basis of Transient Soundings in the Near Zone, Akad. Nauk, SSSR, Siberian Branch, Novosobirsk, USSR.

Kaufman, A.A., Exploration Methods based on Nonstationary Electromagnetic Fields in the Near Zone, 1973, Izv., Earth Physics, No. 11, pp. 43-53.

Keller, G.V. and Crewdson, R.N. and Daniels, J.J., 1978, Time-domain electromagnetic survey in Black Rock Desert - Hualapi Flat area of northwestern Nevada; Quar. Colo. School of Mines, V.73, No. 4, pp.46-56.

Koziar, A. and Strangway, D.W., Permafrost Mapping by Audio Frequency Magnetotelluric. Canad J. of Earth Sci. V.15, pp. 1535-1546, 1978.

Nabighian, M.N., 1979, Quasi-static transient response of a conducting half-space - an approximate representation; Geophysics, V. 44, No. 10, pp. 700-1705.

Rabinovich, B.I. and Surkov, V.S., 1978, Results of the Use of the ZSB Method in the Siberian Platform: in Theory and use of electromagnetic fields in exploration geophysics, Akad. Nauk, SSSR, Novosibirsk, pp. 3-18.

Rabinovich, B.I., Surkov, V.S., and Mandel'baum, M.M., 1977, Electrical prospecting for porous reservoirs with oil and gas in the Siberian platform: Sovetskaya Geologiya, No. 2, pp.51-78.

Rabinovich, B.I., 1973, Handbook of Interpretation of Transient Soundings in The Near Zone, Akad. Nauk, SSSR, Novosobirsk, 1973.

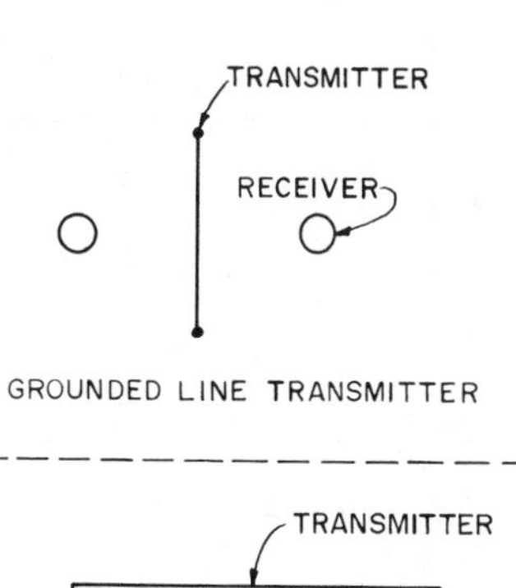

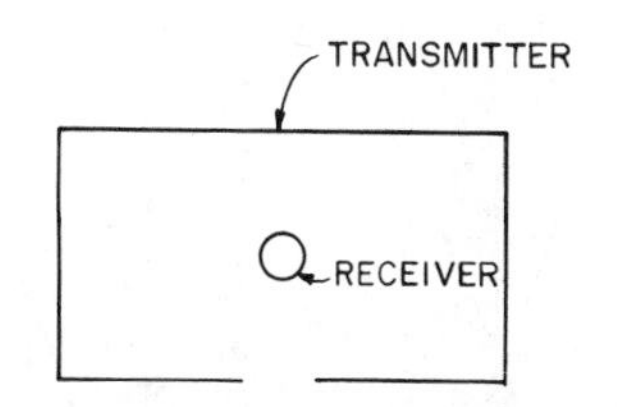

FIGURE 1 Arrays of transmitter-receiver

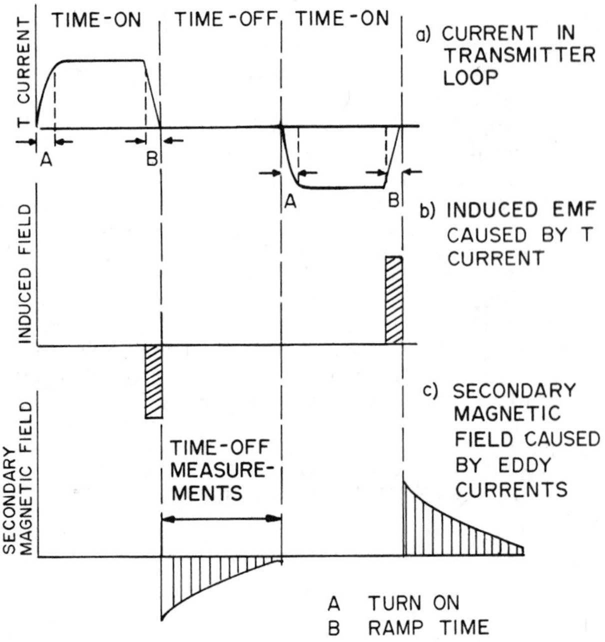

FIGURE 2 System waveforms.

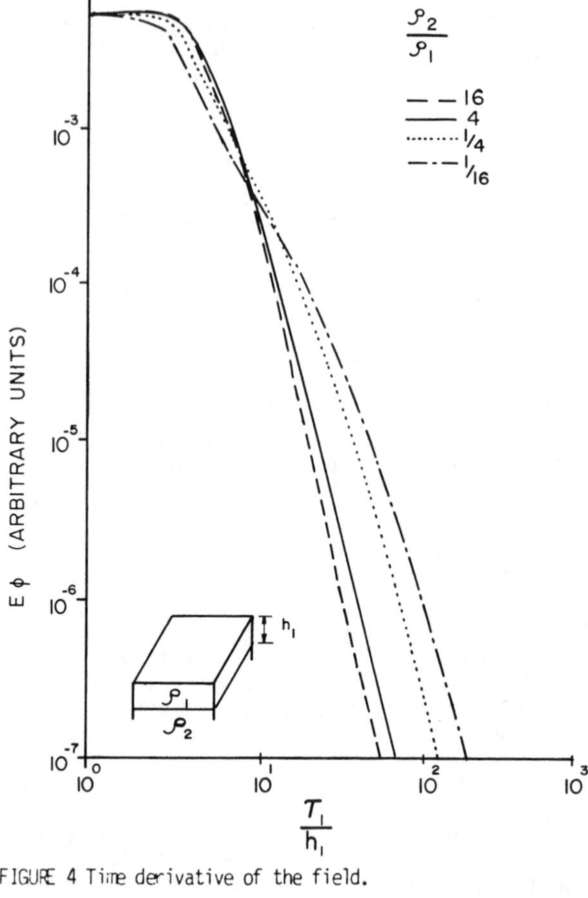

FIGURE 4 Time derivative of the field.

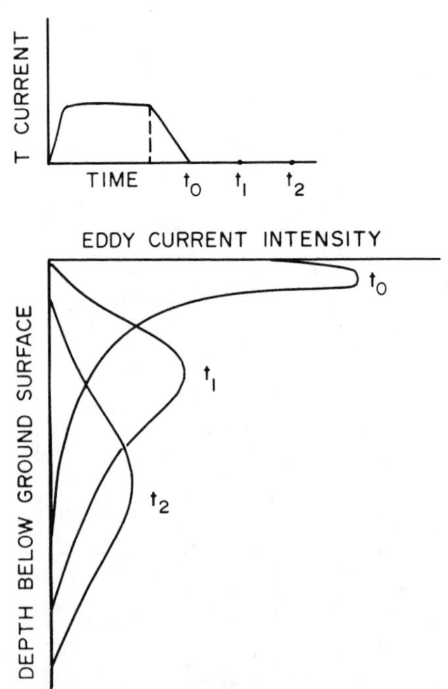

FIGURE 3 Maximum current intensity in vertical plane.

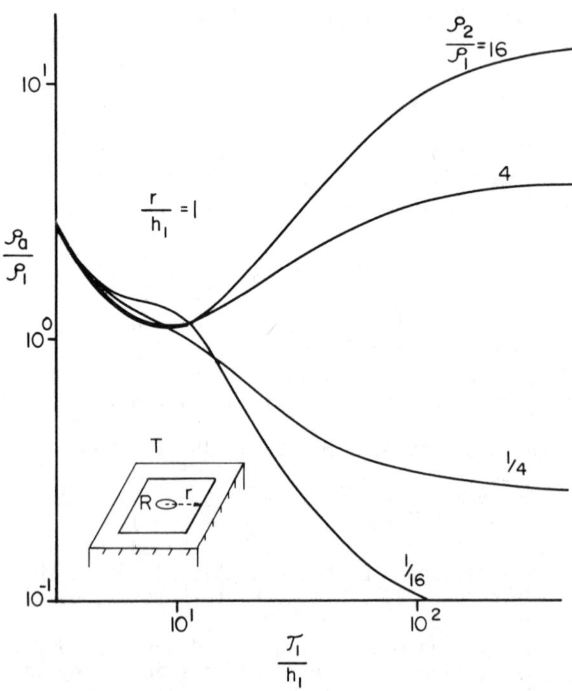

FIGURE 5 Apparent resistivity curves.

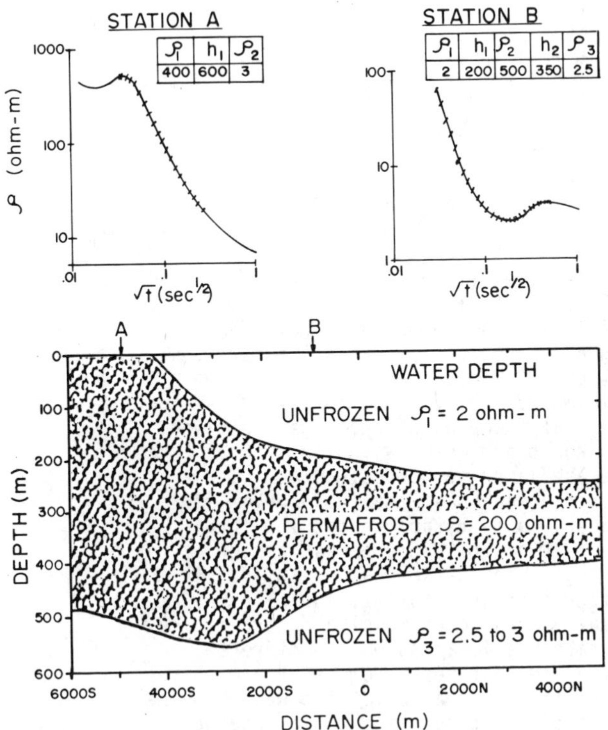

FIGURE 6 Permafrost profiles.

OBSERVATIONS OF SOIL AND GROUND ICE IN PIPELINE TRENCH EXCAVATIONS IN THE SOUTH YUKON

J. R. Ellwood[1] and J. F. Nixon[2]

[1]Foothills Pipe Lines (Yukon) Ltd., Calgary, Alberta, Canada T2P 2V7
[2]Hardy Associates (1978) Ltd., Calgary, Alberta, Canada T2E 6J5

During the winter of 1980-1981, Foothills Pipe Lines (Yukon) Ltd. excavated several long trenches in permafrost in order to install sections of 1.22 m diameter pipeline at the Quill Creek test facility in the South Yukon. The authors directed a program of detailed logging of the different soil strata and ground ice formations encountered. A total of 900 linear meters of trench was excavated mostly by wheeled ditcher to a minimum depth of 2 m. The ditch was logged by an experienced technician, and several examples of the logs are given. The variability in surface peat cover, sand and gravel strata, and a seam of volcanic ash are clearly demonstrated. The ash layer was deposited at a reasonably well-defined date in the recent geological past and therefore allows rates of deposition or growth of surface layers to be established. The large area exposed by the trench wall allowed observations of differences between visible ice and excess ice to be made for logging in coarse-grained soils. It is concluded that in these soil types it is often difficult to distinguish pore ice in the large pore spaces, from what is normally considered massive or excess ice. Several photographs are presented to illustrate the location of the different soil strata and the shapes of several massive ice features. Some of the massive ice forms observed in the trench walls are believed to be relict ice wedges. The trench allowed a three-dimensional impression of the ground ice to be obtained, which is not normally possible from boreholes. Some additional observations are made relating to the success of wheeled ditchers in trenching the different soil types at the Quill Creek test facility.

INTRODUCTION

The Quill Creek test site (Figure 1) is located approximately 30 km northwest of Kluane Lake in the southwestern Yukon. The facility was constructed in 1981 by Foothills Pipe Lines (Yukon) Ltd. to test a variety of pipeline designs, construction procedures, and slope stabilization methods in permafrost terrain. The main portion of the test site (including the area where trenching was conducted) is located on a fine-grained, peaty alluvial plain adjacent to a gravel alluvial fan formed by Quill Creek as it exits from the Kluane Range Mountains into the broad Shakwak valley. The entire test site area is underlain by permafrost, approximately 18 m in thickness.

SITE DESCRIPTION

The site is located in the southern discontinuous zone of permafrost, in a region of thin permafrost and mean ground temperatures in the range of 0° to -1°C. The air freezing index at the nearby localities of Burwash Landing and Destruction Bay is about 3000°C-days, and the thaw index is about 1260°C-days. Preliminary borehole drilling indicated the presence of icy subsoils that would permit significant thaw settlement of the near-surface strata. The ground temperature and climatic data confirm the discontinuous nature of permafrost at the site.

During construction of the test facility, 900 m of ditch was opened using either backhoe or a

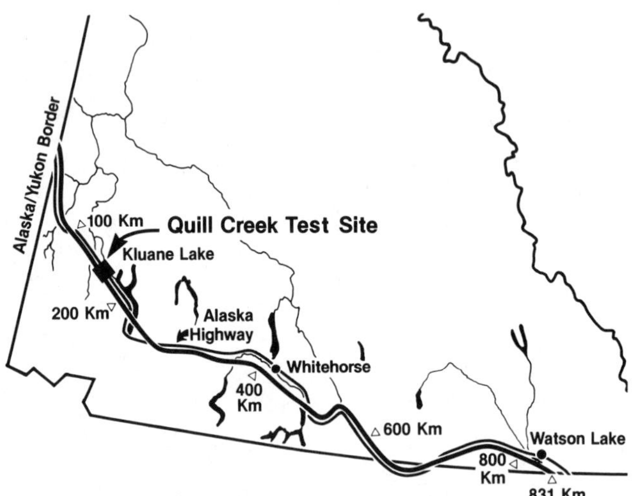

Figure 1: Quill Creek Test Site Location

rotary bucket wheel trenching machine capable of excavating a trench 3 m deep and 2.1 m wide. These wide, deep trenches required for test sections of 1.22 m diameter gas pipeline provided an ideal opportunity for observations and detailed logging of subsurface soil and ice conditions. The soil and ice types were logged according to the Unified Soil Classification System and the National Research Council of Canada Ice Classification System (National Research Council

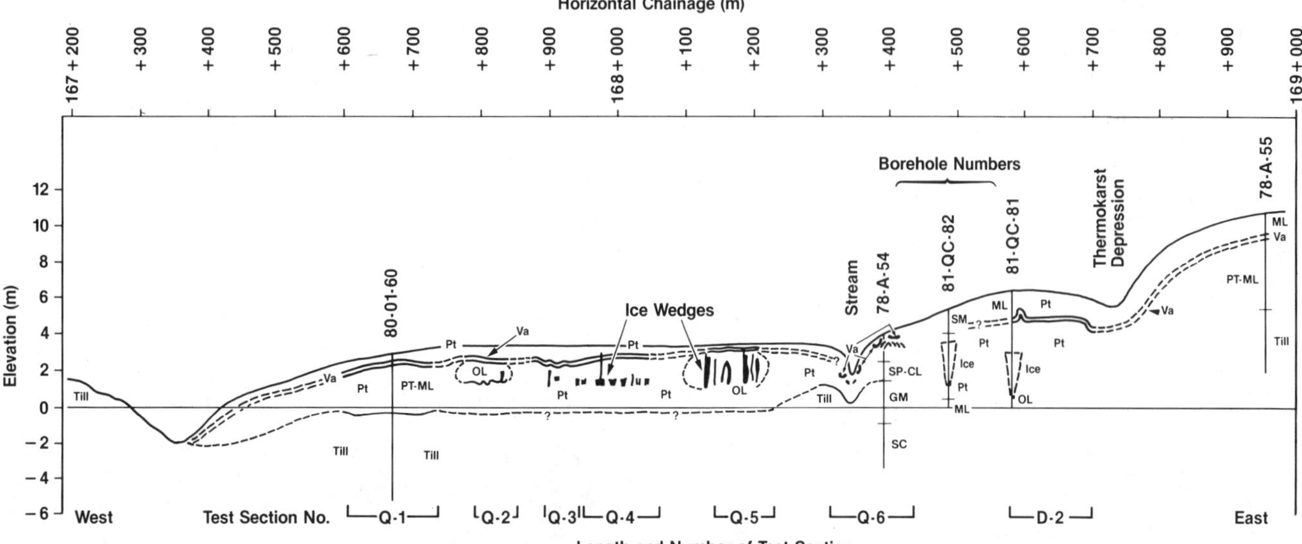

Figure 2: Generalized Stratigraphy of the Quill Creek Test Site

of Canada 1955, 1963). A two-dimensional log of depth versus horizontal distance was prepared showing significant changes in stratigraphy and ice type or content along one wall of the ditch. Representative samples were taken from the wall of the trench during logging and delivered to a commercial soils laboratory for classification and moisture content testing. A generalized stratigraphic profile for the site is shown in Figure 2 along with the specific areas trenched. Several boreholes which were drilled during preliminary investigations to confirm the suitability of the soils for test purposes are also shown plotted in Figure 2.

STRATIGRAPHY OBSERVED IN TRENCH EXCAVATIONS

Volcanic Ash Layers

Two volcanic ash layers were identified in the trench walls. The more recent White River Ash which was desposited about 1220 years B.P. (Hughes et al. 1972) blankets the entire test site area and in this region is 120-180 mm in thickness. On the extreme eastern edge of the test site this ash is overlain by approximately 2.5 m of sands and gravels, with a thin peat cover forming the present-day surface. These sands and gravels are part of the currently inactive western edge of the Quill Creek alluvial fan.

Across the central and western portion of the test site this ash is overlain by peat and organic silt to an average depth of 500 mm. In the area of the trench designated Q-5, however, the ash is only 100-200 mm below the ground surface.

The earlier layer of ash was deposited between 1850 and 1900 years B.P. (Hughes et al. 1972) and is much thinner, averaging only 10-20 mm in thickness. The lower ash unit was present only as a discontinuous band in some of the trench walls (see Figure 3). Where present, this unit is separated from the White River Ash by 50-100 mm of peat. It would appear that the growth rate of the peat during the 650 year interval between the ash layers was significantly slower than it is at present. A detailed examination of the visible ice content descriptions for the ditch walls from areas Q-1 and D-2 resulted in the ice (moisture) content profiles shown in Figure 4. Although the absolute value of the ice content profile is different in the two trench walls, the shape of the profile is similar in each case. It clearly shows the position of the White River Ash relative to significant changes in the moisture and thermal regime of the developing soil profiles. The earlier and thinner ash layer does not coincide with any noticeable effect on the ice content profile.

Figure 3: Ancient stream channel in-filled with sand, showing two layers of volcanic ash.

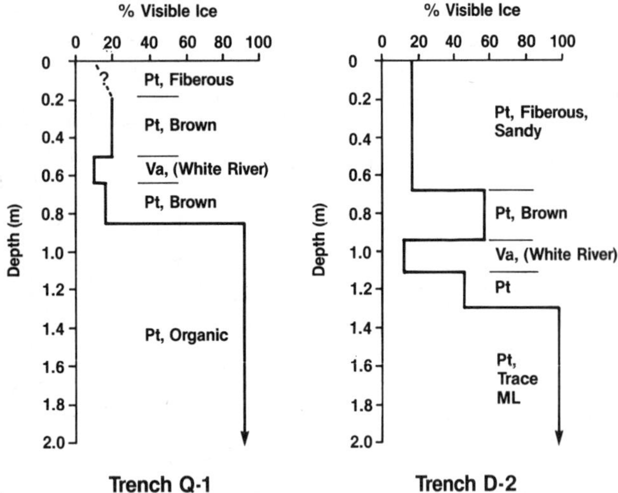

Figure 4: Ice content versus depth for two areas of the site showing similarity of ice profile above and below the White River ash.

Massive Ice Features

Twenty-one rectangular, flat-topped ice wedges were exposed in the trench walls during the ditching operations (Figure 5). All of these wedges were located in a 370 m stretch of the central portion of the test area. Nine of the wedges were greater than 1.5 m wide, four were between 0.5 and 1.0 m wide and the remaining eight were less than 0.3 m in width. The flat tops of most of these ice wedges were 1.5 $\pm$ 0.25 m below the ground surface while the active layer depth in this region is between 0.5 and 0.75 m. Except in a very few cases the White River Ash layer above the ice wedges is undisturbed, indicating that the period of major growth of the ice was prior to 1220 years B.P. Four of the ice wedges show signs of recent growth and have smaller wedges or irregular ice masses extending from the top of the main wedge to the base of the current active layer, as shown in Figure 6.

Figure 5: Rectangular ice wedge in Trench Wall.

Trench Q-6 was excavated across an intermittent stream course, and a portion of the log at the top of the east bank is shown in Figure 7. This area contained an infilled stream channel (also shown in Figure 6) as well as a number of irregular ice lenses and massive ice features which were not found at any other location in the test site area.

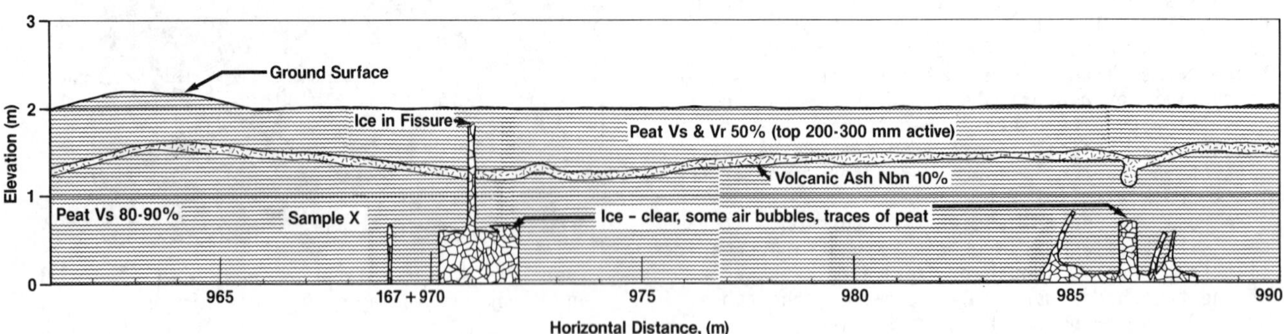

Figure 6: Trench Wall Log - Area Q-4

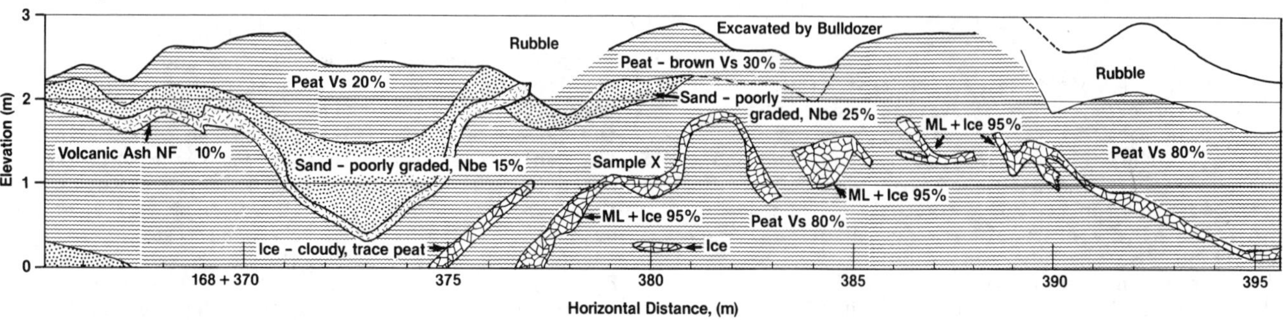

Figure 7: Trench Wall Log - Area Q-6

78-A-54			78-08
USC	Dp	PfC	MC
Pt aVa	1.40	UF Vs Vx	
SP	1.65	VxVs	34 20
CL	2.55	F	
GM aSP tM	4.55		8
SC sML SG		Vx ?	11 8

81-QC-82			81-01
USC	Dp	PfC	MC
Pt	0.10	Nbn	
SM tVa	1.30		
Pt/OL	1.70	Nbe	
ICE + Pt	4.40	ICE +	
Pt	4.85	Nbe	
ML			

METRES 0 1 2 3 4 5 6 7

Figure 8: Geotechnical Logs from two boreholes at site.

COMPARISON OF BOREHOLE RESULTS WITH DITCH WALL LOGS

Several exploratory boreholes were drilled prior to trenching operations, and it is of interest to compare the findings of a conventional drilling investigation with the observations made from a freshly exposed vertical trench wall. Figure 8 presents geotechnical logs from two boreholes (78-A-54 and 81-QC-82) drilled prior to ditching. Although these boreholes were not located on the ditch centerline, they were close enough to illustrate the comparison between the two investigation methods.

Borehole 78-A-54 in Figure 7 is typical of the general stratigraphy in the area and shows a 1.0 m veneer of icy peat and organics overlying a layered sequence of sands and gravels with occasional silty clay layers. This sequence was observed in many of the ditchwall logs, although little appreciation of the three-dimensional nature of subsurface ice could be attained prior to ditching. However, the overall estimate of soil types and ice content in most soil layers was later generally verified by ditchwall logging.

Referring to borehole 81-QC-82, drilled in the same area (location shown in Figure 2), massive ground ice is shown from a depth of 1.8-4.4 m. Clearly, this borehole log indicates an extreme subsurface ice condition, and generalizations based on this log would prove to be completely erroneous. This becomes very apparent when the ditchwall logs are consulted. Wide, massive tabular ice bodies were not discovered at any point across the terrain where the test pipeline trenches were excavated to depths of around 3 m. However, localized ice wedge features of the order of 1 m wide were delineated at many locations. It is believed, then, that the second borehole selected here for illustrative purposes encountered an ice wedge which penetrated to a depth of a least 4.4 m. This comparison illustrates the hazards of generalizing subsurface ice properties based on a few widely spaced boreholes. Interpretations and generalized stratigraphy should rather be based on a combined appreciation of geological information, other exposures, and a knowledge of geological processes in the area as well as the available borehole data base.

Another important comparison emerged in logging the presence of ice in very coarse grained soils. In gravels containing a significant fraction of cobble size particles, the vertical ditchwall provided an excellent appreciation of the three-dimensional structures of the ground ice. However, on site observers at the trenching operations showed a strong tendency to overestimate the excess ice content of the coarser-grained soils. This resulted from the apparently large extent of ice exposed in the vertical section, which was found to be pore ice on closer inspection. That is, ice in large pores might appear to be excess ice, rather than pore ice, due to the large size of the soil pore spaces.

In summary, some advantages and disadvantages of the two methods of subsurface exploration should be mentioned. Boreholes have the advantage of greatly reduced expense per unit depth investigated but fail to provide a full appreciation of the three-dimensional ground ice structure. Ditchwall logs are extremely expensive if used solely for the purposes of investigation and can only penetrate a limited depth. However, they are of great value in assisting the engineer and geologist in appreciating the overall structure of ground ice features and can help in explaining borehole anomalies and more correctly generalizing soil and ice stratigraphy.

DITCHING

Part of the operations at the test site involved an examination of available ditching methods in permafrost. Therefore, it is of interest to briefly outline the methods used and the success attained. One of the selection criteria used for the site was that it should exhibit a wide variety of permafrost soil and ice conditions, so that ditching success could be monitored in differing subsoil conditions.

A Banister Model 710 wheeled ditcher was used for the pipe trench excavations. The ditch dimensions were approximately 2 m wide and 2-3 m deep depending on test requirements. The ditch sidewalls were clean and vertical, and the trench base generally reflected the topography of the surface right-of-way. Most of the ditches were excavated to partial depth and then completed to full depth by subsequent passes. Some additional excavation tests using drill and blast-assisted backhoe excavation methods were also carried out.

Wheel speeds of the ditcher were maintained around 5 rpm, and the maximum available power at the ditcher wheel was 650 kW (870 HP). Average rates of progress varied from about 30 m/h in predominantly frozen granular soils to 40-100 m/h in predominantly fine-grained or organic icy permafrost. These values are very dependent on the degree of right-of-way preparation, the local permafrost soil properties, ground temperatures, the layout and metallurgy of cutting teeth, etc., therefore the rates quoted here should not be extrapolated for use in other areas where these factors are not well established.

It was observed that soils with excess ice in the form of ice lenses tended to fail through the ice during the excavation process. Soils with substantial excess ice content were therefore easier to excavate. When boulders or cobbles were present, the rock particle fractured in many cases before it became dislodged from the in situ frozen soil matrix. Frozen sands caused higher rates of abrasive wear on the cutting teeth than occurred in frozen silts, clay, or peats.

DISCUSSION AND CONCLUSIONS

The ditching, logging, and field observations of near surface permafrost deposits at the Quill Creek test facility have provided a valuable insight into the soil and ice stratigraphy in this area of the South Yukon. Of particular interest is the presence of ice wedge ground ice features. The site is located in a discontinuous permafrost area, where ice wedges are not generally considered to be actively forming at the present time. Ice wedges actively form in continuous permafrost areas (Washburn 1973), or where mean ground temperatures are -2°C or colder (Romanovskii, referenced by Mackay et al. 1978) or where mean air temperatures are colder than -6° to -8°C (Pewe 1973). In addition, Mackay et al. (1978) state that ice wedges grow on recent alluvial islands in the Mackenzie Delta where mean ground temperatures are -2° and -3°C. Finally, Pewe (1966) concluded that ice wedges in Alaska were most abundant in areas of continuous permafrost where the mean annual ground temperature is -5°C or lower.

None of the conditions outlined above are realized for present-day conditions at the Quill Creek site, and the ground ice features are indicative of a significantly colder climate in the geological past. This is not surprising, as Mackay (1975) and others have inferred a warming trend in recent geological times from available evidence. Mackay et al. (1978) have summarized the evidence of ice wedges in many countries, and Fukuda (1981) has reported the presence of ice wedge casts in Hokkaido, Japan. The ice wedge features delineated in the trench wall logs in the South Yukon appear to support ideas that the climate has been significantly colder in this area at some time in the past.

Observations of the volcanic ash layer in this area have allowed an estimate of the growth of surface peat. This appears to be generally in the range of 0.5 m in 1,200 years, with a somewhat slower rate prior to this time.

REFERENCES

National Research Council of Canada, 1955, Guide to the field description of soils: Tech. Memo 37, NRC 3813.

National Research Council of Canada, 1963, Guide to a field description of permafrost: Tech. Memo 79, NRC 7576.

Fukuda, M., 1981, Field observations of ice wedge cracking in the permafrost area near Tuktoyaktuk, N.W.T., Canada: Contribution No. 2424, Institute of Low Temperature Science, Sapporo, Japan, p. 45-60.

Hughes, O. L., Rampton, V. N., and Rutter, N. W., 1972, Quarternary geology and geomorphology, southern and central Yukon (northern Canada): XXIV International Geol. Congress Guidebook A-11, Montreal.

Mackay, J. R., 1975, The stability of permafrost and recent climatic change in the Mackenzie Valley, N.W.T.: Geological Survey of Canada, Paper 75-1, Part B, p. 173-176.

Mackay, J. R., Konischev, V. N., and Popov, A. I., 1978, Geologic controls on the origin, characteristics and distribution of ground ice, Proc. 3rd Intl. Permafrost Conf., v. 2, p. 1-18. Edmonton, NRCC No. 16529.

Pewe, T. L., 1983, Ice wedge casts and past permafrost distribution in North America: Geoforum 15, p. 15-26.

Pewe, T. L., 1966, Ice wedges in Alaska-classification, distribution, and climatic significance: Proc. First Intl. Permafrost, Layfayette, Ind., p. 76-81.

Washburn, A. L., 1973, Periglacial processes and environments: London, Arnold, p. 320.

EVALUATION OF EXPERIMENTAL DESIGN FEATURES FOR ROADWAY CONSTRUCTION OVER PERMAFROST

David C. Esch

STATE OF ALASKA DEPARTMENT OF TRANSPORTATION AND PUBLIC FACILITIES
RESEARCH SECTION - 2301 Peger Road, Fairbanks, Alaska 99701

The side slopes of roadway embankments in warm permafrost areas often cause severe problems from long-term thaw related settlements. Soils underlying the snow-covered slopes do not totally refreeze each winter, as they normally do beneath the cleared portion of the roadway. This results in progressive settlements of the outer edges of the roadway and longitudinal cracking of the roadway surface. Experimental installations of air convection ducts, in conjunction with insulation layers and embankment toe berms, were made during 1973 and 1974 on a newly constructed 7 m high roadway embankment 40 km west of Fairbanks. Performance has been monitored since that time. Results through the thawing season of 1982 are presented. No combination of features tested was totally effective in preventing long-term thaw and settlements of snow-covered embankment slopes. Results indicate that annual refreezing beneath snow-covered side slopes might be achieved by use of the air-duct method. Insulation layers were also installed at different depths in roadway cut sections in ice-rich permafrost at sites approximately 13 km west of Fairbanks. These installations have demonstrated the benefits of placing a 100 mm thick insulation layer at a depth of 1.2 m in a 3 m deep subcut, as compared to installing a similar insulation layer at a greater depth.

INTRODUCTION

Previous studies by the Alaska Department of Highways of roadways constructed over permafrost have shown that progressively deeper thawing generally occurs beneath annually snow-covered roadway side slopes, even in areas where the embankment thickness is adequate to prevent thawing into the underlying permafrost beneath the cleared roadway (Esch 1973, 1978). This side slope thawing results in consolidation and slope settlements in thaw unstable permafrost. Any settlement in the slope areas results in roadway cracking (Figure 1). To study the benefits of different embankment slope designs in controlling thaw beneath side slopes and to evaluate alternative insulated roadway designs for cut sections where the vegetation must be removed, a research project was initiated in 1973. Ten different combinations of insulation layers, toe berms, and air ducting systems were installed on newly constructed highway sections and have been monitored since 1974.

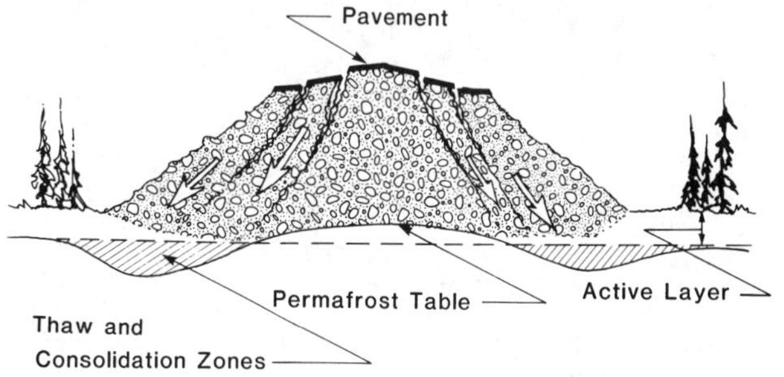

FIGURE 1 Typical roadway distress from thaw beneath side slopes.

SITE LOCATIONS AND DESCRIPTIONS

The two sites selected are located at Alder Creek and Bonanza Creek, approximately 13 and 40 km west of Fairbanks on the Parks Highway. In these areas, new roadway segments were routed across undisturbed terrain, underlain by ice-rich silt permafrost soils. Alder Creek site freezing and thawing indices and Fairbanks snowfall records during the term of this study are shown in Figure 2.

In the Alder Creek valley the new roadway alignment required short cut sections near the valley bottom at two points approximately 180 m east and 360 m west of the creek crossing.

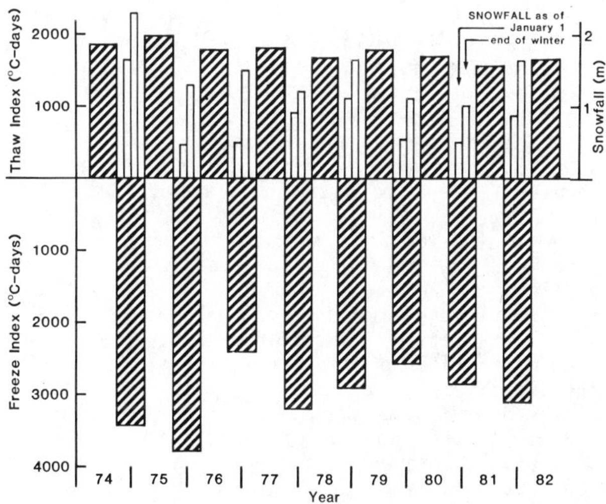

FIGURE 2 Climatological data from Alder Creek Site.

Soils were organic colluvial silts of loessial origin, with a black spruce and sphagnum moss vegetation. Permafrost was generally present beneath a 0.6 m thick active layer. Moisture contents ranged from 30% to 90%. Massive ice was encountered in nearly all borings between depths of 4.5 and 14 m.

In the vicinity of Bonanza Creek the new roadway alignment required an embankment ranging from 6.7 to 7.6 m in height over frozen muskeg. Overlying the permafrost at this site is a 0.3-0.6 m thick peat active layer covered with a surface layer of sphagnum moss, and scattered black spruce. Frozen organic silts were present beneath the surficial peat layer. Frozen water contents of the silt ranged from 30 to 380%, averaging around 100% by weight. No massive ice was encountered.

DESIGN OF EMBANKMENTS

Thermal analyses of the proposed embankments using the "Modified Berggren" calculation approach (Department of the Army 1966), indicated that roadway area thawing could reach to a depth of approximately 4.5 m. The insulation utilized in this study was Styrofoam HI-35, an expanded polystyrene foam from Dow Chemical Company. This product is essentially the only roadway insulation used in North America (Williams 1968) and has a thermal conductivity of 0.029 W/m°C.

At Alder Creek the final roadway grade was only 0.6-1.3 m above the original ground surface, necessitating some excavation and replacement of the underlying soils to prevent settlement problems. Utilizing a fixed depth of subcut of about 3 m below final roadway grade, two alternative insulation depths were selected for field evaluation. The first design type involved placing a 102 mm thick insulation layer as close to the roadway surface (1.2 m) as permitted by construction equipment wheel loadings, and is referred to as the "top-insulated" cut (Figure 3). This design resulted in a thickness of 1.7 m of fill between the insulation layer and the permafrost at the bottom of the subcut, and was used from 326 to 426 m west of Alder Creek.

For comparison purposes, insulation was also installed in a similar cut area located from 326 to 426 m east of Alder Creek. In this area the 102 mm thick insulation was placed at the bottom of the subcut, directly on permafrost at a depth of 3.3 m below final grade (Figure 4).

At Bonanza Creek, a 7 m embankment height was required because of roadway grade restrictions, thermally more than adequate to prevent seasonal thaw penetration into the underlying permafrost. However, problems of slope and shoulder area settlements were anticipated. Three different types of slope modifications were selected for field evaluation to retard this side slope thawing, including insulation layers, toe berms, and air convection cooling ducts.

Toe berms were designed for a thickness of 1.8 m of silt waste, the calculated thickness required to prevent the first year's thaw from penetrating beneath the berm. The full berm width was set at 6.1 m, with a 30.5 m transition length to permit studies of the effect of berm width.

The air convection ducts were included in this study to determine their benefits in removing heat and refreezing the soil beneath the snow cover. A detailed discussion of air ducts for building foundations is presented by the Department of the Army (1966). The ducts function in convection only in winter. The adjacent soil warms the air inside the duct, causing it to expand and rise up the exhaust stack, at the same time drawing in cool air at the inlet. Dampers were included, and are manually closed in summer. Two different layouts for the ducts were utilized on the north and south sides of the embankment, as shown in Figure 5. The ducts were constructed from 0.2 m diameter corrugated metal pipe, installed with the inlet ends above the depth of maximum snow cover, and with 3 m high exhaust stacks.

CONSTRUCTION OPERATIONS

Contract specifications required that excavations into permafrost and the placing of initial 0.9 m thick gravel working pads to be performed during the fall season of 1973. Ground cooling and total refreezing were assured by periodically removing the snow cover from the working pads in the study areas during the 1973-1974 winter.

The insulation layer for the bottom of the subcut area at Alder Creek was also placed during this fall construction period, followed by placement of the initial 1.5 m backfill layer to complete the working pad construction. In all other study areas, insulation placement and berm construction were done early in the spring of 1974, followed by completion and paving in 1975.

Materials utilized at Bonanza Creek included an initial 0.9 m thick working pad of weathered rock placed at 1,790 kg/m³ and 16% moisture, and covered by a 3.0-3.6 m thickness of silt at 1,760 kg/m³ and 10% moisture. The silt soils in the berms had a final density of approximately 1,600 kg/m³ and a moisture content of 15%.

INSTRUMENTATION DETAILS

The thermal performance of the various design features is evaluated from monthly subsurface temperatures measured with a system of 528 thermocouples and 27 thermistors installed in 14 horizontal strings and in 24 vertical borings made through the embankment. Three borings were also made and instrumented in adjacent forest areas to provide comparisons with natural conditions.

Site air temperatures are obtained by battery operated recorders housed in Weather Bureau type shelters placed near each site. Surface temperatures of north and south facing embankment slopes and of the roadway surface are recorded at the Bonanza Creek site.

To provide a basis for repeated measurements of side slope settlements and lateral spreading, cross-section reference hubs and nails were placed at approximately 3 m intervals from the centerline to the toe of the embankment slopes on all study sections, and referenced to a non-heaving benchmark. Thaw depths in the toe of slope and berm areas are measured by means of hand or power assisted probing with 12.7 mm diameter steel rods.

PERFORMANCE COMPARISONS-INSULATED CUTS

The relative performance of similar shallow roadway cut sections made into ice-rich permafrost, and insulated with polystyrene foam layers placed at different depths, was determined by examination of the Alder Creek insulated cuts. Data on thaw depths, settlements, and temperature changes in the underlying permafrost have been obtained over the seven year period since construction.

The section designed and constructed with 102 mm of insulation placed at a depth of 1.2 m beneath the pavement, reached an apparent state of thermal equilibrium within 5 years after construction. Thaw depths beneath the roadway ranged from 3.1 to 3.3 m during the 1979-1982 summers. This annual thawing, which reaches from 1.7 to 1.9 m beneath the insulation, has been totally confined to rock fill placed within the subcut. Full annual refreezing has always occurred beneath the cleared roadway. However, some progressively deeper thawing has occurred each year beneath the side slope and berm areas, and 0.5 m thick talik zones existed beneath these slope areas by 1982. Cross--section settlement profiles since 1975 (Figure 3) indicate that progressive annual settlements are occurring in slope and berm areas, exceeding 0.25 m and 0.33 m, respectively in the 7 years since construction. However, only minor settlements, totaling 58 mm over 7 years, have been observed at the roadway surface. Some longitudinal shoulder area cracks have occurred as a result of the sideslope and berm areas settling much more than the roadway itself.

The bottom insulated cut section, constructed with a thickness of 102 mm of insulation placed directly on the permafrost at a depth of 3.2 m, demonstrated very excessive roadway area settlements (Figure 4) as compared to the nearly stable "top-insulated" cut section discussed above. The relative performance of these two cuts is compared in Table 1, along with data from an adjacent undisturbed forest site. The top-insulated cut design resulted in long-term settlements only 8% as great as adjacent uninsulated roadway areas, while the bottom insulated cut settlements was 48% of the uninsulated area settlements.

The excellent roadway performance of the top-insulated section indicates the benefit of placing insulation layers as near to the surface as practical and providing a thickness of compacted sand or gravel beneath the insulation sufficient to contain the annual thaw zone. Both cut sections design have annually refrozen to the full depth thawed beneath the pavement and shoulder areas.

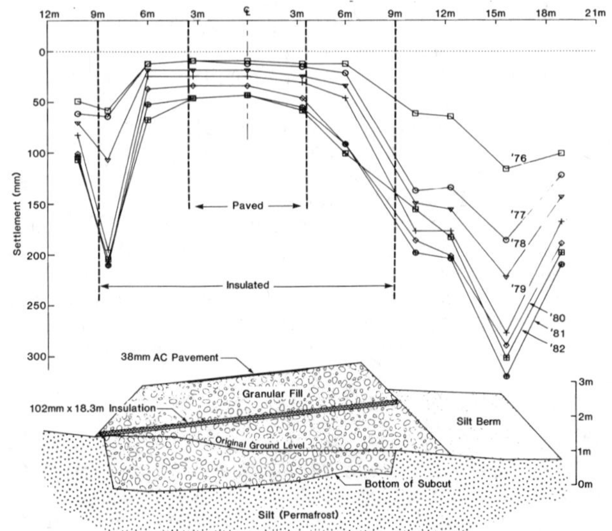

FIGURE 3 Cross-Section detail and settlement plot for top-insulated cut at Alder Creek.

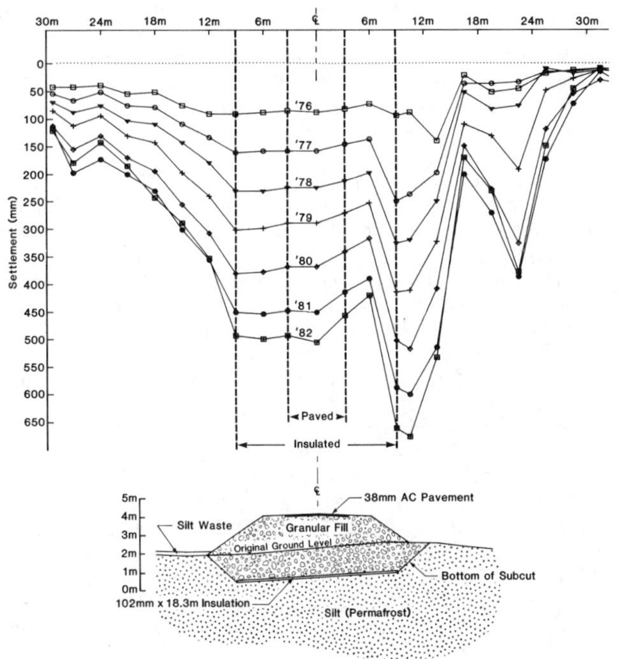

FIGURE 4 Cross-section details and settlement plots for bottom-insulated cut at Alder Creek.

TABLE 1 Comparative Performance of Cut Sections Between 1975 and 1982.

Condition or Insulation Depth (m)	Max. Thaw Depth (m)	Thaw into Permafrost (m)	Centerline Settlement (m)	Permafrost Temp. Change (°C)	Permafrost Temp. Jan. 1982(°C)	Temp. Reference Depth (m)
102 mm @-1.2 m	3.3	0.00	0.06	+0.26	-0.49	-9.4
102 mm @-3.2 m	4.2	0.70	0.38	+0.12	-0.55	-10.0
No Insulation	---	---	0.79	---	---	---
Forest	0.7	0.00	---	+0.01	-0.46	-9.1

However, taliks have developed beneath all sideslope and berm areas, ranging in thickness from 0.5 to 1.0 m after 7 years. These taliks are growing in thickness in nearly equal annual increments.

Temperature trends at depths of about 10 m below the original permafrost surfaces indicate very minor warming for the subroadway permafrost, ranging from 0.12-0.26°C over 7 years. Temperatures at this depth are still slightly colder than beneath the undisturbed forest. This indicates the roadway to have an average surface temperature very close to the original undisturbed forest condition, although the annual temperature variations and thaw depths for the roadway are much greater.

Embankment Performance--Bonanza Creek

The road routing at Bonanza Creek was essentially east-west, and the slope facing south was most heavily instrumented for temperatures. Major differences in vegetation density have been noted, with much better growth on the south slope.

Continuous recorders were installed in 1975 to measure the surface temperatures of the pavement surface and the north and south facing 2:1 sideslopes. During 4 years of observations the average surface temperatures were 0.5°C for the pavement, and 1.9°C and 5.0°C for the north and south-facing slopes, respectively. Surface thawing season n Factors for these points were 1.73, 1.02, and 1.72, respectively, while freezing season n Factors were 1.05, 0.40, and 0.45. The mean air temperature over this period was -2.7°C.

Maximum thaw depths observed beneath the roadway centerline at three locations in the study embankment ranged from 4.1 to 4.9 m, averaging 4.6 m. This was roughly about 2.5 m above the original ground surface. There were no indications of progressive increases in thaw depth with time.

Thaw depths at a constant distance of 2.4 m inside the normal position of the toe of slope, as if no berms had been used, are presented for comparison in Figure 6. Thawing of the embankment permafrost foundation soils at this location is considered to be very detrimental to sideslope stability. This figure demonstrates a progression of thawing in spite of all combinations of treatments, as well as progressively greater benefits in reduced thawing from the addition of berms, air cooling ducts, and berm insulation layers. The embankment sections which featured the combination of toe berms and insulation layers were much more effective than uninsulated berms.

Comparisons of the relative benefits of the different sideslope treatments can be made on the basis of several factors, including thaw depths, surface settlements and lateral slope movements. Thaw depth changes beneath the side slopes over the 7 years following construction are shown in Figure 7. Data from years not shown on these plots were intermediate between the years plotted, and were omitted for clarity.

Air Ducts

Air convection cooling ducts were of significant benefit in retarding thaw of the foundation soils, particularly when used in conjunction with

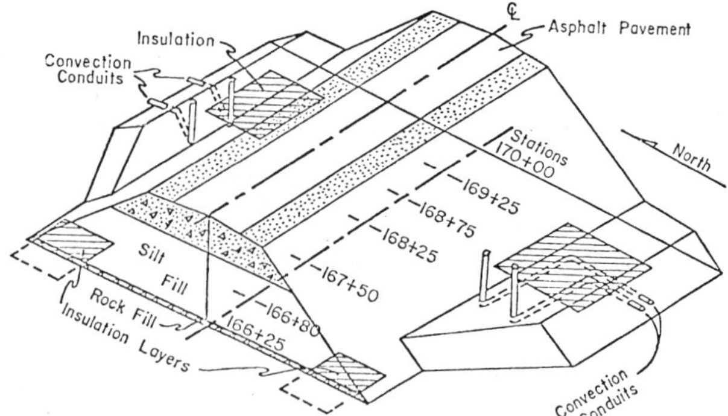

FIGURE 5 Overall view of experimental embankment details at Bonanza Creek.

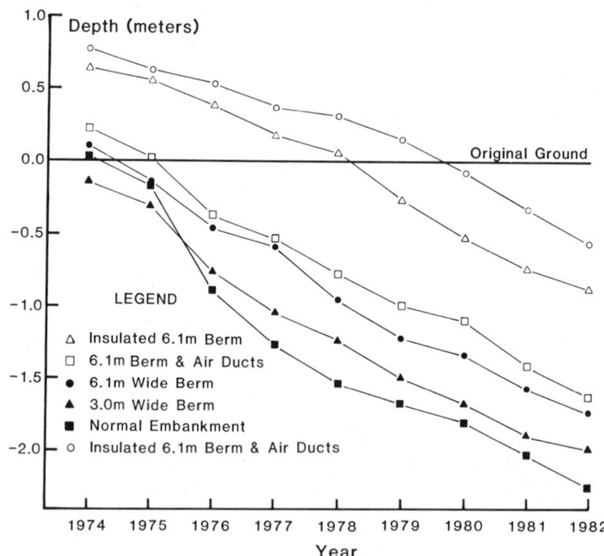

FIGURE 6 Thaw depths for different embankment designs, at 2.4 m inside normal embankment toe.

insulation layers placed to reduce summertime heat gains in the soils cooled by the duct system. Experimental studies and subsequent analyses of these ducts indicated that a larger diameter duct would have been more effective, and a duct design methodology has since been developed. Roadway surface and shoulder movements indicated that these ducts were located too far from the center of the embankment to control those movements.

Embankment Toe Berms and Insulation Layers

The use of toe berms constructed of silt or waste materials showed only very minor benefits on thaw beneath the sideslopes, retarding the progression of thaw to any given depth by 1-3 years. A 6 m berm width was slightly superior to a 3 m wide berm on a short-term basis, but neither berm type prevented long-term thaw progressions. This might be expected, as berms present increased areas of permafrost surface disturbance, which are elevated, well drained, and poorly vegetated, and therefore warmer than the original muskeg surface covered by

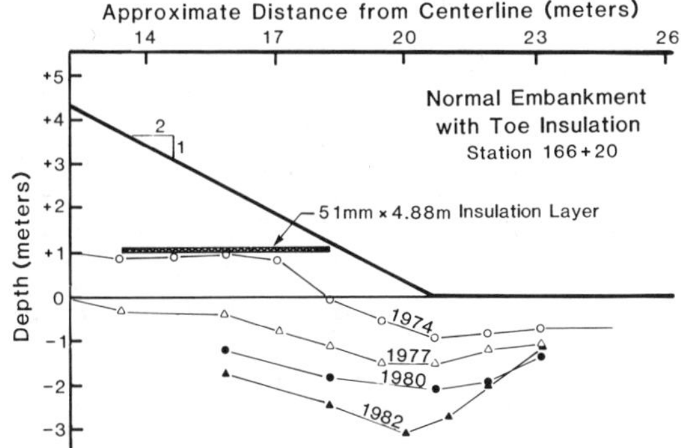

the berm. Berms were marginally effective in reducing settlements and lateral shear-type movements of the lower and mid-slope areas due to their supporting and counterbalancing effects, but had little or no benefit in reducing movements of the critical roadway surface area.

Insulated toe berms provided a major reduction in thawing, with thaw progression into the underlying permafrost delayed by about 4 years by use of 51 mm of insulation as compared to similar uninsulated berms. However, the long-term thawing trends beneath berms and slopes were not arrested by the insulation usage. The placement of 51 mm thick x 4.9 m wide toe insulation layers at a height of 1 m above the toes of a normal embankment without berms, proved of no significant benefit in retarding thaw or related movements.

Observations on Embankment Movements

All embankment measurement points were in a state of progressive movement downward and outward during the 7 years of observations. Vertical settlements of the roadway centerline averaged 29 mm/yr, with little variation between treatments or years. Because observations show no permafrost thawing beneath the center of the embankment, the possibility of these movements being the result of stress-related creep of the underlying warm (-0.6°C) permafrost must be considered.

Sideslope vertical and lateral movements have been very great at this site, with up to 1 m of settlement and lateral movement of some reference points in middle to lower slope areas. Some of these slope movements appear to be related to annual frost action and solifluction type mechanisms, rather than to permafrost thaw and consolidation. Frost action related movements appeared particularly intense on the north-facing slope, perhaps because this was the upstream side of the embankment and had more water availability due to ponding from the dike effect of the frozen embankment. Vertical settlements and lateral movements of the reference hubs are shown by Figures 8 for a normal embankment section and for the full treatment section which combined berms, air ducts, and insulation layers. Intermediate levels of movement were generally noted for other treatments or combinations thereof. At the roadway surface, vertical and lateral movements resulted in the formation of cracks and 20-40 mm deep rutting in the outer wheelpaths, which required patching and levelling of all sections in 1981.

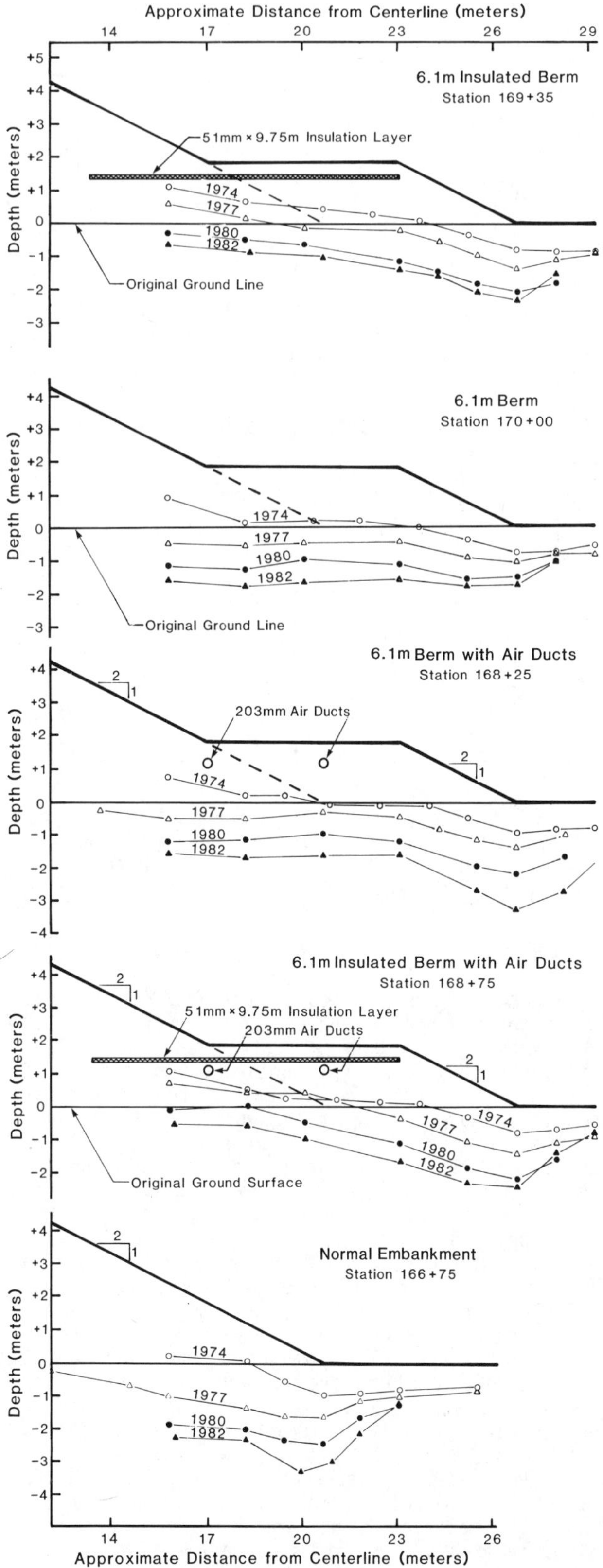

FIGURE 7 Thaw depths beneath lower embankment slopes for various embankment sections.

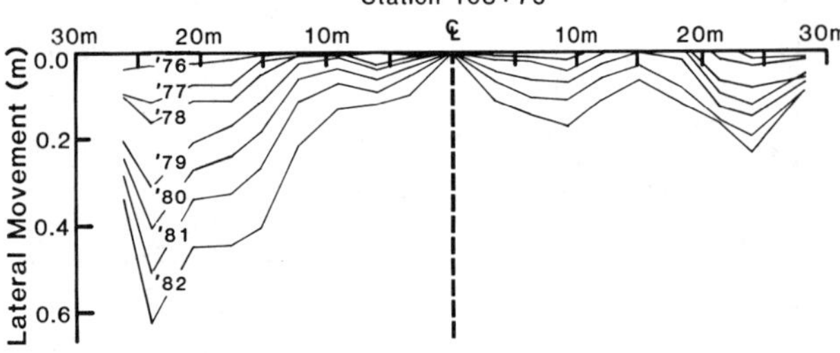

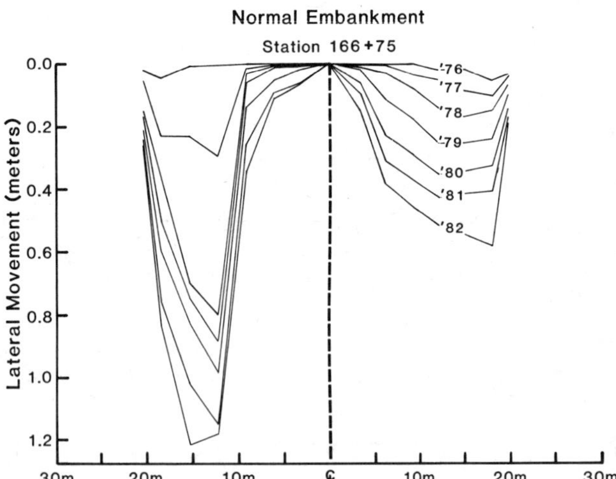

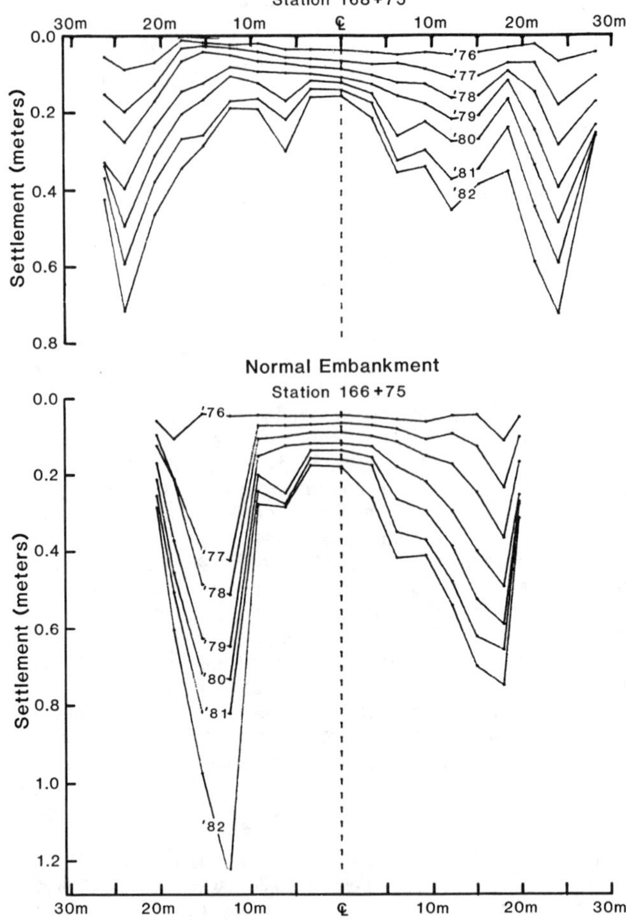

FIGURE 8 Lateral and vertical movements of normal and bermed, insulated, and ducted embankments, from reference points at various distances left and right of roadway centerline.

SUMMARY

Studies of roadway cuts and embankment movements and thermal changes have been made on 10 instrumented roadway study sections near Fairbanks, Alaska, with continuous records kept from 1974 to 1982. In all cases, the permafrost condition has been maintained beneath the cleared roadway and shoulder areas. Excessive thaw-related settlements were observed in a cut section insulated at a depth of 3.2 m, while a similar section with the insulation placed at a 1.2 m depth has had negligible settlements. Roadway sideslopes have developed residual thaw zones or taliks in all areas, and annually deepening thawing beneath the slopes has resulted in progressive deformation and cracking of slopes and roadway shoulder areas.

The benefits of a combination of insulation layers, air cooling ducts, and embankment toe berms were apparent from these studies. However, no combination of treatments was adequate to prevent the long-term progressive degradation of permafrost beneath embankment slope areas, and the eventual distortion of the roadway surface of the 7 m high road embankment. There is some hope that a combination of insulation, embankment shape, and larger air cooling ducts may be sufficient to offset the annual heat gains of sideslope areas and to provide a thermally stable embankment, even for warm (-0.6°C) permafrost conditions.

ACKNOWLEDGMENTS

Funding for this work was provided by the Federal Highway Administration. The assistance of the many research personnel of the Alaska Department of Transportation and the University of Alaska has been essential in maintaining continuous long-term records and data processing.

REFERENCES

Department of the Army (1966) "Calculation Methods for Determination of Depths of Freeze and Thaw in Soils", Technical Manual TM 5-852-6.

Esch, D.C. (1973) "Control of Permafrost Degradation Beneath a Roadway by Subgrade Insulation," Proceedings, Second International Permafrost Conference, pp. 608-621.

Esch, D.C. (1978) "Performance of a Roadway with a Peat Underlay Over Permafrost," Research Report 78-2, Alaska Department of Transportation.

Williams, W.G. (1968) Development and Use of Plastic Foam to Prevent Frost Action Damage to Highways--a Summary of Experience in United States, International Conference on Highway Insulation, Wurtzburg, Germany.

SOLAR SYSTEM ICES AND MARS PERMAFROST

Fraser P. Fanale and Roger N. Clark

Planetary Geosciences Division, Hawaii Institute of Geophysics,
University of Hawaii, Honolulu, Hawaii 96822

Infrared reflectance measurements have resulted in identification of CH_4 ice on Pluto and Triton, H_2O ice on satellites of Uranus, Saturn and on three Jovian satellites (one of which may harbor a liquid H_2O zone in its crust). Frozen SO_2 on Io and bound H_2O on the largest asteroid have also been identified. Such diversity is explicable in terms of accretion conditions, degassing history and atmospheric evolution. Ground H_2O ice appears abundant in ubiquitous Mars permafrost but its distribution is controversial. A key issue is how long ice which is unstable with respect to the atmosphere can be preserved by soil cover. We calculate thermal histories for the regolith at all latitudes, depths and obliquities throughout Mars´ history. Neither seasonal nor obliquity driven freeze-thaw of systems that are stable with respect to global conditions can occur unless the systems have eutectics <235°K. Either this process does not occur on Mars or occurs rarely owing to unusual brines, episodic H_2O emplacement or atypical regional subsurface thermal regimes.

ORIGIN OF SURFACE AND NEAR SURFACE ICES: PRINCIPLES

Formation of solid planets involved accretion of condensates from a gaseous nebula. The condensation temperature determined planetary bulk volatile endowment (Lewis 1974). Objects formed from high temperature (>1000°K) material are almost volatile free. If condensation occurred <200°K, the final objects are icy because the oxygen abundance is more than needed to make silicates or oxidize metal, and combines with hydrogen to make ice equal to twice the silicate plus metal mass. Then formation of $NH_3 \cdot H_2O$ occurs by reaction of NH_3 with ice, and $CH_4 \cdot 8H_2O$ is formed. If the H_2O is hidden in an already accreted planetary body, then the NH_3 condenses as NH_3 ice and the CH_4 as CH_4 ice (Lewis 1972). The Jovian satellites Ganymede and Callisto have ρ ~1.6 g cm^{-3}, or two parts ice and one part silicate. Io and Europa, the two innermost large satellites of Jupiter, are rocky because the early star-like Jupiter disallowed ice condensation near it. Saturn´s larger satellites have densities, ~1.0 g cm^{-3}, because of condensation of other ices below 100°K and dynamic fractionation which depleted silicates. This ratio (ice > rock) is also relevant to comets, though these objects include CO_2 and CO, suggesting kinetic barriers at very low T and P.

If planets form from materials condensing between 1000° and ~200°K, they are "rocky" but have <u>some</u> volatiles: H_2O from hydration of silicates, and P, S, N., etc. from condensed non-icy materials or reactions with them (Fegley and Lewis 1980). This is how volatiles were acquired by Mars, the Earth, Venus, asteroids, Io and Europa.

In large bodies, decay of ^{40}K, ^{238}U, ^{235}U, and ^{232}Th is typically sufficient to induce melting (e.g., Taksoz et al. 1978) causing volcanism and degassing. Roles for accretional heating in initial melting and degassing (cf. Fanale 1971, Wetherill 1976) and for heat from decay of short lived nuclides such as ^{26}Al (Lee et al. 1977) are recognized. The latter is important because some sublunar-sized objects are differentiated, yet are too small to melt by accretion and lack the thermal lag to accumulate the heat from U, Th, and K decay. Tidal heating causes current volcanism on Io (Peale et al. 1979, Smith et al. 1979).

Degassing of a body with some water of hydration can produce an object such as Earth, with its hydrosphere a few km deep, or Europa. Although Europa´s crust is icy--surface temperatures being ~100°K--it may have a liquid zone beneath the first few km or tens of km (Fanale et al. 1977, Soderblom and Luchitta 1983, Squyres et al. 1983). The most likely source of energy to keep the H_2O from freezing is tidal (Squyres et al. 1983). Io, closer to Jupiter, receives vastly more tidal heating and has heat flow similar to geothermal areas in New Zealand (Matson et al. 1981). Volcanism, atmospheric ionization, sputtering, and other loss to space resulted in virtual dehydration (Pollack and Fanale 1982) and Io´s surface currently contains frozen SO_2 (Fanale et al. 1979, Smythe et al. 1979) and sulfur allotropes (Soderblom et al. 1980, Sagan 1979). Icy satellites have melting points so low that a small input of energy should trigger core formation. Ganymede and some Saturnian satellites show tectonic features attributed to volume changes on freezing (Smith et al. 1979, Shoemaker et al. 1982). Liquid H_2O zones may be absent because (1) Ganymede and Callisto (being icy) receive less radioactive heat per gram than Europa, (2) receive far less tidal heating, and (3) their (>1000 km) ice shells are amenable to convection (Cassen et al. 1980).

Condensation models yield a range of H_2O abundances for Mars (Anders and Owen 1977, McElroy et al. 1979, Pollack and Black 1979). Combining these with estimates of degassing efficiency based on atmospheric rare gas analysis yields a global H_2O supply to the surface of 10 to 100 m with most estimates closer to the latter. Much must be present as ground ice (Leighton and Murray 1966, Fanale 1976, Rossbacher and Judson 1981). Thus Mars is like the Earth and Europa, a rocky body with a outer H_2O shell. Most of Mars experiences net H_2O sublimation in the top meter, so the H_2O is largely invisible (see Mars Permafrost section below).

Volatiles are subject to removal by weathering and exospheric escape (see review by Pollack and Yung 1980). Water destruction by dissociation, H escape and oxidation of igneous rock, can destroy a hydrosphere, as it may have on Venus and Io. Ice loss by particle sputtering also occurs on icy satellites. Thus, diversity of surface volatile regimes is a product of diversity in bulk volatile content, degassing histories, and volatile loss processes.

TECHNIQUES FOR IDENTIFICATION OF SURFACE VOLATILES

Planetary densities indicate whether a planet is rocky or icy. Infrared thermal measurements indicate which ices could be stable at any site. Since ices produce landforms, imaging plays a role. Ultimately, γ-ray measurements from orbit and differential scanning calorimetric (and radiometric) measurements from landers, rovers, and penetrators will provide direct analyses of ices. Imaging spectrometers can provide images with high resolution spectra to several microns for each picture element, enabling phase distribution mapping. Meantime, the key means of remote identification of ices is telescopic planetary surface reflectance measurements in the infrared. Sometimes (e.g., the outer solar system) the planet is a 'point of light' yielding only hemispheric phase identification. In other cases spatial resolution is obtained. Combinations of molecular vibrational overtones in ices result in infrared absorption bands allowing phase identification as in Figure 1 where spectra of H_2O, CO_2, and CH_4 ices are compared. The position of an overtone is controlled by molecular bonds and band strength by the probability of photon absorption per unit path length and the total photon path length, which is controlled by scattering. Scattering occurs when there is a change in index of refraction, as at an ice-vacuum boundary. Mean path length is controlled by grain size and absorption coefficient. The smaller the grain size, the more scattering and the smaller the photon path; hence weaker absorptions. The greater the absorption coefficient, the less photons will be able to penetrate, inhibiting band depth. Higher overtones are weaker absorptions, and probe deeper. Rock minerals generally have higher absorption coefficients than ices; thus their presence in ices is easily detected. In some cases, clathrates are detectable, but since the trapped gas has a low abundance compared to the ice, its features are weak (Smythe 1979). Using radiative transfer models, abundances of phases in mixes can be estimated from the reflectance spectrum. Clark (1983) reviews spectral properties of ice-soil mixtures, and techniques for analysis of spectra.

RESULTS

Application to the solar system has revealed a variety of surface ices. Pluto is covered with CH_4 ice (Cruikshank and Salvaggio 1979). Methane gas may contribute but if there were that much, at the ambient temperature of 50^oK, there would be CH_4 ice anyway. This applies to Triton (a large satellite of Neptune) as well. Long photon path lengths are indicated consistent with large grains, as expected from kinetic theory of ice grain growth (Clark et al. 1983). Water ice is the only ice detected on the Uranian satellites (Brown and Cruikshank 1983). Here the optical path is small, since H_2O grain growth rates are slow at $<90^oK$. These satellites have low albedo (~20%), indicating other minerals mixed with ice. Water ice dominates surfaces of all Saturnian satellites whose spectra were measured (Fink et al. 1976, Clark and Owensby 1981). Here also grain size is small (Clark et al. 1983). Saturn's rings are mostly water ice--countless 'frost balls' with other impurities containing Fe^{3+} (Clark 1980).

Before the satellites of Jupiter were seen by spacecraft, H_2O ice was identified on Europa (J2), Ganymede (J3), and Callisto (J4), and its rarity or absence on Io was indicated (Pilcher et al. 1972). Clark (1980) showed J3 has about 90 wt% water ice on its surface (Figure 2), J4 has 30-90 wt%, and J2's surface is mostly water ice with some impurities containing Fe^{3+}, e.g., carbonaceous meteoritic material (see Sill and Clark 1982). The dominant volatile on Io is SO_2 frost and adsorbate (Fanale et al. 1979, Smythe et al. 1979). The 1.04-μm band depths in the H_2O ice on the surfaces of J2, J3, and J4 appear to be (if L = leading hemisphere in the satellite orbit and T = trailing) J2(T) > J3(T) > J2(L) > J3(L) ~ J4(T) ~ J4(L). This is consistent with the hierarchy of particle fluxes and the idea that atomic particle sputtering destroys small grains created by micrometeorites (Clark et al. 1983).

Jupiter's satellites J6-J14 and Saturn's small satellites have broadband reflectances like carbonaceous meteorites. Adequate data have not been obtained. These objects are low in reflectance, but low albedo surfaces could be 99% or more ice (e.g., Clark, in the volume).

The only asteroid with deep H_2O bands is the largest, Ceres (Lebofsky et al. 1981). The absorption near 3-μm is due to bound H_2O. An additional weak absorption was seen also which Lebofsky et al. (1981) interpreted as due to a small amount of H_2O ice. This seems inconsistent with short sublimation times for H_2O ice in the asteroid belt where mean T's are $~190^oK$ and maxima are $~250^oK$ (Watson et al. 1961). The spectrum, however, shows the "ice" feature at a slightly shorter wavelength than ice, and too narrow, so ice is questionable.

Mars is spectrally similar to X-ray amorphous weathering products of mafic volcanic glass found on Mauna Kea, Hawaii (e.g., Singer 1982). Bright region soils show weak H_2O ice bands (McCord et al. 1982), but since frosts are unstable at the time the spectra were obtained, the ice must be in pore structure of weathering products. However, Mars' residual north cap contains approximately 2 x 10^6 km^3 of ice, and this is H_2O, not CO_2, based on thermal measurements (Kieffer et al. 1977).

MARS PERMAFROST

Even volatile-poor models of Mars suggest enough degassed H_2O, to produce a global layer of ground

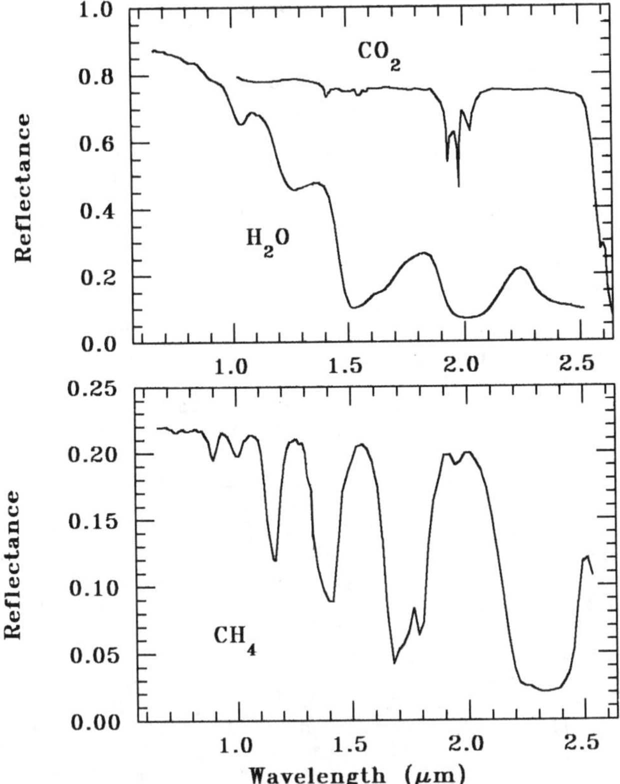

FIGURE 1 Spectral reflectance of ices. Band positions identify the material. Strength of an absorption depends on fundamental strength of transition, photon path length, and abundance. Thus, grain size affects the appearance of the spectrum. The CO_2 spectrum is from Smythe (1979), and the H_2O spectrum is shown in Clark (1983) with frost spectra of other grain sizes. The CH_4 spectrum is from Brown and Clark (personal communication, 1983).

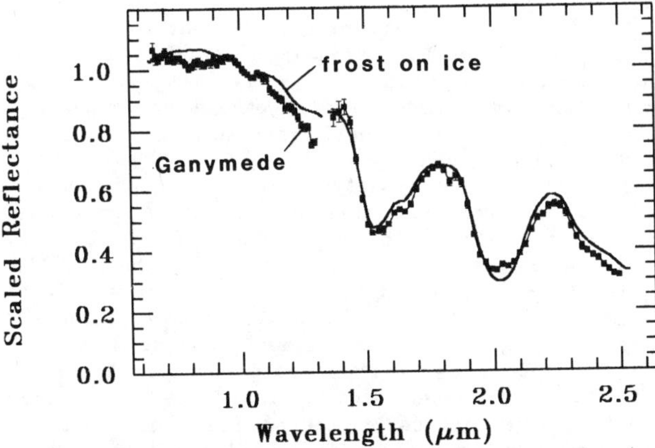

FIGURE 2 The spectrum of Ganymede is compared to frost on ice. The frost layer is 1 mm deep over ice, which increases the 1.04 μm ice band relative to optically thick frost with the same 1.5 and 2.0 μm bands. The strong ice bands indicate purity (~90 wt%). Differences, such as the broad absorption, are due to minerals with Fe^{3+}, typical of carbonaceous chondrite meteorite spectra. Spectra are scaled to 1.0 at 1.02 μm (Clark, 1980).

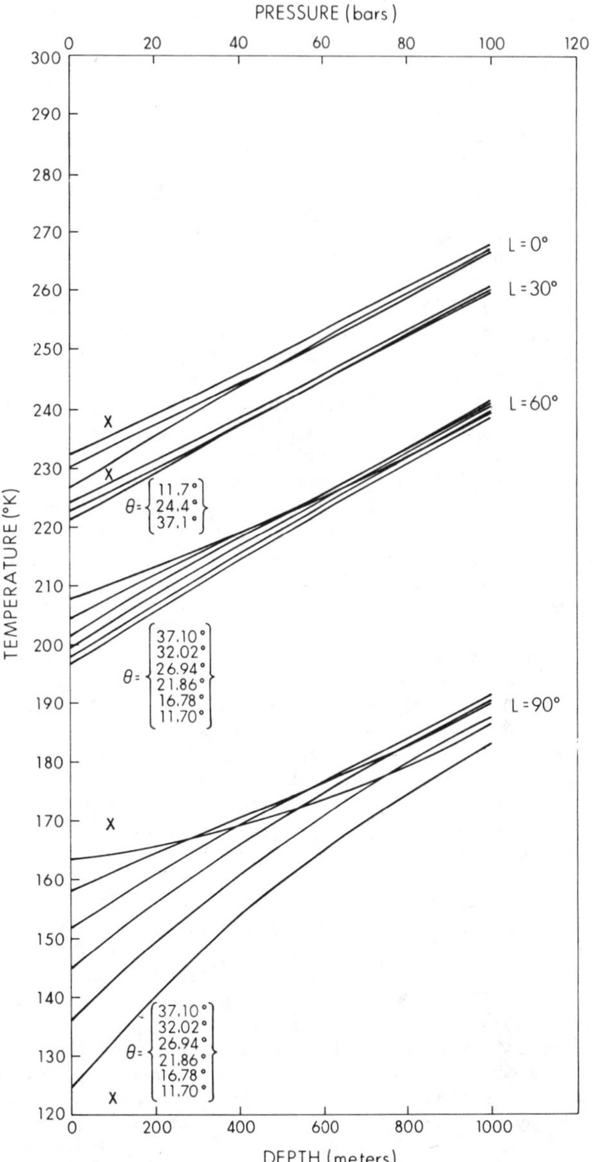

FIGURE 3 The Mars regolith thermal regime below 2 m. Temperature is given as a function of latitude, depth, and obliquity. Assumptions and associated uncertainties are discussed in the text.

ice of 10 to 100 m (Rossbacher and Judson 1981). Its regional presence seems necessary to produce the larger outflow channels (Carr 1979). Also, "nonlunar" ejecta blankets are evident, which suggest fluidization (Carr and Schaber 1977, Mouginis-Mark 1979). A latitudinal dependence for these has been suggested (Johansen 1978), but many variables still need to be controlled (Mouginis-Mark 1979). There are thermokarst-like features, (Gatto and Anderson 1975) polygonal patterned ground, curvilinear features, and craters with central pits and debris flows (Carr and Schaber 1977). The distribution of these, and the fluidized ejecta, has been discussed by Rossbacher and Judson (1981) who show some pitted terrain, aligned pits, and curvilinear features in near equatorial zones.

Ground ice is unstable with respect to the atmosphere at latitudes <40° (Leighton and Murray 1966, Fanale 1976, Farmer and Doms 1979). Thus to explain morphology, it was suggested that "fossil" ice exists near the equator (cf. Smoluchowski 1968). The role of soil in preserving buried ice was reviewed by Clifford and Hillel (1983). The distribution of pores in real soils provides a much greater effective pore size than if, as in Smoluchowski (1968) and Toon et al. (1980), the pore size is equaled to that of the smallest grains. Even a 100 m layer of overlying soil would be unable to prevent sublimation of 200m of ground ice. Thus "outliers" of ground ice in low latitudes. This suggests supply of H_2O from depth rather than preservation of initial ice. Study of temporal and spatial relationships among channels, ground-ice-related features and regional volcanism will test this.

The huge scale of some of the Mars patterned ground features may suggest driving by long term astronomically induced insolation variations, (Carr and Schaber 1977, Rossbacher and Judson 1982, Coradini and Flamini 1979). We will first evaluate the chances of periodic freeze-thaw on seasonal and astronomical (>10^4 yr) timescales. Figure 3 shows variation of Mars regolith temperature with depth, latitude, and obliquity. Here we assumed (1) surface insolation variations with obliquity cycle given by Ward et al. (1974), (2) thermal conductivity of 8 x 10^4 ergs cm^{1-} s^{-1} ^{o}K^{-1} (for hard frozen soil), and (3) bulk heat generation rate of 2.6 x 10^{-7} ergs cm^{-3} s^{-1} (chondritic). We calculated curves in Figure 3 using computational procedure given by Fanale et al. (1982a). The model is sensitive to assumptions: ice free soils range in conductivity from x3 greater to x3 less than assumed. Mars´ K content could be lower (Anders and Owen 1977) which could lower the gradient (angle between bundles and abscissa) by x2.

Still, we can draw conclusions from Figure 3. Coradini and Flamini (1979) assumed an approximately 20° global rise in temperature could result from obliquity rise, causing widespread periodic thaw. Our treatment shows that predicted variations based on Ward et al. (1974) are almost that on the pole, x4 less at 60° latitude, negligible at 30° latitude, and of opposite sign at the equator. Where variation is maximum, near poles, maximum temperature in the first 200 m is under -100°C. Conversely, at low latitudes, where the mean annual temperature is -60° to -40°C, the obliquity cycle does little to vary it. Thus quantitative analysis of Mars´ regolith thermal regime suggests obliquity driven freeze-thaw cycles are implausible.

Analogous argument can be made against seasonal freeze-thaw cycles. The top of any ice choked permafrost on Mars is unlikely to be where the <u>average</u> temperature ~200°K because there and above net annual sublimation is positive, resulting in ice depletion (Leighton and Murray 1966, Fanale 1976, Farmer and Doms 1979). The Viking infrared thermal mapper showed mean diurnal T´s (applicable ~3 cm) to be ~210°-230°K at the equator with an annual mean of ~220°K. At mid latitude they range from ~170° to 230°K with an annual mean of ~200°K, and at polar latitudes, from ~150° to ~220°K with an annual mean of ~170°K (Kieffer et al. 1977). Hence, hard frozen permafrost cannot exist in equilibrium with the atmosphere in low latitude; it must be buried to damp out seasonal variation at mid-latitude, and it can exist in the diurnal zone only at latitudes >75°. Since the diurnal average at seasonal <u>maximum</u> is <230°K wherever frozen permafrost exists in equilibrium with the atmosphere and since even that cannot be achieved at the top of the ice owing to thermal blanketing, we conclude seasonal melting, like diurnal melting, must occur (if at all) at <230°K. Obviously, periodic freeze-thaw on <u>any</u> timescale is a submarginal process for pure water on Mars.

It is also marginal for brines. A brine can be stable to 210°K, with a water concentration of 70%. A more likely $MgCl_2$-$CaCl_2$-H_2O system or $CaCl_2$-alkali chloride system would have a eutectic of 218°K, and for $MgCl_2$-alkali chloride systems it is 238°K. While such systems (barely) satisfy our requirements for freeze-thaw processes they all require equilibrium H_2O vapor pressures about the same as ice at any temperature. This rules out melting at >235°K, since that PH_2O is two orders of magnitude higher than Martian atmospheric PH_2O´s. This is critical for the (shallow) seasonal case. Finally, ubiquitous $MgSO_4$ would react with any $CaCl_2$ solution to produce $MgCl_2$ and $CaSO_4$ solutions far less effective at depressing the freezing point (Clark and Van Hart, 1981).

The effect of "circumscribed" sources such as lava flows in melting permafrost on Mars has been investigated by Coradini and Flamini (1979), who found them of limited effectiveness in space and time. Also, they are not periodic. A convection cell of near global proportions (e.g., associated with Tharsis) <u>in conjunction</u> with obliquity driven brine freeze-thaw may suffice. Magmatism may also emplace ice at shallow depths out of equilibrium with the atmosphere. Also, prior to Tharsis the obliquity variation might have been greater (Ward et al. 1979). We plotted points (x´s) on Figure 3 corresponding to the enhancement in the T variations at 100 m depth that would result. These do not change our conclusions. Also, the solar constant was significantly <u>lower</u> prior to Tharsis. Other possibilities include periodic warming due to continuous dust storms caused by obliquity driven desorption of CO_2 from the regolith (Fanale et al. 1982a). Also, in earlier Mars history some process seemingly produced a different H_2O regime, since highly dissected terrain seems restricted to ancient Mars history and some process akin to sapping seems involved (Pieri 1976). A chronology of terrain types and a distinction between those which require periodic freeze-thaw and those which do not, is called for. Our difficulty in rationalizing periodic freeze-thaw on Mars, when it is so common on Earth, reminds us that comparative planetology is a study of differences as well as similarities.

ACKNOWLEDGMENT

Planetary Geosciences Division publication 371. This work was performed under NASA contract 7-100. We thank D.M. Anderson, R.H. Brown, and J. Salvail for helpful discussions.

REFERENCES

Anders, E., and Owen, T., 1977, Mars and Earth: Origin and abundance of volatiles: Science, v. 198, p. 453-465.

Brass, G. W., 1980, Stability of brines on Mars: Icarus, v. 42, p. 20-28.

Brown, R. H., and Cruikshank, D. P., 1983, The Uranian satellites: Surface compositions and opposition surges: Icarus, in press.

Carr, M. H., 1979, Formation of Martian flood features by release of water from confined aquifers: Journal of Geophysical Research, v. 84, p. 2995-3007.

Carr, M. H. and Schaber, G. G., 1977, Martian permafrost features: Journal of Geophysical Research, v. 82, p. 4039-4054.

Cassen, P., Peale, S. J., and Reynolds, R. T., 1980, On the comparative evolution of Ganymede and Callisto: Icarus, v. 41, p. 232-239.

Clark, B. C., and Van Hart, D. C., 1981. The salts of Mars: Icarus, v. 45, p. 370-378.

Clark, R. N., 1980, Ganymede, Europa, Callisto and Saturn's rings: Compositional analysis from reflectance spectroscopy: Icarus, v. 44, p. 388-409.

Clark, R. N., and Owensby, P. D., 1981, The infrared spectrum of Rhea: Icarus, v. 46, p. 354-360.

Clark, R. N., Fanale, F. P., and Zent, A., 1983, Frost metamorphism: Implications for remote sensing planetary surfaces: Icarus, submitted.

Clifford, S. M., and Hillel, D., 1983, Stability of ground ice in the equatorial region of Mars: Journal of Geophysical Research, in press.

Coradini, M., and Flamini, E., 1979, A thermodynamical study of the martian permafrost: Journal of Geophysical Research, v. 84, p. 8115-8130.

Cruikshank, D. P., and Silvaggio, P. M., 1979, The surface and atmosphere of Pluto: Icarus, v. 41, p. 96-102.

Fanale, F. P., 1971, A case for catastrophic early degassing of the earth: Chemical Geology, v. 8, p. 78-105.

Fanale, F. P., 1976, Martian volatiles: Their degassing history and geochemical fate: Icarus, v. 28, p. 179-202.

Fanale, F. P., Brown, R. H., Cruikshank, D. P., and Clark, R. N., 1979, Significance of absorption features in Io's infrared reflectance spectrum: Nature, v. 280, p. 761-763.

Fanale, F. P., Salvail, J. R., Banerdt, W. B., and Saunders, R. J., 1982, Mars: The regolith-atmosphere-cap system and climate change: Icarus, v. 50, p. 381-407.

Farmer, C. B., and Doms, P. E., 1979, Global and seasonal variation of water vapor on Mars and the implications for permafrost: Journal of Geophysical Research, v. 84, p. 2881-2888.

Fegley, B., Jr., and Lewis, J. S., 1980, Volatile element chemistry in the solar nebula: Na, K, F, Cl, Br and P: Icarus, v. 41, p. 439-455.

Fink, U., Larson, H. P., Gautier, T. N., III, and Treffers, R. R., 1976, Infrared spectra of the satellites of Saturn: Identification of water ice on Iapetus, Rhea, Dione and Tethys, The Astrophysical Journal, v. 207, p. L63-67.

Gatto, L. W., and Anderson, D. M., 1975, Alaskan thermokarst terrain and possible terrain and possible martian analog: Science, v. 188, p. 255-257.

Johansen, L. A., 1978, Martian splash cratering and its relation to water: in Proceedings of the Second Colloquium on Planetary Water and Polar Processes.

Kieffer, H. H., Martin, T. Z., Peterfreund, A. R., Jakosky, B. M., Miner, E. D., and Palluconi, F. D., 1977, Thermal and albedo mapping of Mars during the primary mission: Journal of Geophysical Research, v. 82, p. 4249-4291.

Lebofsky, L. A., Feierberg, M. A., Tokunaga, A. T., Larson, H. P., and Johnson, J. R., 1981, The 1.7- to 4.2-μm spectrum asteroid of 1 Ceres: Evidence for structural water in clay minerals: Icarus, v. 48, 453-459.

Lee, T., Papanastassiou, D. A., and Wasserburg, G. J., 1977, Aluminum-26 in the early solar system: Fossil or fuel?: The Astrophysical Journal, v. 211, p. L107-L110.

Leighton, R. B., and Murray, B. C., 1966, Behavior of carbon dioxide and other volatiles on Mars: Science, v. 153, p. 136-144.

Lewis, J. S., 1972, Low temperature condensation from the solar nebula: Icarus, v. 16, p. 241-252.

Lewis, J. S., 1974, The temperature gradient in the solar nebula: Science, v. 186, p. 440-443.

Matson, D. L., Ransfod, G. A., and Johnson, T. V., 1981, Heat flow from Io (J1): Journal of Geophysical Research, v. 86, p. 1664-1672

McCord, T. B., Clark, R. N., and Singer, R. B., 1982, Mars: Near infrared spectral reflectance of the surface regions and compositional implications: Journal of Geophysical Research, v. 87, p. 3021-3032.

McElroy, M. B., Kong, T. Y., and Yung, Y. L., 1977, Photochemistry and evolution of Mars' atmosphere: A Viking perspective: Journal of Geophysical Research, v. 82, p. 4379-4388.

Mouginis-Mark, P., 1979, Martian fluidized crater morphology: Variations with crater size, latitude, altitude, and target material: Journal of Geophysical Research, v. 84, p. 8011-8022

Peale, S. J., Cassen, P., and Reynolds, R. T., 1979, Melting of Io by tidal dissipation: Science, v. 203, p. 892-894.

Pieri, D., 1976, Martian channels: Distribution of small channels on the martian surface: Icarus, v. 27, p. 25-50.

Pilcher, C. B., Ridgway, S. T., and McCord, T. B., 1972, Galilean satellites: Identification of water frost: Science, v. 178, p. 1087-1089.

Pollack, J. B., and Black, D. C., 1979, Implications of the gas compositional measurements of Pioneer Venus for the origin of planetary atmospheres: Science, v. 105, p. 56-59.

Pollack, J. B., and Fanale, F. P., 1982, Origin and evolution of Jupiter satellites system: in Morrison, D., ed., The Satellites of Jupiter, p. 872-910, University of Arizona Press.

Pollack, J. B., and Yung, Y. L., 1980, Origin and evolution of planetary atmospheres: Annual Review of Earth and Planetary Science, v. 8, p. 425-487.

Rossbacher, L. A., and Judson, S., 1981, Ground ice of Mars: Inventory, distribution, and resulting landforms: Icarus, v. 45, p. 39-60.

Sagan, C., 1979, Sulfur flows on Io: Nature, v. 280, p. 750-753.

Shoemaker, E. M., Luchitta, B. K., Plescia, J. B., Squyres, S. W., and Wilhelms, D. E., 1982, The geology of Ganymede: in Morrison, D., ed., Satellites of Jupiter, p. 435-521, University of Arizona Press.

Sill, G. P., and Clark, R. N., 1982, Composition of the surfaces of the Galilean satellites: in Morrison, D., ed., The Satellites of Jupiter, p. 174-212, University of Arizona Press.

Singer, R. B., 1982, Spectral evidence for the mineralogy of high albedo soils and dust on Mars: Journal of Geophysical Research, v. 87, Third International Colloquium on Mars Special Issue, p. 10159-10168.

Singer, R. B., McCord, T. B., Clark, R. N., Adams, J. B., and Huguenin, R.L., 1979, Mars surface composition from reflectance spectroscopy: A summary: Journal of Geophysical Research, v. 84, p. 8415-8426.

Smith, B. A., and the Voyager Imaging Team, 1979, The Jupiter system through the eyes of Voyager 1: Science, v. 204, p. 951-972.

Smoluchowski, R., 1968, Mars: Retention of Ice: Science, v. 159, p. 1348-1350.

Smythe, W. D., 1979, The detectability of clathrate hydrates in the outer solar system: Ph.D. Thesis, University of California, Los Angeles.

Smythe, W. D., Nelson, R. M., and Nash, D. B., 1979, Spectral evidence for SO_2 frost or adsorbate on Io: Nature, v. 280, p. 766.

Soderblom, L. A., and Luchitta, B. K., 1983, The geology of Europa: in Morrison, D., ed., Satellites of Jupiter, p. 521-555, University of Arizona Press.

Soderblom, L. A., and the Voyager Imaging Team, 1980, Spectrophotometry of Io: Preliminary Voyager 1 results: Geophysical Research Letters, v. 7, p. 963-966.

Squyres, S. W., Cassen, P., Peale, S. J., and Reynolds, R., 1983, Europa: Tidal heating and crustal evolution: Science, in press.

Taksoz, M. N., Hsui, A. T., and Johnson, D. H., 1978, Thermal evolution of the terrestrial planets: The Moon and the Planets, v. 18, p. 281-320.

Toon, O. B., Pollack, J. B., Ward, W., Burns, J. A., and Bilski, K., 1980, The astronomical theory of climatic change on Mars: Icarus, v. 44, p. 552-607.

Ward, W., Murray, B. C., and Malin, M. C., 1974, Climatic variation on Mars 2: Evolution of carbon dioxide and polar caps: Journal of Geophysical Research, v. 79, p. 3387-3395.

Ward, W. R., Burns, J. A., and Toon, O. B., 1979, Climatic variations on Mars, 2: Past obliquity oscillations of Mars: Role of the Tharsis uplift: Journal of Geophysical Research, v. 84, p. 243-249.

Watson, K., Murray, B. C., and Brown, H., 1961, Stability of volatiles in the solar system: Icarus, v. 1, p. 317-327.

Wetherill, G. W., 1976, The role of large bodies in the formation of the earth and the moon, in Proceedings of the Lunar and Planetary Science Conference, 7th, p. 3245-3287.

SUMMER WATER BALANCE OF A HIGH ARCTIC CATCHMENT AREA WITH UNDERLYING PERMAFROST IN OOBLOYAH VALLEY, N-ELLESMERE ISLAND, N.W.T., CANADA

Wolfgang-Albert Flügel

Geographisches Institut, Universität Heidelberg, Im Neuenheimer Feld 348, D-6900 Heidelberg, Federal Republic of Germany

The water balance of Oobloyah Valley, N-Ellesmere Island, N.W.T., Canada, a high arctic catchment area underlying by permafrost was investigated during the arctic summer 1978. The following results will be presented: (1) three main hydrogeological areas with different sediments, thawing depths, and soil drainage were seperated; (2) most of the winter snow cover melted until July 1, thereafter soil thawing and its drainage began; (3) the periglacial streams never had a measurable sediment load, not even during snow melt; (4) considering daily discharges, climatic changes and soil water balance the hydrologic regimes of the three investigated streams were characterized; and (5) summer water balance must include frozen soil water stored in the year(s) before; snow and rain add up to 51 % of the total balance, glaciers contribute up to 48 % and actual evapotranspiration is only 1 %.

INTRODUCTION

During the "Heidelberg-Ellesmere-Island-Expedition" in 1978 to Oobloyah Valley (81.5° N;83.5° W) in N-Ellesmere Island, N.W.T., Canada (Barsch and King, 1981), hydrologic studies were done with the following topics: (1) hydrogeological areas and their physical parameters; (2)climate during field work; (3) hydrological dynamic during snow melt; (4) water balance in representative stations; (5) discharge of streams; and (6) summer water balance of 1978. A measurement method developed for temperate climates (Flügel 1979) based on daily measurements of climatic parameters, soil water balances computed out of tensiometer readings and river discharges. The importance of soil water drainage for arctic water balances is known (Nagel 1979) but often not considered in hydrological studies (Ryden, 1977; Marsh and Woo, 1981). The application of calibrated tensiometers shortly reported by Fügner (1966) in Spitzbergen was till that time not tested under high arctic conditions. By using this method the results of the causal interpretations of the dependency between the components of the water balance were related to the eastern part of Oobloyah Valley, which could not be investigated because of technical reasons.

INSTALLATIONS, MEASUREMENTS, CALIBRATIONS

The following installations were done: (1) two meteorological stations to register precipitation, air humidity and temperature, direction and velocity of wind; (2) four tensiometer stations on three different slopes (Table 1) with tensiometers and temperature cells at 15-,30-,60-, and 90-cm depth and each with 3 evaporimeters 5 cm above the surface; (3) one water level gauge in the periglacial Peri Creek with hourly registrations.

Three streams, the periglacial Peri Creek, the glacial Nukapingwa River, and the mixed periglacial Heidelberg River were investigated. Measurements of soil moistures and discharges were done daily as hourly registration was not possible because of technical reasons and because all measurements had to be done only be the author. The streams showed turbulend flow; therefore the distribution of flow velocity was measured all over the cross section (if possible) with a flow meter out of which the daily discharges were computed.

TABLE 1 Inclination and exposure of the stations

Station	Slope	Inclination	Exposure
Ten A	I	46 %	WSW
Ten B	I	26 %	W
Ten C	II	13 %	S
Ten D	III	6 %	E

To calibrate the tensiometers, undisturbed soil samples from the sediments of each station were drained after saturation in a pressure apparatus. With the resulted exponential calibration equations the measured soil suction in mbar, were converted to adequate soil water contents in volume percent.

HYDROGEOLOGICAL AREAS AND THEIR PHYSICAL PARAMETERS

Oobloyah Bay is a geosyncline developed in Heiberg Sandstone (Thorsteinsson, 1971) situated in the region of continuous permafrost. Therefore common groundwater bodies are non existent. The hydrogeological overview in Figure 1 shows:
(1) sandy debris of Heiberg sandstone covering highly inclined slopes between Carl Troll Glacier and Nukapingwa Glacier; (2) moraine sediments

north of Access Lake with vegetation covered drainage ways among them; and (3) silty sediment between Access Lake, Peri Creek and Heidelberg River

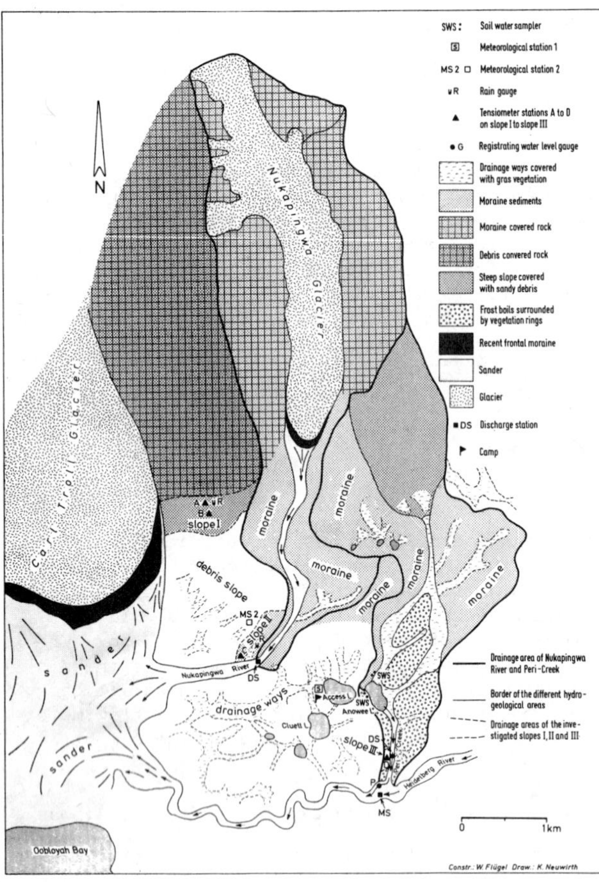

FIGURE 1 Hydrogeological overview

covered with frost boils (Tedrow, 1962, p. 342). Each of them was representatively studied in slope I,II and slope III with mean grain size distributions shown in Table 2.

TABLE 2 Grain size distributions of the slopes

Parameter	Slope I	Slope II	Slope III
Gravel (%)	26.6	27.4	0.5
Sand (%)	35.9	44.6	3.8
Silt (%)	28.5	17.1	62.5
Clay (%)	9.0	10.9	33.2

The sediments of slope I and II had comparable values, but different morphological geneses. Weathering of Heiberg Sandstone is responsible for the debris cover of slope I, glacial transport of moraines for this of slope II (Mäusbacher, 1981). The high amounts of clay and silt of slope III at the valley bottom indicate an aquatic deposition reported by Barsch (1981).

CLIMATE DURING FIELD WORK

The course of the meteorological parameters in Figure 2 monitors little precipitation for July, only 3.7 mm during the whole month; but high precipitations for the first days of August (35.4 mm) with a maximum high of 26.2 mm on August 2. Air temperatures never sank below 0° C and were influenced together with air humidities by the degree

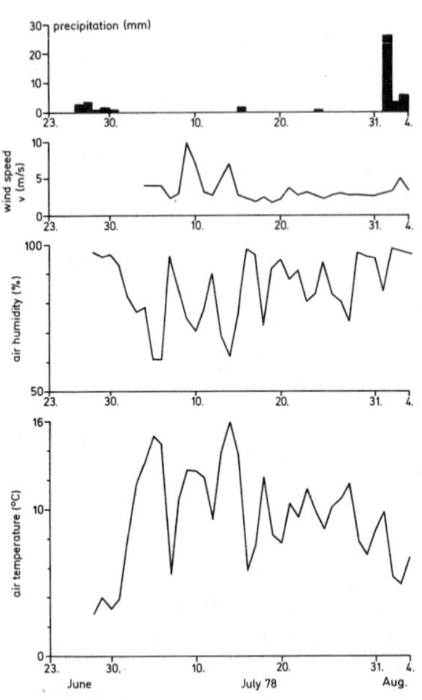

Figure 2 Plot of meteorological parameters

of cloudiness. The following periods will be seperated:

Radiation periods: July 2-6; July 9-10; July 11-15; July 18; July 27-28; July 30-31.

Cloudy periods: June 25- July 1; July 7-8; July 16-17; July 20-22; July 24-25; and August 1-8.

They regulated the thawing of glacier ice and by that the discharge of melt water. Between June 24 and August 4 we computed

Mean air temperature, T = 9.4 °C
Mean air humidity, h = 85.4 %
Mean wind speed, v = 3.5 m/s
Sum of precipitation, P = 50.4 mm

HYDROLOGIC DYNAMIC DURING SNOW MELT

We arrived June 18 in Oobloyah Valley shortly after snow melt began. The expedition area was covered with a nearly 50-cm thick snow layer that had a water content of 37 volume percent, equal to 185 mm rain. This value corresponds to the mean anual precipitation for north Ellesmere Island of 200 mm reported by Hattersley-Smith (1964). Within two weeks, till July 2 nearly all of the snow melted and discharged. Afterward thawing and drainage of the frozen soil began. Although the hydrological dynamic during snow melt was very complex, the three hydrogeological areas showed significant differences, which can be described as (1) surface runoff with solifuction on steep inclined slopes covered with sandy debris of Heiberg Sandstone; (2) thermokarst phenomena in slopes covered with coarse grained moraines and in older terraces of the Nukapingwa River resulted by infiltrated melt water; (3) surface runoff of melt water in broad drainage ways at the valley bottom which also collected water from the moraines.

The periglacial streams never showed sediment transport, not even suspended load, because melt water flew on the surface of the frozen soil, which was partly protected by a thin ice layer. Washout of fine sand by interflow (Flügel, 1979) was observed downslope of the debris cover north of Nukapingwa River. Measurable solifluction was limited to the period of snow melt and mostly occured on steep inclined slopes with little vegetation cover.

SOIL MOISTURE BALANCE

Soil tensions were measured with calibrated tensiometers in four stations, representative for the three investigated hydrogeological areas regardind their grain size distributions (see Table 2). Soil moisture in the four stations showed dependency to slope inclination, grain size distribution, and to depth and relief of the permafrost table. Soil suctions had certain characteristics as (1) the beginning values on July 2 were between 0 and 20 mbar, went up during the dry July to maxima between 40 and 250 mbar, but never reached field capacity at 300 mbar; (2) during the high precipitations on August 2-4 (35.4 mm) the instruments of station C and D fell down to 0 mbar indicating water saturation of their active layers; (3) although water input was high enough to saturate also the sediments of station A and B, soil suctions reached only 20 through 50 mbar, indicating that drainage occured during rain fall resulted by steep inclinations. The soil suctions in each depth were used to compute the avergae soil moisture content of the active layers of each station as geometric means, shown in Figure 3.

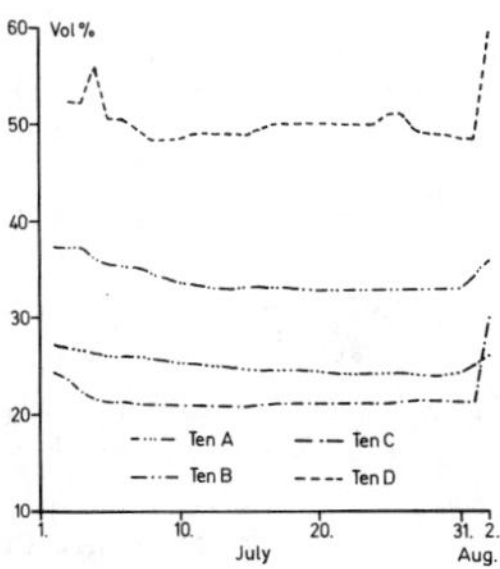

FIGURE 3 Daily soil moisture in station A to D.

Out of Figure 3 and regarding the measured soil suctions the following results will be presented: (1) Drainage of each active layer was almost finihed at July 10. (2) The little precipitation that fell in the middle of July compensated evapotranspiration and low drainage. (3) Only the active layers of slope II and III were saturated by the high precipitation on August 2, meanwhile slope I drained also during the rain fall. (4) As field capacity was never reached soil water always moved by gravity, and was not forced by suction gradients.

The sediments lost water by drainage and evapotranspiration. The latter seemed to be limited to the upper 15 cm of the profiles and was computed out of the daily soil moisture balance regarding also the measured soil temperatures up to 0.1 mm per day. This was only 20 % of the mean potentiel evaporation measured with the evaporimeters.

Most of the drained soil water flew downslope to the receiving streams or drainage ways (Figure 1), but areas covered with frost boils showed a different dynamic. Because under each frost boil a depression developed in the relief of the permafrost table, soil water drained out of the uplifted sediment above flew to the bottom of this mould and couldn't discharge. These areas could not support their streams with drained soil water, what is important for the computation of the yield factor. In Table 3 the soil water balance of each active layer is listed.

TABLE 3 Soil water balance July 2 - August 2.

Parameter		Ten A	Ten B	Ten C	Ten D
Pore volume	(%)	40.2	37.2	38.5	59.3
Field capacity	(%)	23.2	31.3	18.8	44.1
Active layer	(cm)	100	100	80	35
Drainage in July	(mm)	40.0	43.0	28.8	13.7
Refill in August	(mm)	19.0	28.0	70.4	38.5

Table 3 shows, that soil water drainage during July was much higher than the little precipitations of 3.7 mm indicating that older frozen soil water stored in the year(s) before discharged too and had to be calculated as an additional water input.

DISCHARGE OF STREAMS WITH DIFFERENT REGIMES

Three stream (Peri Creek, Nukapingwa River, and Heidelberg River) were investigated. Their drainage basins were planimetered and those of Peri Creek and Nukapingwa River were mapped in detail during field work (see Figure 1). They can be subdivided as shown in Table 4. Their daily discharges (Figure 4) varied significantly corresponding to their regimes because they had different discharge components (melt water, soil water, and precipitation runoff). Regarding the soil water balances (Table 3) of their hydrogeological areas (Table 4) each discharge regime will be quantified.

Peri Creek

The periglacial Creek showed an exponential falling of daily discharges during the dry July with a little increase on July 16 (2.1 mm precipitation). The falling values are best fitted (r=0.99) by the

TABLE 4 Catchment areas or the three streams.

Sediment Area (km^2)	Peri Creek	Nukapingwa River	Heidelberg River
Glacierized	0.00	3.20	38.80
Sandy debris	1.70	4.35	following
Moraines	2.90	2.00	areas all
Silt covered by			together
frost boils	0.90	0.00	135.00
Drainage ways	0.50	0.00	
Lakes	0.06	0.00	
Total area	6.06	9.55	173.80

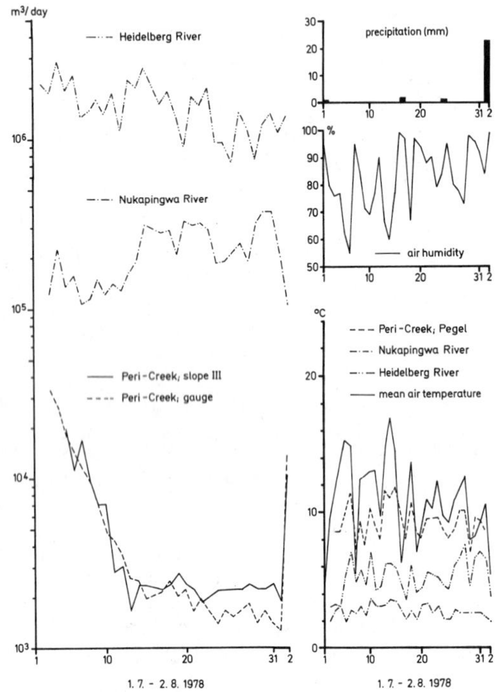

FIGURE 4 Daily discharges of the studied streams.

"Dry Weather Discharge Line" (DWL) in Figure 5 and describes the relation between discharge and duration of dry weather without rain. Baseflow discharge did not limit the DWL, because there was no outflowing groundwater body in this basin with underlying permafrost. By comparing the sum of the computed discharge components with the measured value 10.3 % difference was found, explained by evapotranspiration. Related to the basin area this

Soil water drainage	:	179299 m^3	=	88.7 %
Precipitation runoff	:	22900 m^3	=	11.3 %
Sum of discharge components	:	202100 m^3	=	100.0 %
Measured discharge	:	181200 m^3	=	89.7 %
Difference	:	20300 m^3	=	10.3 %

was about 0.1 mm/day and square meter and corresponds to that value calculated out of the soil water balance.

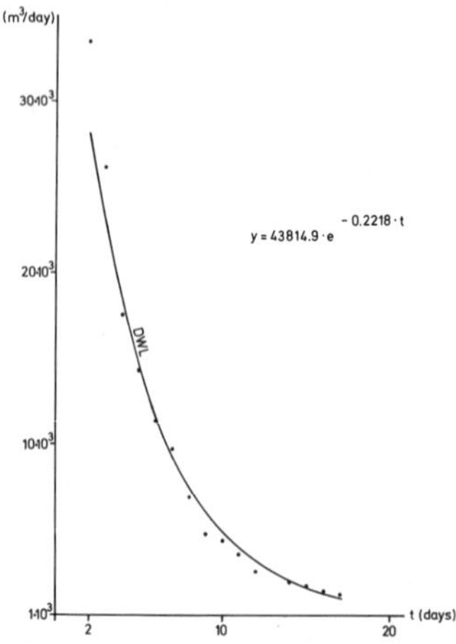

FIGURE 5 DWL of the Peri Creek

Nukapingwa River

The course of the discharge of this glacial stream is mainly influenced by the climatic changes between periods of radiation and cloudiness. They regulated the thawing of the glacier ice and firn. Radiation periods resulted in an exponential increase of daily discharges, the latter in a corresponding decrease. In Figure 6 they are called "Radiation Weather Discharge Line" (RWL), and "Cloudy Weather Discharge Line" (CWL). RWL and CWL

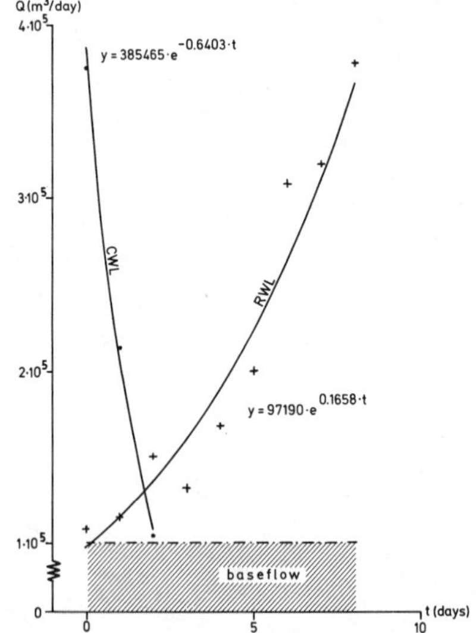

FIGURE 6 RWL and CWL of Nukapingwa River

describe, how daily discharges went up or sank down in dependency of the duration of radiation or cloudiness. Baseflow out of the glacier storage (clefts filled with water; saturated firn) limited the CWL and supported a minimum discharge from which the RWL begins. Baseflow was calculated out of the low water discharges during cloudiness up to 0.1 million m^3/day. By using the soil water balances of the hydrogeological areas (see Figure 1 and Table 4) the following discharge components were determined:

Run off from snow melt	: 0.255 million m^3	=	4.0 %
Soil water drainage	: 0.172 million m^3	=	2.7 %
Precipitation runoff	: 0.032 million m^3	=	0.5 %
Baseflow discharge	: 3.100 million m^3	=	48.3 %
Radiation discharge	: 2.859 million m^3	=	44.5 %
Total discharge	: 6.418 million m^3	=	100.0 %

More than 92 % of the total summer discharge was contributed by melt water from the glacier, which holds water temperatures low.

Heidelberg River

The mixed periglacial-glacial stream drains the eastern part of Oobloyah Valley and a considerable glaciered area of the Krieger Mountains shown in Figure 7. The hydrological regime includes periglacial and glacial components. The periglacial infuence stems from the discharge contribution of the periglacial tributaries and is shown by an exponential decrease of low water discharges during July corresponding to the DWL of the Peri Creek. The glacial components are indicated by an exponential increase and decrease of discharges during radiation and cloudiness. Corresponding to the glacial Nukapingwa River a RWL and CWL were defined, both limited by a baseflow of 0.69 million m^3, which was melt water comming out of the storage capacity of the glacierized areas. In order to calculate the different discharge components the following random condition was defined, that distribution and soil water balance of the hydrogeological areas of the Heidelberg River corresponds to that of the Peri Creek. This condition can be proved by interpretations of air photographs of the valley and is also supported by reports of Mäusbacher (1981). Employing this random condition soil water discharges of the Heidelberg River were calculated out of the daily discharges of Peri Creek in relation of both land areas. The quotient between the remaining melt water discherge and the glacierized area of Heidelberg River then is approximately the same as that value computed for the Nukapingwa River. This gave an additional support to the prior calculation. The following results were computed:

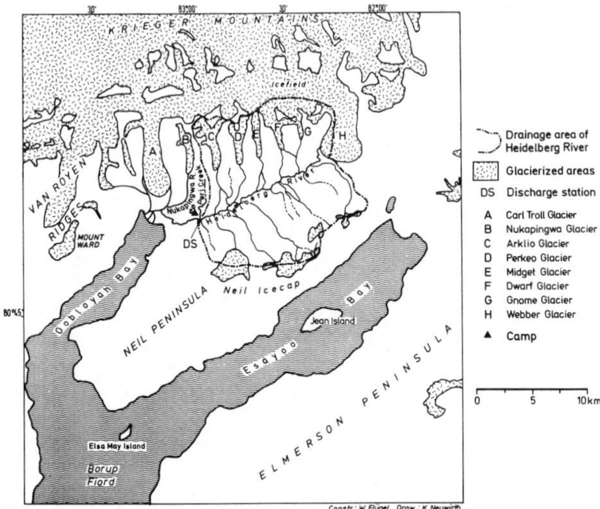

FIGURE 7 Drainage basin of Heidelberg River

Runoff from snow melt	: 5.22 million m^3	=	10.6 %
Soil water runoff	: 3.74 million m^3	=	7.6 %
Precipitation runoff	: 2.12 million m^3	=	4.3 %
Baseflow discharge	: 21.01 million m^3	=	42.7 %
Radiation discharge	: 17.16 million m^3	=	34.8 %
Total discharge	: 49.25 million m^3	=	100.0 %

Melt water from the glaciers is about 77 %, less than in the glacial Nukapingwa River.

SUMMER WATER BALANCE OF 1978

The computation of the summer water balance is based on the results of the single water balances of the three studied streams under the random conditions: (1) distribution of snow cover and summer precipitation is nearly the same all over Oobloyah Valley; and (2) evaporation from the glaciered are areas is comparable to the evapotranspiration estimated from the land areas. Due to a lack of data for the eastern part of the valley and the glaciered areas of the Krieger Mountains (see Figure 7) both had to be defined. My own measurements only supported both of them for the studied region with slope I, II, and III (see Figure 1). The balance had to be subdivided in the duration of snow melt and the following summer period, when the sediments thawed and drained. Input was marked with a plus, output with a minus.

1. Snow melt (20.6.-1.7.1978):	
Snow cover, SC	: + 35.04 million m^3
Precipitation, P	: + 2.14 million m^3
Evaporation (0.1 mm/day), E	: - 31.49 million m^3
Discharge, D	: - 31.49 million m^3
Balance computation	: + 5.46 million m^3

The resulted plus was unmelted snow, which was brought into the second balance.

2. Water balance July 2 through August 2:	
Unmelted snow cover, SC	: + 5.46 million m^3
Frozen soil water, FSW	: + 4.09 million m^3
Precipitation, P	: + 5.66 million m^3
Snow melt discharge, SWD	: - 5.46 million m^3
Soil water discharge, SWD	: - 4.09 million m^3
Precipitation discharge, PD	: - 2.19 million m^3
Baseflow discharge, BD	: - 24.11 million m^3
Radiation discharge, RD	: - 20.02 million m^3
Evapotranspiration, E	: - 0.61 million m^3
Soil water storage, S=P-PD-E	: - 2.86 million m^3
Balance computation	: - 44.13 million m^3

This balance shows a minus, resulted by the melt water contribution of the glaciers, which brought an additionally output as a minus factor. Considering the total balance volume input parameters

SC, P, and FSW add up to 38.5 %, 8.5 % and 4.5 %. As evapotranspiration is only about 1.0 %, and soil water storage adds up to 3.1 %, nearly all of the water input discharged. Additionally, the glacierized regions contributed 48.5 % of melt water discharge to the summer water balance.

ACKNOWLEDGEMENTS

The investigations were supported by the DFG (Deutsche Forschungs Gemeinschaft) and the Polar Continental Shelf Project.

REFERENCES

Barsch,D., 1981. Terrassen, Flussarbeit und das Model der exzessiven Talbildungszone im Expeditionsgebiet Oobloyah Bay, Ellesmere Island, N.W. T. Heidelberger Geographische Arbeiten, H. 69, p. 163-202.

Barsch,D. and King,L., 1981. Zielsetzung und Ablauf der Heidelberg-Ellesmere-Island-Expedition 1978. Heidelberger Geographische Arbeiten, H. 69, p. 1-14.

Flügel,W.-A., 1981. Untersuchungen zum Problem des Interflow. Heidelberger Geographische Arbeiten, H. 56, 170 p.

Fügner,D., 1966. Hydrologen in der Arktis. Wissenschaftliche Zeitschrift der TU-Dresden, 5, p. 933-940.

Hattersley-Smith,G., 1964. Rapid Advance of Glacier in Northern Ellesmere Island. NATURE, 201, p.176

Mäusbacher,R., 1981. Geomorphologische Kartierung im Oobloyah-Tal, N-Ellesmere Island, N.W.T., Kanada. Heidelberger Geographische Arbeiten, H. 69 p. 413-440.

Marsh,P. and Woo,M.-K., 1981. Snowmelt, glacier melt, and high arctic streamflow regimes. Canadian Journal of Earth Sciences, Vol. 18, 8, p. 1380-1384.

Nagel,G., 1979. Untersuchungen zum Wasserkreislauf in Periglazialgebieten. Trierer Geographische Studien, Sonderheft 2.

Ryden,B.E., 1977. Hydrology of Truelove Lowland. in Bliss,L.C. ed., Truelove Lowland Devon Island Canada: A High Arctic Ecosystem, p. 107-136.

Thorsteinsson,R., 1971. Geological Map Greely Fiord West, District of Franklin, Map 1311 A, 1:250000, Geol. Surv. Canada, Ottawa.

Tedrow,J.C.F., 1962. Morphological evidence of frost action in arctic soils. Builetyn Peryglacialny, 11, p. 343-352.

PHYSICAL AND ECOLOGICAL CHARACTERISTICS OF ALEXANDRA FIORD, A HIGH ARCTIC OASIS ON ELLESMERE ISLAND, CANADA

B. Freedman,[1] J. Svoboda,[2] C. Labine,[3] M. Muc,[4]
G. Henry,[2] M. Nams,[1] J. Stewart,[1] and E. Woodley[2]

[1]Department of Biology, Dalhousie University, Halifax, Nova Scotia Canada B3H 4J1
[2]Department of Botany, University of Toronto, Erindale Campus, Mississauga, Ontario Canada L5L 1C6
[3]Department of Geography, University of Alberta, Edmonton, Alberta Canada T6G 2E9
[4]Department of Biology, Camrose Lutheran College, Camrose, Alberta Canada T4V 2R3

Physical and ecological features of an atypically lush high arctic lowland, on Ellesmere Island, are described. Site topography, moisture supply, drainage patterns and radiation input combine favorably to create a highly productive biological oasis amidst a polar desert setting. Growing season radiation and energy balance parameters were higher than those of sites at similar latitudes. Thirteen distinctive lowland plant communities were recognized and divided into a xeric-mesic to mesic lichen-heath-cushion plant dominated series associated with a seasonally discontinuous moisture supply, and a mesic to hydric sedge dominated series associated with a relatively continuous moisture supply. Community species composition, standing crop, and primary production reflected changes in habitat moisture conditions.

GENERAL

This contribution highlights some initial data from a team oriented three year ecological study of a high arctic lowland oasis. The project's objective was to determine and describe the underlying environmental factor(s) contributing to the lowland's diverse and extensive vegetation cover and high productive capacity. The surrounding landscape is characteristically a polar desert/glacier mosaic, with only 46% of the regional study area (ca. 19000 km^2) unglaciated (Freedman and Svoboda 1981). Extensively vegetated oases cover only 1-2% of these glacier free sites, a cover percentage comparable to other high arctic regions (see also Murray 1978; Bliss 1981; Aleksandrova 1977). The studies focused on: (1) meso- and micro climate; (2) soil and active layer features; (3) vascular plant community composition, distribution and productivity; (4) autecology of dominant lowland species; (5) vascular plant responses to nutrient, moisture and microclimate enhancement, and (6) breeding bird and invertebrate population dynamics. This paper focuses specifically on meso and microclimatic features and corresponding vascular plant production.

The lowland study site is located adjacent to Alexandra Fiord (78°53'N, 75°55'W), on the eastern coast of central Ellesmere Island, Canada. The site is a relatively flat, roughly triangular periglacial outwash plain covering 8.0 km^2. The lowland is enclosed by a glacier at its southern extreme; by steep talus slopes and cliffs of a 500 m high eastern and western upland complex; and by the fiord waters along its northern margin.

PHYSICAL CHARACTERISTICS

Total annual precipitation in the region is less than 6 cm and typically less than 1 cm during the growing season (July - August). The relatively insignificant annual moisture contribution, coupled with rapid runoff of snowmelt, creates xeric growing conditions on the bordering upland desert sites. The lowland has a large influx of meltwaters in the early part of the growing season, the magnitude of which is proportional to the annual variation in winter snowpack accumulation. In 1981, there was little snowpack accumulation in winter, and the lowland was almost snowfree by the first week of June. In 1982, with a more typical snowpack regime, only 6% of the lowland was snow free by May 31; 12% by June 7; 47% by June 13; and 90% (following a chinook) by June 16. Isolated snowbanks persisted until June 25. During the remaining course of the growing season a relatively regionalized but continuous moisture supply of glacial and upland meltwater origin prevails. Thus, the lowland shows characteristic features of a hydrologic oasis in a polar desert setting.

Based on a number of energy budget parameters, the lowland also shows evidence of being a thermal oasis. During the growing season both mean daily incoming shortwave and longwave radiation were considerably higher at Alexandra Fiord than in two comparable regional locations (Table 1). This is partly due to consistently clear skies associated with the presence of a persistent summer high-pressure zone over northern Greenland and northeastern Ellesmere Island. A small indirect lowland short-wave radiation component was measured from the glacial and fiord surfaces. Similarly, the dark cliff walls of the adjacent uplands contribute an additional long-wave radiation component to the lowland's energy budget. This "oven effect" has been observed and described from a similar lowland setting on northeastern Devon Island (Courtin and Labine 1977).

Net all-wave radiation was consistently higher in the lowland than on either an adjacent upland barrens site or a site (Eureka) 220 km northwest (Table 2). The differences between the Alexandra Fiord lowland and upland are largely due to differences in albedo, while variations in both incoming radiation and albedo contribute to the Alexandra

Fiord and Eureka differences.

Growing season air temperatures were higher in the lowland than at the unvegetated upland plateau and adjacent coastal lowland sites (Table 3). Lowland monthly air temperatures, with the exception of August, were higher than those at Eureka. Lowland soil surface temperatures were higher during the growing season, than in the upland site, but subsurface (-10 cm) temperatures were similar (Table 4).

Measurements were taken of significant energy budget parameters in two dominant lowland communities - mesic heath and hydric sedge meadow. Use of a roving meteorological station (Campbell Scientific CR-5) prevented simultaneous monitoring of both sites. It is therefore appropriate to compare the relative partitioning of net radiation (Q*) into soil heat flux (G), latent heat flux (LE) and sensible heat flux (H), rather than the absolute values of these parameters. Analysis of the data show G to account for 15% of Q* in the heath and 11% in the sedge meadow, while LE accounted for 26% and 22% at the two sites respectively, and H accounted for 59 and 67% respectively. The ratio of H and LE (= Bowen ratio) is typically high for tundra areas (Barry et al. 1981; Ohmura 1982), and was 2.3 for the mesic heath and 3.0 for the hydric sedge meadow.

PLANT ECOLOGICAL CHARACTERISTICS

The general characteristics of the thirteen distinctive lowland plant communities are summarized in Table 6. A similar series of communities has been described for another high arctic oasis on Devon Island (Barrett 1972; Muc and Bliss 1977). Spatially, these communities are segregated into a complex mosaic according to interrelated patterns of microtopography, drainage, snow persistence, and substrate characteristics. The majority (70%) of the lowland is covered by communities dominated by lichens, heaths, cushion plants, and dwarf shrubs. Soil moisture conditions are typically xeric to mesic, and mid-July active layer development (ca. 70% of peak season depth) averaged 40 cm. The dominant species are woody, long lived (>100 years), slow growing, vegetatively spreading, and possess high proportions (69% of total biomass in *Cassiope tetragona*, and 81% in *Dryas integrifolia*) of attached dead tissue. The latter component may help maintain microhabitat moisture, and facilitate *in situ* nutrient cycling (Svoboda 1977). Total vascular plant standing crop averaged 412 g m^{-2}, with net production averaging 18 g $m^{-2}y^{-1}$. Species net production: total biomass ratios (P/B) were quite low: 0.02 for *C. tetragona*, 0.02 for *D. integrifolia*, 0.05 for *Saxifraga oppositifolia* and 0.07 for *Vaccinium uliginosum*.

Approximately 20% of the lowland is covered by sedge dominated communities, and mosses are an integral but minor component in these. Sites range from mesic to hydric, and mid-July active layer depths averaged 37 cm but thawed as deep as 80 cm by late August. Dominant species are *Carex stans*, *C. membranacea*, *C. misandra* and *Eriophorum angustifolium*. These perennial, shorter lived, herbaceous plants maintain the majority (81%) of their biomass below ground. Total aboveground standing crop averaged 337 g m^{-2} and net production averaged 27 g $m^{-2}y^{-1}$. Graminoid P/B ratios were low: 0.18 in *C. stans*. 0.17 in *C. membranacea*, 0.15 in *C. misandra* and 0.18 in *E. angustifolium*. Although maintaining less aboveground biomass than drier tundra communities, wet sedge meadows are typically more productive (Babb and Bliss 1974, Muc 1977, Brown et al. 1980, Bliss et al. 1981).

The remaining 10% of the lowland is almost equally divided between aquatic (rivers, streams, ponds) and ruderal (human, water or ice modified) habitats. Aquatic habitats possess little to no vascular vegetation. The ruderal sites are generally mesic to mesic-xeric in nature and develop mid-July active layer depths of 27 cm. The diverse vegetation is dominated by short-lived, herbaceous species in which standing crop averages 12 g m^{-2} and production averages 5 g $m^{-2}y^{-1}$. Corresponding P/B ratios are high: 0.71 in *Cochlearia fenestrata*, 0.50 in *Cerastium alpinium*, and 0.35 in *Draba groenlandica* (Maessen et al. 1983).

In conclusion, our study demonstrates that within the overall expanse of high arctic desert, relatively lush oases can exist. This appears to be caused by an environmental amelioration in terms of hydrology and meso- and microclimate.

ACKNOWLEDGEMENTS

This work was funded by the World Wildlife Fund (Canada), and by the Natural Sciences and Engineering Research Council of Canada, operating grants to B.F. and J.S. Logistic support was provided by the Polar Continental Shelf Project of the Department of Energy, Mines and Resources and by the Royal Canadian Mounted Police.

REFERENCES

Aleksandrova, V.D., 1977, The Arctic and Antarctic: their division into geobotanical areas, (Transl. 1980 Doris Love). Cambridge University Press. New York.

Anonymous, 1982, Canadian climate normals. Volumes 1 and 2. Radiation and temperature. Atmospheric Environment Service. Environment Canada, Ottawa.

Babb, T.A. and Bliss, L.C., 1974, Susceptibility to environmental impacts in the Queen Elizabeth Islands. Arctic, 27, p. 234-237.

Barrett, P.E., 1972, Phytogeocoenoses of a coastal lowland ecosystem, Devon Island, N.W.T. Ph.D. thesis, University of British Columbia. Vancouver.

Barry, R.G., Courtin, G.M. and Labine, C., 1981, Tundra climates, *in* Bliss, L.C., Heal, O.W. and Moore, J.J., ed., Tundra Ecosystems: a Comparative Analysis. Cambridge University Press, New York.

Bliss, L.C., 1981, North American and Scandinavian tundras and polar deserts, *in* Bliss, L.C., Heal, O.W. and Moore, J.J., ed., *Tundra Ecosystems: a Comparative Analysis*. Cambridge University Press, New York, p. 8-24.

Brown, J., Everett, K.R., Webber, P.J., Mac Lean, S.F., and Murray, D.F., 1980, The coastal Tundra at Barrow, in Brown, J., Miller, P.C., Tieszen, L.L. and Bunnell, F.L., ed., An Arctic Ecosystem - The Coastal Tundra at Barrow, Alaska. Dowden, Hutchinson and Ross, Inc., Stroudsburg, Pa., p. 1-29.

Courtin, G.M. and Labine, C.L., 1977, Microclimatological studies of the Truelove Lowland, Devon Island, N.W.T. in Bliss, L.C., ed., Truelove Lowland, Devon Island, Canada: a High Arctic Ecosystem. University of Alberta Press, Edmonton, p. 73-106.

Maessen, O., Freedman, B., Nams, M.L.N. and Svoboda, J., 1983, Resource allocation in high arctic vascular plants of differing growth form. Canadian Journal of Botany (in press).

Muc, M., 1977, Ecology and primary production of sedge-moss meadow communities, Truelove Lowland, in Bliss, L.C., ed., Truelove Lowland, Devon Island, Canada: A High Arctic Ecosystem. University of Alberta Press, Edmonton, p, 157-184.

Muc, M. and Bliss, L.C., 1977, Plant communities of Truelove Lowland, in Bliss, L.C., ed., Truelove Lowland, Devon Island, Canada: A High Arctic Ecosystem. University of Alberta Press, Edmonton, p. 143-154.

Murray, D.F., 1978, Vegetation, floristics, and phytogeography of northern Alaska, in Tieszen, L.L., ed., Vegetation and Production Ecology of an Alaskan Arctic Tundra. Springer-Verlag, New York, p. 18-36.

Ohmura, A., 1982, Climate and energy balance on the arctic tundra. Journal of Climatology, 2, p. 65-84.

Svoboda, A., 1977, Ecology and primary production of raised beach communities, Truelove Lowland, in Bliss, L.C., ed., Truelove Lowland, Devon Island, Canada: A High Arctic Ecosystem. The University of Alberta Press, Edmonton, p. 185-216.

TABLE 1 Comparison of Mean Daily Incoming Short-wave and Long-wave Radiation for Several High Arctic Locations.

		Mean Daily Short-wave Incoming (MJ m^{-2} d^{-1}		Mean Daily Long-wave Incoming ($MJ\ m^{-2}\ d^{-1}$)			
		July	Aug.	May	June	July	Aug.
Alexandra Fiord	(79°N,76°W)	19.4	12.0	26.9	31.2	32.4	32.4
Eureka(1)	(80°N,85°W)	18.8	10.5	--	--	--	--
Expedition Fiord(2)	(79°N,92°W)	15.4	12.1	18.9	23.9	25.7	24.9

(1) After Anonymous (1982).

(2) After Ohmura (1982).

TABLE 2 Comparison of Net All-Wave Radiation Between the Alexandra Fiord Lowland, a Site on the Nearby Upland Barrens, and Eureka Normals ($MJ\ m^{-2}\ d^{-1}$).

	1980		1981		
	July	Aug.	June	July	Aug.
Alexandra Fiord lowland	13.6	6.7	9.7	11.9	9.2
Alexandra Fiord upland polar desert	9.0	5.5	9.0	9.5	6.5
Eureka climatic normals(1)	-	-	12.6	9.9	4.5

(1) After Anonymous (1982)

TABLE 3 1982 Mean Monthly Air Temperatures (1.5 m) for Several Sites in the Vicinity of the Alexandra Fiord Lowland (°C).

Station	June	July	August
Central Lowland	6.0	6.4	2.9
Unvegetated Low-land	2.5	4.9	-
Upland Polar Desert	3.8	6.1	1.7
Eureka normals(1)	1.8	5.4	3.3

(1) After Anonymous (1982).

TABLE 4 1982 Mean Monthly Soil Temperatures for Two Sites.

Site	Depth	June	July	Aug.
Central lowland	surface	8.0	9.5	6.0
	-10 cm	5.8	7.7	3.9
Upland polar desert	surface	8.0	8.8	3.2
	-10 cm	5.9	7.9	3.6

TABLE 5 Major Energy Budget Parameters for the Two Most Significant Plant Community Types in the Alexandra Fiord Lowland.

Parameter	Mesic Heath	Sedge Meadow
Absorbed net all-wave radiation, Q*	573	748
Soil heat flux, G	86	85
Latent heat flux, LE	149	166
Sensible heat flux, H	336	497

NOTE: Q*, G, E, and LE were measured, while H was calculated by difference. $KJ\ m^{-2}\ hr^{-1}$, average of several diurnal measurements per site.

TABLE 6 General Characteristics of the Major Plant Communities of the Alexandra Fiord Lowland.

Community Type and Vascular Dominants (1, 2)	Active Layer ($\bar{x}$, cm, mid-July)	Organic Layer (range, cm)	Soil Moisture Index (3)	Areal Coverage (% of lowland) (4)	Cover		Live Biomass (g/m^2)	Total Biomass (g/m^2)
					Vegetated	Unvegetated		
lichen-cushion plant Di, So, Cn, Ct, Sa	65	0	2	4.5	15%	85%	47	148
lichen-cushion plant-sedge Di, So, Ct, Cn, Cmi, Sa	54	2-5	3	7.0	84	16	97	496
lichen-dwarf shrub-heath Di, Ct, So, Cn, Sa	28	3-10	2	19.5	35	65	67	137
spotted dwarf shrub Di, Ct, Sa, So, Cr	42	0-12	3	6.8	67	33	159	544
evergreen dwarf shrub Di, Sa, So, Ct, Cmi, La	40	2	3	8.9	95	5	238	609
mesic heath Ct, Di, Sa, So, La	36	5-6	4	10.3	83	17	224	643
wet heath Ct, Vu, Sa, Di, La	33	2-5	5	8.4	90	10	149	305
deciduous dwarf shrub Sa, Di, Ct, Pv, Pc	34	5-6	3	2.8	97	3	112	161
deciduous dwarf shrub-graminoid Sa, Lc, Di, Pa, Fb	32	2-9	2	1.0	77	23	56	133
sedge meadow Ea, Cs, Cme, Di, Sa	36	2-14	6	12.5	98	2	39	149
dwarf shrub-sedge meadow Di, Cs, Ct, Vu, Ea	38	3-7	5	7.0	99	1	165	525
herb-dwarf shrub El, Sa, So, Di, Pl, La	35	0	3	5.3	9	91	6.8	18.3
salt marsh Pp, Sh, Cma	19	0	5	0.1	6	94	4.2	5.2

(1) Cma = *Carex maritima*; Cme = *C. membranacea*; Cmi = *C. misandra*; Cn = *C. nardina*; Cr = *C. rupestris*; Cs = *C. stans*; Ct = *Cassiope tetragona*; Di = *Dryas integrifolia*; Ea = *Eriophorum angustifolium*; El = *Epilobium latifolium*; Fb = *Festuca brachyphylla*; La = *Luzula arctica*; Lc = *L. confusa*; Pa = *Poa arctica*; Pc = *Pedicularis capitata*; Pl = *Papaver lapponicum*; Pp = *Puccinellia phryganodes*; Pv = *Polygonum viviparum*; Sa = *Salix arctica*; So = *Saxifraga oppositifolia*; Sh = *Stellaria humifusa*; Vu = *Vaccinium uliginosum*.

(2) Species are ranked in descending order of dominance.

(3) Scale of 1-7; 1 = xeric, 4 = mesic, 7 = hydric.

(4) The remaining 5.9% of areal coverage is comprised of rivers, streams, and ponds.

A STUDY OF THE EVOLUTIONARY HISTORY OF PERMAFROST IN NORTHEAST CHINA BY A NUMERICAL METHOD

Fu Liandi, Ding Dewen, and Guo Dongxin

Lanzhou Institute of Glaciology and Cryopedology, Academia Sinica
People's Republic of China

Based on the essential principles of the unity of formation and the difference of existing conditions of permafrost, the comprehensiveness of freezing processes and the relative independence of each subprocess, the continuity (irreversibility and relative stability) and rhythm of the freezing process, the particularity of contradiction and the main contradiction of freezing processes, and the selection of parameters and mathematic models for the description of freezing processes, and taking contemporary permafrost in large blocks in Northeast China as an example, the authors quantitatively reconstruct the evolutionary history of permafrost in China. Results indicate that the average thickness of permafrost in large blocks was 120 m in the Gu Xiangtun periglacial age. Except in the northern regions, permafrost disappeared during the high-temperature period. The average thickness of contemporary permafrost is approximately 75 m. The data obtained from analyses of the paleoclimate and paleoglacial relics are identical to the data obtained from current field explorations. This method may also be used for other types of permafrost and regions. As long as the change of air temperature is given precisely, the development of permafrost can be successfully predicted.

THE ESSENTIAL PRINCIPLES FOR ANALYZING THE EVOLUTIONARY HISTORY OF PERMAFROST

From the physical point of view, permafrost is a geological-geographical system that exists on the surface of the earth's crust and keeps in frozen state. In permafrost, the process of heat motion, including phase change, is in a special stage that has an inherent form of energy conservation and a distributive law of quantity. The change of energy within certain limits evolves the history of permafrost and correspondingly the change of its regional distribution, including the change of its thickness, and occurrence and change of its phenomena.

According to our understanding, analyses of the evolutionary history of permafrost should be based on the following essential principles (Ding Dewen and Guo Dongxin 1982).

UNITY OF FORMATION AND THE DIFFERENCE OF EXISTING CONDITIONS OF PERMAFROST

When the interaction of energy and substance between soils (or rocks) and surroundings reaches a critical state, or when the annual mean ground temperature reaches the freezing point, permafrost forms. On the other hand, the annual mean ground temperature varies with geological, geographical and geophysical conditions, even in the same region and within different periods, or in different regions and within the same period.

CONTINUITY (IRREVERSIBILITY AND RELATIVE STABILITY) AND RHYTHM OF FREEZING PROCESSES

In permafrost, each substance in different grade and its motion have an evolutionary history of their own. The interaction between factors or processes (including feedback) evolves irreversibly. But freezing processes have a relatively steady stage (correspondingly the stage of quantitative change) and a certain change of rhythm regularity (non-periodicity) because of the change of the main energy source and the contradictory feature in the inner system.

PARTICULARITY OF CONTRADICTION AND THE MAIN CONTRADICTION OF FREEZING PROCESSES

Freezing action is one of the changes occurring on the surface of the earth's crust. It becomes a special lithification in permafrost regions, and produces specific permafrost (geomorphological) phenomena and characterizes a specific geographical view (landscape) in which heat motion, including phase change, plays a leading role. Therefore, heat motion can be considered to be a physical process (especially the transportation) that evolves under a certain natural-historical condition.

SELECTION OF PARAMETERS AND MATHEMATICAL MODELS TO DESCRIBE FREEZING PROCESSES

Because freezing processes are a chain of continuous changes of state, choosing the thermodynamic state parameters to describe them is reasonable. Under the case of no external electromegnetic field, choosing the parameters of pressure (P), volume (V), temperature (T), and humidity (W) is enough. Furthermore, because both processes of temperature and phase play a leading role in the evolution of permafrost, and the process of phase change can be objectively described by isotherm, freezing processes can be described by temperature and humidity fields. In the case, the four parameters: the hydraulic conductivity (K), the thermal conductivity (λ), the thermal diffusivity (a) and the volumetric heat capacity (C) are the main parameters. The mathematical models of processes in one dimension can be expressed as follows:

$$\frac{\partial t^i}{\partial \tau} = a^i \frac{\partial^2 t^i}{\partial x^2} \quad (1)$$

$$\frac{\partial W^i}{\partial \tau} = K^i \frac{\partial^2 W^i}{\partial x^2} \quad (2)$$

Where X is the coordinate; τ is time; i represents phase state: "-" freezing, "+" thawing.
Then,

$$X=h, \quad t^- = t^+ = t^*, \quad W_H = W_0 - W' \quad (3)$$

$$\lambda^- \frac{\partial t^-}{\partial x} - \lambda^+ \frac{\partial t^+}{\partial x} = L\gamma_0 (W_h + \Delta W) + \frac{1}{2} C \; t_0 \frac{dh}{d\tau} \quad (4)$$

$$K^+ \frac{\partial W^+}{\partial x} - K^- \frac{\partial W^-}{\partial x} = \Delta W \frac{\partial h}{\partial \tau} \quad (5)$$

Where γ_0 is the dry density of soils; W_0 is the initial water content; W is the quantity of moisture migration per volume, and L is the latent heat.

The initial and boundary conditions can be given out according to features of concrete problems (to set up the condition of the determinative solution).

A DETERMINATIVE SOLUTION FOR THE EVOLUTIONARY HISTORY OF PERMAFROST IN NORTHEAST CHINA

According to our data and those of our predecessors, strata in Northeast China could be divided into two combinations of spore pollen for the periods of the early, middle and late Pleistocene and three for the Holocene. It showed that the climate in this region varied alternatively many times, from cold to warm, from dry to humid, and vice versa and that the frequency of the climate change and the general tendency were the same in the whole region. Permafrost in this region evolved on the background of the climate conditions and formed a general pattern of evolution.

Under influences of the structural system and structural motion within periods of Yian Shan and Himalaya, not only did the local geological, geographical, geomorphological and geophysical conditions tend to be complicated, but also the evolution of permafrost became complicated, among which magma activity and basalt eruption had tremendous influence resulting in higher terrestrial heat, different motion of the earth's crust, and the migration and deposit of the river system, etc.

The component, structure, temperature and thickness of permafrost and its phenomena are representative indexes or marks of freezing processes. According to the data, including the change of palaeoclimate reflected by change of combinations of spore pollen, the relics of paleo-periglacial structures and the change regularity of air temperature with latitude, we can outline the evolution of permafrost as starting from the Gu Xiangtun glacial age in the late Pleistocene, passing through the high temperature period and minor-glacial age in the Holocene, and still developing up to now. In Northeast China, frozen ground can be divided into four types: continuous permafrost in large blocks, permafrost in island-like taliks, island-like permafrost and seasonally frozen ground, all controlled mainly by solar radiation and reflecting obviously the regularities of the horizontal zonation (Guo Dongxin and Li Zuofu. 1981).

The evolution of permafrost in Northeast China can be divided into the following four stages:

First, the late Pleistocene, in which the south limit of continuous permafrost was located near that of contemporary permafrost and reached the San Jiang plain. The south limit of island-like permafrost was situated to the north of Qian Guoangtu in the town of Ao Han.

Second, the end of the high temperature period, in which the permafrost all over the region thawed except for sporadic permafrost remaining in the bottom of valleys or wet low-lying land north of Man Gui.

Third, the minor-glacial age, in which the permafrost formed again and reached the south of Man Gui, and superimposed on old permafrost in the North of Man Gui.

Finally, now-a-days, in which contemporary permafrost formed gradually.

Taking contemporary continuous permafrost in large blocks as an example, we reconstruct the evolutionary history of permafrost quantitatively by using numeric method and equations of temperature field, since the thickness of Quaternary strata is very thin and permafrost belongs to the epigenetic type. Conditions of the determinative solution are as follows:

Geometric and Physical Conditions

The section of strata consists of sandy clay from 0 to 10 meters and granite from 10 to 350 m (the depth of the lower boundary is determined by the transmitted depth of temperature wave in the Gu Xiangtun periglacial period). Physical indexes are as below*:

$\gamma_0 = 1600$ $\quad \lambda^- = 1.66$ $\quad C^- = 504$

$W = 0.25$ $\quad \lambda^+ = 1.47$ $\quad C^+ = 600$

$W_H = 0.025$

$\gamma_0 = 1800$ $\quad \lambda^- = 2.63$ $\quad C^- = 477$

$W = 0.02$ $\quad \lambda^+ = 1.88$ $\quad C^+ = 666$

$W_H = 0$

$Q = L\gamma_0\ (W - W_H)$

*Dimensions are in the system of meter, kilogram, hour, and kilocalorie.

Initial and Boundary Conditions

According to the data of the absolute age of strata and of the paleontological fossils, permafrost was well developed in the Gu Xiangtun periglacial age. After that, it disappeared in many places in the interglacial period between the Gu Xiangtun and Dong Gan glacial ages. Therefore, the starting point of computation is 30,000 years ago. The initial contitions of temperature are:

$t\ (0,0) = 0^\circ C$

$$t\ (X,0) = t\ (0,0) + gx \qquad (6)$$

Where g was the geothermal gradient in 1/30 degree per meter.

From equation (6), the temperature of the lower boundary was $t_H = gH - 11.7^\circ C$.

According to the data of paleontology of relics of paleo-permafrost and of contemporary differences between ground-surface temperature and air temperature, conditions of the upper boundary can be estimated as follows:

$$t\ (0,\tau) = \begin{cases} -6.0^\circ C & 30\ \tau\ 12 \quad \text{thousand} \\ -3.0 & 12\ \tau\ 8 \quad \text{years B.P.} \\ 2.0 & 8\ \tau\ 2.5 \\ -3.0 & 2.5\ \tau \end{cases}$$

COMPUTATIONAL METHODS AND ANALYSES OF RESULTS

The problem of heat conduction with one dimension mentioned above is computerized by the following format of all-implicit difference:

$$T_j^{n+1} = \frac{}{h^2 C}(T_{J+1}^{n+1} - T_J^{n+1} + T_{J-1}^{n+1}) + T_J^n \qquad (7)$$

Where τ is time step; h is space step; n is time of points in the computional net and j is position of points in the computional net.

Cutting-off error is $E = 0(\tau) + 0\ (h^2)$

Although the accuracy of the format has only one grade, the choice of time step τ is not limited by the stability because the state of the temperature field needs to be computerized for 30,000 years. Furthermore, for such a diffusive problem, chosing the implicit format agrees more with the physical model, and for solution of linear equation, three pairs of the angle matrix also have the effectiveness of the tracing method with no requirement for computational time and storage spaces. Therefore, the all-implicit format is suitable for solving the problem of heat condition in one dimension.

In computation, the treatment of parameters is as follows: the thermal conductivity λ and the volumetric specific heat C are constants in the frozen and thawed regions respectively, but vary with soil (or rock) properties. In the phase change region, we have $\lambda^* = \frac{1}{2}(\lambda^- + \lambda^+)$ and $C^* = A + \frac{1}{2}(C^- + C^+)$, where A is the amount of latent heat in the range of $-1^\circ C$ t $0^\circ C$. Consequently,

$$\lambda = \begin{cases} \lambda^- & t\ -1^\circ C \\ \frac{1}{2}(\lambda^+ + \lambda^-) & -1^\circ C\ t\ 0^\circ C \\ \lambda & t\ 0^\circ C \end{cases}$$

$$\begin{cases} C^- & t\ -1^\circ C \\ A + \quad (C^- + C^+) & -1^\circ C\ t\ 0^\circ C \\ C^+ & t\ 0^\circ C \end{cases}$$

Computational results are shown in Table 1.

From Table 1, we can see that in the Qu Xiangtun periglacial age, the thickness of continuous permafrost in large blocks was 120 m on the average and more than 150 m in northern regions. In the high temperature period, permafrost disappeared except for the northern regions, in which permafrost remained in the thickness from 0 to 20 m. Permafrost has developed again since the minor glacial age. The thickness of contemporary permafrost is approximately 75 m on the average and 100 m for individual cases in northern regions. The results obtained from the analyses of paleoclimate and paleoglacial relics is identical with the data obtained from present-day field exploration.

The freezing process can be expressed by the following formula:

$$H = \left[\left(\frac{q\tau}{2a}\right)^2 + \frac{\lambda\tau t_0}{2a} \right]^{\frac{1}{2}} - \frac{q\tau}{2a} \qquad (8)$$

where $\quad a = LW\ \gamma_0 + \frac{1}{2}\ C\ t_0$

TABLE 1 Frost Penetration Depth With the Change of Time

Time, years B.P.	29,900	29,500	29,000	28,000	25,000	12,000	10,000	8,000	3,000	2,000	Present
Frost depth, m	10	50	95	105	115	120	75	70	0	55	75

From formula (8), we can see factors affecting the development of permafrost and relationships between them. When heat flow caused by the change of ground-surface temperature inverts, or when the heat flow changes the direction from output to input, permafrost will be thawed in two directions.

The heat in the interior earth plays a leading role in the thawing process of permafrost.

Permafrost can not be developed unless the condition of $\frac{\tau\lambda t_0}{2a} \quad (\frac{q\tau}{2a})^2$ is satisfied. The greater the absolute value of negative temperature on ground-surface is, the faster it will develop.

If the change of the heat flow is not great, the evolution of permafrost will be mainly controlled by the air temperature and soil (or rock) properties. If the soil (or rock) properties are the same, air temperature will become the main factor and the distribution of permafrost has the obvious regularity of zonation.

In conclusion, the method can also be used for other types of permafrost and regions, and as long as the change of air temperature is given precisely, the prediction of the development of permafrost can be done successfully.

REFERENCES

Ding Dewen and Guo Dongxin, 1982, A Study on the Evolutionary History of Permafrost on Qinghai-Xizhang Plateau, in Proceedings of the First National conference on Glaciology and Cryopedology, Science Publishing House.

Guo Dongxin and Li Zuofu, 1981, A Primary Study on the Evolutionary History of Permafrost in Northeast China and its formation age, Journal of Glacialogy and Cryopedology, Vol.3, No.4.

ULTRASONIC VELOCITY IN FROZEN SOIL

Fu Rong, Zhang Jinsheng, and Hou Zhongjie

Lanzhou Institute of Glaciology and Cryopedology, Academia Sinica
People's Republic of China

The propagation velocities of ultrasonic dilatational and shear waves in frozen Lanzhou medium sand and frozen Tibet clay samples were measured at different water contents and temperatures using the SYC-2 sonic wave detector. The variation of sonic velocity with water content and unit weight substantially depends upon the type of soil. In coarse-grained frozen soil, the sonic-wave velocity consistently increases with increasing water content, while in frozen clayey soil, it decreases with increasing water content in the low water content range, then starts to increase with a fluctuation, and finally increases steadily up to the ultrasonic velocity in ice. Tests indicate that the ultrasonic velocity increases with decreasing temperature. Based on this investigation, the authors consider that ultrasonic detection techniques can be used to determine the physical and mechanical properties of frozen soils, such as water content, density, elastic modulus, and Poisson's ratio.

INTRODUCTION

Temperature, moisture (including ice and unfrozen water) and soil structure directly affect various characteristics of frozen soils. To study how these various fectors influence properties of frozen soils, we used ultrasonic measurement techniques. An understanding of how water content influences the variability of ultrasonic velocities in frozen samples, was based on tests performed on red clay of Qinghai-Xizang and medium sand of Lanzhou.

INFLUENCE OF MOISTURE ON CHARACTERISTICS OF FROZEN SOIL

A high enough temperature, but still below 0°C, when a large part or all of moisture is unfrozen in soil, characteristics of the soils have no apparent difference in comparison with its under normal atmospheric temperature. As the total water content exceeds the amount of unfrozen water, ice crystals will grow in the frozen soils, bonding the soil particles together. The ice forms a firm soil skeleton structure that causes a great change in characteristics of soils after freezing. Therefore, the strength and elasticity are related to the amount of ice present. In general, the quantity of ice depends upon the total amount of water contained in frozen soils. Therefore, the physical-mechanical properties of frozen soils relate directly to water content.

Pore water can exist in different forms, as either free water or bound and osmotic (loosely bound) water in unfrozen soil. In freezing process, the water plays different role, with its form different, and the role change with different grain size of soil. As a result, a change of the soil structure occurs. The finer the soil particles, the larger the effect (Thomas et al. 1971). In fact, the smaller the grain size of soil particles, the larger the surface area, and also, the larger its capacity of maintaining unfrozen water (Anderson and Tice 1972). Apparently, as the grain size of soils changes same water content has different influences upon the physical characteristics of frozen soils.

Using a SYC-2 sonic wave detector, we determined ultrasonic velocities and obtained sonic wave propagation speed (Fu Rong et al. 1980) for frozen soils with different properties (temperature, water content and density). In combination with previous measurements (Zhang Jinsheng and Fu Rong 1981) of unfrozen water content, we indirectly established a relationship between sonic wave speed, ice and unfrozen water content.

Appropriate quantities of distilled water were added to dried specimens of sandy soils to obtain samples with various water content as high as the saturated state. Clay samples with the water contents below the liquid limit will be drier than specimens with water content near liquid limit in the shade, hence they had similar dry density regardless of moisture content. Those with water content above liquid limit are directly made up in the similar way.

The samples were prepared in a cylindrical form with a diameter of 5 cm and a high of 5 cm. They were then frozen in a cooler maintained between -20°C and -30°C. After freezing, prior to testing, they were

TABLE 1 Particle Diameter Analyses for Kinds of Soils.

Kind of soil	Content (%) of various particle diameters (mm)									
	5-2	2-1	1-0.5	0.5-0.25	0.25-0.1	0.1-0.05	0.05-0.01	0.01-0.005	0.005-0.002	0.002
Red clay of Qinghai-Xizang			2.70	2.02	5.00	1.68	26.30	11.59	17.41	33.30
Medium sand of Lanzhou for construction	0.02	1.10	31.81	45.00	17.14		4.93			

TABLE 2 Basic Indexes for Two Type of Soils.

Kind of soil	Specific weight	Liquid limit W_L (%)	Plastic limit W_p (%)	Loose unit weight (T/M^3)	Specific surface area S (m^2/g)	Salt content (%)
Red clay of Qinghai-Xizang	2.76	22.0	15.0		31.8	0.08
Medium sand of Lanzhou for construction	2.65			1.36	6.1	0.05

brought to the desired temperature in a thermostatically controlled chamber (degree of accuracy ±0.1°C).

The main analytical results of the Medium Sand of Lanzhou and Red Clay of Qinghai-Xizang are shown in Tables 1 and 2.

In the course of measurement, test sample is put on the loading frame specially made, and a pressure about 0.5 kg/cm^2 is exerted to guarantee close contact between sample face and sensor or transducer. couplant is water. In low temperature, the water will freeze and become ice, and couple the sensor and transducer with the sample face together.

ULTRASONIC WAVE VELOCITY IN FROZEN SANDS

The grain size of sands is comparatively large, so the effect of the unfrozen water content can be neglected.

Under a comparatively dry state, increasing moisture in sands does not change its soil skeleton structures: Therefore, the density increases with the increases of water content. When water content is rather high, near saturated state, freeze will result in bulk expansion and displacement among sand grains. If the water content increases unlimitedly, the density will decrease and finally approach that of the ice. At lower moisture contents in the sands, the influence of ice volume on sonic velocity is small. The sonic wave speed chiefly depends on the sand and its grain contact conditions. With increase in ice content, not only does the density of frozen sand increase due a larger volume of pores being ice filled, but also the structure becomes more solid due to cementation by the ice. These two aspects tend to influence sonic wave propagation. As a result, sonic wave velocities increase with increased density (Figure 1).

For the single factor of water content, when the increment of moisture results in the ice fill-degree in frozen sand approaching to one, its action for increasing sonic wave speed comes to a fixed value. Hence the increasing rate of wave speed gradually decreases. The relationship between them can be expressed as a parabolic curve (Figure 2). In logarithmic coordinates, they are two groups of parallel straight lines. Since the acoustic impedance is equal to the product of the density and the sonic wave speed, the impedances of dilatational and shear waves ρV_p and ρV_s are mutual parallel straight lines, too, in logarithmic coordinates.

Because the density includes the two factors of moisture and structure, the relative lines between the sonic wave speed and the density and between the sonic wave speed and the water content differ greatly.

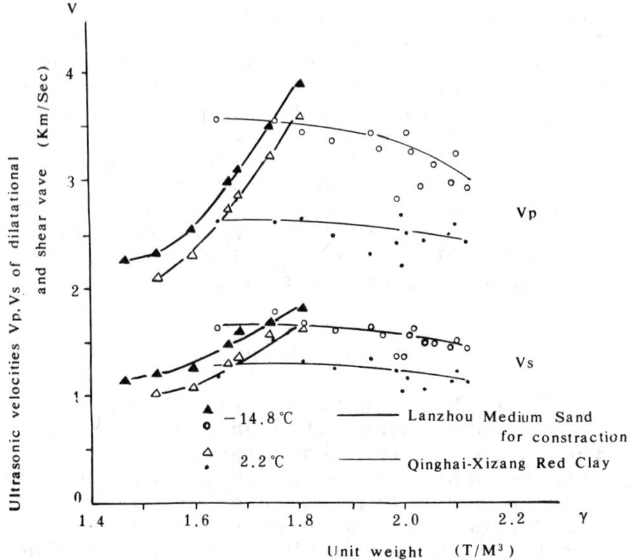

FIGURE 1 Relations between ultrasonic velocities and density, in the frozen consolidated samples of two frozen soils.

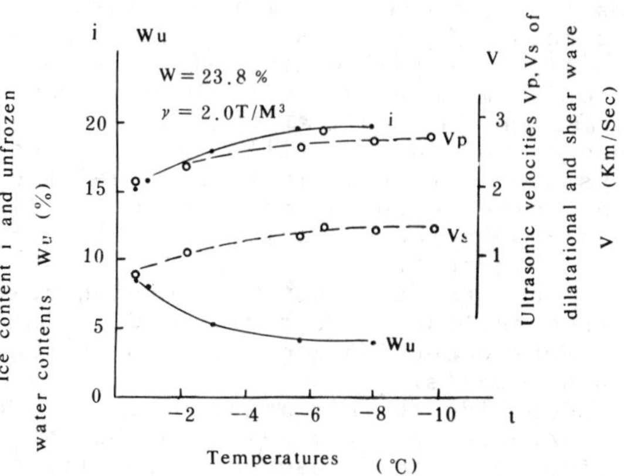

FIGURE 3 Changes of ice content, unfrozen water content and ultrasonic velocities versus temperatures.

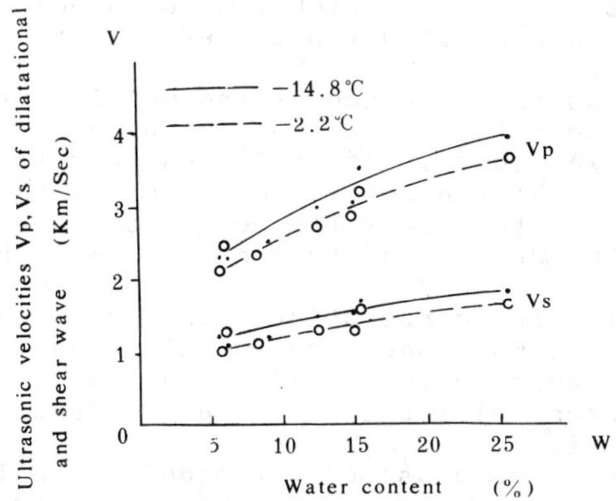

FIGURE 2 Relations between ultrasonic velocities and water content in the frozen consolidated Medium Sand of Lanzhou for construction.

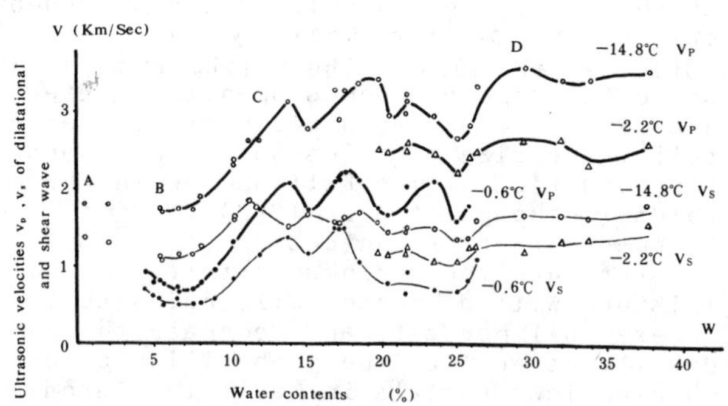

FIGURE 4 A sketch shown the changes of ultrasonic velocities versus water contents in the Red Clay of Qinghai-Xizang.

SONIC WAVE VELOCITY IN FROZEN CLAYEY SOILS

Differing from coarse-grained soils, the moisture in fine-grained soils can disperse soil grains, so that with the increase of water content, the density of soils decreases gradually, and there is a part of unfrozen water existing in the soil after freezing (Figure 1). Since ice crystals strengthen the integrity of the soil skeleton structure and tend to propagate sonic waves, and the unfrozen water content is in relation to temperature, the sonic wave speed and the ice content as well as the unfrozen water content change with negative temperature (Figure 3).

Under dry status, the sonic wave speed of fine-grained soils depends mainly upon the integrity of the structure, and the change of temperature has almost no influence on it. When the water content is very low, there are no ice crystals in soils under negative temperature. Increasing moisture can only thicken the unfrozen water film in soils, so that the soil skeleton structure is loosened. In addition, the propagation speed of sonic wave in water is comparatively low (about 1.5 km/sec). Thereby, with the increase of water content, the sonic wave speed decreases.

Within the whole changing process of sonic wave speed versus water content, the sonic wave speed is the lowest in the stage indicated by section AB in Figure 4.

When the water content is greater than the amount of unfrozen water, ice crystals will appear in soils. With the increment of water content, ice crystals fill-degree in voids will also increase gradually. Since the total amount of water content is smaller, the change of moisture does not result in the displacement of relative positions of soil grains, and the soil porosity maintains a certain constant value, and the sonic wave speed depends only on the fill-degree of ice crystals in voids. In this stage, the sonic wave speed increases with the increase of water content, shown by section BC in Figure 4.

It can be seen that the turning point B from section AB to section BC is just the mark occurring ice crystals in fine-grained soils.

According to this principle, the amount of unfrozen water in soils can be roughly estimated by the measurement results of sonic wave speed.

Surpassing section BC in Figure 4, the water content continues to increase and partial moisture in soils begins to crystallize in voids with the lowest pressure. When ice crystals reach a certain extent, the film water will shrink back or occupy a new position along with the frozen front. At this time, soil particles become crumby granular and are surrounded by ice (Orlov et al. 1977). The magnitude of sonic wave speed depends upon the state and size of crumby-granulary structures of soils, the size of voids among them, and the condition of ice fill-degree in the voids (Fu Rong et al. 1982).

When the water content of soils is greater than the shrinkage limit, the moisture will disperse soil grains to increase soil porosity and decrease the density as well as the probability of occurring ice crystals in soils increased in freezing with the increment of water content, the structure among ice-wrapped bodies tends toward loosening. In other words, with the increase of the density of ice crystal distribution, the ties among soil crumbs are gradually weakened and the sonic wave speed decreases, too. When it arrives at a certain limit, continuing to increase water content will increase both the ice fill-degree in voids and the sonic wave speed. Therefore, with the increase of water content, values of the sonic wave speed change from high to low and back again to high. Nevertheless, after the sonic wave speed reaches a certain high value in this picking-up process, the increment of water content provides a more advantageous condition for ice formation. But the increase of the density of ice crystal distribution in soils results in further weakening of the bonds between soil crumbs. Then the sonic wave speed again tends to descend, and the above mentioned process reappears. However, at this time the size of the soil crumbs (ice-wrapped bodies) is much smaller than that of the last period.

Obviously, with the increase of water content, values of sonic wave speeds fluctuate. The soil crumbs gradually become smaller until the soil grains are separated by ice crystals and suspended in ice. After that the change of water content will not lead to the fluctuation of sonic wave speed, and with the decrease of density of soils after freezing, the sonic wave speed approaches that of ice.

SEVERAL VIEWPOINTS

Under the case of adopting the pulse method to measure ultrasonic velocity in frozen soils, the instantaneous energy exerted upon samples is very big. But the loading and unloading rates are great, and the frozen soil is neither able to yield plastic flow, nor to have the needed time for developing plasticization, hence it is in a pure elastic status. On the basis of the sonic wave velocity, according to elastic theory, the basic parameters of mechanics, elasticity modulus, and Poisson's ratio can be obtained. These parameters are very difficult to be precisely determined by traditional methods of statics.

Acoustic impedance is another important index to reflect the fundamental qualities of frozen soils. If it is possible to measure the sonic wave speed and at the same time to survey the acoustic impedance, then the density of frozen soils will be known.

Utilizing correlation course between the sonic wave speed and density and moisture, can give the density (unit weight) of frozen soils and the water content in some soils.

How to make use of ultrasonic velocity to test precisely various physical and mechanical indexes and fundamental parameters in frozen soils is still an important subject in the future. However, this method undoubtedly has its bright prospect in the study of frozen soils and on practice of engineering.

CONCLUSION

1. In frozen soils, sonic wave speed relates to the state of moisture, and its correlative law varies with the diameter of the soil particles.
2. Comparing the change regularity of sonic wave speeds with the unit weight (or density) and moisture can analyse conditions of the frozen soil structure.
3. Measuring ultrasonic velocity can determine elasticity modulus and Poisson's ratio.
4. Utilizing the correlative law of sonic wave speeds can approximately determine the unit weight of frozen soils and water content in coarse-grained soils.

5. The complicated changing of the ultrasonic velocity with water content in frozen fine-grained soils reflects the multi-variation character of frozen soil structures.

ACKNOWLEDGMENT

The authors gratefully acknowledge associate professor Zheng Dahui, who is working in Changchun Geological Institute, for his instructive advice.

REFERENCES

Anderson, D. M., and Tice, A. R., 1972, Predicting unfrozen water content in frozen soils from surface area measurements, Highway Research Board Record, No. 393.

Fu Rong, Zhang Jinsheng, and Hou Zhongjie, 1980, Preliminary research on ultrasonic measurement of frozen soil and ice: Chi. Journal, Glaciology and Cryopedology, v. 2, Special Issue.

Fu Rong, Zhang Jinsheng, and Hou Zhongjie, 1983, Influence of moisture in frozen soils on the ultrasonic velocity: Chi. Journal, Glaciology and Cryopedology, v. 5.

Orlov, V. O., Dubunov, U. D., and Meilinkov, N. D., 1977, Frost heave of soil and its affected to structure, STRUCTURE Press, Moskow.

Thomas, I. Csalhy., and David, L. Townsed, 1971, Pore size and field frost performance of soils: U.S. Highway Research Board Bull. 331.

Zhang Jinsheng and Fu Rong, 1981, Analysis of several methods for determining the unfrozen water content in frozen soil, _in_ Proceedings of the Second National Conference on Permafrost of China (Selection).

ANALYSIS AND CHARACTERISTICS OF CORES FROM A MASSIVE ICE BODY IN MACKENZIE DELTA, N.W.T., CANADA

K.Fujino, K.Horiguchi, M.Shinbori[1] and K.Kato[2]

[1]Institute of Low Temperature Science, Hokkaido University, Sapporo, 060 Japan
[2]Water Research Institute, Nagoya University, Nagoya, 464 Japan

Distinctive ground features such as polygons, pingos and involuted hills are commonly observed in the arctic regions. Their origins and formation processes remain in question. A massive ice body at Tuktoyaktuk was analyzed to determine its origin. Stratigraphic studies of fabrics, chemical components such as oxygen isotopes, conductivity and pH of core samples of the massive ice body, were investigated.
The results obtained suggest that most parts of the massive ice body originate from superimposed ice by congelation of water in which a snow pack is submerged, and do not support the hypothesis that it is segregated ice. The mechanism of the growth of a massive ice body is also not identical to that of the growth of a pingo in which segregated ice constitues the core.

INTRODUCTION

Distinctive ground features such as polygons, pingos and involuted hills are commonly observed in the arctic regions, especially around Tuktoyaktuk, Mackenzie Delta, N.W.T., Canada.

Their origins and formation processes remain in need of analysis,[1 2] but the fact that large ice bodies are found underneath them suggests that water supply from various sources related to long- and short- term climatic conditions plays an important role in their formation.

Stratigraphic studies of ice fabrics and chemical components such as oxygen isotopes, were carried out to find the origin, the formation mechanisms and characteristics of the massive ice body at Tuktoyaktuk, using core samples which were obtained throughout the massive ice body.

SAMPLING

The massive ice body is located, as shown by ✪ in Fig.1, along a seacoast about 4.5 km southwest of Tuktoyaktuk, Mackenzie Delta, N.W.T., Canada. It was investigated during the joint research expeditions of 1977, 1980 and 1981.[3 4 5]

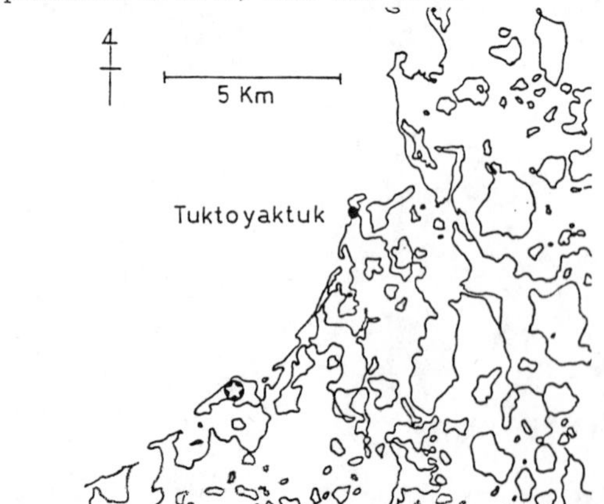

FIGURE 1 Location map of Tuktoyaktuk area. ✪: Sampling site.

Using a newly designed electromechanical core-drill, three holes were drilled into the massive ice body from near the top and continuous cores were obtained in March 1982. One of the holes which were cored throughout the massive ice body, was about 23 m in total length including a 40-cm-thick sand layer of the bed. Cores were arranged and labeled in serial order, put in refrigerated containers and transported to the cold laboratory of the Institute of Low Temperature Science, Hokkaido University, where the following preparations were made.

Stratigraphic features of the cores were described along their entire length, such as bandlike structure, bubble distribution, impurity inclusion, etc. Each piece was cut into halves along the vertical axis. One half was cut perpendicular to the axis at an interval of 10 cm, from which horizontal and vertical thin sections were prepared for the investigation of crystal orientations. Meanwhile, the other half was cut perpendicular to the axis at an interval of 2.5 cm, from which samples were prepared for the determination of oxygen isotopes, conductivity, pH and bubble pressure. Moreover, soil between strata which formed a boundary of the bandlike structure, as well as any impurity inclusions were also prepared for qualitative analysis.

ANALYSES

Characteristics of Stratigraphy and Crystallography

A schematic description of the entire length of core obtained from a single hole is shown in Fig.2. As is clear from the figure, many bandlike structures, which were separated by sediment, were observed. A stratification defined by bubble-rich layer and bubble-free, clear layers were observed along the entire length of the core, although the inclination and thickness of each layer were not consistent.

Sediment contained in layers between two adjacent ice strata varied and consisted of sand, clay, silt or their mixtures. Also, the configurations of the sediment had various forms, one looked to be suspended in the ice (Fig.5), another looked like thick sediment settling down on a concave-convex surface (Fig.4), a third looked to be trapped on the surface of a snow pack (Fig.3).

Crystals in each stratum were clearly different in size, shape and c-axis orientation. Strata could be divided roughly into the following three types: (1), strata of granular ice crystals which are commonly observed in a snow pack, having the nearly homogeneous grain size averaging 0.5 cm and being arranged with random orientations (Fig. 6). (2), those of medium size ice crystals having nearly rectangular shapes averaging 2-3 cm on the long axis and being arranged with some orientation (Fig.7). (3), those of large ice crystals of more than 10 cm, the whole section of the stratum of being occupied by only one or two large crystals (Fig.8). Distributions of the c-axis orientation of crystal in strata are now being measured. But from the results obtained so far, they seem to be comparatively random except within strata of large crystals.

Distributions of air bubbles in the strata were not uniform. Air bubbles in most strata were spherical, although the sizes and distribution densities of bubbles were different in each stratum. But several strata abounded in elongated bubbles that were arranged parallel to the vertical axis of the core (Fig.11). Some of those elongated bubbles were intertwined each other like a three-dimensional network and also coexisted with spherical bubbles.

Bubble pressure was measured using a volume difference before and after melting[7] (Fig.12). As is clear from the figure, the pressure of those bubbles near the surface is the same as the atmospheric pressure, but increases linearly with increasing depth. Those at the bottom of the massive ice body reach nearly twice the atmospheric pressure. High-pressure bubbles like those in glacier ice were not observed in the cores.

Characteristics of Chemical Components

Ice samples at an interval of 2.5 cm along the entire length of the core were prepared for chemical analyses of the oxygen isotopes, conductivity and pH. Determination of oxygen isotopes is continuing at the Water Research Institute, Nagoya University. Results obtained to date are shown in Fig.13. As is clear from the figure, the vertical profiles of $\delta^{18}O$ values abounds in distinctive discontinuities.[8] They seem to correspond to crystallographic and stratigraphic changes in the core. For example, a discontinuity of the profiles, which occurs at about 13 m depth, corresponds to the presence of sedimentary layer shown in Fig.2.

Conductivity and pH of the core were also measured on the same samples used for determinating the oxygen isotope ratios. Results obtained are shown in Fig.14. Values of pH remain reasonably consistent with depth. However, values of conductivity show distinctive spikelike projections in many portions of the profile with depth. These distinctive projections seem to correspond to clear band, bubble-free layers of ice in the core.

DISCUSSION

There are three hypotheses concerning the origin of massive ice bodies. The first one is the snow-glacial hypothesis, which involves the burial of ice, such as stagnant glacial ice, compact snow drift, lake, sea and river ice. The second hypothesis is ground ice such as well developed vein ice. The third is segregated ice from the freezing of drawn up pore water.

Concerning the vein ice hypothesis, there is no evidence favouring it. That is, in the western Arctic, there is no giant syngenetic vein ice as is found in Russia and the conditions which form syngenetic ice wedges have not occurred in the Mackenzie Delta area.[2] The buried glacial ice hypothesis is also not generally accepted but difficult to completely disregard. That is, the presence of large foliated patterns in an outcropping ice wall of the massive ice body seems to support deformation by a glacier. Also the glacial limits of the glaciations in this area are not completely known. The ice segregation hypothesis which has been illustrated by Mackay[1] is that pore water in soil is drawn up and frozen as segregated ice and such freezing within a closed system of high water content would result in the expulsion of water and form the positive relief features. An explanation is still needed, however, to explain the ice fabrics and the foliation in the massive ice body.

From a crystallographic view point, there are some particular distinctive features in glacier ice, that is, in the ice fabrics, a strong concentration of the arrangement of c-axis are commonly observed. The distribution of ice grains and soil inclusions in each stratum are modified and arranged by shear within glacier, and also the highly pressurized air bubbles are observed in the core even after being kept at an annealing stage for a fairly long period.

The present authors' findings on ice fabrics of the core obtained from the massive ice body are not similar to those of glacier ice. That is, there could not be observed such concentration of c-axis orientation in most of strata as in glacier ice. There also could not be observed pressurized air bubbles in the core (Fig.12).

These facts indicate that a previous hypothesis that the massive ice body originates from the remnant glacier is probably wrong.

The hypothesis that the massive ice body is segregated ice is also doubtful based upon our observation of ice fabrics.

More than one half of the core samples consisted of layers of granular ice which abounded in spherical air bubbles of various sizes and homogeneously dispersed (Fig.6). The size distribution, arrangement of c-axes of the ice crystals and the distribution of air bubbles are similar to those within ice formed in a snow pack that freezes after it is immersed in water.

This crystallographic feature is commonly observed in superimposed ice layers such as temperature glacier ice, sea, lake and river ice. Ice fabrics and air bubble distributions of such ice is not uniform because of differences in the amount of water supplied to the snow pack, that is, pores in a snow pack are connected in a three dimensional network and forming a skeleton structure of the snow pack. When the snow pack submerges in water, the pores are isolated and each of them forms an air bubble. The shape and distribution of each air bubble depends upon how much the snow pack is saturated by water in which it is submerged. If the saturation rate is comparatively high, then the shape of a trapped bubble becomes spherical, otherwise, it becomes irregular, keeping the original, elongated shape of the pore.

It may be interpreted that the granular ice layers which consist more than one half of the core are formed by superimposed ice on the bed of ice, considering ice fabrics, shapes and distributions of air bubbles in the strata. Especially, the particular shapes of elongated air bubbles in many strata support the superimposed ice. These elongated air bubbles can not be formed by segregation (Fig. 11).

There are many facts other than ice fabrics that support superimposed ice formed by congelation of water in which a snow pack is submerged.

Stratigraphic changes in chemical components of the core also support this model.

Spikelike projections which were observed in the profile of conductivity against depth (Fig. 14) were formed as follows: Impurities which were contained in the submerged water rejected as ice formed from the water during freezing and enriched the remaining water with chemical components and complete freezing formed layers of high conductivity.

In the profile of $\delta^{18}O$ against depth, changes in value were comparatively small and linear profile was observed except for several portions (Fig. 13). In successive snow strata, glacier ice and ground ice, distinctive periodic changes in value of $\delta^{18}O$ in the profile are commonly observed, reflecting short- and long-term changes in isotopic contents of precipitation which are related to changes in climate.[6] A linear profile without such periodic changes is often observed in superimposed ice in temperate glaciers.[8] It develops when the snow or firn was submerged in melt water which, because of mixing, becomes relatively homogeneous in composition and formed superimposed ice.

Formation of large crystals of bubble-free layers may be explained from the concept of superimposed ice as developing from a snow pack layer which is saturated fully by submergence under water or a fairly thick layer of water which is perched upon a bed of ice below the snow pack, which then freezes under a low cooling rate as commonly observed in the case of lake or pond ice (Fig. 10).

The shapes and other stratigraphic features of some soil sediments support the perching of a large amount of water upon a bed of ice.

A thick soil sediment observed at a depth of 12-13 m (Fig. 4) looks like that formed when the sediment is deposited in water on the bed of ice like a lacustrine sediment and frozen. The sediment looked to be suspended in the ice (Fig. 5) is that formed when the snow pack layer is saturated by melt water and the particles are suspended in that water and frozen in the skelton of the snow pack.

Also, the large foliation bands which were observed in the outcropping ice wall can be explained in terms of superimposed ice layers which preserved accumulation layer of a snow pack or debris of a snow bank.

It may be concluded from ice fabrics, geochemical and stratigraphic analyses that most parts of the massive ice body originate from superimposed ice formed by congelation of water in which it was submerged. The mechanism of its growth is not identical to that of the growth of a pingo in which segregated ice constitutes the core.

An explanation is still needed, however, to explain the preferred location of massive ice bodies beneath positive relief features such as hills and plateaus and the occurrence of beneath a fairly thick cover of till.

Thus, the formation process of the massive ice body might be assumed as follows. At the initial stage, a large mass of superimposed ice was formed, originating from a snow pack or debris of snow bank on a bed of ice, perhaps produced in a lake or pond. While in the process of aging, new snow or debris of drifting snow is deposited on the ice mass. Then, it attains an ablation stage and the upper part of accumulated snow layers on the ice mass melted away. The melt water supplied from surface perched on the ice mass and submerged in the depositing snow layers remaining there, thereby freezing to become superimposed ice. Through repetition of this process, the ice mass develops recognizable strata containing a soil layer at the boundary of each stratum and distinctive discontinuities in the ice fabric as shown in Figs. 3, 4, 8 and 10. Chemical components such as oxygen isotopes and conductivity develop values as shown in Figs. 13 and 14 in this fashion also. After growing to some extent by these intermittent formation processes, the ice mass is buried by reworked till or glaciofluvial sediment which has been formed by the outwash of a river.

Still, some points remain arguable, for instance, preferred location, an overriding till layer, quantity of water in which a snow pack or snow debris are submerged and time of formation of the ice mass.

More detailed information and investigations on massive ice bodies are needed for clarifying the origin and formation processes of them, together with geological conceptions including stratigraphy and dating.

ACKNOWLEDGMENT

Authors wish to express their hearty thanks to Dr. G.D. Hobson, Director of Polar Continental Shelf Project, Energy, Mones and Resources, Canada; to the Camp Manager and his staff of Tuktoyaktuk Base Camp for helping them carry out field work; and also to Dr. J.D. Ostrick of Inuvik Scientific Resource Centre, Indian and Northern Affairs, Canada for his kind arrangements for field and laboratory work. They are also indebted to Dr. Y. Hiratsuka of Northern Forest Research Centre and Mr. Y. Akao of Marubeni, Canada, LTD. and Messrs. T. Okada, K. Kobayashi of Marubeni Reizo LTD. for their kind arrangements for transporting ice core samples.

REFERENCES

1) Mackay, J.R., 1971, The origin of massive ice bed, Western Arctic, Canada: Proc. 2nd Internat. Conf. on Permafrost, North American Contrib. Yaktsk, U.S.S.R., p.223-228.

2) French, H.M., 1976, The Periglacial Environment: Longman Group LTD London., 309pp.

3) Fujino, K. and Kato, K., 1978, Determination of oxygen concentration in the ground ice of a tundra area, Joint Studies on Physical and Biological

Environments in the Permafrost, Alaska and North Canada, July to August 1977: Inst. Low Temp. Sci. Hokkaido Univ., p.77-83.

4) Kato, K. and Fujino, K., 1981, Oxygen isotopic concentration of massive ice at Tuktoyaktuk, North Canada, Joint Studies on Physical and Biological Environments in the Permafrost, North Canada, July-August 1980 and February-March 1981: Inst. Low Temp. Sci. Hokkaido Univ., p.13-20.

5) Kinosita, S., 1978, An outline of research project. Joint Studies on Physical and Biological Environments in the Permafrost, Alaska and North Canada, July to August 1977: Inst. Low Temp. Sci. Hokkaido Univ., p.1-16.

6) Epistein, S. and Mayeda, T., 1953, Variation of O^{18} content of water from natural sources: Geochim. Cosmochim. Acta, 4, p.436-468.

7) Langway Jr., C.C., 1958, Bubble pressures in Greenland glacier ice: Symp. of Chamonix, Physics of the Movement of the Ice, p.16-24.

8) Macpherson, D.S. and Krouse, H.R., 1969, O^{18}/O^{16} ratios in snow and ice of the Hubbard and Kaskawulsh glacier: Ice Field Ranges Research Project. Sci. Results, American Geog. Soc. & Arctic Inst. of North America, Vol.1, p.63-73.

FIGURE 2 Stratigraphic diagrams of the entire length of a core obtained throughout the massive ice body.

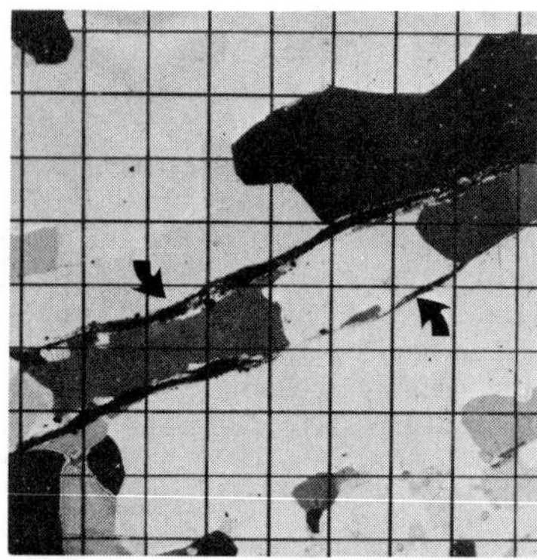

FIGURE 3
Soil layers where ice fablics of each adjacent stratum are clearly discontinuous.

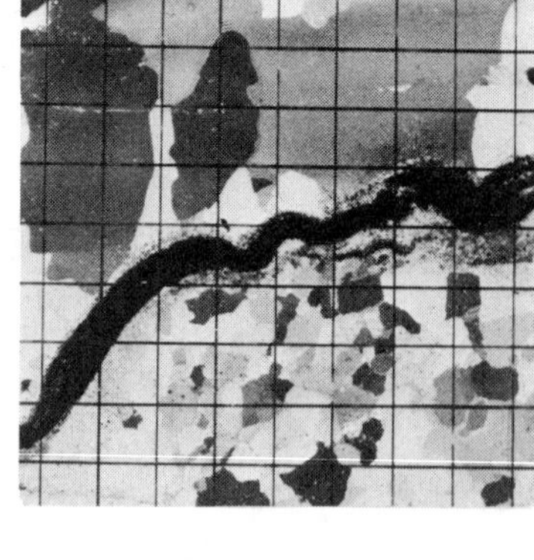

FIGURE 4
Soil layers looking like a sediment having a concave-convex surface plane.

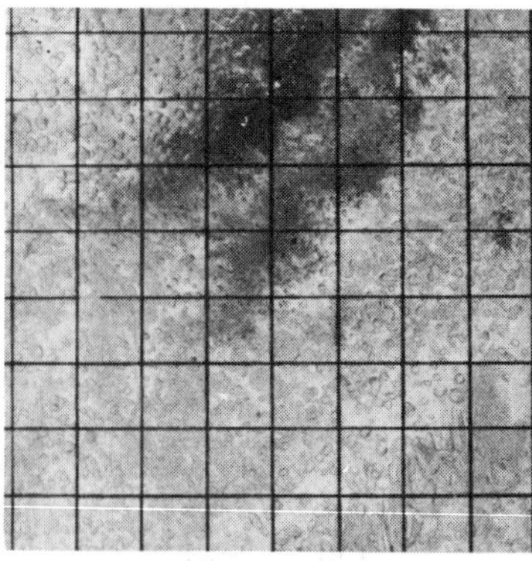

FIGURE 5
Soil particles looking like suspension.

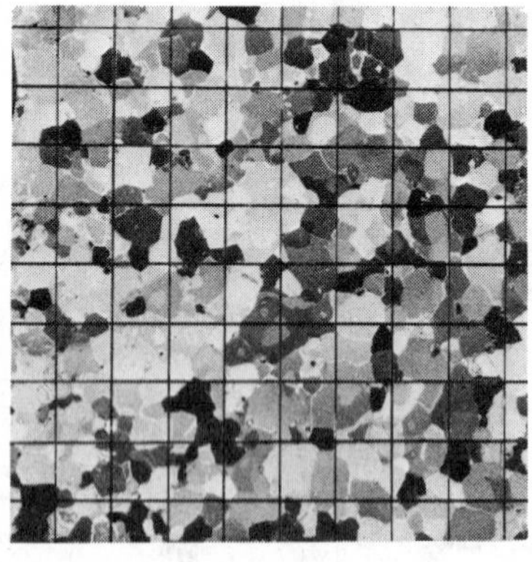

FIGURE 6
A layer of granular ice.

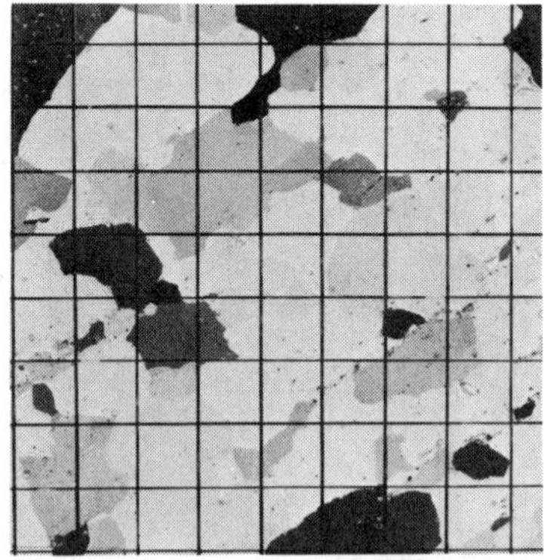

FIGURE 7
A layer of medium-sized ice crystals.

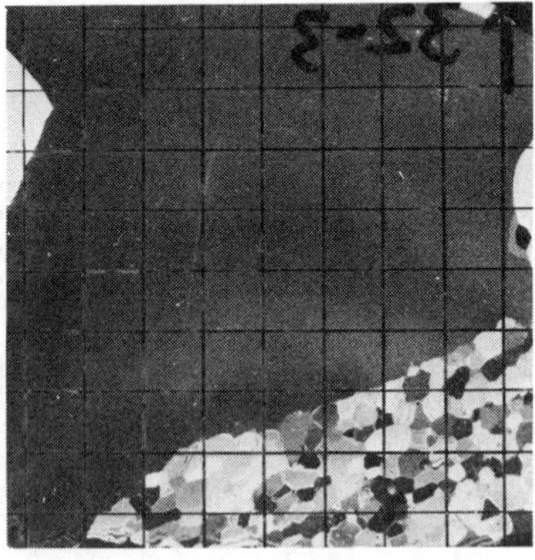

FIGURE 8
A layer of large ice crystals.

FIGURE 9
Vertical section of a bubble layer having various spherical shapes and a soil layer looking like sediment.

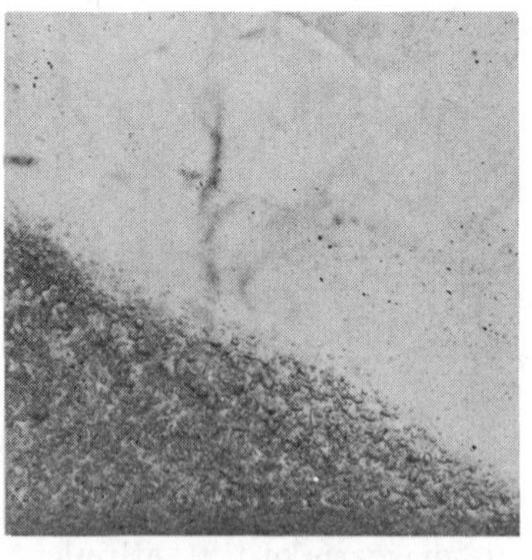

FIGURE 10
Vertical section of a connection between a bubble-free layer and a bubble-rich layer.

FIGURE 11
Vertical section of a mixed layer where elongated bubbles and spherical bubbles coexist.

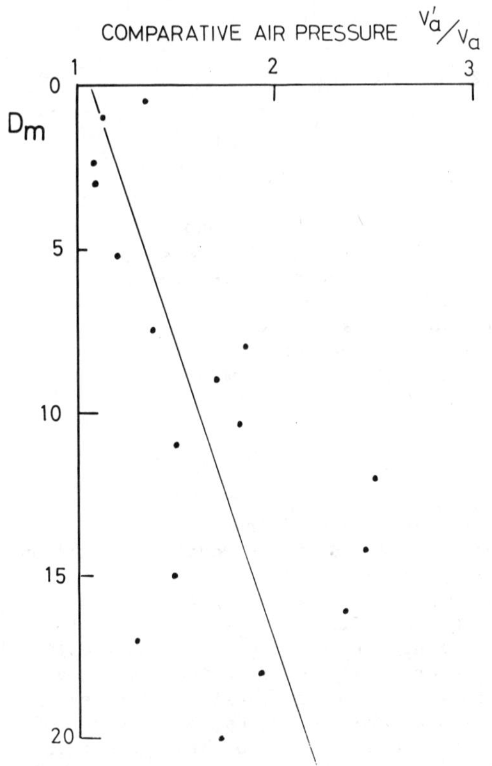

FIGURE 12 Profiles of comparative air pressure of bubbles against depth.

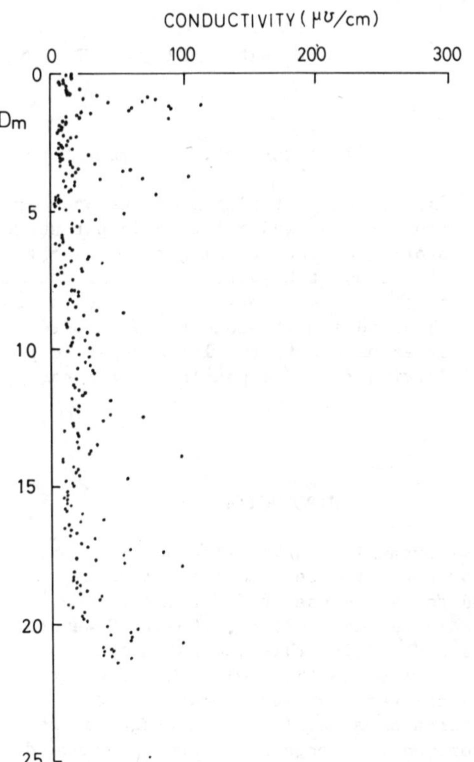

FIGURE 14 Profiles of conductivity against depth.

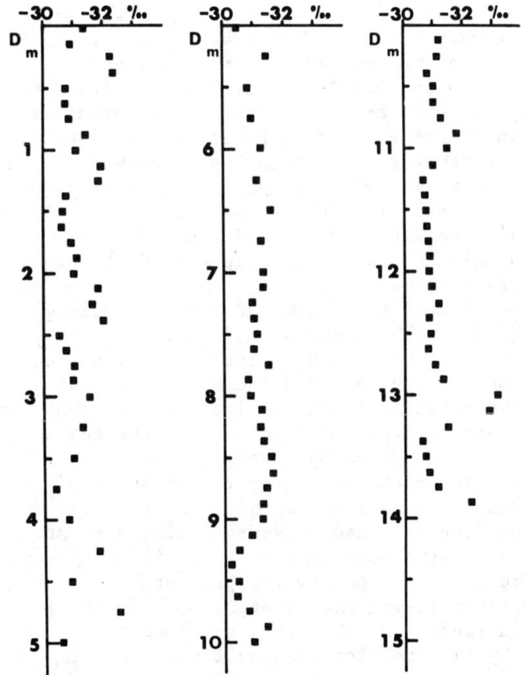

FIGURE 13 Profiles of $\delta^{18}O$ against depth.

THE PORE WATER PRESSURE PROFILE IN POROUS ROCKS DURING FREEZING

Masami Fukuda

Institute of Low Temperature Science, Hokkaido Univ., Sapporo, JAPAN

The frost shattering of rocks was explained by a frost heave mechanism with coupled flow of heat and moisture in porous rocks during freezing. The pore-water pressure profile of the freezing rock was obtained using tensiometers. Pore-water pressures of porous rocks during freezing were monitored by the transducers connected to tensiometers. Both the profiles of temperature and pore-water pressure were obtained simultaneously. From pore-water pressure measurements ahead of the freezing front, the large negative pore-water pressure is suggested as the major driving force responsible for water uptake in porous rocks.

INTRODUCTION

Volume expansion, which accompanies a phase change from water to ice, has traditionally been considered to be a cause of frost shattering of rocks in cold regions (Ollier, 1969). However, conditions under which this mechanism can exist are fairly limited in the field. Thus, other hypotheses are necessary to explain their disintegration as a result of freezing. Everett (1961) proposed the concept of water flow based upon the thermodynamics of water-ice equilibrium. Recently, the process of frost heaving of soils has been studied in detail, and an improved understanding of moisture migration and freezing processes is evolving at both the theoretical and experimental level. The present author has attempted to apply this concept of frost heaving in soils to the frost shattering of rocks. Freezing experiments have been conducted in an effort to verify the application of the concept.

EXPERIMENTS

Neogine Tertiary tuff from central Japan was used in the experiment. It has a high porosity (36%) and is frost susceptible. Column-shaped rock samples (5 cm in diameter and 15 cm in height) were prepared. A sample was first wetted so that the pores were fully saturated with water. Tensiometers were inserted at points 3, 5 and 7 cm from the surface of the rock column. A brass circulation plate through which refrigerant was circulated was placed on the top of the rock column to control the temperature. The rock column was mounted on a water reservoir and maintained hydraulic contact with the water in the reservoir. The temperature of the water in the reservoir was maintained at +2°C with a fluctuation of 0.1°C. Pore-water pressures were measured by pressure transducers, which were connected to the tensiometers by the tubes (Figure 1). The measurements had a resolution of 2 cm H_2O. Fine thermocouples were inserted at intervals of 1 cm in depth in the rock sample. Both read-outs of pore water pressure and temperature were collected by an automatic data logging system controlled by a small computer.

Two different rates of freezing were established under the different cooling conditions on the surface (-3 and -5°C).

At the beginning of each experiment the surface of the rock sample was supercooled, to initiate freezing. If the degree of supercooling exceeded -2°C or below, the freezing front moved down very rapidly. Such a transitional freezing process had to be avoided. Thus, a few drops of liquid nitrogen were applied to the rock surface to prevent supercooling at the beginning.

RESULTS AND DISCUSSION

Changes in pore-water pressure during freezing

More than 20 freezing experiments were conducted. Changes in pressure and temperature were measured every 5 minutes. The frost depths were estimated from the temperature profiles. The initial freezing temperature of pore water was estimated to be -0.08°C. A typical result of the experiments with the slower rate of freezing is shown in Figure 2. During freezing, the rate of advance of the freezing front was 2 mm/hour. The pore water pressure decreased as the freezing front penetrated into the rock. At a depth of 3 cm the pore water pressure decreased continuously. However, at the 7 cm depth no decreases were measured.

The case of the faster advancing freezing front (4 mm/hour) is shown in Figure 3. The pore-water pressure at the depth of 3 cm decreased for 5 hours, then ceased decreasing. At that time, the freezing front was located near the 3 cm depth. At the depth of 5 cm similar changes were noted. At both freezing rates pore-water pressures at locations nearer to the surface were lower than at locations farther away from the freezing front. These pressure differences at different depths suggest that water flows upward.

The result of pore-water pressure measurements during the freezing of Andesite is shown in Figure 4. This rock specimen showed no damage by the freezing-thawing test. In this case, no pressure changes in the rock specimen were observed. This means that there was no water migration in the rock specimen. The pore water pressure at a 3 cm depth increased after 13.8 hours had elapsed. This indicated that the water in the tensiometer had begun to freeze.

The changes in pore-water pressure with depth in the Neogine Tertiary tuff specimens are shown in Figure 5. A small pore water pressure gradient at the beginning of the experiment (0 hour) was due to both drying of the upper portion of the column and the gravitationl potential. Once the freezing started, an increase was observed in pore-water pressure difference between 3 and 5 cm depth. Along this gradient, water must migrate upward. The gradient between 3 and 5 cm in depth increased for 4 hours, but between 4 and 6 hours the increase in the gradient was very small. Additional experimental results, which also illustrate this behavior of the hydraulic gradient between the depth of 3 and 5 cm are shown in Figure 6. The hydraulic gradient increases with time. Each case in Figure 6 indicates that an increase in gradient ceases at a certain time. This time corresponds to the time at which the tensiometer fails due to freezing and depends on the rate of advance of the freezing front. Similar results in the case of soil freezing were reported by Fukuda and Luthin (1980). As the relationship between pore water pressure and water content could not be obtained, it was not possible to estimate the flux of water in a freezing rock from these data.

Estimation of pore water pressure near the freezing front

The hydrodynamic model for soil freezing assumes that the two transport equations, one for heat and the other one for water, are coupled by the relationship between pore water pressure and temperature below 0°C (Taylor and Luthin, 1978). Considering the similarity between the unsaturated soil and the rock, one may apply this model to the rock freezing process. The freezing point depression of pore water in a rock specimen was determined experimentally. The result is shown in Figure 7. The water content is expressed as the degree of water saturation in pores in the rock. The ordinate indicates the freezing point depression, which is the difference in the temperature of freezing of pore water in rocks and bulk water under atmospheric pressure. Thus, if the pores of the rock are not completely filled with water, that water freezes at temperatures below 0°C. Let us apply the Clausius-Clapeyron equation to this situation. If the content of free solutes is negligible, then:

$$dT = (T/L)Vwdpw - Vidpi)$$

where T = temperature, Kelvin
L = heat of fusion
pi, pw = pressure of ice and water, respectively
Vi, Vw = specific volume of ice and water, respectively

When there is no load on the rock, and the rock matrix does not constrain the ice, the pressure of the ice in the pores of the rock is equal to atmospheric pressure. Thus, the equation can be modified as follows (Offenbacher 1981):

$$\Delta T = 0.082p$$

where ΔT is the freezing point depression and p is the pore water pressure in bars. This large negative pore water pressure is the driving force for water to move to the freezing front. The equations given here are based upon the thermodynamics of an equilibrium water-ice situation and are assumed to be applicable although the actual freezing process is a nonequilibrium process.

The author attempted to estimate the pore water pressures near the freezing front from experimental results. The decrease in pore water pressure as a function of distance between the freezing front and the tensiomenters is shown in Figure 8. As the freezing front descends, the pore water pressure decreases gradually. The pore water pressure at the freezing front is calculated through extrapolation to be more negative than 1 m H_2O. This value was obtained by fitting simple regression curves to the data shown in Figure 8 for the three different tensiometer locations. If this value is substituted in the above equation, 0.08°C is obtained as a freezing point depression at the freezing front. This value corresponds to the actual temperature which was used to locate the freezing front.

Potential gradient near the freezing front

In order to estimate the flow rate of water to the freezing front through the unfrozen part of rocks, hydraulic gradients at various locations were calculated. Typical curves of potential gradients between 3 and 5 cm as a function of time are drawn in Figure 6. At the beginning of freezing, the hydraulic gradient increased very sharply. The hydraulic gradient continued to increase until the pore water pressure at the 3 cm depth reached a minimum; then the increase in the gradient ceased. The maximum hydraulic gradient which was observed between 3 and 5 cm and the mean pore-water pressure across this depth, were recorded for each experiment. All the data are plotted in Figure 9. A simple regression having an exponential form was performed. This curve gave the best fit, compared with other simple regressions. The following empirical relation was obtained between pore-water pressures and potential gradients in the unfrozen part of the rock during the freezing:

$$\partial\phi/\partial Z = A.\exp(B//\psi/)$$

where $\partial\phi/\partial Z$ = total hydraulic gradient, cm/cm
$/\psi/$ = absolute value of pore water pressure, cm H_2O
A, B = constants
Z = distance, cm

Evidence of water migration during freezing

The migration of water to the freezing front through unfrozen rocks was implied by the measurements of pore water pressures. However, the flux of water cannot be calculated from these results. The water flow in porous rocks takes place under the unsaturated condition, and the functional relation between the hydraulic conductivity of unsaturated rocks and the water content is not known. As an alternative the author attempted to measure the change in water content of samples which froze under conditions which were identical to those discussed above. At first, rock specimens with exactly the same shape as the ones used in the experiments previously mentioned, were prepared. The rock specimens were set as shown in Figure 1 without tensiometers in them. The top surface of specimens were exposed to -5°C. As the freezing front descended downward, the temperature profile of the specimen was measured once every hour and plotted. The position of the isothermal line of -0.082°C, which was estimated as the freezing front, was determined at given times. Once the freezing front reached specific locations the rock specimens were frozen at -30°C. Then the samples were cut into a discs 1 cm in thickness. The weight of each disk was measured before and after drying in the electric oven. The procedure was based upon previous studies involving freezing soils (Dirksen and Miller 1966).

The results of the experiments are shown in Figure 10. Both temperature and water content profiles are drawn. Arrows in the figure indicate the locations of the freezing front. The bulk density and porosity of these rock specimens were measured. Thus the water content on a weight base (weight of water/weight of dry rock) was converted into volumetric water content. The water content profile after a lapse of 18 hours indicates that the water content is low near the freezing front and high in the frozen part (upper portion). This discontinuity in the water content profile at the freezing front is due to the existence of large negative pore water pressures near the freezing front. After a lapse of 21 hours the freezing front descended, and the zone with a depleted water content also descended. The volumetric water content at a 4 cm depth after a lapse of 18 hours was 34.4 %, and increased to 38.14 % after 21 hours. The water content in the frozen part exceeded the saturation level. Water segregated as ice in the pores. It is suggested that the foregoing flow of water in porous rocks in the process of freezing may explain their frost shattering.

CONCLUSION

From pore-water pressure measurements ahead of the freezing front it is estimated that the negative pore water pressures at or near the freezing front exceeds 1 m H_2O. A steady-state freezing front will result in constant movement of water through porous rocks from the water table and will cause a continuous increase in the amount of segregated ice in pores. Such an excess of water causes the disintegration of rocks. Using a tuff rock with a porosity of 36%, the flux of water through the unfrozen part of each sample was illustrated while it was subjected to freezing. This process of water migration in porous rocks during the freezing is found to be similar to the frost heaving processes of soils.

REFERENCES

Dirksen, C. and Miller, R.D., 1966, Closed-system freezing of unsaturated soil: Soil Sci. Soc. Amer. Proc., V.30, p.168-173.

Everett, D.H. 1961. The thermodynamics of frost damage to porous solids: Transactions of the Farady Society, V.57, p.1541-1551.

Fukuda, M. and Luthin, J.N. 1980. Pore-water pressure profile of a freezing soil: Frost 1 Jord, Nr. 21, p.31-36.

Offenbacher, E.L. 1981. What is the temperature change in pressure melting? Cold Regions Science and Technology, V.4, p.155-156.

Ollier, C.D. 1969. Weathering: Edinburgh, Olver & Boyd.

Taylor, G.S. and Luthin, J.N. 1978. A model for coupled heat and moisture transfer during soil freezing: Canadian Geotechnical Journal, V.15, p.548-555.

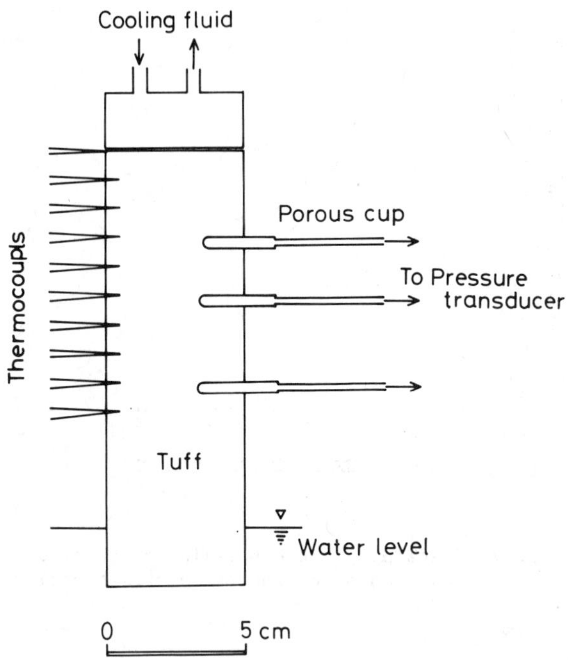

Fig. 1 Diagram of apparatus used to measure pressures in a freezing rock.

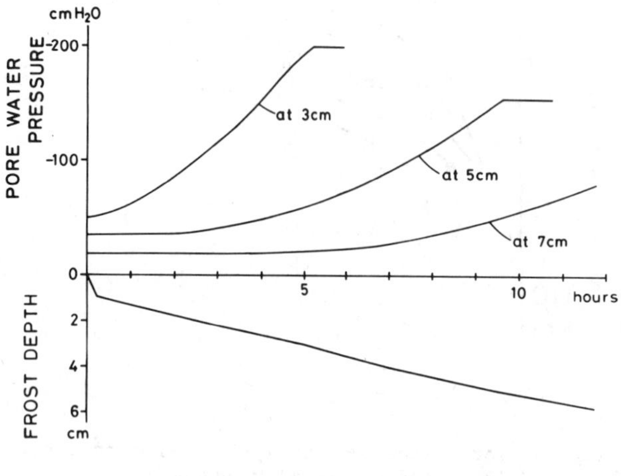

Fig. 3 Typical result of experiment with a high speed of freezing advancing.

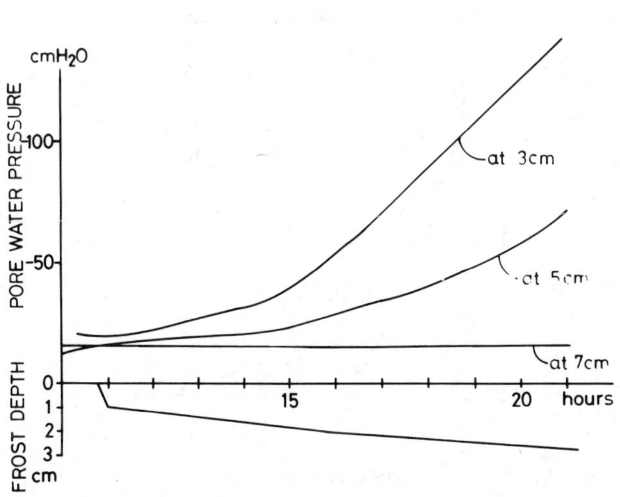

Fig. 2 Typical result of experiment with a low speed of freezing advancing.

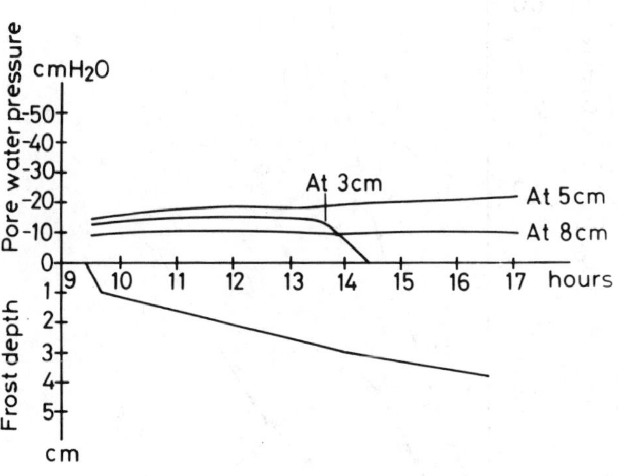

Fig. 4 Typical result of experiment using Andesite specimen.

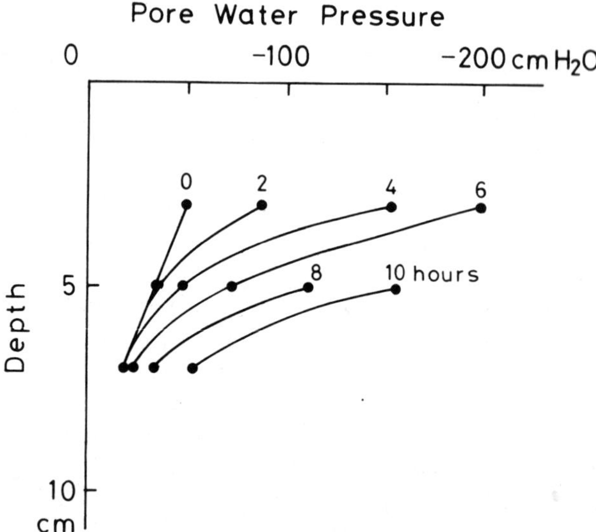

Fig. 5 Pore-water pressure profile for case of low speed of freezing advancing.

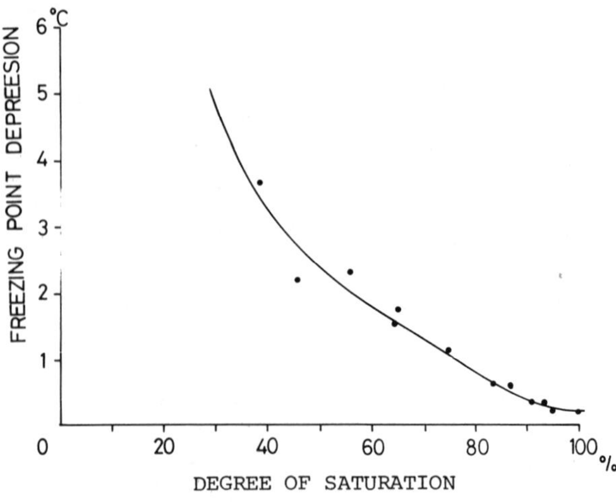

Fig. 7 Freezing point depression of tuff as a function of degree of water saturation.

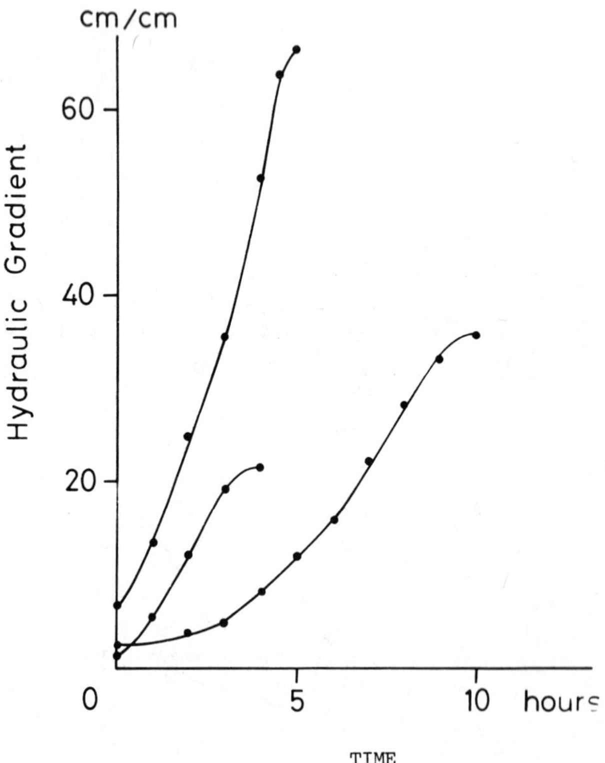

Fig. 6 Rate of change in hydraulic gradients.

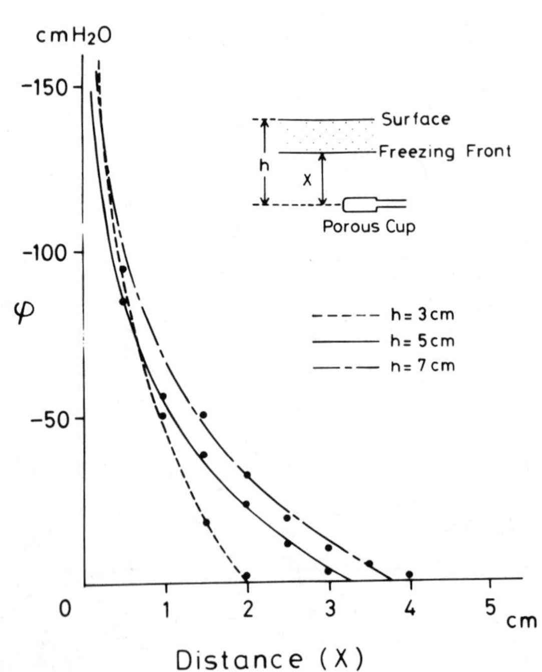

Fig. 8 Increase in pore-water pressure as a function of distance from freezing front.

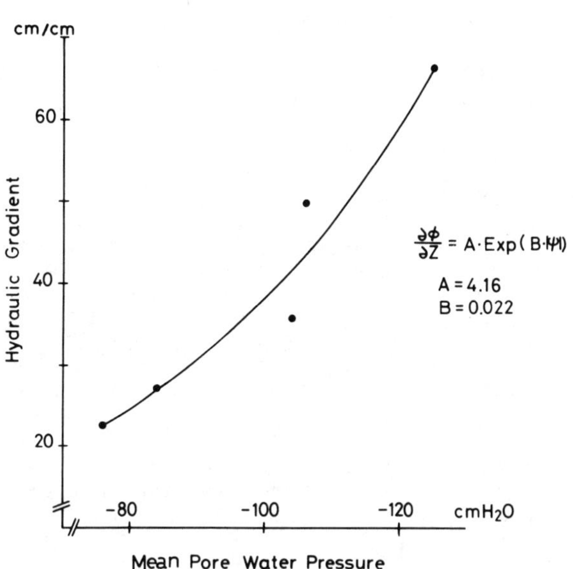

Fig. 9 Empirical relationship between pore-water pressure and hydraulic gradient.

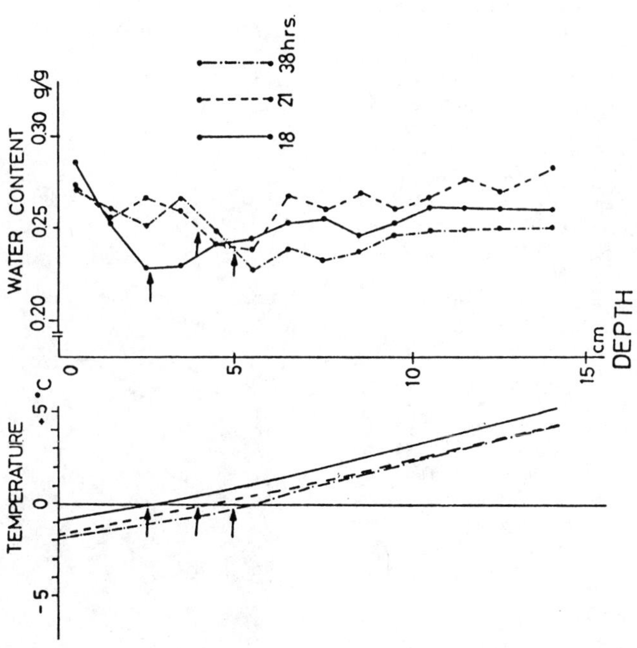

Fig. 10 - Simultaneous changes in profiles of temperature and water content during freezing porous rocks.

CONTROLS AND RATES OF MOVEMENT OF SOLIFLUCTION LOBES IN THE EASTERN SWISS ALPS

Martin W. Gamper

Department of Geography, University of Zürich, 8057 Zürich, Switzerland

Rates and processes of present-day movements of solifluction lobes in the Swiss National Park in places lacking permafrost are compared to Holocene movements established by radiocarbon dating of buried soils in solifluction lobes in similar localities. At a depth of 50 cm, corresponding to the depth of most of the buried soils, movement below vegetation was less than the limit of detection (<0.1 cm/yr). Under vegetation free surface, movements varied from 0.0-0.7 cm/yr (average 0.05 cm/yr). In contrast to present-day rates of movement at depth the rates of movement calculated using radiocarbon dates of buried soils along the length of solifluction lobes range from 0.5 to over 3.0 cm/yr in periods of peak activity during the last 5,000 years. The results of the study of present-day solifluction at the two test sites show that movement is controlled by vegetation cover and by the depth to which the soil freezes during the fall and winter. Thus the comparatively larger rates of movement in past periods of peak activity could be explained by a change in the climate with colder temperatures and winters with less snow cover, which would reduce vegetation cover and promote gelifluction.

INTRODUCTION

Movement rates of solifluction lobes during the Holocene can be reconstructed by radiocarbon dating of buried soils (Benedict 1966, 1976, Worsley and Harris 1974, Alexander and Price 1980, Gamper 1982). For a climatic interpretation of changes in movement rates, a comparison with present-day movement rates of solifluction lobes is required. These lobes have to be at the same altitude and over the same bedrock and have to face in the same directions as the lobes used to reconstruct past movement rates. The controls of present-day movements need also to be studied. Because mass movement by solifluction is slow but highly rate-variable, only a long term survey using numerous data points can provide a reasonable base for this comparison. Similarly, the data base for past movement rates has to be quite broad.

Measurements of present movement rates of solifluction lobes have been made primarily in arctic regions with permafrost (Washburn 1979, Harris 1981). Data for alpine regions and especially for the Alps themselves are rare and cover only short periods (Furrer 1972, Stocker 1973). Data for subsurface movements in the Alps are lacking.

The present study was initiated in 1975. Since then, surface and subsurface movements of solifluction lobes in the Swiss National Park (eastern Swiss Alps) have been continuously measured. Since 1978, soil and air temperatures have also been continuously recorded. The data including 1979 were published by Gamper (1981).

At the same time, over 120 radiocarbon dates of buried soils in solifluction lobes mainly in the eastern Swiss Alps have been made. These dates (Röthlisberger 1976, Schneebeli 1976, Beeler 1977, Furrer 1977, Gamper 1981, 1982) were used for this study together with some unpublished dates.

AREA SETTING

The Swiss National Park is located in the southeastern corner of Switzerland, about 200 km southeast of Zürich. The mountains have an average elevation of about 2,800 m. Bedrock is mainly Triassic dolomite. The timberline (larch and Swiss pine) occurs at about 2,100 m. Solifluction lobes are located on all slopes at an elevation of 2,200-2,800 m depending on the directions in which the slopes face. They are most common at the upper margin of closed vegetation cover.

FIGURE 1 Solifluction lobes in the Swiss National Park (test site 2). Surface movement and soil and air temperature measurements were carried out here from 1977 to present.

On the west slope of Munt Buffalora (test site 1) and on the southwest slope of Munt Chavagl (test site 2, Figure 1), two sites for measurements of solifluidal movements have been selected, both at an elevation of 2,400 m. The fronts and sides of the lobes are covered by vegetation. The backs of the lobes are mostly vegetation free. The soil is not perennially frozen. A detailed geomorphological study of those lobes has been made by Furrer et al. (1971).

On test site 2 (Figure 1), measurements have been carried out from 1977 to the present. The soil and air temperature measurements are also made in this area. Test site 1 was surveyed from 1975 to 1979. In the overlapping period, measured movement rates at both sites have been identical (Gamper 1981).

METHODS

Surface movements were measured using 102 wooden cones which are 10 cm in height and are mounted on 10 cm long wooden dowels which were inserted in the soil (Washburn 1960). In addition, 45 test pillars were inserted to a depth of 90 cm (Rudberg 1958) and 118 aluminum foils were inserted to a depth of 40 cm (Gamper 1981).

Thirty test pillars and 49 aluminum foils were used at the first test site, and all the other measurement devices were used at the second site. Measurements were made 2-5 times during the snow free season during July to October of each year, using a theodolite stationed on bedrock. The test pillars and aluminum foils at the first test site were excavated in 1979 to obtain subsurface movement rates (Gamper 1981). Measurements at the second test site are still in progress. All subsurface movement rates are therefore taken from the first test site and cover only the period 1975-1979.

Temperature measurements were started in September 1978 at one of the lobes at the second test site, using 22 temperature sensors (accurate to 0.1° Celsius). Air temperatures at 1, 50 and 150 cm above surface and soil temperatures from a depth of 5-100 cm were measured every 3 hours. Measurements were made on the back of a solifluction lobe in vegetation free area (Figure 1).

CONTROLS AND RATES OF SOLIFLUCTION MOVEMENT

The measurements show that detectable surface movement is restricted to the backs of solifluction lobes. The front and steep sides of the tongues which are covered by vegetation did not move downslope.

Average surface movement in vegetation free areas on the backs of the lobes was much higher than around the edges of the lobes, which are partly or completely covered by vegetation (Table 1). Solifluction was also restricted to the time during and after snow melting in early summer. During summer no detectable movement occurred.

TABLE 1 Average Surface Soil Movement Rates 1975-1982 at Test Sites 1 and 2 in the Swiss National Park (cm/yr).

Year (Oct. 1- Sept. 30.)	No. of Measured Points	Minimum Rate	Maximum Rate
1975/76	75	0.0	8.9
1976/77	75	0.0	11.0
1977/78	140	0.0	15.0
1978/79	155	0.0	40.3
1979/80	99	0.0	15.8
1980/81	175	0.0	48.2
1981/82	179	0.0	37.6

Year (Oct. 1- Sept. 30.)	Average Downslope Movement Rates			
	All Points	Veg.* Covered Areas	Partly Veg.* Covered Areas	Veg.* Free Areas
1975/76	1.5	0.2	1.6	2.9
1976/77	1.8	0.6	1.8	3.0
1977/78	5.1	1.4	5.2	7.4
1978/79	4.7	1.3	4.1	7.0
1979/80	3.8	2.4	3.3	4.8
1980/81	5.0	1.6	2.7	9.1
1981/82	6.5	2.4	4.8	11.1

*Veg. = Vegetation.

The values in Table 1 show remarkable differences in the average movement rates for different years. The changes in the distribution of slow (0-2 cm/yr), medium (2-6 cm/yr), and high (over 6 cm/yr) movement rates for these years are shown in Figure 2. In years with small average movements, during 1975/76 or 1979/1980, over 80% of the measured points showed only low or medium movements (Figure 2A). In years with high average movements, during 1977/1978 or 1981/1982, almost 50% of the points showed high movement rates. This increase in points showing high movement rates is most evident in the vegetation free areas (Figure 2D) and to a lesser extent in the partly vegetation free areas (Figure 2C) on the lobes. This indicates that vegetation cover is the most important factor in controlling solifluction.

Soil and air temperature measurements on a vegetation free back of a solifluction lobe show (Table 2) that in years with large surface movements (1978/1979, 1981/1982, and to a lesser extent 1980/1981) the soil froze to a depth of close to 1 m or more.

In early summer in those years, the soil thawed from the surface downward and stayed saturated with water as long as the ground in deeper horizons remained frozen. During this time (Table 2), gelifluction, saturated flow of debris, could occur. As soon as the impermeable frozen subsoil is completely thawed, the soil dries and becomes stable (Harris 1973). Therefore, during summer no detectable movement could be measured.

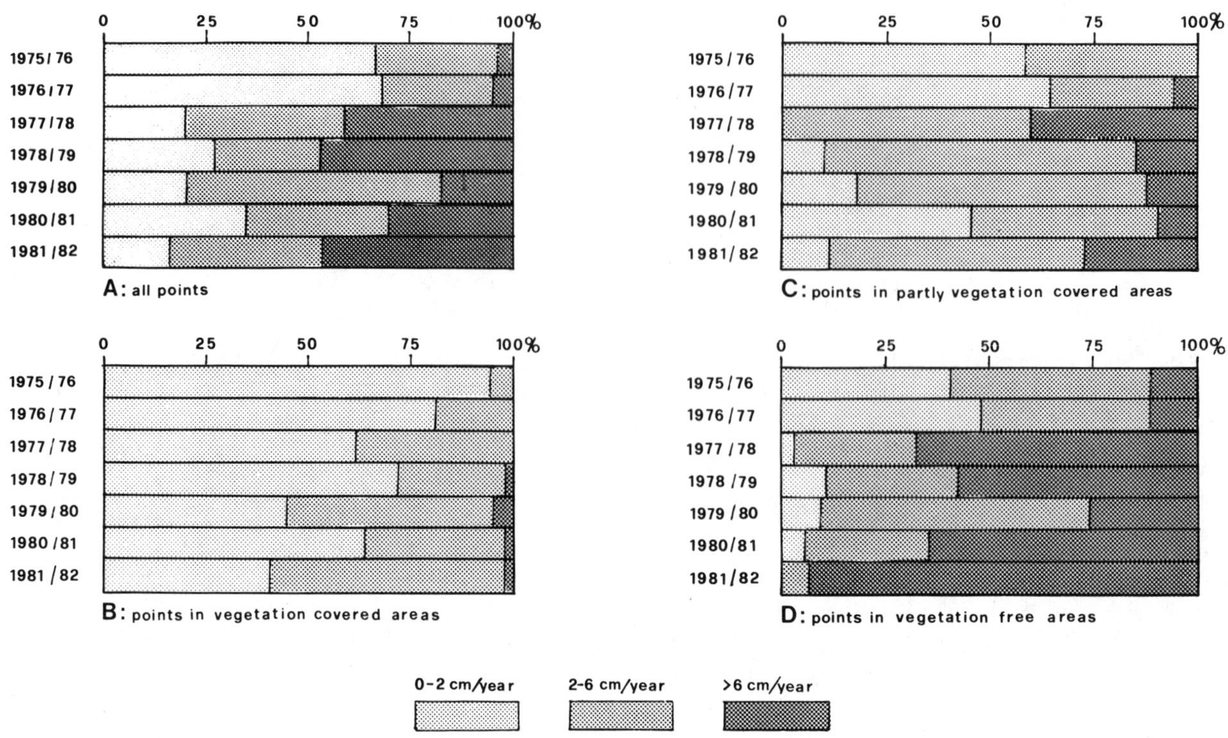

FIGURE 2 Histograms showing the relative distribution of surface movement rates on solifluction lobes at test sites 1 and 2. The figures show that solifluction is mainly controlled by the vegetation cover. Distribution and annual changes of movement rates are controlled by the timing and the duration of the snow cover in winter.

TABLE 2 Soil and Air Temperature Data From Test Site 2 in the Swiss National Park 1978-1982.

Year	Maximum Depth of Frozen Ground (cm)	Days With Gelifluction	Days With Continuous Snow Cover	First and Last Day With Continuous Snow Cover
1978/79	>100	22	186	11/25/78-5/30/79
1979/80	40	0	240	10/17/79-6/13/80
1980/81	85	14	217	10/14/80-5/18/81
1981/82	>100	18	171	12/ 1/81-5/20/82

Year	Mean Annual Air Temperature	Mean Winter Temperature (October 1-May 31) (°C)				
		150 cm Above Ground	2 cm Above Ground	5 cm Below Ground	50 cm Below Ground	80 cm Below Ground
1978/79	-0.5	-4.2	-1.7	-1.4	-0.4	0.2
1979/80	-1.2	-5.0	0.1	0.2	0.6	1.0
1980/81	-0.8	-4.3	-0.7	-0.3	0.2	0.7
1981/82	-0.6	-4.6	-0.9	-0.6	-0.1	0.4

In 1979/1980, a year with small movements, the soil only froze to a depth of 40 cm and thawed from below before the snow melted. Therefore the soil never became water saturated. Only frost creep occurred.

The frost depth itself is controlled by the timing of the onset of continuous snow cover and by the duration of winter snow cover (Table 2). The later in the year that snow falls, the deeper the soil freezes. The shorter the snow cover stays, the better are the chances for a long period of gelifluction. The mean annual air temperature (Table 2) therefore shows no correlation with the average rates of downslope movement (Table 1). The mean temperatures for the winter season (October to May) show that although the air temperature 150 cm above the surface was between -4.2° and -5.0°C, the temperature 2 cm above the ground (the temperature at the base of the snow cover) varied between +0.1° and -1.7°C, because of the presence or absence of an insulating snow cover during fall and early winter.

For a comparison of present-day movement rates with rates of movement during the Holocene as reconstructed by radiocarbon dating of buried soils in solifluction lobes, subsurface movement rates in a depth corresponding to the depth of the buried soils have to be used (Table 3). Present-day surface movement slows down to almost zero at the end of the lobes, because of increasing vegetation cover. The maximum depth of movement also decreases simultaneously. As a consequence, debris from the vegetation free areas above the lobes accumulates in the lobes. The lobes in toto do not move downslope.

TABLE 3 Average Subsurface Soil Movement Rates 1975-1979 at Test Site 1 in the Swiss National Park (cm/yr).

Depth (cm)	All Points	Vegetation Covered Areas	Partly Vegetion Covered Areas	Vegetation Free Areas
0	3.3	0.9	3.2	5.1
5	1.5	0.2	2.1	2.8
10	1.2	0.1	1.9	2.3
20	0.8	0.0	1.2	1.8
30	0.6	0.0	0.8	1.3
40	0.5	0.0	0.6	1.0
50	0.0	0.0	0.0	0.05

The rates of movement at a depth of 50 cm, corresponding to the mean thickness of the overburden on buried soils in solifluction lobes, was generally below the limit of detection (<1 mm/yr). Only 3 of 30 test pillars indicated movements of 0.3-0.7 cm/yr.

Altogether the results of the study of present-day solifluction at the test sites show that movement is generally controlled by the distribution of the vegetation cover. Annual changes of movement rates in vegetation free areas are controlled by the timing and the duration of snow cover in winter.

RATES OF MOVEMENT DURING THE HOLOCENE

Previous studies of buried soils in solifluction lobes in the Swiss Alps (Furrer 1977, Gamper 1981, Gamper and Suter 1982) identified during the past 5,000 years times of solifluction activity and times of dominant soil development on lobes. The results of these studies are based on radiocarbon dating of soils in interbedded solifluction lobes. More than 120 radiocarbon dates (see Furrer 1977, Gamper 1981, Röthlisberger 1976, Schneebeli 1976) have been used to determine these times of peak solifluction activity, shown in Table 4.

By dating buried soils at several places along the length of the covering lobe, rates of lobe advance can be determined by calculating linear regressions for the relationship between age of the soil and distance of the dated soil sample from the lower end of the lobe (Alexander and Price 1980). In some lobes these time-distance profiles indicate fluctuating rates of soil burial as in the example given in Figure 3. In such cases, rates of lobe advance for different times during the Holocene can be calculated. The radiocarbon dates for the buried soil (Figure 3) between 3.2 and 8.4 m distance from the lower end of the lobe are not significantly different in age. The rate of lobe advance of 3.0 cm/yr is therefore only a minimum estimate for the movement rate around 2000 years B.P. In addition, the radiocarbon dates in Figure 3 show only insignificant differences between the ages of the humic acids and the corresponding ages of the organic residues (not NaOH soluble fraction of soil organic matter). This indicates that the samples are not contaminated by recent organic carbon.

Movement rates were calculated for six more lobes at three locations in the eastern Swiss Alps (Swiss National Park, Albulapass 30 km west of National Park, and Berninapass 30 km southwest of National Park). Two lobes were active during two periods, so that altogether nine average values of past lobe advance were determined. All lobes are on west or north facing slopes, and in each case, bedrock is Triassic dolomite. For these calculations, 35 radiocarbon dates were

TABLE 4 Past Rates of Front Advance of Solifluction Lobes Calculated Using Radiocarbon Dates of Buried Soils.

Times of Peak Solifluidal Activity in Years Before Present*)	Average Movement Rates for Individual Lobes in the Eastern Swiss Alps (cm/yr)
4,700-4,300	0.6/>3.0
3,300-2,900	1.5
2,800-2,200	-
2,000-1,400	>3.0/3.0
1,100- 800	0.7/0.5/>2.0/2.7
350- 150	-

*After Gamper and Suter (1982).

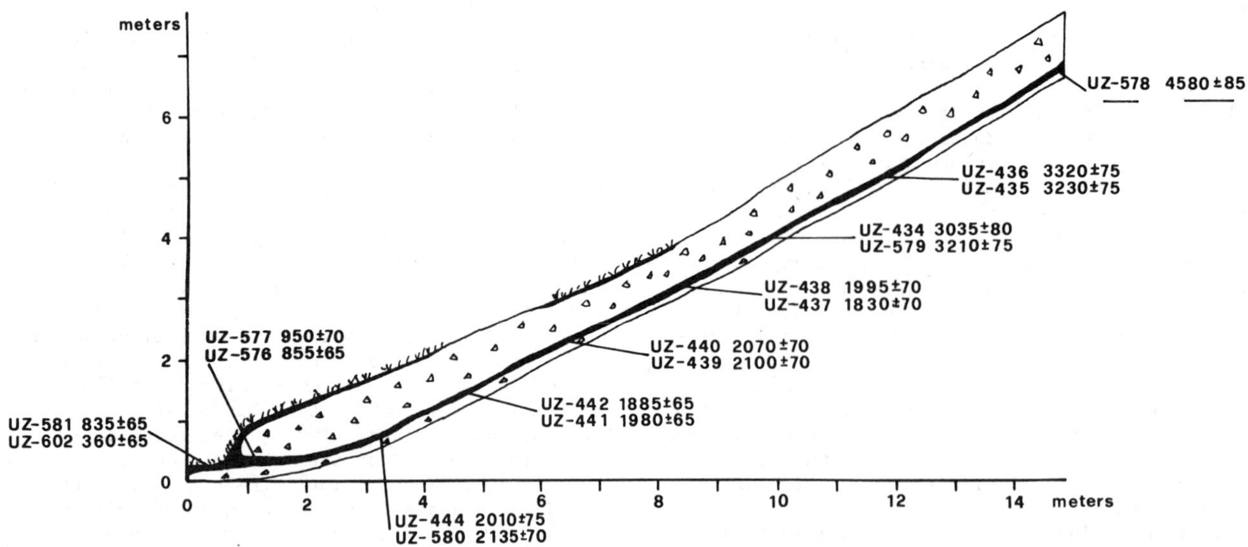

FIGURE 3 Profile of a solifluction lobe (Albulapass, eastern Swiss Alps). The upper radiocarbon date of each pair corresponds to the age of the humic acids, the lower to the age of the organic residues of the buried soil. The radiocarbon dates indicate high rates of lobe advance of 1.5 cm/yr around 3200 years B.P. and of >3.0 cm/yr around 2000 years B.P.

used (Furrer et al. 1975, Beeler 1977, Gamper 1981 and 1982). The rates as shown in Table 4 range from 0.5 to over 3.0 cm/yr. In two lobes, rates for times in between peak activity periods were also calculated (Gamper 1982). They range from 0.1-0.3 cm/yr. Compared to present-day subsurface movement rates in a depth of 50 cm, even these average rates seem to be high. For two periods of solifluction activity (2800-2200 and 350-150 years B.P.) rates could not be calculated, because of insufficient data.

CONCLUSIONS

The rates of movement in past periods of high solifluction activity can be compared with present-day movements. Solifluction lobes show no detectable downslope movement at their fronts. Movement rates at a depth of 50 cm, corresponding to the depth of buried soils, are below the limit of detection (<0.1 cm/yr) even under vegetation free areas.

The high rates of front advance (0.5 to more than 3.0 cm/yr) in former periods of greater solifluction activity may therefore be related to a change in paleoclimate. Present-day solifluction rates in the studied areas with comparable climate are mainly controlled by the vegetation cover and in vegetation free areas by the depth of freezing as influenced by the timing and duration of the snow cover in winter. Former higher movement rates and an advance of lobes could be explained by winters with less snow cover and colder temperatures. Both would reduce vegetation cover and promote solifluction by increasing the depth of freezing and therefore the duration and rates of gelifluction in early summer.

ACKNOWLEDGMENTS

I thank G. Furrer, B. Gamper, and G. Kasper for their help in the field and many discussions. Fieldwork was supported by grants from the Swiss National Science Foundation and the Kommission der Schweizerischen Naturforschenden Gesellschaft zur wissenschaftlichen Erforschung des Nationalparks. This paper was written while I was a research associate at Indiana University sponsored by the Swiss National Science Foundation. R. V. Ruhe critically reviewed the manuscript and made helpful suggestions. S. J. Wintsch and R. P. Wintsch helped prepare the manuscript.

REFERENCES

Alexander, C. S. and Price, L. W., 1980, Radiocarbon dating of the rate of movement of two solifluction lobes in the Ruby Range, Yukon Territory: Quaternary Research, v. 13, p. 365-379.

Beeler, F., 1977, Geomorphologische Untersuchungen am Spät- und Postglazial im Schweizerischen Nationalpark und im Berninapassgebiet (Südrätische Alpen): Ergebnisse der wissenschaftlichen Untersuchungen im Schweizerischen Nationalpark, v. 15, p. 131-276.

Benedict, J. B., 1966, Radiocarbon dates from a stone-banked terrace in the Colorado Rocky Mountains, USA: Geografiska Annaler, v. 48A, p. 24-31.

Benedict, J. B., 1976, Frost creep and gelifluction features: A review: Quaternary Research v. 6, p. 55-76.

Furrer, G., 1972, Bewegungsmessungen auf Solifluktionsdecken: Zeitschrift für Geomorphologie N. F., Supplementband 13, p. 87-101.

Furrer, G., 1977, Klimaschwankungen im Postglazial im Spiegel fossiler Böden: Ein Versuch im Schweizerischen Nationalpark, in Frenzel, B., ed., Dendrochronologie und postglaziale Klimaschwankungen in Europa: Erdwissenschaftliche Forschung, v. 13, p. 267-270.

Furrer, G., Bachmann, F., and Fitze, P., 1971, Erdströme als Formelemente von Solifluktionsdecken im Raum Munt Chavagl/Munt Buffalora (Schweizerischer Nationalpark): Ergebnisse der wissenschaftlichen Untersuchungen im Schweizerischen Nationalpark, v. 11, p. 189-269.

Furrer, G., Leuzinger, H., and Ammann, K., 1975, Klimaschwankungen während des alpinen Postglazials im Spiegel fossiler Böden: Vierteljahresschrift der Naturforschenden Gesellschaft in Zürich, v. 120, p. 15-31.

Gamper, M., 1981, Heutige Solifluktionsbeträge von Erdströmen und klimamorphologische Interpretation fossiler Böden: Ergebnisse der wissenschaftlichen Untersuchungen im Schweizerischen Nationalpark, v. 15, p. 355-443.

Gamper, M., 1982, Postglaziale Solifluktionsphasen am Albulapass (östliche Schweizer Alpen): Physische Geographie, v. 1, p. 171-186.

Gamper, M., and Suter, J., 1982, Postglaziale Klimageschichte der Schweizer Alpen: Geographica Helvetica, v. 37, p. 105-114.

Harris, C., 1973, Some factors affecting the rates and processes of periglacial mass movements: Geografiska Annaler, v. 55A, p. 24-28.

Harris, C., 1981, Periglacial mass-wasting: A review of research, British Geomorphological Research Group, Research Monograph Series, v. 4.

Röthlisberger, F., 1976, Gletscher- und Klimaschwankungen im Raum Zermatt, Ferpècle und Arolla: Die Alpen, v. 52, p. 59-152.

Rudberg, S., 1958, Some observations concerning mass movements on slopes in Sweden: Geologiska föreningens i Stockholm forhandlingar, v. 80, p. 114-125.

Schneebeli, W., 1976, Untersuchungen von Gletscherschwankungen im Val de Bagnes: Die Alpen, v. 52, p. 5-57.

Stocker, E., 1973, Bewegungsmessungen und Studien an Schrägterrassen an einem Hangausschnitt in der Kreuzeckgruppe (Kärnten), in Beiträge zur Klimatologie, Meteorologie und Klimamorphologie, Festschrift für Hanns Tollner zum 70. Geburtstag: Arbeiten aus dem Geographischen Institut der Universität Salzburg, v. 3, p. 193-203.

Washburn, A. L., 1960, Instrumentation for mass-wasting and patterned-ground studies in northeast Greenland: Biuletyn peryglacjalny, v. 8, p. 59-64.

Washburn, A. L., 1979, Geocryology: London, Edward Arnold.

Worsley, P., and Harris, C., 1974, Evidence for neoglacial solifluction at Okstindan, North Norway: Arctic, v. 27, p. 128-144.

GERMINATION CHARACTERISTICS OF ARCTIC PLANTS

B. L. Gartner

Institute of Arctic Biology,
University of Alaska, Fairbanks, Alaska USA

Germination traits of many arctic plants are similar to those of Eriophorum vaginatum. Eriophorum seeds have the following characteristics: they are wind-dispersed, have weakly developed or non-existent dormancy mechanisms, show optimal germination at 20-30°C when tested at constant temperatures, germinate over a wide range of temperatures in the light, germinate only at greater than 15-22°C in the dark, and remain dormant while buried in organic soil to depths of 0-30 cm. These characteristics probably contribute to the colonizing ability of many arctic species.

The conditions controlling seed germination of various species are among the many factors that determine the composition and relative abundance of species in plant communities. Where and when a seed can germinate determine where and when a seedling will have a chance to establish, which in turn determine where and when a plant will have a chance to produce seeds. Because all species in one ecosystem are faced with the same environment, one might expect the successful species to share some characteristics. For example, all arctic tundra plants have low stature. In this paper I analyze the germination traits that are common in arctic plants.

The "weediness" of arctic vegetation and the ability of many tundra species to colonize disturbances have long been recognized (Griggs 1934, Polunin 1934, Savile 1960). Because landscape disturbances are common in arctic areas, I examined whether the seed germination traits may be adaptive for invading disturbed areas. If so, with further research into arctic seed and seedling biology, more options will become available for revegetating disturbed arctic sites.

Eriophorum vaginatum is mentioned throughout this paper, not because it is necessarily typical of arctic plants, but because its seed and adult biology has been studied from several sites around the world, it is present in numerous vegetation assemblages, and it appears as a colonizer of many kinds of sites.

SEED SIZE AND DISPERSAL

Many northern tundra plants produce small, light, wind-dispersed seeds (Porsild 1951, Savile 1972, Densmore 1979, Chester and Shaver 1982). For example, three native grasses used for revegetation (Calamagrostis canadensis, Arctagrostis latifolia, and Poa glauca) have extremely small seeds, averaging 0.12, 0.25, and 0.34 mg respectively (Mitchell 1979). In a study of 54 Alaskan woody species, a higher percentage of the tundra than taiga species produced small or wind-dispersed seeds (Densmore 1979). Among the eight species making up the majority of vascular plant cover (Chester and Shaver 1982), the average seed weight ranged from 0.02 mg for Ledum palustre to 8.0 mg for Rubus chamaemorus. In between these values were 0.2 mg for Betula nana and Vaccinium vitis-idaea, 0.5 mg for Eriophorum vaginatum, and 1.0 mg for Carex bigelowii and Empetrum nigrum.

Seed dispersal in arctic plants relies on a number of mechanisms (Savile 1972). Plumed seeds and fruits are common and are found conspicuously in the genera Eriophorum, Salix, Dryas, Epilobium, Erigeron, and Taraxacum. Small, light seeds are found in Diapensia, Pyrola, some Ericaceae and some of the monocotyledonous species. Fleshy animal-dispersed fruits are also found, including Rubus chamaemorus, Arctostaphylos alpina, A. rubra, Vaccinium vitis-idaea, and V. uliginosum. Winged seeds and fruits, adapted to wind travel, are scarce, but are found in Betula, Rumex, and Oxyria digyna. Hooked fruits are almost absent in the arctic. Dispersal by wind is probably extremely important for arctic seed (Savile 1972). In the winter, small seeds can travel along with blowing snow over the hard-packed snow base. The relatively poor development in the arctic of temperate zone wind-dispersal devices such as wings or plumes may result from their lack of value for arctic winter dispersal (Savile 1972).

GERMINATION

Enforced Dormancy

Enforced or easily broken intrinsic dormancy mechanisms more commonly control germination of arctic plants than do complex intrinsic dormancy mechanisms (Billings and Mooney 1968). Enforced seed dormancy as defined by Harper (1977) is an inability to germinate due to an environmental restraint--shortage of water, low temperature, poor aeration, and so forth. Intrinsic dormancy includes the following conditions (Harper 1977): (1) seeds leaving the parent plant in a viable state but prevented from germinating when exposed to moist aerated conditions by some property of the embryo, endosperm, or maternal structures, and (2) seeds acquiring a condition of inability to germinate caused by some experience after ripening.

Enforced dormancy is certainly in effect in the winter, but may be in effect during the growing season as well. In a study of 54 species of woody tundra and taiga plants in Alaska, Densmore (1979) found that germination of all tundra plants that disperse their seeds in the fall was delayed until the following growing season by some temperature-controlled mechanism. Enforced dormancy appears to operate commonly in alpine species as well. Amen (1966) found no intrinsic dormancy mechanisms for 60% of 62 species of alpine plants; the remaining 40% belonged primarily to only three genera, Carex, Trifolium, and Salix.

The optimal temperatures for germination of many arctic species are high (20-30°C), so dormancy in these species may be enforced throughout most of the year and in most microsites. Papaver radicatum, Saxifraga cernua, and Oxyria digyna require high temperatures to germinate (Bell and Bliss 1980), as does Eriophorum vaginatum (Wein and MacLean 1973, Gartner 1982). For Eriophorum, the optimal temperature varied with population. For three Alaskan and Canadian sites the optimal temperature was about 30°C, but for seed from Scotland the optimal temperature was about 20°C (Wein and MacLean 1973). However, all these tests were made at constant temperatures, and may not reflect germination characteristics under field conditions, where temperature, moisture, and light fluctuate, both during a seed's dormant period and during germination.

Germination in seeds from some populations of Eriophorum vaginatum may be controlled by enforced dormancy only, whereas in other populations there are intrinsic mechanisms. For Eriophorum seeds from three Canadian sites, viability was highest after dispersal and declined progressively when tested 1, 6, 16, and 19 months later (Wein and MacLean 1973). The viability of one seed lot was reduced by 50% after the first six months of storage at room temperature. In contrast, most seeds from Kuparuk Ridge required a period of continued maturation before they germinated (Gartner et al. 1983). This requirement for additional maturation time was related to color: light brown seeds germinated more readily immediately after dispersal than did black seeds, but only black seeds germinated in experiments with 5-year-old seed (Gartner et al. 1983).

Some species, including summer-dispersing willows, germinate immediately after dispersal and lose viability quickly (Bliss 1958, Densmore 1979). For these species, high temperature thresholds restricting germination would only lessen the overall chances of seed germination.

Intrinsic Dormancy

Intrinsic dormancy mechanisms in arctic plants commonly consist of elapsed time, chilling, scarification, or a combination of temperature and light (Billings and Mooney 1968). Both chilling and scarification are effected through elapsed time, the former during winter and the latter probably through freeze-thaw activity during the spring and/or fall. The importance of elapsed time was shown above for Eriophorum, and reviewed for several alpine species in Amen (1966). Two Trifolium, two Carex, and one Luzula species are among the alpine species for which scarification is known to be a germination requirement (Amen 1966). Species of Leguminoseae are among the arctic species known to require scarification (Klebesadel 1971).

In many arctic species, light is required for or greatly enhances germination. The presence of light cues seeds to the fact that they are uncovered, and thus may be located in a site where (1) they will have little competition from established vegetation or (2) cotyledons will be able to reach above the surface to receive light. The latter should be particularly important for species bearing small seeds, which have small energy reserves and thus limited ability to reach the surface from depths. In Densmore's study (1979), nine species having extremely light-weight seeds, $\leq$0.15 mg/seed, required light for germination, regardless of temperature. Overall, 23 of the 52 species she examined required light to germinate, and 26 of the remaining 29 species were in one genus, Salix. Eriophorum vaginatum seeds do not germinate in the dark at temperatures below 15°-22°C (Bliss 1958, Wein and MacLean 1973, Gartner 1982) (Figure 1). These data suggest that in the field where seeds are buried, such as by litter, mosses, or lichens, or where seeds are dispersed to microsites where the temperatures are low, the seed may become dormant and incorporated into the buried seed pool.

Bliss (1958) found that a higher percentage of species characteristic of arctic river bars germinated in both light and dark at about 22°C than did those common to arctic tundra soils with a shallow active layer. This is in keeping with the hypothesis that species characteristic of upland sites may require light to cue germination, but it suggests a separate hypothesis for river bar sites. At these sites, which are subject to frequent deposition by flooding, seeds may germinate

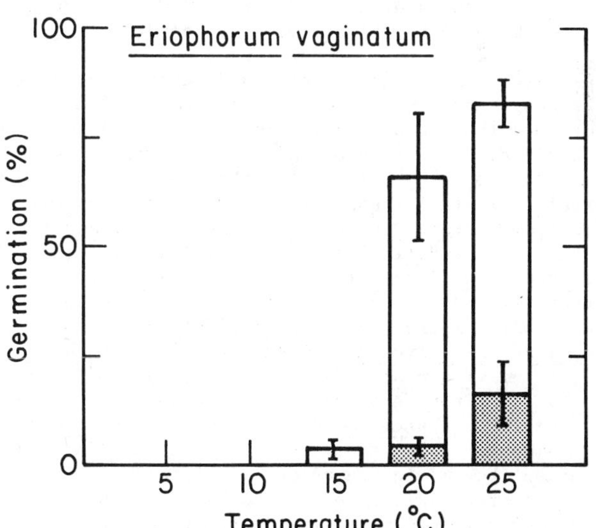

FIGURE 1 Germination of filled Eriophorum vaginatum seeds under controlled light and temperature conditions after 22 days. Mean ± SE (n = 4 groups of 50 seeds). Light treatments consisted of 20 hr light/4 hr dark or 24 hr dark (shaded bars). All treatments were tested at five temperatures. Seeds were stored at 5°C in the dark for 5 yr before start of the experiment (from Gartner 1982).

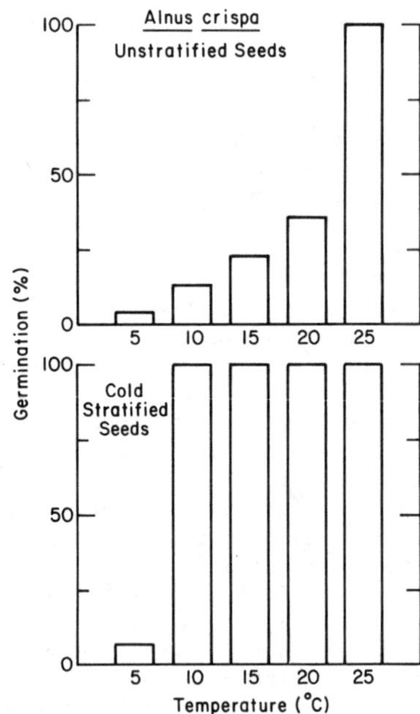

FIGURE 2 Germination of filled *Alnus crispa* seeds in the light (13 hr light/11 hr dark) under controlled temperature conditions for 21 days. Cold-stratified seeds were incubated in the light after 72 days of cold stratification in the light (from Densmore 1979).

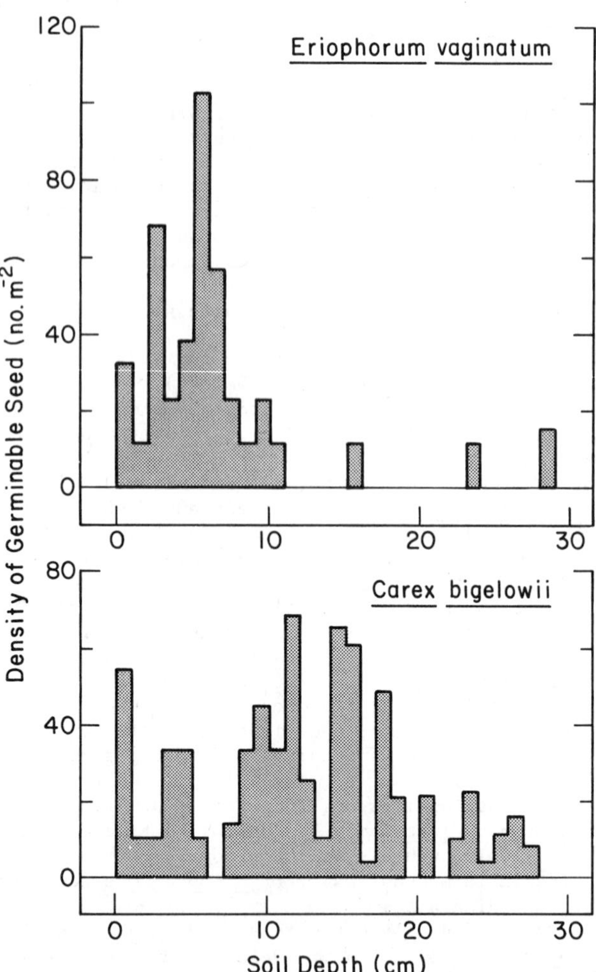

FIGURE 3 Depth profile of germinable buried seed in tussock tundra, calculated from depth profile of seeds in cores from four microhabitats weighted by the relative area of each microhabitat. A total of 118 seeds germinated from 45 cores having a total tundra surface area of 0.14 m^2 (from Gartner et al. 1983).

regardless of the light regime if other physical conditions are met.

Chilling the seeds prior to germination greatly increases the proportion that germinate at low constant temperatures. Densmore (1979) conducted tests on cold-stratified (wet and then chilled) and unstratified seeds of woody tundra and taiga species. Cold stratification substantially lowered the temperature requirements (Figure 2) and broke dormancy in most of the species studied. After cold stratification, seeds of some species were able to germinate at 5°C, a condition that would occur soon after snowmelt. These included the arctic species *Empetrum nigrum*, *Potentilla fruticosa*, *Spiraea beauverdiana*, *Salix glauca*, *S. arctica*, *S. phlebophylla*, *Diapensia lapponica*, and *Dryas octopetala*. A second group could germinate when mean soil temperatures reached 10°C: *Alnus crispa*, *Chamaedaphne calyculata*, *Oxycoccus microcarpus*, *Vaccinium uliginosum*, *Cassiope tetragona*, and *Ledum palustre*. Other species would require temperatures around 15°C: *Vaccinium vitis-idaea* and *Loiseleuria procumbens*. Cold-stratification tests are important because they mimic the effect of winter-chilling on seeds that are dispersed in the late fall but have enforced dormancy mechanisms that prevent fall germination.

BURIED SEED

Buried seed pools exist in all four Alaskan tundra sites where they have been sought. The ages of buried seeds are unknown, but longevity may be greater in cold permafrost soils than in soils of the temperate zone. At three of the sites, sedges were the predominant members: *Carex bigelowii*, *C.* cf. *aquatilis*, and *Eriophorum vaginatum* (McGraw 1980, Gartner et al. 1983, Roach 1983). Buried dicotyledonous seeds were common at two of the tussock tundra sites (McGraw 1980, Roach 1983), including *Ledum palustre*, *Saxifraga punctata*, *Polygonum viviparum*, *Empetrum nigrum*, and *Betula* spp. In a *Carex aquatilis* meadow on the northern coastal plain of Alaska the most abundant buried seed belonged to the dicot *Chrysosplenium tetrandum* (Leck 1980).

The distribution of different kinds of seeds varies with depth. Most dicotyledonous seeds were in the top 5-10 cm of soil (Leck 1980, McGraw 1980, Gartner 1982, Roach 1983), whereas sedges were also abundant down to 30 cm (Gartner et al. 1983). At Kuparuk Ridge, *Eriophorum* made up 38% of the germinating seeds and was found primarily throughout the top 11 cm of soil (Figure 3) (Gartner et al. 1983). This distribution is in contrast to *Carex* bigelowii seeds (58% of all germinating seeds), which were common at depths of 0-29 cm. In addition, all of the buried seeds were found in the organic soil layers. There were no seeds buried in the mineral soil on this upland slope site, even though the mineral soil was at the surface in frost boils. This suggests that they were buried through deposition and/or growth of organic matter on top of the seeds.

DISCUSSION

In some arctic communities there is a high incidence of natural disturbance, caused mostly by frost activity. Because of this disturbance, different parts of the communities are simultaneously in states of disturbance or regeneration such that a mature climax assemblage cannot be identified (Churchill and Hanson 1958). Fires, floods, animal activity, and other disturbances are probably frequent enough in many communities to create bare areas available for colonization on a time scale that allows perpetuation of certain tundra species by seed.

The germination traits of arctic plants may be adaptive for colonization of unvegetated or sparsely vegetated sites. Densmore (1979) concluded that the germination patterns of arctic woody plants are often the same as for the temperate species of a genus and that these patterns were apparently selected in ancestral species. Nonetheless, a discussion on the ways in which these adapted or pre-adapted traits may benefit a species is still valid.

The presence of seeds at a site, whether due to dispersal or storage in the soil, is crucial for a species' colonization. Thus wind-dispersal, a very effective mechanism, and consignment of some seeds to the buried seed pool are important characteristics of colonizers. The same concept is true in the temperate zone, where weedy plants typically have small, wind-dispersed seeds. Generally, most of the seed in buried seed pools in the temperate zone are of colonizing species, even in mature forests and mature pastures (Chippindale and Milton 1934, Sauer and Struik 1964, Livingston and Allessio 1968, Whipple 1977).

By having no dormancy mechanisms or very simple ones, seeds can germinate whenever basic requirements of temperature, moisture, and light are met. The requirement for high temperatures prevents germination in sites that may be temporally or spatially unsuitable for seedling establishment. Seeds with short longevity, however, would gain nothing from high temperature requirements. The requirement for light, like temperature requirements, helps cue a plant as to the suitability of its local environment. This trait has also been shown for colonizers in the temperate zone (Sauer and Struik 1964, Grime and Jarvis 1975).

Cold-stratification alters a seed's germination requirements so that it can have one set of requirements the year it is produced and a second set the following year. Environmental conditions that are correlated with germination success in the fall undoubtedly differ from those correlated with germination success in the spring or summer, so this adaptation can add considerably to the species' germination flexibility.

The general seed traits of arctic plants are similar to those of colonizers elsewhere. Grime (1981) distinguished four categories of colonizing species on the basis of their seed size and shape and their germination requirements (Table 1). The seed sizes range from very small to relatively large. Buried seed banks characterize two of the categories, A and B. Enforced and intrinsic dormancy mechanisms are the same as those listed for tundra plants. It is not surprising that arctic seeds have germination traits similar to those of colonizers elsewhere, for arctic disturbances are probably not uncommon over the long life spans of arctic plants.

TABLE 1 Common characteristics of seed morphology and dormancy for plants that colonize disturbed areas (from Grime 1981).

Group	Seed morphology	Buried seed	Dormancies and germination requirements
A	Small	Yes	No after-ripening, no innate dormancy. Germination in light over broad temperature range. Germination in dark with diurnal temperature fluctuations.
B	Very small	Yes	No dormancy. Germination in light at high or fluctating temperatures. Germination in dark rare.
C	Small, or large with elongated structure	No	No after-ripening, no innate dormancy. Germination in light or dark over a broad temperature range.
D	Relatively large	No	After cold-stratification, germination in the light or dark over a broad temperature range.

CONCLUSIONS

Seeds of Eriophorum vaginatum and many other arctic plants have the following traits:

(1) Seeds are wind-dispersed.

(2) Intrinsic dormancy is non-existent or weakly developed. Seeds from some populations are viable immediately after dispersal, whereas only a small proportion of the seeds from other populations can germinate immediately after dispersal.

(3) The optimal constant temperature for germination is variable from population to population, and is between 20° and 30°C. Some seeds germinate at lower above-freezing temperatures in the light.

(4) In the dark, seeds do not germinate at temperatures below 15° to 22°C.

(5) Some viable seeds are stored in the upper layers of soil. They have been found from 0 to 30 cm deep, but most seeds are probably in the top 10-15 cm. No viable seeds are in the mineral soil.

The data on arctic seeds and the syntheses relating the ecology of species to their germination traits indicate that arctic species have traits similar to those of colonizers of the temperate zone.

ACKNOWLEDGMENT

The research was supported by the U.S. Army Research Office.

REFERENCES

Amen, R. D., 1966, The extent and role of seed dormancy in alpine plants: Quarterly Review of Biology, v. 41, p. 271-281.

Bell, K. C., and Bliss, L. V., 1980, Plant reproduction in a high arctic environment: Arctic and Alpine Research, v. 12, p. 1-10.

Billings, W.D., and Mooney, H.A., 1968, The ecology of arctic and alpine plants: Cambridge Philosophical Society Biological Reviews, v. 43, p. 481-529.

Bliss, L.C., 1958, Seed germination in arctic and alpine species: Arctic, v. 11, p. 180-188.

Chester, A. L., and Shaver, G. R., 1982, Reproductive effort in cottongrass tussock tundra: Holarctic Ecology, v. 5, p. 200-206.

Chippindale, H. G., and Milton, W. E. J., 1934, On the viable seeds present in the soil beneath pastures: Journal of Ecology, v. 22, p. 884-886.

Churchill, E. D., and Hanson, H. C., 1958, The concept of climax in arctic and alpine vegetation: Botanical Review, v. 24, p. 124-191.

Densmore, R., 1979, Aspects of the seed ecology of woody plants of the Alaskan taiga and tundra: Durham, North Carolina, Duke University, Ph.D. thesis (unpublished).

Gartner, B. L., 1982, Controls over regeneration of tundra graminoids in a natural and a man-disturbed site in arctic Alaska: Fairbanks, University of Alaska, M.S. thesis (unpublished).

Gartner, B. L., Chapin, F. S., and Shaver, G. R., 1983, Demographic patterns of seedling establishment and growth of native graminoids in an Alaskan tundra disturbance: Journal of Applied Ecology, in press.

Griggs, R. F., 1934, The problem of arctic vegetation: Journal of Washington Academy of Sciences, v. 34, p. 153-175.

Grime, J. P., 1981, The role of seed dormancy in vegetation dynamics: Annals of Applied Ecology, v. 98, p. 555-558.

Grime, J. P., and Jarvis, B. C., 1975, Shade avoidance and shade tolerance in flowering plants II. Effects of light on the germination of species of contrasted ecology, in Evans, G.C., Bainbridge, R., Rackham, O., eds., Light as an Ecological Factor II: Oxford, Blackwell, p. 525-532.

Harper, J. L., 1977, Population Biology of Plants: New York, Academic Press.

Klebesadel, L.J., 1971, Native Alaskan legumes studied: Agroborealis, v. 3, p. 9-11.

Leck, M. A., 1980, Germination in Barrow, Alaska tundra cores: Arctic and Alpine Research, v. 12, p. 153-175.

Livingston, R. B., and Allessio, M. L., 1968, Buried viable seed in successional field and forest stands, Harvard Forest, Massachusetts: Bulletin of Torrey Botanical Club, v. 95, p. 58-69.

McGraw, J. B., 1980, Seed bank size and distribution of seeds in cottongrass tussock tundra, Eagle Creek, Alaska: Canadian Journal of Botany, v. 58, p. 1607-1611.

Mitchell, W. W., 1979, Three varieties of native Alaskan grasses for revegetation purposes: Agricultural Experiment Station, University of Alaska, Circular 32.

Polunin, N., 1934, The vegetation of Akpatok Island, Part I: Journal of Ecology, v. 22, p. 337-395.

Porsild, A. E., 1951, Plant life in the arctic: Canadian Geographical Journal, v. 42, p. 121-145.

Roach, D. A., 1983, Buried seed and standing vegetation in two adjacent tundra habitats: Oecologia, in press.

Sauer, J., and Struik, G., 1964, A possible ecological relation between soil disturbance, light-flash, and seed germination: Ecology, v. 45, p. 884-886.

Savile, D. B. O., 1960, Limitations of the competitive exclusion principle: Science, v. 132, p. 1761.

Savile, D. B. O., 1972, Arctic adaptations in plants: Canada Department of Agriculture, Monograph No. 6.

Wein, R. W., and MacLean, D. A., 1973, Cottongrass (Eriophorum vaginatum) germination requirements and colonizing potential in the Arctic: Canadian Journal of Botany, v. 51, p. 2509-2513.

Whipple, S. A., 1977, The relationship of buried, germinating seeds to vegetation in an old-growth Colorado subalpine forest: Canadian Journal of Botany, v. 56, p. 1505-1509.

METHODOLOGY FOR USING AIR PHOTOS AND SATELLITE IMAGERY IN PERMAFROST SURVEYS

A. V. Gavrilov, K. A. Kondratyeva, Ye. I. Pizhankova, and Ye. N. Dunayeva

Geology Department,
Moscow State University, Moscow, USSR

The basic principles of a method for using air photos and satellite imagery in permafrost surveys are presented. It is based on the principles developed by V. A. Kudryavtsev and his school for studying rock conditions, the temperature regime and other permafrost characteristics. The method was tested in the compilation of geocryological maps for various permafrost areas in the USSR. A section of a permafrost map for an area in the Central-Yakutian lowland is presented; it was compiled on the basis of laboratory interpretation of multiband satellite imagery taken from the "Soyuz-22" spacecraft. The efficacy of the method is assessed with regard to the north of Western Siberia, and Central and Southern Yakutia.

Intensive economic development of permafrost zones in the USSR requires reliable and quick methods for surveying conditions and characteristics of the natural environment. Aircraft and satellite imagery is especially useful for such surveys of remote areas. The imagery can be used:

1) to analyze the geologic structures and permafrost conditions over large areas since one image may cover areas corresponding to 1:200,000-1:500,000 scale maps;

2) in various spectral bands which provide complementary information on the geologic structures and permafrost conditions;

3) in a series from small-scale satellite imagery to aircraft photography which furnish information on the importance of permafrost-geological features;

4) to extrapolate interpretations over large distances because similar ground features produce similar imagery patterns on synoptic satellite imagery;

5) to make interpretations quicker and more unbiased with optical, optico-electronic and computer techniques;

6) to monitor changes in the natural environment and permafrost conditions, and to follow the development of permafrost-geological processes provided by tele-scanning systems.

Satellite imagery is divided into global, continental, regional, local and detailed (Yeryomin et al., 1978) based on its scale, ground coverage, and resolution. Resolution is a major constraint in permafrost surveys. It drastically limits the application of satellite imagery which is most useful for small- to medium-scale surveys. Local-scale satellite imagery has medium to high resolution, and the utility of medium resolution imagery has already been shown (Haugen et al., 1972). These images, together with the regional and continental photographs, are better for identifying block and fractured structures, for analyzing the geological structures in a region and for permafrost zoning.

Other factors determining the suitability of satellite imagery for permafrost surveys are spectral bands and the number of bands (single- or multiband). The analysis of landscape and geocryological features with aircraft and satellite photographs (ASP) (Gavrilov, 1979, 1980) has shown that multiband and multispectral ASP, obtained in the early summer or early autumn, are most suitable for permafrost surveys. Late autumn is the best time for taking isopanchromatic photographs.

Winter photography is also useful because it shows information on unique winter objects, e.g., water openings in ice. Secondly, winter photography with summer ASP is very useful for distinguishing icings formed by year-round sources from those formed by sources diminishing in a water-critical period. Thirdly, in some cases, winter photography shows more detail on rock lithology, fractures, etc. than some summer photographs.

Geocryological interpretation of ASP is based upon the methodology developed by V. A. Kudryavtsev and his colleagues (Kudryavtsev, 1979) for studying the thermal state and temperature regime of rocks. More precisely, it is based on the identification of cryo-forming factors and the qualitative and quantitative evaluation of the importance of the factors in determining the mean annual temperatures of rocks ($t_{m.a.}$) and the depth of their seasonal thawed or frozen zones (ξ). The methodology is helpful in explaining the existence of permafrost conditions and in analyzing their spatial and temporal changes in light of spatial inhomogeneity and dynamics of natural environment.

It is important to note that information on cryo-forming factors may be obtained through ASP interpretations supplemented by ground truth and the analysis of meteorological data. Photography is useful in identifying cryogenic formations which are often a direct indicator of rock state, and in judging the permafrost series evolution and tendencies in rock temperature regime dynamics. Therefore, the permafrost-temperature interpretations of

ASP depend on the linkage of $t_{m.a.}$ with the elements of the natural environment which govern permafrost formation (cryo-forming factors) and depend on the rock state and temperature regime (cryogenic formations).

Calculation schemes which use parameters that can be determined through ASP interpretations supplemented by ground truth and meteorological data make up the framework of Kudryavtsev's methodology (Kudryavtsev, 1974). His methods define $t_{m.a.}$ by introducing a correction factor for an average annual temperature and an annual air temperature amplitude in light of the effect of natural factors. The ASP analysis is done with the following purposes: 1) interpretation of photos on the basis of field data provides information required for qualitative and quantitative assessment of the influence of the natural factors; 2) the results of the interpretation of cryogenic formations, as indicators of rock state and approximate $t_{m.a.}$, considered in conjunction with the data of thermometry and geophysical prospecting by electrical means and others, while analyzing the calculation results.

This methodology is applied in small-scale permafrost surveys because it is needed to ascertain and further interpret the natural environmental factors that play a key role in determining permafrost conditions. In the tundra environment, key elements are the snow cover distribution and the latitudinal-zonal differentiation of heat exchange. Mapping signs of local permafrost characteristics include such factors as a relative hypsometric location of an area, sizes of relief mesoforms, height and density of vegetation, microrelief, and slope exposure which is responsible for local changes in the snow-cover zonal characteristics.

Factors contributing to the formation of a rock temperature regime in the forest-tundra and taiga zones are more diverse, e.g., snow cover, surface water, rock composition, moisture content and seepage properties, relief, and surface shading.

The complex of cryo-forming factors (and cryo-indicators) in low-mountain areas and plateaus in the southern part of the cryolithic zone with a continental type of heat exchange in the near-ground air layer show the most variety. These are (in respect to the Aldan Plateau and Stanovoi range) absolute altitudes, surface morphology, position in a mesorelief, composition and properties of deposits and surface and ground waters which have maximum effect on a temperature regime in rocks when the rocks have extensive jointing.

In addition to the above, the snow cover distribution, slope exposure and vegetation are also important factors in the formation of $t_{m.a.}$ in low-mountain areas and plateaus. The cryo-indicators in such regions are not individual landscape elements but are composed of complexes of all the external features of the natural environment including relief, vegetation, hydrography, permafrost-geological formations and economic projects (types of earthen roads, in particular).

The above principles of geocryological interpretation of ASP determine the specific nature of comprehensive geocryological surveys at all levels—pre-field, field and laboratory. Experience of small-scale permafrost surveys has indicated that the use of aircraft and satellite imagery requires more extensive pre-field efforts. As a result, the information from pre-field interpretations rises dramatically. This is shown in a portion of the permafrost map for the Central-Yakutian lowland (Figure 1).

This map was compiled from laboratory-based interpretations of multiband Soyuz-22 satellite imagery taken with an MKF-6 camera and data on natural and permafrost conditions of the region (Interpretation ..., 1982). Two methods of photo-interpretation were employed: the visual method of interpreting synthesized images obtained by a multi-channel MC-4 projector with a five-fold magnification (up to a scale of 1:400,000), and a visual-measuring method ensuring the comparison of zonal photos and establishing spectral images of natural objects (Kravtsova, 1980). The resolving capacity of these photos ranges from 16 m (in the 3rd and 4th zones) to 40 m (in the 5th zone). This enables identification of such cryogenic formations as thermokarst lakes and depressions, thermoerosional polygonal block relief, areas affected by eolian and fluvial activity, as well as some 20 types of plant associations (Interpretation ..., 1982; Kravtsova, 1980) which serve as indicators of permafrost on accumulation plains in the northern and central taiga subzone. The map legend describes all the characteristics used in preparing the final map: geological-genetical complexes and rock composition, their thermal state, type and time of freezing, cryogenic textures and ice content, mean annual rock temperatures and temperature amplitudes on the ground surface, depth of seasonal thawing and freezing of rocks and ground surface features (Table 1). Also very essential is the comprehensiveness in characterizing permafrost conditions and lithological rock types.

Therefore, laboratory methods provide useful pre-field data which reduces the amount of and changes the character of field studies, i.e. the number of key plots and job volume are reduced, and to proceed (while selecting keys) from the need of settling down disputable questions and checking results of a pre-field interpretation.

The methodology developed was evaluated in terms of its effectiveness for compiling engineering-geocryological maps for the north of Western Siberia, Central and Southern Yakutia. These maps were compared with the traditional maps. It was determined that ASP-based maps surpass traditional ones by 1.5-5 times (depending on territorial changeability of geocryological conditions) in terms of comprehensiveness in describing composition, moisture content and state of rocks, extension of frozen series, taliks and newly-formed permafrost areas, the development of permafrost-geological phenomena.

The following conclusions are presented based on the above:

1) ASP is useful for permafrost surveys over vast regions, for obtaining supplementary information on the natural environment, for ranking permafrost-geological features by importance, for improving the reliability of interpretations and extrapolation of results, for studying the dynamics of natural and permafrost features and

conditions.

2) Multiband, local satellite imagery of high resolution is the best for doing small-scale and occasionally medium-scale permafrost surveys. The visual interpretation of synthesized photos of a method based on the visual-instrumental measurements of zonal density of images to determine feature patterns are two basic interpretation methods.

3) The permafrost-temperature interpretation of ASP rests upon the Kudryavtsev methodology for studying the state and temperature regime of rocks which includes the identification of cryo-forming factors and the quantitative assessment of the contribution each of them makes to the $t_{m.a.}$ and ξ formation. The information on cryo-forming factors may be obtained through the analysis of ASP, meteorological and field data for key plots. Also very essential is the fact that cryogenic formations are easily discernible on local ASP since the information on them as an indicator of rock state and approximate $t_{m.a.}$ may be taken into account while analyzing estimates.

4) This methodology is employed for interpreting ASP in the course of small-scale permafrost surveys with an aim to identify and assess the effect with ensuing interpretation of environmental elements being decisive in forming t, as well as to discern cryogenic formations on photos.

5) The methodology described provides essential pre-field level information on permafrost conditions which reduces the amount of costly field work. The field work can then be done to resolve problems which cannot be solved with ASP, and to check the results of laboratory interpretations.

6) The assessment of ASP maps indicates that they show more data than the traditional maps.

REFERENCES

Gavrilov, A. V., 1980, Geocryological informative value of local space photos (in respect to the Aldan-Stanovoi range), "Merzlotnye issledovaniya", vol. XIX, Moscow: Izd-vo Mosk. universiteta.

Gavrilov, A. V., 1979, On geocryological informative value of spectro-zonal aerial and space photographs (on the example of the Chulman Region), "Merzlotnye issledovaniya", vol. XVIII, Moscow: Izd-vo Mosk. universiteta.

Interpretation of multiband aerial and space photographs, 1982, Moscow-Berlin: "Nauka", "Academie-Verlag".

Kravtsova, V. I., and Kamyshan, O. L., 1980, Mapping of northern- and central-taiga forests on the example of Central Yakutia, "Kosmicheskaya syomka i temticheskoye kartografirovaniye", Moscow: Izd-vo Mosk. universiteta.

Kudryavtsev, V. A. (Ed.), 1979, Methodology of permafrost surveys, Moscow: Izd-vo Mosk. universiteta.

Kudrayvtsev, V. A. (Ed.), 1974, Foundations of permafrost forecast in engineering-geological investigations, Moscow: Izd-vo Mosk. universiteta.

Yeryomin. V. K., Bryukhanov, V. N., Makhin, G. V., and Mozhayev, B. N., 1978, Present state and perspectives in the application of aerial and space photography in geology, "Izv. vuzov. Ser. geol. i razvedka", no. 10.

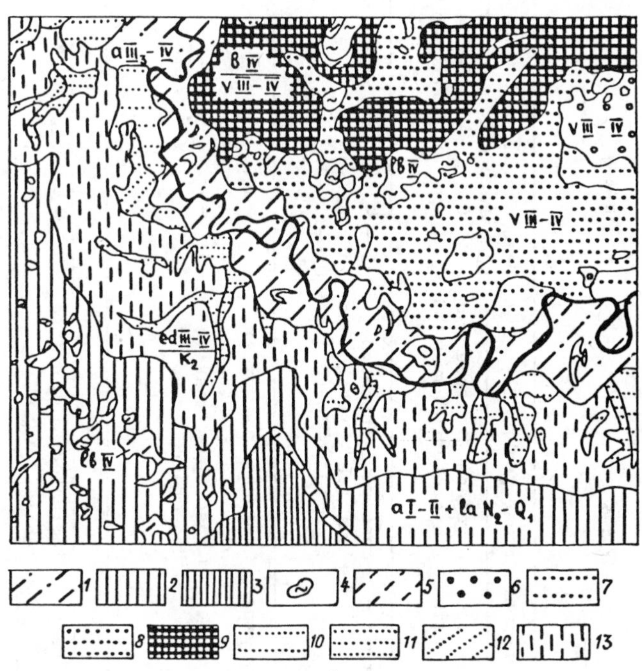

FIGURE 1 Fragment of permafrost map of a region in the Central Yakutian Lowland. (The denotations for the figure are given in the table of the text.)

Table 1

No.	Characteristics of seasonally-frozen and permafrost rocks				Mean annual temperature °C	Amplitude of temperature changes on ground surface, °C	Depth of seasonal thawing and freezing of rocks, m	Exogenous phenomena
	Geologo-genetic complexes	Composition	State, type and time of freezing	Cryogenic texture and ice content (humidity) %				
1	a III_3-IV	fine sand, loamy sand, clay loam	frozen syngenetic III_4-IV	massive, lense-type, laminated, reticulate from 10 to 50-60	-1 - -2 in places 0 - -1, -2 - -3, -5 - -8	13.5-17	1.5-2.6	ice wedges, sandy-icy and ground veins, modern
2	a I-II +	loam and fine loamy sand	frozen, syngenetic, I_2-II	laminated 1st order, ataxitic, laminated, lense-type, massive 2nd order, up to 90	-2 - -3	11-13.5	1.0-1.7	relict ice wedges to 50 m thick with pseudomorphism of surface parts, polygonal-block relief due to thermokarst by ice wedges
3	la N_2-Q_1				-3 - -5		0.8-1.5	
4	lb IV	silt	thawing	-	0	-	-	-
5		silty loamy sand, loam	frozen epigenetic, IV_3	laminated, reticulate, massive, 20-30	0 - -2	13.5-17	1.5-2.6	bulgunyakhi, seldom ice wedges
6			thawing	1-3	0 - +0.8	-	4.0	dunes, modern primary-sand veins in interdune space
7	vIII-IV	sand	frozen, syngenetic, III_2-IV_3	contact, 5-7	0 - -1		2.7-3.2	occasionally primary sandy relict and modern veins, dunes
8				massive, up to 20	-1 - -2		2.6-3.0	
9	bIV / vIII-IV	peat and peaty silt deposits, thin overlying sand		lense-type, laminated, reticulate, up to 80	-3 - -5	11-13.5	0.6-0.9	occasionally modern ice wedges
10		sand	frozen, epigenetic, III_2 - IV_3	contact, 5-7	0 - -1	17-21	2.7-3.2	—
11					-1 - -2		2.3-2.9	
12	edIII-IV,dIV					13.5-17	1.8-3.2	
13	K_2	sand, loamy sand, loam		massive, laminated, reticulate, from 10 to 60	-2 - -3	11-13.5	1.1-1.8	occasionally ice wedges with rare pseudomorphism

MAPPING OF ARCTIC LAND COVER UTILIZING LANDSAT DIGITAL DATA

Leonard Gaydos[1] and Richard E. Witmer[2]

[1]U.S. Geological Survey, Ames Research Center, 240-8
Menlo Park, California, 94035 USA
[2]U.S. Geological Survey, 521 National Center,
Reston, Virginia, 22092 USA

The U.S. Geological Survey, with the assistance of several other agencies and institutions, has been mapping vegetation and other land cover in arctic Alaska as a base for assessing development issues. LANDSAT earth resources satellite imagery in digital form were processed by using algorthims that partitioned the data into spectral classes. Fieldwork supported the later association of these spectral classes with actual land cover types and the construction of a classification system compatible with LANDSAT data and responsive to the needs of resource managers. Problems of classification incompatibility between adjacent LANDSAT images were solved by careful designation of classes and editing procedures.

After an initial investigation within the National Petroleum Reserve in Alaska, further development of the techniques was achieved in the region around the Prudhoe Bay oil fields and in the adjacent Arctic National Wildlife Refuge. LANDSAT was initially proposed as a mapping data source for this region because of cost and time factors, and research has demonstrated that it is extremely well adapted to the task. Not only are the mapped classes comparable to what could be obtained through conventional interpretation of aerial photographs; the resulting data are digital and consequently directly usable in geographic information systems.

THE USGS LAND USE AND LAND COVER MAPPING PROGRAM

In 1974, the U.S. Geological Survey (USGS) received funding to begin the first nationwide program of land use and land cover mapping, data production, and analysis. For several years prior to 1974, the USGS had been developing a land use and land cover classification system that could be applied nationwide, which would provide a comprehensive balanced set of categories useful to planners and land resource managers, and which could be employed by using remotely sensed data as primary source material. Research had also been directed toward other aspects of an inventory that needed resolution before a nationwide mapping program could begin, such as mapping techniques, suitable remotely sensed material, product types and user needs, and a geographic information system able to capture, store, and manipulate the resulting information in digital form.

The final version of the land use and land cover classification system (Anderson et al., 1976) consists of nine general level I classes, that are further subdivided into 37 more detailed level II classes. In the nationwide mapping effort, these level II classes are mapped by conventional photointerpretation methods. High-altitude aerial photographs and other source materials available to the USGS are the typical kinds of remotely sensed data used. USGS 1:250,000-scale and 1:100,000-scale topographic maps are used as base map materials. Areas as small as 10 acres (approximately 4 hectares) are mapped in urban situation(s), for water bodies, and other areas of typically high target/background contrast. For land use and land cover categories having broader extent, such as forests, rangelands, or tundra, an area of 40 acres (approximately 16 hectares) is the smallest unit mapped.

The final products are as follows: (1) hardcopy maps at 1:250,000-scale, with several areas mapped at 1:100,000-scale. In addition to the land use and land cover maps covering each quadrangle, associated maps showing political boundaries, hydrologic units, census county subdivisions, and areas of Federal land ownership were also compiled; (2) statistical tables giving areas of the land use and land cover category for each of the types of regions shown on the associated maps; and (3) various products such as tapes and plots resulting from the geographic information system.

The classification system, techniques, and products described above have been applied consistently since the program began. Nearly two-thirds of the Nation has been mapped; nearly half of those maps have been converted into digital format, and statistical reports for 16 states are now in production.

LAND USE AND LAND COVER MAPPING IN ALASKA

During the first year of the mapping phase of the program, the Fairbanks, Alaska, 1:250,000-scale quadrangle was chosen as one of a dozen maps scattered throughout the United States to be compiled to provide early experience in applying the classification system and mapping techniques in a wide variety of situations. After the map was compiled and reviewed, many interpretation problems were found and the map was not published. These problems were due entirely to the compilers' lack of familiarity with the area. Suitable source materials for other parts of Alaska were not available until 1981, when compilation began on the Valdez 1:250,000-scale quadrangle. Several other quadrangles are now being compiled, and the USGS is presently investigating the possibility of entering into a cooperative agreement with the State of Alaska under which land use and land cover maps conforming to USGS specifications would be compiled as part of the State's Integrated Terrain Unit mapping effort.

LAND COVER MAPPING RESEARCH USING LANDSAT

As part of its continuing research into the capabilities of various remote sensors and their data, the USGS has been experimenting with the extracting of land cover information from LANDSAT multispectral data since 1973. The synoptic coverage, availability, and digital format of these spectral reflectance data held early promise as an economical substitute for aerial photographs where such conventional sources were lacking, for remote areas, and for ready adaptation to digital geographic information systems.

During the first few years of this research, land cover maps were constructed by computer-assisted classification of LANDSAT multispectral data for areas centered on Washington, D.C., Phoenix, Indianapolis, San Jose, and for various areas in the Pacific Northwest. These research efforts gave an early appreciation of the relative advantages and disadvantages of this technology for land cover mapping. Apparent advantages included the availability of data after processing by Goddard Space Flight Center and the EROS Data Center, resolution of data near the 1-acre level, and the digital format of the data allowing economical computer manipulation, including area measurement of each category and use of various digital graphic output devices. Disadvantages included early problems with geometric and radiometric fidelity, lack of optimum coverage for specific seasons, an inability to duplicate many of the categories being used in the nationwide land use and land cover mapping program, a lack of success in extending spectral classes from one LANDSAT scene to another, as well as the persistent problem of producing different sets of categories in different areas.

FIRST OPPORTUNITY--NPRA

A specific need for land cover data of a large remote region was expressed in 1977 when the Bureau of Land Management (BLM) and USGS became responsible for compiling a resource inventory and a land use study for the 97,000 square-kilometer National Petroleum Reserve in Alaska (NPRA). On the basis of the foregoing discussion of perceived advantages and disadvantages of LANDSAT relative to photointerpretation of high-altitude aerial photographs, it was decided to test the hypothesis that analysis of LANDSAT digital imagery could provide an inventory and map of land cover consistent with planning needs.

Full source material coverage of NPRA was obtained from ten summer-season cloud-free LANDSAT scenes, each measuring 185 km on a side. Acquisition dates of the scenes ranged from early July to late August over a three-year period, 1975-1977. Each LANDSAT scene consists of a matrix of 79 x 57-m cells (pixels) sensed by the satellite in each of four wavelength bands, green, red, and two near infrared. The 7.56 million pixels that made up each scene were partitioned into spectral classes by using a clustering algorithm. Each class was defined by pixels with similar spectral characteristics. After defining these spectral classes, the entire scene was classified by assigning each pixel to the spectral class it most nearly resembled.

After classification, each spectral type was identified as a specific land cover by an analyst viewing the classified scene on a color display. Aerial photographs of specific sites and field notes were used as reference material for identifying the spectral classes.

Two visits to the field were made. The first, in 1977, acquainted the analysts with the tundra landscape and helped in preparing a preliminary classification of land cover. In 1978 the analysts returned to investigate problem areas, verify the validity of land cover classes, and sharpen their photointerpretation skills.

Throughout the process, a land cover classification system for the tundra that could be mapped with LANDSAT data was being devised. It was realized at the outset that the LANDSAT data source might not be perfectly compatible with the land use and land cover classification system used by USGS nationwide and that some modification might be desirable in mapping tundra classes. The final system was based on plant communities that have similar lifeform and contained ten classes. Compatibility with the basic USGS classes was shown.

Once the land cover classes were established, it was necessary to ensure consistency in interpretation among all ten LANDSAT frames. Some areas were reclassified following establishment of new spectral classes based on analysis of patterns noted on aerial photographs. In other areas, misclassifications, such as misclassifying shadows on mountain slopes as water, a situation caused by the spectral similarity of water with shadow, were corrected by defining geographic regions where certain classifications did not make sense (Morrisey and Ennis, 1981).

A control network established between topographic base maps and the LANDSAT data enabled correction to the Universal Transverse Mercator (UTM) projection. Original pixels were resampled to form square cells 50 m on a side. After the frames had been corrected, they were mosaicked into two data sets corresponding to the two UTM zones covered by the data. Data were also formatted for each of 11 map quadrangles (1 degree of latitude by 3 degrees of longitude) and made available to BLM.

Land managers at BLM have registered each land cover data set to digital terrain data available from USGS and have used the combination to help monitor potential impacts of mineral exploration (Spencer and Krebs, 1982). The same digital land cover data are also available to the public through the USGS National Cartographic Information Center.

REFINEMENT OF TECHNIQUE

While work was proceeding on the NPRA, contacts were made with the U.S. Army Cold Regions Research and Engineering Laboratory (USA CRREL) where vegetation mapping was in progress at Prudhoe Bay and along the Dalton Highway. USA CRREL and their contractor, the Institute of Arctic and Alpine Research, University of Colorado (INSTAAR) were mapping detailed vegetation by using extensive ground surveys. They are able to achieve excellent detail in environmental mapping but were severely limited in the areas they could cover.

A partnership was forged that used the tundra classification expertise of USA CRREL and INSTAAR and the LANDSAT mapping capabilities of USGS. Detailed maps, already prepared by INSTAAR, were used as reference data for interpreting the LANDSAT classification, thus minimizing new fieldwork. The land cover maps from LANDSAT were used by USA CRREL and INSTAAR to extend their vegetation classification system to the vast regions surrounding their original sites, albeit at less detail.

The Prudhoe Bay region, heart of the petroleum development, was the first target for this joint effort. "The Geobotanical Atlas of Prudhoe Bay" (Walker et al., 1980) was used to differentiate land cover classes within spectral classes resulting from clustering of one LANDSAT frame. The detailed reference data made it possible to tailor spectral classes by focusing on specific known sites, something not possible in the NPRA work. A technique known as reclustering--splitting one spectral class into several--was also employed successfully in refining the classification.

It was recognized that the key to developing a consistent classification using LANDSAT was a better understanding of the capabilities and limitations of the LANDSAT sensor. Spectral classes mapped by LANDSAT were found to be defined by complex reflectances from the ground associated with vegetation, soils, and land forms. These three types of information were mapped separately by Walker et al., (1980) and then combined into master maps. With LANDSAT, components of all three contributed to the spectral reflectance that was sensed, though vegetation was seen to dominate. As work progressed, the arctic tundra region seemed ideally suited to using LANDSAT data. Spectral differences were strongly influenced by moisture and shrub component of the plant community, two factors already considered prominently in the classification system used by Walker et al., (1980). That system was extended and modified to accomodate the units mappable with LANDSAT (Walker et al., 1982).

Work at Prudhoe Bay was extended south by using detailed maps previously compiled by INSTAAR along the Dalton Highway corridor. One LANDSAT frame centered over the Sagavanirktok quadrangle was analyzed by using techniques similar to those used for the Prudhoe Bay scene. Field trips over two seasons were used, as in NPRA, to investigate problem areas or to help in the intrepretation of aerial photographs.

In 1981, the U.S. Fish and Wildlife Service requested a land cover map of the coastal portion of the Arctic National Wildlife Refuge (ANWR) for use in an Environment Impact Statement (EIS). Since the EIS was due within a year, timing was a critical factor. A seven-day reconnaisance trip to the Refuge in August 1981 was the beginning of the project. Three LANDSAT scenes were needed, each treated separately by using cluster analysis. As before, field data were used to identify the spectral classes.

With the lessons learned previously, the analysis went smoothly and a digital data set corrected to the UTM projection was ready within six months. These data were smoothed to enhance cartographic presentation by generalizing tiny mapped units into their surroundings and then used to produce printing plates for lithography using a color laser plotter (Wray and Gaydos, 1982). The finished map (U.S. Geological Survey, 1982) also contained a table showing land cover areal measurement for survey townships located within the study area, a useful byproduct of the LANDSAT analysis.

LESSONS LEARNED

The experiences gained in using LANDSAT data for mapping arctic tundra have been most rewarding. Although LANDSAT was originally proposed as a data source because of cost and time factors, research has demonstrated that it is extremely well adapted to the task. Not only are the mapped classes comparable to what could be obtained through conventional interpretation of aerial photographs, but the resulting data are in digital form and consequently directly usable in geographic information systems. The area covered by each land cover unit for tracts of interest, such as survey townships, are easily obtained by simply digitizing the boundaries of tracts and counting pixels within each. Furthermore, the digital data can be used by a laser plotter to create color separation plates for map reproduction.

The classification system found optimal for use with LANDSAT data has the desireable attributes of being useful for both large-scale and small-scale mapping and consistently applying the same criteria and method for naming communities and vegetation complexes. This system has evolved with our mapping experience and is easily compared with other systems in general use.

REFERENCES

Anderson, J.R., Hardy, E.E., Roach, J.T., and Witmer, R.E., 1976, A land use and land cover classificaton system for use with remote sensor data: U.S. Geological Survey Professional Paper 964, 28 p.

Morrissey, L.A., and Ennis R.A., 1981, Vegetation mapping of the National Petroleum Reserve in Alaska using LANDSAT digital data: U.S. Geological Survey Open File Report 81-31525 p.

Spencer, J.P., and Krebs, P.V., 1982, A digital data base for the National Petroleum Reserve in Alaska: Auto-Carto V, Crystal City, Virginia, August 22-28, 1982, Proceedings, p. 461-469.

U.S. Geological Survey, 1982, Vegetation and land cover--Arctic National Wildlife Refuge coastal plain, Alaska: USGS Miscellaneous Investigations Map I-1443, scale 1:250,000.

Walker, D.A., Acevedo, William, Everett, K.R., Gaydos, Leonard, Brown, J., and Webber, P.J., 1982, LANDSAT-assisted environmental mapping in the Arctic National Wildlife Refuge, Alaska: U.S. Corps of Engineers Cold Regions Research and Engineering Laboratory, CRREL Report 82-27, 59 p.

Walker, D.A., Everett, K.R., Webber, P.J., and Brown, J., 1980, The Geobotanical Atlas of Prudhoe Bay: U.S. Corps of Engineers Cold Regions Research and Engineering Laboratory, CRREL Report 80-14, 69 p.

Wray, J.R., and Gaydos, Leonard, 1982, Vegetation and land cover map and data for an Environmental Impact Statement, Arctic National Wildlife Refuge, Alaska: Multistate demonstration of automation in thematic cartography: Auto-Carto V, Crystal City, Virginia, August 22-28, 1982, Proceedings, p. 283-286.

PERMAFROST CONDITIONS AT THE WATANA DAM SITE

James D. Gill[1] and Michael P. Bruen[2]

[1]Ertec Northwest, Anchorage, Alaska 99503 USA
[2]Harza Engineering, Anchorage, Alaska 99501 USA

The proposed Watana development, consisting of a 270 m (885 ft) high dam and underground powerhouse, is the first phase of a two dam scheme on the upper Susitna River, located in Alaska, approximately 225 km (140 mi) south of Fairbanks. Extensive geotechnical investigations were conducted at this site to confirm the feasibility of the project. The extent of permafrost throughout the reservoir areas was mapped through the use of air photos, and the effect of thawing on reservoir slope stability was evaluated. Permafrost conditions are an important consideration at the Watana site. Lying in a broad glaciated region, the Susitna River at the Watana site flows east to west in a deep valley cut into the surrounding plateau, which lies above elevation 700 m (2300 ft). The river valley is more than 244 m (900 ft) deep and approximately 150 m (500 ft) wide at river level. At the site the north facing slope contains permafrost to depths exceeding 38 m (125 ft), while the south facing slope with favorable exposure is relatively permafrost free. The permafrost in the rock lies within 1°C below freezing. Temperature data suggest that the average annual air temperature is close to the freezing isotherm, suggesting that permafrost, although extensive at the site, is in a state of delicate equilibrium. On the relatively flat terrain above the valley rim the permafrost is more sporadic with extensive but discontinuous permafrost to varying depths. Although most of the permafrost in the overburden is relatively ice free, there exists at one location several miles downstream of the Watana site, a large, flay-lying, lense of ground ice in the valley slope in excess of 3.0 m (10 ft) thick.

INTRODUCTION

The Watana development, a proposed hydroelectric project on the upper Susitna River, is the first phase of what will ultimately be a two dam scheme with a total power generation output of approximately 1600 mw. The Watana site lies some 200 km (125 mi) northwest of Anchorage and 225 km (140 mi) south of Fairbanks, Alaska. The proposed Watana development includes a high earthfill dam over 250 m (820 ft) high with an underground powerhouse. Between 1980 and 1983 extensive geotechnical investigations were conducted at the site to assist in the assessment of the feasibility of the overall development and to provide a basis for submission of a license to the FERC for construction of the project. Previous investigations had been conducted at the site as early as 1975 and as recently as 1978 by the U.S. Army Corps of Engineers. The site location is shown in Figure 1.

CLIMATIC CONDITIONS

The Watana dam site is situated in a valley with an elevation of 442 m (1450 ft) at river level and 686 m (2250 ft) on the abutments. Elevation has a significant influence on the existence of permafrost at the site due to the fact that the mean annual temperature is several degrees colder than locations in the lower Susitna river valley much further to the west but only slightly further south. Initial temperature data from existing sites in the lower valley indicated that the mean annual temperature was close to the freezing isotherm, suggesting that the permafrost at the site was in a very delicate state of equilibrium. A more detailed review of actual temperatures at the Watana site indicates a freezing index of 1873°C days and a thawing index of 1212°C days, with a mean annual temperature of -1.8°C (29°F) averaged over the past 3 years. Although only 3 years of data are available from the newly established climatic station at the Watana site, it is apparent that the site lies within the generally accepted zone of discontinuous permafrost. (Ferriano, 1965)

The southern boundary of discontinuous permafrost is found to coincide with the -1°C annual air temperature isotherm (Gold and Lachenbruch, 1973). Since the Watana site has a mean temperature approaching -2°C, it is apparent that the thermal regime at the site is in equilibrium with the present climatic conditions in most instances. It is possible that in some areas large ice lenses and ground ice were formed at a time when the climatic regime was more conducive to the growth of permafrost and that such occurrences would not form under present conditions. Examples of these conditions are deep ice lenses in the overburden area to the north of the site and a large flat lying lense of ice, estimated to be at lease 3.0 m (10 ft) thick near the Fog Creek area. The location of the ground ice is shown in Figure 2.

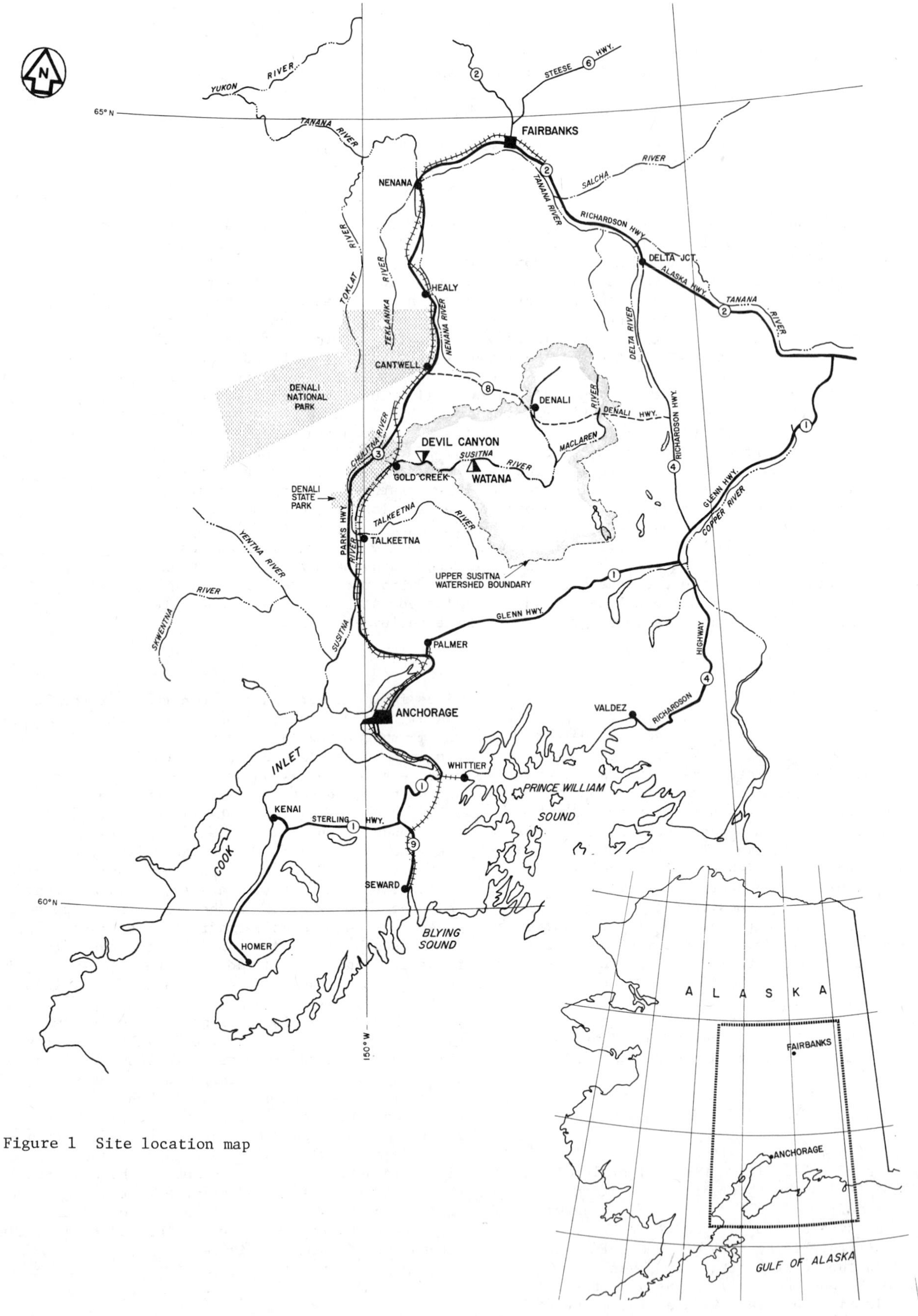

Figure 1 Site location map

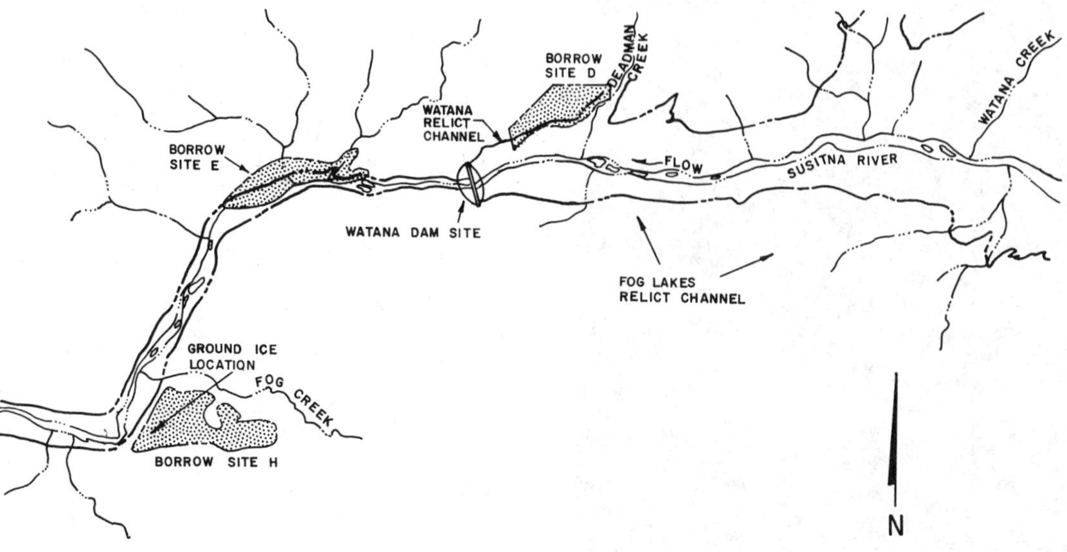

Figure 2 Site area

SOIL AND OVERBURDEN CONDITIONS

The depth of alluvium at the dam site within the river channel is a minimum of 27 m (90 ft) and consists of sand, silt, coarse gravel, and boulders. On the lower valley slopes, the overburden consists primarily of talus. The upper areas of the abutments near the top of the slope are comprised of glacial till, alluvium, and talus deposits. Subsurface investigations show the contact between the overburden and the bedrock to be relatively unweathered.

Just to the north of the Watana dam site a deep relicit channel exists. The location of this buried, preglacial channel, is also shown in Figure 2. The maximum depth of overburden in the thalweg channel is approximately 137 m (450 ft). The channel stratigraphy is compromised of a series of glacial outwash and till and alluvial deposits. Within the glacial deposits frozen ground has been detected in borings to depths of 34 m (110 ft) and has been verified to depths of 18 m (60 ft) in instrumented borings.

In addition to the area above the channel, designated as borrow area D, a number of other borrow areas near the site were investigated and the permafrost conditions were defined in these areas. They include borrow site E near the mouth of Tsusena Creek, and borrow site H near Fog Creek, both of which are shown in Figure 2. Borrow site E contains alluvial gravels, while H consists primarily of glacial till. The alluvial gravels contained no permafrost due primarily to the courseness of the material and the southerly exposure of the area. Borrow site H contains permafrost to over 8 m (25 ft) depth due to its location at a higher elevation and, the flat lying terrain, and the thermal properties of the material.

PERMAFROST IN ROCK

Since overburden depths are very shallow on the valley walls, permafrost depths have penetrated well into the underlying rock on the south abutment. Although no thick ice lenses were observed, it is possible that lenses do exist on preexisting joints in the bedrock. Drill holes in the south abutment indicate that the depth of permafrost exceeds 60 m (200 ft). This depth is unusually deep for the climatic conditions at the site but it is primarily the result of slope aspect. The valley runs almost exactly east to west at the site and hence the south abutment, or north facing slope, is not exposed to solar radiation for many months of the year. Microclimatic effects have been known to influence permafrost distribution (Johnston 1981). Beaty (1977) addresses similar microclimatic effects on slope stability. An air photo of the site is shown in Figure 3.

EFFECT OF SLOPES AND GROUND COVER

The most dramatic and pronounced characteristic of permafrost at the Watana site is the assymetrical distribution at the proposed dam site. Although ground cover has some influence on the depth of permafrost, permafrost distribution in the valley is a consequence of aspect, a characteristic observed by Johnston (1981). Figure 3 shows the Watana site area oriented with north at the top of the figure. The south facing, or north slope, with a minimum of soil cover, direct exposure to the summer sun, and hence, a shorter duration of snow cover, has very little evidence of permafrost. The north facing slope, with a slightly greater depth of insulating overburden and poor exposure to solar radiation, has evidence of permafrost in rock down

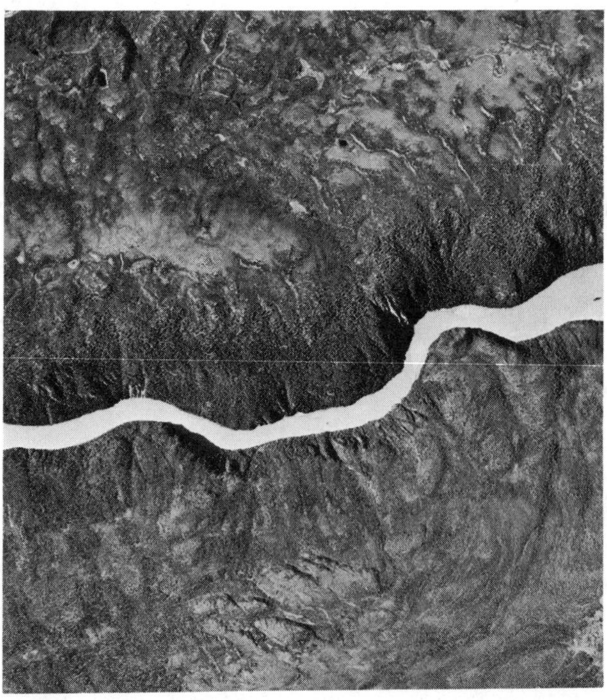

Figure 3 Air photo of site area

to depths of more than 38 m (125 ft).

The exact distribution of permafrost near the river's edge is not known. Although it is normally expected that the influence of a river, or body of water, is to create thaw bulbs below and adjacent the water body, the occurrence of anchor ice during freeze-up has been observed at several locations, suggesting that permafrost may be very close to the water's edge on the south side of the river. This ice develops below the water, firmly attached to the stream bed.

TEMPERATURE OBSERVATIONS

Temperature measurement instrumentation was installed in both soil exploration holes in the overburden and core holes in rock. For holes in rock, thermistor strings were attached to the outside of casing used to install deep piezometers or grouted internally inside 50 mm (2 in) diameter pvc pipes to protect leads from the deep piezometers. Open standpipe-type instrumentation pvc pipes were installed in most vertical soil drill holes. These were filled with an antifreeze fluid permitting the use of a temperature probe to measure the thermal properties of the surficial deposits. The temperature profile from BH6, a deep borehole in rock on the north side of the river, and angled beneath the river, is shown in Figure 4. It is interesting to note that the temperature throughout the length of the hole is at or near freezing, while the annual or seasonal amplitude reaches down to 24 m (80 ft), presumably because of the influence of the river. Although it is considered that the hole is in permafrost, it is likely that interstitial fluids, such as groundwater in the fractures, are not frozen. This was evident when examining the permeability test results which indicated that fractured areas took water during the test, although it may be frozen in some locations. This hole and the area covered by its depth are low enough down in the valley to be affected by a screening effect from the south abutment cliffs, i.e. shade factor reducing incoming solar radiation (Wendler and Ishikawa, 1974).

Typical south abutment temperature profile in the overburden to the north of the dam site are shown in Figure 5.

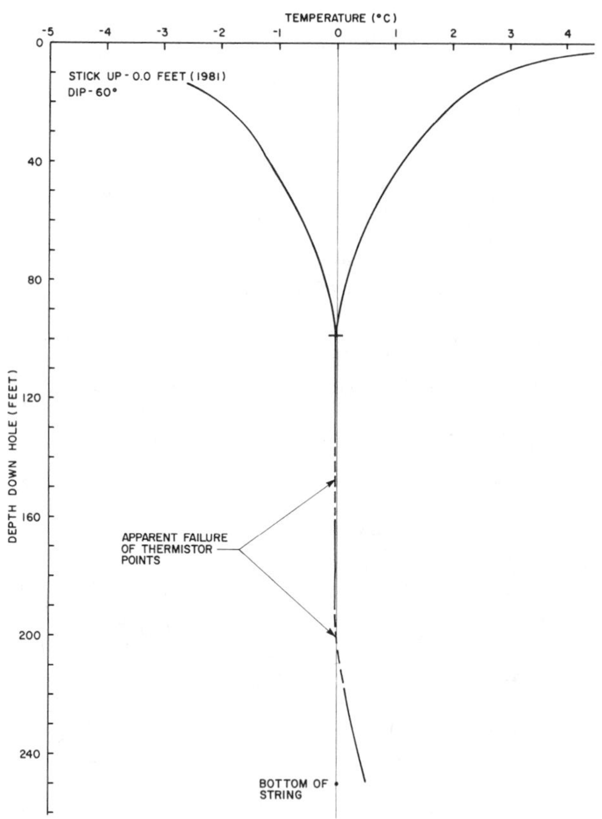

Figure 4 Temperature profile BH6

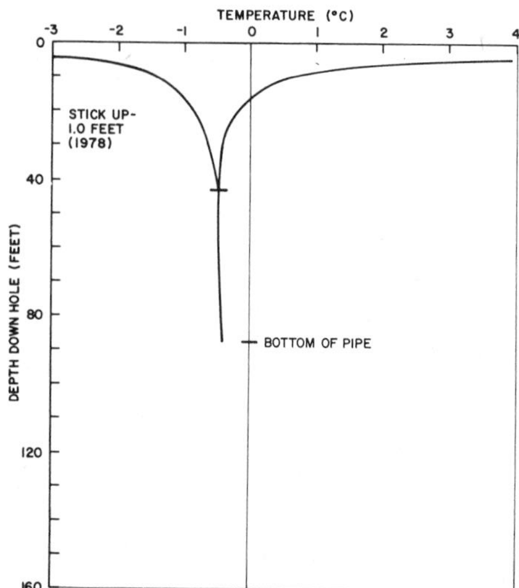

Figure 5 Temperature profile south slope

STABILITY OF RESERVOIR SLOPES

The existence of permafrost throughout the reservoir area has an influence on the future stability of reservoir slopes during the ultimate impoundment of the reservoir. A study was made of reservoir rim stability during the course of the Susitna Feasibility Studies. To aid in the determination of thaw-susceptible slopes, an assessment of permafrost areas was made utilizing aerial photographs and limited field reconnaissance.

As a generalization, a high percentage of the glacial till and lacustrine deposits bordering the reservoir rim are frozen. The south facing slopes based on vegetation characteristics and limited field reconnaissance are better drained and because of slope aspect, permafrost is not as widespread or where found, as deep. Current slope instability in these materials indicates that aspect is important for the distribution of permafrost but aspect relative to incoming solar radiation, is not the key factor for instability and all slopes are susceptibility to mass movement.

The instability of thawing slopes in permafrost has been addressed by McRoberts and Morgenstern (1974). They indicate that bimodal flows, solifluction and skin flows are characteristic of slope instability due to thawing permafrost. Shoreline erosion in permafrost areas has been studied by Newbury et al, (1978). A study was made on the potential for slope instability due to reservoir-induced thawing of permafrost slopes in the Watana reservoir. Most of the problems to be encountered will occur on the sheltered slopes, particularly on the south side of the valley.

Thawing permafrost can generate excess pore pressure in ice rich soils. In cases where the soils are potentially liquefiable such as sands and silts, liquefaction during earthquakes is possible, resulting in additional landslides during seismic activity.

Historically, numerous small slides have occurred around the reservoir rim due to thawing permafrost. One of the largest slides of this type occurred in 1982, and is shown in Figure 6. Although the exact cause of this instability, which is approximately 320 m (1050 ft) long and contains 1.4 million m^3 (1.81 million yd^3) of material, is not known, it was found that the blocky material in the slide is frozen and contains ice lenses. The slide occurred on a northwest facing slope and has a very distinct failure plane sloping at an angle of 15°. Since a large icing appears on the backscarp of the slide it seems that groundwater played a role in the slide, either by inducing thaw along a weak plane and/or creating excess pore pressure along the sliding plane.

Figure 6 Watana Creek slide area

THAWING EFFECTS AND SEEPAGE PROBLEMS

The existence of permafrost in the borrow areas for the proposed dam does not present a problem. The material, although frozen in places, does not contain a great deal of ice and the temperature lies within 1°C below freezing, hence frozen areas can be thawed prior to removal.

The thawing of permafrost in the overburden areas to the north of the dam site does not present serious engineering problems. During excavation of construction materials in borrow site D, the insulating vegetation cover should be stripped during the summer months, to allow the materials to thaw. Proper orientation of the cut face, to a semi-circular south facing slope, will enhance permafrost thaw during development of the borrow source. Some settlement will eventually occur in disturbed areas and design consideration must be given under the freeboard dike required at elevation 2210. However, design techniques for such structures are well established (Macpherson et al 1970).

The thawing of permafrost in rock below the dam foundation requires the greatest attention. The design implications of the thawing of the foundation of dams has been addressed by Johnston et al. (1982) and was carefully considered in the preliminary design of the Watana dam.

The principal results of permafrost thaw in the

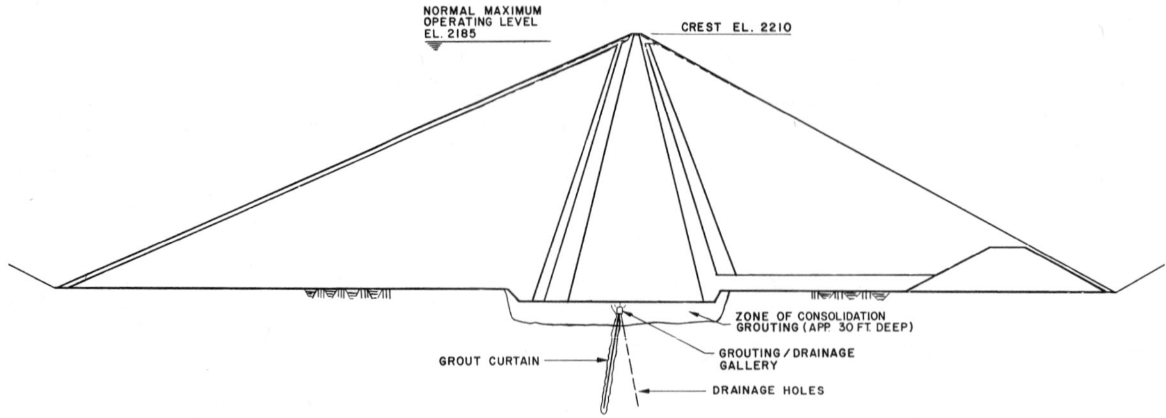

Figure 7 Proposed dam cross section

rock foundation of the dam will be increased seepage through previously ice-filled fractures. One solution to this problem is an effective group curtain. However, in order to be effective, the zone around the grout curtain must be thawed prior to grouting. This can be accomplished by chosing a close arrangement of boreholes in a single or double row curtain. The spacing of the boreholes would depend on the time available for thawing, the temperature of the thaw-water and the economic constraints of additional holes vs. heating and time (Sayles, 1978). The circulating of water into the open boreholes prior to grouting, coupled with a series of holes to monitor temperature will allow for improvements in the thawing process as the operation proceeds. In addition to these precautions during the grouting operations, the preliminary design of the dam includes a gallery in rock below and in the abutments of the dam from which drain holes, as well as the grouting operation, will be undertaken. This tunnel permits observation of the increased in seepage which may still occur after thawing in the abutment. Additional drain holes or remedial grouting may be performed from the gallery. The arrangement of the proposed drainage and grouting tunnel is shown on Figure 7.

ACKNOWLEDGEMENT

The studies for the feasibility and licensing of the Susitna Hydroelectric Project were conducted by Acres American Incorporated for the Alaska Power Authority. The authors wish to thank the Alaska Power Authority for permission to present this paper.

REFERENCES

Beaty, C.B. 1972. Geographical Distribution of Post-Glacial Slumping in Southern Alberta. Canadian Geotechnical Journal, Vol. 9, Number 2.

Ferrians, O.J., Jr. 1965. Permafrost Map of Alaska. United States Geological Survey, Misc. Field Studies Map I-445, 1:250,000.

Gold, L.W. and A.H. Lachenbruch 1973. Thermal Conditions in Permafrost: A review of North America Literature. North American Contribution; Permafrost, Second International Conference, Yakutsk USSR.

Johnston, G.H. 1981. Permafrost Engineering Design and Construction, Associate Committee and Geotechnical Research, National Research Council of Canada.

Johnston, T.C.: E.C. McRoberts and J.F. Nixon 1982. Design Implications of Subsoil Thawing. Presented to the ASCE Spring Convention, Las Vegas, Nevada.

Macpherson, J.D., Watson, and A. Koropatnick 1970. Dikes on Permafrsot Foundations in Northern Manitoba. Canadian Geotechnical Journal, Vol. 7, Number A.

McRoberts, E.C. and M.R. Morganstern 1974. The Stability of Thawing Slopes, The Canadian Geotechnical Jounral, Vol, 11, Number 4.

Newbury, R.W., K.G. Beaty, and G.K. McCullough 1978. Initial Shoreline Erosion in a Permafrost Affected Reservoir, South Indian Lake, Canada. North American Contribution to the Third International Conference on Permafrost, Edmonton, Alberta, Canada. National Academy of Sciences, Washington, D.C.

Sayles, F.H. 1978. Procedure for Estimating Borehole Spacing and Thaw-Water Pumping Requirements for Artificially Thawing the Bedrock Permafrost at the Watana Dam Site. CRREL Technical Note. 8 p.

Wendler, G. and N. Ishikawa, 1974. The Effect of Slope, Exposure and Mountain Screening on the Solar Radiation of McCall Glacier, Alaska: A Contribution to the International Hydrological Decade. Journal of Glaciology, V. 13, No. 68, pg. 213-226.

THERMAL PERFORMANCE OF A SECTION OF THE MACKENZIE HIGHWAY

L. E. Goodrich

Division of Building Research, National Research Council of Canada, Ottawa, Ontario K1A 0R6

The thermal performance of a shallow road embankment constructed on permafrost in the discontinuous zone of the Mackenzie Valley was monitored over a 5-year period. The shallow (1.2 m) embankment consisted of clean borrow material end dumped on a right-of-way which had previously been cleared of trees. The road was never opened to general traffic; consequently snow removal was minimal. Measurements included hourly ground temperatures recorded with a data logger operating unattended. Settlements were monitored and thermal conductivity probes were installed to yield information on seasonal trends within the embankment and in the various natural materials comprising the active layer. The data indicate that degradation of the permafrost is under way and provide information on the relation between road surface temperatures and climatic factors for this region.

INTRODUCTION

In 1972 the Mackenzie Highway Environmental Working Group was established by the Department of Indian Affairs and Northern Development to monitor the environmental impact of the proposed Mackenzie Highway joining Fort Simpson to Inuvik and the Arctic coast. The Division of Building Research of the National Research Council of Canada was invited to carry out a limited study of the thermal aspects. At that time no case studies of shallow embankments constructed on permafrost had been carried out in the Fort Wrigley region. It was expected that although degradation would not pose serious problems in the zone south of Fort Wrigley, the data collected could be used to anticipate problems in the more difficult soils to the north. Although, as originally planned, the road was to be a graveled, all-weather road, construction was halted and during the final two winters of this study there was only minimal snow removal.

OBJECTIVES

The specific objectives of this study were:

1. to observe the changes in temperature regime in the ground underlying the road embankment as well as in the surrounding right-of-way in order to assess the thermal disturbance caused by the embankment;
2. to collect data on temperatures within the embankment for design purposes;
3. to assess the settlements of the embankment;
4. to obtain basic soil information for the site, including profiles of ice content, soil density, and composition; and
5. to measure thermal conductivities both in the natural ground and in the embankment and to assess seasonal and year-to-year variations.

FIELD INSTRUMENTATION

Field work was begun in March 1976 at mile 419.5, near the northern end of the proposed new construction on the Mackenzie Highway (Figure 1). Earlier reconnaissance had indicated the presence of permafrost. This location also offered important logistic advantages. The Hire North camp at River-Between-Two-Mountains could serve as base of operations, and in addition, a winter road facilitated access to Fort Simpson, the nearest point with land connections to the south. For these reasons this site was selected, even though more sensitive terrain could be found to the north.

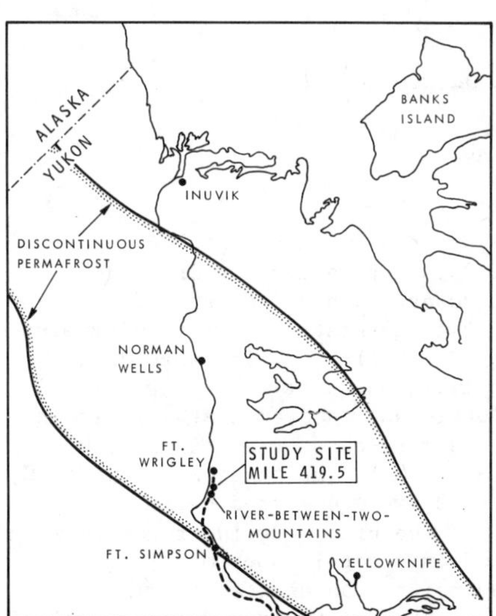

FIGURE 1 Location map.

To compare thermal conditions under the future embankment centerline, the shoulder, the previously cleared right-of-way, and the original undisturbed terrain, four holes were drilled approximately 8 m deep (the maximum achievable with equipment at hand). These locations correspond to profiles A, B, C, and D in Figure 2. Nearly continuous core was recovered from hole A (centerline), and grab samples were collected from holes B and D. The soil type, moisture content, and density were determined on site. In addition, representative core samples were shipped to the laboratory for further evaluation, including thermal conductivity measurements. The holes were instrumented with Yellow Springs #44033 thermistors mounted on multiconductor cable to a depth of 7 m. The cables were cased in PVC plastic tubing for protection from mechanical damage, and the boreholes were backfilled with dry sand. A prefabricated instrument hut made of fiberglas-styrofoam sandwich panels was erected at the site.

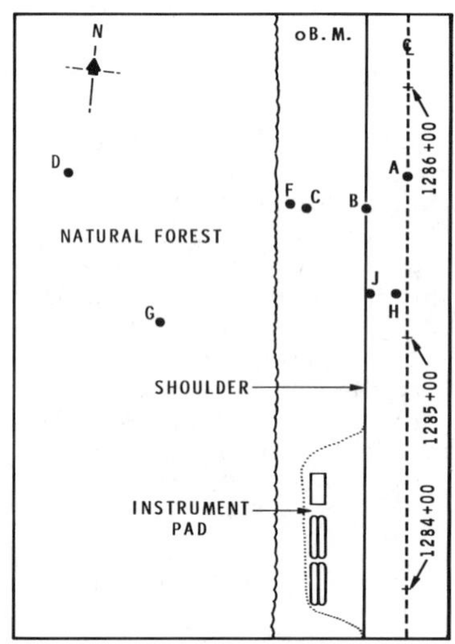

FIGURE 2 Site layout.

A Monitor Systems 9400 data logger using a lower power Digi-Data tape drive was installed in September 1976. Power was provided by a 3M propane-driven thermoelectric generator which simultaneously charged a bank of Gel-Cell batteries serving as a backup power supply. A propane furnace was installed to maintain ambient temperatures within the hut at levels adequate for the data-logging equipment. At this time also, three small pits (E, F, and G in Figure 2) were dug to a depth of approximately 1 m. The pits were instrumented with thermistors potted into metal tubes mounted horizontally on vertical wooden stakes. These installations were intended to provide more reliable information on near-surface temperature conditions than that available from the deep cables A-D. The pits were also instrumented with thermal conductivity probes (Goodrich 1979) to measure the seasonal variation of conductivity for the different materials comprising the active layer.

In June 1977 the final grade was completed past the instrumented section. A trench was dug across the road and stake-mounted thermistor probes, similar to those used in the shallow pits, were installed in the trench wall (Cables H and J in Figure 2). The addition of four thermal conductivity probes completed the thermal instrumentation. Finally, settlement plates were installed on the centerline and near the shoulder to monitor total settlement of the embankment.

The test site was operated, with some interruptions, from September 1976 until March 1981. During this time the data-logging equipment and thermoelectric power supply were left to function unattended for periods of 3-4 months. Approximately 80 channels, including 3 used to verify the equipment, were recorded every hour. Thermal conductivity measurements and level surveys were made whenever possible during site visits.

RESULTS

Figure 3 shows the soil profile for the centerline borehole (profile A). Conditions at the three remaining boreholes were similar, as inferred from split spoon sampling. The profiles consist primarily of silts and sands with some layers of gravel. Fine-grained materials are prevalent in the upper part of the profile, while the gravel layers are found only below 4 m. Correspondingly, the total volumetric moisture content varies from nearly 80% in the upper levels to less than 40% at depth. While some small ice lenses (<5 mm) are present in the silt layers, most of the ice is dispersed, and the soil is well bonded with ice throughout.

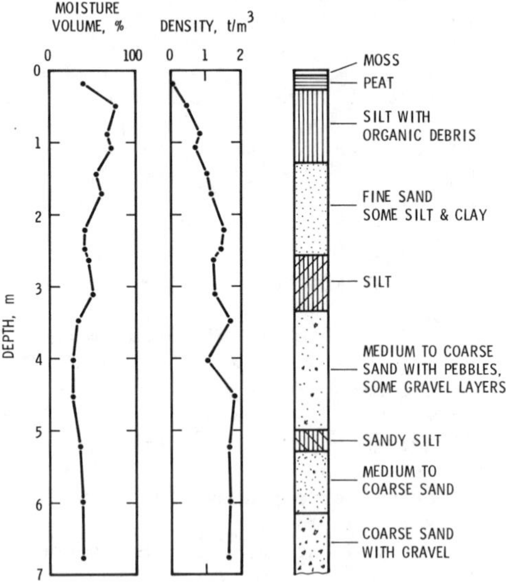

FIGURE 3 Soil conditions: mile 419.5.

Daily mean temperatures for all channels were computed from hourly data recorded on the original tapes. Data processing included a check of each scan for instrument drift, channel sequence, range and sign bit, and clock errors. If more than five unsatisfactory scans occurred in a daily period, the entire daily record was rejected. Since the manufacturer claimed an interchangeability error of less than ±0.1°C, thermistor linearization was done using the same formula for all channels. Examination of Figure 4 reveals, however, that overall errors may be considerably greater. The temperatures for channels 8-10, all of which are located at levels within the active layer (also channel 11 from 1979 onwards), should be near or at the freezing point during the autumn freezeback period (zero-curtain effect). Although this is well verified for channels 8 and 10, the error for channel 9 is rather more than 1°C, and that for channel 11 is nearly as great. It can also be seen that the errors increase with time. Errors of similar magnitude were found in other profiles for several channels in the active layer where a well-defined zero-curtain existed; nearly a third of these channels required corrections of more than 0.2°C. For channels below the active layer there is no certain reference temperature. In a few cases, corrections could be inferred from an examination of the annual mean profiles along with observation of the temporal behavior. In most instances, however, the data for these points were rejected. Typical corrected data are shown for the centerline profile in Figure 5.

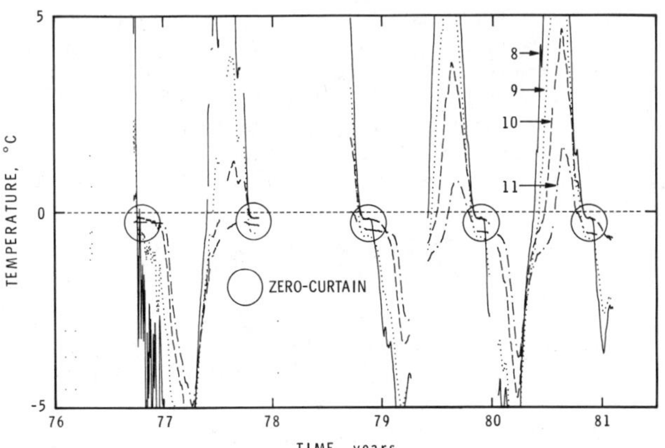

FIGURE 4 Errors in raw data of profile A C/L.

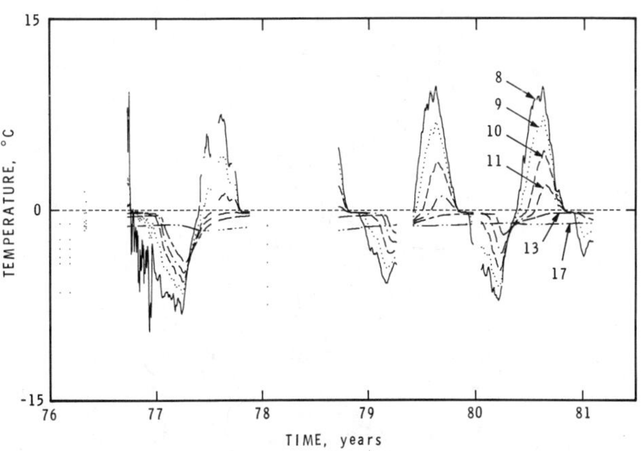

FIGURE 5 Corrected profile A C/L.

As regards equipment reliability, Figures 4 and 5 indicate that a nearly complete record was obtained, with two major exceptions. The long gap from mid-November 1977 to mid-September 1978 was the result of three separate circumstances. The data logger had to be removed for repairs in November 1977. All the equipment, except the defective data logger, was lost when fire destroyed the site in March 1978. Further delay resulted when plans to abandon construction of the road were announced. The second major gap, during early 1979, was purely the result of human error and not of equipment malfunction.

Thermal conductivity measurements were made whenever possible during routine site visits. Fifteen probes were eventually emplaced in a variety of materials, including the natural lichen mat and the decomposed peat layer, as well as at several depths in the natural soil and in the embankment. Figure 6a shows results obtained both in natural peat and in the peat layer underlying the embankment. Data for all years have been plotted on the same axes. The data can, in both cases, be reasonably well represented by constant frozen or thawed values. The thermal conductivity of the compressed peat beneath the embankment is 2-3 times that of the natural peat. In addition, the ratio of frozen to thawed conductivity is reduced from approximately 4 for the natural peat to approximately 2 for the compressed peat. Figure 6b shows thermal conductivities measured near the surface and near the base of the embankment. The data suggest that for the level near the surface the thawed conductivity is remarkably similar from one year to the next. Frozen values are distinctly greater but also show more scatter. It is interesting to note that for the lower level no distinct seasonal pattern is evident. The greatest year-to-year variation is, however, again found in the winter values.

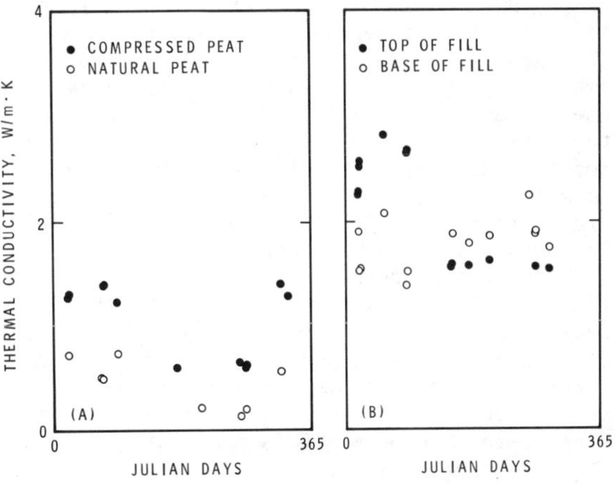

FIGURE 6 Thermal conductivity variation.

Some level survey results for the embankment centerline are summarized in Figure 7. The curves suggest similar trends at the embankment surface and base. (The increase in surface level in mid-1979 reflects road maintenance operations.) Total settlement from construction in mid-1977 to the final survey in mid-1980 was approximately 0.3 m. At many stations in the 200 ft (~60 m) section surveyed, there was considerably more settlement near the shoulders than on centerline. By mid-1980 the road surface had become a series of bumps and hollows with elevation differences of approximately 0.15 m over 10 m intervals.

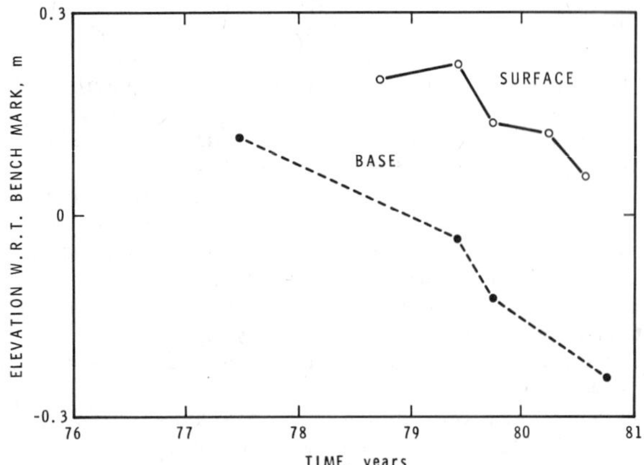

FIGURE 7 Settlement: C/L STN 1285 + 17.

DISCUSSION

Hand probing carried out during the initial reconnaissance (autumn 1975) indicated that the active layer in adjacent undisturbed terrain at the site was typically about 50 cm thick. There was, however, evidence that significant thermal degradation was under way following removal of the black spruce cover some 2 years earlier. Near the eastern side of the cleared right-of-way, advanced degradation was evident along the route of the former Canadian National Telegraph line. Here the ground was thawed to depths of 2 m or more.

The annual progress of the zero isotherm within the embankment and in the underlying ground is shown for the centerline profile in Figure 8. Although the term "zero isotherm" is used here and elsewhere for convenience, the graph, in fact, shows the isotherm for -0.15°C, this value being consistent with the zero-curtain "calibration" of Figure 5. The isotherm histories were computed from the adjusted daily mean temperatures with no temporal or spatial smoothing. This explains some of the apparent scatter in the curves. Note also that the cuspoidal shapes, seen especially in the advancing thaw fronts, are merely artefacts of the method (linear interpolation) used to calculate the position of the isotherm. This procedure will necessarily overestimate isotherm depth, except when the isotherm traverses a measurement level.

It can be seen from Figure 8 that prior to embankment construction the subgrade thawed to a depth of less than 1 m. Since construction, the thaw has progressed to successively greater depths each year so that by 1980 the total thaw reached 1.4 m.

Annual mean temperatures computed from the 3 years of nearly continuous records are shown for the embankment and subgrade centerline profile in

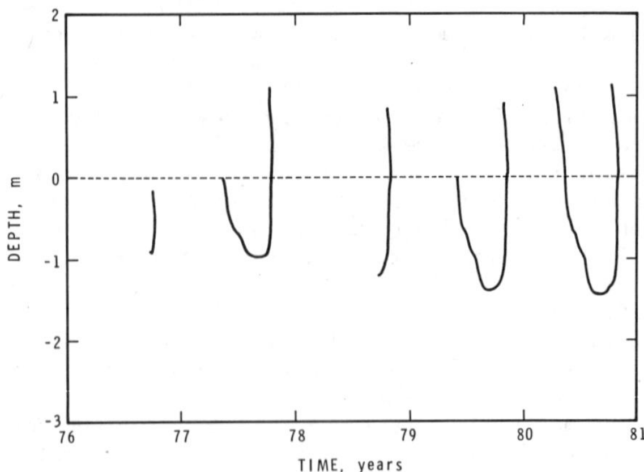

FIGURE 8 Centerline zero isotherm.

Figure 9 and shoulder profile in Figure 10. In 1979 and 1980, annual mean temperatures at the embankment surface were approximately +1.5°C on centerline and +3°C at the shoulder. A substantial increase (about 1°C) has occurred in the upper 1-2 m of the subgrade since the embankment was built. A similar trend is seen in the embankment shoulder profile (Figure 10). While shallow subgrade temperatures under the shoulder are not more than 0.3°C warmer than those at the centerline, near the surface the difference is substantially greater. The reversed curve in the upper levels of the 1977 profiles seen in both Figures 9 and 10 reflects the fact that construction activities took place during the first half of that year.

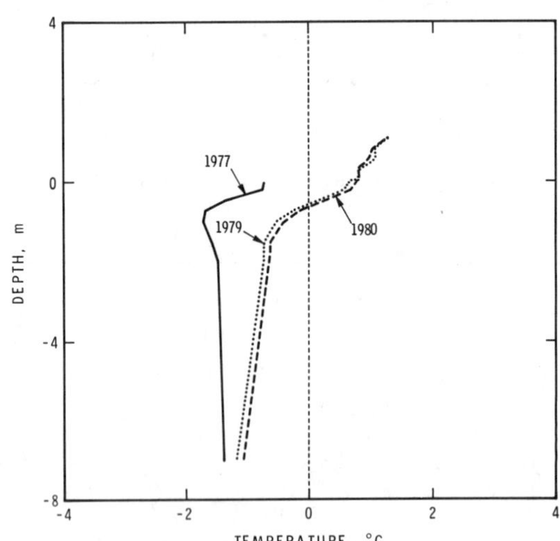

FIGURE 9 Annual means: centerline.

Figure 11 shows the annual mean temperatures for the deep datum cable located in the nearby partially cleared bush. It should be noted that the

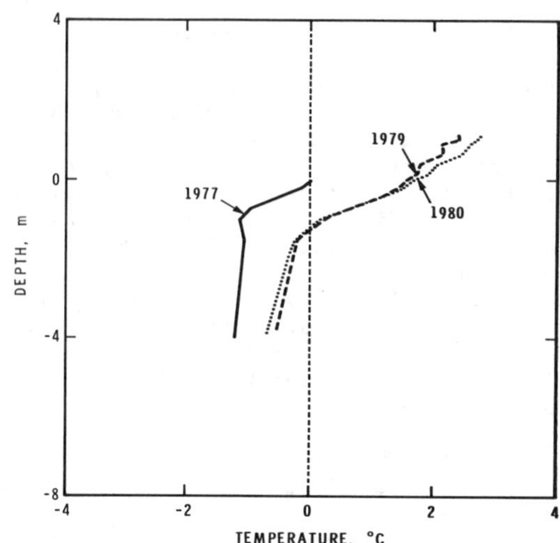

FIGURE 10 Annual means: shoulder.

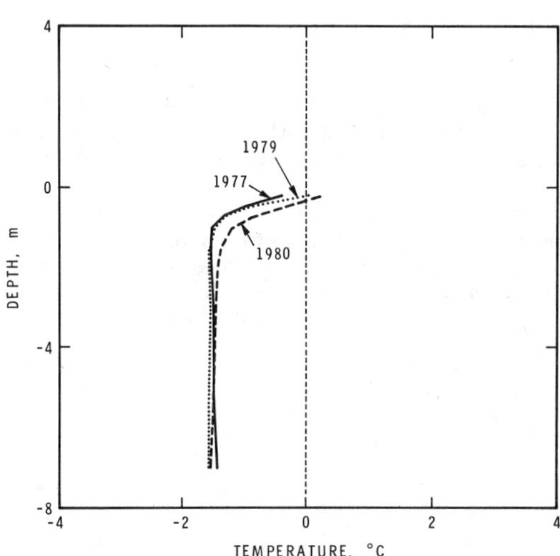

FIGURE 11 Annual means: deep datum.

results for 1979 are nearly identical to those for 1977. This may be interpreted as evidence that the large differences between the temperatures for these years in the embankment profiles are indeed caused by the presence of the embankment and are not merely a reflection of different weather conditions. Note, too, in Figure 11 that in successive years the near surface temperatures are a fraction of a degree warmer. In addition, there is some tendency for the active layer to deepen, as seen in Figure 12.

At all three sites, annual mean temperatures exhibit a strong tendency to decrease sharply through the active layer. This may be, at least in part, a manifestation of the thermal rectification effect resulting from temperature-dependent thermal

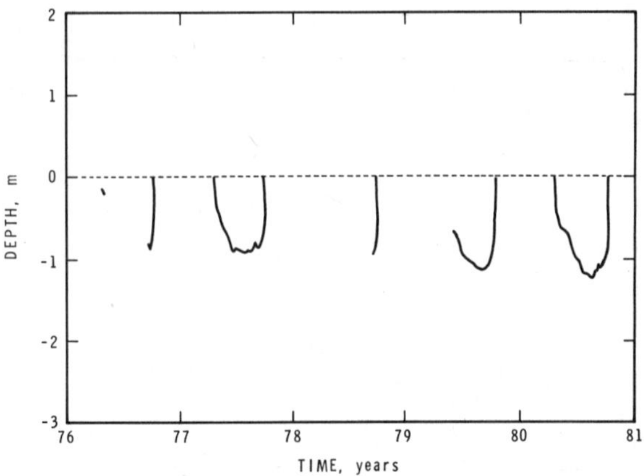

FIGURE 12 Deep datum zero isotherm.

conductivity. When the frozen conductivity is greater than the thawed conductivity, as it is here, annual mean temperatures will decrease with depth. The fact that profile distortion is confined primarily to the active layer suggests that thermal rectification is indeed the principal mechanism at these sites.

Profile curvature may, in addition, indicate that deep temperatures are not yet in equilibrium with long-term conditions at the surface. The 1980 profile at the deep datum site, Figure 11, suggests that mean temperatures will probably continue to rise somewhat and that, not unexpectedly, this datum site is far from being "undisturbed."

Figure 13 shows a comparison of temperatures at 1 m depth below the original ground surface for the embankment centerline, the shoulder, and the deep datum site. From the shape of the curves it can be inferred that thawing did not occur at the deep datum site during the first 2 years, but did occur in 1979 and was more intense in 1980. The data also indicate that some slight thawing occurred at the shoulder profile even before construction. There was no thawing prior to construction at the centerline 1 m level. Finally, it should be noted that the subgrade temperatures are warmer than the datum values not only in winter but also in summer; the shallow embankment enhances summer heating while reducing winter cooling.

Table 1 presents approximate values of "n-factors" for the site. The n-factors are computed as the ratio of the thawing or freezing index at the road surface (that is, beneath the snow cover) to that of the air. Surface temperatures were estimated by extrapolation from measurements starting at a level 20 cm below the surface. Since the air temperature records at the site were incomplete, data were extracted from the published records for Fort Simpson (200 km south of the site), the nearest regularly reporting standard meteorological station. The values in Table 1 should therefore be considered with caution. They can, however, be used for relative comparisons. Table 1 implies that the gravel surface had a thawing n-factor only 16%-19% greater than that of the original terrain. The difference is smaller than expected, and this may, in part, reflect disturbance at the datum site.

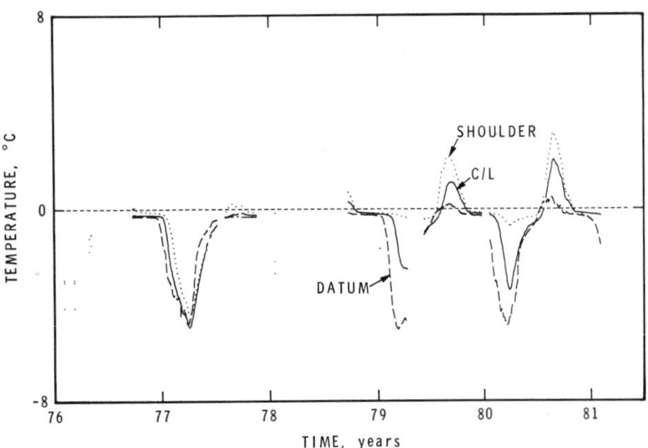

FIGURE 13 Comparison of temperatures at 1 m below original surface.

TABLE 1 Estimated "n-factors".

Year	Thawing Index (°C·day)	Freezing Index (°C·day)	C/L n_t	C/L n_f	Datum n_t	Datum n_f
1978-79	1970	3250	1.08	0.50	0.93	0.56
1979-80	2080	2920	1.02	0.56	0.86	0.64

CONCLUSION

The data suggest that for the climatic and soil conditions in the Fort Wrigley area, road construction results in degradation of the permafrost in spite of special precautions taken to minimize disturbance of the organic cover. For the present conditions of road operation (winter road) with infrequent traffic and limited snow removal, annual mean road surface temperatures are up to 3°C above the freezing point. Since there may, however, be a significant thermal rectification effect associated with temperature-dependent thermal properties at shallow levels, it is possible that mean temperatures at depth will remain negative.

The maximum thaw penetration is about 50% greater beneath the embankment than at the datum site. During the study period there was an evident tendency for the active layer to deepen with time. This tendency, however, in addition to implying a long-term increase in deep-ground temperatures, also mirrors the increase observed in summer air temperatures; consequently, it cannot be stated unequivocally that the active layer beneath the embankment will continue to deepen significantly. Further monitoring of the site would be required to establish the long-term trends with certainty.

Not all the data have been examined at this time. Further analysis, including numerical model simulation, may shed light on the ambiguities noted in this paper and provide useful guidance in designing future embankments in the Mackenzie Valley region.

ACKNOWLEDGMENTS

Prior to 1978 this study was financed by the Mackenzie Highway Environmental Working Group. After suspension of construction activities, financing was continued by the National Research Council of Canada. The author gratefully acknowledges the contribution of G. H. Johnston, whose help with site selection, and initial planning and field work was indispensable. A special thank-you is also due to T. L. White, J. C. Plunkett, W. G. Cooke, and T. Ireson for their unstinting assistance in the field and with essential chores in the laboratory. This paper is a contribution from the Division of Building Research, National Research Council of Canada, and is published with the approval of the director.

REFERENCE

Goodrich, L. E., 1979, Transient probe apparatus for soil thermal conductivity measurements, in Proceedings of a Symposium on Permafrost Field Methods, 3 Oct. 1977, and Permafrost Geophysics, 4 Oct. 1977, NRCC Technical Memorandum No. 124, Ottawa, June 1979.

BIBLIOGRAPHY

Brown, J., and Berg, R.L. (Editors), 1980, Environmental engineering and ecological baseline investigations along the Yukon River-Prudhoe Bay Haul Road, CRREL Report No. 80-19, Sept. 1980, 187 pp.

Judge, A.S., 1973, The thermal regime of the Mackenzie Valley: observations of the natural state, Environmental Social Committee, Northern Pipelines Task Force on Northern Oil Development, Report No. 73-38, Dec. 1973, 177 pp.

ROCK GLACIERS OF THE MOUNTAINS OF MIDDLE ASIA

A. P. Gorbunov

Permafrost Institute, Siberian Branch,
Academy of Sciences, Yakutsk, USSR

About 10,000 active rock glaciers exist in the mountains of Middle Asia. Their total area is about 2,000 km². There are 20 to 30 km³ of detritus and almost 20 km³ of ice concentrated in them. The altitude range of active rock glaciers is from 2,300 to 3,500 m in the northern part of the area, and from 3,300 to 4,500 m in the southern part. The largest rock glacier is 4.4 km long and the rate of advance in certain years is up to 1.3 m per year. The ice content of lobate rock glaciers is 50 to 75% by volume. Waste material from huge excavations on the Rusvumcherr Plateau was dumped on slopes of 30 to 40°. Lower down, the slopes decrease to 15 to 20° and the ice-rich waste material in this permafrost region formed "rock glaciers" with typical form and motion. Locally, an increase downslope to 30° inclination caused the artificial rock glaciers to rupture and a huge body of debris (4,000,000 m³) to catastrophically advance. Such rapid motion in an artificial rock glacier may also have occurred in natural rock glaciers in the past if such changes of slope are present. It is thought that such a catastrophic advance is recorded by a rock glacier of the Uzunvulak Valley, perhaps caused by an earthquake.

In the USSR, rock glaciers are known to exist in the Caucasus, Pamir, Tien Shan, and Dzhungarskiy Alatau, in Altai, Sajany, in the mountains of Transbaikal, in Dzhugdzhur and Cherskiy Ranges in the Karyak upland. However, they are the most common in the mountains of Middle Asia.

The mountains of Middle Asia are Saur Tarbagatay, Dzhungarskiy Alatau, Tien Shan, Pamir-Alai, and Kopetdag. There are no rock glaciers in the first and in the last regions because of either low altitude or the absence of permafrost.

The rock glaciers of Middle Asia have not yet been closely studied. There is more or less detailed information on the rock glaciers of Dzhungarskiy and Zailiyskiy Alatau, Kungey and Terskiy Ala Too, Kichik Alai Range, Ob'-Khingou basin and some other areas (Brukhanda 1976, Glazovskiy 1978, Gorbunov 1970, 1979, Iverova 1950, Palgov 1957, 1961, Suslov 1966, Shultz 1947). For most of the territory, however, there is little or no information on the rock glaciers. Therefore, only an approximate quantitative assessment is possible, although detailed studies of some regions permit generalizations at this time.

A preliminary estimate indicates that there are about 10,000 active rock glaciers in the total area of 2000 km² of the mountains of Middle Asia. They compose about 20-30 km³ of rock material and approximately 20 km³ of ice.

The altitudinal range of the active rock glaciers extends from 2300 to 3500 m in the Dzhungarskiy Alatau in the north and from 2500-3850 m in the Zailijskiy Alatau to 3300-4500 m in the Pamirs in the south. Thus, the maximum altitudinal distribution is about 1000 m.

Inactive rock glaciers are much less common and occur only by 100-200 m lower than the active ones. In Zailijskiy Alatau they total about 15% of all rock glaciers of the Ridge. Some of them have been destroyed, and others are buried under the recent active rock glaciers.

The author (Gorbunov 1979) has suggested that the term "specific density" of these formations should be used as a comparative index of the degree of development of the active rock glaciers. This is the relationship between the total area of active rock glaciers, in m², and the area of the basin of a given valley above the lower limit of distribution of these forms in km². For Zailijskiy Alatau this index is equal to 14,550 m²/km².

From the reports of Barsch (1977), Kershaw (1978), Higuchi et al. (1978) it is possible to determine the approximate specific density for the Switzerland Alps—3333, for the Seluins (Canada)—13,772, and for the Himalayas (Everest region)—1430 m²/km². Thus, the increase of mountain climate continentality creates better conditions for the formation of rock glaciers. This relationship has been known before, but "specific density" enables a quantitative assessment.

Besides climate, conditions affecting the development of rock glaciers and their location are relief features, lithology, and avalanche processes. The most favorable conditions for rock glacier formation, besides climate control, are deeply dissected mountain massifs with steep talus and rock slopes; slightly fissured solid rocks that form large amounts of debris; moderate snow cover; and moderate avalanche activity. Thick snow cover creates perennial freezing of talus and hinders its transformation into rock glaciers. Large avalanches suppress the processes of formation of the classic rock glaciers.

Local factors (relief, lithology, snow cover and avalanches) vary from one river basin to another within the same mountain country, thereby considerably altering the specific density of rock glaciers. Thus, within the Zailijskiy Alatau it varies from 41,610 in the Issyk River basin to 6,500 m²/km² in the Kargaul'dinka River basin. Rock glacier occurrence substantially depends on slope position. In Zailijskiy Alatau 72.5% of all active rock glaciers are confined to the north-facing slopes, and 15.6% to the south.

Such correlation is determined by the fact that permafrost is more common and extensive on north-facing slopes. A similar situation exists in other mountains of the Northern Hemisphere. In the Rocky Mountains of Canada, the former comprise 79.8%, the latter 5.9% (Luckman and Crockett 1978), in the Selwyn Mountains (northwest of Canada) these values are equal to 63.4% and 14.6% respectively (Kershaw 1978). In some instances, however, this regularity breaks in the mountains of Middle Asia. Thus, in Ob'Khingou basin (Darvas, Western Pamir) rock glaciers prevail on the south-facing slopes (Brukhanda 1976).

Rock glaciers have been recently divided into two major groups: those next to talus slopes and those next to end moraines of present glaciers (Barsch 1977). The former, according to Krasnoslobodtsev (1971), are referred to "near-slope," and the latter to "near-glacier" rock glaciers. These terms imply not only genesis of rock glaciers but also their location. In the papers of foreign authors the former are usually called "lobate," and the latter are termed "tongue-shaped" rock glaciers.

Those rock glaciers which look like a continuation of end ice moraines and gradually flow together with them, belong to tongue-shaped rock glaciers. It is this kind or rock glacier that is commonly taken for end moraines. But the active rock glaciers differ from the moraines by their mobility, steep active frontal slope, arched barriers and depressions, as well as by inner structure, about which in most cases we can only guess. The identification of tongue-shaped rock glaciers often becomes difficult during periods of glacier advances, which commonly results in temporary smoothing of the "corrugated" surface of the active rock glacier and superposition of end moraines on it. Besides, the inner structure of rock glacier alters as well, i.e. buried massives of glacial ice can appear to be inside.

It commonly happens that a typical lobate rock glacier formally becomes a tongue-shaped one due to glacier advance. Such a phenomenon was observed when a large lobate rock glacier moved down the valley; then later a glacier advanced and pushed against the rock glacier. Such a lobate rock glacier destroyed by a glacier will be accepted as a typical tongue-shaped one. Hence, the rock glacier position near the glacier edge does not allow one to solve a problem of its genesis. Only petrographical analysis of rudaceous rock and its correlation with surrounding slopes and glacial moraines can clarify the origin of the given rock glacier.

Tongue-shaped rock glaciers are not as common in the mountains of Middle Asia as are the lobate rock glaciers. Only 172 active rock glaciers out of 429 in Zailijskiy Alatau are tongue-shaped ones. The average length, width and area of the tongue-shaped rock glacier in Zailijskiy Alatau are 630 m, 280 m and 0.1897 km^2 respectively.

The largest tongue-shaped rock glacier in Zailijskiy Alatau, and the largest one in the area under discussion, is Prjamoy shcheli (right influx of the Issyk River). It is 4400 m long, the largest width 1320 m, the least width 220 m, mean width 400 m, and area 1662 m^2. The total volume of the rock glacier is estimated to be no less than 8.3×10^7 m^3.

The rock glacier of Prjamoy shcheli is in a trough valley inclined to the north. Its upper part is at 3500 m, the frontal bench foot at 2500 m. The rock glacier is tree-shaped in cross-section.

At present, the area of rocky slopes, providing source material is nearly 7 km^2; in periods of maximum glacier retreat it could reach 11-12 km^2.

Active lobate rock glaciers include forms that originate at the foot of talus slopes of valleys and empty cirques.

In Zailijskiy Alatau the average size of an active lobate rock glacier is 380 m long, 180 m wide, with an area of 0.0667 km^2. One of the largest rock glaciers of the region is situated in the valley of the Kebin River (Chilik basin). Its length is 1750 m, the largest width is 200 m, average width of 150 m, and an area of 0.2775 km^2. The height of the frontal bench is about 40 m. The upper part of the rock glacier lies at the altitude of 3600 m and the frontal bench at 3050 m.

Lobate and tongue-shaped rock glaciers in Zailijskiy Alatau comprise 59.9% and 40.1% of the population respectively. By comparison, in the Swiss Alps they comprise 60.6% and 39.4% (Barsch 1977); in the Rocky Mountains of Canada (Jasper and Banff National Parks) 45.5% and 54.6% (Luckman and Crockett 1978).

An active rock glacier is so named because it shows current motion. The motion is expressed by down valley movement and by vertical rising and lowering of separate parts of the surface.

The usual rate of rock glacier advance is 20-30 cm per year, though in some years it may be substantially higher on some rock glaciers. The central part of the rock glacier of the Gorodetskiy glacier (Zailijskiy Alatau) advanced at a rate of 1.1 m per year from 1923 until 1977. In some years the rate is likely to be considerably more.

Vertical motions of rock glaciers usually do not exceed a few centimeters per year. According to Palgov (1957), rising and subsequent lowering of points on the surface of the "Nizkomorennyy" rock glacier (Jungarskiy Alatau) was 1-2 cm per year. The average rate of vertical motion of the Murtel rock glacier in the Alps was 2-6 cm per year (Barsch and Hell 1975).

It is necessary to emphasize that the horizontal and vertical motions in the rock glaciers are notable for changing over the years. The rate of the front advance of the Gorodetskiy rock glacier during the period 1923-1946 ranged from 0.4 to 0.9 m/yr; from 1946 to 1960, 0.9 m/yr, and during the period from 1960 to 1977 the rate was 0.7 m/yr (Titkov 1979). According to Palgov (1961), the rate of the advance of the "Nizkomorennyy" rock glacier slowed from the period 1953 to 1959 about 36-37% of the 1948-1953 period.

Progressive (horizontal) motion of the rock glacier is controlled by the plastic and ruptural deformations of the contained ice. Vertical motion is generated due to different reasons. Among them are thawing or formation of ice, as well as formation of lateral swells and hollows on the surfaces of rock glaciers due to a difference in the rate of motion between the center and margins of the rock glaciers.

The least known part of a rock glacier is the

inner structure including the amount and type of ice. At present there exists two main concepts for the origin of rock glaciers: (1) a rock glacier is part of an ice glacier buried under rock waste and still has the ability for motion; and (2) a rock glacier is a concentration of detritus, cemented with ice of a different origin and moves by plastic and ruptural deformations within the ice.

Data on the inner structure of rock glaciers are extremely rare, although in the past few years drilling has been undertaken to obtain such information. Two holes, 10.4 and 7.0 m deep were drilled in rock glaciers in the Alps (Barsch, Fierz and Haeberli 1979). These rock glaciers appeared to be composed of perennially frozen rudaceous material containing segregated ice lenses with thicknesses up to 30 to 60 cm. Maximum ice lense thickness is thought to reach 150-170 cm. Segregated ice lenses were found in permafrost of a rock glacier in the South American Andes (Wayne 1981). Such ice lenses also occur in the rock glaciers in Middle Asia.

The studies carried out in the Zailijskiy Alatau reveal that the rock glaciers are composed of perennially frozen rudaceous rocks. The ice content is up to 50-70% by volume. Ice cement, segregated ice in the shape of lenses (up to 15 cm thick and possibly more), and congelation ice that fills fractures in permafrost are the most typical types of rock glaciers. The fractures may be due to permafrost fracture during rock glacier motion and during earthquakes.

Investigations of the structure, dynamics, and physical-mechanical properties of rudaceous debris containing ice, carried out on the slopes of the Khibiny, are of an extreme interest for understanding of the inner structure and motion mechanism of rock glaciers. Great amounts of debris containing ice formed during formations of huge artificial excavations on the Rusvumcherr Plateau and the accumulation of the detritus dumped on slopes attaining angles generally between 30° to 40°. With the slopes decreasing to 20-15° the talus-shaped piles formed into "rock glaciers" with typical horseshoe bars and frontal benches (Otvaly ... 1975, Semekhin, Mjagkov and Vashchalova 1979).

These "technogenic" rock glaciers of the Khibiny originate only on the northern slopes of the mountains where permafrost exists. Perennial ice occurs in the rudaceous material. The volumetric ice content commonly exceeds 50% and occasionally reaches 77%. Ice lenses up to 1.5 to 2.0 m thick occur. This high ice content leads to an abrupt qualitative change of ice-rock mass strains, i.e. the overall mass may behave as a plastic under the stress of 1.6 kg/cm^2 at temperature -1.2°C. Forward motion rates of 120 cm per day are known.

Change in slope steepness from 10 to 15° to 20 to 30° for the artificial rock glaciers caused a sudden increase in rate of flow and rupture of the frozen body. A body of about 4,000,000 m^3 of debris came down, in a mass. The debris form resembles something between an earth flow and a rock slide.

From this observation it follows that under certain conditions the "technogenic" piles are strikingly similar in their appearance and interior structure to natural rock glaciers. Disastrous advance of the artificial rock glacier indicates that such an event can occur in natural conditions as well. Although such rapid advances of rock glaciers have not been registered up to now, it does not mean that they are impossible. Perhaps they occur under certain conditions when the moving rock glacier reaches an abrupt increase in steepness of a slope.

Analyses of aerial photographs of different areas of the Tien Shan and Jungarskiy Alatau let us surmise that some rock glaciers have been subjected to the aforementioned rapid advances.

One of the best examples of a past disastrous advance concerns the rock glacier of Uzunbulak Valley (the Jungarskiy Alatau). A typical near-slope rock glacier of about 1 km long and 150 m wide formerly extended from 2800 to 2600 m. The rock glacier was supplied with detritus from steep slopes.

For some reason, perhaps an earthquake, the rock glacier advanced rapidly. There formed a zone of rupture 600 m long, which looked like a trench confined by swells on both sides. Downslope from the trench exists a typical rock glacier 800 m long and 80-100 m wide. It extends to the low altitude of 2100 m; this is abnormally low for the Jungarskiy Alatau where rock glaciers usually do not descend lower than 2400 to 2500 m.

Active and inactive rock glaciers represent two generations of late Holocene time. The first one (forms that are inactive now) started to develop during the period of mountain glacier retreat approximately 2500 to 3000 years ago, when the firn line was no more than 100 m lower than at present. The beginning of the active rock glaciers formation falls on the last stage of glacier retreat (1500 years ago) when the firn line was near the present level. Absence of rock glaciers more than 2500 to 3000 years old indicates that when the firn line is 300 to 400 m lower than present, the glaciers and snowpacks are more widespread and prevent rock glacier formation because ice covers rock slopes. As the firn line rises, more and more bare rock becomes available and become sources of detritus that form rock glaciers. Rock glacier evolution is then asynchronous to the stages of mountain glaciation. The recent advance of rock glaciers occurs in the period of general retreat of the glaciers in the region. Therefore, rock glaciers are to some extent the antagonists to mountain glaciers and knowledge of their history permits us to know more about retreat of mountain glaciers.

REFERENCES

Barsch, D., 1977, Nature and importance of mass-wasting by rock glaciers in Alpine permafrost environments: Earth surface processes, v. 2, p. 231-245.

Barsch, D., Fierz, H., and Haeberli, W., 1979, Shallow core drilling and bore-hole measurements in permafrost of active rock glacier near the Grubengletscher, Wallis, Swiss Alps: Arctic and Alpine Research, v. 11, no. 2, p. 215-228.

Barsch, D., and Hell, G., 1975, Photogrammetriesche Bewegungsmessungen am Block-gletscher Murtel 1, Oberengadin, Shweizer Alpen: Zeitschrift für Gletscherkunde und Glaziologie, v. XI, no. 2, p. 111-142.

Brukhanda, V. I., 1976, Kamennye gletchery Kavkaza i Pamiro-Altaja i ikh svjaz s pulsacijami lednikov, in Materialy glaciologicheskikh issledovaniy, v. 27: Khronika. Obsuzdenija, p. 63-70.

Glazovskiy, A. F., 1978, Kamennye gletchery basseina r. Bolshoy Almatinki (Zailijskiy Alatau), in Kriogennye javlenija vysokogoroy: Novosibirsk, "Nauka," p. 85-92.

Gorbunov, A. P., 1970, Merzlotnye javlenija Tjan-Shanja: Moscow, Gidrometeoizdat, 265 p.

Gorbunov, A. P., 1979, Kamennye gletchery Zailijskogo Alatau, in Kriogennye javlenija Kazakhstana i Sredney Asii: Yakutsk, izd. Instituta merzlotovedenija, p. 5-34.

Higuchi, K., Fushimi, H., Ohata, T. et al., 1978, Preliminary report on glacier inventory in the Dudh Kosi region: Journal of the Japanese Society of Snow and Ice, SEPPYO, v. 40, p. 78-84.

Iveronova, M. I., 1950, Kamennye gletchery Severnogo Tjan-Shanja: Trudy Instituta geografii AN SSSR, v. 45, Moscow-Leningrad, p. 69-80.

Kershaw, G. P., 1978, Rock glaciers in the Cirque lake area of the Yukon-Northwest Territories in Alberta Geographer, p. 61-68

Krasnoslobodzev, I. S., 1971, O kamennykh gletcherakh Bolshogo Kavhaza: Vestnik MGU. Geografija, v. 1, p. 95-97.

Luckman, B. H., and Crockett, K. J., 1978, Distribution and characteristics of rock glaciers in the southern part of Jasper National Park, Alberta: Canadian Journal of Earth Sciences, v. 15, no. 4, p. 540-550.

Otvaly na gornykh sklonakh, 1975: Leningrad, "Nauka," p. 149.

Palgov, N. N., 1957, Nabludenija nad dvigeniem odnogo iz kamennykh gletcherov khrebta Jungarskogo Alatau, in Voprosy geografii Kazakhstana. Alma-Ata, v. 2, p. 195-207.

Palgov, N. N., 1961, Novye nabludenija nad dvigeniem kamennogo gletchera Nizkomorennogo v Jungarskom Alatau, in Voprosy geografii Kazakhstana. Alma-Ata, v. 8, p. 200-204.

Semekhin, Yu. V., Mjagkov, S. M., and Vashchalova, T. V., 1979, Tekhnogennye geomorfologicheskie processy nivalno-glacialnoy zony (na primere Khibin): Materialy glaciol. issled. Khronika. Obsuz., no. 35, p. 192-195.

Shultz, S. S., 1947, O gravitacionnykh (massovykh) dvigenijakh v Tjan-Shane: Izvestija Kirgizskogo filiala AN SSSR, v. 4, p. 85-96.

Suslov, V. F., 1966, Kamennye gletchery Kichik-Alaja: Trudy Sredneaziatskogo NIGMI, v. 27 (42). Voprosy glaciologii Sredney Azii, Leningrad, p. 13-17.

Titkov, S. N., 1979, O dvigenii nekotorykh gletcherov Zailijskogo Alatau, in Kriogennyje javlenija Kazakhstana i Sredney Azii., Yakutsk, izd. Instituta merzlotovedenija, p. 34-42.

Wayne, W. J., 1981, Ice segregation as an origin for lenses of non-glacial ice in "ice-cemented" rock glaciers: Journal of Glaciology, v. 27, no. 97, p. 506-510.

A THEORETICAL MODEL FOR PREDICTING THE EFFECTIVE THERMAL CONDUCTIVITY OF UNSATURATED FROZEN SOILS

Fabio Gori

Department of Energetics, University of Florence
Via S.Marta 3, 50139 Florence, Italy

The model assumes the soil as a cubic space with a centered cubic solid grain. The hypotheses of parallel isotherms and heat flux lines in the determination of the effective thermal conductivity are discussed. The comparison with several experimental results on saturated two-phase media indicates that the assumption of parallel isotherms gives a good agreement. The unsaturated frozen soils are investigated taking into account phenomena of adsorption and capillarity. The predictions of the present model are finally compared with experimental data on unsaturated frozen soils available in the literature. The model gives better agreement to the data than that found using the expressions of Johansen and Frivik (1980) and the electrical conductivity analogy of De Vries (1963). Some characteristic results are presented for soils studied by Kersten (1949) and Penner et al. (1975)

INTRODUCTION

The prediction of the effective thermal conductivity of unsaturated frozen soils is fundamental for heat transfer engineering in frozen areas. Previously published works on this topic do not seem to resolve the choice of a prediction method.

The work of Johansen and Frivik (1980), which incorporates the features of the Kersten one (1949), is a combination of empirical and interpolating rules. It needs several empirical constants that depend on the kind of soil investigated.

The method of De Vries (1963) uses the Maxwell equation or the analogy with the electrical conductivity. Although it contains some empirical constants, its use is generally acceptable for saturated two-phase media. Unsaturated frozen soil is a porous medium with phenomena of adsorption and capillarity and the electrical conductivity analogy is no longer applicable.

The present work investigates the effective thermal conductivity of unsaturated frozen soils considering phenomena of adsorption and capillarity. The analysis takes into account the conductive heat transfer only, i.e., convective and radiative heat transfer are neglected.

THEORETICAL ANALYSIS

Saturated Frozen or Dry Soils

The soil is represented by a cubic grain inside a cubic space as in Figure 1. The ratio ℓ_t/ℓ_s varies with the porosity of the soil and is given by

$$\beta = \frac{\ell_t}{\ell_s} = \sqrt[3]{\left\{\frac{1}{1 - \varepsilon}\right\}} = \sqrt[3]{\left\{\frac{\rho_s}{\rho_d}\right\}}. \tag{1}$$

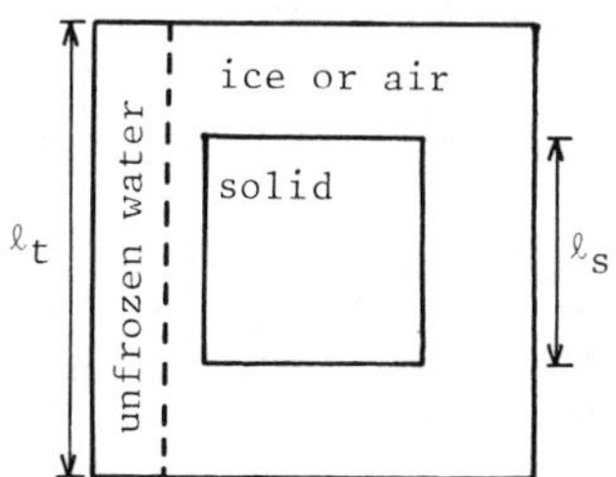

FIGURE 1 Saturated soil.

The effective thermal conductivity of the soil can be determined with the hypothesis that the isotherms are horizontal or the heat flux lines are vertical. These two cases are the extremes for the geometrical assumption adopted (Crane and Vachon 1977).

Unfrozen water, eventually present in the frozen soil, is dispersed along the length of the cell, ℓ_t, as shown in Figure 1. Hence, if u_w is the ratio of the unfrozen water volume to the total volume of the cell, the section of unfrozen water is

$$S_u = u_w \ell_t^2 = u_w \beta^2 \ell_s^2 \tag{2}$$

The effective thermal conductivity for horizontal parallel isotherms is

$$\frac{1}{k_T} = \frac{\beta - 1}{k_c \beta (1 - u_w) + k_w \beta u_w} + \frac{\beta}{k_c \{\beta^2(1 - u_w) - 1\} + k_s + k_w \beta^2 u_w}, \tag{3}$$

while for vertical parallel heat flux lines it is

$$k_Q = \frac{1}{\frac{\beta(\beta - 1)}{k_c} + \frac{\beta}{k_s}} + k_c \frac{\beta^2(1 - u_w) - 1}{\beta^2} + k_w u_w, \quad (4)$$

where k_c is equal to k_i for frozen soils and to k_a for dry soils.

The hypothesis of horizontal parallel isotherms is finally chosen because of the good agreement with the experimental data (see Table 1).

Unsaturated Frozen Soils

Unsaturated frozen soils are divided into two groups depending on whether the ratio V_w/V_v is lower or greater than a certain amount of adsorbed and dispersed water. The adsorbed water increases with the specific surface of the soil; hence, a different value should be used for the various soils. In order to simplify the analysis, an empirical constant value, equal to $V_{wA}/V_v = 0.083$, is used in this work.

Unsaturated Frozen Soils, $V_w/V_v < V_{wA}/V_v$

In this case the ice is supposed to be distributed on the six lateral surfaces of the cubic grain, as shown in Figure 2. Because of the hypothesis of horizontal parallel isotherms, the position of the ice in the cell is not significant but the important assumption is the absence of ice-bridges among the adjacent grains.

The unfrozen water is supposed to be distributed along the length of the cell, as shown in Figure 1. The thickness of ice on the grain surface, ℓ_{ia}, is given by

$$\ell_{ia} = \frac{\delta}{6} \ell_s = \frac{\ell_s}{6} \frac{V_i}{V_s}, \quad (5)$$

where $\delta = V_i/V_s$ and

$$\frac{V_i}{V_s} = \frac{w \rho_d - u_w \rho_w}{(1 - \varepsilon)\rho_i}; \quad (6)$$

the effective thermal conductivity is

$$\frac{1}{k_T} = \frac{\beta - 1 - \frac{\delta}{3}}{\beta\{k_a(1 - u_w) + k_w u_w\}} + \frac{\beta\delta}{3\{k_a(\beta^2 - 1 - \beta^2 u_w) + k_i + k_w \beta^2 u_w\}} + \frac{\beta}{k_s + \frac{2}{3}\delta k_i + k_a(\beta^2 - 1 - \frac{2}{3}\delta - \beta^2 u_w) + k_w \beta^2 u_w} \quad (7)$$

Unsaturated Frozen Soils, $V_w/V_v > V_{wA}/V_v$

If the water content is higher than V_{wA}/V_v, the liquid water accumulates at the points of contact of the grains in the form of biconcave lens (or liquid cups). The curvature of the lateral surfaces and the surface tension between liquid water and air contribute to the capillary pressure of the liquid. The liquid cups coalesce when the liquid water, V_{wf}/V_v, is in the range 0.183 to 0.226, respectively, for $\varepsilon = 0.4764$ and $\varepsilon = 0.2595$ (Luikov 1966). For soils with different porosities the liquid water, accumulated at the points of contact among the grains, is assumed proportional to the porosity of the soil and is evaluated by the following interpolation:

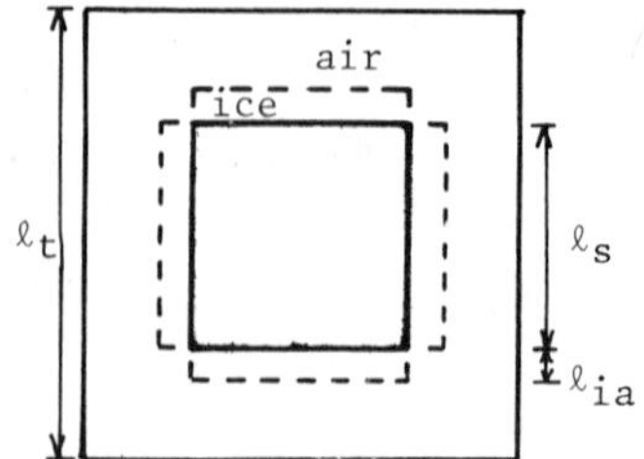

FIGURE 2 Unsaturated frozen soil, $V_w/V_v < V_{wA}/V_v$.

$$\frac{V_{wf}}{V_v} = 0.183 + (0.226 - 0.183)\frac{0.4764 - \varepsilon}{0.4764 - 0.2595}. \quad (8)$$

The interfacial tension between ice and water, σ_{iw}, is reduced, in relation to that between water and air, σ_{wa}, in the proportion

$$|\sigma_{iw}| = 0.052\,|\sigma_{wa}| = |\sigma_{wa}|\,\phi, \quad (9)$$

where $\phi = 0.052$ according to Walden's rule (Perry 1963). Hence the capillary pressure that maintains the ice among the grains is reduced in the same proportion.

The ice accumulated among the grains, V_{if}/V_v, is a fraction of the liquid water content

$$\frac{V_{if}}{V_s} = \frac{V_{wf}}{V_s}\phi = \frac{V_{wf}}{V_v}(\beta^3 - 1)\phi, \quad (10)$$

and the remaining ice is disposed around the grain, as in Figure 3. Unfrozen water is again distributed along the cell, and its section is given by equation (2). The ice volume around the grain is given by

$$\ell_i^3 = \ell_s^3 \gamma^3 = \ell_s^3 \left(\frac{V_i}{V_s} - \frac{V_{if}}{V_s} + 1\right), \quad (11)$$

where V_i/V_s is given by equation (6), from which it is determined $\gamma = \ell_i/\ell_s$. The ice disposed among adjacent grains is distributed on the six surfaces of the cube, yielding a surface of contact given by

$$\ell_{if}^2 = \ell_s^2 \gamma_f^2 = \frac{V_{if}/V_s}{3(\beta - \gamma)}\ell_s^2. \quad (12)$$

The ratio $\ell_{if}/\ell_s = \gamma_f$ can be lower or greater than unity giving two different expressions for the effective thermal conductivity. For $\gamma_f < 1$

$$\frac{1}{k_T} = \frac{\beta^2 - \beta\gamma}{k_a(\beta^2 - \gamma_f^2 - \beta^2 u_w) + k_i \gamma_f^2 + k_w \beta^2 u_w} +$$

$$\frac{\beta\gamma - \beta}{k_a(\beta^2 - \gamma^2 - \beta^2 u_w) + k_i \gamma^2 + k_w \beta^2 u_w} +$$

$$\frac{\beta - \beta\gamma_f}{k_a(\beta^2 - \gamma^2 - \beta^2 u_w) + k_i(\gamma^2 - 1) + k_s + k_w \beta^2 u_w} +$$

$$\frac{\beta\gamma_f}{k_s + k_i(\gamma^2 - 1 + 2\beta\gamma_f - 2\gamma\gamma_f) + A} \quad (13)$$

with

$$A = k_a(\beta^2 - \gamma^2 - 2\beta\gamma_f + 2\gamma\gamma_f - \beta^2 u_w) + k_w \beta^2 u_w$$

and for $\gamma_f > 1$

$$\frac{1}{k_T} = \frac{\beta^2 - \beta\gamma}{k_a(\beta^2 - \gamma_f^2 - \beta^2 u_w) + k_i \gamma_f^2 + k_w \beta^2 u_w} + \quad (14)$$

$$\frac{\beta\gamma - \beta\gamma_f}{k_a(\beta^2 - \gamma^2 - \beta^2 u_w) + k_i \gamma^2 + k_w \beta^2 u_w} +$$

$$\frac{\beta\gamma_f - \beta}{k_a(\beta^2 - \gamma^2 - 2\beta\gamma_f + 2\gamma\gamma_f - \beta^2 u_w) + B} +$$

$$\frac{\beta}{k_s + k_i(\gamma^2 + 2\beta\gamma_f - 2\gamma\gamma_f - 1) + C}$$

with

$$B = k_i(\gamma^2 + 2\beta\gamma_f - 2\gamma\gamma_f) + k_w \beta^2 u_w$$

and

$$C = k_a(\beta^2 - \gamma^2 - 2\beta\gamma_f + 2\gamma\gamma_f - \beta^2 u_w) + k_w \beta^2 u_w.$$

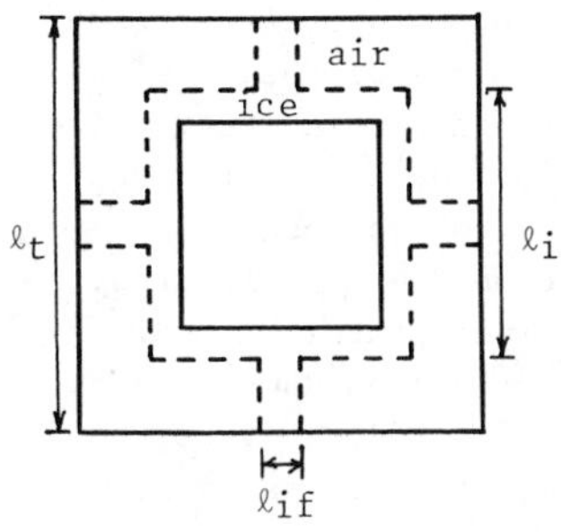

FIGURE 3 Unsaturated frozen soil, $V_w/V_v > V_{wA}/V_v$.

Previous Works

Among the previous works on unsaturated frozen soils, those of De Vries (1963) and Johansen and Frivik (1980) are considered for comparison.

The analysis of De Vries (1963) has been applied to frozen soils by Penner (1970) and Farouki (1982), and it is reported in detail,

$$k_e = \frac{u_w k_w + F_i x_i k_i + F_a x_a k_a + F_s(1 - \varepsilon)k_s}{u_w + F_i x_i + F_a x_a + F_s(1 - \varepsilon)} \quad (15)$$

where

$$F_s = \frac{1}{3}\{\frac{2}{1 + (k_s/k_c - 1)0.125} +$$

$$\frac{1}{1 + (k_s/k_c - 1)0.75}\}, \quad (16)$$

$$F_i = \frac{1}{3}\{\frac{2}{1 + (k_i/k_c - 1)0.125} +$$

$$\frac{1}{1 + (k_i/k_c - 1)0.75}\}, \quad (17)$$

$$F_a = \frac{1}{3}\{\frac{2}{1 + (k_a/k_c - 1)g_a} + \frac{1}{1 + (k_a/k_c - 1)g_c}\}, \quad (18)$$

$$g_c = 1 - 2\, g_a, \quad (19)$$

$$g_a = 0.333 - x_a \frac{0.333 - 0.035}{\varepsilon} \quad (20)$$

for $0.09 \leq x_c \leq \varepsilon$

$$g_a = 0.013 + 0.944\, x_c \quad \text{for } 0 \leq x_c < 0.09, \quad (21)$$

and x_c and k_c are relative to the continuous medium. For unsaturated frozen soils, $k_c = k_w$ if $u_w \neq 0$ and $k_c = k_i$ if $u_w = 0$. For dry soils, $k_c = k_a$.

The expression of Johansen and Frivik (1980) for unsaturated frozen soils is

$$k_e = k_d + (k - k_d)\, S_r, \quad (22)$$

where

$$k_d = 0.034\, \varepsilon^{-2.1} \quad (23)$$

$$k = k_c^{\varepsilon}\, k_s^{(1 - \varepsilon)} \quad (24)$$

$$S_r = \frac{w\, \rho_d - u_w(\rho_w - \rho_i)}{\varepsilon\, \rho_i} \quad (25)$$

$$\text{with} \quad k_c = k_i \qquad \text{if} \quad u_w = 0 \quad (26)$$

$$\text{and} \quad k_c = (k_w - k_i)\frac{u_w}{\varepsilon} + k_i \qquad \text{if} \quad u_w \neq 0 \quad (27)$$

For saturated two-phase media, the expression recently found by Muzzio and Solaini (1982) is used as comparison,

$$k_e = \frac{k_s}{a\{1 + A(k_s/k_c - 1)\}}\{a - \ell n\left[1 + A\left(\frac{k_s}{k_c} - 1\right)(1 - e^a)\right]\} \quad (28)$$

with

$$A = \frac{\varepsilon}{1 - (e^a - 1)/a} \quad (29)$$

and

$$a = 3.3 + \ell n\left[(1 - \varepsilon)^{7/3}\left(\frac{k_s}{k_s - k_c}\right)^2\right]. \quad (30)$$

The thermal conductivity of the solid grain, k_s, used in the above expressions is given by

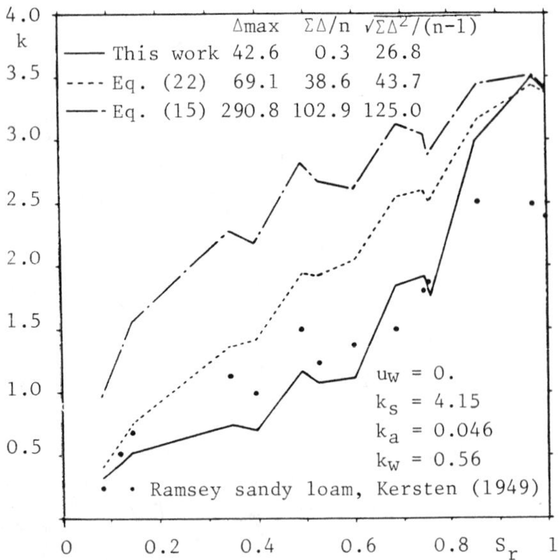

FIGURE 4 Experimental data and theoretical predictions versus degree of saturation.

$$k_s = k_q^q \, k_r^{(1 - q)} \qquad (31)$$

where q is the quartz content of the soil and k_q and k_r are the thermal conductivities of quartz and rest of the soil respectively. For air the value $k_a = 0.046$ is used as proposed by De Vries (1963).

Several purely empirical constants are recognizable in the expressions of De Vries (1963) and Johansen and Frivik (1980), while the empirical expression of a, equation (30), is obtained from the exact analysis of Muzzio and Solaini (1982), reported in the same paper.

A recent analysis of the author of the present work (Gori 1983) can predict the effective thermal conductivity of saturated two-phase media with ratios k_d/k_c lower than 40, i.e., it is to be used for saturated frozen and unfrozen soils but not for dry soils.

RESULTS AND DISCUSSION

Saturated and Dry Soils

The predictions of the effective thermal conductivity are compared with experimental data on saturated and dry soils. Because for dry soils the ratio k_s/k_a can be equal to 160 and in order to test the general validity of the equations used, the comparison is extended to saturated two-phase media with ratios of discontinuous to continuous thermal conductivities, k_d/k_c, ranging from 0.38 to 191.1.

The experimental data used, which amount to 158, have been taken from the review of Farouki (1982) and from the works of Crane and Vachon (1977), Okasaki et al. (1982), and Muzzio and Solaini (1982).

The maximum, Δmax, the average, ΣΔ/n, and the mean quadratic errors, $\sqrt{\Sigma\Delta^2/(n-1)}$, given by the expressions (3), (4), (15) and (28) are reported in Table 1.

Equations (3) and (4) predict results that are generally lower than the experimental data, but the lowest maximum error is found with equation (4) and the lowest mean quadratic error with equation (3). From the comparison of Table 1, the hypothesis of horizontal parallel isotherms is chosen to predict the effective thermal conductivities of unsaturated frozen soils.

TABLE 1 Comparison with 158 Experimental Data on Saturated Two-Phase Media

	eq.(3)	eq.(4)	eq.(15)	eq.(28)
Δmax	80.6	-63.9	167.1	148.4
ΣΔ/n	-5.9	-16.2	8.4	4.5
$\sqrt{\left(\frac{\Sigma\Delta^2}{n-1}\right)}$	19.8	26.9	25.3	24.7

Unsaturated Frozen Soils

Experimental data (115) on unsaturated frozen soils have been considered; some of them are presented in Figures 4 through 6. To justify the present work, the predictions obtained with the analysis of De Vries (1963) and Johansen and Frivik (1980) are reported on the same figures.

Figure 4 presents the experimental thermal conductivities of Kersten (1949) on Ramsey sandy loam, as reported in Farouki (1982), versus the saturation degree, S_r. The present analysis, equations (7), (13), (14), gives effective thermal conductivities that are in good agreement with the experiments up to $S_r = 0.75$; for nearly saturated frozen soil the predictions are higher than the experiments with a maximum error of 42.6%. As evident from Figure 4 the results of De Vries (1963) are the highest,while those of Johansen and Frivik (1980) are intermediate.

Figure 5 shows the experimental data of Kersten (1949) on a Fairbanks silty clay loam, as reported in Farouki (1982). This kind of soil can have some unfrozen water, which has been evaluated as ranging from 0.022 to 0.06 per cent of water volume to total soil volume. In the analysis of De Vries (1963) and Johansen and Frivik (1980) the presence of unfrozen water produces a decrease in the effective thermal conductivity at every degree

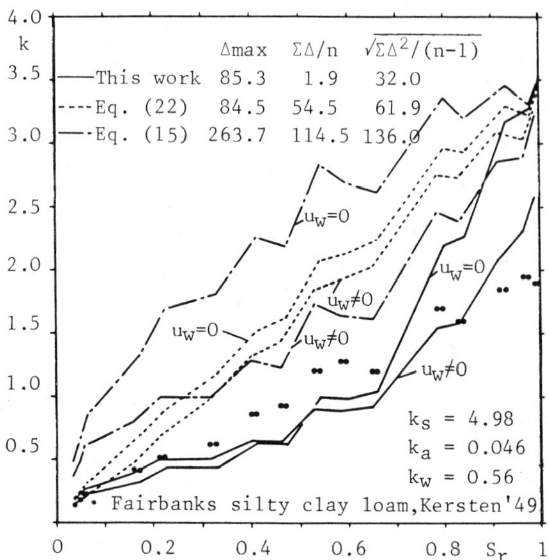

FIGURE 5 Experimental data and theoretical predictions versus degree of saturation.

of saturation. Such a decrease is in the right trend with respect to the experimental data, but both the predictions remain much higher than the experiments.

The results of this work for $u_w = 0$ are lower than the experiments for $S_r < 0.65$ and higher for $S_r > 0.75$, with a better agreement with the experimental data. The introduction of some unfrozen water in the unsaturated frozen soil decreases the effective thermal conductivity for $S_r > 0.5$ and increases for $S_r < 0.5$. These two effects depend on the assumption that unfrozen water is dispersed on the whole length of the cell. At very low saturation degrees the bridge of unfrozen water increases the effective thermal conductivity of the soil because k_w is greater than k_a. Near the saturation the same bridge of unfrozen water decreases the effective thermal conductivity of the soil because k_w is smaller than k_i. The increase of k_T for $S_r < 0.5$ and the decrease for $S_r = 0.5 - 0.65$ are not decisive, while its reduction for $S_r > 0.8$ produces a remarkably better agreement with the experiments.

Figure 6 presents the experimental thermal conductivities of Penner et al. (1975) for three soils, classified as 8, 9, 10. The conclusions of Figure 6 are similar to those of Figure 5 with a generally better agreement between the theoretical predictions and the experiments.

The comparison between the theoretical predictions and 115 experimental results are finally reported in Table 2.

TABLE 2 Comparison with 115 Experimental Data on Unsaturated Frozen Soils

	Δmax	$\Sigma\Delta/n$	$\sqrt{\Sigma\Delta^2/(n-1)}$
This work	81.5	-1.15	24.99
Johansen and Frivik (1980)	86.7	31.44	41.81
De Vries (1963)	291.6	82.96	106.68

The analysis presented in this paper gives average and mean quadratic errors that are lower than those obtained with the expressions of De Vries (1963) and Johansen and Frivik (1980).

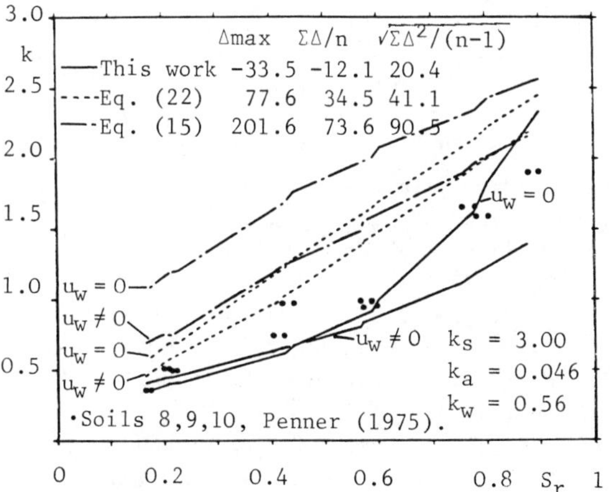

FIGURE 6 Experimental data and theoretical predictions versus degree of saturation.

NOMENCLATURE

k	thermal conductivity, W/mK
ℓ	cubic root of the cube volume
S	surface or section
S_r	degree of saturation
$u_w = \frac{V_{uw}}{V_t}$	unfrozen water
V	volume
w	water content
$x = \frac{V}{V_t}$	volumes fraction

Greek symbols

ε	porosity
ρ	density
σ	surface tension
$\Delta = \frac{k_e - k_{exp}}{k_{exp}}$	percent error

Subscripts

A	relative to adsorbed and dispersed water
a	relative to air
c	relative to continuous phase
d	dry
e	effective
f	funicular
exp	experimental
i	relative to ice
max	maximum
q	relative to quartz
Q	relative to parallel heat flux lines
r	rest of the solid grain
s	relative to solid
T	relative to parallel isotherms
t	total
u	relative to unfrozen water
v	relative to voids
w	relative to water

REFERENCES

Crane, R. A., and Vachon, R. I., 1977, A prediction of the bounds on the effective thermal conductivity of granular materials, International Journal of Heat and Mass Transfer, v. 20, pp. 711-723.

De Vries, D. A., 1963, Thermal properties of soils. In Physics of plant environment (W. R. Van Wijk, ed.), Amsterdam, North Holland Publishing Co.

Farouki, O., 1982, Evaluation of methods for calculating soil thermal conductivity, CRREL Report 82-8, March.

Gori, F., 1983, Prediction of effective thermal conductivity of two-phase media (including saturated soils), Accepted for presentation at XVIth International Congress of Refrigeration, Paris, 1983.

Johansen, Ø., and Frivik, P. E., 1980, Thermal properties of soils and rock materials, The Second International Symposium on Ground Freezing, Tronheim, Norway, pp. 427-453.

Kersten, M. S., 1949, Laboratory research for the determination of the thermal properties of soils, Artic Construction and Frost Effects Laboratory Technical Report 23, AD712516.

Luikov, A. V., 1966, Heat and mass transfer in capillary-porous bodies, Pergamon Press.

Muzzio, A., and Solaini, G., 1982, Effective thermal conductivity of unconsolidated granular media. Theoretical and experimental research. In Italian. XXXVII A.T.I. Congress, Padova (Italy), pp. 679-694.

Okazaki, H., Yamasaki, T., Ninomiya, T., Nakauchi, H., and Toei, R., 1982 Prediction of effective thermal conductivities of wet porous solids, Seventh International Heat Transfer Conference, München (West Germany), v. 6, pp. 93-97.

Penner, E., 1970, Thermal conductivity of frozen soils, Canadian Journal of Earth Sciences, v. 7, pp. 982-987.

Penner, E., Johnston, G.H., and Goodrich, L. E., 1975, Thermal conductivity laboratory studies of some Mackenzie Highway soils, Canadian Geotechnical Journal, v. 12, no. 3, pp. 217-288.

Perry, J. H., 1963, Chemical engineers' handbook, 4th ed., New York, McGraw Hill.

A GEOCRYOLOGICAL ASPECT OF THE PROBLEM OF ENVIRONMENTAL PROTECTION

N. A. Grave

Permafrost Institute, Siberian Branch,
Academy of Sciences, Yakutsk, USSR

A system is proposed for mapping areas of permafrost in the USSR on the basis of surface stability following the removal of vegetation and soil. The lowest degree of stability is typical of areas of fine-grained deposits with high ice contents and substantial moisture at the surface; here minimal disturbance will lead to the development of thermokarst. The levels of surface stability in permafrost areas are: 1A - very unstable; 1B - very unstable; 2 - unstable; 3 - slightly stable; 4 - relatively stable. Some recommendations for monitoring permafrost are made.

The zone of seasonal freezing and permafrost is an important part of the biosphere. The permafrost phenomenon determines some substantial features of northern landscapes, including their great vulnerability to development.

The development of cryogenic processes within permafrost affects the stability of the surface and landscape.

Cryogenic processes that deform the surface arise and intensify particularly dramatically in the presence of technogenic disturbances. Technogenic impacts can be mechanical, chemical, hydrological, pyrogenic, acoustic, or climatic in nature. In the area under development, mechanical impacts are the most widespread: disturbance of the plant, snow and soil covers; excavation and redeposition of earth; and disturbance of rock masses. Hydrological disturbances, i.e. change in moisture content of the surface, are almost always associated with mechanical ones.

Maximum destruction of the surface and landscape is due to thermokarst, which, under favorable conditions, spreads over large areas and extends deep within the earth. Slope processes, frost heaving and fracturing of soil have a substantial influence upon the stability of the structures built, but do not cause such strong landscape transformations as thermokarst does. Therefore, when speaking about the stability of the surface and landscape in permafrost areas, we are referring mainly to the process of subterranean ice thawing and related topographic deformation.

In order to evaluate surface stability in permafrost areas one should primarily have knowledge of thermally induced settlement of permafrost, depending on its ice content and the lithological composition of ice-bearing materials.

If the ice content of the soil and its lithological composition determine the potential for thermokarst-induced topographic deformation, then the actual degree of decay will depend on the intensity of soil thawing and the speed of the thermokarst process itself. With a variation in surface conditions, for example, when the vegetation is removed, the thawing depth depends on a number of physiographical and geocryological factors.

Feldman (1977) has most completely demonstrated this dependence in the expression that includes relationships between the ice content of the soil, the thermal properties of the soil and surface, the air temperature and radiation balance, evaporation, the permafrost temperature, and the properties of the snow cover.

The actual amount of permafrost settlement that occurs when plant and ground cover is removed is generally less than the potential amount, and under some conditions it can be insignificant. If, after removal of vegetation, the thawing depth reaches ice-bearing levels in the permafrost, then further thawing occurs due to the development of thermokarst, thermal erosion and thermal abrasion. These processes can develop only under certain hydrological and topographical landscape conditions.

Prediction of the appearance and development of thermokarst, the most frequently occurring of the processes just mentioned (Methodological recommendations 1981), depends on the geocryological environment where it develops. In regions with ice streaks, the thermokarst manifests itself in a gradual settlement of the surface without forming "funnels" or hollows. In regions with large ice masses (sheet ice, wedge ice, etc.), the thermokarst is more sharply manifested, with shallow depressions, hillocky microrelief with cemetery mounds, and cave-in lakes. The main condition necessary for the development of thermokarst is the accumulation of a sufficient amount of water on the surface so that seasonal soil thawing depth exceeds the depth of seasonal freezing. The thermokarst process within deposit-formed ice is able to proceed until the ice mass thaws out completely. However, as the thermokarst process goes on, it is accompanied by slope processes, solifluction and ground slip, with sediments accumulating on the surface of the thawing, ice-containing soils. This may be the reason for cessation of the thermokarst process at various stages of its development.

After stabilization of cryogenic processes at some stage, the depth of seasonal thawing of soil can decrease and become fixed at some level corresponding to the existing thermal balance on the surface. As indicated by previous studies

(Sukhodrovsky 1978, a.o.), cryogenic morphological processes arising from or aggravated by technogenic impacts on the surface gradually decay following vigorous development during the first 2 or 3 years, and terminate within about 10 years. This is true in cases where disturbance of the surface is halted. If they continue, then the process is able to progress rather deep into the earth, sometimes to the point of complete thawing of subterranean ice.

Thus, observations and analyses of natural and technogenic changes of the parameters mentioned above can be made (Feldman 1977). And on the basis of data on the ice content and lithological composition of permafrost, it is possible not only to gain insight into the causes of variations in thawing depth, but also to predict the future direction and intensity of development of the process.

All the pertinent studies are unified in the problem of permafrost zone monitoring. Permafrost zone monitoring includes observation, analysis, prediction and control of cryogenic processes in the permafrost region during natural and technogenic changes in the geophysical condition of the permafrost.

The object of monitoring in this case is the permafrost zone and cryogenic processes, which are closely associated with natural landscapes and biogeocenoses, and their natural and man-determined development. Therefore, permafrost zone monitoring should be regarded as a part of biosphere monitoring. Scientific investigations in this direction should be carried out with close attention paid to biospheric changes, and, it seems, on common biospheric monitoring sites.

In order to gain a general idea of the degree of stability of the surface in different permafrost zones, necessary in planning monitoring studies, we can use a mapping system devised for the USSR permafrost region (Grave 1982).

The system (Figure 1) categorizes surface stability by generalizing the data reported in the literature on cryogenic structure, ice content, and thermal settlement of soil at the time of thawing; experimental and calculated data on soil thawing following disturbance of plant or snow cover (depending on the factors of physiographical environment and geocryological conditions); and data on the development of cryogenic processes caused by terrain development.

Least stable are regions with soils of high ice content, where minimal surface disturbance (removal or damage of plant cover) is enough to initiate the thermokarst process, with subsequent deterioration of the relief. The maximum potential surface disturbance is determined by a complete thawing-out of subterranean ice. Therefore, territories with thick soil-forming subterranean ice are most vulnerable to exploitation.

In this connection, evaluation of the degree of stability of the surface under technogenic impacts should be based on the ice content of the underlying permafrost. The response to minimal disturbance will depend on climatic, hydrological, geobotanic, lithological and general geocryological conditions in the area.

The surface stability is intimately related to the stability of the landscape. However, changes in the landscape caused by mechanical disturbance of the surface are not always associated with surface failure. They can be produced by technogenic changes of the plant cover in comparatively stable areas.

The following regions are delineated on the map of USSR permafrost regions (Figure 1) according to the degree of vulnerability to the disturbance or removal of surface material.

1A: Regions with a very unstable surface. These are found on the lowland plains of Siberia—the Yamal and Gydana Peninsulas, the North Siberian lowland, the Lena River delta, the littoral lowlands of Yakutiya, the Chaunskaya depression, and the Anadyr lowland. Widely distributed are shallow, thick masses of wedge and streak ice enclosed within suglinok deposits of high ice content of post-Pliocene, Holocene and contemporary ages. A cold climate, subarctic tundra and forest tundra, prevails. The most stable surfaces in these regions occur on floodplains and low terraces within river deltas, formed by sand deposits and deposits of low ice content, as well as on boulder-pebble moraines and ledge rock outcrops.

An example is the middle and southern part of the littoral lowlands of Yakutia (Kuznetsova 1980, Tomirdiaro 1980). The majority of the surface consists of "edoma" (eroded terraces), with post-Pliocene suglinok soils containing thick ice (15-25-40 m or more). The region is forest tundra with permafrost temperatures of -6° to -8°C. A smaller area consists of lake-bog plains with a cover of ice-bearing lake-bog deposits of Holocenic and present-day age. A naturally recurring thermokarst that is due to thermal abrasion and thermal erosion periodically destroys the "edoma." Technogenic impacts intensify these processes, gradually destroying the landscape of a relatively well drained forest tundra and changing it into barely passable lake-bog plains and badlands with ravines, cemetery mounds, and decaying and growing wedge ice.

1B: Regions with a very unstable surface. These occur on lowlands in the northernmost parts of Taimir, the North Siberian Plain, the littoral lowlands of Yakutiya, the Novosibirsk and Severnaya Zemlya islands, and in the central parts of the Yakutsko-Vilyuiskaya depression. The structure and composition of the underlying surfaces of the deposits are similar to those of category 1A. In the northern regions the climate is very cold, with a short summer (Arctic desert, Arctic and subarctic tundra). The permafrost temperature is -12° to -15°C. In Central Yakutiya the climate is very dry and relatively cold (taiga). The permafrost temperature is -3° to -5°C. The cold in the north and the aridity in Central Yakutiya impose some constraints on the development of thermokarst following deterioration or removal of vegetation, and it either subsides in the early stages or does not arise at all. In the north, most stable following plant removal are surfaces typical of edoma, deltas, and portions of river terraces and coasts formed by sandy deposits, as well as bedrock outcrops. On sunken lake-bog plains the thermokarst process developing through thermal abrasion and thermal erosion contributes to a decrease of the surface stability during technogenic impacts. In Central Yakutiya the most stable surfaces accompany alasses and river terraces, formed by sands. Removal of forests during agrarian development leads to the appearance of barren "badlands" and to xerophytization of landscapes and salinization of soils.

2: Regions with an unstable surface. These occur both in lowlands and in hills at about 200-500 m above sea level, in which subterranean ice, as a rule, is not as thick or widespread as in the previously mentioned regions. They occur in different climatic and geobotanical zones and are characterized by different conditions of subterranean ice occurrence. In the mountains, such areas accompany inter-mountain hollows and depressions intersected by river valleys with their beds at about 400-600 m above sea level. Technogenic thermokarst develops everywhere, but as a rule does not exceed the boundaries of the disturbance contour nor critically change the landscape. In the northern part of Western Siberia the landscapes remaining stable during technogenic disturbances include bogs, meadows, and swampy tundras with underlying deposits of low ice content. According to data reported by Moskalenko (Grave, Melnikov and Moskalenko 1983), a deteriorated landscape here recovers within 10 years, and along long-ago-constructed gas lines the natural shape of disturbed bogs and the original geocryological conditions have returned within 20 years. In eastern Siberia, where the composition of the upper part of the permafrost in these regions is coarser, surface settlement deformations are not deep.

3: Regions with a slightly stable surface. These are found in low mountains and elevated plateaus at about 500-600 m to 1000-1200 m above sea level and at lower levels in the Arctic. Thin deposits, mainly deluvial (talus) and eluvial and generally of coarse composition, frequently contain vein and wedge ice. Exploitation of these regions leads to an intensification of thermokarst and (mainly) slope processes, particularly when slopes are cut (Gravis 1976).

4: Regions with a relatively stable surface. These are mainly in mid- and high-altitude areas. In the lowlands they occur where there are deposits of low ice content and bedrock outcrops. In the mountains, when vegetation is removed and slopes are cut, thermokarst intensifies on relic ice veins and rock streams, at whose bases there is buried glacier ice as well as slope processes (talus, landslides, crumbling and solifuction) (Gravis 1976).

5: Regions with permafrost islands and sporadic permafrost. These are mainly in a broad belt along the southern boundary of the permafrost zone. Taiga and forest tundra occurs in the lowlands and mountains, and tundra and sparse forests at high altitudes. The permafrost temperature is high, 0° to -2°C, and decreases in the mountains. Permafrost islands occur in hillocky peatlands and dense, dark coniferous forests with a thick moss cover, while in the Far East they occur with "maris." A large role is played by the snow cover in the freezing and thawing of this permafrost.

In Western Siberia, according to Moskalenko's observations, the surfaces that are most unstable under technogenic impacts are hillocky peatlands and hilly surfaces covered by forests. Here removal of vegetation intensifies settlement phenomena and erosive processes, and accumulations of snow within these settled areas further contribute to the formation of ravines and shallow depressions. Deforestation in places with no permafrost, or where permafrost occurs deep within the earth, leads to an intensification of freezing of the surface, frost fracturing, and heaving.

In the northern part of the Amur region, after removal of the moss-peat cover on the maris there is deep (sometimes complete) thawing of the permafrost, and thermokarst develops. The snow cover in areas under development becomes dense and disappears earlier than usual, which contributes even more to permafrost thawing (Balobaev et al. 1979).

In nature, the proposed gradation system lacks well-defined boundaries. Evaluating large territories from this point of view, we can speak only of the predominance in a region of surfaces with a certain degree of stability; any territory can have areas of almost all degrees of stability in different spatial proportions. Some examples are provided by large-scale classification systems described by Kuznetsova (1980) and Parmuzin and Sukhodolsky (1980).

It is advantageous to select areas in the permafrost zone for systematic long-term observations. This zone is the most susceptible to the impact of man, as cryogenic processes, particularly thermokarst, can develop the most intensively and can extend very deep into the earth. Such areas might, for example, include the banks near the mouths of the Yana and Kolyma rivers, where one should expect a combination of thermokarst and thermal erosion and abrasion. Observations are also pertinent in the northern part of Western Siberia (Yamal, Gydan), and in Central Yakutiya. In the southern part of the permafrost zone an area could be chosen in the eastern part of the BAM zone, possibly within shallow depressions in the North Amur region.

The areas selected for monitoring of the permafrost zone should have large-scale maps of their surface stability (or permafrost stability) prepared. And every 2 to 3 years they should be checked for landscape changes using sequential satellite or aerial photography. In places where processes are proceeding particularly rapidly, repeated topographic surveys would be useful. Monitoring of fluctuations in the temperature of the permafrost zone requires systematic long-term observation of the dynamical behavior of thermal and cryogenic processes, both within the permafrost itself and on the surface, under natural and disturbed conditions. For that purpose, stations should be set up with deep boreholes that pass completely through the permafrost. Unfortunately, such systematic geocryological observations are lacking; and occasional temperature measurements fail to produce a reliable picture of long-term developments in the permafrost zone.

In areas under development, it is advisable to observe geomorphological processes using benchmarks. Of special importance here are observations of the soil conditions and the formation of "kriopegs" (bodies of liquid saline water below 0°C associated with permafrost, or ice-free permafrost with saline pore water) in soils and earth materials as a result of the development activity.

Analysis of data from observations in both undisturbed and developed regions will form a reliable basis for predicting cryogenic changes and for working out protective measures. Therefore, the development of an area should be preceded by a geocryological prediction in the form of maps of the stability of the surface or permafrost during technogenic impacts.

Medium- and small-scale maps of surface stability in some parts of the permafrost zones of the USSR, Canada and Alaska have been generated using various techniques. They show that as a rule, highly unstable surfaces occupy about 30-50% of the area in the most vulnerable regions of the Arctic belt. Preliminary zoning of development areas according to the stability of the surface and landscape will permit a reduction in expenditures for maintaining rational land usage and accomplishing revegetation of deteriorated landscapes.

In summary, when decisions are made that affect the northern biosphere, it is necessary to take into account the geocryological properties of the landscape.

REFERENCES

Balobaev, V. T., Zabolotnik, S. I., Nekrasov, I. A., Shastkevich, Yu. G., and Shender, N. I., 1979, Dynamics of soil temperature of the northern Amur region in reclamation of the territory, in Technogenic Landscapes of the North and Their Recultivation: Novosibirsk, Nauka, p. 74-88.

Feldman, G. M., 1977, Prediction of temperature condition of soils and cryogenic processes: Novosibirsk, Nauka, p. 189.

Grave, N. A., 1982, Surface stability to mechanical disturbances in development of the North: Izv. AN SSSR, Ser. Geograf., in press.

Grave, N. A., Melnikov, P. I., and Moskvalenko, N. G., 1983, Geocryological questions of landscape preservation in developing oil and gas deposits of the permafrost zone: Vestnik MGU, in press.

Gravis, G. F., 1976, Permafrost processes in the BAM zone and evaluation of their development in connection with the economical exploitation of the area, in Problems of Engineering Geology in Relation to Rational Usage of Geological Environment: Leningrad, Proceedings of the All-Union Conference, p. 97-100.

Kuznetsova, I. P., 1980, Engineering-geocryological conditions and stability of permafrost of coastal lowlands of Yakutiya to disturbances of the natural native situation, in Stability of the Surface to Technogenic Impacts in the Permafrost Region: Yakutsk, IM SO AN SSSR, p. 75-107.

Methodological recommendations on prediction of cryogenic physico-geological processes in reclamation areas of the Far North: Moscow, Mingeo. SSSR, 78p.

Sukhodrovsky, V. L., 1979, Exogenic relief formation in the permafrost zone: Moscow, Nauka, 280 p.

Tomirdiaro, S. V., 1980, Loess-ice formation of Eastern Siberia in late Pleistocene and in Holocene: Moscow, Nauka, 184 p.

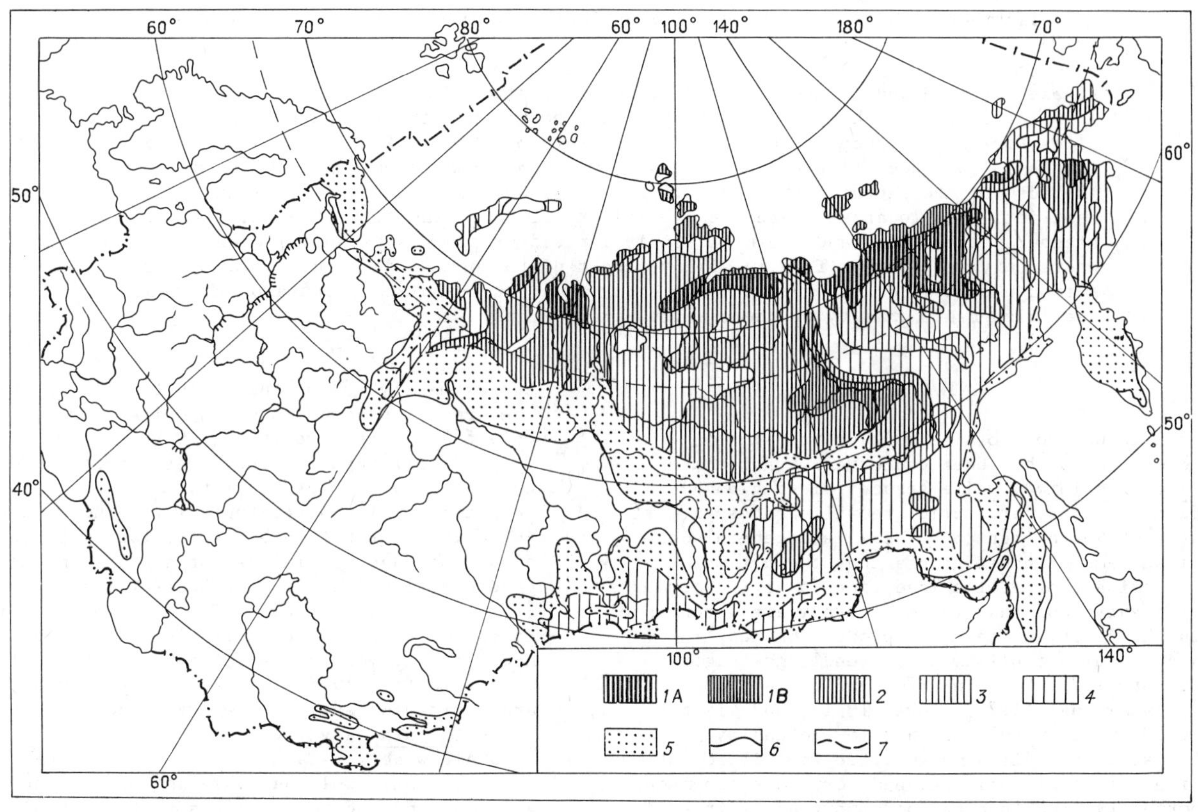

FIGURE 1 System for mapping the USSR permafrost region for surface stability during disturbance.

1-4: Regions with different degrees of surface stability predominating:
- 1A - very unstable
- 1B - very unstable
- 2 - unstable
- 3 - slightly stable
- 4 - relatively stable

5: Regions with permafrost islands and sporadic permafrost

6: Southern boundary of the permafrost region

7: Southern boundary of the region of continuous permafrost and massive permafrost islands.

PERMAFROST, FIRE, AND THE REGENERATION OF WHITE SPRUCE AT TREE-LINE NEAR INUVIK, NORTHWEST TERRITORIES, CANADA

David F. Greene

Department of Geography, University of Calgary, Calgary, Alberta, Canada, T2N 1N4

Adjacent burned and unburned upland stands were studied at the southwest corner of the 1968 Inuvik burn. Shallow active-layers are poorer habitat than are deep active-layer sites for white spruce because of a slower rate of thaw (effective growing season length), lower seed dispersal capacity, lower cone production, and poor seedbed creation (because of the depth of the Of) by fire. Fire delimits the range of white spruce by restricting it to areas where the probability of surviving fire is high, e.g., the steeper valley sides and the most deeply incised or sinuous perennial creeks. Areas where survival through fire is likely also tend to possess deep active layers. Although such sites are not uncommon, many remain uncolonized because of the poor seed dispersal capacity of these white spruce stems.

INTRODUCTION

Although most of the work on tree-line determination has been concerned with black spruce (Picea mariana), white spruce (Picea glauca), forms the conifer arctic tree-line across almost half of North America. There is a general consensus among investigators that the ultimate determinant of arctic tree-line is arctic air-mass frequency, and that the more proximate determinant is the rarity of sexual reproduction and/or successful establishment of seedlings (e.g., Elliott 1979, Nichols 1976, and Larsen 1980).

Van Wagner (1979), Rowe (1971), and Viereck (1979) have speculated that low intensity forest-tundra fires may provide too little optimal (i.e., mineral soil) seedbed, or cause too much mortality, and thus may be important in tree-line determination. Viereck and Schandelmeier (1980) speculated that low seed dispersal capacity and, therefore, survival of seed sources through fire, may be important.

Van Cleve et al. (1981) have shown that soil temperature and rate of soil thaw may be crucial determinants of vegetative vigour in black spruce in the zone of discontinuous permafrost. The stems on a permafrost-free site were more vigorous than those on a nearby permafrost site.

The present investigation focused on the relative importance of fire (and fire-created seedbeds), active-layer depth, and rate of soil thaw in determining the distribution and demography of white spruce near tree-line in the Inuvik area. Additionally, the potential importance of seed dispersal capacity and survival through fire of seed sources were examined.

STUDY AREA

The main study area was in Boot Creek Valley, 2 km east of Inuvik, N.W.T. (68°21' N, 133°40' W). For a general introduction to the Mackenzie Delta area, see Mackay (1963) and Kerfoot (1972).

The uplands around Inuvik are in the zone of continuous permafrost. The climate is cold and dry. Average yearly precipitation is 27 cm; approximately 75% of this amount falls as snow. The area experiences short, cool summers, with only 665 degree-days above 5°C (Burns 1973).

Surficial deposits above permafrost in Boot Creek Valley consist of silty loams. In this area, deep active-layers are typically found on steep (usually greater than 8°) south-facing slopes and along the edges of incised or highly sinuous perennial creeks. At Boot Creek Valley, the steepest slopes are 20°. The Of layer is thin or absent on deep active-layer sites, and the soils are Static Cryosols (Canadian System of Soil Classification 1978). At Boot Creek, such sites are dominated usually by white spruce or paper birch (Betula papyrifera).

Shallow active-layer sites include north-facing slopes, gently rolling tablelands above the valleys, and the more or less flat alluvial and colluvial areas between creeks and valley sides. These sites tend to be dominated by black spruce and dwarf ericaceous shrubs. The Of layers are typically thick and the soils are usually Organic or Turbic Cryosols.

A secondary study site was located 30 km north of Inuvik in an unnamed valley in the Caribou Hills escarpment. The soils at this site are sandy loams. The composition and distribution of vegetation is similar to Boot Creek Valley. The slopes at the Caribou Hills site are, however, much steeper than at Boot Creek. Slope angles can be as great as 30°. The south-facing slope is much more deeply gullied than in Boot Creek Valley. The sandy inter-gully spurs are frequently unvegetated. Both Boot Creek Valley and the Caribou Hills site are asymmetrical valleys with the north-facing slope steeper than the south-facing slope.

Reconnaissance indicated that the great majority of white spruce stems in the uplands around Inuvik are found on mesic to xeric sites. They are especially common where the terrain is well-dissected (i.e., steep south-facing slopes) or along the edges of lakes and perennial streams. Northeast and east of Boot Creek Valley, aerial reconnaissance indicated comparatively few vigorous stands of white spruce. North of Boot Creek Valley, however, the deeply dissected

Caribou Hills escarpment has vigorous, dense white spruce stands extending as much as 70 km north of Inuvik.

For an introduction to the vegetation of the Mackenzie Delta area, see Lambert (1972). Vegetation recovery in the 1968 Inuvik burn is discussed by Wein (1975). Soils in the Inuvik area are discussed by Tarnocai (1973) and Day and Rice (1964).

METHODS

The main study site consisted of adjacent burned and unburned areas at the southwest corner of the 1968 Inuvik burn. Two subjectively chosen transects were run from the creek bottom upslope to the rolling tableland above the south-facing slope in the unburned and burned areas. A fireguard lay between the two sites. Plots 10-m^2 were sampled every 20 m along each transect for slope angle, altitude, and percent cover of each plant species present. Four 1-m^2 quadrats were located at the corner of each plot. The Of, Om, Oh, or LFH depths, and the depth of the permafrost were recorded for each quadrat. All permafrost measurements were made from 3 August to 6 August 1982.

At the unburned transect, all white spruce stems taller than 3 m were cored, and their diameter at breast height (dbh), height, and evidence of cone production or chlorosis recorded. Sections of the unburned transect (10 x 30 m along the flat, and 10 x 20 m on the steep south-facing slope) were intensively sampled for white spruce. In these intensively sampled areas, all white spruce stems, regardless of height, were aged. For white spruce taller than 3 m, cores were taken 30 cm above the root-collar. Stems less than 3 m but greater than 30 cm in height were cut at the root-collar. The very smallest stems were aged by counting the terminal bud scars.

In the burned area, a 50-ha section adjacent to the fireguard was studied for the distribution of white spruce survivors and seedlings. This was necessary because no seedlings were found in the burned transect. For the seedlings, 12 equidistant 10-m-wide transects were walked parallel to the creek. Each of these transects was walked simultaneously by two investigators. The location of seedlings and their distance from a tall, potential seed source were recorded. Survivors taller than 5 m were mapped with binoculars from the opposite slope. Seedlings within the burn were randomly sampled for age determination using a random numbers table. The sampling sites were 10-m^2. Additionally, seedlings from the southeast corner of the burn, from the fireguard, and from recent mud-flows were aged.

Six soil pits were dug in early August. The methods employed in the soils analyses are detailed by Greene (1983).

At the Caribou Hills site, a single transect was used. Stands and quadrats were located on the transect in the same manner as at Boot Creek. The vegetation data that were collected are not included in this paper (but see Greene 1983). Active layer depth measurements were made in the quadrats on 25 June and 6 August 1982. The snow had melted throughout most of this valley by approximately 3 June.

For convenience, south, west, and southwest-facing slopes are referred to as south-facing slopes. All other slopes are called north-facing slopes. A deep active-layer is one that was more than 50 cm deep as of 6 August 1982. Shallow active-layers were less than 50 cm deep. Seed dispersal capacity is defined as that distance from a seed source within which the great majority of the seed rain occurs; magnitude of the seed rain is not implied, and wind direction or velocity is ignored. Fire severity or intensity is defined solely in terms of the depth of the organic layer remaining immediately after a burn. Seedling establishment is defined arbitrarily as the survival of a seedling beyond 4 years of age.

RESULTS AND DISCUSSION

Seedling Establishment

Figure 1 depicts the ages for smaller stems from the unburned transect that were cut at the apparent root-collar. Because of extreme variability in early growth in tree-line white spruce, these stems more closely approximate exact ages than do stems cored at 30 cm. However, because of problems such as adventitious rooting and missing or false rings, the data in Figure 1 should not be interpreted as exact ages. They should be, however, typically within a year or two of the exact age.

As can be seen in Figure 1, seedling establishment events have been common during the last 70 years. The youngest age classes (say, less than 4 years) were undoubtedly underestimated, as these youngest seedlings are extremely difficult to find in deep moss. There is no evidence in the literature of a seed bank in white spruce (Johnson 1975). Further, assumptions of perfect synchrony in flowering, or of increasingly long seed production intervals with increasing latitude are perhaps unwarranted (Harper 1977). The age structure for the older stems cored at 30 cm is (allowing for the increased mortality) similar, but less persuasive because of the tremendous variability in growth to 30 cm.

Of the 185 seedlings younger than 13 years old in the burned area, 94.1% are from the years 1971, 1972, and 1973. This temporal distribution was not a local artifact, but was similar for burned areas at the southeast and northwest corners of the burn, and for unburned areas adjacent to the southwest (Figure 1) and northwest corners.

These tree-line stands are not the "reproductive cripples" one might have expected. Seed production, at least among the more vigorous stems, can be inferred to be adequate for good establishment.

The Relationship of Active-layer Depth to Vegetative Growth and Seed Dispersal

The relationship of active-layer depth to height of stems for the unburned transect is shown in Table 1. These ages are uncorrected, i.e., they represented the age at 30 cm above the apparent

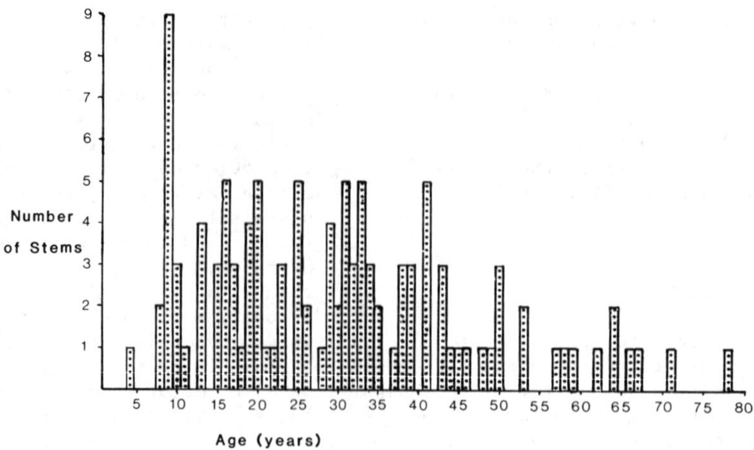

FIGURE 1 Ages of stems along the unburned transect. Only those stems which were cut at the root-collar or aged by terminal bud scars are included.

root-collar. Stems cut at the base are not included. Only the stems in the age classes 50 to 95 years are used for the comparison because the steep slope and footslope experienced a fire in approximately 1880 whereas the rest of the transect was unburned.

The greater height achieved by the stems on the steep slope cannot be merely an artifact of increased insolation, because stems along the creek have equivalent stature. Taking the original observations (n=53), active-layer depth predicts tree height with an r^2 of 0.473, whereas slope angle predicts tree height with an r^2 of 0.321. If a greater number of creekside stems in these age classes could have been included in the transect, the disparity in the correlation coefficients may well have been greater. As with height, differences in dbh are marked between deep and shallow active-layer sites (Greene 1983).

Suppression of growth by shade should only be a factor along the creek where the old-growth (older than 300 years) stand has a cover value of 80%. The stunted stems on the flat and footslope achieve cover values of only a few percent. The stems on the steep slope seeded in after a fire dated at approximately 1880.

Seed dispersal capacity in white spruce is roughly related to tree height; the majority of the seed rain should be limited to within a horizontal distance approximately twice the height of the tree (Viereck and Schandelmeier 1980). Because of low stature, the seed dispersal capacity of white spruce is severely curtailed in these tree-line stands. The tallest trees are 15 to 17 m in height. Examination of the seedling distribution in the 1968 burn showed that 96% of the 470 seedlings enumerated were within 30 m of a tall survivor, and that approximately 80% were within 15 m. No seedlings were found on the burned transect (no survivors were nearby). Seedling density tended to be greatest near the greatest densities of tall survivors.

The vast majority of the seedlings found in the burn were clustered on sites that had exposed mineral soil or retained only a thin veneer of an organic layer. Local seedling densities could be much greater in the burned area (e.g., 0.9 per m^2 for seedlings 8 to 10 years old) than in the unburned area. In the burn, white spruce seedlings were rare where the post-fire vegetation was dominated by asexually reproducing survivors, i.e., where a thick Of layer was maintained.

Survival Through Fire

Given the low seed dispersal capacity of these trees, survival of fire appears to be crucial for maintaining occupancy of a tree-line valley. It is however, probably rare that fire ever kills the entire population of a steep-sided valley, because fires at tree-line burn extremely unevenly (Wein 1975).

In the western third of the burned portion of Boot Creek Valley, 4 out of 5 south-facing gullies had tall surviving white spruce. Approximately half the length of the creek in this area had survivors. All of the 245 surviving tall stems (i.e., taller than 5 m) were clustered along the gullies (12%) or creek (88%). The distribution of the survivors helps explain the tendency of white spruce in this area to be restricted to or near steep south-facing slopes and along the edges of water bodies. The steeper the south-facing slope, the more deeply it tends to be gullied. Additionally, the very steepest slopes make it difficult for a fire to burn downhill, and the more xeric spurs (often unvegetated in coarse soil material) can act as fire-breaks.

Reconnaissance indicated that this pattern of survival was common throughout the 1968 burn. Examination of a new burn (late June 1982) about 30 km southeast of Inuvik showed the same pattern: steep south-facing slopes and the edges of water bodies seldom burned. Such sites were often dominated by white spruce. It would appear that a steep north-facing slope will burn more readily than an equally steep south-facing slope, perhaps because of the deeper and more continuous Of layer. Dessiccated cryptogams, if the cover is continuous,

TABLE 1 Selected Tree Mensuration Data from the Unburned Transect for the Age Classes 50 to 95 Years.

	Landscape Element				
	Creek Edge	Alluvial Flat	Colluvial Flat	Footslope	Steep Slope
Avg slope (^{o})	0	.75	4.3	9	18
Avg depth of Of or LFH (cm)	3	20	27	24	13
Avg active-layer depth 4 August 1982 (cm)	50	27	30	28	58
Number of trees (50 to 95 yr)	3	6	3	23	18
Avg height of stems (m) (50 to 95 yr)	9.8	2.5	5.0	6.3	11
Density of stems per m^2 (all ages)	0.02	0.015	0.004	0.03	0.02
Evidence of recent burning	None	None	None	Surficial charcoal	Surficial charcoal
% of trees (>2 m) with no evidence of cones on branches or near base (all ages)	8	68	100	62	0
% of cone-bearing stems with long axis of cone >2.5 cm (all ages)	92	19	0	50	96
Chlorotic needles (all ages)	No	Common	Common	Uncommon	No

have an important role in supporting a fire in the northern forest-tundra and low arctic (Wein 1975).

It is commonly assumed that fire frequency is extremely low in the northern forest-tundra and low arctic (Wein 1975). The present author found no fire-scarred trees, yet evidence for old fires was common. For example, a soil pit in a white spruce stand that did not burn in 1968 showed 3 buried charcoal horizons.

Approximately 500 m east of the fireguard, the south-facing slope becomes more gentle, there is no gullying, and there is no evidence (burnt stems) of white spruce having recently been on these slopes prior to the 1968 burn. These gently sloping south-facing slopes have tundra vegetation similar to that found on the flatter areas. Mineral soil was seldom exposed, as the fire typically burned only the desiccated upper portion of the thick Of. Thus, white spruce should only be found in steep-sided valleys or valleys with well-incised or highly sinuous perennial creeks. These sites are more optimal because of the high likelihood of some stems surviving through fire and the presence of deep active-layers.

The Effect of Shallow Active-layers on Cone Production

As there was virtually no cone crop during the investigation, conclusions about differential reproductive capacity for stems on deep versus shallow active-layers must rely on indirect evidence and be regarded, therefore, as very tentative. However, the local distribution of stems leads to the conclusion that the more vigorous stems appear to be responsible for a disproportionately large share of the seedling establishment.

The taller trees (i.e., on the deep active-layers) have many more old, opened cones evident on their branches or around their bases than do the more stunted trees on the shallow active-layer sites. The stunted trees commonly show chlorotic needles, smaller needles, and much smaller cones (Table 1). Dead or dying stems younger than 150 years are common on the shallow active-layer sites.

Although the seed-crop in conifers is usually negatively correlated with vegetative growth during and for some time after the crop, the relationship between vegetative vigour and long-term reproductive activity is usually positive (Kozlowski 1971). Environmental conditions that favour rapid vegetative growth in conifers tend to induce earlier flowering (Schmidtling 1969). The more vegetatively vigorous individuals tend to have earlier reproduction (Harper 1977) and larger seed crops (Larson and Schubert 1970, Fowells and Schubert 1956).

In Boot Creek Valley, differences between sites in pH and texture were small. The forested sites and burned deep Of sites tended to have much more exchangeable P, K, NH_4, and NO_3 than did the unburned deep Of sites. This may explain the widespread chlorosis observed in the leaves of the stunted stems on the unburned deep Of.

At the Caribou Hills site, thawing was much more rapid on the steep slope (shallow Of) than on the more flat areas (deep Of). By 25 June, the steep slopes averaged 22.6 cm of thaw, while the gentle to flat slopes averaged 9.8 cm. Similar results have been reported by Van Cleve et al. (1981) and Bliss and Wein (1971). Fine roots of seedlings on the steep slopes tended to be concentrated in the upper few cm, whereas fine roots of seedlings in the deep Of sites were most densely concentrated at about 7 cm below the surface. This difference may be due to the frequent dessiccation of the upper few cm of the deep Of substrates. Consequently, the effective growing season for shallow Of sites may be a few weeks longer than for deep Of sites. Additionally, the poorly aerated, perched water table immediately above the degrading ice may retard growth early in the season (Lees 1972).

If stems on the shallow active-layer sites produced seeds as readily as those on the deep active-layer sites, then it follows that isolated stunted stems (say, more than 100 m from a deep active-layer site) should show local densities and/or age structures similar to those found on deep Of sites with nearby tall stems. Nothing like this appears to occur in the Inuvik area. Invariably, the isolated white spruce stems (e.g., on the tableland or north-facing slope) have no younger stems nearby. Dense yet stunted stands were only found in close proximity to tall trees located on nearby deep active-layer sites, e.g., on narrow sections of alluvium/colluvium between a steep south-facing slope and a creek.

In summary, it would appear that differences in vegetative vigour and (perhaps) seed production on shallow versus deep active-layer sites may be attributed to rate of thaw and, to some degree, nutrient availability. Aeration immediately above the degrading ice may also be important. The rate of thaw should be governed primarily by potential insolation and ice content. In non-gravel substrates, a deep Of layer is more important than either mineral soil texture or slope angle in determining both surficial ice content and relative active-layer depth (Greene 1983). The Of depth, then, is of tremendous consequence in both the demography and spatial distribution of white spruce in the northern forest-tundra.

CONCLUSIONS

Tentatively, it would appear that active-layer depth and fire frequency are the proximate determinants of white spruce tree-line in the uplands around Inuvik. Shallow active-layers delimit the range of white spruce because of a slower rate of June thaw (i.e., effective growing season length), lower seed dispersal capacity, lower cone production (probably), and poor seedbed creation (because of the depth of the Of) by fire. Fire delimits the range of white spruce by restricting it to areas with relatively low fire frequency such as the steeper valley sides and the more deeply incised or highly sinuous perennial creeks. Deep active-layer sites tend to coincide with areas of low fire frequency. In the most heavily dissected terrain in the Inuvik area, such sites are not rare but their distribution is often very disjunct. If fire removes white spruce from an optimal site, the site may remain uncolonized for very long periods because of the low seed dispersal capacity of these stems.

ACKNOWLEDGEMENTS

Funding was provided by the University of Calgary and the Arctic Institute of North America. Logistical support was provided by the Inuvik Scientific Resource Centre and the Department of Forestry in Inuvik. I was assisted in the field-work by D. Privette and M. Bosch. I thank S.A. Harris, E.A. Johnson, and A. Limbird for their criticism of the thesis design and the manuscript.

REFERENCES

Bliss, L.C., and Wein, R.W., 1971, Changes to the active layer caused by surface disturbance, in Proceedings of a Seminar on the Permafrost Active Layer, 4 and 5 May, 1971: National Research Council of Canada, Technical Memorandum Number 103, Ottawa, p. 37-48.

Burns, B.M., 1973, The climate of the Mackenzie Valley-Beaufort Sea, Volume 1: Climatological Studies Number 24, Environment Canada, Ottawa.

Canadian System of Soil Classification, 1978: Resource Branch, Canadian Department of Agriculture, Publication Number 1646.

Day, J.H., and Rice, H.M., 1964, The characteristics of some permafrost soils in the Mackenzie Valley, N.W.T.: Arctic, v. 17, p. 223-236.

Elliot, D.L., 1979, The current regenerative capacity of the northern Canadian trees, Keewatin, N.W.T., Canada: some preliminary observations: Arctic and Alpine Research, v. 11, p. 243-251.

Fowells, H.A., and Schubert, G.H., 1956, Seed crops of forest trees in the pine region of California: United States Department of Agriculture, Technical Bulletin 1150.

Greene, D.F., 1983, Permafrost, fire, and the regeneration of white spruce at treeline near Inuvik, N.W.T.: unpublished Master of Science Thesis, Department of Geography, University of Calgary, Calgary, Alberta, Canada.

Harper, J.L., 1977, Population biology of plants: Academic Press, New York.

Johnson, E.A., 1975, Buried seed populations in the subarctic forest east of Great Slave Lake, Northwest Territories: Canadian Journal of Botany, v. 53, p. 2933-2941.

Kerfoot, D.E., ed., 1972, Mackenzie Delta area monograph: 22nd International Geographic Congress, Brock University, St. Catharines, Ontario, Canada.

Kozlowski, T.T., 1971, Growth and development of trees, 2 Volumes: Academic Press, New York.

Lambert, J.D.H., 1972, Vegetation patterns in the Mackenzie Delta area, N.W.T., in Kerfoot, D.E., ed., Mackenzie Delta area monograph: 22nd International Geographic Congress, Brock University, St. Catharines, Ontario, Canada.

Larsen, J.A., 1980, The boreal ecosystem: Academic Press, New York.

Larson, M.M., and Schubert, G.H., 1970, Cone crops of Ponderosa Pine in central Arizona, including the influence of Albert squirrels: United States Department of Agriculture, Forest Service Research Publication Number RM-58.

Lees, J., 1972, Soil aeration and sitka spruce seedling growth in peat: Journal of Ecology, v. 60, p. 343-349.

Mackay, J.R., 1963, The Mackenzie Delta area, N.W.T.: Canadian Department of Mines, Technical Survey, Geographical Branch Memorandum 8.

Nichols, H., 1976, Historical aspects of the northern Canadian treeline: Arctic, v. 29, p. 38-47.

Rowe, J.S., 1971, Spruce and fire in northwest Canada and Alaska, in Proceedings, Tall Timbers Fire Ecology Conference: Tall Timbers Research Station, Tallahassee, Florida, p. 245-255.

Schmidtling, R.C., 1969, Reproductive maturity related to height of loblolly pine: United States Department of Agriculture, Forest Service Research Note SO-94.

Tarnocai, C., 1973, Soils of the Mackenzie River area: Department of Indian and Northern Affairs, Ottawa, Report Number 73-26.

Van Cleve, K., Barney, R., and Schlentner, R., 1981, Evidence of temperature control of production and nutrient cycling in two interior Alaska black spruce ecosystems: Canadian Journal of Forest Research, v. 11, p. 258-273.

Van Wagner, C.E., 1979, Fire behavior in northern conifer forests and shrublands, in Symposium on Fire in Northern Circumpolar Ecosystems: University of New Brunswick, Fredericton, New Brunswick, Canada.

Viereck, L.A., 1979, Characteristics of treeline plant communities in Alaska: Holarctic Ecology, v. 2, p. 228-238.

Viereck, L.A., and Schandelmeier, L.A., 1980, Effects of fire in Alaska and adjacent Canada--a literature review: Bureau of Land Management, Alaska, Technical Report 6.

Wein, R.W., 1975, Vegetation recovery in Arctic tundra and forest-tundra after fire: Arctic Land Use Research Program, Department of Indian and Northern Affairs, Ottawa, Report 74-75-62.

AN INVESTIGATION OF MIDLATITUDE ALPINE PERMAFROST ON NIWOT RIDGE, COLORADO ROCKY MOUNTAINS, USA

L. A. Greenstein

Department of Geography, University of Colorado, Boulder, Colorado USA 80309

A broad ridge in the Colorado Front Range is the site of a detailed permafrost investigation based on freezing and thawing indices. A 30-year record of temperature data from a station at 3,750 m above sea level is used along with a set of representative lapse rates to determine mean daily temperatures for 12 hypothetical stations on the ridge at different altitudes and with different aspects. Freezing and thawing indices computed for these sites indicate that continuous permafrost may be found down to 3,600 m on south-facing slopes and 3,550 m on north-facing slopes. The discontinuous-sporadic boundary was found to lie at 3,300 m on south-facing slopes and 3,200 m on north-facing slopes. These boundaries are somewhat lower than those proposed in a previous study. It is possible that the freezing and thawing indices used in higher latitude alpine settings as representative of the lower limits of the permafrost zones may overestimate the presence of permafrost in midlatitudes because of higher daily maximum temperatures.

INTRODUCTION

Permafrost research began in the mountains of North America in earnest with Retzer's 1956 study of alpine soils in the Rocky Mountains. Other early work included Pierce (1961), Retzer (1965), and Johnson and Billings (1962). More recently, ground temperature data have been collected by Ives and Fahey (1971), Ives (1973), and Harris and Brown (1978). Harris (1979, 1981a,b,c,d) has continued with extensive studies of climatic relationships and geomorphic indicators and Péwé (in press) has recently written a review of available data on alpine permafrost in the contiguous United States. The present study is a detailed examination of the lower limits of permafrost occurrence in a small area of the Colorado Front Range.

THE STUDY AREA

The Colorado Front Range is an area of north-south-trending mountains approximately 250 km in length. The Indian Peaks region, roughly in the middle of the Front Range at lat. 40°N, encompasses many individual summits higher than 4000 m, U-shaped valleys, and broad ridges that narrow into arêtes. One such ridge, called Niwot Ridge, was selected as the focus of this study because of the availability of a 30-year climatic record for a high-altitude (3,750 m) site. The ridge itself is 10 km long, oriented perpendicular to the Continental Divide. Timberline lies at approximately 3,300 m with the forest-tundra ecotone extending in places up to 3,450 m (Ives and Hansen-Bristow, 1983).

Niwot Ridge experiences a midlatitude continental climate with long winters and cool summers. At high altitudes, the area is very windy throughout the year. Westerly airflow is strongest in winter, with average wind speeds of 14m/s on the tundra. On a knoll on Niwot Ridge, winds at 6 m above the surface exceeded 18m/s during 50% of a 4-month period. The average daily maximum for this same period was 39m/s, and the extreme gust was 62m/s (Barry 1973, 1981). Consequently the ridge is extremely windswept, allowing for only limited snow accumulation in hollows and on the lee slopes of small hills. The summer months are less windy, with July wind speeds averaging 6 m/s. Wide daily and seasonal temperature variations are characteristic. Freezing temperatures can occur in all months. The daily range decreases with increasing altitude as a result of greater windiness and increased mixing of free air.

TECHNIQUES FOR ALPINE PERMAFROST PREDICTION

It is defensible to base continental- or global scale maps of high-latitude permafrost distribution on a theorized relationship between mean annual air temperature and the presence of permafrost, despite the inherent inaccuracies of such an approach (Ives 1974). With decrease in latitude and increase in altitude, however, mean annual air temperature has less and less value for permafrost prediction. Assuming a constant relationship between mean annual air temperature and the ground temperature regime in a level, relatively homogeneous subarctic setting will result in an expected error of 100 km or more in delineation of the equatorward limit of permafrost for each 1°C error in extrapolation (Ives 1973). In the more heterogenous alpine environment, however, these inaccuracies could be magnified considerably. Large discrepancies result when attempts have been made to define a particular isotherm as the lower limit of alpine permafrost. Furrer and Fitze (1970) found the -1 to -2°C isotherm as the lower limit of permafrost in the Alps, while more recently Haeberli (1978), working also in the Alps, placed it between -7 and -8.5°C. The effects of factors such as soil characteristics, vegetation, snowcover, and wind, all of which play roles in the formation of permafrost, are greatly magnified by the high local relief of the alpine environment. In particular, aspect, degree of slope and solar angle, and wind effects on snowcover become very important factors, as they cause drastic changes in climates over very short horizontal distances.

The winter freeze and summer thaw are the two basic energy flows experienced by the soil in mid- and high-latitude settings with low winter snowcover. These basic flows can be represented by the freezing and thawing indices (the sums of the mean daily temperatures below and above freezing, respectively, expressed in degree centigrade days per year). Harris (1981a,b,c,d) has assembled an extensive collection of freezing and thawing indices from known permafrost and nonpermafrost sites in Canada, Greenland, Iceland, Mongolia, USSR, Scandinavia, Spitzbergen, and Switzerland. By plotting freezing indices against thawing indices, Harris was able to determine clusters of data points indicating the continuous, discontinuous, and sporadic zones. This method was applied to the 30-year temperature record from Niwot Ridge.

This technique is only reliable on sites with low winter snowcover. A very thin snowcover has the effect of cooling the soil by reflecting incoming energy (Kudryavtsev 1959). As the snowcover deepens, however, its capacity to reflect energy changes little while its ability to insulate the ground from winter low temperatures increases until the net effect is a warming of the soil. Based on data collected by Harris (1981a,b, c,d) from numerous sites in southern Alberta, 50 cm is indicated as the critical thickness of winter snowcover above which this technique will yield inconsistent results. Snow density and the mean annual range of the air temperature affect the insulating capacity of snow (Kudryavtsev 1959). Consequently, this critical depth would be expected to vary somewhat from one alpine site to another, particularly from a continental to a maritime regime. Krinsley (1963, p. 146) found 15 to 18 cm of snowcover to significantly affect the ambient air temperature fluctuations at low altitude sites in Alaska's interior. Atkinson and Bay (1940) found 25 cm to be the critical thickness at a low altitude site in Wisconsin. Harris (1981a,b,c,d) has found the data from different areas, representing both continental and maritime regimes to agree very well. Harris (1982, personal communication) has recently tested his findings with data from numerous thermister strings installed in the western Yukon territory.

A detailed ground temperature investigation was undertaken by Ives and Fahey (1971) (Ives 1973) on a section of Niwot Ridge. They installed thermister strings at several sites over a 300-m altitudinal range. They encountered permafrost at 3,750 m at a 3.8-m depth, and at 3,500 m at a 1.8-m depth. Ives (1973) was able to infer the presence of permafrost with relative certainty at two other sites located at 3,800 m and 3,750m and the possibility of its occurrence at 3,485m.

The above-mentioned study used direct measurements to establish the presence of permafrost on Niwot Ridge. It is the aim of the present study to expand on the previous work, using indirect prediction methods to extend the boundary of the study area outward.

METHODS

The first step was to compute the mean daily temperature for each day of the year from the 30-year record available. The specific station used is referred to as D-1. This site was established in 1953 by John Marr (1961). It is located on a ridgetop knoll at 3,750 m, 1.5 km east of the Continental Divide. The site is very exposed and windy with a mean annual air temperature of -3.5°C.

Temperature data for numerous other high-altitude sites on Niwot Ridge are available for limited periods. These data were used by Kiladis (1980) along with the data from D-1 to produce four temperature maps of Niwot Ridge and the surrounding area showing average daily minimum and maximum temperatures in January and July. The maps were in part developed from representative lapse rates computed by Kiladis (1980) from the station data available. These representative lapse rates were further refined by Kiladis and this author for the purpose of this study. One important factor in these adjusted lapse rates was the subtraction of the cold air drainage effect common in this and many other alpine settings (Barry 1981, Harris 1982). This results in the temperature estimates, as well as the permafrost predictions, being conservative in those areas that are prone to cold air drainage.

Six separate lapse rates were used in the present study. Above the elevation of D-1 (3,750 m), conditions are similar to those of the free air because of the consistent windiness. Lapse rates computed from these exposed stations were used for temperature estimates above 3,750 m. The average lapse rate of maximum temperatures above 3,750 m is very close to the dry adiabatic rate of 1.0°C/ 100 m. The average lapse rate for minimum temperatures is slightly less at .8°C/100 m. When estimating temperatures downward from D-1, it was necessary to use separate lapse rates for the north and south slopes as well as for minimum and maximum temperatures. These lapse rates are approximately as follows:

North-facing slopes--maximum .5°C/100 m
minimum .6°C/100 m

South-facing slopes--maximum .6°C/100 m
minimum .8°C/100 m

These six separate lapse rates were then used to estimate mean daily temperatures for 12 hypothetical stations. The mean annual air temperatures of the stations are plotted against elevation in Figure 1. As the lapse rates were calculated with data from the tundra and the forest-tundra ecotone, the estimates become less reliable close to timberline.

Freezing and thawing indices were then computed for each of the 13 sites (D-1 and the 12 hypothetical stations). Figure 2 shows the freezing indices plotted against the thawing indices for each of the sites, as well as the limits of the continuous, discontinuous, and sporadic permafrost zones from Harris (1981a,b,c,d).

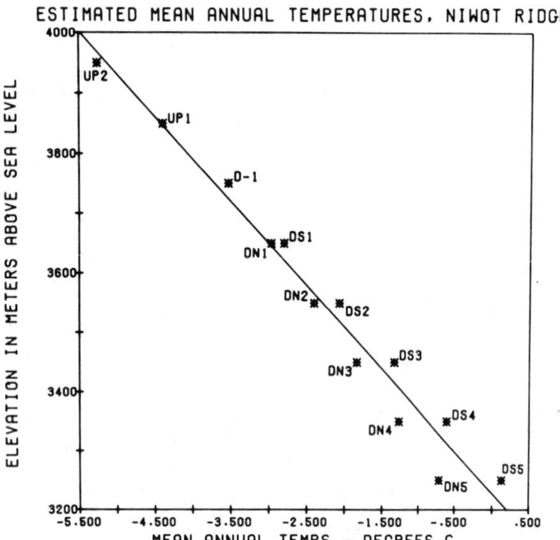

FIGURE 1 Mean annual air temperatures for 13 sites (D-1 and 12 hypothetical stations) on Niwot Ridge, Colorado Front Range. An "N" in the site name indicates a north-facing slope and an "S" indicates a south-facing slope.

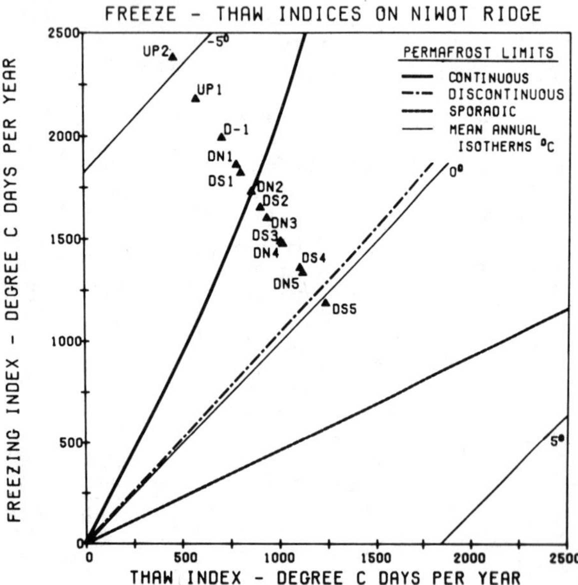

FIGURE 2 Freezing and thawing indices for 13 sites on Niwot Ridge, Colorado Front Range, in relation to the lower limits of the continuous, discontinuous, and sporadic permafrost zones (after Harris 1981a,b,c,d).

DISCUSSION

The lower limit of the continuous permafrost zone is indicated at approximately 3,600 m on south-facing slopes and 3,550 m on north-facing slopes. Above these elevations, permafrost would be expected at most sites. Exceptions would be found in areas with lakes, glaciers, or a substantial winter snowcover. This altitudinal zone, however, is characterized by consistent windiness. On the broad ridges such as Niwot Ridge, very little snow accumulates at these elevations. More snow, and therefore less permafrost, can be found in the cirques and nivation hollows.

The discontinuous permafrost zone appears to extend down to 3,300 m on south-facing slopes and 3,200 m on north-facing slopes. In this zone permafrost occurrence would be patchy in response to the characteristics of the immediate environment. On the ridge itself, some snow accumulates in this zone in areas that are protected from the wind. One such area, referred to as the Saddle (approximately 3,500 m), is the site of a snow study currently under way by J. Halfpenny. Snow depths up to 3.6 m were found in the Saddle in mid-March of 1983 (1983, personal communication). The Saddle would be expected to represent a permafrost-free patch within the discontinuous zone. The thermister studies by Ives and Fahey (1971) and Ives (1973) of three sites in the Saddle confirmed the presence of permafrost at 3,500 m, a possibility of its occurrence at 3,485 m, and a permafrost-free site at 3,490 m.

The lower limit of the sporadic permafrost zone was not determined, as this would have required extending the lapse rate estimates below timberline. In the sporadic zone, small, isolated patches of permafrost would be expected in only the most favorable locations. Frozen ground was encountered late in the summers of 1962 and 1965, 3 to 4 m below the surface during the bulldozing of till for dam fill, on the south side of Niwot Ridge at 3,140 m (Ives and Fahey 1971). Pierce (1961) reported permafrost in a peat deposit at 2,950 m just 5° of latitude north of this study site.

These results indicate a more extensive permafrost cover than that presented in Ives's initial conclusions (1973, 1974, Ives and Fahey 1971). Ives concluded that the lower limits of the continuous and discontinuous zones lie at 3,750 and 3,500 m respectively. Being the first detailed study of permafrost in the Colorado Rockies and based on a limited number of thermister strings, these estimates may have been designedly conservative.

Mention should be made of the possible overestimation of the presence of alpine permafrost using freezing and thawing indices in Colorado. With the exception of the USSR and Mongolia, the data used by Harris (1981a,b,c,d) in defining the indices representative of the three permafrost zones came from significantly higher latitude sites than Niwot Ridge (40°N). While both the Canadian and the Colorado Rockies exhibit cool, continental climatic regimes, incoming energy is temporally allocated quite differently in this more southerly location. While mean annual temperature may be identical between 3,750 m on Niwot

Ridge and a somewhat lower elevation at a higher latitude, Niwot Ridge experiences a more distinct diurnal radiation pattern superimposed on a slightly less severe seasonal regime. Niwot Ridge is affected by higher sun angles and the possibility of higher daily maximum temperatures than a higher-latitude site with the same mean annual air temperature. While the effect of these higher maximum temperatures would be reflected in the freezing and thawing indices, it may be somewhat minimized, as the indices are based on daily mean temperatures. While the use of freezing and thawing indices appears to be valid for midlatitude mountains, the curves representing the lower limits of the permafrost zones may need to be shifted slightly.

ACKNOWLEDGMENTS

The author gratefully acknowledges Jack Ives, Stuart Harris, George Kiladis, and Roger Barry for helpful discussions and suggestions.

REFERENCES

Barry, R. G., 1973, A climatological transect on the east slope of the Front Range, Colorado: Arctic and Alpine Research, v. 5, p. 89-110.

Barry, R. G., 1981, Mountain weather and climate: New York, Methuen.

Black, R. F., 1954, Permafrost--a review: Bulletin of the Geological Society of America, v. 65, p. 839-856.

Furrer, G. and Fitze, P., 1970, Beitragzum permafrostproblem in den Alpen: Vierteljahresschrift der Naturforschenden Gesellschaft in Zürich, v. 115, p. 353-368.

Haeberli, W., 1978, Special aspects of high mountain permafrost methodology and zonation in the Alps, in Proceedings of the Third International Conference on Permafrost, v. 1: Ottawa, National Research Council of Canada.

Harris, S. A., 1979, Ice caves and permafrost zones in southwest Alberta: Erdkunde, v. 33, p. 61-70.

Harris, S. A., 1981a, Climatic relationships of permafrost zones in areas of low winter snowcover: Arctic, v. 34, p. 64-70.

Harris, S. A., 1981b, Climatic relationships of permafrost zones in areas of low winter snowcover: Biuletyn Peryglacjalny, v. 28, p. 227-240.

Harris, S. A., 1981c, Distribution of active glaciers and rock glaciers compared to the distribution of permafrost landforms, based on freezing and thawing indices: Canadian Journal of Earth Sciences, v. 18, p. 376-381.

Harris, S. A., 1981d, Distribution of zonal permafrost landforms with freezing and thawing indices: Erdkunde, v. 35, p. 81-90.

Harris, S. A., 1982, Cold air drainage west of Fort Nelson, British Columbia: Arctic, v. 35, p. 537-541.

Harris, S. A. and Brown, R. J. E., 1978, Plateau Mountain: a case study of alpine permafrost in the Canadian Rocky Mountains, in Proceedings of the Third International Conference on Permafrost, v. 1: Ottawa, National Research Council of Canada.

Johnson, P. L. and Billings, W. D., 1962, The alpine vegetation of the Beartooth Plateau in relation to cryopedogenic processes and patterns: Ecological Monographs, v. 32, p. 105-135.

Ives, J. D., 1973, Permafrost and its relationship to other environmental parameters in a midlatitude, high-altitude setting, Front Range, Colorado Rocky Mountains, in Permafrost--The North American Contribution to the Second International Conference, Yakutsk: Washington, D. C., National Academy of Sciences.

Ives, J. D., 1974, Permafrost, in Ives, J. D. and Barry, R. G., eds., Arctic and alpine environments: London, Methuen.

Ives, J. D. and Fahey, B. D., 1971, Permafrost occurrence in the Front Range, Colorado Rocky Mountains, USA: Journal of Glaciology, v. 10, p. 105-111.

Ives, J. D. and Hansen-Bristow, K. J., 1983, Stability and instability of natural and modified upper timberline landscapes in the Colorado Rocky Mountains, USA: Mountain Research and Development, v. 3, p. 149-155.

Kiladis, G., 1980, Climatic characteristics of the Front Range area in Colorado including temperature maps of Niwot Ridge and case studies: unpublished report, University of Colorado, Institute of Arctic and Alpine Research.

Krinsley, D. B., 1963, Influence of snow cover on frost penetration: United States Geological Survey Professional Paper, 475-B, p. 144-147.

Kudryavtsev, V. A., 1959, Temperature, thickness, and discontinuity of permafrost, in Principles of geocryology, National Research Council Technical Translation 1187: Ottawa, National Research Council of Canada.

Marr, J. W., 1961, Ecosystems of the east slope of the Front Range in Colorado: University of Colorado Studies, Series in Biology, v. 8.

Péwé, T., in press, Alpine permafrost in the contiguous United States: a review: Arctic and Alpine Research, v. 15.

Pierce, W. G., 1961, Permafrost and thaw depressions in a peat deposit in the Beartooth Mountains, northwestern Wyoming: United States Geological Survey Professional Paper 424-B.

Retzer, J. L., 1956, Alpine soils of the Rocky Mountains: Journal of Soil Science, v. 7, p. 22-32.

Retzer, J. L., 1965, Present soil-forming factors and processes in arctic and alpine regions: Journal of Soil Science, v. 16, p. 38-44.

ENGINEERING PROPERTIES AND FOUNDATION DESIGN ALTERNATIVES IN MARINE SVEA CLAY, SVALBARD

Odd Gregersen[1], Arvind Phukan[2], and Torbjørn Johansen[3]

[1]Civil Engineer, Norwegian Geotechnical Institute, Oslo, Norway
[2]Professor of Civil Engineering, University of Alaska, Anchorage, Alaska
[3]Civil Engineer, Norwegian Geotechnical Institute, Oslo, Norway

A new coal mining plant is under planning in Svea, central Spitsbergen, Svalbard. A major part of the plan area is underlain by a marine clay containing ice in vertical cracks. The cracks, 10-20 mm wide, are parallel at a spacing of about 50 mm. This particular structure of the marine clay encountered at the area is related to the geologic history of the region. With a water content of 50% and a high salt content, the engineering properties are extremely poor. At mean annual surface temperature, $-6^{\circ}C$, about 40% of the pore water is unfrozen. The long term strength of the clay is determined by creep tests under unaxial and triaxial conditions. For example, at a temperature of $-4^{\circ}C$, the long term strength is as low as 80 kN/m^2 for a design life of 10 years. Based on field and laboratory data, three viable foundation alternatives are investigated. Spread footings, ventilated gravel pad, and slurried pile foundations are evaluated, and typical design solutions are given.

INTRODUCTION

Svea lies in the inner part of the van Mijenfjord on Spitsbergen, Svalbard, and is one of the three Norwegian communities on the Svalbard Islands. The other two are Ny-Ålesund and Longyearbyen. All three are based on the coal mining industry. Longyearbyen, the largest with a population of about 1,200, is the industrial and administrative center, see Figure 1.

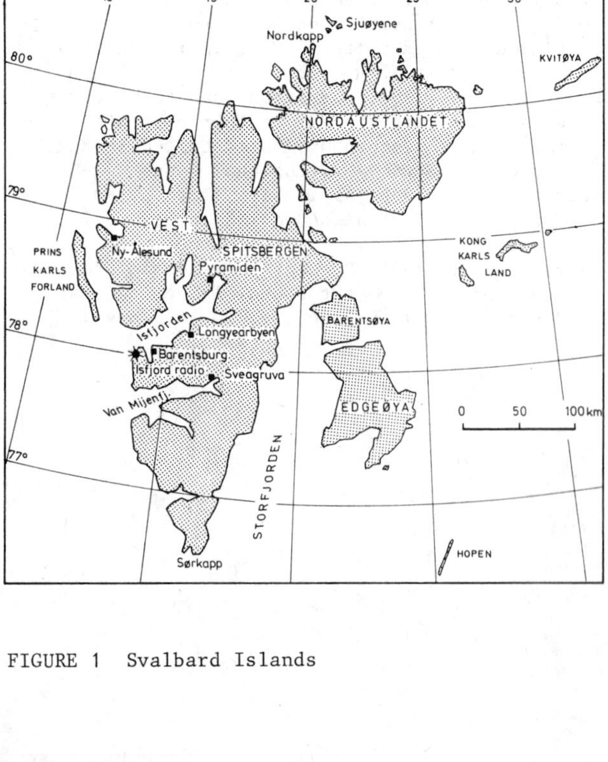

FIGURE 1 Svalbard Islands

Future expansions will take place in Svea. Considerable coal resources have been detected and plans are being made for a large exploitation. This implies extensive construction activities, such as a coal treatment plant, harbour, air field, accomodation, roads, and other service systems.

Svea is located at 78° north and has a mean annual temperature of about $-6^{\circ}C$. The relatively pleasant climate (considering the extreme location) is due to the warm Gulfstream passing along the western shores of Svalbard.

GEOLOGIC HISTORY

A major part of the construction activities will take place within an area of marine clay. This clay was originally sedimented at the bottom of the van Mijenfjord. A surge of the Paula Glacier, from across the other side of the fjord, pushed the fjord-bottom clay in front and onto the Svea Lowland (see Figure 2). This surge is dated to between 600 and 250 years ago (Haga 1978, Pewe 1981).

The clay has a very special structure. A system of parallel, almost vertical, ice filled cracks is present throughout the whole of the deposit. The cracks can be up to 20 mm wide and are more than 5 m (probably 10 m) deep. Investigations done at different sites within a 2 km length indicate that the cracks have the same orientation throughout the deposit, that is, parallel to the shore line (see Figure 3).

The development of the vertical crack system is believed to be related to the glacial surge referred to above. Squeezed between the glacier and the mountain, the clay was subjected to extremely high horizontal pressure. As the glacier withdrew, the horizontal pressure was released, resulting in a lateral expansion of the clay and subsequently giving rise to vertical cracking of the deposit.

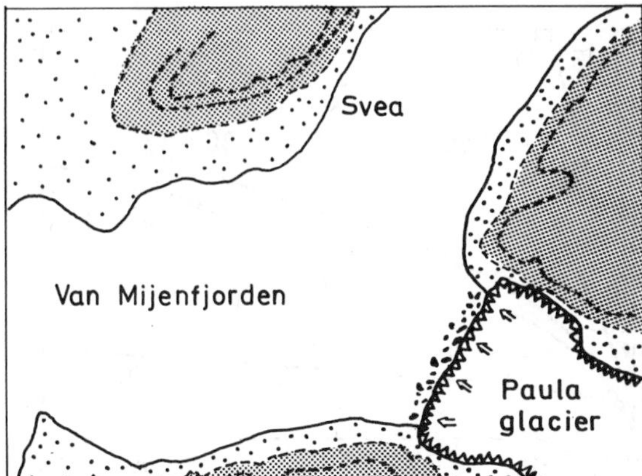

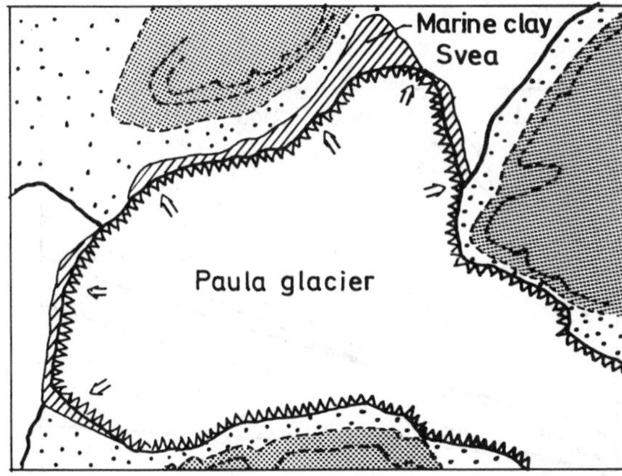

FIGURE 2 Sketch of Paula Glacier surge, 600 to 200 years ago.
a) Presurge stage b) Maximum spread

FIGURE 3 Detailed view of ice-rich marine clay, with parallel ice seams

ENGINEERING PROPERTIES

Basic Data

Figure 4 shows a soil profile of the clay, giving the results of laboratory investigations. It should be noted that the water contents have been determined from the gross sample as well as from the actual clay (excluding the ice cracks). The gross water content is, as shown, above the liquid limit.

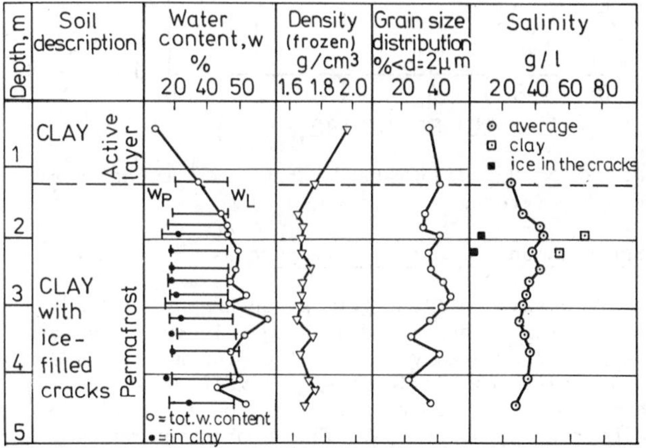

FIGURE 4 Results of laboratory tests

The distribution of salt within the deposit was also evaluated. Salt contents were measured on the gross sample, on the ice in the cracks, and on the actual clay. The gross salt content was found to be 30-40 grams per litre, equivalent to the salinity of saltwater. The ice in the cracks, however, is practically free of salt, while the clay pores have a salt content of about 60 grams per litre. This separation of water and salt is a result of the freezing process. Impurities, in this instance salt, are forced out of the crystals and hence give rise to almost saltfree icelenses.

Grain size distribution tests were made. Figure 4 gives the clay fraction content with depth. As seen, it is about 40% and fairly constant with depth. That is, the material is homogeneous and is defined as a clay. The mineralogic composition is found to be illite (15-40%), clorite (20-30%), and kaoline (10-40%). Small amounts of quartz and feltspar were also detected.

The determination of unfrozen water content as a function of temperature was performed on three samples. The clay specimens were melted and remoulded prior to testing, ensuring a uniformly distribution of water and salt within the test samples. The results from tests of one representative specimen are shown in Figure 5. At -2°C, 100% of the water is unfrozen, i.e. the freezing point depression of this saline clay material is 2°C. At -6°C, the mean annual temperature of Svea, 40% of the water remains unfrozen. The clay in its natural state will show somewhat different behaviour. The clean ice in the vertical lenses will remain frozen up to 0°C, while the saline pore water will be completely melted at above -3°C. These data agree with field observations, that the clay remains soft and can be cut easily with a tool at temperatures a few degrees below 0°C.

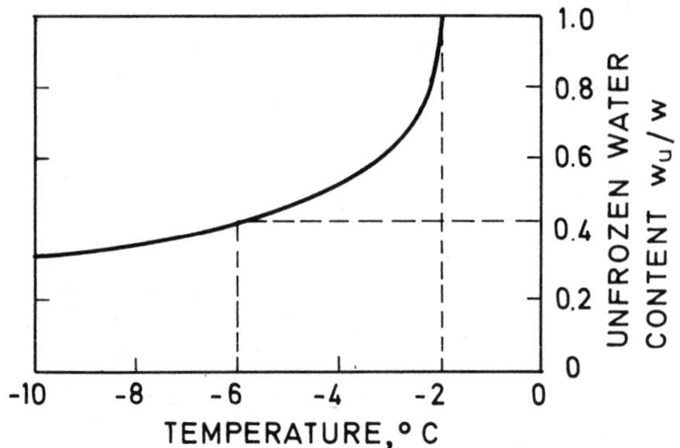

FIGURE 5 Unfrozen water content as a function of temperature

Strength and Deformation Properties

To obtain the long term strength parameters needed for foundation design, a relatively extensive test program was run. The program included short term compression tests as well as unconfined and confined creep tests. Tests were run at two different temperatures. The compression tests served as reference for the creep tests. Preparation of samples and all testing were performed in a cold room at test temperature.

All tests were run on cylindrical soil samples with typical dimensions; diameter = 7 cm and height = 14 cm. The unconfined compression tests were performed with a constant rate of strain of $\dot{e}$ = 1%/min.

The major part of the creep tests were performed in a specially made triaxial creep apparatus. The constant axial load is applied through a Bellofram air cylinder on top of a standard triaxial cell. The apparatus is equipped with electrical transducers. A computer records and processes all the data from the test and produces report-ready plots (Johansen 1981 a). The duration of the creep tests varied from only a few hours up to more than 5 weeks.

A limited number of tests were run with different confining pressures. In Figure 6 the tests with load of 300 kN/m^2 had confining pressures of 0 to 300 kN/m^2. The results from those tests indicated that the Svea clay behaved as a true cohesive material. Therefore, all remaining tests in the program were performed as unconfined creep tests only.

Svea clay was tested at two temperatures, -5°C and -3°C. Most of the tests were performed at -5°C (1°C above the mean annual surface temperature). The latter temperature, -3°C, is the freezing point of the saline clay and was chosen to obtain information on the behaviour of the Svea clay, near melting.

The results of the creep tests are shown in Figures 6 and 7. The scatter in the results of tests at -3°C is a result of different ice structures in the samples, as a large proportion of the deformation is sliding between the ice lenses and the clay. The test results are used to obtain

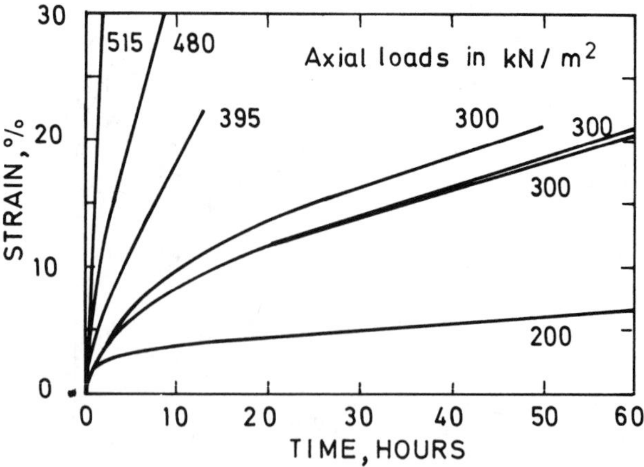

FIGURE 6 Results of creep tests at -5°C

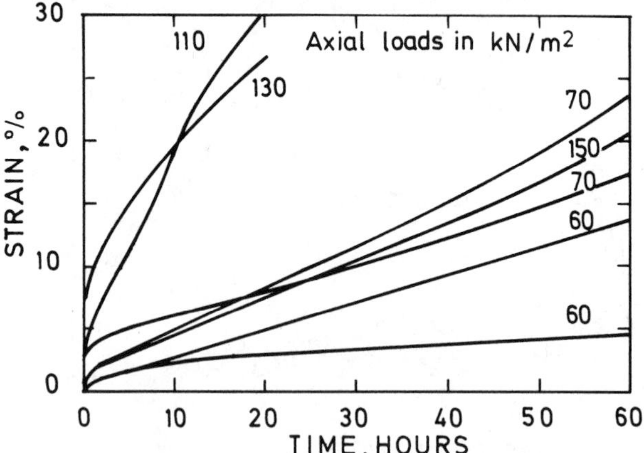

FIGURE 7 Results of creep tests at -3°C

long term strength of the clay by using methods proposed by Vyalov et al. (1966). The method is slightly modified by using a "20% strain criteria" instead of the inflection point between secondary and tertiary creep, in evaluating the creep parameters. This modification is introduced because the tests with low loads did not arrive at the tertiary creep, and often not at the secondary creep, within 20-30% strain. The highly unhomogeneous Svea clay was very difficult to test to a strain larger than approximately 30%.

Figure 8 shows the calculated long term strength for Svea clay based on the creep tests. Each line in the figure represents 1, 10, and 100 years time to failure (defined as 20% strain), respectively. The corresponding short term compressive strength at -5°C and -3°C is 900 kN/m^2 and 380 kN/m^2.

Based on the laboratory test results, the long term strength of the Svea clay is low, even at -5°C. This is mainly due to the high salinity of the solid portion of the clay, and the influence of the thick, vertical and continuous ice lenses.

At -3°C the clay itself is almost totally unfrozen. And in the interface between the clay and the ice, a thin film of unfrozen water is present, strongly influencing the deformation behaviour

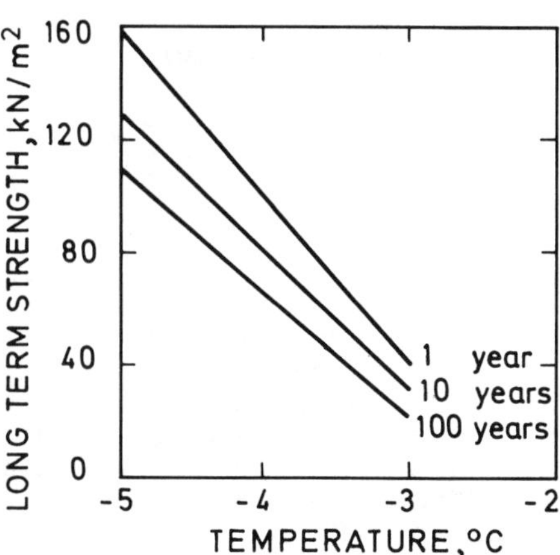

FIGURE 8 Calculated creep strength values as a function of temperature

leading to extremely low creep strength (Berggren 1980, Johansen 1981 b, and Furuberg and Johansen 1983).

Only field tests and monitoring of constructions at Svea can give the necessary feed back of data to correlate laboratory strength parameters and in-situ parameters.

FOUNDATION DESIGN ALTERNATIVES

The assumed subsoil profile consists of clay with engineering properties as described above. The design ground temperature, based on measurements for several years, is shown in Figure 9.

Three most viable foundation types for different structures considered for the Svea clay soil are spread footings with an air gap and directly placed on permafrost, forced air ventilated gravel pads, and slurried pile foundations.

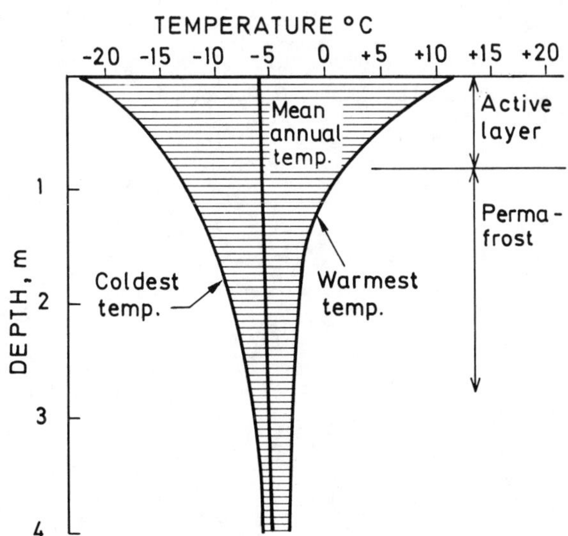

FIGURE 9 Measured temperature profiles from Svea

A typical spread foundation is illustrated in Figure 10. Based on the strength given in Figure 8 and considering a 20-years design period, the allowable soil pressure on the spread footing will be on the order of 100 kN/m^2.

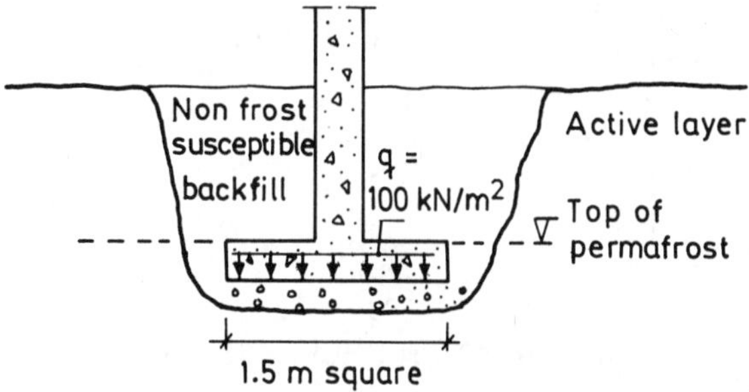

FIGURE 10 Proposed spread footing design

The heated structures with heavy floor loads may be founded on gravel pad with forced air circulation, as shown in Figure 11. The foundation design of such ventilated gravel pads in permafrost zones is principally concerned with two aspects: the pad design for the summer period and the ventilation requirements for the winter periods when the ducts are operative (Phukan and Andersland, 1978). The following data are assumed for a typical case, as shown in Figure 11.

Freezing Index	=	3180°C days
Length of freezing season	=	255 days
Mean annual air temperature	=	-7.5^{o}C
Floor concrete thickness	=	0.2 m
Air duct diameter	=	0.30 m
Gravel density	=	2000 kg/m^3
Gravel water content	=	3%

The methods given by Phukan and Andersland (1978) were used to determine the relationship between the thickness of the gravel pad and insulation for a 40 x 20 m gravel pad with ducts 0.3 m in diameter, and with c/c = 1 meter (see Figure 12). The air velocity required to refreeze the gravel pad was also analysed and is illustrated in the same figure. These relationships show that considerable savings may be obtained by using insulation.

Another viable alternative foundation design will be the slurried pile commonly used in Arctic

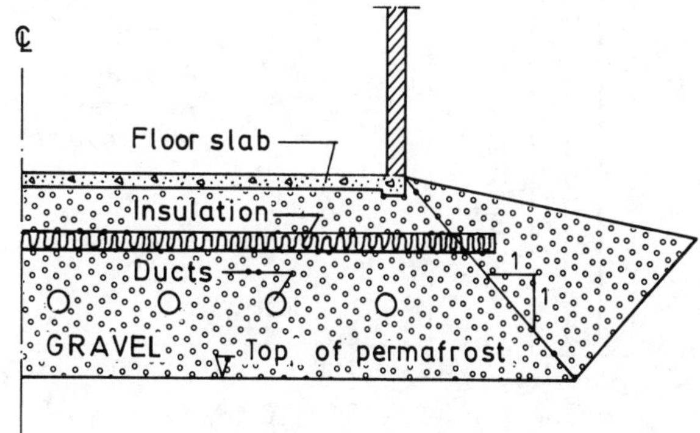

FIGURE 11 Ventilated gravel pad

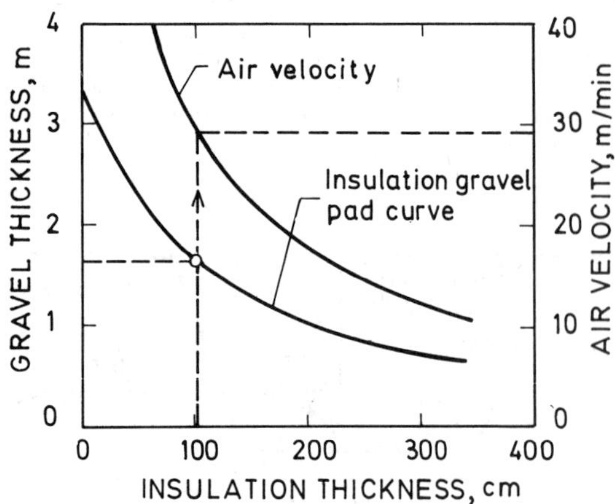

FIGURE 12 Relationships between insulated gravel pad and required air velocity

conditions. It is normally designed based on the adfreeze bond between the pile surface and adjacent frozen slurry, which is placed in the annular space between the drilled hole face and the outside of the pile surface. The design philosophy consists of:

the applied shear stress along the pile surface must be below some acceptable shear strength for the site soil conditions or the imposed loading does not exceed the adfreeze strength for the pile-soil contact, and

the allowable load will maintain the anticipated settlement at the end of the design period within tolerable limits.

A typical slurried pile design for the assumed soil and temperature profiles is presented in Figure 13. Assuming an uplift force of 200 kN/m^2, the required effective length of slurried pile in Svea clay will be on the order of 2.5 m below the active layer (factor of safety used = 2).

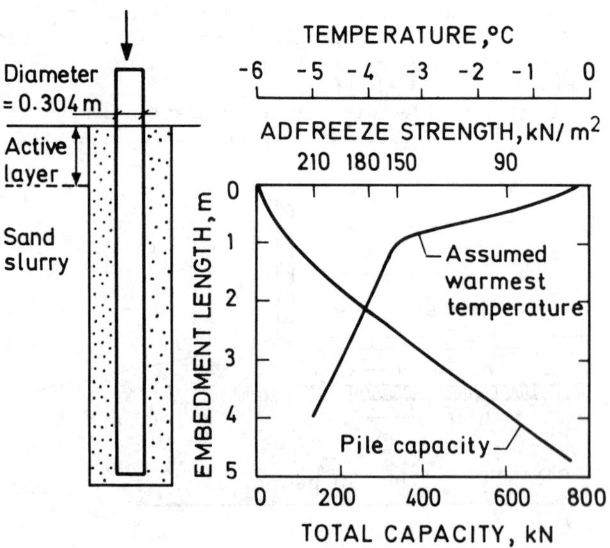

FIGURE 13 Slurried pile foundation design

ACKNOWLEDGMENTS

The research activities concerning the Svea clay were supported by the Norwegian Committee on Permafrost. All the laboratory testing was performed at the Geotechnical Division and at the Division of Refrigeration Engineering at The Norwegian Institute of Technology in Trondheim.

REFERENCES

Berggren, A.L., 1980. Laboratory investigations of a permafrost clay from Svea and an artificial frozen clay from Trondheim (In Norwegian). Geotechnical Division 0.78.06-2, The Norwegian Institute of Technology.

Furuberg, T. and Johansen, T., 1973. Engineering properties of Svea clay - final report (In Norwegian). Geotechnical Division, The Norwegian Institute of Technology.

Haga, Ø., 1978. A study of the effect of The Paula glacier surge on the sediments in the inner part of van Mijen fjord. Thesis (In Norwegian). University of Oslo.

Johansen, T., 1981a. Krypaks - an equipment for running triaxial creep tests at constant deviatoric stress. Users guide (In Norwegian). Geotechnical Division F.81.01-1, The Norwegian Institute of Technology.

Johansen, T., 1981b. Creep tests on Svea clay (In Norwegian). Geotechnical Division 0.81.05. The Norwegian Institute of Technology.

Pewe, T.L., 1981. Engineering geology of the Svea lowland, Spitsbergen, Svalbard. Norwegian Committee on Permafrost, Oslo. 11 p. Frost Action of Soils No. 23.

Phukan, A. and Andersland, O.B., 1978. Foundations for cold regions. Geotechnical engineering for cold regions, N.Y., McGraw Hill. Chapter 6.

Phukan, A., 1982. Design and construction of foundations in permafrost soils related to Svea conditions, Norwegian Geotechnical Institute, Oslo. Report 52701-05.

Vyalov, S.S. et al., 1966. Methods of determining creep, long-term strength and compressibility characteristics of frozen soils, National Research Council of Canada, Technical Translation No. 1364, Ottawa 1969. 109 p.

A HIGH-RESOLUTION TECHNIQUE FOR MEASURING MOTION WITHIN THE ACTIVE LAYER

E. Carrington Gregory[1] and Christopher W. Stubbs[2]

[1]Quaternary Research Center and Department of Geological Sciences
University of Washington, Seattle, Washington 98195 USA
[2]Department of Physics, University of Washington, Seattle, Washington 98195 USA

We have developed an instrument based upon the principle of electromagnetic induction to measure the three-dimensional position of a sensor. This sensor can be buried in the active layer of periglacial areas to study frost heaving, mass movement, and the evolution of patterned ground. The system is sensitive to sensor displacements of 1 mm, a very high resolution relative to anticipated heave rates. The present system is effective at ranges of up to 1 m, and this range could be easily extended. Laboratory tests show the system to be insensitive to variations in temperature, and calibration results are independent of the state of the soil, whether frozen or thawed.

INTRODUCTION

The principal geomorphic processes characteristic of periglacial areas, including frost heaving, solifluction, and the generation of patterned ground, involve the bulk motion of soil and the relative motion of rocks through the soil (Washburn 1980). Valuable information about total subsurface displacements over extended time periods has been obtained from studies of surface displacements (Fahey 1974, Nicholson 1976), from excavations that reveal displaced geologic markers both in the field (Shilts 1978, Washburn 1980) and in the laboratory (Pissart 1970, Burrous 1978, Washburn et al. 1978), and from measurements of pressure and moisture regimes within the active layer (Mackay and MacKay 1976). In spite of this substantial data base, our understanding of patterned ground development and the upfreezing of objects has been limited by the dearth of precise knowledge of the rate, direction, and timing of these subsurface motions.

We have developed an experimental technique that will allow ongoing nondisruptive observation of subsurface displacements in the active layer. The scheme uses electromagnetic induction to measure the three-dimensional position of a buried sensor coil, enabling one to determine its trajectory with high resolution. The sensor can be made to track soil movement or to simulate a rock that may move relative to the soil. Heave rates commonly range up to 50-150 mm yr^{-1} for rocks of a size (diameter $\sim$0.1 m) similar to that of the sensor (Burrous 1978, Washburn 1980), and our system can readily resolve displacements of 1 mm. The present system can be used to a depth of 1 m, which typically covers most of the range of interest in the active layer (Ferrians and Hobson 1973, Washburn 1980). In addition to obtaining precise information about subsurface movements, much could be learned about the underlying driving mechanisms by simultaneously monitoring temperature, pressure, and moisture content with appropriate transducers incorporated in the buried sensor.

APPARATUS

General

The principle of electromagnetic induction is used to measure the position of a buried sensor. The device is shown schematically in Figure 1. We use two wire coils, one as a drive coil on the surface and the second as a sensor that is buried in the active layer, simulating a clast. An ac current in the drive coil produces an oscillatory magnetic field. The component of this field vector normal to the plane of the sensing coil, B_z, induces an electromotive force (EMF) $\mathcal{E}$ in the sensing coil. The EMF is proportional to the time derivative of the magnetic field. For a sensor of fixed area, A (Jackson 1975),

$$\mathcal{E} = -A\frac{\partial}{\partial t}B_z. \quad (1)$$

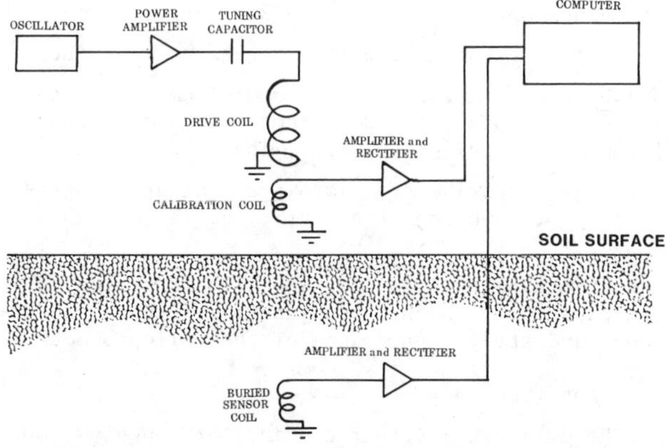

FIGURE 1 Schematic diagram of present system. The drive coil and its associated circuitry lie above the soil surface, while the sensor is buried in the active layer. The physical parameter measured by the system is z, the separation between the drive and sensor coils, once they are coaxial. The horizontal sensing coils are omitted for clarity.

For a sinusoidal drive current of constant amplitude and frequency, when the drive coil is directly above the sensor, the induced EMF is a measure of the separation between the two coils (Jackson 1975). In this configuration the sensor essentially functions as a magnetometer, and the spatial variation of the field amplitude yields vertical position information.

The horizontal position of the sensor is provided by two additional mutually orthogonal accessory coils, wound orthogonal to the main sensor coil already mentioned. When the drive coil is directly above the sensor, no EMF is induced in the accessory coils, since the component of the field normal to each of these coils is zero. The system is very sensitive in this respect, since the EMF scales with sin γ, where γ is the angle between the plane of the coil and the incident field vector. Conversely, the signal in the vertical sensing coil is less sensitive to small horizontal displacements, as sin γ has a broad maximum at $\gamma = \frac{\pi}{2}$. This is to our advantage.

The sensor's position in the horizontal plane is measured on the surface by moving the drive coil horizontally until a null is obtained in both accessory coils. The drive coil is then directly over the sensor, and its horizontal position can be measured. The vertical separation is then determined by measuring the voltage induced across the main sensor coil and comparing it with a calibration curve. To ensure that the sensing coils maintain the proper orientation, the sensor system is mounted on gimbals in a housing like a boating compass.

Electronics

The drive circuitry consists simply of an oscillator and a power amplifier. We use an RC timer (XR 2240) in conjunction with a voltage regulator (μA 7805) to produce a square wave of fixed amplitude. The power amplifier has a variable gain stage (LM 324) and a unity gain buffer (μA 759). The drive system runs on ±12 V and is quite field portable. The drive coil is placed in series with a tuning capacitor, giving an RLC series circuit. This circuit has a resonant frequency, $\omega_o = (LC)^{-\frac{1}{2}}$, at which its impedance is at a minimum. The square wave generator is tuned so that its fundamental frequency coincides with ω_o. Thus the current drawn by the drive coil is sinusoidal at the frequency ω_o, even though the drive voltage is a square wave. This produces a sinusoidally varying magnetic dipole field. We use a 1 μfd series tuning capacitor, which in conjunction with the inductance and stray capacitance of the coil produces a resonant frequency of $f_o = \frac{\omega_o}{2\pi} = 298$ Hz.

The drive coil itself consists of roughly 3,000 turns of #18 magnet wire evenly distributed on a cylindrical plexiglass coil form 0.12 m in diameter and 0.3 m long.

The sensor's coil has about 200 turns of #24 magnet wire wound around an iron core 30 mm in diameter and 120 mm long. The EMF across the sensor is processed through a preamplifier (LM 324) of gain 200, and a rectifier. This dc voltage bears the vertical position information and is carried by coaxial cable to the measurement device on the surface. The horizontal information for the two smaller accessory coils is handled in the same fashion.

A calibration coil at a fixed position relative to the drive coil and similar in construction to the sensor is used to render the system insensitive to small variations in drive current amplitude and frequency. This is achieved by taking the ratio between the EMF induced in the sensor and that induced in the calibration coil. This ratio technique provides good immunity to the circuitry drifts that may occur under harsh field conditions.

Both the sensor signal and the calibration signal are conveniently measured by a microprocessor-based field data acquisition system (Polycorder by Omnidata) which averages each signal for 250 ms and stores the data in memory. It is worth noting that a multimeter will suffice for single measurements.

Mechanical

Ideally the drive coil should be rigidly mounted on a permanent installation for long-term monitoring of the sensor's absolute position, since the system measures the separation between the drive and sensor coils. Periodic adjustments of the drive coil's horizontal position could be made as necessary, followed by vertical position measurements. The most stable configuration would involve a nonmagnetic frame anchored in either bedrock or permafrost.

One drive coil could service an array of sensors, some of which could be outfitted with "fins" in order to determine bulk soil motion. The others would have their density, texture, and thermal characteristics tailored to mimic clasts of interest. Alternatively, sensors could be designed to explicitly investigate any dependence on a given variable. The use of sensors of varying size and shape, for example, would allow comparison with past studies (Burrous 1978 p. 83, 84). Sensor size is somewhat constrained, as sensitivity drops with a reduction in coil area, but sensors of dimension 50 mm on a side seem reasonable.

By mounting the drive coil on the frame the sensor's absolute position is measured. To determine its position relative to the soil surface, surface heave can be measured by using linear potentiometers fixed to the frame and coupled to the ground through rods that move freely with the soil surface. Two adjacent sensors, one with fins and one without, would distinguish between bulk and relative motion. If the vertical position of the sensor relative to the surface suffices, no frame is necessary.

Some disturbance of the soil is unavoidable during the installation of the sensor. This may be minimized, however, by implanting it while the soil is frozen; a soil core could be extracted, and replaced after positioning the sensor in the borehole. Coring at an angle to the anticipated path of the sensor's travel would further minimize the disturbance. After installation, data can be taken at will without further disruption.

Presently, some provision must be made to ensure that the wires do not impede the sensor's motion.

This is probably best achieved by packing a spiral of wire in a grease-filled compartment on the sensor so that it will feed freely. The modifications proposed in the section on future work eliminate external wire entirely.

ELECTROMAGNETIC AND THERMAL CONSIDERATIONS

Magnetic Susceptibility

In general, substances respond to an applied electromagnetic field, and the internal and external fields differ (Jackson 1975). As we will measure the field in the soil, such effects must be considered. The soil's effect is most readily expressed in terms of its magnetic susceptibility, χ, except when a substantial amount of iron is present. Most media perturb a magnetic field only slightly, causing intensity changes of a few parts in 10^4 or 10^5. Table 1 lists the magnetic susceptibility of some substances of interest.

We expect that most soil, frozen or otherwise, in which the sensor is buried will cause a relative field perturbation of less than 0.1%. For a region with a large or inhomogeneous concentration of iron, our technique is of limited use. The actual maximum tolerable field perturbation in such a case will depend upon the desired resolution and accuracy. We made measurements in both air and soil and found no measurable difference between the two (see section on results).

For a particular soil type, a field calibration of one or two data points should establish a scaling factor, if any, to apply to laboratory calibration curves. This scaling factor should be essentially temperature independent and should serve to reduce an already small error.

Eddy Currents

The drive coil's magnetic field generates electrical eddy currents in the earth, which are 90° out of phase with the drive current. These currents have magnetic fields of their own associated with them, which will be detected by our sensor. This is essentially the operating principle of previous permafrost electromagnetic induction work (Hoekstra and McNeill 1973), but is a source of noise for our experiment. Our sensor will be so close to the drive coil, at most 1 m away, that the drive coil's field will dominate by a very large margin. In the unlikely event that this noise source must be actively suppressed to attain higher resolution, synchronous detection of the induced EMF would surely be effective.

TABLE 1 Magnetic Susceptibilities.

Medium	$\lvert\chi\rvert$
Frozen water, 0°C	12.6×10^{-6}*
Liquid water, 0°C	12.9×10^{-6}*
Sedimentary soil	$<10^{-4}$†

*Weast (1977).
†Clark (1966).

The eddy currents also extract energy from the drive coil's field, since the earth is a resistive medium. Hence in earth the magnetic field amplitude diminishes more rapidly with depth than it would in free space. This effect can be quantitatively estimated by taking a simple functional form for the attenuation,

$$B_a(z) = B_o(z)e^{-z/\sigma}, \tag{2}$$

where B_a is the actual field at the position z, B_o is the field in the absence of this effect, and σ is the skin depth. The skin depth is the distance at which the amplitude of a plane wave of electromagnetic radiation drops to $\frac{1}{e}$ of its initial value. The skin depth of the earth at our operating frequency is roughly 1,000 m (Hoekstra and McNeill 1973). For sensor depths of 1 m the exponential can be expanded to obtain

$$\frac{B_a}{B_o} = 1 - \frac{z}{\sigma} + \frac{1}{2}\left(\frac{z}{\sigma}\right)^2 \ldots = 99.9\% \text{ (at } z = 1 \text{ m)}. \tag{3}$$

This energy loss due to eddy current effects is essentially constant in time, but increases with depth and will vary slightly with location. If very high resolution ($>1:10^3$) at long ranges (>1 m) is desired, a few field calibration points should provide a corrected calibration curve; otherwise the effect can be safely neglected.

Temperature

The relevant electromagnetic properties of the earth are effectively constant over the temperature range of interest. Electronic systems, however, are often susceptible to errors introduced by temperature variations. Careful design and component selection (including batteries) are mandatory. Our present drive system functions adequately, but will be modified as outlined in the section on future work.

RESULTS

Data obtained in the laboratory indicate that vertical and horizontal resolution of up to 1 mm is attainable.

The sample data in Table 2 were obtained by suspending the sensor above the drive coil and varying the separation between the two. Both coils were in air, at room temperature. The drive frequency was 298.5 ± 0.2 Hz, and the voltage was 3.000 ± 0.001 V rms. Note the distinct signal change in going from 317 to 318 mm. An average ratio is computed on the basis of five trials at each separation. This information constitutes the calibration curve for the system and is given in Figure 2.

The range is determined by the minimum detectable signal in the sensor and can be extended by increasing the drive amplitude. The vertical resolution of the system depends upon the separation and is governed by the change in B_z with

TABLE 2 Sample Data Obtained in Air.

Z (mm ± 0.5)	Sensor Signal (V)	Calibration Signal (V)	Signal Ratio
317.0	1.246	2.251	0.554
	1.249	2.250	0.555
	1.247	2.250	0.554
318.0	1.239	2.250	0.551
	1.238	2.250	0.550
	1.237	2.250	0.550
320.0	1.226	2.250	0.545
	1.226	2.250	0.545
	1.225	2.250	0.544

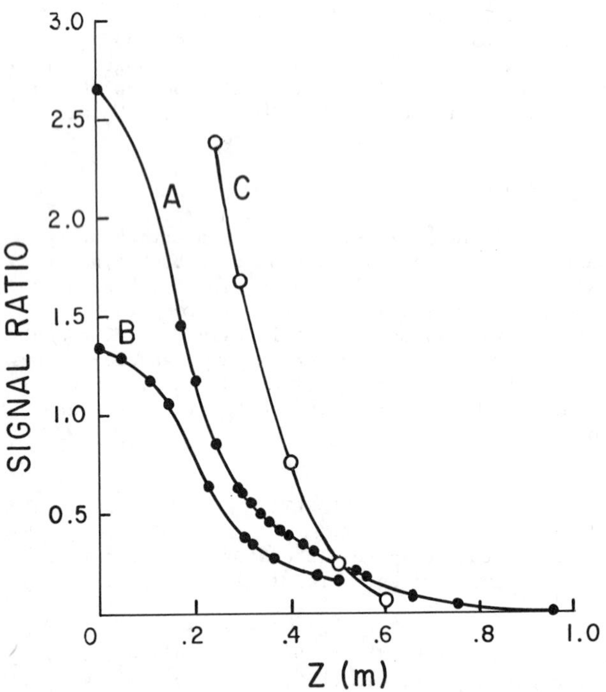

FIGURE 2 Calibration curve obtained in air. Curve A is the calibration curve obtained from a drive of 298.5 Hz at 3.000 V. B was obtained at a lower resonant frequency. C shows the effect of adding an iron core to the drive coil. The uncertainties in the ratio and the position are ±0.001 and ±0.5 mm, respectively, and cannot be resolved on this plot.

separation, $\frac{\partial B_z}{\partial z}$. Hence by optimally shaping the field, the region of maximum sensitivity can be adjusted to coincide with a depth of interest. The field can be modified by adding an iron core, or pole piece, to the drive coil. Figure 2 shows that by adding a laminated iron core we enhanced the sensitivity, which is helpful in probing greater depths. As the addition of the core changes the coil's inductance and therefore the resonant frequency, either the drive frequency or the timing capacitor must be adjusted before calibrating.

To examine reproducibility, we fixed the location of the sensor coil, then dismantled the drive system and reassembled it after several days. The ratio was unaltered within ±0.001.

The effects of soil and temperature were determined experimentally by first taking data in air at an arbitrary separation of roughly 300 mm. The sensor was then surrounded with frost-susceptible soil, including the region between it and the drive coil. The measurement was repeated in the presence of the soil. Water was then added to the soil, and data again taken. Finally, the calibration apparatus, with the wet frost-susceptible soil, was placed in a -3°C environment in the Periglacial Laboratory of the Quaternary Research Center. The soil was allowed to freeze, and the measurements were repeated. Our results were obtained at a drive of 298.5 ± 0.2 Hz and 3.000 ± 0.001 V and are summarized in Table 3.

The data taken at Z_0 show a sizable change in both signal values, but the ratio remains virtually constant. The ratio variation does not jeopardize the resolution at the 1-mm level. The utility of the calibration coil is very apparent in this sequence of measurements. We anticipated some circuitry drifts to be associated with the temperature change, and indeed the resonant frequency dropped by about 1 Hz. Furthermore, the changes in the sensor and calibration signals are consistent with a 5% increase in the magnetic field intensity. We believe that this is due to an impedance change in the RLC circuit, an idea supported by the resonance drift. A current-regulated system would avoid this problem, but in any event the ratio technique very effectively cancels it out.

FUTURE WORK

The most significant single improvement upon the present system would be the elimination of the wires through which the sensor communicates with the surface. This can be achieved using a scheme presented in Figure 3. Essentially, the battery-powered sensor would act like a transponder, transmitting information on command. The use of low power circuitry should allow sensor lifetimes of several years, if transmitting is minimized. A precision voltage reference would be buried with the sensor to provide calibration against drifts in the sensor circuitry. Additional devices, to measure temperature and pressure, for example, could be included in the sensor package.

The revised sensor would continuously monitor the local magnetic field at the drive frequency, ω_d. When the drive coil is turned on, the

TABLE 3 Soil and Temperature Effects.

Temperature (°C)	Medium	Separation	Sensor Signal (V)	Calibration Signal (V)	Signal Ratio
25	Air	Z_o	1.396	2.263	0.617
25	Dry soil	Z_o	1.397	2.264	0.617
25	Wet soil	Z_o	1.397	2.264	0.617
-3	Frozen soil	Z_o	1.473	2.384	0.618
-3	Frozen soil	Z_o + 1 mm	1.462	2.385	0.613
-3	Frozen soil	Z_o + 2 mm	1.449	2.384	0.608

A resolution of 1 mm is readily seen.

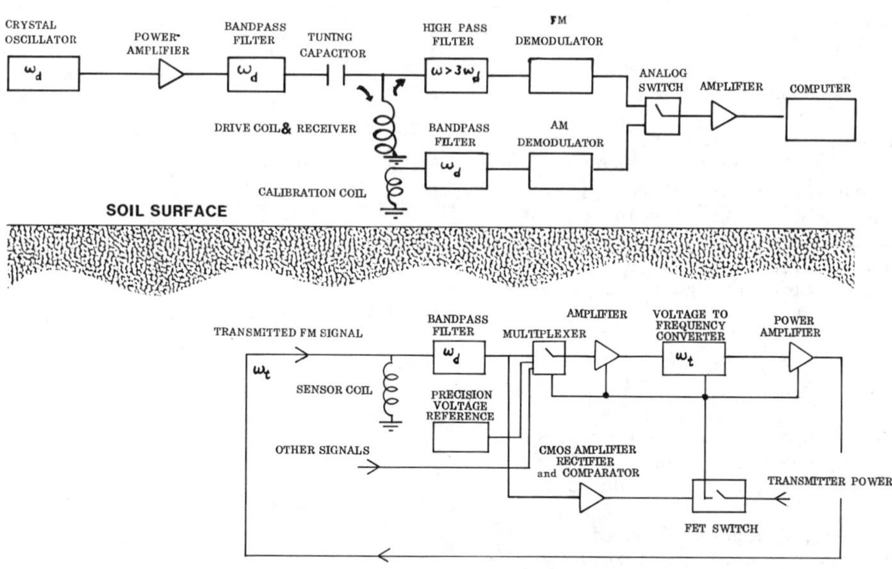

FIGURE 3 Schematic of a proposed wireless version. The underground device would consist of two sections: a stage which is always on and sensing the coil, and a transmitter which would automatically switch on when the drive coil is activated. The position information would consist of a rectified EMF that is transformed into an FM signal, with $\omega_t >> \omega_d$. The FM signal is transmitted to the surface via the sensing coil. A multiplexer would switch between the various transducer signals and a stable reference voltage.

comparator would respond to the increased magnetic field by switching on the transmitter (the multiplexer, amplifier, voltage-to-frequency converter, and power amplifier). The multiplexer would select a signal from the sensor coil, a precision voltage reference, or one of the additional transducers for transmission at a frequency ω_t. Choosing $\omega_t >> \omega_d$ allows decoupling between the incoming and outgoing signals, and thus the sensor coil can serve to both transmit and receive information without interference.

On the surface the drive coil would detect the frequency-modulated (FM) signal from the sensor. The demodulated voltage should reflect the amplified transducer signal in the sensor. The buried voltage reference would provide a fixed point for calibration.

Modifications to the drive circuitry are also in order. Our present drive system functions adequately, but the voltage and frequency must be periodically trimmed by adjusting two variable resistors. A crystal-controlled oscillator with digital counters would provide a stable square wave, with better temperature stability than the present device. A current-regulated system would also be preferable to the voltage-regulated system now in use. This would provide better field amplitude regulation.

CONCLUSION

The current system has the requisite range and resolution to be of use in a quantitative investigation of soil and rock movement in the active layer. Laboratory tests indicate that the electronics operate reliably to our lowest test temperature, -3.0°C, with no signal degradation. High-reliability integrated circuits (Mil Spec) should extend the temperature range to -55°C. Convenient calibration of the system can be performed in air. We found that introducing frost-susceptible soil (dry, wet, frozen) had no appreciable effect. Excluding the support frame, the total instrument package has a mass of 10 kg and is easily field portable.

We plan to install sensors both in sorted circles in the field and in laboratory experiments related to sorted pattern development. We welcome comments and inquiries concerning this system.

ACKNOWLEDGMENTS

Discussions with R. S. Anderson, D. B. Booth, Y. J. Brown, C. A. Gregory, and A. L. Washburn were very helpful. The facilities of the Periglacial Laboratory were invaluable in the development of the apparatus. We are particularly grateful to the Laboratory's director, B. Hallet, for his support and suggestions. Special thanks go to P. Leaming for typing the manuscript.

REFERENCES

Burrous, C. M., 1978, Experimental upfreezing of objects: Effects of object geometry, _in_ Draft report to the Army Research Office, Washburn, A. L. (P. I.).

Clark, S. P., ed., 1966, Handbook of physical constants: Boulder, Colo., Geological Society of America.

Fahey, B. D., 1974, Seasonal frost heave and frost penetration measurements in the Indian Peaks region of the Colorado Front Range: Arctic and Alpine Research, v. 6, p. 63-70.

Ferrians, O. J., Jr., and Hobson, G. D., 1973, Mapping and predicting permafrost in North America: A review, 1963-1973, _in_ North American contribution, Second International Permafrost Conference: Washington, D. C., National Academy of Sciences.

Hoekstra, P., and McNeill, D., 1973, Electromagnetic probing of permafrost, _in_ North American contribution, Second International Permafrost Conference: Washington, D. C., National Academy of Sciences.

Jackson, D., 1975, Classical electrodynamics: New York, John Wiley.

Mackay, J. R., and MacKay, D. K., 1976, Cryostatic pressures in nonsorted circles (mud hummocks), Inuvik, Northwest Territories: Canadian Journal of Earth Sciences, v. 13, p. 889-897.

Nicholson, F. H., 1976, Patterned ground formation and description as suggested by Low Arctic and Subarctic examples: Arctic and Alpine Research, v. 8, p. 329-342.

Pissart, A., 1970, Les phénomènes physiques essentiels liés au gel, les structures périglaciaires qui en résultent et leur signification climatique: Annales de la Société Géologique de Belgique, v. 93, p. 7-49.

Shilts, W. W., 1978, Nature and genesis of mud boils, central Keewatin, Canada: Canadian Journal of Earth Sciences, v. 15, p. 1053-1068.

Washburn, A. L., 1980, Geocryology: New York, Halsted Press.

Washburn, A. L., Burrous, C. M., and Rein, R., Jr., 1978, Soil deformation resulting from some laboratory freeze-thaw experiments, _in_ Proceedings of the Third International Conference on Permafrost, v. 1: Ottawa, National Research Council of Canada.

Weast, R. W., ed., 1977, Handbook of chemistry and physics, 57th edition: Cleveland, Ohio, CRC Press.

ZONATION AND FORMATION HISTORY OF PERMAFROST IN QILIAN MOUNTAINS OF CHINA

Guo Pengfei[1] and Cheng Guodong[2]

[1] 00926 Unit, Xining, Qinghai Province
[2] Lanzhou Institute of Glaciology and Cryopedology, Academia Sinica
People's Republic of China

Alpine permafrost underlies 91,000 km^2 in Qilian Shan, and its distribution shows obvious latitudinal and longitudinal zonation. The lower limit of the permafrost is at 3700-3900 m a.s.l. in the south and at 3494-3650 m in the north. As the latitude increases by 1°N, the permafrost lower limit descends 123.7 m; as the longitude increases 1°E, the permafrost lower limit descends 56.6 m. In the insular permafrost zone, the permafrost is generally 25-35 m thick, with a mean annual ground temperature of 0° to -1.5°C. In the discontinuous permafrost zone, it is 35-95 m thick and the mean annual ground temperature is -15° to -23°C. Temperature inversions, which become stronger northwards, affect permafrost distribution; a maximum thickness of 139.3 m occurs in the Mu Li and Tuo Lei Basins. Permafrost on north-facing slopes is 30-50 m thicker than on south-facing slopes. Qilian Shan has undergone several ice ages, but the existing permafrost was formed around 3000 yr B.P., because during the interglacial period of the Pleistocene and the post-glacial stage at the beginning of the Holocene temperatures were too high for permafrost preservation.

GENERAL DESCRIPTION OF PHYSIOGRAPHY

Qilian Shan is a great peripheral mountain system to the northeast of the Qinghai-Xizang Plateau. It occupies a total area of about 210,000 km^2, and contains seven, five, and three ranges in the western, middle, and eastern sections respectively. All these mountains range from 4000 to 4500 m, and the highest peaks are 5245 to 5826 m. There are perennial snow fields and modern glaciers in the alpine zones above 4300 to 4500 m.

Influenced by the circulation of west wind, the western part of Qilian Shan is dry. The annual rainfall is only 150 to 300 mm, the annual amount of evaporation is 1500 to 2000 mm, belonging to continental desert climate. Affected by the oceanic monsoon, are relatively humid the east and middle sections of the mountains, with an annual precipitation of 400 to 500 mm, belonging to semi-arid alpine steppe climate. Its mean annual temperatures are 0 to -3°C, the temperatures in Jiang Cang, Tuo Lei are -3.2 to -5.7°C.

Qilian Shan belongs to the high altitude periglacial zone. Various periglacial phenomena develop very well, most of them closely related in origin and development to permafrost. These phenomena are also direct indications of modern permafrost.

LAW OF ZONATION IN PERMAFROST DISTRIBUTION

The permafrost in Qilian Shan covers an area of about 91,000 km^2, which is 43% of the total area of the mountains. According to the distributing state of permafrost and the area of taliks, the permafrost can be divided into two main zones: insular permafrost zone and discontinuously blocky permafrost zone (Figure 1). The area of taliks in the insular and discontinuously blocky permafrost zones is 60-80%, and 10-20% respectively.

The Changes of Permafrost Lower Limit and Zonation of the Types of Permafrost

The permafrost in Qilian Shan is of the alpine permafrost type of the middle latitudes. The elevations of its lower limit changes with the latitude and longitude, restricted by solar radiation and the types of climate. The elevations of the south boundary (lower limit) of the permafrost, extending along the southern slopes of La Ji Shan, Qinghai Nan Shan and Qaidam Shan, are 3700, 3800, 3900 m respectively. This boundary coincides with the -2°C isotherm for mean annual air temperature. The altitudes of the north boundary (lower limit) of the permafrost, along the northern slopes of Leng Long Ling, Zou Lang Nan Shan, and Dang He Nan Shan are 3494, 3550, 3650 m respectively. This boundary is identical to the -2.5°C isotherm for mean annual air temperature. The southern boundary is 206 to 250 m higher than the northern boundary. By regression analysis, the decreasing rate of the lower limit of permafrost is 123.7 m per degree of latitude from south to north. The permafrost lower limit falls with the increase of longitude from west to east, decreasing at a rate of 56.6 m per degree of longitude.

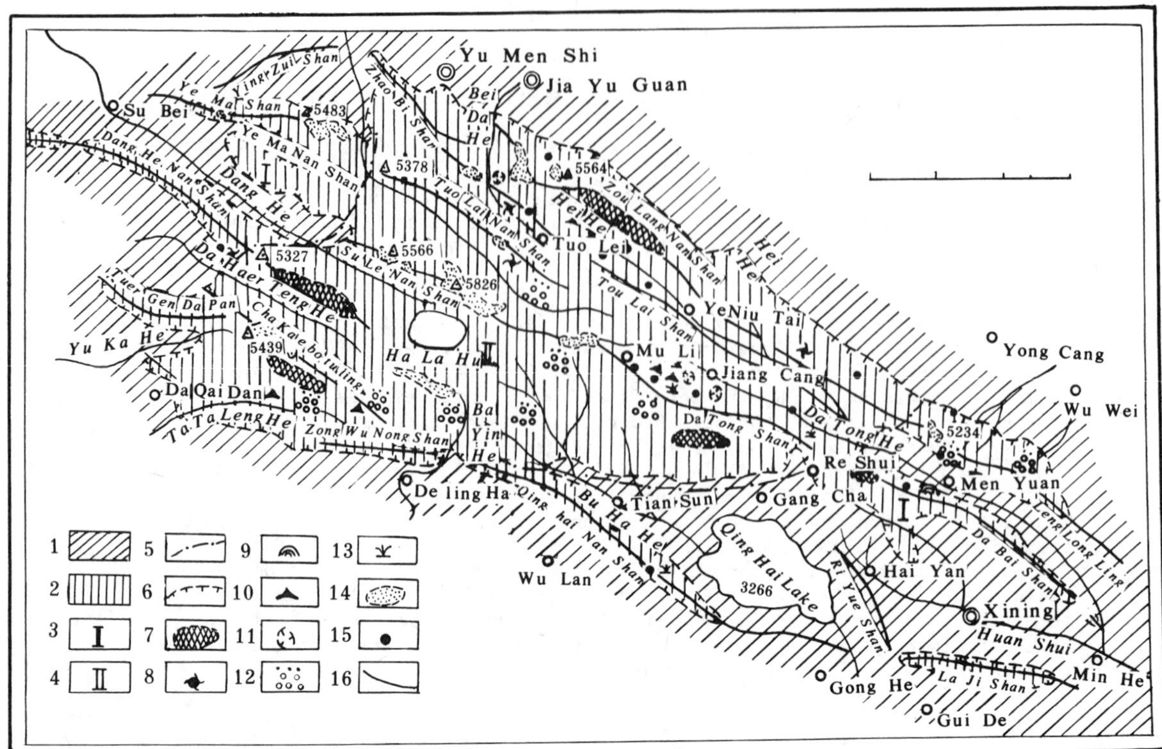

FIGURE 1 Map of permafrost distribution in Qilian Shan.
1. seasonally frozen ground; 2. permafrost; 3. insular permafrost zones; 4. discontinuously blocky permafrost zones; 5. border line between the insular and discontinuously blocky permafrost zones; 6. border line of permafrost distribution; 7. cryoplanation surface; 8. frost heaving mounds; 9. debris flow; 10. icing; 11. thermokarst depression; 12. rock stream slope; 13. grassy marsh in talik; 14. modern glaciers; 15. holes drilled in frozen ground; 16. main mountain ridge.

These changes are contrary to the law that the permafrost lower limit in Europe, America, and Tian Shan of China rises with the decrease of continentality. The restriction of regional climate conditions on the distribution of permafrost is quite apparent. Assuming that the relations between the measured heights of the zones of permafrost and glaciers are taken as the mark dividing the types of permafrost, we then plot a section (Figure 2) in accordance with the relations between Dunde, Yigelige, and Leng Long Ling glaciers and the lower limit of permafrost along 38°N from west to east. The snow line and the permafrost lower limit descend regularly with increasing precipitation from west to east. The distance between the snow line and the lower limit of permafrost is 1350 m in the west part of Qilian Shan, and is less than 1000 m in the east section. The average distance is more than 1000 m, having the characteristics of a continental climate region.

Because of the differences of climatic conditions and corresponding to the types of glaciers, the permafrost in the eastern part of Qilian Shan belongs to subcontinental type and in the western part belongs to extreme continental type (Table 1).

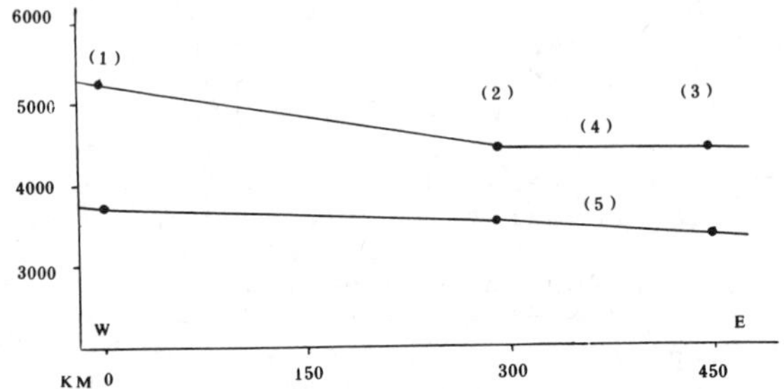

FIGURE 2 Curves of therelations between the snow line and permafrost lower limit in Qilian Shan.
(1) Ka Ke Tu Meng Ke Dun De glacier;
(2) Yi Ge Li Ge glacier;
(3) Leng Long Ling glacier;
(4) Snow line;
(5) Permafrost lower limit.

Changes of Zonation and Thickness of Permafrost

The thickness of permafrost in Qilian Shan varies in different places. In the insular permafrost zones, the permafrost is 25 to 35 m thick, and 1-13 m thick in some areas. The mean annual ground tem-

TABLE 1 Permafrost Types and their Environmental Characteristics in Qilian Shan

Comprehensive characteristics \ Type of permafrost		Extremely continental type (in west part)	Sub-continental type (in east part)
Permafrost features	Permafrost lower limit (m)	3650-3900 (north slope-south slope)	3494-3700 (north slope-south slope)
	Distance between snow line and lower limit (m)	More than 1000	Less than 1000
	Annual average temperature near permafrost lower limit (°C)	-2.5 to -3.0	-1.7 to -2.5
Glacier features	Height of snow line (m)	4830 (Ye Ma Shan glaciers No. 12)	4600-4450 (Zou Lang Nan Shan-Leng Long Ling)
	Annual rainfall near snow line (mm)	400	600-800
	Annual average temperature along snow line (°C)	Lower than -11	Higher than -11
	Type of glaciers	Extremely continental type	Sub-continental type
Main periglacial indications		Frost weathering, debris slope, cryoplanation terraces, frost heave mounds, alpine grassy marshes	Nivation, cryoplanation terraces, rock sea, rock stream, frost heave mounds, solifluction, thaw slumping, stone ring, thaw settlement, thick-layered ground ice
Climate features	Type of climate	Continental arid desert climate	Continental semi-arid alpine steppe climate
	Annual precipitation (mm)	150-300	400-500
Vegetation types		Semi-shrub-desert steppe	Mountain forest, shrub, steppe meadow

perature is 0°C to -1.5°C. In discontinuous blocky permafrost zones, the thickness of permafrost is 35 to 95 m. The mean annual ground temperature is -1.5 to -2.3°C (Table 2).

It is clear that permafrost becomes thicker with the rise in altitude, in response to the vertical decreasing lapse rates (Figure 3, 4). Under the influence of the cold air draining from Qinghai-Xizang Plateau, cold air "lakes" often occur in some small basins in the middle part of the Qilian Shan, and temperature inversions occur. The mean temperature of January in the Mu Li and Tuo Lei basins is -18 to -19.7°C respectively, these being the coldest areas in January. Permafrost in these regions is generally 50 to 95 m thick, with a maximum of 139.3 m. In addition, as the intensity of the temperature inversion becomes stronger from south to north, the permafrost thickness increases at a given altitude from south to north. The thickness of permafrost increases about 42 m in approximately 100 km at a given altitude in the middle section

TABLE 2 The Thickness of Permafrost in Qilian Mountains

Areas	Position No.	Heights (m)	Thickness of permafrost(m)	Permafrost table (m)	Mean annual soil temperature (^{o}C)
Hong Shui Ba	ZK 2	3829	79.3	1.7	lower than -1.5
	ZK 3	4033	139.3	1.7	
Mu Li	CK 2	3989	about 100		-2.3
	CK 3	3968	about 85		-1.8
Jiang Cang	CK 90	3888	86.7	less than 1.0	
	CK 95	3882	50	less than 1.0	
Re Shui	well 40	3696	30	1.1	-0.6
	well 301	3862	more than 60	2.4	-1.5
Mountain area	ZK 11	3950	78.4	1.6	
	Da Ban Shan	3500	25	2.0	
	Zou Lang Nan Shan	3530	33.2	1.8	

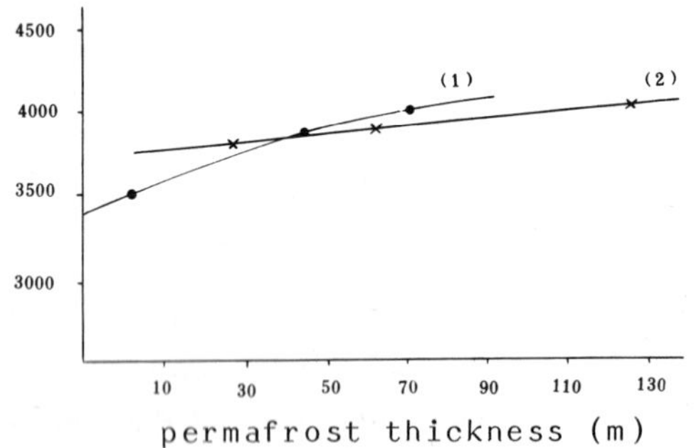

FIGURE 3 Curves of the relations between permafrost thickness and elevations in Mu Li, Hong Shu Ba in the middle part of Qilian Shan.
(1) Mu Li Basin; (2) Hong Shui Ba Basin.

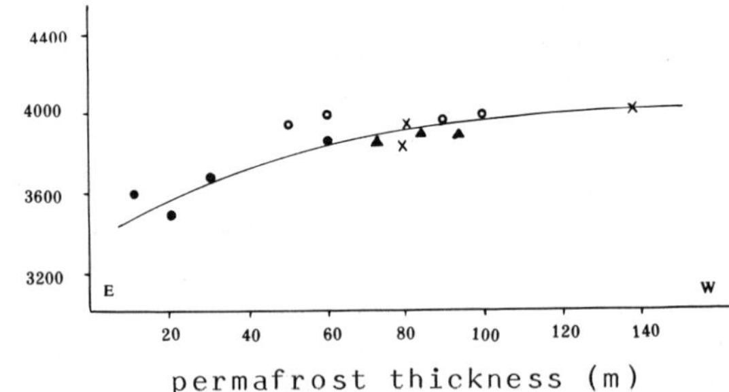

FIGURE 4 Curves of the relations between permafrost thickness and altitudes in the east and west parts in the middle section of Qilian Shan.
• Re Shui, ○ Mu Li, ▲ Jiang Cang, × Hong Shui Ba.

of Qilian Shan. This is a special phenomenon occurring in alpine type permafrost in middle latitudes.

Another feature of the distribution of permafrost thickness is the evident lack of symmetry on the southern and northern sides of the mountains, due to the effect of the duration of the insolation on mountain slopes of different aspects. The thickness of permafrost in the south-facing slopes is 30 to 50 m thinner than that on the north-facing slopes.

Moreover, marshes that absorb solar radiation prevent permafrost from thawing. The thickness of permafrost varies significantly with the differing degree of swamping (Figure 5).

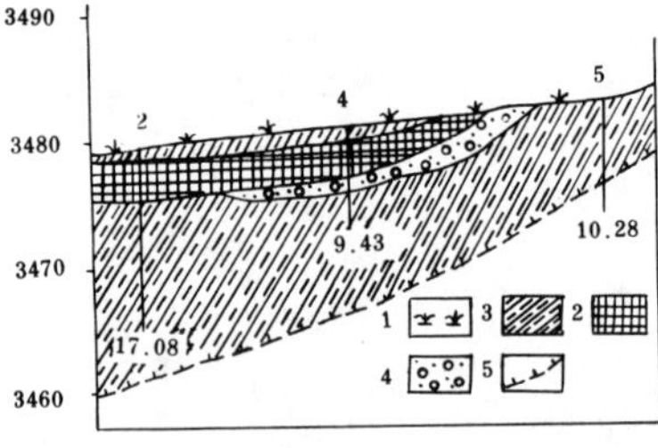

FIGURE 5 Changes of permafrost thickness in marsh regions (according to the data of Re Shui Frozen Ground Team).
(1) Grassy marshland;
(2) Clayey soil;
(3) Ice layer;
(4) Sand and gravel bed;
(5) Permafrost lower limit.

FORMATION HISTORY OF PERMAFROST

It is well-known that the prerequisite conditions for the formation of permafrost are the ancient periglacial climatic environments since the Quaternary. Glaciers and permafrost were produced under cold climatic conditions, and an important way to reveal the history of the formation of permafrost in Qilian Shan is by means of correlations with the division of the glacial periods.

Qilian Shan has witnessed four ice ages and three interglacial periods since the Quaternary (Guo Pengfei 1980). All these can be compared with the glacial ages in the various parts of Qinghai-Xizang Plateau (Table 3). Therefore, fossil perma-

TABLE 3 Comparison of the Glacial Ages in Qilian Shan and Qinghai-Xizang Plateau

Age \ Area	Qilian Shan area, after Guo Pengfei	Jolmolungma area after Zheng Ben-xing, Shi Yafeng	Tanggula Shan area after Zhang Yong et al.	Kunlun Shan area after Zhang Yong et al.
Holocene	Modern glaciers, new moraine	Modern glaciers	Modern glaciers	Modern glaciers
	Gang Shi Ga little ice age	Rong Bu De little ice age	Little ice age	Little ice age
	Warm period	Ya Li period	Warm period	Warm period
Late Pleistocene	San Cha Kou episode	Rong Bu De episode	Ba Si Guo glacial age	Ben Tou Shan ice age
	Intermittent episode	Intermittent episode	Zha Jia Zang Bu/Ba Si Cuo interglacial period	Xi Da Tan/Ben Tou Shan interglacial period
	Dong Gou episode	Ji Long Si episode	Zha Jia Zang Bu ice age	Xi Da Tan ice age
Middle Pleistocene	Leng Long-Da Ban interglacial period	Jia Bu La interglacial period	Bu Qu/Zha Jia Zang Bu interglacial period	Na Chi Tai/Xi Da Tan interglacial period
			Bu Qu ice age	Na Chi Tai ice age
	Leng Long ice age	Nie Nie Xiong La ice age	Bai Duo/Bu Qu interglacial period	Wang Kun/Na Chi Tai interglacial period
			Bai Duo glacial age	Wang Kun ice age
Early Pleistocene	Tuo Lai-Leng Long interglacial period	Pa Li interglacial period	Ke Ma Qu/Bai Duo interglacial period	Shi Zi Shan/Wang Kun interglacial period
			Na Qu cold period	Shi Zi Shan cold period
	Tuo Lai ice age	Xi Xia Bang Ma glacial age	Ke Ma Qu/Na Qu interglacial period	Jing Xian/Shi Zi Shan interglacial period
			Ke Ma Qu ice age	Jing Xian glacial age
Pliocene	Periglacial period, warm climate, no glaciers			

frost must have developed in Qilian Mountains in each ice age. Its area of distribution could have been considerably larger than that of the ancient glaciers. However, the climate over Qilian Shan in every interglacial period in the Quaternary was gradually becoming drier and colder from early to late Pleistocene (Yang Huiqiu and Jiang Dexin 1965). The mean annual air temperatures in the interglacial period were about 8 to 12°C higher than those of today, and the mean annual air temperatures in the hypsithermal were 3 to 5°C (Guo Xudong 1976) higher than now. Furthermore, the influence of the Himalayas, located in the southern part of Qinghai-Xizang Plateau, was not notable in changing the atmospheric circulation and the hydrothermal condition over the Plateau before the middle Pleistocene. As the mountains rose to more than 4500 m in the late Pleistocene, it demonstrates the function of "barriers". The development of glaciers and permafrost were primarily controlled by the vicissitudes of global climate, and the latitude zoning, for both Qilian Shan and the Plateau were under the influence of the Indian Ocean monsoon in early and middle Pleistocene. The ancient permafrost formed in Tuo Lai ice age in early Pleistocene and in Leng Long Ling glacial age in middle Pleistocene must have completely thawed after undergoing a rather long warm climate in Tuo Lai-Leng Long and Leng Long-Da Ban interglacial periods. A careful study should be undertake as to whether the fossil permafrost formed in Da Ban ice age in the late Pleistocene is preserved now or not. Effected by the fluctuation of global climate controlled by solar radiation, the warm climate in Ya Li period in the southern part of the plateau was common on the whole Qinghai-Xizang Plateau, and it corresponds to the Atlantic period in Europe and the Ban Po warm period in China. The ancient permafrost formed during the Da Ban glacial age in Qilian Shan must have melted greatly or totally after experiencing the warm period.

The permafrost distributed in Qilian Shan was then produced in the Gang Shi Ga Neoglacial event. According to Zhu Kezhen (1972), and Qin Dahe (1981), the coldest period in the last one thousand years was during the period 1620-1880 A.D., and lasting two hundred and fifty years. It was a favourable time for permafrost to develop and to be preserved. From the oscillation of radiocarbon records, the climatic fluctuations in Neoglacial periods match reasonably well in southern and northern hemispheres. Consequently, not only a great deal of glaciers developed there but also a huge amount of permafrost was formed. Thus it seems appropriate to confine the age of formation of modern permafrost in Qilian Shan in a period after the Gang Shi Ga Neoglaciation, about 3000 years B.P.

ACKNOWLEDGMENT

Grateful acknowledgment, is made to Interpreter Ran Longde for his help in translating this paper into English.

REFERENCES

Guo Pengfei, 1980, Approaches on Quaternary glacial age in the middle part of Qilian Shan, Journal of Glaciology and Cryopedology, v. 2, No. 1.

Guo Xudong, 1976, Interglacial periods and palaeoclimate in Quaternary in Jolmolungma area, A scientific survey report of the Jolmolungma area, Geology in Quaternary, Scientific Publishing House, Beijing, China.

Qin Dahe, 1981, Primary researches on the glacial changes since Yu Mu ice age of Su Zhu Peak area in Qilian Mounts., Journal of Glaciology and Cryopedology, v. 3, No. 2.

Yang Huiqiu and Jiang Dexin, 1965, The composition of sporo-pollen in Qinghai Lake Basin in Quaternary and its significance, Di Li Xue Bao, v. 31, No. 4.

Zhu Kezhen, 1972, Preliminary studies on the climatic changes in recent 5000 years in China, Archaeological acta.

A COMPARISON OF REMOTELY SENSED SURFACE TEMPERATURE AND BIOMASS ESTIMATES FOR AIDING EVAPOTRANSPIRATION DETERMINATION IN CENTRAL ALASKA

Robert J. Gurney[1], James P. Ormsby[2] and Dorothy K. Hall[2]

[1]Department of Civil Engineering, University of Maryland
College Park, Maryland 20742 USA
[2]Earth Survey Applications Division, NASA/Goddard Space Flight Center
Greenbelt, Maryland 20771 USA

Aerodynamic surface roughness, an important parameter controlling the evaporation rate, is related to the biomass. Previous work indicated that temperature variations in the Heat Capacity Mapping Mission (HCMM) satellite data were related primarily to evaporation differences for similar albedo values. LANDSAT-derived estimates of the biomass were related to HCMM-derived estimates of the surface temperature for an area of the Alaskan sub-Arctic for which contemporaneous LANDSAT and HCMM data existed. It was found that the relationship showed some scatter, the correlation coefficient being only -.62. Thus, land cover data are not good substitutes for remotely sensed thermal infrared data in this area at this time of year, but these data have ancillary uses for situations in which high spatial resolution data are needed to improve coarse resolution thermal infrared data.

INTRODUCTION

Permafrost areas are subject to economic development that may well alter their ecological and environmental equilibria. Unlike in many non-permafrost areas, the tolerance to change is unknown and likely to be small, while the monitoring of changes is very difficult using conventional means because of the remoteness of the areas. Remote sensing techniques provide a set of methods that may allow successful environmental monitoring in these areas. Several techniques may be used: the use of visible and near infrared multispectral data, which are mainly related to vegetation type and biomass, the use of thermal data, which may be related to the evapotranspiration and the use of microwave data, which may be related to the surface roughness and the soil moisture status. The properties to which these systems are sensitive are related. However, the relationships among these properties are not always straightforward. This paper discusses the relationships that may exist, in an area of discontinuous permafrost in central Alaska, among biomass, surface temperature and albedo. Multispectral visible and near-infrared data from the Multispectral Scanner Subsystem (MSS) and a contemporaneous thermal infrared data set from the Heat Capacity Mapping Radiometer (HCMR) are employed. As is discussed, there are good physically-based reasons for thinking that these data types should be related in this area. There is an intricate relationship among vegetation, slope, soil type and presence of permafrost in the study area. For example, interior low brush, muskeg-bog areas and black spruce hardwood associations are generally found in areas that are underlain by continuous permafrost (LUPC, 1973). Bottomland spruce poplar forests are generally found in areas in which permafrost is deep or absent (LUPC, 1973). Thus, a relationship exists between permafrost, evapotranspiration and surface temperature.

BACKGROUND

Previous work has demonstrated that the relationships among the components of the energy balance equation can be modeled using remotely sensed data, e.g. Price (1980). Gurney and Hall (1983) used an HCMM scene of central Alaska to estimate the components of the energy balance equation for 12 May 1978:

$$G = R_N + LE + H \, , \tag{1}$$

where G is the soil heat flux, R_N the net radiation, LE the latent heat flux and H the sensible heat flux. The HCMM data used were obtained just after the ice breakup. At this time, when the ground is wet and the air is still quite dry and cold, evapotranspiration is at or near potential. Evapotranspiration within the study area was found to vary from 2 to 5 mm day^{-1}. This agreed favorably with surface measurements of evapotranspiration taken in previous years in Arctic and Sub-Arctic areas just after the ice breakup.

The latent and sensible heat components of the surface energy balance were modeled as follows:

$$LE = -C_1 \, U_a \, (e_s - e_a) \tag{2}$$

$$H = -C_1 \, \gamma \, U_a \, (T_s - T_a), \tag{3}$$

where T_s is the surface temperature, T_a the air temperature, e_s the surface vapor pressure, e_a the atmospheric vapor pressure, U_a the wind speed, and γ the psychrometric constant. The constant C_1 in Equations (2) and (3) is the bulk diffusion coefficient. This constant, C_1, may be modeled as:

$$C_1 = \frac{\rho c_p k^2}{\gamma \ln^2 \left(\frac{z}{z_o}\right)\left(1 + \frac{r_s}{r_a}\right)} , \quad (4)$$

where ρ and c_p are the density and heat capacity of air, r_a and r_s are the aerodynamic and stomatal resistances, z is the height at which the meteorological data are measured, z_o is the roughness length, and k is von Karman's constant (Brutsaert, 1982).

The bulk diffusion resistance C_1 is thus related to the aerodynamic surface roughness. This surface roughness is in turn related to the vegetation height and thus, at least in part, to the vegetation type (Monteith 1973).

This type of approach has, at least indirectly, been the basis of previous work, such as that of Priestley and Taylor (1972). They define an empirical parameter α that relates potential and equilibrium evaporation rates, E_p and E_e, respectively:

$$E_p = \alpha E_e . \quad (5)$$

Priestley and Taylor (1972) suggest that this parameter has a value of 1.26 for grass, a value supported by subsequent studies, as reviewed by Brutsaert (1982). Other vegetation types with different surface roughnesses yield different estimates of this parameter. For instance, McNaughton and Black (1973) found $\alpha = 1.05$ for a young 8-m-high Douglas fir forest. Thus, it may be seen that the evaporation is often related to the land cover type and, as mentioned previously, vegetation type is related to permafrost. Land cover types may be estimated from multispectral remotely sensed data, such as supplied by the LANDSAT MSS. There are thus at least two ways in which land cover type can be used to improve estimates of evaporation, both involving the use of semiempirical parameters. The surface roughness estimates z_o (Equation 4) can be modified by the type of land cover, or the parameter α (Equation 5) can be changed. Both cases presuppose a simple relationship between biomass and the evaporation rate. This is tested here for this area.

METHOD

Each of the components of the surface energy balance (Equation 1) depends in part on the surface temperature. Gurney and Hall (1983) found that the surface temperature as estimated by the HCMR is a good indicator of evaporation, because the other parameters (such as the albedo) were similar over much of the area. It is thus of interest to compare the surface temperatures estimated by using the HCMR to the land cover types estimated from LANDSAT, and to existing vegetation maps. The temperature estimates are closely related

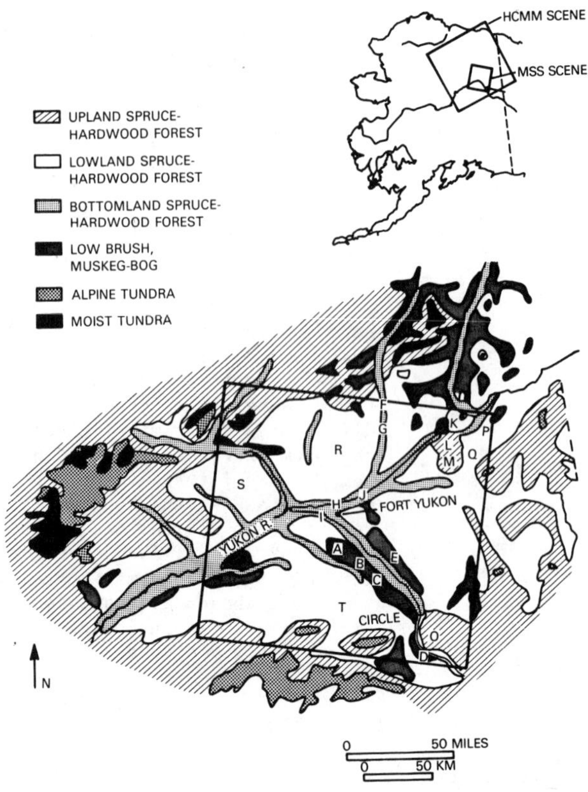

FIGURE 1 Land cover map of the Alaskan study area (after LUPC 1973), showing the major land cover groups and the locations of areas used to extract values for use in this study.

to the actual evapotranspiration, as shown by Gurney and Hall (1983). Gurney and Hall (1983) used an HCMM scene (A0016-22230-2) of the Alaskan sub-Arctic together with a model described by Camillo et al. (1983) to estimate daily evapotranspiration. This same HCMM scene was used in the present study. A LANDSAT scene (30068-20374) was acquired two hours before the HCMM scene of the same area on 12 May 1978 (see Figure 1). The two scenes were geometrically registered (digitally overlaid), as described later, and compared by using the following method.

The bulk diffusion resistance, C_1, is related to the vegetation height and biomass. Holben et al. (1980) have shown that the normalized difference (ND) ratio obtained from MSS data, is also related to the biomass:

$$ND = \frac{R_7 - R_5}{R_7 + R_5} , \quad (6)$$

where R_5 and R_7 are the gray-level counts for LANDSAT MSS Bands 5 and 7, respectively. The temperature for certain areas of known cover type, from a land cover survey (LUPC 1973), was then related to the ND ratio. If there is a relationship between cover type and evaporation, it should appear as a relationship between the ND ratio and the surface temperature.

DATA MANIPULATION

To use the information from the HCMM and LANDSAT data better, the two images were geometrically registered so that the HCMM transformed data lay exactly over the LANDSAT data. The LANDSAT scene was used as the base. The HCMM scene was rotated 45° so that its orientation was approximately the same as that of the LANDSAT scene. Tie points, features that were clearly identifiable in both scenes, were located within the two scenes, a procedure that required two iterations because of the different resolutions of the HCMM and LANDSAT data. A second-order polynomial was fitted to model the geometric errors, and cubic convolution resampling was performed to obtain registered data. The registration root mean square error was estimated at about 250 m, or about half an HCMM pixel. The LANDSAT data were reduced to the HCMM resolution by pixel averaging so that HCMM and LANDSAT pixels were closely overlaid in order to facilitate the extraction of surface temperatures (from HCMM) and ND ratios (from MSS) from precisely the same areas on each image.

An ND ratio image was produced by using Equation 6. This ratio was used as an estimate of the biomass to compare with the HCMM-derived estimates of surface temperature (see Figure 2). Several areas were chosen within each of four cover types, as shown in Figure 1: low brush and muskeg-bog, bottom land spruce-poplar, lowland spruce-hardwood, and upland spruce-hardwood. In each cover type, 10 x 5 pixel areas were selected and mean temperature and the ND ratios estimated. The albedos for all the areas were similar. None was in an area of high relief, and so the surface temperature should be a good indicator of the evaporation rate. The method of estimating surface temperatures from HCMM data is discussed by Gurney and Hall (1983).

RESULTS

The plot of the HCMM estimates of the surface temperature against the estimate of the biomass provided by the ND ratio is shown in Figure 2. The points are divided into the four land cover groups as discussed above. It can be seen that the relationship between temperature and the ND ratio exhibits a lot of scatter. The relationship is basically linear, with a greater surface roughness (implying a larger biomass) being related to a lower surface temperature. The correlation coefficient, r, is -.62, significant at the 99% level. The coefficient of determination is 0.38 and, therefore, there is a large unexplained variance, about 62%. The use of LANDSAT data alone is thus unlikely to produce satisfactory

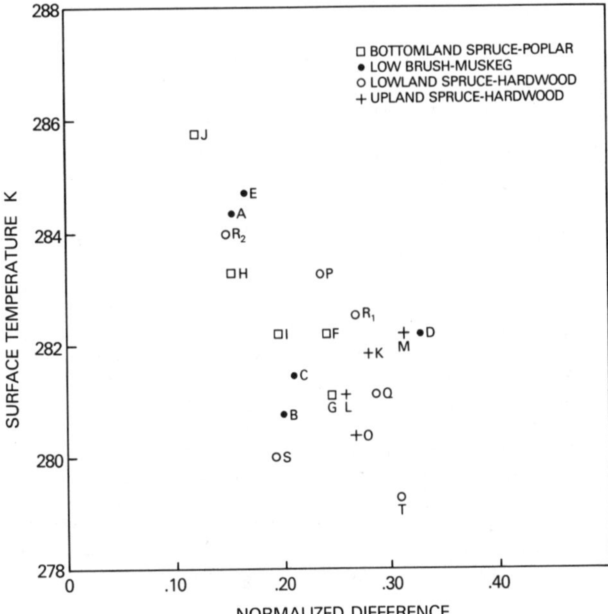

FIGURE 2 Relationship between surface temperature, estimated by using the Heat Capacity Mapping Radiometer, and the Normalized Difference Ratio (ND, Equation 6) estimated by using LANDSAT.

estimates of the evapotranspiration either through the use of a Priestley-Taylor method, or through the estimation of surface roughness from vegetation biomass. However, the use of the thermal data can provide additional information that may be useful in land cover studies using the LANDSAT 4 Thematic Mapper. There is sufficient correlation between LANDSAT and HCMM data for the LANDSAT data to be a useful aid to interpolation in areas where there are mixed pixels on the coarser resolution HCMM or other thermal data. It has previously been shown that the HCMM data provide adequate evaporation estimates in this area for uniform pixels. However, the number of small lakes and areas of bare ground where the modeling assumptions do not hold are sufficient that many pixels of thermal data are contaminated and the model approach cannot readily be used. In these areas, LANDSAT higher spatial resolution data could allow useful evaporation data to be obtained even though the principal method based on thermal data is invalid. It would thus be of use to have suitable multispectral high spatial resolution data co-registered with the low resolution thermal data for environmental monitoring in areas of discontinuous permafrost.

ACKNOWLEDGEMENTS

Dr. R. J. Gurney is funded by NASA Goddard Space Flight Center under Grant NAG5-9.

REFERENCES

Brutsaert, W., 1982, Evaporation into the atmosphere, theory, history and applications, D. Reidel Publishing Company, Dordrecht, Holland.

Camillo, P. J., Gurney, R. J., Schmugge, T. J., 1983, A soil and atmospheric boundary layer model for evapotranspiration and soil moisture studies, Water Resources Research, V. 19, No. 2., p. 371-380.

Gurney, R. J., and Hall, D. K., 1983, Satellite-derived surface energy balance estimates in the Alaskan sub-Arctic, Journal of Climate and Applied Meteorology, Vol. 22, No. 1, p. 115-125.

Holben, B. N., Tucker, C. J., and Fan, C. J., 1980, Spectral assessment of soybean leaf area biomass, Photogrammetric Engineering and Remote Sensing, V. 46, No. 5, p. 651-656.

LUPC (Land Use Planning Commission), 1973, Major ecosystems of Alaska 1:2,500,000 scale map, Joint Federal-State Land Use Planing Commission for Alaska.

McNaughton, K. G., Black, T. A., 1973, A study of evapotranspiration from a Douglas Fir forest using the energy balance approach, Water Resources Research, V. 9, No. 6, p. 1579-1590.

Monteith, J. L., 1973, Principles of environmental physics, American Elsevier Publishing Company, New York.

Price, J. C., 1980, The potential of remotely sensed thermal infrared data to infer surface soil moisture and evaporation, Water Resources Research, V. 16, No. 4, p. 787-795.

Priestley, C. H. B., Taylor, R. J., 1972, On the assessment of surface heat flux and evapotranspiration, Montly Weather Review, V. 106, No. 2, p. 81-92.

ANALYSIS OF INTERACTIONS BETWEEN LINED SHAFTS, PIPES AND THAWING PERMAFROST

I. E. Guryanov

Permafrost Institute, Siberian Branch,
Academy of Sciences, Yakutsk, USSR

Changes in the compressibility properties of permafrost materials were studied with regard to structures in the ground subject to stress-bending, e.g., shaft linings where there is an unfrozen aureole around the shaft, or pipelines built on a thawing foundation. Expressions were derived for the compressibility coefficient of thawing materials in the immediate area of contact depending on the relative thaw radius around the structure. Limits for the applicability of the calculated formulae for types of structures are indicated. Permissible approximations of the calculated models as dictated by the impact of actual uncertainties in the interaction of structure with the ground are determined.

INTRODUCTION

The aim of this paper is to describe the compressibility of thawing soil surrounding vertical structures such as shaft linings supporting surface buildings (head-frames). A similar problem can be formulated for bending of buried pipes whose design lacks distributed vertical loads along their length.

When shaft construction is carried out in permafrost regions, warm operating conditions within a shaft and head-frame give rise to the formation of a thawed zone behind the lining. When near the surface, say within a few tens of meters, there is no firm rock present and foundation curbs are not feasible. In most cases vertical loads due to the lining and head-frame can be transferred to surrounding rock only at great depth. In the lack of such transfer an analysis is to be made of longitudinal bending of the lining in a pliable ground medium, i.e. the thawed zone.

When the lining is deformed under the combined action of horizontal stress near the surface, and vertical loading from above, compression of soil surrounding the lining results from asymmetrical redistribution of soil pressure in each horizontal cross-section. The magnitude of this redistribution, restricted by actual lining deflection, is more than an order of magnitude lower than the general level of soil pressure values. Therefore, with the aid of the method of superposition, the stressed-deformed state can be interpreted as a combined result of axially symmetric rock pressure and unidirectional soil compressibility. Assuming the deformations associated with soil pressure are in equilibrium, compression due to unidirectional movement of the lining can be investigated without taking into account the stressed state that determines soil pressure.

Calculations have shown that horizontal shifts of the lining are small, within the limit of linear deformations. Accordingly, the problem of the compression of soil is formulated here in terms of the theory of elasticity, and deals with the plane deformation of an annulus of thawing soil surrounding the lining. In this case the frozen soil surrounding the thawing zone may be assumed incompressible (Guryanov, 1976). The deformable lining is considered as a cylindrical surface subject to radial displacement (see Fig. 1).

Actual results of construction do not always correspond to expected boundary conditions for the lining-soil interface because the contribution from contact friction is largely indeterminate. To assess possible errors in design, two limiting cases are considered: complete connection of the deformable lining and thawing soil, and complete slip at the interface with separation in an extended zone.

We assume the maximum interfacial strength of the lining and surrounding soil to be equal to the shear strength of the soil which, according to current standards (SNIP, 1974), as well as L. I. Baron's investigations (Baron, 1972), is two orders of magnitude lower than the value of deformation modulus E of thawed soil:

$$\tau_c \sim 10^{-2}E$$

On the other hand, it is easy to show that for tangential stress $\tau_{r\alpha}$ the inequality

$$\tau_{r\alpha} < E\,\frac{u_0}{R_0}$$

is valid in the problem under consideration, where u_0 is the maximum transversal displacement of soil adjacent to the lining (horizontal displacement of a vertical structure), and R_0 is the radius of the structure.

Calculations of lining buckling for a particular problem have shown that maximum horizontal displacements do not exceed 5 mm. At $R_0 \approx$ 3-4 m we have $u_0/R_0 \sim 10^{-3}$. Then, taking into account the preceding expressions we obtain

$$\tau_{r\alpha} < 10^{-1}\,\tau_c$$

Thus, contact tangential stress is at least an order

of magnitude lower than the limiting interfacial shearing strength, i.e. for this particular problem no slip occurs at the contact of shaft lining and surrounding soil.

In the case of a pipeline within a thawing zone having $R_0 \approx$ 20-30 cm, supposed shifts are $u_0 \approx$ 1-2 cm and therefore we have $u_0/R_0 \sim 10^{-1}$. Following the previous reasoning for the case of shaft lining, for pipelines we get in this case:

$$\tau_{r\alpha} \sim 10\tau_c$$

i.e. contact stress is on the average an order of magnitude larger than limiting interfacial shearing strength, and therefore there is virtually full slippage at the contact.

One therefore concludes that the two limiting cases of structural deformation within a thawed zone are potentially observable, many other cases will be intermediate.

ANALYSIS

Stresses and displacements are basic to the definition of compressibility of soil. Using the method of degenerated solutions extended to the system of differential equations of equilibrium for this problem, we apply to the soil the conditions:

$$u = u(r) \cdot u(\alpha) = u(r) \cdot \cos\alpha,$$

$$v = v(r) \cdot v(\alpha) = v(r) \cdot \sin\alpha,$$

where u,v are the radial and tangential displacements, respectively.

In the case of no slippage between the lining and soil it is necessary that $u(r) = -v(r)$ and the degenerated solution follows from the equation of equilibrium:

$$r\,\frac{d^2u(r)}{dr^2} + \left(1 + \frac{1}{1-2\mu}\right)\frac{du(r)}{dr} = 0 \qquad (1)$$

The condition of complete slippage, $v = 0$, occurring despite the entraining action of the deformable lining, corresponds to the condition $dv(r)/dr = 0$ at the interface. In this case the final form of the equation of equilibrium is:

$$r^2\,\frac{d^2u(r)}{dr^2} + r\,\frac{du(r)}{dr} - \frac{3-4\mu}{2(1-\mu)}\,u(r) = 0 \qquad (2)$$

The solutions of equations (1) and (2) subject to the boundary conditions:

$$u(r) = u_0 \quad \text{at} \quad r = R_0,$$
$$u(r) = 0 \quad \text{at} \quad r = R$$

are for equation (1):

$$u(r) = u_0\,\frac{\rho^{\ell} - 1}{\eta^{\ell} - 1}; \quad \ell = \frac{1}{1-2\mu}; \qquad (3)$$

and for equation (2):

$$u(r) = u_0\,\frac{\eta^n(\rho^{2n} - 1)}{\rho^n(\eta^{2n} - 1)}; \quad n = \sqrt{\frac{3-4\mu}{2(1-\mu)}}; \qquad (4)$$

where $\rho = R/r$, $\eta = R/R_0$ are dimensionless parameters and r,R are the coordinate direction radius of the deforming annular soil mass respectively.

Substitution of the solutions (3) and (4) into the known physical equations for stress, expressed in terms of shears, yields the following functions for normal and tangential stress within the thawed soil mass:

according to the solution (3):

$$\sigma_r = -\frac{\eta^{\ell} E u_0 (1-\mu)\cos\alpha}{R_0(1+\mu)(1-2\mu)^2(\eta^{\ell}-1)};$$

$$\tau_{r\alpha} = \frac{\eta^{\ell} E u_0 \sin\alpha}{2R_0(1+\mu)(1-2\mu)(\eta^{\ell}-1)}; \qquad (5)$$

and according to the solution (4):

$$\sigma_r = -\frac{Eu_0(1-\mu)\cos\alpha}{R_0(1+\mu)(1-2\mu)}\left(n\,\frac{\eta^{2n}+1}{\eta^{2n}-1} - \frac{\mu}{1-\mu}\right);$$

$$\tau_{r\alpha} = -\frac{Eu_0\sin\alpha}{2R_0(1+\mu)}. \qquad (6)$$

Tangential stress given by the expression (6) describes the conditions of net shear in a deforming soil mass near the surface of slippage. However, this stress is not of contact but of internal nature and therefore is neglected in analyses of the compressibility of soil mass.

When a structure in contact with soil exerts a unidirectional force (case 1), the resulting compression along a circular contour is the sum of contact stresses projected onto the direction of shear, and is expressed by the function:

$$q = \oint(\sigma_r\cos\alpha - \tau_{r\alpha}\sin\alpha)\,R_0\,d\alpha \qquad (7)$$

where q is a linear load per unit of length of the structure.

According to formula (6), the maximum tensile stress on the structure-soil contact is of the order of

$$\sigma_r \sim -\frac{u_0}{R_0}E \approx -10^{-1}E$$

Since for soils the limiting tensile stress is:

$$\sigma_p \sim 10^{-1}\,\tau_c \sim 10^{-3}E,$$

we get

$$\sigma_r \sim 10^2\sigma_p,$$

that is, for a design scheme without contact friction (case 2), soil resistance to separation is completely overcome. Therefore, in the expression for the compression of soil the circulation is replaced by an integral around the semicircle defining the compression zone, and for tangential contact stress, $\tau_{r\alpha} = 0$. Ultimately we then have:

$$q = \int_0^{\pi} \sigma_r R_0 \cos\alpha \cdot d\alpha. \qquad (8)$$

Taking into account the expressions (5) and (6), formulas for compression (7) and (8) take the form:
for case 1 (no slippage):

$$q = -\frac{\pi\, Eu_0(3-4\mu)\eta^{\ell}}{2(1+\mu)(1-2\mu)^2(\eta^{\ell}-1)}\ ; \qquad (9)$$

for case 2 (complete slippage):

$$q = -\frac{\pi Eu_0(1-\mu)}{2(1+\mu)(1-2\mu)}\left(n\frac{\eta^{2n}+1}{\eta^{2n}-1} - \frac{\mu}{1-\mu}\right). \qquad (10)$$

For an infinitesimal thickness of the thawed zone (at the start of thawing), assuming $u_0 = \xi R_0(\eta-1)$ as $\eta \to 1$, from formulas (9) and (10) we obtain the expressions for initial compression:
for no slippage:

$$q = -\frac{\pi R_0 E\xi(3-4\mu)}{2(1+\mu)(1-2\mu)}\ ; \qquad (11)$$

for complete slippage:

$$q = -\frac{\pi R_0 E\xi(1-\mu)}{2(1+\mu)(1-2\mu)} \qquad (12)$$

It is evident that in the absence of friction, the initial compression according to formula (12) is equivalent to compression along one-fourth of the perimeter of the circle. In comparison with the slippage case, the initial compression in the presence of friction is greater by a factor of 2 at $\mu = 0.5$, and 3 at $\mu = 0$.

Compression of the deforming zone is conveniently expressed in the form of the Fuss-Winckler relation:

$$q = -ku_0\ , \qquad (13)$$

where $k = mE$ is the bed coefficient (proportional to the modulus of soil deformation, E).

The bed coefficient k in formula (13) is related to the conventional expression for bed coefficient k through the relationship:

$$k = 2R_0 k'\ ,$$

since the compressibility q is distributed over the full diameter of the structure. In contrast to the ordinary bed coefficient, k′, with a dimensionality $L^{-2}MT^{-2}(N \cdot m^{-3})$, the coefficient k has a dimensionality of the deformation modulus $L^{-1}MT^{-2}$(Pa) and is independent of the structure diameter under the assumptions since the coefficient k′ is inversely proportional to the size of the lining (Gorbanov-Posadov, 1953). Accordingly, the quantity m in the expression for k is a dimensionless function of soil properties (of the coefficient of transverse deformation, μ) and of compressible zone configuration. Formulas (9) and (10) readily yield expressions for m, which as $\eta \to \infty$, i.e. for unfrozen soil, have, in particular, the form:
for no slippage:

$$m = \frac{\pi(3-4\mu)}{2(1+\mu)(1-2\mu)^2}\ ; \qquad (14)$$

and for complete slippage:

$$m = \frac{\pi[\sqrt{(1-\mu)(1{,}5-2\mu)} - \mu]}{2(1+\mu)(1-2\mu)}. \qquad (15)$$

Plots of m [following formulas (9) and (10)] versus relative depth of thawing, η, are presented in Fig. 2. The figure shows that the limiting influence of boundary conditions leads to a five-fold (and more) change of the compressibility.

The obtained functions of compressibility are intended for use in a linear differential equation of fourth-order which describes the bending of structures supported by an elastic base. The characteristic polynomial involved in this equation has conjugate roots and therefore the general solution of the equation involves sine functions of depth. Therefore for different values of the bed coefficient k the half-wavelength of sinusoids of transverse shears for a structure with rigidity E_kJ is generally inversely proportional to the parameter $\beta = \sqrt[4]{k/4E_kJ}$. Therefore a five-fold increase of k and unchanged loads increases the structure's incline by a factor of about one and a half ($\sqrt[4]{5}$). Similar results are obtained from test calculations. Consequently, in the analysis of a structure's effect on an elastic foundation, the determination of boundary conditions is considerably more important than the refinement of methods of calculation.

Fig. 2 also shows limiting values of the coefficient m for an unlimited increase of the extent of thawing. It is evident that for no slippage of the structure-soil interface and at $\mu \to 0.5$ (for incompressible soil) the compressibility (back pressure) increases without bound. In the case of complete slippage the compressibility remains a finite value, different from zero. This is an important advantage of the equation obtained here over modifications of Flaman's solutions which specify that, for an unlimited compression zone, there are large shears and therefore a zero bed coefficient.

It is interesting that formula (15) for the entire range $0 \leq \mu \leq 0.5$, up to the second significant digit, coincides with the expression:

$$m = \frac{2}{1+\mu} \qquad (16)$$

We substitute (16) into the expression for the bed coefficient,

$$k = mE = \frac{2E}{1+\mu} = 4G \qquad G = \frac{E}{2(1+\mu)}\ ,$$

express the compressibility through normal stress,

$$q = \sigma 2R_0$$

and, introducing these values into formula (13), we obtain:

$$\sigma = -\ 2G\ \frac{u_0}{R_0} \tag{17}$$

Since for this type of stressed-deformed state the mean stress and deformation are zero, from (17) it follows that compressibility properties of a mass of unlimited size can be related only to its shape change.

If we use the solution obtained earlier for the problem (Guryanov, 1976) to estimate the value of compressibility, assuming its distribution along the arc of length 0.5 πR_0, then the relevant expression for the bed coefficient

$$k = \frac{\sqrt{2}\,E\ [1+\eta^2(1-2\mu)]}{(1+\mu)(1-2\mu)(\eta^2-1)} \tag{18}$$

has the same features as function (10), and as $\eta \to \infty$ the expression (17) also follows from (18). Equation (18) also implies the possibility of determining the bed coefficient directly from pressure test results, without using the mechanical characteristics of the soil. In particular, if the working stress σ on the pressure meter of radius R_0 refers to radial shear u_0, the bed coefficient approximating the case of contact with slippage will then be:

$$k = \sqrt{2}\ \sigma\ \frac{R_0}{u_0} \tag{19}$$

Thus, the compressibility of thawing soils interacting with bendible structures depends on boundary conditions no less than on values of soil mechanical characteristics. For actual structures the boundary conditions are generally indeterminate and therefore calculations and design of bendible structures are advantageously done using marginal variations of their size and shears. The combination of a large diameter and small shears of the bendible structure, for example, when shaft lining is being bent, is closer to the case of contact with no slippage and requires that formulas (9) and (14) be used. The combination of a small diameter and large shears, including deformation of pipes in unfrozen soils is closer to the case of complete slippage and requires that formulas (10) and (15) be used. For intermediate cases, calculations for both limiting variant can be used to judge design feasibility.

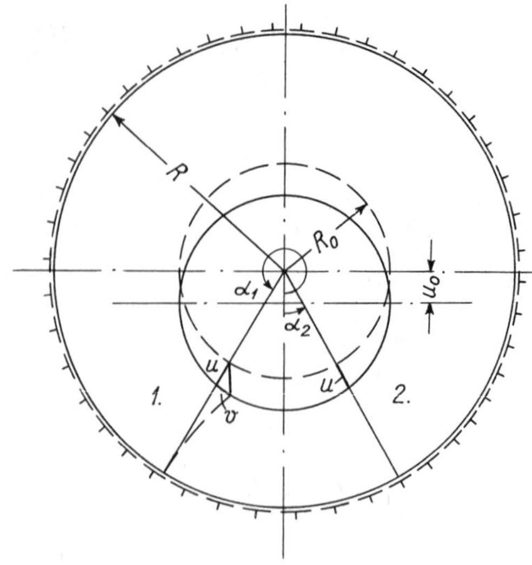

FIGURE 1 Plan view of deforming thawed soil annulus under the action of a cylindrical lining. (1) displacements for no slippage at the structure-soil interface, (2) displacements for complete slippage.

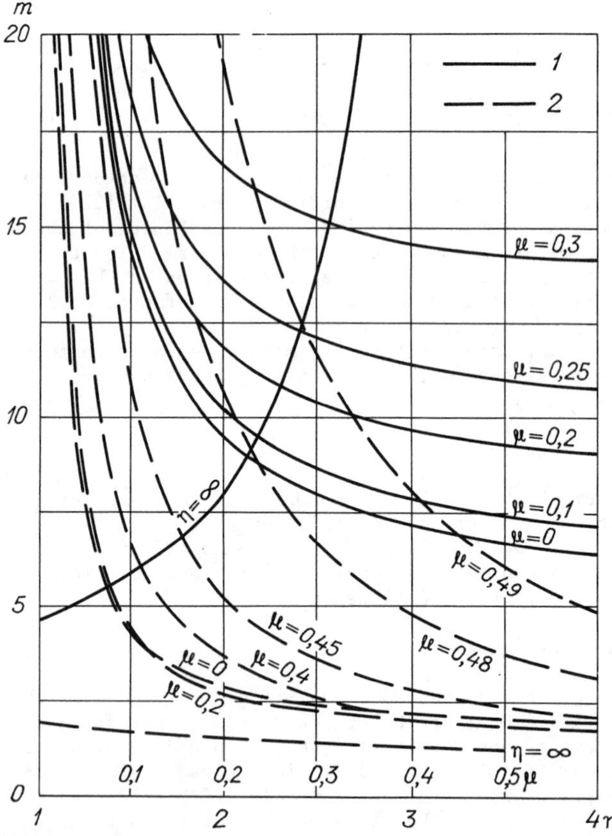

FIGURE 2 Compressibility m for different values of μ, as a function of relative depth of thawing, η. (1) no slippage on the structure-soil interface, (2) complete slippage. The values of μ on the abcissa correspond to two limiting curves for m at $\eta \to \infty$.

REFERENCES

Baron, L. I., 1972, Coefficients of rock hardness, Moscow: "Nauka", 176 p.

Gorbunov-Posadov, M. I., 1953, Design of structures on elastic foundations, Moscow: Gosstroiizdat, 516 p.

Guryanov, I. E., and Malyshev, M. V., 1976, Investigation of the compressibility of thawing soils by means of a pressure meter. Soil Mechanics and Foundation Engineering, vol. 12, no. 4, p. 249-256.

SNiP][-15-74, 1975, Foundations of buildings and structures. Design standards, Moscow: Stroiizdat, 65 p.

FIELD TESTS OF A FROST-HEAVE MODEL

G. L. Guymon[1], R. L. Berg[2], and T. V. Hromadka II[3]

[1]Civil Engineering, University of California, Irvine, California 92717 USA
[2]U.S. Army Cold Regions Research and Engineering Laboratory
Hanover, New Hampshire 03755 USA
[3]U.S. Geological Survey, Laguna Niguel, California 92677 USA

A one-dimensional mathematical model of frost heave based upon a nodal domain integration analog is compared to data collected from a Winchendon, Mass., field site. Air and soil temperatures, pore water pressures, and ground-water level data were collected on test sections containing six different soils during the winters of 1978-1979 and 1979-1980. The soil samples were evaluated in the laboratory to determine soil moisture characteristics, hydraulic conductivity as a function of pore water tensions, density, and other parameters. The parameters were used together with assumed thermal parameters in a one-dimensional model that calculates the distributions of temperature and moisture content as well as the amount of ice segregation (vertically lumped frost heave) and thaw consolidation. Using measured air and soil surface temperatures as input data, the simulated frost heave and thaw consolidation agreed well with measured ground surface displacements that resulted from ice segregation or ice lens melting.

This paper compares measured field data of frost heave and thaw consolidation and simulated frost heave and thaw consolidations using a one-dimensional model that is being developed by USACRREL for a project funded by the Corps of Engineers, the Federal Highway Administration, and the Federal Aviation Administration (Guymon et al. 1981a). This paper reports on continued verification of the mathematical model with some of the field data collected for the tri-agency project.

Beginning in the mid-1970's with Harlan (1973) and Guymon and Luthin (1974), investigators have increasingly studied the use of mathematical models to simulate coupled heat and moisture movement in freezing and thawing soils. In the early 1980's a number of investigators proposed using such models to estimate frost heave as well. Most of these models are reviewed by Guymon et al. (1980) and Hopke (1980). Generally, models are based upon deterministic equations that describe heat and moisture transport; however, the manner of incorporating required ancillary models, i.e. those used to determine latent heat effects, hydraulic properties, and stress distribution in the freezing zone, and the manner of assembling the complex models varies widely between investigators. It is reasonable to state that agreement on either the necessary level of complexity of frost heave models or the formulation of algorithms representing processes in the freezing zone is not wide-spread. Indeed, there is still considerable controversy associated with modeling endeavors. Due to the inherent spatial variability of soil properties, deterministic models will probably never be entirely satisfactory, and we have included probabilistic features in one version of the model used herein (Guymon et al. 1981b).

Probably the best way to show the utility of deterministic mathematical models of frost heave and the problems associated with their use is to compare results from these models with prototype field data and experimental laboratory data. We present such comparisons using a specific model that has been refined over several years.

THE NUMERICAL MODEL

The numerical model has been described elsewhere by Guymon et al. (1980, 1981a, 1981b, 1983) and Hromadka et al. (1982). Consequently, only a brief outline of the model is given here.

Table 1 presents the mathematical basis of the model. The nomenclature used in this table is defined at the end of this paper.

Phase-change processes in freezing or thawing zones are modeled by an isothermal approximation. The phase-change components for discrete times are decoupled using a simple control volume approach that accounts for the total available latent heat for freezing available water or thawing available ice. Overburden and surcharge effects are approximated by computing an equivalent positive pore water pressure, u, and adding it to simulated negative pore water pressures at discrete nodes where ice segregation is occurring. Positive pore water pressures due to consolidation are assumed to be negligible compared to overburden and surcharge stresses.

The equations of state are solved by a nodal domain integration analog (Hromadka et al. 1982), which results in a matrix system similar to the finite element system; that is, for the decoupled heat transport equation:

$$S\ T + C\ \dot{T} = F \tag{1}$$

TABLE 1 Mathematical equation describing frost heave model.

Soil region	Energy state	Liquid moisture transport	Phase change	Sensible heat transport	Ancillary relationships
Surface boundary		$\partial(u-x)/\partial x = 0$		$T_u = T(t)$	
Frozen	$T < T_f$ $u = u(\theta_n) + \delta u_o$ $\delta = \begin{cases} 1, & \theta_i > \theta_o - \theta_n \\ 0, & \theta_i < \theta_o - \theta_n \end{cases}$	$\partial(u-x)/\partial x = 0 = \upsilon$	$\frac{\partial \theta_i}{\partial t} = 0$	$\frac{\partial}{\partial x}[K_T\, \partial T/\partial x] = C_m \frac{\partial T}{\partial t}$	$K_H = K(u) \cdot 10^{-E\theta_i}, E\theta_i > 1$ $\theta_u = \theta(u)$ $\upsilon = -K_H \partial(u-x)/\partial x$
Freezing or thawing	$T = T_f$ $u(\theta_n) < u < 0$	$\frac{\partial \theta_u}{\partial t} = \frac{\partial}{\partial x}[K_H \partial(u-x)/\partial x]$	$-\frac{\rho_i}{\rho_w}\frac{\partial \theta_i}{\partial t}$ $L\frac{\rho_i}{\rho_w}\frac{\partial \theta_i}{\partial t}$	$+\frac{\partial}{\partial x}[K_T\, \partial T/\partial x] - C_w \upsilon \frac{\partial T}{\partial x} = C_m \frac{\partial T}{\partial t}$	$C_m = [C_i\theta_i + C_w\theta_u + C_s(1-\theta_o)]\,/(1+\theta_s)$ $K_T = [K_i\theta_i + K_w\theta_u + K_s(1-\theta_o)]\,/(1+\theta_s)$
Unfrozen	$T > T_f$ $u(\theta_n) < u < u_L$	$\frac{\partial \theta_u}{\partial t} = \frac{\partial}{\partial x}[K_H \partial(u-x)/\partial x]$	$\frac{\partial \theta_i}{\partial t} = 0$	$\frac{\partial}{\partial x}[K_T\, \partial T/\partial x] - C_w \upsilon \frac{\partial T}{\partial x} = C_m \frac{\partial T}{\partial t}$	$\theta_s = \theta_i - (\theta_o - \theta_u),\ \theta_s > 0,\ \theta_u > \theta_n$ $y = \Sigma \theta_s \ell$
Column bottom boundary		$u_L = u(t)$		$T_L = T(t) > T_f$	

where S is a matrix of conductivity parameters and spatial discretization, C is a matrix of capacitance parameters, T and $\dot{T}$ are the unknown state variable and temporal derivative vectors, and F is a vector of specified boundary conditions. The moisture transport equation reduces to an identical form.

The nodal domain integration method results in a C matrix that is also a function of a single mass lumping parameter, η; that is, $C = C(\eta)$ where η can arbitrarily be chosen to represent various numerical analogs. For instance, if $\eta = 2$, a Galerkin finite element scheme results, or if $\eta \to \infty$, an integrated finite difference scheme results. The Galerkin finite element scheme was used in this study.

Governing equations are solved by decoupling them for discrete computational intervals and by holding each parameter constant. The solution is advanced in constant time-steps, Δt, by the Crank-Nicolson approximation. At the end of discrete time periods, nonlinear parameters are updated with ancillary relationships and secondary variables are computed. The computational technique and the sensitivity of results to the choice of numerical analog method are described in detail by Hromadka et al. (1982).

Guymon et al. (1981b, 1983) describe the verification of the model with freezing test data from laboratory soil columns, evaluate the uncertainty relating to boundary condition errors, describe modeling errors in general, and present a detailed analysis of parameter error effects.

FIELD AND LABORATORY DATA

Field data were collected at a test site in Winchendon, Massachusetts, located about 8 km (5 mi.) south of the New Hampshire border and about 32 km (20 mi.) east of the Connecticut River. The site consists of 26 asphalt concrete pavement test sections with 13 different test materials. The test sectons have a 76 mm (3 in.) asphalt concrete pavement underlain by a uniformly compacted subbase material from 0.9 to 1.5 m (3 to 5 ft) deep. Six of the test sections are included in the tri-agency study. Climatic data, ground-water levels, soil temperatures, and soil pore water pressure data were collected in the field during the 1978-1979 and 1979-1980 winters. Undisturbed and disturbed samples were obtained from the test soils and evaluated in the laboratory to determine the physical, hydraulic, and mechanical properties of the various materials. Field observations of frost heave, frost depth, and soil moisture tension were monitored for the following test materials: (a) Ikalanian silt, (b) Graves sandy silt, (c) Hart Brothers sand, (d) Sibley till, (e) Hyannis sand, and (f) dense-graded stone.

Figure 1 shows the mean daily air temperature beginning 10 December 1978 and extending through 15 March 1979. These data are determined from the average of the maximum and minimum daily temperatures taken from a thermograph. Several major freeze-thaw cycles occurred and, due to diurnal temperature variations, there were numerous diurnal freeze-thaw cycles during the winter. Table 2 presents the average monthly diurnal temperature variations.

Table 3 presents the physical data obtained in the laboratory. The soil water characteristics are represented by the equation

$$\theta = \frac{\theta_o}{A_w |u|^n + 1} \qquad (2)$$

where the regression parameters A_w and n are given in Table 3. The freezing soil hydraulic conductivity factors, E, in Tables 1 and 3 were determined by calibration as described below. The E factor (Table 1) is a phenomenological parameter used to adjust unfrozen hydraulic conductivity to represent the hydraulic conductivity in the freezing zone. The version of the model used herein uses a hydraulic conductivity function in tabular form (i.e. hydraulic conductivity as a function of tension).

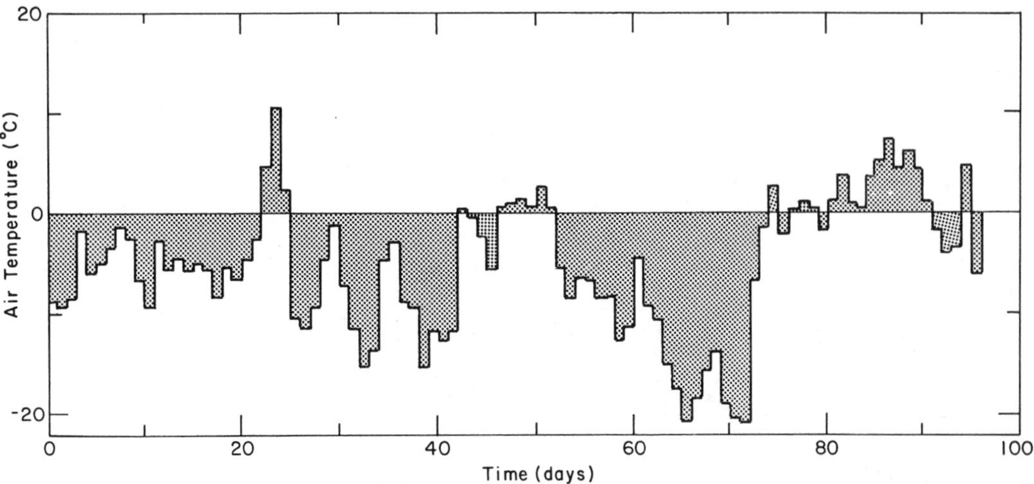

FIGURE 1 Mean daily air temperature, Winchendon, Mass., 10 Dec 1978 (day 0) through 15 Mar 1979.

TABLE 2 Average monthly diurnal temperature variations at the Winchendon test site, 1978-1979.

	Mean daily amplitudes, °C			Max. daily amplitudes, °C			Min. daily amplitudes, °C			Mean daily amplitude coefficient of variation (%)		
Temperature location	Jan	Feb	Mar	Jan	Feb	Mar	Jan	Feb	Mar	Jan	Feb	Mar
Air (1.5 m above g.s.)	4.7	7.4	7.3	10.3	14.4	18.3	0.6	0.7	1.4	51	50	58
Soil surface												
Ikalanian silt	4.6	9.2	7.4	8.5	11.4	16.6	0.6	6.6	2.2	50	20	58
Graves sandy silt	3.2	7.4	5.8	5.0	10.3	17.0	0.1	5.4	1.1	50	30	78
Hart Brothers sand	3.1	7.1	8.1	5.8	10.6	17.4	0.2	1.4	2.0	48	42	58
Sibley till	4.6	10.4	5.9	10.6	14.0	13.2	0.3	5.8	1.8	61	28	69

MODEL SIMULATION

A soil column 1 m long was assumed for all soils except Sibley till, where a 1.3 m column was used. Generally, the water table was a little over 1 m deep in all sections except the one containing Sibley till, which had a lower water table (less than 1.3 m). The soil column was divided into 50 variable- length elements ranging from 0.5 cm at the column top to 10 cm at the column bottom. A Crank-Nicolson time-step of 0.2 hr and a parameter update frequency of 1.0 hr were used. Column bottom boundary conditions were determined from recorded data. Mean daily surface temperature data or air temperature data multiplied by the Corps of Engineers n-factor (Berg 1974) were used for the upper boundary conditions. A sinusoidal diurnal surface temperature variation with a 7°C amplitude was used. Parameters were assumed, measured in the laboratory, or determined by calibration. Table 3 summarizes the parameters used in the simulations.

The results of simulations for the Graves sandy silt and Sibley till during the 1978-1979 winter are shown in Figures 2 and 3 along with the results when mean daily surface temperatures are used without a diurnal variation. Similar results were obtained for other soils studied. In general, errors introduced by using mean daily surface temperature were negligible. The most significant difference observed was for Graves sandy silt (Figure 2). In all cases, the use of mean daily surface temperatures approximated measured thaw penetration more accurately than the 7°C amplitude diurnal variation.

In some cases, it was difficult to predict accurately frost penetration near the end of the season. Measured frost (0°C isotherm) depths, which are subject to some error, are generally deeper than simulated with the model.

The only parameter calibrated was the freezing soil hydraulic conductivity correction factor. Adjusting other parameters, such as the soil water characteristics, might have yielded better overall

TABLE 3 Soil parameters for remolded Winchendon, Massachusetts, test site soils.

Parameter	Ikalanian silt	Graves sandy silt	Hart Bros. sand	Sibley till	Hyannis sand	Dense graded stone
Soil density (g/cm^3)[1]	1.70	1.49	1.690	1.970	1.690	1.870
Soil porosity (cm^3/cm^3)[1]	0.37	0.46	0.391	0.282	0.367	0.334
Soil-water freezing point depression (°C)[2]	0	0	0	0	0	0
Volumetric heat capacity of soil ($cal/cm^3/°C$)[2]	0.20	0.20	0.200	0.200	0.200	0.200
Thermal conductivity of soil (cal/cm/hr/°C)[2]	17.00	17.00	5.000	20.000	17.000	17.000
Unfrozen water content factor (cm^3/cm^3)[2]	0.03	0.12	0.040	0.150	0.010	0.150
Soil water characteristics [Aw,(n)] (unitless)[1]	0.000546 (1.5)	0.0056 (0.9)	0.022 (0.867)	0.062 (3.45)	0.00154 (1.806)	0.053 (0.462)
Saturated hydraulic conductivity (cm/hr)[1]	0.37	1.92	4.080	0.360	1.230	5.540
Freezing soil hydraulic conductivity factor (unitless)[3]	16.00	4.50	5.000	8.000	15.000	15.000

[1] Parameters measured in laboratory from remolded soil samples.
[2] Parameters were assumed.
[3] Parameters were calibrated.

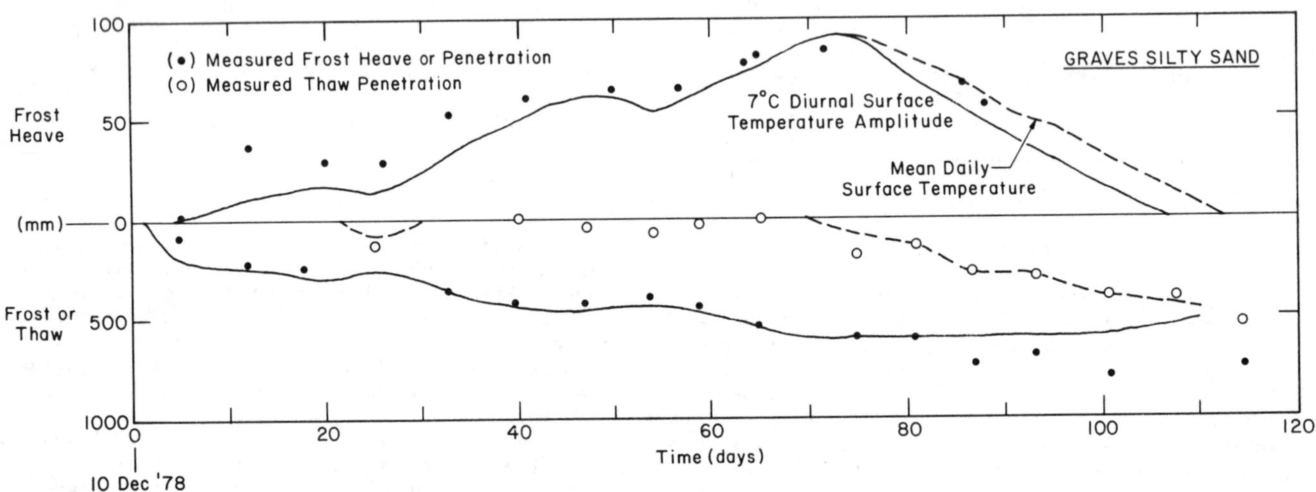

FIGURE 2 Simulated frost heave, thaw consolidation, frost penetration, and thaw penetration for Graves silty sand, 1978-1979 (day 0 = 10 Dec 1978).

results, but this type of calibration is not a wise procedure since it may mask errors in the model. Calibration errors may stem from three sources: (1) a surface moisture flux boundary condition error, (2) the fact that soil parameters vary due to freeze/thaw cycles, and (3) use of pavement surface temperatures instead of soil surface temperatures.

Alternate freezing and thawing occurs near the soil surface in most field prototype situations. It seems appropriate that any complete model of frost heave should include analogs that account for changes in key parameters, such as a hydraulic conductivity due to freeze/thaw cycles. Neither this model nor any other currently available attempts to

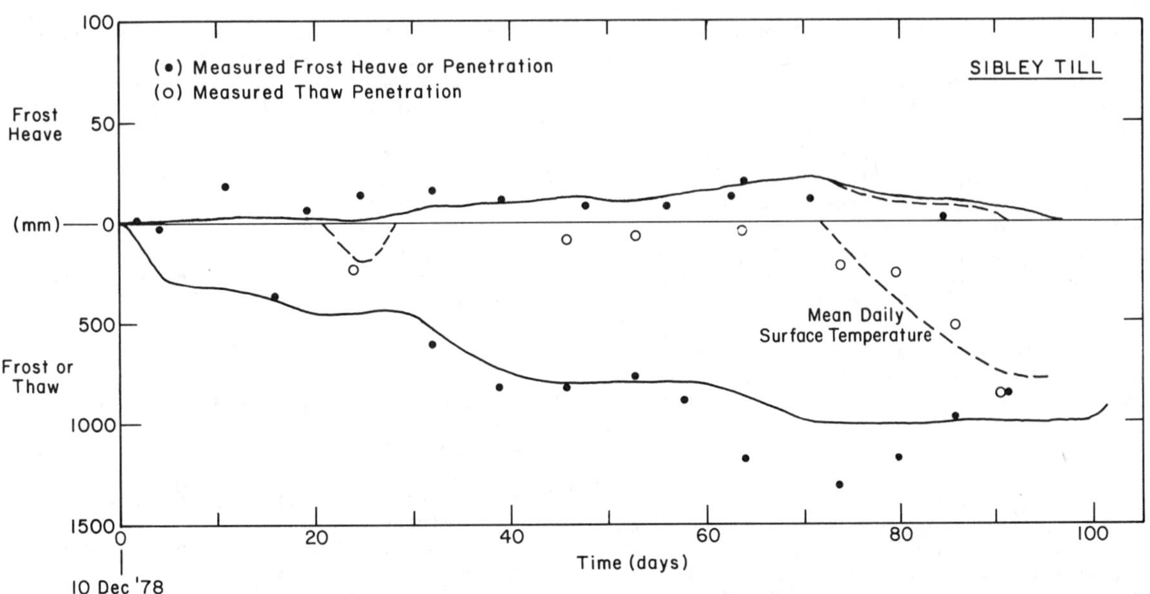

FIGURE 3 Simulated frost heave, thaw consolidation, frost penetration, and thaw penetration for Sibley till, 1978-1979 (day 0 = 10 Dec 1978).

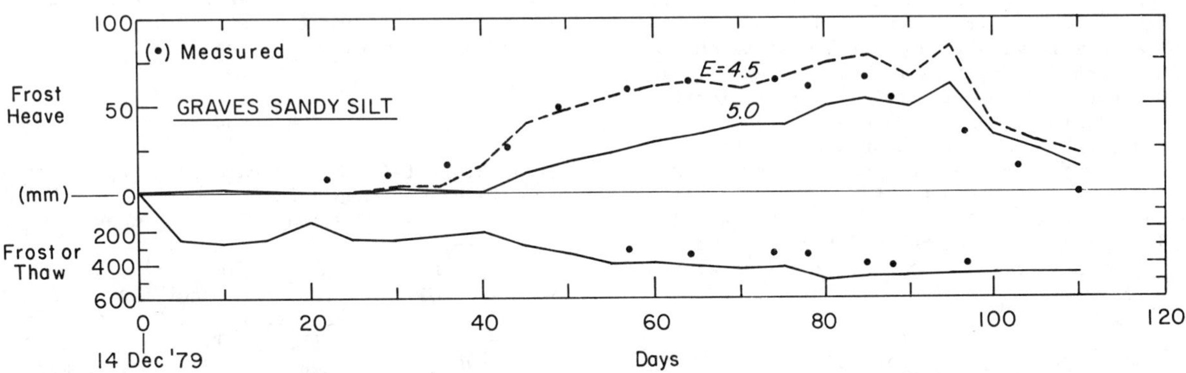

FIGURE 4 Simulated frost heave, thaw consolidation, frost penetration, and thaw penetration for Graves silty sand, 1979-1980.

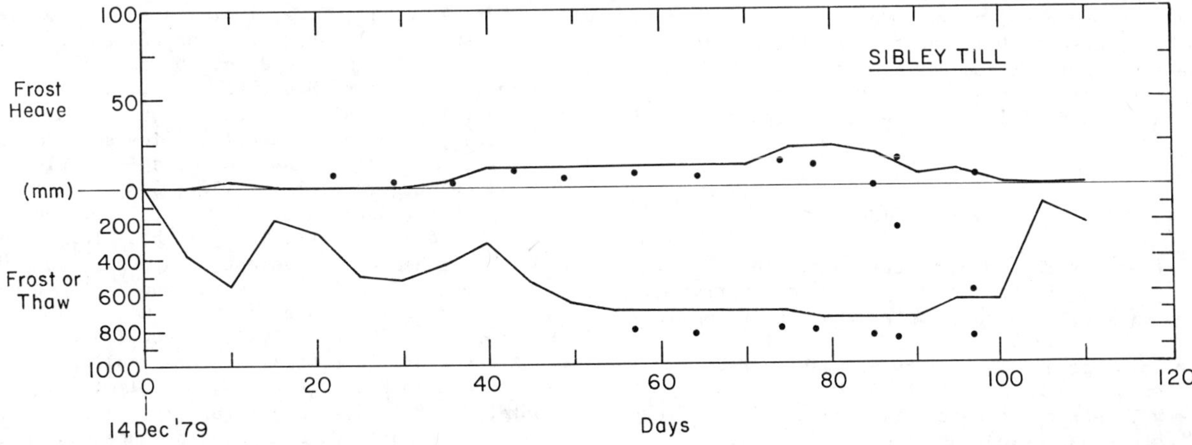

FIGURE 5 Simulated frost heave, thaw consolidation, frost penetration, and thaw penetration for Sibley till, 1979-1980.

include the influence of alternate freezing and thawing.

To further test the model, and particularly the validity of calibrating the E factor, dta for the 1979-1980 winter were used. Air temperature data and frost heave and frost penetration data are only available for (a) Ikalanian silt, (b) Graves sandy silt, (c) Hart Brothers sand, and (d) Sibley till.

Simulated results are compared to measured frost heave and frost penetration for Graves sandy silt and Sibley till in Figures 4 and 5. To match the measured maximum frost heave for the 1979-1980 winter more closely, the E factors are 10.3, 5.0, 9.0, and 8.5 for Ikalanian silt, Graves sandy silt, Hart brothers sand, and Sibley till, respectively. Data in Figure 4 show computed results using both E factors.

CONCLUSIONS

The proposed model can reasonably estimate field frost heave, provided accurate data on boundary conditions and parameters are available. Tests of the model were for field conditions where a number of freeze/thaw cycles occurred as well as significant diurnal temperature variations. Using measured air and soil surface temperatures as input data, the simulated frost heave and thaw consolidation agreed well with ground surface displacements resulting from ice segregation or ice lens melting.

Currently the model requires calibration to determine the E factor for each soil, the phenomenological parameter that accounts for decreased hydraulic conductivity in freezing or thawing zones. The results of the study reported here suggest that it is feasible to calibrate the model for field soils using imperfect input data; however, a long period of data is required with a number of freeze/thaw cycles.

ACKNOWLEDGMENTS

This work was sponsored by the U.S. Army Corps of Engineers, the Federal Highway Administration, and the Federal Aviation Administration. The test site at Winchendon, Massachusetts, was operated by the Massachusetts Department of Public Works. Jonathan Ingersoll and David Carbee, civil engineering technicians, were responsible for the laboratory and field investigations supporting this study.

NOMENCLATURE

T = temperature (primary state variable)
u = pore water pressure (primary state variable)
θ_u = volumetric unfrozen water content (secondary state variable)
θ_i = volumetric ice content (secondary state variable)
θ_s = volumetric segregated ice content (secondary state variable)
x = coordinate vertically downward
ℓ = length of a finite element
y = lumped total frost heave (secondary state variable)
t = time
θ_n = unfrozen water content factor (parameter)
θ_o = soil porosity (parameter)
K_s = soil thermal conductivity (parameter)
C_s = soil volumetric heat capacity (parameter)
T_f = freezing point depression of water (parameter)
K_H = K_H (u) = hydraulic conductivity function (parameter)
E = freezing soil hydraulic conductivity correction (parameter)
K_i = ice thermal conductivity (specified constant)
K_w = water thermal conductivity (specified constant)
K_t = thermal conductivity of soil, water, ice mixture
C_i = ice volumetric heat capacity (specified constant)
C_w = water volumetric heat capacity (specified constant)
C_m = volumetric heat capacity of soil, water, ice mixture
L = bulk water volumetric latent heat coefficient (specified constant)
ρ_i = ice density (specified constant)
ρ_w = water density (specified constant)

Parameter refers to a measured or assumed parameter required in the model. Specific constant refers to a parameter obtained from standard tables. Variables are computed by the model.

REFERENCES

Berg, R.L., 1974, Energy balance on paved surface, Technical Report 226: Hanover, N.H., U.S. Army Cold Regions Research and Engineering Laboratory.

Guymon, G.L., and Luthin, J.N., 1974, A coupled heat and moisture transport model for arctic soils: Water Resources Research, v. 10, p. 995-1001.

Guymon, G.L., Hromadka, II, T.V., and Berg, R.L., 1980, A one-dimensional frost heave model based upon simulation of simultaneous heat and water flux: Cold Regions Science and Technology, v. 3, p. 253-263.

Guymon, G.L., Berg, R.L., Johnson, T.C., and Hromadka, II, T.V., 1981a, Results from a mathematical model of frost heave: Transportation Research Record, v. 809 p. 2-5.

Guymon, G.L., Harr, M.E., Berg, R.L., and Hromadka II, T.V., 1981b, A probabilistic-deterministic analysis of one-dimensioned ice segregation in freezing soil column: Cold Regions Science and Technology. v. 5, p. 127-140.

Guymon, G.L., Berg, R.L., Johnson, T.C., and Hromadka II, T.V., 1983, Mathematical model of frost heave in pavements: Hanover, N.H., U.S. Army Cold Regions Research and Engineering Laboratory, in preparation.

Harlan, R.L., 1973, Analysis of coupled heat-fluid transport in partially frozen soil: Water Resource Research, v. 9, p. 1314-1323.

Hopke, S.W., 1980, A model for frost heave including overburden: Cold Regions Science and Technology, v. 3, p. 111-147.

Hromadka II, T.V., Guymon, G.L., and Berg, R.L., 1982, Sensitivity of a frost heave model to the method of numerical simulation: Cold Regions Science and Technology, v. 6, p. 1-10.

PERMAFROST-GLACIER RELATIONSHIPS IN THE SWISS ALPS - TODAY AND IN THE PAST

Wilfried Haeberli

Versuchsanstalt für Wasserbau, Hydrologie und Glaziologie, ETH Zürich, Switzerland

A conceptual model is proposed to explain the geographical distribution of permafrost and glacier phenomena in high altitude/low latitude mountains. The concept developed shows how the vertical distance between the lower boundary of permafrost distribution and the equilibrium line on glaciers can be related to the mean annual air temperature at the equilibrium line. This temperature term strongly depends on the continentality of the climate and can be empirically linked to thermal conditions in ice and permafrost, and also to glacier activity. Conditions in the Swiss Alps both today and in the past are used to illustrate the concept and to demonstrate its use for paleoclimatic reconstructions. In the humid marginal zones of the Alps, where permafrost is restricted to rock outcrops in the accumulation area of glaciers, active, temperate glaciers are dominant; on the other hand, less active, partially cold glaciers and creep phenomena in perennially frozen sediments (such as rock glaciers and push moraines) are abundant in the sheltered interior regions today. Partially warm-based, cold glaciers of extremely low activity within continuous permafrost were typical for the coldest period of the last ice age. In agreement with palynological reconstructions, this indicates that precipitation was greatly reduced at this time.

INTRODUCTION

Glaciers are the most obvious form of ice on many high altitude/low latitude mountains and have been studied since long before the beginning of the twentieth century. However, high-mountain permafrost is a less striking feature, and research into it has only developed over the last few decades. The growing knowledge about glaciers and permafrost in the Swiss Alps now makes it possible to discuss the relationships between the two phenomena. While the vertical sequence of glacier and permafrost zones in the Alps (discussed earlier by Haeberli (1976, 1978)) mainly depends on the (mean annual) air temperature, the regional distribution of glacier and permafrost features will be shown here to depend primarily on precipitation. Both air temperature and precipitation vary not only in space but also in time, and so does the distribution pattern of permafrost and glacier ice. A greater understanding of the spatial relations between actual glaciers and existing permafrost should therefore also lead to a better interpretation of paleoclimatic conditions.

The purpose of this paper is to propose a conceptual model to explain the relationships between permafrost and glacier phenomena in the Swiss Alps. The concept developed here is mainly based upon ideas and observations by Shumskii (1964), Haeberli (1975), 1978), Barsch (1978), Gorbunov (1978), Kuhn (1980) and Shi and Li (1981).

PERMAFROST-GLACIER RELATIONSHIPS IN THE SWISS ALPS

The concept is a combined consideration of several basic permafrost and glacier phenomena, their interrelationships and their variation with changes in precipitation. It is based upon a model of the cryosphere as depicted in Figure 1; the field, defined by the mean annual air temperature (i.e., "altitude" in mountain regions) and the mean annual precipitation (i.e., "continentality"), is subdivided by the equilibrium line on glaciers (mass balance = 0) and the lower boundary of discontinuous permafrost into three main zones. Zone A, above the equilibrium line, is the accumulation zone of glaciers where no sediments are deposited. Zone B, below the lower boundary of permafrost distribution, is free from permafrost. Zone C, below the equilibrium line of glaciers and above the lower boundary of permafrost distribution, is the zone in which

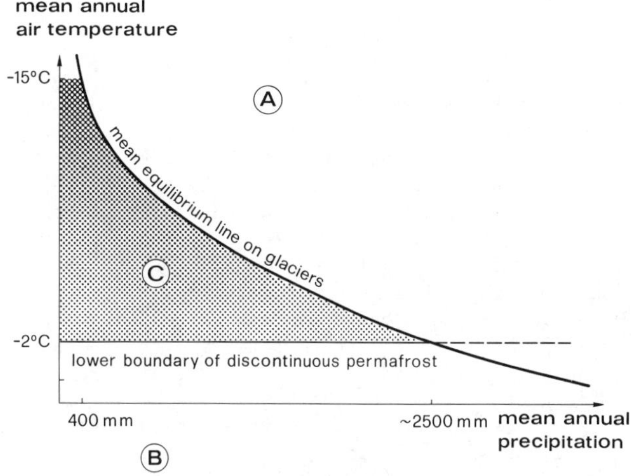

FIGURE 1 Schematic model of the structure of the cryosphere. Explanation in the text.

glacier ice can be cold and unconsolidated sediments can be in a perennially frozen state. The vertical extent of this zone C is a direct expression of precipitation - a principle now used for paleoclimatic reconstructions (Kerschner 1978, 1982, 1983, Haeberli 1982). In a maritime climate, zone C disappears because the equilibrium line descends below the lower boundary of permafrost distribution.

The vertical extent of zone C is related to the mean annual air temperature at the equilibrium line, T_E, via the air temperature gradient (assumed to be constant in this paper, viz., 0.65°C/100 m). As is the case for the mean air temperature during the ablation period, to which it is obviously correlated via precipitation, T_E is a good indicator of precipitation (cf. Ahlmann 1924, 1933, Kuhn 1980, 1982). It is this term, T_E, which may be considered as the central point of the concept, because it is not only related to precipitation, to the vertical extent of zone C (where geomorphological expressions of permafrost such as rock glaciers, push moraines or ice wedges can occur), and to thermal conditions in glacier ice and permafrost, but also to processes of mass exchange at glacier surfaces (mass balance gradient) and glacier activity (ice thickness, shear stress, glacier flow). These terms of the concept and their interrelationships are explained below.

The mean *surface temperature in ice and permafrost* as measured in boreholes is, on average, about 3°-4°C warmer than the mean annual air temperature at the same place (cf. data given by Haeberli 1975, 1976, Markl and Wagner 1978, Barsch et al. 1979). Because of the effect of latent heat exchange by percolating meltwater, the *firn temperature* is 0°C providing the mean annual air temperature is higher than about -8°C. Temperate *firn* bodies (within zone A) therefore occur where $T_E > -8°C$ and are absent in more continental regions, whereas cold *glacier ice* (zone C) exists in regions where $T_E < -3°$ to -4°C but not in regions with a more maritime climate.

As a consequence of the topographical variation of ground temperatures at a given altitude (about ±3°C), the *lower boundary of discontinuous permafrost* follows the -1° to -2°C isotherm, whereas the *lower boundary of continuous permafrost* is close to the -6° to -8°C isotherm (Haeberli 1978). *Rock glaciers* are the morphological expression of creep in perennially frozen, supersaturated sediments under the influence of gravity alone (Barsch 1978, Fisch et al. 1978, Haeberli 1978). On the other hand, the deformation of such sediments through flow of (partially) cold glaciers leads to the formation of *push moraines* in the strict sense (s.s.) (Haeberli 1979). These two phenomena can be used to reconstruct the position of the lower boundary of discontinuous permafrost or the -1° to -2°C isotherm in regions or times without weather station records, if $T_E > -2°C$, where zone C disappears. *Ice wedge* occurrence is limited to the continuous permafrost zone (Péwé 1966, Brown and Péwé 1973) or to regions with $T_E < -6°$ to -8°C (low precipitation).

The equilibrium line on glaciers, when not directly observed in mass balance studies, is inferred from an accumulation area ratio of 0.6 to 0.7 (Gross et al. 1977). With increasing continentality, it moves to colder regions because the decreasing amount of snow accumulation can be balanced by lower amounts of melting (temperature dependent). In these colder regions (lower T_E) the amount of liquid precipitation in the lower reaches of the glacier decreases. This lowers the *mass balance gradient,* the change in snow and ice balance with altitude for a given glacier, often referred to as the "activity index" (Meier 1961). The mass balance gradient not only depends on precipitation (Schytt 1967, Kuhn 1980) and T_E but is also an expression of the mass exchange at the surface of a glacier with a given size. At the same time, it drives the mass flux within a glacier via glacier flow. The latter, involving both ice deformation and sliding at the glacier bed, depends in a nonlinear way on the *basal shear stress in a glacier* (Paterson 1981). The basal shear stress therefore varies with varying mass exchange at the glacier surface; for a given size it decreases with decreasing balance gradient (Maisch and Haeberli 1982) and decreasing T_E, at least for small or continental glaciers. It is related to glacier geometry (thickness, inclination, valley cross section) and may furnish information about glacier activity during past glaciations, even though no mass balance or flow velocity measurements are possible for the vanished glaciers.

Values for balance gradients and basal shear stresses, as quoted in this paper, are averaged over the ablation areas of the corresponding glaciers. It is for these areas only that balance gradients are more or less linear and glacier geometry (for stress calculations) is relatively well documented, both by drillings and geophysical soundings in recent glaciers (e.g., Süsstrunk 1951, Waechter 1981) and by moraines in the case of ice age glaciers (Maisch 1981). *Annual precipitation and mean annual air temperature* are long-term averages estimated from weather station records. Values for precipitation refer to a standard altitude of 2,000 m a.s.ℓ. and can only be rough estimates. Emphasis is placed upon relative differences rather than on absolute numbers. *Permafrost* and *glaciers* are considered to be in equilibrium with climate. This is certainly a simplification, but it probably does not introduce too large an error into the overall picture. The remainder of this paper illustrates the concept outlined above and demonstrates its use for a better understanding of paleoclimatic conditions.

CONDITIONS TODAY

The Swiss Alps are characterized by a transition from maritime to continental climate without showing extreme conditions. An example of each-the "humid" margin" and the "dry interior", respectively-is described here. Many regions in the Alps belong to the transitional zone (Figure 2) between the two examples. More extreme conditions can be found in other low latitude mountain chains of the world (e.g., Asia, South America).

The *"humid margin"* of the Alps, with an annual

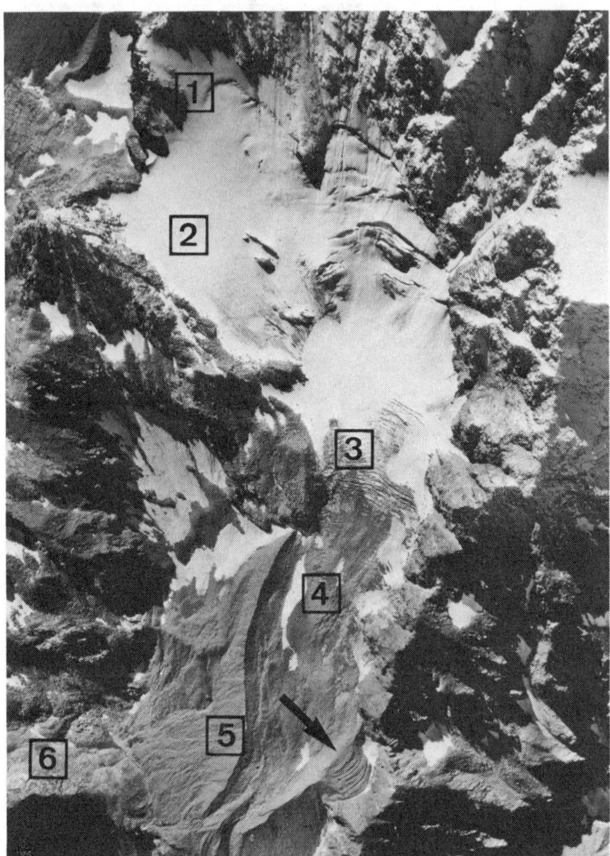

FIGURE 2 Bodmergletscher near the Simplon Pass, an example from the transitional zone between the "humid margin" and the "dry interior" of the Swiss Alps. Typical phenomena discussed in the text are 1, Bergschrund; 2, accumulation area; 3, equilibrium line (accumulation equals ablation); 4, ablation zone (partially debris-covered); 5, lateral moraine of historical and postglacial advances; 6, (small) rock glaciers (creep of perennially frozen, supersaturated sediments). The arrow points to a holocene push moraine (glacially deformed frozen, supersaturated sediments) and at the same time to the north. During postglacial times, T_E was around -4°C, corresponding to an annual precipitation at 2,000 m a.s.ℓ. of 1,000-1,500 mm.

precipitation of about 2,000 mm, is exemplified by the regions of Engelberg in central Switzerland, the upper Reuss valley in the Bernese Alps, and by wide parts of the Alps of Uri and Glarus. A surface trend analysis of snow line altitudes in the Swiss Alps (Müller et al. 1976) shows low values in the regions considered, and the equilibrium line on glaciers indeed descends below 2,600 m a.s.ℓ. T_E is about -1° to -2.5°C, and below the bergschrund, in most cases, the glaciers are temperate throughout. Balance gradients show annual values close to 1 m/100 m (Kasser 1981, p. 123: Silvretta) and basal shear stresses in large glaciers are between 1 and 1.5 bars (Unteraargletscher, Rhonegletscher). As the equilibrium line is very close to the lower boundary of discontinuous permafrost, the presence of permafrost is essentially limited to rock outcrops such as summits, sharp crests, and very steep walls, all within the accumulation area of glaciers. Active rock glaciers and push moraines s.s. are absent because unconsolidated sediments are not in a perennially frozen state. The firn of the accumulation area of glaciers is temperate to an altitude of about 3,500 m a.s.ℓ., at which level the mean annual air temperature is around -8°C and permafrost starts to be continuous. The landscape above the timberline is characterized by active, temperate glaciers, nonfrozen sediments, and bedrock.

Glaciological mountain research in the western world has traditionally concentrated on this "humid margin" type of landscape in which the most spectacular glaciers are found. In the corresponding literature, low-latitude mountains are often generally assumed to be free from permafrost and to be covered with temperate glaciers only. The resulting discrepancies in the identification of rock glaciers and debris-covered glaciers has mainly evolved from this viewpoint.

The Mischabel group within the Wallis Alps or the mountains around the Engadin Valley in the Alps of Grisons are good examples of the relatively *"dry interior"* of the Alps. Annual precipitation is close to, or below, 1,000 mm and the equilibrium line rises to 3,000 m a.s.ℓ. and higher. T_E is about -5° to -7°C. Between the equilibrium line and the lower boundary of discontinuous permafrost (2,400 m a.s.ℓ.), ice temperatures are between 0°C and about -4°C, and many glaciers ending in this belt are probably frozen to their beds at the margins, as is the case for the Grubengletscher (Haeberli 1976). In the driest spots, the equilibrium line approaches the -8°C isotherm of mean annual air temperature, and here firn starts to be cold and permafrost is continuous. Balance gradients are 0.5-1 m/100 m annually and shear stresses in glaciers of considerable size are around 0.8-1.2 bars (Grubengletscher, Griesgletscher: Kasser 1981, p. 123 and author's unpublished data). Large amounts of unconsolidated sediments between the equilibrium line and the lower boundary of discontinuous permafrost are in a perennially frozen state and are supersaturated with ice. Creep of such sediments at a rate which is roughly 1 order of magnitude less than the creep of glacier ice is frequent and leads to the formation of numerous rock glaciers. Push moraines s.s. are abundant in places where partially cold glaciers and perennially frozen sediments are, or have been, in direct contact with one another (many transitions between push moraines and rock glaciers: Figure 3).

For dry, low latitude mountains of the western world there is only scant information available in this context. However, more observations are reported from Asia, where arid mointains are widespread (e.g., Krenke 1975, Gorbunow 1978, Huang et al. 1982, Cheng and Wang 1982, Huang and Sun 1982).

CONDITIONS IN THE PAST

By going back in time, we leave the world of measurements and soundings and we enter the delicate

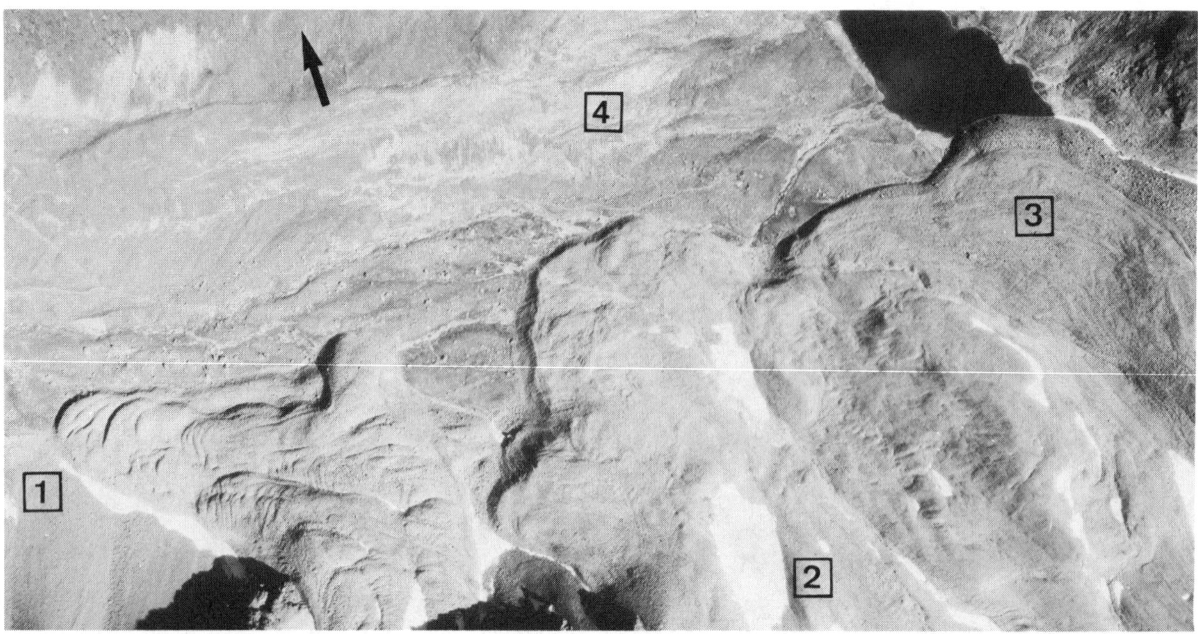

FIGURE 3 Val Muragl, Upper Engadin Valley - an example from the "dry interior". Here 1, active rock glacier; 2, fluted ground moraine, indicating that the perennially frozen sediments were overridden by a partially cold glacier during postglacial times; 3, holocene push moraine (upslope deformation, caused by glacier flow, of perennially frozen sediments during postglacial times) with transitions to rock glacier flow; 4, late glacial lateral moraine showing signs of past permafrost creep. Arrow points to north. T_E in the region is around -5°C, corresponding to an annual precipitation at 2,000 m a.s.ℓ. of about 800-1,000 mm.

field of estimations and analogies. Some features are nevertheless striking and therefore worthy of mention here. Since climatic conditions during *holocene time* probably did not differ much from those of the last few centuries (Patzelt 1977, Gamper and Suter 1982), they need no special attention in this paper.

The picture changes abruptly during the time of the *late-Würm glacier retreat* (10,000-15,000 years B.P., Maisch 1982). The mean annual air temperature during the Younger Dryas cold-phase (10,000-11,000 years B.P.) was lower than that of today by about 2.5°-4°C. As outlined by Kerschner (1982, 1983), precipitation in regions of the "humid margin" described above did not differ too much from today's values but was reduced to about 60-70% of those of today in the "dry interior" (cf. Burga 1982, p. 177). Thus the climatic contrasts seem to have been considerably greater; permafrost boundaries were lower by 400 m or more, whereas the depression of the equilibrium line in the sheltered interior regions was only about 250 m or less. T_E in these interior regions was lower than today by about 1°-2°C, shear stresses in glaciers were slightly reduced (Haeberli 1982, Maisch and Haeberli 1982), the vertical extent of zone C (Figure 1) was enhanced, permafrost phenomena such as rock glaciers were widespread (Maisch 1981), and most of the glacier tongues may have been partially cold and frozen to the bed at the margins. Less information is available at present about earlier stages of glacier retreat (Daun/Gschnitz, Pre-Bölling, 13,000-15,000 years B.P.) when a depression of the mean annual air temperature by at least 4.5°-5°C took place, a fact which is inferred from the depression of the boundary of permafrost and rock glacier occurrence (Kerschner 1983). Shear stresses in glaciers of comparable size were definitely smaller than during postglacial times by about 0.3-0.5 bar in the interior regions. This indicates probably an even more continental climate than that of the Younger Dryas time period (Maisch and Haeberli 1982, p. 124), with even lower T_E values, lower ice temperatures, lower balance gradients, and less active glaciers. Firn or ice covered many steep slopes favourable for rock glacier formation. Since push moraines form independently of local slope conditions, through shearing at the beds of partially cold glaciers, they are probably better indicators of permafrost boundaries during this time. A distinct lack of information exists to date about glacier-permafrost relationships during earlier periods of Oldest Dryas time.

Extreme conditions must have existed during and shortly after the time of *maximum glacier advance within the Würm ice age* (about 18,000 years B.P.). Large push moraines and periglacial ice wedges are the predominant indicators of permafrost (e.g. Schindler et al. 1978, Van der Meer 1980), because only flat lowlands were free from glacier ice; rock glaciers therefore only occurred in exceptional cases (e.g., Van Husen 1981). The lower boundary of continuous permafrost reached beyond Switzerland (Kaiser 1960, Washburn 1979), whereas equilibrium lines on glaciers were around 1,000-1,500 m a.s.ℓ. (Jäckli 1970, Keller and Krayss 1980). The depression of permafrost boundaries by more than 3,000 m corresponds to a lowering of the mean annual air temperature by at least 15°C compared to today's. These boundaries were depressed at least twice as much as the equilibrium lines on glaciers, and T_E is estimated to have been -10° to -15°C. Surface temperatures were well below 0°C, not only in the ice of the ablation areas but probably also throughout the firn of the accumulation areas. Geothermal heat flow and frictional heat production must have nevertheless led to phase equilibrium temperatures at the beds, near the fronts, of piedmont glaciers (Blatter and Haeberli, in preparation). Shear stresses in these piedmont glaciers amounted to only about 0.3-0.5 bar, a stress condition which can be more easily explained by the present concept, than by extraordinary deformation of unconsolidated, nonfrozen sediments at the glacier bed (cf. Boulton and Jones 1979, Haeberli 1981). Balance gradients and ice flow must have been at a very low level. In agreement with palynological reconstructions, precipitation must be assumed to have been in the order of 30% or less of today's values (Frenzel 1980, Haeberli 1982).

IFS AND BUTS

As is the case with every generalization, the proposed concept cannot hope to explain all the intricacies of natural phenomena. It can nevertheless be considered as a guideline for the interpretation of distribution patterns of permafrost and glaciers. These two phenomena are not only related to each other but also to climate in many ways. Thus a combined consideration of all three may furnish improved knowledge about paleoclimates and present ice conditions on earth.

ACKNOWLEDGEMENTS

Through direct contacts, many colleagues have contributed significantly to the clarity of the author's thinking. H. Röthlisberger, head of the Section for Glaciology at the Versuchsanstalt für Wasserbau, Hydrologie und Glaziologie, ETH Zürich, and P. Alean, who assisted in the preparation of the manuscript, receive my thanks in the name of all those not cited in the reference list.

REFERENCES

Ahlmann, H. W, son, 1924, Le Niveau de glaciation comme fonction de l'accumulation d'humidité sous forme solide, Méthode pour le calcul de l'humidité condensée dans la haute montagne et pour l'étude de la fréquence des glaciers: Geografiska Annaler Arg. VI, H. 1, p. 223-272.

Ahlmann, H. W, son, 1933, Present glaciation round the Norwegian Sea: Geografiska Annaler Arg. XV, H. 1, p. 316-348.

Barsch, D., 1978, Rock glaciers as indicators for discontinuous alpine permafrost, An example from the Swiss Alps: Proceedings of the Third International Conference on Permafrost, NRC-Ottawa, V. 1, p. 349-352.

Barsch, D., Fierz, H. and Haeberli, W., 1979, Shallow core drilling and bore hole measurements in the permafrost of an active rock glacier near the Grubengletscher, Wallis, Swiss Alps: Arctic and Alpine Research, V. 11, no. 2, p. 215-228.

Blatter, H. and Haeberli, W., (in preparation), Problems of modelling temperature distribution in Alpine glaciers: submitted to Annals of Glaciology (Symposium on Ice and Climate Modelling).

Boulton, G. S. and Jones, A. S., 1979, Stability of temperate ice caps and ice sheets resting on beds of deformable sediments: Journal of Glaciology, V. 24, no. 90, p. 29-43.

Brown, R. J. E. and Péwé, T. L., 1973, Distribution of permafrost in North America and its relationships to the environment: A review, 1963-1973: Second International Conference on Permafrost, North American Contribution, NAS, Washington D.C., p. 71-100.

Burga, C. A., 1982, Pollenanalytical research in the Grisons (Switzerland), An overview from 1929 to 1981 and some results concerning the history of vegetation since the Late Glacial: Vegetatio 49, p. 173-186.

Cheng, G. and Wang, S., 1982, On the zonation of high altitude permafrost in China: Journal of Glaciology and Cryopedology, V. 4, no. 2, p. 1-17.

Fisch, W., Sr., Fisch, W., Jr. and Haeberli, W., 1978, Electrical D.C. resistivity soundings with long profiles on rock glaciers and moraines in the Alps of Switzerland: Zeitschrift für Gletscherkunde und Glazialgeologie, Bd. XIII, H. 1/2, p. 239-260.

Frenzel, B., 1980, Klima der letzten Eiszeit und der Nacheiszeit in Europa: Veröffentlichung Joachim Jungius-Gesellschaft der Wissenschaft Hamburg 44, p. 9-46.

Gamper, M. and Suter, J., 1982, Postglaziale Klimageschichte der Schweizer Alpen: Geographica Helvetica Jhg. 37, Nr. 2, p. 105-114.

Gorbunow, A. P., 1978, Permafrost investigations in high mountain regions: Arctic and Alpine Research, V. 10, no. 2, p. 283-294.

Gross, G., Kerschner, H. and Patzelt, G., 1977, Methodische Untersuchungen über die Schneegrenze in alpinen Gletschergebieten: Zeitschrift für Gletscherkunde und Glazialgeologie, Bd. VII, H. 2, p. 223-251.

Haeberli, W., 1975, Untersuchungen zur Verbreitung von Permafrost zwischen Flüelapass und Piz Grialetsch (Graubünden): Mitteilung der Versuchsanstalt für Wasserbau, Hydrologie und Glaziologie der ETH Zürich, Nr. 17, 221 p.

Haeberli, W., 1976, Eistemperaturen in den Alpen: Zeitschrift für Gletscherkunde und Glazialgeologie, Bd. XI, H. 2, p. 203-220.

Haeberli, W., 1978, Special aspects of high mountain permafrost methodology and zonation in the Alps: Third International Conference on Permafrost, NCR, Ottawa, Vol. 1, p. 378-384.

Haeberli, W., 1979, Holocene push moraines in alpine permafrost: Geografiska Annaler, V. 61A, no. 1-2, p. 43-48.

Haeberli, W., 1981, Ice motion on deformable sediments: Journal of Glaciology, V. 27, no. 96,

p. 365-366.

Haeberli, W., 1982, Klimarekonstruktionen mit Gletscher-Permafrost-Beziehungen: Materialien zur Physiogeographie, Basel, H. 4, p. 9-17.

Huang, M., Wang, Z. and Ren, J., 1982, Ice temperatures of glaciers in China: Journal of Glaciology and Cryopedology, V. 4, no. 1, p. 20-28.

Huang, M. and Sun, Z., 1982, Some flow characteristics of continental-type glaciers in China: Journal of Glaciology and Cryopedology, V. 4, no. 2, p. 35-45.

Jäckli, H., 1970, Die Schweiz zur letzten Eiszeit: Atlas der Schweiz, Blatt 6, Eidgenössische Landestopographie, Wabern, Bern.

Kaiser, L., 1960, Klimazeugen des periglazialen Dauerfrostbodens in Mittel- und Westeuropa: Eiszeitalter und Gegenwart 11, p. 121-141.

Kasser, P., 1981, Rezente Gletscherveränderungen in den Schweizer Alpen: Gletscher und Klima- Jahrbuch der Schweizerischen Naturforschenden Gesellschaft, wissenschaftlicher Teil, 1978, p. 106-138.

Keller, O. and Krayss, E., 1980, Die letzte Vorlandvereisung in der Nordostschweiz und im Bodensee-Raum (Stadialer Komplex Würm-Stein am Rhein): Eclogae geologiae Helveticae, V. 73/3, p. 823-838.

Kerschner, H., 1978, Paleoclimatic inferences from Late Würm rock glaciers, Eastern Central Alps, Western Tyrol, Austria: Arctic and Alpine Research, V. 10, no. 3, p. 635-644.

Kerschner, H., 1982, Outlines of the climate during the Egesen advance (Younger Dryas, 11,000-10,000 BP) in the Central Alps of the Western Tyrol, Austria: Zeitschrift für Gletscherkunde und Glazialgeologie, Bd. 16, H. 2, p. 229-240.

Kerschner, H., 1983, Late glacial paleotemperatures and paleoprecipitation as derived from permafrost-glacier relationships in the Tyrolian Alps (Austria): Fourth International Conference on Permafrost

Krenke, A.N., 1975, Climatic conditions of present-day glaciation in Soviet Central Asia: Snow and Ice Symposium (Proceedings of the Moscow Symposium, August 1971), IAHS-AISH Publication, no. 104, p. 30-41.

Kuhn, M., 1980, Climate and glaciers: Sea level, Ice and Climate change, IAHS-AISH Publication, no. 131, p. 3-20.

Kuhn, M., 1982, Die Reaktion der Schneegrenze auf Klimaschwankungen: Zeitschrift für Gletscherkunde und Glazialgeologie, Bd. 16, H. 2, p. 241-254.

Maisch, M., 1981, Glazialmorphologische und gletschergeschichtliche Untersuchungen im Gebiet zwischen Landwasser- und Albulatal (Kt. Graubünden, Schweiz): Physische Geographie, Universität Zürich, V. 3, 215 p.

Maisch, M., 1982, Zur Gletscher- und Klimageschichte des alpinen Spätglazials: Geographica Helvetica, Jhg. 37, Nr. 2, p. 93-104.

Maisch, M. and Haeberli, W., 1982, Interpretation geometrischer Parameter von Spätglazialgletschern im Gebiet Mittelbünden, Schweizer Alpen: Physische Geographie, Universität Zürich, V. 1, p. 111-126.

Markl, G. and Wagner, H. P., 1978, Messungen von Eis- und Firntemperaturen am Hintereisferner (Oetztaler Alpen): Zeitschrift für Gletscherkunde und Glazialgeologie, Bd. XIII, p. 261-265.

Meier, M. F., 1961, Mass budget of South Cascade Glacier, 1957-1960: U.S. Geological Survey Professional Paper 424 B, p. 206-211.

Müller, F., Caflisch, T. and Müller, G., 1976, Firn und Eis der Schweizer Alpen / Gletscher inventar: Geographisches Institut der ETH Zürich, Publ. Nr. 57, 174 p.

Paterson, W. S. B., 1981, The physics of glaciers (2nd ed.): Oxford, Pergamon Press, 380 p.

Patzelt, G., 1977, Der zeitliche Ablauf und das Ausmass postglazialer Klimaschwankungen in den Alpen, in Frenzel, B. (ed.), Dendrochronologie und postglaziale Klimaschwankungen in Europa: Wiesbaden, Franz Steiner Verlag, p. 248-259.

Péwé, T. L., 1966, Paleoclimatic significance of fossil ice wedges: Biuletyn Peryglacjalny 15, p. 65-73.

Schindler, C., Röthlisberger, H. and Gyger, M., 1978, Glaziale Stauchungen in der Niederterrassen-Schottern des Aadorfer Feldes und ihre Deutung: Eclogae geologiae Helveticae, V. 71/1, p. 159-174.

Schytt, V., 1967, A study of "ablation gradient": Geografiska Annaler, V. 49A, no. 2-4, p. 327-332.

Shi, Y. and Li, J., 1981, Glaciological research of the Qinghai-Xizang Plateau in China: Proceedings of Symposium on Qinghai-Xizang (Tibet) Plateau (Beijing, China), V. II, p. 1589-1597.

Shumskii, P.A., 1964, Principles of structural glaciology: New York, Dover Publications, 497 p.

Süsstrunk, A., 1951, Sondages du glacier par la méthode sismique: La Houille Blanche, Numéro Spécial A/1951, p. 309-319.

Van der Meer, J. J. M., 1980, Different types of wedges in deposits of Würm age from the Murten area (Western Swiss Plain): Eclogae geologiae Helveticae 73/3, p. 839-854.

Van Husen, D., 1981, Geologisch-sedimentologische Aspekte im Quartär von Oesterreich: Mitteilungen der Oesterreichischen Geologischen Gesellschaft 74/75, p. 197-230.

Waechter, H. P., 1981, Eisdickenmessungen auf dem Rhonegletscher - ein Versuch mit Radio-Echo Sounding: Diplomarbeit am Geographischen Institut der ETH Zürich, 67 p.

Washburn, A.L., 1979, Geocryology-a survey of periglacial processes and environments: London, Edward Arnold Ltd., 406 p.

STRATIGRAPHIC DISTRIBUTION OF PERIGLACIAL FEATURES INDICATIVE OF PERMAFROST IN THE UPPER PLEISTOCENE LOESSES OF BELGIUM

Paul Haesaerts

Koninklijk Belgisch Instituut voor Natuurwetenschappen, Vautierstraat 29, B-1040 Brussels, Belgium

Among the various types of periglacial features known within the loesses of Belgium, only ice-wedge casts and tundra gleys, mainly those associated with an ice-segregation structure, may be considered as indicative of permafrost. The distribution of those features in the upper Pleistocene loesses of Belgium implies that the most widespread episodes of permafrost formation took place at intervals, between 33,000 and 20,000 years B.P., before and during the sedimentation of the late Weichselian loess cover. Permafrost conditions did also occur at the beginning of the middle Weichselian (around 60,000 years B.P.) and during the second half of the late Weichselian (around 15,000 years B.P.).

INTRODUCTION

In Belgium, two loessic deposits, named Brabantian and Hesbayan, were recognized within the upper Pleistocene. Assigned to the late Weichselian and to the middle Weichselian respectively, they are separated by a cryoturbated soil horizon named Kesselt Soil (Gullentops 1954). This has been correlated with the cryoturbated peaty horizon of Zelzate (Flanders) dated 28,000 ± 400 years B.P. (Paepe and Vanhoorne 1967), and subsequently used as a major marker bed for correlations with northwestern France (Paepe and Sommé 1970, Lautridou and Sommé 1974, 1981).

Considering the distribution of desert pavements and large frost wedges within the loessic deposits, those authors concluded that two major cold episodes occurred during the upper Pleistocene. The first has been placed during the beginning of the middle Weichselian, approximately around 55,000 years B.P., and the second one during the late Weichselian, in the interval between 26,000 and 14,000 years B.P. More recently, Kolstrup (1980), having reviewed the main wedge cast localities of northwestern Europe, considered that the major period of continuous permafrost occurred between 23,000 and 19,000 years B.P.

Since 1970 new data have been obtained for the upper Pleistocene loesses of Belgium, especially regarding fossil soils, periglacial features, archeology and tephrostratigraphy (Haesaerts and Van Vliet 1974, 1981, Haesaerts and de Heinzelin 1979, Haesaerts et al. 1981).

From these data it can be considered that the upper part of the Hesbayan loess sequence, as defined by Gullentops (1954), belongs to the late Weichselian, and not to the middle Weichselian as previously stated (Haesaerts et al. 1981). As a result, a new stratigraphic sequence has been developed (Figures 1 and 3) suggesting a more consistent distribution of periglacial features which could be used as markers for long distance correlations between Belgium and northwestern France.

INDICATORS OF PERMAFROST IN LOESS

Various types of periglacial features are known within the upper Pleistocene loesses, namely involutions, large wedges (isolated or in polygons networks), small fissures (isolated or in a row), desert pavements, hummocks, tundra gleys, and solifluction structures. The genesis and the climatic environment of these features were discussed during the field meeting of the I.G.U. Commission for Periglacial Research held in Belgium and in The Netherlands in the autumn of 1978. As far as loess formations are concerned, only tundra gleys, mainly those associated with ice-segregation structures, and ice-wedge casts were considered as permafrost indicators (Haesaerts and Van Vliet 1981).

Tundra gleys

Those soils, also called "Nassboden" or "nanno-podzols" (Gullentops 1954), are well represented within the upper Pleistocene loess sequence.

They usually show a light grey horizon, the thickness of which may vary from 0.1 to 0.5 m, resting in a brown-yellow iron enriched B horizon (see units H.C.6-H.C.3, Figure 3). The lower part of the greyish horizon generally presents a well developed fine foliated platy structure; downwards, this structure often abruptly grades into a more coarse lenticular to subangular blocky structure usually with iron coatings. Sometimes, a coarse prismatic structure is present, starting from the bottom of the greyish horizon.

By comparison with present-day arctic soils (Tedrow 1968), the bleaching of the upper horizon, due to a release of iron, requires moisture and a low surface pH. Considering the high permeability of the loamy or loessic sediments on which those soils developed, this process necessarly implies the presence of a permafrost underneath.

This interpretation is supported by the presence of a laminated structure across the soil profile. It has been demonstrated that the upper platy structure did originate from repeated seasonal frost, while the underlying subangular blocky structure resulted from the slow growth of ice lences into the upper part of the permafrost (Van Vliet 1976, Haesaerts and Van Vliet 1981). Therefore the abrupt transition between both types of structure at the bottom of the iron-enriched horizon most probably corresponds to the mean position

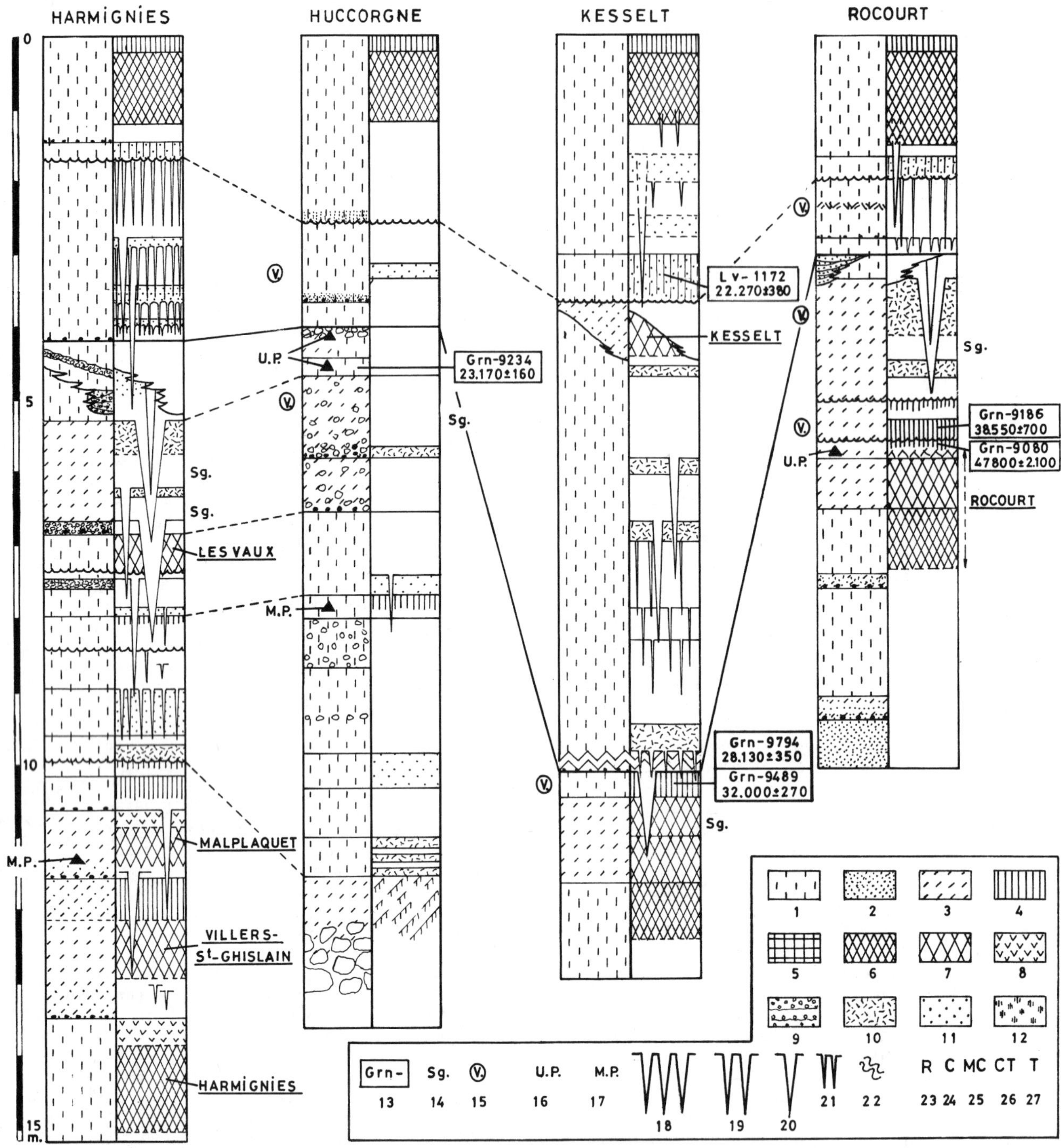

FIGURE 1 Lithostratigraphic correlations of upper Pleistocene type-sections from Belgium. Left part of each column: lithostratigraphy; right part of each column: pedology and periglacial features. Legend : 1, loess; 2, sand; 3, colluviated sediment; 4, humic sediment; 5, peat; 6, Alfisoil (Bt horizon); 7, weakly developed Alfisoil (B horizon); 8, eluvial horizon; 9, chalk or limestone; 10, bleached or reduced sediment; 11, weakly bleached or reduced sediment; 12, iron staining; 13, radiocarbon date; 14, ice-segregation structure; 15, volcanic tuff layer; 16-17, artifacts (U.P.: upper Paleolithic; M.P.: middle Paleolithic); 18, generally occurring polygons; 19, occasional polygons; 20, isolated ice-wedge casts; 21, narrow frost-wedges in a row; 22, solifluction structures; 23-27, paleoclimatic curve based on data derived from fossil soils, periglacial features and from paleontological, palynological and sedimentological analysis: R, rigorous climate (with continuous or discontinuous permafrost); C, cold climate (without active permafrost); MC, moderately cold climate; CT, cool temperate climate; T, temperate climate.

of the permafrost table.

This type of soil profile has been usually observed on loamy sediments, in various topographical environments such as valley-floors, well drained plateaus, as well as on hill-slopes. In very dry environments, as on pure loesses, the structure is usually absent, while the upper greyish horizon may be less developed. In both situations, ice-wedge casts start from the top of the greyish horizon and confirm the relationship between tundra gley and permafrost.

Ice-wedge casts

Most of the large wedge-like structures examined by us within the upper Pleistocene loess sequence have a loessic or loamy filling. Wedges with a sandy or a gravelly infilling are also recorded but are restricted to the upper and the lower part of the sequence.

Almost all the large wedge-like structures observed within the loess possess characteristics generally attributed to ice-wedge casts (cf. Washburn 1979, Black 1976), and therefore may be considered as indicators of former permafrost conditions.

All wedges appear to be epigenetic, although in several localities a cone-in-cone superposition has been observed. In Harmignies (Haine Valley), where one of the most complete upper Pleistocene loess sequence has been recorded (Haesaerts 1974, Haesaerts and Van Vliet 1974), at least nine generations of wedges are present (Figure 1, column 1). The distribution of the wedges, as well as the geometry of the surmounting layers, likely indicate that each generation has been filled with loess or loam before the development of the following generation.

Since the infilling of an ice-wedge by overlying sediments requests a complete melting of the ice-core as well as the degradation of the surrounding permafrost (Black 1976), each generation of ice-wedge casts corresponds to one distinct episode of permafrost development.

In Harmignies, as well as in Rocourt and in Kesselt (Figure 1), a different type of large wedge-like structure has been observed within the late Weichselian loess cover. It results from the coalescence of several long and narrow fissures which form polygons between 0.2 to 0.5 m in diameter (see Figure 2, top of unit K.B.). Most probably originated from dessication (Black 1976), this type of structure is not indicative of permafrost. Nevertheless, in Harmignies, the narrow fissures always start from the lower part of weakly developed tundra gleys which were probably formed under permafrost conditions as suggested by the presence of occasional ice-wedge casts at this level.

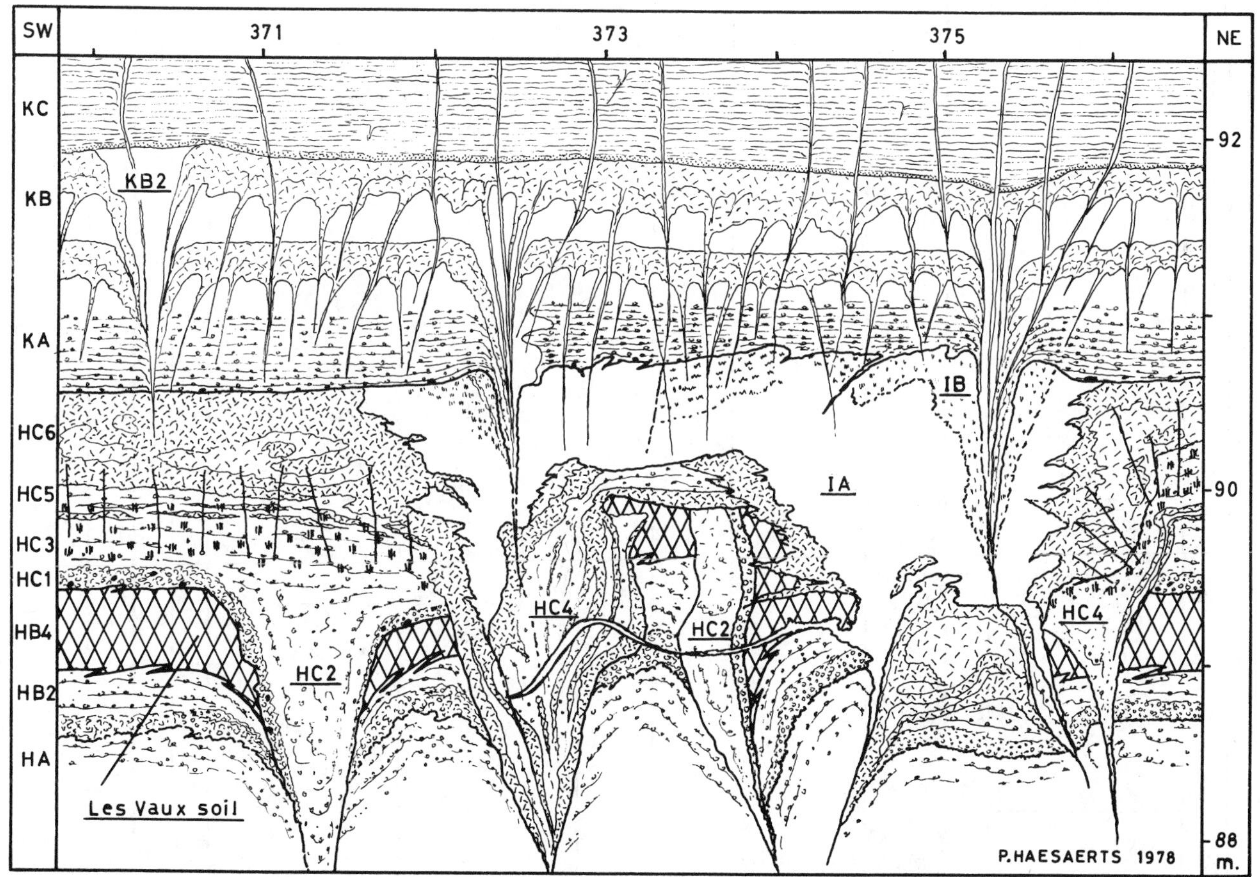

FIGURE 2 Harmignies (Belgium): distribution of ice-wedge casts within the upper part of the Middle Weichselian deposits (same symbols as for Figure 1).

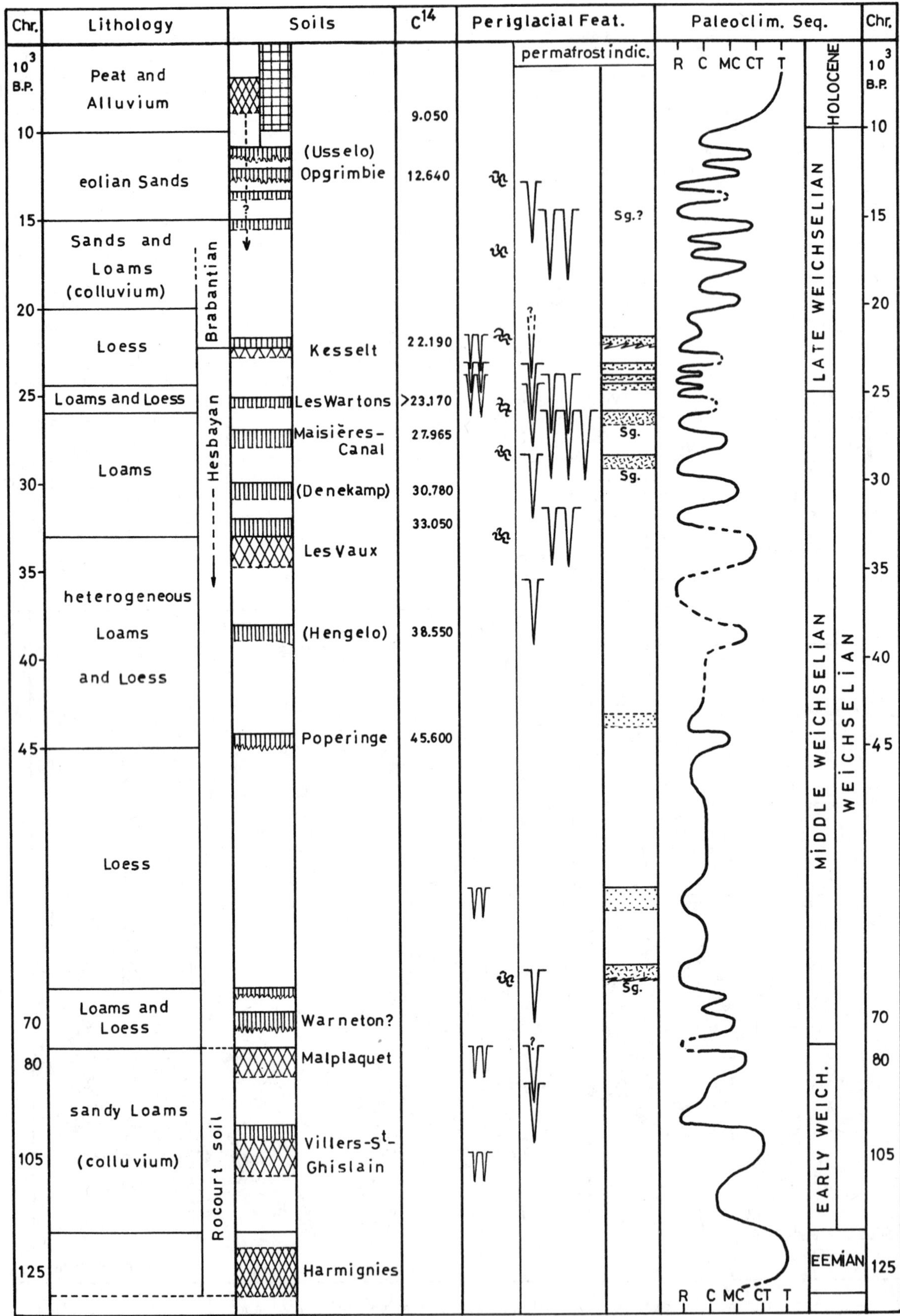

FIGURE 3 Composite section of the Weichselian loess deposits of Belgium (symbols: see Figure 1)

STRATIGRAPHIC AND REGIONAL DISTRIBUTION OF PERMAFROST INDICATORS

Figure 3 shows a lithostratigraphical sequence of the upper Pleistocene loessic deposits of Belgium. Also shown is the distribution of fossil soils, radiocarbon dates, periglacial features indicative of permafrost and proposed paleoclimatic curve. The curve is based on data derived from fossil soils, periglacial features, and from sedimentological, palynological and paleontological analysis (Haesaerts 1974, Haesaerts and Van Vliet 1981). The paleoclimatic curve fluctuates from rigorous climate (left) with discontinuous or continuous permafrost, to temperate climate (right) similar to the present-day climate of northwestern Europe.

In most of the loess sections of Belgium, the early Weichselian is poorly recorded, except at Harmignies where it is represented by two distinct loamy colluvial deposits, each having a grey-brown podzolic soil developed on top (Villers-Saint-Ghislain Soil and Malplaquet Soil, Figure 1, column 1). During this period, humid and rather mild climatic conditions probably prevailed; however, several colder episodes occurred as the small ice-wedge casts present in the lower part of the second colluvial deposit suggest.

During the lower part of the middle Weichselian (approximately 70,000 to 45,000 years B.P.), a first loess cover, locally preserved, was deposited under cold climatic conditions. In Harmignies, two tundra gleys are preserved in the lower part of the loess cover, the lower one, locally associated with small ice-wedge casts, being the most developed. Tundra gleys were also observed, in a similar stratigraphic context, in the Méhaigne Valley at Huccorgne (Figure 1, column 2) and in Normandy (section of Mesnil-Esnard, author's observations); this episode of permafrost development is stratigraphically placed around 60,000 years B.P.

During the central part of the middle Weichselian (approximately 45,000 to 33,000 years B.P.), mild and humid climatic conditions probably existed (Zagwijn and Paepe 1968); this period is poorly represented in the loess sequence because of the dominance of erosional processes

The main period of permafrost development undoubtedly took place during the upper part of the middle Weichselian, between 33,000 and 25,000 years B.P. In Harmignies, at least three distinct networks of ice-wedge polygons, 10 to 30 m in diameter are recorded (Figure 2, units H.C.2, H.C.4 and I.A.). The upper ice-wedge polygons (I.A.), associated with a tundra gley (H.C.6), are the most visible and were stratigraphically situated around 26,000 years B.P., just before the deposition of the late Weichselian loess cover (Figure 2, units K.A. to K.C.).

In a similar stratigraphic position, tundra gleys and large ice-wedge polygons were also recorded in most of the loess sections of Belgium and northwestern France, in plateau position as well as on valley-floor (Figure 4). From north to south, the maximum width of the ice-wedge casts varies from 1.5 to 0.1 m, while the thickness of the active layer is estimated to be between 0.8 and 0.4m.

On this basis, it is likely that around 26,000 years B.P., a continuous permafrost extended across northwestern Europe, probably up to northern France. Considering the present-day climatic environment necessary for ice-wedge polygons to form (Washburn 1979), one may conclude that at that time mean annual air temperatures were lower than -6°C and mean January temperatures were close to -20°C. On the other hand, the maximum width of the ice-wedge casts (1.5 m in Harmignies) suggests that permafrost was present for a considerable period of time, probably more than one thousand years.

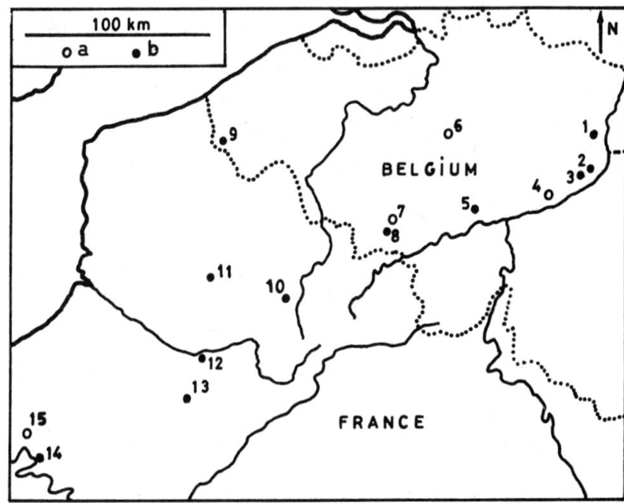

FIGURE 4 Distribution within the loess area of the permafrost indicators reported for the period 26,000-25,000 years B.P. (author's observations). Legend: a, tundra gley; b, tundra gley associated with ice-wedge polygons; 1, Kesselt; 2, Rocourt; 3, Ans; 4, Huccorgne (valley-floor); 5, Tongrinne; 6, Wezembeek-Oppem; 7, Maisières-Canal (valley-floor); 8, Harmignies; 9, Poperinge; 10, Marcoing; 11, Beaumetz-les-Loges; 12, Corbie; 13, Sourdon; 14, Mesnil-Esnard; 15, Roumare.

During the first part of the late Weichselian (25,000 to 20,000 years B.P.), the deposition of the upper loess cover took place; permafrost recurred several times up to Normandy, as shown by the distribution of the tundra gleys within the loess cover. However, climatic conditions were dryer, as suggested by the weak development of the tundra gleys associated with narrow dessication fissures, and the rare occurrences of ice-wedge polygons.

Finally, the last important period of permafrost conditions is situated stratigraphically around 15,000 years B.P.; it is recorded on the valley-floor of the Haine river at Maisières-Canal, near Mons, where a network of ice-wedge polygons, 10 to 20 m in diameter, is present at the base of the late Weichselian coversands (Haesaerts and de Heinzelin 1979, Haesaerts and Van Vliet 1981). Furthermore, evidences of permafrost seem also to be present within most of the Alfisoils developed on top of the late Weichselian loess cover in Belgium (Van Vliet and Langhor 1981).

CONCLUSIONS

The distribution of permafrost indicators within the upper Pleistocene loess sequence of Belgium and surrounding areas, implies that many relatively short episodes of permafrost formation occurred across Belgium and northwestern France during the Weichselian. The most widespread episodes took place at intervals, between 33,000 and 20,000 years B.P., mainly before and during the sedimentation of the late Weichselian loess cover. In Belgium and in northern France at least one episode of continuous permafrost existed around 26,000 years B.P.

Permafrost conditions did also occur at the beginning of the middle Weichselian (around 60,000 years B.P.) and during the second half of the late Weichselian (around 15,000 years B.P.). Furthermore, it seens that, at least in the loess area of Belgium, main periods of permafrost development always precede important episodes of eolian sedimentation.

ACKNOWLEDGMENTS

We thank H.M. French for useful comments and for the critical review of this paper.

REFERENCES

Black, R. F., 1976, Periglacial features indicative of permafrost: Ice and soil wedges: Quaternary Research, v. 6, p. 3-26.

Gullentops, F., 1954, Contribution à la chronologie du Pléistocène et des formes du relief en Belgique: Mémoires de l'Institut de Géologie de l'Université de Louvain, v. 18, p. 123-252.

Haesaerts, P., 1974, Séquence paléoclimatique du Pléistocène supérieur du bassin de la Haine (Belgique): Annales de la Société Géologique de Belgique, v. 97, p. 151-164.

Haesaerts, P., and de Heinzelin, J., 1979, Le site paléolithique de Maisières-Canal: Dissertationes Archaeologicae Gandenses, v. 19: Brugge.

Haesaerts, P., Juvigné, E., Kuyl, O., Mucher, H., and Roebroeks, W., 1981, Compte rendu de l'excursion du 13 juin 1981, en Hesbaye et au Limbourg Néerlandais, consacrée à la stratigraphie des loess du Pléistocène supérieur: Annales de la Société Géologique de Belgique, v. 104, p. 223-240.

Haesaerts, P., and Van Vliet, B., 1974, Compte rendu de l'excursion du 25 mai 1974 consacrée à la stratigraphie des limons aux environs de Mons: Annales de la Société Géologique de Belgique, v. 97, p. 503-516.

Haesaerts, P. and Van Vliet, B., 1981, Phénomènes périglaciaires observés à Maisières-Canal, à Harmignies et à Rocourt (Belgique): Biuletyn Peryglacjalny, v. 28, p. 291-324.

Kolstrup, E., 1980, Climate and stratigraphy in northwestern Europe between 30,000 and 13,000 B.P., with special reference to The Netherlands: Mededelingen rijks Geologische Dienst, v. 32-15, p. 181_251.

Lautridou, J.P., and Sommé, J., 1974, Les loess et les provinces climato- sédimentaires du Pléistocène supérieur dans le Nord-Ouest de la France. Essai de corrélation entre le Nord et la Normandie: Bulletin de l'Association Française pour l'étude du Quaternaire, v. 40-41, p. 237-241.

Lautridou, J.P., and Sommé, J., 1981, L'extension des niveaux-repères périglaciaires à grandes fentes de gel de la stratigraphie du Pléistocene récent dans la France du Nord-Ouest: Biuletyn Peryglacjalny, v. 28, p. 179-185.

Paepe, R., and Sommé, J., 1970, Les loess et la stratigraphie du Pléistocène récent dans le Nord de la France et en Belgique: Annales de la Société Géologique du Nord, v. 90, p. 191-201.

Paepe, R., and Vanhoorne, R., 1967, The stratigraphy and palaeobotany of the Late Pleistocene in Belgium: Mémoires pour servir à l'explication des Cartes géologiques et minières de la Belgique, n° 8, Service Géologique de Belgique, Bruxelles.

Tedrow, J., 1968, Pedogenic gradients in polar regions: Journal of Soil Science, v. 19, p. 197-204.

Van Vliet, B., 1976, Traces de ségrégation de glace en lentilles associées au sol et phénomènes périglaciaires fossiles: Biuletyn Peryglacjalny, v. 26, p. 41-55.

Van Vliet, B., and Langhor, R., 1981, Correlation between fragipans and permafrost with special reference to silty weichselian deposits in Belgium and northern France: Catena, v. 8, p. 137-154.

Washburn, A.L., 1979, Geocryology: Halsted Press, New York.

Zagwijn, W., and Paepe, R., 1968, Die Stratigraphie der weichselzeitlichen Ablagerungen der Niederlande und Belgiens: Eiszeitalter und Gegenwart, v. 19, p. 129-146.

PERIGLACIAL PHENOMENA IN ARID REGIONS OF IRAN

Horst Hagedorn

Geographisches Institut, Universität Würzburg
Am Hubland, D-8700 Würzburg, Federal Republic of Germany

Field work in the Shir-Kuh Mountains (el. 4,055 m) of central Iran included studies of periglacial phenomena as well as of the present and Pleistocene limits of patterned ground and the snow line.

The climate is semi-arid. Average annual precipitation is about 200 mm at 2,000 m and about 400mm at 4,000 m. Precipitation occurs almost exclusively in winter in the form of snow. The summer months are characterized by low cloud cover, and thus high solar radiation.

All of the mountain area possesses Pleistocene glacial landforms down to 1,900 m, as well as present periglacial forms above 2,000 m. Four groups of periglacial features are distinguished:

1. On the Shir-Kuh summit plateau at 4,055 m stone circles, stone stripes and solifluction lobes occur, comparable to alpine forms. Stone circle diameters are 50 - 100 cm. Limestone frost debris is upturned and well oriented.

2. Between 3,400 and 4,000 m, smooth slopes ("Glatthänge") cut across limestone strata. Slope inclinations are between 35° and 40°. They are interpreted as periglacial forms.

3. At the foot of limestone outcrops, between 3,000 and 3,400 m, there are frequent occurrences of vegetation garlands of periglacial origin. Stone stripes are also common and extend down to 2,500 m.

4. Occurrences of patterned ground are found as low as 2,100 m. With the exception of those on very suitable substrates they appear to be convergence forms due to wetting and drying of clays and salts.

There is some evidence for progressive divergence of the snow line and the lower limit of patterned ground towards the arid zones.

INTRODUCTION

There has been a long-standing and controversial discussion on the altitudinal range and lower limit of periglacial phenomena in arid-zone mountains. One opinion, as summarized by Troll (1944, 1947), holds that the lower limit of patterned ground rises and falls in accordance with the tree line and the snow line, thus not culminating in the humid tropics, but in the arid zones of both hemispheres. This view is challenged by Hövermann (1962) who asserts that there is a general rise of the lower limit of patterned ground from the poles to the equator, and that this trend is interrupted, or even reversed, in the arid zones. Since the pattern of altitudinal limits is important for the reconstruction of paleoclimates, there is a need to settle this controversy.

Present periglacial phenomena have been reported from many areas of Iran. Hövermann (1960) attributes an altitude of 2,000 m to the lower limit of patterned ground on the dry southern flank of the Alborz Mountains whereas Bobek (1937, 1952) places the same line 1,000 m higher. Similar figures are given for the Pleistocene depression of altitudinal limits in Iran. Bobek (1955) calculates a rather insignificant depression, interpreting this as evidence that more or less the same forces governed the arid highlands during the last glacial as today.

When members of the Department of Geography of the University of Würzburg did geomorphological fieldwork in the Shir-Kuh Mountains (Figure 1) of central Iran (elevation, 4,055 m; latitude 31° 30'N) from 1972 to 1977, they also collected information upon the present and Pleistocene positions of the lower limit of patterned ground and of the snow line.

GEOLOGY AND GEOMORPHOLOGY

The Shir-Kuh Mountains consist of a granite base topped by as much as 600 m of hard Cretaceous limestones and dolomites. The mountain block is tilted northward. Its sedimentary blanket covers most of the north flank, which is dissected by deep canyons. Limestone forms the highest elevations: 4,055 m at the Shir Kuh summit proper, and about 3,900 m in the Kuh-e-Barfkhaneh. Steep

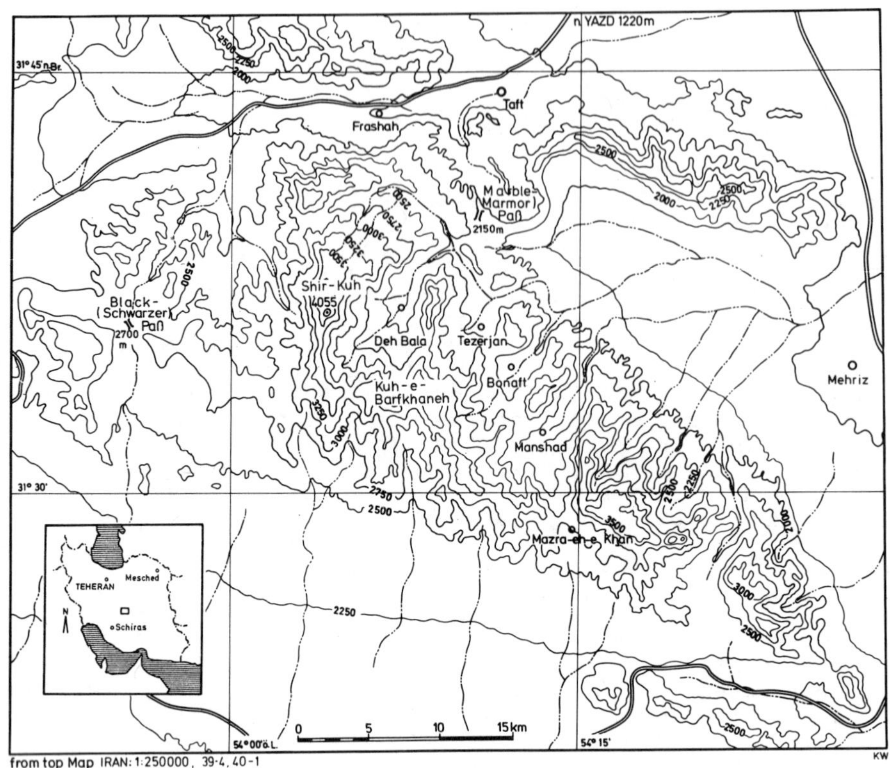

FIGURE 1 Topographic map of the Shir-Kuh Mountains, central Iran, and location of the study area within Iran.

walls plunge from all sides of the summits. The transition to the granitic basement, with steep slopes 30° - 40° in angle, cuts across a series of marls and conglomerates of variable thickness. Below the base of the Cretaceous outcrops V-shaped valleys are cut into the granite. The asymmetric profile of the mountains is further emphasized by the different heights of the foreland plains. On the north side this is the Mehriz plain below 1,900 m; its bedrock deeply buried by alluvial fan deposits. The southern foreland is a wide granitic planation surface, attaining 2,400 m at the foot of the mountain.

CLIMATIC CONDITIONS

There is no weather station within the mountains. The nearest station, Yazd (Figure 2) lies at 1,200 m on the east, or lee, side of the mountains. It receives 124 mm/yr of precipitation. The mountains are likely to be less arid. Incomplete records from a former mountain station, among other evidence, suggest 200 mm for 2,000 m elevation, and further increase to around 400 mm in the summit region. The higher parts of the mountains can thus be called semiarid.

Most of the precipitation falls during the winter, above 2,000 m almost exclusively as snow. There is a continuous snow cover at higher elevations each year from November to April. Snow patches survive until early June: On the north side of the Kuh-e-Barfkhaneh, which means 'snow house mountain', there are perennial firn patches in protected positions at 3,300 m.

Winter snowfalls are accompanied by low temperatures (Figure 2). Even at Yazd, January nighttime temperatures frequently fall below $-5^{\circ}C$. In the high valleys of Deh Bala and Tezerjan (2,200 - 2,500 m), ground temperatures below $-15^{\circ}C$ are not uncommon. For the development of periglacial features the beginning and end of winter are of prime importance. Alternating inroads of cold and warm air masses lead to frequent freeze-thaw cycles in areas above 2,500 m. During the winter the soil remains frozen at these altitudes, with occasional thaws on the south slopes.

The summer months, from June to September, are hot and practically precipitation-free. These are also the months with least cloud cover, and thus with high radiation and evaporation. These conditions result in the rather high elevation of the present snow line.

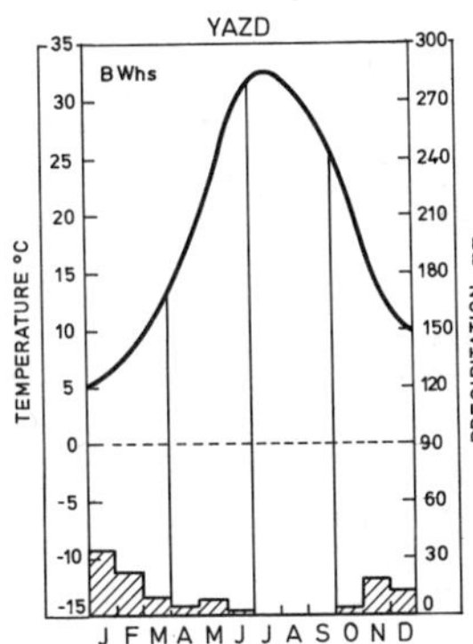

FIGURE 2 Climatic diagram of Yazd, elevation 1,200 m, NE of the Shir-Kuh Mountains (Djavadi, 1966).

PERIGLACIAL FEATURES

1. SHIR-KUH SUMMIT PLATEAU (4,000 - 4,055 m)

The highest elevation of the mountains is a narrow limestone plateau bounded by steep slopes on its south, west and east sides. The rock walls themselves are almost bare. At their base is a fringe of fresh talus cones indicating rapid frost weathering of the heavily jointed limestone walls. Freeze-thaw cycles are frequent and occasionally occur during the summer months.

The dominant periglacial features at this altitude are straight, smooth slopes ("Glatthänge"), that begin at the summit plateau. On S and SW exposures they extend down to about 3,700 m. All limestone strata are uniformly cut at an angle of 35° - 40°. The debris cover is thin, but highly mobile, indicating considerable denudation. According to Klaer (1962), Schweizer (1975) and others, these smooth slopes are typical for the periglacial altitudinal zone of the winter-cold continental mountains of the Near East.

Additional periglacial features occur on the Shir-Kuh summit plateau between 4,000 and 4,055 m. These include stone circles, stone stripes, solifluction lobes, and turf garlands. All are well developed over large areas and compare well with alpine periglacial features (Hagedorn et al. 1978).

The stone circles are 70 - 100 m in diameter. Their coarse angular slabs of limestone frost debris are clearly oriented and pushed into near-vertical positions. The average diameter of the rubble is 15 cm, that of the largest displaced slabs 25 cm. Sorting depth is 40 cm. On slopes of more than 5° the stone circles change to solifluction lobes.

Where the substrate is slightly more loamy, stone stripes a few metres in length occur on slopes of about 10° inclination. The stripes are well sorted into fine and coarse material. The latter have an average particle length of 15 cm and a sorting depth of as much as 30 cm.

The less inclined parts of the summit plateau bear a dense network of vegetation garlands. Strong root lengthening indicates that solifluction is quite efficient. The vegetation garlands are restricted to the windward side of the plateau where snow is blown off and which is thus rather dry. Stone circles concentrate in the lower lee positions and depressions. This clearly indicates the dependence of these features on moisture supply, which relates to differences of snow cover. A few snow patches survive well into summer.

2. KUH-E-BARFKHANEH (3,900 m)

The Kuh-e-Barfkhaneh is an isolated limestone mountain resting, just as the Shir-Kuh summit plateau, on granite. The less resistant marly strata of the basal Cretaceous beds are smoothly cut by slopes 30 - 40° in angle, in contrast to the limestone walls above and the densely dissected granite relief below.

In a few places at the foot of the north wall some snow survives even the hottest summer months. In August and September of 1974 and 1975 there were four perennial snow patches at 3,300 m. The largest was 100 x 30 x 5 m in dimensions and consisted of at least four annual firn layers. The snow patches are flanked by moraine ridges 7 - 10 m high which go down to the granite contact at 3,150 m, and continue into the steep ravines of the granite terrain. They can be traced to a 40 m high moraine in a bowlshaped valley at the foot of the Kuh-e-Barfkhaneh. This follows the valley down to 2,500 m and gives some impression of the remarkable Pleistocene glaciation of the mountains (Hagedorn et al. 1975).

The 30° - 40° slope of the Cretaceous base is covered by loamy debris, in some places by rockfall masses from the limestone wall. On the north side of the mountain there is always a pronounced nick between the wall and its foot, while there is a smooth transition on the south side. Especially in SW exposures periglacial smooth slopes ("Glatthänge") are inclined 35° - 40° across the limestone beds between 3,300 and 3,800 m elevation.

Smaller periglacial features, such as soil stripes and vegetation terraces, are also present. The biggest area of soil stripes lies in a SW exposure underlain by Cretaceous rocks at an elevation of 3,250 m. On an area of 3,000 m^2 stripes of walnut-size limestone rubble with a diameter between 2 and 3 cm alternate with the fines from the underlying basal beds. The stripes are spaced at 15 - 20 cm. Sorting depth is 20 cm; the slope is 25°. They typically bend around larger obstacles, where they get closer to each other, but never unite.

Vegetation terraces occur at the same height (3,100 - 3,300 m) on the slopes of the Cretaceous base, preferably on the sunprotected and thus more humid north side. Aside from some grass the patchy vegetation mainly consists of woody shrubs the roots of which are lengthened downslope. This indicates that considerable debris transport by solifluction still occurs at this height on slopes of 30° - 35°. Severe overgrazing by goats and innumerable goat tracks tend to obscure the periglacial component. Undamaged terraces are preserved in only a few inaccessible places.

3. "BLACK" PASS (2,650 m)

The "Black" Pass, in the western part of the mountains, owes its name to the dark schist debris covering the adjacent slopes. Patterned ground is common above 2,700 m on the east slope. Soil stripes are most frequent, with stone circles on more gentle slopes. Both forms are intermingled with dwarf shrubs.

The biggest area of soil stripes is about 15 m^2. The stripes are 20 - 25 cm apart; sorting depth at a slope inclination of 7° is 12 cm. All particles are oriented with their longer axes downslope. The coarser components have been pushed up subvertically. The solifluction material is splintering dark schist; the particles 3 - 10 cm long are in a matrix of silt and clay. The solifluction horizon is underlain by a reddish-brown fossil soil without any sorting. Debris accumulation on the upslope side of obstacles is evidence of rather strong solifluction. Where the debris has moved around obstacles, it has fully preserved its striped appearance, thus excluding their interpretation as mudflow deposits.

The stone circles on the almost horizontal slope terraces have a diamter of 30 cm; sorting depth is 15 cm. Up to a length of 12 cm the splintered particles are turned up and oriented tangentially. The fine-earth cores, just as in the stone circles, appear to have been squeezed up, thus resembling fine-earth buds. In summer their smooth surfaces are dissected by small dessication cracks.

Similar stone circles occur in other parts of the mountains at comparable altitudes, but mostly only on a suitable substrate. In most cases this is loamy moraine material, as for instance at the so-called hematite pass at 2,650 m, where polygons with a diameter of 20 cm exist on the moraine surface sloping about 2°.

Practically not patterned ground has developed on the grus-size weathering debris of the granite at comparable heights. Poorly developed forms were discovered in only one place, in a shallow depression on the "Pass to Sunij" at 2,800 m. Occasional existence of patterned ground on the slopes would be hard to prove, because of heavy disturbance of the surface by grazing goats.

4. "MARBLE PASS" (2,150 m)

The lowest occurrences of patterned ground are found at 2,000 to 2,200 m in various parts of the mountains. One such location is a broad ridge near the so-called "Marble Pass" at 2,150 m. This gently sloping ridge is covered with fine debris containing much silt. In a few places, small polygons have developed that are transformed into downslope stripes at about 5° inclination. The polygons resemble desiccation cracks. Their diameter is 15 cm, distance between stripes 10 cm, and sorting depth 5 - 7 cm. The latter is clearly visible, as the light-colored sorted soil overlies a red-brown fossil soil. The following profile is typical:

1. A more or less well structured stone pavement 2 - 3 cm thick on top; components up to 5 cm diameter still show signs of upturning and orientation;
2. underneath the pavement a stone-free vesicular layer about 2 cm thick, mostly silty with some clay;
3. occasionally a thin gypsum crust;
4. and below that crusts at 5 - 7 cm the weathered bedrock with debris-rich relicts of a fossil red-brown soil.

These observations raise the suspicion that the patterned ground near Marble Pass is not wholly a frost-dynamic feature but at least partly the result of periodic swelling and shrinking of the clays in the profile. The profile resembles those described by Cooke (1970) from the Chilean and Californian deserts. Cooke attributes the present formation of the stone pavement and the vesicular layer to swelling and shrinking of clay-rich soils (20 - 30 % clay content). He reports that stones were moved up 1 - 2 cm in only 4 - 11 cycles in laboratory tests.

ALTITUDINAL LIMITS AND PRESENT SNOWLINE

Observations suggest an approximate position of the lower limit of "true" periglacial phenomena to be around 2,500 m. Occasionally it may be as low as 2,000 m, but only on suitable loamy substrate and on northerly exposures. The limit should be higher on the south side of the mountains, but evidence is lacking, as the foreland plain rises to 2,400 m, and the granite slopes rising up to 3,000 m are unfavorable to the development of patterned ground.

All patterned ground phenomena below 2,000 m elevation are pseudoforms and not periglacial. This is especially evident where beautiful polygonal ground exists on the hard salt clays in the center of the kavirs, as in the Bafq kavir at 1,000 m. They most likely owe their existence to processes of salt solution and recrystallization together with strong evaporation.

The next question concerns the vertical range of the periglacial zone, i. e., the vertical distance between the "zonal" lower limit of patterned ground as defined by Troll (1944) and Höllermann (1967, 1972) and the present snow line. In the maps published by Bobek (1937, 1952) the present snow line west of Isfahan, at $32^{o}30'$, has been drawn at 4,000 m. Only a short distance to the east the isochione turns sharply westward, leaving an unusually steep rise of the present snow line in the study area to 4,500 or 4,600 m. This hypothetical rise of the snow line toward the arid interior of Iran was extrapolated by Bobek from his observations in the Alborz Mountains.

Assuming the "zonal" lower limit of patterned ground at 2,500 m and the present snow line at around 4,300 m the vertical extent of the periglacial zone of the Shir-Kuh Mountains would be 1,800 m. This supports Hövermann's (1960, 1962, 1972) theory of the wedgelike divergence of the snow line and the lower limit of patterned ground toward the arid zones of the earth.

The Pleistocene snow line may have been depressed by 1,000 m, since the lowest terminal moraines are at 1,900 m on the north side, and of the so-called Turkestan type, i. e., avalanche-fed. This interpretation is supported by the total absence of true cirques and the very rare occurrence of firn fields in the highest parts of the mountains (Hagedorn et al. 1975).

Since the average elevation of the valley floors is 2,500 m, and assuming that the valley glaciers were as much as 300 m thick, the Pleistocene snow line must have occured between the ancient glacier surface at 2,800 m and the enclosing rockwall summits at 3,800 - 4,000 m. By simply halving this interval a snow line between 3,300 and 3,400 m on the north side and between 3,700 and 3,800 m on the south side is suggested.

CONCLUSIONS

Periglacial features diminish in size with decreasing altitude in the Shir-Kuh Mountains, and at the same time become less clear. Similarly the depth of sorting diminishes. On the north side the "zonal" lower limit of periglacial phenomena lies near 2,500 m, and the "extra-zonal" lower limit of patterned ground near 2,000 m. Large-scale solifluction as a factor of slope formation does not occur below 3,000 m. The development of frost structures depends heavily on the suitability of the substrate. Loamy material on slopes of metamorphic schists or on moraines is suitable; granite grus is not suited. Exposure is not an important control over the distribution of periglacial phenomena with the exception of smooth slopes ("Glatthänge") that are best developed on south-facing slopes. Patterned ground as a convergence phenomenon exists below 2,500 m, with only a few genuine forms on highly suitable substrate. For most, a vesicular layer or salt crust indicates a nonperiglacial origin. A zone 300 - 500 m high separates "true" periglacial feature from convergence forms.

ACKNOWLEDGMENTS

H. Hagedorn, D. Busche, J. Grunert, W. Haars, U. Glaser and G. Weise, members of the Department of Geography of the University of Würzburg, who conducted geomorphological field work in the Shir-Kuh Mountains of Central Iran, gratefully acknowledge the financial support of the German Research Council (DFG).

REFERENCES

Bobek, H., 1937, Die Rolle der Eiszeit in Nordwest-Iran: Zeitschrift für Gletscherkunde, 25, p. 130 - 183.

Bobek, H., 1952, Beiträge für klimaökologischen Gliederung Irans: Erdkunde, 6, p. 65 - 85.

Bobek, H., 1955, Klima und Landschaft Irans in vor- und frühgeschichtlicher Zeit: Geographische Jahresberichte aus Österreich, 55, p. 18.

Cooke, R. U., 1970, Stone pavements in deserts: Annals of American Association of Geography, p. 560 - 577.

Djavadi, C., 1966, Climats de l'Iran: Monographie de la Météorologie naturelle 54, 100 p., Paris.

Hagedorn, H., Haars, W., Busche, D. and H. Förster, 1975, Pleistozäne Vergletscherungsspuren in Zentral-Iran: Zeitschrift für Geomorphologie, N. F. Suppl.-Band 15, p. 156 - 166.

Hagedorn, H., Haars, W., Grunert, J. and D. Busche, 1978, Periglazialerscheinungen im Shir-Kuh Massiv (Zentral-Iran): Colloque sur le Periglaciaire d'altitude du domaine mediterranéen et abords. Association Geographique d'Alsace, p. 263 - 278.

Höllermann, P., 1967, Zur Verbreitung rezenter periglazialer Kleinformen in den Pyrenäen und Ostalpen: Göttinger Geographische Abhandlungen, 40, 196 p.

Höllermann, P., 1972, Zur Frage der unteren Strukturbodengrenze in Gebirgen der Trockengebiete: Zeitschrift für Geomorphologie, N. F. Suppl. Band 15, p. 156 - 166.

Hövermann, J., 1960, Über Strukturböden im Elburs (Iran) und zur Frage des Verlaufs der Strukturbodengrenze: Zeitschrift für Geomorphologie, N. F., 4, p. 173 - 174.

Hövermann, J., 1962, Über Verlauf und Gesetzmäßigkeit der Strukturbodengrenze: Biuletyn Periglacjalny 11, p. 201 - 207.

Hövermann, J., 1972, Die periglaziale Region des Tibesti und ihr Verhältnis zu angrenzenden Formungsregionen: Göttinger Geographische Abhandlungen, 60, Festschrift H. Poser, p. 261 - 284.

Klaer, W., 1962, Die periglaziale Höhenstufe in den Gebirgen Vorderasiens: Zeitschrift für Geomorphologie, N. F. 6, p. 17 - 32.

Schweizer, G., 1975, Untersuchungen zur Physiogeographie von Ostanatolien und Nordwestiran: Tübinger Geographische Studien, 60, 145 p.

Troll, C., 1944, Strukturböden, Solifluktion und Frostklimate der Erde: Geologische Rundschau, 34, p. 545 - 694.

Troll, C., 1947, Die Formen der Solifluktion und die periglaziale Bodenabtragung: Erdkunde 1, p. 162 - 175.

THE BREAKDOWN OF ROCK DUE TO FREEZING: A THEORETICAL MODEL

Bernard Hallet

Quaternary Research Center and Department of Geological Sciences,
University of Washington, Seattle, Washington 98195

Frost wedging -- the breakdown of rock due to freezing -- is viewed as a manifestation of slow crack propagation in rocks due to ice growth in cracks. Frost wedging is modeled through a synthesis of Gilpin's (1980a) analysis of freezing in porous media and well-established principles of fracture mechanics. The model predicts most rapid breakdown at temperatures that range from -5° to -15°C for most rocks. At higher temperatures, ice pressures sufficient to produce significant rates of crack propagation are not thermodynamically possible. At lower temperatures the rate of ice growth is greatly reduced because water mobility and, hence, the flux of water necessary to sustain crack growth decrease considerably.

The model clarifies the dependence of frost wedging on lithology, temperature, and moisture conditions. Four important rock properties figure in the analysis: pore size, permeability, average crack length, and fracture toughness. Crack growth rate has a complex dependence on temperature and is proportional to temperature gradient. The moisture content of rock can strongly affect frost wedging rates primarily by controlling the pore water pressure. Solutes are inferred to influence crack growth rates in several distinct ways.

INTRODUCTION

A number of careful studies of the breakdown of rock due to freezing have yielded considerable empirical data (Washburn 1980, McGreevy 1981, Lautridou and Ozouf 1982). However, the actual mechanisms of rock degradation, and the relative importance of lithologic, thermal, and hydraulic parameters have yet to be clearly defined. The theoretical model presented here provides considerable aid in interpreting the numerous and varied experimental results and in evaluating existing concepts of frost-induced rock breakdown. In some cases the model helps reconcile conceptual models that are in apparent conflict. Finally, the model offers guidance for future laboratory, field, and theoretical studies of frost wedging by precisely delineating the important controlling parameters.

MODEL DESCRIPTION

Frost wedging is viewed as resulting from the progressive growth of cracks that are pressurized internally by the freezing process. As in studies of frost heaving in soils (Taber 1929, Gilpin 1980a), the pressure caused by ice growth is not attributed primarily to the volumetric expansion that accompanies the water-ice phase transition. Rather, the induced pressure is assumed to arise thermodynamically because mineral surfaces effectively decrease the chemical potential of water in their close proximity (Gilpin 1979). This effect has two important and relevant consequences: (1) considerable unfrozen water persists in porous media at subzero temperatures (Taber 1929, Anderson and Morgenstern 1973, Miller 1978, Gilpin 1979), and (2) water tends to flow toward mineral surfaces, but at equilibrium this flow is counteracted by an effective pressure increase near surfaces (Gilpin 1979, 1980b). This surface-induced increase in pressure gives rise to a "disjoining pressure" that can be related to the heave pressure in freezing soils (Gilpin 1980a). The disjoining pressure is assumed here to control the pressure generated by ice growth within rock cracks and, as such, it is a form of "crystallization pressure", as used by Taber (1929) and Robin (1978).

Analyses of the thermodynamic equilibrium between ice and water in porous media lead to a modified version of the Clausius-Clapeyron equation (Radd and Oertle 1973, Loch 1978, Gilpin 1980a). The effect of mineral surfaces on the chemical potential of water renders it possible for ice at a certain pressure to exist in contact and in thermodynamic equilibrium with water at a different pressure. When the water is hydraulically connected to a reservoir at reference pressure P_o, the ice pressure and ice-water interface temperature at equilibrium are related by

$$v_s P_s + \frac{LT_1}{T_a} = 0 , \qquad (1)$$

where v_s is the specific volume of ice, L is the heat of fusion of ice, T_a(°K) is effectively the freezing temperature of pure water at pressure P_o; P_s is the ice pressure in excess of P_o, and T_1 is the deviation from T_a of the ice-water interface temperature. For convenience, P_o is taken to be atmospheric; $P_o = 10^5$ Pa and in this case, T_1 is simply the Celsius temperature.

From equation (1) it is evidentthat water at 0°C and atmospheric pressure can be in thermodynamic equilibrium with water in contact with ice at a pressure P_s that increases linearly with the negative Celsius temperature at the rate of $L/(T_a v_s)$. This rate is 1.14 MPa $^oC^{-1}$. If P_s is lower than the equilibrium value given by (1), it would be thermodynamically favorable for water to flow to the ice interface despite adverse pressure gradients. In the context of frost wedging, given sufficiently low temperatures, water would tend to flow to an ice body in a rock crack even if the pressure in the unfrozen water film between the ice and the crack wall is already high. Let us consider, for example, the ice pressure in a crack to be about 10 MPa (100 bars), a value on the order of the tensile strength of sound rock. Water flow into the crack, and hence crack expansion, would be thermodynamically favored as long as the crack temperature is below about -9°C.

The equilibrium condition (1) has been experimentally verified at temperatures down to approximately -18°C (Radd and Oertle 1973, Buil et al. 1981, Takashi et al. 1980), with heaving pressures P_s of up to 20 MPa (200 bars). It is apparent from these studies that freezing could generate pressures on the order of tens of MPa in rock pores and cracks if thermodynamic equilibrium is approached at temperatures on the order of -10°C or lower. As the tensile strength of rocks seldom exceeds 10 MPa, such frost-induced pressures in rock cracks could cause progressive crack propagation.

Crack growth depends on the magnitude (Atkinson 1982) and maintenance of pressure inside cracks. Therefore, it will be controlled both by the ice pressure and the rate of water flow necessary to maintain pressure in an expanding crack. A complex dependence can be anticipated between temperature and crack propagation rate. At very low temperatures, high pressures can be generated in cracks but crack expansion is limited because water mobility decreases sharply with decreasing temperature (Miller 1978, Gilpin 1980a). In contrast, at negative temperatures closer to 0°C, water is considerably more mobile, but pressures inside cracks are limited to low values dictated by the equilibrium condition (1).

Because of space limitations, only the simplest form of the model that contains the essential physics, including these competing effects, is presented. Ongoing work by Joseph Walder and the author has shown that the model can readily be refined and rendered more realistic.

Crack Propagation in Rocks

Modern fracture-mechanics theory focuses attention on the stress intensity factor K_I. This quantity represents the strength of the singularity in the crack-tip stress field that tends to produce opening-mode (tensile) failure (Paris and Sih 1965). For example, neglecting direct interactions between adjacent cracks and idealizing rocks as isotropic linear elastic media,

$$K_I = \left(\frac{\pi l}{2}\right)^{1/2} (P + \sigma), \qquad (2)$$

where l is the length of a two-dimensional crack, P is the pressure inside the crack, and σ is the "applied" normal stress perpendicular to the crack plane (tensile stresses are positive).

In minerals and rocks under low K_I conditions, progressive crack growth--often known as stress corrosion or subcritical crack growth--occurs at rates that are sensitively dependent on K_I, and on the presence of water, water vapor, or other reactive species in the crack-tip environment (Atkinson 1982). One of the most commonly used equations to describe slow crack propagation data is Charles' power law (Atkinson 1982):

$$V_c = V_o \exp\left(-\frac{\Delta H}{RT}\right) K_I^{\,n}, \qquad (3)$$

where V_c is the crack velocity, V_o is a proportionality factor characteristic of the material, ΔH is the activation enthalpy, R is the gas constant, and T is the absolute temperature. n is a material-dependent constant known as the stress corrosion index which is typically in the range of 40 ± 10 for rocks(Atkinson 1982), indicating an extremely sensitive dependence of crack propagation rates on the magnitude of crack loading. Equation (3) adequately describes fracture data for several types of crystalline rocks down to crack velocities of 10^{-9} m s^{-1}, and for Lac du Bonnet granite to crack velocities of 10^{-11} m s^{-1}.

Experimental studies of frost wedging often indicate that significant disaggregation of individual mineral grains occurs in few temperature cycles with periods ranging from hours to days. This would imply that cracks must propagate distances on the order of grain diameters, which typically range from 0.1 to 10 mm, within periods ranging from 1 to 100 days. The corresponding average crack propagation rates range from 10^{-7} to 10^{-11} m s^{-1}. Because the crack velocity is related to K_I raised to a high power (40 ± 10), this 4 order-of-magnitude span for inferred crack velocity collapses to a narrow range of stress intensity factors for a particular rock, with the highest value exceeding the lowest by about 50%.

For illustrative purposes, let us focus on conditions that lead to K_I values corresponding to the midrange of probable crack velocities (i.e., 10^{-9} m s^{-1}). Experimental values for K_I at these crack velocities are presented by Atkinson (1982) for several rock types. He reports approximate K_I values of 1.1, 1.5 and 2.1 MN $m^{-3/2}$ for granite, basalt, and dolerite samples, respectively. According to equation (2) with σ set at zero, these values correspond respectively to pressures P inside cracks of 28, 38, and 53 MPa for 1-mm cracks, and of 9, 12, and 17 MPa for 10-mm cracks. Thus, relatively high pressures in cracks are required to fracture crystalline rocks devoid of large fractures. Although data on relevant stress intensity factor K_I for sedimentary and metamorphic rocks are scarce, it is evident that pressures only a fraction of those necessary for crystalline rocks would suffice to crack many sedimentary and low-grade metamorphic rocks.

Significant sustained crack growth requires that pressure inside cracks be maintained by ice growth as cracks propagate. The rate of ice growth and hence of crack widening is analyzed in the next section.

Ice Growth in Rock Cracks

The model developed by Gilpin (1980a) for prediction of ice lensing and heave in soils offers a particularly suitable theoretical basis for analyzing ice growth in rocks. He derives a simple relationship between the flow rate of water and the driving potential in films of unfrozen water at ice-mineral interfaces. The water flux, which is assumed to be rate limiting, is then related to the rate of ice growth through a continuity relation. The principal governing equations will be presented here (the reader is referred to Gilpin (1979, 1980a) for the complete development). The water flow rate is proportional to the driving potential:

$$q = -k \frac{v_s}{v_l} \frac{d}{dx} \left(P_s + \frac{LT_l}{v_s T_a}\right), \qquad (4)$$

where k is the hydraulic permeability, v_s and v_l are the specific volumes of the solid and liquid phases, and x is the distance normal to the ice lens (perpendicular to the crack). As can be seen from equations (4) and (1), water flow vanishes when thermodynamic equilibrium is established, that is, when $P_s = -LT_l/v_sT_a$. It should also be noted that, in the driving potential term, the effects of both pressure and temperature are likely to be important. For example, a pressure difference of 1 MPa in ice is equivalent to a temperature difference of 0.88°C as far as the potential for driving water flow is concerned.

The mobility of water is a complicated function of temperature because it appears to be largely dictated by the permeability of a partially frozen zone on the warmer side of an ice-filled crack (Figure 1). After a detailed analysis, Gilpin (1980a) showed that the permeability of the partially frozen fringe in non-colloidal soils can be approximated adequately by

$$k_f = k_1 \left(- T_l + T_f\right)^{-2} \qquad (5)$$

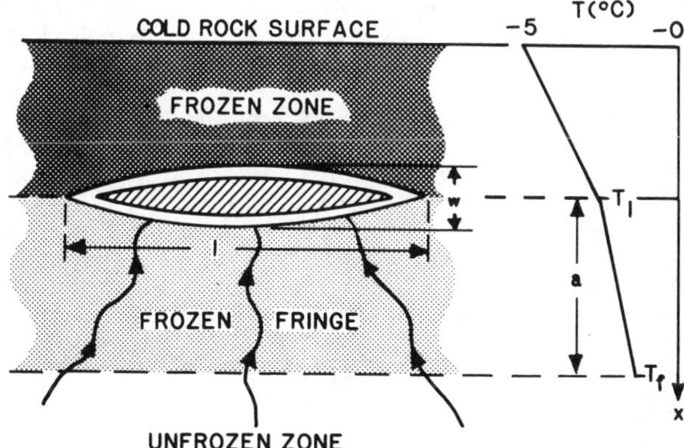

Figure 1 Schematic diagram of ice (oblique stripe pattern) in a rock crack with an intervening water film. Wavy lines represent paths of water flow necessary for sustained ice and crack growth.

where k_1 is a constant dependent on grain size (Gilpin 1980a, Figure 5). T_f, the temperature depression required to form ice in soil pores, is inversely related to pore radius (Gilpin 1980a, equation 18). On the basis of equations (4) and (5), together with conservation of mass, Gilpin (1980a) derived his equation (25) for the rate of heave or ice growth as a function of the effective hydraulic resistance provided by the frozen fringe and by thin water films that envelop growing ice lenses.

For the frost-wedging problem, water influx is simply related to the volume increase of cracks, the shape of which is dictated by the effective internal pressure ($P + \sigma$) and crack length. To streamline this analysis, however, shapewill be taken as invariant by fixing the crack aspect ratio A (crack width/length). This simplification is appropriate for increments of crack growth that are small relative to crack length. The rate of ice growth can then be simply related to the crack propagation rate V_c:

$$V_c = \frac{v_s^2 k_1}{v_l g A (T_f - T_l)^3} \frac{dT}{dx} \left(P_l - P_c - \frac{T_l L}{v_s T_a}\right), \qquad (6)$$

where g is acceleration due to gravity, dT/dx is the temperature gradient, P_c is the pressure exerted on crack walls by ice via the intervening water film and P_l is the pore water pressure outside the frozen fringe. Equation (6) is derived directly from Gilpin's (1980a) equation (25) by assuming that the hydraulic resistance at ice bodies is negligible relative to that of the frozen fringe. This assumption is likely to be valid in rocks for freezing in cracks that are large relative to the spacing of pores and intragranular cracks. Implicit in equation (6) is the hydraulic resistance of the frozen fringe; it scales with the inverse of k_f, approximated by equation (5), and with fringe thickness a, which depends on the temperature gradient: $a = (T_f - T_l)/(dT/dx)$.

The rate of crack propagation, V_c, can be obtained simply from equation (6) by selecting values for the pressure P_c inside cracks that correspond to the broad V_c range of interest for particular rock types, as discussed above. Alternatively, to avoid selecting a fixed value of P_c, the fracture equations (2) and (3) can be combined and solved simultaneously with equation (6) to calculate both V_c and P_c.

RESULTS

From the form of equation (6) it is apparent that crack growth occurs only when

$$\frac{-T_l L}{v_s T_a} + P_l > P_c \qquad (7)$$

For negligible pore pressures P_l, T_l must be less than $v_sT_aP_c/L$. For example, for values of P_c around 10 MPa, crack growth could only occur at temperatures T_l below about -9°C. As the temperature drops below this critical value, V_c first increases steeply but soon falls off because of a rapid decrease in permeability, which scales roughly as T_l^{-3}. This dependence on T_l arises from equation (6) because, as seen in the numerical example below, the contribution of T_f is often insignificant relative to that of T_l. It is also apparent that V_c is directly proportional to the temperature gradient, which controls the thickness of the frozen fringe and, hence, of the effective rock permeability.

A calculation of crack-propagation rate as a function of temperature illustrates many of these relationships quantitatively. The results shown in Figure 2 were obtained using reasonable values for the variable physical parameters: k_1 was taken to be 10^{-13} m s^{-1} $^oC^2$ because, at 1^oC below the freezing temperature T_f, it corresponds to a permeability of about 10^{-13} m s^{-1}, a reasonable value for sound crystalline rocks. The crack length l was assumed to be 100 times its width w, corresponding to A = 0.01. The temperature gradient in the frozen fringe was taken to be 3^oC m^{-1} and the freezing point depression T_f related to pore size was taken to be -1^oC. This T_f corresponds to pores 0.1 μm in diameter given an ice-water interfacial energy of 0.03 N m^{-1} (Gilpin 1980a, equation 18). Crack propagation rates were calculated for internal crack pressures P_c of 5 MPa (appropriate for relatively weak sedimentary rocks) and 10 MPa (for granites). The pore pressure P_l outside the frozen fringe was assumed negligible.

Although V_c values are rather poorly defined primarily because of uncertainties in k_1 and P_c, the form of the temperature depencence of V_c shown in Figure 2 is not expected to vary appreciably. Thus it is evident that frost-induced cracking of rock is expected to occur only below a critical temperature and that cracking is fastest at a slightly lower temperature. Below this the rapid decrease in permeability causes the crack growth rate to diminish with temperature because water flux becomes severely limiting.

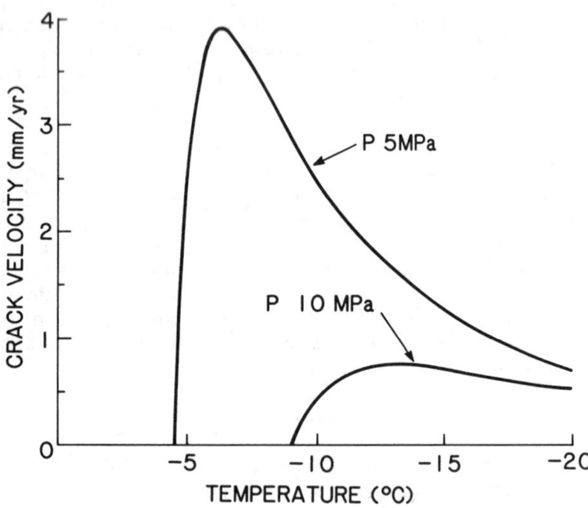

Figure 2 Rate of crack propagation as a function of temperature T_l and ice pressure P_c in cracks. Crack growth is most rapid at temperatures slightly lower than that at which ice under pressure P_c is in thermodynamic equilibrium with water at atmospheric pressure. For lower temperatures, water mobility is so low that ice growth is severely limited by water flux. P_c = 5 and 10 MPa are generally appropriate values for sedimentary rocks and for plutonic rocks, respectively.

DISCUSSION

The analysis shows that freezing can generate sufficient pressure in rock cracks to cause significant cracking over a time scale of interest. It is emphasized that cracks are assumed to be pressurized internally by ice with an intervening film of water. As only water contacts crack walls in the model, this mechanism could be viewed as a form of hydration cracking, which has been perceived by others (e.g., White 1976) as being in direct conflict with frost-shattering concepts. Similarly, the distinction emphasized by Hudec (1974) between "frost sensitive" rocks and "sorption sensitive" rocks now appears largely arbitrary because, according to the model presented here, both "sensitivities" result directly from the reduction of the chemical potential of vicinal water due to mineral surfaces. The thermodynamic properties that are fundamental to the model constitute common ground for a number of apparently distinct frost shattering hypotheses. As a result, this model effectively unifies several hypotheses that commonly figure under separate headings: crystallization pressure, capillarity, and ordered water (Washburn 1980, McGreevy 1981).

The principal implications of the model are subdivided into the effects of lithology, temperature, moisture, and solutes on frost wedging.

Lithologic Controls

Susceptibility to frost wedging is generally recognized to be sensitively dependent on lithology. This is to be expected because lithologic characteristics figure in the analysis of crack-growth rates in several ways: (1) pore size affects the temperature T_f at which ice forms in pores; (2) intrinsic permeability controls water mobility at subfreezing temperatures through k_1; (3) the relevant stress intensity factor K_I of the rock, together with crack geometry A, dictate the value of P_c. These parameters all correspond to rock properties known experimentally to influence the frost susceptibility of rock (Lautridou and Ozouf 1982).

For example, rapid frost wedging in shales and argillites probably results from low K_I values, high permeability at subfreezing temperatures because of high unfrozen water content--a characteristic of frost sensitive rocks (McGreevy 1981)--, and long narrow cracks--often present in rocks with well-developed fissility. In contrast, dolerite with fracture properties similar to those reported by Atkinson (1982 and above) would be very resistant to frost wedging unless long, pre-existing cracks could be exploited.

Thermal Regime

It is clear from equation (6) and Figure 2 that frost-induced breakdown of rock is not expected to occur at appreciable rates at temperatures warmer than a critical subzero value. The instantaneous rate of frost wedging is expected to be highest at a temperature, T_*, slightly below the critical value. The ratio T_*/P_c is approximately 1.3°C/MPa for negligible T_f.

The lack of consensus about thermal conditions that are most favorable for frost wedging (Washburn 1980, McGreevy 1981) is not surprising in view of the complex dependence of crack-propagation rate on both temperature and lithology, as illustrated in Figure 2. The results, however, are compatible with frost wedging being most active in areas where temperatures are substantially below 0° C but not at the extremes found, for example, in Antarctica.

Other aspects of the thermal regime often reported to favor frost wedging, including frequency of freeze-thaw cycles, rate of cooling, minimum temperature, and duration of cold period, can be interpreted with new insight gained from the model. Both the rate of cooling and the cycle frequency affect V_c by controlling temperature gradients dT/dx. The lowest temperatures attained can play a critical role in frost wedging of sound crystalline rock with small cracks that only propagate when stressed highly. The duration of the cold period is clearly important because crack propagation is proportional to time.

Moisture Controls

High initial moisture content and ample water supply are commonly recognized to favor frost wedging (Mellor 1973, McGreevy 1981, White 1976). The effect of moisture content in unsaturated rock cannot be analyzed explicitly in this simple model, which parallels Gilpin's (1980a) heave model for saturated porous media. However, the model presented here does provide considerable guidance for a qualitative assessment of moisture effects. As water freezes in rock pores, the total volume of water in rock expands by about 9%. In saturated rocks freezing results in expulsion of water ahead of the freezing front as long as crack expansion is insufficient to accommodate the volume increase due to the ice-water transition. This expulsion of water, which is characteristic of ice lensing under high confining pressure (Gilpin 1980a), will increase the pore pressure P_l. It is evident from equation (6) that increasing P_l directly increases the crack-propagation rate, and can cause significant cracking at temperatures above those required when $P_l = 0$. In fact, if P_l is sufficiently high ($P_l > P_c$), hydrofractures may form in rock regardless of temperature. This situation is most likely to occur when rock fragments are frozen from all sides, as may be common for talus. Highest pressures would be reached, and hence cracking would be most likely, in the central portion of rocks with high water contents. Such an effect may underlie a causative relationship between the size of rock fragments and climatic variables.

Solute Effects

Current models of freezing in porous media (Miller 1978, Gilpin 1980a) have been developed for pure water systems. Although it is beyond the scope of this paper to treat solutes quantitatively, several observations can help us understand why solutes greatly accelerate the breakdown of rock under freezing conditions in some experiments (White 1976, Williams and Robinson 1981) and have the opposite effect in other experiments (McGreevy 1982).

Solutes are anticipated to affect frost wedging in several ways: 1) the freezing temperature T_f is lowered, thereby increasing the unfrozen water content and water mobility at subfreezing temperatures in accord with the known effect of solutes on phase-equilibria in frozen soils (Anderson and Morgenstern 1973); (2) T_f and the chemical potential of water near freezing interfaces would be selectively lowered by the interfacial solute buildup that results as growing ice preferentially rejects solutes; and (3) the relevant stress-intensity factor K_I, and hence P_c, required to cause particular crack-growth rates may be reduced substantially by high-solute concentrations in cracks (Atkinson 1982). The lowering of T_f effectively shifts the curves in Figure 2 to lower temperatures; this can either increase or decrease frost-wedging rates depending on whether the dominant experimental temperatures are below or above the temperature T_* at which cracking is fastest. The interfacial lowering of the chemical potential would tend to speed up crack growth by increasing water transport to freezing interfaces. This effect is likely to be important for low solute concentrations because, at concentrations approaching the saturation value, the interfacial solute buildup cannot develop fully, as the solute concentration cannot appreciably exceed the eutectic concentration. This is in accord with McGreevy's (1982) experimental finding that for each salt solution the greatest frost-induced damage was associated with the most dilute solutions used. In contrast, however, Williams and Robinson (1981) found that frost wedging in rocks containing saturated salt solutions was considerably more rapid than in the same rock containing water. This result may reflect a strong influence of solute concentration in the vicinity of crack tips on crack-propagation rates. Such an influence is evident in stress corrosion experiments on sandstones in which crack propagation rates increased by an order of magnitude as the NaCl concentration of pore fluids was increased above 0.1 M (D. M. Johnson, personal communication 1983).

CONCLUSIONS

The simple model developed here for frost-induced breakdown of rocks provides considerable guidance in interpreting the large existing data base obtained through laboratory and field studies. Lithology and temperature take on multiple but clearly defined roles in the analysis. The results of the model provide guidelines for future laboratory and field studies and motivate complementary theoretical treatments. Analyses of pore pressure and solute distribution in porous media undergoing freezing would be particularly helpful for improved understanding of the breakdown of rock under such conditions. Some of these analyses are currently underway.

ACKNOWLEDGMENTS

It is a sincere pleasure to acknowledge E. C. Gregory, A. J. Heyneman, and C. W. Stubbs for their truly outstanding assistance without which this paper would not have been completed. I also wish to thank R. S. Anderson for help with numerical analyses, and J. S. Walder and W. M. Bruner for detailed discussions of frost wedging and rock fracture. Critical comments on the manuscript by R. S. Anderson, D. B. Booth, W. M. Bruner, M. P. Muir, M. Sharp, J. S. Walder, and A. L. Washburn are much appreciated. D. L. Gardner and P. S. Leaming's help in preparing the manuscript is acknowledged. This study was supported by NSF grant (EAR 81-09308).

REFERENCES

Anderson, D. M., and Morgenstern, N. R., 1973, Physics, chemistry, and mechanics of frozen ground, in Permafrost--The North American Contribution to the Second International Conference: Washington, D.C., National Academy of Sciences.

Atkinson, B. K., 1982, Subcritical crack propagation in rocks: theory, experimental results and applications: Journal of Structural Geology, v. 4, p. 41-56.

Buil, M., Aguirre-Puente, J., and Soisson, A., 1981, Etude expérimentale de l'équilibre thermodynamique entre deux phases soumises à des pressions différentes. Cas du système eau-glace: Comptes Rendus de l'Académie des Sciences, Paris, Série II, v. 293, p. 653-656.

Gilpin, R. R., 1979, A model of the 'liquid-like' layer between ice and a substrate with applications to wire regelation and particle migration: Journal of Colloid and Interface Science, v. 68, p. 235-251.

Gilpin, R. R., 1980a, A model for the prediction of ice lensing and frost heave in soils: Water Resources Research, v. 16, p. 918-930.

Gilpin, R. R., 1980b, Theoretical studies of particle engulfment: Journal of Colloid and Interface Science, v. 74, p. 44-63.

Hudec, P. P., 1974, Weathering of rocks in Arctic and sub-Arctic environment, in Proceedings of the Symposium on the Geology of the Canadian Arctic: Saskatoon, 1973, Geological Association of Canada and Canadian Society of Petroleum Geology.

Lautridou, J. P., and Ozouf, J. C., 1982, Experimental frost shattering: 15 years of research at the Centre de Géomorphologie du CNRS: Progress in Physical Geography, v. 6, p. 215-232.

Loch, J. P. G., 1978, Thermodynamic equilibrium between ice and water in porous media: Soil Science, v. 126, p. 77-80.

McGreevy, J. P., 1981, Frost shattering: Progress in Physical Geography, v. 5, p. 56-75.

McGreevy, J. P., 1982, "Frost and Salt" weathering: further experimental results, Earth Surface Processes and Landforms, v. 7, p. 475-488.

Mellor, M., 1973, Mechanical properties of rocks at low temperatures, in Permafrost--The North American Contribution to the Second International Conference: Washington, D. C., National Academy of Sciences.

Miller, R. D., 1978, Frost heaving in non-colloidal soils, in Proceedings of The Third International Conference on Permafrost, v. 1: Ottawa, National Research Council of Canada.

Paris, P. C., and Sih, G. C., 1965, Stress analysis in cracks, in Fracture toughness testing and its applications: American Society for Testing and Materials, Special Technique Publication No. 381.

Radd, F. J., and Oertle, D. H., 1973, Experimental pressure studies of frost heave mechanisms and the growth-fusion behavior of ice, in Permafrost--The North American Contribution to the Second International Conference: Washington, D. C., National Academy of Sciences.

Robin, P. F., 1978, Pressure solution at grain-to-grain contacts, Geochimica et Cosmochimica Acta, v. 42, p. 1383-1389.

Taber, S., 1929, Frost heaving: Journal of Geology, v. 37, p. 428-461.

Takashi, T., Ohrai, T., Yamamoto, H., and Okamoto, J., 1980, Upper limit of heaving pressure derived by pore water pressure measurements of partially frozen soil, in Proceedings of The Second International Symposium on Ground Freezing: Trondheim, The Norwegian Institute of Technology.

Washburn, A. L., 1980, Geocryology: New York, Halsted Press.

White, S. E., 1976, Is frost action really only hydration shattering? A review: Arctic and Alpine Research, v. 8, p. 1-6.

Williams, R. B. G., and Robinson, D. A., 1981, Weathering of sandstone by the combined action of frost and salt: Earth Surface Processes and Landforms, v. 6, p. 1-9.

ALASKA HIGHWAY GAS PIPELINE PROJECT (YUKON) SECTION
THAW SETTLEMENT DESIGN APPROACH

A. J. Hanna[1], R. J. Saunders[1], G. N. Lem[1] and L. E. Carlson[2]

[1]Hardy Associates (1978) Ltd., Calgary, Alberta, Canada T2E 6J5
[2]Foothills Pipe Lines (Yukon) Ltd., Calgary, Alberta, Canada T2P 2V7

The Alaska Highway Gas Pipeline will traverse permafrost terrain for much of the route through the Yukon Territory. In the predominantly frozen ground, the pipeline will operate in the "cold" mode. Where the terrain is predominantly unfrozen, the pipe will operate in the "warm" mode requiring mitigative design for thaw settlement in some frozen terrains.

The results of the thaw settlement tests carried out for the project are presented as well as other data available in the literature. Statistical methods were used to arrive at correlations between thaw settlement and moisture content or density. The thaw settlement predictions were based on a computerized Borehole Data Bank containing simplified borehole logs and a computerized thaw settlement program.

The thaw settlement predictions were applied to the pipeline design by assuming that the predicted thaw settlement could occur as a differential settlement over a relatively short length of pipe. The various mitigative designs are presented.

INTRODUCTION

The Alaska Highway Gas Pipeline has been proposed to transport gas from the Prudhoe Bay area on the north shore of Alaska, USA, across Alaska through northwestern Canada to the southern states. Figure 1 shows the Yukon section of the proposed route. The 1.2 m diameter pipeline will traverse permafrost terrain for much of the route through the state of Alaska and the Yukon Territory. The pipeline will be operated continuously at temperatures below 0° C for the entire portion within Alaska and for the first 214 km into the Yukon Territory. Compressor Station 313 at KP 214 is the last point of cold flow, just northwest of Kluane Lake (see Figure 1). The remaining portion of the line is proposed to be operated in a warm mode. Between Kluane Lake and Whitehorse, a high percentage of the route is underlain by permafrost. If significant amounts of excess ice are present in the ground, the thawing created by the warm pipe would cause thaw settlement beneath the pipe over a period of time and lead to possible pipeline stability problems.

This paper summarizes the various components of the thaw settlement design process that were undertaken for the Yukon portion of the pipeline route. These are as follows:
1. investigation of extent and conditions of permafrost along route;
2. assembly of test data and compilation into basic soil groups;
3. establishing thaw settlement correlations;
4. prediction of thaw depths and resulting settlement; and
5. application of the predicted thaw settlements to design.

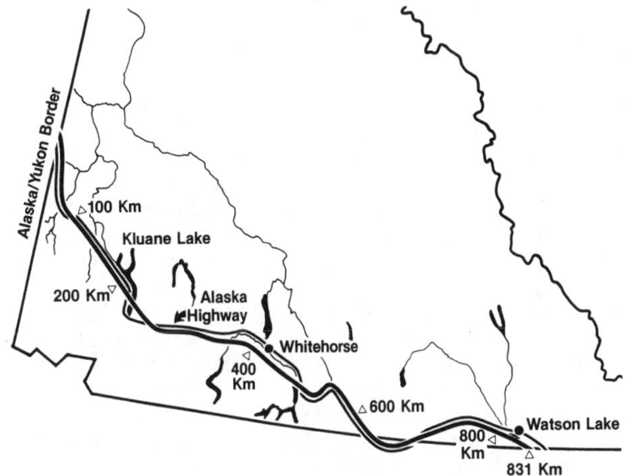

FIGURE 1 The Alaska Highway Gas Pipeline route through Yukon.

EXTENT AND CONDITIONS OF PERMAFROST

Extensive programs of terrain analysis, field reconnaissance, geophysical surveys, and borehole drilling have been carried out in the Yukon. Particular attention was paid to delineating permafrost and determining its properties where present. The extent of permafrost in the warmflow portion of the route is summarized in Table 1 according to different physiographic regions established as part of the overall terrain assessment for the route. Based on airphoto interpretation and considerable borehole data, terrain analysis has made possible the delineation of the potentially ice-rich terrain units along the route. The extent of ice-rich terrain is also given in Table 1.

TABLE 1 Extent of Permafrost South of Kluane Lake.

Physiographic Regions (KP Range)*	Length of Region (km)	Percentage Permafrost	Percentage Ice-Rich Permafrost
226.3-314.5	88.2	46.9	24.3
314.5-439.0	124.5	19.6	13.3
439.0-549.6	110.6	16.9	11.1
549.6-662.0	112.4	18.5	13.4
662.0-769.1	107.1	3.6	0.5
769.1-829.5	60.4	3.1	3.1
226.3-829.5	603.2	18.3	11.2

*KP is the distance in km from the Alaska-Yukon border.

THAW SETTLEMENT TEST DATA

Project Test Data

A total of 185 thaw settlement tests have been performed on frozen core samples obtained during the various drilling programs along the route since 1976. Particular emphasis has been made to define the thaw settlement characteristics for soils with relatively low ice contents.

The development of the basic thaw settlement test has been outlined in Watson et al. (1973). The general test procedure consists of thawing a frozen core under some nominal (5 kPa) pressure, and then two or more subsequent load increments (e.g., 40 and 80 kPa) are applied. This test procedure permits a distinction between the settlement due to thawing of the excess ice alone, from that of the additional settlement attributable to an increase in pressure, as illustrated in Figure 2 (after Watson et al. 1973).

Moisture content and frozen bulk density tests were carried out on the test specimens to permit a correlation between thaw settlement and these more readily available soil properties.

Other Available Thaw Settlement Data

Two main sources of additional thaw settlement data have been used: Mackenzie Valley, N.W.T., data presented by McRoberts et al. (1978) and Alaskan data published by Luscher and Afifi (1973).

The Mackenzie Valley data were compared with the South Yukon data. All the data from the two sources were plotted separately, and a visual comparison was made for each of the soil types represented. The test data for peats differed very considerably, and so the Mackenzie Valley data for peat have not been used. The other soil types were found to compare favorably.

The test data for gravels presented by Luscher and Afifi (1973) represent the only major source of data on granular soil, and therefore, no real comparison with Yukon gravels is possible.

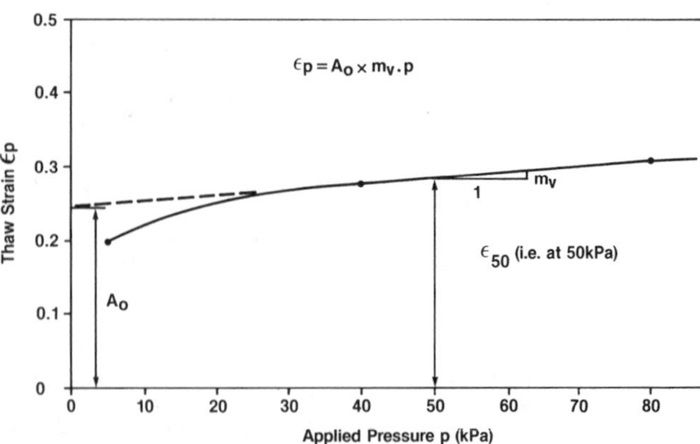

FIGURE 2 Typical thaw settlement test result.

THAW SETTLEMENT CORRELATIONS

Some of the earliest analysis of thaw settlement was carried out by Tsytovich and Sumgin (1937). More recent works include Morgenstern and Nixon (1971), and Crory (1973). In basic form (Figure 2) the thaw strain under any given pressure is expressed by:

$$\epsilon = A_o + m_v \cdot p \qquad (1)$$

where ϵ_p is thaw strain at pressure, p; A_o is initial thaw strain at zero overburden pressure; p is the effective pressure (kPa); and m_v is the coefficient of compressibility (1/kPa).

For the purposes of thaw settlement evaluation over the pipeline route, the thaw strain at a representative effective pressure of 50 kPa has been taken. This pressure conservatively represents the insitu effective pressure in the upper layers of the thaw bulb where most of the thaw settlement occurs. The actual pressure selected is not very critical for general route evaluation, as the major component of the thaw strain value is the initial thaw value, A_o.

Figure 3 shows a typical data plot. Because there is the possibility of error or random variation in both the thaw strain value measured and the moisture content or bulk density determined for each test, there exists a joint probability distribution. Two "best fit" lines have been obtained by linear regression, considering each variable errorfree, respectively, as illustrated in Figure 3. The best estimate of the true correlation lies between these two lines (Lyon, 1970).

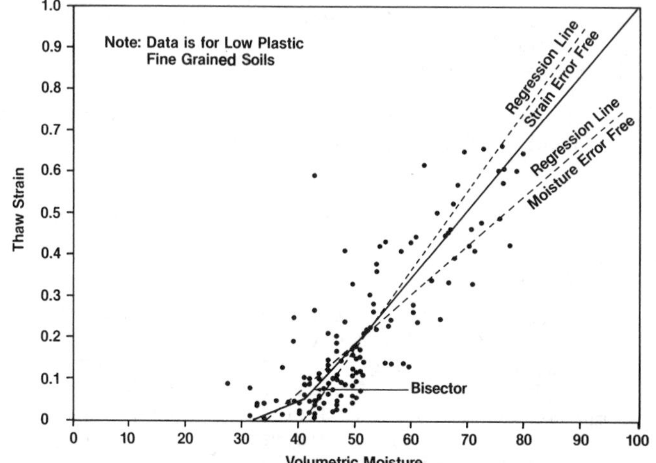

FIGURE 3 Establishing thaw strain-moisture correlation.

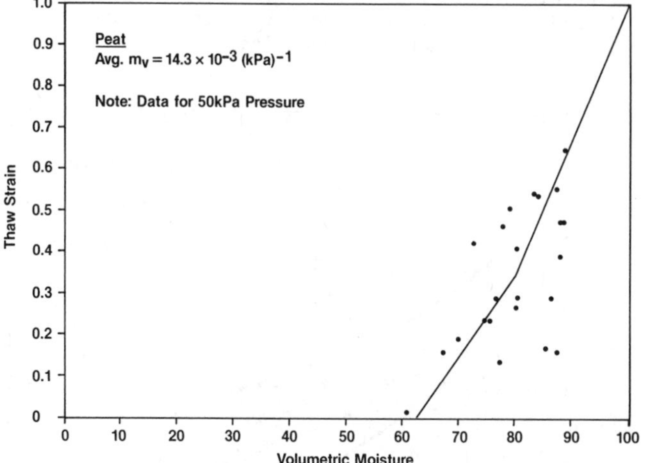

FIGURE 4 Thaw strain-moisture correlation for peat.

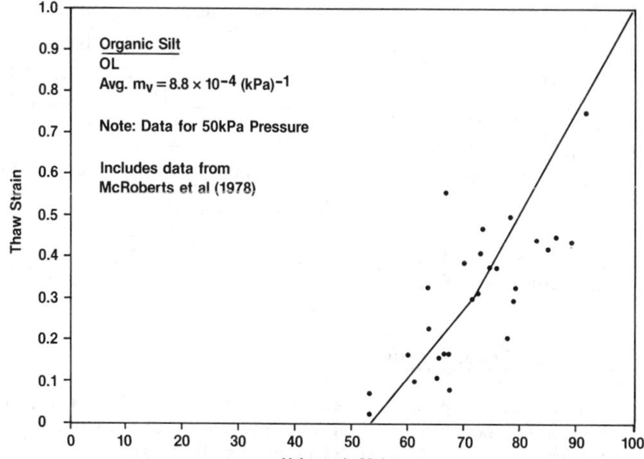

FIGURE 5 Thaw strain-moisture correlation for organic silt.

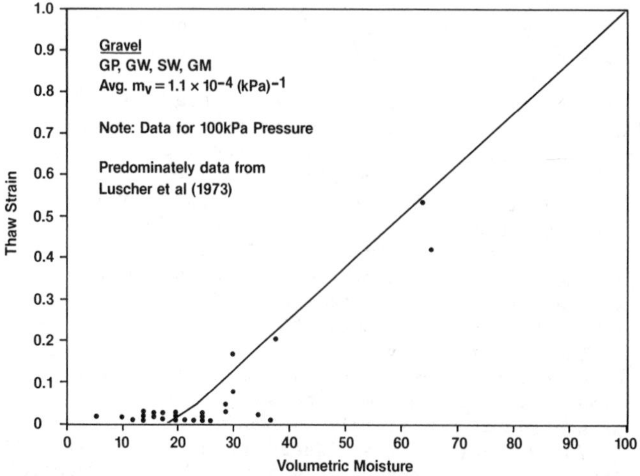

FIGURE 6 Thaw strain-moisture correlation for gravel.

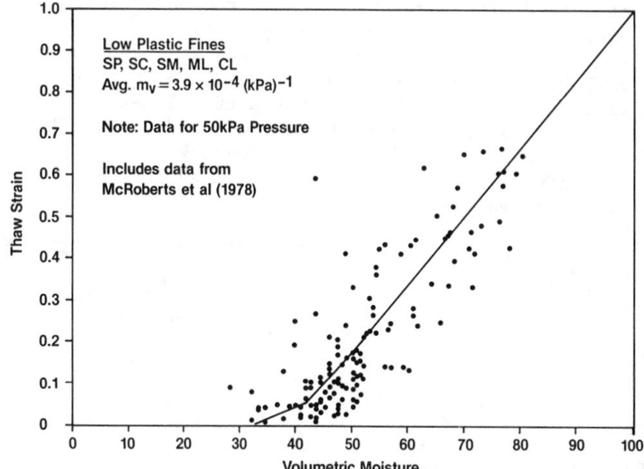

FIGURE 7 Thaw strain-moisture correlation for low plastic fine-grained soils.

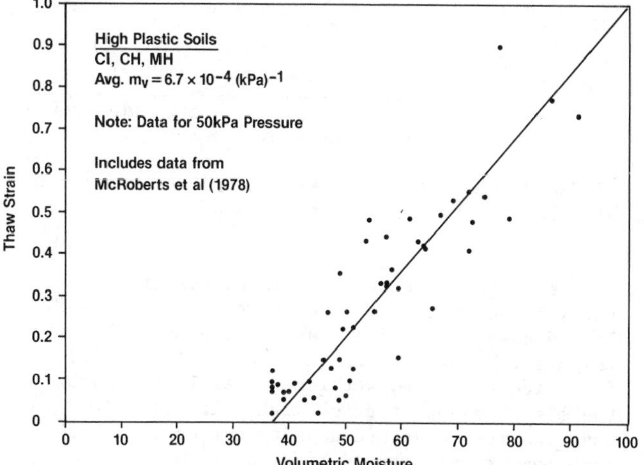

FIGURE 8 Thaw strain-moisture correlation for high plastic soils.

The actual correlations have been obtained as follows:

- For the upper thaw settlement range, the correlation has been defined to provide a thaw strain of 1.0 when the sample becomes pure ice (volumetric moisture content 100%; or frozen bulk density of 917 kg/m^3). This upper limit thaw settlement is joined to the intersection of the two regression lines (see Figure 3).
- The lower portion of the correlation is taken as the bisector of the two regression lines, below the point of intersection.

The thaw strain value ϵ_{50}, at 50 kPa, has been obtained from all the available test data. (The strain at a minimum pressure of 100 kPa is presented for the gravels by Luscher and Afifi (1973).) Based on the visual comparisons of all data, basic soil groups were selected, i.e., peat, organic silts, gravels, non to low plastic silts and clays, and medium to high plastic silts and clays. All of the thaw strain values are presented with respect to volumetric moisture content of the soil groups in Figures 4-8. Volumetric water content was chosen as a dependent variable rather than the more usual gravimetric water content, as the full range of volumetric water content is confined between 0 and 100%. The thaw strain versus frozen bulk density is presented in Figure 9 for the low plastic fine-grained soils only.

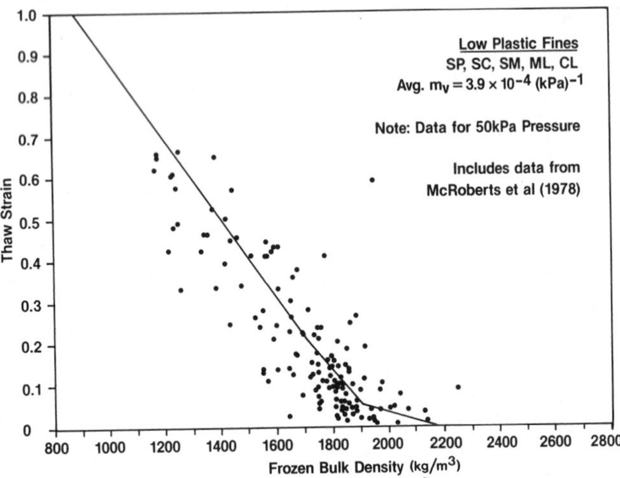

FIGURE 9 Thaw strain-density correlation for low plastic fine-grained soils.

The established thaw strain correlations for the various soil types are shown in Figures 4-9. In the case of the low plastic fine-grained soils (Figure 7), the moisture content (37% by vol.) for which zero thaw settlement was indicated by the bisector appeared too high. It was judged that the moistures and density corresponding to zero thaw strain should be:

- moisture content (dry wt.) = 17.0%; (volumetric) = 31.3%
- frozen bulk density = 2165 kg/m^3

and the lower portion of the correlation for these low plastic soils was modified accordingly.

THAW SETTLEMENT PREDICTIONS

Geothermal Predictions

Thermal simulations of the ground thawing induced by a noninsulated warm buried pipeline were carried out using a computerized geothermal model (Hwang, 1976). Several different pipe temperatures and soil conditions were considered. This study provides the ultimate depth or zone of soil that may eventually be thawed by the operation of the warm pipe. These predicted thaw depths (see Figure 10) have been used for selecting the base of the thaw bulb for thaw settlement calculations. This is primarily dependent on soil type and water content, as these influence the thermal properties.

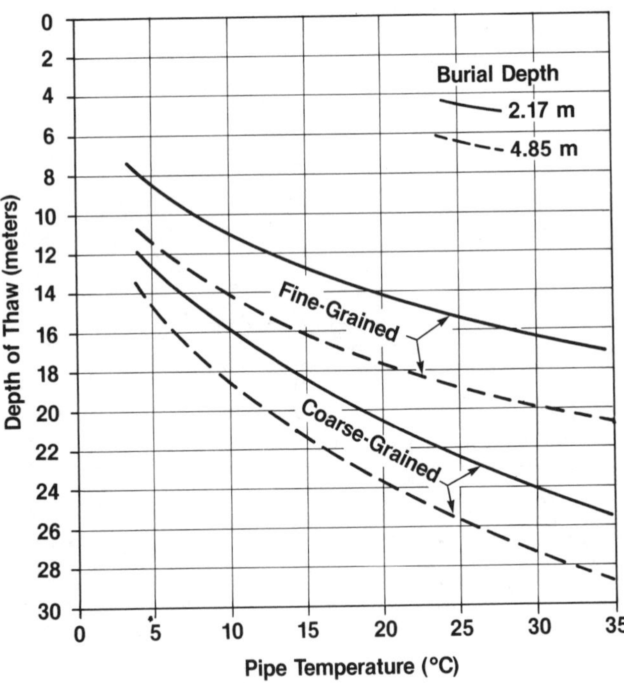

FIGURE 10 Predicted (30 year) thaw depths below the ground surface.

Thaw Settlement Program

A computer program was developed to calculate thaw settlement for both the pipe and the right-of-way (ROW). A brief flow chart in Figure 11 illustrates the major steps in the program. The settlement can be calculated for each individual borehole and for any given depth range by defining the depths to the top and bottom of the thaw zone.

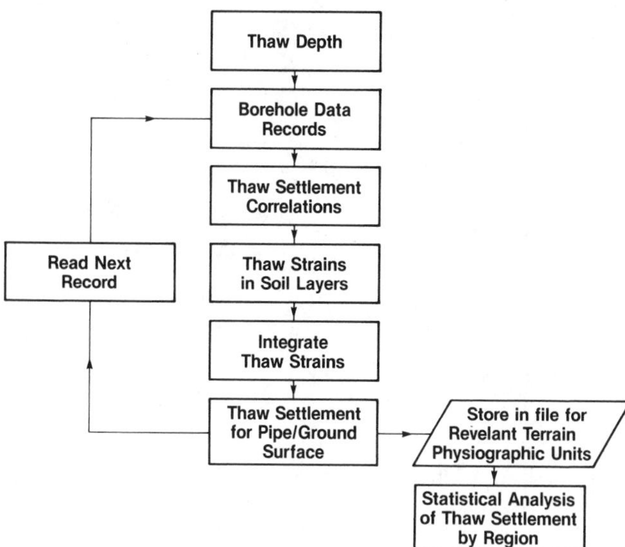

FIGURE 11 Conceptual flow chart for thaw settlement calculations.

The actual thaw settlement calculation was carried out for each individual borehole, layer by layer, using the data stored in a borehole data bank which listed up to eight layers per borehole with moisture and density data. Over 1,000 borehole records were available for this area of the pipeline in the South Yukon, and these were stored on floppy discs and accessed by a Hewlett Packard HP-9845 minicomputer. The layer settlements were then summed to provide the total settlement prediction for the thaw zone defined. The thaw strain values of the individual layers were obtained from the moisture content correlation in preference to the density correlation because moisture data were more abundant.
The mean thaw settlement was calculated for each borehole as

$$S_B = \Sigma(X_i)(\epsilon_i) \tag{2}$$

where S_B is the mean borehole settlement; X_i is the thickness of soil layer i and ϵ_i is the mean thaw strain for the layer, obtained from the correlations.

The standard deviation in the above settlement value was calculated based on the standard error for each soil layer and the layer thickness as follows:

$$\sigma_B = \sqrt{\sigma_1^2 + \sigma_2^2 + \cdots + \sigma_8^2} \tag{3}$$

where $\sigma_1 = (X_1)(s_1)$ etc.
and σ_B is the individual borehole standard deviation; X_1 is the thickness of layer one; and s_1 is the standard error in the mean thaw strain for layer one.

For most design purposes it was required to evaluate the thaw settlement potential for a given terrain unit by examining all the boreholes in that terrain unit. Therefore, the mean value of settlement, $\bar{S}$, for a terrain unit with M boreholes is:

$$\bar{S} = \frac{1}{M}\{(S_B)_1 + (S_B)_2 + \cdots + (S_B)_M\} \tag{4}$$

and the overal standard deviation, $\bar{\sigma}$, for the terrain unit is:

$$\bar{\sigma} = \frac{1}{M}\sqrt{\{(\sigma_B)_1^2 + (\sigma_B)_2^2 + \cdots + (\sigma_B)_M^2\}} \tag{5}$$

For any required confidence level, the thaw settlement can then be stated as

$$S = \bar{S} + (Z_c)(\bar{\sigma}) \tag{6}$$

where Z_c is the number of standard deviations for a given confidence level.
Z_c = 0.0 for 50%; 1.04 for 85%; and 2.33 for 99% confidence level, using a one-sided test.

While these statistically determined data for terrain units provided the major input into final thaw settlement design, thaw settlement prediction for individual boreholes were also used for certain isolated areas.

DESIGN PROCESS

A design confidence level of 85% was adopted for the project. The appropriate predicted thaw settlements were compared to the allowable thaw settlements developed by pipe stress analyses. To arrive at the allowable thaw settlement, it was assumed that the considered settlement could occur suddenly in a stepwise fashion across an interface. This interface could occur anywhere within the terrain unit under consideration (see Johnson et al. 1982). The basic model analyzed is shown in Figure 12. The analysis was performed using a

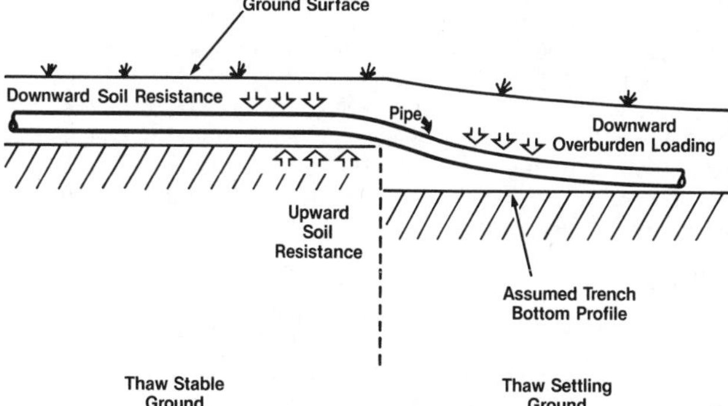

FIGURE 12 Differential thaw settlement at a transition.

commercially available structural pipe stress model (PIPLIN-III, Structural Software Development, Inc., 1981). The allowable thaw settlement was predicted based on established project pipe strain criteria and was dependent largely on the bearing strength of the soil beneath the pipe on the thaw-stable side and the weight of backfill on the pipe on the settling side.

The detailed design of the pipeline has been carried out for two "warm-flow" construction sections. On a kilometer-by-kilometer basis the predicted and allowable settlements were compared considering first of all a standard depth of burial (pipe invert approximately 2 m below original grade). Where the predicted settlement exceeded the allowable settlement, the design proceeded toward selecting one of the alternate design modes to mitigate the thaw settlement problem.

Mitigative Design Modes

The first option was to bury the pipe deeper. This mode was particularly appropriate where icy surface soils extended to depths greater than standard pipe burial depth, but were underlain at reasonable depth by dense, thaw-stable soil types The practical limits to the depth of ditch are normally 3.5 to 4 m, though in special situations up to 5 or even 6 m was considered. Where icy soils extended to depths greater than this, for significant lengths of pipeline, a combination of deep burial with a foamed bedding material was considered. The purpose of the foam (polyurethane was to provide, in effect, a weaker bearing support on the thaw-stable side of the basic model (Figure 12). The allowable settlement could be effectively increased by the amount the foam could compress.

In particularly ice-rich terrain units, such as lacustrine silts and clays and organic deposits, the design selected was invariably an above-grade mode. The concrete-restrained mode evolved during the design process as the preferred above-grade design mode. In this mode, the pipe was essentially placed on an insulated gravel pad laid on the cleared ROW, and specially designed, insulated concrete weights placed over the pipe (see Figure 13). Further details of this design mode are presented by Carlson and Butterwick (1983).

Another special design considered was a buried insulated pipe. This mode was only considered where an above-grade mode would otherwise intersect a highway, and a buried crossing of the highway is required.

FIGURE 13 Construction of above-grade, concrete restrained mode at Quill Creek Test Facility.

ACKNOWLEDGMENTS

The authors wish to acknowledge the assistance of many of their colleagues at Hardy Associates Ltd. and Foothills Pipe Lines (Yukon) Ltd., namely J. F. Nixon, T. Robinson, J. R. Ellwood, and J. R. Wonnacott.

REFERENCES

CARLSON, L. E. and BUTTERWICK, D. E. (1983), Testing Pipelining Techniques in Warm Permafrost, in this conference proceedings.

CRORY, F. E. 1973, Settlement associated with the thawing of permafrost, in Permafrost--The North American Contribution to the Second International Conference, Yakutsk: Washington, D.C., National Academy of Sciences, p. 599-607.

HWANG, C. T., 1976, Predictions and observations on the behavior of a warm gas pipeline on permafrost: Canadian Geotechnical Journal, v. 13, no. 4, p. 452.

JOHNSON, T. C., McROBERTS, E. C. and NIXON, J. F., 1982, Design implications of subsoil thawing: ASCE Spring Convention, Las Vegas, Nevada, April 26-30, 1983.

LUSCHER, U., and AFIFI, S., 1973, Thaw consolidation tests on Alaskan silts and granular soils, in Permafrost - The North American Contribution to the Second International Conference, Yakutsk: Washington, D.C., National Academy of Sciences, p. 325-334.

LYON, A. J., 1970, Dealing with Data: Pergamon Press, New York.

McROBERTS, E. C., LAW, T. C., and MONIZ, E., 1978, Thaw settlement studies in the discontinuous permafrost zone, Proceedings, Third Internation Internatal Conference on Permafrost, Edmonton, Canada, v. 1, p. 700-706.

MORGENSTERN, N. R. and NIXON, J. R. 1971, One-dimensional consolidation of thawing soils: Canadian Geotechnical Journal, v. 8, no. 4, p. 559-565.

PIPLIN-III, Structural Software Development Inc., 1981, Computer program for stress and deformation analysis of pipelines.

TSYTOVICH, N. A. and SUMGIN, M. I., 1937, Principles of mechanics of frozen ground: Cold Regions Research and Engineering Laboratory, Hanover, N.H. Transl. 19, p. 151-161.

WATSON, G. H., SLUSARCHUK, W. A., and ROWLEY, R.K. 1973, Determination of some frozen and thawed properties of permafrost soils: Canadian Geotechnical Journal, v. 10, no. 4, p. 592-606.

VESICLES IN THIN SECTIONS OF PERIGLACIAL SOILS FROM NORTH AND SOUTH NORWAY

Charles Harris

Department of Geology, University College Cardiff, Cardiff CF1 1XL, U.K.

Micromorphological investigations of silty sand soils from a solifluction slope in Okstindan, North Norway, revealed vesicles apparently identical to those encountered in samples from the centers of sorted circles at Gråsubreen, Jotunheimen, Norway. Vesicles appeared in thin section as smooth-walled spherical voids. Walls were clean with no evidence of illuvial silt linings. Other elements of the micromorphology were disrupted by vesicles, including silt coatings on sand grains and silt droplet fabrics. On the basis of simple laboratory experimentation and field observation it was concluded that vesicles resulted from the following conditions: (1) low initial bulk densities during thaw of ice-rich soils; (2) high moisture contents, but air present in larger pores; (3) disturbance of the sensitive soil due to thaw-consolidation, or to the process of sampling. Disturbance caused repacking of grains, reduction in void ratio, increase in pore water pressures and liquefaction. Air in larger voids and pores became bubbles, forming the vesicles observed in thin section.

INTRODUCTION

There have been a number of reports of arctic and alpine soils containing spherical bubble pores or vesicles in their upper layers. The occurrence of vesicles has, for instance, been recorded in the Mesters Vig area of Greenland (Ugolini 1966), the islands of the Canadian High Arctic (Bunting and Jackson 1970, Bunting and Fedoroff 1974, Bunting 1977), Spitzbergen (Chandler 1972), Elephant Island, South Shetlands (O'Brien et al. 1979), northern Norway (Harris 1977, Harris and Ellis 1980) and South-east Iceland (Romans, Robertson and Dent 1980). Vesicles are commonly developed in areas of patterned ground such as sorted and nonsorted circles and gelifluction lobes and terraces where soils are immature, either due to environmental factors or recent exposure of the surface to pedogenesis, and where soils are wet, at least during the thaw period.

In this paper vesicles are described from a soil developed on the tread of a turf-banked solifluction lobe in Okstindan, North Norway, and from the fine-grained centre of a sorted circle below the glacier Gråsubreen in Veodalen, Jotunheimen, southern Norway (Figure 1). On the basis of field observations and simple laboratory experimentation a mechanism of vesicle formation is proposed.

SITE DESCRIPTION

In Okstindan solifluction lobes are developed on an east-facing slope of gradient 5°-15° an altitude of around 710m, some 2km to the east of the outlet glacier Austre Okstindbreen (Figure 1a). Mean annual temperatures are approximately -0.5°C (Harris 1982), and annual freezing has been measured to depths in excess of 1.5m (Harris 1974). Permafrost is absent. At Gråsubreen in the Jotunheim, sorted circles some 3m in diameter are present at an altitude of around 1,860m immedi-

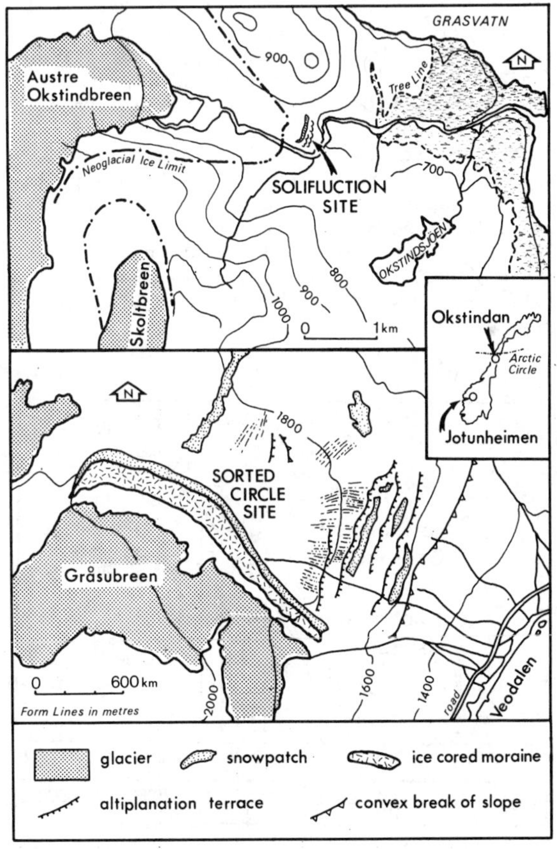

FIGURE 1 (a) Location map Okstindan
(b) Location map Gråsubreen

ately in front of a large ice-cored (Østrem 1964, 1965) terminal moraine (Figure 1b). The circles consist of central islands of fines, surrounded by concentrations of coarse clasts. The coarser margins form an irregular polygonal network which grades laterally into stripes where the gradient increases. Mean annual temperature was estimated as approximately -5°C from the analysis of Green and Harding (1980). It is not known if permafrost is present at this site; none was encountered during sampling, but soil waterlogging prevented deep excavation. At both sites moisture contents were high during sampling, water being supplied by late snow patches immediately upslope in both cases.

MATERIALS AND METHODS

Silty sand tills deposited by the last major Scandinavian glaciation form the substrate at both the Okstindan and the Jotunheim sites, the tills being derived from mica schists and pyroxene granulite gneiss respectively. Grain-size determinations (Figure 2) show both soils to be frost susceptible. Liquid limits and plasticity indices are low (Okstindan site 27% and 5%, respectively; Jotunheimen site 24% and 1%, respectively) and shear strength determination for samples from the Okstindan solifluction slope revealed a peak angle of friction of 35°, with residual angle of friction between 29° and 33°, the cohesion intercept being zero in both cases (Harris 1977). Similar shear strength parameters are to be anticipated for the sorted circle site in view of its similarity in texture and Atterberg properties to the Okstindan soils. Both sites, therefore have nonplastic frictional soils whose consistency is likely to be sensitive to relatively small variations in moisture status. Unit weights for the Okstindan soil are relatively low, with an average of 1.58Mgm^{-3} (Harris 1977).

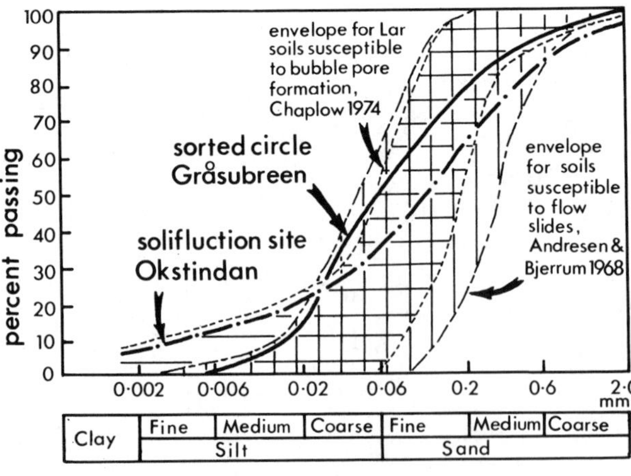

FIGURE 2 Grain size distributions with envelopes of soils susceptible to bubble pore formation

Pedological descriptions of profiles on the Okstindan turf-banked solifluction lobes were given by Harris and Ellis (1980). The thin sections reported here are from the upper part of the slope (profile 1, Harris and Ellis 1980), where late lying snow restricts vegetation to mosses, with some Salix herbacea and Carex bigelowii. Pedogenesis is limited to a 3cm thick A_h horizon which merges into the underlying silty sand soliflucted till. At the Gråsubreen site the vegetation is largely mosses and lichens, and the pedological soil consists of a black organic crust up to 2cm thick above the frost-sorted till substrate.

Undisturbed samples were collected in 6cm^2, 4cm deep open plastic boxes. At the solifluction site two samples were taken from three depths, 2-8cm (nominally 5cm), 22-28cm (nominally 25cm) and 47-53cm (nominally 50cm), and at the sorted circles three samples were collected from the centre of a circle at depths of 5-11cm, soil saturation preventing deeper sampling. Following impregnation, vertical and horizontal thin sections were prepared for microscopic investigation.

VESICLES

At both sites rounded bubble-like pores formed prominent elements of the soil micromorphology (Figure 3). Vesicles were most numerous in the near-surface samples from Okstindan, were less numerous at 25cm, and were not observed in samples from 50cm depth. Samples taken from the center of the Jotunheim sorted circle showed a similar frequency to the near-surface samples in Okstindan. One hundred vesicles were measured for each sample, and results are summarized in Table 1.

TABLE 1 Dimensions of Vesicular Pores

	Okstindan 5cm depth	Okstindan 25cm depth	Jotunheimen sorted circle
mean diameter (μm)	664	390	772
maximum diameter (μm)	4000	1000	2600

At both sites vesicles were simple smooth-walled voids, approximately spherical in shape. Walls were clean with no evidence of illuvial silt linings. These pores appeared to resemble closely the bubbly pores in soils from Elephant Island, South Shetland Islands, and Iceland, described by O'Brien et al. (1979). Other elements of the micromorphology were disrupted by vesicles, including silt coatings on sand grains and lenticular concentrations of silt ("silt droplets", Romans and Robertson 1974) (Figure 3). Clearly the vesicles formed more recently than the grain coatings and silt droplets. Only one generation of vesicles was present, and they had not persisted long enough to acquire illuvial linings.

EXPERIMENTALLY PRODUCED VESICLES

During sampling at Gråsubreen, one undisturbed sample was accidently shaken out of its container and disturbed. Examination of this disturbed sample showed clearly visible bubbly pores, and further pores were found in a wet bulk sample taken by trowel. The possibility arose, therefore, that vesicles observed in thin section did not result directly from the soil freezing process as suggested by Fitzpatrick (1956), but rather from consolidation of the soil under wet conditions following thaw. A simple laboratory experiment tested this hypothesis.

Soil from the Gråsubreen sorted circles was placed in a plastic box 6cm square and 8cm high. The soil was lightly compacted and then thoroughly wetted. The specimen was placed with its base immersed in a reservoir of water, and then frozen from the surface downwards. Ice segregation during freezing gave an excess water content in the frozen soil of $0.3cm^3$ per cm^3 soil. Following freezing the specimen was allowed to thaw from the surface downwards, this being associated with consolidation and the shedding of excess water. The surface was then pressed firmly downwards with a rubber piston. Consolidation led to expulsion of excess water at the surface, and the specimen was transformed into a very viscous slurry. The sample was then air dried for one week, oven dried for 24 hours, and impregnated with resin prior to the preparation of vertical and horizontal thin sections. These revealed abundant vesicles, impossible to differentiate under the microscope from those observed in the field samples (Figure 3(h)). The experimentally produced vesicles had a maximum diameter of 2500μm, and mean diameter 552μm. This experiment was repeated using soil from the Okstindan site and again vesicles identical in form to those in the field samples were produced. A duplicate control specimen was prepared, using Okstindan soil, but not frozen or compacted. No vesicles were observed in thin sections of this control specimen. Finally, dry Okstindan soil was mixed with water to form a paste, placed in the specimen container and compacted with the rubber piston. Liquification and expulsion of water resulted, and when the dried specimen was broken open abundant vesicles up to 3.6mm diameter were found.

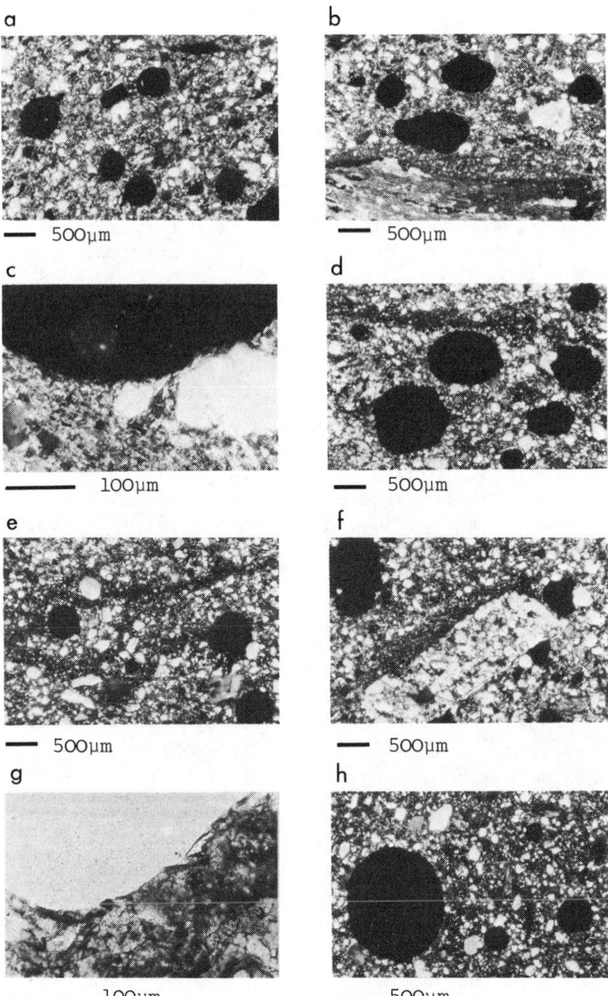

FIGURE 3 Vesicles
(a) 5cm sample, Okstindan
(b) with grain coating, 5cm, Okstindan
(c) wall, 5cm sample, Okstindan
(d) with silt droplet fabric, Gråsubreen
(e) disrupting silt droplets, Gråsubreen
(f) with grain coating, Gråsubreen
(g) wall, Gråsubreen
(h) resulting from freezing experiment using Gråsubreen soil

DISCUSSION

It is unlikely that the vesicles observed following the freezing experiments could have formed during the freezing process in the way proposed by Fitzpatrick (1956), since identical vesicles were produced simply by compacting a wet soil specimen, with no prior freezing. The specimens were also not subjected to rapid drying, such as that which generated vesicles in Devon Island, N.W.T., Canada (Bunting 1977), and initial wetting did not generate air bubbles in the control specimen, as was described by Chaplow (1974) and Romanov (1974). Bunting (1977,p.94) describes vesicular pores in the fine centres of sorted circles as "presumably best related to ejection of moisture and air migration under the stress of cryoturbation". In the experiment described here it seems likely that the observed vesicles developed due to the stresses applied during and after the thawing stage of cryoturbation rather than during the freezing stage.

The results of these simple laboratory experiments suggest the following sequence of vesicle formation:

(1) low bulk density immediately following thaw due to incomplete consolidation. The volumes of the Gråsubreen and Okstindan soil specimens increased by 37.5% and 11.25% respectively due to ice segregation during freezing;

(2) some drainage of excess water immediately following thaw, and prior to compaction, allowing air to enter the larger pores;

(3) compaction under wet conditions leading to a thixotropic reaction as pore space was reduced, excess water expelled, and pore pressures consequently increased. The soil became a viscous

slurry, and pores containing air became bubbles, forming the vesicles observed in thin sections.

The vesicles observed in the field samples are considered to have developed by a very similar process.

Winter freezing in Okstindan is associated with considerable frost heave due to ice segregation (Harris 1972). At the Gråsubreen sorted circle site frost heaving in winter is suggested by the lenticular concentrations of silt observed in thin sections, considered to mark the former locations of ice lenses (Harris and Ellis 1980), and by the ice segregation observed in the laboratory experiment using soil from the Gråsubreen site. The unit weight determinations for the Okstindan site confirmed relatively low values following spring thaw, and Williams (1959) also reported low unit weights immediately following thaw in a soliflucted till in Rondane, Norway, with subsequent increase through the following summer.

During the thaw period moisture contents were high at both field sites, so that any disturbance of the low density soils, either naturally due to consolidation of the thawing soil, or due to the soil sampling procedure, was likely to cause liquefaction as grains resettled into a denser packing, reducing the void ratio and increasing pore pressures as excess pore water was expelled. Harris (1972) recorded settlement of the soil surface at the Okstindan site of up to 8.8cm during the summer thaw of 1971. Air-filled pores would, under these conditions become bubbles. Chaplow (1974) explained bubble voids in borehole samples of sand from the Lar Dam in northern Iran as resulting from liquefaction of partially saturated sand during sampling, and used the grading envelope for soils susceptible to flow slides of Andresen and Bjerrum (1968) as a guide to soils liable to develop bubble structures. The soils at both Norwegian sites described in this paper fall within the Chaplow/Andresen and Bjerrum range of grading curves (Figure 2). Chaplow found that bubble pores developed when the degree of saturation in the initial sample was in excess of 75%, but that as saturation approached 100% the lack of soil air prevented bubble formation.

The fresh appearance of the vesicles described in this paper, their lack of illuvial silt linings, and the absence of traces of former vesicles in the soil micromorphology suggests either that they do not survive from one season to the next, or that they arose purely as a result of disturbance during sampling and did not form under natural conditions. However, at the Okstindan site at least, thaw consolidation has been shown to occur, so that here conditions conducive to vesicle formation are provided naturally.

CONCLUSIONS

On the basis of field observations and simple experimentation the vesicles described in this paper are considered to result from thawing of ice-rich silty sand soils with low liquid limits and plasticity indices. Thaw-consolidation leads to thixotropic behaviour as excess water is expelled, and air present in voids or larger pores becomes bubbles. It is clear that any soil susceptible to liquefaction could form vesicles as a result of consolidation under appropriate moisture conditions. The thawing of ice-rich soils provides particularly favourable conditions for vesicle formation, so that many micromorphological studies of periglacial soils have reported their presence. They are unlikely to greatly affect the hydraulic conductivity of the soil during thaw, for apart from being air-filled, they do not form an interconnecting system of voids.

The presence of bubble pores or vesicles in soils from periglacial patterned ground may therefore indicate that liquefaction occurs during the thaw period. Such liquefaction is likely to form an important component in the process of formation of the patterned ground itself, especially sorted and nonsorted circles, and solifluction lobes and terraces.

ACKNOWLEDGEMENTS

Fieldwork was completed while the author was a member of the University of Reading Okstindan Research Project, and the University College Cardiff Jotunheimen Expedition. Financial support from University College Cardiff and the British Geomorphological Research Group is gratefully acknowledged.

REFERENCES

Andresen, A., and Bjerrum, L., 1968, Slides in subaqueous slopes in loose sand and silt: Norwegian Geotechnical Institute Publ. 81, p.1-9.

Bunting, B.T., 1977, The occurrence of vesicular structures in arctic and subarctic soils: Zeitschrift für Geomorphologie, v.21, p.87-95.

Bunting, B.T., and Fedoroff, N., 1974, Micromorphological aspects of soil development in the Canadian High Arctic, in Rutherford, G.K., ed., Soil Microscopy: Kingston Ont., The Limestone Press, p.350-365.

Bunting, B.T., and Jackson, R.M., 1970, Studies of patterned ground in south-west Devon Island, N.W.T.: Geografiska Annaler, v.52A, p.194-208.

Chandler, R.J., 1972, Periglacial mudslides in Vestspitzbergen and their bearing on the origin of fossil 'solifluction' shears in low angled clay slopes: Quarterly Journal of Engineering Geology, v.5, p.223-241.

Chaplow, R., 1974, The significance of bubble structures in borehole samples of fine sand: Géotechnique, v.24, p.333-344.

Fitzpatrick, E.A., 1956, An indurated soil horizon formed by permafrost: Journal of Soil Science, v.7, p.248-257.

Green, F.H.W. and Harding, R.J., 1980, The altitudinal gradients of air temperature in southern Norway: Geografiska Annaler, v.62A, p.29-36.

Harris, C., 1972, Processes of soil movement in turf-banked solifluction lobes, Okstindan, northern Norway: Institute of British Geographers Special Publ. 4, p.155-174.

Harris, C., 1974, Autumn, winter and spring soil temperatures in Okstindan, Norway: Journal of Glaciology, v.13, p.521-533.

Harris, C., 1977, Engineering properties, groundwater conditions, and the nature of soil movement on a solifluction slope in North Norway:

Quarterly Journal of Engineering Geology, v.10, p.27-43.

Harris, C., 1981, Periglacial mass-wasting: A review of research: British Geomorphological Research Group, Research Monograph Series, No.4, Norwich, Geo Abstracts, 204pp.

Harris, C., 1982, The distribution and altitudinal zonation of periglacial landforms, Okstindan, Norway: Zeitschrift für Geomorphologie, v.26, p.283-304.

Harris, C., and Ellis, S., 1980, Micromorphology of soils in soliflucted materials, Okstindan, northern Norway: Geoderma, v.23, p.11-29.

O'Brien, R.M.G., Romans, J.C.C., and Robertson, L., 1979, Three soil profiles from Elephant Island, South Shetland Islands: British Antarctic Survey Bulletin 47, p.1-12.

Østrem, G., 1964, Ice-cored moraines in Scandinavia: Geografiska Annaler, v.46, p.282-237.

Østrem, G., 1965, Problems of dating ice-cored moraines: Geografiska Annaler, v.47, p.1-38.

Romanov, Y.E.A., 1974, Effect of trapped gas on some processes in soils: Soviet Soil Science, v.4, p.222-228.

Romans, J.C.C., and Robertson, L., 1974, Some aspects of the genesis of alpine and upland soils in the British Isles: in Rutherford, G.K., ed., Soil Microscopy, Kingston, Canada, p.498-510.

Romans, J.C.C., Robertson, L., and Dent, D.L., 1980, The micromorphology of young soils from south-east Iceland: Geografiska Annaler, v.62A, p.93-103.

Ugolini, F.C., 1966, Soils of the Mesters Vig district, Northeast Greenland: Meddelelser om Grønland, Bd. 176, No.2, 26pp.

Williams, P.J., 1959, An investigation into the processes occurring in solifluction: American Journal of Science, v.257, p.481-490.

COMPARISON OF THE CLIMATIC AND GEOMORPHIC METHODS OF PREDICTING PERMAFROST DISTRIBUTION IN WESTERN YUKON TERRITORY

Stuart A. Harris

Department of Geography, University of Calgary, Calgary, Alberta, Canada T2N 1N4.

Drilling and ground temperature measurements along the Dempster Highway demonstrate that continuous permafrost is at least twice as extensive in Western Yukon Territory as previously shown on maps. An extensive belt occurs in the foothills below the St. Elias Range west of Destruction Bay, while continuous permafrost extends south to the North Fork Pass along the Dempster Highway. Between the zones of continuous permafrost are areas of discontinuous permafrost in lowland areas, with 30-80% of the landscape underlain by permafrost. The climatic predictive method is reasonably successful at predicting the permafrost distribution on small scale maps but tends to slightly underpredict the distribution in montane valleys. Cold air drainage is widespread and very marked in these areas, and may cause the anomalies. Groundwater movement is undoubtedly also important locally. The geomorphic method based on the distribution of zonal permafrost landforms is of little use in mapping the permafrost boundaries. It was the evidence used to augment the sparse ground-temperature data available to previous workers.

INTRODUCTION

Permafrost is a temperature condition of the ground where it remains below 0°C for more than 2 years. These conditions can be found occupying some very extensive areas of Canada, Alaska, the USSR, and China, and mapping them presents a considerable problem. Direct methods involving measuring ground temperatures are expensive and time-consuming, and so there is a great need for indirect methods of prediction.

In 1981, a climatic method of mapping or predicting was described based on a correlation between freezing and thawing indices, calculated from daily air temperatures, and the distribution of permafrost zones in areas of low winter snowfall (Harris, 1981). The method is potentially useful where sufficient climatic data are available from representative sites.

Subsequently, it has been shown that there is a general correlation between active zonal permafrost landforms (as defined by Tricart and Cailleux, 1950) and the freezing and thawing indices in areas of low winter snowfall (Harris, 1982a). Since many of these landforms, e.g., ice wedges and pingos, can be identified on aerial photographs taken at suitable scales, it would seem that this method involving photo-interpretation could have considerable value for small scale mapping, if it works adequately.

Recent work in determining the permafrost conditions along the proposed Alcan Gas pipeline route in southern Yukon Territory (courtesy of John Elwood and Doug Fisher, Foothills Pipe Lines (Yukon), Ltd.) and along the Dempster Highway, provides good control for the western Yukon. The purpose of this paper is to compare the actual permafrost distribution in Western Yukon Territory, as determined by drilling and ground temperature measurements, with the predictions based on the climatic and landform methods. This provides a good test of the adequacy of these two methods for prediction of permafrost in constructing small-scale maps.

METHODS USED

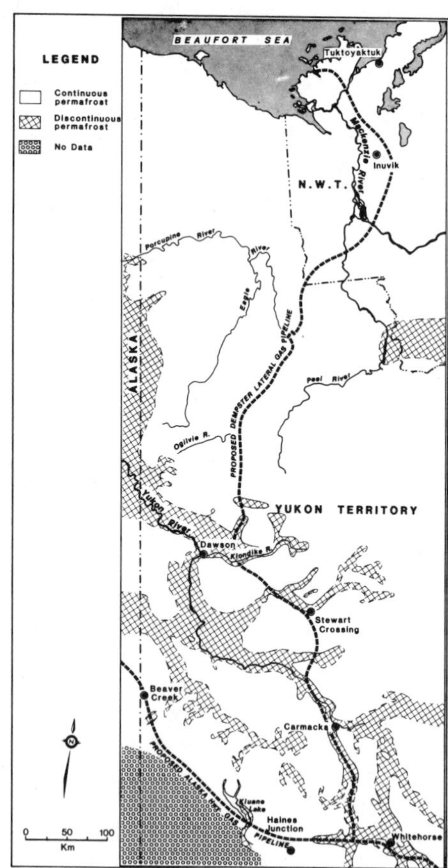

FIGURE 1 Actual distribution of permafrost in the western Yukon Territory based on ground temperature data from our 175 thermistor strings along the proposed Alcan and Dempster/Klondike pipeline routes (data courtesy of John Elwood and Doug Fisher, Foothills Pipelines (Yukon), Ltd.).

The data on permafrost conditions derived from over 50 thermistor cables along the Alcan route west of Whitehorse were used to identify and map the permafrost distribution along the Alcan route in southwest Yukon Territory (Figure 1). The data from the 125 thermistor strings emplaced between Whitehorse and Inuvik along the Klondike and Dempster highways were examined and used. Some extra readings were obtained by the writer from six key cables to check the results. Areas of continuous permafrost (defined as greater than 80% of the thermistor strings in an area proving permafrost) were separated from those with discontinuous or sporadic permafrost and mapped on 1:50,000 scale topographic maps. The boundary was then extrapolated throughout the western Yukon Territory and the adjacent portions of the Northwest Territories, using the topography, vegetation, aspect and active permafrost landforms (e.g. pingos, palsas) as a guide.

The available climatic data for the same area were examined and it was found that the number of Class A weather stations had increased substantially in 1974. All 30 stations showed under 50 cm mean winter snow covers. A check was therefore made to select two subsequent years (post-1974) for which the freezing and thawing indices were close to the long term means of four typical older weather stations. The means for 1975 and 1876 gave roughly comparable results, bearing in mind the rather wide range of individual values from year to year.

The mean daily temperatures (°C) for the year for a given weather station provide the basis for calculation of the freezing index (sum of the negative values) and the thawing index (sum of the positive values). In this way, the freezing and thawing indices for all the Class A weather stations in western Yukon and the adjacent portions of the Northwest Territories for 1975 and 1976 were calculated and averaged. The results (Appendix A) were plotted on a freeze-thaw index diagram and then classified into continuous and discontinuous permafrost by the climatic method (Harris 1981). To provide extra detail, the discontinuous permafrost sector was subdivided into five equal parts (DA, DB, DC, DD, and DE) as in Figure 2. This gave greater precision to mapping the climatic zonation at the margin of the area of continuous permafrost. The latitude, longitude, and altitude of each station were used to check the homogeneity of the climatic parameters within the area, and to establish the relationship of altitude to the indices.

The distribution of landforms usually regarded as being related to permafrost conditions in the western Yukon Territory and adjacent portions of the Northwest Territories is indicated in Figure 3. Considerable detail was found in Zoltai and Tarnocai (1975) and in Hughes (1969), while other information is given in Klassen (1979). The data have been augmented by field observations and study of the aerial photographs. The occurrences of most zonal permafrost landforms were mapped for the western Yukon, although the rock glaciers were omitted since a conclusive separation of tongue-shaped and lobate forms is difficult, while too little information is available concerning which are active and which are fossil.

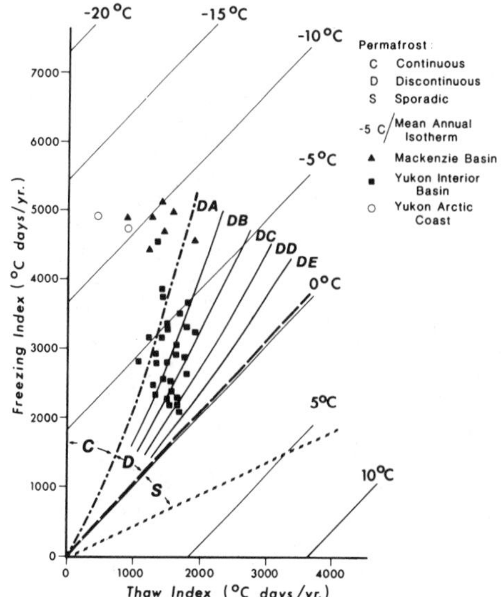

FIGURE 2 Mean annual freezing and thawing indices for 1975 and 1976 for the Class A weather stations in western Yukon Territory and the adjacent areas of the Northwest Territories, plotted on a freeze-thaw index diagram. Also shown are the boundaries of continuous, discontinuous and sporadic permafrost in areas of low winter snow cover. Note the way most stations from the central Yukon Plateau plot near the boundary between continuous and discontinuous permafrost. To provide closer resolution of climate in the discontinuous zone, it is divided into five equal parts labeled DA-DE (warmest).

RESULTS

The data on permafrost distribution based on the boreholes were compiled in Figure 1, and the boundaries extrapolated through the remaining portions of the western Yukon Territory. The results indicate a far larger area of continuous permafrost than previously indicated, except by Brown (1967a). A check with the drill hole logs for sites where no ground temperature cables were installed suggested that the results are correct. In view of the large number of thermistor cables (over 175) and the backup data from over 200 drill hole logs without ground temperature cables, the results should be far more accurate than the maps produced previously (e.g. Brown 1967a, 1978; Washburn 1979), which were based on measurements of about a dozen ground temperature cables, data from mines, and on landforms.

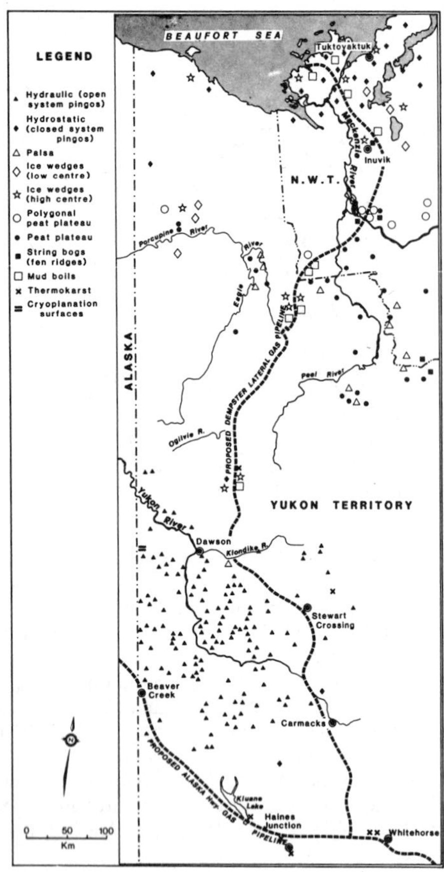

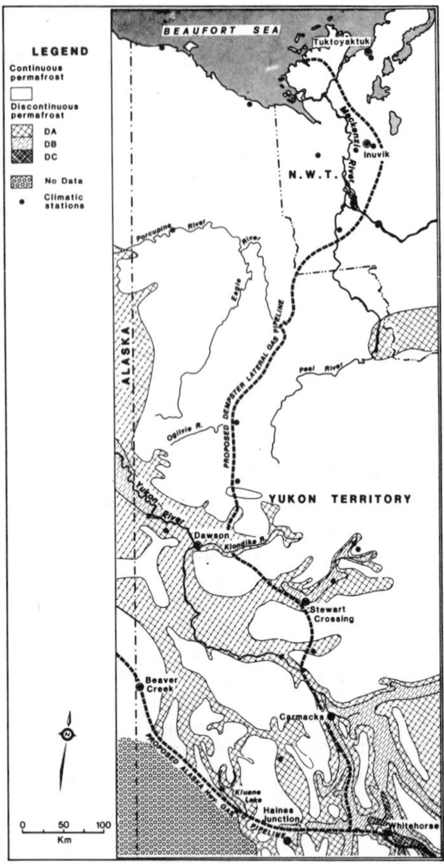

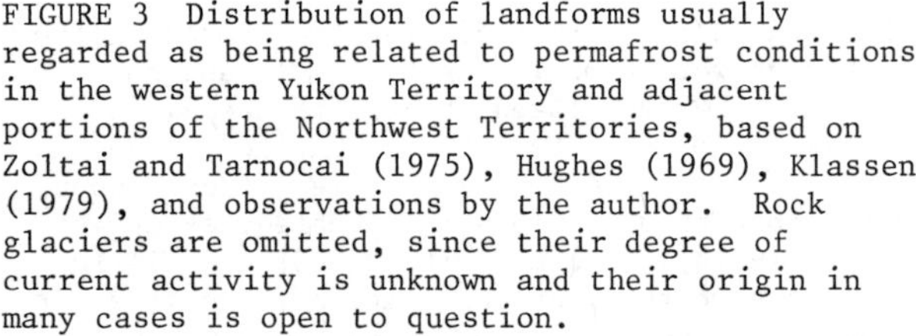

FIGURE 3 Distribution of landforms usually regarded as being related to permafrost conditions in the western Yukon Territory and adjacent portions of the Northwest Territories, based on Zoltai and Tarnocai (1975), Hughes (1969), Klassen (1979), and observations by the author. Rock glaciers are omitted, since their degree of current activity is unknown and their origin in many cases is open to question.

FIGURE 4 Predicted permafrost distribution using the climatic method (see Harris, 1971a). The discontinuous permafrost zone is subdivided into five equal parts as in Figure 2, but only three are found in the area. Note the similarity to Figure 1, apart from underprediction in the valleys below mountain ranges.

Comparison of the Actual Distribution of Permafrost with that Predicted by the Climatic Method

Figure 4 shows the results of (1), calculating the mean freezing and thawing indices for 1975-1976 for the class A weather stations, (2), using these to classify the stations into predicted continuous permafrost or predicted discontinuous permafrost using Figure 2, and (3), then plotting the results on a map of the western Yukon Territory. Boundaries were then interpolated and extrapolated using the topography as a guide. It was assumed that permafrost was more widespread at higher elevations, i.e., no cold air drainage occurred. The boundaries between the predicted continuous and discontinuous permafrost zones show considerable similarity to those in Figure 1.

Thus, in general, the method of climatic prediction described in 1981 seems to work quite well, even though it disagrees with the maps of permafrost published previously (e.g. Washburn, 1979 Figure 3.3; Brown, 1978).

In detail, there are noticeable differences between Figures 1 and 4. From the Alaska border (Snag) to the west end of Kluane Lake (Burwash Landing), along the Alaska Highway, the actual distribution of continuous permafrost is more extensive than would be expected by the results of the climatic method. The same results apply in south-central Yukon. Since this is also the zone of maximum frequency of forest fires, it is probable that the distribution of permafrost in southwest Yukon is in equilibrium with its climatic environment. Disturbance and removal of vegetation would have caused the development of a new thermal profile in the ground, in equilibrium with present-day conditions. Maximum permafrost thickness is commonly about 22 m. Again, the presence of permafrost and active thermokarst subsidence at Haines Junction in an area which was under Neoglacial Lake Alsek in the mid-late nineteenth century (Clague and Rampton, 1982) indicates that permafrost is actively forming in this century. In order to try to determine the cause of these differences, a series of extra checks were used.

Differences of grain size and lithology are known to correlate with different thermal properties (conductivity and diffusivity) (Johnston, 1981). This is a major cause of variations in thickness of permafrost, and can also determine its presence or absence in areas of discontinuous permafrost. However examination of the textures of the materials encountered in the boreholes indicates that there are two main materials present, viz.: river gravels and loess-derived silts (apart from one sample of till), based on the data in Klohn Leonoff Consultants, Ltd., 1979. Both gravels and silts occur in the floors and sides of the valley, so the materials are too homogeneous for this to be the explanation of the differences in Figures 1 and 4.

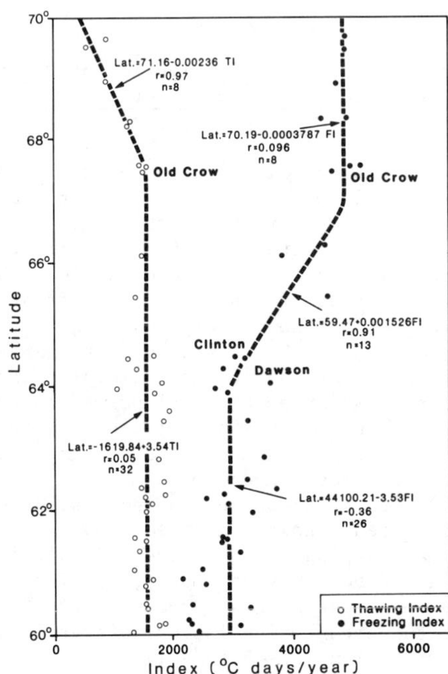

FIGURE 5 Variation of mean freezing and thawing indices for 1975 and 1976 with latitude of the weather stations in western Yukon Territory and the adjacent portions of the Northwest Territories.

Another possible cause could be heterogeneity of the climate across the region. Both freezing and thawing indices may show abrupt change in their trends across a part of a continent, e.g., from Medicine Hat to Inuvik (Harris, 1981). Figure 5 shows the results of plotting the mean freezing and thawing indices for 1975 and 1976 for the weather stations in western Yukon and the adjacent parts of the Northwest Territories against latitude. The r value will be extremely small when the regression line is nearly parallel to an axis, due to the high sensitivity of the factor. As demonstrated on the Medicine Hat/Inuvik traverse, an abrupt change in trend of the thawing and freezing indices occurs in the vicinity of Old Crow and it appears to be present on both sides of the Richardson Mountains. However, south of there, the trend for the thawing index is fairly continuous and indicates a constant relation of index to latitude right up to the foot of the St. Elias Range. This fits with previous work on the climate of southwest Yukon where Taylor-Barge (1969) has claimed that there is an abrupt change in climate between the Maritime influence of the north Pacific Ocean and the more continental interior of Yukon, which lies along the continental divide.

The freezing indices show a break in trend just north of Dawson. The change is marked and corresponds to the transition from dominantly tundra to dominantly forest. Freezing index increases more rapidly with increasing latitude under tundra vegetation. Figure 5 also shows that while the data for the thawing indices fall close to the trend line, those for the freezing index show rather more scatter. This suggests that there could be something complicating the microclimate in winter.

Harris (1982b) has shown that pronounced but sporadic cold air drainage can occur at the latitude of Fort Nelson along the valleys leading from the higher mountains. This can become extremely important in areas where the discontinuous and sporadic permafrost zones lie in the area where the cold air drainage occurs, e.g. east of Steamboat Mountain (west of Fort Nelson, B.C.).

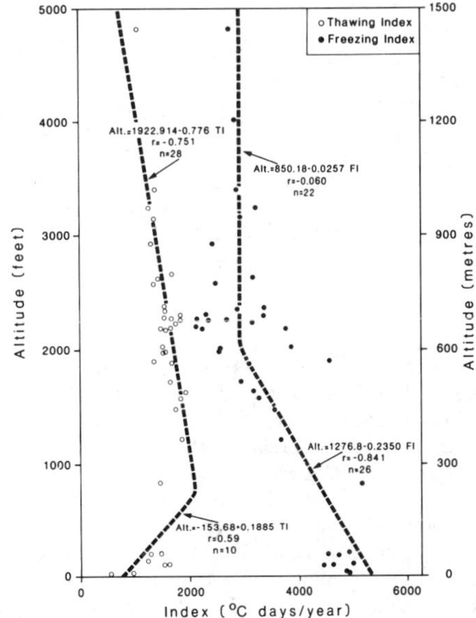

FIGURE 6 Variation of mean freezing and thawing indices for 1975 and 1976 with altitude of the weather stations in western Yukon Territory and the adjacent portions of the Northwest Territories.

Figure 6 shows the results of plotting all the mean freezing and thawing indices for the same stations for 1975 and 1976 against altitude of the climatic station. The thawing index shows a pattern decreasing with altitude. The only exceptions to the trend lie below 350 m and represent the stations along the Arctic coast and the Mackenzie valley.

The trend of the freezing indices is quite different from that found at lower latitudes (see for example Figure 8 in Harris, 1981) in that the index decreases with altitude. There is considerable scatter, but above 700 m altitude, the rate of change is a mere 200 C°days/yr every 600 m. At lower elevations it is over 100°C days/yr every 100 m. This indicates that the winter air temperatures below about 700 m above mean sea level are considerably colder than those at higher elevations. This is consistent with the presence of cold air drainage occurring throughout the region especially during the long, cold winter nights (Marcus, 1974, Harris, 1982b).

The effects of this process on mean annual air temperatures near Stewart Crossing-Keno Hill and Dawson City and vicinity are shown in Figure 7. In both cases, the highest stations have the lowest freezing indices. The effects of the cold air drainage are so great that they even cause a reversal of the direction of the normal lapse rate at lower elevations based on mean annual temperatures (Figure 8), indicating the influence of a strong persistent winter inversion.

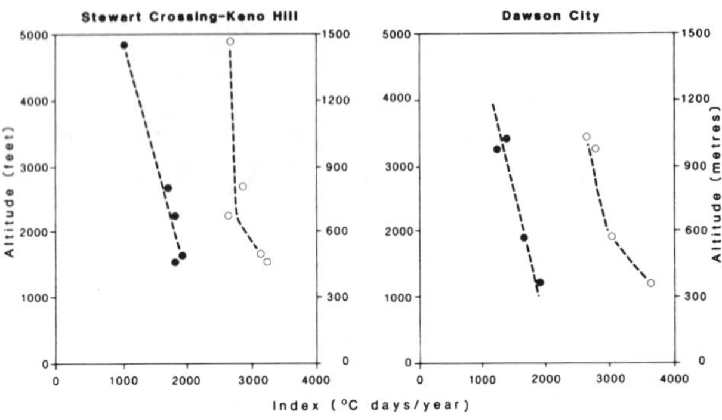

FIGURE 7 Changes in mean freezing (open circles) and thawing indices (solid circles) with altitude around Dawson City and at Keno Hill for 1975 and 1976.

The fact that both the freezing index and the thawing index decrease with altitude (Figure 7) makes detailed prediction of the presence or absence of continuous permafrost by climatic parameters in the Yukon Territory particularly hazardous. Re-examination of Figure 2 shows that the stations from the Yukon Interior Basin, i.e. south of Dawson City, lie straddling the boundary between the discontinuous and continous permafrost zones. This means that any local microclimatic modifications to the temperature regime such as variations in thickness of snow cover and zones of groundwater movement can have a profound effect on the distribution of permafrost. Cold air drainage may cause such a modification and would appear to be a probable contributing factor for the differences between the actual distribution of permafrost (Figure 1) and those predicted by using freezing and thawing indices (Figure 4). The effect of snow cover variability is minimized by the method, but the effects of zones of groundwater movement are obviously also considerable.

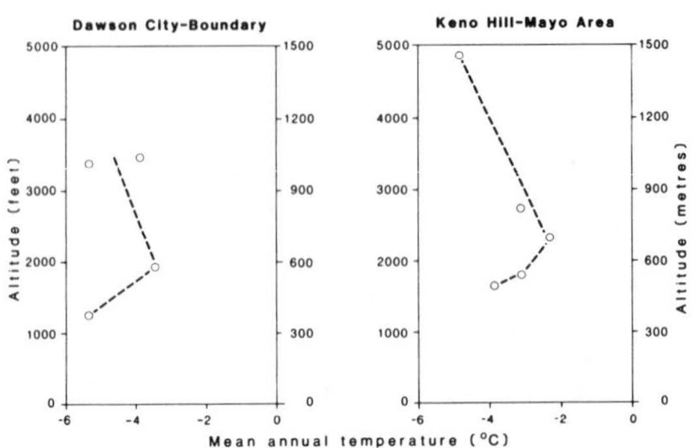

FIGURE 8 Change in mean annual air temperature for 1975 and 1976 with altitude for Class A weather stations around Dawson City and Keno Hill. Note the reversed slope of the lapse rate at lower elevations.

Comparison of the Actual Distribution of Permafrost with that Predicted Using Zonal Permafrost Landforms

Figure 3 shows the distribution of zonal permafrost landforms in western Yukon Territory and in the adjacent portions of the Northwest Territories. When it is compared with the actual distribution of permafrost in Figure 1, it becomes clear that only the major areas of continuous permafrost can be recognized using the variety and frequency of the landforms. Even then, it is only the areas with considerable ground ice that exhibit marked varieties and abundances of zonal permafrost landforms. The better drained alpine permafrost in the central Yukon Plateau lacks appreciable diagnostic permafrost landforms.

The area with both continuous and discontinuous permafrost present shows scattered zonal geomorphic features, mainly pingos, but these are inadequate for mapping the boundary between them. They merely suggest the presence of scattered permafrost throughout the area, which is very different to the actual distribution (see Figure 1).

CONCLUSIONS

Data from thermistor strings in boreholes along the proposed pipeline routes show that permafrost is much more widespread in western Yukon Territory than previously suspected (Figure 1). Continuous permafrost extends throughout the region, although discontinuous permafrost is found in the lower parts of the Yukon River valley and again near Whitehorse.

It is clear from a comparison of Figures 1 and 4

that the climatic method using freezing and thawing indices for weather stations in areas of low winter snowfall works fairly well in predicting the distribution of present day permafrost on small scale maps. However, in southern Yukon, variations in soil moisture and snow cover, and the presence of localized cold air drainage in winter probably cause an extension of the zone of continuous permafrost downslope into valleys below high mountains, e.g. along the foot of the St. Elias range from Snag to Burwash Landing.

As can be seen in Figure 2, the climate of the western Yukon Interior Basin is borderline between continuous and discontinuous permafrost. This explains why the prediction of permafrost is so difficult here. Both the freezing index and the thawing index decrease with altitude, unlike the situation in areas without appreciable cold air drainage, so that the vertical zonation of permafrost is weaker.

The geomorphic method of predicting permafrost works well for the zone of icy continuous permafrost in the north and also for showing that scattered permafrost occurs throughout the area. However, it fails to provide enough data for mapping the boundary between continuous and discontinuous permafrost and also fails to separate dry, well drained continuous permafrost from areas with sporadic icy permafrost. Thus the climatic predictive method is far superior.

ACKNOWLEDGEMENTS

N. M. Waters fitted the regression lines.

APPENDIX A Freezing and Thawing Indices for the Class A Climatic Stations in Western Yukon Territory and the Adjacent Areas of the Northwest Territories

Station	Freezing Index	Thawing Index	Classification
Aishihik A*	2946	1340	DA
Aklavik	4465	1256	C
Arctic Red River	5000E	1600E	C
Beaver Creek	3749	1459	C
Braeburn	2869	1523	DB
Burwash Landing	3164	1417	DA
Carmacks	2950	1640	DB
Clinton Creek	3058	1696	DB
Dawson A	3650	1850	DA
Dempster	3200	1240	C
Dempster 203	3850	1490	C
Dury Creek	2564	1542	DB
Elsa	2905	1702	DB
Faro	2670	1852	DC
Port McPherson	4657	1545	C
Fort Selkirk	3507	1763	DA
Haines Junction	2566	1512	DB
Inuvik A	4913	1284	C
Johnsons Crossing	2318	1552	DB
Keno Hill	2730	1065	C
Kluane Lake	2500	1350	DA
Konakuk Beach	4900	575	C
Mayo A	3156	1945	DB
Mile 34, Boundary Rd.	2830	1392	DA
Ogilvie River	4600	1350	C
Old Crow A	5193	1454	C
Shingle Point	4750	887	C
Stewart Crossing	3256	1819	DB
Tuktoyaktuk A	4906	923	C
Whitehorse A	2174	1664	DC

Based on daily temperatures (°C) in monthly meteorological data (Environment Canada).

E is estimate; C is Continuous Permafrost; DA, DB, DC, DD is Discontinuous Permafrost.

*After Brown (1967b) and Thompson (1963).

REFERENCES

Brown, R. J. E., 1967a, Permafrost in Canada. Geological Survey of Canada Map 1246a: Division of Building Research, National Research Council of Canada, Publication NRC 9769.

Brown, R. J. E., 1967b, Permafrost investigations in British Columbia and Yukon Territory: Division of Building Research, National Research Council of Canada Technical Paper No. 253, 55 p.

Brown, R. J. E., 1978. Permafrost: Plate 32 in Hydrological Atlas of Canada, Ottawa, Fisheries and Environment Canada (34 plates).

Clague, J. J., and Rampton, V. N., 1982, Neoglacial Lake Alsek, Canadian Journal of Earth Sciences, v. 19, p. 94-117.

Harris, S. A., 1981, Climatic relationships of permafrost zones in areas of low winter snowcover, Arctic, v. 34, p. 64-70.

Harris, S. A., 1982a, Identification of permafrost zones using selected periglacial landforms. _In_ the Roger J. E. Brown Memorial volume, Proceedings of the Fourth Canadian Permafrost Conference, H. M. French, Ed.: National Research Council of Canada, Ottawa, p. 49-58.

Harris, S. A., 1982b, Cold air drainage west of Fort Nelson, British Columbia, Arctic, v. 35, p. 537-541.

Hughes, O. L., 1969, Distribution of open-system pingos in central Yukon Territory with respect to glacial limits. Geological Survey of Canada Paper 69-34, 8 p.

Johnston, G. H., 1981, Permafrost engineering design and construction, J. Wiley and Sons, Toronto, 540 p.

Klassen, R. W., 1979, Thermokarst terrain near Whitehorse, Yukon Territory, Geological Survey of Canada Paper 79-1A, p. 385-388.

Klohn Leonoff Consultants Ltd., 1978, Dempster Lateral Drilling Program, v. 1 and 2. Foothills Pipe Lines (Yukon) Ltd.

Marcus, M. G., 1974, Investigations in alpine climatology: The St. Elias Mountains, 1963-1971. _In_ Icefield Ranges Research Project Scientific Results. Edited by V. C. Bushnell and M. G. Marcus, American Geographical Society and Arctic Institute of North America, New York, v. 4, p. 13-26.

Taylor-Barge, B., 1969, The summer climate of the St. Elias Mountains region. _In_ Icefield Ranges Research Project Scientific Results. Edited by V. C. Bushnell and R. H. Ragle: American Geographical Society and Arctic Institute of North America, New York, v. 1, p. 33-50.

Thompson, H. A., 1963, Air temperatures in Northern Canada with emphasis on freezing and thawing indices. Proceedings of the First International Permafrost Conference, Lafayette, Indiana: National Academy of Sciences, National Research Council, Washington, D.C., Publication No. 1284, p. 272-280.

Washburn, A. L., 1979, Geocryology, Edward Arnold, London, 406 p.

Zoltai, S. C., and Tarnocai, C., 1975, Perennially frozen peatlands in the western Arctic and subarctic of Canada. Canadian Journal of Earth Sciences, v. 12, p. 28-43.

THE ORIENTATION AND EVOLUTION OF THAW LAKES, SOUTHWEST BANKS ISLAND, CANADIAN ARCTIC

D. G. Harry[1] and H. M. French[2]

[1]Department of Geography, University of Western Ontario, London, Canada N6A 5C2
[2]Departments of Geography and Geology, University of Ottawa, Ottawa, Canada K1N 6N5

On southwest Banks Island, the melt-out of ice within unconsolidated permafrost sediments has resulted in the formation of numerous thaw lakes. A majority of basins are oriented perpendicular to prevailing winds and possess a D-shaped outline which is in equilibrium with wind-generated geomorphic processes. In particular, a strong relationship exists between lake morphology and the storm wind regime during the summer period of open water conditions. Thaw lakes in this area cannot be interpreted within the traditional "thaw lake cycle" and appear to represent quasi-equilibrium landforms. Shoreline erosion results in asymmetrical expansion rather than a lateral migration of the basin. Lake drainage occurs primarily by catastrophic outflow, following basin capture or truncation by coastal retreat.

INTRODUCTION

Thaw lakes develop by thermokarst subsidence of ice-rich permafrost. Their widespread occurrence has been reported from lowland tundra regions of Alaska, Siberia, and Arctic Canada (e.g., Sellmann et al. 1975, Tomirdiaro and Ryabchun 1978, Bird 1967 pp. 211-216). The conditions under which they evolve are, however, poorly understood and many of the questions raised by Black (1969), in a review of the early literature on the subject, remain unanswered. In particular, the common development of a preferred long axis orientation, first described in North America by Black and Barksdale (1949), is still open to interpretation (e.g., Carson and Hussey 1960, 1962, Rex 1961, Mackay 1956, 1957). Since the development of a preferred orientation is not restricted to thaw lakes (Kaczorowski 1977, Price 1968), it follows that the presence of permafrost serves only to modify the action of essentially azonal processes.

This paper describes the evolution of oriented thaw lakes within the Sachs and Kellett drainage basins of southwest Banks Island (Figure 1). Observations of contemporary lakes and drained lake basins in this area permit an evaluation of the traditional "cycle" of thaw lake evolution (Britton 1967). An alternative model is proposed, based upon the identification of an equilibrium basin morphology which develops in response to currently active lacustrine processes.

LAKE DISTRIBUTION

High concentrations of lakes occur in two major areas of southwest Banks Island. Southeast of Sachs Harbour, the Sachs River lowlands are characterized by a number of large lakes, greater than 1.0 km in diameter, together with numerous smaller tundra ponds (Figure 2). Most of this area has a lake cover in excess of 20%, rising in some places to over 50%. North of Sachs Harbour, lake cover exceeds 20% in parts of the Kellett River valley, with highest concentrations occurring on low terraces above the present floodplain. Numerous small ponds occupy low-centered polygon sites, while larger lakes have formed apparently by basin coalescence following breaching of the polygon ramparts (Figure 3).

Both areas are underlain by outwash sediments, deposited when the late-Pleistocene ice margin lay to the east of Sachs Harbour (Vincent 1982). Permafrost is continuous on Banks Island and probably extends to depths in excess of 500 m (Taylor et al. 1982); however, it is likely that open taliks exist beneath the larger lakes and principal river channels in the Sachs River lowlands (Harry 1982). The surficial sediments are ice-rich, and are characterized by the development of multiple ice-wedge systems (French et al. 1982).

ORIGIN AND AGE

The coastal truncation of thaw lakes in the Sachs River lowlands enables stratigraphic studies to be undertaken of their age and evolution. A thermokarst origin is indicated by the vertical compression of stratigraphic units beneath many basins and by the reduced ice content of the sediments (French and Harry in press). Typically, sediments beneath drained lake basins contain 40-50% less excess ice than equivalent materials outside the basins. Ground ice may constitute nearly 60% by volume of the upper 8.0 m of permafrost in this area, and thawing of this material would result in subsidence of at least 3.0 m. This is consistent with stratigraphic observations beneath drained basins, which suggest an average thaw settlement of 5.0-6.0 m.

Several radiocarbon age determinations relate to the age of thaw lake initiation in the Sachs River lowlands. Organic material collected from the basal horizon of lacustrine sediments beneath one drained basin yields a ^{14}C date of 8560 ± 210 years B.P. (GSC-3292). At a second locality, a ^{14}C date of 8280 ± 140 year B.P. (GSC-2246) has

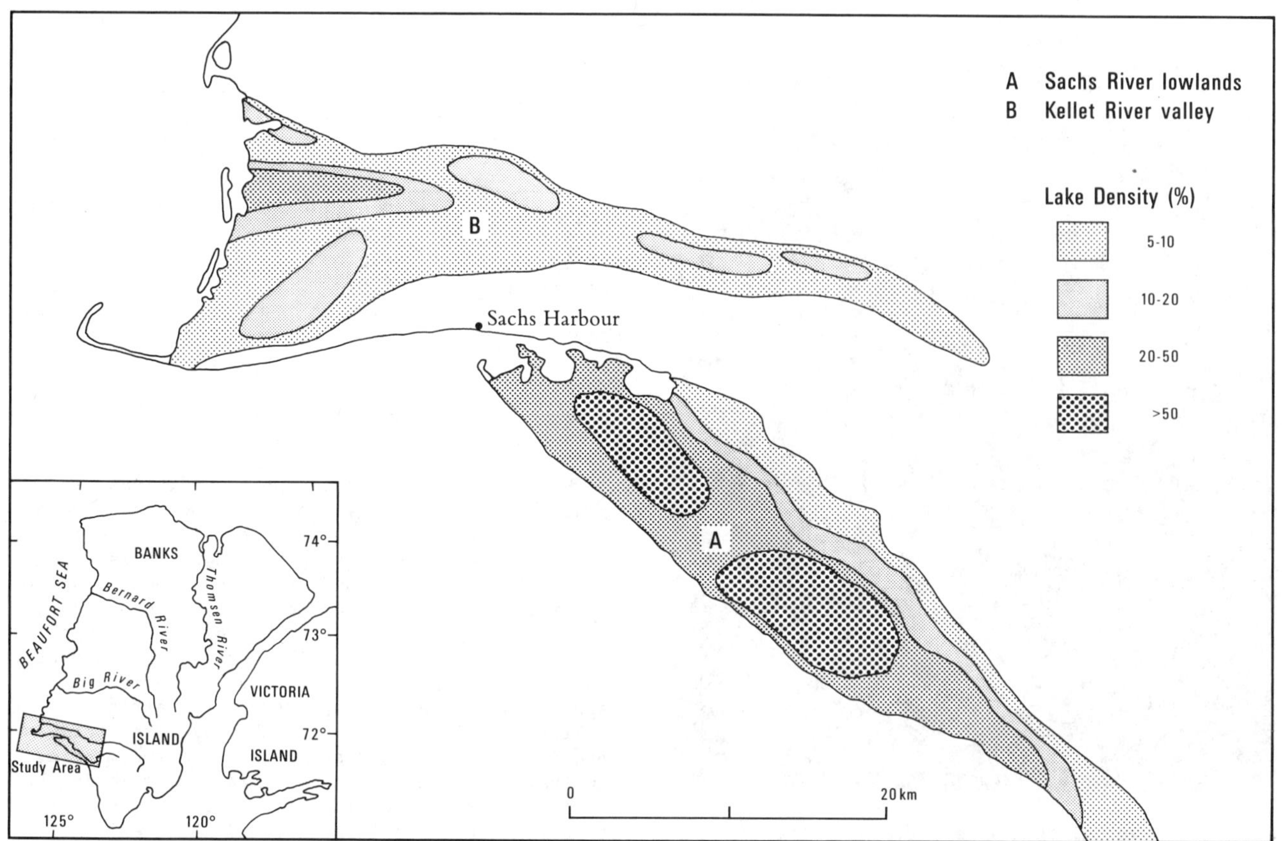

FIGURE 1 Location map of southwest Banks Island, showing distribution of thaw lakes. Density is expressed as percentage lake cover per 25 km^2.

been obtained from peat underlying lacustrine material (Vincent 1980). Third, blocks of fibrous organic material, contained within silts beneath a lake basin 4.0 km west of Sachs Harbour, yield a ^{14}C date of 9490 ± 80 years B.P. (GSC-2364). These dates suggest that many lakes were initiated approximately 8000-9000 years B.P. This period corresponds to an early phase of the mid-Holocene climatic optimum, during which warmer air temperatures may have triggered the melt-out of ground ice.

The age of thaw lake development in the Kellett River valley has not yet been determined. However, the observed influence of ice wedge polygons on basin location suggests that lakes in this area may have been initiated during a period of cold climate conditions. At that time, the formation of a polygon network on the terrace surface would have facilitated the ponding of water bodies and the initiation of self-sustaining thermokarst.

EQUILIBRIUM MORPHOLOGY

Many thaw lakes possess an equilibrium shape which, in its simplest form, is expressed as a preferred long-axis orientation (e.g., Carson and Hussey 1962, Mackay 1956). Random sampling from 1:60,000 scale air photographs indicates that 84% of lakes in the Sachs River lowlands and 97% of lakes in the Kellett River valley have a measurable long-axis orientation (defined by a length-width ratio greater than 1.1). Mean lake elongation is 1.53 in the Sachs River lowlands, as compared to 1.92 in the Kellett River valley. This disparity is partly accounted for by the prevalence of basins in the latter area which have coalesced in the direction of elongation. Lake orientation patterns are similar in the two areas, with a major northeast-southwest component and a minor perpendicular component (Figure 4). Lakes in the Kellett River valley display a particularly strong preferred orientation, with approximately 60% of long axes possessing azimuths between 230° and 240°.

Detailed morphological analysis was based on the identification of commonly occurring lake shape categories. These were defined as circle, oval, ellipse, and rectangle. A D-shaped class was added to describe highly asymmetrical elliptical lakes, and a "complex" category was established to include irregular or multiple basins. Lake shape in each area was determined by visual comparison of these standard categories to 100 lake outlines randomly sampled from 1:60,000 scale air photographs. In the Sachs River lowlands, elliptical or D-shaped morphological types are most common (Table 1). In the Kellett River valley, morphology is frequently complex, reflecting the high percentage of coalescent basins. However, over 50% of lakes possess a D-shaped element.

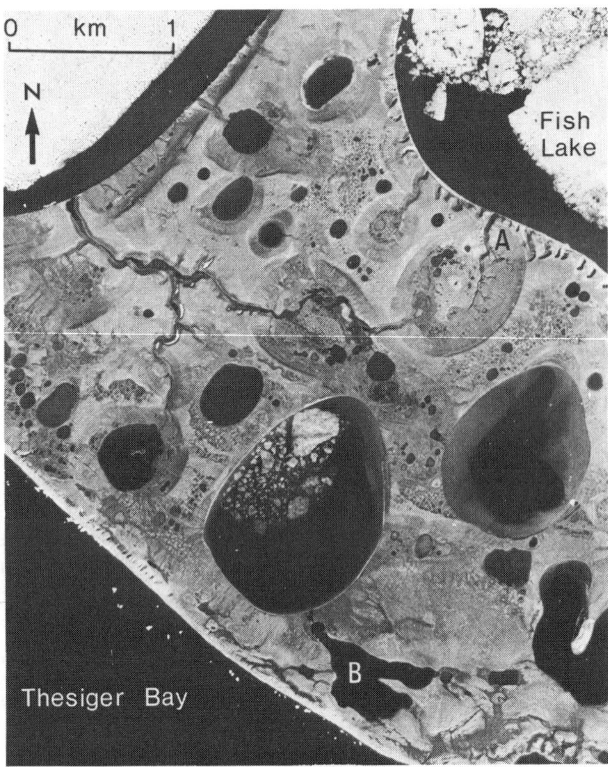

FIGURE 2 Typical thaw lake terrain in the Sachs River lowlands. Note the D-shaped morphology of many lakes and the numerous drained basins. Detail of the outlet channel at locality A is shown in Figure 7. Lake B drained catastrophically in January 1977.

FIGURE 3 Thaw lakes developed on terraces of the Kellett River. Note occurrence of small lakes within ice-wedge polygons and larger basins which have coalesced along a northeast-southwest axis.

TABLE 1 Lake Shape Classification, Southwest Banks Island

Basin Morphology (%)	Kellett River Valley (%)	Sachs River Lowlands (%)	Total (%)
D-shape	52.0	22.0	37.0
Complex	18.0	17.0	17.5
Ellipse	4.0	30.0	17.0
Rectangle	18.0	12.0	15.0
Oval	6.0	13.0	9.5
Circle	2.0	6.0	4.0
Total	100.0	100.0	100.0

Under equilibrium geomorphic conditions, it is probable that a lake basin developed upon a homogeneous surface will possess an outline consisting of a number of smoothly curved bays. This form should be widespread and developed upon a range of surficial materials. On southwest Banks Island, lakes of the D-shaped category most consistently satisfy these criteria. The D-shape is, in fact, curvilinear; the "straight" segment forms approximately half of an extremely flattened ellipse, while the opposing "bow" segment forms a broad semi-ellipse or semicircle. Morphometric analysis of D-shaped lakes of varying size in the Kellett River valley indicates only slight variation from the mean shape shown in Figure 5a. It is suggested, therefore, that this shape represents the equilibrium form of thaw lakes on southwest Banks Island.

Many lakes are oriented perpendicular to the prevailing wind direction during the July-September period of open water (see Figure 4). Directional wind data recorded at the Sachs Harbour meteorological station between 1971 and 1977 display a bimodal frequency distribution, dominated by northerly and southeasterly components. The mean long-axis orientation of lakes bisects the vector resultants of this opposed wind regime. In particular, the long axes of D-shaped lakes, which extend parallel to the "straight" shoreline segment, have an orientation almost exactly perpendicular to the northwest and southeast wind resultants. Similar relationships between lake orientation and wind direction exist elsewhere, especially in parts of northern Alaska and the Mackenzie Delta.

These observations suggest that an understanding of equilibrium lake morphology may be based on models of wind-generated wave and current distributions as first proposed by Livingstone (1954) and Rex (1961). It has been shown that the equilibrium form of a shoreline developed in unconsolidated sediment is a cycloid, within which erosion is greatest in zones oriented at 50° to wave approach (Bruun 1953). Mackay (1963 pp. 46-55) described a simple process-response model of lake morphology, based on the assumption that winds from each compass point tend to develop cycloidal bays. In this way, equilibrium morphology may be regarded as the integrated form of a number of cycloids of different size.

This model was applied to thaw lake morphology

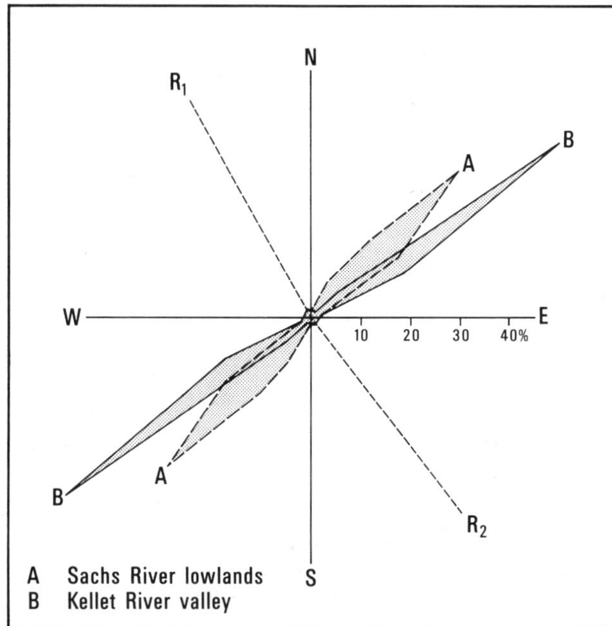

FIGURE 4 Thaw lake orientation, southwest Banks Island. Sample size for each area is 100; R_1 and R_2 are opposed vector resultants of the summer storm wind regime at Sachs Harbour.

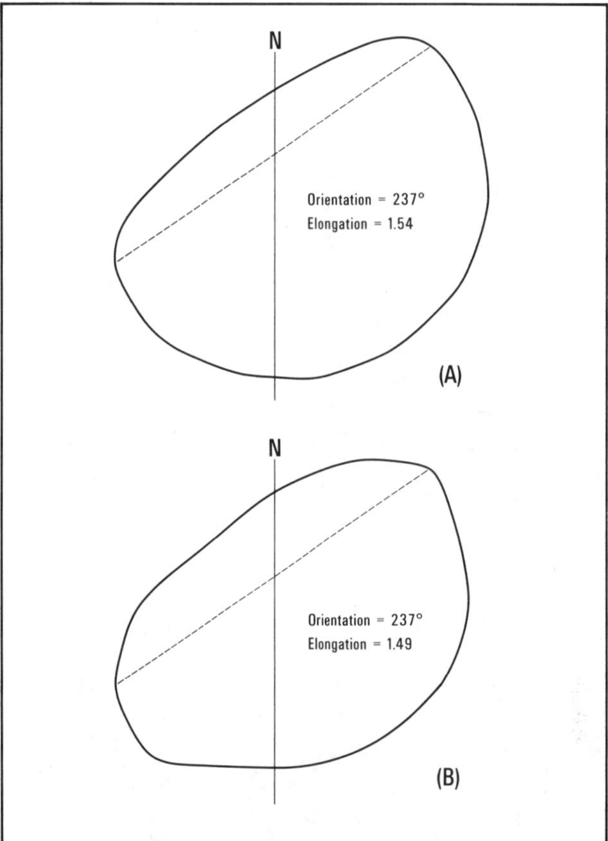

FIGURE 5 Thaw lake morphology, southwest Banks Island. (a) Observed morphology, based on analysis of five D-shaped lakes in the Kellett River valley. (b) Predicted morphology, based on calculation of equilibrium bays related to the summer storm wind regime at Sachs Harbour.

on southwest Banks Island, using seasonal wind data from Sachs Harbour. Predicted lake morphology is elliptical, with a length-width ratio of 1.23 and a long-axis azimuth of approximately 247°. Although this provides a reasonable first approximation to observed lake morphology, a closer correspondence is obtained when storm wind data are used. Storm events are defined, following McCann (1972), as periods during which hourly wind speed is 20 kn (37 km/h) or more for at least three consecutive hours. In this case, the model predicts a D-shaped lake, with a length-width ratio of 1.54 and a long-axis azimuth of approximately 237° (Figure 5b). The significantly better fit suggests that storm events play a major role in shoreline evolution.

Field observations of thaw lakes in the Kellett River valley tend to support this model. Shoreline characteristics may be interpreted with respect to the inferred pattern of wind-generated littoral drift. At the northeast and southwest corners, where maximum rates of littoral drift are expected to occur, shoreline morphology is predominantly erosional. The 10-50 cm high shoreline bluff is undercut and terrace gravels are exposed in the lake bottom. By contrast, the "straight" shoreline corresponds to a zone of minimum littoral drift and is characterized by broad, low-angle depositional flats. Shoreline morphology is therefore consistent with the energy environments predicted by Rex's (1961) model of lake circulation.

LAKE DRAINAGE

Many lake basins, especially within the Sachs River lowlands, are either partly or completely drained. Two models of lake drainage may be considered. The first consists of gradual infilling and segmentation (e.g., Kaczorowki 1977 p. 111), while the second consists of catastrophic outflow, following lake tapping or truncation by coastal retreat (e.g., Mackay 1979 p. 31, Walker 1979, Weller and Derksen 1979). On southwest Banks Island, evidence points strongly to the importance of the catastrophic outflow model.

Catastrophic lake drainage in the Sachs River lowlands has occurred as a result of tapping by headward erosion of streams along ice wedges, coalescence with a basin at a lower elevation, or coastal retreat (Figure 6). Once flow is initiated through the new outlet, drainage may be extremely rapid, resulting in the formation of box-canyon outlet channels. These are characterized by steep lateral bluffs and flat floors, which may be incised into the former lake bed (Figure 7).

Under certain circumstances, buried ice bodies may provide subsurface outlets for lake drainage. For example, one lake located approximately 40 km southeast of Sachs Harbour (see Figure 2, Lake B) experienced catastrophic subterranean drainage in January 1977 (David Nasogaluak, personal communication). While this may be common in areas of thin or discontinuous permafrost, resulting in the

FIGURE 6 Drained basin truncated by sea cliffs, approximately 30 km southeast of Sachs Harbour. Note the deep gully eroded during lake drainage.

FIGURE 7 Box canyon drainage channel, eroded into the lake floor following basin capture (see Figure 2, locality A). Note figures for scale.

formation of thaw sinks (Hopkins 1949, Tedrow 1969), it is rarely reported from regions underlain by thick, continuous permafrost. This lake lies within an area of morainic topography, attributed to the Sand Hills Readvance (Vincent 1982), part of which may be ice-cored (Harry 1982 pp. 88-99). It is possible, therefore, that lake drainage resulted from thermal contraction cracking, allowing water to escape into an ice cave.

DISCUSSION AND CONCLUSIONS

The concept of a thaw lake cycle, consisting of sequential stages of initiation, expansion and drainage is well established (e.g., Billings and Peterson 1980, Britton 1966, Everett 1981, Tedrow 1969). The results of this study suggest, however, that such models are by no means universally applicable. A number of alternative models of thaw lake development are possible, reflecting a delicate equilibrium between process, materials, and morphology.

On southwest Banks Island, thaw lakes appear to have originated either by melt-out of ground ice during a mid-Holocene period of climatic amelioration, or by ponding within ice-wedge polygons developed under cold climate conditions. In the absence of topographic control, subsequent lake expansion is strongly influenced by wind-related patterns of wave and current erosion. This results in the development of a preferred long-axis orientation and D-shaped equilibrium morphology. The sensitivity of this process is demonstrated by the relative rarity of oriented lakes elsewhere on Banks Island. The development of preferred orientation appears dependent on an opposed summer wind regime, unmodified by local topographic effects.

It has been suggested that the processes responsible for thaw lake orientation may also result in migration of the lake basin across the tundra surface (e.g., Tedrow 1969, Tomirdiaro and Ryabchun 1978). No geomorphic or stratigraphic evidence for this is found on southwest Banks Island. Wind-generated patterns of shoreline erosion result in asymmetrical expansion, to form D-shaped lakes, rather than causing an overall translocation of the basin. Oriented thaw lakes in this area appear, therefore, to represent quasi-equilibrium landscape elements.

ACKNOWLEDGMENTS

Geomorphological investigations on Banks Island are being supported by the Natural Sciences and Engineering Research Council (Grant A8367), the Polar Continental Shelf Project, Department of Energy, Mines and Resources (Project 34-73), and the University of Ottawa Northern Research Group.

REFERENCES

Billings, W. D., and Peterson, K. M., 1980, Vegetational change and ice-wedge polygons through the thaw-lake cycle in Arctic Alaska: Arctic and Alpine Research, v. 12, p. 413-432.

Bird, J. B., 1967, The Physiography of Arctic Canada: Baltimore, The John Hopkins Press, 336 p.

Black, R. F., 1969, Thaw depressions and thaw lakes: A review: Biuletyn Peryglacjalny, v. 19, p. 131-150.

Black, R. F., and Barksdale, W. L., 1949, Oriented lakes of northern Alaska: Journal of Geology, v. 57, p. 105-118.

Britton, M. E., 1967, Vegetation of the Arctic Tundra, in Hansen, H. P., ed., Arctic Biology: Corvallis, Oregon State University Press, p. 67-130.

Bruun, P., 1953, Forms of equilibrium of coasts with a littoral drift: University of California, Institute of Engineering Research, v. 3(347), 7 p.

Carson, C. E., and Hussey, K. M., 1960, Hydrodynamics in three arctic lakes: Journal of Geology, v. 68, p. 585-600.

Carson, C. E., and Hussey, K. M., 1962, The oriented lakes of Arctic Alaska: Journal of Geology, v. 70, p. 417-439.

Everett, K. R., 1981, Landforms, in Walker, D. A. Everett, K.R., Webber, P. J., and Brown, J., eds. Geobotanical Atlas of the Prudhoe Bay Region, Alaska: Hannover, New Hampshire, United States Army Cold Regions Research and Engineering Laboratory.

French, H. M., and Harry, D. G., Ground ice conditions and thaw lakes, Sachs River lowlands, Banks Island, Canada: Abhandlungen der Akademie der Wissenschaften in Göttingen, Math.-Phys. Kl. III, in press.

French, H. M., Harry, D. G., and Clark, M. J., 1982, Ground ice stratigraphy and late-Quaternary events, Southwest Banks Island, Canadian Arctic, in French, H. M., ed., The R. J. E. Brown Memorial Volume, Proceedings, Fourth Canadian Permafrost Conference: Ottawa, National Research Council of Canada, p. 81-90.

Harry, D. G., 1982, Aspects of the permafrost geomorphology of southwest Banks Island, Western Canadian Arctic: Unpublished Ph.D. Thesis, University of Ottawa, 230 p.

Hopkins, D. M., 1949, Thaw lakes and thaw sinks in the Imuruk Lake area, Seward Peninsula, Alaska: Journal of Geology, v. 57, p. 119-131.

Kaczorowski, R. T., 1977, The Carolina Bays: A comparison with modern oriented lakes: Coastal Research Division, Department of Geology, University of South Carolina, Technical Report, 13-CRD, 124 p.

Livingstone, D. A., 1954, On the orientation of lake basins: American Journal of Science, v. 252, p. 547-554.

Mackay, J. R., 1956, Notes on oriented lakes of the Liverpool Bay area, N.W.T.: Revue canadienne de géographie, v. 10, p. 169-173.

Mackay, J. R., 1957, Les lacs orientes de la region de la Baie de Liverpool: Discussion: Revue canadienne de géographie, v. 11, p. 175-178.

Mackay, J. R., 1963, The Mackenzie Delta Area, N.W.T.: Ottawa, Department of Mines and Technical Surveys, Geographical Branch, Memoir 8, 202 p.

Mackay, J. R., 1979, Pingos of the Tuktoyaktuk Peninsula area, Northwest Territories: Géographie physique et Quaternaire, v. 33, p. 3-61.

McCann, S. B., 1972, Magnitude and frequency of processes operating on Arctic beaches, Queen Elizabeth Islands, N.W.T., Canada, in Adams, P. W., and Helleiner, F., eds., International Geography, v. 1: Toronto, University of Toronto Press, p. 41-43.

Price, W. A., 1968, Oriented lakes, in Fairbridge, R., ed., Encyclopedia of geomorphology: New York, Reinhold, p. 784-796.

Rex, R. W., 1961, Hydrodynamic analysis of circulation and orientation of lakes in northern Alaska, in Raasch, G. O., ed., Geology of the Arctic, v. 2: Toronto, University of Toronto Press, p. 1021-1043.

Sellman, P. V., Brown, J., Lewellen, R. I., McKim, H., and Merry, C., 1975, The classification and geomorphic implications of thaw lakes on the Arctic coastal plain, Alaska: United States Army, Cold Regions Research and Engineering Laboratory, Research Report, 344, 21 p.

Taylor, A. E., Burgess, M., Judge, A. S., and Allen, V. S., 1982, Canadian Geothermal Series, Number 13: Ottawa, Earth Physics Branch, Department of Energy, Mines and Resources, 289 p.

Tedrow, J. C. F., 1969, Thaw lakes, thaw sinks and soils in northern Alaska: Biuletyn Peryglacjalny, v. 20, p. 337-345.

Tomirdiaro, S. V., and Ryabchun, V. K., 1978, Lake thermokarst on the Lower Anadyr Lowland, in Permafrost: U.S.S.R. Contribution to the Second International Conference, Yakutsk, U.S.S.R.: Washington, D.C., National Academy of Sciences, p. 94-100.

Vincent, J-S., 1980, Les glaciations Quaternaire de l'Ile de Banks, Arctique Canadien: Unpublished Ph.D. Thesis, Université Libre de Bruxelles, 248 p.

Vincent, J-S., 1982, The Quaternary history of Banks Island, N.W.T., Canada: Géographie physique et Quaternaire, v. 36, p. 209-232.

Walker, H. J., 1979, Lake tapping in the Colville River Delta, Alaska, in Proceedings of the Third International Conference on Permafrost, Edmonton, Canada, v. 1: Ottawa, National Research Council of Canada Publication 16529, p. 232-238.

Weller, M. W., and Derksen, D. V., 1979, The geomorphology of Teshepuk Lake in relation to coastline configuration of Alaska's coastal plain: Arctic, v. 32, p. 152-160.

RELATIONSHIPS BETWEEN ESTIMATED MEAN ANNUAL AIR AND PERMAFROST TEMPERATURES IN NORTH-CENTRAL ALASKA

R. K. Haugen[1], S. I. Outcalt[2], and J. C. Harle[3]

[1]U.S. Army Cold Regions Research and Engineering Laboratory
Hanover, New Hampshire 03755 USA
[2]Department of Geological Sciences, University of Michigan,
Ann Arbor, Michigan 48109 USA
[3]Alyeska Pipeline Service Company, Anchorage, Alaska 99512 USA

Mean annual air temperatures (MAAT) are estimated for a transect from central to northern Alaska. The estimated MAAT are compared to mean annual ground temperatures (MAGT) representative of upper permafrost temperatures. The estimation of MAAT for the remote and topographically complex transect area was based on trend surface estimates of numerous short-term (1-7 years) temperature records obtained from climatic stations operated by research projects and longer records from existing National Weather Service stations. The standard error of the estimated MAAT falls within a degree (C) of observed MAAT for stations with long-term records. The MAGT are based on subsurface thermistor measurements made at construction sites and are therefore from disturbed terrain, but data were selected to minimize the effects of disturbance. MAGT measurements ranged from -7.5°C in the north to -0.7°C near Fairbanks. Predicted MAAT ranged from -11.5°C at Prudhoe Bay to -4.5°C in the Fairbanks area. A simple regression relationship showed MAGT to average 3.6°C higher than MAAT. This study suggests that, based on estimated MAAT and MAGT values, the boundary between the zones of continuous and discontinuous permafrost is located at or slightly north of the continental divide at Atigun Pass.

Our understanding of the relationships of permafrost distribution and air temperature in central and northern Alaska is limited by the lack of available direct observations. For the most part, permafrost temperature observations have been confined to populated areas in interior Alaska and in the northern coastal regions. Similarly, long-term observations of air temperature are scarce, and are mostly representative of populated lowland or coastal areas.

Beginning in 1975, climatic data were collected at sites along the route of the Trans-Alaska Pipeline System extending from the Yukon River to the Arctic Ocean. These measurements, together with data from additional remote upland sites in interior Alaska, have been summarized (Haugen 1982) and provide a data base from which mean air temperature values can be established for this poorly known climatic region of north-central Alaska.

Following construction of the Trans-Alaska Pipeline, thermistor strings were installed at selected locations along the pipeline to monitor the performance of its engineering structures. These sites are located in both discontinuous and continuous permafrost zones (Ferrians 1965). Most of these temperature measurements were associated with thermal disturbances. Only a few of the subsurface thermistor arrays were judged to be sufficiently free of the influence of the pipeline or its associated structures to provide subsurface temperatures for comparison with the regional air temperature data. The objectives of this paper are to present a method of estimating air temperature values for a region of diverse terrain and climate and to examine the relationship of mean annual air temperature to ground temperature measurements.

REGIONAL CLIMATE

The Alaskan interior is a zone of temperature extremes and of relatively high precipitation compared to the Arctic (Bowling 1979). During the summer, storms track through the region from the south or southwest. Most summer precipitation is of the convective type and is widely scattered and variable in amount. During the winter, the interior is dominated by relatively dry Continental Polar air masses, and sinking cold air creates high atmospheric pressure. Occasionally, maritime air intrudes into the area from the west or southwest, causing major snowstorms and, rarely, winter rain. The alpine areas within the interior typically have less extreme temperatures but higher precipitation than the forested areas at lower elevations.

North of the continental divide is a region of extremely low winter temperatures, low summer temperatures, and relatively low precipitation. Unlike the continental interior, this is a region where wind is a major environmental factor throughout much of the year. Although winds rarely exceed 17 m/s on the Arctic Coastal Plain, calms are also rare.

Average annual air temperatures range from -11.5°C at Prudhoe Bay to -3.5°C at Fairbanks. The temperatures are some of the most extreme encountered on the North American continent. Extremely low temperatures (<-50°C) recorded at valley stations in the southern foothills of the Brooks Range (e.g. Prospect, Coldfoot, Dietrich) are the result of inversions of the vertical temperature profile, caused by cold air draining downslope. The higher elevations, represented by Gobbler's Knob, Chandalar Shelf, and Atigun Pass, are above the average height of the inversion and have higher average winter temperatures. The lower summer temperatures at the high-elevation sites reflect a normal decrease of temperature with elevation. At least half the stations have summer maximum temperatures over 30°C for an annual range of more than 80°C.

APPROACH TO ESTIMATING TEMPERATURES

The successful estimation of air temperatures over an area of complex topography requires (1) a method for filling in short gaps in the data base which occur at many of the remote sites, (2) a procedure for adjustment of a short record (in this case, 2 to 7 years) to standard 30-year normal values, and (3) the development and testing of methods for extrapolating site-specific air temperature data to locations with differing elevations and ultimately differing slopes and exposures. The initial effort was to estimate mean monthly air temperatures and freezing and thawing degree-day accumulations for the Arctic Slope (Haugen 1982). Although that area is not topographically complex, the sparsity of inland stations created problems. However, simple regression analysis, utilizing distance from the ocean, or multiple regression, using latitude, longitude, and elevation as independent variables, proved reasonably successful for this region. The technique was less successful for the topographically complex areas of central Alaska, primarily because of the lack of data. The present approach treats a 600 x 800 km area extending northward from the latitude of Fairbanks to the Arctic Ocean (SW corner 64.0°N, 158.0°W).

Air Temperature

The stations used in this analysis (Figure 1) were operated by the National Weather Service and

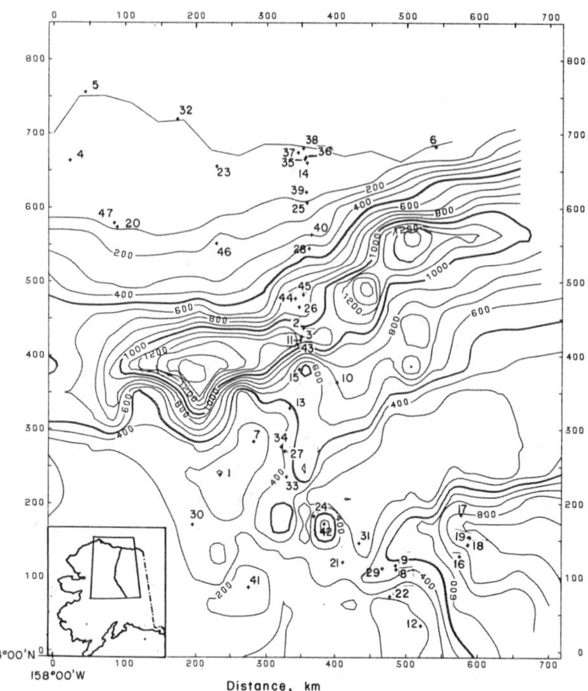

FIGURE 1 Topography of study area. Distance from point of grid origin at 64°N, 158°W used to determine temperature values (Equations 1 and 2). Numbers are keyed to station data, Table 2.

CRREL. Most of the remote stations began operation in 1975 or 1976 and have continued at least intermittently since then. Fewer data were collected in the winter than in the summer because of the difficulty of servicing the instruments in winter. It was decided that the existing data base could be most effectively utilized for comparison with permafrost temperatures by considering only the warmest month (TMAX) and the coldest month (TMIN). TMAX and TMIN are not always for the same month each year. The determination of mean annual temperature is based on the sum of TMAX + TMIN divided by 2 for the period of station record. A comparison of MAAT

TABLE 1 Comparison of mean annual air temperatures based on long-term records and short-term coldest and warmest month.

Period of record	Barrow 1941-70	Barter Island 1948-70	Bettles 1952-70	Fairbanks 1941-70
Long-term average of 12 months	-12.6	-12.7	-5.9	-3.5
1975-81 average of TMAX-TMIN month	-11.7	-12.0	-5.6	-3.8
Difference	+0.9	+0.7	+0.3	+0.3

TABLE 2 Climatic stations grid locations and calculated 1975-1981 mean maximum (TMAX) and minimum (TMIN) monthly air temperatures.

Location (map no.)	Location	Long. X (km)	Lat. Y (km)	Elev. Z (m)	TMAX (°C)	TMIN (°C)
1	Allakaket	235.82	238.90	183.00	14.0	-30.6
2	Atigun Camp	352.90	437.51	1035.00	10.6	
3	Atigun Pass	349.91	425.88	1400.00	9.5	-20.9
4	Atkasook	22.33	663.49	15.25	9.0	
5	Barrow	43.56	756.05	9.20	4.7	-28.6
6	Barter Island	540.04	682.45	12.20	4.8	-28.6
7	Bettles	281.87	283.42	198.00	15.6	26.6
8	Caribou Creek	486.16	110.21	264.00	12.5	-25.2
9	Caribou Peak	485.25	115.69	773.00	12.9	-15.7
10	Chandalar Lake	402.62	363.82	564.00	12.4	-28.6
11	Chandalar Shelf	348.46	421.20	1000.00	11.8	-22.2
12	Chena Dam	520.60	34.82	152.40	15.4	-26.2
13	Coldfoot	332.89	328.36	325.00	15.3	-24.0
14	Deadhorse	358.27	660.80	18.20	7.9	-28.1
15	Dietrich	346.61	381.69	445.00	14.3	-28.3
16	Eagle Creek Lodge	586.66	143.50	690.00	11.3	-28.3
17	Eagle Summit	576.86	184.66	1130.00	11.2	-19.7
18	Eagle Creek Tussock	586.66	143.50	762.00	11.5	-19.1
19	Eagle Summit treeline	587.11	143.59	884.00	10.6	-18.5
20	East Oumalik	88.98	573.91	87.00	11.9	
21	Elliott Highway	411.61	120.10	732.00	12.6	-19.9
22	Fairbanks	477.37	73.82	133.00	17.0	-24.0
23	Fish Creek	230.01	655.81	10.00	9.2	
24	Five Mile Camp	369.28	182.80	145.00	15.4	-29.1
25	Franklin Bluff Camp	358.11	606.78	108.00	13.4	-32.3
26	Galbraith	346.50	465.82	820.00	10.7	-25.7
27	Gobbler's Knob	325.98	270.41	650.00	15.9	
28	Happy Valley	361.44	545.44	290.00	13.0	-29.9
29	Haystack Mt.	467.01	111.59	767.00	12.1	-17.9
30	Indian Mt.	195.29	171.15	372.00	14.7	-22.7
31	Livengood	434.46	145.92	161.00	15.7	-24.0
32	Lonely	172.73	719.98	3.05	4.9	-29.0
33	Old Man Camp	328.75	236.04	406.00	13.8	-25.6
34	Prospect Airfield	320.85	276.53	337.00	14.4	-23.2
35	Prudhoe (Arco)	354.80	665.89	18.50	7.9	-29.9
36	Prudhoe (drill site)	356.32	668.37	14.50	6.8	
37	Prudhoe (Pad F)	344.76	674.50	6.20	6.0	
38	Prudhoe (West Pier)	352.59	680.18	3.00	6.1	
39	Sagavanirktok	357.07	621.24	80.00	11.1	
40	Sagwon Bluff	365.02	563.96	305.00	11.4	-30.6
41	Tanana	275.04	86.16	70.75	15.6	-25.5
42	Yukon River	384.85	172.53	691.00	13.5	-28.6
43	Timberline, Dietrich	344.26	415.02	690.00	13.3	
44	Toolik Lake	340.80	477.39	760.00	13.1	
45	Toolik River	352.80	482.40	850.00	12.1	
46	Umiat	229.75	551.93	81.00	12.7	-31.9
47	West Oumalik	84.91	579.33	55.00	11.6	

Grid origin: 64.00°N, 158.00°W

derived in this manner with MAAT calculated in the usual manner, based on the mean of 12-monthly temperatures for stations with available records, is shown in Table 1. The differences in MAAT according to the two methods are usually less than 1°C. TMIN temperatures during the 1975-1981 period were 1.7°-3.1°C higher than long-term mean temperatures calculated in the same fashion (i.e. based on the coldest month each year). TMAX temperatures were all less than 1°C higher in this comparison.

At least one year of TMIN data was available at 32 stations and TMAX was available at 47 stations (Table 2). Matrices by location consist of maximum (TMAX) and minimum (TMIN) mean monthly temperatures for each year. Missing values are computed according to the average of departures from the regional climate for those years when data does exist. The matrix has the following form:

	1975	1976	•••	1981	Departures for all years
Allakaket	T(1,1)			T(1,NK)	D(1)
•					
•		T(L,K)			D(L)
•					
Umiat	T(NL,1)			T(NL,NK)	D(NL)
Regional means	C(1)	C(K)		C(NK)	

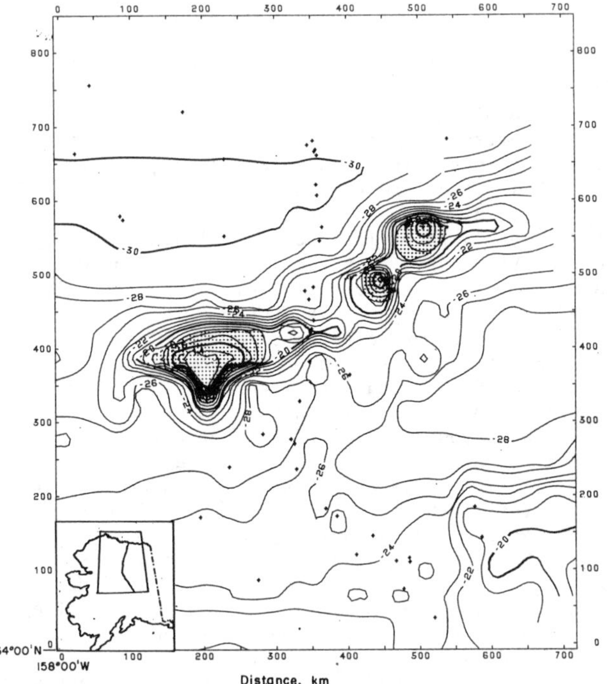

FIGURE 2 Cold month air temperature map. Shaded areas represent elevations above climatic data sample population.

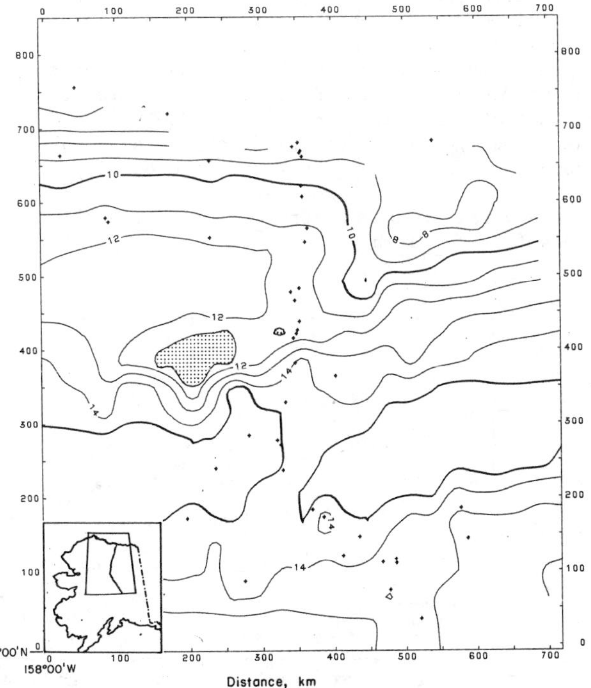

FIGURE 3 Warm month temperature map. Shaded areas as in Figure 2.

The row index is L, representing the location, and the column is K, representing the year. The mean monthly temperature of the region during a given year is the mean value of the entries in each column excluding the missing matrix element. Thus, there is for each year a mean value in what might be termed the regional climate vector C(K). The average deviation D(L) at a climate station is the mean value of [T(L,K) - C(K)] for each station, again excluding the missing entries in the matrix. When replacing a missing entry in the matrix, the best estimate would be the sum of the monthly mean climate for the region C(K) and the average value of the difference between that station value and the regional climate mean when those data were available. The replacement of a missing matrix element is then $T(L,K) \leftarrow C(K) + D(L)$.

A stepwise regression model is used to predict TMIN and TMAX values for any location in the geographic grid. The model that best fitted the data is described by Equations 1 and 2,

$$TMIN = -22.3 - 0.303 \cdot 10^{-1}Y + 0.283 \cdot 10^{-4}Y^2 + 0.752 \cdot 10^{-5}Z^2 \quad R^2=0.65 \; SE = 2.7°C \qquad (1)$$

$$TMAX = 12.9 + 0.265 \cdot 10^{-1}Y - 0.346 \cdot 10^{-2}Z - 0.494 \cdot 10^{-4}Y^2 \quad R^2=0.83 \; SE = 1.3°C \qquad (2)$$

where Y is the distance north (km), Z is the elevation (m), and SE is the standard error of estimate. Distance east was not statistically significant in this model. Elevations within the study area are shown in Figure 1. A spatial representation of the TMIN and TMAX equations is shown in Figures 2 and 3, respectively. Large variations of TMIN and TMAX over the short distances shown in these figures (e.g. the Fairbanks area) are due to elevation differences and should be viewed in comparison with the topographic map (Figure 1).

Ground Temperature

Temperatures representative of the upper permafrost were obtained from thermistors installed at locations between the Tanana River and the vicinity of Prudhoe Bay. These temperature sites were designed to monitor the performance of the oil pipeline and its associated structures in areas of ice-rich permafrost. The thermistor locations utilized in this analysis were selected to minimize the effects of surface disturbance or of the pipeline itself. Unfortunately temperature data are not yet available from areas adjacent to the pipeline route in essentially undisturbed areas. Therefore, the data used probably differ from data for the adjacent undisturbed ground. Temperatures obtained near the buried hot pipeline or near vertical support members (VSM's) that were artificially cooled by thermopiles were not used. At some of the sites with thermopiles, two thermistor strings were installed, one near the VSM and the other some distance (1-10 m) away. In these cases, the temperatures at a distance from the VSM were used.

The depths of the subsurface observations used varied from a minimum of 3.5 m in the north to about 16 m at some of the southern sites. Most temperature observations were obtained bimonthly over a period of 2 or 3 years. For each site, the maximum and minimum temperatures recorded by the deepest thermistor over the entire observation period were averaged to obtain the mean value. The first-year data were not used, to avoid the thermal effects associated with installation. Snow depth was recorded at the time of observation but was not considered in this study.

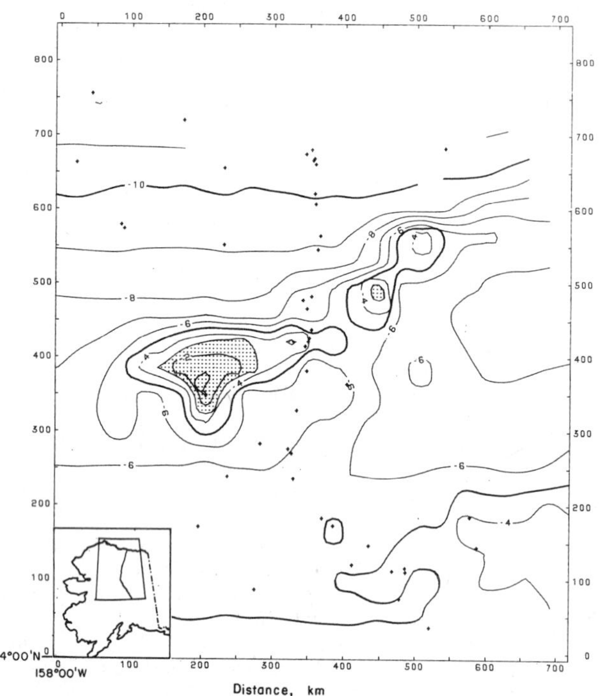

FIGURE 4 Mean annual air temperature (MAAT) map. Shaded areas as in Figure 2.

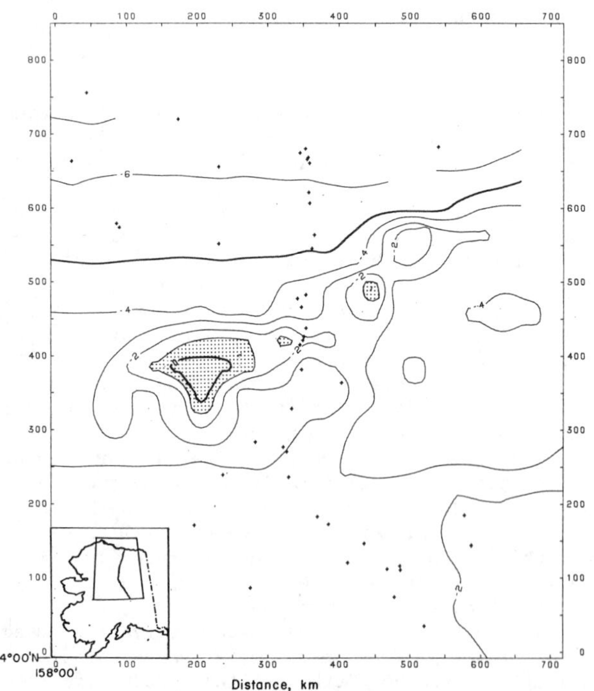

FIGURE 5 Mean annual ground temperature (MAGT) map. Shaded areas as in Figure 2.

RESULTS AND DISCUSSION

Mean annual air temperatures were estimated for locations where permafrost temperatures were available along the pipeline transect. This was done by averaging the TMAX and TMIN values obtained from Equations 1 and 2 for the distance north of 64°50'N latitude and the elevation of the site. The geographic distribution of estimated MAAT in the study area is shown in Figure 4. A simple regression relationship Y = 1.25 + 0.71X between MAAT (X) and the upper permafrost temperature (Y) based on the data shown in Figure 4 indicates an average temperature difference of 3.6°C and a coefficient of determination (r^2) of 0.85. This difference falls within the range of 1° to 5.5°C (average 3.3°C) generally assumed between air and upper permafrost mean annual temperature (Brown and Péwé 1973, Hydrological Atlas of Canada 1978, Plate 32). The mapped values of MAGT are shown in Figure 5. A profile showing the relationship from Fairbanks to Prudhoe Bay is shown in Figure 6.

Subsurface temperature data available for the higher elevations in the Brooks Range were influenced by the buried pipeline: they were above freezing. Therefore, normal temperature measurements from Chandalar Shelf and Atigun Pass were not used in the predictive MAGT equation. Existing maps show the transition from discontinuous to continuous permafrost (>90% on an area basis) south of the continental divide (Ferrians 1965, Brown and Péwé 1973). Recent observations from borings suggest this line more properly belongs in the vicinity of the continental divide (Kreig and Reger 1982). The MAAT isotherm delineating these two zones is in the range of -7°C (Ferrians 1965) to -8.5°C (Brown and Péwé 1973), which corresponds to an approximate subsurface permafrost temperature of -4°C to -5°C. Using Equations 1 and 2, we find an estimated MAAT for Atigun Pass of -6.3°C and, applying the simple regression relationship, we obtain a mean annual ground temperature of -2.7°C. Chandalar Shelf, just to the south of Atigun Pass at an elevation of 1000 m, indicates a MAAT of -5.2°C and MAGT of -1.6°C. This study indicates the position of the -8.5°C MAAT isotherm at or slightly north of the continental divide, as shown in Figure 4.

Several other permafrost temperature observations from comparably shallow depths are available

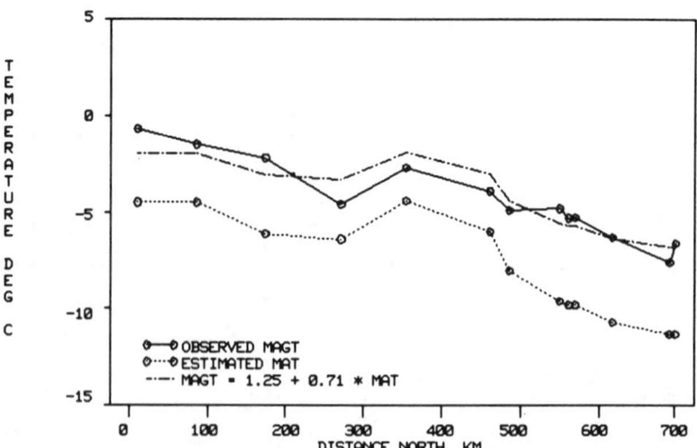

FIGURE 6 Comparison of MAAT, MAGT, and estimated MAGT in profile from Fairbanks to Purdhoe Bay.

for comparison with our approach. At Deadhorse, a MAGT of -8.5°C was obtained over 5 years at 23 m (J. Brown, personal communication), and at Ivotuk, 240 km west of Toolik, a MAGT of -4.7°C was obtained at 7 m (Kachadoorian and Crory 1983). Utilizing the procedures described in this study, we obtain estimated MAGT of -7.9°C for Deadhorse and -3.3°C for Ivotuk. Both of these estimates are on the warm side, probably reflecting the effects of disturbance on our ground temperature data set.

Measurements of permafrost temperatures in undisturbed terrain are rare except for sites along the Arctic Ocean (Lachenbruch et al. 1982) or in some populated areas in the interior (Brown and Péwé 1973). Observed upper permafrost temperatures along the Arctic Coast range from -8.5° to -10.9°C and increase to -6.5°C near Umiat (Brown and Péwé 1973). South of the Brooks Range, permafrost temperatures approach 0°C near Fairbanks. The thickness of the permafrost also decreases southward. Osterkamp and Payne (1981) illustrated this thinning of the ice-rich permafrost between Prudhoe Bay, where it exceeds 600 m, and the Brooks Range, where it becomes less than 200 m thick.

This analysis is intended to provide a broad regional perspective for the relationship of MAAT and MAGT. Direct observations of both values are for the most part lacking in this remote and topographically complex portion of Alaska. Although the MAAT is calculated based on somewhat sparse data and the upper permafrost temperatures are obtained from a disturbed surface environment, we believe that the relationships described are regionally representative. It is possible that using freezing and thawing degree-days to describe air temperatures in relation to permafrost distribution will provide greater precision (e.g. Harris and Brown 1982). However, the approach taken here (the average of TMAX + TMIN) is appropriate for comparison with the available MAGT data and with previous mapping of the region. As more air and ground temperature data become available, these relationships can be further refined.

ACKNOWLEDGMENTS

Collection and analysis of the climatic data were performed under projects of the U. S. Army Corps of Engineers Cold Regions Research and Engineering Laboratory and were partially supported by funding from the U. S. Geological Survey and the U. S. Department of Energy. The ground temperature data were made available through the courtesy of the Alyeska Pipeline Service Company.

REFERENCES

Bowling, S.A., 1979, Alaska's weather and climate, in Alaska's Weather and Climate (Gunter Weller, ed.): Geophysical Institute, University of Alaska, UAGR-269, p. 1-25.

Brown, R.J.E., 1967, Permafrost map of Canada. National Research Council, Division of Building Research, NRC9767.

Brown, R.J.E., and Péwé, T.L., 1973, Distribution of permafrost in North America and its relationship to the environment: A review, 1963-73, in Permafrost: The North American Contribution to the Second International Conference: Washington, D.C., National Academy of Sciences, p. 71-100.

Ferrians, O.J., 1965, Permafrost map of Alaska: U.S. Department of Interior, U.S. Geological Survey, Misc. Geol. Inves. Map 1-445.

Harris, S.A., and Brown, R.J.E., 1982, Permafrost distributions along the Rocky Mountains in Alberta, in Proceedings of the Fourth Canadian Permafrost Conference, Calgary, Alberta: p. 59-67.

Haugen, R.K., 1982, Climate of remote areas in north-central Alaska, 1975-79 summers, CRREL Report 82-35: Hanover, N.H., U.S. Army Cold Regions Research and Engineering Laboratory.

Haugen, R.K., and Brown, J., 1978, Climatic and dendroclimatic indices in the discontinuous permafrost zone of the central Alaskan uplands, in Proceedings of the Third International Permafrost Conference, Edmonton, Alberta, p. 392-398.

Heginbottom, J.A., Kurfurst, P.J., and Larc, J.S.O., 1978, Regional occurrence of permafrost, Mackenzie Valley, Canada, in Proceedings of the Third International Permafrost Conference, Edmonton, Alberta, p. 399-405.

Hydrological Atlas of Canada, 1978, Permafrost: Surveys and Mapping Branch, Department of Energy, Mines, and Resources, Ottawa, Canada, Plate 32.

Kachadoorian, R., and Crory, F., 1983, Temporary airfields in northern Alaska, in Proceedings of the Fourth International Permafrost Conference (this volume).

Kreig, R.A. and Reger, R.D., 1982, Air photo analysis and summary of landform soil properties along the route of the Trans-Alaska Pipeline System: College, Alaska, Department of Natural Resources, Division of Geological and Geophysical Surveys.

Lachenbruch, A.H., Sass, J., Marshall, B., and Moses, T. Jr., 1982, Permafrost, heat flow and the geothermal regimes at Prudhoe Bay, Alaska, Journal of Geophysical Research, v. 87, n. B11, p. 9301-9316.

Osterkamp, T.E., and Payne, M.W., 1981, Estimates of permafrost thickness from well logs in northern Alaska: Cold Regions Science and Technology, v. 5, p. 13-27.

STABILIZATION OF SINKHOLES ON THE HUDSON BAY RAILWAY

Don W. Hayley[1], William D. Roggensack[1], William E. Jubien[2], and Peter V. Johnson[3]

[1]EBA Engineering Consultants Ltd., Edmonton, Canada T5L 2M7
[2]Canadian National Railways, Edmonton, Canada T5J OK2
[3]Canadian National Railways, Saskatoon, Canada S7K OC3

Embankment settlements have presented a long-term track maintenance problem where the Hudson Bay Railway crosses extensive peatlands of northern Manitoba. Localized zones of settlement, termed sinkholes, typically consist of short sections of track that settle about 100 mm each summer. Field studies have verified that subsidence is caused by thaw at the unstable transition between unfrozen fen and permafrost-cored peat plateaus. In 1978, five test sections were constructed on the railway to study techniques for stabilizing the roadbed by stopping permafrost degradation. A total of 40 two-phase heat pipes were installed along the toe of the embankment at four of the sites. Ground temperatures, track movements and surface settlements were monitored from October 1978 through March 1982. A well-defined ground cooling trend was evident at all of the test sites and settlements were substantially reduced after the first year of operation. The research program successfully demonstrated that heat pipes provide a viable means of arresting localized thaw settlement for embankments constructed on peatlands where permafrost is warm and discontinuous.

INTRODUCTION

The Hudson Bay Railway was the first major transportation facility to be built in the permafrost region of Canada. The line provides a shipping route for prairie grain to port facilities located on Hudson Bay at Churchill, Manitoba. The railway runs northeast for 820 km from The Pas to Churchill, along the route shown in Figure 1. During construction, isolated permafrost bodies were observed in peat bogs as far south as Wabowden. The size and frequency of permafrost islands increased as the line advanced toward the north and east.

Track and roadway subsidence, experienced during and immediately following construction, was a precursor to long-term behavior of the roadbed. The most common manifestations of track settlement have been termed sinkholes by Canadian National Railway (CNR) maintenance personnel. Sinkholes consist of short sections of track that typically experience 100 to 150 mm of settlement during a two-month period at the end of summer. These seasonal sags caused by thaw and compression of permafrost soils in the subgrade necessitate frequent track lifting and bank widening. Maintenance records confirm that the sinkholes recur at approximately the same locations year after year.

This paper describes a track stabilization research program undertaken between 1976 and 1982. The aim of the program was to develop practical means for eliminating or reducing the extent of sinkhole development. Tangible benefits to the railway would include reduced operating restrictions, improved safety, and a reduction in track maintenance costs.

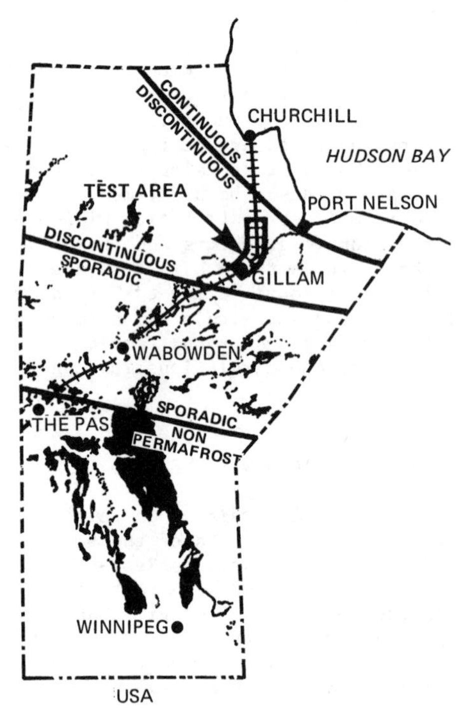

FIGURE 1 Location map, Manitoba Canada.

REGIONAL ENVIRONMENT

The route north of the Nelson River crossing, at Gillam, lies within the Hudson Bay Lowland physiographic region. The topography is subdued, with a maximum elevation about 150 m above mean sea level. The climate varies mainly with latitude but is modified to some degree by the presence of Hudson Bay. The continental climate results in long cold winters and short summers that can be quite warm. Mean annual air temperatures for Churchill and Gillam are -7.1°C and -4.5°C, respectively. Prevailing wind directions are westerly to northwesterly. Between Gillam and Churchill the railway trends in a northerly direction; therefore, drifting results in snow accumulation on either side of the embankment.

Throughout the Hudson Bay Lowland limestone and dolomite bedrock is overlain by a thick sequence of quaternary sediments. Between Gillam and Churchill, the thickness of quaternary sediments ranges from 10 to 50 m. Bedrock outcrops are confined to valleys of deeply incised rivers.

During the Wisconsin glacial maximum, ice occupied the Hudson Bay basin. As the ice retreated, the Tyrell Sea penetrated to the east well beyond the shore of the present bay (Dyke et al. 1982). Glacial retreat was followed by isostatic uplift that eventually forced the Tyrell Sea to retreat toward the present-day shoreline. This retreat exposed a flat plain with a few till-cored knolls protruding above a veneer of fine-grained marine and lacustrine sediments. The climatic and drainage conditions prevailing since deglaciation have resulted in the accumulation of substantial thicknesses of peat throughout the entire lowland. This area exists today as one of the most extensive uninterrupted peatlands in the world.

The internal structure of frozen peat landforms suggests that most of the peat in northern Canada was deposited in a permafrost-free environment (Zoltai and Tarnocai 1975). Permafrost found in peat landforms near the southern boundary of the Hudson Bay Lowlands may be only 600 years old, having been initiated by a brief period of cooler climate at that time (Thie 1974). Once permafrost aggradation has been initiated in peat-covered soil, growth of ice lenses elevates the ground surface, permitting the new plant species to establish. Domed or elevated surfaces then continue to provide the particular thermal characteristics and microclimate required to sustain permafrost (Brown 1968). Most of the relief displayed by frozen peat landforms can be attributed to the formation of ice lenses in the mineral soil beneath the peat (Salmi 1968). Proprietary studies of ground ice distribution conducted adjacent to the railway, along a proposed route for the Polar Gas pipeline, supports this hypothesis.

The terms palsa and peat plateau have been adopted to refer to these elevated landforms. Peat plateaus are the common landforms within the study region, where more than 50% of the area is underlain by permafrost. They plateaus assume a wide range of lateral dimensions and are typically surrounded by unfrozen, water-saturated fens in which ponds and streams are common. The fens are invariably wet and are predominated by marsh sedges and thin non-*Sphagnum* mosses. Conversely, peat plateaus usually have a dry surface and are covered with dense to scattered stands of Black Spruce trees. The ground has a thick forest peat cover consisting of *Sphagnum* mosses and lichens. The peripheries of frozen peat landforms show a distinct response to regional drainage, being molded smooth by the slow flow of ground and surface water in the adjacent fens.

The dimensions of peat plateaus are controlled by climatic conditions, drainage, and local relief. Although they appear to be stable, they respond to both natural and artificial disturbances. Permafrost aggradation and degradation processes can exist simultaneously according to Thie (1974), so the isolated melting of frozen peat landforms does not necessarily reflect a climatic warming trend.

CONSTRUCTION HISTORY

Construction of the Hudson Bay Railway began at The Pas, Manitoba in 1910. The fact that much of this rail line would be built across permafrost terrain was not readily appreciated. Procedures that evolved for coping with the permafrost terrain are described in an historic paper by Charles (1959). Demands on manpower and material imposed by World War I resulted in the project being suspended in 1917. By that time, track had been laid to the Limestone River (north of Gillam), and the right-of-way had been cleared from the end of track all the way to Port Nelson on Hudson Bay. Considerable civil works associated with port development at the mouth of the Nelson River had also been undertaken. For a variety of technical and political reasons, a controversy erupted regarding the selection of Port Nelson as the terminus for the railway. When construction resumed in 1927, the port location had been changed to Churchill, following recommendations of the Palmer Royal Commission.

The first step in resuming construction was to rehabilitate approproximately 320 km of track that had been laid and abandoned 10 years previously. Many embankments and cuts had experienced substantial settlement: ponded water was found lying across the right-of-way in many places, leaving the track and ties suspended in the air. Surveys were conducted to provide a more effective basis for planning drainage, which led to the construction of offtake and parallel ditches in advance of placing the roadbed. However, within the study region shown in Figure 1, the topography is so subdued that it is not feasible to maintain effective drainage control along the right-of-way (J. L. Charles, personal communication, 1978).

Route reconnaissance for the line between the Limestone River and Churchill was carried out on foot during winter. Charles (1959) reports that testholes dug by hand in frozen ground, encountered up to 5.5 m of organics that were usually ice-rich. Recognizing the increasing amount of permafrost encountered north of the Nelson River, it was decided to build the remaining railway almost entirely as an embankment, keeping cuts to a minimum. Initial grading was restricted to that required to permit track to be laid, but occasional cutting into high relief peat plateaus could not be avoided. These cuts were made by using picks and shovels to excavate the frozen peat. The peat was transported by wheelbarrows to adjacent fens where it was used as fill. Track laid on the frozen peat was lifted through train-hauled sand and gravel that originated from offline borrow pits. Figure 2 provides a typical aerial view of the Hudson Bay railway traversing fens and peat plateaus within the test area. The track reached Churchill in March 1929 and has been maintained and operated continuously for more than 50 years.

SINKHOLE DEVELOPMENT

A geotechnical investigation was conducted in 1976, comprising field observations plus drilling and sampling at selected sites. The program's objective was to develop an understanding of the thaw-subsidence mechanism and to rationalize the

FIGURE 2 Hudson Bay Railway crossing peatlands.

distribution of sinkholes along the railway. Abundant segregated ground ice was identified in the peat and mineral soil substrate at each of the sites drilled. Thermistor cables installed in boreholes indicated that maximum ground temperatures within permafrost in the vicinity of Gillam ranged from -0.4°C to 0°C. Field observations supplemented with analysis of the borehole data verified that localized subsidence or sinkhole formation is due to thaw-settlement of the embankment at the transition zone between the nonpermafrost fen and the peat plateau. Track settlement at a typical sinkhole site is shown in Figure 3.

FIGURE 3 Typical sinkhole.

Very little track maintenance has been required on the central portions of peat plateaus. Between the fen and the thermally-stable portion of the peat plateau, however, the track is often bordered by a hummocky actively slumping remnant of the former plateau. These areas of degraded permafrost are usually occupied by thaw ponds 1.5 to 3 m deep.

The hypothesis that was derived to explain development of the sinkholes is illustrated in Figure 4. The rail grade shown in the figure, passes from a substantial embankment across the fen to a small embankment or perhaps a cut across the top of the peat plateau. Disturbance develops at the point where the grade first contacts the boundary of the peat plateau. As small amounts of permafrost degradation accumulate, a transition zone develops and permafrost thaw proceeds along the shoulders before thaw and accompanying settlement occur beneath the railway embankment proper. Gradual degradation along both toes of the embankment eventually coalesce, resulting in a localized sag in the embankment. These sinkholes migrate very slowly along the track into the peat plateau from the original fen-plateau boundary. From year to year, maintenance crews have perceived that settlements recur at the same location, thus leading to the descriptive term sinkhole.

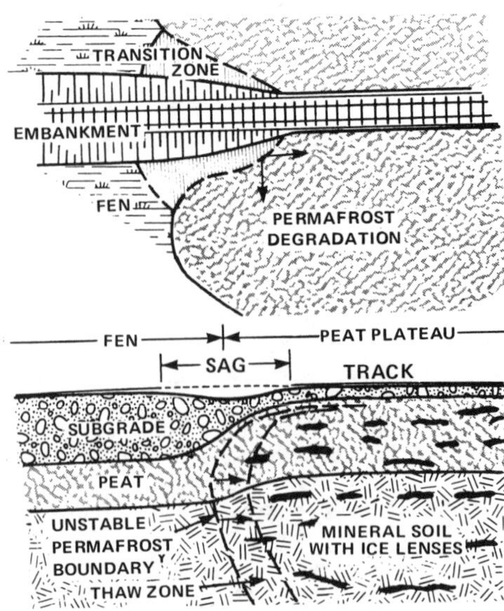

FIGURE 4 Sinkhole development model.

Initiation of permafrost degradation at the transition zone is closely linked to the very warm ground temperatures that are present at the periphery of a peat plateau and the high unfrozen water content of the clayey soils involved in thaw. The degradation is sustained and often accelerated by ponded water and by detrimental snow-drifting patterns. The implication of the thaw-subsidence hypothesis described above is that sinkholes will require maintenance in perpetuity unless progressive lateral shifting of the transition zone can be stopped.

Several alternative remedial measures were proposed and then tested by geothermal simulation to determine their long-term affect. Two-dimensional finite element analyses were first used to investigate the existing section without remedial measures. Formulation and verification of the computer program has been described by Hwang (1976). The simulation predicted an average lateral thaw rate of about 60 mm per year, which reflects a low net positive heat flux into the frozen ground supporting the embankment. To arrest or reverse this process requires that winter heat extraction be increased or that summer heat input be decreased. Embankment modifications that were assessed by further numerical simulation included flattening sideslopes to 6H:1V to limit snow drift accumulation, application of synthetic board insulation on or in the embankment slopes, and insertion of heat pipes into the thawing foundation soils.

Analytical studies of various mitigative measures indicated that a passive system, such as insulation or slope flattening, would probably not be sufficient to stabilize thaw in the fen-plateau transition. However, heat pipes would increase the net heat withdrawal during the winter months by about 25%, and therefore looked promising as a stabilization technique. It was anticipated that cooling beneath the track structure could be achieved with a row of heat pipes installed parallel to the toe of the embankment.

STABILIZATION TEST PROGRAM

A grade stabilization test program was planned to determine the field effectiveness of heat pipes acting both alone and in combination with other measures, such as slope flattening and surface insulation. Five test sites were selected on the basis of their similarity and conformance to a simple and symmetrical pattern of permafrost degradation at the fen-plateau transition. All of the sites were located on the Herchmer Subdivision north of Gillam, Manitoba (Figure 1). This route segment was judged to have the greatest frequency of sinkholes displaying significant settlement magnitudes. The test sites were numbered from north to south and were referenced to railway mileposts. Configurations adopted for the various test sites are summarized in Table 1.

TABLE 1 Test Site Configurations

Site No.	Mile-post	Side-slopes (H:V)	Wood Chip Insulation	Heat Pipes Number	Heat Pipes SPACING
1	368.8	2:1	yes	10	4 m
2	368.8	6:1	yes	0	
3	365.4	2:1	yes	12	4 m
4	363.0	2:1	no	8	4 m
5	340.7	2:1	yes	10	3 m

Model 800 Cryo-Anchor heat pipes, manufactured by the McDonnell Douglas Astronautics Co., were selected for the test program. These two-phase heat pipes afforded a proven history of use on the Trans Alaska Pipeline Project (Heuer 1979) and were also used previously to stabilize a settling foundation on thawing permafrost (Hayley 1982). The units were chosen because of their robust nature and familiarity with their thermal performance characteristics.

A total of 40 units was purchased for installation at 4 of the 5 test sites. The units were 51 mm in diameter and 9.1 m long, with 1.8-m-long extruded aluminum radiator segments. The number of heat pipes installed at each site varied from 8 to 12, as indicated in Table 1. The units were evenly distributed along opposite sides of the track near the embankment toe. Spacing between adjacent heat pipes was 4 m, except for the most southerly site, where the spacing was decreased to 3 m. At 3 of the heat pipe sites, a 150 mm thick layer of wood chips was spread over the embankment sideslopes to provide an insulating cover over the gravel surface. The single site without heat pipes had embankment slopes that were groomed to 6H:1V, then covered with wood chip insulation.

Test site construction and heat pipe installation took place in October 1978. The units were installed in boreholes drilled with a rotary drill rig and grouted into place with a neat cement grout. The extruded aluminum radiator fins were installed on the heat pipes after the borehole grout had set. A typical completed test site is shown in Figure 5.

FIGURE 5 Typical test site.

During the construction period, the following instrumentation was installed at each site: 7 thermistor cables, 1 deep benchmark, 10 surface settlement plates, stations for repeated top-of-rail surveys, and 10 snow depth guages.

Test site performance was monitored for a 3.5-year period. Elevation surveys were repeated at approximately 2 month intervals throughout the 3 thawing seasons. Ground temperature measurements were obtained at 2-month intervals throughout the year with additional readings during the fall freeze-up period.

PERFORMANCE SUMMARY

Meteorological data for both Churchill and Gillam indicated a slight annual warming trend over the 3.5-year monitoring period. The warming resulted chiefly from warmer winters, as reflected by a decrease in air freezing index. Despite this warming trend, the installation of heat pipes has brought about general cooling in the transition zone between the fen and peat plateau. In all cases, track settlement has been substantially reduced. Little, if any, cooling was observed in connection with the flattened sideslopes at Site 2.

Data from two of the thermistor cables installed at Site 3 are shown in Figure 6. The location of the cables relative to the track and the heat pipe array is shown in Figure 7. The data illustrates the modest cooling trend that is typical below the track centerline at all of the transition zone sites where heat pipes were installed. The extent of cooling recorded over the 3.5-year period ranged from 0.3 C° at Site 3 to 0.5 C° at Site 5. The thermistor cable located between adjacent heat pipes shows a stronger cooling trend with a notable decrease in winter ground temperature minima from year to year. This behavior reflects the decrease in latent heat extraction required to fully freeze the soil between heat pipes during successive winters.

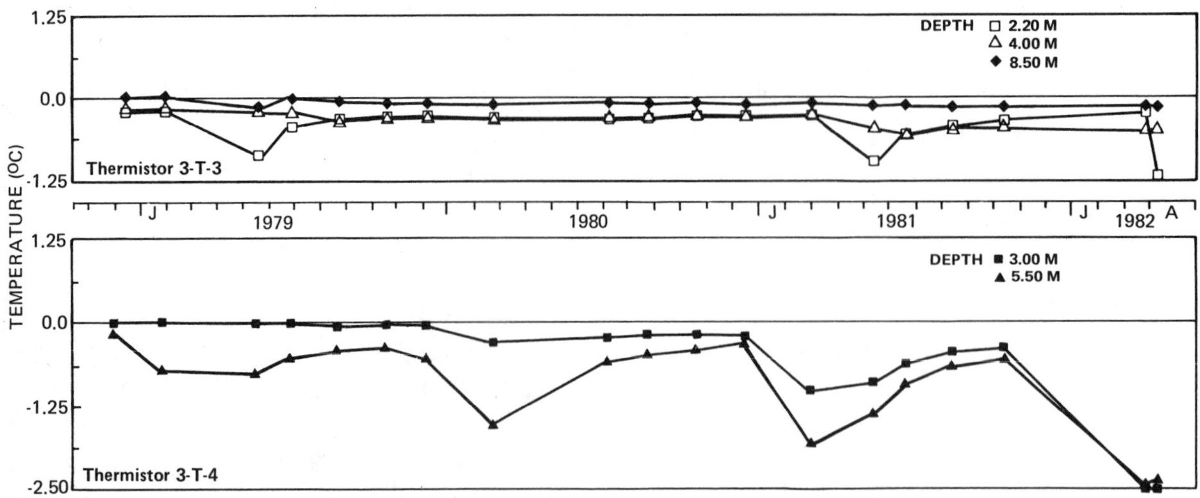

FIGURE 6 Ground temperature variations, Site 3.

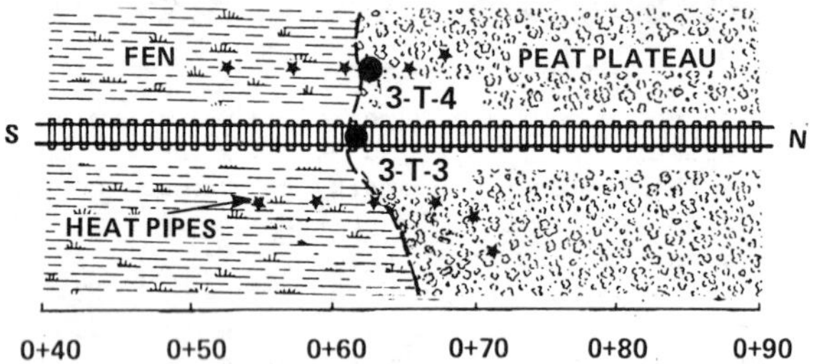

FIGURE 7 Thermistor cable location plan, Site 3.

The performance of individual heat pipes has been evaluated by back analyses of the soil temperature records at the midpoint between adjacent units. The two-dimensional finite-element model described previously (Hwang 1976) was used to predict radial freezing about a vertical line heat sink. The program was modified to allow analyses using polar coordinates to represent cylindrical freezing about a vertical axis of symmetry. Seasonal heat flux from the ground surface was also taken into account in the analyses, with input based on meteorological data from Gillam.

A comparison between predicted and measured ground temperatures at an initially unfrozen location between heat pipes (Site 1) is shown in Figure 8. The predictions have been developed for average winter wind speeds of 5 and 15 km per hour using thermal performance data provided by the manufacturer. The predicted and observed ground temperatures at the depth shown in Figure 8 match very well for wind speeds in the order of 15 km per hour. For the design wind speed of 5 km per hour, the measured maximum summer temperature is cooler than predictions by about 0.2 C°. The excellent overall comparison confirms the reliability of the performance criterion used for the design. The heat pipes have, in fact, performed somewhat better than anticipated.

There have been only minor differences between thermal effects achieved at 3- and 4-m spacings with heat pipes placed along either side of the embankment. A 1-m decrease in spacing does result in quicker freezeback between the heat pipes, but even at 4-m spacings the frost cylinders converge after 2 to 3 years of operation.

Top-of-rail elevation surveys did not provide useful quantitative data pertaining to grade settlements. Periodic resurfacing of the tracks resulted in disparities that could not be adequately evaluated. Settlement plates embedded in the embankment sideslopes within the heat pipe array show negligible settlements. Maximum reported settlements were on the order of 50 mm in 3 years, and these were generally located at the extreme fen end of the test sections. These settlements could be attributed to compression of peat due to placement of new fill. Elevation readings from the most southerly test site show slight heave of the embankment as permafrost conditions have been reestablished in the foundation soils.

CONCLUSIONS AND FUTURE APPLICATIONS

Stabilization of the transition zone by heat pipes has been successful at all of the test sites where the technique was used. This conclusion has been confirmed by operating and maintenance crews who have been pleased with track stability during the monitoring period. The use of sideslope flattening has also been moderately successful during the short period of observation. However, passive methods of stabilization, such as bank widening and slope flattening, are considered to be much less effective than active heat removal. Adverse weather conditions, such as successive years with warmer-than-average temperatures, could initiate irreversible thaw settlements.

The heat pipe array adopted for the test sites appears to be a near optimal configuration. Long-term stabilization has been achieved with a spacing of 4 m. Although the 3-m spacing results in more rapid freezeback, significant improvements in performance have not been realized with the closer spacing. The axisymetric geothermal simulator provides a useful technique for investigating heat pipe effectiveness and spacing at locations outside the immediate test area.

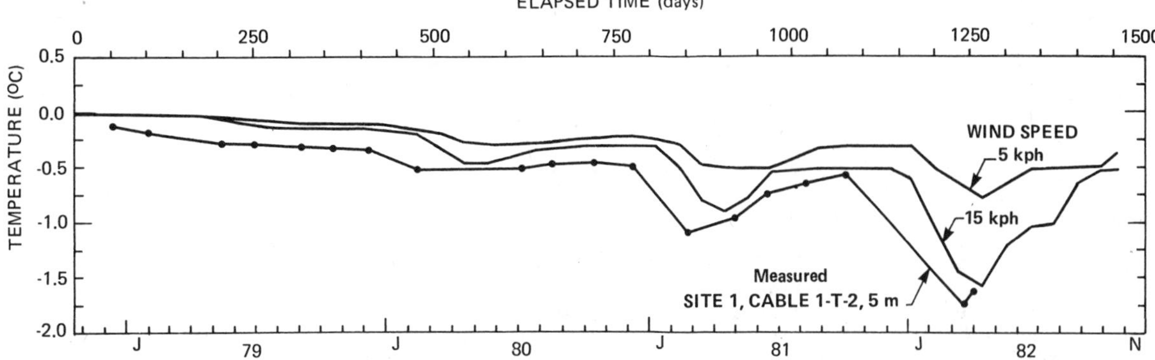

FIGURE 8 Comparison of measured and computed ground temperatures midway between heat pipes.

Heat pipes installed in ground that is presently frozen do not appear to enhance the stabilization process; therefore, accurate definition of the transition zone is a critical component in future large-scale applications of this technique. The total number of heat pipes required to stabilize a localized sinkhole could conceivably be reduced to 4 or 6 if the transition zone can be predefined by geophysical techniques.

Insulating the fill by placing wood chips on the ground surface around the heat pipes improved performance of the test sections only marginally. This minimal improvement in net heat extraction does not appear to justify the difficulties associated with transport and hand placement of surface insulation.

Development of a fully rail-mounted operation has been recommended for future installation of heat pipes to stabilize the railway grade. They could be inserted into a predrilled hole that is sized to ensure a tight fit, thus eliminating the requirements for backfill or grout. The entire operation could be accomplished with either a mobile crane or a crane mounted on a flatbed car. Such an installation technique would be highly efficient in a region where access is available only along the railway. Working from the rails would also minimize handling and placement of additional granular fill in a region where haul distances in excess of 100 km are common.

ACKNOWLEDGMENTS

The authors gratefully acknowledge the support of a number of agencies and individuals who contributed to the program. The early investigative studies on Hudson Bay Railway were funded jointly by the Transportation Development Agency and the Canadian National Railways. Test installation and monitoring were funded by the Prairie Branch Line Rehabilitation Program, administered by Transport Canada. Field support logistics, surveying, and monitoring were provided by CNR from The Pas, Manitoba. Valuable assistance with administering the program was provided by D. J. Horbay of CNR, Saskatoon. The two-dimensional thermal analyses was conducted by D. C. Cathro and Dr. C. T. Hwang of EBA Engineering Consultants Ltd., Edmonton. The Cryo-Anchor heat pipes were supplied and installed by Mobile Augers and Research Ltd. of Edmonton. They also provided valuable assistance with planning and executing the field program.

REFERENCES

Brown, R. J. E., 1968, Permafrost investigations in northern Ontario and northeastern Manitoba, National Research Council - Division of Building Research, Technical Paper No. 21, 75 p. (NRC 10465).

Charles, J. L., 1959, Permafrost aspects of the Hudson Bay Railroad: Journal of the Soil Mechanics and Foundations Division, Proceedings of the American Society of Civil Engineers, v. 85, no. SM6, p. 125-135.

Dyke, A. S., Dredge, L. A., and Vincent, J.-S., 1982, Configuration and dynamics of the Laurentide ice sheet during the Late Wisconsin maximum: Geographic physique et Quaternaire, Vol. XXXVI, nos. 1-2, p. 5-14.

Hayley, D. W., 1982, Application of heat pipes to design of shallow foundations on permafrost, Proceedings of the Fourth Canadian Permafrost Conference (The Roger J.E. Brown memorial volume): Ottawa, National Research Council of Canada.

Heuer, C. E., 1979, The application of heat pipes on the Trans-Alaska Pipeline. CRREL Special Report 79-26, 27 p.

Hwang, C. T., 1976, Predictions and observations on the behavior of a warm gas pipeline in permafrost: Canadian Geotechnical Journal, v. 13, p. 452-480.

Salmi, M., 1968, Development of palsas in Finnish Lapland, Proceedings of the Third International Peat Congress: Quebec, p. 182-189.

Thie, J., 1974, Distribution and thawing of permafrost in the southern part of the discontinuous permafrost zone in Manitoba, <u>in</u> Arctic, v. 27, no. 3, p. 189-200.

Zoltai, S. C. and Tarnocai, C., 1975, Perenially frozen peatlands in the western Arctic and Subarctic of Canada: Canadian Journal of Earth Sciences, v. 12, 28-43.

BUILDING FOUNDATIONS ON PERMAFROST

He Changgeng

The Third Survey and Design Institute of the Ministry of Railways
People's Republic of China

To examine the applicability of various foundations to cold regions, the massive concrete foundation, earth-padded ventilation foundation, reinforced concrete strip foundation, and column and pile foundations were tested at Zhaohui and Jingtao field stations in northeast China, where the mean annual air temperatures are -5° and -5° to -6°C, respectively. The ground temperature fields, thawing settlement, and tangential frost-heaving forces of subsoils were measured from 1972 through 1980. Test results showed that the distribution of ground temperature beneath structures depends upon the type of foundation, and that the thawing depth under foundations is governed by the building's heating system. The formation of cracks in buildings not only depends upon the settlement of foundations, but is also influenced by air temperature. In designing the ground floor of buildings, thermodynamic calculations must be made according to the type of foundation.

Since the type of foundation is not suitable to the floor which settles at thawing and expands at freezing, buildings on permafrost have often sustained damage. In order to study such damage, we have tested buildings on five types of foundations at Zhaohui and Jingtao, where the natural conditions are as follows.

1. Zhaohui Station: mean annual air temperature is about -5°C, original permafrost table 1.6 m. The ground has a pine and birch forest with some moss. Underneath is a peat layer 0.15 m thick, with a sandy loam soil having a 10-25% crushed stone content and 10-80% ice content with a thickness of 4.7-7.3 m. Further down, a sandy pebble layer has thickness of 3.5-5.0 m. Further down is a tuff weathering layer. Constant annual ground temperature is -1.1°C.

2. Jingtao Station: mean annual air temperature is -5 — -6°C original permafrost table 0.4 - 1.0 m. Ground surface has grass mould and moss, with some pine and birch trees. Peat layer is 0.3 - 0.5 m thick, content of ice 65%, ice layer containing soil is about 1.0 m, with ice amounting to 70-80%. Sandy loam soil mixed with gravel is about 1.4 m thick, ice content 20-35%. Pebbles and grit about 4.5 m thick, containing 20-30% ice. Constant annual ground temperature is -2 to -3°C.

The tests began respectively in 1972 and 1976, and on the basis of observations over 5-8 years, we have obtained sufficient data on the thawed settlement and heave of foundation soils ground temperatures, tangential frost heaving forces, as follows.

ZHAOHUI STATION

Explosively Enlarging Pile Foundation of this type of foundation five test buildings were built, using different methods to treat three artificial heat sources (fire stove, space-heated wall, and heated brick bed). When heat from the three sources was suspended on posts or beams, depth of thaw in October each year reached 2.9 m at the maximum, as shown in Fig. 1. The bigger end of the pile foundation is 1.0 m below the thawing plate. Maximum temperature at the bigger end is -0.6°C. At this time, the compressive strength of permafrost is $\sigma = 40t/m^2$, the stress of pile foundation is:

(1) load bearing capacity

$$P = A\,\sigma = (0.4\ m)^2 \pi \times 40t/m^2 = 20.1^t$$

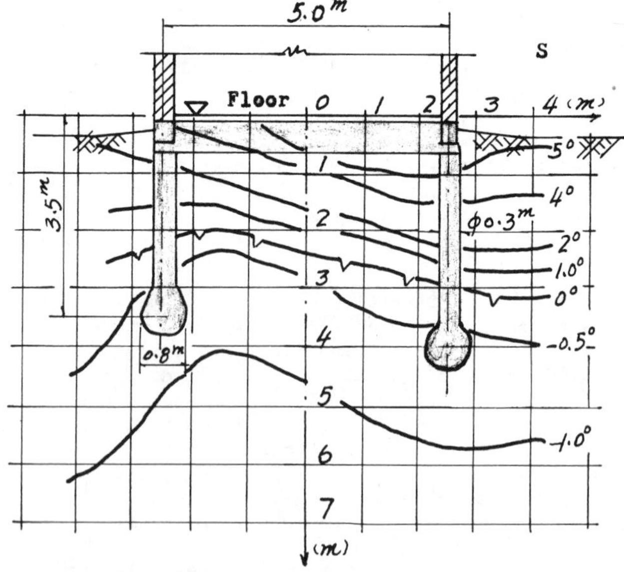

FIGURE 1 Temperature at maximum depth of thaw in subsoil under explosively enlarging pile foundation (3 heat sources on stilts), October 1975.

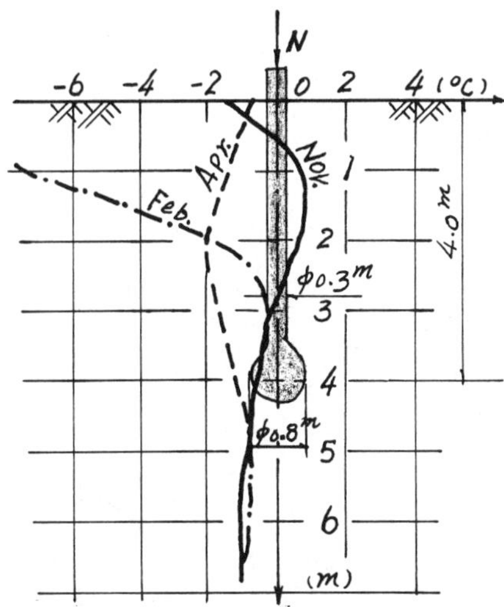

FIGURE 2 Maximum and minimum temperatures explosively enlarging pile foundation (3 heat sources on stilts).

(2) N (weight carried) + G (dead weight of Pile) = $16.3^t < 20.1^t$

When freezing recurs in winter, pile foundations are also subjected to tangential frost heaving forces. By actual measurement of a 0 — 0.8^m section, the tangential frost heaving force 15 t/m^2. Over 2.0 m it was measured as 15 t/m^2; under 2 m as 1/3 of 15 t/m^2. Temperature at the pile foundation is shown in Figure 2.

(3) Frost pulling force $\tau = \sum_{i=1}^{n} A_{ui} \tau i =$

$0.3^m \pi (2^m \times 15t/m^2 + 0.9^m \times 5^{t/m^2}) = 32.5^t$

(4) Antifrost pulling force $T = N+G + \sum_{i=1}^{n}$

$F_{mi} \; t_i = 16.3t + 0.8m \, \pi \times 1.1m \times 6t/m^2 =$

32.8^t

On the basis of a number of observations, maximum foundation subsidence difference is 8 mm, without exceeding the specified value of 10.8 mm. However, a crack has occurred, though this does not affect usage. The big-headed pile has the effect of reactive force on frost heaving, and thus increases the stability of the pile, making it quite safe. But since the floor is built directly on the ground surface, which freezes again in winter, mean floor temperature is only about 5.0°C despite the insulation layer laid on room floor. It is unfit as living quarters.

In similar buildings with heat from stove and heated wall built on stilts, maximum depth of thaw is 3.6 m, the differential in value of foundation settlement is 11 mm, which is in excess of the specified value 6.8 mm, thus producing many cracks.

In houses on the work site with all three heat sources not on stilts, the foot of the northwest wall is filled with a pile of slag for heat insulation. Maximum depth of thaw of the subsoil is 5.7 m, exceeding the pile depth and forming a stable melting plate. The amount of thawing and settlement is 120 mm. The freezeback in winter is faster at the two ends and slow in the middle, thus again causing inhomogeneous freezing expansion, with more cracks in the wall of the house.

From the above-mentioned variations in ground temperature, it may be seen that artificial heat sources on stilts is one of the methods of reducing the depth of thaw. The reason why the depth of thaw is not great is the settlement of subsoil, and beneath the foundation beam crevices are formed resulting in a semiventilated foundation. In winter most of the foundation is frozen, while in summer the depth of thaw is shallow. This type of foundation is stable.

Open Cased Foundation

It is solid as one body with great rigidity, and may rise and fall with the expansion and settlement of the subsoil. It is better adapted to the thawing settlement and freezing expansion of the subsoil. Its maximum depth of thaw is 7.0 m as shown in Figure 3.

In winter, when freezeback occurs, the depth is only 2.1 m, the width of freezeback toward the

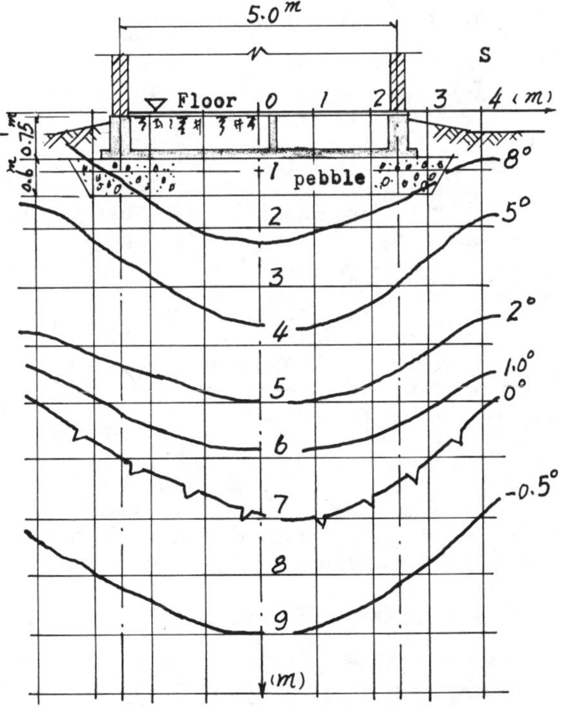

FIGURE 3 Temperature distribution at maximum depth of thaw in subsoil under open-cased foundation.

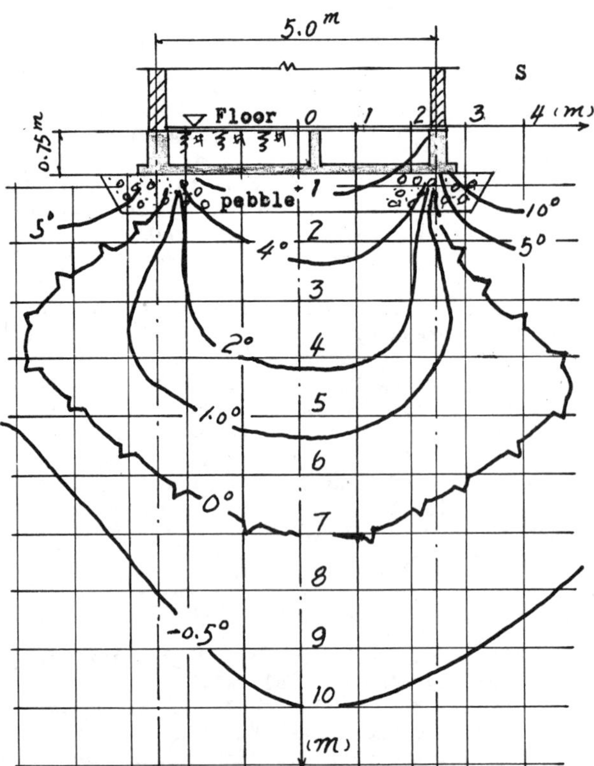

FIGURE 4 Minimum temperature distribution in subsoil under open-cased foundation.

center of the room is 0.3 m, as shown in Figure 4. In the most severe cold room temperature is only 12.6°C, while the floor still retains 8.6°C. It is feasible, therefore, to maintain room floor temperature by the use of thawed subsoil, without ventilation.

The subsoil of this foundation has a maximum settlement as much as 300 mm, while the two ends settle only to a depth of 150-180 mm. In winter, the two ends are entirely frozen and heaved. The inhomogeneous value of settlement and expansion of the foundation at the east and is 100 mm, which is 9 times in excess of the specified value of 10 mm, thus resulting in more cracks at both ends. Some cracks are big enough for the wind to blow through and therefore need repairs. In the mid-section, the depth of thaw is greater but more homogeneous, with a settlement differential of 11 mm, which is in close proximity to the specified maximum value. It has fewer cracks. From this it is seen that to control cracks, the settlement differential value must be controlled.

Soil Filling Cycle-Beam Foundation:

Subsoil filled with sandy gravel with soil has a large amount of stored heat thus deepening the natural permafrost table. With heating provided in the room, the depth of thaw gradually increases. After the station room had been in use for seven years, the depth of thaw reached 9 m, and began to be stabilized. The reason for the greater depth of thaw, apart from the heat provided, is mainly because soil filling on the south is 4 m high, while the three slopes on the east, south and west are all exposed to sunshine. So the subsoil absorbs more heat and thawing becomes deeper. Freezeback in winter is also deeper: 6.0 m under the north wall, and 4.2 m under the south wall. It is evident that the thermal effect of sunshine is considerable. Because the filling soil has a low water content and better conditions for drainage, the extent of settlement is not appreciable, so the room has only two small cracks.

JINGTAO PERMAFROST EXPERIMENT STATION

Built on Stilts Embedded Pile Foundation

This type of foundation is similar to the explosively enlarging pile foundation in nature and condition under stress. After five years of observation, it is found that changes in freezing and thawing in the subsoil are essentially the same each year. Thawing begins at the end of April and reaches in September the maximum depth of thaw of 2.1 m. At this time temperature in the subsoil is -1.4°C, as shown in Figure 5. In October ground surface begins freezing and by January of next year is completely frozen. Temperature of the subsoil in February is -1.4°C, in April -2.8°C, as shown in Figure 6. But when soil temperature drops to -1.8°C, there is some slight freezing expansion. Hence there is a 0-6 mm deformation of the foundation caused by freezing expansion, and also an inhomogeneous deformation of 9 mm, which is less than the specified 13.6 mm. Theoretically, there should be no cracks, but in practice there are several vertical cracks. This is a result of the common effect of freez-

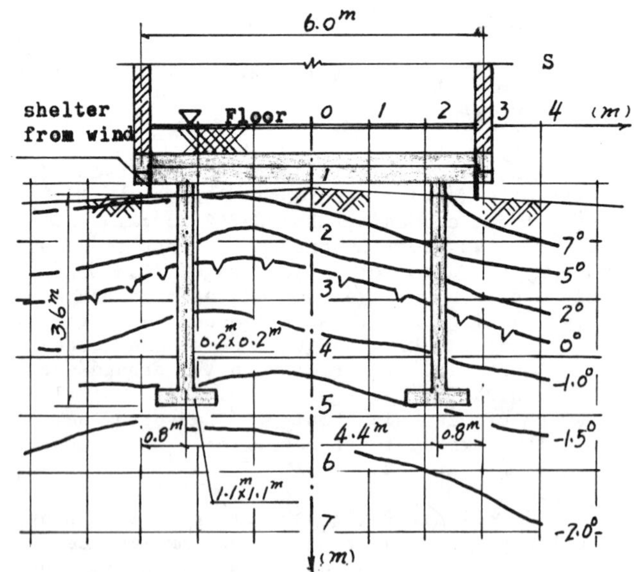

FIGURE 5 Temperature distribution at maximum depth of thaw in subsoil under built-on-stilts ventilation foundation, September 1979.

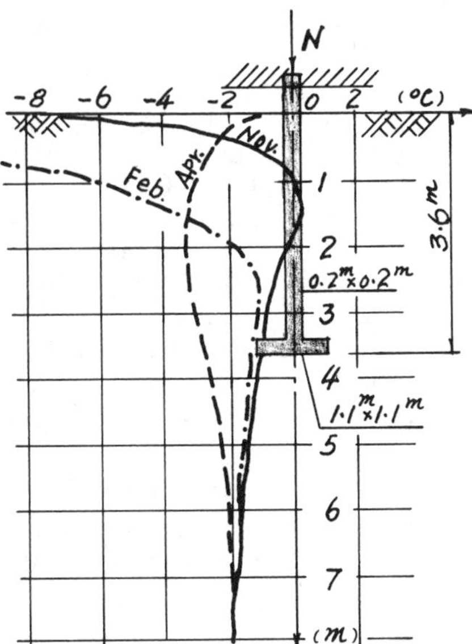

Figure 6 Maximum and minimum temperatures in subsoil under built-on-stilts ventilation foundation.

ing expansion and stress of temperature differential (air temperature difference 34 to -48°C). Usage is not affected.

From Figure 6, it may be seen that there are drastic changes in ground temperature. In this case is the deformation of foundation caused by the tangential frost force? By actual measurement of six foundation piles is obtained the tangential frost force $\tau_i = 13t/m^2$.

Then,

(1) Frost pulling force $\tau = \sum_{i=1}^{n} A_{ui}\,\tau_i =$

$0.2m \times 4 \times 2.1m \times 13t/m^2 = 21.8^t$

(2) Antifrost pulling force $T = \sum_{i=1}^{n} F_{mi}\,t_i +$

$N + G = 1.1m \times 4 \times 1.6\,m \times 7t/m^2 + 16.7^t =$

$66^t > 21.8^t$

It is evident that deformation of the foundation is not due to the effect of tangential frost force. Where possible, the depth of foundation with ventilation on stilts should be further increased, or heat insulation measures be adopted, so that changes in subsoil temperature may not exceed 1.0°C. In short, it is feasible to use embedded pile foundation with ventilation on stilts for buildings in large areas of continuous permafrost soil.

It may be seen from Figure 5 that the foundation pile has moved 0.8 m toward the center of the room, while ground temperature of subsoil in the south drops about 0.2°C, which is favorable to the load bearing capacity of the foundation. If the apron slope at foot of the south wall is made to extend out from the pile it will not only avoid damage from freeze or thaw of subsoil, and also reduce the depth of thaw of subsoil beneath the outer wall.

Drilled and Backfilled Pile Foundations

Two test houses were built on this type of foundation. The foundation pile penetrates the thawing plate, using the freezing force of permafrost for bearing load and anti-frost pulling. In order to reduce the depth of thaw, the outdoor height differential for residence is 1.1 m, that for dormitory is 0.5 m, for the purpose of comparing the effect. After five years of usage, the maximum depth of subsoil is: counting from the outdoor ground surface, 2.0 m for residence, as shown in Figure 7, and 2.0 m for dormitory. The two houses have different height differentials, but the depth of thaw is the same. The main reason is that constant annual ground temperature of the dormitory (-3°C) is 1.0°C lower than that of the residence (-2°C).

Like houses built on the explosively enlarging pile foundation, it too has a ventilation path under the floor beam, so the floor temperature in the room is only 5 to 8°C in the coldest weather.

This load bearing capacity of the drilled and backfield piles, on the basis of ground temperature around the pile as shown in Figure 8,

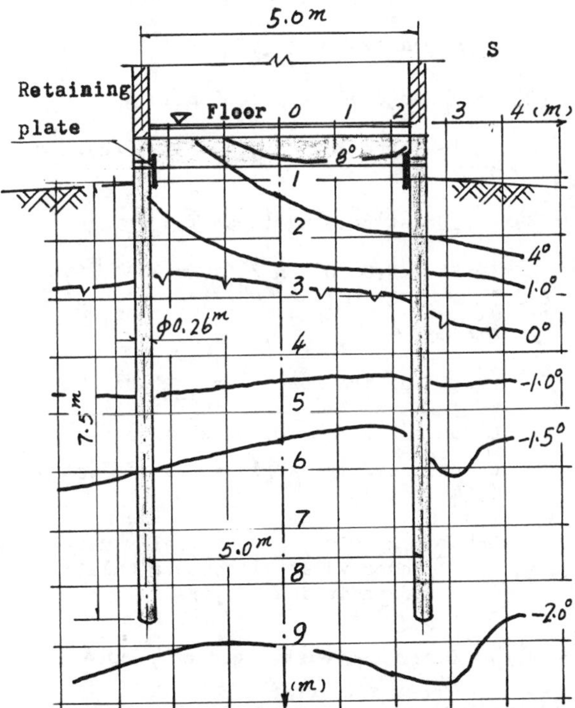

FIGURE 7 Temperature distribution at maximum depth of thaw in subsoil under drilling and filling pile foundation, October 1979.

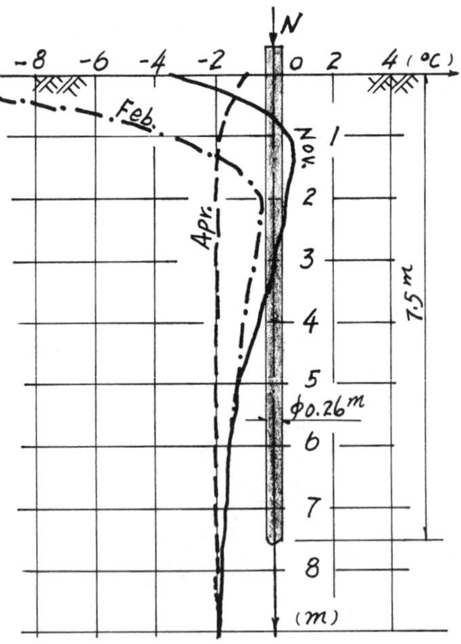

FIGURE 8 Maximum and minimum temperatures at surrounding depth of thaw under drilling and filling pile foundation.

is calculated as follows:

(1) Load bearing capacity $P = \sum_{i=1}^{n} A_{mi} S_i +$ $A\sigma = 0.26m\,\pi\,(1.0m \times 3^t/m^2 + 2.0m \times 8.5^t/m^2 + 1.4m \times 14^t/m^2) + (0.13m)^2\,\pi \times 140^t/m^2 = 39.7^t$

(2) Load and dead weight of pile $N + G = 21.2^t$

Degree of safety $= 39.7^t/21.2^t = 1.87$

(3) Antifrost pulling force $T = \sum_{i=1}^{n} A_{mi}S_i + N + G = 0.26m\,\pi\,(1.0m \times 3^t/m^2 + 2.0m \times 8.5^t/m^2 + 1.4m \times 14^t/m^2) + 9.3t = 32.4^t + 9.3^t = 41.7^t$

(4) Frost pulling force $\tau = \sum_{i=1}^{n} A_{ui}\,\tau_i = 0.26m\,\pi \times 2m \times 20.5\ t/m^2 = 33.5^t$

Overall safety margin = 41.7t/33.5t = 1.24

Note: $\tau_i = 20.5t/m^2$ is actually measured value without calculation of individual layers.

Where: σ — Compressive strength of permafrost soil (t/m^2)

A — Cross section of pile end (m^2)

N — Load on pile (t)

G — Dead weight of pile (including the weight of soil around the pile)(t)

A_{ui} — circumferential area of foundation pile in "i" seasonal frozen soil layer (m^2)

τ_i — tangential frost heaving force around pile in "i" seasonal frozen soil layer (t/m^2)

F_{mi} — circumferential area of soil column on the pile end in "i" permafrost soil layer (m^2)

t_i — Tangential heaving force in "i" permafrost soil layer (t/m^2)

n — number of permafrost soil layers.

A_{mi} — circumferential area of pile in "i" permafrost soil layer (m^2)

S_i — Tangential heaving force of "i" permafrost soil layer (t/m^2)

It is known from this calculation that both the load bearing capacity and antifrost pulling force of the pile are adequate. However, there is still some deformation of the foundation, maximum settlement differential being 8 mm, which is in excess of the permissible value 6.8 mm. Hence there are cracks in the houses. The cracks were found after construction, and they expand and contract with changes in air temperature. It is evident that the stress of temperature differential also has some effect. Besides, there may also be some creep effect produced by permafrost under stress.

From the above analysis and discussions, it may be said that tests on all five types of foundations are basically successful. Each of these, however, has its own optimum conditions for adaptation.

1. Explosively enlarging pile and prefabricated embedded pile: In both cases, the depth of embedment is limited by manual construction, generally to the extent of about 4 m. Therefore, these are suitable for foundations of buildings with underflow ventilation and without heating. Moreover, the maximum subsoil temperature should conform with the requirement for sandy soil at -0.3°C and clayey loam at -0.5°C.

2. Filling pile foundation: This has a greater adaptability and may be used as the foundation for various buildings on different types of permafrost soil. However, permafrost clayey loam with a temperature higher than -0.5°C should not be used because of inadequate load-bearing capacity.

3. Open cased foundation: This is built in the strata of ground surface and easy to construct. It is closely integrated with the subsoil, with a greater depth of thaw. The temperature of the floor is high and suitable for living quarters. But it has a greater settlement and cracks are liable to occur. So

it is suitable for buildings with small areas such as $60m^2$. Preferably, building site should be permafrost soil with negligible settlement at thawing and clayey loam at a temperature over $-0.5^{o}C$.

4. Soil filling cycle-beam foundation: Filling is costly and not easy to ram down, so it is unsuitable for use in ordinary cases. Where production requires, besides having it rammed down, it is necessary to have the height of filling soil plus the depth of natural table greater than the building's maximum depth of thaw, in order to ensure the stability of this type of foundation.

Pile foundations are better types of foundation for buildings on permafrost soil, but because in most cases there are air gaps beneath the subsoil beam thus forming ventilation paths, the temperature of room floor is reduced in winter to the extent that it is uninhabitable. It is recommended, therefore, that buildings on pile foundation, whether ventilated or not, should be built on stilts and provided with ground surface heat insulation in order to make them habitable.

PROBLEMS IN THE CARTOGRAPHY OF GROUND ICE: A PILOT PROJECT FOR NORTHWESTERN CANADA

J.A. Heginbottom

Geological Survey of Canada, Ottawa, Canada, K1A 0E8

A new permafrost map of Canada is being proposed, emphasizing the occurrence and distribution of ground ice. The key element is the extreme variability of ground ice both horizontally and with depth. Problems include the development of a suitable classification scheme and the quality and variability of the source data. Previous work on ground ice classification and mapping is reviewed and a new, morphological classification of ground ice bodies is proposed. The main classes are: ice crystals in voids, equidimensional ice masses, layers or sheets of ice (horizontal, vertical, diagonal) and networks of segregated ice. These classes are subdivided on the basis of the regularity of the shapes of ice masses and whether their occurrences are simple or multiple. A qualitative classification of the amount of ground ice is proposed: low, moderate, high, very high and extreme. Selected geomorphic features, such as pingos, will be shown by symbols. A pilot study map of northwestern Canada, using these classifications is being prepared.

INTRODUCTION

Permafrost is a ubiquitous phenomenon in northern Canada and, as such, is of concern in many aspects of life and in most engineering activities in the area. The engineering problems arise not so much because of the low ground temperature, but because most or all of the water in the ground is in the form of ice. The difficulties are compounded because the ice is unevenly distributed in the ground, commonly occurs in quantities in excess of the void volume, and often exists in large, relatively pure bodies. These attributes result in a soil-ice-water system that is sensitive to disturbance and which has physical properties highly variable over short distances.

Although the existence of permafrost has been known for centuries, the serious engineering problems caused by its existence have been recognized only in the last 50 years. In North America, Muller's (1947) book was a key document in drawing the attention of scientists and engineers to the challenge of permafrost. In Canada, military construction activities gave a number of engineers firsthand experience in coping with permafrost.

Recognizing this, the National Research Council of Canada undertook, in the 1950s and 60s, a survey of permafrost in Canada. This program resulted in a series of reports describing permafrost conditions (e.g. Brown, 1964, 1965 and 1968), and culminated in a map of permafrost in Canada (Brown, 1967).

This map, and the many other maps of permafrost in Canada and North America published in the last 70 years, all have one major deficiency from the perspective of geologists and engineers: they contain essentially no data on the occurrence or distribution of ground ice. Permafrost is largely treated as a climatic phenomenon. The very definition of permafrost, strictly on the basis of temperature, reinforces this approach. And yet, it is the ground ice component of the permafrost that is of most concern to many northern engineering projects.

OCCURRENCE OF GROUND ICE

Ground ice occurs in a variety of forms and a wide range of quantities. At one extreme, ice can comprise 50-70% by volume of the upper 2-3 m of the permafrost over broad areas (J. Brown 1967, Pollard and French 1980). Locally ground ice can exceed 90% of the volume of the ground (Rampton and Mackay 1971). In other areas permafrost contains essentially no ground ice and is termed "dry permafrost". In addition to this gross variability, detailed work along exposures and by means of borings indicates that there is considerable local variation in the distribution of ground ice on a metre-by-metre scale (Heginbottom and Kurfurst 1975). This variability makes the compilation of a ground ice map difficult. Other problems are the selection of a suitable ground ice classification, the availability and the quality of source data and the cartographic difficulties of displaying a complex three-dimensional phenomenon on a map.

Vertical Variability of Ground Ice Conditions

There are relatively few published descriptions of the variation in ground ice content with respect to depth. Examples of two are shown in Figure 1; others from the Mackenzie Delta area have been published by Mackay (1972a, 1973) and Williams (1968). Rampton and Mackay (1971) also analysed the proportion of drill holes which encountered massive ground ice or icy sediments in Richards Island, Tuktoyaktuk Peninsula and adjacent areas, while for the Mackenzie Valley, Heginbottom and Kurfurst (1975) presented detailed borehole logs of ground ice from seven specific localities. The depths of the boreholes described in these sources range from as little as 1 m to over 40 m.

In extreme cases, the ground can consist of essentially pure ice for thickness of over 30 m (Mackay 1973) and the depth to the top of such massive ground ice can be as great as 45 m, although it is commonly less than 20 m (Mackay 1972a, 1973). The profiles shown in Figure 1 seem to be generally representative of the distribution

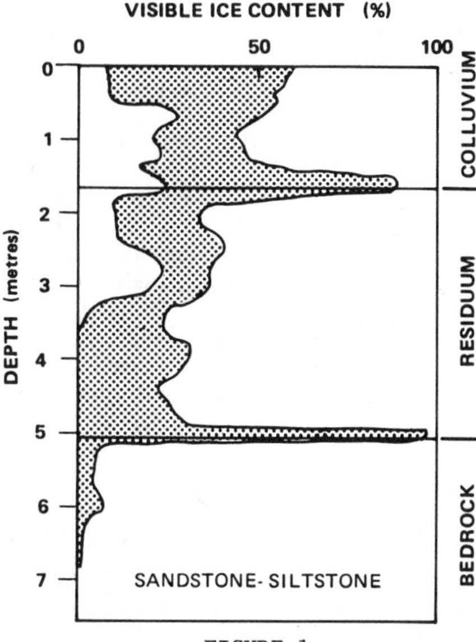

FIGURE 1a

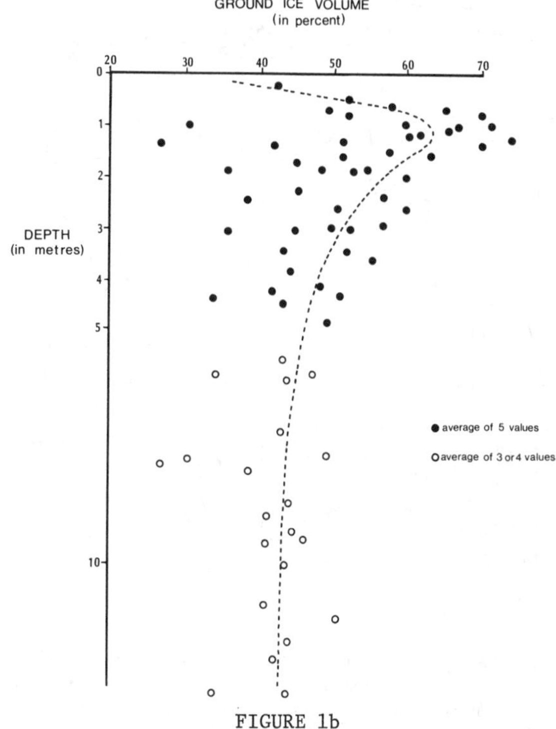

FIGURE 1b

FIGURE 1 Examples of the distribution of ice content with depth for sites in the continuous permafrost zone of Canada. (a) Melville Island, N.W.T. (Stangl et al. 1982). A generalized profile for soils developed on rocks of Mesozoic and Paleozoic age. (b) Richards Island, N.W.T. (Pollard and French 1980). Note: the curve is a visual best fit and not calculated.

of ground ice in soil profiles. The highest ice contents occur at the top of the permafrost profile. The general downward decrease in ice contents is punctuated by local increases, commonly at stratigraphic breaks. Similar observations have been reported from northern Alaska (J. Brown 1967).

Horizontal Variability of Ground Ice Conditions

Information on the horizontal variability of ground ice contents is available locally at either large or small scales and in the form of both plans and profiles. At the large scale, the existence of individual masses of ground ice or, in the discontinuous permafrost zone, of distinct bodies of frozen ground have been described by several workers. Examples from the continuous permafrost zone of Canada are mainly from the western arctic region (Figure 2).

For the zone of discontinuous permafrost, site investigations at Thompson, Manitoba (Johnston et al. 1963), and at Dawson, Yukon Territory (EBA 1977), showed how the existence of frozen ground and ground ice varied on metre-by-metre scale. Both plans and trench profiles are in these reports.

On the broad or regional scale, Mackay (1966) has presented a map of the types of ground ice found in the Mackenzie Delta and on Tuktoyaktuk Peninsula. Other than this example, most reports for Canada are in the form of incidental reports in regional descriptions (e.g. Brown 1972) or annotations to surficial geology maps (e.g. Rampton 1981).

THE CLASSIFICATION OF GROUND ICE

The attributes of ground ice that are of most importance in the compilation of the proposed map are the quantity of ice present and its form. Published classifications do not emphasize either of these characteristics. Those of Shumskii (1959) and Mackay (1966, 1972a) emphasize the genesis of the underground ice rather than its form, while the commonly used engineering classification (Pihlainen and Johnston 1963) is designed to facilitate the production of standardized descriptions of permafrost and ground ice conditions. It does not provide a classification scheme for ground ice. Accordingly a new, simple morphological classification has been developed (Table 1). All of Mackay's (1972a) genetic classes can be related to members of this new classification.

Descriptions of the quantity of ground ice are commonly provided in general terms such as low, moderate and high or unsaturated, saturated and excess. In some cases, percentage values of total moisture content, ice content or excess ice content, on either a gravimetric or a volumetric basis are quoted, often as a range of values. Few sources include both. One is the series of Terrain Classification and Sensitivity Maps of the Mackenzie Valley and Delta regions produced by the Geological Survey of Canada in the early 1970s (e.g. Monroe 1972). The following correlations are used by Monroe for segregated ice by volume.

- very high: 75 or 80%
- high : 40, 50 or 60%
- moderate : 10 or 20%

While these correlations do not lead directly to an acceptable classification, they do indicate the approximate relationship between descriptive and quantitative estimates of ground ice occurrence.

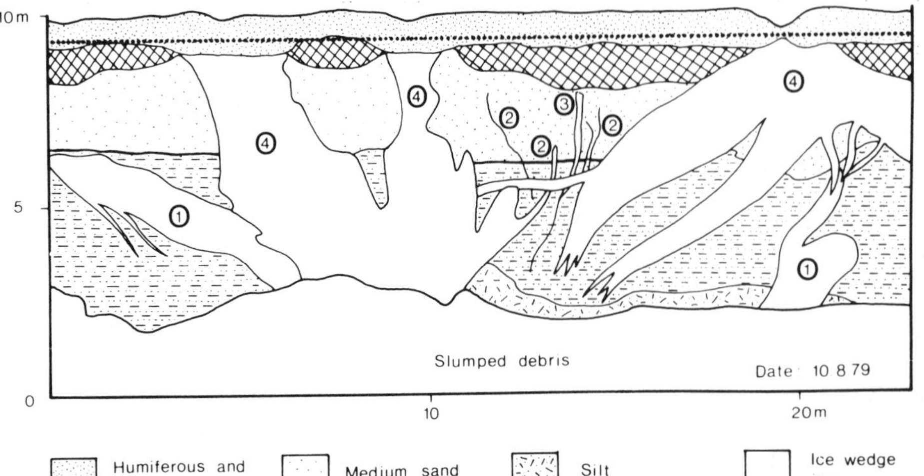

FIGURE 2a

FIGURE 2 Examples of short range horizontal variability of ice content in the continuous permafrost zone of Canada. (a) Ground ice bodies exposed in coastal bluffs, southern Banks Island, N.W.T. (French et al. 1982). (b) Massive ground ice body plotted from seismic shot hole logs, near Tuktoyaktuk, N.W.T. (Mackay 1973).

1 - 4 indicate ice wedge systems

The classification now proposed is given in Table 2.

A special problem arises in the case of permafrost beneath the bed of the sea. The permafrost conditions of the Beaufort Sea are becoming known in some detail (Hunter et al. 1978; Mackay 1972b) and several instances of sub-seabottom ground ice have been recorded from boreholes. There is, however, a major terminological problem (van Everdingen 1976), as the presence of seawater with a temperature below $0^{o}C$ means that the standard definition of permafrost, strictly on the basis of temperature, is not easily applicable. Scientists working in the Beaufort Sea area have been using terms such as ice-bearing permafrost, ice-bonded permafrost and even acoustical permafrost. At this time it is not known if the classifications proposed in Tables 1 and 2 can be applied to sub-seabottom ground ice or if new ones will have to be developed.

MAPPING OF GROUND ICE

There are few published maps showing the occurrence of ground ice in Canada. Mackay (1966) presented a map of types of ground ice in the Mackenzie Delta and Tuktoyaktuk Peninsula areas. Other than this, only maps of pingo distribution have been published (e.g. Mackay 1962; Hughes 1969). In Alaska, the general distribution of ice wedges has been mapped (Péwé 1965) as has the distribution of pingos (Holmes et al. 1965).

For the USSR, the distribution of a wider range of ground ice related features has been depicted, though on small scale maps. These include maps of the general distribution of injection ice and repeated-wedge ice (Shumskii and Vtyurin 1966), of polygons with ice wedges (Popov et al. 1966), of pingos (Frenzel 1967, in Washburn 1980), and of ice wedges (Vtyurin 1975, in J. Brown and Grave 1979). There are also three maps by Vtyurin (1973) showing (i) the total visible ice content

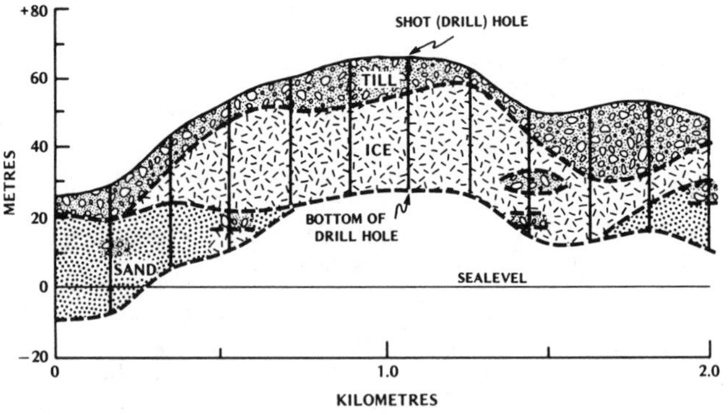

+80
60
40
20
0
−20
METRES
SHOT (DRILL) HOLE
TILL
ICE
SAND
BOTTOM OF DRILL HOLE
SEALEVEL
0
1.0
2.0
KILOMETRES

FIGURE 2b

of the permafrost in the USSR, (ii) three-dimensional macro ice content of the permafrost, as derived from multiwedge ice and (iii) the total amount of visible ice in the upper layers of permafrost. General geocryological maps for Central Siberia (Fotiev, et al. 1973), Yakutia (Melnikov 1966), the Yamal-Gydan area (Yunak 1979) and northeastern USSR (Vtyurin 1980) include some information on the broad distribution of several ground ice related features: injection ice, segregation ice, wedge ice and pingos.

Other published work on the quantity and distribution of ground ice includes estimates on the total quantity of ground ice in specific areas (e.g. J. Brown 1967, Pollard and French 1980), statements of the principles of compiling ice content maps (Lur'e 1972, Trofimov and Badu 1976, Gavrilov and Kondrat'yeva 1978 and Cheng 1981), and regional descriptions of the occurrence of ground ice (e.g. Rampton and Mackay 1971).

PROPOSED METHODOLOGY FOR COMPILING A NEW PERMAFROST MAP OF CANADA

The compilation of a new permafrost and ground ice map of Canada at a scale of 1:5 million is

TABLE 1 Morphologic Classification of Ground Ice

	Forms of occurrence of ground ice bodies	Genetic classes of Mackay (1972a)
1	Individual crystals or small groups of crystals in voids	1 Open cavity ice 5 Closed cavity ice 10 Pore ice
2	Equidimensional masses of ice	
a	- Irregular in shape	
b	- Regular in shape	9 Pingo ice
3	Layers or sheets of ice, more or less continuous	
a	- Horizontal or sub-horizontal layers	7 Aggradational ice
(i)	- Single layer	8 Sill ice
(ii)	- Multiple layers	6 Epigenetic ice
b	- Vertical or sub-vertical layers	2 Single vein ice 3 Wedge ice
(i)	- Single layer	4 Tension crack ice
(ii)	- Multiple layers	
c	- Diagonal layers	
4	Network of segregated ice surrounding masses of soil	

proposed. In preparation for this, a pilot study at a scale of 1:1 million is being undertaken to test the ground ice classifications (Tables 1 and 2) data sources, and map legend and to devise appropriate cartographic techniques for presenting the information.

The area of the pilot study is outlined in Figure 3. It comprises the Yukon, the lower Mackenzie River Valley and part of the Beaufort Sea. This region was selected for the pilot study because: (1) It includes areas of continuous permafrost, and both widespread and sporadic discontinuous permafrost; (2) There is a large body of subsurface borehole data available, resulting from petroleum exploration activities, pipeline proposals and highway construction; (3) It includes the northern end of the Canadian Cordillera, where alpine permafrost occurs; (4) It includes the southern Beaufort Sea, where sub-sea-bottom permafrost occurs; and (5) Recent surficial geology and terrain maps, which include information on permafrost and ground ice conditions, are available for over two thirds of the map area.

Although the topography and Quaternary history of this region are complex, with mountains, plateaus and valleys, and both glaciated and unglaciated areas, they are relatively well known. A wide range of ground ice conditions occur within the map area, including the possibility of relic permafrost in the unglaciated area.

The primary data sources of the pilot study are surficial geology and terrain maps produced by the Geological Survey of Canada over the last 15 years. These maps are available at scales of 1:100 000 to 1:250 000. They include conventional surficial geology maps, terrain inventory maps and terrain sensitivity maps. The map legends include information on permafrost and ground ice conditions, keyed to the main map units. Much of this information has been derived from borehole data, which are from a variety of sources. Most of the boreholes were drilled as either shot holes for petroleum exploration by seismic methods or for geotechnical investigations for pipeline or other construction projects. Less detailed surficial geology, Quaternary geology and bedrock geology maps are available for the remainder of the area. While few or no ground ice data are directly available on these maps, they do give some information on environmental conditions.

In the southern Beaufort Sea area, the primary data sources are geophysical interpretations of marine seismic data collected for petroleum exploration, geocryological and geotechnical purposes.

The primary data sources for both the pilot study map and the final, Canada-wide map provide information on soils, geology and permafrost conditions to a variable depth, generally not greater than 10 m. Since the ground ice conditions in these near surface materials are of greatest consequence for geotechnical purposes, it is proposed to restrict the mapped information to the upper 10 m of ground.

The map legend used for the pilot study is based on the classifications presented in Tables 1 and 2. The compilation approach is to use map units defined on the basis of surface geology, and to label them according to known ground ice conditions. This requires a considerable amount of interpretation and extrapolation in many areas. Nonetheless, it is the only approach currently practicable. Ground ice conditions are shown by using colour for the amount or quantity of ice and a pattern or combination of patterns for its form. In addition, selected geomorphic features, such as pingos, are shown by means of symbols. For the final map of Canada any or all of these elements may be modified.

TABLE 2 Quantitative classification of Ground Ice

Descriptive Term	Equivalent segregated ice content (% by volume)
Extreme	80-100
Very high	60-80
High	40-60
Moderate	20-40
Low	0-20

Any comments, suggestions or information which may assist in the preparation of the final map will be gratefully received by the author.

ACKNOWLEDGMENTS

Thanks are due to Drs. J. Brown, J.G. Fyles, S.A. Harris, J.R. Mackay, T.L. Péwé and

A.L. Washburn for helpful discussions. O.J. Ferrians, Jr., H.M. French, D.A. St-Onge and J.S. Scott kindly read and commented upon the paper. The valuable comments of an anonymous reviewer are gratefully acknowledged.

REFERENCES

Brown, J., 1967, An estimation of volume of ground ice, Coastal Plain, northern Alaska; United States Army Corps of Engineers, Cold Regions Research and Engineering Laboratory, Hanover, NH, Memorandum, 22 p.

Brown, J., and Grave, N.A., 1979, Physical and thermal disturbance and protection of permafrost; in Proceedings of the Third International Conference on Permafrost, v. 2: p. 51-91, Ottawa, National Research Council of Canada.

Brown, R.J.E., 1964, Permafrost investigations on the Mackenzie Highway in Alberta and Mackenzie District; National Research Council of Canada, Publication 7885: 27 p.

Brown, R.J.E., 1965, Permafrost investigations in Saskatchewan and Manitoba; National Research Council of Canada, Publication No. 8375: 36 p.

Brown, R.J.E., 1967, Permafrost in Canada; Geological Survey of Canada, Map 1246A and National Research Council of Canada Publication 9769, 1:7·6 million.

Brown, R.J.E., 1968, Permafrost investigations in northern Ontario and northeastern Manitoba; National Research Council of Canada, Publication No. 10465; 40 p.

Brown, R.J.E., 1972, Permafrost in the Canadian Arctic archipelago; Zeitschrift für Geomorphologie, new series, supplementary volume 13: p. 102-130.

Cheng, G., 1981, Principles for compiling large-scale ice-content maps of permafrost (in Chinese); Bingchuan Dongtu (Journal of Glaciology and Cryopedology), v. 3, p. 53-56.

EBA Engineering Limited, 1977, Geotechnical investigations for utilities design, Dawson City, Yukon; Report submitted to Stanley Associates Engineering Ltd., October 1977, 28 p.

Fotiev, S.M., Danilova, N.S., and Sheveleva, N.S., 1973, Zonal and regional characteristics of permafrost in central Siberia; in Permafrost — the USSR contribution to the Second International Conference, Yakutsk: p. 104-110. Washington, D.C., National Academy of Sciences.

French, H.M., Harry, D.G. and Clark, M.J., 1982, Ground ice stratigraphy and late Quaternary events, southwest Banks Island, Canadian Arctic; in French, H.M., ed., Proceedings of the Fourth Canadian Permafrost Conference, p. 81-90, Ottawa, National Research Council of Canada.

Gavrilov, A.V., and Kondrat'yeva, K.A., 1978, Principles of compiling small scale geocryologic maps; in Permafrost — the USSR contribution to the Second International Conference, Yakutsk: p. 462-466. Washington, D.C., National Academy of Sciences.

Heginbottom, J.A. and Kurfurst, P.J., 1975, Local variability of ground ice at selected sites in the Mackenzie Valley; Geological Survey of Canada, Open File Report No. 476: 56 p.

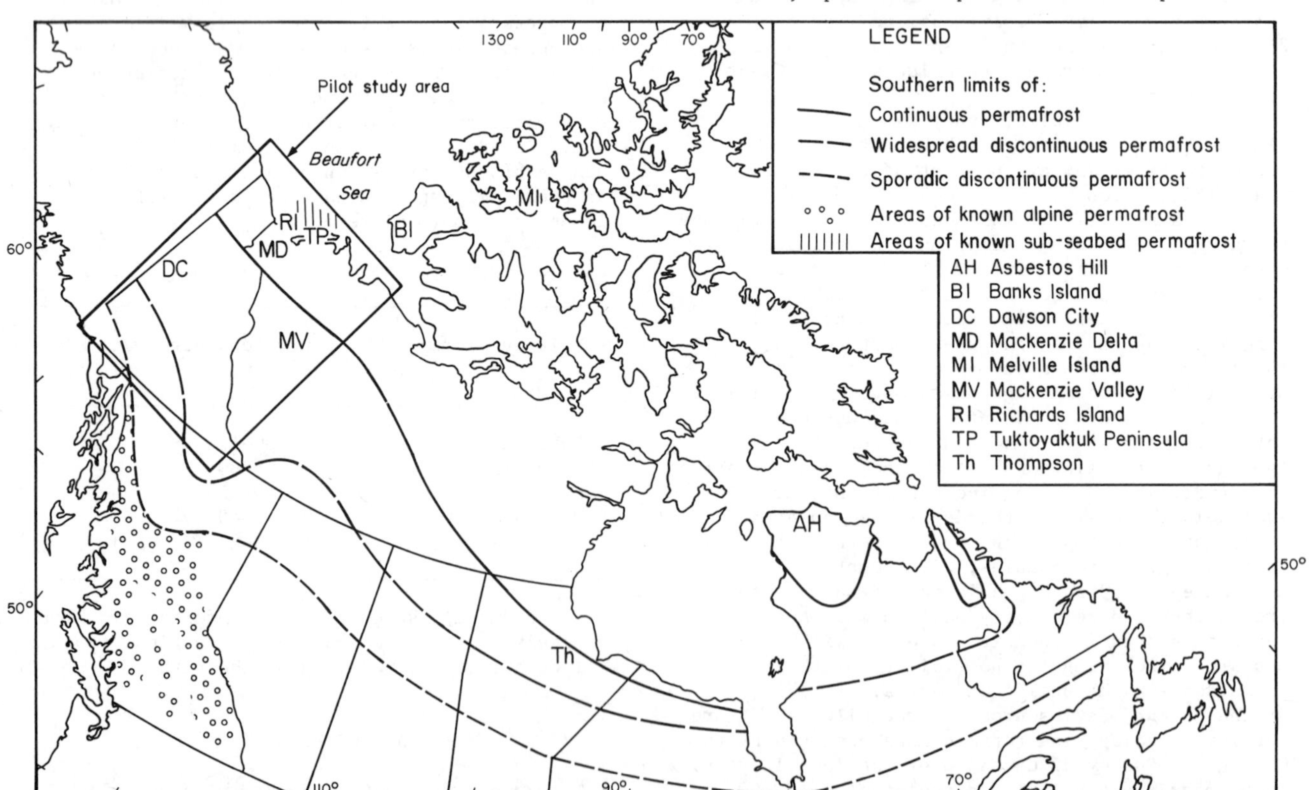

FIGURE 3 Location map. (Permafrost limits modified from Brown 1967).

Holmes, G.H., Foster, H.L., and Hopkins, D.M., 1966, Distribution and age of pingos of interior Alaska; in Proceedings — Permafrost International Conference, N.R.C. Publication 1287: p. 88-93, Washington, D.C., National Academy of Sciences.

Hughes, O.L., 1969, Distribution of open system pingos in central Yukon Territory with respect to glacial limits; Geological Survey of Canada, Paper 69-34: 8 p.

Hunter, J.A., Neave, K.G., MacAuley, H.A., and Hobson, G.D., 1978, Interpretation of sub-seabottom permafrost in the Beaufort Sea by seismic methods. Part I, Seismic refraction methods; in Proceedings of the Third International Conference on Permafrost, v. 1, p. 515-520. Ottawa, National Research Council of Canada.

Johnston, G.H., Brown, R.J.E., and Pickersgill, D.N., 1963, Permafrost investigations at Thompson, Manitoba: Terrain studies; National Research Council of Canada, Publication 7568: 51 p.

Lur'e, I.S., 1972, Mapping the ice content of permafrost in small scale geological-engineering permafrost studies (in Russian); Merzlotnye issledovaniya (Permafrost Studies), No. 12: p. 169-175. Moscow, Moscow University Press.

Mackay, J.R., 1962, Pingos of the Pleistocene Mackenzie Delta area; Geographical Bulletin No. 18: p. 21-63.

Mackay, J.R., 1966, Segregated epigenetic ice and slumps in permafrost, Mackenzie Delta area, N.W.T.; Geographical Bulletin, v. 8: p. 59-80.

Mackay, J.R., 1972a, The world of underground ice; Annals of the Association of American Geographers, v. 62, p. 1-22.

Mackay, J.R., 1972b, Offshore permafrost and ground ice, southern Beaufort Sea, Canada; Canadian Journal of Earth Sciences, v. 9: p. 1550-1561.

Mackay, J.R., 1973, Problems in the origin of massive icy beds, western Arctic, Canada; in Permafrost — The North American Contribution to the Second International Conference, Yakutsk: p. 223-2238. Washington, D.C., National Academy of Sciences.

Melnikov, P.I., 1966, Geocryological map, Yakutskoi ASSR (in Russian); Akademia Nauk, S.S.S.R. (U.S.S.R. Academy of Sciences), 1:5 million.

Monroe, R.L., 1972, Terrain classification and sensitivity maps (97F, 107C, 107D and 107E); Geological Survey of Canada, Open File Report No. 117: four maps and legend, 1:250 000.

Muller, S.W., 1947, Permafrost or permanently frozen ground and related engineering problems; Ann Arbor, Mich., J.W. Edwards, 231 p.

Péwé, T.L., 1965, Ice wedges in Alaska — classification, distribution and climatic significance; in Proceedings — Permafrost International Conference, N.R.C. Publication 1287: p. 76-81. Washington, D.C., National Academy of Sciences.

Pihlainen, J.A., and Johnston, G.H., 1963, Guide to a field description of permafrost; Canada, National Research Council, Associate Committee on Soil and Snow Mechanics, Tech. Memo. 79: 23 p.

Pollard, W.H., and French, H.M., 1980, A first approximation of the volume of ground ice, Richards Island, Pleistocene Mackenzie Delta, Northwest Territories, Canada; Canadian Geotechnical Journal, v. 17, p. 509-516.

Popov, A.I., Kachurin, S.A., and Grave, N.A., 1966, Features of the development of frozen geomorphology in northern Eurasia; in Proceedings — Permafrost International Conference, N.R.C. Publication 1287: p. 181-185. Washington, D.C., National Academy of Sciences.

Rampton, V.N., 1981, Surficial geology, Stanton, District of Mackenzie; Geological Survey of Canada, Map 33-1979, 1:250 000.

Rampton, V.N., and Mackay, J.R., 1971, Massive ice and icy sediments throughout the Tuktoyaktuk peninsula, Richards Island, and nearby areas, District of Mackenzie; Geological Survey of Canada, Paper 71-21: 16 p.

Shumskii, P.A., 1959, Ground (sub-surface) ice; Principles of Geocryology, Part I, Chapter 9, Academy of Sciences, U.S.S.R., V.A. Obruchev Institute of Permafrost Studies, Moscow. Translated by C. De Leuchtenberg, N.R.C., Ottawa, Tech. Trans. 1130: 118 p.

Shumskii, P.A., and Vtyurin, B.I., 1966, Underground ice; in Proceedings — Permafrost International Conference, N.R.C. Publication 1287: p. 108-113. Washington, D.C., National Academy of Sciences.

Stangl, K.O., Roggensack, W.D., and Hayley, D.W., 1982, Engineering geology of surficial soils, eastern Melville Island; in French, H.M., ed., Proceedings of the Fourth Canadian Permafrost Conference: p. 136-147, Ottawa, National Research Council of Canada.

Trofimov, V.T., and Badu, Yu. B., 1976, Mapping the ice content of permafrost (in Russian); Merzlotnye issledovaniya (Permafrost Studies), No. 15, p. 131-139. Moscow, Moscow University Press.

van Everdingen, R.O., 1976, Geocryological terminology; Canadian Journal of Earth Sciences, v. 13, p. 862-867.

Vtyurin, B.I., 1978, Patterns of distribution and a quantitative estimate of the ground ice in the U.S.S.R.; in Permafrost — the U.S.S.R. contribution to the Second International Conference, Yakutsk: p. 159-164. Washington, D.C., National Academy of Sciences.

Vtyurin, B.I., 1980, Principles of cryolithological regionalization of the permafrost zone; Third International Conference on Permafrost, English translation of the forty-nine Soviet papers, the one French paper and the three invited Soviet papers, Part II: p. 1-9. Ottawa, National Research Council.

Washburn, A.L., 1980, Geocryology, 406 p., New York, Halstead Press.

Williams, P.J., 1968, Ice Distribution in permafrost profiles; Canadian Journal of Earth Sciences, v. 15: p. 1381-1386.

Yunak, R.I., 1979, Permafrost in the Yamal-Gydan areas of western Siberia; Polar Geography, v. 3: p. 49-63.

DESIGN OF BURIED SEAFLOOR PIPELINES FOR PERMAFROST THAW SETTLEMENT

C. E. Heuer, J. B. Caldwell, and B. Zamsky
Exxon Production Research Company, Houston, Texas 77001 USA

Computer simulations were performed to model the mechanical and thermal behavior of pipelines buried in the seafloor above ice-bonded permafrost. Conditions typical of the Alaskan Beaufort Sea were used, and the study was directed at water depths greater than 2 m (6 ft). Maximum pipe diameter, operating temperature, and cover depth considered were 61 cm (24 in.), 93°C (200°F), and 3.0 m (10 ft), respectively. A few cases considered two parallel pipelines instead of a single isolated pipeline. Allowable permafrost settlement was determined by modeling the mechanical interaction between the pipeline and settling soil. If the top of ice-bonded permafrost was within about 9 m (30 ft) of the seafloor, then the allowable permafrost settlement was only about 0.3 m (1 ft). However, the allowable settlement increased rapidly as the depth to ice-bonded permafrost increased. For example, doubling the depth to ice-bonded permafrost more than tripled the allowable settlement. The soil thermal regime was modeled to develop guidelines for preventing excessive thaw and settlement. Where the ice-bonded permafrost was below a depth of about 27 m (90 ft), thaw was acceptable for all situations considered. For ice-bonded permafrost depths of about 14-27 m (45-90 ft), insulation was used to limit thaw. If the ice-bonded permafrost was at a depth less than about 14 m, then restrictions on operating temperature or changes in burial geometry were sometimes also necessary. These guidelines can change depending on the specific application but are useful for planning future studies. Salt transfer in the thawed soil was identified as a significant factor in calculating thaw depth because it controls the permafrost thawing temperature.

INTRODUCTION

Within a few years, construction to develop petroleum resources beneath the Alaskan Beaufort Sea could begin, and pipelines buried in the seafloor are likely to be an integral part of any such development. As has been done for exploration, man-made islands will probably be used as platforms for drilling, but they may also serve as locations for large processing facilities. Inter-island pipelines may be needed to connect satellite drilling islands to a central processing island, and island-to-shore pipelines may be needed to deliver the oil and gas to large-diameter onshore trunk lines for transportation to markets. Ice-bonded permafrost is known to be present throughout the potential development areas, and the pipelines may operate at temperatures as high as 93°C (200°F). Therefore, differential settlement due to thawing of permafrost is an important design consideration.

Permafrost thaw settlement has previously been addressed for onshore pipelines, but for offshore pipelines there are three significant differences. First, water depth exerts a controlling influence on seafloor temperature, and even for depths as low as 1 m, the mean annual seafloor temperature is much warmer than the nearby onshore surface temperature. Second, for water depths greater than about 2 m there is no active layer, and the top of ice-bonded permafrost can be tens of meters below the seafloor. The upper layer of thawed soil provides a thermal and mechanical buffer between the pipeline and the ice-bonded permafrost. Finally, pore water salinity decreases the permafrost thawing temperature and makes the frozen soil thermal properties highly temperature dependent. Computer simulations were conducted to study the effects of these differences and other parameters for water depths greater than 2 m. The simulations modeled both the soil thawing and the mechanical interaction between a pipeline and settling soil.

The following assumptions were made in performing the simulations: (1) The design life is 30 years. (2) Differential settlement is the only permafrost-related design problem. (3) Thaw of ice-bonded permafrost is acceptable as long as it does not lead to excessive pipe deformation. (4) The allowable longitudinal compressive or tensile strain is 0.5%, which allows some inelastic pipe deformation. (5) Mapping small areas of ice-rich soil along an entire pipeline route may not be practical, so the worst case length of ice-rich soil is used. (6) A design value for permafrost thaw strain can be developed by testing and correlating a large number of seafloor samples. (7) Maximum differential settlement equals total settlement. (8) Settlement occurs simultaneously with thawing. (9) The thermal and mechanical aspects of the problem can be uncoupled, since allowable settlement is relatively small. (10) Seismic loads may be important in some situations but are not included in this analysis. Additional work may allow relaxation of some of the more restrictive assumptions such as 5 and 7.

The mechanical simulations are described first, and then the thermal simulations are covered. This order is used because the mechanical simulations establish the limits on thaw depth.

MECHANICAL SIMULATIONS

The pipeline mechanical simulations were performed using an earlier version of a commercially available program for two-dimensional stress and deformation analysis of pipelines (Price et al. 1982). The program considers soil settlement, thermal expansion, internal pressure, gravity loads, longitudinal soil friction, and soil deformation. The latter is modeled as a Winkler foundation. Nonlinear effects include inelastic pipe deformation, large pipe displacements, and bilinear soil springs. The program has been used since 1972 in the design of major pipeline projects.

The geometry used to determine soil settlement is shown in Figure 1. This geometry is considered conservative because it assumes that the settled region is infinitely wide perpendicular to the longitudinal axis of the pipeline. Actually, the width is determined by the width of the thaw bulb and can be about the same as the length of the settled region parallel to the pipeline. The resulting three-dimensional settlement would lead to smaller vertical displacement.

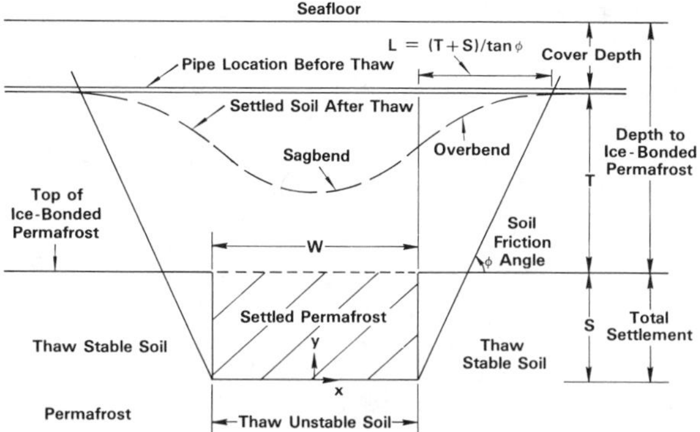

FIGURE 1 Soil settlement geometry.

The soil settlement profile was based on a stochastic model for cohesionless soil (Attewell 1977), which compared well with measured surface settlement above a deformed tunnel. The general shape of the vertical displacement is shown in Figure 2. The hyperbolic tangent function given in Figure 2 is a very close approximation to the results from the stochastic model and is more convenient to use. This function implicitly requires that the volume of settlement directly below the pipe equal the volume of permafrost settlement. This is conservative because in the real case, the former is likely to be smaller than the latter.

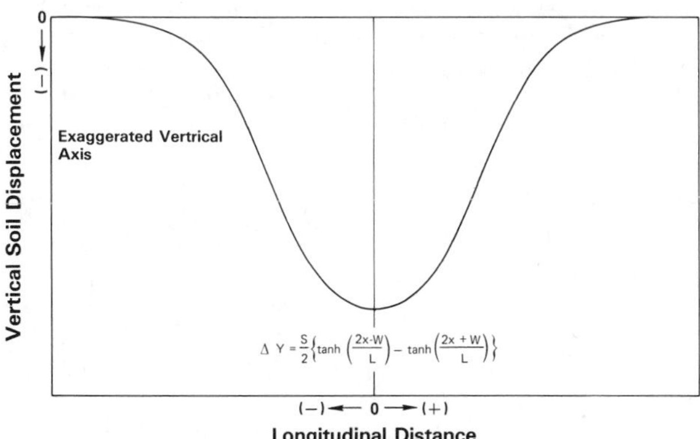

FIGURE 2 Vertical soil displacement.

By assuming an incompressible soil and small strain, the corresponding longitudinal soil displacement can be calculated, and the general shape is shown in Figure 3. In the past, longitudinal displacement has not been

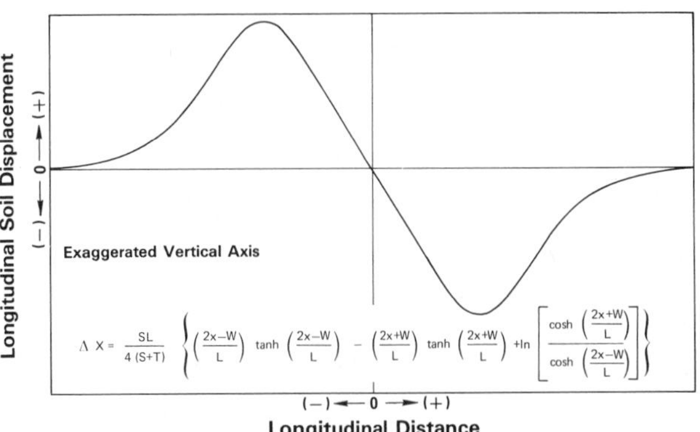

FIGURE 3 Longitudinal soil displacement.

considered. However, for the settlement of deep permafrost, the resulting longitudinal force on the pipeline can be significant.

Figures 1-3 assume that thaw-stable and thaw-unstable soil have the same friction angle, but the thaw-stable soil is likely to have a much higher friction angle. This is because of its higher density. To accommodate the change in friction angle, the following mixing and smoothing functions

$$M = \{\tanh [2(2x + W)/(L_u + L_s)] - \tanh [2(2x - W)/(L_u + L_s)]\}/2$$

$$F_c = M \bullet F_u + (1 - M) F_s$$

were developed. Although arbitrary, they provide for a smooth transition between stable and unstable soil. Subscripts c, s, and u are for composite, stable, and unstable, and F represents either the vertical or the longitudinal displacement functions in Figures 2 and 3. Note that M approaches a maximum value not greater than one as x approaches zero, and that M approaches zero as the absolute value of x increases.

The thaw-stable soil could have smaller latent heat and higher thermal conductivity than the thaw-unstable soil. This would cause the thaw depth in the unstable soil to increase toward the boundary with the stable soil. In view of other uncertainties, modeling this effect was not considered necessary.

Support, uplift and downdrag, and longitudinal soil springs were used to model the force applied to the pipeline when it moved downward, upward, and axially relative to the soil. For each type of spring, the force increased linearly with displacement up to a yield displacement. Above the yield displacement, the force was constant at its ultimate value. The spring constants along with other soil mechanical parameters are given in Table 1 (Bowles 1977, Price et al. 1982).

The pipe steel grade was assumed to be X-65, with the following piecewise-linear stress-strain relation: 0 GPa at 0% strain, 0.36 GPa (52 ksi) at 0.18%, 0.45 GPa (65 ksi) at 0.5%, 0.48 GPa (70 ksi) at 0.8%, and 0.52 GPa (75 ksi) at 3%. Pipeline outside diameters of 30 cm (12 in.) and 61 cm (24 in.) were used with corresponding wall thicknesses of 1.11 cm (0.437 in.) and 1.43 cm (0.563 in.). Cover depth was either 0.9 or 3.0 m (3 or 10 ft).

The gravity load on the pipeline was independent of displacement. It consisted of the buoyant weight of the soil directly above the pipeline plus the buoyant weight of the pipe, insulation, concrete coating, and oil line

TABLE 1 Soil mechanical properties.

Soil Type	Density (Mg/m^3) Dry	Buoyant	Friction Angle (deg.)	Pipe Diameter (cm)	Cover Depth (m)	Support F* (MN/m)	Y* (cm)			
Soft, Unstable (fine-grained)	1.60	1.01	30	30	0.9	0.30	8.1			
					3.0	0.76	8.1			
				61	0.9	0.52	11.9			
					3.0	1.18	11.9			
Stiff, Stable (coarse-grained)	2.08	1.31	45	30	0.9	3.0	2.5			
					3.0	6.9	2.5			
				61	0.9	5.5	3.6			
					3.0	11.0	3.6			
						Uplift & Downdrag F (kN/m)	Y (cm)	Overburden (kN/m)	Longitudinal F (kN/m)	Y (cm)
Ditch fill	1.44	0.91	25	30	0.9	6.7	12.2	8.8	13.4	3.3
					3.0	53.3	12.2	24.2	37.1	3.3
				61	0.9	7.3	17.8	15.2	29.2	4.8
					3.0	61.3	17.8	38.1	80.3	4.8

* F = Ultimate Force
Y = Yield Displacement

fill. To determine the buoyant weight of the pipe, a 0.318 cm (0.125 in.) corrosion allowance was added to the wall thickness. The insulation thickness was 15.2 cm (6 in.), and the concrete thickness was 10.2 cm (4.0 in.) for the 30 cm pipe, and 14.0 cm (5.5 in.) for the 61 cm pipe. The possible increase in overburden pressure due to infilling of a seafloor settlement depression was not modeled.

The pipeline operating pressure was assumed to be 9.31 MPa (1,350 psi), and the installation temperature was assumed to be -1.1°C (30°F). The latter was used to calculate the force due to thermal expansion.

The length of pipeline modeled was over 900 m (3,000 ft). One end was at the midpoint of the settled region. Both ends were restrained in rotation and longitudinal displacement. Before any loads were applied, the pipeline was assumed to be horizontal and stress free. Residual stresses from laying or pressure testing were not considered.

Pipeline diameter, operating temperature, cover depth, and depth to top of ice-bonded permafrost were found to be the major variables controlling allowable permafrost settlement. Friction angle, pipe wall thickness, and soil spring constants were of lesser importance.

Several simulations were needed to determine allowable settlement for a given set of input parameters. For each set, both the depth and the width of the settled region in Figure 1 were varied until two conditions were satisfied: ϵ = 0.5% and $\partial\epsilon/\partial W = 0$, where ϵ is longitudinal strain and W is the width of the settled region. The second condition means that the worst-case value of W was always used.

Figure 4 illustrates the results from a typical series of simulations. For low values of W, the pipeline does not follow the shape of the soil and partially spans the settled region. As W increases, the sag in the pipe increases, the pipe touches down on the soil in the middle of the settled region, and the soil starts to support the pipe sagbend. In Figure 4, the zero slope condition occurs just as the sagbend touches the soil. Consistent with this behavior, the maximum compressive strain occurs at the top of the sagbend for low values of W, but it occurs at the bottom of the overbend for large values of W. Note that the value of W at the design point is only about 12 m (40 ft).

Figure 5 is a plot of allowable permafrost settlement versus depth to top of ice-bonded permafrost for two cover depths, 0.9 and 3.0 m, and two pipeline diameters 30 and 61 cm. The pipe temperature was 66°C (150°F). For shallow ice-bonded permafrost, the allowable settlement is only 0.1-0.4 m (0.4-1.4 ft). The allowable settlement increases rapidly as the depth to ice-bonded permafrost increases because the soil settlement is spread over a longer interval.

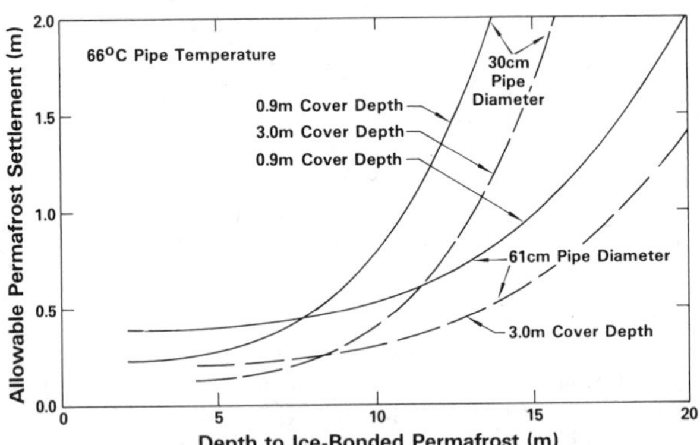

FIGURE 5 Allowable permafrost settlement.

The allowable settlement for the 30 cm pipeline is initially less than that for the 61 cm pipeline. The smaller-diameter pipeline has a lower section modulus, reducing the length of the settled region the pipeline can span. However, for greater depth to ice-bonded permafrost, the smaller pipeline can more easily conform to the shape of the settlement profile.

For both pipe diameters, the allowable settlement is always less for the deeper depth of cover. The

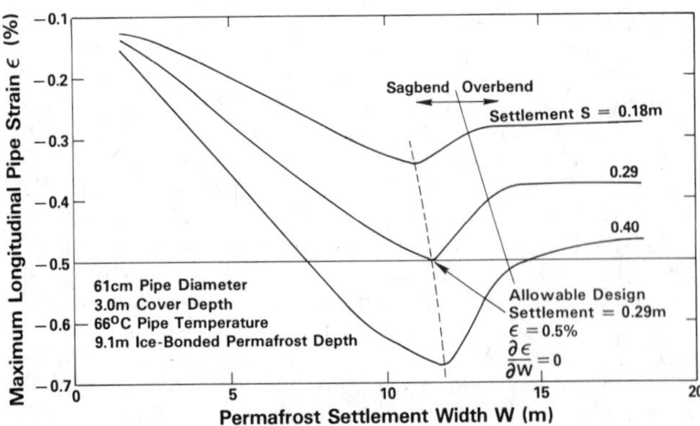

FIGURE 4 Pipe strain.

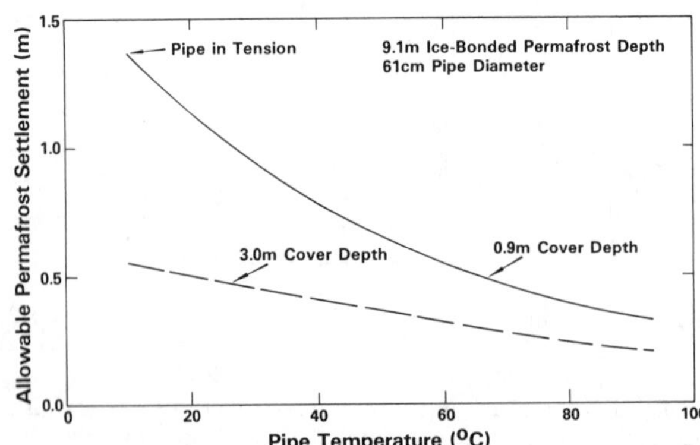

FIGURE 6 Allowable permafrost settlement.

difference increases with increasing depth to ice-bonded permafrost. The greater overburden pressure makes the pipeline follow the settlement profile more closely.

The pipeline operating temperature strongly affects the allowable permafrost settlement. This is shown in Figure 6 for a pipeline diameter of 61 cm. Decreasing the operating temperature from 66°C (150°F) to 10°C (50°F) increases the allowable settlement by a factor of almost 3 for a cover depth of 0.9 m and by a factor of about 2 for a cover depth of 3.0 m. At 10°C and 0.9 m cover, the allowable settlement is controlled by tensile rather than compressive strain.

Figures 5 and 6 are plots of allowable permafrost settlement. To couple these results to the thermal simulations, the allowable settlement must be divided by the design thaw strain.

THERMAL SIMULATIONS

The pipeline thermal simulations were performed using a computer program which models transient, two-dimensional conduction with change of phase (Wheeler 1978). Since 1969, the program has been used to solve a variety of permafrost heat transfer problems and has been compared against analytical solutions, laboratory data, field tests, and other programs.

Some representative soil thermal properties are shown in Figures 7 and 8. The frozen and thawed dry densities were calculated to be consistent with the assumed thaw strain. The values of thawed conductivity and heat capacity were calculated using standard correlations (Kersten 1949) and were not adjusted to account for changes in geometry due to settlement. The corresponding frozen values and latent heat were modified to account for unfrozen moisture due to both salinity (Ono 1975) and specific surface area (McGaw and Tice 1976). In Figure 7, the jump in conductivity for a thawing temperature of -0.3°C (31.5°F) is due to the step increase in dry density upon thawing. The conductivity of the pipe insulation was 0.035 W/m-°C (0.02 Btu/hr-ft-°F).

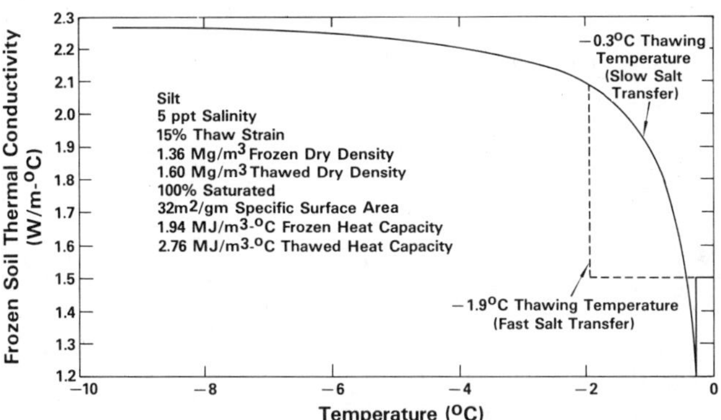

FIGURE 7 Frozen soil thermal conductivity.

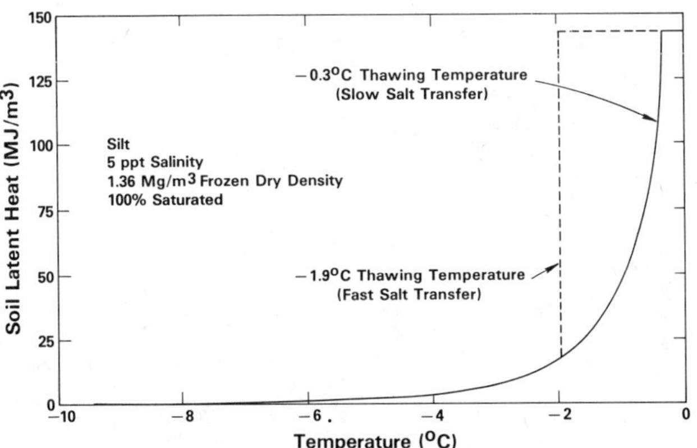

FIGURE 8 Soil latent heat.

Both single and dual pipelines were modeled. When two pipelines were modeled, they had identical dimensions and temperatures. Separation between the pipelines was measured from centerline to centerline.

The finite element grid was rectangular with a width of about 30 m (100 ft) and a depth of about 45 m (150 ft). The top boundary was the seafloor. The right boundary was a plane of symmetry. For a single pipeline, it passed through the pipe centerline. For dual pipelines, it passed midway between the two centerlines. The bottom and left boundaries were far away from the pipeline. The pipeline and insulation were represented by concentric polygons determined by the assumed pipe diameter and insulation thickness. The concrete coating was not explicitly modeled because its conductivity is similar to thawed soil. The cover depth and depth to top of ice-bonded permafrost were used as shown in Figure 1.

The boundary condition at the seafloor was a heat transfer coefficient and a constant seawater temperature. The pipe was a constant temperature boundary. The side and bottom boundaries were adiabatic. The initial temperature was equal to the seawater temperature at the seafloor. It linearly decreased down to the top of the ice-bonded permafrost. There the temperature was equal to the thawing temperature based on the overlying thawed soil salinity. At greater depths, the temperature linearly decreased according to the specified geothermal gradient. All the simulations were run for 30 years.

A sensitivity study determined the possible changes in thaw depth due to variations in the input data. The parameters considered are listed in Table 2, along with the input values used and the corresponding calculated thaw depths. The parameters are listed in order of decreasing variation in thaw depth. They were varied one at a time, while all of the other parameters were set equal to their base case values. The base case

TABLE 2 Thermal sensitivity study.

Parameter	Input Values	Calculated 30-Year Thaw Depth (m)	Variation in Thaw Depth (m)
Insulation Thickness (cm)	0, 5.1, 10.2, 15.2	11.1, 4.7, 2.4, 1.2	9.9
Thawing Temperature (°C)	-1.9, -1.4, -0.8, -0.3	6.2, 5.1, 3.7, 2.4	3.8
Pipe Temperature (°C)	10, 38, 66, 93	0.0, 0.3, 2.4, 3.7	3.7
Ice-Bonded Permafrost Salinity (ppt)	0, 5, 10, 20	2.3, 2.4, 3.2, 5.5	3.2
Pipe Separation (m)	3.0, 6.1, 12, 18, ∞	5.4, 5.0, 3.5, 2.7, 2.4	3.0
Depth of Cover (m)	0.9, 1.8, 3.0	0.0, 0.8, 2.4	2.4
Depth to Top of Ice-Bonded Permafrost (m)	9.1, 18, 27	2.4, 0.0, 0.0	2.4
Pipe Diameter (cm)	30, 61	0.3, 2.4	2.1
Thawed Salinity[1] (ppt)	25, 35, 45	3.2, 2.4, 1.8	1.4
Soil Type	fine, coarse	2.4, 1.5	0.9
Seawater Temperature (°C)	-1.7, -1.1, 0.6	2.0, 2.4, 2.8	0.8
Thaw Strain (%)	0, 15, 30	2.8, 2.4, 2.3	0.5
Geothermal Gradient (°C/m)	0.0091, 0.018, 0.036	2.4, 2.3, 2.1	0.2
Seafloor Heat Transfer Coefficient (W/m-°C)	5.7, 57.0, ∞	2.6, 2.4, 2.4	0.2
Thawed Salinity[2] (ppt)	25, 35, 45	6.3, 6.2, 6.2	0.1

1. Thawed salinity used to determine initial temperature distribution only (slow salt transfer).
2. Thawed salinity used to determine initial temperature distribution and thawing temperature (fast salt transfer).
3. Underlined values are base case conditions.
4. All thaw depths are relative to the top of ice-bonded permafrost.

values are underlined in the table. All the input values were considered possible, but the base case was selected as a typical design situation.

The parameters can be divided into two classes, those that describe environmental conditions and those that describe the pipeline. Only limited control over the environmental parameters is possible when the pipeline route is being selected. The results of the sensitivity study indicate that the primary objectives during route selection should be to minimize thawed soil salinity and permafrost thaw strain and to maximize depth to top of ice-bonded permafrost.

Once the route is selected, the pipeline parameters can be adjusted to control thaw depth. These include insulation thickness, pipe separation, pipe temperature, depth of cover, and pipe diameter. Increasing the first two and decreasing the last three decreases thaw depth. Fortunately, most of these parameters are high on the list in Table 2, so significant control over thaw depth is potentially possible. However, other nonthermal considerations may also affect these parameters.

Thawing temperature is the second most important parameter in Table 2. It is controlled by the rate of salt transfer through thawed soil, which cannot yet be accurately predicted. However, two limiting cases can be considered. "Slow" salt transfer means that the salinity on both sides of the thaw front is equal to the initial ice-bonded permafrost salinity. The thawing temperature is therefore determined by the ice-bonded salinity. "Fast" salt transfer means that the salinity above the thaw front equals the initial thawed soil salinity. In this case, there is a sharp change in salinity at the thaw front. The higher salinity now determines the thawing temperature.

Fast and slow are relative to the thaw front velocity. Based on the allowable permafrost settlement, the allowable thaw front velocity is on the order of 0.1 m/yr, which is an order of magnitude faster than the typical natural ice-bonded permafrost degradation rate of 0.01 m/yr or less. However, the salt transfer rate could be faster than the natural degradation rate. The pore water velocity above the thaw front should provide an upper bound on the salt transfer rate. Field work by Harrison and Osterkamp (1982) for relatively coarse-grained soil indicates that the pore water velocity is probably faster than 0.01 m/yr and may be faster than 0.1 m/yr. The velocity is likely to be lower in fine-grained soil. Although very limited, the available data indicate that there are likely to be areas where the pore water velocity is close to the thaw front velocity. Therefore, the thawing temperature could be below that based on the ice-bonded permafrost salinity. This, in turn, could significantly increase thaw depth.

Decreasing the thawing temperature can increase thaw depth for several reasons: (1) Relative to the thaw front, the seafloor can be a heat source rather than a heat sink. (2) The difference between the pipe temperature and the thawing temperature is increased. (3) The sensible heat input to the thawed soil for a given thaw depth is reduced.

Changes in thaw strain have little effect on thaw depth but have a very big effect on allowable thaw. Doubling the thaw strain from 15 to 30% decreases thaw depth by only about 4%. However, it halves the allowable thaw depth. Therefore, the worst case for thaw settlement occurs for the highest thaw strain, which corresponds to the lowest thaw depth.

An interesting phenomenon occurs when both the overlying thawed soil salinity and the salt transfer rate increase. The former decreases thaw, and the latter increases thaw. However, for the cases modeled (see bottom line in Table 2), the two effects are offsetting because there is essentially no net change in thaw. This may not be universally true.

METHODS TO LIMIT THAW AND SETTLEMENT

Pipeline insulation is considered the primary method for limiting thaw, so Figures 9 and 10 were developed to help determine where and how much insulation is needed. Figure 9 shows the required depth to the top of ice-bonded permafrost as a function of pipe temperature with no insulation. Figure 10 shows the required insulation thickness as a function of pipe temperature for 9.1 m (30 ft) to top of ice-bonded permafrost. In both figures, the pipe diameter is 61 cm, and the cover depth is 3.0 m. Curves are given for fast and slow salt migration. Other conditions are the same as the base case in Table 2. Above the curves, the permafrost settlement is less than allowable, while below the curves, the settlement is more than allowable.

Figures 9 and 10 were obtained by first plotting allowable thaw depth from the mechanical simulations and predicted thaw depth from the thermal simulations versus pipe temperature. The points of intersection for common conditions were then used to plot the curves in the figures.

The curves in Figure 9 have relatively flat slopes. This means that a small increase in depth to top of ice-bonded permafrost allows a big increase in pipe temperature. For example, if the ice-bonded permafrost is

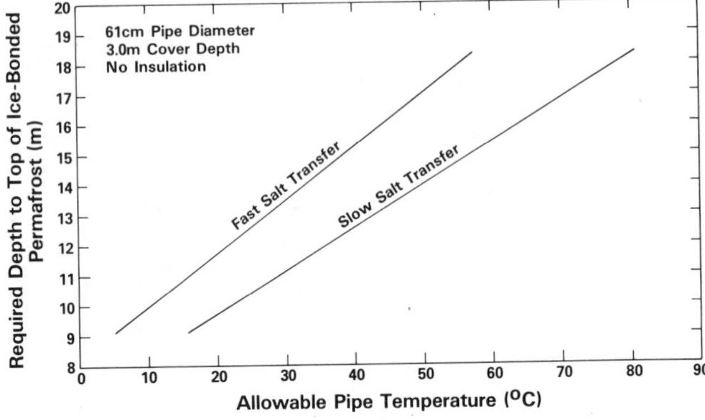

FIGURE 9 Required depth to ice-bonded permafrost.

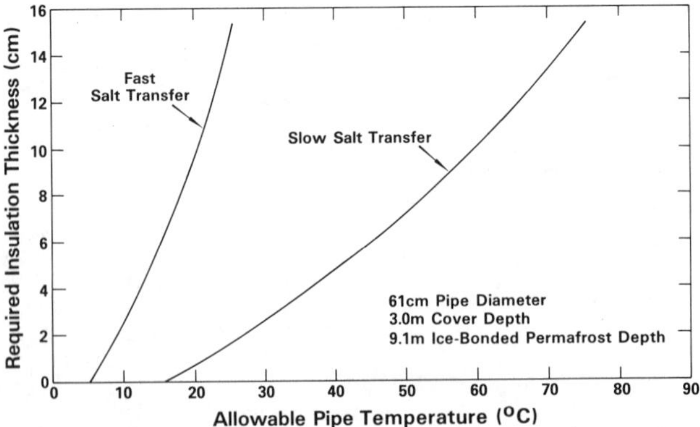

FIGURE 10 Required insulation thickness.

at 9.1 m, the maximum allowable pipeline temperature is 16°C (60°F) with slow salt transfer. If the depth to ice-bonded permafrost is doubled to 18 m (60 ft), then the allowable pipeline temperature increases by more than a factor of 4 to 81°C (178°F). For fast salt transfer, the required depths to top of ice-bonded permafrost are 2-4 m (7-13 ft) deeper.

If the permafrost is too shallow for a given pipeline temperature, then as shown in Figure 10, insulation can be used to reduce thaw depth to an allowable value. The slopes of the curves in this figure are relatively steep. Therefore, small changes in pipeline temperature lead to large increases in required insulation thickness. There is a strong dependence on salt transfer rate. For 15.2 cm (6 in.) of insulation and slow salt transfer, the maximum allowable pipeline temperature is 76°C (168°F). However, for fast salt transfer, the allowable pipeline temperature is only 26°C (78°F).

CONCLUSIONS

Calculation procedures were developed for design of warm pipelines buried in the seafloor above ice-bonded permafrost. Based on relatively limited results, the following guidelines were developed considering a maximum pipe diameter of 61 cm (24 in.), a maximum pipe temperature of 93°C (200°F), a maximum cover depth of 3.0 m (10 ft), a maximum of two pipelines along the same route, and a salt transfer rate intermediate between fast and slow. Where the depth to ice-bonded permafrost is greater than about 27 m (90 ft), no mitigation is needed to reduce thaw. If the depth to ice-bonded permafrost is about 14-27 m (45-90 ft), insulation can be used to limit thaw. For depths less than about 14 m, additional measures to limit thaw may also be required. Possible options are reducing cover depth, reducing pipe temperature, or increasing separation between multiple pipelines. Another option, reducing pipe diameter, will decrease thaw depth, but may also decrease allowable thaw.

These guidelines can be used to develop a preliminary assessment of thaw settlement problems for future applications. However, they are probably overly conservative for some areas of the Beaufort Sea. For example, in the Prudhoe Bay area and further to the east, there is frequently a relatively dense gravel layer at depth. The gravel has a very low thaw strain, and thawing of the gravel would cause little settlement. However, this type of layered soil profile was not considered in developing the guidelines.

There are several areas where additional work is needed to improve the existing design procedure: (1) A quantitative model for salt transfer is needed to evaluate the dependence of thawing temperature on thaw rate. (2) Insulation has never been used for permafrost protection of seafloor pipelines. The potential problems of mechanical damage and moisture absorption must be addressed to ensure long-term reliability of the insulation system. (3) The settlement profile used is reasonable but based on limited information. Model tests and more detailed computer simulations of soil deformation are needed to confirm the adequacy of the current profile. (4) More field data are needed on permafrost thaw strain. Since allowable thaw is inversely proportional to thaw strain, small decreases in strain can lead to much larger increases in allowable thaw. (5) Deformation of the insulation system and concrete coating was not considered in the mechanical simulations, but their deformation may impose a lower limit on allowable settlement. Simulations of the pipe, insulation, and concrete composite are needed. (6) The limit on pipe strain was developed for large diameter, onshore pipelines. The limit should be checked for smaller diameter, offshore pipelines.

The above guidelines are not applicable for water depths less than 2 m. Here the ice-bonded permafrost can be only a meter or so below the seafloor, but the mean annual seafloor temperature can still be close to the permafrost thawing temperature. This combination of very shallow permafrost and warm surface temperature makes burial in the seafloor much more difficult. With additional mitigation, such as overexcavation and backfill with thaw-stable soil, seafloor burial may be possible. Other, more thermally effective methods of avoiding settlement are burying well-insulated pipelines above sea level in a causeway or elevating the pipelines on pier supported trusses. These last two methods can totally prevent thaw of seafloor permafrost.

UNIT CONVERSIONS

°C = 5/9 (°F - 32)	1 MJ/m^3 = 26.8 Btu/ft^3
1°C/m = 0.549 °F/ft	1 MJ/m^3-°C = 14.9 Btu/ft^3-°F
1 cm = 0.394 in.	1 m^3/day = 35.3 ft^3/day
1 km = 0.621 mi	1 MN/m = 68.5 kips/ft
1 m = 3.28 ft	1 MPa = 145 psi
1 Mg/m^3 = 62.4 lb/ft^3	1 W/m-°C = 0.578 Btu/hr-ft-°F

REFERENCES

Attewell, P., 1977, Ground movements caused by tunnelling in soil, in Geddes, J. D., ed., Large ground movements in structures, Proceedings of the Conference at University of Wales, Institute of Science and Technology: London, Pentech, p. 1-167.

Bowles, J. E., 1977, Foundation Analysis and Design: New York, McGraw-Hill.

Harrison, W. D., and Osterkamp, T. E., 1982, Measurements of the electrical conductivity of interstitial water in subsea permafrost, in French, H. M., ed., Proceedings of the Fourth Canadian Permafrost Conference: Ottawa, National Research Council of Canada, p. 229-237.

Kersten, M. S., 1949, Laboratory research for the determination of the thermal properties of soils: Engineering Experiment Station, University of Minnesota.

McGaw, R. W., and Tice, A. R., 1976, A simple procedure to calculate the volume of water remaining unfrozen in freezing soil, in Proceedings of the Second Conference on Soil Water Problems in Cold Regions, Edmonton: Washington, D.C., American Geophysical Union, p. 114-122.

Ono, N., 1975, Thermal properties of sea ice, IV., Thermal constants of sea ice, CRREL Draft Translation 467: Hanover, New Hampshire, Cold Regions Research and Engineering Laboratory.

Price, J., Row, D., and Powell, G., 1982, Structural analysis and design of buried pipelines and piping, seminar notes: Berkeley, California, Structural Software Development, Inc.

Wheeler, J. A., 1978, Permafrost thermal design for the Trans-Alaska Pipeline, in Wilson, D. G., Solomon, A. D., and Boggs, P. T., eds., Moving boundary problems: New York, Academic Press, p. 267-284.

THAW SETTLEMENT AND GROUND TEMPERATURE MODEL FOR HIGHWAY DESIGN IN PERMAFROST AREAS

Eldo E. Hildebrand

Department of Civil Engineering, University of Waterloo,
Waterloo, Ontario, Canada N2L 3G1

Highway roughness produced by the differential settlements of thawing permafrost is probably the major problem for highway engineers in permafrost regions. A one-dimensional finite element model was developed to estimate thaw depths and subsequent settlements of an embankment on permafrost terrain. The computer program accepts basic soil parameters in an interactive manner making its use relatively simple. Soil parameters input to the model are limited to test results normally available to the highway engineer and thermal properties that may be measured or estimated from suitable correlations. Climate data may be input as daily or monthly air temperatures along with suitable n-factors. Results have been compared with theoretical calculations and several measured temperatures under insulated and uninsulated road test sections. Comparisons indicated that the model is capable of giving very good results under a variety of conditions provided that accurate estimates of dry densities and water contents can be obtained.

INTRODUCTION

A one-dimensional finite element model was developed to estimate thaw depths and resulting settlements of a road embankment on permafrost terrain. The computer program was written in a manner allowing the required soil parameters to be supplied in an interactive mode for relative ease of use. The model is intended to form part of a system to estimate highway roughness resulting from the thaw of underlying permafrost, thus performing a vital role in a pavement management system in permafrost areas. This paper will describe the development of the model, some of its features, and results obtained in a comparison with measured soil temperatures.

The number and types of parameters required to define each soil layer were purposely limited to recognize the minimal test data usually available to a highway engineer. Estimates of frozen and thawed thermal conductivities are required, along with the dry density, water contents (initial frozen and final thawed water content), specific heat capacity of the soil matrix, and parameters to define the unfrozen water content. Thaw settlement is estimated from changes in the water content after thawing. Climate data are limited to daily or monthly air temperatures and the Nfactors for thawing and freezing (dependent on local climate and/or highway surface). The rationale for a model requiring minimal input data was to provide a tool for design rather than research.

Many mathematical models have been developed to estimate freeze or thaw depths within layered soil systems (Kawasaki et al, 1982). A number of analytical solutions have evolved from Neumann's original work in the late nineteenth century. However, all of the analytical solutions have limitations in their treatment of material properties and boundary conditions. More recent developments in numerical modeling have produced a range of mathematical models for soil temperature using either finite difference or finite element methods for discretization of the continuous domain. The finite difference approach is conceptually easier but generally less flexible than the finite element method. Finite elements were used in this study because the flexibility of the method made variations in grid spacing and possible expansion to two dimensions relatively straightforward.

A relatively simple one-dimensional finite element model for soil thawing/freezing is described in this paper. HIFIEL (HIghway FInite ELement) was designed to handle large numbers of data points along the alignment of a highway. Variations in temperature within each element are considered linear in space and nonlinear with quadratic variation in the time dimension. A simple approach to thaw settlement has been used with volume changes based on changing of water content after the initial thaw of an ice-rich soil and the change of state from ice to water or vice versa. Heat transfer is by conduction alone with no allowance for heat loss or gain by migrating water. Theoretical justification for this neglect of convective heat transport was established by Nixon (1975) using an analytical solution for one dimensional thaw consolidation. The mathematical structure of HIFIEL and a comparison with measurements at an Alaskan test site are described in this paper.

STRUCTURE OF THE MODEL

In general, the one-dimensional heat-flow problem in a soil considering only conduction is defined by the differential equation:

$$C \frac{\partial T}{\partial t} - k \frac{\partial^2 T}{\partial x^2} = 0 \qquad (1)$$

where C is the volumetric heat capacity, T is the temperature, t is time, k is the thermal conductivity, and x is depth (positive downwards). An analytical solution to the soil freezing/thawing problem derived by Neumann in the nineteenth century has proven useful as a standard against which other methods are evaluated. However, limitations in boundary conditions, initial temperature profiles, and soil homogeneity have made Neumann's approach unrealistic for all but the most simple soil freezing/thawing problems.

Finite Element Formulation

The finite element method was chosen for the research upon which this paper is based to solve the differential equation in a piece-wise manner throughout a portion of the semiinfinite half space of a layered soil system. A one-dimensional linear element was used in the space domain with element size increasing with depth to reduce computational costs. A tridiagonal matrix representing the soil system resulted from the use of Galerkin's version of the method of weighted residuals. Since the thermal properties of the elements are dependent on temperature, a new set of equations is solved for each time step using a solver specifically developed for tridiagonal matrices. The nonlinear heat-flow problem is thus solved as a linear problem with continually varying thermal properties.

The nonlinear component introduced by the variation in material thermal properties with temperature requires either an iterative solution procedure, extremely small time steps, or a three-time-level scheme to ensure accurate results. A three-time-level scheme suggested by Goodrich (1980) was used in the final model after some attempts with other methods. This three-time-level scheme is a straightforward application of the Galerkin weighted residual procedure using quadratic Lagrange polynomials as shape functions. Using fixed time steps the final system of equations for solution developed by Goodrich (1980) reduces to:

$$[C]\cdot[\{T\}^{m-1} - 4\{T\}^{m} + 3\{T\}^{m+1}] + \frac{2\Delta t}{5} \cdot [K]$$

$$[-\{T\}^{m-1} + 2\{T\}^{m} + 4\{T\}^{m+1}] = 0, \qquad (2)$$

where [C] is the heat capacity matrix, [K] is the heat conductance matrix, Δt is the time step, and $\{T\}$ is the appropriate temperature vector at time m-1, m, and m+1. In this particular application, the heat capacity matrix is evaluated for the transition from $\{T\}^{m-1}$ to $\{T\}^{m}$ and [K] is based on the average temperature during the same time step. The temperatures are known or assumed for the first two time levels leaving only $\{T\}^{m+1}$ to be calculated within a tridiagonal system of equations. This procedure has resulted in a reasonably efficient algorithm that has produced good results with time steps as large as three days.

Unfrozen Water Content and Latent Heat

The most critical component of any predictor of freeze/thaw depths in a soil system is the consideration of soil moisture freezing or thawing since the ice and water contents are major factors in the variation of the soil thermal properties. A number of researchers have investigated the factors related to the unfrozen water content of saturated frozen soils (Anderson et al. 1973, Dillon et al. 1966, Tice et al. 1978) and found that fine-grained soils generally have appreciable unfrozen water contents at temperatures as low as -10°C. The neglect of this unfrozen water when considering permafrost at temperatures near 0°C will lead to substantial errors in latent heat calculations. To avoid these errors, a simple method of estimating unfrozen water based on the work of Anderson et al. (1973) has been incorporated in the latent heat calculations.

Latent heat is considered within the model as a component in the apparent specific heat capacity. The apparent specific heat calculation is based on one of two procedures depending on whether an element has changed temperature more than .00001°C during the previous time step. When the temperature change is negligible, the heat capacity is evaluated at the particular temperature based on the slope of the curve shown in Figure 1. However, when the temperature change has been substantial, the heat capacity is evaluated based on the same curve using an apparent heat capacity as required to get from a temperature T_{n-1} to temperature T_n. The alpha and beta values shown in Figure 1 are taken from the work by Anderson et al. (1973). To avoid the difficulty of $W_u \to \infty$ as $T \to 0°C$, the unfrozen water content curve was linearized from -0.5°C to 0°C as shown in Figure 1. In the case of a coarse-grained soil with negligible unfrozen water content below 0°C, the alpha and beta values are fixed such that virtually all of the water changes state in the linear 0.5°C range. This last approach is typical of most freezing/thawing models that consider some range of temperature for the change of state.

One assumption that presents some difficulty within the model is that the apparent specific heat capacity between two time levels t_{n-1} and t_n is constant from t_n to t_{n+1} (the time level that is solved for in each step). This assumption is not generally true. However, it does represent a convenient means of linearizing a nonlinear problem that would otherwise require a more costly algorithm for solution. Within a range of time step sizes from 0.5 to 5 days, the error caused by this assumption seems to be reasonably manageable.

Figure 2 illustrates the variation in the simulated temperature at mid-depth of a 2.0-m-thick soil system subjected to a sudden increase in surface temperature from -10° to 0°C with various sizes of time steps. The temperatures in Figure 2 represent the system after 60 days with the bottom boundary specified as a temperature gradient of .05 C°/m (increasing with depth). The maximum temperature difference from 0.5-day steps to 5-day steps is only 0.129°C which represents a difference in actual heat content of 0.365 kJ/kg for the soil used. Sudden large temperature changes do cause some oscillations. However, as indicated by Goodrich (1980), the quadratic Galerekin time-stepping scheme seems to dampen the oscillations quickly even in large time steps.

Climate Data

One of the primary requirements to drive any soil thermal model is some set of climate data for the time period under consideration. While there are a number of microclimatic models available, a simple model based on the usual n-factor ratio of surface indices to air temperature indices was chosen. This method allows air temperatures to be factored up or down as required to reflect local characteristics or surface conditions. Data requirements are thus minimized while maintaining a reasonable estimate of the surface energy balance. Nfactors have been estimated for several surface types with asphalt pavement ranging from 0.29 to 1.0 for the freezing n-factor and 1.4 to 2.3 for the thawing n-factor (Goodrich and Gold 1981). A specific site near Chitina, Alaska, gave values of 0.86 to 1.18 and 1.58 to 1.66 for the same two factors during three freezing seasons and two thawing seasons (Esch and Livingston 1978). It was thought that estimates of air temperatures and surface n-factors would be available for most northern locations, while more detailed climate information may be unavailable.

Thaw Settlement

The model was altered slightly to provide a means of estimating the settlement if the thaw front advanced beyond the permafrost table. The model was intended for use with relatively light long-term loadings from embankments 1.5 to about 6.0 m high. Various researchers have found that under loadings of less than 100 kN/m (≈6-m embankment), 80 to 90% of the thaw settlement occurs almost instantly as a function of a thaw settlement parameter (Watson et al.1973). The use of a model involving only the instantaneous settlement upon thawing avoids many complications induced by consideration of the consolidation.

The development of the entire model was based on the use of parameters that would normally be measured by a highway engineer. An attempt was made to retain this principle in the thaw settlement portion. McHattie (personal communication 1981) suggested that a practical estimate of thaw settlement could be obtained from estimates of soil water content before and after thawing. An assumption is made that soils undergoing thaw settlement are saturated, so that a void ratio can be estimated based on the water content and the dry density. This information is used by the model to calculate changes in element lengths based on a new void ratio after initial thaw. Once an element has thawed, it will maintain a constant water content and only change size as the water freezes and thaws with the seasons. No attempt was made to account for the water expelled when settlement occurs, nor is the heat in the migrating water considered. Portions of the one-dimensional grid simply expand and contract as elements freeze and thaw. Changes in the geometry allow for the effect of a heat source or disturbance moving closer to the original permafrost table as thaw settlement occurs. When the water content after thawing can be easily established, the model will provide a good estimate of the settlement. The model is not intended for use in situations where rapid thawing from an unusual heat source may cause excess pore pressure leading to possible stability problems, and it is also not intended to model thaw consolidation under loadings greater than 100 kN/m^2.

VERIFICATION

Analytical Versus Numerical Solution

At each stage of development, the subroutines within the Fortran algorithm were checked for their theoretical accuracy, until finally the entire structure required verification. As an initial check of the model performance, the thaw depth of a hypothetical soil structure was evaluated using HIFIEL and the Berggren equation under similar conditions. The soil properties are summarized in Table 1. The temperature at the surface was raised from -10° to +5°C during the first 30 days, and then the surface temperature was held constant at +5°C for the remaining 330 days. Calculations for the equivalent Berggren solution used an average conductivity of 2.5 W/m°C and only 335 days to account for an equivalent 1,675 C degree days. The results of the two simulations are illustrated in Figure 3, with thaw depth plotted against the square root of time. Nearly identical results appear at the end of the time period, with the numerical model predicting 1.44 m of thaw versus 1.45 m for the analytical procedure. Although the result is encouraging, there are some differences that need explanation.

TABLE 1 - Soil Properties for Simulation Check

Soil:	Fairbank Silt
Unfrozen water parameters:	α = .0481 β = -0.326
Water content:	0.55
Dry density:	1.20 t/m^3
Conductivity thawed:	2.0 W/m•K
Conductivity frozen:	3.10 W/m•K
Soil matrix heat capacity:	.710 kJ/kg•K
Initial temperature:	-10°C
Depth of soil:	10 m

First, the initial start of the simulation using finite elements gradually approached the assumed step surface temperature of +5°C during the first 30 days. It is apparent from Figure 3 that this difference due to the startup diminishes after the 100 days as the simulated thaw depth approaches proportionality to the square root of time.

A second contribution to the difference may be the consideration of unfrozen water within the numerical model. The thaw depth has been defined strictly as the point at which the temperature is the 0°C isotherm. The Berggren solution is based partially on the assumption that all water changes state at 0°C. When unfrozen water is allowed to exist below 0°C, there must be some variation from an analytical solution that freezes water at 0°C. The temperature prediction seemed to match the frost depth reasonably well, but further comparison with measured temperatures in other situations seemed necessary.

Chitina Test-Site Comparison

A road section was constructed near Chitina, Alaska, in 1969, to evaluate the performance of embankment insulation and provide needed data concerning the thermal regime under uninsulated road sections. Data from this test site were made available by the Research Section of the Alaska Department of Transportation and Public Facilities (Esch 1972, and personal communications, 1981). The site included a fill section with 5 cm and 10 cm of insulation, uninsulated fill sections, and a cut section. All of the sections were instrumented with a variety of thermocouple strings, including an undisturbed site nearby and a weather station. Data are available from the site for a period of time from October 1969 to the present. However, only three years of data have been used for a verification of the model's accuracy.

Comparisons were made of the model output and the measured temperatures for three different locations under the pavement centerline. The material properties for each location are listed in Table 2. Temperature measurements were made at monthly intervals throughout most of the time period. The results of the simulations and the measured temperatures for the same dates are shown in Figures 4 through 12. The best match between the measured data and the simulations occurred for the section with 10 cm of insulation (thermocouple 3), while in the section with 5 cm of insulation (thermocouple 4) it is apparent that the insulation is approximately 20 cm deeper than Table 2 suggests. The effect of the different thicknesses of insulation is very clear in both the measured temperatures and the simulated temperatures. It appears that over the long term the model may be predicting a cooling trend since all three simulations have colder than measured values at the maximum depth on 25 September, 1972. However, it is difficult to conclude anything from the differences since they are generally less than 0.75°C. Other large differences in the upper layers may have been caused by poor estimates of water content and conductivity. In general the correlation between measured and simulated temperatures appears quite good with no major anomalies except in the extreme upper layers.

For a complete check of the model performance the settlements predicted over a period of time should be matched to those measured at the site. Unfortunately, there were insufficient data at the time of the writing of this paper to make a realistic test of this section of the model. Accurate measurements of the water content of the subgrade soil before thawing and after thaw consolidation are required as input parameters. Settlement of the road has been measured at the site, and a reasonable estimate of the after-thaw water content was used for the uninsulated fill section in an attempt to evaluate the thaw settlement. The measured settlement between 14 July, 1971 and 18 September, 1972 was 0.1 m, while the model predicted 0.126 m during the same time period. While the results cannot be used as a measure of the model's accuracy, they do indicate that reasonable settlement estimates can be expected with the proper soil parameters.

TABLE 2 Layer Properties at Chitina Test Site Used in Simulation

Thermocouple String No.	Material	Thickness (m)	Dry Density (T/m^3)	Water Content	Conductivity Frozen (W/mK)	Conductivity Thawed (W/mk)	Soil Specific Heat Capacity (kJ/kgK)	Unfrozen Water α	Unfrozen Water β
2	Asphalt	.05	2.160	.01	1.300	1.300	1.674	.005	-.333
(uninsulated	Base	.20	2.160	.05	2.750	2.750	.850	.005	-.333
fill)	Subbase	1.30	1.922	.18	3.750	2.750	.850	.005	-.333
	Organics	.30	.256	3.00	1.073	0.430	1.670	.005	-.333
	Silty Peat	6.10	.530	1.50	2.200	0.519	.710	.005	-.333
	Gray Silt	3.05	1.300	.40	2.200	1.200	.710	.050	-.325
3	Asphalt	.05	2.160	.01	1.300	1.300	1.674	.005	-.333
(10-cm	Base	.20	2.160	.05	2.750	2.750	.850	.005	-.333
insulation)	Subbase	1.40	1.922	.05	3.250	2.750	.850	.005	-.333
	Insulation	.10	.034	.005	0.033	0.033	1.000	.005	-.333
	Subbase	.40	1.922	.18	3.750	2.750	.850	.005	-.333
	Organics	.15	.256	3.00	1.073	0.430	1.670	.005	-.333
	Silty Peat	5.55	.530	1.50	2.200	0.519	.710	.005	-.333
	Gray Silt	2.15	1.300	.50	2.200	1.200	.710	.050	-.325
4									
(5-cm	Asphalt	.05	2.160	.01	1.300	1.300	1.674	.005	-.333
insulation)	Base	.20	2.160	.05	2.750	2.750	.850	.005	-.333
	Subbase	1.30	1.992	.05	3.250	2.750	.850	.005	-.333
	Insulation	.05	.034	.005	0.033	0.033	1.000	.005	-.333
	Subbase	.40	1.922	.18	3.750	2.750	.850	.005	-.333
	Organics	.15	.256	3.00	1.073	0.430	1.670	.005	-.333
	Silty Peat	5.55	.530	1.50	2.200	0.519	.710	.005	-.333
	Gray Silt	2.30	1.300	.50	2.200	1.200	.710	.050	-.325

CONCLUSIONS

The model presented in this paper can be used to give estimates of road surface settlements for a variety of designs over permafrost subgrades. It is relatively easy to use with a minimum of soil parameters required as input, and it can include insulating layers or peat material without any special considerations. Thermal modeling has been conducted using a one-dimensional two-node element, but it could be expanded to two dimensions with some increased calculation costs and complexity of the program.

As a design tool for highway engineering, HIFIEL performs as well or better than the currently used modified Berggren equation. In its current form, thermal estimates can be made for a variety of alternative designs in a particular site without great cost (10-year simulation for 15-m depth using 5-day time steps required 96 seconds of CPU time using an IBM 4341). Using a gradient boundary at the bottom of the layers allows long-term warming of the permafrost to be considered in areas where there is potential for complete thaw of the permafrost.

ACKNOWLEDGMENTS

The author would like to thank Dr. R.C.G. Haas for his support as Ph.D. supervisor throughout the development of the computer program. Special thanks also to the Natural Science and Engineering Research Council of Canada, Transport Canada, and the Roads and Transportation Association of Canada for providing scholarships during the author's career as a graduate student at the University of Waterloo.

REFERENCES

Anderson, D.M., Tice, A.R. and McKim, H.L., 1973, The unfrozen water and theapparent specific heat capacity of frozen soils, in Proceedings of the Second International Conference on Permafrost, Yokutsk, U.S.S.R., North American Contribution, National Academy of Sciences.

Dillon, H.B. and Andersland, O.B., 1966, Predicting unfrozen water contents in frozen soils: Canadian Geotechnical Journal, v. 3, no. 2.

Esch, D.C., 1972, Control of permafrost degradation beneath a roadway by subgrade insulation: Alaska Highway Department, Summary Report.

Esch, D.C. and Livingston, H.R., 1978, Performance of a roadway with a peat underlay over permafrost: State of Alaska Department of Transportation and Public Facilities, Interim Report.

Goodrich, L.E., 1980, Three-time-level methods for the numerical solution of soil freezing problems: Cold Regions Science and Technology, v. 3.

Goodrich, L.E., and Gold, L.W., 1981, Ground thermal analysis, in Johnston, G.H., ed., Permafrost Engineering Design and Construction, New York, John Wiley & Sons.

Kawasaki, K., Osterkamp, T.E., and Gosink, J.P., 1982, A preliminary evaluation of numerical models suitable for thermal analysis of roadways and airstrips, Geophysical Institute, University of Alaska, Fairbanks.

Nixon, J.F., 1975, The role of convective heat transport in the thawing of frozen soils: Canadian Geotechnical Journal, v. 12.

Tice, A.R., Burrous, C.M., and Anderson, D.M., 1978, Determination of unfrozen water in frozen soil by pulsed nuclear magnetic resonance, in Proceedings of the Third International Conference on Permafrost, v. 1: Ottawa,National Research Council of Canada.

Watson, G.H., Slusarchuk, W.A. and Rowley, R.K., 1973, Determination of some frozen and thawed properties of permafrost soils: Canadian Geotechnical Journal, v. 10, no. 4.

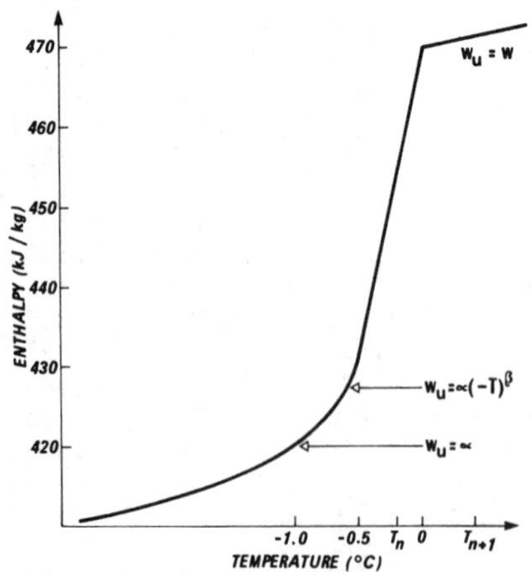

FIGURE 1 Heat content model for a typical soil.

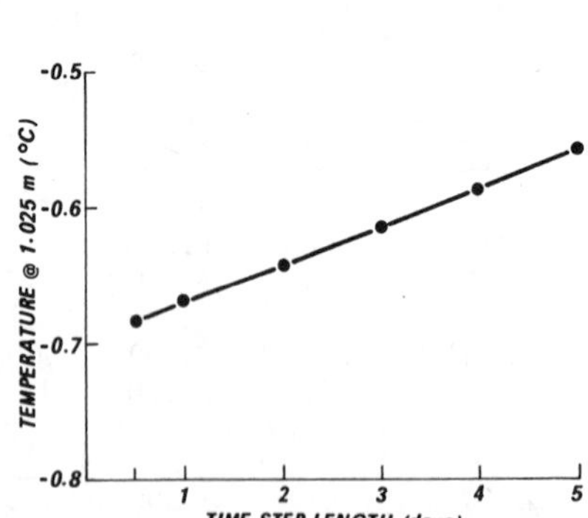

FIGURE 2 Temperature at mid-depth for different time steps.

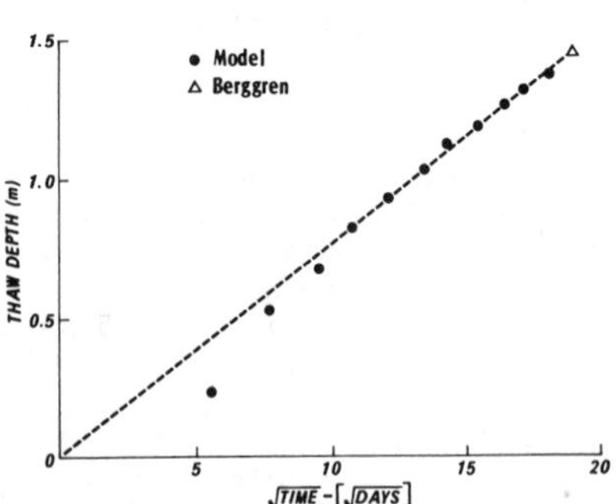

FIGURE 3 Thaw depth versus square root of time.

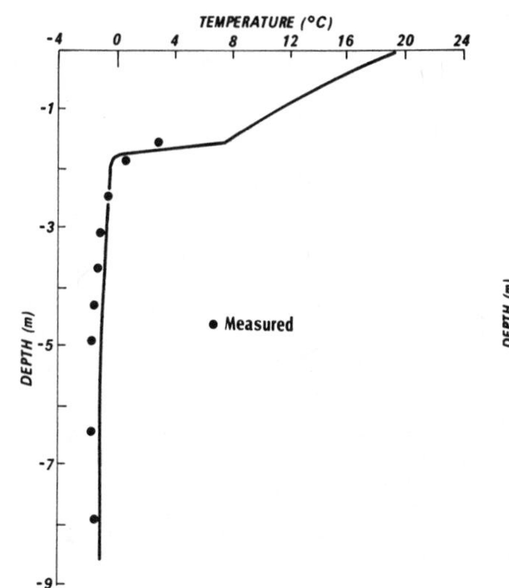

FIGURE 4 Thermocouple 2 temperatures for 7 July, 1970.

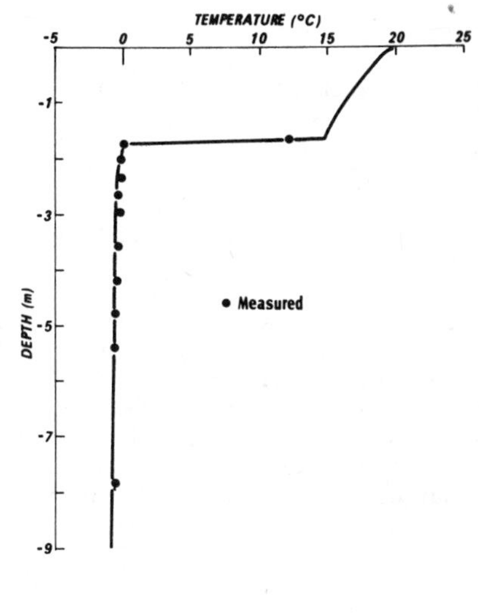

FIGURE 5 Thermocouple 3 temperatures for 7 July, 1970.

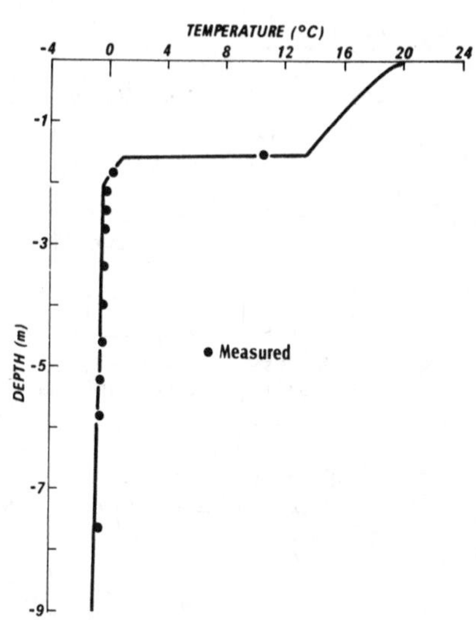

FIGURE 6 Thermocouple 4 temperatures for 7 July, 1970.

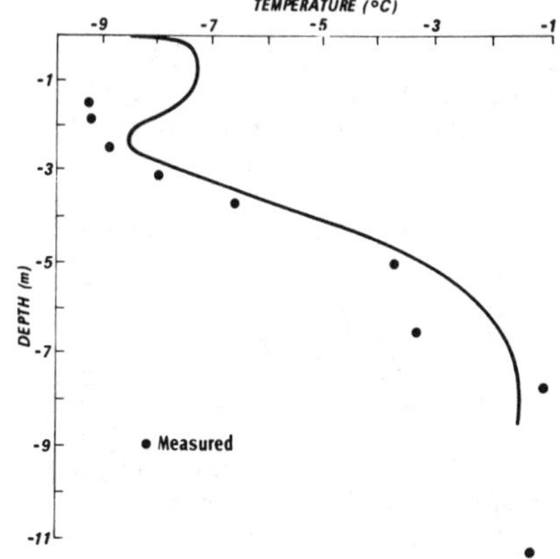

FIGURE 7 Thermocouple 2 temperatures for 4 April, 1972.

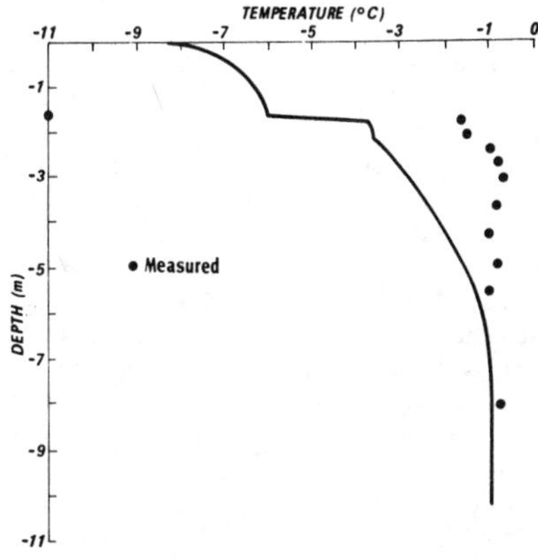

FIGURE 8 Thermocouple 3 temperatures for 4 April, 1972.

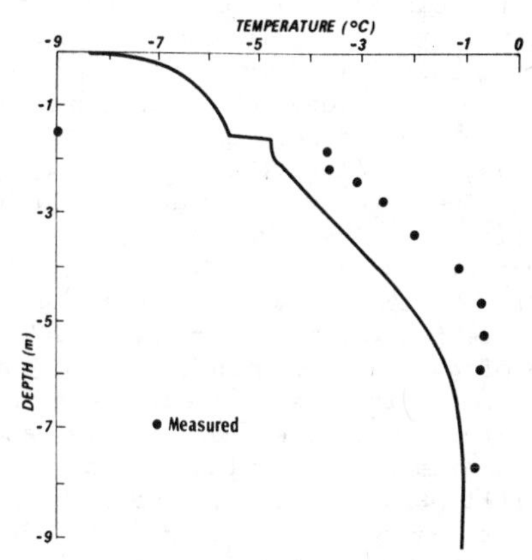

FIGURE 9 Thermocouple 4 temperatures for 4 April, 1972.

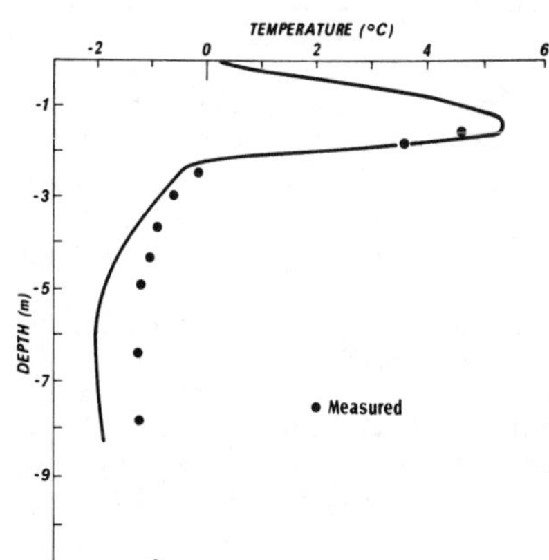

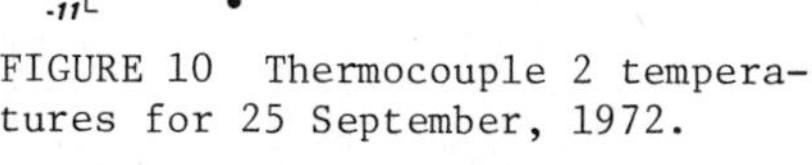

FIGURE 10 Thermocouple 2 temperatures for 25 September, 1972.

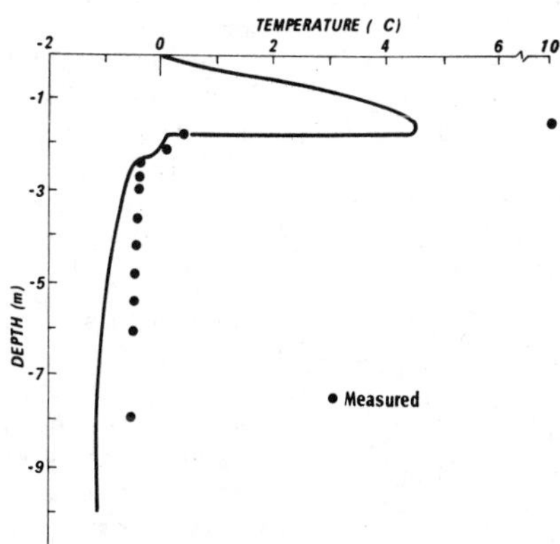

FIGURE 11 Thermocouple 3 temperatures for 25 September, 1972.

FIGURE 12 Thermocouple 4 temperatures for 25 September, 1972.

APPROXIMATE SOLUTIONS FOR MILLER'S THEORY OF SECONDARY HEAVE

John T. Holden

Department of Theoretical Mechanics
University of Nottingham, Nottingham, NG7 2RD, England

Quasi-static approximations for the temperature and pore water pressure profiles are used to describe the one-dimensional freezing of a saturated noncolloidal soil. Using the energy and mass conservation equations the problem is reduced to the solution of two ordinary differential equations. The theory includes a criterion for the initiation of lenses and a simple numerical procedure is used to describe the growth of a lens, the frozen fringe and the initiation of a new lens. Some numerical results are given that are in general qualitative agreement with other studies.

INTRODUCTION

The phenomenon of frost heave has received considerable attention in recent years and various theories and mathematical models have been devised. Theories based on a capillary theory of ice lens growth for non-colloidal soils began with the work of Everett (1961). The earlier models are often termed primary heaving models since they predict the formation of ice lenses at the freezing front, whereas it is found that ice lenses normally form within the already frozen soil. To describe this phenomenon a rigid-ice model of secondary heave was conceived and developed by Miller (1972, 1977, 1978). Miller's model proposed that an ice lens grows behind the freezing front, so that there is a zone of partially frozen soil in front of a lens. This theory is able to predict much higher heaving pressures than the primary models, and it contains a mechanism for the initiation of new lenses in the partially frozen region. The model has been quantified by O'Neill and Miller (1982), wherein the governing partial differential equations are given together with some numerical results. In addition, Miller and Koslow (1980) have examined the predictions of the model when the temperature and water pressure fields are stationary.

This paper considers the one-dimensional freezing of a saturated non-colloidal soil and gives quasi-static approximations for the temperature and pore water pressure profiles. Energy and mass conservation equations that link the parameters introduced by the approximate profiles are formulated. This leads to a pair of coupled nonlinear ordinary differential equations that are solved by using a standard numerical procedure. It is thus possible to follow the development of the frozen region, ice lens and frozen fringe and to determine when and where the next lens forms. This approach is motivated by the slow nature of the freezing process and avoids the much more difficult numerical problem of solving a pair of nonlinear coupled partial differential equations. In addition, the sensitivity to and relative importance of the various physical parameters may be readily investigated.

Numerical results are presented using some typical values for the physical parameters and show the cumulative heave, time lapse between ice lenses, and the thickness of the lenses, as functions of time.

MILLER'S THEORY OF SECONDARY HEAVE

When a column of *saturated* soil is frozen from one end and the column is observed to heave, there are three distinct regions of soil (see Figure 1):

1 a solid frozen region of soil that may contain distinct ice lenses;

2 a partially frozen region of soil (usually known as the frozen fringe) containing soil, ice and water; and

3 an unfrozen region of wet soil.

During freezing, water moves through the unfrozen soil to enter the frozen fringe to feed the growth of the ice lens at the bottom of region 1. The water is sucked toward the lens because in region 2 the ice content increases rapidly with decreasing temperature, and so the water content correspondingly decreases and causes a large suction adjacent to the lens. The model of this process developed by Miller is termed a rigid-ice model, because the ice present in the frozen fringe is assumed to be rigidly connected to the ice of the growing lens. This pore ice moves through the soil pores by a process of microscopic regelation as heave takes place.

To describe the frost heave mechanism within the frozen fringe, a parameter ψ is introduced by the equation

$$u_i - u_w = \sigma_{iw}\psi, \tag{1}$$

where u_i is the ice pressure, u_w the water pressure and σ_{iw} is the ice/water interfacial tension. ψ can be interpreted as a mean curvature of the ice water interface. Using the Clapeyron equation, which relates the equilibrium values of u_i and u_w and the temperature T, the parameter ψ may be written as

$$\psi = Au_w + BT, \tag{2}$$

where A and B are constants. The volumetric ice

content θ_i is assumed to depend only on ψ (Koopmans and Miller (1966)) and the form of the function $\theta_i(\psi)$ used here is the one used by O'Neill and Miller (1982), which was based on data of Horiguchi and Miller (1980).

Another important component of the model is the criterion for the formation of a new lens. The stresses are assumed to be isotropic, and within the frozen fringe the neutral stress σ_n is calculated by the formula

$$\sigma_n = \chi u_w + (1-\chi) u_i, \tag{3}$$

where $\chi(\theta_i)$ is a stress partition factor. In general, σ_n has a maximum in the frozen fringe, and it is when this maximum value reaches the overburden pressure, P, that a new lens is initiated.

APPROXIMATE PROFILES

We consider a semi-infinite mass of soil that initially is at a temperature of T_1. At time t = 0 the surface z = 0 is cooled to a temperature of $T_c < 0$. The freezing front moves down through the soil, ice lenses form and the soil heaves. If z is measured downward and the cumulative heave is H(t), the surface is given by z = -H(t). We denote the position of the bottom of the lowest lens by $z = z_s$, the position of the freezing front by $z = z_f$ and the level of the water table by $z = z_w$. The different levels and associated temperatures are shown in Figure 1.

Because of the magnitude of the latent heat of water, the temperature profiles in the solid frozen region is close to linear and may be written as follows:

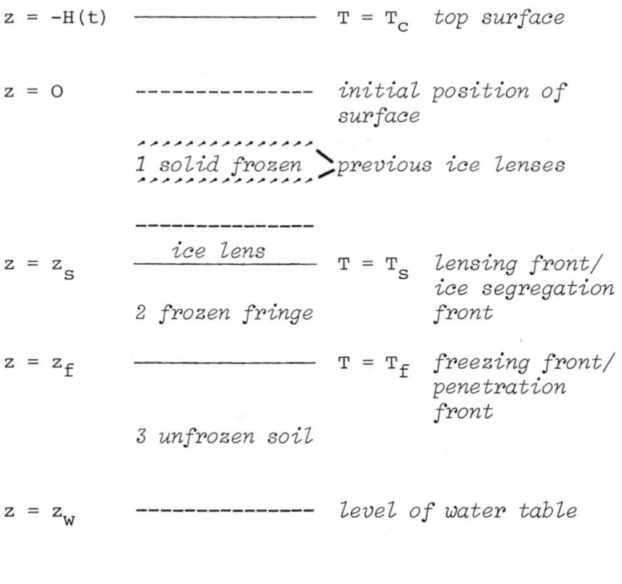

$z = -H(t)$	$T = T_c$	*top surface*
$z = 0$	*initial position of surface*	
	1 solid frozen	*previous ice lenses*
$z = z_s$	*ice lens* $T = T_s$	*lensing front/ ice segregation front*
	2 frozen fringe	
$z = z_f$	$T = T_f$	*freezing front/ penetration front*
	3 unfrozen soil	
$z = z_w$		*level of water table*

FIGURE 1 Section through freezing soil

$$T = \frac{z_s - z}{z_s + H} T_c + \frac{z + H}{z_s + H} T_s, \qquad -H \leqslant z \leqslant z_s. \tag{4}$$

The temperature profile in the frozen fringe is also assumed to be linear and has the form:

$$T = \frac{z_f - z}{z_f - z_s} T_s + \frac{z - z_s}{z_f - z_s} T_f, \qquad z_s \leqslant z \leqslant z_f. \tag{5}$$

In the region below the freezing front, a simple linear profile cannot be used because z tends to infinity. One possibility is to base the approximation on the classical solution:

$$T = T_f + (T_1 - T_f)\,\mathrm{erf}\left(\frac{z}{2\sqrt{\alpha t}}\right), \qquad z \geqslant 0, \tag{6}$$

where α is the thermal diffusivity. This is the solution for a freezing front held stationary at z = 0, and for convenience we replace z by $z - z_f$, thus introducing a simple shift on the temperature profiles to give

$$T = T_f + (T_1 - T_f)\,\mathrm{erf}\left(\frac{z - z_f}{2\sqrt{\alpha t}}\right), \qquad z \geqslant z_f. \tag{7}$$

The effect of this choice is to determine the temperature gradient ahead of the freezing front.

The other profiles needed are for the water pressure in both the frozen fringe and the unfrozen region. The water pressure is zero at the water table level z_w and decreases linearly up to the level of the freezing front. In the frozen fringe the ice content increases rapidly with decreasing temperature, and so the unfrozen water content decreases rapidly. This results in a rapid decrease in the water pressure u_w from the small negative value u_f at level z_f to a much larger negative value u_s just below the lens at level z_s. If we assume the water pressure gradient is continuous at z_f and a quadratic profile in the frozen fringe, we obtain:

$$u_w = \begin{cases} \left\{(z_w - z) - \dfrac{(z_f - z)^2}{z_f - z_s}\right\}\dfrac{u_f}{z_w - z_f} + \left(\dfrac{z_f - z}{z_f - z_s}\right)^2 (u_s - u_f), & z_s < z < z_f, \\[2ex] \dfrac{z_w - z}{z_w - z_f} u_f, \qquad z_f < z < z_w. \end{cases} \tag{8}$$

A quadratic profile was chosen as it serves to illustrate the method. Clearly, other choices of profile are possible.

Two other functions are required for the model: $\theta_i(\psi)$ and $\chi(\theta_i)$. From O'Neill and Miller's (1982) paper an approximation to their curve for θ_i is:

$$\theta_i = \begin{cases} 0.01385 - 6.144 \times 10^{-8}\psi + 3.143 \times 10^{-14}\psi^2, \\ \qquad\qquad 0.26 \times 10^6 < \psi < 3.6 \times 10^6, \\[2ex] 0.4 - \dfrac{0.28 \times 10^6}{\psi - 2.2 \times 10^6}, \qquad \psi > 3.6 \times 10^6 \end{cases} \tag{9}$$

and the form used for the stress partition factor χ is $(1-\theta_i/\theta_0)^{1.5}$.

CONSERVATION EQUATIONS

The equations for this model are derived by using the equations of conservation of heat energy and conservation of mass at the lensing front and at the freezing front. In addition, Darcy's law and an equation involving the load (or overburden) completes the set of equations.

If v_s is the liquid volume flux at the top of the frozen fringe, the energy balance equation at $z = z_s$ is:

$$K_f \left.\frac{\partial T}{\partial z}\right|_{z_s+} - K_s \left.\frac{\partial T}{\partial z}\right|_{z_s-} = \rho_w L v_s, \tag{10}$$

where K_f and K_s are the thermal conductivities of the frozen fringe and solid frozen regions, respectively, ρ_w is the density of water and L the latent heat of fusion of water.

At the freezing front $z = z_f$, the energy balance equation is

$$K_u \left.\frac{\partial T}{\partial z}\right|_{z_f+} - K_f \left.\frac{\partial T}{\partial z}\right|_{z_f-} = -\rho_i L\theta_{if}\frac{dz_f}{dt}, \tag{11}$$

where K_u is the thermal conductivity of the unfrozen soil, ρ_i is the density of ice and θ_{if} is the volumetric ice content at $z = z_f$.

Using the temperature profiles (4), (5) and (7), the equations (10) and (11) become:

$$K_f \frac{T_f - T_s}{z_f - z_s} - K_s \frac{T_s - T_c}{z_s + H} = \rho_w L v_s \tag{12}$$

and

$$K_u \frac{T_1 - T_f}{\sqrt{\pi \alpha t}} - K_f \frac{T_f - T_s}{z_f - z_s} = -\rho_i L\theta_{if}\frac{dz_f}{dt}. \tag{13}$$

The conservation mass equation at $z = z_s$ is:

$$\rho_i(1-\theta_{is})\frac{dH}{dt} = -\rho_w v_s, \tag{14}$$

where θ_{is} is the volumetric ice content at the top of the frozen fringe. If we assume a Darcy law and that at the freezing front there is a jump in the value of the hydraulic conductivity due to the jump in ice content, the mass balance at this level may be written as:

$$\rho_i\theta_{if}\frac{dH}{dt} = -\frac{[k_f]}{g}\frac{u_f}{z_w - z_f}, \tag{15}$$

where $[k_f]$ is the jump in the value of k_f, the hydraulic conductivity at $z = z_f$.

We assume that the Darcy law holds at the top of the frozen fringe so that:

$$v_s = -\frac{k_s}{\rho_w g}\left\{\left(\frac{1}{z_w - z_f} + \frac{2}{z_f - z_s}\right)u_f - \frac{2u_s}{z_f - z_s} - \rho_w g\right\}, \tag{16}$$

where k_s is the hydraulic conductivity at $z = z_s$.

The final equation comes from the assumption that the ice pressure is continuous and equal to the overburden pressure P at $z = z_s$, so that equations (1) and (2) give:

$$P - u_s = \sigma_{iw}(Au_s + BT_s). \tag{17}$$

It is implicit in the above equations that some of the pore water freezes as it moves through the frozen fringe. This is because, as the frozen fringe grows in time, the temperature and water pressure profiles change causing the ice content to increase.

Equations (12) to (17) are six equations in the six unknowns z_f, H, T_s, v_s, u_s and u_f. Elimination of T_f, v_s, u_s and u_f between these equations will give two ordinary differential equations of the form:

$$\frac{dz_f}{dt} = F(z_s, z_f, H, t), \tag{18}$$

and

$$\frac{dH}{dt} = G(z_s, z_f, H, t). \tag{19}$$

Before evaluating the functions F and G, it is advantageous to non-dimensionalize the variables so that a further simplification can be made.

NON-DIMENSIONAL EQUATIONS

The equations are non-dimensionalized using a length scale ℓ_0, time t_0, thermal conductivity K_0, temperature T_0 and pressure P. If we denote the non-dimensional quantities using a bar, the governing equations (12) to (17) may be written in the form:

$$\frac{d\bar{z}_f}{d\bar{t}} = \left(\frac{K_0T_0t_0}{\rho_i L\ell_0^2}\right)\frac{1}{\theta_{if}}\left\{\frac{-\bar{K}_f}{\bar{z}_f-\bar{z}_s}\bar{T}_s + \frac{\bar{K}_f}{\bar{z}_f-\bar{z}_s}\bar{T}_f - \frac{\bar{K}_u(\bar{T}_1-\bar{T}_f)}{\sqrt{(\pi\alpha t_0\bar{t}/\ell_0^2)}}\right\}, \tag{20}$$

$$\frac{d\bar{H}}{d\bar{t}} = \left(\frac{K_0T_0t_0}{\rho_i L\ell_0^2}\right)\frac{1}{1-\theta_{is}}\left\{\left(\frac{\bar{K}_f}{\bar{z}_f-\bar{z}_s} + \frac{\bar{K}_s}{\bar{z}_s+\bar{H}}\right)\bar{T}_s - \frac{\bar{K}_s}{\bar{z}_f-\bar{z}_s}\bar{T}_f - \frac{\bar{K}_s}{\bar{z}_s+\bar{H}}\bar{T}_c\right\}, \tag{21}$$

$$\bar{u}_s = \frac{1 - B\sigma_{iw}T_0\bar{T}_s/P}{1+\sigma_{iw}A}, \tag{22}$$

$$\frac{d\bar{H}}{d\bar{t}} = \left(\frac{k_s P t_0}{\rho_i g\ell_0^2}\right)\frac{1}{1-\theta_{is}}\left\{\left(\frac{1}{\bar{z}_w-\bar{z}_f} + \frac{2}{\bar{z}_f-\bar{z}_s}\right)\bar{u}_f - \frac{2\bar{u}_s}{\bar{z}_f-\bar{z}_s} - \frac{\rho_w g\ell_0}{P}\right\}, \tag{23}$$

$$\frac{d\bar{H}}{d\bar{t}} = \left(-\frac{[k_f]Pt_0}{\rho_i g \ell_0^2}\right)\frac{1}{\theta_{if}}\frac{\bar{u}_f}{\bar{z}_w - \bar{z}_f}. \tag{24}$$

The typical values for the parameters we choose are those of O'Neill and Miller (1982):

$K_0 = 3\text{Wm}^{-1}{}^{\circ}\text{C}^{-1}$,

$\rho_i L = 3.072 \times 10^5 \text{kJm}^{-3}$,

$\rho_i g = 8.99\text{kNm}^{-3}$,

$\rho_w g = 9.8\text{kNm}^{-3}$,

$\alpha = 1.07 \times 10^{-6}\text{m}^2\text{s}^{-1}$,

$A = -2.51\text{mN}^{-1}$,

$B = -3.4 \times 10^7 \text{m}^{-1}{}^{\circ}\text{C}^{-1}$,

$k_s = 8 \times 10^{-12}\text{ms}^{-1}$,

$[k_f] = 10^{-4}\text{ms}^{-1}$,

$P = 150\text{kPa}$,

In addition, if we choose a length scale of 1mm, i.e. $\ell_0 = 10^{-3}$m, a time scale $t_0 = 10^2$s and a temperature scale $T_0 = 1^{\circ}$C, the first of the non-dimensional coefficients is of order unity (see below). The actual values of the coefficients are:

$$\frac{K_0 T_0 t_0}{\rho_i L \ell_0^2} = 0.9766,$$

$$(1 + \sigma_{iw} A)^{-1} = 1.0906,$$

$$B\sigma_{iw} T(1 + \sigma_{iw} A)^{-1}/P = -8.1824,$$

$$\frac{k_s P t_0}{\rho_i g \ell_0^2} = 0.01336,$$

$$-\frac{[k_f]Pt_0}{\rho_i g \ell_0^2} = -1.67 \times 10^5,$$

$$\frac{(\pi\alpha t_0)^{\frac{1}{2}}}{\ell_0} = 18.33,$$

$$\frac{\rho_w g \ell_0}{P} = 6.533 \times 10^{-5}.$$

An examination of the magnitude of the terms in equations (23) and (24) shows that if the heave rate is of order 10^{-8}ms^{-1}, as may be expected (Jones and Lomas 1983), $\bar{u}_f/(\bar{z}_w - \bar{z}_f)$ will be of order 10^{-8}. The result is that the first and third terms in equation (23) will have a negligible effect on the heave rate $d\bar{H}/d\bar{t}$, and equation (23) becomes

$$\frac{d\bar{H}}{d\bar{t}} = \left(\frac{k_s P t_0}{\rho_i g \ell_0^2}\right)\frac{1}{1 - \theta_{is}}\left(-\frac{2\bar{u}_s}{\bar{z}_f - \bar{z}_s}\right). \tag{25}$$

Equation (24) is no longer required as it serves only to calculate $\bar{u}_f$ which is negligibly small.

The system of equations (20), (21), (22) and (25) has the form

$$\frac{d\bar{z}_f}{d\bar{t}} = A_1\bar{T}_s + B_1, \tag{26}$$

$$\frac{d\bar{H}}{d\bar{t}} = A_2\bar{T}_s + B_2, \tag{27}$$

$$\bar{u}_s = A_3\bar{T}_s + B_3, \tag{28}$$

$$\frac{d\bar{H}}{d\bar{t}} = E\bar{u}_s, \tag{29}$$

Elimination of T_s and u_s from equations (26) to (28) gives the formulae for the right-hand sides of the non-dimensional form of equations (18) and (19) as

$$G = E(A_3B_2 - A_2B_3)/(EA_3 - A_2), \tag{30}$$

$$F = A_1(G - EB_3)/EA_3 + B_1. \tag{31}$$

METHOD OF SOLUTION

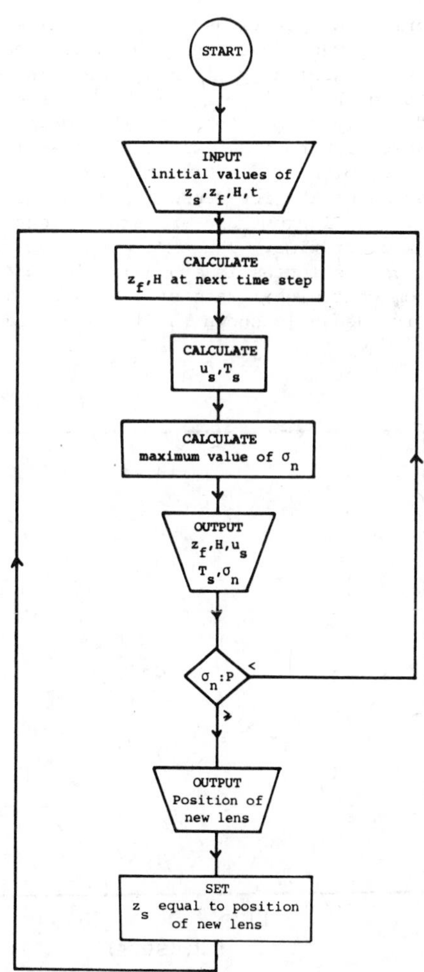

FIGURE 2 Flow diagram

The problem has been reduced to the solution of the two coupled nonlinear ordinary differential equations (18) and (19), where the parameter z_s is a stepwise constant in time. A flow diagram for the computations is shown in Figure 2. The differential equations are solved by using a standard fourth-order Runge-Kutta formula, and a simple search using divided differences is used to find the maximum of the neutral stress σ_n in the frozen fringe. During the formation of a lens, z_s remains fixed; the lens grows at this level until the maximum value of σ_n below the lens reaches the overburden P and a new lens is initiated. At this time, z_s is set equal to the position of the maximum of σ_n for the new lens to form.

In addition to the data quoted above, the values $K_s = 4\text{Wm}^{-1}{}^{\circ}\text{C}^{-1}$, $K_f = 4\text{Wm}^{-1}{}^{\circ}\text{C}^{-1}$, $K_u = 3\text{Wm}^{-1}{}^{\circ}\text{C}^{-1}$, $T_c = -1^{\circ}\text{C}$, $T_f = -0.08^{\circ}\text{C}$, $T_1 = 1^{\circ}\text{C}$, $\theta_{is} = 0.39$ and $\theta_{if} = 0.1$ have been used. Technically, the ice content θ_{is} and the temperature T_f are functions of ψ and thus depends on the temperature and water pressure. However, as the effect of this variation on the coefficients in the equations is relatively slight, for the present work they have been taken to be constant.

RESULTS

Numerical computations using the values of the parameters given above have been performed to evaluate the general performance of this method. It is found that the position of the freezing front z_f is directly proportional to the square root of time, which is the same as for the classical freezing problem in a semi-infinite region. The thicknesses of the ice lenses as a function of time are shown in Figure 3 and are proportional to the square root of time. The time elapsed between successive ice lenses is shown in Figure 4 which has a somewhat higher rate of increase. The cumulative heave is shown in Figure 5. All these

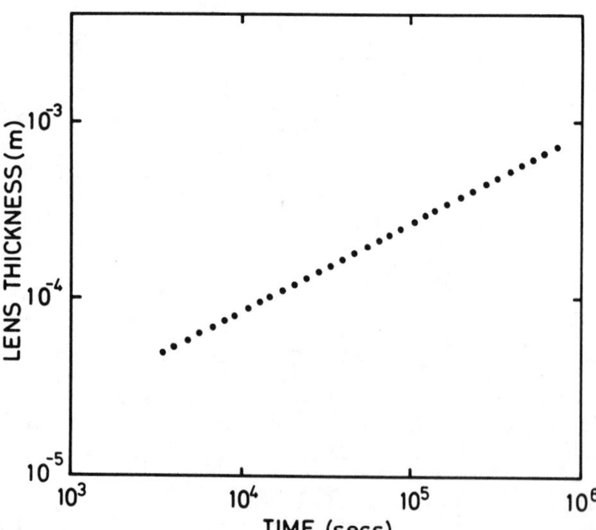

FIGURE 3 Lens thickness versus time

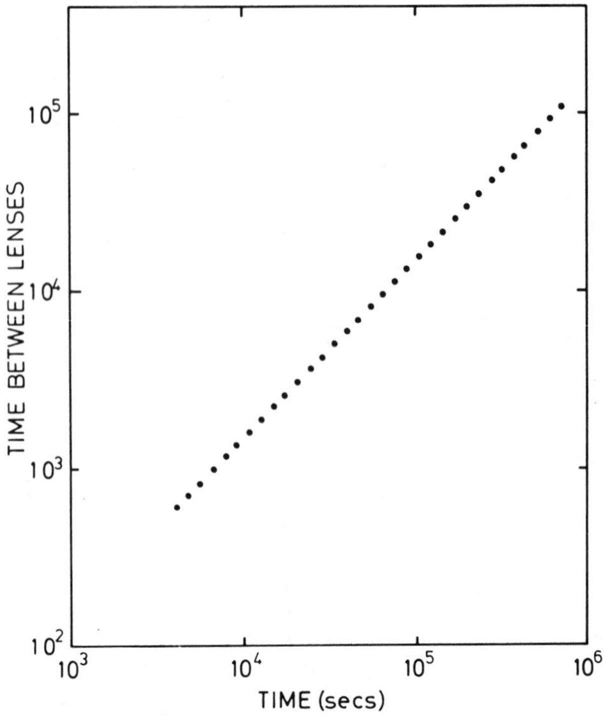

FIGURE 4 Time elapsed between successive lenses versus time

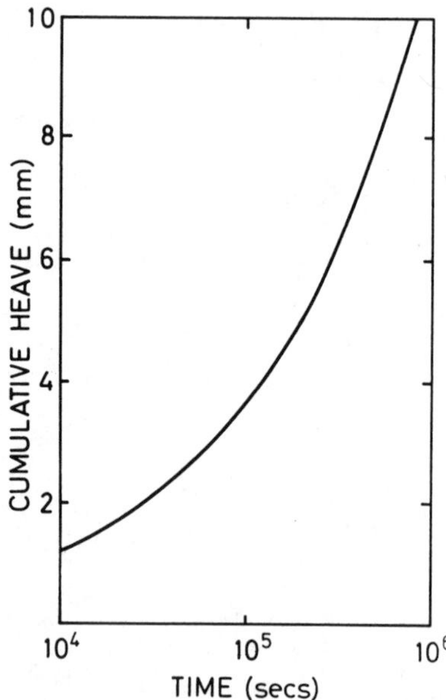

FIGURE 5 Cumulative heave versus time

results are of the same order as those given by O'Neill and Miller (1982). It should be noted that their computations are for a finite column of soil, whereas the present results are for a semi-infinite region. The important difference is that in a finite column the freezing front effectively ceases to penetrate much sooner because of the influence of the lower boundary, and this may account for the differences between the results. The frequency of formation of lenses is also similar.

DISCUSSION

The main advantages of this quasi-static approach are as follows:

- Ease of computation. To solve the full set of partial differential equations for the heat and mass flow in the frozen fringe is difficult. This is because the frozen fringe is a relatively narrow region across which various quantities change very rapidly, so that there are large gradients present. To describe adequately the equations requires a very small mesh size and a correspondingly small time step. Except in the early stages of the computations, when a small time step is required to accommodate the frequent ice lenses, the equations in this paper may be solved using a time step appropriate for the movement of the freezing front. In addition the program is short and uses relatively little store.
- Better understanding of the model. By formulating the equations in this way, it is hoped that it will lead to a greater understanding of the model and a critical evaluation of the assumptions on which it is based.
- Evaluation of parameters. One of the difficulties with theories of frost heave is that the values of some of the parameters are uncertain since they are difficult to measure. With the present theory, we can readily investigate the effects of different values of the parameters and determine the sensitivity and relative importance of the parameters to the heave.

Having shown that this quasi-static approach does predict the formation of ice lenses and heave of the right order, some obvious refinements can be made. One is to allow θ_{is} to vary with ψ and at the same time k_s to vary with θ_{is}. The temperature T_f may also vary with ψ. The choice of temperature profile ahead of the front may be improved in one of two ways: either adapt a more complicated error function solution such as that used for diffusion problems in semi-conductors (Tuck 1974) or compute the profile directly by using a Crank-Nicolson procedure at the same time as solving equations (18) and (19). In addition, other profiles for the water pressure in the frozen fringe may be examined. Meanwhile, it remains only to reiterate O'Neill and Miller's plea for a series of experiments to be established to evaluate the theory and computations.

REFERENCES

Everett, D. H., 1961, The thermodynamics of frost damage to porous solids, Transactions of the Faraday Society, v. 57, p. 1541-1551.

Horiguchi, K. and Miller, R. D., 1980, Experimental studies with frozen soil in an "ice-sandwich" permeameter. Cold Regions Science and Technology, v. 3, p. 177-184.

Jones, R. H. and Lomas, K. J., 1983, The frost susceptibility of granular materials, Proceedings of the Fourth International Conference on Permafrost, Fairbanks, Alaska.

Koopmans, R. W. R. and Miller, R. D., 1966, Soil freezing and soil water characteristic curves, Soil Science Society of America Journal, v. 30, p. 680-685.

Miller, R. D., 1972, Freezing and heaving of saturated and unsaturated soils, Highway Research Record No. 393, Frost Action in Soils, p. 1-11.

Miller, R. D., 1977, Lens initiation in secondary heaving, Proceedings of the International Symposium on Frost Action in Soils, v. 2: Lulea, p. 68-74.

Miller, R. D., 1978, Frost heaving in non-colloidal soils, Proceedings of the Third International Conference on Permafrost, v. 1: Ottawa, National Research Council of Canada, p. 707-713.

Miller, R. D. and Koslow, E. E., 1980, Computation of rate of heave versus load under quasi-steady state, Cold Regions Science and Technology, v. 2, p. 243-252.

O'Neill, K. and Miller, R. D., 1982, Numerical solutions for a rigid-ice model of secondary heave, U.S. Army Cold Regions Research and Engineering Laboratory Report 82-13.

Tuck, B., 1974, Introduction to diffusion in semi-conductors, Institution of Electrical Engineers Monograph Series 16, p. 21.

HYDRAULIC CONDUCTIVITY FUNCTIONS OF FROZEN MATERIALS

K. Horiguchi[1] and R. D. Miller[2]

[1]Institute of Low Temperature Science, Hokkaido University, Sapporo, Japan
[2]Department of Agronomy, Cornell University, Ithaca, New York 14853 USA

A constant-volume permeameter/dilatometer apparatus was used to obtain data for unfrozen water content and hydraulic conductivity for eight materials, including four natural silts, as a function of temperature in the range 0° to -0.35°C with pore water pressure held near zero gauge. After seeding with ice, samples were first chilled and then rewarmed by small steps. Hysteresis effects were extremely large in the first freeze-thaw cycle for all materials but were small in the one natural silt subjected to a second cycle, suggesting possible malfunctions of the seeding procedure in the first round. Typically, hydraulic conductivities fell from values of the order of 10^{-8} m.s^{-1} before freezing to values in the range 10^{-12} - 10^{-13} at temperatures near -0.3°C. After thawing, hydraulic conductivities were higher than before freezing of the thoroughly remolded materials by factors ranging from 1.1 for a natural silt to 60 for a stock sample of Illite.

INTRODUCTION

It is generally accepted that some liquid water persists in frozen soils as a mobile phase at temperatures below 0°C. Many believe that this water has a crucial role in the dynamics of processes induced by freezing and are anxious to have appropriate data for use in mathematical models.

Efforts to measure the apparent hydraulic conductivity of frozen soils by direct methods encounter various problems that compromise credibility of the results, including those to be reported here. The objective in all cases is to measure a coefficient of proportionality (e.g., hydraulic conductivity) presumed to exist between transport and driving force at steady states in which liquid water enters one side of a frozen sample and exits from the other.

Two approaches have been used. In one, solutes are added to the liquid water bathing the two faces, reducing its freezing point to the temperature of the frozen soil (Burt and Williams 1974). Driving forces for transport are imposed by maintaining differences between liquid pressures at the respective faces or differences between their temperatures (Perfect and Williams 1980).

In a second approach, solutes are not used, but ice in frozen soil is prevented from seeding the freezing of adjoining bodies of supercooled water by interposing rigid filters with very fine pores to serve as "phase barriers" that exclude ice but transmit liquid water (Sahin 1973, Miller et al. 1975).

A constant-volume permeameter/dilatometer with phase barriers was used to obtain both unfrozen water content and hydraulic conductivity data for eight materials as functions of temperature when pore water pressure was nominally atmospheric pressure. Results obtained with one of the eight materials were reported previously (Horiguchi and Miller 1980); results obtained with all eight are reported here.

APPARATUS AND PROCEDURE

The constant-volume permeameter/dilatometer and associated apparatus used in these experiments have already been described along with results of tests intended to detect the degree to which flow might pass around the frozen soil or along grain boundaries in the pore ice (Horiguchi and Miller 1980). Results of those tests showed that such flows would normally be negligible. It was also demonstrated that fluxes increased in direct proportion to the imposed hydraulic gradient, in the manner of Darcy's law, using gradients as large as 1500 m/m at low temperatures where conductivity was extremely low.

In use, the permeameter/dilatometer is placed in the dilatometer mode in preparation for each decrement of bath temperature during stepwise chilling of the apparatus immersed in a well-stirred bath, or prior to each increment of rewarming. After equilibration at each step, the system is switched to the permeameter mode. What is reported here as (apparent) hydraulic conductivity is the volume of water emerging from unit area of sample per unit time divided by a hydraulic gradient calculated from the difference in water pressures at the two faces and the thickness of the sample, corrected for head losses across the phase barriers.

In preparation for freezing, the entire apparatus, including sample and seeding tube, was thoroughly flushed with distilled water flowing from an actively boiling reservoir. After a sample had been chilled to a temperature slightly below 0°C, freezing was seeded by way of a port in the side of the sample chamber. The water-filled port led into a length of small-bore water-filled tubing which was induced to freeze by brief exposure to a jet of decompressing CO_2. Ample time (1 - 3 days) was allowed for freezing of all water in the tube and port before the next downstep of bath temperature.

Placement of a sample in its partially

assembled chamber was the most subjective step in the procedure. A thick slurry, thoroughly deaired by boiling at reduced pressure, was spooned into the open water-filled chamber (depth, 3 mm) until it was judged to be exactly filled, whereupon assembly was completed. At the end of an experiment, the sample was recovered, dried, and weighed so that the volume of particles could be calculated. Values reported as porosity, E, represent the ratio: [(Volume of chamber-Volume of particles)/Volume of chamber]. It is obvious that failure to fill the chamber completely, or subsequent consolidation of the sample, would result in ice-filled space that would be reported as if it were a part of the porosity. Such ice would modify pathways for liquid flow. This matter will be discussed further in the light of results obtained.

MATERIALS

Three of the materials used were provided through the courtesy of R. Berg, US Army Cold Regions Research and Engineering Laboratory, Hanover, N.H. These are designated CHENA, NWA, and MANCHESTER. The first two are natural silts from the vicinity of Fairbanks, Alaska, the third is a natural silt from New Hampshire. A fourth natural silt, designated CALGARY, was provided through the courtesy of J. F. Nixon, Hardy Associates, Ltd., and was collected near Calgary, Alberta. V. Snyder, of Cornell University, kindly provided us with a material designated 4-8 μm FRACTION which he had separated from the Manchester sample by repeated decantation. Laboratory stock samples of ILLITE (Grundite) and KAOLINITE (Peerless #2) and a commercial ZEOLITE (Linde 10X, 0.5-5 m) completed the array of materials studied.

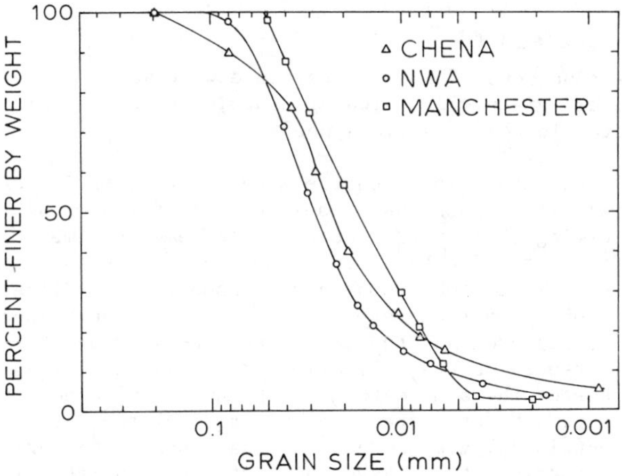

FIGURE 1 Grain analyses of three of the natural silts.

Mechanical analyses of three of the natural silts are shown in Figure 1. Specific surfaces of all but the zeolite were measured at the Institute of Low Temperature Science, Sapporo University, using the BET method and nitrogen gas. Freezing experiments were carried out at Cornell University.

RESULTS

Data from measurements of apparent hydraulic conductivities and specific surfaces are given in Figure 2. Typical data for volumetric unfrozen water content have been plotted in Figure 3.

In Figures 2 and 3, open symbols signify points measured in chilling sequences; solid symbols are for warming sequences. In general, Millipore GSWP (0.10 μm) filters were used as phase barriers. For the 4-8 μm fraction and other examples noted, GSWP (0.22 μm) filters were used. Neither required significant head-loss corrections with the materials used. For one sample of the NWA material, dialysis membrane was used, but its impedance was too high for it to be used in the ice-free range; head loss across the membranes was roughly half of the total even when the soil was frozen and its conductivity was extremely low.

A striking feature of all results for the initial chilling is a very pronounced knee, presumably corresponding to the threshold of ice intrusion. If a similar knee existed during the second freeze-thaw cycle with the Chena material, its location shifted drastically, virtually eliminating the hysteresis that was so pronounced in the first freeze-thaw cycle.

Table 1 gives constants obtained for least-squares fits of exponential equations which approximate data obtained during warming sequences.

TABLE 1 Expressions Describing Results Obtained for Frozen Soils During Warming Sequences.

Unfrozen water content $= AT^B$ m^3/m^3; T = Temperature, °C

Hydraulic conductivity $= CT^D$ m/s

	E	A	B	C	D
CHENA	0.48	$3.6x10^{-2}$	0.63	$8.8x10^{-12}$	3.9
MANCHESTER	0.43	0.112	0.28	$5.1x10^{-11}$	3.7
CALGARY	0.45	0.153	0.13	$1.0x10^{-9}$	1.5
	0.35	---	--	$1.8x10^{-9}$	1.8
4-8 m FRACTION	0.50	$3.3x10^{-2}$	0.61	$7.3x10^{-11}$	3.5
ILLITE	0.67	0.334	0.11	$1.3x10^{-8}$	1.3
	0.65	---	--	$3.4x10^{-9}$	1.2
ZEOLITE	0.62	0.115	0.36	$5.8x10^{-14}$	6.1

The effect of freeze-thaw cycles on the hydraulic conductivity of ice-free soil was very

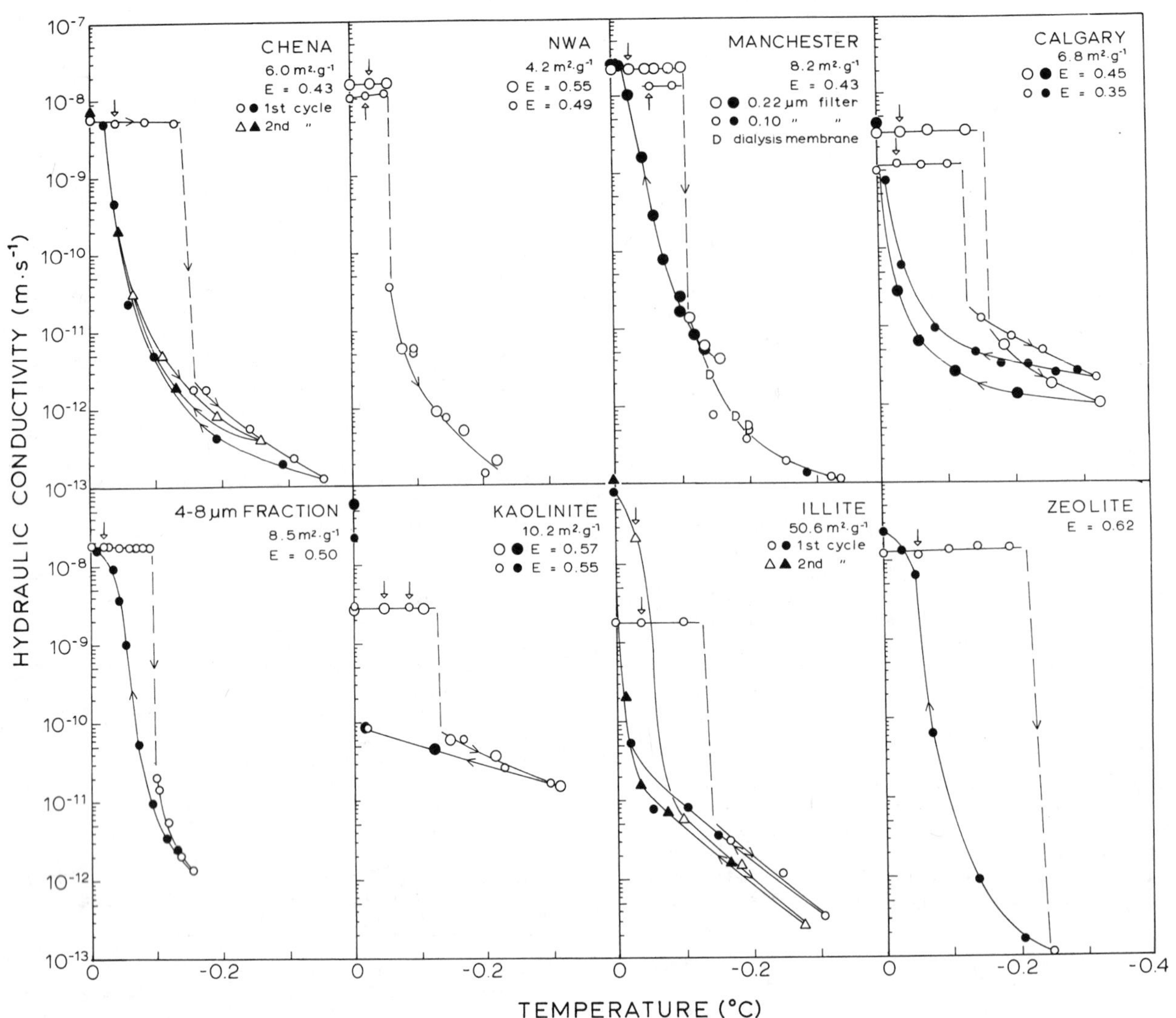

FIGURE 2 Hydraulic conductivities of eight materials as functions of bath temperature when water pressure was near zero gauge. Specific surfaces of each are given. Data measured during chilling are plotted as open symbols. Data measured during warming are plotted as solid symbols.

large for the two clay minerals and was notable, but smaller, in all other materials. Specifically, some (before, after) values, in $m \cdot s^{-1}$, were: Chena (4.1, 7.5) x 10^{-9}; NWA (1.5, 2.4) x 10^{-8} and (1.0, 1.9) x 10^{-8}; Manchester (2.2, 2.5) x 10^{-8}; Calgary, E = 0.45, (3.1, 4.2) x10^{-9}; Kaolinite (2.9, 21.7) x 10^{-9} and (2.6, 60.2) x 10^{-9}; Illite (1.9, 120.) x 10^{-9}; Zeolite (1.3, 2.4) x 10^{-8}.

DISCUSSION

Before the results given in Figures 2 and 3 are accepted at face value it is necessary to examine possible sources of error. Two potential sources of error are obvious. One of these would be a failure of the method of seeding the freezing to work properly. The other would be the appearance of segregated ice within the sample chamber.

The method of seeding was first used in experiments in which freezing characteristics of air-free specimens of soil were compared with drying characteristics of those same specimens after the ice had been melted by warming the samples to room temperature (Koopmans and Miller 1966). In those experiments there was no reason to suppose that the method did not work as intended on every occasion, and this justified use of the same method in the apparatus used here. Unfortunately, examination of the data obtained leaves some doubt that the method worked properly; it may have malfunctioned consistently. Evidence of such malfunction is entirely circumstantial. It has two components, one within the data for the Calgary material, one within the data for the Chena material.

In the two experiments with the Calgary material, one sample was introduced into the chamber at a higher water content (higher porosity) than the other. If the method of seeding brought ice into contact with ice-free soil at a temperature only slightly below 0°C, as

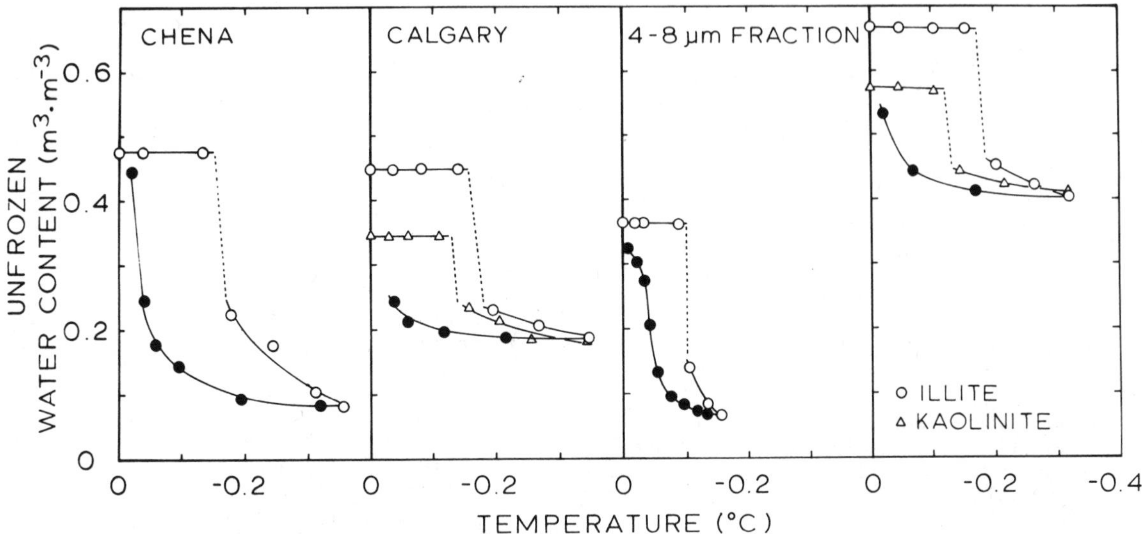

FIGURE 3 Examples of unfrozen water contents deduced from dilatometric measurements.

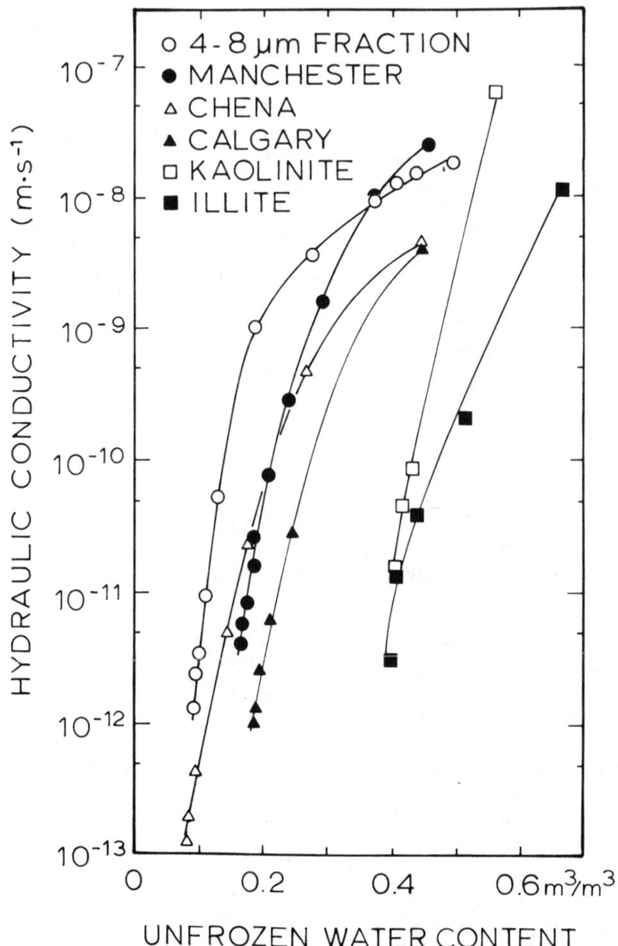

FIGURE 4 Hydraulic conductivities plotted as functions of volumetric water contents.

intended in each case, one would have expected that ice intrusion into the pore system, marked by a knee in the hydraulic conductivity versus temperature curve, would have occurred sooner (at a higher temperature) in the sample with the higher porosity and higher hydraulic conductivity. The reverse seemed to be true. This indication depends, however, upon the validity of a single point on one or the other of the two curves. One is reluctant to give much weight to a single point which has been plotted as it there were no possibility of slight changes in the calibration of the temperature measuring system between the two trials and as if there were no chance of small unrecorded perturbations of bath temperature during the long periods allowed for equilibration of samples before the conductivities were actually measured. If such a perturbation crossed the threshold for ice penetration when the low porosity sample was very close to the threshold, the result could be spurious because of hysteresis as the sample returned to the desired temperature. Hence this inference of possible seeding malfunction is, by itself, of little importance.

There is, however, a peculiarity in the data for the Chena material. This was subjected to two freeze-thaw cycles, but to save time, the sample was not warmed quite to 0°C between cycles, making reseeding unnecessary; the seeding channel presumably remained frozen even if ice disappeared from soil pores. Pore ice had presumably disappeared, to judge from the restoration of hydraulic conductivity, so the second chilling cycle was begun. This time, the pronounced knee that characterized the first freezing was absent. If a less pronounced knee was present, it appears to have been overstepped by the first decrement of temperature.

Is it reasonable to suggest that the knee might be less pronounced in a second freeze-thaw cycle than in the first? The answer must be yes, because if coarse pores are the first to be penetrated by ice, they should tend to enlarge, with a corresponding reduction in the dimensions of smaller pores not yet penetrated by ice. This is what one would expect because the ice pressure is higher than the pressure of water in ice-free pores. If such broadening of the pore size distribution were induced by the first freezing of thoroughly remolded soil, then the hydraulic conductivity of the material after a freeze-thaw cycle ought to be higher than it was before freezing. This was actually observed for all eight of the materials. The induced increase in hydraulic conductivity of the Chena material was not excessive, a factor less than 2, and this is a rather small factor if the same effect were to have virtually eliminated the knee observed in the first cycle. Perhaps the other source of potential error, the appearance of segregated ice, is responsible for camouflaging the presence of the knee in the second cycle.

When a hydraulic gradient is imposed on a soil, it induces a gradient of effective stress and consolidation of soil near the outflow face with the possible appearance of a water-filled void between the upstream face of the sample and the adjoining filter. If the temperature were low enough and ice were present, the void would be ice-filled. It is known that what an observer would perceive to be a stationary ice body can be moving by pressure-induced regelation, making it possible for an observer to record an apparent hydraulic conductivity (Miller et al. 1975). The apparent conductivity would be a complicated function of several properties; the actual hydraulic conductivity of the lens-free soil would be but one of these properties. If, in the course of hydraulic conductivity measurements during the first cycle, such an ice lens were induced between filter and soil at the upstream face by the rather large hydraulic gradients used when conductivity was low, this lens would not melt until temperature reached 0°C unless the sample rebounded strongly during the rewarming steps. It is possible that a residual ice lens induced in this way persisted after the first warming sequence was completed with the Chena material. If so, its presence could have camouflaged the presence of a knee in the second cycle.

If this speculation seems unwarranted, it should be remembered that in an earlier paper (Horiguchi and Miller 1980) this same idea was proposed in the discussion of results obtained with the 4-8 μm fraction. In those studies, it was arranged to reverse the hydraulic gradient from time to time. Each time this was done it was observed that, for a few minutes, thermocouples placed near the inflow and outflow faces of the sample indicated that a temperature differential had been induced but this soon disappeared. It was suggested that this would be consistent with reversal of the location of an induced ice lens. These isolated bits of information do not form a unified and compelling rationale for dismissing the reported hydraulic conductivity values as something other than what they are supposed to be. Neither is it possible to accept those data without reservations.

A completely new and different apparatus, a controlled-stress permeater/dilatometer, has been designed, fabricated, and put into operation at Cornell. It is intended to overcome some of the ambiguities in interpretation of data from the constant-volume permeameter/dilatometer. The new apparatus seems to have its own problems, however, and at this time it would be premature to discuss results being obtained.

CONCLUSIONS

The art of making direct measurements of hydraulic conductivity functions for frozen soils is developing slowly. While measurements of data purporting to be the desired functions can be made, the procedures developed to date require extremely precise and stable control of temperature and long equilibration periods. Even then, it is not certain that the data can be taken at face value until every source of potential error has been recognized and evaluated. The constant-volume permeameter/dilatometer is an attractively simple device, but there are unresolved ambiguities in the interpretation of results. These are probably not serious, but only further work, probably with other forms of permeameters, will resolve the questions that can be raised.

ACKNOWLEDGMENT

We are indebted to R. Berg, USA CRREL, Hanover, N.H., for some of the materials studied and mechanical analyses of the Chena and NWA samples. We are indebted to V. Snyder, of Cornell University, for mechanical analysis data for the Manchester soil and for sharing with us a portion of the 4-8 μm fraction he had isolated from that soil. This work was supported, in part, by National Science Foundation Grant ENG 77-17004 and in part by a subcontract with Battelle-Columbus Laboratories.

REFERENCES

Burt, T. P., and Williams, P. J., 1974, Measurement of hydraulic conductivity of frozen soils: Canadian Geotechnical Journal, v. 11, p. 647-650.

Horiguchi, K., and Miller, R. D., 1980, Experimental studies with frozen soil in an "Ice sandwich" permeameter: Cold Regions Science and Technology, v. 3., p. 177-183.

Koopmans, R. W. R., and Miller, R. D., 1966, Soil freezing and soil water characteristic curves: Soil Science Society of America Proceedings, v. 30, p. 680-685.

Miller, R. D., Loch, J. P. G., and Bresler, E., 1975, Soil Science Society of America, v. 39, p. 1029-1036.

Perfect, E., and Williams, P. J., 1980, Thermally induced water migration in frozen soils: Cold Regions Science and Technology, v. 3, p. 101-109.

Sahin, T., 1973, Transport of water in frozen soil: M.S. thesis, Cornell University, Ithaca, N.Y.

COMPARISON OF TWO-DIMENSIONAL DOMAIN AND BOUNDARY INTEGRAL GEOTHERMAL MODELS WITH EMBANKMENT FREEZE-THAW FIELD DATA

T.V. Hromadka II[1], G.L. Guymon[2], and R.L. Berg[3]

[1]U.S. Geology Survey, Laguna Niguel, California USA
[2]Civil Engineering, University of California, Irvine, California USA
[3]U.S. Army Cold Regions Research and Engineering Laboratory, Hanover, New Hampshire USA

The time- and position-dependent locations of the 0°C isotherm were calculated using two modelling strategies: a domain method and a boundary integral method. Simulations were made for the runway embankment at Deadhorse Airport near Prudhoe Bay, Alaska. The same thermal properties, initial conditions, and boundary conditions were used in both models. Sinusoidal surface temperature variations, dependent upon surface type and exposure, were used in the simulations rather than measured surface temperatures. The positions of the 0°C isotherm determined by the boundary integral method near the time of maximum thaw penetration were essentially the same as those determined by the finite element method, and results from both models agreed closely, within a few centimeters over a total freezing depth of about 2.5 m, with the measured positions. The largest differences between measured and computed positions occurred early in the freezing and thawing seasons. The primary advantage of using the boundary integral method for problems specifically of the type considered herein is that it requires only a few nodal points, so computer simulations can be completed rapidly on a micro computer. If the two-dimensional thermal regime is necessary, the finite element method is most suitable.

During the 1977 and 1978 thawing seasons, the runway at Deadhorse Airport near Prudhoe Bay, Alaska, was improved and paved with an asphaltic concrete pavement. With cooperation from the State of Alaska Department of Transportation and Public Facilities and the Federal Aviation Administration (FAA), USACRREL installed temperature sensors beneath and adjacent to the runway. Subsurface temperatures at some locations are measured manually in liquid-filled access tubes and others are monitored automatically by a data collection platform (Berg and Barber 1982) that transmits information to USACRREL via the ERTS satellite system (McKim et al. 1975). The equipment has been in operation since August 1978.

Two modelling strategies were used to calculate seasonal thaw penetration: (a) a finite element domain method (Guymon and Hromadka 1982) and (b) a boundary integral equation method (Hromadka and Guymon 1982). The time-dependent location of the phase-change isotherm was calculated using both models. In addition, temperature variations at selected positions within the runway embankment were computed using the finite element method. Comparison of calculated temperatures and predicted 0°C isotherm locations with measured data indicate that both numerical modeling approaches are accurate tools in predicting soil thermal response in freezing/thawing environments.

DESCRIPTION OF MODELS

The domain method approximates the well known two-dimensional heat transport equation, which for a freezing or thawing soil is:

$$\frac{\partial(K_x \partial T/\partial x)}{\partial x} + \frac{\partial(K_y \partial T/\partial y)}{\partial y} = C_m \frac{\partial T}{\partial t} - L \frac{\rho_i}{\rho_w} \frac{\partial \theta_i}{\partial t} + C_w \left(v_x \frac{\partial T}{\partial x} + v_y \frac{\partial T}{\partial y}\right) \qquad (1)$$

where x, y = cartesian coordinates
t = time
T = temperature
K_x, K_y = the thermal conductivity of the soil-water-ice mixture
C_m = volumetric heat capacity of the soil-water-ice mixture
L = the volumetric latent heat of fusion of bulk water
ρ_i = ice density
ρ_w = water density
C_w = the volumetric heat capacity of bulk water
v_x, v_y = velocity flux components
θ_i = the volumetric ice content of the soil.

The density parameters are relatively precise for modeling purposes. Although the latent heat parameter is a function of temperature and salinity (Anderson et al. 1973), all of the water is assumed to freeze at 0°C in these simulations. C_w may be regarded as a well defined constant, but the appropriate thermal conductivity and heat capacity of the soil-water-ice mixture are not exactly known and must be estimated. DeVries' (1966) weighting method for estimating these parameters is often employed; i.e.

$$\beta = \Sigma\beta_n P_n \quad (2)$$

where β is the required parameter (K_j or C_m), P_n is the volumetric content of a specific constituent, and n indicates the nth constituent. The velocity flux parameters must be assumed or calculated from a coupled moisture transport model. Since this paper is concerned with heat conduction only, it will be assumed that moisture flux is negligible; consequently, the convective component of Equation 1 is assumed to be zero.

To solve Equation 1, initial and boundary conditions are needed. Initial conditions are of the form:

$$T(t=0) = T(x,y)$$

$$\theta_i(t=0) = \theta_i(x,y) \quad (3)$$

which are usually specified at discrete points in the solution domain. While boundary conditions may be of any form (e.g. a surface energy balance simulation), we will use two types:

$$\frac{\partial T}{\partial n} = 0, \quad t > 0 \quad (4)$$

where n is a unit normal coordinate to the solution domain boundary, and

$$T(s) = N\, T_a(s,t), \quad t > 0 \quad (5)$$

where N is the Corps of Engineers n factor (Berg 1974), s is a tangent coordinate to the solution domain boundary, and T_a is the air temperature, which may be a function of time.

Domain Approach

Commonly used domain approaches include the finite element and finite difference methods. Hromadka et al. (1981) show that an infinity of mass lumped domain numerical analogs may be incorporated into a single matrix expression. The finite element, subdomain, and integrated finite difference schemes are represented as special cases. Depending on the subdomain of integration and the density of the state variable approximation, various domain algorithms may be obtained. These are unified into a single matrix representation called "nodal domain integration," which yields a system matrix similar to the finite element scheme:

$$K\,T + C(\eta)\,\dot{T} = F \quad (6)$$

where K is a square-banded symmetrical conduction matrix that is a function of thermal conductivity and global discretization, $C(\eta)$ is a square-banded symmetrical capacitance matrix that is a function of capacitance parameters and a mass lumping factor (η), $\dot{T}$ and T are vectors of unknown temperatures at discrete points and their temporal derivatives, respectively, and F is a load vector that is a function of the boundary temperatures. Hromadka et al. (1981) give complete details on the derivation of the matrices in Equation 6. Guymon and Hromadka (1982) use this technique to develop a two-dimensional model of coupled heat and moisture transport in freezing or thawing soils. In the analysis presented here, the n factor was set to a value such that a standard Galerkin finite element scheme was used in all domain computations.

Rather than solve Equation 1 in the form shown, the latent heat term is removed and is approximated as an isothermal process (Guymon and Hromadka 1982). Latent heat effects are simulated by a simple control volume approach. A discrete volume of soil is not allowed to reach subfreezing temperatures until the latent heat of fusion of all water in the volume is exhausted. Because this approximation makes it difficult to determine the position of the freezing or thawing isotherm, a pseudo apparent heat capacity approach is used by weighting the diagonal terms of the capacitance matrix. Only the heat transport component of the model was used in the study.

To solve the domain problem, the solution region is discretized into triangular finite elements, as shown in Figure 1. Temperatures are represented by linear shape functions within each triangular finite element.

Boundary Approach

For problems where phase-change effects dominate the solution, the temporal term in Equation 1 may be assumed to be negligible. If one assumes an isotropic, homogeneous medium, Equation 1 reduces to the Laplace equation:

$$\nabla^2 T = 0 \quad (7)$$

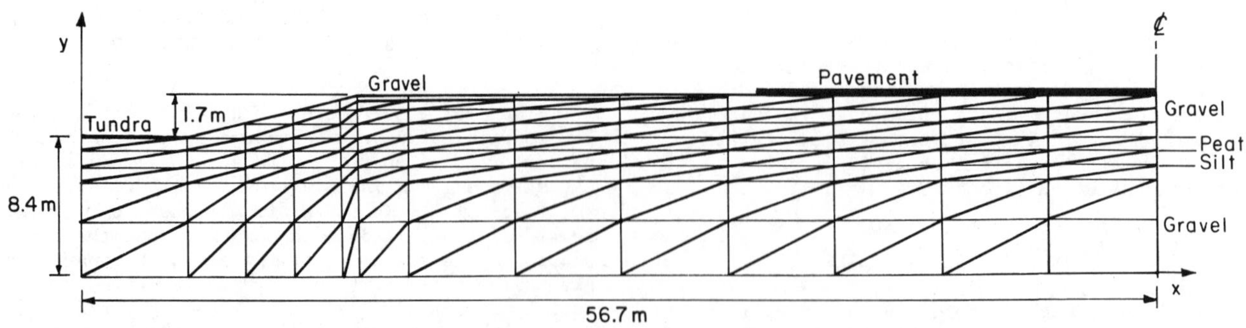

FIGURE 1 Deadhorse runway cross section showing elements and nodes used in the domain solution.

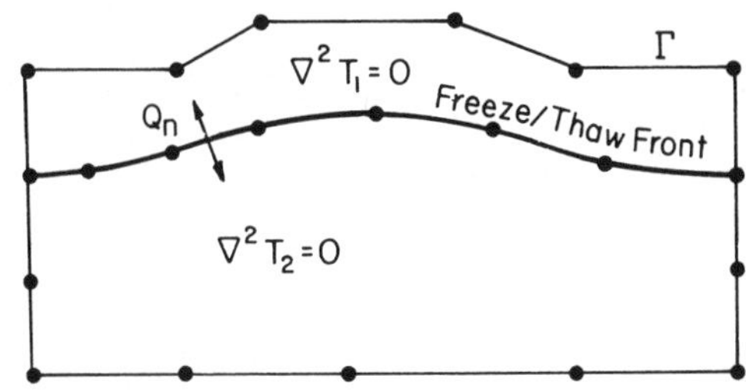

FIGURE 2 Boundaries and nodal locations for a general boundary integral solution.

where again it is assumed that moisture transport is negligible. Equation 7 is assumed to apply in both the frozen and the unfrozen regions, as shown in Figure 2.

Hromadka and Guymon (1982) have shown that boundary integral methods may be applied to geothermal problems involving soil water phase change. They applied the complex variable Cauchy integral theorem to problems where the position of the freezing or thawing isotherm is determined directly. The boundary integral approach estimates heat flux normal to a two-dimensional freezing or thawing surface. The advantage of this method is that it requires much less computer storage and execution time for certain problems than classical domain methods.

The heat flux is given by:

$$Q_x + i\, Q_y = -K \frac{\partial T}{\partial x} - i\, K \frac{\partial T}{\partial y} \tag{8}$$

where Q_x and Q_y are heat flux and $i = \sqrt{-1}$. For any arbitrary closed-contour interior of the solution domain,

$$\int Q_n \, ds = \int Q_s \, ds = 0 \tag{9}$$

where n and s are normal and tangential components of the contour. Equation 9 defines T as harmonic and is related to the harmonic conjugate q by the Cauchy-Riemann equations. The complex temperature is defined by

$$\xi(z) = T(x,y) + i\, q(x,y) \tag{10}$$

where $z = x + iy$. Using Cauchy's theorem, the complex temperature on the boundary may be solved by the contour integral:

$$\xi\,(z_j) = \frac{P_v}{Si} \int \frac{\xi(z)dx}{z - z_j} \tag{11}$$

where z_j is the nth node point on Γ. P_v indicates the Cauchy principle value, and S is the interior line segment angle at node j. Equation 11 is readily solved by assuming $\xi(z)$ is described by straight line segments between each node and by assuming $\xi(z)$ can be represented as a linear polynomial (Hromadka and Guymon 1982). Interior moving freeze/thaw boundaries are located by an isothermal change approximation:

$$L \frac{ds}{dt} = \sum_j (Q_n)_j \tag{12}$$

where $(Q_n)_j$ is a normal heat flux component at a specific location along the phase change front during a time-step Δt. Time-steps of one day or longer may frequently be used because of the time-consuming phase change process in naturally freezing or thawing soils. The solution of Equation 11 requires known temperature or known heat flux conditions on the solution domain boundary, Γ.

The full system of equations is written in the form:

$$K(T,q) \;\; = 0 \tag{13}$$

where K is a fully populated matrix of known coefficients and (T,q) is the vector of the complex temperature ξ_j.

FIELD DATA

Two types of data obtained at Deadhorse Airport were used in this study. The first included properties and dimensions of the soil and pavement layers. The second type included initial and boundary temperatures and the measured subsurface temperatures, which were compared with calculated subsurface temperatures.

Properties and dimensions of the soils and pavement were obtained from boring logs (Division of Aviation 1976) made prior to improving the runway and from test data developed during reconstruction of the airport (Ingersoll et al. 1979). Physical and thermal properties of the materials that were used in the thermal models are shown in Table 1. Parameters shown in Table 1 for density and water content were determined from laboratory samples. Thermal parameters were assumed from USACRREL data.

Upper boundary conditions differed depending upon the horizontal position, but they were developed from measured air temperatures in all situations. Air temperatures were adjusted by n factors (Berg et al. 1978) to obtain the surface freezing and thawing indexes and surface temperatures used in the model simulations. A sinusoidal variation of air temperature coupled with the n-factor approach was used to approximate more closely the type of analysis an engineer usually performs. Consequently, the tests of the two models presented here are conservative.

Approximately 120 thermistors are automatically monitored by a battery-powered Data Collection Platform (DCP) and data are transmitted back to USACRREL via the ERTS satellite. Each thermistor is monitored approximately once every five days, and the temperatures are stored in a computer-accessed file at CRREL. In addition, other subsurface temperature observations were obtained manually three or four times per summer. Temperature observations were plotted in a variety of graphs, i.e. thaw depth (0°C isotherm) vs time, temperatures at specified depths vs time and cross-sections of the runway at various times showing the thermal regime (isotherms).

TABLE 1 Material properties below AC pavement.

Layer	Material	Depth, m	Dry density, g/cm^3	Moisture content, % dry wt.	Thermal conductivity, cal/cm hr °C	Volumetric heat capacity, cal/cm^3 °C	Latent heat of fusion, cal/cm^3
A	Gravel	0.08-0.30	2.00	9.6	30.95	0.354	11.7
B	Gravel	0.30-2.44	1.92	7.0	29.31	0.312	6.8
C	Organic	2.44-3.05	0.80	61.6	18.90	0.400	38.4
D	Silt	3.05-4.11	1.20	50.7	17.71	0.356	46.2
E	Gravel	4.11-10.06	1.68	27.6	29.76	0.368	30.4

MODELING PROCEDURE

Both numerical modeling strategies required discretization of the domain in order to approximate inhomogeneity. In the domain method (Figure 1), 126 nodes and 210 elements were utilized. Included in the element definitions were six parameter groupings (Table 1) that incorporated the various dissimilarities of parameters and initial conditions. The boundary integral solution utilized a rescaled domain so that vertical thermal conductivity was constant above and below the 0°C isotherm (that is, frozen vs. thawed). In the rescaled domain, volumetric latent heat is adjusted to preserve the proper rescaled volumetric properties. The problem chosen is amenable to rescaling. Many heterogeneous domains are sufficiently complicated so that rescaling to arrive at a Laplacian problem is difficult or impossible. The boundary integral solution was based on 28 nodal points with eight nodal points evenly spaced along the phase-change isotherm (Hromadka and Guymon 1982).

Both models used identical specified temperature boundary conditions along the top boundary. The initial temperature distributions were inferred from measured subsurface temperatures. Zero flux conditions were assumed at the bottom and at both sides in both models.

MODEL RESULTS

Results of the domain solution are shown in Figures 3 and 4. Figure 3 shows the computed temperatures (dashed line) at a depth of 4.6 m below the pavement surface and 12.2 m from the runway centerline. Measured temperatures are shown as an envelope (solid lines) for this location. Figure 4 shows the computed thaw depths (dashed line) 12.2 m from the centerline. The measured thaw depths (solid lines) are also shown. The envelope of measured thaw depths results from several temperature assemblies at different locations beneath the pavement. Results shown for the particular embankment regime are typical of results throughout the entire embankment.

Thawing depths predicted by the two modeling approaches are given in Table 2 for modeling day number 155. These depths represent the approximate maximum thawing depths. Table 2 includes two sets of predicted thawing depths from the boundary integral approach, based on step sizes of 6 hours and 3 hours. As can be seen, both modeling approaches agree with each other quite closely. Moreover, both approaches agree well with measured thaw depths. This agreement was achieved in spite of the fact that average sinusoidal air temperature was used as a surface boundary condition in conjunction with the n-factor approach. Soil surface temperatures computed using this approach have a root mean square temperature error of at least 8°C.

In this case, the boundary integral method required significntly less computational effort than the domain method; however, this is not a general

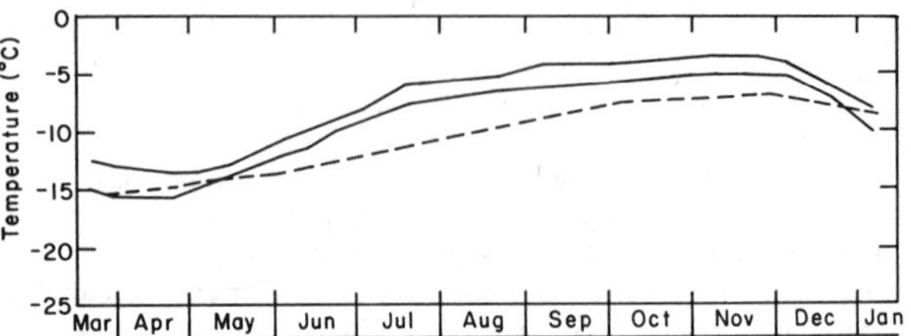

FIGURE 3 Comparison of measured temperatures (solid lines) with those computed in the domain solution (dashed line). Temperatures are approximately 4.6 m (15 ft) below the pavement surface.

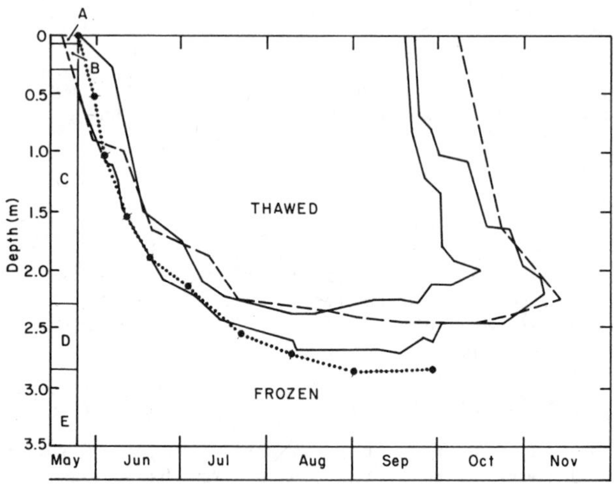

FIGURE 4 Measured (solid lines) and calculated (dashed line represents domain method and dotted line represents boundary integral method) seasonal thaw penetration. Depths beneath the pavement surface. Calculations from the domain solution.

TABLE 2 Thawing penetration depths predicted by numerical models and measured at the airfield (model day 155).

Location (x-coordinate), m	Domain solution, m	Boundary integral solution, m Δt = 6 hrs	Δt = 3 hrs	Measured depths, m
0	0.7	0.7	0.65	0.66
6.1	0.7	0.65	0.64	0.58
7.6	1.1	2.0	2.0	---
9.1	1.7	1.0	1.0	1.09
10.4	2.0	2.3	2.3	1.98
14.0	2.2	2.1	2.2	---
26.2	2.2	2.1	2.2	---
44.5	2.4	2.6	2.7	2.72-3.02
50.6	2.4	2.8	2.7	---
56.7	2.4	2.7	2.8	2.59-3.12

* Δt equals 12.48 hrs; parameter update frequency equals 124.8 hrs.

rule since the boundary integral method has a compact matrix that will lead to less efficient computational effort in some cases. It is concluded that both models can accurately predict the thermal regime of embankments, provided thermal boundary condition and domain solution initial condition information is available.

ACKNOWLEDGMENTS

This research was sponsored by the U.S. Army Corps of Engineers and the Army Research Office under Grant No. DAAG29-79-C-0080 to the University of California, Irvine. Linda Gee, computer assistant, prepared data for the domain simulations.

REFERENCES

Anderson, D.M., Tice, A.R., and McKim, H.L., 1973, The unfrozen water and the apparent specific heat capacity of frozen ground, in Second International Permafrost Conference: Yakutsk, USSR, National Academy of Science, p. 289-294.

Berg, R.L., 1974, Energy balance on paved surface: U.S. Army Cold Regions Research and Engineering Laboratory, Technical Report No. 226.

Berg, R.L., and Barber, M.E., 1982, Prediction of long-term performance of the Deadhorse runway, U.S. Army Cold Regions Research and Engineering Laboratory, Draft Special Report.

Berg, R.L., Brown, J., and Haugen, R., 1978, Thaw penetration and permafrost conditions associated with the Livengood to Prudhoe Bay road, Alaska in Proceedings, Third International Conference on Permafrost: Edmonton, Alberta, National Research Council of Canada, pp. 615-621.

DeVries, D.A., 1966, Thermal properties of soils: Physics of Plant Environment, W.R. Van Wijk, ed.: North-Holland, Amsterdam, p. 210-235.

Division of Aviation, 1976, Engineering geology and soils report, Deadhorse subsurface soils: State of Alaska, Department of Public Works, Anchorage, Alaska.

Guymon, G.L., and Hromadka, T.V., 1982, Two-dimensional numerical model of coupled heat and moisture transport in frost heaving soils, final report: Contract DAAG29-19-C-0080, Army Research Office, Durham, N.C.

Hromadka, T.V. II, Guymon, G.L., and Pardoen, G.C., 1981, Nodal domain integration model of unsaturated two-dimensional soil water flow: development: Water Resources Research, v. 17, p. 1425-1430.

Hromadka, T.V. II, and Guymon, G.L., 1982, Application of a boundary integral equation to prediction of freezing fronts in soil: Cold Regions Sciences and Technology, v. 6, p. 115-121.

Ingersoll, J.I., Collins, C.M., Berg, R.L., and Gardner, C.R., 1979, 1978 field and laboratory studies on the Deadhorse, Alaska airfield: U.S. Army Cold Regions Research and Engineering Laboratory, Technical Note.

McKim, H.L., Anderson, D.M., Berg, R.L., and Tuinstra, R., 1975, Near real time hydraulic data acquisition utilizing the LANDSAT system, in Proceedings of the Conference on Soil-Water Problems in Cold Regions, Calgary, Alberta, p. 200-211.

DETERMINATION OF CRITICAL HEIGHT OF RAILROAD EMBANKMENTS IN THE PERMAFROST REGIONS OF THE QINGHAI-XIZANG PLATEAU

Huang Xiaoming

The Northwest Institute of the Chinese Academy of Railway Sciences
Lanzhou, People's Republic of China

Based on analysis of data obtained from the observation of a railroad embankment in situ, the author presents mathematical formulas for calculating the lower and upper height of railroad embankments. The lower height of the embankment, is the least height that does not affect the natural permafrost table in the subgrade; it can be determined by two empirical formulas: the first expresses the maximum thawed depth in relation to the natural permafrost table, and the second expresses the elevated height of the natural permafrost table in relation to the height of the clay soil embankment. The upper critical height of the embankment is the maximum filling height for the freezing connection of the body of embankment to the natural permafrost table in the first cold season; it can be determined by an empirical formula that expresses the upper height of the embankment in relation to the time of disappearance of frozen ground with temperature of -0.5°C.

"More filling and less outting" has been the main principle used in route selection in permafrost areas, which benefits the stability of foundation for railroad, i.e. less ground surface disturbance and damage might be caused.

The natural permafrost table might be shifted by the disturbance of the thermal equilibrium of the original stratum caused by building embankments. This shift is closely related to different conditions for receiving heat which are relevant to such factors as the height of the embankment, the direction of slope and the construction season. If embankment is too low, the thawing ability of the local climate could lower the natural permafrost table, thus leading to embankment instability, while if the embankment is built in the warm season, the heat carried in by fill material is likely to deplete the remnant freezing ability of the local climate, resulting in a situation where the embankment frost core could not even be formed or could not be formed within a given period. Therefore, in view of thermal stability, there are two critical embankment heights, upper and lower.

The Lower Critical Height of Embankment

The determination of lower critical height, that is the least height that keeps natural permafrost table beneath foundation unchanged, is the key to stabilizing embankment. It depends upon the physical and thermophysical properties of the embankment and the natural stratum of foundation, and the thawing ability of the local climate; it is measured in terms of the thawing rate of the embankment. If the thawing rate of the embankment is higher than that of natural stratum of the foundation, it is worth considering raising the height of embankment or laying a heat insulating layer to prevent the natural permafrost table of foundation from descending.

Effects on Thawing Rate of Fill Material and Cross-section of Embankment

The type of embankment cross-section is advantageous to drain away water content of embankment body itself. As long as the natural permafrost table of foundation rises slightly, water content of soil body beneath foundation will generally be less than that of natural stratum and its annual variation is not significant. According to the data tested, embankment filled up with clayey soil has an annual fluctuation around 3% (Figure 1) and a comparatively stable thermal physical property. Nevertheless, the thawing rate and thawed depth of such embankment are higher than those of natural stratum because of vegetation over it, higher density after ramming, and higher heat conduction property and lower consumption of latent heat of fusion in comparison with those of natural stratum of clayey soil. The estimation made on existing embankments shows that the

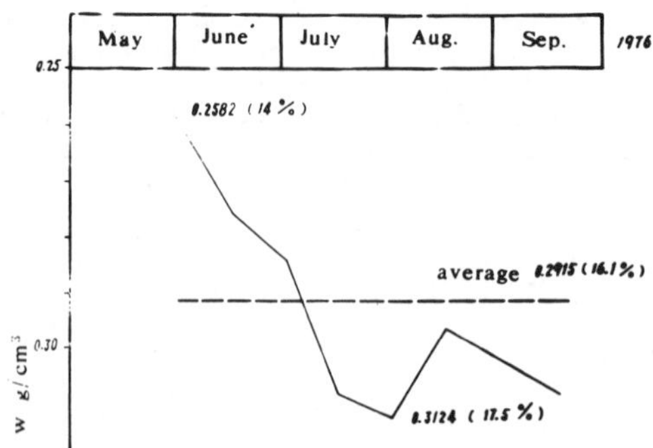

FIGURE 1 The variation curve of average water content in seasonally thawing layer of embankment (0 + 295).

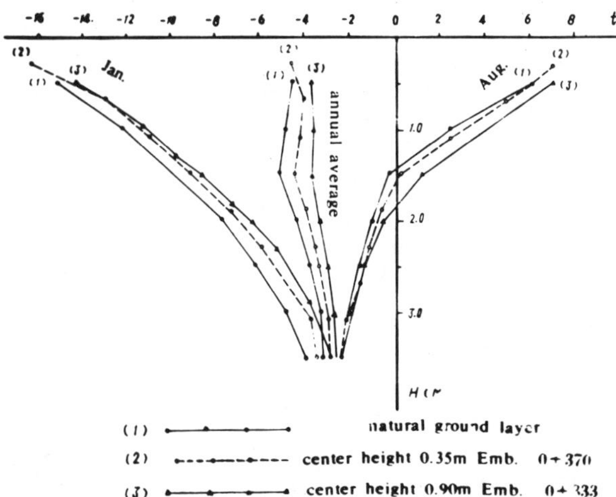

FIGURE 2 A comparison of embankment temperature vs. natural ground temperature.

thawed depth is increased by 20% due to change of thermal physical property.

The additional heating effect on embankment body by horizontal heat flow of embankment slope exposed to the sun makes the slope continually transfer heat towards the center of embankment, thus increasing the thawed depth. For instance, the analysis of the observation data on two low embankments in Mt. Fenghuo region, Qinghai-Xizang Plateau, showed that the annual average temperature of soil body of central plane of embankment—the plane running along the embankment at various depths within 0.5 to 3.5 m—in 1978 was higher than that of natural stratum by about 0.4 to 1.0°C (Figure 2). Corresponding to that of natural stratum, the ground temperature curves of embankment in the coldest and hottest months both shifted toward positive with a thawed depth 29% deeper than the natural permafrost table. Getting rid of the influence caused by physical difference, we found from a rough calculation that the horizontal heat flow would cause the thawing depth of central cross-section of embankment to increase by over 10% (see Figure 2).

The heat effect (mainly the horizontal heat flow of the slope exposed to the sun) on embankment by filling material and type of cross-section again can be reflected by thawing rate. For instance, the thawing rate of embankment center in Mt. Fenghuo region is, in general, $\overline{V}_{emp.}$ = 1.4 to 1.7 cm/day while that of natural stratum is merely $\overline{V}_{nat.}$ = 0.8 to 1.2 cm/day. The former is 1.4 to 1.7 times that of the latter.

For above reasons, it is necessary to determine the minimum filling height of embankment to avoid the thermal degradation of natural permafrost table.

DETERMINATION OF LOWER CRITICAL HEIGHT

Determing H_{min} by Means of Statistics

For a low embankment in center area of permafrost regions of the Qinghai-Xizang Plateau filled with clayey soil, the observed data of embankment height H versus the risen height of natural foundation permafrost table δ are shown in Table 1. The experience equation determined by least square method for calculating the data listed in Table 1 is:

$$\delta = 1.10H - 0.61 \quad (1)$$

provided $H \leq 2.0$ m, where H = height of embankment. The statistical range of natural permafrost table is h_0 = 1.3 to 1.9 m, the relative coefficients being: $\gamma = 0.997$, $\alpha = 0.001$.

By equation 1, when $\delta = 0$, H = 0.55 m. Because of the heating effect of horizontal heat flow at railroad shoulder exposed to the sun, the thawed depth there is deeper than that of foundation for railroad center by 15%. To prevent the ground ice or frozen soil of rich ice beneath the shoulder exposed to the sun from thawing and affecting the stability of founda-

TABLE 1 The Relation Between Filled Height of Low Embankment and Rise or Drop of Natural Permafrost Table

Center height of embankment H (m)	0.00	0.15	0.60	1.13	1.25	1.65	1.75	2.04
Rise or drop of natural permafrost table δ cal (m)	-0.61	-0.45	0.05	0.63	0.77	1.21	1.32	1.63
Rise or drop of natural permafrost table δ test (m)	-0.61	-0.50	0.20	0.61	1.05	1.10	1.10	1.68
Absolute error $\Delta\delta = \delta$ cal $- \delta$ test	0.00	+0.05	-0.15	+0.02	-0.28	+0.11	+0.22	-0.05
Cross-section	Note*	Test Emb. 0+390	Test Emb. 0+370	Test Emb. 0+333	Highway 782	Highway 781	Highway 972	Test Emb. 0+314

Note*: 1. This value would be obtained by taking the mean value of testing embankment in Mt. Fenghuo region 0 + 142, highway embankments; 600 + 000, 628 + 600, 740 + 200, 779 + 600.
2. Test Emb. is the testing foundation for railroad in Mt. Fenghuo region, Highway is the Qinghai-Xizang Highway

tion for railroad, the height of embankment should be controlled by the thawing depth of the mentioned shoulder, i.e.

$$H_{min} \geq 1.15 \times 0.55 = 0.65 \text{ m.}$$

This minimum value is adaptable to foundations for railroad filled with clayey soil along continuous permafrost region with natural permafrost table of 1.3 to 1.9 m.

Determining H_{min} by Conversion

Because the natural permafrost table at the place where embankment is built is the result of weather factors affecting the stratum over a long period, which fully reflects the heat exchange equilibrium relation of weather-ground system, the course of thawing of embankment within the same climate region is bound to be closely related to the thawing process of natural stratum there. The lower critical height of embankment which is adequate for local climate and stratum conditions can be evaluated through converting and simplifying boundary conditions.

Generally, the thawing depth of stratum, h, can be determined by Stephen's equation, i.e.

$$h = \sqrt{\frac{2\lambda \Sigma t\tau}{Q}} \qquad (2)$$

Since heat-absorbing area of low embankment increases only slightly due to its geometrical shape, it is assumed that the heat condition is one-dimensional and the relation between low embankment center and thawing depth of local natural stratum can be expressed by

$$\frac{h'_H}{h_o} = \sqrt{\frac{\lambda_H Q_o}{\lambda_o Q_H}} \sqrt{\frac{\Sigma t\tau_H}{\Sigma t\tau_o}} = \alpha \cdot \beta \qquad (3)$$

or

$$h'_H = \alpha \cdot \beta \cdot h_o \qquad (4)$$

where h'_H is the central thawing depth in meter of embankment on account of one-dimensional heat conduction; h_o is the natural permafrost table; α is revision coefficient (Huang 1979):

$$\alpha = \sqrt{\frac{\lambda_H Q_o}{\lambda_o Q_H}}$$

where λ_H, λ_o, Q_H, and Q_o are the thermal conductivities and the latent heat of volume phase shift of embankment filling material and natural stratum respectively; α is the revision coefficient of surface condition (α = 1.00 as the surface condition of the representative natural table selected coincides with that of embankment, α = 1.06 as the natural ground surface is covered with turf and that of embankment is filled with clayey soil, and α = 1.12 as embankment surface is paved with gravel sand).

Taking account of the horizontal heat flow which actually exists in low embankment slope exposed to the sun, it is necessary to have an additional coefficient of heat flow m_t added to eq 4, then we have

$$h_H = h'_H \cdot m_t \qquad (5)$$

$$h_H = m_t \cdot \alpha \cdot \beta \cdot h_o$$

The value of m_t is generally determined empirically through field data rather than by calculation, which is bound to involve the boundary conditions difficult to be obtained in the design stage. The higher the embankment, the larger the lateral surface absorbing heat, so the larger the ratio between the thawing depth of the shoulder exposed to the sun versus the central thawing depth. For example, the ratio K = 1.25 as H = 2.5 m, while K = 1.40 as H > 6 m. For embankments lower than 1.0 m, K = 1.13 - 1.16, the average value being 1.15, which adequate. The coefficients of equation 4 are listed in Table 2.

Natural permafrost table is kept unaltered provided the maximum seasonally thawing depth of embankment is equal to or less than the sum of the height of embankment and the original natural permafrost table (see Figure 3), i.e.

$$h_H \leq H + h_o \qquad (6)$$

Taking critical condition:

$$h_H = H_{min} + h_o$$

TABLE 2 Coefficients for Calculating Embankment Thawing Depth

Additional coefficient of heat flow, m_t		1.15
Physical revision coefficient		$\frac{\lambda_H Q_o}{\lambda_o Q_H}$
Revision coefficient of surface condition	Embankment surface coincides with natural ground surface	1.00
	Natural ground surface covered with turf, embankment filled with clayey sand	1.06
	Natural ground surface covered with turf, embankment filled with gravel sand	1.12

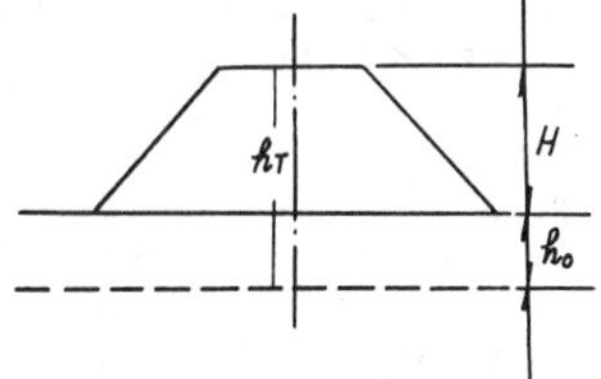

FIGURE 3 Relation between height of embankment and thawing depth. h_T = maximum seasonal thawing depth; H = height of embankment; h_o = natural permafrost table.

Since

$$h_H = m_t \cdot \alpha \cdot \beta \cdot h_o,$$

then

$$H_{min} + H_o = m_t \cdot \alpha \cdot \beta \cdot h_o$$

$$H_{min} = (m_t \cdot \alpha \cdot \beta - 1)\, h_o \qquad (7)$$

The change of physical property of original natural permafrost table caused by shrinkage of foundation soil layer while filling embankment has been cancelled, and the lower critical height of embankment filled in Mt. Fenghuo region H_{min} = 0.63 m. Incidentally, the effects caused by factors as air temperature fluctuation etc. should be taken into account.

THE UPPER CRITICAL HEIGHT OF EMBANKMENT

The maximum filling height of embankment is determined by the local freezing capacity and is closely related to construction technology as well as construction season, etc. For embankment built in cold season, mechanical stability is mainly considered, for fill height is not dominated by the requirements of thermal calculations; while for embankments built in the warm season, the rate of dissipating heat carried into the embankment by fill material must be considered.

The necessary condition for raising the natural permafrost table is that the freezing capacity of climate at the site of project must be higher than the thawing capacity. The remaining freezing capacity after having been consumed in frozen joining of seasonally thawing layer is called remnant (or latent, potential) freezing capacity. The upper critical height tends to be greater with a higher local remnant freezing capacity. Under ordinary conditions, freezing and thawing capacity can be cumulative temperature of ground surface if thawing and freezing depths are considered as functions of it. The annual average temperature t_o is expressed by the following equation:

$$t_o = \frac{\Sigma^+ t\tau + \Sigma^- t\tau}{\tau_w}, \ (^\circ C) \qquad (8)$$

where $\Sigma + t\tau$ and $\Sigma - t\tau$ is the positive and negative cumulative temperature of ground surface respectively, i.e. the sum of average daily positive and negative temperature in a whole year (°C-day) respectively; τ_w is the cycle yearly (365 days); therefore, $t_o > 0^\circ C$ when $|\Sigma^+ t\tau| > |\Sigma^- t\tau|$. At this moment, ground temperature of zero temperature amplitude deep (perennial ground temperature) $t_{cp} > 0^\circ C$ and no permafrost layer is formed. In heat-dissipating or stable-typed permafrost regions, i.e. where the average ground surface temperature annually is less or equal to ground temperature of zero temperature amplitude deep, $t_o = 0^\circ C$ and $t_{cp} \geq 0^\circ C$ when $|\Sigma + t\tau| = |\Sigma - t\tau|$, no permafrost layer is formed or it exists in the form of zero temperature gradient. In this situation no remnant freezing capacity can be utilized. It is only when $|\Sigma + t\tau| < |\Sigma - t\tau|$ that the remnant freezing capacity can be considered for use, and when $|\Sigma + t_\tau| \ll |\Sigma - t_\tau|, t_o \leqslant 0^\circ C$, $t_{cp} \ll 0^\circ C$, i.e. the greater the difference between positive and negative cumulative temperatures of ground surface, the lower the perennial ground temperature, namely, the higher the remnant freezing capacity.

If annual average ground temperature $t_o < 0^\circ C$ is taken as the essential prerequisite for the existence and development of permafrost and since the perennial average ground surface temperature at various ground surfaces in the Qinghai-Xizang Plateau is higher than the annual average atmospheric temperature by 2 to 4°C or more[2], only when the latter is lower than -2 to -4°C, the problem of the embankment's frozen core has to be considered comprehensively concerning water content of soil, orientation and topography. The annual average atmospheric temperature of the continuous permafrost region of Qinghai-Xizang Plateau is, in general, -4 to -7°C. The difference between positive and negative cumulative temperature is fairly great, being 1600-2400°C-day, and the latter is 3-8 times the former. This provides comparatively sufficient remnant freezing capacity, which is able to make newly filled soil layer of certain height frozen and jointed with permafrost.

According to the analysis on the observation data obtained from investigating Qinghai-Xizang Highway and the test embankment, the statistical relationship between the height of embankment filled with clayey soil and the variation of natural permafrost table of the foundation could be expressed by the following equation:

$$\sigma = 0.997H - 0.54 \quad (m) \qquad (9)$$

when the height of embankment H 8 m and the buring depth of natural permafrost table h_o = 1.30 to 1.90 m, eq 9 is suitable.

The critical height of natural table of foundation calculated by the equation that stretches into the embankment body is

$$H > 1.85 - 2.50 \text{ m.}$$

The problem of equilibrium between embankment height and freezing capacity occurs after freezing core stretching into embankment body and it is relevant to construction season.

Effects of Filling Height on Freezing of Embankment

For high embankment built in warm season it requires certain time and condition for heat brought into embankment body with filling material to dissipate. In the course of freezing of embankment built in cold season, which carries on from top to bottom and refreezing from permafrost table upward, the conditions for heat dissipating around foundation is the worst, which frequently results in the forming of frost sandwich of rather high temperature and even thawed sandwich surrounded by permafrost. Statistical material based on tested data shows that the freezing and refreezing rates of embankment are faster (see Table 3) in comparison with those of natural stratum, while the time taken for heat dissipation on the side exposed to the sun near the foundation could be as high as more than 30 months when height of embankment is over 8 m. The relation is roughly:

$$\tau_{month} = 5.89H - 12.1 \text{ month} \qquad (10)$$

Provided

$$8.0 \text{ m} \geq H \geq 3.0 \text{ m}$$

where τ month is the time taken for high temperature frozen soil area (i.e. where ground temperature is higher than -0.5°C) to disappear, in month; H is the central height of embankment filled, in meter.

If the unfavorable effect on freezing by heat brought into embankment body with filling material, the geometrical shape of embankment, as well as the favorable effect on heat dissipation and freezing during contraction of seasonally thawing layer of foundation are not considered, the upper critical height of frozen joining H_{max} of embankment in the first cold season can be converted by the following equation:

$$H_{max} = \left(\frac{\alpha' \cdot \beta'}{1-\mu} - 1\right) h_o' \quad (m) \qquad (11)$$

where α' is the revision coefficient of filling material (Cf Equation 3), i.e.

$$\alpha' = \frac{\lambda_H \, Q_o}{\lambda_o \, Q_H}$$

thermal conductivity adopted is the value under frozen state; β' is the conversion coefficient of frozen depth,

$$\beta' = \frac{\Sigma - t\tau}{\Sigma - t_o\tau_o} \quad ;$$

$\Sigma - t\tau$ is the whole cold season's cumulative temperature ($^{\circ}$C-day) of natural ground surface in the region where embankment is built in; $\Sigma - t_o\tau_o$ is the cumulative temperature of the above natural ground surface ($^{\circ}$C-day) at the moment corresponding to that when frozen joining of seasonally thawing layer is achieved; μ is the coefficient of freezing, namely the ratio of refreezing depth embankment versus the integral frozen depth, which is related to perennial ground temperature and construction season and is taken as 0.25-0.35 according to the data tested; h_o' is

TABLE 3 A Comparison between Ground Temperature and Freezing and Refreezing Characteristics of Embankments of Different Height

Cross-section	Central height filled (m)	Average temperature (Emb.H 1-3 m Nov.15 1975	Average freezing rate of central plane along Emb.(cm/day)	Features of freezing from top to bottom			Features of refreezing from bottom to top (Begin to count in Nov. 15 1975)		
				Freezing depth in this cold season (m)	Starting & ending times of frozen joining (month, date)	Time taken for frozen joining (day)	Refreezing depth (m)	Time taken for refreezing (day)	Rate of refreezing (cm/day)
0+230A	5.4	2.0	3.0	4.5	Oct 3 to Mar 2 next year	150	2.3	107	2.15
0+295A	4.6	1.6	3.3	4.0	Oct 3 to Feb 3 next year	123	1.9	80	2.04
0+314A	3.0	1.2	3.5	3.5	Oct 3 to Jan 12 next year	101	1.1	58	1.55
Natural stratum	0		3.0	1.1	Oct 3 to Nov 11 next year	39	0.32	39	0.82

Note: Isotherm 0°C serves as standard for frozen joining.

the maximum freezing depth of natural stratum in meter, freezing from top to bottom, being about 80% of the depth of seasonally thawing layer.

By the equation the calculated H_{max} for the embankment filled with clayey soil in Mt. Fenghuo region is $H_{max} \leq 7.5$ m, which has been tested and verified by the observation data on testing project.

Effect of Construction Season on Freezing of Embankment

It is proper to build embankment in cold seasons. If construction is carried out in the middle or later stage of warm season, heat exchange of embankment and foundation will be in the worst situation because of the heat stored in the ground as well as the heat brought into it by the filling material. This will unfavorably influence the restoration of the ground temperature and even thaw the ground permafrost (Huang 1980). It is, therefore, imperative to consider the effect on constructing height and stability of foundation caused by thawing compression of foundation permafrost while building embankment in the middle or later stage of warm season. Special attention must be paid to such high embankments beneath whose natural table of foundation there exists ground ice, whose sunward slope lies just on the horizontal slope of the ground surface inclining downward.

Because the heat stored in the ground and that carried into the embankment by fill material are little, the freezing duration is short, and the negative cumulative temperature consumed during joining the upper and lower 0°C isotherms is less, there is more remnant freezing capacity in an embankment built in the early stage of the warm season, which is able to further cool the embankment and its foundation already frozen or favorable for raising the height of embankment. Comparing the data tested, we find that the upper critical height of embankment built in the early stage of warm season is over 1 m higher than that of embankment built in the middle or later stage of the season.

CONCLUSIONS

To prevent the foundation of embankment from thawing subsidence after building it, we have to make the height of embankment higher than the lower critical height in permafrost areas. It is relevant to fill materials and local climate, etc. In center area of permafrost regions of Qinghai-Xizang Plateau, the lower critical height is equal to 0.65 m when the fill materials are clayey soil.

If we want to make the embankment body and natural permafrost table of foundation joint in the first cold season, the height of embankment has to be lower than the upper critical height. It is closely related to the fill materials, local freezing capacity and construction season, etc. The upper critical height for the embankment filled with clayey soil in Mt. Fenghuo region is 7.5 m.

REFERENCES

Huang Xiaoming, 1979, An experience equation for determining the thickness of heat-insulating layer over cuts in permafrost area, Qinghai-Xizang Plateau.

Huang Xiaoming, 1980, Law of variation of man-made permafrost table of embankment built in permafrost region of Qinghai-Xizang Plateau.

Wang Jiacheng, Wang Shaoling and Qui Guoqing, 1979, Permafrost along Qinghai-Xizang Highway: Journal of Geography, v. 84, No. 1.

A GEOPHYSICAL METHOD FOR PERMAFROST SURVEY IN CHINA

Huang Yizhi, Gu Zhongwei, Zeng Zhonggong, He Yixian, and Liu Jinren

Lanzhou Institute of Glaciology and Cryopedology, Academica Sinica, People's Republic of China

Direct-current electric prospecting has been used for permafrost investigations since the end of the 1950's, and remains a major prospecting technique. It has proved to be a useful method for locating the distribution of permafrost and taliks, mapping ground ice, and determining permafrost thickness. Recently, computers have been used to process data from electrical soundings. Experiments in seismic prospecting show that the first arrival refraction can be used to measure the depth of the permafrost table, and shallow reflection can be used to measure the thickness of noncompacted sediments. D.C.-induced polarization, dipole array potential-drop ratio, and radon emanation techniques have been tested and applied to permafrost surveying. The conductivity, acoustical properties, and dielectric constants of frozen soils have also been studied in the laboratory. Future tasks include (1) further research on the characteristics of ground electrical sections of permafrost, to increase the accuracy of quantitative interpretion of V.E.S. curves and to perfect computer data processing, (2) further research on electrical and acoustic properties, and (3) further study of the dynamic characteristics of seismic waves in permafrost.

In China, at the end of the 1950's, D.C. electric prospecting was used for engineering geological surveys on permafrost by engineering departments. At the beginning of the 1960's, the Lanzhou Institute of Glaciology and Cryopedology, Academia Sinica, began their research on geophysical methods for surveying permafrost; D.C. electric sounding was the major prospecting method (Huang Yizhi et al. 1981, 1983). Meanwhile, some colleges and departments of the railroad, highway, and forestry administrations, such as The First Railroad Design Institute, Ministry of Railroad, conducted experiments on geophysical methods for engineering geological prospecting in permafrost areas. In recent years, seismic (Zeng Zhonggong et al. 1982), D.C.-induced polarization (Huang Yizhi et al. 1983), dipole array potential-drop ratio method (Zhou Fucheng 1982), and radon emissions (Yao Baoxing et al. 1982) have been tested and applied to permafrost surveys. The computer has been used for processing of data from electrical soundings (He Yixian 1980). At present, various geophysical methods can be used for locating permafrost, taliks, and thick ground ice layers and determining the position of the top and bottom of permafrost and the thickness of frozen sediments. Preliminary laboratory studies also provided information on the conductivity, acoustical properties, and dielectric constant of frozen soils.

ELECTRIC CONDUCTIVITY, DIELECTRIC CONSTANT, AND ACOUSTICAL PROPERTIES OF FROZEN SOILS

The resistivity of frozen soil is much higher than that of thawed soil because conductive channels are reduced by ice formation. However, even below $0^{\circ}C$ unfrozen water can still play a role as a conductive medium. The increase of resistivity in frozen soil is very steep at temperatures in the 0 to $-2^{\circ}C$ range. A minimum value of resistivity occurs at a water content that corresponds to a materials plastic limit (Gu Zhongwei et al. 1982).

The dielectric constant of frozen soil is considerably lower than that of unfrozen soil, because most of the water that has a high dielectric constant ($\varepsilon=80$) is changed into ice ($\varepsilon=3.2$). The decrease of ε in frozen soil actually depends on the unfrozen water content. The frequency dependence of the dielectric constant for frozen soils is not as strong as that of unfrozen soil, except in the low frequency state. A maximum value of dielectric constant in frozen soil occurs near the thawed materials' plastic limit, which corresponds to the moisture content at which minimum D.C. resistivity would be obtained. Thus, the plastic limit can be a significant turning point for the electrical properties of some frozen soil (Gu Zhongwei and Liu Jinren 1983).

The elastic wave velocity varies with the moisture content. It increases steeply with increasing moisture content in samples when moisture content is less than the plastic limit for fine-grained soils and less than 70% saturation for coarse-grained soils, and then increases slowly with increasing moisture content. When the moisture content goes up to a certain value, between 11% and 20% for the sandy soil, and between the liquid limit and 100% for the fine-grained soil, the wave velocity will remain at a stationary value.

The wave velocity will decrease very slowly with the moisture content increasing and then gradually approaches the wave velocity of pure ice (Liu Jinren and Zeng Zhonggong 1983).

D.C. RESISTIVITY METHOD

D.C. Vertical Electrical Sounding (V.E.S.)

In 1963, V.E.S. was used for permafrost investigation to measure the thickness of permafrost and compare the results with bore hole temperature information at Tumengela, Xizang. This was also done in the Muli, Reshui, Qilian Shan region, and along the Qinghai-Xizang Highway with satisfactory results.

The application of V.E.S. to permafrost surveying covers the following categories (Huang Yizhi et al. 1981).

Locating the distribution of permafrost and taliks. This is done by interpreting the V.E.S. curve types and ρ_s (apparent resistivity) profile maps with the hole data as shown in Figure 1. The V.E.S. curves and the ρ_s profiles in permafrost regions are quite different from those in talik areas.

Investigating the vertical distribution of permafrost. There is a typical V.E.S. curve that sometimes has distortion in the high resistivity part of curve when the permafrost has a so-called double structure in the vertical direction. There is a V.E.S. curve at station 10 on Qumar River along the Qinghai-Xizang Highway. It has been interpreted with the aid of the computer to have seven layers, i.e. ρ_1=69.9Ωm, h_1=0.8 m; ρ_2=152.8 Ωm, h_2=0.46 m; ρ_3=1343 Ωm, h_3=7.39 m; ρ_4=71.2 Ωm, h_4=2.36 m; ρ_5=10 Ωm, h_5=0.28 m; ρ_6=3558 Ωm, h_6=36.85 m; ρ_7=2.5 Ωm. The curve suggests that the fourth and fifth layer form a 2.6 m thick talik.

Investigating taliks below river and lakes. If a complete talik is under a river or lake in permafrost, the V.E.S. curve shows low resistivities, without a high resistivity segment, and an obvious difference from those on permafrost.

Mapping ground ice. Ground ice with its higher resistivity and common position near the upper part of the permafrost section will cause the apparent resistivity at shallow depths to rise steeply. It was found that if the angle between the rising section at the start of the V.E.S. curve and the abscissa is larger than 40°, the curve shows existence of ground ice. A statistical analysis of many V.E.S. curves measured at Fenghuo Shan in Qinghai-Xizang Plateau showed that thickness and buried depth of ground ice can be estimated from an experimental relation developed between the value of electrode spacing at which turning point on the curves occurs and the buried depth of ground ice (He Yixian 1982).

Determining permafrost thickness. Several methods have been used for determining permafrost thickness, namely the so-

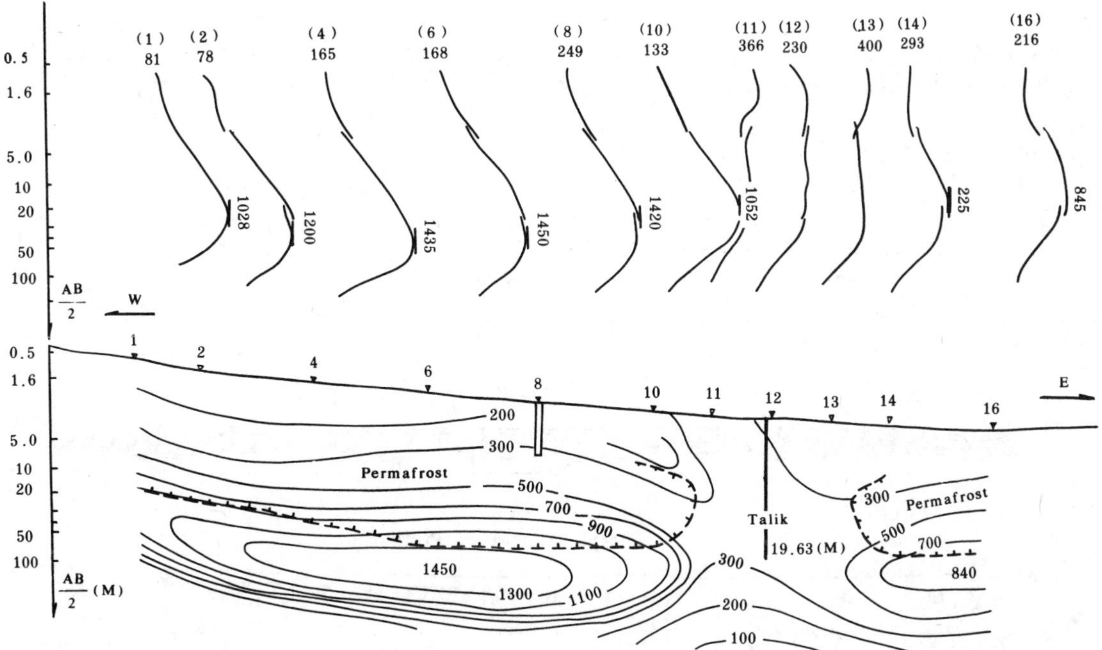

FIGURE 1 V.E.S. profile curve types and ρ_s profile near the 104-II Maintenance Squads of Qinghai-Xizang Highway.

called turning point method, the comparing of master curves, the inverse slope methods, and data processing with the computer. Of these techniques the computer processing method is most efficient due to its automatic approximating method of layer parameters. Many V.E.S. curves measured are matched by the filling program of damping least squares on a DJS-130 computer, made in China, and satisfactory results had been otained. For example, near a hole at Tumengela at the Xizang Plateau, the lower boundary depth of permafrost determined from ground temperature measurement in the hole is 70.3 m, while the depth calculated by computer is 70.7 m.

D.C. Electric Profiling

In 1963, D.C. electrical profiling, usually a double symmetrical four electrode configuration, was used for locating the permafrost and taliks at Tumengela in Xizang. On permafrost, the difference ($\Delta\rho_s$) between the apparent resistivities from both large and small electrode spacings is larger. On taliks, values of $\Delta\rho_s$ are smaller. If $\Delta\rho_s$ is positive and large, permafrost can be considered to exist, and if $\Delta\rho_s$ is negative and small, a talik exists (Huang Yizhi et al. 1981).

Dipole Array Potential-drop Ratio Method

This method has been successfully used for locating islands of permafrost at a mining area in Heilongjiang Province

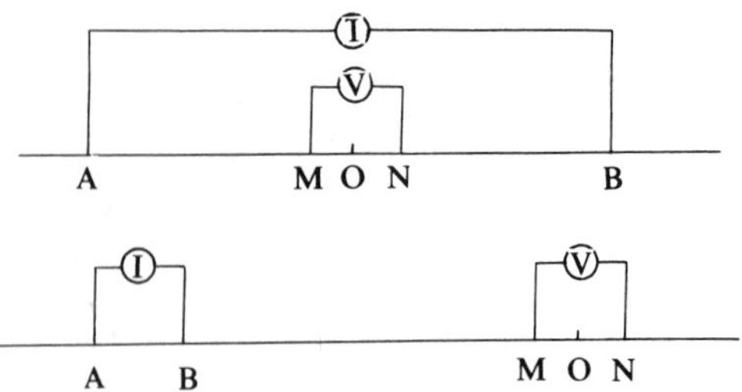

FIGURE 2 The change of AMONB array into ABMON array.

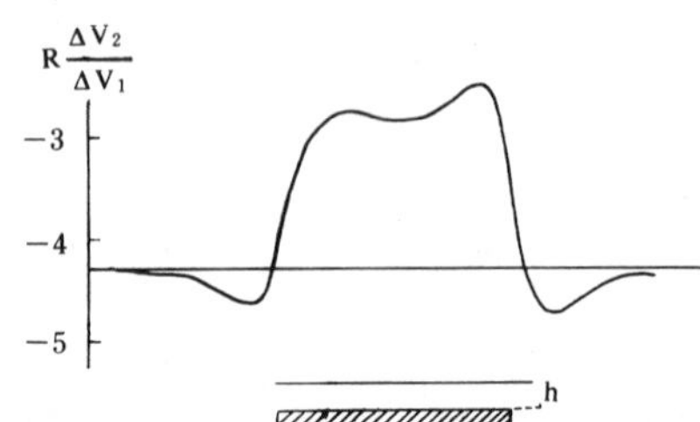

FIGURE 3 A monkcap-shape anomaly due to horizontal wood slab (from model experiment in laboratory).

(Zhou Fucheng 1982). The main points of dipole potential-drop ratio method include: 1) the change of AMONB array of the potential-drop ratio into ABMON array, and then 2) the measuring of potential-drop ratio R between electrodes ON and MO (Figure 2).

This method is excellent for mapping

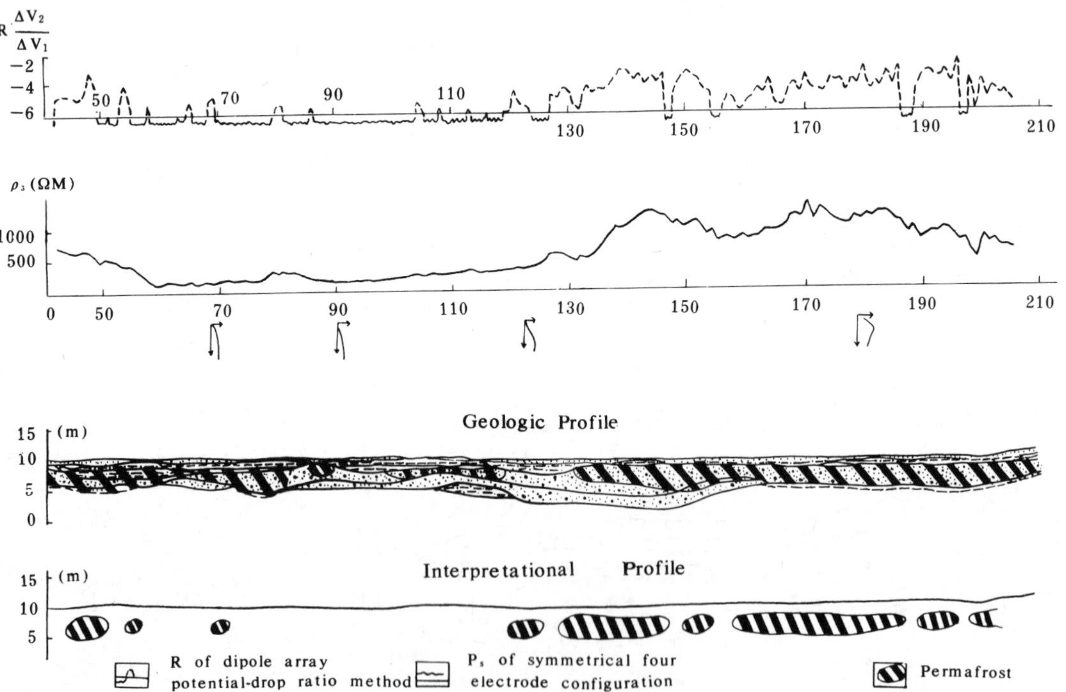

FIGURE 4 A profile of dipole array potential-drop ratio method on island permafrost

by utilizing the inhomogeneity of electrical field. For example, clear monkcap-shape anómaly is obtained on a horizontal wooden plate with high resistivity from a model experiment permafrost performed in the laboratory (Figure 3). A good result has been obtained in lining the distribution of island permafrost (Figure 4). The permafrost boundary shown by the dipole potential-drop ratio method is clearer than that found by electrical profiling. However, the influence of rock inhomogeneity on the dipole potential-drop ratio at the surface is severe, making interpretation of results difficult. Therefore, it is necessary to combine it with electrical sounding.

SEISMIC PROSPECTING

In 1978-79, a seismic prospecting experiment was performed for investigating permafrost at Fenghuo Shan, Qinghai-Xizang Plateau (Zeng et al. 1982, and Zhou and Chen 1980). The experimental results show that the first arrival refraction can be used for measuring the depth of the natural and artifical permafrost table and that shallow reflections can be used for measuring the thickness of sediments. However, attention must be paid to the influence of a seasonal thawed layer on a shallow reflection. On the one hand, the temporary thaw-freeze interface could be mistaken for the top of permafrost because the former is an excellent refractor, through which the wave velocity would change abruptly. On the other hand, due to the large difference of wave impedance at the freeze-thaw interface, most of the energy from the seismic wave is reflected back to the surface, causing the refracted energy from the deeper bedrock surface to be very weak. Thus, the refraction from bedrock surfaces can hardly be recorded by using a hammer as the energy source. It is suggested that this work be done during the period when the earth surface is still frozen, in order to reduce the influence of the thawed material and the shallow reflector.

INDUCED POLARIZATION METHOD

In the summer of 1980, the field experiment of D.C.-induced polarization sounding was undertaken along the Qinghai-Xizang Highway (Huang Yizhi et al. 1983). The experimental results show that IP sounding curves on permafrost are extremely variable. The polarizability of seasonal thaw layers is slightly higher than that of permafrost, and the polarizability curve has little variability in the upper permafrost (which may be related to ground ice) and increases strongly in the lower permafrost section. There is a negative polarizability as the half electrode separation approaches the lower boundary depth of permafrost, and the polarizability of frozen bedrock is generally lower than that of frozen sediments in which the polarizability of coarse-grained soil is generally higher than that of fine-grained soil. The polarizability of coarse-grained soil is more than 2%; finer-grained soil is usually less than 1%. Ground ice has a more steady polarizability of 1.0%-1.2% in general, while pure intrusive ice or the ice-cores of pingos are only 0.2%-0.6%. The difference of D.C. polarizability between permafrost and ground ice is small.

RADON EMISSION SURVEY

On the assumption that permafrost can restrict the migration of radon gas, an experiment to locate local permafrost bodies was undertaken in 1975 (Yao Baoxing et al. 1982). The experimental result shows that values are low in the permafrost area. However, low values were also found on thawed swampy lowland and river taliks due to the absorption of radon gas by water. Concentration at the surface also varies with the lithological characteristics and depth of the bedrock, and the composition and permeability of overlying sediments. Thus, there are two problems to be solved: it is necessary to develop equipment to measure the radon gas concentration of air and water in the field at the same time; and secondly, to study the influence of permafrost properties including variation in ice content on gas.

FUTURE RESEARCH

Permafrost is distributed widely in the alpine regions of China. In order to fully utilize geophysical skills in the study of permafrost, the following future tasks are required.

First, it is necessary to increase the accuracy of quantitative interpretation of V.E.S. curves, and to perfect the computer data processing technique.

Second, research on electrical and acoustic properties should be expanded to offer a broader basis for the interpretation of geophysical data and for the coordination with studies on the phase composition, and physical-chemical process in permafrost.

Finally, the dynamic characteristics of seismic waves in permafrost should be studied, in order to made data interpretation more efficient and accurate.

REFERENCES

Gu Zhongwei et al. 1982, Experimental study for D.C. conductivity of frozen soil, in Proceedings of the Symposium on Glaciology and Cryopedology (Cryopedology), Science Press.

Gu Zhongwei and Liu Jinren, 1983, Preliminary study on dielectric property of frozen soil, in Proceedings of 2nd National Conference on Permafrost, Lanzhou, Gansu People's Press.

He Yixian, 1982, Ground ice investigation with vertical electric sounding at the Fenghuo Shan region in the Qinghai-Xizang Plateau (Abstract) in Proceedings of the Symposium on Glaciology and Cryopedology (Cryopedology), Science Press.

He Yixian, 1980, Numerical interpretation of resistivity sounding data in permafrost regions, in Journal of Glaciology and Cryopedology, v. 2, Special Issue, p. 37-38.

Huang Yizhi et al. 1981, Application of the D.C. vertical electrical resistivity sounding to permafrost surveys, in Collected Papers of Lanzhou Institute of Glaciology and Cryopedology, Academia Sinica, No. 2, Science press.

Huang Yizhi et al. 1983, Application of geophysical method to permafrost survey in China, in Proceedings of 2nd National Conference on Permafrost, Lanzhou, Gansu People's Press.

Liu Jinren and Zeng Zhonggong, 1983, Influence of moisture in frozen soil on velocity of ultrasonic wave, in Proceedings of 2nd National Conference on Permafrost, Lanzhou, Gansu People's Press.

Zeng Zhonggong et al. 1982, Experiment of seismic exploration for permafrost survey (Abstract), in Proceedings of the Symposium on Glaciology and Cryopedology (Cryopedology), Science Press.

Zhou Fucheng, 1982, Geophysical method for locating permafrost region, in Journal of Glaciology and Cryopedology, v. 4, No. 3, p. 41-52.

Zhou Fuchun and Chen Dawei, 1980, Measurement os permafrost table by portable seismograph, in Journal of Glaciology and Cryopedology, v. 2, Special Issue, p. 9-11.

Yao Baoxing et al., 1982, Preliminary application of emanation measurement for permafrost survey (Abstract), in Proceeding of the Symposium on Glaciology and Cryopedology (Cryopedology), Science Press.

SOIL WEDGES ON SPITSBERGEN

Alfred Jahn

Geographical Institute
University of Wroclaw, Poland

In the Van Mijenfjord, in the western part of Spitsbergen, the author found non-sorted polygons built of fine shale debris. Although these polygons were clearly a result of frost cracking, no traces of ice wedges were found down to the permafrost table (depth of 1.2 m). The soil structures found were active-layer soil wedges, similar to the soil wedges known from Siberia. The Spitsbergen findings indicate that such soil wedges can develop in an oceanic periglacial climate and the hypothesis that soil wedges occur only in continental periglacial zones may be incorrect.

Soil wedges (or soil veins) are known chiefly from Siberia. They have been described by Danilova (1956), Katasonov (1973), and Romanovskiy (1973), and are regarded as forms due to frost cracking, especially frost cracking of the active layer (seasonally thawing layer above permafrost). They are commonly termed "ground" or "soil veins", but have also been called "primordial (or primary) soil veins" and "soil-filled veins" in the European literature. One of the first researchers to give an excellent description of ice and soil wedges from Alaska was Leffingwell (1919).

Soil wedges are often confined to the active layer, but perennial ice wedges occur in the permafrost only. [The only exception to this rule is the Antarctic "sand wedge" as defined by Péwé (1959)]. In the Anglo-American literature, based on examples from Alaska or northern Canada, soil that fills the empty space created by a thawing ice wedge becomes an ice-wedge cast in fossil forms but may become part of an active-layer soil wedge if frost-cracking is still active. However, in many cases in the Soviet literature, based on examples from Siberia, the soil or ground wedge (or vein) is synonymous with ice-wedge cast. To avoid confusion, Washburn (1979) distinguished between contemporary and fossil forms of active-layer soil wedges and permafrost soil wedges. The term ice-wedge cast is restricted to replacement of ice wedges and as such is distinct from both contemporary and fossil permafrost soil wedges.

It is not accidental that Siberian researchers have devoted so much attention to soil (ground) wedges or veins. The Siberian summers are warm and the seasonal thaw is deep, whereas in polar climates, soil wedges are rarely identified unless they overlie ice wedges. However, Katasonov (1973) is of the opinion that active-layer soil wedges may exist that have no genetic relationship to ice wedges.

The features observed in the summer of 1978 in the Van Mijenfjord area, Spitsbergen, are of interest and importance (Figures 1 and 2). On a low coastal terrace, soil wedges are found together with the typical polygonal arrangement of frost cracks. The area is underlain by a fine-grained debris of dark Eocene shales originating from frost weathering on the slope and subsequent transport onto the coastal terrace. These flat alluvial fans consist of several generations, from the oldest (A) to the youngest (F), the youngest being on the presently active fan surface (Figure 3). The whole area is a typical "polar desert", almost devoid of vegetation, with single grass tufts in depressed places (Figure 4). The effects of wind action are visible in the form of a stone pavement. The most striking surface features are the regular polygons, with pentagonal and hexagonal forms prevailing (Figure 5). The younger the fan surface, the larger the polygons (Figure 6). The diameter of the youngest forms is 4-6 m, but that of the oldest hardly reaches 1 m. Another regular feature is that the younger forms are flat and the furrows dividing them are shallow, whereas the older forms are convex, resembling small shield-like hills (Figure 7).

Detailed measurements revealed that the mean furrow depth of the youngest polygons is 6 cm, and of the oldest is 18 cm (Figure 3).

The polygons develop as large cells and divide into smaller ones (Figures 8 and 9). Simultaneously, these smaller polygons become convex. The structure of these forms can be traced on the edges that separate the particular fan generations. In places, several cross-sections were made to the permafrost table (Figures 10 and 11), which ranged in depth from 100 to 120 cm, at the beginning of August. It was found that the furrows of the polygons corresponded to the soil wedges, which were situated exactly below the furrows, thus proving their genetic relationship.

The wedge structures are shown in Figure 10, a drawing made in the field, and in the photographs (Figures 11 and 12). It is easy to recognize these structures, since the shale debris consists of flat and elongate particles. The arrangement of the particles gives an exact indication of the direction taken by the forces that formed the soil structures (Figure 12).

Near the soil wedges, the debris is on edge and parallel to the wedge walls, while in the interior of the polygons the clasts lie horizontally against the wedges. No upturned structures due to frost thrusting were noticed anywhere. The soil wedges of this area represent the passive element whose formation is affected only by ice melting and not by frost heaving or thrusting. This observation is important to understanding the development mechanism of these forms.

It is not easy to determine whether the wedge material comes from the polygonal net surface or from the wedge walls, since in both cases it is of the same type, i.e. relatively homogeneous shale debris. There is no difference in color between the wedge and the inner part of the polygon. However, two facts suggest that the greater part of the material within the wedges originates from the polygonal net surface:

(1) There is a marked border between the soil wedge and the inner part of the polygon. The clasts are arranged differently, being on edge rather than horizontal (Figure 11).

(2) The granulometry of the material differs in the two parts of the polygon (Figure 13). That from the polygon interior contains, on average, 40% particles 2-5 mm in diameter, while 60% of the soil-wedge particles are of this size. The cross-sections show clearly that the wedge material is finer than the enclosing material and the material of the polygon surface. This analysis leads to the conclusion that soil-wedge formation here is characterized by a process in which the finer particles are concentrated in the wedges, while the coarser remain on the surface, mainly in the furrows.

It is easy to imagine how this process developed. Soil wedges are started by repeated frost cracking in the winter season. The cracks fill with water during spring and summer thawing, and later freezing gives rise to small ice veins that reach below the active layer and lead to the growth of perennial ice wedges. In summer, however, when the active-layer ice veins have melted, the small cracks trap debris slipping from the polygon surface into the furrow. Only debris with a diameter not exceeding the width of the crack gets into it. The grain-size analyses taken indicate that these cracks were at least 2 mm wide and did not exceed 5 mm; thus they are similar to the generally known Arctic frost cracks. At any rate, I invariably encountered frost cracks of these sizes in early spring on Spitsbergen (Jahn 1975).

Debris up to 5 mm in diameter slips into the cracks and tabular clasts settle edgewise. It would be difficult to imagine this process taking place in a dry environment, so the most reasonable explanation is that the debris is carried into the cracks by meltwater and rainwater.

The soil wedges discussed here have a complex structure. Some are of the "cone-in-cone" type (Figures 10 and 11) and show signs of slight side thrust, such as gentle folds. It is significant that all these features were produced only in the wedges and furrows, whereas the polygons themselves have a simple structure without distortion.

The overall development of the polygons and the soil wedges proceeds as follows. In the first stage large-diameter polygons develop and, as already described, frost cracks result in soil wedges in the surface layer. The regularity of the pattern and the depth of the soil wedges leave no doubt as to their frost origin. Desiccation cracking, which also occurs on the surface of the fans, is entirely different (Figures 10 and 11). The material filling these cracks does not show the regularity that was observed in the soil wedges. The division of the large polygons into smaller pentagons and hexagons is a secondary process (Figures 8 and 9).

A problem that has repeatedly attracted attention is the nature of the effect of ice- and soil-wedge growth on the structure of the enclosing deposits. It is well known that ice wedges act as a thrusting force, even on the frozen soil mass, and cause a characteristic folding or upthrusting of the layers. By contrast, active-layer soil wedges thaw in the summer and lose their dynamic pushing force. In soil wedges, small deformations may occur at wedge walls, but the deformations are considerably smaller than those associated with ice wedges, which tend to produce raised borders running parallel to the furrows (cracks). These raised borders lift the surface of the polygon higher at the edges than in the middle, leading to the concave shape of low-center polygons. The raised borders result from the polygonal block expanding and pressing against the ice and soil wedges as the ground warms. Based on observations made in Alaska and northern Canada, a diagram of the annual dynamic changes of such polygons has been presented (Jahn 1975). This has been confirmed by Mackay (1980), who shows that during the summer, the active layer has a movement away from the center of the polygon toward the furrow (wedge). This movement caused by the thermal expansion of the still-frozen part of the active layer as it warms toward 0°C can distort structures and contribute to the development of low-center polygons.

However, in the Van Mijenfjord high-center polygons, no distortions of the border edges can be found. Any uplift of the ground and surface bulging is related to the center of the polygon. As mentioned above, older polygons are more convex than younger ones, being high-center polygons. Neither the diagram presented earlier (Jahn 1975) or Mackay's (1980) results explain why developing and expanding active-layer soil wedges, which in the Van Mijenfjord area are up to 0.5 m wide in places, do not produce direct distortions in the adjoining soil. The explanation may lie in the type of material and, more importantly, in the lack of vegetation cover. The surfaces of polygon furrows of Siberia, Alaska, and the Canadian Arctic are covered with turf or peat. This vegetation cover seems to retain the effects of horizontal movement in the active layer.

Another cause of the lack of soil wedge impact on the polygon layer is the following: raised borders are known to occur in large polygonal cells up to 20 m in diameter. Even adjoining cells are of different surface shape: small ones are convex and large ones are concave (Jahn 1972). It is very likely that there is a definite limit to the polygon size that determines convexity or concavity. Most of the examples taken from the Van Mijenfjord on Spitsbergen represent the convex nets.

Based on the observations made in the Van Mijenfjord area in the summer of 1978, the following conclusions are drawn:

(1) In addition to ice wedges, which commonly occur under Arctic climatic conditions, there also exist active-layer soil wedges, with widths approaching 0.5 m. They are similar to structures known from cold continental regions, such as Siberia.

(2) These structures are linked to polygonal cells 1-6 m in diameter in fine debris devoid of vegetation.

(3) Active-layer soil wedges need not be dynamic and do not necessarily cause or contribute to thrust distortions in the bordering soil.

(4) The size of frost cracks determines the size of the material deposited in the crack. Water is instrumental in moving debris into the cracks.

(5) Soil wedge structures exist and are developing today in the Van Mijenfjord area. Thus, they are elements of the present Spitsbergen climate. Structures of this type are known from areas of the periglacial Pleistocene zone of Europe and have generally been interpreted as evidence of a climate closer to that of Siberia than to that of Spitsbergen. The Van Mijenfjord area wedges point to a possibly different climatic interpretation of Pleistocene wedge structures.

REFERENCES

Danilova, N. S., 1956, Gruntovyye zhily i ikh proiskhozhdeniye [Ground veins and their origin]. Materialy k osnovam ucheniya o merzlykh zonakh zemnoi kory, v. 3: Moscow, p. 109-122.

Jahn, A., 1972, Tundra polygons in the Mackenzie Delta area: Göttinger Geographisches Abhandlungen, Heff 60, p. 285-292.

Jahn, A., 1975, Problems of the periglacial zone: Polish Scientific Publisher, Warsaw.

Katasonov, E. M., 1973, Present-day ground and ice veins in the region of the Middle Lena. Biul. Peryglacjalny, v. 23, p. 81-90.

Leffingwell, E. K., 1919, The Canning River Region, northern Alaska: U.S. Geological Survey, Prof. Paper 109, 251 p.

Mackay, J. R., 1980, Deformation of ice wedge polygons, Garry Island, Northwest Territories: Geological Survey of Canada, Paper 80-118, p. 287-291.

Péwé, T. L., 1959, Sand-wedge polygons (tesselations) in the McMurdo Sound Region, Antarctica. A progress report: American Journal of Science, v. 257, p. 545-552.

Romanovskiy, N. N., 1973, Regularities in formation of frost fissures and development of frost-fissure polygons. Biul. Peryglacjalny, v. 23, p. 237-277.

Washburn, A. L., 1979, Geocryology. E. Arnold, 406 p.

FIGURE 1 Map of West Spitsbergen showing the location of the areas in which Polish geomorphological and periglacial research was conducted. (1) Research stations. (2) Research areas.

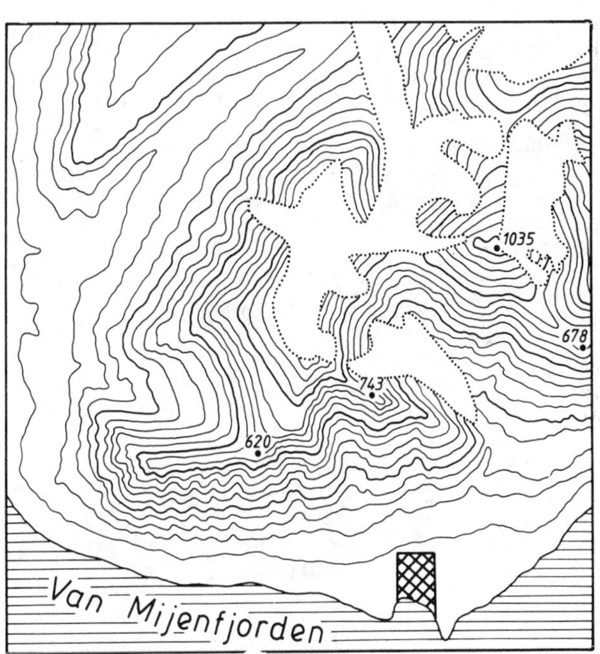

FIGURE 2 Van Mijenfjord research area. Places where detailed studies were carried out are outlined.

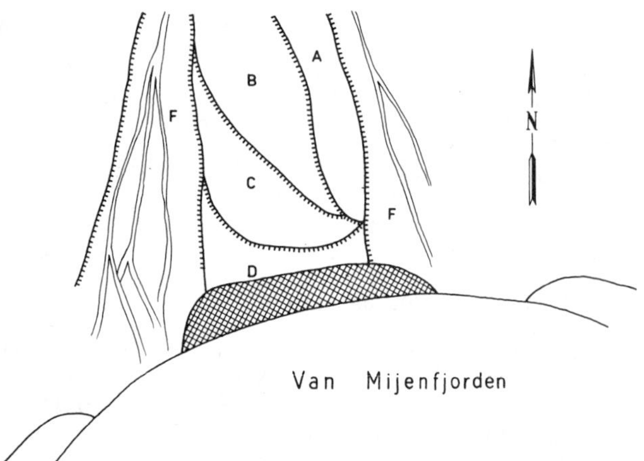

FIGURE 3 Alluvial fan on low Van Mijenfjord terrace. Letters denote successively accumulated fan surfaces, from the oldest (A) to the youngest (F). Coastal rampart marked by checkered pattern.

FIGURE 4 The general view to the polygonal net on the coastal terrace of Van Mijenfjord. In the background is the south coast of the fjord.

FIGURE 5 Oldest part of fan with small polygonal units. Furrows covered with vegetation (surface D).

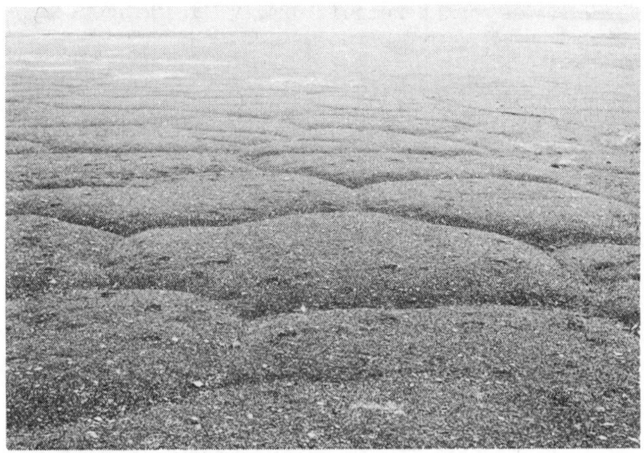

FIGURE 6 Polygon on surface C. Small, convex polygons.

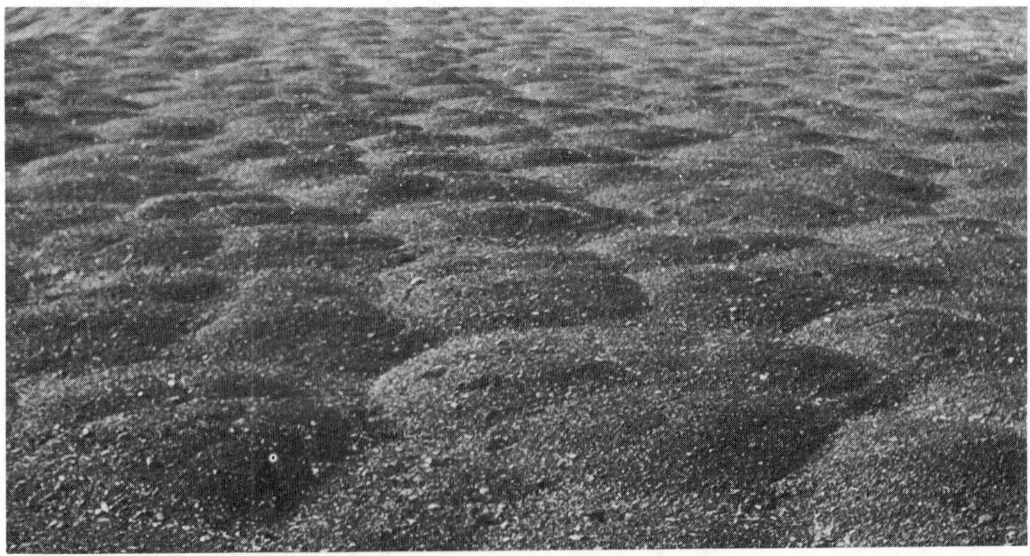

FIGURE 7 Regular, hexagonal polygons, 5 m in diameter.

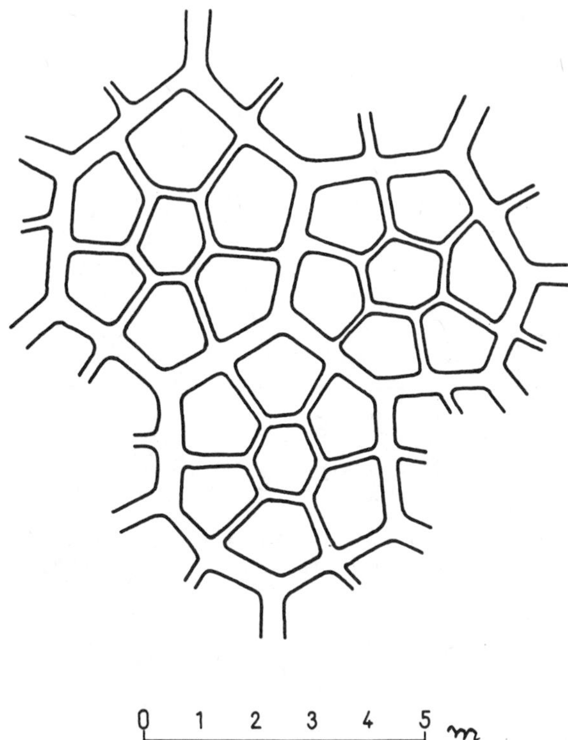

FIGURE 8 Relationship of large polygons (5 m diameter) to small polygons (1 m diameter).

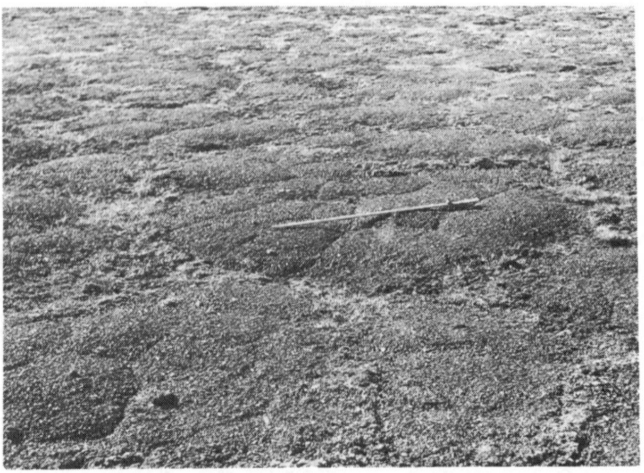

FIGURE 9 Medium-sized polygons dividing into smaller polygonal forms.

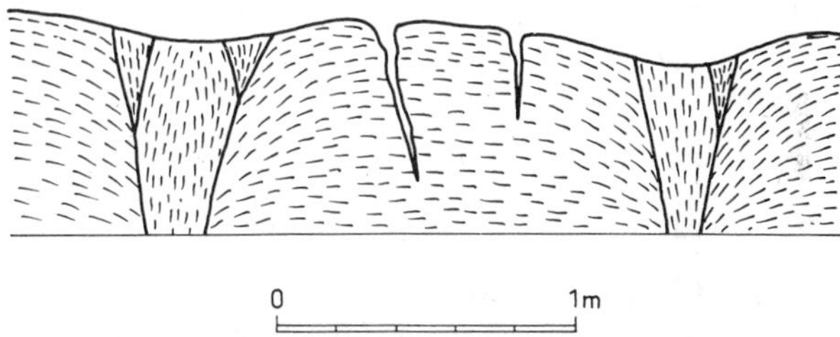

FIGURE 10 Cross-section of polygonal unit and adjoining soil wedges. Empty desiccation cracks at polygon center.

FIGURE 11 Details of structure of polygons and soil wedges. Horizontally stratified debris within the polygons.

FIGURE 12 Cross-section revealing ground wedges. Wedges are dry (better water drainage), so debris is spilling.

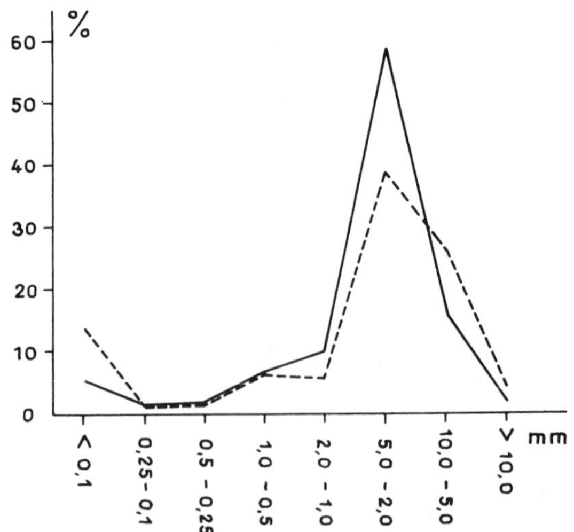

FIGURE 13 Granulometric composition of shale debris from polygon center (broken line) and from soil wedge (solid line).

ACKNOWLEDGMENT

The author wishes to express his appreciation to Dr. A.L. Washburn for his critical reviews and thoughtful comments on the manuscript.

FROST HEAVE MITIGATION AND PERMAFROST PROTECTION FOR A BURIED CHILLED-GAS PIPELINE

H. O. Jahns and C. E. Heuer
Exxon Production Research Company, Houston, Texas 77001 USA

The problems of frost heave and thaw settlement are analyzed for a gas pipeline buried across discontinuous permafrost terrain. First, possible operating temperatures are evaluated, and chilling the gas below the freezing point is concluded to be the best way to minimize geotechnical problems. Next, a reasonable lower limit for the gas temperature is determined. This temperature is then used to calculate freeze bulb sizes for insulated and uninsulated pipe in initially thawed, frost-susceptible soil. The results for insulated pipe demonstrate that the additional heat input caused by the construction surface disturbance can prevent formation of a permanent freeze bulb and, therefore, preclude significant heave. However, if the soil is initially frozen and thaw-unstable, and the gas temperature is near the freezing point, or the pipeline is insulated, then the surface disturbance can also cause thaw settlement. This can be prevented by using a row of heat pipes along both sides of the pipeline. The heat pipes can keep initially frozen ground frozen, and freeze initially thawed ground without causing heave.

INTRODUCTION

In contrast to most crude oil, natural gas can be transported through pipelines at cold temperatures, and power requirements decrease with operating temperature. It is common practice, therefore, that gas pipelines across permafrost terrain are designed to operate at temperatures below 0°C. This allows the pipeline to be buried in thaw-unstable permafrost. Examples are the Tunus-Yakutsk Pipeline in Eastern Siberia and the Alyeska fuel gas line on the North Slope of Alaska. Both of these are in areas of continuous permafrost where the mean annual ground surface temperature is well below the freezing point. Due to the relatively small flow rates in these pipelines, the gas temperature approaches the ambient soil temperature after flowing through only a few kilometers of buried pipe.

The relatively simple concept of an "ambient" temperature gas pipeline in permafrost does not apply to major trunk lines in Alaska and Canada. The reason is twofold: (1) gas transmission lines to southern markets have to cross the zone of discontinuous permafrost where soil conditions below the active layer alternate between frozen and unfrozen at irregular intervals, and (2) the flow rate in a trunk line is usually so large that heat transfer through the pipe wall is no longer sufficient to maintain the gas temperature close to that of the surrounding soil. Significant temperature differences develop due to the Joule-Thomson effect and due to seasonal and climatic variations in ambient soil temperature.

Since both the gas and the soil temperature are changing along the line, and since both frozen and unfrozen ground must be traversed, at least one of the following conditions will of necessity occur: (1) a "warm" pipeline (> 0°C) located in initially frozen ground, or (2) a "cold" pipeline (< 0°C) located in initially thawed ground. Condition 1 is of concern where the permafrost below the pipe is ice-rich, subject to settlement during thawing, while condition 2 is of concern where the soil below the pipe is frost-susceptible, subject to heave during freezing.

Both of these conditions could be avoided by elevating the pipeline above ground. However, despite the potential geotechnical problems, a buried pipeline is preferred because it provides much better security. It also avoids exposure to very cold air temperatures. A winter shutdown could cause condensation of heavier hydrocarbons and could also cause the pipe steel to behave in a relatively brittle manner. Insulation alone cannot prevent these problems.

WARM VERSUS COLD PIPELINE

The gas temperature can be controlled at compressor stations which are located typically 100-200 km (60-120 mi) apart. Thus, the designer can choose between the warm and the cold operating mode for each segment of the pipeline between two compressor stations. The appropriate choice will depend on the extent to which either condition 1 or 2 (discussed above) would occur in the particular pipeline segment, and on the complexity and cost of mitigative measures that would have to be used to prevent excessive pipe deformation due to settlement or heave.

In addition to the cold and warm operating modes, "hybrid" modes have also been proposed with the rationale that, by straddling the freezing point, both thawing and freezing around the pipeline could be minimized. For instance, the gas might be cooled below the freezing point during the summer and warmed above the freezing point during the winter. However, as previously discussed, unavoidable temperature changes in the flowing gas, as well as local climate variations along the pipeline route, make it impossible to achieve a perfectly "balanced" design everywhere along a given pipeline segment. Furthermore, seasonal oscillations of gas temperature around the freezing point may have a very undesirable effect resulting from the repeated formation and disappearance of small temporary thaw bulbs under the pipe in initially frozen ground: where the thaw bulb is water-saturated, the pipe could experience frost jacking due to hydrostatic pressure, even if the soil below the pipe is not frost-susceptible. Thus, the hybrid operating mode does not appear attractive.

There is a similar argument against the warm pipeline mode. A warm pipeline in ice-rich permafrost will have to be insulated to limit thaw settlement. However, where the frozen soil is near 0°C, insulation cannot fully prevent the formation of a small thaw bulb below the pipe during the summer. In order to prevent long-term growth of this thaw bulb, it must be allowed to freeze back, at least partially, during the winter. Thus, the pipe may be subject to seasonal frost jacking.

Another consideration relates to the construction surface disturbance: removal or disturbance of the natural organic layer in permafrost terrain usually causes an increase in the mean annual ground surface temperature. This shift in the thermal balance reinforces the effect of a warm pipeline, but it counteracts the effect of a cold pipeline. Thus, the net thermal disturbance will be smaller for a cold pipeline than for a warm pipeline operating at the same temperature differential relative to the freezing point.

Convective heat transfer by flowing groundwater has a similarly asymmetric effect. The thaw bulb around a warm pipeline in permafrost can be a starting point for thermal erosion unless running water is carefully diverted away from the pipeline. On the other hand, groundwater flow in initally thawed soil will retard or even prevent the formation of a freeze bulb around a cold pipeline.

From these qualitative considerations, it can be concluded that the cold pipeline mode will, in general, be the preferred alternative in discontinuous permafrost terrain; and that the optimum point for the transition from cold to warm pipeline operation will likely be near the southern limit of the occurrence of permafrost along the pipeline route.

Quantitative aspects of the thermal design of a cold pipeline in both permafrost and thawed ground are discussed in the following sections. The first subject addressed is the range of gas temperatures that will be encountered as the gas flows through a hypothetical pipeline segment between two compressor stations. This, then, will provide the boundary condition for an analyşis of the soil thermal regime, and the assessment of special "mitigative" pipeline designs, for both thawed and frozen ground.

In considering special pipeline designs, it should be kept in mind that these will be needed only in certain types of soil and for certain combinations of ground and pipe temperatures. In fact, considerable efforts have been directed in recent years at minimizing the need for mitigative measures in connection with a cold gas pipeline (Myrick et al. 1982). Quantitative methods are being developed to predict pipe deformation due to frost heave and thaw settlement. Over most of a given pipeline route, the predicted deformation should be within allowable limits. Nevertheless, there is a need for special design options that can be used in those areas where the performance of a more conventional pipeline design would be either unacceptable or too difficult to predict.

GAS TEMPERATURE VARIATIONS

Gas temperature variations along the pipeline were calculated using a computer program which models the steady-state thermodynamic and gas dynamic phenomena associated with compressible flow of hydrocarbon mixtures. The program has been compared against field data and has been used to solve many different types of pipeline problems. Using a steady-state program is acceptable because transients within the pipeline are much faster than transients within the soil. The pipeline diameter was 122 cm (48 in.), and the gas properties were based on gas produced from the Prudhoe Bay reservoir.

Heat transfer between the pipeline and the soil was calculated using "effective" heat transfer coefficients and "effective" soil temperatures as input to the pipeline program. Summer and winter values of these parameters were determined with a two-dimensional soil thermal simulator described later. For a range of constant pipe temperatures, the heat flux into the pipe was calculated throughout the year at periodic steady-state. Weather conditions typical of Fairbanks were used for the surface heat balance. The maximum and minimum values of the calculated heat flux were then plotted as functions of pipe temperature, and straight lines were obtained. The effective heat transfer coefficients were equal to the slopes of these lines, and the effective soil temperatures were equal to the abscissa intercepts. All the calculations were done for silt. Depending on the amount of insulation used, the time of year, and whether the silt was initially frozen or thawed, the effective heat transfer coefficient varied from 0.9 to 13.0 W/m-°C (0.5 to 7.5 Btu/hr-ft-°F), and the effective soil temperature varied from -6.7 to 8.3°C (20 to 47°F).

Figure 1 shows some results in the form of idealized gas temperature profiles between two compressor stations 160 km (100 mi) apart. This particular plot was calculated using winter heat transfer coefficients and soil temperatures for initially frozen silt and a flow rate of 56.6 Mm^3/day (2 Gft^3/day) at standard conditions. Initially frozen silt was used to obtain conservatively cool gas temperatures. An operational upper limit of -1.1°C (30°F) on gas temperature was assumed.

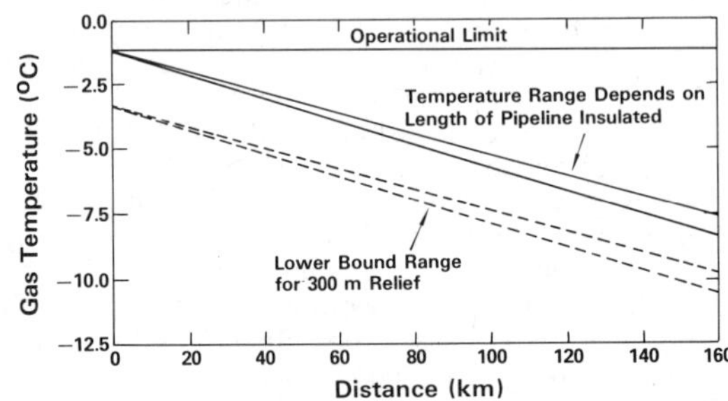

FIGURE 1 Gas temperature profiles.

The solid lines indicate the near-linear temperature drop along the pipeline segment that results from a combination of soil heat transfer and Joule-Thomson effect. On level terrain, the gas temperature would drop from -1.1°C at the compressor station discharge to about -8°C (17.6°F) at the end of the 160 km section, with the small indicated temperature range (-8.3 to -7.6°C) depending upon the fractional length of pipe that is insulated.

Superimposed on this idealized temperature profile will be the effects of adiabatic cooling and warming due to elevation changes on hilly terrain. This effect amounts to about an 0.7°C temperature drop per 100 m elevation increase (0.004°F/ft). The dashed lines in the figure indicate the resulting lower bound gas temperatures for a total relief of no more than 300 m (1,000 ft), which is typical for the bulk of an alignment across interior Alaska. The lower-bound temperatures only occur at the elevation peaks.

During the summer, there will be significant environmental heating of the gas in those portions of the line that are not insulated. This counteracts the Joule-Thomson effect, but gas temperature still decreases with distance. Therefore, the mean annual gas temperature at a particular point will be slightly higher than indicated by the curves in Figure 1, with the amount of the increase depending again on the length of pipe insulated.

It is concluded that at a flow rate of 56.6 Mm^3/day, it should be possible to maintain the mean annual gas temperature above -9.4°C (15°F) along essentially the entire pipeline length in discontinuous permafrost. Exceptions would be a few high passes and ridge crossings. Lower gas temperatures would also result at increased flow rates. But sustained increased flow rates, rather than short-term peaks, are required to significantly affect the mean annual gas temperature. It is this mean annual gas temperature which determines the long-term growth of a freeze bulb below the pipe and the resulting heave rate in frost-susceptible soils.

Improved temperature control can be accomplished by the use of intermediate gas heater stations. For instance, if the gas were heated back up to -1.1°C at the midpoint in this hypothetical segment, the minimum winter temperature with 300 m relief would be raised from about -10.5°C (13°F) to about -6.7°C (20°F). Based on the thermal simulation results reported later in this paper, it does not appear likely that intermediate heating would be necessary at the nominal 56.6 Mm^3/day flow rate. However, heater stations might be considered as a remedial measure in case frost heave problems are encountered unexpectedly, or as a means to allow higher gas throughput with a given set of compressor stations.

In the following sections, it will be assumed that the mean annual gas temperature for design purposes is in the range of -1.1 to -9.4°C, in the absence of intermediate heater stations.

THERMAL SIMULATIONS

Gas pipeline freeze bulb size and heat flux were calculated using a finite-element computer program which models transient, two-dimensional heat conduction with change of phase (Wheeler 1978). Different versions of the program have been used since 1969 on a variety of projects in Alaska and Canada, and the results have been compared against analytical solutions, laboratory data, field tests, and other programs. The program can handle a variety of boundary conditions, including a surface heat balance (Miller 1979).

The pipeline was assumed to be buried in a rectangular ditch below a horizontal ground surface. The pipe diameter was 122 cm (48 in.), and when present, the insulation thickness was 15.2 cm (6 in.). The trench was 2.1 m (7 ft) wide and 2.4 m (8 ft) deep, and pipe centerline was at a depth of 1.5 m (5 ft). The bottom boundary of the finite element grid was 15.2 m (50 ft) below the ground surface, the lefthand boundary passed through pipe centerline, and the righthand boundary was 12.2 m (40 ft) away from pipe centerline.

The native soil was assumed to be either silt or silty sand, and the trench backfill was assumed to be gravel. Thermal properties were calculated using standard correlations (Kersten 1949) and are given in Table 1.

TABLE 1 Soil thermal properties.

Soil	Silt	Silty Sand	Gravel
Dry Density (Mg/m^3)			
Frozen	0.80	1.60	2.08
Thawed	1.60	2.00	2.08
Saturation (% pore volume)			
Frozen	100	100	100
Thawed[1]	100	100	100
Specific Surface Area (m^2/g)	50	10	2.5
Conductivity (W/m-°C)			
Frozen	1.78	3.03	4.15
Thawed	1.52	2.82	3.24
Heat Capacity (MJ/m^3-°C)			
Frozen	1.88	1.85	1.83
Thawed	2.76	2.41	2.34
Heat of Fusion (MJ/m^3)	215	124	69.3
HFUS[2] (MJ/m^3)	200	120	68.2
ALP[2](K)	4.05	3.08	2.43
GAM[2](-)	0.263	0.238	0.249
Thaw Strain (% frozen length)	50	20	0
Heave Strain (% frozen length)	50	20	0

1. If soil initially thawed, thawed saturation reduced to 80%.
2. Unfrozen moisture parameters (Wheeler 1978).
3. Insulation conductivity = 0.0346 W/m-°C.

Unfrozen moisture was based on data reported by McGaw and Tice (1976). Although soil displacement due to freezing or thawing could not be directly accounted for, the thermal properties were adjusted to be consistent with the assumed strains of 50% for silt and 20% for silty sand. These strains were rather arbitrarily selected and were used for both heave and settlement. They were always based on the frozen length and are considered to be relatively high values.

The initial soil temperature distribution was uniform at either 0.06°C (32.1°F) for initially thawed soil or -0.06°C (31.9°F) for initially frozen soil. These temperatures are equal to the "effective" mean annual ground surface temperatures before construction, and they represent worst-case conditions for heave and settlement, respectively. The pipe was a constant temperature boundary, and the side and bottom boundaries of the finite element grid were no-heat-flux boundaries. For most of the simulations, the temperature of the top boundary was assumed to be constant and equal to the "effective" mean annual ground surface temperature after construction. This simplification greatly facilitated the parameter study described below. It was considered adequate to explore long-term pipeline performance and to draw preliminary conclusions. A few simulations used the surface heat balance boundary condition with Fairbanks weather. The results of these simulations were in general agreement with those obtained with the corresponding "effective" constant surface temperature and were used to determine seasonal variations. Additional surface heat balance simulations would be needed to develop the proposed design concepts in more detail.

FREEZE BULB PREDICTIONS

Two typical 30-year freeze bulbs for insulated pipe are shown in Figure 2. The pipeline is buried in silt, and at a temperature of -9.4°C. The larger freeze bulb is for a surface temperature of 0.06°C and has a maximum depth of 3.0 m (9.7 ft) below the bottom of the ditch. The smaller freeze bulb is for a surface temperature of 1.1°C (34°F) and has a maximum depth of 0.6 m (1.9 ft). The lower surface temperature represents an undisturbed surface, while the higher temperature represents a disturbed surface. The figure demonstrates that, by ignoring the surface disturbance, one tends to overestimate the rate of freeze bulb

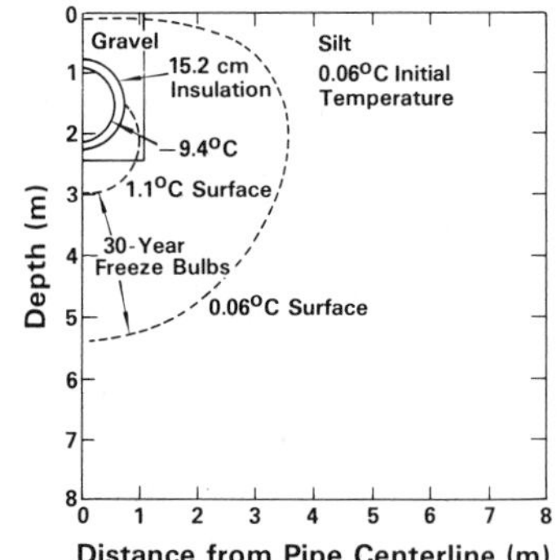

FIGURE 2 Insulated pipe freeze bulbs.

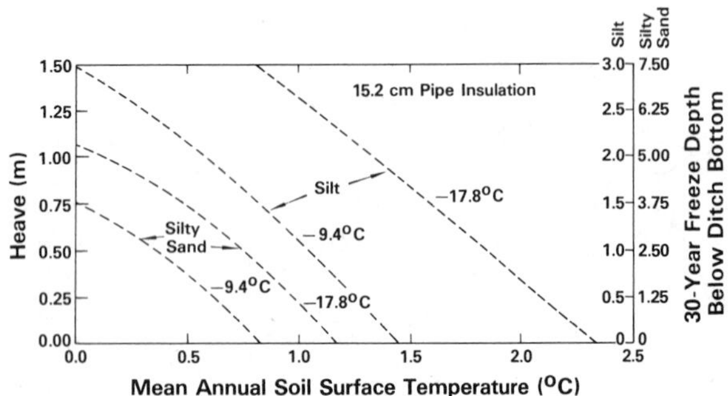

FIGURE 3 Pipeline freeze depth and heave.

growth. Figure 3 illustates this point further.

The plot shows predicted 30-year freeze depth below ditch bottom and the corresponding amount of heave as a function of the mean annual ground surface temperature for an insulated pipe. Heave was calculated by simply multiplying the freeze depth by the assumed heave strain. In the absence of a surface disturbance, the predicted heave is about 1.5 m (5 ft) for silt and a pipe temperature of -9.4°C. But this value decreases rapidly with increasing soil surface temperature, and goes to zero near 1.4°C (34.6°F). This shows the tremendous sensitivity of frost heave predictions for insulated pipe to slight changes in surface temperature.

These relationships will change with local design conditions. Gas temperatures will usually be higher than -9.4°C, reducing the heave potential. Most thawed soils will initially be warmer than 0.06°C. This, too, will reduce the tendency to form a freeze bulb. Soil type is important, too. For instance, in silty sand it appears that a surface temperature increase of only 0.8°C (1.4°F) would be sufficient to eliminate the freeze bulb (see Figure 3).

In contrast, large freeze bulbs will be formed around uninsulated pipe, even in relatively warm soil. For instance, with a pipe temperature of -9.4°C, silt, and surface temperatures of 0.06 and 1.1°C, the predicted 30-year freeze depths below ditch bottom are 9.0 and 8.3 m (29.6 and 27.1 ft), respectively, and are beyond the lower boundary of Figure 2.

The predicted amounts of frost heave are strongly dependent on the assumption regarding the heave strain of the soil. However, it is important to note that predictions for the zero-heave condition are independent of this assumption. The different abscissa intercepts in Figure 3 for silt and silty sand reflect mainly the difference in their thermal conductivity, and not the difference in their heave strain: a no-heave design eliminates the need for accurate prediction of the heave strain in different soil types.

FROST HEAVE MITIGATION

The curves in Figure 3 are an expression of the fact that the heat source represented by the disturbed ground surface can offset the heat sink represented by a well-insulated cold pipeline. The important question then is whether one can count on a mean ground surface temperature increase of 1-2°C (2-4°F) above the freezing point in those situations where the ground is initially thawed. General experience in Alaska, and the results of specific field studies, indicate that this is likely to be the case. For instance, a study by CRREL conducted at a wooded site (Test Section C) near Fairbanks indicated a temperature increase of about 3.0°C (5.4°F) due to complete stripping of the vegetation (Linell 1973). According to Figure 3, an increase this large would prevent a permanent freeze bulb beneath an insulated pipeline for gas temperatures lower than -18°C (0°F). Other measurements by Esch (1982) demonstrate that stripping the organic layer and installing a thin gravel pad can significantly increase heat input to the soil during the summer. Additional field studies, like the above, are needed to quantitatively confirm the effect of the surface disturbance for a wide range of conditions.

Figure 4 is a schematic representation of the envisioned no-heave design with insulated pipe buried in initially thawed silt. To enhance the surface disturbance, the work pad would be extended over the ditch, and the organic layer would be removed a certain distance, perhaps 6 m (20 ft), beyond the ditch and be replaced by gravel to prevent surface erosion. Simulations with the surface heat balance instead of a constant temperature boundary showed that some seasonal freezing below the pipe may occur. However, due to the increased surface temperature, no long-term freeze bulb growth was predicted. For these simulations, the surface heat balance input data were calibrated by matching the 7.0 m (23 ft) thaw depth measured at the CRREL test site over a period of 26 years.

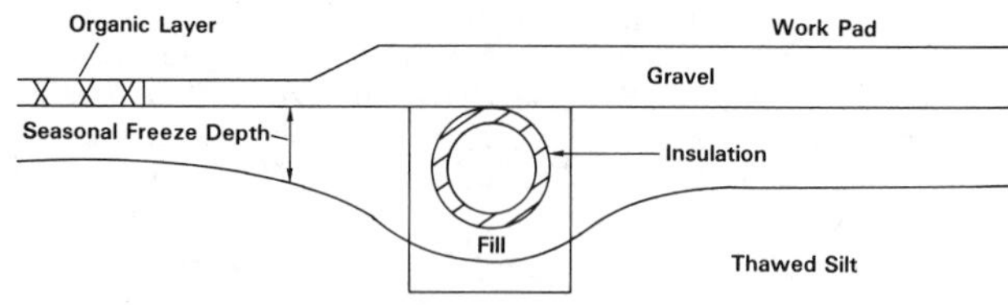

FIGURE 4 No-heave design.

The small amount of seasonal freezing below the pipe could be confined to the ditch backfill by providing perhaps 0.6-0.9 m (2-3 ft) of overexcavation. This design may not eliminate all frost heave everywhere along the line, but the number of exceptions should be limited.

It should be kept in mind that this approach is critically dependent on the quality of the below-ground pipe insulation. Local damage to this insulation could result in the formation of a freeze bulb and local heaving. Reliable methods will be required for field inspection during construction and for repair of damaged insulation.

PERMAFROST THAW PREDICTIONS

The issue of permafrost protection requirements in warm permafrost will now be addressed. Two situations need to be considered: uninsulated pipe in continuous frozen ground, and insulated pipe at transitions between frozen and thawed ground. In both cases, a layer of boardstock insulation could be placed above the pipe to help limit thaw. A 15.2 cm (6 in.) insulation layer was assumed in the calculations.

Figure 5 illustrates the situation for uninsulated pipe. The upper curve indicates the predicted position of the thaw front in degrading permafrost after 30 years with a pipe temperature of -9.4°C. Initial soil temperature was assumed to be -0.06°C, and the mean annual ground surface temperature was assumed to be 2.8°C (37°F) as a result of the construction disturbance. This surface temperature is warmer than that used for the freeze bulb calculations, but it is still slightly less than that indicated by the Linell data for a stripped surface. Taking credit for revegetation by using a lower temperature was considered unconservative. The assumed soil type is silt. A potential dormant period between pipeline construction and startup was not modeled.

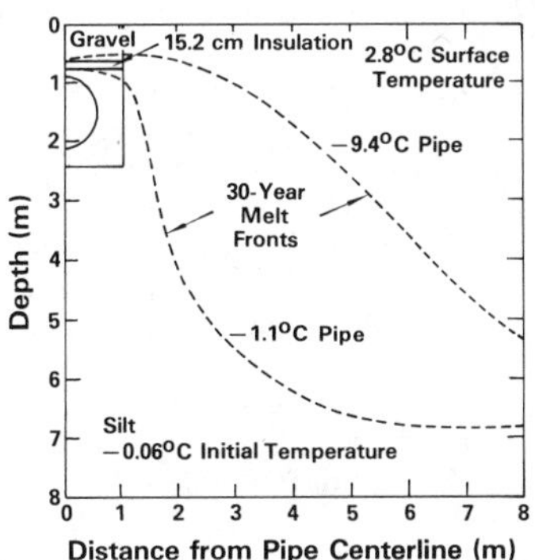

FIGURE 5 Permafrost degradation with uninsulated pipe.

As the figure shows, significant thawing occurs, down to about 4 m (12 ft) below the ground surface at a distance of 6 m (20 ft) from pipe centerline. However, the cold pipeline is effective in maintaining a substantial frozen region around the pipe, assuring its long-term stability.

The size of the frozen region is reduced as the pipe temperature is increased. At a pipe temperature of -1.1°C the remnant freeze bulb would extend less than 2 m (6 ft) from pipe center, which might not be large enough in view of the fact that the actual freeze bulb configuration may be smaller than the calculated one due to groundwater flow and soil settlement. Therefore, the pipeline operating temperature may need to be maintained below about -2°C (28°F).

Figure 6 shows the corresponding situation for insulated pipe. The pipe is now essentially eliminated as a heat sink in the ground and has very little effect on permafrost degradation. For a pipe temperature of -1.1°C, the 30-year melt front is almost 7 m (23 ft) below the surface. This is close to the thaw depth measured at CRREL Test Section C. With the assumed 50% thaw strain, the predicted thaw subsidence would be on the order of 3.5 m (11 ft). Furthermore, in this case, there is not much benefit to be derived from cold gas temperatures, as shown by comparison between the curves for -1.1 and -9.4°C pipe temperatures. This means that insulating the pipeline to prevent heave could lead to settlement where permafrost is present.

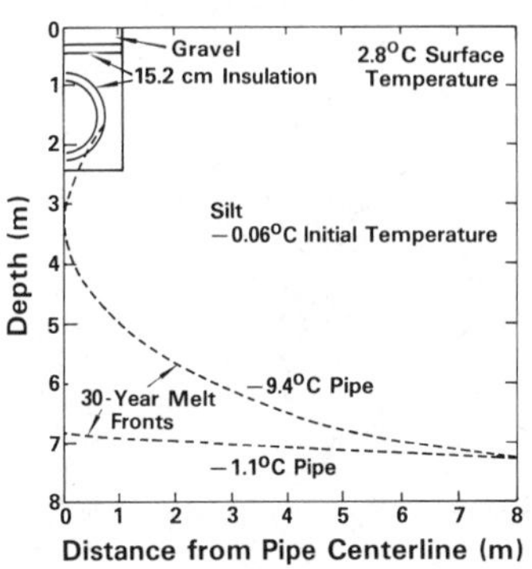

FIGURE 6 Permafrost degradation with insulated pipe.

PERMAFROST PROTECTION

The problem of permafrost degradation was anticipated in the design of the Alyeska pipeline. It was solved there by the use of thermal piles. These are piles equipped with heat pipes which remove large amounts of heat from the ground during the winter to counteract the gradual warming of the soil due to the surface disturbance (Heuer 1979). A similar approach could be used for the gas pipeline.

Figure 7 shows a schematic representation of the proposed mitigative mode for permafrost protection. A row of free-standing heat pipes is placed on either side of the pipeline and embedded to a depth of perhaps 6 m or more. Permafrost degradation will occur under the work pad as it does now under the Alyeska work pad, but near the pipeline the ground would remain frozen, including the ditch backfill below the boardstock in-

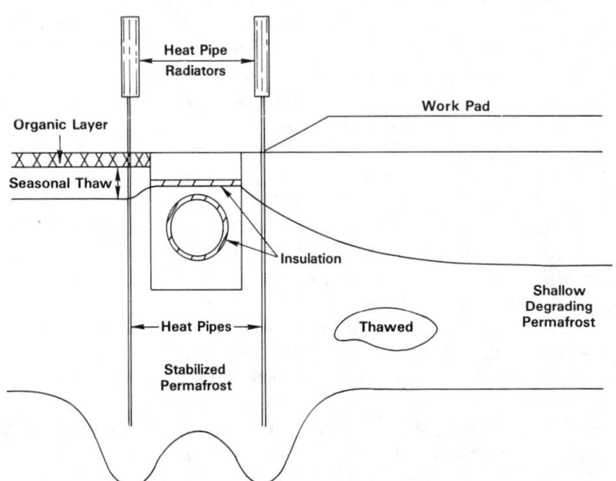

FIGURE 7 Heat pipes for permafrost protection.

sulation. In marginal permafrost, the heat pipes could also be counted on to freeze thawed ground in the vicinity of the pipe to prevent the possibility of frost heave from a freeze bulb around the pipe. Experience with Alyeska and other applications of thermal piles indicates that the heat pipes themselves are generally not subject to heaving, even in thawed ground.

The rationale for the effectiveness of heat pipes in counteracting frost heave is based mainly on geometrical considerations. The long, vertical freeze bulbs around the heat pipes are inherently less conducive to vertical heave than the freeze bulb below a horizontal gas pipeline. Also, before the individual freeze bulbs coalesce, they would reduce the supply of water for ice lens formation directly beneath the pipeline. Thus, heat pipes may also be considered as a means for frost heave mitigation in areas where they are not needed for permafrost protection. For instance, heat pipes might be used as a remedial measure to arrest unexpected frost heave during pipeline operation.

The appropriate spacing of heat pipes along the pipeline may be about one pair every 3 m (10 ft). This spacing would allow the individual heat pipe freeze bulbs to coalesce after one or two winters. Thus, about 300 pairs of heat pipes may be required per kilometer of pipeline. This is about three times the number for Alyeska's thermal pile design, which was four heat pipes per pipeline support every 18 m (60 ft). However, the length of gas pipeline requiring this special protection is expected to be small, so that the total number of heat pipes should be much less than Alyeska's 120,000.

CONCLUSIONS

The optimum operating temperature of a large, buried gas pipeline crossing discontinuous permafrost terrain containing both thaw-unstable frozen soil and frost-susceptible thawed soil is slightly below the freezing point, just cold enough to prevent permafrost degradation in the vicinity of the uninsulated pipe. Calculations for a 122 cm (48 in.) diameter pipeline with 56.6 Mm^3/day (2 Gft^3/day) throughput and 160 km (100 mi) spacing between compressor stations indicate that it should be possible to maintain a mean annual gas temperature between -1.1°C (30°F) and -9.4°C (15°F). For this temperature range, the increased heat input resulting from the construction surface disturbance can be the same order of magnitude as the cooling effect of the pipeline. Where the soil is initially frozen, the surface disturbance can cause thaw settlement problems. However, where the soil is initially thawed, the surface disturbance can be beneficial. It can help prevent heave of the pipeline.

If the pipeline is well insulated in initially thawed soil, then only a slight increase in mean annual ground surface temperature is needed to prevent formation of a permanent freeze bulb. Without a permanent freeze bulb, no significant heave can occur. To take advantage of this effect, the pipeline design must enhance and guarantee the surface disturbance for the life of the pipeline: the organic layer should be removed from a strip perhaps 6 m (20 ft) wide on either side of pipe centerline and be replaced with gravel to prevent erosion and revegetation. One advantage of this approach is that a detailed understanding of frost heave and the mechanical interaction between a pipeline and heaving soil does not have to be developed. The design can be based on thermal considerations alone.

Thaw settlement can occur in initially frozen soil if the gas temperature is relatively warm, or if the pipeline is insulated to prevent heave. The latter is likely at transitions between frozen and thawed soil, and it may not be practical to accurately map all these transitions. Heat pipes could be used in the general areas where transitions are expected. The heat pipes would be arranged in two rows, one on either side of the pipeline. They can keep initially frozen ground frozen, and freeze initially thawed ground without causing heave. Therefore, the design is not sensitive to the initial thermal condition of the soil.

The above results indicate that an enhanced surface disturbance, thick insulation, and heat pipes are promising design concepts for a chilled-gas pipeline. However, additional study is needed to develop these concepts in more detail. The three major areas are: (1) field tests to compare the soil thermal regime for several pairs of adjacent disturbed and undisturbed surfaces, (2) field and laboratory tests to develop a reliable pipeline insulation system, and (3) field tests and computer simulations to study the performance of an array of heat pipes. Work in these and other areas is currently underway (Myrick et al. 1982).

REFERENCES

Esch, D. C., 1982, Thawing of permafrost by passive solar means, in French, H. M., ed., Proceedings of the Fourth Canadian Permafrost Conference: Ottawa, National Research Council of Canada, p. 560-569.

Heuer, C. E., 1979, The application of heat pipes on the trans-Alaska pipeline, CRREL Special Report 79-26: Hanover, New Hampshire, Cold Regions Research and Engineering Laboratory.

Kersten, M. S., 1949, Laboratory research for the determination of the thermal properties of soils: Engineering Experiment Station, University of Minnesota.

Linell, K. A., 1973, Long term effects of vegetative cover on permafrost stability in areas of discontinuous permafrost, in Permafrost--The North American Contribution to the Second International Conference, Yakutsk: Washington, D. C., National Academy of Sciences, p. 688-693.

McGaw, R. W., and Tice, A. R., 1976, A simple procedure to calculate the volume of water remaining unfrozen in freezing soil, in Proceedings of the Second Conference on Soil Water Problems in Cold Regions, Edmonton: Washington, D. C., American Geophysical Union, p. 114-122.

Miller, T. W., 1979, The surface heat balance in simulations of permafrost behavior: Journal of Energy Resources Technology, v. 101, p. 240-250.

Myrick, J. E., Isaacs, R. M., Lin, C. Y., and Luce, R. G., 1982, The frost heave program of the Alaska Natural Gas Transportation System: 103rd Winter Annual ASME Meeting, Phoenix.

Wheeler, J. A., 1978, Permafrost thermal design for the trans-Alaska pipeline, in Wilson, D. G., Solomon, A. D., and Boggs, P. T., eds., Moving Boundary Problems: New York, Academic Press, p. 267-284.

Unit Conversions: °C=5/9(°F-32), 1°C/m=0.549°F/ft, 1cm=0.394in., 1km=0.621mi, 1m=3.28ft, 1m^3/day=35.3 ft^3/day, 1Mg/m^3=62.4lb/ft^3, 1MJ/m^3=26.8Btu/ft^3, 1MJ/m^3-°C=14.9Btu/ft^3°F, 1W/m-°C=0.578Btu/hr-ft-°F

AN ANALYSIS OF PLANT SUCCESSION ON FROST SCARS 1961-1980

Albert W. Johnson[1] and Bonita J. Neiland[2]

[1]Department of Biology, San Diego State University
San Diego, California 92182 USA
[2]School of Agriculture and Land Resource Management, University of Alaska
Fairbanks, Fairbanks, Alaska 99701 USA

Plant succession on frost scars was studied at Ogotoruk Creek, Alaska. In 1961, 326 frost scars were marked for long-term study. The initial physical and biological measurements were repeated several times. In 1965, four transects were established and later revisited to learn if new frost disturbance is occurring. Frost scars at Ogotoruk Creek are convex, primarily oval areas of fine-grained soil with a modal size of about 1 m in diameter. They occur on surfaces of from 1-3° and cover up to 50% of the ground surface. Many frost scars show surface activity due to frost action, but repeated mapping of them indicates that they neither expand nor contract laterally. Repeated measurements along fixed lines suggest that new frost scars are not being formed at present. Plants invade bare areas at Ogotoruk Creek by seeds or vegetative reproduction. Plants growing on frost scars are subject to frost heaving, uplift and disruption from needle ice formation, and wind erosion and desiccation. A direct relationship exists between plant cover on frost scars and soil moisture surrounding the scar. During the 20 years of the study both positive and negative changes in plant cover were recorded. A consistent pattern of plant succession was not detected.

INTRODUCTION

On the wet, relatively flat bottomlands of the Ogotoruk Creek Valley, northwest Alaska (68°06'N, 165°46'W), two major kinds of tundra vegetation occur. On the wettest soils, sedge meadow vegetation consisting primarily of two species, *Carex aquatilis* var. *stans* (Drejer) Boott and *Eriophorum angustifolium* Honck, occurs. On better drained soils a sedge tussock meadow consisting mostly of *Eriophorum vaginatum* L. is extensive. The two types are often mixed where microtopographic differences produce a mosaic of moisture conditions. The two kinds of vegetation occupy approximately 55% of the Ogotoruk Creek watershed (Johnson et al. 1966), an area of about 100 km^{2}. The two vegetation types at Ogotoruk Creek are described more completely in Johnson et al. (1966).

A third kind of vegetation occurs between the upland vegetation of the slopes and the vegetation types of the valley floor. *Carex bigelowii* Torr & Schwein. is the most characteristic plant of several diverse communities which all together comprise about 4.6% of the area, or 458 h. In Johnson et al. (1966) they are referred to as "ecotone communities."

Scattered on the surface in these vegetation types are predominantly oval areas of frost disturbed soil commonly called frost scars or frost boils, or in the formal classification of Washburn (1956), nonsorted circles. They are common in the Ogotoruk Creek Valley, especially in sedge tussock tundra, where up to 50% of the surface may be covered by this patterned ground feature. Frost scars typically measure 1-3 m in diameter; their centers are usually convex, but they may be flat or depressed. Permafrost is deepest under the center of the scar and shallowest at its margins. They are sometimes covered by plants; but in the majority of cases they are sparsely covered or are nearly bare.

At Ogotoruk Creek, frost features are important landscape features and probably play an important role in the dynamics of the plant communities. Nonsorted circles of this type are known from other areas of the Arctic (see especially Hopkins and Sigafoos 1951). Because of the absence of long -term observations, relatively little is known of their genesis, growth, activity, relationship to environmental variables, or other details of their history or future. The colonization of frost scars by plants has been observed, but not followed long enough to determine if plant succession in the traditional sense occurs. Inasmuch as the usual longitudinal method utilized in successional studies does not work well in conditions of soil instability, it was decided to engage in a long-term study of frost scars and their vegetation in order to investigate the following questions:

1. Are frost scars forming under the present climate?
2. Do frost scars exhibit cycles of activity, i.e., initiation, growth, equilibrium, and senescence?
3. Does plant succession on frost scars follow regular and predictable courses?

Preliminary studies were conducted by Johnson (1960) who listed the kinds of patterned ground features present in the Ogotoruk Creek Valley and described the relationships between vegetation and frost features, emphasizing the three vegetation types in which frost action is most important.

In 1961, Neiland permanently marked 326 frost scars located in 24 one acre plots that had been previously established for the study of the vegetation of the Ogotoruk Creek Valley (Neiland et al.

1962, Johnson et al. 1966). She documented the physical details of each scar, mapped its dimensions on a 2 m grid, and described the kinds, amounts, and locations of plants on each scar. Each scar was photographed from a fixed point for future reference.

In 1962, 1963, and 1965, Johnson rephotographed 240, 242, and 299, respectively, of the scars established by Neiland. In 1965, 60 of the scars in 4 plots were remapped in order to determine if physical changes in the scars were detectable. In 1972, 15 of the plots were redescribed, mapped, and photographed, and in 1980 this was repeated for 13 of the plots.

In 1965, in an effort to identify any new areas of disturbance that may have occurred, four lines measuring 640, 305, 457, and 213 m, respectively, were established. Along each line, the areas of disturbance intersecting the line were measured and recorded. Two of the four lines were relocated in 1972 and the measurements repeated.

This paper will concentrate on the differences observed in 189 of 326 frost scars between 1961 and 1980. The detailed descriptions of frost scars made in 1961 will not be given here. Rather, changes, if any, in the physical dimesions of the scars and in the kinds and amount of plant cover associated with them will be emphasized.

OCCURRENCE AND SIZE OF FROST SCARS

Are New Frost Scars Forming Today?

The conditions under which frost scars originate are a matter of speculation, but it is assumed by some authors (e.g., Hopkins and Sigafoos 1951) that they do so under the conditions of the existing climate. This belief is difficult to evaluate, because few, if any, studies have demonstrated it. If frost scars are forming today at Ogotoruk Creek, one should expect to find new scars appearing in a 20 year period. This is not the case. In the study plots at Ogotoruk Creek, no new frost scars have been detected from 1961 to 1980. Once formed, frost scars tend to be self perpetuating because of differences between the temperature-moisture relations of the scar and its better insulated surroundings. It is possible, as has been proposed for sorted circles by Ballantyne and Matthews (1982), that following a period of initial activity under somewhat colder, wetter periods than those of today, nonsorted circles became more or less stable.

In an effort to establish an additional baseline against which the occurrence of new frost scars might be measured, four long lines were established in 1965. The areas of bare soil were mapped along each line in order to determine (1) if any new bare areas occurred and (2) if existing bare areas changed in extent. In 1972, two of the lines were again measured (Table 1). Five new bare areas were recorded along transect 3 in 1972, and 78 of the areas of existing bare soil remained essentially unchanged. The transects were not measured in 1980, but they should be examined again to determine if the 1972 observations are measuring errors, represent newly formed bare areas, or are areas on which renewed activity is taking place.

It is also stated in some of the literature on frost scars that they grow or become reduced in area over a period of years. Hopkins and Sigafoos (1951), for example, suggest that a frost scar can be enlarged under favorable years by marginal needle ice formation, and they speak of the "net growth of many scars."

Are Frost Scars Expanding or Contracting Today?

In order to test the idea that frost scars grow (or contract) in one or more directions, maps of frost scar area were compared among 1961, 1965, 1972, and 1980. An example of these comparisons is given for one scar as it was mapped in 1961, 1965, 1972, and 1980 (Figure 1). A potentially significant error occurs in making these determinations. The precise margin of a frost scar is a matter of interpretation--along any edge it may grade into undisturbed tundra without a clear distinction between it and its surroundings.

What is apparent from this example is that over the periods measured, relatively little change in the shape or areal extent of this frost scar has taken place. Of course, the time periods involved are short and the number of frost scars measured is rather small, but all of the data available support a conclusion that the conditions that occurred at Ogotoruk Creek between 1961 and 1980 did not favor the growth or contraction of frost scars.

The belief that new frost scars are arising or growing is encouraged by the appearance of new frost activity within the margins of the scar. At times the combined action of frost heaving due to the presence of subsurface segregated ice and of surface needle ice gives the appearance of intense activity on the scar surface. Neither form of cryogenic activity implies expansion of the areal extent of the scar. Scars that have been inactive for some time may show a burst of renewed surface activity. For example, at several locations along the Trans-Alaska Oil Pipeline haul road, renewed frost scar surface activity occurred where construction activities blocked drainages and apparently increased the amount of soil moisture in the scar surroundings.

PLANT SUCCESSION ON FROST SCARS

Plants invade the bare soils of frost scars by direct seeding and by vegetative growth from their surroundings. A direct relation exists between the degree to which plant invasion is successful and cover established and the average wetness of the site. Scars occurring in sedge meadow vegetation, the wettest of the three being considered here, show the greatest cover of vegetation (Figure 2); those in ecotonal plant communities, the least.

The results of the plant succession study are summarized in Table 2 and in Figure 3. Two major conclusions are apparent. First, the proportion of frost scars showing changes in plant cover between 1961 and 1980 is highest in the sedge meadow vegetation type and lowest in the ecotone communities. This, together with the data summarized in Figure 2, suggests that the availability of moisture may be a controlling factor in frost scar plant succession. Frost scars occurring in the wettest habitats show the greatest degree of change in plant cover--both positively and negatively. The presence of moisture during the growing season favors plant growth but in the cold periods favors the formation of more soil ice which in turn disrupts plants.

Second, the proportion of frost scars in any vegetation type that show changes in plant cover increases with time. In 20 years the direction of change in plant cover has not been consistent on any single scar.

The discussion that follows describes only the gross changes that occur on frost scar surfaces. Detailed and quantitative descriptions are not given here.

The scars of the sedge meadow show the greatest fluctuations in plant cover of any of the three types (Figures 1 and 4). Plant cover on a few of these scars has changed by nearly 100% over the 20 year period. The large changes that occur in this type are nearly always the result of the establishment or the loss of one or two major species, particularly *Eriophorum angustifolium* or *Carex aquatilis*. The former is particularly susceptible to being uprooted (or detached from its rhizomes) by the effects of needle ice. At least temporary stability with complete plant cover can occur on scars of this type if they can maintain their cover long enough to produce an organic layer over the base mineral soil.

On scars occurring in *Eriophorum* tussock vegetation, changes in plant cover generally involve different species than is the case in sedge meadow vegetation. Some of the most noticeable changes on scars of this type involve tussock forming species, especially *Eriophorum vaginatum* and *Deschampsia cespitosa*. The latter, especially, seeds into the bare surface of the scar and forms small tussocks. These tend to persist and enlarge over several years, but they eventually decompose and are replaced by other tussocks in different locations on the scar. *Eriophorum vaginatum* shows the same kind of behavior, but it is not as abundant as an invader as *Deschampsia cespitosa* (Figure 5). In addition, once established it persists longer and grows to a much larger size. Because the pattern of change on these scars is more like a loss and replacement cycle than an invasion and subsequent elimination, changes in total cover tend to be small. Plant cover in general is lower on these scars than on those in the sedge meadow.

Plant cover on scars of the ecotone is very low, and changes are not obvious. Occasionally plants become established, ususally by seeds, on these base gravelly surfaces, but they do not persist for more than a few years. Invasion of plants from the surroundings of the scar is very slow, if it occurs at all. The surface of these scars tends to be dry, and the substrate coarse and, during the growing season, very hard. Although little evidence of needle ice formation is seen in any of the scars of this type, they show substantial heaving from subsurface ice during the freezeup period. Of the three types of scars, these ecotonal ones as a group show the least changes over the 20 year period.

SUMMARY

This review of a very large number of observations on frost scars at Ogotoruk Creek emphasizes changes that have taken place in the physical and biological characteristics of 188 frost scars between 1961 and 1980. In general, the evidence suggests that the physical dimensions of the scars have changed little in the 20 year period. Although surface activity is high on many frost scars, the lateral dimensions of the scars do not change, at least within our capability of consistent measurement. Neither is there conclusive evidence that new frost scars are forming at present.

Plant succession occurs on frost scars, but not in a directional or progressive sense. Rather it appears that on most frost scars, plant cover waxes and wanes as year-to-year climatic changes occur. Our data and observations suggest that plant cover is likely to be highest on frost scars occurring in the wettest habitats and least on those of the driest habitats. Although high levels of soil moisture encourage plant growth, they also create the most favorable conditions for the formation of soil ice, so it is also true that these frost scars show the greatest and most rapid changes in plant cover. Likewise, frost scars in the driest habitats change least over time. The idea of cycles of frost scar genesis, activity, invasion, and stabilization by plants and ultimate disappearance is not supported by this study. Indeed, the most striking feature of frost scars is their persistence.

The activities on frost scar surfaces involve an interplay of physical and biological phenomena. Only rarely do the biological forces achieve a transient dominance over what are fundamentally the powerful physical effects of the freezing and thawing of soil water.

REFERENCES

Ballantyne, C. K., and Matthews, J. A., 1982, The development of sorted circles on recently deglaciated terrain, Jotunheimen, Norway: Arctic and Alpine Research, v. 14, p. 341-354.

Hopkins, D. M., and Sigafoos, R. S., 1951, Frost action and vegetation patterns on Seward Peninsula, Alaska: U.S. Geological Survey Bulletin no. 974-C, p. 51-100.

Johnson, A. W., 1960, Preliminary studies on the influence of frost action and related phenomena on the vegetation of the Ogotoruk Creek Valley, Progress Report (unpublished) Phase III, Project Chariot: University of Alaska, 60 p.

Johnson, A. W., Viereck, L. A., Johnson, R. E., and Melchior, H., 1966, Vegetation and flora of the Cape Thompson-Ogotoruk Creek Area, Alaska, *in* Wilimovsky, N. J., ed., Environment of the Cape Thompson region, Alaska: U.S.A.E.C. Division of Technical Information Extension, p. 277-354.

Neiland, B. J., Johnson, A. W., and Johnson, R. E., 1962, Influence of frost action on the vegetation of Ogotoruk Creek Valley, Final Progress Report, (unpublished) Project Chariot.

Washburn, A. L., 1956, Classification of patterned ground and review of suggested origins: Bulletin of the Geological Society of America, v. 67, p. 823-866.

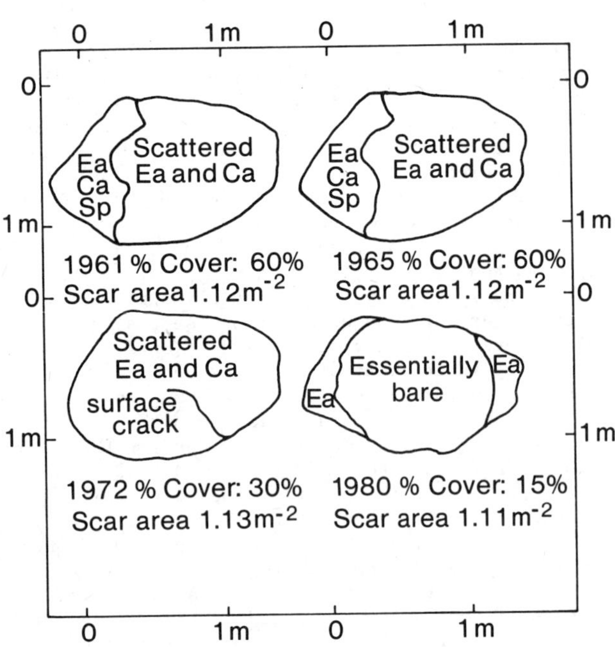

FIGURE 1 Plan of frost scar 1-1 in 1961, 1965, 1972 and 1980. Vegetation cover for each year is expressed in percent cover. The area of each frost scar is expressed in m^{-2}.

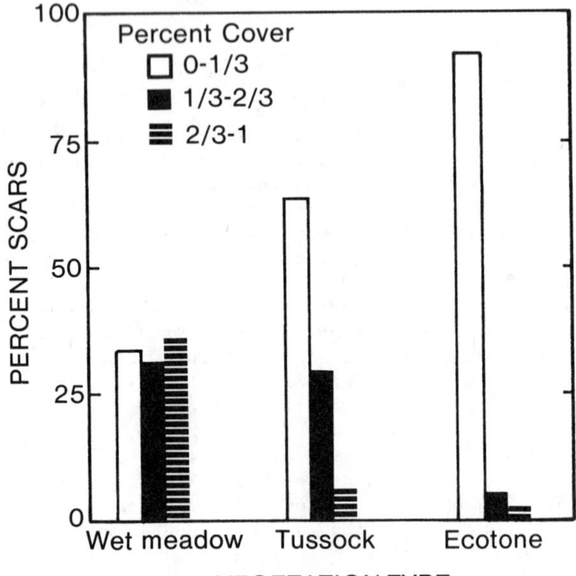

FIGURE 2 Percentage of scars in three vegetation cover classes. Surveyed and staked scars included.

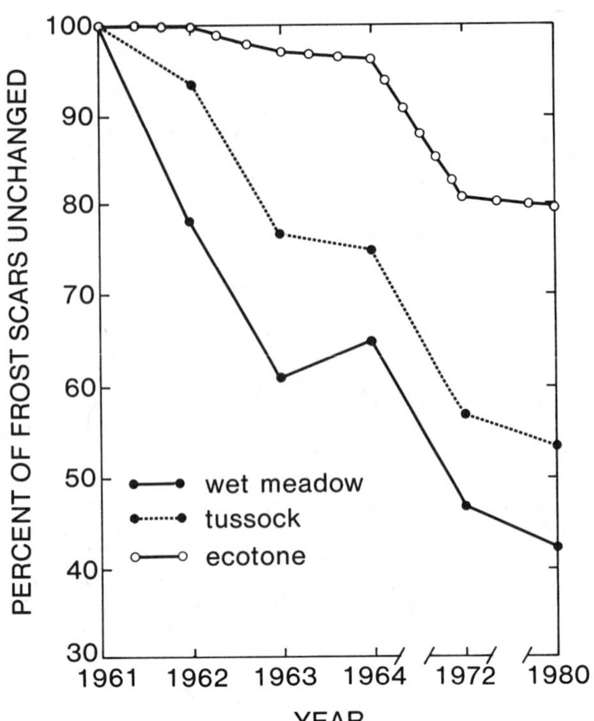

FIGURE 3 Percent frost scars remaining unchanged in percent plant cover from 1961 to indicated year.

Fig 4a

Fig 4b

Fig 4c

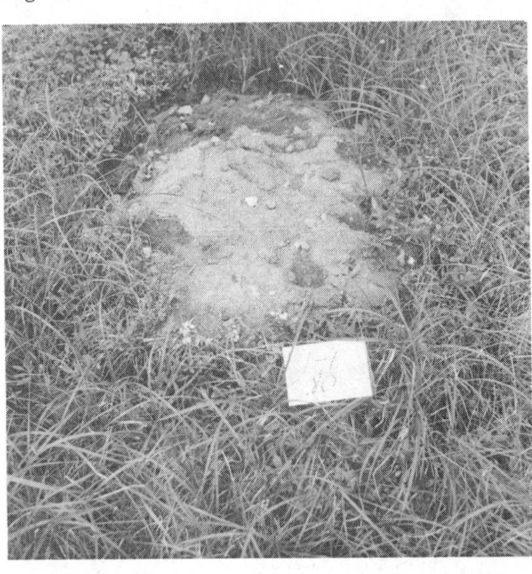

Fig 4d

FIGURE 4 Comparison of plant cover of frost scar 1-1 in the years (4a) 1961 (labeled 1-1), (4b) 1964, (4c) 1972, and (4d) 1980. After relatively little change until 1964, the open area present in 1964 expanded in 1972, and by 1980 almost the entire scar surface is bare of plants. Graminoid plants are primarily a mixture of *Eriophorum angustifolium* and *Carex aquatilis*. For comparison see also Figure 1.

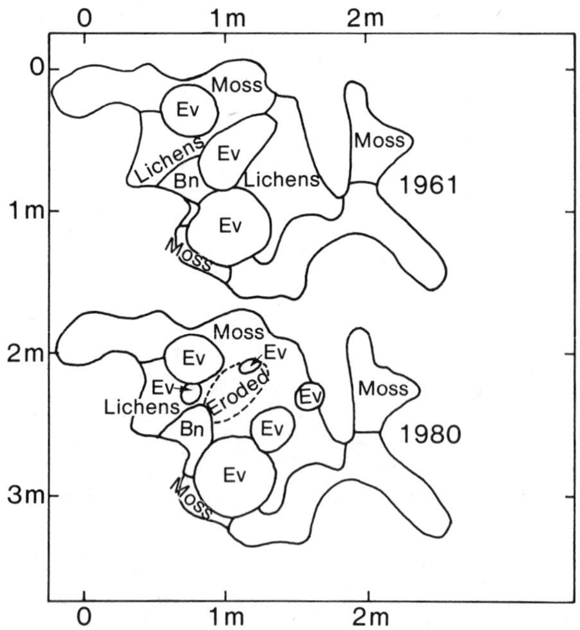

FIGURE 5 Comparison of frost scar 36-7 between 1961 and 1980. One large tussock present in 1961 has almost completely eroded away. Three new tussocks have appeared in 1980. Ev is Eriophorum vaginatum; Bn is Betula nana.

TABLE 1 Comparison of Occurrence of Disturbed Areas Along Two Transects

Transect No.	Length of transect	Number of disturbed Areas along lines in both years.	Areas found in 1965 but not in 1972.	Areas found in 1972 but not in 1965.
1	640 m.	9	1	0
3	457 m.	69	1	5*

*A more detailed examination of these 5 areas suggest that they may be areas of renewed activity, rather than newly formed frost scars.

TABLE 2 Summary of Changes in Plant Cover on Frost Scars of 3 Vegetation Types in the Ogotoruk Creek Valley From 1961 to 1980 (1962) is First Year of Comparison)

	1962					1963					1964					1972					1980				
	N	o	+	-	?	N	o	+	-	?	N	o	+	-	?	N	o	+	-	?	N	o	+	-	?
Wet Meadows	83	65	17	3	4	56	36	17	2	1	102	67	17	16	2	78	37	17	22	2	80	34	10	33	3
Tussock	110	103	2	2	3	115	89	4	12	11	155	116	15	23	1	96	55	21	9	11	78	42	12	17	7
Ecotone	27	27	0	0	0	42	41	1	0	0	54	52	2	0	0	27	22	4	0	1	30	24	0	0	6

NOTE: N = number of scars in sample; o = no change; + = greater plant cover; - = less plant cover; ? = direction of change, if any, uncertain.

RECOVERY AND ACTIVE LAYER CHANGES FOLLOWING A TUNDRA FIRE IN NORTHWESTERN ALASKA

L. Johnson[1] and L. Viereck[2]

[1]U.S. Army Cold Regions Research and Engineering Laboratory, Fairbanks, Alaska 99701 USA
[2]Institute of Northern Forestry, U.S. Forest Service, Fairbanks, Alaska 99701 USA

An upland tundra fire, started by lightning, burned 48 km^2 near the Kokolik River (69°30'N, 151°59'W) in northwestern Alaska during late July and early August 1977. Permanent plots were established to monitor recovery of severely, moderately, and lightly burned areas as well as unburned tundra. During the following 5 years the original permanent plots and other portions of the burn were observed annually. Vegetative recovery was most rapid and active layer effects were least on the moist sedge-shrub tundra. Recovery was slower on a high-centered polygonal area and on severely burned tussock tundra. By August 1979 the sedge-shrub vegetation had largely recovered while both the polygonal ground and the tussock tundra were still readily recognizable as burned areas. Accelerated hydraulic and thermal erosion had occurred on some slopes, resulting in exposures of massive bodies of ground ice. Active layer thicknesses averaged 27 cm in the unburned areas and 35 cm within severely burned areas in August 1977 and reached a maximum at all but one site in August 1979. Depth of thaw decreased between 1979 and 1982 in the sedge-shrub tundra and in the lightly burned shrub tundra and remained at the same increased level through 1982 at all other sites.

INTRODUCTION

There has been increasing recognition of the importance and influence of tundra fires in the last decade (Viereck and Schandelmeier 1980). For example, numerous fires burned more than 200,000 ha of tundra in northwestern Alaska during the dry, hot summer of 1977 (Racine 1979, 1981).

This paper reports on a 1977 tundra fire, started by lightning, that burned 48 km^2 (4800 ha) of upland tundra located southwest of the Kokolik River in northwestern Alaska (69°30'N, 151°59'W). The fire was exceptionally large for a tundra fire north of the Brooks Range. Wein (1976) found that most arctic tundra fires are less than 1 km^2. Of five Arctic Slope tundra fires reported by Barney and Comiskey (1973) only one (1620 ha) exceeded 30 ha. The fire burned from approximately 26 July to 7 August 1977. Early observations of the fire and remote sensing analyses were reported previously (Hall et al. 1978, 1980).

SITE DESCRIPTION

The Kokolik River fire site lies on the boundary between the Arctic Coastal Plain and the northern hills of the Brooks Range. Maximum topographic relief in the area is approximately 100 m. Most of the rolling to flat upland areas are covered with sedge tussock tundra (Viereck and Dyrness 1980) dominated by *Eriophorum vaginatum* tussocks with scattered shrubs of *Ledum decumbens*, *Betula nana*, and *Salix planifolia* subsp. *pulchra*. Shallow drainage depressions and drained thaw lake beds are covered with a wet sedge and sedge-shrub tundra, *Carex aquatilis*, *Eriophorum angustifolium*, and *Salix planifolia* subsp. *pulchra* being the important species. In one very localized area of exposed ridges there are small patches of mat and cushion tundra that were largely unburned by the 1977 fire. Along the streams that border the burned area are herbaceous snow-bed communities and shrub patches dominated by *Salix planifolia* subsp. *pulchra* and *Salix lanata* subsp. *richardsonii*. Much of the actual burned area was sedge tussock (*Eriophorum vaginatum*) tundra with smaller areas of wet sedge-shrub tundra.

The burn lies within the zone of continuous permafrost in northern Alaska. Active layer depths north of the Brooks Range average 30-50 cm with permafrost extending up to 625 m. Permafrost temperatures below the active layer range as low as -11°C, and the upper 1-10 m of soil may include large volumes of ice (Brown and Grave 1979, Osterkamp and Payne 1981).

This fire site is particularly interesting because of its extreme northern location. Most other tundra fire studies have been conducted at locations further south or in warmer climatic areas (Mackay 1977; Racine 1979, 1981; Wein and Bliss 1973; Weber 1975).

Therefore the Kokolik site offers an interesting comparison for both vegetation and permafrost changes following fire.

METHODS

The burn was visited by one or both of the authors yearly during late July to late August from 1977 to 1982. Due to the remoteness of the site, access was by float plane or helicopter, and not all study areas could be revisited each year. (See

Table 2 for areas visited.) In the summer of the fire, 1977, only the southeastern portion of the burn was observed. At that time three 1-m^2 quadrants were established in each of the following types of areas: severely burned tussock tundra, moderately burned tussock tundra, lightly burned shrub tundra adjacent to a stream at the fire's edge, and unburned tussock tundra. The estimates of fire severity were based upon visual observations of the percentage of biomass removed by the fire. All sites are within tussock area 1. Similar plots were established in August 1978 on high-centered polygonal ground, wet sedge-shrub tundra, and another burned and control (unburned) tussock tundra (tussock area 2) in the northeastern portion of the burn.

Our investigations were concentrated in the tussock tundra areas because it is the most widespread vegetation within the burned area. Controls were established only in this type. The unidentified vegetation type that was part of the polygonal complex seems to have a high density of shrubs. The polygonal and the wet sedge-shrub tundra were sampled because they represented the extremes (severe and light) of fire severity.

The following data were collected at each site unit: (1) depths of thaw, 5 depths/plot on burned plots and 5 depths within tussocks plus 5 depths between tussocks on unburned (control) plots; (2) total vascular plant cover, (3) total moss and liverwort cover; and (4) soil temperatures at 5, 10, 20, and 50 cm or until permafrost was reached. In addition, in 1978 the density of living and dead shrubs and tussocks on burned tussock tundra, the areal extent of frost boils in unburned vs. burned tussock tundra, and an estimate of relative biomass volume of cottongrass tussocks in unburned vs. burned areas were recorded.

RESULTS

Although vegetation regrowth on all parts of the burned area was rapid, the extent of change from the original vegetation after 5 years varied greatly depending upon the severity of the fire and the original vegetation. Data from tussock area 1 indicate the severity of the fire in this area. In August 1977 it was visually estimated that 80-90% of the tussock biomass had been consumed in the fire (Hall et al. 1978). In August 1978 tussock volumes were estimated in 5 m^2 of unburned (control) and 5 m^2 of burned tundra within tussock area 1. Burned areas averaged 30% of the tussock biomass of the control. This includes both residual (unburned) and new growth within burned areas.

Within tussock area 2 in August 1978 only 16% of the cottongrass tussocks (Figure 1) and none of the shrubs were regrowing on severely burned plots, while 78% of tussocks and 18% of shrubs were regrowing in moderately burned plots (Table 1). In contrast, all shrubs survived on the lightly burned wet sedge-shrub tundra site, and by August 1978 only some singed mosses and a few burned shrub leaves indicated that the area had even been burned.

Another indication of the fire severity is the increased areal extent of frost boils on burned vs. unburned areas of tussock area 1. Exposed areas of mineral soil averaged 52% greater in severely burned

FIGURE 1 Cottongrass tussocks one year after the burn (August 1978). Only 16% of the tussocks regrew in the severe burn in tussock area 1 after at least 70% of the tussock biomass was consumed by the fire.

TABLE 1 Status of Shrubs and Cottongrass Tussocks One Year After the Burn (August 1978). Based on Three 4-m^2 Plots.

Vegetation type	Severity of burn	
	Severe	Moderate
Tussocks/m^2		
Live	0.8 (16%)	6.4 (78%)
Dead	3.9 (84%)	1.8 (22%)
Shrubs/m^2		
Live	0	0.9 (18%)
Dead	4.6 (100%)	4.2 (82%)

FIGURE 2 An area of massive ground ice near tussock area 1 exposed by hydraulic erosion and thawing 5 years after the burn (July 1982).

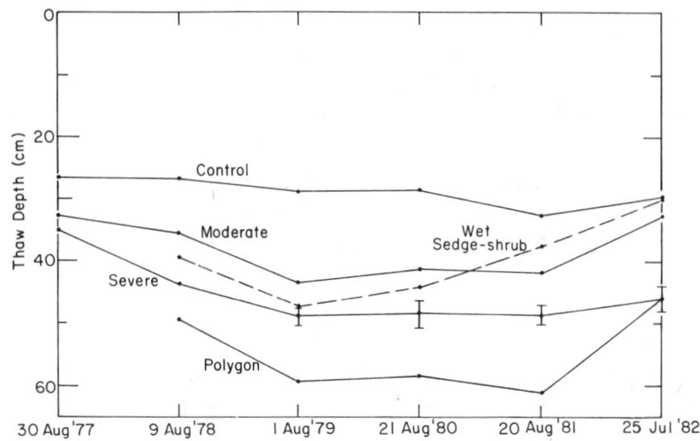

FIGURE 3 Post-fire active layer thicknesses from August 1977 to July 1982. Control, moderate, and severe burn intensities are in tussock area 1 with predominantly sedge tussock tundra. Points shown are means (n = 30 for control and 15 for others) with standard errors greater than 1.5 denoted by I.

areas than in unburned tussock areas, based upon measurements of comparable 30 m^2 sections. This increase was due, at least in part, to combustion of the thin layer of litter and living plants covering the edge of frost boils. However, the fire may also have intensified frost action (Hall et al. 1978), which could increase the areal extent of frost boils.

At least two areas of the burn had massive ice exposed as a result of hydraulic and thermal erosion on slopes. One of these was observed in August 1978 near the eastern shore of the northernmost lake. This area seemed to be stabilized by August 1981, but another exposure of massive ground ice was observed in July 1982 near tussock area 1 (Figure 2).

Active layer thickness increased at all burned sites for at least 2 years after the fire (Fig. 3). Maximum thicknesses at burn sites were recorded in August 1979 except for the maximum overall measurement of 61.0 cm (August 1981 polygon). In contrast, active layer thicknesses in the control plots remained constant with less than 3 cm of annual variation for four of the years and a 6-cm annual variation for 1981. There were unburned (control) plots for comparison only in the tussock tundra areas at tussock areas 1 and 2. The active layer in severely burned plots in tussock area 1 increased by an average of 13.6 cm between August 1977 and August 1979 and was 20 cm thicker than the comparable control plots. These differences persisted through August 1982.

Although no quantitative data are available, melting and subsidence of individual ice wedges did occur following the fire. This was especially noticeable in the polygon study units where surface relief increased for at least the initial 2-3 years of the study. Surface relief also seemed to increase within the severely burned plots in tussock area 1.

Soil temperatures in the active layer were also affected by the fire and its intensity. Although soil temperatures could be measured only once each summer, the measurements should reflect the degree of summer warming of the profile since we visited the sites near the end of the short summer season.

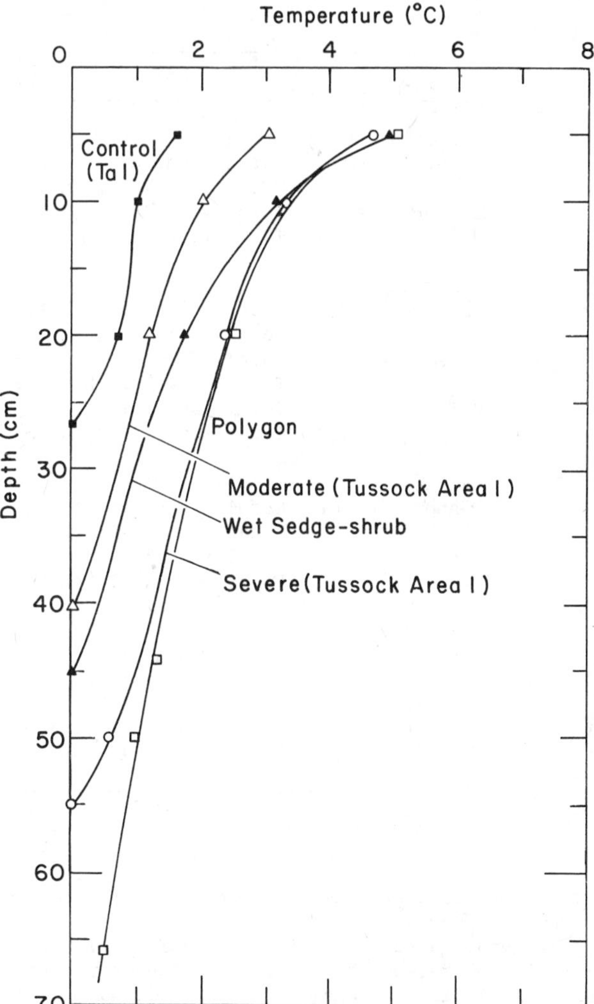

FIGURE 4 Soil temperature profiles in unburned tussock tundra, moderately and severely burned tussock tundra, severely burned polygonal area, and lightly burned sedge-shrub tundra. Temperatures were recorded at the Kokolik fire site on 20-21 August 1980 (n=3) (TA1 = tussock area 1).

Figure 4 shows the results of the soil temperature measurements taken on 20-21 August 1980. These soil temperature profiles are similar to those obtained in other years. The unburned tussock tundra soil-temperature profile for 20 August 1980 is representative of this vegetation type in this area. At 5 cm the temperature was only 1.6°C and temperatures decreased with depth, reaching 0°C at the active layer-permafrost boundary at only 27 cm. For the tussock type and the polygonal area, the severe burn has resulted in soil temperatures that average 2° to 3°C warmer throughout the thawed zone compared with unburned tussock tundra. This warming effect occurs to depths of at least 50 cm.

The moderately burned tussock tundra has soil temperatures intermediate between those of the control and the severely burned sites. The wet sedge-shrub tundra, which was only lightly burned and recovered rapidly, had higher temperatures at 5 and 10 cm (similar to those of the severely burned

TABLE 2 Percent Plant Cover on Burned and Control (Unburned) Areas, Kokolik River, Alaska.

Burn severity and area		Year observed 1977	1978	1979	1980	1981	1982
Control plots							
Tussock area 1	Total vascular	80	80	80	73	70	--
	Total moss & lichen	50	50	50	67	60	--
Tussock area 2	Total vascular	--	80	--	--	--	--
	Total moss & lichen	--	30	--	--	--	--
Severe plots							
Tussock area 1	Total vascular	<10	10	30	30	30	33
	Total moss & lichen	0	10	20	27	40	33
Tussock area 2	Total vascular	--	10	--	35	35	45
	Total moss & lichen	--	30	--	75	80	70
Polygon	Total vascular	--	†	†	13	20	27
	Total moss & lichen	--	70	100	100	100	97
Moderate plots							
Tussock area 1	Total vascular	30	40	50	50	57	--
	Total moss & lichen	†	10	20	33	47	--
Tussock area 2	Total vascular	--	40	--	65	45	70
	Total moss & lichen	--	30	--	70	75	70
Light plots							
Tussock area 1	Total vascular	40	60	60	60	57	67
	Total moss & lichen	80	90	90	90	90	87
Wet sedge-shrub	Total vascular	--	70	80	93	90	90
	Total moss & lichen	--	10	20	93	90	90

† = present, less than 1% cover.
-- = no data collected that year.

sites), but at depth they were lower (similar to the moderately burned site).

Vegetation recovery (Table 2) was rapid for the first 2-3 years, but by August 1979 recovery rates had dramatically decreased or stopped on most study areas. For example, severely burned plots on tussock area 1 had 30% total vascular plant and 20% total moss and lichen cover in August 1979, and only 33% cover for each category in late July 1982. Vegetation cover largely paralleled changes in active layer thickness. Interesting exceptions were the sedge-shrub area, which had rapid recovery of mosses between August 1978 and August 1980, and the polygon area, which, although totally revegetated by mosses and the liverwort *Marchantia polymorpha* by August 1979, also had some recovery of vascular plant cover by 1982.

DISCUSSION

The Kokolik fire offers several interesting comparisons with other tundra fire studies. Although there was limited exposure of massive ice, most of the impact upon the permafrost seems to have been chronic, rather than catastrophic. Active layer thicknesses continued to increase for only 2 years, but remained deeper on more severely burned portions of the fire at least 3 more years. It is uncertain how long these increased thicknesses will persist. The reported increases in the active layer thicknesses are a minimum indication of changes in permafrost. The observed subsidence due to ice melting was not measured, although Mackay (1977) observed subsidence of 50 cm following a fire near Inuvik. All or an unknown portion of the organic layer at Kokolik was consumed, particularly in the severely burned areas. Racine (1981) estimated that at least 5-15 cm of organics were consumed by fires on the Seward Peninsula.

More southerly fires reported thawing that continued to increase for longer periods. At Inuvik (Mackay 1977) and near Fairbanks (Viereck 1982) thaw depths were still slowly increasing 8 years and 9 years, respectively, after fire. These results are consistent with the hypothesized cooler climatic conditions at the Kokolik site, although the paucity of climatic data makes correlations impossible.

Vegetation recovery was quite distinct, at least on severely burned portions of the Kokolik fire. For example, shrub recovery in severely burned areas (tussock areas 1 and 2, polygon) was still very limited 5 years after the fire (less than 10% by August 1982), although controls in tussock tundra averaged 30-40% shrub cover. The dramatic and rapid recovery of the liverwort, *Marchantia polymorpha*, on the severely burned polygonal area is also noteworthy. Since no liverwort was observed in unburned tundra, it either regenerated from buried spores or else spores are readily and widely dispersed in tundra. In contrast, Racine (1981) reports a very low total cover (1% or less) of *Marchantia* in burned areas.

Racine (1981) also reported low recovery rates for shrubs, but this was only one year after the burn. In contrast, tundra fires in four more southerly or warmer areas averaged 55% of their original shrub biomass only 2-3 years after the burn (Wein and Bliss 1973).

CONCLUSIONS

The Kokolik River fire shows the following:

1. Massive thermal and hydraulic erosion can continue at specific sites for at least 5 years after a tundra fire.
2. Increased depths of thaw can persist for at least 5 years on severely burned portions of tundra, although no new increases were seen after 2 years.
3. Vegetation recovery, although initially quite rapid (2-3 years), dramatically slows or halts on many parts of the burn between years 3 and 5. In addition, the species composition of burned areas may remain quite different (especially shrubs and lichens) for at least 5 years after a tundra fire.
4. Severe tundra fires can in fact kill many shrubs and cottongrass tussocks, presumably by elevating soil temperatures during combustion of the organic mat.

ACKNOWLEDGMENTS

The authors would like to acknowledge the support of the U.S. Army Corps of Engineers, Cold Regions Research and Engineering Laboratory projects, the Bureau of Land Management, and the U.S. Geological Survey's National Petroleum Reserve Alaska program for providing logistical support and funding for this project.

REFERENCES

Barney, R.J., and Comiskey, A.L., 1973, Wildfires and thunderstorms on Alaska's north slopes, Research Note PNW-212: Portland, Oregon, USDA Forest Service, Pacific Northwest Forest and Range Experimental Station.

Brown, J., and Grave, N.A., 1979, Physical and thermal disturbance and protection of permafrost, in Proceedings of Third International Conference on Permafrost, v. II: Ottawa, National Research Council of Canada, p. 51-91.

Hall, D.K., Brown, J., and Johnson, L., 1978, The 1977 tundra fire in the Kokolik River area of Alaska: Arctic, v. 31, p. 54-58.

Hall, D.K., Ormsby, J., Johnson, L., and Brown, J., 1980, Landsat digital analysis of the initial recovery of burned tundra at Kokolik River, Alaska: Remote Sensing of the Environment, v. 10, p. 263-272.

Mackay, J.R., 1977, Changes in the active layer from 1968 to 1976 as a result of the Inuvik fire, in Report of Activities, Part B, Geological Survey Conference, Paper 77-1B, p. 273-275.

Osterkamp, T.E., and Payne, M.W., 1981, Estimates of permafrost thickness from well logs in northern Alaska: Cold Regions Science and Technology, v. 5, p. 13-27.

Racine, C.H., 1979, The 1977 tundra fires in the Seward Peninsula, Alaska: effects and initial revegetation, Alaska Technical Report 4: Washington, D.C., USDI Bureau of Land Management.

Racine, C.H., 1981, Tundra fire effects on soils and three plant communities along a hillslope gradient in the Seward Peninsula, Alaska: Arctic, v. 34, p. 71-84.

Viereck, L.A., 1982, Effects of fire and firelines on active layer thickness and soil temperatures in interior Alaska, in Proceedings of the Fourth Canadian Permafrost Conference: Ottawa, National Research Council of Canada, p. 123-135.

Viereck, L.A., and Dyrness, C.T., 1980, A preliminary classification system for vegetation of Alaska, Research Note PNW-332: Portland, Oregon, USDA Forest Service, Pacific Northwest Forest and Range Experimental Station.

Viereck, L.A., and Schandelmeier, L.A., 1980, Effects of fire in Alaska and adjacent Canada - a literature review, Alaska Technical Report 6: Washington, D.C., USDI Bureau of Land Management.

Weber, M.G., 1975, Nutrient budget changes following fire in arctic plant communities, in Arctic Land Use Resource Program, ALUR Report 74-75-62: Ottawa, Ontario, Department of Indian Affairs and Northern Development, p. 32-115.

Wein, R.W., 1976, Frequency and characteristics of arctic tundra fires: Arctic, v. 29, p. 213-222.

Wein, R.W., and Bliss, L.C., 1973, Changes in arctic *Eriophorum* tussock communities following fire: Ecology, v. 54, p. 845-852.

PERFORMANCE OF AN INSULATED ROADWAY ON PERMAFROST, INUVIK, N.W.T.

G.H. Johnston

Division of Building Research, National Research Council Canada, Ottawa, Ontario, Canada K1A 0R6

In 1972 four insulated and two uninsulated test sections were installed and instrumented on the Mackenzie Highway (a gravel-surfaced, all-season road) near Inuvik, N.W.T., to evaluate the use of extruded polystyrene board insulation to control permafrost degradation. Two test sections were insulated with a 90-mm layer of insulation, another with a 50-mm layer, and a fourth with 115 mm. Preliminary analysis had indicated that an optimum insulation thickness of about 90 mm would be required. Ground temperature and settlement observations over a 6-year period showed that permafrost aggraded under all insulated test sections, whereas under the uninsulated (control) sections it receded by as much as 60 cm. Approximately 60 cm of settlement was experienced by the control sections, about 30 cm by the 50-mm insulated section, and little or none by the sections having 90 and 115 mm of insulation. The 90 mm of insulation was shown to provide adequate protection against thawing; 50 mm of insulation was not sufficient.

INTRODUCTION

A layer of insulation placed in an embankment constructed on permafrost can prevent or control thaw penetration and settlement and reduce the quantity of fill required. In 1972 a field study was undertaken by the Canadian Government Interdepartmental Committee on Highway Insulation to evaluate the performance of an insulated, gravel-surfaced roadway constructed on permafrost. Cooperating with the Division of Building Research, National Research Council Canada, were the Departments of Indian and Northern Affairs, Public Works, and Transport Canada, and Dow Chemical of Canada Ltd. Several insulated and uninsulated test sections were constructed on the Mackenzie Highway about 22 km east of Inuvik in April and September 1972. Extruded polystyrene foam boards were placed in the road embankment and ground temperatures and settlements were measured. Design and construction of the test sections and the results of observations made during the period 1972-1978 are summarized.

SITE CONDITIONS

Climate

The mean annual air temperature at Inuvik for the 24-year period 1958-1981 is -9.7°C. For the same period the mean air freezing and thawing indices are 4,690 and 1,210 degree-days Celsius (DDC). Air temperature observations were made at the test site from July 1974 to July 1978. Some mean annual air temperatures and the freezing and thawing indices for Inuvik and the test site are given in Table 1. It is somewhat colder at the test site than at Inuvik, due mainly to more severe winters, as indicated by the freezing indices; thawing indices for both sites are similar. The number of days having mean temperatures below and above freezing were almost identical at both sites for the periods noted.

Terrain

The test site is relatively flat, but the entire area is covered with large hummocks. Depth of thaw in undisturbed areas varies from 15 to 120 cm. Permafrost extends to depths exceeding 100 m in the Inuvik area, which lies within the continuous permafrost zone. Site investigations showed that the soil profile consists of a layer of organic material 15-60 cm thick underlain by grey, silty clay to a depth of 3-4.5 m. Beneath the silty clay is a dense grey, sandy silty till extending to at least 18 m. The organic layer contains typical random inclusions and thin lenses of ice; the layer of silty clay is ice-rich, containing as much as 50% ice by volume, mainly in the form of lenses from hairline to 2.5 mm thick. The underlying till matrix is generally well bonded with ice, but does contain occasional lenses up to 10 mm thick.

DESIGN OF TEST SECTIONS

Because of the complex nature of the layered embankment/subgrade system and the variable surface thermal conditions, numerical methods for ground thermal analysis were used to predict the optimum thickness of insulation to be used and the changes that would occur in the ground thermal regime (Goodrich 1974a,b). The basic profile consisted of 60 cm of gravel, a layer of insulation, and 45 cm of gravel over an ice-rich silt clay subgrade. Several computer runs were made using assumed values for ground temperature distribution, physical and thermal properties of the subgrade and fill materials, and seasonal fluctuations in air and ground surface temperatures. The ground surface temperature was assumed to be either the same as the air temperature or several degrees warmer and to vary sinusoidally during the year. Various thicknesses of extruded polystyrene foam insulation were considered, as well as construction under winter (subgrade completely frozen and fill placed on a snow cover) and summer (active layer thawed) conditions. The road surface was assumed to be free of snow during the winter. Observations at Inuvik indicated that the mean annual ground temperature at a depth of 15-30 m was about -3.3°C.

The results of these computations indicated that an optimum thickness of about 90 mm of extruded polystyrene insulation would be required to prevent thawing of the subgrade below the original ground surface at the center of the roadway. Edge effects from the side slopes were not included in the calculations, nor were changes in geometry that might occur as a result of settlement caused by thawing of the subgrade. The analysis also indicated that major changes in the ground thermal regime would occur during the first 5 or 6 years following construction, and that equilibrium conditions would not be reached until some time later. It showed that if insulation were placed during the summer or fall, the thawed layer underlying it would not refreeze for several months, so that the full advantage of insulation would be lost during the first summer. At least 2.5-3 m of gravel fill would be required to prevent thawing below the original ground surface under an uninsulated gravel embankment. This has been verified at the Inuvik airstrip (Johnston 1982).

CONSTRUCTION OF TEST SECTIONS

Test sections containing different thicknesses of insulation were installed under both winter and summer conditions. Two 38-m insulated test sections separated by a 38-m uninsulated (control) section were constructed at each of two test sites. The "winter" site was constructed in April 1972 and the adjacent "summer" site in September 1972. The width of the layer of insulation at all test sections was 11 m. No special measures were taken to insulate or protect the slopes.

The right-of-way had been cleared by hand of trees and brush early in 1970. Construction of the Mackenzie Highway south from Inuvik (Huculak et al. 1978) began in late summer 1971 and by late November a 0.6-1-m thick "pioneer" fill had been placed across the test area. To construct the test sections about 0.6 m of this fill was removed by bulldozer and ripper. The insulation boards (each 600 by 2,400 mm) were then placed by hand and secured by 150-mm-long steel spikes driven into a 15-cm thick bedding layer of relatively dry silty sand. All joints between boards were staggered horizontally and vertically (where more than one layer of boards was used). About 0.6 m of relatively dry sandy gravel was backfilled over the insulation by bulldozer and carefully rolled on the boards to prevent displacing them. A grader was used to shape the embankment to the desired section. A plan showing the layout of the test sections is given in Figure 1 and construction operations at the sites are shown in Figure 2.

INSTRUMENTATION

Thermocouples were installed to measure temperatures in the fill above and below the insulation to depths of about 6 m below the roadway and 15 m below the ground surface in undisturbed areas at both sites. The location of all cables is shown in Figure 1 and a typical arrangement of sensors at one location is shown in Figure 3. Temperatures were read manually two to four times each month until June 1974, when a specially designed automatic data acquisition system was installed to record temperatures at 360 points every 1-2 h on magnetic tape. Observations were terminated in August 1978 when the equipment was damaged by vandals. Four brass plates were installed in each of the six test sections to monitor settlements (Figure 1). Once or twice each year surveys were made to determine the plate elevations and changes in the shape of the embankment. A special deep bench mark was installed to ensure that a reliable datum was available for elevation survey control.

OBSERVATIONS AND DISCUSSION

The annual thaw pattern and settlement under the center of the road at four of the test sections are shown in Figure 4. In September 1973, 1977, and 1982, trenches were excavated across the road at each of the insulated sections. Three cross-sections showing conditions in 1972 (as built) and in September 1982 are plotted in Figure 5. Differences in the positions of the insulation layers indicate actual movements that have occurred in the 10-year period; differences in the shape of the roadway do not reflect the total settlement since fill was added to maintain a suitable road surface.

Changes in the ground thermal regime under the center of the 90-mm insulated section at the winter site are illustrated in Figure 6. Isotherms show the variation in ground temperature under and adjacent to the winter site insulated and uninsulated test sections (Figure 7). They represent fall and late winter conditions, when ground temperatures are nearing their warmest and coldest values.

Considerable settlement occurred in the control sections (Figure 4). At the winter site total settlement was about 55 cm; approximately 36 cm of this occurred during the first thaw season, due partly (9 cm) to melting of the compacted snow layer under the pioneer fill. The remaining settlement is the result of compression of the thawed fill and subgrade materials caused by the weight of the embankment and, to a lesser extent, traffic loading. Total settlement measured at the summer site control section was much less, averaging about 30 cm. A great deal of settlement (not measured) had occurred during the summer prior to construction of the test sections in the fall of 1972. Thus, observed settlement is due mainly to compression of subgrade soils as thawing of the permafrost occurred.

About 30 cm of settlement occurred at the center of the 50-mm insulated section; again, most of it took place in the first thaw season. Settlement of the 90- and 115-mm insulated sections was negligible. About half of the 15 cm of total settlement experienced by the 90-mm section at the winter site occurred in the first two thaw seasons. Little or no settlement of the two insulated sections occurred at the summer site.

At all test sections settlement of the embankment near the shoulders was from 3-15 cm greater than that at the center of the road. Surveys since 1978 indicate that settlement of the control sections and of the 50-mm insulated section is still continuing, but at a greatly reduced, almost negligible rate. No further significant settlement of the other insulated sections has been observed.

As predicted, freezeback of the subgrade under the summer site insulated sections did not occur until 3 to 4 months after the active layer at the control section and undisturbed areas had refrozen (Figure 4). Between 1972 and 1978 the permafrost table beneath the uninsulated control sections was lowered from 30 to 60 cm. Beneath the insulated sections, however, it rose almost to the bottom of the embankment (i.e., to the original ground surface) under the 50-mm insulation, and where the insulation was thicker it was raised and contained within the embankment.

The foregoing applies to the center of the road and much of the embankment width (Figure 5). As expected, thaw penetration was significantly greater under the slopes. This resulted in considerable sloughing and cracking of the slopes, and owing to the irregular and sharply dipping thaw plane the insulation settled differentially, cracked, broke, and moved outward. The major movements occurred during the first 3-4 years when substantial cracking occurred inside the shoulders. Since then the shoulders and slopes have essentially stabilized in most areas, although some minor cracking and sloughing still occurs.

The marked influence of the insulated and uninsulated embankments on the ground thermal regime is illustrated in Figures 6 and 7. In general, the initial effect was a lowering of ground temperatures to a depth of about 5 m and an increase in temperature amplitude. At all test sections a return to natural conditions is clearly taking place and for most practical purposes equilibrium appears to have been reached by the end of the 6-year study period.

Snow cover plays an important role with respect to the ground thermal regime. During the winter, snow is ploughed from the road surface and deposited on the shoulders and slopes, substantially retarding removal of heat from the underlying ground (Figure 7).

CONCLUSIONS

Observations over a 6-10 year period at several insulated and uninsulated test embankments constructed on ice-rich permafrost on the Mackenzie Highway near Inuvik, N.W.T., confirm that insulation can be used to great advantage to prevent or control thawing of frozen subgrade soils and reduce the quantity of fill required. The predicted 90-mm optimum thickness of extruded polystyrene insulation has given satisfactory protection against thawing at this site; 50 mm of insulation is not adequate. Although no special measures were incorporated to prevent anticipated problems with cracking and sloughing of the shoulders and slopes, it is clear that they are needed, especially if the roadway is to carry any volume of heavy vehicles. On the other hand, it appears that relatively stable conditions will prevail within 5-6 years of construction. During that period only limited maintenance was required, although admittedly traffic volumes and loads were very light.

Construction procedures will have to be modified somewhat if synthetic insulation is used extensively. For satisfactory long-term performance a suitable levelling course must be placed on the ground surface and the backfill carefully placed on top of the insulation to prevent damage and displacement. It is apparent that the most opportune time for construction of an insulated embankment is the early spring, when the snow cover has just disappeared and the ground is still completely frozen. Unless suitable dry materials have been stockpiled in advance or are available in easily excavated borrow pits, however, procurement of good fill may be a problem at that time of the year.

ACKNOWLEDGMENTS

Many people, too numerous to mention individually, participated in the project at various times, and the author is grateful to them all for their cooperation and assistance. He is indebted particularly to J.C. Plunkett for his unflagging assistance and enthusiasm through all stages of the project. This paper is a contribution from the Division of Building Research, National Research Council Canada, and is published with the approval of the Director of the Division.

REFERENCES

Goodrich, L.E., 1974a, A one-dimensional numerical model for geothermal problems: National Research Council Canada, Division of Building Research, NRCC 14123, 28 p.

Goodrich, L.E., 1974b, Fortran IV program for general one-dimensional geothermal problems: National Research Council Canada, Division of Building Research, Computer Program No. 39, 89 p.

Huculak, N., Twach, J.W., Thomson, R.S. and Cook, R.D., 1978, Development of the Dempster Highway north of the Arctic Circle, in Proc. Third International Conference on Permafrost, Edmonton, Alberta, National Research Council Canada, Vol. 1, pp. 798-805.

Johnston, G.H., 1982, Design and performance of the Inuvik, N.W.T., airstrip, in The Roger J.E. Brown Memorial Volume, Proc. Fourth Canadian Permafrost Conference, Calgary, Alberta, March 1981, National Research Council Canada, Associate Committee on Geotechnical Research, pp. 577-585.

TABLE 1 MEAN ANNUAL AIR TEMPERATURES AND FREEZING/THAWING INDICES, INUVIK WEATHER STATION AND INSULATED ROAD TEST SITE

Period	Mean Annual Air Temperature (°C)		Freezing Index DDC		Thawing Index DDC	
	Airport	Road Site	Airport	Road Site	Airport	Road Site
1971	-10.9	-			1,295	-
1971-1972			4,885	-		
1972	- 9.7	-			1,205	-
1972-1973			4,430	-		
1973	- 8.6	-			1,370	-
1973-1974			4,760	-		
1974	-11.9	-			1,120	-
1974-1975			5,010	5,344		
1975	-10.6	-11.2			1,290	1,220
1975-1976			4,950	5,311		
1976	- 8.9	-10.0			1,320	1,330
1976-1977			4,460	4,678		
1977	- 8.9	- 9.5			1,290	1,330
1977-1978			4,185	4,464		
1978	- 8.3	-			1,150	-

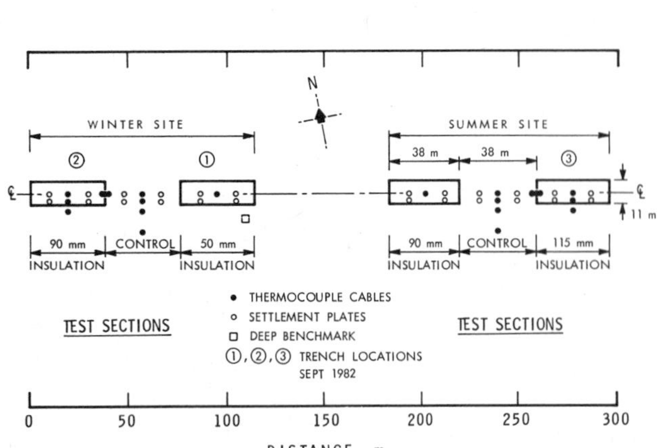

FIGURE 1 Layout of test sections and location of instrumentation.

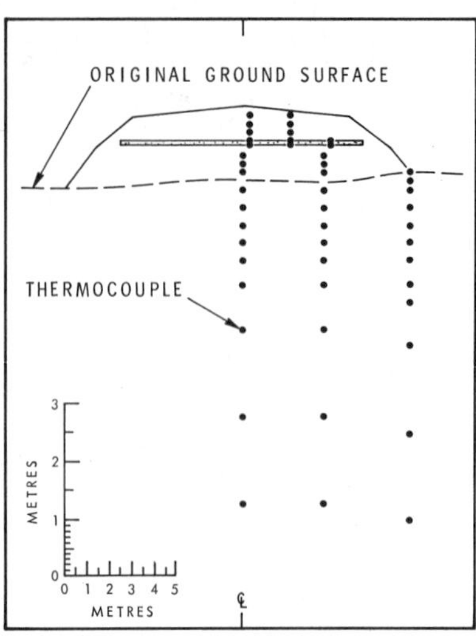

FIGURE 3 Arrangement of thermocouples at center of 90-mm insulated section, winter site.

(a)

(b)

FIGURE 2 (a) Winter site--50-mm insulation and control sections ready for backfilling. Completed 90-mm insulated section in background. (b) Summer site--placing insulation on 115-mm test section; control and 90-mm sections next in line. Completed winter site in distant background.

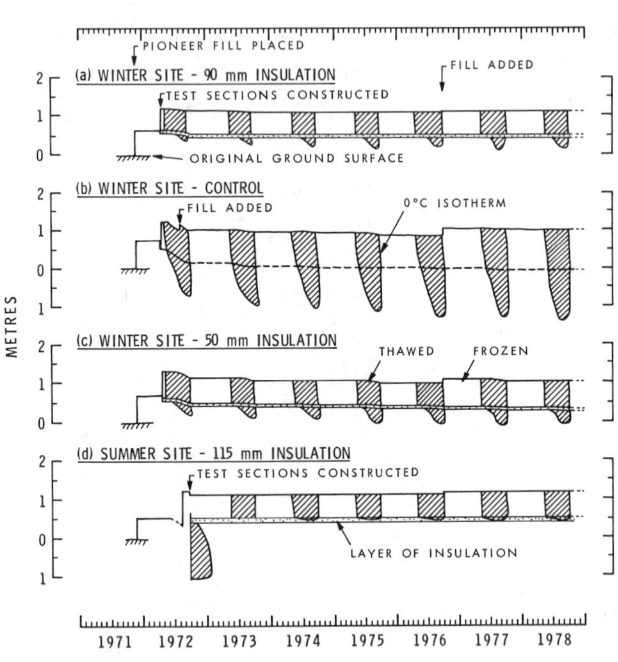

FIGURE 4 Thaw depths and settlement under center of road at several test sections.

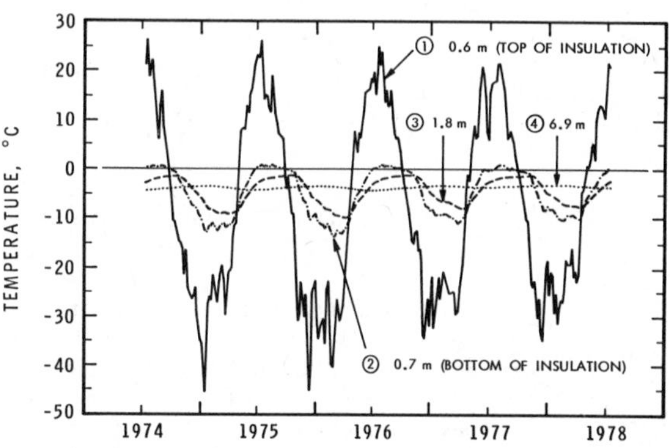

FIGURE 6 Temperature variation at selected depths under centerline of 90-mm insulated section, winter site.

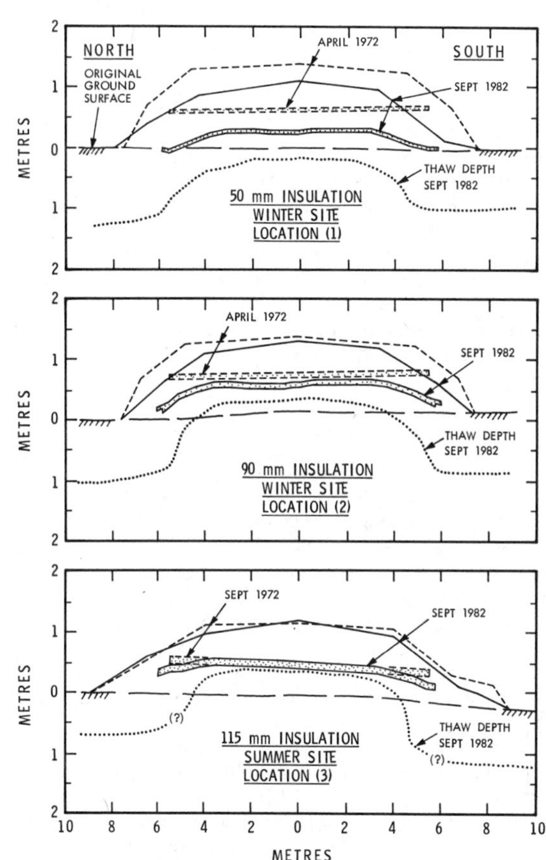

FIGURE 5 Cross-sections at three trenches excavated in September 1982 at locations noted on Figure 1.

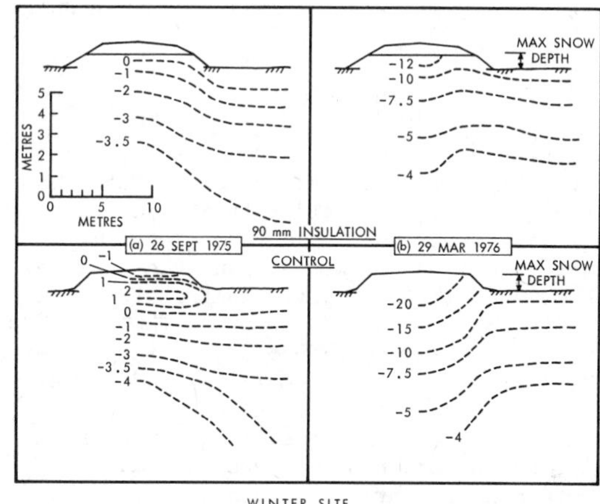

FIGURE 7 Isotherms showing the variation in ground temperature under and adjacent to an insulated and uninsulated embankment in fall and late winter, winter site.

THE FROST SUSCEPTIBILITY OF GRANULAR MATERIALS

R.H. Jones[1] and K.J. Lomas[2]

[1]University of Nottingham, Nottingham, U.K.
[2]Loughborough University of Technology, Loughborough, U.K.

The frost susceptibility and suction characteristics of fifteen granular materials ranging from silty soils to crushed rock sub-base aggregates were determined by the British Transport and Road Research Laboratory (TRRL) test and an osmotic suction technique, respectively. The test specimens were prepared at a number of standardized gradings and compacted to their maximum density at optimum moisture content. The heaves after 250 hours ranged from 2 to 50 mm so that both frost susceptible and non-frost susceptible materials were included. The results for all but one of the materials lay on a single curve relating frost heave to $\theta_{2.5}$, the total volumetric water content at a suction of pF 2.5 (31 kPa). Using the segregation potential approach it appears that there is a characteristic suction that controls the average permeability of the frozen fringe. For most of the materials considered, this suction approximates to pF 2.5. Materials with $\theta_{2.5}$ less than 9% were non-frost susceptible, but tests are required on many more materials to establish whether this relationship has general validity.

INTRODUCTION

The British Transport and Road Research Laboratory (TRRL) frost susceptibility test has been revised recently; self-refrigerated units (SRU's) have superseded cold rooms, specimen preparation has been improved, and the test has been shortened (TRRL 1981). The results of a study carried out under contract to TRRL in which testing in the SRU and the cold room was compared have been reported elsewhere (Lomas and Jones 1981). During that investigation the frost susceptibility of four naturally occuring sandy gravels and two crushed rock sub-base aggregates were determined. Earlier or associated investigations provided data about six more crushed rock sub-bases, a silty soil and mixtures of sand and limestone filler, which had gradings equivalent to those of silty soils. These materials varied from non-frost susceptible to very frost susceptible in the TRRL test.

Previous work had shown that the frost susceptibility of porous limestones could be related to the rock suction characteristics (RSCs) determined by an osmotic technique (Hurt 1976, Jones and Hurt 1978). In the present paper the frost susceptibility is related to the aggregate suction characteristics (ASC). (The ASC reflects the distribution of pores both within and between the particles, unlike the RSC, which characterizes only the within particle pores).

MATERIALS

The authors were concerned jointly with series 100 (the first six aggregates listed below) and series B tests. Hurt (1976) and Thompson (1981) undertook series (H) and (T), respectively. Brief descriptions of the materials are given below and in Table 1.

Ashton Keynes (102): A sand and gravel consisting of weathered fragments of Jurassic Limestone.

Spencers Farm (103): A flint hoggin (gravel-sand-fines mixture).

Stanley Ferry (106): A flood plain sand and gritstone gravel.

Woodhall Spa (114): A sandy hoggin containing flint and chalk particles.

Croft (105): A crushed granite with some honey-combed particles.

Dene (119) (B) (T): A crushed carboniferous limestone.

Whin 2 (B) (T): A crushed fine grained dolerite containing some vesicles up to 5 mm across and phenocrysts.

OOL2 (B): A fine grained calcitic mudstone with shell fragments.

OOL1 (H): Oolitic and shelly limestone.

DOL (H): Buff dolomitic limestone.

MIC (H): Yellow microcellulitic calcite/ dolomite.

CAL (H): Grey or yellow dolomitic limestone.

Sand limestone filler mixtures (SF): Standard Leighton Buzzard sand (600 to 300-μm fraction) was mixed with various proportions of limestone filler to provide reference materials corresponding in grading to silty sands.

Attenborough silt: A non-plastic silt obtained from the washings at a local gravel pit.

Gradings

The materials of series 100 were washed on a 63-μm sieve and separated into component fractions using the 20-, 10-, 5-mm, 600-μm, and, for Woodhall Spa only, the 300-μm sieves. The individual fractions were recombined in an attempt to give the initial gradings F1 and F2 for Woodhall Spa and M for the remainder. (Figure 1). These gradings are typical of sub-bases in Britain. Some difficulty was experienced in achieving the target gradings as the washed materials tended to

TABLE 1 Properties of Materials and Results.

1 Material	2 ACV or (TPF)	3 G_s	4 SMC %	5 MDD Mgm^{-3}	6 OMC %	7 H_{250} mm	8 N	9 $\theta_{2.5}$ %	10 k_{sat} 10^{-6} ms^{-1}	11 v_o 10^{-8} ms^{-1}	12 grad T $°Ccm^{-1}$	13 z_u cm	14 SP_o
(a) Sandy Gravels													
Ashton Keynes (102)	29	2.72	4.0	2.11	7.3	15.6	24	9.5	10	4.8	0.73	7.4	66
Spencers Farm (103)	12	2.64	2.7	2.14	6.6	6.5	24	7.2	46				
Stanley Ferry (106)	21	2.74	4.0	2.12	7.5	7.5	24	8.7	9				
Woodhall- (114 F1)	18	2.68	3.4	2.13	6.9	24.6	24	10.6	11	5.3	0.59	7.0	90
Spa (114 F2)	18	2.67	2.4	1.96	8.9	17.3	24	9.3	32				
(b) Crushed Rocks													
Croft (105)	17	2.69	1.6	2.15	6.9	2.6	24	4.3	19	0.5	0.63	6.1	8
Dene (119)	22	2.70	2.1	2.17	5.4	6.6	6	5.8	10	1.6	0.67	7.1	24
Dene (T)	22	2.70	2.3	2.04	4.8	4.5	3	6.1	10				
Whin 2 (B)	19	2.80	2.8	2.22	8.0	19.4	24	9.6	7	4.2	0.89	7.8	47
Whin 2 (T)	19	2.84	2.8	2.26	8.0	22.0	3	12.2	0.2				
OOL 2 (B)	(60)	2.72	4.1	2.05	7.4	21.9	18	12.5	9	4.0	0.78	7.7	51
OOL 1 (H) M	(75)	2.71	5.4	2.10	9.0	50.0	12	20.0	1				
F1				2.11	9.0	54.0	4	21.3	0.2				
DOL (H) M	30	2.84	4.2	2.23	7.5	13.4	12	13.4	1				
F1				2.25	7.8	19.5	4	14.7	0.1				
MIC (H) M	(40)	2.83	13.9	1.85	13.5	26.8	12	25.9	8				
F1				1.87	14.5	35.8	4	23.6	1				
CAL (H) M	28	2.71	4.6	2.20	7.5	42.0	12	15.4	2				
F1				2.21	8.0	43.5	4	17.3	0.5				
(c) Silty Sands													
SF4/80	-	2.65	-	2.00	9.00	22.9	18	12.8	0.88				
SF4/60 (T)	-	2.65	-	2.03	9.50	44.5	3	17.2	0.06				
Attenborough (T)	-	2.69	-	1.61	19.0	30.5	3	27.3	0.01				

Notes: 1. See text for description of materials. 2. Aggregate crushing value (or Ten Percent Fines/kN) BS812:1975. 3. Apparent specific gravity. 4. Saturation moisture content. 5,6. Maximum dry density and optimum moisture content, test 14, BS1377:1975. 7. Heave at 250 hours. 8. Number of specimens tested. 9. Volumetric water content at pF 2.5. 10. Coefficient of permeability (saturated). 11,12,13&14 apply at onset of terminal lens; v_o = velocity of water, grad T = temperature gradient, $SP_o/10^{-5}$ mm^2 s^{-1} $°C^{-1}$ = segregation potential, z_u = unfrozen depth.

be short of fines. For the purposes of most comparisons this was not a serious drawback.

Series B and H materials were prepared in a similar fashion except that generally they were not washed. In series H, the proportions were adjusted to achieve the M and F1 gradings after compaction.

Sand and limestone filler were mixed to give the gradings SF4/80 and SF4/60 (Figure 1).

Specimen Preparation for Frost Heave Tests

The materials were compacted into a mould in three layers to their maximum dry density at the optimum moisture content by using a vibrating hammer. In the final stage the end plugs were vibrated home to ensure that the required density was achieved. In series H, the final compaction was achieved by static loading of the end plugs after initial compaction with a vibrating hammer, except for the softest aggregate, MIC, which was compacted on a vibrating table.

After compaction and extrusion, specimens were weighed and placed in an apparatus (see below) for 24 hours to allow them to imbibe water prior to the onset of freezing.

THE TRRL FROST SUSCEPTIBILITY TEST

In the TRRL test, nine cylindrical specimens 102 mm in diameter and 152 mm high are placed in an apparatus (Lomas and Jones, 1981) so that the air temperature above them can be maintained at -17 ± 1°C. The specimens rest on porous ceramic discs within copper carriers. The discs are in contact with water, which is maintained at a constant level and at +4 ± 0.5°C. The sides of the specimens are wrapped with waxed paper, and the intervening space is filled with loose, dry sand (5 to 2.36 mm fraction). Push rods bearing on caps placed on top of the specimens enable the heave to be measured. Heave and water intake is

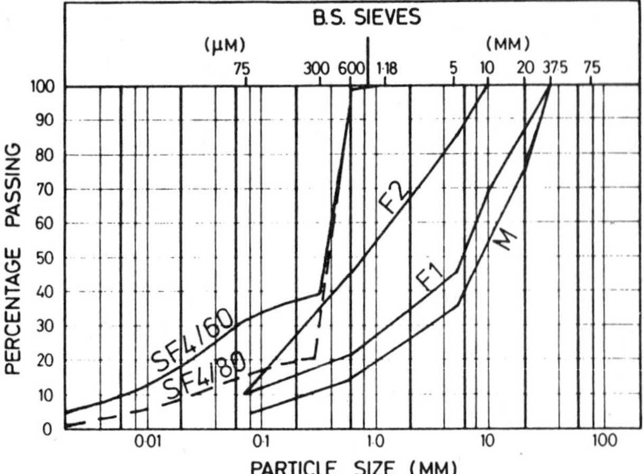

FIGURE 1 Grading Curves

recorded every 24 hours. Thermocouples are used to monitor the boundary and internal temperatures of the specimens.

Tests are normally done in sets of either three or nine. Until very recently, freezing continued for 250 hours and materials were judged frost susceptible, in England and Wales, if they heaved more than 13 mm. The freezing period has now been reduced to 96 hours (TRRL 1981) with a limit of 10 mm set by the Department of Transport.

Although the SRU is now the specified testing facility, properly conducted tests in which an insulated trolley containing the specimens was pushed into a cold room, gave similar results at least within the working range (Lomas and Jones 1981). Series H tests were done entirely in the cold room, but both facilities were used for the remaining series and the heave recorded was the average of all the relevant data.

To facilitate comparison with earlier data, only the 250 hour heaves are considered in this paper. (See Table 1).

The TRRL test is one of a number of direct freezing tests used in various parts of the world. There is some evidence that other tests, e.g., the CRREL test, would rank materials in the same order (Jones 1980).

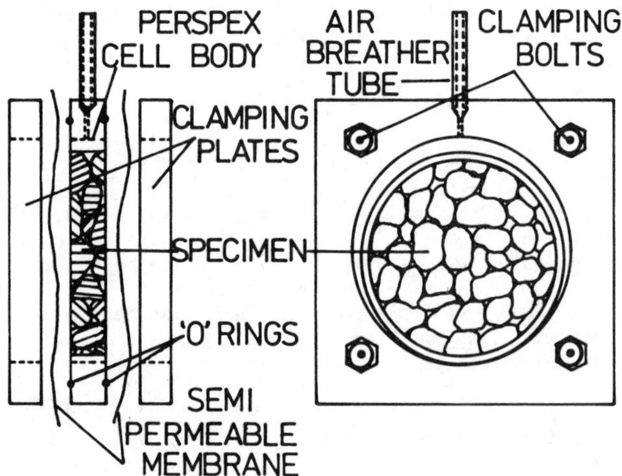

FIGURE 2 Osmotic Cell (Section and Elevation)

DETERMINATION OF SUCTION CHARACTERISTICS

The technique of applying suction by osmosis to the pore water in granular materials, (Jones and Hurt 1978) was further developed with a number of detailed improvements particularly in relation to specimen preparation.

The osmotic cell, figure 2, consisted of a central specimen chamber 110 mm diameter in which a slice of compacted aggregate 15 to 20 mm thick was flanked on either side by a semi-permeable dialysis membrane. The cells were immersed in a tank of polyethylene glycol (carbowax 6000) solution that was maintained at a constant temperature. Water then flowed from the specimen across the membrane, which was impermeable to the long chain carbowax molecules. Equilibrium was established (in five to seven days) between the suction in the pore water and the osmotic suction of the carbowax solution. A calibration curve was determined for each batch of carbowax by using an osmometer. Thereafter the suction could be found simply by measuring the density of the solution with an hydrometer. The effective range of the technique is pF 1.5 to about 4.4 (4 to 2500 kPa). A capillary rise method was used for lower suctions.

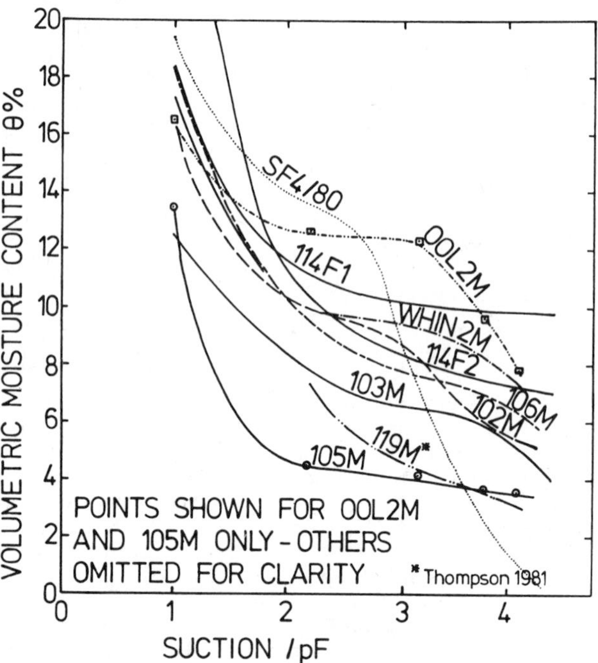

FIGURE 3 Suction Characteristics

Specimens of the harder aggregates were formed by compacting the material into a plastic pipe 102 mm diameter, 100 mm high that was sealed at its base with a disc to form a water-tight container. The entire assembly was frozen in the cold room at -17°C for 24 hours. While still frozen, it was sliced into discs 15 to 20 mm thick with a water cooled diamond studded saw, and the rough edges of the pipe smoothed with a file. The disc of aggregate, still contained within its plastic collar was then placed into the osmotic cell and clamped up. During the suction test, air could easily be drawn into the specimen between the collar and the flexible membrane.

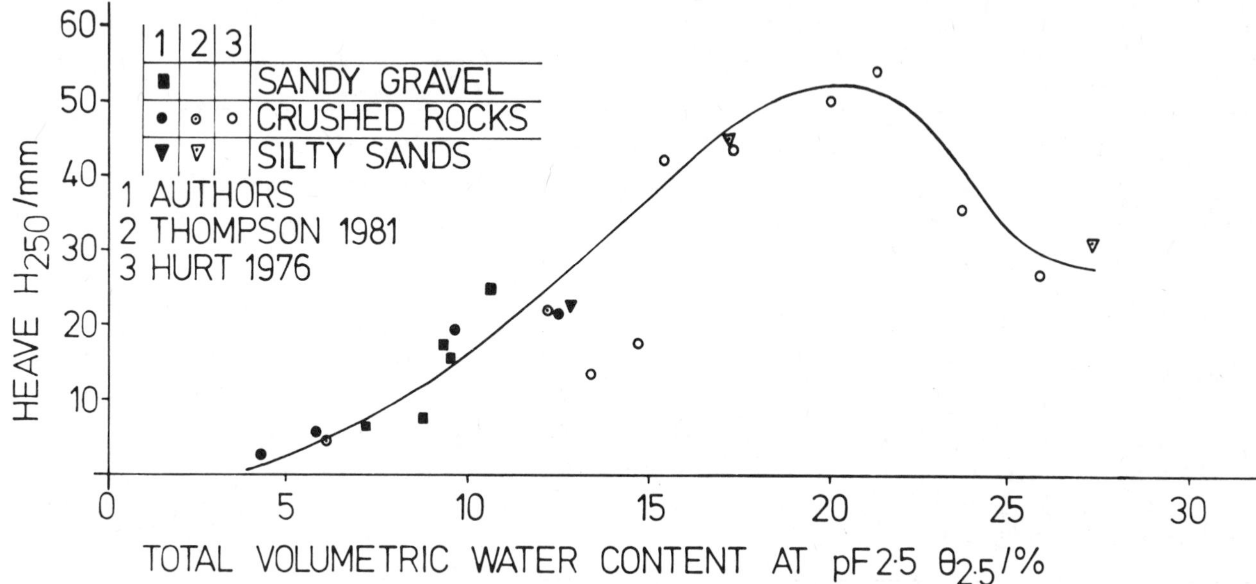

FIGURE 4 250 Hour Heave vs $\theta_{2.5}$

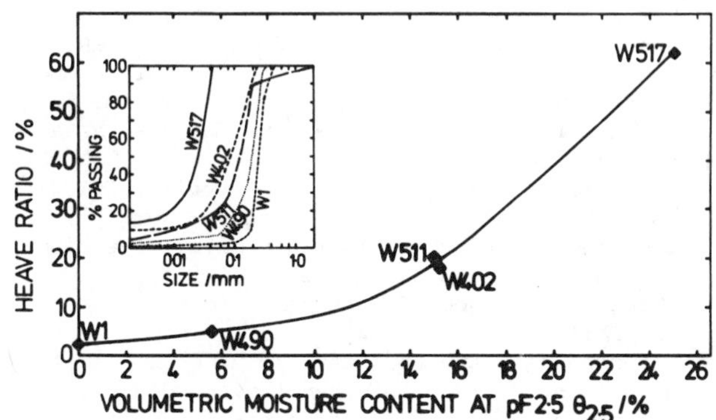

FIGURE 5 Heave Ratio vs $\theta_{2.5}$ (After Gorle, 1980)

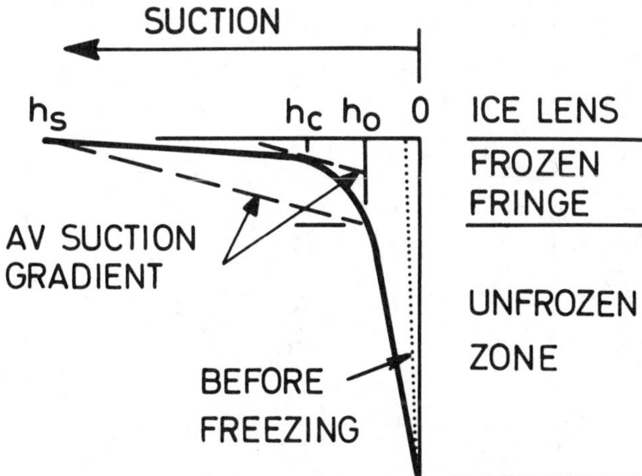

FIGURE 6 Suctions before Freezing

Specimens of sand/filler and of the softer limestones used in series H were prepared without freezing and were extruded from the collars prior to the suction test.

Typical aggregate suction curves are shown in figure 3. Each point was determined by using a separate specimen.

RELATIONSHIP BETWEEN SUCTION CHARACTERISTICS AND FROST SUSCEPTIBILITY

Comparison of figure 3 and column 7 of Table 1 showed that at about pF 2.5, (31 kPa) the volumetric moisture contents, θ ranked the materials in the same order as the frost heave. Figure 4 shows a plot of H_{250} against $\theta_{2.5}$. With one exception (DOL from the earlier series of tests in which slightly different techniques were used) the points lie fairly closely on a smooth curve. For $\theta_{2.5}$ less than 20, the heave increases with H_{250} but thereafter there is a decrease.

A similar dependence between heave behaviour and $\theta_{2.5}$ was observed in results published by Gorle (1980). A plot of heave ratio (heave/depth of frost penetration) against $\theta_{2.5}$ is shown in figure 5.

Significance of $\theta_{2.5}$

In both the secondary heaving theory (Miller 1978) and the segregation potential theory (Konrad and Morgenstern 1980, 1981), the crucial zone is the frozen fringe between the ice lens and the freezing front (Figure 6); the segregation potential theory is simpler and can be readily applied to the interpretation of the results presented earlier; in this theory, frost heave is viewed as a problem of impeded flow to the lens due to the low permeability associated with the high suctions present in the frozen fringe. Salient equations are presented in Table 2.

Estimates of the variation of permeability with suction were made by Thompson (1981) for two of

TABLE 2 Essential Equations of Segregation Potential Theory

Equations		Notation
$T_s = 8.04 \times 10^{-5}\, h_s$	(1)	$-T_s$ = temperature at the lensing front (°C)
		h_s = suction at the lensing front (cm water)
$v = \frac{1}{1.09}\frac{dH}{dt}$	(2)	h_o = suction at the freezing front (cm water)
		v_o = velocity of flow (ms^{-1})
$= (\frac{h_s - h_o}{h_s})\, \bar{k}_{ff}\, \frac{T_s}{z_{ff}}$	(3)	$\frac{dH}{dt}$ = rate of heave (ms^{-1}); grad T = temperature gradient (°C/m)
$= 10^{-4}$ SP grad T	(4)	$\bar{k}_{ff}$ = average coeff. of permeability in the frozen fringe (ms^{-1})
		$\bar{k}_u$ = average coeff. of permeability in the unfrozen zone (ms^{-1})
$= \bar{k}_u \frac{z_u}{h_o}$	(5)	z_{ff} = thickness of frozen fringe (cm)
		z_u = thickness of unfrozen zone (cm)
$\bar{k}_{ff} = \frac{8.04 \times 10^{-5}\, v}{\text{grad } T} (\frac{h_s}{h_s - h_o})$	(6)	SP = segregation potential ($mm^2\ s^{-1}\ °C^{-1}$)

the materials by using series parallel pore iteration theory (Kunze et al 1968) (Figure 7). Other data relevant to the onset of the terminal lens in these materials is given in Table 1. The velocity of flow was calculated from the heave/time curves by using equation (2). The segregation potentials are lower than 100×10^{-5} $mm^2\ s^{-1}\ °C^{-1}$ observed for Devon silt at h_o = 100 cm by Konrad and Morgenstern (1981).

Calculations using Figure 7 and equations (5) and (6) indicated that h_o was of the order of 100 cm or less and h_s was of the order 1000. Similar values for h_o were noted for a compacted glacial till by Garand and Ladanyi (1982). Assigning an average value of 1.05 to $h_s/(h_s - h_o)$ enabled $\bar{k}_{ff}$ to be found from equation (6). The corresponding pF values, of 2.6 for Dene and 2.8 for Whin, were read directly from Figure 7.

These values are in reasonable agreement with pF 2.5, which was shown earlier to be the suction at which the suction characteristics ranked the frost susceptibility. However, this agreement needs to be confirmed by further comparisons with directly measured permeability values at the appropriate suctions.

Nevertheless, it would seem reasonable to associate the average permeability $\bar{k}_{ff}$ with a characteristic suction h_c. This is the suction at the point where the suction gradient is equal to its average value (Figure 6). The characteristic suction h_c will not normally be the average suction. A graph of H_{250} against θ_c, the volumetric water content corresponding to h_c would be expected to show a monotonic increase. Thus pF 2.5 appears to approximate to the characteristic suction of all the materials with an OMC less than 10%.

For finer materials, for example the Attenborough silt or clay soils, the characteristic suction is likely to be higher so that the θ_c would be lower than $\theta_{2.5}$. Similarly, the high internal porosity of the MIC particles could lead to $\theta_{2.5}$ being an overestimate of θ_c. This behaviour is reflected by the higher OMC values in the materials which plot on the decreasing limb of Figure 4.

USE OF $\theta_{2.5}$ AS A FROST SUSCEPTIBILITY INDEX

The five materials with $\theta_{2.5}$ less than 9% were all non-frost susceptible according to the TRRL test. If further testing on a considerable number of other granular materials were to show similar behaviour, then this criteria could be used as a method of assessing frost susceptibility. However, the time taken and the operator effort involved in assessing a single material is similar for the suction method and for a four-day direct freezing test. The latter is therefore likely to remain as the preferred method of assessing frost susceptibility.

Nevertheless, the apparatus required by the suction method is simpler than that needed for direct freezing tests and it is easily duplicated. Thus the suction method could be a useful adjunct

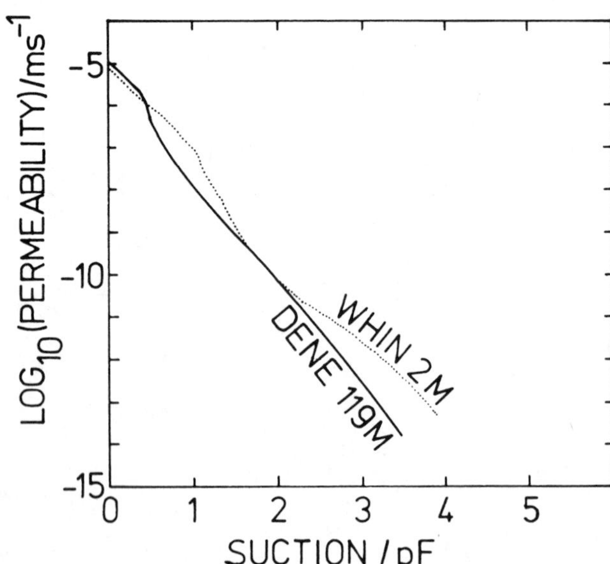

FIGURE 7 Estimated Permeability Characteristics (After Thompson, 1981)

especially when only small samples were available, or when large numbers of materials needed to be assessed simultaneously, e.g., at site investigation or tender (bidding) stages.

CONCLUSIONS

1. For a range of granular materials, frost heave and $\theta_{2.5}$ (the volumetric water content at a suction of pF 2.5, 31 kPa) were related by a single curve.
2. The average permeability in the frozen fringe is associated with the average suction gradient. The suction corresponding to the average gradient, can be considered as the 'characteristic suction' for the fringe.
3. The characteristic suction approximates to pF 2.5 for a wide variety of granular materials with an optimum moisture content less than 10%.
4. Materials for which $\theta_{2.5}$ was less than 9% were non-frost susceptible according to the TRRL test.
5. Further testing is required on a considerable number of materials to establish whether a limit based on suction characteristics is of general applicability.

ACKNOWLEDGEMENTS

We wish to thank TRRL for a contract and the British Quarrying and Slag Federation for grants that provided the main funding for this work, and Drs. K.G. Hurt and J.D. Thompson for their contributions to the overall research program.

REFERENCES

BS812:1975 Methods for sampling and testing of mineral aggregates, sands and fillers: Pt 3 Mechanical properties. British Standards Institution London, 18 pp.

BS1377:1975 Methods of test for soils for engineering purposes. British Standards Institution London, 143 pp.

Garand, P., and Ladanyi, B. (1982) Frost susceptibility testing of a compacted glacial till. Proceedings 3rd International Symposium on Ground Freezing, USA CRREL, p. 277-284.

Gorle, D. (1980) Frost susceptibility of soils: influence of the thermal variables and depth to water table. Proceedings 2nd International Symposium on Ground Freezing, Trondheim, Norway, p. 295-309.

Hurt, K.G. (1976) The prediction of frost susceptibility of limestone aggregates with reference to road construction. Ph.D. Thesis University of Nottingham, 307 pp.

Jones, R.H. (1980) Developments and applications of frost susceptibility testing. Proceedings 2nd International Symposium on Ground Freezing, Trondheim, Norway. Also in Engineering Geology vol. 18 (1981), p. 269-280.

Jones, R.H., and Hurt, K.G. (1978) An osmotic method for determining rock and aggregate suction characteristics with applications to frost heave studies. Quarterly Journal of Engineering Geology, vol. 11, no. 3, p. 245-252.

Konrad, J.M. and Morgenstern, N.R. (1980) A mechanistic theory of ice lens formation in fine grained soils. Canadian Geotechnical Journal, vol. 19, p. 472-486.

Konrad, J.M., and Morgenstern, N.R. (1981) The segregation potential of a freezing soil. Canadian Geotechnical Journal, vol. 18, p. 482-491.

Kunze, R.J., Uehara, I., and Graham, K. (1968) Factors important in the calculation of hydraulic conductivity. Soil Science Journal of America, vol. 26, p. 421-426.

Lomas, K.J., and Jones, R.H. (1981) An evaluation of a self refrigerated unit for frost heave testing. Transportation Research Record 809, p. 6-13.

Miller, R.D. (1978) Frost heaving in non-colloidal soils. Proceedings 3rd International Conference on Permafrost, Edmonton, vol. 1, p. 708-713.

Thompson, J.D. (1981) Subgrade effects on the frost heave of roads. Ph.D. Thesis, University of Nottingham, 325 pp.

TRRL (1981) Specification for the TRRL frost heave test. Materials Memorandum, MM64, 42 pp.

DEFORMATION BEHAVIOUR OF FROZEN SAND-ICE MATERIALS UNDER TRIAXIAL COMPRESSION

S.J. Jones[1] and V.R. Parameswaran[2]

[1]Snow and Ice Division, National Hydrology Research Institute, Environment Canada, Ottawa, Ontario, K1A 0E7
[2]Division of Building Research, National Research Council of Canada, Ottawa, Ontario, K1A 0R6

Deformation behaviour and strength of frozen sand-ice materials containing 0.1 to 0.6 weight fraction of sand were studied by testing the samples at -11°C in compression at a strain-rate of 7.7×10^{-5} s^{-1}, under hydrostatic pressures varying between 0.1 and 85 MPa. The peak strength increased with increasing hydrostatic pressure, reached a maximum for confining pressures around 25 MPa, and decreased for further increase in pressure. The results are compared with those obtained earlier in pure polycrystalline ice and frozen saturated sand.

INTRODUCTION

The effects of hydrostatic pressure on the deformation behaviour, and strength, of frozen saturated sand of uniform grain size containing about 20% moisture by weight of sand, were reported earlier (Parameswaran and Jones, 1981). As a sequel to that work, the authors continued to study the behaviour of frozen sand-ice mixtures of various sand to ice ratios, under triaxial conditions. The present paper reports results on these new materials.

The objective of this project is to obtain an overall picture of the deformation behaviour of sand-ice materials with increasing percentages of sand dispersed in ice, starting from pure ice at one end of the spectrum (Jones, 1978) to saturated sand at the other end (Parameswaran, 1980). Such a study will enable us to understand the physical processes involved in the deformation of frozen earth materials under triaxial conditions, and will also be helpful in evolving design criteria for foundations in ice rich permafrost soils. Also, the sliding of glaciers on their beds is controlled by the deformation of such ice-rich materials (debris-laden ice) at the bottom of the glaciers, and subjected to hydrostatic pressures due to the overburden.

Very little work has been reported on the study of mechanical behaviour of sand-ice materials under high confining pressures. Besides Chamberlain et al. (1972), Alkire and Andersland (1973) and Sayles (1974), the only others are the work of Simonson et al. (1974), and our own work reported earlier (Parameswaran and Jones, 1981).

EXPERIMENTAL PROCEDURE

(a) Specimen Preparation

The sand used for preparing the test specimens was from Ottawa, Illinois, corresponding to specification ASTM C-109, and having a uniform grain size between 0.2 and 0.6 mm. Cylindrical sand-ice test specimens with different compositions were prepared in split acrylic molds (51 mm i.d., 180 mm deep) having provision for evacuation from the top and suction of water from the bottom. Figure 1 shows a mold and a sample made in it.

In the present experiments the compositions of the samples used were 10 to 20% sand by weight dispersed in ice, and sand-ice containing 37 to 50% ice. Test specimens were prepared in a cold room kept at -6°C. A specimen of the required composition was made as follows: The volume fractions of sand and ice for a specimen of desired composition were calculated, snow and sand in the proper weight fractions were mixed to obtain the volume fraction calculated above, the acrylic mold was filled with the snow sand mixture (without compacting, but with light tapping), the mold was evacuated to some extent, and distilled water, precooled to just above 0°C, was drawn into the mold through a port at the bottom, with the vacuum pump running, until the mold was full. The water froze immediately, with the snow particles and the cold sand particles acting as nucleating agents. In this procedure it was impossible to eliminate the air bubbles completely, but the test specimens obtained had a fairly uniform distribution of sand throughout their volume. This was confirmed by cutting a sample perpendicular to its axis into several discs, and determining the moisture content of each by drying.

After allowing the sample in the mold to sit in the cold room overnight to attain temperature equilibrium, the specimen was removed from the mold, 25 mm length was cut off from either end, and the ends faced in the lathe in the cold room. Stainless steel end caps were mounted at either ends of the specimen using precooled distilled water. The finished test specimens had a nominal gauge length of about 108 mm between the end caps, so that the length to diameter ratio was greater than 2. Prior to testing, the diameter and length of each sample were accurately measured.

(b) Test Procedure

The frozen cylindrical sand-ice specimens were tested in a high pressure triaxial cell capable of withstanding hydrostatic pressures up to 200 MPa (2 k bar). Figure 2 shows a schematic diagram of the pressure vessel. The pressure vessel was kept inside a cold box maintained at -11°C, on the base of a universal testing machine (instron Model 1116, having a capacity of 250 kN). All tests were carried out at a constant rate of crosshead motion of 0.5 mm/min, which amounts to a nominal strain rate of about $7.7 \times 10^{-5}\ s^{-1}$ on sample. The hydrostatic pressure was varied for each test. The pressurizing medium was Dow Corning silicone fluid 200. Uniaxial compressive force was applied by a piston (P) sealed by two O-rings. The hydrostatic pressure inside the cell, as a sample was deformed, was constantly monitored by a Bourdon type of pressure gauge connected at the port H of the pressure vessel. Thus the exact hydrostatic pressure inside the cell at the peak load could be known for each test. The load on the specimen at any instant was obtained by subtracting the load due to friction and the hydrostatic pressure from the total load indicated by the load cell of the testing machine. The "nominal strain" shown on the various figures was calculated from the speed of crosshead motion and elapsed time divided by the original sample length. The temperature inside the pressure vessel was monitored by a thermocouple connected to the port T. During a test, the variation in temperature was usually less than 0.1°C. After a test, the sample was taken out of the pressure vessel, drained of silicone fluid, photographed and dried in an oven to determine the moisture content.

RESULTS AND DISCUSSIONS

Figure 3 shows typical deformed sand-ice samples. Figure 4 shows some representative stress-strain curves from samples of various compositions. The deformation behaviour of these samples and their stress-strain curves were more similar to pure polycrystalline ice samples tested under high pressure (Jones, 1978), than the saturated sand samples reported earlier (Parameswaran and Jones, 1981). Unlike the latter, which showed an initial yield point, post yield work hardening behaviour and a peak stress, the present sand-ice samples showed a peak stress coinciding with the yield point at the end of the pseudo elastic region. After this peak the stress dropped rather rapidly at first, and later on more slowly, but continuously. Table I shows the data of peak stress, hydrostatic pressure, strain at peak, etc. for the various samples. The samples tested could be divided into 4 groups with the following sand concentrations: (a) 11.7 ± 1.0%, (b) 16 ± 2%, (c) 48 ± 2%, and (d) 63 ± 3%. The results of these are shown in Figure 5, where the peak stress is plotted against hydrostatic pressure for various samples.

All the curves in Figure 5 show that the peak stress increases with increasing confining pressure and reaches a maximum at a pressure of about 25 MPa, (12 MPa in case of 50:50 sand ice) and then decreases for further increase in hydrostatic pressure. The behaviour is similar to that observed earlier in frozen saturated sands (Parameswaran and Jones, 1981). Chamberlain and others (1972), Alkire and Andersland (1973), and Sayles (1974) have also reported the increase in strength with increasing hydrostatic pressure as observed by us and shown by the early part of Figure 5. This behaviour could be due to the closure of voids and microcracks in the samples. The decrease in strength with further increase in hydrostatic pressure, after the maximum, is thought to be due to increasing amounts of unfrozen water in the samples due to pressure melting around the sand grains.

Simonson et al. (1974) have reported testing ice and frozen soils at -10°C under confining pressures of up to 200 MPa. Their results, at first sight, seem to be the opposite of earlier observations quoted above. They reported that the peak strength in ice decreased monotonically with increasing pressure, whereas in frozen saturated soils (Ottawa sand and kaolinite), the peak strength at first decreased with increasing hydrostatic pressure, reached a minimum at a confining pressure of 120 MPa, and then increased again for further increase in confining pressure. This latter increase in strength was attributed to the frictional effect arising from interparticle contact. A review of their test data (given in Appendix A of their paper), however, shows that all their tests were done at atmospheric pressure (1 bar of 0.1 MPa), or at pressures exceeding 50 MPa (0.5 k bar), except for one sample of polycrystalline ice tested at 40 MPa, and two samples of saturated Ottawa sand tested at 46 and 48 MPa, respectively. From our tests we saw that the strength reaches a peak value at a confining pressure of about 25 MPa for ice and sand-ice materials, and hence Simonson et al. (1974) could have missed this in their tests.

In Figure 6 the present results on sand-ice materials are compared with our earlier results on frozen saturated sand and on ice (Parameswaran and Jones 1981, and Jones 1978, respectively). Curves (a), (b) and (c) are the same as those of Figure 5. Curve (d) is the expected behaviour of ice at a strain rate of $7.7 \times 10^{-5}\ s^{-1}$ (as used in the present tests),calculated from the results of Jones (1978), using the formula for the dependence of strain rate $\dot{\varepsilon}$ on the strength σ:

$$\dot{\varepsilon} \propto \sigma^n$$

where the exponent n was found to have a value of 5.4 at atmospheric pressure (0.1 MPa) and 3.9 in high pressure tests. Curve (e) in Figure 6 shows the variation of peak strength with hydrostatic pressure, for frozen saturated sand containing 20% moisture by weight of sand (this corresponds to a sand/ice ratio of 83/17 by weight, or a volume fraction of 63/37).

These results seem to indicate that the macroscopic strength and deformation behaviour of ice is not considerably affected by the addition of sandy material up to 20% by weight. The strength of the sand-ice material was only slightly higher than that of pure ice. (Similar results on frozen sand-ice materials were reported earlier by Goughnour and Andersland (1968) from uniaxial compression and creep tests, and by Alkire and Andersland (1968) from uniaxial compression and creep tests, and by Alkire and Andersland (1973) from triaxial tests.) This would seem to support the criterion of the application of the flow law for ice for the design of piles in ice-rich soils (Morgenstern et al. 1980). In the sand-ice material containing 48-63% sand, the effect of interparticle soil grain friction also comes into play and causesan enhancement of strength. These samples showed a higher strength than the samples with 11-16% sand, but considerably lower than the frozen saturated sand (containing about 83% sand). For all these materials the strength after the peak drops with increasing hydrostatic pressure, presumably due to increasing amounts of unfrozen water caused by pressure melting at the ice-soil interface. (In polycrystalline ice the pressure melting presumably starts at the grain boundaries).

Further tests on sand-ice materials containing sand fractions between 0.2 and 0.8 are underway and these will be reported elsewhere.

CONCLUSION

Sand-ice materials containing about 10 to 60% sand dispersed uniformly in ice were tested in compression in a high pressure triaxial cell under hydrostatic pressures varying from 0.1 to 85 MPa, at a temperature of -11°C and a compressive strain-rate of $7.7 \times 10^{-5}\ s^{-1}$. The macroscopic deformation behaviour and strength of those samples which contained 11-16% sand were very close to the behaviour of polycrystalline ice tested under similar conditions. For samples containing 48-63% sand, the strength was higher than the previous samples, but much lower than the frozen saturated sand (containing about 17% moisture by weight) tested under similar conditions.

For all these materials the peak strength increased hydrostatic pressure, reached a maximum at a particular confining pressure (about 30 MPa for polycrystalline ice, about 40 MPa for frozen saturated sand, and about 25 MPa for the present sand-ice materials), and then decreased with further increase in confining pressure. The decrease of strength with increasing pressure after the maximum is presumably due to the increase in unfrozen water content caused by pressure melting round the sand particles.

ACKNOWLEDGEMENTS

The authors would like to thank Colin Hubbs (DRB/NRC) and Hemming Chew (NHRI/DOE) for their help in sample preparation and testing.

REFERENCES

Alkire, B.D. and Andersland, O.B., 1973. The effect of confining pressure on the mechanical properties of sand-ice materials. Journal of Glaciology, Vol. 12, No. 66, pp. 469-481.

Chamberlain, E., Groves, C. and Perham, R., 1972. The mechanical behaviour of frozen earth materials under high pressure triaxial test conditions. Géotechnique (London), Vol. 22, No. 3, pp. 469-483.

Goughnour, R.R. and Andersland, O.B., 1968. Mechanical properties of a sand-ice system. Proc. ASCE. J. Soil Mech. and Foundn. Dn., Vol. 94, No. SM4, PP. 923-950.

Jones, S.J., 1978. Triaxial testing of polycrystalline ice. Third International Conference on Permafrost, Edmonton, 1978. Proceedings published by National Research Council of Canada, Ottawa, Vol. I, pp. 670-674.

Morgenstern, N.R., Roggensack, W.D. and Weaver, J.S., 1980. The behaviour of friction piles in ice and ice rich soils. Canadian Geotechnical Journal, Vol. 17, No. 3, pp. 405-415.

Parameswaran, V.R., 1980. Deformation behaviour and strength of frozen sand. Canadian Geotechnical Journal, Vol. 17, No. 1, pp. 74-88.

Parameswaran, V.R. and Jones, S.J., 1981. Triaxial testing of frozen sand. Journal of Glaciology, Vol. 27, No. 95, pp. 147-156.

Sayles, F.H., 1974. Triaxial constant strain rate tests and triaxial creep tests on frozen Ottawa sand. U.S. Army Cold Regions Research and Engineering Technical Report 253, 32 p.

Simonson, E.R., Jones, A.H. and Green, S.J., 1974. High pressure mechanical properties of three frozen materials. Proceedings of the Fourth International Conference on High Pressure, Kyoto, Japan, pp. 115-121. (Also, TerraTek Report No. TR 74-15, 1974, prepared for U.S. Army Natick Laboratories).

TABLE I: Details of all the tests reported.

Sample No.	Temperature °C	Strain Rate $10^{-5}s^{-1}$	Peak (True) Stress $(\sigma_1-\sigma_3)$ MPa	Strain At Peak Stress %	Hydrostatic Pressure At Peak (σ_3) MPa	Sand Content %
20	-11.0	7.69	4.83	1.05	0.1	12.5
21	-10.4	7.68	8.10	1.84	18.0	11.6
22	-11.0	7.66	7.73	1.61	40.0	12.7
23	- 9.8	7.70	6.84	1.66	62.8	11.7
24	-10.2	7.71	6.84	1.85	70.0	12.2
25	-10.9	7.71	7.96	1.71	11.2	11.7
26	-10.4	8.21	4.79	1.72	0.1	11.2
27	-10.6	8.19	8.28	1.96	30.3	11.3
28	-10.8	8.13	7.59	2.19	48.5	11.1
29	-10.6	8.20	7.56	2.95	22.2	10.7
30	-10.8	8.27	7.87	2.23	18.2	12.4
31	-10.8	8.15	5.17	2.69	0.1	11.2
5	-10.1	7.72	4.78	0.75	0.1	-
6	-10.0	8.63	4.94	1.10	0.1	-
7	- 9.6	7.61	3.89	0.75	0.1	-
8	-10.3	7.45	5.50	2.75	36.0	-
9	-10.5	7.52	7.81	1.85	33.5	-
10	-10.5	7.63	5.08	2.00	0.1	-
11	-10.0	7.61	7.60	2.05	28.7	-
12	-10.2	7.67	7.06	1.75	18.0	-
13	-10.0	7.62	6.64	1.85	10.7	-
14	-10.0	7.71	7.54	1.60	54.0	14.2
15	- 9.8	7.70	6.44	1.65	67.5	16.7
16	- 9.5	7.70	4.98	1.70	81.4	14.4
17	- 9.7	7.62	7.01	1.35	40.7	17.0
18	- 9.7	7.69	5.03	1.15	0.1	18.6
19	- 9.4	7.73	6.92	1.50	47.5	18.2
33	-11	8.20	7.82	2.07	0.1	50.5
34	-11	8.23	10.14	2.72	45.6	52.1
35	-11	8.18	9.94	2.31	20.55	49.7
36	-11	8.18	9.47	1.84	61.50	53.7
37	-11	8.24	10.41	1.98	14.95	51.5
38	-11	7.76	10.25	1.63	6.10	44.9
40	-11	7.77	9.97	1.63	15.85	40.6
41	-11	7.76	10.93	1.63	11.40	49.6
52	-11	7.74	7.90	-	0.1	54.3
42	-11	7.67	10.99	1.38	3.65	58.5
43	-11	7.65	12.70	1.52	22.95	62.4
44	-11	7.71	12.25	2.36	28.10	65.3
45	-11	7.70	12.17	2.24	35.20	65.9
46	-11	7.68	9.74	3.00	38.40	66.4
48	-11	7.67	11.61	1.89	43.0	61.5
49	-11	7.69	9.40	2.45	26.4	62.6
50	-10.8	7.68	8.36	-	49.4	64.7
51	-11	7.72	10.26	1.71	23.9	62.2
53	-11.3	7.72	9.26	1.48	0.1	63.5

FIGURE 1 Split acrylic mold and finished test specimen.

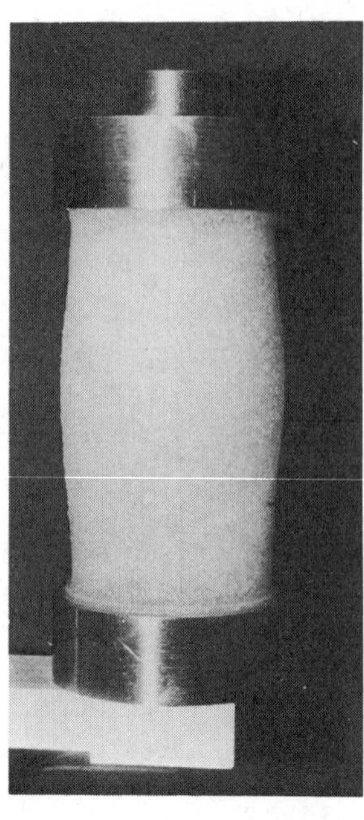

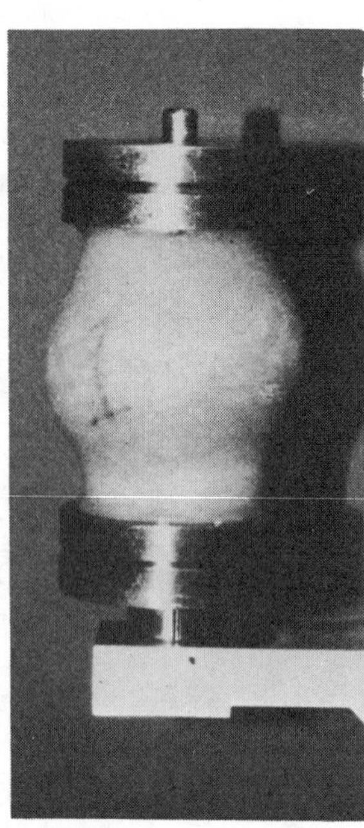

FIGURE 3 Sand-ice samples deformed under high pressure.

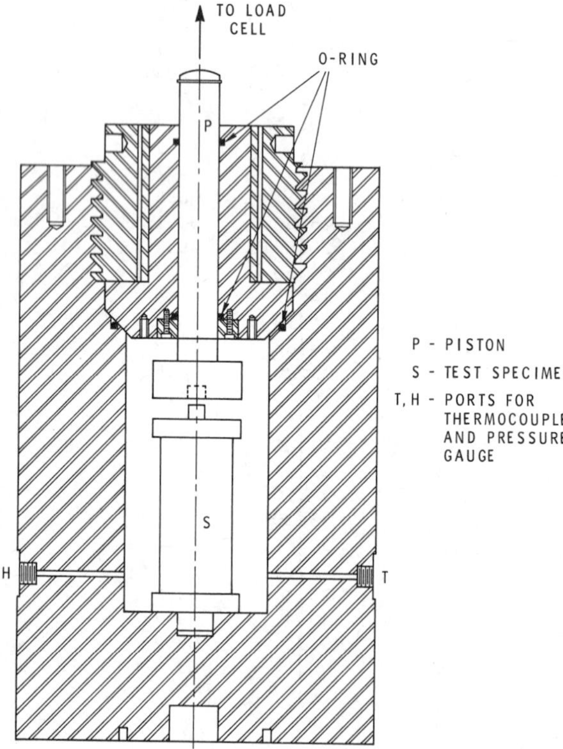

FIGURE 2 Schematic diagram of the high pressure cell.

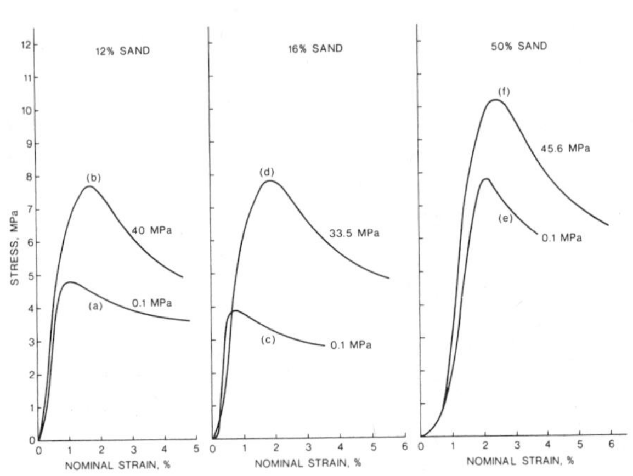

FIGURE 4 Typical stress strain curves for sand-ice materials tested under high pressure. (a, b - 12.% sand; c, d - 16% sand; e, f - 50% sand.).

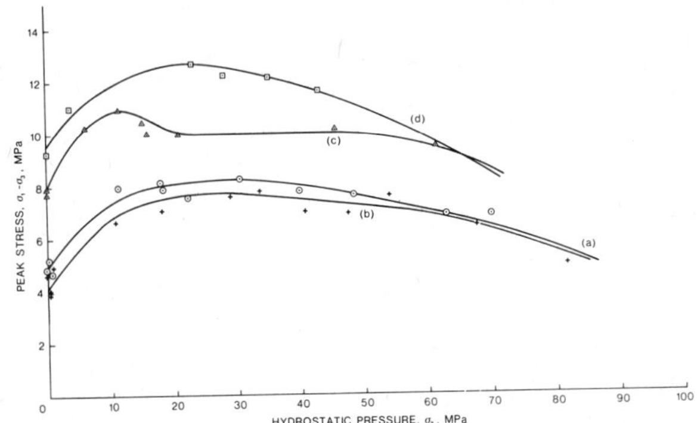

FIGURE 5 Variation of peak strength with confining hydrostatic pressure for various sand-ice materials. (a): 11.7 ± 1% sand, (b) 16 ± 2% sand, (c): 48 ± 2% sand, (d): 63 ± 3% sand.

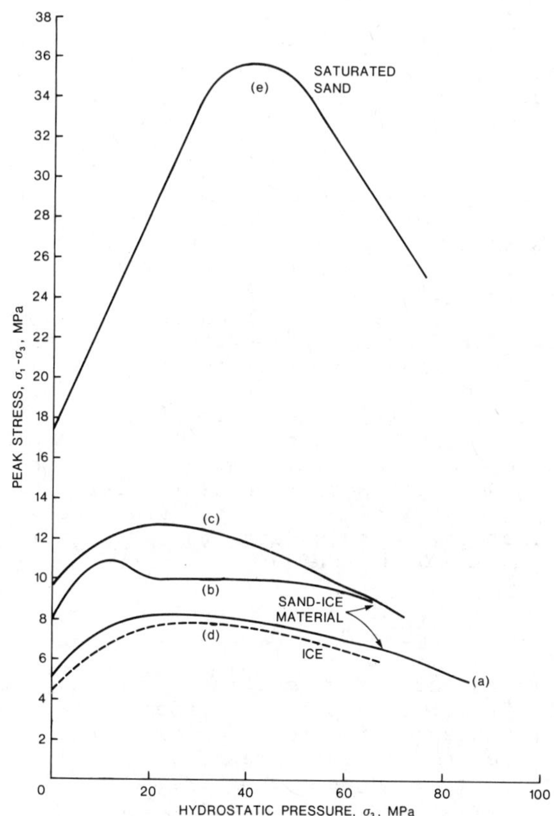

FIGURE 6 Peak strength vs Hydrostatic Pressure: present results (a - 12-16% sand, b - 48% sand, c - 63% sand) compared with those for polycrystalline ice (curve d) and frozen saturated sand (curve e).

PERMAFROST-RELATED TYPES OF LARGE-SCALE DISSECTION, DEGRADATION, AND DEFORMATION OF MARTIAN LANDSCAPE

Heinz - Peter Jöns

Geologisches Institut, Technische Universität Clausthal, Leibnizstrasse 10, 3392 Clausthal-Zellerfeld, Federal Republic of Germany

Aureoles, mega-aureoles and chaotic terrains are among Mars' most significant permafrost-related surface structures. In areas of high relief the melting of a proposed layer of underlying permafrost led to downslope sliding of overlying rigid material. The results were aureoles with a maximum diameter up to about 1,600 km which surround the largest volcanoes (Olympus Mons, Elysium Mons). Large-scale downslope sliding on the flanks of very large updomings led to the development of mega-aureoles with a diameter up to about 6,000 km.
In areas of low relief the melting of underlying permafrost led to the origin of much smaller arcuately bordered depressions with maximum diameters up to about 300 km, the chaotic terrains. Most depressions of this type are related to the Chryse Planitia by large outflow channels. All these features seem to have no counterparts on Earth.

INTRODUCTION

Chaotic terrains and aureoles are members of a larger family of permafrost-related features of the Martian surface. Chaotic terrains were initially recognized and described from Mariner 6 frames (Sharp et al. 1971) and later observed in detail from Mariner 9 (Sharp 1973). Their existence has been confirmed by Mariner 9 and Viking pictures. The origin of chaotic terrains has commonly been attributed to undermining or sapping mechanisms or ground ice deterioration (Sharp et al. 1971, Sharp 1973) and was recently attributed to melting of underlying ice deposits and/or permafrost (Soderblom and Wenner 1978, Jöns 1982a). Carr (1977) mentiones chaotic terrains "as additional supporting evidence for extensive ground ice in the cratered uplands". The role of the chaotic terrains during the development of the Mars surface relief has been explained by Jöns (1982a).

The aureole material which surrounds Olympus Mons was first described from Mariner 9 pictures (Carr 1973). The aureole extending from the volcano contains a series of overlying lobated thin units with arcuate boundaries. Olympus Mons itself is bordered by a scarp which stands 2-8 km high (Wu et al. 1981). Among the numerous theories which have been proposed to explain the origin of the aureole are postulates that the aureole has been caused by the weight of Olympus Mons on bedded deposits (Harris 1977) and subglacial volcanic deposits (Hodges and Moore 1980). Morris (1982) has reinterpreted the aureole materials to be the products of at least six pyroclastic eruptions prior to the construction of Olympus Mons. Lopes et al. (1980, 1982) proposed a mass movement origin for the aureole. They interpreted the scarp as the scar left by the removal of material. Lopes and Guest (1981) have considered "that the emplacement of these rockslides was probably aided by melting of permafrost".

Jöns (1982b, 1983a,b) has pointed out that very large aureole-like features are existing in the surroundings of the Tharsis-Noctis Labyrinthus-Valles Marineris (TaNoVa) updomings and in the surroundings of the Elysium dome. According to the very large diameter of those structures he has called them mega-aureoles.

In this paper it is proposed that both groups of features, the different types of chaotic terrains as well as the aureoles and the mega-aureoles are the result of (?partial) melting of underlying permafrost, with reference to Anderson (1982) who has pointed out that "Calculations based on the Viking Mission Data results indicate that permafrost thicknesses range from about 3.5 kilometers at the equator to approximately 8 kilometers in the polar regions".

AREAS WITH A HIGH RELIEF

If one assumes a melting of underlying permafrost, a reduction in shear strength in the area of the future sliding would produce a high fluid pressure and a low shear stress. Under such conditions slides could occur under their own weight on slopes of 1% (Hubbert and Rubey 1959, Lopes et al. 1982). Therefore it seems to be possible that downslope sliding of overlying rigid- mainly volcanic -material could form aureoles of different scales on Mars in areas with a sufficient degree of relief.

Aureoles

Aureoles have been described from the sur-

FIGURE 1 Fully developed aureole around Olympus Mons (OM).

roundings of some large volcanoes (Olympus Mons, Elysium Mons). The best example of a fully developed aureole is situated around Olympus Mons (Figure 1). It cosists of many overlapping lobate units which indicate a multiple phase origin. Because the aureole around Olympus Mons has been discussed in detail elswhere, it will not be discussed futher here.

The author is of the opinion that it is quite possible that the numerous concentric features which surround Elysium Mons represent a special type of an aureole the development of which came to a halt in a very early stage. Lobate units do not exist but Christiansen and Greeley (1982) suspect mega-lahars in this region (Figure 2).

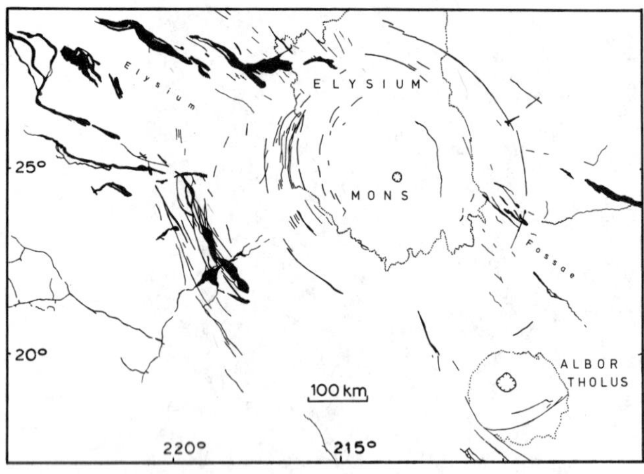

FIGURE 2 Aureole around Elysium Mons.

Mega-Aureoles

Mega-aureoles which surround the TaNoVa updomings and the Elysium dome have been identified by Jöns (1983a,b). In the case of the TaNoVa updomings large parts of the surrounding landscape became instable- perhaps due to melting of underlying permafrost -and formed a special series of narrow landscapes during downslope sliding of overlying rigid material (for detail see Jöns 1982b). All those features run concentrically with respect to Syria Planum -the highest part of the TaNoVa updomings- and have been interpreted as shove structures (Jöns 1982b, 1983a,b).

East and Southeast of the Elysium dome remnants of a former mega-aureole are suspected (Jöns 1983b). The relief of this area is characterized by wrinkle ridges and numerous erosional outliers which run concentrically with respect to the center of the Elysium dome. On the Northwest flank of this dome mega-lahars have been suspected by Christiansen and Greeley (1982), (Figure 3).

AREAS WITH LITTLE RELIEF

Chaotic terrains are characterized by polygonal fracture systems, angular mesas and rounded hummocks arranged disorderly in irregular hollows, circular depressions, or at the base of scarps (Lucchitta, 1982). Further characteristics are arcuate steep escarpments which stand up to about 3,000 m high and deeply incised straight running outflow channels which connect many chaotic terrains with the Chryse Planitia. Most chaotic terrains are found in the Margaritifer Sinus and in the Oxia Palus quadrangle. Sharp (1973) ascribed the origin of the chaotic terrains to possible collapse through the degradation of ground ice. Lucchitta (1982) is of the opinion that some chaotic terrains are associated with fluvial depositional centers and suggests that chaotic terrains may have developed preferentially on sedimentary deposits. In a preliminary classification of the chaotic terrains Jöns (1982a) has distinguished three types: the nonprogressive and the progressive chaotic terrains and the linearly delineated collapse structures. Carful investigations have shown that at least five types can be distinguished.

Nonprogressive Chaotic Terrains

These structures are encircled by arcuate escarpments which stand up to about 3,000 m high. Probably due to degradation of ground ice (Sharp 1973) the overlying rigid material sank and cracked. This led to the origin of numerous linear shallow grabens which transsect the remnants of that material within the chaotic terrains. Most of these depressions are related to the Chryse Planitia by large and deeply incised straight outflow channels (e. g. Ares-, Tiu-, Simud Vallis), (Figure 4).

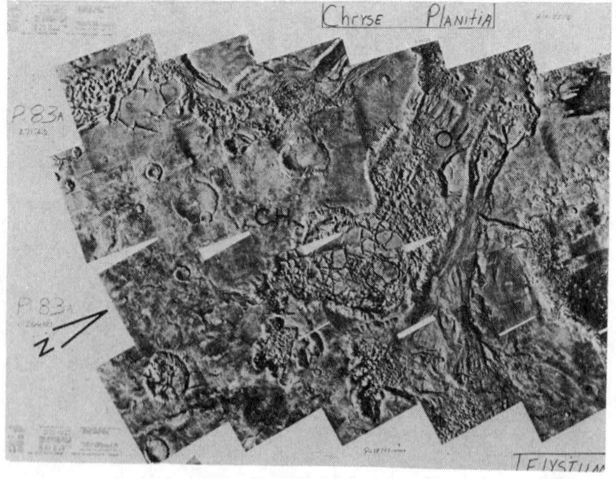

FIGURE 4 Nonprogressive chaotic terrain (CH) with large outflow channel (O).

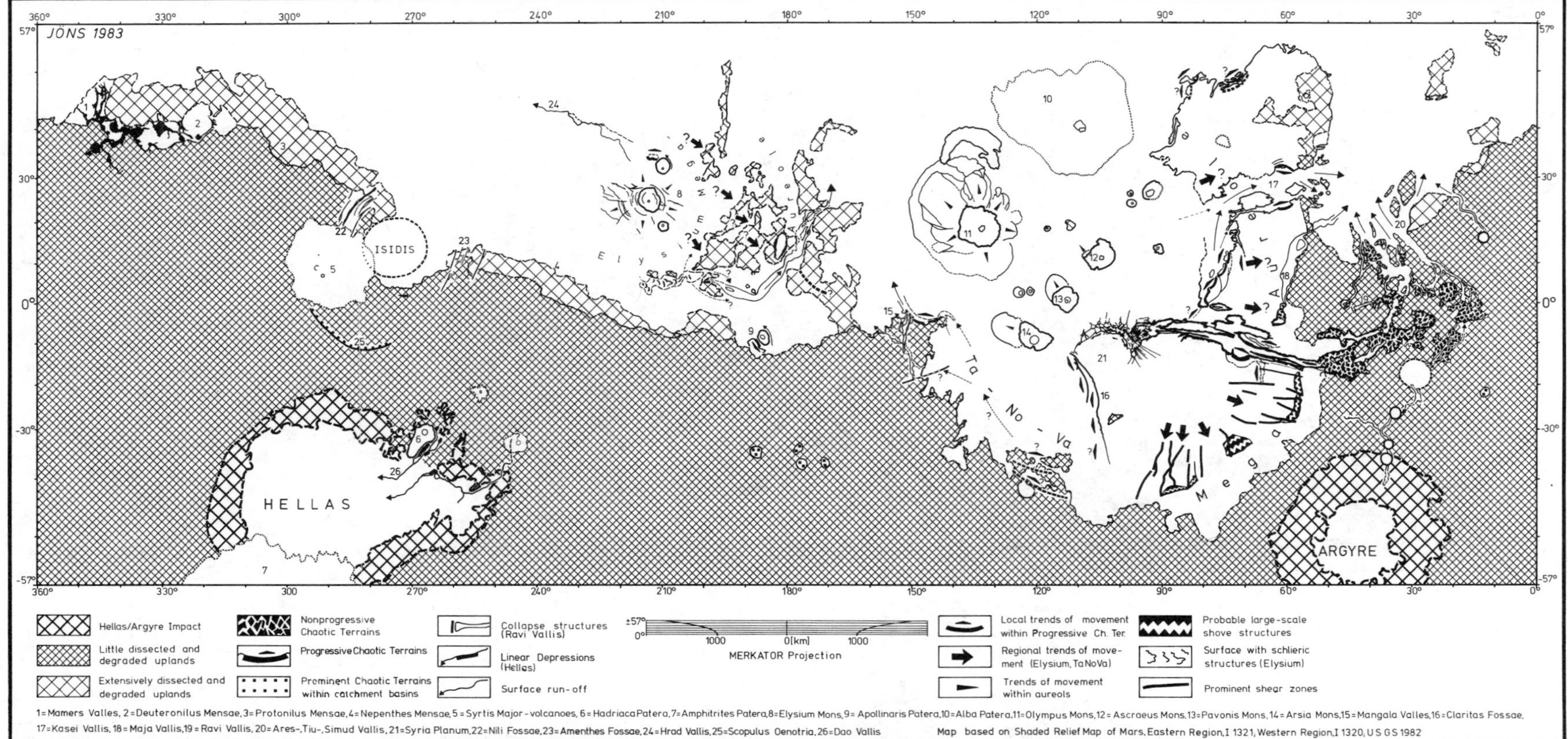

FIGURE 3
Permafrost-Related Types of Large-Scale Dissection, Degradation, and Deformation of Martian Landscape.

Progressive Chaotic Terrains

Chaotic terrains of this type dissect the landscape by cutting off sickle-shaped parts of the adjacent upland. Hence the remnants of the overlying material are characterized by series of arcuate shallow grabens. As no true outflow channels have been formed, the degradation of ground ice probably started with wide fronts along the flanks of preexisting channels (e. g. Echus Chasma) and/or shallow depressions. The best example of a progressive chaotic terrain is situated near Hecates Tholus/North. Lowlands (Figure 5).

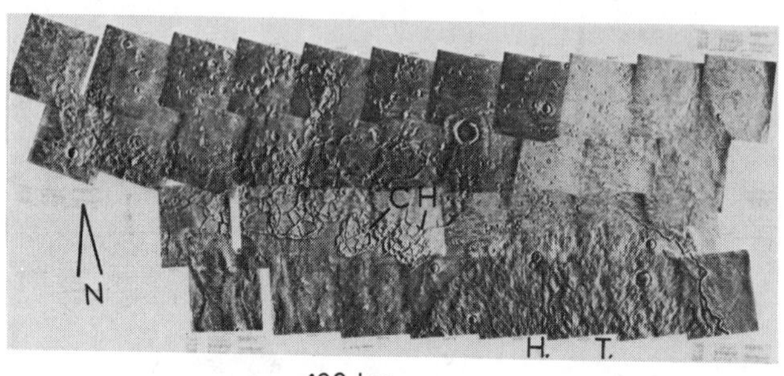

FIGURE 5 Progressive chaotic terrains (CH) near Hecates Tholus (HT).

Chaotic Terrains Within Catchment Basins

Recently, chaotic terrains in apparent shallow depressions (ancient crater scars?) have been described (Lucchitta 1983a). These depressions probably became filled with sediments from local runoff which seem to thin out near their border. Hence many chaotic terrains of this type are characterized by a complex drainage pattern in their surroundings and by the absence of encircling escarpments. They have been identified on the Southern Uplands as well as in the lowest areas in Candor Chasma/ Valles Marineris (Figure 6).

Linear Depressions

Linear depressions seem to exist only in the vicinity of Hadriaca Patera/Southern Uplands. They are characterized by linear escarpments which run more or less parallel to each other forming box-shaped depressions. In contrast to the chaotic depressions described above the average trend of these structures seems to correspond with the main trends of the planets tectonics. The depressions of this type are related to the Hellas Planitia by shallow outflow channels (Figure 7).

FIGURE 6 Chaotic terrain within catchment basin (Southern Uplands).

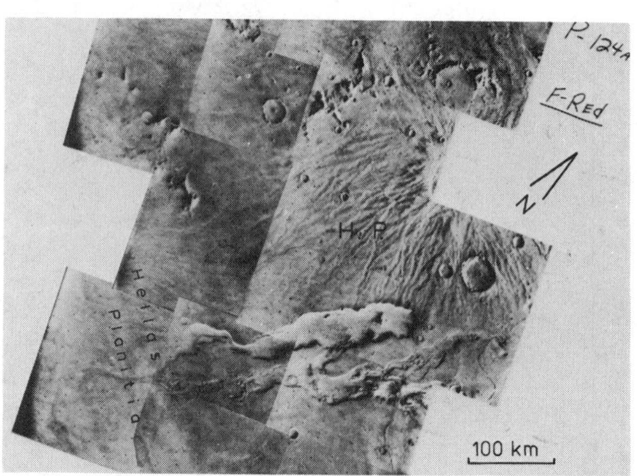

FIGURE 7 Linear depressions near Hadriaca Patera (HP).

Collapse Structures

Collapse structures seem to have been created by release of water from probably confined aquifers (Carr 1979). They consist of broad shallow outflow channels which emerge full born along linear escarpments. The best example of this type is Ravi Vallis. It contains a chaotic terrain of medium size, Aromatum Chaos, which is situated in front of the linear escarpment from which Ravi Vallis emerges (Figure 8).

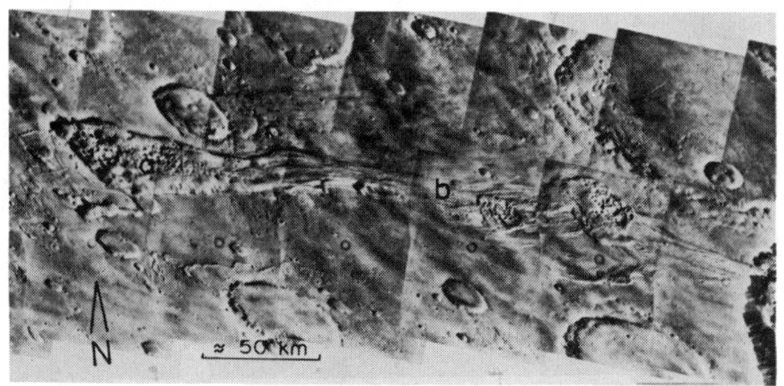

FIGURE 8 Collapse structure with Aromatum Chaos(a) and outflow channel (b), Ravi Vallis.

CONCLUSIONS

As most of the features described in this paper are situated within a broad transition zone between the Southern Uplands and the Northern Lowlands, the possibility can not be ruled out that at least parts of the Northern Lowlands came into existence by ground ice deterioration and/or melting of permafrost.

Consequently it seems to be possible that large-scale permafrost-related dissection, degradation, and deformation of landscape were among Mars' most important relief generating processes.

ACKNOWLEDGEMENTS

The author wants to appreciate the efforts of two anonymous reviewers. This study was supported by the DFG.

REFERENCES

Anderson, D. M., 1982, Physical and Mechanical Properties of Permafrost on Mars, in NASA Technical Memorandum 85127, p. 264.

Carr, M. H. and Saunders, R. S., 1973, A Generalized Geologic Map of Mars: Icarus, 78, p. 4031-4036.

Carr, M. H. and Shaber, G. G., 1977, Martian Permafrost Features: Journal of Geophysical Research, vol. 82, no. 28, p. 4039-4054.

Carr, M. H., 1979, Formation of Martian Flood Features by Release of Water From Confined Aquifers: Journal of Geophysical Research, vol. 84, no. B6, p. 2995-3007.

Christiansen, E. H. and Greeley, R., 1981, Mega-Lahars(?) in the Elysium Region, Mars, in Lunar and Planetary Science XII, part 1, p. 138-139.

Harris, S. A., 1977, The Aureole of Olympus Mons, Mars: Journal of Geophysical Research, vol. 82, p. 3099-3107.

Hodges, C. A. and Moore, H. J., 1979, The Subglacial Birth of Olympus Mons and its Aureoles: Journal of Geophysical Research, vol. 84, no. B14, p. 8061-8074.

Hubbert, M. J. and Rubey, W. W., 1959, Role of fluid pressure in mechanics of overthrust faulting, Mechanics of fluid-filling porous solids and its application to overthrust faulting: Geological Society of America Bulletin, 70, p. 115-166.

Jöns, H. P., 1982a, Die Rolle der Chaotischen Terrains bei der Entstehung des Reliefs der Marsoberfläche: Zeitschrift der Deutschen Geologischen Gesellschaft, 133, p. 339-358.

Jöns, H. P., 1982b, Permafrost Melting and Dissolution of the Landscape of Mars, in The Planet Mars, ESA SP-185, p. 89-105.

Jöns, H. P., 1983a, Large-Scale Permafrost Melting and the Origin of Mega-Aureoles on Mars, in Lunar and Planetary Science XIV, part 1, 355-356.

Jöns, H. P., 1983b, Permafrost-Related Types of Large-Scale Dissection, Degradation and Deformation of Martian Landscape, in Lunar and Planetary Science XIV, part 1, p. 357-358.

Lopes,R. M. C. et al., 1979, Origin of the Olympus Mons Aureole and Perimeter scarp: The Moon and the Planets, vol. 22, no. 2, p. 221-234.

Lopes, R. M. C. et al., 1981, Olympus Mons Aureole: Mechanism of Emplacement, in Third International Colloquium on Mars, presented papers, p. 136.

Lopes, R. M. C. et al., 1982, Further Evidence for a Mass Movement Origin of the Olympus Mons Aureole: Journal of Geophysical Research, vol. 87, no. B12, p. 9917-9928.

Lucchitta, B. K., 1982, Preferential Development of Chaotic Terrains on Sedimentary Deposits, Mars, in NASA Technical Memorandum 85 127, p. 235-236.

Lucchitta, B. K., 1983, Chryse Basin Channels: Low-Gradients and Ponded Flows, in Proceedings of the Thirteenth Lunar and Planetary Science Conference, part 2, Journal of Geophysical Research, vol. 88, Supplement, p. A553-A568.

Morris, E. C., 1982, Aureole deposits of the Martian volcano Olympus Mons: Journal of Geophysical Research, vol. 87, p. 1164-1178.

Sharp, R. P., et al., 1971a, The Surface of Mars 2: uncratered Terrains: Journal of Geophysical Research, vol. 76, p. 331-342.

Sharp, R. P., 1973a, Mars: fretted and chaotic terrains: Journal of Geophysical Research, vol.78, p. 4073-4083.

Soderblom, L. A., 1978, Possible Fossil H_2O Liquid-Ice Interfaces in the Martian Crust: Icarus, 34, p. 622-637.

Wu, S. S. C. et al., 1981, Topographic Map of Olympus Mons, in Third International Colloquium on Mars, presented papers, p. 287-289.

FIELD EVIDENCE OF GROUNDWATER RECHARGE IN INTERIOR ALASKA

Douglas L. Kane and Jean Stein

Institute of Water Resources, University of Alaska,
Fairbanks, Alaska 99701 USA

Increased utilization of groundwater resources in discontinuous permafrost regions requires that we develop a better understanding of the system and its processes. This paper discusses the mechanics of one of the processes, groundwater recharge. Two distinctly different soil conditions exist; generally permafrost inhibits drainage and these areas are poorly drained, whereas nonpermafrost areas are better drained for the same soil type. Field studies near Fairbanks, Alaska, show that the infiltration and hydraulic conductivity properties for frozen and unfrozen soils vary substantially. Freezing of a soil reduces the infiltration rate because of the existence of ice in soil pores. Further, frozen soils with high ice contents have lower infiltration properties than frozen soils with low ice contents. Frozen but relatively dry soils behave in a manner similar to unfrozen soils. Therefore, substantial infiltration can occur in seasonally frozen soils from snowmelt. Our field studies have shown that most of the groundwater recharge occurs during the snowmelt period in nonpermafrost areas. This is partly because of the soil condition and partly because of the large quantity of water available. Recharge during the summer is limited by the pattern of summer precipitation coupled with the ongoing process of evapotranspiration.

INTRODUCTION

Various uncertainties exist in some regions as to the long-term availability of groundwater resources. This problem is compounded in permafrost areas by the lack of data on groundwater availability as related to permafrost distribution. Excessive groundwater withdrawals may exceed natural recharge rates, and thus increase both pumping and development costs due to the water level declines. The potential for groundwater recharge is primarily controlled by the geologic conditions near the ground surface. In cold regions, the hydraulic properties of near-surface soils can vary substantially, depending upon the moisture regime and whether the soil is frozen or unfrozen.

This paper presents a discussion of the potential for groundwater recharge in areas of permafrost and extensive seasonal frost.

GROUNDWATER SYSTEM

In permafrost regions, a delineation should be made between areas of continuous and discontinuous permafrost. This division is very critical when examining groundwater flow in cold regions. In continuous permafrost areas, there is essentially no hydraulic connection between groundwater in the active layer and subpermafrost groundwater. Soil temperatures are colder in continuous permafrost areas, so the permafrost is thicker. Hydraulic properties are also temperature-dependent, and this dependence is more critical at temperatures below 0°C. In discontinuous permafrost areas, a hydraulic connection exists between subpermafrost groundwater and the surface through unfrozen zones in the permafrost. These zones are generally found on south slopes and beneath most lakes and large rivers.

Groundwater zones in permafrost regions are classified as either suprapermafrost groundwater or subpermafrost groundwater. Suprapermafrost groundwater is the saturated zone above the permafrost, and subpermafrost groundwater is the saturated zone below the permafrost. In areas of discontinuous permafrost, there is a transition from the subpermafrost groundwater to the groundwater in the permafrost-free areas. It should be noted that permafrost acts as a relatively impermeable confining layer, so subpermafrost groundwater can behave as water in a confined aquifer.

Because of the shallow nature and limited volume of water, suprapermafrost groundwater cannot be used directly for domestic water. However, drainage of water through this layer to streams and lakes can result in a potential water supply. Many engineering problems are associated with drainage into and out of the active layer.

Development of subpermafrost groundwater has been quite successful in many regions of discontinuous permafrost. In areas of heavy groundwater usage, substantial declines in groundwater levels have been observed, but not explained.

For understanding or estimating groundwater recharge, the mechanics of water infiltration and movement into the unsaturated zone is of the greatest importance. It is through this layer that all groundwater recharge must flow. This layer can be less than 1 m thick in the active layer above permafrost or more than 100 m thick in permafrost-free areas near ridgetops. The unsaturated zone may consist of primarily organic soils in the active layer. The hydrologic role of these organic soils should not be underrated because they act as an important buffer, both hydraulically and thermally, between the atmosphere and mineral soil (Slaughter and Kane 1979). Soil moisture levels can vary, but generally high moisture levels exist in the active layer above permafrost. Better-drained soils with lower water contents are found in permafrost-free areas (Kane et al. 1978a). The permafrost is responsible for impeding drainage and for the development of high moisture levels. In turn, the high moisture levels dictate the type and density of vegetation and subsequent evapotranspiration rates.

The climatic variables of interest in groundwater recharge are the intensity, duration, and timing of the precipitation event. Low-intensity events allow infiltration to occur without producing runoff. High-intensity events may exceed infiltration rates and produce runoff. Duration of the event is important because longer events have greater potential for groundwater recharge. Timing of a hydrologic event is important because the source of water can be either snowmelt or rainfall, or a combination of both. In Alaska, except at high altitudes and along the coast, areas of continuous and discontinuous permafrost are classified as arid climates, based on annual precipitation. Both rainfall and snowmelt intensities are rather low. But the duration can be quite long, especially for snowmelt events. Whereas the occurrence of rainfall events are much less predictable, it is certain that a snowmelt event will occur every spring and that a potential for groundwater recharge exists because the snowpack can represent 40-80% of annual precipitation.

FIELD EVIDENCE OF GROUNDWATER RECHARGE

Reported here are selected results from numerous studies performed by the authors that help provide knowledge of the process of groundwater recharge in cold regions. Some comparative data on hydraulic properties of frozen soils that were reported by other authors are included in the latter part of this section.

Field data were collected near Fairbanks, Alaska, in an area of discontinuous permafrost. Permafrost is found on north-facing slopes and in valley bottoms; south-facing slopes are generally permafrost-free. The predominant soil type in this forested area is wind-deposited silt loam overlain with organic material. The thicknesses of the organic soils vary (generally 0-50 cm, thicker in permafrost areas) as do the mineral soils (0-60 m, greatest thickness in valley bottoms). Permafrost thicknesses can exceed 60 m. We collected field data with two primary goals (1) to assess the potential for groundwater recharge from rainfall and snowmelt events, and (2) to evaluate the potential for groundwater recharge in permafrost and nonpermafrost areas.

Recharge from Rainfall

Numerous studies (Kane, Seifert, and Taylor 1978, Kane et al. 1978b) have been carried out in which tensiometers installed at various depths were used to monitor soil moisture conditions. With a characteristic curve (pore pressure--moisture content relationship), one can estimate the moisture content from tensiometer readings, as well as determine the hydraulic gradient. From this, the direction of flow is determined. Tensiometer readings indicated that for nonpermafrost sites, the soils were very dry for much of the summer. Even during rainfall periods, very little reduction in the soil tensions occurred. The general trend was for the soils to become drier over the summer season. In unsaturated soils, lower moisture content yields lower hydraulic conductivity. For significant groundwater recharge to occur, strong downward vertical gradients are required. These did not occur. We concluded that most of the soil water was lost to evapotranspiration. Braley (1980) examined evapotranspiration losses from barley and rapeseed near Delta, Alaska. He concluded that summer precipitation was equivalent to measured evapotranspiration.

Precipitation events have already been described as being of low intensity and of variable duration, resulting in low quantities of annual precipitation. From the tensiometer data (Luthin and Guymon 1974, Kane et al. 1978a), we concluded that little, if any, recharge occurs during the summer months. The only exception would be periods of sustained precipitation (> 3 cm), which are rare. We then looked at snowmelt as a source of groundwater recharge.

Snowmelt Lysimeters

The first attempt to quantify snowmelt infiltration into seasonal frozen soils was to construct several lysimeters that only allowed vertical movement of meltwater. These one-dimensional lysimeters are described by Kane et al. (1978a). Numerous lysimeters were built so that evaporation losses, moisture retention of organic soil, and moisture retention of mineral soil could be determined.

Some lysimeters collected and measured just snow. The next series of lysimeters collected snowfall, but the snowmelt had to flow through a 10-cm layer of organic material. The last series of lysimeters collected snowfall, and the snowmelt water flowed through the organic layer, plus a 20-cm layer of mineral soil. The system, associated processes, and type of results are shown in Figure 1. Water leaving the bottom of the lysimeters at base of snowpack, organic layer, and mineral layer was collected and measured. The data shown in the figure are for 1976, and the maximum water content prior to ablation of the snowpack was 10 cm. We measured between 17 April and 27 April that 2.1 cm of water was retained in the organic layer, 2.7 cm was retained in mineral soil, and 3.5 cm of water passed through the lysimeters. From this

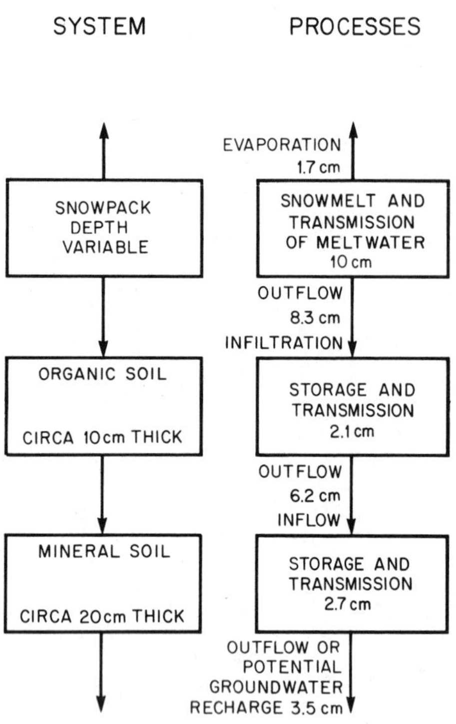

FIGURE 1 One-dimensional field model showing the path of snowmelt.

data we concluded that a maximum of 1.7 cm was lost to evaporation. The combined soil depth of 30 cm in the last series of lysimeters represents the root zone where evapotranspiration would occur. Infiltration water reaching greater depth was considered to be potential groundwater recharge. In this case, 3.5 cm of water (35%) represented potential groundwater recharge. Similar data were collected for 1977.

Because drainage of the soils was allowed from the bottom of the lysimeters, these conditions were only representative of well-drained soils found on nonpermafrost sites. Also, no water was allowed to leave the lysimeter except by evapotranspiration or drainage through the bottom of the lysimeters.

It was concluded (1) that significant groundwater recharge can occur annually from snowmelt in well-drained, although seasonally frozen soils; but (2) that it is necessary to examine the influence that higher moisture contents have on the infiltration rates of seasonally frozen soils to assess the potential for groundwater recharge on sites with other than well-drained soils.

INFLUENCE OF MOISTURE CONTENT ON INFILTRATION

We used two types of infiltrometers: auger hole and double ring. The auger-hole method used a hole 11.1 cm in diameter and 20 cm in depth that was drilled into the seasonally frozen soil. Infiltration rates were obtained and compared for a variety of moisture contents in both frozen and unfrozen conditions (Kane 1980). The major advantage of this method is that one can obtain samples for determination of moisture content and presence of ice lenses from the collected soil core. The main disadvantage of this method is that true vertical infiltration is not obtained, because lateral flow is not prevented. So the measured infiltration rates are greater than actual values. From these data, relative values for a wide range of soil conditions showed that high infiltration rates existed for dry, frozen soils and that infiltration declined as the moisture content of the frozen soil increased.

Double-ring infiltrometers were used to overcome the lateral-flow problem. To minimize the disturbance of the seasonally frozen soil during installation, we decided to install the infiltrometers prior to the development of seasonal frost. To minimize the effect of the infiltrometer on the thermal regime, the surface organic layer was replaced around and inside the infiltrometer until the testing began at the onset of ablation. The diameters of the rings were 18 cm and 36 cm. The ring was installed 9 cm into the mineral soil, and a constant head of 6 cm was maintained. Moisture contents must be measured outside the infiltrometer; soil cores were collected near each infiltrometer to define the moisture profile.

Data from these double-ring tests are shown in Figure 2. These tests were carried out over a longer period of time to observe the infiltration rate variation as the frozen soil thawed. In most cases, the initial infiltration rate was high, but rapidly declined to equilibrium conditions (equivalent to saturated hydraulic conductivity). On the experimental plots, however, we observed occasional increases in the infiltration rate for short periods of time. Then the infiltration rate decreased rapidly to approximately the previous equilibrium value. We attribute this to thawing of an ice-rich layer of soil and vertical variation in total moisture content (ice and water).

We found, from these infiltration tests that very high infiltration rates could be obtained in seasonally frozen soils of low moisture content. In fact, the saturated hydraulic conductivity at equilibrium conditions was reduced only by a factor of 2 when compared to unfrozen Fairbanks silt loam (at 14.8°C). Most of this reduction can be explained by changes in the fluid viscosity that are brought about by lower temperatures (Kloch 1972). For an ice-rich soil, however, a reduction of 2 orders of magnitude occurred at equilibrium conditions.

Both Burt and Williams (1976) and Horiguchi and Miller (1980) have shown that the hydraulic conductivity of frozen soils is further reduced as the temperature is lowered below 0°C because the ice content increases. The magnitude of this reduction is quite significant, because the reduction in the hydraulic conductivity is several orders of magnitude for a small decrease in temperature (< 0.5°C). As the temperature is lowered, the unfrozen-water-to-ice ratio is reduced. This means that there is more ice in the pores and the thickness of the unfrozen water films is reduced. Research on the unfrozen-water-to-ice ratios versus temperature has shown the relationship to be hysteretic. Therefore, one would expect the temperature and hydraulic conductivity relationship also to be hysteretic, as was shown by Horiguchi and Miller (1980).

Daily snowmelt rates measured in this area have varied from zero to a maximum near 2.5 cm. These data are interesting, because when one compares these data with the data in 2b and 2c of Figure 2 for equilibrium rates, it is obvious that somewhere in between the two equilibrium infiltration rates the snowmelt rate exceeds the infiltration rate, and runoff should be produced. If the drier frozen soil is representative of the nonpermafrost areas, groundwater recharge can occur; and if permafrost sites are poorly drained with high moisture contents, the potential for groundwater recharge is reduced, and the potential for runoff is enhanced.

Daily infiltration as calculated from equilibrium rates (Figure 2) is 17.2 cm/day for dry frozen soil and 0.35 cm/day for poorly drained frozen soil. This assumes that infiltration can occur for 24 hours. It is difficult to assess this because of the presence of surface organic material and the capability of this soil to retain some soil water. However, during periods of high snowmelt, saturated conditions should develop in the surface organic layer for soils with low infiltration rates. This should produce near-surface runoff as water moves downslope along the organic-mineral soil interface. A number of runoff plots were designed for studying the influence of moisture content on runoff from seasonally frozen terrain.

INFLUENCE OF MOISTURE CONTENT ON RUNOFF

Prepared runoff plots were 6.1 m wide and 12.2 m long. The plots were situated on a south-facing, permafrost-free forested slope with 15% slope, with approximately 10 cm of organic material over the Fairbanks silt loam, and seasonally frozen depths ranging from 150 to 200 cm. A collection system was constructed at the base of each plot to collect all near-surface runoff moving downslope through the surface organic layer. There were two runoff plots during the winter of 1979/80, three runoff plots during the winter of 1980/81, and four runoff plots during the winter of 1981/82. To alter the soil moisture levels, some plots were irrigated by water sprinklers in the fall 3-4 weeks before the onset of seasonal frost. Large areas outside the plots were irrigated for collecting soil moisture cores throughout the winter and during ablation.

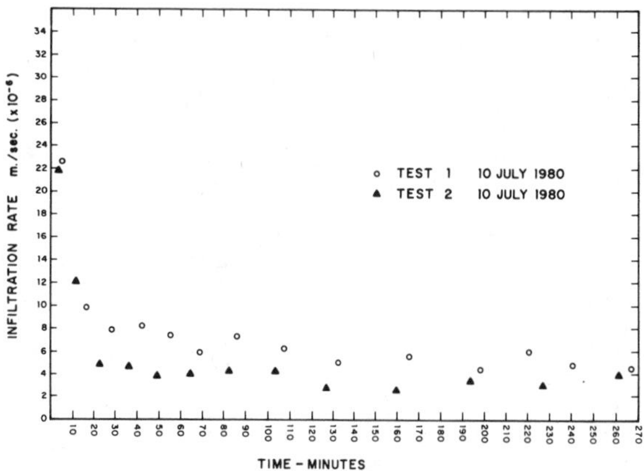

2a. Unfrozen Fairbanks silt loam.

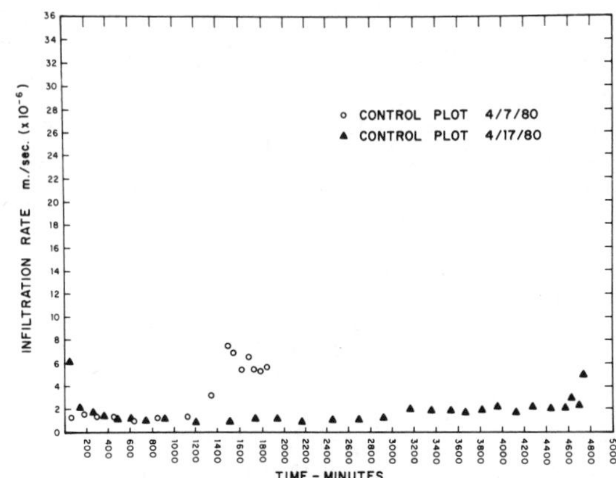

2b. Frozen Fairbanks silt loam (dry case).

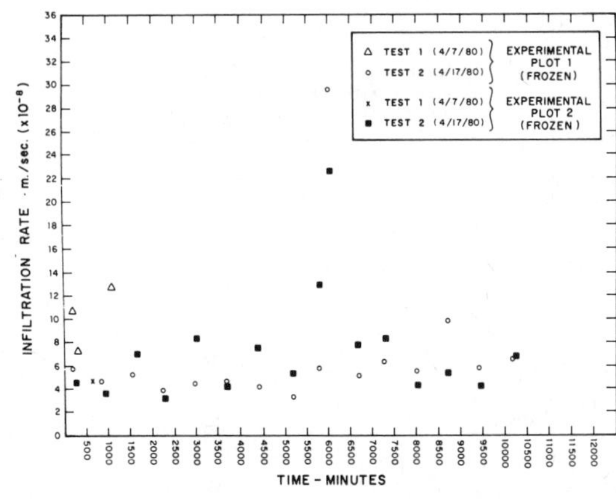

2c. Frozen Fairbanks silt loam (ice-rich case).

FIGURE 2 Infiltration rates for various soil conditions for both frozen and unfrozen conditions.

The infiltration tests described earlier were also performed in this same area.

Data from the runoff plots revealed that for very dry, seasonally frozen soils, no runoff existed. This is the same conclusion that we reached from the infiltration test. From the irrigated runoff plots, measurable runoff always occurred during snowmelt. The ratio of the volume of runoff to total water content of the snowpack varied from 17% to 68%. No clear relationship exists between infiltration rate and runoff volume because of the variation in climate and snowpack from year to year. We found in 1982 that runoff from a control plot (not irrigated) occurred for the first time. This was attributed to rainfall just prior to freeze-up (3.7 cm). It should be noted that 1982 was the first year since 1973 that we observed surface runoff in the natural depressions around this site.

The maximum allowable daily melts that can occur without runoff for various infiltration rates (at 0°C) are shown in Figure 3. In addition, superimposed on this figure are laboratory and field measurements under frozen conditions with the exception of unfrozen Fairbanks silt loam. With the added data of actual measurements, one can see how the infiltration rate relates to soil type and soil condition.

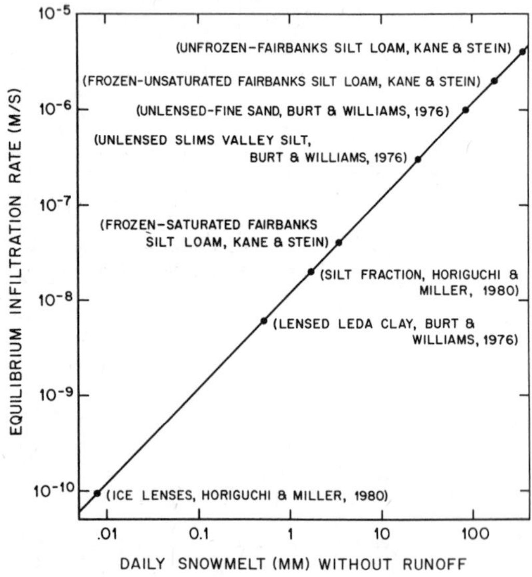

FIGURE 3 Maximum allowable daily melts occurring without runoff for various infiltration rates (at 0°C).

If runoff is generated, snowmelt rate exceeds infiltration rate, and this water is lost as potential groundwater recharge. Our field plots had daily snowmelts from 0 to 25 mm. Figure 3 shows that runoff should be generated if the mineral soil has a high ice content (high moisture level). However, no runoff would be predicted in a well-drained soil. This graph assumes that uniform melt occurs during a 24-hour period, and uniform infiltrations occurs throughout this period. The melt actually occurs over a shorter period of time, but the organic layer can provide temporary storage to minimize the effect of nonuniform melt.

CONCLUSIONS

It appears that the most consistent and predictable period of groundwater recharge in cold regions is during spring snowmelt. Field data demonstrate that most summer precipitation is used for evapotranspiration. The only exceptions occur during long periods of summer precipitation, but these hydrologic events do not occur every year. After 6 months of accumulation, ablation of the snowpack represents a significant annual hydrologic event. For seasonally frozen soils that are well drained and therefore have a low moisture content at the time of ablation, all meltwater that does not evaporate can enter the soil system. Since evapotranspiration is low at this time of year, the potential for groundwater recharge is quite good. As the ice content in the soil pores increases and ice lenses develop, the potential for near-surface runoff increases and the potential for groundwater recharge decreases. Areas underlain by permafrost are generally poorly drained and represent areas of runoff generation rather than groundwater recharge. Some meltwater may enter the active layer, but further downward at the permafrost table, movement is hindered by the presence of ice with much lower transmission properties. So in discontinuous permafrost zones, we conclude that the majority of groundwater recharge occurs in permafrost-free areas during snowmelt, and no significant recharge occurs in permafrost areas of either continuous or discontinuous zones.

ACKNOWLEDGMENTS

The Natural Sciences and Engineering Research Council of Canada and the Quebec General Direction of Higher Education supported in part the work of J. Stein. We also thank Ellen Kane who collected field data.

REFERENCES

Braley, W.A., 1980, Estimates of evapotranspiration from barley and rapeseed in Interior Alaska. M.S. Thesis, University of Alaska, Fairbanks.

Burt, T.P., and Williams, P.J., 1976, Hydraulic conductivity in frozen soils, Earth Surface Processes, v. 1, p. 349-360.

Horiguchi, K., and Miller, R.D., 1980, Experimental studies with frozen soil in an "Ice Sandwich" permeameter, Cold Regions Science and Technology, v. 3, p. 177-183.

Kane, D.L., 1980, Snowmelt infiltration into seasonally frozen soils, Cold Regions Science and Technology, v. 3, p. 153-161.

Kane, D.L., 1981, Groundwater recharge in cold regions, The Northern Engineer, v. 13, N. 3, p. 28-33.

Kane, D.L., Seifert, R.D., and Taylor, G.S., 1978, Hydrologic properties of subarctic organic soils, Institute of Water Resources, University of Alaska, IWR-88.

Kane, D.L., Fox, J.D., Seifert, R.D., and Taylor, G.S., 1978a, Snowmelt infiltration and movement into frozen soils, in 1: Ottawa Proceedings of the Third International Conference on Permafrost, National Research Council of Canada.

Kane, D.L., Fox, J.D., Seifert, R.D., and Taylor, G.S., 1978b, Snowmelt-frozen soil characteristics for a subarctic setting, Institute of Water Resources, IWR-84.

Kloch, G.O., 1972, Snowmelt temperature influence on infiltration and soil water retention, Journal of Soil and Water, v. 27, N. 1, p. 12-14.

Luthin, J.N., and Guymon, G.L., 1974, Soil moisture--vegetation--temperature relationships in central Alaska, Journal of Hydrology, v. 23, p. 233-246.

Slaughter, C.W., and Kane, D.L., 1979, Hydrologic role of shallow organic soils, in Proceedings of the Canadian Hydrology Symposium, National Research Council of Canada.

CONTINUOUS GEOPHYSICAL INVESTIGATION FOR MAPPING PERMAFROST DISTRIBUTION, MACKENZIE VALLEY, N.W.T., CANADA

Anthony E. Kay[1], Andrea M. Allison[2], W.J. Botha[1], and William J. Scott[1]

1 Hardy Associates (1978) Ltd., Calgary, Alberta, T2E 6J5 Canada
2 Canterra Energy Limited, Calgary, Alberta, T2P 2K7 Canada

A continuous inductive conductivity survey using a 10 m station interval was carried out over a distance of 868 km along the Interprovincial Pipe Line route in the Mackenzie River Valley, N.W.T. The work reported herein was done as part of a contract for Interprovincial Pipe Line (NW) Ltd. The study provides a case history for delineating frozen ground along a route that traverses a zone of scattered discontinuous permafrost and a zone of widespread discontinuous permafrost. Terrain conductivity meters were used to measure apparent conductivities along the entire proposed pipeline route. Air photo interpretation was used to map the terrain type distribution along the pipeline route. This information was correlated with vegetation, topographic, and geophysical data to delineate the boundaries of frozen ground. This interpretation was checked by boreholes drilled at an average spacing of 1 borehole every 4 km. The interpreted results provide information useful for the design of the pipeline and substantially reduced the number of boreholes that would normally be required. Statistical analyses were obtained between the occurrence of frozen ground and different terrain types. The results of the analyses indicate that soil texture is a major controlling factor in the occurrence of frozen ground.

INTRODUCTION

Information on the presence and distribution of frozen ground is necessary for the proper planning and construction of engineering projects in permafrost regions. The use of electromagnetic induction methods has become routine for shallow subsurface investigations in permafrost environments (Arcone et al. 1978, Rennie et al. 1978). This investigation reports on a geophysical survey performed along the Interprovincial Pipe Line route from Norman Wells in the Mackenzie River Valley to Zama in Northern Alberta (Figure 1). The survey was carried out during the winter-spring seasons of 1981 and 1982. The geophysical work consisted of electrical conductivity mapping using an electromagnetic induction technique. The purpose of the geophysical work was to delineate the presence of permafrost along the pipeline right-of-way. Since the pipeline route crosses many frozen/unfrozen boundaries, locating and documenting these transitions greatly affects costs and design considerations dealing with frost heave in unfrozen ground and thaw settlement in frozen ground. For the purposes of this study frozen ground is defined as the presence of at least 3 m of frozen ground anywhere in the subsurface down to a depth limit of 10 m. This criterion was chosen on the basis of design considerations for the pipeline.

TECHNICAL APPROACH FOR MAPPING PERMAFROST

The technical approach used in the survey relies on the differences in electrical conductivity (or resistivity) of frozen and unfrozen soils.

Figure 2, after Hoekstra and McNeill (1973), indicates the changes in resistivities of several

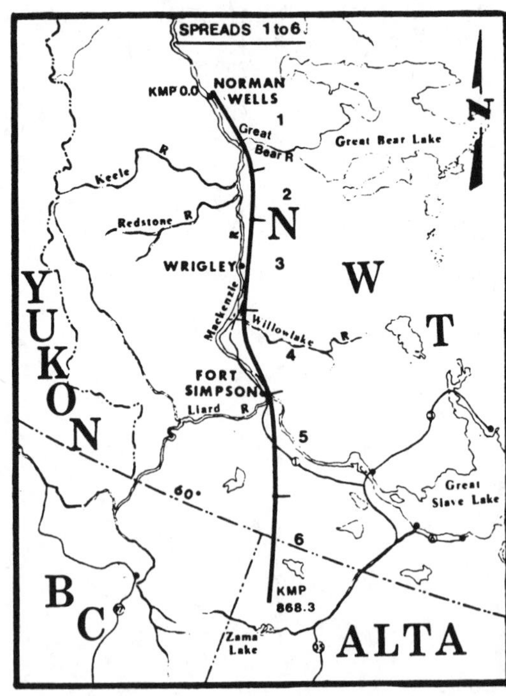

FIGURE 1 Location of geophysical survey route, kilometer post 0 to 868.3.

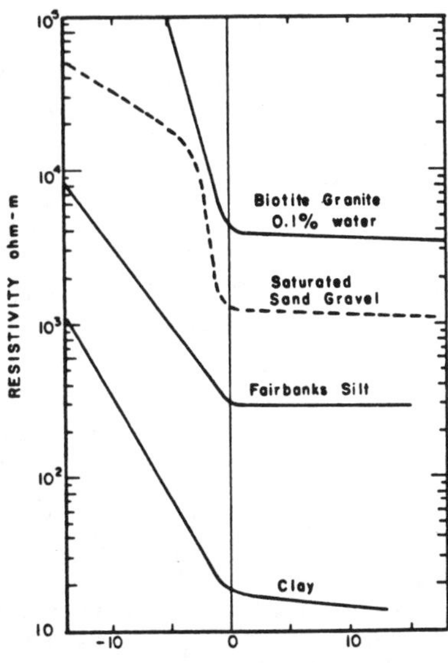

FIGURE 2 The resistivity of several soils and one one rock type as a function of temperature (Hoekstra and McNeill, 1973).

soil types and one rock type as a function of temperature. Resistivity is shown to increase when the ground freezes. Clay-rich soils demonstrate a gradual change in comparison with coarse-grained soils, which undergo an abrupt change. This difference results from the gradual change of pore water to ice in fine-grained soils as temperature drops below 0°C. In general, however, frozen and unfrozen soils can usually be distinguished by electrical resistivity or conductivity measurements, particularly when the soil type is known.

Inductive measurements of conductivity require no electrical contact with the ground and can be made much more rapidly than galvanic measurements. Inductive techniques use an electromagnetic transmitter, which induces localized eddy currents in the ground beneath the system. A receiver senses the strength of the eddy currents, which is proportional to the conductivity of the ground. Apparent conductivity is measured in units of siemens/meter (or 1/ohm-meters) or more conveniently in millisiemens/m (mS/m).

The apparent conductivity values measured at the borehole sites were interpreted by computer forward modelling to obtain the true conductivities of the different soil types where the layer thicknesses were known. This analysis assumed horizontal layering. Then this information was correlated with the vegetation, terrain type and other data to enable the delineation of frozen/unfrozen transitions along the pipeline route. These transitions were assumed to be vertical boundaries since there were generally sharp changes in the apparent conductivity profiles across these boundaries.

SURVEY PROCEDURE

Apparent conductivity readings were taken along the entire survey route from kilometer post (kmp) 0 to kmp 868.3. Measurements were obtained at 10 m station intervals; stations were flagged every 100 m.

The geophysical instruments used in the survey were the Geonics EM 31 and EM 34-3 terrain conductivity meters. The EM 31 was used in two modes of operation. The first mode was in an "up" position, that is, with the coils in a horizontal position. The second mode of operation was in a "side" position, that is, with the coils in a vertical position. The two readings provided information to depths of about 7 m and 3 m respectively (McNeill 1980). The EM 34-3 was used with a 10 m cable and vertical coil orientation to provide information to an approximate depth of 10 m (McNeill 1980). Since two instruments are used at each station, an independant check on the quality of the data is always available.

The use of different coil orientations and spacings with the EM 31 and EM 34-3 allows considerable variation in depth of investigation. The instruments are calibrated to measure apparent conductivity. For a homogeneous half-space this apparent conductivity equals the true conductivity of the subsurface material. For an electrically layered earth, the measured apparent conductivity represents the sum total of responses of subsurface conductivities down to depths somewhat greater than the coil separation.

FIELD PROFILES

A representative set of field profiles is presented to illustrate a variety of conditions encountered during the geophysical survey. These conditions include variations in soil type, frozen/unfrozen boundaries, bedrock influence, cultural disturbances, and effects of north- and south-facing slopes on river banks.

Figure 3 shows the geophysical response observed during a transition from a clay veneer over till soil to a till. The higher apparent

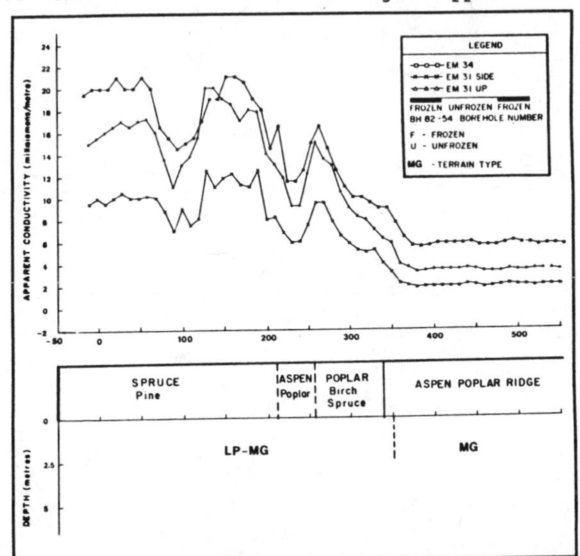

FIGURE 3 Effect of soil type transition on apparent conductivity profiles.

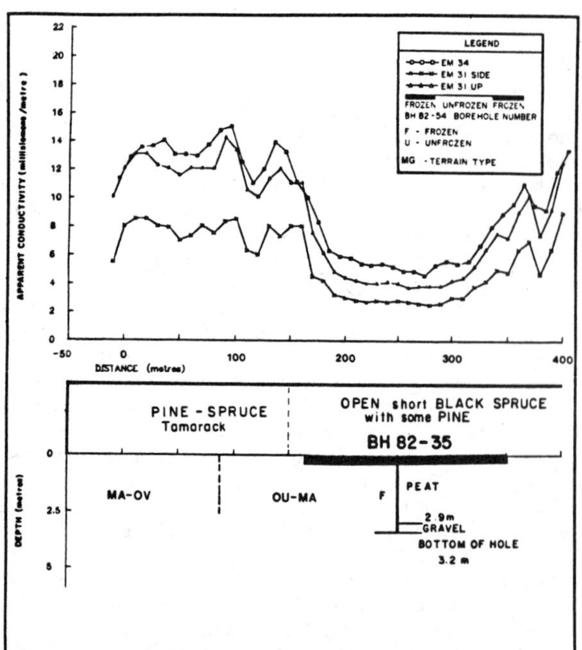

FIGURE 4 Frozen/unfrozen ground transitions in organic terrain.

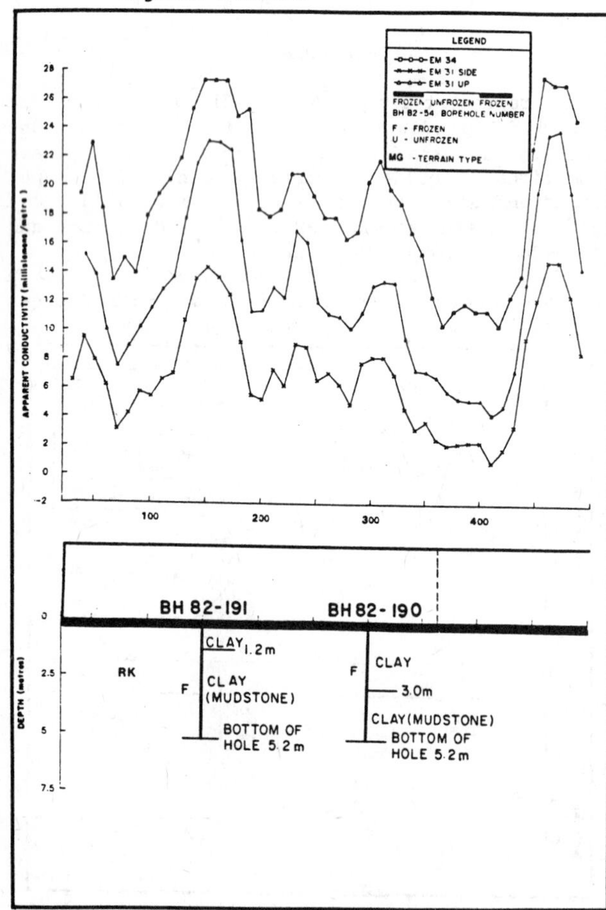

FIGURE 5 Influence of conductive, near surface bedrock on apparent conductivity measurements.

conductivities in the first part of the profile are due to the presence of the conductive clay veneer. Nearby borehole information indicated that there was no frozen ground along this profile.

A frozen/unfrozen transition in an area of discontinuous peat is shown in Figure 4. This type of response is very common in areas with appreciable amounts of peat landforms, where the boundary between frozen peat bogs and unfrozen adjacent terrain can be very sharp. The unfrozen ground in this profile was confirmed by a borehole drilled approximately 1 km away from the vicinity of this profile.

The effect of near-surface bedrock on the apparent conductivity values can be quite dramatic. If the bedrock is resistive, such as a sandstone, then the apparent conductivity values are quite low (< 3 mS/m). If, however, the bedrock is a shale, then the apparent conductivity values are quite high (Figure 5) even when the rock is frozen. In the absence of any borehole control the high peaks on the apparent conductivity curves would probably be interpreted as unfrozen, but the borehole logs clearly indicate frozen ground and a highly variable depth to bedrock.

Cultural influences such as cutlines and winter roads that intersect the proposed pipeline right-of-way are usually reflected on the geophysical profiles (Figure 6). In most cases an increase in depth of thaw to the frozen ground occurs beneath the disturbed zone.

Studies have shown (Heginbottom et al. 1978, for example) that north-facing slopes of river banks are generally wetter and hence are more likely to be frozen than south-facing slopes. Factors other than slope orientation that can

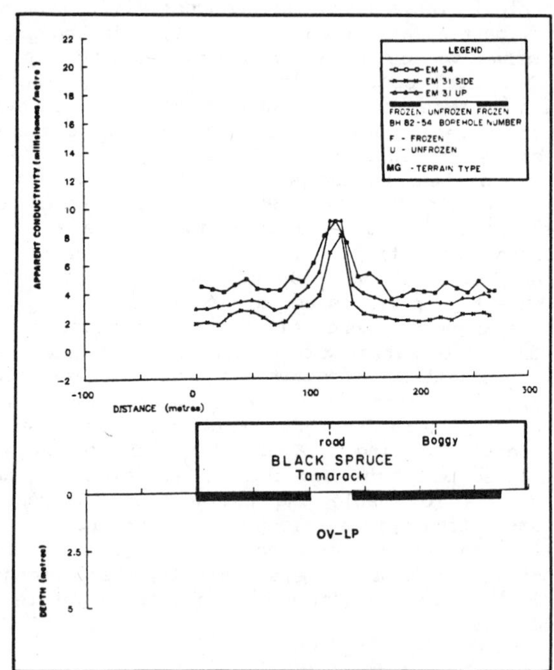

FIGURE 6 Indication of permafrost degradation on apparent conductivity profiles.

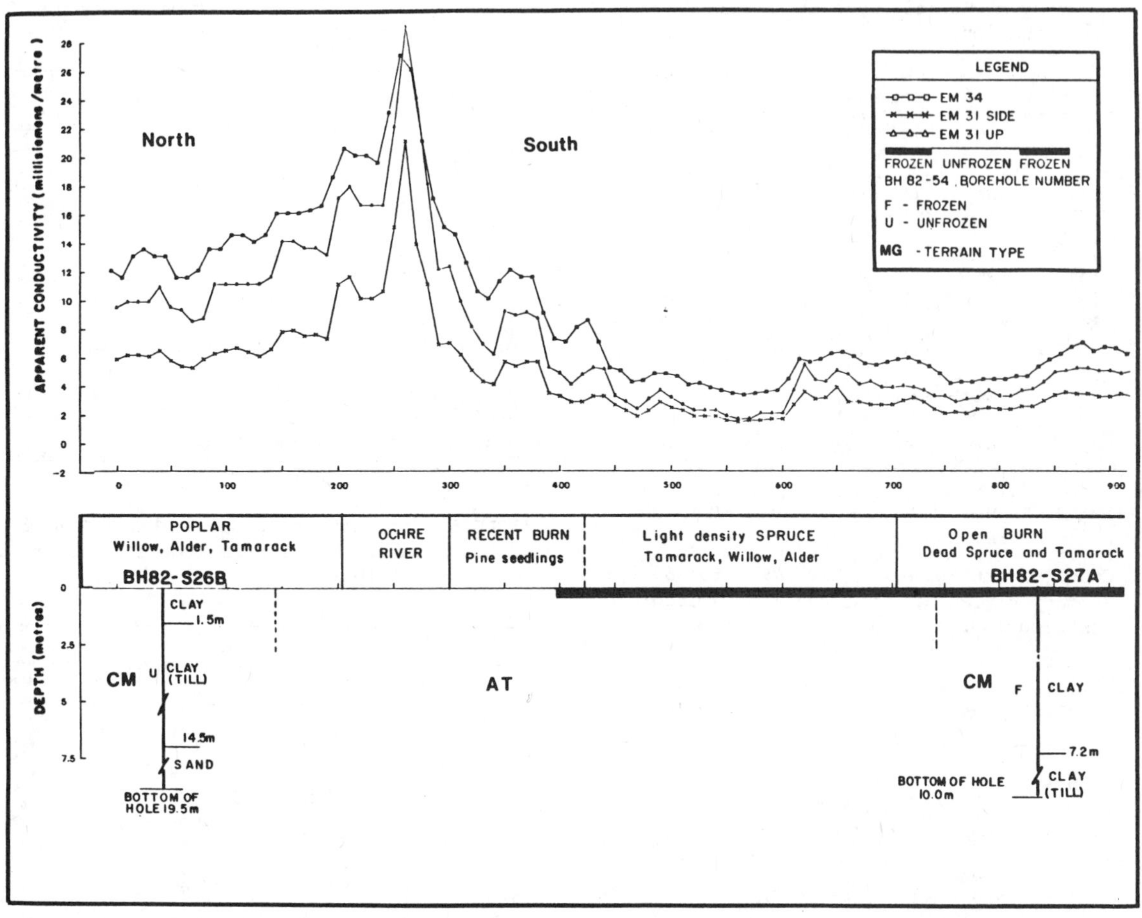

FIGURE 7 Apparent conductivity profiles indicating a frozen north-facing slope and an unfrozen south-facing slope.

influence the presence of frozen ground include lithology, slope aspect, and slope angle. An example of a river valley with a frozen north-facing slope and an unfrozen south-facing slope is illustrated in Figure 7. An elevation profile for the Ochre River indicates that the north and south slopes are very steep. The south slope, which faces north, is probably not exposed to sunlight most of the year and thus remains frozen.

STATISTICAL ANALYSES OF FIELD DATA

The distribution of frozen ground interpreted from the geophysical profiles was digitized and compared with the terrain typing. A statistical breakdown of percentage of frozen ground in each terrain type is important in determining design criteria for the pipeline. Thaw settlement and frost heave considerations will vary depending on terrain type and on the percentage of frozen ground in a particular terrain type.

The data are divided into six spreads with the following kmp boundaries, starting from Norman Wells (Figure 1):

kmp 0 to kmp 149.5	Spread 1
kmp 149.0 to kmp 269.3	Spread 2
kmp 269.3 to kmp 380.4	Spread 3
kmp 380.4 to kmp 528.8	Spread 4
kmp 528.8 to kmp 695.8	Spread 5
kmp 695.8 to kmp 868.3	Spread 6

Tables 1-6 summarize all the statistical information for spreads 1 to 6, respectively. They list the percentage of each terrain type in the spread and the percentage of frozen ground within that terrain type. The percentage contribution of frozen ground in a particular terrain type to the total frozen ground for the entire spread is indicated as percent of total km frozen.

The terrain types are grouped according to divisions established for the Norman Wells Pipeline Project borehole data bank. These groupings are divided on the basis of genetic origin and provide useful information for the design of the pipeline. The terrain-type codes for this data bank are given in Table 7.

A total of 83.7% of the ground is frozen in spread 1, 66.3% in spread 2, 69.2% in spread 3, 31.6% in spread 4, 16.0% in spread 5, and 34.7% in

TABLE 1 Construction Spread 1 (KMP 0.0 to 149.5): Summary of Terrain Type and Frozen Ground Distribution.

Terrain Type	Distance (km)	% of Construction Spread	km Frozen (% of Terrain Type Frozen)	% of Total km Frozen	Avg. No. of Frozen/Unfrozen Interfaces/km
AP	0.51	0.34	0.09 (19.4)	0.08	7.8
AT	1.64	1.10	1.04 (63.6)	0.83	4.9
CM	1.96	1.31	1.65 (84.2)	1.32	5.1
ED	2.03	1.36	0.88 (43.3)	0.71	0.5
GO	4.31	2.88	2.71 (62.9)	2.17	1.4
LB	0.97	0.65	0.39 (40.0)	0.31	7.3
LP	41.66	27.86	33.05 (79.3)	26.42	1.3
LP-MG	14.66	9.82	11.45 (78.1)	9.16	1.9
MG	23.30	15.64	21.09 (90.5)	16.92	1.5
OU	1.05	0.70	0.49 (47.4)	0.40	4.8
OU-LP	8.48	5.67	6.77 (79.8)	5.41	2.1
OV-LP	25.02	16.74	23.58 (94.3)	18.86	1.1
OV-MG	3.12	2.02	2.93 (93.8)	2.26	1.7
RK	17.22	11.52	15.86 (92.1)	12.67	2.6
OV-SP	0.12	0.08	0.01 (9.8)	0.01	8.2
MG-RK	3.45	2.31	3.08 (89.5)	2.47	1.5
TOTAL	149.5	100.00	125.07 (83.7)*	100.00	259**

* Percentage of construction spread 1 frozen.
** Total no. of frozen/unfrozen interfaces in construction spread 1.

TABLE 2 Construction Spread 2 (KMP 149.5 to 269.3): Summary of Terrain Type and Frozen Ground Distribution.

Terrain Type	Distance (km)	% of Construction Spread	km Frozen (% of Terrain Type Frozen)	% of Total km Frozen	Avg. No. of Frozen/Unfrozen Interfaces/km
AF	4.86	4.06	3.82 (94.3)	5.70	1.4
AP	0.92	0.77	0.41 (44.2)	0.51	9.8
AT	1.58	1.32	1.44 (91.2)	1.79	1.9
CM	3.80	3.17	2.48 (65.4)	3.04	4.2
CM-AP	0.31	0.26	0.31 (100.0)	0.38	0
ED	0.41	0.34	0.24 (59.3)	0.30	4.9
GO	14.87	12.41	5.96 (40.1)	7.42	2.9
LP	46.99	39.24	34.74 (73.9)	43.52	4.3
LP-MG	26.98	22.51	18.33 (67.9)	22.77	5.1
MG	2.04	1.70	1.38 (67.7)	1.71	4.9
OU	2.24	1.86	1.38 (61.7)	1.71	10.8
OU-LP	1.31	1.09	0.86 (65.7)	1.06	6.9
OV-LP	10.62	8.87	7.55 (71.1)	9.38	4.2
OV-SP	2.87	2.40	0.53 (18.3)	0.66	2.1
TOTAL	119.80	100.00	79.43 (66.3)*	100.00	513**

* Percentage of construction spread 2 frozen.
** Total no. of frozen/unfrozen interfaces in construction spread 2.

TABLE 3 Construction Spread 3 (KMP 269.3 to 380.4): Summary of Terrain Type and Frozen Ground Distribution.

Terrain Type	Distance (km)	% of Construction Spread	km Frozen (% of Terrain Type Frozen)	% of Total km Frozen	Avg. No. of Frozen/Unfrozen Interfaces/km
AF	6.28	5.65	1.65 (26.2)	2.14	5.1
AP	0.74	0.67	0.16 (21.2)	0.21	2.7
AT	1.89	1.70	1.13 (59.7)	1.47	2.1
CM	3.71	3.34	3.00 (80.9)	3.90	3.2
CM-AP	0.22	0.20	0.02 (10.5)	0.03	4.6
CT	1.40	1.26	1.25 (89.5)	1.63	4.3
ED	0.32	0.29	0 (0)	0	0
GO	7.42	6.68	5.53 (74.5)	7.19	0.7
LP	37.34	33.60	30.39 (81.4)	39.53	4.2
MG	28.63	25.77	15.81 (55.2)	20.56	1.9
OU	1.60	1.44	0.79 (49.7)	1.03	10.6
OV-LP	10.81	9.73	9.83 (90.9)	12.78	2.6
OV-MG	6.26	5.63	4.69 (74.9)	6.10	3.7
RK	1.86	1.68	0.15 (8.0)	0.19	0.5
OV-SP	2.62	2.36	2.48 (94.8)	3.24	0.4
TOTAL	111.10	100.00	76.88 (69.2)*	100.00	342**

* Percentage of construction spread 3 frozen.
** Total no. of frozen/unfrozen interfaces in construction spread 3.

TABLE 4 Construction Spread 4 (KMP 380.4 to 528.8): Summary of Terrain Type and Frozen Ground Distribution.

Terrain Type	Distance (km)	% of Construction Spread	km Frozen (% of Terrain Type Frozen)	% of Total km Frozen	Avg. No. of Frozen/Unfrozen Interfaces/km
AP	0.56	0.38	0.06 (10.4)	0.13	5.3
AT	0.22	0.15	0 (0)	0	0
CM	0.39	0.26	0.12 (30.3)	0.25	13.2
ED	5.13	3.46	0.31 (6.0)	0.66	1.0
GO	7.13	4.80	2.07 (29.1)	4.43	2.8
LB	0.22	0.15	0 (0)	0	0
LP	7.83	5.28	5.17 (66.0)	11.04	8.2
LP-MG	14.73	9.93	6.65 (45.1)	14.18	3.8
MG	62.75	42.27	15.48 (24.7)	33.01	3.3
OU	9.89	6.67	4.40 (44.5)	9.41	6.9
OV-LP	24.89	16.77	7.81 (31.4)	16.67	4.5
OV-MG	14.02	9.45	4.74 (33.9)	10.12	3.9
OV-SP	0.64	0.43	0.05 (7.2)	0.10	3.1
TOTAL	148.40	100.00	46.86 (31.6)*	100.00	592**

* Percentage of construction spread 4 frozen.
** Total no. of frozen/unfrozen interfaces in construction spread 4.

TABLE 5 Construction Spread 5 (KMP 528.8 to 695.9): Summary of Terrain Type and Frozen Ground Distribution.

Terrain Type	Distance (km)	% of Construction Spread	km Frozen (% of Terrain Type Frozen)	% of Total km Frozen	Avg. No. of Frozen/Unfrozen Interfaces/km
AF	0.07	0.04	0 (0)	0	0
AP	0.65	0.39	0.39 (60.4)	1.47	6.1
AT	2.92	1.75	1.20 (41.1)	4.47	2.4
CM	1.44	0.86	0.62 (43.3)	2.32	4.9
ED	7.35	4.40	0.37 (5.0)	1.36	0.5
GO	4.58	2.74	0 (0)	0	0
LP	15.55	9.30	4.02 (25.9)	14.98	3.2
LP-MG	3.00	1.80	0 (0)	0	0
MG	31.89	19.08	2.33 (7.3)	8.68	2.4
MA	1.55	0.93	0.37 (23.9)	1.39	6.4
MH	0.46	0.28	0 (0)	0	0
ME	0.45	0.27	0 (0)	0	0
OU	41.27	24.69	9.49 (23.0)	35.36	4.3
OU-LP	10.69	6.40	1.89 (17.8)	7.08	5.1
OV-LP	5.36	3.21	1.78 (33.3)	6.65	4.7
OV-MG	38.30	22.91	4.33 (11.3)	16.08	3.0
OU-MG	0.50	0.30	0.04 (8.5)	0.16	5.9
OV-SP	1.07	0.64	0 (0)	0	0
TOTAL	167.10	100.00	26.83 (16.0)*	100.00	532**

* Percentage of construction spread 5 frozen.
** Total no. of frozen/unfrozen interfaces in construction spread 5.

TABLE 6 Construction Spread 6 (KMP 695.9 to 868.3): Summary of Terrain Type and Frozen Ground Distribution.

Terrain Type	Distance (km)	% of Construction Spread	km Frozen (% of Terrain Type Frozen)	% of Total km Frozen	Avg. No. of Frozen/Unfrozen Interfaces/km
LP	15.12	8.77	0.77 (5.1)	1.29	2.2
MG	22.97	13.33	4.82 (22.0)	8.04	6.1
MA	2.70	1.55	0.23 (8.7)	0.39	2.6
OU	64.93	37.67	32.57 (50.2)	54.42	13.0
OU-LP	0.20	0.12	0 (0)	0	0
OV-LP	6.73	3.91	0.28 (4.2)	0.47	1.8
OV-MG	40.32	23.39	11.49 (28.5)	19.21	9.5
OU-MG	18.47	10.71	9.43 (51.1)	15.76	15.1
OU-AP	0.23	0.13	0 (0)	0	0
OV-SP	0.73	0.42	0.25 (34.8)	0.42	12.3
TOTAL	172.40	100.00	59.84 (34.7)*	100.00	1708**

* Percentage of construction spread 6 frozen.
** Total no. of frozen/unfrozen interfaces in construction spread 6.

TABLE 7 Terrain Type Codes.

Computer Coded Terrain Type	Description
AF	Alluvial fan
AP	Alluvial floodplain
AT	Alluvial terrace
CM	Colluvial slopewash
CM-AP	Combination of Cm and AP
CT	Colluvial talus slope
ED	Eolian dunes
GO	Glacial deposits
LB	Lacustrine postglacial basin
LP	Lacustrine plain
LP-MG	Combination LP and MG
MG	Ground Moraine
MA	Ablation Moraine
MH	Hummocky Moraine
ME	End Moraine
OU	Organic Undifferentiated
OU-LP	Combination
OV-LP	Organic Veneer over LP
OV-MG	Organic Veneer over MG
OU-MG	Combination
OU-ED	Combination
OU-AP	Combination
OU-GO	Combination
OU-RK	Combination
RK	Bedrock
OV-SP	Organic veneer over granular material of different origin
MG-RK	Combination
OV-RK	Organic veneer over bedrock

spread 6. The percentage of frozen ground generally increases in a northerly direction.

Sands, clays, and till are the predominant soil types in the northern sections in spreads 1, 2 and 3, while tills are dominant in spread 4. A significant portion of spreads 5 and 6 is over boggy terrain as seen by the high percentages of organic terrain types in Tables 5 and 6.

In spread 3 (Table 3) the two largest terrain groups are LP (clays) and MG (tills). Terrain type LP has a higher percentage of frozen ground (81.4%) than tills (55.2%) in this spread.

In spreads 5 and 6 approximately 50% of the total frozen ground occurs in organic terrain types. A large part of the frozen ground in this southern section is thus associated with organic landforms.

DISCUSSION

The entire geophysical survey profile lies within the discontinuous permafrost zone (Brown 1967). The profile section north of Fort Simpson lies in the widespread permafrost region, while the section south of Fort Simpson lies in the southern fringe of the permafrost region. Both the geophysical data and statistical information indicate that the amount of frozen ground generally decreases from north to south. The actual values of percentage frozen ground in spreads 1 to 6 were calculated to be approximately 83.7%, 66.3%, 69.2%, 31.6%, 16.1%, and 34.7%, respectively. The exception is spread 6, but this appears to be a topographic effect, since the elevation averages 300-400 m higher than the elevation in spread 5. In addition, the amount of frozen ground in spread 5 is quite low because of the presence of very wet terrain at a relatively lower altitude.

A major controlling factor for the occurrence of frozen ground is the soil distribution along the pipeline route. Sands and clays are predominant soil types in the northern sections in spreads 1, 2, and 3 while tills and organic terrain are dominant in spreads 4, 5, and 6. Previous studies in the literature (Heginbottom et al. 1978) have shown that finer textured soils have poorer drainage and therefore are more likely to allow ice accumulations. Areas with thick peat cover generally contain frozen ground because of the insulating property of the peat.

The most difficult problem to resolve when mapping the presence of frozen/unfrozen ground with electromagnetic methods is whether a change in apparent conductivity values is due to a change in the frozen/unfrozen state of the ground or due to a change in the soil composition.

Thus it is important to incorporate all available information in the interpretation of geophysical data for delineating frozen/unfrozen ground. This information includes terrain typing, vegetation, borehole logs, and topographic data. The use of this information with the geophysical data has demonstrated that it is possible to delineate the occurrence of permafrost along the pipeline route and supply valuable information for the design considerations of the pipeline. It has also significantly reduced the number of boreholes that would normally be required for such a detailed investigation.

ACKNOWLEDGEMENTS

The authors would like to express their appreciation to Interprovincial Pipe Line (NW) Ltd. for permission to publish this material. The work reported herein was performed as part of a contract for Interprovincial Pipe Line (NW) Ltd.

REFERENCES

Arcone, S.A., Sellmann, P.V., and Delaney, A.J., 1978. Shallow electro-magnetic geophysical investigations of permafrost. Proceedings of the Third International Conference on Permafrost, Vol. 1, pp. 501-507.

Brown, R.J.E., 1967. Permafrost map of Canada. Geological Survey of Canada, Map 1246A, 1:76 million.

Heginbottom, J.A., Kurfurst, P.J., and Lau, J.S.O., 1978. Regional Occurrence of Permafrost, Mackenzie Valley, Canada. Proceedings of the Third International Conference on Permafrost, Vol. 1, pp. 399-405.

Hoekstra, P., and McNeill, J.D., 1973. Electromagnetic Probing of Permafrost. Permafrost: North American Contribution, Second International Conference, p. 517.

McNeill, J.D., 1980. Electromagnetic terrain conductivity measurement at low induction numbers. Geonics Limited, Technical Note TN-6.

Rennie, J.A., Reid, D.E., and Henderson, J.D., 1978. Permafrost extent in the southern fringe of the discontinuous permafrost zone, Fort Simpson, N.W.T. Proceedings of the Third International Conference on Permafrost, Vol. 1, pp. 438-444.

THE REDISTRIBUTION OF SOLUTES IN FREEZING SOIL: EXCLUSION OF SOLUTES

B.D. Kay and P.H. Groenevelt

Department of Land Resource Science, University of Guelph
Guelph, Ontario, Canada, N1G 2W1

The redistribution of "non-reative" solutes during ground freezing has been studied. The combination of field studies, laboratory experiments and theoretical analyses lead to the conclusion that exclusion of a solute by an advancing freezing front may not be a particularly significant mechanism for solute redistribution. However substantial redistribution due to convective transport during the process of ice lens formation does occur.

INTRODUCTION

The redistribution of solutes in a porous medium which is freezing is significant for the engineer attempting to control frost heaving in a marine environment or attempting to use solutes to control frost heaving in non-marine environments. In addition the redistribution of solutes under conditions of seasonal freezing is important to the pedologist studying soil development and the agronomist wishing to optimize the use of agricultural chemicals. This paper summarizes the processes which may influence the redistribution of solutes when ground freezing occurs. Field data on the redistribution of solutes due to seasonal freezing are presented. Finally the dynamics of solute exclusion by a rapidly advancing freezing front is treated theoretically and related to a laboratory study involving coarse sands.

THE REDISTRIBUTION OF SOLUTES BY GROUND FREEZING

Processes which influence the redistribution of solutes in frozen and unfrozen soils are similar. Differences in the relative significance of the different processes in frozen and unfrozen soils relate to the role that the ice phase exerts on the processes. Therefore the extent of reaction with soil colloids, precipitation, physical or chemical transformation, and to a degree biological immobilization or transformation, depends on the degree of exclusion of solutes from the ice phase and the amount of ice present. Likewise diffusion, dispersion and convective transport are influenced by concentration in the liquid phase. In addition, the presence of ice at subzero temperatures may substantially alter the potential of unfrozen water which will influence the convective transport of solutes.

Isolated studies have been carried out to illustrate the influence of the ice phase on the reaction of solutes with colloid surfaces, on chemical transformations, and on biological transformations (eg Nelson and Romkens, 1972; Lahav and Anderson, 1973; Bremner and Zantua, 1975). Exclusion of solutes to thin unfrozen films next to colloid surfaces was considered by these authors to provide the medium whereby such transformations occured in frozen soils. These data apply to the behaviour of solutes within the frozen zone. However these studies provide little insight into the redistribution which solutes may experience in the vicinity of the freezing front as a freezing front is progressing into soil. The advancing freezing front would be expected to exclude solutes with a resultant accumulation of solutes in unfrozen films of water just behind the freezing front and in the unfrozen soil ahead of the freezing front. In the absence of unfrozen water and complete exclusion of solutes by ice, one might speculate that an advancing freezing front would effectively remove or "leach" all of the solutes from the frozen zone (note review by Hallet, 1973). However studies to confirm such speculation have not been published.

The behaviour of water and the behaviour of solutes in freezing soils are linked through processes such as the influence of solutes on the freezing point of the water and the convective transport of solutes within the frozen zone. The influence of solutes on the freezing point depression of water in porous media appears to be quantitatively similar to that in normal solutions (Banin and Anderson, 1974).
The flow of water within the frozen zone due to temperature-induced gradients in liquid pressure will result in solute redistribution (Cary and Mayland, 1972; Campbell et al., 1970; Cary et al., 1979). Solutes transported within the frozen zone would be expected to accumulate in the vicinity of ice accumulation. This in turn will modify the freezing point of the water in this zone, the amount of unfrozen water and consequently the hydraulic conductivity of this zone.

FIELD STUDIES ON SOLUTE REDISTRIBUTION

The field site was located at the Elora Research Station (43°39' N and 80°35' W), which is about 23 km northwest of Guelph, Ontario, Canada. Instrumentation was installed on the site for the purpose of measuring soil temperature profiles, soil moisture profiles, depth to the water table, and displacement of the ground surface. Further details regarding site characterization and instrumentation employed are provided by Sheppard et al. (1981).

The solute selected for investigation was the nitrate ion. Soil cores were obtained from positions that were randomly selected on the site each week during the winter of 1975-1976. At least three completely intact cores were obtained weekly. The cores were sectioned at 1 cm or 2 cm intervals depending on the depth and/or presence of ice lenses. The gravimetric water content was determined by oven drying. The samples were then rewet with excess water (a 1:2 soil:water mixture), shaken, and the nitrate concentration in the supernatant fluid determined. Nitrate analyses were expressed on an oven-dry soil weight basis.

Nitrate profiles for five representative weeks are shown in Figure 1. Two features of these data are salient: the nitrate level appears to fall slightly with the onset of freezing in the zone near the surface (0-10 cm), and the nitrate level appears to increase progressively with time at greater depth (15-35 cm).

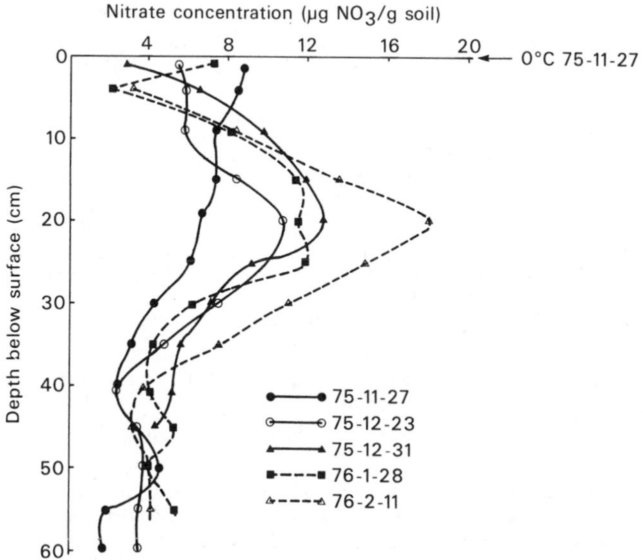

Figure 1: The redistribution of NO_3^- under field conditions due to ground freezing.

Changes in water content due to redistribution in the profile are shown in Figure 2. These data were obtained for the same samples referred to in Figure 1. These data show a progressive increase in water content near the surface (0-10 cm) and at greater depth (15-25 cm). The increase in water content can be attributed to discrete ice lenses formed by the upward movement of water. Near the surface the ice lenses were microscopic in thickness; at depth the ice lenses were often 1-2 mm in thickness. Laboratory studies (Loch and Kay 1978) suggest that the zone of maximum ice accumulation should occur just above the 0°C isotherm when minimum overburden is involved.

A comparison of Figures 1 and 2 illustrates contrasting behaviour in the 0-10 cm zone and the 15-25 cm zone. In the near surface zone the fall in nitrate level coincides with an increase in water content, whereas at greater depth both nitrate and water contents progressively increase.

In the near-surface zone, part of the decrease in nitrate levels may be due to the exclusion of solutes during repeated advances of the freezing front as a consequence of a number of freeze-thaw events between 75-11-27 and 75-12-23. Denitrification and leaching may also have occurred during this period.

At the 15-25 cm depth the increase in nitrate level coincides with the increase in water content. The increase in nitrate level between 75-11-27 and 75-12-23 must reflect the cumulative effect of solute exclusion by an advancing freezing front, and the accumulation of solutes resulting from the convective transport of solution into this zone as a consequence of ice lens formation. However, between 75-12-23 and 76-1-28 the rate of advance of the 0°C isotherm diminished substantially, with visible ice lenses accumulating in the 15-25 cm zone. The depth of maximum water content appeared to move from 20 cm to 25 cm between 75-12-81 and 76-2-11. These data are remarkably similar to data obtained by Dirksen and Miller (1966) during laboratory studies. The maximum in the nitrate profile appears to remain at 20 cm. This observation would suggest that the advancing freezing front is not totally effective in excluding solutes.

The results of the field study demonstrated that significant redistribution of solutes occurred as a consequence of ground freezing. Convective transport of solutes arising from water being drawn into the freezing zone appeared to be a significant mechanism in the redistribution process. The contribution of exclusion of solutes by an advancing freezing front could not be confirmed in either the near-surface zone (where rapid and repeated freezing occurred) or at greater depth (where the rate of advance of the freezing front was slower).

LABORATORY STUDIES

Laboratory studies were initiated in order to clarify the role of solute exclusion by an advancing freezing front on the redistribution of solutes.

A coarse sand (45% w/w 250-500 μm, 33% 500-1000 μm) was used as the soil material. This material is not frost susceptible and would be

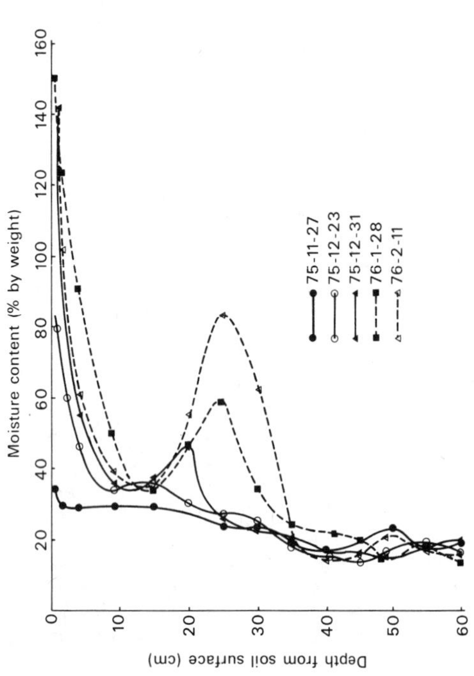

Figure 2 The redistribution of water under field conditions due to ground freezing.

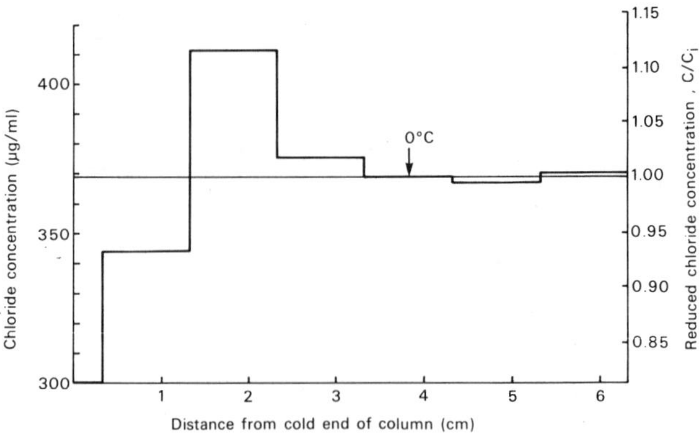

Figure 3 Redistribution of Cl^- during the unidirectional freezing of a saturated sand.

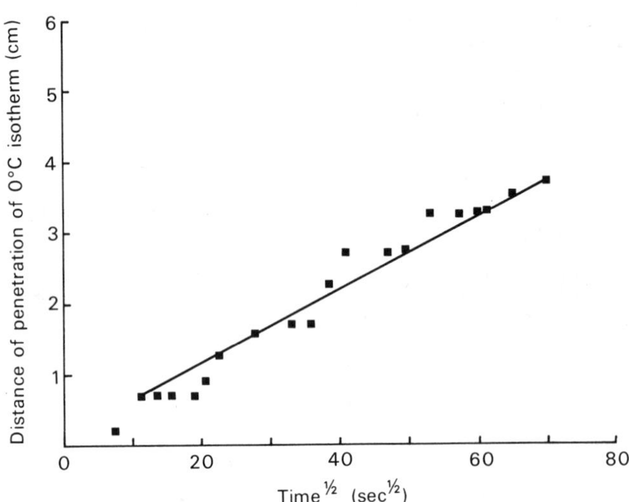

Figure 4 Rate of penetration of 0° isothern into saturated sand.

expected to have minimal unfrozen water contents at subzero temperatures. Therefore any redistribution of solutes during the freezing of this medium would be expected to be due to exclusion at the advancing freezing front.

The solute selected for study was Cl^-. The sand was packed into Plexiglass columns (length 8.0 cm, I.D. 3.73 cm) that were segmented in 1 cm sections. Columns were vacuum saturated with a KCl solution (369 ppm Cl^-).

Unidirectional freezing was achieved using the equipment described by Loch and Kay (1978). Columns were equilibrated at +0.5°C, a 4°C step change in temperature was introduced at the lower boundary, and the columns were then frozen for different lengths of time. At the termination of a freezing experiment the Plexiglass column was segmented and the contents of each segment extracted and chemical analysis of the pore solution carried out.

An example of the redistribution of solute observed at the conclusion of a freezing event is illustrated in Figure 3. This type of distribution is generated by a 0°C isotherm, which advanced upward through the column at rates illustrated in Figure 4. Two features are significant from these figures. First, there is no evidence of effective exclusion of the solute from the frozen zone. The ratio of concentration of solute in the frozen zone to that in the unfrozen zone varies from .81 to 1.12. Second, the 0°C isotherm has not moved uniformly through the sample but has moved in "leaps".

It is hypothesized that exclusion of solutes does occur on a local scale. As ice is forming, the freezing front advances slowly because of the release of latent heat. However, with this advancing freezing front is associated an accumulation of solute excluded from the ice. Increasing the concentration of solute results in increasing the freezing point depression and decreasing the tendency to form ice. As the rate of ice formation diminishes, the rate of advance of the 0°C isotherm begins to rapidly increase until nucleation occurs at some point just beyond the zone of solute accumulation. The process is then repeated. An analysis of Figure 4 suggests that at least three bands of solute accumulation may exist within the zone whose temperature is less than 0°C in this column. Sectioning the frozen zone in 1 cm increments provides concentrations averaged across microregions of solute rejection and solute accumulation. However, the net effect is that solute rejection does not occur on a macroscale, and therefore models designed to predict the redistribution of solutes should take this into account when freezing occurs at the rates used in this study.

THEORY

Theory is developed to describe the temperature profiles and the rate of freezing in a saturated medium. Predictions based on these equations are then compared to the rates of freezing observed in the laboratory studies. Theory describing the rates of solute accumulation at the freezing front is presented and then related to the conditions of the laboratory study in order to evaluate why solute exclusion does not occur on a macroscale.

The temperature profiles and the rate of advance of a freezing front in a saturated medium are calculated using a form of solution to the heat flow equations that is similar to the Neumann solution. Adjustments are made to account for the flow of fluid away from the freezing front due to the differences in volumes occupied by the liquid and ice phases when the matrix is rigid. It is assumed that all of the ice formation occurs at the freezing front. Under these conditions, the temperature profile can be described using the equation:

$$T = T_i - \frac{(T_i - T_f)}{\mathrm{erfc}\,\{(\lambda_f - 2F)/2\sqrt{K_u}\}}\,\mathrm{erfc}\{(\lambda - 2F)/2\sqrt{K_u}\} \quad (1)$$

where $F = \frac{(\rho c)_1}{(\rho c)_m} \times 0.04\ \theta_i \lambda_f$,

where ρc is the volumetric heat capacity and subscripts l and m refer to the liquid and matrix respectively, Θ_i is the volumetric ice content, T is the temperature, subscripts i and f refer to initial temperature and temperature at the freezing front, K_u is the termal diffusivity in the unfrozen zone, $\lambda = x/\sqrt{t}$ where x is the distance coordinate, t is time, and $\lambda_f = \lambda$ at the freezing front ($x = x_f$).

The value of λ_f is found from the heat balance at the freezing front, ie:

$$\frac{k_f}{\sqrt{K_f}}\frac{(T_b - T_f)}{\mathrm{erf}(\lambda_f/2\sqrt{K_f})}\exp[-\lambda_f^2/4K_f] + \frac{\sqrt{\pi}}{2} H_f \rho_i \theta_i \lambda_f +$$

$$\frac{k_u}{\sqrt{K_u}}\frac{(T_i - T_b)}{\mathrm{erfc}\{(\lambda_f - 2F)/2\sqrt{K_u}\}}\exp[-(\lambda_f - 2F)^2/4K_u] = 0 \quad (2)$$

where T_b is the cold side boundary temperature, k_f is the thermal conductivity of the frozen zone, K_f is the thermal diffusivity of the frozen zone, and H_f is the latent heat of fusion. The first term of Eq. (1), represents the amount of heat flowing from the freezing front towards the cold end. The second term, represents the source of heat due to freezing. The third term, represents the amount of heat flowing from the warm end to the freezing front, this term is very small in comparison to the first two terms, at least for the present boundary conditions. This means that the convective heat flow due to the expansion of water upon freezing has a negligible effect on the position of the frost front. The value of λ_f and the temperature profiles for the case that all the water freezes can now be calculated. In this case $F = 0.023\ \lambda_f$. In the experiments it was found that the temperature at the soil surface was -3.3°C. By setting T_b = -3.3°C and $T_f = 0$ it is found from Eq. (2) that $\lambda_f = 0.0337$ cm/$\sqrt{\text{sec}}$

The temperature profiles for the experiment illustrated in Figures 3 and 4, are given in Figure 5. Each point is also replotted as a function of λ (cm/$\sqrt{\text{sec}}$). It appears that for the frozen part, the points fall reasonably well on one line. This means that the similarity principle (scaling) holds for the frozen part. From these graphs it is found that $\lambda_f \simeq 0.06$. An analysis of the data in Figure 4 gives a similar value for λ_f. This large value of λ_f indicates that at the frost front ($T_f = 0$) only part of the water freezes. It further appears that the T(λ) line is not straight. This indicates that ice is continuously formed in the frozen zone (behind the frost front). Thus, there are "heat sources" in the frozen zone. This causes the temperature gradient to be the steepest near the boundary ($\lambda = 0$). The sand used in this experiment is expected to have minimal amounts of unfrozen water at subzero temperatures when the pore water does not contain solutes. Therefore

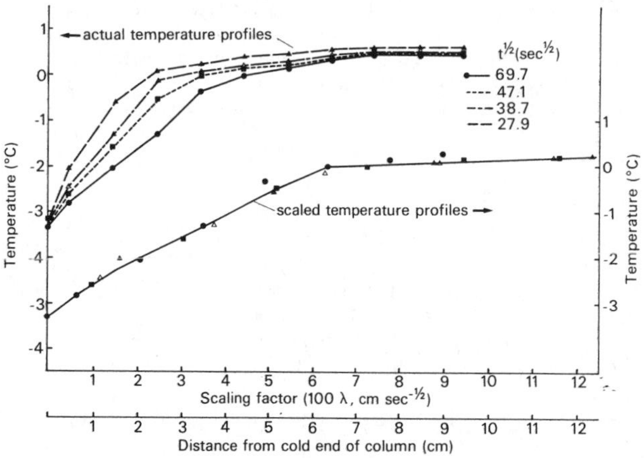

Figure 5 Actual and scaled temperature profiles observed during the unidirectional freezing of a saturated sand.

this will not provide a heat source behind the freezing front. Heat sources in the frozen zone must be associated with brine pockets or bands in which the solute is concentrated.

Ahead of an advancing frost front, i.e. in the not yet frozen part of the soil, the solute present in the liquid phase moves by diffusion and convection. Under these conditions, the concentration profile can be described as:

$$(C - C_i) = A \text{ erfc } [(\lambda - 2B)/2 \sqrt{D_o^1}] \qquad (3)$$

where C (λ) is the solute concentration, C_i is the initial solute concentration, D_o^1 is the effective diffusion coefficient, A is the integration constant, $B = 0.04\ \theta_i \lambda_f/\theta$, and θ is the liquid water content ahead of the freezing front. The value of A may be found from the mass balance of the solute.

The initial amount of solute (before freezing) per unit bulk volume is θC_i. We assume that a certain (constant) amount of solute remains after the frost front passes by. This amount is a fraction of initial amount and is equal to $(1 - k)\ \theta C_i$, where k is the segregation constant. If we assume that no solute is built into the ice lattice, then $(1 - k) = (\theta C)_f/\theta C_i$ where $(\theta C)_f$ is the product of the volumetric liquid content in the frozen zone and the concentration of solute in that liquid.
Using a mass blance approach by equating the amount of solute removed from the frozen zone to the amount of solute added to the unfrozen zone, it is found that:

$$A = k\ C_i\ (\lambda_f/2 \sqrt{D_o^1})/\text{ierfc } [(\lambda_f - 2B)/2 \sqrt{D_o^1}] \qquad (4)$$

The concentration at the frost front follows from Eq. (3):

$$C_f = C_i + A \text{ erfc } [(\lambda_f - 2B)/2 \sqrt{D_o^1}] \qquad (5)$$

Equations (3) and (4) now provide the opportunity to calculate the salt concentration profiles ahead of the freezing front. For reasonable values of the different parameters one finds from Eq. (5) the following approximate relation between C_f and C_i:

$$C_f = C_i + 80\ k\ C_i$$

This indicates that if there is no salt inclusion in the frozen zone (k=1) the concentration at the frost front would rise to about 80 times the initial concentration. The resulting freezing point depression would cause the frost front to "leap" over the zone of high solute concentration. This, in turn, would cause a considerable reduction in the k-value on a macroscopic scale.

ACKNOWLEDGEMENTS

Financial support for this study was provided by the Natural Sciences and Engineering Research Council of Canada and the Inland Water Directorate of Environment Canada.

REFERENCES CITED

Banin, A. and D.M. Anderson. 1974. Effects of salt concentration changes during freezing on the unfrozen water content of porous materials. Water Resour. Res. 10: 124-128.

Bremner, J.M. and M.I. Zantua. 1975. Enzyme activity in soils at subzero temperatures. Soil Biol. Biochem. 7: 383-387.

Campbell, C.A., Ferguson, U.S. and Warder, F.G. 1970. Winter changes in the soil nitrate and exchangeable ammonium. Can. J. Soil Sci. 50: 150-162.

Cary, J.W. and Mayland, H.F. 1972. Salt and water movement in unsaturated frozen soil. Soil Sci. Soc. Amer. Proc. 36: 549-555.

Cary, J.W., R.I. Pappendick, and G.S. Campbell. 1979. Water and salt movement in unsaturated frozen soil: principles and field observations. Soil Sci. Soc. Amer. J. 43: 3-8.

Dirksen, C. and R.D. Miller. 1966. Closed system freezing of unsaturated soil. Soil Sci. Soc. Amer. Proc. 30: 168-173.

Hallet, B. 1978. Solute redistribution in freezing ground. Proc. 3rd Int'l Permafrost Conf. Edmonton, Alta., Can. National Research Council, p. 86-91.

Lahav, N. and Anderson, D.M. 1973. Montmorillonite-benzidine reactions in the frozen and dry states. Clays and Clay Minerals 21: 137-139.

Loch, J.P.G. and B.D. kay. 1978. Water redistribution in partially frozen, saturated silt under several temperature gradients and overburden loads. Soil Sci. Soc. Amer. J. 42: 400-406.

Nelson, D.W. and Romkens, M.J.M. 1972. Suitability of freezing as a method of preserving runoff samples for analysis for soluble phosphate. J. Environmental Qual. 1: 323-324.

Sheppard, M.I., B.D. Kay, J.P.G. Loch. 1981. Coupled transport of water and heat in freezing soils: a field study. Can. J. Soil Sci. 61: 417-429.

LATEGLACIAL PALEOTEMPERATURES AND PALEOPRECIPITATION AS DERIVED FROM PERMAFROST: GLACIER RELATIONSHIPS IN THE TYROLEAN ALPS, AUSTRIA

Hanns Kerschner

Institut für Geographie, Universität Innsbruck, A-6020 Innsbruck, Austria

Numerous fossil rock glaciers in the Alps, situated 450-550 m below the presently active rock glaciers, can be correlated with the Egesen Stadial cold phase (11,000 to 10,000 BP). Fossil rock glaciers of Daun and Senders age (probably older than 13,000 BP) are situated 650 m below the active rock glaciers. From the lowering of the Egesen rock glaciers, a lowering of the mean annual air temperature in the order of -3.5°C against modern values is inferred. The lowering of the summer temperature, as inferred from the lowering of timberline, was in the order of -2.5°C during this period. For the Daun and Senders Stadials, a lowering of the MAAT in the order of -4.5°C is estimated. With the help of a permafrost-glacier-precipitation relationship, paleoprecipitation is estimated. During the Egesen Stadial, precipitation was reduced by 30% in the sheltered valleys of the Tyrolean Alps, whereas precipitation was about equal to present-day values along the northern slope. During the Daun and Senders Stadials, precipitation was reduced by 40% and 25% respectively in some valleys of the northern slope.

INTRODUCTION

In the past paleoclimatic studies in the Alps were concerned with the inference of summer temperature fluctuations from snowline and timberline fluctuations (Patzelt 1975, 1980). A first attempt to derive fluctuations of the annual temperature from rock glacier altitudes was made by the present author (Kerschner 1978). Statements on paleoprecipitation had been qualitative, relying mainly on paleobotanical information. For the last cold phase of the Alpine Lateglacial, the Egesen Stadial (Younger Dryas cold phase, 11,000 to 10,000 BP), a quantitative estimate of the precipitation could be given on the basis of equilibrium line altitudes and timberline altitudes (Kerschner 1982a). As paleobotanical information on timberline fluctuations is limited for various reasons, the distribution of lateglacial fossil rock glaciers is used as a source of paleoclimatic information in this paper. This attempt is largely based on the permafrost-glacier-precipitation model developed by Haeberli (1982, 1983), which relates the mean annual air temperature at the equilibrium line of glaciers with precipitation.

STRATIGRAPHY

The Alpine Lateglacial covers the period of the final recession of the large valley glacier systems

Table 1 Lateglacial glacier and rock glacier data, Tyrolean Alps, Austria

AGE	STADIAL	D ELA (m) M	D ELA (m) C	GLACIERS	BASAL SHEAR STRESS (GLACIERS)	ROCK GLACIERS, PUSH MORAINES M	ROCK GLACIERS, PUSH MORAINES C	D PF(m)	PRECIPITATION % M	PRECIPITATION % C	D MAAT °C	D MST °C
10.000 B.P.												
YOUNGER DRYAS	KROMER	-200	-180	MANY CIRQUE GLACIERS SOME VALLEY GLACIERS	LESS THAN 1 BAR	ABUNDANT, BUT HARD TO DIFFERENTIATE FROM OLDER DEPOSITS		?	POSSIBLY LESS THAN 50 %		-3,0- -3,6 ?	-2,5- -3,0
	LATE EGESEN	-300	-200	MANY CIRQUE GLACIERS SOME VALLEY GLACIERS	LESS THAN 1 BAR	ABUNDANT	ABUNDANT	-450- -550	75 %	50 % OR LESS	-3,0- -3,6	-2,5- -3,0
	EGESEN MAXIMUM	-400	-300	MANY CIRQUE GLACIERS SOME LARGE VALLEY GLACIERS	CA. 1 BAR	ABUNDANT	ABUNDANT	-450- -550	100 % OR MORE	LESS THAN 70 %	-3,0- -3,6	-2,5- -3,0
11.000 B.P.	BØLLING - ALLERØD - INTERSTADIAL COMPLEX											
13.000 B.P.												
OLDEST DRYAS	DAUN	-450- -500	-300- -400	VALLEY GLACIERS MORE FREQUENT	0,6-0,8 BAR	SOME	RARE	-650	60 %	LESS THAN 60 %	-4,5	-3,5- -4,0
	SENDERS	-600	-500- -400	VALLEY GLACIERS	CA. 1 BAR	RARE	?	-650	75 %	LESS THAN 75 %	-4,5	-4,5 ?
	GSCHNITZ	-750	?	LARGE VALLEY GLACIERS	CA. 1 BAR	?	?	?	?	?	?	?
	STEINACH	-750- -800		LARGE VALLEY GLACIERS	?	?	?	?	?	?	?	?
	BÜHL	-1200 OR MORE		VALLEY GLACIER SYSTEMS ("EISSTROMNETZ") INTACT	?	?	?	?	?	?	?	?

D ELA: LOWERING OF THE EQUILIBRIUM LINE ALTITUDE AGAINST MODERN VALUES, D PF: LOWERING OF PERMAFROST (ROCK GLACIERS), D MAAT: LOWERING OF THE MEAN ANNUAL AIR TEMPERATURE, D MST: LOWERING OF THE MEAN SUMMER TEMPERATURE, C: CONTINENTAL AREAS, M: MARITIME AREAS from approximately 16,000 BP until 10,000 BP. By 9,500 BP, glaciers had already readvanced within their mid-19th century limits (Patzelt 1973). This period of general downwasting of the glaciers was interrupted several times by periods of readvance ("stadials"). These stadials, which are named after type localities near Innsbruck (Tyrol), have been described in detail by Mayr and Heuberger (1968), Heuberger (1966) and Kerschner and Berktold (1982).

During the past decade, extensive mapping by several authors showed that the stratigraphical system developed in the Tyrolean Alps is equally applicable in other areas of the Austrian and Swiss Alps (bibliographies in Müller et al. 1981, Maisch 1981). These readvance periods were usually not single phased but consisted of a series of advances and retreats, each one smaller than the preceding. The correlation with the lateglacial stratigraphy of northwestern Europe still presents some problems, mostly for the older stadials. The most likely correlation is given in Table 1.

In the course of mapping lateglacial deposits, a large number of fossil rock glaciers of lateglacial age was found. These rock glaciers developed from talus as well as from morainic debris. Their average length is in the order of 300 to 500 m, but some are more than 1,000 m long and several hundred metres wide. From the collapse structures, a former ice content in the order of 50 to 70% is estimated. A transition from rock glaciers to push moraines (Haeberli 1979a) and vice versa can be observed at numerous localities. Rock glaciers with a core of glacier ice are presently uncommon in the Alps. This is also true for lateglacial rock glaciers. In most cases the transition between the former glacier and the rock glacier can be clearly delineated in the field or from aerial photographs. The period of activity of these rock glaciers can at least be partially fixed with correlation to nearby moraines of known age.

Most fossil rock glaciers at lower altitudes were active during the Egesen Stadial (11,000 to 10,000 BP). Many of them cover large areas which were glacierized during the Daun Stadial as can be deduced from nearby moraines (Kerschner 1978, Maisch 1981). They indicate that a major warming period preceded the Egesen Stadial, which led to an extensive deglaciation. This warm phase is probably identical with the Allerød-Bølling Interstadial complex (13,000 to 11,000 BP; Welten 1982, Eicher 1979, Eicher et al. 1981). As the Younger Dryas cold phase lasted for about 1,000 years and as sufficient debris was available from intense frost shattering and moraines, it is assumed that these rock glaciers closely approached their lowermost position as controlled by topography and climate. In places with appropriate topography, they extend to altitudes between 1,900 and 2,000 m a.s.l., about 450-550 m below presently active rock glaciers. The available data show that there is no significiant spatial variability of the altitude of Egesen rock glacier termini in the study area. Similar altitudes can be observed in central and southwestern Switzerland. Rapid warming after 10,000 BP, as it can be inferred from numerous paleobotanical and glaciological studies (e.g. Welten 1982, Patzelt and Bortenschlager 1982) suggests that the activity of these rock glaciers came to an end at the Lateglacial/Postglacial boundary around 10,000 BP.

While there exists already a fair amount of data on the distribution of rock glaciers of Egesen age, information about rock glaciers of older age is still limited. This is mainly due to the fact that large valley glaciers were more frequent during older phases of the Alpine Lateglacial. These glaciers occupied large areas suitable for rock glacier formation and ended at altitudes that were obviously too low for the existence of permafrost. Rock glaciers of Daun age in the Tuxer Voralpen, 30 km south of Innsbruck, are 620-680 m below the present-day lower boundary of permafrost(Patzelt 1983). Push moraines of Daun age in the Karwendel to the north of Innsbruck can be observed 150-200 m below Egesen rock glaciers. Rock glaciers which can be correlated with the Senders Stadial seem to have been at similar altitudes as the Daun rock glaciers (Kerschner and Berktold 1982).

PALEOCLIMATIC INTERPRETATION

Numerous measurements show that the surface velocity of rock glaciers in the Alps rates from approximately 0.1 to 1.0 m/yr. A surface velocity of 0.3 m/yr is chosen as a reasonable average by Barsch (1977a,b). From this it is deduced that the formation of a rock glacier of considerable length (i.e. several hundred metres) demands at least several centuries (Vietoris 1972). Possible former periods of fossilization and reactivation can usually not be recognized in the field. From this it follows that fossil rock glaciers only provide a general picture of climatic changes on a time scale of several centuries.

About the same level of resolution can be attained for the inference of timberline fluctuations from lateglacial pollen diagrams, which have also been used as a source of paleoclimatic information (Kerschner 1982a). Thus, paleoclimatic models with rock glacier data as input variables should be comparable with models which use timberline fluctuations as one source of information. Both models cannot make full use of the much more detailed, but unrevealed paleoclimatic information which is present in the large number of lateglacial morainic systems.

Mean Annual Air Temperature (MAAT)

The spatial pattern of the distribution of active rock glaciers--as permafrost phenomena--reflects the pattern of annual isotherms in the study area very well (Fliri 1975, Barsch 1977a, Haeberli 1979b). The same should hold true for fossil rock glaciers which had sufficient time to develop. Hence, as a first approximation, the vertical difference between active and fossil rock glaciers of identical exposure, comparable extent and comparable morphologic setting is proposed as a measure of the lowering of the MAAT (Kerschner 1978). This vertical difference is converted into the lowering of the MAAT with the help of a temperature lapse rate of $-0.65^{\circ}C/100$ m. From the lowering of the Egesen rock glaciers (-450 to 550 m) a lowering of the MAAT in the order of -3.0 to

-3.6°C can be inferred for the period 11,000 to 10,000 BP. This is more than the lowering of summer temperature, which can be calculated from the lowering of timberline (approximately -400 m throughout the study area) as about -2.5° to -3.0°C. These results suggest an increase in seasonal contrasts of temperature between summer and winter during the Egesen Stadial. The same figures seem to apply for the late Egesen Stadial, when glaciers were already much smaller than during the maximum.

From the lowering of the Daun and Senders rock glaciers and push moraines (approximately -650 m) a lowering of the MAAT in the order of -4.0° to -4.5°C is inferred. Due to a lack of data, no direct estimates can be given for the summer temperature during these periods.

The figures for the lowering of the MAAT during the Egesen Stadial seem to be rather reliable due to the long duration of the cold phase and the large sample size. The figures calculated for the Daun Stadial and the Senders Stadial are less well supported due to the small sample size and our still insufficient knowledge of the duration of these cold phases.

Paleoprecipitation

Active rock glaciers are abundant in the rather dry valleys of the central Alps which are sheltered from the precipitation bearing air masses from the north and south. Precipitation at 2,000 m a.s.l. is in the order of 750 to 1,200 mm/yr in these areas. The equilibrium line (EL) of glaciers there is high compared with the humid northern slope, where precipitation is in the order of 1,600 to 2,000 mm/yr (Gross 1983, Kuhn 1983). Active rock glaciers are almost absent in the humid northern areas of the Alps where the EL is close to the lower boundary of permafrost, thus leaving no space for the developement of rock glaciers. The vertical distance between the termini of active rock glaciers and the equilibrium line altitude (ELA) increases markedly with decreasing precipitation. From this it follows that the MAAT at the ELA in dry areas is lower than in humid areas (Haeberli 1979b, 1982, 1983). Hence, the MAAT at the ELA can be taken as an indirect measure for the precipitation in a given area. The relationship between the MAAT at the ELA and the precipitation at 2,000 m a.s.l. in the Alps (Figure 1) can be expressed as

$$\text{Precipitation}_{2,000} = 3613.(-\text{MAAT}_E)^{-0.8679} \quad (1)$$

from the data of Haeberli (1982). A standard altitude of 2,000 m a.s.l. is chosen for practical reasons. It is also used in other models for the determination of paleoprecipitation (Kerschner 1982a).

Equation (1) serves as the basis for the quantitative estimation of paleoprecipitation during some of the lateglacial stadials. The MAAT at the ELA is calculated from the vertical difference between the altitude of the termini of fossil rock glaciers and the ELAs of nearby glaciers of similar age. Following the lines of Haeberli (1982), it is assumed that the MAAT at the termini of rock glaciers varies between -2.0° and -4.0°C, depending on aspect and topographic setting of the respective rock glaciers.

Most of the available data come from glaciers and rock glaciers of the Egesen Stadial cold phase (11,000 to 10,000 BP). They show that at the humid northern margin of the study areas the MAAT at the ELA was not below -2.5°C, from which a precipitation in the order of 1,600 to 2,000 mm/yr can be calculated with the help of equation (1). This is about equal to the present-day precipitation in this area. In the sheltered valleys of the central Alps, the MAAT at the ELA was about -6.0° to -7.0°C indicating a precipitation in the order of 650 to 800 mm/yr, which is about 25-33% less than present-day values.

During the late stages of the Egesen Stadial, glaciers were much smaller and their EL was at higher altitutdes than during the maximum. In some areas rock glaciers developed within the area which was glacierized during the maximum advance (Müller et al. 1981, Maisch 1981, Kerschner 1982b). Estimates show that the MAAT at the ELA was of the order of -3.5°C along the northern slope but -8.0°C in the sheltered valleys, thus pointing to a reduction of precipitation by 25 and 55% respectively.

These results can be cross-checked with a stochastic glacier-climate model (Figure 1) which relates the summer temperature at the ELA and the precipitation at 2,000 m a.s.l. (Kerschner 1982a) and with a glacial-meteorological model, which relates ablation and accumulation at the EL through a system of equations defining heat and mass exchange at the glacier surface (Kuhn 1980, 1982). In these models, data on the change of the altitude of tim-

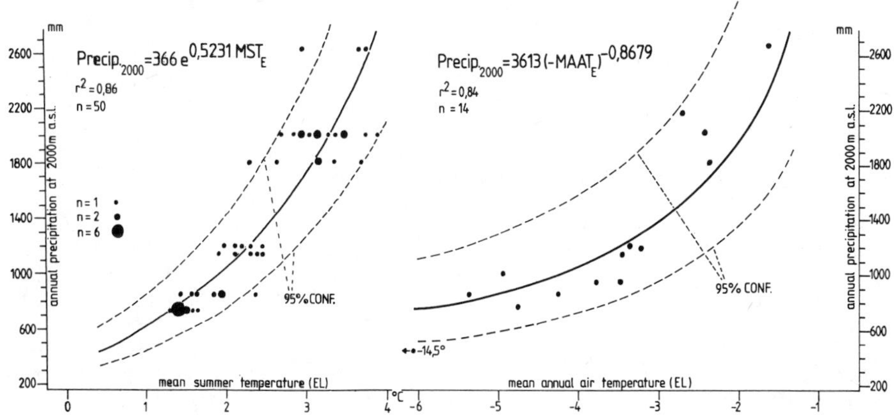

Figure 1 Relation among mean summer temperature at the equilibrium line (MST_E) and precipitation at 2,000m a.s.l. ("glacier-climate-model", Kerschner (1982a)) and relation among mean annual air temperature at the equilibrium line ($MAAT_E$) and precipitation at 2,000 m a.s.l. ("glacier-permafrost-model" after Haeberli (1982), slightly modified).

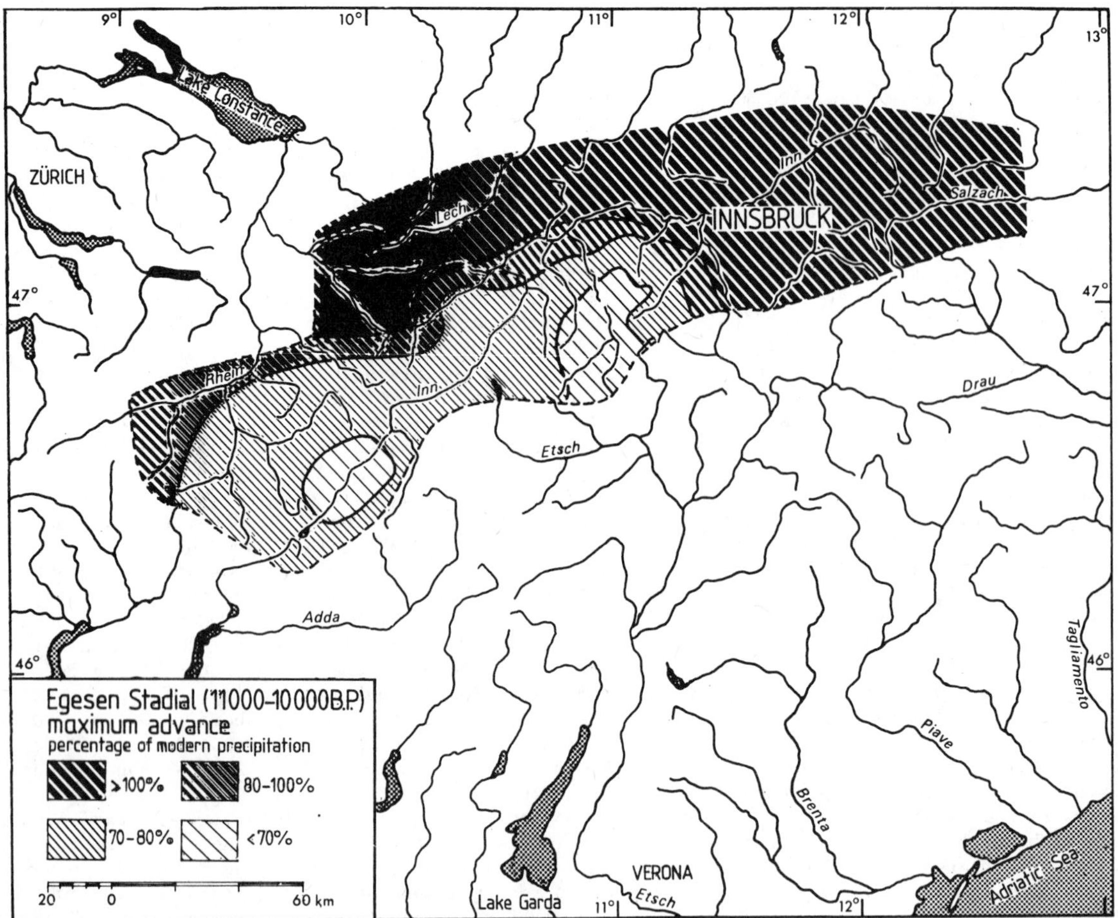

Figure 2 Distribution of precipitation in the central Alps during the Egesen maximum advance, expressed as percentage of present-day precipitation at 2,000 m a.s.l.

berline are used as an independent variable to determine the change of summer temperature at the EL. The use of summer temperature in a glacier-climate model is physically more correct than the use of the MAAT, because summer temperature is an important factor governing ablation. The use of both models is limited to periods for which estimates of the timberline altitude can be made from palynological data. This precludes their application to stadials which preceded the Allerød-Bølling Interstadial complex (13,000 to 11,000 BP). They give similar values of paleoprecipitation during the Egesen Stadial as the permafrost-glacier model discussed in this paper. Differences do not exceed 5% on the average.

A combined application of the glacier-climate model and the permafrost-glacier model allows the plotting of a map of the change of precipitation from present-day values to the precipitation during the maximum advance of the Egesen Stadial (Figure 2) The map covers an area of approximately 250 x 100 km of the central and northern Alps in western Austria and eastern Switzerland. It shows the rapid decrease of precipitation from the northern slope towards the central valleys, which was much more pronounced during the Egesen Stadial than it is today. The transition between humid and dry areas during this period was sharp and concentrated along mountain chains with altitudes of more than 2,700 m a.s.l. to the south of the Inn valley.

Data on the Daun and Senders Stadials are still limited, so only a few remarks on the former climatic conditions can be made. These data come from valleys which were part of the humid northern area during the Egesen Stadial with 100% of the present-day precipitation. They indicate that precipitation in this area was about 25% lower during the Daun Stadial. During these Stadials, the ELA was much lower than during the Egesen Stadial (Table 1). By employing the glacial-meteorological model (Kuhn 1982) the lowering of summer temperature can be calculated. The precipitation values determined with the permafrost-glacier model are serving as input data. During the Senders Stadial, the lowering of summer temperature was in the order of -4.5°C or more. During the Daun Stadial, summer temperatures were approximately -3.5°C to -4.0°C lower than at present-day. These changes of summer temperature cannot be tested quantitatively against other information and thus must be regarded as orders of magnitude only.

The overall impression of the climate during these phases of the Alpine Lateglacial is that of a rather cold and dry mountain climate which seems to be readily comparable with the present-day climate in the high mountain regions of central Asia. It seems that the stadials started with cool and comparably moist conditions which changed to cold-dry conditions within a time span of a few centuries or less. This pattern, which was obtained from glaciological and geomorphological information, agrees favourably with the paleobotanical record of the Alpine Lateglacial. This record shows a cold and dry climate favouring a cold steppe vegetation in the upper reaches of the valleys above timberline (Welten 1972, 1982, Markgraf 1973; extensive bibliographies in Burga 1980, Küttel 1979). Results of $^{18}O/^{16}O$-investigations on lake marls in Switzerland agree also well with this pattern (Eicher 1979).

CONCLUSIONS

Fossil rock glaciers in the Alps seem to be a valuable source for paleoclimatic information. Fluctuations of the MAAT can be inferred from the difference in altitude between fossil and active rock glaciers, if data are derived from rock glaciers of identical exposure comparable extent and comparable morphologic setting. The permafrost-glacier model discussed in this paper seems to provide reliable values of the paleoprecipitation during certain cold phases of the Alpine Lateglacial. It seems to be a useful model for the determination of paleoprecipitation for periods, for which timberline data are absent but rock glacier data are available.

The use of rock glacier data in paleoclimatic considerations allows to extend quantitative paleoclimatic studies in the Alps to the northern Alps, where timberline altitude is not so much controlled by climate, but by relief and the limestone substratum. With some precautions, paleoclimatic estimates can be made for cold phases older than the Allerød-Bølling Interstadial complex (13,000 to 11,000 BP), when timberline had not yet reached its highest possible position. Thus, rock glacier studies allow to broaden the scope of quantitative paleoclimatic studies in the Alps both in space and in time.

ACKNOWLEDGEMENTS

I would like to express my gratitude to Dr. Wilfried Haeberli (Versuchsanstalt für Wasserbau, Hydrologie und Glaziologie, ETH Zürich), who provided valuable encouragement during all stages of the preparation of this paper and to Dr.Klaus Frantz, who assisted in the final stages of the preparation of the manuscript. I greatly appreciate many valuable suggestions for the improvement of the paper by the reviewers of a first draft of the manuscript.

REFERENCES

Barsch, D., 1977a, Alpiner Permafrost - Ein Beitrag zu seiner Verbreitung, zu seiner Charakteristik und seiner Ökologie am Beispiel der Schweizer Alpen: Abhandlungen der Akademie der Wissenschaften in Göttingen, Mathematisch-Physikalische Klasse 3, Folge 31, p. 118-141

Barsch, D., 1977b, Nature and impact of mass-wasting by rock glaciers in alpine permafrost conditions: Earth Surface Processes, v.2, p. 231-245

Burga, C.A., 1980, Pollenanalytische Untersuchungen zur Vegetationsgeschichte des Schams und des San Bernardino (Graubünden, Schweiz), Vaduz: Cramer, 165 p. (=Dissertationes Botanicae 56)

Eicher, U., 1979, Die $^{16}O/^{18}O$ und $^{12}C/^{13}C$-Isotopenverhältnisse in spätglazialen Süßwasserkarbonaten und ihr Zusammenhang mit den Ergebnissen der Pollenanalyse: PhD.thesis, Bern, 205 p.

Eicher, U., U.Siegenthaler, and S.Wegmüller, 1981, Pollen and oxygen isotope analyses on Late- and Postglacial sediments of the Tourbière de Chirens (Dauphiné, France): Quaternary Research v.15, p. 160-170

Fliri, F., 1975, Das Klima der Alpen im Raume von Tirol: Monographien zur Landeskunde von Tirol I, Innsbruck: Wagner, 454 p.

Gross, G., 1983, Die Schneegrenze und die Altschneelinie in den österreichischen Alpen: Innsbrucker Geographische Studien, v.8, p. 59-83

Haeberli, W., 1979a, Holocene push moraines in alpine permafrost: Geografiska Annaler, v.61A, p. 43-48

Haeberli, W., 1979b, Special aspects of high mountain permafrost methodology and zonation in the Alps, in Proceedings of the Third International Conference on Permafrost, v.1: Ottawa, National Research Council of Canada, p. 378-384

Haeberli, W., 1982, Klimarekonstruktionen mit Gletscher-Permafrost-Beziehungen: Materialien zur Physiogeographie, v.4, p. 9-17

Haeberli, W., 1983, Permafrost-glacier relationships in the Swiss Alps - today and in the past: Fourth International Conference on Permafrost, Fairbanks, Alaska

Heuberger, H., 1966, Gletschergeschichtliche Untersuchungen in den Zentralalpen zwischen Sellrain und Ötztal: Wissenschaftliche Alpenvereinshefte 20, 126 p.

Kerschner, H., 1978, Paleoclimatic inferences from Late Würm rock glaciers, eastern Central Alps, western Tyrol, Austria: Arctic and Alpine Research, v. 10, p. 635-644

Kerschner, H., 1982a, Outlines of the climate during the Egesen advance (Younger Dryas, 11,000 - 10,000 BP) in the central Alps of the western Tyrol, Austria: Zeitschrift für Gletscherkunde und Glazialgeologie, v. 16, p. 229-240

Kerschner, H., 1982b, Zeugen der Gletschergeschichte im oberen Radurschltal - Alte Gletscherstände und Blockgletscher in der Umgebung des Hohenzollernhauses: Alpenvereinsjahrbuch 1982/1983 ("Zeitschrift" 107), p. 23-27

Kerschner, H., and E. Berktold, 1982, Spätglaziale Gletscherstände und Schuttformen im Senderstal, nördliche Stubaier Alpen: Zeitschrift für Gletscherkunde und Glazialgeologie, v. 17, p.125-134

Kuhn, M., 1980, Climate and glaciers, in Proceedings of the Canberra Symposium, Dec.1979: IAHS-AISH Publ. 131, p. 3-20

Kuhn, M., 1982, Die Reaktion der Schneegrenze auf Klimaschwankungen: Zeitschrift für Gletscherkunde und Glazialgeologie, v. 16, p. 241-254

Kuhn, M., 1983, Die Höhe der Schneegrenze in Tirol, berechnet aus Fliris klimatischen Profilen: Innsbrucker Geographische Studien, v. 8, p.85-91

Küttel, M., 1979, Pollenanalytische Untersuchungen zur Vegetationsgeschichte und zum Gletscherrückzug in den westlichen Schweizer Alpen: Berichte der Schweizerischen Botanischen Gesellschaft, v. 89, p. 9-26

Maisch, M., 1981, Glazialmorphologische und gletschergeschichtliche Untersuchungen im Gebiet zwischen Landwasser- und Albulatal, (Kt.Graubünden): Physische Geographie, Universität Zürich, v. 3, 215 p.

Mayr, F., and H. Heuberger, 1968, Type areas of lateglacial and postglacial deposits in Tyrol, Eastern Alps, in Richmond, G.M., ed., Glaciation of the Alps: University of Colorado Studies, Series in Earth Sciences, v. 7, p. 143-165

Markgraf, V., 1973, Paleoclimatic evidence derived from timberline fluctuations: Colloque International CNRS Paris, v. 219, 15 p.

Müller, H.N., H. Kerschner, and M. Küttel, 1981, Gletscher- und vegetationsgeschichtliche Untersuchungen im Val de Nendaz (Wallis) - Ein Beitrag zur alpinen Spätglazialchronologie: Zeitschrift für Gletscherkunde und Glazialgeologie, v. 16, p. 61-84

Patzelt, G., 1973, Die postglazialen Gletscher- und Klimaschwankungen in der Venedigergruppe (Hohe Tauern, Ostalpen): Zeitschrift für Geomorphologie N.F., Supplementband 16, p. 25-72

Patzelt, G., 1975, Unterinntal-Zillertal-Pinzgau-Kitzbühel - Spät- und postglaziale Landschaftsentwicklung: Innsbrucker Geographische Studien, v. 2, p. 309-329

Patzelt, G., 1980, Neue Ergebnisse der Spät- und Postglazialforschung in Tirol: Österreichische Geographische Gesellschaft, Zweigverein Innsbruck, Jahresbericht 1976/77, p. 11-18

Patzelt, G., 1983, Die spätglazialen Gletscherstände im Bereich des Mieslkopfes und im Arztal, Tuxer Voralpen, Tirol: Innsbrucker Geographische Studien, v. 8, p. 35-44

Patzelt, G., and S. Bortenschlager, 1978, Zur Chronologie des Spät- und Postglazials im Ötztal und Inntal (Ostalpen, Tirol), in Frenzel, B., ed., Führer zur Exkursionstagung des IGCP-Projektes 73/1/24 "Quaternary Glaciations in the Northern Hemisphere", 5-13 September 1976: Bonn-Bad Godesberg, p. 185-197

Vietoris, L., 1972, Über den Blockgletscher des äußeren Hochebenkars: Zeitschrift für Gletscherkunde und Glazialgeologie, v. 8, p. 169-188

Welten, M., 1972, Das Spätglazial im nördlichen Voralpengebiet der Schweiz - Verlauf, Floristisches, Chronologisches: Berichte der Deutschen Botanischen Gesellschaft, v. 85, p. 69-74

Welten, M., 1982, Pollenanalytische Untersuchungen zur Vegetationsgeschichte des Schweizerischen Nationalparks: Ergebnisse der wissenschaftlichen Untersuchungen im Schweizerischen Nationalpark XVI/80, p. 3-43

SOME ABIOTIC CONSEQUENCES OF THE CANOL CRUDE OIL PIPELINE PROJECT, 35 YEARS AFTER ABANDONMENT

G. Peter Kershaw, Department of Geography, University of Alberta, Edmonton, Alberta Canada, T6G 2H4

The Northwest Territories portion of the CANOL No. 1 crude oil pipeline has been abandoned since 1945. A number of terrain disturbances were initiated by the construction and operation of this pipeline that are comparable to those associated with contemporary developments in northern environments. Abiotic alternations include the removal, compaction, or burial of the often peaty soil surface layers. In all cases there was less organic matter at the surface than in undisturbed areas. Disturbance subsurface temperatures were warmer than controls and it was the amelioration of this ecologically limiting factor that led to a rapid rate of organic matter accumulation (e.g. up to 98% of control values). Subsurface moisture was less for most disturbance types except where depressions resulted from compaction or thermal subsidence in Organic Cryosols. Biological and geomorphological processes continue to respond to the initial surface disruptions and have as yet not stabilized, although the degree of "recovery" varies with the type of disturbance and the terrain in which it occurs.

INTRODUCTION

The CANOL Pipeline Project, 35 years after abandonment, provides an example of how a major, large-scale development has altered the local environment in this mountainous, discontinuous permafrost zone. The Project affected a diversity of environments over a broad geographical area. However this study only reports on the abandoned section within tundra areas in the Northwest Territories, Canada. This area extends over 1.5° of latitude and 3° of longitude between Road Mile Post (R.M.P.) 56.6 near the Mackenzie River at 64°41'N, 127°10'W, and R.M.P. 231.3 on the Continental Divide at 63°14'N, 130°02'W. Elevations of the tundra affected by the CANOL Project range from 775 m near the Mackenzie River to 1,740 m on the Plains of Abraham.

The term "substrate", as it is used in this paper, refers to the surface and near-surface layers of the ground (Agriculture Canada 1976). It includes "soils" of both disturbed and control sites.

Within the study area the pipeline, road, and associated telephone system were constructed between 10 October 1942 and 12 March 1944 and abandoned by 31 May 1945. During the 39 years since the initial disturbances were created, no artificial rehabilitation was attempted and this area still remains closed to auto traffic.

METHODS

Field sampling was carried out between 23 June and 29 September 1979 on those sites where vegetation information had been previously collected. Two criteria were used in selecting representative disturbance and control sites for sampling. These insured that valid comparisons with control conditions could be made.

1. A variety of sites had to be located in close proximity to one another.
2. Each set of sites had to be contained within an area of homogeneous vegetation (i.e., in terms of structure and species composition), considered representative of larger regions.

Layers were numbered and each measured to the nearest 0.5 cm. All subsurface temperatures (i.e., 10 cm depth) for a set of disturbances and their control were determined over a maximum of a few hours time on the same day. At this depth temperatures were expected to vary little from day to day (Geiger 1959).

Samples were thawed in the laboratory and the moisture content determined gravimetrically (McKeague 1976 and Kalra 1971). Moisture content was expressed on an oven dry weight basis. Organic matter was determined by weight loss-on-ignition (McKeague 1976 and Kalra 1971) and expressed as a percentage of precombustion weight. Sieving separated four soil fractions: 1) gravel, 2) very coarse to medium sand, 3) medium to very fine sand, and 4) very fine sand, silt, and clay (McKeague 1976).

Control or reference sites were selected from nearby undisturbed areas. The control soil pit was dug within the plant community in which the initial disturbance was effected.

Discussion is based upon control-corrected values (CC) which are the means of the differences between disturbances (D) and associated controls (C)

$$CC = [(D_1 - C_1) + (D_2 - C_2) \dots + (D_n - C_n)]/n$$

If the mean of the differences between the control-disturbance pairs was positive, then this indicated that disturbances, on average, had greater amounts of a particular characteristic. Negative values indicated that, on average, disturbances were lower than their controls.

In total, 132 substrate descriptions were made from 33 sampling localities with 33 control soil descriptions. The substrate sampling followed two years of investigations during the growing season and winter and selection of sites was based on these studies. The substrates described represent 43% of the 310 stands from which plant community information was collected (Kershaw 1983).

RESULTS AND DISCUSSION

Within the study area, eight distinct types of disturbances were recognized - road, false start road, bladed trail, camp yard, bulldozer track, gravel pit, gravel pit access road and oil spill. In all cases, the road had an aggregate surface, but particle size varied with the source material because no crushing or washing facilities were used at the time of construction. The road substrates remain highly compacted as a result of vehicular use.

False start roads included graded winter roads, poor-weather alternate roads, and partly constructed road reroutings. They were infrequently, if ever, used and as a consequence were less compacted than the Canol Road.

The pipeline right-of-way was bladed to create a level surface and to remove shrubs, large stones, or hummocks (Figure 1). In many cases, low spots would be left intact while hummock tops were planed off. The Resulting bladed trail was generally a shallow depression dominated by mineral exposures.

Camp yards were temporary or permanent areas used as parking, storage or building sites. Generally they were bladed level and seldom infilled but were driven over by tracked and tired vehicles. Relatively undisturbed sites were found immediately adjacent to highly altered, bladed surfaces.

Bulldozer tracks were identified by the presence of two tread furrows that resulted from compression of the organic horizons and therefore were most apparent in areas with peaty surface layers (Figure 1).

Gravel pits were excavated in glacial outwash, till, and modern alluvium as well as in blockfields (Figure 1). They often contained ponded water year round as a result of local drainage into the closed depression or excavation below the local water table (Figure 1). The gravel pit access roads were generally no more than a buildup of loosely compacted, rutted material that fell from trucks as they left the pit (Figure 1).

Oil spills were readily identified by their odor and blackened appearance, and usually lacked vegetation. In some cases, these substrates had high organic content but in others, deflation and running water had left only a mineral surface. A thick, waxy or tarry coating was also present on some sites. The parameters chosen for study do not necessarily reflect the degree of substrate disturbance associated with oil spills. The addition of oil has altered the substrate by introducing substances toxic to plants and thereby dramatically affecting rates of revegetation. Oil spills therefore will be discussed separately.

An initial evaluation of the results indicated that reaction to a particular disturbance varied among the major plant communities in the study area. Further discussion will therefore be grouped by plant community (Figure 1).

Erect Deciduous Shrub Tundra

This type of tundra was found to generally occupy well-drained landforms of alluvium and till throughout the study area. However, the thick surface covering of lichens reduced moisture losses and therefore these soils were also moist (Figure 1). The sporodically distributed patterned ground appeared to be active. However, to a large extent this was the result of needle ice activity and deflation processes, both of which prevent or restrict plant colonization.

Ice-rich permafrost was restricted to wet lands containing isolated palsas, where organic layers were relatively thick. These features were small in area and the soil here was classified an Organic Cryosol.

Dystric Brunisols were most common in this plant community and were typically found beneath *Betula gladulosa-* and *Salix planifolia-* dominated sites. *Salix alaxensis* was the dominant shrub in riparian plant communities with Humic Gleysol the most common soil. Bare ground generally accounted for less than 15% of the total area and was restricted to fine materials in areas of patterned ground.

Twenty-eight samples were taken in this plant community, including six control sites and representatives of all eight disturbance types (Figure 1). A number of relationships were apparent. The large absolute differences in moisture content between controls and disturbances probably resulted from the much higher gravel content and lower organic matter content on the disturbances (Figure 1). The lower moisture content, combined with less organic matter, resulted in disturbances having warmer subsurface temperatures. With the exception of false start roads, rooting depths were shallower on all disturbances. This could be a response to the relatively short period (34-36 years) that roots have had to penetrate the ground. Alternatively, the plant species growing on the disturbances may be more shallow rooted than those on undisturbed terrain.

It is apparent from Figure 1 that substrates of bulldozer tracks and oil spills were most similar to the controls, a result of the limited initial mechanical disturbance of these sites. The gravel pit access roads had 98% of the organic matter on their associated controls. Haag (1973) and Babb and Bliss (1974) reported that the organic surface layer was important in retarding subsurface moisture loss and heat gain. The gravel-dominated disturbances provide drier, warmer substrates. Even those disturbances that left the preexisting soil relatively intact have warmer, drier substrates (Figure 1).

Initially, all of the disturbances removed the shrub canopy. Even without excavation or the addition of gravel, increased exposure to sunlight would result in greater surface warming and greater advection of these areas with a consequent removal of moisture from the surface. When combined with improved drainage (both surface and internal), these factors resulted in dramatic substrate differences between controls and their associated disturbances in this plant community

(Figure 1).

Decumbent Shrub Tundra

This plant community was most common on landforms composed of colluvial and alluvial deposits. Species of *Dryas*, *Salix*, and *Potentilla* were the dominant plants and total cover varied from 60 to 100%. Areas with active needle ice, deflation, or seasonal frost churning processes were devoid of plants. The nine control soil samples were classified as Humic Regosols. These soils were well-to excessively-drained as a result of their high gravel content.

A total of seven disturbance types were described in this plant community and 35 substrate samples were taken. It is noteworthy that over 34-36 years, some disturbances which buried or removed the preexisting soil (i.e., roads, gravel pits) have since accumulated from 55 to 67% of the amount of organic material found in adjacent control areas. This relatively rapid rate of organic accumulation may be a result of enhanced production on disturbances or a reduced rate of decomposition. In this environment, where temperature is a biological limiting factor, the warmer substrates of disturbances must be a positive factor in plant growth. The shallower rooting

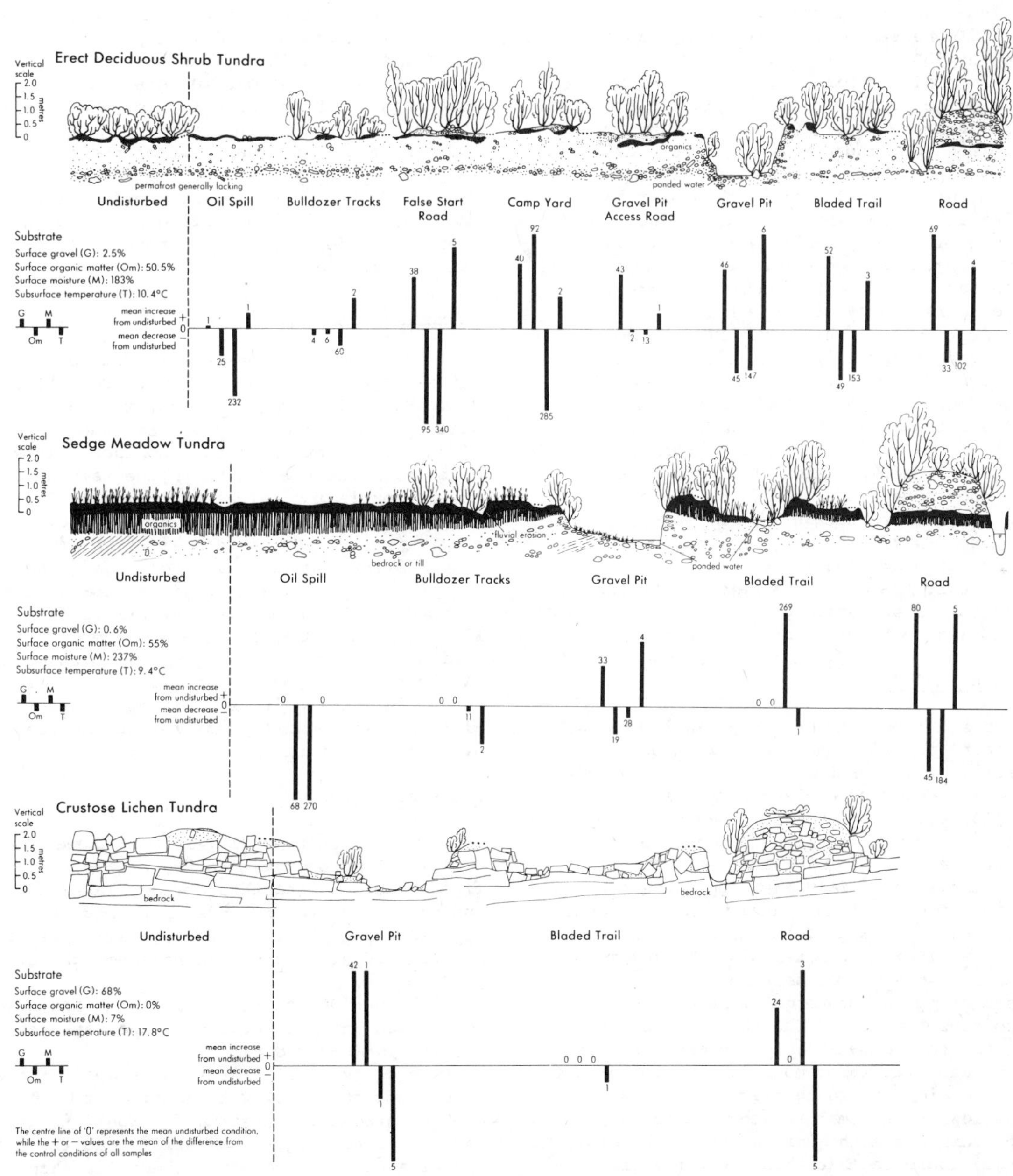

FIGURE 1. CANOL Disturbance Substrate Characteristics in Three Mackenzie Mountain Plant Communities.

depths on disturbances may indicate that plants are capable of taking advantage of these warmer substrate conditions. This also suggests that in this plant community moisture is not a significant limiting factor since the organic accumulations produced by the plants on the site have developed despite drier substrate conditions.

Sedge Meadow Tundra

Sedge dominated wetlands contained Organic Cryosols and Humic Gleysols. The Organic Cryosols had organic layers more than 50 cm thick and a late August thaw layer of 50-60 cm. Sedges in *Carex* and *Eriophorum* were the dominant plants with underlying mats of *Sphagnum* spp. and *Hylocomium splendens*. Plant cover on Organic Cryosols was 100% whereas on the Humic Gleysols it was as low as 75%.

Five control and nine disturbed sites were sampled in Sedge Meadow Tundra, including examples of five disturbance types (Figure 1). Surface layer thickness on bladed trails and bulldozer tracks was reduced by excavation and/or compaction. As a result, bulldozer track substrates were wetter and cooler than their associated controls (Figure 1). Despite the 40% gravel content, in the gravel pit, it had accumulated 81% of the organic matter found in its control (Figure 1). Rooting depth is approximately equal to that of the associated control while subsurface temperature was higher (Figure 1). These conditions have enhanced organic matter production and a rapid rate of accumulation. The roads have warmer but drier substrates and have accumulated 55% of the organic matter found in their controls (Figure 1). If decomposition is also accelerated by warmer substrate temperatures, then the rates of organic matter production must be very high indeed. Loré (1977) has shown that enhanced annual production of standing crop occurred on 50-year-old cart tracks. Thus rates of organic matter accumulations can be greater than those of controls within 50 years of disturbance.

Lichen Heath Tundra

This plant community occurs primarily on landforms composed of colluvium. No permafrost was encountered in this community. Here, the Dystric Brunisol soil supported a plant cover of 90-100%. Dominant plants included *Cassiope tetragona* and *Vaccinium uliginosum* with lichen species of *Cladonia* and *Cetraria* forming mats at ground level. In addition, *Polytrichum* spp. were common.

One set of four disturbances with their associated control were sampled in Lichen Heath Tundra. These well-drained materials with low moisture retention were warmer because of the removal of the insulating surface organic layer and the reduced moisture content. The thick lichen carpet which graded gradually into the organic layer on undisturbed sites was removed and no longer acted as a barrier to ground heat and moisture flux. The low organic matter accumulation over the 34-36 years since the initiation of the disturbance may indicate that moisture was a limiting factor in this plant community, with the higher substrate temperatures (twice as high as those of the control) causing periods of drought and subsequent mortality of seedlings and/or adult plants.

Fruticose Lichen Tundra

This type of tundra was found on well-drained, gravelly landforms composed of alluvium or colluvium with Humic Regosols and Dystric Brunisols. Lichen species of *Cladonia* and *Cetraria* were the dominant plants and generally formed over 80% of the normally 100% plant cover.

Six disturbances were described with four control and 14 disturbed sites sampled. Where a disturbance affected both soil types in this plant community, the differences in organic and moisture content relative to the controls were greater in Humic Regosols, despite the fact that control soil conditions were similar. Rooting depth in Humic Regosols was greater on all disturbances (with the exception of gravel pits) whereas in Dystric Brunisols the depth of rooting was generally less than that of the controls. It is evident that warmer substrates result from the removal or burial of organic surface layers and the insulating lichen surface covering in addition to the deposition or exposure of coarse, well-drained materials.

Cushion Plant Tundra

This plant community was most common on colluvial deposits where soils were Regosolic with Humic Regosols under plant cover. These soils were well to excessively drained and gravelly, often with a shallow lithic contact. This, combined with their location in the rainshadow of the Mackenzie Mountains, has produced dry soils. Plant cover values ranged from 3 to 20%. Dominant plants included *Dryas octopetala*, *D. integrifolia*, *Silene acaulis*, and *Salix dodgeana*.

Five controls and 10 sites representing four types of disturbances were described from Cushion Plant Tundra. Many substrate parameters were similar on disturbed and control sites because of the naturally coarse soils. In comparison with other plant community substrates, those of Cushion Plant Tundra had little difference between controls and disturbances.

Crustoce Lichen Tundra

This plant community is most common in blockfields. Some of these landforms frequently had finer material beneath a surface layer of blocks and/or in isolated islands dispersed throughout (Figure 1). All of these areas were classified as rockland since they did not meet the criteria of Regosols.

The 90-100% plant cover was dominated by long lived and slow growing lichen species in *Rhizocarpon* and *Lecanora*.

Three types of disturbances were studied in this plant community with a total of five samples and three controls described. Most disturbances had more fine-textured materials and therefore had greater moisture storage ability than did their controls (Figure 1). Disturbances were lighter in color than the black lichen covered controls and with a higher albedo, had lower surface temperatures (Figure 1). These conditions offered a much less harsh environment for vascular plants on disturbances than on controls. The high temperatures and low moisture of control substrates would create water stress conditions for most vascular plants, and disturbances would provide a more habitable environment. The fine materials found on disturbances also provided a rooting medium for vascular plants that was absent in undisturbed areas.

Oil Spill Disturbances

The type and amount of oil residue remaining in the soil after 35 yrs. varied from one oil spill to the next (Kershaw 1983). Oil spills were deposited on the ground surface and the soil structure was little altered by the addition of oil. Surface vegetation and organics usually absorbed the oil and have undergone some consolidation over the past 35 years due to the weight of the oil. Moisture content was less than controls, as a result of the sealing of the surface by oil and the filling of mineral soil spaces to prevent water infiltration (Bliss and Wein 1972). In organic layers, oil was absorbed by the litter which acted as a sponge. This was also noted by McCown et al. (1972). Most oil spills produced a low surface albedo that persists and produced warmer subsurface temperatures (Figure 1).

SUMMARY

On road, false start road, bladed trail, gravel pit, and gravel pit access road disturbances the loss of organic matter most affected moisture, and subsurface temperatures. These characteristics control seed germination and survival success as is reflected by rooting depth and organic matter accumulations (on initially mineral dominated substrates) 34-36 years after the initial disturbance. As expected in control plant communities where organic matter is lacking or limited in the surface layers (e.g., Cushion Plant Tundra, Crustose Lichen Tundra) and on disturbances where surface layers have remained relatively intact (e.g., some bladed trails, bulldozer tracks) the overall substrate differences between controls and disturbances were relatively small.

Many surfaces were denuded of plants when they were disturbed. The current accumulations of organic matter on disturbances in several plant communities were a surprisingly high proportion of what was found on the associated controls. Accumulation of organic matter must be accelerating with time as plants reestablish (unless it is now in an equilibrium with decomposition). This implies that substrate conditions on some disturbances are not especially limiting to plant production in general.

Several disturbances in all plant communities had substrates with different particle size composition than was found in their controls. The most significant difference was the greater gravel content found on several disturbances. The absence of fine fractions, combined with the removal of surface organics on several disturbances increased through flow, decreased internal storage, and increased the effectiveness of evaporation. All these conditions have been described for short-term disturbances (Haag and Bliss 1974). Roads, false start roads, camp yards, gravel pits, and gravel pit access roads were generally drier and warmer than controls (Figure 1). Important exceptions to this occurred in Cushion Plant and Crustose Lichen Tundra where the controls lacked significant organic layers. Despite increased or unchanged gravel content, disturbance substrates here had more fine material and were able to retain more moisture. Consequently subsurface temperatures were generally cooler than in control areas.

CONCLUSION

Undisturbed soils in the study area were generally not well developed with thin surface organic layers (except Cryosols) and weakly developed mineral layers. This was the case despite the extreme age of sections of the study area lying within the unglaciated portion of the Mackenzie Mountains. Soils were generally cold and in areas with thick accumulations of organics, permafrost was often present. Regosols, Brunisols, Gleysols, and Cryosols were described from the region.

With the exception of oil spills, there are numerous cases where disturbed substrates have responded positively after 34-36 years. Certainly, differences exist between controls and disturbances. However, it is difficult to equate or compare a substrate that is 34-36 years of age with a soil that is thousands or, in the unglaciated area, even a million years old. Even those substrates on such devastating disturbances as roads and gravel pits have accumulated organic matter and offer warmer substrates (Figure 1) to colonizing plants in an environment where temperature is a major ecologically limiting factor. However, the removal or reduction of organic matter (primarily in the L-H horizon) and the lower fine particle content on many of the disturbances has created drier conditions (Figure 1) that can limit plant recolonization (Johnson 1981 and Johnson et. al. 1981).

No efforts were made to minimize the initial CANOL disturbances. The project proceeded with virtually no preplanning, was executed without restrictions on construction and operation practices and no rehabilitation program was instituted following its abandonment. However, the CANOL disturbances provide confirmation that rehabilitation would have resulted in much greater recovery than has been observed. For example, false start roads where the predisturbance soil was removed had highly irregular surface topography, no drainage ditches, slight compaction, and included some organic matter. These disturbances were frequently more similar to controls than were roads, which provided quite different substrates (Figure 1). It is also possible that attempts to rehabilitate some types of disturbances (e.g., bulldozer tracks) might result in greater disruption than if left to recover naturally.

ACKNOWLEDGEMENTS

Financial support for this research was provided by the Boreal Institute for Northern Studies, the Arctic Institute of North America, and the National Wildlife Federation. Some logistical support was provided by Imperial Oil Limited and AMAX Northwest Mining Co. Figure 1 was drafted by the Reprographics Section of the Department of Geography, University of Alberta. Linda Kershaw contributed and collaborated during all phases of this research.

REFERENCES

Agriculture Canada. 1976. Glossary of terms in soil science. Agriculture Canada Publication 1459. Ottawa.

Babb, T.A. and Bliss, L.C. 1974. Effects of physical disturbance on Arctic Vegetation in the Queen Elizabeth Islands. Journal of Applied Ecology, V. 11, p. 549-562.

Bliss, L.C. and Wein, R.W. 1972. Plant community responses to disturbances in the western Canadian Arctic. Canadian Journal of Botany, V. 50, p. 1097-1109.

Geiger, R. 1959. The Climate Near the Ground. Harvard University Press, Cambridge.

Haag, R.W. 1973. Energy budget changes following surface disturbance to two northern vegetation types. In Botanical Studies of Natural and Man-Modified Habitats in the Mackenzie Valley Eastern Mackenzie Delta Region and the Arctic Islands. L.C. Bliss ed. Environmental-Social Program Northern Pipelines, Report No. 73-43, p. 6-26.

Haag, R.W. and Bliss, L.C. 1974. Energy budget changes following surface disturbance to upland tundra. Journal of Applied Ecology, V. 11, p. 355-374.

Johnson, L.A. 1981. Revegetation and selected terrain disturbances along the trans-Alaska pipeline, 1975-1978. CRREL Report 81-12, Hanover.

Johnson, L.A. Rindge, S.D. and Gaskin, D.A. 1981. Chena River Lakes Project revegetation study, three year study. CRREL Report 81-18, Hanover.

Kalra, Y.P. 1971. Methods used for soil, plant, and water analysis at the soils laboratory of the Manitoba-Saskatchewan region 1967-1970. Northern Forest Research Centre, Information Report NOR-X-11, Edmonton.

Kershaw, G.P. 1983. Long-term Ecological Consequences in Tundra Environments of the CANOL Crude Oil Pipeline Project, N.W.T., 1942-1945. Unpublished Ph.D. thesis, University of Alberta, Edmonton.

Loré, J. 1977. Tundra Disturbance Study: Burwash Uplands, Yukon Territory. Unpublished M.Sc. Thesis, University of Alberta, Edmonton.

McCown, B.H., Deneke, F.J., Rickard, W.E. and Tieszen, L.L. 1972. The response of Alaskan terrestrial plant communities to the presence of petroleum. In Proceedings of the Symposium on the Impact of Oil Resource Development on Northern Plant Communities, Occasional Publications on Northern Life, Number 1, Institute of Arctic Biology, University of Alaska, Fairbanks, p. 34-43.

McKeague, J.A. 1976. Manual on Soil Sampling and Methods of Analysis. Subcommittee of Canada Soil Survey, Committee on Methods of Analysis, Ottawa.

SITE INVESTIGATION AND FOUNDATION DESIGN ASPECTS OF CABLE CAR CONSTRUCTION IN ALPINE PERMAFROST AT THE "CHLI MATTERHORN," WALLIS, SWISS ALPS

H. R. Keusen[1], W. Haeberli[2]

[1]Geotest AG, Zollikofen Bern, Switzerland
[2]Versuchsanstalt für Wasserbau, Hydrologie und Glaziologie, ETH Zürich, Switzerland

The construction of the uppermost section of the cable car leading from Zermatt (1,600 m a.s.l.) to the "Chli Matterhorn ("little Matterhorn", 3,884 m a.s.l), Wallis, Switzerland, posed several geotechnical problems related to the presence of alpine permafrost. One important pylon had to be designed to withstand the creep pressure from a perennially frozen historical moraine, rich in ice. The site was investigated by geoelectrical D.C. resistivity soundings, core drillings, and geodetic measurements of creep displacements. Pits excavated for the foundation of the pylon were filled with pure ice. This ice has first to be squeezed out by creep of the frozen moraine before the full creep pressure can be exerted on the pylon. In the rocks of the summit station (with a mean temperature of about -12°C), the anchorage of heavy cables had to be effected in a purely mechanical way, in tunnels which were specially excavated for this purpose. Attention was given to the thermal insulation of buildings in order to avoid any melting of ice in cracks and joints. Deformation of the rock mass is being measured periodically and has remained small.

INTRODUCTION

High mountain construction work related to tourism and to scientific and military activities has often encountered permafrost conditions in the Swiss Alps (Salomon 1929, Barsch 1969, Furrer and Fitze 1970, Haeberli et al. 1979). However, the phenomenon of high altitude permafrost in low latitude mountains has only relatively recently appeared in the scientific and technical literature. Thus the occurrence of permafrost at high altitude construction sites was often, and sometimes still is, somewhat of a surprise, causing unexpected technical and financial problems of considerable extent. In contrast, permafrost aspects were taken into account from the very beginning of construction for the cable car to the "Chli Matterhorn" (i.e., little Matterhorn) near Zermatt. This paper discusses some of these aspects, especially questions of site investigation and foundation design in perennially frozen sediments and anchorage of heavy cables in negative temperature bedrock.

Today the tourist can travel easily and quickly from Zermatt at 1,600 m a.s.l, where permafrost is sporadic, to the "Trockener Steg" at 2,939 m a.s.l, at the border of the "Oberer Theodul-

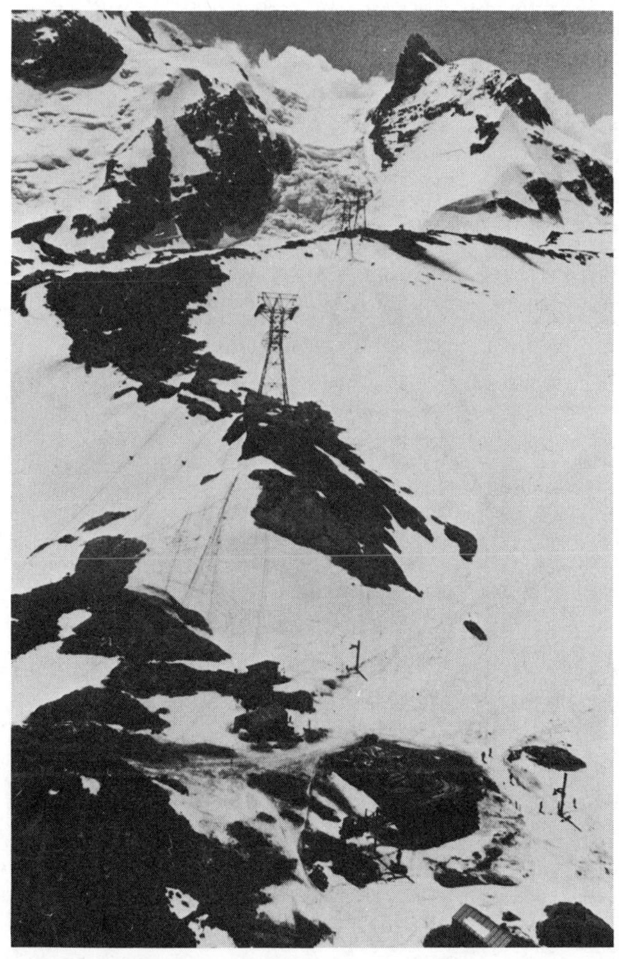

Figure 1:
View from the station of "Trockener Steg" to pylons 1, 2, and 3 and the "Chli Matterhorn." Pylon 2 is situated on a historical lateral moraine of the "Oberer Theodulgletscher" (off the right of the picture). This moraine is perennially frozen, contains a large amount of ice, and creeps downhill at a speed of a few centimeters per year. An icefall of the "Unterer Theodulgletscher" is visible in the background.

gletscher" within discontinuous permafrost, up into the belt of continuous permafrost at the spectacular summit of the "Chli Matterhorn" at 3,884 m a.s.l. Permafrost-related engineering problems were restricted to the section leading from the "Trockener Steg" to the "Chli Matterhorn" (Rieder et al. 1980).

PYLON 2: CREEP OF PERENNIALLY FROZEN SEDIMENTS

Before crossing the "Unterer Theodulgletscher" with a cable approximately 3 km in length, the cable car passes over three pylons, situated between about 2,950 and 3,050 m a.s.l. (Figure 1). As is also the case for the lower station of the cable car at "Trockener Steg," two of these pylons (pylons 1 and 3) have their foundations in bedrock, and geotechnical problems specifically related to permafrost were not encountered at these positions. Pylon 2, however, had to be installed at a place where the bedrock (serpentinite) was covered by a layer of lateral moraine, still occupied by the "Oberer Theodulgletscher" during the early twentieth century. Based upon experience from a study of permafrost distribution in alpine regions with comparable climate (Haeberli 1975, p. 121), it was estimated that this moraine was in a perennially frozen state, and adequate site investigation was necessary.

Geoelectrical resistivity soundings (Röthlisberger 1967) using the Schlumberger configuration were carried out across the moraine in August 1974. The results are summarized in Figure 2. Below the thin, low resistivity, active layer (1,200-3,000 Ωm, about 2 m thick), frozen sediments with a very high resistivity were encountered (145,000-156,000 Ωm). Similar resistivity values have been reported from rock glacier permafrost by Fisch et al. (1978). No low resistivity layer was observed between these sediments and the bedrock-the latter being a massive serpentinite with a resistivity of about 18,000-25,000 Ωm. The absence of a low resistivity layer indicated that the sediments were frozen throughout and that permafrost conditions had also to be expected in the underlying bedrock. Dead glacier ice with resistivity values of the order of many million Ωm (Röthisberger 1967, Röthlisberger and Vögtli 1967, Fisch et al. 1978) was either too thin to be detected by the resistivity soundings or was not embedded in the frozen morainic sediments et all. Because of the rather unexpected existence of a local depression in the topography of the bedrock surface, the thickness of the frozen moraine at the proposed site for pylon 2 reached about 30 m. As a consequence, the construction site for pylon 2 was shifted to a place were bedrock depth was expected to be shallower. Three core drillings furnished supplementary information about the thickness and the properties of the frozen sediments (Figure 3). A double-wall core drill was used to avoid the melting of ice in the cores as much as possible. As was expected from the resistivity soundings, bedrock was encountered here between 6 m and 11 m below the surface. Core drilling C penetrated a snow and ice patch of limited thickness. All three drillings recovered frozen, silty sands and gravels

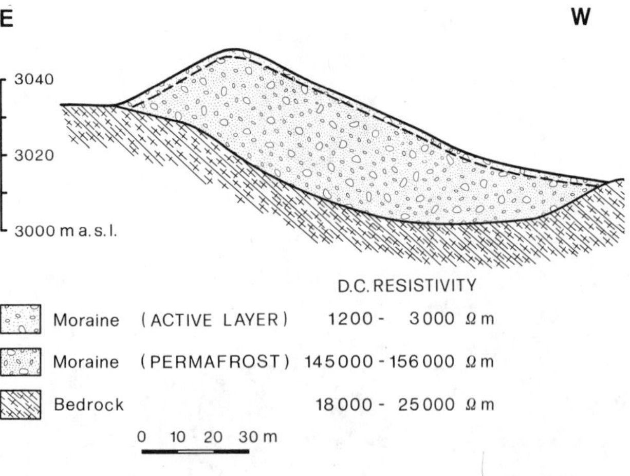

Figure 2:
Cross section through the perennially frozen moraine near pylon 2, as determined by geoelectical resistivity soundings.

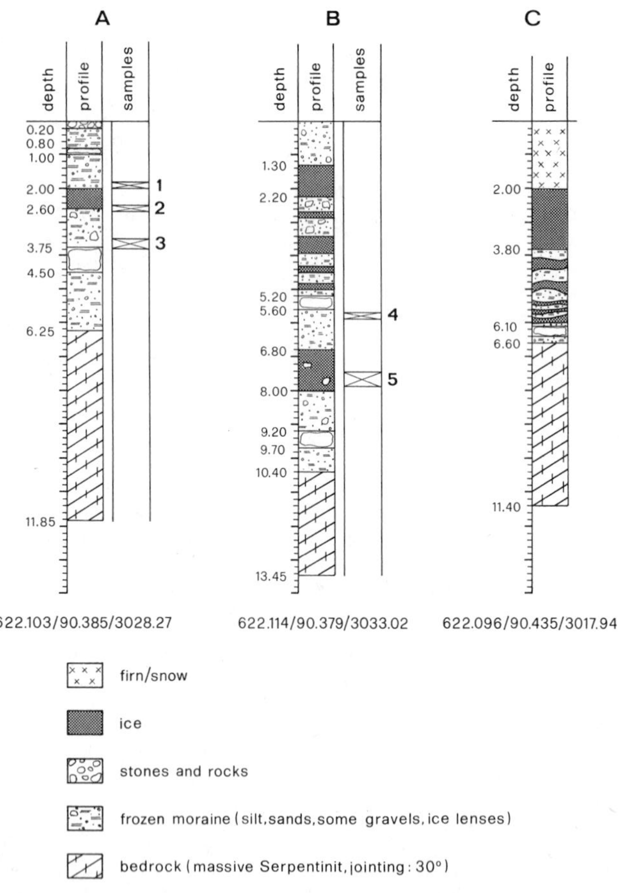

Figure 3:
Three core drillings taken at the site of the foundation for pylon 2. 1-5: samples for laboratory analysis.

with some larger rocks, and ice lenses. More than 30% of the frozen moraine consisted of relatively pure ice. Five samples were analyzed in the laboratory. All of them contained very limited amounts of air-only 1-2% by volume on average. Ice lenses (samples 2 and 5, Figure 3) were almost completely free from air and gave density values of 1-1.2 Mg m^{-3}, corresponding to an ice content of about 85-95% by volume. Samples 1, 3, and 4 (Figure 3), taken from sections of frozen moraine without macroscopically visible ice, had densities between about 1.4 and 1.6 Mg m^{-3} and ice contents above 60% by volume. The overall ice content (around 75% by volume) in the obviously supersaturated frozen sediments was therefore higher than the porosity of comparable nonfrozen sediments by a factor of 2-3. These observations were in good agreement with the results of scientific core drillings in rock glacier permafrost (Barsch 1977, Barsch et al. 1979).

Supersaturated frozen sediments with reduced particle-to-particle contacts were known to creep following a stress-strain relationship quite similar to that of ice (e.g., Vyalov 1962, Thompson and Sayles 1972, Anderson and Morgenstern 1973, McRoberts and Morgenstern 1974, Sayles and Haynes 1974, cf. also Huder 1979). The creep velocity of the frozen sediments at the construction site of pylon 2 was therefore geodetically measured at 5 poles. These were anchored by concrete within the permafrost (below the active layer). However, rough estimates of creep displacements were already needed before the end of the observation period (1975-1978) in order to design the foundation for planned pylon 2. On the basis of Glen's flow law for pure ice (Glen 1955) and taking into account results from photogrammetric movement studies (Messerli and Zurbuchen 1968, Barsch and Hell 1975, Haeberli et al. 1979) on alpine rock glaciers (which represent natural large-scale experiments on the long-term creep behavior of frozen supersaturated sediments), 1-10 cm yr^{-1} seemed to be a reasonable order of magnitude for the considered 11°-18° slope, with the given geometry and density of the frozen moraine. The direct observations finally gave 5-8 cm yr^{-1} with one exception (23 cm). It is worthwhile noting that the displacement vectors deviated considerably from the line of maximum slope.

Three approaches were used to determine creep pressure against the pylon within the frozen moraine: (1) the method of "passive earth pressure" after Haefeli (in Mohr et al. 1947) which considers the combined effect of passive earth pressure and shear stresses acting at the sides of an obstruction with cylindrical cross-section; for the case under consideration, internal friction was assumed to be zero and a long-term shear strength had to be introduced (1 t/m^2 for ice and 2 t/m^2 for frozen sediments were chosen to give upper limits for the estimation of creep pressure); (2) the use of an empirical relationship for the calculation of creep pressure against cylindrical piles, after Haefeli; and (3) the use of an equation for piles of circular cross section, by Wenz (Schenk and Smoltczyk 1966)

$$Q_c \simeq 2.5\ \gamma(z + d)d \qquad (2)$$

$$Q_c = 1.7(1 + \pi)dc_u \qquad (3)$$

where Q_c is creep pressure (per m pile height); γ, specific weight of the creeping mass; z, depth below surface; d, pile diameter; and c_u, "long-term shear strength" of the creeping material. Because of the geometrical and structural complexity of the creeping mass, all pressure determinations are necessarily rather rough approximations. Safety considerations were based upon all three approaches.

A scheme of the system selected for the foundation of pylon 2 is shown in Figure 4. Anchors had to be installed within the bedrock, below the foundation, because rock resistance, even with fixation depths of 3.8 and 4.8 m, was not high enough for the required safety. Since all three

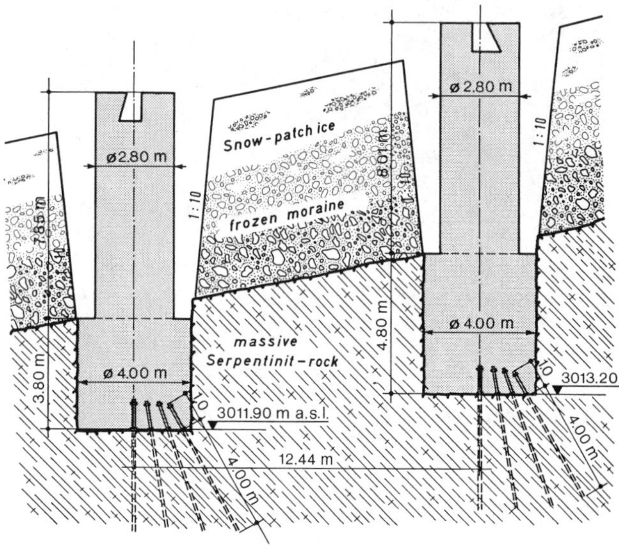

Figure 4:
Foundation design for pylon 2.

approaches used for the estimation of creep pressure gave smaller pressures for ice than for frozen sediments, the pits excavated for the foundation were refilled with ice instead of morainic material. This ice has first to be squeezed out by the creeping moraine before the full creep pressure is exerted on the pylon by the frozen sediments. During this time, not only the pylon can be observed but also the nearby glacier. If the present glacier advance in the Alps continues (Kasser 1981), the ice of the "Oberer Theodulgletscher" could build up irresistable pressures against pylon 2 within a time period of a few decades.

SUMMIT: ANCHORING HEAVY CABLES IN NEGATIVE TEMPERATURE BEDROCK

The summit station (Fig. 5) was anticipated to be within the alpine belt of continuous permafrost and to have a mean rock temperature of around $-10^{o}C$ (cf. Haeberli 1976, p. 211). During the excavation of a tunnel from the steep north side to

Figure 5:
Concrete transportation to the construction site of the summit station at 3,820 m a.s.l. Concrete was prepared at "Trockener Steg" to avoid frost problems. Mont Blanc (French Alps), the highest peak of the Alps, is visible in the background. (Photograph taken by Air Zermatt.)

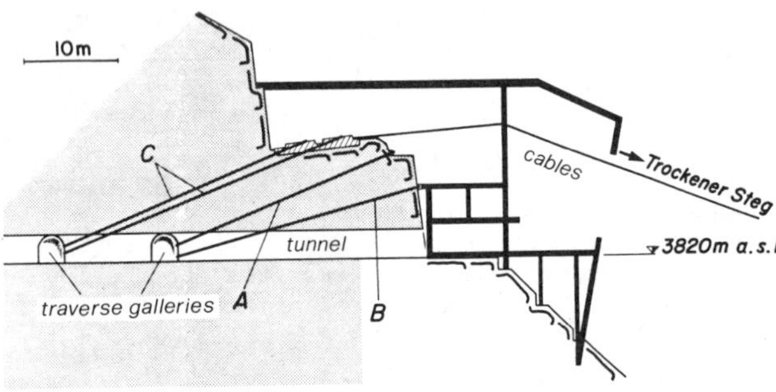

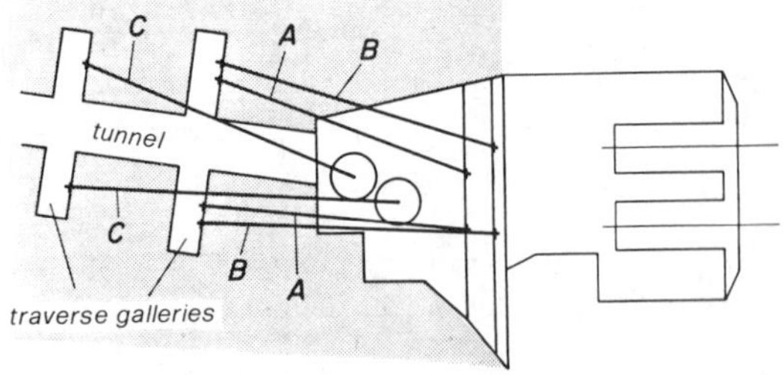

Figure 6:
Scheme of the anchor arrangement in the summit rocks of the "Chli Matterhorn." A, B, C: anchor groups.

the glaciers on the south side of the mountain, rock temperatures were measured by infrared thermometry of tunnel walls in the section which was barely disturbed thermally by the excavation. However, in the section where heating was inevitable because water was employed during drilling, rock temperature was determined by thermistor readings in shallow boreholes near the tunnel front. The observed -10° to -12°C temperature values represent thermal conditions within the nearby west side of the mountain. Ice-filled cracks were noted at both ends of the tunnel.

The existence of this tunnel made it easy to avoid frost problems normally associated with injections of concrete in anchors for the steel construction and cables of the cable car were, in fact, fixed on blocks installed in small transverse galleries which were specially excavated for this purpose (Figure 6). Each group of anchors was designed with a safety factor of $F = 1.5$, to cover the case that one anchor in the group should fail. This arrangement permits periodic control and even, if necessary, replacement of anchors without major interruptions to the cable car operation. Anchor force at each group is observed with permanently installed strain gauges. Relaxation caused a decrease in anchor force of about 1.5% of the initiated tension during the first year. Attention was given to thermal insulation of the summit station to avoid any melting of ice in cracks and thus any related destabilization of the negative temperature bedrock around the anchors.

Rock deformation is measured with an extensometer arranged parallel to the anchors and by convergence measurements between the anchor blocks and the opposite tunnel walls. The extensometer did not give measurable deformations while putting the anchors under stress, whereas the convergence measurements showed displacements of the anchor blocks toward the rock wall of up to 0.5 mm, followed by long-term reverse movements of about 1.6 mm, probably as a consequence of rock relaxation.

Cable car operation started at the end of 1979.

ACKNOWLEDGEMENTS

The authors are grateful to the Burgergemeinde Zermatt for giving permission to publish this study, to many colleagues for scientific and technical advice, and to P. Alean for helping to prepare this manuscript.

REFERENCES

Anderson; D. M., and Morgenstern, N. R., 1973, Physics, chemistry and mechanics of frozen ground: a review: Permafrost Second International Conference - North American Contribution, NAS-Washington D.C., p. 257-288.

Barsch, D., 1969, Permafrost in der oberen subnivalen Stufe der Alpen: Geographica Helvetica, 24. Jhg., Nr. 1, p. 10-12.

Barsch, D., 1977, Ein Permafrostprofil aus Graubünden, Schweizer Alpen: Zeitschrift für Geomorphologie, NF Bd. 21, p. 79-86.

Barsch, D., and Hell, G., 1975, Photogrammetrische

Bewegungsmessungen am Blockgletscher Murtèl I, Oberengadin, Schweizer Alpen: Zeitschrift für Gletscherkunde und Glazialgeologie, B. XI, H. 2, p. 111-142.

Barsch, D., Fierz, H., and Haeberli, W., 1979, Shallow core drilling and borehole measurements in the permafrost of an active rock glacier near the Grubengletscher, Wallis, Swiss Alps: Arctic and Alpine Research, v. 11, no. 2, p. 215-228.

Fisch, W., Sr., Fisch, W., Jr. and Haeberli, W., 1978, Electrical D.C. resistivity soundings with long profiles on rock glaciers and moraines in the Alps of Switzerland: Zeitschrift für Gletscherkunde und Glazialgeologie, Bd XIII, H. 1/2, p. 239-260.

Furrer, G., and Fitze, P., 1970, Beitrag zum Permafrostproblem in den Alpen: Vierteljahreszeitschrift Naturforschende Gesellschaft Zürich, Jhg. 115, H. 3, p. 353-368.

Glen, J. W., 1955, The creep of polycrystalline ice: Proceedings of the Royal Society, Ser. A, 228, p. 519-538.

Haeberli, W., 1975, Untersuchungen zur Verbreitung von Permafrost zwischen Flüelapass und Piz Grialetsch (Graubünden): Mitteilung der Versuchsanstalt für Wasserbau, Hydrologie und Glaziologie der ETH Zürich, Nr. 17, 221 p.

Haeberli, W., 1975, Eistemperaturen in den Alpen: Zeitschrift für Gletscherkunde und Glazialgeologie, Bd. XI, H. 2, p. 203-220.

Haeberli, W., Iken, A., and Siegenthaler, H., 1979, Glaziologische Aspekte beim Bau der Fernmelde-Mehrzweckanlage der PTT auf dem Chli Titlis: Mitteilung der Versuchsanstalt für Wasserbau, Hydrologie und Glaziologie der ETH Zürich, Nr. 41, p. 59-75.

Haeberli, W., King, L., and Flotron, A., 1979, Surface movement and lichen-cover studies at the active rock glacier near the Grubengletscher, Wallis, Swiss Alps: Arctic and Alpine Research, v. 11, no. 4, p. 421-441.

Huder , J., 1979, Geotechnische Untersuchungen für das Gefrierverfahren: Mitteilung der Versuchsanstalt für Wasserbau, Hydrologie und Glaziologie der ETH Zürich, Nr. 41, p. 77-98.

Kasser, P., 1981, Rezente Gletscherveränderungen in den Schweizer Alpen: Gletscher und Klima-Jahrbuch der Schweizerischen Naturforschenden Gesellschaft, wissenschaftlicher Teil, p. 106-138.

McRoberts, E. C., and Morgenstern, N. R., 1974, Stability of slopes in frozen soil, Mackenzie Valley, N.W.T: Canadian Geotechnical Journal 11, no. 4, p. 554-573.

Messerli, B., and Zurbuchen, M., 1968, Blockgletscher im Weissmies und Aletsch und ihre photogrammetrische Kartierung: Die Alpen 1968, H. 3, p. 139-152.

Mohr, C., Haefeli, R., Meisser, L., Waltz, F., and Schaad, W., 1947, Umbau der Landquartbrücke der Rhätischen Bahn in Klosters: Schweizerische Bauzeitung 1947, Nr. 1-3, p. 5-8, 20-24, 32-37.

Rieder, U., Keusen, H. R., and Amiguet, J.-L., 1980, Geotechnische Probleme beim Bau der Luftseilbahn Trockener Steg-Klein Matterhorn: Schweizer Ingenieur und Architekt 1980, H. 18, p. 428-431.

Röthlisberger, H., 1967, Electrical resistivity measurements and soundings on glaciers: Introductory remarks: Journal of Glaciology, v. 6, no. 47, p. 599-606

Röthlisberger, H., and Vögtli, K., 1967, Recent D.C. resistivity soundings on Swiss glaciers: Journal of Glaciology, v. 6, no. 47, p. 607-621.

Salomon, W., 1929, Arktische Bodenformen in den Alpen: Heidelberger Akademie der Wissenschaft, Sitzungsberichte Nr. 5.

Sayles, F. H., and Haines, D., 1974, Creep of frozen silt and clay. Technical Report 252: Hanover, N.H. CRREL 51 p.

Schenk, W., and Smoltczyk, H. U., 1966, Pfahlroste, Berechnung und Ausbildung: Grundbautaschenbuch, Band 1, Wilhelm Ernst, Berlin/ München, p. 658-715.

Thompson, E. G., and Sayles, F. H., 1972, In situ creep analysis of room in frozen soil: Journal of Soil Mechanics and Foundation Division, Proc. ASCE 98, p. 899-916.

Vyalov, S. S., 1962, The strength and creep of frozen soils and calculations for ice-soil retaining structures. Translation 76, AD 484093: Hanover, N.H., CRREL.

SELECTION OF PRINCIPLES FOR THE USE OF PERMAFROST MATERIALS AS A FOUNDATION FOR BUILDINGS AND STRUCTURES

L. N. Khrustalyov

Research Institute of Foundations and Underground Structures
Vorkuta, USSR

In designing buildings and structures in the North two principles are applied in using permafrost materials as a foundation (the permafrost materials are used either in a frozen state or in a thawed state). The choice of one of these construction principles represents a sophisticated technical and economical problem. It is made on the basis of optimization of the reliability of the foundation in terms of cost. Calculations of the reliability of the foundation and of the costs involved were made by computer using an algorithm developed by the author and construction of a design nomogram that interrelates foundation reliability, costs involved, and the controlling parameters of the foundation. Then the nomogram was used to select one of the principles for using permafrost materials, the foundation depth and the depth of preliminary thawing or freezing of the foundation materials.

STATE OF THE PROBLEM

Either of the following two construction principles of permafrost ground utilization as soil bases is currently in use for designing soil bases and foundations for building and structures to be built in northern areas.

Construction Principle 1 — The permafrost ground of soil bases is utilized in the frozen state, which is preserved over the construction period and throughout the entire service life of a building or structure.

Construction Principle 2 — The permafrost ground of soil bases is utilized in the thawed state (its thawing is permitted to the design depth prior to the structure use, in the course of the structure use, with the upper permafrost boundary stabilized).

In selecting the construction principle of permafrost ground utilization, the following rules should be observed:

1) For sites where the seasonal thawing layer is in contact with the permafrost strata over the entire area of a structure under design, it is recommended to utilize soil bases in the frozen state (construction principle 1). This rule does not cover those cases when special technological or structural features of the building or structure make the arrangement of a cooling system either impossible or economically disadvantageous.

2) For sites where the seasonal freezing-thawing layer is not in contact with the permafrost strata at all or is in contact with it not over the entire area of a structure under design, soil bases may be utilized both in the frozen state (construction principle 1) and in the thawed state (construction principle 2). Principle selection will be made on the condition that the minimum construction costs are provided.

3) For sites with deep permafrost ground occurrence it is more advantageous to use construction principle 2.

Such a small body of information affords ample opportunities for adoption of peremptory decisions that cannot be put up with in the engineering practice.

Let us now consider this problem in terms of the reliability theory.

DETERMINATION OF THE SOIL BASE RELIABILITY

The reliability of a soil base is understood as its capacity to take up external loads and effects (thermal and mechanical) while maintaining serviceability of the erected building or structure. The reliability of a soil base is thought to be ensured if the ultimate conditions for the bearing capacity and the deformation (settling) of soils are met (design in terms of ultimate stressed state of soils). Nevertheless, the answer to what extent it is ensured and what guarantee there is for the conservation of reliability over a specified period of time profiting by the deterministic approach adopted now in the engineering practice cannot be given yet. But the probabilistically statistical approach provides a means for getting it. This approach consists in coherent understanding of the fact that thermal and mechanical interaction of a building with its base is a time-developing stochastic process that is dependent on a set of natural climatic and technogenic factors that are accidental by their nature. According to the general concepts of the reliability theory a stochastic process is developed within the system quality space (Bolotin, Gol'denblat, & Smirnov, 1972). A set of the system states to be allowable in terms of quality forms a region of allowable states (Ω_0) within the quality space (V). If a stochastic process intersects the boundaries of this region, then the system quality will be lost and a failure will occur. The probability of quality conservation over a specified period of time is system reliability.

The quality space of the system "base" is formed

by its elements which are as follows: the bearing capacity of base and the quantity of simultaneous deformation (settling) of the base and building or structure. The boundaries of the allowable states region are the load imposed by the building or structure on its base and the allowable limiting quantity of simultaneous deformation (settling) of the base and building or structure. Proceeding from the methodology of the ultimate states we shall think that the base quality is conserved if the bearing capacity of base exceeds the load imposed by the building on its base and the simultaneous deformation (settling) of the base and building is less than the allowable limit. The probability of this event throughout the in-service life of a building will be termed base reliability and the entire flow of values of this time-allowable probability will be termed base reliability function.

The determination of the reliability function comprises three stages. The first stage consists in schematization of the system and the external effects it is exposed to, as well as in selection of the quality space and the allowable states region. The second stage involves determination of the stochastic behaviour of the system being exposed to accidental effects. The problem comes therefore to solving the following stochastic equation (1):

$$v = L \cdot u \tag{1}$$

where v = the component of the quality space V;
u = the component of the entry parameters space U; and
L = the system operator.

The third stage incorporates determination of the reliability function P(t) as equation (1's) complement of a probability of the process accidental fall-out from the allowable states region.

The entry parameters space U is understood as a set of natural climatic and technogenic factors, which determine conditions of thermal and mechanical interaction between the building or structure and its base.

The system operator L represents the algorithm for sequential solving of the following problems:

- Temperature calculation for soil surface under building as well as beyond its area.
- Temperature and freezing-thawing depth calculation for soil layer under building.
- Calculation of ground base on the first ultimate stressed state of soils (bearing capacity).
- Calculation of ground base on the second ultimate stressed state of soils (settling).

In virtue of extreme complexity of the operator L we recommend to compute reliability functions by the method of statistical tests (the Monte Carlo method). This method is based on running a mass experiment on a computer, which allows one to apply multiple solving (testing) of a determinate problem on a computer instead of solving a statistical problem (Ermakov, 1971). The required number of tests is evaluated by the formula:

$$N = \frac{4P(1-P)}{\Delta^2} \tag{2}$$

where P = the expected probability; and
Δ = the given accuracy of calculation.

After completion of tests the statistical estimate of the reliability function is computed:

$$P(t) = P(\Delta t \cdot i) = 1 - \frac{n_i}{N} \tag{3}$$

where n_i = the number of failures within the time interval ranging from 0 to $\Delta t \cdot i$;
Δt = the time spacing;
i = the ordinal number of time interval (a total of m time intervals is all that can be expected from the condition $m\Delta t = t_{use}$; and
t_{use} = the in-service period.

A block-diagram of the program for solving the problem of base reliability estimation by the Monte Carlo method is shown in Fig. 1. It consists of 19 operators.

With the known operator L and the specified boundaries of the allowable states region Ω_0 the form of the reliability function P(t) will be completely determined by the input parameters, i.e. by elements of the population U. A part of them, namely constructive elements, included in the group of technogenic factors, has a particularly strong effect on the function value and unlike the rest of these elements they may be arbitrarily specified by a designer, i.e. they are the governing elements.

If the structure was predetermined, the basic governing parameters will be as follows: the thickness of the thawed layer left under the building or structure H; the depth of the foundation, h, and the type of the foundation; the number of ventilation openings (ducts) in the basement of the building or structure (ventilation modulus of ventilated basement μ).

A combination of these governing parameters completely determines the construction principle of permafrost ground utilization as soil bases of buildings or structures. For example, if $h > H$ and this inequality is met during the entire service life of a building or structure, we use the first construction principle of permafrost ground utilization as soil bases; if $h < H$, we use the second construction principle and if, in addition, $\mu \neq 0$ the principle of stabilization is used for the building or structure construction.

The task is to select and to apply such a combination of the governing parameters which will ensure the optimal reliability of the system.

OPTIMIZATION OF THE SOIL BASE RELIABILITY

There are no systems with the absolute reliability in Nature. All systems have reliability less than one. As it was mentioned above, the reliability is understood as a minimum of the reliability function, i.e. it is its value at the end of the in-service period. As a consequence, the failures are bound to occur causing collapses of some built-up structures and appearance of a detriment. The higher the reliability, the less the detriment. On the other hand, the high reliability requires high costs to develop a structure. It is evident that the system must be designed in such a way as to ensure its optimum reliability.

The problem of determining optimum reliability has been formulated more than once (Khotsialov, 1929; Rzhanitsyn, 1961), but in doing so the time factor has been disregarded.

A more correct formulation of this problem (Bolotin, 1971) is that for every point of the quality space V, for every time moment t, there is a definite value of the payoff function C(v,t) equal to the detriment over a unit time provided the system is to be found in the point of the quality space V involved. If the system is within the region of the allowable states Ω_0, the value of the payoff function C(v,t) is taken as zero. Mathematical expectation of the total detriment (the value of risk) is determined by the foluma:

$$C_1 = \int_0^{t_{use}} dt \int_V C(v,t)\ P(v,t)\, dV \tag{4}$$

where P(v,t)dV = the probability of finding in the time moment t of the component v existing in volume dV of the quality space V.

The soil base, foundation, and frame of the building are generally considered to be nonrecoverable systems. In systems of that kind any type of failure results in the collapse of the structure. Therefore, in the case considered, the value of the payoff function will only depend on whether the system is within the region of the allowable states or not.

Analytically it is represented by:

$$C(v,t) = \begin{cases} 0 & \text{with } v \in \Omega_0 \\ C(t) & \text{with } v \ni \Omega_0 \end{cases} \tag{5}$$

The payoff function C(t) is in turn introduced for the reasons which are as follows.

If a building fails before its in-service period expires, the detriment will be equal to the depreciated cost of the failed building, which is determined as a difference between the renewal cost and the depreciation cost. Besides, the building's failure will cause an indirect detriment: either underproduction of products or reduction in transport service operations, or reduction in housing facilities. The indirect detriment is generally expressed as a fraction of the building's renewal cost. In order that the time-differentiated costs could be compared, they should be reduced to a common time. The start-up of the building use is generally taken as a reference date for comparison. Based on the statement above, the payoff function may be described as follows:

$$C(t) = \frac{C_0(1 + k_{econ.} - 0.01k_{renewal} \cdot t)}{(1+E)^t} \tag{6}$$

where C_0 = the acquisition cost of a building or structure;

E = the standard reduction coefficient for the time-differentiated costs;

t = the period of reduction;

$k_{renewal}$ = the charge rate for complete renewal, it is taken according to the capitality class of a building or structure; and

$k_{econ.}$ = the coefficient of economical responsibility, it is taken as ratio between the indirect detriment and the acquisition cost of a building or structure.

Subject to expression (6), for determination of the value of risk, expression (4) may be redescribed as:

$$C_1 = C_0 \int_0^{t_{use}} \frac{(0.01k_{renewal} \cdot t - k_{econ.} - 1)\ P^1(t)}{(1+E)^t} dt \qquad \ldots (7)$$

where $P^1(t)$ = the first-order derivative of the reliability function.

There are discerned systems characterized by economical detriment and systems characterized by moral detriment. In the case of systems with economical detriment the minimum of the reduced costs serves as the optimization criterion:

$$C_0 + C_1 = \min. \tag{8}$$

In the case of systems with moral detriment it is the minimum of the reduced costs with the given reliability:

$$\left| C_0 + C_1 \right|_{P \geq P_1} = \min. \tag{9}$$

where P_1 = the reliability with which safety of a building or structure use is provided.

The problem of finding an optimum solution involves the selection of the best combination of the governing parameters at which the conditions of expression (8) or expression (9) will be met. As it was stated above, the combination of the governing parameters determines completely the construction principle of permafrost ground utilization as soil bases of buildings or structures.

For design purposes it is necessary to calculate the reliability P as well as the reduced cost C as functions of the soil base parameters H, h, and μ. Determination of these relations is connected with a large volume of calculations which can be performed on the computer only. The algorithm for these computations is shown in Fig. 2. It comprises 9 operators.

The acquisition cost of a building or structure is computed as the sum of the superstructure cost, $C_{superstructure}$, costs of foundations and the cost of soil base preparation. The value of risk C_1 is computed by formula (10) which interprets formula (7) in terms of finite differences:

$$C_1 = C_0 \sum_{i=1}^{m} (1 + k_{econ.} - 0.01k_{renewal} \cdot i\Delta t)\ \frac{P[(i-1)\Delta t] - P(i\Delta t)}{(1+E)^{i\Delta t}} \tag{10}$$

The reduced cost C is determined as a percentage of the superstructure cost:

$$C = \frac{C_0 + C_1 - C_{superstructure}}{C_{superstructure}} \cdot 100\% \qquad (11)$$

Calculation results serve as a basis for drawing the nomogram which illustrates the relationship between 5 parameters (P, C, h, H, μ) as well as for selection of the construction principle.

EXAMPLE OF THE CONSTRUCTION PRINCIPLE SELECTION

An example of the design nomogram drawn for a 5-storeyed large-panel building on a pile foundation is shown in Fig. 3. The foundation depth, h, is plotted on the ordinate of the nomogram, on the abscissa—the thickness of the thawed layer H to be left in the base of the building by the beginning of its use. The line h = H is the boundary between the first and the second construction principles of permafrost ground utilization as soil bases. On the left side of the nomogram field the reliability levels fitting the minimum value of the reliability function cut off a region where preparation of a base defined by the given reliability is impossible. On the nomogram there are also drawn the isolines of the reduced costs of the building's substructure works expressed as a percentage of the superstructure cost. The isolines of the reduced costs outline the region of the optimum reliability.

Let us illustrate how the construction principle of permafrost ground utilization as a soil base and the values of the governing parameters are selected through the use of the nomogram. If the building is classed with the systems featuring a purely economical detriment, then it is evident that on the nomogram the design point is to be found in the region of the minimum reduced cost. In our example it will be the point with the parameters h = 8 m and H = 10 m. This point is positioned in the region associated with the second construction principle; to be more precise, in the region associated with the method of stabilization $\mu = 5.9 \cdot 10^{-4}$. It fits the reduced cost of the substructure works amounting to 12% of the superstructure cost and the minimum value of the reliability function 0.78.

In the case referred to, reliability has an indirect effect on the selection of the governing parameters via the reduced cost; to be more precise, via the detriment function. It is not assigned beforehand, but it is obtained from expression (8) as a consequence.

The picture will be quite different if the building is classed with the systems featuring the moral detriment. In such a case the reliability is the governing factor in the selection of the construction principle. Let us assume that the value of the reliability function should not be less than 0.99 to ensure safety. In our specific example we shall have two isolines P = 0.99. The first one is to be found in the region associated with the first construction principle and the other—in the region associated with the second construction principle. Any point on these two isolines will assure the specified reliability and, hence, may be taken as a design one. However, the reduced cost is dependent in many respects on what specific region of the nomogram this point is positioned in. It goes without saying that the optimum solution will be that at which the specified reliability will fit the minimum of the reduced cost. Moving along the isolines P = 0.99, we shall find this solution. It will fit the point with the coordinates h = 12 m; H = 4 m. This point is positioned in the region associated with the first construction principle and fits the reduced cost of the substructure works amounting to 40% of the superstructure cost.

Thus, in the first case the building should be constructed on the method of stabilization. The depth of pile sinking into the thawed ground should be put equal to 8 m and the minimum thickness of the thawed layer equal to 10 m. This means that in those areas where the permafrost ground occurs at higher depths, it is necessary to prethaw it down to a depth of 10 m. In the second case the building should be constructed on the first construction principle. The depth of pile sinking into the frozen ground should be put equal to 12 m and the maximum thickness of the thawed layer equal to 4 m. This means that in those areas where the permafrost ground occurs deeper than 4 m, it is necessary to prefreeze the thawed ground layer.

In the design practice, application of the procedure set forth in this paper enables one to perform a scientifically grounded selection of the construction principle of permafrost ground utilization as soil bases which secures a building or structure against the untimely failure due to an unfavourable combination of factors and offers therewith the least reduced construction costs.

REFERENCES

Bolotin, V. V., 1971, Application of methods of probability and reliability theories for design of structures, Stroyizdat.

Bolotin, V. V., Gol'denblat, I. I., and Smirnov, A. F., 1972, Structural mechanics. Up-to-date state and development, Stroyizdat.

Ermakov, S. N., 1971, Monte Carlo method and contiguous problems, Nauka Press.

Khotsialov, N. F., 1929, Safety margins. Construction industry, no. 10.

Rzhanitsyn, A. P., 1961, Determination of safety characteristics and margin factors from economical considerations, *in* Problems of plasticity theory and structural strength, Stroyizdat.

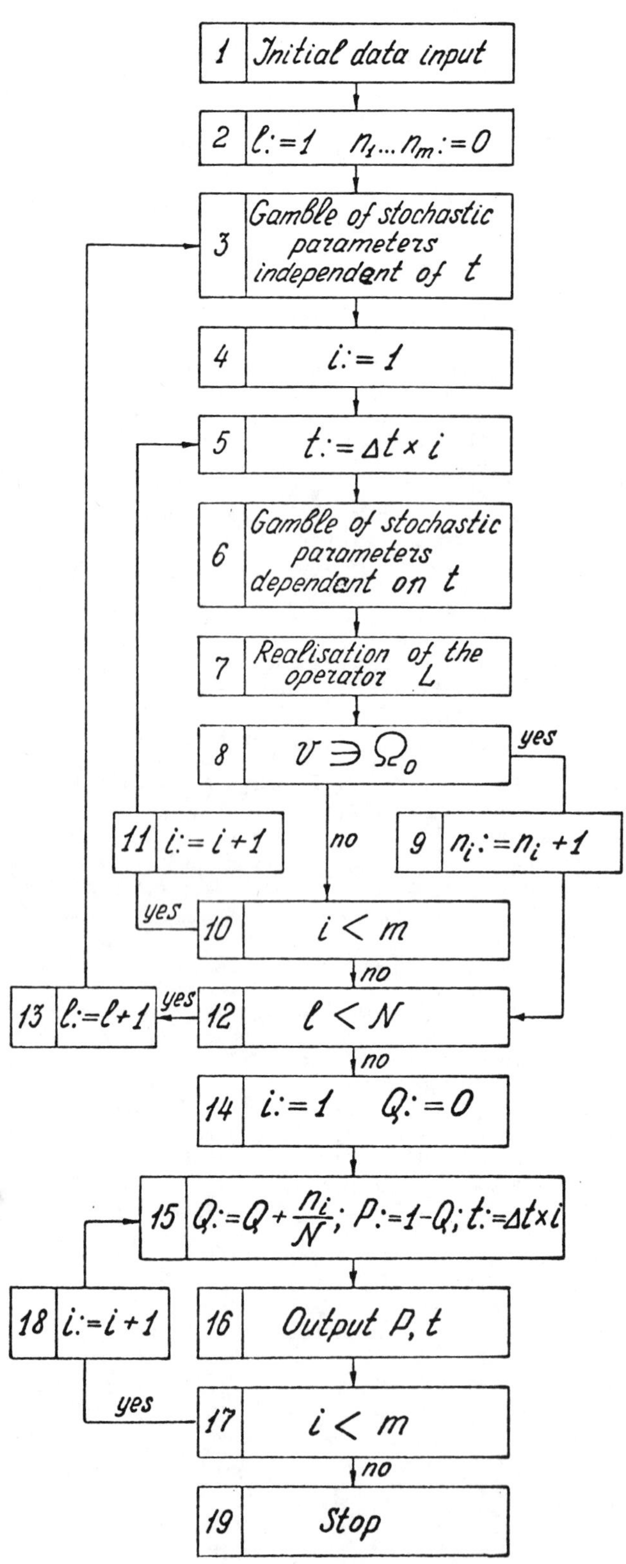

FIGURE 1 Block diagram of the program for estimation of soil base reliability by the Monte Carlo method.

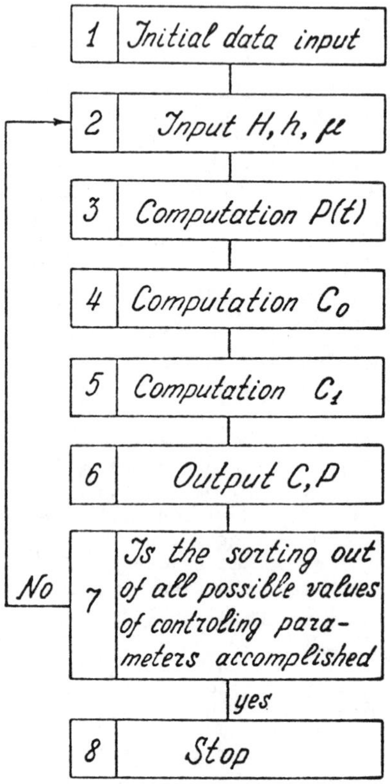

1 Initial data input

2 Input H, h, μ

3 Computation $P(t)$

4 Computation C_0

5 Computation C_1

6 Output C, P

7 Is the sorting out of all possible values of controling parameters accomplished

No

yes

8 Stop

FIGURE 2 Block diagram of the program for computation of the quantities P and C as functions of the parameters H, h, and μ.

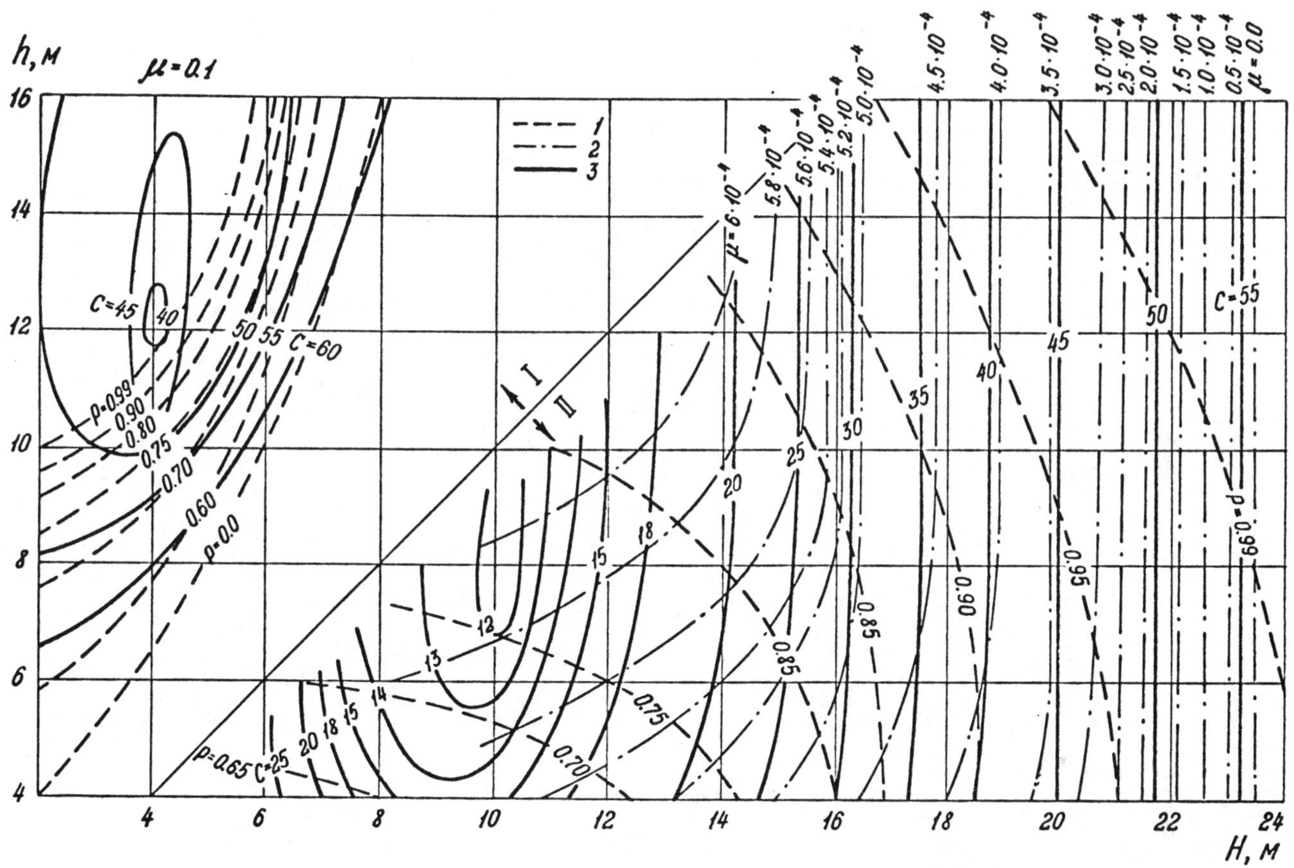

FIGURE 3 The nomogram for selection of the construction principle.
1 - reliability level
2 - ventilation modulus
3 - reduced cost

HIGH MOUNTAIN PERMAFROST IN SCANDINAVIA

Lorenz King

Geographisches Institut der Universität Heidelberg, Im Neuenheimer Feld 348, D-6900 Heidelberg, West Germany

In the vicinity of the highest elevations of northern Europe, the Kebnekaise Massif in northern Sweden (peak of Kebnekaise is 2,100 m a.s.l.) and the Jotunheimen area in southern Norway (Galdhöpiggen and Glittertind reach about 2,470 m a.s.l.), extensive permafrost occurences could be proved. Two test areas, the Tarfala Valley (67°55'N, 18°35'E) and the Leirvassbu/Juvasshytta area (61°40'N, 8°20'E), have been investigated in detail by means of ground temperature measurements, hammer seismic soundings, dc-geoelectric soundings and studies of snow depth and snow temperature (BTS values). In both areas a permafrost thickness of more than 100 m could be proved above 1,500 and 2,200 m, respectively. A continuous alpine permafrost zone seems to exist here. Additional information concerning the Sognefjell and Rondane/Dovre area and the Lyngen Alps is given, and a model of alpine permafrost zonation for the central parts of the Scandinavian high mountains is derived from these findings. Cross sections through the high mountains of Scandinavia show that large areas underlain by widerspread and patchy discontinuous permafrost must exist and help explain the rich periglacial morphology well known from Scandinavia.

INTRODUCTION

Knowledge of permafrost distribution in Scandinavia is still limited and rather restricted to two very special features: palsas and ice-moraines. Since the study of Fries and Bergström (1910), the number of studies on palsa formation and distribution in Scandinavia has increased rapidly (cf. references in Åhman 1977), and further studies are still being conducted (e.g. Seppälä 1982). Some papers even state that permafrost in Scandinavia is restricted mainly to these specific features (cf. Vorren 1967, Wramner 1973). The study of ice-cored moraines requires sophisticated methods, and Østrem (1964, 1965) after thorough studies, was able to present an inventory of ice-cored moraines covering large parts of Scandinavia. Barsch (1971) discusses the assumption that in general ice-cored moraines are rock glaciers (cf. also Østrem 1971). In so far, Østrem's inventory of ice-cored moraines represents a collection with information about high mountain permafrost in Scandinavia, although this was not recognized by many researchers. Besides these papers, there are only a few punctual findings of permafrost (see Rapp and Annersten 1969, Ekman 1957). This paucity of detailed work on alpine permafrost is characteristic of many high mountain areas in the world and has been regretted by scientists such as Haeberli (1978) and Harris and Brown (1978). In Scandinavia, until very recently, this lack is especially due to the absence of engineering work. In 1977, Åhman could still state with full justification that in Scandinavia "permafrost in the ground outside palsas exists in different parts, (but) little is known of its distribution."

Our studies began in 1976, when we could prove that in general, permafrost exists also in ridges located in the vicinity of ice-cored moraines (King 1976). This discovery, together with climatic studies, led us to the assumption that large areas of the Scandinavian high mountains must be underlain by permafrost. Systematic studies on permafrost in Scandinavia were started in 1977, and we are now able to present some highlights of permafrost distribution. The most important results have been summarized recently (King 1982 and King unpublished manuscript).

INVESTIGATED AREAS

In order to give a valid picture for many areas of the Scandinavian high mountain belt, four widely separated regions were selected: Jotunheimen and Dovre/Rondane in the southern part, and the Lyngen Alps and Kebnekaise/Abisko north of the Arctic Circle (Figure 1). The important environmental features of these areas are given by Sømme (1960), Østrem et al. (1973), Østrem and Ziegler (1969) and King (unpublished manuscript). The purpose of the project was to obtain a pattern of permafrost and lower limits of permafrost distribution that may be correlated with the climatic parameters along an east-west gradient and a north-south gradient. The intention was to obtain a picture of the permafrost distribution that should be valid for most areas of the Scandinavian high mountain belt.

STUDY METHODS AND METHODOLOGICAL RESULTS

In all four test regions, investigations were carried out in summer and in winter. The methodological results obtained during our studies

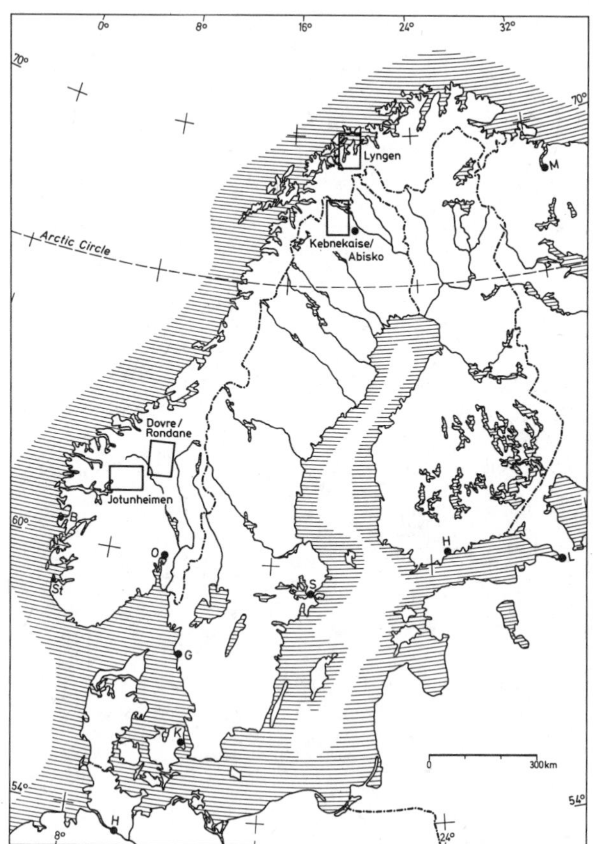

FIGURE 1 Location of investigated areas Jotunheimen, Dovre/Rondane, Kebnekaise/Abisko, and Lyngen.

are presented in detail in King (Manus.) and can be summarized as follows: For the detection of permafrost, four methods prove to be most efficient: (1) hammer seismic field work that reveals qualitatively the existence of permafrost in debris (p-wave velocities are 2,200 - 3,800 m/s) and gives the thickness of the active layer; (2) dc-geoelectrical soundings; (3) ground temperature measurements over several years, the latter two methods give an indication to the total permafrost thickness; and (4) measurement of the basal temperatures of the snow cover (BTS) at the end of the winter, permitting very efficient mapping of permafrost on a larger scale, provided that there is a mean minimum snow thickness of about 80 cm. The electrical resistivity in unfrozen debris usually amounts to values between 100 and 10,000 Ohm-m, and that in frozen debris to values between 10,000 and 900,000 Ohm-m. Mainly in frozen rocks the resistivity values increase significantly with lower temperatures, and at a temperature of about -6°C the resistivity of schistose plutonic rock may be 5 times higher than at a temperature of -1°C (cf. Bogolyubov et al. 1978, Séguin 1978). Geoelectrical and seismic soundings in permafrost near the freezing point may give equivocal results, since the resistivity and p-wave velocity values of frozen and unfrozen ground are overlapping (cf. Dzhurik et al. 1978). Accurate ground temperature measurements and diggings will then have to prove permafrost. However, dc-geoelectrical sounding provides a very effective tool for detecting larger ice cores or ice lenses, as these show electrical resistivities of several tens of MOhm-m. The method as developed by Haeberli (1973, 1978) of measuring the basal temperature of a snow cover in the Alps yields good results also in northern Europe. The limiting temperature values for distinguishing between areas with unfrozen ground from those with questionable and reliable permafrost occurences are -2° and -3°C at a snow depth of 150 cm. The limiting values are dependent on the snow depth, the minimum snow cover being 80 cm. BTS values furthermore may give a good indication of the mean annual ground temperature at the depth of the Zero Annual Amplitude (King unpublished manuscript). Figure 2 gives an idea of the number of soundings in our main test area, Tarfala, just east of the Kebnekaise peak, Sweden's highest point. BTS sites are not shown, but 172 BTS values have been recorded at the end of the winter of 1979/1980, and many more snow depths were measured, too.

REGIONAL RESULTS

In the Kebnekaise/Abisko area (cf. Figure 1), where mainly the Tarfala Valley has been intensively studied, a permafrost thickness of more than 100 m is probable at altitudes of more than 1,500 m a.s.l. (mean annual ground temperature is about -4°C; thickness of active layer is about 130 cm). Widespread discontinuous permafrost is encountered between 1,500 and 1,200 m a.s.l. (mean annual ground temperature is in the range of -4° to -2°C; thickness of the active layer attains 200-400 cm). Active permafrost is mostly missing on the valley floor due to a thick snow cover in winter and probably hydrological heat transport (cf. Figures 2 and 3), but inactive permafrost seems to occur here. Sporadic permafrost exists in palsas down to the lowest places along the shores of Lake Torneträsk (342 m a.s.l.).

In the Lyngen Alps the valleys Veidalen and Gjerdelvdalen have been investigated. The lower limit of discontinuous permafrost was found at about 960 m a.s.l. in block tongues exposed to NW. The lower limit of widespread permafrost is reached on Falsnesfjellet (1,200 m a.s.l.). Theoretically, areas with continuous permafrost should exist only at the highest elevations of the Lyngen Peninsula (peak of Jiekkevarre is 1,833m a.s.l.).

In Jotunheimen, where the Juvasshytta area has been studied in detail, a permafrost thickness of 100-200 m occurs at 2,200 m a.s.l. (estimated mean ground temperature is about -6°C; thickness of active layer is about 110 cm). Permafrost is still very widespread between 1,900 and 1,600 m a.s.l. (mean ground temperature is about -3°C; active layer is about 150 cm). In the Leirvassbu area 15 km SW of Juvasshytta, scattered permafrost may be found between 1,600 and 1,400 m a.s.l. As at the test sites in the Lyngen Alps, it seems that the limit of permafrost distribution is lower in the Sognefjell area and the area of permafrost

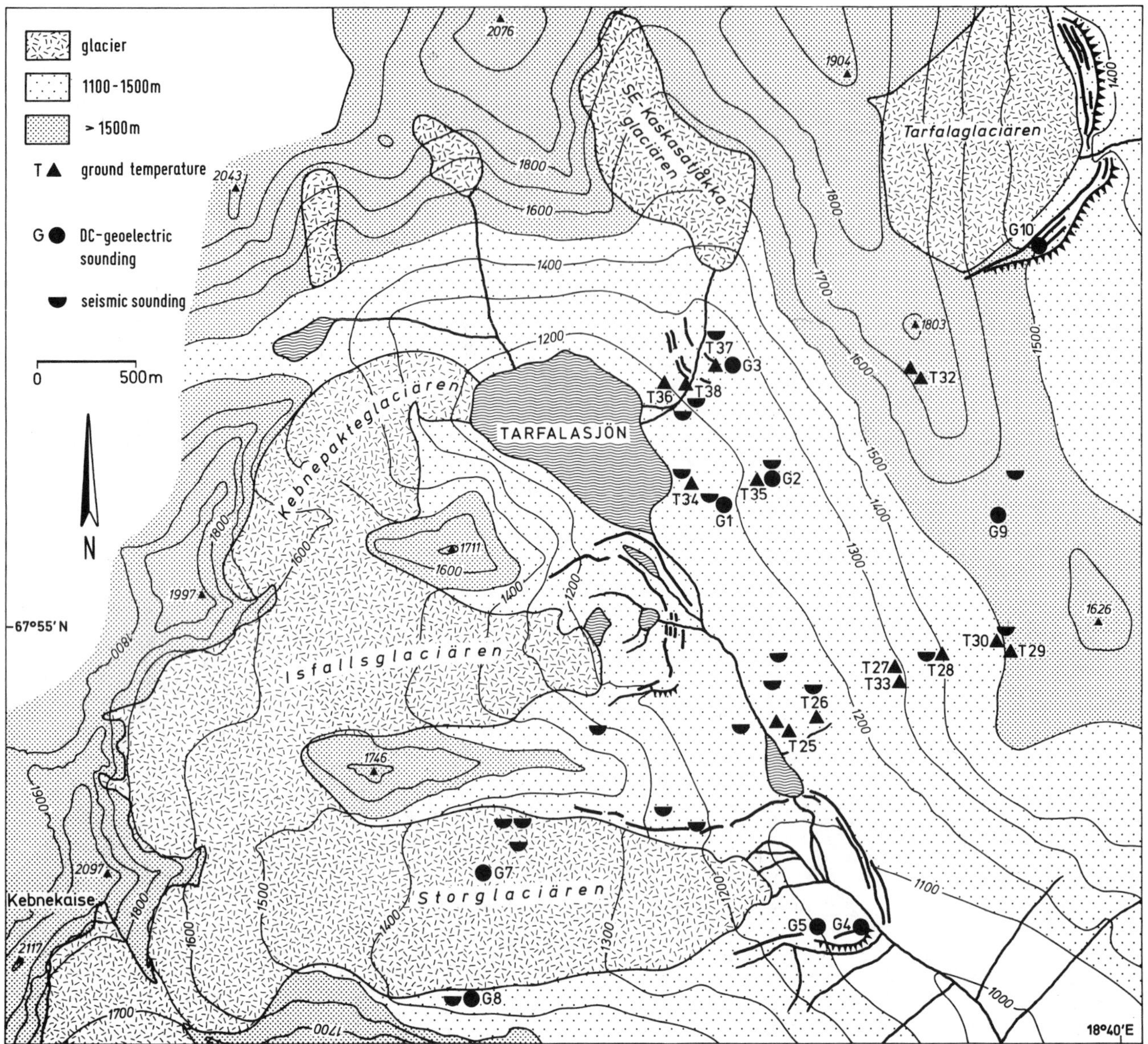

FIGURE 2 Map of the test area Tarfala (Kebnekaise region) with location of ground temperature, geoelectric, and seismic sites. In addition, 172 BTS values and many more snow depths have been measured in the Tarfala Valley.

occurrences is increased due to the existence of perennial snow fields.

In Rondane/Dovre, sporadic permafrost is often encountered in palsa fields down to 940 m a.s.l. (Melöya, Einunndalen), and relict permafrost seems to exist here in windswept terraces. The lower limits of discontinuous permafrost in slopes exposed to the NW could be found in the Rondane area at 1,330 m a.s.l. Measurements at the top of Simlepiggen suggest that the lower limit of continuous permafrost is not much higher than 1,720 m. The highest peaks of the Rondane area should reach into the continuous permafrost zone but have not been investigated to date.

CONCLUSIONS AND ZONATION

Permafrost distribution in alpine areas is primarily controlled by relief (altitude, angle, and slope) and secondarily by the duration and thickness of the snow cover (cf. Gorbunov 1978). Other factors such as vegetation, thermal conductivity of the ground, and hydrology may be of importance for local permafrost occurrences but are of minor interest for the regional distribution. The lower limit of discontinuous permafrost in the alpine areas investigated may be defined as the lower limit in slopes exposed to the north. There, angles of not more than 45° still allow geoelectrical, seismic or temperature soundings in an economical way. Steeper slopes could not be investigated

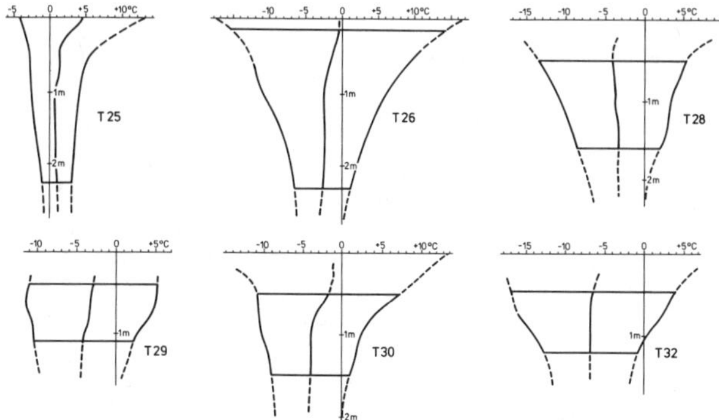

FIGURE 3 Ground temperatures of selected sites (cf. Figure 2). The temperature envelopes mark minimum and maximum temperatures registered between August 1977 and September 1980.

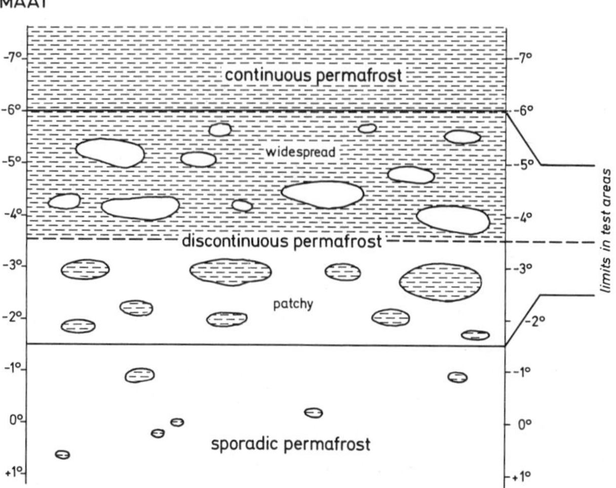

FIGURE 4 Zonation of alpine permafrost. The corresponding altitudes for the investigated areas are given in Table 1

safely, especially in winter time (avalanches!). The same principle applied to the lower limit of continuous permafrost and corresponding soundings in slopes exposed to the south. The limit between widespread and patchy permafrost separates areas where more or less than 50% of the ground, respectively, is frozen (Karte 1979). It corresponds to the altitude of those valley floors where snow free ridges contain permafrost but where the ground in flat areas is only partly frozen, depending on the thickness of the snow cover.

The lower limits of continuous, discontinuous widespread, and patchy discontinuous permafrost in the four investigation regions correspond closely to the mean annual air temperature (MAAT) of -5°, -3,5°, and -2,5°C for most of the investigated places (slopes up to 45°). If the lower limit of continuous permafrost (walls exposed to the south) is fixed at the altitude of the -6°C isotherm of the MAAT and the lower limit of discontinuous permafrost (walls exposed to the N) at the altitude of the -1,5°C isotherm, sites that cannot be investigated economically by our methods should be included in our model, too (Figure 4).

We realize that the -6°C isotherm is a rather high value compared with -8,5°C in the zonation of high latitude permafrost (Brown and Péwé 1973). Haeberli (1978) mentions field observations that suggest that the lower boundary of continuous permafrost may be somewhat lower in the Swiss Alps (-7°C) and states a limit of perhaps -6° oder -7°C in his most recent report (Haeberli 1982). We would like to leave this question open for further research and think that this might not only be a matter of permafrost findings but also a matter of definition of the term "continuous".

The lower limits of permafrost occurrences seem to be lower due to the maritime influence of unfrozen fjords in the western part of our investigation regions, and similar differences might also be found toward more continental areas of Finnish Lapland . Research in these areas is certainly needed and also intended.

The fact that the altitude of the lower limit of permafrost coincides with the altitude of a certain MAAT allows us to propose an extrapolation for the noninvestigated areas. A permafrost map of Lapland and southern Scandinavia is in preparation (King, unpublished manuscript). Four cross sections through the Scandinavian high mountain range are shown in Figure 5. Forest limits (taken from topographical maps) and glaciation limits (cf. Østrem et al. 1973, Østrem and Ziegler 1969) are also marked. It is noteworthy that contrary to the permafrost limits, the forest limit and the glaciation limit are rising with increasing continentality. The uppermost cross section shows that large potential forest areas rise above the altitude of the -1,5°C isotherm in Finnish Lapland. These are the areas where the most impressing palsas in Scandinavia exist favored by climate, vegetation, and relief. They belong to the discontinuous permafrost zone (cf. King 1979). In the other cross sections the forest limit hardly reaches up to the altitude of the -1,5°C isotherm, and minute palsa forms belonging to the sporadic zone are reported from there (Sollid and Sørbel 1974).

Further permafrost research is urgently needed especially in the areas outside the main mountain range. Models with freezing and thawing indices are developed by the author (cf. Harris 1982). We trust that already now the model for permafrost distribution in Scandinavia presented above will bring new aspects to other research topics. We especially think of the study of periglacial processes and forms by Scandinavian scientists in several classical studies (Rapp and Rudberg 1964, Rudberg 1977). Geomorphological surveys in the investigated areas have been done by the author. They suggest that "the spatial extent of periglacial regions not underlain by permafrost of one sort or another is limited" (French 1979). Apparently, this seems to be true for many alpine regions, too.

ACKNOWLEDGMENTS

Research was carried out with the support of the Deutsche Forschungsgemeinschaft, Bonn. Special thanks are due to V. Schytt (Stockholm), director of the Tarfala Research Station, and to D. Barsch (Heidelberg), who both supported the studies with logistic help and discussions. The author is very grateful to many other persons who helped him to obtain the results that form the background of this paper. Although the long list of friends and colleagues cannot be published here, the author realizes that these studies could not have been done without their support and will keep them well in mind.

REFERENCES

Åhman, R., 1977, Palsar i Nordnorge, En studie av palsars morfologi, utbredning och klimatiska förutsättningar i Finnmarks och Troms fylke: Meddelanden från Lunds Universitets Geografiska Institution Avhandlingar, LXXVIII, 165 pp.

Barsch, D., 1971, Rock glaciers and ice-cored moraines: Geografiska Annaler 53 A, 3-4, p. 203-206.

Bogolyubov, A. N., 1978, Investigation of dependence of specific resistance of unconsolidated frozen deposits in original bedding on cryogenous textures and temperatures, in Permafrost--USSR Contribution to the Second International Conference, Yakutsk, July 13-28, 1973: National Academy of Sciences, Washington, D.C., p. 482-485.

Brown, R. J. E., and Péwé, T. L., 1973, Distribution of permafrost in North America and its relationship to the environment: A review, 1963-1973, in Permafrost--North American Contribution to the Second International Confer ence, Yakutsk, 1973: National Academy of Sciences, Washington, D.C., p. 71-100.

Dzhurik, V., and Leshchikov, F. N., 1978, Experimental investigations of seismic properties of frozen soils, in Permafrost--USSR Contribution to the Second International Conference, Yakutsk, July 13-28, 1973: National Academy of Sciences, Washington, D.C., p. 485-488.

Ekman, S., 1957, Die Gewässer des Abisko-Gebietes und ihre Bedingungen: Stockholm, Kungl. Sv. Vetenskapsakademiens Handlingar 4, 6: 6.

French, H. M., 1979, Periglacial geomorphology: Progress in Physical Geography, p. 264-273.

Fries, T., and Bergström, E., 1910, Några iakttagelser öfver palsar och deras förekomst i nordligaste Sverige: Stockholm, Geologiske Föreningen Förhandlingar 32, 1, p. 195-205.

Gorbunov, A. P., 1978, Permafrost investigations in high-mountain regions: Arctic and Alpine Research 10, 2, p. 283-294.

Haeberli, W., 1973, Die Basis-Temperatur der winterlichen Schneedecke als möglicher Indikator für die Verbreitung von Permafrost in den Alpen: Zeitschrift für Gletscherkunde und Glazialgeologie, IX, 1-2, p. 221-227.

Haeberli, W., 1978, Special aspects of high mountain permafrost methodology and zonation in the Alps: in Proceedings of the Third International Conference on Permafrost, July 1o-13, 1978- Edmonton, Alberta, Canada: National Research Council of Canada, Ottawa, Vol 1, p. 379-384.

Haeberli, W., 1982, Klimarekonstruktionen mit Gletscher-Permafrost-Beziehungen: Basel, Materialien zur Physiogeographie 4, p. 9-17.

Harris, S. A., 1982, Identification of permafrost landforms, (The Roger J. E. Brown Memorial Volume). Proceedings Fourth Canadian Permafrost Conference: Ottawa, National Research Council of Canada, p. 49-58.

Harris, S. A. and Brown, R. J. E., 1978, Plateau Mountain: A case study of alpine permafrost in the Canadian Rocky Mountains, in Proceedings of the Third International Conference on Permafrost, July 10-13, 1978, Edmonton, Alberta, Canada: National Research Council of Canada, Ottawa, Vol. 1, p. 385-391.

Karte, J., 1979, Räumliche Abgrenzung und regionale Differenzierung des Periglaziärs: Bochumer Geographische Arbeiten 35, 211 pp.

King, L., 1976, Permafrostuntersuchungen in Tarfala (Schwedisch Lappland) mit Hilfe der Hammerschlagseismik: Zeitschrift für Gletscherkunde und Glazialgeologie 12, 2, p. 187-204.

King, L., 1979, Palsen und Permafrost in Quebec: Trierer Geographische Studien, Sonderheft 2, Kanada und das Nordpolargebiet, 141-156.

King, L., 1982, Qualitative und quantitative Erfassung von Permafrost in Tarfala (Schwedisch-Lappland) und Jotunheimen (Norwegen) mit Hilfe geoelektrischer Sondierungen: Zeitschrift für Geomorphologie, N. F., Supplement-Band 43, p. 139-160.

King, L., unpublished manuscript, Permafrost in Skandinavien - Untersuchungsergebnisse aus Lappland, Jotunheimen und Dovre/Rondane: Habilitationsarbeit, Universität Heidelberg, 225 pp.

Østrem, G., 1964, Ice-cored moraines in Scandinavia: Geografiska Annaler 46, p. 282-337.

Østrem, G., 1965, Problems of dating ice-cored moraines: Geografiska Annaler 47, p. 1-38.

Østrem, G., 1971, Rock glaciers and ice-cored moraines, A reply to D. Barsch: Geografiska Annaler 53 A, 3-4, p. 207-213.

Østrem, G., Haakensen, N., and Melander, O., 1973, Atlas over Breer i Nord-Skandinavia: Meddelanden nr. 22 fra Hydrologisk Avdeling, Norges Vassdrags og Elektrisitetsvesen, Oslo, 315 pp.

Østrem, G., and Ziegler, T., 1969, Atlas over Breer i Sør-Norge, Atlas of Glaciers in South Norway: Meddelanden nr. 20 fra Hydrologisk Avdeling, Norges Vassdrags- og Elektrisitesvesen, Oslo, 207 pp.

Rapp, A., and Annersten, L., 1969, Permafrost and tundra polygons in northern Sweden, in Péwé, T. L., ed, The periglacial environment past and present: Montreal, McGill-Queen's Univ. Press, p. 65-91.

Rapp, A., and Rudberg, S., 1964, Studies on periglacial phenomena in Scandinavia 1960-1963: Biuletin Peryglacialny 14, p. 75-89.

Rudberg, S., 1977, Periglacial Zonation in Scandinavia: in Poser, H. (Hrsg.), Formen, Formengesellschaften und Untergrenzen in den heutigen periglazialen Höhenstufen der Hochgebirge Europas und Afrikas zwischen Arktis und Äquator: Abhandlungen der Akademie der Wissenschaften in Göttingen, Math.-Physikal. Klasse, 3. Folge, Nr. 31, p. 92-104.

Séguin, M. K., 1978, Temperature-electrical resistivity relationship in continuous permafrost at Purtuniq, Ungava Peninsula: in

Proceedings of the Third International Conference on Permafrost, July 10-13, 1978, Edmonton, Alberta, Canada: National Research Council of Canada, Ottawa, Vol. 1, p. 137-144.

Seppälä, M., 1982, An experimental study of the formation of palsas, Proc. 4th Can. Permafrost Conf. Calgary (The Roger J. E. Brown Memorial Volume): Ottawa, National Research Council of Canada, p. 36-42.

Sollid, J. L. and Sørbel, L., 1974, Palsa bogs at Haugtjørnin, Dovrefjell, South Norway: Norsk geografisk Tidsskrift 28, 1, p. 53-60.

Sømme, A., ed., 1960, A geography of Norden - Denmark, Finland, Iceland, Norway, Sweden: Oslo, Cappelens, 364 pp.

Vorren, K. D., 1967, Evig Tele i Norge: Ottar 51, 26 pp.

Wramner, P., 1973, Studies of palsa bogs in Taavavuoma and the Laiva Valley, Swedish Lapland: Göteborg, Akademisk Avhandling, 7 pp.

TABLE 1 Altitude of MAAT Key Values (1931-1960), Temperature Gradient = 0.53°C/100m

region: area:	Jotunheimen Leirvassbu	Rondane Simlepigg	Kebnekaise Tarfala	Lyngen Gjerdelvdal
-6.0°	2,060	1,960	1,600	1,720
-5.0°	1,870	1,770	1,420	1,530
-3.5°	1,580	1,490	1,130	1,250
-2.5°	1,400	1,300	940	1,060
-1.5°	1,210	1,11o	750	870

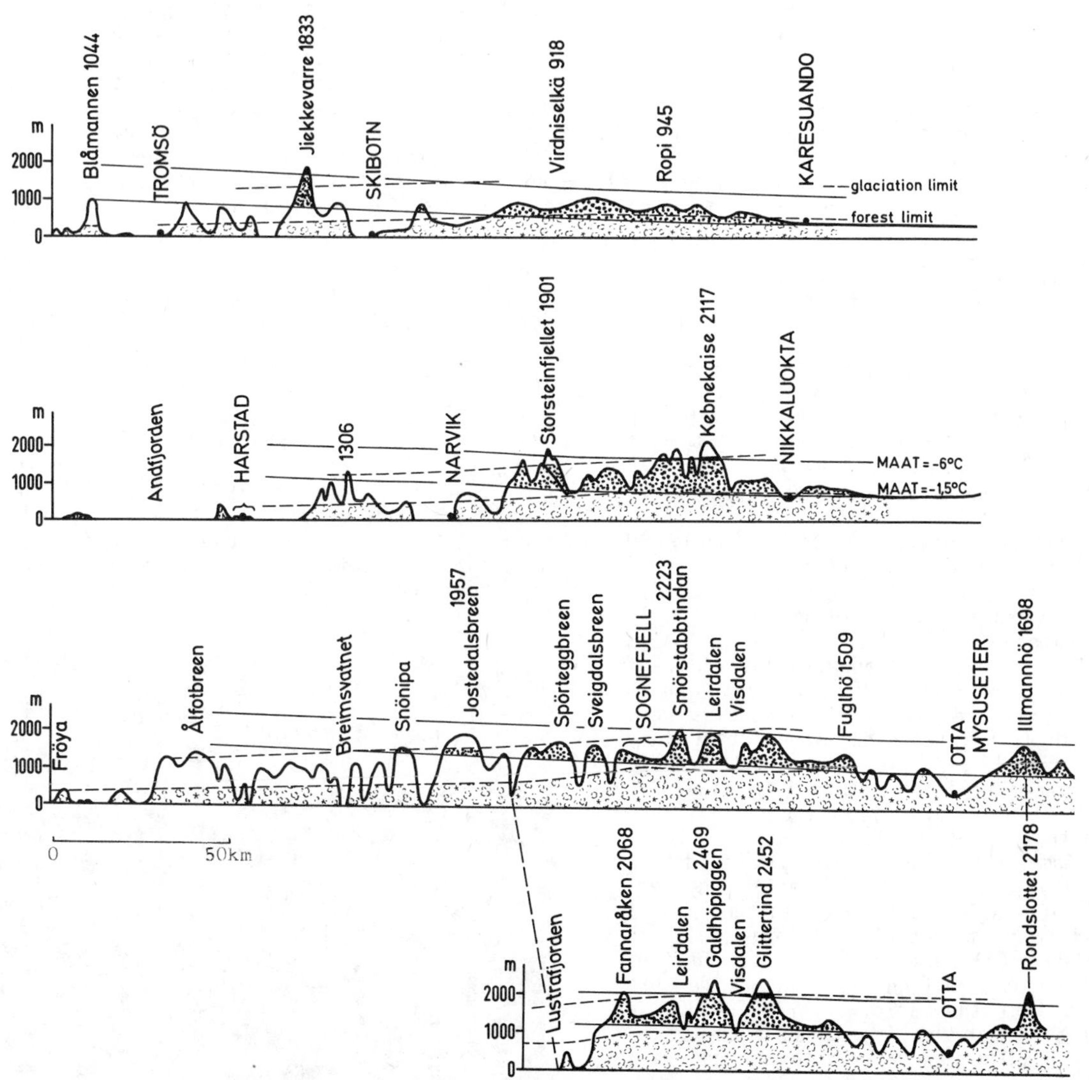

FIGURE 5 Cross sections through the Scandinavian high mountain belt with position of lower limits of permafrost zones, glaciation limit, and tree limit.

FOUNDATION STABILIZATION OF CENTRAL GAS INJECTION FACILITIES, PRUDHOE BAY, ALASKA

Thomas C. Kinney[1], Barry W. Santana[2], D. Michael Hawkins, Erwin L. Long, and Edward Yarmak Jr.[3]

[1]Shannon & Wilson, Inc., Fairbanks, Alaska 99707, USA; [2]ARCO Alaska Inc., Prudhoe Bay, Alaska, USA; [3]Arctic Foundations, Inc., Anchorage, Alaska, 99502, USA

The importance of the Central Gas Injection Facilities at Prudhoe Bay, Alaska, prompted a detailed study of its adfreeze pile foundation when differential settlements were noticed in some piles. As a result, an extensive data gathering program was initiated to monitor pile movement, subsoil temperatures, and microclimate temperatures beneath the facility. The data gathered indicated that pile settlement was occurring and that settlement was not related only to the magnitude of applied load. Studies indicated that settlement was more related to soil temperature and the creep characteristics of soil near the pile. A theoretical analysis successfully predicted observed settlement rates on those piles showing the most settlement. The theoretical model assumed the piles were embedded in ice or ice-rich soils. Remedial measures were undertaken to control and limit the pile movement. The ground surface beneath the entire facility was insulated, many piles were converted to thermosyphons, exposed pile surfaces were painted white to increase their reflectivity, and an impervious membrane was installed over the insulation to control moisture and contaminant infiltration. These remedial measures appear to have controlled pile movement to an acceptable level and to within the accuracy of survey measurements.

INTRODUCTION

The Central Gas Injection Facilities at Prudhoe Bay, Alaska, are located on a 1.8 m thick, 13 ha gravel pad on the east side of the Prudhoe Bay field. The gravel pad was placed directly on the tundra mat which covers about 5 m of silt overlying sands and gravels. The area contains polygonal ground, and massive ice was encountered in many of the borings.

The facilities consist of approximately 40 modules having a total first level floor area of nearly 15,000 m^2. The modules are elevated 2.5 m above the pad and are supported on approximately 1200 steel pipe piles. The piles are 46 cm in diameter, about 6.5 m long and are slurried and frozen in the silt and ice permafrost which overlies the sand and gravel. The modules were constructed in Tacoma, Washington, and barged to Prudhoe Bay in 1975, 1976, and 1977. Field start-up of the facilities occurred in mid-1977. Photograph 1 is an aerial view of the facilities.

The Central Injection Facilities consist of the Central Compressor Plant (CCP), the Field Fuel Gas Unit (FFGU), and the gas injection system. The function of the CCP is to compress the natural gas produced from the entire field for injection into the Prudhoe Oil Pool gas cap. The function of the FFGU is to condition produced gas for use as fuel throughout the field. Nearly 254,000 KW of compression equipment inject as much as 57 million m^3 STP of gas per day into the gas cap. The equipment enclosed in this facility is critical to the production of oil from the entire Prudhoe Bay field.

This paper discusses the history of a differential foundation settlement problem at the Central Injection Facilities. The foundation stabilization procedure used and its effectiveness are described.

HISTORY

In the summer of 1976, Central Injection Facilities module foundations were suspected of movement because of observed architectural, structural, piping, and equipment alignment problems. Door frames in the facilities were

PHOTOGRAPH 1 Central Gas Injection Facilities

racking causing doors to bind, intermodular connections were showing distress, instances of unsupported piping were occurring, and rotating equipment realignment requirements were excessive. A pile monitoring program was implemented at this time to track and verify suspected differential pile movement.

By the end of 1979, 20 piles had shown relative settlements greater than 3 cm. One pile had settled 8.5 cm, while other piles had negligible movement. In general, the rate of settlement was greatest during the late summer, fall, and early winter when subgrade temperatures were maximum, and nearly stopped during the spring when the ground temperature was at a minimum.

Several hypotheses were developed to explain the settlements. Subgrade temperatures may have been increased due to elevated ambient temperatures below the turbine modules and solar radiation at the south and west module perimeters. Glycol spilled in one area may have penetrated the soils and depressed the freezing point. Ponding water under and around the modules may also have resulted in an increased ground temperature. There was a lack of quality control on pile slurrying during construction which may have resulted in poor workmanship, and the original pile adfreeze bond strength design criteria were optimistic.

In the spring of 1981, a program was undertaken to stabilize the foundation of the Central Injection Facilities. The subgrade below 22 modules was insulated by placing a 10 cm thick layer of polystyrene board insulation on the ground surface and anchoring it with a shallow layer of gravel. This was intended to reduce the active layer thickness. In addition, nearly 500 of the piles exhibiting high settlement rates were converted to two-phase thermosyphons to provide additional cooling to the ground surrounding the pile during winter months.

INSTRUMENTATION AND MONITORING

Vertical movement of selected piles at the Central Injection Facilities has been monitored since 1976. A comprehensive and sophisticated pile monitoring program was implemented in September 1981 to carefully track and analyze foundation movement on a larger scale. The pile monitoring program is designed to obtain first order level accuracy or better. Pile elevations are referenced to four arctic bench marks, which are referenced to a U. S. Geological Survey monument.

Accuracy is a primary concern when dealing with the small vertical movements of piles frozen in permafrost. The environmental extremes of the arctic affected survey procedures and the accuracy of the elevation data collected. Large temperature variations and ground movement caused by seasonal freezing and thawing cycles affected the reference bench marks and caused apparent cyclic movement of the piles. Extreme low temperatures, long hours of darkness, and whiteout conditions caused hardships on men and equipment during winter surveys. Bright sunlight and radiation heat waves caused distortions and difficult sight conditions in the summer months.

Special techniques were used in the pile monitoring program to maintain errors of closure on level loops to within ±0.100 cm and individual pile elevation readings to within ±0.015 cm. Seasonal movement between the arctic bench marks over several years has been approximately 0.02 cm. It is estimated that pile elevation differences between studies are accurate to ±0.090 cm, neglecting the effects of thermal expansion and contraction. Thermal effects on piles between studies can be as great as 0.100 cm. Correction for this condition is possible but has generally not been attempted on this project.

The subgrade thermal regime below the Central Injection Facilities has been monitored on a monthly basis since July 1981. Twenty-eight thermistor strings with over 250 thermistor points were placed below the insulation (approximately 40 cm below existing grade) to monitor temperatures in the horizontal plane. Eighteen vertical thermistor strings with 25 thermistors each were installed throughout the site at locations such as at the ends of the facility, beneath the facility, at locations having unique shading conditions, and spaced radially from selected piles converted to thermosyphons. All but one vertical string extend 9 m below grade. One 30.5 m vertical string was installed to obtain temperature information below depths affected by seasonal temperatures. Three thermistor strings were installed within steel piles which were converted to thermosyphons in an attempt to record internal pile wall temperatures.

Early in the project localized climatic conditions were observed at various locations beneath the turbine modules. In addition, weather at the Central Injection Facilities site is notably different from that of the nearest meteorological station, the Prudhoe Bay airstrip. Instrumentation was placed below two typical turbine modules to collect and record ambient air temperature, surface soil temperature, pile wall contact temperature, and wind speed and direction. Maximum, minimum, and average values were recorded on the hour. The system has been somewhat unreliable and limited usable data have been collected to date.

REMEDIAL DESIGN DEVELOPMENT

Large differential settlements of many piles prompted the search for a design that would limit additional pile settlements to within tolerable limits. In order to formulate a design to minimize settlements, it was first necessary to investigate the settlement history and determine potential causes for pile settlement.

Generally, the piles exhibiting the greatest settlements were placed during the late winter of 1975-1976. The specification that governed pile placement for many of those piles called for warm water to be placed in the annulus between the pile and the drilled hole. Sand was then dumped into the water and vibrated to cause densification. Piles were placed in position without pile caps or covers. Freezeback of the slurry occurred naturally.

Seasonal thaw penetration at the site varied

from 1.2 m to 2.3 m, depending upon the location, the local depth of gravel fill, and the microclimate. Pile settlement measurements revealed that the piles with excessive settlement settled most during those portions of the year when the ground temperature throughout the depth of the pile was relatively warm and portions of the gravel pad were thawed.

A structural analysis that incorporated settlement data showed that settlement was not proportional to loading. The greatest settlement was measured on a pile with a calculated axial load of less than 23 metric tons; while analysis showed that other piles were loaded to more than 135 metric tons and showed comparatively little movement.

Pile settlement rates were analyzed using techniques based on those presented by Nixon and McRoberts (1976), Weaver and Morgenstern (1981a), and Morgenstern et al (1980). Data from full-scale proprietary pile tests conducted by British Petroleum/SOHIO at Prudhoe Bay in 1971, as well as other data from Weaver and Morgensten (1981b), and Johnston (1981), were used to estimate the creep rate relationships for the piles. Back-calculated creep relationships from pile settlement rate data during periods of rapid creep on those piles experiencing settlement indicated that adfreeze stresses were on the order of those for ice below the active layer. This indicated that support along those piles showing appreciable settlement was primarily by ice.

Glycol spills and the subsequent solution of ice within the soil probably increased the settlement rates in those areas where the glycol spills occurred. Heat conduction into the subgrade and along the pile walls may have been increased by solar heat gain in exposed areas and by exhuast heat in the vicinity of exhaust stacks. Water from the roof condensers and snowmelt may have caused increased subgrade temperatures.

The remedial measures that were used to limit future pile settlement consisted of the following four components: the ground surface beneath all modules was insulated with 10 cm of expanded polystyrene insulation extending a minimum of 5 m beyond the structure, an impervious membrane was installed above the insulation, all exposed portions of piles were painted white, and selected piles were converted to two-phase thermosyphons. The remedial measures were designed to reduce the depth of the active layer and to reduce the maximum ground temperatures throughout the depth of the piles, thereby providing additional adfreeze area and a lower creep rate within the supporting soils. The impervious membrane was installed to inhibit further glycol and water infiltration into the subgrade.

Piles were selected for conversion to thermosyphons based on the following order of importance: (1) piles with the highest settlement; (2) piles directly supporting turbines, compressors and exhaust stacks; (3) piles in the glycol spill area; and (4) additional piles as needed to provide bulk heat removal from the subgrade.

THERMOPILE CONVERSION PROCEDURE

Each pile selected for conversion was de-iced, pressure tested, evacuated, and charged with refrigerant. The piles at the site contained various quantities of snow and ice. The snow apparently entered the piles during the interval of time between pile placement and pile cap attachment. A layer of ice capped the snow in the piles whenever snow levels were as high as the bottom of the preconversion active layer adjacent to the piles.

Refrigerated methanol was used to de-ice pile interiors to prevent warming of the material around the pile. Chilled methanol was sprayed into the piles, coating the walls with a thin film to ensure thorough de-icing. The methanol remained in the piles until all of the snow and ice was in solution with the methanol. The normal time period required for complete solution of the snow and ice was between 1 and 2 days.

Once the de-icing procedure was completed, the piles were pressure tested for fitting leaks to 310 kPa. The piles were then evacuated to a vacuum of 4 kPa. After evacuation the piles were charged with Refrigerant-12 (R-12). R-12 was selected because it was readily available at a reasonable cost and it would operate at a low positive pressure.

After approximately 2 months of operation, the thermosyphons were checked for operation and noncondensing gases were purged from the units. Most commercially available fluorocarbon refrigerants contain small quantities of impurities. After thermocycling has taken place, the noncondensing impurities effectively cut off the available condensation area, thus reducing the heat transfer capability of the system.

Snow and ice were not the only items inside of the piles. Sand slurry, pieces of wooden boards, a coffee cup, glycol, and a section of 40.64 cm O.D. steel pipe were some of the items discovered inside the piling. Many times in the field, ingenious techniques were devised to side-step problems caused by debris in the piles.

Thermosyphon pressures are monitored annually prior to cold weather activation. Temperature of the liquid/vapor interface within the pile is derived from the measured pressures using published data for the refrigerant. Monitoring has shown that the average maximum pile wall temperature at a depth of 4 m decreased approximately 1°C during the first year of operation.

DATA ANALYSIS

Pile Settlement

Only a few piles were initially surveyed, and as a result complete sets of settlement data do not exist for most of the piles. Figure 1 shows a typical settlement trend exhibited by many piles. Initially, settlements occurred during late summer and fall, with little settlement, and in some instances apparent heave, occurring during the winter and spring. The pile represented in Figure

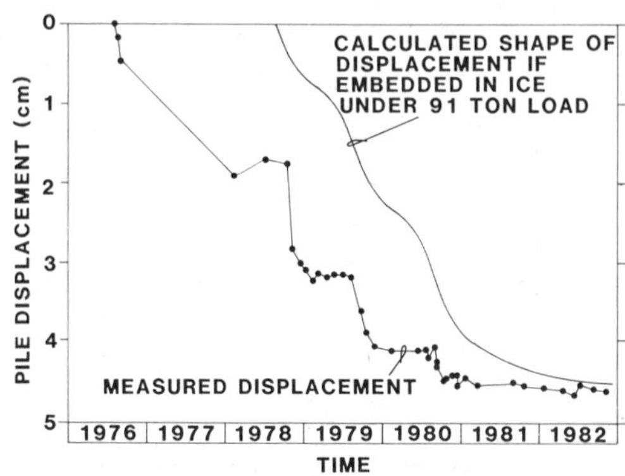

FIGURE 1 Typical pile displacement

1 was converted to a thermosyphon in early 1981 with ground surface insulation being placed shortly thereafter. Negligible settlement has been observed since the thermosyphon conversion and ground insulation placement. Analyses indicate that the reduction in settlement rate was caused by the remedial measures and not by the natural reduction in secondary creep rate with time.

Recent survey measurements taken throughout 1982 show many small heave and settlement fluctuations, many of which do not tend toward overall heave or settlement of a pile. These fluctuations in surveyed elevations are of a greater magnitude than expected heave or settlement, making prediction of future performance based on these settlement measurements difficult. It appears that the fluctuations are predominantly a result of survey inaccuracy and thermal effects above the ground surface, and are not real pile movement within the ground.

Microclimate

Microclimate conditions beneath two compressor train modules were monitored with 42 thermistors placed to record: air temperature 30 cm above ground, ground surface temperature, and pile surface temperature. Several trends were observed in the microclimate data.

During warm periods, pile temperatures were generally observed to be cooler near the center of the modules. Temperatures 1°C to 3°C colder than ambient air were noted. Pile temperatures near the exhaust stacks and along the perimeter of the modules generally followed the ambient air temperature at the CCP during periods when the thermosyphons were expected to be inactive. As expected, active thermosyphons had surface temperatures warmer than the ambient air temperature.

Microclimate air temperatures beneath the modules follow the same trend as pile surface temperatures, however, the temperature extremes and fluctuations are greater. Ground surface temperatures appear to vary closely with the microclimate air temperature at a given location.

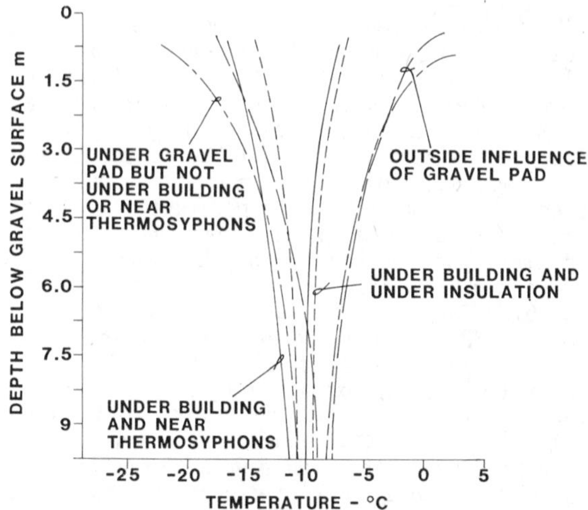

FIGURE 2 Minimum and maximum temperatures

Soil Temperatures

Vertical thermistor strings were placed at numerous locations around the facilities to gather data that would be representative of different conditions. Profiles of the minimum and maximum temperature versus depth are shown for various conditions on Figure 2. Placing a gravel pad on the surface increased the width of the temperature range at the surface but decreased the temperature range at the surface of the organic layer. The average temperature at a depth of 9 m was decreased about 1°C.

The temperature ranges recorded under the insulation and under the building were smaller, and the average temperature at a depth of 9 m was about 1°C colder than in those areas without the insulation. Near piles converted to thermosyphons, the average temperature at a depth of 9 m was approximately 1°C colder than at greater distances from the converted piles.

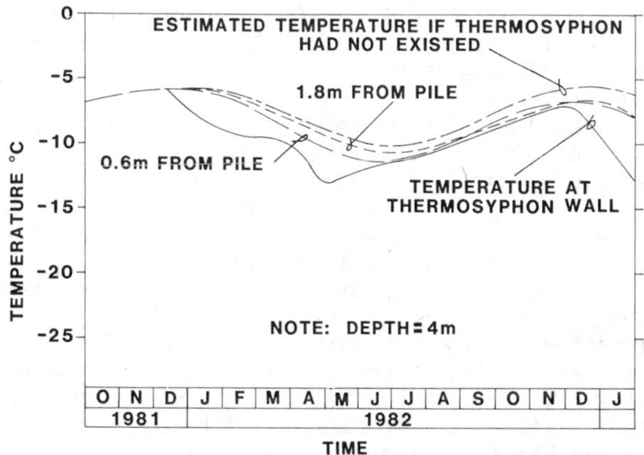

FIGURE 3 Effect of thermosyphon on temperature

Vertical thermistor strings were placed inside the pile and at 0.6 m and 1.8 m radial distances from converted piles at three locations. A typical relationship between temperature and time is shown at the 4 m depth in Figure 3. As expected, the thermosyphon wall became colder than the surrounding soil when the air temperature dropped below the ground temperature. Once the air temperature became warmer than the pile wall, the thermosyphon ceased to function and the pile soon warmed to the same temperature as the surrounding soil. Between November 1981 and November 1982, the net decrease in temperature of the soil surrounding the converted piles at a depth of 4 m was approximately 1°C.

Based on data from the thermistors placed immediately below the insulation, the warmest temperatures under the module were on the order of -3°C and were about 1°C colder in the summer of 1982 than the summer of 1981. No significant difference in the change in warmest temperatures could be detected between those areas immediately adjacent to thermosyphons and those areas at least 6 m from any thermosyphon indicating that the cooling effect is widespread. The data does not indicate whether the cooling is caused by climatic changes or the thermosyphons. The thermosyphons should have caused cooling, however, the climate was also getting colder. The thermistors do not show isolated hot spots such as under exhaust stacks.

PILE ANALYSIS TECHNIQUE

The settlement rates of the piles were calculated assuming that the piles exhibiting the most settlement were embedded in ice. Using the measured temperature-depth profile at various times of the year it was possible to determine the creep rate versus stress dependency of the ice around the pile at any depth and at any time of year. Since the pile displacement rate (ice creep rate) is uniform throughout the length of the pile, it was possible to determine the stress level on the pile at any depth and time, given the pile displacement rate and the temperature. Since the sum of the pile wall stresses is the total load on the pile, it is possible to calculate the pile displacement rate throughout the year under any total load.

As shown in Figure 1, the calculated pile displacement rates throughout the portions of the year when settlement rates are greatest qualitatively agree with the displacement rates measured on those piles that have undergone the most settlement (excluding the piles in the glycol spill area). During those portions of the year when ground temperatures are coldest, the predicted settlement rate is greater than the measured settlement rate. During this portion of the year, the gravel pad is frozen, providing high shear stresses on the pile near the top, within a material that is subject to low creep rates. The theoretical model used for the predictions shown on Figure 1 did not take this into account.

Accurate theoretical analyses of pile creep rates are not possible because of the uncertainty of the load on a given pile and of the actual material surrounding each pile. The load on a given pile may not be the same as the design load due to redistribution of the load caused by differential pile movements.

CONCLUSIONS

Pile Settlement

The remedial measures undertaken to decrease or stop the settlement of piles have appeared to work well. Pile movement over the first year since they were implemented has been, on the average, less than ±0.1 cm, in contrast to earlier years when as much as 2 cm was observed in a single year. The magnitude of pile movement recorded in the first year appears to be within the limit of survey accuracy and therefore, it is difficult to accurately predict future pile performance based on trends developed in pile movement over the one year.

Subgrade Temperatures

Within the first year after remedial measures were completed, the maximum temperature at the base of the pad had been reduced from about -1°C to about -7°C, and the gravel pad has remained frozen throughout the year. The thermosyphons reduced the ground temperature within about 0.6 m of the pile at the 4 m depth by about 2°C by June. The temperature near the pile-soil interface comes to equilibrium with the surrounding soil within a few weeks after the air temperature rises above the pile wall temperature.

ACKNOWLEDGMENTS

The authors wish to acknowledge the assistance of their colleagues who have been involved with the field work, data collection, data reduction, and computer programming necessary to accomplish a project of this magnitude. This paper is presented with the permission of the management of ARCO Alaska, Inc., and ARCO Oil and Gas Company.

REFERENCES

Nixon, J.F., and McRoberts, E.C., 1976, A design approach for pile foundations in permafrost: Canadian Geotechnical Journal, V.13, p. 40.

Weaver, J.S., and Morgenstern, N.R., 1981a, Pile design in permafrost: Canadian Geotechnical Journal, V. 18, p. 357-370.

Weaver, J.S., and Morgenstern, N.R., 1981b, Simple shear creep tests on frozen soils: Canadian Geotechnical Journal, V.18, P. 217-229.

Johnston, G.H., 1981, Permafrost engineering design and construction: New York, John Wiley & Sons.

Morgenstern, N.R., Roggensack, W.D. and Weaver, J.S., 1980, The behavior of friction piles in ice and ice-rich soils: Canadian Geotechnical Journal, V.17, p. 405-415.

THE PROCESS OF NALED (ICING) FORMATION IN THE TROUGHS OF THE PACIFIC MOUNTAIN BELT

I. V. Klimovskiy

Permafrost Institute, Siberian Branch,
Academy of Sciences, Yakutsk, USSR

The formation and existence of belts of icings are pre-determined by the geology and evolutionary history of the regions during the Pleistocene and Holocene. These belts are related geomorphologically to the activities of former and contemporary glaciers. The action of the latter has led to the formation of moraine-and-riegel complexes, series of which have resulted in the icing belts. Icing formation occurs within the moraine-and-riegel complexes. A definite interrelationship exists between the icings and sub-icing taliks. The dimensions of the moraine-and-riegel complexes and of the icing complexes regularly decrease with altitude. There may be several separate taliks, or a single extensive one, but they are always smaller in area than the area occupied annually by icings. The number of icings corresponds to the number of talik basins, which in turn coincides with that of the erosional valleys draining the sides of the troughs. Freezing of the taliks and icing formation always begin from the upper part of the area affected by the icing and the freezing barrier migrates from higher to lower altitudes.

In many mining developments in East Siberia, within the so-called mobile Pacific belt, traces of stadial contraction of Sartan glaciation have been revealed to date. In different regions the number of stadial formations, because of their different conditions of preservation and variations in the de-glaciation process, does not remain constant.

Marginal moraines forming part of the above-mentioned stadial complexes had formed mainly along the edges of glaciers as these discontinued their advance, or originated as the advancing ice was pushing debris material, thus fixing the maximum advance of the glaciers at that particular stage. The thickness of moraines reached considerable values; their fragments, still well preserved, are arch-shaped hills 150 to 200 m high. In front of such obstacles many large moraine-dammed lakes formed. Their breakthrough took place outside the former bed of valleys and the river network was then reshaped. Such a moraine-dammed lake, well preserved to date, is exemplified by Tabanda Lake in Chibagalakhsky Ridge (Chersky Mountain Range) which represents a trough scoured during the Sartan glaciation and later flooded for a distance of more than 11 km. An example of a sharp curve in a river in front of an obstacle such as a terminal moraine is in turn presented by the valley of the Buordakh River near its confluence with the Lyunkide River (Ulakhan-Chistaisky Ridge). The rivers just mentioned had been north-flowing, but once blocked by the huge Sartan moraines, they first filled large depressions and produced lakes. The lake in the valley of the Lyunkide River drained at the eastern termination of the moraine dam, and it was this that altered the original direction of the river. At the present time, the Lyunkide River, meandering on the bed of the drained lake, discharges into the Buordakh River and abruptly turns eastward in front of the aforementioned terminal morainal ridge.

Simultaneously with the formation of terminal moraines, stadial riegels were forming in the valleys. They were produced by the combined action on the valley floor of a glacier trough and the water flow from beneath it. Somewhat upstream from the termination of the glacier tongue, there was an area of intense glacial erosion (Markov, 1941; Avsyuk, 1948; Evteev, 1964; et al.). Right beneath the snout of the glacier, the valley floor practically does not erode, while downstream from it the floor is strongly affected by fluvial erosion. After the glacier melts away, a riegel with well-defined relief remains in such a place, forming a ledge in the bedrock floor of the valley shaped by the glacier, on which a ridge of the stadial terminal moraine is superimposed. The "choking" effect on the valley due to a moraine-riegel complex is further intensified because at the time of glacial stagnation, the trough edges above the stadial moraine continue to be affected by intense glacial erosion whereas downstream in the valley there is no such impact and here the river erodes predominantly downwards into its bed. After the river overcomes the moraine-riegel obstacle, there normally arises a canyon- or ravine-shaped valley with narrow precipitous sides.

Under severe continental climatic conditions in the mountains of East Siberia, the contraction of Sartan glaciation seems to have been accompanied only by stoppages and thereafter only by minor re-advance of glaciers. It was the duration of the still-stands of the retreating Sartan glaciers that contributed to the formation of well-defined moraine-riegel complexes, whose unconsolidated sediments, in the presence of a perennially frozen substrate within the sides of the valleys, changed into a frozen state which further enhanced the preservation of the formations in question (Sheinkman, 1978).

After the Sartan glaciers had melted away, the valley floors of that region were affected by cryogenic processes. The most active of them is the formation of icings which not infrequently was pre-determined by the dynamic evolution of Sartan glaciers. Thus, most of the major perennial icings in the troughs of the region are associated with moraine-riegel complexes. This is largely due to the alternation of extended and contracted areas, to the presence within their floors of a large thickness of glacial and alluvium-sandr sediments, as well as due to the stepped profile of the valleys (Sheinkman & Nekrasov, 1976, 1978). In addition, present-day streams commonly flow on broad, even valley floors and meander strongly, which is not typical of mountain rivers; on such places at shoals there arise fairly large (but only seasonal) icings.

In general, in the middle and lower parts of the troughs within the moraine-riegel complex, an icing first develops on an area of the bedrock riegel which is intersected by a river. This is due to the fact that at such places the alluvium layer within the valley floor is of small thickness and within it there are no water-bearing taliks below its bed. An icing is produced owing to the waters coming from the upper portions of the valley.

After the formation of an icing on a portion of the canyon, the groundwater collector that exists upstream from the riegel in an erosional basin filled with morainal material, becomes dammed. As the pre-riegel basin is frozen, there is an increase in pressure in the groundwater, concentrated within ground-seepage taliks within these sediments (according to N. N. Romanovsky's classification, 1972). Ultimately, these waters dig through the frozen roof, forming an icing, the size of which is determined by the configuration and water-bearing capacity of the taliks (Klimovskiy, 1980). An icing can also form downstream from the moraine-riegel obstacle when there exist glaciofluvial sediments of considerable thickness which store groundwater.

It is interesting to note that if under the best conditions the number of stadial complexes in Sartan troughs of the region under consideration is as large as eight, systems of associated perennial icings occur only in some cases. This is explained by the limited amount of water available to form icings. A perennial icing complex, including the three types of icing mentioned, on occasion forming part of a single ice mass, forms only in areas with the best conditions for feeding groundwater. Generally, this occurs in middle parts of Sartan troughs, where the flow of icing-forming waters is sufficient to form a large perennial icing. Besides, the moraine-riegel obstacles in the middle parts of Sartan troughs have already been somewhat destroyed, and the flow of icing-forming water, escaping out of the pre-riegel collector, is able to transit the region of the canyon-shaped valley, and to contribute to feeding groundwater within the accumulation of glaciofluvial deposits associated with the riegel.

Downstream in the valley, the icing-forming groundwater flow is sufficient to form large perennial icings only toward the lower reaches of the troughs, especially as the formation of an icing here will be facilitated by a thinning of the glacial deposits (Preobrazhenskiy, 1963). In this case, however, the oldest moraine-riegel obstacle is usually largely destroyed, and the groundwater flow from the pre-riegel collector will largely traverse this section freely. In such a situation no icing will form upstream from the riegel.

As far as the upper reaches of Sartan troughs are concerned, the dynamics of the icing process is rarely the same, as discussed above, but more often has a quite different character. This is particularly true in cases where troughs are incised into the middle or upper terraces (for example, the upper reaches of the Garbynya River).

From the engineering-geological point of view, the formation of icings in such areas involves two aspects: hydrological and geocryological. The former implies the determination of basic trends in the association of icings with individual elements of the relief, the detection of the sources that feed them, and the delineation of thawed basins and the establishment of the nature of taliks. The latter involves the study of the geomorphological activity of an icing as a distinctive cryogenic phenomenon.

In this connection, it is interesting to elucidate some questions concerning permafrost and geohydrological behavior that reflect the main properties of the icing process within the youngest moraine-riegel complexes, and to show the ratio between icing areas and unfrozen areas.

It is known that upstream from each bedrock riegel ridge along the axis of main troughs, there occurs a pre-riegel basin. Like the entire trough bed, even in the upper reaches of troughs, this basin is always filled with morainal debris with generous porosity and permeability. Upstream in the valley the thickness of these deposits declines rather sharply, and present-day alluvium has not yet formed. Hence it is also evident that the largest amounts of water below a river bed can concentrate only within a pre-riegel basin.

However, as is shown by geomorphological research, ridges of lateral moraines between two neighboring riegels occur as a rule on the sides of the trough, and are intersected by present-day erosional valleys, for which the bottom of the main trough is a local base level of erosion. In this connection, at the mouth of the valleys there are large alluvial fans composed of a thick layer of boulder-shingle deposits. These fans not infrequently merge to form bajadas and thereby can have substantial influence not only upon redistribution of loose coarse debris, but also upon the hydrogeological regime of the waters below a river bed. This, in turn, gives rise to changes in the dynamics of the icing process and in the regime of the taliks below river beds, determining their morphology (both in horizontal and vertical sections) and icing duration.

A vertical section shows that within through floors the bedrock is covered by young alluvium while moraine suglinoks* with a large number of boulders predominate toward the sides. It should be noted that moraine deposits may lie directly on bedrock.

---

* Editor's Note: *Suglinok* is loam or loamy soil.

The flows of streams that drain glacial terraces and moraine ridges have high energy owing to steep gradients. As a result, they wash out and remove practically all finely dispersed matrix material. Therefore, alluvial fans are formed of shingly uncemented deposits containing boulders. The zone of fine washed out material exceeds substantially the width of the channels of these streams, and extends deep into the masses of sediment. Its maximum depth is observed at the upper parts of the alluvial cones, and it can vary from 3 to 10-15 m. This value increases at the mouth of each subsequent downstream tributary along the valley of the trough. This is associated with increasing depth of the erosional incision. Such wash-out zones exhibit very good collector properties, and contribute to intensive deep penetration of surface waters which warm the sediments deposited on the sides and floors of the valleys of tributary streams and of the main trough and they thereby contribute to the formation of taliks. Taliks, initially having the form of isolated capsules, grow in size and merge together, forming a talik of complex configuration extended downstream along the river valley: expansions of the talik alterate with narrowing. A maximum cross-section of collector taliks occurs within trough floors, opposite the mouth areas of erosion valleys. Downvalley the size of the talik gradually decreases, reaching a minimum near the mouth of the next tributary, i.e. where the flow velocity drops off and the alluvium of the floodplain and terraces is saturated with a sufficient amount of fine-grained material. This decreases the permeability properties of the unconsolidated deposits.

The above-mentioned morphological features of collector-taliks regularly repeat themselves downstream along the axis of the trough, with the only difference being that the size of the talik increases in both expansions and contractions. The total number of talik basins corresponds to the number of tributary erosional valleys that drain the slopes of the trough or erode the terrace surface.

Thus, the vertical section along the longitudinal axis of trough valleys shows that the lower and upper boundaries of the talik below the river bed are not patterned after the relief and inclination of the bedrock floor, but behave in a wavelike manner: the lower talik boundary rises and the upper one descends near mouths of tributary erosional valleys, i.e. at the site of alluvial fans.

As far as the morphology of a talik below a river bed as seen in plan view is concerned, a bead-like structure is clearly marked here: expansions usually correspond to wash-out zones and are fixed on the trough bottom by sections where alluvial fans are well developed, while abrupt contractions occur where fragments of above-floodplain terraces have remained. These are mainly the low interfluves of neighboring erosional tributary valleys.

From the above analysis we can draw the following conclusions.

1. In the upper reaches of troughs, within one moraine-riegel complex and icing area, there can exist one common talik or several separate taliks with a complex configuration as viewed horizontally and vertically. The area of these taliks is everywhere smaller than that of icing areas and they do not repeat their shape in a plan view.

2. Icing masses form owing to discharge of groundwater stored in several talik basins, coinciding with one icing area. Therefore, within one moraine-riegel complex an icing consists of several merging icings, the number of which is determined by that of the erosion valleys within the trough sides.

3. The formation of an icing, as a specific hydrocryogenic process, starts in the upper part of the icing area where the first frozen barrier forms with the beginning of cold weather because, in this case, the supply of water in the talik below a river bed is the smallest and the cross-section of the supply collectors is also small.

The icing formation extends gradually downvalley and reaches the riegel. The freezing taliks merge to form a single frozen body. Hydrodynamical stresses arising in the process contribute to the formation of icing frost mounds in areas upstream from the riegel.

4. The outwash zones extending far beyond the boundaries of present-day alluvial fans, being good collectors for groundwater below a river bed moving at depth through the sediments considerably warm the adjacent areas of moraine relief and of Holocene terraces. This can lead to the formation of single masses of "non-convergent permafrost".

The above mechanism of icing processes in the upper reaches of Sartan troughs does not, however, rule out formation of rather large perennial icings. This occurs when bedrock hills delineating troughs reach high absolute altitudes, and accumulate on their slopes a considerable amount of precipitation which decreases abruptly for mounttains of more moderate altitude. This situation favours an ample supply of surface water and ground water which contribute to the formation of icings. The best conditions for icing-forming water supply exist when contemporary glaciers have been preserved in the upper reaches of Sartan troughs. Here icings are steadily fed by glacier melt water and coincide with the youngest, and therefore, best-preserved moraine-riegel formations. Icings, formed upstream from a riegel and in the area of a down-cutting canyon, practically entirely intercept the icing-forming ground water flow. As a result, no icing forms downstream from the riegel, especially because glaciofluvial sediments associated with young glacial stadial formations do not reach large thicknesses, and are unable to accumulate considerable reserves of ground water.

Because in the mountains of East Siberia the clastic material in valleys (under conditions of neo-tectonic uplifting of rock structures) is of considerable thickness, particularly within Sartan troughs, these are the processes of talik formation that determine the formation of icings. But this icing formation, as shown above, is then to a considerable extent due to the dynamics of Sartan glaciers and current denudation.

As for tectonics, however, the contribution of which to the formation of icings is unquestioned (Simakov, 1961; Tolstikhin, 1966; et al.), we will remark here that Sartan troughs occur mostly within the region of deeply circulating ground water, and only the lower reaches of valleys can intersect the flow lines of this water discharge. These situations frequently occur in the near-margin parts of inter-mountain depressions, the floors of which, in turn, had commonly been the source for propagation of Sartan glaciers. Thus, on the one hand, part of the waters from the ground water collectors, produced by the activity of Sartan glaciers in the upper and middle parts of these troughs, can go to supply the deep-circulation waters along tectonic fractures that break the integrity of permafrost. On the other hand, in the lower reaches of troughs, such collectors may be intersected by deep water circulation, and therefore the formation of icings there will be still more intense.

Thus, icings in Sartan troughs of mountain regions of the Pacific belt represent part of a regulated drainage system, corresponding to a given stage of evolution of nature. Any changes in the regulation of this system will immediately entail an increase in the icing effect in some places, and conversely, a decrease of icing activity in others. In the end, this will cause the icings to migrate to benches of moraine-riegel complexes, and the migration will go on until the entire system is in equilibrium with the new conditions. One of the most important regulators of this process is the deglaciation that caused a stadial contraction of the Sartan glaciation (Sheinkman & Nekrasov, 1976, 1978).

As is known (Alekseev, 1973; Romanovsky, 1973), icings contribute much to the transformation of valleys. Therefore, when the icing system, as deglaciation develops, migrates along the valley, the earlier positions of icings are marked by relict sites of annual icings, yet are free from ice and well defined in the relief. As a result, the bench-like profile of valleys in areas of moraine-riegel complexes, as well as the whole appearance of troughs, becomes more sharply expressed.

After the site of annual appearance of an icing changes into a relict state, taliks within the valley floor deposits start to freeze up. For some time thereafter, ground water still moves through them but they are usually no longer capable of breaking the frozen talik roof, because most of the icing-forming flow is intercepted by newly emerged icings. In this case within unconsolidated deposits, there arise only layered deposits of injection ice.

From the aforesaid, we can conclude that along with general regularities of the formation of fluvial valleys, in glaciated regions of the Pacific belt, this process will exhibit features of its own. These features are determined by the peculiarities of interaction of tectonic, glacial, and cryogenic processes. Icing formation and icing glaciation are the most active external agents determining the present-day appearance of the valleys. The most important regulator of development of both phenomena proved to be the stadial process, developing throughout the Holocene.

The above considerations confirm once more that numerous icings and sources feeding them are closely related, not only to the area of a water collecting basin and the river flow rate, but also primarily related to the geological and geomorphological structure of regions, and the position of various sections in relief (Shvetsov, 1968).

To date we have sufficient observational data to state that in high-altitude regions of the Verkhoyansk-Kolyma Mountain Range, as well as in the mountains of the North Trans-Baikal, the distribution of icings allows us to identify an "icing belt" (according to O. N. Tolstikhin, 1974), confined to an altitude range with an upper boundary 400-900 m lower than the altitude of drainage basin divides, and a lower boundary observed about 50-100 m below the regional base level of erosion.

The formation and existence of an "icing belt" is due not only to the properties of the geological structure of the region, but also due to its evolutionary past, especially during the Quaternary epochs, i.e. Pleistocene and Holocene, for which development of glaciation over considerable areas was typical.

The formation of "icing belts" is closely linked geomorphologically with the activity of Pleistocene glaciers, and in some regions (Ulakhan-Chistai Range, Kodar Range) of present-day glaciers as well. Consequently, geomorphological activity of glaciers implies not only glacial erosion and general planation of the relief, but also the formation of a great variety of moraine-riegel complexes. It is a series of such complexes that produces an "icing belt".

REFERENCES

Alekseev, V. R., 1973, Icings as a factor of valley morpholithogenesis, in Regional geomorphology of Siberia, Irkutsk, p. 99-135

Alekseev, V. R., 1974, Icings of Siberia and Far East, in Siberian Geogr., v. 8, Novosibirsk: "Nauka", p. 5-68.

Avsyuk, G. A., 1948, On some questions of glaciology, in Problems of physical geography, v. 8, Moscow-Leningrad, izd. AN SSSR, p. 122-144.

Evteev, S. A., 1964, Geological activity of glacier cover of Eastern Antarctic region, Moscow: "Nauka", 120 p.

Kalabin, A. I., 1960, Permafrost and hydrogeology of the North-East of the USSR, Magadan, 472 p.

Klimovskiy, I. V., 1980, Icing formation as a periglacial process. Proc. of the All-Union Meeting on Marginal Formations, Kiev, p. 15-17.

Markov, K. K., 1940, Erosion of glaciers and mountain relief, in Problems of physical geography, v. 10, Moscow-Leningrad, izd. AN SSSR, p. 75-86.

Nekrasov, I. A., Maksimov, E. V., and Klimovskiy, I. V., 1973, The last glaciation and cryolithozone of the Southern Verkhoyansk, Yakutsk, 152 p.

Preobrazhenskiy, V. S., 1963, Icings and ancient glaciation of Stanovoi Upland—West Transbaikal. Division of the Geogr. Soc. USSR, v. 22, p. 131-134.

Romanovskiy, N. N., 1972, Taliks in the permafrost zone and schemes of their classification—Vestn. Mosk. un-ta Geol., no. 1, p. 23-24.

Romanovskiy, N. N., 1973, On geological activity of icings, in Permafrost research, v. 8, Moscow, izd. Mosk. un-ta, p. 66-89.

Sheinkman, V. S., 1978, The influence of climatic fluctuations on the relief of Sartan troughs in mountainous regions of the North-East of the USSR, in Climatic factor of relief formation, Kazan, izd, Kazanskogo un-ta, p. 89-90.

Sheinkman, V. S. and Nekrasov, I. A., 1976, Near-glacier icing phenomena of Cherskiy Range, in Hydrogeological conditions of permafrost zone, Yakutsk, p. 86-96.

Sheinkman, V. S. and Nekrasov, I. A., 1978, Icings of the near-glacier zone of East Siberia mountains, in General geocryology. Contributions to the 3rd International Conference on Permafrost, Novosibirsk: "Nauka", p. 151-156.

Shvetsov, P. F., 1968, Some regularities of hydrogeothermic processes in the Extreme North and North-East of the USSR, Moscow: "Nauka", p. 110.

Simakov, A. S., 1961, Icings of the North-East of the USSR, in Third Meeting on Ground Water and Engineering Geology of Siberia and the Far East, v. 2, Irkutsk, p. 117-119.

Tolstikhin, O. N., 1966, Icings and geotectonics of the North-East of Yakutia. Sov. Geol., no. 8, p. 106-119.

Tolstikhin, O. N., 1974, Icings and ground water of the North-East of the USSR, Novosibirsk: "Nauka", 164 p.

THE EFFECTS OF GRAVEL ROADS ON ALASKAN ARCTIC COASTAL PLAIN TUNDRA

Lee F. Klinger, Donald A. Walker, and Patrick J. Webber

Institute of Arctic and Alpine Research
and Department of Environmental, Population, and Organismic Biology
University of Colorado, Boulder, Colorado 80309 USA

Gravel roads on the Alaskan Arctic Coastal Plain may induce considerable habitat alteration through snowbanks, flooding, and dust. An integrated sampling approach is taken to delineate the magnitude and extent of these impacts on the tundra communities along the West Road at Prudhoe Bay, Alaska. After 1 year of monitoring, minor vegetation changes are noted. Snowbanks within 10 m of the road are 5-10 times deeper than the natural snowpack. Unlike roads with heavy winter traffic where dust on the snow surface decreases albedo which hastens snowmelt, the West Road receives no winter traffic and snowbanks persist for many weeks after natural snowpack melt. These snowbanks block culverts which causes extensive early summer flooding which is the predominant impact of the road, inducing greater depth of thaw and greening of the vegetation. The level of dust fallout along the road is relatively low, only about one-third of that measured along other roads in the area. A matrix of predicted impacts is presented and certain mitigative measures are suggested for gravel road construction and maintenance.

INTRODUCTION

The present approach of building gravel roads in the Arctic is to build the roadbed thick enough to prevent thawing of the underlying tundra. In northern Alaska, 2-m thick roadbeds are generally used. Although this design is successful in maintaining the integrity of the roadbed, there are certain consequences which may drastically alter the adjacent physical and biological environments. The three main effects of these roads on the adjacent tundra are: a deepening of snowpack, the impedance of surface drainage with resultant flooding and dust fallout. These effects have been noted for several years by numerous authors (Benson et al. 1975, Everett 1980, Walker and Werbe 1980, Spatt and Miller 1981). There is still, however, a lack of understanding of the integrated system of road-tundra interactions. This paper integrates the observations from earlier studies with provisional results of a multiyear, monitoring study documenting the abiotic and biotic changes adjacent to a gravel road on wet Arctic Coastal Plain tundra at Prudhoe Bay, Alaska. This work was performed for the Department of the Army, Alaska District, Corps of Engineers, Anchorage. The monitoring program is sponsored by ARCO Alaska, Inc. and Sohio Alaska Petroleum Company and is required to fulfill permitting stipulations. The program was managed by Woodward-Clyde Consultants in 1981-1982, and by Envirosphere Company in 1982-1983.

Studies of this kind are useful in understanding the cumulative impacts of roads on tundra communities and in determining reliable mitigative measurements for road construction and maintenance. The additive effects of road-induced snowbanks, flooding, and dust on the vegetation result in notable losses and gains in wildlife species through habitat alteration. By examining road usage and design in the context of habitat changes, sound recommendations can be made in route selection, culvert location, and road maintenance.

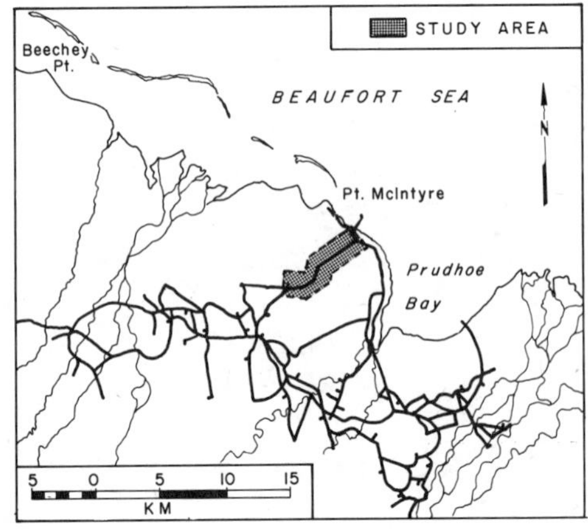

FIGURE 1 Location of the West Road study area (lat. 70° 22', long. 148° 36') at Prudhoe Bay, Alaska. Heavy lines denote gravel roads.

The new results were gathered along the West Road (Figure 1). It was constructed in the winter of 1980-1981 and extends from the Arctic coast for 7 km southwest, traversing an area of continuous permafrost. The area is flat with numerous shallow lakes and drained lake basins. The vegetation, soils, and landforms of the region are described by Walker et al. (1980).

Monitoring of the West Road impacts is being

accomplished with four principal methods: mapping, road transects, dust measurement, and permanently marked study plots. The mapping approach utilizes vertical aerial photographs to delineate the extent of snowbanks and flooding (Figures 2a and 2b, respectively). A dust distribution map (Figure 2c) was constructed by plotting isolines of equal dust fallout using the results of the dust measurement. Road transects are used to monitor changes in snow, water, and thaw depths adjacent to the road. Dust quantity and quality measurements are being made on material collected in dust settlement traps located at each plot. Traffic and meteorological monitoring are being performed as well. Permanent study plots are being sampled nondestructively for vegetation cover, community composition, and numerous soil and environmental variables. Plots are distributed on both sides of the road evenly between impacted, experimental and nonimpacted control areas and among the three main tundra moisture regimes: dry, moist, and wet.

MAJOR EFFECTS OF THE ROAD

Snowbanks

Benson et al. (1975) have reported on snowbank formation along roads in the Prudhoe Bay region. They observed that roads oriented perpendicular to the prevailing winter winds generated deep and extensive snowbanks, whereas roads running parallel to the winds did not cause a deepening of the natural snowpack.

Changes in snow depth with distance to the road are characterized by transect measurements (Figure 3) which show a deeper, more extensive snowbank on the west (downwind) side. The winds that built the 1982 snowbanks apparently prevailed from an easterly direction with strong secondaries from the west. This accumulation pattern is somewhat different from the results of Benson et al. (1975), who found deeper snowbanks on the east side of the roads. During the winter of their observations, most of the storms with significant snowfall had winds from the west and southwest. There is considerable variation among years with respect to wind patterns and total snowfall, and both of these factors are important with regard to the size and position of the road snowbanks.

The snow distribution map for 1982 (Figure 2a) indicates that the majority (over 80%) of the study area is under patchy snow cover or is entirely snow free just prior to breakup. It is the snow free areas that are most utilized by early breeding bird species. Snowbank formation due to the road has clearly diminished the extent of this habitat in the early summer. Road-augmented snowbanks overlie 0.65 km^2 of the 22.21 km^2 study area which may be better expressed as 9.2 ha km^{-1} (37 acres mi^{-1}) of road.

One of the main effects of snowbanks is the blockage of culverts. This effect is reduced by heavy winter traffic when dust accumulation on the snow surface near a road decreases albedo and snowmelt is hastened. Even though snowbanks next to the elevated roads may be 5-10 times deeper than the average winter snowpack as a result of drifting and/or plowing, these are often the first areas to melt (Benson et al. 1975). Conversely, roads with little or no winter traffic have insignificant dust fallout and snowbanks may persist for many weeks after natural snowpack meltout and completely block culverts. This situation can cause the deepest and most extensive flooding following spring runoff.

Certain plants are likely to respond favorably to increased snow accumulations. Prostrate willows, lichens, horsetails, and numerous forbs flourish in snowbanks of the Prudhoe Bay region (Walker 1981). It should be noted, however, that snowbanks associated with roads are quite different than most natural snowbanks in the Prudhoe Bay region. Most of the larger natural snowbanks are associated with pingos, creek bluffs, and other areas of high relief which are well drained. These natural snowbank areas are relatively well drained, and the vegetation has a mesic or xeric character. Road-associated snowbanks, in contrast, commonly occur in the very wet areas since this is the most common moisture regime on either side of the road.

A notable consequence, so far, of snowbanks is a significant suppression of graminoid cover in wet areas. Snow is naturally deeper on wet plots and additional snow is likely to delay vascular plant phenology to a measurable extent. Since graminoids make up over 95% of the vascular plant cover in wet plots, this is the growth form in which the change is manifested most.

Flooding

Flooding is the predominant impact of the West Road both in terms of areal extent and observable and measurable change. Extensive flooding can occur during snowmelt in areas where a drainage basin is traversed by an elevated gravel roadbed. Normally, culverts in the roadbed at the tundra surface can prevent deep ponding if they are designed, installed, and maintained properly. However, numerous problems render many culverts inoperable. Roadway compaction can deform culverts, causing the ends or middle to bow (Berg 1980). Ends of culverts can clog with gravel from road shoulder grading (Berg 1980). Early in the summer, snow and ice may block the passage of water through the culverts. Walker's (1981) study of the vegetation of the Prudhoe Bay region noted that thermokarst was associated with road-induced flooding. Ponding caused by the Dalton Highway (formerly called the TAPS Haul Road) has been well documented by Berg (1980). He also discussed some possible effects of ponding on permafrost and roadbed degradation.

Along the West Road, 99% of the road-related flooding occurs on the east side. This is due to the main axis of the road running obliquely to the northerly direction of surface water flow. The area of flooding shown in Figure 2b is 1.34 km^2 or 18.2 ha km^{-1} (75 acres mi^{-1}) of road. In 1982, additional culverts were properly placed along sections of the road which brought about more rapid drainage of the impoundments than in 1981.

Early season water levels are very high in areas next to the east side of the road, but drop off rapidly by mid-season to depths slightly above

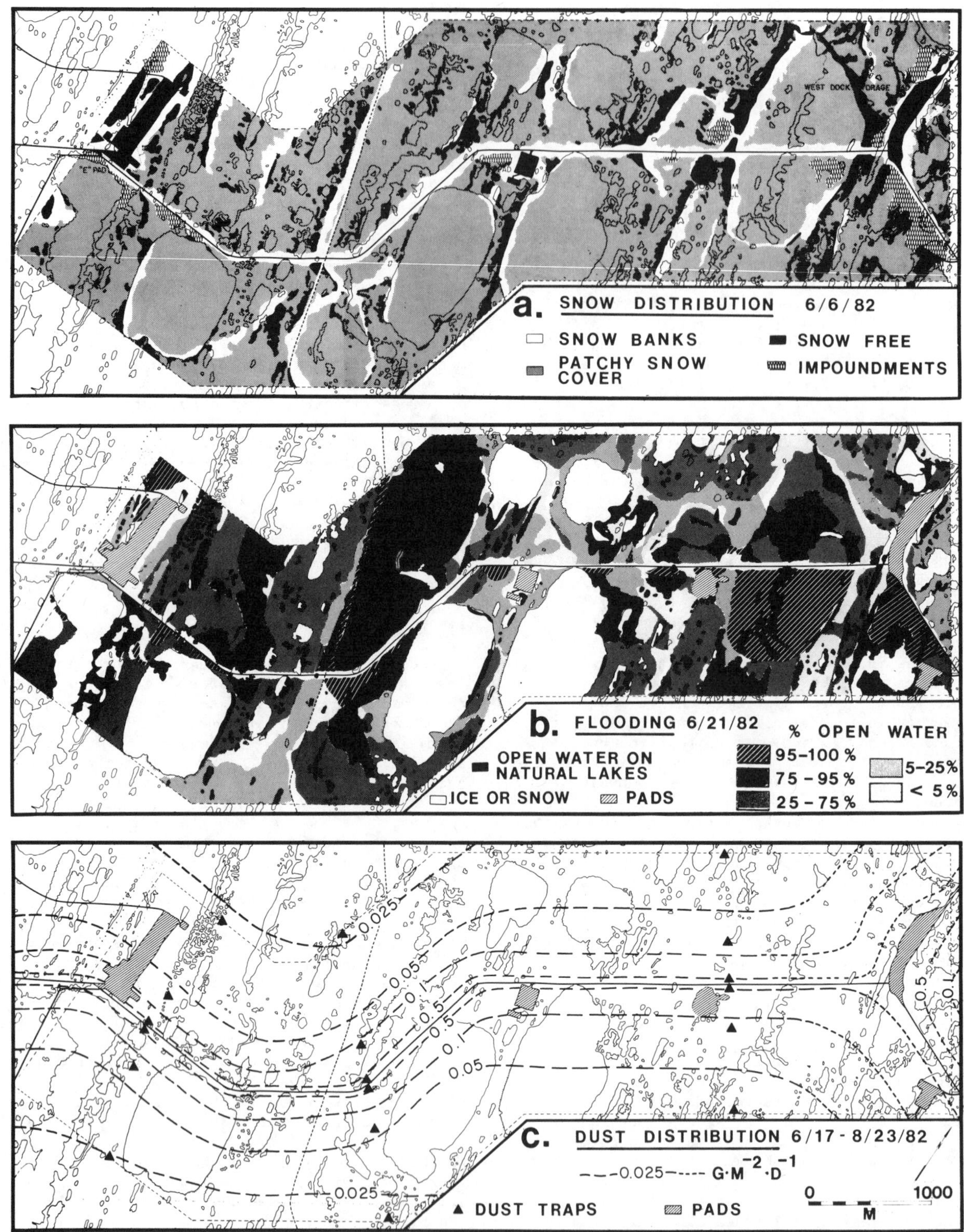

FIGURE 2 West road impact maps. (a) Snow distribution shows extensive snowbanks adjacent to roads and pads. (b) Flooding maps shows flooding concentrated on the southeastern side of the road. (c) Dust distribution map delineates isolines of dust fallout.

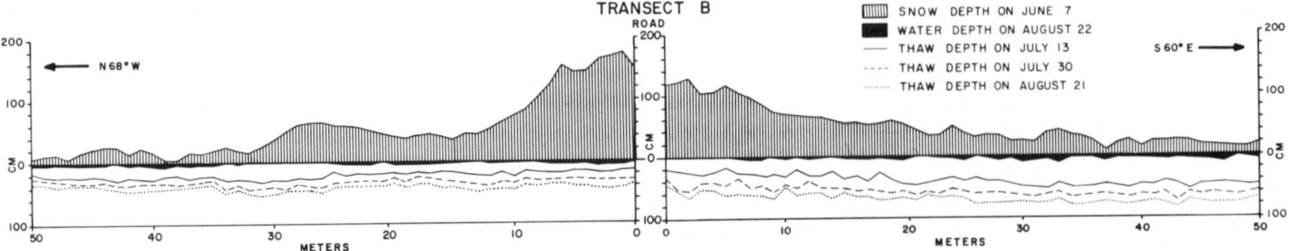

FIGURE 3 Snow, water, and thaw depths at Transect B in 1982. Note the deeper snowbank on the northwest (downwind) side, and the greater thaw depths on the southeast (flooded) side of the road.

those in control areas (Figure 4). Flooding also induces greater thawing of the active layer. A flooding index was calculated for each permanent plot. This index is based on the depth of flooding, the duration of flooding, and the normal (preroad) water levels. End of season active layer thickness is highly correlated ($p < 0.001$) with this index. Similar increases in depth of thaw due to flooding have been noted by other authors (e.g., Berg 1980).

The most apparent effect of flooding on plants is a deep greening of the vegetation evident later in the summer when water levels are at a minimum. This probable increase in graminoid production is likely caused by several factors including warmer temperatures associated with ponds, increased nutrient availability and changes in phenological patterns. The assumed increase in plant productivity in greening areas has not yet been demonstrated; however, changes in cover of certain plant categories can be correlated with the flooding index. In moist plots, which are not regularly subject to persistent standing water, lichens and shrubs are responding unfavorably to flooding, whereas bryophytes, which often flourish in a water-saturated environment, appear to be benefiting from an enhanced water regime.

Road Dust

Everett (1980) has studied the physical and chemical properties of road dust and its distribution patterns at Prudhoe Bay and along the Dalton Highway. He has demonstrated clearly the log-log relationship between the mass of dust fallout and distance to the road. The effects of road dust on the vegetation along the Dalton Highway have been reported by Walker and Werbe (1980), and Spatt and Miller (1981). They have shown that heavy dust loads are detrimental to many mosses and lichens and certain vascular plants, such as *Cassiope tetragona* and *Lycopodium annotinum*.

Dust fallout along the West Road in 1982 was only one-third of that measured along other roads in the Prudhoe Bay region by Everett (1980). This is the result of restricted access to the West Road throughout most of the year. Most of the dust accumulating along the road was produced during July and August when traffic was unrestricted. Dust was distributed fairly evenly on both sides of the road (Figure 2c) and followed the log-log relationship mentioned above.

Walker (1981) has described the two distinct vegetation regions at Prudhoe Bay. The first is

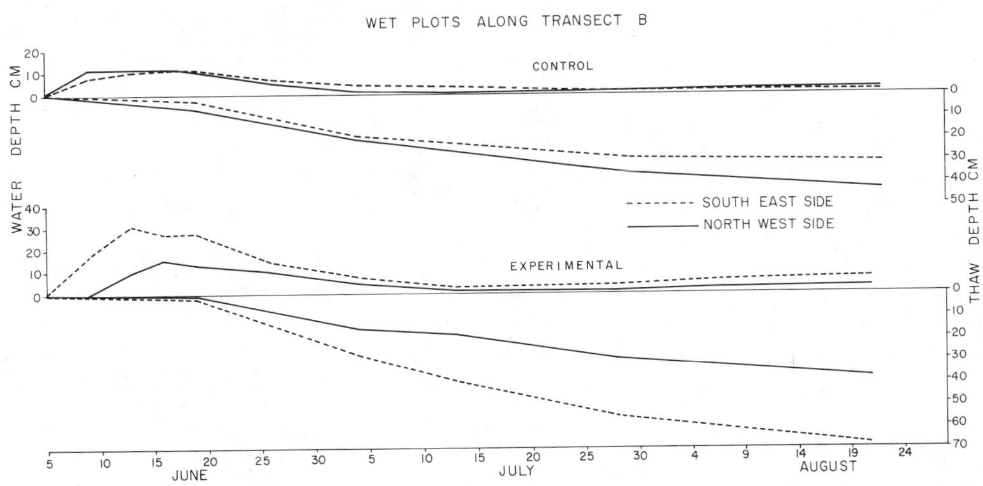

FIGURE 4 Seasonal progression of water and thaw depths at the wet plots along Transect B in 1982. Southeast side experimental plots show greater water and thaw depths due to flooding.

TABLE 1 West Road Vegetation Impacts Observed After 2 Years and Expected Impacts After 10 Years.

Vegetation Type		Impact Type: Snowbanks	Impact Type: Flooding	Impact Type: Dust[a]
Dry Prostrate Shrub Tundra	Impact after 2 years:	No negative effects Increase in lichens. (?)	None.	Little or no effect.
	Expected impact after 10 years:	Will probably be difficult to detect snow effects because of overriding impact of dust.	Areas of high-centered polygons will have thermokarst in the troughs. Tops of polygons will be eroding into the troughs.	Near the road most mosses and lichens will be smothered by dust. Low growing forbs will be doing poorly. Sedges and low shrubs will be doing relatively well.
Moist Graminoid Tundra	Impact after 2 years:	No negative effects Possible increase in bryophyte and forb productivity.	Decrease in lichens and *Dryas* Increase in bryophyte production.	Possible increase in forbs.
	Expected Impact after 10 years:	Will probably be difficult to detect snow effects because of the overriding impact of dust. Willows near the road may increase in stature due to winter protection afforded by the snow bank.	Most moist areas will not be impacted. Where they are, *Dryas integrifolia*, lichens and many mosses will be eliminated. Sedges and hydrophilous mosses will increase.	Less tolerant taxa that will decrease include *Cassiope tetragona*, *Tomenthypnum nitens*, *Catascopium nigritum*. These may be eliminated near the road. More tolerant taxa include *Dryas*, *Drepanocladus* spp., *Campylium stellatum* willows and sedges.
Wet Graminoid Tundra	Impact after 2 years:	Decrease in standing crop algae and live vascular plants. Difficult to separate snowbank effects from flooding effects.	Decrease in graminoids in the deepest flooded areas, also decreases in algae, live vascular plants and standing crop. Greening of graminoids and possible increase in production in some flooded areas.	Increase in algae, possible increase in standing crop.
	Expected impact after 10 years:	Will probably be difficult to detect snow effects because of the overriding impact of flooding. Species composition of unflooded wet areas with deep snow will change very little.	Deepest flooded areas will increase in depth due to thermokarst. Graminoid production will probably decrease in these areas. In less deeply flooded areas, many forbs and mosses will decrease or be eliminated. Areas flooded only during the early season melt will not change much.	Some species of moss such as *Catascopium nigritum* will be eliminated near the road. Other dominant mosses such as *Drepanocladus* and *Scorpidium* are relatively tolerant of dust.
Wet Graminoid Tundra	Impact after 2 years:	Decrease in standing crop algae and live vascular plants. Difficult to separate snowbank effects from flooding effects.	Decrease in graminoids in the deepest flooded areas, also decreases in algae, live vascular plants and standing crop. Greening of graminoids and possible increase in production in some flooded areas.	Increase in algae, possible increase in standing crop.
	Expected impact after 10 years:	Will probably be difficult to detect snow effects because of the overriding impact of flooding. Species composition of unflooded wet areas with deep snow will change very little.	Deepest flooded areas will increase in depth due to thermokarst. Graminoid production will probably decrease in these areas. In less deeply flooded areas, many forbs and mosses will decrease or be eliminated. Areas flooded only during the early season melt will not change much.	Some species of moss such as *Catascopium nigritum* will be eliminated near the road. Other dominant mosses such as *Drepanocladus* and *Scorpidium* are relatively tolerant of dust.
Aquatic Graminoid Tundra	Expected impact after 10 years:	No effects.	Generally no effects, except in deeply flooded areas that fail to drain because of lack of culverts. Here all plants are likely to be eliminated and permanent lakes formed. Areas with *Arctophila fulva* will be unaffected.	No effects except immediately adjacent to the road where some aquatic mosses may be smothered by dust.

[a] Projected dust impacts after 10 years are based on observations from along the Prudhoe Bay Spine Road and the Dalton Highway haul road (Walker and Werbe 1980).

an area of alkaline tundra associated with loess from the Sagavanirktok River, and the second is from acidic tundra that is outside the main area of loess deposition. The study area lies entirely within the region of acidic tundra. This situation is important when considering the effects of road dust. The roadbed material is quarried from calcareous alluvial deposits. The high pH of the dust released from traffic across this roadbed ultimately can be expected to change the acidic tundra by reducing or eliminating many of the acidophilous plant taxa.

Certain species in the region of alkaline tundra are preadapted to a more calcareous environment and experience no change or even an increase in productivity next to the road. Conversely, acidophilous species, such as *Sphagnum* spp. and *Dicranum* spp., are likely to be eliminated near the road. Along older roads in northern Alaska, many cryptogams have been eliminated due to smothering and toxicity effects of road dust (Walker and Werbe 1980). By removing this important insulating layer, thermokarst potential is greatly enhanced.

The impacts of dust from the West Road on the vegetation at this time are slight. The only noteworthy effect of dust to date is an increase in the alga *Nostoc commune* which is apparently responding to a changed nutrient regime near the road. We expect that with time, other dust-related changes will be apparent.

CONCLUSIONS

The monitoring of the road impacts has shown that flooding is the major impact both in terms of magnitude and extent. The most important abiotic effect noted at this time is a significant increase in the active layer of flooded areas. Snowbanks and dust are currently relatively minor impacts along the West Road.

It is important to note that the West Road is very different in many respects from other roads at Prudhoe Bay. The restricted vehicle access is causing low levels of dust fallout. The road is located in an area of tundra that is somewhat wetter and more acidic than most of the region. Furthermore, the road runs obliquely to the drainage patterns, causing more flooding than other roads in the area which parallel the direction of drainage. Due to these high levels of flooding, considerable attention has been paid to the number and placement of culverts in the West Road.

Based on the preliminary results of the West Road study along with many years of observations around the Prudhoe Bay region, a predictive impact matrix is presented which summarizes the short-term and long-term effects of gravel roads on tundra vegetation (Table 1). Measurable changes in the vegetation, so far, have been slight. Most tundra plants are long-lived perennials with large below-ground food reserves and are likely to persist for many years after the habitat has changed. However, road effects are likely to be cumulative over many years, and vegetation changes may occur very quickly once a critical level of impact is reached.

To reduce the level of impact along gravel road, certain mitigative measures are recommended. First, culverts should be installed during the summer thaw season when drainage patterns can be more readily observed, rather than during the winter which is the current practice. Second, culverts should be kept free of snow and ice during snowmelt to prevent extensive flooding. This can be accomplished by removing snow from culvert openings and by using steam pipes to prevent ice clogging in the culvert. Third, dust suppression techniques should be used on a regular basis.

Future plans for development on the Arctic Coastal Plain call for many hundreds of kilometers of gravel roads. The potential impacts of this extensive road system on wildlife and habitat are immense. With a good understanding of road-tundra interactions derived from integrated studies on existing roads, well-devised mitigative measures can be applied to minimize future impacts.

REFERENCES

Benson, C., Holmgren, B., Timmer, R., Weller, G., and Parish, S., 1975, Observations on the seasonal snow cover and radiation climate at Prudhoe Bay, Alaska during 1972, *in* Brown, J., ed., Ecological investigations of the tundra biome in the Prudhoe Bay region, Alaska: Biological Papers of the University of Alaska Special Report No. 2, p. 12-50.

Berg, R. L., 1980, Road performance and associated investigations, *in* Brown, J. and Berg, R.L., eds., Environmental engineering and ecological baseline investigations along the Yukon River-Prudhoe Bay Haul Road: U.S. Army Cold Regions Research and Engineering Laboratory, Hanover, N.H., CRREL Report 80-19, p. 53-100.

Everett, K. R., 1980, Distribution and properties of road dust along the northern portion of the Haul Road, *in* Brown, J. and Berg, R. L., eds., Environmental engineering and ecological baseline investigations along the Yukon River-Prudhoe Bay Haul Road: U.S. Army Cold Regions Research and Engineering Laboratory, Hanover, H.H., CRREL Report 80-29, p. 101-128.

Spatt, P. D. and Miller, M. C., 1981, Growth conditions and vitality of *Sphagnum* in a tundra community along the Alaska Pipeline Haul Road: Arctic, v. 34, p. 48-54.

Walker, D. A., 1981, The vegetation and environmental gradients of the Prudhoe Bay Region, Alaska: Doctoral thesis, University of Colorado, Boulder, CO, 484 p.

Walker, D. A., Everett, K. R., Webber, P. J., and Brown, J., 1980, Geobotanical atlas of the Prudhoe Bay region, Alaska: U.S. Army Cold Regions Research and Engineering Laboratory, Hanover, H.H., CRREL Report 80-14, 69 p.

Walker, D. A. and Werbe, E., 1980, Dust impacts on vegetation, *in* Brown, J. and Berg, R. L. eds., Environmental engineering and ecological baseline investigations along the Yukon River-Prudhoe Bay Haul Road: U.S. Army Cold Regions Research and Engineering Laboratory, Hanover, H.H, CREEL Report 80-19, p. 126-127.

ALPINE PLANT COMMUNITY COMPLEX ON PERMAFROST AREAS OF THE DAISETSU MOUNTAINS, CENTRAL HOKKAIDO, JAPAN

Takeei Koizumi

Department of Geography, Tokyo Gakugei University, Koganei City, Tokyo, 184 Japan

On the summit areas of the Daisetsu Volcanic Mountains, the existence of permafrost is recognized. In this area an alpine plant community complex is distributed widely. It is composed of alpine windblown herb-heath, an alpine stony desert plant community, and two kind of Salix (Salix pauciflora and Salix yezoalpina), which are thought to grow in a humid habitat. To clarify the cause of this complex, the depth of the active layer in mid-July, vertical changes of soil temperature, and the depth of the water table were examined. It became clear that the depth of the water table was the only factor controlling the distribution of the two species of Salix, though there were no relations between the distribution of windblown vegetation and three conditions. Salix is limited to the humid places of the shallow water table, less than 35 cm in mid-July. They do not emerge in the dry places of the low water table. Owing to the rise of the water table caused by the permafrost, two species of Salix are thought to have mixed in windblown vegetation, which is ordinarily a dry type plant community. And as a result, the alpine plant community complex is formed.

INTRODUCTION

There are some investigations on the influences of permafrost on vegetation. Johnson (1966) suggests the following effects of permafrost; the impeder of the drainage of water, the cooling of the surficial soil, the making of the surficial irregulalities and the provision of an impervious substrates which reject the penetration of the plant root. Ronning (1969) investigates the relationships between microtopography, permafrost and vegetation in Spitzbergen. He showes that the Dryas community occupy the dry ridges with the low groundwater level, though the Deschmpsia community grows in the wet depressions with the high groundwater level. Dingman and Koutz (1974) studies the relationships between the forest types and the distribution of permafrost, in the discontinuous permafrost areas of Alaska. Brown (1970) makes clear that the terrain formed by permafrost decides the vegetation in palsa areas. In any case, the groundwater level is considered to be very important(Ito 1978, Peterson and Billings 1980, etc.).

In Japan the distribution of permafrost is recognized in only two places: one is the summit area of Mt. Fuji and the other is the summit areas of the Daisetsu Mountains (Fukuda and Kinoshita 1974). However, the influence of permafrost on the alpine vegetation has not been studied yet. This paper discusses the influences of permafrost on the alpine vegetation, especially on the composition of plant communities. The studied area is the summit area of Mt. Koizumitake (2,158 m) in the eastern part of the Daisetsu Mountains.

OUTLINE OF STUDIED AREA

The Daisetsu Mountains are located in the central part of Hokkaido, the northernmost part of Japan. It is a plateaulike large volcano composed of many volcanic cones and pyroclastic or lava flow plateaus. The altitude of these cones and plateaus is about 2,100-2,300 m a.s.l.(Figure 1). Gentle slopes are conspicuous (Plate 1).

The forest line runs at about 1,600 m and the above belongs to the alpine zone. In the alpine

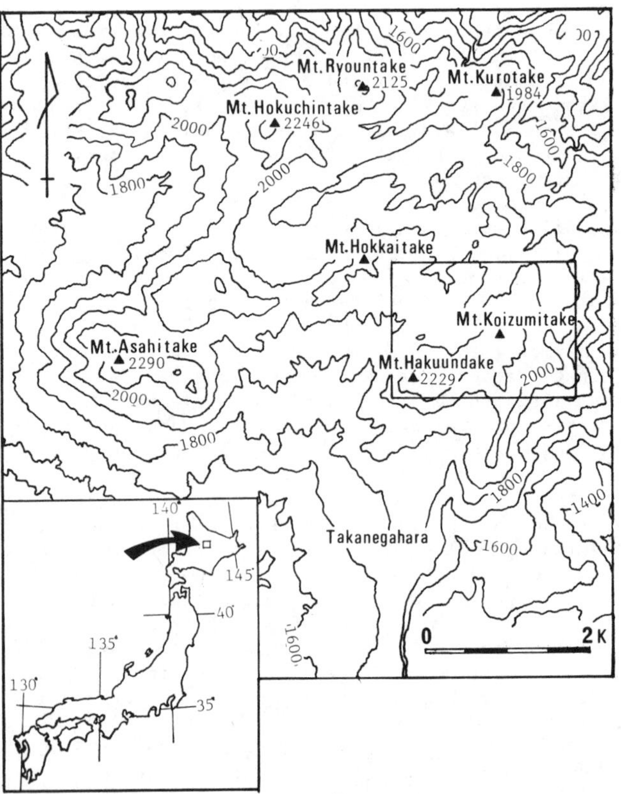

FIGURE 1 Location of studied area.

PLATE 1 View of the Daisetsu volcano.

zone Pinus pumila scrubs are eminent. However, in summit areas or areas along the main rounded ridges there are periglacial debris fields and windblown herb-heath and alpine stony desert plant communities. They emerge as summit phenomena. Because of strong northwest monsoons, there is less snow in winter on summit areas. So Pinus pumila cannot grow in such places and windblown herb-heath and alpine stony desert plant communities emerge instead. There is frost action and many kinds of patterned ground, for example, sorted stripes, solifluction lobes, or turf-banked terraces, are distributed (Koaze 1965).

Mt. Koizumitake is one of the rounded peaks of the Daisetsu volcanos (Plate 2).

PLATE 2 View of the Mt. Koizumitae from Hokkaidaira
distant view: Mt. Koizumitake
foreground: Hokkaidaira (pyroclastic flow plateau)

COMPOSITION OF PLANT COMMUNITIES OF MT. KOIZUMITAKE

The author set the 21 quadrant on the eight points for vegetation surveying from the top of the Mt. Koizumitake to the gentle col between Mt. Koizumitake and Mt. Hakuundake (Figure 2). One quadrant is 2 x 2 m width. Results of vegetation

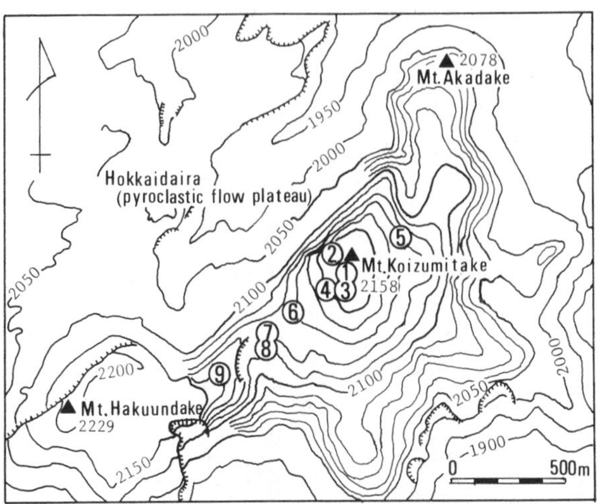

FIGURE 2 Topography of Mt. Koizumitake and the location of surveying point.

surveying are shown in Table 1.

Generally speaking, the vegetation cover is poor and the periglacial debris fields are extensive (Plate 3). The main plants are Diapensia lapponica var. obovata, Carex flavocuspis, and Loiseleuria procumbens, etc., which make up the windblown herb-heath. However, reflecting the complexity of the habitat, the components of alpine stony desert plant community are also present. These species, for example, Viola crassa, Dicentra peregrina var. pusilla, etc. grow on unstable habitats as the part of the fine-size gravels of solifluction lobes.

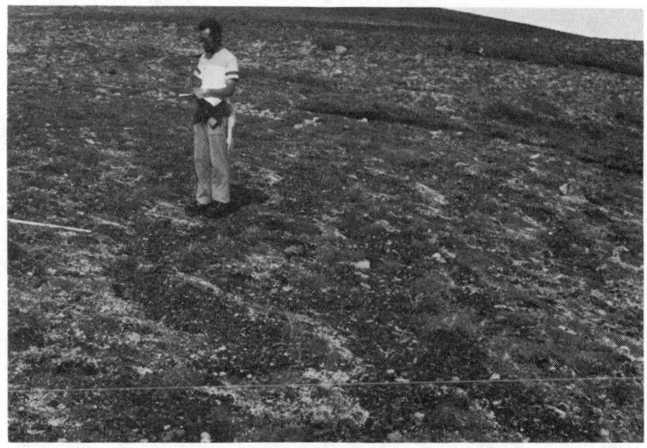

PLATE 3 Plant communities and periglacial debris fields on Mt. Koizumitake.

In addition, two species of Salix (Salix pauciflora and Salix yezoalpina) mix in these plant communities. They are distributed on points 2, 4, 5, 6, and 8. Usually Salix is thought to grow in a rather humid habitat. In the Japanese Alps they are not distributed on the windblown areas, which are usually dried. So this distribution phenomenon means that the plants of humid habitat get mixed with the dry-type windblown plant communities.

TABLE 1 Composition of Plant Communities on Permafrost Areas of Mt. Koizumitake.

point number	1		2		3			4			4-A		5		6			7	8	9	
altitude	2,159		2,155		2,155			2,145					2,130		2,125			2,121	2,119	2,140	
direction	-		-		SSW			SW					SE		W			S	S	E	
inclination	flat		flat		7			9					8		4			7	10	3	
height of herbs (cm)	2-10		2-15		2-5			2					2-10		5			2-8	2-8	2-8	
vegetation coverage (%)	20		25		40			20			75		35		20			50	50	60	
quadrant	a	b	a	b	a	b	c	a	b	c	d	e	a	b	a	b	c	a	a	a	b
number of species	9	10	15	16	12	13	12	11	11	14	10	13	16	14	13	7	11	13	16	16	15
Diapensia lapponica var. *obovata*	10%	30	10	5	30	30	30	20	20	20	60	50	30	25	10	5	2	30	20	40	50
Carex flavocuspis		2	1		15	5	10	1	1	4	40	30	6	6	2	2	1	1	2		1
Loiseleuria procumbens		1			2	2	2	2	2	2	2	2	1	2	5		2	20	30	4	4
Carex blepharicarpa var. *dueensis*	15	10		1	2	1	5	1	2	2		20	1	2	2	10	4	6	6	25	6
Minuartia macrocarpa var. *minutiflora*			10	8																	
Saussurea yanagisawae	2		1	5	1	2	2	1	1	1	2	1	6	4	5	5	5	2	4	1	2
Potentilla matsumurae	1	1	1	2	1	1	1	1	1	1	1	1	1		1	1		1	1	1	1
Arcterica nana	1	2	1		1	1	1	1		1	2	1					1			1	1
Artemisia trifurcata var. *pedunculosa*	1		2	2							6	2	2	1	1	2	1		1	6	6
Bryanthes gmelinii													1	10				6	2		
Deschampsia flexuosa				1	1	1	1			1		1	1	1					1		
Oxytropis japonica var. *sericea*																				10	10
Cassiope lycopodioides					1	1		1		1			2	1						1	
Artemisia arctica f. *villosa*																		2	4		
Luzula wahlenbergii			2	2							1									2	1
Polygonum ajanense			1	1																1	
Arctous alpinus var. *japonicus*														4							
Pedicularis oederi																				1	1
Spiraea betulifolia var. *ameliana*													1								
Saxifraga laciniata													1					1			
Stellaria pterosperma															1		1			1	
Lagostis stelleri var. *yesoensis*			10	20																	
Viola crassa			1	1																	
Dicentra peregrina		1		1		1	1														
Salix pauciflora			5	1				1	1	1	6	2	4	1	2				2		
Salix yezoalpina									2	1		1			3	10	4		1		
Thamnolia vermicularis	2	10	2	1	25	10	10	30	30	6	2	1	2	2	3		2				
Cetraria	2	10	2	1	10	30	20	25	40	1		1	2	1	1					2	1
Polytrichum	1	1	2	1	1	1	1		2	1			1		2					2	1
Cetraria cucullata														2						2	2
Cladonia																					1

PLATE 4 Salix pauciflora

(Photo by H. Shinsho)

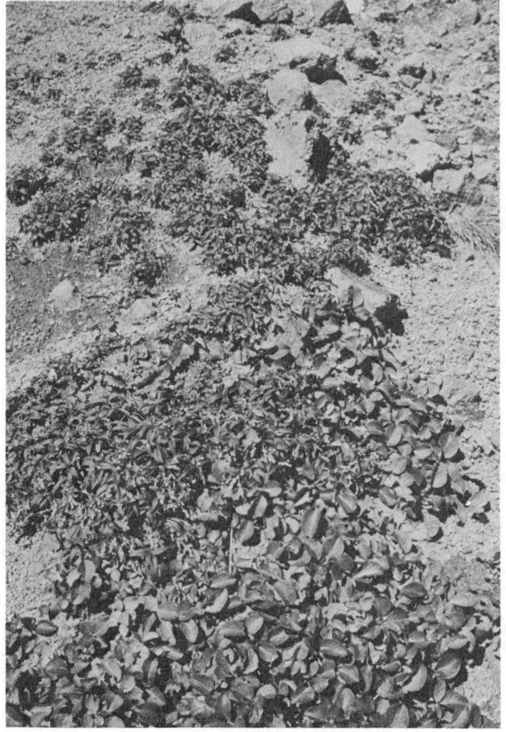

PLATE 5 Salix yezoalpina

(Photo by H. Shinsho)

Therefore, the alpine vegetation consists of a plant community complex here.

This plant community complex is reported already (Ito and Nishikawa 1977, Ito and Sato 1980). But the reason of the distribution of this complex in not been clarified.

INFLUENCE OF PERMAFROST ON VEGETATION

The distribution of such a plant community complex is limited on the permafrost areas of the Daisetsu Mountains. To clarify the cause of this complex, the influences of permafrost on vegetation

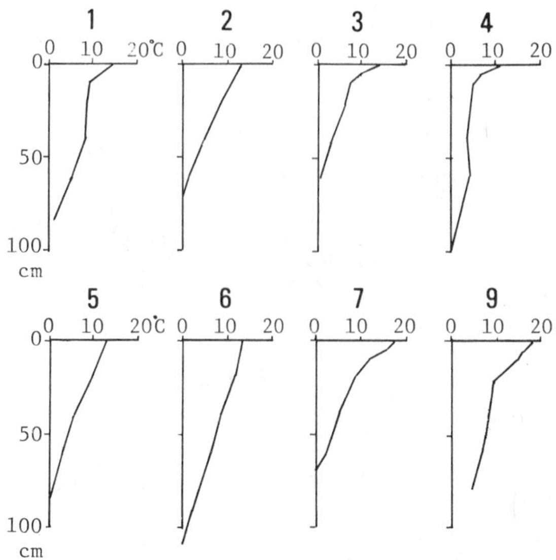

FIGURE 3 Vertical change of soil temperature in each point

were checked, specifically, the cooling of surface soil or the change of a habitat to a humid state caused by permafrost. So the distribution of permafrost, the depth of the active layer in mid-July, the vertical changes of soil temperature, and the depth of the water table were examined.

In these conditions, permafrost is recognized at all points on Mt. Koizumitake. Therefore, the existence of permafrost itself has no relation to plant distribution. Cooling of surface soil is not recognized in each point. The soil temperature of the earth's surface rises to 15-20 C or more in mid-July (Figure 3). So it is improbable that the cooling of soil limits the growth of plants. For the same reason, the the depth of the active layer has no relation to plant distribution (Table 2).

But it becomes clear that the depth of the water table has a close relation to the distribution of the two species of Salix. In this area the water table exists generally in shallow places (Table 2, Figure 4). The reason for such a high water table is that the permafrost prevents the penetration of groundwater.

The water table is especially shallow on the gentle mountain slope (point 4-6), but around the col it becomes rather deep. The two species of

TABLE 2 Depth of Water Table and Depth of Active Layer at Each Point.

point number	1	2	3	4	4A	5	6	7	8	9
depth of water table (cm)	deep	20-26	42	0	25	35	30	50	19	deep
depth of active layer in mid-July (cm)	85	73	62	104	104	85	110	68	58	deep

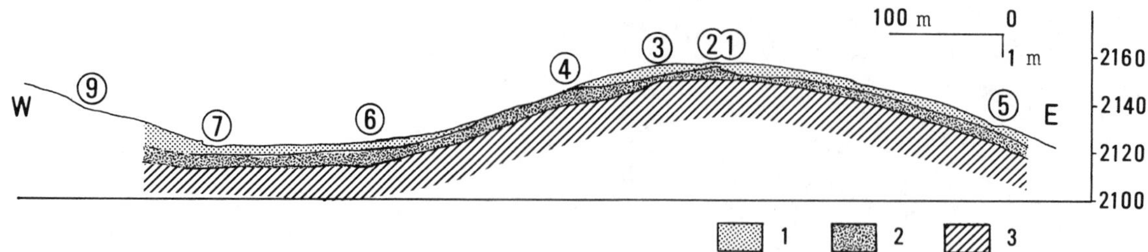

FIGURE 4 Schematic profiles of frozen ground and the water table on Mt.Koizumitake in mid-July.
1: active layer above water table
2: active layer below water table
3: frozen ground
Numbers show the surveyed points.

Salix emerge in the habitat with a shallow water table but not in the dry places of the low water table. The borderline is at -35 cm in mid-July.

As mentioned already, the surveyed areas are located on windy places, so the mixture of the two Salix is an abnormal phenomenon in a sense. But the presence of permafrost enables such a distribution to exist. And the formation of a plant community complex is also the result of such a distribution. Since such a habitat does not exist other high mountains, the plant community complex in this mountain is unique to Japan.

ACKNOWLEDGEMENTS

The author expresses grateful thanks to Mr. H. Shinsho and Mr. O. Yanagimachi for the help in the field study and also to Mr. M. Tamura for the aid with illustrations.

REFERENCES

Brown, R. J. E., 1970, Permafrost as an ecological factor in the Subarctic, in Ecology of the Subarctic Regions: UNESCO, p. 129-140.

Dingman, S. S., Koutz, F. R., 1974, Relation among vegetation, permafrost, and petential insolation in Central Alaska: Arctic and Alpine Research, v. 6, p. 37-42.

Fukuda, M., Kinoshita, S., 1974, Permafrost at Mt. Taisetsu, Hokkaido and its climatic environment: Quaternary Research (Japan), v.12, p. 192-202. (Japanese with English summary)

Ito, K., 1978, Plant communities of permafrost, in Kinoshita,S., ed., Joint Studies on Physical and Biological Environments in the Permafrost, Alaska, and North Canada, July to August 1977: Sapporo, Institute of Low Temperature Science, Hokkaido University, p. 105-147.

Ito, K., Nishikawa, T., 1977, Alpine communities of the northern Taisetsu mountain range (2); Syntaxonomy of plant communities: Reports of the Taisetsuzan Institute of Sciences, Asahikawa College Hokkaido University of Education, No.12, p. 1-26.

Ito, K., Sato, K., 1981, Outline of the vegetation map of the Taisetsu mountain range: Sapporo, Hokkaido Government, 23p. (Japanese)

Koaze, T., 1965, The patterned grounds on the Daisetsu Volcanic Group, Central Hokkaido: Geographical Review of Japan, v. 38, p. 179-199. (Japanese with English summary)

Johnson, A. W., 1966, Plant ecology in permafrost areas. in Proceedings of the Permafrost International Conference (1963): NAS-NRC Publication, No. 1287, Washington, D.C.

Peterson, K. M., Billings, W. D. 1980, Tundra vegetational patterns and succession in relation to microtopography near Atkasook, Alaska: Arctic and Alpine Research, v. 12, p. 473-482.

Roning, O. I., 1969, Features of the ecology of some arctic Svalbard (Spitsbergen) plant communities: Arctic and Alpine Research, v. 1. p. 29-44.

Yoshie, F., Fujino, K. 1982, Vegetational changes in relation to topography and soil environments in the arctic tundra, in Joint Studies on Physical and Biological Environments in the Permafrost, North Canada, July-August 1980 and February-March 1981: Sapporo, Institute of Low Temperature Science, Hokkaido University, p. 73-93.

COVER SANDS IN SOUTHERN JUTLAND (DENMARK)

Else Kolstrup

Danmarks Geologiske Undersøgelse, Thoravej 31, DK-2400 København NV, Denmark

Investigations in southern Jutland (Denmark) have revealed the presence of Older and Younger Cover Sands. By means of pollen analysis and stratigraphic comparisons with neighbouring countries, it is suggested that the Upper Pleniglacial is represented by the Older Cover Sand I and II, with a phase of fluvial activity, deflation, and frost action in between. There is a Late-Glacial succession of Younger Cover Sand I, peat of Allerød age, and Younger Cover Sand II on top of the Older Cover Sand II. Holocene peat is found on top in some areas. Within the area glaciated during the Weichselian, diamicton is locally overlain by cover sand, while west of the glaciated area a period of cold climate is reflected within the cover-sand sequence. It is suggested that the conditions for the accumulation of cover sands are: a flat or very gently sloping relief in the area of deposition, (short?) vegetation to catch and retain the sand and silt, winds, and a sandy-silty source area without protective vegetation cover.

INTRODUCTION

The geology of large parts of Denmark was mapped during the early half of this century. The maps (including descriptions) of western and southern Jutland (e.g., Hansen 1965, Milthers 1925) show large areas covered by eolian deposits, most of which constitute a thin, usually only a few decimeters thick, cover on the landscape. This cover of eolian sand very often lies on and in the vicinity of so-called "fresh-water" deposits and "layered sand", deposits which were thought to be water-lain, but descriptions and further suggestions concerning their environment(s) of deposition are scanty.

A gas pipe line system has been under construction in Denmark since 1981. The locations of the gas pipe line trenches, Frøslev-Egtved (S-N), and Egtved-Lillebælt (W-E), are indicated on the map of southern Jutland in fig. 1 together with an outline of the geology of that area. The gas pipe lines are buried in continuous trenches approximately two meters deep. These trenches made it possible to study the vertical succession of sediments and the lateral changes of layers. In the areas with a top sheet of eolian sand, the top layer as well as the underlying sediments were investigated in vertical section with regard to structure, texture, and stratigraphic succession.

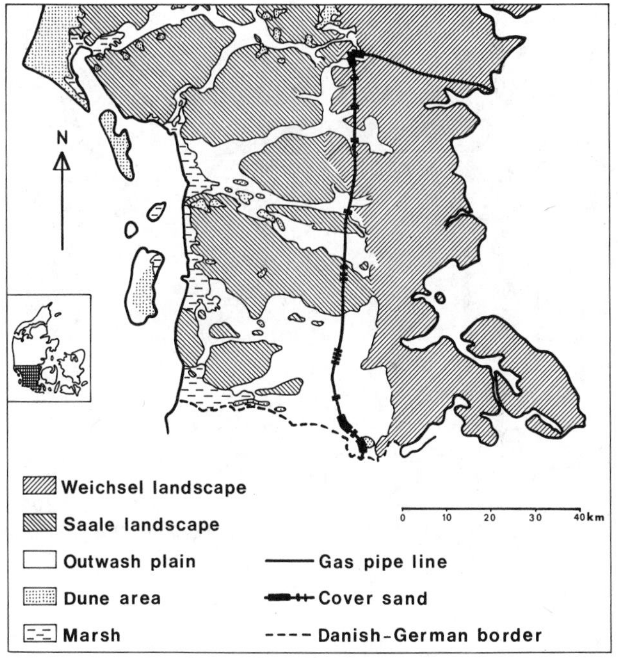

Figure 1. Map of southern Jutland with locations of cover-sand occurrences indicated along the gas pipe line trenches Frøslev-Egtved (S-N) and Egtved-Lillebælt (W-E). The geological boundaries are based on P. Smed, 1982, Landskabskort over Danmark, Blad 3, Sønderjylland, Fyn. Geografforlaget, Brenderup.

SEDIMENTS

Two types of sediments were found to be prevalent in the areas with a top sheet of eolian sand (Kolstrup & Jørgensen 1982).

Type 1

Type 1 corresponds to the eolian sand sheets of the Danish geological maps. This type consists of horizontal to subhorizontal layers from mm to a few centimeters thick (fig. 2). Consecutive layers are well-sorted and normally they have approximately the same grain size, with a maximum usually between 150 µm and 300 µm (dashed lines to the left in fig. 3), but layers dominated by grains about 2 to 4 mm in diameter may be found locally, as well as single pebbles up to 1.5 cm in size. The composition of the sand is strongly dominated by quartz grains which are rounded to well rounded and polished. In some localities this unit is divided

Figure 2. Photograph of Older Cover Sand II (lower. half of profile) with a gradual transition upwards to Younger Cover Sand. The ruler is two meters long.

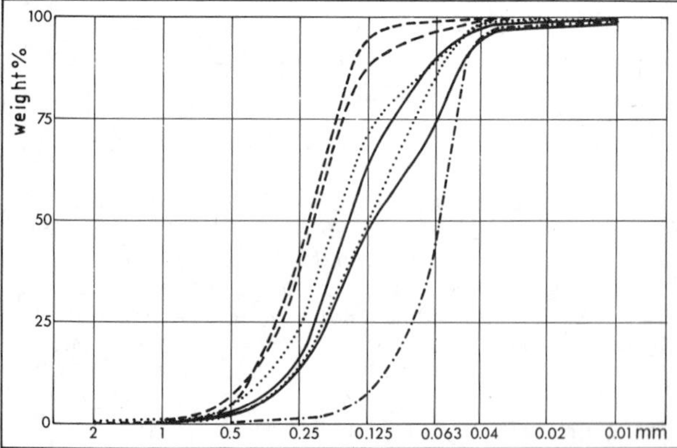

Figure 3. Grain-size distribution of Younger Cover Sand (dashed lines), Older Cover Sand I (full lines), Older Cover Sand II (dotted lines), and a fine-grained layer in the Older Cover Sand (dashed-dotted line).

by a peat layer or by a somewhat finer horizon which may possess bioturbation. Although type 1 may be up to 1.5 m thick, its thickness is usually less than one meter. It makes up the top layer of the profiles where it is present, and the original layering in its upper part is therefore often disturbed by surface processes or has disappeared altogether. In some depressions type 1 is overlain by peat. The good sorting and the polished grains of this unit, together with its appearance as a morphologically inconspicuous surface layer covering flat and slightly undulating areas, makes an eolian origin probable.

Type 2

Type 2 is included in the "fresh-water" deposits and "layered sand" of the geological maps. Also, this sediment normally consists of horizontal to subhorizontal from mm to a few centimeters thick layers (fig. 2); however, there is a vertical alternation of fine and medium sand layers in this deposit. In some places the fine fraction, usually dominated by particles ranging between 63 µm and 105 µm, may be dominant in vertical profile; in others, the coarse fraction prevails. The dotted-dashed line in the right-hand part of fig. 3 represents a decimeter-thick layer of the fine fraction, and it can be seen that individual layers are well sorted. Field observations reveal that the medium sand layers of type 2 are similar in texture and sorting to the sand of type 1. Bulk samples of type 2, therefore, are made up of various mixtures of the pure populations (dotted and full lines in fig. 3). The larger grains of this unit are well rounded and polished whereas the smaller grains are more angular and shiny.

In the southern part of the Frøslev-Egtved stretch, type 2 is divided into two subunits by a gravel layer with a horizontal to subhorizontal layer of wind-polished and faceted pebbles and stones on top. The upper subunit, which rests upon this pebble layer is usually undisturbed, whereas the part underlying the pebble layer is usually strongly involuted and includes somewhat coarser sand and gravel.

In the northern part of the above-mentioned stretch, and in the western part of the Egtved-Lillebælt trench, undisturbed type 2 sediment is found on top of a diamicton, predominantly till, with grain sizes ranging from clay to boulders with wind-polished and faceted stones on top. The upper (undisturbed) subunit of type 2 is usually between 0.5 and 1 meters thick, although in some areas it may be more than two meters thick. This deposit either grades upwards into type 1 sediment, or there may be an up-to-10-cm-thick layer of fine sand and loam separating the two. Due to the texture of the individual sand grains, the sorting, and the polished and faceted pebble layer, the type 2 deposit is also thought to be of eolian origin.

POLLEN ANALYSIS

The peat within type 1 contains fairly high percentages of Betula pollen including both larger and smaller grains. Pollen of Gramineae, Cyperaceae and herbs are also fairly well represented, as are

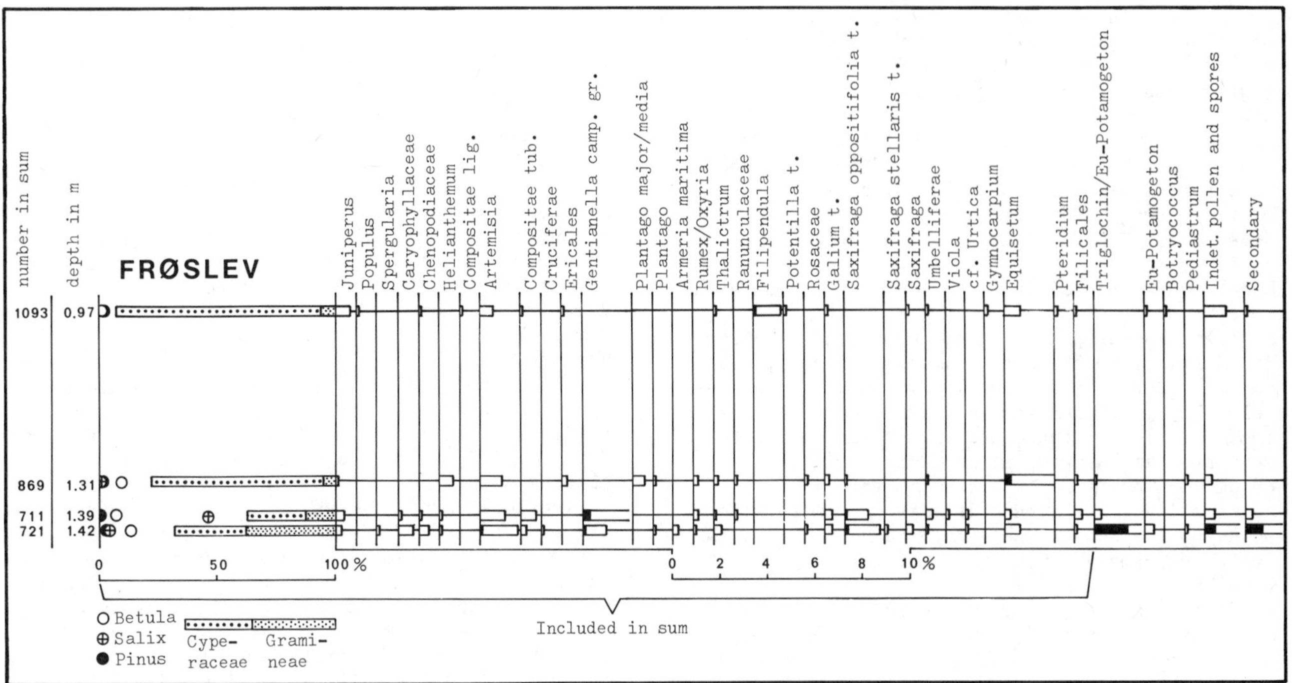

Figure 4. Pollen diagram representing organic layers within the Older Cover Sand II near Frøslev. The black parts in the spectra of the resolved diagram are of the same scale of percentage as that of the main-diagram to the left; the scale of the open bars is indicated below the resolved diagram.

spores of Gymnocarpium. Pollen from trees other than Betula, Pinus, and Salix are absent. This pollen composition points to a vegetation of short plants, with scattered occurrences of birch (Betula pubescens and B. nana) and willow.

Within the upper, undisturbed subunit of type 2, four approximately 1-cm-thick layers containing organic material could be followed laterally for more than 350 meters in the southern part of the Frøslev-Egtved trench. The composition of the pollen from the organic layers is given in fig. 4. The pollen is strongly dominated by Cyperaceae and Gramineae, and a number of herb pollen types are present, notably Gentianella and Saxifraga in the lower spectra. Betula is almost exclusively represented by small pollen, viz., pollen deriving from Betula nana, and the pollen of Salix probably also represents short forms rather than trees. The spectra in fig. 4, therefore, indicate that short plants were growing in a moist environment.

LITHOSTRATIGRAPHY			CHRONOSTRATIGRAPHY
	SOUTH	NORTH	
	SOIL OR PEAT		HOLOCENE
	YOUNGER COVER SAND II		LATE DRYAS
	LOAMY SAND OR PEAT		ALLERØD
	YOUNGER COVER SAND I		EARLIER DRYAS
	LOWER LOAMY BED		BØLLING
	OLDER COVER SAND II		PLENIGLACIAL
FRØSLEV	FRØSLEV GRAVEL BED		
COMPLEX	(periglacial processes) SAND AND GRAVEL	TILL	
	OLDER COVER SAND I	?	
	WIND POLISHED PEBBLES SAND AND GRAVEL		

Figure 5. Outline of the stratigraphy in the cover-sand areas in the southern and northern part of the area of investigation.

STRATIGRAPHY

A comparison of type 1 and 2 deposits with deposits from The Netherlands, western Belgium, northern Germany, and Poland shows that type 1 and 2 are similar to the Younger and Older Cover Sands respectively (compare, e.g., van der Hammen 1951; Nowaczyk 1976).

Lithostratigraphy

Two types of stratigraphic sequences were exposed in the trenches (fig. 5). In the southern part of the Frøslev-Egtved trench, the lower part of the sequence consists of a water-lain deposit of sand and gravel, sometimes covered by a layer of wind-polished gravel. Above this is type 2, viz., Older Cover Sand deposit, which in turn is covered by another water-lain deposit of sand and gravel. This part of the sequence is usually strongly involuted or undulating, and, locally, frost-wedge casts extend downward from within the involuted layer. In the northern part of the Frøslev-Egtved trench, and in the western part of the Egtved-Lillebælt trench (i.e., within the area glaciated during the Weichselian), the lower part of the sequence usually consists of a diamicton, mostly till.

On the surface of both the involuted sequence and the diamicton, a layer of wind-polished and faceted pebbles and stones is found, which is overlain by an undisturbed deposit of type 2, viz., Older Cover Sand. The upper part of both sequences consists of type 1 deposit, i.e., Younger Cover Sand, except for sites where peat has accumulated later.

The stratigraphic sequence in southern Jutland is identical to the Dutch, western Belgian, and northern German sequences (e.g., Dücker & Maarleveld 1957, van der Hammen 1971, Zagwijn & Paepe 1968) and it can therefore be suggested that the sequence of changes in the environmental conditions are similar. Chronostratigraphic correlations have been made between the cover-sand sequences in western Belgium, The Netherlands, and northern Germany (Dücker & Maarleveld 1957, Zagwijn & Paepe 1968).

In the northern part of the investigated area in Denmark, only that part of the column overlying the diamicton is comparable to others, i.e., undisturbed Older Cover Sand overlain by Younger Cover Sand.

Chronostratigraphy

Investigations of peat found upon the Younger Cover Sand in Denmark show that the accumulation of organic material started in early Holocene time, i.e., around or shortly after 10,000 y.B.P. (e.g., Milthers 1925; compare also Iversen 1973).

A comparison of the pollen composition of the organic layer from within the Younger Cover Sand with palynological data from other localities suggests an Allerød age (i.e., 11,000 to 11,800 y. B.P.), as is also the case in the neighbouring countries (e.g., van der Hammen 1951).

A layer of bleached fine sand and loam separates the Older and the Younger Cover Sands in some localities. A similar layer is found in the same stratigraphic position in western Belgium, northern Germany, and The Netherlands, where it is called the Lower Loamy Bed and dated to the Bølling time (12,000 to ca. 12,400 y.B.P.). It is possible that this fine-grained layer in Denmark may also be of Bølling age.

It is also possible to subdivide the Younger Cover Sand into two units in Denmark, a Younger Cover Sand II which was predominantly deposited during Late Dryas time (i.e., predominantly between 11,000 and 10,000 y.B.P.), and a Younger Cover Sand I, which was deposited before the Allerød period and possibly after the Bølling.

The pollen composition of the organic layers in the upper, undisturbed Older Cover Sand unit (three lower spectra of fig. 4) corresponds with deposits of Upper Pleniglacial age (i.e., between ca. 27,000 and ca. 13,000 y.B.P.) in The Netherlands (van der Hammen et al. 1967, Kolstrup 1980), and therefore, the Danish deposit is also thought to be of Upper Pleniglacial age.

In southern Jutland the involutions, which are locally more than 1.5 meters deep, and the frost-wedge casts connected to them, indicate a period of cold climate. Complete Dutch sequences reveal that the deposition of Older Cover Sand was temporarily interrupted by a period with permafrost, fluvial activity, and deflation, some time between ca. 22,000 y.B.P. and ca. 14,000 y.B.P. (The Beuningen Complex: van der Hammen et al. 1967, Kolstrup 1980). Since the stratigraphic sequences in neighbouring countries and Denmark are identical, and since a cold phase is reflected in all countries at the same stratigraphic level, it seems reasonable to accept that the involutions and frost-wedge casts were formed during the same cold spell. Following the Dutch nomenclature (van der Hammen et al. 1967), the Danish Older Cover Sand is consequently divided into an upper, undisturbed, Older Cover Sand II, deposited after the cold period, and a lower, involuted (cryoturbated) Older Cover Sand I, deposited before the period of maximum cold.

The local name of Frøslev Complex is introduced here for the Danish deposits and phenomena which are probably equivalent to the Beuningen Complex. The layer of wind-polished and faceted pebbles and stones forming the top layer of the Frøslev Complex is accordingly termed the Frøslev gravel bed.

Within the area glaciated during the Weichselian the Older Cover Sand II rests upon a diamicton with wind-polished and faceted stones on top.

A comparison between the stratigraphies of the two types of areas in Denmark (fig. 5) suggests a parallel development from the time of wind-polishing and onward, whereas before that time, the northern part was glaciated while periglacial processes were active in the southern, unglaciated area.

ORIGIN OF COVER SAND

As long as cover sands have been known in northwestern Europe the environmental conditions for their formation, especially those of the Older Cover Sand, have been a puzzle which has not been solved yet.

Significant studies describing eolian deposits of a similar appearance in northern latitudes where accumulation is taking place today have been made by e.g., Pissart et al. (1977) and Trainer (1961).

Source Area and Appearance

It has been suggested that the exposed sea floors of the North Sea and the Channel were the source areas for the sands and silts (loesses) in Europe (e.g., Vink 1949). This suggestion has been queried by, among others, Crommelin (1964) who advocated the idea of a more "local" origin based upon three Dutch heavy mineral provincies being closely related to the mineralogy of the underlying sediments.

It might be suggested that the alternating grain sizes in the Older Cover Sand represent winds blowing from different directions, thus representing different source areas, but the investigations by Pissart et al. (1977) and Trainer (1961), and also, indirectly, that of Crommelin (1964) show that this is not a necessary requirement.

The alternation of grain sizes in the Older

Cover Sand has been explained as being the result of transportation and deposition by snow, possibly reflecting differences in mode of deposition during summers and winters (van der Hammen 1951). Others, e.g., Krumbein (1937) attribute the formation of interbedded eolian sand and silt to the transportation of sand by traction and dust in suspension. On the other hand, investigations by de Ploey (1977) reveal that both silt and fine sand may be transported as a basal flow along the ground surface, and he suggested that the two fractions may be deposited by simultaneous dumping.

Following Trainer's (1961) investigations and considerations, it may be suggested that the fine-grained layers were deposited during slackened deposition and/or were transported over a longer distance than the coarser fraction. Trainer believes that the characteristics of the eolian sediment in Alaska can be related to changes in the character of the source area of the eolian material and puts forward an elegant solution. He suggests that the changes may be related to formation and destruction of stabilized flood-plain surfaces and "it is apparent that any considerable change in stream regimen must be dependant upon changes in the glacier, in sea level, or in both, and hence ultimately upon changes in climate" (Trainer 1961, p. 30).

The periods during which the cover sands were deposited in northwestern Europe (i.e., some time before and some time after, but not during, the maximum ice extension) were periods of changing climate, with fluctuations of ice caps and glaciers and accompanying changes of sea level. Sea level changes may therefore have altered the stream regimen of northwest European watercourses, thus giving rise to incisions and floodings which in turn might have resulted in the exposure of unvegetated deposits of sand and silt. The changes in stream regimen may also, together with changes in temperature and precipitation, have altered the moisture balance of higher ground. Although the drying out of the surface sediment is not a necessary requirement for the start of deflation (de Ploey 1977), changes in sea level may thus indirectly have contributed to the erosional effect of the wind, so that sand could relatively easily be picked up in sparcely vegetated regions and become deposited in neighbouring areas.

Vegetation

Both Pissart et al. (1977) and Trainer (1961) consider vegetation cover an important factor for catching and retaining the wind-blown sands which come from neighbouring barren, sandy areas, particularly flood plains.

The pollen content of the organic layers in the cover sands in The Netherlands (Kolstrup 1980) and Denmark show that parts of the landscape were, at least periodically, covered by vegetation during the deposition of the sands. It may therefore be proposed that vegetation, in parts of the northwest European cover-sand areas, acted as a trap for the wind-blown sands and silts as well as a protection against erosion. It is difficult to envisage the extension of vegetation during the cover-sand accumulation. It was probably much more extensive than the scattered findings in northwestern Europe suggest, but since the accumulation and preservation of organic material is dependant on a non-aerated, non-oxidized environment, organic remains have only become preserved in a few places with favourable conditions.

Furthermore, the presence of pollen from perennial plants within single, thin, organic layers of the Older Cover Sand may suggest that deposition of sand and silt only took place occasionally, thus yielding a relatively slow accumulation rate of the sand sheets.

Air Temperature

The air temperature conditions during the deposition of the cover sands in northwestern Europe may be deduced from palynological records.

The presence of pollen of *Armeria maritima* var. *planifolia* found in the Older Cover Sand I and II in The Netherlands (Kolstrup 1980), in the Older Cover Sand II in Denmark, and in Late-Glacial deposits (e.g., Iversen 1954, 1973) point to a mean January temperature, at least periodically, of -8°C or warmer (Iversen 1954). The mean July temperature may have been different during the deposition of the various units, but generally speaking, it may have been on the order of +8°C to +13°C (Kolstrup 1980) during the deposition of some of the layers in the Older Cover Sand, and between ca. +12°C and +15°C (Iversen 1954, 1973, Kolstrup 1982) during the deposition of the Younger Cover Sand.

Relief

Within the area glaciated during the Weichselian in Jutland, it can sometimes be seen that the layering of the cover sand conforms with underlying undulations of the diamicton, and in some sites the horizontal layers of the Younger Cover Sand can be followed laterally into more than 2-meters-thick eolian deposits with steep foresets. This change is thought to reflect a change in the depositional environment caused by a change from a flat, or gently undulating area, to a more accentuated relief. It is, however, unknown whether the relief alone could account for the change or whether it was caused by a combination of steeper relief, associated hydrological change, and an accompanying change in vegetation.

CONCLUSIONS

The present investigation demonstrates that parts of Denmark belong to the northwest European cover-sand area and that the sequence of changes in environmental conditions have been similar within the region. Observations within and outside the area glaciated during the Weichselian make it possible to establish a probable link between the glacial and the periglacial stratigraphies.

The environmental conditions for the formation of the cover sands are still insufficiently known, but further investigations in northern areas where similar deposits accumulate today will undoubtedly add information.

ACKNOWLEDGEMENTS

The author wants to thank J. Stockmarr (Copenhagen) and I. Sørensen (Århus) for their valuable criticism of the manuscript, D. Blom (Kolding) for the correction of the English text, and I. and C. Torres (Copenhagen) for their help with the photographical work. The Danish Oil and Natural Gas Company A/S (D.O.N.G. A/S) kindly gave permission to follow the trenching. The Geological Survey of Denmark, The Danish Natural Science Research Council (S.N.F.), and The Natural Agency for the Protection of Nature, Monuments, and Sites (Fredningsstyrelsen) financed the investigation.

REFERENCES

Crommelin, R.D., 1964, A Contribution to the Sedimentary Petrology and Provenance of Young Pleistocene Cover Sand in The Netherlands. Geologie en Mijnbouw, 43(9), p. 389-402.

Dücker, A. & Maarleveld, G.C., 1957, Hoch- und spätglaziale äolische Sande in Nordwestdeutschland und in den Niederlanden. Geologisches Jahrbuch, 73, p. 215-234.

Hansen, S., 1965, Kortbladet Tinglev, 1:100.000. Danmarks geologiske Undersøgelse, I Rk., 23A.

Iversen, J., 1954, The Late-Glacial Flora of Denmark and its Relation to Climate and Soil. In Iversen, J., ed., Studies in Vegetational History in Honour of Knud Jessen: Danmarks geologiske Undersøgelse, II Rk., 80, p. 87-119.

Iversen, J., 1973, The Development of Denmark's Nature since the Last Glacial. Danmarks geologiske Undersøgelse, V Rk., 7-C, 125pp.

Kolstrup, E., 1980, Climate and Stratigraphy in Northwestern Europe Between 30.000 B.P. and 13.000 B.P., with Special Reference to The Netherlands. Mededelingen Rijks Geologische Dienst, 32-15, p. 181-253.

Kolstrup, E., 1982, Late-Glacial pollen diagrams from Hjelm and Draved Mose (Denmark) with a suggestion of the possibility of drought during the Earlier Dryas. Review of Palaeobotany and Palynology, 36, p. 35-63.

Kolstrup, E. & Jørgensen, J.B., 1982, Older and Younger Coversand in southern Jutland (Denmark). Bulletin of the Geological Society of Denmark, vol. 30, p. 71-77.

Krumbein, W.C., 1937, Sediments and Exponential Curves. The Journal of Geology, XLV, 6, p. 577-601.

Milthers, V., 1925, Beskrivelse til Geologisk Kort over Danmark (1:100.000). Kortbladet Bække. Danmarks geologiske Undersøgelse, I Rk., 15, 175pp.

Nowaczyk, B., 1976, Eolian Cover Sands in Central-West Poland. Quaestiones Geographicae, 3, p. 57-77.

Pissart, A., Vincent, J.-S. & Edlund, S.A., 1977, Dépôts et phenomènes éoliens sur l'île de Banks, Territoires du Nord-Ouest, Canada. Canadian Journal of Earth Sciences, 14, 11, p. 2462-2480.

de Ploey, J., 1977, Some Experimental Data on Slopewash and Wind Action with Reference to Quaternary Morphogenesis in Belgium. Earth Surface Processes, 2, p. 101-115.

Trainer, F.W., 1961, Eolian Deposits of the Matanuska Valley Agricultural Area Alaska. U.S. Geological Survey Bull., 1121-C, p. 1-35.

van der Hammen, T., 1951, Late-Glacial Flora and Periglacial Phenomena in The Netherlands. Leidse Geologische Mededelingen, XVII, p. 71-183.

van der Hammen, T., 1971, The Upper Quaternary stratigraphy of the Dinkel valley (with an annotated list of radiocarbon dates). In van der Hammen, T. & Wijmstra, T.A., eds., The Upper Quaternary of the Dinkel valley: Mededelingen Rijks Geologische Dienst, N.S. 22, p. 59-72.

van der Hammen, T., Maarleveld, G.C., Vogel, J.C.& Zagwijn, W.H., 1967, Stratigraphy, Climatic Succession and Radiocarbon Dating of the Last Glacial in The Netherlands. Geologie en Mijnbouw, 46(3), p. 79-95.

Vink, A.P.A., 1949, Bijdrage tot de kennis van loess en dekzanden, in het bijzonder van de zuidoostelijke Veluwe. Thesis, Wageningen.

Zagwijn, W. & Paepe, R., 1968, Die Stratigraphie der weichselzeitlichen Ablagerungen der Niederlande und Belgiens. Eiszeitalter und Gegenwart, 19, p. 129-146.

RECOVERY OF PLANT COMMUNITIES AND SUMMER THAW AT THE 1949 FISH CREEK TEST WELL 1, ARCTIC ALASKA

Vera Komarkova

Institute of Arctic and Alpine Research
and Department of Environmental, Organismic, and Population Biology,
University of Colorado, Boulder, Colorado 80309 USA

The Fish Creek Test Well 1 site on the Arctic Coastal Plain (70°18'36"N, 151°52'40"W) has been abandoned since its construction in 1949. The plant communities that developed on the disturbed site are all successional. The degree of community recovery has differed with the habitat/vegetation type and corresponds to the degree of recovery of the habitat. The average depth of thaw was greater in recovering than in undisturbed habitats, particularly in mesic uplands. Marshes have recovered much faster than mesic uplands. Mesic uplands are more resistant to surface disturbance, are less resilient, and show less fluctuation and narrower ecological range on some environmental gradients, such as depth of thaw. The time of recovery of mesic upland ecosystems at the Fish Creek site is estimated to be 600-800 years and that of marshes, 100-200 years. In most general features the resulting communities will probably resemble the surrounding undisturbed complex communities.

Like other test oil well sites in arctic Alaska from the 1944-1953 oil and gas exploration period, the Fish Creek Test Well 1 site is an excellent subject for studies of arctic tundra recovery because it has been relatively undisturbed since its construction in 1949. The site was initially studied in 1977 after 28 years of recovery (Lawson et al. 1978). This paper examines the hypothesis that in permafrost areas in arctic Alaska the degree of plant community recovery after surface disturbances is closely related to the degree of recovery of the disturbed physical environment, particularly its thermal regime.

Fish Creek Test Well 1 is located at 70°18'36"N and 151°52'40"W on the Arctic Coastal Plain in the region of continuous permafrost. The landforms, soils (Everett 1979), and vegetation are characteristic of the landscape of oriented lake basins and streams active in consolidated dunes (Carter 1981). The original vegetation habitats at the site included wet polygon troughs, ponds and marshes, elevated dry ridges, snow patches, mesic uplands, and marshy lowlands; today the site is surrounded by vegetation that is probably very similar to that which was originally disturbed (Lawson et al. 1978).

During the construction and operation of the test well in 1949, vegetation and much of the surface organic soil layer on the site were removed. In 1979, solid waste was removed from the site by the U.S. Geological Survey under contract with Husky Oil NPR Operations, Inc. (Husky Oil 1981). The recovery of the vegetation cover at Fish Creek is almost complete, but ground subsidence, thermokarsting, thermal erosion, and drainage modification caused by the disturbance may still be continuing (Lawson 1982). The new vegetation is composed of 30-40 successional plant communities that developed during primary and secondary succession. The diversity of the successional communities is high, but the taxa diversity is lower on the disturbed site than in undisturbed tundra; almost all colonizing taxa occur in the undisturbed area surrounding the site (Lawson et al. 1978).

METHODS

Vegetation recovery is defined here as a gradual process of colonization and composition changes leading to the reestablishment of a vegetation cover and, at a later stage, to a mature plant community, which may be similar or dissimilar to the original community.

For the overall study, 339 plots were sampled according to the Braun-Blanquet method of vegetation analysis (Westhoff and van der Maarel 1978). The environmental characteristics of each plot, including disturbance, were estimated on subjective gradient scales, and soils were collected and analyzed. Several depth-of-thaw measurements were taken at each plot and averaged. The thaw data (Table 1) were collected in the summers of 1978, 1979, and 1980, in June and early July at the beginning of the growing season, and in August at the end of the growing season.

The samples selected for this analysis were collected on topographically equivalent sites within each habitat/vegetation category; for example, both the recovering and undisturbed snow patches were well-drained depressions with late-melting snowcover. One to three representative samples were selected for both undisturbed and recovering vegetation at the main disturbed site in each of the following common habitat/vegetation categories: upland, snow patch, lowland, marsh, and ridge. Two pairs of ridge samples were included to test the validity of the categories for floristically different communities. Vascular plant data averaged for each category were analyzed by numerical methods, including ordination and numerical classification.

TABLE 1 Means and standard errors of thaw for the investigated habitat/vegetation categories, in cm.

	Beginning of growing season				End of growing season			
	Undisturbed		Recovering		Undisturbed		Recovering	
	Mean	Standard error	Mean	Standard error	Mean	Standard error	Mean	Standard error
Upland	4.6	0.39	14.9	0.90	25.7	2.19	40.1	2.43
Snow patch	11.7	2.72	23.0	4.78	48.0	0	55.8	8.42
Lowland	16.0	2.97	15.5	2.45	44.8	5.57	46.0	5.38
Marsh	14.0	1.89	17.9	1.88	47.0	5.41	61.6	6.47
Ridge	36.4	3.12	34.4	3.53	84.4	6.07	69.4	3.83
Average	16.4		21.1		50.0		54.6	

Recovery habitats show deeper thaw in all categories except for ridges, where further differentiation and recovery will result in deeper thaw depths. Shallower thaw depths will result during further differentiation and recovery in all other habitat categories.

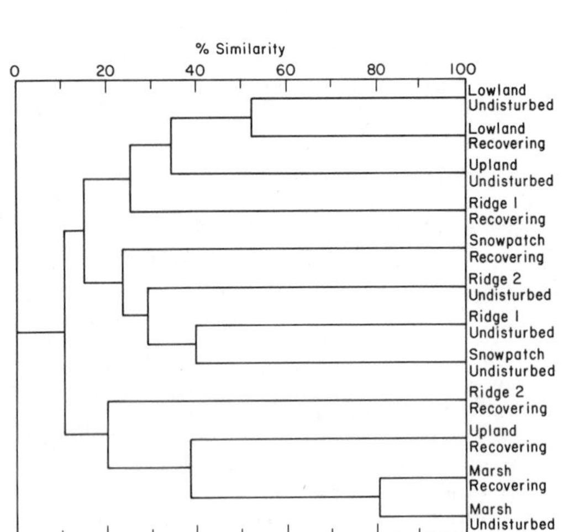

FIGURE 1 Polythetic, hierarchical dendrogram of composite plant community samples representing the selected habitat/vegetation categories. Based on percent-cover taxa data and constructed by the average linkage method using Pearson's product-moment correlation coefficient as the clustering criterion. Each category includes both undisturbed and recovering communities. The level of similarity at which the undisturbed and recovering communities were paired decreases in the following order: marsh (82%), lowland (52%), snow patch (26%), ridge 1 (16%), ridge 2 (14%), and upland (14%).

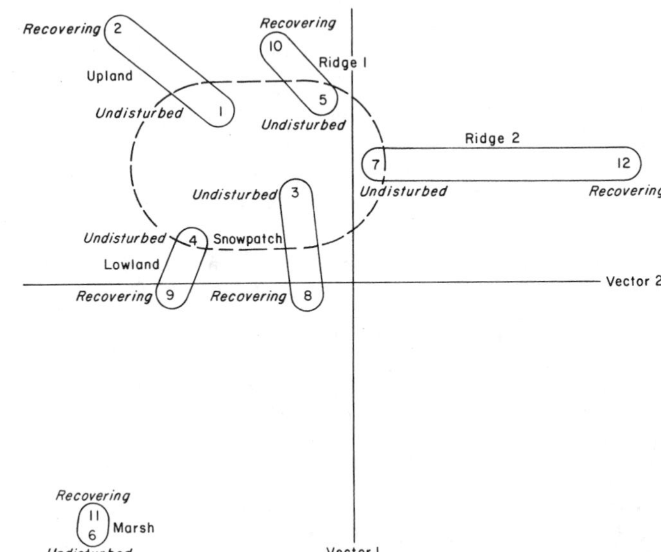

FIGURE 2 Detrended correspondence analysis ordination of quantitative vascular taxa data for the same plant communities as in Figure 1. Undisturbed and recovering plant communities were paired in the outlined habitat/vegetation categories (solid line). A group of undisturbed communities was formed in the center of the ordination space (dashed line), indicating that disturbance is one of the principal factors ordering the communities on the ordination planes. Moisture, duration of snow cover, and depth of thaw were other important environmental factors ordering the communities.

RESULTS

The degree of recovery of the original plant communities at the Fish Creek site was determined by computing the floristic similarities between undisturbed and recovering communities. The average similarity of recovering communities (22%) was slightly lower than the average similarity of undisturbed communities (25%), but they are not clearly separated in the dendrogram (Figure 1). Undisturbed and recovering communities in marshes and lowlands were highly similar, but the pairs of upland and ridge communities clustered at very low levels of similarity. These results are consistent with field observations and indicate that the rate of community recovery at Fish Creek differs with the habitat/vegetation type (cf. Billings 1973, Hernandez 1973).

In the ordination space (Figure 2), communities were ordered according to both human disturbance and the most important natural environmental factors controlling community composition, including moisture, duration of snow cover, and possibly depth of thaw. The relationship of these factors with the ordination axes was not linear. Undisturbed and recovering communities that belonged to the same habitat/vegetation category usually appeared close to each other. A group of undisturbed communities formed in the center and appeared on all the first three ordination planes (only the first plane is shown); marshes were the only category not included in this group. The distance between the undisturbed and recovering communities in the ordination space represents the environmental differences between their habitats. This distance is well correlated with the preceding results as it is longest between undisturbed and recovering ridges and uplands and shortest between marshes.

The degree of recovery of the original plant communities and the degree of recovery of their habitats thus appear to correspond, just as the degree of uniformity of communities undisturbed by humans and the degree of uniformity of their habitats correspond (Komarkova 1980). Given the original vegetation type, it should be possible to predict the degree of vegetation recovery from the degree of physical environment recovery, and vice versa. The distance between the undisturbed and recovering communities in the ordination space will decrease as recovery progresses.

Brown et al. (1969) reported that human surface disturbances, especially those that result in removal of the surface organic layer, significantly affect the thermal regime in permafrost terrain. Lawson et al. (1978) and Lawson (1982) related the recovery of a disturbed site on permafrost to the recovery of the thermal regime. The average depth of thaw at the Fish Creek site was greater at recovering sites than at undisturbed ones, particularly toward the end of the growing season (Table 1). The greatest difference was in the uplands, which also had the least seasonal thaw increases. The average depth of thaw was 33 cm for all the undisturbed habitats and 37 cm for all the recovering habitats, indicating that most of the disturbed site at Fish Creek is still thermally dissimilar, but that thermal recovery is taking place in some habitats. While the overall mean thaw depth was greater for recovering habitats, the extremes were greatest in some of the undisturbed habitats. This indicates that the development of specialized thaw depth conditions on the Fish Creek site takes more than 34 years (since 1949). Except for marshes, the differences in thaw between undisturbed and recovering habitats correlated positively with the degree of floristic difference between undisturbed and recovering communities.

DISCUSSION

Intensity, type and frequency of disturbance, and ecosystem resistance (ability to remain unaffected by disturbances) and resilience (ability to recover to a more or less persistent state) determine the speed, degree, and direction of recovery from partial or total disturbance (Boesch and Rosenberg 1981).

In the Alaskan arctic tundra, marshes appear to be more responsive and resilient and less resistant to partial surface disturbance than mesic uplands. Although the same vehicle will cause more severe damage in a marsh, where the surface organic matter is considerably more compressed, than in a well-drained environment where the resistance to compression is higher (Bliss and Wein 1972, Walker et al. 1980), recovery is faster in a marsh where, under saturated conditions, compressed fibrous organic material regains its original structure faster (Abele et al., 1983).

Marshes are also more resilient after total surface disturbance. A major man-inflicted surface disturbance overcomes the resistance threshold beyond which the ecosystem does not resist displacement from its initial state. At Fish Creek, the initial disturbance resulted in a new, relatively homogeneous, bladed surface that was later differentiated by newly triggered disturbance processes such as the continuing thaw of ground ice, thermokarsting, and erosion; some live plant propagules may have been preserved at the site. Colonizing plants and successional communities slowed the new disturbance processes, contributing to the differentiation of the newly forming landforms and habitats. A few sloping, eroding sites may never recover. In 30 years marshes that resemble undisturbed marshes have developed in thermokarst depressions, but the uplands on elevated sites between the depressions do not resemble undisturbed uplands. Nevertheless, the resilience thresholds (at which the ecosystem returns to a different persistent state, characterized by altered taxa composition) of the disturbed uplands have probably not been exceeded; they may only be recovering at a very slow rate.

The resilience of the marshes at Fish Creek after the total surface disturbance is greater due to both abiotic and biotic circumstances. On disturbed sites, rapid differentiation of marshes is aided when organic matter and sediments from thawing ice wedges are deposited on thermokarsting surfaces (Lawson 1982). Marshes are strongly controlled by local natural disturbance factors, water, and the freeze-thaw cycle; in particular, water makes the pre- and post-disturbance environments similar, which may speed up colonization by the dominants of undisturbed marshes. It also speeds up colonization of organic matter left after the disturbance, which under dry conditions does not conduct enough moisture for plant establishment (Deneke et al. 1975).

The dominants of undisturbed marshes, *Carex aquatilis* Wg. ssp. *stans* (Drejer) Hultén and *Eriophorum angustifolium* Honck. ssp. *subarcticum* (V. Vassil.) Hultén, colonize and dominate newly available marsh habitats at Fish Creek. As observed in small patches bared by the 1979 cleanup, both taxa spread rapidly by rhizomes and, at a slower rate, by seedlings. The recovering marsh communities are similar to the undisturbed ones, but they lack a number of floristically important taxa that would be present were the community fully recovered. On naturally disturbed surfaces, such as wet stabilizing dunes and natural thermokarst, developing marsh communities with similar composition are common.

Alaskan arctic uplands recovering after surface disturbance contain no ice; due to compression, soil pore space and moisture content are generally reduced, and the soil bulk density is increased (Lawson 1982). Changes in soil characteristics, particularly decreased moisture, may hinder germination and establishment of *Eriophorum vaginatum* L.

ssp. spissum (Fern.) Hultén, which dominates the undisturbed mesic upland habitats but has failed to colonize the small upland patches bared by the 1979 cleanup, nor does it dominate the successional upland communities that developed at Fish Creek.

As on many human and natural disturbances throughout the Arctic, the successional upland communities at Fish Creek are formed primarily by grasses (Arctagrostis and Poa spp.) that disperse, grow, and reproduce rapidly. These grasses represent a very low percentage of undisturbed upland vegetation, but they occur even on very small disturbed patches (a few m^2) and persist wherever they are periodically disturbed. Along with willows, these taxa may be enrichment opportunists, and their dominance may be promoted by the greater availability of nutrients in disturbed habitats (Chapin and Shaver 1981). Nutrients are released when permafrost containing organic matter thaws in both marsh and upland habitats (Brown et al. 1969), which is what happened at Fish Creek. On sites where the top organic layer is underlain by mineral matter, the removal of the surface depletes the nutrient capital (Van Cleve 1977).

Another factor contributing to the success of grasses over Eriophorum vaginatum during the colonization stage at Fish Creek may be their rapid advance into newly available bare surfaces by rhizomes; Eriophorum vaginatum reproduces only by seeds, and the seedlings have a high mortality rate (Gartner 1982). Eriophorum vaginatum appears to be less well adapted for both colonization and dominance than the dominants of both recovering and undisturbed marshes: Carex aquatilis ssp. stans and Eriophorum angustifolium ssp. subarcticum, which spread rapidly by rhizomes. Carex aquatilis also has larger below-ground storage reserves and slower root turnover than Eriophorum vaginatum (Chapin and Slack 1979).

Eriophorum tussocks accumulate organic matter, which insulates the ground, holds moisture, and raises the permafrost table to shallower depths that change very little throughout the growing season. In marshes, fluctuations in the depth of thaw are considerably greater, which may contribute to the better colonizing abilities of the marsh dominants. In fluctuating environments in general, taxa adapted for rapid dispersal, growth, and reproduction predominate and make the community highly resilient, while in nonfluctuating environments, taxa adapted for high competitive ability are dominant and make the community less resilient (Grime 1979).

Eriophorum vaginatum is a successful colonizer elsewhere, such as in wet, 5- to 10-year-old disturbed sites along the forested parts of the Yukon River-Prudhoe Bay road. Chapin and Chapin (1980) observed that within 5 years of a disturbance, Eriophorum vaginatum revegetated an organic disturbed site in interior Alaska that had been successfully seeded with commercially available nonnative grasses. This taxon also colonizes frost boils at Cape Thompson and on the Seward Peninsula (Hopkins and Sigafoos 1951, Johnson et al. 1966), and Gartner (1982) observed its good colonizing ability and early dominance in the Toolik Lake area. These sites are all south of Fish Creek, however; Eriophorum vaginatum may be less efficient at the northern limits of its occurrence and dominance.

At Fish Creek, the change from successional grasses and sedges to Eriophorum vaginatum will be very slow. Permanent plots established in successional communities in 1980 showed no noticeable composition changes by 1982, although there is some indication that the replacement is beginning: Eriophorum tussocks appear in several places on the disturbed site, including a few ripped trenches. The development of tussock communities at Fish Creek is apparently inhibited by the Arctagrostis, Poa, and Carex communities that precede them in the successional sequence; introduced grasses delay invasion by relatively slow-growing native taxa (Johnson and Van Cleve 1976). No such inhibition exists in marshes: once the dominant achieves high cover, the subordinate taxa establish themselves more easily, even in different tussock communities (Hopkins and Sigafoos 1951, Chapin and Chapin 1980).

On the Arctic Coastal Plain, complex Eriophorum vaginatum communities grow on river meanders that are 600-800 years old (Everett 1979, 1980; Komarkova and Webber 1980). The development of complex Eriophorum vaginatum communities at the Fish Creek site may take at least that long. Propagule availability, organic-rich substrate, and moisture conditions are more favorable in the small disturbed patch at Fish Creek, but the development of the tussock community is hindered by the successional upland community of grasses and sedges.

Marsh communities appear on somewhat younger, 400-600-year-old meanders and in relatively young drained lakes; in 100-200 years they may develop on thermokarsting sites where man has disturbed the surface. During the several thousand years between large landscape disturbances, the complexity of both marsh and tussock upland communities gradually increases, particularly through the development of complex microtopography; this can be seen on a series of old river meanders at Atkasook, Alaska.

Most of the communities resulting from the recovery of the Fish Creek site will probably resemble the current, undisturbed complex communities in most general features. It is likely that the Eriophorum vaginatum tussock communities are in equilibrium with the present climate and will recover once disturbed, contrary to Webber and Ives (1978) and Lawson et al. (1978). Tussock communities can be seen developing on relatively young surfaces in the Fish Creek area and after recent disturbances in southern locations in arctic Alaska. However, very recent climate change could have lowered the recovery potential of Eriophorum vaginatum communities at their northern distribution limit.

Human surface disturbances, such as the Fish Creek Test Well 1 site, represent a relatively small patch, and landscape changes triggered by the original disturbance are very localized compared to the natural landscape changes caused by the thaw lake cycle, meandering streams, and wind. Because these natural disturbances overwork most surfaces on the Arctic Coastal Plain in cycles of several thousand years, it can be expected that the small sites disturbed by humans will eventually enter this disturbance-and-recovery cycle and disappear.

CONCEPTUAL MODEL

Figure 3 shows a conceptual model of the recovery of upland and marsh communities at Fish Creek. Mesic uplands have a narrower ecological range, show less fluctuation in environmental

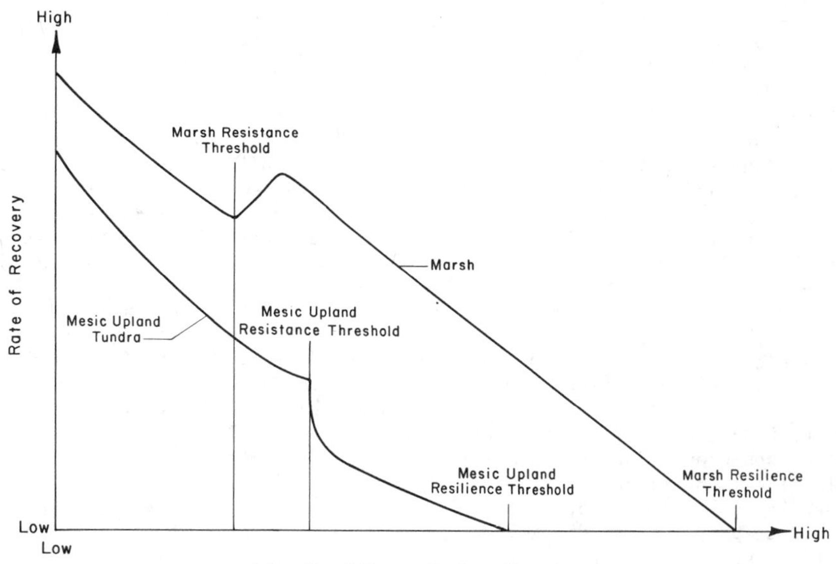

FIGURE 3 Conceptual model of recovery rates of Fish Creek communities related to community resistance and to the intensity of human surface disturbance. Community resistance is independent of disturbance. More resilient communities have higher overall recovery rates, which decrease slower than in less resilient communities. In highly resilient communities, such as marshes, the rate of recovery decreases with the increasing intensity of disturbance until it reaches the threshold of community resistance, and then increases when the accumulated organic matter is removed. In less resilient communities, such as upland tundra, the rate of recovery drops considerably once the community resistance threshold is passed, and the space is occupied by a successional community composed of successful colonizers, which are slowly replaced by the dominants of undisturbed communities. The communities are presumed to be at their ecological optimum, i.e. at the most favorable conditions under which they occur.

factors such as depth of thaw, and recover more slowly. This is consistent with the hypotheses that (1) complex, mature, local-climate-controlled systems are stable (resistant and resilient) only within a comparatively narrow range of conditions, whereas simple, pioneer systems controlled by factors other than local climate are stable in a wider range of environmental conditions (May 1973, Goodman 1975); and (2) frequently disturbed communities or fluctuating environments are more resilient after catastrophic disturbance than similar communities in less fluctuating environments. According to Haber (1979), the persistent, resistant, diverse ecosystems occur in relatively constant environments and can develop a rich and constant assemblage of taxa; the dynamic, resilient, simple ecosystems occur in fluctuating environments and can only develop a poor and fluctuating assemblage of taxa (van der Maarel 1980).

The tundra on the Alaskan Arctic Coastal Plain is strongly controlled by low temperature and ground moisture and by periodic disturbances of the thaw-lake cycle, rivers, and wind. It represents an "azonal" ecosystem that is taxa-poorer and has a higher degree of dominance than less-stressed ecosystems. Because these ecosystems have evolved under drastic periodic landscape changes, they are probably well equipped to cope with small-scale human surface disturbances such as occurred at Fish Creek, and they may be more resilient to disturbance than less-stressed ecosystems. Because of the slow dynamics of the system, however, the recovery times of complex communities may be very slow.

ACKNOWLEDGMENT

This study was funded primarily by the U.S. Geological Survey's National Petroleum Reserve, Alaska program through the U.S. Army Cold Regions Research and Engineering Laboratory. I would like to thank the reviewers for valuable comments.

REFERENCES

Abele, G., Brown, J., and Brewer, M.C., 1983, Effects of offroad vehicle tests on tundra terrain in northwestern Alaska, in Proceedings of the Fourth International Conference on Permafrost, Fairbanks, Alaska (this volume).

Billings, W.D., 1973, Arctic and alpine vegetations: Similarities, differences, and susceptibility to disturbance: BioScience, v. 23, p. 697-704.

Bliss, L.C. and Wein, R.W., 1972, Plant community responses to disturbances in the western Canadian Arctic: Canadian Journal of Botany, v. 50, p. 1097-1109.

Boesch, D.F., and Rosenberg, R., 1981, Response to stress in marine benthic communities, in Stress effect on natural ecosystems (Barrett, G.W., and Rosenberg, R., eds.): New York, Wiley, p. 179-200.

Brown, J., Rickard, W., and Vietor, D., 1969, The effect of disturbance on permafrost terrain, CRREL Special Report 138: Hanover, N.H., U.S. Army Cold Regions Research and Engineering Laboratory.

Carter, L.D., 1981, A Pleistocene sand sea on the Alaskan Arctic Coastal Plain: Science, v. 211, p. 381-383.

Chapin, F.S., III, and Slack, M., 1979, Effect of defoliation upon root growth, phosphate absorption and respiration in nutrient-limited tundra graminoids: Oecologia, v. 42, p. 67-79.

Chapin, F.S., III, and Chapin, M.C., 1980, Revegetation of an arctic disturbed site by native tundra species: Journal of Applied Ecology, v. 17, p. 449-456.

Chapin, F.S. III, and Shaver, G.R., 1981, Changes in soil properties and vegetation following disturbance of Alaskan arctic tundra: Journal of Applied Ecology, v. 18, p. 605-617.

Deneke, F.J., McCown, B.H., Coyne, P.I., Rickard, W., and Brown, J., 1975, Biological aspects of terrestrial oil spills, USACRREL Oil Research in Alaska, 1970-74, CRREL Research Report 346: Hanover, N.H., U.S. Army Cold Regions Research and Engineering Laboratory.

Everett, K.R., 1979, Evolution of the soil landscape in the sand region of the Arctic Coastal Plain as exemplified at Atkasook, Alaska: Arctic, v. 32, p. 207-223.

Everett, K.R., 1980, Distribution and variability of soils near Atkasook, Alaska: Arctic and Alpine Research, v. 12, p. 433-446.

Gartner, B.L., 1982, Controls over regeneration of tundra graminoids in a natural and a man-disturbed site in arctic Alaska, M.Sc. thesis, University of Alaska, Fairbanks, Alaska.

Goodman, D., 1975, The theory of diversity-stability relationship in ecology: Quarterly Review of Biology, v. 50, p. 237-266.

Grime, J.P., 1979, Plant strategies and vegetation processes: New York, Wiley.

Haber, W., 1979, Theoretische Anmerkungen zür ökologischen Planung: Verhandlungen der Gesellschaft für Ökologie, Freising-Weihen-Stephan, v. 7, p. 19-30.

Hernandez, H., 1973, Natural plant recolonization of surficial disturbances, Tuktoyaktuk Peninsula region, Northwest Territories: Canadian Journal of Botany, v. 51, p. 2177-2196.

Hopkins, D.M. and Sigafoos, R.S., 1951, Frost action and vegetation patterns on Seward Peninsula, Alaska. U.S. Geological Survey Bulletin, v. 974-C, p. 51-100.

Husky Oil NPR Operations, Inc., 1981, Environmental cleanup of solid waste abandoned during the early years of exploration activities on the National Petroleum Reserve in Alaska, 1979 and 1980 summer season.

Johnson, A.W., Viereck, L.A., Johnson, R.E., and Melchior, H., 1966, Vegetation and flora, in Environment of the Cape Thompson region, Alaska (Wilimovsky, N.J., and Wolfe, J.N., eds.): U.S. Atomic Energy Commission, Washington, D.C., p. 277-354.

Johnson, L.A., and Van Cleve, K., 1976, Revegetation in arctic and subarctic North America: A literature review, CRREL Report 76-15: Hanover, N.H., U.S. Army Cold Regions Research and Engineering Laboratory.

Komarkova, V., 1980, Classification and ordination in the Indian Peaks area, Colorado Rocky Mountains: Vegetatio, v. 42, p. 149-163.

Komarkova, V., and Webber, P.J., 1980, Two low arctic vegetation maps along the Meade River near Atkasook, Alaska: Arctic and Alpine Research, v. 12, p. 447-472.

Lawson, D.E., 1982, Long-term modifications of perennially frozen sediment and terrain at East Oumalik, Northern Alaska, CRREL Report 82-36: Hanover, N.H., U.S. Army Cold Regions Research and Engineering Laboratory.

Lawson, D.E., Brown, J., Everett, K.R., Johnson, A.W., Komarkova, V., Murray, B.M., Murray, D.F., and Webber, P.J., eds., 1978, Tundra disturbances and recovery following the 1949 exploratory drilling, Fish Creek, Northern Alaska, CRREL Report 78-28: Hanover, H.H., U.S. Army Cold Regions Research and Engineering Laboratory.

May, R.M., 1973, Stability and complexity in model ecosystems: Princeton, N.J., Princeton University Press.

Van Cleve, K., 1977, Recovery of disturbed tundra and taiga surfaces in Alaska, in Recovery and restoration of damaged ecosystems (Cairns, J., Jr., Dickson, K.L., and Herricks, E.E., eds.): Charlottesville, Va., University Press of Virginia, p. 422-455.

van der Maarel, E., 1980, Towards an ecological theory of nature management, in Haber, W., ed., Verhandlungen der Gesellschaft für Ökologie (Freising-Weihenstephen 1979), v. 8, p. 13-24.

Walker, D.A., Everett, K.R., Webber, P.J., and Brown, J., 1980, Geobotanical atlas of the Prudhoe Bay region, Alaska, CRREL Report 80-14: Hanover, N.H., U.S. Army Cold Regions Research and Engineering Laboratory.

Webber, P.J., and Ives, J.D., 1978, Recommendations concerning the damage and recovery of tundra vegetation: Environmental Conservation, v. 5, p. 171-182.

Westhoff, V., and van der Maarel, E., 1978, The Braun-Blanquet approach, in Classification of plant communities (Whittaker, R.H., ed.): The Hague, Junk, p. 287-399.

GEOCRYOLOGICAL MONITORING OF PIPELINES

V. G. Kondratyev

The Chita Polytechnical Institute
Chita, USSR

Long-term geocryological-geological investigations along the route of an arctic pipeline revealed environmental disturbances during both construction and operation and also corresponding changes in the permafrost condition. The impact of cryogenic processes on pipeline operation was studied; causes of breakdowns were revealed; and possible ways of improving operation were pinpointed. The conclusion was reached that geocryological monitoring must be included in the design construction and operation of pipelines in permafrost zones, as well as permafrost surveys and forecasts. The term monitoring implies systematic field observations of the dynamics of permafrost conditions along the route of the pipeline as the natural environment evolves under the technogenic influences imposed on it. The scientific and procedural aspects of geocryological monitoring are described. Its aims and practical applications at the various stages of exploration, construction and operation are defined. It is noted that the monitoring data are necessary for revealing trends in the formation and development of permafrost conditions along the pipeline, and for developing, checking and streamlining forecasts of changes and means of modifying permafrost conditions to ensure optimum conditions for pipeline operation consistent with environmental protection.

In modern literature devoted to the problems of environmental protection the term "monitoring" is more and more often used. It means, according to the Merriam-Webster Dictionary (1974), "watching or observing for a special purpose." However, in some scientific literature this term includes a broader sense, though different authors use it in various meanings. Thus, in the booklet "Global environmental monitoring", published in 1971 by a special committee of the Scientific Commission on Problems of the Environmental of the International Council of Scientific Unions the comment is made that a system of monitoring includes systematic observations on the state of the environment, determination of possible changes (especially due to human influence), control of such changes, and measures to regulate (control) the environment (Ananichev 1974). In a similar definition Gerasimov (1975) describes it as "monitoring is a system of observations, control and environment conduction, which are carried out on different scales and on a global one as well." Israel (1974) says, "it is more correct to call the system of observations which helps to distinguish changes of the biosphere state under the influence of the human activity—monitoring." Mann (1973) considers monitoring an estimation of repeated observations of one or some more elements of environment; in space and time, with the definite aims and program prepared beforehand. Thus the environment observations should be considered as the main thing in monitoring, though some authors also include in it the environment estimation and control.

Many countries have set about creating a global environment monitoring system on the initiative of the United Nations. In the USSR this is carried out by the State Service of Observation and Control over the pollution level of the environment. Recently the question arose about the necessity of environmental monitoring as a consistent of a global one. In particular, Konoplyantsev and other authors (1982) suggest organizing an observation system that would watch all processes arising with respect to underground water under the influence of the anthropogenic changes, to accumulate necessary series of comparable information, to evaluate the present situation, and to carry out the prediction of the regime, resources of the underground water and their quality for far and near perspective (underground water monitoring). There are also suggestions about organizing hydrogeological monitoring during exploitation of mineral resources (Korobeynikov et al. 1982). Grave (1980) has reported on the monitoring of the cryolithozone and the control of development in the geocryological environment as related to natural factors and technogenic operations.

Frozen rocks (Kudryavtsev 1978) are closely interrelated with the complex of the geological-geographical factors of the natural environment. Their changes cause natural changes of the permafrost conditions in the course of natural evolution and especially under technogenic operations. They are expressed in terms or the changes of temperature of the rock regime, rate and depth of their seasonal freezing or melting, the development of the cryogenic processes and phenomena. In this connection essential change of the engineering geological conditions of the territory are possible and may have an unfavorable influence on the stability of structures. For example, pipelines, being linear structures with variable heat regimes, can exert considerable influence (mechanical and thermal) upon the environment. Consequently, the permafrost situation can be changed considerably along the pipeline. The latter, in turn, acting on the pipeline may cause additional stresses and deformations, which may result in its destruction.

Thus, because of the specific characteristics of the permafrost situation which is constantly developed and if taking into account its functional interconnection with the complex of natural factors changing both in natural conditions and during the pipeline construction and operation, along with the permafrost survey and forecast geocryological monitoring prototype systematic observations of changes of permafrost conditions along the pipeline in the course of natural evolution and under the technogenic influence should be fulfilled. It is necessary to do monitoring constantly from the moment of prospecting, during construction and after operation of the pipeline, because of the continuous changes of permafrost conditions along the pipeline.

During prospecting, prototype observations are carried out, together with the permafrost survey, and should be aimed at solving the main task of exposing general and specific regularities of forming and developing seasonal rocks and permafrost, cryogenic processes and phenomena connecting with them. Geocryological monitoring is organized in the early stages of the technological and economic calculations of the pipeline construction. For this purpose observation areas are designated in all types of terrain, for studying the temperature regime of soils, their seasonal freezing and melting, development of cryogenic processes and phenomena depending upon the main geological-geographical factors and their natural dynamics. In the subsequent site selection, monitoring should be continued on the chosen variant of the pipeline. Essential increase of the number of the observation areas and the regime observation program becomes more complicated. In addition to more detailed observations of natural permafrost situations at control areas, observations at experimental areas, with different types of technogenic operations, i.e. at "technogenic" areas are organized. These are, for example, areas with the vegetative cover taken off, areas kept free of snow, drained ponds, etc. The monitoring data during engineering prospecting along the pipeline are used for the engineering-geological estimation and making recommendations concerning the control of the permafrost situation to ensure optimum conditions for the pipeline construction, operation and environment protection.

In connection with the possibility of unfavorable development of cryogenic processes along the pipeline during construction, it is advisable to create special geocryological services (geocryological monitoring service, geocryological station, etc.) attached to the management of the pipeline under construction (or to some research institution). Such groups would carry out: (1) constant permafrost control during the pipeline construction and (2) operation of the fulfilment of measures ensuring its stability and geocryological environment protection; (3) regime observations of changes of permafrost conditions in the course of natural evolution and in connection with the disturbance of the complex of natural conditions during the pipeline construction and its heat influence during the pipeline operation; and (4) systematic study of the influence of the cryogenic processes on the pipeline operation and working out effective measures, ensuring better geocryological conditions for its operation and environmental protection.

During the pipeline construction the fulfilment of all measures concerning the control of the permafrost situation suggested in the project are controlled, with the purpose of ensuring stability of construction and environment protection. Special attention should be paid to the quality of the earthwork, surface planning, maximum preservation of the turf-mantle of soil and to the fulfillment of measures concerning earth recultivation and decreasing the development of undesirable cryogenic processes. During this period geocryological monitoring continues at the areas equipped during prospecting (they characterize, mainly, natural conditions of work) and at additional areas which are made along the pipeline during its construction. Observation areas should characterize every type of locality, at the most complicated parts of the pipeline, for example, in areas crossing the valleys of large rivers. Observations are also made at locations where the pipeline laying, structural peculiarities (diameter, thermal insulation, etc.) and the operational regimes differ. Regime observations of the temperature dynamics, seasonal freezing and melting of soil, development of the cryogenic processes are conducted in natural conditions (undisturbed), along the construction area (in disturbed conditions) and near the pipe directly. All disturbances of the complex of natural factors occurring as the result of construction operations are fixed to maximum details. Systematic observations of the relaxation of these disturbances and regeneration of the vegetative cover are carried out; simultaneously, conditions of snow accumulation and surface structure of the radiation-thermal balance and, hence, the permafrost situation along the pipeline will change.

During pipeline operation observations are carried out at the areas equipped during construction and at those left after prospecting. They are: (1) near the pipe—observations of the temperature field change, the aureole formation of seasonal soil melting and freezing, the vertical displacement and stressed state of the structure, the thermoerosion development, winter-killing, cracking, and other cryogenic processes; (2) within the construction zone, and (3) in natural conditions—of the temperature field changes, soil dynamics of seasonal freezing and melting, new formation of permafrost and freezing or, on the contrary, of the formation of non-confluent permafrost areas, development of cryogenic processes and phenomena. Special attention should be paid to the study of influence of the cryogenic processes on the pipeline operation, efficiency of the project measures for the construction stability and environment protection. All cases of pipeline ruptures should be thoroughly analyzed from the point of view of the influence of the cryogenic processes. Based upon the regime observations it is necessary to test and to improve the accuracy of geocryological forecasting. In accordance with this to make new recommendations for improvement of the geocryological conditions of the pipeline operation. During this period it is necessary to carry out systematic and strict control over fulfilment of the measures ensuring the pipeline stability and environment protection, particularly over limitations of the movement of crawler tractors near the pipeline and in the adjacent area.

Scientific and procedural aspects of the geocryologic pipeline monitoring described above may be illustrated by the long-term data of the geocryological-geological investigations conducted by the Geocryological Department of Moscow University at one of the northern gas pipelines (Poltev et al. 1979, Kondratyev et al. 1979). Experimental investigations were made at the most complicated and important section of the pipeline; the part of the pipeline crossing a large river valley.

With respect to climate this region is characterized by a low average annual air temperature (-10.6°C), a large amplitude of annual fluctuation (44.3°C), and 44.3 cm of precipitation annually. Due to snow-storm winds thickness of snow cover often is irregular in valleys, at the foot of the slopes and in areas covered with thick shrubs it reaches 1.5-3 m and more, while in elevated areas, especially from the bare ones, snow is blown off and its thickness is 0 to 0.5 m. In this connection the river valley and the adjacent areas of the alluvial plain are characterized by the very complicated permafrost situation. The areas of continuous (low flood plain, first terrace above the flood plain, watershed surface), discontinuous permafrost (high flood plain), and also the areas without permafrost (river-bed, bottom of large lakes) are observed. Thickness of the permafrost mass varies from 3-10 m at the bottoms of the drained lake potholes to one hundred metres at the water divide surface. The upper 10-15 m layer of the permafrost consists of rocks of the Middle and Upper Pleistocene and also Holocene, mainly, of loamy and sandy composition. The development of the epigenetic as well as syngenetic and polygenetic permafrost is marked. In accordance with the genesis of the permafrost masses, and their composition, the character of the cryogenic texture and their ice-content is changed. Stratified and reticulate structures prevail. At some areas the ice-content in the soil is 50% and more. Average annual temperature of soils at the depth of annual zero amplitudes varies from 3 to 4°C under large lakes to -4 to -6°C on the watershed surface. Typical temperatures of the soils in the river valley range from -0.2 to 2°C. The temperature regime in rocks is influenced considerably by the snow cover, increasing the average annual temperature of soils by 2-12°C. Ponds (by 0.1-6°C), bog formation (by 0.1-1°C) and summer precipitation infiltration (0.1-0.5) give heat effect as well. Vegetation cools soils by 0.1 to 1.5°C, however, retaining the snow (especially in shrubs) increases the temperature of rocks. Temperature shift in annual thawed layer reduces temperature by 0.2-1.1°C. Thermal influences by other factors is not significant. In the territory under study seasonal melting as well as seasonal freezing of soils is observed, the former process spreading much wider than the latter one. Thickness of annual thawed layer ranges, depending on the complex of natural factors, from 0.6 to 0.75 m to 1.8-2.5 m and that of an active layer from 0.4-0.5 to 1.3-1.5 m.

In natural conditions practically all main geocryological-geological processes and phenomena take place here, however, the degree of their development is very different. Solifluction, heaving, thermokarst are widespread. Such processes as frost cracking and ice wedge formation are not common. Spreading the geocryological processes and phenomena along the area is not constant. High flood plain (heaving, thermokarst) and right-bank valley side (solifluction, ice formation) are subjected to their influence to a great extent. On the right bank within the lake-alluvial plain different age forms of the thermokarst and heaving hillocks are also met. Trench-laying of tunnels was made because of the danger of the pipeline destruction by the spring ice drift in the river floodplain. During the pipeline construction natural conditions were considerably disturbed. So, under each tunnel in the floodplain a strip from 30 to 50 m wide was cleared away by bulldozers and then the trench from 1 to 5 m deep and from 1.5 to 10 m wide, was dug by excavating machines (with the preliminary ripping of soil by blasting). It was accompanied by partial change of microrelief, shrub destruction, considerable moss-herbage disturbance and also by changes of the runoff conditions of the surface, intrapermafrost and subpermafrost water, partial or complete drainage of some swamps or lakes. Later on (2-3 years) recovery of herbage took place within the construction strip and in 5-6 years, cuttings were covered with meadow herbage 0.2 to 0.8 m in height of different density. In some areas even small thin willow shrubs (to 0.2-0.4 m) appeared. However, in the areas of the intensive thermoerosion development, e.g. on the slope of the terrace above the floodplain, herbage did not recover. Shrub felling changed considerably the conditions of the snow cover formation. Thus, in the high flood-lain area the depth of snow decreased from 2 to 2.5 m (density, 0.25 to 0.35 g/cm^3) to 0.12 to 0.90 m (density, 0.3 to 0.4 g/cm^3).

Everything mentioned above, together with the variable pipeline thermal regime, caused great changes in permafrost conditions along the pipeline, in comparison with natural situation. Experimental studies of the dynamics of the permafrost situation, in connection with the pipeline construction and operation, was made on 18 specially equipped observation areas which made it possible to carry on investigations, in practically all micro-districts of the territory having different operation regimes of the pipeline (Kondratyev et al. 1976). Observations of the soil temperature dynamics, using response thermometers and thermistors, in holes of 7 to 11 m in depth, proved that the considerable changes within the construction strip occurred in areas where shrubs were destroyed and the depth of the snow cover essentially decreased. Thus, average annual temperature of soils at the base of the annual thawed layer decreased by 2-5°C in high floodplain areas and, at the depth of the annual zero amplitudes, by 0.3 to 2.5°C. In areas where the disturbance of the complex of natural conditions was not critical or it recovered rapidly (low floodplain), the soil temperature within the construction strip did not change significantly. As the average annual temperature of gas at the starting point of the underground area is almost the same as the air temperature and the temperature of rocks changes from 0° to -4°C, the pipe during the average annual cycle causes a cooling effect on the pipeline section to the riverbed equal to 9°C at the pipe entrance into the soil; 3.5°C at the 0.5 km distance, and only 2.5°C at the 1.5 km distance. Thermal effect on the pipe occurs beyond the river, the riverbed being 2.5 km wide there. The pipeline thermal influence zone decreases rapidly downwards and laterally from the pipe. When the pipeline operates all year round, and is buried at a depth of 1.0 to 1.5 m, the thermal influence of the pipe downwards is not more than 8 to 10 m, and to the side only 3 to 4 m.

Changes in temperature regime of soils and their thermophysical properties caused the depth changes of their annual thawing. The greatest increase in the depth of the annual thawed layer (by 25-50%) occurs within the construction zone in areas such as the high floodplain and on the terrace cusp, where the shrub and the moss cover were destroyed. In the areas where there were no essential vegetative changes, such as low floodplains and the bottom of the lake basin, the depth of annual thawing of soils did not, for practical purposes, change. Due to the high moisture content of soils in the river floodplain and the decreased average annual temperature within the construction zone, because of the reduced snow cover, there was a comparatively small increase in the depth of annual thawing within the construction zone, in comparison with natural conditions away from the pipeline. A sharp increase in the depth of freezing was observed in the areas of non-confluent permafrost, where linking of the new permafrost layer with the permafrost table at greater depths took place during the first 1-3 years.

The deepest layer of soils with depths of 1.2 to 2.5 m thaw in the trench. Such deep thawing takes place both because of the day-surface and the pipe in which gas transporting in summer has a positive temperature. Thawing of soils under the pipe in the river valley varies from 0.6 m at the lake basin bottom to 1.6 m on the terrace slope, the increase being connected in the latter case mostly with the greatest thermal effect of the pipeline, since this area is the nearest one to the point of the pipe entrance into the soil and also to the suprazone water run-off along the trench. The least waste of heat on the phase water transition is observed there, as ice content of sandy loam in the trench is not essential. The least thawing layer is formed in those areas with shallow burial of the pipe at the lake basin bottom, where the soils are characterized by large ice content and low average annual temperature at the base of the annual thawed layer.

Destruction of the vegetative cover, and the sod-mantle of soil, changes in the temperature soil regime, depth and rate of their annual thawing and freezing caused considerable active geocryological-geological processes. Thus, it stimulates soil solifluction on the slopes. On many slopes and ravines, thermoerosive rain rills and gullies (Figure 1) are intensively developed because of the sod cover destruction and the formation of the concentrated run off along the technogenic microrecesses, as a rule, along the crawler tractor tracks. Their formation rate is rather great with gullies 3 to 4 m deep and 5 to 20 m wide, forming 1-2 years after construction. Especially intensive growth of gullies occurred during rainfalls. The increase in depth of the annual thawed layer, as well as the change of its moisture regime causes an increase in seasonal heaving and settlement of soils within the construction zone 1.1 to 2.0 times as great (up to 4-12 cm), in the trench 2 to 5 times as great (up to 8-30 cm).

The variety of permafrost conditions of the region under study causes rather dissimilar effect of the permafrost on the pipeline. This effect becomes sharply complicated by the changes of the permafrost conditions because of the destruction of the natural situation in the course of the pipeline construction and its subsequent thermal effect during operation. Everything said above together with variable thermal regime of the pipeline causes the development of pipe tensions of different values that often result in the breaking of the pipe (Kondratyev et al. 1979). It is characteristic that the overwhelming number of breaks occurred in the river valley, during the period from November until February in contact zones of permafrost and thawed rocks (Figure 2) and in places of large curves as well (Poltev et al. 1979).

The analysis of the permafrost situation of soils in places of the pipeline breaks and of the dynamics of freezing of aureoles of seasonal thawing of soils around the pipe in the areas of the permafrost development and of new formation of permafrost masses, temperature observations of the pipeline sides helped to explain the pipe breakdowns. Soil in the aureoles of seasonal thawing around the pipe in November-December and sometimes later because of the cold gas transportation freezing over, causes the pipe in the permafrost to be squeezed. In the areas of thawed rocks (under fluvial, under lakebed taliks, and in regions of non-confluent permafrost) during the whole winter, new formation of permafrost takes place under the pipeline with the intensive ice formation and soil heaving. Thus, the pipe section squeezed at the ends in permafrost masses, undergoes the ever-growing pressure on the freezing soils. Curved pipeline sections, with a small radius of bend and large temperature amplitudes at the pipe wall, causes severe stresses in the pipe. Everything mentioned above results in breaking the pipe at the weakest points, as a rule, along welded joints. That is why the following precautions should be taken into consideration: (1) avoid, if possible, laying the pipe in taliks; (2) decreasing the size and unevenness of soil heaving in a trench; (3) preventing the pipe squeezing and permitting its movement under the influence of heaving, temperature, and other tensions; (4) utilize insulation of the pipe at certain areas, with the purpose of decreasing the amplitude of temperature fluctuations, and (5) laying the pipe, if possible, without curves.

On the right-bank area of the river crossing pipeline breakdowns were not encountered. However, the development of solifluction and thermoerosive processes is of great danger for the pipeline stability (Figure 1).

Thus, the research carried out shows that rather essential changes of permafrost conditions along the pipelines take place. These changes may be quite different in various sections of the pipelines depending on the complex of the geological-geographical factors. Together with variable thermal regimes of the pipeline they cause the pipe tensions, often resulting in its breakage. In this connection geocryological monitoring should be constantly carried out with the purpose of studying the dynamics of permafrost situation, in natural conditions, and under the influence of the pipeline construction and operation. Thus, it is possible to work out effective measures concerning the control of the permafrost situation along the pipeline construction with the purpose of ensuring its stability and environmental protection.

REFERENCES

Ananichev, K. V., 1974, Problems of environment, energy and natural resources, Moscow, USSR.

Gerasimov, I. P., 1975, Scientific fundamentals of modern environment monitoring, Izvestia, Academy of Sciences of the USSR, Geography serial, no. 3, p. 13-25.

Grave, N. A., 1980, Position and direction of geocryological investigations in the problem of environmental protection and rational use of nature in permafrost regions, in Surface stability against technogenic influence in permafrost regions: Yakutsk, USSR, p. 6-16.

Israel, U. A., 1974, Global system of observations. Forecast and estimation of changes of natural environmental state, Fundamentals of monitoring: Meteorology and Hydrology, no. 7.

Kondratyev, V. G., Zontov, M. N., Chushkina, N. I., and Chalyavina, N. S., 1976, Natural study of buried pipeline and permafrost rocks interaction, in Proceedings of the Third Scientific Conference of Young Scientists and Postgraduates of Moscow University: Geocryology serial, Moscow, USSR, Izdatelstvo "Moscow University," p. 16-24.

Kondratyev, V. G., Zamolotchikova, S. A., and Poltev, N. F., 1979, Pipeline crossing projecting through the valleys of large rivers in arctic regions: Pipeline Construction, no. 8, p. 30-31.

Konoplyantsev, A. A., Lapshova, L. P., and Semyonov, S. M., 1982, Underground water monitoring and environmental protection of Siberia and the Far East, in Proceedings of the All-Union Conference on Underground Water of the East of the USSR, Irkutsk, p. 75-76.

Korobeinikov, V. A., Buchkin, M. N., and Ustinova, Z. G., 1982, Hydrogeological monitoring of the mining complex, in Proceedings of the All-Union Conference on Underground Water of the East of the USSR, Irkutsk, USSR, p. 82.

Kudryavtsev, V. A., ed., 1978, General geocryology: Izdatelstvo "Moscow University," Moscow, USSR.

Mann, R. E., 1973, Global environment monitoring system, CKOPE, rep. 3, Toronto, Canada.

Poltev, N. F., Kondratyev, V. G., Zamolotchikova, S. A., Zontov, M. N., and Chushkina, N. I., 1979, Geocryological geological investigations in connection with pipeline construction and operation in permafrost regions, in Engineering Geocryology: Novosibirsk, USSR, Izdatelstvo "Nauka," p. 213-222.

Webster, M., 1974, The Merriam-Webster Dictionary: New York, Pocket Books.

FIGURE 1 Thermoerosive gully development in connection with turf vegetative cover disturbance during the pipeline construction.

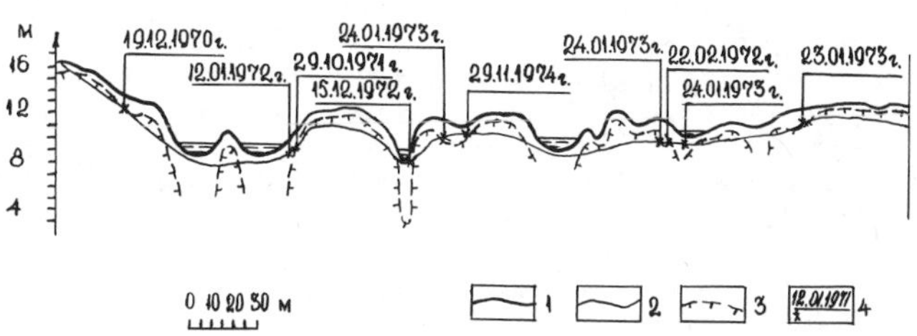

FIGURE 2 Places of pipeline breaks in the river floodplain marked in contact zones of permafrost and thawed rocks. 1) earth surface, 2) pipe, 3) boundary of permafrost rocks, 4) place and date of the break.

THE CRYOGENIC EVOLUTION OF MINERAL MATTER
(AN EXPERIMENTAL MODEL)

V. N. Konishchev and V. V. Rogov

Geography Department
Moscow State University, Moscow, USSR

An analysis of existing hypotheses as to the interaction between water and minerals in connection with their crystal and chemical properties allows one to present a theoretical model for the existence of minerals in the zone of cryogenic weathering. According to this model, the destruction threshold depends on the peculiarities of the surface energy of the minerals, which is a function of the crystallo-chemical properties of the surface, the characteristics of the ground solution, and the degree of dispersion. The threshold of cryogenic destruction was established experimentally for different groups of minerals and for different temperature regimes. It was established that the degree of mineral disintegration was two to three times higher in the case of temperature oscillations across the 0°C threshold than in the case of oscillations entirely beneath that threshold.

One of the main processes of the transformation of mineral matter in the zone of cryolithogenesis is that of destruction of primary minerals and rocks in the course of water↔ice phase transitions. By its mineralogical essence, this process is the simplest out of the known forms of hypergenesis. The mechanism of cryogenic disintegration, however, is of a rather complicated nature and refers to the physical-chemical type. The latter statement is based on the assumption that in fine-grained cryogenic crushing of minerals, their specific surface energy (SSE) has a determining role. The concrete form of the SSE effect on the process of cryogenic disintegration is realized through the protective function of the film of unfrozen water. Thus the different categories of bound water of dispersed systems play a dialectically contradictory part in the process of their freezing-thawing with respect to the resistance of mineral particles. The metastable thermoactive film of bound water inserts a destructive effect, in the main, and the stable film of unfrozen water, on the contrary, raises particle resistance, since the whole range of the outer effects on them occurs mediately, through the film of water unfrozen at the given concrete temperatures. Thus, the resistance of mineral grains is not only determined by their strength, but also depends on the thickness and properties of the adsorbed unfrozen water.

The above-given considerations were assumed as a basis of a theoretical model of cryogenic resistance of different, mainly rock-forming minerals, or rather of their relative correlations, and, in the end, of an ideal scheme of the mineralogical composition of cryogenic eluvium (Konishchev, 1977; Konishchev & Rogov, 1978).

A theoretical study of the process of cryogenic disintegration has resulted in the conclusion on the reverse distribution, over the granulometric spectrum, of the maxima of contents of individual rock-forming minerals in the deposits of the zone of cryolithogenesis, as compared with dispersed deposits formed under the conditions of the warm and temperate climatic zones. One of the particular corollaries of this general proposition has been an inference of the fact that quartz grains in the zone of cryolithogenesis are less resistant than feldspar grains unchanged previously by some other processes.

The need for evaluation of cryogenic resistance in the "particle + unfrozen water film" system is a fundamentally important aspect in substantiating the theoretical model of cryogenic resistance of particles of different mineralogical composition. Thus it follows that in order to verify the validity of the proposed theoretical scheme of cryogenic eluvium, possibilities should be sought that would allow determining not only the total volume of unfrozen water in a dispersed system, but also its real structural correlations with the mineral skeleton. Unfortunately, the widely accepted methods of estimating the amount of unfrozen water do not give a direct answer to the question of its protective functions for different components of the skeleton of frozen ground.

One of the promising approaches to the solution of this complicated problem is related to the microscopic investigation into the structure of frozen dispersed systems.

Studies of the microstructure of mineral suspensions frozen and cooled to low temperatures with the use of electronic microscopy have shown that basal cement-ice that originates with the freezing of suspensions subdivides clearly into two types. One of these, named volumetric, forms from free and loosely bound water, and the other, the near-contact type, forms from firmly bound water. In frozen suspensions consisting of particles of different mineralogical composition with the sizes of >0.1 mm, it was recorded that the volume of the near-contact ice around feldspar grains exceeded that of quartz particles by many times. The value of the volume of the near-contact cement-ice can serve as a measure of SSE of particles of different mineral composition, and it reflects the thickness of unfrozen water films on particles at higher negative temperatures real for the natural environmental conditions. Therefore, the correlation between the volumes of the near-contact cement-ice on quartz and feldspar grains demonstrates quite objectively and obviously the higher resistance of feldspar grains than those of quartz with respect to the factors of the outer cryogenic effect on the ground system.

The laboratory modelling of cryogenic destruction of different minerals with dimensions of the granulometric fractions widely used in the USSR, viz., 1 to 0.5 mm, 0.5 to 0.25 mm, 0.25 to 0.1 mm, 0.1 to 0.05 mm, and 0.05 to 0.01 mm, has made it possible to determine the cryogenic resistance of each of the monomineral fractions. The cryogenic resistance was estimated by a simple index of:

$$Hcr = \frac{a - b}{a \times n},$$

where a = the amount of granulometric monomineral fraction in the specimen prior to the effect of the cryogenic factor on it;
b = the amount of the same fraction after this effect; and
n = the number of cycles of freezing and thawing.

The lower the value of Hcr, the higher is the resistance of fraction. The index suggested is very convenient to generalizing the data on laboratory modelling of cryogenic crushing and allows proceeding from the theoretical scheme to an experimental model of cryogenic disintegration. An analysis of the experimental data has shown that cryogenic resistance is a function of the mineralogical composition, degree of dispersion, humidity, and temperature conditions.

A series of cryogenic resistance of minerals was obtained for different conditions, which turned out to be most original. Thus, for example, the following sequence was observed for 1 to 0.5 mm-fraction with complete moisture saturation and the temperature conditions of freezing-thawing at t^0 = -10; +20°C: apatite < chlorite < quartz < limestone < limonite < magnetite < garnet < microcline < muscovite (the resistance grows from left to right).

A system of such a series has made it possible to assert that quartz grains in all the fractions proved to be less resistant than feldspar grains. We also succeeded in establishing experimentally the limits of cryogenic disintegration of the most widespread minerals. This limit for quartz, amphiboles and pyroxenes corresponds to grain sizes of 0.05 to 0.01 mm, and for fresh feldspars, 0.1 to 0.05 mm.

The dependence of cryogenic stability of minerals on the degree of their erosion and freshness of grains is also clearly traced. Resistance of quartz grains may vary by several times depending on their prehistory, eg., grains from a pegmatite vein are four times more resistant than quartz grains from alluvial sand that underwent several cycles of lithogenesis.

As for feldspars, their grains, even with a low degree of pelitization, reveal a fundamentally different trend in their behaviour as compared with fresh grains, viz., they are characterized by a high degree of disintegration in fine fractions.

The effect of humidity on cryogenic destruction of different minerals is not simple-valued. Grains of minerals with low surface energy (quartz, feldspars, amphiboles, pyroxenes, magnetite and apatite) disintegrate more intensively in a humid medium than in an air-dry state.

The degree of dispersion of minerals with high surface energy (muscovite, biotite and limonite) change fundamentally in a different way depending on the degree of humidity. Fine grains (0.1 to 0.05 mm) of biotite and muscovite are considerably less resistant in an air-dry state than with full saturation with water. These facts illustrate convincingly the protective role of adsorbed water films with respect to the factors of cryogenic influence, although the degree of the protective effect and the grain sizes with which this effect starts playing the leading role are different for different minerals.

The character of changes in the cryogenic destruction depending on dispersion is also most specific for individual minerals. For quartz, increasing resistance of grains is rather clearly traced as their sizes diminish right to the ultimate 0.05 to 0.01 mm.

The same character of dependence is also inherent in other groups of minerals, such as feldspars, amphiboles and pyroxenes. As for the grains of magnetite, apatite and carbonate, their resistance as a whole lowers, though with some divergences, as grain sizes diminish. The dependence of resistance on dispersion for minerals with high surface energy is more complicated, as there is usually a definite size of grains with maximum resistance, while particles coarser and finer than this size disintegrate. Thus, for example, maximum resistance for limonite is observed with its particles of 0.5 to 0.25 mm and 0.25 to 0.1 mm which aggregate under moderate temperature conditions of the experiment (t^0 = -10°; +20°C) coarser and finer particles of limonite disintegrate markedly. For biotite, maximum resistance is found with 0.25 to 0.1 mm grains; and for muscovite, two maxima of resistance are revealed, viz., 0.1 to 0.05 mm and 1 to 0.5 mm.

Under natural environmental conditions, cryogenic disintegration proceeds at different temperature parameters of freezing and thawing of the ground systems and temperature variation amplitudes. Two series of experiments were carried out to study the influence of temperature conditions on the cryogenic transformation of dispersed systems of different mineral compositions.

In the first series, granulometric monomineral fractions of standard sizes and of different minerals were subjected to cyclic freezing and thawing under four regimes (t^0 = -5°, +5°C; t^0 = -10°, +20°C; t^0 = -20°, +20°C; t^0 = -40°, +20°C). A factor common for these regimes was that of transition of temperature through 0°C.

Minerals are rather clearly divided into two groups with respect to these freezing-thawing regimes. The first group is composed of minerals with high surface energy (biotite and limonite). The cryogenic resistance of these minerals is minimal under the most rigorous regime (t^0 = -40°, +20°C).

The second group is comprised of minerals with relatively low surface energy (quartz, magnetite and apatite). The maximum destruction of grains of this group occurs under the conditions of complete moisture saturation and under the optimum regime of freezing and thawing ($t^0 = -10°$ to $-20°$, $+20°C$). The resistance of grains of these minerals grows under milder and more rigorous regimes. The groups of minerals that differ in cryogenic resistance with respect to the regimes of freezing-thawing correspond fully to the mineral groups singled out above depending on the influence of the degree of humidity on changes in their dispersion.

In the second series of the experiments, a spectrum of granulometric fractions of different minerals in the moistened state was frozen and then subjected to the cyclic effects of temperature variations ($t^0 = -1°$, $-20°C$). The results of this series were to answer the question whether cryogenic transformation did take place and to what degree with the phase transitions of unfrozen water and with temperature variations within the negative spectrum of temperatures. Such methods helped to achieve reduction, to a possible minimum, of the effect of the disjoining action of ice on the particle destruction. This series of experiments simulates the conditions of cryogenic transformation of matter in the cold period of the year, after the layer of seasonal thawing freezes, and the upper layer of soil-eluvium can also change its temperature rather sharply due to temperature variations at the surface during the winter period.

The most important and general result of this series of experiments is the inference that the cryogenic disintegration of all the minerals, with the exception of hornblende and magnetite, is considerably lower with temperature variations within the spectrum of negative temperatures than with temperature variations involving the transition through 0°C.

The difference is particularly great for coarse fractions (1 mm to 0.25 mm), viz., cryogenic resistance for these differs by 2 to 3 orders. In finer fractions (0.1 mm to 0.05 mm), the values of cryogenic disintegration of different minerals have considerably lower variations. At the same time, it is necessary to emphasize that the disintegration of minerals with variations of temperatures within the spectrum of their negative values does nevertheless take place, though with lower intensity than under the regime involving the transition through 0°C. And individual minerals disintegrate within the spectrum of negative temperature values even more intensively than with the transition through 0°C. A theoretically important result of this experiment is that of ascertaining the dependence of the degree of destruction on the degree of dispersion. Under the temperature regime of $t^0 = -1°$, $-20°C$, the cryogenic disintegration of minerals tends rather clearly to grow with the diminishing of the sizes of their grains. This dependence is contrary to the pattern that was revealed under the freezing-thawing regimes with the transition through 0°C. A comparison with the first group of experiments shows that the destructive effect of the cleaving action of ice is the highest for coarse fractions, while the cryohydratational weathering is the most effective for finer fractions, with the former factor being absolutely far more intensive.

Under the temperature regimes of $t^0 = -1°$, $-20°C$ (as under other regimes), different minerals are characterized by individual features according to the nature of disintegration. The general reduction of the resistance of mineral grains as their sizes diminish can be asserted to be only a tendency. The concrete curves of resistance of separate minerals depending on the degree of dispersion under the given regime of temperature variations have a more complicated character. At the same time, the fundamental difference between the most widespread rock-forming minerals of quartz and feldspars with respect to the limits of cryogenic destruction takes place also under the regime of $t^0 = -1°$, $-20°C$. And it can be seen here that in the fine-dispersed portion of the granulometric spectrum, quartz grains of 0.05 mm to 0.01 mm sizes are distinguished by maximum resistance. This maximum for feldspars is found to be shifted to the area of coarser particles.

The results of the experiments carried out allow understanding rather comprehensively the role of different factors and mechanisms of the cryogenic transformations of mineral matter. The different behaviour of the basic groups of minerals under different regimes of humidity and freezing-thawing is based on the fundamental differences in their structure and crystallochemical properties which are expressed in different values of the surface energy.

The process of cryogenic destruction of minerals occurs most actively under temperature variations involving the transition of temperature through 0°C, when the major part of water↔ice phase transitions takes place. In this case, the disjoining effect of ice and the cryohydratational destruction act simultaneously. Under temperature variations within the negative spectrum of its values, the degree of destruction of minerals due to the reducing of the cleaving action of ice decreases, as a whole, by more than one order, though it is not extinguished. Hence it follows that under such conditions, it is the disjoining action of water that is displayed most fully. This process manifests itself most vividly in fine-dispersed particles for which a tendency to intensification of the cryohydratational destruction begins to show.

REFERENCES

Konishchev, V. N., 1977, General regularities of the cryogenic disintegration of minerals, *in* Frozen rocks and the snow cover, Moscow: Nauka, p. 48-62.

Konishchev, V. N., and Rogov, V. V., 1978, An experimental model of cryogenic resistance of the basic rock-forming minerals, *in* Problems of cryolithology, vol. VII, Moscow: University Press, p. 189-198.

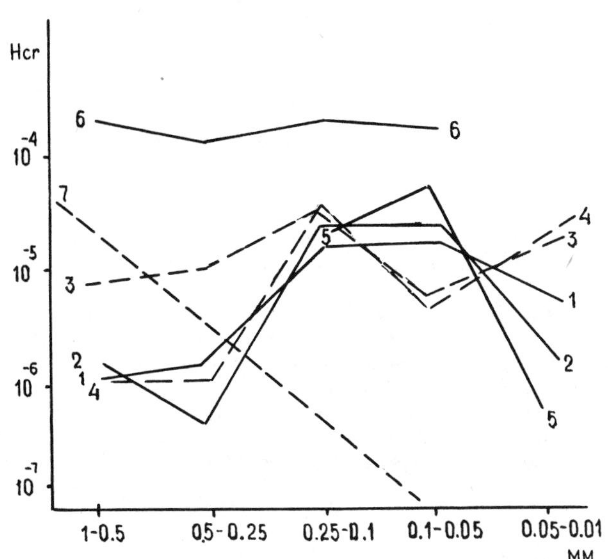

FIGURE 1 Cryogenic resistance of light minerals. 1 - Quartz from a pegmatite vein; 2 - Quartz (the lower reaches of the Kolyma); 3 - Albite; 4 - Labradorite; 5 - Quartz (lyuberetsky); 6 - Quartz; 7 - Feldspars.

1 to 5: $\frac{-20°}{-1°}$ temperature regime;

6 and 7: $\frac{-10°}{+20°C}$ temperature regime.

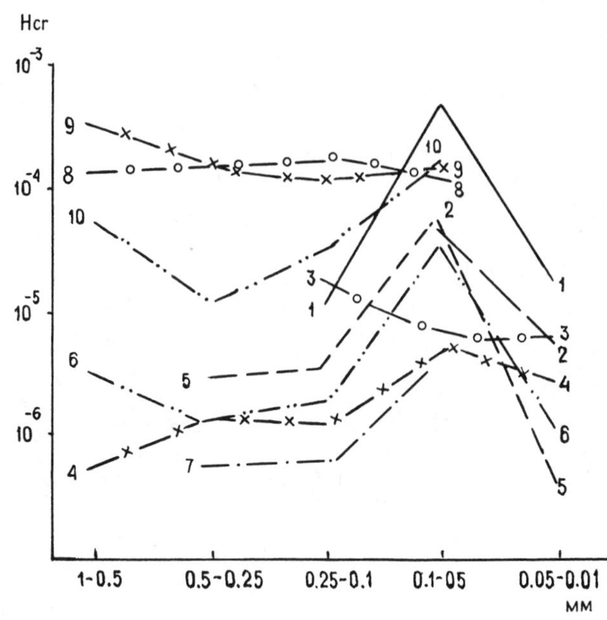

FIGURE 2 Cryogenic resistance of heavy minerals. 1 - Hornblende; 2 - Muscovite; 3 - Magnetite; 4 - Apatite; 5 - Diopside; 6 - Garnet; 7 - Cassiterite; 8 - Magnetite; 9 - Apatite; 10 - Garnet.

1 to 7: $\frac{-20°}{-1°}$ temperature regime;

8 to 10: $\frac{-10°}{+20°C}$ temperature regime.

FROST SUSCEPTIBILITY OF SOILS IN TERMS OF THEIR SEGREGATION POTENTIAL

J. M. Konrad[1] and N. R. Morgenstern

Department of Civil Engineering, University of Alberta, Edmonton, Alberta, Canada T6G 2G7
[1]Presently at the Division of Building Research, National Research Council Canada, Ottawa, Canada K1A 0R6

Recent studies have shown that the segregation potential, which is the ratio of water intake velocity to temperature gradient near the frost front, provides a basis for coupling mass flow with heat flow in a freezing soil. Since the segregation potential depends upon the composition of the soil system and the externally applied load, it appears to be an effective parameter for classifying frost susceptible soils. Moreover, it can be obtained readily for various applied pressures from a one-dimensional freezing test and can be used directly to estimate frost heave in the field versus time, provided that the solution to the corresponding thermal problem can be obtained. Data is presented and practical cases are described that illustrate the use of the segregation potential for determining frost susceptibility with respect to various project requirements.

FROST ACTION IN SOILS

The problem of frost action damage is widespread; it occurs in all temperate zones wherever there is seasonal soil freezing as well as in the active layer of more northerly permafrost regions. Moreover, recent resource development in arctic environments have brought up novel freezing situations such as long-term freezing induced by chilled buried gas pipelines. For unheated buildings, cold-storage facilities, and other structures, deformation of the structure by heaving rather than by thaw softening is the principal frost problem. Usually most of these structures do not heave uniformly, nor do they settle to the original elevation after the thawing period. Thus uplift due to cyclic freezing and thawing tends to be cumulative. When a chilled pipeline is buried in discontinuous permafrost, differential frost heave will occur that could lead to unacceptable strains in the pipe. Dams constructed in a northern environment may be subjected to severe frost action, and ice lensing may develop in the upper part of the impervious core. Upon thawing, preferential seepage paths may lead to serious internal erosion problems in the crest of the dam.

This paper presents the concept of segregation potential, which provides a new basis for evaluating the degree of frost susceptibility. The concept is tested on various case histories, illustrating the different types of problems associated with frost action in soils.

PHYSICAL PROCESSES ASSOCIATED WITH FREEZING OF SOILS

When wet soil is subjected to freezing, frost heave occurs as a result of two simultaneous processes: pore water in the soil freezes in situ and water supplied from the unfrozen soil, or possibly an external source, is sucked to the freezing front where it freezes, thereby leading to the development of ice lenses. The amount of heave by ice lensing depends on the frost susceptibility characteristics of the soil.

As the frost front progresses, ice nucleates in water-filled pores, but some water remains unfrozen as adsorbed films on the surface of soil particles. Water from the unfrozen soil migrates through the unfrozen water films into the frozen zone under the action of a temperature-induced suction gradient. Since the thickness of the unfrozen water films decreases as the temperature of the frozen soil decreases, the migratory water will be essentially stopped in a zone of extremely low permeability within the frozen soil where it freezes. When sufficient water accumulates, an ice lens forms (see Figure 1).

The zone between the frost front corresponding to the warmest isotherm at which ice exists in the soil pores and the segregation front where the ice lens is growing has been referred to as the frozen fringe (Miller 1972). Direct experimental evidence for the existence of the frozen fringe has been given by Konrad and Morgenstern (1982b).

Moisture transfer in the frozen soil behind the warmest ice lens is strongly reduced owing to very low frozen soil permeabilities and the presence of continuous ice. The contribution to the final heave from this region has been found to be negligible under both laboratory (Hoekstra 1969) and field conditions, at least over several years (Slusarchuk et al. 1978).

From a phenomenological point of view, the mechanics of frost heave can be regarded as a problem of impeded drainage to an ice-water interface at the segregation front. Substantial suctions are generated at this interface, but the reduced permeability of the frozen fringe impedes the flow of water to the segregation front, particularly for fully saturated soils. Thus, it appears that the frost heave characteristics or the frost susceptibility of freezing soils are related to the characteristics of the frozen fringe that, in turn, depend on the unfrozen water content. It is well known that the relationship between unfrozen water content (and hence frost susceptibility) and temperature for soils of different textures depends mainly on (1) specific surface of the soil, (2) gradation, grain size, amount of fines, type of minerals, (3) quantity and type of exchangeable cations, (4) pressure in ice and water phase, (5) solute concentration, and (6) density of the unfrozen soil.

LIMITATIONS OF EXISTING FROST SUSCEPTIBILITY CRITERIA

Once the physical processes associated with the freezing of wet soils are understood, the relevance of the criteria commonly utilized in existing frost susceptibility classifications can be evaluated.

The survey by Chamberlain et al. (1982) identified in general 3 levels of sophistication in estimating the frost susceptibility of soils used in roads and airfield construction. The first level, type I, is primarily based on the percent of soil finer than a specified particle size, commonly 0.074 mm or 0.02 mm. In some cases, a uniformity designation is an added requirement. Since its criterion is based on a single particle size, it cannot directly assess frost susceptibility because the controlling factors of the unfrozen water content are not addressed solely by the particle size. In addition, it is too conservative because non-frost susceptible soils are frequently classified as frost susceptible.

The second level, type II, is generally based on soil type or classification of particle size curves, and supplementary tests for Atterberg limits, capillarity, pore size distribution, or permeability of unfrozen soil. Type II methods, although more time consuming and costly, have limitations as well, since the parameters, which solely characterize the unfrozen soil, do not address the main factors affecting the frozen fringe.

The third level, type III, requires a laboratory freezing test and observations of frost heave and/or thaw weakening. However, the time required for most specified freezing tests, about 10 days, imposes serious restraints on their use. Freezing test procedures tend to differ from agency to agency; as a result, it is not always possible to compare their results directly. Moreover, existing classification work using a type III approach has been hampered by a lack of clear transfer to field conditions. In general, two freezing modes are used either with a constant cold plate temperature or a constant rate of frost penetration and often only one freeze-thaw cycle. Cold plate temperatures vary significantly, ranging between -3 and -20 °C and even -25 °C. This results in substantial temperature gradient variations in the various tests and thus affects significantly the water migration rate as shown by Konrad and Morgenstern (1980, 1981). Hence, freezing a given soil sample with different thermal boundary conditions may result in different degrees of frost susceptibility, which in turn may create an inadequate basis for decision making for a variety of field conditions.

SEGREGATION POTENTIAL OF FREEZING SOILS

If one considers that water migration occurs through a layered medium composed of saturated unfrozen soil and the frozen fringe, the freezing soil would be best characterized by suction at the segregation-freezing front and permeability of the frozen fringe (Figure 1). Both parameters are temperature and pressure dependent. Although these parameters provide a better insight into the physical processes involved during freezing, from a practical point of view they are difficult to measure.

As discussed by Konrad and Morgenstern (1980, 1981, 1982a, b) the characteristics of a freezing soil can also be represented by the segregation potential, SP, defined as the ratio of the rate of water migration, v, and the overall temperature gradient in the frozen fringe, grad T_{fr}. The use of SP is advantageous because it can be determined directly from a freezing test in which water intake rate and temperature profile with time are measured. Since the physical size of the frozen fringe is relatively small, its temperature gradient is approximated by the measured overall temperature gradient in the frozen soil near the 0 °C isotherm, grad T_f.

Both theoretical analyses and experimental results reported by Konrad and Morgenstern (op.cit.) indicate that SP is dependent on the suction at the frost front, P_u, the pressure applied on the warmest ice lens, P_e, and the rate of cooling of the frozen fringe, $\dot{T}_f$. $\dot{T}_f$ includes both geometrical and thermal boundary conditions and is defined as the change in average temperature of the frozen fringe per unit time. Its value can be approximated by the change of temperature at the 0 °C isotherm per unit time from:

$$\dot{T}_f \simeq \text{grad } T_f \cdot \dot{X} \qquad (1)$$

where $\dot{X}$ is the rate of frost advance.

The rate of segregation frost heave, $\dot{h}_s$, in a freezing soil can therefore be expressed as:

$$\dot{h}_s = 1.09 \cdot v = SP(P_u,\ P_e,\ \dot{T}_f) \cdot \text{grad } T_f \qquad (2)$$

To successfully predict frost heave in the field using equation (2), SP must be determined from laboratory tests using representative values of P_u, P_e and $\dot{T}_f$.

For field conditions, frost penetration rates are usually small (for example, in Figure 3 they are only about 0.07 cm/h), and the average temperature gradients in the frozen zone are also small, say approximately 0.07 °C/cm. Because equation (1) results in an extremely small $\dot{T}_f$ (field), about 0.005 °C/h, it is argued that SP (field) may be approximated by SP obtained at the formation of the final ice lens using freezing tests with constant temperature boundary conditions. For these conditions, $\dot{T}_f$ as the final ice lens is initiated is of the order of 0.01 °C/h.

In many field conditions, it may be safely assumed that P_u will be small enough to ignore owing mainly to high mass permeability of the unfrozen soil and small migration rates. Consequently, equation (2) reduces simply to:

$$\dot{h}_s \text{ (field)} \approx 1.09 \text{ SP } (P_e) \text{ grad } T_f \qquad (3)$$

The relationship between the segregation potential and applied pressure is best expressed as (Konrad and Morgenstern, 1982b):

$$SP = SP_0 \; e^{-aP_e} \qquad (4)$$

where SP_0 corresponds to the formation of the final ice lens for zero applied load and for P_u close to atmospheric pressure, and "a" is a constant determined from tests for various P_e.

Finally the change in water content, (ΔWC), with depth due to moisture migration is evaluated from:

$$\Delta WC(\%) = 100 \; \frac{V}{\dot{X}} \frac{\gamma_w}{\gamma_d} \qquad (5)$$

where γ_w and γ_d are the unit weight of water and soil, respectively.

The results of a literature survey on well-documented cases of frost heave data are shown in Figure 2. The segregation potential for various applied pressures can be expressed by using equation (4) for many soils ranging from sandy silts to clays, thereby giving strong support for the general validity of this relationship to characterize any freezing soil.

Figure 2 also reveals the importance of including the overburden pressure in any frost susceptibility classification. In general, the segregation potential of soils with low unfrozen water contents (silty sands) will be very sensitive to small changes in overburden pressures; this is not the case for soils with higher unfrozen water contents such as clays.

In conclusion, it appears that only a limited number of well-controlled freezing tests may be required to characterize adequately the segregation potential of any soil for field freezing conditions over a wide range of overburden pressures. In practice, these freezing tests are performed with constant temperature boundary conditions (the warm end at +1 °C) and different applied surcharges covering the expected stress range in the field. Using a cold end temperature of about -5°C to -8°C and 10 cm high samples shortens the time required to reach the maximum frost penetration, and hence the growth of the final ice lens, to less than 1 day.

A NEW APPROACH FOR FROST SUSCEPTIBILITY EVALUATION

The existing frost susceptibility criteria, either index tests or empirical relationships relating laboratory investigations to field performance, are usually effective for regional seasonal freezing problems only. Extension of these criteria to new types of problems such as sustained freezing beneath a chilled gas pipeline is therefore very limited. Furthermore, these criteria lead only to the acceptance or rejection of soils without specifying the degree of damage resulting from the freeze-thaw cycles.

The authors are suggesting a new approach for determining the relative frost susceptibility of soils. This method considers that the freezing conditions, the soil properties, the project requirements, and the acceptable risks should be taken into account in assessing the degree of frost susceptibility. Frost heave and water content changes in the field during seasonal or long-term freezing can be estimated by using the segregation potential measured in the laboratory as specified in the previous section. Frost susceptibility is then evaluated by comparing these parameters with allowable parameters that ensure adequate safety of the project.

It is proposed to distinguish between problems where the thaw weakening effect is of prime importance (class I), problems where frost heave and/or differential frost heave can lead to damage (class II), and problems where the combined effects of frost action (i.e., lensing and thaw weakening) is a concern (class III).

Class I Problems

These problems are generally associated with the performance of roads and airstrips. Surface heave due to ice lensing does not usually affect trafficability during the freezing period. However, the loss of strength by thaw weakening is of paramount importance, particularly for cases where dynamic loading is involved.

Class I problems, (e.g., evaluating the adequacy of a particular soil for construction purposes) can be dealt with as follows. According to climatologic factors at a given locality, the depth of frost is determined either from a geothermal analysis with appropriate boundary conditions or from existing solutions for relevant problems. The amount of heaving and change in water content with depth are then calculated under the most adverse conditions by using equations (3) to (5). The relationships between the change in water content and the loss of strength upon thawing should be obtained from laboratory and/or field tests. These tests are not standard practice to date, but efforts are being made in this direction (Chamberlain 1982). From such relationships, a thaw weakening factor could be obtained for a given soil subjected to given freezing conditions, and an allowable increase in water content could be determined for adequate safety.

The following situation illustrates frost susceptibility evaluation for a class I problem. The till at the La Grande hydro electric project in northern Quebec (James Bay) is a well-graded, gravelly, silty sand (60-80% passing #4 sieve, 20-40% passing #200 sieve) (Paré et al. 1978). Both the Casagrande and Beskow frost susceptibility criteria classify the till as being frost susceptible. Townsend and Csathy (1963), using the concept of limiting pore-size ratio, also classified similar tills as frost susceptible. Cyclic freeze-thaw tests were performed for various surcharges on the matrix material from the till by the U.S. Army Cold Regions Research and Engineering Laboratory (USA CRREL) and reported by Paré et al. (1978). The results of these tests have been reanalysed in terms of segregation potential and are presented in Figure 2. Frost penetration, segregational frost heave and increase in water content have been calculated for this till at two different locations characterized by freezing indexes of 1000 °C days and 2800 °C days using the suggested segregation potential relationship. These climatic conditions are representative of the Montreal-Ottawa and James Bay areas, respectively. The results of the analysis are presented in Figure 3 and compare relatively well with field observations reported by Paré et al. (1978) for the James Bay area and by Penner (1976) for the Ottawa area. Penner indicated that frost heaving in Leda clay (70% clay size and 30% silt size) is approximately 9 to 12 cm during most winters in snow-cleared areas. It should be noted that Leda clay has higher SP values than James Bay till (Figure 2).

The analysis reveals that a significant increase in water content due to moisture migration will occur in the Montreal-Ottawa region at shallow depths. Furthermore, because the frost front remains almost stationary for about one month, a discrete ice lens may grow approximately 0.6 m below ground surface, resulting in serious road damage when it thaws. At James Bay, frost penetrates much deeper and the increase in water content per unit depth is smaller than for the warmer Montreal-Ottawa area. Moreover, the overburden pressure reduces significantly the moisture migration at larger depths, hence reducing the occurrence of ice lenses as the frost front becomes stationary.

This analysis clearly shows that the degree of frost susceptibility of the till at James Bay is significantly smaller than for the same till in Montreal or Ottawa. For specific projects, the design requirements will determine the suitability of the till with respect to performance associated with varying degrees of frost action.

Class II Problems

These problems are associated with projects such as building a pipeline, that involve soils subjected to long-term freezing (more than one season). Two important new design considerations arise. First, how much frost heave (restrained or unrestrained) will occur over the lifetime of the project? In addition, where the pipeline crosses from permafrost to unfrozen ground and back to permafrost, how much differential heave will occur and will it lead to unacceptable strains in the pipe? Konrad and Morgenstern (1983) describe a procedure for calculating the amount of heave under a chilled gas pipeline based on a finite difference formulation of the heat and mass transfer in saturated soils and on the segregation potential concept. Good agreement is found between the predictions of heave obtained with this procedure and that of long-term, full-scale experiments at a test site in Calgary, Canada. Figure 4 shows the predictions of frost heave and frost penetration over 20 years for a hypothetical pipeline with a temperature of -8 °C, buried in soil with an initial ground temperature of +2 °C at a nominal depth of about 2 m below ground surface. The frost heave characteristics, i.e., the SP versus P_e relationship, were assumed to be those of the Calgary silt shown in Figure 2. The maximum pipe heave after 20 years is nearly 0.8 m, and the frost penetration below the original position of the pipeline is about 6 m. The combined influence of small temperature gradient and significant overburden pressure at depths greater than 4 m explains why the heave rate is only 0.012 m/yr after 10 years of operation. The maximum differential heave that might occur after 20 years is a function of the depth of the shallow permafrost. Deciding whether this soil is considered acceptable with respect to frost susceptibility will mainly depend on the allowable strains for the pipeline. If no remedial measures are considered during operation of the pipeline, two approaches can be used. A maximum differential heave could be specified for adequate safety or, alternatively, a more sophisticated analysis of frost heave pipeline interaction may be performed such as that presented by Nixon et al. (1983). A pipeline situated in the warmer discontinuous permafrost could present a problem of differential heave if variable soil conditions are encountered. In these conditions, for a ground temperature of about +5 °C, the maximum differential heave would be approximately 1.0 m, which would occur for a layer of nonheaving soil thicker than 4 m. The previous example illustrates clearly that the degree of frost susceptibility of a given soil depends essentially on the project requirements that can be expressed in terms of maximum total and/or differential heave and related allowable pipe strains.

Class III Problems

Problems associated with dam construction in colder areas (such as James Bay) fall into this category. When frost penetrates into the till core of an embankment, ice lensing may occur in its upper part. The thawing of the ice lenses may cause preferential seepage paths to form parallel to the water flow, which, in turn, may induce internal erosion and lead to the destruction of the top part of the dam.

Frost heave and water content increases with depth may be calculated for a two-layered medium composed of a top protection layer of sand and gravel (non-frost susceptible) and the till core. A simulation of frost action in a typical crest of James Bay embankments is shown in Figure 5. It is assumed that there is no frost heave in the upper sand and gravel protection and that the water table is 3 m below the crest surface. Segregational frost heave in the till core is significantly reduced owing to the effect of overburden pressure; hence the change in water content with depth is also negligible, as shown in Figure 5. These results have been confirmed by a test pit in a till embankment at the LG 3 Project, which revealed no visible ice lenses below a depth of 1.5 m (Paré, 1982). One can conclude that for James Bay freezing conditions, minor heaving without any real ice lens formation will occur in till cores placed under a sand and gravel cover of about 1.5 to 1.8 m. Since the increase in water content after freezing is negligible, damage of till core with time due to freeze-thaw cycles is therefore not likely to be expected in the dams and dykes at the La Grande complex although the accepted standards classify the till as frost susceptible.

CONCLUSIONS

Existing procedures are not very effective in distinguishing between differing degrees of frost susceptibility in soils. The segregation potential SP, defined as the ratio of the velocity of migration water and overall temperature gradient in the frozen soil adjacent to the frost front, provides a new basis for a quantitative evaluation of frost susceptibility. A standard freezing test can be devised to determine SP for different applied surcharges. SP is therefore a parameter that includes the influence of mineralogy, pore water solutes, and other compositional factors known to influence frost heave susceptibility. Furthermore, it has been established that the laboratory-based SP associated with a "quasi" stationary frost front corresponds reasonably well with values of SP deduced from field frost heave observations.

Rejection or acceptance of a given soil with respect to its frost heave potential should be determined by comparing calculated heave and ice content distribution using the SP concept with an allowable water content increase per unit depth for problems where thaw weakening is of paramount importance. Comparisons with an allowable heave and/or differential heave should be made where vertical movements could lead to structural damage. Finally, where freeze-thaw effects are of concern, the frost susceptibility should be determined with respect to both allowable parameters. Specific project requirements and acceptable risks will predicate the maximum values of frost heave and water content increase per unit depth in order to ensure adequate safety. The SP-based approach applies to any climatic or design condition and thus facilitates an extension of the experience gained on regional problems to new problems.

ACKNOWLEDGEMENT

J.M. Konrad gratefully acknowledges the assistance provided by the Division of Building Research, National Research Council Canada.

REFERENCES

Chamberlain, E.J., Gaskin, P.N., Esch, D., and Berg, R.L., 1982, Identification and classification of frost susceptible soils, ASCE Spring Convention, Las Vegas, Nevada, April 26-30.

Hoekstra, P., 1969. Water movement and freezing pressures, Proceedings, Soil Science Society of America, v. 33, pp. 512-518.

Konrad, J.-M., and Morgenstern, N.R., 1980, A mechanistic theory of ice lens formation in fine-grained soils, Canadian Geotechnical Journal, v. 17, pp. 473-486.

Konrad, J.-M., and Morgenstern, N.R., 1981, The segregation potential of a freezing soil, Canadian Geotechnical Journal, v. 18, pp. 482-491.

Konrad, J.-M., and Morgenstern, N.R., 1982a, Prediction of frost heave in the laboratory during transient freezing, Canadian Geotechnical Journal, v. 19, pp. 250-259.

Konrad, J.-M., and Morgenstern, N.R., 1982b, Effects of applied pressure on freezing soils, Canadian Geotechnical Journal, v. 19, pp. 494-505.

Konrad, J.-M., and Morgenstern, N.R., 1983, Frost heave of chilled pipelines buried in unfrozen soils, Canadian Geotechnical Journal (in press).

Miller, R.D., 1972, Freezing and heaving of saturated and unsaturated soils, Highway Research Record, No. 393, pp. 1-11.

Nixon, J.F., Morgenstern, N.R., and Reesor, S.N., 1983, Frost heave - pipe-line interaction using continuum mechanics, Canadian Geotechnical Journal (in press).

Paré, J.J., 1982, Personal communication.

Paré, J.J., Lavallée, J.G., and Rosenberg, P., 1978, Frost penetration studies in glacial till on the James Bay Hydroelectric Complex, Canadian Geotechnical Journal, v. 15, pp. 473-493.

Penner, E., 1976, Insulated road study, Transportation Research Record 612, pp. 80-83.

Slusarchuk, W., Clark, J., Nixon, J.F., Morgenstern, N.R., and Gaskin, P., 1978, Field test results of a chilled pipeline buried in unfrozen ground, Proceedings, 3rd International Conference on Permafrost, Edmonton, Alta., pp. 878-890.

Townsend, D.L., and Csathy, I.I., 1963, Soil type in relation to frost action, Dept. of Civil Engineering, Queen's University, Kingston, Ont., Ontario Joint Highway Research Project Report No. 15.

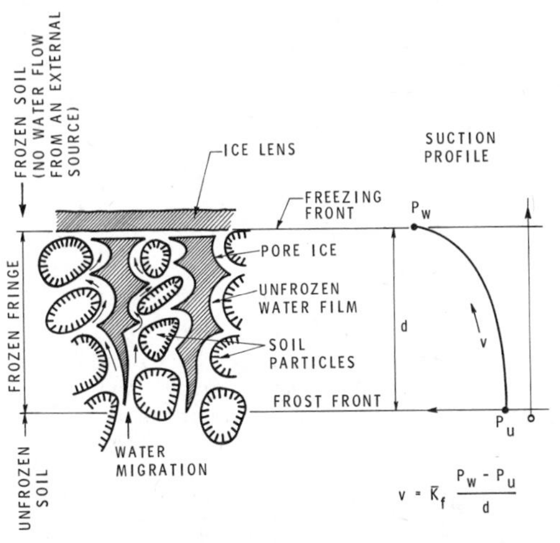

FIGURE 1 Schematic of frozen fringe in freezing soils

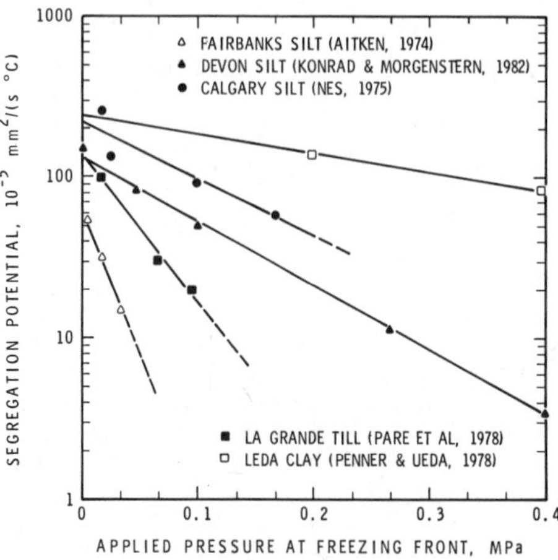

FIGURE 2 The segregation potential of various soils

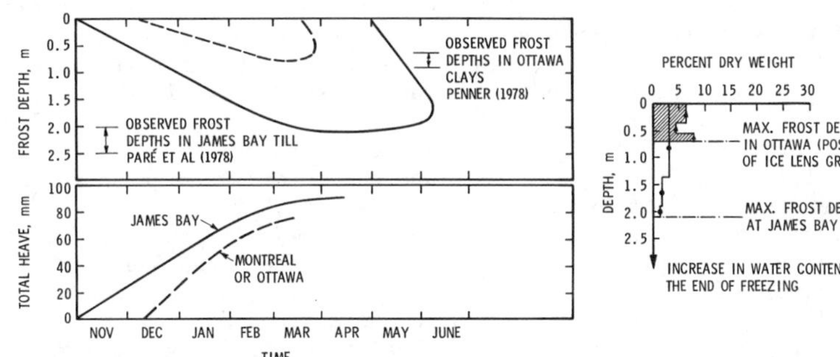

FIGURE 3 Typical example of class I problems

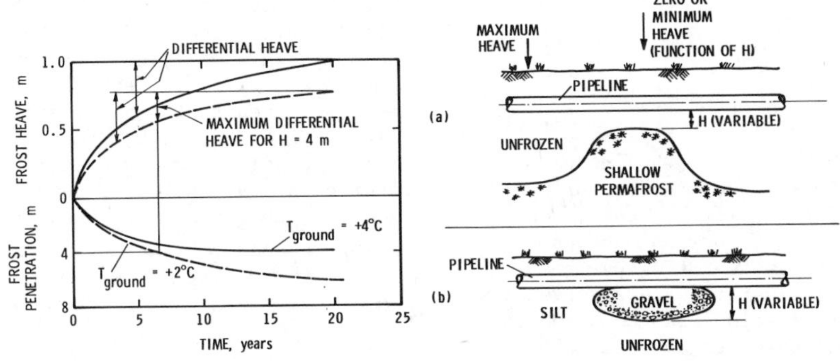

FIGURE 4 Typical example of class II problems

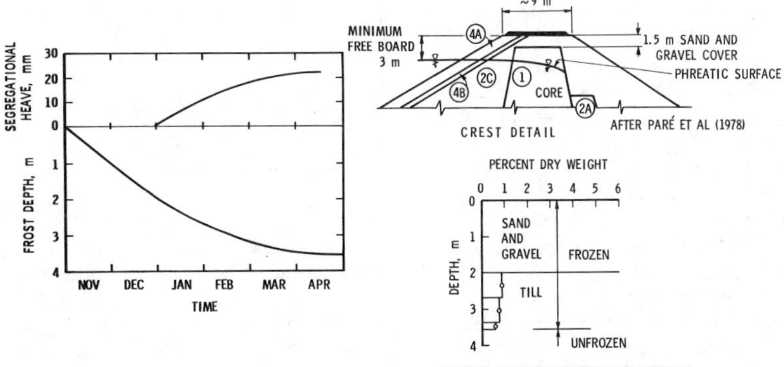

FIGURE 5 Typical example of class III problems

HYDROTHERMAL PROPERTIES OF PERMAFROST SOLONCHAKS IN THE CENTRAL MONGOLIAN ARID-STEPPE ZONE

A. Kowalkowski

Soil and Fertilization Division, Forest Research Institute, Raszyn, Poland

A three-layer, unstable system exists within the permanently frozen solonchak profiles of the dry-steppe zone in the southern part of the Central Gobi Peneplain. It is subject to deformation due to hydrostatic pressure and low temperature. The action of the latter factors results in the formation of characteristic continental mud boils on the lower parts of the slopes. These are perennial features and lack frost segregation.

INTRODUCTION

The well known theory of the cryostatic genesis of tundra "mottled soils," "mottled tundra," or "tundra craters" in the periglacial zone is open to criticism. These phenomena are now interpreted as resulting from gravitational processes in liquefied and thixotropic soil materials over a frozen or low permeability substrate. The studies by MacKay and MacKay (1976) and French (1977), as well as earlier studies by Jahn (1948) indicate the possibility of horizontal and vertical subsurface shifts of semi-fluid materials under periodically increased hydrostatic pressure within the soil profiles. These processes are initiated after the saturated layers immediately above the permafrost table have thawed. The processes in question result in diapiric uplifts and open muddy craters.

AIMS AND METHODS

During the period of 5-7 July 1978, in the vicinity of Augutuin (Nuur) Lake, the author observed a cluster of mud boils (frost scars, non-sorted circles) in permafrost turf solonchaks composed of rhyolite-granite debris with a basalt admixture. Similar boil clusters were observed in the semi-desert permafrost solonchaks composed of basalt debris with a volcanic-ash admixture in the vicinity of Shagshuraga (Nuur) Lake (Figure 1). These two saline lakes are in the northern part of the Central Gobi Peneplain (Korzhnev 1982) in the southern reaches of the insular Central Asian permafrost region (Gavrilova 1981).

Data recorded at the nearby physico-geographical Gurvan Turuu Station (lat. 47°03'N, long. 07°38'E, elevation = 1371 m) indicate that the winter period, with temperatures lower than 0°C, lasts for 187 days, while summer, with temperatures higher than 15°C, lasts for only 22 days. January is the coldest month with temperatures averaging -24°C, while July is the warmest with average temperatures of +16°C; the mean annual temperature is -2.7°C. Mean precipitation totals 250 mm, of which 43% falls in July (Hess et al. 1980). Groundwater flow and precipitation usually result in the formation of intermittent lakes in summer in thermokarst-ablation alluvia filled basins. Soils of these depressions are takyr solonchaks (sorovyi solonchak) according to Nogina et al. (1977) with permafrost.

As a result of the arid climate, the subsurface waters are highly mineralized, with dissolved solids ranging from 0.1 to 0.7 g/ℓ. Chemically these waters are included in the calcium bicarbonate and potassium bicarbonate category (Chelmicki and Tserev 1980).

Low snowfall results in the freezing of soil to depths greater than 3.5 m and in the formation of discontinuous, insular permafrost. In permafrost solonchaks, an active layer 0.4 to 4.5 m thick is usually underlain by permafrost in which temperatures range from -0.1°C to -0.4°C (Kowalkowski et al. 1981).

Soil samples taken from mud boils and from soil profiles excavated at distances of 0.5-1.0 m from the boils were analyzed granulometrically using a combination of sieve methods. Bulk density (BD) was determined by the pycnometric method in vacuo, volumetric density (VD), actual humidity (AH), maximal capillary water capacity (MCWC), and field water capacity (FWC) were determined for 100 cm³ samples with the natural structure preserved.

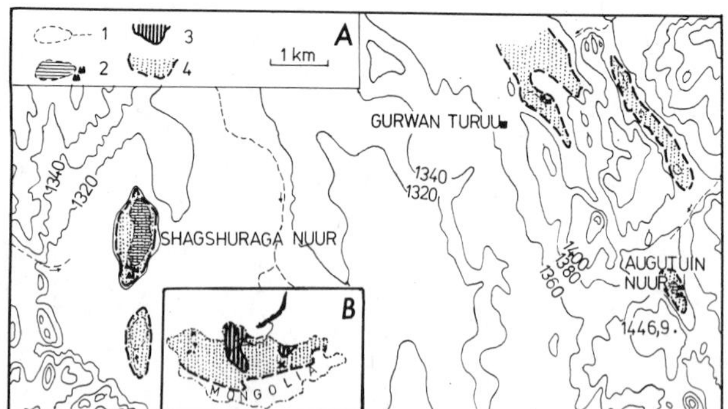

FIGURE 1 Location and topography of the solonchaks under study and of the Physico-Geographical Gurwan Turuu Station. 1 - ephemeral (intermittent) saline lakes and streams; 2 - ephemeral saline lakes with takyr solonchaks and fields of mud boils under study; 3 - southern boundary of discontinous permafrost; 4 - southern boundary of sporadic and insular permafrost.

Wilting moisture (WM) was determined by multiplying the maximum hygroscopicity by the coefficient of 1.34 determined for Mongolian dry-steppe soils by Umarov and Yakushkina (1978). Total porosity (TP) was calculated from both the specific and the volumetric density. Soil temperature was measured under field conditions with an accuracy of 0.1°C.

RESULTS

In the basins of two permafrost solonchaks located 9.5 km apart, there were accumulated slightly segregated alluvia of different origin and grain size (Figure 2). These alluvia determined the forms of the mud boils (Figures 3 and 4). The frost boils form irregular clusters along the eastern, southeastern, and southern banks of the intermittent lakes that occur on the lower parts of slopes inclined 0.5° to 2.5° and underlain by permafrost at depths of 0.6-1.4 m (Figure 5).

Within the areas of mud boils the soil surface is characterized by a great variety of morphogenetic forms. On turf solonchaks, among the turf tussocks, the soil surface has patches of salt crusts 0.2-0.4 cm thick in the form of polygons (Figure 3). The surface of semidesert solonchaks, which have a vegetation cover of 2 to 4%, is mosaic of deflation-pavement on a thin desert crust with salt efflorescence (Figure 6), cracked silt alluvia (Figure 3), and forms of eolian accumulation and surface and furrow flushing. The upper parts of the slopes, which are underlain by permafrost at depths of 0.4-0.6 m are dominated by fields of recent cryogenic peat thufurs, which are saturated, particularly during the rainy season.

Materials that comprise the profiles of the soils under study form an unstable layered system where hydrostatic pressure is the main factor that stimulates disturbances, without, however, the frost segregation processes that are typical of the periglacial environment. The stratigraphy consists of three units of different density (Figure 7) and satisfies the conditions defined by Cegła and Dzhulynski (1970) for a geologically unstable layered system. The respective layers have the following characteristics:

(1) Compact soil layer VD 1.73-1.28 g/cm³, TP 37-54%.

(2) Thixotropic layer, periodically liquified, VD approximately 0.35 g/cm³, TP approximately 85%.

(3) Permafrost layer, with lamellar ice, VD

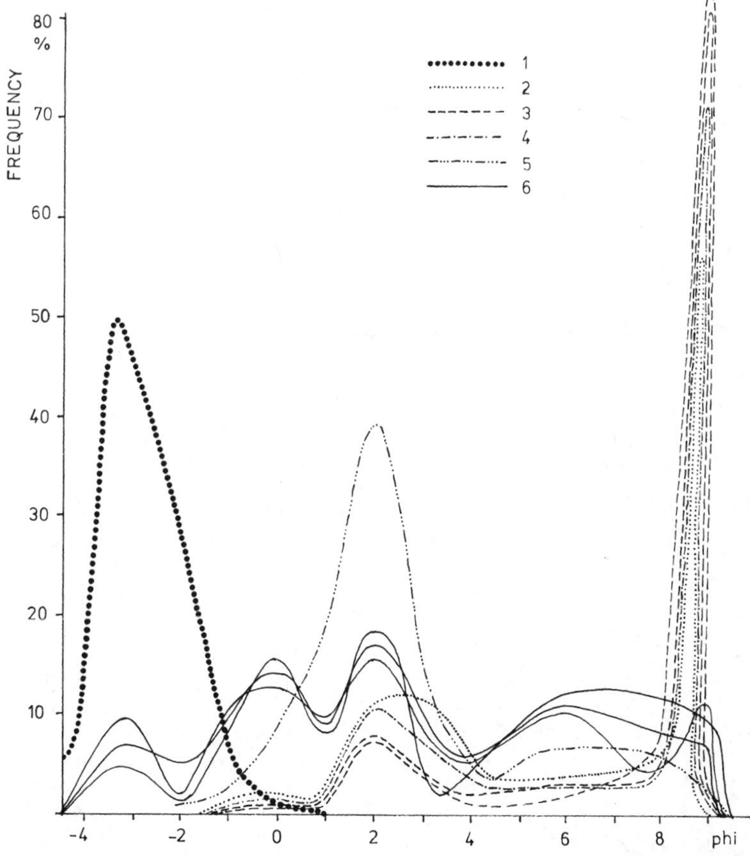

FIGURE 2 Grain size curves of permafrost semidesert solonchak (profile 107) and of permafrost turf solonchak (profile 109). Profile 107: 1 - deflation cover; 2 - compact soil layer; 3 - thixotopic layer; 4 - dried-out crust on top of mud boil; 5 - sand lenses in compact soil layer. Profile 109: 6 - compact soil layer.

FIGURE 3 Landscape of permafrost semidesert solonchaks with mud-boil craters 60-350 cm in diameter and heights up to 60 cm at the south of the Shagshuraga Nuur Lake. At the center are the mud boils under study, at profile 107.

FIGURE 4 Surface of the permafrost turf solonchaks covered with salt crust, and mud boils with crater diameters up to 60 cm and 20 cm high with a fresh outflow of liquefied material, to the southeast of Augutuin Nuur Lake, at profile 109. To the left of the boils are soil clods pressed out onto the surface, while to the right is dried-up mud related to several outflow phases.

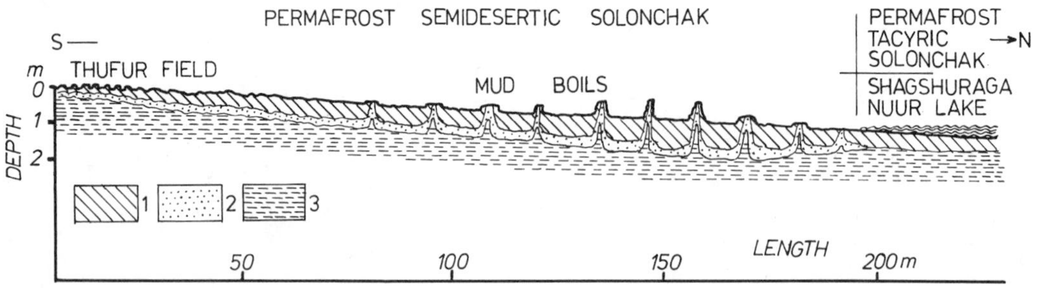

FIGURE 5 Schematic vertical cross-section through the soil profile in thufur and mud-boil fields on the southern bank of the Shagshuraga Nuur Lake. 1 - compact soil layer; 2 - thixotropic layer; 3 - permafrost layer.

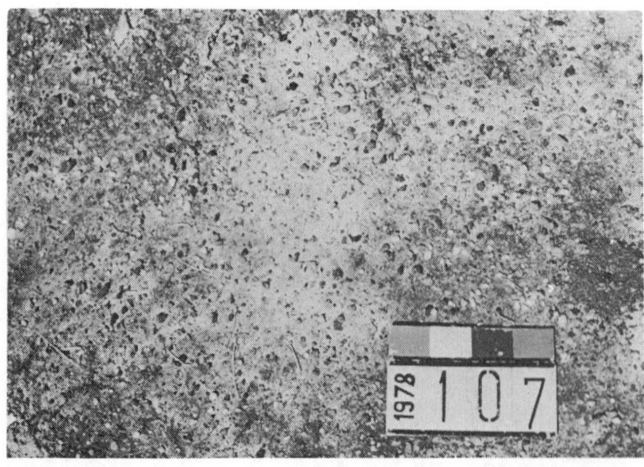

FIGURE 6 Deflation surface of the permafrost semidesert solonchaks with a gravel cover and desert crust showing a white salt efflorescence at profile 107. Location is within the 8-10 m section of Figure 9.

approximately 0.65 g/cm^3, TP approximately 75%.

The above layers have high water contents ranging from 72% to 100% (by volume) in the compact soil layer and 98% to 100% in the thixotropic layer, in the mud-boil craters, and in the underlying permafrost (Figure 8).

The formation of peat thufurs and mud boils depends upon increased hydrostatic pressure in the 20-40-cm-thick cryogenic thixotropic layer that extends along the slopes above the permafrost. This pressure increases during spring thaw on adjacent plateaus and mountain ridges and in the rainy season. The pressure magnitude depends on four factors: the distance between the point of water inflow and the support of the thixotropic layer by compact materials of takyr solonchaks, the thickness of compact soil above the thixotropic layer, the slope length and inclination (Figure 5), and the geological origin of the soil material (Figures 3 and 4).

If most of the permafrost water exists in the form of horizontal, 1-4-mm-thick ice lamellae, then in the thixotropic layer or in the mud-boil craters it will form highly durable coagulation structures

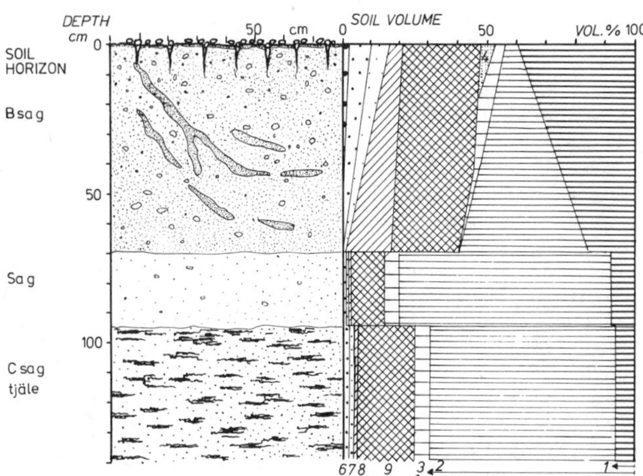

FIGURE 7 Stratigraphy and physical properties of the permafrost semidesert solonchaks measured in profile 107. BSAG - saline gleyed compact soil layer; Sag - saline thixotropic layer; CsaG tjale - saline permafrost parent rock; 1 - wilting moisture, 2 - field water capacity, 3 - maximal capillary water capacity, 4 - air capacity, 5 - >1.0 mm fraction, 6 - 1.0-0.1 mm fraction, 7 - 0.1-0.02 mm fraction, 8 - 0.02-0.002 mm fraction, 9 - fraction less than 0.002 mm.

FIGURE 8 Cracked surface of crusted silt alluvia on the permafrost semidesert solonchaks, at profile 107. Location is within the 22-24 m section of Figure 9.

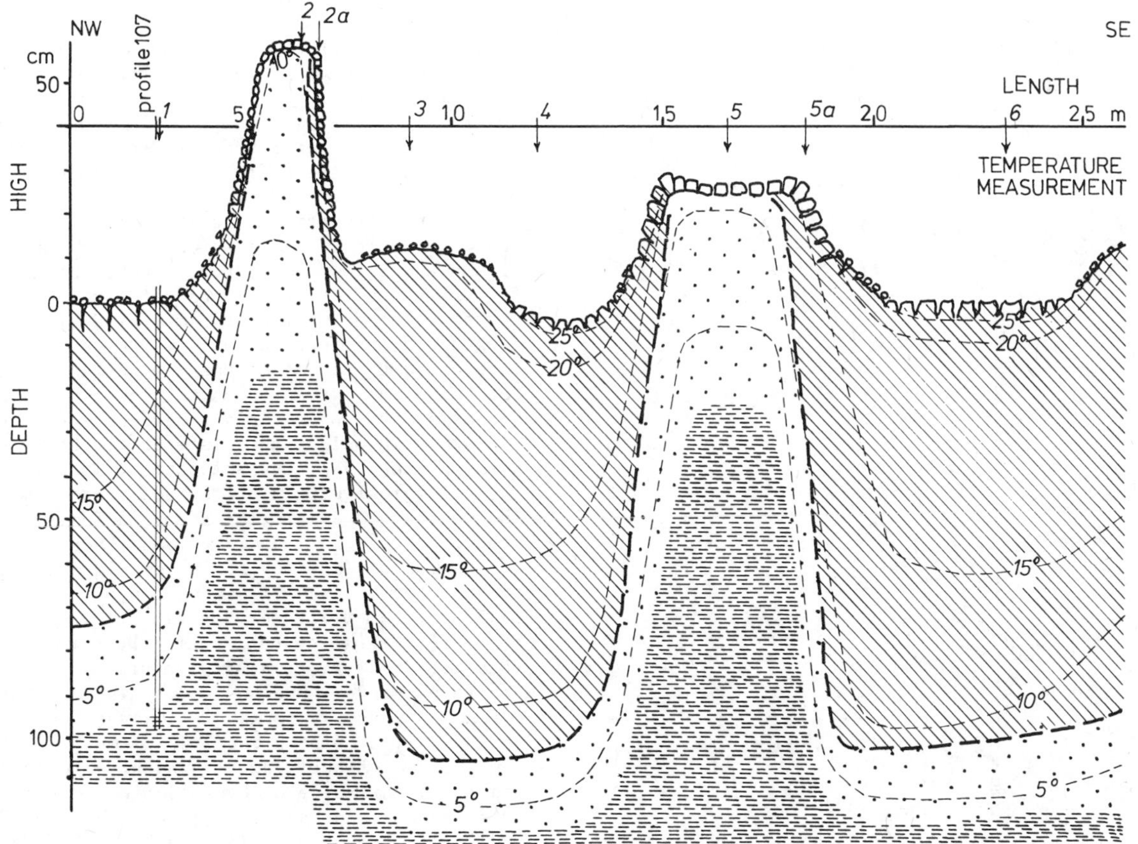

FIGURE 9 Hydrothermal conditions in continental mud boils on the Shagshuraga Nuur Lake bank (profile 107), as recorded on 5 July 1978 at 0100 h. (See Figure 5 for legend)

that are resistant to the action of mechanical factors. Within this structure, two directions of capillary motion predominate--toward the cold front of the permafrost at the bottom, and toward the warm soil surface, especially in the open mud-boil craters (Figure 9).

Transition of water into ice lamellae and the resulting increase in permafrost volume is usually accompanied by liquefaction of the thixotropic mass immediately above the permafrost table. By contrast, within the mud-boil craters, a dense suspension, high-molecular gel forms, containing permanently suspended rock and mineral fragments up to 4 cm in diameter. Water is lost by evaporation into the atmosphere through a cracked dry crust less than 5 cm thick (Figure 9). As a result of this process a coagulation structure is formed that remains static under uniformly changing pressure and undergoes syneresis. At the same time the freezing point of the soil structure is depressed and, consequently, ice can aggrade but not veins or lamellae.

The ascending capillary motion of water above the permafrost table and the loss of heat through evaporation result in a temperature drop within the crater. One consequence of this process is the growth of a permafrost cone over 100 cm high in the crater. In some cases, the cone apex is 10-15 cm lower than the slope surface. In general, however, it occurs at a depth of 40-80 cm beneath the top of the mud boil. There is also an increase in crater diameter of up to 2.5 m as a result of liquefaction of soil material under the influence of the aggrading permafrost cone.

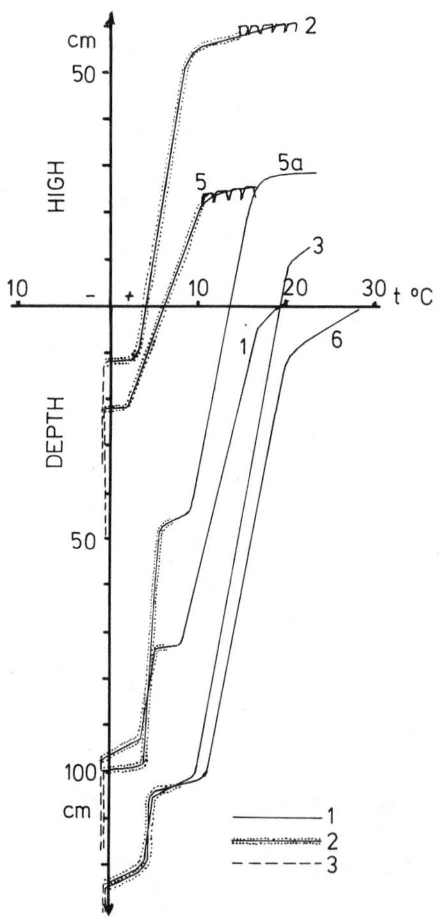

FIGURE 10 Geothermal profiles of permafrost semi-desert solonchaks and mud boils, as recorded on 5 July 1978, at 0100 h. 1 - compact soil layer, 2 - thixotropic layer, 3 - permafrost layer.

The most important factors responsible for deformation and transformation of the unstable layered system are the dynamics of temperature change with time and local temperature variation in the vertical and horizontal soil planes (Figure 9). From the temperature depth profiles shown in Figure 10, it can be concluded that the greatest vertical temperature range within the thixotropic layer occurs above the apexes of permafrost cones. Here, temperature differences of 15-20°C occur in the 50-75 cm thick layer. Between the permafrost cones, the thixotropic layer is 20-30 cm thick and the vertical temperature range is 7°C. The degree of exposure influences both the temperature distribution in the boils as well as the form of the permafrost cones (Figure 9).

Figure 10 illustrates the characteristic decrease in the temperature jump with the depth of the contact between the thixotropic layer and the permafrost layer.

SUMMARY

The feature that distinguishes the islands of permafrost solonchaks in the northern dry-steppe area of the Central Gobi Peneplain is a system of layers with variable density. The basic deformation conditions determined by Cegła and Dzhulynski (1970) include the difference in volumetric mass between the upper and lower layer, the degree of soil water saturation, and the soil liquefaction characteristics. In addition, the hydrostatic pressure and low temperature conditions are major factors stimulating disturbance of the unstable system. These two factors, modified by slope length and inclination, extent of the thixotropic layer, and thickness of the compact soil layer, determine the formation on the lower parts of slopes of the characteristic clusters of mud boils. Variation in height and diameter of these forms, between the respective solonchaks, is a function of the geological origin of the soil parent material. The occurrence of similar forms in Iceland has been described by Stingl and Herrmann (1976).

The mud boils in question come into being as a consequence of local liquefaction of the compact soil-cover in vertical channels and the pressing through of these channels onto the surface of the thixotropic layer remaining constantly under hydrostatic pressure. Mud boils in this area constitute perennially active forms, in contrast to the periodically repeated outflows of mud in the polygon systems of periglacial tundra, which also occur under the influence of hydrostatic pressure (Jahn 1948, Mackay and Mackay 1978, French 1979, Wojcik 1981). This major contrast reflects the continental Central-Asiatic peculiarity of hydrothermal conditions of the permafrost solonchaks that contain fields of cryogenic mud boils.

REFERENCES

Chelmicki, W., and Tserev, O., 1980, Variability of ground water total mineralization in steppe around Gurwan Turuu, Bulletin de L'Academie Polonaise des Sciences de la Terre: v. 38, no. 2-3, p. 189-198.

Cegla, J., and Dzhulynski, S., 1970, Uklady niestatecznie warstwowane i ich wystepowanie w srodowisku peryglacjalnym, Acta Universitatis Wratislawiensis: 124, p. 17-42.

French, H. M., 1979, Periglacial geomorphology: Progress in physical geography: 3, London.

Gavrilova, M. K., 1981, Sovremennyi klimat i vechnaya merzlota na kontinentnakh: Novosibirsk, Nauka.

Hess, M., Kowanetz, L., and Suchanek, R., 1980, Some results of climatological research in Mongolia, Bulletin de L'Academie Polonaise des Sciences de la Terre: v. 38, no. 2-3, p. 157-164.

Jahn, A., 1948, Badania nad struktura i temperatura gleb w Zachoe niej Grenlandii: Rozprawy Wydzialu Matematyczno-Przyrodniczego PAN, v. 72, Krakow, Poland.

Korzhnev, S. S., 1982, Central Gobi Peneplain: Geomorphology of Mongolian Peoples Republic: Moscow, Nauka, p. 150-165.

Kowalkowski, A., Borzyszkowski, J., and Charaldamba, G., 1981, Soil moisture dynamics of arid steppe in the growing season of 1978: Soil Science Annual, v. 32, no. 1, p. 3-17.

Mackay, J. R., and Mackay, D. K., 1976, Cryostatic pressures in nonsorted circles (mud hummocks), Inuvik, NWT: Canadian Journal of Earth Sciences, v. 13, Ottawa, p. 889-897.

Nogina, N. A., Jevstifeyev, Yu. G., and Ufimtseva, K. A., 1977, Pochvy niskogornykh i ravninnykh stepei i pustyn Mongolii: Aridnye pochvy, ikh genesis, geokhimya, izpolzovaniye: Moscow, p. 165-195.

Stingl, H., and Herrmann, R., 1976, Untersuchungen zum strukturbodenproblem auf Island: Geländebeobachtungen und statistische Auswertung: Zeitschrift für Geomorphologi, NF, 20.2, Berlin-Stuttgart, p. 205-226.

Umarov, K. U., and Yakushkin, G. N., 1978, Dinamika vlazhnosti i temperatury kashtanovykh pochv: Pochvennyi pokrov osnovnykh prirodnykh zon Mongolii: Moscow, Nauka, p. 167-176.

Wojcik, C., 1981, Deformations of unstatically layered gravel-clay system on a chosen example from the Kaffiöyra Lowland (Oscar II Land, Spitsbergen): Polish Polar Research, v. 2, no. 1-2, PWN Warsaw, p. 63-72.

ERODIBILITY OF UNCONSOLIDATED MATERIALS IN PERMAFROST

E. Z. Kuchukov and D. V. Malinovsky

Geology Department,
Moscow State University, Moscow, USSR

Mechanisms and laws governing the erosion of unconsolidated rocks in permafrost are examined. Four types of erosion, distinguished on the basis of the interaction of water flow with the rock, are identified. The most typical for the erosion of frozen rocks are the thermoerosional and extreme-thermoerosional types. It was established that the intensity of erosion increases with increasing stream discharge (in the case of the erosional and thermoerosional types) and with increasing water temperature (in the case of the extreme-thermoerosional type). It was shown that the composition, cryogenic structure, ice content, and density of the rocks are basic factors in dictating their erodibility.

Indices of erodibility of permafrost materials, which reflect the mechanical, thermal, and physical aspects of erosion as well as methods of determining them are proposed. A scheme is developed for subdividing permafrost materials on the basis of erodibility, which might be used for assessing and mapping permafrost and seasonally-frozen materials within the Yamal Peninsula. Much of the peninsula is composed of deposits susceptible to catastrophic and rapid erosion with the percentage area of such deposits increasing from south to north.

Stream activity in permafrost conditions emerges as one of the most important factors of cryolithogenesis and is essential for the development of surface denudation, transportation and accumulation of weathering products. The specific nature of this process, also called thermoerosion, is the simultaneous thermal and mechanical effects of streams on permafrost. The present work, carried out under the guidance of Professor E. Yershov at the University's Permafrost Studies Department, limits itself to discussing the scouring of permafrost dispersed rocks by intermittent streams, as the most hazardous effects of man's economic activity, and of scourability of permafrost reflecting its sensitivity to concurrent thermal and mechanical impact of flowing water. As a result, the degree of scouring is dependent on the thermal-physical and stability properties of permafrost (Yershov, Kuchukov et al., 1980). The former determine the speed of rock thawing, and the latter determines the destruction intensity and the follow-up material removal, i.e., the scouring intensity (I), which may be defined as a rock amount carried off with flow from a unit area in a unit time, or, for the flat flow, as a rock layer scoured in a unit time.

Laboratory and field studies have shown that permafrost is scoured while removing individual particles (for sandy rocks) and microaggregates and aggregates (for clay). Basically, the character and size of jointings teared off are specified by the cryogenic texture.

Three essentially differing patterns of this process may be ascertained with regard to the correlation of the thermal and mechanical factors, and accordingly, the scouring types identified by the stream-rock relationship.

1) Scouring of permafrost proper, when erosion capacity of a stream surpasses the scouring resistance of permafrost is the permafrost-erosional type (Fig. 1, curve 1). The pivotal point here is the high strength of permafrost rocks, and their scouring forestalls their thawing. Under natural conditions this type of scouring is improbable since actual streams in permafrost areas do not possess adequate mechanical energy to destroy frozen rocks. Therefore, many specialists have difficulty depicting the whole permafrost area (except for its most southern part) on many gully-erosion maps as gullyless and not prone to erosion.

2) Thawing of permafrost and instantaneous removal of thawed material while the stream is in contact with a permafrost basement is the maximum-thermoerosional type (Fig. 1, curve 2). It is characterized by an equal intensity of thawing and scouring; while the flow rate reaches its critical point (N_{Kr_1}), the erosion front inevitably catches up with the thawing line. The scouring intensity then becomes a function of the thawing speed, and more important, of a water temperature, dispersity and thermal-physical properties of thawing rocks. The linear relationship was traced between the critical scouring intensity (N_{Kr_1}) for permafrost rocks at a given temperature, and the stream temperature (Fig. 2). The scouring of clear ice goes on slower than that of sand but much quicker than of clay. And loamy sand is more easily scoured than loam. What is important here is that the ice content of permafrost determines the heat level at phase transitions and the cryogenic texture determines movement of the scouring front. The scouring of clay having a laminated or reticulate texture develops in two ways. If the size of mineral aggregates is smaller than the geometrical parameters (depth and width) of jet-type water flow, then the scouring acquires a discrete nature because mineral aggregates, due to their scouring resistance, are removed only after they thaw. If the size of

mineral aggregates is comparable to the water flow parameters (e.g., at scouring of mineral aggregates over 3 cm in size), then scouring proceeds through selective erosion along ice schlierens. In this case the removal of frozen aggregates is possible. The critical scouring intensity may be calculated as follows:

$$I_{Kr} = K_{tp}\, t_w \quad m/s \,, \tag{1}$$

where t_w = water temperature, °C:
K_{tp} = temperature index of scouring, m/s• degree, describing an increment of the critical scouring intensity per one degree Centigrade. It is determined by resolving the problem of a thawing line movement for the case of a complete "instant" removal of any thawed rock:

$$K_{tp} = \frac{\delta + \Delta}{\sigma[L_{vol} \cdot \delta\left(\frac{\delta}{2\lambda} + \frac{1}{\alpha_0}\right) + \frac{\Delta}{\alpha}} \quad m/s \cdot degree \,, \tag{2}$$

where δ and Δ = the sizes of mineral and ice layers, respectively, m;
σ = the ice-thawing heat, J;
L_{vol} = the ice volume in mineral interbeds;
α_0 and α = coefficients of a turbulent exchange between water and permafrost rock, water and ice, J/sq.m•degree;
λ = the thermal conductivity of a mineral interbed, J/m.s•degree.

This solution is valid only for the condition when the temperature of a permafrost rock series is close to zero. Observations of the rock temperature in a one-meter layer and under a stream bed, carried out at permanent stations in the north of Western Siberia show that it does not drop below -3°C in the spring-summer period. But taking into account the heat amount used for heating permafrost rocks under the stream bed, the proposed relationship (2) may be employed with adequate reliability.

3) Forestalled rock thawing and a stream contact with thawing or thawed deposits: Two types of scouring are found here: a) thermoerosional (Fig. 1, curve 3)—erosion of thawing rocks having a post-cryogenic structure and a low mechanical strength due to the increased humidity and structural linkages not fully restored during thawing; b) erosional (Fig. 1, curve 4)—scouring of thawed deposits, not differing from the traditional erosion of non-frozen rocks. In both cases, when scouring is done by a jet-type flow, the relationship between the scouring intensity and the water flow power may be described as follows:

$$I = K_{sc}\, N \quad m/s \,, \tag{3}$$

where K_{sc} = the scouring resistance factor, cu.m/J, being equal to the scouring intensity at a unit water flow power, calculated by the formula:

$$N = 0.5\rho \cdot V^3 \quad W/sq\ m \,, \tag{4}$$

where ρ = water density, kg/cu.m;
V = flow velocity, m/s.

The value of the resistance factor and, correspondingly, the scouring intensity, is determined by the stability of thawing rocks. We have established a close correlation (a correlation factor being 0.95) between the scouring resistance factor and the maximum rock displacement stress measured by micropenetrometer MB-3 (Yershov, Malinovsky et al. 1982) for multiple dispersed thawing rocks of various composition, found in the north of Western and Eastern Siberia. The relationship obtained may be presented as follows:

$$K_{sc} = 10^{-15} \frac{1}{R} \quad cu.m/J \tag{5}$$

where r = the maximum displacement stress, Pa. Research has indicated that there is actually no relationship between the maximum displacement stress and the angle of internal friction. On the other hand, a close relationship was traced between the R value and cohesion, taking up a linear form for non-frozen rocks and a parabolic form for thawing rocks. Using equation 3, it may be easily stated that the admissible non-scouring velocity (ANV) for sands does not exceed 0.1 m/s, while for clays with a laminated or reticulate texture it ranges from 0.1 to 1.0 m/s and exceeds 1.0 m/s for rocks with a cryogenic texture. The established relationship between the scourability of permafrost rocks and the stability of thawing rocks is quite reliable and in accord with the data of Ts. Ye. Mirtskhulava (1970), ascertained the dependence of ANV on cohesion of water-saturated rocks.

Thus, we may define the likely type and intensity of permafrost scouring relying upon the scouring indices, when external conditions (N and t_w) are known. If the potential scouring intensity exceeds the critical value, then the maximum-thermoerosional type of scouring is to be anticipated, and if below critical, then thermoerosional or erosional. Usually deposits having the maximum displacement stress less than 0.03 MPa are subject to thermoerosional scouring.

The likely intensity of scouring of a thermoerosional type is of utmost importance for erosion forecasts in permafrost. This is so because the resistance factor may change by two orders depending on the rock structure and texture, while temperatures, although different in different rocks studied, are still rather close to each other. Besides, since many varieties of dispersed permafrost show high stability to the effects of natural stream erosion (e.g., heavy clays), the maximum-thermoerosional type of scouring cannot be obtained. Then the subdivision of permafrost rocks by scouring is done using the resistance factor with regard to their composition, ice content, cryogenic composition and stability (Table 1). The first three categories differ by the ice content and the cryogenic rock texture. The fourth and fifth categories comprise ice-rich rocks, having differing cryogenic texture and low stability at thawing. The increased dispersion generally results in the diminished scouring in view of the growing cohesion. Should the structure of permafrost rocks be disturbed, their susceptibility to scouring grows; readily thawing rocks are easy to scour. But this is not always valid because of differences in thawing and scouring patterns. Hence the data on the speed and character of thawing may be taken (conventionally) to describe the scouring intensity. The presence of plant roots in permafrost rocks is very essential for diminishing the rock scouring.

The scouring assessment of major geological-genetical types of rocks on the Yamal Peninsula was carried out making use of the described classification of permafrost rocks by scouring. It is based on the analysis of permafrost-engineering geological peculiarities of upper horizons (to a depth of 5 m) within the Peninsula. The results obtained have indicated that the rocks of the fourth category (highly susceptible to scouring) are most widely spread over the Peninsula; and the scouring here is of the maximum-thermoerosional type because in most typical conditions their potential scouring intensity goes beyond $1 \cdot 10^{-4}$ m/s. These are syn- and epigenetic ice-rich sands of a marine, lagoon-marine and alluvial geneses possessing low structural stability in a water-saturated state. The scouring resistance factor ranges within (0.3-10) 10^{-6} cu.m/J. Widely met also are rocks of the fifth category, including syngenetic alluvial ice-rich clay with an ataxitic cryogenic texture of disastrous scouring, behaving in this respect as clear ice. They are scoured by the maximum-thermoerosional type. Their stability after thawing is nearing zero. Their scouring resistance factor is always above 10^{-7} cu.m/J. Many researchers (Kosov and Konstantinova 1969) claim that it is the presence of deposits of the fourth and fifth categories that are conducive to a disastrously rapid scouring. Less developed on the Yamal Peninsula are deposits of the third and second categories, in which scouring intensity is equal to (0.1 to 1.0) 10^{-4} m/s following the thermoerosional type. These rocks are distinguished by the lack of cohesion between aggregates and, usually, the post-cryogenic structure at scouring. The third category includes syngenetic clay of the marine, lagoon-marine and alluvial geneses. They are given to thixotropy and have low stability (not more than 0.01 MPa) after thawing. In typical conditions the scouring may sometimes be of the maximum-thermoerosional type. The scouring resistance factor makes up (0.1 to 0.3) 10^{-6} cu.m/J which refer them to easily-scoured rocks. The second category encompasses medium-scoured syngenetic clay of the marine, lagoon-marine and glacial-marine origin. What set them apart is the fact that the tearing of particles is attributable not only to textural but structural peculiarities which emanate from lithological, physical, chemical, biological and other factors. Their stability when thawing varies from 0.01 to 0.03 MPa and, in typical conditions, the scouring is of the thermoerosional, and more seldom, erosional type. The resistance factor is estimated to be (0.03 to 0.10) 10^{-6} cu.m/J. The first category of slightly scourable rocks includes epigenetic clay deposits in which scouring (erosional type) intensity is less than $0.1 \cdot 10^{-4}$ m/s. These rocks are characterized by specific scouring along schlierens, but, in general, erosion tends to decrease. Stability of mineral aggregates does not exceed 0.03 MPa and the scouring resistance factor is not more than $0.03 \cdot 10^{-6}$ cu.m/J.

Slope-type deposits (alluvial, deluvial, solifluctious, proluvial) may enter any of the above categories with regard to their composition and structure (Yershov, Malinovsky et al. 1982).

Studies of permafrost dispersed rock scouring have indicated that syngenetic permafrost series which area increases northward, are characterized by less resistance to scouring. This explains, in our view, the tendency to intensified gully development to the north of the so-called gullyless belt in the USSR, which is dwelt on in many publications (Kosov and Konstantinova 1969). This belt is confined to the southern part of the permafrost zone typical for which are a maximum depth of seasonal freezing and thawing and a lack of syngenetic permafrost. It is the latter that, in view of their high ice content and, mainly, due to the development of small-meshed and thinly-laminated cryogenic textures, show the lowest erosion resistance.

Generally the scouring properties of permafrost rocks tend to decrease in the following order: syngenetic alluvial - marine - lagoon-marine - epigenetic lacustrine - marine - lacustrine-glacial - water-glacial. Therefore, the genesis of permafrost series (the freezing type being inclusive) gives an idea about the deposits' scouring.

The results obtained laid the basis for zoning the scour potential of Yamal Peninsula at a scale 1:5,000,000 (Yershov, Malinovsky et al. 1982). Identified on the map were regions, microregions, etc. of deposits having similar factors contributing to permafrost scouring. Permafrost surveys, providing ample material for ascertaining these factors, were used. For the purposes of zoning, taxonomic units distinguished for the West-Siberian map by V. T. Trofimov in 1977 were employed. Reflecting the changeability of engineering-geological rock properties in space, they are also helpful in tracing the scouring variability in particular. Indeed, while dividing a region into provinces on the basis of rock classes (with or without cementation), there are distinguished, first of all, areas composed of practically unscoured and scoured (dispersed) rocks. The thermal effect of streams on rock scouring (thermoerosion or traditional erosion) becomes clear when subdividing the provinces into zones and subzones by the present state of rocks.

Most important for scientific and practical purposes is zoning of territories (for assessing rock scouring) composed, firstly, of deposits without cementation and, secondly, rocks in a permafrost state. This somehow narrows the problem and allows application of the zoning pattern, described for the Yamal Peninsula, to predominantly permafrost areas.

The zoning of the Yamal Peninsula was carried out using the system of taxonomic units inherent in the engineering-geological zoning. The largest taxonomic units—districts—are identified by the morphological specifics of a territory and the genesis of the upper horizons in permafrost series: (1) district of predominantly Late Quaternary and modern lagoon-marine, marine, lacustrine and alluvial terraces, with a surface layer of syngenetic permafrost; (2) district of predominant Middle and Late Quaternary marine, glacio-marine, glacio-lacustrine and glacial plains in which the surface is epigenetic permafrost. The next taxonomic unit, in a decreasing order, is a region or a part of a district distinguished by bedrock in the upper part of a profile. Subregions are differing by the cryogenic structure of deposits. A definite rock category for scouring corresponds to each subregion. Disregarding composition and structure, deposits with well-developed ice wedges are referred to the fifth category. The assessment undertaken on Yamal Peninsula has indicated that much of the territory is composed of disastrously and rapidly scoured deposits, and their area increases northward.

REFERENCES

Kosov, B. F., and Konstantinova, G. S., 1969, Gully-erosion intensity in the newly-developed regions in the North, "Vestnik Moskovskogo universiteta," Geographical series, no. 1, p. 22-31.

Mirtskhulava, Ts. Ye., 1970, Engineering methods of calculation and forecast of water erosion, Moscow: Kolos.

Yershov, E. D., Kuchukov, E. Z., Malinovsky, D. V., et al., 1980, Methods of studying the permafrost rocks' scouring, in Engineering and construction investigations, no. 1, Moscow: Stroiizdat, p. 26-34.

Yershov, E. D., Malinovsky, D. V., Kuchukov, E. Z., et al., 1982, Thermal erosion of dispersed rocks, Moscow: Izd-vo Mosk. universiteta.

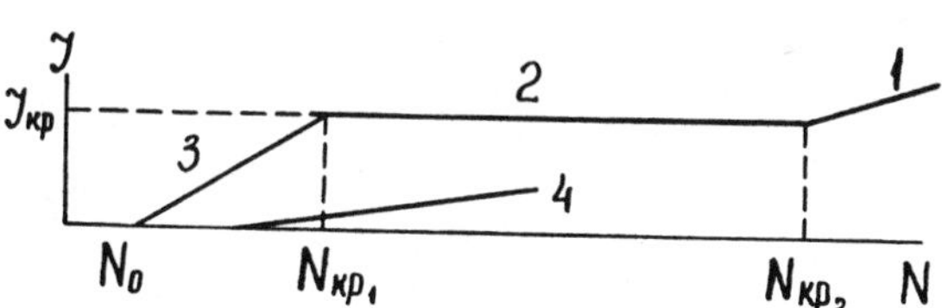

FIGURE 1 Relationship between the scouring intensity of permafrost rocks and water flow power at a constant temperature.

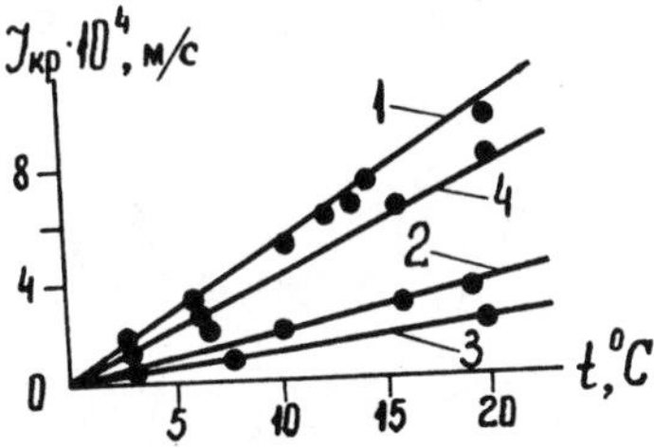

FIGURE 2 Relationship between the critical scouring intensity of dispersed rocks and the water flow temperature: 1 - sand; 2 - ice; 3 - loamy sand; 4 - loam.

TABLE 1 Classification of Dispersed Permafrost Rocks by Scouring

Category	Scouring	Scouring resistance factor, K_{sc} 10^6, cu.m/J	Admissible non-scouring velocities, V m/s	Rocks	Ice content in rocks, π_{vol} %	Cryogenic textures	
						Type	Variety
I	slight	0.03	1.0		low 0–25	massive	rare and coarse
II	medium	0.03–0.10	1.0–0.3	clay, loamy sand, loam	average 25–50	laminated and reticulate	average
III	rapid	0.10–0.30	0.3–0.1				fine
IV	intensive	0.30–10.0	0.1–0.003	sand	high >50	massive and reticulate	-
V	disastrous	10.0	0.003	sand and clay		ataxitic	

DILATOMETER TESTING IN THICK CYLINDERS OF FROZEN SAND

B. Ladanyi[1] and H. Eckardt[2]

[1]Northern Engineering Centre, Ecole Polytechnique, Montréal, H3C 3A7, Canada
[2]Züblin AG, Stuttgart, West Germany

The borehole dilatometer method for in situ determination of creep properties of frozen soils, introduced initially by Ladanyi and Johnston (1973), has since that time been used with reasonable success in several field investigations in frozen soils and ice, the results of which have been reported in several previous papers. However, since, the method has not yet been tested under controlled laboratory conditions, it was decided to carry out a series of dilatometer creep tests in confined thick cylinders of frozen sand, using the conventional field equipment. This paper presents and discusses the results of a series of stage-loaded medium- and long-term creep tests carried out in these cylinders with the Menard pressuremeter and gives the whole theoretical interpretation of such tests, which appear to be an interesting alternative method for the determination of creep parameters of frozen soils under plane strain conditions.

INTRODUCTION

After the first experience with field testing of creep properties of frozen soils, the results of which were published by Ladanyi and Johnston (1973), several additional field studies have been carried out in the North, both in frozen soils and in sea ice. Most of these studies were designed to improve the understanding of the borehole expansion testing method, in which two kinds of borehole dilatometers were used for performing short- and long-term borehole creep and relaxation tests (Ladanyi 1982).

This paper presents some results from a recent study of borehole expansion testing in frozen sand, in which the tests were performed with the conventional field equipment but under cold room conditions (Eckardt 1981). Such a study, carried out in thick cylinders of frozen sand under controlled temperature and lateral stress conditions, was considered necessary to be able to better evaluate and understand the effects of stress level and time on the creep behavior of frozen soil. Although the main purpose of this investigation was to simulate the field testing, the thick cylinder expansion method has proved to be an interesting alternative to other types of laboratory tests with frozen soils, especially when large scale plane strain information is required.

Special problems discussed in this paper are the interpretation of such thick-cylinder tests, the shape of stress-strain curves deduced from short-term tests, the different creep behavior of the soil at low and high stress levels, the extrapolation of creep curves to longer times, and the effect of stress redistribution on the observed creep behavior in stage-loaded borehole creep tests.

TEST EQUIPMENT

Although Eckardt (1981) used both the Menard pressuremeter and the CSM cell for performing borehole creep and relaxation tests, because of the space limitation only the results obtained with the pressuremeter will be described in this paper. The pressuremeter used in the tests was a Menard type GC with a volume capacity of about 800 cm^3 and a pressure range of up to 10 MPa. The probe had a minimum diameter of 7.1 cm, and its total length was 45.5 cm.

A sketch of the test chamber, specially constructed for sample preparation and test performance, is shown in Figure 1. The chamber, a steel tank with an outer diameter of 90 cm and a height of 50 cm, was designed to contain cylindrical hollow samples of frozen soil that were about 50 cm in diameter and up to 45.5 cm long. A rigid construction of the bottom and the top of the tank furnished essentially plane strain conditions for the sample. The bottom of the tank was perforated with 16 holes connected by plastic tubes to a water container for saturating the samples. The lateral confinement of the sample was supplied by two superimposed air-filled rubber cushions (actually inner tubes of large truck tires) placed between the sample and the steel tank wall, making it possible to apply a confining pressure of up to 0.3 MPa.

The test chamber was located in a cold room in which the temperature was controlled with an accuracy of ±1°C. In general, the temperature measured within the chamber and on the sample did not show much variation, remaining around -2.5^{o}C in all the tests.

TEST MATERIAL

A natural medium-grain-size sand was used in all the tests. Its grain size distribution obtained by a periodic ckeck of different batches, was about 90% between 0.1 and 1.0 mm, with less than 2% below 0.1 mm and less than 1% above 2 mm, giving a coefficient of uniformity of about 3.0. The maximum and minimum dry densities of the sand, determined by the ASTM standard method, were 1,810 and 1,510 kg/m^3, respectively. Constant-cell-pressure triaxial tests on the dry sand gave peak strength angles of internal friction of 45° at the maximum density, and 39° at a medium density of 1,680 kg/m^3.

The residual friction angle was about 35° at both densities.

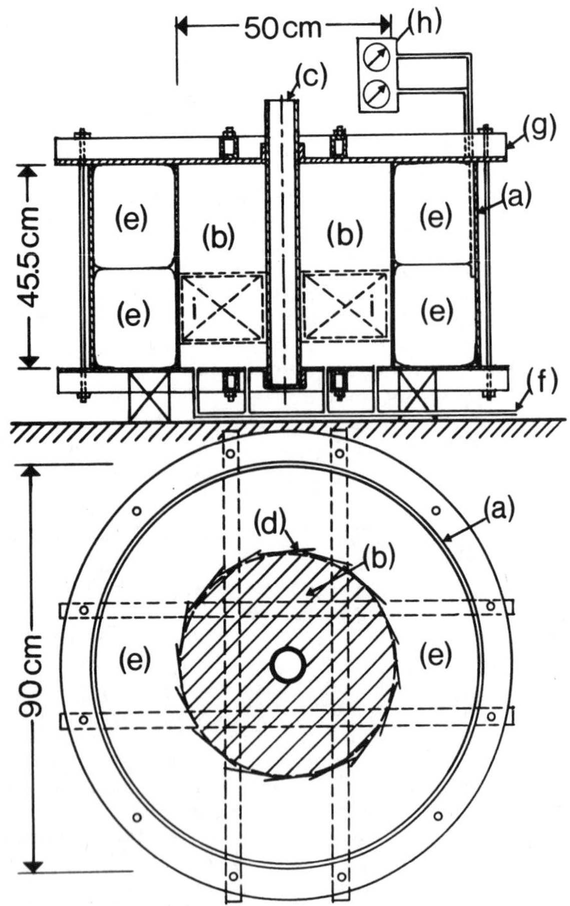

FIGURE 1 Thick-cylinder test setup for pressumeremeter tests in frozen sand. (a) steel tank, (b) sample, (c) central tube during sample preparation, (d) steel strips for sample preparation, (e) rubber cushions for lateral confining pressure, (f) saturation water supply line, (g) reinforced lid, (h) lateral pressure gages, (i) auxiliary frame.

PREPARATION OF SAMPLES

Before each sample preparation, the bottom of the steel tank was greased with a cold-resistant lubricant and then covered with a plastic sheet, perforated to permit sample saturation. Next, the rubber cushions for lateral pressure application were inserted and slightly inflated. Then 16 steel strips, each 13 cm wide and 45.5 cm long, were installed vertically against the inner side of the cushions, to form the outer sample envelope. The hole for the test was provided by installing in advance a central steel tube with a 7.4 cm diameter. The tube, which was greased and covered with a plastic sheet, was extracted after the sample was frozen. Lateral steel sheets and the central tube were fixed in the empty container by an auxiliary mobile frame, as shown in Figure 1. During the filling of the dry sand into the tank, this frame was pulled continuously upward. The sand was compacted by a vibrator to a medium density by moving the vibrator around the sample and inserting it in four different cycles. After the whole sample was prepared, the chamber was closed with a reinforced lid. Then the rubber cushions were inflated up to 100 kPa lateral pressure, and the sand was saturated from the bottom up with distilled and deaerated water.

The sample was frozen at a cold room temperature of -10°C. During freezing, the steel tank was insulated on the top and the sides to force the freezing front to move upward, which minimized freezing strains and stresses. After 3 days, the sample was frozen, and the room temperature was increased to the desired test level of -2.5°C. The sample was allowed to attain a homogeneous temperature distribution for three more days without insulation. Thus, altogether, the preparation of each sample took between 6 and 8 days.

The densities and water contents of the test samples were determined by taking six specimens after each tests, two from the top, two from the bottom, and two from the center.

THEORY OF DILATOMETER TESTS IN THICK-WALLED CYLINDERS

Short Term Tests

In the borehole dilatometer and pressuremeter practice (e.g., Ladanyi and Johnston 1973, 1978), the results of a short-term borehole expansion test are usually first plotted as a corrected pressure-expansion curve. The "true pressure-expansion curve" used for interpretation is obtained by shifting the pressure origin of that first curve to p_o, which denotes the average original ground pressure acting normally to the borehole axis. The "true pressure-expansion curve" represents then a relationship of the form $\Delta V = f(p)$, where $p = p_c - p_o$ is the net pressure acting on the borehole wall, and ΔV is the corresponding volume increase of the measuring section of the borehole, whose underformed volume was V_o.

For evaluation of test results, it is still more convenient to plot, instead of ΔV, the dimensionless quantity $\Delta V/V$, which represents in fact the shear distortion strain γ at the cavity wall. Here $V = V_o + \Delta V$ denotes the current borehole volume at a net pressure p.

As shown by Ladanyi (1982) , the classical theory of an expanding elastic thick cylinder in plane strain, enables to determine from the initial slope of the true pressure-expansion curve the value of the shear modulus of the soil.

$$G = \frac{E}{2(1+\nu)} = \lambda \frac{\Delta p}{\Delta \gamma} \tag{1}$$

where E and ν denote the Young's modulus and the Poisson's ratio of the soil, respectively, $\gamma = \Delta V/V$, and,

$$\lambda = \frac{1 + (1 - 2\nu)\,\alpha^2}{1 - \alpha^2} \tag{2}$$

where $\alpha = r_i/r_e$ is the ratio between the inner and the outer radius of the cylinder.

As far as the short-term strength is concerned, a method enabling one to deduce the whole stress-strain curve from a pressure-borehole expansion curve, developed originally by Ladanyi (1972), can easily be modified to include the tests performed in thick cylinders. It can be shown that in such a case, from any two consecutive points i,i+1 on the pressure-expansion curve, the value of the corresponding principal stress difference $q_{i,i+1} \equiv (\sigma_1 - \sigma_3)$, can be calculated from

$$q_{i,i+1} = \frac{2(p_i - p_{i+1})}{\ln(\gamma_i/\gamma_{i+1}) - (\alpha^2/\gamma_f)(\gamma_i - \gamma_{i+1})}, \quad (3)$$

where $\gamma_f = q_f/2G$ is the shear strain at failure of the soil and q_f is its unconfined compression strength, all valid for a plane strain case.

For the same interval, the average value of shear strain is

$$\gamma_{i,i+1} \equiv (\varepsilon_1 - \varepsilon_3) = \tfrac{1}{2}(\gamma_i + \gamma_{i+1}) . \quad (4)$$

All these formulas remain valid only until radial cracking of the cylinder occurs. It follows from Lamé's theory that in a linear-elastic material, radial fissures start to appear as soon as the internal pressure p_c in the borehole attains the value

$$p_c = \frac{2p_o + |T_s|(1 - \alpha^2)}{1 + \alpha^2}, \quad (5)$$

where p_o is the external pressure applied to the cylinder at $r = r_e$ and $|T_s|$ is the absolute value of the tensile strength of the material.

Creep Tests

For an ice-rich frozen soil or ice, the total strain attained after a given time under a constant stress is given by

$$\varepsilon_e = \varepsilon_e^{(i)} + \varepsilon_e^{(c)}, \quad (6)$$

where $\varepsilon_e^{(i)}$ is the instantaneous (elastic and plastic) portion of the total strain and $\varepsilon_e^{(c)}$ is the creep strain, which may be expressed by a power law equation

$$\varepsilon_e^{(c)} = (\dot{\varepsilon}_c/B)^B(\sigma_e/\sigma_c)^n\, t^B . \quad (7)$$

Here the subscript e denotes the von Mises equivalent stress and strain, σ_c is the reference stress at the reference strain rate $\dot{\varepsilon}_c$, t is the time, and B and n are creep exponents.

Following Ladanyi and Johnston (1973, 1978) and Ladanyi (1982) , it will be seen that the creep portion of the borehole expansion in a thick cylinder at a constant stress $p = p_c - p_o > 0$, applied during stage i, is given by the equation

$$\ln(V/V_{i-1}) \approx (\Delta V_c/V) \approx \gamma_c = 2D_n p^n t^B, \quad (8)$$

where V is the current volume of the cavity, $V_c = V - V_{i-1}$ is its creep increment with respect to the volume V_{i-1} attained in the preceeding stage, and D_n is defined by

$$D_n = (\sqrt{3}/2)(\dot{\varepsilon}_c/B)^B(\sqrt{3}/n\,\bar{\sigma}_c)^n , \quad (9)$$

with

$$\bar{\sigma}_c = m\sigma_c , \quad \text{and} \quad m = 1 - \alpha^{2/n} . \quad (10)$$

As shown in the above-mentioned references, the values of creep parameters B, n, and $\bar{\sigma}_c$ can be obtained from a series of borehole creep tests or from one multistage creep test, by plotting the results in a double-log plot, first the creep strain γ_c versus time and then $(2D_n p^n)$ versus the applied pressure p. The value of creep parameters deduced from such tests is, however, subject to a certain error which is due to the apparent primary creep, resulting from the stress redistribution around the hole, as shown later.

TEST RESULTS

As mentioned before, only some typical results of tests carried out with the Menard pressuremeter will be presented here. A more complete report on these tests and on the CSM cell tests performed in this investigation can be found in Eckardt (1981) and in a companion paper by Eckardt and Ladanyi (1983).

Altogether 15 pressuremeter tests were carried out for determining the short- and long-term parameters of frozen sand in this study. Eight of these tests were conducted as stage-loaded tests with eight to ten 30-min-long stages at constant load. The results of these tests were used not only for determining the deformation modulus G and the short-term strength parameters of the soil but also for calculating its medium-term creep parameters. The remaining seven tests, designed to study long-term behavior of frozen sand, were conducted with two to three creep stages, one of which was at a low stress level. The longest creep time reached in one of the tests was about 8 days. The tests were performed at different confining pressures (p_o = 0, 0.1, 0.2, and 0.3 MPa) and at an average soil temperature of -2.4° (-2.2° to -2.6°C).

Short-Term Parameters

The short-term parameters of the frozen sand were determined from the results of pressuremeter tests by the usual method, described before (e.g., Ladanyi 1972), but by using equations (1)-(5), valid for the tests in thick-walled cylinders. The procedure consisted in calculating the corrected pressuremeter curve $\Delta V/V = f(p)$ for the end-of-stage (30 min) readings of each test, and in determining the corresponding stress-strain curves. Five of such stress-strain curves, relating the principal stress difference $q = \sigma_1 - \sigma_3$ to the shear strain $\gamma = \sigma_1 - \sigma_3$ are shown in Figure 2. As they represent a plane strain information, the corresponding axial symmetry curves can be obtained through the von Mises relationships:

$q_{ax} = (\sqrt{3}/2)q_{ps}$ and $\varepsilon_{1ax} = (\sqrt{3}/3)\gamma_{ps}$. The slope through the origin of these curves gives the value of the shear modulus $G = \frac{1}{2}\ \Delta q/\Delta\gamma$.

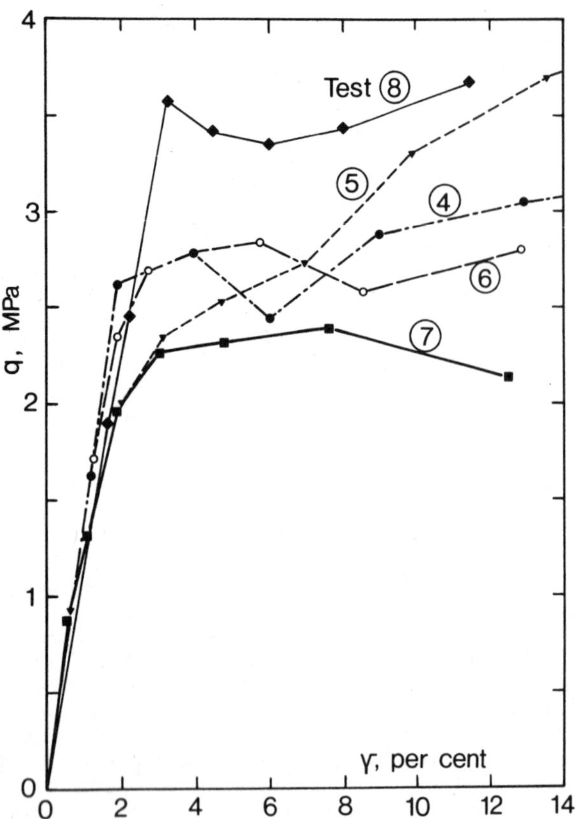

FIGURE 2 Stress-strain curves from 30-min-per-stage creep tests.

In addition, a set of Mohr circles was also plotted for each test (Eckardt 1981), from which the minimum value of the tensile strength T_s of the soil was determined. Although the tests were conducted at external confining pressures varying from 0 to 0.3 MPa, no systematic effect of that pressure on test results could be detected. In general, in spite of the great care in preparing the tests, the results were rather variable, which may be due to the great sensitivity of the sand behavior to small variations of its density and ice saturation.

Table 1 shows the main data of the six most consistent short-term tests 3-8.

The rather variable shape of the stress-strain curves in Figure 2, some with two peaks and others showing strain hardening, may be explained by the fact that, on one hand, radial tensile cracks started propagating from the central hole rather early in the tests, which might have caused the first peak, while, on the other hand, the ever-increasing strain rate, which is typical in such stage-loaded tests, combined with the mobilization of friction, resulted in an apparent strain hardening.

Theoretically, according to equation (5), with $|T_s| \approx 1$ MPa and $\alpha^2 \approx 0.02$, the radial tensile cracks should have started propagating at $p_c \geq 1$ MPa with $p_o = 0$, and at $p_c \geq 1.6$ with $p_o = 0.3$ MPa. In fact, they were observed only at nearly 4 times higher p_c values, which may be explained by the nonlinearity of stress-strain behavior of frozen sand due to creep (Ladanyi and Gill 1981). In other words, although the tests invariably finished with a complete failure of the cylinders, during the major portion of the tests the radial cracks were insificant and did not affect seriously the creep behavior.

TABLE 1 Results of Short-Term (30 min/stage) Tests.

Test	ρ_d (kg/m^3)	w (%)	S (%)	T (°C)	p_o (MPa)	$\|T_s\|$ (MPa)	q_{ps} (MPa)	γ_f (%)	G (MPa)
3	-	-	-	-2.5	0.1	1.5	3.7	2.5	90
4	1,560	17.5	66.8	-2.1	0.2	0.9	2.8	4.0	70
5	1,540	18.1	66.5	-2.5	0.3	0.8	3.7	3.0	70
6	1,590	17.9	70.6	-2.3	0	1.0	2.8	3.0	70
7	1,640	15.1	65.0	-2.5	0.3	1.0	2.4	3.0	50
8	1,590	17.6	70.1	-2.3	0	0.8	3.6	3.2	60
Average	1,584	17.3	67.8	-2.37	-	1.0	3.17	3.1	68.3

Note: ρ_d = dry density; w = water content; S = degree of saturation; q_{ps} = plane strain compression strength; γ_f = shear strain at failure.

Creep Behavior: Medium Term

As described by Ladanyi and Johnston (1973), the medium term creep parameters can be determined from a stage-loaded pressuremeter test by plotting the successive creep curves in a log-log plot of creep strain $\gamma \approx \ln(V/V_{i-1}) \approx \Delta V/V$ versus time t, such as shown in Figure 3 for test 4. However, as noted by Ladanyi and Johnston (1978), a disadvantage of such a plot is to include not only the creep strains, but also the instantaneous plastic strains, which makes its use for parameter determination difficult and inaccurate, especially at low pressures. At higher pressures, i.e., at $p > 2$ MPa, it was nevertheless possible to determine from the tests 3-8 average values of creep parameters B, n, and σ_c in equation (7). These values and their ranges were B = 0.80 (0.72 - 0.84), n = 2.40 (1.80 - 3.60), and σ_c = 1.23 MPa (0.82 - 2.00) at $\dot{\varepsilon}_c = 10^{-5}$/min. In particular, for test 4 (Figure 3), the values were B = 0.81 (0.5 - 1.1), n = 2.26, and σ_c = 0.94 MPa.

A better way of interpreting the same creep data is to subtract from the measured strains the pseudo-instantaneous ones read from the first plot at a given short time interval, say, 1 min, and replot the remaining strains, as suggested by Ladanyi and Johnston (1978).

The plot in Figure 4 is the result of such a manipulation of data for test 4, where the intercepts at 1 min in Fig. 3 were subtracted from the total strains and the rest was replotted at the times reduced by 1 min. It will be seen that a much more regular plot is obtained in such a manner. In addition, two distinct regions of creep behavior can clearly be discerned: one at low stresses, in which B strongly increases with stress, and

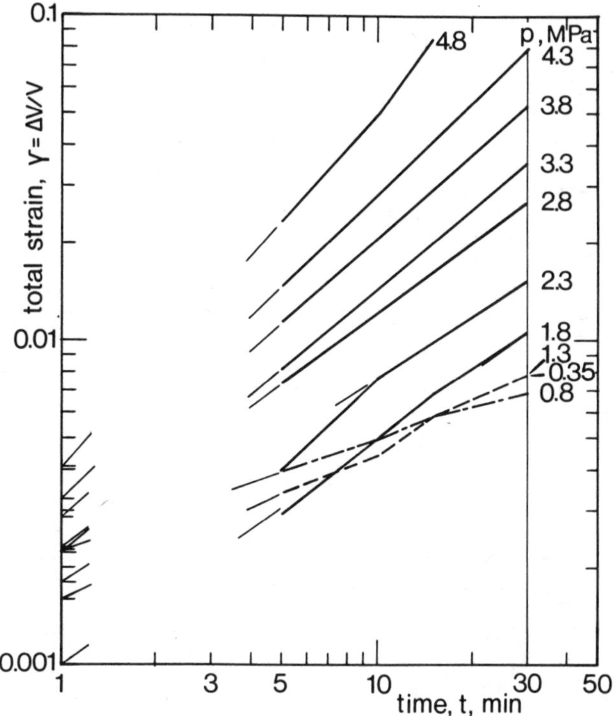

FIGURE 3 Test 4: Total strain versus time curves.

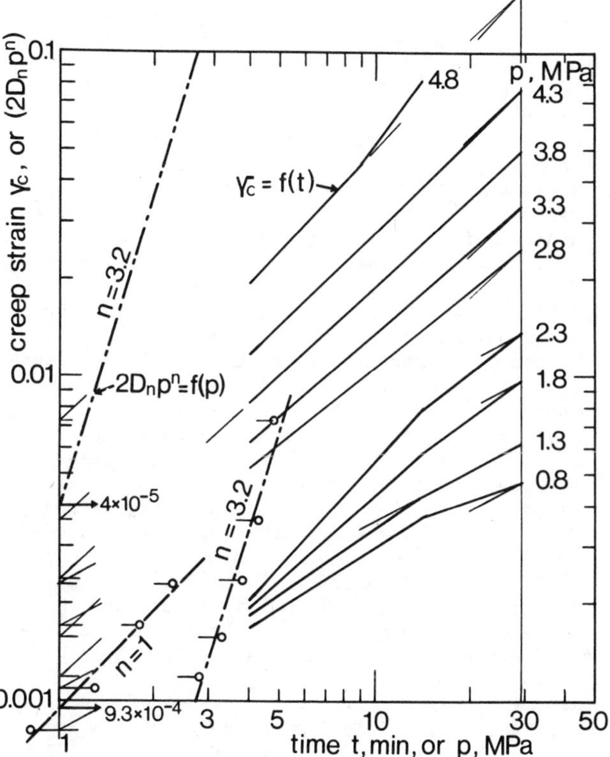

FIGURE 4 Test 4: Creep strain versus time, and ($2D_np^n$) versus pressure curves.

another at high stress, where B remains practically constant. In test 4, for $t > 15$ min, these values were B = 0.34 - 0.77 (average 0.52) for p between 0.8 and 2.8 MPa, and B = 0.8 - 1.0 (average 0.90) for $p > 2.8$ MPa. Using the average B values, the corresponding n and σ_c values obtained for the two regions were n = 1, σ_c = 11.63 MPa for $p \leq 2.8$ MPa, and n = 3.2, σ_c = 0.878 MPa for $p > 2.8$ MPa (all at $\dot{\varepsilon}_c = 10^{-5}$/min). In addition, by treating the cumulative 1 min pseudo-instantaneous strains in a manner similar to the creep strains, it was possible to determine also the first term in equation (6) for both regions.

The final result of such interpretation were the following two complete stress-strain-time equations, describing the behavior of frozen sand during a medium-term creep at -2.1°C: (1) for $p \leq 2.8$ MPa,

$$\varepsilon_e = \sigma_e(0.00117 + 0.00030\ t^{0.52})$$

and (2) for $p > 2.8$ MPa,

$$\varepsilon_e = 0.000874\ \sigma_e^{1.5} + 5.277 \times 10^{-5}\sigma_e^{3.2}\ t^{0.90}\ ,$$

where the stresses are in MPa and the time in minutes.

Creep Behavior: Long Term

As creep time intervals of more than 24 hours have been difficult to realize in the field, one of the main tasks of this study was to see whether the creep rate remains unchanged after longer periods of time. Several tests were made for that purpose with two to three loading stages, the first one being up to 10 days long. The results of two of such tests are shown in Figure 5. The figure shows the creep curves from which the 1 min portion of strains has already been subtracted.

In Figure 5, for test 10, it appears that B remained constant during the first 4 hours but increased with the stress level from 0.38 at 0.9 MPa to 0.64 at 3.0 MPa. There was some decrease in slope after 4 hours, which could have been caused by small variations in the soil temperature and in the confining pressure during this test.

A similar trend of B increasing with pressure was observed in test 12 (Figure 5), as well as in several other similar tests not shown here. In test 12, the first stage at p = 0.92 MPa was held for 8 days and showed a fairly constant average slope of B = 0.32. The second stage at 3.96 MPa manifested a much faster creep rate with B = 0.82, ending with the cylinder failure after 3.5 hours.

The variation of the time exponent B with pressure is considered to be quite an important phenomenon, which should be taken into account when using the pressuremeter-determined creep data for the design of structures in permafrost. Based on the results of pressuremeter tests performed in this study, it was found that for the frozen sand at about -2.5°C, the variation of B could be described on the average by $B = 0.45p^{0.5}$, with an upper bound of $B = 0.52p^{0.4}$ and a lower bound of $B = 0.38p^{0.6}$, valid for $0.8 < p < 5.0$ MPa.

It is noted that the values of creep exponents n and B, reported in the literature for various frozen sands, $1.28 < n < 2.63$ and $0.45 < B < 0.63$, do not differ much from those found here (Sayles 1968).

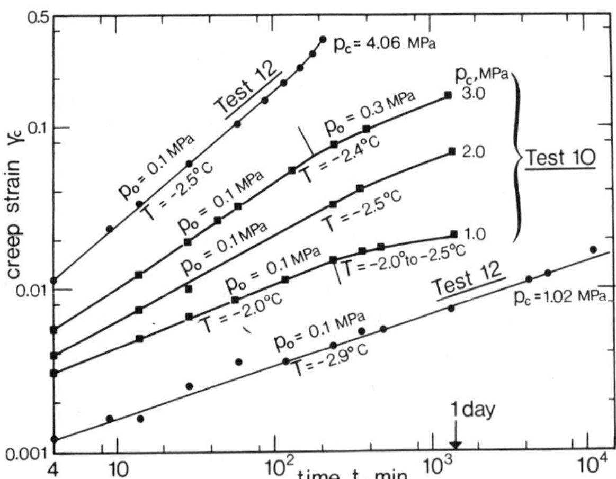

FIGURE 5 Test 10: 24 h creep curves.
Test 12: a two-stage creep test, the first stage being 8 days long.

Stress Redistribution Time

In a stage-loaded dilatometer test, a certain time is necessary after load application for the stresses to redistribute from the initial linear-elastic state to that corresponding to the nonlinear stationary creep. As during this time the measured creep parameters may be in error, it is interesting to see how long this time would be under the conditions prevailing in this study. While the problem of stress redistribution in such a case can be solved only numerically, an approximate answer may be found by using the method proposed by Calladine (1969). Using that approach, the authors have developed the following equation, valid for a thick-walled cylinder in plane strain under a stage load p.

$$\tau_R = t_R^B = \frac{(m\,\sigma_c)^n \ln R}{G\,n\sqrt{3}(\sqrt{3}/n)^n\,(\dot{\varepsilon}_c/B)^B\,p^{n-1}} \quad (11)$$

Here, $_R$ denotes the "time" to achieve redistribution of stress to within a fraction 1/R of the stationary state, or for the creep strain to become equal to (1/n)ln R times the elastic strain.

Substituting in equation (11) the average values of creep parameters found in tests 3-8, i.e., B = 0.80, n = 2.40, G = 70 MPa, σ_c = 1.23 MPa at $\dot{\varepsilon}_c = 10^{-5}$/min, and m = 0.80, it is found that the redistribution time is given by $t_R = 476.4/p^{1.75}$, with t_R in minutes and p in MPa. This equation gives for t_R about 27 hours at 0.5 MPa, 8 hours at 1 MPa, over 2 hours at 2 MPa, and about 1/2 hour at 5 MPa. Although these times vary with the values of creep parameters, it is interesting to note that stress redistribution times are much longer at low stresses than at high stresses.

An independent Finite Element Method study, designed for checking these predictions, is presently under way.

CONCLUSIONS

This cold-room study of the pressuremeter test as a means for determining certain basic stress-strain and creep parameters of frozen soil has revealed the potential of such a test, not only as a field test, but also as an alternative method for large-scale testing of frozen soils under plane strain conditions. In particular, this study, carried out in large thick cylinders of frozen sand, has made it possible to attain creep times much longer than under field conditions. The results show that the assumed power law formulation for the creep behavior of frozen sand can be retained and used for long-term extrapolation, provided one takes into account the fact that the creep parameters n and B both tend to decrease with decreasing deviatoric stress.

REFERENCES

Calladine, C.R., 1969, Time scales for redistribution of stress in creep of structures, Proceedings, Royal Society of London, A, v.309, p. 363-375.

Eckardt, H., 1979, Tragverhalten gefrorener Erdkörper, Publications, Institute for Soil and Rock Mechanics, University of Karlsruhe, Heft 81, 141 p.

Eckardt, H., 1981, Laboratory borehole creep and relaxation tests in thick-walled cylinder samples of frozen sand, Report 222, Northern Engineering Centre, Ecole Polytechnique, Montréal, 125 p.

Eckardt, H., and Ladanyi, B., 1983, Dilatometer-Laborversuche mit gefrorenem Sand, Die Bautechnik, v. 60, 2, p. 56-65.

Ladanyi, B., 1972, In-situ determination of undrained stress-strain behavior of sensitive clays with the pressuremeter, Canadian Geotechnical Journal, v. 9, p. 313-319.

Ladanyi, B., 1982, Borehole creep and relaxation tests in ice-rich permafrost, Proceedings of the 4th Canadian Permafrost Conference (The Roger J.E. Brown Memorial Volume), National Research Council of Canada, Ottawa, p. 406-415.

Ladanyi, B., and Gill, D.E., 1981, Determination of creep parameters of rock salt by means of a borehole dilatometer, Proceedings, First Conference on the Mechanical Behavior of Salt, Pennsylvania State University (in press).

Ladanyi, B., and Johnston, G.H., 1973, Evaluation of in-situ creep properties of frozen soils with the pressuremeter, in Permafrost, The North American contribution to the Second International Conference, Yakutsk: Washington, D.C., National Academy of Sciences, p. 310-318.

Ladanyi, B., and Johnston, G.H., 1978, Field investigations of frozen ground, in Andersland, O.B. and Anderson, D.M., ed., chapter 9, Geotechnical engineering for cold regions: McGraw-Hill, New York, p. 459-504.

Sayles, F.H., 1968, Creep of frozen sands, Technical Report 190, Cold Regions Research and Engineering Laboratory, Hanover, New Hampshire, 31 p.

THE EXTENSION OF PERMAFROST IN WESTERN EUROPE IN THE PERIOD BETWEEN 18,000 AND 10,000 YEARS B.P. (TARDIGLACIAL) : INFORMATION FROM SOIL STUDIES

R. LANGOHR

Department of General Pedology, State University of Ghent
Krijgslaan 281, 9000 Ghent, Belgium

Soils developed since the beginning, or during the Tardiglacial (±18,000 to 10,000 years B.P.), are compared with "Holocene" soils. The area covers the loess and coversand belts of NW France, Middle Belgium, and The Netherlands. Eight major kinds of Tardiglacial soils are recognized. All of these except one have a consolidated horizon which is present regardless of the texture and drainage class. This horizon has a particular structure and porosity fabric, shows relict pseudogley properties, and is not observed in the Holocene soils. The formation of the consolidated horizon either by freeze-thaw cycles or by permafrost is discussed. It is concluded that only the permafrost hypothesis can explain the complete set of properties. Consequently, permafrost would have been at least once continuous during the Tardiglacial in the studied area.

INTRODUCTION

There is no agreement whether permafrost conditions occurred in NW France, Belgium, and The Netherlands during the Weichsel Tardiglacial age (±18,000 to 10,000 years B.P.). Many authors mention traces of periglacial processes like cryoturbations and small frost wedges but do not specify whether permafrost was present or not (e.g., Zagewijn and Paepe 1968, Paepe 1969, Paepe and Sommé 1970). Others mention permafrost (e.g., Haesaerts and Bastin 1977, Haesaerts and Van Vliet 1981). Our purpose is to investigate this controversy on the basis of the surface soil characteristics (upper 2-3 m of the regolith).

SELECTION OF THE AREAS AND FIELD SURVEY

In studying the traces left in the surface soils by the Tardiglacial environment the areas of interest are severely reduced because of the following requirements.

1. The soils must be formed in sediments deposited just before or during the Tardiglacial. For this reason the investigation is limited to (1) the upper meters of the Weichsel Pleniglacial B loess and coversands, deposited after 23,000 years B.P. and probably before 18,000 years B.P. (Haesaerts et al. 1981) and (2) the recent coversands deposited during some of the colder periods of the Tardiglacial. The loess deposits are dominant in NW France and Middle Belgium, the coversands occur mainly in N Belgium and in The Netherlands.
2. The influence of man on the soil characteristics must be avoided as much as possible. Consequently, the field survey is limited to only those forested areas which as verified by historical data have not been cleared and plowed since at least the eleventh or twelfth century.
3. For the study of complete toposequences, including the water divide, the slope, and the valley bottom positions, only the large forests are suitable.
4. Only those toposequences are selected where soils are developed in thick (more than 1 m) sediments of the same litho-chronostratigraphic unit.

The above mentioned requirements reduced the area of interest for this investigation to less than 3% of the loess belt and less than 15% of the coversand belt.

The soil descriptions were made using a methodology developed by the author, combining procedures from the soil survey as well as the archeology and the field survey made for lithostratigraphy. Special attention was paid to the combination of both vertical and horizontal sections of the soil profile and to the relative position of each soil feature. When useful, samples were collected for physicochemical analysis or for the preparation of thin sections.

In total, some 60 profiles have been described in loess and coversands. These data have been compared with those from soils developed in Holocene sediments, particularly sand dunes, marine and alluvial deposits, and colluvial sediments in areas under cultivation.

RESULTS

For reasons of simplicity the soils developed since ±10,000-18,000 years will be called "Tardiglacial soils" and the younger soils will be called "Holocene soils". Figure 1 represents the main soil kinds (SK) observed. SK1-SK6 are Tardiglacial soils developed in the upper meters of Pleniglacial B loess and coversands. SK7 and SK8 are the most common Tardiglacial soils in the recent coversands. SK9-SK11 represent the most common types of Holocene soils.

A striking feature of Tardiglacial soils (SK2 up to SK8) is the presence of a particular consolida-

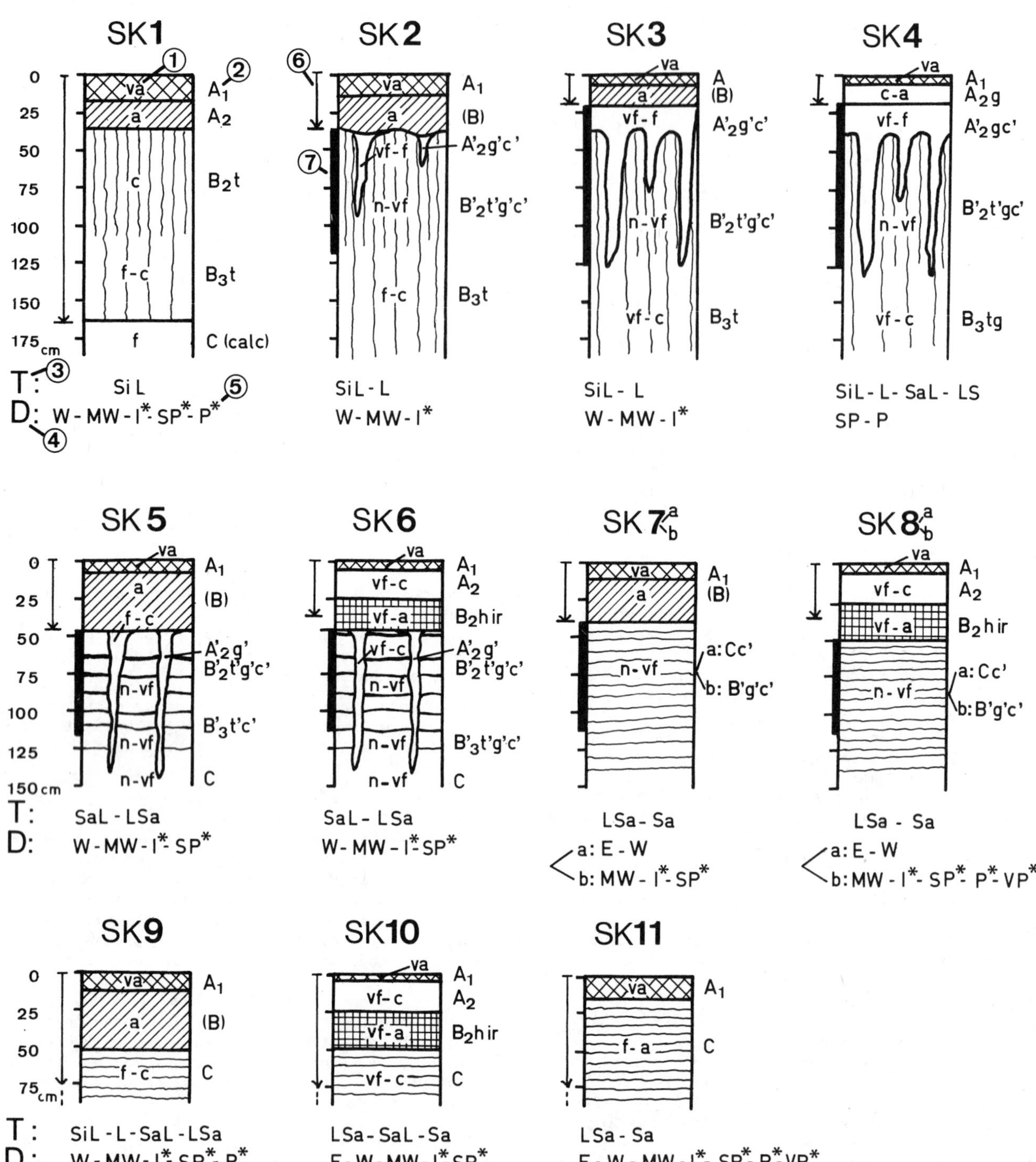

FIGURE 1 Main soil kinds (SK) developed in sediments deposited within the last 23,000 years. SK1 indicates soil kind 1. 1 : abundance of roots, n, no; vf, very few; f, few; c, common; a, abundant; va, very abundant. 2 : soil horizon symbols, mainly after Soil Survey Staff (1975), except c (consolidated horizon). In B'2t'g'c', the accents stand for all the processes that are thought not to be active any more. 3 : T = texture classes, after Soil Survey Staff (1975), SiL, silt loam; L, loam; SaL, sandy loam; LSa, loamy sand; Sa, sand. 4 : D = drainage class, after Soil Survey Staff (1951), E, excessively; W, well; MW, moderately well; I, imperfectly; SP, somewhat poorly; P, poorly; VP, very poorly. 5 : with these drainage classes, the symbol g (active gleying) should be added to the horizon symbols. 6 : biologically most active part of the profile. 7 : total thickness of the consolidated horizon.

ted horizon or pan indicated in Figure 1 by the horizon symbol c and by the thick black line on the lefthand side of the profiles. This layer was neither observed in the Holocene soils nor in SK1 of the Tardiglacial soils. In latter soils a calcareous C horizon is observed relatively close to the soil surface (mostly at 1.5-2.2 m depth). These soils are formed in particular landscape positions of the loess belt, namely (1) relatively steep slopes (gradient mostly of more than 30%) and (2) in some poorly drained smooth upland depressions.

The main characteristics of the consolidated horizon are the following.

1. The upper limit of the compact horizon always represents an abrupt increase in bulk density (see Table 1). The density remains more or less constant throughout the pan and drops slightly in the underlying B3.

TABLE 1 Average Bulk Density in Soils Developed in Pleniglacial B Loess (Ring Samples Collected at Field Capacity)

Horizon	Depth of samples (cm)	Average Bulk density	Number of horizons analyzed
Root-active (B)	10-25	1.25	11
Consolidated A'2g'c'	25-35	1.59	7
Consolidated B'2t'g'c'	40-110	1.60	19
Root-active B3t	130-180	1.53	10

2. All the consolidated horizons, regardless of the texture and drainage class, have a particularly high penetration resistance. Field measurements made at field capacity always indicate a penetration resistance of more than 50 Newton/cm^2 (upper limit of many field instruments). These high values are obtained in all parts of the horizon, including the bleached vertical streaks and the porous central part of the prisms. In the underlying nonconsolidated B3 horizon the resistance drops to values ranging mostly between 10 and 20 Newton/cm^2.

3. No evidences of recent or present-day activity of burrowing animals such as earthworms or larvae of beetles is seen.

4. Few if any roots pass through the pan. Where they do, they are always oriented downwards and situated along the faces of a coarse prismatic structure (see point 10 below). The only biological activity inside these prisms is limited to hyphae of fungae along some pores and secondary ped faces.

5. The upper boundary is at 20-50 cm depth and is abrupt and roughly parallel to the soil surface. Its topography is in places smooth, but commonly the overlying much more porous earth tongues down to several decimeters depth. The lower limit is mostly at a depth of 80-150 cm, yet depths up to 180-230 cm have been observed (mainly in the sandy soils), this boundary is gradual and smooth to wavy.

6. The soil material is not cemented, and air dried fragments slake in water.

7. The consolidated layer overlaps different soil genetic horizons such as bleached clay-eluvial horizons (A'2g'c' of SK3 and SK4), brown clay-illuvial horizons (B'2t'g'c' of SK2-SK6), color and/or structure B horizons (B'g'c' of SK7b and SK8b) and even C horizons (Cc' of SK7a and SK8a). In the B'2t'g'c' horizons most of the pores and ped faces are devoid of clay coatings. However, thin sections show numerous disrupted fragments of clay-illuvial bodies in the interior of the peds. In contrast, below the compact horizon (B3t) the clay coatings are covering pore walls and ped faces.

8. The present-day drainage class of these soils ranges from excessively drained up to poorly drained. In the loamy and silt loamy soils with a fluctuating groundwater table, the water saturation is in fact limited to the bleached streaks in between the prisms (see point 11 below); because of the low hydraulic conductivity of the prism border, the large pores inside the prisms remain empty of water, even when the whole horizon is below the groundwater table level.

9. All soils show at least in the upper part of the pan morphological evidences of a fluctuating groundwater table. Bleached and iron oxide enriched mottles occur along the pores and ped faces; the manganese oxide enriched areas are found mainly inside the aggregates. In at least the excessively-, the well- and the moderately well-drained soils (the latter two soils represent more than 50% of the studied area), this hydromorphic characteristic is relict. In fact, all the observed soils with compact horizon have pseudogley characteristics, as the most intense mottling is situated in the upper part of the pan, whereas it decreases in the lower part. In the soils without a groundwater table today, the mottling disappears completely in the underlying B3. On the other hand, in soils with a fluctuating groundwater table the mottling increases again in depth.

In some profiles the lower limit of the pseudogley is abrupt and contains a few centimeter thick layer with, successively from top to bottom, an accumulation of iron and manganese oxides. These profiles are mainly found in very flat plateau positions or in depressions. In the latter profiles the accumulation horizon is situated deeper in the profile (80-130 cm versus 35-60 cm).

10. Where the compact horizon starts nearest to the soil surface (around 20 cm, SK3 and SK4 in Figure 1,and Figure 2) the following structures sequence is observed. (1) The structure is composed of very thin plates (1-3 mm thick) in the upper 5-10 cm of the pan. The packing density of the plates is very high, and no roots pass in-between the peds. Yet discontinuous pores and fissures occur in-between the plates. (2) At a depth of 30-40 cm the structure becomes gradually coarser and lenticular. (3) From 25-35 cm depth this structure type overlaps with the upper part of coarse prisms. These prisms vanish toward the lower limit of the consolidated horizon. The prisms are largest (up to 4-5 m) in the excessively and in the well-drained sandy soils of SK7a and

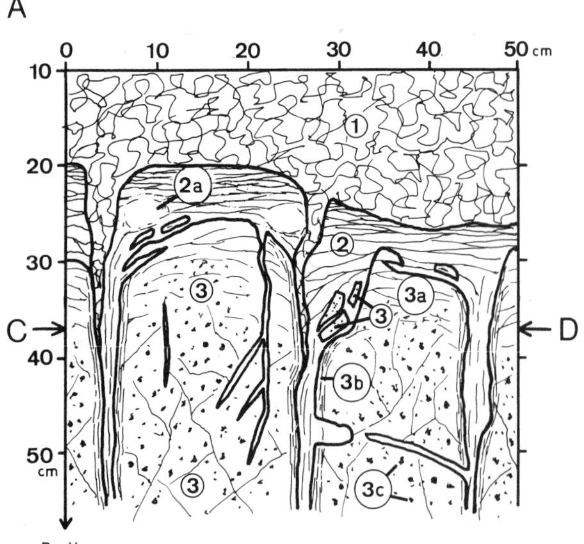

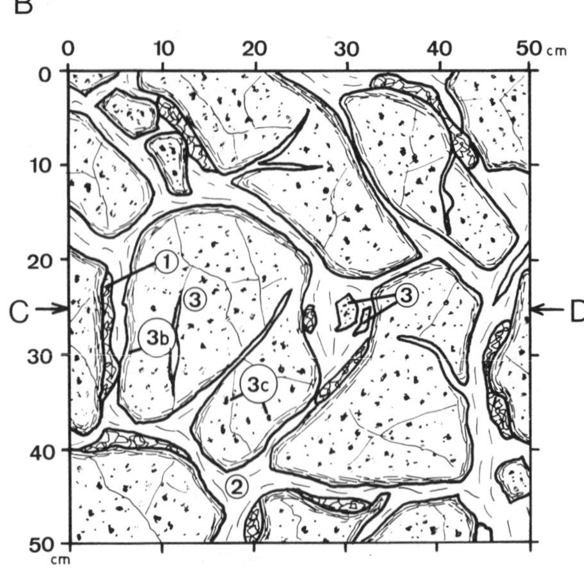

FIGURE 2 Detail of the upper part of a consolidated horizon in an imperfectly drained soil that was under forest and developed in a ±20,000 year old loess (SK3 of Figure 1). A : Vertical section-- 1 : biologically very active (B) horizon, with downward tongues; 2 : consolidated bleached A'2g'c' horizon with platy to lenticular structure and tonguing down along continuous vertical streaks, (2a, very compact massive earth nodules). 3 : consolidated B'2t'g'c' horizon with coarse prismatic structure, (3a, upper part of the prisms, with a coarse lenticular structure; 3b, particularly compact rim of the prisms, with weak vertical platy structure; 3c, coarse irregular discontinuous pores inside the prisms). B : Horizontal section at 37 m depth-- 1 : vertical tongues with material of the (B) horizon, all roots are situated here; 2 : vertical streaks with consolidated material similar to the one of the A'2g'c' horizon; 3, 3b, 3c : see A; CD : line of intersection between A and B.

SK8a. They are smallest (5-10 cm) in the somewhat poorly and the poorly drained silt loamy soils. Present-day well-drained soils located in depression also can have relatively small prisms. The outer rim of the prisms, with a width of 0.5-2 cm, has only pores smaller than 0.2 cm and has a weak vertical platy structure oriented parallel to the prism walls. The structure inside the prisms varies from profile to profile and can even vary within the same profile. The most common types are (1) structureless (the stratification of the sediment is still present in SK7 and SK8). (2) moderately fine to coarse angular blocky, and (3) weak very irregularly fine up to coarse prismatic. The in-ped porosity of the prisms is also variable; in the sandy soils few or no pores are visible with a lens; in the medium textured soils, 0.5 up to 2 pores of 0.5-3 mm diameter are present per square centimeter. These relatively large pores have an irregular shape and are discontinuous.

11. A particular feature of most compact horizons is the presence of bleached streaks from a few millimeters up to 5-8 cm wide in-between the prisms (see point 8). They are widest in the upper part of the horizon and gradually become thinner with depth; the largest ones reach a depth of 150-230 cm. The material in these streaks is the same as that in the overlying A'2g'c', whenever this horizon is still present. This was confirmed in at least two soil kinds, (1) in soils where only the A'2g'c' and the streaks contained small charcoal fragments and (2) in sandy loam soils with a graded texture becoming more silty downward and where the streaks contained the same coarser material as that in the upper part of the pan. The soil material in the streaks is also rather compact, particularly in the upper part of the horizon, where close examination shows only very few pores larger than 0.5 mm and very locally a weak vertical platy structure. The lower part of the streaks can contain larger pores which are irregularly shaped and discontinuous. When roots pass through the consolidated horizon, they do so at the contact between the streaks and prism walls. The largest roots are flat shaped.

12. Locally angular fragments of the brown B'2t'g'c' are included in the bleached A'2g'c' (Figure 2). Inversely, intrusions of the A'2g'c' and of the bleached streaks penetrate along fine fissures into the brown prisms of the B'2t'g'c'.

13. Chemical analysis, total analysis, and clay mineralogy do not show any significant difference between profiles with a consolidated horizon as in SK2 and very similar soils without this horizon (SK1) (Lozet and Herbillon 1971, Guidotti Pena 1976, Van Ranst 1981).

DISCUSSION

Among the Tardiglacial soils, only those of soil kind 1 (SK1 in Figure 1) have no consolidated horizon. In these SK1 soils the presence, relatively close to the soil surface, of a still calcareous parent material is probably not a coincidence and the absence of a consolidated horizon can be explained by the following mechanism. When the upper part of the soil became compacted, the soil at that depth was still calcareous; as during further pedogenesis the calcium carbonate was gradually leached, the compaction disappeared, and biological activity and roots could easily penetrate; consequently the consolidated horizon was not conserved (Langohr and Van Vliet 1981).

From the data described previously it can be concluded that the formation of the consolidated horizon was mainly a physical process, creating a particular aggregate and porosity fabric associated with or immediately followed by a perched groundwater table. The process had a gradient parallel to the soil surface and was active in all soils, regardless of texture, drainage, and previous profile development conditions. However, parallel to these three soil attributes, differences in morphology of the compact horizon are observed. The process was not active during the Holocene or during the beginning of the Tardiglacial. The latter statement is concluded from the fact that the compact horizon formed after the development of other pedogenetic soil horizons.

Also, the high packing density of the platy and lenticular aggregates in the upper part of the horizon indicate a pedogenetic process that has been active in a period when there were no roots and no worm activity at these depths (± 20-50 cm). Such an environment does not correspond to what we know of the Holocene in Western Europe.

The question arises whether this consolidation developed during freeze-thaw cycles or whether it is indicative of permafrost conditions.

In soil science literature, more or less similar compact horizons have often been called "fragipan". The author previously used this term (Langohr and Van Vliet 1981). However, in view the data described in previous chapter, he prefers not to use this term any more. Indeed, as it is defined in USDA Soil Taxonomy (Soil Survey Staff 1975), fragipans are restricted to compact horizons situated below an eluvial horizon. Several of the compact horizons represented in Figure 1, or part of some of these horizons, do not fit in this definition.

Fragipan horizons have been described in many countries (Van Vliet and Langohr 1981). There is no general agreement concerning the genesis of this horizon, and it is impossible to summarize here all the theories that have been proposed. Literature reviews have been made by Grossman and Carlisle (1969) and recently by Smalley and Davin (1982). The latter authors recorded five approaches to fragipan formation. One of these is by periglacial processes. In western Europe, unspecified periglacial processes have been proposed as the cause for the formation of this compact horizon in The Netherlands by Sevink and Vink (1969-1970), and Vink and Sevink (1971) and in Belgium by Lozet and Herbillon (1971). For Scotland, Fitzpatrick (1956) has proposed a link with permafrost conditions and so have Langohr and Van Vliet (1981) for Belgium and northern France.

Ice segregation in lenses, creating a platy or lenticular structure has fequently been observed in active layers (e.g., Pissart 1975, Mackay and MacKay 1976) and in soils with annual freeze-thaw cycles (open system) (e.g., Allan et al. 1969 , Fedorova and Yarilova 1972, Bunting and Fedoroff 1974). Prismatic and blocky structures, on the other hand, are more often described at the level of permafrost layers (e.g., Kuznecova 1973, Svenson 1964, Mackay 1974). The smallest prisms and blocks are mainly observed where permafrost developed in clayey and water saturated sediments.

Several authors mention the particular consolidation of the aggregates as a result of freezing (Mackay and MacKay 1976, Mackay 1974, Pissart 1970). This consolidation, which remains after thawing, seems to be associated with the extreme dessiccation and consequent collapsing of the pores during the freezing process.

CONCLUSIONS

The observations from areas with active permafrost today permit to explain the very particular set of characteristics of the consolidated horizons.

1. The thin plates in the upper part of the horizon can correspond to the lower part of the active layer, and the prismatic and blocky aggregates below would then coincide with the upper part of the permafrost.

2. The compact spherical nodules in the bleached A'2g'c' probably are those cores of the lower part of the active layer that freeze last in autumn. From such spots, water can be extracted toward the downward and upward freezing fronts starting from the soil surface and from the permafrost table respectively. This water extraction can promote the consolidation of the soil (Mackay and MacKay 1976).

3. The locally observed thin layer of iron and manganese oxide accumulation is very similar to the layer described by Black (1976) as one of the soil properties indicative of the limit between the active layer and the permafrost.

4. Permafrost conditions can explain the presence, starting from a depth of a few decimeters, of wedge shaped cracks which reach a depth of 1-2 m and a final width of several centimeters in the upper part. Whether these cracks were filled with ice or with soil material, is not yet known. The local presence of a weak vertical platy structure is an argument for a soil wedge; the presence of coarse pores in the lower part of the streaks and the compaction of the outer rim of the prisms points more to ice segregation. Possibly both, soil material and ice, were present.

5. The hydromorphic mottling, coinciding with at least the upper part of the compact horizon, even in landscape positions where today no groundwater table occurs, can be explained by the water saturated layer perched on top of the permafrost table. When the permaforst gradually thawed, this water saturated layer gradually moved downward until the whole frozen layer vanished. This mechanism permits to understand why this mottling occurs along the pores and aggregate faces of the compact horizon.

Considering that the consolidated subsurface horizon is observed in all those Tardiglacial soils where no biological, chemical or physical turbation occurred at that soil depth during the Holocene, we can conclude that in the Tardiglacial, at least once, permafrost was continuous in NW France, Middle and N Belgium, and The Netherlands.

REFERENCES

Allan, R., Brown, J., and Rieger, S., 1969, Poorly drained soils with permafrost in interior Alaska: Soil Science Society America Proceedings, v. 33, p. 599-605.

Black, R. F., 1976, Features indicative of permafrost: Annual Review of Earth and Planetary Sciences, v. 4, p. 75-94.

Bunting, B. T., and Fedoroff, N., 1974, Micromorphological aspects of soil development in the Canadian High Arctic, in Soil Microscopy, Proceeding 4th International Meeting on Soil Micromorphology, Ontario, p. 350-365.

Fedorova, N. N., and Yarilova, A., 1972, Morhology and genesis of prolonged seasonally frozen soils in western Siberia: Geoderma, v. 7, p. 1-13.

Fitzpatrick, E. A., 1956, An indurated soil horizon formed by permafrost: Journal of Soil science, v. 7, p. 248-257.

Grossman, R. B., and Carlisle, F. J., 1969, Fragipan soils of the eastern United States: Agronomy, v. 21, p. 238-279.

Guidotti Pena, J. C., 1976, Bodemkartering in het zuidelijk deel van de gemeente Ellezelles (België) en vergelijkende studie van gronden met en zonder fragipan: Unpublished Master thesis, State University of Ghent.

Haesaerts, P., and Bastin, B., 1977, Chronostratigraphie de la fin de la dernière glaciation, à la lumière des résultats de l'étude lithostratigraphique et palynologique du site de Maisière Canal (Belgique): Geobios, v. 10, p. 123-127.

Haesaerts, P., Juvigné, E., Kuyl, O., Mücher, H., and Roebroeks, W., 1981, Compte rendue de l'excursion du 13 juin 1981, en Hesbaye et au Limbourg Néerlandais, consacrée à la chronostratigraphie des loess du pleistocene supérieur: Annales de la Société Géologique de Belgique, v. 104, p. 223-240.

Haesaerts, P., and Van Vliet, B., 1981, Phénomènes périglaciaires et sols fossiles observés à Maisières-canal, à Harmignies et à Rocourt: Biuletyn Peryglacjalny, v. 28, p. 291-324.

Kuznecova, T. P., 1973, Special features of cryolithogenesis in the alluvial plains, Central Yakutia: Biuletyn Peryglacjalny, v. 22, p. 221-231.

Langohr, R., and Van Vliet, B., 1981, Properties and distribution of Vistulian permafrost traces in today's surface soils of Belgium, with special reference to the data provided by the soil survey: Biuletyn Peryglacjalny, v. 28, p. 137-148.

Lozet, J. M., and Herbillon, A. J., 1971, Fragipan soils of Condroz (Belgium) : Mineralogical, chemical and physical aspects in relation with their genesis: Geoderma, v. 5, p. 325-343.

Mackay, R. J., 1974, Reticulate ice veins in permafrost, Northern Canada: Canadian Geotechnical Journal, v. 2, p. 230-237.

Mackay, J. R., and MacKay, D. K., 1976, Cryostatic pressures in nonsorted circles (mud hummocks), Inuvik, Northwest Territories: Canadien Journal of Earth Sciences, v. 13, p. 889-897.

Paepe, R., 1969, Les unités litho-stratigraphiques du Pleistocène supérieur de la Belgique, in La stratigraphie des loess d'Europe: Bulletin de l'Association Française des Etudes du Quaternaire, Supplément, p. 45-51.

Paepe, R., and Sommé, J., 1970, Les loess et la stratigraphie du Pleisocène récent dans le Nord de la France et en Belgique: Annales de la Société Géologique du Nord, v. 40, p. 191-201.

Pissart, A., 1970, Les phénomènes physiques essentiels liés au gel, les structures périglaciaires qui en résultent et leur signification climatique: Annales de la Société Géologique de Belgique, v. 93, p. 7-49.

Pissart, A., 1975, Glace de ségrégation, soulèvement du sol et phénomènes thermokarstiques dans les régions à pergélisol: Bulletin de la Société Géographique de Liège, v. 11, p. 89-96.

Sevink, J., and Vink, A. P. A., 1969/70, Some banded soils in the southern Veluwe (The Netherlands): Geoderma, v. 3, p. 157-173.

Smalley, I. J., and Davin, J. E., 1982, Fragipan horizons in soils : A bibliographic study and review of some of the hard layers in loess and other materials: New Zealand Soil Bureau Bibliographic Report 30, Department of Scientific and Industrial research New Zealand.

Soil Survey Staff, 1951, Soil Survey Manual, U.S. Department of Agriculture Handbook, no. 18, U.S. Government Printing Office, Washington, D.C.

Soil Survey Staff, 1975, Soil Taxonomy, U.S. Department of Agriculture Handbook no. 436, U.S. Government Printing Office, Washington, D.C.

Svensson, H., 1964, Structural observations in the minerogenic core of a palsa: Svensk Geografisk Årsbok, v. 40, p. 138-142.

Van Ranst, E., 1981, Vorming en eigenschappen van lemige bosgronden in midden- en hoog België: Unpublished Ph. D. thesis, State University of Ghent.

Van Vliet, B., and Langohr, R., 1981, Correlation between fragipan and permafrost -with special reference to Weichsel silty deposits in Belgium and northern France: Catena, v. 8, p. 137-154.

Vink, A. P. A., and Sevink, J., 1971, XII. Soils and paleosols in the Lutterzand: Mededelingen Rijks Geologische Dienst, Nieuwe serie, no. 22, p. 165-185.

Zagewijn, W., and Paepe, E., 1968, Die Stratigraphie der weichselzeitlichen Ablagerungen der Niederlande und Belgiens: Eiszeitalter Gegenwart, v. 19, p. 129-146.

NUIQSUT AIRPORT DREDGE PROJECT

Craig C. LaVielle, Scott C. Gladden, Alvin R. Zeman

Rittenhouse-Zeman & Associates, Inc.
Anchorage, Alaska

The unique Nuiqsut Airport Dredge Project is the first dredged fill placed over permafrost on the Alaska North Slope. The specially designed 45.7 cm diameter suction dredge capable of being air or truck transported, was used to unify the steps of mining, transporting and placement of the runway embankment fill material. This technique has the advantage of minimizing damage to the fragile tundra and ice-rich permafrost. The need for large on-shore borrow pits and long haul roads was eliminated. Approximately 152.911 cubic meters of sand and gravel was hydraulically placed along the 1554.0 meter long runway right-of-way. The predicted depth of thaw beneath the fill and the amount of thaw consolidation in the natural soils compared well with measured values. The direct placement of dredged fill material on permafrost with controlled and limited thawing of the subgrade has proved to be a rapid, ecologically sound, and cost-effective technique.

Design and construction of the Nuiqsut, Alaska runway was completed in the spring and summer of 1981. This unique project used the direct placement of pumped dredged fill to construct a 1.8 m to 3.66 m (6- to 12-foot) thick runway embankment on an ice-rich permafrost subgrade. Direct dredged-fill placement on permafrost is a new technique to the North Slope. The process had previously been considered too risky for use on thaw-sensitive, ice-rich permafrost because of the large volume of relatively warm river water that is required. However, preliminary analysis indicated that, depending on the thermal regime of the specific site, dredge placement procedures may not only be feasible, but cost-effective in comparison to conventional dragline mining and pit excavations.

Our involvement in the project as geotechnical consultants for the North Slope Borough, included conducting exploration for submerged fill material source areas and detailed sampling of permafrost soils along the proposed runway alignment. In addition, engineering studies were undertaken including a complete thermal analysis concerning the effects that a pumped hydraulic fill would have on thaw-sensitive permafrost. Periodic construction monitoring was maintained throughout the construction process.

PROJECT DESCRIPTION

The Inupiaq Village of Nuiqsut, Alaska is located on the North Slope of the Brooks Range near the head of the Colville River Delta. This isolated village is approximately 402 km (250 miles) east of Barrow and 129 km (80 miles) west of Prudhoe Bay (Deadhorse) Alaska. Air access to Nuiqsut in the past has been limited to a small 457 m (1500-foot) gravel runway and emporary winter landing strip on the ice of Nechelik Channel. During breakup, the runway in the channel is lost. In addition, breakup in the channel often results in the short gravel runway being flooded by the river which may rise 12 m (40 feet) or more above normal channel levels. During the month of June, and at times into July, accesss to the village is limited to helicopter and light aircraft which are forced to land on the main streets of the village. Even when the small strip is repaired, it is far too short to accommodate fully loaded large cargo aircraft such as the Hercules C-130 transport planes. For this reason, a large permanent runway located close to the village was deemed necessary. Such a runway would allow transportation of needed construction materials and fuel to the village at any time during the year.

The Nuiqsut runway as planned would be 1554.5 m (5100 feet) long, 45.7 m (150 feet) wide, consisting of approximately 152,911 m^3 (200,000 cubic yards) of material. The runway would also have a taxiway, parking apron and a shipping/receiving terminal building. The runway core embankment would have to be constructed of high-quality NFS (non-frost susceptible) fill of sufficient thickness to protect the ice-rich subgrade. The surface course of the runway would also have to be designed for easy maintenance and not be subjected to extensive frost-heave and rutting. In addition, construction of the runway should take place within the short Arctic summer when thawed fill soils could be placed and properly compacted.

Existing borrow sources in the village contained large amounts of silt and could not be considered as non-frost susceptible. In order to construct the runway from locally available sand and gravel, a system was needed for excavation, washing, transportation and placement of clean fill material. Equipment to complete this project must be minimal, since transportation of equipment to the village would be costly, having to be done during the winter across a temporary ice road from Prudhoe Bay. For this reason, the decision was made to utilize a suction-dredge system. The dredge unifies all the steps for mining, transportation and placement of fill along the runway alignment, along with the help of two locally available bulldozers and a grader. Costly haul-roads and trucking equipment would not be needed. In addition, a specially designed dredge could be disassembled into easily transported units and then carried to the site by truck or Hercules C-130 air transport.

Our involvement in the Nuiqsut Runway Project included conducting an exploration for suitable submerged sites of high-quality sand and gravel which were thawed. This material would have to be within the Nechelik Channel of the Colville River and within a two-mile radius of the runway alignment to allow for pumping. In addition, subsurface sampling must be conducted along the runway alignment in order to determine the nature of the underlying permafrost. This information in turn would be utilized for the thermal analysis of the site to determine the thaw-sensitive nature of the permafrost. Finally, during construction phases of the runway, we would monitor fill procedures to insure quality control. At this time, temperature probes would be installed both in the center of the runway and along its perimeter to monitor thaw depths during construction and to determine the long-term outcome of dredge-placed fills of this size.

SUBSURFACE EXPLORATION

Material Source:

A subsurface exploration for suitable submerged high-quality sand and gravel for the Nuiqsut Runway was conducted during the summer of 1980 and winter of 1981. As previously discussed, the dredge system together with its booster pump, had a practical discharge limit of two miles. For this reason, material had to be located within these limitations and in sufficient quantities to complete the project.

The first phase of exploration was to conduct a summer surface-deposit reconnaissance within the Nechelik Channel. This was done in order to observe fluvial deposition characteristics of the channel as well as locating strand and bar deposits of sand and gravel. Likely areas of accumulation were denoted on a topographic map of the area. These locations were then used to delineate areas for subsurface sampling. The subsurface exploration within the Nechelik Channel took place during the winter of 1981. The frozen condition of the channel and upland tundra allowed access to sites of expected sand and gravel accumulation delineated during the summer reconnaissance.

In general, the ice thickness at the time of exploration on the channel ranged from 1.8 m to 2.4 m (6 to 8 feet) underlain by approximately 3 m (10 feet) of water followed by thawed sediments in the thalweg of the channel. Where the channel water was less than 1.8 m (6 feet) deep, the water and sediment were completely frozen except for remnant talik zones within the sediments at depth. In order to cut through the ice for bottom sampling, a 15.2 cm (6-inch) solid-stem auger was advanced as a pilot hole, utilizing a Nodwell-mounted Acker Mark IV drill rig. Once the pilot hole was completed, an 8-inch hollow-stem auger was attached and advanced down through the pilot hole and underlying water to the thawed sediments below. At this stage, samples were taken of the thawed sediments through the hollow-stem auger at selected intervals. These samples were used to determine quality and quantity of the deposit as well as to detect the permafrost boundary if encountered. This procedure proved useful for acquiring relatively undisturbed soil samples. However, at temperatures below -40°C (-40°F), freezing of water within the hollow-stem auger caused a slowdown in production. These problems were not encountered when ambient air temperatures were above -32°C (-25°F). Exploration of the channel was successfully completed upon delineating approximately 49,699 cubic meters (65,000 cubic yards), pit-run grade sand and gravel and 168,212 cubic meters (220,000 cubic yards) of fine to medium sand, (see Figure 1 for range in grain size distribution).

Runway Alignment:

The second phase of exploration involved taking continuous core samples at 61 m (200-foot) intervals down the proposed runway alignment. This was done in order to determine the subsurface soil and ice conditions to facilitate thermal analysis of the thaw-sensitive permafrost. Each test boring location was cored to a depth of at least 1.5 m (5 feet). These samples were extruded on-site, classified, photographed, and returned undisturbed to our Anchorage laboratory for thaw consolidation tests. Drilling fluids were not used for this coring procedure.

Typically, the subsurface soils along the runway alignment consisted of a 15 cm (6-inch) tundra mat over sandy, ice-rich organic silts to a depth of .61 m to 1.5 m (2 to 5 feet). Underlying the silts were generally found silty medium sands having a high ice content. Ice formations generally occurred as pore ice, veins, lenses, and in massive formations.

DREDGE, HYDRAULIC FILLING PROCESS and RUNWAY CONSTRUCTION

The technique of direct placement of pumped dredged fill was used to construct the airport runway embankment at Nuiqsut, Alaska. The direct placement technique has several distinct advantages over more traditional techniques of fill embankment construction. The advantages include unifying the project operations of mining, transporting, placing, washing and compacting the fill material. By unifying these operations, significant construction cost savings are realized. The direct placement technique also proved to be more environmentally acceptable than the traditional open-pit mining and trucking techniques.

The Nuiqsut Airport project used an 946 cm (18-inch) suction dredge to mine the thawed sands and gravels from the sources located in the river channel (see Figure 2). The special unitized dredge, designed by Dredging Consultant, Michael Weston, is approximately 12.8 m (42-feet) long. A 9.1 m (30-foot) ladder with the suction pipe and a revolving cutter head allowed material to be mined from as deep as 7.9 m (26 feet) below the water surface. Two large pumps, one onboard the dredge and a second acting as a booster pump at the half-way point, pumped the sands and gravels up a 6.1 m (20-foot) rise and out onto the alignment. Sand and gravel was pumped 3.2 km (2 miles) to the farthest end of the runway alignment utilizing a 46 cm (18-inch) diameter, 4.45 cm (1.75 inch) thick polyethelene pipe as a discharge line. Approximately 3823 cm (5000 cubic yards) per shift was delivered to the site.

The excavation and transportation of the sand and gravel to the site required only two dredging pumps, as opposed to the traditional fleet of loaders and trucks. For this reason, no haul-roads were needed from the borrow source area. Therefore, all fill material could be used on the project itself. The traditional maintenance headaches of pit operations in the arctic such as frozen stockpiles, thawing and draining stockpiles, pit drainage, and blasting were avoided.

The construction of the Nuiqsut Airfield required some 152,920 cubic meters (200,000 cubic yards) of select fill with frost-susceptible silt and clay materials removed. Once staking of the runway limits was finished, a small dike approximately .91 m (3 feet) high and 3.65 m (12 feet) wide was constructed around the perimeter of the proposed runway (see Figure 3). These dikes as well as the parking apron and taxiway were constructed of silty sand scraped and truck-hauled from small channel bar deposits. The dredge located approximately two miles distant in the Nechelik Channel of the Colville River, was maneuvered over a known, submerged sand and gravel deposit.

As the saturated dredged material was deposited at the end of the outfall line, two small bulldozers (Caterpillar D5 and D7) graded the fill across the runway alignment building up the dikes as needed see Figure 4. The winch-equiped dozers could move the flexible discharge line laterally across the runway. When enough material had been deposited at any one spot, sections of the plastic pipeline were quickly cut and removed, shortening the line. If extension of the pipe was required, a small portable fusion weld machine was used. As the water settled out of the fill material, consolidation of the soil took place, leaving a well-compacted NFS material averaging 90 to 95 percent of the laboratory maximum density (ASTM:D-1557). The millions of gallons of water used to carry the sand and gravel was allowed to drain from the fill to a settling pond at the middle of the runway and then pumped back to the river. The rapid draining of the water acted to wash the sands and gravels and carry the silt and clay fines down to the settling pond. In addition, the washed fill material was densified due to active water settlement. For this reason, no compaction equipment was required for the embankment core to meet compaction specifications.

The above described filling process was continued until the eastern third and western third of the runway was completed. At this point, a large stockpile of material was collected on the borders of the settling pond located in the middle of the runway. The return pump was dismantled and frost-susceptible silt accumulation removed. The stockpiled fill was then placed into the location of the excavated silt pond completing the core embankment of the runway. The entire fill project was completed within 75 days during the short Arctic summer. The following summer, stockpiles of substantial quantities of gravel for other village improvement programs and for the gravel topping course of the runway was collected and crushed. At this time, the 1555 m (5100-foot) runway can support a fully loaded C-130 Hercules transport aircraft. The permanent runway surfacing consisting of approximately 25 cm (10 inches) of compacted crushed gravel will allow all-season operation and is currently being installed.

By combining the mining transporting, washing, placing and compacting operations, significant construction cost savings was achieved. The savings came as a result of reductions in the required fuel, labor and capital costs.

CONSTRUCTION MONITORING

Thawing beneath the fill as a result of the hydraulic filling process was monitored at five points in the runway alignment. Monitoring points were placed at center line STNs 14+50, (feet) 23+50, 31+50, 38+50 and 45+00. The temperature of the soil was measured by means of thermistors placed at depths ranging from 0.3 to 2.1 m (1 to 7 feet) below the top of the tundra. Measurements were made at random intervals not exceeding two weeks, throughout the construction period. A plot of the temperature beneath the fill from July 1981 through October 1983 are shown in Figure 5.

Maintaining the monitoring stations and thermistor wires proved to be difficult. Construction equipment moving along the containment dikes while moving the discharge pipe frequently cut the wires. In addition, ground squirrels chewed through several of the buried wires.

Thirty-three compaction tests, using a nuclear densometer were also taken along the full-length of the runway near the top of the completed fill surface. The test results ranged from 90-95 percent of the laboratory maximum using ASTM:D-1557 as the standard. Observation wells were installed in early September at five locations along the right-of-way; STNs 25+00, 30+00, 35+00, 40+00 and 45+00. The reading obtained during the early September period indicate water depths averaging just over 1 m (3 feet) below the finished runway surface. The compaction test results and the temperature readings over a one-year period show the fill is performing very well and that a total freeze-back was achieved and a second-season thawing did not penetrate the base of the fill.

A significant benefit of suction dredging versus open-pit mining was that washing of the sand and gravel fill material occurred during placement. Grain size analysis of the source materials prior to dredging indicated a silt and clay content of of up to 10 or 12 percent finer than .07 mm. Samples obtained of the completed runway showed that 90 percent of the silt had been removed. A technically non-frost susceptible (NFS) material with less than 1 percent finer than .02 mm was produced without expensive screening.

LABORATORY TESTING

The laboratory phase of our thermal study included a series of tests designed to determine the material parameters pertinent to this project. These material parameters included grain size distribution, total moisture content and dry unit weight. The thermal properties of soil thermal conductivity and heat capacity which are used in the thermal analyses could be derived from these parameters.

The results of our laboratory testing program, have shown that the natural soils along the runway alignment tend to display high ice contents and low dry unit weights. Moisture contents of the tundra ranged from 41% to 237% and averaged 172%.

Moisture values for the silt unit ranged from 46% to 368% and averaged 127%. Moisture values for the sand unit ranged from 19% to 188% and averaged 42.5% by weight. Dry unit weights averaged 309 kg/m (19.3 pcf), 637.5 kg/m (39.8 pcf), and 1401.8 kg/m (87.6 pcf) for the tundra, silt and sand units respectively.

In addition, a series of thaw consolidation tests were performed on the frozen, undisturbed core samples taken from the runway alignment. The thaw consolidation tests consisted of cutting 5 cm (2-inch) long sections from the 6.4 cm (2-1/2 inch) diameter, undisturbed core samples, and loading the thawing sample with 147 to 295 kg per square meter (720 to 1440 psf for 48 hours). This surcharge is approximately equivalent to the load which would be expected to be imposed by the completed runway fill. The samples were kept saturated and confined within a bronze ring 6.4 cm (2-1/2 inch) in diameter during consolidation. Sample drainage was allowed through both top and bottom of the sample, by means of filter paper and porous stones. The mean percent consolidation for tundra unit and the silt unit was 36 and 23 respectively. The range of consolidation values was 15% to 55% for the tundra and 17% to 32.5% for the silt unit.

Although only three basic soil types were involved in our study, a wide range of soil thermal conductivities were found to apply. Thermal conductivities were determined from curves published by Kersten (1949). It was noted that soil thermal conductivities can change significantly with changes in moisture content, soil dry unit weight, soil temperature and changes from the frozen to the thawed state. Of the variables mentioned above, all were expected to effect the soils in the runway alignment. The greatest variations in thermal conductivity would be due to the wide variation in soil moisture content. As previously mentioned, the moisture contents in the tundra unit varied from 41% to 237% by weight. According to the charts published by Kersten for peat, this variation in moisture content represents a variation in soil thermal conductivity of .1038 to 1.038 W/(m·°C) (0.06 to 0.6 BTU/hr ft °F) in frozen peat and .086 to .43 W/(m·°C) (.05 to .25 BTU/hr ft °F) in frozen peat. Conductivity values for the silt unit and sand unit averaged 1.21 W/(m·°C) (0.7 BTU/hr ft °F and 3.07 W/(m·°C) (1.78 BTU/hr ft °F) respectively.

The variations in thermal conductivity caused by temperature and the changes from the frozen to the thawed state were dealt with by averaging the frozen conductivity with the thaw conductivity (k_u+k_f/2).

In order to perform a meaningful thermal analysis of the entire runway alignment, it was necessary to assume mean soil properties. By taking this approach in our analyses, we realized we would loose some accuracy in our depth-of-thaw estimates in localized areas. However, a large overall view of the scale of potential thawing was produced and, therefore, the project's feasibility could be presented. An accounting of the possible consequences of deeper thawing in areas where the deviation from our mean soil properties occurred was also made. The deviation was then used along with experience to temper the estimates presented.

THERMAL ANALYSES

The thermal analyses consisted of determining by mathematical methods the estimated depth of thaw to be expected in the natural frozen ground as a result of the hydraulic filling operation; the estimated time required for freeze-back of the thawed natural soils and the fill itself. These estimates are based upon the assumed conditions and parameters of an undisturbed tundra mat, subsurface soils, the mean Nuiqsut area climatic conditions, and thermal conditions imposed by the hydraulic filling operation. The soil parameters were determined experimentally.

The climatic condition assumptions were obtained from data published by the Environmental Atlas of Alaska, U of A, Fairbanks. The conditions assumed for this study are listed below.

Nuiqsut Townsite Area	Parameters
Thawing Index	500 degree-days
Length of Thaw Season	95 days
Freezing Index	8500 degree-days

The length of the thaw season has been defined as the number of days that have a mean daily air temperature (M.A.T.) above 0°C (32°F). Typically, the months of June, July, August, compose the thaw season, at Nuiqsut, although a variation of one-half month on either end is possible. The initial part of our thermal analysis consisted of constructing as complete a picture as possible, of the events, time tables and temperatures anticipated for the filling operation.

The model of the hydraulic filling operation consisted of millions of gallons of a sand and gravel slurry being quickly placed on the frozen soil. The sand and gravel slurry pumped from the river would tend to initially act as a constant temperature heat source. The temperature of the slurry was estimated at 7°C to 10°C (45°F to 50°F). This constant temperature heat source would allow heat to flow into the soil causing melting and thaw consolidation. The temperature of the slurry fill presumably then would tend to cool near it's base as heat was lost to the frozen soil. The surface of the slurry/ fill would either warm or cool in response to changes in air temperature and solar radiation. The rate of temperature change in the slurry/fill would depend strongly on how quickly the fill drained. The drained fill would tend to reach an equilibrium more rapidly with the frozen subgrade and the air.

The heat capacity of the river water is far greater than the soil particles. Therefore, the water in the fill was acting as the constant heat source. By draining the water away, thawing could be limited. If the water could be drained away quickly, only the climatic influences of air temperature, wind speed, and solar radiation on the dark fill surface would promote thawing. The completed thickness would tend to insulate the natural soils from the climatic influences and limit thawing beneath the fill.

Our analysis assumed a maximum of two days would be required to place the full thickness of fill over the new area. The surface of the fill was expected to drain within hours. The base of the fill, however, was anticipated to require three months or more to drain. Complete drainage of the fill would not be required and would not be advantageous. Additional insulation against climatic influences would be available if most of the fill retained a moisture content equivalent to 10 percent of the dry soil weight and the lower 15 cm (6 inches) of fill remained saturated.

Our study of the subsurface thermal aspects of the placement of hydraulic fill indicated the project was feasible if proper drainage could be maintained. Our conclusions covering potential thawing and thaw consolidation are presented below with a discussion of what actually occurred.

The estimated average depth-of-thaw beneath the existing tundra surface was 1.1 to 1.2 m (3½ feet to 4 feet). Variations in the depth of thaw were of as much as ± 0.3 m (1 foot) could be expected depending on local conditions. This estimate was based again on a 95-day filling period. In actuality, approximately 75 days were required to complete the runway fill. Measured depths-of-thaw averaged 0.76 m (2½ feet) with as little as 0.6 m (2 feet) of thaw occurring in some areas and as much as 1.5 m (5 feet) of thaw occurring in a small localized area. The maximum depth of thaw that actually occurred (0.76 m) was within the range anticipated. The total filling period was approximately 25% shorter than assumed and a proportionately reduced amounts of thawing were found to have occurred.

The total percent consolidation of the thawed soils was estimated at 25% of the newly thawed thickness or approximately 0.3 m (1 foot) of the estimated 1 to 1.2 m (3½ feet to 4 foot) thaw depth. Considering the wide range of ice and soil conditions, actual ground settlements were expected to range from 0.23 to 0.46 m (3/4 to 1½ feet). The actual thaw consolidation that occurred averaged approximately 0.30 m (1 foot). The range of settlement that occurred at the points measured was as much as 0.46 m (1½ feet) and as little as 0.15 m (½ foot). The correlation between the estimated and the actual consolidation are good. Actual consolidation was slightly overestimated, possibly because lesser amounts of ice were present than assumed. The estimated date of freeze-back of the total thickness of fill and thawed natural soil was mid-February. A frozen thickness sufficient to allow for heavy aircraft traffic was estimated to exist by mid-October. Actual total freeze-back of the fill and the natural ground was complete by early January. Our estimate proved to be conservative but within an acceptable range. Although sufficient data is not available for any theoretical analysis, we do expect that the raised profile of the fill mass allowed greater exposure to the wind and more rapid cooling than initially were anticipated.

SUMMARY

Until quite recently, sand and gravel for construction purposes has been mined from large upland pits. These huge pits become a permanent feature in the Arctic landscape along with their attendant haul roads. Even so, at many locations across the North Slope of Alaska there is a lack of quality sand and gravel which could be considered as a suitable Arctic construction material. Where present, these sands and gravels contain high percentages of silt or clay which renders them marginally suitable at best for use in roads, pads and runways. Other methods of mining and improving the quality of the borrow material, such as by washing or screening, have normally not been considered economically feasible in the short construction season.

As an alternative to pit mining and truck hauling, the runway at Nuiqsut represents the first major use of a pumped dredge fill on upland permafrost. A special 46 cm (18-inch) suction dredge, capable of being broken down into truckable units, was designed and constructed for this project as well as future projects planned by the North Slope Borough. The use of a dredge for this major fill allowed for the mining, washing, transportation and placement of the material by essentially a single piece of equipment.

The use of a dredge for mining and placing fill material has been hampered in the past by concerns about thawing and thermal degradation of ice-rich permafrost. This project has demonstrated that with proper engineering and construction techniques, the use of a dredge for material extraction and placement offers a viable option to open-pit mining for sand and gravel. In the future, we expect to see dredges such as this used for the extraction and placement of materials for the many proposed offshore oil drilling islands.

Aside from the advantages associated with extracting and placing fill materials, the use of a dredge can have important advantages from an environmental standpoint. Rather than being left with the scars of an open-pit, the deepend channel in this case study is expected to provide an improved over-wintering habitat for fish in the river. At the same time, there has been little or no damage to the surrounding tundra or beaded streams since an extensive system of haul roads was not required.

ACKNOWLEDGMENTS

We wish to acknowledge the important participation of Michael Weston, Dredging Consultant, on this project. Mr. Weston performed the early feasibility studies as well as being in charge of the design, construction and operation of the dredge. Our studies were sponsored by the North Slope Borough, the governmental unit responsible for dredge funding and runway construction. Particular credit is due Mayor Eugene Brower and Public Works Director, Irving Igtanloc, who actively supported this innovative construction technique.

REFERENCES

Andersland, O.B. and Anderson, D.M. (ed.) 1978. Geotechnical engineering for cold regions. McGraw-Hill, New York, NY, 576 p.

Goodrich, L.E. 1973. Computer simulation. Appendix to: Thermal Conditions in Permafrost: A review of North American Literature, L.W. Gold and A.H. Lachenbruch. Proc. 2nd International Conference on Permafrost, Yakutsk, U.S.S.R.,

North American Contribution, U.S. National Academy of Sciences, pp. 23-25.

Goodrich, L.E. 1976. A numerical model for assessing the influence of snow cover on the ground thermal regime. Ph.D. Thesis, McGill Univ., Montreal, Que., 538 p.

Goodrich, L.E. 1978. Some results of a numerical study of ground thermal regimes. Proc. 3rd International Conference on Permafrost, Edmonton, Alta., Canada, National Research Council, Vol. 1, pp. 24-34.

Kersten, M.S. 1949. Thermal properties of soils. Univ. of Minnesota, Engineering Experiment Station, Bull. 28, 227 p.

McRoberts, E.C., Fletcher, E.B. and Nixon J.F. 1978. Thaw consolidation effects in degrading permafrost. Proc. 3rd International Conference on Permafrost, Edmonton, Alta., Canada, National Research Council, Vol. 1, pp. 693-699.

Morgenstern, N.R. and Smith, L.B. 1973. Thaw consolidation tests on remoulded clays. Can. Geotech. J., Vol. 10, No. 1. pp. 25-40.

Nixon, J.F. and McRoberts, E.C. 1973. A study of some factors affecting the thawing of frozen soils. Can. Geotech. J., Vol. 10, No. 3, pp. 439-452.

Nixon, J.F. and Morgenstern, N.R. 1974. Thaw consolidation tests on undisturbed fine-grained permafrost. Can. Geotech. J., Vol. 11, No. 1, pp. 202-214.

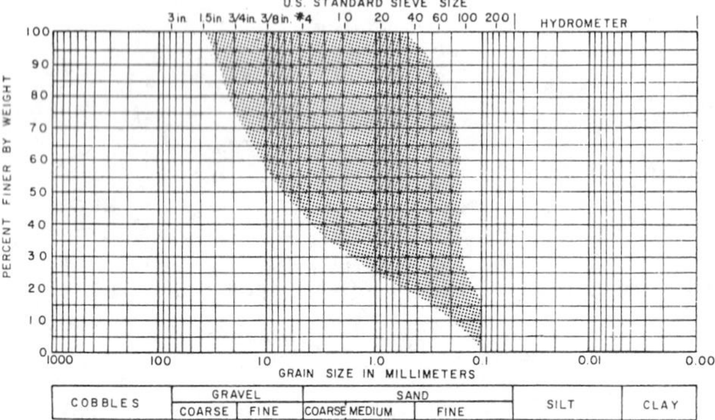

FIGURE 1 Range of grain size distribution of fill material source sampled from the Nechelik Channel.

FIGURE 2 Dredge used to construct the Nuiqsut runway as designed by Michael Weston, Dredging Consultant for North Slope Borough, and built by Kenner Marine, Baton Rouge, Louisiana.

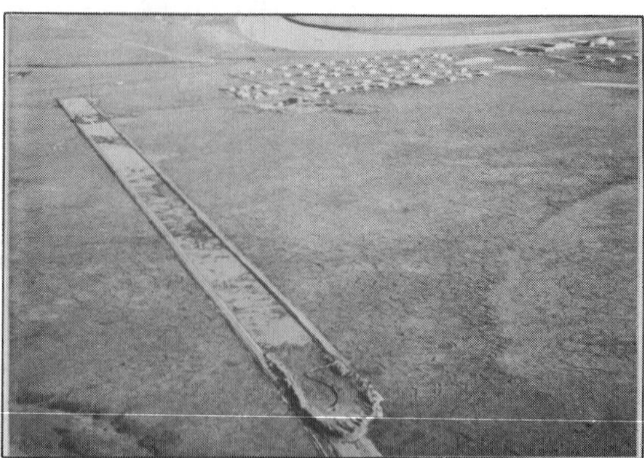

FIGURE 3 Nuiqsut runway alignment outlined by the perimeter dikes during the first phase of dredge placement.

FIGURE 4 Bulldozer grading dredge fill within airfield perimeter dikes as it is deposited from discharge line.

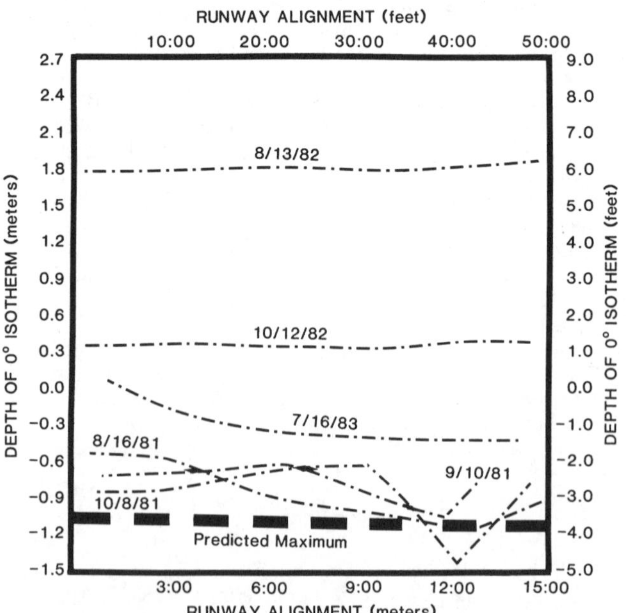

FIGURE 5 Depth of 0°c isotherm along runway alignment.

GROUND ICE IN PERENNIALLY FROZEN SEDIMENTS, NORTHERN ALASKA

D. E. Lawson

U.S. Army Cold Regions Research and Engineering Laboratory
Hanover, New Hampshire 03755 USA

The distribution and volume of ice in perennially frozen sediments beneath three unglaciated sites in northern Alaska vary with the grain size and depositional origins of the sediment, thermal history (permafrost aggradation and degradation), and age of the terrain and deposits. Substantial lateral variation in near-surface ice volume exists between and within each site, but reasonably consistent trends in ice content with depth were measured beneath individual landforms. Primary deposits, those deposited and frozen without postdepositional thermal or sedimentologic modification, contain the highest volume of ice at each locality. Sediments that have undergone thawing or resedimentation typically contain much less excess ice. Thaw lake, slope, or fluvial processes modify ice contents and produce sedimentary sequences with a spatial distribution of ice determined by these depositional processes and the subsequent thermal history.

The volume, distribution, and configuration of ice in perennially frozen ground are extremely important parameters that affect its physical response to disturbance as well as its geotechnical properties (e.g. Muller 1947, Terzaghi 1952, Ferrians et al. 1969, Mackay 1970, French 1975, Johnston 1981, Lawson 1982). Although knowledge of the occurrence and types of ground ice has substantially increased (e.g. Mackay and Black 1973, Mackay et al. 1978), few studies have analyzed the volume and distribution of ground ice in North America in relation to geologic parameters. There are few published measurements or estimates of ice volume for North American regions or sites (e.g. Livingstone et al. 1958; Carson and Hussey 1962; Black 1964; Hussey and Michelson 1966; Brown and Sellmann 1973; Williams and Yeend 1979; Pollard and French 1980).

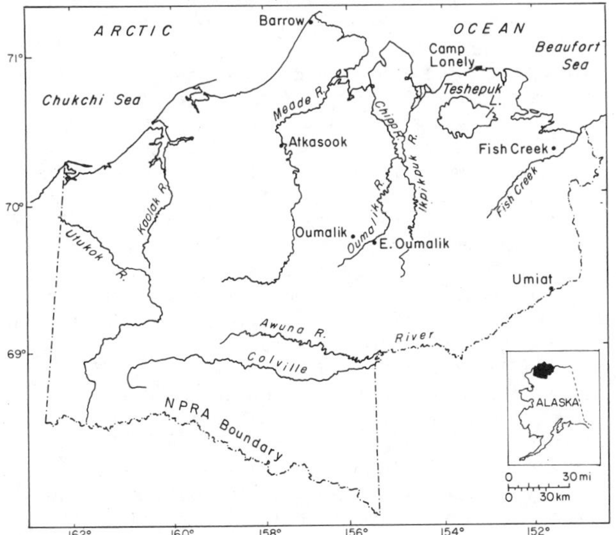

FIGURE 1 Location of Oumalik, East Oumalik, and Fish Creek in northern Alaska.

From 1978 to 1981, I examined ground ice in perennially frozen sediments at three unglaciated localities in northern Alaska: Oumalik, East Oumalik, and Fish Creek (Figure 1). Each site is the location of an exploratory well drilled about 30 years ago by the U.S. Navy in Naval Petroleum Reserve No. 4 (now designated National Petroleum Reserve, Alaska) (Reed 1958). This paper summarizes the data on ice volume and its relationship to each site's physical attributes and recent sedimentary history. A paper in preparation will discuss the results of the analyses in detail.

METHODS

The geology and occurrence of ground ice were defined by drilling and coring 375 holes ranging in depth from about 2.5–40 m. An additional 450 holes were augered to depths of 1.5–11 m. Drill holes were generally located along transects across the primary landforms at each site.

After logging the geology of each drill hole, frozen samples were selected, usually at an interval of 1 m or less (max. 2-m spacing), for laboratory analyses that included grain size distribution, moisture content by weight, ice content by volume, organic content by loss-on-ignition, and bulk/dry density following standard techniques. The undisturbed frozen cores were 7.5 cm in diameter, with samples cut to a 5-cm length. Degree of saturation, porosity, void ratio, and textural statistics were calculated from the analytical data.

GROUND ICE OCCURRENCE

Fish Creek

The Fish Creek site lies on the Arctic Coastal Plain in a region of gently rolling terrain with low relief and numerous oriented lakes and drained lake basins that are commonly separated by ridges of inactive eolian dunes (Sellmann et al. 1975, Williams et al. 1977). Sediments beneath Fish Creek, as well as at Oumalik and East Oumalik, are part of the Gubik formation of the Quaternary Age (Black 1964). They range from about 10-20 m thick over bedrock at this site. The landforms examined were drained lake basins, an intervening upland surface between those basins, and stream valley floodplains and slopes (Figure 2).

The upland surface is composed of 8-14 m of eolian sand that overlies a second sand that is of fluvial, or possibly marine, origin. The eolian deposits are stratified to massive, well-sorted, fine-grained sand, with some layers of medium-grained or silty sand. Some sections exhibit high angle cross-bedding. Organics are generally sparse (1-3%), but stringers of peaty material are interspersed randomly throughout. The underlying sands are medium- to fine-grained and stratified, with horizons rich in polished pebbles, shell fragments, and disseminated organic materials mixed with silt. Radiocarbon analysis of six in situ peats in the upper 3 m yielded ages ranging from 2200-3500 yr B.P., while an organic-rich horizon at the approximate contact of the two sand units yielded an age of 22,750 ± 1700 yr B.P. These dates suggest deposition of the eolian sands in late Wisconsinan time; they are correlative with sand sea deposits described by Carter (1981).

Ground ice is mainly interstitial, filling voids between sand grains. Less frequently, irregular lenses (5-10 mm across) and thin veins (0.1-3 mm thick) are present. Ice volume was fairly consistent to depths of 15-18 m, ranging generally from 40-50%, with volumes of up to 68% measured in strata containing silt or organic material (Figure 3a). Ice volume was most variable near the contact of the two sand units. Most samples were unsaturated.

Wedge ice was encountered only near the ground surface in association with irregular, polygonal troughs of 5-15 m spacing. Wedges range from 1.5-2.4 m in width at the top and extend 2.3-3.2 m below the surface. They compose about 10-15% of the upper 10 m of sediment.

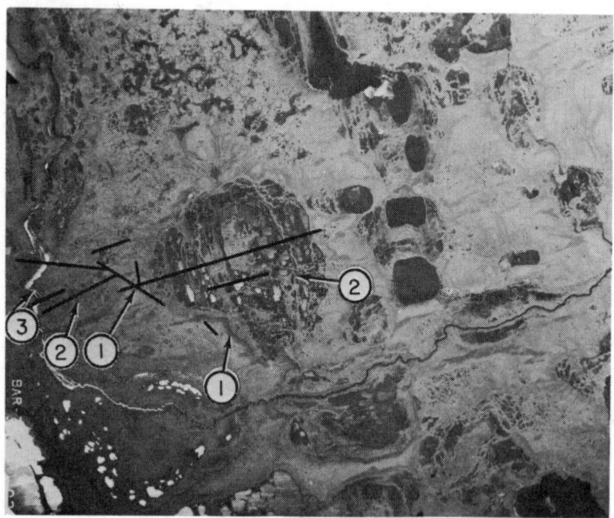

FIGURE 2 Aerial photograph of Fish Creek (BAR 741-030, 7/10/48). Lines indicate general position of drilling transects. Landforms examined include (1) upland sand surfaces, (2) drained lake basins, and (3) stream valley floodplain (scale 1:20,000). North to top.

Lacustrine deposits in the drained lake basins consist of interstratified silt, sandy silt, and fine sand, with peaty horizons (40-60% organics by weight), organic-rich zones, and shelly zones common. Alternating light silt and dark sandy silt layers (2-3 mm thick) commonly form a varve-like stratification. These deposits average 3-4 m thick in the large basin east of the upland (5.5 m in a mound near basin center), and 1-1.5 m thick in the smaller basin to the west (Figure 2). Radiocarbon analyses of peat horizons indicate deposition in the eastern basin began about 10,700 ± 230 yr B.P. and ended near 8900 ± 180 yr B.P., while activity in the western basin took place after 3500 yr B.P.

The lacustrine deposits are underlain by two stratified sands that appear laterally continuous with upland deposits. The physical properties of these sands are the same as the upland sand, except that the upper unit is only 3-6 m thick and both units contain much less ice in veins and lenses.

Ice volume of basin sediments is a highly variable 50-90% in the lacustrine deposits but drops sharply to a consistent 30-40% in the lower sands (Figure 3b). A comparable downhole trend in ice volume was measured by Hussey and Michelson (1966) in sediments beneath drained lake basins near Barrow. Ice in the lake deposits occurs as lenses (2-30 mm thick), veins (0.1-3 mm thick, of vertical to horizontal orientation), and thin layers (10-15 mm). Ice wedges in the eastern basin are similar in dimension to those of the upland, while those in the western basin are smaller (1.5-2.4 m deep, 1.1-1.7 m wide). Larger segregated ice bodies (up to 15 cm thick) form mounds located near the center of the basins (Figure 3b). Whereas lake deposits are generally saturated or oversaturated, underlying sands are unsaturated. Further, these sands are less saturated with ice than those of the upland and have a higher bulk density and lower void ratio.

The trend in ice volume with depth can be explained as the direct result of the thaw lake cycle (Britton 1967), with the lower ice volume of the sand a result of its thaw during lake formation and expansion. Thawing of sediments beneath lakes that are more than 2 m deep will melt the ground ice, thereby inducing thaw subsidence and consolidation. The higher density and lower void ratios of the sand in comparison to upland deposits suggest that compaction has taken place. Refreezing after lake drainage then "resets" the content and distribution of ice in the sand. During freezeback, water may migrate toward the freezing front, particularly into the lacustrine deposits, but ice in the underlying sand would be derived mainly from interstitial water present in its unfrozen state.

Indirect support for this process is that upland sands, which thawed because of disturbance by activities on the old drill site, then subsequently refroze, show a similar change in physical properties, with ice volume decreasing to a relatively consistent 35-40% (Lawson 1982). In addition, similar ice contents are found in the colluvial deposits mantling valley slopes and the sand underlying the alluvium in the stream valley, which also presumably thawed during fluvial activity (Figure 3c).

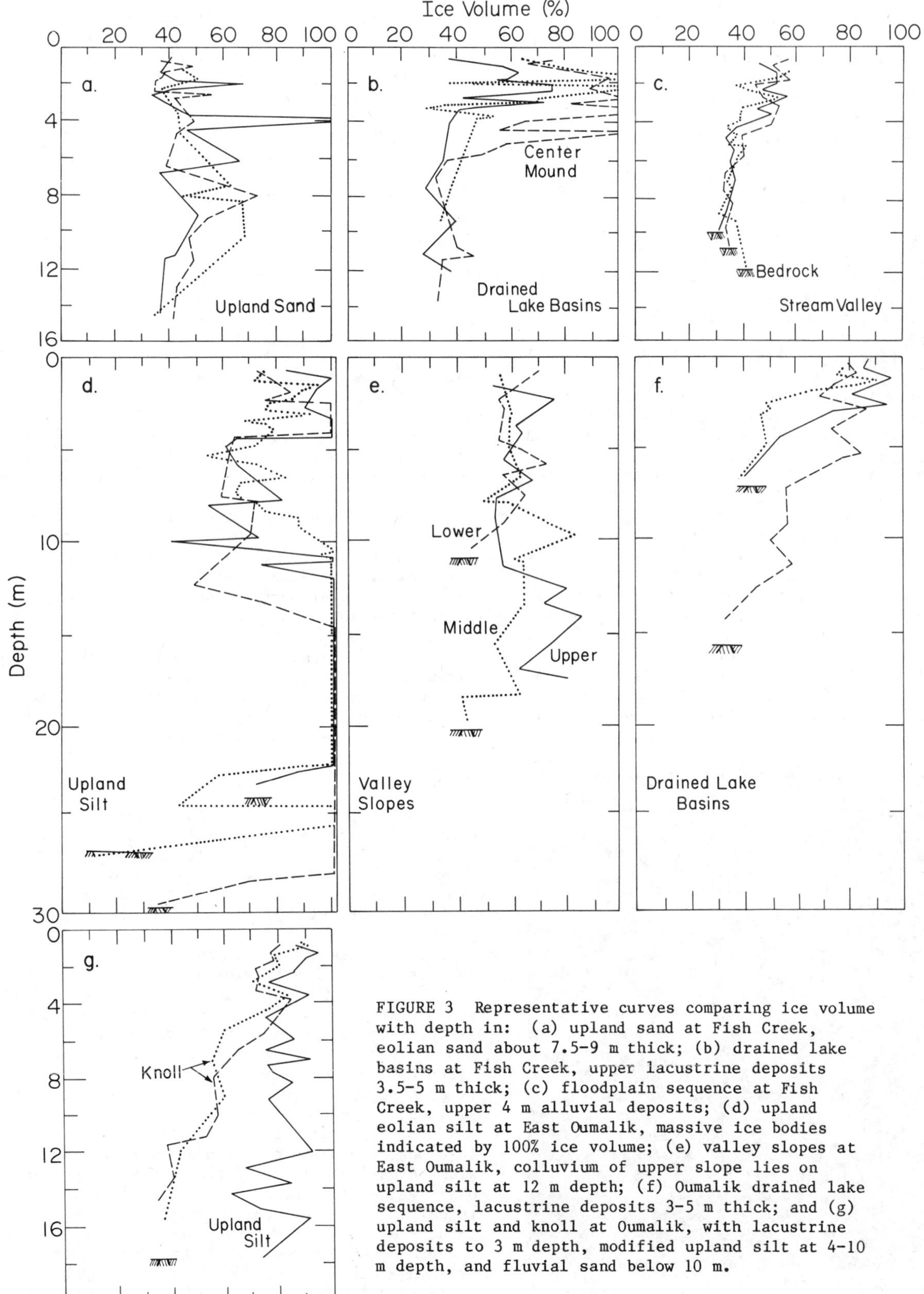

FIGURE 3 Representative curves comparing ice volume with depth in: (a) upland sand at Fish Creek, eolian sand about 7.5-9 m thick; (b) drained lake basins at Fish Creek, upper lacustrine deposits 3.5-5 m thick; (c) floodplain sequence at Fish Creek, upper 4 m alluvial deposits; (d) upland eolian silt at East Oumalik, massive ice bodies indicated by 100% ice volume; (e) valley slopes at East Oumalik, colluvium of upper slope lies on upland silt at 12 m depth; (f) Oumalik drained lake sequence, lacustrine deposits 3-5 m thick; and (g) upland silt and knoll at Oumalik, with lacustrine deposits to 3 m depth, modified upland silt at 4-10 m depth, and fluvial sand below 10 m.

East Oumalik

East Oumalik, which lies near the southern boundary of the Arctic Coastal Plain, is located in an area of upland surfaces cut by meandering streams and active, often deeply incised, thaw lakes (Figure 4). The maximum relief is about 20 m.

The upland at East Oumalik (Figure 4) is composed of relatively homogeneous silt that overlies sedimentary bedrock at a depth of 24-39 m. Approximately 80% or more of the upland silt ranges in size from 0.01-0.06 mm. It is structurally massive or faintly to strongly laminated, with occasional strata of fine sandy silt. Some strata exhibit low angle cross-bedding. Organics are sparse, with organic-rich horizons (10-30%) limited to the upper 2-3 m, and only small fragments of wood (2-3 mm) or fibrous material randomly disseminated at depth.

The upland silt is interpreted as eolian in origin. Its age is not well defined. Twelve radiocarbon analyses of organics within 2-3 m of the surface yielded ages ranging from 10,550 ± 150 to 5980 ± 115 yr B.P., while fibrous material and wood at depths of 6 and 12 m were older than 37,000 yr B.P.

Extreme variability in ice volume, with rapid shifts from 50-100%, characterizes all holes in the upland (Figure 3d). Ice occurs as irregular lenses (0.1-20 cm thick), discontinuous layers (generally 5-10 cm thick, but exceptionally to 2 m), veins in near-horizontal or near-vertical orientation (0.1-5 mm thick), and a massive sill (tabular body).

The ice sill ranges from 11-13 m thick and may compose up to 50% of a vertical section above bedrock. It is laterally continuous over at least 8 km^2, occurring in the lower part of each upland hole and separated from bedrock by a 0.5-3.5-m-thick layer of silt and weathered rock. Debris may be suspended randomly within the ice, or it may occur in veins that parallel or cross-cut ice foliation. Fracture fillings are also common. Alternating bubble-rich and bubble-poor ice in bands of varying thickness (1-10 cm) contain mostly spheroidal bubbles. Analyses of the ice sill data are continuing, but the similarity of the ice to the segregation ice of Mackay (1971) and Gell (1976) suggest that it is also of a segregated origin.

Downhole variability in ice volume and form does not relate to changes in grain size distribution or other sedimentary features, suggesting cryogenic processes were more important in determining ice occurrence. The increase in the ice content of sediments just above the ice sill, and the decrease below it, may relate to these processes.

Ice wedges in the upland are from 5-10 m wide at the top and extend to a depth of 7-15 m. Wedge ice may compose up to 60-70% of the upper 10 m of the upland. Two sets of ice wedges, one without surficial expression and apparently inactive, are present. The top surface of the "inactive" wedges occurs at a depth of about 0.9-1.2 m, whereas the other wedges lie at 0.3-0.7 m. The present active layer thickness of the upland is 0.3-0.6 m.

Colluvial sediments of the stream valley slopes, the lower parts of which have been reworked by solifluction, show less spatial variability in ice volume, generally in the range of 50-70% (Figure 3e). Thin veins (2.1-5 mm thick) or small, irregular lenses (1-10 mm thick) are common, but larger bodies of ice are not. These deposits form a wedge that mantles valley slopes to a thickness of more than 15 m at some locations.

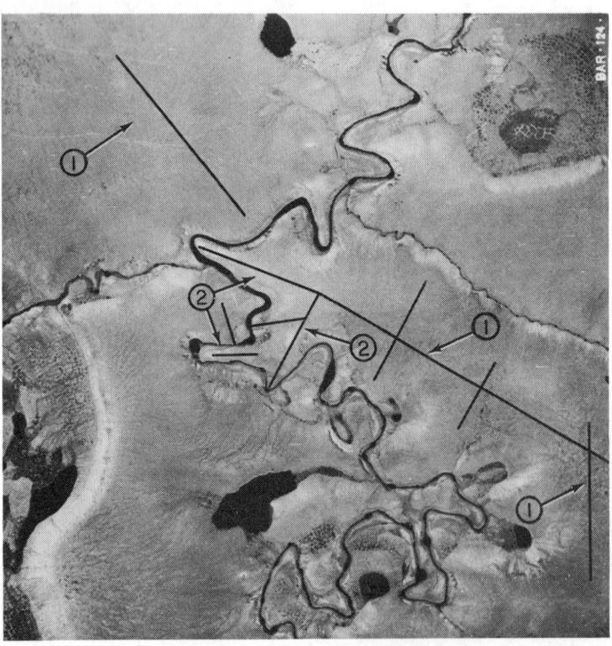

FIGURE 4 Aerial photograph of East Oumalik (BAR 124-147, 9/6/48). Lines indicate general location of drilling transects. Valley slope (1) and upland silt (2) are located (scale 1:20,000). North to top.

Oumalik

Oumalik, located about 18 km west of East Oumalik (Figure 1), is an area principally of coalesced, drained thaw lake basins, many of which contain or are separated by remnants of uplands like those at East Oumalik (Figure 5). The lake basins are poorly drained and are covered by polygonal ground.

Silt beneath the uplands at Oumalik (Figure 5) is physically and mineralogically similar to that at East Oumalik. The radiocarbon ages of near-surface peat horizons (upper 5 m) are correlative. The silt is therefore believed to be contemporaneous windblown material derived from the same source. The distribution and volume of ground ice are the same as at East Oumalik (Figure 3g), except that an ice sill was not encountered.

The upper lacustrine deposits beneath drained lake basins consist of 4-6 m of silt or silty clay, with occasional strata of fine sand. They possess a well-defined horizontal stratification (0.5-2 cm thick) including varve-like couplets. Sedimentary structures, including flame structures, load casts, and graded bedding, occur within some strata. Organic material is disseminated throughout the deposits, with peat and woody horizons common in the upper 3 m. The lake deposits lie directly on bedrock or on a thin layer (<2 m) of fine to medium sand over bedrock. Radiocarbon dating of six peat samples suggests lake activity was established by

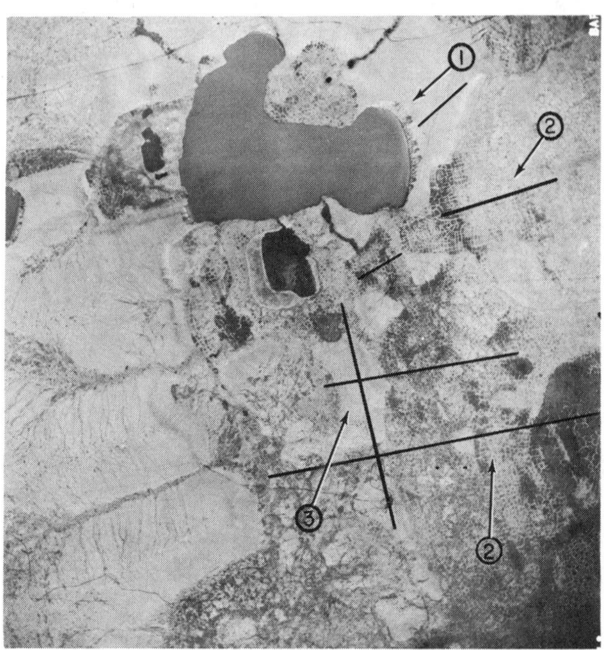

FIGURE 5 Aerial photograph of Oumalik (BAR 106-117, 9/4/48). Lines indicate general position of drilling transects. Upland (1), drained lake basins (2), and knoll (3) are located (scale 1:20,000). North to top.

7500 yr B.P. and continued for at least another 1500 years.

Ice occurs as small lenses, veins, and irregular layers (max. 20 mm thick) within the lake deposits, composing 60-90% of the upper 4 m (Figure 3f). Within basinal slopes, lake deposits overlie silt, like that of the uplands, but with an ice content of from 50-60%. Melting of excess ice in the silt during thawing beneath the active lake accounts for the lower ice content. Ice occurs as sporadic, small lenses (1 mm) or thin veins (1-3 mm thick) in this modified silt.

The small knoll adjacent to the drained lake basin (Figure 5) is an upland remnant that has been modified by lacustrine and subaerial slope processes. The uppermost 3 m are laminated lacustrine deposits that contain substantial amounts of ice (Figure 3g). They are comparable in age to lake deposits in the adjacent basin. Beneath these deposits is silt of a composition similar to that of the uplands but containing distinctly less excess ice (Figure 3g). Stratified, medium sands with pebbly horizons and cross-bedding indicative of fluvial deposition lie at the base of the sequence (lower 4 m). The ice volume of the sand is comparable to the thawed and refrozen sand at Fish Creek (Figure 3c). The ice content of the middle eolian silt and lower fluvial sand reflects thaw and refreezing during thaw lake activity and drainage. Colluvial silt on the knoll slopes contains 50-65% ice, lacks structure, and forms a basinward thickening wedge onto the adjacent lake deposits.

DISCUSSION

The total ice volume of perennially frozen sediments at each locality results from the combined effects of geological and cryogenic depositional and postdepositional processes. Stratigraphic variability in ice occurrence relates to the origin and composition of the sedimentary sequence and thermal events that accompanied sedimentary activity. Trends in volume with increasing depth as well as lateral near-surface variability at each site reflect this sedimentary and thermal history. Shifts in ice volume are thus indirectly correlated with the relative age of deposits and landforms at a given site.

Primary deposits--those not reworked by post-depositional processes, which are fine-grained or organic-rich--contain significantly more ice than coarse-grained deposits and possess the greatest downhole variability in ice volume. This relationship is not necessarily valid if sediments subsequently thaw; ice content then depends upon the processes of resedimentation or refreezing events that follow.

This concept is illustrated by the effects of thaw lake processes. Thawing and reworking of sediments during lake formation and expansion mixes and concentrates fine-grained sediments and organic material in deposits on the lake bed. Following drainage, freezing of the lacustrine deposits results in ice segregation, with moisture migrating from the underlying, still-thawed sediments. Their ice content equals or exceeds surrounding primary deposits. Sediments that are thawed by lake activity refreeze to an ice content lower than before lake formation, with segregated ice less common. Ice wedge dimensions are smallest in basins drained most recently.

Additional detailed analyses of ground ice occurrence in relation to geologic parameters are clearly needed to examine further the relationships of spatial variability found in this study. An ability to predict the general distribution and volume of ice based upon limited borehole data, for example, would be particularly useful in selecting sites that are least susceptible to disturbance, or in assessing the variability in geotechnical properties of frozen ground during presiting studies.

ACKNOWLEDGMENT

This study was supported by the U.S. Geological Survey's National Petroleum Reserve, Alaska, program. I thank Bruce Brockett for working with me on the sampling program and Drs. Max Brewer and Jerry Brown for encouraging and supporting this research.

REFERENCES

Black, R.F., 1964, Gubik formation of Quaternary age in northern Alaska: U.S. Geological Survey Professional Paper 302-C.

Britton, M.E., 1967, Vegetation of the arctic tundra, in Hansen, H.P., ed., Arctic Biology: Oregon State University Press, p. 67-130.

Brown, J. and Sellmann, P.V., 1973, Permafrost and coastal plain history of arctic Alaska, in Britton, M.E., ed., Alaskan Arctic Tundra: AINA, p. 31-47.

Carson, C.E. and Hussey, K.M., 1962, The oriented lakes of arctic Alaska: Journal of Geology, v. 70, p. 417-439.

Carter, L.D., 1981, A Pleistocene sand sea on the Alaskan arctic coastal plain: Science, v. 211, p. 381-383.

Ferrians, O.J., Kachadoorian, R., and Greene, G.W., 1969, Permafrost and related engineering problems in Alaska: U.S. Geological Survey Professional Paper 678.

French, H.M., 1975, Man-induced thermokarst, Sachs Harbor airstrip, Banks Island, Northwest Territories: Canadian Journal of Earth Sciences, v. 12, p. 132-144.

Gell, A.W., 1976, Underground ice in permafrost, Mackenzie Delta, Tuktoyaktuk Peninsula, N.W.T.: Ph.D. dissertation, University of British Columbia, 258 p.

Hussey, K.M., and Michelson, R.W., 1966, Tundra relief features near Point Barrow, Alaska: Arctic, v. 19, p. 162-184.

Johnston, G.H., ed., 1981, Permafrost: engineering design and construction: New York, John Wiley and Sons.

Lawson, D.E., 1982, Long-term modifications of perennially frozen sediments and terrain at East Oumalik, northern Alaska, CRREL Report 82-36: Hanover, N.H., U.S. Army Cold Regions Research and Engineering Laboratory.

Livingstone, D.A., Bryan Jr., K., and Leahy, R.G., 1958, Effects of an arctic environment on the origin and development of freshwater lakes: Limnology and Oceanography, v. 3, p. 192-214.

Mackay, J.R., 1970, Disturbances to the tundra and forest tundra environment of the western arctic: Canadian Geotechnical Journal, v. 7, p. 420-432.

Mackay, J.R., 1971, Origin of massive icy beds in permafrost, western Arctic coast, Canada: Canadian Journal Earth Sciences, v. 8, p. 397-422.

Mackay, J.R., and Black, R.F., 1973, Origin, composition and structure of frozen ground and ground ice: a review, in Permafrost--The North American Contribution to the Second International Conference: Washington, D.C., National Academy of Sciences Publication 2115, p. 185-192.

Mackay, J.R., Konishchev, V.N., and Popov, A.I., 1978, Geologic control of the origin, characteristics and distribution of ground ice, in Proceedings, Third International Conference on Permafrost, v. 2, p. 1-18: Ottawa, National Research Council of Canada.

Muller, S.W., 1947, Permafrost or permanently frozen ground and related engineering problems: Ann Arbor, Michigan, J.W. Edwards, Inc.

Pollard, W.H., and French, H.M., 1980, A first approximation of the volume of ground ice Richards Island, Pleistocene MacKenzie Delta, Northwest Terrorities, Canada: Canadian Geotechnical Journal, v. 17, p. 509-516.

Reed, J.C., 1958, Exploration of Naval Petroleum Reserve No. 4 and adjacent areas, northern Alaska: U.S. Geological Survey Professional Paper 301.

Sellmann, P.V., Brown, J., Lewellen, R.I., McKim, H., and Merry, C., 1975, The classification and geomorphic implications of thaw lakes on the Arctic Coastal Plain, Alaska, Research Report 344: Hanover, N.H., U.S. Army Cold Regions Research and Engineering Laboratory.

Terzaghi, K., 1952, Permafrost: Journal of the Boston Society of Civil Engineers, v. 39, n. 1, p. 1-50.

Williams, J.R., and Yeend, W.E., 1979, Deep thaw basins of the inner Arctic coastal plain, Alaska: The U.S. Geological Survey in Alaska, U.S. Geological Survey Circular 804-B, p. B35-B37.

Williams, J.R., Yeend, W.E., Carter, L.D., and Hamilton, T.D., 1977, Preliminary surficial deposits map of National Petroleum Reserve-Alaska: U.S. Geological Survey Open File Report 77-868.

EROSION BY OVERLAND FLOW, CENTRAL BANKS ISLAND, WESTERN CANADIAN ARCTIC

Antoni G. Lewkowicz

Department of Geography, University of Toronto
Toronto, Ontario, Canada, M5S 1A1

Process studies carried out at runoff plots on Banks Island emphasize the importance of solute erosion by overland flow, rather than suspended sediment removal. Rates are low, despite the influence of permafrost in promoting surface runoff. Observations of feedback within the slopewash system suggest that these trends may not be universal. In tundra zones, vegetation is thought to be more protective than in the transitional zone of Banks Island. In semidesert and polar desert zones, where vegetation cover in snowbed locations is limited, removal of sediment particles by overland flow may be more significant.

INTRODUCTION

Overland flow (i.e. surface wash) is an azonal process not restricted to permafrost regions. The presence of permafrost, however, favours saturation of the active layer by preventing deep percolation of water, while the rapidity of snowmelt in early summer further increases the probability of overland flow (Washburn 1980 p. 244). The role played by overland flow in eroding and transporting material in permafrost regions is speculative. French (1976 pp. 141-142) hypothesizes that erosion by overland flow is most effective in areas transitional between tundra and polar desert. On Spitsbergen, Jahn (1961) measured rates of erosion due to overland flow of up to 16 $g/m^2/yr$. Apart from these, and more limited observations by Wilkinson and Bunting (1975), there is an absence of data available on the erosional capability of surface wash.

This paper summarizes data collected during 3 years of direct field observations of overland flow in the Western Canadian Arctic. Measurements were made in the Thomsen River lowlands of north central Banks Island (Figure 1; latitude 73°14'N, longitude 119°32'W), an area regarded as transitional between the tundra and semidesert and polar desert zones. Four instrumented runoff plots (67-525 m^2) were located on typical low angled slopes, and standard hydrologic and geomorphic monitoring techniques were employed to collect data (Lewkowicz 1981). Earlier papers provide descriptions of the frequency and magnitude of overland and subsurface flow, and the relevance of hillslope hydrologic theory to this part of the Arctic (Lewkowicz and French 1982a, 1982b). This paper deals with the erosional capability of overland flow and consequently with the rapidity and nature of slope evolution under periglacial conditions (see French 1976 pp. 165-166).

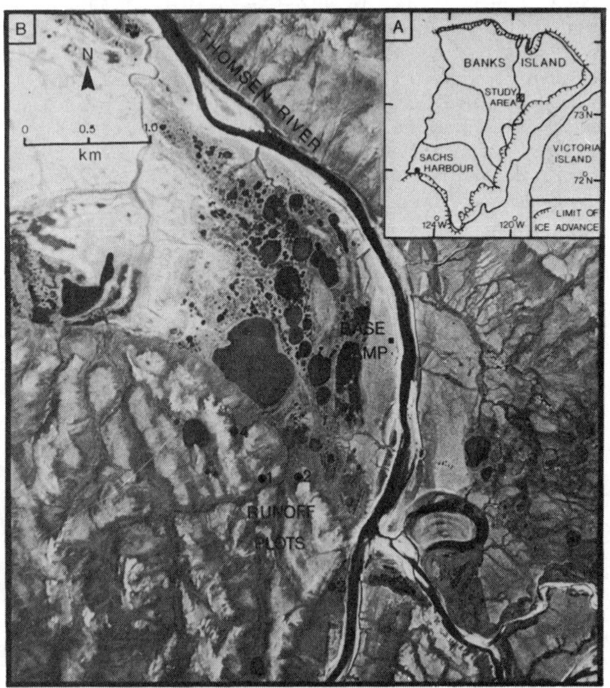

FIGURE 1 Location of study: A (inset)--Banks Island, N.W.T., Canada, showing Early Wisconsinan glacial limit (after Vincent 1982); B--Air photograph of study area (part of A-17379-40).

METHODS

The methodology of the study involved independent measurements of overland flow volume, solute concentration, and suspended sediment concentration. These data were then used to calculate net rates of erosion. Standard procedures were used to monitor overland flow and other hydrologic parameters at the runoff plots (see Figures 2 and 3). A full description is presented in Lewkowicz and French (1982b). Samples of overland flow were taken manually at the runoff collectors. Solute concentrations were estimated by field titration for total, calcium and magnesium hardness (Schwarzenbach method) and/or by a measurement of specific conductance.

FIGURE 2 Runoff plot 1, the interfluve site, June 3, 1977. Note small hummocks, sparse vegetation cover and breakup of thin snowpack into discrete pieces.

FIGURE 3 Runoff plot 2, a snowbank site, June 30, 1978. Note tricycle for scale at top of plot, widespread overland flow, and extensive vegetation cover. Weir and water-level recorder provide a continuous record of overland flow.

Conductance values in µS/cm were multiplied by 0.65 (the middle of the range given by Rainwater and Thatcher (1960)) to obtain solute concentrations in mg/l. One litre samples for suspended sediment were filtered through predried and preweighed filter papers using an Østrem filter pump, and were reweighed in a laboratory after drying at the end of the field season. Control experiments revealed an accuracy of approximately ±4 mg/l.

SOLUTES

A wide range of solute concentrations was determined in overland flow samples collected at the runoff plots (Table 1). Minimum values of 2 mg/l were recorded in the vicinity of the snow edge, but concentrations increased as the water moved downslope. The low values confirm that basal ice (e.g. Lewkowicz 1981 p. 197, Woo et al. 1982) prevents meltwater from coming into contact with the ground prior to leaving the snowbank. Consequently, no wash erosion takes place beneath the snow.

Variability in solute concentrations can largely be explained by (1) the distance between the snow edge and the collector and (2) the overland flow discharge. These are related to the residence time and the degree of mixing of the flow, and can be considered at diurnal and seasonal time scales.

On any particular day, changes in solute concentrations (Figure 4) relate mainly to velocity variations associated with diurnal hydrographs. The relationship between discharge and solute concentration is inverse and best described for plots 3 and 4 by semilogarithmic curves. Hysteresis, thought to be the result of snow edge retreat (0.5 m at plot 3 and 0.9 m at plot 4 on July 4-5, 1978), is also present. On a longer time scale, solute concentrations increase as the melt season progresses. This is partly the result of higher flow temperatures and vegetation growth (linked to biogenic CO_2 production (Woo and Marsh 1977)), but is mostly due to the greater distances between the snow and the collector, the result of snowbank ablation.

TABLE 1 Variability of Solute Concentration in Overland Flow and Best Fit Multiple Regression Equations

Plot No.	Years	Number of samples	Mean solute concentration in samples (mg/l)	Standard deviation (mg/l)	Regression Equation[a]	r
1	1977 & 1979	46	217	95	C=122.1-14.0Q+12.3D	+0.90
2	1977-1979	58	58	34	C=-3.8Q+4.4D	+0.90
3	1977-1979	121	93	86	C=64.4-37.4 lnQ+8.2D	+0.87
4	1978-1979	28	52	32	C=13.2-10.6 lnQ+9.0D	+0.98

[a] Symbols in regression equation: C is solute concentration (mg/l); Q is overland flow (l/min); D is distance between the snow edge and overland flow collector (m).

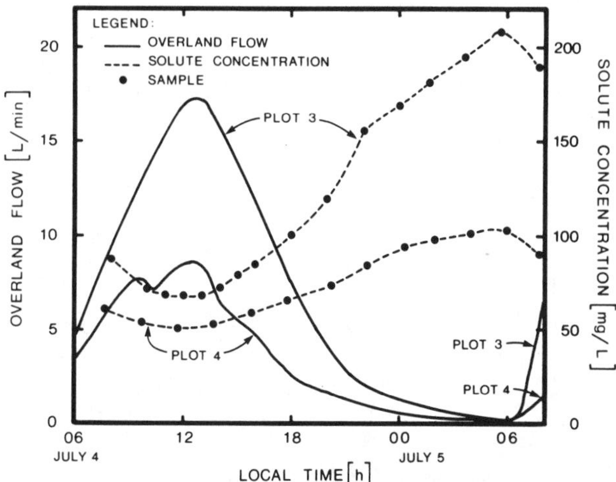

FIGURE 4 Diurnal variation of solute concentration in overland flow, plots 3 and 4, July 4-5, 1978.

Multiple regression of solute concentration against overland flow discharge and distance travelled by the flow gives correlation coefficients (r) between +0.87 and +0.98 for the four plots (Table 1). Since these values are high, the multiple regression equations in Table 1 can be used as predictive rating curves for their respective plots. The total weight of dissolved solids can be calculated by multiplying concentrations derived for each 30-min period of the discharge record, by the discharge over that time. The only case where it is not possible to derive totals in this way, is plot 1 for 1977. The breakup of the snow cover into a number of irregular masses during that year prevented assessment of the distance to the snow edge. A sufficient number of conductivity measurements exist, however, for direct use to be made of the solute concentration data. Since the regression equations provide imperfect explanation of the data variation, the totals produced should be regarded as "best estimates".

The mean solute concentration at the interfluve site (plot 1) is considerably higher than those at the snowbank sites (plots 2, 3, and 4) (Table 2). This may be the result of contact between water and the ground surface within the snow-covered area due to a poorly developed basal ice layer in the thin snowpack. Best estimates of solute erosion vary by 2 orders of magnitude, from a 3-year average of 1.3 $g/m^2/yr$ at plot 2 to 49.6 $g/m^2/yr$ at plot 3. The average for the interfluve is higher than that for plot 2, primarily because the mean solute concentration is 4 times greater. These rates represent erosion averaged over the complete plot. Because of the presence of basal ice, however, the true picture is one of zero erosion over that portion of the plot upslope of the final runoff-producing position of the snowbank, and enhanced erosion on the lower portion of the plot - the "affected area". The latter constitutes between 30 and 54 % of the total plot area (calculated from snow edge retreat records). If weight losses are distributed over the affected areas, rates increase 2-3 times, reaching a maximum of 165 $g/m^2/yr$ at plot 3 (Table 2).

The values of weight loss by solution can be converted into denudation rates using a bulk density factor for the eroding surface of 0.67 Mg/m^3 (from Pearce 1976) (Table 2). Average rates for the plots range from 2 to 74 mm/1,000 yr, and for the affected areas from 6 to 246 mm/1,000 yr. Data available for comparison with these figures are extremely limited, since most measurements of net solution loss have been made by examination of the stream solute load. One of the few comparable measurement rates was obtained by Thorn (1974), working in the alpine environment of the Colorado Front Range. A denudation value of 6.7 mm/1,000 yr was attributed to solution by overland flow downslope of a large snowbank. This value falls within the range measured at the Thomsen River runoff plots but is an order of magnitude less than the maximum recorded.

TABLE 2 Best Estimate Rates of Solute Erosion by Overland Flow

Plot No.	Year	Mean solute concentration (mg/l)	Weight loss over all the plot ($g/m^2/yr$)	Area of plot affected by overland flow[a] (%)	Weight loss over affected area ($g/m^2/yr$)	Denudation over all the plot[b] (mm/1,000 yr)	Denudation over affected area[b] (mm/1,000 yr)
1	1977	182	1.2	40	2.9	1.8	4.4
	1979	252	4.8	54	8.9	7.2	13.3
2	1977	44	2.2	30	7.4	3.3	11.1
	1978	63	1.4	37	3.8	2.1	5.7
	1979	51	0.4	31	1.2	0.6	1.9
3	1978	94	49.6	30	164.8	74.1	246.0
4	1978	70	21.1	35	60.6	31.5	90.4

[a] Represents area of plot downslope of final position of snow edge.
[b] Dry bulk density assumed to be 0.67 Mg/m^3.

SUSPENDED SEDIMENT

Suspended sediment concentrations in overland flow were lower and less variable than solute concentrations. A total of 78 samples were collected at the plots in 1979, and more than half of these possessed concentrations less than 4 mg/l. Moreover, with one exception, no diurnal or seasonal cycles in concentrations were identified. Since the transport capacity of the overland flow varied on an hourly basis, these observations suggest that the concentrations were limited by the availability of sediment ("weathering limited").

The single exception to the lack of concentration variation occurred at plot 1 on June 30, 1979. More than half of the total season's overland flow was generated in 14 hours, and a maximum discharge of 10.7 l/min was recorded. The concentration of suspended sediment reached 70 mg/l, but peaked prior to the maximum flow (Figure 5). Similar cycles of sediment concentrations in overland flow have been observed in nonpermafrost areas in response to natural or simulated rainfall (e.g. Emmett 1970). They indicate the removal of easily transported particles in a relatively short time. The absence of any discernible cycles at the other three plots, in spite of unit discharges greater than those at plot 1, is attributed to the stabilizing presence of vegetation downslope of the snowbanks.

Despite problems with the measurement of suspended sediment, resulting from concentrations being close to the level of discrimination of the technique, it is still possible to produce estimates of the rates of weight loss and denudation. In order to perform the calculations, a number of assumptions are necessary. First, because of variability in the weight of filter papers during drying, any concentration apparently less than 4 mg/l was assumed to be 2 mg/l. Second, values greater than 4 mg/l were used unchanged. Third, after calculation of total suspended sediment lost during the melt season by multiplying flow volumes by their appropriate concentrations, the residue of sediment trapped in the collector and weir was added. When divided by

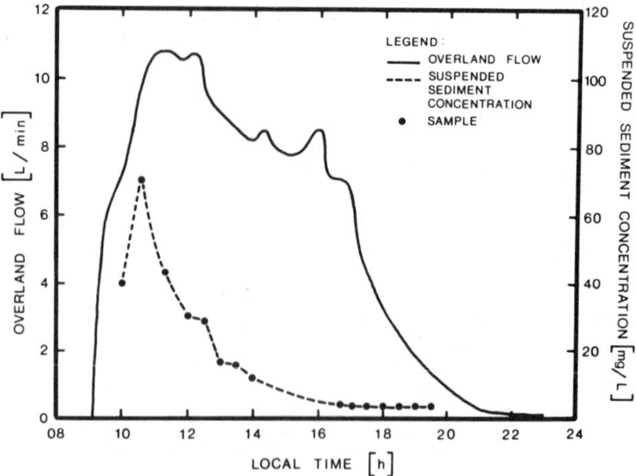

FIGURE 5 Diurnal variation of suspended sediment concentration in overland flow, plot 1, June 30, 1979.

the flow volume, this total gave the mean concentration. It was assumed that this concentration, obtained from 1979 data, was applicable to the 1977 and 1978 field seasons.

The results of the suspended sediment calculations are given in Table 3. Best-estimate mean concentrations are less than 4 mg/l at the snowbank plots and 21 mg/l on the interfluve. Average sediment weight losses from the plots vary from <0.1 to 1.7 g/m^2/yr. These values can be compared with Jahn's (1961) measurements on unvegetated slopes on Spitsbergen. His maximum value of 16 g/m^2/yr exceeds that at Thomsen River by a factor of 10, and his minimum value of 1 g/m^2/yr is close to the maximum recorded in the study area. It is probable that these differences result from the absence of a relationship between sustained moisture supply and vegetation growth on the coarse regolith slopes in Spitsbergen. A further contrast between the results is the positive correlation between snowbank size and erosion rate that exists on Spitsbergen but not on Banks Island. This also indicates the protective

TABLE 3 Best Estimate Rates of Suspended Sediment Erosion by Overland Flow

Plot No.	Year	Mean suspended sediment concentration[a] (mg/l)	Weight loss over all the plot (g/m^2/yr)	Area of plot affected by overland flow[b] (%)	Weight loss over affected area (g/m^2/yr)	Denudation over all the plot[c] (mm/1,000 yr)	Denudation over affected area[c] (mm/1,000 yr)
1	1977	21	0.1	40	0.3	0.2	0.5
	1979	21	0.4	54	0.8	0.6	1.1
2	1977	3	0.1	30	0.4	0.2	0.7
	1978	3	0.1	37	0.2	0.1	0.2
	1979	3	<0.1	31	0.1	<0.1	0.1
3	1978	3	1.7	30	5.8	2.6	8.6
4	1978	4	1.2	35	3.5	1.8	5.2

[a] Inferred from 1979 results and sediment collected from weirs.
[b] Represents area of plot downslope of final position of snow edge.
[c] Dry bulk density assumed to be 0.67 Mg/m^3.

influence of vegetation, resulting in a lower average erosion rate at plot 2 than at the interfluve plot (Table 3).

Conversion of the weight losses to denudation rates (Table 3) gives low values, with a maximum of 2.6 mm/1,000 yr for the whole of plot 3, and 8.6 mm/1,000 yr for the area affected by wash. These values can be compared with the global range of denudation rates by surface wash of 0.1 to >230,000 mm/1,000 yr (Young 1974 p. 69). The estimated losses from the runoff plots are among the lowest in the world and somewhat lower than those obtained by Thorn (1974) of 3.5-8.4 mm/1,000 yr.

DISCUSSION AND CONCLUSIONS

Comparison of the denudation rates shown in Tables 2 and 3 reveals that solute removal by overland flow in central Banks Island is 8 to 30 times greater than removal of sediment in suspension. This dominance is most pronounced at the site of the largest snowbank and least at the interfluve site.

Problems arise, however, if attempts are made to extrapolate from these results to other permafrost environments. These values may not be representative of overland flow erosion in all areas north of the tree line due to variations in (1) snow amounts and distribution, (2) rates and distribution of snowmelt runoff, and (3) vegetation cover. Observations made at Thomsen River, however, do permit the formulation of a hypothesis concerning the spatial variability of overland flow erosion in permafrost areas. The details of the hypothesized trends (Figure 6) are as follows:

(1) A simple decrease in interfluve solution may occur between the tundra zones and the semidesert and polar desert zones, resulting from decreases in overland flow runoff, flow temperature, and biogenic CO_2 production (Line A).

(2) Interfluve suspended sediment removal may increase between the tundra zones and the semi-desert and polar desert zones because the positive erosional effects of vegetation decline outweigh the negative influences of lower runoff production (Line B).

(3) Snowbank solution may decrease between the tundra zones and the semidesert and polar desert zones as overland flow runoff, flow temperature, and biogenic CO_2 production all decline (Line C).

(4) Snowbank suspended sediment removal may be more complex, with little change expected in erosion rates between the tundra and transitional zones, since vegetation is protective in both. A substantial increase may occur between the transitional and semidesert and polar desert areas, however, as locations downslope of snowbanks are subjected to high unit discharges without the presence of a completely protective vegetation cover (mainly due to a shortened growing season) (Line D).

In conclusion, therefore, the generalization that surface wash erosion increases northward from the tree line (French 1976 p. 142) is

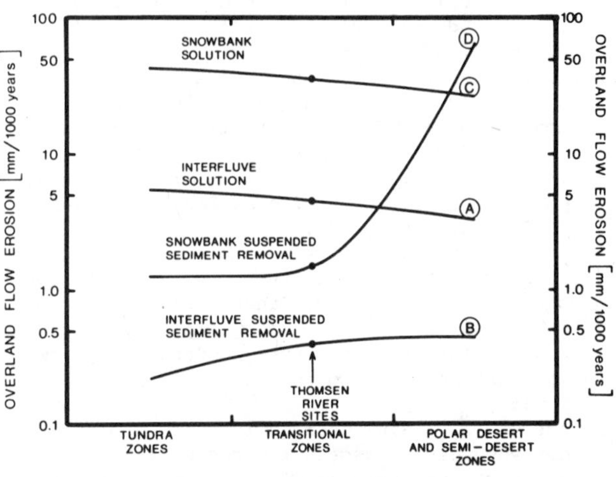

FIGURE 6 Inferred variation of overland flow erosion in areas of continuous permafrost.

supported by the trends summarized in Figure 6. The maximum erosion rate, however, is thought to occur in the polar desert zones rather than in the tundra-polar desert transition zones. On the other hand, the measurements reported in this paper suggest that rapid landscape modification by surface wash processes is unlikely, even in the polar desert regions of the High Arctic.

ACKNOWLEDGEMENTS

Fieldwork on Banks Island was supported by the Natural Sciences and Engineering Council (Grant A-8367), the University of Ottawa Northern Research Group, and the Polar Continental Shelf Project, Energy, Mines and Resources (Project 34-73). Equipment was kindly loaned by the Terrain Sciences Division, Geological Survey of Canada, Inland Waters Directorate, Environment Canada, and the Rideau Valley Conservation Authority. Use of the Terrain Sciences Division's tundra hut at Thomsen River is gratefully acknowledged. Excellent field assistance was provided by G. Champagne, T. Colosimo, J. M. C. Lalonde, A. LeHenaff and R. Spencer. Illustrations were drawn by B. Nakrosius using financial aid from a University of Toronto grant. Helpful comments on draft of this paper were provided by H. M. French and J. M. C. Lalonde.

REFERENCES

Emmett, W. W., 1970, The hydraulic of overland flow on hillslopes: United States Geological Survey Professional Paper, no. 662-A, 68 p.

French, H. M., 1976, The periglacial environment: London, Longmans.

Jahn, A., 1961, Quantitative analysis of some periglacial processes in Spitsbergen: Nauka O Ziemi II, Seria B, v. 5, p. 3-34.

Lewkowicz, A. G., 1981, A study of slopewash processes in the continuous permafrost zone,

Western Canadian Arctic: Unpublished Ph.D. thesis, Department of Geography, University of Ottawa, 269 p.

Lewkowicz, A. G., and French, H. M., 1982a, Downslope water movement and solute concentrations within the active layer, Banks Island, N.W.T., in French, H. M., ed., Proceedings of the Fourth Canadian Permafrost Conference: Ottawa, National Research Council of Canada, p. 163-172.

Lewkowicz, A. G., and French, H. M., 1982b, The hydrology of small runoff plots in an area of continuous permafrost, Banks Island, N.W.T., in French, H. M., ed., Proceedings of the Fourth Canadian Permafrost Conference: Ottawa, National Research Council of Canada, p. 151-162.

Pearce, A. J., 1976, Geomorphic and hydrologic consequences of vegetation destruction, Sudbury, Ontario: Canadian Journal of Earth Sciences, v. 13, p. 1358-1373.

Rainwater, F. H., and Thatcher, L. L., 1960, Methods for collection and analysis of water samples: United States Geological Survey Water Supply Paper, no. 1454, 297 p.

Thorn, C. E., 1974, An analysis of nivation processes and their geomorphic significance, Niwot Ridge, Colorado Front Range: Unpublished Ph.D. thesis, Department of Geography, University of Colorado, 351 p.

Vincent, J.-S., 1982, The Quaternary history of Banks Island, N.W.T., Canada: Geographie physique et Quaternaire, v. 36, p. 209-232.

Washburn, A. L., 1980, Geocryology: New York, Halsted Press.

Wilkinson, T. J., and Bunting, B. T., 1975, Overland transport of sediment by rill water in a periglacial environment in the Canadian High Arctic: Geografiska Annaler, v. 57A, p. 105-116.

Woo, M.-K., Heron, R., and Marsh, P., 1982, Basal ice in High Arctic snow-packs: Arctic and Alpine Research, v. 14, p. 251-260.

Woo, M. K., and Marsh, P., 1977, Effect of vegetation on limestone solution in a small High Arctic basin: Canadian Journal of Earth Sciences, v. 14, p. 571-581.

Young, A., 1974, The rate of slope retreat, in Brown, E. H., and Waters, R. S., eds., Progress in geomorphology--Papers in honour of David L. Linton, Institute of British Geographers Special Publication, no. 7, p. 65-78.

PERMAFROST STUDY AND RAILROAD CONSTRUCTION IN PERMAFROST AREAS OF CHINA

Li Yusheng[1], Wang Zhugui[2],
Dai Jingbo[3], Cui Chenghan[3],
He Changgen[3], and Zhao Yunlong[4]

[1]Chinese Academy of Railway Sciences (CARS)
[2]Northwest Institute of CARS
[3]Third Railroad Survey and Design Institute
[4]Qiqihar Railway Administration
People's Republic of China

Field investigation, laboratory experiments, and engineering practice during the past 30 years has given China a great deal of information on railroad construction in permafrost regions. The thermal and mechanical properties of soils both in frozen and thawed states were determined, several formulas for predicting the natural and artificial permafrost tables were developed, four distribution maps of permafrost at different scales were compiled, and a principle for selecting railroad routes was presented. In the field of foundation design for railroads, bridges, and culverts, methods for constructing foundations in sectors of thick-layer ground ice have been studied, along with measures to prevent freezing of blind drain outlets and avoid thaw damage to embankment foundations. Raft and wood-reinforced culvert foundations were successfully used in the permafrost regions of the Great Hinggan Mountains, reinforced concrete pile foundations were tested, and a method of rapid construction of foundation pits for small bridges and culverts was developed. In tunnel design and construction, a sluiceway drainage system was used. A set of design principles and methods has been developed for buildings, pipelines, and pavements. In addition, two kinds of low-temperature construction materials have been developed for reinforced concrete-poured pile foundations and culverts.

In the mid-1950s, with the development of our socialist economic construction, railroad survey and design crews entered the Great Hinggan Mountains to exploit forest resources. The permafrost regions there presented unfamiliar geological, geographical, and climatic conditions. As neither the nature nor the distribution of permafrost were known, conventional methods of railroad design and construction were used. A number of problems inevitably resulted, emphasizing the need to study railroad survey, design, construction, and maintenance in cold regions. This was the beginning of our research into permafrost and its effect on railroad engineering.

Gradually, through research, testing, and more than 20 years of practical experience, we have mastered some aspects of building railroads in permafrost areas. So far, the railroad built through the Great Hinggan Mountains of northeast China is more than 2000 km long, of which nearly 450 km runs through continuous permafrost. To meet the needs of China's economic development and to establish the prerequisites for building the Qinghai-Xizang Railroad in the near future, some essential experiments and studies have been conducted in the permafrost region of the Qinghai-Xizang Plateau.

The following is a brief introduction to various aspects of our research and experience.

ENGINEERING GEOLOGY IN PERMAFROST AREAS

In cold regions, the controlling factor for construction projects is engineering geology; this is especially true for railroad construction in permafrost areas. For this reason, much effort has been concentrated on this problem, and the Lanzhou Institute of Glaciology and Cryopedology of the Academia Sinica has made a major contribution.

The subjects under study are: (1) the essential properties of permafrost, (2) the classification of permafrost for engineering purposes, (3) permafrost distribution, (4) physical and geological phenomena in permafrost areas, (5) special features of surveying and mapping the engineering geology of permafrost areas, and (6) principles for route selection in permafrost areas.

Satisfactory results have been achieved in both field and laboratory work. The results have been incorporated into the "Railroad Design Standards" and the "Detailed Rules and Regulations of Railroad Survey and Design."

China's progress in the study of cold regions includes achievements in permafrost mechanics, the study of the thermodynamics of frozen ground, mapping of permafrost distribution, and selection of railway routes in permafrost regions.

Permafrost Mechanics

Our achievements in permafrost mechanics include developing a method for calculating permafrost bearing capacity, a distribution law of frost-heave pressures on foundations and retaining walls, and indexes of shear strength of frozen soils.

The Lanzhou Institute of Glaciology and Cryopedology has played a principal role in developing the mechanical indexes of typical soils and calculating building stability. The indexes are the basic bearing capacity of permafrost (First Railroad Survey and Design Institute et al. 1972), tangential frost-heaving pressure on concrete and wooden foundations from seasonally thawing or freezing soil during freezing (Wu et al. 1979b), and the long-term adfreezing force between permafrost and concrete or wooden foundations (Wu et al. 1979a). These have all been included in "Railroad Design Standards" and "Detailed Rules and Regulations of Railroad Survey and Design."

The Third Railroad Survey and Design Institute has defined a distribution law for tangential frost-heaving pressures in the thawing-freezing layer on building foundations in the Great Hinggan Mountains, and has developed a technique for determining appropriate values for foundation design (Cui and Zhou 1981). In addition, empirical formulas to determine the thawing and compression coefficients of frozen clayey soil in the area were obtained from regression analysis of many of the experimental data (Wu et al. 1979c).

The Northwest Institute of the Chinese Academy of Railway Sciences (the Northwest Institute, for short), through analysis of L-shaped, model retaining walls at the Mt. Fenghuo test site, has defined not only the distribution law of horizontal frost-heaving pressure acting on vertical retaining structures, but the relation between horizontal frost-heaving pressure and the deformation of a retaining structure, and the maximum horizontal frost-heaving pressure (Ding 1983). Recently, the long-term resistant force of anchor rods in permafrost has been studied (Ding and Zhang 1983).

To obtain indexes of the shear strength of soil so that the stability of the slope and ground base of railroad embankments can be determined, the Northwest Institute has conducted a series of direct shear tests in the laboratory and large-scale shear tests in the field to study the nature of sandy and clayey soils in the permafrost areas of the Qinghai-Xizang Plateau. The results show that the shear strength at the freeze-thaw interface is greater than that of fully thawed soil (Tong 1983). This conclusion is valuable in computing soil stability in all permafrost areas.

Thermodynamics of Frozen Ground

Based on laboratory experiments made on peat, clayey soil, sandy soil, gravel-clayey soil, and sandy gravel along with a number of field tests, the Lanzhou Institute presents the relationships between five thermal indexes (specific heat, volumetric heat capacity, thermal conductivity, thermal diffusivity, and latent heat of fusion when thawing and freezing) and two physical indexes (dry bulk density and water content) of each of the above materials as well as a list of thermal parameters (Xu et al. 1979b). In addition, in the field of thermal engineering, the Lanzhou Institute has developed an empirical formula to determine the depth of the permafrost table at different elevations and latitudes (Xu et al. 1979a), an approximation equation to evaluate the critical height of an embankment in a permafrost area (Qinghai-Xizang Highway Research Group 1981), and a method for calculating the required thickness of the thermal insulation layer within a railroad embankment using a one-dimensional model (Ding et al. 1979).

The Northwest Institute has studied the thermodynamics of the permafrost areas of the Qinghai-Xizang Plateau to determine burial depths for building foundations and the design of insulating layers. They have also developed a method for determining the depth of the natural and artificial permafrost tables for railroad, bridge, and culvert foundations based on:

- the seasonal freezing and thawing characteristics of soil layers in the permafrost areas of the Qinghai-Xizang Plateau,
- a method for determining the permafrost table, and its application to cut engineering,
- an empirical formula to determine the necessary thickness of the heat insulation layer in the area,
- the law of variation of the artificial table of an embankment in the area,
- an empirical formula for evaluating the artificial table of a culvert foundation and an equation to calculate the thaw depth of the ground under culvert foundations in permafrost, and
- the artificial table of a bridge foundation in permafrost.

Some of these achievements represent significant progress, e.g. evaluating the latent freezing capacity of the soil layer in view of the surplus freezing index; using the period from October to September as the calculation year, which is more precise and logical than the calendar year used previously; presenting a quantitative estimate of the variation of the natural permafrost table in sandy and clayey soil layers (1.7% when the variation in average temperature of the calculation year is 1°C), and then making a 100-year forecast (Huang 1981). The statistical equation to determine the natural permafrost table of the peat and clay soils in the Mt. Fenghuo region is based on the average August temperature in the Wudaoliang region, and is verified by data from the preceding seven years with a maximum calculation error of no more than 7%. The forecast conclusion for the first 20 years, based on the 100-year forecast, proved to be identical with the data obtained from the test observations in those 20 years (Liu 1983).

Permafrost Distribution

Our research on the distribution of permafrost is so extensive that only those achievements relevant to railroad engineering are introduced in this report.

A map "The Distribution of Permafrost in Northeast China" (1:2,000,000) was compiled by the Yakeshi Institute of Forest Survey and Design, the Lanzhou Institute, and the Third Railroad Survey and Design Institute.

A map of the distribution of permafrost along the railroad within the jurisdiction of the Qiqihar Railway Administration (1:200,000) was compiled by its regional bureau from information collected by the Third Railroad Survey and Design Institute and the Geological Bureau of Heilongjiang Province, as well as that collected by the bureau.

An abridged map "The Distribution of Permafrost in the Great and Minor Hinggan Mountain Regions"

(1:4,660,000) was compiled by the Third Railroad Survey and Design Institute from "The Distribution of Permafrost in Northeast China."

A map "The Distribution of Permafrost along the Qinghai-Xizang Highway" (1:600,000) was drawn by the Lanzhou Institute.

The Third Railroad Survey and Design Institute has studied ground temperature characteristics in the northern permafrost areas of the Great Hinggan Mountains and has determined the ground temperature curve, annual variation depths of ground temperature, factors affecting perennial ground temperature, and laws governing ground temperature variations in this area to complete the ground temperature data base for the region (Dai and Li 1981). These data will be valuable in future research and engineering.

In addition, the Institute has studied the close relationship between various plant communities and permafrost distribution and ice content in the area (Lin 1981). It offers a feasible method for speeding investigation and mapping for geological engineering.

Railway Route Selection in Permafrost Regions

This work was undertaken by the First and Third Railroad Survey and Design Institutes and the Northwest Institute, based on analysis of geological, topographical, and physical-geological phenomena as well as other geological engineering conditions for each sector of the permafrost region.

FIGURE 1 Road foundation test projects.

SEVERAL PROBLEMS IN THE DESIGN OF RAILROAD FOUNDATIONS

Methods for designing railroad foundations in sectors of thick layer ground ice have been studied, along with measures to prevent freezing of blind drain outlets and means to avoid settlement of embankment foundations due to thawing. The primary achievements in this area are as follows.

Designing Road Foundations in Sectors of Thick-Layer Ground Ice

In the area from the Kunlun Shan to Tanggula Shan on the Qinghai-Xizang Plateau, thick-layer ground ice is widely distributed on mountain and hill slopes as well as on the high plains, with single-layer thicknesses up to 4 m (Qinghai-Xizang Highway Research Group 1983). Road foundations in these sectors frequently subside and collapse due to thawing. Through in-situ tests (Figure 1) and laboratory experiments, the Northwest Institute and the First Railroad Survey and Design Institute have successfully addressed the following subjects as they affect the design of railroad foundations in sectors of thick-layer ground ice (Shu and Huang 1983): (1) the principles of design, (2) the sectional form of cut and engineering measures, (3) design and preparation of minimum cuts and fills and low embankments, (4) methods for calculating the thickness of the thermal insulation layer and the artificial table of an embankment, and (5) drainage system design.

FIGURE 2 Heat-preserving outlets.

FIGURE 3 Wide, shallow, large-sectioned drainage ditch.

FIGURE 4 Thermal-insulating revetment road without soil covering.

FIGURE 5 Thermal-insulating revetment road covered with soil and tatau grass.

Heat-Preserving Water-Discharge Outlets

To control the formation of ice mounds in the permafrost areas of the Great Hinggan Mountains, blind drains are often used to discharge water. Since the outlets were inevitably subject to freezing, in the past a small cabin containing a burning furnace was built as a shield. The Third Railroad Survey and Design Institute designed and built seepage blind drains with cone-shaped, heat-preserving outlets that do not freeze (Figure 2). Recently, the original design has been further improved (He and Li 1983). This achievement not only reduces project costs, but facilitates transportation and management. It has been widely and successfully used in the design of railroad foundations, tunnels, and water-supply and drainage outlets.

Comprehensive Treatment of Embankment Subsidence

Embankment subsidence is a major hazard for railroad foundations in permafrost regions, directly affecting the quality and safety of the roadbed. The Qiqihar Railway Administration built a wide, shallow, large-section drainage ditch near an embankment on the Ya-Lin Railroad (Figure 3), and constructed a thermal-insulating, single-slope revetment road covered with tatau grass between the railroad and the drainage ditch. The permafrost table under the embankment and the road isolated the water and heat sources of the thawing trough of the embankment foundation from ground water (Figures 4 and 5). This gradually reduced the thickness of the weak layers of the foundation and prevented the embankment from sinking (Wang 1980).

DESIGN AND CONSTRUCTION OF BRIDGE AND CULVERT FOUNDATIONS

In China, smaller bridges and culverts in permafrost regions are subject to damage, so the study of their design and construction is stressed. Our major achievements are described below.

Raft and Wood-Reinforced Culvert Foundations

There are many smooth valleys in the permafrost regions of the Great Hinggan Mountains that are covered by tatau grass and brush; they store water all year round. Not far beneath the ground surface there is a layer of peat 0.4-1.0 m thick, below which is a layer of clayey soil mixed with gravel up to 5-6 m thick. Culverts built in such areas are frequently subject to such hazards as crushing of the end and side walls, staggering or pulling of pipe sections, and subsiding and bowing. To overcome these hazards, the Third Railroad Survey and Design Institute has used raft or wood-reinforced continuous foundations (Third Railroad Survey and Design Institute 1971). The raft foundation preserves the permafrost and adapts to slightly uneven subsidence, and wood-reinforced foundations can adapt to uneven subsidence if thawing ground has been allowed for in the design. The culverts constructed according to these principles have been successfully tested for 7 or 8 years.

Reinforced Concrete Pile Foundations

A pile foundation is the optimal type of foundation for bridges and buildings in permafrost regions, because it not only causes minimal damage to the frozen state of the underlying soil, overcoming frost heaving as well as deformation caused by thawing, but it also can speed construction, lower costs, and improve working conditions. Reinforced concrete pile foundations appear to suit Chinese conditions better than any other type, so the Northwest Institute, the First Railroad Survey and Design Institute, and the Low-Temperature Construction Institute of Heilongjiang Province have studied several subjects (Chen 1982): (1) the construction conditions, technology, and equipment necessary to emplace inserted, reinforced concrete, driven, and poured piles, (2) a cost-effective method of calculating a beam's capacity for use in field tests, and (3) a method for determining horizontal and vertical loads (bearing capacity and thrust) on pile foundations in permafrost (Figure 6).

Rapid Construction of Foundation Pits for Small Bridges and Culverts

To maintain the underlying soil in the frozen state, foundations for small bridges and culverts must be excavated as quickly as possible. After conducting a number of tests on various types of excavation blasts, the Railroad Construction Institute of CARS developed techniques for designing foundation excavation blasts in permafrost (Qinghai-Xizang Railroad Research Group of Testing Project 1976).

FIGURE 6 Pile foundation test site in Kunlun Shan.

FIGURE 8 Drilled pile foundation test structure.

FIGURE 7 Sluiceway blocked by ice.

FIGURE 9 Buried pile foundation test structure.

Live demolition blasts on the Qinghai-Xizang Plateau showed that the goal of rapid construction was achieved. Upon evaluation, the method was approved for use in the excavation of foundation pits for small bridges and culverts in permafrost regions.

DRAINING WATER AND MELTING ICE IN TUNNELS

A major problem of tunnel design in permafrost areas is water drainage. Ground water leaks into the tunnel and freezes, causing the liner to crack, break, and peel off, and ice accumulates on the foundation and hangs from the ceiling. This not only damages the tunnel, but also impedes and even endangers trains. The Third Railroad Survey and Design Institute has used various devices, such as deeply buried, shallowly buried, and frost-heave-protected ditches and sluiceway drainage systems to eliminate hazards to tunnels in permafrost areas (Nie 1981). The sluiceway drainage system has proved to be the most efficient. It uses a small gallery beneath the tunnel along with shafts, water-discharging borings, and natural fissures to collect and drain ground water from around the tunnel. The complete drainage system includes an inspection well and a cone-shaped, heat-preserving outlet.

In permafrost areas ground water often freezes inside sluiceways (Figure 7). For example, an emergency occurred in February 1973 in a tunnel in the Great Hinggan Mountains---freezing had blocked the sluiceway, causing freezing in the tunnel and endangering traffic. Various methods were used to try to clear the blockage--manually shaving the ice and melting it with electricity, with locomotive steam, and by burning coal--but they were unsuccessful because of the large amount of heat required. In 1977 tests, the Qiqihar Railway Administration succeeded in melting ice by mechanical ventilation. Expanded experiments were conducted in 1978; over three separate periods of time, a cumulative ventilation of 23 days and nights successfully melted as much as 1400 m^3 of ice. This has proved to be an economical and effective way to treat freezing in tunnels.

BUILDING DESIGN AND CONSTRUCTION

Subsidence due to thawing and deformation due to frost heaving frequently affect buildings constructed in permafrost areas because of poor choice of location or inappropriate design of the bases and foundation. Research has been conducted to determine the suitability of various kinds of foundations (Figures 8, 9, 10, and 11) to permafrost regions and to develop a method for determining the shape and dimension of the thawing plate beneath heated buildings (Ma and Wang 1983, Zhao and Wang 1983). On the basis of engineering practice as well as several years of testing and research, the First and Third Railroad Survey and Design Institutes, the Qiqihar Railway Administration, and the Northwest Institute offer a set of design principles and methods: (1) proper choice and layout of construction site, (2) proper base and foundation design, (3) protecting foundations from frost heaving, (4) proper design of structures above ground, and (5) proper design of aprons.

FIGURE 10 High-filled, ventilated foundation test structure.

FIGURE 11 Girth foundation test structure.

FIGURE 12 Water-supply system with heat-preserving pipes on ground and outside cistern.

FIGURE 13 Joining culvert foundations.

FIGURE 14 Joining culvert box pipes.

DESIGN AND CONSTRUCTION OF WATER-SUPPLY PIPELINES

Since the founding of the People's Republic of China, 25 railroad water stations have been established in the northeastern permafrost region. More than half of these are located in discontinuous and continuous permafrost areas. At first, varying degrees of freezing or pipeline breakage occurred at almost every station. So as early as the mid-1960's, the Third Railroad Survey and Design Institute began to search for a new method of constructing water-supply pipelines by using pipes of various materials, various heat preservation materials, and various methods of laying pipelines (Figure 12) (He 1983). After 20 years of continuous engineering practice, testing, and research, the problem of design and construction of water-supply pipelines in permafrost areas of the Great Hinggan Mountains has been solved.

LOW-TEMPERATURE CONSTRUCTION MATERIAL

The success of trial production of low-temperature, early-strengthened concrete makes it possible to adapt reinforced poured-concrete pile foundations to permafrost regions.

Many methods for producing low-temperature and negative-temperature concrete have been used in China, but there is little complete or systematic information, especially about freeze- and thaw-resistant indexes. The Railroad Construction Institute of CARS has succeeded in developing three types of low-temperature early-strengthened concrete, whose low-temperature and negative-temperature strength, compared with those of ordinary concrete, are 30-67% stronger over the same time using the same amount of cement; cement consumption can be reduced by 100 kg/m^3 without reducing the strength. The freeze- and thaw-resistant indexes when the concrete is soaked in fresh water or salt water solutions of 0.1%, 1.0%, and 5.0% are all over M-200, which is 4 to 11 times those for ordinary concrete without additional chemicals. The predicted lifetime of low-temperature concrete structures is about 50 years. Another advantage of this concrete is that the temperature of the mixture can be controlled as low as 0°C so that its thawing effect on permafrost is greatly reduced (Railroad Construction Institute 1979, Shi 1978). This concrete is suited for use in regions where ground temperatures range from 0 to 3°C, so it may be used in frigid zones as well as in permafrost regions. It has been recommended that the technical construction specification be incorporated into the "Standards of Construction of Railroad Engineering Concrete and Reinforced Concrete."

An inexpensive and easy-to-use negative-temperature cement has been developed to cement joints, so that culverts can be constructed at temperatures below -25°C (Li and Wu 1981).

One of the subjects of our research into railroad construction in frigid and permafrost areas is the use of prefabrication techniques in bridge and culvert construction (Figures 13 and 14). Constructing prefabricated, reinforced concrete structures in regions with temperatures below -25°C requires a cementing material (for joints) that has the advantages of being fast developing, is long lasting, has strength at negative temperatures, shows little shrinkage, and has no corrosive effect on reinforcing bars. The Railroad Construction Institute of CARS was commissioned to undertake this work and has fulfilled its charge. The material has a compressive strength that approximates 300 kg/cm^2 in three days, it cannot be damaged even when subjected to freeze (-18 to -25°C) shortly after it is poured, and it has a freeze- and thaw-resistant strength up to M-150. Its other features are: (1) low drying shrinkage, about one-fourth that of concrete containing ordinary dry-to-harden cement, (2) no corrosive effect on reinforcing bars, (3) resistance of sulphate corrosion, (4) long-term stability and strength, and (5) cementing strength, tensile strength, and modulus of elasticity with reinforcing bars to meet the railroad design standards.

The material was first used in a test bridge in the northwestern frigid zone in August 1976, and then in two prefabricated bridges in the same zone in December 1977. The construction quality was excellent and these three bridges have been used safely for 4 and 5 years. Since 1978, the material has been used in prefabricated winter projects throughout northern China. The material has passed inspection and is approved for use in frigid-zone railroad engineering and similar projects. According to the evaluation, it is suggested that the temporary detailed construction specifications for using the material be incorporated into the next revision of the "Standards of Construction of Railroad Engineering Concrete and Reinforced Concrete."

SUMMARY

This has been a brief description of research into railroad construction in the permafrost regions of China. Some of the work has been successfully completed, but a number of technical problems remain to be studied and solved, including the prevention and treatment of frost heaving of road and building foundations, the design of light and pliable culverts, further improvement of the design of reinforced concrete pile foundations, and the design of shallow ditches with frost-heave protection for tunnels and buildings constructed on warm permafrost ground.

REFERENCES

Chen Zhuohuai, 1982, Experimental research on pile foundation in permafrost, in Symposium of Glaciology and Cryopedology Research Conference: Scientific Publishing House.

Cui Chenghan and Zhou Kaijiong, 1983, Study of various laws of tangential frost heaving pressure, in Symposium of Second National Conference on Permafrost: Gansu People's Publishing House.

Dai Jingbo and Li Enying, 1981, Law of temperature variation of permafrost in the northern part of the Great Hinggan Mountains.

Ding Dewen, Ma Xiaowu, and Lou Xuebao, 1979, A method for calculating the thickness of the thermal insulation layer using a one-dimensional model, in Collected Publications of Lanzhou Institute of Glaciology and Cryopedology, no. 1: Scientific Publishing House.

Ding Jingkang, 1983, Research on horizontal frost heaving pressure, in Symposium of Second National Conference on Permafrost: Gansu People's Publishing House.

Ding Jingkang and Zhang Luxin, 1983, Research on ultimate long-term resistant forces of anchor arm in permafrost, in Symposium of Second National Conference on Permafrost: Gansu People's Publishing House.

First Railroad Survey and Design Institute, Northwest Institute of Chinese Academy of Railway Sciences, and Lanzhou Institute of Glaciology and Cryopedology, 1972, A study on constructing railroads in the permafrost of the Qinghai-Xizang Plateau.

He Jie, 1983, Experimental research on methods of laying water supply and drainage pipes in permafrost regions, in Symposium of Second National Conference on Permafrost: Gansu People's Publishing House.

He Jie and Li Enging, 1983, Cone-shaped, heat-preserving outlet and analysis of its thermodynamic regime, in Symposium of Second National Conference on Permafrost: Gansu People's Publishing House.

Huang Xiaoming, 1981, The seasonal freezing and thawing characteristics of soil layers in the permafrost region of the Qinghai-Xizang Plateau: Northwest Institute of Chinese Academy of Railway Sciences.

Li Qili and Wu Suhua, 1981, Material for cementing structures under negative temperature construction: Railroad Science and Technique, Maintenance Engineering, no. 10.

Lin Fengton, 1981, Characteristics of phreatic water of river valleys in permafrost regions of Great and Minor Hinggan Mountains and methods for seeking ground water.

Liu Tieliang, 1983, A method for determining the depth of the permafrost table and its application to cutting engineering, in Symposium of Second National Conference on Permafrost: Gansu People's Publishing House.

Ma Zonglong and Wang Ziyuan, 1983, Studies of adaptability and design principles of building foundations in permafrost regions of Qinghai-Xizang Plateau, in Symposium of Second National Conference on Permafrost: Gansu People's Publishing House.

Nie Fengming, 1981, Tunnel engineering in cold regions: Third Railroad Survey and Design Institute.

Qinghai-Xizang Highway Research Group, 1983, Distribution laws and recognition marks of permafrost with rich ice content along Qinghai-Xizang Highway, in Symposium of Second National Conference on Permafrost: Gansu People's Publishing House.

Qinghai-Xizang Highway Research Group, 1981, The height of embankment with asphaltic pavement in permafrost regions of Qinghai-Xizang Highway.

Qinghai-Xizang Railroad Research Group of Testing Project, 1976, Experimental research on excavating foundation pits for bridges and culverts with demolition in permafrost regions.

Railroad Construction Institute of CARS, 1979, Technical specifications for designing and constructing bridges and culverts in permafrost regions of the Qinghai-Xizang Plateau.

Shi Renjun, 1978, Research on low-temperature early-strengthened concrete for bridges and culverts in permafrost regions of the Qinghai-Xizang Plateau.

Shu Daode and Huang Xiaoming, 1983, Design and construction of cuttings in massive ground ice sectors, in Symposium of Second National Conference on Permafrost: Gansu People's Publishing House.

Third Railroad Survey and Design Institute, 1971, Design summary of bridges and culverts on Nun-Lin railroad.

Tong Zhiquan, 1983, Experimental research on large-scale direct shear test for the freeze-thaw interface and thawed soil in field, in Symposium of Second National Conference on Permafrost: Gansu People's Publishing House.

Wang Wenbao, 1980, Preliminary research on method to prevent road foundations from subsiding in permafrost regions: Railroad Science and Technique, Maintenance Engineering, no. 2.

Wu Ziwang, Shen Zhongyan, Wang Yaqing, and Zhang Jiayi, 1979a, Experimental research on adfreezing strength between permafrost and foundation, in Collected publications of Lanzhou Institute of Glaciology and Cryopedology, no. 1: Scientific Publishing House.

Wu Ziwang, Shen Zhongyan, Zhang Jiayi, and Wang Yaqing, 1979b, Experimental research on tangential frost heaving pressure acting on foundations during freezing, in Collected Publications of Lanzhou Institute of Glaciology and Cryopedology, no. 1: Scientific Publishing House.

Wu Ziwang, Zhang Jiayi, Shen Zhongyan, and Wang Yaqing, 1979c, Preliminary research on thawing compression property of permafrost, in Collected Publications of Lanzhou Institute of Glaciology and Cryopedology, no. 1: Scientific Publishing House.

Xu Xiaozu, Fu Liandi, and Zhu Linnan, 1979a, An empirical formula to determine the depth of the permafrost table using elevation and latitude values, in Collected Publications of Lanzhou Institute of Glaciology and Cryopedology: Scientific Publishing House.

Xu Xiaozu, Tao Zhaoxiang, and Fu Sulan, 1979b, Research on thermodynamic properties of typical thawing-freezing soils, in Collected Publications of Lanzhou Institute of Glaciology and Cryopedology, no. 1: Scientific Publishing House.

Zhao Yunlong and Wang Jianfu, 1983, Computation of the thawing depth for foundations of heated buildings in permafrost regons, in Symposium of Second National Conference on Permafrost: Gansu People's Publishing House.

PERIGLACIAL SLOPEWASH AND SEDIMENTATION IN NORTHWESTERN GERMANY DURING THE WÜRM (WEICHSEL-) GLACIATION

Herbert Liedtke

Geographisches Institut, Ruhr-Universität Bochum,
D-4630 Bochum, Federal Republic of Germany

The older morainic landscape of the Northwest German Lowland has a very subdued relief. This was caused by Weichsel periglacial processes, partly of a solifluidal but mostly of an ablual kind. The latter encompasses all eroding sheetwash processes on slopes and accumulation of fine sandy material in basins. Abluation and solifluction were processes during cold periods in the foreland of central European glaciations; erosion and sedimentation occurred mainly during the end of interstadial phases. The most impressive incision happened at the very end of the last interglacial.

THE SUBDUED RELIEF OF CENTRAL EUROPE'S OLDER MORAINIC LANDSCAPE

In the Northwest German Lowland two geomorphologically different types of glacial landscape can be distinguished: young morainic and older morainic landscape.

The young morainic landscape is characterized by fresh forms of relief which clearly bear the marks of the about 20 000 years old Weichsel glaciation and whose topography is so distinct that it is east of Kiel called "Holsteinian Switzerland"; the older morainic landscape shows, deposited during the Saale (Riss) glaciation, a totally subdued relief, which is characterized by low gradients of slope and wide, almost flat plains underlain by Pleistocene deposits (Figure 1). Large differences in altitudes like those so conspicuous in the young morainic landscape are absent, and the endmoraines originally existed are so subdued that different investigators disagree as to their positions. The boundaries of ice-marginal channels (Urstromtäler) against neighbouring terrain are also difficult to recognize. Even the boundaries of outwash plains have to be deduced and their connection with ice-marginal positions is in most cases not clear. On the whole one can speak of a terrain in which glacial deposits are still present, but glacial relief is almost completely extinguished (Liedtke 1981b).

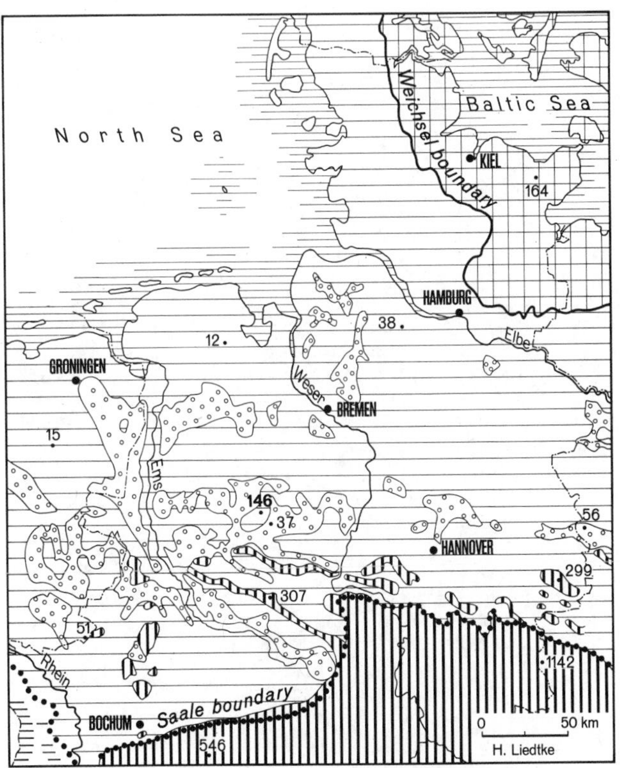

FIGURE 1 Northwestern Germany's young and older morainic landscapes and the major basins filled up by ablual sediments during the Weichsel glaciation

Young morainic landscape (Weichsel = Würm)
Outcrop of hard rock
Old morainic landscape (Saale = Riß)
Area of Rhine terraces
Major basins filled up mainly by ablual sediments during the Weichselean
146 Damme Hills

Why have the main features of glacial relief disappeared in the older morainic landscape? Why are glacial channels and tonguelike basins no longer recognizeable? Why is the landscape so extraordinarily flat? The answer is easy: it was the periglacial climate with permafrost and lack of vegetation which created these conditions and caused a flattening of relief in the region adjacent to the inlandice of the last glaciation.

STRATIGRAPHIC EVIDENCE FOR INCISION AND ABLUAL ACCUMULATION

In general, solifluction is made responsible for this erosional process. But, as known from areas where hard rock is underlying a thin glacial cover, the effectiveness of solifluction is limited and the main features of relief have been altered comparatively little. Apart from that there are large areas underlain by sandy-gravelly sediments which, due to their lack of fine material, were hardly liable to solifluction. The "sanding up" of areas with Saale ground moraine shows that here not only a simple transport of material occurred but also sorting: blocks and coarse sand remained, whereas finer sand, silt, and clay were carried away. Therefore, slopewash and not solifluction played the main role in flattening the glacial relief in older morainic landscapes. Fine materials such as clay, silt and fine sands were washed off the land surface and carried away during the melting period every year whereas the coarser material remained. For this process I propose the term "abluation" (from Latin "abluere", to wash off), which is understood as a periglacial aquatic process of wash on slopes, with deposition of this material at the bottom of such slopes where the inclination decreases. For northwestern Germany this process explains the filling up of wide glacial basins through which scattered small islands of ground moraine protrude. The sediments were transported in periglacial-fluvial valleys, which can be subdivided into three parts: the upper reaches with denudational abluation, the middle reaches where material was transported through, and the lower reaches and foreland with depositional abluation.

Evidence on which these statements are based, were found at the Damme Hills in Northwestern Germany southwest of Bremen, which belong to the typical older morainic landscape. These Damme Hills, 20 km long and 10 km broad, consist of probably pre-Saale pushed glacial sands and gravels (Stauchmoräne) which were formerly covered with a thin layer of Saale till or ground moraine. Due to the glacial push, the relief was very hilly and has still today a relief of about 100 m between the highest point (146 m) and the lowest (38-50 m). During the entire Weichsel glaciation, this area was exposed to a periglacial climate, which was the reason that the covering Saale ground moraine was mostly removed and only erratic boulders were left. In the push area selective erosion took place; valleys were cut in the sandy parts whilst the gravelly structures formed ridges, because they acted as a resistant layer. Geomorphological mapping shows that eroded material was transported partly along former glacial channels (Galbas, Klecker & Liedtke 1980).

These valleys with a graded longitudinal profile have a flat-floored cross-profile and show a very uniform stratigraphy:

0.5-0.8 m sandy-gravelly blanket, unsorted (Geschiebedecksand);

2.0-3.0 m well-sorted fine sands in the upper reaches, sandy-gravelly sands in the middle reaches;

0.5-1.0 m basal conglomerate;

sharp unconformity;

Saale layers;

The basal conglomerate contains flat blocks of Bunter Sandstone up to a diameter of 0.5 m in a thalweg that is 5-10 m wide. This shows that a rapid discharge occurred during the time of incision even in the upper reaches of small valleys. All coarse clasts, bigger than the transport capacity allowed sank down and were moved only occasionally. The basal conglomerate became packed so densely that it can be exposed only by a pickaxe.

The erosion phase is followed by a sedimentary accumulation of several meters. On the top we find an unsorted sandy-stony cover, called "Geschiebedecksand", which is thought to have been formed under periglacial conditions (Kopp 1970, Lembke 1965).

The material that reached the valleys was partly sedimented in valley floors and mostly in basins of the lowlands, where often more than 5-10 m were accumulated. It is typical and remarkable that nearly always fine or medium sands were deposited, as an example in a large sand pit at Holdorf station shows. Within the uppermost 5 m of horizontally stratified sandy layers I found only three beds with scattered small gravels up to 3 cm in diameter, though the edge of the Damme Hills is nearby (1 km). Organic layers are totally lacking (Liedtke 1981a). Several syngenetic frost cracks are proving an accumulation under cold periglacial conditions (Karte 1979).

Such very common accumulation plains, consisting mostly of medium to fine sands without gravels, demonstrate an activity of water, but are not the typical deposit of a river. The slight gradient of 2-3 ‰ allowed only the transport of fine sand on the top of the permafrost during the seasonal melting of winter snow when the water ran down in a kind of sheet discharge. Coarse material was left behind, sands were deposited in the basins, and only silt and clay could reach the North Sea. According to these processes, all the basins were filled up and sediment spread more and more, drowning the adjacent flat glacial landscapes. Geologists, not recognizing these processes, mapped these sandy areas as "Talsand" (valley sands). The borders to the adjacent higher areas are often very uncertain, especially where during the end of the Weichsel cold period eolian coversands or dunes were sedimented or where Holocene low fens or raised bogs developed.

THE AGE OF THE MOST ACTIVE PROCESSES SUBDUING THE TYPICAL GEOMORPHOLOGICAL FEATURES OF THE SAALE GLACIATION

As the original constructional topography of this area was formed during the Saale (Riss or Illinoian) glaciation, we now have to solve the question whether reduction of the relief and aggradation in the basins already began during the end of the Saale glaciation. We cannot expect considerable denudation at that time, because in the young morainic landscape we find a merely changed rugged glacial relief with a lot of typically closed depressions still visible today. On the other hand, we have direct evidence that the aggradation in depressions did not start earlier than the beginning of the Weichsel glaciation, as Menke (1976) showed in

Schleswig-Holstein and Lembke et al. (1970) in Brandenburg (Figure 2).

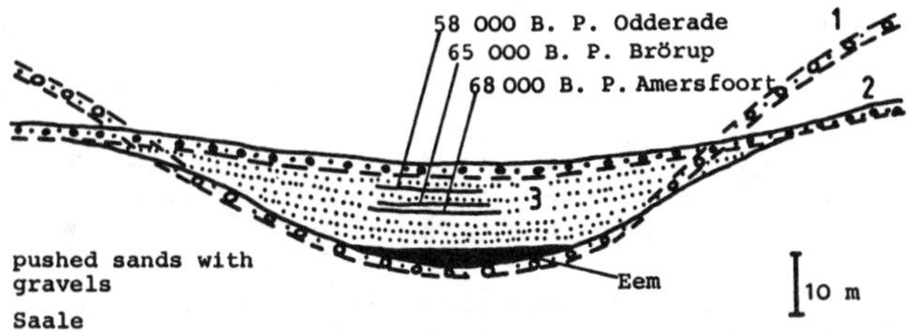

1 sandy-gravelly blanket, unsorted, Saale
2 sandy-gravelly blanket, unsorted, Weichsel
3 ablual sandy sediments, well stratified, Weichsel

Figure 2 Schematic representation of Weichsel ablual filling-up of closed depressions.

For in the older morainic landscape nearly all glacial lakes have disappeared and most areas have become drained by valleys, it is surprising that thousands of former kettle holes have survived. There are more than 1000 of them on the Fläming (south of Berlin, Liedtke 1960), and a special mapping showed that in Lower Saxony more than 10 000 closed depressions are still existing; partly they were explained as pingos (Garleff 1968). Most of them are glacial depressions that were not totally filled up by post-Saale or Weichsel periglacial processes. They are recognizable in the landscape as depressions with a depth of 1-2 m and a diameter of up to 250 m. Sometimes they are filled with water. On topographical large-scale maps they are identified by a special sign (arrow).

Why are we very sure that these depressions were filled up not earlier than during the Weichsel cold period? There are 3 indications:

1. Many of these depressions contain an interglacial Eem (Sangamon) layer at their bottom (Figure 2).

2. The Eem layer covers the typical late Saalean "Geschiebedecksand", a kind of sediment which we can find on all young morainic deposits as well.

3. The upper parts of the aggradated sands in some of the depressions contain Weichsel interstadial layers. They allow to fix the time of aggradation closely, and they demonstrate that about 80 % of the filling up happened during the early Weichsel (Figure 3).

So we can fix two processes very accurately: The dominant aggradation happened during the cold stages of early Weichsel by ablual sheetwash and was preceded by strong erosive incision. Without any doubt, it was the geographical milieu that offered the best conditions for these two processes: it was the transitional phase from the last interglacial to the last glacial. At that time everywhere in central Europe an erosion phase took place and was connected or immediately followed by aggradation in basins and lowlands. It is now proved by recent research in periglacially formed areas of central Europe (Menke 1976) that this erosive phase introduced the beginning of the Weichsel cold period. This idea can also be applied to former cold periods of the Quaternary, although no stratigraphic evidence has been found that would demonstrate this.

We now know that the filling up of depressions

STRATIGR. (LIEDTKE 1981)	C^{14}- TIME	JULY-AVERAGE (°C)	INTERSTADIAL (10^3)	VEGETATION IN ADJACENT NETHERLANDS (WIJMSTRA 1978, ZAGWIJN 1974)	
MID - WEICHSEL	50 000	6°			tundra, treeless
EARLY					subarctic with pine & birch
		14°	Odderade ca. 58		pine-birch forest with oak, hazel
	60 000	8°			subarctic with pine & birch
WEICHSEL		16°	Brörup ca. 65		spruce-pine-birch forest, alder
		10°			subarctic with pine & birch
		15°	Amersfoort ca. 68		pine forest with oak, alder & spruce
		11°			subarctic parklandscape, pine & birch and heather
EEM	>70 000	16°			pine-birch forest with spruce
					pine-spruce forest
		18°			spruce forest with elm, hazel & hornbeam

tundra · birch · pine · oak · hazel · spruce · alder · elm

Figure 3 Vegetation succession during the early Weichsel.

began with the first cold forest climate. That means that the main erosive phase took place during the very end of the Eem interglacial, and we can conclude:

> The very end of an interglacial or a warm interstadial started strong erosion in hilly areas and accumulation in basins.
>
> During a stadial, accumulation filled up the valley bottoms according to solifluction and (or) slope wash.

During every end of the early Weichsel interstadials, after the vegetation had returned and the area was reforested, a phase of steady erosion occurred. To cause this erosion a special fluvial process is needed: a seasonal sudden heavy discharge produced by melting of winter snow and favored by winter frost in the surface layers. There was no continous permafrost needed, but it was in formation and expansion; and there was still a thick mollisol that together with a large amount of running water allowed erosive incision, which was increased by the eroded material as an attacking weapon. Big blocks remained more or less stationary at the bottom; they were only projected downwards to form the basal conglomerate.

During the cold tundra periods accumulation took place in valleys and basins. The material came from the slopes; on sandy-gravelly slopes abluation prevailed, on sandy-silty slopes solifluction was usual. Both of the processes supplied the valleys with fine or coarse material. The wetter the conditions were the more material reached the valleys. During the extremely dry High (Main) Weichsel the loess cover could develop.

Unfortunately, sand pits with interglacial or interstadial layers are very rare. Here was shown the ablual filling up of a glacial depression during the begin of the early Weichsel. But also during the middle Weichsel abluation proceded, and sandy material was ablually deposited in the basins. A temporary sand pit at Lemförde, 8 km southeast of the Damme Hills, showed more than 8 m of horizontally stratified sandy layers with four cryoturbated beds. Two of these beds were recently C^{14}-dated (Nr. 11770, 11771 by Niedersächsisches Landesamt für Bodenforschung, Hannover, April 6, 1983); their age of 35 060 ± 870 B. P. and 36 030 ± 650 B. P. lies between the Hengelo and Denekamp interstadials and indicates that also during the middle Weichsel ablual processes were active under periglacial conditions with permafrost.

So, ablual processes happened under full periglacial permafrost conditions as well as under conditions when the formation of permafrost began and a long duration of seasonal frost existed.

We must keep in mind that during the very end of the Eem interglacial, erosion took place, which was accompanied and followed by a mighty accumulation during the first tundra periods of the early Weichsel. There were only about 10 000 years of tundra climate in which this accumulation occurred during early Weichsel (Liedtke 1981b) and filled up nearly most closed depressions, left from the Saale glaciation; at that time also began a mighty sedimentation in the lower parts of Saale-aged glacial basins.

REFERENCES

Galbas, P. U., Klecker, P. M. and Liedtke, H., 1980, Erläuterungen zur geomorphologischen Karte der Bundesrepublik Deutschland 1 : 25 000, GMK 25, Blatt 3415 Damme: Berlin, 48 p.

Garleff, K., 1968, Geomorphologische Untersuchungen an geschlossenen Hohlformen ("Kaven") des niedersächsischen Tieflandes: Göttingen, Göttinger Geographische Abhandlungen, Heft 44, 142 p.

Karte, J., 1979, Räumliche Abgrenzung und regionale Differenzierung des Periglaziärs: Paderborn, Bochumer Geographische Arbeiten, Heft 35, 211 p.

Kopp, D., 1970, Kryogene Perstruktion und ihre Beziehung zur Bodenbildung im Moränengebiet.: Gotha, Petermanns Geographische Mitteilungen, Ergänzungsheft Nr. 274, p. 269-279.

Lembke, H., 1965, Probleme des Geschiebedecksandes im Jung- und Altmoränengebiet: Berlin, Berichte der Geologischen Gesellschaft der DDR, 10, p. 721-726.

Lembke, H., Altermann, M., Markuse, G. and Nitz, B., 1970, Die periglaziäre Fazies im Alt- und Jungmoränengebiet nördlich des Lößgürtels: Gotha, Petermanns Geographische Mitteilungen, Ergänzungsheft Nr. 274, p. 213-268.

Liedtke, H., 1960, Geologischer Aufbau und geomorphologische Gestaltung im Fläming: Bad Godesberg, Berichte zur deutschen Landeskunde, 26, p. 45-81.

Liedtke, H., 1981a, Führer für die Exkursion in das Gebiet des Dümmers am 9. - 10. Oktober 1980: Paderborn, Bochumer Geographische Arbeiten, Heft 40, p. 97-137.

Liedtke, H., 1981b, Die nordischen Vereisungen in Mitteleuropa. Erläuterungen zu einer farbigen Übersichtskarte 1 : 1 000 000: Trier, Forschungen zur deutschen Landeskunde, Band 204, 307 p.

Menke, B., 1976, Neue Ergebnisse zur Stratigraphie und Landschaftsentwicklung im Jungpleistozän Westholsteins: Öhringen, Eiszeitalter und Gegenwart, Band 27, p. 53-68.

Wijmstra, T. A., 1978, Palaeobotany and climatic change, in Gribbin, J., ed., Climatic change: Cambridge, p. 25-45.

Zagwijn, W. H., 1974, Vegetation, climate and radiocarbon datings in the Late Pleistocene of the Netherlands, Part II Middle Weichselian: Haarlem, Mededelingen Rijks Geologische Dienst, N.S. 25 (3), p. 101-110.

A COLLOCATION ANALYSIS FOR SOLVING PERIODIC THAWING AND FREEZING PROCESSES IN ACTIVE LAYERS OF PERMAFROST

Sui Lin and Shimao Ni

Department of Mechanical Engineering, Concordia University
Montreal, Quebec, Canada H3G 1M8

The periodic variation of the ground surface temperature causes alternative thawing and freezing processes in the active layer of the permafrost. The ground surface is an important location of the present problem, which is selected as the collocation point of the analysis. Analytical results are compared with those obtained from the zeroth-order and the first-order quasi-steady approximations, and the first-order perturbation method. The accuracy of the collocation analysis is not less than that obtained from the first-order perturbation approximation.

INTRODUCTION

Due to the seasonal variation of temperature, alternating thawing and freezing processes take place in the active layer of permafrost. When the ambient temperature increases above the melting point of the permafrost, melting starts at the ground surface and propagates downward into the permafrost. There co-exist two phase regions: liquid and solid. When the ambient temperature decreases to below the melting point of the permafrost, the liquid phase solidifies starting also from the ground surface and propagates downward. There co-exist three phase regions: solid, liquid, and solid. When the solidification process reaches the end of the liquid phase, the liquid region disappears, and it becomes one solid phase region again.

The prime characteristic of the melting and solidification processes is the existence of moving boundaries between the solid and liquid phases, whose locations are unknown and which are to be determined as a part of the solution. The moving boundary problem is nonlinear and solutions involve considerable mathematical difficulties. Exact solutions to such problems are presently restricted to only a few idealized situations [1]. With regard to the nonlinear problem, solutions have been obtained by analytical approximations and numerical methods (see, for example, [2] and [3]).

For the periodic thawing and freezing processes, we consider that the ground surface is subject to a temperature variation with a harmonic function varying around the melting temperature of the permafrost. The surface temperature variation causes steady periodic thawing and freezing processes in the active layer of the permafrost. For solving such a nonlinear problem, the first-order perturbation approximation and a numerical analysis were conducted by Lock et al. [4], the zeroth-order quasi-steady approximation was used by Seban [5], and the first-order quasi-steady approximation was used by Lunardini [1]. In the present paper, an approximate analysis utilizing the method of collocation is developed for solving the steady thawing and freezing processes in the active layer of the permafrost. The result is compared with results obtained by Lock et al., Seban, and Lunardini.

MATHEMATICAL FORMULATION AND ANALYTICAL APPROXIMATION

Consider a permafrost region having its melting point T_m. Initially the active layer is at the melting temperature and in solid state. After a definite time the surface at x = 0 is subject to a periodic temperature variation, $T_{x=0} = A_0 \sin \omega t + T_m$, where A_0 is the amplitude and ω is the frequency. We wish to find the response of the active layer to this periodic temperature disturbance for large values of time, that is, the steady periodic solution of the problem, for which the initial condition may be ignored. During the time period

$$\frac{2n\pi}{\omega} < t < \frac{(2n + 1)\pi}{\omega} \quad (n = 1, 2, 3, \ldots)$$

because the surface temperature is higher than the melting temperature of the permafrost, a melting process propagates from the surface downward into the active layer of the permafrost. In this case, there co-exist two phase regions: liquid and solid. During the time period

$$\frac{(2n + 1)\pi}{\omega} < t < \frac{(2n + 2)\pi}{\omega}$$

because the surface temperature is lower than the melting temperature of the permafrost, a soldification process propagates from the surface downward into the liquid phase of the active layer of the permafrost. There co-exist therefore three phase regions: solid, liquid, and solid. When the solidification process reaches the end of the liquid phase, the active layer becomes solid. At that time the three phase regions disappear and return to one phase region.

For mathematical formulation of the problem, we designate the initial complete solid phase as region 0, the liquid phase formed during the melting process as region 1, and the solid phase formed during the freezing process as region 2. As described above, after the liquid phase disappears, region 2 becomes region 0 as shown in Figure 1, where s_1, s_2 represent the locations of the phase boundaries between regions 0 and 1, and between regions 1 and 2 , respectively. To simplify the problem, it is assumed that material properties in each phase are constant, but they may be different in different phases and that density change across the phase boundary may be neglected. Furthermore, it is assumed that the latent heat of fusion of the permafrost is much larger than its sensible heat. Under this assumption, the following approximations may be made:

1. After the time, $t = t^*$, when the liquid phase disappears, or when $s_2 = s_1 = s^*$, it is assumed that the temperature at the position $x = s^*$ is maintained at the melting point.
2. The heat flux in the solid phase during the melting process, and the heat flux in the liquid phase during the solidification process, may be ignored.

The following is the system of equations describing the one-dimensional periodic melting and freezing processes with the melting temperature of the permafrost as a reference temperature, that is, $T_m = 0$:

$$\frac{\partial T_i}{\partial t} = a_i \frac{\partial^2 T_i}{\partial x^2} \tag{1}$$

$$T_i\ (o,t) = A_o \sin \omega t \tag{2}$$

$$T_i\ (s_i,\ t) = 0 \tag{3}$$

$$\rho L \frac{ds_i}{dt} = (-1)^i k_i \frac{\partial T_i(s_i,\ t)}{\partial x} \tag{4}$$

where $i = 1$ refers to the melting process for the time period

$$\frac{2n\pi}{\omega} < t < \frac{(2n+1)\pi}{\omega}$$

for $n = 0, 1, 2, \ldots$, and $i = 2$ refers to the freezing process for the time period

$$\frac{(2n+1)\pi}{\omega} < t < \frac{((2n+2)\pi}{\omega}$$

For solving the problem, the method of collocation will be used [6], in which the heat conduction differential equations are satisfied only at preselected points (collocation points). For the present problem, the periodic melting and solidification processes are caused by the periodic variation of the surface temperature. Therefore the most important location of the present problem is at the surface $x = 0$. Hence, we select the point $x = 0$ as the collocation point.

The temperature profile is assumed as

$$T_i = A_i e^{-\lambda_i x/\sqrt{2}} \sin\,(a_i \lambda_i^2 t - \lambda_i x/\sqrt{2}) + B_i x \tag{5}$$

which satisfies the differential equation, Equation (1), at the collocation point $x = 0$, where A_i and λ_i are constant, and B_i is a function of time. By making use of the boundary conditions, Equations (2) and (3), A_i, λ_i, and B_i can be determined as follows:

$$A_i = A_o \tag{6}$$

$$\lambda_i = \sqrt{\frac{\omega}{a_i}} \tag{7}$$

$$B_i = -\frac{1}{s_i} A_o e^{-\lambda_i s_i/\sqrt{2}} \sin\,(a_i \lambda_i^2 t - \lambda_i s_i/\sqrt{2}) \tag{8}$$

The temperature profile can then be written as

$$T_i = A_o \Big[\exp\Big(-\sqrt{\frac{\omega}{2a_i}}\,x\Big) \sin\Big(\omega t - \sqrt{\frac{\omega}{2a_i}}\,x\Big) - \frac{x}{s_i} \exp\Big(-\sqrt{\frac{\omega}{2a_i}}\,s_i\Big) \cdot \sin\Big(\omega t - \sqrt{\frac{\omega}{2a_i}}\,s_i\Big)\Big] \tag{9}$$

The derivation of the temperature T_i, at the position $x = s_i$, can be expressed by

$$\left.\frac{dT_i}{dx}\right|_{x=s_i} = -A_o \exp\Big(-\sqrt{\frac{\omega}{2a_i}}\,s_i\Big)\Big[\Big(\sqrt{\frac{\omega}{2a_i}} + \frac{1}{s_i}\Big) \sin\Big(\omega t - \sqrt{\frac{\omega}{2a_i}}\,s_i\Big) + \sqrt{\frac{\omega}{2a_i}} \cos\Big(\omega t - \sqrt{\frac{\omega}{2a_i}}\,s_i\Big)\Big] \tag{10}$$

Substitution of Equation (10) into Equation (4) yields

$$\frac{d\bar{s}_i}{d\tau} = (-1)^{i+1} \frac{St_i}{2} \frac{\exp(-\bar{s}_i)}{\bar{s}_i} \big[(\bar{s}_i + 1) \sin(\tau - \bar{s}_i) + \bar{s}_i \cos(\tau - \bar{s}_i)\big] \tag{11}$$

where $\bar{s}_i = \sqrt{\frac{\omega}{2a_i}}\ s_i$ (12)

$$\tau = \omega t \tag{13}$$

$$St_i = \frac{A_o c_i}{L} \tag{14}$$

For a seasonal temperature variation, the value of ω and, in turn, the value of $\bar{s}_i$ are very small. For a small value of $\bar{s}_i$, Equation (11) may be approximated, after the expansion of the functions $\exp(-\bar{s}_i)$, $\sin(\tau - \bar{s}_i)$, and $\cos(\tau - \bar{s}_i)$, by neglecting the terms having $\bar{s}_i$ of order higher than two, as follows:

$$\frac{d\bar{s}_i}{d\tau} = \frac{St_i \sin\tau}{2\ \bar{s}_i} \tag{15}$$

Integration of Equation (15) yields

$$\bar{s}_i = \sqrt{2\ St_i}\ \sin\frac{\tau}{2} \tag{16}$$

RESULTS AND DISCUSSION

In order to compare the location of the moving phase boundary obtained from Equations (11) and (16) with those obtained by Lock et al. [4], Seban [5], and Lunardini [2], the coordinates of the moving boundary must have the same scale for all cases. With the following transformation,

$$\bar{s}_i = \sqrt{\frac{St_i}{2}}\ \xi_i \tag{17}$$

Equations (11) and (16) become

$$\frac{d\xi_i}{d\tau} = (-1)^{i+1}\exp\left(-\sqrt{\frac{St_i}{2}}\ \xi_i\right)\left[\left(\sqrt{\frac{St_i}{2}}\ \xi_i + 1\right)\right.$$
$$\sin\left(\tau - \sqrt{\frac{St_i}{2}}\ \xi_i\right) + \sqrt{\frac{St_i}{2}}\ \xi_i$$
$$\left.\cos\left(\tau - \sqrt{\frac{St_i}{2}}\ \xi_i\right)\right] \tag{18}$$

and

$$\xi_i = 2\sin\frac{\tau}{2} \tag{19}$$

The results obtained in [4], [5], and [2] are, respectively, as follows:

$$\xi_i = 2\sin\frac{\tau}{2} - \frac{St_i}{3}\sin\frac{\tau}{2}\sin\tau \tag{20}$$

$$\xi_i = 2\sin\frac{\tau}{2} \tag{21}$$

and

$$\xi_i = 2\sin\frac{\tau}{2} - \frac{St_i}{2}\left[\frac{2}{3} + 2\cos\frac{\tau}{2} - \frac{8}{3}\cos^3\frac{\tau}{2}\right] \tag{22}$$

It is seen that the simplified Equation (19) is identical with Equation (21) obtained from the zeroth-order quasi-steady approximation. By using the fourth-order Runge-Kutta method, Equation (18) can be integrated. Figures 2-4 show the comparison of the results calculated from the collocation method (Equation (18)), the first-order perturbation approximation(Equation (20)), the zeroth-order quasi-steady method (Equation (21)), and the first-order quasi-steady approximation (Equation (22)) with St_i as a paprameter. For a small Stefan number, $St_i = 0.01$, there is practically no difference among these results as shown in Figure 2. However, deviations among these results increase with an increase of the Stefan number as shown in Figures 3 and 4. From the comparison, it can be concluded that the accuracy of the collocation analysis is between those obtained by the first-order perturbation approximation and the first-order quasi-steady method.

NOTATION

a	thermal diffusivity
A_o	amplitude
c	specific heat
k	thermal conductivity
L	specific latent heat of fusion
s_1, $\bar{s}_1$	phase boundary and dimensionless phase boundary between regions 0 and 1, respectively
s_2, $\bar{s}_2$	phase boundary and dimensionless phase boundary between regions 1 and 2, respectively
s*	location of the maximum depth of phase boundary
St	Stefan number defined in Equation (14).
t, τ	time and dimensionless time, respectively
t*	time at which $s_1 = s_2 = s^*$
T	temperature
T_m	melting temperature
x	space coordinate
ξ	dimensionless phase boundary
ρ	density
ω	frequency

Subscripts

0	solid phase, region 0, in which without phase change.
1	liquid phase, region 1, created by melting process
2	solid phase, region 2, created by solidification process
i	i=1, melting process; i=2, solidification process

ACKNOWLEDGMENT

The present work is being supported by the Natural Sciences and Engineering Council of Canada, under grant A7929.

REFERENCES

1. Carslaw, H.S. and Jaeger, J.C., Conduction of Heat in Solids, Second Edition, Oxford University Press, London, 1959, Chapter Xl.
2. Lunardini, V.J., Heat Transfer in Cold Climates, Van Nostrand Reinhold Company, 1981, Chapter 8.
3. Wilson, D.G., Solomon, A.D., and Boggs, P.T., (Editors), Moving Boundary Problems, Academic Press, New York, 1978.
4. Lock, G.S.H., Gunderson, J.R., Quon, D., and Donnelly, J.K., A Study of One-Dimensional Ice Formation with Particular Reference to Periodic Growth and Decay, International Journal of Heat Mass Transfer, v. 12, 1969, p. 1343-1352.

5. Seban, R.A., A Comment of the Periodic Freezing and Melting of Water, International Journal of Heat Mass Transfer, v. 14, 1971, p. 1862-1864.
6. Rohsenow, W.M., and Hartnett, J.P., Handbook of Heat Transfer, McGraw-Hill, 1973, section 4, p. 4-72.

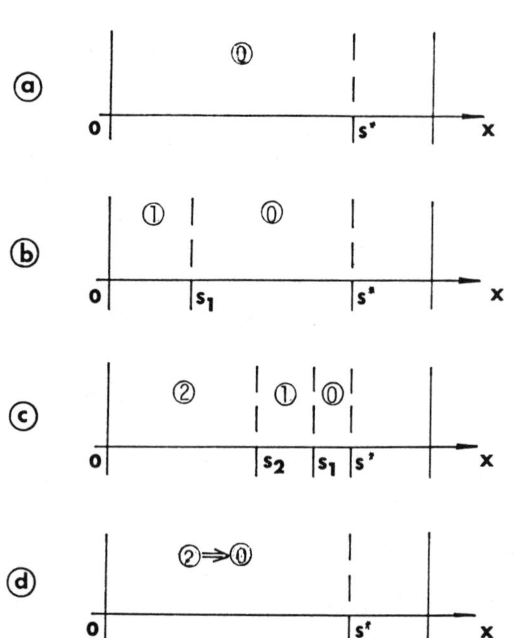

FIGURE 1 Schematic diagram of steady periodic melting and solidification processes (a) after disappearance of the liquid phase and before the start of the melting process, (b) during the melting process, (c) during the solidification process, and (d) end of the solidification process. (0) Solid phase, (1) liquid phase, and (2) solid phase.

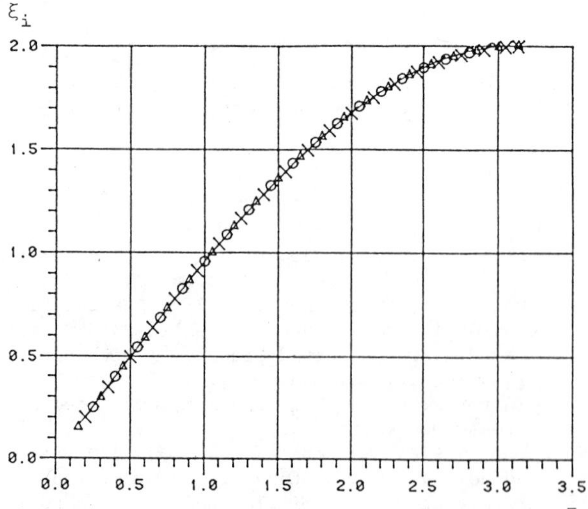

FIGURE 2 Comparison of different analytical results with $St_i = 0.01$. Circles represent zeroth-order quasi-steady approximation [5]; crosses, first-order quasi-steady approximation [2]; triangles, first-order perturbation approximation [4]; and solid line, present analysis.

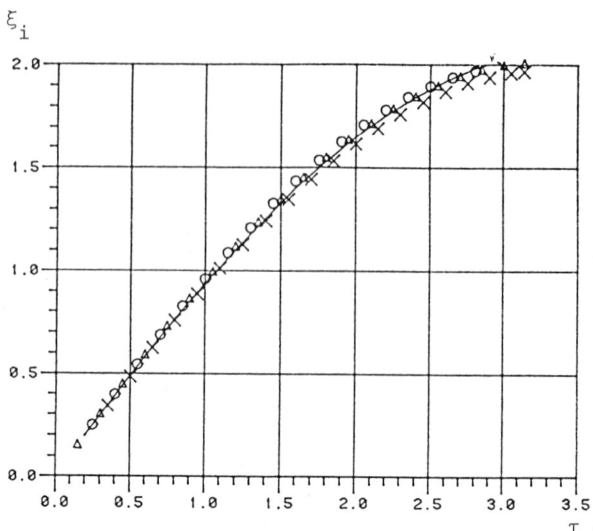

FIGURE 3 Comparison of different analytical results with $St_i = 0.1$. Circles represent zeroth-order quasi steady approximation [5]; crosses, first-order quasi-steady approximation [2]; triangles, first-order perturbation approximation [4]; and solid line, present analysis.

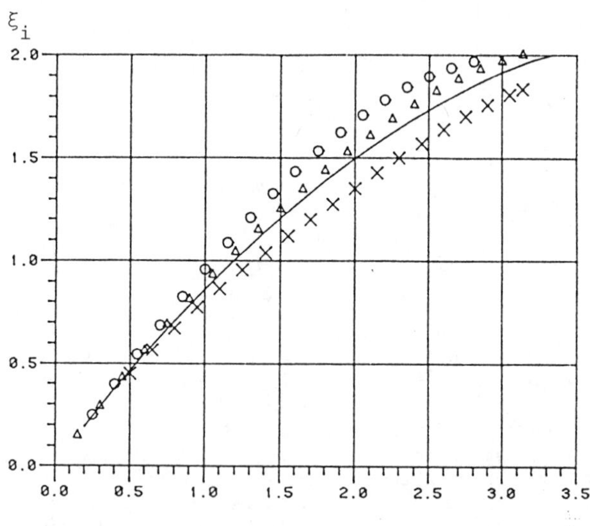

FIGURE 4 Comparison of different analytical results with $St_i = 0.5$. Circles represent zeroth-order quasi-steady approximation [5]; crosses, first-order quasi-steady approximation [2]; triangles, first-order perturbation approximation [4]; and solid line, present analysis.

EFFECT OF SURFACE-APPLIED PRUDHOE BAY CRUDE OIL ON VEGETATION AND SOIL PROCESSES IN TUSSOCK TUNDRA

A. E. Linkins[1] and N. Fetcher[2]

[1]Biology Department, Virginia Polytechnic Institute and State University, Blacksburg, Virginia 24061 USA
[2]Duke Phytotron, Department of Botany, Duke University, Durham, North Carolina 27706 USA

Prudhoe Bay crude oil was applied as a 20 liter m^{-2} surface spray on Eriophorumus vaginatum tussock tundra at Eagle Summit, Alaska, in August 1979. Oil caused a significant reduction in mycorrhizal root numbers and root respiration rates in Betula nana, but not in Eriophorum vaginatum root tips which had grown through the 5- to 15-cm-deep oil-contaminated soil. Significant changes did occur in leaf senescen patterns of B. nana and the tillering index of E. vaginatum. Soil cellulase and phosphatase enzyme activities both declined in the oiled soil horizons but were unaffected in horizons immediately below visibly contaminated organic matter.

INTRODUCTION

The impact of oil on soils and vegetation in arctic and subarctic regions has been studied in association with the development of oil exploitation and transportation in these areas. The effect of oil on soils and vegetation has recently been reviewed by Linkins et al. (1983). This review points out that the toxic effects of direct oil-vegetation contact are relatively straightforward (leaf death occurring from direct oil contact, with regrowth from protected buds (Walker et al. 1978, Wein and Bliss 1973) and vigorous regrowth occurring where the soil is water saturated and oil does not penetrate the soil.) However, where oil penetrates the soil into the plant rooting zone, the plant responses are more complex, ranging from rapid death, or gradual losses in growth and reproduction capability, to complete or limited revegetation (Deneke et al. 1975, Johnson 1981, Johnson et al. 1980, Linkins et al. 1978, Linkins and Antibus 1978). All these responses must be due to oil-related altered soil water potential, structure, cation exchange capacity, aeration, as well as direct toxic effects of the oil (Everett 1978, Johnson, et al 1980, Pesek 1981, Sparrow et al. 1978). The complexity of these oil-soil-plant interactions makes it quite difficult to establish a successful revegetation/restoration plan in oil-contaminated arctic-subarctic soils (Linkins et al. 1983).

It is the purpose of this paper to present information on the responses of two plant types to a surface oil spill of 20 liters m^{-2}--Eriophorum vaginatum and Betula nana. The decomposition of cellulose and organic phosphorous is also discussed as it relates to overall site recovery potential.

MATERIALS AND METHODS

Oil Application

Prudhoe Bay crude oil was applied as a uniform surface spray to give a final coverage volume of 20 liters m^{-2} as a uniform surface spray (Everett 1978) on a 3 x 20 m plot of E. vaginatum tussock tundra at Eagle Creek, Alaska, on August 14, 1979. Oil was applied at ambient temperature. A control plot of 3 x 20 m was delineated adjacent and upslope of the oil spill plot.

Plant Studies:

Xylem pressure potential was measured on a clear day (July 16, 1980) at midday for five species of tussock plants, Eriophorum vaginatum, Carex bigloweii, Rubus chamaemorus, Betula nana, and Vaccinium uliginosum. Measurements were taken on unshaded stems or blades following the recommendations of Ritchie and Hinckley (1975). Readings were taken using nonoiled current year's vegetation from plants in the oiled plot. Similar vegetation was used from plants in the control plot. Data were analyzed using a two-way analysis of variance.

Plant demography studies were conducted with E. vaginatum. Demographic analysis of new tillers (Vo), 1-year tillers (V1), and mature tillers (Vm) was done according to Goodman and Perkins (1968). Demographic analysis was conducted using two to three tussocks per sample, with all tillers in a harvested tussock categorized.

Viable root tips from the above tussocks were suspended immediately upon harvesting in a 5 mM $CaCl_2$ solution. Respiration rates of these root tips were measured at 10°C using an Orbisphere polarographic electrode (Model 2709).

Leaves of the deciduous shrub B. nana were categorized as follows: live, dark green and turgid; senescent, light green colored and turgid; or dead, brown or dull red and dry. Determinations were made on a minimum of five main stems of at least three separate plant clumps.

Ectomycorrhizal root tips from the above plant clumps were harvested and quantified on a soil volume basis as described by Antibus and Linkins (1978). Root tip respiration rates were measured as described above.

Soil Studies

Potential for cellulose hydrolysis in oiled and control soils was measured by the exocellulase and endocellulase assays as described by Linkins et al. (1978). Potential for organic phosphomonoester hydrolysis in the soil was measured using the phosphomonoesterase enzyme assay with p-nitrophenylphosphate (PNPP) (Sigma) substrate as described by Tabatabi and Bremner (1969) and modified to 0.3M PNPP as described by Herbein (1981). All enzyme assays were conducted at 20°C with appropriate controls as described by Linkins and Neal (1982) and Herbein (1981).

RESULTS AND DISCUSSION

The effect of crude oil on the water potential of representative plants in tussock tundra is shown in Figure 1.

FIGURE 1 Water potential in megapascals for E. vaginatum (E.V.), C. bigloweii (C.B.), Rubus chameomorus, B. nana (B.N.,) and Vaccinium uliginosum (V.U.) in control 0 and 20 liter m^{-2} oil spill 0 plots. Data are given as means with standard deviation.

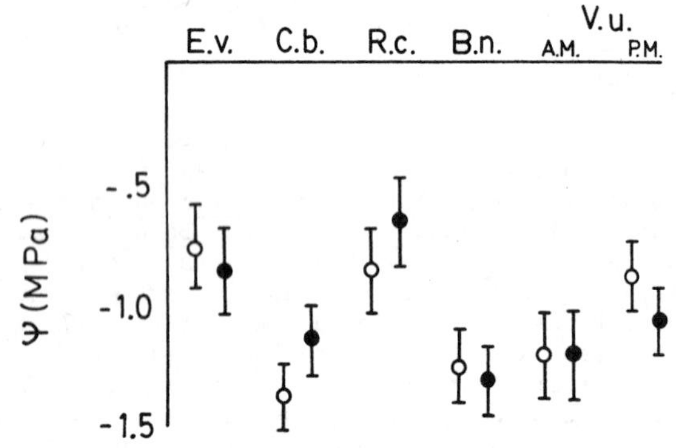

In no case was there any significant difference in the water potential of any plants tested. This was even true for plants which have the majority of their root biomass in the 1-15 cm soil layer visibly contaminated with the oil (e.g. B. nana, V. uliginosum, C. bigloweii (Miller et al. 1982)). There were not even any water potential differences for known ectomycorrhizal plants (B. nana), even though Antibus and Linkins (1978) and Linkins and Antibus (1978) showed significant structural and functional changes in oil contaminated ectomycorrhizal roots of Salix rotundifolia at Barrow, Alaska.

The ectomycorrhizal roots of B. nana, many of which were found to reside in the 1- to 15-cm oil contaminated soil depth, did show a significant reduction in both numbers and the respiration rates of those remaining (Table 1). These responses were similar to those noted for S. rotundifolia at Barrow following a 5- or 12-liter m^{-2} Prudhoe Bay oil spill (Antibus and Linkins 1978, Linkins and Antibus 1978).

This reduction in mycorrhizal root biomass and activity seemed to be balanced by a reduction in the number of leaves per stem (Table 2). There was a trend in the reduction of leaves per stem during the season for plants in oiled soils, although it was not always significant. This concomitant reduction in both absorptive root surface and leaf surface area could explain the lack of observed water stress in these plants (Figure 1).

However, there are other differences in the leaves on the stems of oiled and nonoiled plants. There is a significant shift in the proportion of leaves which are green, senescing, or dead. For 3 years there was a significant increase in the proportion of leaves in a senescing or dead class for oiled plants (Table 2). This shift was independent of the time of the growing season. These data suggest that there is an effect of oil on these plants which continues to cause not only an overall reduction in viable roots and leaves but also an accelerated aging and death in leaves.

TABLE 1 Effect of Prudhoe Bay crude oil (20 liters m^{-2}) on Root Tip Numbers (cm^{-3} soil) and Respiration Rates (μl O_2 h^{-1} g dry wt root^{-1}) of the Mycorrhizal Root Tips of B. nana and on Root Tip Respiration (μl O_2 h^{-1} g dry wt root) of E. vaginatum in E. vaginatum tussock tundra at Eagle Creek, Alaska.

	1980		1981	1982
	17 July	20 August	16 August	21 August
B. nana Mycorrhizal Root Tip Numbers				
Control	22.4(2.4)^a	21.5(2.6)^a	25.6(1.9)^a	25.8(2.2)^a
Oiled	17.7(2.1)^b	10.8(2.5)^c	7.3(0.9)^c	5.2(1.7)^{cd}
B. nana Mycorrhizal Root Tip Respiration Rates				
Control	19.5(1.5)^a	18.0(1.8)^a	17.5(1.5)^a	21.0(1.2)^a
Oiled	8.2(0.9)^b	5.2(0.9)^c	5.7(0.9)^c	6.5(1.3)^c
E. vaginatum Root Tip Respiration Rates				
Control	22.6(2.1)^a	24.5(0.9)^a	22.0(5.0)^a	22.9(2.1)^a
Oiled	24.3(1.5)^a	23.2(1.0)^a	19.2(3.5)^a	25.9(2.9)^a

Values given as means with standard deviation in parentheses. N = 6. Values in a row or column within a section followed by a different superscript are significantly different at the 0.05 level.

Whether these changes in the dynamics of roots and leaves are due to direct toxic effects or altered edaphic factors unfortunately cannot be determined from these data. Similar data on the dynamics of growth allocation root respiration rates, water potential, and root location of E. vaginatum may help in part to clarify this.

TABLE 2 Effect of Prudhoe Bay Crude Oil (20 liters m^{-2} on the Proportion of Live (Green, Turgid), Senescent (Light Green Colored, Turgid) or Dead (Brown or Dull Red, dry) Leaves on the Stems of B. nana in E. vaginatum Tussock Tundra at Eagle Creek, Alaska.

	17 July		20 August	
	Control[1]	Oiled[1]	Control[1]	Oiled[1]
1980				
Live(L)	98	70	47	10
Sen(S)	2	21	51	40
Dead(D)	0	8	2	50
Total leaves per stem	15(3)[a]	8(2)[b]	14(3)[a]	5(2)[a]
1981	20 July		16 August	
Live(L)	90	60	52	15
Sen(S)	10	30	43	45
Dead(D)	0	10	5	40
Total leaves per stem	20(3)[a]	6(2)[b]	15(3)[a]	3(1)[b]
1982			22 August	
Live(l)			35	18
Sen(S)	ND		60	32
Dead(D)			14(3)[a]	3(2)[b]

Data expressed as percentage of total leaf number present on a stem. Total leaves per stem presented as means with standard deviation in parentheses. Values followed by a different superscript are significantly different at the 0.05 level.

[1]Distributions of leaf types are all significantly different at 0.05 level except SEN on August 16, 1981.

Table 2 shows that the root tip respiration rates of E. vaginatum are not affected by the oil in the top 1-15 cm of soil, probably because the root tips exhumed for all experiments were 18-30 cm deep and not in the oiled soil horizon. The roots also looked normal, as noted by Johnson et al. (1980) for E. vaginatum in oiled tundra at the Poker-Caribou Creek oil spill. Aboveground biomass looked normal, with no apparent accelerated senescence, or unusual water stress (Figure 1).

Determination of the tillering index ($Ti = \frac{(Vl + Vo)}{Vm}$) for E. vaginatum, however, shows that there is a significant change in the tillering index of plants in oiled soil (Table 3).

TABLE 3 E. vaginatum Tillering Index ($Ti = \frac{(Vl + Vo)}{Vm}$) for E. vaginatum After a Surface Oil Spill of 20 liters m^{-2} Prudhoe Crude in August 1979.

	1981		1982	
	Control	Oiled	Control	Oiled
Vo	7.5[a]	20.5[b]	8.9[a]	32.5[c]
Vl	17.2[d]	21.6[d]	18.8[d]	23.8[d]
Vm	75.3[e]	57.9[f]	72.3[e]	43.7[g]
Ti	0.32[h]	0.73[i]	0.38[h]	1.29[k]

V_1 = first-year tillers, Vo = new tillers, Vm = mature tillers. Values in a row followed by a different superscript are significantly different at the 0.05 level.

The tillering index increases for the 1981-1982 seasons and is always significantly greater than that of the control plants. The shift in growth allocation patterns of the plants in the oiled soils is toward a predominance in Vo tillers and a loss in Vm tillers. The proportion of Vl tillers seems constant regardless of the time or treatment, but survival to Vm is significantly reduced in all plants in oiled soils. This response to oil is seemingly independent of a direct water potential or root tip activity interaction, so perhaps it is due to an oil-meristem or nonroot tip-root interaction.

Regardless, it will be important to understand the nature of this oil/E. vaginatum interaction as it may undermine long-term restoration of oil-contaminated tussock tundra by E. vaginatum.

Soil Enzymes

The enzyme complex cellulose hydrolyzes cellulose and similar 14 linked polymers into soluble carbohydrates which can be assimilated by soil microorganisms (Eriksson 1978, Linkins et al. 1978, Linkins and Neal 1982, Reese 1977). Two major functional groups of this complex are the endocellulases, which hydrolyze internal bonds in the cellulose chain, and exocellulases, which hydrolyze bonds from the nonreducing end of the cellulose chain (Eriksson 1978, Reese 1977). This latter group is most directly associated with the production of soluble molecules for microbial uptake.

A ratio of the exocellulase to endocellulase activity has been shown in natural ecosystems to be related to the quantity (degree of lignification) or availability of cellulose in plant litter (Linkins et al. 1978, Linkins and Neal 1982, Linkins et al. 1980, Shaver et al. 1983, Sinsabaugh et al. 1981). The higher the ratio the more cellulose is hydrolyzed for microbial uptake and oxidation. Data in Table 4 show that oil-contaminated soils lose their ability to support the hydrolysis of cellulose to products taken up by microorganisms.

TABLE 4 Effect of Prudhoe Crude Oil (20 liters m^{-2}) on Cellulase Activity in Oil-Contaminated Oi-Oe Horizons of Tussock and Intertussock Soils in E. vaginatum Tussock Tundra at Eagle Creek Alaska.

	1979		1980	
	Exocellulase	Endocellulase	Exocellulase	Endocellulase
Tussock				
Control	83^{a}	158^{a}	104^{a}	151^{a}
Oiled	74^{a}	152^{a}	35^{b}	142^{a}
Intertussock				
Control	89^{a}	147^{a}	100^{a}	163^{a}
Oiled	60^{a}	129^{b}	35^{b}	127^{b}

	1981		1982	
	Exocellulase	Endocellulase	Exocelulase	Endocellulase
Tussock				
Control	100^{a}	162^{a}	98^{a}	150^{a}
Oiled	20^{c}	100^{b}	10^{a}	60^{c}
Intertussock				
Control	95^{a}	148^{a}	100^{a}	165^{c}
Oiled	19^{c}	80^{c}	12^{d}	43^{c}

Endocellulase activity is given as unit h^{-1} g dry wt $soil^{-1}$; Exocellulase activity is given as μg glucose equivalents released h^{-1} g dry wt $soil^{-1}$. N = 18. Activities given as means. Value in an enzyme type followed by a different superscript are significantly different at the 0.05 level.

The loss in cellulase activity is initially in the exocellulase component, but eventually both components' activities are reduced. In addition to an absolute loss in cellulase activities there is a reduction in the efficiency of the enzyme complex to produce microbially utilized products. The exocellulase to endocellulase ratio shifts from the 0.50-0.62 range in control soils to 0.27-0.16 in 1982 oiled soil horizons. These patterns are similar to those seen for cellulase activities in oil-contaminated soils at Barrow, Alaska, where cellulose was lost as an energy source in oiled soils (Linkins et al. 1978).

Acid phosphomonoesterase is an enzyme that hydroloyzes inorganic phosphorous off organic phosphomonoesters. In soils, these enzymes, like most other extracellulase soil hydrolases, are physically adsorbed to soil particles (Spier and Ross 1978). The effect of oil on acid phosphomonoesterase activity in the oil contaminated soil Oe-Oi horizons is given in Table 5. Unlike the exocellulase activity, phosphatase activity did not show a significant reduction in activity until the third season, 1981. This reduction remained through the following season. Thus, the potential for generation of inorganic phosphorous, if organic phosphorous was available to the enzyme in these oiled soils, was unaffected for 2 years after the oil spill. This maintenance of activity could be due to the tolerance of the extant enzyme or production of the extracellular enzyme by oil tolerant organisms. The ultimate decline in activity is probably due to loss of the enzyme as a result of loss of suitable available substrate and/or enzyme binding sites in the oil-coated soil material (Everett 1978, Linkins et al. 1978, Pesek 1981).

The influence of oil in a particular soil horizon (Oe-Oi in this study) seems to primarily be limited to that horizon. Tables 5 and 6 show that soil enzyme activity in the Oe-Oa soils material immediately under the visibly contaminated Oi-Oe horizons was unaffected. These data suggest that those oil components, if any, that leach down in the soil do not adversely effect certain enzymatic aspects of decomposition.

TABLE 5 Effect of Prudhoe Bay Crude Oil (20 liters m^{-2}) on Phosphomonoesterase Activity in the Oe-Oi or Oa Horizon of Tussock and Intertussock Soils in E. vaginatum Dominated Tussock Tundra at Eagle Creek, Alaska.

	1979		1980	
	Oe-Oi	Oa	Oe-Oi	Oa
Tussock				
Control	1,725[a]	1,000[a]	1,677[a]	1,300[a]
Oiled	1,650[a]	1,250[a]	1,496[a]	1,275[a]
Intertussock				
Control	2,000[a]	950[a]	2,218[a]	1,200[a]
Oiled	2,115[a]	1,075[a]	2,590[a]	1,350[a]

	1981		1982	
	Oe-Oi	Oa	Oe-Oi	Oa
Tussock				
Control	1,952[a]	1,250[a]	2,101[a]	1,275[a]
Oiled	752[b]	1,300[a]	572[b]	1,100[a]
Intertussock				
Control	1,979[a]	1,179[a]	2,278[a]	1,020[a]
Oiled	679[b]	1,220[a]	452[b]	2,257[a]

Phosphatase activity expressed as µg PNP release h^{-1} g dry wt $soil^{-1}$. Activities given as means. N = 18. Values in a soil horizon column followed by different superscript are significantly different at the 0.05 level.

TABLE 6 Cellulase Activities in the Oe-Oa horizons Immediately Below the Oe-Oi Oil-Contaminated Horizons.

	1981		1982	
	Tussock			
	Exocel	Endocel	Exocel	Endocel
Control	85[a]	36[a]	88[a]	40[a]
Oiled	80[a]	32[a]	92[a]	43[a]
	Intertussock			
Control	65[a]	10[a]	62[a]	12[a]
Oiled	62[a]	12[a]	65[a]	15[a]

Data given as means. N = 10. Values followed by a different superscript are significantly different at the 0.05 level, as described in Table 4.

CONCLUSIONS

Limited observations on the water potential of plants in oiled soils suggest that oil in the soil may not induce significant water stress in plants. The concomitant reductions in both leaf area and absorptive root surface area in B. nana (a plant with the bulk of its roots in the oiled soil horizon) undoubtly played a role in the maintenance of observed plant water potentials (Oberbauer and Miller 1982). The consistent premature senescing of leaves in B. nana and reduction in mycorrhizal root tips (Antibus and Linkins 1978, Linkins and Antibus 1978), plus the altered tillering index in E. vaginatum, suggest that the persistence of crude oil in the soil significantly alters plant growth and carbon-nutrient allocation patterns. These changes which were observed for 3 years after a 20 liter m^{-2} crude oil spill, provide useful additional information in planning cleanup and revegetation/restoration efforts in tussock tundra. Johnson (1981) and Linkins et al. (1983) suggest that an important option for consideration in planning cleanup and revegetation/restoration efforts should be no action (passive). However, the plant response data in this paper show that even after 3 years the presence of oil still alters growth and nutrient allocation patterns such that successful regrowth and restoration may not occur, or at least be very slow. Perhaps a passive approach should occur only on small or contained spills where acceptable revegetation can occur by invasion from adjacent tundra by plants with some porportion of their root system in uncontaminated tundra.

Data on soil processes in this paper show that the effect of oil is limited to areas visibly contaminated by the oil. The organic soils immediately adjacent to or below the contaminated area are unaffected. McGraw and Shaver (1982) and Shaver et al. (1983) have shown that the organic mat is an important natural seed source for seedling establishment of many tussock tundra plants. Therefore, in oil spills with limited vertical movement where some uncontaminated organic material remains, removal of only the contaminated soil may leave organic material with normal nutrient cycling and decomposition potential (enzyme activity) and a viable native seed source. This may promote the most acceptable long-term cost-effective restoration of the site.

ACKNOWLEDGMENTS

The authors wish to thank M. Bryhan for her technical assistance in this work and Drs. F. S.

Chapin and G. R. Shaver for their help in the demographic studies. This work was supported by grants from the U.S. Dept. of Energy and U.S. Army Research Office.

REFERENCES

Antibus, R. K., and A. E. Linkins, 1978. Ectomycorrhizal fungi of Salix rotundifolia. I. Impact of surface applied Prudhoe Bay crude oil on mycorrhizal structure and composition. Arctic 31:366-380.

Deneke, F. J., B. H. McCown, P. I. Coyne, W. Richard, and J. Brown. 1975. Biological aspects of terrestial oil spills. USA CRREL Oil Research in Alaska, 1970-1974. USA-CRREL Res. Rpt. 346 ADA 047365.

Eriksson, K. E. 1978. Enzyme mechanisms involved in cellulose hydrolysis by the white rot fungus Sporotrichum pulverulentum. Biotech. Bioeng. 20:317-332.

Everett, K. R. 1978. Some effects of oil on the physical and chemical characteristics of wet tundra soils. Arctic 31:260-276.

Goodman, G. T., and D. F. Perkins. 1968. The role of nutrients in Eriophorum communities. III. Growth response to added inorganic elements in two E. vaginatum communities. J. Ecol. 56:667-683.

Herbein, S. B. 1981. Soil phosphatases: Factors affecting enzyme activity in Arctic tussock tundra and Virginia mineral soils. M.S. thesis. Biology Dept., VPI & SU, Blacksburg, VA. p. 101.

Johnson, L. A. 1981. Revegetation and selected terrain disturbances along the trans-Alaskan pipeline, 1975-1978. U.S. Army CRREL Report 81-12. U.S. Army CRREL, Hanover, N.H.

Johnson, L. A., A. E. Sparrow, T. F. Jenkins, C. M. Collins, C. V. Davenport, and T. T. McFadden. 1980. The fate and effects of crude oil spilled on subarctic permafrost terrain in interior Alaska. EPA Report, E.P.A.-600/3-8-400 and CRREL Report 80-29, U.S. Army CRREL, Hanover, N.H. p. 128.

Linkins, A. E. and R. K. Antibus. 1978. Ectomycorrhizal fungi of Salix rotundifolia Trantu. II. Impact of surface applied Prudhoe Bay crude oil on mycorrhizal root respiration and cold acclimation. Arctic 31:381-391.

Linkins, A. E., R. M. Atlas, and P. Gustin. 1978. Effect of surface applied crude oil on soil and vascular plant root respiration, soil cellulase and aryl hydrocarbon hydroxylase at Barrow, Ala. Arctic 31:355-365.

Linkins, A. E., and J. L. Neal. 1982. Soil cellulase, chitinase, and protease activity in Eriophorum vaginatum tussock tundra at Eagle Summit, Ala. Holarctic Ecol. 5:135-138.

Linkins, A. E., L. A. Johnson, K. R. Everett, and R. M. Atlas. 1983. Oil spills-damage and recovery in tundra and taiga. In: Oil spill restoration and recovery J. Cairns and A. L. Buikema (eds.). Ann Arbor Press, Ann Arbor, MI.

McGraw, J. B., and G. R. Shaver. 1982. Seedling density and seedling survival in Alaskan cotton grass tussock tundra. Holarctic Ecol. 5:212-217.

Miller, O. K. 1982. Mycorrhizae, mycorrhizal fungi and fungal-biomass in subalpine tundra at Eagle Summit, Ala. Holarctic Ecol. 5125-134.

Miller, P. C., R. Mangan, and J. Kummerow. 1982. Vertical distribution of organic matter in eight vegetation types near Eagle Summit, Ala. Holarctic Ecol. 5:117-124.

Oberbauer, S., and P. C. Miller. 1982. Growth of Alaskan tundra plants in relation to water potential. Holarctic Ecol. 5:194-199.

Pesek, E. A. 1981. Oil-cellulose interactions in Arctic tundra soils. Master's thesis, Biology Dept., Univ. of Louisville, Louisville, Ky. p. 128.

Reese, E. T. 1977. Degradation of polymeric carbohydrates by microbial enzymes. pp. 311-367. In: The structure, biosynthesis, and biodegradation of wood. F. A. Loewus, and B. C. Runeckles (eds.), Plenum Press, N.Y.

Ritchie, G. A. and T. M. Hinckley. 1975. The pressure chamber as an instrument for ecological research. Adv. Ecol. Res. 9:165-254.

Shaver, G. R., F. S. Chapin III, and A. E. Linkins. 1983. Revegetation of Arctic disturbed sites by native tundra plants. 4th Int. Conf. on Permafrost. Fairbanks, Ala.

Sinsabaugh, R. L., E. F. Benfield, and A. E. Linkins. 1981. Cellulase activity associated with the decomposition of leaf litter in a woodland stream. Oikos 36:184-190.

Sparrow, E. B., C. V. Davenport, and R. C. Gordon. 1978. Response of microorganisms to hot crude oil spills on a subarctic taiga soil. Arctic 31:339-347.

Speir, T. W., and D. J. Ross. 1978. Soil phosphatase and sulphatase. In: Soil Enzymes. R. A. Burns (ed). pp. 197-250. Academic Press, London.

Tabatabi, M. A., and J. M. Bremner. 1969. The use of p-nitrophenyl phosphate for assay of soil phosphatase activity. Soil Biol. Biochem. 1:301-307.

Walker, D. A., P. J. Webber, K. R. Everett, and J. Brown. 1978. Effects of crude and diesel oil spills on plant communities at Prudhoe Bay, Ala. and the derivation of sensitivity maps. Arctic 31:242-259.

Wein, R. W., and L. C. Bliss. 1973. Experimental crude oil spills on Arctic plant communities. J. Appl. Ecology 10:671-682.

CALCULATION OF FROST-HEAVING FORCES IN SEASONALLY FROZEN SUBSOILS

Liu Hongxu

Heilongjiang Provincial Low-Temperature Construction Research Institute
People's Republic of China

The mechanical properties of frozen soil differ considerably from those of unfrozen soil. The author proposes that the soil beneath foundations on seasonally frozen ground be considered as a double layer, composed of a layer of frozen soil and a layer of unfrozen soil. Based on the linear elastic theory, the author proposes a method for calculating stress distribution at the freeze/thaw interface in soils under foundations. Analyzing the characteristics of stress distribution during the frost process in a double-layer subsoil, the author has presented a method for estimating the normal and horizontal frost-heaving forces in frost-susceptibile subsoils. According to field test results, a nomograph for determining normal and horizontal frost-heaving forces in various frost-susceptible soils is also presented. To demonstrate the application of double-layer calculations to foundation design, several examples for examining foundation stability are given.

Ordinary foundation soil is of one layer in seasonal frozen soil districts. With the freezing and expansion of the foundation soil, its stress state gradually and finally changes to that of the two-layer soil. When thawing takes place in the spring, the mechanical properties the thawing soil are much less than those of frozen soil, thus forming a three-layer foundation soil. This paper analyzes the force imposed on building foundations during the process of frost heaving by frost-heaving soil using the calculation result of the two-layer soil.

Based on the relation curve of normal frost-heaving with respect to freezing depths actually measured at the frozen soil observation station, by using a finite element method under a principle of non-linearity during a certain load history, we can obtain the frost-heaving stress distribution in different depths of the foundation soil.

When the stress is comparatively small, numerical calculation should be done by the finite element method based on the linearity concept. Similarly, the shearing frost-heaving force can also be developed by using the calculation of frost-heaving for a two-layer soil.

Using the non-linear analyses of the force imposed on two-layer foundation soil, the bearing capacity of the frozen soil can also be calculated.

The strength of a frost-heaving soil usually becomes lower when it is thawing. By using an analysis of the stress for a three-layer foundation soil, the stability of a foundation during thawing process can be checked.

CALCULATION OF NORMAL FROST-HEAVING FORCE

Before freezing, the distribution of additional stress induced by loading is uniform (in a single layer) in the foundation soil. When freezing develops to a depth under the base of the foundation, the foundation soil will turn into two layers because of the difference of mechanical properties between the frozen soil and the unfrozen soil. If the foundation soil is not expansive in freezing, distribution of stress in the soil also remains uniform, although the foundation soil has turned into two layers. But if the foundation soil is of frost-heaving soil, then the stress in it will undergo a series of changes or redistributions as frost-heaving force occurs and increases continuously. If the additional stress is σ_z at the intersecting point of freezing interface, depth of which is h counted from the foundation base, and the symmetric axis (z-axis) of the foundation base is n, considering the foundation soil as two layers, and the heaving stress on the interface σ_t is m, the additional load $\frac{m}{n}$ on the foundation will be equal to the stress in the two-layer foundation soil, when $m < n$, while the remaining part $(1-\frac{m}{n})$, is equal to the stress in the unchanged one-layer foundation soil. When $m = n$, the stress state of the foundation soil becomes that of two-layer soil completely. If the soil expands only a little during freezing, its stress state may not attain that of the two-layer soil at all.

Loads on buildings with their foundations buried within the range of freezing are comparatively small. The stress in the soil, as a homogeneous body linearly deformed, is usually calculated by using the elastic theory. After the soil is frozen, its mechanical properties will be greatly improved. It is reasonable to analyze the stress in foundation soil as a linearly deformed body in a two-layer semi-infinite space.

Deformation modulus of the frozen soil is closely related to temperature. In winter, distribution of temperature in the seasonal frozen layer along its depth can be taken as linear, when the thickness of the frozen layer does not exceed 3/4 of the maximum frozen depth, i.e. when the temperature increases again. According to actual measurements in Northeast China, where the frozen depth is about 2m (about in the region of 46° N, 120° E), the mean vertical temperature drop in the soil is 0.1° c/cm. Temperature at various points under the ground surface could be calculated by the following formula:

$$t = f(z) = -0.1(h-z) \quad (1)$$

where h is thickness of the frozen layer counted from the base of the foundation, in cm; and z is the vertical coordinate of a certain point in the frozen layer under the base of the foundation. The relation between the deformation modulus of the frozen soil and the soil temperature is

$$E_1 = f(t) = E_0 + Kt^{\alpha} = 100 + 430t^{0.865} \quad (2)$$

Where E_0 is the deformation modulus of frozen soil at 0° C.
Substituting eq1 into eq2, we have

$$E_1 = 100 + 59(h - z)^{0.865} \quad (3)$$

Figure 1 shows a schematic diagram for the calculation of two-layer foundation soil. For foundations in the form of circular disc with diameter of 50 cm, calculation is made by computer with parameters somewhat different from those in Figure 1, before freezing and at different frozen depth respectively. Results are listed in Table 1. When h is 25 cm, the stress coefficient σ_{zx} at the point a reduces from the original value, before freezing 0.6659 to 0.3313, i.e. 50% lowered. When h reaches 50 cm, σ_{zx} reduces from the value before freezing 0.2835, to 0.770, i.e. 73% lowered. It is evident that within the range of bearing, layer stress after freezing is much lower than before freezing, indicating that during the process of freezing and subsequent expansion, stress attenuates continuously, and the smaller the foundation disc and thicker the frozen layer, the larger the attenuation. When h is 25 cm, the stress reduces to 50%. If part of the foundation soil expands nonuniformly during freezing and stiffness of the whole building is comparatively good, then internal stress in the building will

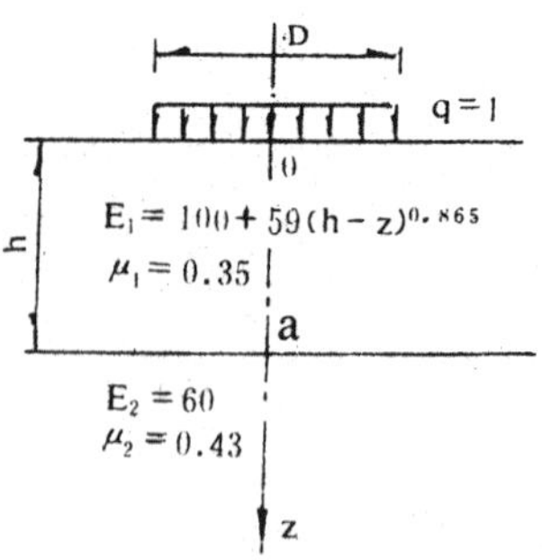

FIGURE 1 Schematic diagram showing calculation for two-layer foundation soil.

TABLE 1 Stress coefficient on interface between two layers and that of one layer.

Thickness of frozen layer(cm)	Foundation in form of circular Disc (50 cm dia.)	
	Homogeneous Single Layer	Frozen Interface between Two Layers
25	0.6659	0.3313
50	0.2835	0.0770
75	0.1380	0.0299
100	0.0799	0.0116
125	0.0536	0.0061
150	0.0349	0.0033

redistribute as frost heaving occurs. The soil near the foundation will unload to the foundation. At that time, if the load increases twice the compression stress in the interface is just equal to that before freezing. At this moment, ordinary buildings with lower stiffness and strength will crack essentially. When h is 50 cm, the compression stress on the interface will be equal to that of one-layer foundation soil before freezing, only when load on the foundation increases 4 times approximately. Therefore, in the process of frost heaving, no compression will apply to the underlying unfrozen soil.

Measurement of foundation models located on ground surface and at different depths under ground surface has been performed for many years at the Yanjiagang Obervation Station in Harbin (frost heaving soil of type IV), and at the Longfeng Observation Station in Daqing (frozen heaving soil of type III and type II). Anchored load cell was used in the measurement. Stress distribution on the frozen interface was obtained, using the measured frost heaving force at different frozen depths by foundation models of different shapes and sizes. The maximum value on the symmetrical points will be the frost heave force on the frozen interface at the given depth.

Frost heaving stress is the upward expanding force per unit area while frost-

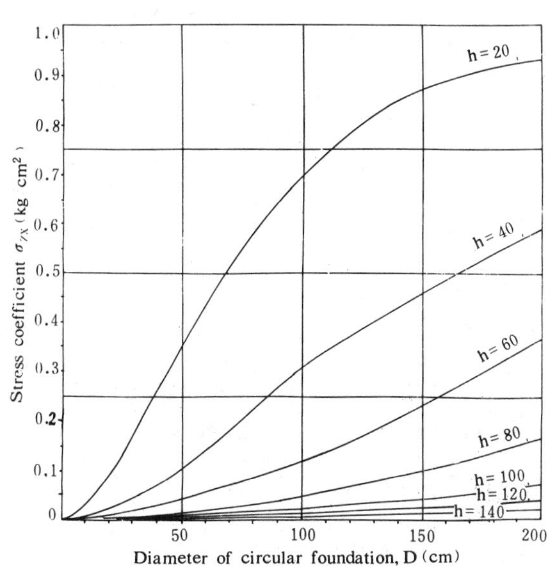

FIGURE 2 Relation curve between diameter of circular foundation D and stress coefficient σ_{zx}.

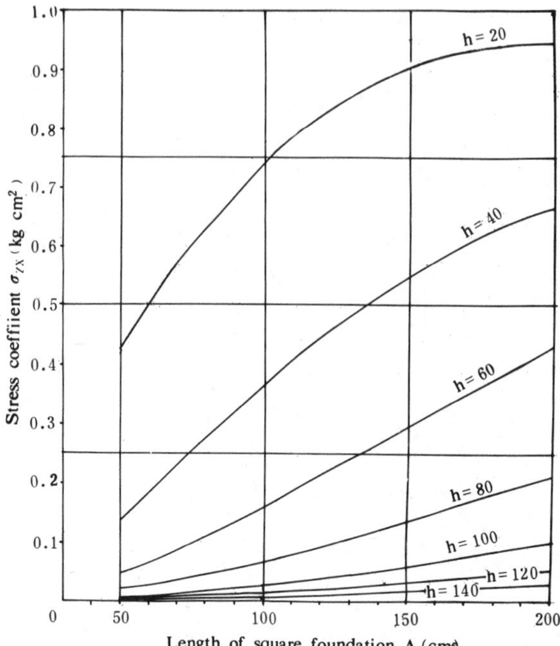

FIGURE 3 Relation curve between length of square foundation A and stress coefficient σ_{zx}.

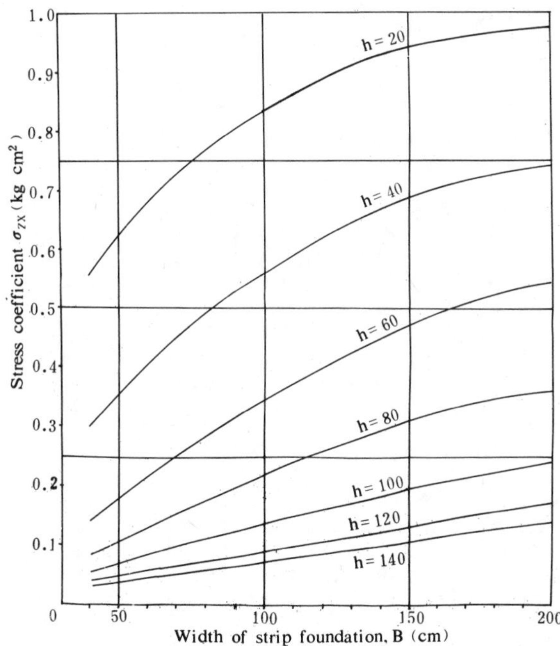

FIGURE 4 Relation curve between width of strip foundation B and stress coefficient σ_{zx}.

heaving occurs at various depths. It is the inherent mechanical property of the frost heaving foundation soil, independent of the foundation itself. In finding the frost heaving stress, when the contact pressure on the base of the foundation is very great, the plastic-viscous property of frozen soil under higher stress should be considered based on the concept that non-linearity exists during a certain load history. If the contact pressure is too great, the frost heaving force should be revised by stress relaxation. When the stress is comparatively small, numerical calculation should be done by the finite element method based on the linearity concept.

Figures 2, 3, and 4 show the calculated stress coefficient σ_{zx} on the frozen interface between two layers for circular, square and strip foundations, using the parameters shown in Figure 1. Relation of the frost heave stress versus frozen depth curve, based on experimental data, and the mean frost heaving rate η for different frozen depths is shown in Figure 5.

The mean frost heaving rate η is defined as the ratio of the amount of frost heave to the thickness of frozen layer. In the calculation of frost heave force, the external load imposed on the foundation should be used as the actual load during the long winter period in which frost heave develops. Live loads, which do not always exist, must not be included in the calculation. For example, during the

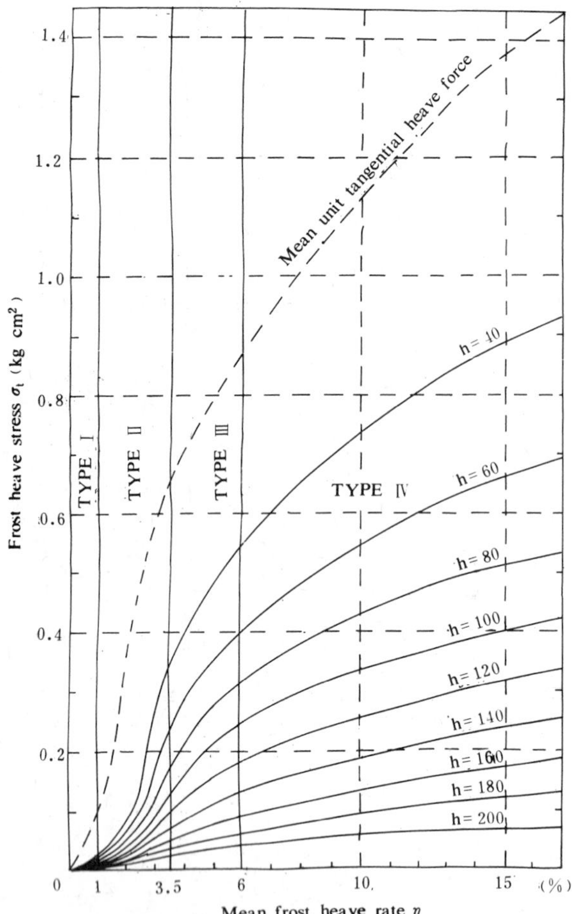

FIGURE 5 Relation curve between mean frost heaving rate η and frost heaving stress σ_t in soil.

season of acute frost heaving, there would be no snow load or only a thin layer of snow on the roof; the live load on the floor of a school classroom during winter vacation is practically zero; though there are activities in meeting rooms and clubs usually, there will be no live load in the spare time. Under these conditions, it is unsafe if these buildings are considered to have full loads in the calculation for balancing the frost heaving force.

Example 1. Consider a square independent foundation of 100 cm X 100 cm, buried depth of foundation 100cm, additional load (permanent load in winter) on the foundation p = 2.0 kg/cm^2, the foundation soil is of type IV frost heaving soil, frost heave rate η = 10%. Maximum frozen depth 200 cm. Tangential frost heave force on the lateral wall of the foundation has been removed.

When the frozen depth reaches 150 cm, the thickness of the frozen layer under the foundation h is 50 cm. From Figure 3, the vertical downward additional stress σ_z is 0.255 X 2 = 0.53 kg/cm^2. From Figure 5, when h = 150 cm, the upward frost heaving stress of soil σ_t is 0.16 kg/cm^2 (while finding the frost heaving stress in soil, the weight of the foundation itself has been considered already). Because $\sigma_z > \sigma_t$, the foundation is stable. At a frozen depth of 200cm, σ_z is 0.06 kg/cm , σ_t is 0.05 kg/cm, σ_z is still larger than σ_t, the foundation is safe.

CALCULATION OF TANGENTIAL FROST HEAVING FORCE

For pile or column foundations and the like, before freezing of the foundation soil, the load (not for supporting piles) is balanced mainly by friction and stress-induced distributions in one layer. As soil freezes, frozen strength is produced by the freezing of soil on the lateral surface of the foundation. If the foundation soil is not of frost heaving soil, distribution of stress at different locations after freezing remains the same as in one-layer soil, despite the foundation soil becomes two-layer soil for the frost heaving soil, with the freezing and expansion of soil, internal force undergoes redistribution, i.e. freezing force gradually replaces friction. As result of redistribution, the component of stress in the two-layer foundation soil gradually and finally replaces that of the single layer soil at all. Therefore, the calculation method of two-layer foundation can be applied to tangential frost heave also.

At present, it is more convenient and practical to use average tangential frost heave force per unit area along the frozen depth in calculation for engineering purpose. The relation between average tangential frost heave force per unit area and the frost heave rate, according to the result of measurement at two Frozen Soil Stations, is shown by the dotted line in Figure 5. In calculation, if the actual frost heave rate of the local foundation soil is given, its corresponding tangential frost heave force per unit area will be obtained. Multiply this force by the lateral surface area of the foundation in frozen soil layer, and the product is the tangential frost heave force.

Example 2: Consider an independent square foundation of 100 cm x 100 cm, buried depth of the foundation is 100 cm, additional load on the foundation p = 2.0 kg/cm^2, the foundation soil is of type IV frost heaving soil, frost heave rate η = 10%, maximum frozen depth is 200 cm, measures are taken to eliminate normal frost heave force on the base of foundation and maintain the bearing force.

When the frozen depth reaches the base of the foundation, the frost heave force of soil σ_t at this depth is 0.335 kg/cm^2, lateral surface area is 40,0000 cm^2 mean height of lateral surface from the frozen surface is 50 cm. From Table 2 the stress coefficient for a square foundation with side length 100 cm and frozen depth 50 cm is 0.27. Take the force acting on the lateral surface as that acting on the lateral side of the cross-section at the mean height, the force on such assumed area is 0.335 / 0.270 X 10.000 kg = 12407kg force per lateral unit area is 12,407 / 40,000 = 0.310 kg/cm^2.

When the frozen depth reaches 150 cm, the frost heave force of soil σ_t = 0.16 kg/cm^2, the distance between the lateral surface and the frozen surface is 100 cm, stress coefficient σ_{zx} = 0.03, the maximum tangential frost heave force developed at this moment is 0.16 / 0.03 X 10.000 kg = 53,333 kg. According to actual measured data of foundation model, it can be approximately taken as the maximum unit tangential frost heave force as shown by dotted line in Figure 5, or the frozen strength between soil and lateral surface of foundation. When frost heaving rate is 10%, the maximum tangential frost heave force is 1.13 kg/cm^2. Because of restriction by the maximum unit tangential frost heave force, the actual value is only 1.13 X 40,000 = 45,200 kg, while the remaining force induces shear displacement between the soil and lateral side of the foundation. Because the existing load is only 20,000 kg and the tangential frost heave force could not be balanced, this foundation would be raised by frost heaving and lose its stability.

According to the roughness shown by the dotted line in Figure 5, the lateral surface of the foundation is an ordinary plane. For especially smooth or especially rough ones, it should be multiplied by a modified coefficient either smaller than 1 or larger than 1.

CALCULATION OF COUPLED ACTION OF NORMAL AND TANGENTIAL FROST-HEAVE FORCES

When the normal frost heave force and the tangential frost heave force act simultaneously, the resulting force is not a simple superposition of such two individual forces. When the frozen depth is h, either the tangential frost heave force sustained by the lateral surface of the foundation or the normal frost heave force sustained by the base of the foundation is induced by the frost-heaving stress acting on the frozen interface.

Example 3: Consider a strip foundation of width b=100 cm, buried depth of foundation is 50 cm, additional load p=2.0 kg/cm^2, the foundation soil is of type IV frost heaving soil, frost heave rate is 10%, maximum frozen depth is 200 cm.

When frozen depth reaches 100 cm, the depth of the frozen layer under the base of the foundation is 50 cm, the distance between the centre of the lateral surface of the foundation and the frozen interface is 75 cm. At the depth of 100 cm, the frost heaving stress of soil σ_t is 0.335 kg/cm^2. From Figure 5, the unit tangential frost heave force is 1.13 kg/cm^2 and the total tangential frost heave force is 50 X 100 X2X1.13 kg = 11,300 kg. From Figure 4, the stress coefficient for the cross-section at the centre of the lateral surface of the foundation to the frozen interface is 0.255 kg/cm^2, the stress on the cross-section is 1,13 kg/cm^2, the frost heave
stress of soil consumed in the generation of 11,300 kg of tangential frost heave force is 0.255 X 1.13 = 0.288 kg/cm^2, the remaining frost heave stress of soil on the frozen interface is 0.335 - 0.288 = 0.047 kg/cm^2. From Figure 4, the stress coefficient for a frozen layer of 50 cm thickness is 0.45. At the same time, the existing normal frost heave force is 0.047 / 0.45 = 0.1044 kg/cm^2, and the resultant normal frost heave force is 1044 kg. The total sustained frost heave force is 11,300 + 1044 = 12,344 kg, smaller than external load 20,000. So it is safe. If the frozen depth increases to 150 cm, the distance from the centre of the lateral surface of the foundation to the frozen interface is 125 cm, its stress coefficient is 0.085. When the distance from the bottom of the foundation to the frozen interface is 100 cm, its stress coefficient is 0.135, the frost heave stress of soil at the depth is 0.16 kg/cm^2. The frost heave stress consumed in the generation of tangential frost heave force is 0.085 X 1.13 = 0.096 kg/cm^2, the normal frost heave force generated from the remaining frost heave stress is 0.064 / 0.135 = 0.474 kg/cm^2, and the resultant force is 0.474 kg/cm^2 X 10,000 = 4,740 kg, and the total sustained frost heave force is 11,300 kg + 4,740 kg = 16,040 kg, smaller than 20,000 kg. The foundation is safe. At frozen depth of 200 cm, using the same way of calculation, the frost heave force obtained is still smaller than the external load, and the foundation is safe.

The above calculations are not the stability of foundations, i.e. whether it would be raised in the process of frost heaving, and as long as the frost heave force is not larger than the external load, stability is guaranteed. But if frost heave force exceeds the external load not too much, it is also allowable, because each building is allowed to sustain certain deformation according to its stiffness, except redistribution of internal force. Within the range of allowable deformation, no crack or failure would occur. With the occurrence of frost heave deformation, the frost heave force attenuates continuously. Further studies will be needed for the relation between frost heave force and deformation, as well as the calculation of frost force from soil deformation. At present, testing data in these respects are not too much, so data accumulation should be carried on.

CHECK OF FOUNDATION STABILITY DURING THAWING

If deformation occurs in the frozen foundation soil during thawing, a series of changes (redistribution of internal force) also occur in its interior, which could be found through the analysis of internal force of the linearly deformed body in the three-layer semi-infinite space.

Ordinary foundation soil is of one layer. With the freezing and expansion of the foundation soil, its stress state gradually and finally changes to that of the two-layer soil. When thawing takes place in the spring, the upper part of the soil mass begins to thaw. The mechanical proties of the thawing soil are much less than those of frozen soil, thus forming a three-layer foundation soil. Since the stiffness of the frozen soil in the second layer is great and its compressibility is relatively small, the concentration of stress appears in the first layer, i.e. the stress in the thawing soil suddenly increases. The larger the foundation and the thinner the thawing layer, the more obvious is the concentration of stress. Under loading, the frost heaving foundation soil freezes first and then melts; change of its stress is shown in Table 2.

Taking the soil of the Longfeng Frozen Soil Station as an example, when the water content in the upper part of the soil layer increased 5% during freezing, the plastic state of the soil will change to the soft plastic state, thus lowering 10% of its allowable bearing strength. During thawing in spring, stress concentrates on the thawing interface, concentration is more serious, generally increasing more than 20%, especially when the thawing depth is within the range of the radius of the foundation. For foundation, the strength of which is just enough for use, it is necessary to check whether it is in an allowable condition during thawing.

For piles located in the range of frozen layer, when the foundation soil is of a strong frost heaving soil, relative sliding generally occurs between the soil and the lateral side of the foundation. For soil subjected sliding during thawing, the bearing capacity owing to friction before freezing turns into negative friction, which plays the role of bearing during thawing. At this time, it is also necessary to check the bearing capacity of the pile.

TABLE 2 Variation of Stress in Foundation Soil During Freezing and Thawing.

Depth (cm)	Circular foundation of (50 cm diameter, q=1 kg/cm^2)			
	Single homogeneous layer	At the frozen interface of two layers soil	The thawed interface of three-layer soil	Increment during thawing %
25	0.6659	0.3313	0.8042	21
50	0.2835	0.0770	0.3556	25
75	0.1380	0.0299	0.1782	29
100	0.0799	0.0116	0.1240	55
125	0.0536	0.0061	0.0766	45

Remarks: The 2nd interface located at depth of 200 cm.

CALCULATION OF THE BEARING CAPACITY OF FROZEN SOIL

When seasonal frozen layer freezes to a certain depth, it has a much higher bearing capacity than that before freezing. With the increase of frozen depth and the decrease of soil temperature in the upper part, its bearing capacity still increases. Before weather becomes warm and the soil temperature increases, it is quite possible to utlize such bearing capacity in this short period, for instance, in the emergent repairing of bridges, framework foundation construction for engineering practice, roads construction for heavy trucks in war time, runway construction for airplane, etc. during winter.

Bearing capacity of the frozen soil is calculated according to the nonlinearity of the two-layer foundation soil. Based on the assumed load, find out the stress and also the soil temperature at each point according to the distribution of soil temperature at different depths. Then, from soil temperature, find out the corresponding allowable compressive strength. It would be safe when the calculated compressive stress is not greater than the allowable value, but it would be unsafe if the compressive stress exceeds the allowable value.

CONCLUSION

This calculating method is based on the theoretical analysis of layered deformed body in semi-infinite space. The physical meaning of this method is quite explicit, and the stress state obtained is very clear. In this method, using the finite element program, numerical computation can be made by electronic computer, and a lot of diagrams, graphs, and tables can be compiled to make calculations easy and convenient.

FROST SUSCEPTIBILITY OF SOILS

C. W. Lovell

School of Civil Engineering, Purdue University
West Lafayette, Indiana 47907 U.S.A.

The fabric of soil significantly influences all aspects of its behavior including frost susceptibility. While the arrangement of solids in a soil is very difficult to quantify, the size distribution of pores may be simply accomplished by techniques of mercury intrusion porosimetry. The resulting frequency distribution may be characterized in any convenient statistical manner, and related quantitatively to other variables. A conventional technique of relating frost susceptibility to grain size distribution has a serious intrinsic deficiency in that a soil has a single grain size distribution, but a multitude of pore size distributions. It is, of course, the pore size distribution which controls the movement of fluids under all gradients. This has been demonstrated experimentally by producing a wide variety of frost heaving rates for a single soil; in all cases, the heaving rates were related to the pore size distributions in the soil. Pore size distribution has also been related to suction and hydraulic conductivity values. It is an appropriate substitution for (1) direct determination of the water characteristics curve and (2) measurement of permeability, in fine grained soils, since it requires but a fraction of the time required for these tests. Examples of all correlations cited above are contained in the paper.

INTRODUCTION

The fabric of soil significantly influences all aspects of its behavior, including frost susceptibility. This fact is illustrated by Figure 1, where such characteristics as grain size and shape, density, and total porosity are constant between the two parts of the picture. Fabric is seen to be strongly contrasting, whether evaluated in terms of the distribution of solids, or in terms of the size distribution of pores. Many susceptibility ratings would predict identical effects of frost action for the two soil conditions, although such can obviously not be true with strong differences in fabric.

The author supports the use of pore size distribution to quantitatively characterize fabric and differences in fabric. These characterizations have been successfully related to soil stress-strain (Ahmed et al. 1974), to hydraulic conductivity (Garcia-Bengochea et al. 1979), to name but a few responses.

PORE SIZE DISTRIBUTION

Successful application of mercury intrusion porosimetry to characterize soil and rock fabric required several research studies, culminating in the reports by Ahmed (1971) and Bhasin (1975). Limited and largely unsuccessful experience with the method prior to this time is summarized by Ahmed and Bhasin. A major stumbling block was the need to dry specimens, prior to pore size measurements, without change in the fabric. Two techniques were identified for fine-grained soils, viz., critical region drying and freeze drying. The latter is much more simple and less expensive than the former, and is the preferred technique. Freeze drying can be accomplished within a working day (Reed 1977). Porosimetry for soils and rocks has been sufficiently debugged to be the subject of a draft standard in Committee D18 of the American Society for Testing and Materials. Laboratory porosimeters are available "off the shelf" from a number of commercial equipment companies.

In the pore size distribution determinations, the nonwetting fluid (mercury) is forced into the pores by pressure. The absolute pressure (p) - pore diameter (d) relationship is given by the capillary equation,

$$p = -\frac{4\,\sigma_m \cos\theta}{d}$$

where σ_m = surface tension of mercury and θ = contact angle of mercury with the intruded material.

As the pressure is increased, smaller pore diameters are intruded. By precisely measuring the volume of mercury pushed into the pores by an increase in pressure, a cumulative curve of pore space vs. pressure (pore size) is generated. The intrusion and all calculations can be accomplished within a working day.

The evidence of a great number of experimental measurements on soil is that all (or essentially all) of the pore spaces can be intruded by mercury, which means that these spaces are interconnected (Bhasin 1975, Garcia-Bengochea 1978,

Reed et al 1980). These same measurements provide explicit proof that the total porosity of the soil is unchanged by the mercury penetration, and implicit evidence that the pore size distribution is also unchanged. In addition, it has been shown (Reed et al. 1979, Garcia-Bengochea et al 1979, Juang 1981) that the coarser portions of the porosity control the engineering properties of greatest interest. These spaces are penetrated at modest values of pressure.

Figure 2 shows both cumulative and differential representations of the pore size distribution of a compacted sand-clay mixture. The bimodal distribution is commonly observed with clayey soils. The larger mode is interpreted to represent the pores between grains or aggregations of particles (Garcia-Bengochea et al. 1979). The magnitude of the small pore mode is likely a "signature" of the clay minerals present, while the large one varies with the details of soil placement. For example, both the magnitude and frequency of the large pore mode can be changed by varying the compaction water content and the roller used.

Frost Heaving Rate

The importance of fabric to rate of frost heaving has been clearly demonstrated by Reed et al (1979). Figure 3 shows a wide range in frost heave rates for a single soil (90% silt-10% clay). The two compaction pressures are represented by three water contents each. The lower and upper values of water content represent equal porosities (densities) on the dry side and on the wet side of optimum, respectively. The dry side samples have a fabric more conducive to frost heaving than the wet side ones. Lower compaction pressures also allow a more open fabric in the fine-grained soil, and result in a higher frost heave rate.

Using the rating scale shown in Table 1, Reed (1977) clearly showed that any of three silt-clay test mixtures could be prepared to produce any level of frost susceptibility, simply by varying the compaction energy and water content. Reed also produced statistical euqations which related frost heaving rates to pore size distributions of the compacted soils. These relationships are available in the open literature (Reed et al. 1979 and 1980) and are too lengthy to be presented and explained here. The statistical credentials of these regressions are excellent...statistical equations of course require statistical input, as opposed to single values input, to produce reliable values.

Determination in Sands

Juang (1981) extended the porosimetry techniques to sandy soils, including clean sands. He accomplished the retention of fabric in cohesionless soils by mixing in a trace of thermally sensitive polymer prior to sample preparation. A subsequent heating of the sample melted the polymer and achieved a sufficient cohesion to allow handling and pore size distribution determination. The amount of polymer was insufficient to change the total sample porosity.

Changes in fabric are more subtle in sands than in fine-grained soil, but porosimetry is a feasible way of detecting these. Thus, the effects of sand fabric on frost susceptibility can be experimentally determined. Juang (1981) did not accomplish this, although he did produce excellent correlations between pore size distributions and hydraulic conductivity, for a range of soil textures from clay to sand. The analytical techniques developed to relate pore size distribution and hydraulic conductivity included: the Kozeny hydraulic radius model, a variable diameter capillary approach, and a simple (Marshall) probabilistic model (Garcia-Bengochea 1978, Garcia-Bengochea et al. 1979 and 1980, Garcia-Bengochea and Lovell 1981). The model ultimately used by Juang (1981) is a reasonably versatile probabilistic one, and its statistical reliability is illustrated in Figure 4. When the time and expense of direct permeability measurements are considered, the potential value of pore size distribution correlations becomes more obvious.

It is unlikely that the mercury intrusion porosimetry technique can be employed for soils of coarser texture than sands. There are two major reasons for this (Juang 1981): sample size must be rather small (approximately 1 cm cube), and penetration of pores larger than about 300 micrometers is difficult to control. Other techniques (e.g., optical) are needed to quantify the distribution of such macro-porosity in both intergranular and extended forms.

Determination of Moisture Characteristics

Wood et al. (1982) report that pore-size distribution data can be used to produce a soil moisture characteristic curve. This curve relates the water content and its associated capillary pressure.

The capillary pressure (p_w) equation for water has the same form as that previously stated for mercury, viz.,

$$p_w = \frac{4\,\sigma_w \cos\theta_w}{d}$$

$$d = \frac{-4\,\sigma_m \cos\theta}{p} = \frac{4\,\sigma_w \cos\theta_w}{p_w}$$

and

$$p_w = \frac{+\,\sigma_w \cos\theta_w\, p}{-\,\sigma_m \cos\theta}$$

where σ_w = surface tension of water, θ_w = contact angle of water with the soil, and the other terms have been previously defined.

The water content (w) for the partially saturated condition which matches this p_w may be calculated from

$$w = \frac{V_w \gamma_w}{W_s}$$

where V_w = volume of water, γ_w = unit weight of water, and W_s = weight of dry solids, and

$$V_w = V_v - V_{vm}$$

where V_v = total volume of voids and V_{vm} = volume of voids intruded by mercury at absolute pressure (p).

Direct determination of the soil moisture characteristics curve would normally require much more time than that required to define the pore size distribution. Therefore, the above correlations have important practical value.

Use as a Susceptibility Index

In his comprehensive reivew of the subject, Chamberlain (1982) identified several tests which are particularly promising for the prediction of frost susceptibility. These include grain size distribution, moisture-tension, hydraulic conductivity, and frost heave tests. Limited experimentation with pore size distribution on soil textures from clay to sand indicate that it too has great promise as an indicator test. To date, experiments have:

(1) illustrated the intrinsic limitations of the grain size criteria (Reed et al. 1979);

(2) shown that the frost heave is statistically related to pore size distribution (Reed et al. 1979);

(3) defined the relationships between hydraulic conductivity and pore size distribution (Garcia-Bengochea et al. 1979, Juang 1981); and

(4) demonstrated the determinations of moisture tension curves from pore size measurements (Wood et al. 1982).

In many cases, the pore size determination is less time consuming than direct determination of the index property with which it is correlated.

CONCLUSIONS

The mercury intrusion method of determination of pore size distribution in soils and rocks has been developed and improved over the past decade. The procedure is presently being standardized in Committee D18 (Soils and Rocks) of the American Society for Testing and Materials.

The pore size relations generated by this technique for soil textures from clay to sand have been statistically correlated with the following frost susceptibility indices: frost heaving rate, hydraulic conductivity, and moisture-tension relationships. In each case the pore size determination is expected to be less time consuming than the cited index quantity. The author concludes that the pore size method deserves more attention and application for definition of frost susceptibility than it has received in the past.

ACKNOWLEDGMENTS

The research summarized and interpreted in this paper was funded by the Federal Highway Administration of the U.S. Department of Transportation and the Indiana Department of Highways, and administered through the Joint Highway Research Project, School of Civil Engineering, Purdue University.

REFERENCES

Ahmed, S., 1971, Pore Size and Its Effects on the Behavior of a Compacted Clay, Master of Science in Civil Engineering Thesis, Purdue University, West Lafayette, Indiana, June, 200 pp.

Ahmed, S., Lovell, C. W. and Diamond, S., 1974, Pore Sizes and Strength of Compacted Clay, Journal, Geotechnical Division, American Society of Civil Engineers, V. 100, No. GT4, April, pp. 407-425.

Bhasin, R. N., 1975, Pore Size Distribution of Compacted Soils After Critical Region Drying, Doctor of Philosophy Thesis, Purdue University, West Lafayette, Indiana, May, 222 pp.

Chamberlain, E. J., 1982, Frost Susceptibility of Soils; Review of Index Tests, Report No. FHWA/RD-82/081, Federal Highway Administration, U.S. Department of Transportation, Washington, D.C., August, 110 pp.

Garcia-Bengochea, I., 1978, The Relation Between Permeability and Pore Size Distribution of Compacted Clayey Silts, Joint Highway Research Project Report No. 78-4, School of Civil Engineering, Purdue University, West Lafayette, Indiana, April, 179 pp.

Garcia-Bengochea, I., Lovell, C. W. and Altschaeffl, A. G., 1979, Pore Distribution and Permeability of Silty Clays, Journal, Geotechnical Engineering Division, American Society of Civil Engineers, V. 105, No. GT7, July, pp. 839-856.

Garcia-Bengochea, I., Lovell, C. W. and Altschaeffl, A. G., 1980, Closure to Pore Distribution and Permeability of Silty Clays, Journal, Geotechnical Engineering Division, American Society of Civil Engineers, V. 106, No. GT10, October, pp. 1168-1170.

Garcia-Bengochea, I. and Lovell, C. W., 1981, Correlative Measurements of Pore Size Distribution and Permeability in Soils, Permeability and Groundwater Contaminant Transport, American Society for Testing and Materials, Special Technical Publication 746, ASTM, October, pp. 137-150.

Juang, C. H., 1981, Pore Size Distribution of Sandy Soils and the Prediction of Permeability, Joint Highway Research Project Report No. FHWA/IN/JHRP-81/15, School of Civil Engineering, Purdue University, West Lafayette, Indiana, August, 109 pp.

Reed, M. A., 1971, Frost Heaving Rate of Silty Soils as a Function of Pore Size Distribution, Joint Highway Research Project Report No. 77-15, School of Civil Engineering, Purdue University, West Lafayette, Indiana, September, 116 pp.

Reed, M. A., Lovell, C. W., Altschaeffl, A. G. and Wood, L. E., 1979, Frost-heaving rate predicted from pore-size distribution, Canadian Geotechnical Journal, V. 16, No. 3, August, pp. 463-472.

Reed, M. A., Lovell, C. W., Altschaeffl, A. G. and Wood, L. E., 1980, Reply to Discussion on Frost-heaving rate predicted from pore-size distribution, Canadian Geotechnical Journal, V. 17, No. 4, November, pp. 639-640.

Wood, L. E., Altschaeffl, A. G. and Lovell, C. W., 1982, Final Report: Effects of Pore Size Distribution on Permeability and Frost Susceptibility of Selected Subgrade Materials, Joint Highway Research Project Report No. FHWA/IN/JHRP-82/17, School of Civil Engineering, Purdue University, West Lafayette, Indiana, October, 30 pp.

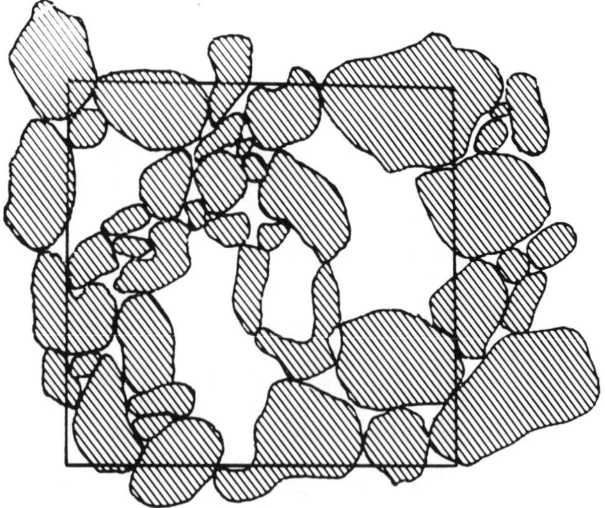

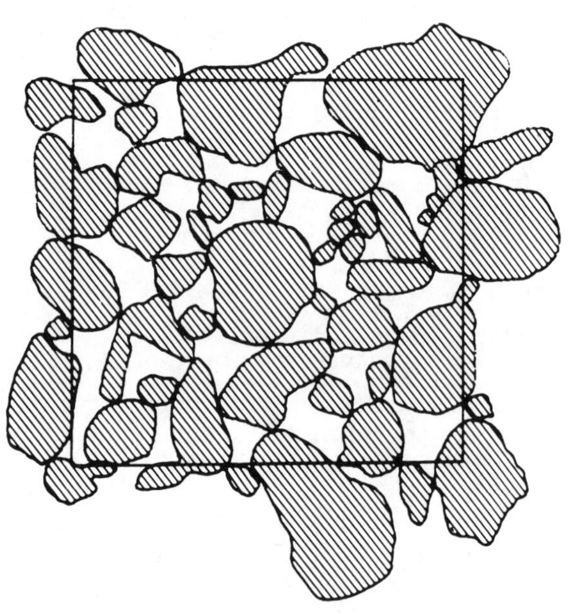

FIGURE 1 Schematic diagram of soil fabric.

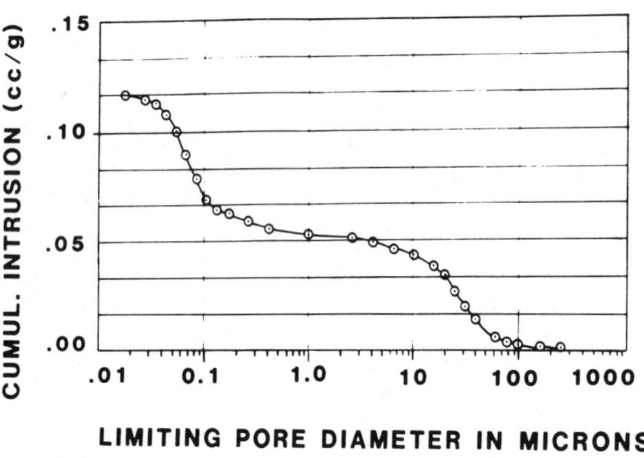

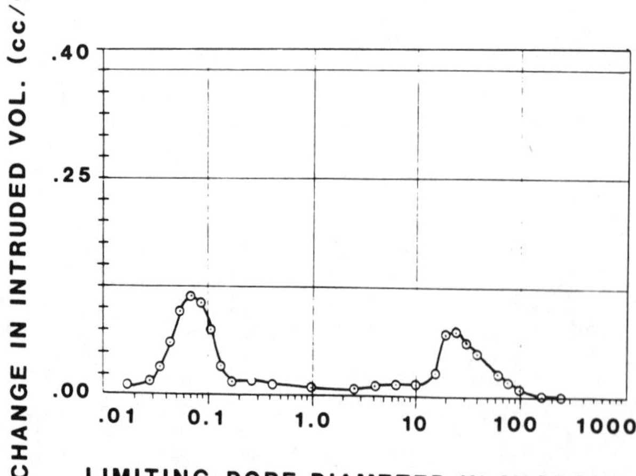

FIGURE 2 Pore size distribution curves for compacted sand-clay.

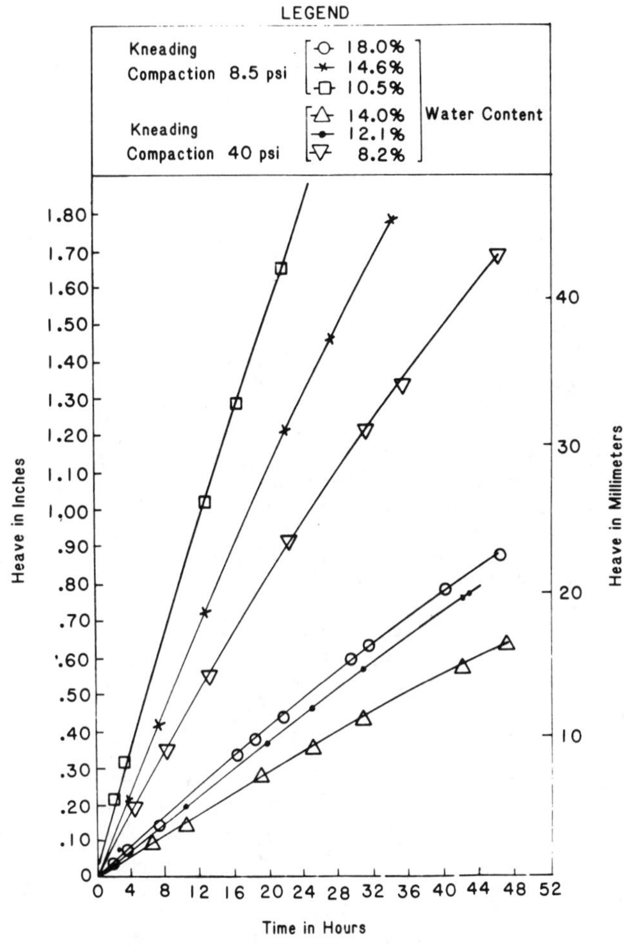

Figure 3 Typical frost heave curves for 90% silt-10% kaolin.

TABLE 1 Frost Susceptibility Ratings of Test Soils.

24 Hour Frost Heave (mm)	Frost Susceptibility Rating
< 6	Very low
6 - 11	Low
11 - 20	Medium
20 - 30	High
> 30	Very High

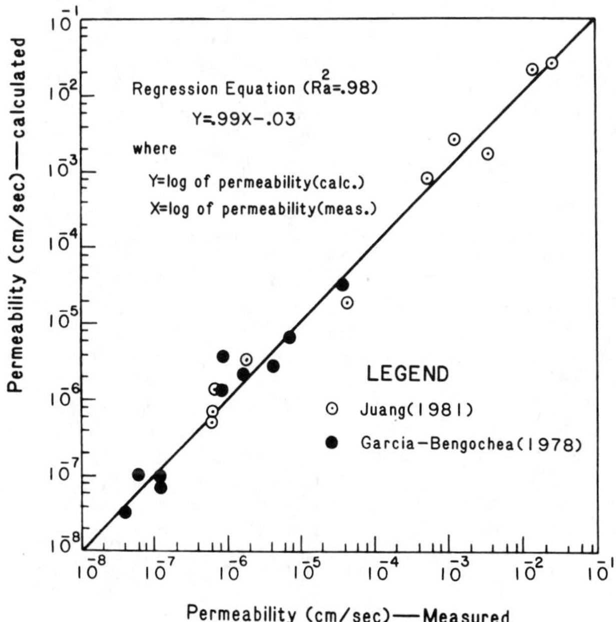

FIGURE 4 Prediction of permeability from pore size distribution.

BASIC CHARACTERISTICS OF PERMAFROST IN NORTHEAST CHINA

Lu Guowei[1], Guo Dongxin[2], and Dai Jingbo[3]

[1] Yakeshi Survey and Design Institute of Nei Monggol Ministry of Forestry
[2] Lanzhou Institute of Glaciology and Cryopedology, Academia Sinica
[3] Third Survey and Design Institute, Ministry of Railways
People's Republic of China

There are two kinds of permafrost in northeast China: high-latitude permafrost and alpine (high-altitude) permafrost. The high-latitude permafrost is distributed mainly in the Da-Xiao Hinggan Ling and occupies an area of about 38,000 km^2, while the alpine permafrost is only sporadic. From south to north in Northeast China, the mean annual air temperature drops from 0°-1°C down to -5° or -6°C, and the landscape changes from forest-steppe, through mixed coniferous and broad-leaf forest, to coniferous forest, and the permafrost varies correspondingly in continuity, temperature, and thickness. In the south, there is insular permafrost, 5-10 m thick, with a mean annual ground temperature of 0°C. In the north, the large-patch continuous permafrost has a mean annual ground temperature of 0° to -2.5°C (and in some places even lower than -4.2°C) and is 80-100 m thick. Between these two extremes there is a transitional permafrost zone with insular taliks. The southern limit of permafrost descended to the upper course of the Liao He River during the last glacial period. The existing permafrost is the sum of relic permafrost formed in the Late Pleistocene and younger permafrost formed in the last 3000 years.

There are two types of permafrost in Northeast China, i.e. high-latitude and alpine permafrost. The former mainly occurs in the Da and Xiao Hinggan Ling and occupies an area of about 380,000-390,000 km^2. The alpine permafrost is scattered in Changpai Shan Mountain and Huanggang Liang Mountain situated in the southern end of Da Hinggan Ling.

HIGH-LATITUDE PERMAFROST

Natural Conditions Forming High-Latitude Permafrost

Both the Da and Xiao Hinggan Ling are in the northernmost part of China, and stretch northeastward and northwestward through the eastern and western parts of the region respectively, separated by the Nen Jiang Valley. Thus, the landforms of the region are higher in the eastern and western parts, but lower in the middle. The Da and Xiao Hinggan Ling form the shape of "λ" on a plane. The elevation in the middle section of Da Hinggan Ling is 1000-1400 m and descends gradually to the north to 500-600 m northward. Da Hinggan Ling is an asymmetrical mountain. The west-facing slope consists of undulating hills, but the east-facing slope is steep and is deeply cut by numerous river valleys. However, the Xiao Hinggan Ling has more gentle slopes with dendritic type of valleys. The mean elevation of this mountain is 500-600 m with some peaks reaching up to 800 m.

Since the Late Tertiary, the Da Hinggan Ling Mountain has risen unevenly parallel to an old tectonic system undergoing planation and denudation for a long time. On the top and along the slope of the mountains, 1-2 m of colluvium, eluvium and slope wash occur consisting of crushed stones, angular clasts or silty clay with crushed stones. The loose deposits in pediments, intermontane valleys and basins are relatively thicker and consist mainly of the solifluction, slope wash, and pluvial sediments with peat and humus soil above them, and clayey silt, silty clay with crushed stones, sand, and gravel below them, usually with a total thickness of less than 10 m.

The climate in this region belongs to the continental monsoon of the cold-temperate zone. From south to north and southeast to northwest, the mean annual air temperature (MAAT) drops from 1°-0°C to -5° or -6°C, the annual precipitation decreases from 500-600 mm to less than 200 mm, while the amplitude of mean daily air temperature increases from 35°C to 50°C.

The main feature of climate in the region is severe cold, especially in winter, because of the influence of the high pressure over Siberia and Mongolia. The strong cold air mass current often intrudes into this region, greatly lowering the air temperature, so that in January the mean air temperature is 5°-10°C lower than that in Xinjiang at the same latitude.

In summer, this region is affected by the Pacific high pressure cell. The southeast monsoon is strengthened, warm air with moisture from south meets the cold air from north in this region, producing heavy rains so that 80-90% of the annual precipitation is concentrated in the summer. The associated cloud keeps the air temperature cool, so that the air temperature in summer is 1-3°C lower than that in Xinjiang at the same latitude. Thus, the low mean annual air temperature constitutes not only the basic condition for the existence and growth of permafrost, but also one of the main causes of the southern boundary of Eurasian permafrost protruding southward in this region.

Under the control and influence of Siberian and Mongolian high pressures, the winter temperature inversions are widespread and stable, and have an important effect on both the regional differences of permafrost and the process of permafrost development.

Comparison and analysis of the air temperature at the 850 mb (1500 m) elevation indicate that this region is located in the southern margin of the inversion layer controlled by Siberian high pressure. In the Middle Siberian Mountains the thickness of the inversion layer is 1200-1500 m, the inversion gradient is 1.5-2.0°C/100 m (Fotiyer, S. M. et al. 1974), but in China the former is 500-1000 m, and the latter is 1.0°C/100 m in Mohe, 0.8°C/100 m in Nunjiang, and 0.5°C/100 m in Hailar.

Most of the region is covered with forest, shrub and sedge, and is known as the "green ocean". The northern part belongs to the cold-temperate montane coniferous forest consisting of Hinggan larches, and southward it gradually becomes coniferous-broad leaf mixed forest and forest-steppe. To the southern edge of the region, crops can be planted.

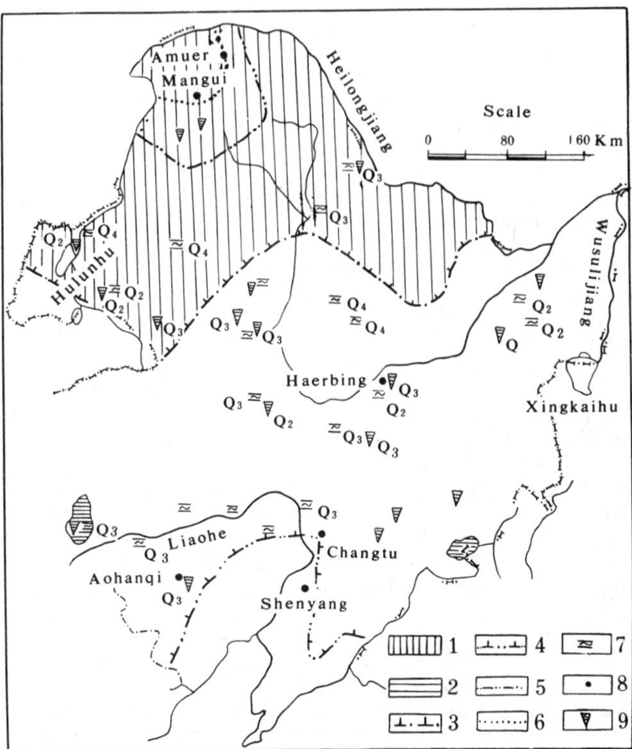

FIGURE 1 The distribution and south boundary change of permafrost from late Pleistocene to now in northeast China. 1) existing high latitude permafrost, 2) island alpine permafrost, 3) existing south boundary of permafrost, 4) south boundary of permafrost during late Pleistocene, 5) existing south boundary of continuous permafrost with patch taliks, 6) inferring south boundary of permafrost during Hypsithermal interval, 7) involution layers, 8) town, 9) sand wedge.

Distribution, Temperature and Thickness of High-Latitude Permafrost

The formation of frozen ground is the result of heat exchange between the atmosphere and the lithosphere. The above mentioned natural factors, which act jointly on heat exchange, exert great influence on the features of permafrost. Therefore, the latitudinal zonation of natural conditions from south to north and from southeast to northwest as mentioned above should surely reappear in the features of permafrost (distribution, temperature, thickness, ground ice, etc.). Along the directions mentioned above, from south to north, the area of permafrost increases from 5-20% to 70-80%; the ground temperature drops from 0° -1.0°C to -2.0°C with the lowest being -4.2°C. The average thickness of permafrost increases from 5-20 m to 70-80 m, with extreme cases with more than 100 m (Guo Dongxin 1981). The type of permafrost changes gradually from insular permafrost into insular talik and large patch continuous permafrost (Figure 1).

Because of the differences in landform, temperature inversion, vegetation and thickness of loose deposits, and of the combined influence of various natural factors, such as surface water, the above-mentioned principal regularities of the permafrost features are made less obvious and on the background of latitudinal zonation, the distribution of permafrost also shows the regional and local differences.

Because of the influences of landform and elevation (Figure 1), the permafrost southern boundary does not trend parallel to the latitude, but presents a "W" shape. The southern boundary of permafrost obviously stretches out southward along the limbs of the Da and Xiao Hinggan Ling but extends northward in the Nunjiang plain. The difference of latitude between them is nearly two degrees. Besides, as mentioned above, the lower mean annual air temperature in this region due to the influences of Siberian and Mongolian high pressure cells is another important factor causing the southern boundary of Eurasian permafrost to be extended southward.

The northern part of the region belongs

to the large patch continuous permafrost zone. The distribution, temperature, and the thickness of permafrost, as well as the growth of ground ice at different landform units of the same region are quite different owing to the influences of the above mentioned factors.

In the shady intermontane valleys, in marshlands and lower terraces covered with dense moss and lichen, as well as beneath thick loose deposits (8-10 m), the permafrost develops very well and is up to 60-80 m thick, occasionally reaching more than 100 m in places, with ground temperatures of -2° to -4.2°C and layered ground ice. In contrast, on sunny and semi-sunny slopes with sparse trees and thinner loose cover, generally the permafrost develops less well and may be absent, its thickness is less than 20 m with ground temperatures of 0° -1.0°C, and there is no ground ice. According to the data of adjacent region, when the slope is 20°-22°, the value of direct solar radiation is 92 kcal/cm^2.yr, on the sunny slope, but 42.6 kcal/cm^2.yr on the shady slope (Shpelynskaya, N. A., 1978). The thickness of loose deposits on the shady and semi-shady slopes is as large as that on the sunny slope, but the shady slope has more dense vegetation and less solar radiation than sunny one. Therefore, the growth of permafrost on shady slope is interposed between the valley bottom, depression, low terrace and sunny slope.

In the northern part of the region, under large river beds and within 10-30 m on either side of a large river, the permafrost is usually absent and open talik exists. Under smaller river beds, the permafrost table generally descends and closed talik forms.

The southern part of the region belongs to the insular permafrost region. The permafrost features vary according to the natural environment also. In the high plain of Hulun Buir, the intermittent marsh zone evolved from paleochannels, and the pools in the centre of this marsh are the outstanding characteristics of the landscape. Marsh and pool are closely related to island permafrost. Generally speaking, in the margin of marsh and pool, there is always scattered insular permafrost with its area less than 10% of total area, its thickness 3-5 m to 10-15 m and its temperature 0° to -1°C. In the undulating ground on the east-facing and west-facing slopes of Da Hinggan Ling in the southern part of the region, the permafrost islands are distributed along the flood plains and low terraces in intermittent bands. Northward from the southern boundary the permafrost increases from several dozens of meters and 100-200 m to several hundred meters in size, covering 10% to 30% in area. The dominant permafrost thickness is 5-10 m with a ground temperature of 0° to -1°C. In Xiao Hinggan Ling Mountains, the permafrost islands occur only in the intermontane valleys, flood plains land and low terraces with dense vegetation and marsh. The area of permafrost is about 20% of the total area, with permafrost thicknesses of 5-10 m, ground temperature colder than -1.0°C. In the southern part of the region, the permafrost in profile is quite complex (some connected, some disconnected, and some multi-layered.) due to different soil types and the influence of ground water.

On the whole, either in the south or in the north of the region, the distribution and growth of permafrost is better in the low land than in the high land, having greater thickness and lower temperature, and is obviously different from the permafrost existing in the western mountains and plateau, so it is called "Hinggan Ling-type permafrost". The characteristics of this type of permafrost are not only reflected in the distribution and growth, but also exert an influence on the process of permafrost degradation. As early as the 1960's, it was found that the permafrost degradation occur first on the south-facing slopes, high land and mountains, and only later on the north-facing slope, low lands and valley bottom. Although the formation of such a characteristic of Hinggan type permafrost is the result of comprehensive action of various natural factors, the wide and stable inversion in winter, thick unconsolidated deposits, and dense vegetation in the valley bottom and depression exert a dominant control.

ALPINE PERMAFROST

The permafrost found south of the southern boundary of contemporary permafrost in the Da and Xiao Hinggan Ling and above a designated elevation is called alpine permafrost. This type of permafrost mainly occurs in the Changpai Shan Mountain and Huanggang Ling Mountain which is located in the southern part of Da Hinggan Ling. The natural conditions for growing permafrost and the lower limit in these two mountains are also different.

Huanggang Ling Mountain situated in the western part of Songliao plain and the eastern part of Nei Monggol high plain, is the highest mountain at the southern end of Da Hinggan Ling, with elevations exceeding 1400-1500 m. The climate in this region is semi-arid continental and is influenced by the Mongolia high pressure cell. The mean annual air temperature at an elevation of 1500 m is -2.9°C. The landscape has an obvious vertical zonation with the semi-arid steppe and desert zone below 1500 m; sparse coniferous-broad leaf mixed forest above 1500 m, and marshes on the gentle hill top and north-facing slope. The lower limit of permafrost on the mountain is 1600 m. The permafrost islands mainly occur on the gently sloping hill top and in the wet, marsh areas. The permafrost thickness is less than 10 m and most of it is 3-5 m thick.

Changpai Shan is the highest mountain in Northeast China. There are 16 peaks all over 2500 m high. The montane climate and landscape possess an obvious vertical zonation: coniferous-broad leaf mixed forest occuring below 1000 m, montane dark coniferous forest at 1000-1600 m, dark coniferous forest at 1600-1800 m, birch forest at 1800-2000 m, and alpine tundra above 2000 m. The mean annual air temperature at an elevation of 2670 m is -7.3°C.

Up to now there have been no studies of location of the lower limit of permafrost. Taking the 0°C mean annual ground surface temperature as a criterion for existing permafrost and considering the higher value of geothermal heat flow in this region, we suggest that the lower limit of alpine permafrost should be at an elevation of 1800 m (mean annual air temperature of -3° to -4°C), roughly coincident with the upper limit of dark coniferous forest.

HISTORY OF PERMAFROST SINCE THE LATE PLEISTOCENE

During the Quaternary Period, there were several occurrences of aggradation and degradation of permafrost with cold climate alternating with warm. According to the periglacial remnants and other evidence, the history of permafrost since the Late Pleistocene may be divided into three parts (Guo and Li 1981).

The Periglacial Conditions in the Late Pleistocene

The large number and wide extent of periglacial remains belonging to this stage indicate that permafrost was much more extensive. According to the paleo-air temperature calculated from fossil plant and pollen, it is estimated that within the Liao He Plain, the 0°C isotherm of mean annual ground surface temperature at that time stretched roughly along the northern part of Aohanqi and Changtu, and coincided with the southern boundary of relic permafrost at that time. Moreover, both of them roughly coincided with the 7°-8°C isotherm of the present mean annual air temperature. Therefore, we have estimated the southern boundary of permafrost in the Late Pleistocene in regions lacking relic permafrost as having lain at the present 7°-8°C isotherm.

The Hypsithermal Interval

During the Hypsithermal warm period lasted from 3000-7000 years B.P. According to the paleo-air temperature calculated from spore-pollen analysis, it is estimated that the air temperature at that time was about 0°C in the northern part of the existing large patch continuous permafrost zone (the region to the north of Mangui). On the other hand, the analyses of light and heavy minerals of unconsolidated sediments in drill hole No. 50 at Armer indicate that humid and warm climate was absent there at the Hypsithermal Interval. According to this, it is suggested that the permafrost in the northern half part of existing large patch continuous permafrost zone might not have been entirely destroyed during the Hypsithermal Interval.

Post-Hypsithermal Permafrost

Beginning around 3000 years B.P., the thawed permafrost regenerated as the temperature dropped. In the northern part of the existing large patch continuous permafrost region, the ground temperature of relic permafrost formed during the Late Pleistocene decreased, its thickness increased and the permafrost has been developing to present state. In the southern part, new permafrost formed. In the 17-18th century, the young permafrost reached its maximum. The alpine insular permafrost in Huanggang Ling Mountain above 1600 m and Changpai Shan Mountain above 1800-2000 m is the product of that stage.

REFERENCES

Fotiyev, S. M., Darnilova, N. S., and Saviliva, N. S., 1974, Geocryological conditions in Middle Siberia, Moscow: Science Press.

Guo Dongxin, 1981, Division of permafrost regions in Daxiao Hinggan Ling Northeast China, Journal of Glaciology and Cryopedology, v. 3, No. 3, p. 1-9.

Guo Dongxin and Li Zuofu, 1981, Preliminary approach to the history and age of permafrost in Northeast China, Journal of Glaciology and Cryopedology, v. 3, No. 4, p. 1-15.

Shpelynskaya, N. A., 1978, Permafrost in Outbaykal. Moscow: Science Press.

PERMAFROST ON MARS: POLYGONALLY FRACTURED GROUND

Baerbel K. Lucchitta

U.S. Geological Survey, 2255 North Gemini Drive, Flagstaff, Arizona 86001 USA

Newly discovered martian polygons of 50-300 m diameter have random orthogonal patterns and are similar in size and shape to terrestrial unsorted polygons in permafrost terrain. The martian polygons occur at lat 47°N, long 346°W, where conditions are favorable for ice-related fracturing processes; ice is in equilibrium with the atmosphere and could remain in the ground for extended periods. Desiccation of ground ice due to astronomically forced climatic cycles, or thermal contraction due to seasonal cycles could have fractured the ground. The desiccation hypothesis has uncertainties concerning the temperature regime and the exchange of water between the atmosphere and the ground. An origin by seasonal thermal contraction, however, is compatible with the size and shape of the polygons, with a seasonal cooling of -60°C, and with the availability of seasonal frost. Both sand wedges and ice wedges could have filled the polygonal fractures. Crispness of the fractures suggests that they are only slightly degraded and thus relatively young, but local ground collapse indicates that they are probably not forming actively today.

INTRODUCTION

Mars is the only explored body in our Solar System that has many landforms similar to terrestrial ones, especially those associated with permafrost or cold climate. Permafrost, 1-3 km thick, probably occurs over the entire planet because the mean temperature is -80°C (Fanale 1976). Rossbacher and Judson (1981) calculated this thickness, assuming the heat flow values of Toksöz and Hsui (1978) and a surface layer composed of hard-frozen limonitic soil with a high (60%) ice content. Near the equator, midday temperatures may reach 15°C (Michaux and Newburn 1972), but nighttime temperatures remain far below 0°C, and, even though freeze-thaw processes are possible, the formation of a thick active layer is unlikely. Although liquid water cannot exist at the surface because of the thin (6.1 mbar) atmosphere, the presence of fluvial features, such as channels, implies that water once flowed on the surface of Mars. Remote-sensing measurements suggest that water presently exists in the atmosphere and in the polar caps (Kieffer et al. 1977, Farmer and Doms 1979). At the Viking landing sites water in the soil was detected from instrumental readings (Anderson and Tice 1978), and in late winter a thin frost layer was observed by the cameras on the Viking 2 Lander (Wall 1981). Ground ice would be in equilibrium with the atmosphere north and south of lat 40° (Farmer and Doms 1979), and even in the equatorial areas, ice could remain in the ground if blanketed by fine debris (Smoluchowski 1968). Thus, conditions on Mars are conducive to permafrost processes, and permafrost features can be expected on its surface. Such features were illustrated in the summary reports by Carr and Schaber (1977) and Lucchitta (1981).

LARGE POLYGONS

Polygonal-fracture patterns on the martian surface were discovered on Viking Orbiter images; these patterns are abundant throughout the northern latitudes. The polygons are 2-20 km in diameter, much larger than those of known patterned ground on Earth (Washburn 1980). That martian polygonal ground may be related to thermal contraction was suggested by Carr and Schaber (1977), but Black (1978) objected because the size of the polygons is incompatible with the thermal-contraction process as observed on the Earth. Alternative hypotheses include contraction by cooling lava (Morris and Underwood 1978) or tensional tectonism (Pechmann 1980). Lava cooling, however, should form small polygons, and the tensional tectonism should not be restricted to northern latitudes.

Helfenstein and Mouginis-Mark (1980) pointed out that thermal-contraction polygons 2-3 times the size of terrestrial ones could have formed under Martian lithostatic stresses and suggested that desiccation later preferentially enlarged and coalesced some polygons to form the network presently observed. Also, cooling cycles on Mars may be 10^5-10^6 years long and related to changes in the obliquity and eccentricity of Mars' orbit rather than seasonal. Using secular cycles of 1.2×10^5 years, Coradini and Flamini (1979) calculated that Mars would have an "active layer" of 100 m; isotropic tensile horizontal stresses penetrating to such depths could form polygons of kilometer size (Helfenstein 1980). Nevertheless, the origin of the 3-7-km diameter polygons, most common in the northern plains of Mars, remains enigmatic.

SMALL POLYGONS

Several new observations show that some patterned ground on Mars has polygon diameters similar to those of ice-wedge polygons on Earth (generally a few meters to more than 100 m, Washburn 1980). Evans and Rossbacher (1980) noticed networks of polygons outlined by ridges in the smooth plains material of Lunae Planum (lat 17°N, long 56°W). These polygons range in diameter size from 400-1,000 m; considering martian conditions, those dimensions are comparable to those of ice-wedge polygons on the Earth. Brook (1982) has discovered light-colored polygonal networks on a layered tableland (lat 23°N, long 35°W). The polygons range from 100-1,000 m in diameter and are concentrated near the rims of several flat-floored, steep-walled depressions that were previously interpreted as alases (thermokarst collapse depressions) by Carr and Schaber (1977) and Lucchitta (1981). Brook hypothesized that the association of polygons and alases indicates ground ice and that the network is composed of ice-wedge polygons.

A recently discovered area underlain by polygonal ground occurs on a platform in Deuteronilus Mensae, near lat 47°N, long 346°W (figs. 1,2). The polygons measure 50-300 m in diameter, but smaller ones may also exist locally; faint networks can be detected down to the limit of resolution of the pictures. The interiors of these polygons are dark, similar to the polygons observed by Brook (1982). The polygons occur mostly on level ground but are also found on the sloping rim of a crater (8 km diam.), where the light-toned network can be clearly identified as linear depressions or cracks. Fractures parallel to the slope are emphasized, similar to what is observed on some terrestrial polygons (Washburn 1980). No upturned margins are visible; it appears that most of the polygons are flat or high centered. Apparent influence on the shape of the polygons by subparallel linear structures resulted in a network pattern most compatible with Lachenbruch's (1962) random-orthogonal class. Even though some small (about 500 m diam.), subdued old craters have polygonal fractures on their rims, most other small craters and their ejecta are devoid of cracks. The absence of such features may indicate that the craters are superposed and thus younger than the polygonal ground, or that the craters are older and the ejecta have physical properties which do not support formation of polygons. Some craters are, indeed, older than the polygons because they are surrounded by light radial lines that appear to merge with the polygonal ground pattern; these polygons apparently developed along preexisting radial impact fractures. Other radial lines may be rays. Figure 2a shows that polygons become less well defined northeastward; the light-toned networks widen, the dark centers shrink, and the polygonal patterns are more difficult to trace. Eventually, the patterned ground dissolves into a field of dark spots or low hummocks on a lighter-toned background.

Terrestrial patterned ground may be sorted or unsorted (Washburn 1980). Sorted patterned ground has circles, polygons, or nets that would be below the resolution of available images on Mars. Most large polygons on the Earth are unsorted; they may be more than 100 m in diameter, have cracks as much as 30 m deep, and occur in well-sorted fines, sand, gravel, or poorly sorted glacial deposits (Washburn 1980). Patterns may be dominated by orthogonal intersections (random or oriented orthogonal systems) or by triradial 120° intersections (nonorthogonal system) (Lachenbruch 1962). Orthogonal systems, which are common, develop mostly in nonhomogeneous or plastic media, or wherever loci of low strength or high stress predisposed certain areas to cracking; and some fractures predate others. Pure nonorthogonal systems are rare (Kerfoot 1972); they develop in homogeneous materials under isotropic tensional-stress regimes, and cracks propagate from point sources simultaneously. Kerfoot's study showed that on close inspection, most apparent triradial 120° intersections are orthogonal or at 60°. The polygons on Mars are similar in shape to unsorted random-orthogonal polygons on the Earth. Figures 1 and 2 show predominantly random-orthogonal systems that locally appear to be influenced by structural lines of weakness; these lines include fractures radial to craters, on slopes, and in subparallel sets of unknown origin.

ORIGIN OF POLYGONS

Unsorted polygons can be formed by desiccation, thermal contraction, or phase changes (Lachenbruch 1962, Washburn 1980). Desiccation polygons, for example, occur in mud and are generally less than 1 m in diameter (Washburn 1980). Phase-change polygons occur in cooling lava and thermal-contraction polygons in ice or salt. Salt polygons, on the Earth, are restricted to a few hot deserts. Polygons due to thermal contraction of ice are common at high latitudes, where they are known as tundra polygons or ice-wedge polygons. Fundamental to their formation are ice-rich ground, temperatures below -6°C (Péwé 1973), and rapid cooling with a large temperature differential. The fractures that form when the ground cools and contracts become filled with ice (ice-wedge polygons), or with dirt (sand-wedge polygons) where moisture is scarce. The polygons are not restricted to the active layer, but form within the permafrost layer, owing to seasonal temperature changes. Active polygons (low centered) have upturned edges caused by expansion of material during the warm season; inactive polygons (high centered) are surrounded by troughs from melting of ice wedges. Upon further degradation of wedges and ground ice, troughs widen, polygon centers slump into troughs, fields of hills (baydjarakhs) develop, and eventually the terrain disintegrates into thermokarst depressions.

An origin of the martian polygons by contraction of cooling volcanic materials is implausible because of the age relations of cracks and impact craters. Lavas or ash flows would develop cracks immediately upon cooling, and all ensuing impact craters would be superposed on the cracks. However, the pictures show that some polygons developed on the rims of small craters and that, locally, crater and polygon fractures apparently interacted. Therefore, it is more likely that some time elapsed between the cooling of possible flows and the formation of polygons.

An origin by desiccation, though possible for the martian polygons, presents some difficulties. On the Earth, desiccation polygons are commonly small, whereas the polygons on Mars are large. However, the common occurrence of desiccation polygons on the Earth in thin layers of mud predisposes cracks to be limited to shallow depths and polygons to small sizes. By contrast, desiccation on Mars may have affected materials to considerable depth, the desiccation being caused by seasonal or secular climatic cycles. Since thick layers may thus have been affected by desiccation on Mars, the polygons might have grown commensurately larger than those on the Earth.

The areas on Mars most severely desiccated lie in the equatorial zone between lat ±40° where water is in disequilibrium with the atmosphere and depleted from the ground (Fanale 1976). A wealth of very high resolution images fail to show polygonal ground in those regions. The fracture patterns observed by Evans and Rossbacher (1980) lie in the equatorial area, but the patterns are composed of ridges rather than fractures, and so contraction due to desiccation may not be the appropriate mechanism. The pattern detected by Brook (1982) is a better candidate for desiccation: It also lies equatorward of lat 40° and the polygons are associated with well-developed thermokarst depressions (Lucchitta 1981). Brook has interpreted this relation to indicate disintegration of ice wedges, although the loss of ice could also have been by pervasive desiccation of the ground. The distinction between polygons formed by desiccation or by thermal contraction may be difficult on Mars on the basis of observational evidence alone because both processes require the existence of ice-rich ground, which in both processes may generate features of ground collapse.

Recent seasonal desiccation to the depths required to form polygons 50-300 m in diameter is unlikely at the latitude of Deuteronilus Mensae, where the newly discovered patterned ground is situated, because water there is in equilibrium with the atmosphere year round and may remain in the ground below about 1 m depth (Farmer and Doms 1979). During astronomic cycles of 10^5-10^6 years, however, resulting in a temporary high obliquity of Mars' rotational axis, the ground might have possibly warmed to sufficient depth for large-scale desiccation. Fanale and Jakosky (1982), however, have considered that water exchanged between the regolith and the atmosphere would be greatly inhibited by carbon dioxide pore gas and surface absorption, so that the depth of desiccation should not exceed a few meters. Helfenstein (1980) calculated that these climatic cycles have large effects near the poles but only negligible effects near lat 45°, and so fractured ground near and equatorward of this latitude would probably not be affected much by secular climatic cycles. These considerations indicate that, near the poles, desiccation due to climatic cycles may have created polygons there, but also indicate that secular desiccation is not a likely mechanism for the origin of the patterned ground in the latitudes discussed here and by Brook (1982).

An origin of the martian polygons as thermal-contraction cracks in ice-rich ground appears to be most compatible with the observations. Ground ice is a prerequisite for thermal-contraction polygons. As pointed out above, theoretical considerations suggest that ice is currently stable in the martian ground below about 1 m depth at the latitudes of the patterned ground here discussed. According to Coradini and Flamini (1979), only water ice would be able to withstand the seasonal thermal wave into the ground; clathrates and carbon dioxide ice would evaporate too quickly.

Observational evidence also suggests that ice exists in the martian ground. As Figure 2 shows, the polygonal network appears to disintegrate into a field of hummocks northeastward across the picture. This feature is similar to the collapse of ice-rich ground on the Earth, where polygons disintegrate into fields of baydjarakhs. The destruction process on Mars, however, is more difficult to explain. As discussed above, evaporation would not occur under present climatic conditions below about 1 m depth, but ice at the surface could be seasonally depleted, and any disturbance, such as slumping, could initiate exposure of deeper ice to the atmosphere and trigger progressive loss of ice and collapse. Also, desiccation to a few meters depth during a past warmer period of the secular climatic cycle, as proposed by Fanale and Jakosky (1982), may have caused loss of ice and partial disintegration of the terrain. This process, however, as mentioned above, is less likely.

The seasonal temperature change in the area discussed is substantial. The mean diurnal temperature ranges from -50°C in the summer to -110°C in the winter during a martian year, a drop of 60°C (Kieffer et al. 1977). The amount of this drop is more than adequate to initiate thermal contraction in ice-rich ground. A decrease in the linear coefficient of expansion of ice with decreasing temperature (from 40×10^{-6}/°C at -50°C to 30×10^{-6}/°C at -110°C; Hobbs 1974) would lessen the thermal contraction at the lower temperatures but would not prohibit the formation of thermal-contraction polygons at the cold temperatures of the martian surface.

If the polygons on Mars are, indeed, thermal ice-contraction cracks, the question remains whether the fractures are filled by ice wedges or sand wedges. The climate on Mars is extremely dry; only very few to about 30 precipitable microns of water exist at this latitude on Mars in winter and summer, respectively (Jakosky and Farmer 1982). Wind-driven material, however, is ubiquitous on Mars and would readily fill cracks. The lighter tone of the polygonal network also is compatible with infilling by drift: sand wedges, therefore, are plausible. On the other hand, as seen at the Viking 2 Lander site, waterfrost is precipitated in winter on Mars (Wall 1981), and crystals may grow in cracks. On the Earth, ice wedges form mostly by percolation of meltwater into previously opened cracks (Leffingwell 1919, p. 205-214), although windborne snow may also contribute (Black 1976). Windblown movement of frost is possible on Mars in winter and could contribute to the infilling of cracks and the growth of ice wedges. Furthermore, the polygonal ground observed is transitional with the hummocky ground. This observation suggests that disintegration occurred along the polygonal-fracture lines, which, in turn, suggest that ice was contained in the fractures. Overall, the evidence is compatible with ice-wedge polygons, although sand-wedge polygons cannot be ruled out; the fractures probably contain a mixture of dirt and ice.

It is difficult to ascertain whether the polygons are actively forming today or are relics from the past. The crispness of some cracks suggests a recent origin, but the absence of recognizable upturned edges (indicating actively forming ice wedges), the locally disintegrating ground, and the possible superposition of some rayed craters suggest that the polygons may be old. The polygons may belong to a former part of the climatic cycle, when the temperatures were colder or when air currents favored supply of moisture to the region.

The composition of the material in which the polygons are developed is not known; it could be regolith on old volcanic surfaces, windblown deposits, or other sediment. The proximity of several channels, including a previously unknown outflow channel, suggests that the surface might be covered by fluviatile or glacial (Lucchitta et al. 1981, Lucchitta 1982) materials. Corte (1963) noted that the largest and best developed ice-wedge polygons occur on somewhat elevated, relatively well-drained and coarse outwash from glaciers. Deposition of similar sediment in this area on Mars may have enabled the polygonal ground to become exceptionally well developed there.

CONCLUSIONS

Newly discovered polygonal ground on Mars occurs in Deuteronilus Mensae at lat 47°N, long 346°W, near the northern margin of the martian highlands. The polygons measure 50-300 m in diameter and form cracks in random orthogonal patterns. The size and shape of these polygons are analogous to those of unsorted polygons on the Earth, which are most common at high latitudes and form by thermal contraction of ice-rich ground. On Mars, ice is in equilibrium with the atmosphere at the latitudes of this patterned ground, and ice may remain in the ground for extended periods. The presence of ice is also indicated by the disintegration of some polygonal patterns to fields of hummocks, which suggests loss of material. Therefore, the polygons on Mars are probably related to the presence of ground ice in this region.

Both desiccation due to climatic cycles and thermal contraction due to seasonal cycles could have formed the polygonal fractures. The desiccation hypothesis has uncertainties concerning the temperature regime and atmospheric exchange of water during climatic cycles. The thermal-contraction hypothesis is favored because it is compatible with the size and shape of the polygons, with theoretical and observational evidence for ground ice, with seasonal cooling of -60°C, adequate to initiate thermal contraction of ice-rich ground, and with the availability of seasonal water vapor or frost. The fractures may be filled by sand wedges or ice wedges, or, most likely, by a combination of the two. The dry martian climate and abundant supply of windblown materials favor the formation of sand wedges, whereas the disintegration of the ground along polygonal patterns favors the formation of ice wedges. The crispness of some fractures suggests that they are little degraded and thus relatively young, although local destruction of the ground indicates that presently they are not forming actively.

ACKNOWLEDGMENTS

This study was supported by the Planetary Geology Program of the U.S. National Aeronautics and Space Administration.

REFERENCES

Anderson, D. M., and Tice, A. R., 1978, The Viking GCMS analysis of water in the martian regolith, in Proceedings Second Colloquium on Planetary Water and Polar Processes, Hanover N. H., 16-18 October, 1978, p. 55-61: Hanover, N. H., U.S. Army, Cold Regions Research and Engineering Laboratory.

Black, R. F., 1976, Periglacial features indicative of permafrost: Ice and soil wedges: Quaternary Research, v. 6, p. 3-26.

Black, R. F., 1978, Comparison of some permafrost features on Earth and Mars: Some cautions and restrictions, in Proceedings of the Second Colloquium on Planetary Water and Polar Processes, Hanover, N. H., 16-18 Oct. 1978, p. 127-130: Hanover, N. H., U.S. Army, Cold Regions Research and Engineering Laboratory.

Brook, G. A., 1982, Ice-wedge polygons, baydjarakhs, and alases in Lunae Planum and Chryse Planitia, Mars, in Reports of Planetary Geology Program, 1982, p. 265-267: U.S. National Aeronautics and Space Administration Technical Memorandum 85127.

Carr, M. H. and Schaber G. G., 1977, Martian permafrost features: Journal of Geophysical Research, v. 82, no. 28, p. 4039-4054.

Coradini, M. and Flamini, E., 1979, A thermodynamical study of the martian permafrost: Journal of Geophysical Research, v. 84, no. B14, p. 8115-8130.

Corte, A. E., 1963, Relationship between four ground patterns, structure of the active layer, and type and distribution of ice in the permafrost: Biuletyn Peryglacial, no. 12, p. 8-86.

Evans, N. and Rossbacher, L. A., 1980, The last picture show: Small-scale patterned ground in Lunae Planum, in Reports of Planetary Geology Program, 1980, p. 376-378: U.S. National Aeronautics and Space Administration Technical Memorandum 82385.

Fanale, F. R., 1976, Martian volatiles: The degassing history and geochemical fate: Icarus, v. 28, p. 179-202.

Fanale, F. P. and Jakosky, B. M., 1982, Regolith atmosphere exchange of water and carbon dioxide on Mars: Effects on atmospheric history and climate change: Planetary Space Science, v. 30, no. 8, p. 819-831.

Farmer, C. B. and Doms, P. E., 1979, Global seasonal variation of water vapor on Mars and the implications for permafrost: Journal of Geophysical Research, v. 84, no. B6, p. 2881-2888.

Helfenstein, P., 1980, Martian fractured terrain: Possible consequences of ice-heaving, in Reports of Planetary Geology, 1980, p. 373-375: U.S. National Aeronautics and Space Administration Technical Memorandum 82385.

Helfenstein, P. and Mouginis-Mark, P. J., 1980, Morphology and distribution of fractured terrain on Mars, in Lunar and Planetary Science XI, p. 429-431: Houston, TX., Lunar and Planetary Science Institute.

Hobbs, P. V., 1974, Ice Physics: Oxford, Clarendon Press, 837 p.

Jakosky, B. M., and Farmer, C. B., 1982, The seasonal and global behavior of water vapor in the Mars atmosphere: Complete global results of the Viking atmospheric water detector experiment: Journal of Geophysical Research, v. 87, no. B4, p. 2999-3019.

Kerfoot, D. E., 1972, Thermal contraction cracks in an arctic tundra environment: Arctic, v. 25, no. 2, p. 142-150.

Kieffer, H. H., Martin, T. Z., Peterfreund, A. R., Jakosky, B. M., Miner, E. D., and Palluconi, F. D., 1977, Thermal and albedo mapping of Mars during the Viking primary mission: Journal of Geophysical Research, v. 82, no. 28, p. 4249-4271.

Lachenbruch, A. H., 1962, Mechanics of thermal contraction cracks and ice-wedge polygons in permafrost: Geological Society of America Special Paper 70, p. 69.

Leffingwell, E. de K., 1919, The Canning River Region, Northern Alaska: U.S. Geological Survey Professional Paper 109, 251 p.

Lucchitta, B. K., 1981, Mars and earth: Comparison of cold-climate features: Icarus, v. 45, p. 264-303.

Lucchitta, B. K., 1982, Ice sculpture in the martian outflow channels: Journal of Geophysical Research, v. 87, no. B12, p. 9951-9973.

Lucchitta, B. K., Anderson, D. M., and Shoji, H., 1981, Did ice streams carve martian outflow channels?: Nature, v. 290, no. 5809, p. 759-763.

Michaux, C. M., and Newburn, R. L., 1972, Mars scientific model: Pasadena, California Institute of Technology, Propulsion Laboratory Document 606-1.

Morris, E. C., and Underwood, J. R., 1978, Polygonal fractures of the Martian plains (abs.), in Reports of Planetary Geology Program, 1977-1978 p. 97-99: U.S. National Aeronautics and Space Administration Technical Memorandum 79729.

Pechmann, J. C., 1980, The origin of polygonal troughs on the northern plains of Mars: Icarus, v. 42, p. 185-210.

Péwé, T. L., 1973, Ice-wedge casts and past permafrost distribution in North America: Geoforum, v. 15, p. 15-26.

Rossbacher, L. A. and Judson, S., 1981, Ground ice on Mars: Inventory, distribution, and resulting landforms: Icarus, v. 45, p. 39-59.

Toksöz, M. N., and Hsui, A. T., 1978, Thermal history and evolution of Mars: Icarus, v. 34, p. 537-547.

Smoluchowski, R., 1968, Mars: Retention of ice: Science, v. 159, p. 1348-1350.

Wall, S. D., 1981, Analysis of condensates formed at the Viking 2 Lander site: The first winter: Icarus, v. 47, no. 2, p. 173-183.

Washburn, A. L., 1980, Geocryology: A survey of periglacial processes and environments: New York, John Wiley and Sons, 406 p.

FIGURE 1 (a) - Polygonally fractured ground in Deuteronilus Mensae, Mars. Polygons occur on flat terrain and on slopes of an ancient impact crater rim. Illumination is from bottom. Center of picture is at lat 46.7°N, long 346.1°W. Viking Orbiter frame 458B67. Orientation and scale same as in (b).
(b) - Sketch map of polygons shown in (a). Polygons are about 50-300 m in diameter. Stippled areas denote impact craters and their ejecta, dashed lines crater rim crests of large subdued craters.

FIGURE 2 (a) - Polygonally fractured ground in Deuteronilus Mensae. Polygons disintegrate to hummocky terrain northeastward, toward top of picture. Illumination is from bottom. Center of picture near 46.9°N, 346.0°W. Viking Orbiter frame 458B69. Orientation and scale same as in (b).
(b) - Sketch map of polygons in (a). Stippled areas denote craters and their ejecta.

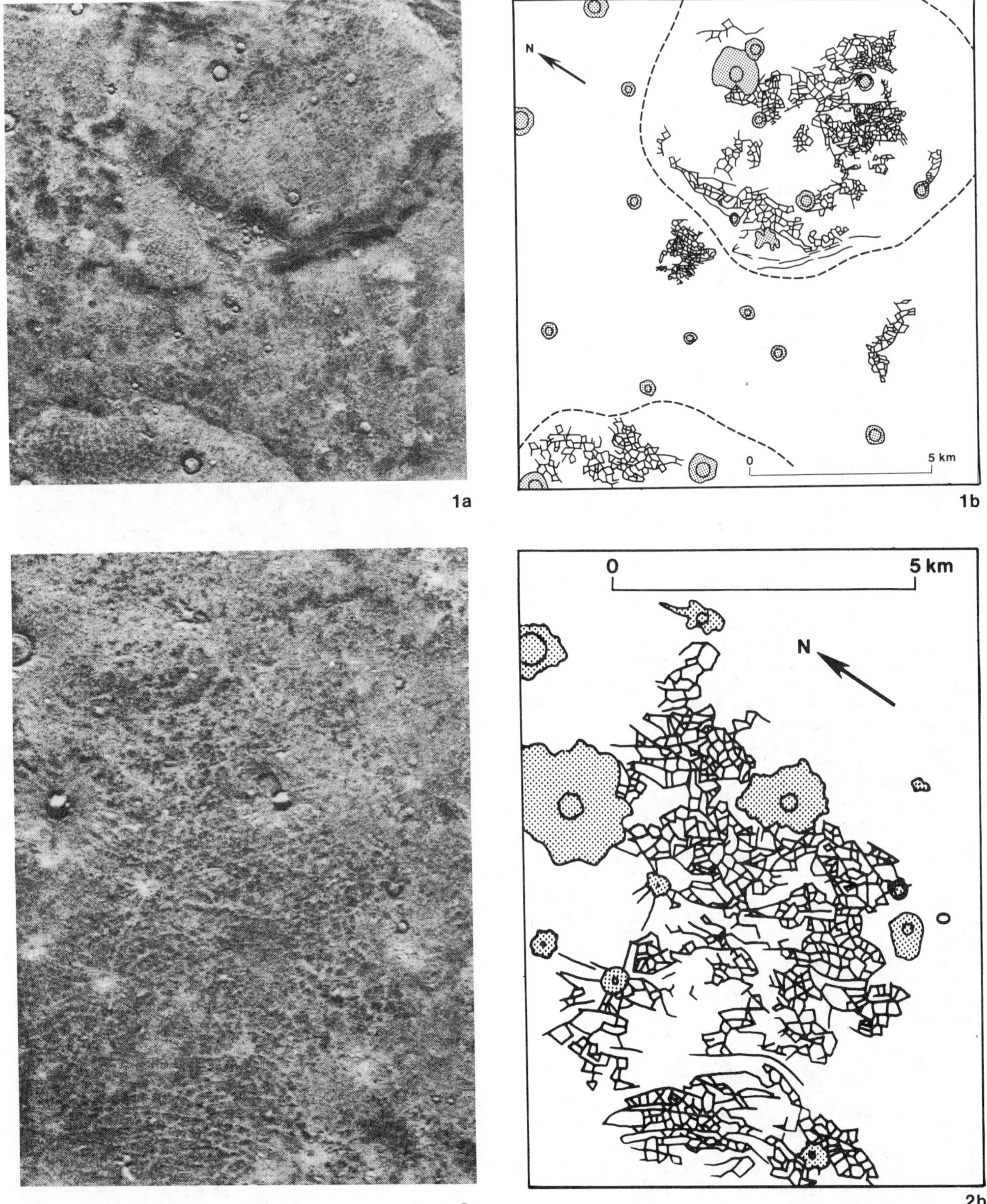
N
0
5 km
1a
1b
0
5 km
N
2a
2b

THAWING BENEATH INSULATED STRUCTURES ON PERMAFROST

V.J. Lunardini

U.S. Army Cold Regions Research and Engineering Laboratory
Hanover, New Hampshire 03755 USA

The problem of thawing beneath heated structures on permafrost (or cooled structures in nonpermafrost zones) must be addressed if safe engineering designs are to be conceived. In general there are no exact solutions to the problem of conduction heat transfer with phase change for practical geometries. The quasi-steady approximation is used to solve the phase-change problem for insulated geometries, including infinite strips, rectangular buildings, and circular storage tanks. Analytical solutions are presented and graphed for a range of parameters with practical importance.

Large-scale exploration and development of the high latitudes of the northern hemisphere have stimulated interest in a number of thermal problems. Not the least of these is the effect of heated structures on underlying or surrounding permafrost. This involves the study of conduction heat transfer in media that can undergo freezing or thawing. Lachenbruch (1957a,b, 1959) and Jumikis (1978) applied linear conduction theory to the effect of heating on permafrost since no phase change was considered. However, the phase change conduction problem is nonlinear and few exact solutions exist (Lunardini 1981a).

Geometries of practical interest do not allow exact solutions of the thermal problem. Closed form solutions, as opposed to numerical evaluations, have relied on approximate methods such as the heat balance integral method and the quasi-steady method. The heat balance integral method gives excellent accuracy and has been applied successfully to simple geometries such as semi-infinite systems (Goodman 1958, Lunardini 1981c, Lunardini and Varotta 1981, Yuen 1980) and pipes buried at infinite depth (Lunardini 1980, Bell 1978).

The quasi-steady method is not as rigorous as the heat balance integral method, but it can be applied to a wide variety of geometries. Applications have included uninsulated buried pipes (Hwang 1977, Thornton 1976, Porkhayev 1963), insulated buried pipes (Lunardini 1981b,d; Seshadri and Krishnayya 1980], and three-dimensional structures (Porkhayev 1970). Probably the most widely used calculated results are those of Porkhayev (1970). New quasi-steady relations are derived and applied to insulated geometries including semi-infinite strips (roads, rivers), rectangular buildings, and circular storage tanks.

QUASI-STEADY METHOD

The quasi-steady approximation assumes that the temperature field changes slowly from one steady state to another. Consider an infinite strip, as shown in Figure 1, with different properties in the thawed and frozen zones. Initially, the temperature is uniform at T_o and the insulated surface of the strip jumps suddenly to T_p. The equations for the thawed and frozen regions are

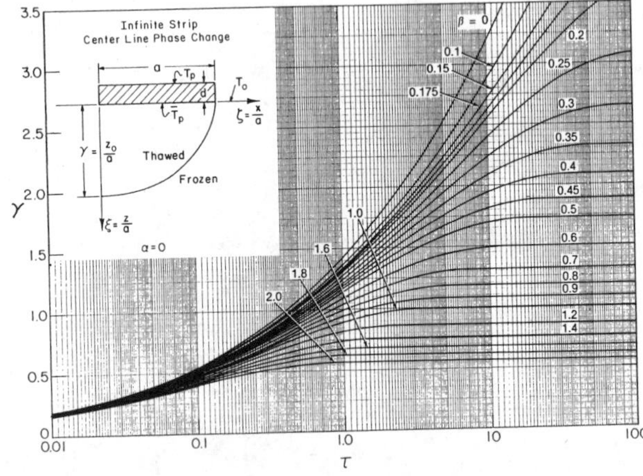

FIGURE 1 Phase change beneath center of an uninsulated infinite strip.

$$\frac{\partial^2\phi_1}{\partial\zeta^2} + \frac{\partial^2\phi_1}{\partial\xi^2} = \frac{2}{\pi} S_T \frac{\partial\phi_1}{\partial\tau} \quad (1)$$

$$\phi_1(\zeta,o,\tau) = 1 \qquad -1 \le \zeta \le +1 \quad (1a)$$

$$\phi_1(\zeta_o,\xi_o,\tau) = 0 \quad (1b)$$

$$\frac{\partial^2\phi_2}{\partial\zeta^2} + \frac{\partial^2\phi_2}{\partial\xi^2} = \frac{2S_T}{\pi\kappa_{21}} \frac{\partial\phi_2}{\partial\tau} \quad (2)$$

$$\phi_2(\zeta,o,\tau) = 0 \qquad \zeta < -1 \text{ or } \zeta > +1 \quad (2a)$$

$$\phi_2(\zeta,\xi,o) = 0 \quad (2b)$$

$$\phi_2(\zeta_o,\xi_o,\tau) = 1 \quad (2c)$$

$$\lim_{\zeta,\xi\to\infty} \phi_2(\zeta,\xi,\tau) = 0 \quad (2d)$$

The energy balance over the surface S' separating the frozen and thawed zones is

$$\int_{S'} \left(\frac{\overline{T}_p - T_f}{T_p - T_f}\frac{\partial\phi_1}{\partial m'} - \beta\frac{\partial\phi_2}{\partial m'}\right)_{\zeta_o,\xi_o} ds' = -\frac{2}{\pi}\frac{dV'}{d\tau} \quad (3)$$

Equations 1 through 3 cannot be solved exactly, but if S_T is small then the diffusion equations reduce to the steady-state case that is more easily solved. If the properties of the frozen and thawed media are identical, then the quasi-steady system is

$$\frac{\partial^2\phi}{\partial\zeta^2} + \frac{\partial^2\phi}{\partial\xi^2} = 0 \qquad (4)$$

$$\phi(\zeta,o) = \begin{cases} 1 & -1 \leq \zeta \leq 1 \\ 0 & \zeta > 1 \text{ or } \zeta < -1 \end{cases} \qquad (4a)$$

$$\lim_{\zeta,\xi\to\infty} \phi(\zeta,\xi) = 0 \qquad (4b)$$

Notice that the energy boundary equation is solved by merely using the quasi-steady solution for ϕ. Lunardini (1981a) showed that Equations 4 through 4b are the zeroth order of a perturbation solution of Equations 1 and 2. The quasi-steady method consists of solving the steady-state problem and locating the phase change boundary with an equation similar to Equation 3. The accuracy of the method depends upon the magnitude of the Stefan number. For systems with a large latent heat relative to the sensible heat, the approximation will be reasonably good. This includes soil systems and many latent heat storage materials.

GENERAL QUASI-STEADY RELATIONS

General equations can be derived that will be valid for a class of important phase change problems. Assume that the solutions to the quasi-steady forms of Equations 1 and 2 are

$$\phi_1 = A_1(\tau) + B_1(\tau)\ f(\zeta,\xi) \qquad (5)$$

$$\phi_2 = A_2(\tau) + B_2(\tau)\ f(\zeta,\xi) \qquad (6)$$

When combined with the boundary conditions, these give

$$\phi_1 = (f-f_o)/(1-f_o) \qquad (7)$$

$$\phi_2 = f/f_o \qquad (8)$$

The function f is the solution to the equivalent steady-state equations, and f_o is its value on the phase-change surface.

The heat flow through the insulation will be equated to the heat flow from the thawed soil at ξ = 0, a concept used by Seshradi and Krishnayya (1980) and by Lunardini (1981b). Using Equation 7, the relation for the insulation/ground temperature is

$$\frac{\overline{T}_p - T_f}{T_p - T_f} = \frac{1}{1 - \dfrac{2\alpha}{1-f_o}\dfrac{\partial f(o,o)}{\partial\xi}} \qquad (9)$$

The general interface equation, Equation 3, is now

$$\left(\frac{1}{f_o - 1 + 2\alpha\dfrac{\partial f(o,o)}{\partial\xi}} + \frac{\beta}{f_o}\right)\left(\frac{\partial f}{\partial\xi}\right)_{\zeta_o,\xi_o} = \frac{2}{\pi}\frac{d\xi_o}{d\tau} \qquad (10)$$

The steady-state (limiting) solution occurs when $d\xi_o/d\tau \equiv 0$. This is

$$f_{o\infty} = \frac{\beta}{1+\beta}\left[1 - 2\alpha\frac{\partial f(o,o)}{\partial\xi}\right] \qquad (11)$$

The heat flux into the thawed soil at the center of the heated surface is

$$q = \frac{-k_1(T_p - T_f)}{a\left[1 - f_o - 2\alpha\dfrac{\partial f(o,o)}{\partial\xi}\right]}\frac{\partial f(o,o)}{\partial\xi} \qquad (12)$$

The effect of the insulation can also be accounted for by considering an excess layer of soil, applied to the thawed zone temperature relations, with a thermal resistance equal to that of the actual insulation. The concept was introduced by Porkhayev (1963, 1970), however the method used here gives better results (Lunardini 1982).

SEMI-INFINITE STRIP

The semi-infinite strip can represent roads, shallow rivers, or very long rectangular buildings. The steady-state solution is given by Lunardini (1981a) as

$$f(\zeta,\xi) = \frac{1}{\pi}\tan^{-1}\left(\frac{2\xi}{\zeta^2+\xi^2-1}\right) \qquad (13)$$

Equation 10 can be integrated to obtain

$$\tau = \int_o^{\xi_o} \frac{[(\zeta^2+u^2-1)^2 + 4u]\ du}{(\zeta^2-u^2-1)\left[\dfrac{1}{f(\zeta,u)-1-\dfrac{4\alpha}{\pi}} + \dfrac{\beta}{f(\zeta,u)}\right]} \qquad (14)$$

Equation 14 was evaluated numerically, along the axis of symmetry at ζ = 0, with the values for α = 0 plotted as Figure 1. Correction factors for nonzero α values are defined as

$$\gamma = \gamma(\alpha{=}0) - K_i \qquad (15)$$

and plotted in Figures 2 and 3. The single-phase solution (β = 0) is

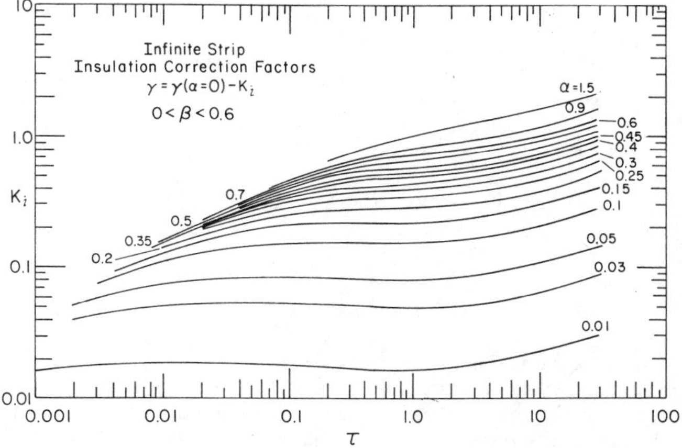

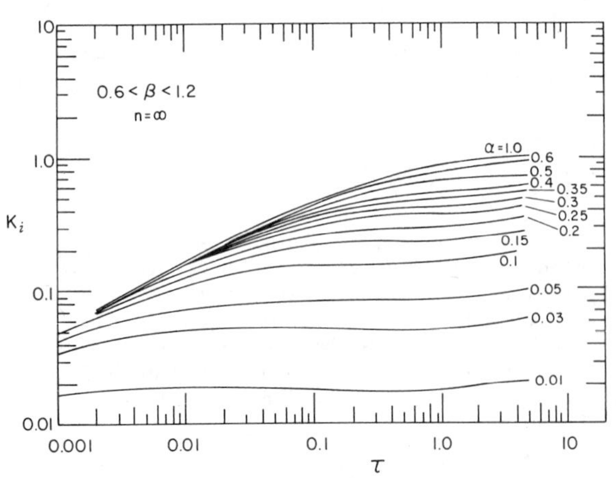

FIGURE 3 Insulation correction factor, infinite strip, $0.6 < \beta < 1.2$.

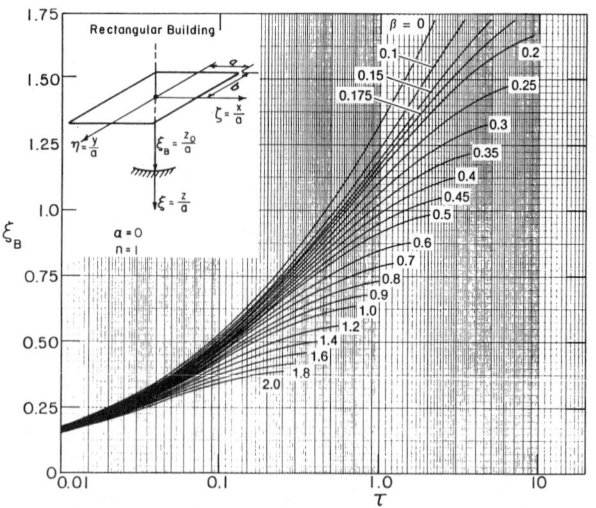

FIGURE 4 Phase change beneath center of rectangle, $n = 1$, $\alpha = 0$.

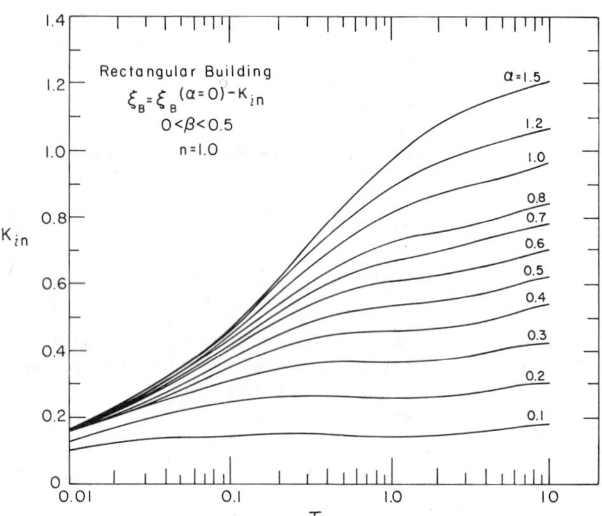

FIGURE 5 Insulation factor, $n = 1$, $0 < \beta < 0.5$.

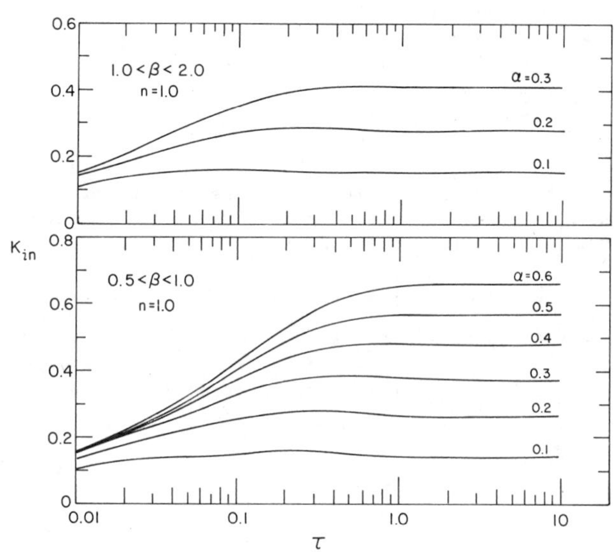

FIGURE 6 Insulation factor, $n = 1$, $0.5 < \beta < 2.0$.

$$\tau = \left(1 + \frac{4\alpha}{\pi}\right)\left(\frac{\gamma^3}{3} + \gamma\right) - \frac{2}{\pi}\left\{\left(\frac{\gamma^3}{3} + \gamma\right)\text{ctn}^{-1}\gamma + \frac{1}{3}\ln(1+\gamma^2) + \frac{1}{6}\gamma^2\right\} \quad (16)$$

The temperatures, in the thawed and frozen zones, can be found with Equations 7 and 8.

RECTANGULAR BUILDING

The semi-infinite strip can represent a building with a very large length-to-width ratio, but solutions for small aspect ratios are needed. The general steady-state solution is given by Lunardini (1981a) and the geometric factor along the axis of symmetry, for $\zeta = \eta = 0$, can be written

$$f(\xi) = \frac{2}{\pi}\tan^{-1}\left(\frac{n}{\xi\sqrt{\xi^2+1+n^2}}\right) \quad (17)$$

The phase-change interface is

$$\tau = -\frac{1}{n}\int_0^{\xi_B} \frac{[u^2(u^2+1+n^2)+n^2]\sqrt{u^2+1+n^2}\,du}{(2u^2+1+n^2)\left[\dfrac{1}{f(u) - 1 - \dfrac{4\alpha\sqrt{1+n^2}}{\pi n}} + \dfrac{\beta}{f(u)}\right]} \quad (18)$$

The steady-state solution is

$$\xi_{B\infty}^2 = \sqrt{\left(\frac{1+n^2}{2}\right)^2 + n^2\cot^2\left[\frac{\beta}{1+\beta}\left(\frac{\pi}{2} + \frac{2\alpha\sqrt{1+n^2}}{n}\right)\right]} - \frac{(1+n^2)}{2} \quad (19)$$

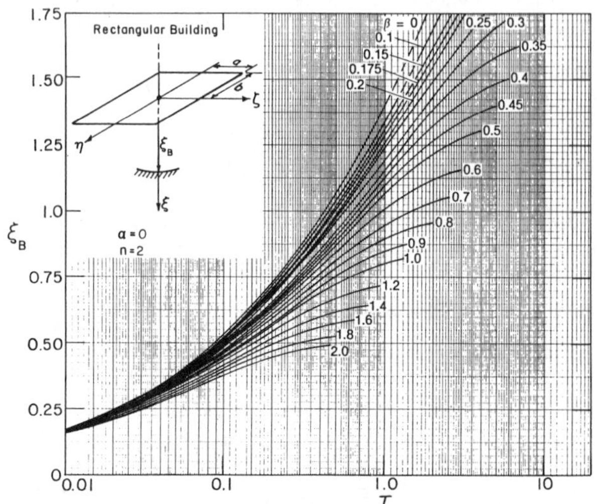

FIGURE 7 Phase change beneath center of rectangle, n = 2, α = 0.

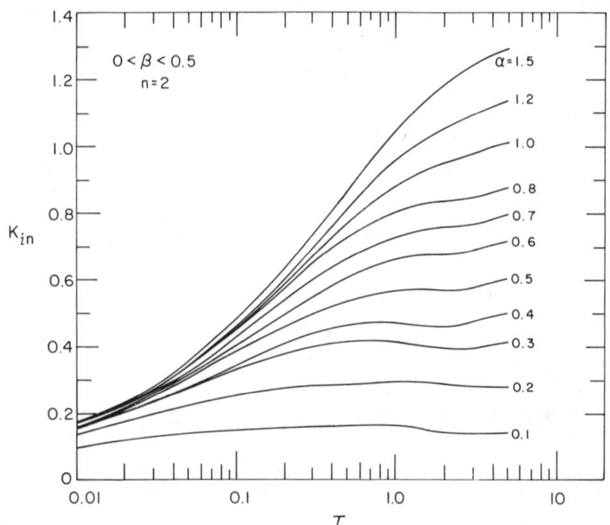

FIGURE 8 Insulation factor, n = 2, 0 < β < 0.5.

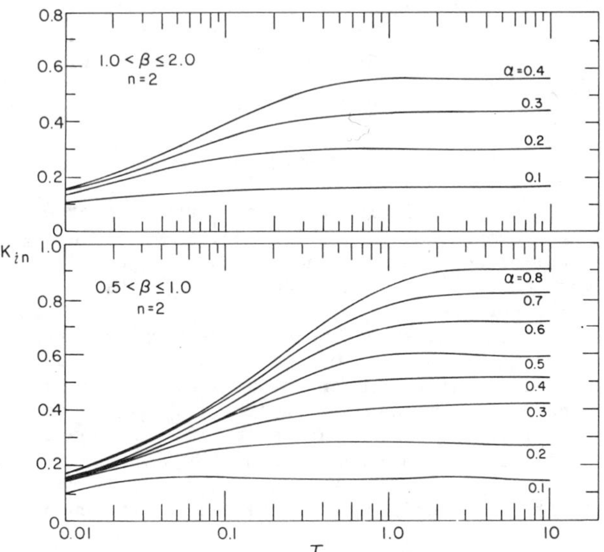

FIGURE 9 Insulation factor, n = 2, 0.5 < β < 2.0.

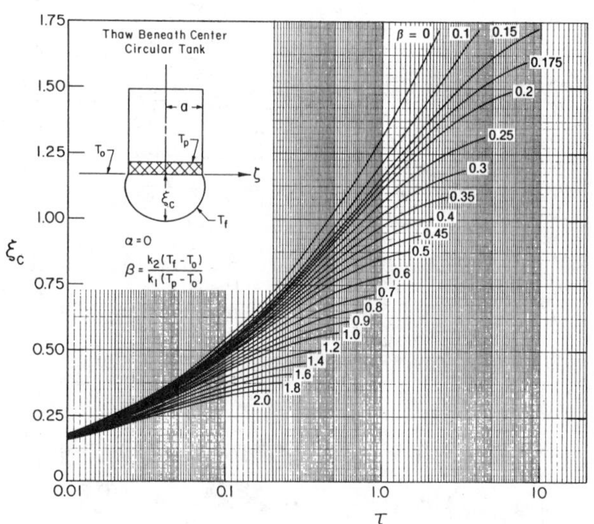

FIGURE 10 Phase change beneath center of uninsulated circular tank.

Equation 19 is valid for the infinite strip if the limit is taken as $n \to \infty$.

Numerical quadrature of Equation 19 was carried out and a correction factor was defined as

$$\xi_B = \xi_B(\alpha=0) - K_{in} \tag{20}$$

Figures 4 through 9 can be used for rectangular structures.

CIRCULAR TANK

The transient solution for the temperature in a semi-infinite medium initially at T_o, after a surface area S_1 is disturbed with a temperature T_p, is given by Lachenbruch (1957). The steady-state solution is

$$\frac{T-T_o}{T_p-T_o} = \frac{z}{2\pi} \iint_{S_1} \frac{dA}{\left[(x-x')^2+(y-y')^2+z^2\right]^{3/2}} \tag{21}$$

Equation 21 cannot be integrated in closed form for the circle, but the temperature along the z-axis can be written immediately, and the geometric function is then

$$f(\xi) = 1 - \frac{\xi}{\sqrt{\xi^2+1}} \tag{22}$$

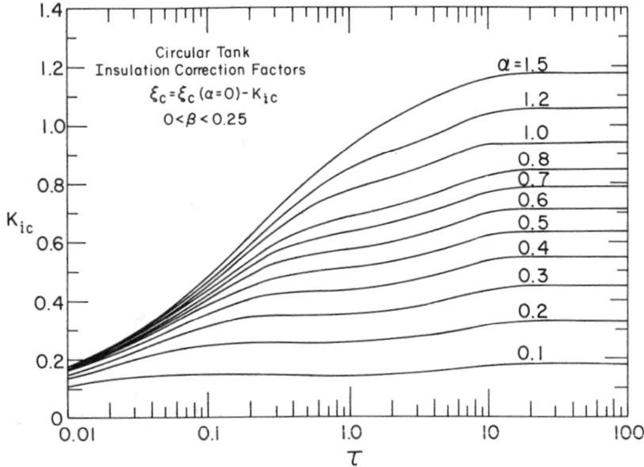

FIGURE 11 Insulation factor, $0 < \beta < 0.25$.

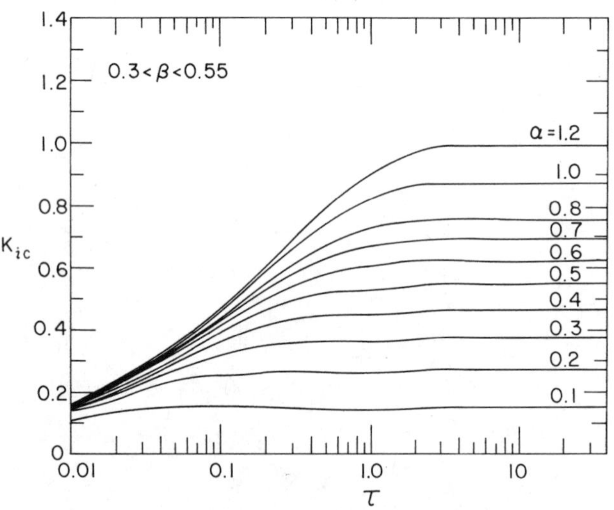

FIGURE 12 Insulation factor, $0.3 < \beta < 0.55$.

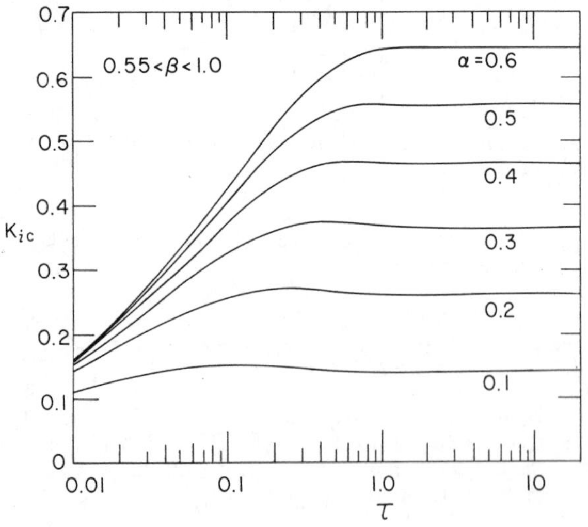

FIGURE 13 Insulation factor, $0.55 < \beta < 1.0$.

The phase-change interface depth along the axis of symmetry is

$$\tau = \frac{2}{\pi}\int_0^{\xi_c} \frac{(u^2+1)\,du}{\left[\dfrac{1}{u+2\alpha\sqrt{u^2+1}} - \dfrac{\beta}{\sqrt{u^2+1}-u}\right]} \tag{23}$$

The limiting solution is

$$\xi_{c\infty} = \left(\left[\frac{1+\beta}{1-2\alpha\beta}\right]^2 - 1\right)^{-1/2} \tag{24}$$

The single-phase solution can be written as

$$\tau = \frac{1}{\pi}\left\{\frac{\xi_c^4}{2} + \xi_c^2 + \frac{\alpha}{2}\left[\xi_c(2\xi_c^2+5)\sqrt{\xi_c^2+1} + 3\ln\left(\xi_c + \sqrt{\xi_c^2+1}\right)\right]\right\} \tag{25}$$

The circular tank correction factor is

$$\xi_c = \xi_c(\alpha=0) - K_{ic} \tag{26}$$

Numerical values are given in Figures 10 through 13.

CONCLUSIONS

The problem of conduction heat transfer with phase change has been solved approximately using the quasi-steady method. The method is expected to be more reliable than the widely used graphs of Porkhayev (1970).

The quasi-steady method is extremely useful since it can be applied quite easily to a number of practical cases, and the results can be presented in a compact form, as shown by the graphs. These graphs are felt to be acceptable for engineering estimates with maximum errors of ±20%. For insulated systems, where the phase change is more limited, the graphs should be even more accurate.

ACKNOWLEDGMENT

This study was funded by the U.S. Army Corps of Engineers.

NOTATION

2a,2b	- width and length of structure
C	- volumetric specific heat
d	- insulation thickness
k	- thermal conductivity
k_{12}	$\equiv k_1/k_2$
K	- insulation correction factor
L	- volumetric latent heat
m'	= m/a, outward normal on phase-change interface
n	= b/a
q	- heat flux at bottom of insulation

s' - s/a, distance along phase change interface
S' - S/a^2, area of phase change interface
S_T = $C_1(T_p - T_f)/L$, Stefan number
t - time
T_f, T_o - freezing, initial temperatures
T_p - temperature of bottom of structure
$\overline{T}_p$ - insulation-ground temperature
V' = V/a^3, volume of region changing phase
x,y,z - cartesian coordinates
x_o, y_o, z_o - coordinates of phase change interface
α = $k_{1i}\ d/2a$, insulation parameter
β = $k_{21}\ (T_f - T_o)/(T_p - T_f)$
γ - value of ξ_o at center of semi-infinite strip
ζ, η, ξ = x/a, y/a, z/a
κ - thermal diffusivity
ξ_o = z_o/a
ξ_c - value of ξ_o beneath center of circular tank
ξ_B - value of ξ_o at center of rectangle
τ = $2\kappa_1 S_T t/(\pi a^2)$, dimensionless time
ϕ = $(T-T_o)/(T_p-T_o)$
ϕ_1 = $(T_1-T_f)/(\overline{T}_p-T_f)$
ϕ_2 = $(T_2-T_o)/(T_f-T_o)$

REFERENCES

Bell, G.E., 1978, A refinement of the heat balance integral method applied to a melting problem: International Journal of Heat Mass Transfer, v. 21, p. 1357-1362.

Carslaw, H.S., and Jaeger, J.C., 1959, Conduction of heat in solids, 2nd ed.: Oxford at the Clarendon Press.

Goodman, T.R., 1958, The heat-balance integral and its application to problems involving a change of phase: ASME Transactions, v. 80, p. 335-342.

Hwang, C.T., 1977, On quasi-static solutions for buried pipes in permafrost: Canadian Geotechnical Journal, v. 14, p. 180-192.

Jumikis, A.R., 1978, Graphs for disturbance temperature distribution in permafrost under heated rectangular structures, <u>in</u> Proceedings of the Third International Conference on Permafrost: Ottawa, National Research Council of Canada, v. 1, p. 589-598.

Lachenbruch, A.H., 1957a, Three dimensional heat conduction in permafrost beneath heated buildings, U.S. Geological Survey Bulletin 1052-B: Washington, D.C., U.S. Government Printing Office.

Lachenbruch, A.H., 1957b, Thermal effects of the ocean on permafrost: Bulletin of the Geological Society of America, v. 68, p. 1515-1530.

Lachenbruch, A.H., 1959, Periodic heat flow in a stratified medium with applications to permafrost problems: U.S. Geological Survey Bulletin 1083-A.

Lunardini, V.J., 1980, Phase change around a circular pipe: Hanover, N.H., U.S. Army Cold Regions Research and Engineering Laboratory, CRREL Report 80-27.

Lunardini, V.J., 1981a, Heat transfer in cold climates: New York, Van Nostrand Reinhold.

Lunardini, V.J., 1981b, Phase change around insulated buried pipes: quasi steady method: Journal of Energy Resources Technology, v. 103, no. 3, p. 201-207.

Lunardini, V.J., 1981c, Application of the heat balance integral to conduction phase change problems: Hanover, N.H., U.S. Army Cold Regions Research and Engineering Laboratory, CRREL Report 81-25.

Lunardini, V.J., 1981d, Approximate phase change solutions for insulated buried cylinders, ASME Paper 81-WA/HT-50, presented at 1981 ASME Winter Annual Meeting: Washington, D.C.

Lunardini, V.J., 1982, Conduction phase change beneath insulated heated or cooled structures: Hanover, N.H., U.S. Army Cold Regions Research and Engineering Laboratory, CRREL Report 82-22.

Lunardini, V.J., and Varotta, R., 1981, Approximate solution to Neumann problem for soils systems: Journal of Energy Resources Technology, v. 103, no. 1, p. 76-81.

Porkhayev, G.V., 1963, Temperature fields in foundations, <u>in</u> Proceedings of Permafrost International Conference: Washington, D.C., National Academy of Sciences, NRC Publication 1287, p. 285-291.

Porkhayev, G.V., 1970, Thermal interaction between buildings, structures, and perennially frozen ground: Moscow, Nauka Publisher.

Seshadri, R., and Krishnayya, A.V.G., 1980, Quasi-steady approach for thermal analysis of insulated structures: International Journal of Heat Mass Transfer, v. 23, p. 111-121.

Thornton, D.C., 1976, Steady state and quasi-static thermal results for bare and insulated pipes in permafrost: Canadian Geotechnical Journal, v. 13, no. 2, p. 161-170.

Yuen, W.W., 1980, Application of the heat balance integral to melting problems with initial subcooling: International Journal of Heat Mass Transfer, v. 23, p. 1157-1160.

RESULTS OF LOAD TESTS ON TEMPERATURE-CONTROLLED PILES IN PERMAFROST

Ulrich Luscher,[1] William T. Black,[2] and James F. McPhail[3]

[1]Woodward-Clyde Consultants, 100 Pringle Ave., Walnut Creek, California 94596 USA
[2]Office of the Federal Inspector, Alaska Natural Gas Transportation System, 2222 Martin Drive, Irvine, California 99715 USA
[3]Alyeska Pipeline Service Company, P. O. Box 2220, Houston, Texas 77001 USA

Eight temperature-controlled pile load tests have been made at three sites representing typical interior Alaskan permafrost conditions. The loads were applied in increments, and accurate displacement histories were obtained for each increment. Direct results of these tests and evaluations based on primary creep and ultimate strength are presented in this paper. It is believed that these tests represent an extremely valuable data base to the profession, to evaluate the load-carrying and settlement characteristics of slurried piles in permafrost.

INTRODUCTION

This paper presents data obtained during a significant load test program on piles in permafrost. An earlier description of these pile load tests was published by Black and Thomas (1978).

As of the early 1970s, piles had been installed in Alaska at several locations, principally in connection with Air Force facilities and other national defense programs. Also, the U.S. Army's Cold Regions Research and Engineering Laboratory had tested many piles in permafrost conditions (Sanger 1969). This previous work represented the starting point for the development of the design criteria for the piles used by Alyeska Pipeline Service Company to support the elevated portion of the Trans-Alaska Pipeline System (TAPS). Alyeska was required to conduct a substantial amount of independent research and criteria development for its design purposes, because these prior projects and research differed in many important respects from the pipeline project. Basically, prior pile projects had been installed at single sites covering a small geographic area, generally on projects requiring only 100 or so piles, and most were installed in cold permafrost. Alyeska, in contrast, needed to develop criteria which were suitable for approximately 78,000 piles installed along a 1,280-km route, in soils which varied widely in composition and thermal state. In addition, the design criteria which were developed for Alyeska's purposes needed to ensure the integrity of the pile system for a 30-year useful life, with allowance only for limited maintenance. The end result was several significant differences between the piles and the criteria developed for Alyeska and the research and projects which had preceded that program. For example:

- Many of the prior projects used timber piles and mechanical refrigeration systems for initial freezeback, which were uneconomical or impractical for use by Alyeska. Alyeska used steel pipe piles, many of which were corrugated and contained self-refrigerating thermal devices.
- Most prior installations used in situ soil, which frequently was ice-rich silt, or even water, to backfill the holes around the piles. Alyeska, in contrast, engineered a carefully controlled sand-water slurry which was used to backfill the annulus around the pile and which was vibrated during placement for densification.
- While other projects had typically been designed on the basis of the exact temperature of the soils, which often were substantially below freezing, this was impractical over the long route of the pipeline. So a specific upper design temperature was established (-0.3°C for most of the line) for the pile and the adjacent frozen ground.

The pile tests discussed herein were confirmation tests, made by Alyeska as one condition of acceptance of Alyeska's pile design criteria by the federal government review agencies in late 1974. The purpose of the tests was to confirm previously established criteria, not to serve as the basis for setting new criteria. The initial design criteria for the piles had been established on the basis of earlier criteria and performance results, extensive laboratory creep tests on frozen slurries and natural soils, and thorough analytical modeling of the thermal and mechanical performance of the piles.

TEST CONDITIONS AND DETAILS

Test Plan

The basic test plan was to install and test three identical piles at each of three test sites representing different soil and permafrost characteristics. The piles were configured and installed as similar to pipeline prototype piles as was possible. Each pile was tested by applying loads in increments, at 3-day intervals, and pile top displacements were measured very accurately during the 3-day period after each incremental load application.

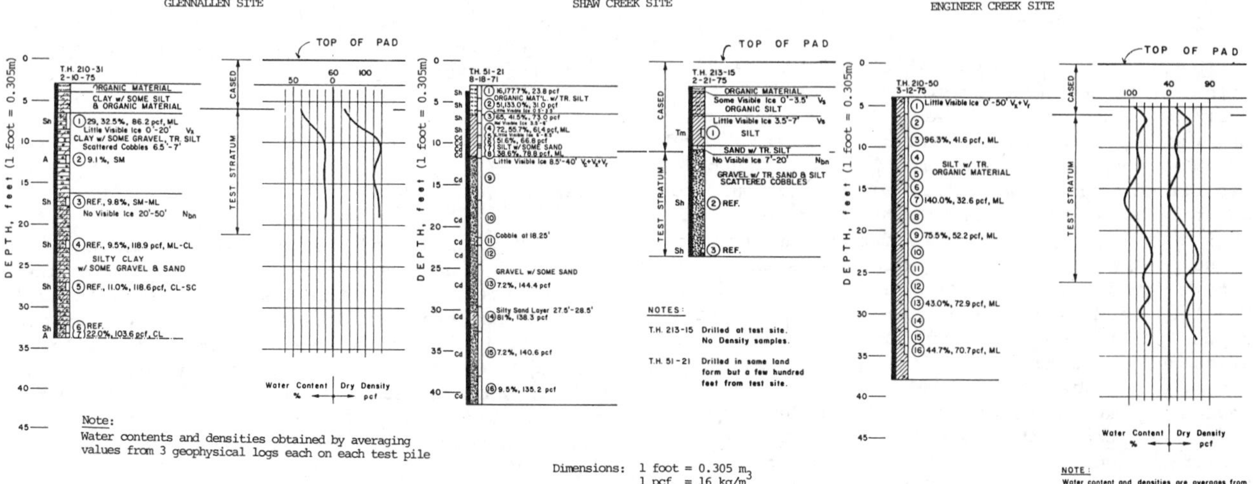

FIGURE 1 Test Sites Soil Profiles.

Sites and Site Geotechnical Conditions

Three permafrost sites were selected in south-central Alaska with widely varying soil conditions:

- Glennallen (Copper River Basin): stiff clays;
- Shaw Creek Flats (north of Delta Junction): dense alluvial sands and gravels; and
- Engineer Creek (north of Fairbanks): ice-rich colluvial silts.

The sites were selected to be representative of significant lengths of the pipeline alignment. Typical soil profiles at each of the three test sites are shown in Figure 1.

Site Development

At each of the three sites an insulated gravel pad was placed in early 1975, prior to spring thawing of the active layer. Four test piles were installed in a line at each site at 3.6-m spacing. Each test pile was straddled by two reaction piles, spaced 2.3 m from the test pile. Thus a total of 12 piles was installed at each test site. The plan was to test three of the four test piles, with the fourth test pile being retained as redundant in case of some future need. A building was subsequently placed over the test site to maintain a closely controlled temperature around the test installation. This temperature control was needed to allow the required very accurate displacement measurements of the pile heads and to facilitate the temperature control in the piles.

Prototype Test Pile Installation

In as many respects as possible the test piles (see Figure 2) were identical to the piles being installed at the same time along the pipeline alignment. Hence the piles were 450-mm (18-in.) diameter steel pipe piles with 12.7-mm (1/2-in.) wall thickness. The piles were installed in a 600-mm (24-in.) diameter hole, which was auger-drilled in the dry. After pile placement the annulus between the pile and the hole wall was backfilled with a sand-water slurry, which had a water content selected to give about 90% water saturation after placement and densification. The piles were vibrated during slurry placement to densify the slurry. The piles were then corrugated in the bearing layer at 300-mm spacing with 12- to 18-mm-high corrugations, which are circumferential ridges created by an expansion tool placed inside the pile after installation but before slurry freezeback. The tested lengths were representative of bearing lengths in the respective three soils and were 4.6 m (15 ft) in the stiff clay, 3.7 m (12 ft) in the gravel, and 6.1 m (20 ft) in the ice-rich silt. Just like the prototype piles, the test piles were intended to transfer part of the applied load through end bearing, so they had end plates or a structural equivalent.

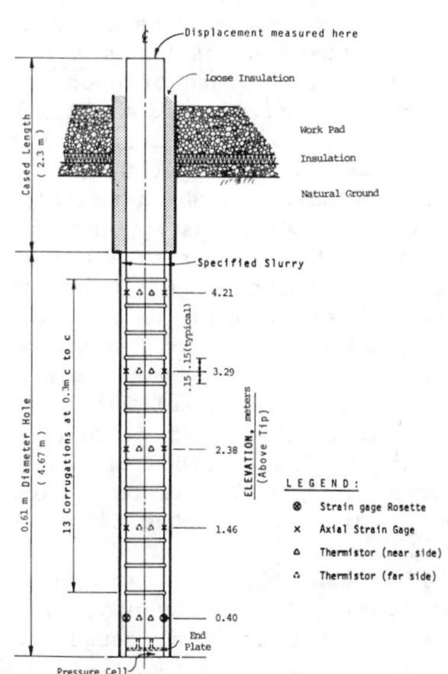

Figure 2 Typical As-Built Test Pile - Glennallen Site

Special Test Conditions

Special test conditions which made these test piles different from prototype piles were as follows. The test piles were cased down to the bearing layer, a distance varying between 1.8 and 3.6 m at the three sites. The inside of the piles was filled with a temperature-controlled fluid to maintain a constant-temperature environment of -0.3°C in and near the piles during the loading phase of the test; this temperature was the highest design temperature for the prototype piles. The test loads were applied in discrete increments, which were held constant for 3 days to allow very accurate creep displacement measurements. The size of the load increments was selected to lead to failure in about 10 increments.

Further the test piles were subjected to a "thermal shock" loading (Heuer 1979) of -30°C temperature for a period of 6 hours prior to the load test, to simulate a rapid ambiant temperature drop in fall.

Controls and Measurements

The test load was applied to each pile through a hydraulic ram which was furnished pressure by an air-actuated hydraulic pump. An electronic load controller was programmed to maintain the desired load within 1%. A calibrated load cell monitored the load and provided feedback to the load controller. This system provided load control well within the specified 1%.

The pile top displacement was measured by both dial gauges and displacement transducers. The displacements were measured with respect to an independently supported reference frame. A reading sensitivity of 0.0025 mm was desired, and a precision of about 0.025 mm was obtained.

The maximum prototype pile design temperature of -0.3°C was selected as the test temperature. It was considered impossible to select test sites where natural permafrost conditions would provide a constant test temperature at that level. Therefore, separate control units were provided to control the temperature of each test pile. Heat was removed by circulating a refrigerant in coils inserted within the test length of each test pile. The coils were thermally coupled to the pile wall by a glycol solution which was also circulated. The heat withdrawal was opposed by heating elements, to meet a temperature criterion of -0.3°C with plus 0°C and minus 0.1°C tolerance.

Though load transfer along the pile and end bearing will not be discussed herein, the corresponding measurement systems are noted. They included a Gloetzl load cell at the pile tip to measure end bearing, and pairs of strain gauges placed diametrically opposite each other at five locations along each pile to measure loads transferred at different depths along the pile (see Figure 2).

A 128-channel data acquisition system was used to monitor all instrumentation at each site, except for the dial gauge readings, which were manually read and recorded. Output from the data recorder included a typed tape used for hourly and daily monitoring of the results and for adjustments to the test plan, and a computer-coded punched tape which furnished the record copy of all test data.

TESTS AND TEST RESULTS

Tests Accomplished

The planned three tests were implemented at the Glennallen and Shaw Creek sites. At the Engineer Creek site, two of the test piles were damaged during construction and were considered unsuitable for testing. Thus one of the originally designated test piles, plus the redundant test pile, were tested at this location. The latter had a steel end plate instead of a bottom load cell, otherwise it was equivalent to the other test piles.

In all cases the test length had been selected to give a design load of about 267 kN (60 kips) and anticipated failure at approximately 890-1,340 kN (200-300 kips) of applied load. Correspondingly, the reaction frame was designed for a maximum load of 1,780 kN (400 kips). The test load was applied in all cases in 89-kN (20-kip) increments up to 890 kN (200 kips) applied load and in 178-kN (40-kip) increments above that load. The maximum applied load, if no prior failure occurred, was 1,780 kN (400 kips). This load was reached in all tests except those at Engineer Creek, where the maximum applied load was 1,600 kN (360 kips).

Test Results

The results of a typical load test at the Glennallen site are presented in Figure 3, which shows a displacement-time history and a load-displacement curve for the entire test. The reported displacement is that at the top of the load-bearing portion of the pile; i.e., elastic shortening of the pile above that location was subtracted from the measured displacements. The displacement-time curve illustrates the occurrence of displacement in each of the 15 applied load increments. The load-displacement curve is similar to that of a typical pile load test.

Additional detail on the development of displacement under several selected load increments is given in Figure 4. The figure shows that even at relatively small loads an equilibrium condition was not reached within the 3-day duration of the increment; rather, the curves are indicative of increasing displacements with a decreasing creep rate. This shape, which is frequently termed "primary creep," was typical of the vast majority of the results. Only in three of the eight tests was a constant or increasing creep rate, i.e., secondary or tertiary creep, observed, and then only at the highest applied load level. These were Pile 2 at Shaw Creek at 1,780 kN (400 kips) load and both piles at Engineer Creek at 1,600 kN (360 kips) load.

Figures 3 and 4 give results for Pile 3 at the Glennallen site. This pile had intermediate strength results of the three tests at Glennallen.

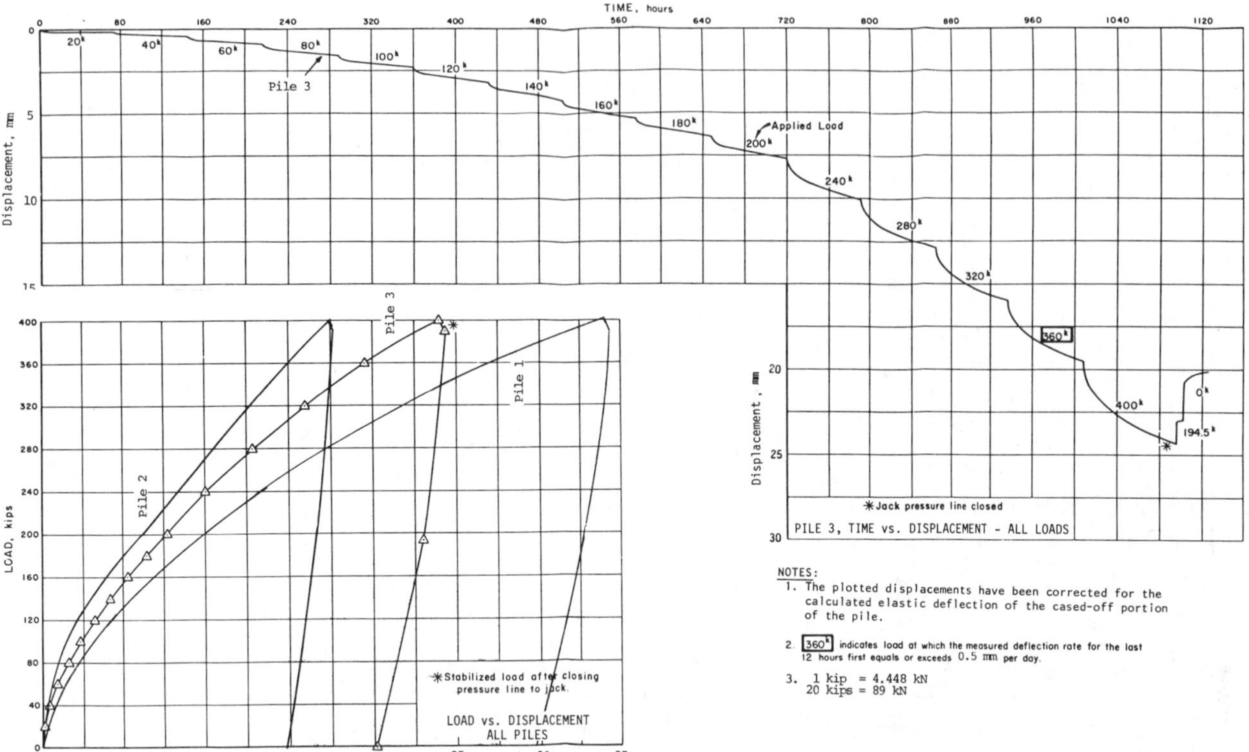

FIGURE 3 Pile Test Results--Glennallen Site.

In Figure 3 the load-displacement curves of the other two tests at Glennallen are also shown. At the Shaw Creek site, Pile 1 gave the intermediate strength results of the three piles there, and the corresponding load-displacement-time results are presented in Figure 5, together with load-displacement curves for the other two tests at Shaw Creek. At Engineer Creek, the two test piles gave very similar results. The displacement-time curve of Pile 4 and the load-displacement curve for both tests are presented in Figure 6.

EVALUATION OF RESULTS

General Observations

The two or three tests at each site showed reasonable repeatability of tests, as seen in Figures 3, 5, and 6. One exception is Pile 2 at Shaw Creek, which showed a significantly greater displacement than the other two piles at the same site. On the other hand, systematic differences between the three soil types are quite marked.

The pile tests in the gravel at the Shaw Creek site had the smallest displacements and the highest ultimate loads. All three piles reached the 1,780 kN (400 kip) maximum test load, with only one showing any signs of weakening. It is believed that considerable end bearing was mobilized in addition to the resistance along the vertical sides in these tests at Shaw Creek.

The piles in the stiff Copper River basin clays at the Glennallen site showed intermediate strength. All three piles reached the 1,780 kN (400 kip) maximum load, though the displacements were considerably larger than at Shaw Creek at all loads.

The test piles at the Engineer Creek site, though they had the longest bearing length, both showed large displacements and displacement rates at 1,600 kN (360 kips) of load and accordingly were not taken to the 1,780 kN (400 kip) maximum load. These piles were relatively stiff at low loads but then showed a dramatic increase in displacements above 530 kN (120 kips).

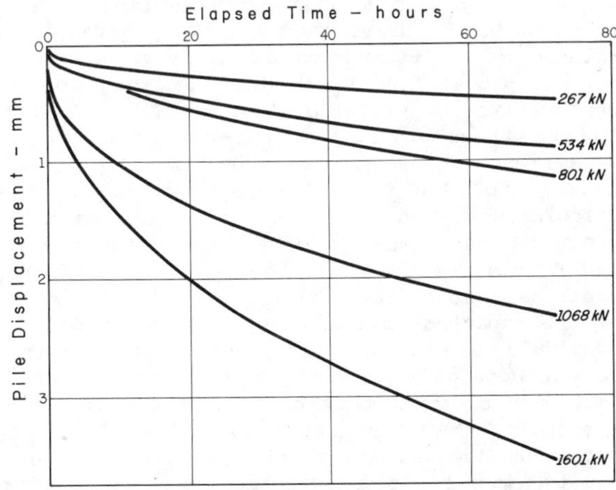

FIGURE 4 Typical Creep Curves--Pile 3, Glennallen Site

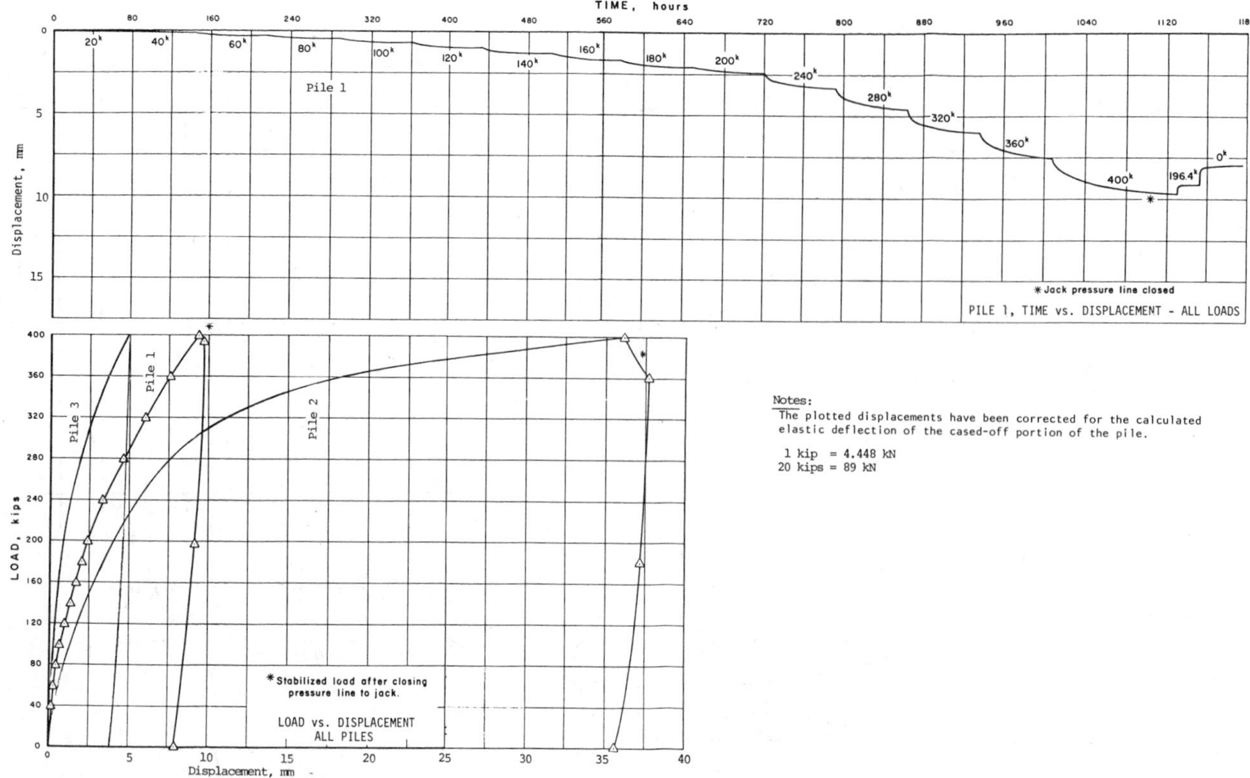

FIGURE 5 Pile Test Results--Shaw Creek Site

Specific Evaluations

Two evaluations of the pile load test results are presented below as examples of what can be done with these data. These include an evaluation based on primary-creep data and another based on indications of ultimate strength.

Evaluation of primary creep can be done utilizing a plotting method introduced by Sayles (1968), who suggested that primary creep is described by a creep rate power function of the form

$$D = D_1 \left(^1/t\right)^{1/M},$$

where D is a creep rate at some time, t is time, and D_1 and M are constants for a given creep relationship. Evaluation of the data by this method is best done by plotting the displacement rate versus reciprocal of time on log-log paper. To make these plots, the time-displacement data were smoothed, and the corresponding displacement rates were calculated. The curve fitting and rate calculations were done using a simple computer program. Determination of the two constants D_1 and M from the plots (for the later times in the usual case of bilinear plots) allows calculation of the expected primary-creep settlement for the particular load at any time, for instance over a 30-year design life. These 30-year expected settlements can then be plotted versus the load (presuming that the settlement under each load increment is independent of the prior smaller loading increments), and a reasonable and conservative relationship established. This relationship allows determining, for any selected design load, the expected 30-year primary-creep settlement as indicated by the particular test. An example of the two plots is given by Black and Thomas (1978) in their Figures 7 and 8.

This evaluation was made for all eight test piles in this program. Results from all eight tests for a 276 kN (60 kip) design load sustained for 30 years are summarized in Table 1. They show a maximum predicted 30-year primary-creep settlement of 40 mm (1.6 in.). Displacements of this magnitude could easily be accommodated by Alyeska's elevated pipeline design.

The second evaluation is analogous to load test evaluations that might be made for pile tests in

TABLE 1 Results of Evaluations

	Prim. Creep	Ultimate Capacity		
Site and Pile No.	30-year Settlement for 267 kN mm	Ultimate Capacity kN	Design Load (F=2.5) kN	Design Stress (F=2.5) kPa
GA 1	38	1,250	500	77
GA 2	10	>1,780	>710	>110
GA 3	40	1,600	640	96
SC 1	<3	>1,760	>710	>134
SC 2	8	1,600	640	120
SC 3	<3	>1,760	>710	>134
EC 1	8	1,070	430	48
EC 4	8	1,070	430	48

1mm = 0.04in., 1kN = 0.22kips, 1kPa = 0.021ksf

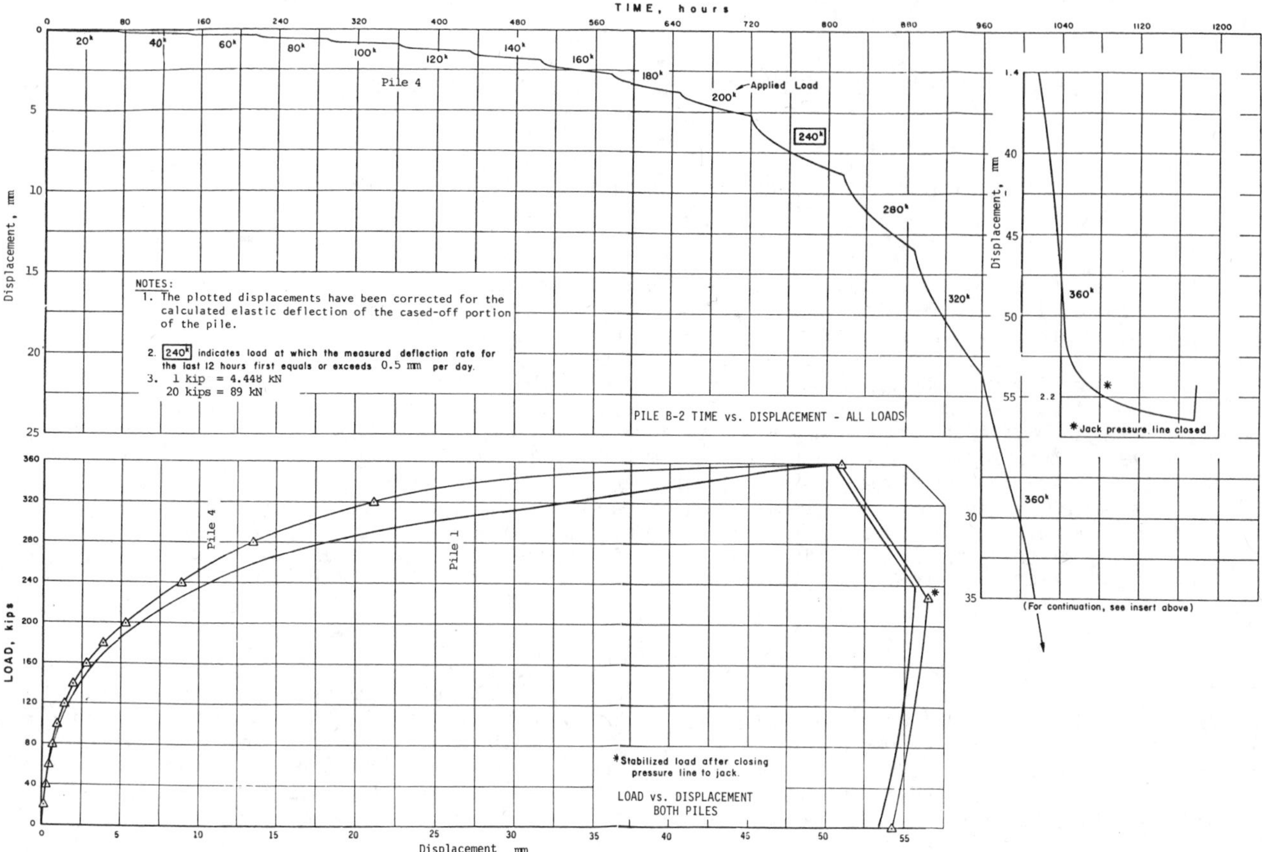

FIGURE 6 Pile Test Results--Engineer Creek Site

unfrozen soils; it is based on the pile's ultimate capacity, as evidenced by the creep rate at the end of each load increment. Specifically, it has been suggested that the load at which a creep rate of 0.5 mm (0.02 in.) per 24 hours was occurring at the end of the 3-day duration of each load increment was a reasonable representation of the pile's ultimate capacity. The highest load increments at which the displacement rate was just below 0.5 mm/day are indicated in Figures 3, 5, and 6 for the three highlighted piles, and in Table 1 these ultimate capacities are tabulated for all eight test piles. If a factor of safety of 2.5 is used, as suggested by Crory (1974) for piles in permafrost when pile load tests have been made (3.0 without load tests), the design load and the design adfreeze stress levels are also presented in the last columns of the table. The design loads are 427 kN (96 kips) or higher for all tests, well in excess of the 267 kN (60 kips) intended design load.

Thus both types of evaluations of the tests in terms of primary-creep displacement and of ultimate load confirmed the adequacy of the previously adopted pile design criteria.

ACKNOWLEDGMENTS

The permission of the owners of TAPS and their agent, Alyeska Pipeline Service Company, to publish this paper is gratefully acknowledged. The owners of TAPS are Amerada Hess Pipeline Corporation, Arco Pipe Line Company, BP Pipelines, Inc., Exxon Pipeline Company, Mobil Alaska Pipeline Company, Phillips Alaska Pipeline Corporation, Sohio Pipe Line Company, and Union Alaska Pipeline Company. Recognition is given to J. H. Wilson, who designed most of the complex control and measurement systems employed in these tests.

The contents of this paper reflect the views of the authors and do not necessarily reflect the official views of Alyeska.

REFERENCES

Black, W. T., and Thomas, H. P., 1978, Prototype pile tests in permafrost soil, in Pipelines in adverse environments, Proceedings of ASCE Pipeline Division Specialty Conference: New York, American Society of Civil Engineers.

Crory, F. E., 1974, Adfreeze strength for a steel pile embedded in a saturated sand slurry, Memorandum to Alaska Pipeline Office, November.

Heuer, C. E., 1979, The application of heat pipes on the Trans-Alaska Pipeline, CRREL Special Report 79-26: Hanover, N.H., Cold Regions Research and Engineering Laboratory.

Sanger, F. J., 1969, Foundation of structures in cold regions, CRREL Monograph 111-C4, June: Hanover, N.H., Cold Regions Research and Engineering Laboratory.

Sayles, F.S., 1968, Creep of Frozen Sands, CRREL Technical Report 190: Hanover, N.H., Cold Regions Research and Engineering Laboratory.

PINGO GROWTH AND SUBPINGO WATER LENSES, WESTERN ARCTIC COAST, CANADA

J. Ross Mackay

Department of Geography, University of British Columbia
Vancouver, B.C., Canada, V6T 1W5

Field surveys of hydrostatic (closed) system pingos which are growing in the bottoms of drained lakes have been carried out along the western arctic coast, Canada from 1969-1982. The water pressures beneath a pingo are derived from pore-water expulsion in saturated sandy lake-bottom sediments during permafrost aggradation. A subpingo water lens develops when the pore water pressure uplifts the pingo overburden and intrudes water beneath it. Three growth patterns have been identified for the pingos with subpingo water lenses. (1) Constant year-to-year growth suggesting a balance between water supply and freezing of the water lens; (2) decreasing year-to-year growth suggesting a freeze-through of the sub-pingo water lens; (3) erratic year-to-year growth with periods of increasing height followed by pingo rupture, water loss, and pingo subsidence. The freezing point of the water in a subpingo water lens is depressed slightly below 0°C by a combination of hydrostatic pressure and the concentration of solutes rejected during freezing.

INTRODUCTION

A pingo is an ice-cored hill which is typically conical in shape and can only grow and persist in permafrost. There are about 1,450 pingos along the western arctic coast of Canada, and of these, about 1,350 occur on Richards Island, Tuktoyaktuk Peninsula, and the coastal area just south of the Eskimo Lakes and Liverpool Bay (Figure 1). Pingos can be divided into two main types: (1) hydraulic (open) system pingos where the water is derived from a source higher in elevation than the pingo and (2) hydrostatic (closed) system pingos where the water is derived from a hydrostatic head resulting from pore water expulsion during permafrost aggradation (Mackay 1962, 1979). The pingos under discussion are of the hydrostatic type and have grown in residual ponds left by rapid lake drainage. The lake bottom sediments are sandy. The pingo cores are composed mainly of segregated ice, "injection ice" frozen from bulk water in a subpingo water lens, or any combination of the two types. Although the term "injection ice" is a misnomer, because it is the water which is injected and not the ice, the term is used here because it is well established in the permafrost literature. Field surveys have been carried out on many growing pingos (Figure 1) for the period 1969-1982. The focus of this discussion is on pingos 13, 14, and 17. The mean annual air temperature in the area is about -11°C, the mean annual ground temperature about -8° to -9°C, and permafrost is about 500-600 m thick. The purpose of this paper is to discuss aspects of pingo growth affected by the presence of subpingo water lenses.

PINGOS 13, 14, AND 17

Pingos 13, 14, and 17 have formed in the bottom of a large lake which drained prior to 1890. The lake was about 1 km wide, 6.5 km long, and deeper than the maximum winter ice thickness over most of its area, so that a large talik would have existed beneath the lake bottom prior to lake drainage. The tops of each of the three pingos rise to a uniform height of about 12 m above the lake bed and the position of the three tops forms the vertices of a triangle the sides of which are about 450 m long.

Pingo 14

A section across the width of the drained lake basin with pingo 14 in the center is shown in Figure 2. The 1971-1982 changes in height of a bench mark installed into permafrost on the top of the pingo, and referenced by precise survey to a stable datum beyond the former lake shore, are plotted in Figure 3. Holes were drilled through ice-bonded permafrost in the pingo down to the subpingo water lens in 1973, 1976, and 1977. All holes produced artesian flow (Figure 4). In 1973 and 1976, the holes were allowed to freeze in. In 1977, two drill holes were kept open for 2 weeks, and two pressure transducers were installed in the subpingo water lens where it was 1.2 m thick (Figure 2). Because the projections of the growth trends for the pingo top (Figure 3) were parallel for 1971-1972, 1973-1976, and 1978-1982 when there was no water loss, the rate of recharge to the subpingo water lens seems to have been constant. The water pressure in the subpingo water lens for 1977-1982 is plotted in Figure 5. Within a month following freezing of the drill holes in 1977 the pressure stabilized at 370 kPa where it has remained essentially constant ever since.

Pingos 13 and 17

Increase in elevation of the tops of pingos 13 and 17 is plotted in Figure 6. Although pingos 13, 14, and 17 undoubtedly share a common subpermafrost groundwater system, drilling in pingo 14 seems to have had no detrimental effect upon the growth of pingo 13. Even though pingo 17 has not been

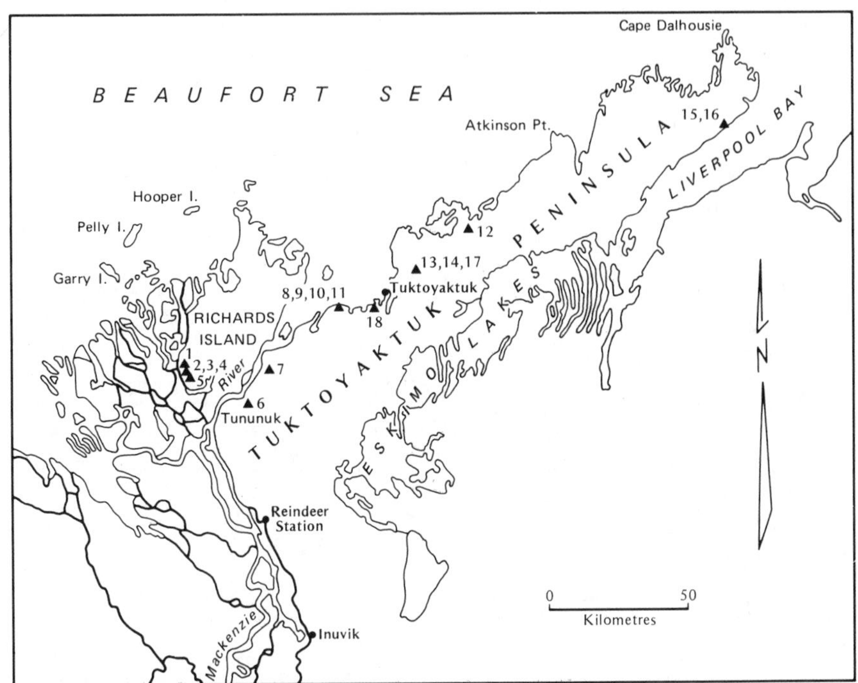

Figure 1. Location map. Pingos are assigned numbers, because most are unnamed.

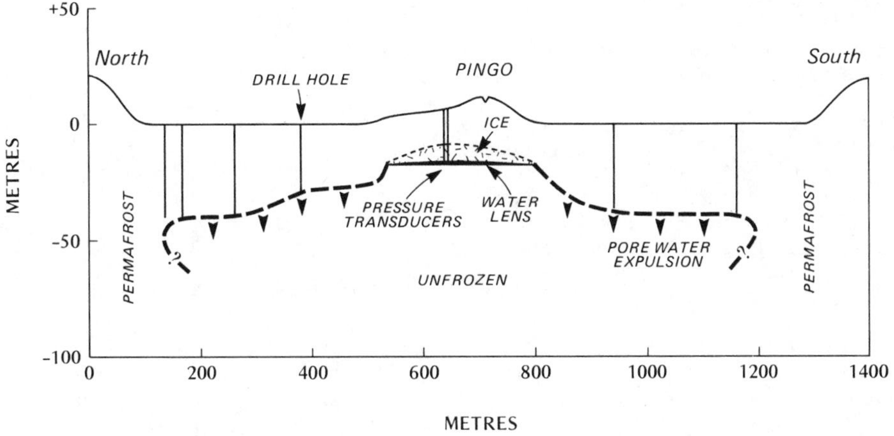

Figure 2. Cross section of pingo 14 and the drained lake.

drilled, there is abundant evidence from periodic spring flow, peripheral pingo faulting with water loss, and subsidence of the top to show that pingo 17 is underlain by a large and deep water lens (Mackay 1977, 1979). Figure 6 shows a variable growth pattern for pingo 17 between 1974 and 1977 when there was frequent spring discharge and pingo rupture. From 1977-1979, following extensive water loss from drilling in nearby pingo 14, the growth has been steady. No spring discharge has been observed for the 1977-1982 period.

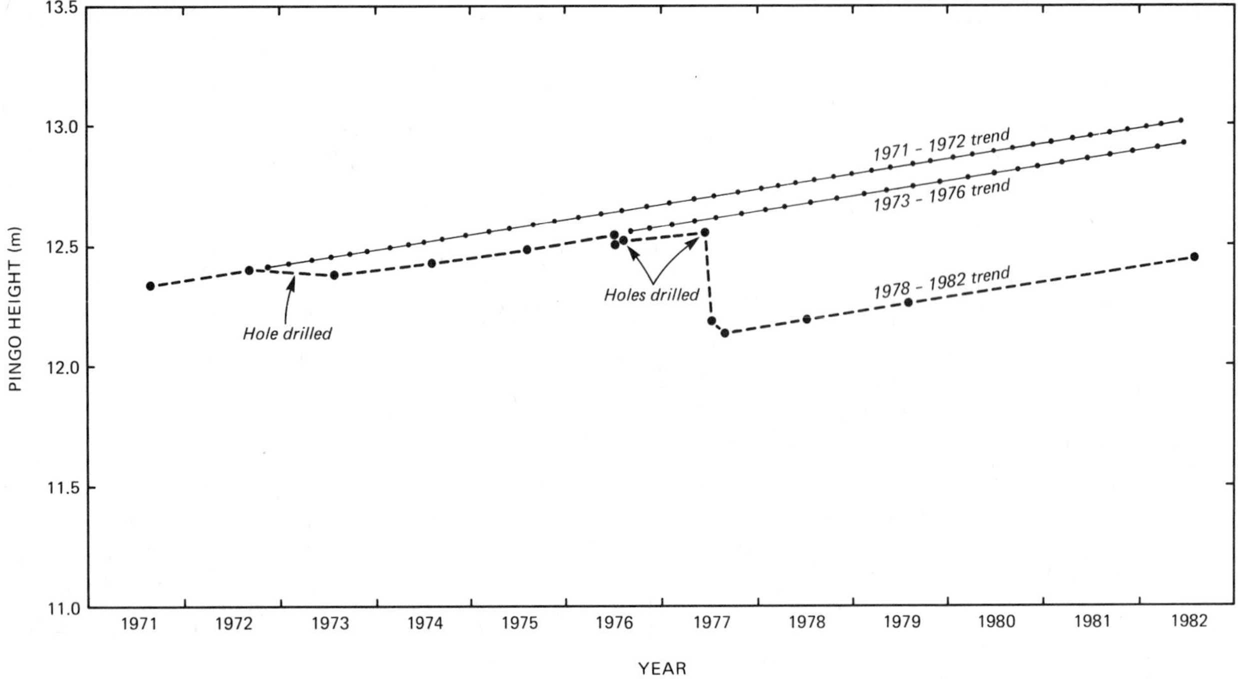

Figure 3. Graph showing the height of a bench mark on the top of pingo 14 for the 1971-1982 survey period. Note the loss of height following each period of drilling.

PERMAFROST GROWTH AND PORE-WATER EXPULSION

When downward freezing occurs in saturated sand, such as is present beneath the drained lake of pingos 13, 14, and 17, pore water will be expelled in front of the freezing surface if drainage is permitted, but if drainage is blocked, the pore-water pressure will rise and may approach the overburden pressure (e.g. Arvidson and Morgenstern 1977; Khakimov 1957; McRoberts and Morgenstern 1975). It is permafrost growth accompanied by pore-water pressure which provides the mechanism for pingo growth. Furthermore, it is obvious that if a pingo has a subpingo water lens, the water pressure is high enough to uplift the superincumbent pingo with its ice core and overburden. In the case of pingo 14, the hydrostatic head of the subpingo water lens lies about 11 m above the top of the pingo and about 23 m above the lake flat.

There are numerous methods which can be used to estimate the rate of growth of permafrost in thawed sediments underlying a drained lake bottom, a widely used one being Stefan's solution:

$$z = \sqrt{\frac{-2Tkt}{Q_L}} = b\sqrt{t}, \qquad (1)$$

where

z = depth of permafrost;
k = thermal conductivity;
T = temperature, °C;
t = time;
Q_L = the volumetric latent heat of fusion of the soil;
b = a constant for Stefan's solution.

The rate of permafrost growth is

$$\frac{dz}{dt} = \frac{b}{2\sqrt{t}} \ . \qquad (2)$$

From equations 1 and 2,

$$\frac{dz}{dt} = \frac{b^2}{2z} = \frac{z}{2t} = \frac{b}{2\sqrt{t}} \ . \qquad (3)$$

If the porosity is η, the volume expansion of water to ice is 9%, all of the 9% excess volume is expelled as pore water, and the thickness of the layer of expelled pore water is H, then

$$\frac{dH}{dt} = .09\eta\frac{dz}{dt} = \frac{.09\eta b^2}{2z} \ . \qquad (4)$$

The value of b for four recently drained lakes in the Tuktoyaktuk Peninsula area is about 3 m/yr$^{\frac{1}{2}}$ (Mackay 1979). If this value of b is used with a porosity (η) of 0.3 and a mean depth of permafrost (z) of 40 m (Figure 1), from equation 4 the expelled pore water would amount to about a 3 mm thick layer of water in a year. The cumulative addition of water to pingos 13, 14, and

Figure 4. Drill hole flow in 1977 from pingo 14. The water came from a depth of 22 m and jetted up 2.6 m high from a 7.5 cm diameter hole. The clear mineralized water had an orifice temperature of about -0.1°C.

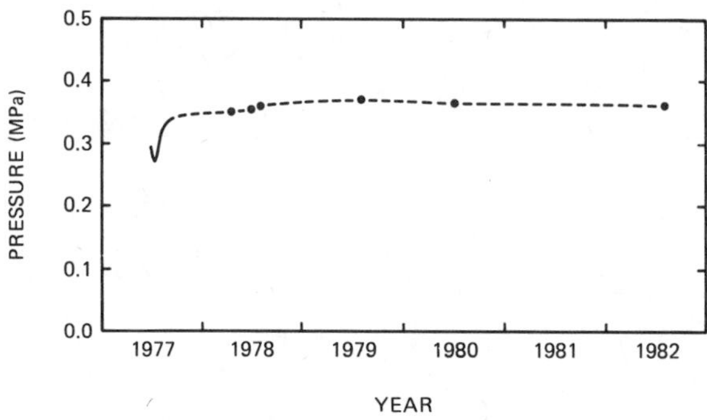

Figure 5. Water pressure of pingo 14 as measured with a pressure transducer in the subpingo water lens (Figure 2). The drop in pressure in 1977 was associated with drilling and transducer installation.

17 is about 8,000 m^3/yr, the estimate being based upon the pingo areas and growth patterns. From the preceding, the area of permafrost pore water expulsion required to supply the 8,000 m^3/yr of water at the rate of 3 mm/yr is then about 2.7 km^2, or about half the total area of the drained lake bottom. This seems to be of the right order of magnitude. The water which accumulates in a subpingo water lens, as sampled from spring and drill hole flow, has a much higher concentration of solutes than the original pore water because of the rejection of solutes during freezing (Mackay 1979). Consequently, the 0°C isotherm is typically many meters below the bottom of the ice-bonded permafrost. The "unfrozen" zone in Figure 1, as an example, is based upon the lack of ice-bonding, not upon a positive temperature.

GROWTH PATTERN

A pingo underlain by a subpingo water lens is in an unstable growth situation. At one extreme, the pingo will eventually rupture, if the yearly addition of water to the subpingo water lens exceeds that which is frozen; at the other extreme, the subpingo water lens will freeze through, if the yearly amount added is less than that which is frozen. Figure 7 illustrates three different growth types for pingos with subpingo water lenses. Type 1, constant rate of growth: Pingo 9, which had a 2.2 m thick water lens in 1978 (Mackay 1978), has a constant rise in elevation for the pingo top. If all of the 2.2 m deep water lens were to freeze solid in one winter, a thermal impossibility, without any addition or loss of water, the 9% volume expansion could only produce a summit growth of 20 cm/yr in contrast to the actual growth of about 30 cm/yr. Consequently, there must be a net annual addition of water to the subpingo water lens. Type 2, decreasing rate of growth: The growth rate of pingo 8 is decreasing and presumably growth will soon cease. Pingo 8 was drilled in 1980, and although a "cavity" was penetrated beneath the summit, there was no artesian flow. The water lens appears to be freezing through without addition of more water. Type 3, erratic growth: Pingos 4 and 15 show episodic periods of growth from water accumulation, followed by subsidence from pingo rupture and water loss. The existence of a subpingo water lens is self-evident from the erratic growth pattern of Type 3, but the growth patterns of Types 1 and 2 could occur for pingos without subpingo water lenses.

CONCLUSION

Some pingos are underlain by a subpingo water lens whose freezing results in nearly pure "injection ice." The water pressure is derived from pore water expulsion which accompanies permafrost aggradation in saturated sandy

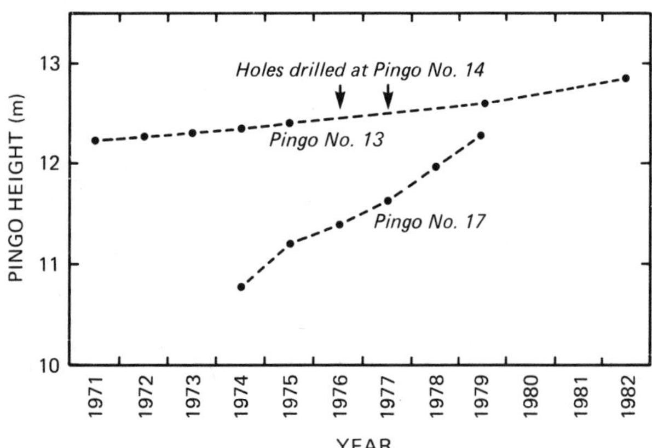

Figure 6. Graph showing the heights of bench marks on the tops of pingos 13 and 17 for the respective survey period.

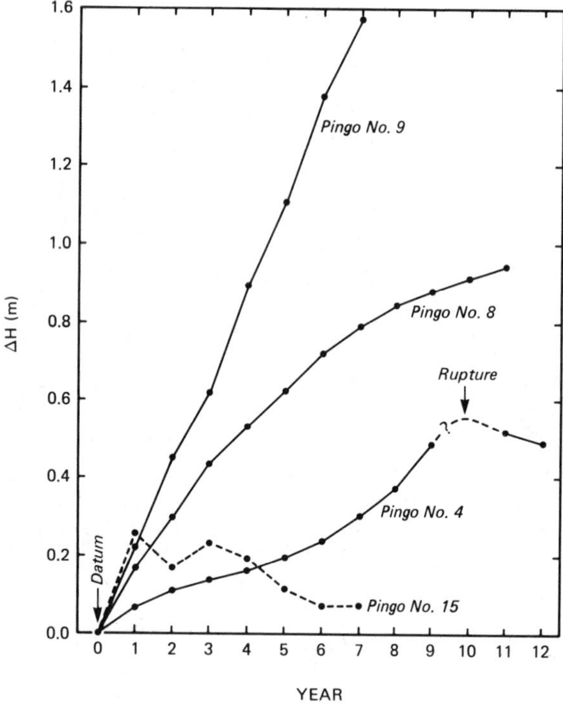

Figure 7. Graph showing the changes in height of four pingos relative to a datum year. For pingos 4, 8, and 15, the datum year was 1969; for pingo 9, it was 1970.

sediments. The water pressure in a subpingo water lens is sufficient to uplift the superincumbent pingo. A pingo with a subpingo water lens is in an unstable situation and subject to rupture, water loss, and subsidence. Pingos with subpingo water lenses exhibit a variety of growth patterns.

ACKNOWLEDGEMENTS

The fieldwork has been supported by the Geological Survey of Canada, the Polar Continental Shelf Project, Canada, and the Natural Sciences and Engineering Research Council of Canada.

REFERENCES

Arvidson, W. D., and Morgenstern, N. R., 1977, Water flow induced by soil freezing: Canadian Geotechnical Journal, v. 14, p. 237-245.

Khakimov, Kh. R., 1957, Problems in the theory and practice of artificial soil freezing, (in Russian), Moscow, Publishing House, Academy of Sciences, U.S.S.R., Translated by Cold Regions Research and Engineering Laboratory, Hanover, N.H., 1970, 178 p.

Mackay, J. R., 1962, Pingos of the Pleistocene Mackenzie Delta area: Geographical Bulletin, no. 18, p. 21-63.

Mackay, J. R., 1977, Pulsating pingos, Tuktoyaktuk Peninsula, N.W.T.: Canadian Journal Earth Sciences, v. 14, p. 209-222.

Mackay, J. R., 1978, Sub-pingo water lenses, Tuktoyaktuk Peninsula, Northwest Territories: Canadian Journal Earth Sciences, v. 15, p. 1219-1227.

Mackay, J. R., 1979, Pingos of the Tuktoyaktuk Peninsula area, Northwest Territories: Géographie Physique et Quaternaire, v. 33, p. 3-61.

McRoberts, E. C., and Morgenstern, N. R., 1975, Pore water expulsion during freezing: Canadian Geotechnical Journal, v. 12, p. 130-141.

FROST CELL DESIGN AND OPERATION

Daniel W. Mageau[1] and Mark B. Sherman[2]

[1]R & M Consultants, Inc., Fairbanks, Alaska USA
[2]City Engineering Office, Fairbanks, Alaska USA

This paper presents the design and recommended operation of a frost heave cell for use in conducting laboratory frost heave experiments. Specific features of the design include split barrel construction for easy assembly, controlled top and bottom temperatures, well-insulated sides, simple thermistor installation for temperature measurement, and new heat exchanger details for top and bottom plates. Tests can be performed on undisturbed samples, remolded cohesive soil, or remolded cohesionless soil. Details of sample preparation are presented for each sample type. Test duration depends on soil type and amount of heave, and generally ranges from 1 to 3 days. Tests can be conducted at various overburden pressures. Test results are interpreted using the Konrad-Morgenstern segregation potential concept of frost heave.

INTRODUCTION

Many researchers over the past several decades have attempted to relate frost heave to soil index properties, the most common of which has been grain-size, (Penner 1968, Kaplar 1974, Rieke et al. 1983). Rieke et al. (1983) performed a series of laboratory frost heave tests in which the amount of silt and clay and the type of clay mineral were varied from test to test. They found that while relationships between frost heave and soil index properties do exist, the sensitivity of frost heave to even small changes in clay content (1-2%) resulted in significant changes in observed heave rate under the same testing conditions. It is concluded from these tests and from the lack of correlation between grain size and results from frost heave tests performed by others (e.g., Kaplar 1974, Leary et al. 1968) that soil index properties alone are insufficient to adequately assess the frost heave potential of a soil. Furthermore, it is our contention, and that of others (Konrad and Morgenstern 1982b), that laboratory frost heave tests are necessary to accurately determine the frost susceptibility of soil.

One drawback of laboratory frost heave testing in the past has been the lack of uniformity of test equipment, test procedures, and techniques to interpret results. In this paper, we present the design of a frost cell which, although similar in design to frost cells used by others, incorporates special features developed by the authors. Procedures for testing either fine-grained or coarse-grained soils are presented with special attention given to problems commonly encountered during frost heave testing. Finally, a method is presented to interpret frost heave test results based on a frost heave theory proposed by Konrad and Morgenstern (1980).

The purpose of this paper is not to present frost heave test results but rather to present a proven approach for performing consistent laboratory frost heave tests on a variety of soils along with details for obtaining the necessary parameters to analyze test results.

FROST HEAVE TEST EQUIPMENT

The cell described in this paper (Figure 1) uses a 150-mm high soil sample of either granular or fine-grained material. The test apparatus consists of 150-mm I.D. by 200-mm O.D. split nylon cylinder, instrumented along the sides with thermistors. The cylinder rests on a copper heat exchanger base plate. The soil sample (either remolded or undisturbed) is placed in the cylinder within a latex membrane, and a copper piston-like heat exchanger is situated on top of the soil. Lines in the top heat exchanger allow water from a reservoir (burette) to flow into the sample and eventually to the freezing front as it progresses through the sample. Freezing is initiated at the base of the sample, and as the sample heaves, the top piston moves upward. A vertical load is applied to this piston via a dead weight system to simulate field overburden conditions.

The cell is placed in a temperature controlled environment slightly above freezing. External coolant circulating baths supply a glycol-water mixture to the heat exchangers at adjustable preselected temperatures, allowing any desired gradient to be applied. Thermistors in the top and base plates permit the plate temperatures to be monitored during the test. Figure 2 is a schematic diagram of the frost cell experimental setup.

Design Considerations

The heat exchanger plates at the top and bottom of the sample must be capable of transferring a sufficient quantity of heat in or out of the sample in order to maintain the desired thermal gradient for the duration of the test. In turn, the cooling mechanism attached to the plates must be able to remove or supply this heat to maintain isothermal conditions in the plates within very close tolerances. Copper has been selected as the material for the heat exchanger plate, although aluminum has also been used successfully (Mageau and Morgenstern 1980, Rieke et al. 1983). The actual configuration of the top and base heat exchangers is shown in Figures 3 through 6. Our design minimizes thermal gradients across the plate and between passages. Assuming a 50% water/50% ethylene glycol mixture at a minimum flow rate of 1.2 l/min, the total temperature drop in the fluid between the inlet and outlet should be on the order of 0.02°C, once a steady-state thermal gradient in the sample has been achieved.

The coolant baths should be capable of maintaining a temperature stability on the order of 0.02°C. Our experience has shown that overall temperature stability is improved if the coolant baths are located externally from the cell environment and coolant lines are well insulated. In order to minimize radial heat flow into the sample, the cells are insulated with 100 mm of polyurethane foam. The cell environment itself should be maintained as close to +1.0°C as possible. It has been found that a large cold room can be difficult to maintain at a constant temperature and that a small cold room is more desirable. Rieke et al. (1983) have successfully used a standard refrigerator modified to accept frost cell equipment.

The basic parameters measured during the frost heave test are temperature, water intake, and vertical heave. For temperature measurement, we recommend waterproof and pressure resistant thermistors, which are calibrated prior to each test using a carefully prepared ice bath. An accuracy of ±0.01°C is desirable in order to properly estimate thermal gradients and frost penetration. Water intake to the sample is measured with a burette accurate to ±0.2 ml. Vertical heave is measured using an LVDT (or dial gage) accurate to ±0.0025 mm. The accuracy of vertical heave measurement is especially critical for coarse-grained samples which may show little heave over the duration of the test. An automated data collection and storage system is recommended.

SAMPLE PREPARATION

Preparation of a soil sample for frost heave testing depends on the type of soil and on whether the sample is remolded or undisturbed. We recommend testing an undisturbed sample, as the soil structure and density would more closely resemble insitu conditions than would those associated with a remolded sample. However, for many engineering projects, it may not be feasible or, in the case of coarse-grained soils, it may not be possible to obtain undisturbed samples, and laboratory tests must be conducted on remolded samples.

Sample preparation for remolded cohesive soil differs from that for remolded cohesionless soil. The basic difference is that remolded cohesive soil is consolidated in a separate consolidometer, while remolded cohesionless soil is compacted in lifts inside the frost cell. Many natural soils are borderline between cohesive and cohesionless, and there is no absolute rule for determining to which category a soil should belong. If clay mineral is present, a soil with as little as 20% passing the #200 sieve may be considered cohesive. In general, external consolidation yields a more uniform and consistent sample and is used by the authors for any soil which has enough cohesion to remain intact after removal from the consolidometer.

This section for sample preparation is divided, therefore, into three categories based on the type of sample: undisturbed, remolded cohesive, and remolded cohesionless.

Undisturbed Soil

The required diameter of a sample to be tested in the frost cell is 150 mm (after extrusion from the sampler). A sample with a diameter smaller than 150 mm would allow water to migrate between the sample and the membrane and therefore is not used. A sample with diameter larger than 150 mm can be used after trimming, provided the trimming process does not result in voids along the sides of the sample. The sample length can range from 80 to 150 mm.

Remolded Cohesive Soil

Even in cohesive soil, material greater than 30 mm may be present and can lead to difficulties in sample preparation. Therefore, all material greater than 30 mm is removed from the sample and replaced with an equivalent weight of materal larger than the #4 sieve and smaller than 30 mm. This procedure is expected to have little effect on frost heave test results since it is the silt and clay fraction (i.e., specific surface) which controls the heave process (Rieke et al. 1983).

A sufficient quantity of soil is then thoroughly mixed with enough distilled water to produce a slurry with a water content in excess of the liquid limit of the soil. The soil is allowed to soak for at least 12 hours. The slurry is then consolidated in a consolidometer (approximately 150 mm I.D.) at the pressure at which the frost heave test will be conducted. If this pressure is not sufficient to consolidate the sample so that it can be handled without deforming, the consolidation pressure is increased to a level which yields a sample that will not easily deform under zero confining pressure. This minimum consolidation pressure depends primarily upon the amount of fines (particularly the clay fraction) in the soil and typically ranges from 50 to 100 kPa.

If the sample is undisturbed or has been prepared using the consolidometer, the next step is to place the test specimen on the base plate and place the top cap (with a filter paper) on the sample. The latex membrane is installed around the sample and secured at the top and base plates using "O" rings. This membrane is allowed some slack to provide for stretching as the soil heaves vertically during the test. The membrane is coated with a silicone lubricant and the two barrel halves are placed around the sample and fastened. At this time the top cap is checked to ensure it is not binding.

Remolded Cohesionless Soil

As with cohesive soil, all material larger than 30 mm is removed and replaced with an equivalent weight of material larger than the #4 sieve and smaller than 30 mm. The soil is completely covered with distilled water and allowed to soak for at least 12 hours. The frost cell is partially assembled with the membrane attached to the base plate by means of an "O" ring and stretched over the sides of the bottom portion of the nylon barrel. A low-viscosity silicone lubricant is applied to the inside walls of the barrel prior to attaching the membrane to reduce friction between the barrel and the membrane during the test.

The cohesionless soil is placed in the frost cell in 20-mm lifts and compactd to the desired density using a hand-held tamper 40-mm in diameter. Distilled water is added to each lift prior to tamping to ensure that the soil remains saturated. The top cap is placed on the sample (with a filter paper between the porous stone and the sample), and the membrane is attached to the top cap with two "O" rings. The top portion of the nylon split-barrel is placed around the top cap and secured.

FROST CELL PREPARATION

The remainder of the frost cell is assembled, including the insulation, load ram guide bar, and clamps, and the entire assembly is placed in the controlled-temperature cold room set at a temperature of approximately +1°C. The drainage lines are attached and de-aired, and the sample is consolidated to the desired pressure by placing appropriate weights on the hanger assembly. The sample is kept in the cold room overnight with the drainage lines open to allow equilization of both pore pressure and soil temperature. For most soils 12 hours should be adequate to dissipate excess pore pressures. For clay soils complete excess pore dissipation may take longer. A standard constant or falling head permeability test (without back pressure) can be performed on the sample at this time if desired. Otherwise, the sample is ready for testing.

TEST PROCEDURE

Coolant lines from the warm temperature bath (Tw=+0.3°C) are attached to both top and bottom plates, and fluid is circulated for several hours in order to induce a uniform temperature in the sample.

In fine-grained soil, pore spaces are generally too small for initial nucleation of ice crystals to occur at the cold-side test temperature of -1°C. Therefore, initial nucleation must be induced by briefly supercooling the base of the sample. This procedure is usually not necessary for cohesionless soils. First, the coolant lines from the warm refrigeration bath are disconnected from the base plate (circulation through the top cap should be continued). Second, coolant lines from a separate refrigeration bath which has been set at -15° to -20°C are connected to the base plate. The change in height and base plate temperature readings are continually monitored after circulation is initiated. Nucleation, and subsequent change in height, should commence within 1 min. for most fine-grained soils. As soon as any change in height is observed, the coolant lines from the nucleation refrigeration bath are disconnected, and coolant lines from the bath set at -1°C are attached to the base plate. Fluid circulation is immediately initiated.

Readings for temperatures (top and bottom plates, sidewalls, and room), change in height and water intake (or expulsion) are recorded at the following times: 0, 1 min, 5 min, 10 min, 20 min, 30 min, 1 hr, 2 hrs, 4 hrs, 8 hrs, 16 hrs (optional), and 24 hrs. Readings are recorded at least twice daily thereafter.

The test is continued for a minimum of 12 hours after the rate of frost penetration becomes approximately zero. The time for zero frost penetration to occur depends on the soil type and the amount of water intake. For cohesionless soils, which usually intake little water during the test, the total length of a frost heave test may range from 1 to 2 days. For silty material, the length of a test may range rom 2 to 4 days. For soils with significant clay content, the length of a test may range from 3 to 5 days.

After completion of the test the sample is removed from the cell and examined. The location of the final ice lens (if visible) with respect to the bottom of the sample (cold-side) is measured. The sample is divided into frozen and unfrozen parts, and water content determinations are made on each. Frost heave (change in height), water intake, and frost penetration are then plotted as a function of time.

INTERPRETATION OF TEST RESULTS

Results from laboratory frost heave tests are interpreted using a mechanistic theory of frost heave developed by Konrad and Morgenstern (1980) which allows prediction of field frost heave behavior using parameters obtained from laboratory frost heave tests. The theory is derived from thermodynamic processes within the frozen fringe of soil between the frozen-unfrozen interface and the last ice lens. Simply stated, frost heave is controlled by the temperature gradient within the frozen fringe as expressed by:

$$v = SP(Pe) \text{ grad } T, \tag{1}$$

where v is the velocity of water intake (mm/s), grad T is the temperature gradient across the frozen fringe (°C/mm) and SP(Pe) is the segregation potential ($mm^2/s°C$) as a function of overburden pressure, Pe (kPa).

The amount of segregational heave (versus insitu heave) is thus given by:

$$hs = \sum 1.09 \ v\Delta t \tag{2}$$

where hs is the total segregational heave (mm) and Δt is the time interval (sec).

From equations (1) and (2) it can be seen that segregational heave is directly related to SP. However, SP is dependent on several variables. Because of the many unknowns associated with using laboratory test results for predicting field behavior, it is considered prudent to determine the highest segregation potential for a soil. Konrad and Morgenstern (1981) have observed that SP is at its maximum when the suction at the frozen-unfrozen interface (Pu) is zero. Moreover, they observed that Pu decreases as the warm-side temperature, Tw, decreases. By conducting the laboratory tests at Tw as close to 0°C as practical, the SP value obtained from test results can be maximized. SP is also affected by the rate of cooling at the frost front. Konrad and Morgenstern (1982a) found that maximum SP value occurs when the rate of cooling is near zero (0.1°C/hr for Devon silt). This rate of cooling corresponds to the formation of the final (warmest) ice lens. Therefore, it is conservative to evaluate v and grad T from equation (1) at the time the final ice lens begins to develop, t_o. Finally, SP is dependent on overburden pressure. As overburden pressure increases, the value of SP decreases. Therefore, a minimum of two frost heave tests is required to define a relationship between SP and pressure for a soil similar to that shown in Figure 7 for Devon silt.

To obtain values of v and grad T for use in equation (1), time t_o must first be determined. If the final ice lens is visible, t_o can be estimated by measuring the distance from the bottom of the final ice lens to the bottom of the sample at the end of the test and then scaling off this distance on the frost penetration versus time plot, as shown in Figure 8a. The velocity of water intake, v is then determined at time t_o from the water intake versus time curve, as shown in Figure 8b. The value for grad T is determined from a plot of sidewall temperatures recorded at (or near) time t_o, as shown in Figure 8c.

If no final ice lens is visible, such as with coarse-grained soils that exhibit little heave, time t_o must be estimated by assuming that the rate of cooling at the frost front which corresponds to the maximum SP value occurs as the frost penetration rate approaches zero. Once this point on the frost penetration versus time curve has been established, values for v and grad T are determined in the same manner described above for samples when the final ice lens is visible.

It is acknowledged that there is a certain amount of error associated with determining SP values for a soil. It is the authors' opinion, however, that the test procedures and method of analysis described herein result in consistent and reproducible segregation potential values for a soil within an acceptable margin of error for most engineering projects.

The segregation potential concept is relatively new (1980), and its application to engineering projects has seen limited use. Nixon (1983), and Konrad and Morgenstern (1983) have reported good correlations between field behavior predicted from laboratory frost heave tests (similar to that described in this paper) and observed field behavior. It is hopeful that the use of a laboratory frost heave test such as that presented in this paper will eventually become standard practice for projects where prediction of frost heave is required.

CONCLUSIONS

The design of a frost cell and recommended procedures for testing undisturbed, remolded cohesive and remolded cohesionless soil samples have been presented. The cell is designed to minimize friction and radial heat flow and is relatively easy to assemble and disassemble. The test procedures involve applying constant warm and cold temperatures (+0.3° and -1.0°C) to the top and bottom plates, respectively, and recording heave, temperature, and water intake until after the frost penetration rate becomes zero. The duration of the test can range from 1 to 5 days but typically ranges from 1 to 3 days. Test results are interpreted using the segregation potential concept of frost heave introduced by Konrad and Morgenstern (1980). This method appears to be the most reliable for assessing the frost susceptibility of a soil and should be used whenever prediction of field frost heave behavior is required.

REFERENCES

Kaplar, C. W., 1974, "Freezing test for evaluating relative frost susceptibility of various soils", Technical Report 250, U.S. Army Cold Regions Research and Engineering Laboratory, Hanover, N.H.

Konrad, J. M., and Morgenstern, N. R., 1980, "A mechanistic theory of ice lens formation in fine-grained soils", Canadian Geotechnical Journal, 17, pp. 473-486.

Konrad, J. M., and Morgenstern, N. R., 1981, "The segregation potential of a freezing soil, Canadian Geotechnical Journal, 18, pp. 482-491.

Konrad, J. M., and Morgenstern, N. R., 1982a, "Prediction of frost heave in the laboratory during transient freezing", Canadian Geotechnical Journal, 19, pp. 250-259.

Konrad, J. M., and Morgenstern, N. R., 1982b, "Effects of applied pressure on freezing soils", Canadian Geotechnical Journal, 19, pp. 494-505.

Konrad, J. M., and Morgenstern, N. R., 1983, "Frost heave prediction of chilled pipelines buried in unfrozen soils", Proceedings from Fourth International Conference on Permafrost, Fairbanks, Alaska (this volume).

Leary, R., Sanborn, J., Zoller, J., and Biddiscombe, J., 1968, "Freezing tests of granular materials", Highway Research Record No. 215, pp. 60-74.

Mageau, D. W., and Morgenstern, N. R., 1980, "Observations on moisture migration in frozen soils", Canadian Geotechnical Journal, 17, pp. 54-60.

Nixon, J. F., 1982, "Field frost heave predictions using the segregation potential concept", Canadian Geotechnical Journal, 19, pp. 526-529.

Penner, E., 1968, "Particle size as a basis for predicting frost action in soils", Soils and Foundations, 8 (4), pp. 21-29.

Rieke, R., Vinson, T., and Mageau, D. W., 1983, "The role of specific surface area and related index properties in the frost susceptibility of soils", Proceedings from Fourth International Conference on Permafrost, Fairbanks, Alaska (this volume).

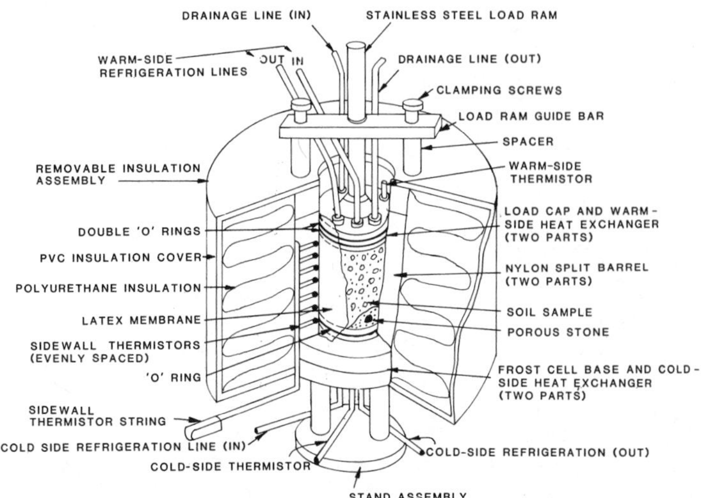

FIGURE 1 CUTAWAY VIEW OF FROST CELL.

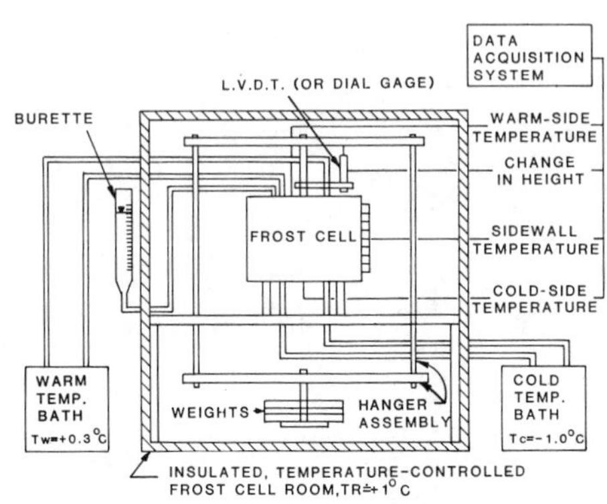

FIGURE 2 FROST HEAVE EXPERIMENTAL SET-UP.

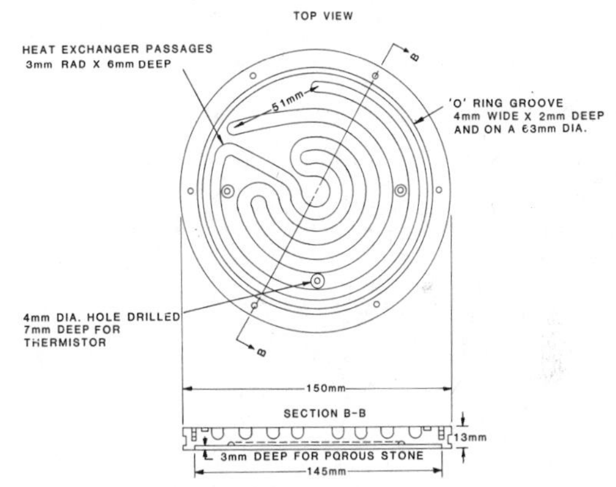

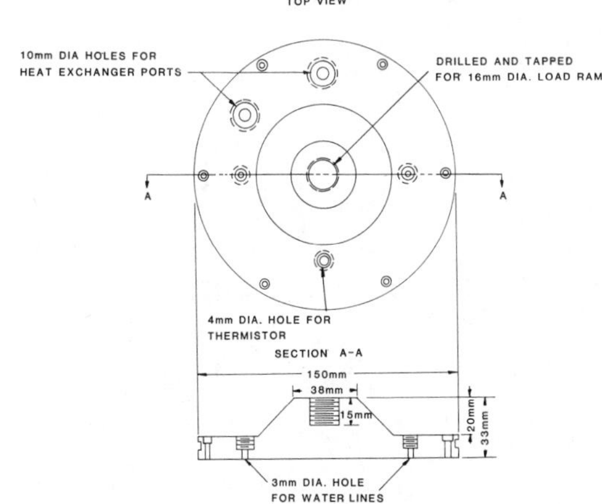

FIGURE 3 PART 1 OF TOP CAP.

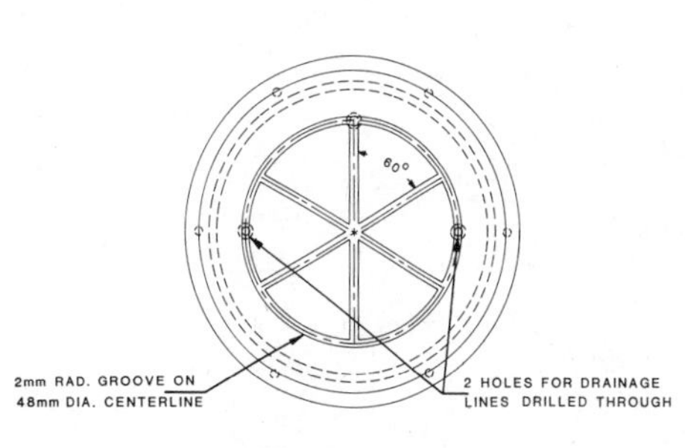

FIGURE 4 PART 2 OF TOP CAP.

FIGURE 2 Insulation correction factor, infinite strip, $0 < \beta < 0.6$.

TOP VIEW

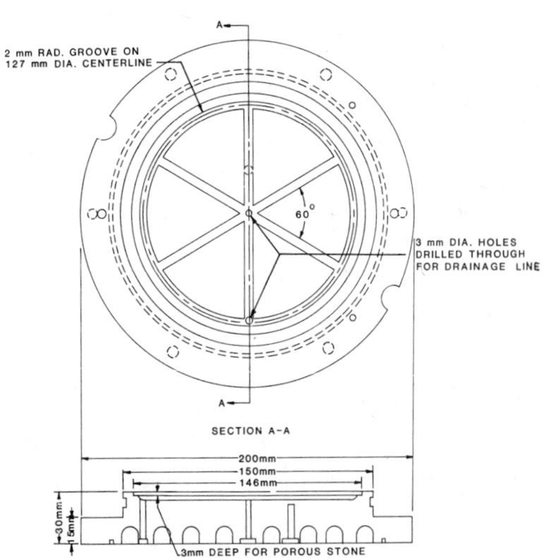

SECTION A-A

BOTTOM VIEW

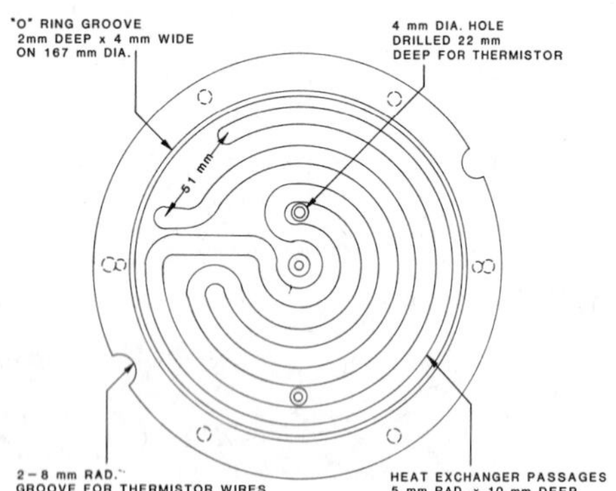

FIGURE 5 PART 1 OF BASE PLATE.

TOP VIEW

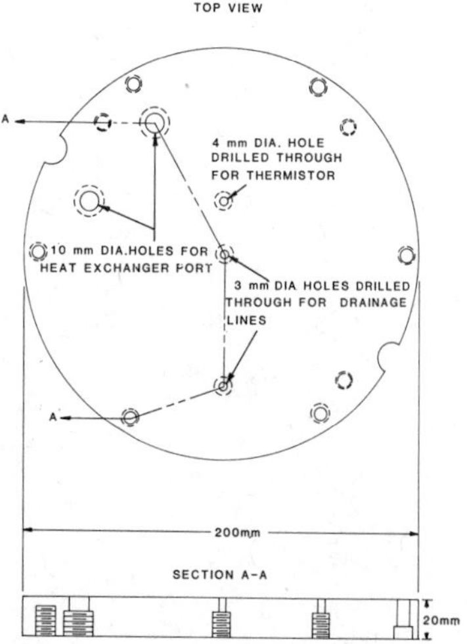

FIGURE 6 PART 2 OF BASE PLATE.

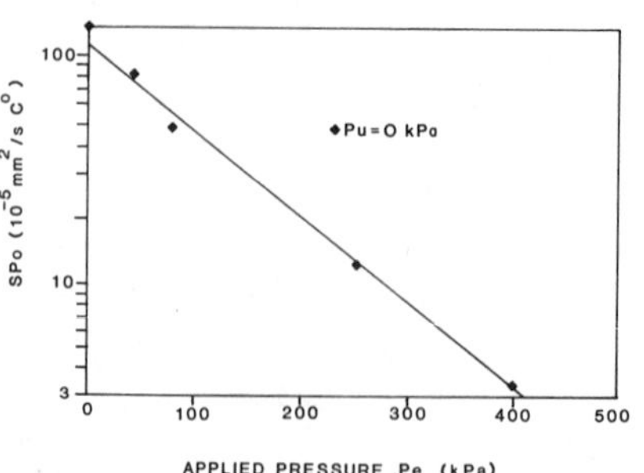

FIGURE 7 SEGREGATION POTENTIAL VERSUS APPLIED PRESSURE FOR DEVON SILT (FROM KONRAD AND MORGENSTERN 1982b).

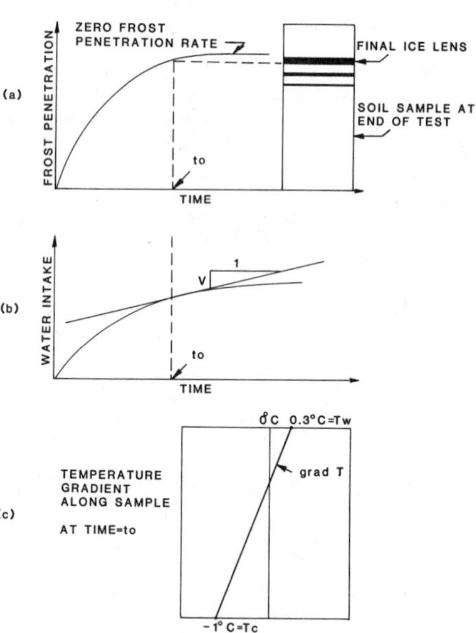

FIGURE 8 EXAMPLE OF FROST HEAVE TEST RESULTS USED TO DETERMINE SEGREGATION POTENTIAL.

PHYSICAL AND NUMERICAL MODELING OF UNIAXIAL FREEZING IN A SALINE GRAVEL

Larry J. Mahar[1], Ralph M. Wilson[2], and Ted S. Vinson[2]

[1]Ertec Western Inc., Long Beech, California 90807 USA
[2]Department of Civil Engineering, Oregon State University
Corvallis, Oregon 97331 USA

Laboratory model studies were conducted to study the freezing process and the distribution of salt during advancement of a freezing front through a seawater saturated column of granular soil. The freezing front in a saline soil may be described as a zone of partially frozen soil. The strength of the frozen soil in the freezing zone is visually weaker than for soils above the freezing zone. The salt concentration and the unfrozen water content in the partially frozen zone decreases toward the colder temperature. It is shown that conventional analytical modeling of freezing front penetration will underpredict the rate of freezing when no account is made for the effects of salinity. The rate of advance to a given depth of freezing increased with increasing salinity.

The paper presents a discussion based on the results of 17 laboratory freezing column tests in which the effects of salinity on the freezing front advance and the redistribution of solutes was observed. Comparisons between observed behavior and computer simulations of three tests are given to demonstrate difficulties associated with analytical modeling freezing front penetration in a saline soil.

INTRODUCTION

The design of artificial islands for arctic waters requires a detailed thermal analysis to determine the extent of soil freezing that will occur during the design life of the structure. This analysis is important because the unfrozen/ frozen interface at any time may represent the location of a potential failure plane through the island. For the long-term design of production islands, additional thermal analyses are required to determine the growth of a strenghtening permafrost core within the island, thaw degradation around well conductors, thermal disturbance beneath production facilities, disturbance to subsea permafrost and thermal effects of pipelines and oil storage vessels.

There are many numerical computer models available to solve two-dimensional heat transport problems with the porewater phase change modeled at the frozen-unfrozen interface. Many of these models have been shown, through field and laboratory measurements, to accurately predict thermal regimes and the location of a freezing interface for soils with non-saline porewater. However, in the marine environment, where a significant concentration of sea salts are present in the porewater, the thermal regime during freezing is altered owing to the effects of freezing point depression, salt exclusion and solute transport.

In recognition of the need to predict the thermal regime in offshore artificial islands, computer simulations were performed of uniaxial freezing in saline and non-saline saturated gravels. The results from the computer simulations were compared to the results of laboratory freezing column tests performed on saline and non-saline saturated gravels. The computer comparisons demonstrate the difficulties associated with modeling uniaxial freezing of a saline saturated soil.

In addition, data is presented from the loboratory freezing column tests to demonstrate the redistribution of solutes during uniaxial freezing. The results are used to develop the relationship of solute concentration to the unfrozen water content profile. Finally, the effects of solute transport in a soil-water system as it relates to freezing of saline soils is discussed.

LABORATORY MODEL TESTS

A series of 17 large-scale tests were conducted in the laboratory to study the effects of salinity on uniaxial freezing of seawater saturated gravel. Detailed descriptions of the test apparatus and procedures are given by Mahar et al. (1982) along with typical results. Briefly, a 46 cm diameter soil column (see Figure 1) saturated with a seawater solution of a known salinity was frozen uniaxially from the top of the column downward. The initial porewater salinities ranged from 0 to 80 parts per thousand (ppt) and various freezing rates were applied. During freezing, the temperature profile at the center of the column was monitored at regular time intervals using a string of closely spaced thermistors. Porewater samples were extracted at various locations along the column during the test for salinity determinations. It was possible in certain tests, to extract porefluid in the partially frozen soil after the freeze front had passed.

The results from thirteen tests are summarized by Vinson et al. (1983) and demonstrate the general effect of salinity on the rate of freezing front penetration and the changes in unfrozen porewater salinity with time at discrete points in the soil column (for selected tests). It was concluded, based on the laboratory tests, that the nature of the freezing interface was quite different for the distilled porewater and saline porewater cases. The freezing interface in the distilled porewater column was planar with clear ice filling the pore spaces. The location of the interface at any time was easily identified on the temperature profile by a distinct break at the zero degree (celsius) isotherm.

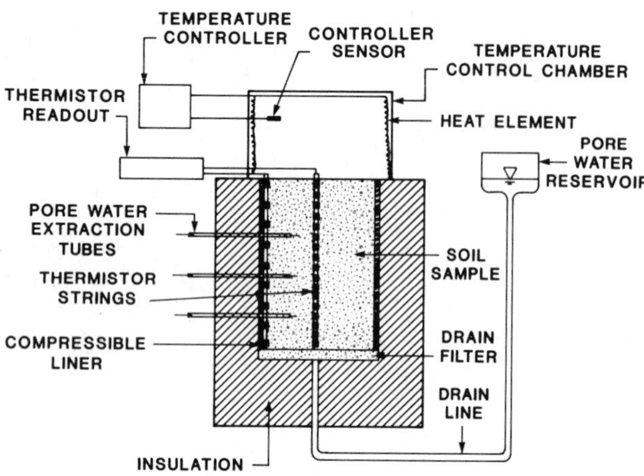

FIGURE 1 Schematic of freeze front penetration test system.

For the saline porewater cases, it was found that a distinct freezing interface did not exist. Instead, a zone of partially frozen soil with a high unfrozen water content beginning at the depressed freezing point isotherm (defined herein as the freeze front) was observed. The unfrozen water content was found to decrease with distance toward the colder temperatures from the freeze front. In addition, the location of the freeze front was not clearly identified by a break in the temperature profile.

Salinity measurements in the partially frozen zone, reported by Vinson et al. (1983) indicated very high values, reaching concentrations in excess of 100 ppt. The salinities were found to increase with decreasing temperature. The significance of these measurements, along with a presentation of additional tests, where more extensive salinity measurements were made, is presented later in this paper.

THERMAL ANALYSIS

In order to assess the potential of existing computer models to predict the thermal regime in soils with saline porewater, computer simulations of two of the freezing column tests were performed. The computer code selected for the simulation is based on a finite element model developed at the University of California at Irvine (Guymon et al. 1981). Computer code FROST 2B represents a two-dimensional model of coupled heat and moisture transport in freezing/thawing soil with an isothermal soilwater phase change approximation. The numerical solution used to solve the governing flow equations is the nodal domain integration method with a linear trial shape function for triangular elements. Time domain solutions are obtained with the Crank-Nicolson method and fully implicit method for the heat transport and the moisture transport models, respectively. The computer code and model are similar to other thermal models presently used in design practice.

The two freezing column tests simulated with FROST 2B were identical with respect to test conditions with the exception that one was saturated with distilled water and the other with a seawater having a salinity of 28 ppt. In both cases, thermal properties for the soil were selected for input based on recommended values from the literature (Kersten 1949). No attempt was made to adjust the thermal properties to account for salinity except to use a depressed freezing point of -1.6°C for the saline porewater case.

The results of the two computer simulations are presented in Figures 2 and 3. The simulations for the distilled porewater case compare favorably with the laboratory measurements. Both the temperature profiles and the location of the freezing front were modeled accurately, indicating the adequacy of the computer method for non-saline soils.

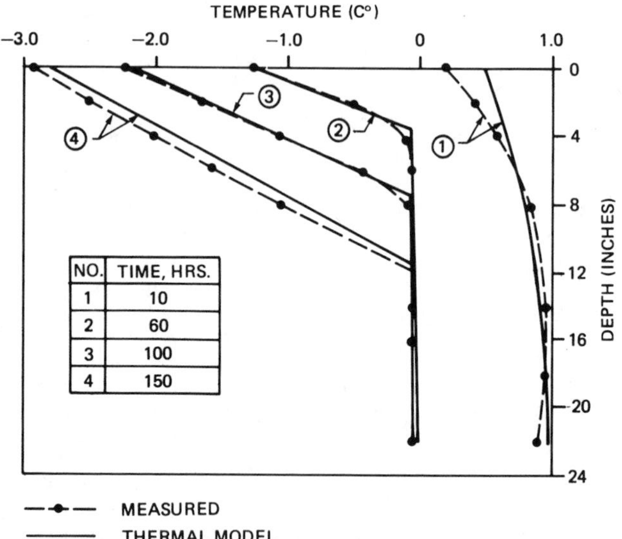

FIGURE 2 Comparison of thermal model results to laboratory measurements- distilled porewater case.

In contrast, for the saline porewater case, (Figure 3) the simulations compared poorly with the laboratory measurements. The depth to the freeze front was significantly underpredicted based on the computer solution and the temperature profiles were inaccurate. The apparent effect of salinity on the thermal behavior of the soil-water system and, particularly, on the phase change phenomena is strongly evident.

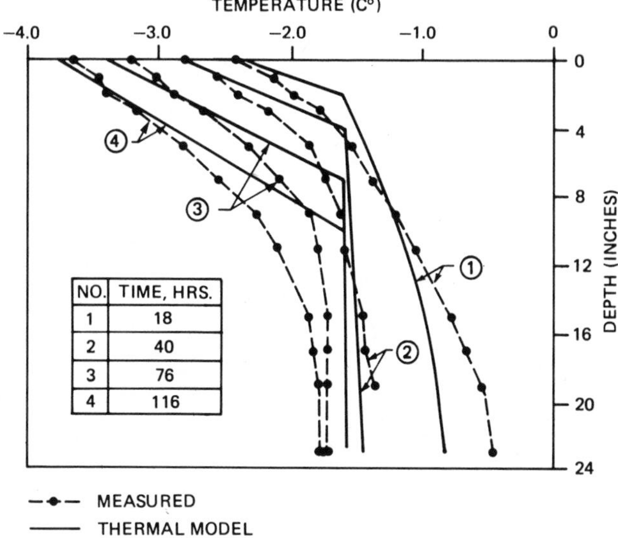

FIGURE 3 Comparison of thermal model results to laboratory measurements - saline porewater case (salinity = 28 ppt).

The primary source of error in the saline porewater case is found in the method of modeling the phase change. In the distilled porewater case, the phase change occurs at the freezing point isotherm (at the freeze front) where nearly all the porewater freezes, releasing its latent heat. This results in the sharp break in the temperature profile at the freeze front in both the computer simulation and laboratory test results.

In the saline porewater case, no clearly distinguishable freeze front exists. Freezing occurs over a range of temperatures and depths due to a changing solute concentration in the unfrozen porewater as a consequence of salt exclusion. During the phase change, latent heat is released gradually over the entire depth of the partially frozen region as temperatures are lowered. Since the computer model only allows latent heat release to occur instantaneously along a plane, significant error results.

Clearly, a better understanding of the effects of salinity on the freezing of soils is needed to develop accurate thermal modeling methods. The following sections of this paper addresses solute redistribution as it pertains to the soil freezing problem, based on a laboratory model of uniaxial freezing in saline gravel.

FREEZING OF SALINE SATURATED SOIL COLUMNS

In a saline saturated soil column subjected to freezing temperatures, ice crystal growth begins only as the freezing point depression is reached. The crystals grow in a pure (or nearly pure) water ice form while excluding solute into the remaining unfrozen porewater (Sheeran and Yong 1975). As this occurs, the concentration of the unfrozen porewater increases, further depressing the freezing point. If the heat transfer is slow enough, the solute concentration in the unfrozen porewater will be maintained in equilibrium with the pore ice. Thus, the soil temperature at any point will be equal to the equilibrium solidification temperature for the enriched solute concentration of the unfrozen porewater. Further lowering of the temperature results in more ice growth and increasing the porewater salinity to maintain the equilibrium balance (Hallet 1978).

The effect of salt transport due to diffusion or natural convection downward through the column will cause this freezing process to occur more readily. A solute concentration gradient occurs above the freeze front in which higher solute concentrations and porewater densities occur at the colder temperatures (near the top of the soil column). This condition results in solute transport downward toward the lower concentrations due to convection associated with the density gradient and diffusion associated with the concentration gradient. As solute transport occurs, the salinity at a given point has a tendency to decrease. However, the decrease in salinity results in a reduction in the freezing point depression, thus stimulating further ice growth. The further ice growth, in turn, increases the salinity maintaining the concentration gradient in the unfrozen porewater. Given sufficient time and a sustained thermal gradient, it is believed that continued solute transport will occur until nearly all the solutes are leached out of the soil leaving the bulk porewater salinity near zero parts per thousand.

SOLUTE DISTRIBUTION

Three freezing column tests were conducted in which it was possible to extract porewater samples at discrete locations along the gravel column at any given time. The three test columns were prepared with initial porewater salinities of 16, 30, and 50 ppt. Typical salinity profiles at various times during the 50 ppt test are given in Figure 4. The salinity profiles verify the presence of high solute concentrations in the unfrozen porewater near the top of the column where the soil was at its coldest temperatures. The dashed line on Figure 4 represents the final salinity of the bulk porewater after thawing. The reduction in the bulk salinity from the initial value of 50 ppt may be attributed to solute transport due to the high concentration gradients in the partially frozen zone.

For each of the unfrozen porewater salinity measurements made during the three tests, the corresponding soil temperature was recorded. The salinity measurements for the 30 and 50 ppt tests are plotted against temperature in Figure 5. The results indicate that the porewater salinity for the 30 ppt case remained in equilibrium with the soil temperature as compared to the equilibrium phase diagram for a pure NaCl solution (Jumikis 1966). However, for the 50 ppt porewater case, the measured salinities were in excess of the equilibrium concentration shown on Figure 5. The result suggest the freezing rate was rapid enough to prevent the solutes from diffusing away from the fréezing zone, despite the high concentration gradients. If the heat transfer rate was reduced, it would be expected that the excess solute concentrations would decrease to concentrations given by the equilibrium phase diagram.

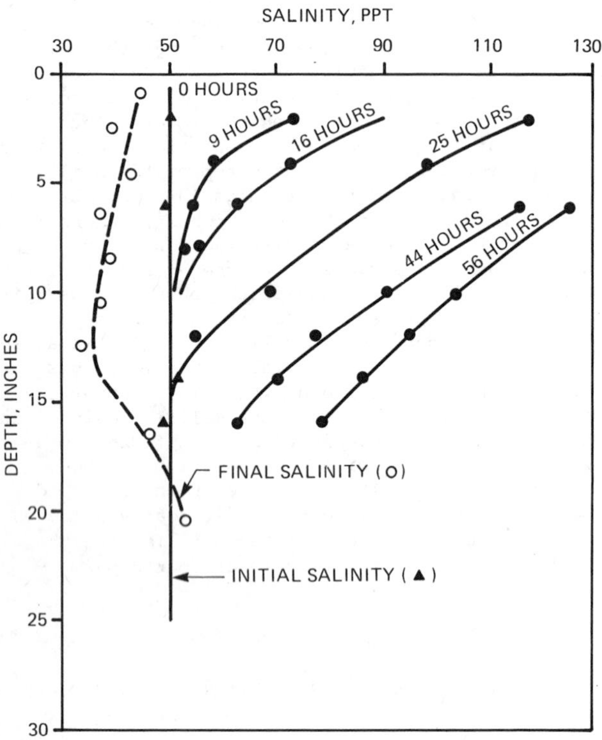

FIGURE 4 Salinity profiles at various times during a typical freezing column test.

UNFROZEN WATER CONTENT

In modeling heat transport with porewater freezing, the release of latent heat during the phase change dominates the solution. In the uniaxial freezing column case, latent heat becomes the primary source of heat conducted upward through the column. The amount of latent heat generated at any given point in the column is a direct function of the volume of water freezing out.

For a closed system in which a saline solution is permitted to freeze and, for simplicity, assuming that solute transport is negligible, the amount of water freezing out is a function of temperature and initial salinity and may be calculated from the results shown on Figure 5 (Sheeran and Yong 1975). The calculation is based on the principal that, for a given volume of solution having a known initial salinity, the amount of water remaining unfrozen will be inversely proportional to its enriched solute concentration at the desired temperature. For example, when the solute concentration of the unfrozen porewater doubles, the amount of unfrozen porewater remaining is one-half of the original volume.

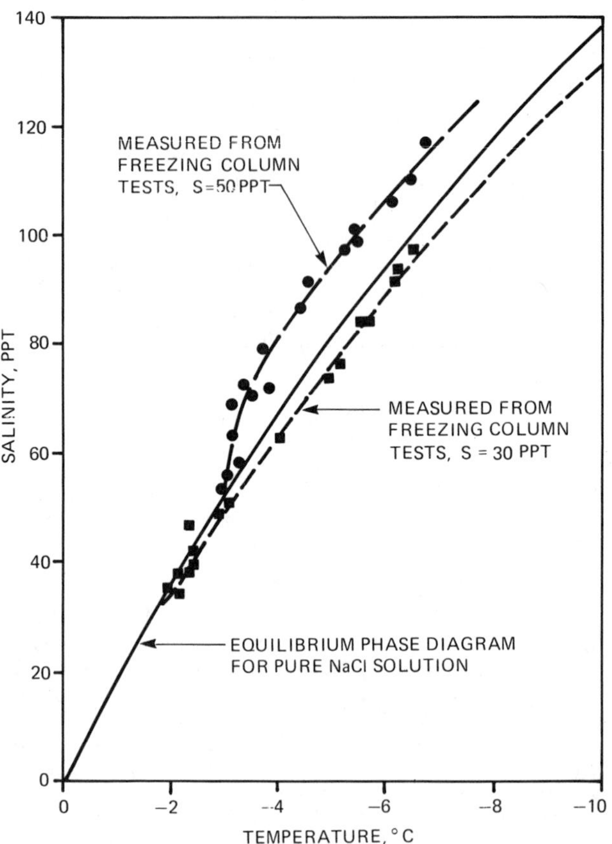

FIGURE 5 Measured salinity versus temperature relationship in unfrozen porewater for two freezing column tests.

Based on Figure 5, unfrozen water content versus temperature relationships for the three freezing column tests were derived as presented in Figure 6. Similar relationships were derived by Yong et al. (1973) based on phase equilibria for bulk saline solutions. From these relationships, the unfrozen water content profile for a soil column may be estimated at any time if the temperature profile and initial porewater salinity is known as shown in Figure 7 for the 30 ppt porewater case. Although there is some error associated with neglecting solute transport, the results presented demonstrate that throughout the soil column, above the freezing front, a significant unfrozen water content remains.

It has been shown that a significant percentage of the free water does not freeze in the freezing column tests. Thus, the lower measured temperature shown on Figure 3, compared to the computer simulation, may be explained by the fact that a smaller amount of latent heat is released during the partial freezing of the porewater owing to solute exclusion. Further, the non-linear nature of the temperature profile above the freeze front is the result of the gradual release of latent heat as the temperatures are lowered at any given point. Therefore, in order to mathematically model the freezing of saline saturated soils, one must couple the heat flow equations with a constituative relationship for the unfrozen water content as a function of temperature and initial salinity applied to all partially frozen zones.

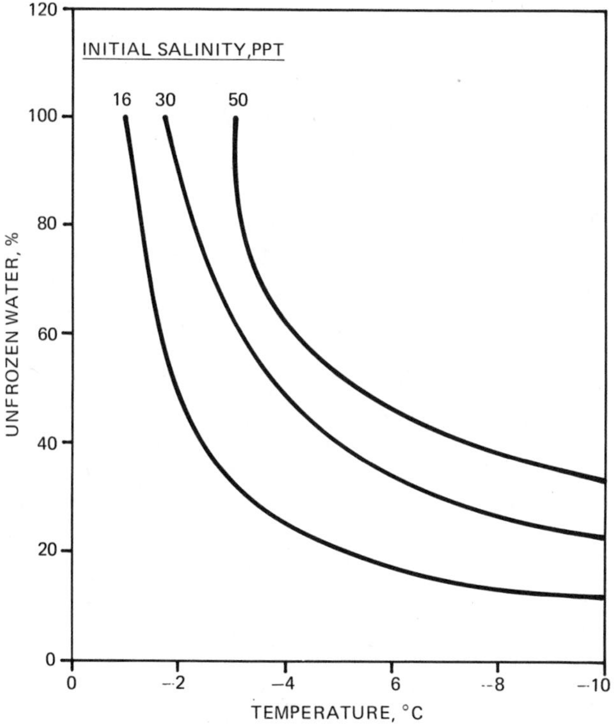

FIGURE 6 Temperature versus unfrozen water content for three freezing column tests (derived from Figure 5).

CONCLUSIONS

The results from physical and numerical model studies of uniaxial freezing of a saline saturated gravel soil column have been presented. Emphasis was placed on identifying parameters affecting heat transport, soil porewater phase change and solute distribution. Based on the results of the model studies, the following may be concluded:

1. Existing numerical methods do not adequately model freezing of saline soils due to the effects of solute exclusion during the phase change.

2. The solute concentration in the unfrozen porewater of a freezing column of saline saturated gravel increases toward the colder temperatures.

3. The unfrozen water content in the partially frozen soil above the freeze front decreases with decreasing temperature. Unfrozen water contents as high as 20 percent were observed at the top of the soil column in one test in which the soil temperature was measured at -8°C.

4. Salt transport occurs in a freezing soil column if a thermal gradient is maintained for an extended period of time.

5. To develop a computer model for the freezing of saline saturated soils, it is necessary to develop constituative relationships between unfrozen water content, diffusivity and hydraulic conductivity as a function of temperature and initial salinity.

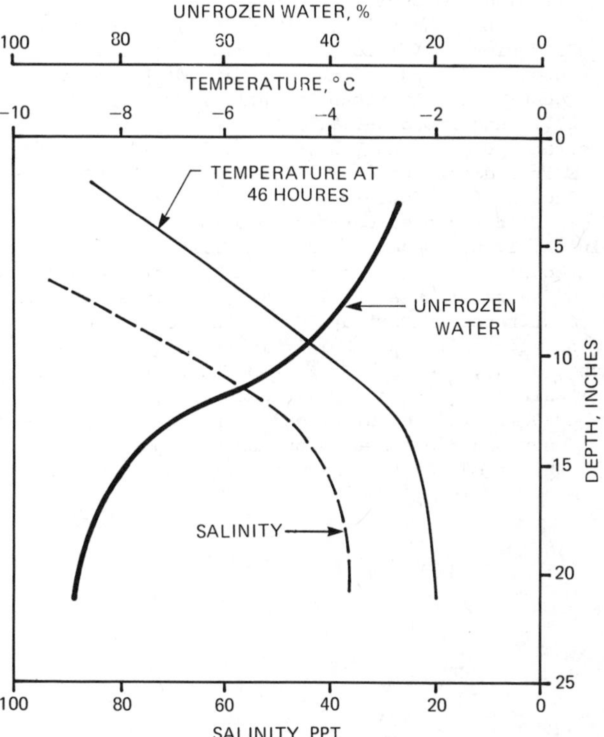

FIGURE 7 Temperature, salinity and unfrozen water content profiles for a freezing column test with initial salinity of 30 ppt.

ACKNOWLEDGMENTS

The research presented in this paper was supported by Ertec Western, Inc. The generous support of Ertec is greatfully acknowledged. Many helpful suggestions were provided by Mr. Brian Stuckert.

REFERENCES

Broecker, W. S., and Oversby, V. M., 1971, Chemical Equilibria in the Earth, McGraw Hill Book Company, New York, Chapter 5, pp 99 - 129.

Guymon, G. L. Berg, R. L., Johnson, T. C., and Hromadka, T. V., 1981. "Results from a Mathematical Model of Frost Heave" Transporation Research Record 809.

Hallet, B., 1978, "Solute Redistribution in Freezing Ground, Proceedings of the Third International Conference on Permafrost, V.1: Ottawa, National Research Council of Canada.

Jumikis, A.T., 1966, Thermal Soil Mechanics, Appendix 12, Rutgers University Press, New Brunswick, New Jersey.

Kersten, M.S., 1949, "Laboratory Research for the determination of the Thermal Properties of Soils", University of Minnesota Engineering Experiment Station, Final Report.

Mahar, L. J., Vinson, T.S., and Wilson, R., 1982, "Effects of Salinity on Freezing of Granular Soils", International Symposium on Ground Freezing, Hanover, New Hampshire.

McDuff, R.E., and Ellis, R. A., 1979, "Determining Diffusion Coefficients in Marine Sediments: A Laboratory Study of the Validity of Resistivity Techniques", American Journal of Science, Vol. 279, June, pp 666-675.

Sheeran , D. E., and Yong, R. N., 1975, "Water and Salt Redistribution in Freezing Soils", Proceedings: First Conference on Soil-Water Problems in Cold Regions, Canada, pp 58-69.

Vinson, T. S., Mahar, L. J. and Wilson, R., 1983, "Model Study of Freezing Front Penetration in Offshore Granular Fill Soils", 7th International Conference on Port and Ocean Engineering Under Arctic Conditions, Helsinki, April.

Yong, R. N., Sheeran, D. E. and Janiga, P. V., 1973, "Salt Migration and Frost Heaving of Salt Treated Soils in View of Freezing and Thawing", Symposium on Frost Action on Roads, Report II, Organization for Economic Cooperation and Development, Paris.

A MECHANISTIC APPROACH TO PAVEMENT DESIGN IN COLD REGIONS

Joe P. Mahoney[1] and Ted S. Vinson[2]

[1]Department of Civil Engineering, University of Washington
Seattle, Washington 98195 USA
[2]Department of Civil Engineering, Oregon State University
Corvallis, Oregon 97331 USA

A mechanistic approach to pavement design in continuous permafrost areas is examined. Typical material, environmental and loading conditions are evaluated for Barrow, Alaska. A presentation of typical pavement response parameters is made and used to evaluate several possible pavement structural sections.

INTRODUCTION

A mechanistic approach to pavement design involves (1) predicting stress, strains, and deflections in a pavement structure owing to a specified geometry and magnitude of wheel loading conditions and (2) adjusting the properties and thicknesses of the elements in the pavement structure to insure the predicted stresses, strains, and deflections are within allowable limits. The mechanistic approach provides flexibility in modeling a pavement structure owing to the fact that measured material properties associated with a range of thermal and drainage conditions in the field may be incorporated in the analysis.

To illustrate the mechanistic design approach, an analysis was performed for several conceptual pavement structures subjected to representative arctic environmental and loading conditions (specifically the Barrow, Alaska, area). Limiting pavement response parameters included (1) pavement surface deflection (associated with fatigue cracking and/or rutting), (2) horizontal tensile strain at the bottom of an asphalt concrete surface layer (associated with fatigue cracking), (3) vertical compressive stress at the top of rigid board insulation (associated with the design compressive strength of the material), and (4) vertical compressive strain at the top of the subgrade (associated with rutting). Pavement structural sections with various thicknesses of asphalt concrete, base, insulation, and subgrade were considered under three environmental conditions: (1) early winter, (2) late winter, and (3) spring thaw. The design loading condition selected was a fully loaded "DJB" dump truck and 100,000 repetitions in a 20-year design period. A performance evaluation rating system was established for limiting the pavement response parameters considered.

In general, pavements located throughout the state of Alaska or any cold region may be classified into three general surface types. These are (1) gravel surfaced, (2) flexible - bituminous surfaced, and (3) rigid - portland cement concrete (PCC) surfaced. The Alaska Department of Transportation and Public Facilities (Alaska DOTPF) maintains a total centerline network of about 8000 km (5000 mi) of which about 3520 km (2200 mi) are paved. Essentially all of the paved roads are bituminous surfaced (asphalt concrete, bituminous surface treatment, seal coat, etc.).

Several types of pavement structures may be possible for continuous permafrost areas. Two basic types of pavement structures have been constructed in the Alaskan arctic over continuous permafrost areas. Most common are gravel surfaced roads. Other pavements include stabilized layers such as asphalt concrete or cement treated surfacing and are located primarily at Alaska DOTPF owned and operated airports.

For example, at Barrow, Alaska, asphalt concrete surfaced pavements constitute the runway, associated taxiways, and primary parking apron at the Wiley Post - Will Rogers Memorial Airport. The asphalt concrete is 229 mm (9 in) thick, overlying various unbound embankment materials (minimum total embankment thickness of about 1.4 m (4.5 ft)). Cement treated pavement is located in the general aviation apron (east end of aircraft parking area).

DESIGN CONSIDERATION AND METHODS

This section overviews some of the design considerations and methods associated with the development the pavement concepts specifically for Barrow, Alaska. Included are the following subsections:

1. Site conditions;
2. Current pavement design practice;
3. Layered elastic theory;
4. Development of typical pavement cross sections.

Site Conditions

A brief review of the existing site conditions at a specific arctic location (Barrow) as they relate to a pavement structure follows. This includes a review of existing soil borings, aggregate sources, and other relevant materials.

The primary indications of various soil investigations include the following:

1. Permafrost underlies the Barrow area to a depth of about 400 m (1300 ft).
2. The Barrow area topography is flat to gently rolling and generally less than 8 m (25 ft) above mean sea level.

3. Seasonal variation of ground temperatures occurs in the upper 12 m (40 ft) of soil (the temperature changes decrease rapidly with increasing depth).
4. Ground ice can compose up to 100% of the volume (primarily silts, ice lenses, and wedges) in the upper 3.0 to 4.6 m (10 to 15 ft) of the soil profile.
5. Polygonal ground is common in the area (although with a variable density), thus indicating extensive ice.
6. Existing roads and streets are gravel surfaced. Gravel fill is generally 0.9 to 1.5 m (3 to 5 ft) thick, originally having been placed directly over the tundra. The thickness of the active layer beneath the existing roads in Barrow averages about 1.0 m (3 ft) (measured during August 1978). The maximum active layer depth measured in Barrow was 3.5 m (11.5 ft) (at a culvert). The depth of thaw in relatively undisturbed tundra extended 200 to 305 m (8 to 12 in) below the ground surface.
7. The fill materials composing the existing roads and streets can be broadly classified as gravels or sands. The dominant materials in these fills, according to the Unified Classification System, are sands (SM, SP-SM, SP and SW) followed by gravels (GM, GP and GP-GM).
8. The upper 200 to 305 mm (8 to 12 in) of some of the existing gravel surfaced roads has enough "binder" to reduce rutting in the surface layer.

Of major importance, as well, is the type and amount of available aggregates from borrow sources. Aggregate sources are limited and are of a gradation which precludes the production of significant amounts of crushed material.

Current Pavement Design Practice

Since World War II, the Corps of Engineers has developed pavement design procedures (street and airfield) that can be used to develop structural design requirements. The available design procedures for pavements subject to freezing and thawing in the underlying soils are based on either of two basic concepts:

1. Control of surface deformation resulting from frost action (or thaw).
2. Provision of adequate bearing capacity during the most critical climatic period.

Based on the above considerations, three separate design approaches can be used:

1. Complete protection method: Sufficient thicknesses of pavement and non-frost susceptible base course are provided to prevent frost penetration into the subgrade.
2. Limited subgrade frost penetration method: Sufficient thicknesses of pavement and non-frost susceptible base course are provided to limit subgrade frost penetration to amounts that restrict surface deformation to within acceptable limits.
3. Reduced subgrade strength method: The amount of frost heave is neglected and the design is based primarily on the anticipated reduced subgrade strength during the thaw.

Hennion and Lobacz (1973) recommend that seasonal thawing and freezing should be restricted to the pavement (surfacing and non-frost susceptible base course) in continuous permafrost regions. The concept is comparable to the complete protection method wherein the critical factor is the depth of thaw rather than the depth of frost penetration.

The Alaska DOTPF (1982) has recently issued pavement design guidelines based on research begun in 1976. The primary objective was to study the various relationships that controlled the performance of flexible pavements. Approximately 120 pavement sections were selected on the existing state maintained road network for the research program. The results of that study (McHattie et al., 1980) have significantly influenced the new design procedure.

The Alaska DOTPF design guidelines focus on the fact that increased fines (No. 200 minus material in the unbound layers of a pavement structure) lead to increased thickness of asphalt concrete surfacing. In those portions of Alaska where asphalt concrete is readily available, increased fines in the underlying layers may be acceptable. However, in remote arctic regions the production of asphalt concrete is expensive and should be minimized. Thus, it is prudent to specify non-frost susceptible (NFS) materials in the underlying layers to the extent possible to minimize the following:

1. The requirements for asphalt concrete (if not eliminate its need, altogether);
2. The potential for frost heave;
3. The potential for other types of pavement distress such as alligator cracking and rutting.

Layered Elastic Design Approach

The use of layered elastic theory in evaluating on-grade pavement design concepts has several attractive features. First, estimated (or measured) material properties in the form of elastic moduli and Poisson's ratio for each layer are used. This provides flexibility in "modeling" the pavement structure thus reducing the need (or use) of design procedures not strictly suitable for cold regions. Second, the calculation process can be used to estimate the pavement response for any combination of material properties, layers and loading conditions.

The determination of the elastic material parameters for a mechanistic design approach (resilient or elastic modulus and Poisson's ratio) is difficult due to the complex and variable nature of the materials in each pavement layer and associated subgrade soils. To evaluate the initial pavement designs considered in this paper, all elastic parameters were estimated based on reasonable assumptions or published data. In practice, final pavement designs must be based on laboratory derived elastic material parameters determined on the actual materials to be used in the pavement structural section and/or field correlations.

Modeling a pavement structure as a layered elastic system is a complex process if calculations of pavement response are made manually. In fact, layered systems with greater than three layers are difficult to solve without a computer program. Thus, various computer programs have been developed that can be used to accomplish the necessary calculations to generate the required stresses, strains, and deflections due to a given loading condition. A computer program was selected for use in this study that can treat multiple pavement layers and several individual wheel loads simultaneously.

The pavement responses for the cases studied can be compared with appropriate failure criteria to determine the potential performance of an in-service flexible pavement structure. Critical pavement response parameters include:

1. Pavement surface deflection;
2. Horizontal tensile strain at the bottom of the asphalt concrete layer;
3. Vertical compressive stress at the top of the rigid board insulation;
4. Vertical compressive strain at the top of subgrade (or permafrost).

The pavement surface deflection (if excessive) can be associated with fatigue cracking and/or rutting. The horizontal tensile strain at the bottom of the asphalt concrete layer can be used to estimate fatigue cracking. The vertical compressive strain at the top of the insulation is used to determine if the design compressive strength will be exceeded. Finally, the vertical compressive strains at the top of the subgrade can be used to estimate the potential for rutting. In the mechanistic approach to pavement design it is necessary to control or limit pavement responses to develop suitable pavement designs. Somewhat unique to cold regions, the importance of these pavement responses will vary greatly throughout the year. For example, the complete structural section will be in a totally frozen condition during the winter months. With all pavement layers frozen, pavement responses generally will be quite low resulting in little or no pavement distress. During the spring thaw, the pavement structure will experience its weakest condition resulting in the greatest amount (if not the majority) of pavement damage for the year.

Extensive work accomplished in various laboratories worldwide has resulted in tensile stress or strain relationships as a function of the number of loads to failure for asphalt concrete fatigue characterization. It is difficult to obtain a consensus among engineers as to how well laboratory derived relationships duplicate actual in-service fatigue for asphalt concrete pavements. In an attempt to avoid possible pitfalls associated with laboratory derived relationships, a fatigue relationship was chosen for this study that was developed from AASHO Road Test data. This criterion was developed by Finn et al. (1977) and is as follows:

$$\log N_f = 15.947 - 3.291 \log \left(\frac{\varepsilon}{10^{-6}}\right) - 0.854 \log \left(\frac{E}{10^3}\right)$$

where N = Number of loads of constant stess to failure;
ε = Tensile strain at the bottom of asphalt concrete layer;
E = Modulus (psi).

To estimate the potential for rutting, the criterion developed by Dorman and Metcalf (1964) was adopted. As for fatigue, this criterion was based on data derived from the AASHO Road Test. The vertical compressive strain at the top of the subgrade is used to determine the number of loads to failure as follows:

Load Repetitions (Design Loads of DJB Truck)	Vertical Compressive Strain
100,000	1050×10^{-6}
1,000,000	650×10^{-6}
10,000,000	420×10^{-6}
100,000,000	260×10^{-6}

The allowable surface deflections for a pavement are based on criteria developed by Bushey et al. (1974) in California. The criteria are a function of the type of surface course (asphalt concrete or bituminous surface treatment), thickness and the anticipated number of traffic loadings. To convert the DJB design wheel load to a 22.2 kN (5,000 lb) wheel load (as used by the California DOT), a multiplier of 105 was used; i.e., a 71.2 kN (16,000 lb) wheel load for the DJB truck is 105 times more damaging than a 22.2 kN (5,000 lb) wheel load. The surface deflection criteria used are as follows:

1. Asphalt concrete (63.5 mm (2.5 in) thick):
   (a) acceptable : $\leq$0.58 mm (0.023 in);
   (b) marginal: $\simeq$0.58 mm (0.023 in);
   (c) unacceptable: $\gg$0.58 mm (0.023 in).
2. Asphalt concrete (127 mm (5.0 in) thick):
   (a) acceptable: $\leq$0.43 mm (0.017 in);
   (b) marginal: $\simeq$0.43 mm (0.017 in);
   (c) unacceptable: $\gg$0.43 mm (0.017 in).
3. Bituminous surface treatment:
   (a) acceptable: $\leq$1.01 mm (0.040 in);
   (b) marginal: >1.01 and $\leq$1.27 mm (>0.040 and $\leq$0.050 in);
   (c) unacceptable: >1.27 mm (0.050 in).

The surface deflection criteria should be used only in a general way since the surface deflection is in part dependent on the assumed elastic modulus of the asphalt concrete layer.

The number of allowable load repetitions to failure (fatigue) were used to set the following criterion for asphalt concrete surfaced pavements:

1. Acceptable: $\gg$100,000;
2. Marginal: $\simeq$100,000;
3. Unacceptable: $\ll$100,000.

In a similar manner, the allowable vertical compressive stress for the rigid insulation board was conservatively set at 80.7 kPa (11.7 psi). The following criterion resulted:

1. Acceptable: <80.7 kPa (11.7 psi);
2. Marginal: $\simeq$80.7 kPa (11.7 psi);
3. Unacceptable >80.7 kPa (11.7 psi).

The final criterion selected was for rutting that is a function of the vertical compressive strain at the top of the subgrade (or permafrost):

1. Acceptable: <500 x 10^{-6};
2. Marginal: ≃500-800 x 10^{-6};
3. Unacceptable: >800 x 10^{-6}.

The failure criteria for fatigue, rutting, and surface deflection identified above were used to examine several possible pavement structural sections. It must be recognized that the failure criteria presented were originally developed for non-permafrost areas. Thus, the combination of the estimated elastic material properties in conjunction with the available failure criteria may result in less than desirable estimates of pavement life, although such estimates should be suitable for the development of pavement concepts.

Pavement Cross Sections

The design loading condition selected for analysis was a "DJB" dump truck that is commonly used in the Barrow area. This truck has a total of six wheels as shown in Figure 1. Fully loaded, each front wheel carries 72.6 kN (16,320 lbs) and each rear wheel 70.5 kN (15,840 lbs) (all at 345 kPa (50 psi) tire pressure). The design number of loaded DJB repetitions applied to a road was assumed at 20 per day. This results in approximately 100,000 repetitions in a 20-year design period. This estimate represents a "worst case" condition. There are few streets in the Barrow area that will experience this level of traffic; however, the estimate provides some latitude in accounting for other heavy traffic loadings (truck, heavy construction equipment, etc.).

Table 1 contains a brief description of the initial structural sections evaluated. The latter sections (cases 7 through 12) are analogous to gravel or a bituminous surface treatment surfaced road section. Figure 2 contains a sketch of the cross sections described above.

The elastic moduli used in the analysis were estimated. For the permafrost (subgrade), three moduli were used (6895, 689.5, and 6.9 MPa) (1,000,000, 100,000 and 1,000 psi). Permafrost with a modulus of 6895 MPa (1,000,000 psi) represents a condition well below freezing. A modulus of 689.5 MPa (100,000 psi) approximates "strong" permafrost but warming. Finally, a modulus of 6.9 MPa (1,000 psi) represents permafrost during thawing conditions. Actually, the effective modulus of thawed ice-rich ground approaches zero, although the modulus used represents a less than desirable condition.

The elastic moduli for the non-frost susceptible materials were estimated to be 140 MPa (20,000 psi) (assuming some lateral confinement). This may be somewhat unconservative for these materials in the unfrozen state but is conservative for the frozen condition. The pit-run material, if properly thawed and compacted, has the potential for a somewhat higher modulus (240 MPa (35,000 psi)) due to slightly increased fines (binder); however, lower moduli values also were used (140 and 70 MPa (20,000 and 10,000 psi)). Additionally, the insulation (Dow HI-35) was assumed to have the same modulus as the surrounding non-frost susceptible material.

Asphalt concrete is a viscoelastic material and its strength and deformation properties vary with temperature. The higher the temperature, the lower the asphalt concrete elastic modulus. The necessity of using a "soft" asphalt at Barrow or any arctic site is apparent. Additionally, it is highly desirable to use crushed coarse aggregate in the mixture. Assuming the use of a "soft" asphalt, crushed coarse aggregate and a design surface temperature of 21°C (summer condition), a modulus of 1725 MPa (250,000 psi) was selected for the analysis. Only a complete mix design could allow the proper determination of the value.

The results of the pavement response to the design loads and structural section cases are summarized in Table 2. Also contained in this table is a summary of the failure criteria applied to the selected pavement response measures. Ratings for each case and criterion were made on the basis of the following:

1. A: criterion is within acceptable limits;
2. B: criterion is within marginal limits;
3. C: criterion is within unacceptable limits.

Additionally, the overall rating for each case is based on the lowest rating for any one of the four (or three) criteria.

On the basis of structural considerations alone, case nos. 1, 9, 10a, 11 and 12a are fully acceptable (refer to Table 2). Case 1 is surfaced with 63.5 mm (2.5 in) of asphalt concrete and cases 9, 10a, 11 and 12a are the same general cross sections except without the asphalt concrete (however, the layer moduli vary). Therefore, it may be an attractive alternative to construct a bituminous surface treatment over structural sections such as cases 9, 10a, 11, and 12a. The possible advantage of using asphalt concrete is a higher level of service but with the risk of increased maintenance and construction costs.

EVALUATION OF SELECTED DESIGN CONCEPTS

Initial construction costs are only one (but important) cost to consider. Additionally, the costs associated with the maintenance of a pavement structure need to be evaluated. Historically, on state maintained highway networks, asphalt concrete pavements may need to be overlayed with additional asphalt concrete every 10 years (although this period tends to vary significantly and is a function of the initial design and other site specific conditions). Generally, such overlays are about 40 to 75 mm (1.5 to 3.0 in) thick. For bituminous surfaced pavements, thermal cracks should be expected at a transverse spacing of about 12 m (40 ft) or less. These cracks should be sealed early each summer (minor maintenance). Tradeoffs between asphalt concrete and bituminous surface treatment flexible pavements are:

1. The asphalt concrete surface provides extra load capacity for the pavement structure; the bituminous surface treatment does not.
2. The asphalt concrete surface should provide a longer, reduced maintenance life; however, if thermal stability is not achieved, maintenance will be extensive and expensive.
3. The bituminous surface treatment is less expensive initially, provides for reduced

METHODS FOR DETERMINING THE LONG-TERM STRENGTH OF FROZEN SOILS

R. V. Maksimjak, S. S. Vyalov and A. A. Chapaev

Research Institute of Foundations and Underground Structures,
Moscow, USSR

Various methods are available for determining the strength characteristics of soils; and as numerous investigations have demonstrated, the values of the parameters being determined depend on the conditions and type of testing. The article discusses the problems of the impact of the test conditions and type of test on the value of the long-term strength characteristics of frozen soils. The tests were carried out to determine long-term strength and to compare the experimental data from uniaxial compression tests with constant loading, with those derived with a constant rate of deformation and a constant rate of loading and under conditions where the operating stress was relaxed. These investigations revealed the mechanisms whereby the soil strength decreased under various types of tests and demonstrated the possibility of determining long-term values for the strength of frozen soils from the various tests utilized.

Values of ultimate long-term strength (time-dependent failure) of frozen soils are the principal design characteristics in calculations of bearing capacities of foundations of structures.

At the present time the following methods for determining the strength characteristics of frozen soils are used as a standard practice in laboratory studies: (a) creep tests at a constant load (σ = const); (b) tests at constant rate of deformation ($V = \frac{d\varepsilon}{dt}$ = const) or at constant rate of loading ($V = \frac{d\sigma}{dt}$ = const); or (c) the dynamometric test method $[(\sigma = f(\varepsilon), \sigma = f(t)]$. Due to the large variety of methods used, the test results sometimes differ, causing different interpretation of the results. At the same time, the long-term strength is a physical characteristic of soil, therefore this should be an invariant independent of the test method.

To compare the test results and to determine the values of the long-term strength, the specimens of a given frozen soil, were tested using each of the above methods. Particular attention was paid to the dynamometric method, which in our opinion seems to be very promising.

The soil selected for investigation was frozen Chegan (neogenic) clay featuring a disturbed structure and a massive cryotexture. All the tests were performed under axial compression at a temperature $\theta = -5°C$.

Creep tests under permanent loads are the most common types. According to this method, a conventional instantaneous value of the soil strength was determined first by applying a quick loading (20-30 s duration). Upon finding the instantaneous strength (σ_{inst} = 3.75 MPa), identical specimens were tested under different constant stresses at a percentage of the conventional instantaneous strength (Vyalov et al. n.d.). The results of the tests in the form of creep curves are illustrated in Figure 1. The criterion for creep strength for a given time to failure of the sample was taken as the transition point on the creep curve where the tertiary stage of creep (progressive flow) begins. Then, in compliance with the commonly accepted method, a long-term strength curve was plotted. The analytical expression takes the following form:

$$\sigma = a \cdot t^{-\lambda} \qquad (1)$$

where a and t = the parameters of the strength of the frozen soil.

The correctness of the selected relationship is confirmed by the resulting straight-line plots of the long-term strength curve in logarithmic coordinates.

The ultimate long-term strength was taken as the soil strength corresponding to a 50-year service life (t_{calc}) of the structures. In this case equation 1 gives the value σ_∞ = 0.52 MPa.

In the step-loading regime, the load was applied by steps equal to a certain percentage of the conventional instantaneous strength. The load at each step was held until the moment of conventional damping of deformations. The tests were carried out until, at some successive step of loading, the deformation tended to continue until failure occurred. The last step of loading, when a damping of deformations was observed, was taken as the ultimate long-term strength of the soil (σ_∞ = 0.75 MPa).

The subsequent series of tests were performed at a constant rate of loading. The loading rates ranged from $\dot{\sigma}$ = 0.375 MPa to 9.0 MPa. The test results (Figure 2) showed that the strength of the soil decreases with decreasing loading rate. For example, at $\dot{\sigma}$ = 7.5 MPa per hour the value of σ_{fail} = 2.4 MPa, while at $\dot{\sigma}$ = 0.75 MPa per hour the parameter σ_{fail} = 1.9 MPa.

The parameters of the long-term strength equation 1 were determined for constant rate of applied stress ($\dot{\sigma}$ = const) according to the criterion of linear summation of defects. This criterion enabled the researchers to determine the analytical relationship between the time to failure and the loading rate. On the basis of the obtained parameters the strength for a period of 50 years was calculated (σ_∞ = 0.7 MPa).

When testing the soils at a constant rate of deformation, the peak (σ_s) and residual (σ_r) strengths must be determined. The peak strength approaches the conventional instantaneous strength when the rate of deformation is high, while under it, it tends to approach σ_∞ when the rate of deformation is low.

As the result of the tests, a family of stress-time curves (Figure 3) was obtained, which make it possible to characterize the change of the soil resistance to failure at a constant rate of deformation. The analysis of these curves indicates that the peak strength tends to decrease as the deformation rate decreases. For instance, the decrease of the deformation rate from $\dot{\varepsilon}$ = 15 hr^{-1} to 0.03 hr^{-1} reduces the value of the peak strength by 2.3 times (σ_{fail_1} = 2.65 MPa, σ_{fail_2} = 1.17 MPa).

The parameters of the long-term strength equation 1 were determined with due regard for the loading procedure ($\dot{\varepsilon}$ = const) according to the criterion of linear summation of defects together with the deformation equation expressed according to the theory of stress hardening. On the basis of the obtained parameters the 50-year strength of the soil was calculated to be σ_∞ = 0.75 MPa.

The dynamometric method of testing frozen soils for long-term strength consists of subjecting the soil specimens to a variable load that is gradually reduced in the deformation process. The method is described below.

The soil specimen is loaded through a dynamometer. Upon reaching the predetermined load, the dynamometer is retained in a fixed position. The stress imparted to the soil specimen through the dynamometer causes the development of creep deformations which in turn reduces the load applied by the dynamometer, thereby reducing the stress. In the test specimen* the process of decreasing the stress in the dynamometer lasts until the deformation of the specimen is stabilized and the load applied to the soil specimen and the internal forces of resistance of the soil reach equilibrium (Vyalov and Maksimyak 1970).

The tests showed that, in the range of the initial stresses (σ_{init} = 0.1 to 0.5 σ_{inst}), there is a dependence between the final stress (σ_{final}) and the initial stress (σ_{init}). At initial stress $\sigma_{init} \geqq 0.6\ \sigma_{inst}$, the stress decreases practically to σ_{final}. The latter is taken as the value of the ultimate long-term strength of soil. In the experiments under consideration, it proved to be equal to σ_∞ = 0.54 MPa for the frozen soil tested.

In the present study, the dynamometric test method was improved by using a step-loading regime in combination with a dynamometer (Vyalov et al. 1980). In this method, the load was applied to the specimen in steps, as in the stepped tests under constant load. Each subsequent step of the load was applied only after the deformation of the preceding step had stabilized. Experiments with the specimens of the Chegan clay using step loading show that when the total stress imparted to the specimen does not exceed 0.7 σ_{inst}, the stress drops to the final value, which depends upon the initial load at each step. When the total load reaches or exceeds 0.7 σ_{inst} the stress drops to a final constant value, which appears to be independent of any further increase of the load. The results of such experiments are presented in Figure 4. The initial load in these experiments was σ_{init} = 0.1 σ_{inst}, while the value of each step was $\Delta\sigma$ = 0.1 σ_{inst}. It can be seen from the illustration that the load at the first step has dropped to σ_{final} = 0.4 MPa, at the second step to σ_{final} = 0.58 MPa, at the third step to σ_{final} = 0.64 MPa, at the fourth step to σ_{final} = 0.68 MPa, at the fifth step to σ_{final} = 0.7 MPa, and, beginning with the sixth step, the final stress remains unchanged (σ_{final} = 0.7 MPa) in spite of the fact that the total load is still rising.

To develop an optimum test technique, a special study was undertaken on the effect of each step's load value on the final stress value (Vyalov et al. 1980). The tests were conducted at the following loading steps: 0.1, 0.16, 0.20, and 0.33 of σ_{inst}. The tests indicated (Figure 4) that with the step value of $\Delta\sigma$ = 0.16 σ_{inst}, the stress at the first four steps drops to 0.2 MPa, 0.4 MPa, 0.5 MPa, and 0.7 MPa at the first, second, third, and fourth steps respectively. The initial load in this case increased from σ_{init} = 0.56 MPa to σ_{init} = 2.5 MPa, which corresponds to 0.7 σ_{inst}. Further load increases reduces the stress to σ_{final} = 0.7 MPa, irrespective of the value of the initial load. However, with such a value of the step load, each test takes a long time, since each step is to be held until the deformations are stabilized, and several loading steps are necessary to reach the critical load (σ = 0.7 σ_{inst}).

With the increase of the step value, the number of steps for reaching the critical load diminishes. However, the increase of the value of the steps not only reduces the duration of the test but also impairs the accuracy of the results. Thus, for example, with a step value of $\Delta\sigma$ = 0.16 σ_{inst}, the value of the critical load was σ = 2.5 MPa, but with a step equal to $\Delta\sigma$ = 0.2 σ_{inst} this load amounted to σ = 2.66 MPa. Moreover, with a value of $\Delta\sigma$ = 0.33 σ_{inst}, the researchers failed to note the critical stress: before reaching a load of σ = 0.66 σ_{inst} the final stress had not yet stabilized, and the application of the next step of the load caused the failure of the soil specimen. The experiments suggested that the most favorable value of the step is $\Delta\sigma$ = 0.1 σ_{inst}.

To reduce test duration, a study was undertaken to determine the effect of the loaded time periods of the steps and of the initial load upon the value of the ultimate stress (Vyalov et al. 1980). With this object in mind, the relevant tests were performed under different initial loads, i.e., 0.1, 0.5, 0.6, 0.7, and 0.8 of the conventional instantaneous strength of the soil. The further loading operations were subjected to steps identical and equal to $\Delta\sigma$ = 0.1 σ_{inst}. An optimum value of the step was selected to determine more accurately the value of the final stabilized stress. The experiments showed that regardless of the value of the initial stress, by gradually increasing the loading of the specimen in steps of $\Delta\sigma$ = 0.1 σ_{inst} starting from the initial load of 0.7 σ_{inst}, the specimen-

*This test is often called a "relaxation" test.

dynamometer system tends to relax at a constant final stress σ_{final} = 0.72 MPa.

Thus, changing the value of the initial load has no effect on the value of the final stabilized stress, while decreasing the initial load increases the test duration. For instance, reduction of the tests by one step cuts the length of the test time by 72 hours.

The results of these experiments made it evident that the best variant is a test under an initial load lying within 0.5-0.6 of the conventional instantaneous strength of the soil under test.

These tests were conducted with every subsequent step being held until a predetermined stabilization of the deformations was attained. To reduce the duration of the tests, the effect of the time the applied load was held upon its final stabilized value was investigated. To this end, tests were performed in which the loading steps were held for 12 and 24 hours, as well as until the predetermined stabilization of the deformations was obtained. The investigations showed that, for 12- and 24-hour durations, the value of the final stress at each step is higher than at the previous step, up to the application of the last step of σ_{init} = 0.9 σ_{inst}, after which the soil specimen failed. Thus, the experimenters failed to succeed in obtaining the stabilized final stress.

Thus, the studies showed that the best variant is testing under an initial load of 0.5-0.6 σ_{inst} with a gradual step-by-step rise of the load in increments of $\Delta\sigma$ = 0.1 σ_{inst}, with a predetermined stabilization of the deformation at each step of the loading.

Table 1 gives a comparison of the characteristics of long-term strength of frozen soil in various types of tests.

bonds, etc. (Vyalov et al. n.d., Vyalov et al. 1973).

Thus, when selecting the method for determining strength characteristics of frozen soils it is necessary to take into consideration the loading conditions and the state of the soil in situ.

However, it should be emphasized that to determine the long-term strength of the soil by testing at σ = const, as well as at constant stress and constant rates of loading and deformation, it is essential to obtain a family of creep curves. This requires the testing of a series of identical specimens. As a rule, this is accompanied by a large scatter of test results, and calls for an additional number of tests. The other shortcoming of these methods is the ambiguity in determining the time of failure of the soil specimens. Thus, despite the fact that these methods reflect correctly the state of the soil under constant and time-variable loads, they are time-consuming. The method of step loading at a constant stress allows the test to be performed with the use of one or two specimens. This method had the disadvantages characteristic of the previous methods in determining the moment of failure of the soil, and, on the other hand, requires too much time to accomplish the tests.

The most promising are the dyanmometric test methods, i.e. the improved method of single loading and the revised method of stepped tests. They offer a number of advantages. First, they allow the value of the ultimate long-term strength σ_∞ to be determined by testing only one or two specimens; second, they materially reduce the duration of the tests; and third, the values σ_∞ are determined directly on the basis of the results of the experiments and require no preliminary mathematical processing. The particular dynamometric test method should be selected on the basis of the expected stress state of the soil in situ in the foundations of the structures to be built on permafrost.

TABLE 1 Ultimate long-term strength of frozen soil σ_∞ under various types of tests

	Test conditions					
	σ = const				Dynamometric method	
σ_∞, MPa	Single loading	Stepped loading	$\dot{\sigma}$ = const	$\dot{\varepsilon}$ = const	Single loading	Stepped loading
	0.52	0.75	0.70	0.75	0.54	0.72

Analysis of these data demonstrates that all the obtained values could be divided into two groups: (a) a group comprising σ_∞ = 0.5 MPa for the case of σ = const and σ_∞ 0.54 MPa for the case of single dynamometric tests, and (b) a group including σ_∞ = 0.75 MPa in the case of stepped tests at σ = const, σ_∞ = 0.7 MPa whenever $\dot{\sigma}$ = const, σ_∞ = 0.75 MPa for $\dot{\varepsilon}$ = const, and σ_∞ = 0.7 MPa in the case of stepped dynamometric tests.

Comparison of the foregoing data indicates that in all cases when the test specimens were loaded gradually the ultimate long-term strength turned out to be higher than the constant load. This is due to the fact that a gradual increase of the load strengthens the soil due to the rearrangement of the soil particles, development of new structural

REFERENCES

Vyalov, S. S., and Maksimyak, R. V., 1970, Long-term strength of clay soils and accelerated method of its determination (Trans.), 7th European Conference on Mechanics of Soils and Construction of Foundations: Briton.

Vyalov, S. S., Maksimyak, R. V., and Chapaev, A. A., 1980, Accelerated method of determination of the value of ultimate long-term strength of frozen soils (Trans.), 3rd Symposium on Rheology of Soils: Erevan.

Vyalov, S. S., Pekarskaya, N. K., and Maksimyak, R. V., n.d., Change of soil strength properties in the process of creep.

Vyalov, S. S., Zaretsky, Yu. K., and Maksimyak, R. V., 1973, Kinetics of structural deformations and failure of clays (Trans.), 7th International Congress on Mechanics of Soils: Moscow, Stroiizdat.

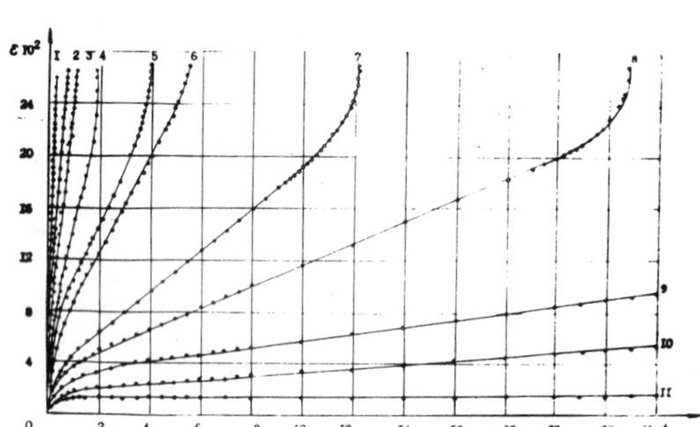

FIGURE 1 Unconfined compression creep curves ε - t.

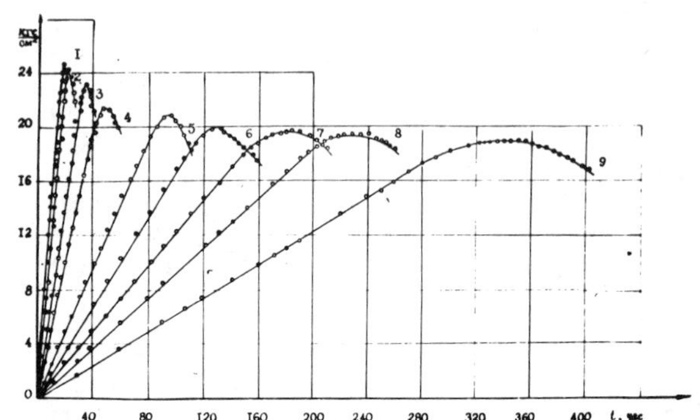

FIGURE 2 Compression strength curves σ - t under different strain rate $\dot{\varepsilon}$.

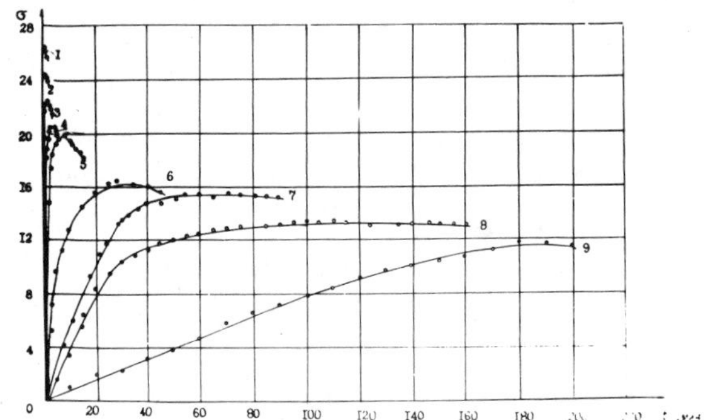

FIGURE 3 Compression strength curves σ - t under different rate of loading.

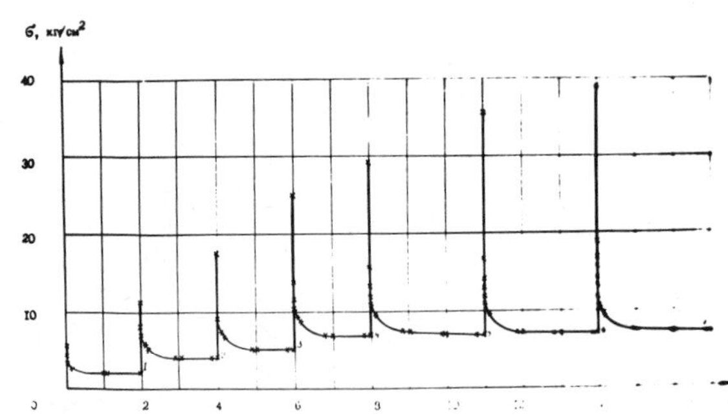

FIGURE 4 Strain curves under step loading on dynamometric apparatus.

OBSERVATIONS OF MARTIAN FRETTED TERRAIN

Michael C. Malin and Dean B. Eppler

Department of Geology, Arizona State University, Tempe, Arizona 85287

Fretted terrain is one of several lowland landscapes found on Mars that suggest intense geomorphic activity. Fretted terrain was formed initially during a period of heavy cratering and possibly fluvial activity early in martian history. Subsequent modification processes have been active in the "recent" past, and may still be active today. Evidence of the process(es) that formed the fretted terrain is generally lacking because of subsequent modification. The location of the terrain along the boundary between heavily cratered uplands and sparsely cratered lowlands that extends around the planet, and the orientation of portions of the terrain along structural trends associated with the boundary, suggest a tectonic origin in part. Subsidence through ground ice melting or sublimation may then have led to progressive deterioration of the relatively incompetent rock units. Modification of the surfaces has been widespread; most severely altered are valley floors and walls, and aprons surrounding mesas. Lineations on the surfaces of aprons resemble leveed channels, whereas those on valley floors, though superficially similar to medial moraines and other indicators of longitudinal flow, most likely reflect differential degradation by subsidence and eolian action. Measurements of the aprons show some similarities to certain terrestrial gravity-induced and gravity-driven mass flows.

INTRODUCTION

Fretted terrain is the name applied to a complex physiographic province marking a portion of the boundary between heavily cratered uplands and sparsely cratered lowlands on Mars. The terrain is characterized by steep, abrupt escarpments, and outliers of high-standing terrain surrounded by relatively smooth, flat-lying plains. The zone of most highly developed fretted terrain spans approximately 3500 km between 30°N,28°W and 50°N,350°W and is about 600 km wide. The importance to martian geology of fretted terrain lies in its spatial and temporal relationships to features suggesting profound changes in martian geological evolution: the terrain is developed only along the above mentioned boundary between "ancient" cratered terrains and "younger" sparsely cratered plains. This boundary probably reflects major tectonic events early in Mars' history (Mutch et al. 1976, Malin et al. 1978). Concurrent with or subsequent to these tectonic events, major erosional processes re-shaped portions of the martian surface (Malin 1976a, Mutch et al. 1976). Fretted terrains were formed, and have been modified, by these processes, and fretted landforms provide constraints on the nature and duration of the erosion.

Although early studies by Sharp (1973), Coradini and Arvidson (1976), and Malin (1976a) described fretted terrain morphology and age relationships from examination of Mariner 9 images, it was not until the receipt of Viking orbiter images, beginning in 1976, that more detailed study of the "fretting process(es)" could be conducted. Carr and Schaber (1977), Squyres (1978,1979), Schultz and Glicken (1979), and Breed et al. (1980) suggested several alternative mechanisms for the formation and modification of the fretted terrain, based on these Viking data. The papers by Carr and Schaber, and Squyres, addressed principally the modification of the fretted terrain. Features discussed by these investigators included lobate aprons beneath scarps and around mesas, and lineated valley floors. Carr and Schaber suggested that the features resembled those that are formed by frost creep or gelifluction, whereas their appearance suggested to Squyres that they formed by rockmass creep in a manner analogous to rock glaciers. Both Schultz and Glicken, and Breed and her co-workers, considered the initial formation of fretted terrain to require the action of liquid water, either as mudflows generated by subsidence over melted ground ice, as catastrophic floods, or as surface run-off.

MORPHOLOGY

No area of fretted terrain looks exactly like another. However, over a 3500 km distance two or three general assemblages of features occur. At the south and east end of the zone of principal occurrence, in Nilosyrtis, the assemblage of forms includes a ridge-and-valley topography strongly oriented parallel to the upland-lowland boundary, with flat or slightly concave valley floors that characteristically display longitudinal lineations (Figure 1). The valleys are typically a few kilometers across and several tens of kilometers in length. The ridges have subdued, rounded, undulating crests, some relatively narrow and others much broader (to 10-15 km). Valley floors occur at several discrete topographic levels; tributaries reach concordancy through gentle, ramp-like longitudinal slopes descending from higher to lower levels. Straight to arcuate or slightly sinuous lineations (<100 m wide) within many of these valleys exhibit both positive and negative relief (probably less than a few tens of meters).

At the northwest end of the fretted zone (in Deuteronilus), the terrain is characterized by a progressive transition from undulating, heavily cratered surfaces indented by occasional channels to sparsely cratered lowland surfaces surrounding a few remnant outliers of high-standing terrain. These mesas typically have flat-tops and sharp brinks; their planimetric shapes vary widely, suggesting that there was little regional structural control of outline form. The lowland surface displays considerable variation in texture. However, most prominent of the characteristics of this area of fretted terrain are aprons surrounding each mesa and extending from the bases of other cliffs in the region (Figure 2). These aprons, reaching up to several tens of kilometers from the base of scarps, are faintly lineated in the direction generally perpendicular to the superjacent cliff. The lineations, usually less than a kilometer across, resemble shallow grooves (perhaps only a few tens of meters deep), and in some cases, leveed channels.

A third general pattern of fretted terrain occurs between the two areas discussed above (i.e., in Protonilus) and consists of a combination of their features (Figure 3). This pattern includes both mesas and ridges, separated by relatively narrow (<10 km), flat-floored valleys similar to those in Nilosyrtis; the pattern exhibits some regional tectonic alignment. Valley floors are coarsely lineated, but in places the lineations merge with pits and other irregular depressions, producing a chaotic and rough texture.

AGE RELATIONSHIPS

Estimates of relative age relationships between various elements of the fretted terrain, and between fretted and surrounding terrains, are derived from counts of impact craters (the number of superimposed craters is directly related to the amount of time a surface has been exposed to cratering) and from other examples of superposition.

Crater counts show that, in general, the highland surfaces are more heavily cratered and hence older than the lowland surfaces (Coradini and Arvidson 1976). The dearth of impact craters on the aprons and lineated valley floors, even in moderately high-resolution images, implies "extreme" relative youth. Crater counts by Carr and Schaber (1977) set an upper limit of about one crater larger than 1 km diameter per 2000 km^2, compared to surfaces on the superjacent highlands with values typically an order of magnitude higher. Assuming a relatively uniform flux of cratering objects, there are three possible explanations for this observation: (1) the aprons are young and unmodified, (2) the aprons are young and experiencing contemporary modification, or (3) the aprons are old but their surfaces have been "recently" modified. In all three cases the processes that crated the surfaces must have acted on Mars in the not-too-distant past. The mechanisms that crated the fretted terrain (as opposed to those that have modified the surface), however, appear to have operated substantially earlier (as indicated by the number of large craters superimposed on the fretted terrain valleys and ridges), perhaps under conditions that no longer persist on Mars.

MEASUREMENTS OF FRETTED TERRAIN APRONS

Measurements of apron length perpendicular to superjacent scarp, scarp height above the apron (determined from measurements of shadow lengths), and scarp azimuthal orientation were made on aprons surrounding mesas in Deuteronilus. The results of these measurements are shown in Figures 4, 5, and 6. Figure 4 shows the relationship between scarp height and apron length for 50 aprons with similar morphology and occurrence. This type of plot is related to that used to compare landslides on the Earth, Moon, and Mars (e.g., Howard 1973, Lucchitta 1978), inasmuch as the scarp height is realted to the potential energy available and the apron length is related to the efficiency of conversion of potential to kinetic energy. Our inability to measure volumes (owing to poor topographic control over most areas) prevents calculation of potential energies and flow efficiencies. No statistically significant trend is seen in the data, although there is a visual similarity to the landslide relationship that clearly derives its features from conversion of potential energy. It is not possible to demonstrate that aprons form on Mars as a result of gravity-induced movements, although their morphology suggest such an origin. Figure 5 shows the relationship of average apron length as a function of azimuth. A peak in this curve centered on N30E is apparent. Also plotted on Figure 5 are the azimuths of structural trends in the local area, measured on linear crater wall segments, scarps, and mesa cliffs. Figure 6 shows that scarp height is also a slight function of azimuth, but not in a manner consistent with that of apron length on azimuth.

The observation of an azimuthal dependence of apron length is a possible constraint on the nature of the fretted terrain modification process. Factors that might contribute to an azimuthal variation include structural and environmental variables (such as temperature, atmospheric pressure and H_2O content in solid or liquid form). As there is little correspondence between structural "grain" in the region and apron length (i.e., longest aprons are not at azimuths perpendicular to regional structure, nor are the most aprons perpendicular to the structural fabric), then the environmental factors may be responsible for the observed relationship.

ENVIRONMENTAL CONSIDERATIONS

Current conditions do not permit liquid water to occur in the near surface (<1 km deep) materials of the fretted terrain (Kieffer et al. 1977, Palluconi and Kieffer 1981). Diurnal variations of as much as 80°C occur, with maximum theoretical summer temperatures under current insolation conditions of -10°C. The average temperature is close to -73°C, and the winter low is about -120°C, at which point precipitation of the principal atmospheric gas (CO_2) acts as a buffer to prevent appreciably colder surface temperatures. Highest rate of change of temperature occurs during morning illumination, when rates as much as 20°C/hour are observed. The eastern orientation of the longest aprons might indicate some influence of the rate of heating on the materials of the fretted terrain.

Variations in the obliquity (angle between the spin axis and solar equator) induced by gravitational perturbations of the other planets cause significant variations in martian surface temperatures and atmospheric pressure (Toon et al 1980). The temperature changes at many latitudes are quite large, but at 45°N the effects are small. The effects of obliquity variations on the magnitude and phase of the diurnal and seasonal temperature cycles at fretted terrain latitudes has not yet been computed. It is possible, and perhaps even probable, that such cycles will enhance the insolation received by eastern slopes, thus enhancing the effect seen in apron-length distribution.

The isotopic composition of nitrogen on Mars suggests that that planet once had significantly more nitrogen in its atmosphere. Studies of the isotopic composition of the rare gasses, oxygen, and hydrogen, and arguments made on the basis of cosmochemical abundances of these gases, seem to indicate that a denser atmosphere occurred early in martian history (e.g., Owen et al. 1977). Evidence of fluvial processes similarly date from that period (Malin 1976b, Pieri 1980). Such evidence is consistent with a warmer, "moister" environment conducive to many familiar processes of erosion. The exact nature of this environment (i.e., diurnal and seasonal temperatures, atmospheric pressure, and variations in it, insolation as a function of latitude, obliquity and eccentricity, and so on) is unknown, and thus consideration of it has not yet provided insight into surface evolution, although such studies are currently underway.

SPECULATIONS

Owing to the severe modification of valley wall slopes and floors, and the apparent retreat of scarps and the concurrent production of aprons, it is not possible to infer from available data on its present morphology the origin of fretted terrain. Its physical association with a boundary between old-and-high and low-and-young lands that extends around the planet, as well as the apparent control of some of its elements by structures related to this boundary, implies at least in part a tectonic genesis. Channels and evidence for collapse within the cratered uplands suggest some fluid activity, thought by some to involve ground ice melting and subsequent fluvial activity. Such evidence of erosion is lacking farther into the transition zone to lowland plains (as in Deuteronilus), and in areas where valleys are typically closed at one or both ends (as in portions of Nilosyrtis). Ground ice sapping (Sharp 1973), a process of ground deterioration occurring as ice sublimes without melting, releasing otherwise poorly consolidated materials, remains a viable and intriguing alternative.

Modification of fretter terrain appears to have occurred in the relatively recent past. Several different materials, and differences in their responses to different processes, are probably responsible for the variations in form seen in the fretted terrains. Aprons appear to form at the base of scarps in Deuteronilus; they are most likely formed of debris shed from these cliffs and transported as debris avalanches, debris flows, or debris creep masses. Each of these mechanisms is suggested by some observations, although none is consistent with all observations. The straited valley floors in Nilosyrtis and Protonilus show little evidence of flowage (despite the initial impression of longitudinal flowage evoked by the lineations) either off the valley walls or down the valley floors. Rather, the lineations appear to be formed by differential degradation of floor material by unknown processes (possibly ground ice deterioration and eolian action). Why such processes act to produce the organized pattern seen on the valley floors is similarly unknown, but the pattern may reflect structural control (for example, joints formed by cooling or differential compaction after emplacement) by the material being eroded. It is these unknown processes and material controls that present a great challenge to our understanding of Mars.

ACKNOWLEDGEMENTS

This research was supported by the National Aeronautics and Space Administration's Planetary Geology Program, through grant NAGW-1.

REFERENCES

Breed, C. S., McCauley, J. F., and Grolier, M. J., 1980, Evolution of inselbergs in the hyperarid western desert of Egypt--Comparisons with martian fretter terrain, in Holt, H. E. and Kosters, E. C., (eds.), Reports of the Planetary Geology Program--1980: Washington, D.C., NASA Technical Memorandum 82385, p. 307-311.

Carr, M. H., and Schaber, G., 1977, Martian permafrost features, Journal of Geophysical Research, v. 82, p. 4039-4054.

Coradini, M., and Arvidson, R. E., 1976, Age and formation of martian fretted terrain, Geologica Romana, v. 15, p. 377-381.

Howard, K., 1973, Avalanche mode of motion: Implications from lunar examples: Science, v. 180, p. 1052-1055.

Kieffer, H. H., Martin, T. Z., Peterfreund, A. R., Jakosky, B. M., Miner, E. D., and Palluconi, F. D., 1977, Thermal and albedo mapping of Mars during the Viking primary mission: Journal of Geophysical Research, v. 82, p. 4249-4291.

Lucchitta, B. K., 1978, A large landslide on Mars: Bulletin of the Geological Society of America, v. 89, p. 1601-1609.

Malin, M. C., 1976a, Studies of the martian surface (Ph.D. Dissertation): Pasadena, California Institute of Technology, 176 p.

Malin, M.C., 1976b, Age of martian channels: Journal of Geophysical Research, v. 81, p. 4825-4845.

Malin, M. C., Phillips, R. J. and Saunders, R. S., 1978, The nature and origin of the martian planetary dichotomy (is still a problem), in Holt, H. E. (ed.), Reports of the Planetary Geology Program--1977-1978: Washington, D.C., NASA Technical Memorandum 79729, p. 83-85.

Mutch, T. A., Arvidson, R. E., Head, J. W. III, Jones, K. L., and Saunders, R. S., 1976, The geology of Mars: Princeton, Princeton University Press, 400 p.

Owen, T., Biemann, K., Rushneck, D. R., Biller, J. E., Howarth, D. W., and LaFleur, A. L., 1977, The composition of the atmosphere at the surface of Mars: Journal of Geophysical Research, v. 82, p. 4635-4639.

Palluconi, F. D., and Kieffer, H. H., 1981, Thermal inertia mapping of the surface of Mars from 60°S to 60°N: Icarus, v. 45, p. 415-426.

Pieri, D., 1980, Martian Valleys: Morphology, distribution, age, and origin: Science, v. 210, p. 895-897.

Schultz, P. H., and Glicken, H., 1979, Impact crater and basin control of igneous processes on Mars, Journal of Geophysical Research, v. 84, p. 8033-8047.

Sharp, R. P., 1973, Mars: Fretted and chaotic terrains: Journal of Geophysical Research, v. 81, p. 4073-4083.

Squyres, S. W., 1978, Martian fretted terrain: Flow of erosional debris: Icarus, v. 34, p. 600-613.

Squyres, S. W., 1979, The distribution of lobate debris aprons and similar flows on Mars: Journal of Geophysical Research, v. 84, p. 8087-8096.

Toon, O. B., Pollack, J. B., Ward, W., Burns, J. A., and Bilski, K., 1980, The astronomical theory of climate change on Mars: Icarus, v. 44, p. 552-607.

FIGURES

Figure 1. Viking Orbiter 1 Image 084A73, showing a portion of the fretted terrain in Nilosyrtis. The photograph shows an area approximately 45 km on a side.

Figure 2. Viking Orbiter 2 Image 058B55, showing a portion of the fretted terrain in Deuteronilus. The photograph shows an area approximately 90 km on a side.

Figure 3. Viking Orbiter 2 Image 058B65, showing a portion of the fretted terrain in Protonilus. This photograph shows an area approximately 95 km on a side.

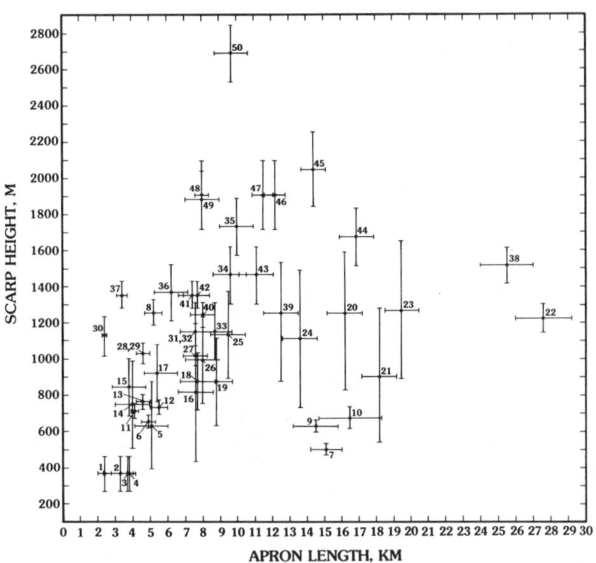

Figure 4. Plot of scarp height as a function of apron length perpendicular to superjacent scarp for 50 aprons on the Deuteronilus fretted terrain.

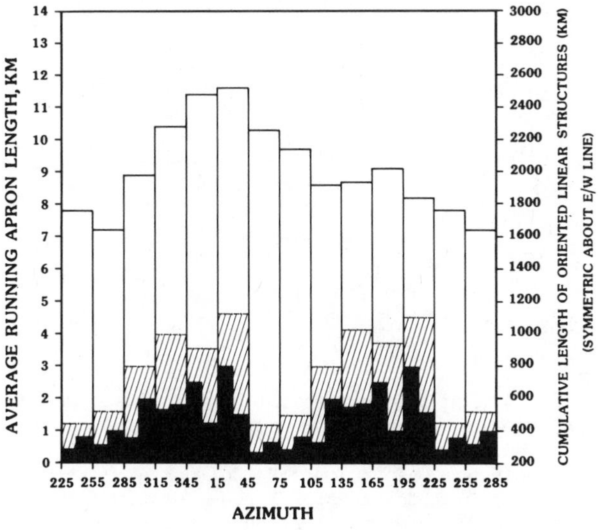

Figure 5. Plot of average apron length as a function of azimuth for 79 aprons in the Deuteronilus fretted terrain, using a three-sample running average. Note the peak at N30E. Also plotted is total length of linear structural features (e.g., crater wall elements, scarps, etc.) as a function of azimuth.

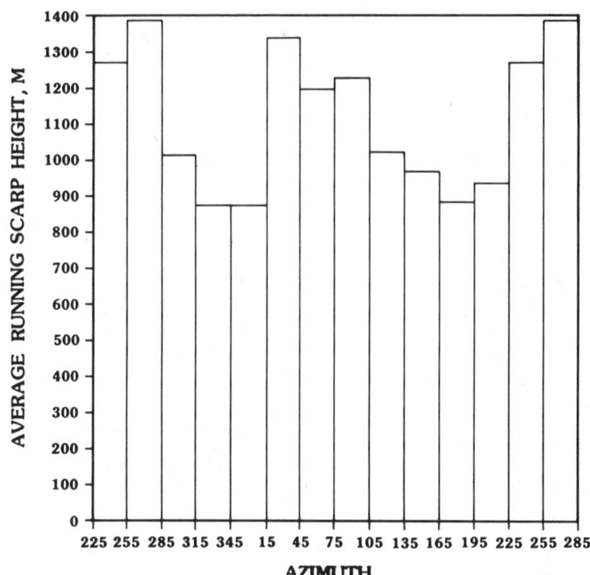

Figure 6. Plot of average scarp height as a function of azimuth for 50 samples shown in Figure 4.

REVEGETATION OF FOSSIL PATTERNED GROUND EXPOSED BY SEVERE FIRE ON THE NORTH YORK MOORS

E. Maltby and C.J. Legg

Department of Geography, University of Exeter
Amory Building, Rennes Drive, Exeter, Devon, England EX4 4RJ

Severe fire in 1976 ignited large areas of blanket peat on the North York Moors. Direct combustion, conversion to ash, and subsequent erosion have exposed a dense system of patterned ground at Rosedale. Disorientated mineral material and exhumed, sorted stone hummock-hollow terrain indicate periglacial activity dating to the Late Devensian or possibly immediate postglacial times. Geometric and chemical characteristics of the patterned ground are outlined. Revegetation by a succession of bryophyte species has been observed since 1976. Early colonization by Ceratodon purpureus and lichens (Lecidea spp.) has given way to a clear pattern adjusted to moisture conditions involving Polytrichum piliferum (dry hummock microsites) and Polytrichum commune (moist hollow microsites). Factors influencing colonization on the fossil hummocks are examined and compared with patterns of mature vegetation in contemporary periglacial environments.

INTRODUCTION

Thaw-freeze processes mold both the landscape detail and small-scale vegetation patterns in high latitudes, high altitudes, and at the margins of ice masses. There has been considerable interest in the relationships between active periglacial conditions, frequently producing distinctive patterned ground, and vegetation responses (e.g., Billings and Mooney 1959, Bryant and Scheinburg 1970, Hopkins and Sigafoos 1950, Warren Wilson 1952). Much of the early work has not considered the process and rates of plant colonization of distinctive periglacial features. The investigations of Heilbronn and Walton (in press) fill this gap for sorted features in the Subantarctic.

Although sorting processes are still active in Britain in some high mountain areas (e.g., Tinto Hill, Lanarkshire (Miller et al. 1954)) the present temperate climate and modern vegetation cover are generally inimical to patterned ground formation. Abundant evidence exists, however, for fossil periglacial features, including patterned ground elements (Fitzpatrick 1956). In some cases, distinctive vegetation responses have been related to complex fossil subsoil structures which cause large variations in textural, hydrological, and parent material (nutrient) conditions over small distances (Williams 1964). Often, however, there is no obvious vegetation response, and no information exists on how fossil patterned ground exposed by removal of later soils or deposits might influence vegetation development.

STUDY AREA

The North York Moors are an isolated, compact Jurassic upland of some 550 mi^2 (1,425 km^2) in northeast England (Figure 1). The area is dominated by an extensive plateau surface comprising mainly Middle Jurassic grits, sandstones, and shales which attains a maximum altitude of 454 m.

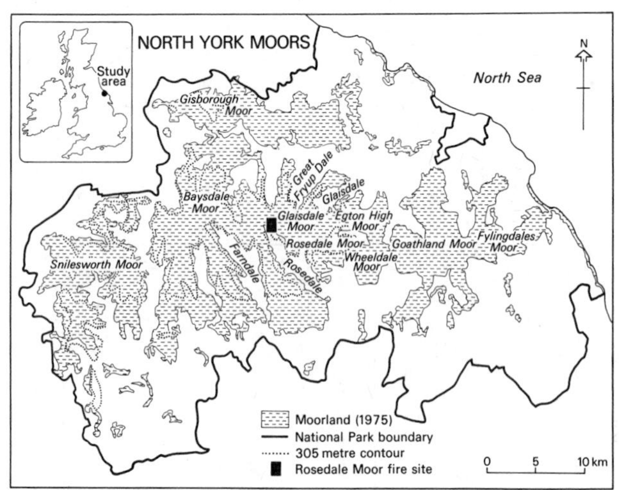

FIGURE 1 Location map.

Ice overran the area in all but the last (Devensian) glaciation and removed most pre-Pleistocene soils. Some evidence for the survival of earlier palaeosols has been described (Bullock et al. 1973), but such relict features are rare and do not occur on the central plateau moorland watershed. The upland block has been subjected to strong periglacial modifications. This has resulted in marked congelifraction, disorientation, and downslope mixing of bedrock strata. The evidence for frost riving is apparent when removal of the postglacial peat cover exposes a litter of angular blocks, many of which have been tipped into a vertical position by downslope movement or by ground ice activity or by reorientation in the mollisol (Plate 1). The exhumed landscape is often strongly reminiscent of fell-field.

PLATE 1 Irregular and disorientated sandstone fragments resembling fell-field, typical of the exhumed periglacial surface exposed by peat and soil removal from plateaus on the North York Moors.

The evidence for patterned ground development elsewhere on the North York Moors has been largely inferential based on air photographs and vegetation contrasts. Polygonal networks of cotton grass (Eriophorum spp.) occupying debris-filled cracks in bedrock in the otherwise heather (Calluna vulgaris) dominated vegetation of the upland have been described by Dimbleby (1952). Infill of the 3 m deep gaps is thought to have occurred when wedges of ground ice melted with amelioration of the periglacial climate. It is by no means certain, however, in which cold period the features formed. Much of the plateau is dominated by blanket peat, which in places exceeds 3.5 m depth. The peat generally comprises Eriophorum and Sphagnum remains, and the succession to Calluna almost certainly indicates some recent drying out of the surface (Tansley 1968). Type soil and peat profiles are described in Bendelow and Caroll (1976). Postglacial vegetational history, and possible cause of change, peat development, and soil alteration have been thoroughly investigated (e.g., Dimbleby 1962, Simmons 1969, Cundill 1971, Simmons and Cundill 1974, Atherden 1976).

Rainfall totals for the moorland plateau are relatively low compared with other upland areas in Britain. Annual amounts exceeding 1,140 mm (45 ins) are exceptional. Winters are cold by oceanic temperate standards, with a January mean of 1.7°C. Snow may lie for 45-80 days depending on altitude, and on average there are 75 days with frost.

Rosedale Moor Site

Severe 'accidental' fire persisting for 4 weeks affected 32 ha of moorland at the head of Rosedale in August 1976. The site occupies mainly gently sloping land, generally 1°-3° on the southern moorland flank of the central plateau within an altitudinal range of 380-403 m. Vegetation cover over the entire area was eliminated, but underlying peat and organic horizons were altered to a varying extent depending on depth, moisture, and presence of structural features allowing localized entry of oxygen (Maltby 1980). In the northwest part of the burnt area on the most gentle gradients and at the highest elevation, the thin blanket peat/peaty horizons originally averaged 300 mm depth and were associated with both stagnohumic gley and iron pan stagnopodzol profiles. The fire burnt most intensively in this area and ignited most of the organic materials to a fine powdery red-yellow ash. Only a small proportion of the original peat was incompletely combusted, and such residual material was charred into hard fragments. Deflation and erosion of the ash by surface wash rapidly exposed a series of stone hummocks and intervening hollows comprising a distinctive patterned ground complex (Plate 2).

PLATE 2 Hummock with residual charred peat surrounded by ash 1 year after the fire. (Scale is 1 m.)

This hummock-hollow complex occupies a SSW facing upper slope facet and seems to be restricted to this part of the site. The features may extend in elongated form downslope, but the intact moorland surface makes the geometry of stony areas difficult to establish with certainty. The progress of erosion, differential recolonization patterns elsewhere in the fire-site, and a general account of biological, chemical, and physical alteration of substrate conditions are given in Maltby (1980).

Methods

The position of over 50 distinctive stony hummocks has been recorded. For each feature, measurements were made of diameter (including for random examples long axis (D_1) and short axis (D_2) measurements where these could be clearly defined), height, and stone length for five individuals from each hummock.

Permanent markers were established shortly after the fire, and changes in surface morphology and vegetation colonization have been monitored through repeated analysis of point quadrat frames and at 0.5 m intervals along a line transect. Photographic records have been maintained throughout the postburn period, and detailed maps of the mosaic of colonizing bryophyte species made for selected sites. Soil samples have been taken from the surface 65 mm in representative areas of each surface type for routine chemical analysis. Cations were extracted with 1 M ammonium acetate, organic carbon determined by a wet oxidation procedure (Walkley and Black 1934), nitrogen using a Carlo Erba nitrogen analyzer (model 1400), and pH measured at sticky point.

Detailed examination of a representative zone of the patterned ground was made in 1981, 5 years after the fire. A transect was established extending between two ash filled hollows and over the long axis of a stony hummock with a small area of disturbed ash caused by damage to the bryophyte turf. Vegetation, soil, and topographic characteristics were recorded in 0.25 x 0.25 m quadrats. Frequency of exposed stones is much increased on the hummocks; hence the percent cover of species is expressed as percent stone free area of the quadrat.

RESULTS

Patterned Ground Complex

The striking regular spacing of the hummocks is illustrated in Plate 3.

PLATE 3 The patterned ground complex at Rosedale with colonization of the ash filled hollows by *Ceratodon purpureus* (light grey) and residual charred peat remaining on the hummocks (dark grey) 3 years after the fire.

For further analysis the features have been stratified into three groups according to slope position and angle. Amplitude of the hummock-hollow microtopography is greatest on the lowest slope angle (<1°) nearest the plateau summit. Height decreases, and the hummocks become more elongated with increasing distance downslope and the associated increase in gradient (Table 1). There is no statistically significant change in size, but variability increases downslope. The hummocks downslope have become fully revegetated by a dense moss carpet; eventually, they are impossible to measure accurately. It appears that they are more subdued because of original genesis rather than because of differential exhumation of the stone hummocks upslope by greater contemporary erosion.

TABLE 1 Geometric Characteristics of Exposed Hummocks.

Slope Position Angle (Sample n)	Hummock Statistics			
	Diameter (m)	D_1/D_2[1]	Height (mm)	Stone long axis[2]
Upper (28) <1.0	2.31 ±0.08[3]	1.47	156.3 ±07.3	152.0 ±09.4
Middle (15) 1-2°	2.05 ±0.11	1.58	136.0 ±08.3	113.5 ±06.9
Lower (10) >2°	1.99 ±0.22	1.88	107.0 ±05.6	95.8 ±04.2
All (53)	2.20 ±0.07	-	141.2 ±05.3	128.2 ±06.9

[1]Hummock long (D_1)/short (D_2) axis.

[2]Mean of samples, each comprising five individuals.

[3]Standard error.

The hummocks comprise a convex surface of angular sandstone blocks above a circle of distinctly imbricated stones which dip toward the centre of the features. Blocks and fragments are generally moderately weathered and possess a characteristically bleached outer skin where acids from the peat have removed iron. In the intervening hollows, the pedogenic eluvial (E(g)) horizon of the ignited and eroded soil contains abundant small stones, but the larger blocks which make up the hummocks are absent.

Higher combustion temperatures are inferred in the hollows compared with the hummocks because most of the original peaty cover of the hollows was reduced completely to ash, whereas on the hummocks a much larger proportion of material remained as charred peat (Plate 2). This may have been related to the differential thickness of the peat, which prior to the fire gave no surface expressions of the buried pattern.

Microenvironment for Plant Colonization

In contrast to controlled fires associated with moorland management, the severe burn resulted in (1) radical alteration of physical and chemical substrate properties (Figure 2) and (2) destruction or dramatic reduction of the viable seed bank and root stocks in burnt surface horizons, delaying establishment of vascular plants (Table 2).

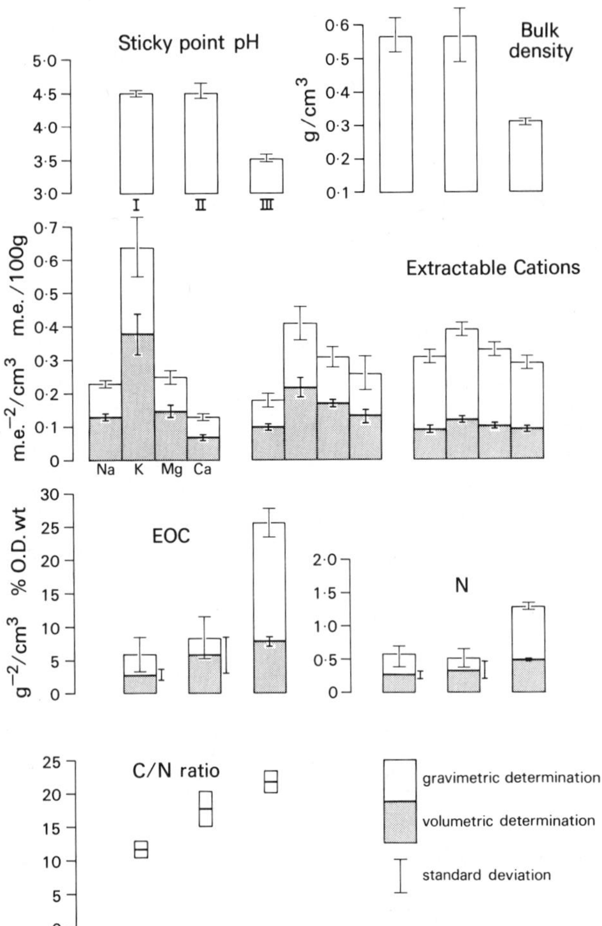

FIGURE 2 Analytical characteristics of surface 65 mm of ash-dominated surfaces above subdued microrelief (I), exhumed patterned ground with distinctive microrelief (II), and intact peat surfaces (III). EOC is easily oxidizable organic carbon (Walkley and Black 1934).

The mosaic of residual charred peat and ashed material found at the patterned ground site provides a juxtaposition of contrasting habitats with soil pH ranging from less than 4 to more than 6.5 and parallel changes in nutrient status (Table 3). In the absence of vegetation and original organic cover, surface runoff increases and water retention declines, resulting in a change from wet and often waterlogged conditions to a frequently xeric environment. Exposure of the surface to frost action, wind, and surface water leads to sustained surface instability and erosion. Colonization is therefore not by the usually dominant heath and moorland species but by a succession of algal, lichen, and bryophyte species, whose distributions are related primarily to the substrate and microclimatic variations caused by the exhumed microtopography.

The Pattern of Colonization

Primary colonization of the site during the first year following the fire was by algae. These were predominantly filamentous forms over the ash surface, though gelatinous and coccoid types were present on the surface of charred peat. The latter were accompanied by the ubiquitous presence of *Lecidea uliginosa* on crusted peat surfaces and numerous conspicuous patches of *Lecidea granulosa*. The surface film produced by these species has been shown to inhibit the germination of seeds of *Calluna vulgaris* and may be a significant factor affecting further colonization.

TABLE 2 Number of Points Represented by Each Surface or Vegetation Type Recorded in 1979 and 1982 for the Whole Area of Patterned Ground

	Number of Point Observations	
	1979	1982
Granular peat	31	0
Charred peat	6	6
Ash	134	60
Stone	43	32
Ceratodon purpureus)		7
Polytrichum piliferum)	313	258
P. commune)		151
Aira praecox	0	7
Poa annua	0	7
Agrostis tenuis	0	2
Juncus effusus	0	1
Calluna vulgaris	0	1

TABLE 3 Surface Characteristics of the Hummock-Hollow Complex and Adjacent Peat

	Analysis of Surface 0-20 mm					
	Extractable nutrients mg/kg				pH	Moisture content
	Na	Mg	K	P		
Charred peat (hummock) 1977	70	64	73	25	3.6	60-135%
Ash 1976	156	349	945	60	5.8-7.2	50-140%
1977 (hollow)	72	119	240	73	5.0-6.8	
Intact peat 1977	109	142	56	44	3.2-3.4	400-450%

The ashed material in the hollows was particularly susceptible to frost disturbance during the first and second winters, and pipkrake formation was observed on numerous occasions. However, during the second summer following the fire *Ceratodon purpureus* rapidly covered and stabilized ashed surfaces. *Ceratodon purpureus* is an acrocarpous moss having low (10-20 mm) "short turf" morphology (Gimingham and Birse 1957). This species then also spread to the charred peat of the stony hummocks, establishing particularly in the cracks created by shrinkage of the charred peat. The occurrence of Ceratodon purpureus seems to be related to the chemical quality of the ash deposits and thin residual peat, as it is totally absent from neighboring sites with a relatively intact peat surface or deeper residual peats, where Dicranella heteromalla and Pohlia nutans are abundant.

During the third season following the fire, Polytrichum piliferum became widely established. The somewhat taller (20-30 m) "tall turf" morphology of this species with a spreading underground rhizome-like system enabled it to gain immediate dominance over the Ceratodon purpureus. Though covering the ground more rapidly in the hollows, this xeromorphic species quickly established on the hummocks in sheltered microsites provided by desiccation cracks in the residual peat and in the shelter of stones.

Polytrichum commune, which is morphologically dominant over the other two species, has also established in the area, growing through and over the Polytrichum piliferum turf such that by 1982 it represented almost 40% of all plant cover. It is, however, less xeromorphic than Polytrichum piliferum and is largely restricted to the hollows on deeper ash, and areas downslope where soil moisture is higher. The situation six seasons after the fire is therefore a mosaic of Polytrichum commune, and Polytrichum piliferum reflecting soil moisture. Ceratodon purpureus is abundant but at this stage is rarely more than a degenerate or dead turf beneath the Polytrichum piliferum (Figure 3).

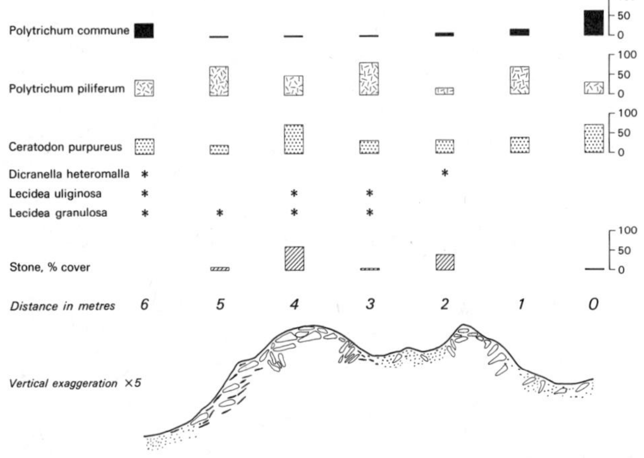

FIGURE 3 Distribution of different species along a scale transect across typical hummock and hollow features. (Stones are indicated schematically only.)

Where the protective surface of mosses has been broken the secondary recolonization process is slow, and open patches of unstable, frost-heaved ash remain for some time, with an open mixed community of the three bryophyte species developing over a 2 year period.

On the patterned ground, vascular plants are restricted to a limited growth of the small annual grass, Aira praecox, at the margins of frost-disturbed sites, and the occasional plants of Agrostis tenuis, Poa annua, and Festuca ovina, predominantly in the hollows, originating probably from sheep droppings.

DISCUSSION

The microrelief described here is a somewhat unusual example of periglacial patterned ground since it combines hummocks and rings of imbricated stones more characteristic of stone circles. The stone hummocks contrast markedly with the fossil polygons described elsewhere on the North York Moors (Dimbleby 1952) and perhaps bear greatest morphological similarity to earth hummocks (Washburn 1973). These are widespread south of the tree line, and typical examples 1-3 m in diameter and 0.40-0.50 m high have been described in the northern Mackenzie Valley and North Yukon (Zoltai and Pettapiece 1974). In contrast, these modern microrelief counterparts are generally not described as stony. However, excavation of similar surviving features on Dartmoor has revealed that a stone core may have segregated at some depth due to ground ice (D.J. Miller personal communication). The occurrence of earth hummock features elsewhere in Britain has been reviewed by Pemberton (1980), who discounts a periglacial origin in favor of much more recent development either during the "Little Ice Age" or in some cases even within the last 150 years. He has reasoned that the microrelief of original periglacial hummocks would not have survived postglacial events. This would not necessarily contradict a true periglacial origin for the Rosedale features, however, since they have been protected by a peaty cover for a significant part of the postglacial.

The dominant features of the vegetation colonizing the burnt area are the abundance of bryophyte species but absence of vascular plants. Although similar successions of moss species may be observed following fire in other heath vegetation (e.g., Clement et al. 1980), the scarcity of vascular plants is generally unusual for sites in the north temperate zone. Similar patterns, however, are described for fire on patterned ground at Kokolik River, Alaska, where a tundra fire over "high centred polygons" results in 70% cover of Marchantia polymorpha and Ceratodon purpureus but only 5% cover of vascular plants (Hall et al 1980). Following controlled burning for moorland management, the establishment of vascular plants from seed or buried rootstocks occurs during the first season after fire, and subsequent spread is by vegetative growth. Where initial colonization is delayed by severe fire which destroys the seed bank and potential regeneration centers, establishment is further inhibited by changes in the substrate quality, ecesis resistance of the bryophyte communities, and grazing pressure. Establishment of vascular plants is therefore extremely slow and depends upon the availability of suitable microsites. These may occur either on dung or where the moss turf has been broken to expose the subsurface ash.

Frost lift is undoubtedly important in reducing the rate of colonisation during the first two winters and subsequently in patches where the moss turf is broken. These areas, however, are not obviously related to the pattern of fossil hummocks and therefore frost and exposure are not the prime factors in determining the vegetational pattern associated with the hummocks. This contrasts with the observations of Heilbronn and Walton (in press) in which the establishment of vascular plants on patterned ground is directly related to particle size and frost movement. The distribution of plants in mature vegetation on patterned ground, and a complex of interactions between exposure, frost action, and the insulating and stabilizing properties of vegetation have been described elsewhere (e.g. Webb 1972, Warren Wilson 1952). The early colonisation patterns developed on the Rosedale site, however, appear to reflect the controls of substrate quality and microclimate.

CONCLUSION

Examination of active features in modern periglacial environments can assist greatly in the interpretation of fossil landforms. However, the reverse may be true also under special conditions. The vegetation succession on the patterned ground microrelief exposed by a major peat fire on the North York Moors parallels examples of postfire development on sorted ground features in the Arctic. Continued monitoring will establish the full extent to which the exhumed periglacial microrelief influences progressive vegetation development in a climate less severe than that normally associated with active hummock formation or contemporary stone sorted features.

REFERENCES

Atherden, M. A., 1976, Late Quaternary vegetation history of the North York Moors, III, Fen Bogs: Journal of Biogeography, v. 3, p. 115-124.

Bendelow, V. C., and Caroll, D. M., 1976, Soils in North Yorkshire, III: Soil Survey Record No. 35, Harpenden.

Billings, W. D., and Mooney, H. A., 1959, An apparent frost hummock-sorted polygon cycle in the Alpine Tundra of Wyoming: Ecology, v. 40, p. 16-20.

Bryant, J. P., and Scheinburg, E., 1970, Vegetation and frost activity in an alpine fellfield on the summit of Plateau Mountain, Alberta: Canadian Journal of Botany, v. 48, p. 751-771.

Bullock, P., Carroll, D. M., and Jarvis, R. A., 1973, Palaeosol features in northern England: Nature, v. 242, p. 53-54.

Clément, B., Forgeard, F., and Touffet, J., 1980, Importance de la vegetation muscinale dans les premiers stades de recolonisation des landes après incendie: Bulletin d'Ecologie, v. 11, p. 359-364.

Cundill, P. R., 1971, Ecological history and the development of peat in the central watershed of the North Yorkshire Moors: Ph.D. thesis, University of Durham.

Dimbleby, G. W., 1952, Pleistocene ice wedges in North-East Yorkshire: Journal of Soil Science, v. 3, p. 1-19.

Dimbleby, G. W., 1962, The development of British heathlands and their soils: Oxford Forestry Memoir, No. 23, Oxford.

Fitzpatrick, E. A., 1956, Progress report on the observations of periglacial phenomena in the British Isles: Biuletyn Peryglacjalny, v. 4, p. 99-115.

Gimingham, C. H., and Birse, E. M., 1957, Ecological studies on growth-form in bryophytes, I. Correlations between growth-form and habitat: Journal of Ecology, v. 45, p. 533-545.

Hall, D. K., Ormsby, J. P., Johnson, L., and Brown, J., 1980, Landsat digital analysis of the initial recovery of burned tundra at Kokolik River, Alaska: Remote Sensing of Environment, v. 10, p. 262-272.

Heilbronn, T. D., and Walton, D. W. H., in press, Plant colonisation of actively sorted stone stripes in the subantarctic: Arctic and Alpine Research.

Hopkins, D. M., and Sigafoos, R. S., 1950, Frost action and vegetation patterns on Seward Peninsula, Alaska: U.S. Geological Survey Bulletin, v. 974-C, p. 51-100.

Maltby, E., 1980, The impact of severe fire on Calluna moorland in the North York Moors, Bulletin D'Écologie, v. 11, p. 683-708.

Miller, R., Common, R., and Galloway, R. W., 1954, Stone stripes and other surface features of Tinto Hill, Geographical Journal, v. 120, p. 216-219.

Pemberton, M., 1980, Earth hummocks at low elevation in the Vale of Eden, Cumbria: Transactions Institute of British Geographers New Series, v. 5, p. 487-501.

Simmons, I. G., 1969, Pollen diagrams from the North York Moors: New Phytologist, v. 68, p. 807-827.

Simmons, I. G., and Cundill, P. R., 1974, Pollen analysis and vegetational history of the North York Moors, I, Pollen analysis of blanket peat: Journal of Biogeography, v. I, p. 159-169.

Tansley, A. G., 1968, Britain's green mantle, 2nd ed.: London, George Allen and Unwin.

Walkley, A., and Black, I. A., 1934, An examination of the Degtjareff method for determining soil organic matter, and a proposed modification of the chromic acid titration method: Soil Science, v. 37, p. 29-38.

Warren Wilson, J., 1952, Vegetation patterns associated with soil movement on Jan Mayen Island, Journal of Ecology, v. 40, p. 249-264.

Washburn, A. L., 1973, Periglacial processes and environments: London, Edward Arnold.

Webb, R., 1972, Vegetation cover on Icelandic thúfur: Acta Botanica Islandica, v. 1, p. 51-60.

Williams, R. B. G., 1964, Fossil patterned ground in eastern England: Biuletyn Peryglacjalny, v. 14, p. 337-349.

Zoltai, S. C., and Pettapiece, W. W., 1974, Tree distribution on perennially frozen earth hummocks: Arctic and Alpine Research, v. 6, p. 403-411.

ULTIMATE LONG-TERM STRENGTH OF FROZEN SOIL AS THE PHASE BOUNDARY OF A VISCOELASTIC SOLID-FLUID TRANSITION

Chi-Sing Man

Department of Civil Engineering, University of Manitoba, Winnipeg, Manitoba, Canada R3T 2N2

Herein we explore the possibility of using wave propagation methods to determine the ultimate long-term strengths of frozen soils. Our discussion is based on the assumptions that the creep behavior of frozen soils (under conditions of non-failure) could be modeled as nonlinear viscoelastic, that the ultimate long-term strengths of a frozen soil in fact constitute the phase boundary of a viscoelastic solid-fluid transition, as well as the lack of smoothness of such a transition. The general discussion is illustrated by a detailed analysis of one particularly simple example, namely, why we could likely locate by experiments of wave propagation the phase boundary of a saturated sand-ice material under low-stress conditions. Finally we discuss the possibility that for frozen soils the transition from solid to fluid behaviour be strain-controlled rather than stress-controlled, and explain briefly how wave propagation experiments could still be used to determine the transition strain which demarcates between, for each given temperature, the solid and fluid behavior of frozen soils.

INTRODUCTION

Materials which creep under a sustained load are susceptible to loss of strength in time. Soils, frozen and unfrozen alike, are no exception. The importance of this fact has long been recognized in geotechnical engineering (see, for example, Casagrande and Wilson 1951). To characterize the reduction of strength in time, the concept of "long-term strength" is introduced. Under uniaxial compressive stress conditions the long-term strength or "creep failure stress" $\sigma_f(t_f, \theta)$ of a soil for the "elapsed time to failure" t_f and temperature θ is defined as the stress level under the sustained isothermal loading of which will lead to failure of the soil after a load period of length t_f; σ_f is a function of t_f and θ. (For simplicity the discussion in this paper will be restricted mainly to uniaxial compressive stress conditions, and we shall adopt the sign convention that compressive stress and compressive strain are positive.) The limit of the function $\sigma_f = \sigma_f(t_f, \theta)$ for a constant θ as t_f approaches infinity is called the "ultimate long-term strength" at the temperature θ. Methods for the determination of ultimate long-term strength are discussed in the literature (see, for example, Vyalov 1979, chap. 9). Although we have used the temperature as the parameter in the definitions above, in general it could be replaced by other physical variables. For example, the volume fraction of sand could be used as the scalar parameter for saturated sand-ice mixture in random packing when other factors which affect long-term strength are held constant.

Everyone who has had the experience of running constant-stress creep tests on frozen soil must be familiar with the following difficulty: Given a creep curve which attenuates over a relatively short interval of time to a very low rate of strain, it is practically impossible to determine without prejudice whether the curve should be extrapolated to the long-term as exhibiting attenuating creep or as approaching stationary creep at a low strain rate. The implication of this difficulty is that no conventional method for the determination of ultimate long-term strength could deliver that quantity for sure, because all such methods rely on extrapolation into a regime in which we literally do not know how we should extrapolate.

Here it may be argued that in practice what we are really interested in is the loss of strength during the life time of a structure, not the "ultimate" strength. But, to determine strength reduction in the long run (say, in time units of decades), the conventional experiments themselves may take a rather long time to complete, if we do not accept ready-made formulae like Vyalov's "law of strength reduction" as applicable *a priori*.

The unsatisfactory situation described above raises the question whether we can have other experiments that could be completed in a relatively short time and will deliver the ultimate long-term strength of frozen soils without ambiguity. Once such new experimental methods become available, studies on the improvement of frozen soils to enhance their ultimate long-term strength would become feasible.

In the classical theory of linear viscoelasticity, a material is characterized by its creep compliance J or its relaxation modulus G. A linear viscoelastic material is called a solid (e.g., "the standard linear solid") if the time derivative of its creep compliance $\dot{J}(t)$ approaches zero as the time t approaches infinity, and is called a fluid (e.g., "the Maxwell fluid") if $\dot{J}(t)$ approaches a positive constant. Under constant stress a linear viscoelastic solid will exhibit only attenuating creep, and a fluid will ultimately undergo stationary creep. Linear viscoelastic solids and fluids, indeed, have very different mechanical behavior under many circumstances (Pipkin 1972, chap. 3). Thus far in this section we have restricted our discussion to one given temperature. Should we consider a range of temperatures, then a material could be a linear viscoelastic fluid, say, for temperatures θ above some fixed temperature θ_o and be a solid for θ below θ_o. If we use a straight line segment to represent the range of temperatures under consideration, each temperature will correspond to one phase of the material---either solid or fluid. Thus we have a "phase diagram", and θ_o is the "phase boundary" which divides those temperatures at which the material is solid from those temperatures at which the material is fluid.

The foregoing discussion can be generalized for nonlinear viscoelastic materials which manifest solid-fluid transitions. For nonlinear materials whose transitions are temperature and stress controlled, the phase diagram will be defined on a domain D in the θ-σ or temperature-stress plane. Each point (θ, σ) in D corresponds to a phase of the material---either solid or fluid, and we call each (θ, σ) a phase point. If the material is solid at the phase point (θ, σ), then the creep curve defined by the temperature θ and the stress level σ exhibits attenuating creep; otherwise the creep curve will ultimately manifest stationary creep. The phase boundary which separates the solid and the fluid phase points is now a curve in the domain D. Consider the straight line defined by a fixed temperature θ. Suppose this straight line meets the phase boundary curve at the point $(\theta, \sigma_\infty(\theta))$. Suppose, without loss of generality, that for all $\sigma > \sigma_\infty(\theta) > 0$, the (θ, σ)'s are fluid phase points, and that for $0 < \sigma < \sigma_\infty(\theta)$, the (θ, σ)'s are solid phase points. Suppose further that for the material in question the nonlinear viscoelastic model is valid under conditions of non-failure, and that stationary creep will invariably lead to ultimate failure. If a series of constant-stress creep tests were performed on samples of such a material at the same temperature θ but at various stress levels, and if we could determine ultimate long-term strength from such experiments, we should conclude that $\sigma_\infty(\theta)$ be the ultimate long-term strength of that material. (See Figure 1.)

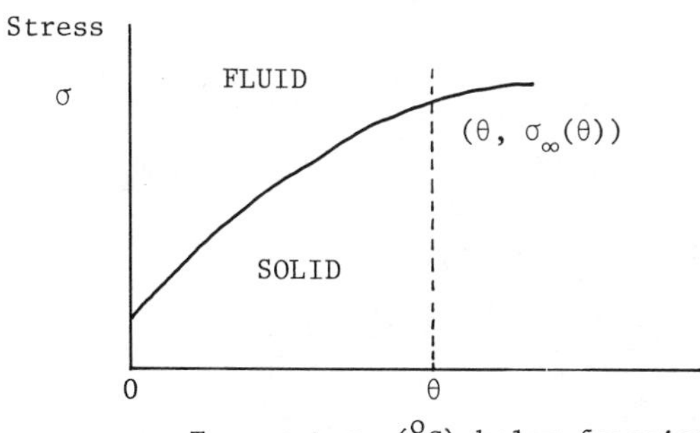

FIGURE 1 Phase diagram of a typical nonlinear viscoelastic material. The curve $\sigma = \sigma_\infty(\theta)$ defines the phase boundary of the solid-fluid transition.

One basic premise of this paper is that the foregoing discussion can be applied to frozen soils. It has been said that under non-failure conditions the behavior of a frozen soil is similar to that of a "non-linear Maxwell body" (Ladanyi 1981). As far as creep under constant stress is concerned, the preceding statement will be correct only if the stress applied is above the ultimate long-term strength. For stress levels lower than the ultimate long-term strength, it would be more appropriate to say that a frozen soil behaves like a nonlinear viscoelastic solid. Indeed, for a fixed temperature, if the ultimate long-term strength at that temperature demarcates between solid and fluid behavior, it is the phase boundary of the viscoelastic solid-fluid transition.

The main object of this paper is to study the possibility of locating that solid-fluid phase boundary by wave propagation methods. Since the solid-fluid transition of frozen soil might be strain-controlled rather than stress-controlled, we shall round off our discussion by explaining briefly how wave propagation studies could be used to clarify the important question whether ultimate long-term strength is really a constitutive quantity for frozen soils.

NOTATION AND PRELIMINARIES

Hereafter we shall assume that all samples of the frozen soil under consideration are homogeneous and they undergo homogeneous deformations in creep tests. In particular, we assume that each sample is prepared under the same standardized procedure at some fixed temperature, brought to the test temperature and cured for a fixed length of time before test at that temperature. Let L_o and $L(t)$ be the original length of a sample and its length at time t after loading, respectively. Then $F(t) \equiv L(t)/L_o$ is the deformation gradient at time t and $\varepsilon \equiv 1 - F$ is the compressive strain. In what follows we shall explicitly distinguish a function and the values it assumes. For example, while $f(\cdot)$ denotes a function with the dot $\cdot$ standing for the running variable, $f(x)$ will be the value of the function $f(\cdot)$ when the variable in question takes the specific value x.

Let D be a domain in the θ-σ plane which defines the range of temperature and stress level within our interest. Each constant-stress creep test is defined by a point (θ, σ) in D; here θ is the temperature of the test and σ the stress applied. To each (θ, σ) is associated a creep curve. We shall denote by $\varepsilon(\cdot; \theta, \sigma)$ the creep curve parametrized by the control variables (θ, σ); here the dot $\cdot$, as usual, stands for the running variable, which for the present instance is time. And we shall denote by $\varepsilon(t; \theta, \sigma)$ the strain assumed at the time t by the creep curve $\varepsilon(\cdot; \theta, \sigma)$. If a creep curve has a point of inflection, there is tertiary creep and the creep curve will be defined only for some finite interval of time. On the assumptions that frozen soils are nonlinear viscoelastic materials under conditions of non-failure, and that it is prolonged stationary creep that leads to failure, we replace the tertiary portion of each creep curve with a straight line segment which extends to $t = \infty$ and has its slope equal to the minimum strain rate. After this surgery and reconstruction, every (θ, σ) in D will correspond to a creep curve which as a function of t is continuously differentiable and is defined on the interval $[0, \infty)$. Henceforth when we use the symbol $\varepsilon(\cdot; \theta, \sigma)$ or refer to a creep curve, unless otherwise stated we shall mean the one after reconstruction. Let

$$\dot{\varepsilon}(\infty; \theta, \sigma) \equiv \lim_{t \to \infty} \frac{\partial \varepsilon}{\partial t}(t; \theta, \sigma). \tag{1}$$

A point (θ, σ) in D with $\sigma > 0$ is called a solid phase point if $\dot{\varepsilon}(\infty; \theta, \sigma) = 0$; it is called a fluid phase point if $\dot{\varepsilon}(\infty; \theta, \sigma) > 0$. The intersection of the frontier of the set of solid phase points and that of the fluid phase points is called the phase boundary. Hereafter we shall denote the phase boundary and a generic point on it by $\sigma_\infty(\cdot)$ and $(\theta, \sigma_\infty(\theta))$, respectively. (Here we assume that for each θ, there is only one $\sigma_\infty(\theta)$.) It is our basic assumption that $\sigma_\infty(\theta)$ is indeed the ultimate long-term strength at the temperature θ. When another physical variable replaces the temperature as the parameter, the foregoing discussion and definitions would still apply after obvious modifications.

When the temperature θ is the parameter, what follows is a familiar empirical fact: When the points (θ_1, σ_1) and (θ_2, σ_2) are sufficiently close to each other, the corresponding reconstructed creep curves $\varepsilon(\cdot; \theta_1, \sigma_1)$ and $\varepsilon(\cdot; \theta_2, \sigma_2)$ become virtually indistinguishable. (This empirical fact can be stated in a precise way after we endow the set of creep curves with a suitable topology---then it simply says that the mapping $(\theta, \sigma) \mapsto \varepsilon(\cdot; \theta, \sigma)$ is continuous. A topology to that effect was presented by Man (1982). In what follows we shall assume that a suitable topology has been chosen, but will leave the chosen topology implicit. Suffice it to say here that the set of creep curves is embedded in the space E of continuously differentiable functions $f(\cdot)$ defined on $[0, \infty)$ with $\dot{f}(\infty) < \infty$; a suitable topology is defined on E and the set of creep curves is given the induced subspace topology.) Now suppose (θ, σ_1) is a solid phase point, (θ, σ_2) is a fluid phase point, and $|\sigma_2 - \sigma_1|$ is small. Then the creep curves $\varepsilon(\cdot; \theta, \sigma_1)$ and $\varepsilon(\cdot; \theta, \sigma_2)$ are practically indistinguishable. As a result, unless we wait so long that the creep test at temperature θ and stress level σ_2 brings the sample in question to failure, there is no way to tell whether σ_2 exceeds the ultimate long-term strength $\sigma_\infty(\theta)$. Moreover, by means of creep tests alone we can never say for sure that σ_1 does not exceed $\sigma_\infty(\theta)$. This explains why it is difficult to determine ultimate long-term strength by creep tests.

In what follows we explore the possibility of locating the phase boundary $\sigma_\infty(\cdot)$ by means of wave propagation methods.

A SIMPLE EXAMPLE

The best way to illustrate why we could likely locate the phase boundary $\sigma_\infty(\cdot)$ by wave propagation experiments is perhaps by way of a simple example.

Consider saturated mixtures of ice and a fine sand (in random packing). The discussion below will be restricted to one given temperature and such other experimental conditions that the single parameter which affects the ultimate long-term strength of the sand-ice material could be taken as the volume fraction of sand ν. As ice has zero ultimate long-term strength, sand-ice mixtures with low values of sand concentration presumably would behave like a viscoelastic fluid. If mixtures with sufficiently high sand concentration have non-zero ultimate strengths, the phase boundary $\sigma_\infty(\cdot)$, which is now a curve in the ν-σ plane, will meet the $\sigma = 0$ axis at some point $(\nu_o, 0)$. (See Figure 2.) In other words, under the above assumptions, those mixtures with $\nu < \nu_o$ always behave like viscoelastic fluids in creep tests at the given temperature; those with $\nu > \nu_o$ will behave like solids when the compressive stress σ does not exceed $\sigma_\infty(\nu)$, and like fluids when $\sigma > \sigma_\infty(\nu)$.

According to Stevens (1975), for the propagation of infinitesimal sinusoidal progressive waves (peak dynamic stress 0.1-0.5 psi, frequency 500-10,000 Hz), "the theory of linear viscoelasticity appears to be adequate for frozen soils". This suggests that if the stress has always been maintained at a sufficiently low level and the strain has always been small, the sand-ice material can be adequately described by the constitutive relation

$$\varepsilon(t) = J(0; \nu)\sigma(t) + \int_0^t \dot{J}(\tau; \nu)\sigma(t - \tau)d\tau, \tag{2}$$

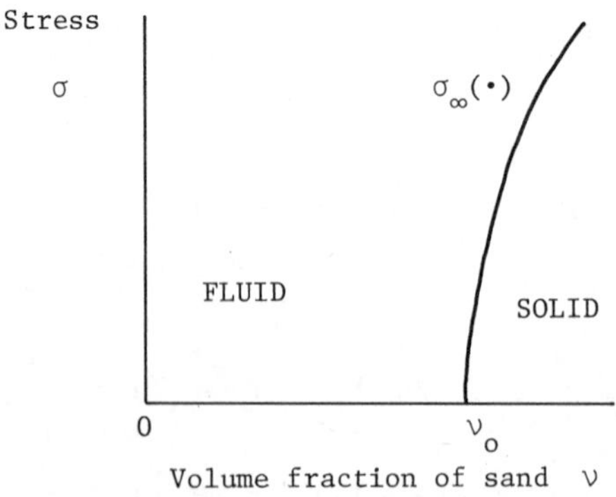

FIGURE 2 Phase diagram of a saturated sand-ice mixture (in random packing).

for a range of ν that contains ν_o; here $J(\cdot;\ \nu)$ is the creep compliance when the volume fraction of sand takes the value ν, $J(0;\ \nu)$ the initial value of $J(\cdot;\ \nu)$, $\dot{J}(\cdot;\ \nu)$ the time derivative of $J(\cdot;\ \nu)$, and τ is a dummy variable which runs from $\tau = 0$ to $\tau = t$. Since the point $(\nu_o,\ 0)$ is on the $\sigma = 0$ axis, to locate it we can restrict ourselves to experimental situations under which Equation (2) is valid. Henceforth we shall base our further discussion in this section on Equation (2).

In Equation (2), the creep compliance $J(\cdot;\ \nu)$ is that of a fluid for $\nu < \nu_o$, and is that of a solid for $\nu > \nu_o$. Since $J(\cdot;\ \nu)$ has drastically different characteristics for $\nu < \nu_o$ and for $\nu > \nu_o$, it is convenient to express $J(\cdot;\ \nu)$ in terms of two functions. Let $J^f(\cdot;\ \nu)$ and $J^s(\cdot;\ \nu)$ be defined for $\nu < \nu_o$ and $\nu > \nu_o$, respectively, such that $J^f(\cdot;\ \nu) \equiv J(\cdot;\ \nu)$ for $\nu < \nu_o$, and $J^s(\cdot;\ \nu) \equiv J(\cdot;\ \nu)$ for $\nu > \nu_o$. For each ν, $J^f(\cdot;\ \nu)$ and $J^s(\cdot;\ \nu)$ are in fact creep curves for unit step stress. Let $\dot{J}^f(\cdot;\ \nu)$ and $\dot{J}^s(\cdot;\ \nu)$ be the corresponding functions of time that define the creep rate per unit stress. We assume that $J^f(\cdot;\ \nu)$, $\dot{J}^f(\cdot;\ \nu)$, $J^s(\cdot;\ \nu)$, and $\dot{J}^s(\cdot;\ \nu)$, as functions of ν, are continuously differentiable. This assumption of smoothness seems to be consistent with the experimental findings of Goughnour and Andersland (1968) and Hooke et al. (1972). Under this assumption, for instance, the secondary creep rate $\dot{J}^f(\infty;\ \nu)$ is a continuously differentiable function of ν. Let $J^f_\nu(\cdot;\ \nu)$ and $J^s_\nu(\cdot;\ \nu)$ denote the derivatives of $J^f(\cdot;\ \nu)$ and $J^s(\cdot;\ \nu)$ with respect to ν, respectively. By the assumption of smoothness they are continuous as functions of ν. (The words "continuous" and "continuously differentiable" above can be given a precise meaning when the topology on the space E is made explicit.)

Now let us consider what happens to $J^f(\cdot;\ \nu)$ and $J^s(\cdot;\ \nu)$ when $\nu \to \nu_o^-$ and $\nu \to \nu_o^+$, respectively. From what I can gather from the reported experimental findings, the best guess is $J^f(\cdot;\ \nu_o^-) = J^s(\cdot;\ \nu_o^+)$. Should it be true, then the creep curve $\varepsilon(\cdot;\ \nu_1,\ \sigma) = J^f(\cdot;\ \nu_1)\sigma$ for $\nu_1 < \nu_o$ will be practically indistinguishable from the creep curve $\varepsilon(\cdot;\ \nu_2,\ \sigma) = J^s(\cdot;\ \nu_2)\sigma$ for $\nu_2 > \nu_o$, when $|\nu_2 - \nu_1|$ is sufficiently small, and thence creep tests are not suitable for the determination of ν_o.

Even if $J^f(\cdot;\ \nu_o^-) = J^s(\cdot;\ \nu_o^+)$, however, $J^f_\nu(\cdot;\ \nu_o^-)$ need not be equal to $J^s_\nu(\cdot;\ \nu_o^+)$. In other words, even if two functions take the same value at a given point, their derivatives need not take the same value there. Indeed the best uninformed guess is that the derivatives take different values, i.e., $J^f_\nu(\cdot;\ \nu_o^-) \neq J^s_\nu(\cdot;\ \nu_o^+)$. (An argument by "genericity" can be given here. Such an argument was presented by Man (1982) when he discussed a similar problem.) In what follows we shall assume that $J^f(\cdot;\ \nu_o^-) = J^s(\cdot;\ \nu_o^+)$, $J^f_\nu(\cdot;\ \nu_o^-) \neq J^s_\nu(\cdot;\ \nu_o^+)$, and in particular $J^f_\nu(0;\ \nu_o^-) \neq J^s_\nu(0;\ \nu_o^+)$, unless stated otherwise.

For a material which has its constitutive relation defined by Equation (2) for a given ν and has density $\rho(\nu)$, it is well known (Coleman et al. 1965, Coleman and Gurtin 1965) that the value of $(\rho(\nu)J(0;\ \nu))^{-\frac{1}{2}}$ is equal to (i) the intrinsic velocity of acceleration waves, (ii) the intrinsic velocity of shock waves, and (iii) the "ultrasonic speed" of infinitesimal sinusoidal progressive waves propagating in a homogeneous sample of the material. Let $U \equiv (\rho(\nu)J(0;\ \nu))^{-\frac{1}{2}}$. For our sand-ice mixtures,

$$U = \begin{cases} (\rho(\nu)J^f(0;\ \nu))^{-\frac{1}{2}}, & \text{if } \nu < \nu_o, \\ (\rho(\nu)J^s(0;\ \nu))^{-\frac{1}{2}}, & \text{if } \nu > \nu_o. \end{cases} \tag{3}$$

Since ρ is a smooth function of ν, U is a continuously differentiable function of ν both for $\nu < \nu_o$ and for $\nu > \nu_o$ by the assumption of smoothness on J^f and J^s. Moreover, $U(\nu_o^-) = U(\nu_o^+)$ because $J^f(0;\ \nu_o^-) = J^s(0;\ \nu_o^+)$. But

$$\frac{dU}{d\nu}(\nu_o^-) \neq \frac{dU}{d\nu}(\nu_o^+), \tag{4}$$

because $J^f_\nu(0;\ \nu_o^-) \neq J^s_\nu(0;\ \nu_o^+)$ and $dJ^f(0;\ \nu)/d\nu = J^f_\nu(0;\ \nu)$, $dJ^s(0;\ \nu)/d\nu = J^s_\nu(0;\ \nu)$. (Proof that the two preceding equalities necessarily hold can be given when the topology on the space E is stated explicitly.)

If we determine $U(\nu)$ by appropriate wave propagation experiments for the sand-ice material at various sand concentrations, the graph we obtain by plotting $U(\nu)$ against ν should be smooth for $\nu \neq \nu_o$ but has a discontinuous change in slope at $\nu = \nu_o$. Thus the critical volume fraction of sand ν_o in princip could be determined by wave propagation experiments, provided that $J^f_\nu(0;\ \nu_o^-) \neq J^s_\nu(0;\ \nu_o^+)$, which is very likely to be true. Theorems on the attenuation of waves (Coleman and Gurtin 1965) provide other possible ways of determining ν_o by wave propagation methods.

It should be clear that the discussion above is based in part on the assumed lack of smoothness of the solid-fluid transition.

DETERMINATION OF ULTIMATE LONG-TERM STRENGTH

Although the preceding example concerns only linearly viscoelastic materials, small strains and low stresses, wave propagation methods are by no means so restricted; indeed the foregoing argument can be generalized so that it also pertains to non-linear materials under finite deformations and arbitrary levels of stress.

Henceforth we shall restrict our discussion to one given temperature θ and for simplicity shall suppress the symbol θ from all expressions. For instance, the ultimate long-term strength $\sigma_\infty(\theta)$ at the temperature θ will be denoted by σ_∞.

We assume that under conditions of non-failure the frozen soil in question is described by the constitutive relation

$$\sigma(t) = G(F_r^{(t)}; F(t)), \tag{5}$$

i.e., the stress σ at a body point at time t is determined jointly by the past history $F_r^{(t)}$ of the deformation gradient up to time t and the present deformation gradient $F(t)$ at that body point; here the past history $F_r^{(t)}$ is a function defined on the interval $(0, \infty)$, and $F_r^{(t)}(s) = F(t - s)$ for $0 < s < \infty$, where the positive quantity s is defined through the present time t and the past time t' by $s \equiv t - t'$. The pair $(F_r^{(t)}, F(t))$ together defines the history of the deformation gradient up to time t. It should be emphasized that in Equation (5) the functional G is generally nonlinear, and its domain of definition would include histories for which the corresponding strain $\varepsilon \equiv 1 - F$ need not be small, either for the present time t or for any time t' in the past.

Consider a constant-stress creep test in which a compressive stress σ is applied to a sample of the frozen soil at $t = 0$. Since the sample has always been in the "original" configuration before the load is applied, $F_r^{(0)}(s) = 1$ for all s in $(0, \infty)$, and we write $F_r^{(0)} = 1_r^{(0)}$. If $F_0 \equiv F(0)$ describes the instantaneous response to the loading, then

$$\sigma = G(1_r^{(0)}; F_0) = \sigma(F_0). \tag{6}$$

If Equation (6) could be inverted, we obtain $F_0 = F_0(\sigma)$. Since $\varepsilon \equiv 1 - F$, we have $\varepsilon(0; \sigma) = 1 - F_0(\sigma)$.

Suppose for the frozen soil in question the transition from solid to fluid behavior is determined by the level of stress σ in the sample. Suppose also, without loss of generality, that the compressive stress σ corresponds to a fluid phase point if $\sigma > \sigma_\infty$ and to a solid phase point if $\sigma < \sigma_\infty$. It is convenient to introduce two new functions ε^s and ε^f defined through the expressions $\varepsilon^s(\cdot; \sigma) \equiv \varepsilon(\cdot; \sigma)$ for $\sigma < \sigma_\infty$, and $\varepsilon^f(\cdot; \sigma) \equiv \varepsilon(\cdot; \sigma)$ for $\sigma > \sigma_\infty$. We assume that ε^s and ε^f are continuously differentiable as functions of σ. Furthermore, empirically we have $\varepsilon^s(0; \sigma_\infty^-) = \varepsilon^f(0; \sigma_\infty^+)$. But, the preceding equality notwithstanding, by the same argument as in the last section, in general it is most likely that

$$\frac{d\varepsilon^s}{d\sigma}(0; \sigma_\infty^-) \neq \frac{d\varepsilon^f}{d\sigma}(0; \sigma_\infty^+). \tag{7}$$

As defined by Coleman et al. (1965), the "instantaneous tangent modulus" E corresponding to the history of deformation $(F_r^{(t)}, F(t))$ determines the instantaneous response to a small strain impulse superimposed at the present time t on the given history of (possibly finite) deformation. Corresponding to history of the form $(1_r^{(0)}, F_0(\sigma))$, E is a function of σ only and is given by

$$\begin{aligned} E(\sigma) &\equiv E(1_r^{(0)}, F_0(\sigma)) \\ &= -\lim_{\Lambda \to 0} [G(1_r^{(0)}; F_0 + \Lambda) - G(1_r^{(0)}; F_0)]/\Lambda \\ &= -(d\sigma/dF_0) = [d\varepsilon(0; \sigma)/d\sigma]^{-1}; \end{aligned} \tag{8}$$

here Λ stands for the strain impulse superimposed at time $t = 0$ on the history $(1_r^{(0)}, F_0)$, for which the strain ε_0 corresponding to the deformation gradient F_0 need not be small. If we can determine $E(\sigma)$ for various values of σ, the graph we obtain by plotting $E(\sigma)$ against σ is continuous for $\sigma < \sigma_\infty$ and $\sigma > \sigma_\infty$. By Equation (7), however, the same graph will suffer a discontinuous jump at $\sigma = \sigma_\infty$. Thus the ultimate long-term strength σ_∞ could be determined from the plot of $E(\sigma)$ versus σ.

For the class of materials described by Equation (5), it is well known that those with fading memory behave as perfectly elastic materials in sufficiently fast processes. While frozen soils in general do not behave as materials with fading memory, experience with seismic studies does seem to corroborate the contention that they also behave as elastic materials in sufficiently fast processes. Granted this, if our sample, which has been in the original configuration until $t = 0$, is subjected to a stress σ that produces a homogeneous strain $\varepsilon_0 = 1 - F_0(\sigma)$ at time $t = 0$, and we let an acceleration wave of arbitrary amplitude propagate through the sample immediately after it has developed the homogeneous strain (which need not be small), the average velocity of propagation of the wave through the sample should give a good approximation to the value $(E(\sigma)/\rho_0)^{1/2}$, from which $E(\sigma)$ could be calculated after we substitute in the density of the original configuration ρ_0. We use the word "approximation" in the preceding assertion because the wave takes, however short, a finite time to propagate through the sample, which is creeping during this time under the stress σ. Nevertheless we expect this approximation to be good enough for the evaluation of $E(\sigma)$. Instead of acceleration waves, we can also let infinitesimal sinusoidal progressive waves of higher and higher frequency propagate through copies of our sample after they have developed the initial homogeneous strain under the constant stress σ. Infinitesimal progressive waves of frequency ω has a propagating speed $v(\omega)$, and the "ultrasonic speed" (i.e., the limit of $v(\omega)$ as $\omega \to \infty$) is equal to $(E(\sigma)/\rho_0)^{\frac{1}{2}}$. This gives an alternate way to determine $E(\sigma)$ as a function of σ, but this second method, unlike the first, requires extrapolation.

CLOSING REMARKS

In the preceding section an attempt is made to convince the reader that we should try using wave propagation methods to determine ultimate long-term strength of frozen soil. The reader may not be convinced by the argument above, especially if he believes that for frozen soils the transition from solid to fluid behavior occurs only after a certain amount of strain has developed, i.e., if he believes that the solid-fluid transition of frozen soil is strain-controlled rather than stress-controlled. Indeed, such a belief can easily find indirect support from the literature. (For example, the role of strain is underlined by Ting (1983) when he attributes the correlation of minimum creep rate and the time to reach that rate to "the existence of an approximately constant strain at the minimum strain rate".) But this belief, if true, implies that ultimate long-term strength is not a constitutive quantity for frozen soils and it has no significance independent of the specific process in question. On the other hand the strain which demarcates between solid and fluid behavior would be a constitutive quantity and it would merit our attention and study.

In this regard the argument presented above remains valid if we consider phase diagram in the ε-θ plane instead of the σ-θ plane and modify our argument accordingly. To determine the transition strain at temperature θ, we subject similar samples at the given temperature to various constant homogeneous strains, let acceleration waves and infinitesimal sinusoidal progressive waves propagate through the samples, plot the intrinsic velocity of the acceleration waves or the ultrasonic speed of the infinitesimal progressive waves against the strain and look for discontinuity in the graph of the plot. By performing wave propagation experiments at various times d after the constant homogeneous strain in question has been imposed, we can also investigate whether the duration d has any appreciable effect on the transition strain. If the phase transition is determined purely by the amount of strain developed, the duration d should have no effect on the transition strain. It should also be pointed out that even if the solid-fluid transition for each given temperature is purely strain-controlled, we shall still obtain a transition stress when we perform the experiments suggested in the preceding section; the transition stress thus obtained, however, is not related to the ultimate long-term strength.

No physical conjecture could win the respect of the sceptical empiricist before it is corroborated experimentally. Being a theorist, I lack the means to test what I believe. Here I do not claim that ultimate long-term strength could indeed be determined by wave propagation methods, for that concept itself may not even be a constitutive quantity and thence lacks physical significance. But I do strongly believe that wave propagation methods are useful in general in the study of the solid-fluid transition of frozen soils, and in particular for ascertaining whether ultimate long-term strength is really a constitutive quantity. The present paper would have served its purpose if it could draw the attention of some experimentists to the possible fruitfulness of using wave propagation methods in the study of ultimate long-term strength.

REFERENCES

Casagrande, A., and Wilson S.D., 1951, Effect of rate of loading on the strength of clays and shales at constant water content: Geotechnique, v. II, p. 251-263.

Coleman, B.D., and Gurtin, M.E., 1965, Waves in materials with memory, II. On the growth and decay of one-dimensional acceleration waves: Archive for Rational Mechanics and Analysis, v.19, p. 239-265.

Coleman, B.D., Gurtin, M.E., and Herrera R., I., 1965, Waves in materials with memory, I. The velocity of one-dimensional shock and acceleration waves: Archive for Rational Mechanics and Analysis, v. 19, p. 1-19.

Goughnour, R.R., and Andersland, O.B., 1968, Mechanical properties of sand-ice system: Journal of the Soil Mechanics and Foundations Division, Proceedings of the American Society of Civil Engineers, v. 94, p. 925-950.

Hooke, R.L., Dahlin, B.B., and Kauper, M.T., 1972, Creep of ice containing dispersed fine sand: Journal of Glaciology, v. 11, p. 327-336.

Ladanyi, B., 1981, State of the art: determination of creep settlement of shallow foundations in permafrost, paper presented at the Annual Convention of the American Society of Civil Engineers, October 26-30, 1981, St. Louis, Missouri, USA.

Man, C.-S., Solid-fluid transitions in uniaxial creep tests of nonlinear viscoelastic materials, paper presented at The 19th Annual Meeting of the Society of Engineering Science, Inc., October 27-29, 1982, Rolla, Missouri, USA.

Pipkin, A.C., 1972, Lectures on viscoelasticity theory: New York etc., Springer-Verlag.

Stevens, H.W., 1975, The response of frozen soils to vibratory loads, U.S. Army, CRREL, Tech. Rpt. 265: Hanover, New Hampshire, USA.

Ting, J.M., 1983, On the nature of the minimum creep rate-time correlation for soil, ice, and frozen soil: Canadian Geotechnical Journal, v. 20, p. 176-182.

Vyalov, S.S., 1979, Excerpts from rheological fundamentals of soil mechanics, U.S. Army, CRREL, Draft Transl. 716: Hanover, New Hampshire, USA.

PILE DRIVING AND LOAD TESTS IN PERMAFROST FOR THE KUPARUK PIPELINE SYSTEM

Victor Manikian

ARCO Alaska, Inc., Anchorage, Alaska 99510 USA

Drilled and slurried piles have been used at the North Slope of Alaska. These tests were conducted for the determination of a faster pile placement method in permafrost. Ultimately this refined, thermally-modified pile driving method was selected by ARCO Alaska, Inc., in 1980 in its construction of the aboveground Kuparuk Pipeline.

INTRODUCTION

Placement of structural foundation support piles on Alaska's North Slope generally has been performed using drilled holes and slurry backfill techniques. The process of drilling, placing, aligning and slurrying piles is time consuming and requires considerable manpower and equipment. A method of accomplishing the same results more quickly, using fewer men and less equipment, and the early success of specially-modified driven H-pile structural shapes in permafrost (Crory 1973, Crory 1975, Nottingham 1981), prompted ARCO Alaska, Inc., to initiate a series of driving and load tests starting in February 1980.

These field tests were designed to verify the practicality of driving piles in cold permafrost on a production basis using thermally modified predrilled pilot holes. This technique employs a small drilled pilot hole, generally 15.2 to 21.6 cm (6 to 8.5 in.) in diameter depending on pile diameter, filled with water, heated to a minimum of 66°C (150°F). The pilot hole is filled with the hot water 30 min before driving in frozen granular soils and 60 min before driving in frozen fine-grained soils, in order to raise the surrounding soil temperature and facilitate the pile placement. Thermal modification and pile driving procedures were refined to suit the arctic pile driving. Ultimate adfreeze strengths of these piles were determined by load tests after isolating and eliminating the effects of end bearing and the active layer. In Prudhoe Bay, heaving forces from silty soils in the active layer have controlled the design of lightly-loaded piles (Crory 1965).

Driving tests were conducted in three phases, with a total of 45 piles being driven. In Phase 1, 27 pipe piles were driven to a depth of 4.6 m (15 ft) in frozen silty sand using a Delmag D15 diesel hammer and predrilled, thermally-modified pilot holes. In Phase 2, eight reinforced H-piles were driven to depths ranging from 6.7 to 7.6 m (22 to 25 ft) in frozen silts and sands, using thermally-modified pilot hole procedures similar to those used in Phase 1. Phase 2 also included the installation of two drilled and slurry pipe piles. In Phase 3, ten pipe piles with cast steel tips were driven into frozen river gravelly sands, using the Phase 1 procedures.

Short-term pile load tests were performed on seven piles and long-term (creep) tests were performed on five piles. For comparative purposes the two slurried piles were subjected to short- and long-term (creep) load tests under conditions similar to those used for driven piles. All piles were loaded until an apparent failure occurred. In the long-term load tests, the piles were loaded until the load-deformation plots indicated that irreversible progressive failure, greater than 0.5-in. (1.3 cm) downward movement, had occurred.

The following paragraphs describe each of the three testing phases and results obtained from them.

PHASE 1 - PILE DRIVING AND LOAD TESTS FOR PIPE PILES IN FROZEN ICE-RICH SILTY SAND SOILS

Test Equipment and Installation

The test site near the ARCO airstrip represented typical Prudhoe Bay soil conditions with ice-rich frozen silts overlying frozen sands and gravels. The selected site was the center of a 30 m (100 ft) diameter polygon surrounded by ice wedges.

The test piles were set on a 1.5 m (5 ft) centers grid to allow use of a 3 m (10 ft) long test beam at any location. Vertical load testing equipment was sized to develop a minimum of 3.52 kg/cm^2 (50 psi) adfreeze working stress and 7.0 kg/cm^2 (100 psi) adfreeze ultimate stress for a 45.7 cm (18 in.) diameter pile over the embedded test length. For the vertical tests, 273,000 kg (600 kips) capacity jacks were used.

Previous experience with driving in cold weather showed that, if kept warm, single-acting diesel hammers would operate satisfactorily. A decision was made to use a D15 Delmag, single-acting diesel hammer, with a maximum rated energy of 3,750 kg-m (27,100 ft-lb), which turned out to be too large for the smaller 27.3 cm (10.75 in.)

diameter piles. The Delmag D15 hammer was mounted on a Marion 36,000 kg (40-ton) crane with semi-fixed leads. In addition, the test driving experiment required a Nodwell track-mounted Mobile B-61 drill.

Pile Driving - General Approach

The general approach was to establish a method using hot-water-filled, drilled pilot holes that would provide the quickest pile driving without affecting pile integrity. Various pile wall thicknesses and tip reinforcing were used, along with variable thaw times for water in the pilot holes. Pile placement proceeded in the following general manner:

1. Drill pilot hole.
2. Fill pilot hole with hot water.
3. Align driving equipment and pile.
4. Allow for sufficient time for ground around pilot hole to thaw or warm.
5. Drive pile.

Pilot holes were drilled using a Mobile B-61 drill. The pilot holes were 15.2 cm (6 in.) diameter for the 27.3 cm (10.75 in.) and 32.4 cm (12.75 in.) diameter piles, however, 21.6 cm (8.5 in.) diameter holes were used for the 45.7 cm (18 in.) diameter piles. Each pilot hole was visually logged by a field geologist, and site soil samples were recovered for laboratory analysis. Following pilot hole drilling, and for varying periods before driving, the holes were filled with hot water of different temperatures in order to modify soil thermal conditions and determine their effect upon the subsequent driving effort.

Driving Operations and Results

Driving proved extremely easy, when using the hot-water-filled pilot holes, if the water remained in place for sufficient time. In this project, rated hammer energy was rarely developed, except occasionally during the final stages of pile driving. With continuous, uninterruped driving and proper pilot hole thermal modification, an average driving time of about 5 min was required for the 4.6 m (15 ft) penetration. The diesel impact hammer provided a very important gauge in the pile driving experiment, between success and failure, of thermally-modified piles against piles driven in a drilled pilot hole without hot water. Thermally-modified pile No. 4 recorded 22 blows per 30.5 cm (22 blows/ft) near its tip penetration while pile No. 5 with its dry-bored hole recorded 263 blows per 30.5 cm (263 blows/ft) near its tip penetration.

After the pipe test piles were driven, their interior was reamed and the soil inside and below the test pile tips was removed for inspection purposes. Removal of inside soil 15-30 cm (6-12 in.) below the pile tip allowed pile load testing to measure adfreeze frictional strengths directly without end-bearing effects. Reaming also provided a convenient way to install thermistors to measure pile wall temperatures. The reaming operation was specified for load test research only and is not to be construed as part of normal construction operations.

After the pile tests were completed, two piles, Nos. 7 and 16, were pulled for further examination of their tip failures. Their progressive tip failures were caused by uneven lateral tip pressures. This type of failure usually occurs during light driving and would not be produced by the axial buckling condition commonly caused by overdriving.

From a structural point of view, pipe piles can be successfully driven into permafrost if certain precautions are taken. Driven into water-filled pilot holes they try to follow the path of least resistance, and if not properly centered in the thermally-modified zone or if they encounter a discontinuity, they will experience variable tip pressures. These in turn are resisted primarily by pile bending and lateral soil pressure. If the tip is not strong enough and thus owing to ring bending cannot resist uneven lateral pressures, a type of progressive tip failure will result. During these tests, four pipe piles (with thinner walls) were driven with tips, however, they were not the hard cast steel types nor were they as thick. Piles with these tips, fabricated from mild steel plates, did prove more successful when tested against similar piles with open ends. The design of the pile tips were flush on the outside to ensure continuous contact to the pile surface and avoid air gaps.

Of the 27 piles driven in Phase 1, using the parameters indicated above, 10 piles were driven without deformation. From the results of various methods tried, Table 1 has been prepared to show the relationship of parameters affecting driving piles in permafrost. Based on these test results, installation of driven pipe piles in frozen silty sands, require pilot holes and, prior to driving, a thermal modification time between 40 and 60 min to prevent tip damage.

In general, for these tests, planned position tolerances of less than 5 cm (2 in.) and plumbness of less than 2% from the vertical were achieved without difficulty. Minor leveling adjustment of piles was accomplished during the pile driving operation by gently pushing the pile into alignment with heavy equipment. Generally, with preparation and care, driving tolerances can be made as accurate as may be required for design.

Short-Term Vertical Load Tests

Purpose and Procedures: Since little adfreeze data exist on piles driven in permafrost and there are virtually no data on piles driven in Prudhoe Bay permafrost conditions, short-term tests were planned to establish starting points for better definition of long-term load and creep rate tests.

Two short-term tests, Pile Nos. 2 and 17 were conducted. Because of the length of time required for pile load distribution, short-term load tests do not produce usable data other than for short-term loading. Additional tests, called "reheal tests", were run on the two test piles after their initial significant deformation (failure), to determine pile load resistance when allowed a short recovery time of approximately 1 day.

After the driving operations, piles were reamed out below the tip to relieve pile tip support capacity in order to determine adfreeze frictional values after the isolation and elimination of the end bearing effects. Test procedures for Phase 1 included a downward vertical pile load application. Excavation of the top meter (3 ft) of soil adjacent to each test pile assured an effective test length of 3.7 m (12 ft) below the frost heave zone.

Initial readings were taken and loads were applied in increments of 17,850 kg (39,270 lbs). A minimum time interval of 15 min was provided between load applications to allow for pile settlement stability. Periodic deformation readings were taken, and ground temperatures were recorded.

Results: In general, each pile began to show signs of rapid failure after about ± 0.38 cm (0.15 in.) of settlement. Tests were carried out to settlement of 0.64 cm (0.25 in.) or more to ensure that maximum resistance conditions had been encountered. Results of the short-term load tests are summarized in Table 2.

Pile No. 17 was driven primarily into massive ice and sustained a short-term adfreeze stress of 5.1 kg/cm^2 (72 psi) before failure. This was considerably lower than the 7.0+ kg/cm^2 (100+ psi) values of test pile Nos. 2, 19 and 23 in Table 2, though still significant. Normally, allowable design adfreeze stress considers creep effects for sustained loading conditions and this effect becomes critical when ice is used as structural support.

Reheal tests were run on Pile Nos. 2 and 17 after their deformation (failure) and the short recovery period of 1 day. As indicated in Table 2, the fast recovery of the adfreeze stress of these piles provides interesting data that could be used for short-term pile design use under emergency conditions.

Long-Term Vertical Load Tests

Purpose and Procedure: The purpose of these long-term tests was to add information to the developing data base on design criteria. As experience is gained from long-term sustained pile loading in permafrost, it becomes apparent that, in certain designs, creep-related settlements control rather than adfreeze strength (Nixon, 1976). The design of pipeline supports, with large short-term live loads, has adfreeze strength criteria which will limit the creep to a design limit.

After completion of the short-term tests, the two test beams and equipment were moved to Pile Nos. 19 and 23. Tests were performed by applying sustained loads of 17,850 kg (39,270 lbs) in 2-week load increments. Jack pressure readings, dial gauge readings, survey levels, and temperature readings were taken every day at night, and before and after load changes.

Results: Figure 1 presents the results of the load test on pile No. 19 during the long-term test. Of particular interest in this diagram is the adfreeze and deformation curve slope, which are indicated for curves within usable design ranges only. It should be noted that, over these usable creep ranges, the average ground temperature remained fairly stable, with only minor deviations during the testing period.

At the conclusion of long-term testing, which concluded after 103 days with a load of 125,000 kg (275 kips) and adfreeze stress of 3.4 kg/cm^2 (48 psi), pile No. 19 was failed in a manner similar to the previous short-term tests (i.e., every 15 min, jack pressure was increased by loads of 17,850 kg until pile failure). Results of long-term load tests on pile Nos. 19 and 23 are summarized in Table 3.

The short-term tests which were conducted at the conclusion of the long-term testing of pile Nos. 19 and 23 indicated very high short-term adfreeze strengths in permafrost. These values were 8.2 kg/cm^2 (116 psi) for pile No. 19 and 9.1 kg/cm^2 (129 psi) for pile No. 23.

PHASE 2 - PILE DRIVING AND LOAD TESTS FOR SLURRY PIPE AND DRIVEN H-PILES IN FROZEN ICE-RICH SILTY SAND SOILS.

Test Plan

Subsequent to the Phase 1 tests in early 1980, Phase 2 tests were initiated to accomplish the following:

1. Establish feasibility and requirements for driving H-piles through a gravel workpad into underlying frozen fine-grained soils.
2. Determine comparative adfreeze strengths, both short- and long-term, for slurry pipe piles and driven H-piles.
3. Determine the feasibility of driving longer pile embedment lengths. Phase 2 piles were driven at least 6.7 m (22 ft) below grade compared to 4.6 m (15 ft) for Phase 1 piles.

Little literature is available regarding the strength of driven piles or the comparative strength of driven versus slurry piles in cold permafrost. For this reason, in Phase 2 it was planned to drive two test H-piles 6.7 m (22 ft) below grade, and place two slurry test pipe piles 6.7 (22 ft) below grade. Six H-piles were also driven 7.6 m (25 ft) below grade for anchor piles. H-piles were structurally designed to resist variable tip loading and localized buckling under driving conditions.

Test Equipment and Implementation

Contrary to the previous pile testing in which the piles were loaded vertically downward, tests in this series were planned to pull the piles upward, in tension. In this manner, a realistic test of adfreeze strength without pile end-bearing effects could be realized for driven H-piles and slurry pipe piles. For every 1,000 kg of load on the jack, 500 kg of load was transmitted to the test pile.

Pile load jacks, load beams, and equipment were sized for ultimate loads of 7.0 kg/cm^2 (100 psi) adfreeze, applied to a 45.7 cm (18 in.) pipe pile frozen 4.9 m (16 ft) into the ground. All test piles were cased with a 61 cm (24 in.) diameter pipe down 1.8 m (6 ft), with the annulus filled with an oil-sand mixture and the pile wrapped with plastic in this upper zone to prevent adfreeze bond stresses in the active layer.

The driven piles were placed in accordance with the following procedures:

1. Drill a 15.2 cm (6 in.) pilot hole to the tip elevation.
2. Position pile leads initially with Delmag D15 pile hammer.
3. Fill pilot hole with water heated to a minimum of 66°C (150°F).
4. After waiting 40-60 min for thermal modification of hole in silty sands, drive pile.
5. Keep installation records of blow counts, drive time, water temperature, etc. (See for example, the summary shown in Table 1.)

The slurry test piles were placed in 61 cm (24 in.) diameter holes using a premixed sand slurry with a consistency similar to 15.2 cm (6 in.) slump concrete. Slurry piles were accurately positioned and firmly held in place until freezeback was assured.

Placement and driving proceeded as expected, except at certain locations where pilot holes were drilled out of plumb. Since driven piles follow the pilot hole closely, this resulted in piles being driven out of plumb. Immediately after driving, piles that were driven out of plumb in excess of test requirements were adjusted into close tolerances by pushing with a Caterpillar 966 loader. This method did not seem to bend or otherwise damage piles treated in this manner, if performed before freezeback occurred.

In summary, the eight H-piles were driven using production driving techniques in an interrupted manner. Actual setup and driving time for individual piles ranges from 15 to 30 min. All piles (except where pilot holes were drilled out of plumb) were driven without difficulty.

Short-Term Vertical Load Tests

Test pile No. 2 was a driven HP 10 x 57 with angle reinforcement. This pile was rapidly loaded in increments of 35,700 kg (78,540 lbs) every 15 minutes. The results are summarized in Table 2.

Test pile No. 9 was a slurry 45.7 cm (18 in.) diameter pipe pile. Testing proceeded with this pile in a manner similar to that used for pile No. 2 up to the adfreeze value of 6.3 kg/cm^2 (89.3 psi). At this stress level the pile began to show signs of significant creep, a 15.2 cm (6 in.) diameter pin, which was part of the load beam connecting system, broke at a load of 441,000 kg (970 kips) and the test was terminated. The projected load to failure was estimated at about 455,000 kg (1,000 kips) or near the ultimate equipment capacity. The results are summarized in Table 2.

Long-Term Vertical Load Tests

Test pile No. 4 was a driven HP 10 x 57 with angle reinforcement. The pile failed at 2.1 kg/cm^2 (30.5 psi) during long-term adfreeze tests, conducted in accordance with the previously outlined procedures. The short-term loading completed at the end of long-term tests indicated maximum adfreeze values of 7.9 kg/cm^2 (111.8 psi) using the encompassed perimeter and 5.0 kg/cm^2 (70.7 psi) for the contact perimeter area. The results are summarized in Table 3.

Test pile No. 7 was a 45.7 cm (18 in.) diameter slurry pipe pile. The pile failed at an adfreeze bond stress value of 2.5 kg/cm^2 (36.2 psi) during long-term tests. Short-term loading completed at the end of long-term tests indicated a maximum adfreeze value of 6.6 kg/cm^2 (94.1 psi). The results are summarized in Table 3.

PHASE 3 - PILE DRIVING AND LOAD TESTS FOR PIPE PILES IN FROZEN GRAVEL

Test Plan

Ten 27.3 cm (10.75 in.) diameter by 1.27 cm (0.5 in.) wall pipe piles were driven using APF 0-14001 tips in frozen gravel and 15.2 cm (6 in.) diameter pilot holes, drilled 0.61 m (2 ft) deeper than the nominal tip embedment.

Concerns had been expressed regarding driving piles in frozen gravelly soils. As a result, these tests were planned primarily to drive pipe piles with cast steel tips to depths of 4.6 m (15 ft) in frozen gravelly materials. The test site was adjacent to the Sagavanirktok River, where typical riverbed soils included frozen sandy gravels. Following driving, load tests were performed to establish data concerning adfreeze strengths in these soils. Tests were initially conducted during the summer, when ground temperatures are generally warmest and adfreeze values are lowest.

Adjustments were made to the hot water thaw time after initial driven tests indicated high blow counts. It was determined that a 20-min minimum waiting period produced best results before driving in the gravelly soils. Pilot holes drilled in thawed active layer gravels had top-caving problems that interfered with pilot hole thawing. For this reason, pilot holes drilled 0.61 m (2 ft) deeper than the pile tip embedment helped contain the surface caved-in gravels. A summary of driven piles for Phase 3 is shown in Table 1. The vertical load tests were performed by extracting the piles upward in tension, similar to the Phase 2 load tests.

Short-Term Vertical Load Tests

Short-term testing on pile No. 4 was begun in August 1980 but upon applying 36,000 kg (78.5 kips) tension on the pile, the pile suddenly failed and the test was terminated. Results of ground temperature monitoring indicated that this pile was only embedded in 2.1 m (7 ft) of frozen soil. The calculated adfreeze stress over this region of frozen contact indicates a value of 1.9 kg/cm^2 (27.7 psi). The calculated adfreeze stress over the 4.6 m (15 ft) of thawed and frozen soil contact equals 0.9 kg/cm^2 (12.9 psi), as shown in Table 2.

Testing of pile No. 7 began in November 1980 after the previously thawed active layer had refrozen. Results indicated an increase in adfreeze strength over test pile No. 4. The maximum adfreeze bond stress yielded approximately 3.6 kg/cm^2 (51.7 psi), see Table 2.

Long-Term Vertical Load Tests

Long-term load test procedures were conducted in accordance with the Phase 2 test program. Pile No. 2 failed at approximately 3.4 kg/cm^2 (48.5 psi) over the embedded pile test length. Pile No. 9 failed at an adfreeze bond stress value of approximately 2.3 kg/cm^2 (32.3 psi). The results on pile Nos. 2 and 9 are summarized in Table 3.

For piles driven into frozen gravelly sand, the adfreeze values achieved were not as high as those achieved for the piles in frozen fine-grained soils. One contributing condition may have been warmer ground temperatures at this test pile location, which were due to the effect of the river floodplain and time of year.

CONCLUSIONS AND RECOMMENDATIONS

The results of these field tests to verify the practicality of diesel pile driving in cold permafrost, on a production basis using thermally modified, predrilled pilot holes, are encouraging. Subsequent to these tests the use of vibratory hammer for driving production piles has also proven successful. The technique of employing a small drilled pilot hole filled with water heated to a minimum of 66°C (150°F) prior to placement is effective. Thaw time before driving was approximately 30 min for frozen granular soils and 60 min for frozen fine-grained soils in order to raise the surrounding soil temperature and improve the pile driving. A summary of the pile driving tests is given in Table 1.

The results of the pile load tests conducted during this program provided useful information for development of pile design criteria. The pile load testing equipment, especially designed for arctic conditions, worked well. The development of design criteria should recognize permafrost peculiarities such as loading conditions and creep. For purposes of comparison, Tables 2 and 3 are provided as a summary of short-term and long-term (as compared to short-term) load test data for all load tests.

Generally, differences between soil types were more important than installation techniques in the determination of adfreeze strengths. Driven pile placement methods produced comparable strengths to typical slurried piles in ice-rich silty sandy soils. However, driven piles in frozen gravelly soils indicate lower adfreeze values than ice-rich silty sands due to their location near the rivers and the warmer ground temperature effects.

Some significant factors affecting pile performance are soil temperature, pile diameter, and creep. For design, two conditions should be considered, short-term loading and long-term creep.

Based on results of these tests, it was determined that pipe and H-piles could be driven in permafrost with acceptable accuracy, providing ARCO Alaska, Inc. with a technically sound alternate pile placement method. The thermally-modified pile driving method was ultimately selected in 1980 for the aboveground Kuparuk Pipeline systems design.

ACKNOWLEDGEMENTS

The permission of the owners of the Kuparuk Pipeline System to publish this paper is gratefully acknowledged.

The consultant selected to conduct the test program, under guidelines determined by ARCO engineering, was Mr. D. Nottingham of Peratrovich and Nottingham, Inc. The field construction for the installation of these testing facilities and the surveying services was conducted by VECO, Inc. Recognition is given to the above for their efforts.

REFERENCES

Crory, F. E., 1973, "Installation of driven test piles in permafrost at Bethel Air Force Station, Alaska," U.S. Army Cold Regions Research Engineering Laboratory, Technical Report 139.

Crory, F. E., 1975, "Bridge foundations in permafrost areas," U.S. Army Cold Regions Research Engineering Laboratory, Technical Report 266.

Crory, F. E., and Reed, R. E., 1965, "Measurement of frost heaving forces on piles," U.S. Army Cold Regions Research Engineering Laboratory, Technical Report 145.

Nixon, J. F., and McRoberts, E. C., 1976, "A design approach for pile foundations in permafrost," Canadian Geotechnical Journal, Volume 13, No. 1, pp 40-57.

Nottingham, D., 1981, "Method and H-piles in permafrost," U.S. Patent 4,297,056 October 27.

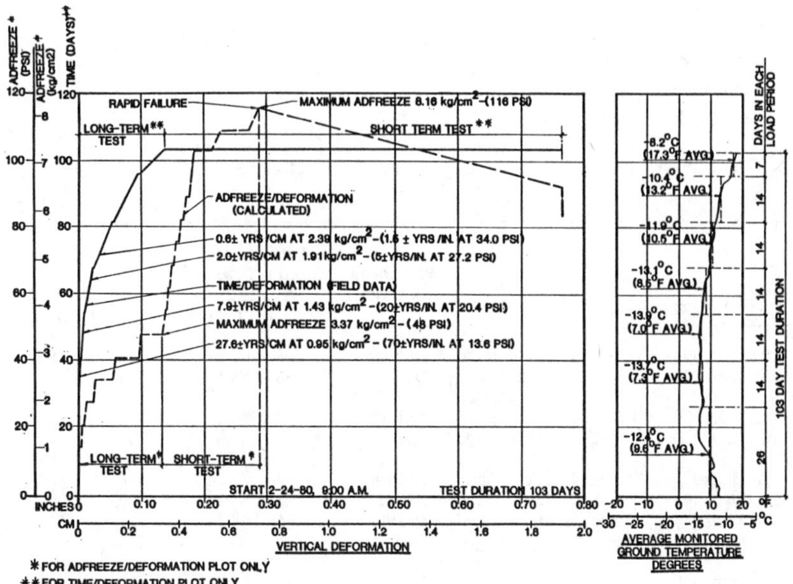

FIGURE 1 Phase 1 long-term load test for pile no. 19, 32.4 cm diameter x 1.27 cm wall.

TABLE 1 Summary of Driven Piles

Pile No.	Size Dia. & Wall (cm)	Size Dia. & Wall (in.)	Tip Size (cm)	Approx. Pilot Hole Fill Time (min)	Approx. Pilot Hole Thaw Time (min)	Driving Time* (min)	Pile Condition After Driving (0 Sound, x Failed)	Initial Pilot Hole Water Temp. (°C)	Initial Pilot Hole Water Temp. (°F)	Thaw Remarks Just Prior to Driving
Phase 1										
1	27.3x1.27	10.75x0.500	None	10	60	15	0	54	130	Reheated to 21°C(70°F)
2	27.3x1.27	10.75x0.500	None	10	18	19	0	54	130	No reheat
3	27.3x1.11	10.75x0.438	None	10	40	13	0	34	94	Cooled to 4°C(39°F)
4	27.3x0.93	10.75x0.365	1.9(0.75")	10	15	16	0	37	98	No reheat
5	27.3x0.93	10.75x0.365	None	0	0	55	x	-	-	Dry hole
6	27.3x0.93	10.75x0.365	None	10	20	15	x	43	110	No reheat
7	27.3x0.93	10.75x0.365	None	10	15	9	x	40	105	No reheat
8	32.4x1.27	12.75x0.500	None	10	42	9	0	32	90	Cooled to 7°C(45°F)
9	32.4x1.03	12.75x0.406	None	10	12	8	x	49	120	No reheat
10	32.4x1.27	12.75x0.500	None	10	18	24	x	14	58	No reheat
11	32.4x0.95	12.75x0.375	None	10	19	7	x	38	100	No reheat
12	32.4x1.27	12.75x0.500	None	10	17	9	x	27	80	No reheat
13	32.4x1.03	12.75x0.406	None	10	19	88**	x	43	110	No reheat
14	32.4x0.95	12.75x0.375	None	10	20	17	x	43	110	No reheat
15	32.4x1.27	12.75x0.500	None	10	20	12	x	44	112	No reheat
16	32.4x1.27	12.75x0.500	None	10	15	7	x	49	120	No reheat
17	45.7x1.27	18.00x0.500	None	20	78	14	0	43	110 Est.	Cooled to 0°C(32°F)
18	45.7x0.95	18.00x0.375	None	20	30	36	x	43	110 Est.	No reheat
19	32.4x1.27	12.75x0.500	2.54(1.0")	0	70	10	0	18	64	Reheated to 21°C(70°F)
20	32.4x1.03	12.75x0.406	None	0	41	9	0	31	88	Reheated to 17°C(62°F)
21	32.4x1.27	12.75x0.500	None	0	58	11	x	23	74	Reheated to 32°C(90°F)
22	32.4x0.95	12.75x0.375	None	0	53	7	x	23	74	No reheat
23	27.3x1.27	10.75x0.500	None	0	67	4	0	18	64	Reheated to 21°C(70°F)
24	27.3x1.27	10.75x0.500	None	0	43	6	x	14	58	Cooled to 3°C(38°F)
25	27.3x1.27	10.75x0.500	None	0	41	5	x	6	43	No reheat
26	-	-	-	-	-	-	-	-	-	-
27	27.3x0.93	10.75x0.365	1.9	0	19	5	x	-	-	No temp. measurement
28	27.3x1.11	10.75x0.438	1.9	0	20	4	0	-	-	No temp. measurement
Phase 2										
1	HP 10x57		APF 75750	10	35	14	0	82	180	No reheat
2	HP 10x57		APF 75750	10	26	8	0	66	150	No reheat
3	HP 10x57		APF 75750	10	61	12	0	71	160	No reheat
4	HP 10x57		APF 75750	10	60	15	0	93	200	No reheat
5	HP 10x57		APF 75750	10	64	12	0	72	162	No reheat
6	HP 10x57		APF 75750	10	81	25	0	93	200	No reheat
7	Slurry Pile		-	-	-	-	-	-	-	-
8	HP 10x57		APF 75750	10	59	19	0	93	200	No reheat
9	Slurry Pile		-	-	-	-	-	-	-	-
10	HP 10x57		APF 75750	10	50	13	0	88	190	No reheat
Phase 3										
1	27.3x1.27	10.75x0.500	1.9	10	-	13	0	-	-	-
2	27.3x1.27	10.75x0.500	1.9	10	20	11	0	71	160	No reheat
3	27.3x1.27	10.75x0.500	1.9	10	15	12	0	66	150	No reheat
4	27.3x1.27	10.75x0.500	1.9	10	29	12	0	80	175	No reheat
5	27.3x1.27	10.75x0.500	1.9	10	60	11	0	85	186	No reheat
6	27.3x1.27	10.75x0.500	1.9	10	52	15	0	66	150	No reheat
7	27.3x1.27	10.75x0.500	1.9	10	20	10	0	66	150	No reheat
8	27.3x1.27	10.75x0.500	1.9	10	20	11	0	66	150	No reheat
9	27.3x1.27	10.75x0.500	1.9	10	20	7	0	66	150	No reheat
10	27.3x1.27	10.75x0.500	1.9	10	20	8	0	66	150	No reheat

* Driving time includes intermittent hammer firing. Piles 23 through 28 drove without significant interruption.
** Values not reliable due to hammer malfunction.

TABLE 2 Summary of Short-Term Vertical Load Test Results: Phases 1, 2, and 3.

Test Phase	Pile No.	Size Dia. & Wall (cm)	(in.)	Soil Type	Placement Method	Monitored Soil Temp. at Failure (°C)	(°F)	Maximum Adfreeze Strength (kg/cm²)	(psi)	One-Day Reheal Test Adfreeze Strength (kg/cm²)	(psi)	Test Duration (min)
1	2	27.3x1.27	10.75x0.500	Ice-Rich Silty Sand	Driven	-8	18	7.4	105	4.0	57	536+121[3]
1	17	45.7x1.27	18.00x0.500	Massive Ice and Ice-Rich Silty Sand	Driven	-6	21	5.1	72	4.8	68	429+162[3]
1	19	32.4x1.27	12.75x0.500	Ice-Rich Silty Sand	Driven	-8	18	8.2	116[1]		-	275
1	23	27.3x1.27	10.75x0.500	Ice-Rich Silty Sand	Driven	-8	18	9.1	129[1]		-	124
2	2	HP 10x57 w/Angles		Ice-Rich Silty Sand	Driven	-6	22	7.0	100[2]		-	150
2	9	45.7x1.27	18.00x0.500	Ice-Rich Silty Sand	Slurry	-6	21	6.3	89		-	197
2	4	HP 10x57 w/Angles		Ice-Rich Silty Sand	Driven	-5	23	7.9	112[1,2]		-	140
2	7	45.7x1.27	18.00x0.500	Ice-Rich Silty Sand	Slurry	-5	23	6.6	94[1]		-	106
3	4*	27.3x1.27	10.75x0.500	Gravelly Sand	Driven	-3	27	2.0	28		-	2
3	7	27.3x1.27	10.75x0.500	Gravelly Sand	Driven	-6	22	3.7	52		-	90

1 Short-term loading at completion of long-term test.
2 Encompassed perimeter area is considered rather than actual contact area, since creep will tend to occur on that basis.
3 Initial test plus reheal test duration.
* Load test attempt failed 2 min after load application. Pile was partially frozen (2.1 m, or 7 ft, from tip) and ground temperatures may have been higher than initially measured due to thermal modification.

Phase 1 load tests were conducted as downward tests.
Phase 2 & 3 load tests were conducted as uplift tests.

TABLE 3 Summary of Long-Term Vertical Load Test Results: Phases 1, 2, and 3.

Test Phase	Pile No.	Size Dia. & Wall (cm)	(in.)	Soil Type	Placement Method	Monitored Soil Temp. at Failure (°C)	(°F)	Maximum Adfreeze Strength (kg/cm²)	(psi)	Test Duration (days)
1	19*	32.4x1.27	12.75x0.500	Ice-Rich Silty Sand	Driven	-8	18	3.4	48[2]	103
1	23	27.3x1.27	10.75x0.500	Ice-Rich Silty Sand	Driven	-8	18	4.0	57[2]	103
2	4	HP 10x57		Ice-Rich Silty Sand	Driven	-5	23	2.1	31[1,2]	49
2	7	45.7x1.27	18.00x0.500	Ice-Rich Silty Sand	Slurry	-5	23	2.5	36[2]	62
3	2	27.3x1.27	10.75x0.500	Gravelly Sand	Driven	-11	12	3.4	49	28
3	9	27.3x1.27	10.75x0.500	Gravelly Sand	Driven	-9	15	2.2	32	20

1 Encompassed perimeter area is considered rather than actual contact area, since creep will tend to occur on that basis.
2 Failed during short-term testing subsequent to long-term tests (see Table 2).
* Failed during long-term testing. Results are documented graphically in Figure 1.

GEOMORPHOLOGICAL MAPPING AND ENVIRONMENTAL PLANNING IN PERMAFROST REGIONS

R. Mäusbacher

Department of Geographie, University of Heidelberg
Heidelberg, Germany

With the help of the geomorphological map of Oobloyah Bay, Ellesmere Island, N.W.T., Canada, the applicability of the legend, developed in a priority project of the German Research Foundation, in permafrost regions is demonstrated. In addition, possibilities for the practical applicability of geomorphological information like this are presented. For the user who is informed about the terminology used in geomorphological maps, different levels of information (e.g., slope angles, subsurface material, etc.) can be printed separately. For those not informed about the terminology, the geomorphological information can be prepared with ecological evaluation on the basis of a suitability test, which allows direct use for nongeomorphological problems.

INTRODUCTION

With the financial aid of the German Research Foundation (DFG), there was a research project established in the Federal Republic of Germany in 1976, creating geomorphological maps at scales of 1:25,000 (GMK 25) and 1:100,000 (GMK 100) (Barsch and Liedtke 1980). Focusing on these two different scales, guidelines for the field work activities and the legends of maps were to be developed, applicable to all regions (lowland, mountains, and high mountains) of the FRG. The field work guidelines for mapping the exemplary sheets are based on the legends developed by Leser and Stäblein (1975) for the GMK 25 and Fränzle et al. (1979) for the GMK 100, employing up-to-date international experience in this field (Leser 1981).

The geomorphological map 1:25,000 Oobloyah Bay, Ellesmere Island, N.W.T., Canada, prepared during the Heidelberg/Ellesmere Island Expedition is also based on the legend developed for the GMK 25. The aim of this mapping program was to test the applicability of the legend in a permafrost region.

Considering the ever increasing human activity in permafrost regions, environmental planning in these areas is becoming more and more necessary (Robitaille 1960, St-Onge 1965, Kreig and Reger 1976, Everett et al. 1978). Therefore, the problem of practical applicability of geomorphological information on environmental planning in these regions will be discussed also.

EXPEDITION AREA AND BASE MAP

The expedition area is located in the northwestern part of Ellesmere Island, near Oobloyah Bay at 81.5°N and 83.5°W (Figure 1). Geologically, it is a synclinal valley of about 15 km fringed by two anticlines: in the north the so-called Krieger Mountains reaching hights of more than 1,000 m, and in the south the Neil Peninsula, climbing up to 600 m above sea level. This association with the geological structure, however, is only reflected in major landforms. The landforms mapped at 1:25,000 scale are the product of glacial, fluvioglacial, cryogenetic, and marine processes (Mäusbacher 1981).

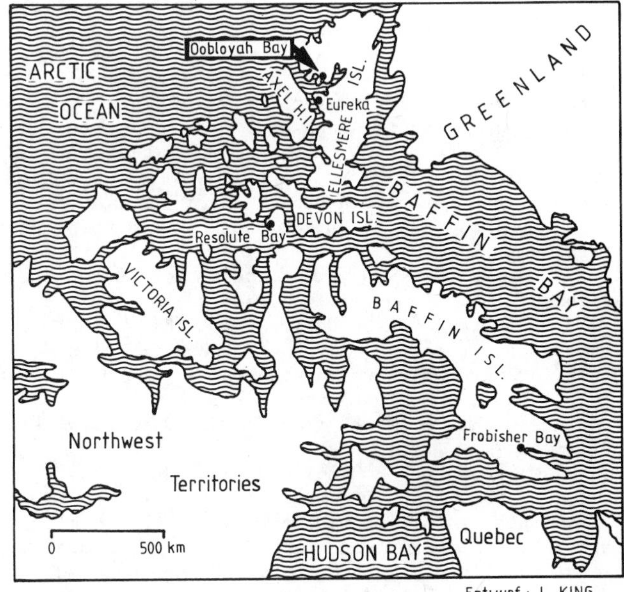

FIGURE 1 Location of study site.

For the field survey and the cartographical presentation the author used orthophotomaps at 1:25,000 with 25 and 50 m isolines. These were produced from aerial photographs of the National Air Photo Library, Ottawa, Canada (Hell 1981).

THE LEGEND

Unlike many other legends, this one does not contain a catalog of individual landforms. It is developed on a system of basic elements combined so that the cartographical presentation of the relief is given by a combination of different layers of information. These levels are differentiated in the following manner:

. slope angles (divided into three groups: lowland, mountains, and high mountains)

. axes of curved slope and crest segments and curvatures of hillocks and depressions (basic line >100 m)

. steps and breaks of slope, valleys, minor landforms, and roughness (basic line <100 m)

. subsurface material (bedrock and unconsolidated deposits); classification of unconsolidated deposits is given by the German Industrial Standard (DIN) 18196

. hydrography

. morphogenetic process areas (only those processes are mapped that are responsible for the latest significant shaping of an area)

. individual actual processes and traces of processes

. topography and supplementary information (e.g., drilling site)

The cartographical means of expressions are symbols, colors, and screens.

The advantages and disadvantages of this legend for the preparation of the sheet on Oobloyah Bay is shown in the following remarks:

. The mapping of slope angles is only possible if large scale aerial photos or maps with contour lines are available. As for all detailed investigations, one of these mapping bases are necessary, the mapping of slope angles does not represent any problem.

. Due to thin vegetation cover, morphographical elements such as axes of curvatures, breaks of slope, etc., could be derived from the aerial photographs. Therefore only the morphometrical differentiation (radius of curvatures, hights of steps, etc.) had to be studied in the field. This is a great advantage if the time for field mapping is limited.

. Most of the probings and trenchings had to be carried out at the end of the summer because then the depth of the active layer was thickest. With the aid of information (morphography, hydrography, etc.) previously mapped, it was possible to select more representative probing sites. In addition, these probings made it possible to determin the approximate maximum thickness of the active layer in different soils.

. For the display of the hydrographical situation, the legend had to be extended. This indicates that the legend for hydrography depends more on climatic conditions than does the other information.

. The hierarchical system of decisions for the registration and delimitation of the geomorphogenetic process areas developed for the GMK 25 (Barsch and Liedtke 1980) could not be used. Due to these guidelines, all areas with active processes had to be presented in the same color (orange). Using this system it would not have been possible to differentiate glacial, fluvioglacial, and the other processes. Therefore, active process areas could not been shown. All actual processes are presented with symbols of "individual active processes."

These statements concerning the different items of the legend indicate that the legend developed first of all for geomorphological mapping in Central Europe is also adoptable to permafrost regions.

APPLICABILITY TO ENVIRONMENTAL PLANNING

Geomorphological maps such as geological and soil maps are, first of all, information systems prepared by scientists for scientists. According to the aim of geomorphology is to explain landforms and the processes which create them, geomorphological maps contain a global view of the relief forms, as well as evidence of present and past morphogenetic processes that have contributed to the evolution of the landscape. That means that these maps are very complex with a high data concentration, which makes their reading by a nongeomorphologist quite difficult. Due to the importance of the relief parameters on environmental problems including planning, the content of geomorphological maps must be presented in a form, which is easily understood by nongeomorphologists. With the legend, used in the Oobloyah Bay region, it is possible to derive special maps.

Derivative Maps

Due to the conception of the legend, all layers of information can be printed separately and all combinations are possible (Figure 2). Thereby the

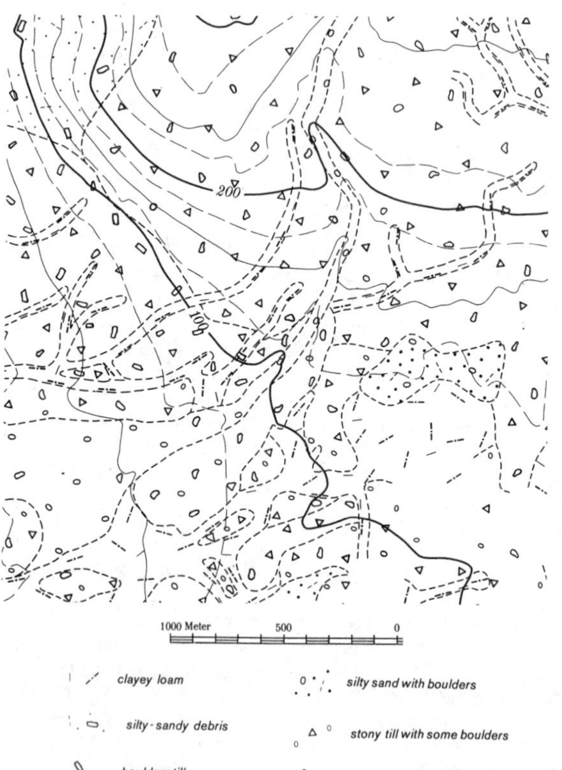

clayey loam

silty-sandy debris

bouldery till

silty sand with boulders

stony till with some boulders

gravel with boulders

FIGURE 2 Derivative map with only one level of information; substrate near the surface.

density of information can be diminished without loss of information. This is important, as for nongeomorphological problems often only parts of the whole array of information are necessary. Prerequisite to the use of this "derivative map" (Barsch and Mäusbacher 1980) is that the user is well informed about the terminology of geomorphological maps (classification of bedrock and unconsolidated material, definitions for the geomorphological processes, etc.). Therefore, these products will be used first of all by scientists in the neighboring sciences of geomorphology (geology, applied geology, and soil sciences). Several possible applications of the different levels of information are shown in the following examples.

The actual processes, the substrate near the surface, and the slope angle are of great importance for planning infrastructure. Using this information, the estimation of actual and potential danger is possible.

Additionally, by employing the slope angle and the subsurface material, the traffic limitations (load capacity of the active layer, maximum angle of climb for vehicles, and traversing of drainage ways) can be estimated.

In the framework of planning transport arteries (oil pipelines) the encroachment on the environment by defective pipes has to be taken into account. Since the degree of the encroachment is first of all dependent on the distribution (transport by water or gravity) of the damaging material, morphography and hydrography are of considerable influence.

All parameters of the geomorphological map, and additionally the potential radiation derived from slope angle and exposure, are important for soil science and botany because soilgenesis and the distribution of plants in areas without human impact are dependent on the form and genesis of relief (Everett et al. 1978).

For all these examples the neccessary transfer of the GMK information for different problems had to be made by the user.

Interpretative Maps

For the potential user who is not versed in geomorphological map terminology, the geomorphologist has to prepare his information in a way which will allow its direct use for nongeomorphic problems. For this preparation, ecological evaluation on the basis of a suitability test, developed for environmental planning purposes, would be a good solution. With this procedure of a land systems classification, different ecological parameters are evaluated with regard to human use and impact (Leser 1974, Seibert 1975). According to the information given by geomorphological map, the land systems classification is realized with a "multiple factor analysis" dominated by relief parameters. That means that the delimitation of the units is carried out with several parameters (morphography, subsurface material, hydrography, etc.), the majority of which are relief parameters. An extract of the land systems map of Oobloyah Bay is shown by Figure 3. These maps can also be used as base maps for soil and vegetation mapping (Everett et al. 1978).

In the second step the parameter used for the delimitation of the land systems will be classified. But this step is only possible if these para-

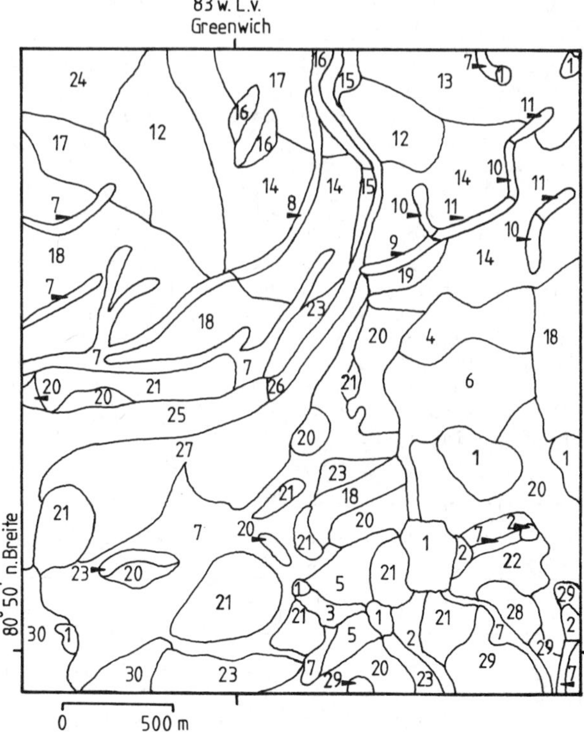

Legend for the ecological units

Code	Identifying field characteristics
1	lake
2	plains; clayey loam; area flooded during summer
3	plains; silty sand; area flooded during summer
4	plains; silty sand
5	plains; silty sand; with vegetation hummocks
6	slopes 2-7° S; clayey loam with variable proportion of material 2 mm; periglacial drainage way
7	slopes 2-7° SE/SW; clayey loam with variable proportion of material 2 mm; periglacial drainage way
8	slopes 7-15° SE/SW; clayey loam with variable proportion of material 2 mm; periglacial drainage way
9	box-shaped valley with NW/SE slopes; clayey loam with variable proportion of material 2 mm; periglacial drainage way
10	saucer-shaped valley with W/E slopes; clayey loam with variable proportion of material 2 mm; periglacial drainage way
11	saucer-shaped valley with NW/SE slopes; clayey loam with variable proportion of material 2 mm; periglacial drainage way
12	slopes 15-35° SE/SW; bouldery till
13	slopes 2-7° N; bouldery till
14	slopes 7-15° S; bouldery till
15	slopes 7-15° E/W; bouldery till
16	slopes 7-15° NE/NW; bouldery till
17	slopes 7-15° SE/SW; bouldery till
18	slopes 2-7° SW/SE; bouldery till
19	slopes 7-15° N; bouldery till
20	slopes 2-7° SE/SW; stony till with boulders
21	slopes 2-7° W/E; stony till with boulders
22	plains; stony till with boulders
23	slopes 2-7° NW/NE; stony till with boulders
24	slopes 15-35° SE/SW; silty sandy debris
25	box-shaped valley with N/S slopes; gravel with boulders; fluvioglacial drainage way
26	box-shaped valley with NW/SE or NE/SW slopes; gravel with boulders; fluvioglacial drainage way
27	plains; gravel with boulders
28	plains; silty sand and gravel
29	slopes 2-7° S; silty sand and gravel
30	plains; bouldery gravel over silty sand

FIGURE 3 Excerpt from the land systems map (ecological units) Oobloyah Bay with legend.

meters are already quantified and if threshold values are defined. The latter procedure, at the present time is difficult for many factors because measurements which show the relations between the different parameters (e.g., dependence of slope stability on slope angle and subsurface material) are not available. This problem is also shown by the two following examples.

A very important factor is potential solar radiation because it influences snowmelt, depth of thaw, and distribution and growth of vegetation. These values are not shown directly in the geomorphological map; however, they can be calculated with trigonometrical and astronomical equations using the values of slope angle and exposure (Gessler 1925, Heyne 1969). The values calculated with $\varphi = 80°$ (latitude of the expedition area) and with the slope angles of the GMK 25 Oobloyah Bay are shown in Figure 4.

slope angle \ exposure	S	SW/SE	E/W	NW/NE	N
0°	290				
2°	294	293	290	287	286
7°	301	298	288	265	247
15°	302	294	294	241	193
35°	219	209	205	187	150

FIGURE 4 Daily means (only for the days from 22.2 to 19.10) of potential solar radiation in W/m² for 80°N (expedition area) and the slope angles of the GMK 25 Oobloyah Bay.

According to other authors (Knoch 1963), a scale of five parts is used for the classification of these values, and the threshold values are adapted to the local values. That means that the classes are formed starting with the maximum and minimum values. This is reasonable as long as absolute threshold values are not available, because for most problems the local differentiation is more important. The corresponding classification is shown in Figure 5.

The second example contains the potential danger by actual geomorphological processes, particularly by landslides and mud flows. For this example the mapped actual processes, the slope angle, and the subsurface material are used as basic values. Threshold values are derived from values of slope angles and the subsurface materials of these areas showing actual geomorphological processes. Due to limited data for this classification, a scale in three parts was selected (Figure 6).

class	potential solar radiation	valuation
I	300 - 350	very high
II	250 - 300	high
III	200 - 250	moderate
IV	150 - 200	low
V	100 - 150	very low

FIGURE 5 Key for the classification of values of solar radiation.

class	slope angle and subsurface material: mud flow	slope angle and subsurface material: land slides	potential danger
I	flaggy debris with sandy matrix >15° silty sandy debris >15° stony till >15°	silty sandy debris >15° silty sand and gravel >15° debris over sandstone >35°	high
II	flaggy debris with sandy matrix 7-15° silty sandy debris 7-15° stony till 7-35°	silty sandy debris 7-15° silty sand and gravel 7-15° debris over sandstone 7-35°	moderate
III	flaggy debris with sandy matrix silty sandy debris 0- 7° stony till	silty sandy debris 0- 7° silty sand and gravel 0- 7° debris over sandstone 0- 7°	low

FIGURE 6 Key for the classification of actual geomorphological processes of the sheet on Oobloyah Bay.

In the third step the classified parameters are evaluated by associating suitability values to the different classes. Prerequisite for this type of allocation, however, is knowledge of the degree of influence (positive or negative) which each of these factors display concerning the different problems. Therefore, the evaluation of each problem has to be done separately.

For example, solar radiation controls the process of snowmelt and the depth of active layer. Therefore, the evaluation of these two processes is a precondition for evaluating solar radiation. An example is shown in Figure 7.

The second example monitors the evaluation of the geomorphological processes concerning man-made buildings (houses, roads, pipelines, etc.). In this connection it is postulated that an area with high risk is unsuitable for buildings (Figure 8).

In the last step the factors evaluated for the same problem are combined (e.g., for thickness of the active layer, solar radiation is combined with soil moisture as a function of grain size distribution, and with hydrographical information). This total value is the result of calculating the arithmetical means of the values of suitability for each factor. These values, besides others, form the content of the interpretation map. The cartographical presentation of an interpretation map like this is discussed in Mäusbacher (1983).

class	evaluation	suitability	suitability values
I	very high	very good	1
II	high	good	2
III	moderate	moderate	3
IV	low	bad	4
V	very low	very bad	5

FIGURE 7 Evaluation of solar radiation, if a thick active layer is considered positive.

class	hazard level	suitability	suitability values
I	high	good	1
II	moderate	moderate	2
III	low	bad	3

FIGURE 8 Evaluation of the potential risk caused by geomorphological processes.

CONCLUSIONS

The legend developed for geomorphological mapping in Central Europe is also applicable in permafrost regions as shown by the Oobloyah-Bay map (Mausbacher 1981). In addition, interpretation of the geomorphological information for non-geomorphological problems is possible. The array of interpretations at this time depends on the threshold values, which are available for the different factors.

ACKNOWLEDGEMENTS

This research was made possible by a Grant from the National Research Foundation of the Federal Republic of Germany (DFG). The writer's sincere personal thanks are extended to Prof. Dr. Dietrich Barsch, Scientific Leader.

REFERENCES

Barsch, D. & Liedtke, H., 1980, Principles, scientific value and practical applicability of the geomorphological map of the Federal Republic of Germany at the scale of 1 : 25 000 (GMK 25) and 1 : 100 000 (GMK 100): Z. Geomorph. Suppl. Bd. 36, p. 296-313.

Barsch, D. & Mausbacher, R., 1980, Auszugs- und Auswertungskar- ten als mogliche nutzungsorientierte Interpretation der geomorphologischen Karte 1 : 25 000 (GMK 25): Berliner Geogr. Abh. 31: p. 31-48.

Everett, K.R., Webber, P.J., Walker, D.A., Parkinson, R.J. & Brown, J., 1978, A geoecological mapping scheme for Alaskan coastal tundra, in Proceedings of the Third International Conference on Permafrost, v.l.: Ottawa, National Research Council.

Franzle, O., Barsch, D., Leser, H., Liedtke, H. & Stablein, G., 1979, Legendenentwurf fur die geomorphologische Karte 1 : 100 000 (GMK 100): Heidelberger Geogr. Arb. 65, pp. 18.

Gessler, R., 1925, Die Starke der unmittelbaren Sonnenstrahlung und ihre Abhangigkeit von der Auslage: Abh. d. Preussischen Meteorol. Inst. 8, 1, p. 3-25.

Hell, G., 1981, Geodatische und photogrammetrische Arbeiten an der Oobloyah-Bay, N-Ellesmere-Island, N.W.T., Kanada, im Rahmen der Heidelberg Ellesmere Island Expedition 1978: Heidelberger Geogr. Arb. 69, p. 35-45.

Kreig, A.R. & Reger, R.D., 1976, Preconstruction terrain evaluation for the Trans-Alaska-Pipeline Project, in Coates, D.R. ed., Geomorphology and engineering, New York, Halsted Press.

Knoch, K., 1963, Die Landesklimaaufnahme, Wesen und Methodik: Ber. d. Dt. Wetterdienstes, Bd. 12, 85, p. 1-60.

Leser, H., 1974, Nutzflachenanderung im Umland der Stadt Esslingen am Neckar und ihre Konsequenzen fur Planungsarbeiten zur Landeserhaltung aus landschaftsokologischer Sicht: Tubinger Geogr. Stud. 55, p. 65-101.

Leser, H., 1981, Die geomorphologische Kartierung in der Schweiz: pp. 1-25, Basel.

Leser, H. & Stablein, G., 1975, Geomorphologische Kartierung. Richtlinien zur Herstellung geomorphologischer Karten 1 : 25 000. - 2. verand. Aufl.: Berliner Geogr. Abh., Sonderheft, pp. 39.

Mausbacher, R., 1981, Geomorphologische Kartierung im Oobloyah-Tal, N-Ellesmere Island, N.W.T., Kanada: Heidelberger Geogr. Arb. 69, p. 413-440.

Mausbacher, R., 1983, Die geomorphologische Detailkarte der Bundesrepublik Deutschland (GMK 25) - ein nutzbarer Informationstrager auch fur Nicht-Geomorphologen: Regio Basiliensis (im Druck).

Robitaille, B., 1960, Presentation d'une carte geomorphologique de la region de mould bay, ile du Prince-Patrick, Territoires du Nord-Ouest: Canadian geographer, 15, p. 39-43.

Seibert, H., 1975, Versuch einer synoptischen Eignungsbewertung von Okosystemen und Landschaftseinheiten: Forst-archiv 46, p. 89-87.

THE INFLUENCE OF SURCHARGE LOADS ON FROST SUSCEPTIBILITY

E.Y. McCabe and R.J. Kettle

Department of Civil Engineering, University of Aston in Birmingham, U.K.

When a frost susceptible soil is frozen and the heave is restrained, either partially or completely, heaving pressures will also be generated. The work reported in this paper describes a laboratory study of the interactions between heave, heaving pressure, and surcharge pressure. The tests were performed on a highly frost susceptible mixture of sand and ground chalk to which various amounts of coarse aggregate were also added to produce a range of granular mixtures. The frost heave and heaving pressures were determined for these materials. To quantify the effects of surcharge, several levels of loading were applied to the freezing specimens, with either dead loads or an air-diaphragm system, to produce surcharges between 3 and 120 kN/m^2. Freezing behaviour was clearly modified by the application of surcharge loads, and for each material a hyperbolic relationship was established between frost heave and surcharge. Although the complete elimination of frost heave was not achieved, the application of low surcharges produced large reductions in heave. The surcharge required to render each material non-frost susceptible was dependent on heaving pressure rather than frost heave, as demonstrated by the ratio between surcharge pressure and heaving pressure. Typically, a 5% ratio was sufficient to achieve a 50% reduction in heave.

INTRODUCTION

Reductions in heave, following increases in the overburden pressure, have been widely reported in literature, and this approach has been suggested as one of the possible ways of controlling frost action in frost susceptible soils. It is unlikely that heave could be completely eliminated in many engineering locations, such as highways or airfield construction, because of the high surcharges that could be required. However, with structures such as pipelines and LNG tanks, the weight and stiffness of the structure could provide significant restraint to the heave forces developed in the soil. The associated pressures will act on the structure and must be allowed for during design so that the inter-relation between heave, heaving pressure and overburden pressure is of considerable interest.

The influence of surcharges, or overburden pressures, were reported some fifty years ago (Taber 1930, Beskow 1935) and since then others have demonstrated that surcharges can reduce the heave exhibited both in the laboratory and the field (Aitken 1974). It has been suggested (Hill and Morgenstern 1977, McRoberts and Nixon 1975) that there is a critical "shut-off pressure" at which moisture transfer to the freezing front ceases so that subsequent heave may be controlled. More recently, however, it has been demonstrated (Penner and Ueda 1977) that no "shut-off pressure" exists for soils, providing that freezing is continued for a sufficiently long period, although marked reductions in heave rate are produced when an external load is applied. Under such conditions, frost heave will coexist with the developed heaving pressures.

Takashi et al (1971) derived an empirical equation connecting heave and surcharge load and again underlined that the complete elimination of heave would be difficult. The majority of the investigations have been concerned with the relationship between heave and surcharge stress. However, this investigation has been concerned with the inter-relationships between heave, heaving pressure, and surcharge stress. Of particular interest has been the level of surcharge required to render the soils non-frost susceptible, as defined by the U.K. criterion (Croney and Jacobs 1967).

MATERIALS

These have been considered (McCabe 1982) as mixed granular materials produced by artificial blending. The basic matrix consisted of a highly frost susceptible mixture of sand and ground chalk (ratio 4:1). Three types of coarse particles were selected -- blastfurnace slag, basalt, and limestone -- each presieved to the 3.35-20mm group. Each aggregate was blended with the matrix to produce various mixtures containing up to 50% coarse particles.

Outline of the Investigation

Initially a series of frost heave and heaving pressure tests were performed on mixtures containing up to 50% coarse particles. The frost heave was measured in the cold room following the procedure of the TRRL frost heave test (Croney and Jacobs 1967, TRRL 1977). The heaving pressures were determined when the heave was restricted by a stiff reaction frame. The specimen was supported in a multi-ring cell and frozen from the top by using a thermoelectric device (TED) or a cryostat system. Detailed results are reported elsewhere (Kettle and McCabe 1981, McCabe and Kettle 1982, McCabe 1982). From these results a limited number of mixtures were selected for the study of the influence of the surcharge. These were made on the basis of the following conditions:

1. Materials with similar heaves, but different heaving pressures (30% slag, 30% basalt).
2. Materials with similar heaves and similar heaving pressures (30% basalt, 20% limestone).
3. Materials with different heaves, but similar heaving pressures (30% basalt, 10% limestone).

The data for these materials are collected in Table 1.

TABLE 1. Heave and Heaving Pressure Results.

Materials	Total Heave (mm)	Heaving pressure (kN/m^2)
matrix	25.0	300
30% slag	16.0	192
30% basalt	16.4	250
20% limestone	16.4	246
10% limestone	18.4	254

SURCHARGE TEST PROGRAMME

Low Level Surcharge

It was considered (McCabe 1982) that low-level surcharges could be produced by placing steel weights on top of the frost heave specimens, with the self refrigerated unit (SRU) being the most suitable facility for such tests. The specimens were prepared following the standard procedure for the TRRL frost heave test (Croney and Jacobs 1967, TRRL 1977). The specimens are wrapped in waterproof paper in the standard test, but this was not sufficiently strong to provide adequate lateral support to the surcharged specimens. Oversize Tufnol rings were, therefore, slid over the paper-wrapped specimens, and this system was shown (McCabe 1982) to produce similar results to those of the standard specimens. The conditioning and freezing procedure followed that of the standard TRRL frost heave test, apart from the application of the surcharge. The steel weights were applied to the top of the specimen when it had just started to freeze with a heave of 0.1 mm. This occurred approximately 10 hours after the start of the test. Application of the surcharge before frost action had commenced could produce consolidation in the material and so modify the effects of frost action. It was considered that if the steel weights were placed directly on the Tufnol cap they could modify the boundary conditions since they would rapidly cool to -17°C. It was, therefore, decided to place the weights in a slightly elevated position, on a wooden support table, so that air could circulate above the specimen as in the standard test.

High Level Surcharge

It was not possible to apply surcharges greater than 16 kN/m^2 with the dead weights, and so a Bellofram air diaphragm was employed. This was used in conjunction with a controlled heave unit (CHU) (McCabe 1982). The specimens were produced as for the Low Level Series and positioned in the CHU as shown in the arrangement detailed in Figure 1. The specimen was conditioned for 24 hours at +4°C and frozen from the top by using a circulating cryostat. The air diaphragm was not connected to the air supply until freezing had commenced, so as to, again, prevent pre-test consolidation. The frost heave was continuously monitored on a trace recorder. The test was normally stopped when the rate of increase in heave was less than 0.01 mm/hour (McCabe 1982), and at least two tests were performed on each material.

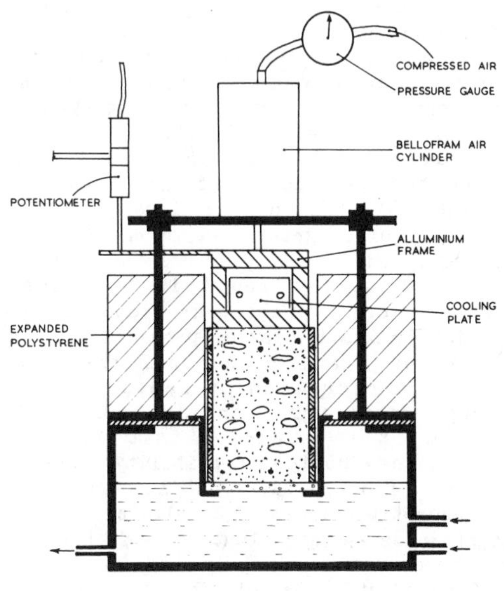

FIGURE 1. Modified CHU for frost heave test with surcharge.

DISCUSSION OF THE RESULTS

Low Level Surcharge

The results for the sand-snowcal matrix are given in Table 2, and those for the selected aggregate-matrix mixtures are given in Table 3. It can be seen from Table 3 that, although the 30% slag and basalt mixtures have similar frost heaves without surcharge, the basalt mixture requires a much higher surcharge to reduce the heave to the non-frost susceptible value of 9.75 mm. It is interesting to note that the basalt mixture developed a higher maximum heaving pressure than the slag mixture. The sand/snowcal matrix, with even higher values of frost heave and heaving pressure, also required significantly higher surcharges to reduce the heave to values below the 13 mm criterion for frost susceptibility as is apparent from the results in Table 2.

TABLE 2. Frost Heave Values for Sand/Snowcal Matrix Under Different Levels of Surcharge.

Surcharge (kN/m^2)	Heave (mm)	Surcharge / Heaving Pressure (%)
0	25.01	0
3.25	20.79	1.1
7.86	18.23	2.6
12.48	13.91	4.2
15.25	12.25	5.1

The 30% basalt and 20% limestone mixtures, as indicated in Table 3, required similar surcharges to reduce the frost heave to approximately the same value. However, the 10% limestone mixture which initially had a higher frost heave but a similar heaving pressure to 30% basalt and 20% limestone mixtures, required a similar level of surcharge to achieve similar heaves to those developed by other mixtures. It appears, therefore, from the above results that the heaving pressure dictates the behaviour of a particular material under surcharge. Those materials that had developed higher heaving pressures required greater surcharge to reduce the frost heave and, hence, the frost susceptibility.

In order to examine this influence, the surcharges were expressed as proportions of the appropriate maximum heaving pressure. It is clear from Tables 2 and 3 that all the materials exhibited a reduction in frost heave as the surcharge ratio was increased. Of particular interest, it is apparent that with all the materials the frost susceptibility classification was reduced from highly frost susceptible to non-frost susceptible (below 13 mm) with comparatively low surcharges -- equivalent to 4-5% of the maximum heaving pressure.

To investigate this relationship, the percentage reduction in heave was calculated and is plotted in Figure 2 against the level of surcharge, expressed as a percentage of the maximum heaving pressure. A regression analysis revealed a linear relationship between the two parameters (r = 0.96) and this regression line can be expressed as:

$$Y = 8.81X$$

where Y is the reduction of heave (%) and X is the surcharge/heaving pressure ratio (%).

Thus, for all the materials tested only a relatively low surcharge was required to reduce frost susceptibility, and this may be calculated if the frost heave and heaving pressure are known.

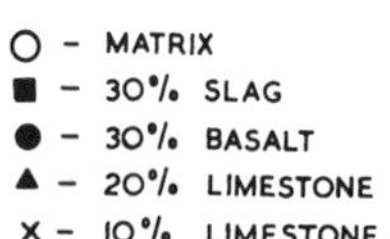

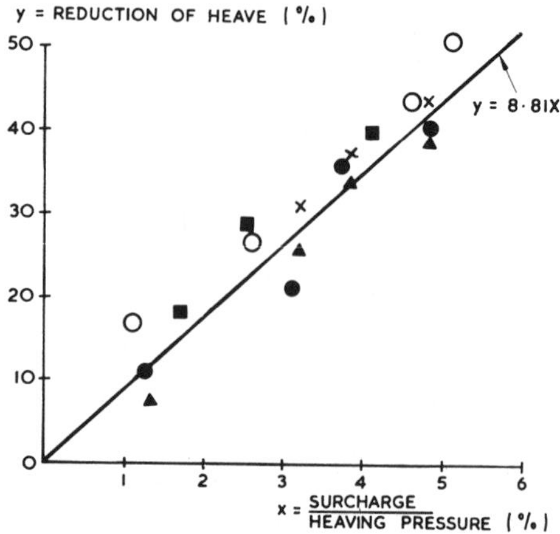

FIGURE 2. Relationship between reduction of heave and surcharge ratio.

High Level Surcharge

The results are summarized in Table 4 and are presented in Figure 3 together with the data from the previous section for the low-level surcharge. The relationship between frost heave and surcharge is hyperbolic, which agrees with other experimental work (Nakazaka and Ueda 1976) and theoretical considerations (Kaplar 1970). It can clearly be seen that in order to reduce the frost heave to below 10 mm, only relatively low surcharges are necessary, although limited movement is still apparent at high surcharges. Indeed, movements of the order of 0.1 mm, corresponding to the stiffness of the load cell (McCabe 1982), accompanied the maximum pressures recorded in the heaving pressure tests. It would, therefore, appear that for the materials tested shut-off pressures were not achieved. Thus, while it was not possible to eliminate heave completely, it is reduced through an increase in the surcharge.

TABLE 3. Heave Values Under Surcharge.

Surcharge kN/m^2	30% Slag		30% Basalt		20% Limestone		10% Limestone	
	Heave (mm)	S(%) / H.P	Heave (mm)	S(%) / H.P	Heave (mm)	S(%) / H.P	Heave (mm)	S(%) / H.P
0	16.0	0	16.4	0	16.4	0	18.4	0
3.25	13.10	1.7	14.62	1.3	15.20	1.3	-	-
4.80	11.37	2.5	-	-	-	-	-	-
7.86	9.60	4.1	12.90	3.1	12.10	3.2	12.64	3.1
9.35	-	-	10.50	3.7	10.80	3.8	11.60	3.7
11.90	-	-	9.75	4.8	9.97	4.8	10.50	4.7

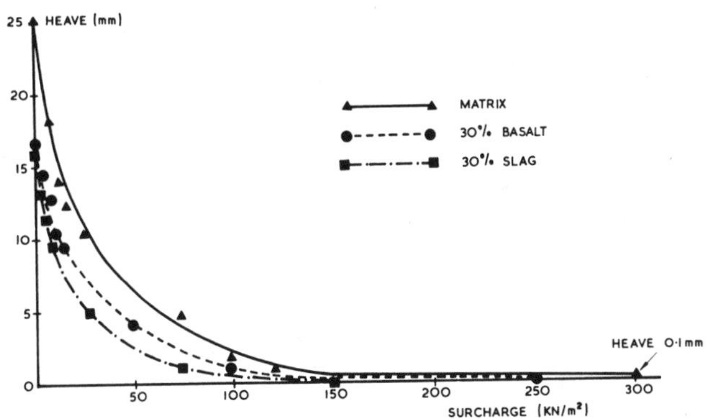

FIGURE 3. Frost heave against surcharge.

TABLE 4. Heave Values Under Surcharge.

Surcharge kN/m^2	Sand/snowcal Heave (mm)	Sand/snowcal S(%) H.P	30% Slag Heave (mm)	30% Slag S(%) H.P	30% Basalt Heave (mm)	30% Basalt S(%) H.P
0	25.01	0	16.0	0	16.4	0
24	10.5	8	5.2	12	-	-
48	6.7	16	-	-	4.2	19.0
72	4.5	24	1.1	38	2.5	29.0
97	1.7	32	-	-	1.0	38.0
120	1.0	40	-	-	-	-

The surcharge is not uniquely related to frost heave. It can be seen from Figure 3 that different surcharges are necessary for different materials to produce a given value of heave. Those materials with higher frost heaves and, especially, heaving pressures required higher surcharge for a given reduction in frost heave and hence frost susceptibility.

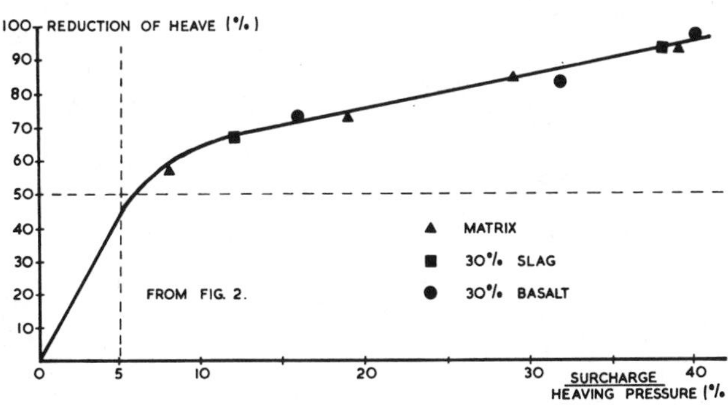

FIGURE 4. Reduction of heave against surcharge ratio.

In Figure 4 the reduction in heave is plotted against the ratio of surcharge to heaving pressure, including data from Figure 2. It appears that there are two separate relationships between the reduction in heave and the surcharge ratio. At low levels of surcharge, up to 5% of the heaving pressure, there is a very clear relationship such that quite small changes in the surcharge produced marked differences in the reduction of heave. A surcharge ratio of approximately 5% is sufficient to reduce the heave by 50%. However, significantly higher levels of surcharge are required to achieve larger reductions in heave, so that for the complete elimination, if possible, of heave would require extremely high surcharges.

From these experimental results it seems that the magnitude of heaving pressure dictates the behaviour of a soil when frozen under surcharge loads. Typically, for two materials with similar frost heaves but different heaving pressures, the material that developed the higher heaving pressure required a higher surcharge to achieve the same reduction in heave. Thus, the magnitude of any reduction in heave depends on the heaving pressure rather than the initial heave value.

This investigation is currently being extended to study the heat flow through the surcharged specimens when subjected to controlled freezing. These measurements are being coupled with the water intake rate so that linkage may be developed between the thermal parameters and the soil properties to predict frost heave and heaving pressures.

CONCLUSION

1. Relatively low surcharges produced reductions in frost heave and hence frost susceptibility.
2. The frost heave was related to surcharge by a hyperbolic curve that depended on the type of material.
3. For all the materials, the surcharge required to render a material non-frost susceptible was dependent on the heaving pressure rather than the frost heave.
4. The ratio of surcharge pressure to heaving pressure can be used to predict the reduction in heave. A ratio of approximately 5% was sufficient to achieve 50% reduction in heave and so considerably modified the frost susceptibility.
5. The complete elimination of the frost heave was not achieved, even at high levels of surcharge, although frost heave was considerably reduced.

ACKNOWLEDGMENTS

The authors wish to thank the University of Aston for making available facilities and for providing the opportunity to undertake the research reported in this paper. An acknowledgement is also extended to the Science and Engineering Research Council for assistance in funding the continued investigation.

REFERENCES

Aitken, G.W., 1974, Reduction of frost heave by surcharge stress. Technical Report 184. U.S. Army Cold Regions Research and Engineering Laboratory, Hanover, New Hampshire.

Beskow, G., 1935, Soil freezing and frost heaving with a special application to roads and railways, Swedish Geological Society, 26th Year Book, No.

3, Series C No. 375, 145 pp. With a special supplement to the English translation of progress between 1935-1946. Translated by J.O. Osterberg, Technical Institute, Northwestern University, Evanston, Illinois.

Croney, D. and Jacobs, J.C., 1967, The frost susceptibility of soils and road materials, Road Research Laboratory Report, LR 90, Crowthorne, England.

Hill, D., and Morgenstern, N.R., 1977, Influence of load and heat extraction on moisture transfer in freezing soils, in Proceedings of the International Symposium on Frost Action in Soils, University of Lulea, Sweden, Vol. I, pp. 76-91.

Kaplar, C.W., 1970, Phenomenon and mechanism of frost heaving, Highway Research Record No. 304, pp. 1-13, Washington, U.S.A.

Kettle, R.J., and McCabe, E.Y., 1981, Heaving pressures as a means of assessing frost susceptibility, in Proceedings of International Seminar on the Prediction of Frost Heave, University of Nottingham, England, pp. 51-58.

McCabe, E.Y., 1982, Frost action in granular materials, Ph.D. Thesis, University of Aston in Birmingham, England, 278 pp.

McCabe, E.Y., and Kettle, R.J., 1982, Heaving pressures and frost susceptibility, in Proceedings of the Third International Symposium on Ground Freezing, U.S. Army Corps of Engineers, CRREL, Hanover, N.H., U.S.A. Special Report SP 82-86, pp. 285-294.

McRoberts, E.C., and Nixon, J.F., 1975, Some geotechnical observations on the role of surcharge pressure in soil freezing, in Proceedings of First Conference, Calgary, Canada, pp. 42-57.

Nakazaka, H., and Ueda, T., 1976, Experimental study on frost heaving characteristics of soils, Takenaka Technical Research Report No 15, part 1, Japan.

Penner, E., and Ueda, T., 1977, The dependence of frost heaving on load application, in Proceedings of International Symposium on Frost Action in Soils, University of Lulea, Sweden, Vol. I, pp. 92-102.

Taber, S., 1930, Freezing and thawing of soils as factors in the distruction of road pavements, Public Roads, Vol. 11, No. 6, pp. 113-132, U.S.A.

Takashi, T., et al, 1971, An experimental study on the effect of loads on frost heaving and soil moisture migration, Journal of Japan Society of Snow and Ice, Vol. 33, No. 3, pp. 109-119.

Transport and Road Research Laboratory, 1977, The LR90 Frost Heave Test - interim specification for use with granular materials, TRRL Supplementary Report, SR318, Crowthorne, England.

AN INVESTIGATION OF TRANSIENT PROCESSES IN AN ADVANCING ZONE OF FREEZING

R. W. McGaw, R. L. Berg, and J. W. Ingersoll

U.S. Army Cold Regions Research and Engineering Laboratory
Hanover, New Hampshire 03755 USA

Studies have indicated a relation between subfreezing temperature in a fine-grained soil and pressure (moisture tension) in the film water adjacent to an ice lens. During the experiments reported here, concurrent measurements were obtained of temperature and pressure in the liquid water phase of a freezing silt soil. Freezing was from the top down utilizing an open system, with the water table held at the base of a specimen 30 cm long. The freezing front advanced into the specimen at a generally decreasing rate from 20 mm/day to 5 mm/day. The tests utilized a special tensiometer developed at CRREL that continues to measure moisture tension below a temperature of 0°C as long as continuity with the unfrozen water is maintained. Moisture tensions were registered continuously up to 75 kPa (0.75 atm), after which the tension remained constant or decreased slightly. Temperature corresponding to the maximum tension was -2.0°C; the tensiometers were unfrozen at this temperature. Atmospheric pressure was indicated in the soil water until the local temperature was colder than about -0.45°C. Combining the pressure-temperature data with data on unfrozen water vs. temperature resulted in an apparent moisture characteristic for the unfrozen water phase that is dependent on temperature.

Taber (1918) first showed that the major cause of volume increase when a moist soil freezes is the formation of segregated ice, i.e., ice composed of water in excess of that which was already present in the unfrozen soil. Subsequent investigators are still seeking a rational explanation for the drawing-up of excess water from lower depths in response to freezing.

Beskow (1935) and others have demonstrated that a gradient of negative pressure (i.e., suction) develops within the moisture phase of a fine-grained soil during freezing. It is known that liquid water can withstand tensions equal to several hundred atmospheres owing to its intrinsic molecular cohesion. However, the presence of dissolved air typically reduces the measurable tension to the order of one atmosphere, inasmuch as the nucleation and expansion of bubbles locally interrupts the continuity of the water phase in the soil.

Although moisture tensions are known to exist below an advancing freezing front, and are suspected to exist within the freezing zone itself, the seat of these tensions has not yet been adequately explained. Recent attempts by Derjagin and Churaev (1978), Loch (1978), Miller (1978), O'Neill and Miller (1980), and others to develop comprehensive theories of frost heaving that could rationalize this and other occurrences during freezing have not been fully successful. Such theories have required critical assumptions in their development that pertain to the levels of temperature and pressure existing within the freezing zone itself. These assumptions are at present supported by little or no substantive experimental information.

In an effort to begin to remedy this situation, the tests reported here were designed to provide concurrent measurements of temperature and pressure in the moisture phase of a freezing silt soil.

MATERIALS AND METHODS

Two silty soils were each frozen unidirectionally by applying increasingly colder temperatures to the upper surface with the lateral surfaces insulated. The two soils used were Graves sandy silt from western Massachusetts and Northwest silt from the vicinity of Fairbanks, Alaska. These soils were frozen in a saturated condition at densities of 1.60 and 1.39 x $10^3 kg/m^3$, respectively. Selected physical properties are listed in Tables 1 and 2.

Experimental results for the Graves sandy silt are presented in this paper. Analysis of the results for the Northwest silt is presently in progress and will be reported in a subsequent paper, together with a comparison of the behavior of the two soils.

A special apparatus was designed and fabricated to contain the 30-cm long soil specimen during freezing and to support the devices for measuring temperature and moisture tension. Top and bottom temperatures were maintained to a tolerance of 0.01°C by separate fluid baths; the upper temperature was adjusted daily to cause the advance of a freezing front into the specimen at a generally decreasing rate between 20 mm/d and 5 mm/d over a 24-day period. The specimen was subsequently allowed to thaw over an 8-day period.

De-aired water was supplied continuously to the bottom of the soil through a porous stone connected to an external water supply, which was maintained at a temperature of 4°C. The external water table was held level with the bottom of the specimen.

Temperatures along the length of the soil column were measured using closely spaced thermistors mounted within the wall of the containing vessel. It had been determined that this arrangement registers actual soil temperatures within 0.05°C.

TABLE 1 General properties.

Soil type	Specific gravity	Amount finer than (%) 2 mm	0.074 mm	0.020 mm	0.002 mm	Liquid limit	Plastic limit
Graves sandy silt	2.73	100	46	13	2	23.2	-
Northwest silt	2.65	100	99	36	5	32.7	30.4

TABLE 2 Compaction properties (optimum @ CE-55* procedure).

Soil type	Max. density 10^3 kg/m^3 (pcf)	Water content (%)	Degree saturation (%)	Volume solids	Volume voids	Volume water	Volume air
Graves sandy silt	1.76 (110.0)	15.5	75	0.64	0.36	0.27	0.09
Northwest silt	1.86 (115.5)	10.5	63	0.70	0.30	0.19	0.11

*Similar to U.S. modified AASHO procedure.

Moisture tensions were measured in the unfrozen and the frozen portions of the soil column by means of special tensiometers designed at CRREL to respond to moisture tensions both above and below 0°C. These tensiometers have been described in detail in McKim et al. (1976). Briefly, they differ from conventional tensiometers in two respects: a) the sensing element is a strain gauge mounted on an internal metal diaphragm that is vented to atmospheric pressure on the side away from the soil; soil moisture tension is applied to the forward side of the diaphragm in the conventional way through a porous ceramic tip; and b) the porous tip in contact with the moist soil is filled with a 10% solution of ethylene glycol and water, which remains unfrozen to a temperature of approximately -4°C (25°F).

During the development of these devices at CRREL it had been determined that a) the antifreeze does not diffuse into the soil water (because the ethylene glycol molecule is both large and hygroscopic); b) in 10% aqueous solution, the density of the antifreeze is 1% greater than that of water, but its viscosity is twice that of water; c) within experimental error the calibration with soil moisture tension is found to be the same for a 10% solution as for water (surface tension of the aqueous solution being only 5% greater than that of water). The physical properties referred to here are from data given by Dow Chemical Co. (1956).

Experimentally, the use of a strain gauge transducer (rather than a manometer) provided a very fast response to rapidly changing moisture tensions in the transient freezing situation. This effect was the result of the extremely small volume changes required to deflect the diaphragm and thereby to register the amplitude of the moisture tension.

RESULTS AND DISCUSSION

During the freezing tests, temperatures to the nearest 0.01°C and moisture tensions to the nearest 0.1 kPa (0.001 atm or approximately 1 cm water) were recorded once daily along the full length of the soil specimen. Immediately following each reading the temperature at the surface was reduced by 0.5°C to maintain a continuously advancing frost front during the subsequent 24-hr period. By this procedure each set of readings represented the response of the freezing specimen 24 hours after a change in applied surface temperature, essentially ensuring that the measured tensions were in phase with the measured temperatures.

Figure 1 is an example of the data obtained daily for temperature and moisture tension in the Graves sandy silt, plotted as a function of depth. Figure 2 is a graph of moisture tension vs. temperature in the freezing zone for the entire test period; data are shown for depths of 5, 10, and 15 cm below the original surface.

Several occurrences are notable in the record of moisture tension at temperatures below 0°C. Not only was tension registered continuously by the tensiometers, but the pressure differed significantly from atmospheric only when the local temperature of the soil was colder than about -0.5°C. The level of moisture tension then increased rapidly as temperatures became lower, finally reaching an apparent maximum tension between 70 and 80 kPa (0.7 to 0.8 atm). Thereafter, the tension remained constant or decreased slightly. The temperature associated with the apparent maximum moisture tension was approximately -2.0°C for the Graves sandy silt.

In these tests, no attempt was made to monitor the location or amount of ice formation behind the freezing front. Ice was clearly present because heaving was measured at the surface of the specimen. Nevertheless, the continuous recording of moisture tension by the tensiometers indicates that an unfrozen (liquid) water phase remained that was continuous with the liquid in the tensiometer.

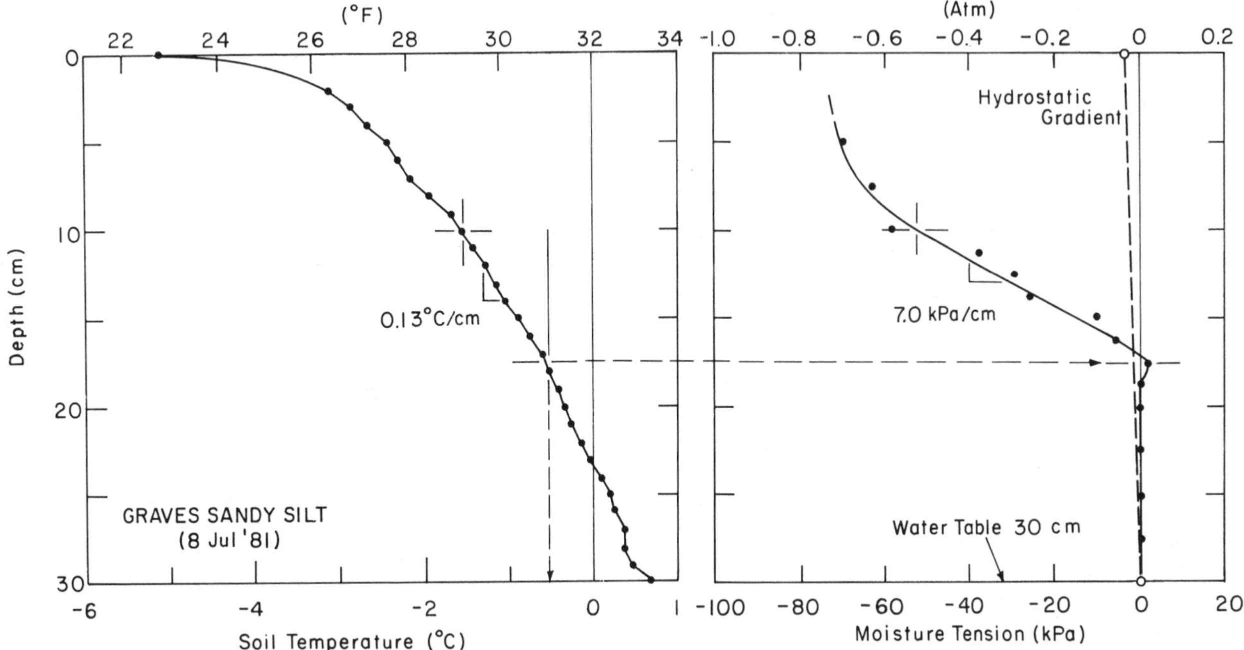

FIGURE 1 Concurrent temperature and moisture tension vs. depth in freezing zone of Graves sandy silt.

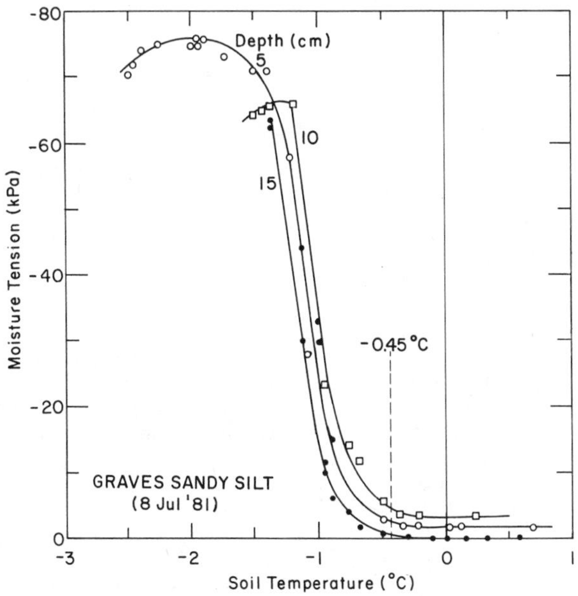

FIGURE 2 Moisture tension vs. temperature in freezing zone of Graves sandy silt.

APPARENT MOISTURE CHARACTERISTIC FOR UNFROZEN WATER IN A FREEZING ZONE

Figure 3 is a graph of the unfrozen water content for this soil as a function of the below-freezing temperature; this plot is a transcription of unfrozen-water data measured by a nuclear magnetic resonance (NMR) procedure described in Tice et al. (1982). At the time of this writing, the data shown were unpublished (Tice, personal communication).

Combining the data of Figures 2 and 3 provides an important new step in the assessment of frost heaving, and results in an apparent moisture-tension characteristic for the water remaining unfrozen behind the freezing front (Fig. 4). The combining of pressure and temperature information in this way rests on the assumption that the pressure measured by the tensiometers is that of the unfrozen water itself, i.e., that the state of water in the active freezing zone is the same as it is in Tice's steady-state tests.

An observed correspondence in the two types of test supports this assumption: the temperature of initial nucleation found by Tice on cooling when he seeded with ice (-0.45°C) is very nearly the same temperature at which moisture tension first exceeds the hydrostatic level in the advancing freezing front. Moreover, a steep gradient of moisture tension develops at temperatures colder than -0.45°C, giving rise to the probability that it is the pressure of the unfrozen water that is being measured.

It is important to note that the moisture characteristic presented in Figure 4 represents the pressure of liquid water measured relative to the atmosphere external to the specimen. This is because the reference gas in the tensiometer (i.e., on the non-wetted face of the diaphragm) is the atmosphere of the room; a tube in the tensiometer assembly provides a positive vent to the ambient air. Thus, the tensiometer reading is the actual pressure in the liquid water behind the freezing front. This test does not determine what interfaces exist within the soil in conjunction with the water.

Unlike a moisture characteristic for an unfrozen soil, the freezing-zone moisture characteristic is not independent of temperature. Each

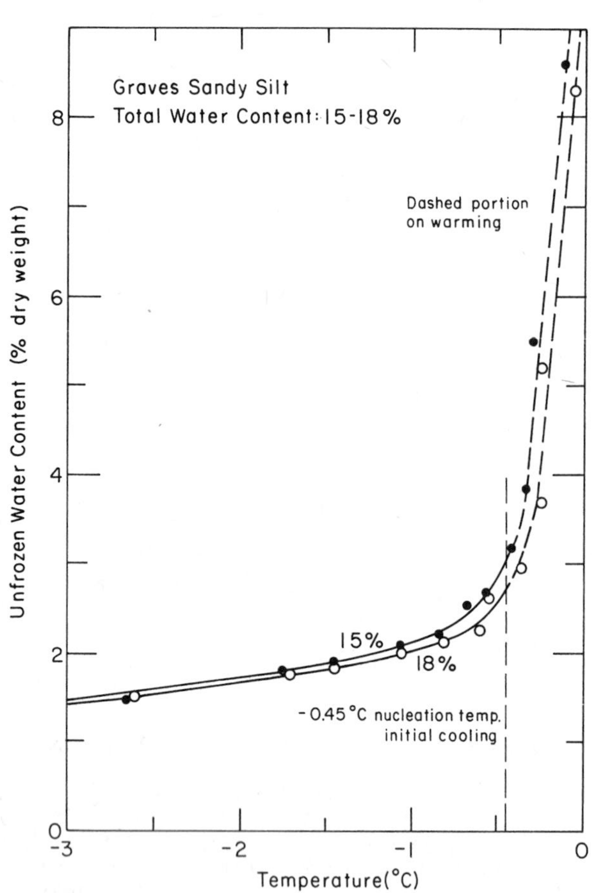

FIGURE 3 Unfrozen water content vs. temperature for Graves sandy silt (data by Tice).

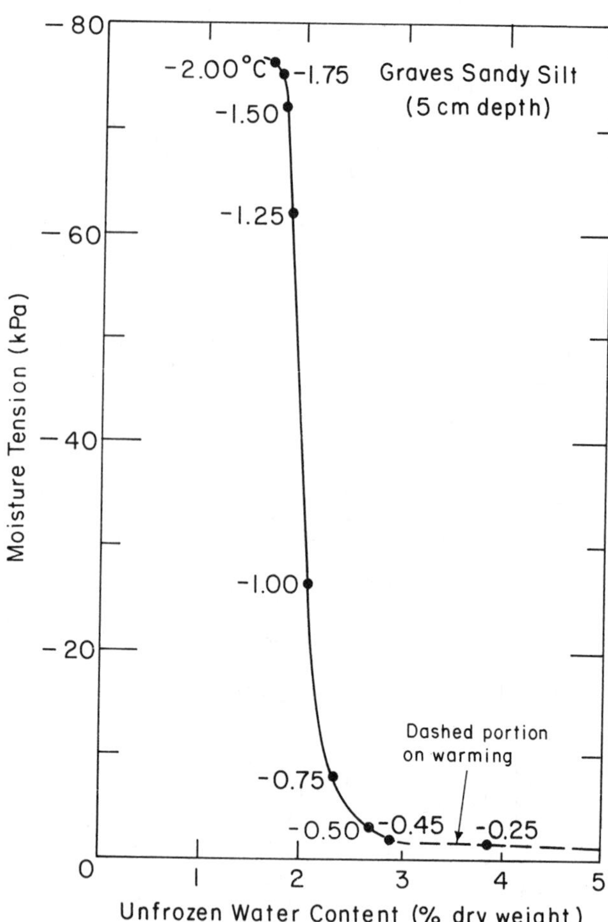

FIGURE 4 Freezing-zone moisture characteristic between unfrozen water and external atmosphere for Graves sandy silt.

intersection of moisture tension and unfrozen water content on a plot such as Figure 4 corresponds to a narrow range of temperatures below 0°C. The observed range of temperature is most likely the result of local variations in gradation, density, or water content in the vicinity of the tensiometer. In addition, a normal moisture characteristic corresponds to an air/water interfacial system within the soil; the freezing-zone moisture characteristic probably conforms both to a water/ice interfacial system with a lower intensity of surface energy, and an air/water interfacial system within the soil somehow coupled to the former.

SUMMARY AND CONCLUSIONS

In the tests reported here, a silty soil was frozen unidirectionally using specially designed apparatus that allowed the measurement of moisture tensions and temperatures in the freezing zone. The highest (warmest) temperature at which significant moisture tension developed was approximately -0.45°C, which corresponded to the warmest nucleation temperature for ice measured independently in steady-state tests for unfrozen water on the same soil.

The measured variation of moisture tension took place in a zone approximately 7.5 cm (3 in.) in vertical dimension bounded by the temperatures of -0.45°C and -2.0°C. The level of moisture tension observed ranged from nearly zero (hydrostatic) to 75 kPa (0.75 atm).

Water flowed through the unfrozen soil below the freezing zone under a moisture tension gradient that was barely detectable, consistent with the measured water intake and the estimated hydraulic conductivity of the soil tested. Moisture tension gradients in the freezing zone reached a maximum of approximately 7.0 kPa/cm (1 atm/15 cm).

Utilizing measured values of temperature, moisture tension, and unfrozen water, a freezing-zone moisture characteristic was derived for the soil tested. For moisture tensions from 0 to 75 kPa the unfrozen water content in the freezing zone was in the range 2% to 3% by dry weight.

The capability of deriving the freezing-zone moisture characteristic for a frost-susceptible soil represents an important development in the prediction of the amount of segregated ice in a freezing soil. The resulting correspondence among the factors of temperature, moisture tension, and unfrozen water content will provide a better calibration of existing simulation models of the frost heaving process and may lead to computer algorithms for the freezing zone that are more nearly physically based than those presently in use.

ACKNOWLEDGMENTS

This work was supported by Corps of Engineers projects. The authors wish to thank Drs. A. Assur and V. Lunardini, both of CRREL, for their sincere interest, encouragement, and technical suggestions during the course of this work.

REFERENCES

Beskow, G., 1935, Soil freezing and frost heaving, with special reference to highways and railroads: Stockholm, Sweden, Statens Vaginstitut, Meddelande 48, Swedish Geological Society, Series C, No. 375.

Derjagin, B.V., and Churaev, N.V., 1978, The theory of frost heaving: Journal of Coll. International Science, v. 67, p. 391-396.

Dow Chemical Co., 1956, Properties and uses of glycol: Midland, Mich.

Loch, J., 1978, Thermodynamic equilibrium between ice and water in porous media: Soil Science, v. 126, p. 77-80.

McKim, H., Berg, R., McGaw, R., Atkins, R., and Ingersoll, J.W., 1976, Development of a remote reading tensiometer transducer system for use in subfreezing temperatures: Proceedings, Second Conference on Soil Water Problems in Cold Regions, Edmonton, Alberta.

Miller, R.D., 1978, Frost heaving in non-colloidal soils: Third International Conference on Permafrost, Edmonton, Alberta, p. 707-713.

O'Neill, K., and Miller, R.D., 1980, Numerical solutions for rigid ice model of secondary frost heave: Second International Symposium on Ground Freezing, Norwegian Institute of Technology, June 24-26, p. 656-669.

Taber, S., 1918, Surface heaving caused by segregation of water forming ice crystals: Engineering News-Record, v. 81, p. 683-684.

Tice, A.R., Oliphant, J.L., Nakano, Y., and Jenkins, T.F., 1982, Relationship between the ice and unfrozen water phases in frozen soil as determined by pulsed nuclear magnetic resonance and physical desorption data, CRREL Report 82-15: Hanover, N.H., U.S. Army Cold Regions Research and Engineering Laboratory.

BENEFITS OF A PEAT UNDERLAY USED IN ROAD CONSTRUCTION ON PERMAFROST

Robert L. McHattie, Senior Research Engineer and
David C. Esch, Chief of Highway Research

STATE OF ALASKA DEPARTMENT OF TRANSPORTATION AND PUBLIC FACILITIES
DIVISION OF PLANNING AND PROGRAMMING-RESEARCH SECTION
2301 Peger Road, Fairbanks, Alaska 99701

In 1973, two adjacent experimental road sections incorporating peat underlays were constructed on a permafrost foundation site near Fairbanks, Alaska. The project's objective was to determine if the theoretical advantage of such a design technique would in fact prevent thawing of the thermally unstable foundation. Because the thermal conductivity of peat is much lower in a thawed condition than when frozen, it should act as a summertime thermal barrier while allowing considerable winter refreeze to occur. Field studies have continued through 1982 in order to estimate the long term consequences of this construction method.

Results indicate that a 0.75 m thick layer of consolidated peat can provide significant protection against permafrost degradation under at least the central portion of a road embankment in a warm (-1°C) permafrost area. However, side slope and ditch area thawing and settlement problems, which result primarily from the insulating effects of snow cover, were not prevented by the installation of the peat underlay.

INTRODUCTION

Thaw consolidation due to the melting of ground ice is a costly problem associated with road construction on permafrost. Embankment thicknesses necessary to prevent thaw and differential settlement are often economically impractical. This is especially true in discontinuous permafrost areas where permafrost temperatures approach 0°C. Under such conditions, highway construction methods are usually employed which avoid disturbance of the natural ground cover and in particular, cut sections. This report summarizes a 9 year field study of experimental roadway cut sections which utilized peat as an underlay material in an attempt to reduce or prevent differential settlements. These sections were placed within a new roadway alignment traversing warm permafrost and located approximately 97 km southeast of Fairbanks, Alaska.

Because of its high moisture content and, therefore, high thermal mass, peat has been used in Norway since 1903 as a soil replacement layer to control frost heaving under railways and roads (Skaven-Haug 1959) (Solbraa 1971). However, with a frozen conductivity nearly twice its thawed value, peat should be better suited to preserving a frozen state than a thawed one. This behavior suggested placement of peat into road embankments where a warm permafrost foundation must be maintained.

PHYSICAL SETTING

The study site is located on the Richardson Highway, about 106 km southeast of Fairbanks. The field site is about 1.6 km north of and 76 m above the Tanana River in an area of rolling hills. A cut of up to 9 m in frozen silt was required as a part of a realignment constructed in 1972 and 1973.

The roadway cut intersects a contact between a weathered and frozen schist bedrock and frozen aeolian silt. The natural ground surface is composed of a 0.3 m layer of miscellaneous organics topped by 0.2 to 0.3 m of sphagnum moss. The insulating quality of the surficial organic mat has resulted in a local active layer thickness of 0.5 to 0.8 m. Indigenous black spruce trees have minimized summer's solar heating and the winter's insulative snow layer.

Local silts were characterized by organic contents between 3 and 7% by weight, with 90% minus 0.074 mm, 10% minus 0.01 mm particle size fractions. Excluding the active layer, silt moisture contents decreased with depth and ranged between 30 and 87%, averaging 47%. All samples were above their liquid limit values of 23 to 30%.

Moisture contents within the schist bedrock ranged between 16 and 22%, indicating some potential for thaw-settlement within this material.

Climate conditions at the site are average for interior Alaska with cold dry winters and warm dry summers. The mean site air temperature averaged approximately -3.9°C with short duration extremes of about -51°C and +35°C. Fairbanks weather records compiled since 1931 indicated annual freezing and thawing indices near -3,080 and +1,830°C-days respectively.

DESCRIPTION OF THE RESEARCH SITE

A location was selected for the peat study between Stations 3246+30 and 3252+17 on a Richardson Highway reconstruction project. Figure 1 shows a plan view of the research site. Note that stationing locations described in this report are in English units (feet).

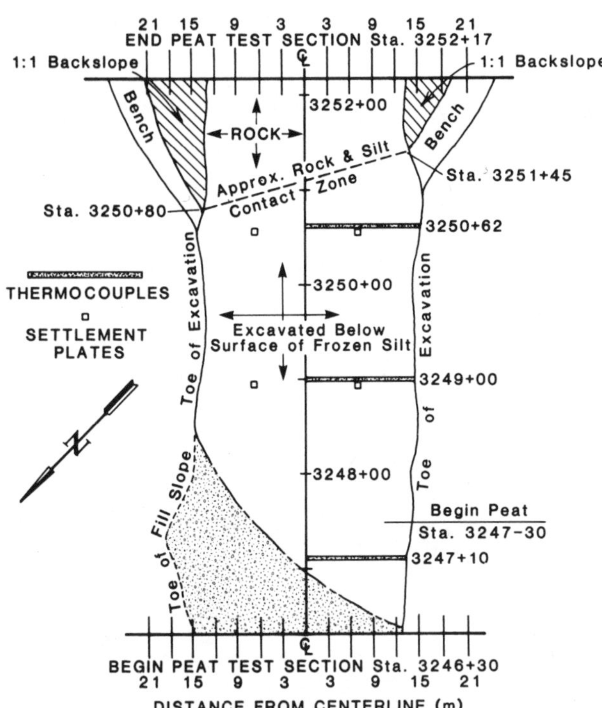

FIGURE 1 Plan view of research site

GENERAL DESIGN

Calculations of seasonal thaw depths used for design of the study sections were based on the Modified Berggren method (U.S. Army 1966). Assumptions made for the in-place peat materials included an equilibrium moisture content, after thawing, of 125% and an unfrozen moisture content of 40% in frozen peat. Frozen peat was used because thawed material was not readily available for construction purposes. Calculations indicated that maximum seasonal thawing would advance 0.6 to 1.0m into the consolidated peat, covered by a 1.5 m gravel layer and an asphalt pavement. For peat placed in an initially frozen state, it was apparent that consolidation would occur in any portion subsequently thawed. Thaw consolidation tests indicated that the peat would compress to about 60% of its original thickness under 1.5 m of fill.

Only a thin granular overlay was initially placed above the peat to carry construction traffic and to promote maximum thaw and consolidation of the peat during construction. A one year waiting period was specified prior to fill completion and pavement placement, to ensure consolidation of the peat layers. Initial (frozen) peat layer thicknesses of 1.2 to 1.5 m were selected, to give consolidated thicknesses of approximately 0.7 and 0.9 m.

For purposes of comparison, a 30 m long control section, similar to the 1.2 m peat embankment design but containing no peat, was also placed within the construction cut. It was to be built using only granular fill and calculations indicated that a maximum yearly thaw of about 4.6 m could be expected. The control, 1.2 m and 1.5 m experimental sections will be referred to as sections A, B and C respectively.

INSTRUMENTATION

One cross-section located in each of the three test areas was selected for temperature monitoring instrumentation. A system of 204 transducers was installed for monitoring thaw depths and temperature changes with time. An undisturbed forest location 27 m left of the road centerline was also instrumented for purposes of general comparison.

Settlement plates were installed at the top, middle and base of the peat layers in sections B and C so that consolidations within and beneath the peat could be measured. Air temperatures were recorded continuously, as were pavement and side slope temperatures at a depth of 2.5 cm. Temperatures along all the transducer strings were recorded monthly.

CONSTRUCTION

Hand clearing of the test area began July 9, 1972. No cut excavation or removal of the insulative moss layer was allowed until daily average temperatures approached freezing. All excavation and placement of frozen peat materials was accomplished between October 16 and 27, 1972, under subfreezing air temperatures. Minor thawing occurred only on the south facing cut slope prior to winter. A total of 4,770 cubic meters of frozen peat was placed in a 1.2 m thickness (sta. 3248+10 to 3249+80) and 1.5 m thickness (sta. 3250+00 to 3251+65).

Construction of the test sections was done in two stages. The 1972 construction included peat placement followed by a 0.9 to 1.2 m covering of decomposed granite. This stage ended November 1, 1972 and normal traffic was allowed over the section during the 1972-1973 winter. Work resumed on the study sections on June 24, 1973. The embankment was raised to final subgrade profile with alluvial gravel, followed by placement of 15 cm of subbase, 11.4 cm of base course and a 3.8 cm thick asphalt concrete surface. The test section was completed July 24, 1973, about a year ahead of schedule. Early completion was problematical, because a year of peat thaw and consolidation was part of the initial research design strategy to prevent excessive postconstruction settlements.

MATERIALS

Properties of embankment materials used in this study are indicated in Table 1. Thermal properties were estimated based on moisture and density data.

Frozen peat used as embankment fill ranged in structure from coarse to fine fibrous. Moisture and organic contents varied from 124 to 300% and 39 to 54% respectively. Compacted wet density was approximately 945 Kg/m^3. Thaw consolidation tests indicated that under 1.5 to 1.8 m of fill, the peat would compress to about 60% of its initial thickness. Moisture contents after thaw consolidation testing averaged 95%.

Materials	Avg. Dry Density (kg/m³)	Moist. Avg. (%)	Moist. Range (%)	Thermal Conductivity (W/mK*) Froz.	Thaw.
Gravel	2,400	4.3	3-6	4.0	3.8
Common Fill	2,062	6.7	3-9	2.8	2.8
Peat Layer (as Placed)	309	204.0	124-299	0.9	---
Peat Layer (Consolidated)	543	95.0	81-133	1.0	0.5

* Note: Heat flow properties from Kersten (1949)

TABLE 1 Embankment materials-physical properties

Surface water flows were locally persistent during the time of this study and resulted in a continually saturated condition in the south side roadway ditch.

TEMPERATURE AND SETTLEMENT OBSERVATIONS

General Temperature Observations

Annual temperature and snowfall data are summarized in Table 2. Site air temperatures averaged -3.2 degrees C during the period of study. Air freezing and thawing indices averaged -2,900 and 1,940°C-days, respectively.

Surface temperatures and their relationships to air temperatures are presented in Table 3. The N factors calculated for these freezing and thawing

Year & Season	Freezing or Thawing Indices (°C-Days) Air	Pavement	Sideslope	N Factors Pavement	Sideslope
74-75 Winter	-3,561	-3,056	-1,000	0.86	0.28
1975 Summer	1,469	2,444	2,000	1.66	1.36
75-76 Winter	-3,611	-3,111	-1,000	0.86	0.28
1976 Summer	1,656	2,722	2,278	1.64	1.38
76-77* Winter	-2,167	-2,444	-1,333	1.13	0.62

* Instrumentation used for measuring surface temperature variations was removed during the summer of 1977.

TABLE 2 Climatological data table

seasons refer to the ratio of ground to air freezing and thawing indices. Average air and surface temperatures, calculated over the same 2½ year period, indicate average air, pavement and side slope surface temperatures to be -4.6, -1.1 and 2.9°C, respectively.

The subsurface thermal regime was greatly affected by surface temperature differences between centerline and side slope/ditch areas. Annual refreeze at the centerline of both peat sections extended through the seasonal thaw zone during all years of the study. In contrast, a perennially thawed zone or "talik" developed

Year & Season	Yearly Avg. Air Temp. (°C)	Season Lengths Thawing (Days)	Freezing (Days)	Snowfall** Total cm	Dist. Factor (%)*	Freezing or Thawing Index for Air (°C-days)
72-73 Winter			191	145	73	-2806
1973 Summer	-4.4	161				1403
73-74 Winter			187	147	62	-3367
1974 Summer	-5.5	166				1722
74-75 Winter			215	251	68	-3561
1975 Summer	-6.2	159				1469
75-76 Winter			195	104	65	-3611
1976 Summer	-3.6	174				1656
76-77 Winter			191	165	41	-2167
1977 Summer	-3.1	177				1722
77-78 Winter			176	94	68	-3185
1978 Summer	-1.0	205				2675
78-79 Winter			199	201	63	-3350
1979 Summer	-2.7	182				2602
79-80 Winter			180	94	54	-1278
1980 Summer	-0.7	177				2482
80-81 Winter			192	112	46	-2697
1981 Summer	-1.3	167				1734
81-82 Winter			204	152	47	-3002
1982 Summer	-3.8	--				

* % of total snowfall before January 1st.

** Snowfall data is from Eielson A.F.B., located 48 km northwest of site.

TABLE 3 N Factors

under the control section centerline after the unusually warm winter of 1976-1977. Maximum seasonal thaw depths have continued to increase under sections A and B. Only section C provided enough thermal barrier to actually stabilize and in fact contain annual thaw penetration.

Ditch and side slope areas experienced very little wintertime cooling due to heavy wintertime accumulations of snow. Progressively deeper thawing under shoulders and side slopes of all test sections, due to the lack of seasonal cooling, produced taliks that continued to grow through the duration of the study.

Temperature versus depth

Seasonal lag is the response time necessary for seasonal surface temperature extremes to be reflected at a given depth. An assessment of the data indicates that seasonal temperature lag was virtually uneffected by the presence of the peat. Temperature extremes within the permafrost layer shifted by approximately 16-20 days per meter of depth.

Temperatures at depths below 9 m are indicated by the deep thermistor data plotted on Figure 2.

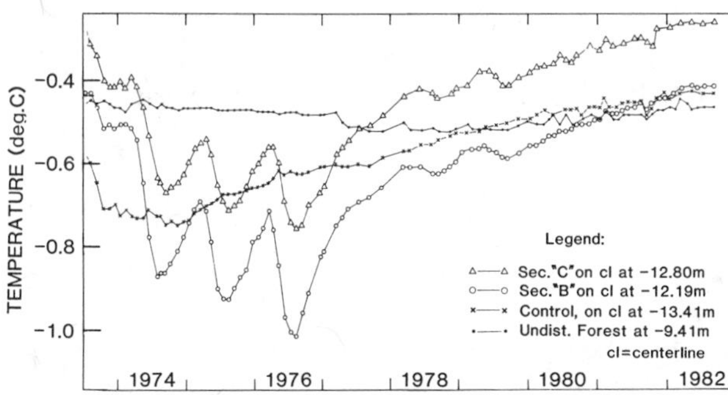

FIGURE 2 Temperature history of deep thermistors

Prior to 1977, the peat insulated sections were characterized by temperatures which were yearly variable but generally decreasing. No such seasonal variations occurred in the non peat control section, which has continually warmed since 1974.

Bottom thermistor temperatures of all sections have risen since 1977. This trend has been strongest in the peat sections (B and C), which have become warmer at depth than the control section. A slight, local warming trend is further evidenced by the natural ground thermistor, which has indicated slight warming since about 1979.

Heat Flow Analysis

The test sections can be compared on the basis of vertical heat flows into and out of the underlying permafrost. Cumulative heat flows were estimated based on vertically spaced thermocouple pairs located within the permafrost, just below the depth of maximum thaw. Since the soil between thermocouples remained frozen, a constant conductivity of 2.1 W/mK, estimated from known soil moisture and density conditions (Kersten 1949), was assumed for the analysis.

Cumulative heat flows in terms of kJ/m² are plotted on Figure 3. Positive values indicate a net accumulation of heat going into the ground (net warming), while negative values indicate a net heat loss or net cooling. All heat flow data are cumulative from the beginning of the project.

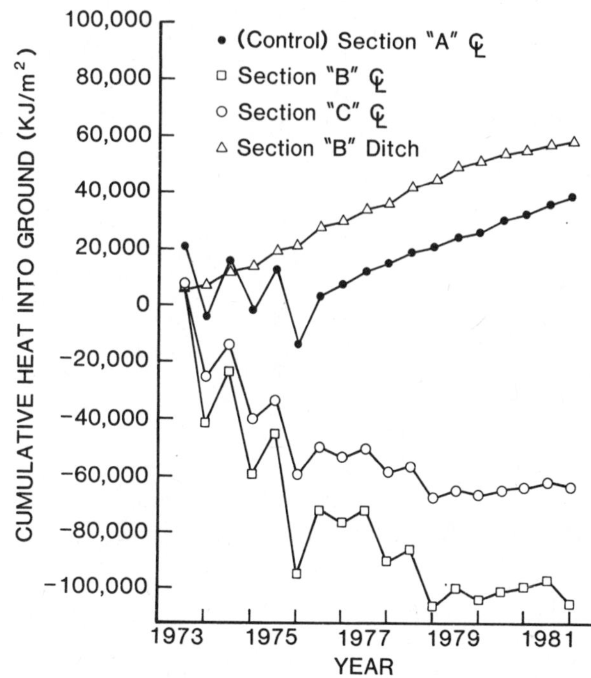

FIGURE 3 Cumulative heat flow into ground

Figure 3 provides a fairly clear demonstration that a thermal advantage in terms of vertical heat flow is gained beneath the cleared roadway as a result of the peat layering, while the snow-covered ditch continued to warm by vertical heat flow throughout the study. It also indicates, however, that a slight but continual heat gain commenced in 1979, even at the centerline of the peat sections. This plus some lateral heating from the ditch/shoulder talik areas would account for the temperature increases at all centerline, deep thermistor locations.

Recent trends in ground heat flow and temperature rise appear to have followed the obvious rise in recent air temperature averages indicated in Table 2. Air warming was reflected more quickly by the non peat control section, where a steadily positive heat accumulation began in 1977.

Seasonal Thaw Depths

Maximum seasonal thawing was found to occur under the test sections in late September or October. The changes in maximum thaw depth from 1973 to 1982 (Figures 4, 5 and 6) illustrate that thawing has progressively increased to the present time. Only the centerline area of section C has shown an apparent stabilization, with a maximum seasonal thaw depth to date extending only slightly below the original cut surface. Ditch areas of section C were subjected to the deepest

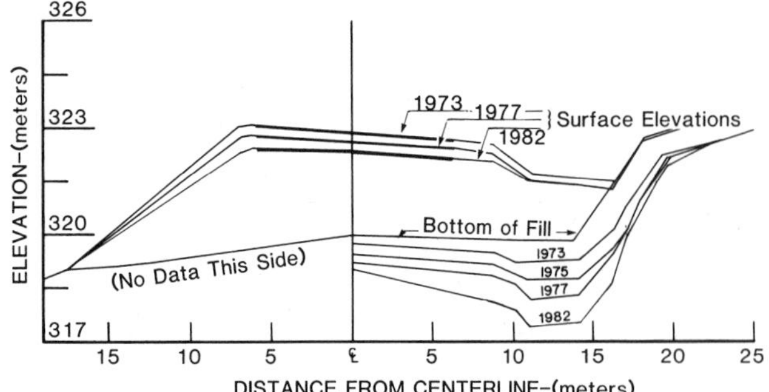

FIGURE 4 Section A--Maximum thaw depths

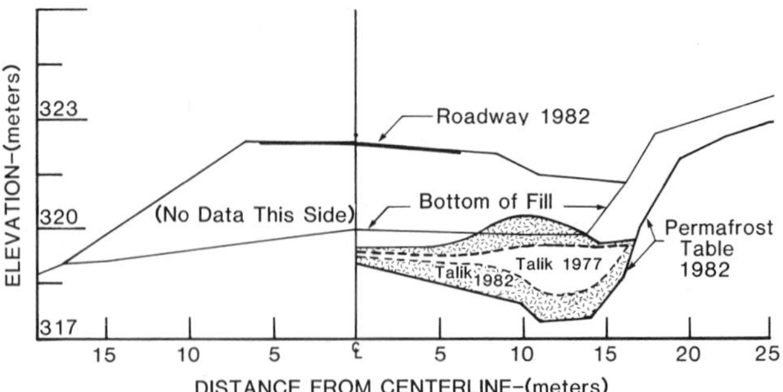

FIGURE 7 Section A--Residual thaw zones

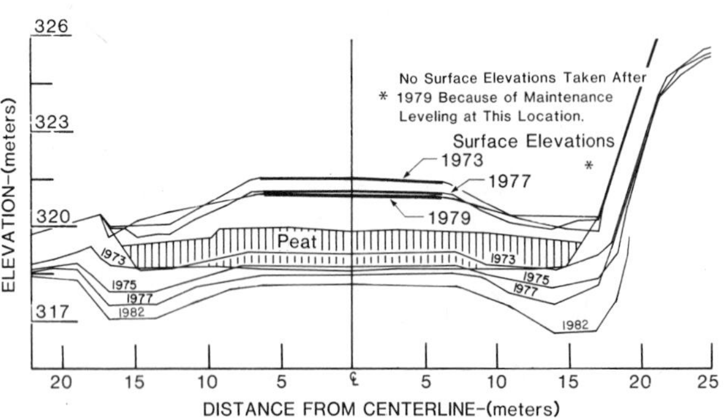

FIGURE 5 Section B--Maximum thaw depths

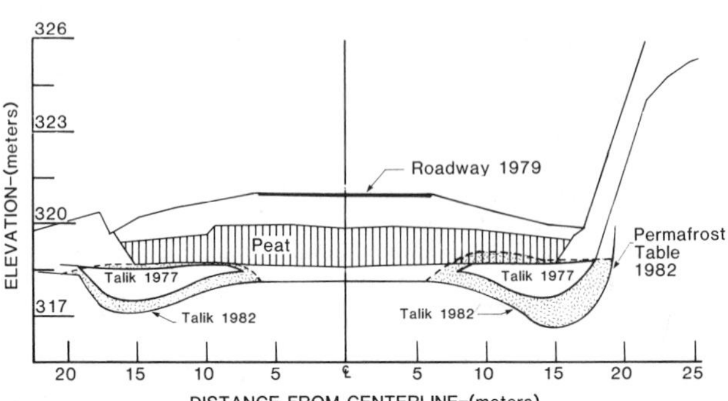

FIGURE 8 Section B--Residual thaw zones

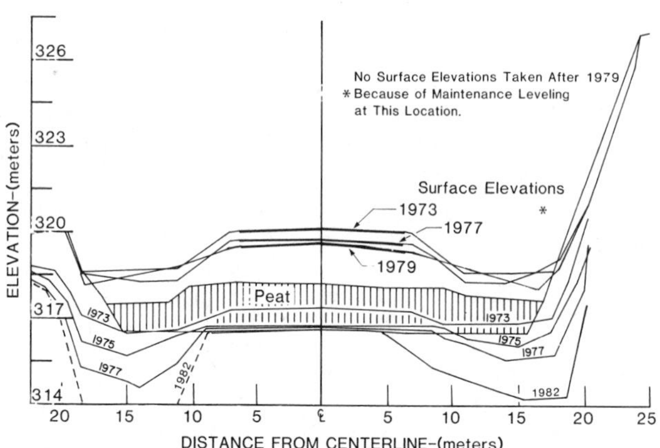

FIGURE 6 Section C--Maximum thaw depths

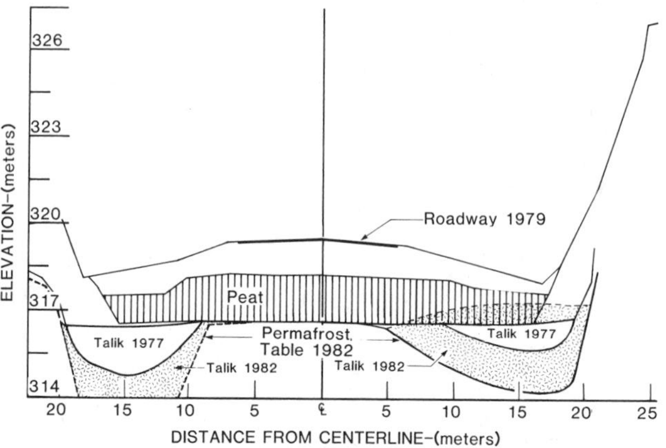

FIGURE 9 Section C--Residual thaw zones

measured seasonal thaw.

Residual Thaw Zones

As indicated by Table 3, the roadway side slope and ditch areas had an average annual surface temperature considerable above the average air temperature. This extreme warming has resulted because the unvegetated gravel surfaces are exposed to the intense warming effects of the summer sun while protected by an insulating snow cover during winter. These areas have thawed more deeply each summer than the maximum depth of refreezing. Talik growth created by this thermal condition is illustrated in Figures 7, 8 and 9. Taliks warm quickly after summer thawing of the overlying seasonal frost layer and progressively deeper annual thawing of the underlying permafrost therefore readily occurs. This results in continuing settlements in the ditch and side slope areas even though an equilibrium thaw depth may have been reached in a more central portion of the paved roadway. Side slope settlements can eventually lead to differential settlements, and cracking of the roadway shoulder. Such settlements are evidenced in Figures 8 and 9 by the irregularly shaped side slopes and ditches.

Settlement Observations

Consolidation histories for the peat layers in both test sections are shown in Figure 10. Essentially no elevation change was measured for any of

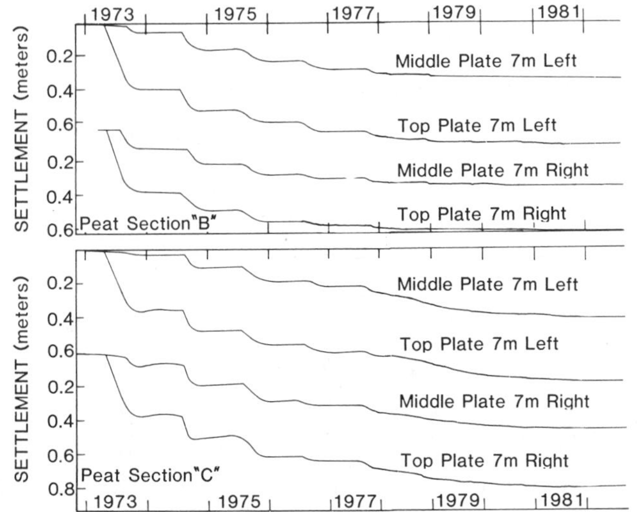

FIGURE 10 Settlement plate evelvation changes

the bottom settlement plates.

Centerline thermocouples indicate that maximum seasonal thawing within section B progressed from the mid point to the bottom of the peat layer between September 1973 and 1977. By 1977 the original 1.2 m thickness had decreased to 0.61 m, and to 0.55 m by October 1982. Similarly, maximum thaw depths reached the mid depth point of the peat layer of section C during September of 1977 and have since appeared to stabilize at the base of the peat. The original 1.5 m thickness had decreased to 0.91 m by 1977, and to 0.73 m by October 1982.

Conclusions

The thermal advantage of peat layering was demonstrated very clearly in the 1973-1977 ground temperature data. A rise in average air temperature since 1977 has caused a generally correlative ground temperature increase and has somewhat obscured the long term effects of peat placement. In general, however, peat improved thermal conditions under the road surface but demonstrated no positive effect in the side slope and ditch areas.

Stabilization of the maximum annual thaw depth has occurred only under the central roadway portion of section C, which exhibits a 1982 consolidated peat thickness of 0.73 m. All other disturbed areas within the research site have shown an unattenuated thaw depth increase since 1973.

Ditch taliks formed in each experimental section apparently without regard to the presence or absence of peat and caused side slope and shoulder cracking. It was obvious that the peat layers did not completely alleviate differential settlement related problems at the research site. The exaggeration of subsurface temperature variations due to a combination of peat layering and warm ditches may, in fact, actually tend to accentuate differential settlements in the case of a uniform permafrost foundation. Conversely, where massive ice exists in the foundation, peat should protect the central portion of the roadway from localized thermokarst effects and reduce necessary maintenance. Combined use of peat construction and wintertime removal of snow from roadside areas should result in a relatively low maintenance road structure for the warm permafrost environment.

ACKNOWLEDGMENTS

This study was supported by the Federal Highway Administration under the Highway Planning and Research Program.

REFERENCES

Kersten, M., 1949, Laboratory research for the determination of the thermal properties of soils, Engineering Experiment Station, University of Minnesota, Eng. Exp. Stn., Final Report.

Skaven-Haug, S., Sept. 1959, Protection against frost heaving on the Norwegian railways, Geotechnique, Vol. 9, No. 3.

Solbraa, K., Dec. 1971, The durability of bark in road construction, Royal Norwegian Council on Science and Industry, No. 5.

U.S. Army, Jan. 1966, Calculation methods for determination of depths of freeze and thaw in soils, Department of the Army Technical Manual, TM-5-852-6.

USE OF SEASONAL WINDOWS FOR RADAR AND OTHER IMAGE ACQUISITION AND ARCTIC LAKE REGION MANAGEMENT

Jack C. Mellor

U.S. Department of the Interior, Bureau of Land Management
P.O. Box 1150
Fairbanks, Alaska 99707 USA

This paper introduces regional remote-sensing data and methods that can provide arctic lake basin characterizations for economical site-specific management decisions. It also illustrates conditions and time periods, hereafter referred to as "windows," that can reduce natural-resource/human-activity conflicts and enhance remote-sensor acquisition of aquatic resource inventory data. Most oil exploration activity can be completed in a cost-effective manner with the least environmental disturbance during the winter surface use window. Lakes, rivers, and wetlands can be managed much like a terrestrial environment during the long arctic winter. Lakes, ponds, and hygroscopic soils cover most of the Arctic Coastal Plain and support an active flora and fauna that are susceptible to disturbance during the short summer. Some seasonal land management factors to be considered are overland and ice-road travel, avoidance of critical habitats during periods of high wildlife use, and water withdrawal from lakes. Data useful for regional inventory, classification, and management of arctic lake resources are April radar images, ice thicknesses, and summer aerial photographs and satellite images. April radar images of arctic lakes may be used to interpret isobaths separating three ranges of water depth (approximately 0-2 m, 2-4 m, and >4 m).

INTRODUCTION

Most of the Alaska Arctic is being studied, explored, and/or leased for energy resource development. Permafrost, ice thicknesses, and surface water present physical conditions that affect oil and gas field operations and wildlife. Seasonal windows can favor the gathering of data for resource inventories and industrial field operations during specific, limited time periods on the Alaskan Arctic Coastal Plain (Figure 1). Knowledge of specific lake depths and regional ice thicknesses aids land managers, who assess resource conflicts, and land users, who need supplies of winter water and heavy overland transport.

From June to September, the Alaskan Arctic Coastal Plain is a wetland continuum covered with tens of thousands of lakes and ponds. Regional inventory or comprehensive lake surveys do not exist. No lake-specific data are available for the vast majority of Arctic Coastal Plain lakes. The only regional data available are thematic maps, such as land cover from computer-classified LANDSAT digital data (Morrissey and Ennis 1981, Walker et al. 1982), which provide only generalized information. Most surface land management decisions are based on exiguous field data and past experience. Site-specific assessments are accomplished when necessary but are kept to a minimum in light of expensive field logistic costs.

Water is a predominate surface feature, which supports a very active flora and fauna, during the summer on the Arctic Coastal Plain. Winter brings changed and reduced faunal distributions and a frozen surface substrate from which lakes, rivers, and wetlands are difficult to discern. Surface water can be evaluated using summer photographs and satellite images.

The bathymetry and potential uses of lakes can be evaluated with winter radar images. Past investigators (Sellmann et al. 1975, Elachi et al. 1976, Arcone et al. 1979, Weeks et al. 1977, 1978, 1981, Mellor 1982a) have identified the potential use of radar images to separate lake areas into two categories, those areas that freeze to the lake bottom versus those that do not. The radar image studies reported here were aimed at applying these findings by quantifying and regionally mapping this radar interpreted 1-2 m isobath. During the course of this research a second mappable isobath ($\approx$4.0 m) was discovered, and the mechanism proposed by the above authors, for radar signal returns through ice containing columnar gas bubbles, was strengthened (Mellor 1982a).

METHODS

Data were acquired to test the hypothesis that significant aquatic resource inventory and assessment could be accomplished economically by applying knowledge of arctic lake ecosystems to regional remote sensing image interpretations. Figure 1 identifies study areas A, B, and C, where lakes were sampled and the stippled transect from Barrow to the Brooks Range where repetitive Side-Looking Airborne Radar (SLAR) images were acquired at a frequency of 9.1 to 9.4 GHz ($\approx$3 cm wavelength) to study the use of radar images for determining lake depths. This paper highlights potential use of airborne radar, photographic, satellite multispectral scanner, and lake ecosystem data that can aid in the management of Arctic Coastal Plain resources. The detailed verification of regional remote-sensing data acquisition and the interpretation methods is provided in Mellor (1982a) and is summarized in Mellor (1982b).

The use of such terms as "operating windows" and "design solution" has only recently crept into arctic resource management (U.S. Department of the Interior, 1983), but the concepts and practices to which they apply have been evolving over the past decade. Biologists and Arctic engineers have identified time periods and conditions when wildlife and the tundra surface can be significantly disturbed. Stipulations developed for field operating windows by identifying those time periods outside the critical periods of impact (i.e., summer thaw, caribou calving, and peregrine falcon and waterfowl nesting) to address that season when activities could be conducted without adverse effect. Design solution and window concepts are provided to the petroleum industry in lease stipulations before exploration field planning begins. This encourages their cooperation and effort in predesign planning of environmentally sound operating procedures.

RESULTS AND DISCUSSION

Ice Thicknesses and Lake Depths

Human activity in the Arctic and the physical presence and activity of fish and wildlife are intimately tied to the freeze/thaw cycle. Most of the surface water volume on the Arctic Coastal Plain is frozen by late April as the ice approaches 2 m in thickness. Lake basins typically have a small percentage area where depths exceed 2-3 m.

Figure 2 illustrates ice accretion on arctic lakes based upon 1978-1979 data taken from nine lakes located within lake study areas A, B, and C within the transect in Figure 1 (Mellor 1982a). Ice accretion continued into May 1979 in northern "A" and mid-coastal "B" plain lakes, while some surface and perimeter melt occurred in foothill "C" lakes. Ice grew most rapidly during October, immediately after freezing over, and was thicker on northern than on foothill lakes throughout the winter.

As ice thickens the winter biota must either move to deeper water or become inactive where frozen to the bottom. The deeper areas are often of small volume with extreme temperature and dissolved oxygen and solute conditions. The amount of water remaining under winter ice cover is critical to aquatic life and is also the only freshwater source for human/industrial activity in this permafrost region lacking in ground water.

Regional ice thickness data in combination with SLAR images have been used to interpret lake isobaths. Regional ice thickness and the 3 April SLAR image in Figure 3 were used to classify the larger lake areas labeled with three ranges of water depth: (a) 0 to 1.5 m, (b) 1.5 m to 4.0 m, and (c) greater than 4.0 m. This image shows an area of deep mid-coastal plain lakes ("B" area in Figure 1) approximately 100 km south of Barrow, Alaska. By 3 April 1979, ice thickness was approximately 1.5 m. The black perimeters of lake images marked "a" depict where they are frozen to the bottom from the shoreline out to the 1.5 m

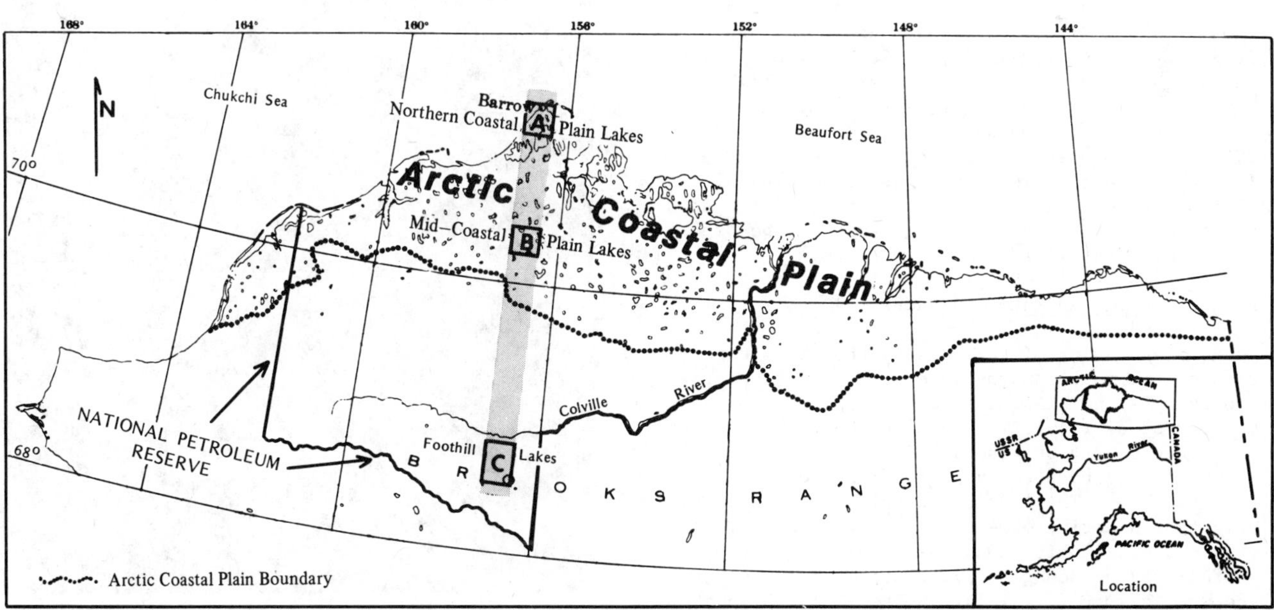

FIGURE 1 Location of the Arctic Coastal Plain, National Petroleum Reserve in Alaska, and study areas within the Alaskan Arctic.

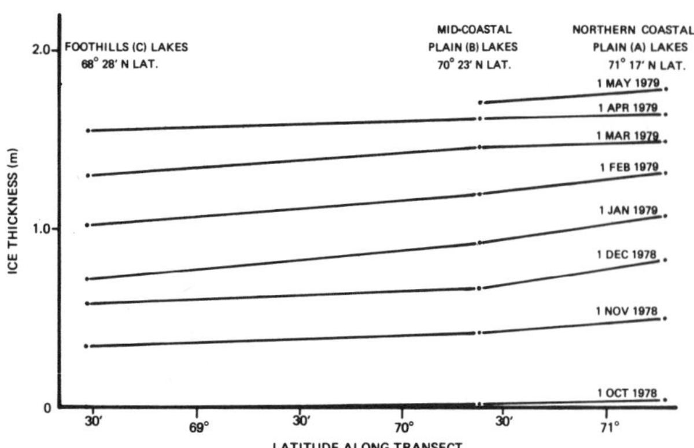

FIGURE 2 Average ice thickness at 1-month intervals as a function of latitude along the study transect.

isobath. One lake in the southwest corner of the image is frozen to the bottom across the entire 2 km long basin. The white areas marked "b" are shallow areas where liquid water exists below the ice from 1.5 m to approximately 4 m. The gray areas over the central portions of four imaged lake basins marked with a "c" have been interpreted to be greater than 4 m deep.

Although devoid of overwintering fish, lake basins shallower than 1.5 m thaw early in the summer, thus attracting early waterfowl arrivals. Lakes with maximum depths in the second category (1.5-4.0 m) may have only marginal amounts of water for winter fish survival. Lakes with water depths greater than 4 m ("c" category) have the greatest potential for a well-established fishery. The areal extent of each depth category compared with the proposed water withdrawal from a lake identifies potential industry/fishery use conflicts. SLAR images acquired in April 1980 over 8 million ha of the National Petroleum Reserve in Alaska (NPR-A) are being used to inventory the three depth ranges for thousands of basins on the 12 USGS, 1:250,000 scale quadrangles covering NPR-A. The mapped depths are used to assess potential conflicts between industry and fish and wildlife uses and to determine the best lakes that may be used as winter water sources and ice runways.

Remote-Sensing Inventory Windows

Remote-sensing of lakes using the visible portion of the electromagnetic spectrum (i.e., aerial photographs and LANDSAT images) is plagued by a variety of difficulties. During the winter, skies are often clear, but the sun is low, on or below the horizon, and the lakes are frozen over and snow covered. The Arctic Coastal Plain, to the uninitiated aerial observer, might appear like a flat, ethereal white sand desert lacking anything to relieve the homogeneity. During the 3 months of ice-free summer (i.e., July-September), while the sun is high and the lakes discernible, incessant ground fog plagues efforts to image the surface synoptically. Completely cloud- and fog-free synoptic, summer coverage of the Arctic Coastal Plain does not exist from either the National Aeronautics and Space Administration (NASA) high-altitude aerial photographic program or the past 10 years of LANDSAT Multispectral Scanner (MSS) images, but these data are sufficient and helpful for updating the 1955 photographic data used for surface water identification on U.S. Geological Survey maps.

Computer-aided characterization of the many Arctic Coastal Plain basins could aid lake inventory and lake data management. For example, a computer software system has been developed (Mellor 1982a) that manipulates summer LANDSAT digital data features (i.e., area, perimeter, crenulation, and centroid). The latitude and longitude of each lake centroid identifies each lake in the data files, providing for retrieval by user-specified geographic area. Lack of LANDSAT images with ice-free basins and cloud-free skies have, among other things, impeded testing and application of this automated system. Lake surface areas are frequently measured manually from aerial photographs for site-specific assessments, but this method is expensive to apply regionally.

Winter ice cover provides a fortuitous medium for radar image interpretation for the ranges of lake depths discussed earlier. April is the best

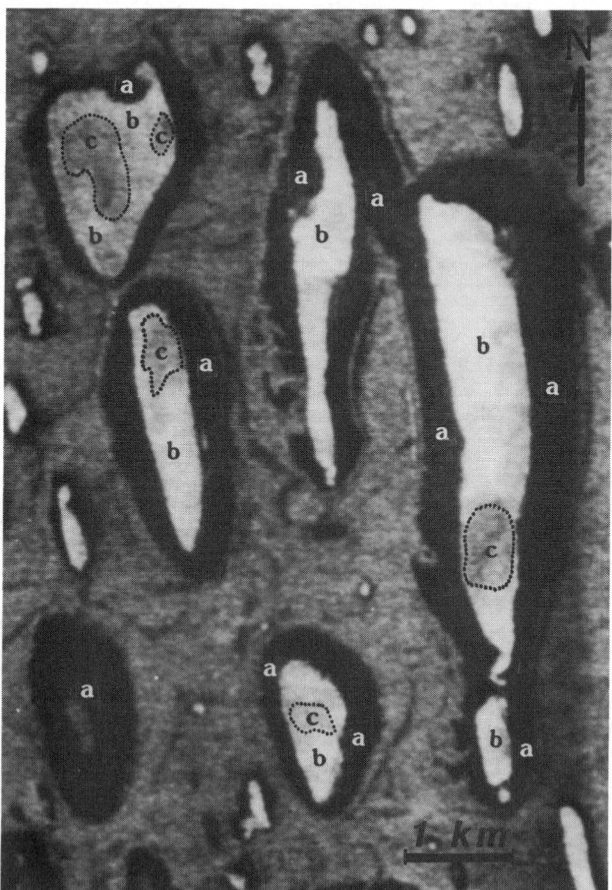

FIGURE 3 SLAR image acquired over mid-coastal plain "B" lakes, 3 April 1979 with depth interpretations annotated (i.e., (a) 0-1.5 m, (b) 1.5-4.0 m, and (c)>4.0 m).

month to record frozen lakes in the Alaskan Arctic because the ice has reached its greatest thickness in the southern foothill lakes before any surface melting occurs. Moisture at the surface or within the ice matrix absorbs the radar signal preventing a strong return upon warming that occurs during May in the Arctic.

The depth interpretation window results from Arctic lake conditions that are different from lower-latitude lakes that lack the strong radar signal returns. The unique aspect of Arctic lake SLAR images is the strong return (Figure 3 with white image "b" areas) over the intermediate 1.5-4.0 m depth range. This depth is a dominant characteristic of Arctic thaw lakes.

The mechanism hypothesized for the strong SLAR signal return is the existence of multitudes of columnar gas bubbles (1-15 cm high) observed in ice cover over lake areas from 0-4 m depth (Mellor 1982a). The gas bubbles are incorporated into the ice matrix after the dissolved gases reach saturation within the limited volume of water beneath the ice. Apparently the columnar bubbles scatter the incident signal radially and forward (downward) without changing the vertical angle of incidence. After reflecting off the ice/water interface the radar signal is forward (upward) and radially scattered and emerges from the ice and snow cover in the shape of an inverted cone. Since the surface of the cone maintains the angle of the incident radar signal, the return signal is strong and produces the bright image over the lake areas less than 4 m deep but deep enough to have water beneath the ice cover. Fewer bubbles provide less radial scattering in ice over lake areas with >4 m depth and produce subtle gray tones ("c" areas) signifying less signal return. Areas frozen to the lake bottom, that lack a finite ice/water interface to reflect the signal, produce a black image ("a" areas). Subtle changes in SLAR image gray tones over lakes can be caused by factors other than >4 m lake depth (i.e., lakes almost frozen to the bottom and ≥ 2 o/oo dissolved salts). More field verification of these hypotheses and criteria (i.e., depth $\gtrsim 4$ m with reduced bubble content in ice cover and increased SLAR image gray scale) are warranted.

The conditions favoring the occurrence of bright SLAR images acquired over Arctic Coastal Plain lakes versus ice-covered, lower-latitude lakes that lack the bright images are as follows. Persistent winds mix the Arctic Coastal Plain lakes so that they are nearly saturated. Most area of even the large deep basins is within the 0-4 m depth range, in which the bright image returns can occur. The long freezing season both concentrates the gases below the ever-thickening ice and incorporates the columnar bubbles within it. Lack of terrain relief allows for image acquisition with little or no shadows. Finally, throughout the arctic winter and spring, lack of solar energy and persistent freezing temperatures cause dry ice and snow conditions conducive to radar signal penetration. Radar signal reflection observations relative to different natural Arctic snow conditions show little correlation with snow roughness characteristics observed.

Additional information can be obtained from sequential radar images acquired throughout the freezing season. For example, images from November, January, and March may be used to determine 0.5, 1.0, and 1.5 meter isobaths, respectively (Mellor 1982a, 1982b). Future satellites may acquire repetitive regional radar images that will provide an economic means for further water depth inventory.

Activity Management Windows

With the warmth of April and May, arctic wildlife activity increases significantly. Raptors, such as the endangered peregrine falcon, arrive in April to begin nesting. During May, caribou begin calving in large, concentrated numbers, and migratory waterfowl arrive in search of ice-free lakes and ponds in which to feed, molt, and nest. The thawing of the active layer above permafrost begins coincident with snow-melt during May. Cross-country transport of heavy equipment that would disturb the tundra surface and could also disturb calving and nesting wildlife must be curtailed.

Figure 4 illustrates these windows of critical wildlife use and some human activities during the calendar year. Since most human and wildlife activities are intimately associated with the presence, absence, or thickness of ice cover, a curve for mid-coastal plain lake average ice thickness measured during 1978-1979 (Mellor 1982a) is included. The active layer and water surfaces begin to refreeze in September, initiating the following progression of thickening ice and increased use by heavy equipment until breakup begins in May and June. Cross-country tractor trains begin tundra travel about the first of November when the active layer has frozen down to approximately 30 cm and there is an average of 15 cm of snow cover. By then, lake ice of about the same thickness (30-50 cm) is safe for light aircraft and surface vehicle use. SLAR images can be used to locate shallow basin areas that freeze to the bottom early in the winter, providing early yet safe trails and landing sites for heavy transport. Snow and water have been used successfully to construct winter ice roads, pads, and airports on the tundra to avoid permanent construction with gravel. January lake surfaces are thick enough to support heavy tractor train transport. February ice thickness on lakes can attain 1.2 m, providing natural landing strips for heavy aircraft, such as the Hercules C-130. Freshwater sources dwindle in both number and volume as the ice thickens. By March water basins greater than 2 m deep must be found to provide a reliable source of water, and care must be taken to avoid dewatering critical fish habitat. April SLAR image depth determinations can help locate sources and assess potential use conflicts with fish such as dewatering and polluting. For example, lake areas with depths from 0.5 to 1.5 m have unfrozen water from late June to February (Figure 4). Cross-country use of the tundra surface is terminated during May snowmelt. Breakup progresses rapidly in June, precluding travel except by boat or improved roads and airstrips. In June and July breakup coincides with the single most critical period of wildlife use. Critical raptor and waterfowl uses continue through most of August. Caribou calving occurs from mid-May to mid-July, and some important summer migration continues through mid-September. Although some potential summer wildlife disturbance

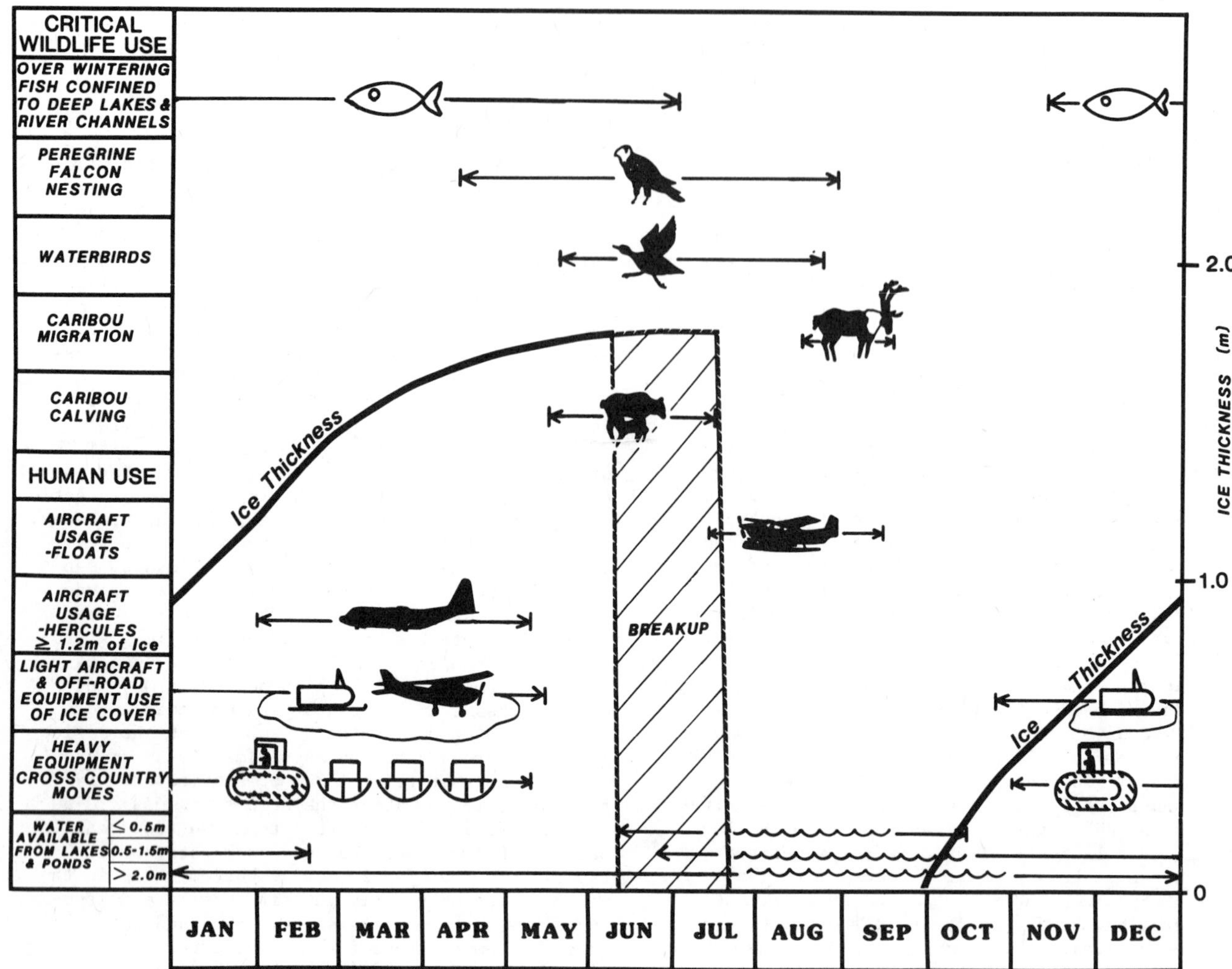

FIGURE 4 Calendar of wildlife/human activity windows relative to mid-coastal plain 1978-1979 ice thicknesses.

exists from aircraft use and improved all-season facilities, breakup and an unfrozen active layer limit vehicle surface travel and construction during these critical wildlife use windows. Infrequent floatplane use occurs while the lakes are ice-free.

The natural separation between human and wildlife use shown in Figure 4 may change with Arctic oil and gas development. Human activity in summer will increase in direct proportion to the construction and maintenance of permanent all-season facilities and must be analyzed carefully to minimize conflicts in use between industry and fish/wildlife. During oil and gas exploration, use conflicts can be reduced by recognizing the natural separation of use windows; encouraging the use of temporary winter ice roads, ice pads, and ice strips; and by discouraging construction of all-season roads and airstrips. This will conserve migratory concentrations of fish and wildlife and populations with low productivity that are sensitive to disturbance by human activity.

SUMMARY

Arctic conditions permit only a narrow summer window for acquisition of visible spectral, remote-sensing data of lake surface characteristics. Active radar systems can provide a reliable means for acquiring both Arctic lake surface and depth characteristics, because they are not dependent upon solar energy or clear skies absent during much of the year. Other physical and climatic conditions provide fortuitous radar interpretation windows for Arctic lake basin characterizations. April radar images and ice thicknesses are being used to interpret three ranges of Arctic lake depth.

Arctic conditions cause natural use windows

that separate critical summer wildlife use periods from most present human activity including winter oil and gas exploration in NPR-A. Permanent all-season facility development needs careful planning to minimize fish and wildlife use conflicts. Lake depth determinations from April SLAR images can be useful in evaluating winter water supplies and also avoiding conflicts with fisheries.

REFERENCES

Arcone, S. A., A. J. Delaney and P. V. Sellmann, 1979, Detection of arctic water supplies with geophysical techniques. U.S. Army Cold Regions Research and Engineering Laboratory, Report 79-15, Hanover, N.H., 30 p.

Elachi, C., M. L. Bryan and W. F. Weeks, 1976, Imaging radar observations of frozen arctic lakes. Remote Sensing of Environment, v. 5, p. 169-175.

Mellor, J. C., 1982a, Bathymetry of Alaskan Arctic Lakes: a key to resource inventory with remote-sensing methods. Ph.D. Thesis, University of Alaska, Fairbanks, AK, 342 p.

Mellor, J. C., 1982b, Remote-sensing inventory of Alaskan lakes, in Proceedings of the Symposium on acquisition and utilization of aquatic habitat inventory information, October 28-30, 1981. Portland, OR. American Fisheries Society, pp. 312-318.

Morrissey, L. A. and R. A. Ennis, 1981, Vegetation mapping of the National Petroleum Reserve in Alaska using LANDSAT digital data. U.S. Department of the Interior Geological Survey, Open File Report 81-315, Reston, Virginia.

Sellmann, P. V., W. F. Weeks and W. J. Campbell, 1975, Use of Side-Looking Airborne Radar to determine depth on the Alaskan North Slope, U.S. Army Cold Regions Research and Engineering Laboratory, Special Report No. 230, Hanover, N. H.

U.S. Department of the Interior, 1983, Final environmental impact statement on leasing in the National Petroleum Reserve in Alaska. U.S. Department of the Interior Bureau of Land Management, Anchorage, AK.

Walker, D. A., W. Acevedo, K. R. Everett, L. Gaydos, J. Brown and P. J. Webber, 1982. LANDSAT-assisted environmental mapping in the Arctic National Wildlife Refuge, Alaska. U.S. Army Cold Regions Research and Engineering Laboratory, Report 82-37, Hanover, N.H.

Weeks, W. F., P. Sellmann and W. J. Campbell, 1977. Interesting features of radar imagery of ice-covered North Slope lakes. Journal of Glaciology, v. 18, p. 129-135.

Weeks, W. F., A. G. Fountain, M. L. Bryan, and C. Elachi, 1978, Differences in radar return from ice-covered lakes. Journal of Geophysical Research, v. 83, p. 4069-4073.

Weeks, W. F., A. J. Gow and R. J. Schertler, 1981, Ground-truth observations of ice-covered North Slope lakes imaged by radar, Cold Regions Research and Engineering Laboratory, Report 81-19, Hanover, N.H.

PERMAFROST DYNAMICS IN THE AREA OF THE VILUY RIVER HYDROELECTRIC SCHEME

P. I. Melnikov and B. A. Olovin

Permafrost Institute, Siberian Branch
Academy of Sciences, Yakutsk, USSR

During 14 years of operation of the hydroelectric installation the permafrost in the flooded zone has thawed to a depth of 15-17 m and to a depth of 20-50 m in the base of the dam. During the construction phase and during the first few years of operation freezing of the taliks below the river bed to a depth of 10-20 m was observed as a result of the drainage of the construction site, the systematic disturbance of the snow cover and to intense air convection in the rock fill. The permafrost layer which had formed had thawed almost completely by 1982. The temperature field of the dam is characterized by extremely low mean annual temperatures; however, the spatial characteristics of the zone of negative temperatures differ significantly from those forecasted using a stationary model. The formation of temperature anomalies within the body of the dam is due to use of highly gas-permeable materials ($k = 1 \cdot 10^{-7}$ m^2) in its construction.

In water development projects fundamental changes occur in the conditions of the external heat exchange of permafrost formation owing to (1) artificial regulation of the river run-off and construction of water storage reservoirs, (2) removal during the construction of considerable volumes of earth material, (3) driving mining workings, and (4) construction of dikes and dams. The variation of the hydrological regime of rivers is characterized by an increase in winter run-off, intensification of filtration processes at the base and shore contiguities of dams, and a rise in the mean annual temperatures of the rock surface in the flowage zone.

For hydroelectric projects, the study of the temperature dynamics of the strata in the zone of thermal effect is necessary to predict filtration losses from water reservoirs and to evaluate the stability of structural elements of engineering projects. The prediction of the nonstationary temperature field, taking into account water filtration, may currently be highly approximate due to the low reliability for determination of the filtration properties (permittivity and piezoconductivity) of thawing permafrost, carried out most frequently by indirect methods. Water development projects in the extreme North are marked by the extreme complexity of permafrost-hydrogeological conditions, by a large number of factors that influence the temperature field at the base of the project during construction and operation, and by the nonpredictability of the many physical properties of the strata near the project. This paper presents some results of long-term temperature observations around one of the largest hydroelectric stations in the USSR, constructed in an area of continuous permafrost. The structure, construction methods, and primary changes in the development of the temperature field inside and at the base of the Viluyskaya Hydro structures have been reported in a series of papers (Biyanov 1975, Kamensky 1973, Olovin and Medvedev 1980).

ENGINEERING-GEOLOGICAL AND GEOCRYOLOGICAL CONDITIONS OF THE REGION

The Viluy Hydro dam, 75 m high and 600 m long, was erected in West Yakutiya on a stratum intrusion of weak-jointed dolerites (diabase), existing to a depth of more than 100 m. The dolerite permeability in the region of the river bed is, according to field test measurements, about 1 m per day. Before construction, the permafrost permeability of the dam base after thawing was not specifically investigated. On the basis of visual estimations in developmental work, the rough value of the permeability of the thawing ground was assumed to be 10 m per day. Deep pressure grouting was planned for areas with higher permeability.

The frozen rock temperature on watersheds varies from -1° to -3°C and averages -2.5°C in areas with undisturbed soil-plant cover. The frozen rock temperature at the bottom of the annual temperature variation zone varies from -1 to -9°C, depending on the distance from the river bank, exposure, mean slope angle, surface altitude, and lithology. Statistical treatment of geothermal observations on slopes has shown that outside the thermal influence zone of the Viluy River ($L \geq 100$-200 m) the frozen rock temperature varies in accordance with:

$$t = -6.6^\circ + \Delta t(z) + 0.037\beta - 1.4\cos(63 \pm \alpha) + \Delta t(\alpha_m),$$

where t = the temperature at a depth of 20 m;
$\Delta t(z)$ = the empirical correction for the difference in surface altitudes;
β = the angle between the drop line and north ($\beta = 0$-180°);
α = the incidence angle of the slope (positive value α is taken for slopes of north and near-north exposure, the negative value for south);
$\Delta t(\alpha_m)$ = the temperature correction, taking into

account the presence on the slope of a suffusion-destruction subfacing of a rock stream slope.

The influence of the various factors on permafrost temperature decreases in the following order:

- Slope exposure, $\Delta t(\beta) \leq 5.6°C$;
- Absolute elevation, $\Delta t(z) \leq 2.9°C$;
- Presence of abrupt slopes, $\alpha > 40°C$, with an extension of steep areas $l > 10\text{-}15$ m, $\Delta t(\alpha_m) = 2.1\text{-}2.5°$;
- Incidence angle of the slope at $\alpha < 25°$, $\Delta t(\alpha) \leq 1.0°$.

The thickness of the permafrost bench on the watershed areas near the hydro project is 600-800 m. Ice in porous spaces is found only in the upper levels: in a layer 80-90 m thick in the river bed and 300-320 m on watersheds. Below this, water shows increased mineralization. The temperature field in the river valley has been traced down to 100 m. In the interval of the depths studied the rocks are in a frozen state. Taliks (unfrozen ground within permafrost) occur everywhere beneath the river bed, but since all the temperature curves for bore holes driven in the river bed are characterized by negative gradients, one may surmise that the river bed talik in areas of bed narrowing is not an open one.

The permafrost water content outside the weathering zone, due to the presence of crack-vein ice, does not exceed several percent and increases toward the ground surface and toward the valley edges by the most disturbed process of frost weathering.

PERMAFROST THAWING IN THE FLOW ZONE

Permafrost thawing dynamics within the reservoir bed have been studied for 15 years in an experimental area about 1 km from the dam. The rocks in this area are represented by jointed dolerites, weathered from the surface down to 1 m to form grus and crushed rock. The area had been entirely stripped of trees, and the soil-plant layer was completely removed from a considerable portion. The extent of the study area extended for 600 m perpendicular to the valley. The surface slope is 7.9°. The depth of seasonal thawing, which under natural conditions did not exceed 1.5-2.0 m, increased to 3.4-4.0 m after the soil-plant layer had been removed. The thermal diffusivity K of the frozen dolerite as measured by in situ methods at depths of 20-50 m is $28 \cdot 10^{-4} m^2/hr$ at a variation coefficient of less than 5%; from 10-20 m, K increases up to 34 or $35 \cdot 10^{-4} m^2/hr$; heat conductivity coefficient λ is 2.27 W/m·K at 20-50 m and at 2.76-2.84 W/m·K at shallower depths. The permafrost temperature in the area is -4 to -4.5°C. The mean annual water temperature in the reservoir is close to 4°C.

Statistical treatment of observations from three bore holes in the flow zone over a 12-year period has yielded empirical relationships between the thawing depth ξ (m) and time τ (years) of the form

$$\xi = C_1 + C_2 \sqrt{\tau},$$

notably for bore hole 25:

$$\xi = -1.21 + 4.83 \sqrt{\tau}\text{ , m} \quad (1)$$

bore hole 26:

$$\xi = -1.57 + 5.03 \sqrt{\tau}\text{ , m} \quad (2)$$

bore hole 27:

$$\xi = 2.51 + 3.83 \sqrt{\tau}\text{ , m} \quad (3)$$

The relationship between ξ and τ is characterized by the following correlation coefficients: dependence (1) 0.976;
(2) 0.955;
(3) 0.973.

During the last 5 years the dolerite's mean rate of thawing has been 0.6-0.7 m per year. The obtained data characterize the mean rates of thawing of dolerite, due solely to conductive heat transfer.

The permafrost thaw rate at the base of the dam is influenced largely by convective heat transfer, due to water filtration through the jointed rocks of the base. The mean rate of permafrost thawing at the base of the grouting gallery (Figures 1 and 2) varies from 0.8 to 1.2 m per year on the right-hand bank (bore holes 2-4), up to 7-8 m per year (bore holes 9 and 10) on the left-hand bank. The extremely large thaw rates observed in 1973 and subsequent years in the vicinity of bore hole 1 are due to an additional influence of horizontal heat flow on the side of the right-bank water supply channel. The high thaw rates in the upper 10-m layer observed at the time the reservoir was filled (1968-1969) in the vicinity of bore holes 3 and 4 (Figure 3) are due to the high permeability of the upper levels of the cryogenic structure. As the reservoir was filled, the permafrost thawed very rapidly around the tunnel conduits that supplied water from the canal to the hydroelectric units of the right-bank station. The permafrost thaw rate was about 5 m per year radially from the water duct walls.

Permafrost thawing at the base of the left-bank structures occurred with a significant time lag despite the higher permeability of the rocks. In 1975 the talik was only 2 m thick, about 10 m wide, and it was located at the base of the left-bank supply canal beneath the upper cofferdam 24 m high (the maximum water pressure on the side of the upper reach was 0.17 mPa), indicating a very insignificant discharge of filtration flow in the left-bank contiguity and an extremely insignificant thawing of the base of the river diversion canal.

The talik below the river bed began to freeze while the dam was under construction, after a short time blocking of the bed in 1964 and full blocking in 1965. During that period the soil-plant layer was almost completely removed and the conditions of snow cover formation, as compared to natural ones, were substantially violated. In the river bed this resulted in the formation of a permafrost layer up to 8-9 m thick in the extended area of the bed and 10-12 m thick in the narrowed area due to the decrease of mean annual temperature from 3-4° to -3° to -4°C. In the winter of 1964-65, a 40-m-high retaining wall of rock fill was constructed, causing the mean annual temperatures of the base surface near the geometric axis of the dam to decrease to -10° and -15°C. Toward the downstream wall the low mean annual temperatures were from -20° to -30°C.

An estimate of the freezing rate of the base can be made by the known formula $\xi = (2\lambda \cdot \Delta t \cdot \tau / q_o W)^{0.5}$, in the form

$$\xi = (\mathit{f} \cdot \Delta t \cdot \tau)^{0.5} \quad (4)$$

convenient for using the data on the thermal physical properties of dolerites, influencing the value of the empirical coefficients in formulas 1 through 3). Taking into account that permafrost thawing in the flow zone occurred as a result of an increase in the surface mean annual temperature by $\Delta t = 8°C$, the mean value of ξ for dolerites of the study area can be determined by formulas 1 through 3, assuming $C_1 = 0$, thence

$$\xi = 2.65 \frac{M^2}{°C/yr}.$$

Thus, for rough calculations of freezing rate in the absence of water filtration in base rocks one may use a semi-empirical formula

$$\xi = (2.65 \cdot \Delta t \cdot \tau)^{0.5} m, \quad (5)$$

where τ is measured in years; Δt in °C.

Formula 5 can be used to determine the freezing depth of the dam base in the river bed which at a single temperature jump for the surface $\Delta t = 20°C$ should have been 16 m within 5 years of the start of construction (1970) and 23 m 10 years later, in 1975.

As the dam filled, rock freezing occurred, most intensively in regions distant from direct contact with the water reservoir. The main reasons for deep freezing were earth material fill at negative air temperatures and intense inter-pore air convection in large fragmented rocks of the retaining wall. In the construction stage the dam-retaining wall and the loam shield core were almost completely frozen up to 20 m in height. Due to the high air permeability of the rock fill, on the order of $1 \cdot 10^{-7} m^2$ (10^5 darcy) and a substantial difference in the air densities of the atmosphere and the interpore space throughout the year, there is air motion through the surface of a dry slope with periods of 24 hours and 1 year. There is a close correlation between air flow Q, through the dam surface, and ground layer temperature t_B:

$$r(Q, t_B) = 0.3 - 0.6.$$

There are 3 circulation patterns—winter, summer, and mixed. In winter the air motion is predominantly ascending and in summer, it is descending. Mixed circulation may be observed at any time of the year. It is a circulation pattern in which atmospheric air enters the middle part of the retaining structure and divides into two flows, ascending and descending. The total air flow, as measured with flowmeters with a receiving surface area of 22.5 m^2 (1.5 m wide and 15 m long along the slope drop), varies throughout the year in the direction with an amplitude of 50 m^3 per hour. Fluctuations of instantaneous flow values during the day are several times smaller.

As a result of air convection in the retaining structure, there are anomalously large annual temperature variations, reaching several °C even 50-60 m away from the surface of a dry slope. The mean annual temperatures of the rock fill are lower by 10-20°C than those characteristic of West Yakutiya. The depth of seasonal freezing-thawing near the dam crest reaches 20 m (Figure 3). In the 14 years since the beginning of construction the mean annual temperature of rocks within the dam body increased by 5-10°C, the amplitude of annual variations decreased substantially, and frozen loams in the shield thawed completely. In the river bed, rapid thawing of rocks occurred where the dam base comes into contact with the retaining wall. The increase in the rock temperatures might be due to a lowering of their permeability caused by collapse of earth material that was weakly pressurized during construction and to the formation of intraground ice due to infiltrating atmospheric precipitates and condensation of water vapor from the air moving through the pores of the rock filling.

Despite the continual decrease of the area with negative rock temperatures, the spatial distribution of mean annual temperatures is far from being a relevant stationary state as predicted in terms of an analog model (Kudryavtsev et al. 1963) that took into account the possibility of heat transfer only by the mechanism for molecular heat conduction. Evidently, it will take several decades to establish a stationary state for the dam at Viluy Hydro, in the course of which the porous space in the negative temperature zone can be completely filled with ice.

The main source of water for the formation of intraground ice is seepage. The mean flow of water through the dam's surface is about 100 kg/m^2 per year.

The lowest temperatures in the rock fill were observed under the road on the berm at the 16 m mark (Figure 3a). The berm is constructed where the river bed is narrowest, not exceeding 100-200 m at normal water level; therefore at the berm base there are favorable conditions for deep freezing of the dam and formation of a continuous ice-ground curtain. The temperature variations in bore holes 14, 17, 18, and 19 (Figure 4b), as measured in 1972-1975, confirm the deep freezing of the talik below the river bed at the base of the dam. Observations of the intensity of air filtration through the dry slope of the dam in the range of 0-16 m and a study of pit walls driven in rock filling from 16 m showed that the amount of intraground ice in the pore space does not exceed 50% of its volume, therefore despite the low temperatures of the rock fill this was not an essential obstacle to the water flow, filtered through the shield and base and moving toward the tail race on the same surface as the dam base. The flow of filtered water during 1967-1974 seems to have been in the region of point A (Figure 4b). The short-term permafrost formed earlier has now practically entirely thawed, while the permafrost boundary approaches what it was before construction (Figure 4c).

CONCLUSION

Several years of investigations and observations at Viluy Hydro, constructed under the climatic conditions of the extreme North and the massive propagation of rock permafrost, yielded unique data characterizing the heat and mass exchanges within the body of a rock-filled dam, at its base, and in the river bed, as well as an understanding of the influence of the water reservoir in thawing the underlying permafrost. The construction of the hydro project has greatly influenced the thermodynamic characteristics of the permafrost. This influence is propagated tens of meters away from the boundaries of the structures erected. With a general trend toward permafrost degradation occurring at a rate of from 0.6 to 7-8 m per year, in individual areas there is rapid freezing of taliks. This must be taken into account in the design of drainage systems.

During the past 14 years of project operation, the heat and mass exchange within the body of the dam, at its base, and in the river bed has not stabilized. The formation of a temperature field continues on and its dynamics do not yet permit an accurate prediction of the thermodynamic cryogenic process that is occurring. The investigation and observation at the Viluy hydro project must be continued and extended.

REFERENCES

Biyanov, G. F., 1975, Dams on permafrost: Moscow, Energiya.

Kamensky, R. M., 1973, Thermal regime of the base and shield of the dam of the Viluy Hydro, in II International Conference on Permafrost, v. 7: Yakutsk, p. 228-235.

Kondratyeva, K. A., 1963, On the influence of topography on the formation of the temperature regime of rocks in the presence of temperature inversion in the region of the Viluy Hydro, in Permafrost investigations, v. 3: Moscow, Izd-vo Moscow University, p. 73-82.

Kudryavtsev, V. A. and Bakulin, V. P., 1963, Prediction of the temperature regime established in the process of exploitation of the Viluy Hydro of the body and base of the dam: Vestnik MGU, Ser. geolog., no. 5, p. 70-77.

Olovin, B. A., and Medvedev, B. A., 1980, Dynamics of the temperature field of the dam of the Viluy Hydro: Novosibirsk, Nauka.

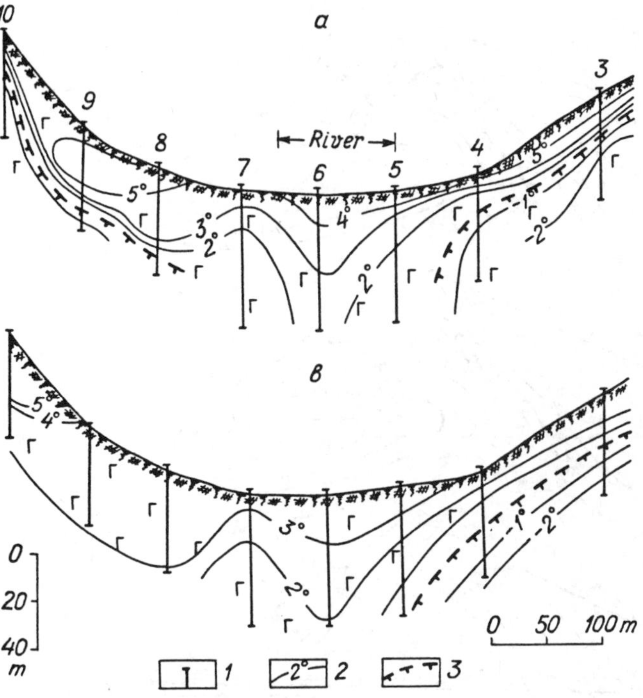

FIGURE 1 Temperature field at the base of a grouted gallery, (a) November 1968 and (b) November 1981. (1) geothermal bore holes; (2) isotherms, (3) isotherm t = 0°C.

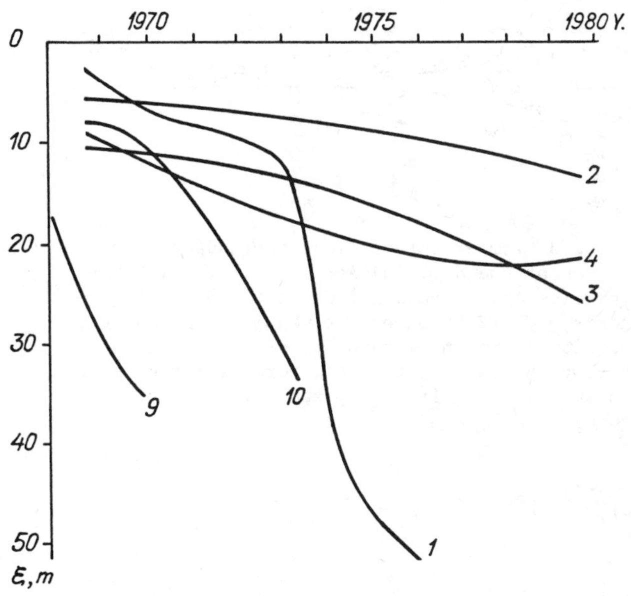

FIGURE 2 Dynamics of thawing of the base of a grouted gallery in the vicinity of bore holes 1-4, 9-10.

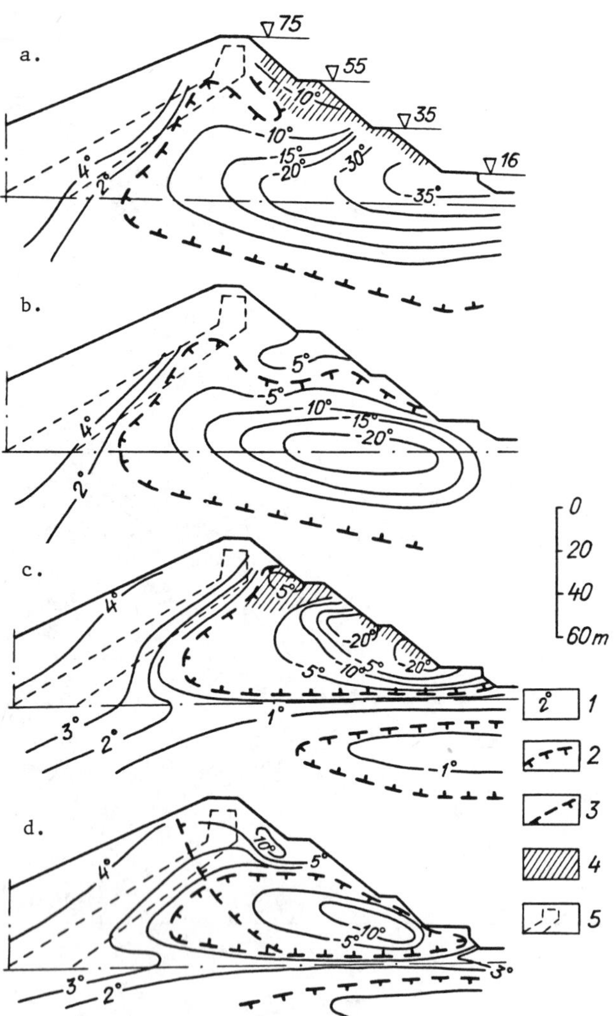

FIGURE 3 Temperature field throughout the cross-section of the dam; (a) April 1969, (b) October 1969, (c) April 1981, (d) October 1981. (1) isotherms, °C; (2) isotherm, t = 0°C; (3) calculated position of isotherm t = 0°C, corresponding to a stationary state of the structure; (4) region of seasonal thawing-freezing; and (5) loam shield (core).

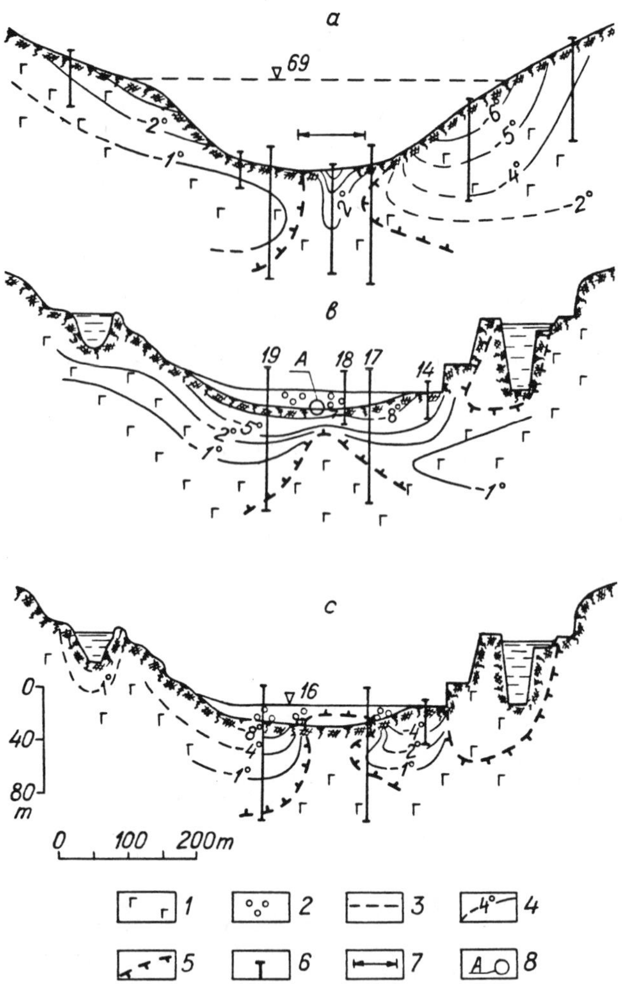

FIGURE 4 Transverse temperature cross-sections through the Viluy River valley near the tail race, on the berm with a 16 m mark, constructed from the observations made in (a) 1960, (b) 1970, (c) 1981. Data from Kondratyeva (1963) were used in generating the cross-sections. (1) dolerite (dam's base); (2) large-fragmented rocks (dam); (3) water; (4) isotherms, °C; (5) isotherm, t=0°C; (6) geothermal bore holes; (7) position of the Viluy River bed prior to construction; (8) anticipated position of talik, where water flow filtrated through structures.

ISOTOPE VARIATIONS IN PERMAFROST WATERS ALONG THE DEMPSTER HIGHWAY PIPELINE CORRIDOR

F. A. Michel

Department of Geology, Carleton University, Ottawa, Ontario, Canada K1S 5B6

Oxygen-18 contents of permafrost waters have been determined for representative core samples from the proposed Dempster Highway pipeline route. Many samples contain water with an isotopic content ($\delta^{18}O$ = -20 to -23°/oo) similar to modern precipitation which suggests significant groundwater migration near surface. Samples from the unglaciated northern Yukon have ^{18}O contents ranging from -14 to -27°/oo which are suggestive of climatic changes since the late-Wisconsinan. Superimposed on these variations are 2 - 2.5°/oo fluctuations formed by isotope fractionation during freezing.

INTRODUCTION

Information about the history of permafrost, and the water or ice contained within the permafrost zone at any site, can be obtained through the study of environmental isotopes. Mackay (1972) demonstrated through the use of oxygen isotopes that relic land-based permafrost exists within the Beaufort Sea, while Stuiver et al. (1976) were able to distinguish zones of marine and glacial water within permafrost in Antarctica. Detailed isotope investigations within the Mackenzie Valley and the Mackenzie Delta (Michel 1982, Michel and Fritz 1978, 1982a) have been successful in documenting the existence of age variations for water (ice) in the permafrost, and correlating these isotope variations to paleoclimatic conditions. In this paper, variations in the isotopic composition of permafrost waters collected along the proposed Dempster Highway pipeline route are described. The term 'permafrost waters' is used, herein, to refer collectively to H_2O in all 3 states of existence within the permafrost, most of which will usually be as ice.

STUDY SITES

The proposed route for the Dempster Highway pipeline closely follows the Dempster and Klondike Highways from Inuvik to Whitehorse. The major physiographic regions traversed by the route include the Mackenize Delta, Mackenzie Valley, Peel Plain, Richardson Mountains, Eagle Plain, Ogilvie Mountains, Tintina Trench, and Pelly Mountains. Permafrost has been encountered along the entire route with the largest ice-rich sections located in the unglaciated Eagle Plain of the northern Yukon (Foothills 1978). A large number of core samples, from boreholes located along the entire route, were provided by Foothills Pipe Lines (Yukon) Ltd. from their 1978 drilling program.

On the basis of the borehole logs provided to the author by Foothills Pipe Lines (Yukon) Ltd. and a preliminary examination of samples from a number of boreholes, a detailed investigation was undertaken on core material from the Eagle Plain - northern Ogilvie Mountains region. All available core samples from boreholes 78-141, 78-142, 78-148, and 78-165, were processed for isotopic analysis. The location of each borehole is shown in Figure 1.

ANALYTICAL PROCEDURE

All samples were shipped frozen from the field to Calgary where they were stored in freezers for geotechnical testing. Samples which were not tested were sent frozen to the author at the University of Waterloo in 1981. Upon arrival, the samples were stored in a walk-in freezer at -10°C.

Each sample processed was unwrapped in the freezer where the outer layer of soil and ice were removed and discarded. The innermost core of each undisturbed sample was described and then sealed in a thick-walled plastic bag. The sample was first allowed to thaw completely and equilibrate at room temperature. Each sample was then placed within a steel jacket and subjected to a controlled hydraulic pressure in order to squeeze water from the soil. The procedures for water extraction are described by Patterson et al. (1978).

To determine the ^{18}O content of a sample, 8 ml of water were equilibrated for at least 4 hours with a standard CO_2 gas. The oxygen isotope ratio of the equilibrated gas was measured relative to a laboratory working standard. To measure the ^2H content of selected samples, 20 µl of water were circulated over hot uranium metal (800° - 850°C) and converted to hydrogen gas for analysis. Both the oxygen and hydrogen isotopes were measured using a Micromass 602D mass spectrometer at the Isotope Laboratory, Department of Earth Sciences, University of Waterloo. The results are expressed as parts per thousand (permil, °/oo) difference between the sample and the reference: Standard Mean Ocean Water (SMOW). These differences are referred to as δ values and a negative value describes a concentration less than that of the reference standard. The analytical error in δ for ^{18}O is less than 0.2°/oo, while the error for ^2H is less than 2.0°/oo.

DISCUSSION OF RESULTS

Shallow Groundwater

Throughout the area of study, shallow groundwaters have been sampled for isotope analyses. ^{18}O contents are consistently within the range of -20 to -23°/oo (Michel and Fritz 1982b). During the preliminary examination of core material by Michel and Fritz (1982b), the water within the uppermost permafrost was generally found to have $\delta^{18}O$ values in the range of -19 to -22°/oo. The uniformity of $\delta^{18}O$ values within a relatively narrow range for groundwaters and upper permafrost waters is indicative of the average annual composition of modern precipitation and this provides a reference with which to compare the isotopic composition of other permafrost waters.

Eagle Plain - Northern Ogilvie Mountains Region

Previous studies by Michel (1982) and Michel and Fritz (1978) were undertaken in areas which were glaciated during the late-Wisconsinan. As a result, it is impossible to determine whether the $\delta^{18}O$ values of -29 to -31°/oo reported in these studies represent waters which infiltrated from melting glaciers, or directly as precipitation. Glacier ice could have formed from snow which fell tens or hundreds of kilometres away from the point of melting and, therefore, would not be representative of the local climate at the time of melting and recharge.

The Eagle Plain, lying between the Richardson Mountains to the northeast and the Ogilvie Mountains to the south, is an area which was not glaciated during the Wisconsinan. The northern edge of the Ogilvie Mountains was subjected to localized alpine glaciation, but the cores examined during the study were from beyond the limit of glaciation. Therefore, it is possible within this region to examine isotope profiles with groundwater recharge which was not derived from glacier ice. However, when examining cores from low-lying areas, such as valley floors, it is necessary to recognize the possible existence of meltwaters from alpine glaciers. Four cores from the region have been examined during the present study.

Borehole 78-165 is located adjacent to the Dempster Highway just south of where the highway crosses the Eagle River. The borehole was drilled approximately 10 m from the edge of the bank along the Eagle River valley. This bank rises several tens of metres above the valley which served as a major glacial spillway during the late-Wisconsinan period. Work by Hughes et al. (1981) indicates that deglaciation of the Peel River - Bonnet Plume Basin area to the south was underway by 16,000 years ago.

The ^{18}O profile shown in Figure 2 reveals the presence of two distinct water masses. Within the silt and clay units below 10 m, the ^{18}O contents are slightly less negative than modern groundwater values ($\delta^{18}O$ = -21.5°/oo) for the area. The waters are considered, therefore, to be representative of a slightly warmer time period. The ^{18}O contents of water within the organic silts vary between -26 and -27.5°/oo and are approximately 6°/oo more negative than modern waters in the area. A temperature estimate based on these data would indicate that these waters were recharged during a period when the annual air temperature was at least 9°C lower than today. If such an estimate is correct, then the waters would have to be late-Wisconsinan in age. The deeper permafrost waters are considered to be part of a groundwater flow system active during the Holocene. Crampton (1979) has reported the existence of a thick interpermafrost talik within the Eagle River valley near this site and water encountered in the talik was under pressure.

Further to the southwest, borehole 78-148 was drilled on gently sloping ground approximately 3 km east of the Ogilvie River. The stratigraphy at this site consists of peat and organic silt to a depth of 2 m underlain by 4.5 m of silt which overlies at least 3.3 m of gravel. Core samples available for study do not include the gravel unit at depth.

As can be seen in Figure 3, the ^{18}O profile displays a shift from modern values near surface (-20 to -22°/oo) to more negative values (-26 to -27°/oo) at depth. The $\delta^{18}O$ values in the range of -26 to -27°/oo are similar to those in the organic silts of borehole 78-165. These values would yield a similar temperature estimate as previously described for borehole 78-165 and these waters are assigned, therefore, to a late-Wisconsinan age.

It is unlikely that the shift to less negative values toward the surface is the result of mixing between modern groundwaters and late-Wisconsinan groundwaters, since such an effect would tend to produce a smoothed profile. Variations formed as a result of freezing could have a magnitude of up to 2.8°/oo for ^{18}O, if the isotopic fractionation occurred under equilibrium conditions (Michel 1982). In the profile of Figure 3, variations at a depth of 4 m exceed the 2.8°/oo value and cannot be explained solely by fractionation during freezing. Therefore, this gradual shift is probably representative of changing climatic conditions since the late-Wisconsinan.

The lowermost sample analysed in this borehole has a much higher ^{18}O content than the sample immediately above and is several permil less negative than modern water in the area. Examination of the ^{2}H vs. ^{18}O plot in Figure 4 reveals that this sample has not been altered by secondary processes. The thick gravel unit only 0.5 m below this sample could represent a former interpermafrost talik through which isotopically heavier Hypsithermal waters have migrated. Exchange with such waters could result in the present isotopic composition. Unfortunately, without a sample from the gravel it is impossible at present to determine if this was the case.

Boreholes 78-141 and 78-142 are located further upstream along the Ogilvie River (see Figure 1). Core material recovered from borehole 78-141 consists primarily of organic silt with some segregated massive ice zones. The borehole is located near the river on gently sloping ground; a mountain is located immediately to the north. The isotope profile for this core (Figure 5) is devoid of any major shifts throughout the 6 m section. The minor fluctuations which do exist are characteristic of fractionations resulting from freezing as described by Michel (1982). The lack of major variations in this core could be due either to the formation of permafrost in this setion during a period of constant climatic conditions, or it could reflect the destruction of previously existing isotopic variations by groundwater migration.

Borehole 78-142 is located on a hillside sloping 3° westward toward the Ogilvie River. During drilling, a total of 16.45 m of massive segregated ice was encountered between a depth of 1.25 and 19.0 m. Peat and organic silt overlie most of the massive ice (Figure 6). Between the two main ice sections, a 0.9 m zone of silty sand was encountered. Drilling was terminated after 0.5 m of silt was cored beneath the largest ice section. Unfortunately, no samples of the ice were retained during the drilling program. Therefore, all isotopic determinations reported are for waters from the sediment-rich horizons.

In Figure 6, the organic-rich sediments above the massive ice can be seen to have a relatively constant oxygen-18 composition similar to modern water. The silty sand layer at a depth of 5.2 - 6.1 m contains ice with much lower ^{18}O concentrations (approximately -29°/oo). The lowermost silt at a depth of 19.0-19.5 m has a $\delta^{18}O$ value of -14.7°/oo which is much less negative than present-day values.

Without isotope data for the ice sections, it is very difficult to interpret these fluctuations. The lowermost silt sample is characteristic of a warmer climatic period than currently exists and is isotopically more positive than would be expected for water from the Hypsithermal. Examination of the ^{2}H-^{18}O relationship for this core in Figure 4 reveals that the deep sample lies to the right of the meteoric water line. If this is a result of evaporation during storage, the corrected $\delta^{18}O$ value would be approximately -18°/oo, which would then make a Hypsithermal age reasonable. However, since there was no evidence of evaporation when the core sample was examined during processing, it is doubtful whether such an adjustment can be made legitimately to the isotopic value. It is also possible that the offset from the normal trend could be due to a minor fractionation during freezing causing a decrease in the slope of the ^{2}H-^{18}O line. Regardless of how the $\delta^{18}O$ value is manipulated, the water still must be from a climatically warmer time period than currently exists in the area.

The low concentration of ^{18}O in the silty sand samples could be explained in two ways. First, the water may be representative of conditions during the glacial maximum of the late-Wisconsinan. It should be noted that the ^{18}O contents of these samples are roughly 2°/oo less than the samples from boreholes 78-148 and 78-165. This could be explained either by local variations in the average annual isotopic composition of the precipitation or by time differences between the water masses.

Secondly, the contact between the lower massive ice and the silty sand unit may represent a former permafrost table. The organic-rich sediments accumulating above the silty sand would have been seasonally frozen and water saturated. During freezing from the surface downward, the silty sand layer could have acted as a confined aquifer through which groundwaters migrated. In this situation, a closed system environment of freezing, similar to that for the formation of frost blisters as described by van Everdingen (1978), could develop once part of the aquifer was completely frozen. This would explain both the 3.3 m of massive ice above the silty sand layer and the very low concentration of ^{18}O in the ice of the silty sand. To confirm such a hypothesis would require analysis of the massive ice above the silty sand.

CONCLUSIONS

Four cores collected from the un-glaciated Eagle River-Ogilvie Mountains region of the northern Yukon have been examined for their ^{18}O contents. Two of the cores contain large fluctuations in isotopic composition which are suggestive of paleoclimatic variations during the past 20,000 years. Permafrost waters interpreted as being late-Wisconsinan in age have $\delta^{18}O$ values of -26 to -28°/oo and suggest that temperatures were on average 9°C lower than today. These $\delta^{18}O$ values are slightly less negative than waters in the Mackenzie Delta region which have also been interpreted as late-Wisconsinan in age. This discrepancy may be the result of localized variations in the isotopic contents of precipitation or it might reflect the presence of some glacial meltwater in the Mackenzie Delta samples.

One core contained permafrost waters with $\delta^{18}O$ values similar to modern precipitation and shallow permafrost waters ($\delta^{18}O$ = 21.5°/oo). This has been interpreted as indicating either formation of the permafrost during a period of constant climatic conditions or destruction of ^{18}O evidence for past climatic variations due to groundwater migration.

The fourth core encountered large quantitities of massive ice, but only samples from the sediment-rich sections were available for study. At depth, the permafrost waters have an isotopic composition which is interpreted as indicating a warmer climatic interval. Above the main massive ice zone $\delta^{18}O$ values of -29 to -30°/oo were encountered. To determine whether these values are related to the late-Wisconsinan period or to a closed system fractionation effect would require analysis of the massive ice zones.

ACKNOWLEDGEMENTS

My thanks are extended to Dr. P. Fritz of the University of Waterloo for his helpful discussions during this study and for the analytical work conducted at his laboratory. I would also like to thank Dr. A. Judge of the Department of Energy, Mines and Resources for his continued support of this research. Finally, I would like to express my appreciation to Foothills Pipe Lines (Yukon) Ltd. for kindly providing the core material and borehole logs, and to Mr. D. Fisher of Foothills for his support of the program.

REFERENCES

Crampton, C.B., 1979, Changes in permafrost distribution produced by a migrating river meander in the northern Yukon, Canada: Arctic, v. 32, p. 148-151.

Foothills Pipe Lines (Yukon) Ltd., 1978, Dempster lateral drilling program: prepared by Klohn Leonoff Consultants Ltd., 2 volumes.

Hughes, O.L., Harrington, C.R., Janssens, J.A., Matthews, Jr., J.V., Morlan, R.E., Rutter, N.W. and Schweger, C.E., 1981, Upper Pleistocene stratigraphy, paleoecology, and archaeology of the northern Yukon interior, Eastern Beringia. 1: Bonnet Plume Basin, Arctic, v. 34, p. 329-365.

Mackay, J.R., 1972, Offshore permafrost and ground ice, southern Beaufort Sea, Canada: Can. J. Earth Sci., v. 9, p. 1550-1561.

Michel, F.A. 1982, Isotope investigations of permafrost waters in northern Canada: unpublished Ph.D. thesis, University of Waterloo, 424 p.

Michel, F.A. and Fritz, P., 1978, Environmental isotopes in permafrost related waters along the Mackenzie Valley corridor: Proc. Third International Conference on Permafrost, Edmonton, Alta., Canada, National Research Council, v. 1, p. 207-211.

Michel, F.A. and Fritz, P., 1982a, Significance of isotope variations in permafrost waters at Illisarvik, N.W.T.: Proc. Fourth Canadian Permafrost Conference, Calgary, Alberta, p. 173-181.

Michel, F.A. and Fritz, P., 1982b, Laboratory and field studies to investigate isotope effects occurring during the formation of permafrost, Part 4: report on Project No. 606-12-06 for Canada Department of Energy, Mines and Resources, Serial No. OSU81-00076, Waterloo Research Institute, University of Waterloo, 76 p.

Patterson, R.J., Frape, S.K., Dykes, L.S. and McLeod, R.A., 1978, A coring and squeezing technique for the detailed study of subsurface water chemistry: Can. J. Earth Sci., v. 15, p. 162-169.

Stuiver, M., Yang, I.C. and Denton, G.H., 1976, Permafrost oxygen isotope ratios and chronology of three cores from Antarctica: Nature, v. 261, p. 547-550.

van Everdingen, R.O., 1978, Frost mounds at Bear Rock, near Fort Norman, Northwest Territories, 1975-1976: Can., J. Earth Sci., v. 15, p. 263-276.

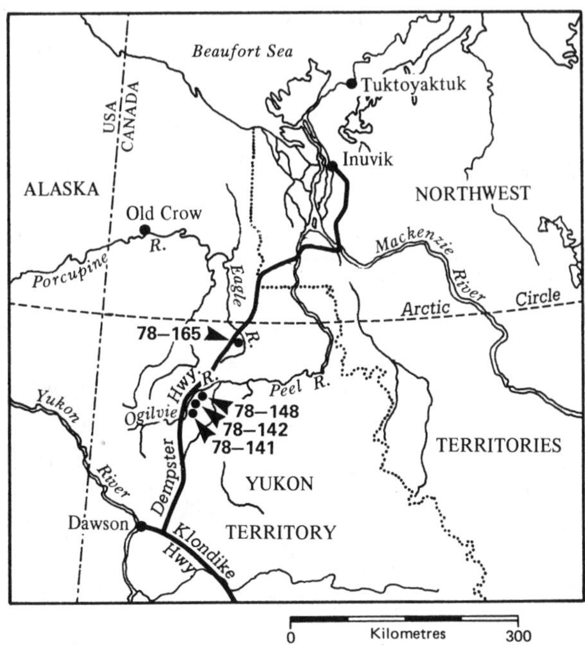

Figure 1 Location map of borehole sites along the Dempster Highway.

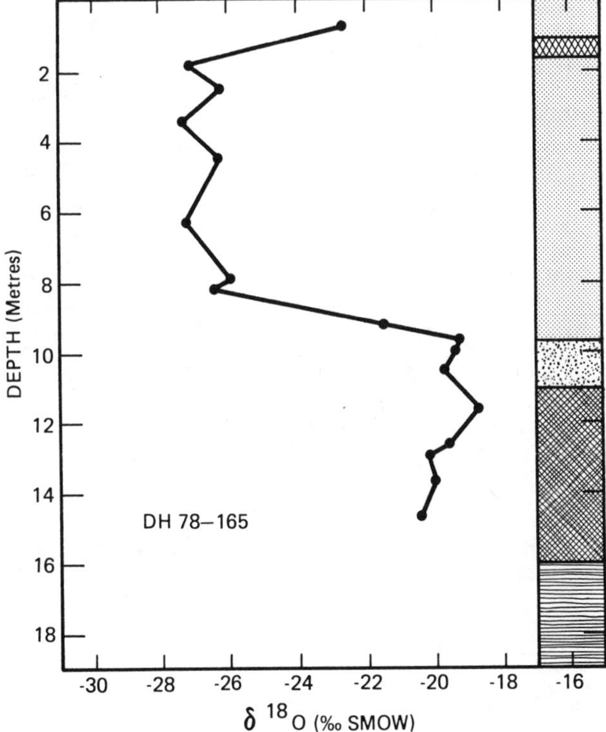

Figure 2 Variation in ^{18}O content with depth for core 78-165.

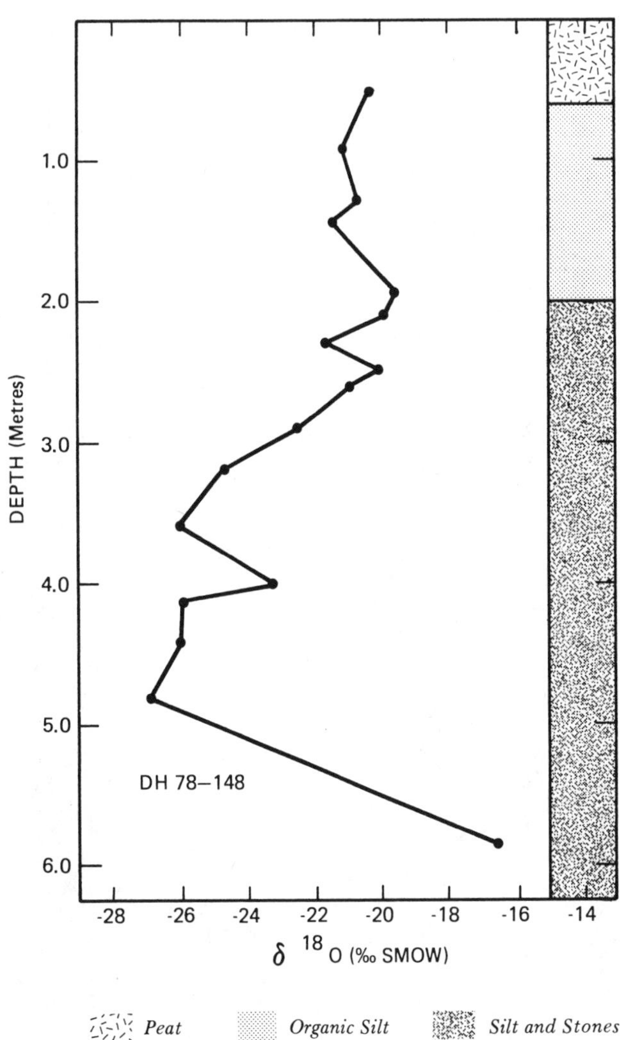

Figure 3 Variation in ^{18}O content with depth for core 78-148.

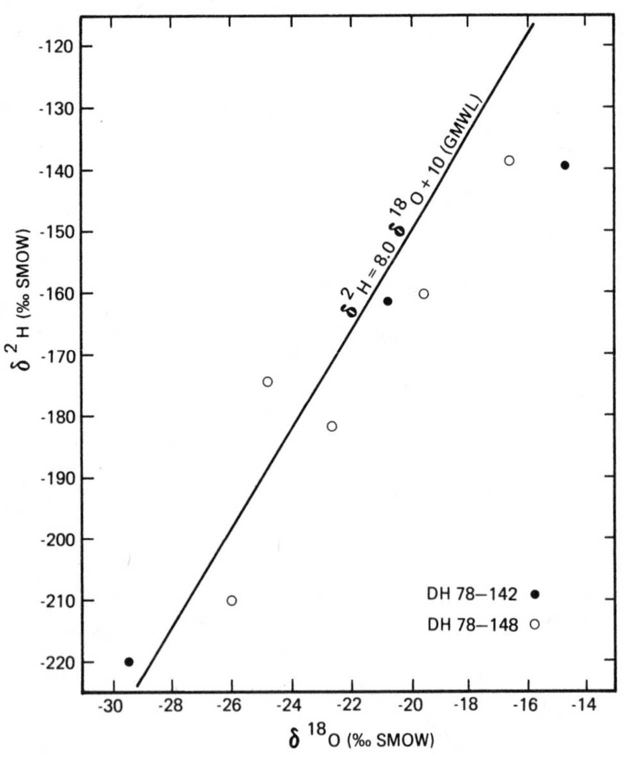

Figure 4 Relationship between ^2H and ^{18}O contents for cores 78-142 and 78-148.

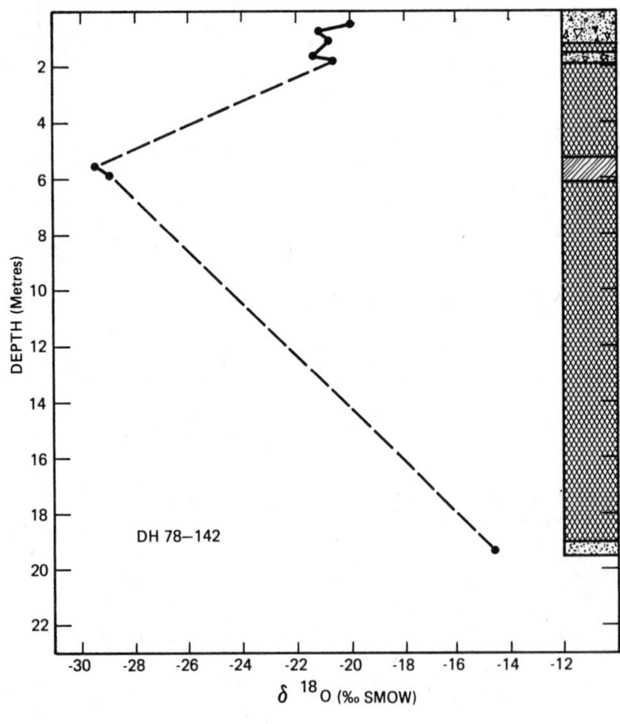

Figure 6 Variation in ^{18}O content with depth for core 78-142.

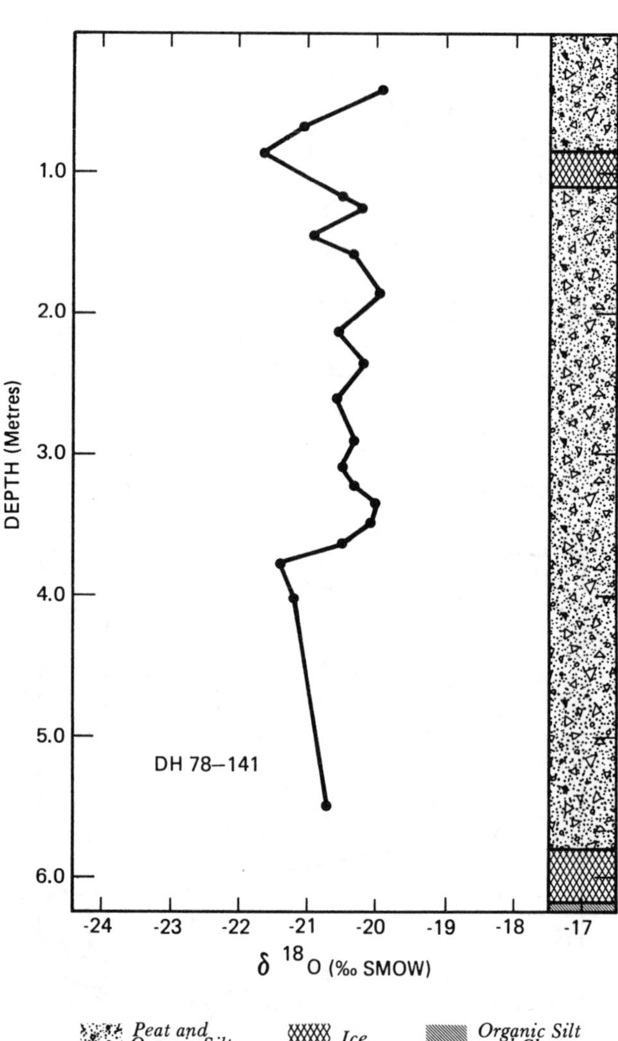

Figure 5 Variation in ^{18}O content with depth for core 78-141.

BEHAVIOR OF CONCRETE STRIP FOUNDATIONS ON THAWED LAYER OF CLAYEY PERMAFROST

Alexey Mindich

Design and Construction Department, Portland Cement Association, Skokie, Illinois 60077 USA

This paper reports on laboratory and field investigations into the behavior of thawed layers of clayey permafrost under strip plate loads. The aim was to study the load-settlement relationships of unfrozen as well as thawed soil layers. Such relationships were found to be nonlinear and were expressed in terms of a maximum load and a parameter dependent on the soil layer thickness, the foundation shape, and soil properties.

INTRODUCTION

Two-layer soil systems with a weak thawed upper layer underlain by a stiff frozen base are often encountered in permafrost regions. Thawing of the upper layer of clayey permafrost is accompanied by a decrease in its bearing capacity and by its transformation into a semi-liquid state. Whether a structure can be founded on such soil depends on the deformability as well as the bearing capacity of the soil. The latter is important because even a small load can cause the thawed clayey soil to be squeezed out from beneath the foundation, causing the collapse of the supported structure.

An extensive study into the behavior of two-layer soil systems consisting of a weak soil layer underlain by a rigid base was carried out in the laboratory as well as at an outdoor site. The aim was to determine the settlement of the weak soil layer considering its nonlinear stress-strain characteristics, to estimate the maximum load, and to ascertain the influence of soil layer thickness on both the settlement and the maximum load.

TEST SETUP

Laboratory Experiment

The laboratory research was carried out on soil contained in a mold 1940 mm long, 820 mm wide and 600 mm deep. Load was applied through strip test plates 150, 225, and 300 mm wide. The length of each plate equaled the width of the mold. Each segment was loaded independently. The load on the soil was kept concentric. The load on and the settlement of only the middle segment of the test plates were considered as test results, to avoid the influence of the mold walls (Figure 1). The load on the test plate was increased in steps of 15 kPa until an average pressure (strip plate load divided by plate area) of 120-180 kPa was attained. The value of load attained at the end of a load step was kept substantially constant (within ± 0.25 kPa) during measurements following that step. Settlement of the test plate as well as vertical and horizontal displacements of the six surface markers (three on each side of the strip plate) were measured with an accuracy of 0.01 mm 5 minutes after a load increment, 10 minutes thereafter, and every 20 minutes after that, until stabilization of test plate settlement was accomplished. Strip plate settlement was assumed to have stabilized when it increased by no more than 0.02 mm in 20 minutes and by no more than 0.05 mm in 1 hour of observation. Only the final value of plate settlement corresponding to each load increment was considered in analyzing the test results. Consideration of creep beyond the stabilization of plate settlement was excluded from this study. Such consideration would have interfered with the purpose of this investigation, which was to gather primary data on the behavior of clayey soil layers from large-scale tests (Mindich and Vyalov, 1973).

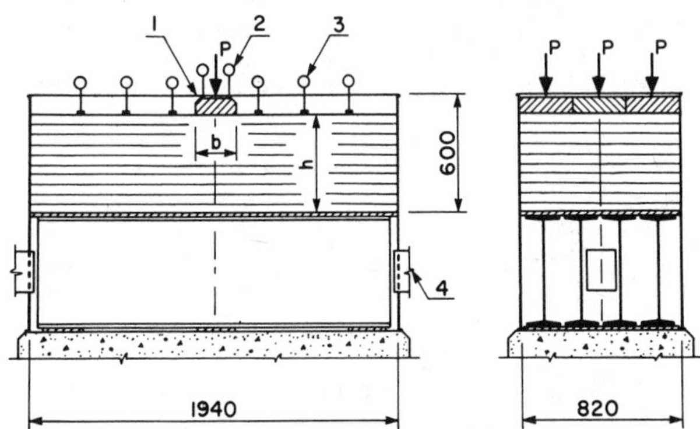

FIGURE 1 Laboratory test setup. 1, strip plate; 2, plate settlement dial gages; 3, surface markers; 4, cold air duct.

The thickness of the weak soil layer was 75, 150, 300, and 450 mm in tests with unfrozen soil layers underlain by a rigid base. Tests with artificially thawed soil overlying frozen soil were carried out in the same mold. A 500 mm thick soil mass was frozen by means of cold air driven through a cold air duct under the bottom plate of the mold. Then the upper part of that frozen soil was artificially thawed. The thickness of the thawed layer was 35-40, 60-70, 130-180, 260-280, and 430-440 mm. Thus, the ratio (λ) of the thickness of the weak soil layer (h) to the width of the loading plate (b) was varied from 0.25 to 3.

Field Experiment

In addition to the laboratory tests described above, on-site experiments were carried out on an artificially thawed 0.5 m thick layer of loam permafrost. The 1000 x 2400 mm test plate consisted of three 800 x 1000 mm sections. The horizontal and vertical displacements at six points, three on each side of the test plate, were measured (Figure 2). Horizontal displacements inside the thawed soil layer were recorded by means of colored sand poured into vertical cuts that extended through the layer from each measuring point on the surface. Preparatory work consisted of removal of vegetation, leveling of the test site, and installation of devices and equipment. Two rods with electrothermometers mounted at a close regular interval of 50 mm were embedded into soil on two sides of the test plate location. The depth to the boundary between frozen and unfrozen soil (i.e. the depth to the thermometer reading 0°C) was thus known at all times. This depth was verified by poking pointed

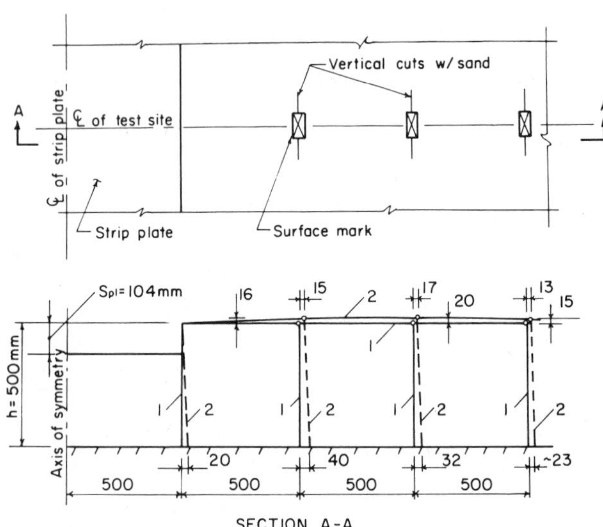

FIGURE 2 Field test. 1, positions of horizontal surface and of vertical strips before test; 2, the same after test.

steel rods into the soil at random locations. At the time when the test plate was ready to be installed in position, the thickness of the thawed soil layer was 0.3 - 0.35 m. Special heating plates were used to increase the thickness of the thawed layer. After a few hours of intensive preheating of a 3.5 x 5.0 m area, the thickness of the thawed layer equaled 0.42 - 0.45 m, and at the beginning of test, that value was $h_{ave} = 0.45 \pm 0.01$ m. 200 hours after the test began, the average thickness of the thawed layer was 0.48 m, and at the end of the test, 0.60 m. Considering that the final test plate settlement was about 0.10 m, it can reasonably be assumed that the thickness of the thawed layer equaled a constant average value of 0.5 m. The settlement of the middle section of the test plate as well as the surface marker displacements were measured with an accuracy 0.01 mm.

Plate settlement was considered to have stabilized when it did not increase by more than 0.05 mm in 2 hours of observation. The test plate settlement and the displacements of the surface markers were registered every 15 minutes during the first hour after each load increment of approximately 50 kPa, every 30 minutes during the second hour, and every hour thereafter up to the stabilization of plate settlement. Plate settlement during the test is illustrated in Figure 3.

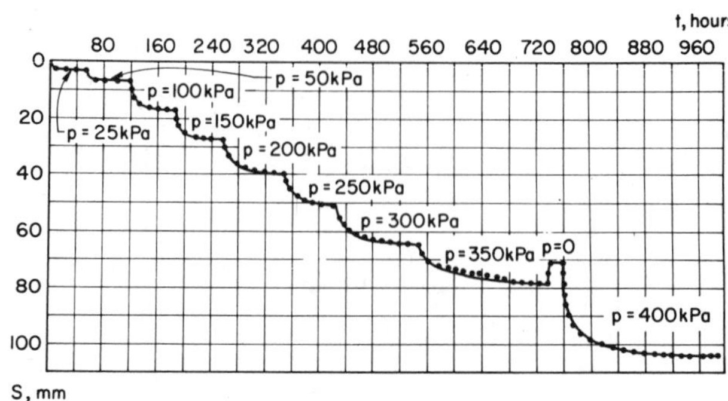

FIGURE 3 Graph of strip plate settlement with time.

SOIL PROPERTIES

The soil chosen for laboratory research was similar to the weak thawing clayey soil of the Vorkuta region of the USSR. Such strain and stress properties are commonly possessed also by "young" and "light" clays such as muds of littoral zones of seas, on flood plains of rivers, and so forth. The properties of the laboratory research soil (kaolin) and of the field test soil are listed in Table 1.

The stress-strain relationships of the test soil were investigated using Bishop's triaxial test apparatus and also on compression and shear devices. The relationships were found to be nonlinear and describable by a hyperbolic function (Vyalov and Mindich, 1974).

Table 1 Soil Properties

Characteristic	Laboratory Test Soil	Field Test Soil
Liquid limit, percent	46.5	32.3
Plastic limit, percent	28.6	17.5
Water content, percent	36.0	21-23
Unit weight, kg/m^3	1850	1990
Specific gravity	2.62	2.68
Porosity, percent	48.0	39.2
Angle of internal friction, deg.	18.0	18.5
Cohesion, kPa	7.0	40.0
Initial stress-strain modulus, kPa	850	2340
Poisson's ratio	0.40	0.35

Settlement and Maximum Load

The results of laboratory strip plate tests for different values of b and h are shown in Figures 4a-d. The results of the field tests are shown in Figure 5a (curve 1).

The load-settlement curves in Figures 4 and 5 can be approximated by the following expression (Popov, 1950):

$$s = (Ap) \cdot \frac{1}{1 - p/p_{max}}, \qquad (1)$$

where A(mm/kPa) is a coefficient that depends on the soil properties, the shape and size of the plate, and thickness of the compressed soil layer. Equation (1) can be written in a linear form:

$$\frac{p}{s} = \frac{1}{A} - \frac{1}{A\,p_{max}}\,p\,. \qquad (1a)$$

Parameters A and p_{max} were determined by fitting linearized graphs of equation (1a) to the test data in coordinates p/s versus p. An example of such curve fitting for data from the field test is shown in Figure 5. Dots plotted in Figure 5a in

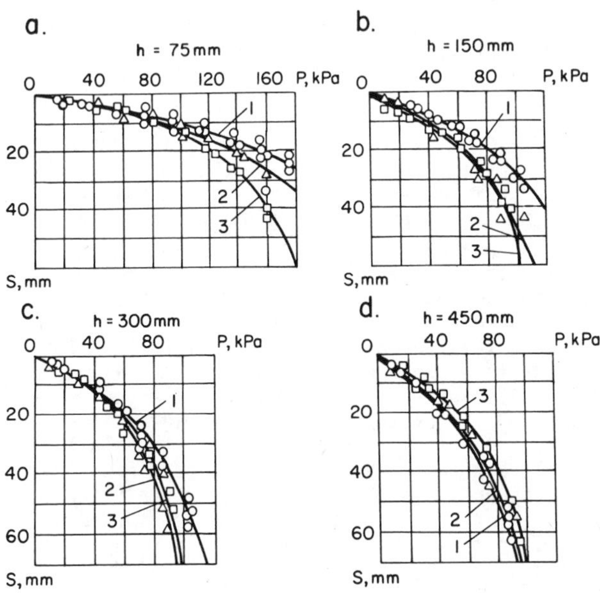

FIGURE 4 Results of strip plate tests for different plate widths and layer thicknesses: (1) b = 150, (2) b = 225, and (3) b = 300 mm.

coordinates p versus s correspond to the values of stabilized settlement for each load increment shown in Figure 3. Then the same dots were plotted in coordinates p/s versus p. Linearization of this series of dots by the least square method yielded the equation

$$p/s = 6.9 - 0.0075p. \qquad (1b)$$

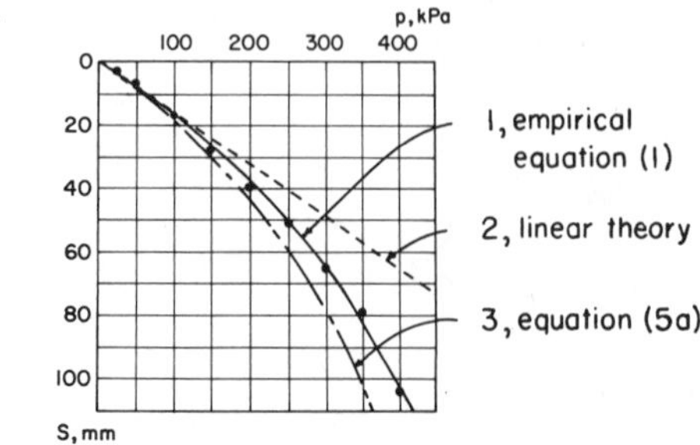

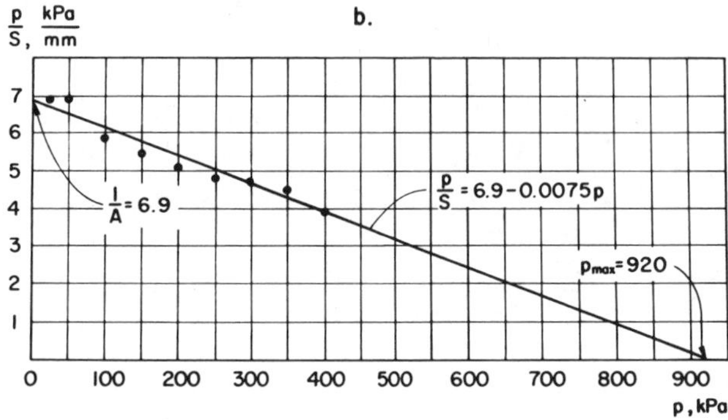

FIGURE 5 (a) Test data and load-settlement relationships; (b) auxilliary graph for determination of parameters of empirical load-settlement relationship (curve 1).

From equations (1a) and (1b), A = 1/6.9 = 0.145 mm/kPa and p_{max} = 6.9/0.0075 = 920 kPa. Parameters 1/A and p_{max} can be determined directly as shown in Figure 5b. Curve 1 in Figure 5(a) is a plot of equation 1, with values of A and p_{max} as determined above. Linear regression analysis in this particular case gave values of residual standard deviation and correlation coefficient of 0.296 and 0.932, respectively.

The values of A and p_{max} for the laboratory tests were determined in the manner indicated above, and are listed in Table 2.

Parameter A

Referring to Equation (1), s = Ap represents a straight line tangent to the initial portion of the s-p curve, and $\frac{1}{1 - p/p_{max}}$ determines the deviation of the s-p curve from that straight line. Assuming that the theory of elasticity is applicable to the initial straight portion

$$A = b\,\frac{1 - \nu_o^2}{E_o}\,C, \qquad (2)$$

where E_o and ν_o are the initial stress-strain modulus and the Poisson's ratio of the soil, respectively, and C is a nondimensional coefficient, dependent on the shape of the loading plate or foundation and on $\lambda = h/b$. Gorbunov-Posadov (1946) computed and tabulated values of C for different values of λ ranging from 0 to 25, based on the linear elastic theory. For strip foundations, these theoretical values can be approximated by

$$C = \frac{1.07\,\lambda}{1 + 0.4\lambda} \; . \qquad (3)$$

Use of the empirical equation (3) not only gives

TABLE 2 Parameters Involved in Load-Settlement Relationship, Equation (1)

Layer Thickness	Plate Width	Relative Layer Thickness	Parameters of Equation (1)		
			P_{max}	A mm/kPa	
				Test Data	by Eq.(2)
mm	mm	$\lambda = h/b$	kPa		
75	300	0.25	420	0.074	0.072
75	225	0.33	302	0.072	0.069
75	150	0.50	226	0.071	0.066
150	300	0.50	206	0.139	0.132
150	225	0.67	177	0.200	0.126
150	150	1.00	155	0.169	0.113
300	300	1.00	172	0.200	0.226
300	225	1.33	135	0.200	0.206
300	150	2.00	134	0.182	0.176
450	300	1.50	163	0.298	0.297
450	225	2.00	145	0.256	0.264
450	150	3.00	140	0.212	0.216
500	1000	0.50	920	0.145	0.167

good correlation with theoretical results (in most cases within 2-3%) but also allows determination of the value of coefficient C when $\lambda \to \infty$ (indicating transition from a finite layer to a semi-infinite half-space). According to equation (3) that value $C_\infty = 2.68$, whereas the theoretical solution yields $C_\infty \to \infty$ and C = 2.66 for the maximum value of λ (= 25).

Ultimate State of Soil Behavior

The ultimate state of a soil layer supported on a rigid base was considered by Vyalov (1966) and Mandel and Salencon (1969). According to Vyalov, a drastic reduction in the bearing capacity of a weak soil layer takes place following the formation of a plastic zone under the loading plate (Figure 6a). This plastic zone acts as a wedge which consolidates and slides the surrounding soil apart. Mandel and Salencon assumed that a weak soil layer loses bearing capacity following the formation of differently shaped plastic zones which are exposed to the surface and correspond to the overflow of soil beyond the plate edges (Figure 6b). It is stated in both papers that the value of maximum load depends on the value of $\lambda = h/b$.

To determine the actual soil behavior, a ditch was dug on the test ground, following the conclusion of the field test. One edge of the ditch was along the centerline of the loading strip and connected the measuring points (Figure 2). The measurement of distances between the vertical strips of colored sand showed that the maximum horizontal displacements had taken place at the interface between the thawed and frozen soils, rather than on the free surface, confirming the sliding apart of the soil. The primary mechanism of the failure of

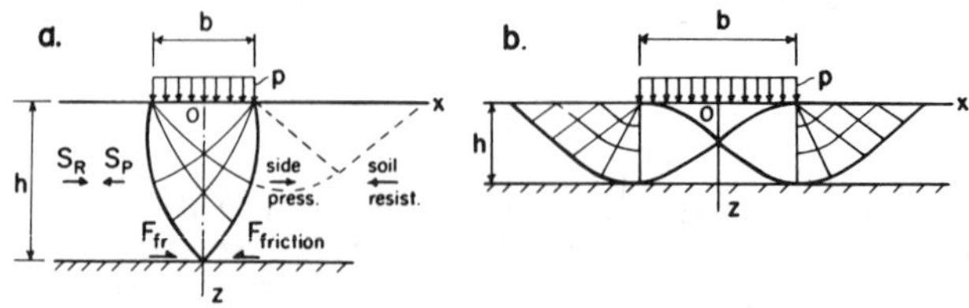

Figure 6 Limit state of a weak soil layer underlain by a rigid base (a) according to Vyalov, (b) according to Mandel and Salencon.

a weak soil layer supported on a rigid foundation bed is the dislocation of soil in different directions caused by the penetration of an elasto-plastic soil wedge that forms under the loading plate. A secondary mechanism is the simultaneous partial overflow of the soil (more exactly, the lifting of the free soil surface) beyond the plate edge.

Maximum Load

Maximum load by definition is supposed to be the load that corresponds to infinite deformation or settlement. Working with a weak soil layer of finite thickness, it is not possible to directly determine that value experimentally, because a decrease in soil layer thickness due to plate settlement leads to an increase in the maximum load which is a function of soil layer thickness. In the laboratory experiments presented, the value of plate settlement was restricted not to exceed 40-60 mm or 20-30 percent of the strip plate width. The experimental data were sufficient to establish the nonlinear character of the load-settlement relationship and to enable graphical or analytical determination of the maximum load value for each test, using the method described earlier.

Values for p_{max} based on the laboratory tests are plotted as a function of λ in Figure 7a, where points 1, 2, 3, etc., correspond to the different h and b combinations listed in Table 2. It is evident that p_{max} decreases with an increase in λ and approaches a minimum value equal to the limit load of a semi-infinite half-space. The test data can be approximately represented by

$$p_{max} = \alpha \frac{\lambda}{\lambda - \beta} \qquad (4)$$

where, for the test soil, α = 131 kPa and β = 0.180. The theoretical solution of Mandel and Salencon (1969) shows dependence between p_{max} and λ in the same form as equation (4). However, as shown by curve II on Figure 7a, for any λ, the theoretical solution yields a lower p_{max} than is given by equation (4), with α and β values as given above.

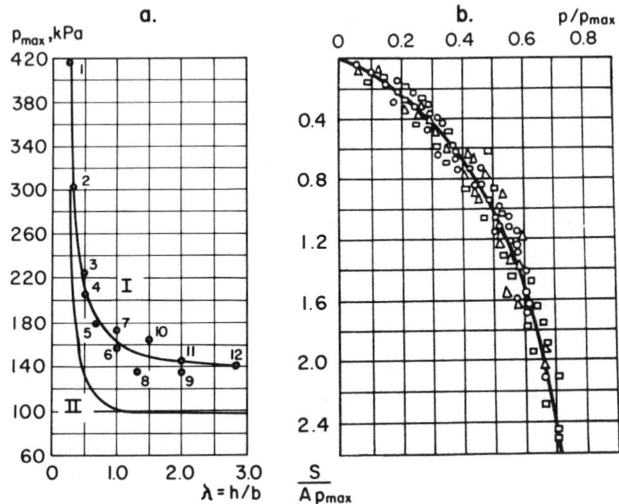

Figure 7 Relationship of P_{max} with λ, and generalized load-settlement curve.

Load-Settlement Relationship

Combining equations (1) - (3), the final s-p relationship becomes

$$s = A\frac{p}{1 - p/p_{max}} = b\frac{1-\nu_o^2}{E_o}\cdot\frac{1.07\lambda}{1+0.4\lambda}\cdot\frac{p}{1-p/p_{max}} \qquad (5a)$$

Equation (5a) applies to strip foundations only. For foundations of other shapes, equation (5a) can be generalized, using equation (4), to:

$$s = A\frac{p}{1-p/p_{max}} = \left(b\frac{Y\lambda}{1+Z\lambda}\right)\frac{p}{1-p(\lambda-\beta)/\alpha\lambda}, \qquad (5b)$$

where Y and Z are parameters to be determined from a test. They depend on soil properties and on the shape of the loading plate or foundation. Referring to equation (5a), when the test data are plotted in the coordinate system $s/(Ap_{max})$ versus (p_{max}/p), the resulting curve is valid for any value of λ. Figure 7b shows such a generalized curve, with the values of E_o and ν_o from triaxial tests. The test data, including those from laboratory tests with artificially thawed clayey soil layers, were correlated with the generalized curve through the use of a computer. The coefficient of correlation was a low 0.93, demonstrating the reliability of equation (5a).

It was earlier found experimentally (Mindich and Mirenburg, 1973) that equation (1) can be used for evaluation of settlement and carrying capacity of driven piles, particularly those driven into plastic frozen soils (with temperatures around -1.5^oC).

CONCLUSION

The necessity of considering the nonlinear properties of thawed clayey soil layers underlain by permafrost is illustrated by Figure 5a, where curve 1 illustrates the experimental load-settlement relationship corresponding to equation (1), straight line 2 represents the relationship s = Ap predicted by the linear theory with the value of A as given by equation (2), and curve 3 shows the results of analytical calculation of settlement by equation (5a). All three curves are based on a value of $\lambda = 0.5$ which corresponds to the field test. Curve 1, because of the way the parameters A and p_{max} of equation (1) were derived, is in the closest agreement with the load-settlement data from the field test. The analytical prediction, curve 3, overestimates strip plate settlement by 10-15 percent. Use of the linear theory for calculation of the possible settlement of a strip foundation on a thawed clayey soil layer can result in significant underestimation of the predicted settlement and in uncertainty as to the relationship between the design load and the bearing capacity of the soil layer.

The settlement of a strip foundation supported on a thawed clayey soil layer and the maximum load allowable on such soil (both quantities are needed for design purposes) can be determined in two ways.

"Field" Method

This method consists in carrying out no less than two strip plate load tests directly at the construction site, with different ratios of the thawed soil layer thickness to the width of the strip plate. The number of load increments should be no less than 7 or 8. Sufficient time for stabilization of settlement should be allowed between successive load increments. Test results (load p and the corresponding settlement s) should be plotted in coordinates p/s versus p. The best fit of equation (1a) to the plotted points would yield the parameters A_i and $p_{max\ i}$ for each test. Having two pairs of values, $p_{max\ i}$ and λ_i, the numerical values of α and β of equation (5b) can be found from equation (4). Data A_i, b_i, and λ_i from two tests can be used to find an average value of parameter Y from equation (5b); parameter Z = 0.4 for strip foundations. This method results in the load-settlement relationship for a strip foundation on a specific construction site.

Analytical (General) Method

Such soil parameters as initial stress-strain modulus E_o, Poisson's ratio ν_o, cohesion c, and angle of internal friction φ^o can be determined from laboratory tests usually accompanying the prospecting of a construction site. With these values, the maximum bearing capacity of a thawed soil layer can be theoretically computed (Mandel and Salencon, 1969). The calculated p_{max} and the experimentally determined values of E_o and ν_o can be substituted in equation (5a) for the prediction of the load-settlement relationship of a strip foundation supported on clayey thawed soil underlain by permafrost.

ACKNOWLEDGMENT

The author would like to express his deep appreciation to Professor S. S. Vyalov whose ideas and scientific advice were of immense value.

REFERENCES

Gorbunov-Posadov, M. I., 1946, Settlement of a foundation on a soil layer underlain by a rock foundation bed; Moscow, Stroyizdat Publishing House (in Russian).

Mandel, J., and Salencon, J., 1969. The bearing capacity of soils on a rock foundation: Proceedings, VII International Conference on Soil Mechanics, V. II, Mexico, p. 157-164.

Popov, B. P., 1950, Application of dimensional analysis to experiments with test loads, in Engineering-Geological Investigations for Water-Power Construction (in Russian), V. 1; Moscow, Gosgeolizdat Publishing House.

Vyalov, S. S., and Mindich, A. L., 1974, Settlement and limit equilibrium of a layer of weak soil underlain by a rigid foundation bed: Journal Soil Mechanics and Foundation Engineering, V. 11, No. 6: New York, Plenum Publishing Corporation, p. 381-386.

Mindich, A. L., and Vyalov, S. S., 1973. An installation for determining stresses and deformations of a layer of clayey soil on a rigid base, in Proceedings of the Eighth International Conference on Soil Mechanics and Foundation Engineering, V. 4.3: Moscow, p. 38

Mindich, A. L., and Mirenburg, Yu. S., 1973, Accelerated method of plate-load and pile tests for estimate of foundation settlement and bearing capacity of foundation bed, in Proceedings of the Lake Baidal Branch of Geographic Society of USSR, Issue XCI; Chita, Geographic Society of USSR Publishing House, p. 63-65 (in Russian).

WELL CASING STRAINS DUE TO PERMAFROST THAW SUBSIDENCE IN THE CANADIAN BEAUFORT SEA.

D.E. Mitchell[1], D.D. Curtis[2], S.W. Laut[1], N.K. Pui[1], A.S. Burgess[2]

[1] Dome Petroleum Limited, Calgary, Alberta, Canada
[2] Acres Consulting Services Limited, Calgary, Alberta, Canada

Permafrost sediments will experience local melting and thaw subsidence with the prolonged extraction of 60° to 85° C hydrocarbons through production wellbores. Design concerns include the settlement of topside production systems and the potential detrimental effects of the displacement-controlled loading of production casings. A total stress analysis, assuming undrained loading, was employed in parametric, single-well studies to assess the influence of the composite thaw subsidence loading mechanisms -- thaw strain and stiffness reduction. The thaw strain results from the potential 9% pore-ice to pore-water volume reduction in ice-bonded permafrost. The stiffness reduction strain is due to the increase in the soil's undrained compressibility with the melting of the pore ice. The validity of the undrained approach to thaw subsidence analysis has been supported by a major field investigation and monitoring program performed during the drilling of the Tarsiut N-44 exploratory well in 1981-82.

INTRODUCTION

In the early 1970s, Exxon, Arco, and BP pioneered research into geotechnical problems related to the extraction of hydrocarbons in the Arctic environment of Prudhoe Bay, Alaska. Specifically, they were concerned with the thermal degradation of permafrost penetrated by "hot" production casings and the resulting soil-displacement-controlled loadings of those casings. The term "thaw subsidence" was coined to describe this phenomenon.

The Prudhoe Bay work was laudable because of its completeness. It encompassed field coring, laboratory testing, definition of constitutive mechanisms, finite element modelling and a full-scale, 5-spot field thaw test. The 5-spot test provided data for the verification of the proposed mathematic model; but, the calibration process was not as satisfying as it might have been. According to Mitchell (1977):

> "The initial model runs matched the field test qualitatively and predicted casing strains that were within +/- 0.02% strain of the field test results. <u>The mechanical properties were then adjusted to match the results at the base of the permafrost and the greatest casing strains in the region of alternating strains.</u> The model tracked the measured strains of these specific joints within the accuracy of the field test measurements without any further adjustments."

In 1978, Dome Petroleum Limited initiated a similar research program preparatory to feasibility level design of offshore production structures for their Canadian Beaufort acreage. Exploratory drilling activities in this region had documented the presence of unbonded (no pore ice) and bonded (pore ice) permafrost sediments up to 600 m in thickness. The basic objective of the program was to extend the Prudhoe Bay work to define a more general, non site-specific and non material-specific approach to the thaw subsidence problem. The "Prudhoe" approach was not felt to be directly applicable to the Beaufort regime because of the difficulty in reliably defining pre and post-thaw pore pressures and the difference in soil types between the two areas. The Prudhoe Bay materials are primarily coarse-grained, while those in the Beaufort are fine-grained.

To this end, both single well and multi-well production configurations were analyzed. A summary of the single well study and the complementary field and laboratory testing programs is presented herein.

THAW SUBSIDENCE MECHANISMS

Two schools of thaw subsidence analysis were reviewed prior to initiation of the single well analytical studies: that of the Prudhoe Bay researchers, Mitchell and Goodman, and that of Morgenstern and Weerdenburg from the University of Alberta. Both teams identified two soil straining mechanisms integral to the thaw subsidence problem. The first relates to phase change and the 9% pore-ice to pore-water volume reduction in ice-bonded permafrost on thaw. This has been addressed under the heading of <u>Thaw Strain</u> by the University of Alberta and <u>Pore Water Pressure Reduction</u> by Mitchell and Goodman. The second soil straining mechanism pertains to the decrease in load-carrying capacity of the soil as the pore material changes from ice to water. This mechanism has been commonly termed <u>Stiffness Reduction</u>.

Phase Change Strain

The phase change of pore ice to water will result in a potential soil volume decrease equivalent to 9% of the pore ice volume. For initially saturated conditions in soils, this can be accommodated by:

(a) a volumetric contraction of the soil skeleton (thaw strain) under free boundary conditions;

(b) a reduction in pore water pressure potentially to desaturation within the pore space, if full volumetric contraction is not permitted;

(c) imbibition of water.

Processes (a) and (b) take place without any flow of water into or out of a representative element of the soil and, therefore, represent "undrained" conditions. In contrast, process (c) would be regarded as "drained" because it involves movement of water from high to low potential.

The thawing of normally consolidated, fine-grained materials under unconstrained conditions will generally induce a volumetric strain because the material is compressible (process (a)). In contrast, dense or coarse-grained materials under the same conditions may not experience the full strain because of the restraining stiffness of the soil skeleton. The result is a decrease in pore pressure that can be permanent (process (b)) or transient (process (c)) if the material possesses a high enough permeability to allow water inflow and satisfaction of the reduced pore water pressure.

Stiffness Reduction

The stiffness reduction loading mechanism can be conceptualized by a two parameter, elastic spring model (Figure 1). The model comprises a series of horizontal beams and vertical springs to represent the permafrost and thawed soil annulus surrounding the well casing.

Prior to thaw, the soil/springs are stiff because of the pore ice and the system is in equilibrium. On thawing, the stiffness of the soil/springs is reduced as the ice changes to water and the beams deflect downward due to their self-weight. The full downward deflection is prevented by the end restraint at the casing and the permafrost, resulting in stress transfer (arching) to these boundaries. This causes a reduction in vertical load at the base of the permafrost and the "springs" in the basal, unfrozen material expand. This results in a compressive straining of the casing within the frozen zone and its tensile straining in the underlying unfrozen material (Figure 3a).

PARAMETRIC SINGLE WELL STUDIES

Basic Approach

The undrained total stress methodology was selected for use in the majority of the Dome analyses on the basis of the predominantly fine-grained nature of the Beaufort materials and concern over the general validity of the pore pressure reduction approach to phase change loading. The pore pressure reduction concept was questioned on the basis that the establishment of pore pressures in the ice and subsequently in the thawed ground on an a priori basis was highly contentious.

Finite Element Modelling

The finite element program NONSAP was chosen for analysis of the thaw subsidence phenomenon because of its linear and non-linear capabilities. The structured model was axi-symmetric with a fixed base and rollered side. A permafrost thickness of 600 m was selected in light of recorded Beaufort conditions, and the external lateral boundary was set at a radial distance of 500 m to meet the zero displacement criterion (Figure 2). The modelled casing was 500 mm in diameter, and any structural contribution of a grouted annulus was conservatively neglected. A no-slip condition at the casing/soil interface was assumed during preliminary analysis.

The thaw strain and stiffness reduction loading mechanisms were treated separately and then superimposed for the linear elastic analyses. The elasto-plastic analyses required that both loading mechanisms be combined and incrementally applied during modelling.

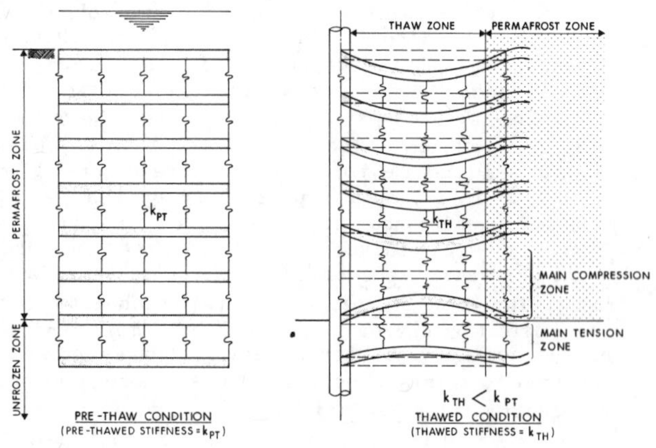

FIGURE 1 : CONCEPTUAL STIFFNESS REDUCTION MODEL

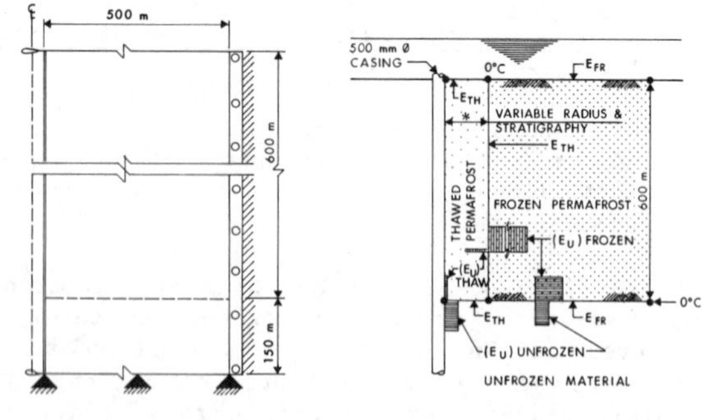

FIGURE 2a : FINITE ELEMENT MODEL CONFIGURATION

FIGURE 2b : SINGLE WELL PARAMETRIC GEOMETRY

The desired thaw strains were input as initial strains into the finite element model. The volumetric thaw strain was distributed equally in three orthogonal directions, i.e., $\varepsilon_r = \varepsilon_z = \varepsilon_\theta = \varepsilon_v/3$. A brief discussion and derivation of initial strain loads is presented in Appendix A.

The casing stresses and strains resulting from the stiffness reduction mechanism were assessed by comparing the pre-thaw and post-thaw regimes (i.e. E_{FROZEN} vs E_{THAWED}). Derivation of the constitutive laws for the stiffness reduction mechanism is presented in Appendix B.

Parametric Runs

Two basic deposit geometries were investigated in the single well parametric study package:

(a) a homogeneous soil stratigraphy with variable thaw radius, and

(b) a layered stratigraphic sequence with a fixed 12 m thaw radius.

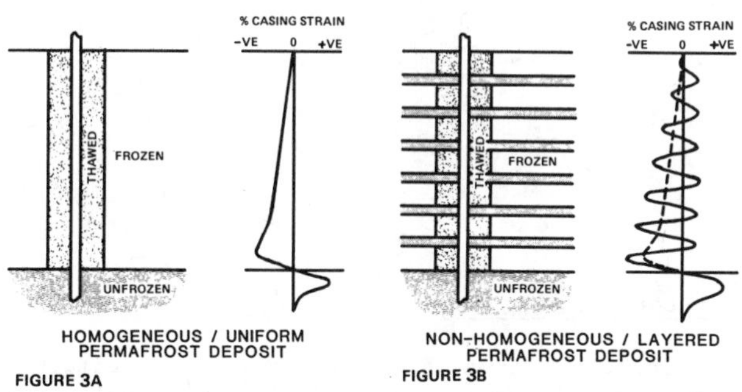

HOMOGENEOUS / UNIFORM PERMAFROST DEPOSIT
FIGURE 3A

NON-HOMOGENEOUS / LAYERED PERMAFROST DEPOSIT
FIGURE 3B

With a homogeneous soil stratigraphy, the peak casing strains occurred at the base of the permafrost. Maximum compressive casing strains occurred adjacent to the thawed soil, while peak tensile casing strains were found directly below and adjacent to the unfrozen soil (Figure 3a). A summary of the homogeneous geometry results is presented in Figure 4, along with the basic design parameters.

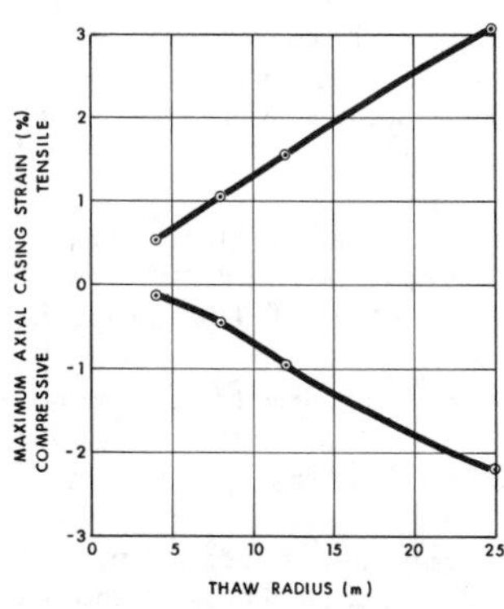

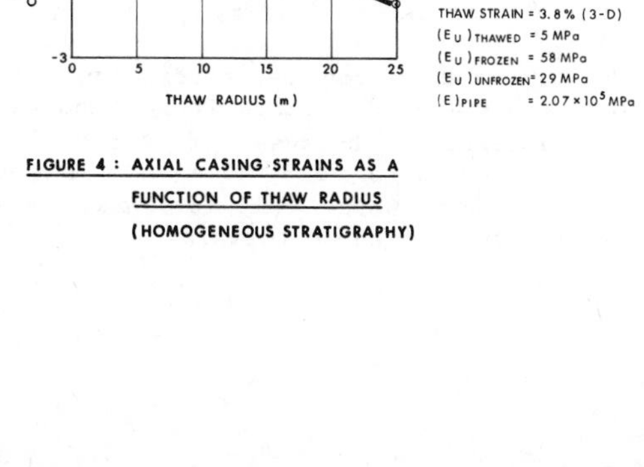

FIGURE 4 : AXIAL CASING STRAINS AS A FUNCTION OF THAW RADIUS (HOMOGENEOUS STRATIGRAPHY)

Deep coring of Beaufort sediments revealed a layered stratigraphy made up of materials with sharply varying mechanical properties; therefore, layering effects were investigated using a repeating stiff and soft material couplet. The "layered" casing strains followed similar mean trends to their homogeneous counterparts, but the interbedding induced minor strain variations that locally amplified the general trends (Figure 3b). The worst-case layer configuration was a thin, soft-soil layer located below a thick-stiff layer. A summary of the layered sequence results is presented in Figure 5.

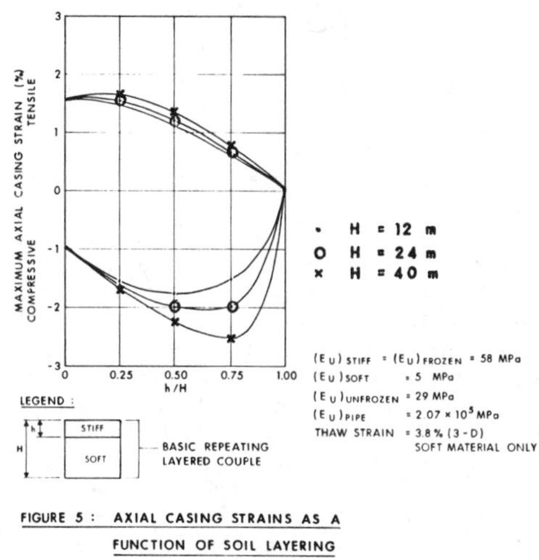

FIGURE 5 : AXIAL CASING STRAINS AS A FUNCTION OF SOIL LAYERING (12 m THAW RADIUS)

The general conclusion of the parametric work was that a layered sequence represented the critical design scenario, ceteris paribus, and that changes in compressibility within interbedded sequences most significantly affected the casing strains. Cursory elasto-plastic analysis marginally reduced the strains predicted from the elastic model, but did not alter this basic conclusion.

TARSIUT N-44 CASE HISTORY

General

A major field investigation and thaw subsidence monitoring program was performed during the drilling of the Tarsiut N-44 well in 1981-82 to latterly aid in model verification. A preliminary calibration run had been made using Prudhoe Bay results, but this, in spite of good agreement between field and model results (Figure 6), did not represent an objective verification of the approach because of the lack of original laboratory data in the literature. The Tarsiut program included field drilling, sampling, logging, and testing; field casing instrumentation and monitoring; laboratory testing; and back-analysis of the measured casing strains through finite element modelling.

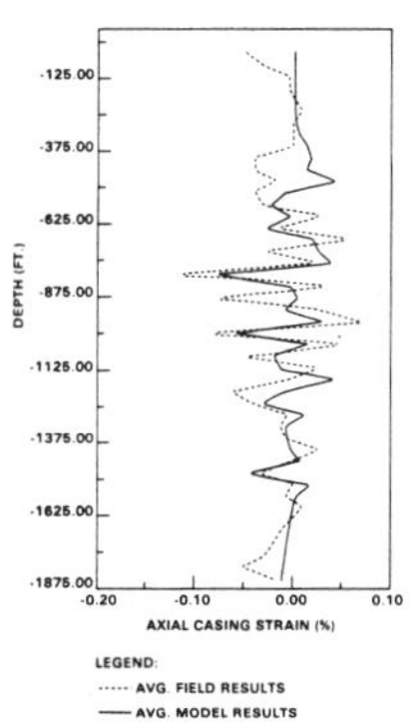

FIGURE 6

Field Investigations

The geotechnical field investigation at Tarsiut "piggy-backed" the well drilling operation to the 650 m depth level. The first phase of the program was carried out between 20 m and 170 m below the island/seabed contact, and was effectively a standard geotechnical investigation with pseudo-continuous sampling of the formation at 1 metre intervals and local in-situ testing. The retrieved samples were subjected to simple index testing in the field (i.e., density, temperature, strength, total and frozen moisture contents, etc.) and then preserved for shipment south in thermostatically controlled freezers.

Operational and schedule constraints precluded geotechnical coring in the lower portion of the permafrost zone. As a result, the second phase of the investigation, which related to the 170 m to 650 m depth range, involved the inference of lithology from drill bit cuttings, drilling performance, geophysical logs, and sidewall cores.

The investigations revealed that the Tarsiut N-44 foundation stratigraphy was characterized by interbedded silts and clays with isolated sand stringers. Bonded permafrost soils existed between 60 m and 410 m below the island/seabed contact and contained little or no visible ice. Temperatures measured downhole and in core samples indicated that minimum ground temperatures approached -3.5° C. The consistencies of cohesive units ranged between firm and very stiff when thawed, and test results indicated that these soils were moderately compressible and normally consolidated.

Casing Instrumentation

In-situ temperatures, pore pressures, and casing strain data were considered to be crucial to the successful verification of a thaw subsidence model, and a downhole instrumentation system was designed to effect the retrieval of such data during the drilling and testing period at Tarsiut N-44. Integral to the system was a monitoring casing joint that housed two temperature transducers, three pore pressure transducers, and nine strain gauges (3 hoop and 6 axial). These instruments were sensed by downhole data acquisition units and, subsequently, a topside wellhead data concentrator and processing unit (Figure 7).

Six, 30-inch diameter and eight, 20-inch diameter monitoring joints were installed within the permafrost zone at Tarsiut . Specific locations were selected to obtain casing strain data representative of material type (i.e. located in the middle of massive deposits) and facies changes (i.e., located at layer boundaries).

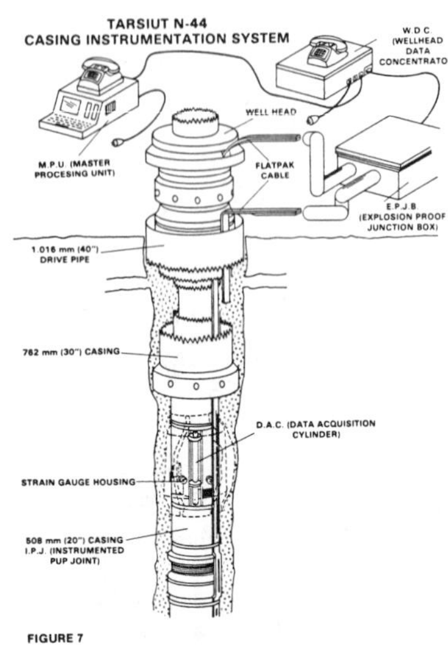

FIGURE 7

Laboratory Testing Program

Previous research showed that the primary design variable in the thaw subsidence analysis was the modulus of deformation ($E\mu$) and that small design changes in elastic properties were more important than increased sophistication in modelling technique. As a result, Dome Petroleum Limited commissioned EBA Engineering of Edmonton to evaluate the frozen and thawed deformation moduli ($E\mu$), Poisson's ratios (ν), and thaw strains (ε_v) of the retrieved Tarsiut cores. A complete description of the temperature and stress path-controlled testing program is documented in the referenced report.

While detailing of the testing program results is not possible herein due to their proprietary nature, the following qualitative statements can be made concerning the program findings:

(a) the measured thaw strains were compatible with the mineralogy, pore water salinity, and in situ temperatures of the samples;

(b) the measured thaw deformation properties agreed well with those of normally consolidated "southern" materials;

(c) the measured frozen deformation properties were generally lower than those reported in the literature because of their "warm" nature, but were compatible with the mineralogy, pore water salinity, and in situ temperatures of the materials.

(d) the thaw strain treatment of phase change loading appears to be justified.

Tarsiut Back-Analysis

Casing strain data for the Tarsiut back-analysis was selected from the January to April 1982 time period. All mechanical drilling- and testing-related strains, such as casing hanging and tensioning, were subtracted from the pup joint gauge data on the basis of drilling records. The remaining strains were felt to be attributable to the thaw-induced, soil displacement-controlled loading of the casing and, therefore, represented the datum for model verification.

The amount of thaw generated by the N-44 exploratory well was obtained through the use of a finite element thermal program. Input parameters included pre-drilling measured formation temperatures, external wellbore temperature variation with time measured by the in situ temperature sensors, and the thermal conductivities, volumetric latent heats, and volumetric specific heats of composite soil layers. The thaw radius for the Tarsiut scenario was thereby predicted to be 1.75 m.

Due to the low measured casing strains and small thaw radius, the thaw-induced strains were modelled using the linear elastic portion of the finite element program. Input moduli and thaw strains were obtained from representative material stress/strain curves as determined by EBA Engineering. The thaw strain and stiffness reduction loading mechanisms were analyzed separately and then superimposed in accordance with the methods outlined previously.

As shown in Figure 8, the predicted casing strains agree well with those measured, in both magnitude and trend. The displayed results pertain to the initial thaw subsidence model run using objectively selected material properties. A subsequent parametric study showed high sensitivity to material properties, thereby lending credence to the laboratory testing procedure and the method of analysis.

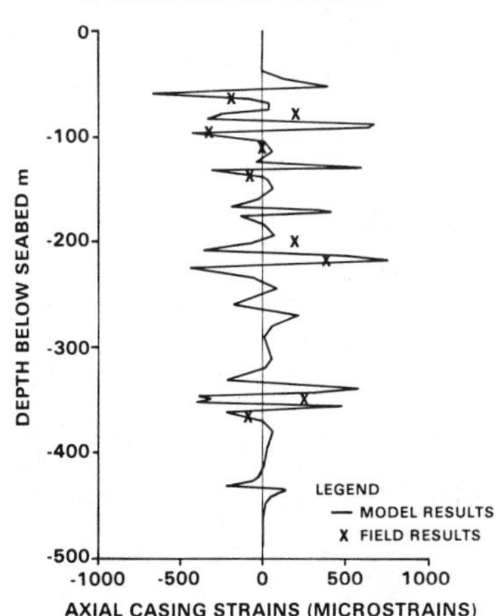

FIGURE 8

FUTURE WORK

On the basis of the Tarsiut N-44 validation study, the undrained thaw subsidence approach is being used to analyze the more realistic multi-well/cluster development scenario. The work will be performed as a joint industry sponsored research project (APOA 209) and will include examination of the influence of production structure-induced consolidation (i.e. negative skin friction) and the stability of deviated well clusters. It is anticipated that this study will be completed by January 1984.

ACKNOWLEDGEMENTS

The writers wish to acknowledge the contributions of personnel from Dome Petroleum Limited (R. Saint, M. Bradshaw), Acres Consulting Services (Dr. A. S. Burgess), EBA Engineering Ltd. (Dr. W. D. Roggensack, Dr. D. Taplin, P. Ruffell, P. Romeril), and the University of Alberta (Dr. N. R. Morgenstern, P. Weerdenburg) to this study. We also thank Shelly Lewchuk for her assistance in preparing this paper and the Tarsiut partners (Gulf, Dome, Norcen, and Canterra) for permission to publish it.

LIST OF REFERENCES

Acres Consulting Services, Influence of Permafrost Thawing on Single Production Wells in the Beaufort Region, report to Dome Petroleum Limited, 1982.

Andersland and Anderson, Geotechnical Engineering for Cold Regions, McGraw Hill Books, New York, 1978.

Arco, Prudhoe Bay Field Permafrost Casing and Well Design for Thaw Subsidence Protection, (authored by Perkins, T.K., Ruedrich, R.A., Wooley, G.R., Ruchon, J.A., Schultz, F.J.), 1975

EBA Engineering Consultants Ltd., "1981 Offshore Geotechnical Site Investigation Tarsiut Deep Core, Beaufort Sea," report to Dome Petroleum Limited, April 1982.

EBA Engineering Consultants Ltd., "Laboratory Investigation of Elastic and Thaw Strain Parameters," report to Dome Petroleum Limited, March 1983.

Goodman, M.A., "Mechanical Properties of Simulated Deep Permafrost," Transactions of the ASME, Journal of Engineering for Industry, 1975, pp. 417-425.

Goodman, M.A., "Loading Mechanisms in Thawed Permafrost Around Arctic Wells," Journal of Pressure Vessel Technology, November 1977, pp. 641-645.

Mitchell, R.F., "A Mechanical Model for Permafrost Thaw Subsidence," Transactions of the ASME, Journal of Pressure Vessel Technology, February 1977, pp. 183-186.

Mitchell, R.F. and Goodman, M.A., "Permafrost Thaw Subsidence Casing Design," Journal of Petroleum Technology, Vol. 30, 1978, pp. 455-460.

Morgenstern, N.R. and Weerdenburg, P.C., Thaw Subsidence Effects on a Well in Permafrost, report to Dome Petroleum Limited, 1980.

Perkins, T.K. and Ruedrick, R.A., "The Mechanical Behaviour of Synthetic Permafrost," Society of Petroleum Engineers Journal, SPE ƒ4057, August 1973, pp. 211-220.

Smith, R.E. and Clegg, M.W., "Analysis and Design of Production Wells Through Thick Permafrost," Proceedings of the 8th World Petroleum Congress, Moscow, June 1971, pp. 379-389.

Tsytovich, N.A., The Mechanics of Frozen Ground, McGraw Hill Books, New York, 1975.

APPENDIX A

Finite Element Treatment of Initial Strains

1 - Finite Element Analysis (Axisymmetric) with Initial Strain Loads

Start with the result of the minimization of potential energy.

$$\{P\} = [K]\ \{\Delta\} \tag{1}$$

This equation applies to the whole continuum

Where $\{P\}$ = global force vector (loads)

$[K]$ = global structure stiffness matrix

$\{\Delta\}$ = unknown nodal displacement.

For an individual element

$$[K] = [B]^T[D][B]dV \tag{2}$$

Where $[K]$ = element stiffness matrix

$[B]$ = relates displacements to strains.

Also

$$\{\varepsilon\} = [B]\ \{\Delta\} \tag{3}$$

$$\{\sigma\} = [D]\ (\{\varepsilon\}-\{\varepsilon_o\}) \tag{4}$$

Where

$[D]$ = constitutive matrix for an axisymmetric solid

$\{\varepsilon\}$ = strain from applied loads

$\{\varepsilon_o\}$ = initial strain.

With the element stiffness matrices formed, the structure stiffness matrix can be assembled for the entire continuum.

The applied loads generated by the initial strains are

$$\{F\}^e = [B]^T[D]\{\varepsilon_o\}dV \tag{5}$$

The element force vector $\{F\}^e$ is assembled into $\{P\}$. Knowing $\{P\}$ and $[K]$, one can solve for the overall displacements

$$\{\Delta\} = [K]^{-1}\ \{P\} \tag{6}$$

The results of (6) are substituted into (3) then (4) to obtain stresses.

APPENDIX B

2 - Stiffness Reduction Loading - Undrained

(a) Initial Stress State

$$\sigma_z = \gamma h;\ \sigma_r = K_o\sigma_z;\ \sigma_\theta = K_o\sigma_z$$

Where γ = total soil density

h = height of soil above element centroid

(b) Initial Strain - Frozen State

$$\varepsilon_{z_F} = \frac{\sigma_z}{Eu_F} - \frac{\nu u_F}{Eu_F}(\sigma_r + \sigma_\theta)$$

$$\varepsilon_{r_F} = \frac{\sigma_r}{Eu_F} - \frac{\nu u_F}{Eu_F}(\sigma_z + \sigma_\theta)$$

$$\varepsilon_{\theta_F} = \frac{\sigma_\theta}{Eu_F} - \frac{\nu u_F}{Eu_F}(\sigma_r + \sigma_z)$$

Where Eu_F = undrained Young's Modulus (frozen)

νu_F = undrained Poisson's Ratio (frozen)

(c) Initial Strain - Thawed State

$$\varepsilon_{z_T} = \frac{\sigma_z}{Eu_T} - \frac{\nu u_T}{Eu_T}(\sigma_r + \sigma_\theta)$$

$$\varepsilon_{r_T} = \frac{\sigma_r}{Eu_T} - \frac{\nu u_T}{Eu_T}(\sigma_z + \sigma_\theta)$$

$$\varepsilon_{\theta_T} = \frac{\sigma_\theta}{Eu_T} - \frac{\nu u_T}{Eu_T}(\sigma_r + \sigma_z)$$

Where Eu_T = undrained Young's Modulus (thawed)

νu_T = undrained Poisson's Ratio (thawed)

(d) To compute the change in initial strain due to stiffness reduction, the frozen initial strains are subtracted from thawed initial strains.

FROST FEATURES IN THE KARST REGIONS OF THE WEST CARPATHIAN MOUNTAINS

Pavol Mitter

Museum of Slovak Karst and Protection of Nature
Liptovsky Mikulas, Czechoslovakia

The action of freezing and thawing on limestone near cave entrances in the high mountains of Czechoslovakia produces modification of rock walls, rock shelters, debris ramparts, and sorted polygons and stripes in the frost-weathered debris of the cave floor. Debris ramparts, about 1 m high and wide, form near cave entrances. They are the result of intense frost action caused by the outside freezing air entering the cave and contacting the moisture-soaked, jointed limestone walls. Frost-weathering debris on the cave floors near the entrances is sorted into small polygons and stripes by frost action.

INTRODUCTION

The occurrence of permafrost is rare in Czechoslovakia. A special type of permafrost exists in the high mountains, in colluvium and mainly in ice caves among which Dobsinská Ladová jaskyna (968 m), Silická Ladnica (503 m), Demänovská Ladová jaskyna (837 m) and Ladová jaskyna na Muráni (860 m) are the bset known localities. The areal extent of permafrost in ice caves is only several 1000 m^2. The occurrence of frost stirred soil forms can be expected on the border of the glacier where the surface part of sediments partly melts in summer and where favorable conditions for typical solifluction arise. During research of the karst in the Západné Karpaty (West Carpathian) Mountains in the territory of Slovakia, we often met forms of relief which are influenced by or depend on frost activity. Some of them are due to periglacial processes without the existence of permafrost (Washburn 1973). In winter, when the air temperature often alternates below and above $0^{\circ}C$, mostly in sunny localities, the weathered mantle of loose sediments alternately freezes and thaws, and gives rise to favorable conditions for frost shattering and sorting of soils. In our country, Sekyra (1960) has evaluated the effect of the frost on the soil and Demek (1972) has suggested the classification and the terminology of cryogenic forms. Some manifestations of frost activity are considered to be due to periglacial processes, but others are due to local ice growth and exfoliation.

We have directed our attention partly to the effect of temperature changes on the rock walls of limestones and dolomites, on the rock walls of shelters, and on the unvegetated entrance areas of caves.

FORMATION OF ROCK WALLS, ROCK SHELTERS, AND CAVE ENTRANCES

The formation of debris ramparts in front of cave entrances is a manifestation of frost action. The debris ramparts are about 1 m high and 1 m wide. They are vertical to the axis of the cave entrances and may extend in front of the entire entrance if it is beneath a great rock wall.

Debris ramparts have been found from a lower elevation of about 600 m to the highest ones at elevations of 800-1000 m; they are the most widespread. They form and are preserved in front of entrances oriented to various cardinal points if cryogenic processes cannot remove the debris rampart, e.g. Malá jaskyna za Kanel'om (997 m) in the Nizke Tatry (Figure 1) and Lòm (862 m) and Biela jaskyna (840 m) in the Vel'ká Fatra.

Weathered material falls from the rock walls mostly in spring, suggesting a direct relationship to frost action. This area is in contact with the outside atmosphere as well as with the air in the cave. In addition, surface water seeping downward through minor joints in the rocks and condensation moisture are very important in frost weathering. Most of the rock shelters in the intensely tectonically shattered limestones may be considered to be chiefly a product of frost action.

The difference between the formation activity of frost action and formation that occurs without frost action is evident in Malá jaskyňa za Kancl'om, a horizontal cave about 43 m long. The entrance is partially filled by debris that had been isolated from the inner part of the cave. The walls of the entrance are modified by frost weathering. The weathered material and powder sinter on the floor of this part of the cave are sorted by frost action. The walls of the inner part of the cave, however, have been separated for a long period of time from the outside atmosphere and they are smooth, with the original form of erosion and karst formation, without traces of frost weathering. Even debris on the floor in the inner part of the cave is not frost-sorted (Figure 2).

PATTERNED GROUND

Patterned ground is a common phenomenon on rock shelter floors and in cave entrances. It forms where debris and powder sinter have accumulated. We use the term "powder sinter" to indicate a modification of secondary $CaCO_3$, which occurs on the walls of rock shelters and in cave entrances in

FIGURE 1 Entrance to the cave Malá jaskyňa za Kancľom with debris rampart. (Photo by P. Mitter)

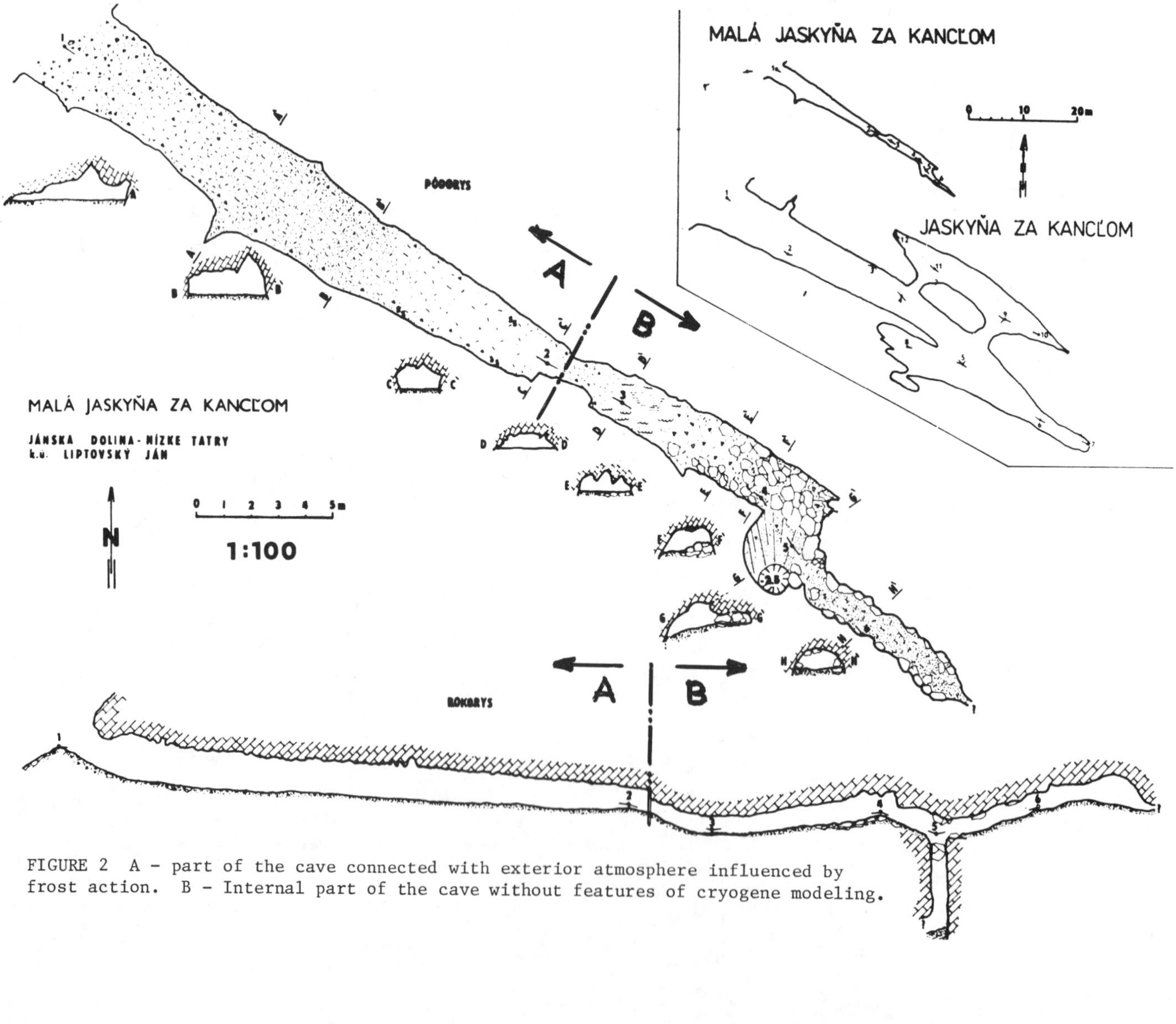

MALÁ JASKYŇA ZA KANCĽOM

0 10 20m

JASKYŇA ZA KANCĽOM

PÔDORYS

A B

MALÁ JASKYŇA ZA KANCĽOM

JÁNSKA DOLINA - NÍZKE TATRY
k.ú. LIPTOVSKÝ JÁN

N

0 1 2 3 4 5m

1:100

A B

BOKORYS

FIGURE 2 A - part of the cave connected with exterior atmosphere influenced by frost action. B - Internal part of the cave without features of cryogene modeling.

FIGURE 3 Patterned ground in the cave Jelenia/Vel'ká Fatra Mountains. (Photo by P. Mitter)

cold climates [Ložek (1964) used the term "pěnitec], forming separate crystals on the walls that float very slowly by percolated water to the floor of the shelter or cave. The unconsolidated material on the floor consists of debris, powder sinter, and the insoluble residue of limestone, which holds water well and creates a very favorable environment for frost action. There is often water there, and during more humid periods this sediment turns almost marsh-like. This type of wet sediment is favorable for the formation of patterned ground (Figure 3). The absence of vegetation and humus substance and the absence of the immediate effects of rainfall exclude the possibility that this phenomenon is a product of processes other than frost action, as Panos (1973) reported from the Belanské Tatry Mountains. We also found small non-sorted polygons in places with no debris. Our findings lower the altitudinal limit of the occurrence of patterned ground considerably from those of Jahn (1970).

SORTED STRIPES

Sorted stripes commonly occur on inclines in the floors of rock shelters and in the entrance areas of caves. The environment and the mechanism by which they are formed are very similar to those in which sorted polygonal forms occur; however, the inclines provide very favorable conditions for sediment movement. We wish to emphasize the importance of these processes in the creation of rock shelters, small caves, and cave entrances.

In the Západné Karpaty Mountains, where the mean annual air temperature ranges from +5°C to -3.7°C from the lower slopes to the top (the highest point in Czechoslovakia with a meteorological station), and the annual precipitation is 900-2000 mm, weathered material and free sinter material are transported out of caves by alternate freezing and thawing, mainly at the beginning and the end of the winter period. The transported material accumulates in front of the caves. In the Jelenia Jaskyňa cave (844 m), which has a spacious entrance and permits easy contact with the outside atmosphere, features of cryogenic transport occur as far as 35 m back in the cave from the entrance. Transported material in front of the Piesková cave in the Vel'ká Fatra Mountains can be seen in Figure 4. In many small caves, we have noted small sorted polygons, unsorted polygons, and sorted stripes.

After 10 years of research on karst in the Vel'ká Fatra and the Nizke Tatry Mountains, examining several tens of small caves and rock shelters, we believe that frost weathering and the transport activity of frost (Figure 5) play the main role forming in these caves and shelters. A specific type of cave or shelter is formed where the solution features are greatly subdued and frost-weathering phenomena are well developed. We term such caves "slope caves." The main features of these caves are: (1) linear form, (2) cave walls that are modelled by mechanical weathering, (3) cave floors that incline to the exit of the cave, and (4) debris on the floor of the cave that exhibits sorted polygons and stripes. These caves can be up to some tens of meters long.

FIGURE 4 Entrance to the cave Piesková with debris transported from the cave by frost action. (Photo by P. Mitter)

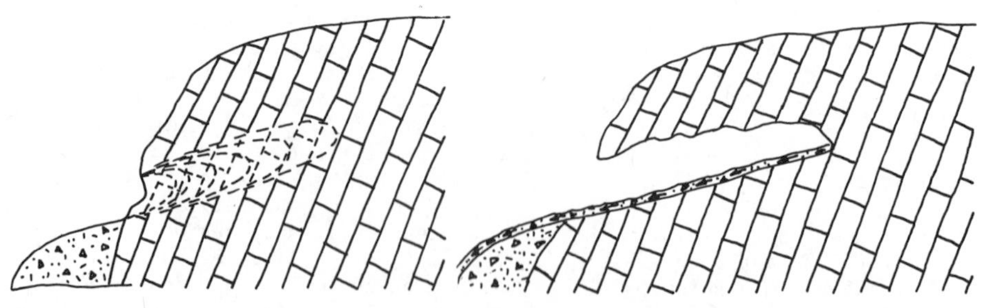

FIGURE 5 Development of slope caves with cryogene transport material from the cave.

CONCLUSIONS

On the exposed parts of the middle and high elevations of the Západné Karpaty Mountains, under the existing central European climate, frost action is one of the most important geomorphic factors of the karst regions. In front of caves, remarkable ramparts of debris form as a result of frost action on the exposed rock walls. In cave entrances, where there is easy, direct contact with the outside atmosphere, and in small caves as well, small sorted polygons and stripes form in the weathered debris on the cave floors. Cryogenic processes transport the loose sediments on the inclined cave floor towards the exit, and special types of caves, e.g. "slope caves," are created.

REFERENCES

Demek, J., 1972, Klasifikace a terminologie kryogénnych tvarů: Sbornik ceskoslovenské spolecnosti zemepisné: Academia, Praha (Prague), v. 77, no. 3, p. 303-309.

Jahn, A., 1970, Najnizsze stanowisko czynnych gruntow strukturalnych w Tatrach i problem dolnej granicy wystepowania zjawisk periglacjalnych w gorach: Acta Geographica Lodźiensiz, no. 24, p. 217-224.

Lozek, V., 1964, Ruzový previs ve Vrátné doline u Turcianské Blatnice: Academia, Praha (Prague), Ceskoslovenský kras, no. 15, p. 105-117.

Panos, V., 1973, The development dynamics of small landscape forms in the weathering and the vegetation mantles of the Belanské Tatry Mountains (Czechoslovakia): Acta Universitatis Palackianae Olomoucensis FRN, no. 42, Geographica-Geologica XIII, SPN, Praha, p. 109-126.

Pulina, M., 1968, Gleby poligonalne w jaskyni Czarnej, Tatry Zachodnie (les soils polygonales dans la grotte Czarna, les Tatres Occidentales): Speleologia, Warszawa, no. 3(2), p. 99-104.

Sekyra, J., 1960, Pusobeni mrazu na pudu: Geotechnica, Praha (Prague), no. 27, p. 164.

Washburn, A. L., 1973, Periglacial processes and environments: London, Edward Arnold Ltd.

GEOPHYSICAL MEASUREMENTS OF SUBBOTTOM PERMAFROST IN THE CANADIAN BEAUFORT SEA

J. L. Morack[1], H. A. MacAulay and J. A. Hunter

Geological Survey of Canada, 601 Booth Street, Ottawa, Ontario, Canada

[1]Currently on sabbatical leave from the University of Alaska, Fairbanks

Early marine seismic refraction interpretations on the Canadian Beaufort Sea shelf have indicated the widespread occurrence of ice-bearing sub-seabottom sediments. A large area of the shelf is underlain by coarse grained high velocity (> 3000 m/s, typical of ice-bonded sand and gravel) Cenozoic sediments. Recent data have reaffirmed the subdivision of this area into two major seismic layers, a deep thick continuous layer with the upper surface generally greater than 100 m below the seabottom, and a thin discontinuous layer at depths of 5-50 m below the seabottom. High resolution multichannel refraction seismic surveying has provided detailed structure of these layers.

INTRODUCTION

The existence of subsea permafrost in the Beaufort Sea shelf was predicted by Mackay (1972). Subsequent studies (Hunter and Hobson 1974, Hunter et al. 1976, 1978, MacAulay and Hunter 1982, Morack and Rogers 1982, 1983, Neave and Sellmann 1982, Osterkamp and Harrison 1982, Rogers and Morack 1980 Sellmann and Chamberlain 1979) in the Alaskan and Canadian Beaufort Seas have shown that ice-bonded sediments exist under large areas of the continental shelf.

Hunter et al. (1978) employed seismic refraction techniques on first arrival events from oil industry records to determine depths and velocities of thick ice-bonded permafrost. These measurements were limited by the large spacing of the hydrophone groups and the preferential attenuation of refraction events. Recently, the oil industry has been collecting high resolution reflection data in some areas of the Beaufort Sea shelf. The shorter hydrophone lines, the closer spacing of the hydrophone groups, and the higher source frequencies used for these data are much better suited to the refraction interpretation of ice-bonded subsea permafrost. This paper presents the analysis of two such lines.

The location of these lines is shown in Figure 1. Line A transects the outer Beaufort Sea shelf and Line B the inner shelf to the east of Mackenzie Bay. Also shown in this figure are areas underlain by shallow and deep high velocity refractors as determined from previous work (Hunter et al. 1978).

SEISMIC VELOCITIES OF ICE-BONDED PERMAFROST

Seismic velocities of water-saturated materials increase with decreasing temperatures below 0°C responding to the increase in interstitial ice content. Ice content varies directly with grain size (i.e., sand generally contains higher ice content than clay) and inversely with pore water salinity (saline interstitial water lowers the freezing point).

Numerous laboratory measurements of seismic velocities of ice-bonded permafrost have been made (Aptikaev 1964, Nakano et al. 1971, 1972, King et al. 1982). Velocity ranges for samples taken in the Canadian Beaufort Sea are discussed in detail in the paper by King et al. Their data show that the increase in ice content, and hence seismic velocity, with decreasing temperature below 0°C is less dramatic for fine grained than for coarse grained materials. Velocities for ice-bonded sand in the Beaufort Sea range from 4,000 m/s down to approximately 2,500 m/s, and may be as low as 1,800 m/s if the materials are only partially ice-bonded. Velocities for ice-bonded silt range from 3,300 to 2,200 m/s, while those for ice-bonded clay range from 2,300 to 2,000 m/s. These velocity ranges may be a few hundred m/s lower if the materials are only partially ice-bonded. These general velocity ranges are used later in this paper for the interpretations given to the refraction sections.

DISCUSSION OF HIGH RESOLUTION DATA

Detailed seismic refraction data have been analyzed along two marine survey lines (A and B in Figure 1) recorded during the summers of 1979 and 1981 for Dome and Gulf Petroleum. An air gun array was used as the source in both cases. Line A data were taken using the near 24 equally spaced hydrophone groups of a longer streamer giving an effective length of 767 m with a 240 m offset from source to near group. The shot points were spaced every 33.3 m for reflection anaysis; however, only every tenth shot point was analyzed for refractors Line B data were collected using 96 equally spaced hydrophone groups along a 600 m long streamer having a 14 m offset. Every eighth shot point was analyzed giving a spacing of 100 m. The first 0.80 s of the Line A data and 0.50 s of the Line B

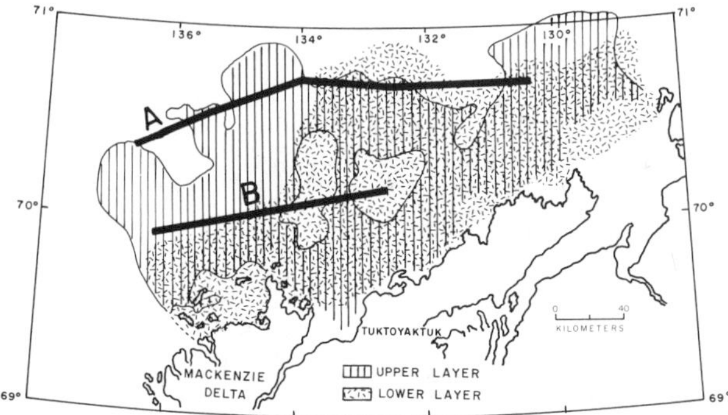

FIGURE 1 Map of the study area showing locations of seismic lines. Shaded areas indicate locations of upper (< 60 m) and lower (> 100 m) ice-bonded sediments as determined from previous studies.

data were made available for refraction analysis. A typical "wiggle trace" record for the Line B data is shown in Figure 2. The first arrival events were manually picked using microcomputer assisted digitizing techniques. The data were analyzed for depth and velocity information using standard refraction techniques (Telford et al. 1976). The different refraction events were easily determined by displaying the data as a reduced travel time plot where arrival times were shifted by the amount x/v, where v is near the refractor velocity.

ISO-OFFSET TIME SECTIONS

The first arrival times for all records were displayed in a format that we call an "Iso-Offset Time Section." This is a plot of shot point locations along the horizontal axis and first arrival time in the downward direction. Arrival times at a given hydrophone group for each of its shot point locations are then connected by straight line segments. The position of each hydrophone group was shifted by one-half its distance from the shot point to more accurately locate the midpoint of the ray path. An Iso-Offset plot for Line A is shown in Figure 3. The vertical spacing of the connected lines decreases as the seismic velocity increases due to the shorter interval time. Thus, a sharp velocity contrast is indicated by a change in the density of the connected lines; areas underlain by high velocity material appear as areas of closely spaced lines on the plot. Figure 3 shows at a glance the locations along the line where high velocity materials exist. Additionally, the vertical displacement of the high density segments gives a qualitative indication of the depth of the velocity discontinuities; those segments appearing later in time occur at greater depth.

For the purposes of discussion here, the shelf area has been divided into several geotechnical divisions suggested by M. J. O'Conner (personal communication, 1983). From the west, these divisions are the Kringalik Plateau, the Ikit Trough, the Akpak Plateau, the Kugmallit Channel, the Tingmiark Plain, the Niglik Channels, and the Kaglulik Plain. Line A traverses the outer shelf

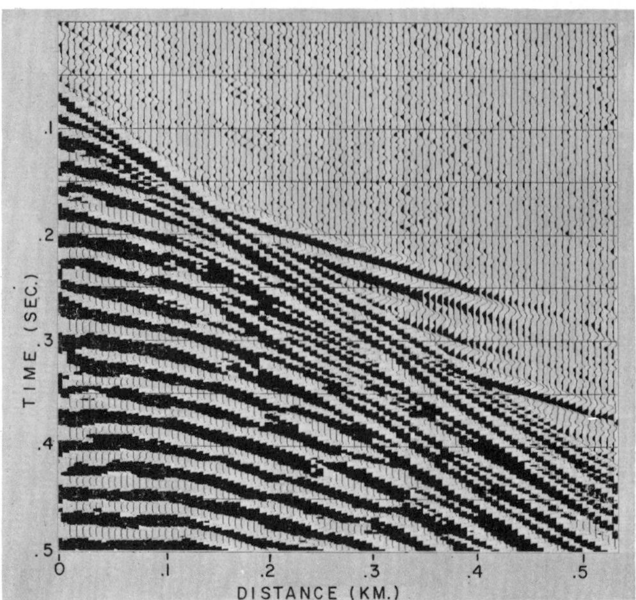

FIGURE 2 Typical "Wiggle trace" record showing high velocity refractors.

area and crosses all of these divisions. The west end of the line beneath the Kringalik Plateau shows a few shallow high velocity events. The section beneath the Ikit Trough shows a depression in the high velocity zones. The area beneath the Akpak Plateau is characterized by nearly continuous shallow high velocity materials. The Kugmallit Channel area shows a shallow layer, having a seismic velocity in the range of 1,800-2,100 m/s, underlain by a deeper high velocity (> 2,500 m/s) layer which is visible for most of the rest of section as a gradually upward sloping layer. East of the Kugmallit Channel this deep layer is overlain by a shallow high velocity layer which shows two definite depressions at the Niglik Channels. All of these high velocity segments have been interpreted as ice-bonded or ice-bearing materials. In general there is a good correlation with the suggested geotechnical divisions.

Figure 4 shows an Iso-Offset plot for Line B which traverses the inner shelf. Because of the large number of shot points examined, this line is shown in three segments. The velocity contrasts are much more apparent than those in Line A because of the increased number of traces per record and because the vertical time scale is only one-half that of Line A. Beginning at the west end of the line, sections B-1 and B-2 up to approximately 135° W longitude show only an occasional high velocity segment. Beneath the Akpak Plateau an almost continuous shallow high velocity layer is present. Underneath this shallow layer, a deeper high velocity layer is apparent along much of the section. Across the Kugmallit Channel the shallow high velocity refractor is only sporadically present, whereas the deeper high velocity layer is essentially continuous. Along segment B-3 the deep high velocity layer continues and short segments of a shallow layer occur. The refraction analysis indicates that the depth from the sea surface to

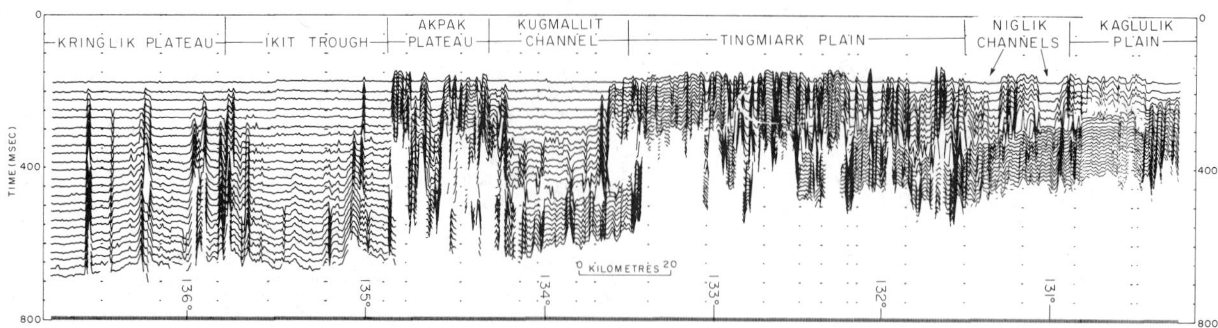

FIGURE 3 Iso-Offset section of seismic Line A.

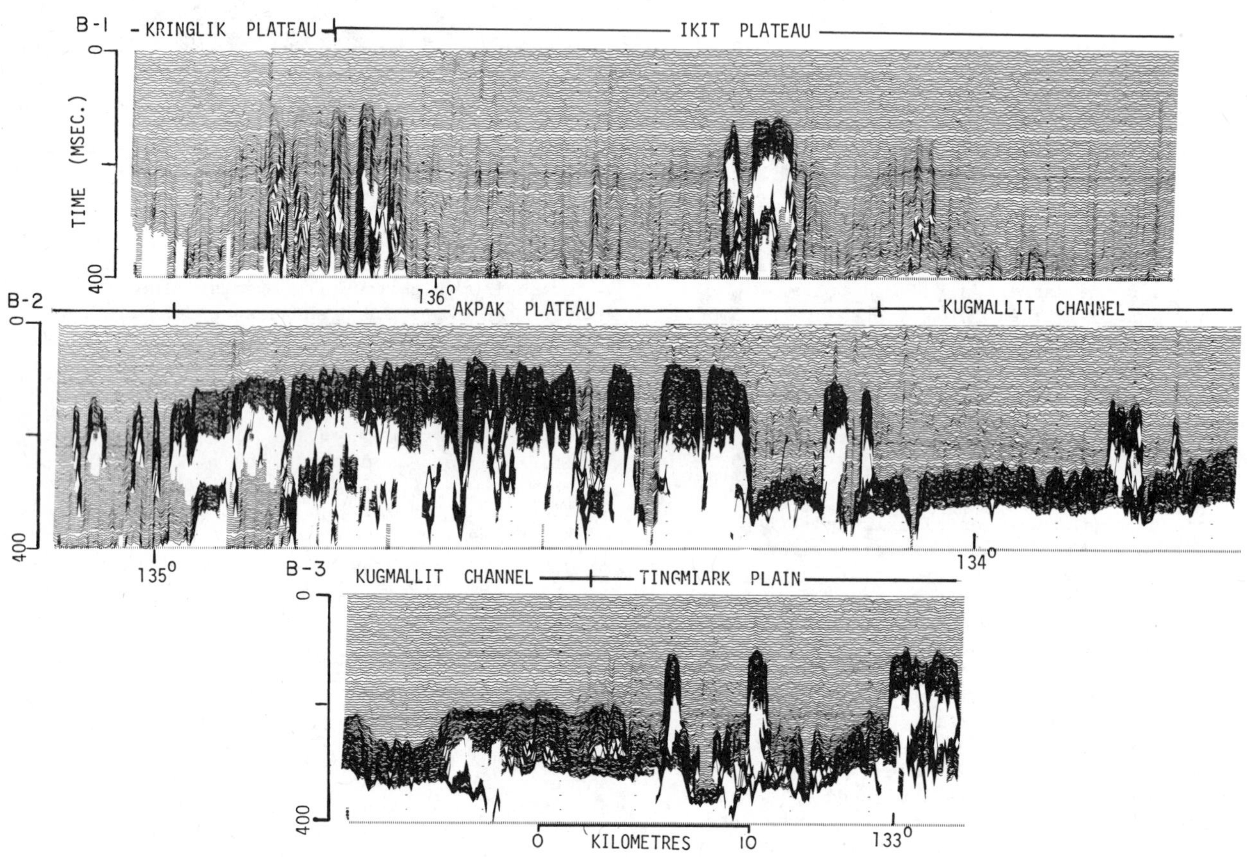

FIGURE 4 Iso-Offset section of seismic Line B shown in three adjoining segments.

the top of the shallow high velocity layer ranges from 20 to 60 m, while the depth to the top of the deep high velocity layer is between 120 and 160 m. The seismic velocities of the high velocity layers in segments B-2 and B-3 are greater than 2,500 m/s indicating ice-bonded materials, while those in segment B-1 are greater than 1,800 m/s indicating ice-bearing materials. In general the Iso-Offset section does not correlate as well with the suggested geotechnical divisions as the previously discussed section farther offshore.

RESULTS OF REFRACTION ANALYSIS

A detailed refraction analysis and interpretation has been carried out on selected portions of both Lines A and B. Figure 5 shows an Iso-Offset section from Line A which crosses the Kugmallit Channel. Beneath this is a plot of the refraction analysis including calculated velocities and depths to refractors. The error bars shown on the velocity data points are the standard deviations of the least squares fits made to the time-distance

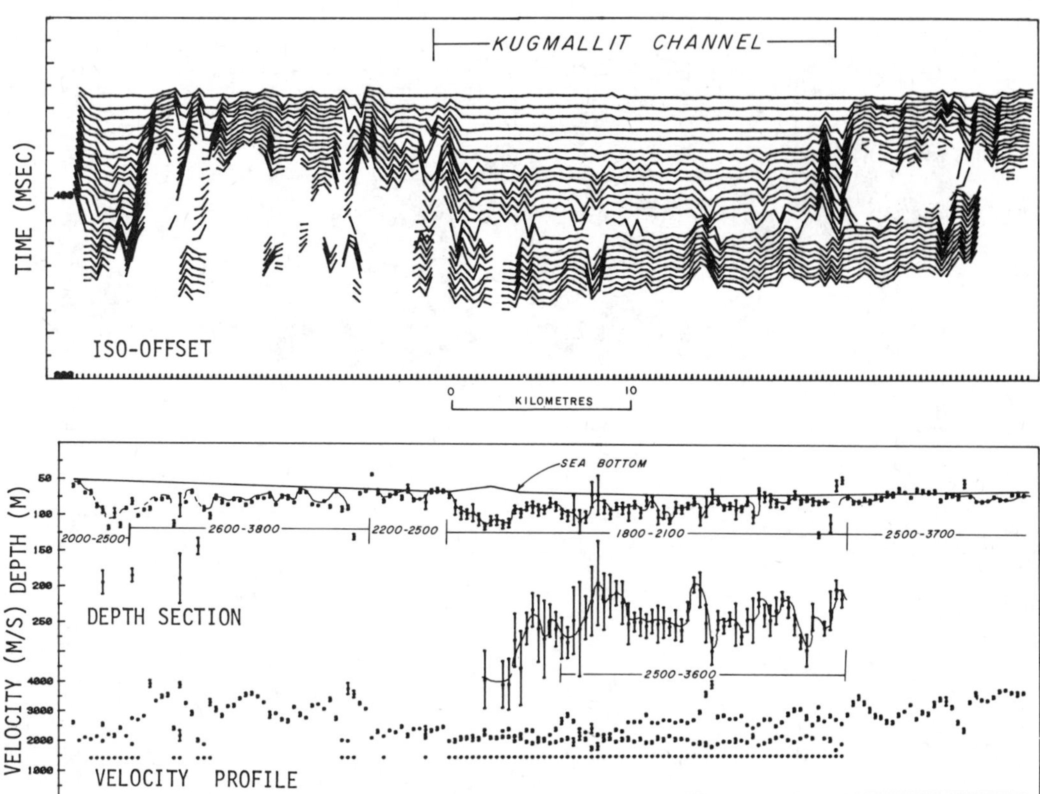

FIGURE 5 Iso-Offset section and velocity-depth section for a portion of Line A. The numbers shown beneath the depths are ranges for the corresponding refractor velocities in m/s.

plots. The errors shown for the depth calculations are the propagated errors of the velocities and time intercepts used in the calculation. A water velocity of 1,420 m/s was used for the depth calculations when a water arrival was not observed on the records. This occurs when the first arrivals are generated by a shallow high velocity layer.

West of the Kugmallit Channel the section shows a number of discontinuous high velocity zones within 50 m of the seabed. These can be divided into three segments exhibiting different velocity characteristics. At the west end is a dipping layer with velocities ranging from 2,000 to 2,500 m/s. This material is interpreted to vary from partially ice-bonded to ice-bonded fine grained materials. To the east of this is a section showing a shallow layer where the velocities are from 2,600 to 3,800 m/s. This layer is interpreted to be ice-bonded sands. Immediately to the west of the Kugmallit Channel is a shallow layer having velocities ranging from 2,200 to 2,500 m/s which are interpreted to be ice-bonded fine grained materials.

The section beneath the Kugmallit Channel indicates a refractor approximately 25 m beneath the seabed with velocities ranging from 1,800 to 2,100 m/s, and a deeper refractor approximately 160 m below seabottom with velocities of 2,200 to 3,500 m/s. The upper layer is interpreted to be partially ice-bonded or ice-bonded fine grained sediments and the deeper layer to be partially ice-bonded or ice-bonded sands.

East of the channel the section shows a high velocity layer lying essentially at the seabed. The layer exhibits velocities in the range of 2,500 - 3,700 m/s indicating ice-bonded fine sands. Due to the high velocity of this refractor, a hidden layer probably exists above it. Its absence in the refraction analysis would make the calculated depth of the high velocity layer too shallow. Beneath this layer the continuation of the deeper layer is evident on the Iso-Offset plot but has not been included in the refraction analysis due to a lack of information of the intervening material velocities. The amplitude attenuation of the seismic signals indicates that the thickness of the shallow ice-bonded layer is probably no greater than 50 m; however, additional data including drill hole data is needed to confirm this. Thus a low velocity zone is probably present beneath the shallow ice-bonded layer in this region. This section illustrates how useful it is to have a high resolution reflection profile in conjunction with the refraction analysis. The reflection profile shows the layering, while the refraction analysis supplies the velocities needed to calculate depths.

A detailed analysis of a section of Line B-2 just west of the Kugmallit Channel is shown in Figure 6. An Iso-Offset section and the refraction analysis are shown. This section is characterized by discontinuous shallow high velocity refractors which lie from 20 to 60 m beneath the seabed and a deep high velocity refractor which is nearly continuous at a depth of approximately 125 m below

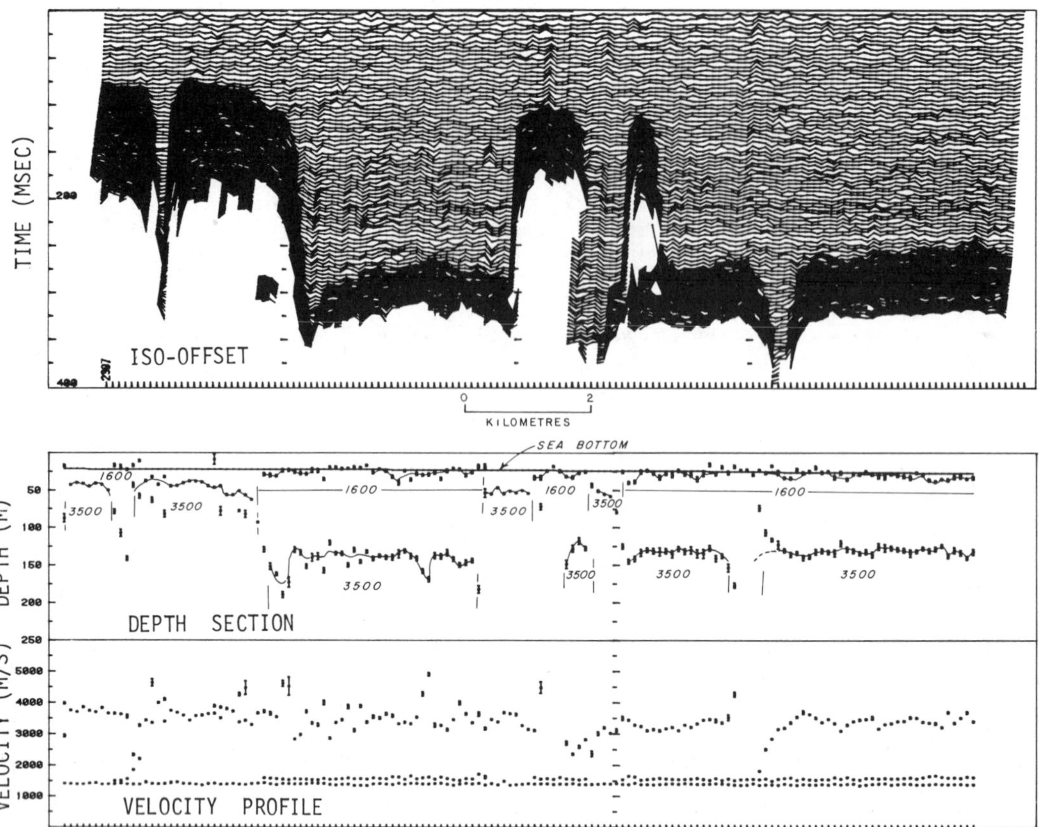

FIGURE 6 Iso-Offset section and velocity-depth section for a portion of Line B. The numbers shown beneath the depths are the corresponding approximate refractor velocities in m/s.

the seabed. The velocity of these refractors is approximately 3,500 m/s and indicates that they are ice-bonded sands. In areas where the shallow high velocity layer is absent, the deep layer is overlain by materials having velocities of approximately 1,600 m/s whose upper surface is essentially the seabed.

A thin hidden layer is undoubtedly present in the case of the shallow high velocity layers, and consequently the calculated depths are slightly shallow. Again from the amplitude behaviour of the seismic signals, the thickness of these shallow ice-bonded layers is estimated to be less than 50 m. A short segment of the deep layer is also apparent on the Iso-Offset section beneath the shallow layer near the west end. It is only slightly "pulled up" above the deep layer lying just to the east, again indicating that the upper layer is thin. An apparent gap in the deep layer is present near the east end of the section. The anomalous velocity values that occur at each end of the gap are explained by recalling that the hydrophone line extends across several shot points. These incorrect velocity values are responsible for the nature of the depths shown.

COMPARISON OF REFLECTION AND ISO-OFFSET SECTIONS

The previous sections have shown that the Iso-Offset display allows a quick determination of the location of high velocity materials and gives a qualitative indication of their depths. Another use of this display is for the identification of shallow high velocity materials on high resolution reflection sections. Figure 7 shows a Common-depth-point stacked reflection section of Line B-2 and the corresponding Iso-Offset section. The high velocity layers have been marked on the reflection section and the correlation between the two sections is generally excellent. The shallow high velocity event on the reflection section usually occurs as a bright spot. Where it is not obvious, the Iso-Offset section confirms its occurrence.

ACKNOWLEDGMENTS

The authors wish to thank Dome Petroleum Ltd. and Gulf Canada Resources Inc. for permission to publish data and for their cooperation and encouragement in this research by freely supplying seismic data tape for reprocessing.

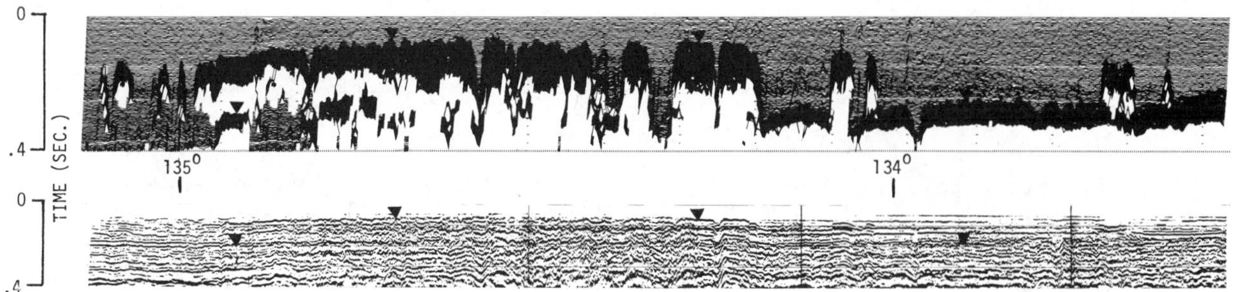

FIGURE 7 Iso-Offset and reflection (first .4 s) section of Line B-2.

REFERENCES

Aptikaev, F. F., 1964, Temperature field effect on the distribution of seismic velocities in the permafrost zone: Akad Nauk SSSR Sibjrskoe otdie. Inst. Merzlotovedeniia. Teplovye protesessy v merzlykh porod. (Transl.).

Hunter, J. A., and Hobson, G. D., 1974, A seismic refraction method of detecting sub-seabottom permafrost: in The Coast and and Shelf of the Beaufort Sea, J. C. Reed, and J. E. Suter, eds., A.I.N.A. pp. 401-416.

Hunter, J. A., Judge, A. S., MacAulay, H. A., Good, R. L., Gagné, R. M., and Burns, R. A., 1976, Permafrost and frozen sub-bottom materials in the Southern Beaufort Sea: Beaufort Sea Project, Tech. Rep. no. 22, Dep. Environ., 174 p.

Hunter, J. A., Neave, K. G., MacAulay, H. A., and Hobson, G. D., 1978, Interpretation of sub-seabottom permafrost: in The Beaufort Sea by Seismic Methods, Part I, Seismic Refraction Methods. Proc. 3rd Int. Conf. Permafrost, Natl. Res. Counc. pp. 515-521.

King, M. S., Pandit, B. I., Hunter, J. A., and Gajtani, M., 1982, Some seismic, electrical and thermal properties of sub-seabottom permafrost samples from the Beaufort Sea: Proc. 4th Can. Permafrost Conf., Natl. Res. Counc. pp. 268-273.

MacAulay, H. A., and Hunter, J. A., 1982, Detailed seismic refraction analysis of ice-bonded permafrost layering in the Canadian Beaufort Sea: Proc. 4th Can. Permafrost Conf. pp. 256-267.

Mackay, J. R., 1972, Offshore permafrost and ground ice, Southern Beaufort Sea: Can. J. Earth Sci., v. 9, pp. 1550-1561.

Morack, J. L., and Rogers, J. C., 1982, Marine seismic refraction measurement of near-shore subsea permafrost: Proc. 4th Can. permafrost Conf., Natl. Res. Counc. pp. 249-255.

Morack, J. L., and Rogers, J. C., 1983, Acoustic velocities of near-shore materials in the Alaskan and Chukchi Seas: in Alaskan Beaufort Sea, Ecosystem and Environment, P. Barnes, E. Reimnitz, and D. Shell, eds., Academic Press. In press.

Nakano, Y., Smith, M., Martin, R., Stevens, H., and Knuth, K., 1971, Determination of the acoustic properties of frozen soils: Prepared for Advanced Res. Projects Agency ARPA order 1525 by Cold Regions Res. Eng. Lab., U.S. Army, 72 p.

Nakano, Y., Martin, R., and Smith, M., 1972, Ultrasonic velocities of the dilatational and shear waves in frozen soils: Water Resour. Res., v. 8, no. 4, pp. 1024-1030.

Neave, K. G., and Sellmann, P. V., 1982, Subsea permafrost in Harrison Bay, Alaska: CRREL Report 82-24, 62 p.

Osterkamp, T. E., and Harrison, W. D., 1982, Temperature measurements in subsea permafrost off the coast of Alaska: in Proc. 4th Can. Permafrost Conf., pp. 238-248.

Rogers, J. C., and Morack, J. L., 1980, Geophysical evidence of shallow near-shore permafrost, Prudhoe Bay, Alaska: J. Geophys. Res., v. 85, no. B9, pp. 4845-4853.

Sellmann, P. V., and Chamberlain, E. J., 1979, Permafrost beneath the Beaufort Sea near Prudhoe Bay, Alaska: Proc. of the Eleventh Off-shore Technology Conference, 3, pp. 1481-1488.

Telford, W. M., Geldart, L. P., Sheriff, R. E., and Keys, D. A., 1976, Applied Geophysics: Cambridge University press, New York, 860 p.

THE UTILITY OF REMOTELY SENSED DATA FOR PERMAFROST STUDIES

Leslie A. Morrissey

Technicolor Government Services, Inc., NASA/Ames Research Center,
Moffett Field, California 94035 USA

The presence of permafrost determines to a large degree, the capability or suitability of land for development. Many environmental factors which influence permafrost can be studied through the analysis of remotely sensed data. This paper discusses the utilization of remotely sensed data in the development of derivative maps and data layers based on associations among vegetation and terrain factors and permafrost conditions. An approach will be presented for deriving data layers from an existing data base to illustrate their potential contribution to predictive permafrost modeling.

INTRODUCTION

Land cover maps derived from satellite data have been utilized extensively by natural resource managers. However, land cover information alone provides only one piece of the puzzle. The utility of such information can be greatly extended when remotely sensed data is integrated with other data to provide resource managers with derivative maps specifically tailored to complex environmental issues. For example, in Alaska the presence of permafrost may determine the capability and suitability of land for development. Many environmental factors which influence or reflect permafrost conditions can be studied through the analysis of remotely sensed data. It is the purpose of this paper to demonstrate the utility of remotely sensed data in the development of derivative maps for integration into predictive permafrost models through illustration with a select number of examples.

Scientists at NASA/Ames Research Center and the Alaska Department of Natural Resources (ADNR) have conducted a cooperative effort to test the feasibility of using LANDSAT digital data in conjunction with topographic data to map forest cover types in a study area northwest of Fairbanks (Morrissey and Ambrosia 1982). The primary objective of this demonstration project was to test and evaluate a technique combining multi-date LANDSAT data with digital terrain data to provide forest resource information -- specifically stand size, crown cover density, and species composition -- in a cost effective and timely manner. A multivariate clustering approach, using multispectral waveband combinations from both August and September (1979) dates, provided the basis for an initial land cover classification. This preliminary product was refined using elevation, slope, and aspect information.

Digital elevation data were acquired and registered to the LANDSAT satellite data. Slope and aspect data were calculated from the elevation data (Figure 1). In addition, statistical tests were utilized to determine significant relationships between forest cover types derived from LANDSAT and terrain. The results of these tests provided the basis for grouping and reassigning spectral classes within specific topographic gradients during the final classification process and resulted in a detailed timber type map encompassing the Fairbanks, Livengood, and Minto Flats regions.

The use of an existing data base, such as the Tanana data base, will be used to illustrate the development of derivative maps based on select vegetation and terrain factors that are associated with permafrost distribution. The potential utility of this existing data base for permafrost mapping will be demonstrated by a number of examples.

Existing maps of permafrost distribution are general, primarily because of a lack of more extensive and detailed information (Pewe 1966). The delineation of permafrost boundaries from conventional aerial photography is difficult and requires extensive and costly work for confirmation. Moreover, photo- interpretation does not lend itself to convenient or accurate mapping of extensive areas. However, an alternative approach that acknowledges the close interaction of landscape factors within a region is based on the development of predictive environmental models. Such

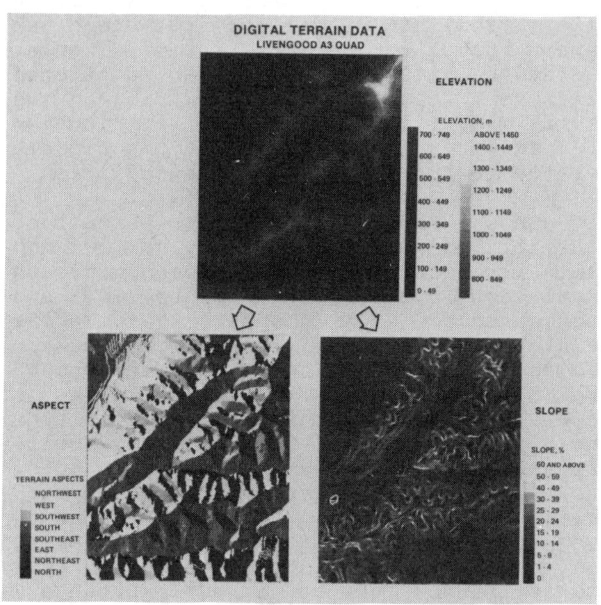

Figure 1 Aspect and slope derivative data layers generated from digital terrain data.

models, based on known relationships among permafrost and biotic and terrain factors, have great potential for providing increased understanding of the character and distribution of permafrost over large regions.

Mapping permafrost is a much more difficult task than mapping vegetation because underlying permafrost interacts with the landscape in a myriad of ways. Understanding the interrelationships among the various factors that affect the distribution and character of permafrost is the key to developing predictive permafrost models. Brown (1969) provides a thorough review of factors which influence discontinuous permafrost. Figure 2 illustrates the development of environmental models designed to predict the occurrence of permafrost based on the integration of derivative maps. Initially, this model would require a systematic collection of ground data relating to a wide range of vegetation and terrain factors, together with their spatial correlation. The correlations would permit the determination of environmental associations that constitute the basis for the predictive model. A data base could be devised which integrates those environmental factors (or surrogate factors) that are highly correlated with the occurrence of permafrost. Three environmental factors, vegetative cover, topographic data and snow cover, and their corresponding derivative maps are shown in Figure 2. This data base could be compiled from available maps, photo-interpretation of low and high altitude and satellite imagery and digital classification of satellite imagery. The resulting predictive model could then be used to generate regional maps for testing and refinement as appropriate. The next section will discuss the environmental factors in Figure 2 in more detail and illustrate their contribution to a permafrost model.

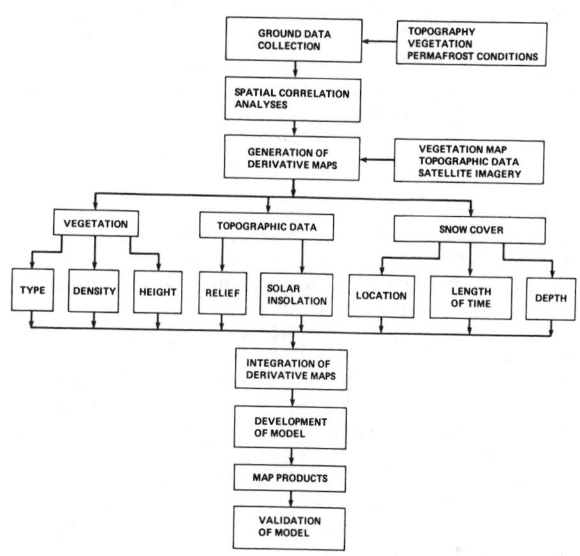

Figure 2 The development of a permafrost model based on the integration of derivative maps.

ENVIRONMENTAL FACTORS

Creation of a permafrost model involves the development of derivative maps to be utilized as surrogates for other factors indicative of the presence or absence of permafrost. For instance, vegetation alone does not indicate the existence of permafrost. However, it is well known that permafrost indirectly affects vegetation by controlling drainage, by maintaining low temperatures in the root zone, and by providing an impervious substrate that restricts the extent of root growth (Price 1972). Consequently, it is possible to develop correlations between vegetation types and the probability of permafrost occurrence. In this way, knowledge of the reciprocal relationship between vegetation types and permafrost can be utilized in the development of a predictive model. Similarly, other vegetative and terrain factors can be integrated and modeled to predict permafrost conditions. A close correspondence between vegetation types and the presence or absence of near surface permafrost in the uplands of the discontinuous-permafrost zone of central Alaska has been noted by numerous investigators (Stoeckeler 1949, Rieger et al. 1963, Péwé 1966, Dingman and Koutz 1974). In general, black spruce scrub forests and shrub bogs are indicative of underlying permafrost close to the surface, while white spruce, birch and aspen forests occur on slopes which are free of near surface but which may have permafrost below the surface. On the basis of these relationships, classes within the detailed timber type map (referred to above) can be reclassified to reflect the probability of permafrost occurrence based on vegetation type.

Forest stand structure indirectly affects permafrost by influencing near surface conditions. Trees shade the ground from solar radiation and intercept some of the snowfall in winter. Furthermore, the density and height of trees influence the micro-climatic effects of ground surface wind velocities (Brown and Péwé 1973), wind speeds being lower in areas of dense growth than in areas where trees are sparse or absent. Tree height and density, as well as other timber factors, have been derived from satellite imagery using multistage sampling techniques (Morrissey and Ambrosia 1982) and could be utilized as individual data layers to further refine permafrost models.

Various terrain conditions have important effects on permafrost growth and thickness (Brown and Péwé 1973). Surface relief directly influences permafrost formation by controlling the amount of solar radiation received by the ground. In the discontinuous zone, permafrost may occur only on north-facing slopes, which receive less solar radiation; in the continuous zone, permafrost on north-facing slopes may be thicker with a thinner active layer than south-facing slopes. Relief also influences snowfall accumulation and vegetative cover, which in turn affect permafrost thickness. Digital topographic data (slope, aspect, and elevation) can provide the basis for modeling permafrost based on terrain conditions -- for instance, in the discontinuous zone, north-facing slopes have the highest probability of permafrost occurrence, south-facing slopes have the lowest probability, with other slopes having intermediate probability values. In addition, solar illumination information can be derived from slope and aspect data to measure incoming solar radiation through time for an additional refinement of the model. Dingman and Koutz (1974) found that the distribution of vegetation and permafrost is closely related to the influx of solar radiation.

Snow cover influences permafrost distribution and thickness by controlling heat transfer to the ground through insulation from the atmosphere (Gold and Lachenbrunch 1973, Grandberg 1973, Thomas et al. 1978). Snowfall conditions and the length of time that snow is on the ground are important factors. While early, heavy snowfall in winter effectively insulates the ground from the severe cold, preventing permafrost aggradation, thick snow cover that lasts late into spring delays thaw and has the opposite effect. Even in the discontinuous zone, snow cover can be a critical factor in the formation and existence of permafrost (Brown and Pewe 1973). The thickest and most extensive permafrost and thinnest active layer exist in areas of minimal snow cover. A comparison of satellite images throughout spring and fall can provide locational information concerning the accumulation of snow, length of time on the ground, and depth (Figure 3). Measurements in the microwave region not only provide an all weather capability but have been shown to have potential for determing snowpack properties, such as snow depth and snow water equivalent, which cannot be determined in other portions of the spectrum (Barnes 1981).

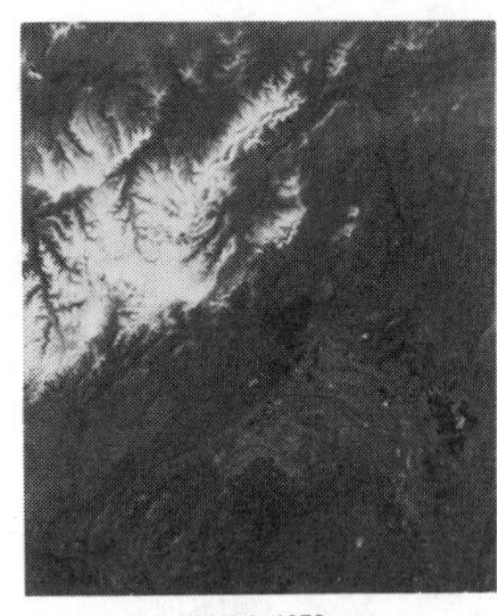

MAY 8, 1979

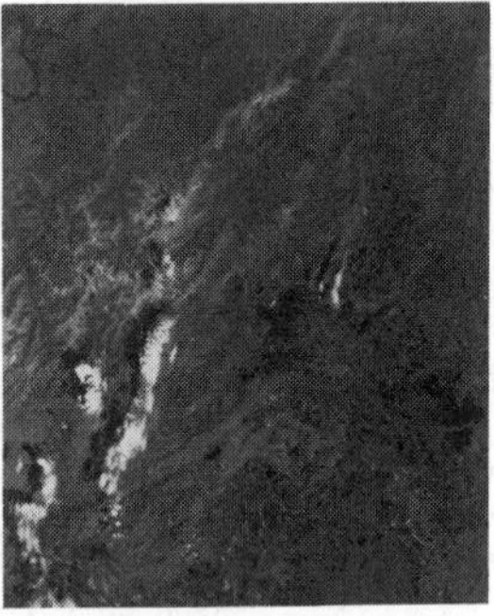

JUNE 5, 1979

SAWTOOTH MOUNTAIN
NORTHWEST OF FAIRBANKS, ALASKA

Figure 3 Snow cover monitoring from sequential LANDSAT satellite imagery (note: the May scene is cloudfree while partially snow covered, while the June scene is free of snow but partially cloud covered).

Permafrost maps based on vegetative cover, terrain conditions, and snow cover can be integrated with other information such as soils, climatic, hydrologic, and geologic data to further refine the model. However, the effect of additional data layers to the models predictive accuracy is not clear. The addition of increasing layers of information in the modeling process may increase the likelihood of error in the model itself. The inaccuracies of each data layer may be compounded when combined with other layers. Conversely, the errors in one data layer may counterbalance those in other layers, increasing the models overall performance. In general, each model must be examined on a case specific basis and subjected to empirical verification.

SUMMARY

The immense size, lack of ground transportation and harsh climatic conditions in arctic and subarctic regions make remote sensing techniques and satellite derived information well suited for reconnaissance mapping of permafrost. The availability of satellite data at various spatial resolutions and in various portions of the electromagnetic spectrum provides both a baseline and a periodic updating mechanism for use in environmental studies. Future satellite sensors, with increased spectral and spatial resolution, will enhance the utility of remotely sensed data for both regional and site specific studies. However, before this data can be fully exploited, two fundamental areas of research must be addressed. Initially, the spatial interrelationships between the various vegetation and environmental factors and the occurrence of permafrost require further study. This type of research has been continuing for a number of years and the complexity of the task requires additional study. Second, systematic research is required to evaluate the utility of present and future remote sensor data for information extraction relevant to arctic and subarctic regions. In addition, there is a need for fundamental research to identify spectral, spatial, and temporal resolution requirements for specific features. For example, the spatial resolution that is required to identify polygonal ground patterns on the Arctic Coastal Plain.

A tremendous potential exists for the utilization of remotely sensed data to provide meaningful resource information for permafrost studies. The integration of this remotely derived data with other ancillary data in a geographic information system context is the next logical step in the process of mapping permafrost in harsh and inaccessible arctic and subarctic regions.

REFERENCES

Barnes, J.C., 1981, Survey paper on remote sensing techniques to map snow cover, in Proceedings of the International Geoscience and Remote Sensing Symposium, Washington, D.C., June 8-10, 1981, v. 1, p. 123-132.

Brown, R.J.E., 1969, Factors influencing discontinuous permafrost in Canada, in the Periglacial Environment, Past and Present, International Association Quaternary Research Congress, McGill Queens University Press, Montreal, p. 11-53.

Brown, R.J.E. and Péwé, T.L., 1973, Distribution of permafrost in North America and its relationship to the environment: a review, in Permafrost--The North American Contribution to the Second International Permafrost Conference, Yakutsk: Washington, D.C., National Academy of Sciences, p. 71-100.

Dingman, S.L. and Koutz, F.R., 1974, Relations among vegetation, permafrost, and potential insolation in central Alaska: Journal of Arctic and Alpine Research v. 6, p. 37-42.

Gold, L.W. and Lachenbruch, A.H., 1973, Thermal conditions in permafrost, a review of North American literature in Permafrost-- The North American Contribution to the Second International Permafrost Conference, Yakutsk: Washington, D.C., National Academy of Sciences, p. 3-26.

Grandberg, H.B., 1973, Indirect mapping of the snow cover for permafrost predictions at Schefferville, Quebec, in Permafrost-- The North American Contribution to the Second International Permafrost Conference, Yakutsk: Washington, D.C., National Academy of Sciences, p. 113-120.

Morrissey, L.A. and Ambrosia, V.G., 1982, Forest timber typing final report, National Aeronautics and Space Administration Report 166391, NASA Ames Research Center.

Péwé, 1966, Permafrost and its effect on life in the north, in Arctic Biology, Hansen, H.P. editor, Oregon State University Press, Corvallis, Oregon, p. 27-66.

Price, L.W., 1972, The periglacial environment, permafrost and man: Association of American Geographers, Resource Paper 14, 88 pages. Rieger, S., Dement, J.A. and Sanders, D., 1963, Soil survey of the Fairbanks area, Alaska, U.S. Soil Conservation Service, series 25.

Stoeckeler, E.G., 1949, Identification and Evaluation of Alaskan Vegetation from Air Photos with Reference to Soil, Moisture, and Permafrost Conditions: U.S. Army Corps of Engineers, St. Paul District, 102 pages.

Thomas, I.L., Lewis, A.J. and Ching, N.P., 1978, Snowfield assessment from LANDSAT, Photogrammetric Engineering and Remote Sensing, v. 44, p. 493-502.

THE ROLE OF PRE-EXISTING, CORRUGATED TOPOGRAPHY IN THE DEVELOPMENT OF STONE STRIPES

Mark P. Muir

Quaternary Research Center, Mail Stop AK-60
University of Washington, Seattle, Washington 98195 USA

An experiment was conducted to explore the role of small-scale topography in the development of stone stripes. The surface of a silt loam with a homogeneous admixture of gravel was shaped into sinusoidal corrugations with wavelengths ranging from 0.05 to 0.8 m. The soil was then subjected to both diurnal and storm-length freeze-thaw cycles. Surface heave and subsurface temperature were monitored electronically. Sorting was measured by displacement of marker stones and changes in gravel concentration in soil samples. The volume concentration of gravel increased in the soil underlying troughs. Marker stones tended to move toward the nearest trough. Stones close to the surface in more steeply inclined regions traveled the greatest lateral distance with several stones moving as far as 40 mm. Sorting related to topography was apparent only under the larger corrugations. Heave records indicate that significant heave occurred in troughs during the thaw portion of the temperature cycle. Residual differences in heave between the crests and troughs of sinusoids on the order of 1 mm per storm-length freeze-thaw cycle tended to accentuate relief through time.

INTRODUCTION

The development of sorted patterned ground requires both lateral and vertical components of sorting of coarse particles in the soil. As a soil freezes, coarse particles tend to move in the direction of local heat flow relative to the soil matrix (Corte 1966). The combination of lateral and vertical components of sorting can only occur then, when a freezing front is nonplanar as it penetrates a soil. This may result from lateral variations in soil-moisture content, sorting, vegetation, or relief (Schmertmann and Taylor 1965, Nicholson 1976). A laboratory experiment was designed to study the sorting involved in the formation of stone stripes in a soil with an initially corrugated topography comparable to rills on a hillslope. Such topography has been implicated in the development of stone stripes (Brockie 1968).

EXPERIMENTAL DESIGN

Silt loam was mixed with about 15% by volume of gravel (volume of gravel/total volume of saturated soil plus gravel). The gravel consists of clasts ranging from 10 to 30 mm in diameter obtained from a quarry in glacial outwash deposits. The soil was then placed in a container with length of 3.6 m and width of 2.5 m to an average depth of about 0.2 m (Figure 1). Underlying the soil and separated from it by a sheet of muslin was a 20 mm thick layer of sand. Perforated pipes running through this layer were connected by pipes (water headers) rising through the soil to an external water supply system. Water could be added to the soil through these pipes at any time. The container is located in a cold room capable of achieving temperatures as low as -50°C and has controls for automatic cycling of temperature.

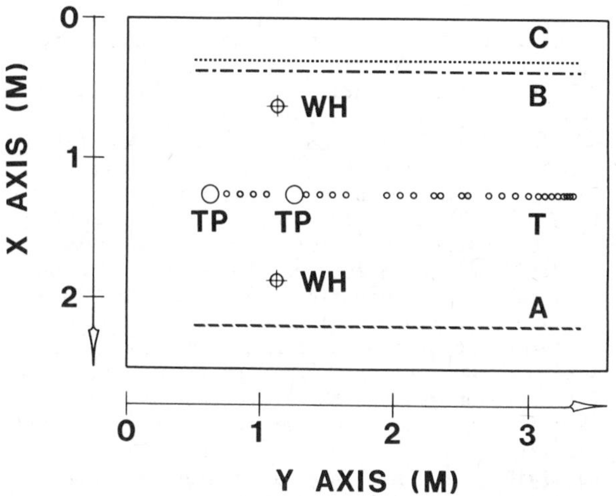

FIGURE 1 Plan view of the experiment showing the locations of thermocouple piles (TP), individual thermocouples (T), and water headers (WH). Also shown are the three marker stone trenches A, B, and C.

Parallel corrugations were formed on the surface of the soil to produce five sets of sinusoids with wavelengths of 0.05, 0.1, 0.2, 0.4, and 0.8 m (Figure 2). The amplitude of each sinusoid was one-tenth the wavelength. Three flat-floored trenches were cut to different depths perpendicular to the trend of the sinusoids (Figure 2). Glitter was spread in a thin layer over the floor of each trench to function as a marker horizon. One hundred twenty-eight painted marker stones were emplaced in the side walls of the trenches. Their positions were surveyed and the trenches were back filled.

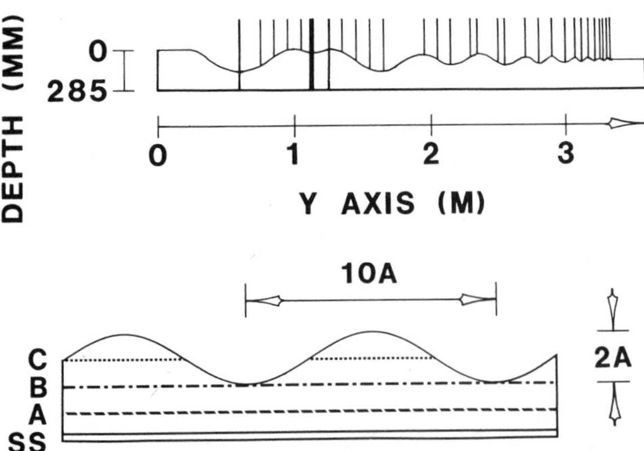

FIGURE 2 (Top) Cross-sectional view of the experiment showing water headers (heavy line), thermocouple piles (light lines extending to the base of the soil), and the wooden structural supports for individual thermocouples (light lines terminating at the soil surface). Note the sinusoidal topography. (Bottom) Cross-sectional view of one set of sinusoids showing the depth to marker horizons A, B, and C relative to the amplitude A of topography. SS marks the 20-mm thick sand substrate.

Soil samples taken prior to shaping the topography had a volumetric moisure content of 0.40. The gravel concentration was 14.0 ± 3.6% by volume (volume of gravel/bulk volume of sample) and showed no significant variation with depth. Additional samples were taken after the topography was created, but only to an average depth of 78 mm. The mean concentration of gravel in these samples was 8.5% by volume. Therefore, the processes by which the original topography was formed caused substantial downward movement of gravel, leaving the near-surface region relatively depleted in gravel at the start of the experiment. The same processes were used to re-create the topography over marker stone trenches after they were back filled and it is inferred that the marker stones were similarly displaced. Independent experiments show that negligible (< 1 mm) lateral movement accompanied this vertical displacement. Thus, at the start of the experiment, the gravel and marker stones had random horizontal distributions, but were concentrated at depth.

The soil was then subjected to a series of freeze-thaw cycles, including 100 diurnal and 12 storm-length (2 to 10 days of freeze) cycles. During the thaw portion of a cycle, the ambient temperature in the cold room was raised from -5° C to between 12° and 20° C in 8 hours. The maximum temperature depended on whether and how long infrared heaters were used to accelerate thawing. The temperature was then lowered in 4 hours back to -5° C, where it remained until the next thaw.

Heave was monitored continuously with linear displacement heave transducers. Soil temperature was recorded by a network of isolated thermocouples and two verticle groupings of thermocouples (thermocouple piles).

RESULTS

Sorting

Following the conclusions of the experiment, the soil was dried to prevent gravel movement during excavation. Twenty-eight sites were excavated with 2 to 3 samples taken every 40 to 50 mm vertically throughout the soil column at each site. Each sample was analyzed for bulk density, porosity, and concentration of gravel.

The mean concentration of gravel by volume for these samples was 23.3 ± 3.3% (volume of gravel/volume of solid constituents of soil). As the minimum sample width was 100 mm, the sampling discriminated between gravel concentrations under crests and troughs only in the three largest sets of sinusoids. The mean concentration of gravel was 22.1 ± 3.8% for crest sampling sites (12 samples) and 24.8 ± 2.6% for trough sites (8 samples). A t test comparing data from crest and trough samples shows that they represent statistically distinct populations with a significance level greater than 97.5%. Thus, lateral sorting occurred in the larger sinusoids, with clasts converging on troughs.

During periods of thaw, solifluction transferred material from the crests to the troughs, particularly in the larger corrugations. As fine material was preferentially transported, the degree of sorting implied by the data understates the actual lateral movement toward troughs.

The lateral displacements of marker stones are shown in Figure 3. Almost all stones showing lateral displacement moved toward the nearest trough. Stones located near the surface in regions of steep slope experienced the greatest lateral displacement, with several stones moving as far as 40 mm.

Throughout the experiment, there was a distinct soil-moisture gradient parallel to the corrugations with soil moisture increasing in the positive x direction. In the larger sinusoids, this resulted in a marked change in small-scale mass-wasting processes along the length of corrugations. Solifluction on the drier end of a corrugation was characterized by debris flows, whereas the relatively wet end commonly experienced thin mud flows. Although differences in soil moisture may have affected the pattern and magnitude of heave along a corrugation, the degree of sorting was unaffected (Table 1). The absence of a distinct change in sorting along this gradient during a period in which significant sorting occurred suggests that topography controlled sorting more effectively than did soil-moisture gradients parallel to the corrugations.

Marker stones in the B and C trenches under corrugations with wavelengths of 0.2 m or less were not systematically displaced toward troughs. These stones all moved in the same lateral direction, perhaps in response to a soil-moisture gradient oriented perpendicular to the corrugations, with increasing moisture in the direction of the larger sinusoids.

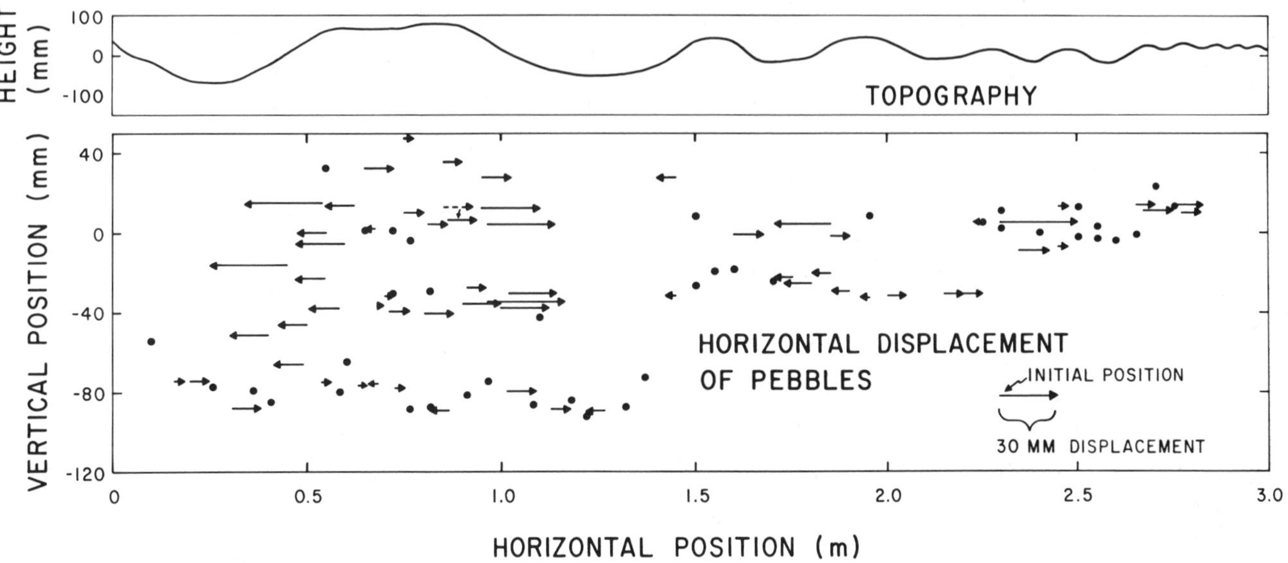

FIGURE 3 The lateral displacements of marker stones relative to topography. The upper graph shows topography without vertical exaggeration. The lower graph has a vertical exaggeration of 3.95. The arrows are displacement vectors for marker stones. Arrow tails mark the original position of marker stones within the soil. Arrows point in the direction of displacement. The scale for displacement is shown on the map. Dots represent stones with lateral displacement less than 3 mm.

TABLE 1 Two Comparisons of Groups of Marker Stones From Different Trenches.

Trench	Range of Y Co-ordinates of Position (m)	Depth (m)	Number of Stones in Group	Mean Δy (mm)	Mean $((\Delta y)^2)^{1/2}$ (mm)
A	0.880-1.420	0.099-0.116	9	3.9 ± 16.3	13.8 ± 10.8
B	0.800-1.500	0.108-0.127	8	3.8 ± 14.9	12.5 ± 9.0
A	0.940-1.365	0.061-0.081	9	-2.8 ± 21.0	16.1 ± 13.7
B	0.850-1.350	0.064-0.092	6	0.8 ± 23.5	19.2 ± 13.0

The groups of stones in each comparison come from equivalent positions with respect to the sinusoidal topography, but from marker stone trenches on opposite sides of the experiment.

These results may be considered using data from this experiment in conjunction with a simplified model of the temperature field in the soil. Consider a homogeneous half-space with a topography described by

$$h = h_o \cos(2\pi y/\lambda) \quad (1)$$

where h_o is the amplitude of the topography, y is a point on the surface, and λ is the wavelength of the topography. The temperature T at a depth z under the point y is a function of the average surface temperature T_o, heat flow q at a distance below the surface, and thermal conductivity k (Turcotte and Schubert 1982) such that

$$T = T_o + qz/k - qh_o/k \cos(2\pi y/\lambda) \exp(-2\pi z/\lambda) \quad (2)$$

This may be rearranged to

$$T - T_o - zq/k = qh_o/k \cos(2\pi y/\lambda) \exp(-2\pi z/\lambda) \quad (3)$$

to isolate the factor dependent on topography. Therefore, subsurface isotherms have the same wavelength as, and are in phase with, topography, but their amplitudes attenuate exponentially with depth.

It is important to note that this model neglects the considerable effects of latent heat and water transport in the freezing of wet soils. Thus, any results derived from it must be considered as purely conceptual. Also, this model is strictly valid only for small slopes. As the amplitude of each sinusoid is one-tenth its wavelength, the experiment is in reasonable compliance with this restriction.

The depth at which the ratio of attenuation R (amplitude of isotherm/amplitude of topography) reaches a particular value is

$$z = (\ln R/2\pi)\lambda \tag{4}$$

Lateral displacement should occur then, at greater depths under larger corrugations than under smaller ones. Depths associated with values of R for the range of wavelengths in this experiment are shown in Figure 4.

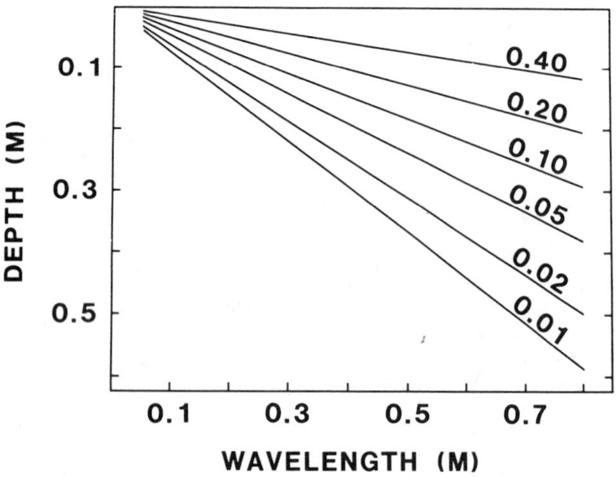

FIGURE 4 Values of the attenuation ratio R with the associated depths for the range of wavelengths in the experiment.

The relative size of clasts to corrugations may also be an imporant factor in reducing the importance of topography in sorting under small corrugations. In this experiment, the size of the average clast was within one order of magnitude of the wavelengths of the 0.1 and 0.05 m sinusoids. Clasts under these sinusoids were subjected to a range of heaving directions that, to some extent, were competing with each other. This probably inhibited lateral displacement related to topography.

Surface Heave

Heaving of the trough during thaw was common (sites T1 and T2 in Figure 5). The largest measured heave was 1.1 mm at site T1. The relative timing, rate, and magnitude of trough heave varied greatly, both geographically and temporally.

Mackay (1983) has measured heave during seasonal thaw at a number of sites in the Canadian Arctic. One site (Garry Island, N.W.T.) heaved 14.4 mm in one summer. He argues that heave during thaw occurs when water flows into a zone of permafrost at sub-zero temperatures and refreezes. Local soil structure seems to be important in increasing the permeability of permafrost.

The rapidity with which heaving began after the onset of thaw (T1 in Figure 5) implies that water flowed into frozen ground utilizing soil structure, probably cracks, to penetrate beyond the thaw front. The variability of trough heave during thaw may reflect strong control of local crack systems.

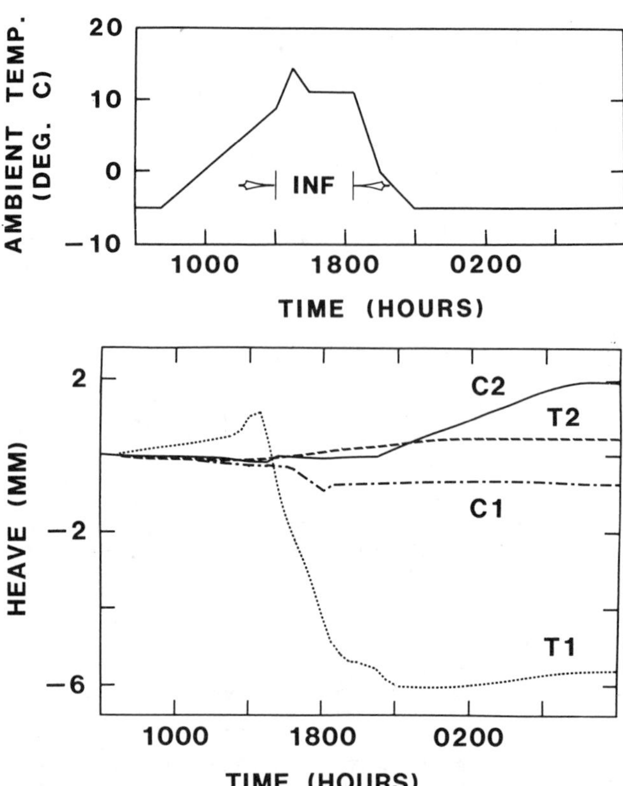

FIGURE 5 (Top) Ambient temperature versus time for October 28-29, 1981. INF denotes the time in which the infrared heaters were turned on. (Bottom) Surface heave versus time for October 28-29, 1981, at four sites. The position coordinates of the measurement sites are as follows: T1 (x = 0.6, y = 1.65), C1 (x = 0.6, y = 1.25), T2 (x = 1.9, y = 1.65), and C2 (x = 1.9, y = 1.25). Data is in meters.

Crests generally heaved only when ambient temperatures were sub-zero (C2 in Figure 5). At the end of storm-length freeze-thaw cycels, there was a residual change on the order of 1 mm in relief between the crests and troughs in the 0.8 m sinusoids. This differential heave tended to accentuate relief through time. Washburn (1980) has summarized the research describing differential heave in patterned ground. This process tends to elevate regions of fine material relative to regions of coarser material.

In this experiment, differential heave seems to have arisen from the flow of water into regions at depth below crests during thaw and its subsequent freezing. The increase in basal temperature beneath a crest to a magnitude approximating that at the surface (between 1900 and 2300 hours at a depth greater than 0.2 m under the crest in Figure 6) probably reflects an influx of melt water from

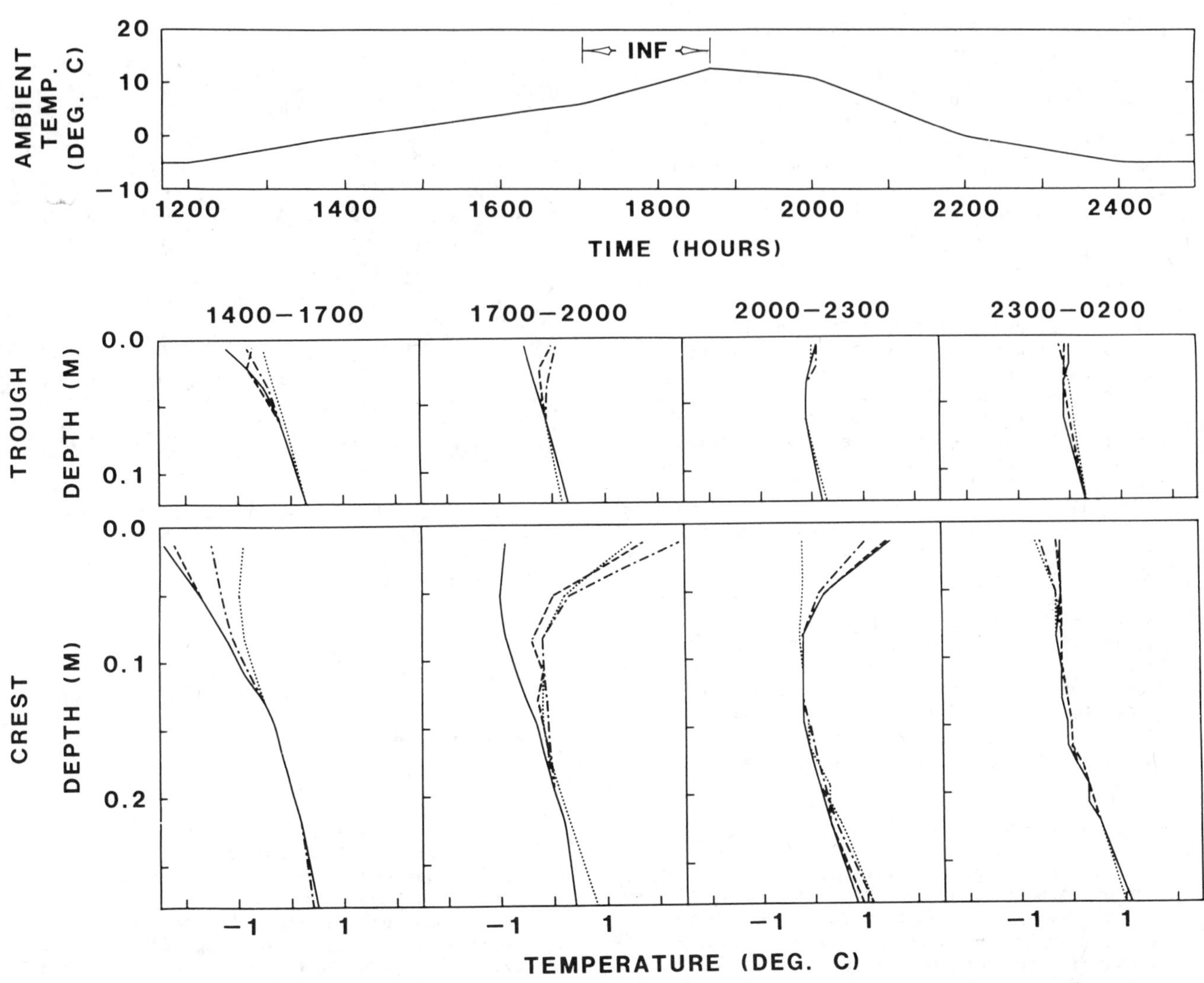

FIGURE 6 The upper graph represents ambient temperature versus time for June 26-27, 1981. INF denotes the time interval in which the infrared heaters were turned on. The lower pair of graphs represents subsurface temperatures through time measured with the two thermocouple piles. The heading (e.g., 1400-1700) above each pair of graphs is the time interval in which the measurements were made. Temperatures were recorded at one-hour intervals. For each group of four temperature profiles, the solid line corresponds to the temperature with depth at the start of the time interval. Succeeding measurements are indicated by successively more broken lines. The dashed line shows the second temperature measurement in the time interval, alternating dots and dashes represent the third measurement, and the dotted line is the fourth.

the surface. The thermal signal of a forming ice lens is likely to be an anomalous local increase in temperature due to the release of latent heat in regions with temperatures of 0° C or lower. The increase in temperature under the crest at depths of 0.1 to 0.17 m between 2300 and 2400 in Figure 6 may be due to this process, though the change in temperature (0.1° C) is at the limit of resolution for the thermocouples employed in this experiment.

Changes in the surface elevation of the soil may have been due not only to frost heave and thaw consolidation, but to other factors as well, such as changes in soil moisture or thermal expansion of the soil. The component of change due to thermal expansion as the soil thawed may be estimated in the following way. The average increase in temperature in the soil over the depth of the soil under the crest was about 2° C (Figure 6 between 1400 and 2000 hours). The soil thickness under the crest is about 0.3 m. Horiguchi (1978) reported a value of 4.8×10^{-5} °C^{-1} for the linear coefficient of thermal expansion of ice. This value is much larger than that for most silicate rocks (Lachenbruch 1966). Using this value to ensure that the magnitude of change of length due to thermal expansion is not underestimated yields a value for the maximum increase in surface elevation because of thermal expansion as soil thawed of 2.9×10^{-5} m, which is negligible. The comparable value for sites in troughs would be as little as one third this number because of the lesser thickness of soil there.

Thermal expansion of the steel heave transducers may be computed as follows. The change in ambient temperature during a temperature cycle was approximately 20° C (-5° C to +15° C). At maximum extension, which occurred when transducers were located in the trough of the 0.8 m sinusoid, the length of the transducers didn't exceed 0.3 m. The linear coefficient of thermal expansion for steel in this temperature range is 1.1×10^{-5}° C^{-1} (Resnick and Halliday 1966). The maximum expansion of a heave transducer during a temperature cycle then, is 6.6×10^{-5} m. This amount is also negligible. Values at other sites range up to 50% less, since the transducers weren't fully extended. The sense of movement, as recorded by the heave transducers, is equivalent to a heave of the soil surface.

A simple experiment was performed to determine whether the hydration of clay minerals by melt water during periods of thaw would produce surface heave. Water was poured on a thawed, dried sample of the soil used in the experiment and the change in surface elevation was measured. Instead of heaving, the surface was lowered about one mm. Apparently, changes occurred in the soil structure, such as the elimination of void spaces, which tended to decrease the soil volume. Thus, measurements of heave during thaw probably underestimate the actual amount of ice lens growth taking place.

Marker Horizons

Some of the horizons were deformed such that they appeared to conform to the topography (Figure 7). This may be a result of structure imposed on the soil by the formation of ice lenses. Any volumetric expansion of the soil will be perpendicular to the local orientation of the freezing front as ice lenses form. After thawing, a remnant porosity may exist as an expression of this, which records the deformation of the soil. This deformation should have the same wavelength as and be in phase with topography. In other words, it "mimics" the topography. Local variations are large enough, though, that most horizons show deformation that has no discernible relationship to topography.

CONCLUSIONS

Freeze-thaw cycles in soils with corrugated topography can induce sorting analogous to that found in stone stripes, if the corrugations are large enough. Small stone stripes probably do not form in this way, because temperature variations induced by the presence of topography die out so quickly beneath corrugations of small wavelength.

Frost heaving under sinusoids is characterized by instances of trough heave during thaw and residual differences in heave that accentuated topography through time.

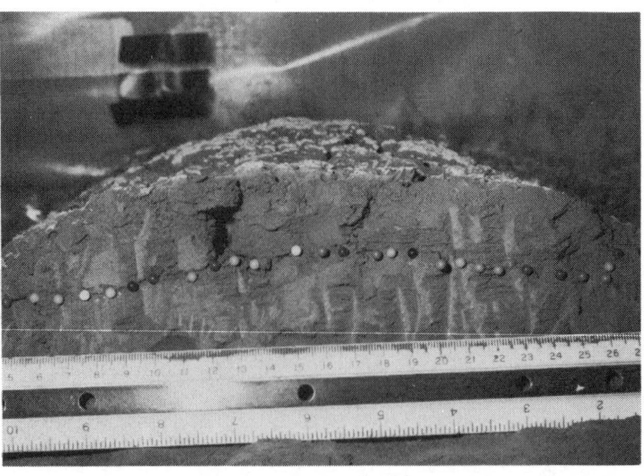

FIGURE 7 This photo shows the crest of a 0.4 m sinusoid in cross-section with the B marker horizon indicated by pin heads. The upper scale of the ruler is in centimeters.

ACKNOWLEDGMENTS

The author gratefully acknowledges numerous helpful discussions with Dr. Bernard Hallet of the Quaternary Research Center.

REFERENCES

Brockie, W. J., 1968, Contributions to the study of frozen ground phenomena -- preliminary investigations into a form of miniature stone stripes in East Otago, in Proceedings of the Fifth New Zealand Geography Conference: Auckland, New Zealand Geographical Society.

Corte, A. E., 1966, Particle sorting by repeated freezing and thawing: Biuletyn Peryglacjalny, no. 15, p. 175-240.

Horiguchi, K., 1978, Thermal expansion of a frozen bentonite-water system, in Proceedings of the Third International Conference on Permafrost, v. 1: Otawa, National Research Council of Canada.

Mackay, J. R., 1983, Downward water movement into frozen ground, Western Arctic Coast, Canada: Canadian Journal of Earth Science, v. 20, p. 120-134.

Nicholson, F. H., 1976, Patterned ground formation and description as suggested by Low Arctic and Subarctic examples: Arctic and Alpine Research, v. 8, no. 4, p. 329-342.

Resnick, R., and Halliday, D., 1966, Physics; Part 1 (2nd ed.): New York, John Wiley and Sons.

Schmertmann, J. H., and Taylor, R. S., 1965, Quantitative data from a patterned ground site over permafrost: United States Army Cold Regions Research and Engineering Research Report no. 96.

Turcotte, D. L., and Schubert, G., 1982, Geodynamics: New York, John Wiley and Sons.

Washburn, A. L., 1980, Geocryology: New York, Halsted Press.

BIOGEOGRAPHIC SIGNIFICANCE OF STEPPE VEGETATION IN SUBARCTIC ALASKA

David F. Murray,[1] Barbara M. Murray,[1] Boris A. Yurtsev,[2] and Randal Howenstein[1]

[1]Institute of Arctic Biology and the Museum, University of Alaska, Fairbanks, Alaska 99701 USA
[2]Komarov Botanical Institute, Leningrad 197022 USSR

Fossils in late-glacial deposits of Beringia suggest that a dry grassland or steppe was a fundamental part of the periglacial vegetation mosaic. This is consistent with the expected effect of the Bering Land Bridge on climate during glacial maxima: colder and drier overall, possibly with shorter and warmer summers when deeper seasonal thaw created conditions suitable for steppe plants but inimical to the dominant taxa found here today. Remnants of that vegetation may persist today in Siberia and the continental interior of Alaska and adjacent Yukon on steep, dry south-facing bluffs and terraces. We reconnoitered the steppe flora and its communities on bluffs and terraces along the Yukon, Tanana, Copper, and Porcupine rivers, examining both glaciated and unglaciated areas and surfaces of Tertiary, Quaternary, and Holocene ages. These steppes are comparable to those of the upper to middle Yana and Indigirka drainages and the upper Kolyma River in western Beringia. Steppe vegetation of Alaska and Yukon is to a significant degree composed of vascular plants that are either the same species or close relatives (vicariants) of Siberian species. Others are American taxa either endemic to eastern Beringia or the same as or vicariants of taxa found in continental western North America. The mosses and lichens are mostly members of a widely distributed, although often disjunct and rare, flora of arid and semi-arid regions. The Asian taxa presumably arrived in eastern Beringia via the Bering Land Bridge, some perhaps as early as late Tertiary. The American contribution to the modern steppes might be both ancient and recent, since the present mix of species could also reflect contributions from a post-glacial warm interval. These taxa became restricted to their present sites during the Holocene by the expansion of taiga and mires.

INTRODUCTION

Numerous fossils of a grazing megafauna dominated by mammoth, bison, and horse (but also including saiga) and pollen indicating a relative abundance of sage (*Artemisia* spp.), grasses, and willows (and an absence of spruce, birch, and alder) have been recovered from late-glacial deposits in eastern Beringia (Alaska and Yukon). Reconstructing a grassland or steppe from this evidence has been considered consistent with and necessary for the support of this fauna. The apparent shift to a vegetation rich in birch, spruce, and alder during the Pleistocene-Holocene transition provides one explanation for the rapid extinction of the grazers (Guthrie 1968). Since the pollen record for roughly the same time period from places widely separated from each other is similar, the concept of a widespread, full-glacial Beringian steppe emerged. Since studies of modern pollen have failed to identify a modern analog for this steppe, some have assumed that both the plant and animal communities were reduced by extinctions in the Holocene (Matthews 1976, cf. Hopkins et al. 1982).

Young (1976a,b) and Yurtsev (1963, 1972, 1981), on the other hand, have suggested that late-glacial steppe remnants persist in areas of northeastern Asia and northwestern America--on steep, dry south-facing bluffs and terraces. Stimulated by the data from studies by Hanson (1951), Gjaerevoll (1958-1967), Shacklette (1966), Young (1976a,b), and Batten et al. (1979), we have begun a multiyear, international project to study the steppe flora in these local, azonal habitats of Alaska and Yukon where the zonal vegetation is taiga. Our objectives are to describe the flora and vegetation and to compare and contrast our results with those for similar landscapes in Chukotka and eastern Siberia. Through a study of floristics and the systematics of the taxa we can obtain indirect evidence with which to assess the validity of relict status for the modern plant community.

VASCULAR PLANTS

The relict nature of the modern steppe is inferred from (1) the general floristic similarity of the present vascular vegetation to the late-glacial, herb-rich pollen assemblage and (2) the specific floristic similarity of the vegetation in two distant areas of the Beringian refugium that today share an extremely continental climate and are isolated by environments inimical to exchange or dispersion. Explanation for the disjunction is the Holocene separation and reduction of a once continuous distribution for these plant communities that developed when Beringia was hypercontinental in climate and the Bering Land Bridge enabled the migration of Asian taxa to America.

Steppes were examined over a wide geographic area, and on surfaces of late Tertiary (Yukon and Tanana rivers), late Pleistocene (Porcupine River), and early Holocene (Copper River) ages. In these river valleys, a fairly predictable set of xerophilous herbs, sage (predominately *Artemisia frigida*), lichens, and mosses occupies the south-facing slopes and terraces over sections of tens of meters up to several kilometers and over a vertical extent of 200-300 m. These steppes are comparable to the montane steppes of the upper to middle Yana and Indigirka drainages of northeastern Yakutia and the upper Kolyma River. Major differences are the absence in Alaska and Yukon of true tufted grass species of *Festuca* and *Helictotrichon* and the lower floristic diversity of the vascular taxa, suggesting that portions of the Asian steppe failed to expand eastward. Nevertheless, to Yurtsev who has experience in both countries the impression of overall similarity was remarkable.

Preliminary analysis shows that the flora is a composite of several elements. Some of the prominent taxa such as *Calamagrostis purpurascens*, *Poa glauca*, and *Potentilla hookeriana* are common subarctic steppe plants of both eastern and western Beringia and also occur in dry habitats in arctic and alpine areas. Other taxa such as *Rosa acicularis*, *Solidago decumbens*, *Galium boreale*, and *Zygadenus elegans* are derived from open woodlands nearby. Some such as *Amelanchier alnifolia* and *Apocynum androsaemifolium* are species of temperate woodlands and grasslands that are restricted at their northern limits to these generally warmer sites. But there are, however, several taxa generally restricted to bluffs, terraces, cutbanks, and outcrops. Some of these can be found also growing on active floodplain gravels, and others can inhabit screes and similar alpine habitats where limestones predominate. Notable are endemics to the continental interior of Alaska and Yukon (*Draba murrayi*, *Penstemon gormanii*, and *Podistera yukonensis*), disjuncts and vicariants from similar continental areas of northeastern Asia (e.g., *Alyssum americanum*, *Plantago canescens*, *Silene repens*, *Artemisia alaskana*), and disjuncts and endemic vicariants from the high plains and continental intermontane valleys of western North America (*Cryptantha shackletteana*, *Erysimum angustatum*, *Eriogonum flavum* var. *aquilinum*, *Townsendia hookeri*).

Table 1 shows the vascular taxa more or less restricted in the subarctic to steppe bluffs. The arctic-alpine cryophyte element is omitted. Data are from Hansen (1951) and Gjaerevoll (1958-1967) for the Copper River, from Shacklette (1966), Young (1976a,b), and Batten et al. (1979) for the upper Yukon River region, and our own observations on the Copper, Tanana, and Yukon rivers (D.F.M., B.A.Y.) and on the Porcupine River (R.H.). Comparative data from steppes in central and southwestern Yukon (YT) are from Porsild (1951, 1966), Johnson and Raup (1964), Hoefs et al. (1975), Nimis (1981a), Hultén (1968), and D. Murray (unpublished).

TABLE 1 Vascular Taxa Generally Restricted in Subarctic Alaska and Yukon to Steep, Dry South-Facing Bluffs, Terraces, Cutbanks, and Outcrops.

Taxon	Porcupine	Yukon	Tanana	Copper	YT
Agropyron spicatum	X	X	X		X
A. yukonense		X			X
Alyssum americanum	X	X			X
Amelanchier alnifolia	X	X	X		
Apocynum androsaemifolium		X	X		
Arabis divaricarpa	X				X
A. holboellii	X	X	X	X	X
Artemisia alaskana		X	X		X
A. frigida	X	X	X	X	X
A. laciniatiformis	X	X	X		X
A. rupestris ssp. woodii					X
Carex duriuscula		X	X		X
C. filifolia	X	X	X		
C. sabulosa					X
C. supina ssp. spaniocarpa	X	X	X		
Chamaerhodos erecta		X			X
Cryptantha shackletteana		X			
Draba murrayi		X			
Erigeron caespitosus	X	X	X		X
E. compositus	X	X		X	X
Erigeron sp. (? pumilus of Hulten)	X				X
Eriogonum flavum var. aquilinum	X	X			
Erysimum angustatum		X			
Eurotia lanata					X
Halimolobus mollis	X	X		X	X
Haplopappus macleanii					X
Juniperus communis	X	X	X	X	X
J. horizontalis				X	X
Lesquerella arctica				X	X
Linum lewisii	X	X		X	X
Oxytropis splendens	X	X			X
Papaver nudicaule ssp. americanum		X			
Penstemon gormanii	X	X	X	X	X
Phacelia sericea		X			
Phlox hoodii	X	X			
Plantago canescens	X	X		X	X
Podistera yukonensis		X			
Potentilla hookeriana	X	X	X	X	X
P. pensylvanica	X	X	X	X	X
Pulsatilla patens	X	X	X	X	X
Selaginella sibirica	X	X	X		X
Silene menziesii ssp. williamsii		X	X		
Silene repens	X	X			
*Townsendia hookeri	X				X

*New to Alaska

The upper Yukon River has the most extensive system of steppe bluffs and also the largest array of endemic and disjunct taxa. The steep slopes in this region were never glaciated or reworked by riverine processes in the Quaternary. The physical isolation of vascular endemics from their closest morphological relatives and the disjunct bryophytes discovered here suggest an ancient origin for these taxa. Clearly some of the disjuncts arrived at their present positions in the Holocene (e.g., Carex filifolia at Chitina), and although associated with arid grasslands, they cannot be interpreted as late-glacial relicts. Eriogonum flavum var. aquilinum, Erysimum angustatum, and Cryptantha shackletteana are endemics only weakly differentiated from species more widespread in arid regions of western North America; therefore it is tempting to imagine the migration of these stocks northward during early to mid Holocene warm intervals via the northern prairies and intermountain dry grasslands into southwestern and central Yukon and adjacent interior Alaska. Isolation was soon effected by development of boreal forest and mires and the concomitant rise in the permafrost table, reducing steppe environments to habitats where strong topographic and geomorphic controls maintain steppe-like conditions. That relatively complex steppe communities can form within postglacial time is evident from the vegetation at Chitina on the Copper River and the extensive well-developed steppes in the region of Kluane Lake in southwestern Yukon.

MOSSES AND LICHENS

The moss and lichen flora studied (by B.M.M.) consists of a large group of xerocontinental species that fully support the notion of present-day steppe. They are part of a widely distributed, though often highly disjunct, arid lands flora of families and genera showing great specialization, presumably to enhance survival in extremely hot and/or dry conditions. In this regard, the composition of the steppe moss flora is particularly striking. Although analysis is preliminary, 30 species in 13 families have thus far been identified (Table 2). Of these, 21, or 2/3 of the total, are members of just two families, Pottiaceae and Grimmiaceae, both of which show many xeric adaptations.

The lichen flora is composed of about 15 families in which the genera and species are more evenly distributed, but many of the genera are characteristic of steppe and other arid landscapes. Especially important on xeric, fine soils are the genera Dermatocarpon, Diploschistes, Endocarpon, Fulgensia, Psora, Toninia, and Squamarina and on coarser soils Xanthoparmelia. Among other important genera of subarctic Alaskan steppes are Acarospora, Aspicilia, Caloplaca, Candelariella, and Collema (Table 3).

Of the twenty-one taxa in the two moss families that are important on the steppes, Pottiaceae and Grimmiaceae, five are new to Alaska, including one that is new to North America, and six represent significant range extensions (Table 2). Five of the forty lichens reported here are also new to Alaska (Table 3) (B. M. Murray unpublished). Thus the cryptogamic steppe flora is of great phytogeographic interest.

The azonal steppe floras of northeast Asia (see Yurtsev 1981) and Alaska are only recently becoming known, but when these floras are compared with each other and with those of zonal steppes of Middle and Central Asia and even deserts of North America, great affinities are apparent.

Several species have a cosmopolitan distribution. Indeed most of the important taxa found on Alaskan steppes are widely distributed, occurring in Europe, Central and Northeast Asia, and other parts of North America; some also occur in North Africa and South America. Some are widespread in these areas; others are very disjunct or occur only in a portion of this general range. For example, the moss, Indusiella thianschanica, found at one unglaciated site on the Yukon River where several endemic vascular plants occur, is known otherwise only from mountain steppes in the Caucasus, Middle and Central Asia, and adjacent Siberia and one locality in North Africa (B. M. Murray unpublished). The lichen, Fulgensia desertorum, found in the Copper River valley, is known otherwise from Eurasia and Colorado (Poelt 1971).

Mosses and lichens both have generally wider distributions than vascular plants and often demonstrate less endemism. Whereas long distance dispersal may account for such patterns, the large, dramatic disjunctions and the presence of a repeated suite of species in steppe sites argue for persistence over a long period of time, and lack of endemism argues for great genetic stability. There is no question of the capability of mosses and lichens to persist. They very well fit Hoffmann's (1981) principles for survival (of mammals) in refugia: (1) be adapted to the right habitat, that is, one that can survive periods of change; (2) be small; (3) live in a large refugium; and (4) avoid competitors. The habitat of most steppe mosses and lichens is the fine soil between the above-ground parts of vascular plants.

It is likely that many of the xerocontinental bryophytes and mosses will be shown to have wider and more continuous ranges as azonal steppes and other arid habitats become better known and comparisons are made with Asian and North American material. Until recently, Caloplaca tominii was known only from Eurasia, but it is now known also from Yukon Territory (Nimis 1981b), and, based on re-identified material, from Northwest Territories and Saskatchewan, Canada; Montana, Utah, S. Dakota, Utah, Washington, and Wyoming, USA; and Peru (Thomson 1982)--and now Alaska.

In addition to the historical (age) differences between sites, edaphic differences determined by lithology affect the distribution of taxa. Coarse soils and rubble slopes tend to have a greater proportion of vascular plants and cryptogams from the arctic-alpine floristic element; thus the bluffs of the Copper River,

TABLE 2 Mosses.

Taxon (Family in order of importance in vegetation)	Yukon	Tanana	Copper
Pottiaceae			
Aloina brevirostris	X		X
*A. rigida	X	X	
Aloina unidentified	X		
Barbula convoluta	X	X	
Bryoerythrophyllum recurvirostrum	X		
*B. recurvirostrum var. latinervia		X	
+Didymodon cf. rigidulus	X	X	
Molendoa sendtnerianum	X		
*Phascum cuspidatum	X	X	
+Pterygoneurum lamellatum			X
+P. ovatum		X	X
+P. subsessile	X	X	X
Tortella fragilis	X		
+Tortula caninervis	X		
Tortula ruralis	X		X
Trichostomum or Weissia	X		
Stegonia latifolia			X
S. pilifera			X
Grimmiaceae			
+Coscinodon cribosus	X		
Grimmia anodon	X		X
*Grimmia teretinervis	X		
**Indusiella thianschanica	X		
Schistidium apocarpum	X		
Bryaceae			
Bryum argenteum	X	X	X
Bryum unidentified	X	X	X
Encalyptaceae			
Encalypta rhaptocarpa	X	X	
Encalypta unidentified	X		X
Amblystegiaceae			
Hypnum vaucheri	X		
Hypnum unidentified	X	X	X
Leskeaceae	X		
Rhytidiaceae			
Rhytidium rugosum	X	X	
Hedwigiaceae			
Hedwigia ciliata	X	X	
Polytrichaceae			
Polytrichum piliferum		X	
Thuidiaceae			
Thuidium abietinum	X		
Fissidentaceae			
Fissidens	X		
Hylocomiaceae			
Hylocomium splendens	X		
Orthotrichaceae			
Orthotrichum speciosum	X		

*New to Alaska
**New to North America
+Significant range extension in Alaska

TABLE 3 Lichens.

Taxon (alphabetical)	Yukon	Tanana	Copper
Acarosporaceae	X	X	X
Aspicilia unidentified	X	X	X
Buellia unidentified	X		X
Caloplaca stillicidiorum	X	X	X
*C. tominii			X
Caloplaca unidentified	X	X	
Candelariella aurella			X
Candelariella unidentified	X	X	X
Cetraria nivalis		X	
Cladonia pyxidata (incl. pocillum)	X	X	
Cladonia squamules	X	X	
*Collema coccophorum			X
Collemataceae	X	X	X
Cornicularia aculeata		X	
Dermatocarpon (incl. hepaticum)	X	X	X
Diploschistes bryophilus	X	X	
Endocarpon pusillum	X		X
*Fulgensia desertorum			X
F. sp. (sterile)	X		X
Lecanora cf. frustulosa	X	X	
L. cf. palanderi			X
Lecanora unidentified	X	X	X
Parmelia sulcata	X		
Peltigera aphthosa	X	X	
P. cf. lepidophora	X		
Peltigera cf. malacea	X		
P. rufescens	X		
P. spuria		X	X
Phaeophyscia constipata	X		
P. sciastra	X		
Physcia caesia	X		
Physciaceae unidentified	X	X	
Physconia muscigena		X	
P. cf. perisidiosa	X		
Psora decipiens		X	X
Psora unidentified	X	X	X
*Psorula rufonigra	X	X	
Rhizocarpon disporum			X
Rhizoplaca chrysoleuca	X	X	
R. melanophthalma		X	
Rinodina roscida			X
Rinodina unidentified	X	X	X
*Spilonema revertens	X	X	
Squamarina lentigera			X
Staurothele perradiata			X
Stereocaulon sp.		X	
Toninia caeruleonigricans		X	X
Toninia unidentified	X	X	X
Verrucaria cf. leightonii			X
Xanthoparmelia taractica	X	X	
X. wyomingica	X		
Xanthoria elegans			X
Xanthoria unidentified	X	X	

*New to Alaska

where the bedrock is deeply mantled by calcareous loess, are differentiated from the upper Yukon and Tanana river regions where carbonate-rich soils can overlie both acid and basic parent material. At Chitina soil lichens characteristic of calcareous crust soils typical of steppes are common: *Caloplaca tominii*, *Fulgensia desertorum*, *Psora decipiens*, *Squamarina lentigera*, and *Toninia caeruleonigricans*. These steppes show a great affinity to those in southwestern Yukon (Nimis 1981a) and to the Canadian prairies (Looman 1964), and several are found on loess on end moraines of present day glaciers in Yukon and Alaska (B. M. Murray unpublished). Since bedrock is abundantly exposed at the Yukon and Tanana sites, saxicolous taxa such as *Lecanora* spp., *Rhizoplaca* spp., *Psorula rufonigra*, and *Spilonema revertens* are found there. The less basic soil and the coarser texture accounts for the presence also of the genus *Xanthoparmelia*. Consequently, more information on soil chemistry will be gathered as field work progresses.

FLORISTICS AND PALEOECOLOGY

What relationship is there between the flora, vegetation, and physical environment of the steppe bluffs today and those of Beringia during late-glacial time? If the steppe communities were in fact widespread and dominant, well-drained soils had to be far more common than they are today. It would follow that the permafrost table was below present levels, yet not to the extent of precluding the growth of ice wedges or the preservation in frozen ground of faunal remains.

If we take the view of Cwynar and Ritchie (1980) that pollen influx was very low for all taxa and that the influx of *Artemisia* did not change during the Pleistocene-Holocene transition, then the protracted drought of an intensely continental climate coupled with broad, more extensive braided floodplains and sand seas might have provided sufficient trophic support for a modest grazing megafauna and allowed for the vegetation mosaic that seems appropriate (Schweger and Habgood 1976).

If we adopt another proposal by Cwynar and Ritchie (1980) that the sage responsible for the pollen in the late-glacial record was *A. arctica*, and not *A. frigida* as has been assumed, then an image of herb-rich communities more similar to fellfields emerges. The permafrost table need not be lower, and as Péwé (1975) has suggested, it might have even been higher. Their points are persuasive, but the abundance of arctic-alpine cryophytes in their pollen spectra is not fatal to the idea of a steppe-like vegetation, for we would expect with the lowering of tree line during cooler times that the mountain and steppe species would mingle. Indeed, cryophytes are found on steppe bluffs today. While it is true that their lists of taxa include genera that are well distributed in the Arctic and do occur on fellfields, certain species of those genera are definitely restricted to temperate dry grasslands and reach their northern limits at steppe sites in Alaska.

Our data show that occurrence of a complex of xerophilous temperate woodland and steppe species and cryophilous arctic-alpine species is determined by topography and soils. Such large gaps in distribution exist between these communities and others in Asia and western America that normal methods of dispersal cannot create them today. Long distance dispersal of such a complex flora is less likely than the formation of these communities by reduction of formerly more widespread ranges. These plants could only have been more widespread under conditions different from today, conditions that better match the cold, dry full-glacial intervals when the Bering Land Bridge was in place. These plants did not arrive simultaneously as a single community, but at intervals since the Tertiary when appropriate conditions prevailed.

The flora of these steppes is ancient, but the communities they form are not necessarily relicts of any one specific period. More likely these steppes have received migrants from Asia beginning in late Tertiary and including arrivals as recently as mid-Holocene. The immigrant taxa share an ability to disperse and a tolerance to drought. Rather than a process of steady accretion of taxa, local extinctions must have occurred through the millennia of climatic changes. We have more than a depauperate version of a late-glacial steppe; we have the current version of what, on a timescale of geologic history, is probably a continually changing community.

Steppes are restricted today to bluffs and terraces where extremes of summer drought occur. These sites also experience extremes of summer warmth, and carbonate-rich soils are common. The steep slopes show surface instability from mass wasting. This combination of physical features results in the formation of open plant communities for which bare ground is an important aspect. An inability to compete with other plants is usually presumed as a reason these plants are not found elsewhere.

The introduction of the late-glacial steppe into paleo-reconstructions was stimulated in large part by the need to provide a suitable trophic base for the grazing megafauna. Uncertainties remain: are the mammalian fossils contemporaneous and do they signify a diverse and abundant fauna? Is a true steppe consistent with such a fauna? Furthermore, although an herb-rich pollen zone is consistently a part of late-glacial pollen spectra taken from many areas of Alaska and Yukon, how geographically prominent was this vegetation? Could the pollen record have been produced if we invoke only the expansion of xeric sites under a hypercontinental climate? It is crucial to know whether or not the amount of *Artemisia* and grass in the vegetation changed dramatically in the Holocene.

REFERENCES

Batten, A. R., Murray, D. F., and Dawe, J. C., 1979, Threatened and endangered plants in selected areas of the BLM Fortymile Planning Unit, Alaska: Bureau of Land Management-Alaska Technical Report 3.

Cwynar, L., and Ritchie, J. C., 1980, Arctic-steppe tundra: a Yukon perspective: Science, v. 208, p. 1375-1377.

Gjaerevoll, O., 1958-1967, Botanical investigations in central Alaska, especially in the White Mountains: Kgl Norske Videnskabers Selskabs Skrifter, v. 1958 nr 5, p. 1-74; v. 1963, nr 4, p. 1-115; v. 1967, nr 10, p. 1-63.

Guthrie, R. D., 1968, Paleoecology of the large mammal community in Interior Alaska during the Late Pleistocene: American Midland Naturalist, v. 79, p. 346-363.

Hanson, H. C., 1951, Characteristics of some grassland, marsh, and other plant communities in western Alaska: Ecological Monographs, v. 21, p. 317-378.

Hoefs, M., McT. Cowan, I., and Krajina, V., 1975, Phytosociological analysis and synthesis of Sheep Mountain, southwest Yukon Territory, Canada: Syesis, v. 8, supplement 1, p. 125-228.

Hoffman, R. S., 1981, Different voles for different holes: Environmental restrictions on refugial survival of mammals, in Scudder, G. G. E., and Reveal, J. L. (eds.), Evolution Today. Proceedings of the Second International Congress of Systematic and Evolutionary Biology: Hunt Institute for Botanical Documentation, Carnegie-Mellon University, Pittsburgh.

Hopkins, D. M., Matthews, J. V., Jr., Schweger, C. E., and Young, S. B., 1982, Paleoecology of Beringia: New York, Academic Press.

Hultén, E., 1968, Flora of Alaska and neighboring territories: Stanford, Stanford University Press.

Johnson, F., and Raup, H. M., 1964, Investigations in southwest Yukon: geobotanical and archaeological reconnaissance, in Investigations in southwest Yukon, Papers of the Robert S. Peabody Foundation for Archaeology, v. 6: Phillips Academy, Andover, Massachusetts.

Looman, J., 1964, The distribution of some lichen communities in the Prairie Provinces and adjacent parts of the Great Plains: Bryologist, v. 67, p. 209-224.

Matthews, J. V., 1976, Arctic steppe--an extinct biome, in Abstracts of the Fourth Biennial Meeting of the American Quaternary Association.

Nimis, P. L., 1981a, Epigaeous lichen synusiae in the Yukon Territory: Cryptogamie, Bryologie et Lichenologie, v. 2, p. 127-151.

Nimis, P. L., 1981b, Caloplaca tominii new to North America: Bryologist, v. 84, p. 222-225.

Péwé, T. L., 1975, Quaternary geology of Alaska: Geological Survey Professional Paper 835, p. 1-145.

Poelt, J., 1971, Uber einige fur Nordamerika neue flechten: Bryologist, v. 74, p. 154-158.

Porsild, A. E., 1951, Botany of southeastern Yukon adjacent to the Canol Road: National Museum of Canada Bulletin, Number 121, p. 1-400.

Porsild, A. E., 1966, Contributions to the flora of southwestern Yukon Territory: National Museum of Canada Bulletin, Number 216, p. 1-86.

Schweger, C. and Habgood, T., 1976, The Late Pleistocene steppe-tundra of Beringia--a critique, in Abstracts of the Fourth Biennial meeting of the American Quaternary Association.

Shacklette, H. T., 1966, Phytoecology of a greenstone habitat at Eagle, Alaska: Geological Survey Bulletin 1198-F, p. 1-36.

Thomson, J. W., 1982, A further note on Caloplaca tominii Savicz in the Americas: Bryologist, v. 85, p. 251.

Weber, W. A., 1965, Plant geography in the southern Rocky Mountains, in Wright, H. E., Jr., and Frey, D. G. (eds.), The Quaternary of the United States: Princeton, Princeton University Press.

Young, S. B., 1976a, The environment of the Yukon-Charley rivers area, Alaska, Contributions from the Center for Northern Studies, v. 9: Wolcott, Vermont.

Young, S. B., 1976b, Is steppe tundra alive and well in Alaska, in Abstracts of the Fourth Biennial Meeting of the American Quaternary Association.

Yurtsev, B. A., 1963, On the floristic relations between steppes and prairies: Botaniska Notiser, v. 116, p. 396-408.

Yurtsev, B. A., 1972, Phytogeography of northeastern Asia and the problem of transberingian floristic interrelations, in A. Graham (ed.), Floristics and paleofloristics of Asia and eastern North America: Amsterdam, Elsevier.

Yurtsev, B. A., 1981, [Relict steppe complexes of northeastern Asia.] Reliktovye stepniye kompleksy severo-vostochnoi Azii: Novosibirsk, Akademiya Nauk SSSR.

SOIL-WATER DIFFUSIVITY OF UNSATURATED SOILS AT SUBZERO TEMPERATURES

Y. Nakano, A. R. Tice, J. L. Oliphant, and T. F. Jenkins

U.S. Army Cold Regions Research and Engineering Laboratory
Hanover, New Hampshire 03755 USA

The soil-water diffusivities of soils containing no ice were determined at -1°C by an experimental method recently introduced. The theoretical basis of the method is presented. The measured diffusivities of three kinds of soils are found to have a common feature in that the diffusivity increases with increasing water content, attains a peak, and increases again as the water content increases. This common feature of the soils at the subzero temperature is discussed in comparison with unfrozen soils. The experimental data appear to indicate that the basic transport mechanism of water in soils containing no ice at the subzero temperature is essentially the same as that in unfrozen soils containing a small amount of water.

It has long been recognized that accurate knowledge of water movement in frozen soil is very important for many practical applications. For example, the freezing of water transported to the freezing part of soil is known to provide the largest contribution to frost heaving (Anderson et al. 1978), and the damage of frost heave is of major concern to engineers.

Introducing a new experimental method, Nakano et al. (1982, 1983a-c) investigated the rate of water transport in unsaturated frozen clay under isothermal conditions and found that the rate of water transport is proportional to the gradient of total water content, which is generally the sum of the unfrozen water content and the ice content. In particular, when ice is absent, the flux of water F in one direction is given as

$$F = - D(\theta)\, \rho \frac{\partial \theta}{\partial x} \qquad (1)$$

where θ is the weight of unfrozen water per weight of dry soil, x the space coordinate, ρ the dry density, and D the soil-water diffusivity.

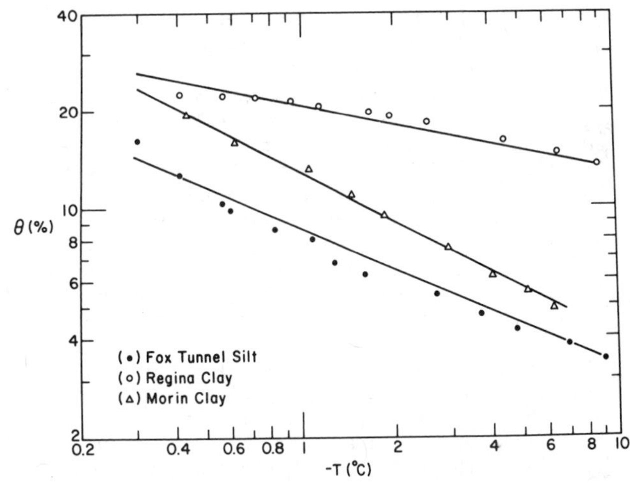

FIGURE 1 Unfrozen water content θ(%) vs temperature - T(°C) in a warming cycle for Fox Tunnel silt (w = 21.0%, ρ = 1.24 g/cm^3), Regina clay (w = 22.6%, ρ = 1.45 g/cm^3), and Morin clay (w = 20.0%, ρ = 1.42 g/cm^3).

EXPERIMENT

The two kinds of soil selected for this study were Fox Tunnel silt, obtained from Fairbanks, Alaska, and Regina clay from Saskatchewan, Canada. The specific surface area of Fox Tunnel silt determined by ethylene glycol retention is 31 m^2/g and that of Regina clay 291 m^2/g. In a previous study (Nakano et al. 1982, 1983b) we conducted experiments similar to this by using a marine-deposited Morin clay from Maine with a specific surface area of 60 m^2/g. We will use the experimental data for Morin clay in this work for comparison.

The unfrozen water contents θ of the three kinds of soils at various temperatures in a warming cycle are presented in Figure 1. Measurements were made by the pulsed nuclear magnetic resonance method described in Tice et al. (1978). The total water content w and the dry density ρ of Fox Tunnel silt were 21.0% and 1.24 g/cm^3, respectively. For Regina clay w = 22.6% and ρ = 1.45 g/cm^3, while for Morin clay w = 20.0% and ρ = 1.42 g/cm^3. The three lines in Figure 1 are the approximate analytic expressions of data points for each soil.

The experiment consisted of connecting two long columns of soil of the same size (0.80 cm in diameter and 20.33 cm long) and the same specified dry density. One of them was uniformly dry with a negligibly small total water content b, while the other was uniformly wet with a specified total water content a. At time t = 0 we connected the two columns to make a single column 40.66 cm in length. While we maintained the column at -1°C, water was transported from the wet part to the dry part across the contact surface between the wet and dry soil columns. After 10 days passed, the soil column was quickly sectioned into a total of 64 equal thin column segments. The water content and the dry

TABLE 1 Experimental Conditions.

Exp. No.	Soil	Initial condition a(%)	Initial condition b(%)	Time duration (day)	Dry density (g/cm^3) Mean	Dry density (g/cm^3) Standard deviation
1	Fox Tunnel silt	6.75	0.29	10	1.24	0.112
2	Fox Tunnel silt	6.80	0.35	10	1.23	0.105
3	Regina clay	18.1	0.49	10	1.45	0.150
4	Regina clay	17.9	0.52	10	1.45	0.141

density of each segment were determined gravimetrically.

We used 10-mL graduated plastic pipettes to enclose the soil columns. The dispensing end of each pipette was cut off and sealed with a stopper. For the preparation of wet columns, we thoroughly mixed water and soil and allowed the mixture to set for at least three days to attain moisture equilibrium prior to the packing. For the preparation of dry columns, oven-dried soil was packed into a pipette with slightly greater compactive force. During the packing, soil in the dry columns usually gained a small amount of water from the air, while soil in the wet columns lost a small amount of water to the air. We weighed each soil column after packing so that any gain or loss of water during the packing was accounted for when the initial water content was determined. All wet columns were frozen to −16°C and then warmed up to −1°C before each experiment so that they were in a warming cycle at t=0.

We conducted four experiments with the initial conditions given in Table 1. Experiments 2 and 4 were repeat experiments under nearly the same initial conditions as Experiments 1 and 3, respectively. Since the equilibrium unfrozen water contents for Fox Tunnel silt and Regina clay at −1°C are 8.62% and 20.6%, respectively, no ice was expected to be present in all these experiments at t = 0. The uniform packing of soil into a pipette was quite difficult. We examined the uniformity of the soil columns by using the measured dry density of all segments after each experiment. The mean and standard deviation of dry density are given in Table 1. Although these values are affected by errors involved in sectioning the soil columns into thin columns with ideally equal length in a relatively short time, they serve as an indicator of uniformity in packing and accuracy in sectioning.

Theory

When Equation 1 holds true, our experiment can be described in mathematical terms by the following initial value problem on the set $S = \{(-\infty,\infty) \times (0,T);\ T>0\}$:

$$\frac{\partial u}{\partial t} = \frac{\partial}{\partial x}\left[D(u)\,\frac{\partial u}{\partial x}\right] \quad \text{on } S \tag{2a}$$

$$u(x,0) = u_0(x) = a - b \quad x < 0; \qquad = 0 \quad x \geq 0 \tag{2b}$$

where $u = \theta - b$, a and b are real numbers with $a > b \geq 0$, and soil columns in the experiment are approximated to be infinite. This approximation should remain accurate until the time when column end effects become significant. We also introduce a function $\phi(m)$ defined as

$$\phi(m) = \int_0^m D(u)\, u^{-1}\, du \tag{3}$$

where m is some positive number. It should be mentioned that the commonly used method (Bruce and Klute 1956) for the experimental measurement of diffusivities utilizes the mixed initial and boundary value problem of Equation 2a.

Equation 2a can be transformed into an ordinary differential equation by introducing a similarity variable $\xi = x(t + c)^{-1/2}$ where c is a positive real number. The similarity solution $u(x,t) \equiv h(\xi)$ should satisfy the equation:

$$[D(h)h']' + \frac{1}{2}\xi h' = 0 \quad , \quad -\infty < \xi < +\infty \tag{4a}$$

where primes denote differentiation with respect to ξ. We will seek a solution of Equation 4a satisfying the boundary condition, given as

$$h(-\infty) = a - b, \qquad h(+\infty) = 0 \tag{4b}$$

Oleinik et al. (1958) investigated the initial value problem of Equation 2a for the case in which initial data $u_0(x)$ is continuous. Introducing a class of weak (generalized) solutions, they proved the existence and uniqueness of solutions in this class. These solutions can be classified into two types depending upon whether or not $\phi(m)$ is finite. When ϕ is infinite the solutions are continuously differentiable. However, when ϕ is finite, Equation 2a is parabolic near points where $u > 0$, but not near points where $u = 0$. Equations of this kind are often called degenerate parabolic. Because of this degeneracy, the transition between a part $u > 0$ and a part $u = 0$ is not smooth, and the solutions satisfy Equation 2a in the sense of distribution.

In soil physics the existence of a steep wetting front with finite propagating speed has been recognized for some time, based upon experimental observations (Swarztendruber 1969). A new theoretical interpretation of a wetting front was recently advanced (Nakano 1980a-c, 1981, 1982, 1983a-b) by applying mathematical theorems of degenerate parabolic equations. A wetting front, defined as the interface between a part $u > 0$ and a part $u = 0$ in the solutions of Equation 2a for finite ϕ, is a singular surface in the sense of Truesdell and Toupin (1965), and propagates with finite speed. It is well established that the flow of water through unsaturated porous media is described accurately by Equation 1 when the inertial force is negligibly small in comparison to the viscous force (Raats and Klute 1968). Recently Nakano (1983a-b) has shown that the finiteness of ϕ defined by Equation 3 is equivalent to a physical condition that the inertial force is negligibly small at $\theta = b$ and $D(b) = 0$. Since in physical problems $D(\theta)$ tends to vanish as θ approaches zero, it is expected that ϕ is finite when b is negligibly small, and that under such conditions a wetting front with finite propagating speed appears.

It has been shown (Van Duyn and Peletier 1977a, Van Duyn 1979) that the problem of Equations 4a and 4b also has two types of solutions, depending upon whether or not ϕ is finite. When ϕ is infinite, $h(\xi) > 0$ for all ξ, and h approaches zero as ξ tends to infinity. When ϕ is finite, there exists a number ξ_0 such that

$$h(\xi) = 0 \quad , \quad \text{for } \xi_0 \leq \xi$$
$$\quad > 0 \quad , \quad \text{for } -\infty < \xi < \xi_0 \tag{5a}$$

We will consider a similarity solution of Equation 2a with a similarity variable of $\eta = xt^{-1/2}$. The similarity solution $u(x,t) \equiv f(\eta)$ should satisfy the equations:

$$[D(f)f']' + \frac{1}{2}\eta f' = 0 \ , \quad -\infty < \eta < \infty \tag{6a}$$

$$f(-\infty) = a - b, \qquad f(+\infty) = 0 \tag{6b}$$

We may consider $f(\eta)$ as a limit of $h(\xi)$ when c approaches zero.

A prevailing proposition states that similarity solutions are not only particular solutions, but, in many instances (Barenblatt and Zel'dovich 1971), are also asymptotic solutions to a wide class of the Cauchy problems of the original partial differential equation. There has been marked interest in the validity of this proposition applied to problems of porous media equations (Peletier 1971, Van Duyn and Peletier 1977b, Gilding 1979, Aronson and Peletier 1981). Van Duyn (1979) showed that similarity solutions of Equations 4a and 4b are indeed asymptotic solutions of the initial value problems of Equation 2a with continuous initial data.

From a strictly mathematical point of view, the discontinuity in initial data such as Equation 2b causes a serious complication. Applying Van Duyn's theorem (1979) and restricting the problem to the case in which the discontinuity in initial data disappears in some finite time, Nakano et al. (1983c) have recently shown that the solution of Equations 2a and b asymptotically converges upon the solution of Equations 6a and 6b. The extension of Van Duyn's theorem (1979) to the case of discontinuous initial data may not be mathematically exact, but should be an accurate approximation, from a physical point of view. We will use the similarity solution of Equations 6a and 6b to analyze the experimental data below.

When the solution of Equations 6a and 6b describes the asymptotic profile, we integrate Equation 6a to obtain

$$D[f(\eta)] = \frac{1}{2}\left[\int_{\eta_0}^{\eta} f(\eta)d\eta - \eta f(\eta)\right] / \frac{d}{d\eta} f(\eta), \quad \eta \leq \eta_0 \tag{7}$$

where $\eta_0 > 0$ is the location of a wetting front when ϕ is finite, and η_0 is infinite when ϕ is infinite. Once $f(\eta)$ is determined experimentally, $D(f)$ can be evaluated by Equation 7.

ANALYSIS AND DISCUSSION

The time duration of each experiment should be long enough for an initial profile to converge upon its asymptotic profile, but should not be excessively long so that column end effects are negligible. According to Van Duyn's theorem (1979), the deviation of a developing profile from the asymptotic profile is proportional to $t^{-1/2}$. Based on experimental data for Morin clay (Nakano et al. 1982, 1983a-c) and some preliminary experiments on Fox Tunnel silt and Regina clay, we set the time duration of 10 days. The results of the experiments are presented in Figure 2 and Figure 3, where the average water contents of the soils from two

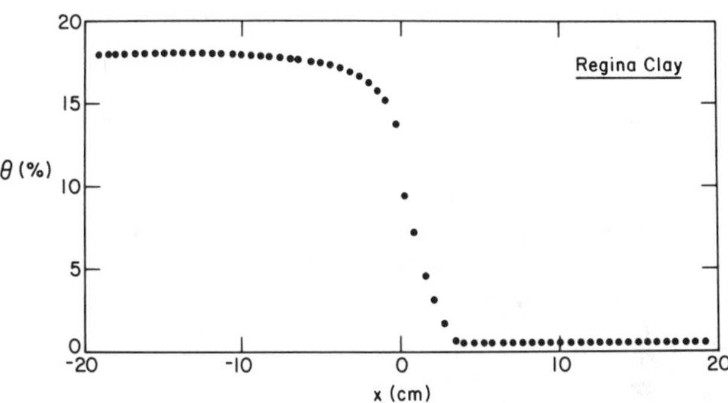

FIGURE 2 Average water contents of two repeated experiments θ(%) vs distance x(cm) for Regina clay.

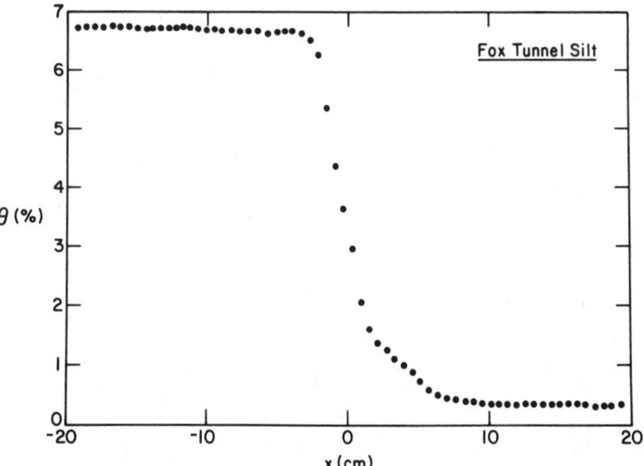

FIGURE 3 Average water contents of two repeated experiments θ(%) vs distance x(cm) for Fox Tunnel silt.

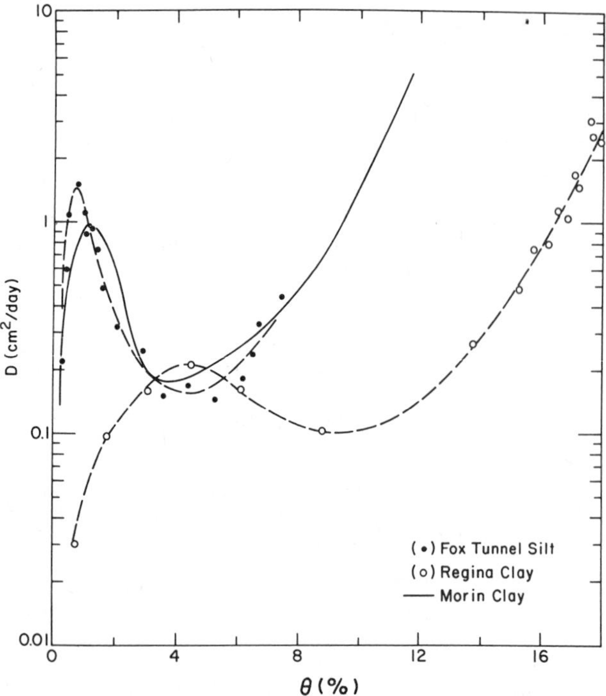

FIGURE 4 Diffusivities D(cm^2/day) vs water content θ(%) for Fox Tunnel silt, Regina clay, and Morin clay.

repeated experiments are plotted versus the distance x (cm).

As discussed in the preceding section, a wetting front is expected to appear when D(b) is negligibly small. A wetting front is clearly identifiable for Regina clay (Figure 2), but it is not easy to pinpoint for Fox Tunnel silt. Fortunately, this difficulty is of no serious concern in evaluating Equation 7 because no significant amount of error is introduced by using a value of η that is larger than the actual location of a wetting front η_0 for a starting point of the integration. Hence, one should select a starting point of η large enough so that θ = b in the neighborhood of this point. Numerical differentiation is generally considered to be a difficult operation. It is desirable to use formulas that include several neighboring data points so that the variation of data points due to experimental errors is smoothed over. Actual calculations of derivatives of f(η) in Equation 7 were made by the standard finite difference method in which six neighboring points were used for interior points and three for end points (Nakano et al. 1983c).

The calculated values of D(θ) were plotted versus θ in Figure 4 where broken curves were drawn by visual inspection to show the trend of data points of Fox Tunnel silt and Regina clay. For the sake of comparison, the diffusivity of Morin clay with ρ = 1.42 g/cm^3 at -1°C (Nakano et al. 1983c) is also presented as a solid curve in Figure 4. The common feature of these three kinds of soil is that D increases with increasing θ, attains a peak, and increases again as θ increases.

It has been recognized that the greatest amount of vapor diffusion occurs near the water content θ_c, when a monolayer of liquid water forms over soil surfaces, and that vapor diffusion levels off rapidly as the water content increases from this critical value (Jackson 1965, Nakano et al. 1982). Assuming the area of a water molecule to be 1.08 x 10^{-15} cm^2, we calculated θ_c to be 0.86%, 1.66%, and 8.05% for Fox Tunnel silt, Morin clay, and Regina clay, respectively. On the other hand, from Figure 4 we find the water content θ_m at which the measured D(θ) attains a peak of about 0.8%, 1.2%, and 4.5% for Fox Tunnel silt, Morin clay, and Regina clay, respectively. Although θ_c does not coincide with θ_m, the soil with the greater specific surface area has the greater value of water content at which D(θ) attains a peak. From Figure 4 we also find that soil with less specific surface area has the greater peak. This is also the case for Jackson's data (1965).

We have found that the specific surface area is an important parameter affecting D(θ). However, the specific surface area does not seem to be directly linked to the mobility of water. Despite the difference in the specific surface area, the diffusivity of Fox Tunnel silt is close to that of Morin clay. In addition, the diffusivity of Regina clay is much less than those of the other two except at water contents ranging from 3% to 5%.

Although our experimental data are limited, the general behavior of D(θ) of soils at the subzero temperature is quite similar to that of unfrozen soils. Our data appear to indicate that the basic transport mechanism of water in soil containing no ice at the subzero temperature is essentially the same as that in unfrozen soil containing a small amount of water, although kinetic factors of transport are affected by the temperature.

ACKNOWLEDGEMENT

This work was supported by several Corps of Engineers projects.

REFERENCES

Anderson, D.M., Pusch, R., and Penner, E., 1978, Physical and thermal properties of frozen ground, in Andersland, O.B., and Anderson, D.M., eds., Geotechnical Engineering for Cold Regions: New York, p. 37-102.

Aronson, D.G., and Peletier, L.A., 1981, Large time behavior of solutions of the porous medium equation in bounded domains: Journal of Differential Equations, v. 39, p. 378-412.

Barenblatt, G.I., and Zel'dovich, Ya.B., 1971, Intermediate asymptotics in mathematical physics: Russian Mathematical Surveys, v. 26, p. 45-61.

Bruce, R.R. and Klute, A., 1956, The measurement of soil water diffusivity: Soil Science Society of American Proceedings, v. 20, p. 458-462.

Gilding, B.H., 1979, Stabilization of flows through porous media: SIAM Journal of Mathematical Analysis, v. 10, p. 237-246.

Jackson, R.D., 1965, Water vapor diffusion in relatively dry soil: IV. Temperature and pressure effects on sorption diffusion coefficient: Soil Science Society of America Proceedings, v. 29, p. 144-148.

Nakano, Y., 1980a, Particular solutions to the problem of horizontal flow of water and air through porous media near a wetting front: Advances in Water Resources, v. 3, p. 81-85.

Nakano, Y., 1980b, Traveling wave solutions of saturated-unsaturated flow through porous media: Water Resources Research, v. 16, p. 117-122.

Nakano, Y., 1980c, Application of recent results in functional analysis to the problem of wetting fronts: Water Resources Research, v. 16, p. 314-318.

Nakano, Y., 1981, A traveling wave solution to the problem of simultaneous flow of water and air through homogeneous porous media: Water Resources Research, v. 17, p. 57-64.

Nakano, Y., 1982, Use of similarity solution for the problem of a wetting front - a question of unique representation: Advances in Water Resources, v. 5, p. 156-166.

Nakano, Y., 1983a, Similarity solutions to the second boundary value problem of unsaturated flow through porous media: adsorption of water by semi-infinite, homogeneous and dry soil under flux boundary conditions: Advances in Water Resources, in press.

Nakano, Y., 1983b, Asymptotic behavior of solutions to the problem of wetting fronts in one-dimensional, horizontal and infinite porous media: Advances in Water Resources, in press.

Nakano, Y., Tice, A., Oliphant, J., and Jenkins, T., 1982, Transport of water in frozen soil: I. Experimental determination of soil-water diffusivity under isothermal conditions: Advances in Water Resources, v. 5, p. 221-226.

Nakano, Y., Tice, A., Oliphant, J., and Jenkins, T., 1983a, Transport of water in frozen soil: II. Effects of ice on the transport of water under isothermal conditions: Advances in Water Resources, v. 6, p. 15-26.

Nakano, Y., Tice, A. and Oliphant, J., 1983b, Transport of water in frozen soil: III. Experiments on the effects of ice content: Advances in Water Resources, in press.

Nakano, Y., Tice, A., and Oliphant, J., 1983c, Transport of water in frozen soil: IV. Analysis of experimental results on the effects of ice content: Advances in Water Resources, in press.

Oleinik, O.A., Kalashnikov, A.S., and Yui-Lin, C., 1958, The Cauchy problem and boundary problems for equations of the type of non-stationary infiltration. Izvestia Akademii Nauk, USSR Ser. Mat., v. 22, p. 667-704.

Peletier, L.A., 1971, Asymptotic behavior of solutions of the porous media equation: SIAM Journal of Applied Mathematics, v. 21, p. 542-551.

Raats, P.A.C., and Klute, A., 1968, Transport in soils: the balance of momentum: Soil Science Society of America Proceedings, v. 32, p. 452-456.

Swarztendruber, D., 1969, The flow of water in unsaturated soils, in DeWiest, J.M., ed., Flow through Porous Media: New York, Academic Press, p. 215-292.

Tice, A.R., Burrows, C.M., and Anderson, D.M., 1978, Phase composition measurement on soils at very high water contents by the pulsed nuclear magnetic resonance technique: Transportation Research Record, no. 675, p. 11-14.

Truesdell, C., and Toupin, R.A., 1965, The classical field theories, Handbuch der Physik: Berlin, Springer-Verlag, v. 3, p. 226-793.

Van Duyn, C.J., 1979, On the diffusion of immiscible fluids in porous media: SIAM Journal of Mathematical Analysis, v. 10, p. 486-497.

Van Duyn, C.J., and Peletier, L.A., 1977a, A class of similarity solutions of the nonlinear diffusion equation: Nonlinear Analysis, Theory, Methods and Application, v. 1, p. 223-233.

Van Duyn, C.J., and Peletier, L.A., 1977b, Asymptotic behavior of solutions of a nonlinear diffusion equation, Architectural Rational Mechanical Analysis, v. 65, p. 363-377.

SEISMIC VELOCITIES AND SUBSEA PERMAFROST IN THE BEAUFORT SEA, ALASKA

K. G. Neave[1] and P. V. Sellmann[2]

[1]Northern Seismic Analysis, Echo Bay, Ontario, Canada
[2]U.S. Army Cold Regions Research and Engineering Laboratory
Hanover, New Hampshire 03755 USA

The distribution of high-velocity material was used as an indicator of ice-bonded permafrost. Observations from ice survey and marine seismic records, coupled with control from a small number of drill holes, suggest that ice-bonded permafrost is extremely widespread in the Beaufort Sea. Large areas of high-velocity material at shallow depths, 10-40 m below the seabed, were observed near Prudhoe and Harrison Bays. In some cases these zones extended up to 35 km from shore. It was also common to find that depths to the high-velocity material increased with distance from the shore. Observed depths were as great as 150-230 m below the seabed. Velocities also commonly decreased with distance from shore to a point where it was no longer possible to identify ice-bonded horizons. Reflection analysis revealed several deep near-horizontal reflectors that recurred along segments of the coastline from Harrison Bay to the Canadian border. Reflectors at approximately 200 m and 450 m could be permafrost-related, although no control exists for an evaluation of this interpretation. Locally these reflectors may extend out to the shelf edge to approximately the 100-m water depth.

Petroleum industry seismic data were used to obtain an improved understanding of the regional distribution of subsea permafrost. Our observations in the U.S. Beaufort Sea are based on data from Harrison Bay east to the Canadian border. The most detailed information is from the coast out to a water depth of approximately 15 m, with the greatest detail from the Prudhoe and Harrison Bay regions (Figure 1).

It has been demonstrated that refraction analysis of the upper part of deep seismic records can be useful in locating high-velocity zones that may be ice-bonded permafrost (Hunter et al. 1976, 1978). However, not all frozen materials can be located, due to their low velocities, which are caused by warm subsea permafrost temperatures, saline pore water, gas in voids, and the coarse resolution of this technique.

Reflection interpretations can also be used to acquire information on permafrost distribution and structure, but the lack of velocity data usually makes these interpretations more speculative than those from refraction analysis. Beyond the 15-m water depth, out to the shelf edge, our only information is from reflection analysis.

METHODS

The early-return data used for this study included nonproprietary records from Western Geophysical Company, Geophysical Service Inc., as well as data released by British Petroleum. Records from the land-sea transition were acquired during the winter over the sea ice. Examples of marine and ice survey records showing high-velocity zones interpreted as ice-bonded permafrost can be found in Neave and Sellmann (1982) and Hunter et al. (1976).

Control for these observations is very limited and it is virtually nonexistent beyond the 15-m water depth and in shallower water outside the Harrison and Prudhoe Bay regions. The existing control is from shallow drilling, penetrometer observations, and high-resolution seismic studies (Osterkamp and Harrison 1976, 1980, 1981; Harrison and Osterkamp 1981; Blouin et al. 1979; Sellmann and Chamberlain 1979; Miller and Bruggers 1980; Rogers and Morack 1980, 1981). Observations in the Canadian Beaufort Sea provide indirect control with a range of permafrost velocities, horizontal limits, and thicknesses (Hunter et al. 1976).

Error estimates were made to verify the analytical methods and measurements (Neave and Sellmann 1982). The consistency of results from duplicate coverage, often acquired during different years and seasons, also helped develop confidence in the results.

A model and appropriate equations are required to interpret the records and obtain depth and velocity data for the construction of maps and sections. The model used was a simple plane homogeneous layer over a half space, and our equations are derived from those provided by Grant and West (1965). Minor modifications to these equations are reported in Neave and Sellmann (1982).

A simplifying assumption was used for the refraction depth determinations: the water layer was combined with the low-velocity bottom sediments to make a single upper layer. An upper-layer velocity of 1.8 km/s was used for all profiles out to the 15-m water depth. In this zone, the upper-layer velocities were observed to range from 1.6 to 2 km/s. In deeper water an upper-layer velocity of 1.5 km/s was used.

The spatial resolution of data from this study is obviously not as great as could be obtained from a seismic investigation specifically designed to study offshore permafrost. In general, the horizontal extent of a feature that can be detected should be a minimum of three detector spacings. This means that the minimum size of a target that can be resolved will vary from 150 to 300 m, depending on the source of the data used in this study. The minimum vertical thickness of a detectable high-velocity layer is usually determined by the wavelength of the refracted single. Resolution is possible to approximately 1/2 wavelength, or about 50 m for most of these data (Sherwood 1967). However, shallow layers less than 30 m thick may also be detectable.

OBSERVATIONS

Our observations are influenced by a number of factors, including the amount of data examined for a region, its quality, the type of analyses that were possible, and the availability of control required to support the interpretations.

In the coastal zone, control has established a correlation between high-velocity zones and ice-bonded permafrost. Nearly all the records in the region have high-velocity refractors (greater than 2 km/s), but they are found at different depths. This great variability in the distribution of high-velocity material prompted us to outline coherent blocks of data that could be analyzed separately (Figure 1). The boundaries of the zones (A^1, A^2, etc.) were established by examining our seismic sections and isolating the segments with similar properties. The following results of our seismic analysis show the validity of these zones. It should be possible to find a geological explanation for the zones.

In area A^1, just north of the Sagavanirktok delta, refractors range from 0-40 m below sea level (Figure 2), with an average velocity of approximately 3 km/s. This area is part of a much larger shallow high-velocity patch, C^1, that extends north and east of Reindeer Island (Neave and Sellmann, in press). Rogers and Morack (1980) were the first to make observations on the southern edge of this zone, based on seismic data. Drilling and penetrometer observations confirm that shallow ice-bonded permafrost is present north of Reindeer Island and at several other locations in area A^1 near the Sagavanirktok delta (Miller and Bruggers 1980, Sellmann and Chamberlain 1979, Blouin et al. 1979).

In area A^2, which stretches from Foggy Island Bay to Flaxman Island with the exception of part of Mikkelsen Bay (Figure 1), high-velocity refractors are found in the range of 60-130 m below sea level. Velocities associated with this refractor are seen in Figure 3. The average velocity along this refracting layer is approximately 3 km/s.

In area A^3, in Prudhoe and Mikkelson Bays, high-velocity refractors occur between 160-230 m below sea level. The refractor velocities in this zone range from 2.5-4.5 km/s with an average of approximately 3.5 km/s (Figure 4). The position of this refractor in Prudhoe Bay is supported by logs from an offshore well (Osterkamp and Payne 1981). The refractor depth in this case corresponds to the top of the ice-bonded permafrost, and the structure appears to continue seaward in some parts of the study area as a reflector.

In addition to the deep refractor that occurs in the 200-m depth range, refractors are also found in area A^3 at depths of 10-70 m (Figure 4). These shallow refractors have an average velocity of approximately 2.5 km/s, and may correlate with Horizon B reported by Reimnitz et al. (1971) around

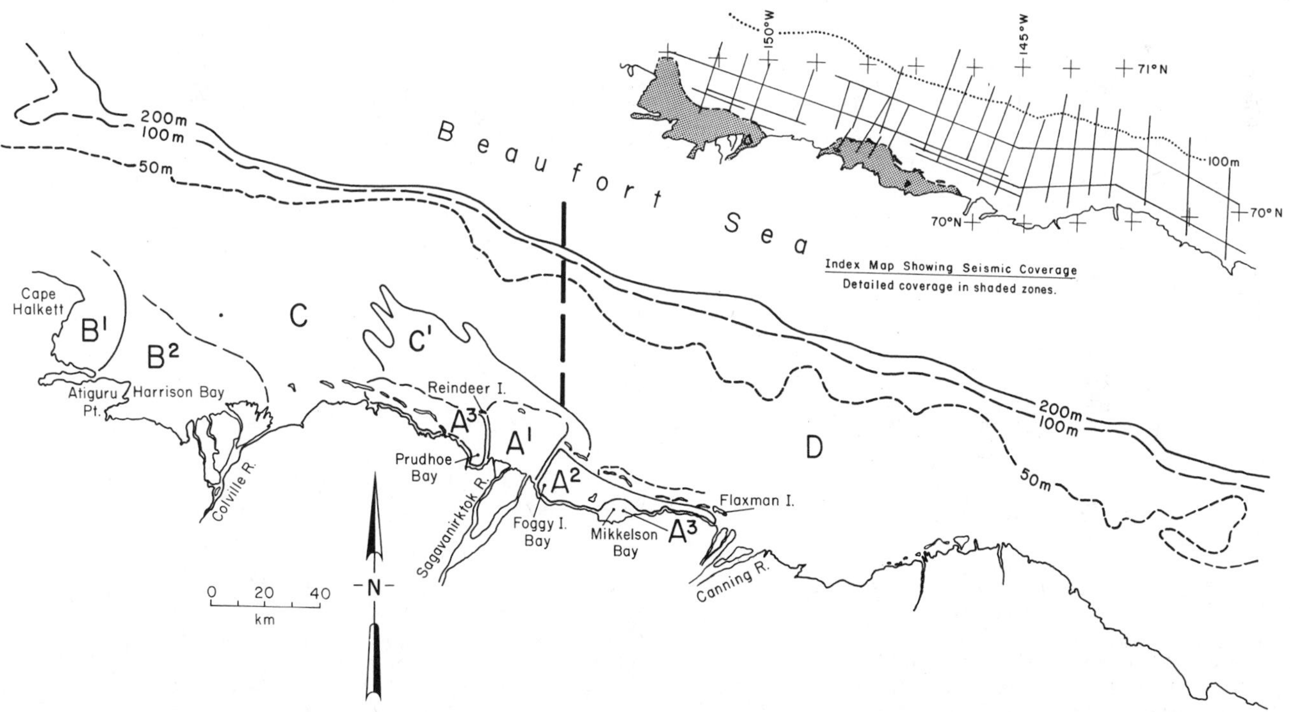

FIGURE 1 Index map showing the study area and zones with similar seismic velocity and depth structure.

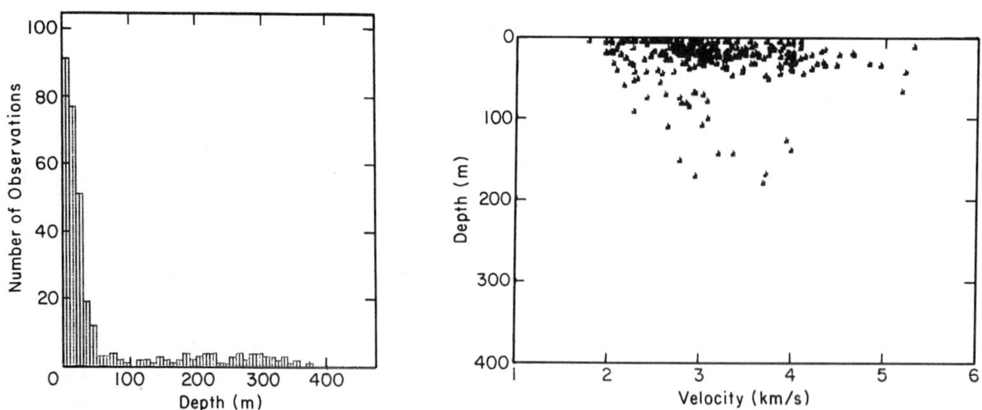

FIGURE 2 Histogram of refractor depths and velocity depth plot for area A[1].

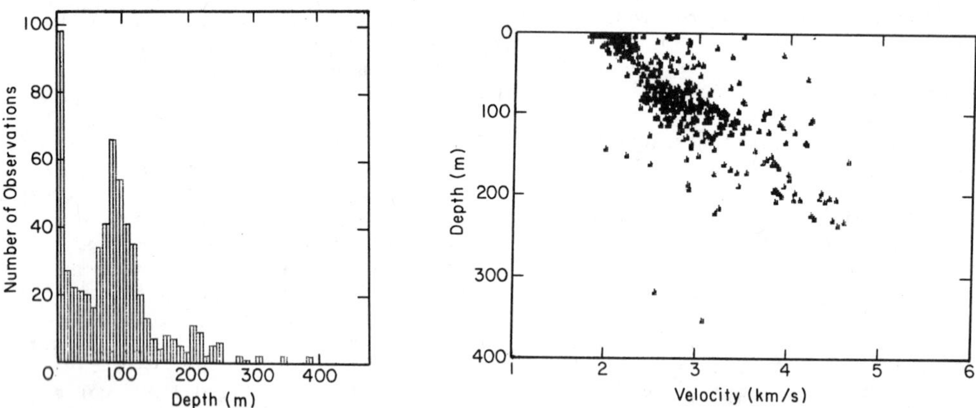

FIGURE 3 Histogram of refractor depths and velocity depth plot for area A[2].

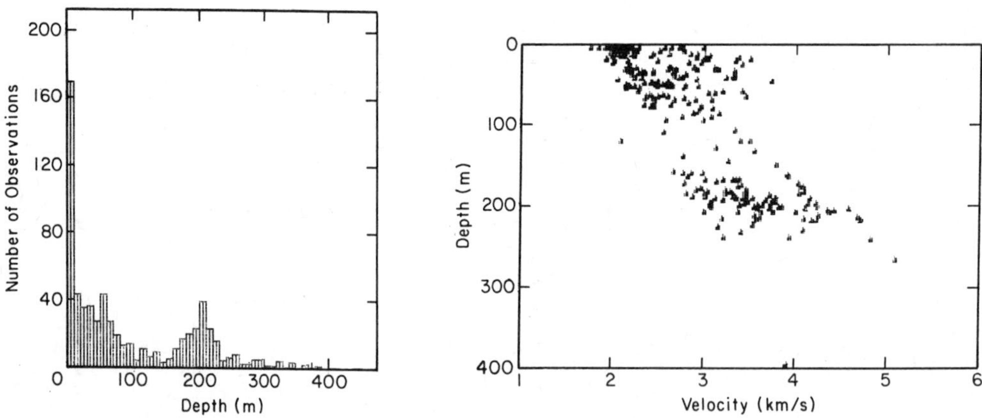

FIGURE 4 Histogram of refractor depths and velocity depth plot for area A[3].

Pingok Island. It was part of their reflection survey, which covered the region between area A^2 and the Colville delta.

The depth histogram for the Harrison Bay data (Figure 5) is very similar to the one for area A^3 in Prudhoe and Mikkelson Bays (Figure 4). There is a peak at the depth of 210 m and a grouping in the 10–70 m range. Most of the shallow high-velocity refractors are west of Atigaru Point in area B^1 (Figure 1). The shallow refractors in this area are lower velocity, approximately 2.2 km/s, in contrast to 3 km/s for the shallow refractors in the Prudhoe area. This characteristic has been reported earlier by Rogers and Morack (1981), and they suggest it may

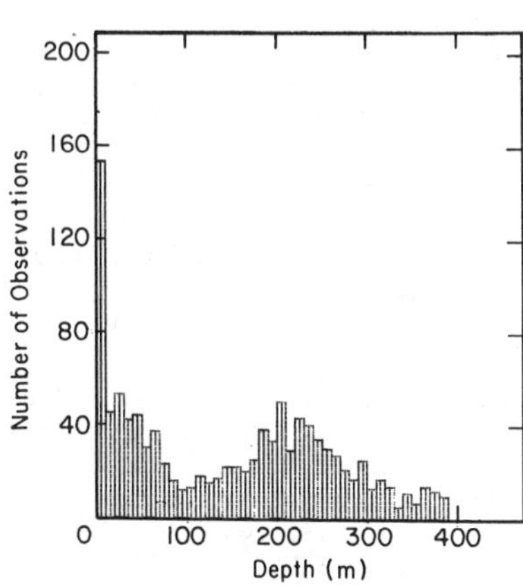

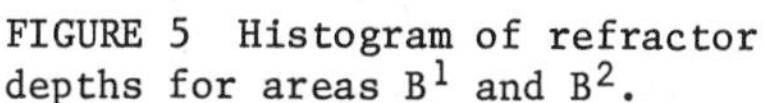

FIGURE 5 Histogram of refractor depths for areas B^1 and B^2.

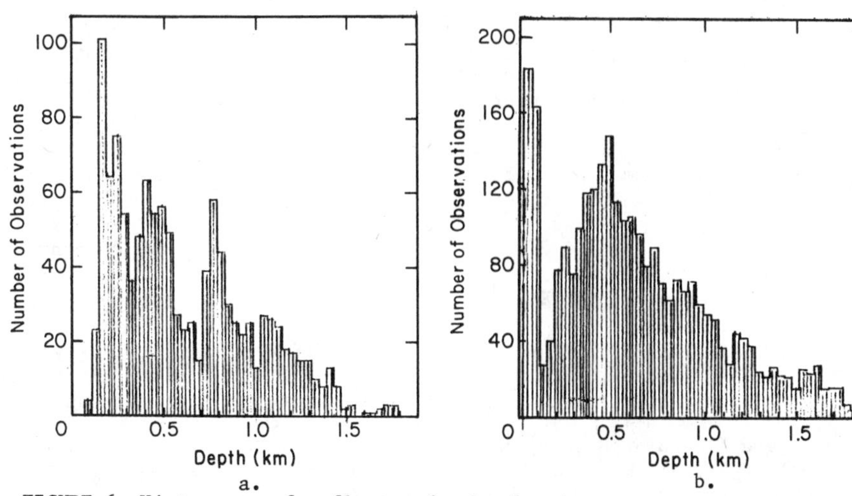

FIGURE 6 Histograms of reflector depths for deep-water data. (a) Area C, west of the dashed line in Figure 1. (b) Area D, east of the dashed line in Figure 1.

be related to the occurrence of finer grained material in Harrison Bay. The deep refractors, at around the 210-m depth, are from the eastern two-thirds of Harrison Bay (area B^2). The average velocity for this refractor is approximately 3.5 km/s.

On the lines north of the barrier islands, the refraction part of the records has a low signal-to-noise ratio along with low velocity readings. The only obvious high-velocity zone is the large area C^1, beyond Reindeer Island that appears to be an extension of area A^1 off the Sagavanirktok delta (Figure 1). This zone is discussed in greater detail in Neave and Sellmann (in press) and Rogers and Morack (1980).

Other zones are restricted to intermittent strips of marginally high velocity, in some cases just greater than 2 km/s. These zones can be found near shore and near the shelf edge. Reflection data for the lines west of Foggy Island Bay in area C (Figure 6a) define three distinct horizons, at 200 m, 450 m, and 750 m. The 200-m reflectors may be correlated with nearshore refractors in area B^2 and both parts of area A^3. A tie seems likely between the 200-m reflector and the nearshore 200-m refractor that can be connected with onshore permafrost. This correlation, as well as that of the 450-m reflector with the bottom of ice-bonded permafrost, is discussed in Neave and Sellmann (in press).

The reflection data from the lines east of Foggy Island Bay (area D) are not of the same quality as the western lines. The resulting histogram for the eastern lines only shows the 450-m peak clearly (Figure 6b), but there are zones within this area that show shallower and deeper horizons. We therefore assume that these horizontal reflectors are common to the shelf from Harrison Bay to the Canadian border.

The most distinct transition observed in these records occurs just beyond the edge of the shelf. North of the 300-m water depth there is a noticeable increase in the quality and resolution of the reflection data.

CONCLUSION

The distribution of high-velocity material and possible related reflectors suggests that ice-bonded permafrost is extremely widespread and in some locations may even extend out to the shelf edge. It also indicates the variability of the materials in the permafrost zone.

The evidence for high-velocity zones in water depths less than 6 m being ice-bonded permafrost is good because of their connection with high-velocity material on shore. Drilling in the Prudhoe Bay region supports the conclusion that extensive shallow high-velocity material in this region is ice-bonded. This high-velocity zone is large (Figure 1) and may extend as much as 35 km from shore. A similar shallow zone (10-40 m below sea level) also exists in the western part of Harrison Bay. Other extremely shallow high-velocity zones may exist, but may not have been detected due to the resolution of this technique, the low velocities associated with warm ice-bonded material, and insufficient coverage. In areas where shallow high-velocity zones were not found, deeper high-velocity zones up to 230 m deep were observed. These zones were also thought to represent ice-bonded sediments, and in some cases the high-velocity material could be traced, with decreasing depth, back to shallow high-velocity material at the coastline.

Information on ice-bonded permafrost beyond the 15-m water depth is speculative, since no control exists beyond this point. However, deep reflectors at depths of approximately 200 m and 450 m may indicate that ice-bonded materials exist well out onto the shelf. The likelihood of the deep structures on the shelf being related to permafrost may be supported by similar observations in the Canadian Beaufort Sea, where deep layers thought to be ice-bonded permafrost persist more than 100 km from shore (Hunter et al. 1976). Additional weak evidence that permafrost may exist out to the shelf edge is suggested by a noticeable improvement in record quality beyond the shelf margin, indicating a considerable change in material properties.

ACKNLOWLEDGMENT

This study was funded wholly by the Bureau of Land Management through inter-agency agreement with the National Oceanic and Atmospheric Administration, as part of the Outer Continental Shelf Environmental Assessment Program. We would like to thank the geophysical companies for their cooperative spirit in providing necessary seismic data.

REFERENCES

Blouin, S.E., Chamberlain, E.J., Sellmann, P.V., and Garfield, D.E., 1979, Determining subsea permafrost characteristics with a cone penetrometer - Prudhoe Bay, Alaska: Cold Regions Science and Technology, v. 1, no. 1.

Grant, F.S., and West, G.F., 1965, Interpolation theory in applied geophysics: New York, McGraw-Hill.

Harrison, W.D., and Osterkamp, T.W., 1981, Subsea permafrost: Probing thermal regime and data analysis: Environmental Assessment of the Alaskan Continental Shelf, Annual Reports: Washington, D.C., National Oceanic and Atmospheric Administration, U.S. Department of Commerce, v. VII, p. 291-349.

Hunter, J.A., Judge, A.S., MacAulay, H.A., Good, R.L., Gagne, R.M., and Burns, R.A., 1976, The occurrence of permafrost and frozen subseabottom materials in the southern Beaufort Sea: Geological Survey of Canada and Earth Physics Branch, Department of Energy, Mines and Resources, Canada, Beaufort Sea Technical Report No. 22.

Hunter, J.A., Neave, K.G., MacAulay, H.A., and Hobson, G.D., 1978, Interpretation of subseabottom permafrost in the Beaufort Sea by seismic methods. Part 1: Seismic Refraction Methods, in Proceedings of the Third International Conference on Permafrost, v. 1, p. 521-526.

Miller, D.L., and Bruggers, D.E., 1980, Soils and permafrost conditions in the Alaskan Beaufort Sea, in Proceedings of the Twelfth Annual Offshore Technology Conference 30 April-3 May, 1979: Houston, p. 325-332.

Neave, K.G., and Sellmann, P.V., 1982, Subsea permafrost in Harrison Bay, Alaska: An interpretation from seismic data, CRREL Report 82-24: Hanover, N.H., U.S. Cold Regions Research and Engineering Laboratory.

Neave, K.G., and Sellmann, P.V., in press, Determining distribution patterns of ice-bonded permafrost in the U.S. Beaufort Sea from seismic data, in Outer Continental Shelf Environmental Assessment Program (OCSEAP), Beaufort Sea Volume: New York, Academic Press.

Osterkamp, T.E., and Harrison, W.D., 1976, Subsea permafrost at Prudhoe Bay, Alaska, Drilling report and data analysis: Fairbanks, Alaska, Geophysical Institute, University of Alaska (UAG-R-245).

Osterkamp, T.E., and Harrison, W.D., 1980, Subsea permafrost: Probing, thermal regime, and data analysis, in Environmental Assessment of the Alaskan Continental Shelf, Annual Reports: Washington, D.C., National Oceanic and Atmospheric Administration, U.S. Department of Commerce, v. IV, p. 497-677.

Osterkamp, T.E., and Harrison, W.D., 1981, Methods and equipment for temperature measurements in subsea permafrost: Fairbanks, Alaska, Geophysical Institute, University of Alaska (UAG-R-285).

Osterkamp, T.E., and Payne, M.W., 1981, Estimates of permafrost thickness from well logs in northern Alaska: Cold Regions Science and Technology, v. 5, p. 13-27.

Reimnitz, E., Wolf, S.C., and Rodeick, C.A., 1972, Preliminary interpretation of seismic profiles in the Prudhoe Bay area, Beaufort Sea, Alaska: U.S. Geological Survey Open File Report 548.

Rogers, J.C., and Morack, J.L., 1980, Geophysical evidence of shallow nearshore permafrost, Prudhoe Bay, Alaska: Journal of Geophysical Research, v. 85, p. 4845-4853.

Rogers, J.C., and Morack, J.L., 1981, Beaufort and Chukchi sea coast permafrost studies, in Environmental assessment of the Alaskan Continental Shelf, Annual Reports: Washington, D.C., National Oceanic and Atmospheric Administration, U.S. Department of Commerce, v. VIII, p. 293-332.

Sellmann, P.V., and Chamberlain, E.J., 1979, Permafrost beneath the Beaufort Sea near Prudhoe Bay, Alaska, in Proceedings of the Eleventh Annual Offshore Technology Conference, p. 1481-1488.

Sherwood, J.W.C., 1967, Refraction along an embedded high-speed layer, in Seismic refraction prospecting (Musgrove, A.W., ed.): Washington, D.C., Society of Exploration Geophysicists, p. 138-151.

PREVENTING FROST DAMAGE TO RAILWAY TUNNEL DRAINAGE DITCHES IN COLD REGIONS

Nei Fengming

The Third Survey and Design Institute of the Ministry of Railways
Tianjing, People's Republic of China

Cold weather and permafrost usually have an adverse effect on the drainage system of a railway tunnel. Based on several years' experience in railway construction and operation in the Great Xingan Mountains of China, the author discusses the effect of drainage ditches, deeply buried seepage ditches below tunnels, thermo-insulated drains, heated drains, etc., on the prevention of frost damage by using actual engineering examples. His conclusion is that, in cold regions where underground water is presumed to exist in winter, the drainage ditch is still preferable, even though it is difficult to construct and increase project costs. The seepage ditch is applicable only in regions where the frost depth is 2.7-4.8 m, because the ditch can affect the stability of the sidewalls if it is buried too deeply below the center of the tunnel. The thermo-insulated ditch and the heating ditch are more convenient and cheaper to construct, but should be used on only an experimental basis until more research can be done.

Because of cold weather and permafrost the drainage system in the railway tunnel usually does not work well. Based on experiences in railway construction and operation in the Great Khingan Mountains for many years, the author has introduced the effect of the drainage ditch, the seepage ditch buried deep in the centre of tunnel bottom, the thermo-isolated drain and the heating drain, etc., on the prevention of frost damage by using actual engineering examples. The detailed structure of the drainage ditch and the seepage ditch and their effect are presented with observed data.

PREVENTING FROST DAMAGE TO RAILWAY TUNNEL DRAINAGE DITCHES IN FOREIGN COUNTRIES

U.S.S.R.

The heating of railway tunnels in some places is provided by electric power using tubular heaters, while heat preservation in the ditch is effected by using modern and efficient thermal insulation materials, such as hard foamed plastics, cellular glass, pottery clay, etc. In some places, the ditch is warmed by the use of hot water or vapor.

Norway

Heating cables are installed in the tunnel drainage system. Some tunnels are equipped with frost-prevention double doors, which automatically open when the train approaches and close after it passes through.

U.S.A.

The primary measure adopted is to check carefully on every drainage ditch before the onset of winter and install heat preservation materials on the outlet. Heat is preserved in the drainage ditch by using foamed materials and heating cables are installed in the drainage pipelines.

Japan

Two in-situ experiments were undertaken in the Kamiwahoro tunnel at Hokkaido, where the method used was to cover tunnel lining surface with thermal insulation materials, so as to reduce release of ground heat, in order to keep the tunnel lining surface at a temperature above the freezing point.

Type A: Use polyethylene corrugated plates installed on lining surface as waterproof plates; a gap of 18 mm between plates is left for drainage. Waterproof plates are coated with a 35 mm thick insulation layer of foamed urethane.

Type B: The concrete surface lining is coated directly with 13 mm thick waterproof mortar to serve as a waterproof layer, and then covered with 35 mm thick foamed urethane as an insulation layer.

Results obtained from the Japan experiments are: 1) In the severe cold region where the lowest temperature reaches -20°C and the yearly cumulative total negative temperature exceeds 510°C, the minimum surface temperature of the concrete lining of Type A can be kept at about 3°C; 2) The minimum surface temperature of the concrete lining of Type B can be kept at about 4°C; 3) The minimum surface temperature of the lining in the region where no heat preservation measures are adopted is -11°C, and the freezing period at the lining lasts 4 months.

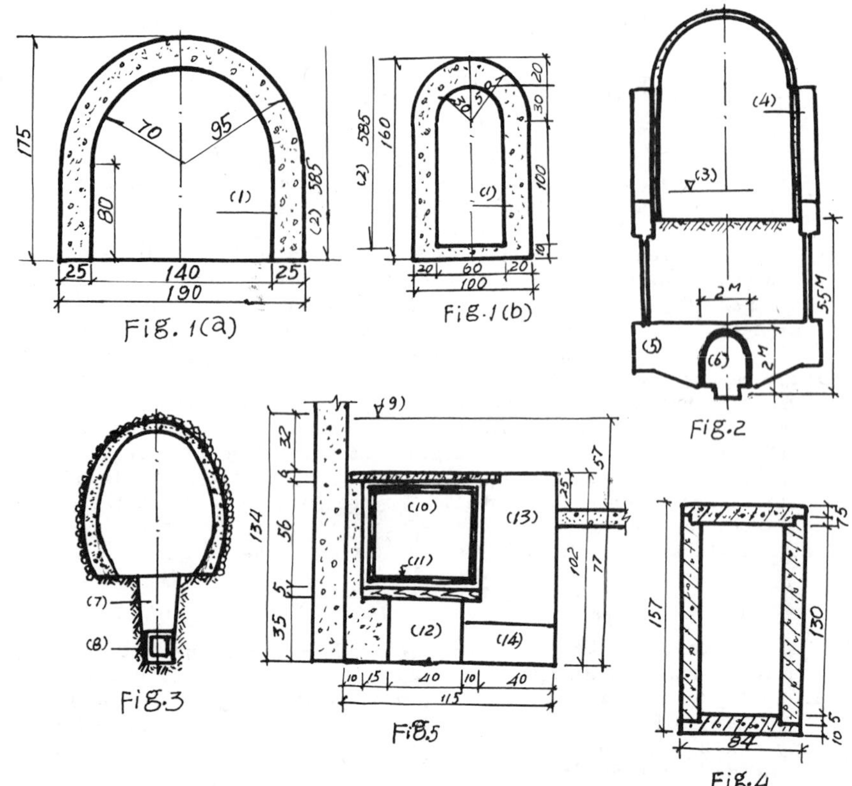

Figure 1 (a). Drainage ditch type A (cm).

Figure 1 (b). Drainage ditch type B (cm).

Figure 2. The drainage system of the drainage ditch (m).

Figure 3. The seepage ditch buried in the bottom.

Figure 4. Structural diagram of the deep buried seepage ditch (cm).

Figure 5. The thermo-isolated drain (cm).

Signs of drawings

(1) grade 140 concrete
(2) to roadbed
(3) rail surface
(4) vertical blind drain outlet
(5) sub-ditch
(6) drainage ditch
(7) back fill ballast
(8) deep-buried seepage ditch
(9) rail surface
(10) glass fibre filling
(11) waterproof layer
(12) water trough
(13) foamed concrete thermo-isolated wall
(14) grade 100 concrete prefabricated block

PREVENTING FROST DAMAGE TO TUNNEL DRAINAGE DITCH IN COLD REGIONS

Drainage ditch

The drainage ditch looks like a small tunnel and is constructed beneath the railway tunnel. The water table around the tunnel may be reduced by the use of the shaft, drilling or cracks, thus preventing lining from the effect of expansion caused by the freezing of the surrounding medium. Experiments show that the drainage ditch is a comparatively good drainage system in cold regions and is particularly suitable for regions where the mean monthly temperature is below -20°C and rock freezing depth is about 5-6 meters. A section of the drainage ditch is shown in Fig. 1. The section of Type A, constructed with close-cut method, is suitable for the sector inside the tunnel; the section of Type B, constructed with open-cut method, is suitable for the sector outside the tunnel.

To make full use of the drainage capacity of the drainage ditch, at the time when it is being built, it is necessary to build an auxiliary drainage system, including a vertical blind drain, drain drilling and sub-ditch, as shown in Fig. 2.

Performance of the Drainage Ditch: Two kinds of drainage ditches are used in the railway tunnels in the permafrost region of the Great Khingan Mountains - those which pass through permafrost, and those which pass through non-permafrost soil. These drainage ditches have been in operation for many years with good effect.

The drainage ditch in the mountain-top tunnel on the Yalin railway line has a section (about 190 m) passing through the permafrost soil (of the linking-up type, with mean ground temperature over many years at -1°C), thus resulting in freezing inside the drainage ditch each year. According to the survey data of July 1977, thickest ice in the drainage ditch is found in the permafrost soil region, and the thickness measures up to 80 cm. The freeze damage caused to the two kinds of drainage ditches which partly pass through permafrost soil is analyzed as follows: The drainage ditch passing through the permafrost soil and that passing through non-permafrost soil are located in different environs. The former is one form of heat source occurring in the frozen soil. Since the water temperature of the thawing region through which the tunnel passes is lower (1-3°C), the inner potential of this heat source is very small. There occurs heat exchange in the form of transmission between water at certain temperature and the surrounding frozen rocks, with the result that the inner potential of water, i.e., temperature drops. Besides, in the course of ground water flowing from the thawing region to the frozen soil region, the water temperature also drops gradually and when the water reaches the freezing point and releases all its potential heat, then freezing will occur.

Cuiling tunnel No. 2, for example, has all its drain ditches on permafrost soil. In February 1973 these were ice-bound, causing freezing inside the tunnel and affecting traffic safety. It was necessary to use a steam locomotive to thaw the ice and manpower to excavate it. In the winter of 1973, though measures had been strengthened, freezing still occurred when ice thickened up to 70 cm. The effect of thawing by electric stove was neglible. In 1977, 75 kW ventilators were used to blow hot air (from outside the tunnel) into the drain ditch, and it took seven days and nights to melt the ice of more than 400 m³ inside it. The fact that freezing occurred in both cases shows that to prevent frost damage it is necessary to take effective measures regardless whether the drain ditches pass wholly or partly through the permafrost soil.

The seepage ditch buried at center of tunnel bottom

The seepage ditch is buried deep in the ditch under the tunnel at a certain freeze depth with the aim of preventing frost damage by the use of ground temperature. The Chaoyang No. 1 tunnel is the first tunnel which has adopted the deep buried seepage ditch. The overall length of this tunnel is 420 m. The lowest mean temperature of the region is -27.9° C (in 1969), and annual mean temperature -3.2°C (in 1969). Its elevation drawing is given in Fig. 3. Its buried depth (from roadbed to flow surface of the seepage ditch) ranges from 3.5m to 4.5 m, and the actual freeze depth inside the tunnel is 5 m (where the rocks are rough-grained granite). Therefore, freezing occurs inside the ditch in winter season. However, since this kind of drainage system has a good drainage effect, frost damage is not liable to occur. In summer, when the covering slab of the inspection shaft is taken off, ice in the seepage ditch will thaw within several days, and normal operation is not affected. The seepage ditch is in Fig. 4.

The thermo-isolated drain

The thermo-isolated drain may generally reduce the ditch depth and raise the foot of the sidewall, thus making the cost cheaper and construction easier. For this reason, the thermo-isolated drain was constructed and tested in the Baikaer tunnel. There the freeze depth of the rock (Anshan rock) outside the tunnel is 7.4 m, and the minimum mean monthly temperature is -32°C (in 1973). Fig. 5 shows the section of the thermo-isolated drain. By testing the Baikaer thermo-isolated drain, we made the following observations: (1) The thermo-isolated drain costs less and is easier to construct. Thus it should be the object of research on the type of tunnel drainage in severe cold and permafrost soil regions; (2) If tunnels are provided with thermo-isolated drains, it is necessary to build such drains on both sides, because one-sided drain is liabole to cause frost damage; (3) It is necessary to have thermo-isolation materials well treated to be dampproof, because the heat conduction coefficient increases with the increase of dampness; (4) It is also necessary to provide thermo-isolated layer at the bottom of the thermo-isolated drain. Only in this way is it possible to protect thermo-isolated drain from frost damage; (5) In construction it is necessary to reduce, as far as possible, the effect of air convection in the ditch, in order to minimize thermal loss due to the effect of air convection.

Heating drain

To prevent frost damage to the drain, in some tunnels different methods are used, such as hot air ventilation, hot wind ventilation, and stove heating. In the Aer Mountain tunnel hot air is used to raise the temperature. This tunnel has a total length of 3215 m and is sloped in the shape of an inverted V. The drain has a buried depth of 0.7 m inside it and 1.5 m at the opening. The mean annual temperature in the region is -3.5°C, and the minimum mean monthly temperature is -25.7°C (in January). Outside the tunnel, the minimum temperature is -48°C and maximum freezing depth is about 2.4 m. About 200 m inside the north end of the tunnel, the drain is provided with heating facilities, placed between the upper and lower covering slabs. The end of the heating pipe has a diameter of 65 mm, without radiator or backwater pipe. Each year, from October 15 to April 20, heating is maintained with steam supplied three times a day. At the time steam is supplied, the temperature in the drain reaches beyond 20°C. Usually the temperature is kept at 7-8°C.

Thermo-isolated door-screen

This is installed at the tunnel opening for the purpose of preventing cool air from entering into the tunnel, preventing the drain from freezing and reducing frost damage. The tunnels which are equipped with thermo-isolated door-screen should have a watchman on duty to communicate with the railway station by means of signals. This method works better but is unfavorable for natural ventilation. It is applicable only when few trains pass through in opposite directions and when the tunnel is comparatively long. The effect of thermo-isolation door-screen used in the Xinan tunnel is as follows: the winter season in this tunnel lasts as long as seven months. The mean temperature in the coldest months is -26°C, and the minimum temperature is -54°C. The wind draft in winter blows directly to the door, at a velocity of 4.5 m/sec. on the average and 10 m/sec. at the maximum. Because it is a single slope tunnel, the difference in height between the two ends of the tunnel is 37 m. Therefore, there is considerable air convection inside the tunnel. When the temperature at the tunnel opening is below -15°C without the thermo-isolation door-screen, ice is formed along the entire 3000 meter length of the tunnel. With the thermo-isolation door-screen, air convection is greatly reduced. The minimum temperature in the middle of the tunnel is 2°C, usually at 5°C. The range of freezing is 500 m at one end, and 800 m at the other end. Sometimes, when the intervals between trains are short, and the door-screen is open for 30 minutes in one stretch, the range of freezing is 800 m and 1000 m respectively. After the thermo-isolated door-screen is installed, the amount of ice excavated is greatly reduced, so that maintenace work can be undertaken inside the tunnel in winter time. The door-screen method has been used for 30 years with good effect. The thermo-isolated door-screen is made of thick canvas, size 6 X 8 m, opened and closed by hand. This tunnel has double-line clearance and single-line traffic. There is no mechanical ventilation.

SELECTION OF THE TYPE OF TUNNEL DRAINAGE

None of the types of drainage discussed is perfect, and any application should be made according to specific conditions in a flexible and comprehensive manner for purpose of economy and safety. Since the thermo-isolation door is questionable in terms of safety, it is seldom used. Even though the drainage ditch has problems such as difficult construction and high cost, it is still advisable to make preferential use of the drainage ditch in cold regions where underground water is presumed to exist in winter. The seepage ditch, if buried too deep in the center, would affect the stability of the sidewalls, and is therefore applicable only in regions where the freezing depth of rocks is 2.7 - 4.8 m. The thermo-isolated ditch and the heating ditch are more convenient and cheaper to construct, and should be used on an experimental basis. More research should be done on these two types of tunnel drainage in the cold regions.

REFERENCES

Katswya Okada, 1979, Analytical tests of Icicle prevention work by Adiabatic Treatment of Tunnel Surface in very cold area, in Quarterly Report, Railway Technical Research Institute, Vol. 20, No. 2, p. 65-69.

Souhihe, E.G. 1964, Prevention of Frost Damage to tunnel drainage ditch, in Experiences of bridge tunnel maintenance, p. 62-64 (in Russian).

DYNAMICS OF THE CRYOLITHOZONE IN THE NORTHERN HEMISPHERE DURING THE PLEISTOCENE

I. A. Nekrasov

Permafrost Institute, Siberian Branch,
Academy of Sciences, Yakutsk, USSR

During a study of deep freezing of the lithosphere in various parts of the Earth, the concept of an extremely old cryolithozone (namely 2 mil. years old) was adopted. It was shown that the last phase of deep freezing of the lithosphere in the mountains of Northern Siberia and Alaska started in the Early Pleistocene (600-700 thousand years BP) and has continued from that time. During the last 50 thousand years four stages can be identified in the development of these frozen strata; during these stages the dynamics of the cryolithozone were closely linked to the progress of glacial events. Summarizing the information on traces of the past distribution of permafrost, the author has compiled a map of its maximum distribution. On the basis of the trends derived from this, a prognosis of the evolution of the cryolithozone for the next 100 years has been attempted. However, it is emphasized that man's industrial activities are starting to exert a critical impact on the course of natural processes.

A problem of geocryology is to determine the beginning of freezing of the upper levels of the lithosphere in various regions of the globe.

It is generally recognized that deep freezing of the Earth's crust has been a recurrent phenomenon. This is suggested by numerous traces of the old glaciations which are found within deposits ranging from Archean to Cainozoic in age. Glacial periods alternated with warmer periods, during which most of the Earth was dominated by subtropical climates. During the latter, the permafrost that had formed in previous glacial ages degraded completely. Present-day permafrost commenced forming during the Cainozoic fall of temperature.

The earliest ideas of the mechanism for deep freezing of the lithosphere are, in principle, correct. As early as 1886 Voeikov formulated the principle of equilibrium thermal state of the Earth's crust by demonstrating that heat losses of a crust layer with yearly varying temperatures are in equilibrium with the heat input from the Sun and from deeper layers of the lithosphere. The equilibrium between soil and atmosphere is, according to his view, represented by the annual average temperature at the depth of zero annual fluctuations (Voeikov 1952, p. 378).

Later, Shostakovich (1916) pointed out that the formation of "permafrost" starts when a soil layer, once frozen during winter, does not thaw out in summer, and a layer with subzero-temperature remains. One year later the phenomenon occurs again and the frozen layer grows in thickness. With every winter, the thickness of a frozen layer will increase until the freezing reaches the depth where the penetrating subzero-temperatures are in equilibrium with the warmth of deeper soil layers. Thus, the formation of permafrost is a current phenomenon, readily accounted for by existing conditions. Permafrost starts to form in conditions of ever-decreasing heat input when the annual average magnitude of the radiation balance decreases to less than 30-35 kcal/cm² per year. It is at such values that the annual average ground temperature falls below 0°C, i.e. annual heat storage takes place at the subzero-temperature at the bottom of a layer of zero annual fluctuations.

General laws governing the permafrost history of the Earth have been subject to study by many researchers. These conclusions are summarized by Nekrasov (1976).

The concept of permafrost has been suggested by Sher (Sher 1971, Arkhangelov and Sher 1973). By identifying deposits of the Lower-Kolyma lowland (Olyorskaya suite) containing pseudo-morphs of wedge ice, and on the basis of faunal remains he attributes their formation to early Pleistocene. Sher believes that "... we feel that all these data taken together remove all doubts that in the early Pleistocene, in the formative stage of Olyorskaya suite deposits there existed on the Kolyma lowland conditions of severe continental climate that initiated the development of permafrost ..." (p. 63). Sher attributes the formation of the permafrost to 2 million years ago.

Another view is proposed by Velichko. By critically reviewing all earlier concepts (including that proposed by Sher), Velichko believes that "the permafrost zone in Eastern Siberia ... formed not at the end of the Pliocene but at least at the beginning of the second Pleistocene glacial stage" (Velichko 1973, p. 108), or about 700 thousand years ago. As far as Europe is concerned, he claims that "the first broad general-climatic wave of development of permafrost" (p. 47) is traceable only back to the beginning of the Würm, i.e. only 60-70 thousand years ago.

Explorers of the "Mamontova Gora" area attribute to approximately the same time, about 600 thousand years ago, the initiation of the permafrost in Central Yakutiya (Agadzhanyan et al. 1973, p. 174).

"Once formed in lower Pleistocene, subterranean glaciation had existed throughout the entire Pleistocene, increasing simultaneously with increasing continentality of climate and vice versa."

Ravsky (1972) attributes the initiation of permafrost in Eastern Siberia to the Samara glaciation, which, according to Ravsky, started about 200 thousand years ago. Later, throughout the Messov inter-glacial period "the permafrost ... degraded completely south of the polar circle but remained, similar to present-day extent, north of it" (Ravsky 1972, p. 293). Then, during the Tazovsky stage permafrost aggraded and in the Kazantsevo interglacial degradation occurred over considerable territory of Inner Asia (p. 295). The present times (Holocene), according to Ravsky remains a period of permafrost degradation (p. 304).

The difference in dating of the start of deep freezing of the lithosphere in Eastern Siberia is nearly two million years. But if we take into account the latitude and altitude variation these differences are largely apparent. It should be emphasized that the data reported by Sher refer to the Kolyma lowland (70° N), those by Velichko and explorers of Mamontova Mountain to Central Yakutiya (63°N) while Ravsky refers to southern Siberia (52-54° N). The difference in latitude reaches nearly 20°.

It is known (Nekrasov 1976) that as one moves from north to south ground temperature increases by 0.7°C by every degree of latitude. Therefore, if we assume that this relationship still remained during the Pleistocene one can calculate that while on the Kolyma lowland the rock temperature had already dropped to -2, -3°C, it was still above +10°C in southern regions of Siberia in areas close to sea level. Exceptions were the high mountains of the southern part of Siberia. Owing to altitude variation the rock temperature decreased with height and therefore at altitudes in excess of 2000 m the permafrost formed simultaneously with that of the Kolyma lowland while at lower altitudes its development was close to that in Central Yakutiya. In depressions of Baikal-type its evolution was comparable to the history outlined by Ravsky.

During the colder epochs, the permafrost spread southward beyond the boundaries of its present position. For example, during the Samara glaciation of Siberia "perennial freezing covered the entire area of the Mongolian People's Republic, including its souther regions, the Gobi" (Gravis et al. 1973, p. 45).

On summarizing all data on the boundaries of maximum development of permafrost in Asia, Europe, and North America (Novoselskaya 1961, Popov and Kostyaev 1962, Flint 1963, Zeiner 1963, Brown and Péwé 1973, Huges 1973, Lorenzo 1969, Williams 1969) we have generated a map, showing the limits of maximum development of early permafrost within the northern hemisphere, by combining it with the one produced earlier (Nekrasov 1971) for present-day permafrost (Figure 1).

We have carried out paleogeographical reconstructions for the entire Quaternary era and more detailed reconstructions for the Upper Pleistocene and Holocene, including rock temperature fluctuations at the surface. The analysis of stratigraphical data on precipitation storage within mountain depressions and an evaluation of neotectonic up-liftings and the modelling of deep freezing of the lithosphere (Nekrasov 1976) permit the following conclusions:

(1) The permafrost of the northern hemisphere started to form about 2 million years ago on lowlands along the Arctic Ocean and in watershed areas of the highest altitudes of the temperate zone. The permafrost of the southern regions originated only 200 thousand years ago, with its repeated complete or partial degradation during warming and new formation during colder periods. The latter was simultaneous with the formation of powerful cover and mountain-valley glaciers.

(2) During colder periods there was an expansion of permafrost so that its maximum advance in Europe reached 45° N, in North America, 35° N, and in Asia, 25° N.

(3) The evolution of permafrost of the northern hemisphere can be reconstructed with maximum reliability only since the Karginskoye interglacial age, i.e. for about the last 50-40 thousand years. During that period one can identify four stages, differing significantly from each other with respect to heat exchange between the lithosphere and the atmosphere:

(a) The first stage (Karginskoye interglacial) was from 50 to 25 thousand years ago. The Zyryanskoe glaciation, which had lasted for over 50 thousand years, terminated about 50 thousand years ago. Of course, by the beginning of the Karginskoe interglacial, permafrost was widespread within Eurasia and North America and probably, its extent exceeded that of today. With the advent of the Karginskoe interglacial, together with a substantial warming, degradation of permafrost started. Permafrost degraded completely over most of Southern Siberia, and remained only on highlands. North of 60° N degradation in Western Siberia contributed to partial thawing of permafrost. In Eastern Siberia, this led to an increase of temperature and a substantial decrease of thickness. The coldness of the permafrost zone in near-polar regions is suggested by mammoth carcasses, found with well-preserved soft tissues, whose age, according to C^{14}, reaches 40 thousand years.

(b) The second stage (Sartanskoe glaciation) was from 25 to 12 thousand years ago. About 25 thousand years ago, a cold period recurred, which, according to Velichko (1973b), was the coldest one in the Quaternary era. Glaciers in the mountains were reactivated. In Siberia that glaciation was presented at high-altitudes while the maximum length of glaciers of northern slopes reached 120 km, with ice up to 500-700 m thick. Uplands and large depressions in between mountains remained free from glaciers.

In the course of the fall of temperature, intense freezing occurred with an increase of thickness of permafrost in the mountains, particularly in watershed areas at high altitudes where its maximum thickness reached 1400-1500 m about 16-18 thousand years ago. Maximum thickness exceeded 600-700 m on uplands and over 1000 m in valleys in Arctic regions.

As the thickness and size of glaciers increased, there was a decrease in the rate of freezing of rocks within the bottoms of glacier trough-valleys. Where the thickness of cover glaciers was at a maximum, partial degradation of permafrost started.

In the period of maximum development of glaciers the high-altitude regions were covered by 70-80% with ice. On tablelands, periglacial steppes, and tundras, sparse growth of birch trees was dominant.

(c) The third stage, from 12 to 5 thousand years ago, was associated with a warming of climate and a retreat of glaciers. Thawing gave rise to discharge carrying huge amounts of suspended materials from high-altitude regions into depressions and valleys of uplands. These processes occurred most intensively along northern slopes of ridges, where glaciers were large. For example, in Trans-Baikalia sediment washed from the foothills of the Upper-Angara, Delyun-Uransky, North-Muisky and Kodarsky Ranges and the Longdor Mountain massif was carried away through the Patomsky and North-Baikal uplands. Outwash material from glaciers on the southern slopes of ridges generally did not move as far from the glacier edge because water discharges were less intensive.

Again, the most essential changes in the extent of the permafrost zone occurred in Europe and the mountains of Southern Siberia. Within the North-Baikal and Patomsky uplands and on the Selenga tableland a complete degradation of the permafrost occurred within river valleys and partial degradation in watershed areas. In Western Siberia deep thawing of permafrost occurred (Baulin et al. 1981).

Within depressions fluvio-glacial deposits from that stage infill the valleys and masses of mid-Pleistocene sands dominate their bottoms. Permafrost within the valleys also degraded partially. In most of the north-eastern part of Siberia, in association with a general warming of climate there was a gradual increase in temperature of the permafrost and a decrease of its thickness.

We associate the termination of this period with the warming maximum during a climatic optimum when glaciers had ablated to nearly their present-day dimensions.

(d) The fourth stage, of about 5 thousand years duration, is characterized by a gradual fall of temperature and establishment of the present-day climate and by an expansion of permafrost.

In high-altitude regions, the taliks within river valleys which were formed in the third period and which attained their maximum dimensions in the period of climatic optimum, start to contract. The higher parts freeze but their area still remains insignificant. There is the commencement of deep freezing in the bottom of glacier valleys (troughs) and small-size ice-collecting hollows. Intensive freezing of deposits also occurs within river valleys that drain depressions between the mountains, with the formation of ice wedges.

The permafrost zone within the tablelands and uplands of the north-eastern part of Siberia remains unaltered. In Southern Siberia the permafrost penetrates again farther south, into regions from which it had disappeared during the climatic optimum. An upper layer of permafrost was forming in Western Siberia at that time.

What are the main features of the present thermal equilibrium? Paleogeographical analysis shows that the present epoch refers to the fall of temperature that started about 5 thousand years ago after the climatic optimum. On the other hand, we are witnessing a warming epoch that started following the "small glacial period" (the period of fluctuations of 1850 AD. During 500 years there has been an irreversible increase of annual average temperatures. This is complicated by short-term fluctuations. The glaciers on a world-wide scale are retreating and such a retreat of glaciers and the warming of climate has continued since at least 2300-2400 with a maximum meso-oscillation at a period of 1850 AD. Of course, owing to the general fall of temperature (of 40,000 years recurrence), the temperatures during the maximum period will be considerably lower than those at a maximum of the preceding stage that occurred in the III-VI centuries AD.

It is these basic trends in the thermal development of the Earth that should be used in evaluating the present-day extent of permafrost. But we should not omit the large inertia of the permafrost in comparison with the variations of thermal balance on the surface. Thus, with climatic fluctuations of a period of 40,000 years a time delay of 6-8 thousand years is present for permafrost 1000 m thick.

At the same time, interference of man in nature can abruptly change the picture presented above (Nekrasov and Balobaev 1982). Very hazardous is the increase of aerosols and carbon dioxide in the atmosphere. According to Budyko et al (1978), who made a prediction of climate changes for the coming 50 years, fuel burning will increase the atmospheric concentration of carbon dioxide by 30-40% by the year 2000 and 200% by 2025.

Similar to the past, this leads to an increase in the annual average air temperature by 0.5°C by 1990 and by 10-15°C by 2025. This means that the Arctic Ocean will be free from ice and its impact on climate will change. The predicted climatic changes are gigantic and at times provoke scepticism.

However, estimations made by American scientists (Energy ... 1977) are close to inferences of Soviet investigators. Objection may be caused only by the under-estimation of the self-regulating role of nature which must activate the mechanism of carbon dioxide absorption by plants and the ocean, which will greatly hinder the thawing of ice of the Arctic Ocean.

If we assume, however, that the predicted changes of climate are possible, then permafrost will be faced with a disastrous situation. In the greater part of its present-day range, an irreversible process of permafrost degradation will start. Thus, at temperate latitudes [the zone of the Baikal-Amur Railroad (BAM)] the thickness of the upper unfrozen layer will be 10-15 m by 2025. Farther to the north and at higher altitudes the thawing will be smaller. Where conditions still remain for the persistence of permafrost, the depth of seasonal thawing of soils will increase by 0.5-1.0 m. In turn, this will lead to an intensification of solifluction and thermokarst. Conditions will exist where structures supported by piles, constructed following the principle of a frozen foundation, will fail.

Deformations will occur in subgrades, and beneath roads and railroads wherever soils with high ice contents and with ice masses are present.

Since the process of phase transitions of ice

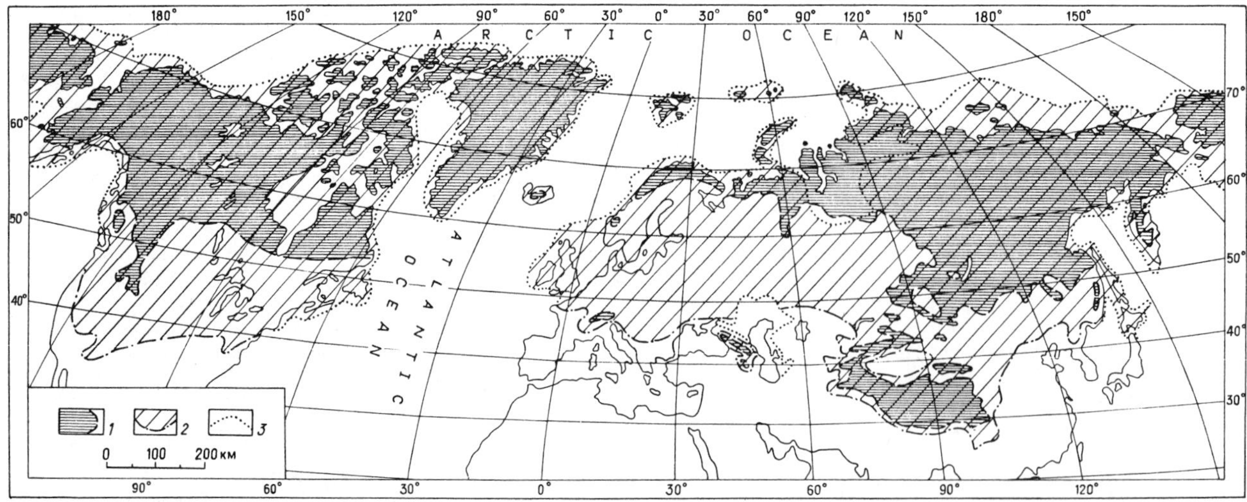

FIGURE 1 Extent of permafrost distribution in the northern hemisphere during different periods of the Quaternary: 1 - region of present-day permafrost; 2 - region of maximum development of permafrost; 3 - boundaries of sea basins during maximum development of permafrost.

into water proceeds slowly, mining structures, except for open cast operations, will not deteriorate in the near future. The principles of extracting deep ore deposits will not be altered within the next one hundred years.

REFERENCES

Agadzhanyan, A. K., Boyarskaya, T. D., Glushankova, N. I., Sudakova, N. G., Faustov, S. S., Khorev, V. S., and Shlyukhov, A. I., 1973, "Mamontova Gora" cross-section of the latest deposits, Moscow: Izd. MGU, 200 p.

Arkhangelov, A. A., and Sher, A. V., 1973, Toward the age of permafrost in the extreme north-eastern part of the USSR, II International Conference on Permafrost (Reports and Communications), v. 3, Yakutsk, p. 5-11.

Baulin, V. V., Chekhovsky, A. L., and Sukhodolsky, S. E., 1981, The main stages of development of permafrost of north-eastern European part of the USSR and Western Siberia, in The evolution of permafrost rocks of Eurasia, Moscow: "Nauka," p. 41-60.

Brown, R. J. E., and Péwé, T., 1973, Distribution of permafrost in North America and its relationship to the environment, a review, 1963-1974. North American Contribution, 2nd International Conference on Permafrost, Washington, p. 71-100.

Energy and Climate, 1977, in Studies in Geophysics, Washington, D.C.: National Academy of Sciences.

Flint, R. F., 1963, Glaciers and paleogeography of the Pleistocene, Moscow: Izd. inostr. lit., 576 p.

Gravis, G. F., Zabolotnik, S. I., Lisun, A. M., and Sukhodrovsky, V. L., 1973, A geocryological characteristic of the Mongolian People's Republic and some features of the development of permafrost in the past, 2nd International Conference on Permafrost (Reports and Communications), v. 2, Yakutsk, p. 37-45.

Huges, T., 1973, Glacial permafrost and Pleistocene ice ages. North American Contribution, 2nd International Conference on Permafrost, p. 213-223.

Lorenzo, Josiz., 1969, Minor periglacial phenomena among the high volcanoes of Mexico. The periglacial environment, Montreal: McGill Queen's University Press, p. 161-175.

Nekrasov, I. A., 1971, Permafrost zone of the northern hemisphere of the Earth, in Geocryological investigations, Yakutsk, p. 146-152.

Nekrasov, I. A., 1976, Permafrost zone of the north-eastern and southern parts of Siberia and some features of its development, Yakutsk: Yakutskoe knizhnoe izd-vo, 248 p.

Nekrasov, I. A., and Balobaev, V. T., 1982, Evolution of rock permafrost in the BAM zone, "Geografiya i prirodnye resursy,", no. 2, Novosibirsk: "Nauka," p. 86-91.

Novoselskaya, N. B., 1961, Traces of the existence of rock permafrost within the European part of the USSR outside the region of their present-day occurrence, in Sketches on regional and historical cryology, Moscow: Izd. AN SSSR, p. 78-93.

Popov, A. I., and Kostyaev, A. G., 1962, Maps of periglacial formations of Asia, at present and during mid-Pleistocene, in Questions of geographical cryology and periglacial morphology, Moscow: IZD. MGU, p. 59-63.

Ravsky, E. I., 1972, Sediment of storage and climate of Central Asia in the Quaternary era, Moscow: "Nauka," 336 p.

Sher, A. V., 1971, Mammalia and stratigraphy of Pleistocene of the extreme north-eastern part of the USSR and North America, Moscow: "Nauka," 310 p.

Shostakovich, V. B., 1916, Permafrost, "Priroda" (May-June).

Velichko, A. A., 1973, Natural process in the Pleistocene, Moscow: "Nauka," 256 p.

Voeikov, A. I., 1952, Selection of works, v. 3, Moscow: Izd. AN SSSR, 502 p.

Williams, R. B. G., 1969, Permafrost and temperature conditions in England during the last glacial period. The periglacial environment, Montreal: McGill-Queen's University Press, p. 399-410.

Zeiner, F., 1963, Pleistocene, Moscow: Izd. inostr. lit., 504 p.

A FROST INDEX NUMBER FOR SPATIAL PREDICTION OF GROUND-FROST ZONES

Fritz Nelson and Sam I. Outcalt

Department of Geological Sciences, University of Michigan
Ann Arbor, Michigan 48109 USA

Recent investigations of the relationship between the minimum and maximum mean monthly air temperatures along the trans-Alaska pipeline route indicate that good estimates of freezing and thawing degree days at a site can be obtained from these two extreme monthly mean values. The dimensionless ratio between the Stefan solution for the annual frost and thaw depths at a site can be obtained from the degree-day estimates, winter snow-cover data, and estimates of site thermal properties. The ratio of Stefan estimates of frost to thaw depths is termed the "frost number." Its value is much larger than unity in regions of continuous permafrost and descends to unity at the equatorward limit of the discontinuous permafrost zone. A preliminary mapping of the frost number in northwestern North America shows a very high spatial correspondence with observed permafrost distribution. The frost number can also be mapped using climate data in successive years, which yields information useful for correlation with such geomorphic processes as palsa growth and thermokarst development.

INTRODUCTION

Although numerous attempts have been made to regionalize permafrost through compilation of field observations and by means of computations involving climatic data, little effort has been made to incorporate calculations involving climatic data, substrate properties, and snow cover into an easily computed "permafrost index." Use of above-surface climate data alone can result in only a crude approximation of permafrost distribution, since the very large influences of snow, microclimate, and the thermal properties of the substrate are ignored. This paper presents a method which can incorporate spatial variations in some of these variables. Using this technique, an initial attempt is made to regionalize permafrost patterns in northwestern North America by combining degree-day estimates, winter-precipitation data, and, as a first approximation, simplified assumptions about subsurface thermal properties, topography, and vegetation cover.

PREVIOUS STUDIES

Many investigators have attempted to estimate and/or regionalize soil-frost conditions and permafrost distribution; these studies span a broad range of methods and scales but share the common objective of determining the nature and spatial or temporal extent of freezing conditions in the ground. Methods can be divided into two broad groupings. The first utilizes direct field-based observations on permafrost obtained from borehole, remote sensing, or geophysical methods, as reviewed by Ferrians and Hobson (1973). Trends can be extrapolated through time, as when fluctuations in a temperature profile are used to infer climatic change (Lachenbruch and Marshall 1969), and through space by delimiting ground-frost zones (e.g., Ferrians 1965), or mapping permafrost thickness (Osterkamp and Payne 1981).

Of more relevance to this paper is a second group, which employs environmental information to make what are essentially "educated guesses" about spatial patterns of frost and/or permafrost. The major problem of such regionalization is treating the complex interactions between various factors such as snow and vegetation cover, terrain climate, substrate heterogeneity, soil moisture, and variations of geothermal heat flow, all of which can be quite variable within small areas. These problems have been addressed in a number of ways, ranging from ignoring most of the complicating factors to statistical analysis of a large number of variables.

Geographical scale is a critical consideration because the influence of local ground-cover, microclimate, and substrate-property variations is less discernible at smaller scales. For example, Brown (1967) used the -1.1° and -8.3°C mean annual air isotherms to delimit the equatorward boundaries of the discontinuous and continuous permafrost zones in areas of Canada where field measurements on permafrost were lacking. The resulting document is very useful for showing spatial variations of permafrost at the continental scale, but at large scales significant deviations from these boundaries appear, as shown by Zoltai (1971).

More recently, Harris (1981a, b) attempted to correlate freezing and thawing indices derived from screen-level climate data with permafrost occurrence and selected periglacial landform elements. His method consists of plotting freezing against thawing degree days and superimposing classes of observed ground-frost conditions or landforms for the same approximate locations on the resulting graphs. Although this procedure has revealed some interesting relationships, it is limited in its usefulness by its neglect of snow-cover effects and because the inherently spatial nature of the permafrost-distribution problem was largely ignored.

Many other investigators have attempted to

predict various frost and permafrost characteristics and their spatial variations by means of calculations involving environmental data, but the topics of many of these papers are only marginally related to those addressed in this investigation and will not be considered here.

DERIVATION OF THE FROST NUMBER

In this paper we develop an easily computed index of frost/permafrost conditions which reflects geographically variable climatic conditions, including snow cover, but which for present purposes is independent of variations in subsurface properties or conditions. The technique we propose could incorporate planimetric variations in the geothermal gradient, substrate thermal properties, and soil-moisture content although these variables are treated as constant in this study. Future studies could consider spatial variations in subsurface variables if reliable data become widely available for locations at which climate information is also collected.

In conjunction with a technique of spatial interpolation known as "lattice tuning," which takes maximum advantage of the spatial autocorrelation inherent in geographically distributed climatic data, the frost number can be used to predict ground-frost conditions over large areas in which common climatic data are available. Information yielded by the frost number pertains to permafrost conditions in equilibrium with the climatic data used as input; relict permafrost cannot be predicted by the method. Application of the frost number is restricted to sites experiencing thermal regimes in which the annual temperature amplitude exceeds that of the diurnal.

The well-known Stefan formula for predicting frost and thaw depths in soils has received wide use by engineers, agronomists, and geomorphologists (see, for example, Jumikis 1977 chapters 14-15). Although in unmodified form it can overestimate the depth of frost penetration because it ignores sensible heat, the Stefan formula provides an easily computed approximation of frost and thaw depths that is especially useful when little is known about site properties. The ratio of Stefan's frost and thaw estmates, modified for the effects of winter snow cover, can be used as an index of permafrost occurrence (cf. Lunardini 1981 p. 551). When used in conjunction with a simple method for abstracting climatic data and an efficient method of spatial interpolation, this frost index has potential for mapping permafrost distribution at small scales.

Analysis of the monthly mean air-temperature records of 19 stations in northern Alaska showed that their thermal regimes can be abstracted by three parameters: 1) mean annual temperature (T_m), 2) annual temperature amplitude (A), and 3) the phase angle (ϕ) in radians. Mean annual temperature and thermal amplitude can be approximated through simple manipulation of the mean temperatures of the extreme months (Haugen et al. 1983):

$$T_m = (T_h + T_c)/2 \quad (1)$$

$$A = (T_h - T_c)/2, \quad (2)$$

where T_h and T_c are the mean temperatures of the "hottest" and coldest months, respectively. These equations preserved 97.3 $\pm$ 1.6% of the variance in the 19 series analyzed. The air temperature T(d) for any day of the year can be estimated as a function of Julian date (d) by

$$T(d) = T_m + A \cos\{(2\pi d/365) - \phi\}. \quad (3)$$

In the absence of phase change, the solution for penetration of a periodic surface thermal disturbance into a homogeneous medium is given by

$$T(Z, d) = T_m + A \exp(-Z/Z_*) \cos\{(2\pi d/365) - \phi + (-Z/Z_*)\}, \quad (4)$$

where Z is depth and Z_* is the damping depth in the substrate (Ingersoll et al. 1954). The latter parameter is a function of the thermal diffusivity (α) of the substrate material and the wavelength (P) of the cyclic disturbance, expressed by

$$Z_* = (\alpha P/\pi)^{\frac{1}{2}}. \quad (5)$$

Decrease of amplitude with depth is given by

$$A(Z) = A \exp(-Z/Z_*). \quad (6)$$

The division between summer and winter occurs at the point on the cosine curve where the temperature crosses 0^oC. This "frost angle" (β) is defined as

$$\beta = \cos^{-1}\{(T_f - T_m)/A\}, \quad (7)$$

where T_f is the freezing point. When the temperature curve is symmetric, the mean summer (T_s) and winter (T_w) temperatures can be estimated by integration over only half the annual cycle (π radians):

$$T_s = T_m + A \frac{1}{\beta} \int_0^\beta \cos\theta \, d\theta = T_m + \frac{A}{\beta} \sin\beta \quad A \quad (8a)$$

$$T_w = T_m + A \frac{1}{\pi - \beta} \int_\beta^\pi \cos\theta \, d\theta = T_m - \frac{A}{\pi-\beta} \sin\beta. \quad (8b)$$

However, the winter temperature amplitude is compressed at the ground surface due to snow cover of "average" thickness Z_s. The winter surface amplitude A_w can be found by

$$A_w = A \exp(-Z_s/Z_{*s}). \quad (9)$$

The mean winter surface temperature must also be modified to reflect snow-cover effects; the mean summer surface temperature is for present purposes taken as the mean summer air temperature. This approximation could possibly be refined in future applications by Lunardini's (1981 p. 559-575) "surface n factor" using a surface-cover category representative of the vicinity of a station. In the equations that follow, the subscript (+) indicates effects attributable to snow cover. The mean winter surface temperature (T_{w+}) is given by

$$T_{w+} = T_m - \frac{A_w}{\pi-\beta} \sin \beta, \tag{10}$$

and the lengths of the summer and winter seasons in days can be calculated as functions of the frost cycle by

$$L_s = 365 \ (\beta/\pi) \tag{11a}$$

$$L_w = 365 \ \{(\pi - \beta)/\pi\}. \tag{11b}$$

A set of degree-day values adjusted for snow-cover effects can now be calculated as

$$DDT = T_s \ L_s \tag{12a}$$

$$DDF = |T_w \ L_w| \tag{12b}$$

$$DDF_+ = |T_{w+} \ L_w| \ , \tag{12c}$$

where the degree days of thaw (DDT) and frost (DDF) are the time-temperature integrals above and below the freezing point and T_s, L_s, T_w, and L_w are the mean summer temperature, length of summer (days), mean winter temperature, and length of winter, respectively.

Finally, by means of a formulation similar to that of Lachenbruch (1959), the mean annual surface temperature can be adjusted for snow cover:

$$T_{m+} = T_s \ (L_s/365) + T_{w+} \ (L_w/365) \tag{13a}$$

$$T_{m+} = T_m + (\sin \beta/\pi) \ A \ \{1 - \exp \ (-Z_s/Z_{*s})\}. \tag{13b}$$

Phase change, which cannot be easily incorporated into equations for periodic thermal disturbances in semi-infinite solids, is a very important consideration in frost/permafrost problems. The depth to which frost extends can be approximated by considering the thermal conductivity (λ) of the substrate, its volumetric water/ice content (x_w), the latent heat of fusion (L), and the degree days of thaw and frost. Frost and thaw depths can be obtained from

$$Z_t = \{(2 \ \lambda \ S \ DDT)/(x_w \ L)\}^{1/2} \tag{14a}$$

$$Z_f = \{(2 \ \lambda \ S \ DDF)/(x_w \ L)\}^{1/2}, \tag{14b}$$

where Z_t and Z_f are the thaw and frost depths, respectively, and S is a scale factor representing the appropriate units of time for the metric system employed.

If the seasonal differences in thermal properties $\{\lambda/(x_w \ L)\}$ are small, the ratio of frost to thaw depths at a site may be calculated with and without snow-cover effects, yielding a dimensionless ratio. These ratios are termed the air (F) and surface (F_+) frost numbers:

$$F = Z_f/Z_t \tag{15a}$$

$$F_+ = Z_{f+}/Z_t. \tag{15b}$$

Note that the surface frost number will always be equal to or less than the value calculated for air temperature because snow cover increases the mean annual surface temperature. If for mapping purposes substrate properties are assumed homogeneous, equations (14) and (15) reduce to

$$F = (DDF/DDT)^{1/2} \tag{16a}$$

$$F_+ = (DDF_+/DDT)^{1/2}. \tag{16b}$$

INTERPRETATION AND MAPPING

Interpretation of the frost number is straight forward. At locations where $F_+ < 1$, only seasonal frost is possible. Where F_+ is approximately unity, equilibrium permafrost should be near its climatic limit, and this value can be used to map the seasonal-frost/discontinuous-permafrost "boundary." Although delimitation of the continuous permafrost zone is always somewhat arbitrary, comparisons with empirical mappings indicate that a reasonable value for its equatorward limit may be taken as 1.5, which corresponds to a ratio of DDF_+ to DDT equal to 2.25.

A preliminary mapping (Figure 1) of the surface frost number F_+ was performed for northwestern North America using climatic data from Alaska (Searby 1978, Rieger et al. 1979), and the Yukon and Northwest territories (Hare and Thomas 1979, Canada 1967). The station records are of variable length but all represent at least part of the period 1941-1970. All but two of the records containing less than 10 years of observations are from DEW line stations and have been edited to eliminate extreme values (Canada 1967).

Four parameters were used as input: mean annual air temperature, the annual temperature amplitude, "average" snow depth, and a binary variable which indicates whether the snow cover is "packed" or "fresh" to facilitate calculation of the snow's thermal influence. The substrate was for preliminary purposes assumed to consist of a homogeneous, moderately wet silt.

The most problematic variable with respect to data availability was "average" snow-cover depth. Since such summary statistics are not readily available for a large number of stations, a surrogate method of obtaining such information was necessary. Our solution was to calculate an average snow depth Z_s by the formula

$$Z_s = 0.67 \ \{\Sigma \ (P_i/D) \ (K - (i-1))\}/K \tag{17}$$

where P is water-equivalent precipitation in those months i (i = 1,2...k) in which the mean temperature is $\leq 0^0$C and D is average snow density at the site (Bilello 1969 Figure 6). This procedure yields snow-depth estimates which are probably too low in the northern part of the study area and too high in the southern part. Nevertheless, the deviations are not extreme in cases for which average snow-depth data were available, and even this crude approximation is an improvement over ignoring snow-cover effects, as has been done in some other studies.

Locations of the data sites are given in Figure 1. The sites are largely restricted to coastal locations and a cluster of points in interior Alaska and Yukon, making some form of spatial interpolation necessary. A 34 x 24 grid (65-km mesh) was set up over the mapped area, and a technique known as "lattice tuning" (Tobler 1979) was employed to create a regular point lattice from the original irregularly spaced data points. Lattice

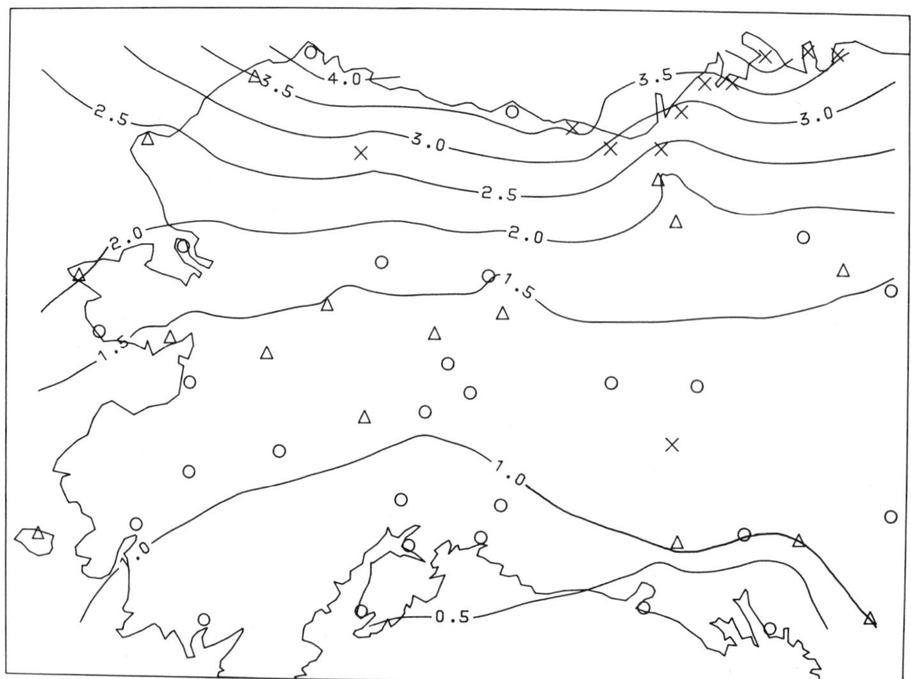

FIGURE 1 Map of the frost number in northwestern North America. Note strong increase from south to north. Locations of climate stations are shown by symbols denoting record lengths as follows: O, 30-year record; Δ, 10-29 years; ×, <10 years.

tuning is an iterative procedure in which estimated values at nodes lacking observations are adjusted to the original observations until successive operations yield no effective changes. The result is a smooth surface that interpolates correctly to the original data points from surrounding lattice nodes; in this respect the procedure is superior to more common gridding techniques which do not strictly preserve the integrity of the original values. The resulting lattice was then contoured and superimposed on a map of the study area (Figure 1).

A very strong, smooth increase in values of the frost number is apparent from south to north, and is consistent with trends of mean annual air temperature in the study area. The southward contour flexure in western Alaska is a response to reduction of thawing degree days near the Bering Sea coast. The southern boundaries of the discontinuous and continuous permafrost zones, delimited by the 1.0 and 1.5 contours, are well placed when compared with empirical delimitations (cf. Brown 1967, Ferrians 1965, Péwé 1975 Figure 23). The slight southward displacement of the continuous/discontinuous "boundary" relative to those of other authors' maps is a reflection of the simplifying assumptions made in this study about substrate properties, surface cover, and the lack of climate data in this part of the study area. The overall smoothness of the map and its lack of local peaks and pits show the importance of regional climate for permafrost distribution. If subsurface information was incorporated, the smoothness might be reduced, but local sources of deviation could be treated analytically by trend-surface or map-comparison methods.

DISCUSSION

In addition to the difficulty of calculating good estimates of "average" snow cover, our method of obtaining degree-day estimates yields values for the thawing season which are somewhat low in northerly coastal locations. Values of the frost number in the northern part of the mapped area are therefore slightly higher than they would be if empirical values of DDT and DDF were used, although comparison with values obtained using data contained in Haugen (1982) indicates that the deviation is not extreme for the stations used here. For obvious reasons, the frost number can become very large when applied to locations with few thawing degree days. In such cases the values of DDT used must be as accurate as possible or gross errors in F_+ can result. An extreme example is Alert, N. W. T.; using the degree-day estimation techniques outlined in this paper, a value of nearly 27.0 is obtained for the surface frost number, whereas the value is about 5.5 if empirical degree-day values are used (data from Taylor et al. 1982). This is nearly as extreme an example as could possibly be found, however, and such large discrepancies are not possible at most lowland locations. However, this example does point out the dangers of indiscriminate application of the estimation techniques employed in this paper. The frost number can be accurately computed only when reliable estimates of freezing and thawing indices are employed.

In addition to the use to which it was put in this paper, the frost number has some potential for use with annual climatic data at individual sites. Annual values of the number can fluctuate in response to variations in a wide range of natural and anthropogenic environmental influences, including snow depth, temperature, and soil moisture. Studies dealing with thermokarst development and palsa formation would be particularly appropriate for such applications. However, the frost number probably achieves its greatest utility in a spatial context; its simplicity and minimal input-data requirements make it an effective tool for mapping permafrost over large areas in which data lack detail or uniform spatial coverage. Mapping its variation in other parts of the earth may point to interesting environmental relationships recognizable only in a spatial context. A possibility not explored in this paper is use of statistical spatial-analytic techniques for comparing different years and/or the various environmental parameters involved in computation of the frost number.

ACKNOWLEDGMENTS

We wish to thank the U. S. National Science Foundation and the U. S. Army's Cold Regions Research and Engineering Laboratory, which provided financial support for this work. We also appreciate the comments of an anonymous reviewer, who made us aware of some important literature overlooked in an earlier draft.

REFERENCES

Bilello, M. A., 1969, Relationships between climate and regional variations in snow-cover density in North America: U. S. Army Cold Regions Research and Engineering Laboratory, Research Report 267.

Brown, R. J. E., 1967, Permafrost in Canada: Geological Survey of Canada, Map 1246A.

Canada, Meteorological Branch, 1967, Temperature and precipitation tables for the north--Y. T. and N. W. T.: Toronto, Department of Transport.

Ferrians, O. J., Jr., 1965, Permafrost map of Alaska: U.S. Geological Survey Miscellaneous Geological Investigations Map I-445.

Ferrians, O. J., Jr., and Hobson, G. D., 1973, Mapping and predicting permafrost in North America: A review 1963-1973, in Permafrost--The North American contribution to the Second International Conference, Yakutsk: Washington, D. C., National Academy of Sciences, p. 479-498.

Hare, F. K., and Thomas, M. K., 1979, Climate Canada: Toronto, Wiley.

Harris, S. A., 1981a, Climatic relationships of permafrost zones in areas of low winter snow-cover: Arctic, v. 34, p. 64-70.

Harris, S. A., 1981b, Distribution of zonal permafrost landforms with freezing and thawing indices: Erdkunde, v. 35, p. 81-90.

Haugen, R. K., 1982, Climate of remote areas in north-central Alaska: U. S. Army Cold Regions Research and Engineering Laboratory Report 82-35.

Haugen, R. K., Outcalt, S. I., and Harle, J. C., 1983, Regional relationships between estimated mean annual air temperature and upper permafrost temperatures in north central Alaska: this volume.

Ingersoll, L. R., Zobel, O. J., and Ingersoll, A. C., 1954, Heat conduction with engineering, geological and other applications: University of Wisconsin Press.

Jumikis, A. R., 1977, Thermal geotechnics: New Brunswick, Rutgers University Press.

Lachenbruch, A. H., 1959, Periodic heat flow in a stratified medium with application to permafrost problems: U. S. Geological Survey Bulletin 1083-A.

Lachenbruch, A. H., and Marshall, B. V., 1969, Heat flow in the Arctic: Arctic, v. 22, p. 300-311.

Lunardini, V. J., 1981, Heat transfer in cold climates: New York, van Nostrand Reinhold Company.

Osterkamp, T. E., and Payne, M. W., 1981, Estimates of permafrost thickness from well logs in northern Alaska: Cold Regions Science and Technology, v. 5, p. 13-27.

Péwé, T. L., 1975, Quaternary geology of Alaska: U. S. Geological Survey Professional Paper 835.

Rieger, S., Schoephorster, D. B., and Furbush, C. E., 1979, Exploratory soil survey of Alaska: Washington, D. C., U. S. Department of Agriculture, Soil Conservation Service.

Searby, H. W., 1978, Climates of the states, v. 1: Detroit, Gale Research Company.

Taylor, A., Brown, R. J. E., and Judge, A. S., 1982, Permafrost and the shallow thermal regime at Alert, N. W. T., in Proceedings of the Fourth Canadian Permafrost Conference: Ottawa, National Research Council of Canada.

Tobler, W. R., 1979, Lattice tuning: Geographical Analysis, v. 11, p. 36-44.

Zoltai, S. C., 1971, Southern limit of permafrost features in peat landforms, Manitoba and Saskatchewan: Geological Association of Canada Special Paper, v. 9, p. 305-310.

THAW STRAIN DATA AND THAW SETTLEMENT PREDICTIONS FOR ALASKAN SOILS

R. A. Nelson[1,4]; U. Luscher[1]; J. W. Rooney[2]; and A. A. Stramler[3]

[1]Woodward-Clyde Consultants, Walnut Creek, California 94596 USA
[2]R&M Consultants, Anchorage, Alaska 99503 USA
[3]ARCO Pipeline Company, Los Angeles, California 90071 USA
[4]Formerly with R&M Consultants, Anchorage, Alaska 99503 USA

To aid design of the trans-Alaska pipeline, a method was developed to predict thaw strain from correlations with simple frozen soil index properties. Laboratory thaw consolidation tests were made on about 1,000 representative soil samples recovered during subsurface exploration programs along the pipeline alignment. The test results were then grouped by sample soil type and landform profile. Various relationships among the soil index properties (frozen dry density and moisture content, specific gravity, gradation) and thaw strain measurements of samples within each category were examined to empirically determine a "best fit" equation by multiple variable regression analysis of the data. A computerized analysis procedure determined the set of regression coefficients to yield a unique equation for each soil category to predict thaw strain of a frozen soil sample with only borehole index property values. These estimated thaw strain values could be multiplied by stratum thickness and summed over an anticipated thaw depth to predict thaw settlement.

INTRODUCTION

Ground settlement resulting from thaw of permanently frozen ground is a problem often associated with civil construction in areas of discontinuous and continuous permafrost. One such construction project exemplifying a variety of thaw-inducing features is the trans-Alaska 1.22-m-diameter oil pipeline, constructed from 1974 to 1976, and in operation since its completion. Heat transfer from hot oil flowing through a buried pipe causes a generally cylindrical volume of permafrost surrounding the pipe to thaw. Disturbance of the natural vegetative regime by stripping of the organic mat from the pipeline right-of-way, construction equipment activity on the mat, and placement of embankment atop the organic mat alters the thermal and heat transfer properties of the ground surface, thereby effecting gradual thermal degradation (thaw) of the permafrost.

This paper presents laboratory thaw consolidation test results that were the basis for predicting thaw settlement along the entire 1,300-km pipeline alignment. These results were used in a multiple variable regression analysis with readily measured frozen soil index properties as the regression variables, ultimately producing a series of empirical equations used to predict thaw strain from those index properties. Once the thaw strain of a particular soil stratum was predicted, it was a simple matter to multiply the stratum thickness by the thaw strain to arrive at the stratum's thaw settlement. A summation of thaw settlements for the various strata encountered in a borehole to the total anticipated depth of thaw yields the total settlement for the borehole.

The analyses and procedures detailed in the following paragraphs were used to make borehole-by-borehole thaw settlement estimates for borings with permafrost from among the approximately 2,000 borings located along the pipeline centerline. As the boreholes were sited at an average spacing of about 0.5 km in permafrost sections of the alignment, settlement estimates for all borings along a short design section within a particular geologic landform (Kreig 1977) were considered together to express the range of anticipated settlements. Thaw settlement estimates were then used in such pipeline design functions as construction mode selection, elevated pipeline support design, and gravel workpad thickness requirements.

LABORATORY TESTING

Samples

Permafrost samples used in the laboratory test programs were collected in the field and handled in transport and preparation with attention to maintaining their frozen state. Borings were made using refrigerated drilling fluid. The samples were cored with various drill bits of approximately 76-mm inside diameter (i.d.). Field geologists classified the samples according to Pihlainen and Johnston (1963) field description of permafrost and the Unified Soil Classification system, and identified the geologic landform in which each boring was located as well. The samples were then transported under refrigeration from the field to testing laboratories in Fairbanks, Alaska, and Oakland, California.

There, sample ends were cut with a diamond saw in a cold room. Index properties of the still-frozen samples--dry density, moisture content, specific gravity, and gradation--were measured for each. Samples were generally 7-8 cm in diameter with a 2-2.5 height-to-diameter ratio, though some fine-grained samples with diameters as small as 4 cm were tested.

Thaw Consolidation Test Procedure

Thaw consolidation testing was done by R&M Consultants (R&M) in their Fairbanks laboratory and by Woodward-Clyde Consultants (WCC) in their Oakland laboratory. Though the test procedures used by these two organizations differed slightly, for the large magnitudes of thaw strain typically measured, the axial strain did not appear to be influenced by test method.

R&M implemented a uniaxial technique in performing thaw consolidation tests, using an 8.9-cm i.d. plastic cylinder to confine the samples. The frozen samples were first sheathed with a rubber membrane and fit with porous stones and filter paper on their ends. As the samples typically were not perfectly cylindrical nor of a single diameter, the plastic confining cylinder was slightly oversized, and any gap between the sample membrane exterior and the interior of the plastic cylinder was filled with rounded clean (Ottawa) sand. An axial stress equivalent to the overburden load at the sample's depth was maintained on the sample during thaw. Axial strain was recorded until all measurable deformation had ceased; however, where post-test diameter measurements denoted any lateral deformation, those values were then used to correct measured axial strain to a value corresponding to zero lateral strain.

WCC tests employed both uniaxial and triaxial techniques. Uniaxial tests, performed on clays, silts, and sands, were made in a standard 5-cm odometer device; triaxial tests, performed on coarse-grained sands and gravels, were made on 4- to 8-cm-diameter samples with height-to-diameter ratios of 2-2.5. Additional details of these tests are presented in Luscher and Afifi (1973).

Laboratory Test Results

The 1,024 thaw consolidation tests for the pipeline project were grouped into the following soil categories with the indicated Unified Soil Classifications:

(1) Upland Silt - Upland Landforms (ML)
(2) Lowland Silt - Lowland Landforms (ML)
(3) Organic Soil (OL, OH, Peat)
(4) High Plasticity Silt and Clay (CH, MH)
(5) Low Plasticity Clay (CL)
(6) Silty Sand (SM, SM-SC, SC)
(7) Clean Sand (SW, SP, SW-SM, SP-SM, SW-SC, SP-SC)
(8) Silty Gravel (GM, GM-GC, GC)
(9) Clean Gravel (GW, GP, GW-GM, GP-GM, GW-GC, GP-GC)

Test results for these nine soil type categories are plotted as axial strain versus frozen dry density in Figures 1-9 in the order of the above listing. Axial strain includes any consolidation due to dissipation of excess pore water pressures in the thawed state; typically, consolidation while in the thawed state was considered a small proportion of the total measured thaw strain. Also shown in the figures are thaw strain prediction equations and the resultant curves, data produced by procedures to be discussed further.

MULTIPLE VARIABLE REGRESSION ANALYSES

Large geotechnical projects often make use of relationships between simple soil index properties and complex engineering property test results for certain design aspects, particularly when extremely variable subsurface soil conditions make prohibitive the time and cost of performing many complex tests. The more than 2,000 borings drilled along the trans-Alaska pipeline centerline showed soil conditions at some locations variable in soil composition and in distribution of moisture content and dry density within a single borehole as well. Use of correlations, then, proved a practicable geotechnical method, exemplified by the correlation of thaw consolidation tests with simple soil index properties.

Thaw Strain Prediction Equation

Thaw consolidation test results and associated frozen index properties were used as input to a multiple variable regression analysis. An equation of the form

$$\varepsilon_t = A_1 n^2 + A_2 n + A_3 \frac{n^2 w}{S} + A_4 \frac{n}{S} + A_5 \frac{n}{w} + A_6 \qquad (1)$$

where

ε_t = predicted thaw strain,
n = porosity of frozen sample (1-(frozen dry density / (specific gravity of soil particles x unit weight of water))),
w = moisture content of frozen sample (weight of water / weight of dry soil),
S = degree of saturation of frozen sample (specific gravity x w x (1-n) / n), and
$A_1 \ldots A_6$ = regression coefficients,

was used as the thaw strain regression equation. After an extensive parametric study of laboratory data and a number of empirical computer trials of different equation forms, Equation 1 was established as that algebraic expression yielding the most representative prediction.

A computerized mathematic analysis procedure determined the set of regression coefficients that made a predicted thaw strain, ε_t, most nearly equivalent to measured thaw strain, ε_{tm}. The mathematic procedure was similar to a least squares fit of a linear relationship with the exception that a nonlinear regression equation (Equation 1) was used, thereby allowing nonlinear data fits of several variables. Thaw strain was then predicted by this equation with only such simple index properties as frozen dry density, frozen moisture content, and specific gravity of a particular soil available.

Display of Thaw Strain Prediction Equation

The thaw strain (ε_t) predicted for any given combination of moisture content, dry density, and specific gravity is not obvious from examination of Equation 1 using the appropriate coefficients for each soil category. Therefore, the results of the regression analysis have been superimposed on the data displays in Figures 1-9 as curves based on a representative specific gravity value and saturation values of 80% and 100%, bounding values for the saturation of most samples recovered from the field. A specific gravity of 2.70 was used for all display curves except those for lowland silt and organic soil, which were plotted for specific gravity of 2.65 and 2.30, respectively. The solid lines are curves calculated from the prediction equations shown in each figure. The short-dashed lines are linear extensions of the regression curves in high and low density ranges; these extensions are based on engineering judgment in the ranges where laboratory data are scarce.

Discussion of Prediction Curves Versus Laboratory Data

Each prediction curve resulting from the multiple variable regression analysis is drawn through a rather wide scatter of data points. Examination of a typical plot, such as Figure 1, then indicates that the thaw strain predictions may be in error by as much as 10% from what was measured in the laboratory. However, as not all data points on the figure have the same specific gravity or saturation range as for the curves drawn on the figure, the actual predicted strain may be closer to some of the outlying thaw strain results than that strain shown by the curves. The standard error of estimate (analogous to the standard deviation of the predicted versus measured error) and the multiple correlation of each of the nine thaw strain prediction equations are listed on Table 1. In the higher density ranges, the standard error of estimate decreases from this value, while, at the lower densities, the standard error of estimate is magnified.

TABLE 1 Prediction Error.

Primary Soil Type	Number of Samples	Multiple Correlation	Standard Error of Estimate (% Thaw Strain)
ML up	336	0.951	5.7
ML low	44	0.949	5.5
OL, OH	67	0.771	13.3
CH, MH	41	0.818	2.9
CL	122	0.886	5.1
SM, SC	196	0.890	5.3
SW, SP	90	0.829	2.1
GM, GC	59	0.921	3.8
GW, GP	69	0.913	3.0

The regression analysis results in a prediction equation that merely "averages" the data shown in the figures. Specific design problems at discrete locations may necessitate a more detailed examination of soil conditions and laboratory thaw consolidation results from a specific site.

Coarse-grained soils are subdivided into clean and silty fractions, for each displays a distinctive thaw consolidation behavior. Clean sands and gravels (less than 12% silt) exhibit some measurable strains, though of generally lower value than those shown by siltier sands and gravels. Because the data are scarce in the low density range for clean soils, the short-dashed line for low density extrapolation is substantial. However, as few samples have low density and the estimated prediction curves roughly correlate with the lower density portion of the silty soil prediction curve, the prediction curves shown in Figures 7 and 9 are considered as representative.

BOREHOLE THAW SETTLEMENT ESTIMATES

The following procedure was then used to calculate thaw settlement. The representative alignment borings were chosen from the complete alignment boring program as those proximate to the pipeline centerline and generally containing the most detailed soils and geologic data. A typical boring and the associated tabulation of its predicted thaw settlements are used in explanation both of the assignment of frozen soil properties to each stratum and of the output from a thaw settlement prediction routine. Figure 10 illustrates typical information available for an alignment borehole, and Table 2 presents the thaw settlement predictions for that boring.

Frozen soil index properties were assigned to each meter of alignment borings according to the following systematic procedure. Though most boreholes were drilled to a depth of 15 m, thaw settlement estimates were extrapolated to a 30 m depth. The total borehole depth was generally subdivided by stratum thicknesses of approximately 1.5 m, corresponding closely to field sampling intervals, or at natural changes in soil type. Index properties were then determined. Laboratory test results were used when possible as the basis for designating index properties; however, as an exhaustive testing program was not feasible, soil properties were sometimes estimated for a stratum (data on Table 2 followed by an "E") by making comparisons with tested samples of similar soils.

The first two columns of Table 2 give the identifying number ("SAMP ID") and sampling depth ("SAMPLE INTERVAL") of a sample tested to determine index properties. Frozen dry density, frozen moisture content, frozen saturation, and specific gravity of soil particles of each stratum are entered to the right following "STRATA INTERVAL." The degree of saturation was calculated from the other three index property values (neglecting the 9% volumetric expansion of ice as discussed by Crory (1973)).

Once the frozen soil properties of dry density, moisture content, and specific gravity were determined and the Unified Soil Classification soil type assigned, these values were entered into a computer subroutine that calculated thaw strain and thawed soil properties. The calculation routine for each soil type includes the thaw strain prediction equation as shown in the appropriate one of Figures 1-9, as well as similar equations predicting thawed water content from the frozen soil index properties.

The thaw strain prediction equation (Equation 1) was not used over an entire range of frozen dry densities; rather, linear extrapolations from the

TABLE 2 Typical Thaw Settlement Prediction, Boring 84-3.

SAMP ID	SAMPLE INTERVAL (m)	STRATA INTERVAL (m)	FROZEN DRY DENSITY (kg/m^3)	FROZEN MOISTURE CONTENT (%)	FROZEN SAT (%)	SPEC GRAV	SOIL CLASS	THAW STRAIN (%)	THAW SETTL (cm)	ACCUM THAW SETTL (cm)
1	0 - 0.6	0.0- 0.9	1492	22.0	71.7	2.75 E	ML	9.1	8	8
4	1.5- 2.4	0.9- 2.4	1591	21.2	86.4	2.61	SM	7.6	11	19
6C	3.5- 3.8	2.4- 4.0	1482	26.7	91.6	2.61 E	SM	9.5	15	34
7B	4.6- 5.3	4.0- 5.5	1436	30.2	96.3	2.61 E	SM	10.4	16	50
8	5.5- 6.7	5.5- 7.0	1337	34.1	93.4	2.61	SM	12.2	18	68
9	7.0- 7.5	7.0- 7.5	1431	30.1	95.3	2.61 E	SM	10.4	5	73
		7.5- 8.5	1218 E	40.0 E	85.7	2.82 E	ML	14.4	14	87
11	8.5- 9.8	8.5-10.1	1029	54.4	88.1	2.82 E	ML	23.7	38	125
		10.1-10.8	1029 E	54.4 E	88.1	2.82 E	ML	23.7	17	142
		10.8-11.3	657 E	105.0 E	89.9	2.82 E	MLICE	50.5	25	167
13A	11.3-11.6	11.3-11.6	1202	32.2	67.4	2.82 E	ML	18.6	6	173
13B	11.6-12.6	11.6-12.6	1631	23.0	89.0	2.82 E	ML	3.8	4	177
		12.6-13.1	2131 E	10.0 E	97.2	2.73 E	WB	0.9	1	178
16	13.4-13.9	13.1-14.6	2228	8.1	98.0	2.73 E	WB	0.1	0	178
17	14.6-15.4	14.6-16.2	2191	8.9	98.6	2.73	WB	0.4	1	179
		16.2-19.2	2452 E	3.0 E	67.8	2.75 E	WB	0.0	0	179
		19.2-30.4	2452 E	3.0 E	67.8	2.75 E	BD-RK	0.0	0	179

low density data to 100% strain at 0.0 kg/m^3 (melting of pure ice) and from the high density data to a maximum dry density for each soil grouping were extended from the prediction curves as shown in Figures 1-9. These extrapolations were necessary, for the data were insufficient at the extremes of the density spectrum. After a stratum's thaw strain had been predicted, thaw settlement and thawed soil properties for that interval were calculated from the thaw strain and thawed moisture content predictions. The thaw settlement value for each stratum (thaw strain multiplied by the stratum thickness) is listed as "THAW SETTL" on Table 2. A cumulative sum of these settlement values, commencing at the existing ground surface, is included in the table as the entry under "ACCUM THAW SETTL"; those values represent the total ground surface settlement when thaw has advanced to each stratum's base.

CONCLUSIONS

1. Complete index properties (frozen dry density, frozen moisture content, specific gravity, gradation) of thaw consolidation samples were used in multiple variable regression analyses to relate measured thaw strain to each sample's index properties. After extensive trials, an equation was developed for each soil category that would predict representative thaw strain for the density range in which data were available. The predictions had a standard error of estimate generally within about 5% strain.

2. Thaw strain prediction equations can be used to estimate thaw settlement over an anticipated thaw depth using simple index properties of the frozen soil.

ACKNOWLEDGMENTS

This work was done for Alyeska Pipeline Service Company; their permission to publish these results is appreciated. The authors could not have completed the studies without the assistance of Dr. W. Wang, California State University, Los Angeles, and of H. Olson and J. Patterson, both formerly with R&M Consultants, and without the oversight of C. Whorton and A. Condo, both with ARCO.

REFERENCES

Crory, F.E., 1973, Settlement associated with the thawing of permafrost, in Permafrost--The North American Contribution to the Second International Conference, Yakutsk: Washington, D.C., National Academy of Sciences, pp. 599-607.

Kreig, R.A., 1977, Terrain analysis for the trans-Alaska pipeline: Civil Engineering, July, pp. 61-65.

Luscher, U., and Afifi, S.S., 1973, Thaw consolidation of Alaskan silts and granular soils, in Permafrost--The North American Contribution to the Second International Conference, Yakutsk: Washington, D.C., National Academy of Sciences, pp. 325-334.

Pihlainen, J., and Johnston, G., 1963, Guide to a field description of permafrost for engineering purposes, Technical Memorandum 79: Ottawa, National Research Council of Canada, 23 p.

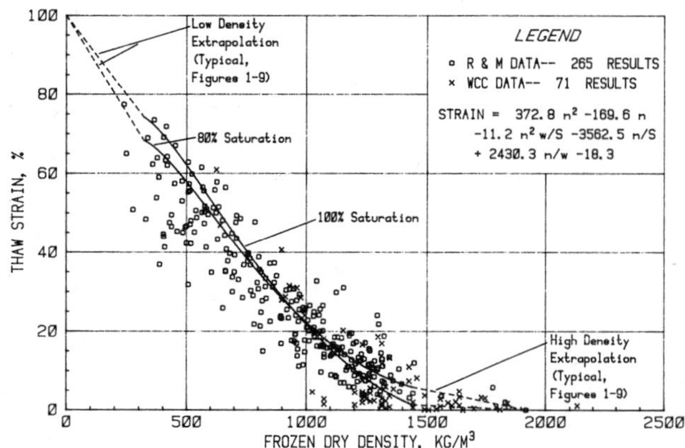

FIGURE 1 Thaw strain of upland silt (ML).

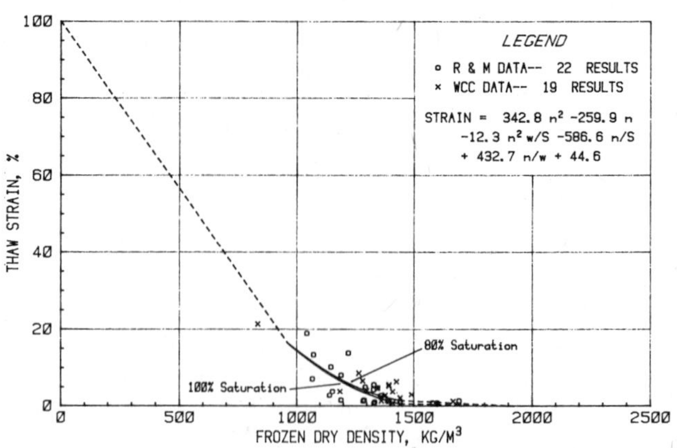

FIGURE 4 Thaw strain of high plasticity clay (CH and MH).

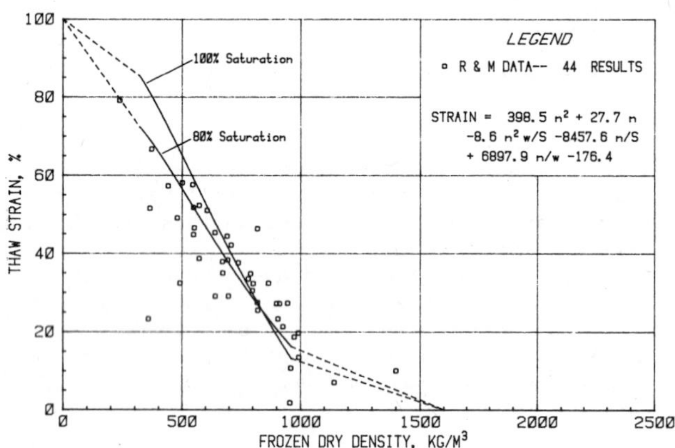

FIGURE 2 Thaw strain of lowland silt (ML).

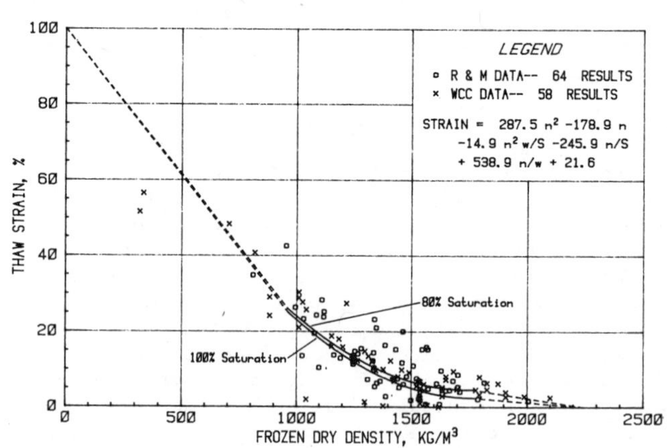

FIGURE 5 Thaw strain of low plasticity clay (CL).

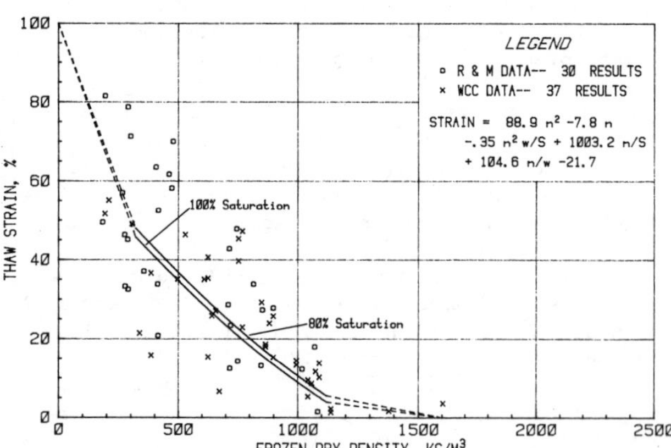

FIGURE 3 Thaw strain of organic soil (OL, OH, and Peat).

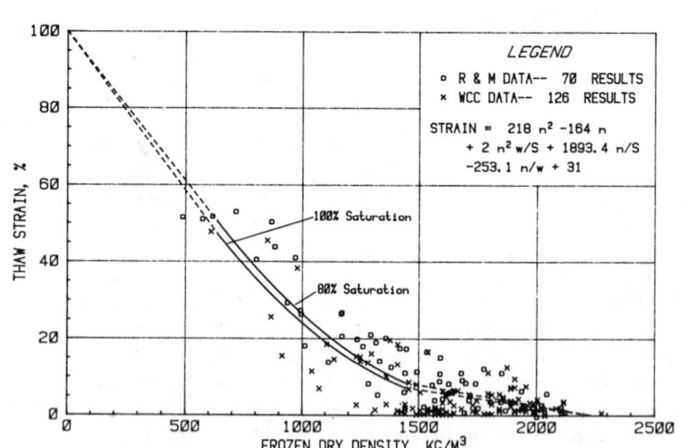

FIGURE 6 Thaw strain of silty sand (SM, SM-SC, and SC).

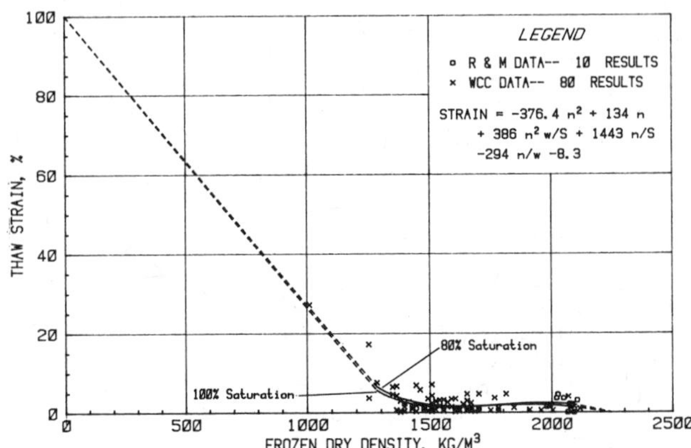

FIGURE 7 Thaw strain of clean sand (SW, SP, SW-SM, SP-SM, SW-SC, and SP-SC).

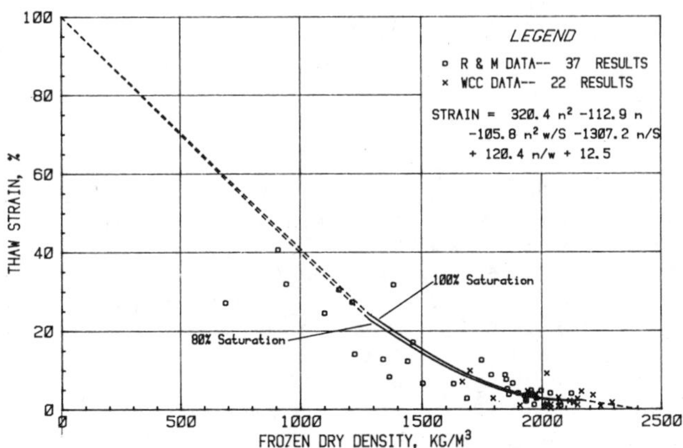

FIGURE 8 Thaw strain of silty gravel (GM, GM-GC, and GC).

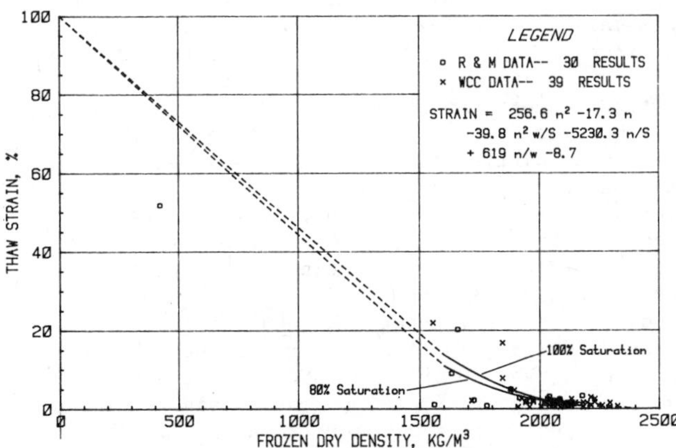

FIGURE 9 Thaw strain of clean gravel (GW, GP, GW-GM, GP-GM, GW-GC, and GP-GC).

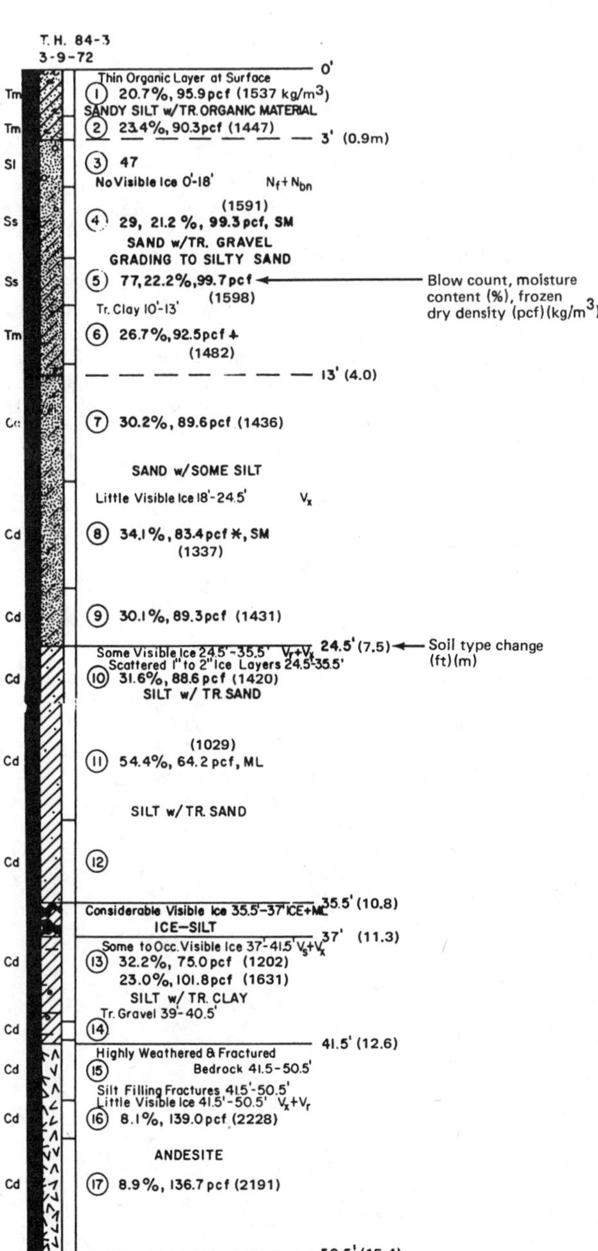

FIGURE 10 Typical boring log 84-3 from 100 km north of Yukon river pipeline crossing.

SHORELINE EROSION AND RESTABILIZATION IN A PERMAFROST-AFFECTED IMPOUNDMENT

R. W. Newbury and G. K. McCullough

Government of Canada, Fisheries and Oceans, Freshwater Institute
501 University Crescent, Winnipeg, Manitoba R3T 2N6 Canada

In 1976, an 850 m^3/s river diversion was constructed through 300 km of permafrost-affected landscape in northern Manitoba. The diversion was accomplished by raising the level of a 1,977 km^2 riverine lake on the Churchill River (Southern Indian Lake) until the water spilled across a terrestrial drainage divide into a series of small valleys tributary to the Nelson River. Over 400 km^2 of permafrost-affected backshore area surrounding the lake were flooded. The mean annual temperature in the Southern Indian Lake region is -5°C. Three repeated phases of shoreline erosion in permafrost materials were observed; melting and undercutting of the backshore zone, massive faulting of the overhanging shoreline, and removal of the melting and slumping debris. At erosion monitoring sites in fine-grained frozen silts and clays, representative of over three-quarters of the postimpoundment shoreline, rates of retreat of up to 12 m/yr were measured. The index of erosion based on the wave energy impinging on the shoreline was 0.00035 m^2/tonne. After 5 years of erosion, restabilization of the shoreline has occurred only where bedrock has been encountered on the retreating backshore. Clearing of the forested backshore prior to flooding did not affect the erosion rates. The rapidly eroding shorelines have increased the suspended sediment concentration in Southern Indian Lake water and have triggered degradation of the commercial fishery.

INTRODUCTION

Southern Indian Lake lies in shallow irregular Precambrian bedrock depressions on the Churchill River in northern Manitoba (latitude 57°N, longitude 99°W) (Figure 1). The climate of the region is continental, with long cold winters and short, cool summers. Average mean monthly temperatures vary from -26.5°C in January to +16°C in July.

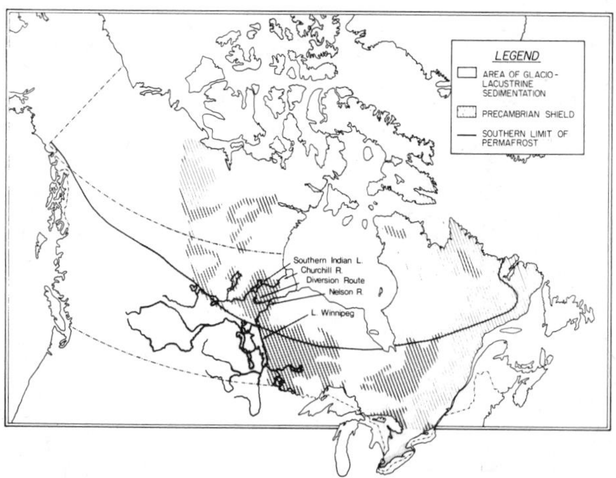

FIGURE 1 The 2,391 km^2 Southern Indian Lake reservoir is located on the Precambrian Shield in the discontinuous permafrost zone of central Canada. The lake lies on the northern boundary of the glacio-lacustrine deposits of Lake Agassiz.

The mean annual temperature is -5°C. One third of the annual precipitation of 430 mm occurs as snowfall during the average 200-day mid-October to late May snow cover period. The average accumulated depth of snow is 60 cm. Black spruce (Picea mariana (Mill.)B.S.P.), jackpine (Pinus banksiana Lamb.), and tamarack (Larix laricina (DuRoi)K. Koch) are the principal tree species of the boreal forest that covers the upland surrounding the lake. A layer of decaying mosses and lichens varying in thickness from a few cm to 5 m has accumulated since the final glacial retreat from the region 7,000 to 9,000 BP. During the deglaciation period, eskers, kames, and other proglacial landforms were deposited on the bedrock surface in the northern third of the region surrounding the lake. The deposits form a rolling upland with a local relief seldom exceeding 20 m. The uplands in the southern two-thirds of the basin lay within the area covered by glacial Lake Agassiz, a large proglacial lake that extended southward to the northern United States. Deposits of laminated silty clays up to 20 m thick occur throughout the region. In the southern region, the upland relief is greater and more abrupt, with exposed knolls and ridges of bedrock separated by poorly drained wetlands.

Permafrost is widespread in all terrain types surrounding the lake with the exception of the proglacial deposits in the northern region. The depth of the active layer varies from 0.5 to 2 m, depending upon local topography and the thickness of peat deposits. The temperature of the permafrost ranges from -0.2° to -0.8°C. Landforms associated with permafrost such as palsas, collapse scars, and peat plateaus occur frequently in the southern glacio-lacustrine region. Although the permafrost exceeds 10 m in depth in the upland areas (Brown 1973), it does not exist under the lake or under the narrow valleys of major tributaries.

In 1966, a program of hydroelectric development began in northern Manitoba to supply electrical energy to southern Manitoba and the central United States. A 927 km direct-current transmission line was constructed from southern Manitoba to hydroelectric dams at Kettle Rapids (1,272 MW capacity) and Long Spruce Rapids (980 MW capacity) on the lower Nelson River, 200 km southeast of Southern Indian Lake. Rather than extending the transmission line to potential dam sites on the Churchill River, a license was granted to Manitoba Hydro to divert 850 m^3/s (about 85%) of the Churchill River waters southwards across a drainage divide, through a long series of small channels and lake basins in the Rat and Burntwood River valleys tributary to the Nelson River above the power dams. In June 1976, a dam was completed across the natural outlet of Southern Indian Lake at Missi Falls, and the mean lake level was raised 3 m to divert the flow across the drainage divide at the southern end of the basin (Figure 2). The area of the lake was increased from 1,977 to 2,391 km^2. Flooding extended beyond the sub-lake thawed zone into the permafrost-affected upland.

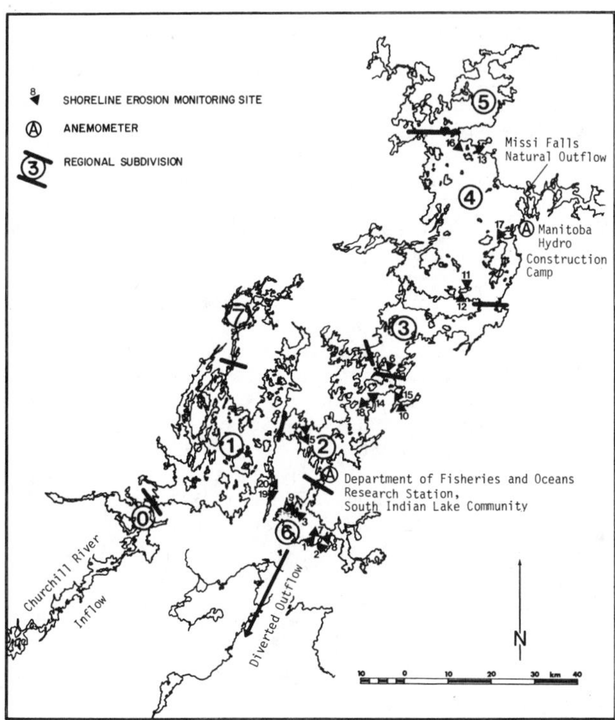

FIGURE 2 Southern Indian Lake is a series of bedrock-controlled, riverine basins on the Churchill River. The basins have been numbered to facilitate limnological studies. Erosion monitoring stations in granular deposits and fine-grained permafrost materials are located at 20 sites of varying exposure throughout the lake.

Preimpoundment studies of the effects of flooding on shoreline stability predicted qualitatively that erosion and solifluction of the shoreline materials would occur (Underwood-McLellan 1970, Lake Winnipeg Churchill Nelson Rivers Study Board 1975). Quantitative estimates of the rates and extent of shoreline erosion were not made, as analogous conditions had not occurred or had not been reported for such a large impoundments in permafrost.

METHODS

The study reported in this paper began in 1975, 1 year prior to the impoundment. Erosion monitoring sites were selected and surveyed at 20 locations having different exposures to wave conditions. The rates of erosion in the initial year of impoundment and the relative resistance to erosion of the permafrost materials have been previously reported (Newbury et al. 1978). The resistance to erosion was based on an index of erosion reported by Kachugin (1966) as a "washout coefficient, K_e," which expressed the volume of backshore material eroded per unit of wave energy dissipated on the shoreline. Kachugin's units of the washout coefficient of m^2/tonne are derived from the quotient of cubic meters of eroded materials per meter of shoreline length divided by the perpendicular component of the wave energy acting on the backshore expressed as tonne-meters per meter of shoreline length $(m^3/m)/(tonne\text{-}m/m)$. Eroded volumes have been surveyed annually since impoundment at erosion monitoring sites. Wave energies have been hindcast from wind speeds and directions recorded at 2 sites adjacent to the lake (Figure 2) using the modified Sverdrup-Munk procedure (U.S. Army Coastal Engineering Research Center 1966).

The lake was divided into eight sub-basins (Figure 2), for which the contribution of shoreline materials to the lake after impoundment was estimated. The washout coefficients determined from the monitoring sites were combined with the hindcast wave energies acting on 331 reaches of shoreline to determine the total erosion in each basin. The actively eroding shoreline reaches were mapped by aerial and boat reconnaissance of the whole lake in 1976 and 1978.

RESULTS AND DISCUSSION

Erosion Processes

Representative textures and ice contents of shoreline deposits at the monitoring sites are given in Table 1. An example of surveyed profiles showing annual erosion and nearshore deposition of lacustrine clay at a relatively high wave energy site is shown in Figure 3. The erosion of frozen fine-grained materials on shorelines surrounding the larger basins of the lake was observed to proceed in a repeated sequence of melting, slumping and removal phases. In the initial phase, melting occurs below and slightly above the water surface. In the second phase, the partially thawed materials flow out to form a silty-clay beach strewn with scattered frozen blocks. In some cases, caverns or melt niches are formed that are up to 1 m in height and extend up to 3 m into and under the frozen backshore materials (Figure 4). In this situation, the overlying cantilevered block splits away from the main land mass and falls onto the foreshore (Figure 5). In the final phase, wave erosion re-

TABLE 1 Texture and Ice Content of Mineral Materials at Shoreline Monitoring Sites. (Dashes indicate samples were not analysed.)

Site	Textural analysis of parent materials (%)			Ice content (% of dry weight)
	Sand	Silt	Clay	
1	1	15	84	43
2	0	16	84	-
3	1	34	65	-
4	0	15	85	64
5	1	26	73	-
6	1	34	65	64
7	0	49	51	-
8	19	16	65	-
9	1	17	82	-
10	35	46	19	-
11	10	45	45	47
12	1	19	80	-
13	8	39	53	56
14[a]	2	36	62	92
15	98	0	2	-
16[a]	9	33	58	-
17[a]	4	37	59	43
18[b]	-	-	-	-
19	1	34	65	-
20	-	-	-	-

[a]Samples analyzed were of backshore lacustrine deposits. To date, erosion at these sites has been predominantly of former sandy beach materials.

[b]Predominantly fine to coarse sand with some silty beds.

moves the fallen debris and the warm lake water is again brought in contact with frozen backshore materials. In the silty-clay glacio-lacustrine materials, the form of the slumping and eroding shorelines does not change substantially as the backshore retreats inland. If bedrock or coarse granular materials are encountered at the eroding face, the inland movement at the water level ceases but erosion of the backshore continues until a wave-washed bedrock zone or a stable beach is established.

At non-permafrost sites in coarse granular deposits (Sites 15 and 18), erosion and deposition processes agreed with those reported by Bruun (1962) for similar materials.

Erosion Rates at Monitoring Sites

The total volume of annual erosion and the washout coefficient, K_e, at each of the 20 monitoring sites surrounding the lake are summarized for 4 years of impoundment in Table 2. Where bedrock was not encountered, the mean washout coefficient for the permafrost materials was generally one-half of that reported by Kachugin (1966) for similar materials in the unfrozen state. Although this suggests that permafrost conditions may retard erosion, no data are supplied with the Soviet Union observations, and the magnitude of the washout coefficient may have been based on a different method of determining an index of the wave energy.

There was a wider range of K_e values observed during the first year of impoundment than in the following 3 years. At sheltered sites exposed to low wave energies, several open water seasons were required to destroy the protective moss and root cover at the water's edge. At exposed sites, large volumes of peat were quickly removed from the flooded foreshore, producing high K_e values in the first year. Because of the variability of early years, general K_e values were calculated using 1978-1980 data only. Also excluded from the general K_e determinations were values at sites after bedrock had been encountered at the eroding face (Sites 6, 8, 10, and 11). Based on the 16 shoreline sites which extend over the range of materials and fetches encountered on Southern Indian Lake, the general K_e value found by linear regression was 0.00035 m^2/tonne (r^2 = 0.71, n = 42, Figure 6).

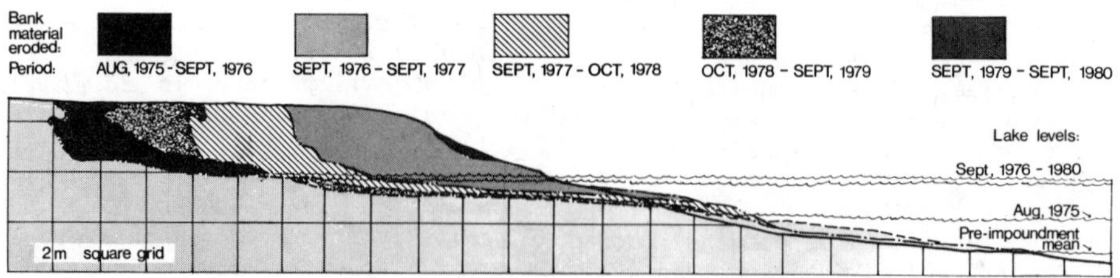

FIGURE 3 A typical consecutive annual sequence of eroding shoreline profiles at Site 1 (Figure 2). Slumped material from the initial 1975-76 period was removed in the following year.

FIGURE 4 The sequence of erosion in permafrost materials begins with the thawing and removal of backshore materials at the reservoir water level. Deep niches and caverns may be formed under the backshore zone as shown here at an island in region 4. The portion of the survey boat protruding into the cavern is 5 m in length.

FIGURE 5 As the thawing and eroding niche proceeds under the backshore zone, the length of the cantilevered block of frozen materials overhead increases until the entire backshore zone splits and tumbles onto the foreshore. The block is then thawed and removed by waves, and the formation of a new niche commences under the freshly exposed frozen bank.

TABLE 2 Total Volumes of Material Eroded Annually From Monitored Shoreline Sites on Southern Indian Lake (m^3 per m of shoreline).

Site	1977	1978	1979	1980
1	23.4 (0.77)	15.7 (0.48)	10.1 (0.61)	15.3 (0.66)
2	1.8 (0.18)	1.7 (0.16)	1.3 (0.27)	0.8 (0.12)
3	8.9 (0.35)	7.5 (0.29)	3.0 (0.24)	8.4 (0.50)
4	7.0 (0.67)	4.3 (0.34)	1.3 (0.15)	2.9 (0.35)
5	0.9 (0.71)	0.6 (0.38)	0.9 (0.74)	0.5 (0.56)
6	21.0 (1.32)	14.4 (0.90)	1.6 (0.14)	4.5 (0.28)
7	9.4 (0.48)	4.4 (0.19)	0.9 (0.04)	5.1 (0.26)
8	1.9 (0.84)	0.6 (0.26)	0.2 (0.13)	0.4 (0.17)
9	0.0 (0.00)	1.2 (0.19)	1.5 (0.38)	3.1 (1.70)
10	0.0 (0.00)	0.2 (0.06)	0.1 (0.07)	0.5 (0.21)
11	17.4 (0.54)	4.3 (0.11)	0.7 (0.03)	not surveyed
12	4.8 (0.98)	2.1 (0.33)	2.6 (0.51)	" "
13	1.4 (0.35)	⟶	2.0 (0.14)	" "
14	2.0 (0.34)	3.5 (0.47)	1.4 (0.27)	3.5 (0.66)
15	0.0 (0.00)	0.1 (0.56)	0.2 (1.32)	0.1 (0.51)
16	8.6 (0.94)	⟶	14.6 (0.32)	not surveyed
17	2.2 (1.35)	2.7 (0.55)	4.2 (0.42)	" "
18	1.0 (0.16)	1.5 (0.16)	6.3 (0.40)	4.3 (0.44)
19	2.1 (0.32)	0.1 (0.03)	0.1 (0.05)	0.2 (0.08)
20	1.3 (0.29)	0.4 (0.14)	0.3 (0.22)	0.4 (0.20)

The erosion index, K_e, is shown in parentheses for each period (x 10^{-3} m^2/tonne). Arrows indicate cumulative erosion at shorelines not surveyed in 1978.

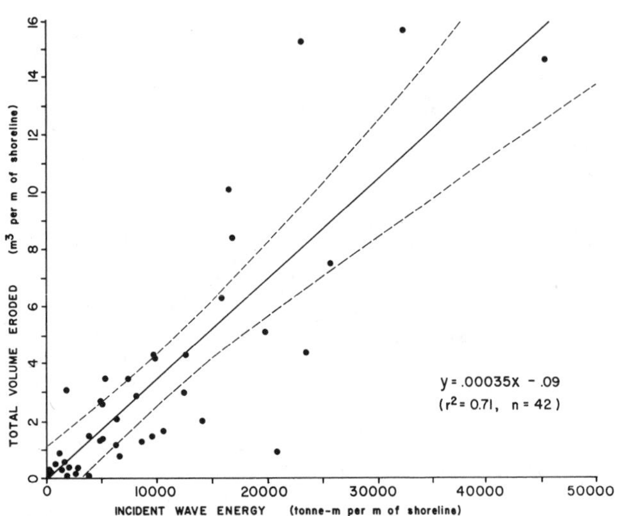

FIGURE 6 The linear relationship between eroded materials and wave energy for the 1978-1980 period on Southern Indian Lake is similar to that proposed by Kachugin (1966) for reservoirs in the Soviet Union. The scattered data points and wide confidence limits (95%) at high erosion and high energy sites suggest that the relationship may be curvilinear.

TABLE 3 Estimated Total Dry Weight[a] of Mineral Materials Eroded From the Shorelines of the Major Basins of Southern Indian Lake for the Period 1976-1978 (10^6 kg)[b]

Region	1976	1977	1978
0	122	177	166
1	528	672	615
2	238	311	290
3	478	668	608
4	1594	2099	1916
5	207	275	229
6	190	273	247
Whole Lake	3357	4475	4071

[a]The volume of dry mineral material eroded was calculated using an average water content of the perennially frozen silty-clays of 58% dry weight and assuming a bulk dry density of 2600 kg/m^3.

[b]After Hecky and McCullough (1983).

Sediment Contributed by Shoreline Erosion

The total dry weight of mineral materials eroded in the years 1976, 1977, and 1978 for each basin excluding the limited exposure shorelines of region 7 of the lake, is summarized in Table 3. Estimates were not extended beyond 1978, as that was the last year in which a reconnaissance survey was undertaken to determine the portions of the total shoreline in overburden and bedrock materials. Postimpoundment bank erosion dramatically increased the turbidity of the lake (Hecky and McCullough 1983). The long-term preimpoundment sediment input to the whole lake, estimated from Churchill River inflows, is 200 x 10^6 kg/yr. The sediment input from eroding shorelines following impoundment exceeds 4,000 x 10^6 kg/yr.

Time Required for Shoreline Restabilization

In the first 5 years of impoundment on Southern Indian Lake, restabilization in permafrost-affected fine-grained materials has occurred only on shorelines where bedrock underlying the backshore zone was exposed at the water's edge. Where bedrock was not encountered, there has been no change in the melting, slumping and eroding sequence of shoreline migration. The annual erosion indices at monitoring sites in fine-grained materials have shown no diminishing trend following the first year of impoundment (Table 2). The clearing of shorelines up to the impoundment level did not affect erosion rates.

Because sub-surface exploration of the bedrock topography surrounding the lake is prohibitively expensive, the time required for shoreline restabilization can be estimated only from the frequency of occurrence of bedrock at the monitoring sites scattered throughout the lake. Eighteen of the monitoring sites occur in fine-grained materials which are representative of over three quarters (2,841 km) of the postimpoundment shoreline. In the initial 4 years of impoundment, bedrock was encountered in the retreating backshore of the 4 most exposed monitoring sites. Assuming that the bedrock distribution is similar at less exposed sites,the rate of bedrock encounters should decrease, because more time will be required to remove the overburden. If the rate decays geometrically, 4/18 of the remaining eroding shoreline will strike bedrock every 4 years until the preimpoundment condition is restored.

Prior to impoundment, 76% of the shoreline was bedrock controlled. Following impoundment, bedrock was exposed on only 14% of the shoreline. Because the wave energy distribution on the lake and the bedrock topography were not changed by the impoundment, the same proportion of shoreline will ultimately be bedrock controlled. At the assumed geometric recovery rate of the sample shorelines, it would take 35 years to restore 90% of the fine-grained permafrost shorelines to their preimpoundment condition. Although this is an approximate estimate, it is likely that the instability of the Southern Indian Lake shoreline will not change for several decades. The discharge of bank sediment into the lake will continue to disrupt the fishery resources upon which the local residents are dependent (Bodaly et al. 1983a,b).

ACKNOWLEDGMENT

For over 20 years, the late R. J. E. Brown of the National Research Council of Canada undertook permafrost research in Manitoba and throughout northern Canada. In 1978, following the Third International Conference on Permafrost, he conducted a memorable tour of the Freshwater Institute project on Southern Indian Lake, which focused international attention on hazards of extensive flooding in permafrost terrain. The research staff of the Southern Indian Lake project wish to acknowledge his support, encouragement, and significant contribution to the understanding of permafrost phenomena.

REFERENCES

Bodaly, R. A., Hecky, R. E., and Fudge, R. J. P. 1983a, Increases in fish mercury levels in lakes flooded by the Churchill River hydroelectric diversion, northern Manitoba, Ottawa: Canadian Journal of Fisheries and Aquatic Sciences (in press).

Bodaly, R. A., Johnson, T. W. D., and Fudge, R. J. P., 1983b, Post-impoundment declines in the grade of commercial whitefish catch from Southern Indian Lake, northern Manitoba, Ottawa: Canadian Journal of Fisheries and Aquatic Sciences (in press).

Brown, R. J. E., 1973, Influence of climatic and terrain factors on ground temperatures at three locations in the permafrost region of Canada in Proceedings of the IInd International Conference on Permafrost, North American Contribution: Washington, D.C., National Academy of Sciences.

Bruun, P., 1962, Sea level rise as a cause of shoreline erosion: Journal of the Waterways and Harbours Division, American Society of Civil Engineers, 88.

Hecky, R. E., and McCullough, G. K., 1983, The effects of impoundment and river diversion on the sedimentary regime of Southern Indian Lake, Ottawa: Canadian Journal of Fisheries and Aquatic Sciences (in press).

Kachugin, E. G., 1966, The destructive action of waves on the water reservoir banks, in Proceedings of the International Association of Scientific Hydrology Symposium Garda 1: Brussels.

Lake Winnipeg, Churchill and Nelson Rivers Study Board, 1975, Summary report: Winnipeg, Queens Printer.

Newbury, R. W., Beaty, K. G., and McCullough, G. K., 1978, Initial shoreline erosion in a permafrost-affected reservoir, Southern Indian Lake, Canada, in Proceedings of the IIIrd International Conference on Permafrost, v. 1: Ottawa, National Research Council of Canada.

Underwood-McLellan, 1970, Manitoba Hydro Churchill River diversion: Study of alternative diversions: Winnipeg, Underwood-McLellan and Assoc. Ltd.

U.S. Army Coastal Engineering Research Center, 1966, Shore protection, planning and design, Technical Report No. 4, 3rd ed.: Washington, D.C., U.S. Government Printing Office.

GEOTHERMAL DESIGN OF INSULATED FOUNDATIONS FOR THAW PREVENTION

J.F. Nixon
Hardy Associates (1978) Ltd., 221-18 St. S.E., Calgary, Alberta, Canada T2E 6J5

As development in the North American Arctic proceeds to more northerly remote regions, the demand for reliable and thermally acceptable foundations for heated structures on permafrost will increase. In subarctic areas, or regions where the permafrost ground temperatures are close to melting, some method of ventilation or heat removal is necessary to provide a stable, frozen subsoil condition for support of the structure. However, in more northerly areas, communities or new energy developments are located where the mean ground temperatures are colder than $-5^{\circ}C$ or so, and another foundation alternative is available to the designer. It is possible using insulation alone, to retain the melting isotherm above the thermally unstable permafrost layer. In some cases, a distinct economic advantage may exist (particularly for smaller structures) to placing the foundation on a thicker layer of insulation, and omitting any active form of heat removal altogether. In this way, the capital costs of an elevated structural floor, or costs for ventilation or other heat removal systems can be omitted. In many remote northern areas, regular maintenance cannot be relied upon, and the use of a pure insulated foundation system does not require any long-term maintenance, and may have more than simply economic advantages. A thermal analysis for rectangular insulated structures is reviewed, and a new solution for a heated circular insulated structure is presented. A convenient chart allows the designer to select the insulation thickness that will prevent the $0^{\circ}C$ isotherm (or any other isotherm) from entering the permafrost subsoil. In addition, an available solution for an insulated buried pipe in permafrost is reviewed for comparison.

INTRODUCTION

Foundations for heated structures on permafrost have tradionally employed ventilation as well as insulation to preserve the frozen condition beneath the structure. (See for example Tobiasson (1973) or Nixon (1978).) This is likely the most economical method of preserving permafrost beneath a building placed on grade in areas where the mean ground temperatures are within a few degrees of the melting point. Industrial and community development has proceeded further north into remoter and colder areas of the North American continent. In such areas, the requirement for essentially maintenance free foundation systems becomes more acute. In addition, the mean ground temperatures may be several degrees colder than some of the subarctic or discontinuous areas where development has been traditionally more active in past decades. Therefore, the use of a foundation system involving insulation alone as a means of permafrost protection becomes more attractive, where it is necessary to place a structure directly on grade. This removes the need for pile placement equipment and a costly structural slab, or the installation and maintenance of costly subfloor ventilating systems.

This paper reviews an existing thermal solution for insulated rectangular structures, and introduces a new solution for circular heated areas placed on grade. It is assumed that thawing of the subgrade must be completely prevented in the long term, and the thaw line maintained within the insulation layer, or layer of stable granular fill beneath the structure. It is stressed again that such foundation systems are only attractive for areas where the mean ground temperature is several degrees below the melting point, and become particularly useful where the lateral dimension of the heated area is not large.

GEOTHERMAL SOLUTIONS FOR THAW PREVENTION

Porkhaev (1963) has presented charts for the calculation of thaw depth beneath a two-dimensional heated strip area on permafrost. Based on a later review by Tsytovich (1975), the solution for zero thaw depth can be written as:

$$\alpha = 0.5 \text{ cotan } \left\{\frac{\pi\beta}{2(1+\beta)}\right\} \qquad (1)$$

where $\beta = k_f(T_f-T_g)/k_u\ (T_s-T_f)$

$\alpha = k_u h_i/B\ k_i$, a dimensionless insulation parameter

(T_s-T_f) = structure temperature in $^{\circ}C$ above melting point

(T_s-T_f) = mean ground temperature in $^{\circ}C$ below melting point

h_i = insulation thickness (m)

B = total width of foundation (m)

and k_i, k_u and k_f are the conductivities of insulation, unfrozen and frozen ground in $W/m^{\circ}C$.

This solution provides the required insulation thickness to prevent thaw beneath the center of a long strip heated area on permafrost of width B. Porkhaev (1963) provides further approximate methods of modifying this two-dimensional solution for square or rectangular shaped structures. The basis for these is not apparent and the following solution handles the three-dimensional case more accurately.

For heated circular areas such as tanks (or approximation to square areas), a similar method to Porkhaev's "quasi-static" method can be used to obtain the temperature field and thaw depth beneath a circular area. Although physically a three-dimensional problem, it is made simpler by considering it as a two-dimensional case in radially symmetrical coordinates. The author has based this solution on the work of Ruckli (1948) and others. It is beyond the scope of this paper to present the complete derivation of this new solution for circular heated structures. For zero thaw depth, the thickness of insulation can be written in non-dimensional form as

$$a = \frac{\pi}{8} \; \frac{1}{\beta} \qquad (2)$$

with the same notation as before.

Equations (1) and (2) can then be plotted as shown in Figure 1, and may be used to obtain the insulation requirements beneath the center of a long rectangular or circular structure.

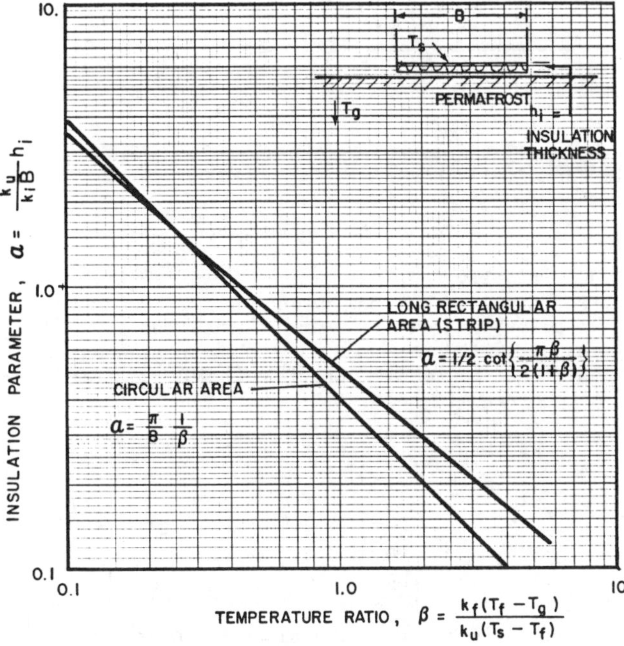

FIGURE 1 Required Insulation Thickness For No Thaw Beneath Heated Building On Permafrost - General Solution

DISCUSSION

It is of practical interest to note that the thickness of insulation increases in direct proportion with the minimum lateral dimension of the structure. The foundation requirements are also heavily dependent on the ratio of the structure temperature to the ground temperature, as one would expect. A circular or square area will require somewhat less insulation than a long heated strip area, other conditions being equal. This is due to the differences between heat flow in two- and three- dimensions.

Figure 2 shows a specific solution for a typical set of soil thermal properties, and for a typical polystyrene insulation. As an example, a tank heated to +15°C continuously, and founded on a permafrost subsoil having a mean ground temperature of -8°C, would require 20 cm of insulation for a tank 20 m in diameter. At first sight this may appear to be a large insulation thickness, but may be economically attractive when one considers the great reduction in complexity during construction, and the significantly reduced maintenance effort during operation, as compared with a ventilated or piled foundation.

The insulation requirements toward the edge of the heated area might be reduced somewhat, and more detailed thermal analysis using a two-dimensional thermal simulator could be carried out to optimize the insulation thickness across the structure (see Nixon and Halliwell (1982) for a description of a two-dimensional thermal program operating in radial symmetry).

Figure 2 shows a specific solution for circular and rectangular heated areas, where typical or representative values for the thermal conductivity of frozen soil and insulation have been selected. The insulation conductivity is a recommended design value for rigid polystyrene insulation used in such applications. When the ratio of the structure temperature to the ground temperature becomes greater than about 2:1, the figure illustrates the sharp rise in insulation thickness that would be required to prevent thaw beneath the centre of the structure.

Experience with thermal analyses of this nature shows that thermal equilibrium is usually approached or reached within about 20 years. However, structures that are designed to function for much less than this might employ less insulation, depending on a more detailed time dependent thermal analysis, i.e., Porkhaev (1963) or Nixon and Halliwell (1982).

The freezing temperature, T_f, given in equations (1) and (2) is normally assumed to be the melting point of bulk fresh water, 0°C. However, for this foundation type, it would be prudent to adopt a somewhat lower temperature (say, -1°C) for the maximum temperature at the underside of the insulated pad. This allows some safety factor to be incorporated in the design, and reduces some of the frozen soil strength or creep problems that might result if the subsoil remained too close to its melting point for much of its design life. Design in saline soils would require the use of a lower freezing temperature, T_f.

Cross-drainage and the flow of meltwater through the granular pad beneath the insulation pose a major hazard for the success of this foundation type, and steps must be taken to limit or prevent this completely using surface grading or placement of impervious layers.

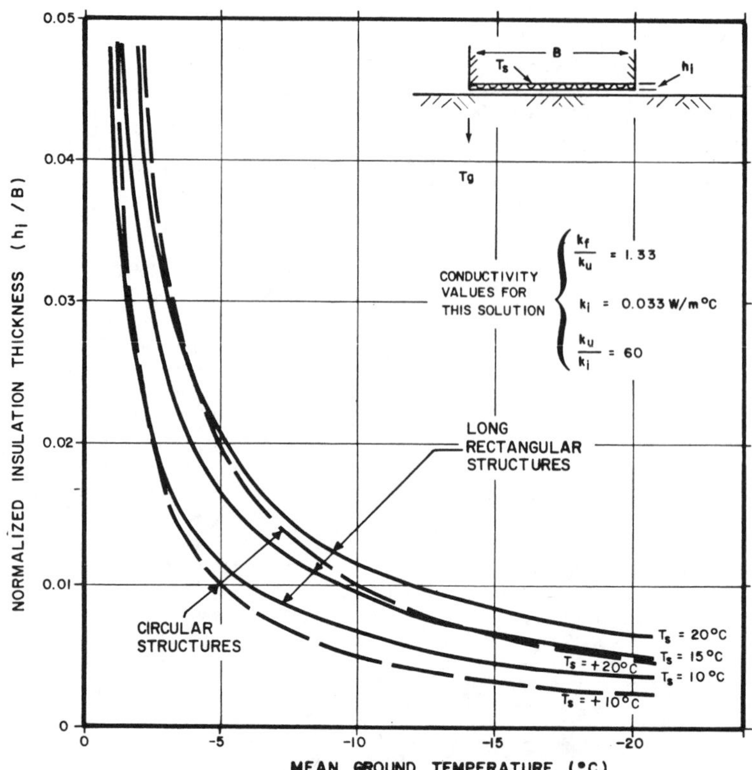

FIGURE 2 Specific Solution for Required Insulation Thickness to Maintain Permafrost Frozen

This form of foundation design allows a large depth of soil beneath the loaded areas of the structure to warm up close to the melting point. Large, persistent structural or live loadings may introduce time-dependent bearing capacity or soil creep problems. Granular pads should normally underlie the insulation layer to distribute concentrated surface loadings, and should be of a sufficient thickness that soil creep, in particular, does not progress on a continuing basis. Normally, surface loads imposed by a hanger, tank, or warehouse on a ventilated or piled foundation might not induce stresses that would cause creep or time-dependent bearing capacity problems. For these insulated foundation types, a large "bowl" of permafrost has been subjected to increasing temperatures in the long term. Unfortunately, the region of highest stresses coincides with the region of warmest temperature, and the designer must be very conscious of the foundation design issues (such as creep settlement) that remain.

No documented case histories of this foundation type are available to the author, and it will become a high priority to instrument foundations constructed using this design method to obtain some field verification of the design approach presented here. However, the above relationships will be of value to the design engineer as no straightforward chart exists, to the author's knowledge, that summarizes the geothermal design for heated foundations using insulation alone to prevent thaw.

INSULATED PIPES

Finally, a solution given originally by Thornton (1976) has been reworked to obtain the insulation thickness for no thaw beneath a buried pipe. His solution can be rearranged into a similar format as previous solutions for heated structures to become:

$$a = \frac{k_u}{k_i}\,\frac{h_i}{B} \tag{3}$$

$$= \tfrac{1}{2}\,\frac{k_u}{k_i}\left\{\exp\left[\left(\frac{k_u}{k_i}\right)^{-1} \beta^{-1}\ \ln\left(\frac{D}{r_o} + \sqrt{\left(\frac{D}{r_o}\right)^2 - 1}\right)\right] - 1\right\}$$

where B is now the pipe diameter, D is the depth of burial to the pipe axis, and r_o is the radius of the insulated pipe.

The insulated pipe radius is related to the pipe diameter and insulation thickness by

$$r_o = 0.5\ B + h_i \tag{4}$$

The function expressed by equation (3) is plotted in Figure 3.

The solution is only weakly dependent on the ratio D/r_o, and the insulation thickness can be obtained explicity from equation (3) as long as the approximate ratio D/r_o is known.

For β greater than about 0.3, equation (3) can be greatly simplified to

$$a = \tfrac{1}{2} \ln \left(\frac{D}{r_o} + \sqrt{\left(\frac{D}{r_o}\right)^2 - 1} \right) \cdot 1/\beta \qquad (5)$$

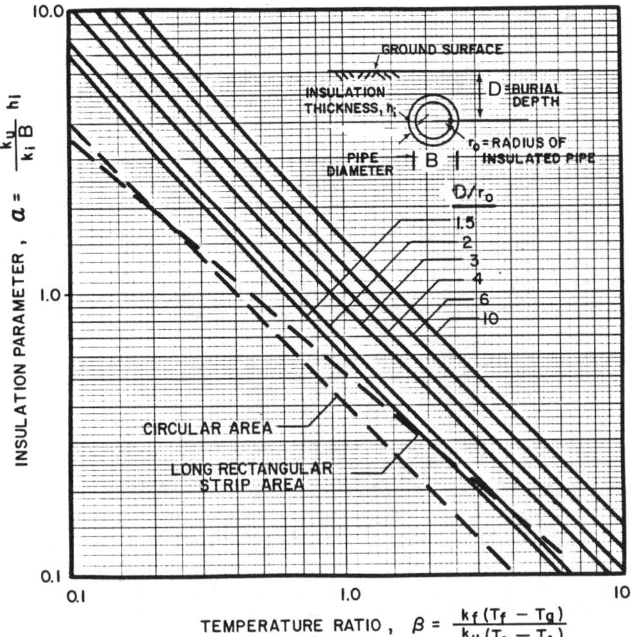

FIGURE 3 Required Insulation Thickness for No Thaw Beneath Heated Pipe in Permafrost-General Solution

For example, with a burial depth ratio, D/r_o of 3, the natural log function is 1.763, and the insulation thickness is simply obtained from

$$a = 0.881/\beta \qquad (6)$$

This function has therefore a similar form to the circular heated area on the ground surface.

Replacing a and β by their usual representation, the equation for insulation thickness is

$$\frac{h_i}{B} = 0.881 \frac{k_i}{k_f} \left(\frac{T_s - T_f}{T_f - T_g} \right) \quad \text{for } D/r_o = 3 \qquad (7)$$

If $k_i/k_f = 1/65$, then

$$\frac{h_i}{B} = 0.014 \left(\frac{T_s - T_f}{T_f - T_g} \right)$$

For a 60°C pipe buried in -6°C permafrost, the required insulation thickness is about 14% of the pipe diameter.

It appears, as for the heated structures considered earlier, that insulation on its own may provide a viable design option, provided the permafrost is cold, and the structure is not excessively hot. It is also worth noting that with all other things being equal, the required insulation thickness goes up in direct proportion to the pipe diameter.

Finally, it should be noted as before that these sections should be used only as a rough guide to insulation requirements, and detailed numerical analysis accounting for actual soil stratigraphy and surface heat transfer conditions should be undertaken to refine these preliminary predictions.

It is further noted that all of the solutions presented here can be used equally well for determining insulation requirements for chilled structures on unfrozen ground, provided the subscripts for frozen and unfrozen soil are interchanged.

REFERENCES

Nixon, J.F., 1978. Geothermal aspects of ventilated pad design. Proc. Third International Permafrost Conference, p. 841. Edmonton.

Nixon, J.F. and Halliwell, D., 1982. Practical Applications of a versatile geothermal simulator. Proc. Winter Annual Meeting of American Society of Mechanical Engineers, Phoenix, Ar. November, 1982.

Porkhaev, G., 1963. Temperature fields in foundations. Proc. First Internation Permafrost Conference, Lafayette, Indiana. pp. 285-291.

Ruckli, R., 1948. Two and Three-dimensional ground water flow towards ice lenses formed in the freezing ground. Proc. Second International Conference on Soil Mechanics and Foundation Engineering, Vol. II, p. 282.

Thornton, D., 1976. Steady state and quasi-state results for bare and insulated pipes in permafrost. Can. Geotech. J., 13, pp. 161.

Tobiasson, W., 1973. Performance of the Thule hangar soil cooling systems. Proc. Second International Permafrost Conference, Yakutsk.

Tsytovich, N.A., 1975. The mechanics of frozen ground. Trans. from Russian by Swinzow/ Tschebotarioff. Scripta Book Co., Washington, D.C.

DRIVEN PILES IN PERMAFROST: STATE OF THE ART

Dennis Nottingham and Alan B. Christopherson

Peratrovich, Nottingham & Drage, Inc., Anchorage, Alaska 99503 USA

Pile driving techniques in permafrost have developed rapidly over the last 5 years, to a point where the driven pile method using thermally modified drilled pilot hole now offers a fast and economical solution to many frozen foundation problems. Most types of piles, including H-shape, pipe, and sheet piles, now can be driven, assuming that there is a knowledge of soil conditions and that the correct procedure is followed. Presented in this paper are the latest research, testing, and production pile driving developments in Alaska. Current proposed pile design approaches will be discussed as related to short-term loading and long-term creep. The scope of this paper is based largely on actual practical experience with about 5,000 piles recently driven in cold and warm permafrost soils; as well as recent laboratory work and field testing experience on a large number of driven piles. The proposed criteria presented in this paper are primarily addressed to the practicing design engineer, including design and construction considerations. As more research and experience accumulate, factors presented in this paper may change.

INTRODUCTION

In the past, piles used as structural support in permafrost have consisted primarily of the drilled and slurry backfill type. Early pile driving efforts in Alaska by the U.S. Army Corps of Engineers, the State of Alaska, oil companies, and others had varying success. Although piles had been driven in permafrost, the methods used were not entirely reliable or economical. Conversely, development of slurry backfill pile placement was advanced through extensive use on the Trans-Alaska Pipeline and North Slope oilfield projects.

Pile driving in permafrost using thermally modified pre-drilled pilot holes has now developed to the point where predictability, reliability, and economy make it a viable method in most applications. Current research efforts have helped define characteristics associated with the general soil types encountered. Contractors also are becoming aware of these methods and advantages, particularly after recent experience with piles driven into permafrost on the North Slope and other locations in Alaska.

Driven pile placement requires less equipment and support. When properly planned, installation placement rates of possibly 2 to 3 times the rate for slurry piles can be achieved. Driven piles can be loaded after placement much sooner than slurry backfill piles. Principal limitations of driven piles are location tolerances and present lack of refined driving equipment, although many good components do exist.

This paper is directed primarily toward design engineers who must apply the results of knowledge gained to date in a practical manner. A design concept is presented that uses short-term loading criteria to define maximum adfreeze limits under any condition including long-term, followed by long-term loading to establish long-term pile adfreeze strength limits based on creep deflection. This approach clarifies a past area of confusion to many engineers concerning the appropriate values to use for long-term strength.

DRIVEN PILE PLACEMENT METHODS

Piles, including H-shape, pipe and sheet, can be, depending on soil conditions, driven with both impact and vibratory hammers and the more sophisticated sonic hammers. Where extremely hard driving is encountered, an impact hammer and cast-steel pile driving tips are necessary. Experience has shown that even with relatively easy driving, pile tips should probably be used with an impact hammer to prevent tip damage. Pipe piles are particularly subject to tip ovaling and flattening during impact driving into pilot holes. Pile tips on pipe piles should be of the open, flush exterior type, and H-pile tips should be flush on the exterior. Vibratory hammers are particularly good in fine-grained saturated thawed soils or weak frozen soils, such as those produced by the thermally modified pilot hole method. They have been used for slow driving in warm frozen silts without the use of pilot holes. Vibratory hammers may have difficulty driving piles into cold dense frozen soils or where there is a predominance of coarse gravels and cobbles, or hard layers. It should be noted that driven piles made of mild steel (i.e., A36, A252, etc.) have not been observed to fail due to fracture while driving with impact hammers in extremely cold environments; however, they may fail from improper structural design considerations or driving methods. To more clearly identify various hammer types suitable for use in cold weather and permafrost, the following discussion is presented.

Impact Hammers

Impact hammers rely on falling mass to produce energy and have many forms and types. Practice in Alaska now centers primarily on diesels, with air hammers and hydraulic impact hammers also in use. Diesel hammers work well if kept warm, and meet resistance to driving to assure ignition. When used with pilot hole thermal modification and short piles, driving is often too easy for efficient diesel operation. Air hammers offer very controllable driving, but during cold weather line deicers or heaters may be needed to prevent freezeup. Hydraulic impact hammers are small, fast-hitting devices that offer tremendous potential for driving small piles, such as for remote building foundations. Mounted on tracked vehicles, they are highly efficient and mobile machines. As with all hydraulic systems in cold weather, attention must be given to use of proper fluids and to keeping components warm and protected from the environment.

Typical driving rates in permafrost for properly sized impact hammers are 300 mm (1 ft) per minute in warm fine-grained soils, 300 mm (1 ft) per minute in dense warm granular soils with the use of a pilot hole, and up to 1524 mm (5 ft) per minute by use of thermally modified pilot holes in most soil types and temperatures.

Vibratory Hammers

Vibratory hammers are either hydraulic or electric and operate on a principle which uses two counter-rotating eccentric weights. Even the largest vibratory hammer has driving energy only equivalent to a small impact hammer, and will perform the same should difficult driving be encountered due to coarse granular or dense material. They are particularly good in fine-grained saturated soils or under conditions where soil particles can be displaced. As a result, vibratory hammers are highly efficient when used with thermally modified pilot holes.

Properly sized vibratory hammers have achieved driving rates in permafrost of less than 150 mm (0.5 ft) per minute at best in some warm fine-grained soils, but approximately 6 m (20 ft) or more per minute in most cold permafrost soils when the thermally modified pilot hole method is used properly.

Sonic Hammers

Often confused with vibratory hammers, sonic hammers and drills are inherently capable of tremendous driving rates. These high frequency devices offer great potential, but at the present time they are expensive, few in number and have many significant operational problems, particularly in cold weather. In most frozen fine-grained soils without pilot holes, driving rates comparable to vibratory hammers using thermally modified pilot holes have been achieved. To date, frozen granular soils have presented difficult driving for this type of pile hammer and thermally modified pilot holes have been used under these conditions to speed pile installation. Without the use of thermally modified pilot holes, voids have been noted around the pile near the ground surface, and typically pile embedment is increased to account for loss of strength in these upper sections.

DRIVEN PILE INSTALLATION TOLERANCES

Designers specifying driven piles must recognize that placement tolerances are to be expected, and plans must be detailed accordingly. Horizontal tolerances of piles installed with an impact hammer can be ± 50 mm (2 in), with an extreme of ± 76 mm (3 in) in plan, while variation from plumb may be up to 2%. Vibratory hammers are somewhat better in this regard, and can usually drive piles to within a 12 mm (0.5 in) horizontal tolerance and virtually plumb. This is because piles can be vibrated up and down the thawed pilot hole until desired tolerances are achieved. Piles driven by impact hammers cannot be adjusted in this manner. An important factor in achieving specified design tolerances if pilot holes are used is to drill an accurate pilot hole, since the pile follows hole alignment during driving. At times it may be desirable to drill the pilot hole 0.5 - 1 m (1 - 2 ft) deeper than pile tip elevation, particularly if driving to close vertical tolerances.

To reduce potential accumulations of soil and water pressures within driven pipe piles, placement of a small diameter weep hole in the pile wall prior to driving just above final ground line elevation is recommended. From the authors' experience, this hole need not be greater than 25 mm (1 in) in diameter. It has been noted on several driving jobs that water will spray out of these weep holes many feet as the pile nears grade; when welded pile cap is attached to pile top.

EXPERIMENTAL TESTING

Split Spoon Frozen Soil Penetration

Experiments using standard penetration tests (SPT) in frozen ground are shown in Figure 1. Analysis of these data and subjective comparison to past pile driving and soil sampling experience show that suitably designed piles probably can be driven in fine-grained soils as cold as -3°C (26.6°F) and in coarse-grained soils of possibly -1°C (30.2°F) using pilot holes. Piles probably cannot be driven efficiently without pilot holes in frozen soils much colder than -0.5° to -1.0°C (31.1° to 30.2°F), depending on soil type. With this information, parameters are established for potential driven pile foundations in permafrost. Obviously, in permafrost colder than these temperatures, some method such as pilot hole thermal modification must be used to achieve suitable soil temperature in the immediate pile area during driving.

Pilot Hole Thermal Modification

The technique of modifying permafrost temperature by use of a small pilot hole and non-circulated hot water allows easy pile driving where previously it seemed impossible. A pilot hole is a drilled hole generally extending to the desired pile tip depth. Pilot hole thermal modification has proven to be a more reliable and controllable method than other methods of thawing, including steaming, and if used properly affects the soil regime significantly less than drill and slurry methods.

Use of water-filled pilot holes in permafrost may have other important implications. Water offers a noncompressible media which when subject to shock tends to transfer vibrations, causing soil particles to temporarily loosen and then densify against the pile thereby promoting good adfreeze bond between the pile and soil. Water also fills all voids in material around the pile, thus assuring a strong pile/soil/ice bond. Regardless of the exact mechanism, water-filled pilot holes in general allow piles to be driven easily and improves placement accuracy.

Drilled pilot hole diameter can be determined for any size pile by equating the pilot hole diameter to the pile diameter and subtracting 2d. Where d is the desired isotherm distance from the edge of the pilot hole in mm (in). For H-piles, an equivalent diameter equal to slightly larger than the section depth may be appropriate.

Using concepts taken from Nottingham (1981) concerning early discoveries relating to pile driving with water-filled pilot holes, and curve fitting techniques, the following approximate relationship was established:

$$d = k(T)^{1/2} \qquad (1)$$

here: d = isotherm distance from the pilot hole edge in mm (in)

k = constant for various soil types and isotherm desired

T = ground thaw time in minutes

Considering a -3°C (26.6°F) isotherm for silty soils and a -1°C (30.2°F) isotherm for gravelly soils, k will be approximately 0.3 - 0.5 for most conditions, using British units. For a thaw time of 60 minutes, d from the equation above will be approximately 50 - 101 mm (2 - 4 in).

In practice, vibratory pile driving tends to cause soil to be vibrated from the sides of the hole and be displaced to the pile tip. This will cause the pilot hole water to rise along or in the pile, depending on type and results in the following:

1. Soil refreezes near the pile tip faster because of lower heat requirements, as verified by field measurements.

2. The water-soil mixture created by driving action acts to effectively slurry the pile into place.

To date, field observations in -5°C (23°F) soils indicate freezeback in less than 2 days and in -7°C (19.4°F) soils in about 1 day. Structural strength for most load applications is achieved after this period. To the authors' knowledge, no significant frost jacking during the refreeze process has been observed or measured in the Prudhoe Bay area.

Based on laboratory tests and field experience, initial temperature of pilot hole water does not appear to be critical. Water temperatures between 15-27°C (60-80°F) may be suitable for warmer permafrost, while water temperatures between 65-100°C (150-212°F) appear to be appropriate for cold permafrost. Generally, water in actual installations has been placed in pilot holes 40-60 minutes before pile driving with good success.

PILE LOADING CRITERIA

Presented in Figure 2 is an idealized mode of pile action in ice-rich permafrost under constant loading. After a load increment is applied, for a period of a few hours to a few days and depending on pile length, a load adjustment period will be required for stresses to be uniformly distributed over the pile surface. This period is often described in literature as primary creep. Steady state creep is of interest for the low long-term adfreeze stresses normally used in design, and is often referred to as secondary creep. For most structural applications, it is usually necessary to limit long-term creep of piles to less than 12 - 25 mm (0.5 - 1 in); thus pile failure within the tertiary creep region is of less interest to the design engineer.

Four specific conditions are of importance to the design engineer:

1. short-term vertical loading

2. long-term vertical loading

3. frost jacking loading

4. lateral loading

Conditions 1 and 2 are discussed in this paper.

Short-Term Vertical Loading

Short-term pile tests in cold permafrost have demonstrated tremendous adfreeze resistance values, but values rapidly decrease near 0°C (32°F). Short-term has been conservatively taken in this report to be loads of generally less than 5 hours' duration. This loading group contains the following categories:

1. building live loads (other than permanent loads such as furniture, files, etc.)

2. wind loads

3. earthquake loads

4. moving vehicle loads

5. ice impact forces

6. other loads applied for short duration

By accurately assessing these loads, the design engineer can reduce pile lengths where short-term loads are a significant factor because of potentially higher allowable adfreeze stress in most cases, and where creep is not an important factor. Figure 3 depicts the authors current recommended design adfreeze stresses with an approximate safety factor of 3 for the short-term condition. It is important to note that for driven piles the adfreeze of coarse-grained soils to steel is much lower than for fine-grained soils. The thermally modified pilot hole approach helps to assure a kind of slurry bond; however, high allowable adfreeze stresses are not advisable in coarse soils. Tests on slurry piles indicate little difference in strength compared to similar size driven piles in fine-grained soils, but slurry piles exhibit greater short-term strength in coarse-grained soils.

Design charts shown in this paper do not account for end bearing, which may be significant for some conditions, such as gravel. However, end bearing is ignored in favor of a more conservative approach at this time.

When designing piles, engineers should disregard pile embedment in the active zone as contributing to pile strength.

The graphs used here do not account for saline soil conditions. The design engineer confronted with this situation should perform additional tests, which are beyond the scope of this paper.

Long-Term Vertical Loading

Piles in permafrost are subject to creep-related settlement when loads are of a sustained nature. This cannot be overemphasized in designing significant structures such as water tanks, heavy machinery supports, or other critical structures. The phenomenon is not unlike creep, which engineers commonly consider for design of concrete or timber structures.

Conversely, failure of the designer to separate short-term loads from long-term loads may lead to uneconomical foundation solutions.

Long-term creep is sensitive primarily to these aspects:

1. sustained loading
2. soil temperature
3. pile diameter
4. soil type and moisture content
5. vibrations
6. adfreeze stress

Long-term deformation of ice and ice-rich soils may be approximately represented by steady-state creep where the flow law for ice provides the upper limit for ice-rich soil. Limited numbers of long-term tests on driven piles in various soil types and temperatures have produced a series of approximate creep rates for various adfreeze values.

Pile displacement rate is very dependent upon the applied shaft shear stress. However, the effect of temperature is also clearly important, with pile displacement rates changing by as much as one order of magnitude for soil temperature changes of a few degrees.

The effect of pile diameter is also important, as found by several other researchers (Nixon and McRoberts 1976, Weaver 1979). Increasing pile diameter appears to lower allowable stress for equal creep rates. In other words, under equal adfreeze stress, a large diameter pile will settle faster than a smaller pile.

In piles supporting sustained loads, long-term deformation is highly sensitive to adfreeze stress, and design loads must be held to low levels for supports which are critical with respect to deflection.

Figure 4 was constructed from the best data currently available for ice-rich soils. The chart is based mainly on soil temperature and pile size, and only generally considers soil type and moisture content. This chart can be used conservatively for most soil types, including granular soils, but pure ice will have greater creep rates. Soils with ice content greater than 50% should be viewed with consideration given to potentially greater creep rates and reduced adfreeze strength.

Long-term creep calculations should consider average soil temperature over the pile length, excluding the active layer. Where average temperature varies with the season, it may be desirable to evaluate time increments to reflect these variations. Where unusual conditions such as heavily long-term loaded closely spaced piles are used, group creep action should be given special consideration; this aspect is beyond the scope of this paper.

Driving Methods

The authors recommend the use of driven piles in thermally modified pilot holes for the following reasons:

1. Foundation soil can be logged and examined for exact conditions at each pile.

2. Difficulties in achieving driving tolerances are reduced.

3. Driving stresses and pile installation time are reduced.

4. Soil slurry is developed around the pile thus assuring a more complete adfreeze bond.

5. Impact on the permafrost thermal regime around the pile is minimized, compared to slurry methods.

6. Discontinuous permafrost, taliks, and perched water tables present few construction problems when compared to slurry piles in similar conditions.

CONCLUSIONS

Driven piles in permafrost can offer an attractive alternative to many foundation systems. Suitable design procedures recognizing permafrost peculiarities must be used, with attention to loads, creep, and frost jacking over the design life. Currently driving procedures and techniques presented in this paper are being implemented during production driving tests on the North Slope.

For more detailed information the reader is referred to the paper by Nottingham and Christopherson (1983). Additional reference data can be found in the paper by Victor Manikian (1983).

ACKNOWLEDGMENTS

The authors are grateful to ARCO Oil and Gas Company for their cooperation and dedication to the further development of driven pile techniques in permafrost on the North Slope. Also the authors wish to thank Todd Nottingham for his technical assistance in experiments for this project.

REFERENCES

Crory, F.E., 1973, Installation of driven test piles in permafrost at Bethel Air Force Station Alaska, U.S. Army Cold Regions Research Engineering Laboratory, Technical Report 139.

Crory, F.E., 1975, Bridge foundations in permafrost areas, U.S. Army Cold Regions Research Engineering Laboratory, Technical Report 266.

Manikian, V., 1983, Pile Driving and Load Tests in Permafrost for the Kuparuk Pipeline System, in Proceedings of the Fourth International Conference on Permafrost, V. 1: Fairbanks, Alaska USA, Polar Research Board, National Academy of Sciences.

Nixon, J.F., 1978, Foundation design approaches in permafrost areas, Canadian Geotechnical Journal, V. 15(1), P. 96-112.

Nixon, J.F., and McRoberts, E.C., 1976, A design approach for pile foundations in permafrost, Canadian Geotechnical Journal, V. 13(1), P. 40-57.

Nottingham, D., 1981, Method and H-pile tip for driving piles in permafrost, U.S. Patent 4,297,056.

Nottingham, D., and Christopherson, A.B., 1983, Design criteria for driven piles in permafrost, Research report, State of Alaska Department of Transportation and Public Facilities.

Parameswaran, V.R., 1978, Adreeze strength of frozen sand to model piles, Canadian Geotechnical Journal, V. 15(4), P. 494-500.

Parameswaran, V.R., 1981, Adreeze strength of model piles in ice. Canadian Geotechnical Journal, V. 18(1), P. 8-16.

Rooney, J.W., Nottingham, D., and Davison, B.E., 1976, Driven H-pile foundations in frozen sands and gravels, Proceedings--Second International Symposium on Cold Region Engineering, Fairbanks, Alaska, pp. 169-188.

Tsytovich, N.A., 1975, The mechanics of frozen ground, McGraw-Hill Book Company.

Weaver, J.S., 1979, Pile foundations in permafrost, Doctoral Thesis submitted to Department of Civil Engineering, Edmonton, Alberta.

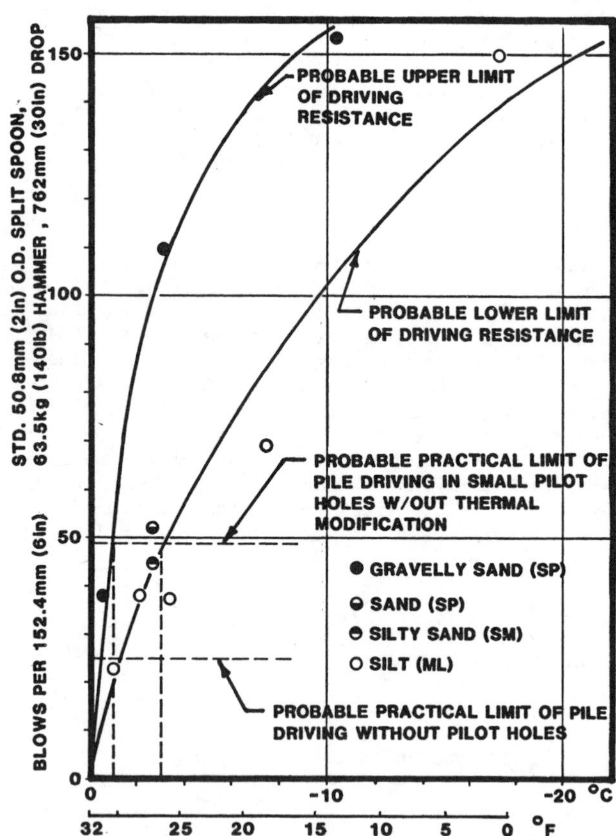

FIGURE 1 Split-Spoon Soil Sampling Resistance Versus Frozen Soil Temperature.

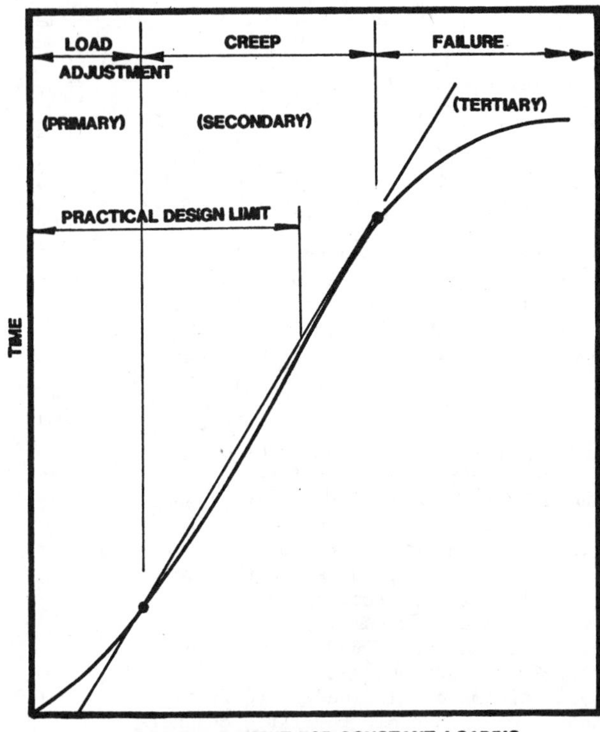

FIGURE 2 Idealized Modes Of Pile Settlement in Ice-Rich Permafrost.

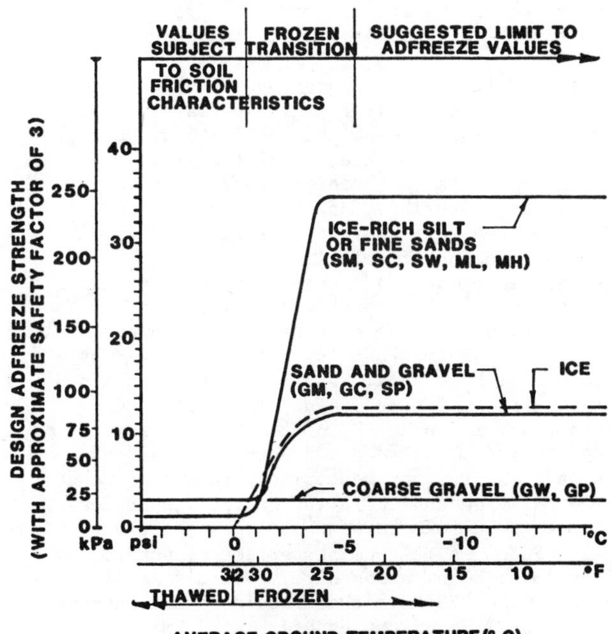

FIGURE 3 Proposed Maximum Short-Term Design Adfreeze To Steel Piles.

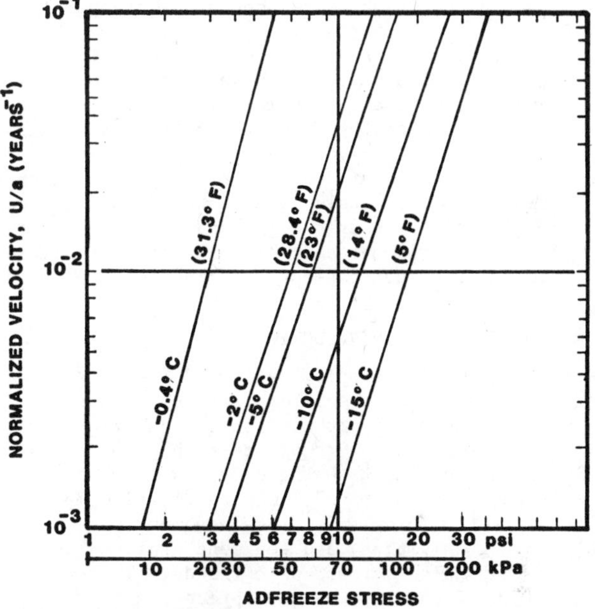

FIGURE 4 Proposed Design Pile Settlement for Ice-Rich Permafrost.

PERMAFROST ON MARS: DISTRIBUTION, FORMATION, AND GEOLOGICAL ROLE

Dag Nummedal

Department of Geology, Louisiana State University,
Baton Rouge, Louisiana 70803-4101 USA

Temperature data for Mars suggest the existence of global permafrost. Due to the low pressure of water vapor in the Martian atmosphere the frost point lies at about 198K. Consequently, ground ice can only be in diffusive equilibrium with the atmosphere at relatively high latitudes (poleward of about 40°). Slow diffusion rates through fine-grained regolith, however, may permit ground ice to have existed for billions of years anywhere within the permafrost zone.

The climatic changes on Mars, on time-scales of millions of years, appear to be directly controlled by astronomical variables. Climatic changes on time-scales of billions of years are probably of geological nature. It appears likely that an early phase of intense outgassing could have produced global temperatures considerably in excess of those at present.

Probably more than 100 m of precipitable water was released during this early phase of outgassing. The water may have charged a global ground water system which subsequently froze during a period of cooling climate. The downward progression of the freezing front may have increased the pressure of the underlying ground water system. Locally, the artesian pressure could have exceeded the lithostatic pressure, fractured the ground ice seal and released water at enormous discharges. Alternatively, the pressure build-up in the confined aquifers could have liquefied subsurface unconsolidated materials leading to fracturing of the overlying ground ice layer. Both mechanisms would explain the large channels as results of catastrophic discharge. Smaller-scale morphologic features indicative of deterioration of near-surface ground ice occur over vast areas on Mars.

INTRODUCTION

The morphology of channels, valleys, chaotic and fretted terrains and many smaller features on Mars, is consistent with the hypothesis that localized deterioration of thick layers of ice-rich permafrost has been a dominant geologic process on the Martian surface. Such ground ice deterioration may have given rise to large-scale mass movement, including sliding, slumping, and sediment gravity flowage, perhaps also catastrophic floods. In contrast to Earth, such mass movement processes on Mars lack effective competition from erosion by surface runoff. Therefore, Martian features due to mass movement have grown to reach immense size without being greatly modified by secondary erosional processes.

The Viking mission to Mars in 1976 provided adequate measurements of the relevant physical parameters to constrain models for Martian permafrost. The temperatures on Mars suggest thick global permafrost and models for planetary evolution indicate abundant water. Consequently, the most reasonable conclusion based on the probable physics of the Martian regolith is that it does contain a thick layer of ice-rich permafrost, a conclusion consistent with the observed surface geology.

GLOBAL DISTRIBUTION OF PERMAFROST

Permafrost, as originally defined by Muller (1947), refers to "a thickness of soil or other superficial deposit, or even of bedrock, at a variable depth beneath the surface of the earth in which a temperature below freezing has existed continually for a long time (from two years to tens of thousands of years). Permanently frozen ground is defined exclusively on the basis of temperature, irrespective of texture, degree of induration, water content, or lithologic character". Unfortunately, some confusion exists in the planetary literature regarding the use of "permafrost", hence the need to explicitly state the above definition.

Muller's definition appears consistent with the view that 0°C is the criterion for permafrost. Most workers currently use this definition (Washburn 1980), rather than the actual freezing point which varies with a number of factors including water content, lithology, and salinity.

Permafrost on Mars plays a potentially major role in the cycling of volatiles and in its effects on crustal geological processes. The mean annual surface temperature on Mars is everywhere below 273 K. Thus, permafrost is global. The lower boundary of the permafrost layer is controlled by the geothermal heat flux. Fanale (1976) determined the present rate of heat production on Mars by assuming that, in bulk, the planet had the same ^{40}K, ^{232}Th, ^{238}U, and ^{235}U concentrations as chondrites (a type of stony meteorite with small rounded bodies of olivine and enstatite called chondrules). Based on this he found a near-surface heat flux of 29.3 ergs/cm^2/s. Assuming perfect temperature equilibrium with the atmosphere, Fanale's model showed that the 273 K isotherm would be encountered at a depth of 1 km at the equator, at 2 km at 40° latitude and at greater depths at higher latitudes (Figure 1). These numbers probably give the maximum thickness of the global permafrost. Toksoz and Hsui (1978) calculated a heat flux of 35 ergs/cm^2/s. This higher value would correspond to a steeper temperature gradient and somewhat thinner permafrost

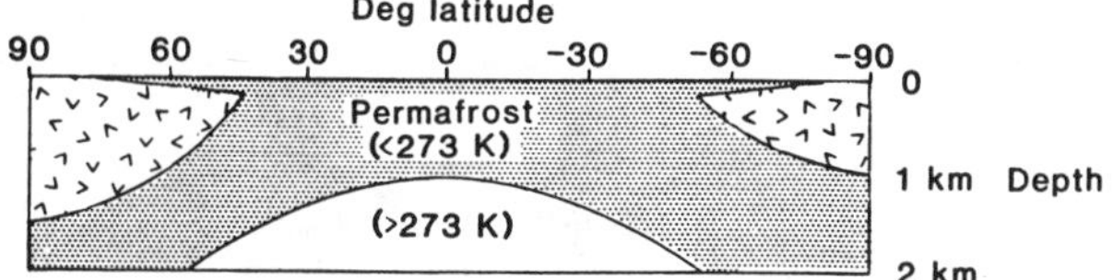

FIGURE 1 Theoretical distribution of ice and subfreezing (< 273 K) temperatures on Mars. Ground ice is in equilibrium with respect to the present atmosphere in lenticular regions extending from the poles equatorward to about 40° latitude. Ice may also be present in the remaining permafrost zone because of slow water vapor diffusion through fine-grained regolith. Modified from Fanale (1976).

zone than the one calculated by Fanale (1976). Rossbacker and Judson (1981) have computed the permafrost distribution based on this revised heat flux value.

The zone in which ground ice could exist in equilibrium with the atmosphere is much smaller than the permafrost zone, because of sublimation of ice in diffusive contact with the atmosphere wherever the temperature exceeds the frost point (the temperature at which ice will begin to precipitate out of atmospheric water vapor). The upper boundary of the zone containing permanent ice intersects the ground surface at the edge of the permanent ice cap. The base is at a depth of 1 to 1.5 km at the poles. The zone of equilibrium ground ice narrows to zero at about 40° latitude (Figure 1).

The pattern depicted in Figure 1 is highly idealized. It assumes perfect diffusive contact between the ground ice and the atmosphere. The soil, however, is probably not in perfect diffusive contact with the atmosphere at all depths. Due to slow diffusion rates of water vapor, ground ice could exist for billions of years at nearly any depth within the permafrost layer (Smoluchowski 1968).

Measurements of the atmospheric water vapor by the Mars atmospheric water detector (MAWD) onboard the Viking orbiters, demonstrated diurnal and seasonal variability within a range of 5 to 40 precipitable μm in equatorial and mid-latitudes (Farmer et al.1977). Using the equatorial average of 12 pr. μm, and assuming a well-mixed atmosphere, Farmer and Doms (1979) calculated the associated frost point temperature to be 198 K. Figure 2 illustrates the results of their calculations of the depth below the surface at which temperatures during the diurnal cycle never exceed this frost point temperature. In this figure the regions labeled "cold" refer to regolith which never rises above the frost point temperature any time during the year, i.e these are the zones of equilibrium ice. The central hatched area, labeled "hot" is characterized by temperatures which do exceed the atmospheric frost point at some time during the day in all seasons. Farmer and Doms' (1979) investigation shows that between latitudes 35°S and 46°N the near-surface temperatures are always above the frost point, so that ice can never be present in equilibrium with the atmosphere. The near surface regolith at these latitudes is therefore

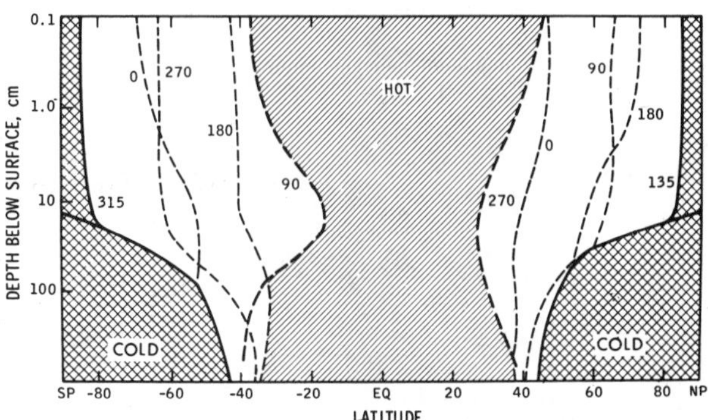

FIGURE 2 Stability of water ice in the Martian regolith as a function of latitude. The results are based on assumed perfect diffusive exchange between the ground ice and an average well-mixed Martian atmosphere containing 12 μm of precipitable water. Such an atmosphere gives a frost point of 198 K. Regions labeled "cold" may retain permanent ground ice, that labeled "hot" will never retain equilibrium ice. In intervening areas water ice is stable during part of the year. Numbers on the vertical boundaries (270, 0, 90 ...) refer to areocentric longitude (L_s). L_s = 0 corresponds to northern vernal equinox. From Farmer and Doms (Journal of Geophysical Research, v. 84, p. 2881-2888, 1979, with permission).

"desiccated", or free of ice and water, at least within layers which maintain diffusive contact with the atmosphere. At latitudes higher than about 45° the regolith at a depth of one meter never reaches the frost point at any time of the year; this is the region of potential permanent ice. The upper boundary of the zone of permanent ice intersects the surface at the edge of the ice caps (~80°).

CLIMATIC CHANGE

The global permafrost distribution discussed in the preceding chapter is based on the currently observed Martian climate. The Martian temperature regime, however, is variable. Kieffer et al's. (1977) surface temperature model for Mars calls for wide diurnal and seasonal temperature fluctuations. At equatorial latitudes the daily maxima range from 290 K to 260 K. Daily minima range from 180 K to 170 K (Figure 3). The equatorial zone is warmest during southern hemisphere summer. Surface temperatures observed by the infrared thermal mapper (IRTM) during the Viking primary mission ranged from 130 to 290 K (Kieffer et al. 1977).

After decades of controversy the idea of primary astronomical control of periodic fluctuations in the Earth's climate now appears to have been firmly established (Berger 1977, Imbrie and Imbrie 1980). The Earth's climatic history, however, is exceedingly complex because the signature of the astronomical variations is overprinted by factors controlled by oceanic circulation and global tectonics. Also, of the orbital parameters responsible for climatic change, axial obliquity and orbital eccentricity vary relatively little for the Earth.

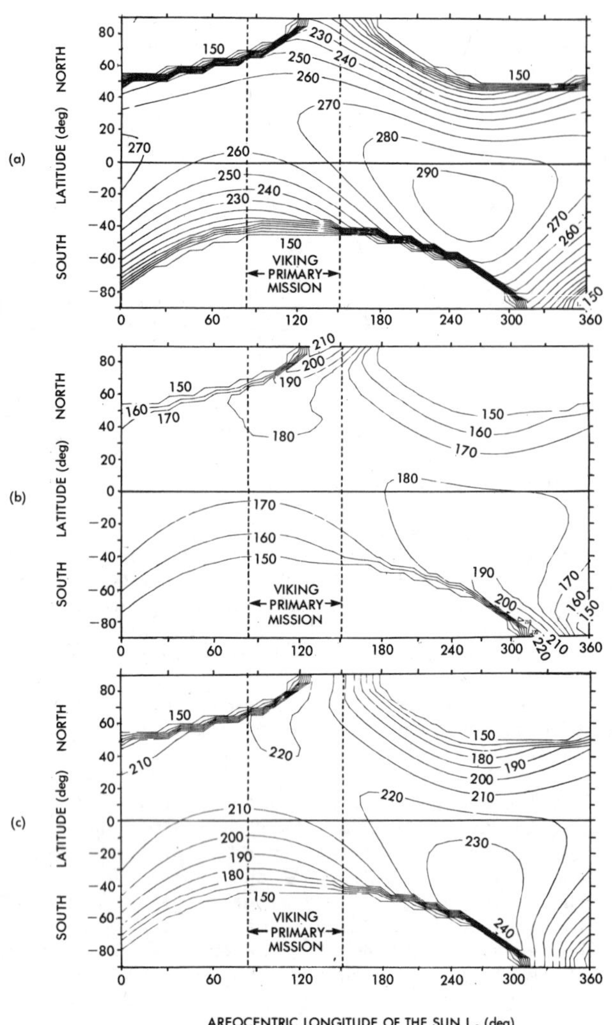

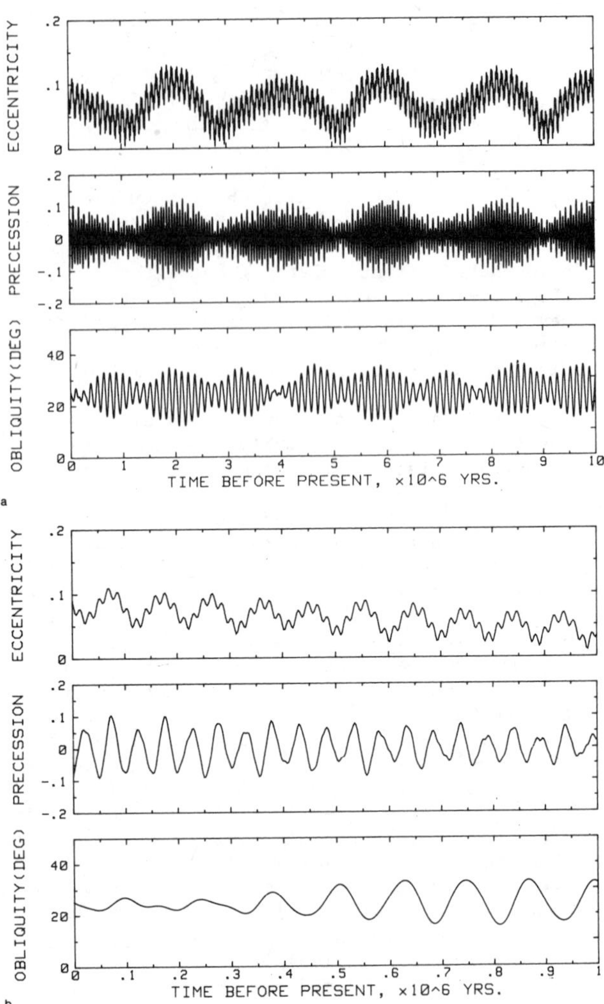

FIGURE 3 Diurnal surface temperature means and extremes for the primary Viking thermal model. (a) maximum temperatures, (b) minimum temperatures and (c) mean temperatures. Modified from Kieffer et al. 1977.

FIGURE 4 Calculated time series for the variation in eccentricity, precession index and obliquity for Mars. (a) presents results for the last 10 million years, (b) presents, with greater resolution, the variations within the last 1 million years. Based on analysis by Ward (1974). Figure reproduced from Cutts and Lewis (Icarus, v. 50, p. 216-244, 1982, with permission).

For Mars the situation seems to favor a much stronger astronomical control on climatic change. The absence of oceans will probably produce a much more stable global circulation pattern, and orbital parameters are subject to great change. The obliquity of the Mars spin axis varies on a time scale of 10^5 years (Ward 1974). The amplitude of this behavior is itself modulated on a 10^6 year time scale by changes in orbital inclination (Figure 4). Together with orbital eccentricity variations and changes in the heliocentric longitude of perihelion of Mars' orbit, the obliquity variations will produce great changes in the intensity and distribution of insolation on Mars. The climatic response is likely to be great and geological evidence should be discernible. It is commonly thought that the layered terrain near the Martian poles preserves a record of cyclic aeolian sedimentation probably caused by periodic changes in global climate (Pollack and Toon 1982, Carr, 1982, Cutts and Lewis, 1982).

As on Earth, longer-term climatic change driven by geological factors has probably also occurred on Mars. One such factor is possible polar wandering in response to a reorientation of the Mars spin axis following the growth of the Tharsis volcanic province.

The climate on Mars early in that planet's geologic history may have been dramatically different from the present because of possible "greenhouse effects" in the early atmosphere. Mars today has an atmosphere which does not lead to appreciable self-enhanced warming. However, if a primordial reducing atmosphere occurred, greenhouse warming could have been substantial (Sagan, 1977). Pollack (1979) proposed that abundant water was available in the primordial source materials, and that Mars had been subject to a brief period of intense outgassing early in its history. The resulting atmosphere would have been rich in both oxidized gas species (carbon dioxide) and reduced ones (ammonia). Such an atmosphere would have produced both high atmospheric pressure and high temperatures.

The depth penetration of the thermal wave corresponding to these temperature variations depends on the wave frequency. The daily thermal wave probably affects only the upper few centimeters, the annual changes the upper 1-10 m. These fluctuations produce a layer which is subject to alternating ice formation and desiccation on an annual basis. In mid-latitudes this layer is from 10 cm to 1 m thick, in the equatorial zone it extends down to 10 m (Figure 2). Temperature changes on time scales of 10^5-10^6 years should penetrate to depths of a few hundred meters to 1 km. Temperature changes over geologic time scales of billions of years may affect the entire regolith (Table 1).

TABLE 1 Climatic changes on Mars: Factors, Time Scales, and Depths of Penetration of the Associated Thermal Wave

Time Scale	Cause	Penetration Depth of Thermal Wave
Daily	Planetary Rotation	cm
Year	Orbital revolution	1-10 m
10^5-10^6 years	Obliquity and eccentricity variations	100 m-1 km
1-4 x 10^9 years	Polar wandering Greenhouse effect Solar luminosity	km

The depth-porosity relationship of the Martian upper crust is still subject to conjecture because of the absence of relevant seismic data. Based on comparisons with Earth and the moon, Carr (1979) suggested that the base of the porous layer on Mars should be in the 10-20 km range. Pore space for the containment of interstitial ice would, therefore, be expected to exist throughout the permafrost zone. Pore fluids could exist down to considerable depth below the permafrost boundary.

FORMATION OF ICE-RICH PERMAFROST

Prior to the Mariner 9 mission in 1971-1972 Mars was considered a planet greatly impoverished in water. Morphological evidence for flowing water in the form of large channels, discovered by Mariner 9 (McCauley et al. 1972), completely changed that perception. As stated by Pollack (1979), "Water is the only plausible, cosmically abundant substance that meets all the morphological constraints imposed by the many channels on Mars." Water is probably present in frozen interstitial form in the permafrost zone, some is tied up in the polar caps, and some is tied up in hydrated minerals or adsorbed onto clay particles in the regolith.

Fanale (1976) estimated that for reasonable porosities and thickness of the impact-generated regolith one could easily accomodate the equivalent of 400 m of water evenly spread over the surface of the planet. Anders and Owen (1977), basing their analysis on the atmospheric abundance of the primordial nobel gas isotope ^{36}Ar, concluded that the outgassing of Mars should have produced the equivalent of 9.5 m of water, distributed globally. This is the lowest estimate available. Pollack and Black (1979) predicted a Mars water inventory in the range of 80-160 m.

One commonly held model for planetary accretion implies an increasing volatile abundance with distance from the early sun. Presumably, this was controlled by the temperature gradient in the solar nebula. Accordingly, one should expect more primordial water on Mars than on Earth. As we have seen, however, available estimates imply that Mars has outgassed much less water than Earth. Ringwood (1978) has addressed this problem in a comprehensive study of water in the solar system. He argues that Mars probably did have a much greater amount of water-rich condensates in its primordial composition, but due to a low accretion energy which prevented complete differentiation of that planet, the water remained trapped in the mantle. The differentiated Earth with its active tectonic system has a mechanism for effective release of mantle water to the surface. Such a mechanism does not exist on Mars. As stated by Baker (1982), "The question is not why we see an abundance of water-related surficial features on Mars. Rather, the question is why there is not much more evidence for water on the planet's surface". The answer, according to Ringwood (1978), lies in the incomplete degassing of the planet.

The temperatures on Mars suggest thick global permafrost. Models for planetary evolution indicate abundant water. Consequently, a most reasonable conclusion is that the Martian regolith contain large amounts of ice-rich permafrost. In all probability, additional water is available in the liquid phase in pore spaces beneath the ice-rich zone.

This layer of ice-rich permafrost could have formed, as in the terrestrial case, from downward percolation of meteoric water precipitated as the planet's atmosphere cooled after an initial high-temperature stage (Carr 1979). This would be the most likely scenario if the planet experienced an early phase of massive outgassing, with consequent "greenhouse" warming, as assumed in Pollack's (1979) model.

Instead of producing an ocean of liquid water, as on Earth, the outgassing of the planet Mars may have generated a subsurface reservoir of water and ground ice. Although this reservoir may play the role as both source and sink in the Martian water cycle, it does not drive a regular hydrologic system as on Earth. The geological effects of the Martian water reservoir differ greatly from those on Earth; the sculpted surface features differ accordingly.

INFERRED GEOLOGICAL CONSEQUENCES

Excellent papers by Carr and Schaber (1977) and

Lucchitta (1981) and books by Carr (1981) and Baker (1982) present an abundance of pictorial evidence for morphological features on Mars strongly indicative of ice-rich permafrost. Small-scale features include patterned ground, and a range of thermokarst features. Larger-scale features include areas of collapsed ("chaotic") terrains (Sharp 1973), abundant in the "headwater" regions of the large outflow channels (Sharp and Malin 1975) of the Chryse Basin, and many regions of dissected ("fretted") terrain along the edge of the Martian uplands (Sharp 1973, Gatto and Anderson 1975). Sapping, due to ground ice deterioration, is probably a global phenomenon; the process may have been particularly important in development of the valley networks of the old Martian terrains (Pieri 1980).

Soderblom and Wenner (1978) considered some of the geological implications of global permafrost. If, indeed, a thick ice-rich permafrost is underlain by a layer where water does reside in liquid form, a global discontinuity in regolith strength may have developed. Diagenetic processes would have progressed much more rapidly within this liquid zone than within the permafrost, perhaps producing tightly cemented rocks below the interface. This discontinuity subsequently controlled the development of scarps bounding the chaotic and fretted terrains. The lower zone would be much more resistant, allowing the overlying impact breccia and volcanic deposits to easily erode down to the common base level of that diagenetic interface. The escarpments along the fretted terrain margins have heights in the 1 to 2 km range with very slight regional variations (Sharp 1973).

Carr (1979) suggested that venting of subsurface water reservoirs through fracture of the ground ice seal might be a mechanism for the release of vast floods. Morphologic evidence for such floods are the large outflow channels. Carr's (1979) proposal calls for a regional system of confined aquifers. These were created by the progressive downward freezing of ground ice during a period of cooling climate. Liquid water between this freezing front and the base of the porous regolith would gradually increase in pressure. In regions downslope of high potential head differences the artesian pressure could have exceeded the lithostatic pressure, fractured the ground ice seal, and caused water to rush to the surface at enormous discharge rates. Carr's calculations yield discharge rates commensurate with the hydraulic geometry of the Chryse outflow channels.

Nummedal (1978) called on liquefaction of subsurface unconsolidated materials to explain the Chryse channels and associated chaos. This scenario was developed because of striking morphological similarities between the Martian chaos-channel systems and liquefaction slide scars in Norwegian quick-clay terrain (Bjerrum 1971) and submarine slides subsequently discovered on the Mississippi delta front (Nummedal and Prior 1981). If the Martian features are due to liquefaction, then most of the channel volume is due to overburden collapse rather than erosion by running water. The apparent headward growth of many outflow channels is consistent with headscarp retrogression, a common mechanism in terrestrial liquefaction (Prior and Suhayda 1979).

Liquefaction in unfrozen terrain is normally limited to very fine-grained materials where internal pore water pressure can reach the vertical confining pressure without substantial fluid leakage. In permafrost terrain, however, an ice-rich cap may provide the necessary seal such that even gravels may become liquefied (Finn et al. 1978). In fact, Carr's (1979) proposed system of confined aquifers is likely first to reach an internal pressure adequate for liquefaction of unconsolidated materials before the artesian pressure can rupture the overlying seal. Once such a highly sensitive subpermafrost layer was formed, a seismic shock (internal or impact) could have triggered the liquefaction, with consequent collapse and fluid release.

Terrestrial morphologic features formed by the collapse of ice-rich permafrost are generally of small scale. This appears to be due to their rapid modification by surface runoff. On Mars, on the contrary, such modification would be insignificant as it probably has not rained since the initial formation of ground ice collapse features. Consequently, the surface morphology on Mars has evolved through the continued action of mass movement. The related features have grown to reach immense size. Channels, valleys, chaos, and fretted terrains on Mars have grown to assume scales and morphologic characteristics similar to those found in the only terrestrial morphogenetic province where mass-movement is dominant: the submarine continental slope and rise (Nummedal and Prior 1981, Nummedal 1982).

CONCLUSIONS

Observed temperatures and calculated internal heat flux for Mars suggest the existence of global thick permafrost. The layer is probably 1 km thick near the equator and substantially thicker in higher latitudes.

Changes in the thermal regime of the Martian regolith have occurred in response to cyclic astronomic perturbations (time scales of a day to 10^6 years) and the geologic evolution of the planet ($\sim 10^9$ years). The consequent distribution of ice within the permafrost zone may be very complex.

Existing models for Martian outgassing yield an adequate amount of water to have produced a global ice-rich permafrost. Water may have entered the regolith either as meteoric water during a planetary warm phase or directly as juvenile water from below.

This ice-rich permafrost is commonly thought to be the cause of abundant "cold-region" morphologic features seen in Viking images. It is proposed that some of the large-scale channels and chaos areas derive from catastrophic failure of a permafrost seal above deep artesian aquifers or liquefaction of unconsolidated materials beneath such a seal.

REFERENCES

Anders, E., and Owen, T., 1977, Mars and earth: origin and abundance of volatiles: Science, v. 198, p. 453-465.

Baker, V. R., 1982, The Channels of Mars: University of Texas Press, Austin, Texas, 198 p.

Berger, A., 1977, Support for the astronomical

theory of climatic change: Nature, v. 296, p. 44-45.

Bjerrum, L., 1971, Kvikkleireskred; et studium av årsaksforhold og forebygnings-muligheter: Norwegian Geotechnical Institute, Publ. no. 89, 14 p. (in Norwegian).

Carr, N. H., 1979, Formation of Martian flood features by release of water from confined aquifers: Journal of Geophysical Research, v. 84, p. 2995-3007.

Carr, M. H., 1981, The Surface of Mars: New Haven, Conn., Yale University Press, 232 p.

Carr, M. H., 1982, Periodic climate change on Mars: Review of evidence and effects on distribution of volatiles: Icarus, v. 50, p. 129-139.

Carr, M. H., and Schaber, G. G., 1977, Martian permafrost features: Journal of Geophysical Research, v. 82, p. 4039-4054.

Cutts, J. A., and Lewis, B. H., 1982, Models of climate cycles recorded in Martian polar layered deposits: Icarus, v. 50, p. 216-244.

Fanale, F. P., 1976, Martian volatiles: Their degassing history and geochemical fate: Icarus, v. 28, p. 179-202.

Farmer, C. B., Davis, D. W., Holland, A. L., LaPort, D. D., and Doms, P. E., 1977, Mars: Water vapor observations from the Viking orbiters: Journal of Geophysical Research, p. 4225-4248.

Farmer, C. B., and Doms, P. E., 1979, Global seasonal variations of water vapor on Mars and the implications for permafrost: Journal of Geophysical Research, v. 84, p. 2881-2888.

Finn, W. D. L., Yong, R. N., and Lee, K. W., 1978, Liquefaction of thawed layers in frozen soil: Journal of Geotechnical Engineering Division, American Association of Civil Engineers, v. 104, p. 1243-1255.

Gatto, L. W., and Anderson, D. M., 1975, Alaskan thermokarst terrain and possible Martian analog: Science, v. 188, p. 255-257.

Imbrie, J., and Imbrie, D. C., 1980, Modeling the climatic response to orbital variations: Science, v. 207, p. 943-953.

Kieffer, H. H., Martin, T. Z., Peterfreund, A. R., Jakosky, B. M., Miner, E. D., and Palluconi, F. D., 1977, Thermal and albedo mapping of Mars during the Viking primary mission: Journal of Geophysical Research, v. 82, p. 4249-4291.

Lucchitta, B. K., 1981, Mars and Earth: Comparison of cold-climate features: Icarus, v. 45, p. 264-303.

McCauley, J. F., Carr, M. H., Cutts, J. A., Hartman, W. K., Masursky, H., Milton, D. J., Sharp, R. D., and Wilhelms, D. E., 1972, Preliminary Mariner 9 report on the geology of Mars: Icarus, v. 17, p. 289-327.

Muller, S. W., 1947, Permafrost or permanently frozen ground and related engineering problems: Ann Arbor, Mich., J. W. Edwards, 231 p.

Nummedal, D., 1978, The role of liquefaction in channel development on Mars: NASA Technical Memorandum 79729, p. 257-259.

Nummedal, D., 1982, Continental margin sedimentation: its relevance to the morphology of Mars: NASA Technical Memorandum 85127, p. 256-257.

Nummedal, D., and Prior, D. B., 1981, Generation of Martian chaos and channels by debris flows: Icarus, v. 45, p. 77-86.

Pieri, D. C., 1980, Martian valleys: Morphology, distribution, age and origin: Science, v. 210, p. 895-897.

Pollack, J. B., 1979, Climate change on the terrestrial planets: Icarus, v. 37, p. 479-553.

Pollack, J. B., and Black, D. C., 1979, Implications of the gas compositional measurements of Pioneer Venus for the origin of planetary atmospheres: Science, v. 205, p. 56-59.

Pollack, J. B., and Toon, O. B., 1982, Quasi-periodic climate change on Mars: A review: Icarus, v. 50, p. 259-287.

Prior, D. B., and Suhayda, J. N., 1979, Submarine mudslide morphology and development mechanisms: Proceedings 11th Offshore Technology Conference, p. 1055-1061.

Ringwood, A. E., 1978, Water in the solar system, in McIntyre, A. D., ed., Water, planets, and people: Australian Academy of Sciences, Canberra, p. 18-34.

Rossbacker, L. A., and Judson, S., 1981, Ground ice on Mars: Inventory, distribution and resulting landforms: Icarus, v. 45, p. 39-59.

Sagan, C., 1977, Reducing greenhouses and the temperature history of Earth and Mars: Nature, v. 269, p. 224-226.

Sharp, R. P., 1973, Mars: Fretted and chaotic terrains: Journal of Geophysical Research, v. 78, p. 4073-4083.

Sharp, R. P., and Malin, M. C., 1975, Channels on Mars: Geological Society of America, Bulletin, v. 86, p. 593-609.

Smoluchowski, R., 1968, Mars: Retention of ice: Science, v. 159, p. 1348-1350.

Soderblom, L. A., and Wenner, D. B., 1978, Possible fossil H_2O liquid-ice interfaces in the Martian crust: Icarus, v. 34, p. 662-637.

Toksoz, M. N., and Hsui, A. T., 1978, Thermal history and evolution of Mars: Icarus, v. 34, p. 537-547.

Ward, W. R., 1974, Climatic variations on Mars, 1, Astronomical theory of insolation: Journal of Geophysical Research, v. 79, p. 3335-3386.

Washburn, A. L., 1980, Geocryology: New York, John Wiley and Sons, 406 p.

PRACTICAL APPLICATION OF UNDERSLAB VENTILATION SYSTEM: PRUDHOE BAY CASE STUDY

William B. Odom
Sohio Alaska Petroleum Company, Pouch 6-612, Anchorage, Alaska 99502 USA

When designing buildings at grade on permafrost, consideration must be given to maintaining the frozen subgrade. One foundation system cools by ventilating ambient air through ducts in the subgrade. This paper responds to the need for documented results correlating performance of underslab ventilation systems to theoretical thermal analysis for operating facilities at Prudhoe Bay, Alaska. Experiences from the design and construction of a 7200 sq. ft. Fabrication Shop in Prudhoe during the 1981 summer construction season are the basis for the discussion. Design criteria recommendations are presented for slab-on-grade underslab ventilation systems for thermal analysis and system design. Aspects of construction specifications which contribute to ensuring good system performance are highlighted. The article concludes with a case study for Sohio's Fabrication Shop. Thermal analysis calculations to determine ventilation rates which contrast F. J. Sanger's method to J. F. Nixon's approach using the Fabrication Shop's design criteria are appended. Instrumentation data is tabulated for intake and exhaust duct temperatures, and subgrade temperatures over an eighteen (18) month period covering initial freeze-back, winter operation, and summer thaw. From information presented, correlations are made on how well theoretical analysis approximates actual performance of a slab-on-grade underslab ventilation system.

INTRODUCTION

When designing buildings at grade on permafrost, consideration must be given to maintaining the frozen subgrade to prevent excessive differential settlement. One method of ventilating a foundation system is by forced ventilation using ambient air through ducts in the subgrade.

This paper responds to the need for documented results correlating performance of underslab ventilation systems to theoretical thermal analysis for operating facilities at Prudhoe Bay, Alaska. Design criteria and aspects of construction specifications are presented. The article concludes with a case study of an underslab ventilation system for Sohio's Fabrication Shop.

ANALYSIS

Two thermal analysis methods are commonly used to determine subgrade ventilation rates. One method was authored by Sanger (1967). A more current and conservative method was presented by Harlan and Nixon (1978); the basis for this method was originally published by Nixon (1978) and includes some charts and two case histories. After applying each method to the Fabrication Shop's design criteria (see section on case study) and using a 3°C temperature difference between inlet and outlet duct temperatures for both methods, Harlan and Nixon's approach provided approximately a factor of 3 greater ventilation than that computed using Sanger's method. Furthermore, for our design criteria, Sanger's method suggested using a temperature difference between inlet and outlet duct temperatures of 15.4°C. When strictly applying Sanger's method with a 15.4°C design temperature difference, and using a 3°C design temperature difference with Harlan and Nixon's method, Harlan and Nixon provided a factor of 21 greater ventilation than Sanger.

In reviewing the two methods, Harlan and Nixon's approach is based on peak heat flux consisting of heat from the structure and heat gain from the pad subjected to a sinusoidal change in surface temperature, maintaining a small temperature difference between inlet and outlet air, and shutting ventilation off when ambient temperatures are above -4°C (25°F). Sanger's method is based on average heat flux consisting of heat gain from the structure and heat that must be removed to refreeze pad, and considers larger temperature differences between inlet and outlet air. Although Sanger's method does not consider a ventilation shutoff temperature, it does consider ducts closed during the summer thaw season. Harlan and Nixon's method appears to yield more conservative results because peak heat flux is used to design ventilation rates, and the heat gain from the pad is significant.

SYSTEM DESIGN

Subgrade Ducts

Helical, corrugated metal pipe (culvert material), hot dipped in aluminum was used for the Fabrication Shop subgrade ducts. Although an economical product, it is difficult to field weld should the seams leak. The Fabrication Shop corrugated ducts failed a 0.25 N/m^2 hydrotest, causing a 2 week delay while the system was reexcavated and repaired. Mild, thin-walled steel pipe would have allowed for field welding of joints and repair of seams. Another advantage of continuous mild steel pipe would be the ability to electrically thaw any plugged ducts; however, the high cost of steel pipe generally limits its use. Ducts should be sloped toward a collection point where access can be made to pump out any water which may infiltrate the system.

System Controls

As a safety precaution, it is desirable to supercool the foundation subgrade to insure that the summer thaw below the slab does not propagate into the tundra. Ideally, to take advantage of the annual cooling cycle, the fan system should operate when subgrade temperature is warmer than ambient air. This requires a subgrade thermal probe, ambient air thermal probe, and comparator. Location of the subgrade probe should be evaluated to determine a representative depth. This system will achieve maximum super cooling of subgrade. The Fabrication Shop, however, only uses an ambient air thermal probe; the system is set to operate when ambient air is -4°C (25°F) or colder. Although simple in operation, this system does not use the entire annual cooling cycle and can actually cause warming of the subgrade.

The most desirable system from a maintenance standpoint is a simple system such as one with a single fan for all ducts, access to each duct, and an on/off switch. Because we have experienced electrical component failures, controller malfunctions, and nonfunctioning dampers, careful selection of system components is required.

Intake and Exhaust Ducts

Location of intake/exhaust ducts should be a minimum of 3.7 m above grade. Systems with intake/exhaust ducts at low elevations are experiencing drifting snow problems. Intake/exhaust ducts located at roof line appear to perform well.

The exhaust ducts for the Fabrication Shop have a gooseneck profile and are detachable for inspection and maintenance. Dampers were installed to allow for balancing the system, and bird screens cap both intake and exhaust ducts.

Insulation

Extruded polystyrene is recommended over expanded polystyrene, or any of the urethane board insulations for below grade applications under permanent structures. Studies have shown that extruded polystyrene maintains its thermal characteristics better than urethane when subjected to freeze/thaw cycles in saturated soils conditions; in addition, urethanes tend to disintegrate. From a cost standpoint, extruded polystyrene is generally competitive with urethane when comparing cost per "R" value (thermal resistance coefficient). Where polystyrenes are used beneath buildings when spilled solvents such as diesel fuel could come in contact with the insulation, an impervious membrane should be used to protect the polystyrene.

Instrumentation

The subgrade should be instrumented in order to monitor the thermal performance of the subgrade. The Fabrication Shop used 3 nonreplaceable thermistor strings evenly spaced along the longitudinal centerline of the building. Each string had 4 nodes at 1.5 m on center with the first node located 1.5 m below grade. Each string was placed in a small diameter plastic pipe which was filled with a water and glycol mixture. Thermistors were chosen because of their high degree of accuracy, stability, and simplicity in obtaining readings. If thermistor strings cannot be replaced, a minimum of three thermistor strings should be located beneath any major building to cross check data. A digital ohm meter is required to read thermistor resistance, which is converted to temperature with a simple equation. Thermocouples were not used for the Fabrication Shop because of the difficulty in obtaining the proper instruments to read the sensors and the lack of knowledge of the parameters that affect the accuracy of the readings.

Radiant Heat

Typically, the heat loss computations for a ventilation system are based on slab temperature of approximately 15°C (60°F). Monitoring of the Fabrication Shop floor confirmed that this was a reasonable assumption for that building. The use of radiant heat can increase the slab temperature to 27°C or greater. If the slab temperature is not accurately determined, the long-term performance of the foundation system could be jeopardized. Radiant heat is not recommended as a method of heating buildings with subgrade ventilation systems. During the summer months, the ventilation system is inactive and the abnormally high heat input caused by radiant heat may thaw the subgrade below the theoretically computed depth.

Duct Spacing

Underslab ducts for the Fabrication Shop were arbitrarily spaced 10 times the diameter of the duct and ventilation rates were computed based on this duct spacing. A more rational approach to determine duct spacing would be to evaluate the maximum freeze radius of the duct subjected to the anticipated thermal load.

Thawing Provisions

Should the subgrade ducts become plugged with frozen water, provision should be made to easily thaw the plug. For one North Slope building, we used a steam probe and small pump which worked well; it is imperative that there be access to each duct. Heat tracing or installing a continuous small diameter pipe to inject steam would also work.

CONSTRUCTION SPECIFICATIONS

Installation Specifications

Specifications should explicitly state "how" and "when" the contractor is to install the ventilation ducts. The time of year is a significant parameter in the construction of an underslab ventilation system. During construction of the Fabrication Shop, the duct trenches filled with water, and because construction duration was not specified for excavation and backfill, the contractor was not required to expedite his work. Duct trenches were left open during the warmest season and for a longer period than anticipated. Ideally, the best procedure is to install the ducts during one construction season and to construct the building the subsequent construction season. However, as an alternate, we recommend a stringent installation specification which will require the contractor to perform in a short period of time. This will limit thaw bulb propagation.

Pressure Test

The ventilation system should be pressure tested prior to placement of backfill to eliminate the problem of water intrusion into the ducts during the summer thaw. The Fabrication Shop specifications required an air pressure test of 0.25 N/m^2 (one-in. water gauge) for 1 hour without an allowance for bleed-down. However, the contractor had difficulty in capping the ducts and in connecting the air pressure equipment to the ducts. Consequently, an equivalent hydrotest was substituted.

Temporary System Operation

Performance of a subgrade ventilation system assumes a full season of operation. As the construction of a major building may require 6 months, we recommend that provisions be made in the construction contract to require temporary operation of the underslab ventilation system when ambient air temperature is -4°C (25°F) or colder. Temporary heat in the building will be approximately the same as normal operating temperature. Therefore, it is essential that the ventilation system operate as soon as practical to ensure long term performance of the foundation system.

CASE STUDY

Synopsis

The Sohio Fabrication Shop is an 18 x 37 m preengineered building with a slab-on-grade foundation in conjunction with an active subgrade ventilation system. Figure 1 shows building plan dimensions and subgrade duct locations.

Figure 2 shows a typical section through floor slab. The existing gravel pad was approximately 1.2 m thick. In order to maintain access to adjacent buildings and roads, pad elevations could not be raised. Foundation construction started in June 1981, and Figure 3 illustrates water intrusion and sloughing of trenches during placement of subgrade ducts. The subgrade ventilation system was commissioned in November 1981.

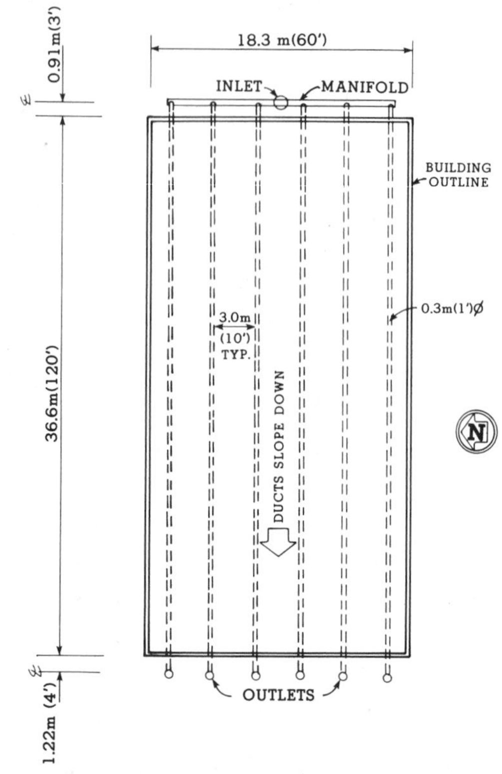

FIGURE 1 Subgrade Ducts for Sohio Fabrication Shop, Prudhoe Bay, Alaska

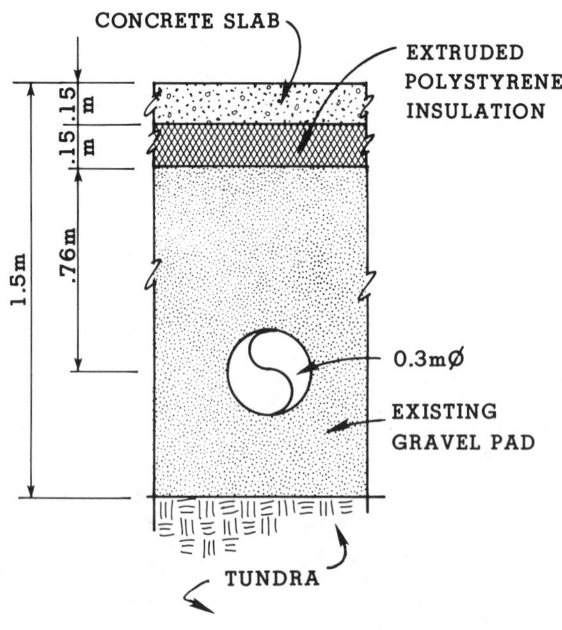

FIGURE 2 Typical Section Through Subgrade Duct

FIGURE 3 Subgrade Ducts During Summer Construction. Note water intrusion.

DATE	INTAKE (ambient) °C	AVE. OUTLET °C	△ T
11/5/81	-18	-4	14
11/19/81	-10	-3	7
12/1/81	-18	-6	12
12/9/81	-19	-7	12
12/16/81	-23	-9	14
12/22/81	-14	-8	6
12/30/81	-4	-12	8
1/19/82	-31	-15	16
2/2/82	-20	-15	5
1/10/83	-29	-23	6

FIGURE 5 Air Duct Temperatures

Building and system performance has been good to date. As our thermal analysis indicated that thaw bulb depths were marginally safe, the total subgrade ventilation rate was increased from 0.189 m^3/s (400 ft^3/min) to 1.89 m^3/s (4,000 ft^3/min). During freezeback, 1.6 mm cracks developed in the slab and were thought to be a combination of concrete shrinkage and expansion of the saturated subgrade during freezing. However, the cracks did not propagate, and the floor's serviceability remains intact. Figure 4 shows thermal profiles over a 1 year period. Figure 5 presents inlet and outlet duct temperature readings taken during the first and second year of operation. Figure 6 illustrates system simplicity using a single fan on the intake manifold. Figure 7 shows the problem of drifting snow over duct outlets. This problem was later rectified by raising the duct outlet elevations to approximately 3.7 m.

FIGURE 6 Intake Manifold
Note single fan system.

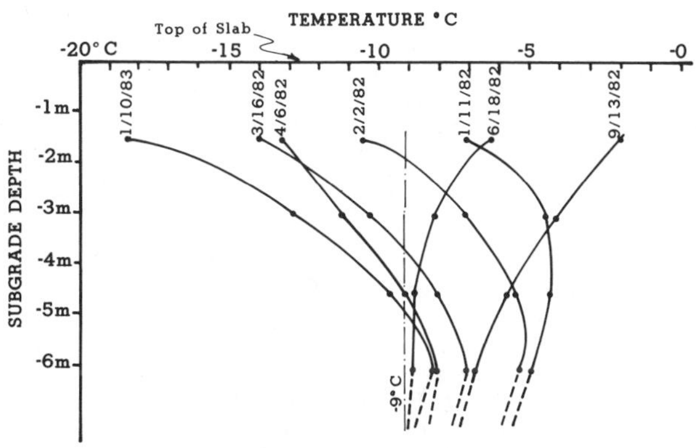

NOTE: Undisturbed Tundra
Mean Annual Temperature -9°C

FIGURE 4 Temperature Profiles
Center of Building

FIGURE 7 Exhaust Ducts
Note drifting snow.

Design Criteria

Freezing index (F.I.)	4166°C days (7500°F days)
Thaw index (T.I.)	555°C days (1000°F days)
Mean annual air temperature	-12.2°C
Mean annual temperature variation	17.8°C
Freezing season	265 days
Building temperature (Ts)	15.6°C
Insulation Thickness	0.15 m
Thermal conductivity insulation (K_1)	0.0311 W/mK
Thermal conductivity gravel pad (K_2)	1.80 W/mK
Dry density gravel pad	2,020 kg/m^3
Moisture content pad (w)	3%
Volumetric heat capacity air (C_a) @ -12.2°C	1.355 kJ/M^{3}K
Time ambient air above -4°C	126 days
Temperature rise through ducts	3°C (Harlan and Nixon or 15.4°C (Sanger)
Mean annual ground temperature (zero amplitude @ -14 m below existing grade)	-9°C

Thermal Analysis--Harlan and Nixon's Method

Substituting the Fabrication Shop's design criteria into formulas presented by Harlan and Nixon, and using a temperature difference between inlet and outlet duct temperatures of 3°C, the following thermal quantities are computed.

Thaw depth @ 126 days	1.31 m
Peak heat flux to ducts occurs	@ 237 days
Heat gain from structure	7.9 J/m^{2}s
Heat gain from pad	13.9 J/m^{2}s
Maximum heat flux	21.8 J/m^{2}s
Total required ventilation rate	3.6 m^3/s (7622 CFM)

Thermal Analysis--Sanger's Method

Substituting the Fabrication Shop's design criteria into formulas presented by Sanger and using a temperature difference between inlet and outlet duct temperatures of 15.4°C, the following thermal quantities are computed.

Thaw depth (modified Berggren equation)	1.06 m
Average heat flow from pad	0.103 J/m^{2}s
Average thaw index	77.8°C days
Average surface temperature	-0.3°C
Average air inlet temperature	-15.7°C
Average temperature rise in ducts	15.4°C
Average heat flow from floor	0.432 J/m^{2}s
Average total heat flux	0.535 J/m^{2}s
Duct velocity	0.391 m/s
Total required ventilation rate	0.171 m^3/s (363 CFM)

CONCLUSION

Thermal analysis methods are most sensitive to the temperature difference between inlet and outlet ducts. Air temperature measurements at inlet and outlet ducts indicate an average temperature difference during the first year of operation around 10°C based on random readings over a 4 month period. During the second year, one reading taken when ambient air temperature was -29°C indicated a temperature difference between inlet and outlet ducts of 6°C. Therefore, use of a design temperature difference of 3°C in Harlan and Nixon's method provides a conservative ventilation rate for the first year of operation, and may be a realistic number after several years of operation. Increasing ventilation rate simply requires an increase in fan size, which insignificantly affects the total system cost. Although the Fabrication Shop's actual ventilation rate was approximately 10 times greater than that computed by Sanger, and 50% less than that required by Harlan and Nixon, the system has performed well. The success of the Sohio Fabrication Shop is attributed to the high ventilation rate which supercools the subgrade, the low mean annual air temperature (-12.2°C), and the relatively cold permafrost temperatures (-9°C @ -14 m depth).

REFERENCES

Harlan, R. L., and Nixon, J. F., 1978, Ground thermal regime, in Andersland and Anderson Geotechnical engineering for cold regions: New New York, McGraw-Hill.

Nixon, J. F., 1978, Geothermal aspects of ventilated pad design, in, Proceedings of the Third International Conference on Permafrost, v.1: Ottowa, National Research Council of Canada, p. 841-846.

Sanger, F. J., 1969, Foundations of structures in cold regions, U.S. Army CRREL Monograph-III-C4: Hanover, N. H., Cold Regions Research and Engineering Laboratory.

UNIAXIAL COMPRESSIVE STRENGTH OF ICE SEGREGATED FROM SOIL

Takahiro Ohrai, Tsutomu Takashi,
Hideo Yamamoto and Jun Okamoto

Technical Research Laboratory, Seiken Co., Ltd.
2-11-16 Kawaraya-machi, Minamiku, Osaka, Japan

Segregated ice (ice lens), which was columnar grained, was subjected to uniaxial compression tests. Specimens of such ice (5 cm in diameter and 9 cm in height) were prepared by freezing a soil composed of a frost-susceptible clay collected in Tokyo. The main factor thought to influences the strength of frozen clay is either unfrozen water or segregated ice. Therefore, the strength of segregated ice constitutes an important element in the clarification of the strength of the frozen clay. When a load is applied in the direction of ice growth, the uniaxial compressive strength increases with decreasing temperature, giving values of 18.6 Mega Newton/square meter at -10 °C and 24.5 MN/m^2 at -20 °C. When it is applied perpendicular to the growth direction, the uniaxial compressive strength, which is about 4.9 MN/m^2 at -20 °C, is approximately equal to the strength of commercial ice. When the loading angle ranges between 20° and 90° to the growth direction, the strength is nearly constant and decreases at -10 °C to one-fifth of the strength in the growth direction.

INTRODUCTION

The artificially frozen wall has come to be widely applied to the technique of supporting earth pressure as well as of preventing water from seeping into an area to be excavated. In recent years the ground-freezing technique has become more important due to increased number of large scale works and increased depth of work sites, e.g., 50 m depths. The need to save energy and to reduce the amount of frost heaving calls for the frozen soil wall to be made as thin as possible. Accordingly, knowledge of the bearing strength of the frozen soil wall is required.

The uniaxial compression test is relatively simple for obtaining data on the mechanical properties of frozen soils so that we have studied them consistently, limiting our study to the uniaxial compressive strength (Takashi et al. 1980, 1981a). The reason the uniaxial compressive strength of frozen clay is lower than that of sand is considered to be the existence of segregated ice, the so-called ice lens, as well as unfrozen water, which is one characteristic of frozen clay.

The authors have succeeded in forming large specimens of segregated ice, which contain no trace of soil particles, by applying the knowledge that the frost heave ratio increases with decreases in both effective stress and frost penetration rate (Takashi et al. 1978).

The experimental results, presented in this paper, were obtained systematically with respect to the dependence of the uniaxial compressive strength of segregated ice on temperature, loading direction, and strain rate, which are discussed below.

The results thus obtained will serve as an attempt to understand the mechanical behavior of frozen clay.

PREPARATION OF SPECIMENS

Soil Collection

The soil, from which specimens of segregated ice were later prepared, was collected in Tokyo. The soil is a Manaita-bridge clay, undisturbed and saturated, markedly frost susceptible, and diluvial (Takashi et al. 1981b). The grain-size distribution curve and properties of the soil collected are shown in Figure 1 and Table 1, respectively.

Technique

Shown in Figure 2 is the schematic diagram of the apparatus used to prepare specimens of segregated ice. Uniaxial upward freezing was achieved by lowering an acrylite cylinder, which contains the soil collected, into a chilled brine bath, described as an antifreeze bath in the figure. The lowering speed was about 2 mm/hr. This arrangement made it possible to provide nonload freezing in an open system because only the weight of the unfrozen part of the soil above the freezing front exerted pressure on the frozen part. The portion of sand above the bottom of the cylinder was designed to eliminate the shock of violent freezing, which comes about the moment that the cylinder bottom touches the surface of the chilled bath.

Thus, the thickness of sand was selected to be more than 40 mm. When the soil was thick, rhythmic ice banding (Martin 1959), which produced frost heaving in the soil, was found, as shown in Figure 3. With the frost line penetrating further up, the thickness of the unfrozen part of the soil decreased; and then, with the resultant decrease in resistance to water movement in the unfrozen soil, segregation of ice

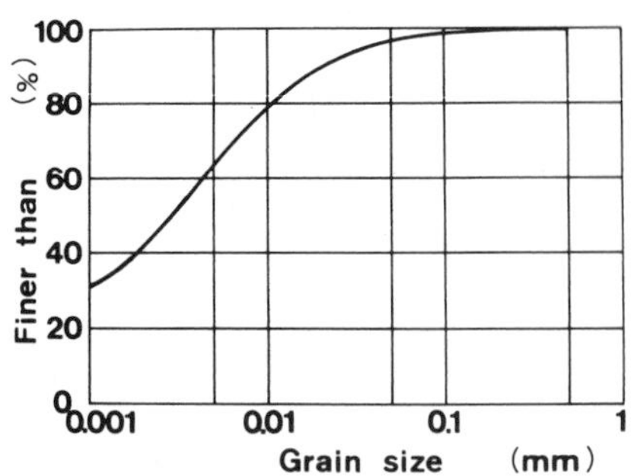

FIGURE 1 Grain-size distribution curve.

TABLE 1 Properties of Soil Collected.

Specific gravity of soil particles	G_s	2.662	
Specific surface area	S_g	39.0	m^2/g
Liquid limit	w_L	84.8	%
Plastic limit	w_P	43.7	%
Water content	w	84.4	%
Wet density	ρ_t	1.496	g/cm^3
Porosity	n	0.695	
Degree of saturation	S_r	98.5	%
Hydraulic conductivity	k	7.0×10^{-8}	cm/s

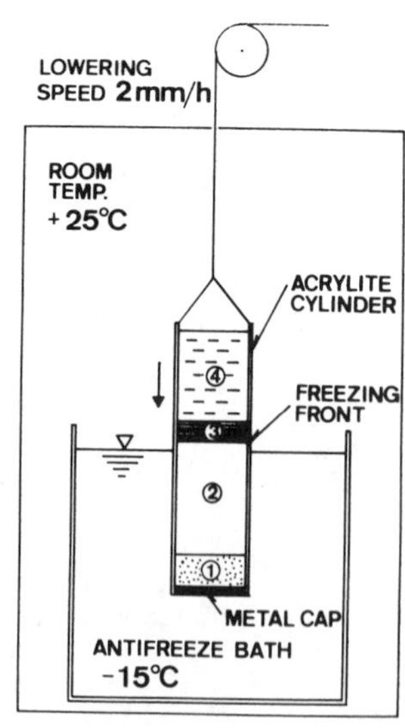

FIGURE 2 Apparatus for preparing specimens.

increased gradually. Finally, the segregation reached a stage of so-called perfect ice segregation (Arakawa 1966), in which the formation of segregated ice continued, being free from soil particles, as shown in Figure 4. Once perfect ice segregation started, it occurred continuously in the space between the unfrozen part of the soil and the sand at the bottom in the cylinder by repeating the same process as seen in Figures 2 and 4.

FIGURE 3 Vertical section of rhythmic ice banding.

FIGURE 4 Exterior appearance of a cylinder in the process of perfect ice segregation.

Structural Characteristics of Ice Specimens

Photographs of a vertical thin section and a horizontal thin section obtained from a segregated ice specimen (5 cm in diameter and 9 cm in height) are shown in Figure 5. Segregated ice was columnar

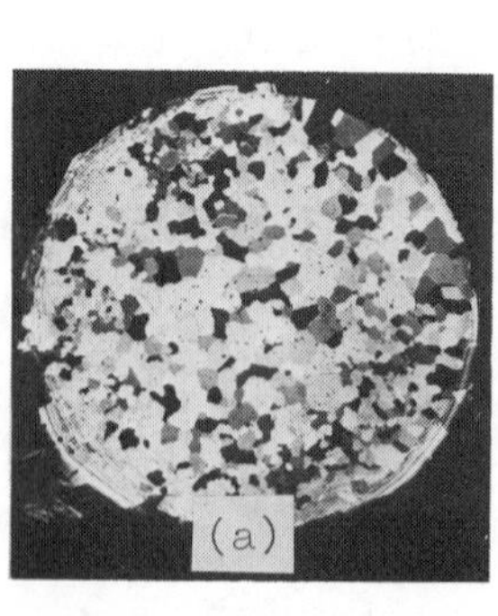

FIGURE 5 Photographs of (a)horizontal and (b)vertical thin section of a specimen ($5\ cm^{\phi} \times 9\ cm^{H}$).

grained ice, and its grain diameter on a plane perpendicular to the growth of the columnar grains was approximately 1 mm.

Usually, columns of bubbles as fine as a thread were found throughout the specimen in parallel with the growth direction, while small bubbles about 1 mm in diameter were found at the center of the specimen. The density of the specimens ranged from 0.877 to 0.904 g/cm^3, the average for 35 specimens being 0.893 g/cm^3.

No trace of soil particles was detected in the segregated ice through a magnifying glass of 50 magnifications.

TEST PROCEDURE

The shape of the specimens was cylindrical, and their standard dimensions were 5 cm in diameter and 9 cm in height.

The uniaxial compression apparatus used was capable of applying a load for compression at a constant strain rate ranging from 1.0 x 10^{-5} to 6.7 x 10^{-3} s^{-1}, the maximum loading capacity being 98 Kilo Newton. The apparatus was set up in a walk-in room with a temperature range of -0.2° to -30 °C (about ±0.2 °C accuracy).

A loading platen was a rigid acrylite disk. Specimens were exposed to the test temperature for at least 24 hours before testing. Axial load and displacement during the test were recorded by an X-Y recorder.

RESULTS AND DISCUSSION

Stress-Strain Curve and Uniaxial Compressive Strength q_u

Stress-strain curves which are typical at different temperatures and strain rates are shown in Figures 6 and 7, respectively.

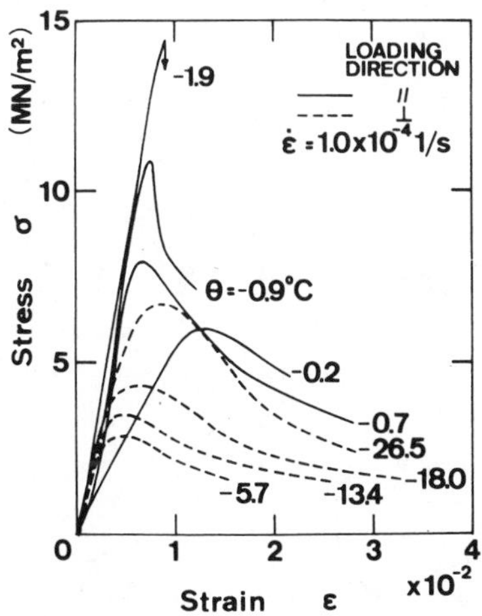

FIGURE 6 Stress-strain curves at various temperatures.

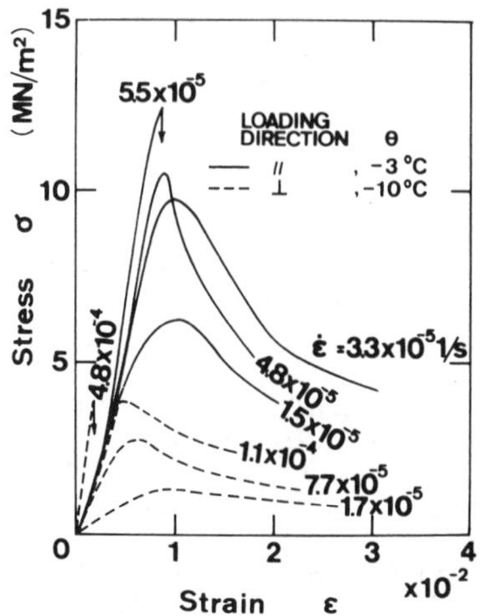

FIGURE 7 Stress-strain curves at various strain rates.

The peak of the curves becomes sharp with decreasing temperature or increasing strain rate. When the temperature was low or the strain rate was large, the specimen collapsed into fine pieces with explosive rupture. This mode of failure, shown by arrows in both figures, is called "brittle failure" in this paper. On the other hand, the mode of failure in which the specimen remains whole without rupture after the stress reaches its maximum, is called "ductile failure."

The uniaxial compressive strength is defined as the maximum stress q_u under each condition. In this paper, $q_{u//}$ represents the q_u when the load is applied in the direction of segregated ice growth and $q_{u\perp}$ when applied perpendicular to the direction.

Temperature Dependence of q_u

Shown in Figure 8 is the relation between the uniaxial compressive strength q_u and temperature θ under conditions of two strain rates and two loading directions. In all cases, q_u increases with decreasing temperature under the same conditions.

Values of $q_{u//}$ at a strain rate of 1.0 x 10^{-4} s^{-1} are fairly large, giving values of 9.8 MN/m^2 at -1.5 °C, 18.62 MN/m^2 at -10 °C, and 24.5 MN/m^2 at -20 °C. These values are larger than those of various types of ice. At temperatures close to freezing, it seemes that the relation between the value of $q_{u//}$ and temperature agrees with the relation between the pressure and freezing point calculated from the Clausius-Clapeyron equation. On the other hand, the values of $q_{u\perp}$ are much smaller than those of $q_{u//}$ and are equal to the strength of commercial ice, ranging from 1.96 to 6.86 MN/m^2 (Butkovich 1954).

Referring to the gradient of q_u/θ, that of $q_{u//}/\theta$ is larger than that of $q_{u\perp}/\theta$.

It is found that the shape of stress-strain curves varies with the temperature as seen in Figure 6. In the case of $q_{u//}$, the mode of failure was brittle at -1.9 °C, whereas even at -26.5 °C it was ductile in the case of $q_{u\perp}$.

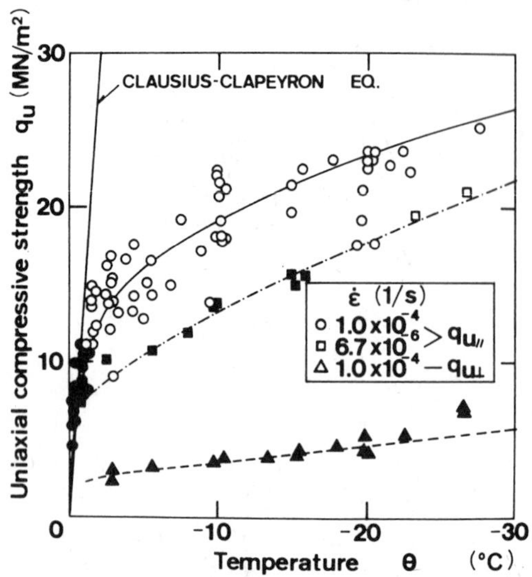

FIGURE 8 Uniaxial compressive strength q_u vs. temperature θ. Open symbols represent brittle failure; solid symbols, ductile failure.

Loading Direction Dependence of q_u

A great difference exists between values of $q_{u//}$ and $q_{u\perp}$ under the same temperature and strain rate, as seen in Figure 8. It is considered from photographs in Figure 5 that the value of q_u may depend on the loading direction.

The relation between uniaxial compressive strength and loading direction at -10 °C is shown in Figure 9, in which the abscissa represents angles ψ between loading direction and ice growth direction. In the range 0° - 10°, the value of q_u decreases markedly with increasing ψ. In the range 20° - 90°, its value is nearly constant and is equal to that of $q_{u\perp}$. The mode of failure was ductile except when ψ = 0°. In this case it was presumed from the state of deformation and how cracks formed, which is described later, that the dominant cause is the gliding of the grain boundary.

The value of q_u is larger at an angle of 10° than in the range 20° - 90°. The main reason is presumed to be differences in angle, or friction of grain-boundary glides, or both.

Strain Rate Dependence of q_u

Shown in Figure 10 is the relation between the uniaxial compressive strength q_u and the strain rate $\dot{\varepsilon}$ at two tempera-

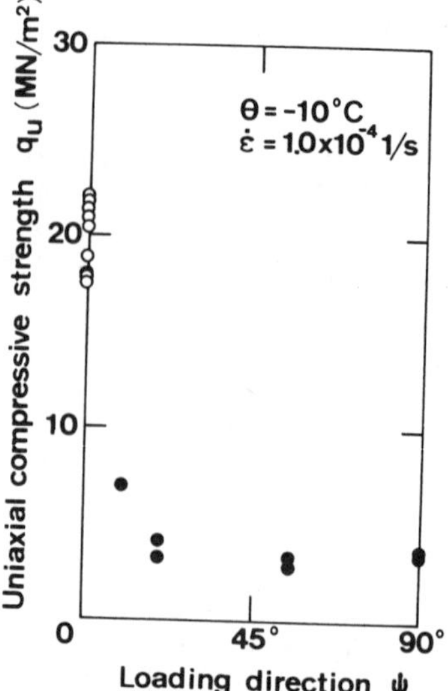

FIGURE 9 Uniaxial compressive strength q_u vs. loading direction ψ. Open circles represent brittle failure; solid circles, ductile failure.

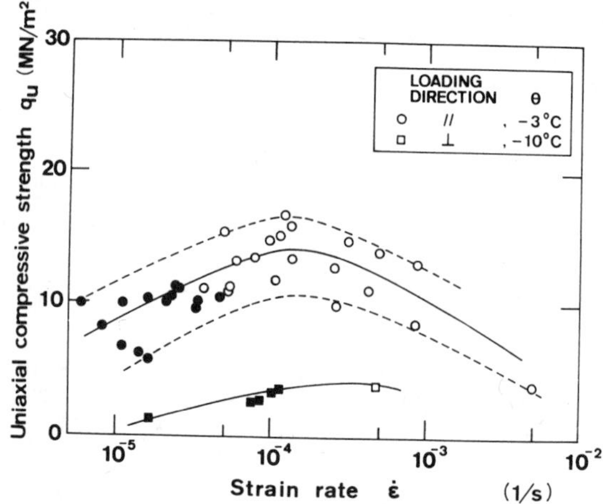

FIGURE 10 Uniaxial compressive strength q_u vs. strain rate $\dot{\varepsilon}$. Open symbols represent brittle failure; solid symbols, ductile failure.

tures, -3 °C in the case of $q_{u//}$ and -10 °C in the case of $q_{u\perp}$, chosen on the basis of expected occurrences of both brittle and ductile failures. In the figure, maximum and minimum $q_{u//}$ values at each strain rate are shown by two broken lines because of wide scatter in the data.

It is apparent that a maximum value of q_u exists at a certain strain rate. It can be seen that the compressive strength has a maximum at a strain rate between 8.3×10^{-5} and 3.3×10^{-4} s^{-1} in the case of $q_{u//}$ and between 1.2×10^{-4} and 5.0×10^{-4} s^{-1} in the case of $q_{u\perp}$. In two series of previous tests, a strain rate of 1.0×10^{-4} s^{-1} was used, at which the maximum measured strengths were consequently observed.

It was found that the shape of stress-strain curves varies with the strain rate, as seen in Figure 7. The variations are similar to those with the temperature. The mode of failure was brittle at 5.5×10^{-5} s^{-1} at -3 °C in the case of $q_{u//}$; and at 4.8×10^{-4} s^{-1} at -10 °C in the case of $q_{u\perp}$.

Characteristics of $q_{u//}$ at near 0 °C

Shown in Figure 11 is the relation between uniaxial compressive strength $q_{u//}$ and strain rate $\dot{\varepsilon}$ at about -0.4 °C. Room temperatures recorded during the test varied by no more than ±0.05 °C. Temperatures at points in contact with both ends of the specimen were measured by means of copper-constantan thermo couple measurements that have a sensitivity of ±0.003 °C.

In Figure 11, with an increase in strain rate the temperature of the specimen is lowered and the value of $q_{u//}$ increases.

It is considered that part of the specimen melts and that the specimen temperature is lowered by latent heat. The freezing-point depression of an ice-water system is expressed by the Clausius-Clapeyron equation as follows:

$$\Delta P = - \frac{L}{V_i - V_w} \frac{\Delta T}{T} = -13.41 \Delta T \text{ (MN/m}^2\text{)} \qquad (1)$$

where P, T, L, V_i , V_w are pressure, absolute temperature of freezing, latent heat of fusion, and specific volume of ice and of water, respectively; ΔT is the freezing-point depression. The value of ΔP calculated from equation (1) using the temperature that was lowered during the compression is shown by triangle symbols in Figure 11. The value of ΔP coincides with the value of $q_{u//}$, approximately.

Only near 0 °C did the phenomenon of decreasing temperature occur; at -5 °C and

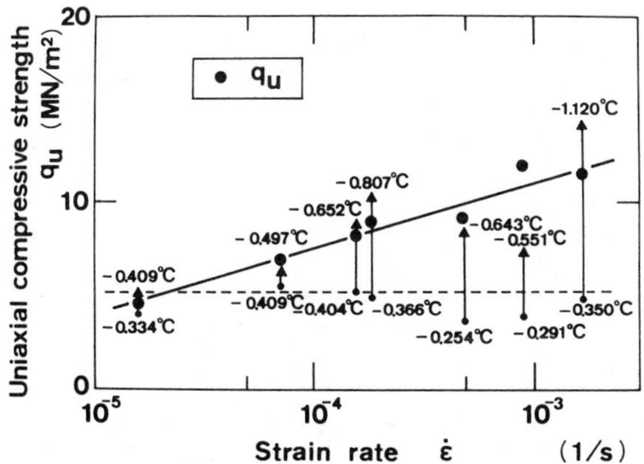

FIGURE 11 Uniaxial compressive strength $q_{u//}$ vs. strain rate $\dot{\varepsilon}$ at about -0.4 °C.

-10 °C, the temperature rose. The cause remains unexplainable at present.

Transition from Ductile to Brittle Failure

Shown in Figure 12 is the dependence of the strength of each of the brittle and the ductile failure, $q_{u//}$, on temperature , for which the data have different strain rates. Open and solid circles represent brittle and ductile failure, respectively. Half-open circles represent the brittle failure in which the specimen collapsed immediately after the peak in the stress-strain curve. Data near 0 °C and at large strain rates such as 5.0×10^{-3} s^{-1} were omitted in the figure. It is clearly seen from the figure that regions of brittle and ductile failure are separated by a straight line. The transition from ductile to brittle failure occurred at strain rates of 1.0×10^{-4}, 3.3×10^{-5}, and 6.7×10^{-6} s^{-1} at constant temperatures of -1.7°, -3°, and -18 °C, respectively.

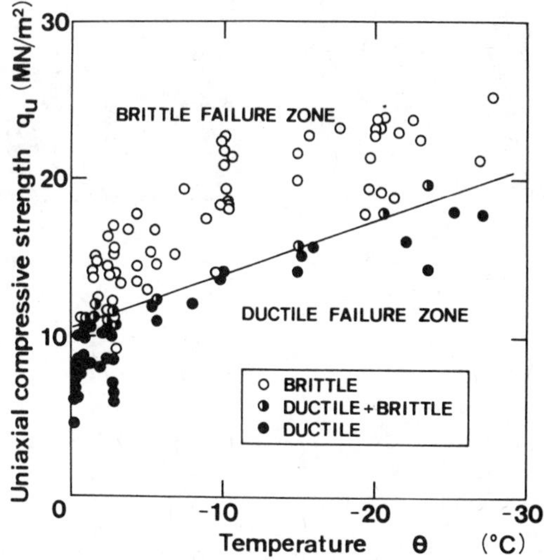

FIGURE 12 Zones of ductile and brittle failure in the case of $q_{u//}$.

Cracks

In the case of brittle failure, visible crack formation did not occur until the specimen collapsed into pieces with explosive rupture.

In the case of ductile failure, however, visible cracks occur when the stress is near maximum, when the strain is about 1×10^{-2}, as seen in Figures 6 and 7. Typical cracks are shown in Figure 13. These cracks appear perpendicular to the direction of segregated ice growth.

In the case of data adjacent to the straight line in Figure 12, the cracks appear regularly in the growth direction and perpendicular to it, as seen in Figure 14. Near 0 °C, cracks in the growth direction dominated.

When the load was applied perpendicular to the growth direction, failure was mostly ductile. In this case, the horizontal section of the specimen after compression was in the shape of an ellipse, which had a short axis in the growth direction, and the majority of cracks appeared in the growth direction.

As mentioned above, it was characteristic of segregated ice that visible cracks appeared regularly in the growth direction, or perpendicular to it, or both.

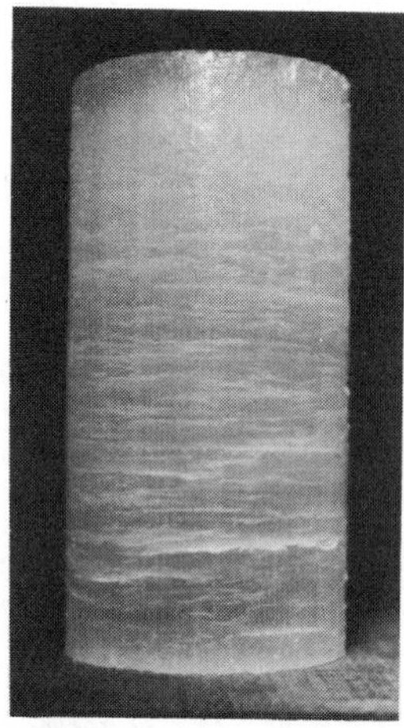

FIGURE 13 Photograph of typical cracks (θ=-0.6 °C, ε=1%, $q_{u//}$=7.25MN/m^2).

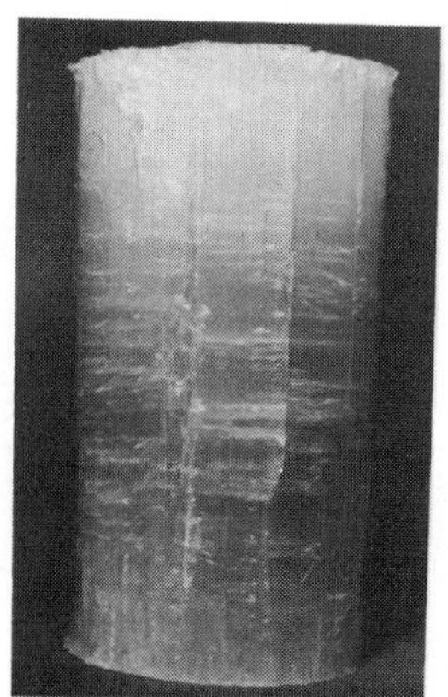

FIGURE 14 Photograph of cracks in transition zone from ductile to brittle failure (θ=-16 °C, ε=1%, $q_{u//}$=15.4MN/m^2).

CONCLUSIONS

The experimental results with respect to the uniaxial compressive strength q_u of segregated ice are as follows:

1° When the load is applied in the direction of ice growth, the longitudinal uniaxial compressive strength $q_{u//}$ increases with decreasing temperature, giving values of 18.62 MN/m^2 at -10 °C and 24.5 MN/m^2 at -20 °C. These values are larger than those of various types of ice.

2° When the load is applied perpendicular to the growth direction, the lateral uniaxial compressive strength $q_{u\perp}$, which is about 4.9 MN/m^2 at -20 °C, is approximately equal to the strength of commercial ice.

3° In the case of a loading direction in the range 0° - 10°, which is the angle ψ between the loading and the ice growth direction, the value of q_u decreases markedly with increasing ψ. In the range 20° - 90°, its value is nearly constant.

4° The compressive strength has its maximum at a strain rate between 8.3×10^{-5} and 3.3×10^{-4} s^{-1} in the case of $q_{u//}$, and between 1.2×10^{-4} and 5.0×10^{-4} s^{-1} in the case of $q_{u\perp}$.

5° Near 0 °C, part of the specimen melts and the specimen temperature is lowered by the latent heat corresponding to the melting. With an increase in strain rate, the value of $q_{u//}$ increases with a resultant fall in temperature.

6° The transition from ductile to brittle failure occurred at strain rates of 1.0×10^{-4}, 3.3×10^{-5}, and 6.7×10^{-6} s^{-1} at constant temperatures of -1.7°, -3°, and -18 °C, respectively. Visible cracks appeared regularly in the direction of ice growth, perpendicular to it, or both.

REFERENCES

Arakawa, K., 1966, Theoretical studies of ice segregation in soil, J. Glaciol., 6, 44, 255-266.

Butkovich, T. R., 1954, Ultimate strength of ice, SIPRE, Res. Paper 11, 12pp.

Martin, R. T., 1959, Rhythmic ice banding in soil, Highway Res. Board, Bull., 218, 11-23.

Takashi, T. et al., 1978, Effect of penetration rate of freezing and confining stress on the frost heave ratio of soil, Permafrost, 3rd Int. Conf., 1, 737-742.

Takashi, T. et al., 1980, Experimental study on uniaxial compressive strength of frozen sand, Proc. of JSCE, 302, 79-88 (In Japanese).

Takashi, T. et al., 1981a, Experimental study on uniaxial compressive strength of homogeneously frozen clay, Proc. of JSCE, 315, 83-93 (In Japanese).

Takashi, T. et al., 1981b, Upper limit of heaving pressure derived by pore-water pressure measurements of partially frozen soil, Eng. Geol., 18, 245-257.

WATER MIGRATION DUE TO A TEMPERATURE GRADIENT IN FROZEN SOIL

J. L. Oliphant, A. R. Tice, and Y. Nakano

U.S. Army Cold Regions Research and Engineering Laboratory
Hanover, New Hampshire 03755 USA

Closed soil columns at an initially uniform total water content were subjected to a nearly linear and constant temperature gradient along their length. At various times, the columns were sectioned and water content as a function of position was determined gravimetrically. Unfrozen water content vs. temperature curves were also determined with a nuclear magnetic resonance technique on separate samples of the same soil at the same dry density. It was found that the water migrated from the warm to the cold end and two zones developed in each of the tubes, one that contained only liquid water and the other containing ice and water. The boundary between the two zones also migrated toward the cold end as the experiment progressed, and the water content of the zone containing only water fell while that of the zone containing ice and water increased. The free energy of the liquid water was calculated as a function of position, assuming the validity of a form of the Clausius-Clapeyron equation. Hydraulic conductivity coefficients were then calculated from the free energy gradients and water migration data. Values ranged from 4.5×10^{-12} m/s at -0.3°C to 3.5×10^{-13} m/s at -2°C for the Morin clay soil tested.

It has been known for many years that a portion of the water in soils remains unfrozen at temperatures below 0°C. Methods of measuring the unfrozen water content as a function of temperature have been discussed in a recent paper by Oliphant and Tice (1982). Details of the nuclear magnetic resonance technique developed at USA CRREL for accurately measuring unfrozen water contents are given in a report by Tice et al. (1982). The existence of unfrozen water in frozen soils allows for significant rates of ice and water migration in frozen soils along water free-energy gradients. This causes ice lensing, frost heave, breakup of roads and foundations, and loss of soil strength when the soil thaws. Thus, it is important to understand the mechanisms of water migration and ice formation in frozen and freezing soils.

Free energy gradients of water in frozen soils can be caused by temperature gradients, water or dissolved salt concentration gradients, or pressure gradients. To understand the movement of water due to each of these causes, we are carrying out a series of experiments in which the gradients causing water migration are carefully controlled and separated so that, as nearly as possible, only one type of gradient acts on the free energy of the water in a given experiment. The data from these experiments can then be used to validate models concerning the migration of water in frozen soils. An understanding of both the driving forces for water migration and the responses to these driving forces, or conductivities, is needed to model water migration in soils under natural conditions.

The results of experiments on water migration in frozen soils due to water concentration gradients have been reported by Nakano et al. (1982, 1983). The purpose of this paper is to report the initial results of an experiment that measures water migration due to a temperature gradient. Water migration due to a temperature gradient in porous materials containing ice has previously been measured by Dirksen and Miller (1966), Hoekstra (1966), and Williams and Perfect (1980). In the experiments by Dirksen and Miller and by Hoekstra, a closed moist soil column was subjected on one end to a temperature below freezing and on the other end to a temperature above freezing. The soil column itself was used to conduct heat from the warm end to the cold end. The soil columns, initially unfrozen, would begin to freeze at the cold end, and the temperature gradient along the column would change as the experiment proceeded. This changing temperature gradient was caused by the initial constant temperature of the soil column and by changing thermal diffusivities in the column as water migrated and ice rearranged. Although this type of experiment is good for obtaining a qualitative idea of what happens when moist soil is subjected to a temperature gradient, it is difficult to obtain quantitative data on moisture migration as a function of temperature and the temperature gradient in the soil.

In the more recent experiment by Williams and Perfect, a saturated frozen soil column was connected on either end to water reservoirs containing a lactose solution. Semipermeable membranes were placed between the lactose solution and the ends of the soil column to help slow the migration of lactose into the soil. The two reservoirs were kept at two different temperatures and the rate of water movement from the reservoirs to the soil column was monitored. Again, in this experiment heat was conducted from the warm reservoir to the cold reservoir through the soil column, so a completely linear temperature gradient could not be established. After a long time a steady-state temperature gradient and water flow rate through the column were established, and an average value of water migration through the column due to the overall temperature gradient was obtained.

Williams and Perfect (1980) discuss how to calculate hydraulic conductivity coefficients from the water migration data by using a form of the Clausius-Clapeyron equation to convert temperature gradients to effective pressure gradients in frozen soil. The coefficients they calculate show considerable scatter but agree on an order-of-magnitude basis from sample to sample. More importantly, the calculated coefficients seem to be compatible with data obtained from the same apparatus using a pressure gradient instead of a temperature gradient between the two reservoirs (Burt and Williams 1976). This seems to indicate that the applied form of the Clausius-Clapeyron equation is valid for ice and water in a frozen soil. If this is the case, then accurate measurements of water migration due to temperature gradients will also yield accurate values of the hydraulic conductivity of frozen soils.

The experiment described herein was designed to allow accurate measurement of moisture migration in soil columns due to carefully controlled temperature gradients. The assumptions involved in deriving hydraulic conductivity coefficients from these data are discussed. In addition, methods of comparing the coefficients with those derived from direct pressure and moisture content gradient experiments are given.

EXPERIMENTAL PROCEDURE

The apparatus shown in Figure 1 was used to maintain a constant and nearly linear temperature gradient along the soil column. It consists of two aluminum blocks A, 30.5 cm long by 11.2 cm across by 2.54 cm thick. Five semicircular grooves 0.95 cm in diameter were machined along one face of each block. These grooves were spaced 1 cm apart across each block so that when the blocks are bolted together, five holes 0.95 cm in diameter are formed. The soil is packed into 10-mL plastic serological pipettes that fit snugly into the holes.

Two aluminum blocks, B, bolt to either end of the combined blocks, A, as shown in the figure. These blocks are 5.08 cm by 5.08 cm by 11.2 cm. Each has a 2.54-cm diameter hole drilled into it through which an antifreeze mixture from a temperature-controlled bath is circulated. The four aluminum blocks, when bolted together, are then surrounded by a box of 5.08-cm thick fiber-impregnated plastic, which acts as thermal insulation and damps out short-term temperature changes in the laboratory room.

Aluminum has a thermal conductivity about 90 times higher than that of ice and thus about two orders of magnitude higher than a partially frozen soil-water mixture. When antifreeze mixtures at two different temperatures are circulated through the B blocks, heat is tranferred at a high rate through the A blocks, from the warm end to the cold end. Most of this heat is transferred through the aluminum blocks and only a small amount through the embedded soil columns. The diameter of the soil columns is small compared with their length, so the temperature profile that obtains along the aluminum blocks is impressed on the soil columns, keeping them at the same constant and nearly linear temperature gradient as the surrounding aluminum. Temperatures are monitored by a row of copper-constantan thermocouples placed through the bottom aluminum block, A, flush with the inside surface of the middle groove.

The soil selected for the initial experiment, called Morin clay, was a marine-deposited clay obtained from the Morin brickyard, Auburn, Maine.

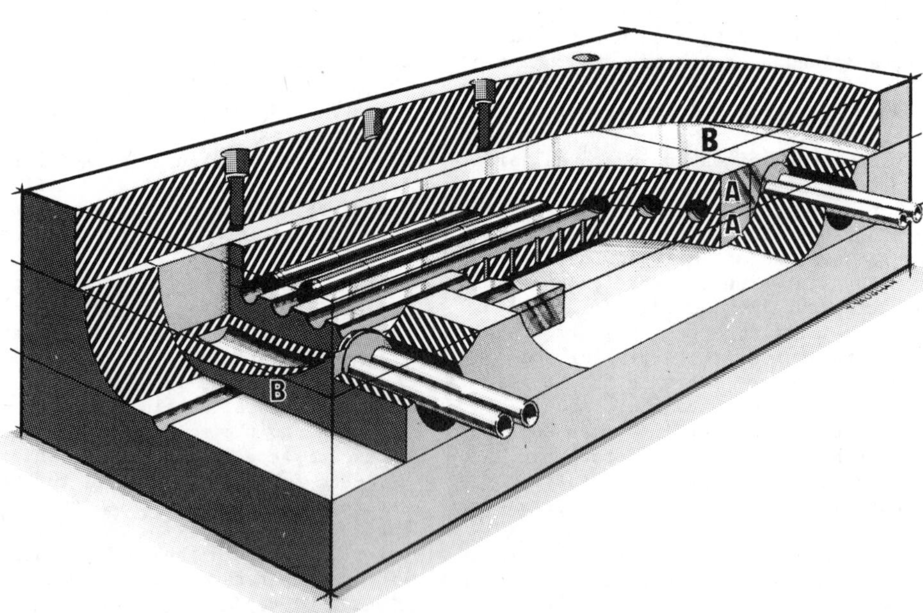

FIGURE 1 Apparatus for obtaining a constant temperature gradient along a soil column.

The area was uplifted and original solutes were leached by natural seepage. The clay is non-swelling and has a specific surface area of 60 m^2/g. It is the same soil used in the experiments reported by Nakano et al. (1982, 1983).

For the initial experiments, the ends of five plastic serological pipettes were trimmed, leaving plastic tubes 27.8 cm long with 0.95 cm o.d. and 0.79 cm i.d. A rubber stopper was placed in one end of each tube filling 0.5 cm of the tube. Morin clay with a nominal water content of 20% of the dry weight was then packed into each tube in layers approximately 1 cm thick. This packing was done carefully to achieve the same average dry density in each tube. The average dry densities in tubes 1 through 5 were 1.46, 1.50, 1.46, 1.47, and 1.48 g/cm^3. These dry densities are almost the same as those used in the water migration experiments reported by Nakano et al. (1982, 1983). A rubber stopper was then placed in the last 0.5 cm of the tube, leaving a packed soil column 26.8 cm long.

The five packed tubes were set aside for several days at room temperature to allow moisture equilibration in the soil. They were then placed in a -20°C cold room and allowed to freeze quickly. Meanwhile, a temperature gradient was established along the apparatus shown in Figure 1 by circulating antifreeze at two different temperatures through the end blocks. Five less carefully packed soil columns were placed in the apparatus while the temperature gradient was being established. The thermocouple temperature readings obtained for the first experiment are shown graphically in Figure 2. They range from -0.25°C at the warm end to -2.05°C at the cold end of the soil columns. It can be seen that the temperature gradient is almost linear except near the warm end. After a stable temperature gradient was established, the five less carefully packed tubes were quickly removed and the five soil columns frozen at -20°C were placed in the apparatus. It was noted that in about 45 minutes the thermocouple readings along the apparatus returned to those obtained for the first experiment.

Tubes 1 through 5 were removed after being subjected to the temperature gradient for 2, 4, 8, 16, and 32 days, respectively. As each tube was removed it was replaced by one of the blank tubes.

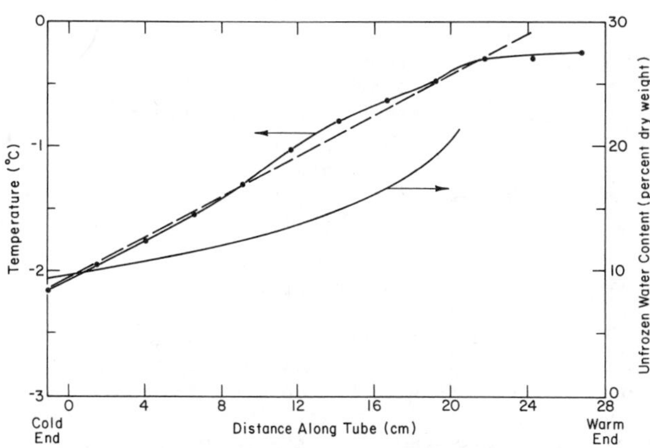

FIGURE 2 Temperatures and unfrozen water contents along the soil column.

It took about 30 minutes for the original temperature readings to be reestablished after each tube was removed. The removed tubes were quickly placed on their side back in the -20°C cold room and allowed to freeze for 30 minutes. Each soil column was then sectioned into 44 segments and the total water content of each segment was determined gravimetrically.

RESULTS AND DISCUSSION

The measured water content of each of the 44 segments of each tube is shown in Figure 3. The total water content seems to divide qualitatively into two sections for each of the tubes. On the warm end there is a section of relatively flat and even water content that is below the average for the tube. The length of this section increases and the water content falls slightly as time goes on. On the cold end there is a section, which initially covers most of the tube, that has a relatively uneven total water content. The total water content of this section is above the average for the tube and it appears that, as time goes on, two peaks are formed in the total water content curve, one close to the cold end and one close to the boundary between the even and uneven sections. The curves shown in Figure 3 look qualitatively similar to those reported by Dirksen and Miller (1966) and Hoekstra (1966). Both of these reports show curves that are relatively flat on the warm end, dipping slightly toward the cold end, and eventually forming two peaks in the cold end section.

A qualitative explanation of the shape of the curves shown in Figure 3 can be given. The smooth section toward the warm end is made up of segments that contain only unfrozen water, while the section toward the cold end is made up of segments that contain both ice and unfrozen water. The unevenness of the cold end section is probably due to the inherently uneven distribution of small ice lenses forming in the soil. The peak that forms at the far cold end is due to the closed boundary condition of

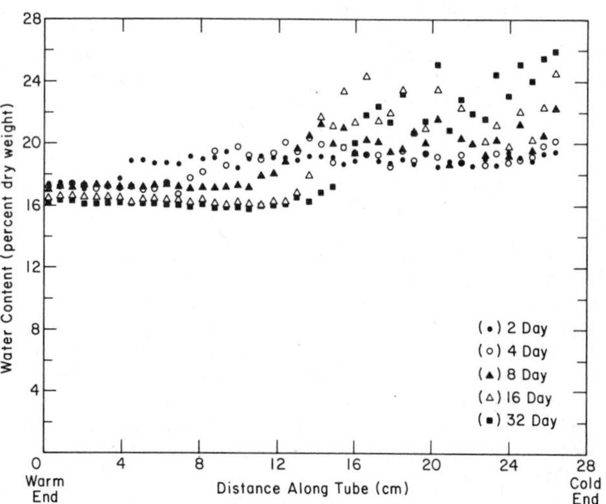

FIGURE 3 Final total water contents along the soil column.

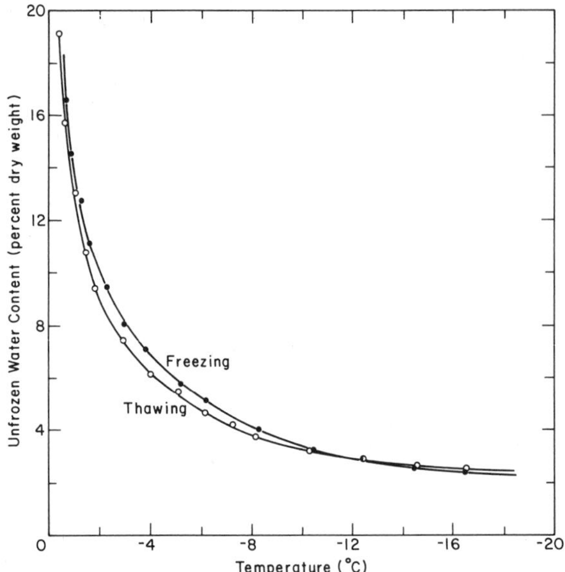

FIGURE 4 Unfrozen water contents as a function of temperature for Morin clay at a dry density of 1.45 gm/cm^3 and a total water content of 21.8%.

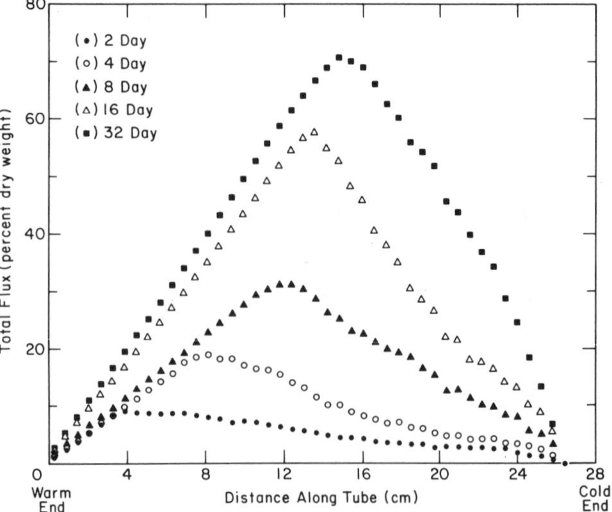

FIGURE 5 Total fluxes of water across segment boundaries for the soil column.

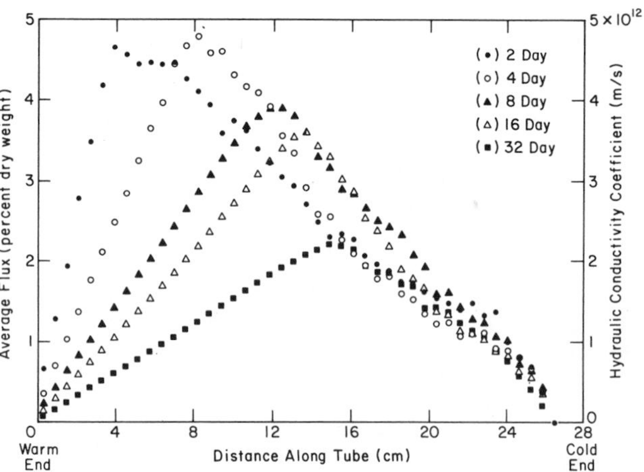

FIGURE 6 Average fluxes of water across segment boundaries and calculated hydraulic conductivity coefficients for the soil columns.

the tube at this end. The peak that forms at the boundary between the two sections probably forms because the hydraulic conductivity is higher close to the boundary than it is farther along in the tube where the soil is colder and there is less unfrozen water.

The unfrozen water content vs. temperature curves determined by NMR on a separate sample of the Morin clay at the same dry density as the other samples are shown in Figure 4. Both the freezing and thawing curves are shown; the hysteresis between them is evident. Using the curves in Figure 4 and the temperature measured along the soil column as shown in Figure 2, the expected unfrozen water content as a function of position along the tube can be obtained. The values obtained from the freezing curve in Figure 4 are plotted in Figure 2. It can be seen from Figure 2 that if the boundary between ice-containing and ice-free parts of the tube is to move gradually toward the cold end, as it does in Figure 3, then the total water content, which is equal to the unfrozen water content in the ice-free sections, must decrease in these sections. This is seen to happen on a qualitative basis in Figure 3, although the decrease is less than that expected from Figure 2.

The data shown in Figure 3 can be analyzed quantitatively by first calculating the water fluxes across the boundaries of adjacent segments. The average water content of all the segments was calculated and then the individual mass balances were calculated for each segment. Water fluxes away from the warm end and toward the cold end were counted as positive. For example, if the first segment on the warm end had a total water content less than the average, then the amount of water making up the difference has moved across the boundary from the first to the second segment. A mass balance around the second segment shows that the flux from the second to the third segment must be equal to the flux from the first to the second segment plus the difference between the average water content, which is assumed to be the initial water content of all the segments, and the final measured water content. In general, for the i'th segment, the flux from that segment to the next, f_i, can be calculated from the equation

$$f_i = f_{i-1} + (W_i - W_{ave}) \tag{1}$$

where W_i is the water content of the i'th segment and W_{ave} is the average or initial water content. Flux values calculated according to Equation 1 are shown for the boundaries between each of the segments of each of the five tubes in Figure 5. These values were then divided by the number of days that each tube was exposed to the temperature gradient to obtain an average flux per day at each segment boundary. These values are shown in Figure 6.

Comparing Figure 3 with Figure 5, it can be seen that the position of maximum flux corresponds to the boundary between the sections of each tube that contain ice and water and only water. Figure 5 also indicates that fluxes through the section that contains only liquid water increase only gradually with time. Comparing Figure 5 and Figure 6, it can be seen that most of the flux calculated for the segments in the section that contained only liquid water at the end of the experiment came at a time when there was both ice and water in the segments, before the boundary between the two sections had passed these segments. After the boundary passed, only a small amount of flux moved through the ice-free segments. This is especially evident when comparing data on the 2-, 4-, and 8-day tubes (Figure 5).

In Figure 6, it is evident that in the cold end section, where both ice and water were present in the soil, the flux at a given position in the tube and thus at a given temperature and temperature gradient remained relatively constant for all of the tubes as the experiment progressed. In other words, a frozen soil containing both ice and unfrozen water that is subjected to a temperature gradient acts as a pump, pumping water toward the cold end. The rate of pumping can depend on the type of soil, the temperature, and the temperature gradient, but as long as ice is present it does not depend on the total water content. Similar conclusions were reached by Williams and Perfect (1980).

A line has been drawn in Figure 2 that approximates the temperature profile over most of the length of the temperature gradient apparatus. This line has a slope of 0.082°C/cm. If we assume, as do Williams and Perfect (1980), that the form of the Clausius-Clapeyron equation that applies is

$$H = \frac{L}{Tg} \Delta T, \tag{2}$$

where H is suction in cm of water, L is the latent heat of fusion of pure water equal to 3.33×10^9 ergs/g, T is temperature in °K, ΔT is the freezing point depression or degrees below 0°C, and g is the gravitational acceleration equal to 981 cm/sec^2, then an effective pressure gradient can be calculated from the temperature gradient. Evidence that Equation 2 does apply is given by Williams (1967). Using Equation 2 and the temperature gradient given above, we can calculate that the driving force for water movement in the part of the soil column that contains both ice and water is 1020 cm of H_2O/cm. The fluxes in Figure 6 are given in percent of dry soil wt/day. To convert these values to cm^3/day, multiply by the dry density of the soil, then by the volume of an individual segment (0.30 cm^3), and divide by 100%. Hydraulic conductivity coefficients, K, can then be found by solving a form of Darcy's law

$$q = - K A \frac{dP}{dX}, \tag{3}$$

where q is the flux of water in cm^3/day, A is the cross-sectional area in the tube available for flow (0.49 cm^2), dP/dX is the effective driving force calculated above to be 1020 cm of H_2O/cm, and K has units of cm^2/cm of H_2O/day. To convert K to the more usual units of m/s, divide by the number of centimeters in a meter (100) and by the number of seconds in a day (86,400).

Carrying out all of the above calculations, we find that if the numbers given for the flux values in Figure 6 are multiplied by 1.0×10^{-12}, they are equal to the hydraulic conductivity coefficient, K, in m/s. By comparing Figures 2 and 6, we can see that the hydraulic conductivity coefficients range from about 4.5×10^{-12} m/s at -0.3°C to about 3.5×10^{-13} m/s at -2°C. Williams and Perfect (1980) report values of the average hydraulic conductivity coefficients on several silts that range from 4×10^{-12} m/s at an average temperature of -0.23°C to 6×10^{-14} m/s at an average temperature of -0.40°C. If it is remembered that the Morin clay is a finer material and thus has a higher surface area and a higher unfrozen water content at a given subfreezing temperature, then the higher hydraulic conductivity coefficients at a given temperature reported in this paper are not surprising. It is also not surprising that, as temperature decreases, the hydraulic conductivity coefficient values decrease at a slower rate for the Morin clay than for the silts. What is remarkable about the conductivity values shown in Figure 6 is the reproducibility of the values from sample to sample. In addition, values could be calculated over a broad temperature range using just one soil column. It appears that by using the experiment given in this paper, values of hydraulic conductivities down to at least -10°C can be measured.

If, in the future, values of hydraulic conductivity coefficients in frozen soils can be directly measured using some type of direct pressure gradient apparatus, such as that used by Burt and Williams (1976), then those values could be compared with values obtained from temperature gradient experiments such as those reported in this paper, and the assumption that the correct form of the Clausius-Clapeyron equation is being used could be checked. At present, there are no direct methods available accurate enough to allow a quantitative comparison. Harlan (1973) has suggested that hydraulic conductivity values obtained for unsaturated, unfrozen soils could be used for frozen soils at like unfrozen water contents. The experiments reported by Nakano et al. (1982, 1983) give values for the soil-water diffusivity, $D(\theta)$, for Morin clay at various water contents. The form of Darcy's law used was

$$q = - D(\theta) A \frac{\partial \theta}{\partial X} \tag{4}$$

where θ is the water content and $D(\theta)$ is the diffusivity. If the soil-water characteristic curve, i.e. suction or pressure in the soil as a function of water content, is known, then Equation 4 can be written as

$$q = - D(\theta) A \frac{\partial \theta}{\partial P} \frac{\partial P}{\partial X} \tag{5}$$

where P indicates soil-water suction. Equation 5 is analagous to Equation 3 with K equal to $D(\theta)\ \partial\theta/\partial P$. Thus, a comparison can be made between the temperature gradient and water concentration gradient experiments. We do not yet have good soil-water characteristic curves on the Morin clay, but these can be obtained. We will make these comparisons in the future.

CONCLUSIONS

In this paper we have presented a method for measuring water migration due to a temperature gradient in frozen soils. The important part of the experimental technique that is new is that long and thin soil columns are embedded in an aluminum block and the temperature gradient created in the block is impressed upon the soil columns. This allows the temperature gradient in the soil to be controlled even though water and ice migrate. Hydraulic conductivity coefficients were calculated for a Morin clay sample down to a temperature of -2.0°C. This calculation involves the assumption that a specific form of the Clausius-Clapeyron equation is valid for relating the pressure or chemical potential of the water in the soil to temperature as long as both ice and liquid water phases are present.

It appears that experiments can be done down to at least -10°C to determine hydraulic conductivity coefficients. We are presently doing further experiments with Morin clay, using steeper temperature gradients and going to lower temperatures, and we plan to run other soils in the future. We hope that the method becomes a quick and convenient way to determine the hydraulic conductivities of frozen soils over a wide temperature range.

ACKNOWLEDGMENTS

This work was conducted under sponsorship of the U.S. Army Corps of Engineers Cold Regions Research and Engineering Laboratory.

REFERENCES

Burt, T.P., and Williams, P.J., 1976, Hydraulic conductivity in frozen soils: Earth Surface Processes, v. 1, p. 349-360.

Dirksen, C., and Miller, R.D., 1966, Closed-system freezing of unsaturated soil: Soil Science Society of America Proceedings, v. 30, p. 168-173.

Harlan, R.L., 1973, Analysis of coupled heat-fluid transport in partially frozen soil: Water Resources Research, v. 9, p. 1314-1323.

Hoekstra, P., 1966, Moisture movement in soils under temperature gradients with the cold-side temperature below freezing: Water Resources Research, v. 2, p. 1314-1323.

Nakano, Y., Tice, A.R., Oliphant, J.L., and Jenkins, T.F., 1982, Transport of water in frozen soil: I. Experimental determination of soil-water diffusivity under isothermal conditions: Advances in Water Resources, v. 5, p. 221-226.

Nakano, Y., Tice, A.R., Oliphant, J.L., and Jenkins, T.F., 1983, Transport of water in frozen soil: II. Effects of ice on the transport of water under isothermal conditions: Advances in Water Resources, v. 6, p. 15-26.

Oliphant, J.L., and Tice, A.R., 1982, Comparison of unfrozen water contents measured by DSC and NMR, in Proceedings of the Third International Symposium on Ground Freezing, 22-24 June: Hanover, N.H., U.S. Army Cold Regions Research and Engineering Laboratory.

Tice, A.R., Oliphant, J.L., Nakano, Y., and Jenkins, T.F., 1982, Relationship between the ice and unfrozen water phases in frozen soil as determined by pulsed nuclear magnetic resonance and physical desorption data: Hanover, N.H., U.S. Army Cold Regions Research and Engineering Laboratory, CRREL Report 82-15.

Williams, P.J., 1967, Unfrozen water content of frozen soils and soil moisture suction, in Norwegian Geotechnical Institute Publication No. 72, Oslo, Norway.

Williams, P.J., and Perfect, E., 1980, Investigation of thermally actuated water migration in frozen soils: Final report to the Department of Energy, Mines and Resources, Earth Physics Branch: Ottawa, Canada, Geotechnical Science Laboratories, Dept. of Geography, Carleton University.

Hydrologic Characteristics of Small Arctic-Alpine Watersheds Central Brooks Range, Alaska

Lawrence J. Onesti and Steven A. Walti

Indiana University, Bloomington, Indiana 47405 USA

A hydrologic investigation of six watersheds was conducted during three summer runoff seasons (1978, 1979, and 1980) in the central Brooks Range. The orographic effect produced by the Brooks Range significantly affects both summer and winter precipitation patterns and consequently stream runoff characteristics. Summer precipitation north of the continental divide is normally greater than that to the south, which is reflected in the volume of runoff per surface area for the small watersheds near the divide. The largest percentage of seasonal runoff takes place from breakup to early summer and is attributed to snowmelt and to the significant proportion of summer precipitation which occurs during this period. Precipitation discharge calculations suggest that the major portion of annual precipitation becomes stream discharge representing a large water yield in proportion to total precipitation. These yields are attributed to low evaporation rates and to the presence of continuous permafrost.

INTRODUCTION

The hydrologic characteristics of six arctic-alpine watersheds ranging in size from 4.8 to 125 km^2 were monitored during three consecutive runoff seasons (1978, 1979, 1980) at Atigun Pass in the central Brooks Range. Atigun Pass, having an elevation of 1,400 m, is located in the Endicott Mountains of the Phillip Smith quadrangle (68°N latitude, 150°W longitude) and crosses the east-west trending continental divide of the Brooks Range (see Figure 1). Three of the watersheds under investigation drain south to the Yukon River system, while the remainder flow to the north. Relatively small watersheds were selected in an attempt to limit the variability in geologic, morphometric, and climatic factors which could affect the stream's hydrologic regime.

SURFICIAL CHARACTERISTICS OF WATERSHEDS

The Brooks Range is an arctic-alpine periglacial environment located within the zone of continuous permafrost (Péwé, 1975). The active permafrost layer is highly variable in the study area. Ellis (1979) documented a lower limit of the active layer which rarely exceeds 1.5 m during the thaw period.

Pleistocene glacial deposits are present along all the valley floors, interrupted only by recent fluvial deposits associated with the present streams. The valley walls both north and south of the continental divide are covered with more recent talus cones, aprons of colluvium and/or alluvial fans, slushflow deposits, protalus ramparts, and lobate rock glaciers.

VEGETATION

Arctic-alpine tundra is the characteristic vegetation throughout the watersheds of the study area. The low relief valley floors are host to a wide variety of tussock, sedge, and grass communities. Inhabiting the stream banks of the larger valleys are extensive stands of dwarf birch and shrub willows. This shallow rete of vegetation dominates valley floors, and low relief fans, providing adequate protection against surficial runoff.

Recent glacial activity, frost action, prolific mass wasting, and rapid erosion of the steep mountainsides have left the valley hillslopes unstable and constantly changing. The shallow soil and coarse debris found on the valley floors, create drastic changes in vegetation. Lichens, ferns, grasses, small herbs, and saxifrages are scattered among the coarse debris, offering little resistance to surficial runoff and erosion.

CLIMATE

Weather conditions in this alpine region are influenced primarily by the radiation balance and the movement of air masses, both of which are highly variable during the runoff season. Air temperature normally remained above freezing; however, intrusion of arctic air masses during the runoff period would inhibit snowmelt runoff.

Four Belfort Universal weighing bucket-type recording snow and rain gages were installed at sites N-1, N-2, S-1, and S-2 (see Figure 1) to record summer precipitation. Although this network of gages could be considered rather thin, it was determined to be representative for the purposes of this study.

The orographic effect produced by the Brooks Range gives this area the highest annual precipitation values along the trans-Alaska pipeline from the Yukon River to Prudhoe Bay (Haugen, 1979).

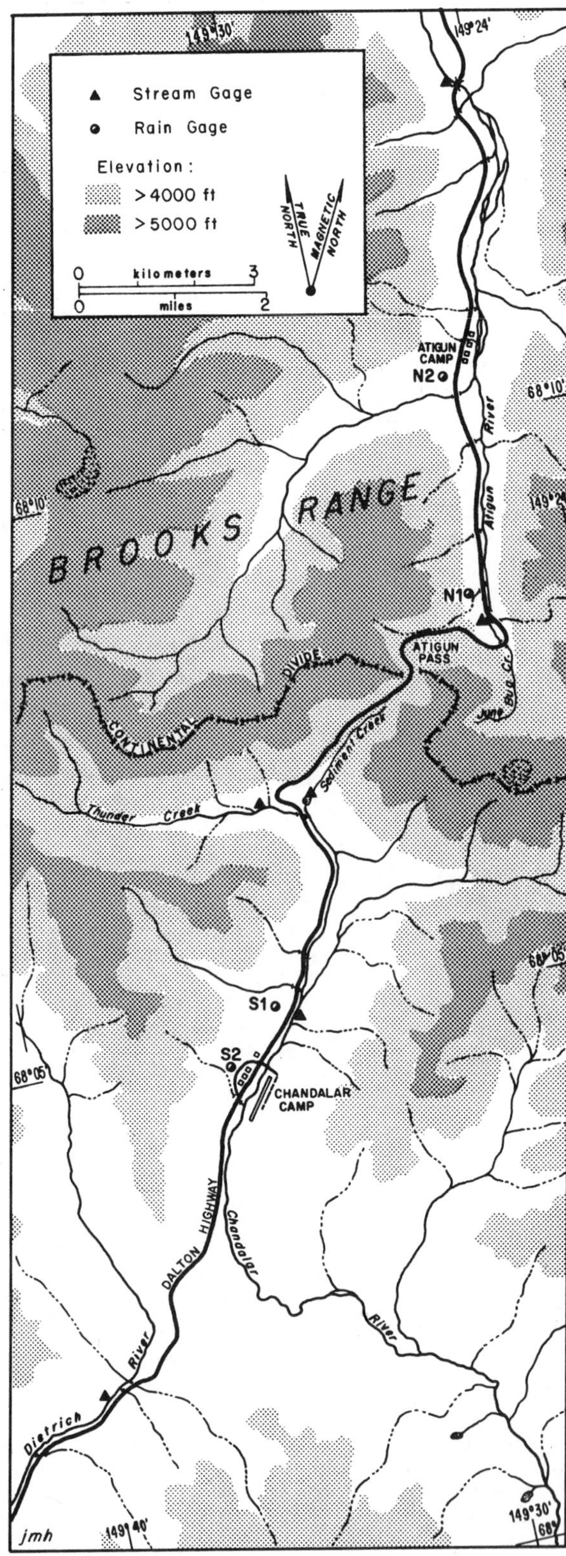

FIGURE 1 Study area and location of gages.

Precipitation data from a network of gages distributed over a distance of 60 km both North and South of Atigun Pass for the period of record suggest that summer rainfall varies considerably from year to year as well as north and south of the continental divide. This network is not shown in Figure 1. Precipitation north of the divide is normally greater than that to the south.

The largest total summer rainfall recorded during the study period was 332.74 mm at Station N-1 in the Atigun Pass during the 1980 field season and the least 74.00 mm at station N-2 in 1978. However, the 1978 value may be somewhat misleading, since this period of record is much shorter than the 1979 and 1980 seasons (see Table 1). The average total summer precipitation for the 1979 and 1980 seasons was 227.72 mm for stations north of the divide and 186.06 mm for stations to the south (see Table 1). Considering that all the periods of record terminate in mid-August and freezeup normally occurs in late September or early October, these rainfall values would be higher if precipitation for the latter half of August and September were included.

Unlike the reported significant influence of evaporation in the middle and low latitudes, in arctic regions it has been reported to account for only a 0.5-2.0% reduction in total runoff (Hövind 1965). Thus evaporation has been disregarded in the calculations pertaining to the study.

TABLE 1 Precipitation Values North and South of Continental Divide (in millimeters).

Year	Length of Record	Stations (North) N-1	N-2
1978	6/20-8/12	97.00	74.00
1979	5/13-8/18	245.13	151.63
1980	5/26-8/13	332.74	181.37
Average 227.72			

Year	Length of Record	Stations (South) S-1	S-2
1978	6/20-8/12	102.7	102.7
1979	5/14-8/17	166.89	166.13
1980	5/26-8/10	185.17	226.06
Average 186.06			

FIELD METHODS

Six continuous recording gages were used to monitor the stage of the streams under investigation. Stage recorders were operational within a few days of the initial spring flow; however, icing and slushflow conditions as well as winter snow accumulation in the channels inhibited early installation. A Marsh-McBernie velocity meter was used to record flow velocities. Velocity measurements were used to establish a representative rating curve for stream discharge. Six seasonal hydrographs were derived from stage values taken at 2 hour intervals from each stage recorder chart.

NATURE OF THE DATA

The period of record for both precipitation and discharge does not extend through the entire runoff season. Data collection normally ended in mid-August due to university commitments (see Tables 1 and 2 for length of records). In the Atigun Pass area freezeup normally occurs in late September or early October (G. Elliot, personal communication). Since the 1978 record does not include the breakup period, no conclusions will be drawn for this period but the data will be reported.

The 1979 and 1980 period of record are rather similar in length and do include most of the breakup period; however, approximately 1.5 months of the precipitation and discharge data are missing. Nevertheless the authors feel the major portion of the seasonal runoff was recorded and only late summer low flows and perhaps a few precipitation events are missing. Due to the paucity of hydrologic information for this area, these data merit comment; however, it is recognized that the record is incomplete.

DISCHARGE PATTERNS

The patterns of precipitation and runoff for the period of record are highly skewed toward late spring and early summer in the central Brooks Range. On the average, approximately 30% of the recorded summer precipitation occurred during the first 30 days of this period. The large proportion of summer precipitation which occurs in the early portion of the runoff season in combination with snowmelt contributes to the exceptionally large runoff experienced.

Seasonal hydrographs for the streams under investigation indicate that 50% of the stream flow was discharged in the first 30 days of the 1980 runoff season, while 85% of the recorded flow occurred in the seven week period after June 1 in both 1979 and 1980. These values correspond well to values documented by Pissart (1967), Church (1972), McCann and Cogley (1973), and Collins (1979) for other arctic watersheds.

Daily stream flow fluctuations seen in June are somewhat more uniform than those experienced in the latter portion of the runoff season and may be attributed to snowmelt response to diurnal fluctuation in air temperature (see Figure 2). Snowmelt rates increase during the early portion of the runoff period. This acceleration of snowmelt is brought on by discontinuous or patchy snow cover characteristic of south facing slopes, which receive a disproportionate amount of radiation during the early spring. Gray and O'Neill (1974) established that during a 6 day interval, 44% of the energy supplied to an isolated snow patch was by sensible heat transfer. During a period of continuous snow cover the corresponding figure for the same location was 7%.

Diurnal peak flows on the average occur from approximately 7:00 P.M. to 10:20 P.M., and minimum flows are recorded between approximately 9:30 A.M. and 12:30 P.M. depending upon basin size and location within their individual drainage networks. The degree of diurnal flow variation ranges from 0.35 to 3.15 m^3/s again depending primarily on basin size. A rapid and short-term decline in the hydrograph is normally related to an intrusion of arctic air with temperatures at or below freezing (see Figure 2). The gradual decline in the seasonal hydrograph reflects the melting and final disappearance of practically all of the snowpack by mid-July, with a few large late summer frontal or convectional showers producing peak discharge events which are in some cases larger than events experienced in the earlier part of the runoff season but are of shorter duration (see Figure 2). The highest peak discharges recorded occurred during the 1978 period of record in the larger basins, while in the smallest basins June Bug and Sediment Creek, the peak discharges occurred in 1979 and 1980 respectively. Since both peak flows recorded in the latter two basins are associated with precipitation events, this would appear to suggest that smaller basins are somewhat more sensitive to individual rainfall events.

TABLE 2 Total Recorded Basin Runoff.

Streams	Year	Length of Record	Seasonal Runoff (m^3/ record x 10^3)	Peak Discharge (m^3/ s)
Dietrich River	1979	5/22-8/17	14,778	286.5
	1980	5/29-8/14	19,718	239.0
Chandalar River	1978	6/26-8/13	10,678	644.0
	1979	5/28-8/17	17,114	437.0
	1980	5/30-8/14	23,499	459.0
Thunder Creek	1978	6/15-8/13	5,413	166.5
	1979	5/30-8/18	5,790	127.5
	1980	5/29-8/14	5,645	134.0
Sediment Creek	1978	6/13-8/13	888	65.6
	1979	5/25-8/18	1,784	58.9
	1980	5/26-8/14	1,800	91.8
June Bug Creek	1978	6/18-8/13	1,484	44.0
	1979	5/29-8/18	2,962	106.5
	1980	5/29-8/14	3,287	62.5
Atigun River	1978	6/14-8/13	50,504	1020.0
	1979	6/4-8/18	42,684	850.0
	1980	5/30-8/14	53,957	932.8

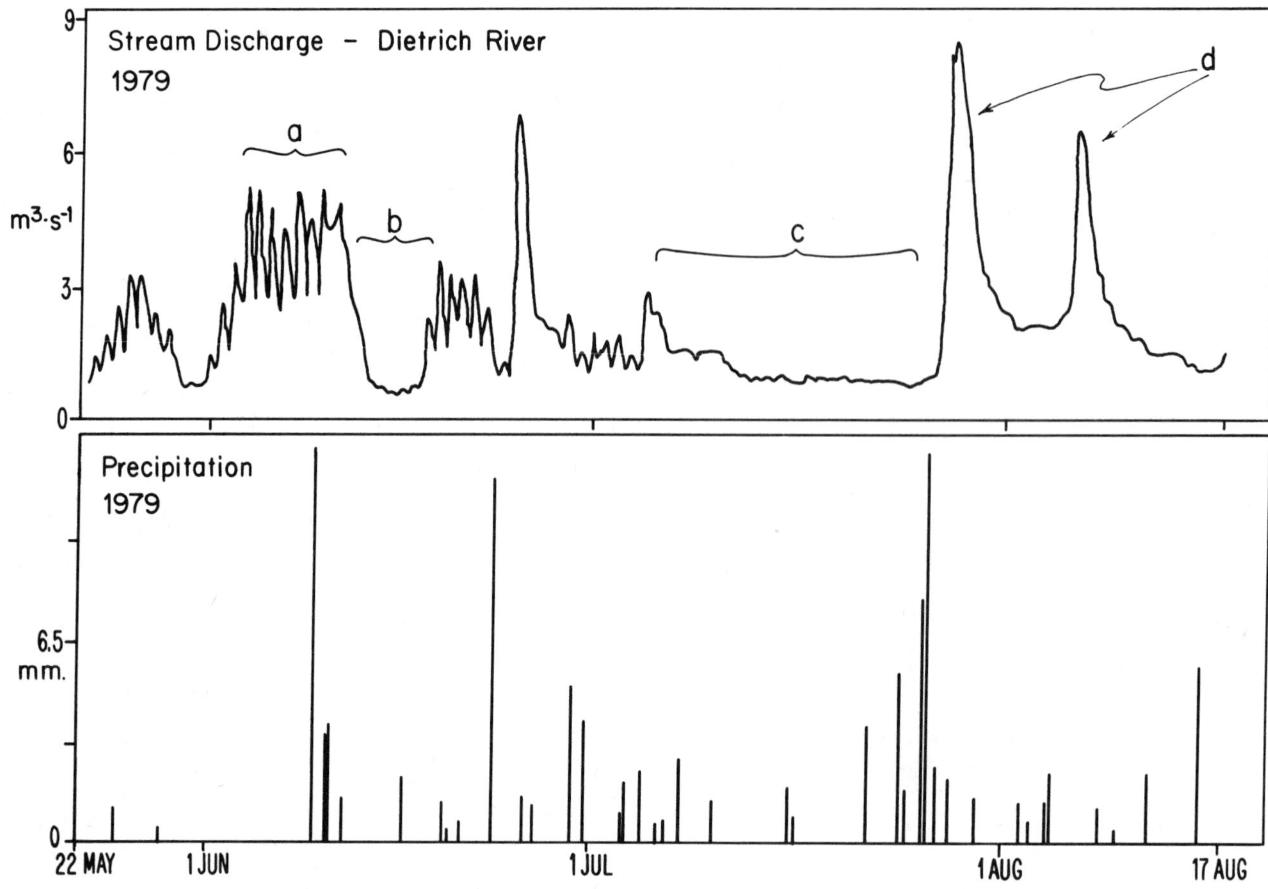
Stream Discharge - Dietrich River
1979
m3·s-1
9
6
3
0
a
b
c
d
Precipitation
1979
6.5
mm.
0
22 MAY
1 JUN
1 JUL
1 AUG
17 AUG
a) Nival freshet with remarkable diurnal flow.
b) A cold spell with a dramatic decline in flow.
c) Late summer period, little precipitation and little runoff.
d) Large peaks due to frontal precipitation.

FIGURE 2 An example of a seasonal hydrograph and associated precipitation.

TOTAL RUNOFF CALCULATION

Precipitation-runoff calculations represent the contribution of rainfall and snowfall to stream runoff (Langbein 1949). The equation used to calculate total annual runoff (Rp) is as follows:

$$R_p = P_t \times (A_b \times C_f),$$

where (P_t) is seasonal precipitation in mm, (A_b) is basin area in km^2, and (C_f) the correction factor for time. Calculated values are expressed in cubic meters per season.

For the period of record summer rainfall accounts for approximately 42% of stream discharge in the study area, suggesting that approximately 58% of the seasonal discharge is attributable to nonrecorded snow and ice melt as well as to groundwater contribution.

The above percentages are in fair agreement with annual frozen versus unfrozen precipitation values indicated by Haugen (1979). However, since discharge and precipitation for late August and September were not recorded by this study, precipitation-runoff values attributed to summer rainfall may be considered underestimates.

TABLE 3 Precipitation/Runoff Calculations.

Stream	Area (km^2)	Year	Seasonal Precip. (mm)**	Total Runoff (m^3/ record x 10^3)	Runoff per km^2 (m^3/ km^2 x 10^3)
Atigun River (N-2)*	125.2	1978	74.0	8,989	71
		1979	151.6	13,792	110
		1980	181.4	23,553	188
June Bug Creek (N-1)*	5.9	1978	97.0	1,152	195
		1979	245.1	1,237	210
		1980	332.7	1,939	329
Thunder Creek (S-1)*	12.3	1978	102.7	1,259	102
		1979	166.9	1,636	133
		1980	185.2	2,387	194
Sediment Creek (S-1)*	4.8	1978	102.7	495	103
		1979	166.9	644	134
		1980	185.2	939	195
Chandalar River (S-1)*	45.0	1978		4,636	103
		1979	166.9	6,026	134
		1980	185.2	8,790	195
Dietrich River (S-2)*	44.9	1978	102.7	4,617	103
		1979	166.1	6,944	155
		1980	226.1	8,166	182

*Rain gage site used in calculation.
**See Table 2 for length of record.

Total water yield for the period of record indicates that the seasonal volume of discharge is directly related to drainage basin size, as would be expected; however, water yield per km^2 does not show this relationship. Small watersheds near the continental divide have greater volume yields per surface area than larger watersheds (see Table 3). The precipitation gage N-1 normally records the highest summer rainfall values (see Table 1) while Calkin and Ellis (1980) reported greater depth of snowpack with decreasing distance from the continental divide; both of these factors most likely contribute to this variation in volume discharge per unit area.

SUMMARY

Four distinct runoff periods characterize stream flow in the arctic-alpine environment of the central Brooks Range. Spring breakup is short and associated with penetration of warm air masses into the study area in combination with high inputs and longer durations of radiation as the summer solstice approaches. The largest volume of discharge occurred during the snowmelt runoff period.

Low stream discharge characterizes the late summer period; however, sporadic intense rainfall events develop peak flow conditions in all streams. The major portion of all rainfall and snowmelt becomes runoff due to the continuous nature of the permafrost and to low evaporation rates. Precipitation-runoff values for the period of record indicate that 42% of stream runoff is attributed to summer precipitation and the remaining 58% to snowmelt and baseflow.

ACKNOWLEDGEMENTS

This study was supported by NSF Grant DDP 789982 and in part by the U.S. Army Cold Regions Research and Engineering Laboratory, Hanover, New Hampshire, through the Indiana University Foundation. The authors thank Dr. John R. Williams and the anonymous reviewers for their comments.

REFERENCES

Calkin, P. E., and Ellis, J. M., 1980, A lichenometric dating curve and its application to holocene glacier studies in the central Brooks Range, Alaska: Arctic and Alpine Research, v. 12(3), p. 245, 264.

Church, M., 1972, Baffin Island sandurs--a study of arctic fluvial processes: Geological Survey Canadian Bulletin, v. 216, 208 p.

Collins, D. N., 1979, Hydrochemistry of meltwaters draining from an alpine glacier: Arctic and Alpine Research, v. 11(3), p. 307-324.

Ellis, J. M., 1979, Neoglaciation of the Atigun Pass area, east-central Brooks Range, Alaska: Unpublished M.A. thesis, State University of New York at Buffalo.

Gray, D. M. and O'Neill, A. D. J., 1974. In S. H. Santeford and J. L. Smith, eds., Advanced concepts and techniques in the study of snow and ice resources: Washington, D.C., National Academy of Sciences, p. 108-1180.

Haugen, R. K., 1979, Climatic investigations along the Yukon River to Prudhoe Bay haul road, Alaska, CRREL Technical Note (unpublished): Hanover, N.H., Cold Regions Research and Engineering Laboratory, 23 p.

Hövind, E. L., 1965, Precipitation distribution around a windy mountain peak: J. Geophysical Research, v. 70, p. 3271-3278.

Langbein, W. B., 1949, Annual runoff in the United States: U.S. Geological Survey Circular, 52.

McCann, S. B., and Cogley, J. G., 1973, The geomorphic significance of fluvial activity at high latitudes: Research in Polar and Alpine Geomorphology, 3rd Guelph Symposium on Geomorphology, p. 118-35.

Péwé, T. L., 1975, Quaternary geology of Alaska: U.S. Geological Survey Professional Paper 835, 145 p.

Pissart, A., 1967, Les modalités de l'écoulement de l'eau sur l'île Prince Patrick (76° lat. N, 120° long. O, Arctique Canadien): Biuletyn Peryglacjalny, no.16, p. 217-24.

FIELD MEASUREMENTS OF ELECTRICAL FREEZING POTENTIALS IN PERMAFROST AREAS

V. R. Parameswaran[1] and J. R. Mackay[2]

[1]Division of Building Research, National Research Council Canada, Ottawa, Ontario, Canada K1A 0R6
[2]Department of Geography, University of British Columbia, Vancouver, British Columbia, Canada V6T 1W5

Electrical potentials developed during freezing of natural soils have been measured in two permafrost areas: one in a saturated sand in a drained lake (Illisarvik) on Richards Island where permafrost was aggrading downward 5.65 m below ground; the other, in a mud hummock in the active layer at Inuvik. Peak potentials of up to 1350 mV were measured on electrodes located on the advancing freezing front at Illisarvik. At Inuvik, maximum freezing potentials of up to 700 mV were measured as the active layer froze in winter. There was also evidence of downward water migration and freezing as the ground started to thaw at the surface in spring.

INTRODUCTION

The phenomenon of electrical potentials generated during freezing of aqueous solutions and other systems such as moist soils has been known for some time, although the exact mechanism of development is not fully understood. Several plausible explanations have been proposed for pure water and dilute aqueous solutions, but no such theory has been suggested for soils.

When water freezes on the surface of a solid, the dipole field of the oriented water molecules in the initial ice layer causes rotation and reorientation of the molecules in the liquid adjacent to the ice. An electrical double-layer of a few nanometers is formed at the interface between the frozen and unfrozen parts. It behaves like a semipermeable membrane, allowing one kind of charge to pass through more readily than the other, and causes selective incorporation of an excess charge of one sign in the frozen part and an equal and opposite charge in the unfrozen part. The potential developed in this way is commonly called the freezing electrical potential. During freezing, there is a tendency for water to migrate to the freezing front, and thus several electrokinetic effects (such as electrophoresis, electro-osmosis, streaming potential, and sedimentation potential) may also arise as a result of the relative motion between water and solid phase.

Early measurements of freezing potentials were carried out in water and in dilute aqueous solutions containing different kinds of ions, for example, alkali metals, ammonia, halogens, and hydroxyl ions (for early references, see Parameswaran 1982). The early investigations showed that the magnitude and sign of the freezing potential depend upon various factors, including the type of ion in solution, type and distance between electrodes, concentration of electrolyte solution, and rate of cooling.

Reports on the observation of freezing potentials in soils and rocks are meagre and inconclusive. Jumikis (1958) measured potentials of 40-120 mV during freezing of Dunellen silty soil (a glacial till) and suggested that such potentials could enhance moisture transport in the material. Korkina (1975) studied the freezing electrical potentials in suspensions containing particles of less than a micron and saturated with different kinds of ions such as Fe^{3+}, Ca^{++}, Na^{+}, and NH_4^{+}. The magnitude and polarity of the induced voltage depended on the density of the suspensions, type of particles, and type of ions. Under laboratory conditions, Borovitskii (1976) measured freezing voltages of 150-200 mV in argillaceous rocks, and Yarkin (1974, 1978) measured freezing voltages of up to 325 mV in powdered sand containing about 23% moisture and smaller freezing voltages in other soils. More recently, Hanley and Ramachandra Rao (1980, 1981) studied the effects of various cations on the generation of freezing potentials in a bentonite clay. The maximum potentials were in the range of 30-50 mV, and they concluded that the development of freezing potential and migration of moisture are closely associated phenomena, each influencing the other. Parameswaran (1982) has also reported laboratory measurements of electrical freezing potentials in water and soils under various rates of cooling. Ice-positive potentials of 4-12 V developed in pure water under rapid cooling; and in moist sands and soils ice-negative potentials of 100-200 mV were measured. Under slow cooling at -2.2°C reconstituted natural soils from permafrost areas showed freezing voltages of 200-300 mV. The freezing potentials in water could be due to a combination of those related to the phase change of water and those arising out of charge separation and ion incorporation from traces of impurities (known as Workman-Reynolds effect). In addition to polarization of water molecules and alignment of dipoles across the potential barrier at the freezing boundary, charges inherently present at the surface of mineral particles in soils could also contribute to the development of freezing potentials by exchange absorption processes.

The development of contact electrical potentials during freezing is of considerable importance in many construction problems in northern North America, the USSR, and Europe, where vast areas are underlain by perennially and seasonally frozen ground. Under the influence of an electrical gradient such as that caused by freezing, water is transported through frozen clay and silt towards the freezing front where segregated ice can form (Verschinin et al. 1949, Hoekstra and Chamberlain 1963). Such movement of water under the influence of geoelectric fields in the presence of a thermal gradient could also cause electro-osmotic drainage of unfrozen and frozen soils (Nersesova and Tsytovich 1966). The additional heave caused by such processes could necessitate special construction techniques for pipelines, foundations for buildings, and other structures in permafrost areas. Geoelectric fields generated near buried pipelines could also affect their cathodic protection.

FIELD STUDIES

Very few measurements of electrical freezing potentials under natural conditions have been published. The only documented report to the authors' knowledge is that of Borovitskii (1976), who measured the inherent electrical fields developed in the active layer between frozen and thawed parts in a permafrost region of the USSR. The measured values did not exceed 50 mV, although the value of the potential varied with depth in the ground. This he attributed to differences in quantity and direction of moisture movement. Mackay has also carried out measurements of the potentials developed in electrodes embedded to different depths in the active layer and in perennially frozen ground in three different areas of the Northwest Territories. When electrodes were placed in the active layer, he observed a potential change as the freezing front passed the electrodes. Values of potentials measured in the field were of the same order of magnitude as values measured in the laboratory on natural soils (Parameswaran 1982). This similarity of the freezing potentials measured in the laboratory and in the field prompted the authors to install field electrode assemblies in two permafrost areas and to measure freezing potentials in an attempt to understand the physical processes that occur during freezing of the ground. One chosen area was the middle of an artificially drained lake (Illisarvik) on Richards Island, N.W.T., where permafrost was aggrading from the top down (Mackay 1981). The other area was at Inuvik, N.W.T., where the active layer in a mud hummock was monitored. This paper describes the equipment and some of the results.

Illisarvik Lake Site, Richards Island

An electrode probe consisting of five plexiglas tubes, each 1.93 m long, 38 mm OD and 25 mm ID, was installed at the site, the middle of a drained lake, on 20 June 1981. Twenty-two copper electrodes in the form of circular bands (12.5 mm wide) were installed on the probe in suitable grooves 1 mm deep. The bottom electrode (numbered zero) was 89 mm from the tip of the plug closing the first tube. All the others, numbered 1 to 21 from the bottom upwards, were set 150 mm apart. A shielded coaxial cable (RG 174/U) was connected to each electrode through a slanted hole in the wall of the tube and the soldered joints were protected by epoxy resin. The tubes were designed to be joined with plexiglas couplings as they were installed.

Wires from the electrodes were connected to a rotary switch having 24 terminals (Figure 1b). The bottom electrode was connected to the common point (-); the others were connected to switch points 1 to 21. The central switch contact point was made the (+) terminal. The shields from all the cables were connected to the ground terminal, which was in turn connected to a grounding steel post at the lake site.

Two thermistor cables were installed near the electrodes for temperature measurements. The first cable had 12 thermistors (YSI 44033, calibrated to ±0.01°C at 0°C), each inside a soil salinity sensor (Soilmoisture No. 5100-A). These were embedded in a 30 mm diam PVC rod at 500 mm intervals. The rod (with salinity sensors) was installed in a hole drilled by water-jet 0.4 m from the electrode assembly (Figure 2). A second temperature cable with 26 thermistors (YSI 44033) spaced 150 mm apart was installed 4 m from the electrode assembly.

Depth of permafrost was 5.65 m when the electrodes were installed in June 1981. The uppermost electrode (21) was at a depth of 6.11 m and the bottom electrode (0) was 9.31 m below ground level (lake bottom). This placement ensured that the 0°C isotherm would soon reach the uppermost electrode. The material in which the electrodes were installed was a saturated sand with medium-to-fine grain size.

Inuvik Site

This site is about 3 km north of the town of Inuvik in an area of colluvium that has developed a pattern of mud hummocks. The electrode probe consisted of a PVC tubing (20.6 mm OD, 12.7 mm ID, 3 m long, closed at one end) on which 12 copper electrodes in the form of bands or rings (12.7 mm wide and about 1 mm thick) were installed in suitable grooves 150 mm apart. Coaxial cables were soldered to the electrodes and the wires were led out through a central hole and connected to the terminals of a rotary switch. The outer shields of the cables were grounded to a steel post at the site. A thermistor cable with 13 thermistors (YSI-44033) was also made. The electrode probe and the thermistor cable (Figure 3) were installed (by hand drilling) in August 1981 in the centre of a mud hummock about 2 m in diameter.

RESULTS AND DISCUSSIONS

Illisarvik Lake Site, Richards Island

Owing to the thermal disturbance of the ground caused by drilling and the resulting freeze-back, the readings taken in August 1981 are questionable and are not reported here. The Illisarvik site is relatively inaccessible and frequent readings were impossible, so that the next readings were taken 23 March 1982, 8 June 1982, and 12 August 1982. The temperature and freezing potential profiles for these dates are plotted in Figure 4(a-c). None of the three temperature profiles shows any change in gradient in crossing 0°C, from positive to negative temperatures. In Figure 4(a) there is a gradual change in the profile at about -0.1°C; and in Figure 4(b, c) a definite inflection point at -0.1°C. With little doubt this marks the freezing point depression so commonly observed in soils. The Illisarvik sediments consist mainly of sand, and the freezing point depression is slightly lower than might be expected. Since the sub-permafrost pore water has an increased salinity from ion rejection during permafrost aggradation, however, a freezing point depression below that of the original sands is to be expected. To illustrate, in the summer of 1982 the specific conductance of the pore water just below permafrost, as measured with the soil salinity sensors, was about 300 $\mu ohm^{-1}\ cm^{-1}$; salinity had increased appreciably since the previous summer. Nearby measurements of heave of the ground surface suggest that most of the ice formed during the downward growth of permafrost was pore ice and not segregated ice.

Figure 4(a-c) shows plots of peak potentials and temperature gradients. On 23 March 1982 there was a very pronounced peak in the potential at 1.08 V at a depth just 15 cm below the 0 deg isotherm. As the temperature measurements were taken 0.4 m from the electrodes, the agreement seems good. By way of contrast, on 8 June 1982 the peak potential (about 1.35 V) occurred at a ground temperature of -0.1°C, the peak coinciding closely with the inflection point on the temperature curve. By 12 August 1982, however, the peak potential was unchanged at the same depth as on 8 June 1982, although permafrost had aggraded 50 cm and the ground temperature had decreased to about -0.4°C. In summary, the results from the Illisarvik site indicate that freezing potentials develop on electrodes located at the freezing front, but these results are by no means conclusive.

Inuvik Site

The Inuvik site contrasts with the Illisarvik site in soil type and the placement of the electrodes. Here the soil is a silty clay, with particles about 50 % finer than 0.002 mm and a specific surface area of about 120 m^2/g. The amount of unfrozen pore water in the clay, using the specific surface area method of Anderson and Tice (1972), is estimated at about 9 % at -5°C and 7 % at -10°C. Voltage and temperature readings were taken every two weeks by the staff of the Inuvik Scientific Resource Centre. Variations of voltage and temperature with depth below surface were much more predictable and systematic than those at the Illisarvik Lake site. Figure 5(a-e) shows the variations of voltage and temperature at various depths as the ground froze and thawed in the 1981-1982 season. At depths of 0.05 and 0.20 m, respectively, a freezing potential developed as the temperature passed through 0°C in the month of October 1981 (Figure 5(a,b)). The maximum freezing potentials developed were as high as 650 mV. As the ground temperature at these locations dropped in winter, the potential dropped too and remained at the lower level. In spring (May 1982), as the ground started to thaw and the temperature rose above 0°C, the potentials again rose to values of up to 700 mV. The interpretation of this potential is uncertain, but there is considerable evidence that during the summer thaw period water may migrate from the thawed active layer into the still-frozen active layer to freeze and increase the ice content (Cheng 1982, Mackay, in press; Parmuzina 1978, Wright 1981). A minor freezing potential may thus develop during the summer thaw period. This behaviour of meltwater, possibly percolating downwards and freezing to generate a freezing potential, was not observed at the electrodes at lower levels. For example, in Figure 5(c) the freezing potential on the electrode at a depth of 0.51 m below surface level rose to 700 mV as the ground temperature passed through -0.1°C. With further decrease in ground temperature the freezing potential remained unchanged. A slight drop in potential was noted when the temperature passed through 0°C in June 1982 as the ground thawed, but essentially the voltage was constant around 700 mV even when the ground started to freeze again in October 1982. The same type of behaviour was observed 0.96 m below ground level (Figure 5(d)).

Figure 5(a,b) shows the development of pronounced freezing potentials at depths of 0.05 and 0.20 m, respectively, in the upper part of the active layer where ice lensing commonly takes place. In addition, potentials at 0.51 and 0.96 m remained at 600-700 mV during the winter, after temperatures dropped below 0°C. Measurements carried out at Inuvik for many years have shown a prolonged and very gradual frost heave in the winter months long after the temperature has dropped below 0°C (Mackay et al. 1979). This suggests that, in winter, upward migration of unfrozen water from the freezing front into the frozen active layer could cause a minor freezing potential to develop. At levels close to the permafrost table, where ground temperature is maintained at or below 0°C, no freezing potentials developed on the electrode 1.42 m below ground level (Figure 5(e)). Figure 5(a-e) also shows that as depth increases below ground level the minimum temperature, as expected, becomes higher. For example, at 0.05 m the lowest ground temperature measured was -10°C, whereas at 0.96 and 1.42 m the lowest soil temperatures measured were -3.5 and -2.5°C, respectively.

CONCLUSION

Freezing potentials appear to have developed on electrodes installed in saturated sand at Illisarvik where the 0°C isotherm was that of an aggrading lower permafrost surface, but the results are inconclusive. Peak potentials of up to 1350 mV were measured on electrodes located on the advancing freezing front. At Inuvik, the electrodes were placed in the active layer on top of permafrost, in a silty clay. The electrodes in the upper part of the active layer where lens ice forms in the freeze-back period registered substantial increases in potential as the freezing front passed the electrodes. The maximum freezing potential measured here was about 500 mV when the ground temperature was about -0.1°C. There was also evidence of downward water migration and freezing, as evidenced by a rise in freezing potential at levels below the surface as the ground thawed at the surface. These field measurements indicate that by suitably modifying electrode probes and improving measuring techniques it is possible to locate and study the advancing freezing front and water migration at different levels below the surface as ground freezes and thaws.

ACKNOWLEDGEMENTS

The authors would like to express their sincere thanks to the following: C. Hubbs, J.C. Plunkett, and G.H. Johnston; the DBR/NRCC Workshop staff, and the staff of the Inuvik Scientific Resource Centre. Field support for J.R. Mackay from the Geological Survey of Canada, the Polar Continental Shelf Project (Department of Energy, Mines and Resources) and the Natural Sciences and Engineering Research Council is gratefully acknowledged. This paper is a contribution from the Division of Building Research, National Research Council Canada, and is published with the approval of the Director of the Division.

REFERENCES

Anderson, D.M. and Tice, A.R., 1972, Predicting unfrozen water contents in frozen soils from surface area measurements: Highway Research Record No. 393, p. 12-18.

Borovitskii, V.P., 1976, The development of inherent electrical fields during the freezing of rocks in the active layer and their role in the migration of trace elements: Journal of Geochemical Exploration, v. 5, p. 65-70.

Cheng, C., 1982, The forming process of thick layered ground ice: Scientia Sinica (Series B), v. 25, No. 7, p. 777-788.

Hanley, T. O'D. and Ramachandra Rao, S., 1980, Freezing potentials in wet clays: I. Early results, and II. Specific systems: Cold Regions Science and Technology, v. 3, p. 163-168, 169-175.

Hanley, T. O'D. and Ramachandra Rao, S., 1981, Freezing potential studies in wet clay systems: Presented ASME (Heat Transfer Division), 16-21 Nov. 1980, Chicago (Preprint No. 80-WA/HT-8), 5 p.

Hoekstra, P. and Chamberlain, E., 1963, Electro-osmosis in frozen soils: Nature, v. 203, p. 1406.

Jumikis, A.R., 1958, Some concepts pertaining to the freezing soil systems: Highway Research Board, Special Report 40, p. 178-190.

Korkina, R.I., 1975, Electrical potentials in freezing solutions and effect on migration. U.S. Army, Cold Regions Research and Engineering Laboratory, Draft Translation 490, 15 p.

Mackay, J.R., 1981, An experiment in lake drainage, Richards Island, Northwest Territories: Progress Report: Current Research, Part A, Geological Survey of Canada, Paper 18-1A, p. 63-68.

Mackay, J.R., In press, Downward water movement into frozen ground, Western Arctic coast, Canada.

Mackay, J.R., Ostrick, J., Lewis, C.P. and MacKay, D.K., 1979, Frost heave at ground temperatures below 0°C, Inuvik, Northwest Territories: Current Research, Part A, Geological Survey of Canada, Paper 79-1A, p. 403-405.

Nersesova, Z.A. and Tsytovich, N.A., 1966, Unfrozen water in frozen soils: Permafrost, Proceedings of the First International Conference, National Academy of Sciences, Washington, D.C., p. 230-234.

Parameswaran, V.R., 1982, Electrical freezing potentials in water and soils: Presented, Third International Symposium on Ground Freezing, Hanover, N.H. (To be published in Proceedings, Vol. 2, 1983)

Parmuzina, O. Yu., 1978, [Cryogenic texture and some characteristics of ice formation in the active layer] (in Russian): Problemy kriolitologii, No. 7, p. 141-164. (Translated in Polar Geography and Geology, July-September 1980, p. 131-152.)

Verschinin, D.V., Deriagin, B.V., and Virilenko, N.V., 1949, The non-freezing water in soil: U.S. Army, Cold Regions Research and Engineering Laboratory, Translation No. 30.

Wright, R.K., 1981, The water balance of a lichen tundra underlain by permafrost: McGill University, McGill Subarctic Research Paper No. 33, Climatological Research Series No. 11, 110 p.

Yarkin, I.G., 1974, Physico-chemical processes in freezing soils and ways of controlling them: Collected papers, No. 64, N.M. Gersevanov Foundation and Underground Structure Research Institute, Moscow, p. 59-81.

Yarkin, I.G., 1978, Effect of natural electrical potentials on water migration in freezing soils: Permafrost, Proceedings of the Second International Conference, National Academy of Sciences, Washington, D.C., p. 351-361.

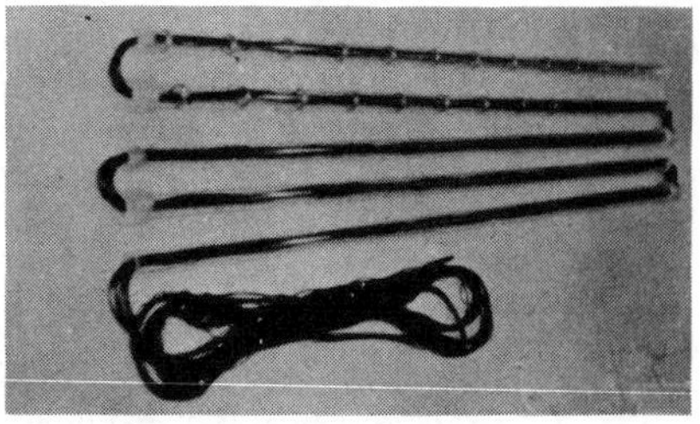

FIGURE 1(a) The Illisarvik electrode assembly.

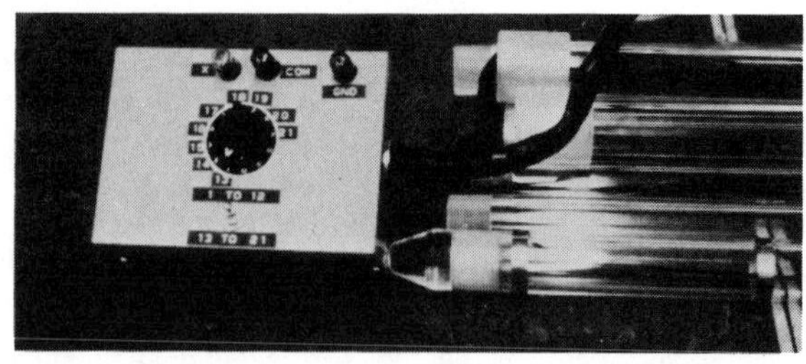

FIGURE 1(b) Switching box.

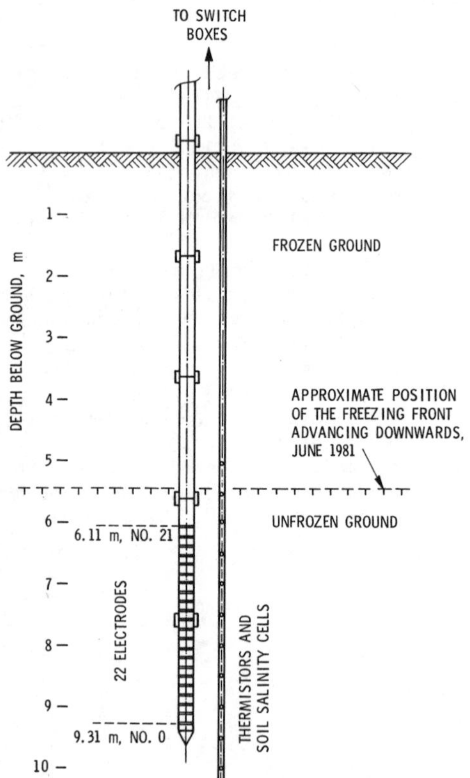

FIGURE 2 Schematic diagram showing the location of electrodes and thermistors after installation, Illisarvik site.

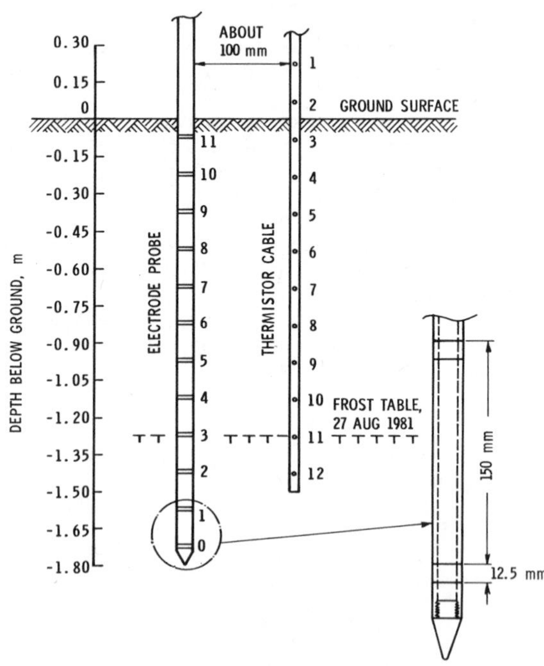

FIGURE 3 Schematic diagram showing the location of electrodes and thermistors after installation at the Inuvik site.

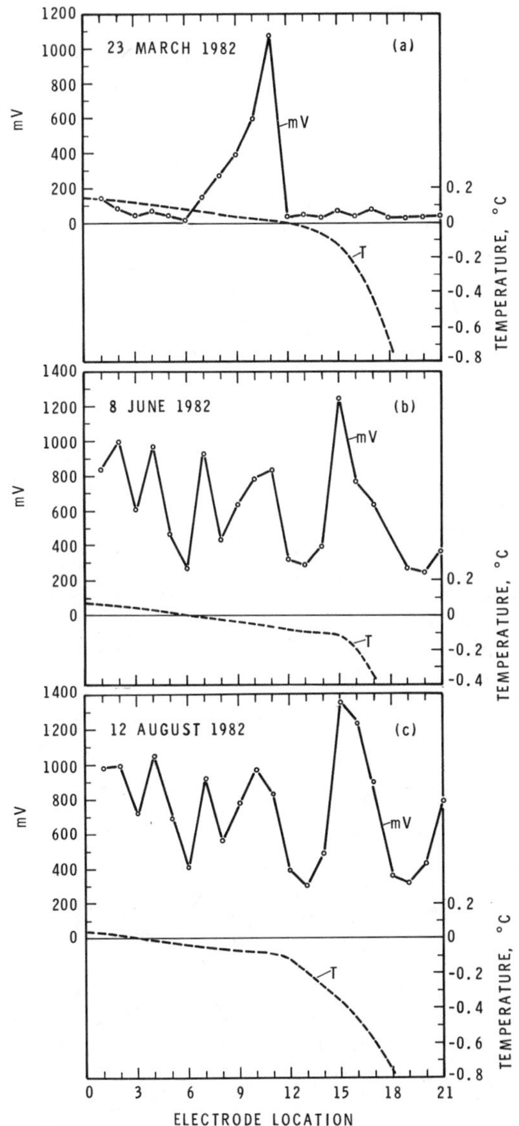

FIGURE 4 Variation of voltage and temperature with depth below ground surface at the Illisarvik site, (a) 9 months, (b) about 1 year, and (c) 14 months after installation.

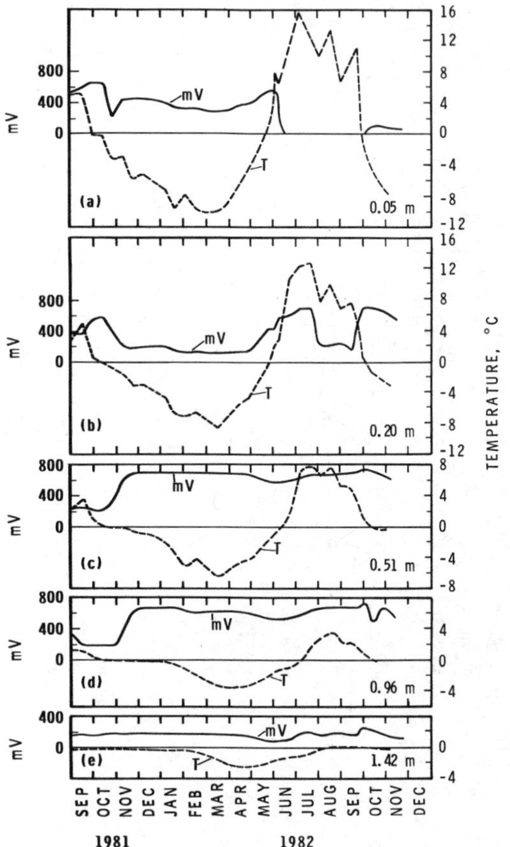

FIGURE 5 Variation of voltage and temperature with time at different depths below the surface, Inuvik site.

MEASUREMENT OF UNFROZEN WATER CONTENT IN SALINE PERMAFROST USING TIME DOMAIN REFLECTOMETRY

Daniel E. Patterson and Michael W. Smith

Geotechnical Science Laboratories, Geography Department, Carleton University, Ottawa, Ontario Canada, K1S 5B6

Time Domain Reflectometry has been previously used to determine volumetric unfrozen water contents, θ_u, in frozen soils. This paper presents laboratory results which indicate that the technique can be extended to use in saline permafrost samples. TDR determined values of θ_u for samples at the same salinity, are generally reproducible to within 2.5%. Measured θ_u data, for the soil at its natural salinity, were used to calculate values at other salinities, by accounting for the freezing point shift. The measured and calculated values generally agree to within 2-5% in θ_u. Finally, some field data are presented.

INTRODUCTION

Ice-bonded permafrost is found extensively beneath the floor of the Beaufort Sea, and its presence requires special attention in offshore engineering. To reliably predict the behaviour of permafrost for engineering purposes, accurate data on the volumes of ice and water are required. Since seabed permafrost may not be in a state of thermodynamic equilibrium, it is not possible to simply calculate the proportions of ice and water present in the soil based on temperature, salinity and pressure. The phase composition can only be determined by direct measurements on actual core samples, carried out immediately upon recovery. This paper presents preliminary results on the use of Time Domain Reflectometry (TDR) to determine the volumetric unfrozen water content in saline soils. If total water content is also known (e.g. by gravimetric means or by TDR on the thawed sample), the complete phase composition of the sample can be determined.

THE TDR TECHNIQUE

Previously we have reported on the use of TDR to determine volumetric unfrozen water contents of soils, θ_u, from the measured dielectric constant, K_a (Patterson and Smith 1981). Background information on the TDR technique can be found in the literature (e.g. Topp et al. 1980, Patterson and Smith 1981, Smith and Patterson 1982). Briefly, the technique is used to determine the apparent dielectric constant (K_a) of a soil, from measurement of the travel time (t) of the TDR's step-voltage along a transmission of known length (L). Parallel rod or coaxial lines are the two types of transmission line most commonly used, where the former is embedded in the sample and the latter is used to contain the sample. Figures 1 and 2 show sample TDR traces for each type respectively. The A point on these figures is the start of the line in the soil, and the B point (where recognizable) is the response of the open circuit termination. K_a is determined from the following equation:

$$K_a = \left[\frac{c \times t}{L}\right]^2 \quad (1)$$

where c is the free space velocity (3×10^8 m/s) and t is measured on the horizontal axis of the TDR trace.

Topp et al. (1980) determined an empirical relationship between K_a and volumetric water content, θ_v, for unfrozen mineral soils, which is largely insensitive to soil type, temperature or other physical properties:

$$K_a = 3.03 + 9.3\,\theta_v + 146.0\,\theta_v^2 - 76.7\,\theta_v^3 \quad (2)$$

This relationship has also been found to apply to the determination of volumetric unfrozen water content in frozen soils (Patterson and Smith 1981). Figure 3 presents a comparison of θ_u data determined via the TDR technique, and by other methods for the same soils. In general, the data agree to within +/-2.5% in θ_u. In addition, the authors have also shown that TDR determined freezing characteristic data agree well with published data for soils of similar texture and physical properties (Patterson and Smith 1981, Smith and Patterson 1981, 1982). An example is shown in Figure 5, for two Kaolinite soils.

The use of TDR in highly saline soils (salinity > 5 g NaCl/l) can be difficult due to the large electrical losses associated with the conductive pore water. Under certain conditions, the losses can make it difficult to distinguish the open circuit response (B point), making K_a measurements impossible. The ability to determine K_a will depend upon water content, pore water salinity, temperature and the length of the transmission line. Some of these aspects are illustrated in Figures 1 and 2.

The traces shown in Figure 1 are for an unfrozen saline sand at two temperatures using parallel rod lines. At 0^oC, the open circuit response for the 5 and 10 cm lines is quite clear, but it cannot be located for the 20 cm line. At 25^oC, the B point is still clear for the 5 cm line, however, it is somewhat uncertain for the 10 cm line. This example shows that the conductive losses, at any salinity, are temperature dependent and that the ability to determine K_a is a function of line length.

Figure 2 shows TDR traces for frozen Kaolinite samples, at initial salinities of 10 and 20 g NaCl/1, in 5 cm coaxial lines. In these examples, liquid water content decreases and pore water salinity increases as the soil freezes. As this occurs, the open circuit response improves, even though the pore water salinity increases as more water turns to ice.

When using TDR in saline soils, the combined effect of pore water salinity, water content and temperature will dictate the line length to be used. To obtain K_a measurements over a wide range of freezing temperatures, several line lengths can be used. As a guide, we recommend using the longest line that gives a recognizable B point (see Smith and Patterson 1982). Further, Figure 1 suggests the possibility of using a dual length probe to determine B points in very lossy materials.

PROCEDURES

Sample Preparation

A silty clay (Allendale) and a clay (Kaolinite #3) were used in laboratory experiments with precision 50 ohm coaxial lines serving as sample containers, of lengths 5-15 cm (depending upon initial pore water salinity). To prepare the samples, a coaxial line was placed in a low temperature chamber and a soil slurry, at the desired pore water salinity, added in small incremental amounts and frozen until the line was completely filled. In this manner, a sample without ice lensing or variations in salinity can be obtained. The sample containers were then covered in an impermeable latex membrane, and immersed in a controlled-temperature bath.

TDR traces were recorded on an x-y recorder at various temperatures on warming cycles. K_a and θ_u were determined from equations (1) and (2) respectively.

Determining the Freezing Point Shift

If freezing characteristic data (θ_u versus temperature) for a soil at its natural salinity are known, then values at other salinities can be calculated using the method described in Banin and Anderson (1974). The assumptions of this analysis are: 1. all the salts are in solution, 2. they are uniformly distributed in the pore water, and 3. the salts do not precipitate out during freezing. For samples containing only NaCl in the pore water, and with unfrozen water contents expressed on a volumetric basis, the method of Banin and Anderson can be simplified as follows:

$$T_n = T_i + \frac{S_o \cdot \Delta}{\theta_i/\theta_o} \quad (3)$$

where T_n is the new temperature at which an unfrozen water content θ_i will occur, θ_i is the unfrozen water content at temperature T_i for the soil at its natural salinity, θ_o is the water content of the thawed sample, S_o is the pore water salinity (g NaCl/1) of the thawed sample and Δ equals -5.8675×10^{-2}°C/(g NaCl/1). This formulation can be used if $S_o/(\theta_i/\theta_o)$ is less than about 90 g NaCl/1 (see Wolf et al. 1978).

Sources of Error

In these experiments, travel times in the samples were between 0.5 and 3.0×10^{-9}s (depending upon θ_u and line length). The authors have verified that in this range, travel times measured in coaxial lines filled with low loss materials (air, dielectric powders and liquids of known properties) are within +/-2% of expected values, as given by the time base accuracy of the TDR used. In saline soils, the open circuit response (B point on Figures 1 and 2) may become rounded, creating uncertainties in travel time measurements (hence K_a and θ_u determinations). The authors estimate that the travel time can be measured to within +/-2.5% of the available time window. This implies an uncertainty in θ_u estimates of +/-3.5% (from equations (1) and (2)).

The error in calculating the temperature shift in a soil freezing characteristic curve is a function of the certainty in water content and pore water salinity. Ignoring uncertainty in the pore water salinity, the error in calculating the freezing point shift will be due to uncertainties in the θ_i/θ_o ratio in equation (3). For example, if θ_i/θ_o is assumed to be 0.5, +/-5%, then the uncertainty in the freezing point shift will be about 5%. The error in the calculated freezing point shift increases with the salinity of the thawed sample. Depending upon S_o, θ_i, θ_o and T_i, calculated values for T_n can be in error by several hundredths to one or more degrees.

RESULTS AND DISCUSSION

Figure 4 presents the freezing characteristic data for the Allendale silty clay soil, measured at salinities from 0.6 g NaCl/1 (natural salinity) to 10.6 g NaCl/1. The curves at various salinities were calculated from the data at 0.6 g NaCl/1, accounting for the freezing point shift. At salinities of 5.6 g NaCl/1 or less, the agreement between measured and calculated θ_u is within ± /-2%. At 10.6 g NaCl/1, the measured values are between 1 to 5% lower than calculated.

Figure 5 shows freezing characteristic data for Kaolinite #3 determined from four samples. The data generally agree within 2.5% in θ_u with the largest difference of 5% at -0.2°C. This reproducibility in TDR results has been noted previously in Patterson and Smith (1981). The dashed line in Figure 5 is a power curve fit to the freezing characteristic data for Kaolinite #7, from Anderson and Tice (1973) (see also Anderson et al. 1973 and Tice et al. 1978). The agreement in the mean unfrozen water content data (either expressed gravimetrically or volumetrically) is good for these soils of similar properties.

Figure 5 shows the comparison between measures and calculated θ_u values for Kaolinite

#3 at initial salinities of 10 and 20 g NaCl/l. The calculated values were derived from the data for the sample at its natural salinity, by the method discussed previously. The right hand axis denotes calculated pore water salinity, which increases as θ_u decreases.

The measured and calculated values agree well, although the measurements tend to be about 2-5% higher, with the largest differences found for the 10 g NaCl/l sample. This aspect was not present in the 10.6 g NaCl/l Allendale silty clay results; however, data in Yong et al. (1979) for a Kaolin, show that measured values also were higher than calculated.

It is estimated that the error in the measured θ_u data for the saline samples is about +/-3.5% (based upon possible errors in K_a), although data from different runs at any given temperature are within about 2.5%. Other factors, such as the effects of salt on soil hydrologic properties, or modification to the K_a-θ_u relationship, cannot be ruled out as contributing to the observed differences. The authors have noted some effect of salinity on travel time when salinities are large. This aspect is presently under investigation using a combined TDR-dilatometer experiment to simultaneously obtain phase composition data for saline samples (cf. Patterson and Smith 1981). These experiments should provide information for refinement, if necessary, to the use of the TDR technique in saline permafrost materials.

FIELD OBSERVATIONS

The TDR technique has recently been used in the field, during a core logging program in the Beaufort Sea. θ_u data were obtained using a 10 cm parallel rod line which was inserted into the core within minutes of recovery. The probe could generally be pushed into the sample; however, if there was sufficient ice-bonding, pilot holes could be made using a hand drill. An x-y recorder trace of the TDR display was made for data analysis. Values of the apparent conductivity, σ, were also determined via TDR (see Smith and Patterson 1982). The results for one core are shown in Table 1. The data in the table indicate two distinct layers in the profile. The values of θ_u in the 15-48 m layer, determined from the thawed sample and via TDR, agree to within 5%. Below this, the differences are between 8 and 26%. There is also a three-fold difference in the apparent conductivity between these layers. In view of this, the differences in θ_u in the 60-90 m layer could be interpreted as ice contents.

CONCLUSIONS

The TDR technique has been used to obtain phase composition data in saline frozen soils. TDR determined values of θ_u for different samples at the same salinity are generally reproducible to within 2.5%. This reproducibility has been noted previously for soils at low salinity (Patterson and Smith 1981).

For the silty clay samples, measured and calculated θ_u values generally agree to within +/-2%, for salinities less than 5.6 g NaCl/l. At 10.6 g NaCl/l, the measured θ_u data are 1 to 5% lower than calculated. For Kaolinite #3 samples, the measured θ_u data are up to 5% higher than calculated but with most differences in the 2 to 3% range.

TABLE 1 Summary of Field Observations

Depth (m)	Temp. (°C)	PWS	θ_u Grav.	θ_u TDR	$\Delta\theta$	σ (S/m)
15.0	-	-	39.0	43.2	-4.2	0.40
23.5	-	-	46.1	41.2	4.9	0.48
29.8	-1.7	-	46.2	43.2	3.0	0.48
35.7	-2.1	-	42.0	41.2	0.8	0.48
38.7	-2.1	-	49.5	43.9	5.3	0.45
48.0	-	11.5	45.0	41.8	3.2	0.45
66.3	-2.3	10.7	36.1	18.6	17.5	0.10
72.5	-	8.3	34.8	19.9	14.9	0.11
78.6	-	7.5	38.3	18.6	19.7	0.12
81.6	-	-	46.3	20.2	26.1	0.16
87.8	-2.9	7.6	38.3	26.8	11.5	0.16
90.6	-2.9	-	31.2	22.6	8.6	0.12

$\Delta\theta$ = difference between θ_u gravimetric, and θ_u TDR
PWS = pore water salinity (g NaCl/l)
σ = apparent conductivity from TDR
\- indicates no data

ACKNOWLEDGMENTS

The authors wish to thank Charles Lewis for his help in the laboratory and for collecting the field data. The laboratory work formed part of a research program supported by the Natural Sciences and Engineering Research Council and the Earth Physics Branch, Department of Energy, Mines and Resources. The continued interest and support of Dr. Alan Judge is gratefully acknowledged. Finally we wish to thank Dome Petroleum Ltd for their permission to use the field data, and EBA Engineering (Edmonton) for their continuing interest in using the TDR technique in the field.

REFERENCES

Anderson, D. M., and Tice A. R., 1973, The unfrozen interfacial phase in frozen soil water systems, in Ecological Studies, Analysis and Synthesis, A. Hadas et al. eds., v. 4, p. 107-124.

Anderson, D. M., Tice, A. R., and Banin, A., 1973, The water-ice phase composition of clay-water systems: 1. the kaolinite water system: Soil Science Society of America Proceedings, v. 37, n. 6, p. 819-822.

Banin, A., and Anderson D. M., 1974, Effects of salt concentration changes during freezing on the unfrozen water content of porous materials: Water Resources Research, v. 10, n. 1, p. 124-128.

Patterson, D. E. and Smith M. W., 1981, The measurement of unfrozen water content by time domain reflectometry: results from laboratory tests: Canadian Geotechnical Journal v. 18, p. 131-144.

Smith, M. W. and Patterson, D. E., 1981, Investigation of freezing soils using time domain reflectometry, Report to the Dept. of Energy, Mines and Resources, Earth Physics Branch.

Tice, A. R., Burrous, C. M., and Anderson, D. M., 1978, Determination of unfrozen water in frozen soil by pulsed nuclear magnetic resonance, in Proceedings of the Third International Conference on Permafrost, v. 1, Ottawa, National Research Council of Canada, p. 149-155.

Topp, G. C., Davis, J. L., and Annan A. P., 1980, Electromagnetic determination of soil water content: measurements in coaxial transmission lines: Water Resources Research, v. 16, n. 3, p. 574-582.

Wolf, A. V., Brown, M. G., and Prentiss, P. G., 1977, Concentrative properties of aqueous solutions: conversion tables, in Weast, R. C., ed., CRC Handbook of Chemistry and Physics: Cleveland, CRC Press.

Yong, R. N., Cheung, C. H., and Sheeran, D. E., 1979, Prediction of salt influence on unfrozen water content in frozen soils: Engineering Geology, v. 13, p. 137-155.

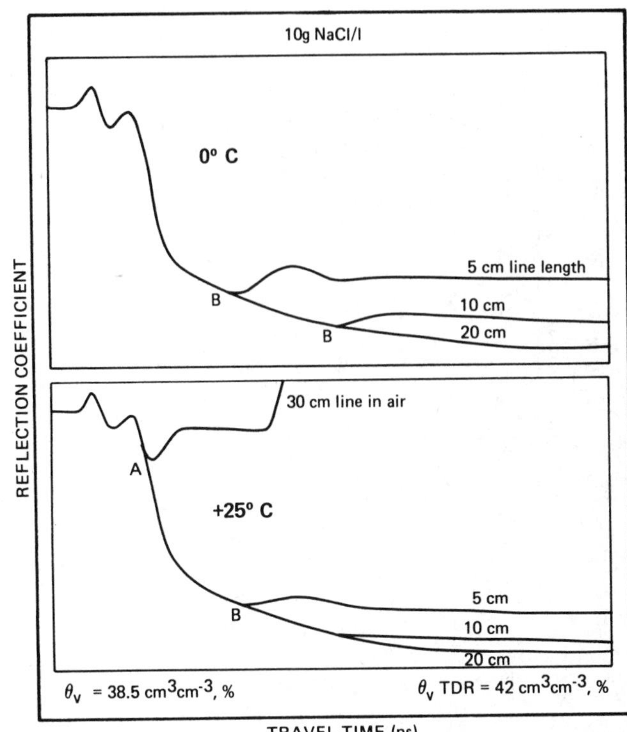

FIGURE 1 TDR traces for various parallel rod lines in saline sand.

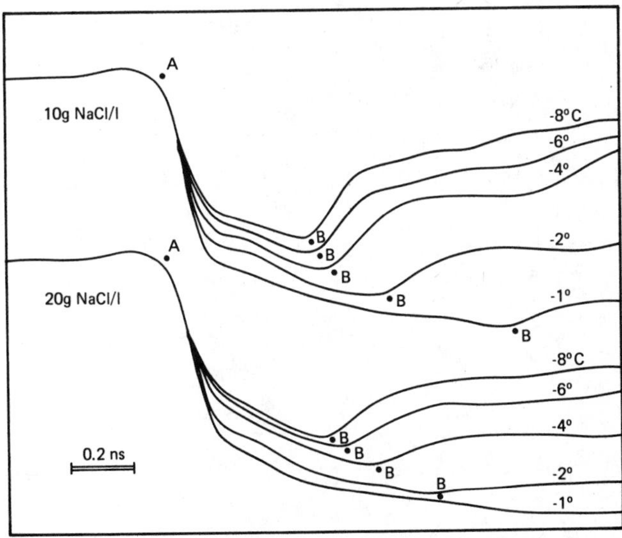

FIGURE 2 TDR traces for coaxial lines containing Kaolinite at different salinities and temperatures (axes same as Figure 1).

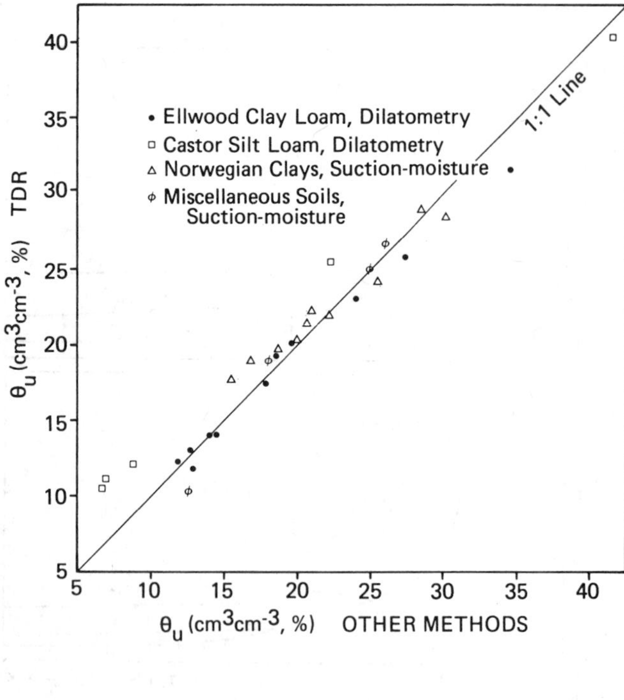

FIGURE 3 Comparison of θ_u determinations by TDR and various other methods.

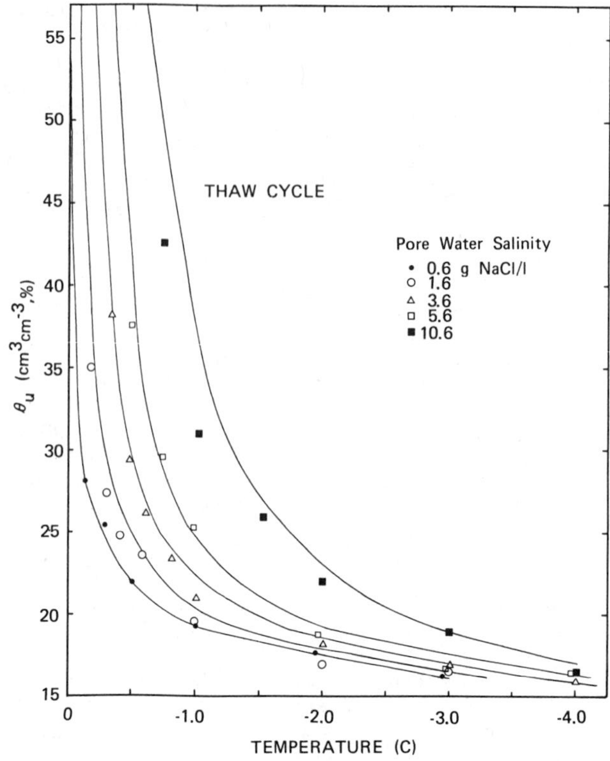

FIGURE 4 Freezing characteristic data for Allendale silty clay at various salinities.

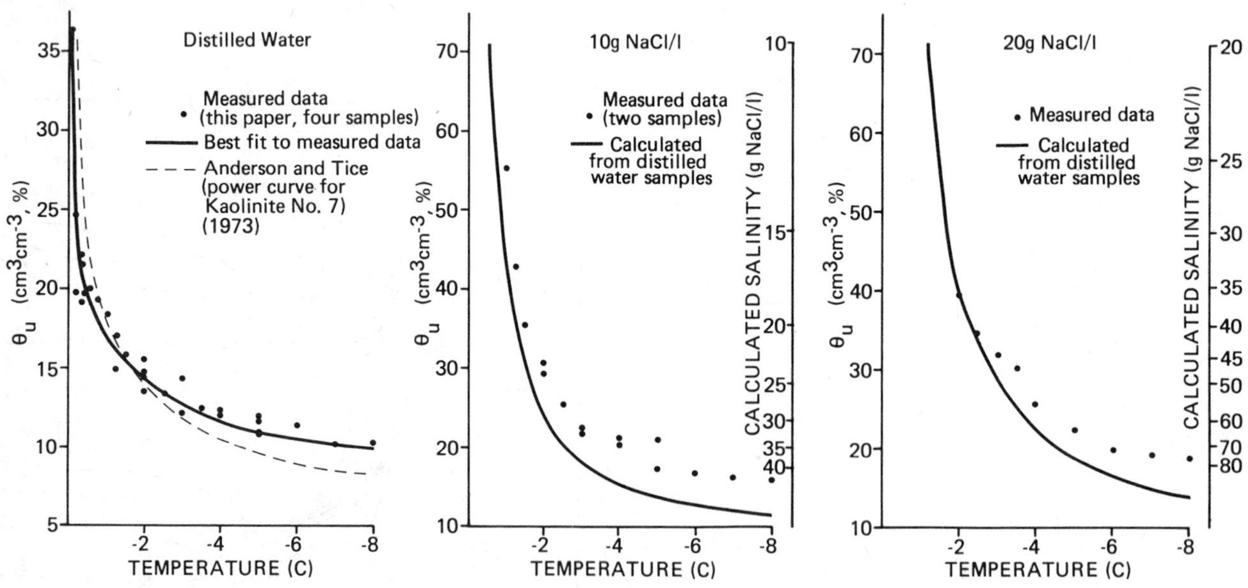

FIGURE 5 Freezing characteristic data for Kaolinite #3 at various salinities.

SONIC AND RESISTIVITY MEASUREMENTS ON BEREA SANDSTONE CONTAINING TETRAHYDROFURAN HYDRATES: A POSSIBLE ANALOG TO NATURAL GAS HYDRATE DEPOSITS

C. Pearson, J. Murphy, P. Halleck, R. Hermes, and M. Mathews

Earth and Space Sciences Division, Los Alamos National Laboratory
Los Alamos, New Mexico 87545 USA

Deposits of natural gas hydrates exist in arctic sedimentary basins and in marine sediments on continental slopes and rises. However, the physical properties of such sediments, which may represent a large potential energy resource, are largely unknown. In this paper, we report laboratory sonic and resistivity measurements on Berea sandstone cores saturated with a stoichiometric mixture of tetrahydrofuran (THF) and water. We used THF as the guest species rather than methane or propane gas because THF can be mixed with water to form a solution containing proportions of the proper stoichiometric THF and water. Because neither methane nor propane is soluble in water, mixing the guest species with water sufficiently to form solid hydrate is a difficult experimental problem, particularly in a core. Because THF solutions form hydrates readily at atmospheric pressures it is an excellent experimental analog to natural gas hydrates. Hydrate formation increased the sonic P-wave velocities from a room temperature value of 2.5 km/s to 4.5 km/s at -5°C when the pores were nearly filled with hydrates. Lowering the temperature below -5°C did not appreciably change the velocity however. In contrast, the electrical resistivity increases nearly two orders of magnitude upon hydrate formation and continues to increase more slowly as the temperature is further decreased. In all cases the resistivities are nearly frequency independent to 30 kHz and the loss tangents are high, always greater than 5. The dielectric loss shows a linear decrease with frequency, suggesting that ionic conduction through a brine phase dominates at all frequencies, even when the pores are nearly filled with hydrates. We find that the resistivities are strongly a function of the dissolved salt content of the pore water. Pore water salinity also influences the sonic velocity, but this effect is much smaller and only important near the hydrate formation temperature.

INTRODUCTION

Hydrates are a crystalline form of water containing voids or cavities that can trap other (guest) molecules which play an important role in stabilizing the hydrate structure. Two types of hydrates (types 1 and 2) are known to form. The first type contains only relatively small cavities (>8.6 Å in diameter) that can trap guest species smaller than ethane, whereas the second type contains a mixture of small and large (9.5 Å in diameter) cavities that can trap molecules as large as isobutane. Most of the common constituents of natural gas form hydrates, usually type 1, but type 2 hydrates can also form if significant amounts of C_3H_8, C_4H_{10}, or CO_2 are present.

Until recently, natural gas hydrates were widely known only as a nuisance that condensed in gas transmission lines. Then Soviet investigators reported natural gas hydrate deposits in the Siberian Arctic. These early reports were confirmed with the discovery of large natural gas hydrate deposits in arctic North America and in marine sediments. Research interest in hydrates increased as hydrates became not only a potential engineering problem in the Arctic, but also a potential energy resource. Although a considerable body of experimental data exists on the electrical properties of pure crystalline hydrate (see Davidson 1973, for a summary) these measurements are usually conducted at frequencies far higher than those used in exploration geophysics and surprisingly little is known about the physical properties of sediments containing hydrates in their pores except for the pioneering work by Stoll and Bryan (1979).

In this paper, we present laboratory sonic and resistivity measurements on Berea Sandstone cores containing tetrahydrofuran (THF) hydrate. Tetrahydrofuran was used as a guest species instead of methane or some other constituent of natural gas because THF hydrate is stable at moderate temperatures (+4°C) and atmospheric pressures, which greatly simplifies experimental procedures. A second major advantage of THF hydrate is that the guest is mixable with water. This eliminates the problem of ensuring complete mixing between the guest species and water, which is a formidable problem inside the pore spaces of sedimentary rock. Because the crystal structure of hydrates is largely independent of the guest species, the physical properties of THF-hydrate-containing samples are probably similar to the physical properties of a natural gas hydrate deposit formed in similar rocks. This is particularly likely if the natural hydrates form from gas-containing molecules large enough to form type 2 hydrates.

This paper focuses on sonic and electrical measurements because preliminary calculations (Pearson 1982) show that sonic velocities and resistivities are more strongly affected by the presence of hydrates than are other physical properties such as densities or thermal conductivities. In addition, seismic and electrical methods are the most commonly applied exploration geophysical techniques. Clearly a detailed understanding of the electrical and acoustic properties of hydrates is necessary to design and interpret geophysical surveys over natural gas hydrate deposits.

During this study we restricted our electrical measurements to the frequency range from 10 Hz to 30 kHz because most exploration geophysical surveys are conducted at these relatively low frequencies. In addition Olhoeft (1977) demonstrates that the electrical properties of permafrost at higher frequencies are very variable. If hydrated sediments are similar in this respect, then even if surveys could be conducted at higher frequencies, the results would be very difficult to interpret.

EXPERIMENTAL METHOD

The samples were cylindrical cores approximately 5 cm long and 2.54 cm in diameter cut perpendicular to the bedding plane from a block of Berea sandstone. The ends were ground parallel to ensure good contact between the ends of the samples and the electrodes or transducers. Samples were saturated with a stoichiometric mixture of THF and water (18 parts water to 1 part THF; Gough and Davidson 1971) under vacuum. As part of the study, various amounts of NaCl were added to the fluid. The concentration of salt is reported by the molarity of the water NaCl solution before THF was added to the mixture. Because temperature is an important variable in our study, all measurements were conducted in a NESLAB RTE-8 constant temperature bath. We left the sample in the bath for 24 hours, well after a temperature change, to ensure the sample had equilibrated with the bath. Because the crystalline water that formed in the pores disassociated near 4°C which is near the disassociation temperature of THF hydrates (Gough and Davidson 1971) and significantly above the melting point of ice, THF hydrate apparently formed in the pore spaces of the rock, not water ice.

The electrical measurements were conducted using a two-electrode system similar to that described by Collett and Katsube (1973). We used a Princeton Applied Research model 5204 lock-in analyzer, which can measure the in-phase and quadrature components of the voltage drop across the precision resistor, and were able to calculate the real and imaginary components of the resistivity, the phase angle, and the complex relative permittivity. The complex electrical properties (i.e., the complex permittivity (K*) and the loss tangent (D) were calculated using the following equations presented by Collett and Katsube (1973).

$$D = \frac{P'}{\sqrt{[P^*]^2-(P')^2}} \text{ , and} \qquad (1)$$

$$K^* = \frac{1}{\omega\,[P^*]\,\varepsilon_0} \text{ ,} \qquad (2)$$

where [P*] and P′ are the magnitude complex resistivity and the real part of the complex resistivity respectively, is the frequency, and ε_0 is the permittivity of free space. The real (K′) part of the relative permittivity can be calculated from K* using

$$K' = \frac{K^*}{1+D^2} \text{ .} \qquad (3)$$

Two-electrode systems of this type are subject to several systematic errors, among which are inductive and capacity coupling between the leads, current leakage around the sample, transfer impedances between the sample and the electrodes and contact polarization (Olhoeft 1975). We minimized coupling between the leads by using shielded wires. Calibration using known resistors showed that coupling and resistance in the leads had a negligible effect on our measurements. Current leakage was minimized by jacketing the samples tightly in shrink tubing before conducting measurements. We minimized transfer impedance between the electrode and the sample by spring loading the electrodes and using a conducting brine as an interstitial fluid. Polarization processes at the electrodes can also cause errors in two-electrode measurements, particularly if the sample is highly resistive or the measurement frequency is high. These effects may cause the small-frequency dependence above 1 kHz in the high resistivity curves in Figures 1 and 2. However we always used 200-Hz measurements when comparing resistivities because, at these frequencies, the

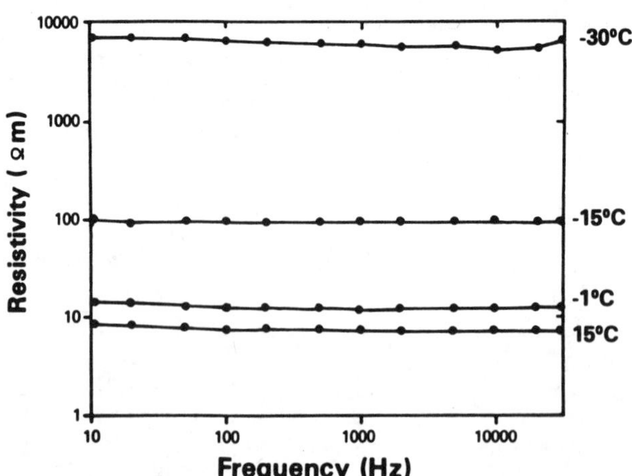

FIGURE 1 Resistivity as a function of temperature and frequency for a Berea core saturated with 0.5 N NaCl THF solution.

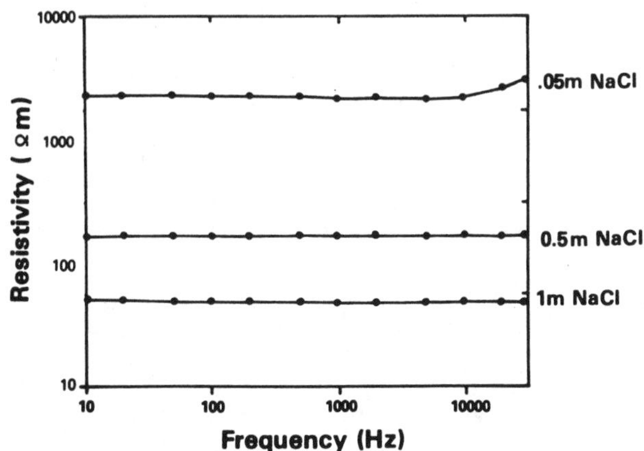

FIGURE 2 Resistivity as a function of frequency and salinity at -24°C.

resistivities were frequency independent. Rust (1952) compared resistivity measurements on porous sedimentary rocks using both two- and four-electrode (which avoid most of the experimental problems described above) techniques. He concluded that, as long as the sample is saturated with a conductive fluid, either method gives an accurate measure of the complex resistivity.

Sonic measurements were conducted using the pulse transmission described by Mattaboni and Schreiber (1967). We used Valpey Fisher LTZ-5 quartz 1-MHz piezoelectric transducers that were attached to the sample by a spring-loading device. All measurements (both sonic and electric) were conducted at atmospheric pressure, except for a small axial load, >0.1 MPa, which was applied to ensure that the electrode or transducer remained in contact with the sample. Olhoeft (1977) found the electrical properties of permafrost to be strongly pressure dependent from 10 Hz to 10 kHz. Thus, care must be taken when using our results to interpret geophysical surveys over natural hydrate deposits because the *in situ* pressure will always be much greater than atmospheric pressure.

ELECTRICAL RESULTS

As shown in Figures 1 and 2, the resistivities of Berea cores containing THF hydrates are functions of both temperature and salinity at which the measurements were made. However, to 30 kHz, the resistivities are nearly independent of frequency. Figures 1 and 2 plot the complex resistivity, but because the imaginary component of the resistivity was always very small, usually less than 10% of the real component, the real and complex resistivities are nearly equal. As a result, loss tangents are very high, often in excess of 100. Both the real relative permittivity and the imaginary part which is proportional to the dielectric loss are linear functions of frequency. The log linear relationship between the dielectric loss and frequency (Figure 3) is particularly important because the dielectric loss is a parameter that describes the motion of electric charge. If the material displays conduction that arises not from the effect of polarization on the displacement current but from actual charge transport, Hasted (1974) shows that

$$\varepsilon = \varepsilon_{dielectric} + \frac{\sigma}{4\pi\omega\varepsilon_o} \quad . \qquad (4)$$

Here $\varepsilon_{dielectric}$ is the component of the dielectric loss associated with loss from polarization currents. With a dielectric loss mechanism such as that from water molecules and in the absence of conductors, will normally show a strong peak when plotted vs. frequency. However, if conduction dominates the polarization effects, ε will be inversely proportional to frequency. Thus, the slope of the line from Figure 3, which is (0.994) calculated using least square regression techniques, implies that conduction is much more important than polarization effects in determining the electrical properties of hydrate-containing Berea sandstone cores.

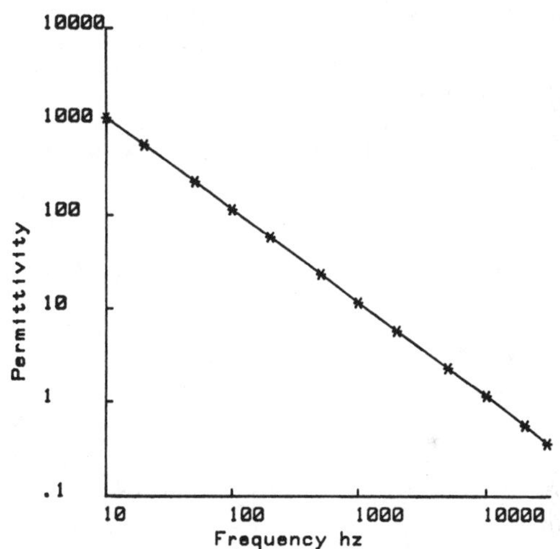

FIGURE 3 Imaginary part of the relative permittivity x 10^{-4} vs. frequency for a Berea core saturated with a 0.5 N NaCl THF solution at -24°C.

The effect of temperature and salinity on the resistivities of the Berea cores also suggests that the electric currents flow because of ionic conduction in an unfrozen brine phase, which is present in the rock even after hydrates start to form. The exponential increase in resistivity with temperature occurs because lowered temperatures cause the proportion of hydrates in the pores to increase, further constricting the brine phase. The decrease in resistivity as the salinity increases (shown in Figure 2) is caused by an increase in the ionic concentration of the brine phase. The additional ions present probably also inhibit the formation of hydrates increasing the amount of brine present in the pores.

The electrical properties of hydrate-containing sediments can be quantitatively understood using Archie's law ($\rho = a\rho_w\phi^{-m}Sw^{-n}$), an empirical relationship between water content and the resistivity of water-saturated sediments. Here ρ is the resistivity of the sediments, ρ_w is the pore water resistivity, Sw is the fraction of the porosity occupied by liquid water, and a, m, and n are empirically derived parameters. This equation also applies to rocks where the pore spaces are partially filled with ice or hydrates. However, as the amount of liquid water decreases, Sw and ρ_w are both reduced, Sw because some of the available pore space is now filled with a solid nonconductor, and ρ_w because the dissolved salts are concentrated in the remaining unfrozen brine. If the brine is not very near saturation, the effect on hydrate or ice formation of ρ_w is relatively easy to quantify because an increase in salt concentration causes a linear decrease in ρ_w. Because hydrates of ice exclude all of the dissolved salts as they form, the salt concentration of the brine inclusions is inversely proportional to the volume fraction of liquid water, if we assume that the sediments were initially water saturated. In addition, the resistivity of aqueous solutions increases exponentially with decreasing temperatures. Including both the temperature and concentration effects, the resistivity of a partially frozen brine at temperature T is thus proportional to $(C)^T$ Sw, where C is a constant. Substituting this relationship into Archie's equation and dividing by the resistivity at 0°C, we find that the ratio of frozen (ρ_f) and thawed (ρ_t) resistivities is

$$\rho_f/\rho_t = C^{-T} \ Sw^{1-n} \quad . \qquad (5)$$

Archie's law accounts for the rapid decrease in resistivity as a function of temperature (see Figure 4). Because N is usually equal to 2, the resistivity is inversely proportional to Sw. As the temperature decreases, the concentration of the brine at equilibrium with hydrates increases, causing Sw to decrease and the resistivity to increase. Increasing the molarity of the salt solution causes the resistivity to decrease because the increased salinity of the pore water inhibits the formation of hydrates, which increases the amount of unfrozen water (Sw) present.

In order to compare the electrical properties of partially frozen sediments with the measurements on hydrate containing sediments described above, we performed a series of electrical measurements on a Berea sandstone sample saturated with an 0.5 N NaCl solution. Because this sample did not contain any THF, ice rather than hydrates formed when the temperature of the sample was reduced below freezing. As was the case with hydrates, the complex resistivities were frequency independent and a linear relationship existed between log (E) and log frequency. The relationship between hydrate and ice containing samples is shown in Figure 4. Note that the curves are similar in that both curves show an exponential decrease in resistivity with temperature. The quantitative difference between the curves is probably caused by differing amounts of crystalline hydrates and ice in the pore spaces. Such differences are not unexpected because ice and THF hydrates have different stability curves and dissociation temperatures. The resistivity of ice reaches a nearly constant value at -4°C, the freezing point of a 0.5 N NaCl solution, whereas the THF-containing sample does not reach plateau until +2°C, presumably the dissociation temperature of THF hydrates in a 0.5 N NaCl solution.

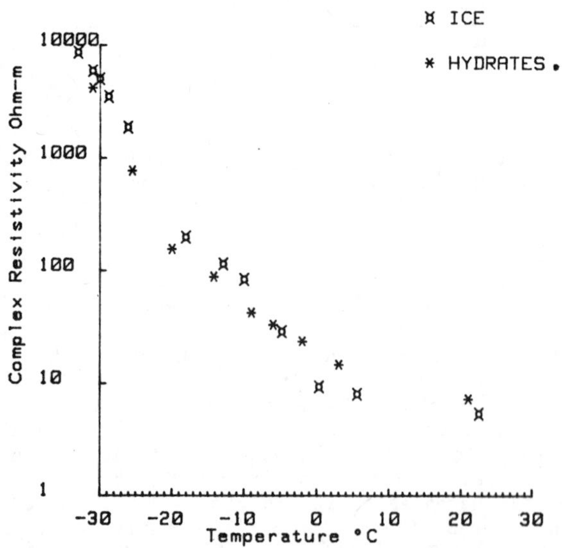

FIGURE 4 Resistivity as a function of temperature for two Berea sandstone samples, one saturated with 0.5 N NaCl THF solution and one saturated with a 0.5 N NaCl solution without THF.

SEISMIC RESULTS

Sonic P-wave velocities, measured on hydrate-containing Berea sandstone cores as a function of temperature, are shown in Figure 5. This figure

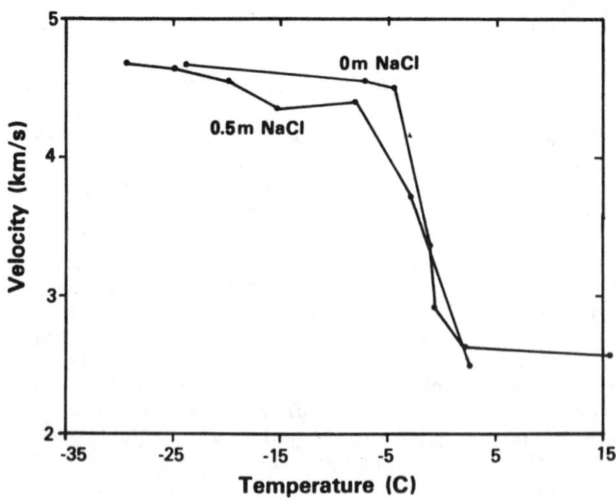

FIGURE 5 Sonic velocities vs. temperature for two Berea cores showing the effect of salinity.

shows results from cores saturated with two different NaCl solutions. Note that in both cases the sonic velocities increase from 2.5 to 4.5 km/s when hydrates begin to form in the pore spaces. Once hydrates form, the velocities reach a plateau where further cooling produces very little change.

As shown in Figure 5 the salinity of the saturating liquid has very little effect on the sonic velocities once hydrates have formed in the cores. However cores saturated with saline water and THF approach the high-velocity plateau more gradually than do samples saturated with pure water and THF. Thus the behavior of sonic velocities as a function of temperature contrasts with the electrical resistivity measurements in that electrical resistivities decrease rapidly as a function of temperature even after the pore spaces of the sample are probably nearly full of hydrates, while sonic velocities rapidly increase when hydrates start to form in the pore spaces. Then the velocities reach a plateau where further cooling produces very little change in sonic velocities. The difference in the temperature dependence of sonic velocities and resistivities illustrates a fundamental difference in the mechanism by which electrical and acoustic signals are transmitted in rock. Electrical signs are transmitted through the brine phase so electrical properties remain sensitive to the amount of brine present, even when the fraction of the pore volume containing brine becomes very small. In contrast, acoustic pulses are transmitted primarily through the solid matrix, so once the pore volume is largely filled with hydrates, a further decrease in the small brine fraction produces only a negligible change in velocity. However the slower asymptotic approach in the brine-rich sample suggests that amount of fluid in the unfrozen brine phase does have some effect on velocity when the brine phase consists of a relatively large amount of fluid.

The compressional velocity of hydrates forming in a sediment can probably be understood using a three-phase time-averaged equation, first proposed by Timur (1968) for partially frozen sediments and since tested by several other authors. The compressional velocity (Vp) is related to the velocity of ice (Vs), the velocity of the brine inclusions (Vb), and the velocity of the solid matrix (Vm) by

$$1/Vp = \frac{\phi(Sw)}{Vb} + \frac{\phi(1 - Sw)}{Vs} + \frac{(1 - \phi)}{Vm} \quad . \qquad (6)$$

Because of the similarities between the seismic velocites of ice and hydrates, this equation can probably be used to calculate the velocity of a mixture of hydrates and brine in sedimentary rock. Note that Equation 6 depends linearly on Sw, in contrast to Equation 5 which, if $n = 2$, is inversely proportional. The difference in electrical and sonic properties as a function of temperature can be explained by the difference in the dependence of Equations 5 and 6 on Sw. The electrical properties are inversely proportional to Sw so the electrical resistivity remains sensitive to changes in Sw even when very little unfrozen water remains in the rock. In contrast, in Equation 6 Sw enters directly as a term added to other quantities. Thus as Sw becomes small it has a negligible effect on the seismic velocity.

We performed no sonic measurements on Berea samples that did not contain a THF hydrate mixture so we could not compare the effect of THF hydrates and ice on the sonic velocities. However because our measurements are very similar to results published by Pandt and King (1979) for partially frozen sediments, we expect ice and hydrates to have nearly the same effect on the sonic velocities.

CONCLUSIONS

Several significant conclusions can be drawn from this study: (1) the resistivities and sonic velocities of Berea sandstone cores are strongly affected by the presence of either ice or hydrates. Resistivities increased by an order of magnitude and continued to increase rapidly as further decreases in temperature reduced the amount of unfrozen brine present in the rock. The sonic velocities, in contrast, rapidly increased when hydrates began to form in the cores but soon approached a limiting value. Further cooling produced only a very small increase in sonic velocities. (2) Ice and hydrates produce very similar changes in the sonic and electrical properties of Berea sandstone. Any differences can probably be ascribed to differing amounts of unfrozen brine present in the pores. This suggests that it may be difficult to distinguish permafrost and hydrate-containing layers using ordinary geophysical techniques. (3) The salinity of the pore water in which the hydrates form has a strong effect on the resistivities but a very small effect on the sonic velocities. We suggest that the effect of temperature and salinity on resistivities and sonic velocities can be explained if the samples obey Archie's law for resistivities and the three-phase rule for velocities. (4) Resistivities of hydrate containing cores are nearly frequency independent in the range from 10 Hz to 30 kHz. However, both the dielectric constant and the dielectric loss decrease rapidly as a function of frequency. The log linear relationship between frequency and dielectric loss suggests that the electrical properties of the hydrate containing samples are controlled by ionic conduction in a frozen brine phase.

Our experimental results show that the presence of hydrates has a strong effect on the acoustic and electric properties of sediments compared to unfrozen or unhydrated sediments. An increase of several orders of magnitude in electrical resistivities can easily be detected using a variety of electrical exploration techniques. Also an 80% increase in sonic velocity is sufficient to produce a very strong reflection in seismic reflection data and can easily be detected in seismic refraction surveys. This very strong velocity contrast may account for strong reflections that are often observed at the bottom of possible hydrate bearing horizons in marine seismic surveys (Shipley et al. 1979).

ACKNOWLEDGMENTS

We thank the ESS-3 Rock Mechanics Group, especially Tom Dey, without whose help this research could not have been undertaken. We would also like to thank Tom Shankland who reviewed this paper and made many helpful suggestions. This research was supported by the U.S. Department of Energy.

REFERENCES

Collett, L. S. and T. J. Katsube, Electrical parameters of rocks in developing geophysical techniques, Geophysics, 38, 76-91, 1973.

Davidson, D. W., Clathrate hydrates in water, a comprehensive treatise, F. Frank, ed., Plenum Press, New York, 2, 115-234, 1973.

Gough, S. R. and D. W. Davidson, Composition of tetrahydrofuran hydrate and the effect of pressure on the decomposition, Canadian Journal of Chemistry, 49, 2691-2699, 1971.

Hasted, J. B., Aqueous Dielectrics, Chapman and Hall, London, 302 pp., 1974.

Mattaboni, P. and E. Schreiber, Method of pulse transmission measurements for determining sound velocities, J. Geophys. Res., 72, 5160-5163, 1967.

Olhoeft, G. R., Electrical properties of permafrost, Ph.D thesis, Dept. of Physics, Univ. of Toronto, Toronto, Canada, 1975.

Olhoeft, G. R., Electrical properties of natural clay permafrost, Canadian Journal of the Earth Sciences, 14, 16-24, 1977.

Pandt, B. I. and M. S. King, A study of the effect of pore water salinity on some physical properties of sedimentary rocks at permafrost temperatures, Canadian Journal of the Earth Sciences, 16, 1566-1580, 1979.

Pearson, C. F., Physical properties of natural gas hydrate deposits, Los Alamos National Laboratory report LA-9422-MS, 1982.

Rust, C. F., Electrical resistivity measurements on reservoir rock samples by the two-electrode and four-electrode methods, Petroleum Transactions, American Institute of Mining Engineers, 195, 217-224, 1952.

Shipley, T. H., M. H. Houston, R. T. Buffler, F. J. Shaub, K. J. McMillen, J. W. Ladd and J. L. Worzel, Seismic evidence for widespread possible gas hydrate horizons on continental slopes and rises, American Association of Petroleum Geologists Bulletin, 63, 2204-2213, 1979.

Stoll, R. D. and G. M. Bryan, Physical properties of sediments containing gas hydrates, J. Geophys. Res., 84, 1629-1634, 1979.

Timur, A., Velocity of compressional waves in porous media at permafrost temperatures, Geophysics, 33, 584-595, 1968.

ADFREEZING STRESSES ON STEEL PIPE PILES, THOMPSON, MANITOBA

E. Penner and L. E. Goodrich

Geotechnical Section, Division of Building Research
National Research Council of Canada, Ottawa, Ontario, K1A 0R6

Studies of uplift forces on steel pipe piles in frost-susceptible varved clays were carried out near Thompson, Manitoba, over three consecutive winters. The installation consisted of pipe piles, 169, 323 and 458 mm in diameter and 3.3 m long. Each pile size was replicated three times on the same site, and all piles had separate reaction frames anchored to bedrock. For the first winter, when the thick gravel layer that was placed over the site was allowed to freeze to the piles, high uplift forces were measured. The pile was isolated from the gravel in the two subsequent winters and the adfreeze values corresponded more closely to those determined previously.

Adfreezing of frost-susceptible soil to foundations and subsequent uplift is an important engineering problem. Special precautions are required for structures that must function in this environment, to avoid destructive displacement forces. The question that arises in the consideration of various design possibilities is the magnitude of the force generated and how it varies with soil type, soil moisture content, climate and foundation material. Crory and Reed (1965) and Vialov (1959) have shown the importance of such uplift forces and that they can be substantial in the active layer of permanently frozen ground. Earlier adfreezing studies in seasonal frost areas were described by Trow (1955), Kinoshita and Ono (1963), Penner and Irwin (1969), Penner and Gold (1971), and Penner (1974). The work of Penner, and Penner et al, involved field measurements on piles of wood, concrete, and steel pipe of various diameters from 7.6 to 30.5 cm.

Displacement of foundations in frost-susceptible soils may occur because the wet soil freezes to the foundation during frost penetration. The formation of ice lenses, the main cause of frost heaving, may lift structures unless special precautions are taken. Preventative measures may simply involve replacing the frost-susceptible soil, preventing freezing by insulating, special heating techniques, isolating the structure from the heaving soil or controlling the water supply. Loading of the structure is rarely used in practice.

The technique commonly used in the measurement of adfreeze forces is to place a force gauge between a suitable reaction frame and the structure, e.g., foundation walls or piles. Normally, characteristic surface displacements develop in the surrounding soil (Penner and Gold 1971).

Primary heaving forces always develop in the same direction as the heat flow, normal to the direction of ice lens growth. Complex thermal patterns sometimes develop due to differences in the thermal properties of the structure and the soil. For example, a steel pile often projects above the ground surface. In areas of seasonal frost, the depth of freezing is greatest at the pile. This pattern is enhanced further when snow accumulates on the site and the pile projects above the insulating snow layer.

A field study was undertaken in 1971 at Thompson, Manitoba, where a Division of Building Research (DBR) field station is located, to extend investigations carried out at Ottawa on the performance and behaviour of various foundation structures. The initial objective of this study was to measure uplift forces on steel piles.

SITE SELECTION AND PREPARATION

The main site requirements were easy accessibility, a nearby power line, shallow bedrock for anchorage of the steel reaction frames, and surface soil characteristics common to the region. Islands of permafrost occur at Thompson, so a site was chosen where no permafrost was present.

The test site was about 3.2 km south of the town. Site preparation consisted of removing the trees and placing a layer of gravel over the ground surface to provide good trafficability for construction equipment with a minimum of soil disturbance. A thin layer of asphaltic concrete was placed over the gravel to improve local working conditions during pile and rock anchor installations. A schematic diagram of the site with installations in place is shown in Figure 1.

Following the development of unusually high forces during the first winter's operations, the gravel layer on the ground surface was isolated from the pile with a metal shield. A layer of grease was placed between the pile and the metal shield to reduce friction and the possibility of adfreezing. The influence and importance of this design change in the profile next to the pile cannot be overemphasized. The experience might well serve as a caution wherever piles are used in cold climates. The effect on the force measurements and adfreeze stresses will be included in the section on results and discussion.

INSTALLATION OF STEEL PIPE PILES AND ROCK ANCHORS FOR REACTION FRAMES AND BENCHMARKS

The surface of the piles was sandblasted to remove the anticorrosion treatment. Piles were stored outside for several months to allow surface rust to form. The ends of the piles were capped with steel plates, 2.54 cm thick on top and 1.25 cm on the bottom. All nine piles, three of each diameter, 169, 323 and 458 mm, were prepared this way. These piles were nominally 6, 12 and 18 in. in diameter and are referred to as S6A, S12A, and S18A, respectively. Copper-constantan thermocouples were placed at 0.3 m intervals on the surface of one pile of each diameter for temperature measurement. Lead wires were placed in a 1.9 cm diameter conduit welded along the length of the pile. To reduce thermal conduction errors along the leads on temperature measurements, the thermocouple wires and tip were embedded with epoxy resin in a 15 cm long horizontal groove around the circumference of the pile at each measurement location. This procedure also ensured good thermal contact between pile and thermocouple.

The piles were randomly located on the site in machine-augered holes 15 cm larger than the pile diameter. Backfilling was done immediately with the augered soil to avoid large moisture changes. The surface gravel layer was replaced to conform with the rest of site, for the first year only.

The steel reaction frame consisted of two steel beams (S150 x 19) 0.91 m long, crossed at 90°, and tied down at the four ends with 1.9 cm diameter high-strength steel anchor rods. Stelco expansion shells (3.2 cm "S" type) were placed at a depth of 0.46 m into bedrock (Figure 2). The strength of each anchor and rod assembly was tested by applying a tensile load of 0.1 MN. Two benchmarks, BM 1 and BM 2 (Figure 1), were installed in bedrock in the same way.

Thermocouple cables were placed to a depth of 3 m in augered vertical holes at various distances (0.15, 0.30, 0.61, 1.22, and 3 m) from the three temperature-instrumented piles. Fine-grained dry mine slag was used for backfilling around the cables.

Snow was allowed to accumulate on the site during the winter. Lightweight elevated wooden catwalks were built along the main row of ground survey stakes, to the nearby benchmark, and to the instrumented weather centre, to avoid snow disturbance by the survey crew. A photograph of a portion of the site is shown in Figure 3.

Other Instrumentation

Dillon force gauges, placed between reaction frame and pile, were used to measure uplift forces. The gauge capacity selected was based on the Ottawa experience, with suitable provisions for the higher forces expected. During the first winter the reaction frames had to be adjusted at critical times, to reduce the forces and avoid damaging the gauges by exceeding their rated capacity; hence the maximum uplift forces were not measured. For the second and third winters (1973/1974 and 1974/1975), vibrating wire force gauges capable of measuring a force of 0.44 MN were installed. The new gauge could be read remotely in addition to having a high capacity. As will be discussed later, the layer of gravel placed over the site was thought to be the main cause of the very high forces measured in the first year, so gravel and pile were isolated for the second and third winters.

Observations

Ground temperatures were recorded at least once a day using a multipoint L & N recorder for the main points; others were recorded by hand. The overall accuracy of the temperature measurements, including installation errors, was believed to be within $\pm\frac{1}{4}$°C. All force gauges were calibrated and checked before installation and read daily. Level surveys were done weekly or bimonthly for the center of the reaction frame, the piles and surface gauges. Air temperatures were measured with a thermocouple inside a Stevenson screen at the site. Temperatures from maximum/minimum thermometers were used to determine the freezing index. Snow depths were measured daily at two locations, using permanently placed snow stakes.

SOIL, CLIMATE, AND PERMAFROST CONDITIONS

The town of Thompson (55°45'N, 97°50'W), a recently developed industrial settlement, lies within the discontinuous permafrost zone (Johnston et al 1963). Although the site did not contain any permafrost, there were patches of it within a few hundred metres. The mean annual air temperature at Thompson, based on very limited data, is about -3.9°C. This value is consistent with that determined for nearby settlements from 7 years of observations. The mean annual precipitation, based on measurements made between 1968 and 1972, was 44.6 cm of water, 33 cm as rain, and 116 cm of snow.

The Thompson area was extensively glaciated during the Pleistocene epoch. Varved clays and glacial tills constitute the surficial deposits throughout this region. At the test site the overburden ranges in thickness from 3.7 m in the NW corner to 6.1 m in the SW corner. It consists of varved clays with the exception of a thin layer of pebbly sandy silt overlying the bedrock. There was a thin layer of peat at the SE corner, through which one pile passed (Figure 1, S6A).

The clay size content determined on augered samples was about 50%, with most of the remaining coarser materials passing the 200 sieve. Moisture contents ranged from 20 to 30% by weight. The grain size at the site falls within the grain size envelope for the Thompson soil (Johnston et al 1963).

RESULTS AND DISCUSSION

Figure 4 shows the 1972/1973 and 1973/1974 maximum heave along a NS line through the temperature-instrumented piles, S6A, S12A, and S18A. The maximum frost penetration in the vicinity of the piles is also shown. This figure illustrates the contrasting results, to which special attention is directed.

The cause for the high adfreeze results was the adfreezing of the gravel layer during the first winter, 1972/1973. The large vertical displacements of the reaction frames are consistent with the high forces that were generated. The frozen wet gravel forms a strong layer over the frost-susceptible soil. It was considered that this layer frozen to the piles transferred these larger than normal forces to the pile as the soil heaved. The reaction frames had to be adjusted several times to release the forces and avoid overstressing the Dillon force gauges; hence the record of the first winter's results does not give the maximum forces that probably developed. Table 1 gives the calculated adfreeze stresses (1972/1973) from the highest forces measured prior to release of the frames. The adfreezing pressures for the next two winters were much lower and are comparable to the results of studies carried out earlier. For comparative purposes, the (D) section in Table 1 gives the results of measurements made at Ottawa on piles of the same diameter. As at Thompson, the Ottawa adfreeze results are the average of three piles for each of two diameters, 169 and 323 mm.

The adfreeze forces measured in Ottawa are about twice as high as those measured at Thompson. This probably reflects differences in the amount of heave in soils at the two sites. The measured heave in Ottawa was in excess of 7 cm, while at Thompson it was less than 2 cm. The Ottawa soil--Leda clay--has a clay size content of 70% and a moisture content of 44%, while at Thompson, the average clay size content of mixed varve material was about 50% with a much lower moisture content (20-30%). One further difference between the two sites was the thick gravel layer at Thompson, which did not exist at the Ottawa site. The response to air temperature changes, therefore, would be much slower at Thompson.

Summarizing Remarks

The most important and unexpected result is the unusual influence of the surface gravel layer. For piles that may be subjected to uplift forces, a surface layer of wet gravel over a heaving soil may have a serious detrimental effect and should be avoided by isolating the pile. When the effect of the gravel layer was removed, the adfreezing forces were consistent with the heave and the nature of the underlying clay soils.

REFERENCES

Crory, F. E. and Reed, R. E., 1965, Measurement of frost heaving forces on piles, TR145; Hanover, N.H., U.S. Army Material Command, Cold Regions Research Engineering Laboratory.

Johnston, G. H., Brown, R. J. E., and Pickersgill, D. N., 1963, Permafrost investigations at Thompson, Manitoba, terrain studies, Technical Paper No. 158, Ottawa: Division of Building Research, National Research Council of Canada.

Kinoshita, S., and Ono, T., 1963, Heaving forces of frozen ground, I, Mainly on the results of field research: Low Temperature Science Laboratory, Teron Kagaku, Serial A 21, p. 117-139 (National Research Council of Canada, Technical Translation 1246, 1966).

Penner, E., 1974, Uplift forces on foundations in frost heaving soils: Canadian Geotechnical Journal, v. 11, p. 323-338.

Penner, E., and Gold, L. W., 1971, Transfer of heaving forces by adfreezing to columns and foundation walls in frost-susceptible soils: Canadian Geotechnical Journal, v. 8, p. 514-526.

Penner, E., and Irwin, W. W., 1969, Adfreezing of Leda clay to anchored footing columns: Canadian Geotechnical Journal, v. 6, p. 327-337.

Trow, W. A., 1955, Frost action on small footings, Highway Research Board, Bulletin 100, p. 22-28.

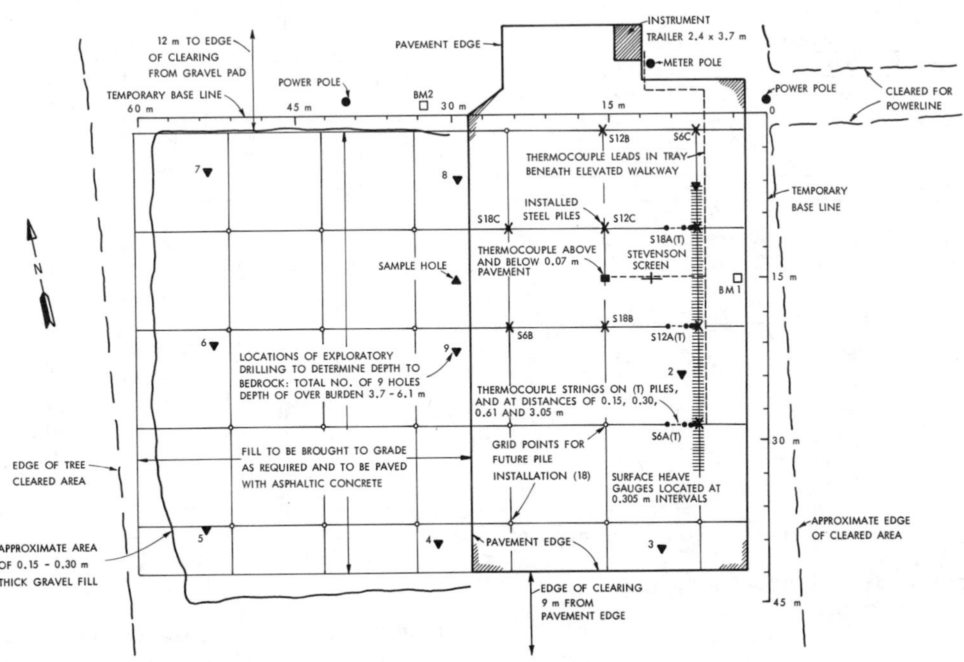

FIGURE 1 Schematic drawings of the frost heave test site, Thompson, Man.

Vialov, S. S., 1959, Reologicheskie Svoistua i Nesushchaia Sposobnost' Merzlykh Gruntov, Izdatel'stov Akademic Nauk SSSR, Moscow, (Rheological properties and heaving capacity of frozen soils) Technical Translation 74 CRREL: Hanover, N.H., U.S. Army Material Command, Cold Regions Research Engineering Laboratory.

TABLE 1 Calculated Adfreeze Stresses *(kPa) on Steel Pipe Piles

Thompson, Manitoba

(A) 1972/73, Fl 2599°C-days, Maximum stress before release--surface gravel layer adfrozen to piles

S6 (dia. 169 mm) - 380**
S12 (dia. 323 mm) - 179**
S18 (dia. 458 mm) - 124**

(B) 1973/74, Fl 3189°C-days, Average of 3 piles for each diameter--gravel layer isolated from pile

	Nov.	Dec.	Jan.	Feb.	Mar.	Apr.	May
S6	35.9	38.9	79.9	59.1	36.1	19.8	13.4
S12	56.9	29.7	47.0	35.1	23.2	14.1	12.0
S18	37.1	22.6	46.2	37.8	29.1	17.6	6.2

Winter Average	Winter Maximum
40.5	117
31.1	76
28.1	69

(C) 1974/75 Fl 2406°C-days, Average of 3 piles for each diameter--gravel layer isolated from pile

	Nov.	Dec.	Jan.	Feb.	Mar.	Apr.	May
S6	32.4	39.2	47.8	39.7	26.1	20.6	**
S12	18.5	20.2	24.2	21.4	12.1	10.4	**
S18	10.3	18.7	29.2	24.1	16.0	12.4	**

Winter Average	Winter Maximum
34.3	69
17.8	55
18.5	48

Ottawa, Ontario

(D) 1971/72 Fl, 1067°C-days, Average of 3 piles of each diameter--no gravel layer

	Nov.	Dec.	Jan.	Feb.	Mar.	Apr.	May
S6	***	137.0	75.8	88.2	73.1	***	***
S12	***	105.5	64.8	62.7	72.4	***	***

Winter Average
93.8
76.5

* Adfreeze stress = measured force/surface area of pile in frozen ground.
** Maximum reached in the Dec.-Jan. period.
*** No frost in the ground.

ACKNOWLEDGMENTS

The authors wish to express their appreciation for the assistance of colleagues in the Section. This paper is a contribution from the Division of Building Research, National Research Council of Canada, and is published with the approval of the Director of the Division.

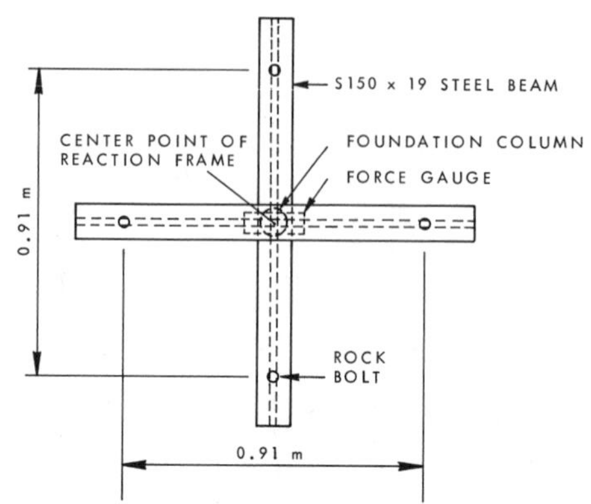

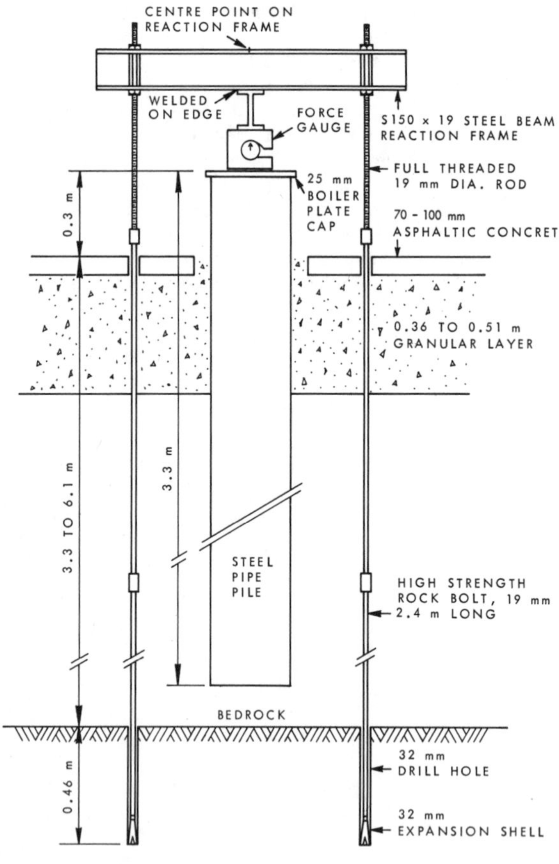

FIGURE 2 Reaction frame for measurement of uplift forces on steel pipe piles.

FIGURE 3 Test site, showing reaction frames with insulated force gauges, catwalk, instrument trailers, Stevenson screen, and thermocouple cables.

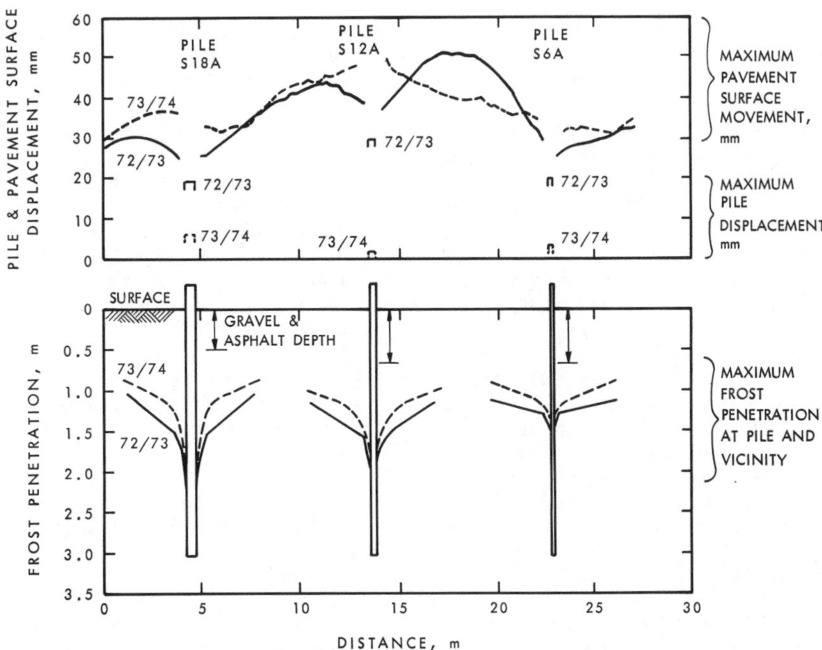

FIGURE 4 Heave profile, maximum frost penetration and maximum vertical pile strain for 1972/1973 and 1973/1974.

LAWS GOVERNING INTERACTIONS BETWEEN RAILROAD ROADBEDS AND PERMAFROST

N. A. Peretrukhin[1] and T. V. Potatueva[2]

[1]Institute of Transport Construction, Moscow, USSR
[2]Institute of Civil Engineering, Tomsk, USSR

In southern areas of permafrost occurrence interactions between railroad roadbeds and permafrost are characterized by changes through time both in the intensity of the thawing of permafrost and settlement of the grade as deformation of the foundation materials occurs. These interactions may be divided provisionally into three periods: those of intense deformation, moderate deformation, and complete stabilization. The first period is characterized by an irregular regime of thawing and settlement of the foundation materials. Initially thawing and settlement are very intense but towards the end of the period this intensity declines markedly. The beginning of the period coincides with that of construction and it ends 1-2 years after completion of the grade. These trends are dictated by the changes in the natural permafrost conditions provoked by construction of the roadbed. The second period is characterized by a relatively low intensity of thawing and settlement which tends to die away with time. The duration of this period is dictated by the degree of disturbance of the natural local conditions and the deformability of the thawing permafrost materials under the influence of the completed grade. The third period is characterized by complete stabilization of the roadbed which corresponds to the needs of normal operation of the line. Depths of thaw and amounts of settlement display no linear correlation. The thaw depth depends on the degree of disturbance of the natural thermal regime, and on hydrological, permafrost, and other natural conditions, whereas total settlement depends on the composition and ice content of the thawing permafrost and also on the compressibility of the permafrost materials under a given operating load.

INTRODUCTION

In past years, construction of railway roadbeds in southern areas of permafrost required relatively complicated engineering decisions, such as removal of soft active layer soils and permafrost to depths of 3 m under the embankment and backfilling the excavation with granular material; increasing the embankment width to account for settlements caused by thawing of the underlying permafrost, and also the construction of berms. These measures are taken to prevent or control the large, differential settlements and other embankment deformations that occur, and result in an increase of 30-100% in the volume of earth work and a complicated and labor-intensive construction operation. Experience has shown, however, that such measures do not result in a stabilized roadbed, and, even when some permafrost has been excavated, the roadbed experiences long-term, large and irregular settlements. It is very difficult to control both permafrost thawing and roadbed settlement. Field experiments were carried out, therefore, to study the mechanisms of roadbed-permafrost interaction, to observe the effectiveness and anti-deformation measures, and to accumulate data characterizing roadbed stability of typical sections (Peretrukhin 1978).

RESEARCH METHOD

Values of the active layer in natural conditions, the depth to the permafrost table under the embankment, roadbed settlements caused by compression of soft active layer soils and thawing permafrost are the main quantitative indices of roadbed-permafrost interaction. These indices are generalized parameters of the process considered. They indicate the overall effect of a great number of variable natural factors (climatic, geomorphological, hydrogeological etc.) on the thermal regime and the condition and properties of the perennially frozen ground, as well as changes in them resulting from roadbed construction. At the same time, these indices characterize the thermal and mechanical interaction of embankments and perennially frozen ground during the construction and operating periods, as well as the consequences of such interaction—the permafrost thaws and embankment settlements occur.

The field research included a detailed study of the natural frozen soil and engineering-geological conditions at typical locations and embankment sections, as well as changes that occurred prior to, during, and following construction of the railway. The engineering-geological work and geocryological survey were conducted using standard methods. In doing so, values of the generalized indices that characterize the roadbed-permafrost interaction were measured. Investigations were conducted at five locations on a railway being constructed in the southern part of the permafrost region.

NATURAL CONDITIONS AND DESIGN DECISIONS

Permafrost occurs in scattered patches along the route. It was found on north-facing shady slopes and in river valleys and dells where the ground was covered by moss and vegetation. The active layer thickness is as much as 0.4-1.2 m in undisturbed areas. The roadbed design was chosen, taking into consideration the operating experience of railways constructed in regions having similar natural conditions. In doing so, it was assumed that all permafrost would thaw and that long-term and irregular roadbed settlements would occur. The total settlement was calculated taking into account the permafrost ice content. The embankment width was increased and berms were constructed on both sides of the embankment as anti-deformation measures (Table 1). In addition, at test section 4 the soft foundation soil (a loam of medium plasticity) was excavated to a depth of 0.5-0.8 m and replaced by gravel.

before the preconstruction period. In places where the permafrost thickness exceeded 5 m, the depth of thaw was somewhat greater than 5 m. The calculated settlement differed from that actually measured by not more than 10% (Table 2). During the first 3 years the embankment settled 0.10 m and in the next 8 years it settled very little (Figure 1, curve 4). At section 4 the permafrost thawed more rapidly (Figure 1, curve 2). This was due to the comparatively higher temperature of the permafrost as well as the warming influence of water that filled a trench excavated in order to replace the soft foundations soils under the embankment. Water filled the trench during the summer construction period. In the first year the permafrost thawed to a depth of 0.7 m under the embankment centerline and to 2.4 m under a berm on the west side. Three years after the embankment was built the permafrost table was at a depth of 4 m below the embankment centerline, at a depth of 5 m under a berm on the east side. During the following 8-9 years the depth of thaw exceeded 11 m (Figure 1, curve 2).

TABLE 1

Section		Embankment (m)		Berms (m)		Earth work time
Number	Length (m)	Height	Increase in width	Height	Width	
1	260	2.0	0.7-1.5	-	-	Winter
2	290	1.8-4.0	-	1.5-3.6	3.0-5.0	Winter & summer
3	150	2.0-1.5	2.0	-	-	Summer
4	123	14.0	2.0	5.0	10.0	Summer
5	50	6.0	5.6	2.0	6.0	Winter

The embankments were to be constructed of fine to medium sands (sections 1-3) and gravels (sections 4 and 5). The upper part of the embankment was constructed of clayey loam with gravelly inclusions (sections 1-3) and of clayey loams with stones and cobbles (sections 4 and 5).

RESEARCH RESULTS

Natural frozen soil conditions began to change during the preconstruction works period. The active layer thickness increased by 2.2-2.7 times compared to that in undisturbed areas and reached 1.2-1.4 m in sections 1, 2 and 3, and 1.0-1.2 m in section 5. During and after construction of the embankment the depth to the permafrost table under the embankment continued to increase at all sections following a definite pattern: the rate and depth of permafrost thaw attenuated with time (Figure 1).

During the first two years of railway operation the rate of permafrost thaw under the embankment centerline was 0.4-0.5 m/yr at sections 1 and 2, constructed in winter, and to 0.8 m/yr in sections 4 and 5, constructed in summer. During the next two years the thaw rate decreased almost by half. In the 4th year following construction the depth to the permafrost table exceeded 3 m and by the 12th year it was almost 5 m (Figure 1, curve 1).

The permafrost thawed completely in places where the permafrost thickness was less than 5 m

The considerable difference between calculated and actual values of the permafrost thaw depth at this section (Table 2) can be explained as follows: the effect of both time (summer) and method of construction, and also of the warming influence of water in the trench, were not taken into account in the calculations. In spite of the rapid, irregular and deep permafrost thaw, the actual embankment settlement proved to be negligible (0.05 m). It occurred only in the first few years after construction of the embankment when compressible soils were thawing. The underlying frozen soils, including shales, did not deform when thawing. As a result, during the 12 years of railway operation, the actual dimension of the roadbed and the ballast layer did not change significantly from that measured before the railway was put into regular operation.

At section 5 the change in depth to the permafrost table followed a regular pattern from the time construction started (Figure 1, curve 3). The rate of thaw in the first two years was greater compared to that in the following years and the thaw rate under the roadbed centerline was approximately half that occurring under berms and embankment slopes. By the end of the 3rd year the permafrost table was at a depth of about 1.2 m under the centerline and 1.6-2.8 m under berms. In this period the embankment settlement due to thaw

TABLE 2

Section number	Permafrost Thickness (m)	Permafrost Temperature (°C)	Depth to permafrost table in 12th year (m) Calculated In design	Depth Calculated More precise	Depth Actual	Embankment settlement by 12th year (m) Calculated In design	Settlement Calculated More precise	Settlement Actual
1	3-10	0.02	complete thaw	-	-	to 1.0	0.4	0.15
2	3-10	1.4	"	5.0	5.0-5.5	1.06	0.4	0.15
3	3-10	1.4	"	5.0	5.0-5.5	0.67	0.4	0.16
4	>9	0.13	"	5.0	to 11.5	0.55	0.15	0.05
5	~20	0.36	"	1.7	1.5-1.7	1.87	0.85	0.75

Note: More precise calculated values of the depth of thaw and settlements were determined using methods specified by the authors (1974).

consolidations of both active layer and permafrost soils about 0.55 m under the centerline (Figure 1, curve 6) and 0.35-0.70 m under berms. After 10 years of railway operation the permafrost table under the embankment centerline was just below the original moss cover. The latter appeared to have settled to a depth of 1.75 m below the original ground surface. In this case, the more precise calculated value of the depth of thaw under the embankment did not differ significantly from that actually observed.

The curve showing the rate of thaw at section 5 indicates that thawing ceased under the embankment (Figure 1, curve 3). At the same time, the high rate of embankment settlement at this section, compared to that occurring at the other sections, is worth noting (Figure 1, curve 6). Moreover, settlement was observed not only in the first two years, when the frozen active layer soils thawed, but also in the following 8 years, when thawing of permafrost had ceased. This was due to compression of the permafrost soils, which originally were in a plastic-frozen state at a temperature of -0.3°C. The actual total settlement of the roadbed at this section was about 2.5 times less than the design value and about 12% less than the more precise value calculated using methods specified by the authors (1974).

The research showed that properties of the foundation soils change significantly as settlement occurs. During 11 years of observation the moisture content of the thawed permafrost under the embankment at sections 1-3 decreased by 5% with a corresponding increase in density. At section 5 the moisture content of the foundation soils decreased by as much as 50% and the density increased 1.3-1.8 times. Present calculation methods do not take these changes into account.

Analysis of the research results shows that permafrost thaw and embankment settlement follow a predictable pattern. The permafrost thaws more rapidly during construction and in the first 1-2 years thereafter. Under certain conditions it decreases by approximately 70-80% in the first 2-3 years and by 20-30% in the following 8-10 years. This pattern is not followed, however, in cases where the moss-vegetative cover is disturbed or removed and especially when soft formulation soils are excavated and replaced under an embankment. The length of the thaw period under such conditions is dependent on the permafrost thickness and temperature, the embankment height and structure and also on the local frozen soil, hydrological and hydrogeological conditions and changes that occur in them prior to and during construction. Embankment settlements may cease long before thawing of permafrost ends, if compressibility of the thawing soil appears to be insignificant. At the same time, settlements of high embankments are possible due to compression of plastic perennially frozen soils, thawing of which has practically ceased.

Differences between actual and calculated settlement values were observed at all sections. These were due partly to the incorrect design assumption that, regardless of its thickness, all permafrost would thaw under railway embankments in areas where high-temperature permafrost occurred in scattered islands. The lack of good data on the ice content and compressibility of the perennially frozen foundation soils also introduced errors in the calculations. The roadbed at all test sections, as well as at other embankment sections constructed on permafrost along this railway route, was as stable as necessary. Most settlements occurring in the construction period were compensated for before putting the railway into regular operation. The small settlements that occurred later were corrected by placing more ballast during maintenance operations.

One factor that deserves special mention is that the effectiveness of berms on improving both the stability and condition of embankments was not determined even though the warming effect on foundation soils was ascertained sufficiently well under all conditions considered. Thus, increasing the embankment width and construction of berms can be considered as expedient anti-deformation measures. However, the increased width at all test sections appeared to be excessive and the use of berms on both sides of the embankment at section 2 was not well founded. Increasing the roadbed width based on calculated settlement values would be more expedient for this section.

CONCLUSION

The results of the field research described in this paper were used in the design of embankments for another railway situated in similar natural conditions. In that case, excavation of peat at "mari" (bogs) was not undertaken when embankments were greater than 3 m thick and the width of embankments and berms were prescribed taking into consideration the condition and properties of perennially frozen soils. The roadbed at such sections remained sufficiently stable and ensured railway traffic could proceed at established speeds.

REFERENCES

Peretrukhin, N. A., 1978, Strength and stability of railway embankments in permafrost regions, in Proceedings of the Third International Conference on Permafrost, v. 1: Ottawa, National Research Council of Canada, p. 847-852.

Peretrukhin, N. A. and Potatueva, T. V., 1974, Influence of railway embankment construction on permafrost conditions, in Transportnoe Stroitelstwo, no. 11, p. 36-37.

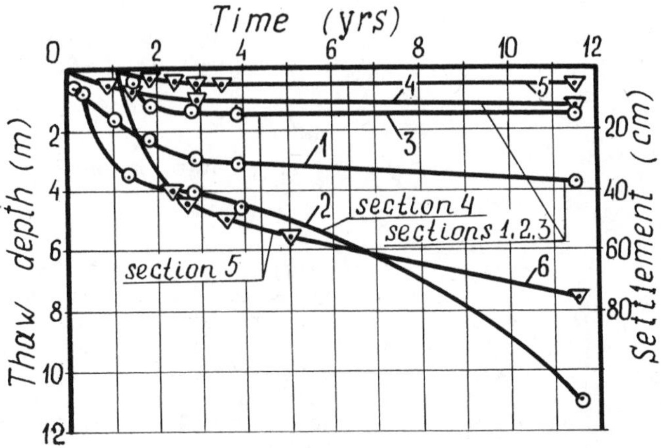

FIGURE 1 Curves showing permafrost degradation (1,2,3) and embankment settlement (4,5,6).

APPLICATIONS OF THE FAST FOURIER TRANSFORM TO COLD REGIONS ENGINEERING

W. F. Phillips

Utah State University, Logan, Utah 84322 USA

The utility of using the fast Fourier transform as a means of consolidating climatic data into a form useful for cold regions engineering design is presented and discussed in this paper. It is shown that a 20-year data set consisting of 175,320 hourly values of a climatic variable can be represented by only 118 Fourier coefficients with little loss of information content. Complete consolidated climatic data sets are presented for Barrow and Fairbanks, Alaska. In addition, some other applications of the Fourier transform to the field of cold regions engineering are discussed and specific examples are presented.

INTRODUCTION

Construction and maintenance of structures in the arctic and subarctic regions are characterized by a wide range of problems in addition to those experienced elsewhere. Most of these problems are directly related to some type of thermal interaction between the structure, the environment, and the underlying permafrost. An engineer must be able to predict such thermal interactions in order to design structures which perform satisfactorily in such an environment. Because the performance of such structures is coupled so closely to climatic conditions, the design engineer often needs data which describe how the ambient temperature and solar radiation vary with time at the design location. Such data are available for several Alaska locations through the National Climatic Center (1978, 1979).

Because of the variable nature of climatic conditions, a very large volume of data is needed to adequately describe these climatic variables. For example, 8,760 data points are needed to describe the hourly variations in temperature for 1 year at a single location. Furthermore, because climatic conditions change somewhat from year to year, 10 - 20 years' data are needed to obtain meaningful statistics. Thus, something on the order of 100,000 data points are needed to describe a single climatic variable at a single location. However, a data set of this volume has only very limited usefulness in routine engineering design.

There is a need for some means by which climatic data sets can be consolidated into a more manageable form. The consolidated data must retain all the information required for design and yet it should be compact enough to be handled with small desk top computers. It is the purpose of the present paper to show that the fast Fourier transform can be used to achieve such a consolidation.

The Fourier transform has long been a principal tool in many diverse fields of science, and with the development of the modern fast Fourier transform, many facets of scientific analysis have been completely revolutionized. The Fourier transform identifies or distinguishes the amplitude and phase of the different frequency sinusoids which combine to form an arbitrary function of time. Mathematically, this relationship is stated as

$$F(\nu) = \int_{-\infty}^{\infty} f(t)\ \exp(-i2\pi\nu t)dt, \tag{1}$$

where f is the function of time, t; $i = \sqrt{-1}$; and F is the Fourier transform expressed as a function of frequency, ν. The function of time can likewise be expressed in terms of its Fourier transform

$$f(t) = \int_{-\infty}^{\infty} F(\nu)\ \exp(i2\pi\nu t)d\nu. \tag{2}$$

If f(t) is a periodic, band-limited function of time, sampled 2N times over some integer multiple of the fundamental period with a sampling frequency that is more than 2 times the largest frequency component of f(t), then the Fourier transform becomes a set of 2N Fourier coefficients which can be calculated from the 2N sampled values of f(t):

$$F_n = \frac{1}{2N}\sum_{k=0}^{2N-1} f(kt_s)\ \exp(-i\pi nk/N);\left\{\begin{array}{l} n = 0,1,\ldots,2N-1 \\ k = 0,1,\ldots,2N-1 \end{array}\right.$$

where t_s is the sample interval. The function f(t) is expressed in terms of these Fourier coefficents as

$$f(t) = \sum_{n=0}^{2N-1} F_n\ \exp(i\pi nt/Nt_s);\quad n = 0,1,\ldots,2N-1$$

Also, in view of the facts that $\exp[i\pi(2N-n)k/N] = \exp(-i\pi nk/N)$ and the Nyquist component, F_N, is zero for the class of functions described above, equation (4) may be rewritten as

$$f(t) = \sum_{n=-N+1}^{N-1} F_n\ \exp(i\pi nt/Nt_s), \tag{5}$$

where,

$$F_{-n} = F_{2N-n}\ ;\qquad n = 1,2,\ldots,N-1 \tag{6}$$

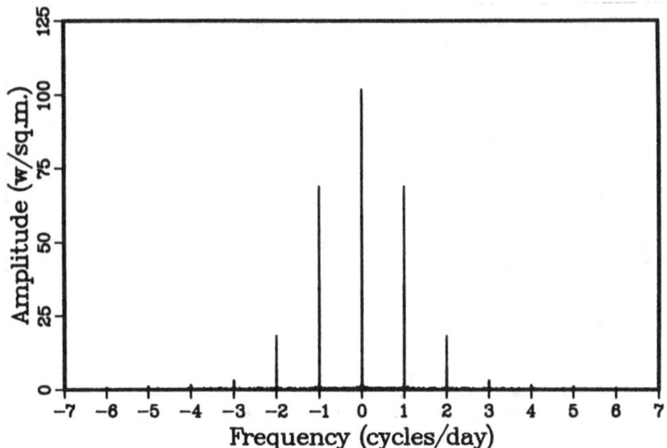

FIGURE 1 Fourier transform of solar flux data for Fairbanks, Alaska.

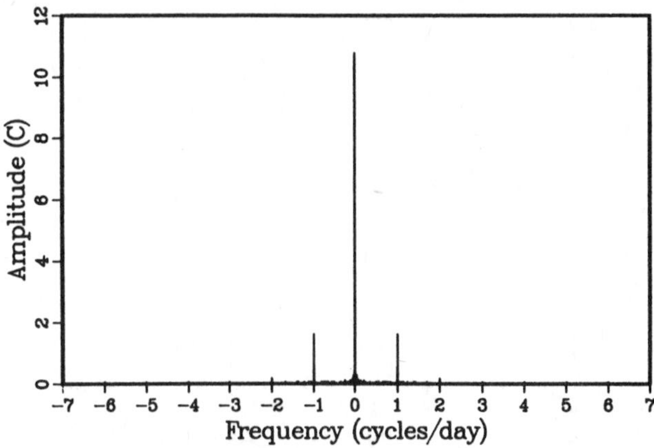

FIGURE 2 Fourier transform of ambient temperature data for Fairbanks, Alaska.

Furthermore, if f(t) is a real valued function, then F_{-n} is the complex conjugate of F_n and f(t) may be equivalently written as

$$f(t) = \sum_{n=0}^{N-1} [A_n \cos(\pi nt/Nt_s) + B_n \sin(\pi nt/Nt_s)], \quad (7)$$

where A_n is twice the real part of F_n and B_n is the negative of twice the imaginary part of F_n, for n greater than zero, and A_0 is the real part F_0.

For a detailed discussion of Fourier transforms, discrete Fourier transforms and fast Fourier transforms, see Brigham (1974).

Meteorologists and climatologists have used the Fourier transform as an analytical tool for many years. The hope of finding some predictable element in the year-to-year variations of climate has long inspired the search for dependable periodicities in climatic data. Harmonic analysis has been a standard tool in this search.

The usual procedure has been to remove the annual and diurnal variations from a data set through averaging or some other filtering process. Fourier analysis is then used to identify any low frequency oscillations which may still be present in the data set. From analysis of this type almost every conceivable period from 1 to 36 years has been suggested by investigations of one observation series or another. For detailed bibliography and description of methodology see Panofsky and Brier (1958) and Lamb (1978).

More recently, spherical harmonic analysis has been used to identify two-dimensional spatial periodicities in global temperature and pressure fields (see Christidis and Spar 1981). Analysis of this type has aided in improvement of the sophisticated space-time models used for short term weather prediction.

In the present paper harmonic analysis is applied to climatic data with a different objective. The Fourier transform is used here as a means of restructuring a climatic data set into a form that is more useful for engineering analysis and design.

FOURIER TRANSFORMS OF HOURLY DATA

To demonstrate the utility of transforming climatic data from the time domain to the frequency domain, transforms for two different Alaska stations will be presented. These two stations are Barrow and Fairbanks.

Fourier transforms were computed for both solar flux and ambient temperature using 20 years of hourly data for both of the above locations. These data were obtained on magnetic tape from the National Climatic Center (1978), Asheville, North Carolina. For each climatic variable at each location, 175,320 hourly readings for the period 1957 - 1976 were used to generate a Fourier transform. These Fourier transforms each contain 87,661 complex pairs, comprising the amplitude and phase information for the 87,661 different frequency sinusoids that combine to exactly form the original data set. It is interesting to note, however, that in each case, at most, 59 frequencies were found to have magnitudes detectable above a low noise level. Furthermore, it was always the same frequencies that were found to have significant magnitudes, independent of the climatic variable or the location for which the transform was computed. The magnitudes of the Fourier coefficients in equation (5), for the Fairbanks transforms, are plotted in Figures 1-2. Similar results were obtained for the Barrow transform.

Further examination reveals that each of the spikes shown in Figures 1-2 are actually formed by a set of closely spaced frequencies and not by a single frequency. The low frequency spike, that appears at zero cycles per day, may consist of up to five separate significant frequencies. These are the zero frequency and the fundamental annual frequency and its first three harmonics (0, ±1, ±2, ±3, ±4, cyc/yr). For example, Figure 3 shows the

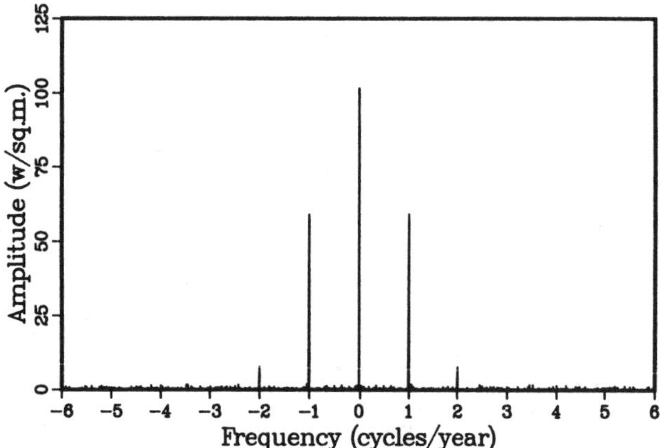

FIGURE 3 The low frequency harmonics of the solar flux transform for Fairbanks, Alaska.

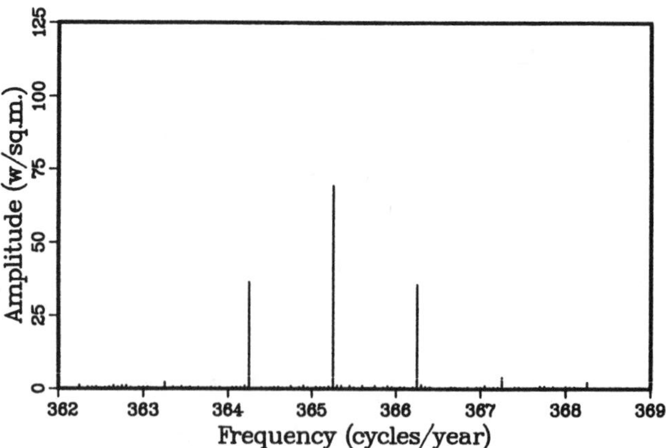

FIGURE 4 The near-daily frequencies of the solar flux transform for Fairbanks, Alaska.

low frequency portion of the same transform plotted in Figure 1. The expanded scale of Figure 3 allows one to distinguish between the many different frequencies, that all appear as a single line in Figure 1. Notice that the transform has a significant component at the zero frequency. This is simply the average value of the climatic variable. There is also a significant component at the annual frequency ±1 cyc/yr. It should be pointed out that a year is taken here to be a solar year, that is, 8,766 hours or 365.25 days. This annual component was found to have a period of exactly 8,766 hours. There is also a small, but definitely detectable, component at the first harmonic of the annual frequency ±2 cyc/yr. In Figure 3 the higher order harmonics are not detectable, however, when the numerical values are examined, these harmonics are found to have magnitudes slightly above the noise level. All of the intermediate frequencies were found to have only very low noise level magnitudes. In Figure 3 the computer has plotted the magnitudes of 240 different frequencies; however, only those mentioned above appear significant.

When the low frequency portions of the other transforms are expanded, the results are very similar to Figure 3. In all cases the only significant frequencies were found to be 0, ±1, ±2, ±3, ±4, cyc/yr. The Fourier coefficients for these frequencies describe the annual periodic variations in the climatic data.

From Figures 1-2 it can be seen that the daily frequency and its first five harmonics (±1, ±2, ±3, ±4, ±5, ±6 cyc/day) are also significant. This is to be expected, since the Fourier coefficients for these frequencies describe the daily variations in the climatic data. What might be less obvious is that there are eight other significant frequencies in the neighborhood of each of the daily harmonics. For example, that portion of the transform shown in Figure 1, consisting of the frequencies very near 1 cyc/day, is shown in Figure 4 on an expanded scale. Notice that the primary frequency occurs at 365.25 cyc/yr. The frequencies nearest to this fundamental harmonic, on either side, are found to have only a very low noise level magnitude, barely detectable in Figure 4. Then two more significant frequencies occur, one on each side of the primary frequency, at 364.25 and 366.25 cyc/yr. Then, at twice this spacing, two more frequencies with small but detectable magnitudes are observed. These frequencies, which flank the fundamental frequency at intervals of 1 cyc/yr, will be called "beat frequencies."

In a similar manner, each of the harmonics of the daily frequency is found to be flanked by four pairs of beat frequencies. The spacing between these frequencies is always exactly one cycle per solar year. Similar results are found for the other transforms. It is the daily harmonics and their associated beat frequencies that contain the information which describes the daily variations in the climatic data and how these daily variations change with the seasons of the year.

PREDICTABLE CLIMATIC VARIATIONS

It seems reasonable to assume that variations in a climatic variable at a particular location can be adequately described by the 175,320 hourly values which make up a 20-year data set. Such data sets contain two different types of variations. There are predictable variations, that result primarily from the motion of the earth relative to the sun, and there are random variations, resulting from random motions in the earth's atmosphere. All of the information about the predictable variations in a climatic data set is contained in the 59 complex pairs which are the Fourier coefficients for the 59 significant frequencies described above.

When these 59 sinusoids are recombined, one obtains the climatic variable as a function of time, excluding all of the random variations. Each of the component frequencies contributes to the temporal distribution in a way that can be related to physical observations. The role of the annual harmonics and the daily harmonics is rather straightforward. However, the importance of the beat frequencies, associated with the daily harmonics, may be less obvious.

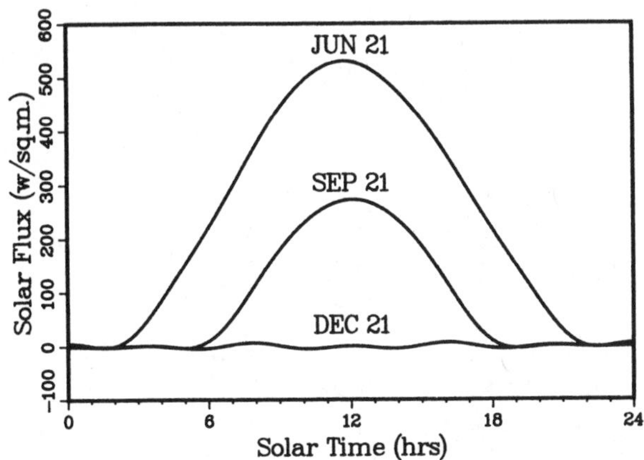

FIGURE 5 The daily variations in Fairbanks solar flux as predicted from the Fourier transform excluding the noise frequencies.

To demonstrate the role which the beat frequencies play in the Fourier transform, consider a function of time, $f(t)$, formed by combining a primary sinusoid of amplitude A and frequency ω with two secondary sinusoids of amplitude B and frequencies $\omega-\delta$ and $\omega+\delta$, respectively,

$$f(t) = B\sin(\omega t-\delta t) + A\sin(\omega t) + B\sin(\omega t+\delta t) \quad (8)$$

Recalling that

$$\sin(\theta\pm\phi) = \sin\theta\cos\phi \pm \cos\theta\sin\phi, \quad (9)$$

equation (8) can be rewritten as

$$f(t) = [A + 2B\cos(\delta t)]\sin(\omega t). \quad (10)$$

Thus it is seen that the function described by equation (8) is an amplitude modulated sine wave. In a similar manner, each of the daily harmonics combines with the associated beat frequencies to form an amplitude modulated sinusoid.

Just as constant amplitude sinusoids can be combined to produce a repeating but nonsinusoidal wave form, these amplitude modulated sinusoids combine to produce daily variations which have a wave form that changes over the year. This can best be demonstrated by recombining the Fourier components shown in Figure 1. This result is shown in Figure 5.

It can be seen from Figure 5 that recombining these harmonics results in the prediction of small nighttime oscillations in the solar radiation, when the flux should remain constant at zero. These small inaccuracies result from the fact that the Fourier series was truncated to contain only five harmonics of the daily frequency.

NUMERICAL RESULTS

For future work it will be convenient to have each transform, expressed in a numerical form. One has the option of expressing any Fourier series either in complex form or in sinusoidal form. Each form has certain advantages and one form can be easily computed from the other. Because many small computers and hand held calculators do not easily handle complex functions, the sinusoidal form will be presented here.

The Fourier transforms of the solar flux and ambient temperature data for Fairbanks and Barrow, Alaska are presented in Tables 1 and 2. Column one is the frequency in cycles/year; columns two and three give the cosine amplitude and the sine amplitude, respectively, for the corresponding frequency in the Fourier series representation of the solar flux data; and columns four and five give the same information for the ambient temperature data.

The average or most probable value of either solar flux or ambient temperature, at any time during the year, can easily be determined from the results given in Tables 1 and 2. Suppose, for example, that the average value of the solar flux in Fairbanks on a particular Julian day, D, at some solar time, t, is desired. This is computed from the equation,

$$q_s = \sum_{n=0}^{58} \{A_n\cos[2\pi\omega_n(D - 1 + t/24)/365.25] \quad (11)$$

$$+ B_n\sin[2\pi\omega_n(D - 1 + t/24)/365.25]\},$$

where ω_n, A_n, and B_n are taken from Table 1, columns 1, 2, and 3, repectively. Thus the average solar flux on June 21, D = 172, at 2:15 p.m. solar time, t = 14.25, is 460 W/m^2. The average ambient temperature at the same time would be 20.5°C.

DISCUSSION

This paper has pointed out some of the advantages of using Fourier analysis on climatic data. It has been shown that a 20 year data set can be consolidated by more than 1,000:1 through use of the Fourier transform. This result alone seems very significant. There are, however, other advantages to Fourier analysis which may be even more significant. Perhaps the biggest advantage to be gained from transforming climatic data from the time domain to the frequency domain is the simplification achieved by using the transformed data for engineering analysis.

Consider, for example, the problem of computing the change in ground temperature which results from a change in surface conditions. In order to simplify this example, the earth will be taken as a homogeneous semi-infinite solid and the effects of any phase change will be ignored. The equation that governs the temperature in the earth is

$$\frac{\partial\theta}{\partial t} = \frac{k}{\rho c}\frac{\partial^2\theta}{\partial x^2}, \quad (12)$$

where θ is the earth temperature as a function of depth x and time t and where k, ρ, and c are the thermal conductivity, density, and specific heat of the earth, respectively. Assuming a ground cover

Table 1 Fourier transform from hourly climatic data for Fairbanks, Alaska.

	SOLAR FLUX (W/sq.m)		AMBIENT TEMPERATURE (C)	
FREQUENCY (Cyc/yr)	COSINE AMPLITUDE	SINE AMPLITUDE	COSINE AMPLITUDE	SINE AMPLITUDE
0.00	101.61	0.00	-3.244	0.000
1.00	-114.35	30.84	-21.300	-3.517
2.00	9.66	-12.27	-0.890	-0.095
3.00	1.79	1.30	0.119	0.823
4.00	1.15	-2.81	0.017	-0.127
361.25	-1.17	-1.87	0.188	-0.135
362.25	-2.09	-1.50	0.161	0.121
363.25	2.57	-3.74	0.035	0.471
364.25	70.30	20.04	0.257	1.090
365.25	-138.05	-0.37	-1.937	-2.615
366.25	68.30	-20.02	0.786	0.559
367.25	5.39	4.61	0.390	0.200
368.25	-4.39	0.95	0.040	0.021
369.25	-0.14	1.62	-0.051	0.117
726.50	1.13	1.34	-0.035	0.006
727.50	-0.81	1.77	-0.076	-0.082
728.50	-12.75	-5.15	0.036	-0.281
729.50	-6.69	-3.33	0.004	0.173
730.50	36.68	-0.03	0.268	0.245
731.50	-4.99	4.11	0.127	0.279
732.50	-15.04	3.67	-0.221	-0.134
733.50	0.97	-1.02	-0.076	-0.051
734.50	0.46	-1.09	-0.016	-0.020
1091.75	0.13	-0.67	-0.021	0.026
1092.75	4.91	1.85	-0.009	0.058
1093.75	2.25	2.24	0.024	-0.032
1094.75	-5.96	-1.59	0.023	-0.087
1095.75	-1.20	0.51	-0.054	0.041
1096.75	-6.39	0.40	-0.048	-0.105
1097.75	2.65	-1.19	-0.025	-0.077
1098.75	4.69	-2.11	0.081	0.046
1099.75	0.13	0.38	0.037	0.021
1457.00	-2.27	-1.49	0.032	-0.019
1458.00	-1.16	-1.04	-0.002	0.014
1459.00	2.85	0.87	-0.039	0.024
1460.00	-0.35	0.39	0.007	-0.003
1461.00	1.57	-0.42	0.001	0.024
1462.00	-0.41	0.25	0.003	-0.024
1463.00	3.14	-1.05	0.010	0.054
1464.00	-1.47	0.90	-0.001	0.025
1465.00	-2.16	1.64	-0.026	-0.022
1822.25	0.82	0.95	-0.005	-0.002
1823.25	-1.70	-0.83	0.002	-0.019
1824.25	0.01	-0.16	-0.008	0.004
1825.25	-0.04	-0.08	0.017	-0.007
1826.25	-0.46	0.15	-0.008	0.006
1827.25	-0.05	0.01	0.018	-0.006
1828.25	-0.16	0.01	0.009	0.009
1829.25	-1.48	0.82	-0.015	-0.022
1830.25	0.73	-0.83	-0.003	-0.004
2187.50	0.93	0.63	-0.008	-0.007
2188.50	0.08	0.16	0.013	-0.010
2189.50	0.08	0.02	0.018	0.021
2190.50	-0.08	-0.02	0.001	-0.001
2191.50	0.31	-0.01	-0.011	0.000
2192.50	-0.02	-0.07	-0.025	0.004
2193.50	0.10	0.09	0.004	0.002
2194.50	-0.01	-0.05	0.000	-0.011
2195.50	1.01	-0.70	0.006	0.005

Table 2 Fourier transform from hourly climatic data for Barrow, Alaska.

	SOLAR FLUX (W/sq.m)		AMBIENT TEMPERATURE (C)	
FREQUENCY (Cyc/yr)	COSINE AMPLITUDE	SINE AMPLITUDE	COSINE AMPLITUDE	SINE AMPLITUDE
0.00	79.01	0.00	-12.671	0.000
1.00	-99.46	28.81	-14.814	-7.968
2.00	17.65	-9.64	1.300	0.121
3.00	-3.91	-9.67	0.596	1.395
4.00	12.66	3.95	-0.041	0.172
361.25	-6.93	2.77	-0.102	0.022
362.25	-2.84	-9.42	0.041	-0.201
363.25	1.83	-3.02	0.143	0.137
364.25	52.50	19.88	0.055	0.358
365.25	-94.85	-2.08	-0.706	-0.576
366.25	51.90	-16.36	0.587	0.030
367.25	2.57	1.02	0.069	0.194
368.25	-3.40	10.05	-0.140	0.055
369.25	-6.66	-2.91	-0.048	-0.097
726.50	3.39	1.70	-0.016	0.048
727.50	0.64	3.18	-0.031	0.012
728.50	-10.48	-2.69	-0.049	-0.096
729.50	-2.95	-5.27	0.058	-0.016
730.50	18.91	1.44	0.092	0.059
731.50	-2.70	2.97	-0.005	0.052
732.50	-10.71	3.60	-0.065	-0.028
733.50	0.82	-2.92	0.009	0.007
734.50	3.24	-2.24	0.009	0.003
1091.75	-0.27	-1.00	0.019	0.018
1092.75	3.12	1.26	0.001	0.040
1093.75	1.08	1.14	-0.025	0.002
1094.75	-2.08	-0.17	-0.015	-0.028
1095.75	-0.77	-0.04	-0.003	-0.007
1096.75	-2.00	0.17	0.001	-0.009
1097.75	0.90	-0.86	0.021	0.005
1098.75	3.25	-1.78	0.024	0.021
1099.75	-0.25	1.55	-0.021	0.008
1457.00	-1.54	-1.12	0.004	-0.008
1458.00	-0.18	-0.32	0.003	0.002
1459.00	0.55	-0.12	-0.009	0.013
1460.00	0.15	0.53	-0.005	-0.006
1461.00	0.10	-0.38	-0.001	-0.003
1462.00	0.05	0.05	0.002	-0.006
1463.00	0.76	-0.20	0.007	0.002
1464.00	-0.34	0.57	-0.001	0.004
1465.00	-1.52	0.80	-0.009	-0.002
1822.25	0.12	0.37	0.000	-0.002
1823.25	-0.38	-0.23	-0.004	0.000
1824.25	0.12	0.08	0.000	0.002
1825.25	-0.12	-0.19	0.002	0.002
1826.25	0.08	0.15	-0.001	0.002
1827.25	-0.09	0.00	-0.002	-0.001
1828.25	-0.02	-0.01	-0.004	-0.003
1829.25	-0.27	0.11	-0.004	0.002
1830.25	0.17	-0.19	-0.001	-0.004
2187.50	0.17	0.07	-0.008	0.002
2188.50	-0.02	0.10	-0.007	-0.006
2189.50	0.02	-0.09	0.006	-0.010
2190.50	-0.01	0.07	0.007	0.002
2191.50	-0.04	0.00	-0.005	0.013
2192.50	0.06	-0.10	-0.008	0.006
2193.50	0.03	0.10	-0.005	-0.003
2194.50	-0.07	-0.02	0.001	-0.006
2195.50	0.16	-0.18	0.004	-0.004

with thermal conductance u, solar absorptance α, and a surface film coefficient h, the boundary condition at x = 0 is

$$h\,\theta(o,t) - \left(\frac{h}{u} + 1\right)k\frac{\partial\theta}{\partial x}(o,t) = \alpha q(t) + h\,\phi(t) \qquad (13)$$

where q and ϕ are the solar flux and ambient temperature. The remaining boundary condition is

$$k\frac{\partial\theta}{\partial x}(\infty,t) = q_\infty, \qquad (14)$$

where q_∞ is the steady heat loss from the earth's core.

Thus, to predict the temperature variations in the earth as a function of surface conditions, it is necessary to express the solar flux and ambient temperature as a function of time and then solve a partial differential equation which involves these forcing functions. However, if we introduce the Fourier transforms,

$$q(t) = \sum_{n=-N}^{N} Q_n \exp(i\omega_n t), \qquad (15)$$

$$\phi(t) = \sum_{n=-N}^{N} \Phi_n \exp(i\omega_n t), \qquad (16)$$

$$\theta(x,t) = \sum_{n=-N}^{N} \Theta_n(X) \exp(i\omega_n t), \qquad (17)$$

then the transform of equations (12)-(14) is

$$i\omega_n\Theta_n = \frac{k}{\rho c}\frac{d^2\Theta_n}{dx^2}, \qquad (18)$$

$$h\,\Theta_n(o) - \left(\frac{h}{u}+1\right)k\frac{d\Theta_n}{dx}(o) = \alpha Q_n + h\Phi_n \qquad (19)$$

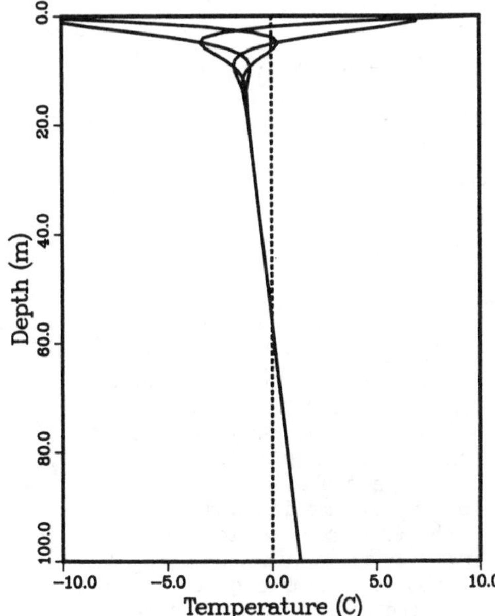

FIGURE 6 Predicted ground temperature in the Fairbanks area with moderate ground cover ($u = 1.5$ W/m^2°C).

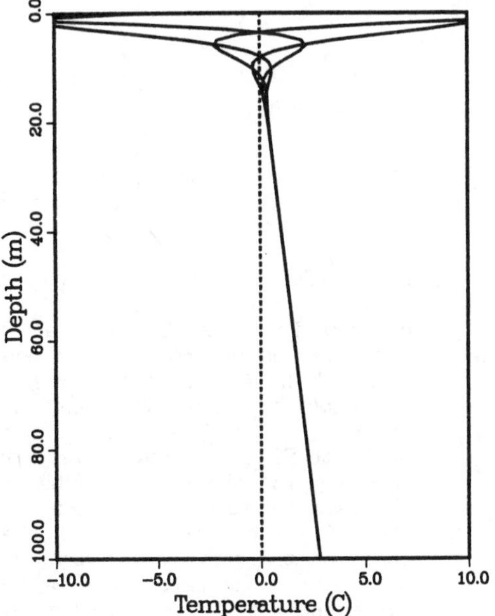

FIGURE 7 Predicted ground temperature in the Fairbanks area with no ground cover ($u >> h$).

$$k \frac{d\Theta_n}{dx}(\infty) = \begin{cases} q_\infty, & n = o \\ o, & n \neq o \end{cases} . \qquad (20)$$

Thus, what was a partial differential equation in the time domain becomes an ordinary differential equation in the frequency domain.

Equation (18) is easily solved subject to equations (19) and (20) to give

$$\Theta_n(x) = \begin{cases} \Phi_o + \dfrac{\alpha Q_o}{h} + \dfrac{(h + u)q_\infty}{hu} + \dfrac{q_\infty}{k} x, & n = 0 \\ \dfrac{(\alpha Q_n + h\,\Phi_n)\exp(-\gamma_n x)\exp[-i(\gamma_n x + \beta_n)]}{\{[h + (h+u)k\gamma_n/u]^2 + [(h+u)k\gamma_n/u]^2\}^{1/2}}, & n \neq 0 \end{cases}$$

where

$$\gamma_n = (\omega_n \rho c/2k)^{1/2} , \qquad (22)$$

$$\beta_n = \tan^{-1}\left[\frac{(h+u)k\gamma_n}{hu + (h+u)k\gamma_n}\right]. \qquad (23)$$

The Fourier transforms given in Table 1 or 2, can be used in equation (21) and the result can be transformed back to the time domain, using equation (17), to give the desired solution.

A typical solution using the Fairbanks data is shown in Figure 6. The four curves shown in this figure are the predicted ground temperature distributions at solar noon on March 21, June 21, September 21, and December 21. For this solution, a ground cover with thermal conductance of 1.5 W/m^2°C was assumed. Note that the permafrost extends to a depth of about 60 m with this ground cover present. Figure 7 shows the solution for the same conditions with the ground cover removed ($u >> h$). Note that removal of the ground cover will completely thaw the permafrost and greatly increase the thickness of the active layer.

This example points out how much of the heat transfer analysis associated with cold regions engineering can be simplified through the use of Fourier transforms. This was a very simple example: however, a similar approach can be used in the analysis of more complex problems that involve thermal interaction with the environment.

REFERENCES

Brigham, E. O., 1974, The Fast Fourier Transform, Prentice-Hall, Englewood Cliffs, New Jersey.

Christidis, Z.D. and Spar, J., 1981, "Spherical Harmonic Analysis of a Model Generated Climatology", Monthly Weather Review 109, V. 215.

Lamb, H. H., 1978, Climate Present, Past and Future, Methuen, London.

National Climatic Center, 1978, SOLMET Manual, "Hourly Solar Radiation - Surface Meteorological Observations", TD-9724, US National Climatic Center, Asheville, North Carolina.

National Climatic Center, 1979, SOLDAY Manual, "Daily Solar Radiation - Surface Meteorological Data", TD-9739, US National Climatic Center, Asheville, North Carolina.

Panofsky, H.A. and Brier, G.W., 1958, Some Applications of Statistics to Meteorology, Pennsylvania State University Press.

LONG-TERM CREEP DEFORMATION OF ROADWAY EMBANKMENT ON ICE-RICH PERMAFROST

Arvind Phukan

Department of Civil Engineering, University of Alaska
Anchorage, Alaska, 99508 USA

Six different combinations of insulation layers, toe berms, and air-ducting systems were installed at an embankment near Bonanza Creek to study thermal degradation at side slopes and toe sections. The performance of these design measures has been monitored since the completion of construction in 1974. Though some of these measures show encouraging results in minimizing long-term thaw settlements, creep settlements are recorded at some of the central sections of the embankment where soil conditions consist of generally frozen ice-rich organic soils to fine-grained soils. In this paper, the long-term creep deformations recorded for the roadway embankment near Bonanza Creek are evaluated. Based on all data recorded for the last 8 years, a long-term creep deformation behavior of the roadway embankment (about 7.6 m height) in ice-rich permafrost is formulated. Based on theoretical analyses, the vertical creep movements (within the central portions of the embankment) are calculated, and they show reasonable qualitative agreement with the measured data. Measured movements indicate that the primary creep movements occurred within the first 3 years after the construction. The analysis presented provides a rational basis for embankment design on ice-rich permafrost.

INTRODUCTION

Roadway embankments built on permafrost have encountered significant deformation with time (Esch , 1973, 1978, Johnston , 1969, Phukan , 1981). Previous studies (Figure 1) by the Alaska Department of Transportation on the roadways embankment constructed over permafrost have shown that relatively deeper thawing beneath the roadway side slopes occurs, even in areas where the embankment thickness beneath the roadway centerline is sufficient to prevent thawing into the underlying permafrost. Because of snow cover, the side slopes will not freezeback the thawed portions near the toe sections, and therefore, annually deeper thawing of underlying permafrost often occurs beneath the roadway side slopes. Those problems are further intensified by increased embankment heights and steeper side slopes which are constructed over ice-rich permafrost. In 1974, the Alaska Department of Transportation installed six different combinations of insulation layers, toe berms, and air-ducting system at a newly constructed highway section near Bonanza Creek to study the embankment slope design in controlling thaw of permafrost beneath roadway side slopes (Esch , 1978). Details of these design measures are illustrated in Figure 1, and measurements have been monitored since the completion of construction in 1974. Though some of the design measures show encouraging results in minimizing long term thaw settlements, creep settlements are recorded at most of the central sections of the embankment where the subsoil conditions consist of generally frozen ice-rich organic soils to silty soils. Limited data are available on the creep deformation behavior of roadway embankment constructed on these ice-rich soils.

This paper evaluates the creep deformations recorded for the roadway embankment constructed near Bonanza Creek, which is located approximately 25 miles west of Fairbanks on the Parks Highway, Alaska, USA. Based on the data, a creep theory is formulated to predict the long-term deformation. Results of the laboratory studies on ice-rich frozen soil samples as well as some of the available information on ice, are used to determine the magnitude of creep parameters. The predicted creep settlement agrees closely with the measured movements.

SITE CONDITIONS AND INSTRUMENTATION

The site located at Bonanza Creek lies within the discontinuous permafrost regions of Alaska, and the thickness of permafrost is of the order of 30-40 m in the regions. Esch (1978) reported the detailed information regarding the construction and design measures for this embankment section.

The embankment was built on about Ø.6 m of ice-rich organic soils underlain by ice-rich silt. The variation of frozen moisture content of the underlying soils is presented in Figure 2. The natural vegetation near the site consists of scattered black spruce, tamarack, willow, and alder, which is typical of the interior muskeg environments in Alaska. The mean annual permafrost temperature recorded in an adjacent undisturbed forest area is in the range of -1° to $-1.5^{o}c$ at a depth of 10 m.

Three hundred and fifty-two thermocouples and 20 thermistors were installed both in horizontal strings and in 18 vertical borings drilled through the embankment. Readings of both thermocouples (accuracy $\pm 0.2^{o}C$) and thermistors (accuracy $\pm 0.03^{o}C$) are made monthly. Details of these temperature readings are given by Esch (1978). For this study, the recorded temperature readings by the Alaska Department of Transportation are evaluated, and the data at two different sections, at sta. 170+00 and sta. 166+75, are shown in Figures 3 and 4.

Survey measurements are made annually in late September to monitor annual settlements at different cross sections. Centerline settlements recorded from 1976 to 1981 at cross sections sta. 166+75 and sta. 170+00 are presented in Figure 5. Deformations recorded at other cross sections of the embankment lie within the shaded portion of the envelope shown in this figure.

THEORETICAL SETTLEMENT ANALYSIS

It can be seen in Figure 5 that the embankment built on ice-rich permafrost shows typical creep behavior with distinct primary and secondary stages. As shown in Figures 3 and 4, the temperature at the midpoints of the embankment was changing and approached steady state conditions after 6 years of construction. The effect of berm at the toe of the embankment on the temperature distribution is reflected (see Figures 3 and 4). The settlement reached the secondary creep stage after 3 years of construction, (2 years after the end of construction) and in practice one is mainly concerned with the deformation during this stage.

The following equation may be written:

$$S(t) = S^{(i)} + \dot{S}^{(c)}t, \qquad (1)$$

Where: $S(t)$ = time-dependent settlement;
$S^{(i)}$ = pseudo-instantaneous settlement;
$\dot{S}^{(c)} = \frac{ds}{dt}$ = steady state settlement rate;
t = time.

At a given frozen soil temperature, both $S^{(i)}$ and $\dot{S}^{(c)}$ are functions of the applied stress.

Based on previous works (Landanyi , 1972, Johnston and Ladanyi , 1972, Hult , 1966, Vyalov , 1965), an approximation of the plane-strain condition equation (1) can be written as

$$S(t) = C_1 S_k \left(\frac{\sigma_1 - \sigma_3}{\sigma_k}\right)^k + C_2 \dot{S}_c \left(\frac{\sigma_1 - \sigma_3}{\sigma_c}\right)^n t \,, \qquad (2)$$

where $C_1 = \left(\frac{(3)^{1/2}}{2}\right)^{k+1}$

$C_2 = \left(\frac{(3)^{1/2}}{2}\right)^{n+1}$;

σ_1 and σ_3 are the major and minor principal stresses, respectively; σ_k, σ_c, K, n are creep parameters dependent on temperature and unique to particular frozen soil; and S_k and $\dot{S}_c$ are arbitrarily selected settlement and settlement rate, respectively, which are introduced into the equation to put it into normalized form. The four experimental parameters σ_k, K, σ_c, and n in equation (2) can be obtained by plotting pseudoinstantaneous deformation and steady state settlement rates, versus the applied stress in a log - log plot.

Equation (2) can be rewritten as

$$s(t) = B_1(\sigma_d)^k + B_2(\sigma_d)^n t \qquad (3)$$

where $B_1 = C_1 S_k/\sigma_k^{\,k}$, $B_2 = C_2 \dot{S}_c/\sigma_c^{\,n}$

Phukan (1982) investigated the creep behavior of ice-rich frozen soil samples obtained from various regions of Alaska.

In the new synthesis presented here, these data as well as some of the available data on ice are reviewed, and the assumed creep parameters for ice-rich silt are summarized in Table 1.

TABLE 1
SUMMARY OF CREEP PARAMETERS FOR ICE-RICH SILT

Temperature Range (^{o}C)	B1 (KPa^{-k} x cm)	K	B2 (KPa^{-n} x $^{-1}$ cm)	n
0 to -1	10^{-2}	1.5	5 x 10^{-6}	2.7
-1 to -3	5 x 10^{-4}	1.9	0.7 x 10^{-6}	2.7

Using equation (3) and Table 1, it is now possible to predict the deformation rates under constant embankment load on ice-rich permafrost.

In order to introduce some cross-check between the theoretical treatment described above and field measurements of embankment settlement, it is important to compare the predictions offered by this preliminary analysis with the recorded data presented in Figure 5.

From the embankment construction data, the average deviator stress below the midsections at sta. 166+75 and 170+00 may be calculated as 71.4 and 88.2 KPa, respectively, and the temperature range is about $-0.5^{o}C$. Using the creep parameters of ice-rich soils (Table 1), equation (3) may be used to predict the settlement of the embankment. The range of settlement calculated for the embankment for different time periods agree with the measured information plotted in Figure 5.

The foregoing analysis for the creep settlements of embankment in ice-rich permafrost employed assumptions concerning the uniformity of soil conditions and temperatures. Moreover, it was assumed that the embankment maintained its geometry (no significant change in height). This was done to preserve the simplicity of the treatment, and it is encouraging to observe the reasonable success of the method in predicting field behavior.

Clearly, the embankment design on ice-rich permafrost will involve limiting the height of embankment and thereby stress to keep the settlement within allowable limits. Obviously, the temperature beneath the embankment must also be maintained below the freezing temperature.

CONCLUSIONS

An analysis has been developed for the prediction of settlement for embankment built on ice-rich frozen soils. As more creep data for frozen soils of varying ice contents become available, the creep parameters for the settlement equation synthesized here may be modified to reflect the properties of the supported soil beneath the embankment. This will result in a more rational basis for the settlement calculation.

From a practical point of view, it is most important that the effects of disturbance of the ground thermal regime (due to construction operations or naturally occurring changes in terrain conditions and climate), which may result in significant settlement with time, be duly considered in the design of embankment in ice-rich permafrost. Moreover, depending on the existing temperature and ice content of founded frozen materials, the height of the embankment (reflecting stress) may be limited to maintain the embankment settlement within an acceptable range. More information on creep parameters of frozen soils at different stress levels is required to limit the maximum height of embankment that can be built so that only negligible deformation with time occurs.

ACKNOWLEDGMENTS

This study has been funded in part by the state of Alaska, Department of Transportation and Public Facilities, Division of Planning and Programming, Research Section. The author acknowledges D. C. Esch for useful field measurement data for this study.

REFERENCES

Esch, D. C., 1973, Control of permafrost degradation beneath a roadway by subgrade insulation: North American Contribution, Permafrost Second International Conference, National Academy of Sciences, p. 608-621.

Esch, D. C., 1978, Road embankment design alternatives over permafrost, in Proceedings of the Conference on Applied Techniques for Cold Environments, Anchorage: American Society of Civil Engineers, v. 1, p. 159-170.

Hult, J. A. H., 1966, Creep in engineering structures: Blaisdell Publishing Company, Waltham, Mass. p. 115.

Johnston, G. H., 1969, Dykes on permafrost, Kelsey Generating Station, Manitoba: Canadian Geotechnical Journal, v.6, no. 2, pl 139-157.

Johnston, G. H., and Ladanyi, B., 1972, Field tests of grouted rod anchors in permafrost: Canadian Geotechnical Journal, v. 9, p. 63-80.

Ladanyi, B., 1972, An engineering theory of creep in frozen soils: Canadian Geotechnical Journal, v. 9(1), p. 63-80.

Phukan, A., 1981, Design guide for roadways on permafrost: Reported to the State of Alaska, Department of Transportation and Public Facilities, Division of Planning and Programming, Research Section, October.

Phukan, A., 1982, Laboratory investigation of the creep behavior of ice-rich frozen soils, School of Engineering, University of Alaska, Fairbanks, interim report to be publishe1.

Vyalov, S. S., Ed., 1965, The strength and creep of frozen soils and calculation for ice-soil retaining structures: U.S. Army CRREL, Translation 76, Hanover, N.H. p. 321.

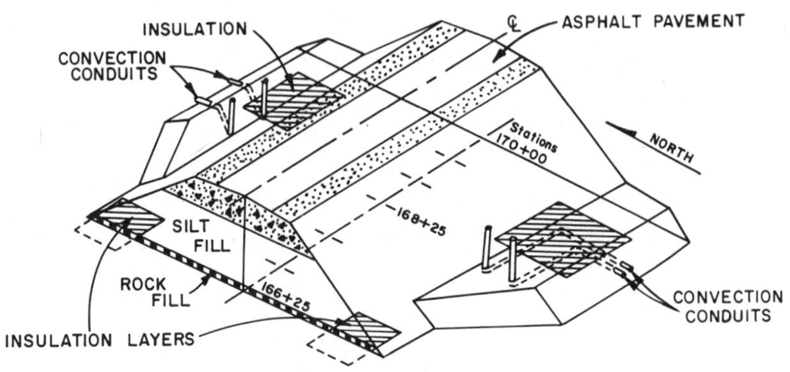

FIG. 1 EMBANKMENT DETAILS
(After Esch, 1978)

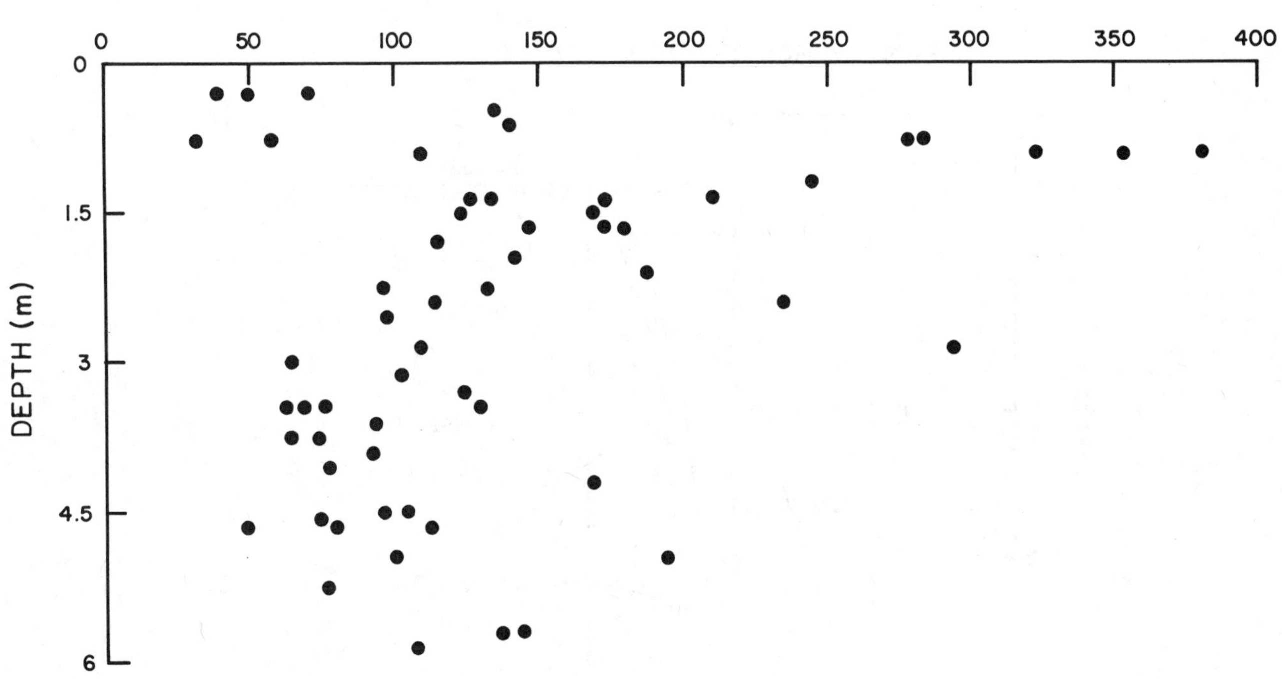

FIG. 2 FROZEN WATER CONTENT PROFILE

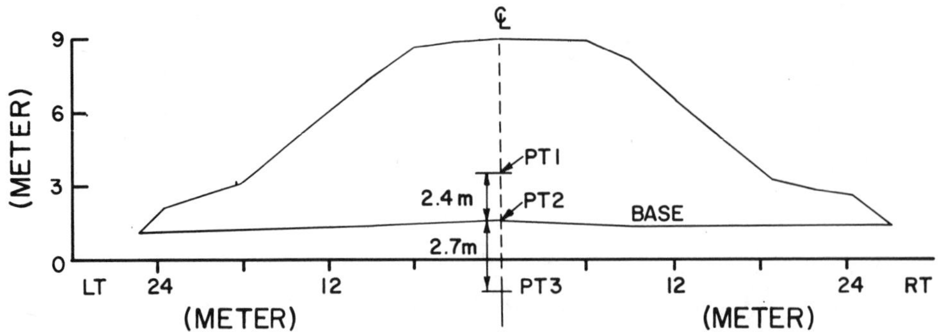

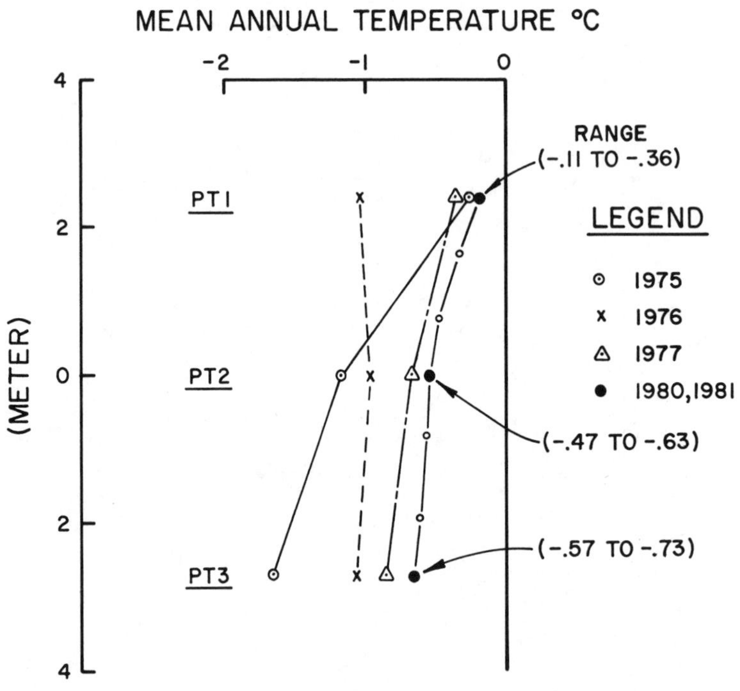

FIG. 3 TEMPERATURE PROFILE AT STA. 170+00

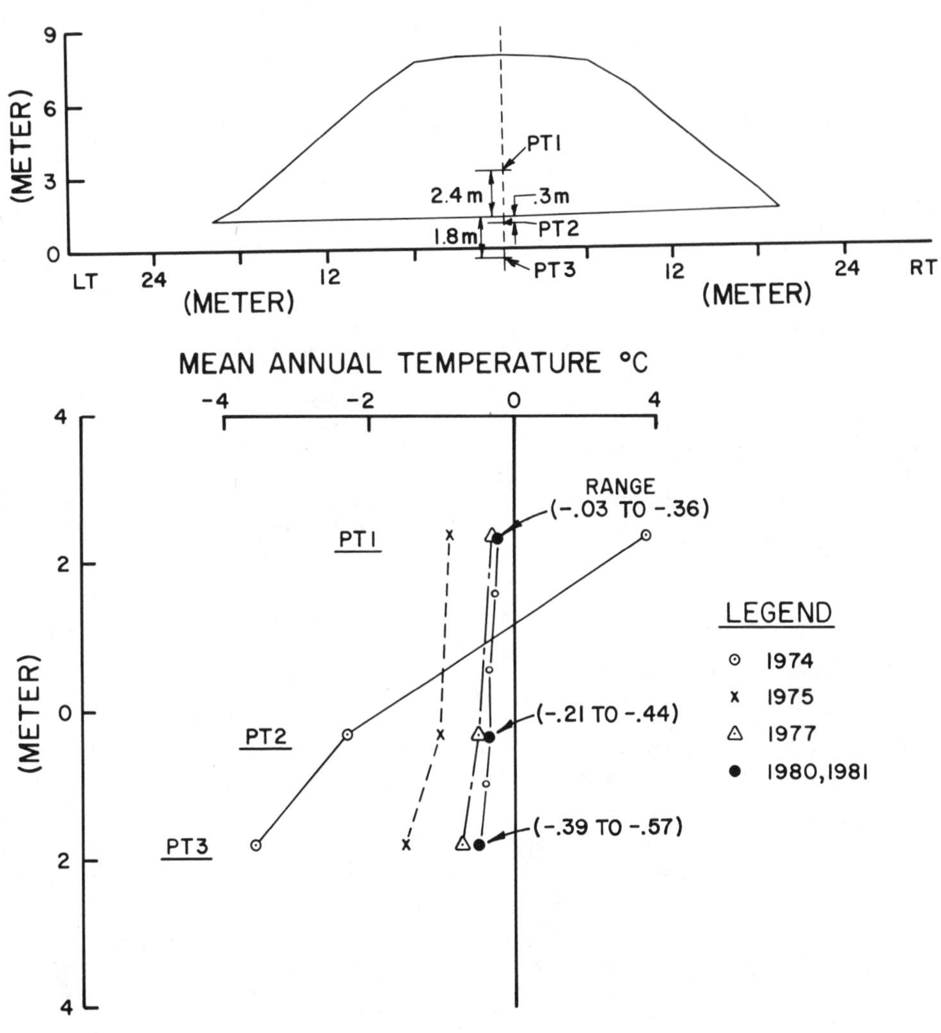

FIG. 4 TEMPERATURE PROFILE AT STA. 166+75

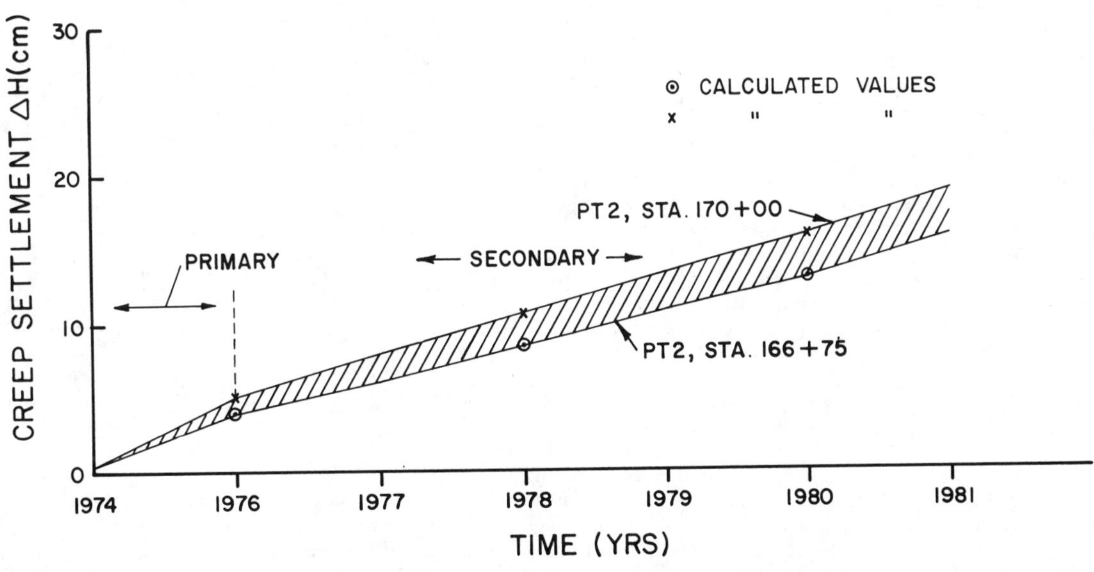

FIG. 5 RELATIONSHIP BETWEEN SETTLEMENT Vs TIME

SEASONAL FROST MOUND OCCURRENCE, NORTH FORK PASS, OGILVIE MOUNTAINS, NORTHERN YUKON, CANADA

W. H. Pollard[1] and H. M. French[2]

[1]Department of Geography, University of Ottawa, Ottawa, Canada K1N 6N5
[2]Departments of Geography and Geology, University of Ottawa, Ottawa, Canada K1N 6N5

Between 1980 and 1982 a total of 65 seasonal frost mounds were observed at several localities in the North Fork Pass, interior northern Yukon. The majority were of the frost blister variety, although icing blisters and icing mounds also occurred. The largest was 3.5 m high. Several persisted for more than 1 year and a few experienced reactivation and further growth in a second winter. Stratigraphic investigations, together with ice fabric analyses, suggest the mounds result from the freezing of suprapermafrost groundwaters during winter freezeback under conditions of high hydraulic potential. Piezometers, installed in areas of active mound formation, indicate pressures ranging from 40 to 80 kPa are associated with mounds 1-2 m high.

INTRODUCTION

Seasonal frost mounds are a groundwater discharge phenomenon resulting from increased hydraulic potentials developed by suprapermafrost groundwater in the active layer during winter freezeback. They form a distinct group of ephemeral and unstable winter landforms.

Although their occurrence has been widely reported in arctic North America (e.g., Brown and Berg 1980, p. 29, 82; Leffingwell 1919, p. 158; Muller 1945, p. 59; Porsild 1938, p. 47; Sharp 1942; Toth 1972, p. 161, 162; Zoltai and Johnson 1978, p. 24, 27, 51), only a few studies describe them in detail (e.g., van Everdingen 1978, 1982; van Everdingen and Banner 1979). In some instances, seasonal frost mounds have been confused with palsas (e.g., Hughes et al 1972, p. 38; Maarleveld 1965, p. 10-11) and they should also be distinguished from other types of short-lived ice-cored mounds which occur in tundra regions (e.g., French 1971). In the USSR, seasonal frost mounds have been the subject of several in-depth investigations (e.g., Shumskii 1964; see Williams 1965, p. 100, 103, 211-212).

Within this context, this paper documents the occurrence and internal structure of seasonal frost mounds in the North Fork Pass area of the Ogilvie Mountains, interior Yukon Territory (64°35'N, 138°18'W). At intervals between September 1980 and July 1982, the growth and decay patterns of more than 65 seasonal frost mounds were observed. The measurement of hydraulic pressures associated with mound activity is also reported.

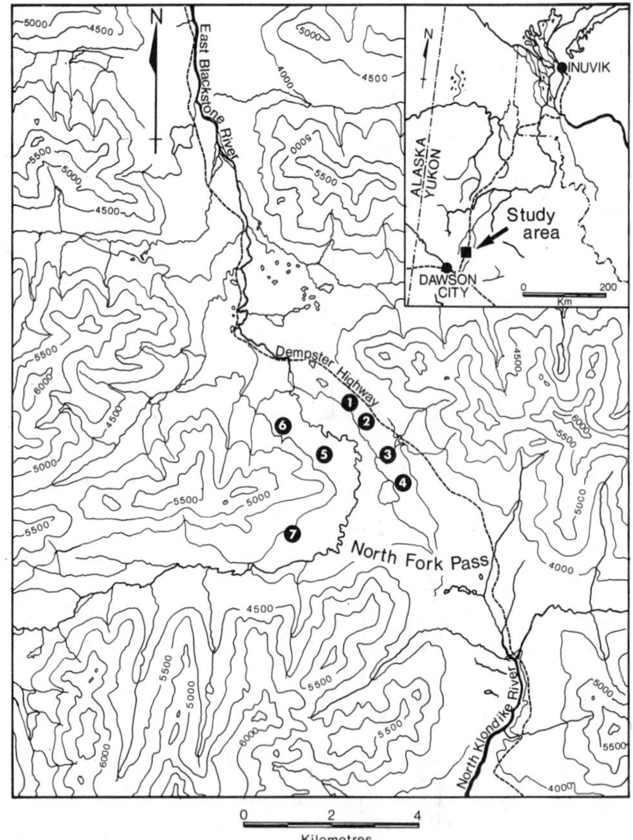

FIGURE 1 Location map of North Fork Pass, interior Yukon, showing sites (1-7) where seasonal frost mounds occur.

STUDY AREA

The seasonal frost mounds described here occur in several locations in the North Fork Pass area between kilometers 82 and 86 of the Dempster Highway (Figure 1). The elevation of the pass is 1,370 m a.s.l. The mean annual air temperature is approximately -8°C and average precipitation is 480 mm. The area is mapped by Brown (1978) as being in the most northern zone of discontinuous permafrost. However, according to borehole data (Foothills Pipe Lines [Yukon] Ltd. 1978), permafrost temperatures immediately north of the North Fork Pass are between -8°C and -5°C below an active layer which varies from 0.3 m to 2.0 m thick depending on site conditions. The boundary between continuous and discontinuous permafrost may be just south of the pass at an elevation of approximately 1,000 m a.s.l. (Harris et al. 1983).

Various landforms diagnostic of permafrost conditions are present in the vicinity of the pass. These include ice wedge polygons, poorly sorted circles and stripes, beaded drainage and small thermokarst ponds. Solifluction is widespread on valleyside slopes above 1,450 m a.s.l. Vegetation consists, for the most part, of shrub and tussock tundra in the valley bottom and alpine tundra on the valleyside slopes.

MORPHOLOGY AND SETTING

At least 7 localities of frost mound activity have been identified (see Figure 1), all of which are associated with cold mineralized perennial springs. The mounds occur at or below the break of slope between the valleyside and the floodplain of the East Blackstone River. The break of slope is hydrologically significant since it approximates the intersection point between the water table and the ground surface. Frost blisters usually form in organic-rich silts covered by a layer of peat and/or tundra vegetation (Figure 2a) while icing mounds and blisters (Figure 2b) form in the icings that result from winter overflow caused by spring discharge. In all localities, alluvial fan deposits, formed where small tributary channels join the main valley, cover the lower valley slopes and serve as primary aquifers. The angles of slope of all sites range between 2° and 10°. Since the mounds occur on both sides of the valley, aspect does not appear to be a critical factor in their distribution. However, aspect may influence the timing of mound activity, since freezeback occurs earlier on northeast facing slopes. This limits the availability of groundwater and may explain why the largest icings occur on the northwest side of the valley (sites 1, 3, and 4; see Figure 1). The largest mounds, by contrast, occur at sites 5 and 6 on the northeast side of the valley (see Figures 1 and 2a), and are associated with the steepest local relief (i.e., highest hydraulic potentials).

FIGURE 2 Photographs illustrating the range and variety of seasonal frost mounds, North Fork Pass, interior Yukon. (a) Frost blister, 3.5 m high on west side of North Fork Pass, March 15, 1982. (b) Icing blister, approximately 200 m downslope from km 86, Dempster Highway, March 13, 1982. The blister is 1.7 m high and associated icings extend to the nearest tributary of the East Blackstone River. (c) Section through frost blister showing ice core and internal stratigraphy, September 25, 1980. (d) Remnants of frost blisters at same locality as Figure 2(b) but observed September 20, 1981.

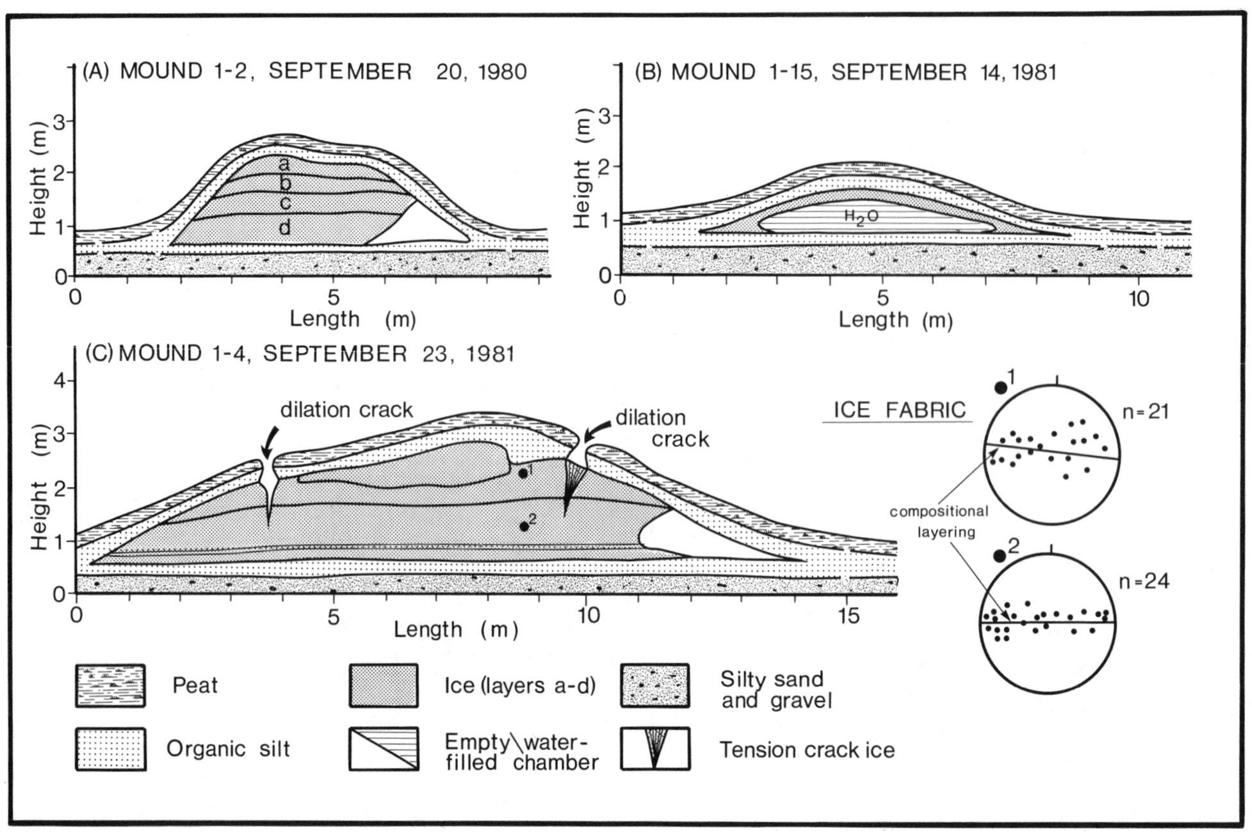

FIGURE 3 Diagram showing the internal stratigraphy of three frost blisters. (A) Frost blister 1-2. (B) Frost blister 1-15. (C) Frost blister 1-4. Ice fabric diagrams from vertical thin sections are plotted on Schmidt equal area, lower hemisphere projections. For description of stratigraphy, see text.

Mound heights range from 0.5 m to 3.5 m while long-axis basal diameters range from 3 m to 70 m, often oriented parallel to the local slope. In general, the seasonal frost mounds of the North Fork Pass area are smaller than those described elsewhere (e.g., Lewis 1962; van Everdingen 1978, 1982). On the other hand, their frequency of occurrence exceeds all others previously cited.

In plan the mounds may be either circular, elongate, lobate, or complex. The mound profile is frequently modified by dilation cracks up to 35 cm wide, and by slumping or collapse caused by melting of the ice core (see Figures 2c and 2d). Scattered lumps of organic material observed incorporated within the icings indicate that explosive rupture may have occurred in some instances. Frost mounds were observed to survive the summer with minimal degradation or thaw. Some were reactivated the following winter; in one instance, reactivation resulted in a doubling of both mound height and length.

INTERNAL STRUCTURE

Stratigraphic investigation, along with detailed process studies, provides one method of distinguishing between seasonal frost mounds and other morphologically similar frost mound phenomena such as palsas.

Frost mound stratigraphy was determined either mechanically using a motor-driven coring unit and light drilling equipment, or manually by excavating along dilation cracks and areas of collapse.

The typical mound stratigraphy is summarized by the following generalized sequence. This is also illustrated in Figure 3 which shows the internal structures of three mounds located at site 1 (see Figure 1). All are covered by a surface vegetation mat which, during winter, is also covered by varying thicknesses of packed snow or icing ice. Beneath the surface organic layer is a thin (10-50 cm thick) layer of peat grading into organic-rich silt. This contact between the ice and silt is sharp. In many mounds, the ice is layered and includes bands of gas bubbles and some sediment inclusions. Gas inclusions range from small and spherical to vertically oriented and elongate bubbles, as much as 1 cm in diameter and 3-4 cm long. In some small blisters less than 1.0 m high, the mound is solid; the ice core forms an epigenetic ice body unconformably overlying frozen ice-rich sediments. By contrast, the larger frost blisters have cores consisting of one or more layers of clear ice which are usually arched over a water-filled chamber. Typically, the thickness of ice varies between 10 and 50 cm and the height of the chamber ranges from 20 to 70 cm. The horizontal dimensions of the ice core are difficult to document as are the sizes and shapes of the interior water chambers. Based on the drilling and sectioning undertaken, it appears that the shape of the ice core resembles the surface expression of the mound, while the water chamber is low and lens shaped. During winter drilling, water encountered in the chamber was found to be clear and sometimes slushy; during the summer the water was murky and dirty. On several occasions in summer drilling an empty chamber was observed, where frost mound rupture the previous winter had presumably resulted in drainage. The sediment located beneath the ice core consists of in situ layers of frozen organic silt and silty sand with gravel. The bottom of the ice core probably correlates with the top of the frost table at the time the mound formed. Frost blisters 1-2 and 1-15 (see Figure 3) are typical of this general stratigraphy.

Frost blister 1-4, however, incorporates at least two ice-rich sediment layers in the ice core (see Figure 3). These separated ice of visibly different crystal and bubble characteristics. This mound was observed to experience reactivation during the 1980-1981 winter, following its initial formation the previous winter. The sequence of sediment-rich and clear ice, present in the lower part of the core, may represent a multiple cycle of mound growth where water with a high suspended sediment content was injected either between existing layers of ice or into a nearly totally frozen ice core.

Ice fabric studies (see Figure 3) also suggest that the mounds result from the injection of water rather than ice segregation, as might be expected if they were in palsas. Ice crystal orientations were measured using a four-axis universal stage (Langway 1958). Two fabric patterns are differentiated: first, a chill zone of small ice crystals with c-axis orientations forms a loose girdle parallel to compositional layering and perpendicular to growth direction; and second, the fabric is characterized at depth by large (0.5-2.0 cm diameter) vertically oriented long columnar crystals formed parallel to their growth direction. A preferred c-axis orientation normal to the growth direction was also observed. This fabric pattern is similar to that described for lake ice or injection ice (Gell 1978; Shumskii 1964) and might be expected in the freezing of a closed water-filled chamber.

HYDRAULIC CONDITIONS

Although the origin of these features is well known in general terms (e.g., Muller 1945, p. 59-60; van Everdingen 1978, p. 273-274), the details of their growth mechanism will normally reflect site-specific hydrologic conditions.

During drilling of several mounds in September 1981, it was noted that when the water-filled chamber was penetrated, trapped gases were released. On at least one occasion, sufficient pressure was present to temporarily force water 15 cm up the borehole. Presumably, these water pressures represented residual pressures associated with the hydraulic potentials necessary for mound growth the previous winter. Accordingly, piezometers were inserted in several actively growing mounds during March 1982. The piezometer consisted of a PVC standpipe containing silicone oil which was closed at the top and had a flexible rubber diaphragm at the bottom. A simple pressure meter was attached to the top. Although any gas pressures contained within the mound were released upon penetration of the cavity by the drill, the insertion of the piezometer enabled measurement of any hydraulic potential that subsequently developed. A foam collar, attached to the piezometer, ensured a complete seal. Maximum values recorded over a 7-day period in four frost mounds, all between 1.0 and 2.0 m high (Figure 4), ranged between 40 and 80 kPa. Highest potentials were recorded in the two icing blisters (1-23 and 1-24) while the two frost blisters (1-25 and 1-26) were significantly lower. This may reflect the different tensile strengths of the confining materials but equally may indicate different hydraulic gradients at the four localities.

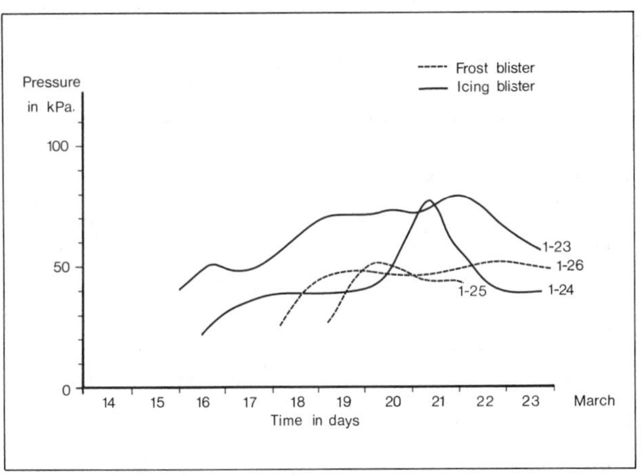

FIGURE 4 Diagram showing hydraulic potentials recorded at four seasonal frost mounds, North Fork Pass, March 16-23, 1982.

DISCUSSION

Similar measurements of hydraulic potentials associated with seasonal frost mounds have not been reported previously in the literature, although V. G. Petrov (see Williams 1965, p. 160) attempted to deduce the pressures involved via the depression of the freezing point of water inside a mound. However, van Everdingen (1982), using assumed densities of confining materials at Bear Rock, N.W.T., inferred that the minimum hydraulic potential necessary to form a 2.9 m high mound was 35.5 kPa, and for a 4.9 m high mound was 56.8 kPa. Although these calculations do not consider the planimetric dimensions of the mounds, if it is assumed that there is a general relationship between mound height and overall dimensions, it would appear that the maximum potentials recorded at the North Fork Pass exceed the minimum values assumed to operate at Bear Rock. At North Fork Pass, using similar values to those of van Everdingen (1982), a frost mound 2.0 m high (e.g., 1-26) would require a minimum pressure of 38.7 kPa. Since we observed pressures as high as 52 kPa at this mound, we assume that the difference reflects the resistance to deformation of the confining materials. We do not think that higher hydraulic pressures exist in the North Fork Pass area, since the maximum dimensions of the mounds and the discharge of the perennial springs are considerably less than at Bear Rock. Instead, the presence of peat and organic-rich silts in the North Fork Pass are not conducive to the retention of high hydraulic potentials because the threshold values for mound rupture will be relatively low. On the other hand, the number of frost mounds exceeds that of the Bear Rock locality. It may be that the generally shallower active layer in the North Fork Pass, the dispersed nature of the springs, and the variability of freezeback at the high elevation all favor the formation of small, but numerous, mounds.

CONCLUSIONS

The occurrence of seasonal frost mounds is relatively common in the North Fork Pass area. They reflect suitable site-specific conditions of groundwater discharge, climate and the presence of permafrost as an impermeable lower boundary to groundwater movement. Stratigraphic studies indicate that the mounds possess an internal structure which differentiates them from palsas. The ice core, as illustrated by ice fabric and isotope analyses, indicates injection and freezing of water under conditions of high hydraulic potentials.

The direct measurement of the hydraulic potentials associated with seasonal frost mound activity is possible using a simple piezometer apparatus. Although more measurements are required from a wider range of localities, the data presented in this paper suggest that potentials of between 40 and 80 kPa are necessary to form mounds 2.0 m high. Large mounds probably require higher potentials.

ACKNOWLEDGMENTS

Research was funded by the Natural Sciences and Engineering Research Council grant A-8367 (H. M. French) and the University of Ottawa Northern Research Group. Drilling equipment was loaned from the Terrain Sciences Division, Geological Survey of Canada, courtesy of O. L. Hughes, and the National Hydrology Research Institute, Environment Canada, courtesy of R. O. van Everdingen. Numerous helpful discussions with R. O. van Everdingen, J. Banner, O. L. Hughes, and F. Michel are much appreciated.

REFERENCES

Brown, J., and Berg, R. L., eds., 1980, Environmental engineering and ecological base investigations along the Yukon River-Prudhoe Bay Haul Road: Cold Regions Research and Engineering Laboratory, Report 80-19, 187 p.

Brown, R. J. E., 1978, Permafrost, Plate 32, Hydrological Atlas of Canada: Fisheries and Environment Canada, Ottawa.

Foothills Pipe Lines (Yukon) Ltd., 1978, Dempster Lateral Drilling Program: Consultants' Report. Klohn Leonoff Consultants Ltd., 2 volumes, 533 p.

French, H. M., 1971, Ice cored mounds and patterned ground, southern Banks Island, western Canadian Arctic: Geografiska Annaler, 53A, p. 32-38.

Gell, W. A., 1978, Fabrics of icing-mound and pingo ice in permafrost: Journal of Glaciology, 20 (84), p. 563-569.

Harris, S. A., van Everdingen, R. O., and Pollard, W. H., 1983, The Dempster Highway: Dawson to Eagle Plain, in H. M. French and J. A. Heginbottom, eds., Fieldtrip B-3, Northern Yukon Territory and Mackenzie Delta--Fourth International Conference on Permafrost, Fairbanks, Alaska, in press.

Hughes, O. L., Rampton, V. N., and Rutter, N. W., 1972, Quaternary geology and geomorphology, Southern and Central Yukon: XXIV International Geological Congress, Montreal, 1972, Guidebook, Field Excursion A-11, 59 p.

Langway, C. C., 1958, Ice fabrics and the universal stage: United States Army CRREL, Technical Report 62, 16 p.

Leffingwell, E. de K., 1919, The Canning River Region, northern Alaska: United States Geological Survey, Professional Paper 109, 251 p.

Lewis, C. R., 1962, Icing mound on Sadlerochit River, Alaska: Arctic 15 (2), p. 145-150.

Maarleveld, G. C., 1965, Frost mounds; a summary of the literature of the past decade: Mededelingen van de Geologische Stichting: Nieuwe Serie, No. 17, 16 p.

Muller, S. W., 1945, Permafrost or permanently frozen ground and related engineering problems: Special report, Strategic Engineering Study No. 62, Military Intelligence Division Office, Chief of Engineers, U.S. Army, 231 p.

Porsild, A. E., 1938, Earth mounds in unglaciated arctic Northwestern America: Geographical Review 28 (1), p. 46-58.

Sharp, R. P., 1942, Ground-ice mounds in tundra: Geographical Review 32 (3), p. 416-423.

Shumskii, P. A., 1964, Ground (subsurface) ice: National Research Council of Canada, Technical Translation 1130, Ottawa, 118 p.

Toth, J., 1972, Properties and manifestations of regional groundwater movement: International Geological Congress, 24th Session (Canada), Section 12, p. 153-163.

van Everdingen, R. O., 1978, Frost mounds at Bear Rock, near Fort Norman, Northwest Territories, 1975-1976: Canadian Journal of Earth Sciences, 15, p. 263-276.

van Everdingen, R. O., 1982, Frost blisters of the Bear Rock area, Fort Norman, N.W.T.: Arctic 35 (2), p. 243-265.

van Everdingen, R. O., and Banner, J. A., 1979, Use of long-term automatic time-lapse photography to measure the growth of frost blisters: Canadian Journal of Earth Sciences 16, p. 1632-1635.

Williams, J. R., 1965, Ground water in permafrost regions--an annotated bibliography: United States Geological Survey Water Supply Paper 1792, 295 p.

Zoltai, S. C., and Johnson, J. D., 1978, Vegetation-soil relationship in Keewatin District: Environmental-Social Program A1-25 Canadian Forestry Service; Fisheries and Environment Canada, Ottawa, 95 p.

CRYOLITHOGENESIS AS A TYPE OF LITHOGENESIS

A. I. Popov

Faculty of Geography,
Moscow State University, USSR

The paper examines a special, climatically controlled type of lithogenesis, namely cryolithogenesis. This zonal type of lithogenesis is dictated by the processes of cryodiagenesis and cryohypergenesis, occurring specifically in the polar and subpolar zones. Typical cryolithogenic products of cryodiagenesis and cryohypergenesis are distinguished, along with the corresponding peculiarities of their geographical distribution. Some critical comments are presented with regard to certain alternate existing concepts of cryolithogenesis.

Recognition of cryolithogenesis as an independent climatic type of lithogenesis was a logical consequence of elucidation of the role of cryogenic processes in the earth's crust (Popov 1967). The variety of interpretations of the concept of cryolithogenesis in the literature, however, requires a return to this subject.

Lithogenesis is not only the formation of sedimentary rocks, but also the formation of weathering crusts; all transformations of primary and secondary rocks under subaerial and subaqueous conditions, i.e., under thermodynamic conditions at shallow depths, characterized by low temperatures and low pressure, should be attributed to it. All the processes of exogenic rock-formation under the effect of sedimentogenesis, diagenesis and weathering are thus processes of lithogenesis. The intensity and the zonal character of the processes of lithogenesis are determined by physical-geographical conditions that vary from one natural zone to another.

Catagenesis, metagenesis and metamorphism correspond to different thermodynamic conditions characterized by high temperatures and higher pressure. Such conditions are caused by endogenic factors and are not subject to natural zonation.

Because cryogenic processes are exogenic processes, cryolithogenesis is a part of general lithogenesis.

The cryogenic processes should be regarded as processes of diagenesis and weathering which develop under the near-surface thermodynamic conditions; they have no continuation under the thermodynamic conditions of catagenesis, the subsequent stage of transformation of earlier-formed rocks.

Cryolithogenesis is active in polar, subpolar and high mountain regions of both the Northern and Southern Hemispheres.

Arid lithogenesis develops under dry conditions and humid lithogenesis under sufficiently wet conditions; cryolithogenesis occurs under the action of ice and water that changes its phase periodically. According to all data available, the ice type of lithogenesis (Strakhov 1960) should be regarded as cryolithogenesis (Popov 1967).

Diagenesis should be understood not only as the physical-chemical equilibration of sediment in a subaqueous medium (Strakhov 1960), but also as the physical-mechanical equilibration in both the subaqueous and subaerial media, viz., compaction, relative dehydration, and squeezing out of free water and migration of loosely-bound water (Shvetsov 1958, Rukhin 1969, Shantser 1966, Lomtadze 1970, and others). Consequently, cryogenic processes mainly responsible for physical-mechanical transformations in sediments and rocks and contributing to their long-term consolidation should be understood as diagenesis (cryodiagenesis). Under cryolithogenesis, the thermodynamic factor of diagenesis, i.e. water crystallization, suppresses the physical-chemical equilibration but does not exclude it altogether.

Cryodiagenesis is the long-term, irreversible freezing of moist sediments and dispersed rocks; ice forms in these substrates as an authigenic mineral. Cryodiagenesis influences the aleuropelitic series of these sediments and rocks, which is expressed in relative dehydration of mineral aggregates during migration of both free and loosely-bound water, in their intravoluminal contraction and the formation, as a rule, of various types of ice streaks which form the characteristic cryogenic textures of a diagenetized perennially frozen rock. No appreciable changes occur in the mechanical makeup of the sediment or rock. Cryodiagenesis of coarsely dispersed rocks (sands and the like) is a simpler process associated with fixation of freezing free water in situ, with forcing it out of the freezing front, with formation of a massive cryogenic texture or centers of ice-formation that do not markedly disturb the structure of coarse sediments, and also with retaining the initial mechanical composition of coarse clastic and other materials. The formation of massive wedge-ice lattices in dispersed rocks during the successive frost cracking should also be referred to the category of phenomena of cryodiagenesis. Also included here should be the formation of the ice cores of pingos and palsas.

As the gravitational compaction of a dispersed rock develops during its subsidence and dehydration, the effects of cryodiagenesis cease at the stage of complete disappearance of loosely bound water whose presence and ability to migrate represents the most important factor of cryodiagenesis. The freezing of a moist aleuropelitic rock that underwent general diagenesis and had not frozen previously, but had become only moderately consolidated, retaining the loosely-bound water capable of cryogenic migration, is a process of diagenesis with all the mentioned cryotextural aftereffects. The freezing of such a rock is not a transition to the stage of catagenesis; it is only a pause in the common diagenesis of the rock, during which originate new, cryodiagenetic processes whose textural features superimpose upon the textural features of the preceding pre-cryogenic diagenesis. Thus, cryodiagenesis can manifest itself as the primary diagenesis when sediments, which did not undergo diagenesis before, freeze, but it can also act as secondary, superimposed diagenesis, when rocks freeze which did undergo general diagenesis but which did not become heavily consolidated and retained loosely-bound water capable of cryogenic migration (Popov 1967).

Cryodiagenesis also manifests itself by similar cryotextural effects in longterm, irreversible freezing of unconsolidated moist clayey and coarser weathering deposits. In this case, cryodiagenesis appears as posteluvial diagenesis.

Clastic formations stand out as products of cryogenic weathering in cold regions of the earth; they result from physical disintegration of crystalline, metamorphic and consolidated sedimentary rocks. Silty formations also form through cryogenic weathering as an ultimate result of disintegration of the sand and coarser fraction and aggregation of the clay fraction.

Cryogenic weathering is manifested most fully by systematic, seasonal or diurnal freezing-thawing of rocks. Two basic mechanisms of cryogenic destruction of rocks are then taking place: (1) cryogenic weathering proper as a result of the wedging action of ice in relatively large fissures and pores, which causes the formation of coarse detritus (cryoclastites, such as blocks, rubble and coarse sand), and (2) cryohydrational weathering (Konishchev 1977) due to the variation in pressure in thin water films in microfissures, which is the result of phase changes which cause the formation of the silty and fine-sandy fraction (cryopelite). Cryopelite is the end product of cryogenic weathering.

Cryogenic weathering resulting in the formation of cryoclastites and cryopelites is restricted to the horizon of seasonal freezing-thawing. The epigenetic freezing of crystalline, metamorphic and compact sedimentary rocks (sandstones, aleurolites, argillites, limestones), when a perennially frozen rock is being formed, is also, in the end, a process of destruction. During this process, mainly free water freezes in fissures due to tectonic processes or denudation that relieve the rock mass of its load. The subsequent possible thawing of these rocks (perhaps in a short period of time or in hundreds or thousands of years) would necessarily lead to their destruction and disintegration. Therefore the process of freezing of compact bedrocks should be regarded as a process of the initial croygenic weathering and also as a process that prepares these rocks for destruction; such perennially frozen rocks should be called "rocks of cryogenic pre-destruction."

The process of nivation is a process of cryogenic weathering. It occurs on rock surfaces under the conditions of diurnal changes of phase (ice-water) in the peripheral, intensively moistened zone in the immediate vicinity of snowpatches and glaciers. Silty nival fine material is the product of nivation.

The regelational type of weathering occurring at the contact of a glacier bed and the underlying rock is an independent type of cryogenic weathering. This type of weathering produces characteristically a so-called regelational, fine silt (Moskalevsky 1978) which constitutes the bulk of the fine material in ground moraines. While glaciers are second, after rivers, in the volume of load (Gurrels and Mackenzie 1974), the cryogenic process should be considered as one of the most important factors in the complex geological activity of glaciers.

Cryogenic weathering (cryohypergenesis) is diverse, and involves at least four types.

On the basis of the preceding discussion, the products of cryodiagenesis are the following: (1) polymineral dispersed ice-rich rocks, i.e., cryolitites, and (2) monomineral ice rocks (veined, injectional, etc.), i.e., cryolites. The products of cryogenic weathering, i.e., of cryohypergenesis, are as follows: (1) rocks of cryogenic pre-destruction, such as perennially frozen masses of crystalline and other compact rocks; (2) cryoclastites, i.e., initial products of cryogenic destruction, such as blocks, rubble and sand; and (3) cryopelites, i.e., end products of cryohypergenesis, namely, silty loessial fine material. The products of cryohypergenesis are components of a distinctive cryogenic crust of weathering.

A number of authors substantiated the delimitation of "polar lithogenesis" (Lapina et al. 1968, Danilov 1978); the conditions of sedimentogenesis in polar seas are the main criterion, while the processes of lithogenesis under continental conditions are considerably less important.

The well-known specificity of sedimentation in the seas of the cold zone takes place more appreciably on the shelf than outside it. However, Strakhov (1976) concluded that there was no single polar type of lithogenesis that would include both polar land and seas. The lithogenesis of sediments over the major part of polar waters is mainly the lithogenesis of an oceanic humid type that does not differ essentially from the oceanic type of lithogenesis at middle latitudes.

In conclusion, only the direct influence of ice or of systematic phase changes on sediments and rocks determines the classification of a territory or a water area under cryolithogenesis. The indirect effects of floating ice on sediments (transport of boulders), the presence of salt brines, and other such phenomena without direct influence of ice on sediments and rocks cannot be regarded as factors of cryolithogenesis.

A definition of cryolithogenesis as a sedimentary process inherent to the cold regions of the earth has been proposed (Gasanov 1976). Under that definition, all cryogenic processes that are not immediately associated with sedimentation are not regarded as components of cryolithogenesis, but are understood as phenomena of cryogenesis.

In my opinion, the restriction of cryolithogenesis to participation of cryogenic factors in the process of sedimentation, i.e., in the formation of syngenetic frozen rock masses, is erroneous because a number of links within the complex process of cryolithogenesis such as cryogenic weathering, freezing of bedrock, cryodiagenesis of unconsolidated or poorly consolidated rocks with epigenetic freezing are omitted.

It is only through consideration of the climatically preconditioned specificity of diagenesis and weathering that one can determine properly the type of lithogenesis, in our case, cryolithogenesis, including cryodiagenesis and cryogenic weathering (cryohypergenesis).

Researchers regard cryolithogenesis as a sedimentary process because of the strict succession between the areas of denudation, transportation of material, and its accumulation in the terminal runoff basins. This approach is permissible when this succession is well developed, i.e., when there is a causal interrelationship between the areas of drift, transportation and accumulation. However, as we know now, such succession and interrelationship is very inextensive within the zone of cryolithogenesis. Also, the manifestation of cryolithogenesis in time is relatively restricted—from the end of the Pliocene to the Holocene inclusive.

Based on sufficiently extensive field data, I have concluded that cryogenic rocks and their complexes are formed under the conditions of a limited group of processes inherent in the areas of predominant denudation, predominant accumulation, and relative stabilization of the surfaces undisturbed by drift or material accumulation (Popov 1958). Each specific area is characterized by its own complex of cryogenic rocks, which may fail to reach into another area. On the Taymyr Peninsula, for example, under the conditions of denudation and downslope material movement, deposition of the displaced products of cryogenic weathering and their local accumulation took place on slopes and at the slope bases, in depressions and in small river valleys during long periods of time corresponding to the major time periods of the Pleistocene. There are pertinent examples referring to the other two mentioned areas, viz., areas of predominant accumulation and of relative stabilization of accumulation and drift of cryogenic material.

Apparently, the conclusion is true that lithogenesis in general and cryolithogenesis in particular can be realized both successively in stages from the drift sources to the terminal runoff basins and, to a considerable degree, through closed development within the areas of predominant denudation, predominant accumulation, and relative stabilization of surfaces without drift or accumulation or complete continuity between them.

Under the conditions of accumulation (in large river valleys, in deltas and in the sea), the self-containment of the type of formation of cryogenic rocks and its relative independence of the adjacent areas of denudation and drift of the cryogenic weathering products are particularly expressive. The following materials enter such accumulation areas: material formed cryogenically on water divides, products of river-washout of the Pre-Quaternary and Quaternary "non-cryogenic" rocks, considerable percentage of the present-day organic materials which are mainly vegetative, part of sediment runoff, and non-volcanogenic and volcanogenic eolian materials. All these become components of the deposits of the terminal runoff basins and major river valleys as a sum of non-cryogenic and cryogenic products, and it is far from being clear which of these components will prevail in each particular case. This is greatly confusing for those researchers who are studying only products of cryolithogenesis in marine sediments of the polar basin.

Observations showed that during both subaqueous and subaerial sedimentation in the Arctic, cryogenic weathering in a thin active layer (0.2 to 0.5 m) only weakly "reprocesses" the accumulated material; cryodiagenesis fixes its initial composition and changes only its mechanical properties and the texture due to ice segregation. Therefore the mixed composition of such a cryogenically diagenetized rock may show few or no traces of previous cryogenic weathering at the stage of denudation or at the place of accumulation. The areas of accumulation and the local areas of cryogenic denudation in the Arctic and Subarctic lack direct and close continuity.

Fully independent cryogenic, mainly cryo-eluvial rock formation proceeds under the conditions of relative stabilization of the surfaces not subject to drift or material accumulation; there is often no input or output of material underway in these areas. This is illustrated by considerable areas of some plateaus, vast drainage divides of the great lowlands, and old terraces of the major river valleys. Under such conditions, the exogenic process at the stage of cryogenic weathering with a thin cryogenic weathering crust is clearly autonomous; an example are loessial formations of loessial mantle loams.

The closure of processes and products of cryolithogenesis within the areas of denudation and drift, and the well-known autonomy of cryogenic exogeny under the conditions of denudation and slope migration were already discussed.

The rate of the processes of formation of the products of cryogenic denudation and cryogenic weathering crust should also be taken into account. Contrary to the rather widely accepted high rates of denudation and, in general, high rates of development of exogenic processes in permafrost and tundra regions, the data available show irrefutably that in reality the areas of stable cooling are characterized by relative retardation of these processes and the high rates are an exception. It must be concluded that one cycle of cryogenic denudation, transportation and accumulation can sometimes extend over a period of time corresponding to the Pleistocene, i.e., over the entire epoch of the existence of permafrost and the climate under which cryolithogenesis occurs. Hence the inertness and the relative conservatism of materials that form are moved and accumulate in the course of cryolithogenesis, and the long time they take to reach the terminal runnoff water basins.

The epigenetic freezing of both Pre-Quaternary and Quaternary rocks should be mentioned; this is a process bearing no relation to either the immediate cryogenic denudation, drift, or accumulation.

As was already shown, this is either a process of cryodiagenesis or of the initial cryogenic weathering, i.e., cryogenic pre-destruction that can operate for a relatively long time.

It was established earlier that cryolithogenesis occurs in the area of stable cooling of the earth; this area coincides with the permafrost regime. But it is known that cryolithogenesis is manifested also in the areas of only seasonal freezing, outside the limits of permafrost occurrence. It is displayed here only as cryogenic weathering. This is the area of predominantly humid lithogenesis, and cryolithogenesis is active here only under certain conditions. This zone with a combination of the processes of humid and cryogenic lithogenesis should be probably named a cryohumid zone and referred to the area of unstable cooling.

The recognition of cryolithogenesis as an independent type of lithogenesis (along with the arid and humid types) is supported by the following circumstance. While in the zones of arid and humid lithogenesis the major lithogenesis agent is water and the product is a solid mineral constituent (transformed sediments and materials of weathering crust), in the zone of cryolithogenesis the major lithogenesis agent is ice, and the main product is also ice which is accompanied by important solid mineral products of cryolithogenesis. This unity of agent and product places the zone of cryolithogenesis into a special position; the cryohumid zone with its exclusive cryogenic weathering is an exception.

Ice formation imparts clearly expressed individual features to sediments and all rocks and preconditions their originality in terms of lithology and facies. Freezing accompanies other geological processes in the cold zone, such as erosion, water accumulation, etc. To a considerable degree, freezing directs all exogenic geological processes along a new, specific path. Besides, freezing fixed as permafrost represents a limit for the usual subsequent processes of rock alteration at positive temperatures, e.g., for diagenesis and catagenesis.

It should be noted that frozen rocks and underground ice do not enter any of the general lithological classifications, which by itself is an indication of their special, exclusive position. Their seeming ephemerality and relative instability under the thermodynamic conditions of the earth cannot be the reason for such exclusion. Under the conditions of the Arctic, in many regions of the Subarctic and in the Antarctic, many underground ice formations remain relatively unaltered during tens and hundreds of thousands of years and have undoubtedly a right to be singled out as a special group of rocks. The cryogenic processes causing most appreciable changes in the zone of hypergenesis of the cold regions of the earth acted during the entire Upper Cenozoic era upon considerable areas of dry land in the cold stages of the Pleistocene.

REFERENCES

Danilov, I. D., 1978, The polar lithogenesis: Moscow, Nedra.

Gasanov, S. S., 1981, A cryolithological analysis: Moscow.

Gurrels, R., and Mackenzie, F., 1974, The evolution of sedimentary rocks: Moscow, Mir.

Konishchev, V. N., 1977, Some general regularities of the transformation of dispersed rocks by cryogenic processes, in Problems of Cryolithology, vol. VI: Moscow University Press.

Lapina, N. N., et al., 1968, Marine sedimentogenesis in the polar regions of the Earth, in Physical and Chemical Processes and Facies: Moscow, Nauka.

Lomtadze, V. D., 1970, Engineering geology, engineering petrology: Leningrad, Nedra.

Moskalevsky, M. Yu., 1978, On the role of cryogenic factor in the formation of ground-moraine deposits (under the conditions of glaciation of the Severnaya Zemlya), in Problems of Cryolithology, vol. VII: Moscow University Press.

Popov, A. I., 1958, The polar surface complex, in Problems of Physical Geography of the Polar Lands, vol. I: Moscow University Press.

Popov, A. I., 1967, Cryogenic phenomena in the Earth's crust (cryolithology): Moscow University Press.

Rukhin, L. B., 1969, The foundations of lithology. The science of sedimentary rocks: Leningrad, Nedra.

Shantser, E. V., 1966, Essays on the science of the genetic types of continental sedimentary formations: Moscow, Nedra.

Strakhov, N. M., 1960, The principles of the theory of lithogenesis, vol. I: Moscow, Academy of Sciences of the U.S.S.R. Publishers.

Strakhov, N. M., 1976, An approach to the problem of the types of lithogenesis in the oceanic sector of the Earth, in Lithology and Mineral Resources, vol. 6: Moscow.

THE EFFECT OF HYDROLOGY ON GROUND FREEZING IN A WATERSHED WITH ORGANIC TERRAIN

Jonathan S. Price

Department of Civil Engineering, University of Saskatchewan
Saskatoon, Saskatchewan, Canada S7N 0W0

This is an investigation into factors affecting seasonal frost penetration in organic and inorganic terrains including the effect of the water table on freezing. Five different terrain types were instrumented with frost gauges and snow stakes. Piezometers were used to record water table fluctuations, and soil thermistors were used to determine temperature gradient and heat flux. Frost penetration among sites, and variation of frost depth within sites was investigated. Where the water table was at the surface and considerable flow occurred through upper layers, circulation of water reduced the rate of freezing but caused cooling to occur more rapidly in deeper layers, but where little flow was observed, cooling of lower layers lagged considerably, resulting in more rapid frost penetration in upper layers. Freezing of positive and negative relief elements was different for bogs and fens. Here again, the elevation of the water table was largely responsible.

INTRODUCTION

An assortment of terrain types both organic and inorganic can occur within one watershed. In this paper the term "terrain" is used to define a geographical area typified by the soil type, vegetation and hydrologic characteristics (for example bog or fen). The characteristics of each terrain type can result in a variety of freezing patterns in response to similar macroclimatic stimuli. During winter, the hydrologic and freezing regimes of a particular terrain type are interdependent. Price and FitzGibbon (1982) have shown that in organic terrain where the water table is at or near the surface when freezing begins, the pattern of subsurface drainage and runoff are affected. They noted in both mineral terrain and organic soils with deeper water tables, freezing has little or no affect on local hydrology. Obviously, the reverse is also true. The presence of the water table at the surface will have a strong effect on freezing. Thermal conductivity is directly proportional to soil moisture. Higher specific heat of moist soils, however, demands greater energy loss for cooling and freezing. Thus peat which is moist but not saturated experiences rapid freezing (Tyrtikov 1959). The role of water overlying permafrost in organic terrain was investigated by Ryden and Kostov (1980) who found freezing rates in depressed (wetter) areas to be 20% slower than in areas which were elevated.

The freezing of sphagnum hummocks proceeds more quickly than freezing of adjacent depressions. Brown and Williams (1972) and Romanov (1961) found higher moisture content of depressions to be partly responsible. Romanov (1961) also found that the increased surface area of hummocks promoted a more rapid cooling. A similar pattern of freezing in ridge-trough microrelief of ribbed fen was noted by Romanov (1961). He found deeper frost in ribs (than in troughs) but with less ice (water) content, hence, equivalent "frozen mass."

In addition to the effect of hydrological conditions on freezing, the physical and mechanical properties of the soil and vegetation are important. Brown and Williams (1972) found mineral soils to have thermal conductivity roughly an order of magnitude greater than dry peat. This results in greater annual amplitude in soil temperature for mineral terrains. This paper is an investigation of the hydrological properties which affect the formation and penetration of ground frost in a watershed containing large areas of organic terrain.

THE STUDY AREA

A 17.4 km^2 watershed located in Nipawin Provincial Park, Saskatchewan (lat. 53° 55' N, long. 104° 45 W) was instrumented for the winter of 1981-1982. Located north of the 0° C mean annual isotherm, it is on the southern fringe of the zone of scattered permafrost. The mean annual precipitation is 440 mm (133 mm snow) (Fisheries and Environment Canada 1978). The area is gently to moderately rolling, and is underlain with a relatively impermeable glacial till which is blanketed by a 120-200 cm sand layer. No permafrost was encountered in this basin.

The watershed (shown in Figure 1) contains five distinct terrain types consisting of mineral terrain (Site 1), wet low-lying bog (Site 2a), sedge fen (Site 2b), dry high-moor bog (Site 3), and ribbed fen (Site 4). Organic terrains account for 35% of the area of the watershed. On well drained mineral terrain (sandy soil) the dominant vegetation is white spruce (Picea glauca) and jack pine (Pinus banksiana) with some aspen (Populus tremuloids). Black spruce (Picea mariana) and occasional balsam poplar (Populus balsamifera) dominate lower lying areas on mineral terrain. There is generally a 70 to 80% canopy cover and a ground cover of feather moss (Pleurozium schreberi). Open areas have scattered jack pine and are covered with a 5 cm mat of lichens.

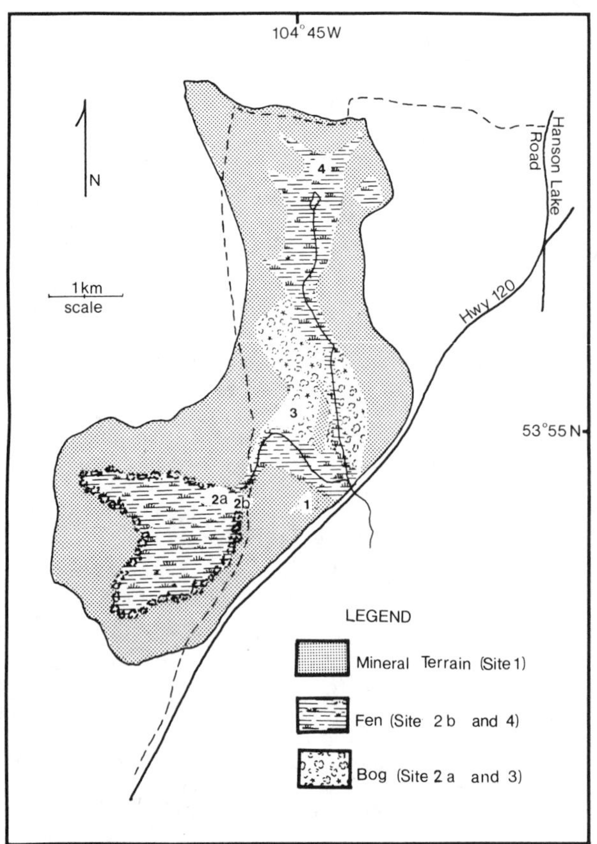

Figure 1 The study area.

The wet bog (Site 2a) has a poorly developed hummocky microtopography formed from sphagnum moss (Sphagnum fuscum) and is covered with Labrador tea (Ledum groenlandicum). It has 20 to 50% canopy cover composed of black spruce and tamarack (Larix laricina). Mean peat thickness is 165 cm. The water table was very close to the surface with small pools in some depressions. The dry bog (Site 3) has a 60-80% hummocky ground cover, comprised primarily of sphagnum mosses, with Labrador tea on the hummocks. Mean peat thickness is 140 cm, and the water table was 20-30 cm below the surface (of the depressions) at the onset of winter.

The open sedge fen (Site 2b) has low relief and is dominated by Carex (sedges) with some willow (Salix) and alder (Alnus) shrubs at the edges. The mean peat thickness is 170 cm and the water table was at or above the surface. Considerable flow was observed over the fen at high water levels during the autumn. Ribbed fen (Site 4) consists of alternating rib-trough microtopography, orientated perpendicular to the general direction of flow. Low Carex (sedges) grow in the troughs, with willow and alder shrubs and sedges growing on the ribs. Ribs are spaced 3-5 m apart and are 20-50 m in length and 10-20 cm in height. Mean peat thickness is 180 cm and the water table was at the surface of the troughs prior to freeze-up.

METHODS

A recording piezometer was installed in all sample areas except for the wet bog (Site 2a). The piezometer tube was constructed from a 10 cm diameter plastic pipe extending 1.5 m below the soil surface, and 0.5 m above. The above ground section of the piezometer was covered by an insulated housing (R 12 fiberglass insulation), and an ethylene glycol - methyl alcohol solution was poured into the tube with a small quantity of light oil (to prevent evaporation of the alcohol). This was largely successful in preventing the piezometer from failing due to freezing. A stage recorder in a wooden housing recorded the water level in the tube.

Each site had 10 frost gauges (Rickard and Brown 1972) and 10 snow stakes along a transect. At each station along the transect one frost gauge and one snow stake were located adjacently. Stations were located from 5 to 10 m apart or as specified below. On mineral terrain (Site 1), five frost tubes and five snow stakes were situated under the forest canopy, and five each in the open. On the high bog (Site 3) they were located alternately on hummock and depression, and similarly for rib and trough in the ribbed fen. A transect of 20 snow stakes and 20 frost gauges extended from the low bog (Site 2a) over the adjoining fen (Site 2b), oriented perpendicular to the direction of flow.

At each site one or two sets of thermistors were implanted in the soil at depths of 10, 60, and 120 cm to measure temperature. A mercury in steel probe and recording thermograph was also used in the sedge fen (Site 2b) with probes at 10 and 60 cm. Air temperature data were recorded at Site 3, and had to be supplemented with records from the LaRonge weather station (125 km to the north west) due to instrument malfunction. Correlation of LaRonge temperatures with on site temperatures was $r^2 = 0.98$.

RESULTS AND DISCUSSION

Depth and standard deviation of frost penetration are shown in Figure 2 for each terrain type. Considerable variation in frost depth was displayed within both the mineral terrain and the high (dry) bog. In both of these terrains, the water table remained below the freezing front (see Figure 3). Much less variability was observed for the fens and the low bog, where the water table elevation was at or above the surface at least during the first few months of winter, and the freezing front was in contact with saturated peat at all times. This is shown for the fens (Site 2b and Site 4) in Figure 3. It was found that movement of water through the saturated zone imported heat to the soil (Figure 4). On 17 October, which was the last day of positive mean daily temperatures for 1981, the temperature profile for the fens differed markedly from those of the mineral and bog terrains (Figure 4). In mineral and bog terrains, the soil was nearly isothermal. In the bogs, cool temperatures existed at depth because of slow heating over the

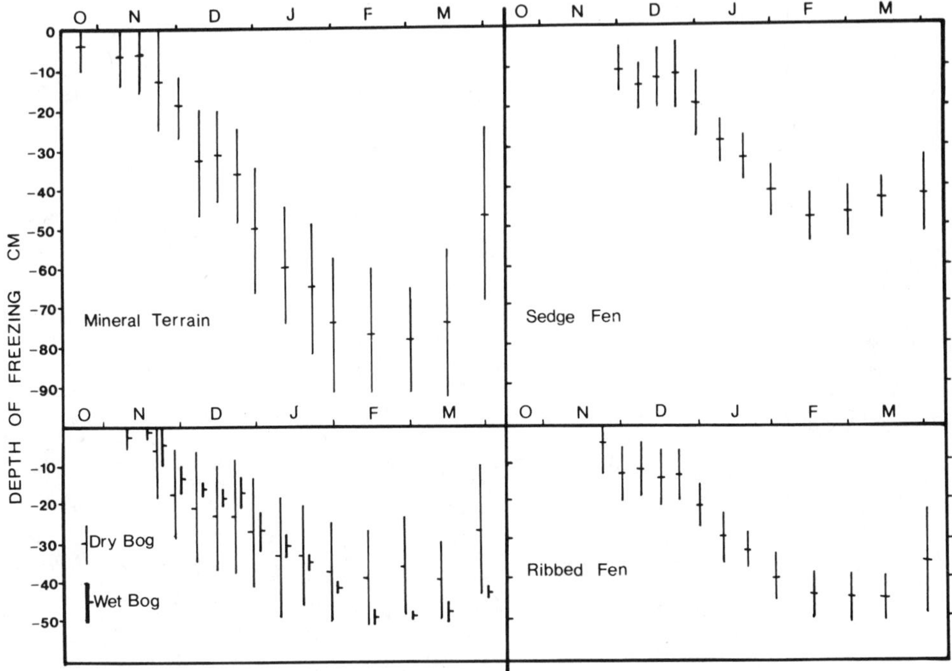

Figure 2 Depth and standard deviation of freezing of five different terrain types.

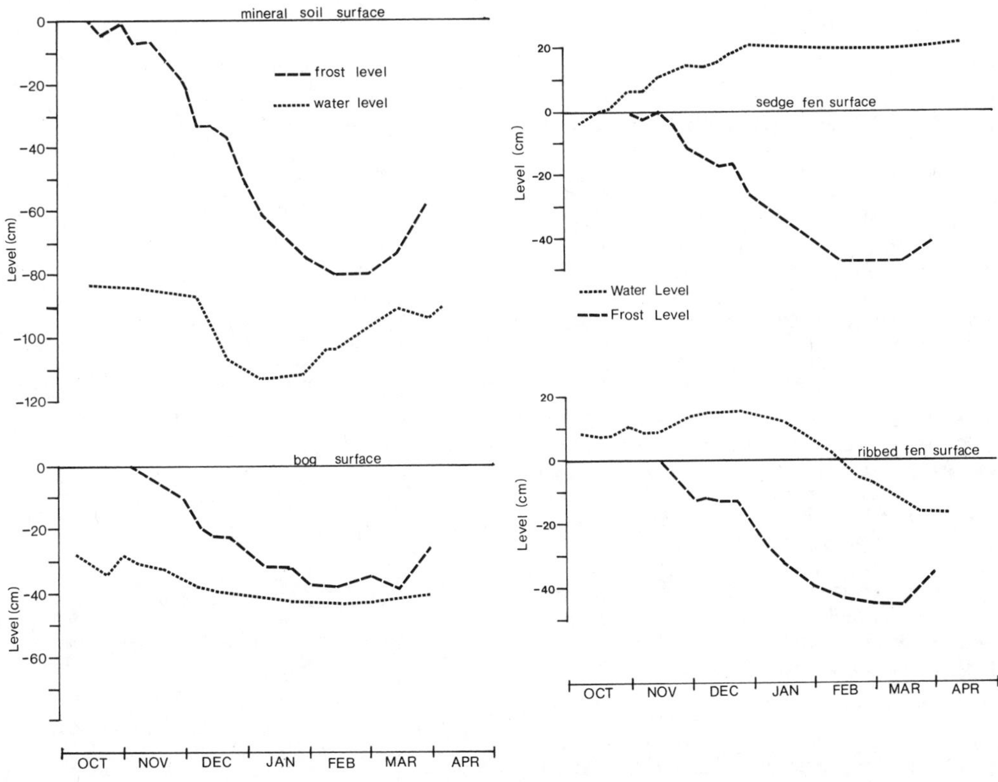

Figure 3 Elevation of water in the piezometer and depth of freezing for four terrain types.

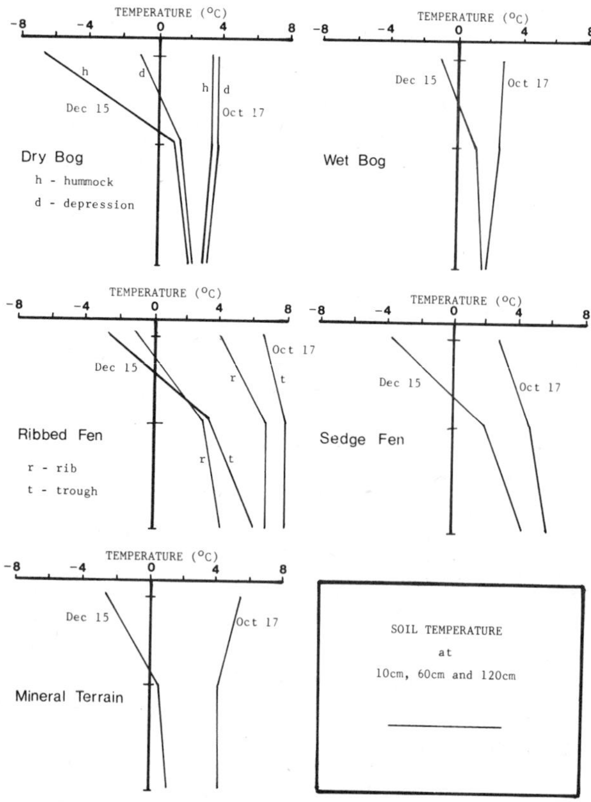

Figure 4 Ground temperature profiles for five terrain types on 17 October and 15 December.

summer. Insulative properties of surface mosses when dry, and slow drainage inherent to bogs minimized heat exchange. Water in the fens, on the contrary, experienced greater volumes of flow on and within surface layers and in the underlying mineral aquifer, allowing greater heat exchange than in the bogs. Additionally, the saturated fen surface provided little insulative cover, and the unforested condition permitted greater direct solar radiation receipt. Consequently, warmer temperatures prevailed in the fens, particularly at depth.

By 15 December, 436 negative degree-days had accumulated. From inspection of the temperature profiles shown for this date (Figure 4), it is evident that in the bogs most of the heat lost was from the surface layers, and cooling of deeper layers was slow, whereas in the fens, flowing water provided the mechanism for more rapid heat exchange. In the mineral site, large changes in surface and basal temperatures for the dates given (in Figure 4) reflect the greater thermal conductivity and lower volumetric heat capacity of mineral soils.

Frost penetration at each site was affected by the variable height of surface features with respect to the water table, vegetation characteristics, and snow accumulation patterns. Discussion of these effects is presented site by site.

Mineral Terrain

On the mineral terrain both frost depth and snow depth were found to be significantly different between forested and open sites. The difference between mean values were tested with t-test at a 95% significance level. Snow accumulation was 50% greater in the open area than under the dense black spruce due to canopy interception. Nevertheless, on mineral terrain, significantly greater frost penetration occurred in the open areas than in the forested location (see Figure 5). This is contrary to Baldwin's (1956) and Tyrtikov's (1959) suggestion that deeper frost occurs under dense coniferous forests. Although snow cover is normally important in reducing the depth of the frozen layer, the effect of the forest microclimate cannot be overlooked. Longwave radiation emitted by forest biomass is an important source of energy for the snowpack and ground in the forest ecosystem (Jeffrey 1968). Additionally, lower wind velocities in the forest reduces sensible heat exchange. FitzGibbon (1976) found windspeed in similar forest conditions near Schefferville, Quebec to be 7.06 km/hr lower than in the open. Here, a combination of these factors was sufficient to result in significantly less frost in the forested area. The difference in snow density between open and forested sites sampled was not significant (n = 10). The moss cover in the forest appeared to have some influence on frost penetration, with less frost forming beneath moss covered soil than in ground covered by lichen.

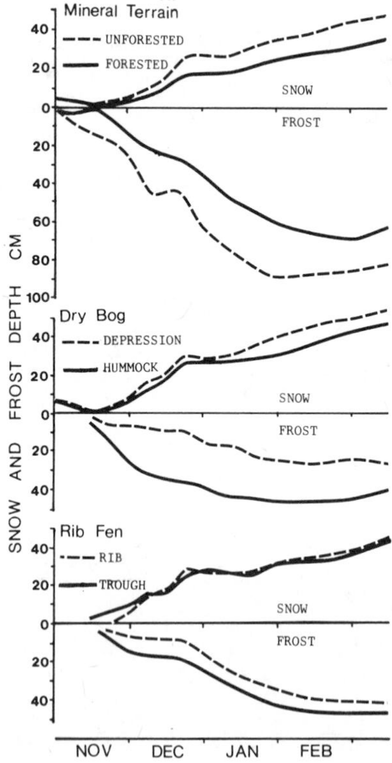

Figure 5 Freezing and snow depth for mineral terrain (top), dry bog (centre) and ribbed fen (lower).

In the mineral terrain the frost depth reached its maximum by 28 February and then began to degrade for the remainder of the winter, even though 43 continuous subfreezing days followed totalling 486 freezing degree- days. This was a result of the rising water table in the mineral aquifer caused by icing of the drainage channel 250 m away. The rising water table can be seen from Figure 3. The water in the aquifer had sufficient heat energy to cause thawing of frost at that location, and raise the temperature at 120 cm depth by 3 C° between 14 February and 15 March.

Wet Bog

In the wet bog (Site 2a) the water table was close to the surface at the time of freeze-up. Because of the relatively low microrelief and slow subsurface drainage, ground freezing showed little variability among sampling stations (Figure 2). Snow depth also showed a very low standard deviation. The average rate of frost penetration was 4 mm/day, as compared to 7 mm/day reported by Ryden and Kostov in a depressed (wet) area underlain by permafrost in Sweden. The lower rates of freezing encountered here are in part due to the fact that permafrost was not present. Initial freezing was slower than in the dry bog (Site 3) because of the high volumetric heat capacity of saturated peat. The rate of frost penetration remained relatively constant when compared to the dry bog (Figure 2), as a result of the physical homogeneity of the site, and the constant saturated moisture conditions experienced at the freezing front. Upward heat transfer from lower layers was slight as reflected by the temperature profiles (Figure 4). The heat flux also remained reasonably constant over the winter, and only dropping slightly in response to thicker snow cover.

Sedge Fen

In the sedge fen (Site 2b) there was little surface microrelief across the transect. The water table was at or above the surface prior to freeze-up. The weekly accumulation of snow was evenly distributed across the transect with standard deviation generally half of that for frost depth. Variability in frost depth, was related to zones of higher subsurface water flow. Shallower freezing was noted to occur where most surface water flow was observed prior to winter. Frost depth in these zones was typically 10-20 cm less than in other parts of the fen. This was of considerable importance to winter drainage because it maintained a subsurface "channel" in the most active hydrological layer (i.e. the upper layers of peat have the highest hydraulic conductivity, whereas lower layers are almost impermeable).

An 11.4 mm (water equivalent) snowfall occuring between 20 and 23 December reduced the temperature gradient of the fen and caused the temperature 10 cm below the peat surface to rise 2.2C° (while mean daily air temperature dropped). In high flow zones, total degradation of the frost layer occurred (depth of frozen layer was initially 10 to 13 cm) and the average thaw for the transect was 4.9 cm. This illustrates the large thermal capacity of the fen.

In the sedge fen the frozen layer not only extended downward, but grew upward from the primary freezing surface as a result of flooding and subsequent freezing of the lower snow layers which became saturated by water. Water seepage onto the surface was caused by a) hydraulic pressure created by the expulsion of water from the peat as it froze, b) increased hydrostatic pressure created by the overburden pressure exerted by the snowpack c) artesian pressure formed as a result of the impermeable frozen layer which constricted inflowing water beneath (Price and FitzGibbon 1982). The rising hydrostatic pressure in the fens is evident from Figure 3.

Dry Bog

In the dry bog there was a deep water table relative to the sphagnum hummocks. Similar moisture content was measured in both positive and negative relief elements, hence specific heat and thermal conductivity were also similar. Under such conditions, Romanov (1961) found freezing intensity to depend exclusively on surface morphology during the early winter before the accumulation of a snow cover. Hummocks, which have greater surface area, experienced deeper and more rapid freezing than depressions (Figure 5). Here, the heat flux ratio of positive to negative relief elements was 2.1:1 between 15 to 29 November. Additionally, the difference between snow depth on hummocks vs. depressions was not significant until after 31 January, which indicates snow cover may not have been as important as surface form or proximity to the water table (no data is available on snow density during this period). Between 31 January and 15 February the heat flux for positive and negative relief elements was nearly equivalent as was freezing intensity.

In the dry bog (Site 3) the mean rate of frost penetration was initially higher than in the wet bog (Site 2a) due to the lower volumetric heat capacity of the drier surface. As the freezing front approached the water table, the increased volumetric heat capacity of peat at the freezing front resulted in slower freezing.

Ribbed Fen

Prior to freezing (17 October), ribs were cooler than troughs (Figure 4), but by 15 December the upper layers of the trough were considerably cooler than ribs, and by 31 January the troughs were cooler throughout the vertical profile. The heat flux for troughs was 3.5 times greater than for ribs between 15 to 29 November. Although the saturated condition of the troughs resulted in a higher volumetric heat capacity than ribs, freezing occurred first in troughs and penetrated more rapidly (Figure 5).

On 15 December, three (of five) frost gauges on ribs were unfrozen while mean frost depth in troughs was 19.4 cm. There was some evidence that the water table level in ribs rose slightly, and that water circulation through ribs might have been important in delaying freezing. It is possible that lateral flow may have occurred in ribs as deeper freezing in adjacent troughs (in addition to their already lower position with respect to the water table) presented a barrier to flow in the normal downslope direction. This might explain the pools of unfrozen water that were observed at the edge of the fen in early winter.

From 15 February onward, the difference in frost depth for ribs and troughs was no longer significant. This coincided with the period of rapidly diminishing water inflows (as evidenced by the falling piezometric water levels shown in Figure 3) and with frost depths in ribs and troughs exceeding 40 cm. With freezing at this depth the hydrologically most active layer described by Romanov (1961) becomes inert, resulting in similar conditions at the freezing front in ribs and troughs. As the deepening snowpack provided greater insulation, and as water circulation at the freezing front was minimized, the rate of frost penetration in positive and negative elements became similar.

CONCLUSIONS

Several generalizations on the effect of the groundwater table on frost penetration can be made. Where the groundwater table is deep, it has little effect on the freezing process, but as the freezing front approaches the water table, the freezing rate slows considerably. Frost which has already formed may degenerate if the water table rises or a thick, low density snowpack accumulates. Where the water table is close to the surface and water experiences little flow, fairly homogeneous freezing occurs. If the water table is at or above the surface and considerable subsurface flow occurs (as in fens), the patterns of freezing can be complex. Water flowing beneath the frozen surface inhibits deep frost penetration. The effect of snow as an insulator is important but can be overshadowed by other local conditions such as forest cover, depth to the water table, or surface morphology of relief elements.

ACKNOWLEDGEMENTS

The financial assistance provided by the Northern Research Grant administered by the University of Saskatchewan for field and travel expense is gratefully acknowledged.

REFERENCES

Baldwin, H.I., 1956, The effect of forest on snow cover: Proceedings -- 13th Annual Meeting of the Eastern Snow Conference, Hanover, N.H., pp. 18-24.

Brown, R.J.E., and Williams, G.P., 1972, The freezing of peatland: N.R.C. Technical paper 381, N.R.C. 12881, 29p.

Fisheries and Environment Canada, 1978, Hydrologic Atlas of Canada: Department of Energy, Mines and Resources, Ottawa.

FitzGibbon, J.E., 1976, Generation of the snowmelt flood in the subarctic, Schefferville, Quebec: Ph.D. thesis, Department of Geography, McGill University, Montreal, Quebec, 207 p.

Jeffrey, W.W., 1968, Snow hydrology in the forest environment, in Snow hydrology, Proceedings -- Workshop Seminar, CNC/IHD. pp. 1-19.

Price, J.S., and FitzGibbon, J.E., 1982, Winter hydrology of a forested drainage basin: Proceedings -- Canadian Hydrology Symposium N.R.C., Fredericton, New Brunswick, pp. 347-360.

Rickard, W., and Brown, J., 1972, The performance of a frost tube for the determination of soil freezing and thawing depths: Soil Science, 113(2), pp. 149-154.

Romanov, V.V., 1961, Hydrophysics of bogs: Isreal Program for Scientific Translations, Jerusalem, 1968.

Ryden, B.E., and Kostov, L., 1980, Thawing and freezing in tundra soils, in Ecology of a subarctic mire, M. Sonesson, ed., Ecological Bulletins, Swedish Natural Science Research Council, Stockholm, pp. 251-281.

Tyrtikov, A.P., 1959, Perennially frozen ground and vegetation, in Principles of geocryology, ed. P.F. Shvetsov, (Academy of Sciences of the U.S.S.R.), vol. 1, pp. 299-421, N.R.C. of Canada Technical Translation Translation 1163, 1964.

PEDIMENTATION AND FLUVIAL DISSECTION IN PERMAFROST REGIONS WITH SPECIAL REFERENCE TO NW-CANADA

Kuno Priesnitz and Ekkehard Schunke

Department of Geography, University of Göttingen,
Federal Republic of Germany

The paper deals with the question, whether under periglacial conditions the destruction of pediments by fluvial dissection or the formation of pediments are prevailing processes. Studies on this problem were carried out mainly in the Mackenzie and in the central Richardson Mountains, in areas, which since the beginning of the Quaternary period are subject to periglacial but not to glacial conditions. Pediplains are common in the whole area. Most of them are thought to be periglacial pediplanation surfaces (cryopediments). The aim of the paper is to report observations about the main processes causing the formation of these pediments by backwearing (intense slope retreat of the adjacent steep slopes by mass wasting and rill-wash, solifluction and sheet wash activity, favoured by the surplus water from the steep relief and nivation on the upper parts of the pediplains, transport of the debris across the pediments into the main rivers). Fluvial action is different in areas not yet affected by the dissection caused by isostatic uplift and in the marginal areas, where this dissection is very intense. Outside the recently dissected parts there is no competition between cryoplanation and fluvial activity.

INTRODUCTION

In climatic geomorphology, the approach exists to make a distinction between the climatomorphic zones dominated by planation (formation of plains) and those dominated by dissection (formation of valleys) (e.g. Büdel 1971, 1977, 1982). With perhaps the exception of the humid tropics, no geomorphic zone generates so much controversy over the prevailing style of erosion as the arctic periglacial zone. The extreme positions are that the dominant geomorphic tendency in the periglacial zone is (1) "excessive valley-cutting" ("exzessive Talbildung", Büdel 1977, 1982) or (2) most intense pediplanation (Dedkov 1965).

Studies on these problems were made in 1978 and 1981 in the Mackenzie and Richardson Mountains, especially in the central Richardson Mountains accessible by the Dempster Highway (an area of continuous permafrost), in the eastern and western foothills of both ranges and around the "Plains of Abraham" in the Canyon Ranges, SW of Norman Wells (an area of discontinuous permafrost). Comparative observations were made in the Brooks Range near Atigun Pass, in several parts of Alaska, including the Arctic Slope, and in the Canadian Archipelago.

The main criterion in selecting these study areas was that they have been subject to periglacial conditions since the beginning of the Quaternary period without being glaciated at any time (cf. Hughes 1969, Denny 1970, Douglas 1972, Prest et al. 1974). Thus, if there exists any "convenient natural test site" (Büdel 1982:63) for meeting pure and mature periglacial features, it is here that we will find it. Many of the difficulties in climatic geomorphology arise from the comparison of forms of different age and maturity and from neglect of the time factor (cf. Brunsdén 1980 and Douglas 1980). Areas in which a recent change in the general modeling style has occured, as, e.g., those subject to periglacial conditions for only a few thousand years since the end of Pleistocene glaciation, may be well suited for process research, for studies on short lived microforms or on modeling tendencies, but they are not so well suited to demonstrate what the typical climatomorphic relief (including mesoforms and macroforms) may be. In this regard the unglaciated regions of NW-Canada and Alaska offer ideal conditions.

Pediplains—in the strictly descriptive sense of gently sloping surfaces (1°–10°) at the foot of higher mountains—are very common in the mentioned area. Their frequency and ubiquity suggests that the non-climatic environmental conditions leading to their formation are common and unspecific:

1. They occur at different altitudes. Many of them start right from the valley bottoms or the lowest terraces of the large rivers. At medium altitudes of 300-900 m, they are most frequent. The highest observed pediplains (not cryoplanation terraces) are situated at 1,300 m in the Mackenzie Mountains. The formal difference between the planation surfaces described here and the better known cryoplanation terraces lies in their sizes and their positions relative to the mountain slope, not in their absolute altitudes (cf. Czudek and Demek 1970, French 1976:155 ff.).

2. A strict dependence on permafrost could not be established: Frequency and appearance as well as evidence of active backwearing (see below) have been observed to be the same both in areas with continuous permafrost (northern Richardson Mountains, about 67°N) and in areas with discontinuous permafrost (Ft. Simpson area, surroundings of the Plains of Abraham in the Mackenzie Mountains near 64°N). Though it appears plausible that the processes causing cryoplanation are favored by permafrost, our observations seem to agree more with the views of Demek (1969:56 ff.) than with those of Reger and

Péwé (1976:107 ff.).

3. A dependence on lithology could not be established either. Pediplains cut rocks of all kinds and ages: Cambrian schists in the southern Richardson Mountains, dolomites of Ordovician and Silurian age, as well as Devonian sandstones and shales in the Plains of Abraham area and sandstones, siltstones, and shales of Devonian, Mississippian, and lower Cretaceous age in the Richardson Mountains foothills.

4. As to the general orographic position, three types of pediplains may be distinguished, piedmont pediments, valley pediments and intermontane pediments (classification by Mensching 1958, 1973)(Figures 1 and 2).

FIGURE 1 Cryopediment with sharp upper limit and inclination of 2°-4°, about 4 km wide; crossed by shallow valleys. Little Keel River Valley, North of "Plains of Abraham", Mackenzie Mountains.

FIGURE 2 Narrow valley-cryopediment showing dissection by box-shaped valley. Bedrock (Upper Devonian sandstone) is cut sharply by pediment. Central Richardson Mountains (66°50'N).

The features which suggest the morphogenesis of the pediplains are described in the following three sections. These treat, first, the steep slopes delimiting the pediplains at the upper end, second, the pediplains themselves and the processes acting on their surfaces and, finally, the valleys to which the pediplains usually run.

FIGURE 3 Mountain slope dissected by steep V-shaped valleys into triangular sections. Pediment interrupted by Carcajou Lake in the course of Little Keel River, a canyonlike incised tributary of Carcajou River, Mackenzie Mountains.

THE STEEP RELIEF AT THE UPPER MARGIN OF THE PEDIPLAINS

The marginal chains of the Mackenzie and Richardson Mountains mostly have the character of hogback mountains at 1,100-1,600 m elevation.

The characteristic traits of the relief are that the mountain sides are deeply cut by V-shaped valleys, independent of the structure. The valleys have steep rectilinear slopes of about 30°, very steep longitudinal profiles and, despite that, often flat floors.

Erosion in these valleys is obviously very intense. On the slopes, mass wasting and rillwash dominate, with episodic runoff and gravel transport by meltwater in the valley bottoms. Valley-erosion cuts down to the local erosion base and then develops widening flat floors without any change in the slope processes and slope profiles. Thus, small and large valleys are characterized by the same slope angles of about 30° and finally the whole mountain slope becomes dissected into very characteristic triangular facets (Figure 3).

The same processes operate on these "triangular slopes" (cf. Büdel 1977:72) as on the valley slopes, namely frost wedging, mass wasting, and rillwash, and cause very intense backwearing.

THE PEDIPLAINS

The important supply of frost debris from above does not lead to accumulation at the foot of the slope for denudational activity at the upper edge of the pediplains is capable of keeping the foot of the steep slopes clean. Commonly the upper limit of the pediplain is marked by a sharp knickpoint, most exceptions being structurally controlled. The processes responsible for this denudational cleaning of the upper edge of the pediplains are solifluction and sheetwash, which were both observed directly and deduced from micro-forms,

FIGURE 4 Oblique view from aeroplane showing sharp upper end of cryopediment and forms giving evidence of active solifluction and sheetwash. NW of the "Plains of Abraham", Mackenzie Mountains.

FIGURE 5 Cryopediments with pediment passes and inselberglike mountain remnants. Vegetation marks now inactive transport-paths directed toward very shallow valleys. NE footregion of the "Plains of Abraham", Mackenzie Mountains.

transport patterns, and sediment structures, e.g.on pediplains at the south side of Little Keel River (Plains of Abraham), and in the upper ends of many valley pediments in the central Richardson Mountains (cf. Figure 4).

The main factor aiding all types of solifluction and sheetwash in this orographic position is the increased water supply, caused by surface runoff from the steep slopes and by meltwater from snow trapped in the mountainfoot concavity.

The upper parts of the pediplains have usually been found to be slightly fan-shaped, with slope angles varying from 6°-12°. This allows solifluction and sheetwash to deliver debris to stream channels in a short distance. We have seldom observed that the entire pediplain is an active "slope of transportation" (cf. French 1976:156); most debris transport is done—after a short distance of nonlinear transport—by the rivers crossing the pediplains (cf. Figure 5). These rivers are:

- small runoff channels not connected to the mountain drainage and not eroding at all;
- rivers coming from the V-shaped or flat-floored valleys of the adjacent mountains, heavily charged with sediment (these are characterized by braided channels, either eroding only weakly or more often flowing above the pediplain level through accumulations of their own sediments);
- some rivers derived from areas in the adjacent mountains which do not supply detritus, e.g., some quarzite areas in the McDermott syncline of the Mackenzie Mountains. These rivers are able to cut shallow box-shaped valleys into the pediplain—5 m being the maximum observed depth cut into solid rock. These valleys expose the sharply cut bedrock core of the pediments, overlain by a fanglomerate of 0.5-3 m thickness.

The lower parts of the pediplains, with slope angles of 1°-5°, are usually inactive with surfaces characterized by patterned ground, lichen-covered stones, or dense and undisturbed tundra vegetation.

Given our observations of structure, genesis, and erosional intensity, as well as the probable duration of periglacial erosion in these areas, most of the pediplains in all probability are cryopediments in the sense of Macar (1969:138) and Czudek and Demek (1970:101; 1973) or cryoplanation forms following Bryan (1946:639 ff.) and Demek (1963). The morphographic and morphogenetic criteria described in these references—most of them are summarized by French (1976:155 ff.)—fit the observed forms surprisingly well.

Since we lack studies over a long period and sediments which would provide independent age estimates for the cryoplanation surfaces, we can only evaluate the probability whether or not periglacial processes would be capable of producing pediplains (including those several kilometers wide) in the time available. The main view against a genesis in cryoplanation is that which considers the pediplains as Tertiary forms which have been preserved or only slightly modified by Quaternary processes (cf. Büdel 1971, Washburn 1979:237 ff.). Arguments to refute such a view include the following facts.

High levels interpreted as Tertiary relic surfaces in several parts of Yukon (cf. Bostock 1972), e.g., in the Wernecke and Ogilvie Mountains, lie clearly at a higher elevation than the cryopediments and are dissected by the V-shaped valleys and the steep frost slopes above the pediments.

The known glaciation limits and the "rather meagre understanding of changing climates of central Yukon during the Quaternary" (Foscolos et al. 1977: 1; cf. Bostock 1966, Hughes et al. 1969, Zoltai and Tarnocai 1974) suggest that, during most of the Quaternary, there must have been some kind of a periglacial climate in the Mackenzie and Richardson Mountains. Consequently, careful extrapolation of present morphodynamic rates to the entire Quaternary may be possible. Extrapolating comparable studies of periglacial denudation (cf. French 1976: 148, 172) and taking into account the age of those planation surfaces which can be dated (cf. Demek 1969, Reger and Péwé 1976, Washburn 1979:240), it may be argued that even the largest pediplains in the non-glaciated parts of NW-Canada and Alaska (with the exception of the North Slope) can be explained by the present cryoplanation rates. Thus, the actual relief of this never-glaciated area may be considered as one of the most pure and mature periglacial reliefs in the world.

Cryoplanation is not only a process of widening of preexisting plains. As demonstrated by the research on cryoplanation terraces (as compiled recently by Karrasch 1972) and our own observations in many periglacial areas, the slightest concavity at the foot of a slope can start (by backwearing effects) the development of a frost cliff/frost slope, and a cryoplanation terrace/cryopediment.

In this context, the main problems which remain to be solved seem to be (1) the reconstruction of the Tertiary relief, (2) the Quaternary history with its fluctuations of the periglacial climate and their geomorphic effects and (3) quantitative studies of periglacial backwearing rates and cryopediment growth.

THE VALLEYS

The valley network of the Mackenzie and Richardson Mountains appears on the ground as well as on air photographs and LANDSAT imagery to consist of two overlapping river systems: an older one with wide shallow valleys originating in the center of the mountain ranges, and a younger, more sharply incised one, which is growing retrogressively at the expense of the older one.

It seems evident, that the intense erosion causing the younger valley system follows recent uplifts, by glacial rebound as well as by tectonics. Recent isostatic uplift is estimated at about 100-200 m for the Richardson Mountains.

The southern Mackenzie Mountains are believed to have been recently uplifted even more: in the Nahanni River area by as much as 485 m in postglacial time (Ford 1973). The upper valleys, not yet affected by retrogressive dissection, are shallow and flat-floored with depths of 2-3 m, 10 m at the maximum, and widths of the broad gravel fills of up to more than hundred meters. The rivers swing in anastomosing channels over the whole width of the valley floor, locally undercutting the banks and locally restricted by the gravel fans of tributary rivers.

The adjacent parts of the pediments are the oldest, inactive sections which suggests that the rivers have been flowing at the same level since the beginning of the pediment formation. They are used to capacity by their gravel load and have no tendency to downward cutting.

The tributary rivers of the Mackenzie and the Peel River which are presently incising have a totally different appearance. Deep, narrow box-shaped canyons up to 100 m deep (note the "Canyon Ranges") are common. The Nahanni Canyon, with rock walls of more than 1,000 m in places, represents a most impressive example of the vertical erosion achieved by an antecedent river.

Fluvial activity is in general limited to short runoff periods during the snowmelt season (lasting several weeks to several months depending on latitude) and exceptional heavy rain events. Glacier and baseflow runoffs are morphologically of secondary importance. Our assessment of fluvial activity in a longterm periglacial environment is that, because the runoff is concentrated into short periods, the rivers are capable of enormous erosion rates. This explains the overrating of periglacial vertical erosion by Büdel (int. al. 1982: "zone of excessive valley-cutting") and others, who have generalized from areas of recent isostatic uplift as, e.g., Spitsbergen.

The "normal case" in the sense of climatic geomorphology seems to be that the energy of a comparatively very steep longitudinal gradient (up to several degrees) is needed for the through-transport of the heavy frost debris load. There seems to exist a long-term equilibrium between different and seemingly antagonistic tendencies (formation of plains versus formation of valleys), which allows the conclusion that under stable conditions the actual surface is to be considered as mature in a climatic-geomorphological sense as well as in a Davisian sense. Only in the steepest headwater regions or after a break in the equilibrium profile does considerable vertical erosion take place.

The threshold value of the gradient of rivers used to capacity by the through-transport and of vertically eroding rivers depends of course not only on climatic but also on other environmental factors as, e.g., petrography and on the age of the actual relief.

REFERENCES

Bird, B., 1967, The physiography of Arctic Canada, Baltimore, 336 pp.

Bostock, H. S., 1966, Notes on glaciation in central Yukon Territory, Geological Survey of Canada, Paper 65-36.

Bostock, H. S., 1972, Physiographic subdivisions of Canada, in Douglas, R. J. W., ed., Geology and economic minerals of Canada, Geological Survey of Canada, Economic Geology Report No.1, p. 10-30.

Brunsden, D., 1980, Applicable models of long term landform evolution: Zeitschrift für Geomorphologie, Neue Folge, Supplementband 36, p.16-36.

Bryan, K., 1946, Cryopedology—the study of frozen ground and intensive frost-action with suggestions on nomenclature: American Journal of Science 244, p. 622-642.

Büdel, J., 1971, Das natürliche System der Geomorphologie (mit kritischen Gängen zum Formenschatz der Tropen): Würzburger Geographische Arbeiten 34, 152 pp.

Büdel, J., 1977, Klima-Geomorphologie, Berlin, 304 pp.

Büdel, J., 1982, Climatic geomorphology, Princeton, 443 pp.

Czudek, T., and Demek, J., 1970, Pleistocene cryopedimentation in Czechoslovakia: Acta Geographica Lodziensia 24, p. 101-108.

Czudek, T., and Demek, J., 1973, The valley cryopediments in Eastern Siberia: Biuletyn Peryglacjalny 22, p. 117-130.

Dedkov, A., 1965, Das Problem der Oberflächenverebnungen: Petermanns Geographische Mitteilungen 87, p. 258-264.

Demek, J., 1963, Hangforschung in der Tschechoslowakei, in Mortensen, H., ed., Neue Beiträge zur internationalen Hangforschung, Nachrichten der Akademie der Wissenschaften in Göttingen, Mathematisch-Physikalische Klasse, 3. Folge, Heft 9, p. 99-138.

Demek, J., 1969, Cryoplanation terraces, their geographical distribution, genesis and development, Rozpravy Ceskoslovenske Akademie Ved, Rad Matematickych A Prirodnich Ved, Rocnik 79, Sesit 4, 80 pp.

Denny, C. S., 1970, Glacial geology of Alaska, in The National Atlas of the United States, p. 76.

Douglas, I., 1980, Climatic geomorphology—present-day processes and landform evolution—problems of interpretation: Zeitschrift für Geomorphologie, Neue Folge, Supplementband 36, p. 27-47.

Douglas, R. J. W., ed., 1972, Geology and economic minerals of Canada, Geological Survey of Canada, Economic Geology Report No.1, 838 pp.

Ford, D., 1973, Development of the canyons of the South Nahanni River, N.W.T.: Canadian Journal of Earth Sciences 10, p. 366-378.

Foscolos, A. E., Rutter, N. W., and Hughes, O. L., 1977, The use of pedological studies in interpreting the Quaternary history of central Yukon Territory: Geological Survey of Canada, Bulletin 271, 48 pp.

French, H. M., 1976, The periglacial environment, London, New York, 309 pp.

Hughes, O. L., 1969, Distribution of open-system pingos in central Yukon Territory with respect to glacial limits: Geological Survey of Canada, Paper 69-34, 8 pp.

Hughes, O. L., Campbell, R. B., Muller, J. E., and Wheeler, J. O., 1969, Glacial map of Yukon Territory, south of 65 degrees north latitude, Geological Survey of Canada, Paper 68-34, Map 6-1968.

Karrasch, H., 1972, Flächenbildung unter periglazialen Klimabedingungen?, Göttinger Geographische Abhandlungen 60 (Hans-Poser-Festschrift), p. 155-168.

Macar, P., 1969, Actions périglaciaires et évolution des pentes en Belgique: Biuletin Peryglacjalny 18, p. 137-152.

Mensching, H., 1958, Glacis, Fußfläche, Pediment: Zeitschrift für Geomorphologie, Neue Folge, Band 2, p. 165-186.

Mensching, H., 1973, Pediment and Glacis, ihre Morphogenese und Einordnung in das System der klimatischen Geomorphologie aufgrund von Beobachtungen in Trockengebieten Nordamerikas (USA und Nordmexiko), Zeitschrift für Geomorphologie, Neue Folge, Supplementband 17, p. 133-155.

Péwé, T. L., 1975, Quaternary geology of Alaska: U.S. Geological Survey Professional Paper 835, p. 1-145.

Prest, V. K., Grant, D. R., and Rampton, V. N., 1968, Glacial map of Canada, Geological Survey of Canada, 1253 A.

Reger, R. D., and Péwé, T. L., 1976, Cryoplanation terraces: indicators of a permafrost environment: Quaternary Research 6, p. 99-109.

Washburn, A. L., 1979, Geocryology, a survey of periglacial processes and environments, London, 406 pp.

Zoltai, S. C., and Tarnocai, C., 1974, Perenially frozen peatlands in the western Arctic and Subarctic of Canada: Canadian Journal of Earth Sciences 12, p. 28-43.

ALPINE PERMAFROST IN TIAN SHAN, CHINA

Qiu Guoqing, Huang Yizhi, and Li Zuofu

Lanzhou Institute of Glaciology and Cryopedology, Academia Sinica
People's Republic of China

The distribution of permafrost in Tian Shan depends chiefly on vertical zonation. Because of local conditions, the lower limit of permafrost varies from place to place: the lowest limit is 2700 m on north-facing slopes and 3100 m on south-facing slopes, about 1000 m below the local snow line. In general, the lower limit descends 171 m with each 1° increase in latitude and 10.6 m with each degree of increase in longitude. Alpine permafrost can be divided into three zones: unstable permafrost, transitional permafrost, and stable permafrost. The map of permafrost distribution in Tian Shan is based on temperature observations in bore holes, field observations of cryogenic phenomena, and interpretation of aerial photographs. The development of ground ice in coarse-grained sediments, both buried glacial ice and segregated ice, is a peculiarity of Tian Shan. Under the stone stripes, rock circles, and frost-heaved rocks, massive ground ice is found. An old moraine with a silt content of less than 15% has proved to be rich in ice.

In China, the Tian Shan stretches for 1700 km west to east and is 100-300 km wide from south to north. There are many ridges and fault basins controlled by the WNW and ENE tectonic lines and other factors. Most of the Tian Shan consists of metamorphic and igneous rocks. Ridge crests generally exceed 4000 m a.s.l; the highest summit, Tomor Peak in the west, is 7443 m. Local relative relief is usually 3000 m or more, so vertical zonation of air temperature, precipitation, and landscape is inevitable.

The mean annual air temperature decreases with elevation. The gradient is 0.33°-0.37°C/100 m between 1830 and 3000 m a.s.l., and 0.61°-0.66°C/100 m between 3000 and 4000 m (Zhang and Xie 1962).

In general, there are two zones with high precipitation: the first is coincident with the forest-steppe zone (about 1800-2500 m), with precipitation of about 600 mm/year; the second lies at the snow line, with precipitation of from 500-700 to more than 1000 mm/year. About 90% of the annual precipitation is concentrated in the period from April to October. Observations show that the precipitation decreases from west to east.

With rising elevation, a series of landscapes is encountered, namely, desert and semidesert, forest-steppe, subalpine-alpine meadow, and cold desert. Their boundaries vary with slope orientation, latitude, and longitude.

All these affect the characteristics of permafrost distribution.

PERMAFROST DISTRIBUTION

The total area of permafrost in the Tian Shan of China is about 63,000 km^2 (Figure 1). The distribution shows both vertical zonation and latitudinal and longitudinal variations.

According to observations and 4 years of ground temperature measurements in the Kuyxian Pass region of the middle Tian Shan, the lower limit of alpine permafrost is at 2700 m on north-facing slopes and 3100 m on south-facing slopes (Qiu and Zhang 1981). The lower limit varies, however, because of differences in relief, surface cover, geologic-lithologic characteristics, and precipitation, and does not coincide with a particular contour line (Table 1). Despite variation, the lower limit always lies in the subalpine-alpine meadow zone, higher than the local tree line, and is 900-1200 m lower than the snow line. These characteristics, together with the distribution of cryogenetic phenomena, are useful aids when mapping permafrost through the interpretation of aerial photographs (Figure 1).

The thickness, temperature, and stability of the permafrost and the depth of the active layer varies with elevation. Usually, the depth of seasonally frozen ground is 1-2 m at 1000 m a.s.l. and becomes 4-5 m adjacent to the lower limit of permafrost, while the active layer in the permafrost region is 4-5 m thick in the vicinity of the lower limit, becoming $\leq$2 m in the stable permafrost area of the high mountains. Generally, there are three zones of alpine permafrost: the unstable zone, the transitional zone, and the stable zone.

Adjacent to the lower limit, the perennially frozen ground has a mean annual temperature of -0.1° to -0.2°C, so it is very unstable. For example, in the Kuyxian Pass region, the rainfall in 1972 was 100 mm higher than normal, and the permafrost table observed in drill hole F5 descended to a depth of 3.5 m. In 1973-1975, annual precipitation was normal, and the permafrost table gradually returned to a depth of 2.7 m (Figure 2).

However, at high altitudes, for example at 3200 m in the Kuyxian Pass region, the permafrost is as

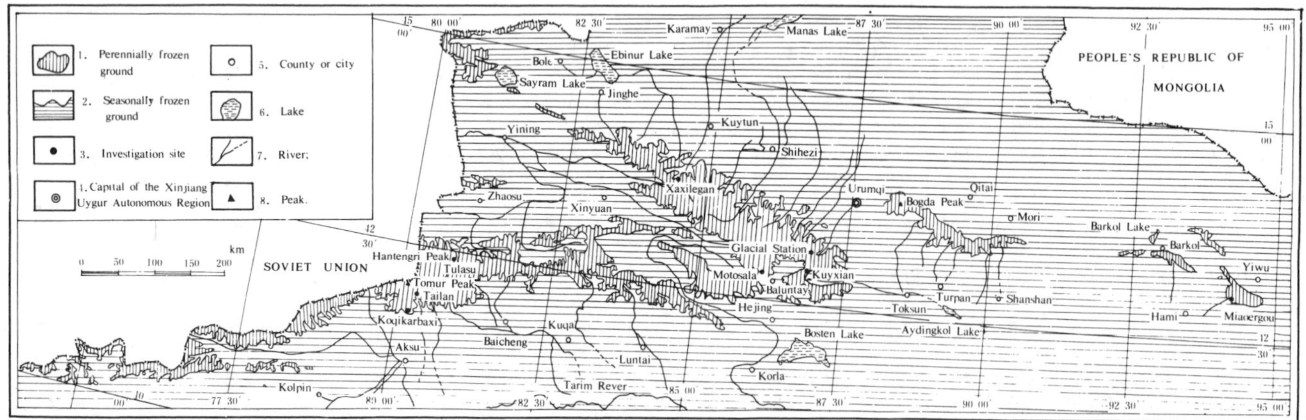

FIGURE 1 Distribution of frozen ground in Tian Shan, West China.

TABLE 1 Selected data on the lower limit of alpine permafrost in the Tian Shan.

Site	Location	Elevation of permafrost lower limit (m), slope facing					Elevation of glacial terminus (m)	Elevation of snow line (m), slope facing		Vertical distance between permafrost lower limit and snow line (m)
		N	NE/NW	E/W	SW/SE	S		N	S	
Miaoergou	43°00'N 94°15'E			2950			3600		4200 (SW)	1250
Kuyxian Pass	42°56'N 86°53'E	2700				3100	3700	3900	4100	1100
Source of Urumqi River	43°07'N 87°00'E	2900				3250	3600	3800	4200	925
Motosala	42°48'N 86°08'E	3250				3390				
Xaxilegan	43°46'N 84°27'E			2920	3050		3250	3600		870
Congtailan	41°54'N 80°18'E				3350					
Tulasu	42°04'N 80°16'E	3000			3250		3140	4000 (NE)		920
Koqikarbaxi	41°42'N 80°00'E		3000		3100		3070		4100	950

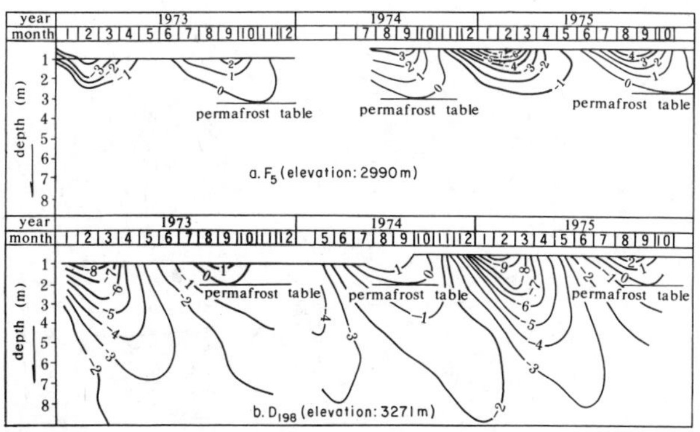

FIGURE 2 Isogeotherm of drill holes F5 and D198 in the Kuyxian Pass region.

much as 100 m thick and has a mean annual ground temperature of -2°C, which can be defined as stable permafrost. As a result, the level of the permafrost table did not change during the same 4-year period no matter how much the rainfall fluctuated (Figure 2). Moreover, in 1981 the temperature at the level of zero annual amplitude remained the same as that measured in 1973. Gorbunov (1978) assumed that at 7000 m the ground temperature might be as low as -20° to -25°C, but the presence of such highly stable permafrost remains to be proven.

In the transitional zone, between the unstable and stable zones, permafrost thicknesses increase 20 m for each 100-m increase in elevation at Miaoergou in the east and 50 m for each 100 m at the source of the Urumqi River in the central region.

Because the lower limit of alpine permafrost varies in altitude and the width of the transitional zone varies from place to place, it is impossible to associate the boundaries of the three zones with specific elevations. Nevertheless, binary regression analysis provides the following equation for prediction of the elevation of the lower permafrost limit:

$$H = 11089.5 - 10.6X - 171.2Y$$

where H = elevation of lower limit (m)
X = longitude (°)
Y = latitude (°)

The correlation coefficient of this equation is 0.915. This equation shows that the latitudinal variation of the lower limit of permafrost is -171.2 m/lat. 1°N, which differs from the data presented by Fujii and Higuchi (1978) and Zhou and Guo (1978). This equation also shows a longitudinal variation of -10.6 m/long. 1°E, which might be caused by the decrease of precipitation from west to east. Since the Tian Shan extends through 21° of longitude (from 74° to 95°E), the longitudinal variation of the limit should not be omitted.

GROUND ICE

The development of ground ice in the Tian Shan depends on geologic-geomorphologic conditions. There are few ice veins in the cracks of the frozen bedrock; permafrost is usually ice-poor on scree slopes and in the gravel and pebbles in well-sorted alluvial fans because they are well drained. Permafrost in clayey soils on gentle slopes, in intermontane depressions, and in hollows at the boundary between alluvial fans is usually rich in ground ice (Qiu and Zhang 1981).

The following discussion is concerned with ground ice in coarse-grained sediments, which may be a peculiarity of the Tian Shan.

Exogenous Ground Ice

At the source of the Urumqi River, buried ice occurs in Little Ice Age moraines. A 10-m-thick buried glacial ice body is exposed in a natural cave of the ice-cored end moraine of Glacier 6. The ice is preserved by the protection of the debris mantle. In addition, the Tomor-type glacier, which has a great deal of avalanche cover and a superglacial moraine, forms similar kinds of buried glacial ice.

Another kind of buried ice is from buried snow. For example, on the south slope of Haxilegan, a snow deposit was covered by artificial waste rocks in May 1977, and in a year it had become a 20-50 cm ice layer. It appears that buried "snow-ice" may be formed in places with thick snow deposit and frequent rock avalanches in west Tian Shan.

Endogenous Ground Ice

(1) Segregated ground ice in sediments rich in fine-grained materials: The deluvial-solifluctional deposits in the Tian Shan are usually poorly sorted, and segregated ice will form if there is an appropriate water supply. For example, in well D-135, on a gentle slope adjacent to a steep cliff in the Kuyxian Pass region, it was found that a deluvial sandy clay layer containing 25-35% gravels and pebbles is underlain by a deluvial debris (slope wash) layer with 21-31% fines. The upper layer has a netted-stratified cryogenetic structure (mineral layers separated by ice veins and horizontal ice layers) with a volumetric ice content of 60%, and the debris layer has a conglomeratic cryogenetic structure (debris cemented by ice lenses) with a volumetric ice content of 55%. This kind of ground ice has been found in many places in the Tian Shan.

(2) Segregated ground ice in glaciofluvial-deluvial deposits with alternative beds of coarse and fine-grained material: This kind of ice occurs in a swampy lowland at the source of the Urumqi River. As a result of syngenetic frost action, a thick ice layer (>50 cm) with a volumetric ice content of 95% has formed beneath the new permafrost table (old earth surface, as indicated by the presence of humic materials and decayed roots). Due to the impermeability of the massive ground ice, the active layer is usually saturated and therefore experiences strong frost heaving. This results in many frost-sorting phenomena on the slopes and the swampy lowland, including upheaving stones (Figure 3), stone stripes (Figure 4), and rock circles. According to pit observations, the sorted stones are standing in the principal frost-heaving zone of the active layer, and between the base of the sorted stones and the permafrost table there is an undisturbed zone. This demonstrates that frost sorting is directly related to the heaving process that occurs in the active layer.

(3) Ground ice in coarse-grained deposits: In the Kuyxian Pass region, a 110-m-thick permafrost layer occurs in the Alaxigongjin Moraine that is rich in ice lenses and ice horizons and has a stratified conglomeratic cryogenetic structure. Based on its depositional characteristics and cryogenetic structure, as seen in drill hole samples, Qiu and Zhang (1981) pointed out that this is a syngenetic permafrost formed in association with sediment aggredation during glacier advance and recession, and the ice layers there differ from buried glacial ice.

There are two logical explanations for how such ground ice could form in a sediment containing such

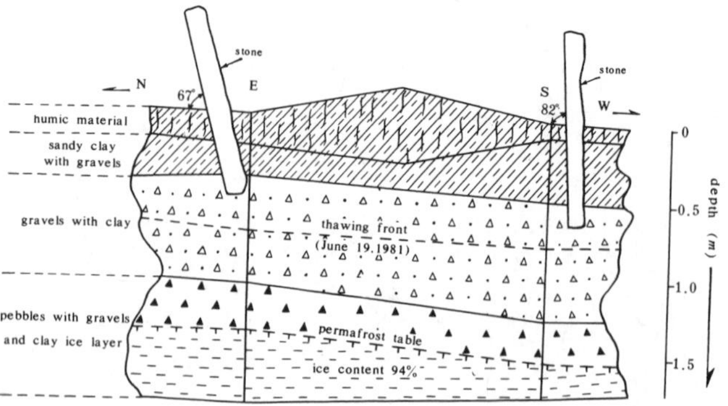

FIGURE 3 Upfreezing stone and the underlying permafrost.

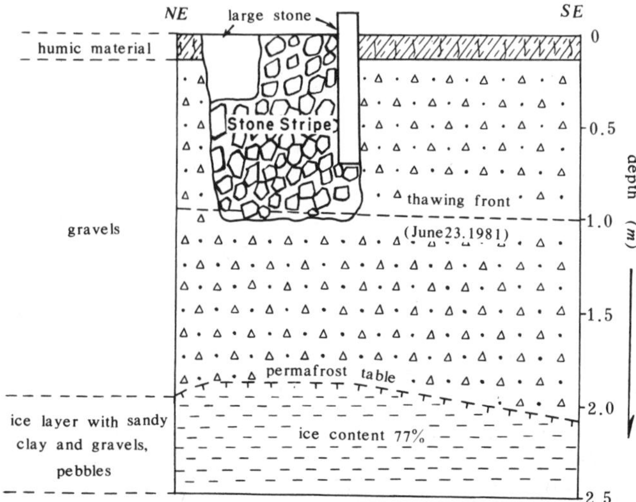

FIGURE 4 Section of large stone stripe.

a small amount of fine-grained material (the percentage of <0.1 mm grains is no more than 10% in general): repeated ice formation over an extended period of time, and the poor permeability of ice-rich frozen ground.

During sediment aggradation and after glacier melting, the sediments remain in a cold climatic environment and undergo frost action with the formation of ice crystals. At the initial stage of the sediment freezing, moisture migration might be insignificant; however, with ice crystal formation, the porosity of the soil may gradually diminish. Then water would migrate from places with higher temperature to those with lower temperature, and from places with thicker unfrozen water films surrounding ice crystals or soil grains to places with thinner unfrozen water films. After a long period of repeated ice segregation, ice lenses and ice horizons would be formed.

Soil permeability may decrease as ice forms at depth. With the aggregation of permafrost, gravitational water might freeze at the base of the active layer and form ice horizons joined to the underlying frozen ground.

CONCLUSIONS

Distribution of permafrost in the Tian Shan depends chiefly upon elevation. With increasing elevation, three permafrost zones may be recognized: an unstable permafrost zone with mean annual ground temperatures t_{cp} of -0.1° to -0.2°C and a thickness less than 20 m; a stable permafrost zone with t_{cp} below -2°C and permafrost >100 m; and a transitional zone between them. The lower limit of alpine permafrost and the boundaries between zones vary and do not coincide with particular contour lines.

The lower limit of permafrost shows latitudinal and longitudinal variations. For each 1° of latitude to the north, the lower limit drops 171 m; and with each 1° of longitude to the east, the lower limit drops 10 m.

Several kinds of ground ice develop in coarse sediments, with ice contents as high as 50-95%, a characteristic that may be a peculiarity of the Tian Shan.

ACKNOWLEDGMENT

The authors wish to express their thanks to Prof. T. L. Péwé for his valuable suggestions on this paper.

REFERENCES

Fujii, Y., and Higuchi, K., 1978, Distribution of alpine permafrost in the Northern Hemisphere and its relation to air temperature, in Proceedings of the Third International Conference on Permafrost, v. 1: Ottawa, National Research Council of Canada.

Geomorphological Study Group of Xinjiang Exploration Team, 1980, Geomorphology of Xinjiang: Beijing, Science Press.

Gorbunov, A.P., 1978, Permafrost investigation in high-mountain regions: Journal of Arctic and Alpine Research, v. 10, no. 2.

Qiu Guoqing, 1980, On engineering geological survey in alpine permafrost regions of West China: Journal of Glaciology and Cryopedology, v. 2, no. 3, p.17-23.

Qiu Guoqing and Zhang Changqing, 1981, Distributive features of permafrost near the Kuyxian Daban in the Tian Shan Mountains, in Memoirs of Lanzhou Institute of Glaciology and Cryopedology, Academia Sinica, no. 2: Beijing, Science Press.

Zhang Jinhua and Xie Yinqing, 1962, Climatic features in the valley of Urumqi River, manuscript.

Zhou Youwu and Guo Dongxin, 1982, Principal characteristics of permafrost in China: Journal of Glaciology and Cryopedology, v. 4, no. 1, p. 1-19.

PERMAFROST THAW ASSOCIATED WITH TUNDRA FIRES IN NORTHWEST ALASKA

Charles H. Racine[1], William A. Patterson III[2] and John G. Dennis[3]

[1]Division of Environmental Studies, Johnson State College, Johnson, Vermont 05656 USA
[2]Department of Forestry and Wildlife Management, University of Massachusetts, Amherst, Massachusetts 01003 USA
[3]Biological Resources Division, National Park Service, Washington, D.C. 20240 USA

Thaw depths, soil temperatures, and vegetation cover were measured between 1978 and 1982 in tussock tundra on burns representing current, 1,2,3,4,5,7,8,9, and 10 year old fires in either the Seward Peninsula or the Noatak River areas of Alaska. Percentage increase of thaw in tussock tundra on flat terrain was 10-40% during the first 5 years following fire with a possible peak at 2 years and a return nearly to prefire thaw depths by the tenth year. Thaw depths following fire were significantly deeper in tussock tundra on slopes (> 5%) than on flatter terrain. Tussock tundra soil temperatures were significantly higher in a 3-5 week old burn than in an unburned control, but were similar to the control in a 10 year old burn. Increases in vascular plant cover following fire in tussock tundra averaged 10% per year during the 10 years following fire and could account for decreases in thawing and soil temperatures over this time. Other factors, such as seasonal time of burning, thickness of the soil organic horizons and tussock density, vary regionally and may affect postfire thawing. Prolonged deeper thawing found in burned white spruce taiga emphasizes the importance of vegetation recovery for ameliorating postfire thawing regimes.

INTRODUCTION

Lightning-caused tundra fires constitute a natural form of arctic vegetation and soil disturbance which potentially can produce active layer changes and permafrost degradation. Individual fires as large as 980 km^2 have been reported (Racine 1979), and more than 3,800 km^2 burned in northwest Alaska in 1977, alone. Although it is clear that permafrost conditions could be affected over large areas, relatively little is known about how fire affects active layer thicknesses and soil temperatures of tundra, particularly with respect to changes over time spans of as many as 10 years (Viereck and Schandelmeier 1980).

Our study reports active layer thickness measurements from burns up to 10 years old in northwest Alaska and attempts to construct a vegetation-soil thaw word model to describe ecosystem changes due to natural fire. Between 1978 and 1982 we made measurements of thaw depths and vegetation cover at a wide variety of sites representing 2 week to 10 year old burns in the Seward Peninsula or Noatak River areas (Figure 1). By surveying a wide range of burned sites in northwest Alaska, we attempted to obtain data that would permit us to propose hypotheses which suggest how quickly and by what mechanisms the permafrost responds to fire.

The available information on fire-permafrost relationships has been summarized recently by Viereck and Schandelmeier (1980) and Brown and Grave (1979). Long term information (8-20 years) for active layer changes following taiga fires has been presented by Viereck (1981) for interior Alaska and by Mackay (1977) and Black and Bliss (1978) for northern Canada at Inuvik. These reviews indicate that there are relatively few long term measurements of thaw depths following tundra fires in Alaska. Much of the research to date has dealt with the first and second year postfire thaw measurements, such as those by Wein and Bliss (1973), Brown et al. (1969), Racine (1981), and Hall et al (1978), which for tussock tundra, generally show increases of 10-40% over controls. Continued monitoring by Johnson and Viereck (in this volume) of thaw depths over 5 years following a 1977 tundra fire on the North Slope represents some of the longest monitoring of postfire thaw patterns in tundra that is available.

STUDY AREAS

The Noatak River and Seward Peninsula study areas are located above and below the Arctic Circle, respectively, and are at or near the boundary between tundra and taiga in northwest Alaska (Figure 1). White spruce (Picea glauca) forests extend into the southeastern base of the Seward Peninsula and into the lower drainages of the Noatak watershed. In all other respects, both areas are arctic in character with continuous permafrost. Mean annual temperatures are about -8° to -5° C and annual precipitation probably ranges between 22 and 50 cm (Racine 1979).

The Noatak River drains over 33,600 km^2 of the western Brooks Range as it courses primarily westerly through a 644 km long, generally broad

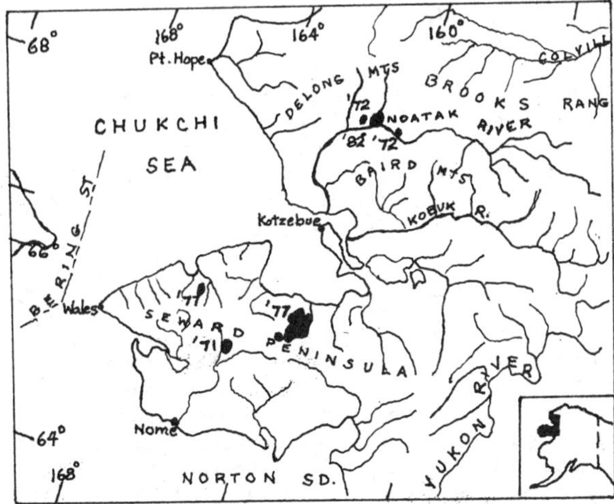

FIGURE 1 Location and year of burn for areas studied between 1978 and 1982 in the Seward Peninsula and Noatak River watershed, northwest Alaska.

TABLE 1 Sites Sampled Between 1977 and 1982 in Northwest Alaska for Study of Tundra Fire-Permafrost Relationships.

Name	North Latitude	West Longitude	Fire Dates [1]	Years Since Burn at Time of Sample
Seward Peninsula				
Coffee Creek*	65° 17'	164° 45'	6/29/71	7,8,10
Utica Creek	65° 66'	163° 0 '	7/09/77	1,2,3
Imuruk Lake*	65° 35'	163° 10'	7/09/77	1,2,3
Arctic River	66° 14'	165° 36'	7/23/77	1
Noatak River Watershed				
Kougururok River*	67° 59'	161° 55'	7/13/72	10
Poktovik Creek	68° 01'	161° 20'	7/13/72	10
Noatak R. sites 1-5,7*	67° 58'	161° 50'	7/14/77	4,5
Uchugrak Hills	68° 04'	161° 35'	7/14/77	5
Kungiakrok Creek*	67° 57'	161° 55'	6/21/82	0.1
Mukachiak Creek	67° 54'	160° 55'	7/12/82	0.05

1. As reported in Bureau of Land Management (BLM) Fire Detail sheets.

* Intensive study sites.

lowland that separates the Brooks Range into two smaller mountain groups. Elevations vary from sea level where the river enters Kotzebue Sound to over 1,830 m in the mountains, with about 66% of the watershed occurring below 610 m elevation. Lowland areas (below 610 m) are generally covered with tussock tundra with patches of white spruce forest along the lower reaches of the river. Soils are slightly acidic and have surface organic layers in the intertussock spaces of 5-14 cm thickness.

The generally lower than 914 m, 320 km by 192 km Seward Peninsula consists of rolling uplands, interior basins, coastal lowlands, and a central mountain range. Deep silty loess deposits cover much of the area, and soils generally are moderately to highly acidic, with organic matter thicknesses in the intertussock spaces of up to 20-30 cm. Sedge tussock tundra is dominant on the rolling gentle slopes, with willow shrub thickets along drainageways and sedge wet meadows on poorly drained flats.

The known fire history for the Seward Peninsula has been reviewed by Racine (1979) and that for the Noatak by Racine et al (in preparation). Tundra fires have been fairly extensive in both areas during the 1970's. An estimated 3,000 km^2 burned in the Seward Peninsula in July 1977 and 640 km^2 burned during 1977 in the Noatak basin. Based on present data, tundra fires appear to be of more frequent occurrence in these two areas than elsewhere in arctic Alaska.

Eight different burns representing 1971 (1), 1972 (2), 1977 (3), and 1982 (2) fires were visited in the Seward, Noatak, or both (Table 1, Figure 1). All of the Noatak River fires visited were located near the river in the lower to middle part of the watershed below an elevation of 305 m. Seward Peninsula sites included Imuruk Lake and Utica Creek on a large (940 km^2) 1977 burn, a 1971 burn at Coffee Creek about 75 miles north of Nome, and a second 1977 burn at Arctic River in the northwestern part of the peninsula. Data from these fires were supplemented with information on unburned tundra vegetation and soils obtained in 1973 by Racine and Anderson (1979) and Holowaychuk and Smeck (1979) in the Seward Peninsula and by Ugolini and Walters (1974) in the Noatak River area.

METHODS

Thaw depths were measured with a steel probe pushed vertically into the ground until stopped by the frost table. Therefore, rather than measuring the distance to the 0° C isotherm in the soil, we measured the distance to an ice-cemented surface. Thaw depths were measured both from the ground surface (active layer thickness) and/or from a horizontal reference string tied tightly between two stakes 5-10 m apart. Also recorded was the distance from the string to the ground surface so that profile cross sections of both the ground surface and the frost table could be constructed to examine interrelationships and the roughness of both. When probing in tussock tundra, the type of surface (live or dead tussock, intertussock space, frost scar, etc.) through which the probe passed was recorded each time. Soil temperatures were measured with a YSI Telethermometer and a long steel thermistor probe.

Probing was carried out in conjunction with plant cover, density, and biomass sampling in 1 m x 1 m plots or along transects extending from unburned into burned tundra. One thaw depth measurement was taken in the center of each 1 m x 1 m plot, and in most plots 4 additional measurements were taken midway between the center and the corners of the plot. At 3 Seward Peninsula sites (Table 1), thaw depths and vegetation were sampled once in each of 3 years following a 1977 fire. In addition, prefire (1973) thaw and vegetation measurements were obtained by Racine (1979) and Holowaychuk and Smeck (1979) at the Seward Peninsula-Imuruk Lake sites. During July and August of either or both 1981 and 1982, several sites in the Noatak area (Table 1) were probed at 10-14 day intervals. Additional one-time depth-of-thaw measurements were obtained at remote sites in both the Noatak and Seward Peninsula, representing both burned and unburned tundra.

RESULTS

Depth of Thaw

Because thaw depths vary seasonally and with surficial conditions of soil texture, exposure, drainage and vegetation cover, we found it necessary to control these variables as much as possible to permit detection of the effects of fire on thawing. We achieved this control by limiting our discussion primarily to tussock vegetation on 2 types of terrain: on flats (slopes less than 5%) and on slopes (angles of 5% or more). Because we could not obtain unburned (control) thaw measurements at all burn sites and because measurements were taken at the different sites in different years and at different stages in the progession of summer thawing, we constructed composite seasonal thaw curves for unburned (control) tussock tundra in the Seward Peninsula (Figure 2) and the Noatak (Figure 3).

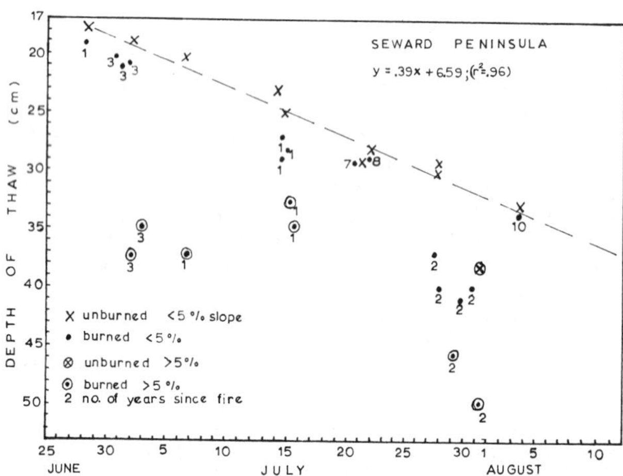

FIGURE 2 Thaw depths in burned and unburned tussock tundra in the Seward Peninsula at different times during the summer, on different slope angles (< 5% vs. >5%), and for varying number of years since fire.

Seward Peninsula. Although the unburned thaw depth measurements were obtained in different years and at different sites, the resultant curve appears to define fairly well (r^2 = .96) the seasonal thaw pattern for unburned (control) tussock tundra on flat terrain in the Seward Peninsula (Figure 2). By plotting thaw depths similarly measured in relatively flat tussock tundra sites in the Seward Peninsula representing 1, 2, 3, 7, 8, and 10 years following fire (Figure 2), we can determine that the depth of thaw increased by a maximum of 30-40% by the second year following fire, after which it returned nearly to control levels by the seventh, eighth, and tenth years.

Burned tussock tundra sites on steeper slopes showed a much greater and probably longer lasting thawing response (Figure 2). Without a reliable control for these sloping sites, we are unable to compute the actual percentage increase in thawing due to fire for these slopes. In 1 case, however, we found increased thawing of over 100% 1 year after a fire on a steep, 12-13% slope.

Noatak River. The unburned (control) thaw curve for the Noatak River area (Figure 3) was constructed from a composite of several thaw curves obtained by monitoring thaw at unburned sites through portions of the 1981 and 1982 summer thaw seasons. We also used 1960 thaw data from Ogotoruk Creek at Cape Thompson (Johnson 1963), west of the Noatak study area. The composite curve is similar to that determined for the Seward Peninsula, with a daily rate of thaw of 0.4-0.5 cm per day for both areas.

Thaw depth measurements in July-August 1982 on a June 1982 tussock tundra burn in the Noatak show a first year increase in thaw of 20-30%. On a nearby 10 year old tussock tundra burn, thaw depths were only 10-15% deeper than the control. Three, 4 and 5 year old burned tussock tundra sites were thawed 16-22% deeper than the control. Other 4 and 5 year old burned tussock tundra sites sampled in the Noatak were located on slopes greater than 5%, for which no

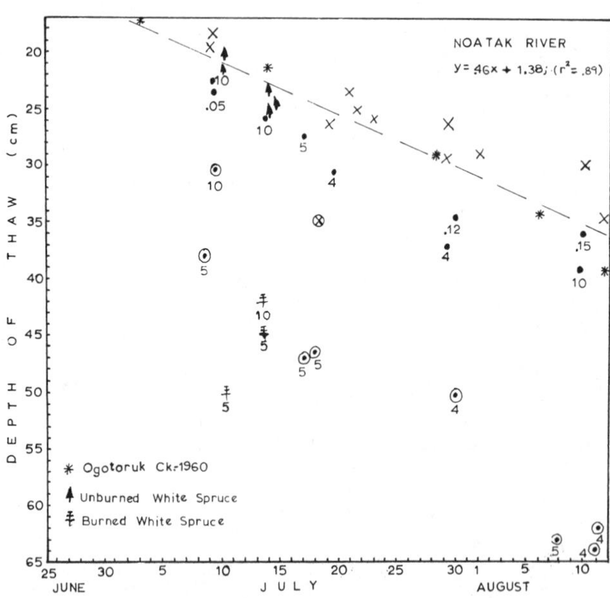

FIGURE 3 Thaw depths in burned and unburned tussock tundra in the Noatak River watershed at different times during the summer, on different slope angles (< 5% vs. > 5%), and for varying numbers of years since fire. For symbols see Figure 2. Unburned 1960 tussock tundra thaw measurements for Ogotoruk Creek Valley, Alaska are from Johnson (1963).

unburned control curve was available. In agreement with the Seward findings, we found that these burns thawed more deeply than burns on flat tussock tundra sites.

Based on our limited measurements and those of Ugolini and Walters (1974), we conclude that thaw depths in unburned white spruce stands in the Noatak are shallow and similar to those in unburned tussock tundra (Figure 3). In contrast, we measured thawing in a spruce area burned 5 years earlier that was 2 times deeper than in an adjacent unburned stand.

Soil Temperatures

To examine the thermal characteristics of the observed initial increase in thawing followed by a return to nearly prefire levels by the tenth year, we obtained measurements of soil temperatures in unburned, recently burned (1982), and 10 year old burned (1972) tussock tundra sites in the Noatak River area (Table 2). Temperatures at depths of 10 cm below the unburned intertussock ground surface occured in the range of 2°-5° C. In contrast, temperatures 10 cm below the tops of surrounding unburned tussocks were 11°-13° C, which indicates a 6°-8° C temperature difference between the tussocks and intertussock spaces.

During a clear warm period in July 1982 we found intertussock soil temperatures were 7°-10° C at 10 cm depths in recently burned (June 1982) tussock tundra. These values were significantly higher by 3°-5° C than in the unburned controls (Table 2, $t = 7.0$, $p \leq 0.001$). Similarly, tussock temperatures at 10 cm below the tussock top also were warmer than in unburned tussocks by 5°-8° C during this period ($t = 6.7$, $p \leq 0.001$). Likewise, soil temperatures during July 1982 at 20 cm depths were significantly lower by 2.7° in unburned intertussock spaces than in recently burned intertussock spaces ($t = 8.9$, $p \leq 0.001$). Lastly, during the warm July 1982 period, temperatures at 10 cm depths in tussocks and intertussock spaces of tundra burned 10 years ago were not significantly higher than the unburned control (Table 2), but were significantly lower by 3°-4° C than temperatures at comparable depths in the recent burn ($t = 5.1$; $p \leq 0.001$).

Such differences were ephemeral, however, and depended on weather conditions. With the onset of rains and cool, cloudy weather in August, we detected little temperature difference between the recently burned area and the unburned control (Table 2).

DISCUSSION AND CONCLUSIONS

Role of Revegetation

The results presented here suggest that, following fire in tussock tundra on level terrain, active layer thicknesses and soil temperatures rise relatively rapidly and then return to preburn levels within as few as ten years. We hypothesize that this relatively rapid recovery is due largely to a rapid reestablishment of graminoid cover following fire (Figure 4).

At one Noatak River tussock tundra site burned in 1977, and probably representing an old drained lake basin, live vascular plant cover had reached levels of 75-90% in five years, due to the rapid regrowth mainly of *E. vaginatum* (36%) and *Betula nana* (26%). Six weeks after a June 21, 1982 tussock tundra fire on the Noatak River (Figure 1), we measured cover of 40% contributed mainly by regrowth of *E. vaginatum* tussocks which had leaf lengths of 20 to 30 cm by the end of July. Such data show how quickly *E. vaginatum* leaves can grow to overhang and shade the intertussock spaces, increasing reflection and reducing the penetration of radiation.

Racine et al. (in preparation) found that about one half of the recorded tundra fires between 1956 and 1981 in the Noatak watershed occurred during June, with the other half occurring during July. For those June burns

TABLE 2 July-August 1982 Tussock Tundra Soil Temperatures (° C) at 10 and 20 cm Depths in Intertussock Spaces (IT) and in *Eriophorum vaginatum* Tussocks (TT) and Depth of Thaw (cm) in Intertussock Spaces in an Unburned Control, a June 21, 1982 Burn and a 10-Year Old Burn (Average ± 1 Standard Deviation, n = 4-10).

	Unburned Control			June 21, 1982 Burn			July 1972 Burn	
	11 July	30 July	9 August	11 July	29 July	10 August	14 July	9 August
Temperatures at 10 cm								
IT	2.4±0.8	4.4±0.8	4.1±0.9	7.1±1.0	8.1±1.0	4.9±1.0	4.3±1.2	4.6±1.0
TT	12.7±1.5	12.4±0.6	10.9±0.3	20.7±0.1	16.7**	11.7±1.3	14.3±1.4	11.1±0.8
Temperatures at 20 cm								
IT	0.0*	2.2±0.4	2.4±0.7	3.3**	4.9±0.3	2.8±0.7	1.0±0.6	3.1±0.7
TT	7.6±2.1	9.1±0.6	9.4±1.0	15.8±0.4	13.9**	8.4±2.0	8.8±2.1	7.9±0.6
Depth of Thaw								
IT	18.4±5.5	27.2±3.8	29.4±5.4	24.3±5.3	35.7±5.3	36.2±5.7	26.4±5.0	40.5±5.2

* Assumed
** n = 1

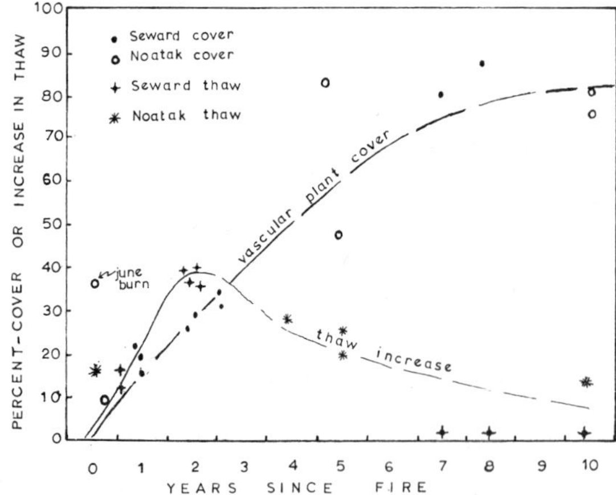

FIGURE 4 Percent cover and percent increase in thaw depth in relation to time since fire in tussock tundra.

followed by favorable growing conditions, the rapid regrowth of sedges as described above could reduce heat absorption and increase surface albedo relatively quickly. In contrast, data from mid to late July burns indicate slower rates of revegetation: in the Noatak by the end of July live vascular plant cover on a mid-July burn had reached only about 10% with E. vaginatum leaf lengths of only 4-9 cm; following the late July 1977 Kokolik River burn on the North Slope of Alaska, Hall et al. (1978) reported only 5-10% vascular plant cover regrowth by the end of August 1977. These observations suggest that vegetation regrowth does not contribute to reduction of heat absorption and increase of albedo in the same year of a mid-summer fire. These differences in system response to timing of fire suggest that there also may be some difference in rate of recovery of the permafrost table.

Role of Organic Thickness

In addition to influences of revegetation on postfire thawing patterns, the thickness of the intertussock organic soil horizon remaining after fire further controls the rate of postfire permafrost recovery. Our measurements suggest that in the Seward Peninsula, tussock tundra generally has deeper organic horizon thicknesses (20-30 cm) and greater sphagnum moss cover than in the Noatak, where tussock tundra organic horizons are only 10-14 cm and sphagnum is less common. In the Seward Peninsula on most tussock tundra sites, less than 5 cm of this organic horizon is lost in a fire (Racine 1977) and, as a result, the increase in thaw 7-10 years after fire is proportionally lower in the Seward Peninsula sites than in the Noatak sites (Figure 4). Based on this finding, we suggest that postfire buildup of the organic horizons is not rapid enough to account for the observed rapidity of postfire stabilization.

Role of Slope

On slopes greater than 5%, postfire thawing was always deeper than in tussock tundra on flats. Unfortunately, without an adequate unburned control for such slopes, we cannot determine percent increases in thaw due to fire. We hypothesize that the increased thawing we found on slopes may relate to greater drainage and insolation which would result in greater drying, leading to greater fire-caused fuel consumption and damage to vegetation and underlying organic soil horizons. Delayed revegetation due to slower plant growth rates where tussocks are absent or sparse may prolong periods of increased susceptibility to thawing while increased insolation at the soil surface may produce greater thawing. We generally have found prefire tussock density and cover to be lower, and the burning to be more severe, on slopes than on flats, so that slopes would not benefit as much from the combination of rapid tussock regrowth and survival of an insulating organic horizon. In addition, the shrubs that are proportionately more abundant in the intertussock spaces of tussock tundra on steeper slopes usually do not contribute significantly to regrowth of cover (25% or more) until at least 7-10 years after fire. Lastly, although bryophytes (e.g., Ceratodon purpurea, Marchantia polymorpha) can develop high cover rapidly on burned sites, we do not believe that they significantly reduce the ability of soil to thaw and may actually slow the development of a vascular plant cover.

Tundra/Taiga Comparison

Viereck (1981) suggests that thaw stabilization following fire may be more rapid in arctic regions of Alaska with low mean annual temperatures (-8° to -9° C) than in taiga areas of interior Alaska having higher mean annual temperatures (-3.5° C). However, our results from burned tundra flats and slopes of the Noatak River and Seward Peninsula areas suggest that recovery of the prefire permafrost table may be delayed significantly where surface revegetation is slow, and that regrowth of the vegetation is a function of the preexisting vegetation, slope characteristics, and timing of the fire.

ACKNOWLEDGEMENTS

Our tundra fire studies were carried out between 1978 and 1981 in the Seward Peninsula through support from the Bureau of Land Management (BLM), U.S. Army Cold Regions Research and Engineering Laboratory (CRREL), National Park Service (NPS) and Earthwatch; and in 1981 and 1982 in the Noatak through support from the U.S. Man and the Biosphere program, NPS, and Earthwatch. We thank the several reviewers of the draft manuscript who provided valuable comments and suggestions.

REFERENCES

Black, R.A. and Bliss, L.C., 1978, Recovery sequence of Picea mariana - Vaccinium uliginosum forest after burning near Inuvik, Northwest Territories, Canada: Canadian Journal of Botany 56, p. 2020-2030.

Brown, J. and Grave, N.A., 1979, Physical and thermal disturbance and protection of permafrost: U.S. Army Cold Regions Research and Engineering Laboratory Special Report 79-5, 42p. Also in 1978, Proceedings of the Third International Conference on Permafrost, 2: Ottawa, National Research Council of Canada, p. 51-92.

Brown, J., Rickard, W., and Vietor, D., 1969, The effect of disturbance on permafrost terrain: U.S. Army Cold Regions Research and Engineering Laboratory Special Report 138, 13 p.

Hall, D., Brown, J. and Johnson, L., 1978, The 1977 tundra fire in the Kokolik River area of Alaska: Arctic, 31, p. 54-58.

Holowaychuk, N., and Smeck, N.E., 1979, Soils of the Chukchi-Imuruk area, in Biological Survey of the Bering Land Bridge National Monument (H.R. Melchior, ed): Revised final report, Alaska Cooperative Park Studies Unit, University of Alaska, Fairbanks.

Johnson, A.W, 1963, Plant ecology in permafrost areas, in Proceedings Permafrost International Conference, National Academy of Sciences-National Research Council Publication 1278, p. 25-30.

Mackay, J.R, 1977, Changes in the active layer from 1969 to 1976 as a result of the Inuvik fire, Report of Activities Part B, Geological Survey of Canada, Paper 77-1B, p. 273-275.

Racine, C.H, 1979, The 1979 tundra fires in the Seward Peninsula, Alaska: effects and initial revegetation: Bureau of Land Management, Alaska, Technical Report 4.

Racine, C.H., 1981, Tundra fire effects on soils and three plant communities along a hillslope gradient in the Seward Peninsula, Alaska: Arctic 34, p. 71-85.

Racine, C.H., and Anderson, J.H., 1979, Flora and vegetation of the Chukchi-Imuruk area, in Biological Survey of the Bering Land Bridge National Monument (H.R. Melchior, ed.): Revised final report, Alaska Cooperative Park Studies Unit, University of Alaska, Fairbanks.

Racine, C.H., Dennis, J.G., and Patterson III, W.A., (in preparation), Analysis of the tundra fire regime in the Noatak River watershed, Alaska: 1956-1981: Arctic (submitted).

Ugolini, F., and Walters, J., 1974, Pedological survey of the Noatak River Basin, in The Environment of the Noatak River Basin, Alaska, (S.B. Young, ed.): Center for Northern Studies, Wolcott, VT.

Viereck, L.A, 1981, Effects of fire and firelines on the active layer thickness and soil temperatures in interior Alaska, Proceedings of the Fourth Canadian Permafrost Conference: Ottawa, National Research Council of Canada, p. 123-135.

Viereck, L.A., and Schandelmeier, L.A., 1980, Effects of fire in Alaska and adjacent Canada - a literature review: Bureau of Land Management, Alaska, Technical Report 6, 124 p.

Wein, R.W., and Bliss, L.C., 1973, Changes in arctic Eriophorum tussock communities following fire: Ecology, v 54, p. 845-852.

DISTRIBUTION AND GEOLOGY OF GROUND ICE ALONG THE YUKON PORTION OF THE ALASKA HIGHWAY GAS PIPELINE

V. N. Rampton[1], J. R. Ellwood[2], and R. D. Thomas[1]

[1]Terrain Analysis and Mapping Services Ltd., Carp, Ontario, Canada K0A 1L0
[2]Foothills Pipe Lines (Yukon) Ltd., Calgary, Alberta, Canada T2P 2V7

Ground ice occurs within specific geologic environments along the Alaska Highway Gas Pipeline in the Yukon Territory. At shallow depths, ice is commonly in the form of ice lenses (aggradational); rarely in the form of ice wedges. Great thicknesses of aggradational ice are present where thick fine-grained postglacial sediments have accumulated. Epigenetic ground ice has formed in fine-grained sediments under poorly drained ground adjacent to streams, lakes, swamps, ponds and drainageways. It has also formed in fine-grained sediments draped over truncated permeable strata, in fine-grained sediments complexly interbedded with coarse-grained sediments and in fine-grained sediments adjacent to bedrock valley walls. Generally, epigenetic ground ice has formed in environments where ample groundwater has been supplied to aggrading permafrost tables, environments characterized by high groundwater gradients.

INTRODUCTION

The Alaska Highway Gas Pipeline parallels the Alaska Highway in the Yukon Territory (Figure 1). It follows mainly valleys and lowlands. All terrain except the first 5 km has been glaciated. However, distribution of permafrost and variations in local geology, most specifically the past and present hydrogeologic environment, governs the distribution of ground ice.

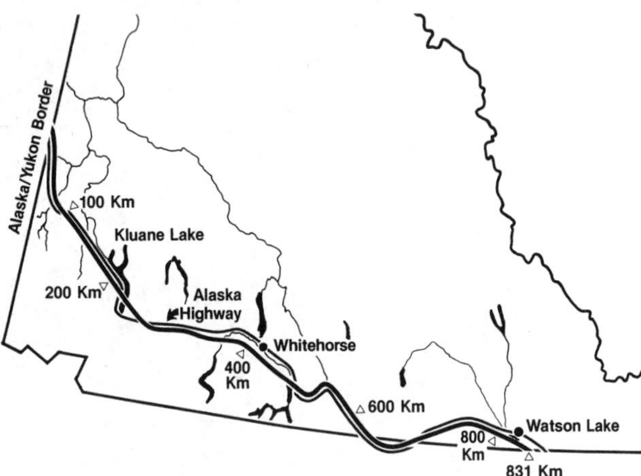

Figure 1: The Alaska Highway Gas Pipeline Route through Yukon.

Investigations conducted along the pipeline right-of-way indicate that more than 80% of the terrain traversed north of Kluane Lake is perennially frozen, that significantly less than one-half of the terrain traversed between Kluane Lake and Takhini River (KP 379) is perennially frozen, that less than 20% of the terrain traversed between Takhini River and KP 653 is perennially frozen; and that less than 5% of the terrain traversed south of KP 653 is perennially frozen (Table 1). Ground ice is found in different concentrations throughout most frozen ground, but it is only present in high concentrations in specific geologic environments.

Ground ice is most abundant in the following permafrost environments: (1) shallow depths--mainly aggradational ice and locally ice wedges (ground ice classification according to MacKay 1972); (2) areas of fine-grained postglacial deposits--alluvium and eolian silt; (3) poorly drained areas adjacent to surface water--mainly swamps, lakes and streams; (4) areas where near surface fine-grained strata receive groundwater recharge from permeable strata; (5) valley bottoms having extensive fine-grained sediments adjacent to bedrock valley walls; and (6) disintegration moraines--complexes of glacial deposits formed on stagnant glacial ice.

SHALLOW GROUND ICE

Ice lenses occur frequently in the upper 2-4 m where permafrost is present (Table 1), especially where fine-grained sediments overlie coarse-textured sediments. Typical areas occur north of Kluane Lake between KP's 133 and 221 (cf. 78-CS-2 in Figure 2) where till and glaciofluvial deposits are adjacent to broad braided channels or outwash plains, which provided moderate amounts of loess upon deglaciation. During deglaciation permafrost began to aggrade and an active layer was established. As loess was deposited, the base of the active layer rose and water perched on the permafrost in the fall eventually became incorporated into the top of the permafrost. In areas where loess has been re-deposited by slopewash, aggradational ice is similarly incorporated into the profile.

Aggradation ice has also been incorporated into the profile where the active layer has thickened and thinned due to effects of climatic changes, forest fires and mass wastage. Peat beds in a sequence of icy colluvium (mainly re-worked loess)

Table 1: Summary of frozen/unfrozen terrain - ground conditions.

	Ground Condition Below 3 Metres Depth (%)				Some Med-High* Ice Content
KPs	Frozen	Probably Frozen	Uncertain	Unfrozen	Upper 2 m (%)
0-55	36.2	30.0	19.9	13.9	81.2
55-110	50.5	32.0	8.1	8.4	65.8
110-165	69.7	17.8	9.2	3.3	96.8
165-219	66.3	15.3	2.6	15.8	58.2
Kluane Lake					
225-300	14.2	3.2	35.0	47.5	29.0
300-379	4.9	1.7	26.2	67.3	12.9
379-451	2.3	—	—	81.3	1.3
451-520	4.0	0.0	2.4	93.6	7.8
520-588	0.6	—	—	83.3	2.4
588-653	2.1	—	—	78.9	16.6
653-718	0.0	—	—	98.0	0.1
718-760	0.0	—	—	99.2	0.5
760-785	0.6	—	—	98.8	3.1
785-800	0.0	—	—	100.0	0.0
800-831	0.0	—	—	99.5	0.0

* Low - <10% excess ice by volume; Medium - 10-20% excess ice by volume; High - >20% excess ice by volume.

78 - CS - 2 (T) 78-08

USC	Dp	PfC	MC
VA	0.25	F	254
OL & Pt	0.80	Nbe	20
SM	1.10		
OL & Pt	1.50	Vx	171
GM sML cbl (till-like)	3.00	Vr Vx	18 18
SC fc aCL sG (till-like)	6.50	4.65 Nbe	28 16 9
Cl (G) (till-like)	9.10	7.10 F	7 10 15
GC aCl lfcS	↓ 11.50		6

15.10 m END OF HOLE

76 - O - 1 76-09

USC	Dp	PfC	MC
OL lS	0.60	UF	61
VA	1.00		33
OL lS	1.37	1.45	86
Pt - OL lyrd	5.03	Nbe; Nbe Vs Vr; Nbe; Vs; Nbe	170 227 185 135 108 134 153
ML lS org incl	5.79	Vr	76
G aS tML few cbl lyrd bldr or bdrk	8.84	N	72 17 17 16 15

8.84m END OF HOLE

METRES

Figure 2: Bore hole logs for 78-CS-2(T) and 76-01-1. Peat (Pt), loess (OL, SM) and volcanic ash (VA) in 78-CS-2(T) show high moisture contents (right-hand column) whereas the underlying glacial drift have low moisture contents except for the upper few metres. Thick organic-rich loess (Pt-OL, ML) in 76-01-1 shows high moisture contents throughout its thickness. USC - Unified Soil Classification; Dp - Depth; PfC - Permafrost Classification; MC - Moisture Content.

Figure 3: Ice wedges buried by organic-rich fine-grained alluvium, volcanic ash and peat near Quill Creek.

and peat on a slope at the Quill Creek test site have been dated at 10,600 ± 170 (BGS-738), 6,340 ± 140 (BGS-739) and 3,670 ± 120 (BGS-751). This suggests that thaw intervals occurred shortly after 10,600 B.P., 6,400 B.P. and 3,700 B.P. allowing material to be eroded downslope. At sites of deposition the active layer rose and ice lenses were incorporated into the profile. Not only has aggradational ice developed here, but it also must have developed in flat areas where the active layer thickened and thinned without sediment accumulation. Climate may have been warmer than present between about 10,000 and 3,500 years B.P. and then subsequently was cooler than present until recently (Denton and Stuiver 1967, Rampton 1970). The White River Ash (deposited about 1,220 years ago (Hughes et al. 1962)) shows no sign of movement at the Quill Creek test Site (KP 155) and throughout the area; this suggests frigid conditions and general landscape stability since its deposition.

Ice wedges are present in some areas north of Kluane Lake. Polygonal patterns (their surface expression) can be recognized on air photos of terrain underlain by fine-grained sediments near the Alaska border. Further south where ice wedges are weakly active or inactive, they are present in peat near White River and in the subsurface at the Quill Creek test site (Figure 3). At the latter location the larger ice wedges are relic as they are buried by alluvium, volcanic ash and organics. Ice wedges are rarely present in coarse-textured fluvial deposits and drift; e.g., Hughes et al. (1962) noted polygonal trenches developing on morainic topography stripped of its organic cover near KP 50.

EOLIAN AND ALLUVIAL ENVIRONMENTS NORTH OF KLUANE LAKE

Postglacial eolian and fluvial activity has resulted in the accumulation of thick fine-grained deposits, especially in areas where sources of wind-blown loess are close at hand (Rampton 1981).

For example, silt blowing off the Donjek River's braided floodplain (KP 140) onto moss-covered substrata has resulted in the buildup of organic-rich and icy loess to depths of over 7 m (cf. 76-0-1 in Figure 2). The continuous rain of loess upon the landscape plus alluvial action and fluctuations in climate leading to mass wastage have also resulted in redeposition of fine-grained alluvium in valley bottoms during postglacial time. As shown by the nature of the permafrost and high moisture contents in borehole log 81-01-060 (Figure 4), ice lenses are very common in the more than 10 m of fine-grained organic sediment that accumulated on an imperfectly drained alluvial plain during the last 8,000 years. Variable thicknesses of icy organic-rich fine-grained alluvium, which cap gently-sloping coarse-textured sediments in other places (e.g. between KP's 25-40, 55-71, 75-125, and 169-213), generally have similar origins.

Ice lenses are also associated with thin beds of fine-grained sediments within coarse-textured alluvial fans. This ice probably has an epigenetic origin following adandonment of the fan by surface and subsurface water flow, which would tend to maintain unfrozen ground conditions.

Of special note is the distal edge of a large alluvial fan north of Beaver Creek where organics and fine-grained alluvium overlie gravel and sand (KP 7-16). Groundwater movement through this fan has maintained relatively high subsurface temperatures. At the toe of the alluvial fan where it abuts against bedrock, springs and taliks have developed because of warmish water being forced to the surface there. Ground ice is restricted to the organics and fine-grained alluvium due to the groundwater flow (cf. 80-11-010 in Figure 4). Ground ice here is probably epigenetic, as the area has been subjected to a number of thermokarst cycles.

PERMAFROST ENVIRONMENTS ADJACENT TO SURFACE WATER

In sporadic permafrost south of Kluane Lake, ground ice is abundant where ample groundwater has been supplied to aggrading permafrost through the subsurface from water bodies or from groundwater moving toward these water bodies. Under normal conditions, the aggrading permafrost table is isolated from a water supply by the initiation of permafrost development, but where water bodies occupy taliks, water can be laterally supplied to the permafrost table (Figure 5). Thus extensive ice lensing and massive ice is often found adjacent to swamps, ponds, lakes, rivers and drainageways.

Massive ice and ice lenses under palsa-type mounds extend to depths of 9 m near Sulphur Lake (KP 258) and to depths of 4-5 m near Jake's Corner (KP 504) and Morely River (KP 614-625) in areas adjacent to ponds or swampy areas in broad poorly drained depressions (Figure 6). Groundwater must have moved from the ponds (commonly of thermokarst origin) or from the taliks underlying the ponds to an aggrading permafrost table in the adjacent terrain. Permafrost and ground ice has probably aggraded and degraded intermittently during postglacial time, as witnessed by active thermokarst at these localities.

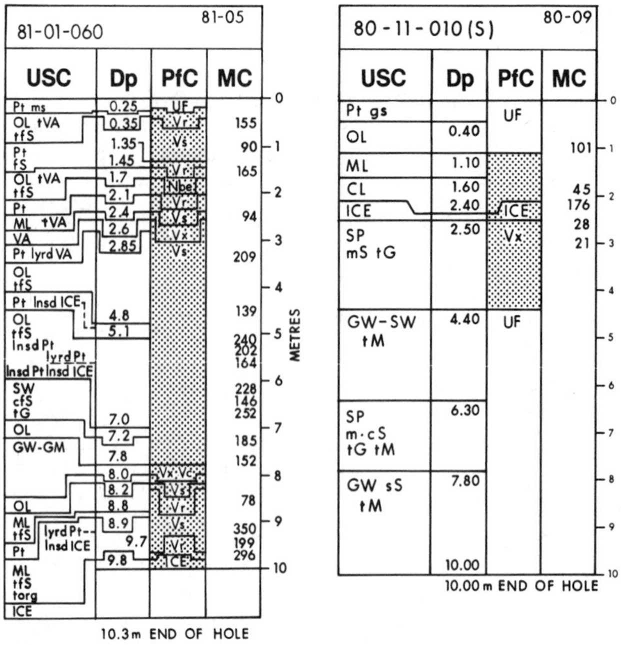

Figure 4: Borehole logs for 81-01-060 and 80-11-010(S). See Figure 2 for explanation of USC, Dp, PfC, MC.

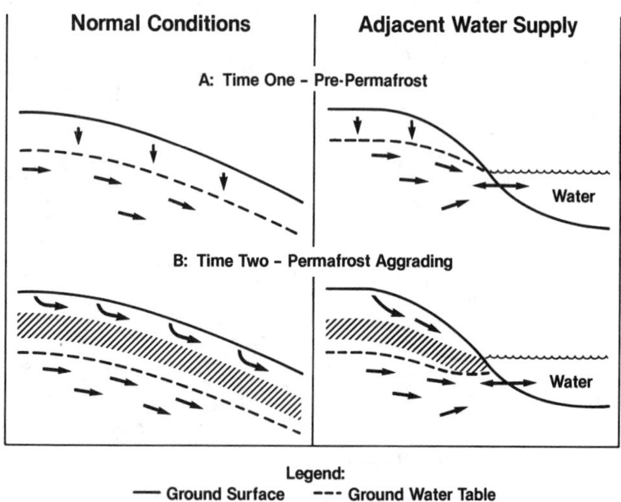

Figure 5: Ground water conditions in areas of aggrading permafrost where no surface water body is present ("normal conditions") and with surface water body present. Note that water may flow to or from water body depending upon position of water table.

Abundant ice lenses and massive ice are present in sediments underlying low terraces adjacent to Pine Creek (KP 288.5), Marshall Creek (KP 300), Aishihik River (KP 315) and McClintock River (KP 466.5). At Marshall Creek and Aishihik River, ground ice is found in sediments underlying flood

80-P/L-C-15B (LAGOON) 80-04

USC	Dp	PfC	MC
ML IS tG	1.20	UF	17
	2.00	F	
ICE + CL	2.20	ICE+	44
	3.00		57
	3.80	Nbe	26
		ICE+	50
			120
			47
			80
			98
			75
CL	7.00	UF	28
SM IM IC	9.00		
	10.00		24

10.00 m END OF HOLE

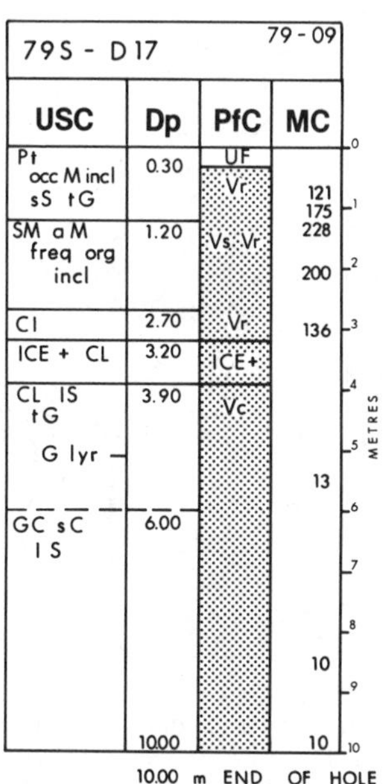

Figure 6: Logs of boreholes drilled adjacent to ponds or poorly drained ground near Sulphur Lake (80-P/L-C-15B) and Morley River (79S-D17). Drill holes on well-drained uplands adjacent to 80-P/L-C-15B are characterized by low ice contents. See Figure 2 for explanation of USC, Dp, PfC, MC.

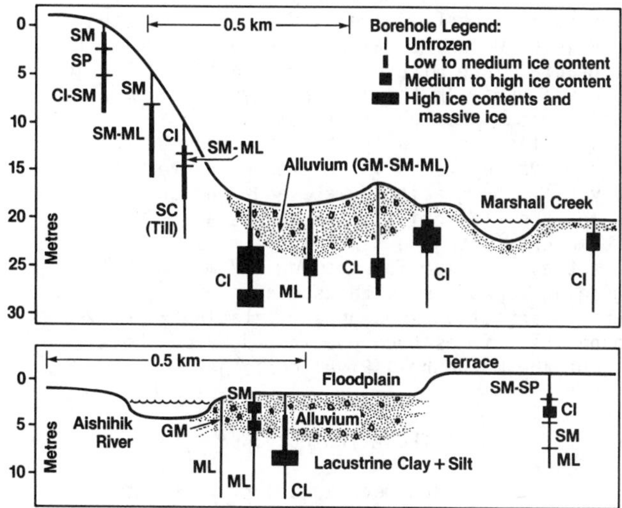

Figure 7: Ground conditions underlying Marshall Creek and Aishihik River terraces and floodplains. See Table 1 for quantification of ice contents.

plains and low terraces to depths of 9 and 6 m, respectively (Figure 7). Permafrost aggradation and the formation of ground ice at these localities has undoubtedly occurred after the formation of the postglacial terraces and floodplain, as the ground immediately under the stream creating these landforms would have been thawed at that time. The ground ice has developed in zones of high groundwater gradients; whether the area is one of

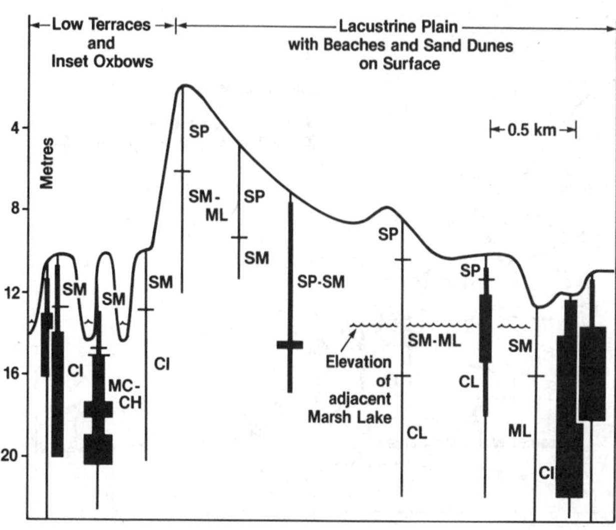

Figure 8: Ground conditions underlying terrain adjacent to McClintock River and Marsh Lake. See Figure 7 for borehole explanation.

discharge to the streams, recharge from the streams or varies according to season is not known. Clearly however, ample water was being provided to the aggrading permafrost table.

Near McClintock River (KP 466.5), ice lenses and massive ice have developed to depths of 9 m in glaciolacustrine clays and silts that underlie lacustrine and eolian sands (Figure 8). Ice underlying the lacustrine plains postdates the drainage of the late-glacial lake; ice underlying the terraces postdates their formation. The abundance of ground ice can be explained by an ample supply of water moving short distances from the water bodies (river or lake) along sand partings in the clays to an aggrading permafrost table in the lacustrine sediments. Ground ice present to depths of 6.5 m under an isolated palsa-like mound in the large alluvial-lacustrine complex near Marsh Lake's outlet (KP 460) probably developed similarly.

PERMAFROST ENVIRONMENTS ADJACENT TO AREAS OF GROUNDWATER DISCHARGE

Ice lenses are common in the lacustrine clays and clayey tills that form the near-surface substrata on the slope west of Pine Creek (KP 288). Here lacustrine sediments blanket a valley bottom (Figure 9). Erosion of unconsolidated deposits by stream and glacier action to form the valley truncated permeable strata and allowed groundwater to be discharged on the surface of the slope as evidenced by springs at the base of scarps along the slope. Thus ample groundwater was supplied quickly to any aggrading permafrost table on the lower slope. Massive ice in sediments east of Cracker Creek (KP 334) may have a similar origin.

Unconsolidated sediments forming the moderately sloping banks of major stream valleys, e.g., the Marshall Creek valley (KP 299) and the Takhini River valley (KP 379), also contain ground with

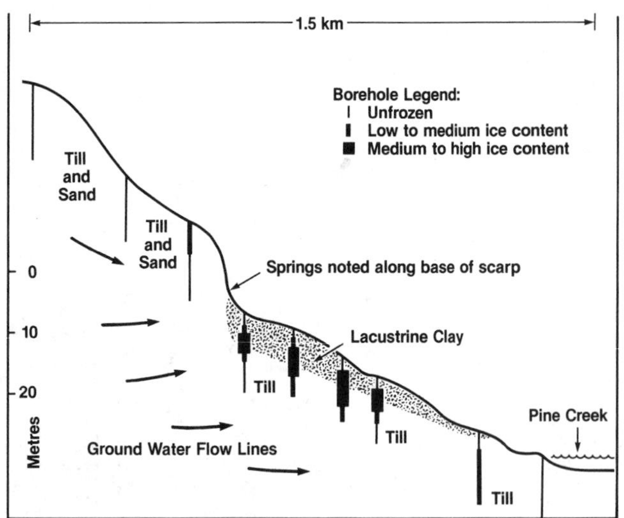

Figure 9: Ground conditions underlying slope to west of Pine Creek. See Table 1 for quantification of ice contents.

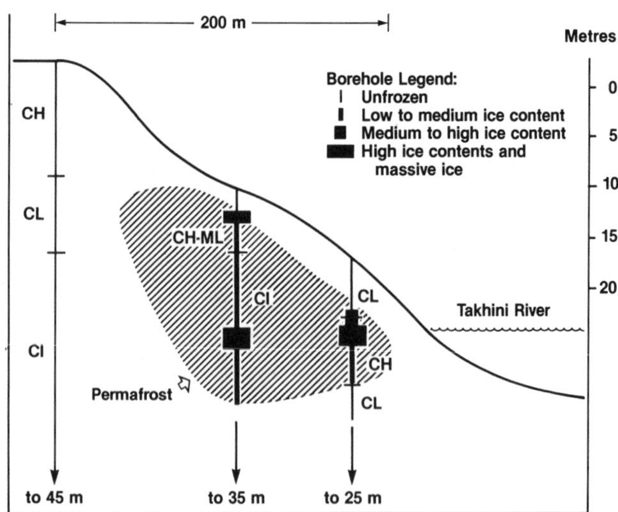

Figure 10: Ground conditions in west bank of Takhini River. See Table 1 for quantification of ice contents.

high ice contents, which may be due to groundwater discharge. At Takhini River, geophysics and drilling outlined a wedge of frozen low-permeability clays with high ice contents (Figure 10). However, exposures along the Takhini River show sand partings and bands of silty fine sands interbedded in the glaciolacustrine sequence. Probably, groundwater moving through these more permeable strata created a damp environment on the river banks. This allowed vegetation favorable to permafrost aggradation to flourish. Groundwater continued to flow to the aggrading permafrost table, and ice lenses and massive ice developed. Obstructions of groundwater flow lines by the permafrost would cause the groundwater to move above and below the initial permafrost pod, and cause icy permafrost to form at some height above stream level and the original level of the groundwater discharge.

PERMAFROST ENVIRONMENTS ADJACENT TO BEDROCK VALLEY WALLS

Large areas near the base of extensive bedrock valley walls are underlain by massive ground ice to depths exceeding 9 m in the Takhini and Medenhall River (KP 359) valley bottoms. These areas are marked by an abundance of thermokarst depressions.

Permafrost probably formed immediately after drainage of the glacial lake, especially near the base of valley walls where moist conditions would prevail. Water falling on the surrounding bedrock-cored uplands entered the groundwater system by seeping directly into the bedrock or through permeable strata on the bedrock surface. This water was probably discharged at the base of the valley walls or flowed through sediments to incised streams as idealized in Figure 11. Thus strata at the base of the aggrading permafrost were saturated by groundwater from the surrounding uplands in a zone of high groundwater gradients, and ground ice formed as the permafrost aggraded.

Areas where permeable sandy beds are interbedded with the silts and clays are most favorable for the accumulation of ground ice, as water can quickly move toward the aggrading permafrost table. This may be the case in the Mendenhall River valley, where currents may have deposited sandy belts when the valley acted as an outlet for the glacial lake within the Mendenhall and Takhini River valleys. In the Takhini River valley, areas of high ground ice are most common near bedrock valley walls, especially where the base of the adjacent valley walls is covered by permeable sandy tills or glaciofluvial sands and gravels. Perhaps the glaciofluvial sediments are interlayered with the glaciolacustrine sediments and act as a medium for groundwater movement to aggrading permafrost. Lacustrine sediments in the central part of valleys appear relatively free of ground ice, as has been documented by boreholes west of the Mendenhall River crossing.

DISINTEGRATION MORAINES

Part of the route between Enger Lakes and White River (KP 40-KP 63) crosses disintegration moraine (Rampton 1971) characterized by discontinuous lenses and beds of interbedded diamicton (till), silt, sand, and gravel (Figure 12) in an area of relatively continuous permafrost. Thick ice lenses and massive ice are common to depths of over 20 m. Many strata are probably inclined and vertical due to melt-out of dead ice containing much englacial debris and underlying much supraglacial debris and to depositional and deformational processes common to the terminal zone of glaciers. The diamicton (till) in the disintegration moraines is commonly loose and coarse-textured. This complex configuration of strata allows groundwater to flow through

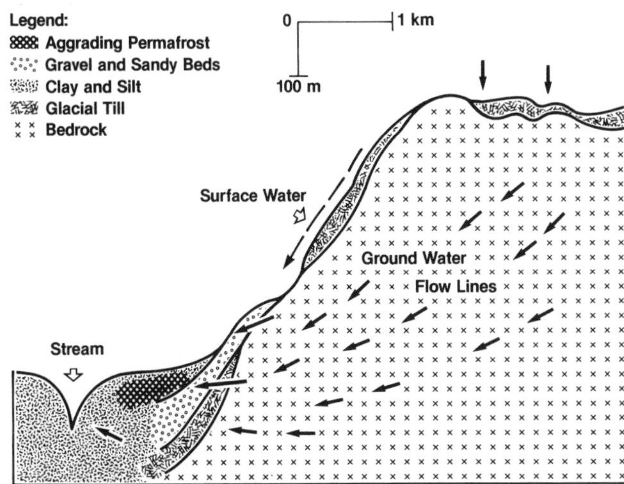

Figure 11: Diagramatic cross-section showing conditions leading to ground ice formation during permafrost aggradation in glaciolacustrine sediments at base of bedrock valley walls.

permeable sediments to focii of ground ice formation during permafrost aggradation. The complex configuration of strata has led to a very irregular distribution of ground ice as illustrated in Figure 12.

Most ground ice is probably epigenetic, although possible occurrences of relict glacier ice cannot be discounted. During glaciation the 0°C isotherm probably rose to the base of the glacier. Permafrost would then degrade subglacially, and melt any ground ice previously present. During deglaciation permafrost would immediately begin to aggrade and water would be conducted to focii of ground ice formation via permeable strata from lakes, ponds or the saturated taliks underlying them. Alternately, subglacial meltwater from basal melting of the decaying glacier may have been driven through permeable strata by the hydraulic gradient created by glacier overburden thickness; much as has been envisaged for the formation of massive ice in the Mackenzie-Beaufort region (Rampton 1974).

CONCLUSIONS

Ground ice has developed along the pipeline route throughout late glacial and postglacial time (approximately the last 13,000 years). Ground ice most commonly occurs as aggradational ice or epigenetic ice. Aggradational ice is most extensive where water is perched upon a rising permafrost table, either where the table is rising due to aftereffects of forest fires, mass wastage or changing climatic conditions or where continuous deposition of fine-grained materials is occurring. Epigenetic ice is most extensive where ample groundwater is available and can be transmitted via permeable strata or under high groundwater gradients to an aggrading permafrost table. This usually occurs adjacent to rivers, lakes, ponds, and permeable taliks or in discharge areas adjacent to truncated unconsolidated sequences or bedrock valley walls.

79 - CS - 1 (79-03)

USC	Dp	PfC	MC
OL aPt	0.10	VrVx	
GM sS sM lC lcbl	3.50		
Cl-CH tfS aM	4.30		32
SW lfG lC+M	5.50		22 12
SW aG lMtC tcbl	9.00		11 10 9
ML sC lS lG	10.00		

22.00 m END OF HOLE

78 - CS - 1 (T) (78-08)

USC	Dp	PfC	MC
Pt/OL	0.3		106
GC sS sCL-ML	1.60	Vx Vr	15 25
aS	3.10	Nbe	15 28 5
CL sS tG	5.00	Vx Vr	95 113 154
		ICE	
SM aML tG few ash layers	5.10 9.10	Vx Vr Vr Vc	302 179 35 103 188 228 110
		ICE	
CL-ML	10.20	Nbe	22

15.50 m END OF HOLE

Figure 12: Logs of boreholes 79-CS-1 and 78-CS-1 (T) showing the high degree of variability of strata and ground ice contents in disintegration moraines. These drill holes are within 100 metres of one another. See Figure 2 for explanation of USC, Dp, PfC, MC.

REFERENCES

Denton, G.H. and Stuiver, M., 1967, Late Pleistocene glacial stratigraphy and chronology, northeastern St. Elias Mountains, Yukon Territory, Canada; Geol. Soc. Am. Bull., V. 78, p. 485-510.

Hughes, O.L., Rampton, V.N. and Rutter, N.W., 1962, Quaternary geology and geomorphology, southern and central Yukon (northern Canada); XXIV International Geol. Congress, Guidebook A11, Montreal.

MacKay, J.R., 1972, The world of underground ice; Annals Assoc. Am. Geograph., v. 62, p. 1-22.

Rampton, V.N., 1970, Neoglacial fluctuations of the Natahat and Klutlan Glaciers, Yukon Territory, Canada; Can. Jour. Earth Sci., v. 7, p. 1236-1263.

Rampton, V.N., 1971, Late Pleistocene glaciations of the Snag-Klutlan area, Yukon Territory; Arctic, v. 24, p. 277-300.

Rampton, V.N., 1974, The influence of ground ice and thermokarst upon the geomorphology of the Mackenzie-Beaufort region, in Fahey, B.D. and Thompson, B.D., ed., Research in Polar and Alpine Geomorphology, Proceedings 3rd Guelph Symposium on Geomorphology, 1973.

Rampton, V.N., 1981, Surficial materials and landforms of Kluane National Park, Yukon Territory; Geol. Surv. Can., Paper 79-24.

A MATHEMATICAL MODEL FOR PATTERNED GROUND: SORTED POLYGONS AND STRIPES, AND UNDERWATER POLYGONS

R. J. Ray,[1] W. B. Krantz,[1] T. N. Caine,[2] and R. D. Gunn[3]

[1]Department of Chemical Engineering, University of Colorado
Boulder, Colorado 80309 USA
[2]Institute of Arctic and Alpine Research and Department of Geography
University of Colorado, Boulder, Colorado 80309 USA
[3]Department of Chemical Engineering, University of Wyoming
Laramie, Wyoming 82071 USA

This paper presents a model in which Rayleigh convection of the water in an active layer is responsible for the regularity, size, and shape of some types of sorted patterned ground. These convection cells result in uneven melting of the underlying ice front during thawing. The resulting undulatory ice front provides a pattern of which sorted stone polygons and stripes are the "fossil mirror." The model predicts the width to depth-of-sorting ratio for sorted patterned ground. Furthermore, the model predicts the existence and width to depth-of-sorting ratio of underwater sorted polygons. These predictions are corroborated with field study data from both "normal" and underwater sorted polygons.

INTRODUCTION

Sorted patterned ground is a periglacial phenomenon which displays many unusual and oftentimes strikingly beautiful characteristics. Near perfectly hexagonal stone polygons (Figure 1) are common in remote locations above timberline in the highest mountains and even at sea level in the arctic. Sorted polygon widths range from a few centimeters up to giant forms of 10 m or more (Washburn 1980 p. 142). Networks of regular, sorted polygons often transform into parallel, sorted stripes as the terrain changes to sloped. Furthermore, the impressive regularity of sorted patterned ground may extend over several square kilometers at a single site (Richmond 1949).

Among the most interesting sorted patterned ground features are the nearly constant width to depth-of-sorting ratios for any size form and the occurrence of sorted polygons on the bottom of shallow lakes or ponds. Meinardus (1912) noticed that the characteristic width of sorted patterned ground increases as the depth-of-sorting increases. Sorted polygons, which presumably have been formed underwater, have been reported (e.g., Jennings 1960) and in fact often occur at the bottom of a pond or lake at lower elevations than "normal" sorted patterned ground.

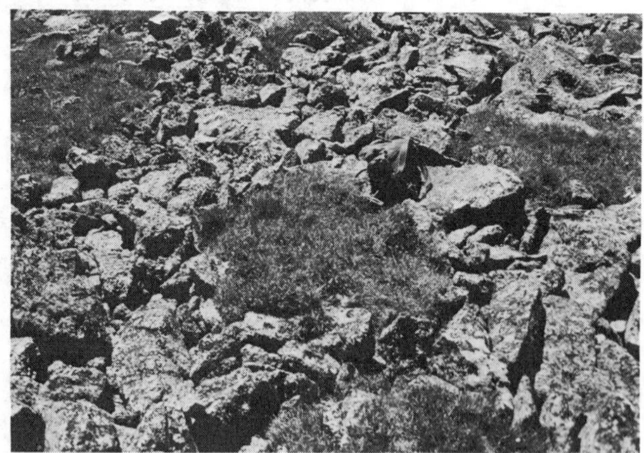

FIGURE 1 Relict sorted polygon, Arapahoe Pass, Front Range of Colorado. Stone polygon is approximately 3 m across.

This paper presents a Rayleigh convection model for the regularity of sorted patterned ground which explains these unusual characteristics including the prediction of the width to depth ratio of a broad class of sorted polygons observed both above and below water. The scope of this paper is as follows. First, a brief review of previous models for sorted patterned ground formation is given. Next, a new model for the regularity of sorted patterned ground is presented. This convection model then is supported with width and depth-of-sorting measurements of both "normal" and underwater sorted polygons. Finally, many interesting characteristics of sorted patterned ground are explained with the use of this model.

It is generally agreed that the origin of patterned ground is polygenetic (Caine 1972). There are many mechanisms which can cause a homogeneous mixture of small and large stones to become sorted or segregated through frost action: cryostatic pressure, differential frost heave, and upfreezing of stones are possible mechanisms (Washburn 1956). Other models for sorted patterned ground formation invoke some sort of cracking by desiccation, or thermal contraction in seasonally frozen ground or permafrost (Washburn 1980 p. 156).

Although all these processes do occur in nature, it is generally agreed that they are inadequate in explaining the regularity of all sorted forms. The noncracking mechanisms require an initial regularity of some kind to be developed in the soil prior to sorting. The cracking mechanisms yield crack networks too small (desiccation cracking) or too large (permafrost cracking) to explain most sorted polygons or stripes. Furthermore, none of these mechanisms can adequately explain the constant width to depth-of-sorting ratio in sorted polygons, the transition from sorted polygons to sorted stripes, the increase in depth-of-sorting with depth of the active layer, the regularity of polygon networks or the occurrence of sorted polygons underwater.

The model presented here suggests that thermally driven convection or "Rayleigh" convection of the water in the active layer can explain the interesting characteristics of a broad class of sorted patterned ground. Rayleigh convection has been advanced as an explanation for sorted patterned ground before. Nordenskjold (1909) suggested that sorted polygons are a result of convection currents in the ground due to the temperature difference between the surface of the frost table and the surface of the thawed ground. Low (1925) and Gripp (1926, 1927, 1929) proposed that currents capable of lifting stones occurred due to the density inversion of water between 273 K and 277 K. However, Elton (1927) showed that the flow caused by this density difference is too weak to carry stones. The model developed in this paper, while involving Rayleigh convection, is substantially different from the convection models discussed above.

RAYLEIGH CONVECTION MODEL FOR SORTED PATTERNED GROUND REGULARITY

This section discusses the implications Rayleigh convection cells in an active layer would have toward sorted patterned ground formation. First, the new model for inducing regularity in the active layer is discussed. Then, a summary of the principal results of the mathematical analysis for the onset of Rayleigh convection cells in the active layer is given. It should be emphasized that this model in no way attempts to explain the entire formation process of sorted patterned ground; rather, it provides a mechanism for the occurrence of regularity in the active layer.

Formation of Undulatory Ice Front by Rayleigh Convection

During the warmer months, the top portion of the permafrost melts to form the active layer. The active layer is thus water-saturated earth bounded below by thawing, but still-frozen, soil at 273 K and bounded above by either air or standing water if it lies below a shallow lake or pond. For modeling purposes, it is assumed that the upper surface is at or near 277 K. This situation often occurs naturally (e.g., Ives 1973, Washburn 1980 p. 59); situations where the surface temperature is different from 277 K are easily incorporated into the theory (see Ray et al. 1983). It is well known that in this temperature range, the density of water increases with temperature, reaching a maximum value at 277 K.

Since the density of the water near the upper surface (at 277 K) is greater than that near the lower ice-liquid interface (at 273 K), the system is potentially unstable and will tend to form convection cells (Figure 2). Lord Rayleigh (Strutt 1916) showed that convection cells do not form in every fluid layer with an inverse density stratification. The onset of convection cells is controlled by opposing forces. The density gradient and corresponding buoyancy forces must be large enough to force the fluid to circulate. The viscous drag, which in an active layer is a function of the viscosity of the water and the permeability of the soil, must be small enough to avoid damping out the circulation cells. This criterion is commonly expressed in terms of a critical value of the Rayleigh number (Ra), a dimensionless parameter defined as the ratio of the buoyancy force divided by the product of the viscous drag and the rate of heat conduction:

$$Ra \equiv \frac{\rho_c^2 C_p \beta \Delta T k g L}{\lambda_s \mu} \qquad (1)$$

where ρ_c is the density of the fluid at its reference temperature (water at 273 K), C_p is the heat capacity at constant pressure, β is the thermal expansion coefficient, ΔT is the temperature difference imposed across the depth L, k is the permeability, g is the gravitational acceleration, λ_s is the thermal conductivity, and μ is the shear viscosity. Any parameter which makes the Rayleigh number large contributes to the formation of Rayleigh convection cells. The critical Rayleigh number, above which the onset of convection cells will occur in the active layer, will be given when the results of the mathematical analysis are summarized.

With the aid of these Rayleigh instability arguments, a sequence of events is now proposed for the formation of sorted polygons. First, it is assumed that the permeability, k, of the soil is initially too low for the density inversion to result in Rayleigh convection cells; that is, the k which appears in the numerator of Equation (1) is so small that the Rayleigh number is below its critical value. The thickness of the active layer is often nearly the same each year (Ives 1973,

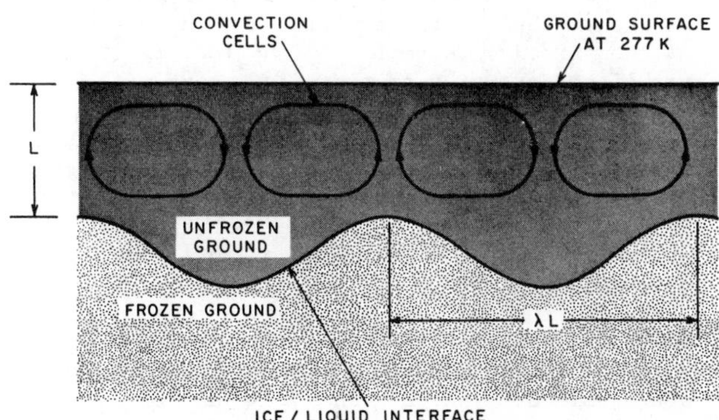

FIGURE 2 Two-dimensional cross-section of hexagonal Rayleigh convection cells in the active layer. Note the resulting uneven melting of the permafrost front.

Washburn 1980 p. 59). Frost heave causes vertical (the primary degree of freedom for movement) expansion of the active layer. On thawing, it does not settle to its original density; that is, it stays slightly lofted. Corte (1962) has shown experimentally that up to 10% increase in volume of the active layer can result from as few as 10 freeze-thaw cycles. Application of the Blake-Kozeny equation (Bird et al. 1960 p. 199) indicates that a 10% volume increase can result in an increase in permeability of nearly 70%. After many such freeze-thaw cycles, the active layer will have lofted enough to reach a permeability which allows the critical Rayleigh number to be reached. If the 4 K temperature difference is present across the active layer, all properties appearing in Equation (1) will be nearly constant except the permeability, k, and the depth, L.

Attaining the critical Rayleigh number then is first possible when the nonconstant parameters in the Rayleigh number reach their extreme values; that is, k is large enough to achieve the onset of convection when L is at its maximum (deepest annual thaw). Note here that the Rayleigh number is largest when the upper surface temperature is at or near 277 K. This implies that convection cells will be initiated at that time of the year when the upper surface temperature is about 277 K.

Earlier models which attempted to explain the regularity of patterned ground by Rayleigh convection were based on the contention that these weak circulatory currents could cause direct movement of the larger stones which form the borders of sorted polygons. The model developed in this paper is not based on any argument that Rayleigh convection cells can move stones directly to effect a sorting process. In contrast, this model contends that Rayleigh convection determines the regularity indirectly through its influence on the shape of the ice front which underlies the water-saturated porous medium.

The influence of Rayleigh convection on the shape of the underlying ice front was ignored in earlier Rayleigh convection models which attempted to explain patterned ground regularity. The Rayleigh convection cells determine the shape of the underlying ice front because they induce uneven heat transfer to this front. In regions of downflow, the transport of warmer water from the upper surface to the ice front will cause increased melting; conversely, in regions of upflow, the transport of colder water from the ice front will retard melting. Figure 2 shows a schematic of the two-dimensional cross-section of the resulting undulatory ice front. Note the ice "peaks" under the upflow regions and the ice "valleys" under the downflow regions. It can be proven that the plan form configuration of the Rayleigh convection cells will be hexagonal.

The presence of hexagonal undulations on the thaw front then influences any sorting processes which are slowly occurring in time such that the sorting mirrors the pattern of the underlying ice front. This could occur by any one or combination of possible sorting mechanisms such as cryostatic pressure, differential frost heave, and primary frost sorting. This paper does not attempt to determine which mechanism or combination of mechanisms causes the sorting process to mirror the patterned ice front. Indeed it is not possible to infer from the field data taken in this study which mechanism caused the sorting. However, it is possible to determine if the geometry and characteristic spacing of the patterned ground correspond to that which would be induced by an undulatory ice front caused by Rayleigh convection. If this model for explaining the regularity of patterned ground is correct, then the ratio of the polygon width to depth-of-sorting should be equal to the width to depth ratio predicted by the mathematical model assuming that the depth-of-sorting is nearly equal to the depth of the active layer when free convection is initiated.

Summary of Principal Results From Mathematical Analysis

In order to determine the critical Rayleigh number above which convection will occur and the corresponding convection cell width to depth ratio, a linear stability analysis was carried out. Although this analysis is straightforward, it is quite lengthy and hence only the principal results can be summarized here. The interested reader desiring more details of the analysis is referred to Ray (1981) or Ray et al. (1983).

The governing equations for fluid flow and heat transfer in the unstably stratified active layer are Darcy's law, the equations of motion, the thermal energy equation, and the equation of continuity. Also, an equation of state is required which describes the increase of water density with temperature between 273 K and 277 K. The boundary conditions needed to solve these equations correspond to specified temperatures of 273 K and 277 K at the lower and upper boundaries, respectively. In addition, in the case of "normal" polygons formed in an active layer bounded above by air, the upper and lower boundaries are assumed to be impermeable. In the case of underwater polygons, this latter condition is relaxed at the upper boundary; this leads to dramatically different predictions. In addition, for both "normal" and underwater polygons, an energy balance condition must be satisfied at the lower ice front boundary which is allowed to be compliant in that it can deform in response to the convection cells. The resulting linear stability theory solution is novel in allowing for this compliant boundary.

This system of equations and boundary conditions is solved using the formalism of linear stability theory. For the case of "normal" polygons, linear stability theory predicts a critical Rayleigh number $Ra_c = 27.1$ for the inception of convection cells and a corresponding convection cell width to depth ratio given by

$$\frac{\lambda}{L} = 3.81 \qquad (2)$$

For the case of parallel stripes on a slope, theory predicts a critical Rayleigh number of $Ra_c = 27.1/\cos\phi$, where ϕ is the angle of inclination. The corresponding width to depth ratio is then in the range $2.68 \leq \lambda/L \leq 3.81$, depending on the slope angle and the magnitude of the subsurface flow in the active layer.

Sorted stone polygons formed underwater are

predicted to have a smaller critical Rayleigh number ($Ra_c = 17.1$) and a correspondingly higher width to depth ratio:

$$\frac{\lambda}{L} = 5.07 \qquad (3)$$

Theoretical considerations suggest that this ratio may depend on the water depth. However, we have insufficient data to warrant refining the theory at this time.

These results allow prediction of the conditions for which convection cells will form. When the physical parameters of an active layer bounded above by air yield a Rayleigh number just above 27.1 (i.e., the permeability is large enough for flow), convection cells should form which have a width to depth ratio of about 3.8. When the Rayleigh number of an active layer below a shallow lake or pond goes just above 17.1, convection cells with a width to depth ratio of about 5.1 should form.

FIELD STUDIES

The field studies were designed to test the hypothesis that the regularity of sorted patterned ground corresponds to that predicted for Rayleigh convection cells. This would suggest that Rayleigh convection in the active layer is responsible for the observed regularity of sorted patterned ground. If Rayleigh convection is involved in sorted polygon inception, it would have occurred long ago in relict sites that are now polygonally patterned; however, sites about to form sorted polygons are very difficult to identify. For this reason, the field studies presented below focused on measurements of existing sorted polygons as a test of the Rayleigh convection hypothesis.

Two sets of data were collected. Figure 3 is a plot of 18 data points taken from a wide range of "normal" sorted polygon sizes, that is, where the active layer was bordered above by air. The dark points are diurnal small polygon data whereas the light points are data for polygons formed over longer time periods. The characteristic width is an average of center-to-center measurements for several adjacent polygons. The depth was found by trenching through the stone border of a polygon, then measuring from undisturbed ground reference level to the bottom of the sorted border, or to the point where the stones were no longer oriented vertically. A linear regression of these data yields a slope of 3.6 and an intercept of 0.041 m with a correlation coefficient of 0.98. This slope of 3.6 agrees reasonably well with the theoretical value of 3.8 given by Equation (2).

The data plotted in Figure 4 are from measurements taken in the manner described above for polygons appearing on the bottom of shallow lakes or ponds. A linear regression fit of these data yields a slope of 5.7 with an intercept of -0.18 m and a correlation coefficient of 0.94. This slope agrees reasonably well with the value of 5.1 given by Equation (3).

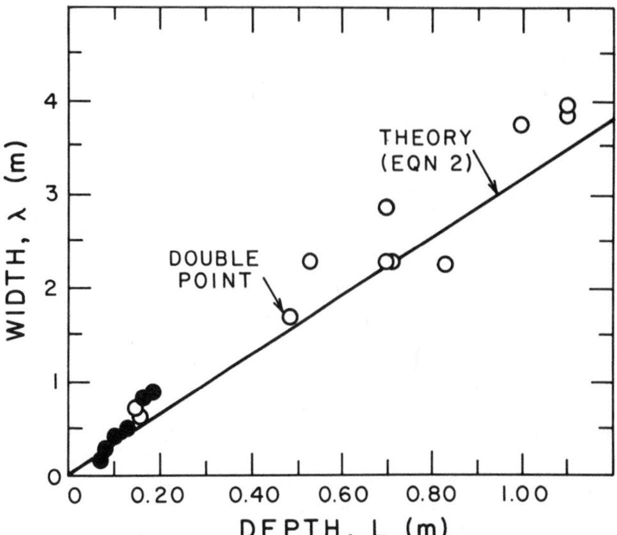

FIGURE 3 "Normal" sorted polygons: Characteristic width versus depth-of-sorting. The dark points are data from diurnal sorted polygons.

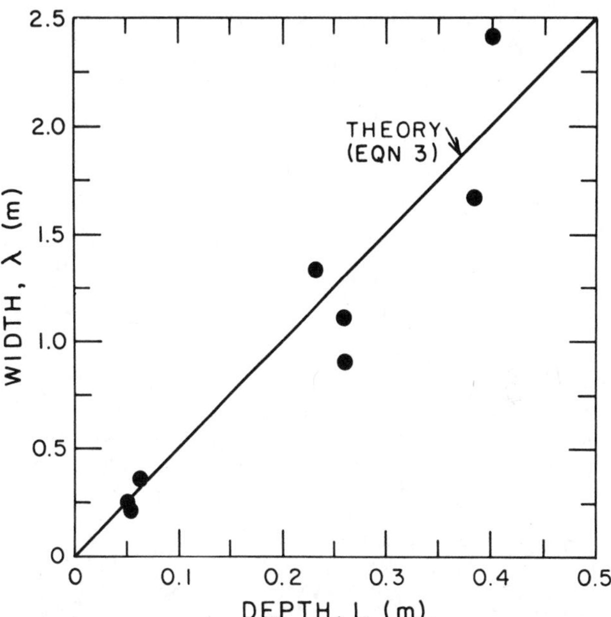

FIGURE 4 Underwater sorted polygons: characteristic width versus depth-of-sorting.

DISCUSSION

The excellent agreement between measured and predicted values of the width to depth ratios for both normal and underwater polygons provides convincing support for the hypothesis that Rayleigh convection in the active layer was responsible for the regularity of the sorted patterned ground studied here. This Rayleigh convection model can also explain many other characteristics of patterned ground. For example, the constant width to depth-of-sorting ratio observed for "normal" sorted polygons follows from the fact that for any depth active layer, the critical convection cell width to depth ratio is 3.81. This then explains the wide range in sorted patterned ground sizes since thinner active layers will produce smaller polygons.

Since theory predicts that Rayleigh convection cells will be hexagonal in a horizontal active layer, this implies that the mirroring stone polygons also will tend toward this shape. Many investigators have noticed this preferred shape of sorted polygons, especially in smaller, diurnal forms (see Ray et al. 1983). The regularity and extent of some sorted patterned ground networks suggest that, when conditions are conducive to Rayleigh convection on any part of a site, they probably prevail throughout the site.

The transition from sorted polygons to sorted stripes is predicted by the Rayleigh convection model as well. It is well known in thermal convection theory that as a saturated porous medium undergoing Rayleigh convection is tilted from the horizontal, the three-dimensional, polygonal convection cells, which occur in the flat configuration, become two-dimensional roll cells oriented downslope (Combarnous and Bories 1975). In a sloped active layer, such two-dimensional roll cells would melt the ice-water interface into troughs and ridges running parallel to the fall line. The resulting sorted forms would then be parallel sorted stripes (Ray et al. 1983). The fact that the critical Rayleigh number is predicted to increase with slope angle may explain in part why stripes are not observed on steeper slopes.

Finally, the occurrence of underwater sorted polygons can be explained in light of the Rayleigh convection model. Figure 5 shows a network of underwater polygons on the bed of a shallow lake in the Snowy Range of Wyoming. Freezing of a shallow pond in a periglacial environment will extend into the bed below it and produce an active layer which is bordered above by standing water during thaw conditions.

Figure 4 shows that the measured width to depth ratio for underwater polygons is seen to agree remarkably well with the value of 5.1 predicted by the Rayleigh convection model. Note that the predicted critical Rayleigh number for underwater polygons is 17.7 as opposed to 27.1 for "normal" polygons. This implies that underwater sorted polygons may be observed under less constraining conditions than "normal" patterned ground (i.e., lower permeability, shallower active layer, reduced temperature gradient). Throughout the Southern Rockies, most large active sorted polygons are on the beds of ephemeral ponds. Commonly these polygons do not continue onto the shore above the high water level and may be found at elevations several hundred meters lower than those of "normal" polygons (Ray 1981 p. 230).

FIGURE 5 Underwater sorted polygons, Snowy Range of Wyoming. Polygons are approximately 0.8 m wide.

CONCLUSIONS

The model presented here hypothesizes that the regularity, size, and shape of some types of sorted patterned ground can be explained by Rayleigh convection cells in the active layer. The model predicts a width to depth-of-sorting ratio of 3.8 for "normal" sorted polygons and a width to depth-of-sorting ratio of 5.1 for underwater sorted polygons which compare favorably with field measurements of 3.6 and 5.7, respectively. Finally, many interesting characteristics of sorted patterned ground can be explained by this Rayleigh convection model.

ACKNOWLEDGMENTS

The authors wish to acknowledge financial support provided for these studies by the National Science Foundation (grant DPP-8210156). A seed grant for the field studies was provided by the Council for Research and Creative Work of the University of Colorado. One of the authors (RJR) was partially supported during this work by a Dean's Graduate School Fellowship from the University of Colorado. Finally, the authors wish to acknowledge Ms. Ellen M. Romig and Mr. Dennis L. Ehmsen for their assistance in preparing the final manuscript and figures and Mr. David Newbold for his help in the field program.

REFERENCES

Bird, R. E., Stewart, W. E., and Lightfoot, E. N., 1960, Transport phenomena: New York, John Wiley and Sons, Inc.

Caine, N., 1972, The distribution of sorted patterned ground in the English lake district: Revue Geomorphologie Dynamique, v. 21, n. 2, p. 49-56.

Combarnous, M. S. and Bories, S. A., 1975, Hydrothermal convection in saturated porous media, in Advances in Hydroscience, v. 10: Academic Press.

Corte, A. E., 1962, The frost behavior of soils: laboratory and field data for a new concept. Part II horizontal sorting: U.S. Army Corps of Engineers, Cold Regions Research & Engineering Laboratory, Research Report 85, part II: Hanover, New Hampshire.

Elton, C. S., 1927, The nature and origin of soil polygons in Spitzbergen: Geological Society of London Quarterly Journal, v. 83, Part I, p. 163-194.

Gripp, K., 1926, Uber frost und Strukturboden auf Spitzbergen, in Zeitschrift der Gesellschaft fur Erdkunde: Berlin 1926, p. 351-354.

Gripp, K., 1927, Beitrage zue geologie von spitzbergen, in Abhandlung der Naturwissenschaftliche Vereinigung: Hamburg. Band 21, Heft 3-4, p. 1-38.

Gripp, K., 1929, Glaciologische und geologische Ergebnisse der Hamburger Spitzbergen--expedition 1927, in Abhandlung der Naturwissenschaftliche Vereinigung: Hamburg Band 22, Heft 2-4, p. 145-249.

Ives, J. D., 1973, Permafrost and its relationship to other environmental parameters in midlatitude, high-altitude setting, front range, Colorado Rocky Mountains, in Permafrost: North American contribution [to the] Second International Conference: Washington, DC, National Academy of Sciences, p. 121-125.

Jennings, J. N., 1960, On an unusual occurrence of stone polygons in the French Alps: Biuletyn Peryglacjalny, no. 7, p. 169-173.

Low, A. R., 1925, Instability of viscous fluid motion: Nature, v. 115, no. 2887, p. 299-300.

Meinardus, W., 1912, Beobachtungun uber detritussortierung und strukturboden auf spitzbergen, in Zeitschrift der Gesellschaft fur Erdkunde: Berlin 1912, p. 250-259.

Nordenskjold, O., 1909, Die polarwelt und ihre nachbarlander: Leipzig, B. G. Teubner.

Ray, R. J., 1981, A Rayleigh convection compliant ice front model for sorted patterned ground: M.S. thesis, University of Colorado, Boulder, Colorado.

Ray, R. J., Krantz, W. B., Caine, T. N., and Gunn, R. D., 1983, A model for sorted patterned ground regularity: Journal of Glaciology, in press.

Richmond, G. M., 1949, Stone nets, stone stripes, and soil stripes in the Wind River Mountains, Wyoming: Journal of Geology, v. 57, no. 2, p. 143-153.

Strutt, J. W. (Lord Rayleigh), 1916, On convection currents in a horizontal layer of fluid, when the higher temperature is on the underside: Philogophical Magazine, series 16, v. 32, p. 529-546.

Washburn, A. L., 1956, Classification of patterned ground and review of suggested origins: Bulletin of the Geological Society of America, v. 67, no. 7, p. 823-866.

Washburn, A. L., 1980, Geocryology: New York, John Wiley and Sons, Inc.

GELIFLUCTION DEPOSITS AS SOURCES OF PALEOENVIRONMENTAL INFORMATION

Richard E. Reanier and Fiorenzo C. Ugolini
College of Forest Resources, University of Washington
Seattle, Washington 98195 USA

A gelifluction lobe above treeline at Walker Lake in the Brooks Range of Alaska was examined for pedological and palynological evidence of the expansion of the boreal forest during the Holocene. Soil pits excavated to the base of the active layer (1 m) revealed organic rich material consisting of ancient vegetation buried by the advancing lobe front. Laminations, platy soil structure, slickensided soil peds, and "plough marks" from pebbles in this organic-rich deposit indicate basal sliding is an important component of downslope movement. Radiocarbon dates obtained from buried organics indicate a relatively constant lobe advance rate of about 3.21 mm yr^{-1} throughout much of the Holocene. A paleosol buried beneath the advancing lobe shows that a degree of soil development comparable to that of modern soils of the area was achieved by about 3,200 years B.P. Pollen spectra from the buried organic layer differ markedly from the regional lake pollen record, but give evidence that the treeline has recently advanced to its modern position during the past 5,000 years, and has not been more expanded in the past. We conclude that gelifluction deposits contain valuable paleoenvironmental data for studies of vegetation history and soil development. Such deposits are particularly important in geomorphic situations where lakes and bogs are absent and site-specific information is required.

INTRODUCTION

Historically, studies of periglacial phenomena have concentrated on geomorphic description and on analysis of physical properties and processes involved in landform development (Washburn 1980). Paleoenvironmental uses of periglacial landforms generally have been restricted to regional inferences about former climates from the presence of relict landforms. Climatic inferences are based upon analogy with climatic parameters associated with similar active landforms. Gelifluction, the gradual downslope flow of water-saturated soil in permafrost affected regions, results in such periglacial landforms as gelifluction sheets, benches, lobes, and streams (Washburn 1980). Benedict (1976) suggested that deposits buried beneath advancing gelifluction features should provide high-resolution palynological and pedological records.

As part of a multidisciplinary study of Holocene treeline changes in the southern Brooks Range of Alaska (Brubaker et al. 1983, Garfinkel and Brubaker 1980, Goldstein 1981, Ugolini et al. 1981, 1982), we examined gelifluction lobes as potential sources of information on treeline fluctuation. If the treeline had expanded beyond its present limit during the Holocene, as it did in northern Canada (Bryson et al. 1965, Ritchie and Spear 1982), we reasoned that gelifluction deposits near the treeline would contain a record of that expansion. We report here the results of palynological and pedological investigations of one lobe located above Walker Lake (67°9'N, 154°22'W) in the south central Brooks Range of Alaska (Figure 1).

STUDY AREA

The Walker Lake study area was selected because of its well-defined treeline and

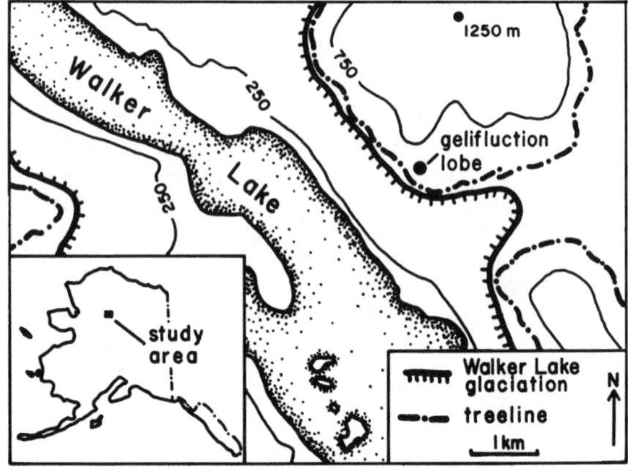

FIGURE 1 Map of study area.

noncalcareous parent material suitable for pedogenic studies and ^{14}C dating. The south facing slope where our study was centered is composed of low grade quartz-muscovite schist of Paleozoic age (Nelson and Grybeck 1981). The slope is mantled to an altitude of about 525 m by drift of the late Wisconsin Walker Lake glaciation (Figure 1), but was overridden by the earlier and more extensive Itkillik and Kobuk glaciations (Hamilton 1982). Walker Lake is near the boreal forest and shrub tundra transition. The altitudinal limit of continuous forest occurs at about 550 m at the study site. White spruce (*Picea glauca*) dominates the forest at treeline and in well-drained areas, while black spruce (*Picea mariana)* dominates the poorly drained areas at lower altitudes. Quaking aspen (*Populus tremuloides*) and paper birch (*Betula papyrifera*) are secondary tree species. Ericaceous shrubs and lichens are common in the forest understory and in the shrub tundra above treeline. Alder

(*Alnus crispa*) and willow (*Salix* spp.) are locally abundant in draws and along stream banks. The area is within the zone of discontinuous permafrost (Brown and Péwé 1973).

The gelifluction lobe examined here is located at an altitude of 620 m, approximately 70 m above the limit of continuous forest (Figures 1 and 2). It is about 35 m in axial length with the axis oriented downslope at 175° azimuth. The surface gradient is 14° and the lobe is from 1 to 1.5 m thick. Active layer thickness in midsummer is about 1 m. It is a turf-banked lobe (Benedict 1976) with a continuous cover of shrub tundra. Although the lobe is located above the continuous forest limit, a solitary white spruce grows on the lobe near its western margin 13 m upslope from the lobe front, and several white spruce seedlings are present.

METHODS

Five trenches were opened along the lobe axis in order to examine the internal structure of the lobe (Figure 2). Soil profiles were described in each trench and one of the profiles was selected for laboratory chemical analyses. Fabric measurements were made in this same profile. Soil description followed standard methods (cf. Ugolini et al. 1981). Soil textural analysis utilized standard sieve techniques for the fraction coarser than 4 ϕ and a Sedigraph 5000D particle size analyzer for the fines after pretreatment for organic matter and iron removal. Soil chemical analyses were made on the air-dry soil fraction finer than -1 ϕ (2 mm). Details of the analytical procedures are given in Ugolini et al. (1981). Samples of the buried organic deposit were taken for pollen analysis and ^{14}C dating (Figure 2). Radiocarbon samples were stored at 3°C, air-dried, and carefully examined under magnification to remove any rootlets prior to submission for dating. The dates were run by Beta Analytic, Inc. after dispersion in hot HCl to remove carbonates and have been normalized to -25 per mil ^{13}C. Pollen samples were processed by standard acetolysis techniques (Faegri and Iverson 1975). Extracted pollen was mounted in silicone oil and counted under 400X magnification until at least 300 grains were identified in each sample.

RESULTS AND DISCUSSION

Structural Analyses

Fabric measurements taken from 100 stones between 25 and 75 cm below the surface at the soil profile location (Figure 2) show long-axis orientation parallel to the lobe axis (Figure 3). The mean fabric dip angle is 26.6° with an angular deviation of 12.0° as compared with a lobe surface gradient of about 14°. The mean fabric azimuth is 178.5° with an angular deviation of 42.0°, while the lobe axis azimuth was estimated to be 175° true. Such fabric orientations are commonly found in gelifluction deposits (Benedict 1976), and have been taken as evidence of active lobe movement.

Although the exact mechanism of gelifluction movement, whether by flow or by sliding along shear planes, remains undetermined (Washburn

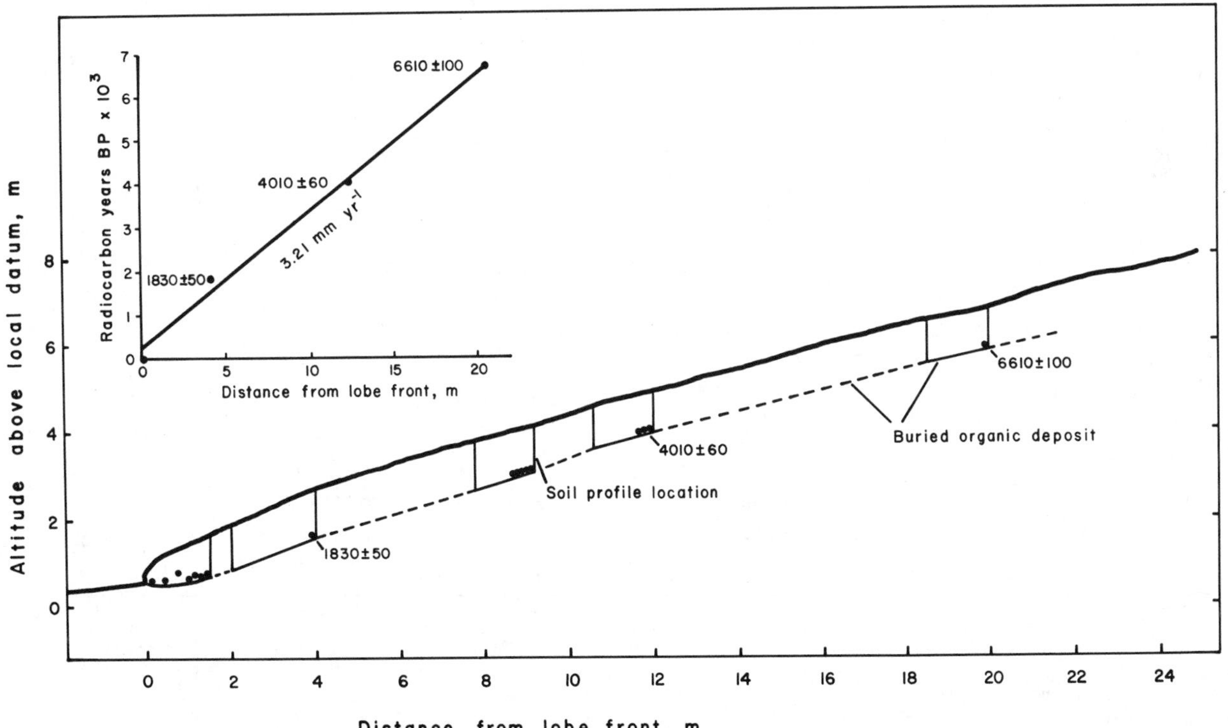

FIGURE 2 Longtudinal section of gelifluction lobe showing sample locations (black dots) and soil profile location, and a plot of radiocarbon age and distance from lobe front.

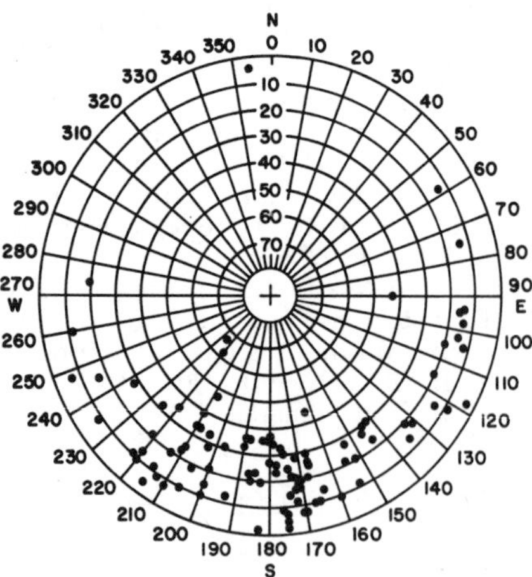

FIGURE 3 Scatter plot of fabric measurements.

1980), basal sliding appears to be an important component of downslope movement of this lobe. The soil structure of the lobe material is dominated by fine to very fine platy structure (Table 1), the development of which is favored by ice segregation that preferentially forms planes parallel to the soil surface (Benedict 1976). Such planes are potential places for sliding to occur as the active layer thaws. Detailed field examination of the still-frozen Alb horizon revealed apparent glide planes separated by thin segregation ice lenses. The soil peds had shiny, slickensided surfaces due to the orientation of fine muscovite plates. Occasional stones protruding into the organic horizon from the overlying C2 horizon had left striae and "plough marks" several cm long in the Alb material.

Rates of Movement

Radiocarbon dates of 1,830±50 years B.P. (Beta-2840), 4,010±60 years B.P. (Beta-2645), and 6,610±100 years B.P. (Beta-2840) were obtained from the buried organic horizon 4.1, 12.4, and 20.6 m from the lobe front, respectively. A plot of these dates (including the present) against distance from the lobe front is linear (r^2=0.99) and indicates a relatively constant rate of movement of about 3.21 mm yr $^{-1}$ throughout most of the Holocene at the time resolution examined here (Figure 2). This compares with other reported northern hemisphere basal movement rates determined by ^{14}C dating ranging from 0.6 to 10 mm yr $^{-1}$ (Benedict 1976, Alexander and Price 1980). No attempt was made to correct for calendar age or for mean residence time of soil carbon as others have done (Benedict 1970) because of differences among calibration curves and the apparent unreliability of ^{14}C ages less than about 500 B.P. (Stuiver 1982). However, the use of reasonable estimates of residence time (300-500 yr) in the regression would tend to somewhat increase both the movement rate and correlation.

Pedological Analyses

Soils of the study area have been characterized by Ugolini et al. (1981). The soils are weakly podzolized; those displaying albic horizons fail to meet the spodic horizon chemical requirements of the Soil Taxonomy (Soil Survey Staff 1975), and are classified as Dystric Cryochrepts together with similar soils lacking albic horizons. Soils of the gelifluction lobe lack albic horizons, except immediately beneath the

TABLE 1 Selected morphological and chemical data from soil profile.

Horizon	Depth (cm)	Color (moist)	Texture	Structure	Consistence	Roots	Boundary	pH 1:1	EC 1:5 μmho cm^{-1}	C %	C_p %	Fe_d† %	Fe_p† %	Al_d† %	Al_p† %
01	2-1	*Cladonia* sp.; *Juniperus communis* and *Empetrum nigrum* needles													
02	1-0	Decomposed organic material													
A1	0-1	5YR 3/2	sil	2vfcr	fr,ss,sp	3vf&f	aw	4.10*	101.3	12.3	5.3	1.64	0.77	0.39	0.39
B2	1-15	10YR 3/3	gl	2f-vfpl	fr,ss,sp	2vf&f	cs	3.77	72.7	2.7	1.5	2.11	0.70	0.32	0.27
C1	15-50	10YR 4/4	gl	2f-vfpl	fr,ss,sp	1vf	aw	4.82	7.47	0.80	0.76	2.13	0.58	0.48	0.39
C2	50-109	10YR 4/3	gl	2f-vfpl	fr,ss,sp	1vf	aw	4.74	6.52	0.93	0.75	2.06	0.66	0.37	0.31
Albf	109-117	10YR 3/2	gsicl	3vfpl	fr,ss,sp	1vf	aw	4.44	13.24	4.73	2.8	2.41	1.24	0.57	0.56
B2bf	117-127	10YR 4/4	gl	2f-vfpl	fr,ss,sp	none	cw	4.68	7.95	1.00	0.92	2.47	0.93	0.36	0.31
Cbf	127-155+	10YR 5/4	gl	2f-vfsbk	fr,ss,sp	none		5.03	5.46	0.50	0.50	1.89	0.42	0.40	0.32

Abbreviations:

Texture: si-silty; c-clay; l-loam; g-gravelly.

Structure: 2-moderate; 3-strong; vf-very fine or thin; cr-crumb; pl-platy; sbk-subangular blocky.

Consistence: fr-moist, friable; ss-wet, slightly sticky; sp-wet, slightly plastic.

Roots: 1-few; 2-common; 3-many; vf-very fine; f-fine.

Boundary: a-abrupt; c-clear; w-wavy; s-smooth.

* 1:2

† Subscript p indicates Na-pyrophosphate extractable form. Subscript d indicates Na-dithionite extractable form.

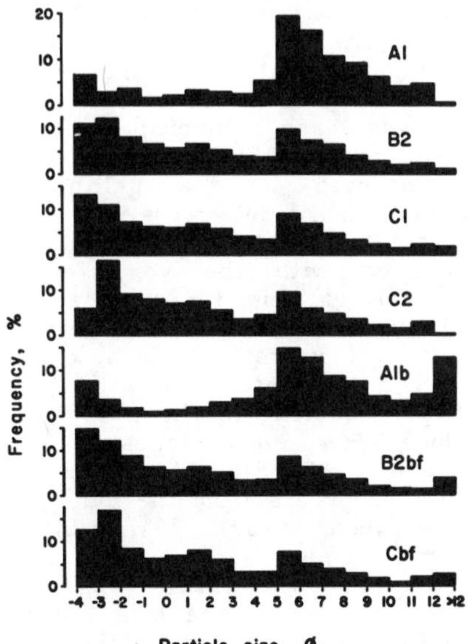

FIGURE 4 Particle size distribution in soil profile.

single white spruce tree. Detailed morphological and chemical analyses of a profile extending through the lobe to the paleosol beneath (Figure 2) are presented in Table 1. The Alb horizon (buried surface horizon) of the paleosol is very dark grayish brown in color and contains an amount of readily oxidizable carbon (C) exceeded only by that of the modern surface horizon. C is lower in the B2bf horizon than in the modern B2 horizon, perhaps as a result of in situ oxidization after burial. A similar trend is seen in the humified carbon fraction (Cp). As measured by electrical conductivity (EC), ionic strength similarly shows a peak in the Alb horizon but is nearly an order of magnitude lower in the buried than in the modern solum. However, both Na-dithionite and Na-pyrophosphate extractable Fe and Al reach peak values in the paleosol, suggesting the preburial soil was at least as well developed as the modern soil. Carbon and extractable sesquioxide values for other soils near treeline in the study area are comparable to levels reported here (Ugolini et al. 1981). The interpolated time of burial of this portion of the paleosol, and therefore the date of cessation of surficial pedogenesis, is 3,200 years ago (Figure 2), by which time the Holocene soils above treeline had evidently reached modern levels of development.

The particle size distributions of the soil horizons for material finer than -4 ϕ show strikingly similar modes in the medium silt (5 to 6 ϕ) and in the pebble (-4 to -2 ϕ) size classes (Figure 4). The bimodal pattern may result from comminution of the quartz-muscovite schist during gelifluction itself, or may be a distribution inherited from till (cf. Dreimanis and Vagners 1971). The particle size distributions generally show the poor sorting characteristic of gelifluction deposits. Despite the overall similarities, the two Al horizons differ from the other horizons in having relatively more material in the silt and clay fractions (5 to 12 ϕ) and less material in the sand and gravel fractions (-4 to 5 ϕ). This result is consistent with greater rates of both physical and chemical weathering in the surface horizons.

Pollen Analyses

At the outset of this investigation it became obvious that the reliability of pollen data from gelifluction deposits needed to be tested. Pollen from lake cores has been "smoothed" by mixing in the water column before deposition on the lake bottom. However, soil pollen may reflect microscale heterogeneity in plant distributions, so that samples of the same age from the same locality may have dissimilar pollen spectra. Accordingly, we examimed paired samples of the buried organic material taken from points about 0.5 m apart on a line normal to the lobe axis to see if contemporaneous samples had similar spectra. The sample pairs were taken from distances of 0.2, 1.5, 4.1, 9.6, 12.6, and 20.6 m from the lobe front (Figure 2). Using a multinomial homogeneity test reported by Mosimann (1965: test A7), we evaluated the null hypothesis that the pairs had the same true proportions of the seven pollen types in Figure 5. In five of the six pairs the hypotheses of homogeneity was rejected at the 0.05 level of significance. Gramineae (grasses) Cyperaceae (sedges), and *Artemisia* typically contribute the highest individual terms to the total chi-square, reflecting the higher heterogeniety of these taxa. This is understandable since these are anemophylous taxa producing large amounts of pollen and all grow locally on the lobe today. The chance proximity of such plants to the portion of soil that eventually becomes a "sample" may explain the higher heterogeneity of these taxa in the pollen spectra. When these three taxa are excluded from the test, the homogeneity hypothesis is not rejected in any case at the 0.05 level. These results indicate that the interpretation of changes in the relative abundances of Gramineae, Cyperaceae, and *Artemisia* must be made with caution.

Pollen preservation in the gelifluction lobe samples is generally poor in relation to normal preservation in lake pollen records. Pollen degradation in the lobe takes the form of both crumpling and corrosion; degradation generally increases with sample age. However, even in the oldest samples some well preserved grains were encountered, and in the youngest samples poorly preserved grains were found. Of the taxa in Figure 5, Cyperaceae and Gramineae seemed most susceptible to corrosion and are possibly underrepresented in the samples, while *Picea* pollen seemed relatively unaffected by corrosion. Fragile *Juniperus* pollen was not preserved in the lobe deposits, although *Juniperus communis* is common on the lobe today. However, a *Juniperus* needle was recovered from the 4,000 year old sample, suggesting that this species was present at the site throughout the record.

Brubaker et al. (1983) have recently completed a detailed study of late Wisconsin and Holocene vegetation changes in the central Brooks Range based on lake pollen records. The major vegetation changes in the region occur prior to the beginning of our gelifluction lobe record at 6,600 years B.P. The exception is the establishment of the boreal forest at the south shore of Walker Lake about 5,000 years B.P. Otherwise, this younger portion of the record is conspicuous by its lack of change.

Pollen spectra from the buried organic layer how systematic departures from those of the regional lacustrine record. In contrast to the regional spectra, *Artemisia* pollen percentages dominate those of other taxa throughout the gelifluction lobe record, reaching levels of 40-60% of the pollen sum (Figure 5). Pollen of Ericaceae and Gramineae are insignificant components of the regional spectra, but reach much higher levels in the gelifluction deposit. *Betula* percentages are perhaps half those of the lake records. Spores of *Lycopodium* and monolete spores of the Filicales (ferns) dominate the total pollen throughout the gelifluction lobe record, exceeding 90% of the total microfossils in the oldest sample (Figure 5). This may result in part from differential preservation due to the high resistance of these spores (along with conifer pollen) to both corrosion and oxidation (Havinga 1967). However these spores are common in surface litter samples from the study area, and increases in these types in the regional record between 8,000 and 10,000 B.P. have been interpreted as increasing abundances of *Lycopodium* and ferns in the vegetation (Brubaker et al. 1983).

In addition to these overall differences between the two records, changes in the pollen percentages of specific taxa are seen in the lobe record which do not occur in the lake pollen record. Gramineae pollen percentages reach a minor peak in the 3,200 year old samples. Both Ericaceae and Cyperaceae percentages increase in the gelifluction lobe samples younger than 1,800 years. *Picea* pollen is absent from the 6,600 year old gelifluction lobe sample, consistent with the very low regional *Picea* percentages at this time. The arrival of the boreal forest at the south shore of Walker Lake at about 5,000 years B.P. is accompanied by lake pollen *Picea* percentages of about 20% that remain relatively constant to the present. In the gelifluction lobe, *Picea* percentages of 1-2% occur in the 4,000 and 3,200 year old samples. Samples younger than 1,800 years from the lobe have *Picea* percentages between 5 and 10%, while modern surface litter samples from the lobe have percentages between 14 and 21%. In the absence of any change recorded in the lake record, these increases in *Picea* percentages in the gelifluction lobe record suggest treeline has advanced recently to its modern position from lake shore altitudes (200 m) where it stood about 5,000 years B.P. Latitudinal treeline in the Alatna River valley (35 km east of Walker Lake) apparently reached its modern position about 1,500 years ago (Brubaker

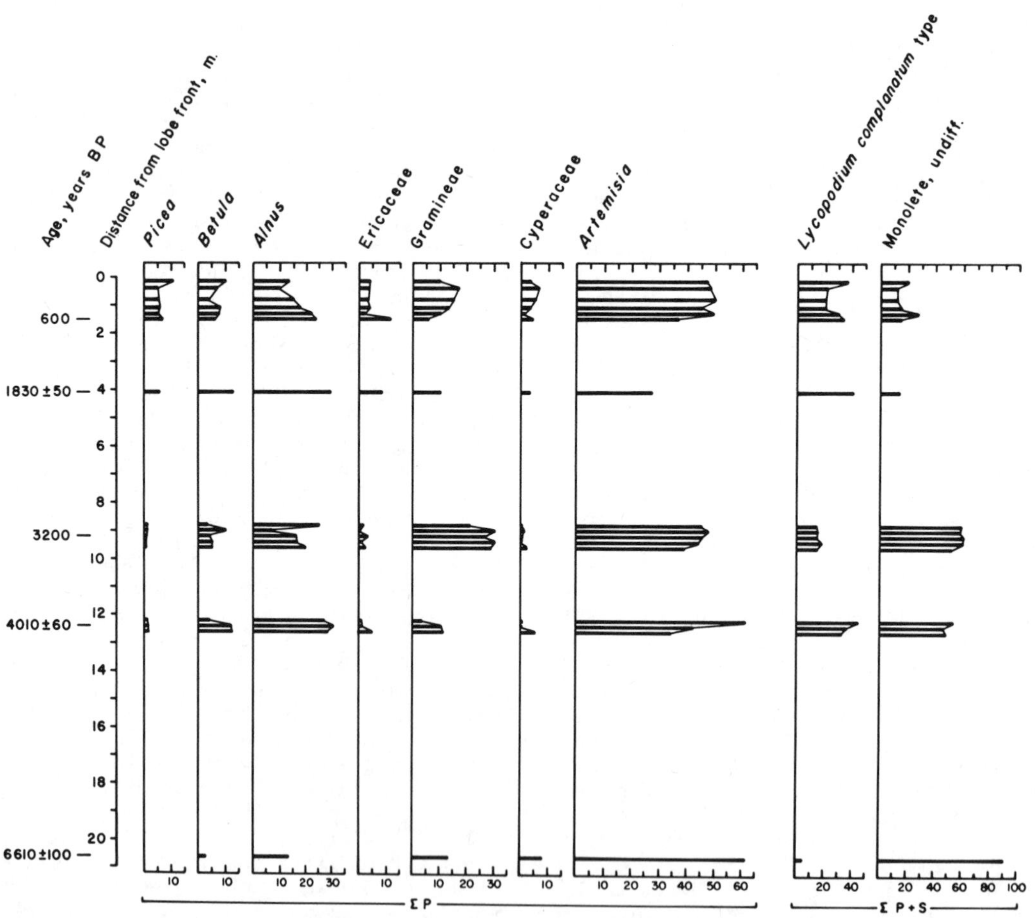

FIGURE 5 Pollen percentage diagram from the buried organic layer of the gelifluction lobe.

et al. 1983). Modern treelines at Walker Lake and in the Alatna valley are both at about 550 m altitude, although the treeline on the Alatna River is about 50 km farther north.

CONCLUSIONS

The degree of soil development exhibited by the buried paleosol and the *Picea* pollen percentages that suggest a recent increase of treeline to its modern altitude both indicate that treeline did not expand beyond its modern position during the Holocene. This conclusion was reached by others solely on the basis of soil (Ugolini et al. 1981) and lake pollen (Brubaker et al. 1983) data. We have also shown that basal sliding is an important component of gelifluction lobe movement at this site. This study has demonstrated that gelifluction deposits are useful sources of paleoenvironmental data. Paleosols preserved beneath advancing lobes contain information about past degrees of soil development. The advancing lobe front buries organic matter which can be used to date lobe activity and to calculate past rates of movement. The buried organic material contains a fossil pollen record useful in documenting past vegetation changes.

Shortcomings in the use of gelifluction deposits in paleoenvironmental studies include the excessive labor required to excavate samples, pollen spectra heterogeneity resulting from the effects of nearby individual plants, and the poor degree of pollen preservation which may bias the record. However, gelifluction deposits are frequently found in locations where lakes are absent, and may be the only available source of paleoenvironmental data. Additionally, gelifluction pollen records can complement lake pollen records by providing information about local vegetation not recorded in the lake records.

ACKNOWLEDGMENTS

We wish to thank J. M. Zachara, G. H. Goldstein, and H. L. Garfinkel for assistance with the field work, D. A. Naslund for laboratory assistance, and L. B. Brubaker, P. M. Anderson, and D. H. Mann for helpful comments on the manuscript. This research was supported in part by NSF grant DPP 76-23041.

REFERENCES

Alexander, C. S. and Price, L. W., 1980, Radiocarbon dating of the rate of movement of two solifluction lobes in the Ruby Range, Yukon Territory: Quaternary Research, v. 13 p. 365-379.

Benedict, J. B., 1976, Frost creep and gelifluction features: a review: Quaternary Research, v. 6 p. 55-76.

Brown, R. J. E. and Péwé, T. L., 1973, Distribution of permafrost in North America and its relationship to environment: a review, *in* Permafrost--The North American Contribution to the Second International Conference, Yakutsk: Washington D.C., National Academy of Sciences.

Brubaker, L. B., Garfinkel, H. L., and Edwards, M. E., 1983, A late-Wisconsin and Holocene vegetation history from the central Brooks Range: implications for Alaskan paleoecology: Quaternary Research, in press.

Bryson, R. A., Irving, W. N., and Larsen, J. A., 1965, Radiocarbon and soil evidence of former forest in the southern Canadian (arctic) tundra: Science, v. 147 p. 46-48.

Dreimanis, A. and Vagners, U. J., 1971, Bimodal distribution of rock and mineral fragments in basal tills, *in* Goldthwait, R.P., ed., Till, a symposium: Columbus, Ohio State University Press.

Faegri, K., and Iverson, J., 1975, Textbook of pollen analysis: New York, Hafner Press.

Garfinkel, H. L. and Brubaker, L. B., 1980, Modern climate-tree-growth relationships and climate reconstruction in sub-arctic Alaska: Nature, v. 286 p. 872-874.

Goldstein, G. H., 1981, Ecophysiological and demographic studies of white spruce (*Picea glauca* [Moench] Voss) at treeline in the central Brooks Range of Alaska: Ph.D. dissertation, University of Washington.

Hamilton, T. D., 1982, A late Pleistocene glacial chronology for the southern Brooks Range: stratigraphic record and regional significance: Geological Society of America Bulletin, v. 93 p. 700-716.

Havinga, A. J., 1967, Palynology and pollen preservation: Review of Palaeobotany and Palynology, v. 2 p. 81-98.

Mosimann, J. E., 1965, Statistical methods for the pollen analyst: multinomial and negative multinomial techniques, *in* Kummel, B. and Raup, D. M., eds., Handbook of paleontological techniques: San Francisco, W. H. Freeman.

Nelson, S. W. and Grybeck, D., 1981, Map showing distribution of metamorphic rocks in the Survey Pass quadrangle, Brooks Range, Alaska MF-1176-C: Washington D.C., U.S. Geological Survey.

Ritchie, J. C., and Spear, R., 1982, Late Pleistocene and Holocene pollen stratigraphy of the Mackenzie region of northwest Canada: American Quaternary Association Abstracts, Seattle, p. 157.

Soil Survey Staff, 1975, Soil taxonomy, a basic system of soil classification for making and interpreting soil surveys, Agriculture handbook 436: Washington, D.C., U.S. Government Printing Office.

Stuiver, M., 1982, A high-precision calibration of the AD radiocarbon timescale: Radiocarbon, v. 24 p. 1-26.

Ugolini, F. C., Reanier, R. E., Rau, G. H., and Hedges, J. I., 1981, Pedological, isotopic and geochemical investigations of the soils at the boreal forest and alpine tundra transition in northern Alaska: Soil Science, v. 131 p. 359-374.

Ugolini, F. C., Zachara, J. M., and Reanier, R. E., 1982, Dynamics of soil forming processes in the arctic, *in* French, H. M., ed., Proceedings of the Fourth Canadian Permafrost Conference: Ottawa, National Research Council of Canada.

Washburn, A. L., 1980, Geocryology: New York, Halsted Press.

INVESTIGATION OF THE AIR CONVECTION PILE AS A PERMAFROST PROTECTION DEVICE

R. L. Reid[1,3] and A. L. Evans[2,3]

1 Mechanical and Industrial Engineering Department
University of Texas at El Paso, El Paso, Texas 79968 USA
2 Mechanical Engineering Department, Cleveland State University
Cleveland, Ohio 44115 USA

3 Formerly, Mechanical and Aerospace Engineering Department, University of Tennessee, Knoxville, Tennessee 37916 USA

Winter performance and summer characteristics were measured for the air convection pile permafrost protection device. Velocity and temperature profiles were measured in the tube and annulus of air convection piles having below ground lengths of 3 m (10 ft) and 6 m (20 ft) containing 0.25 m (10 in) diameter tubes concentric with a 0.46 m (18 in) diameter outer piles. It was found that there was turbulent flow in both the tube and annulus and that the thermal performance could be expressed in terms of Nusselt and Rayleigh numbers by the equation $Nu = 0.21\ Ra^{0.29}$ for temperature differences between permafrost and ambient air temperature exceeding 14°C (25°F). In another 3 m (10 ft) pile, ice formation rates were measured under simulated summer conditions. It was found that a maximum ice thickness of 3.8 cm (1.5 in) is obtained and that the rate of formation is very slow for aboveground heights of about 2 m (7 ft) or more. Although the air is conceptually stagnant during summer, air movement was detected in the device for surface winds exceeding 2.0 m/s (4.5 mph) for the 3 m (10 ft) pile and 4.5 m/s (10 mph) for the 6 m (20 ft) pile.

INTRODUCTION

Thermal protection of permafrost is a well-recognized concern whenever construction of any type is contemplated in arctic and subarctic regions. The removal of naturally insulating vegetation can cause increased melt depths for permafrost in the summer and possible lack of refreezing in the winter. This potential problem was identified early in the planning for the Trans-Alaska Pipeline and a vigorous program was undertaken to find the most effective permafrost protection device (Jahns et al, 1973).

There are two obvious means of protecting permafrost in pipeline projects. These both involve active refrigeration. One method is to cool the fluid before it is pumped into the pipeline and the other method is to refrigerate the outside of the pipeline along its entire passage through permafrost. The first method is feasible for a gas but not a liquid since fluid friction is much higher with the liquid pipeline necessitating periodic and large amounts of cooling along the pipeline route. Refrigerating the pipeline wall is possible with both liquid and gas pipelines, but is extremely costly. This techique was used in the Trans-Alaska Pipeline in sections where pipeline elevation above the permafrost was not possible.

Elevating and insulating a pipeline removes most of the direct heating problem but not the indirect effect caused by the removal of vegetation during construction. This problem also arises whenever buildings, highways, or any other type of construction is attempted in permafrost regions. One method of protecting the permafrost is to install passive devices in the permafrost that will transport heat out of the permafrost in winter, when the permafrost temperature exceeds ambient temperature, to refreeze and possibly subfreeze the permafrost. These devices should cease operation in the summer when the permafrost is colder than ambient air. These devices can operate through transport of a gas, a liquid, or liquid and vapor through a phase change (heat pipes). Several different devices of all three types were tested by Galate, et al (1973) to determine overall heat removal rates. The type of device ultimately selected for the pipeline was a heat pipe although it was among the most costly devices. An air device, commonly known as the air convection pile, was not chosen for the Trans-Alaska Pipeline because of the lack of a detailed understanding of its performance and because of concerns about possible summer problems. Although the heat pipe had higher heat removal rates in the laboratory tests, Galate (1975) showed that both devices exceeded a minimum heat removal rate set by computer simulations for installations in permafrost. Because the air convection pile is so inexpensive compared to heat pipes (about one-tenth the cost) and because it has a potential of a longer service life, research has continued on the performance and characteristics of the air convection pile.

The air convection pile is an open system, tube-in-tube device that is installed in permafrost at depths of 3 - 18 m (10 - 60 ft). As shown in Figure 1, an aboveground head section keeps precipitation out of the pile and allows passage of ambient air through the pile where, in winter, the warmer permafrost heats the air in the annulus causing air to enter the tube, pass down to the bottom, rise up the annulus, and exit back to the environment. The outer tube is typicallly 0.46 m (18 in) diameter and can be made of heavy gauge steel if the pile is to support an above ground load. The inner tube is typically a thin gauge material, 0.25 m (10 in) in diameter, and extends nearly to the bottom of the pile leaving only 0.15 m (6 in) of clearance. A support arm can be attached to the aboveground section if a load is to be supported as in an elevated pipeline.

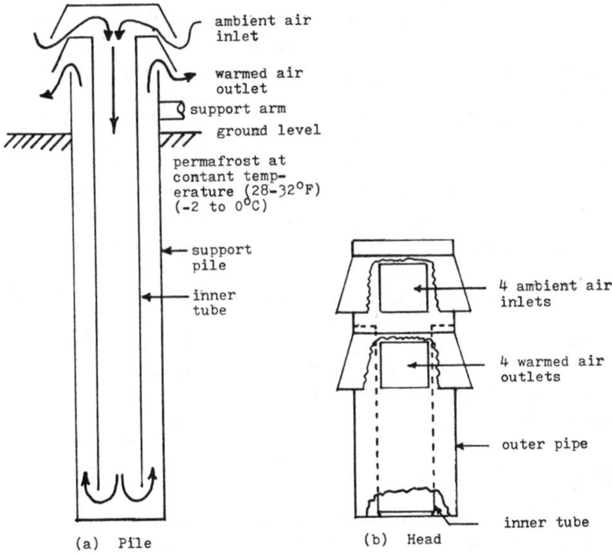

FIGURE 1 The air convection pile.

Even though the construction of the air convection pile is very simple, the heat and mass transfer mechanisms are actually very complicated. Although the driving force is natural convection, there is unequal heating of the annulus because of radiation heat transfer from the outer pile to the inner tube which provides heat for both the tube and the annulus. Also the tube flow acts as a drag on the annular flow and there are entrance effects in both the tube and annulus. The experiments by Galate (1975) were done in a heated calorimeter bath with a refrigerated chamber over the aboveground head and measured only the overall thermal performance in the device. No measurments were made of the velocity or temperature profiles in the device. An early mathematical model developed by Reid et al (1975) predicted temperature and velocity profiles as well as thermal performance based on assumed heat transfer modes in the tube and annulus. One objective of the study reported here was to perform detailed velocity and temperature measurements as well as measure overall performance to better understand the heat and mass transfer mechanisms and to develop correlations based on the observed mechanisms and data.

The concerns relative to the summer behavior of the air convection pile were that (1) because of the very low pressure drops through the pile, it might be susceptible to wind induced forced convection, and (2) moisture might enter the pile by diffusion and/or condensation, forming ice at the frost line, and possibly completely blocking the air flow through the pile in the following winter, when the air temperature falls below the permafrost temperature. Galate (1975) reported some observations of the first effect but provided no data as to a possible threshold velocity or magnitude of the effect. Additional objectives of the study reported here were to investigate possible ice formation and wind induced circulation during summer.

THERMAL PERFORMANCE

A full-scale prototype air convection pile was constructed and instrumented. Since the predominant driving force for the flow is the temperature difference between the permafrost and the ambient air, it was decided to simplify the apparatus by using ambient room air as the source and heating the outer pile wall. The difference in temperature level of the experiments relative to arctic conditions was correlated by the Rayleigh number as discussed below. A simulated below ground section was constructed of galvanized ducting, wrapped with copper tubing, and covered with insulation as shown in Figure 2. This section was built to be expandable and data were taken for both 3 m (10 ft) and 6 m (20 ft) depths. A hot water supply system which could control and measure the temperature and flow rate of the water was constructed. The inner tube and head assembly were also constructed from galvanized sheet and tubing.

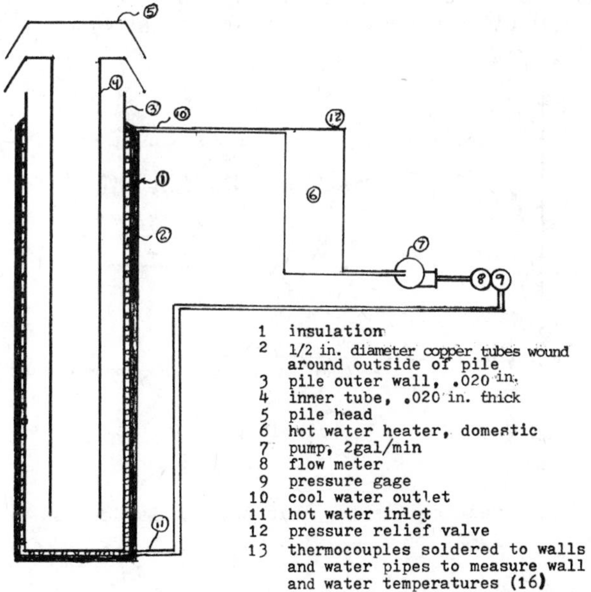

FIGURE 2 Test configuration.

Details of the instrumentation and experimental method were reported previously by Evans and Reid (1982). Velocity and temperature profiles obtained by hot wire anemometer and thermocouple traverses were reported for the 3 m (10 ft) heated length in the same reference. The velocity measurements were complicated by the discovery of low-frequency turbulent flow (4-28 Hz). This required signal conditioning and linearization to obtain the true time average velocity at each point.

Velocity and temperature profiles were measured at three vertical positions in the pile. A typical velocity profile at the bottom of the tube is shown in Figure 3 along with the curve calculated for fully developed turbulent flow from Kays (1966). Profiles at upstream sections were somewhat different, showing the flow progressing to a fully developed state. Figure 4 shows a measured profile at the top of the annulus along with calculated curves for fully developed laminar and turbulent flow in an annulus from Knudsen and Katz (1958). It can be seen that the profile resembles neither laminar or turbulent flow. At lower (upstream) locations, the profiles were basically of the same shape. Typical dimensionless temperature profiles are given in Figure 5.

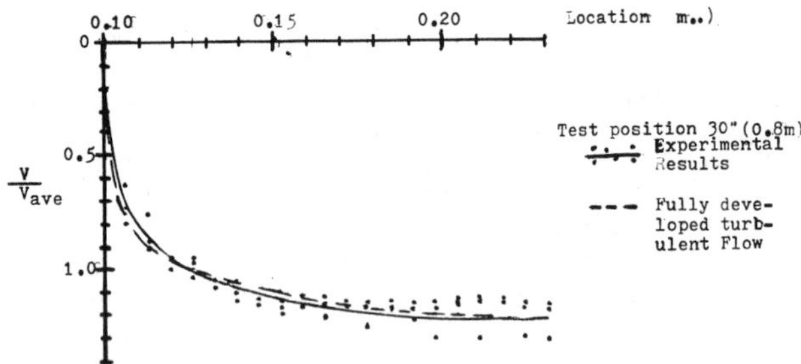

FIGURE 3 Velocity profile in the tube.

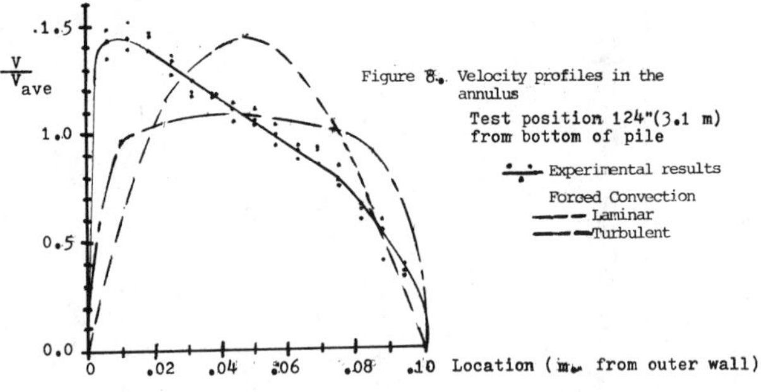

FIGURE 4 Velocity profile in the annulus.

The dimensionless temperature, θ, is defined as

$$\theta = \frac{T - T_i}{T_w - T_i}$$

where T_i = ambient air temperature and T_w = outer pile wall temperature. The dimensionless radial position Y is defined such that Y = 0 is the outer pile wall and Y = 1.0 is the centerline of the pile. (The tube wall is located at Y ∓ 0.44.)

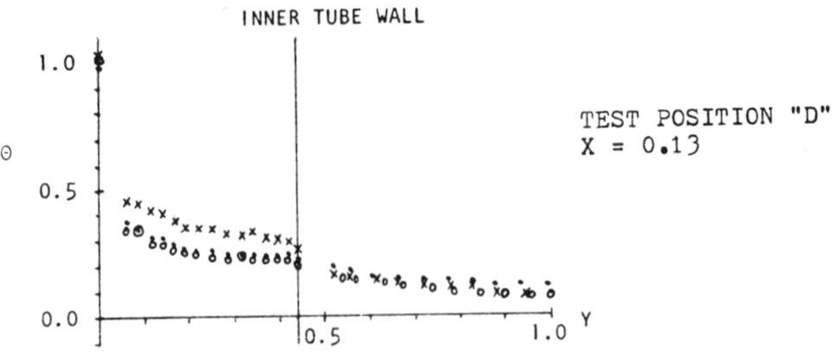

FIGURE 5 Temperature Profiles

The shape of the velocity profile in Figure 4 does not resemble fully developed turbulent flow because of the unequal heating from the walls and the strong coupling of the momentum and energy equation through the buoyancy force. There are three distinct regions in each profile. From the tube wall, at Y = 0.44, is an inner boundary layer developing off the tube wall. From Y = 0.4 to 0.075 then is core type of flow in which the velocity increases linearly with Y. Finally for Y < 0.075 there is an outer boundary layer being generated off the outer pile wall at Y = 0. The maximum velocity occurs at Y = 0.05, 1.1 cm (0.45 in) from the outer wall. The profiles taken at the bottom to the pile showed this general profile is already almost fully developed only a short distance of the annulus. This is due to the high amplitude of the turbulence at this point generated by the turn of the flow at the bottom of the pile.

Average velocities in the tube and annulus as a function of temperature difference between the ambient and outer pile wall are shown in Figure 6. The overall heat removal rates are shown in Figure 7.

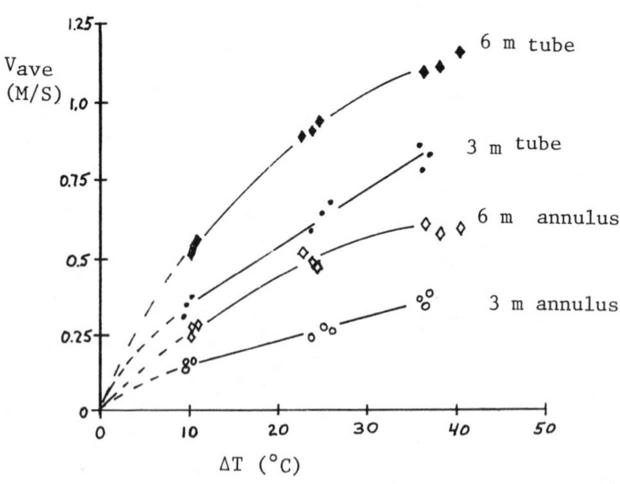

FIGURE 6 Average velocities in the Tube and Annulus.

The data for the 6 m (20 ft) below ground section were not reported in the previous reference, but the velocity and temperature profiles were similar in shape as for the 3 m (10 ft) section. The velocities and overall heat removal rates are also given in Figures 6 and 7, for the 6 m (20 ft) pile.

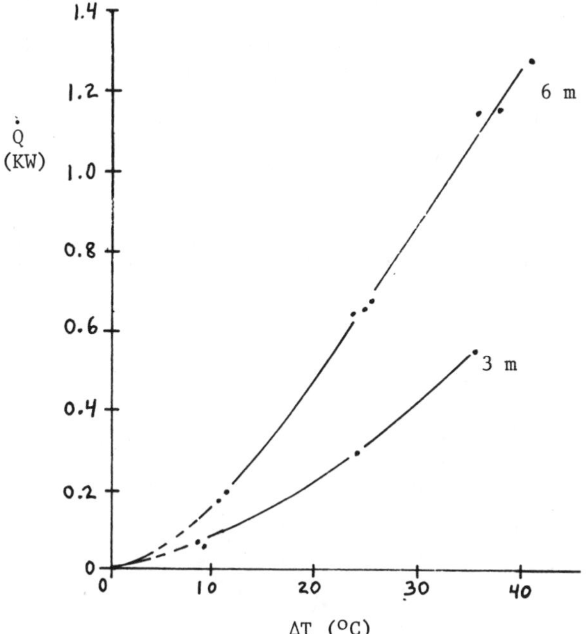

FIGURE 7 Overall heat removal rate for the air convection pile.

Since the measurements of air convection pile performance were made at a higher temperature level than occurs with permafrost, the ΔT between pile wall and inlet air temperature in the laboratory does not correspond directly with the ΔT that would be measured in a field application. The important parameter that governs the driving force for heat transfer is the Rayleigh number, defined as:

$$(Ra)_r = \frac{g\beta\Delta T r^3}{\alpha\nu}$$

where

- g = gravitational acceleration
- β = volume coefficient of expansion
- ΔT = temperature difference between inlet air and outer pile wall
- r = radius of the pile
- α = thermal diffusivity
- ν = kinematic viscosity.

The relationship between the Rayleigh number and arctic ΔT assuming a permafrost temperature of 0° C (32° F) is shown in Figure 8. The results for the heat transfer rate can also be expressed in a dimensionless form by the Nusselt number, which is defined as:

$$(Nu)_r = \frac{\dot{Q}r}{A\Delta T k}$$

where

- Q = heat transfer rate
- A = heated area of the pile (outside wall + bottom)
- k = thermal conductivity.

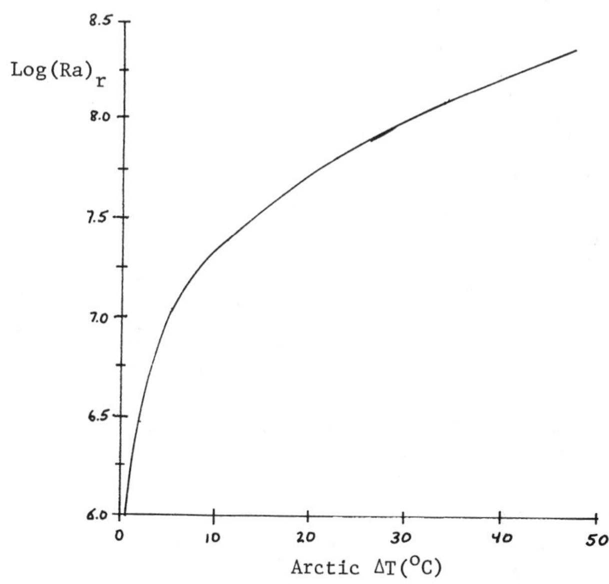

FIGURE 8 Relationship between Rayleigh number and arctic ΔT.

The plot of $(Nu)_r$ versus $(Ra)_r$ for the results of these experiments and the results of Galate et al. (1973) are shown in Figure 9. All the results fall on the same curve since the pile is equally effective per unit length for any total length of pile. For the same ΔT, the total heat transfer will be twice as great for a pile of twice the length because of the heated area factor in the Nusselt number as shown above. Above a Rayleigh number of about 2.5 X 10^7 (log $(Ra)_r$ = 7.4, arctic ΔT = 14°C (25°F)), the curve is approximately a straight line which is determined by the equation:

$$(Nu)_r = 0.21\ (Ra)_r^{0.29} \quad (1)$$

The results for Q predicted from equation (1) and shown in Figure 9 are about 10% lower than given by Galate (1976) for a 3 m (10 ft) air convection pile. As shown in the same reference, the air convection pile has a heat removal rate about 30% lower than that attained by heat pipes. However, as discussed earlier, both types of device exceed a minimum required for permafrost installation. This is because the heat transfer to the permafrost is governed by a series of thermal resistances which include the thermal resistance in the soil and the thermal resistance from the pile to the ambient air. The latter resistance is zero for the air convection pile but not zero for heat pipes. Also the thermal resistance in the soil is large and may dominate the other resistances. The important result is that the less expensive air convection pile provides adequate thermal protection of the permafrost.

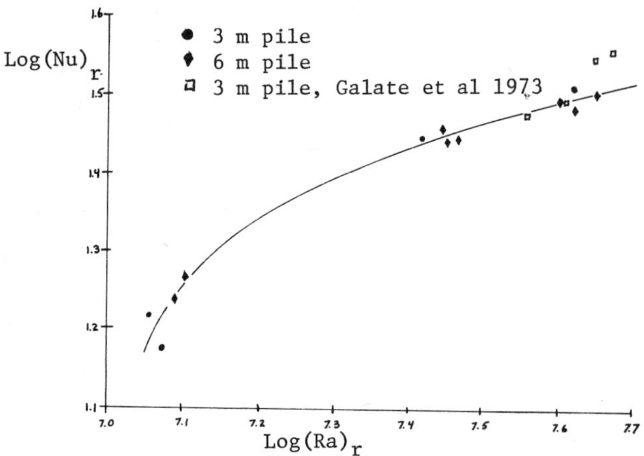

FIGURE 9 Nusselt number versus Rayleigh number.

FROST AND ICE FORMATION

As discussed earlier, one concern about the air convection pile was that ice might form on the walls of the pile during summer. It was postulated that the ice could bridge the annulus of the pile and prevent or delay the start up of the pile the next winter. Therefore, a study was undertaken to determine the degree of frost and ice formation in the annulus of the pile under arctic conditions and to obtain a method or methods to reduce or stop the ice formation.

To determine the effects and the amounts of ice formation another 3.05 m (10 ft) model of an air convection pile and a simple environmental chamber were constructed as described by Reid et al (1982). The pile wall was refrigerated and the environmental chamber temperature controlled to simulate actual arctic conditions.

During a test the following parameters were monitored and recorded: pile wall and inner tube temperature, chamber humidity and temperature, ice thickness, volume and final mass, and the annulus concentration gradient. The ice thickness was measured by a profilometer riding on a track attached to the tube wall. From the profiles, the ice volumes were determined by integrating the cross section and revolving it about an axis. The final ice mass was found by letting the ice melt at the end of a test and weighing the resulting water. The chamber humidity and the annulus concentration gradient were determined using an hygrometer system which gave a reading in percent relative humidity. The percent relative humidity was converted to absolute humidity ratio using the dry bulb temperature and psychrometric chart.

Detailed results from this study were given in the paper by Reid et al (1982). Two important observations can be made from the results. First, the rate of ice formation is relatively insensitive to ambient humidity, but is very sensitive to the length of air convection pile above the ground surface. Vertical humidity traverses in the annulus showed a concentration gradient consistent with this observation since the diffusion path length is larger for longer unheated necks. The second observation is that the maximum ice thickness levels off at 3.8 cm (1.5 in) after 25-30 days. It was postulated that the maximum possible ice thickness is governed by a conduction, convection, and radiation heat transfer balance. That is, after the ice obtains a certain thickness no further increase in this thickness is possible at that point due to the surface temeratures not being low enough for any moisture to freeze to ice. The ice will continue to increase in thickness on levels where the thickness is less than the maximum possible ice thickness. This was verified by the data which showed that the ice volume was still increasing when the maximum ice thickness had leveled out. Another concern with ice formation is whether a partial blockage will prevent the pile from starting to operate during the following winter. A test designed for the thermal performance apparatus using an inflatable bag showed that partial blockage would not prevent winter operation. Approximate calculations have also shown that ice may sublimate out of the pile in 10 - 50 days during the winter depending on the amount of blockage.

The conclusion from this part of the study was that ice formation should not be a problem under stagnant air conditions since the maximum ice thickness will be 3.8 cm (1.5 in). However, added protection can be obtained by always using an aboveground height of at least 2.1 m (7 ft) which includes 1.4 m (4.5 ft) for the neck and 0.7 m (2.5 ft) for the inlet and outlet section.

EFFECT OF SUMMER WIND

As mentioned earlier, another concern during summer was that wind blowing across the pile head might cause circulation of ambient air throught the device causing additional permafrost melting. The same experimental apparatus as was used for simulated winter performance testing was used for this study. However, the hot water supply was replaced by a chiller to cool the pile wall approximately 6.5 C (11.7 F) below the ambient temperature in the laboratory. A large air supply was constructed which included straightener tubes to provide a relatively uniform flow field. The wind speed was slowly increased while the hot wire anemometer probes were being monitored and the wind speed was recorded at which flow just started in the air convection pile. This critical wind speed was 2.0 m/s (4.5 mph) for the 3 m (10.0 ft) pile and 4.5 m/s (10.0 mph) for the 6 m (20 ft) pile. The rate of heat transfer from the air to the pile wall was measured for the 10 ft pile at a wind speed of 4.5 m/s (10.0 mph). A mass flow rate of 0.0133 kg/s (1.75 pound mass/min) and a heat transfer rate of 9.6 W (32.9 BTU/hr) were measured. These measuremnts were not made for the 6 m (20 ft) pile because of the inadequacy of the chiller to cool the pile for the long period of time necessary to make the traverses.

The important observation in this section is that for piles of 6 m (20 ft) in depth, summer induced circulation may not be a problem since winds will seldom exceed 4.5 m/s (10 mph). It appears that this effect may be approximately linear such that piles of 12 m depth or greater may not be wind sensitive because winds above 9 m/s (20 mph) are rare in summer. Further data are needed, however, to fully define the effect of summer winds.

SUMMARY

Prototype 3 m (10 ft) and 6 m (20 ft) air convection piles were constructed and performance was measured as a function of Rayleigh number. These results are given in Figure 9 and equation (1) with the relationship between Rayleigh number and arctic ΔT shown in Figure 8. Measured velocity profiles showed turbulent flow in both the tube and annulus.

An additional 3 m (10 ft) prototype pile was placed in an environmental chamber simulating summer conditions and the rate of ice formation monitored as a function of chamber humidity and uncooled neck length (height that protrudes above ground). It was found that the maximum ice thickness in the 10 cm (4 in) annulus would be 3.8 cm (15 in). The rate of ice formation was only weakly dependent on chamber humidity bui decreased greatly as the uncooled neck length was increased such that very low rates were obtained if the height above ground surface was 2.1 m (7 ft).

The pile was shown to be wind sensitive under summer conditions but this effect was very dependent on below ground length. Since flow did not start until the wind velocity equaled 4.5 m/s (10 mph) in the 6 m (20 ft) pile, wind sensitivity may be negligible for longer piles.

ACKNOWLEDGEMENT

This research was supported by the National Science Foundation under grant ENG 77-2098.

REFERENCES

Evans, A. L., and Reid, R. L., 1982, Heat transfer in an air thermosyphon permafrost protection device, American Society of Mechanical Engineers Journal of Energy Resources Technology, V 104, p 205-210.

Galate, J. W., Jahns, H. O., Miller, T. W., Power, L. D., Rickey, W. P. Scanlan, F. G., and Wheeler, J. A., 1973, Thermal piles for elevated pipelines, Esso Production Research Company, report no. EPR.30PS.73.

Galate, J. W., 1976 Passive refrigeraiton for arctic pile suports, American Society of Mechanical Engineers Journal of Engineering for Industry, V.98, series B, p 695-700.

Jahns, H. O., Miller, T. W., Power, L. D., Rickey, W. P., Taylor T. P., and Wheeler, J. A., 1973, Permafrost protection for pipelines, in Permafrost--The North American contribution to the Second International Permafrost Conference, Yakutsk: Washington, D. C., National Academy of Sciences.

Kays, W. M., 1966, Convective Heat and Mass Transfer, New York, McGraw-Hill.

Knudsen, J. G., and Katz, D. L., 1958, Fluid Dynamics and Heat Transfer, New York, McGraw-Hill.

Reid, R. L., Tennant, J. S., and Childs, K. W., 1975, The modeling of a thermosyphon type permafrost protection device, American Society of Mechanical Engineers Journal of Heat Transfer, v. 97, p 382-286.

Reid, R. L., Hudgins, E. H., and Onufer, J. S., 1982, Frost and ice formation in the air convection pile permafrost protection device, American Society of Mechanical Engineers Journal of Energy Resource Technology, v. 104, p 199-204.

SOME EXPERIMENTAL RESULTS ON SNOW COMPACTION

Eugenio Retamal

Geotechnical Division, Institute of Research & Testing Materials (IDIEM), Universidad de Chile, Plaza Ercilla 883, Santiago, Chile

This paper presents the results of snow field compaction at snow temperatures near 0°C and some stress-strain properties obtained from rigid plate load and unconfined compression tests. The data and results of some load tests performed at outer Antarctica are also presented.

INTRODUCTION

In the most southern part of continental Chile (lat.45°-55°S) and in areas sufficiently separated from the sea, there are large areas covered with snow almost all year round with ambient temperatures characteristically fluctuating between -10°and +5°C. These temperatures gradually diminish toward the inner part of the southern cone of South America and also toward the south. The systematic diminution of temperature means that at the Shetland Islands the average annual isotherm is -2°C and at Margarita Bay (lat. 68°10'S, long. 67°05'W) it is -6°C. During summer the isotherm at the Shetland Islands is 1°C, at Margarita Bay it is 0°C, and in winter it is -7° and -14°C, respectively. Toward the inner part of the Antartic continent, temperatures constantly diminish, and they change abruptly when the coastal areas are left behind.

Therefore human activities in these areas require aerial transportation, and consequently, there is need of methods to increase the bearing capacity of snow and of working techniques for the construction of airfields on deep to not very thick snow. Of special interest is the case of snow temperatures near 0°C since there are very few experimental data dealing with snow in the temperature range -5°C to the melting point, which is of importance for Chile.

This paper adds data obtained from field compaction of snow in the vicinities of the Lonquimay Vulcan in a test field located at 1,400 m above sea level and data for tests performed at the Collins Glacier, King George Island (Shetland Islands).

TESTS AT LONQUIMAY

The Lonquimay Vulcan, 2,860 m high, is located in the southern Andes Mountains (lat. 38°26'S, long.71°14'W). An area was selected near its south slope and 1,400 m high because of temperature similarities with the Antarctic Peninsula.

Two types of test fields were constructed using a Bomag BW 75 S, two steel wheel roller with a gross weight of 928 kg. The first type of test field was built by just passing the roller without vibration over the existing natural snow, which had a total thickness of about 80 cm. For the first five passes the roller was assisted by a system of cables and winches and by a thin rubber carpet 10 m by 1.2 m put over the snow surface. From the fifth pass the roller was able to operate in a self moving way with or without the rubber carpet and without assistance from the cable-winch system. In Figure 1 is presented the experimental curve of the variation of unit weight of snow of the upper part versus the number of passes.

A second type of test field was built using an excavation of about 8 m by 1.5 m in plan area and filling it, from the hard soil surface, with compacted snow. For this purpose the first layer of loose snow was put over the existing soil with a loose thickness of about 20 cm and then compacted using the roller, which passed over the rubber carpet during the first 4-5 passes. In this way a test field 8 m long by 1.2-1.5 m wide with a total height of about 80 cm was built. On the surface of the test field several vertical load tests were performed using a 30 cm diameter rigid load plate. The results of a typical test are presented in Figure 2. In Figure 3 is shown the variation of the tangent deformation modulus E vs. time (Tangent modulus is obtained from the straight parts of the curves of Figure 2 which have similar slopes). To obtain the value of E the classical equation was used:

$$E = \frac{q_s B (1-\mu^2)}{\rho} I_\rho$$

where

q_s = total intensity of contact pressure

B = diameter, 30 cm

μ = Poisson's ratio; 0.35 considered for this case

I_ρ = influence factor, $\pi/4$

ρ^ρ = measured settlement

In Figure 4 the results of unconfined compression tests are shown. The tests were performed with samples taken from the test field and with different degrees of compaction.

TESTS AT COLLINS GLACIER

Collins Glacier is located at King George Island. Tests were performed during winter over a multilayer system of metamorphic snow and ice.

Vertical rigid plate load tests were performed at different bottom levels of two pits during their excavation. These pits were also used for getting the stratigraphy of the snow-ice system shown in Table 1.

TABLE 1 Stratigraphy of the Pits.

PIT	DEPTH (cm)	DESCRIPTION
1	0-33	Loose snow
	33-36	Ice
	36-170	Intercalated thin layers (1-5 cm) of medium to hard snow and ice layers
2	0-37	Loose snow
	37-47	Medium to hard snow
	47-52	Hard ice
	52-85	Medium snow
	85-90	Ice
	90-93	Loose snow
	93-95	Ice
	95-105	Loose to medium snow
	105-108	Ice
	108-127	Medium snow
	127-185	Ice

In Figure 5 are shown the results of a vertical load test performed at a depth of 33 cm in Pit 1. In Figure 6 can be seen the variation of the deformation modulus E, obtained as a secant modulus between the origin and a settlement of 0.1 and 0.2 cm, versus time. In Figure 7 are presented the data of a load test at a depth of 40 cm in Pit 2. In Figure 8 is shown the variation of E versus time if E is obtained as a secant modulus in the curves of Figure 7, and in Figure 9 the variation of the modulus of deformation if E is obtained as a tangent modulus for large settlements from the same curves. Finally in Figure 10 is presented the variation of E in a deeper load test performed in Pit 2 if E is obtained as a tangent modulus for large settlements.

COMMENTS AND CONCLUSIONS

It can be seen that snow, under the conditions described, can be compacted using light equipment and that with 5-10 passes of the roller used it is possible to get a unit weight of about 6.5-7 kN/m^3. Probably with heavy equipment, such as a heavy rubber-tired roller, the results would be better.

The influence depth of compaction for the roller used was between 15 and 20 cm. Vibrations were not useful for compacting snow.

The values of E are highly contaminated with creep, which is a very important factor in snow and ice behavior under load. As a consequence, the load speed is of great importance.

Experimental work at Collins Glacier shows that under a first layer of loose snow of about 35-40 cm, at the time and under the conditions of the field tests, there is a multilayer system of ice and snow and its deformation modulus E, also affected by considerable creep, is comparable to the same modulus as that of a soft to hard clay, depending on time. This level could be a relatively good subbase for upper layers of compacted snow for skiways and probably as a runway for emergencies or restricted use. It is convenient to point out that the temperature of deeper layers of the glacier tend to diminish, and consequently their engineering properties tend to improve.

From the load tests at Collins Glacier and also from the same tests on compacted snow it seems that the applied pressure-settlement behavior could be represented by a bilinear system with E_{sec} and E_{tg} approximately independent of time at least for $t<10-15$ min according to Figure 11. In this figure, q_L would vary with time. If it is supposed that for landings $t \approx 0$, the deformation would be defined for E_{sec}, E_{tg}, and q_L.

ACKNOWLEDGMENTS

The author is grateful to the Department for the Development of Research, University of Chile, for supporting this work and also to the Chilean Air Force for the logistic support at Antarctica.

REFERENCES

Fukue, M. 1979, Mechanical Perfomance of Snow Under Loading, Tokai University Press, Tokyo.

Klokov, V.D. 1982, Air Navigation Support for Scheduled Flights of IL-18 Airplanes to Molodezknaya Station, Symposium on Antarctic Logistics at the XVII SCAR, Leningrad.

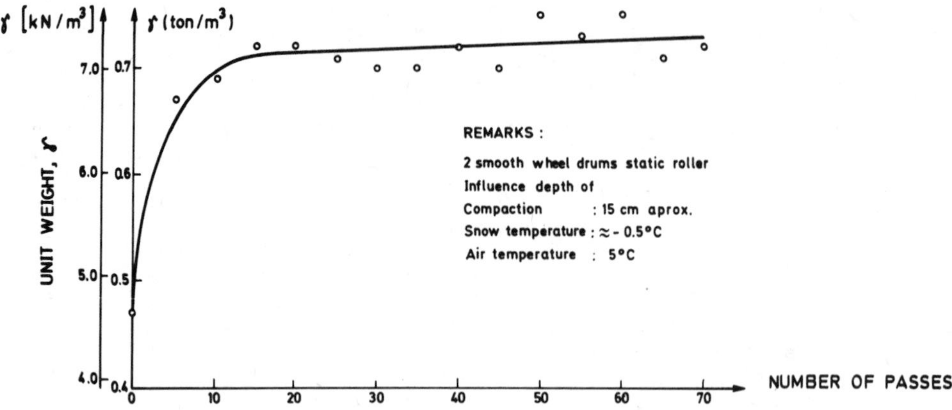

FIGURE 1 Variation of the superficial snow unit weight vs. number of passes using a BW 75 S (928 kg) roller

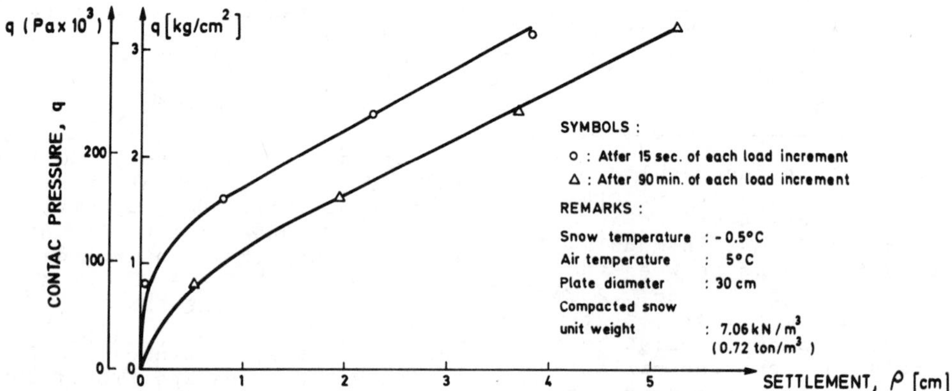

FIGURE 2 Vertical settlements in a compacted test field of 80 cm. in thickness during a rigid plate load test.

FIGURE 3 Variation of E vs. time from the load test of Figure 2 if E is the tangent modulus for large settlements.

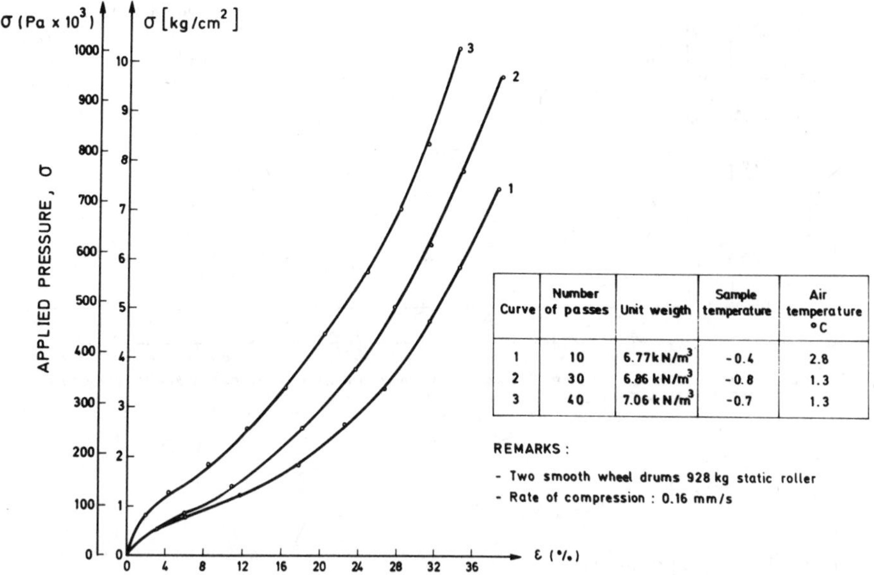

Curve	Number of passes	Unit weigth	Sample temperature	Air temperature °C
1	10	6.77 kN/m³	-0.4	2.8
2	30	6.86 kN/m³	-0.8	1.3
3	40	7.06 kN/m³	-0.7	1.3

FIGURE 4 Stress - strain curves from unconfined compression tests of compacted snow.

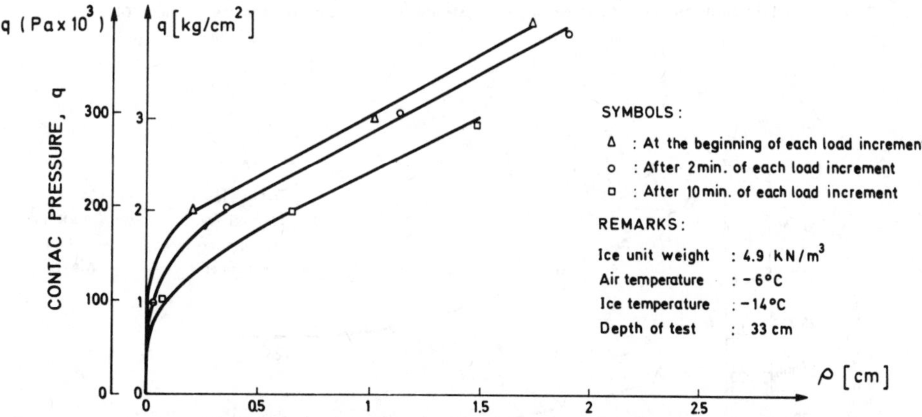

FIGURE 5 Vertical settlements in a layered snow-ice system during a rigid plate load test. Collins Glacier, Pit 1.

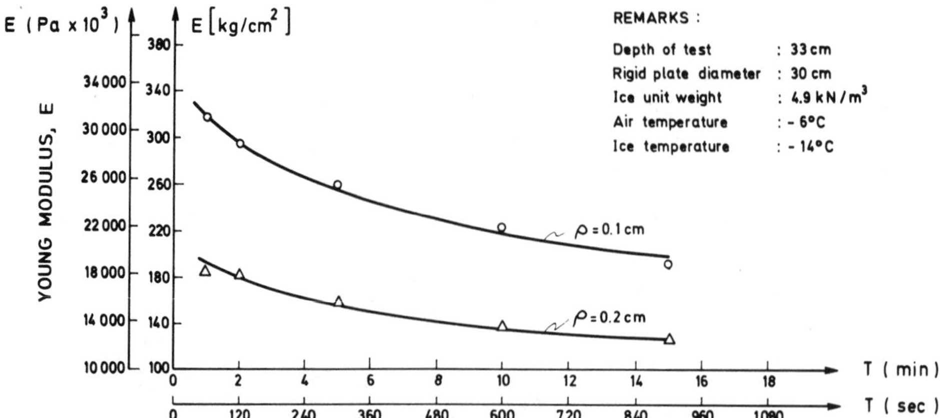

FIGURE 6 Variation of E vs. time from the load test of Figure 5 if E is obtained as a secant modulus for small settlements. Collins Glacier, Pit 1.

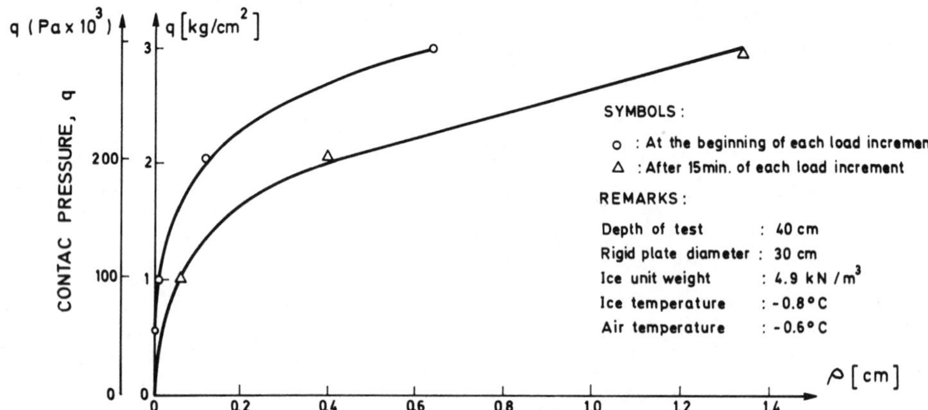

FIGURE 7 Vertical settlements in a layered snow-ice system during a rigid plate load test. Collins Glacier, Pit 2.

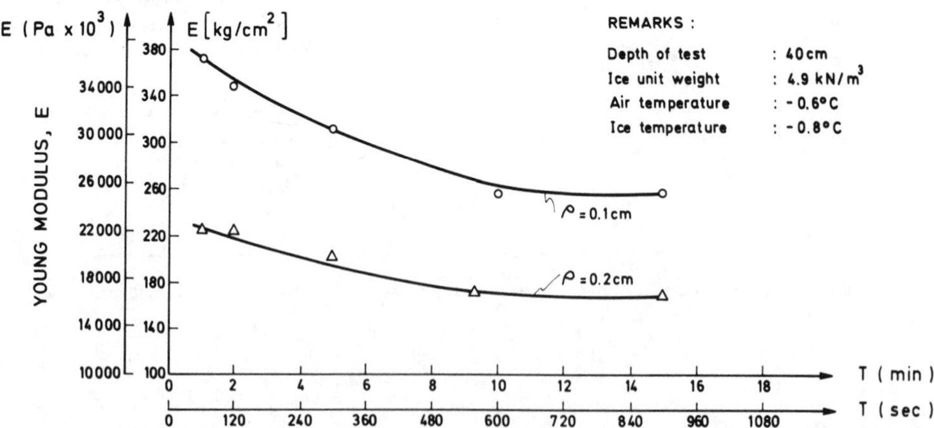

FIGURE 8 Variation of E vs. time from the load test of Figure 7 if E is obtained as a secant modulus for small settlements.

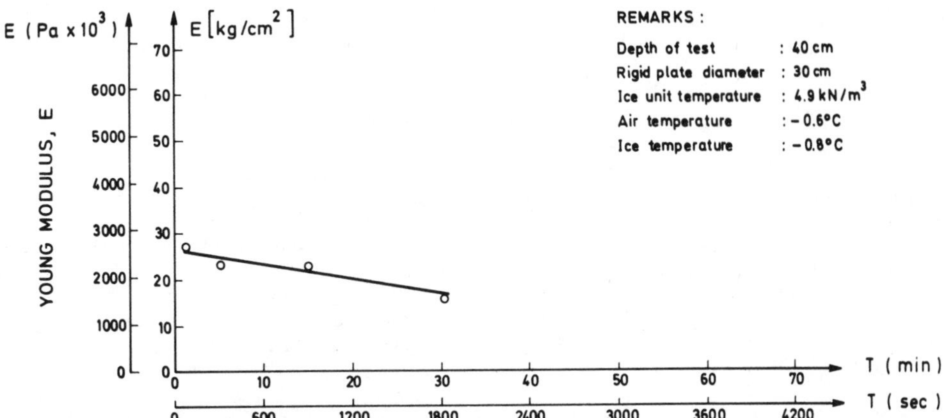

FIGURE 9 Variation of E vs. time from a load test if E is obtained as a tangent modulus for large settlements. Collins Glacier, Pit 2.

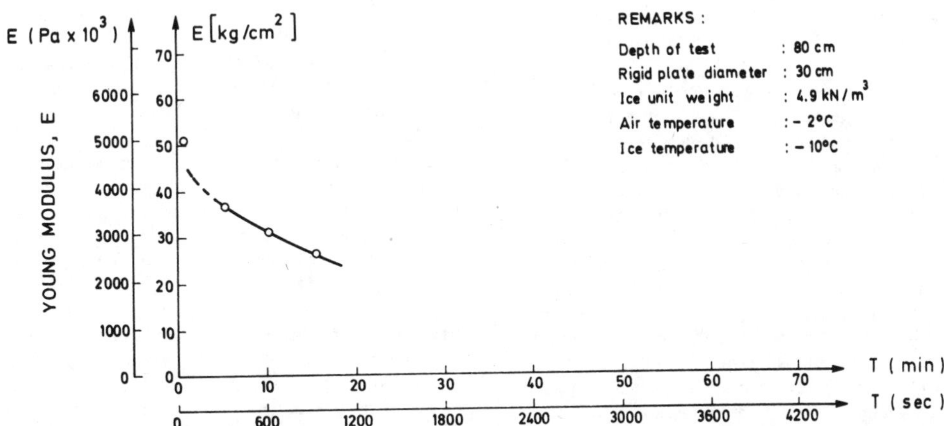

FIGURE 10 Variation of E vs. time in a deeper load test if E is obtained as a tangent modulus for large settlements. Collins Glacier, Pit 2.

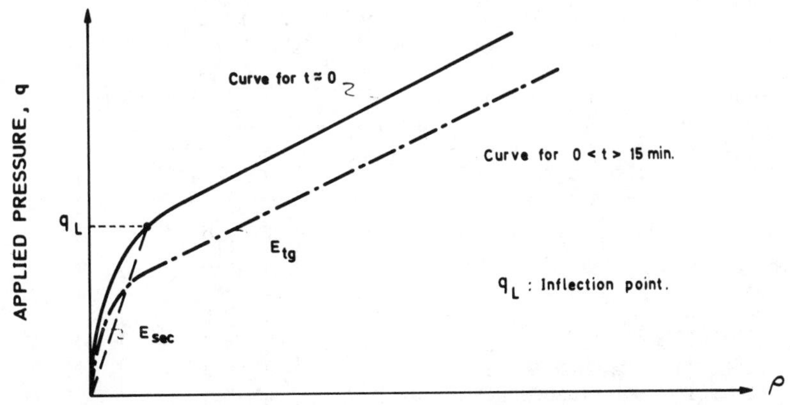

FIGURE 11 Vertical settlements, ρ, in a layered snow - ice system

POWER LINES IN THE ARCTIC AND SUBARCTIC -- EXPERIENCE IN ALASKA

Robert W. Retherford
International Engineering Company, Inc., 813 D Street, Anchorage, Alaska 99502 USA

There is a substantial background of experience on the design, construction, and operation of power lines on and in the frozen earth of Alaska. This paper contains a digest of mostly successful examples of Alaska power line elements operating with the frozen earth. The paper considers footings, anchors, structures, conductor and electric grounding. Examples are taken from operating power systems which serve the small village (50 consumers); the large urban/rural community (50,000 consumers); and a number of federal, state and industry installations.

INTRODUCTION

Ice, snow, wind, heat and cold make and remake the frozen earth of the world's northern latitudes. Much of Alaska is in this region so its residents must deal with those changing elements (Eaton et al 1975). Power line construction and operation has grown with Alaska's development and gained experience in living with this environment.

Long-term weather records do not exist in most of Alaska. Experience has also shown that climates may differ drastically with small changes in location and elevation. Skill in reading the evidence of nature is then important in decision-making which may radically alter the feasibility of projects.

Differences in the weather criteria chosen for the design of electric lines listed herein arise partly from the subjectivity that results from these conditions. Other differences in the terrain, flora and fauna of a line location provide the challenge that encourages innovative solutions to match the variety of Alaska's earth and sea.

The overload capacity design factors chosen to assure reliability in a power line are influenced by the value of the affected product or service. For example: Loss of power to the Prudhoe Bay oil fields could shut down a cash flow of $2 million/hr. while the loss of hydroelectric power to a typical comunity of Southeast Alaska might require replacement by using diesel fuel costing about $1,000/hr.

The experience described is with power lines in Alaska, although not all are located within a strict permafrost regime. Examples are included for footings, anchors, structures, conductors, and grounding.

ALASKA POWER LINES

The key map (Figure 3) shows the approximate location of a selected group of power lines. Table 1 lists the lines, general data, and a key indicator pointing to specific experience described in more detail in the text.

SPECIFIC EXPERIENCE ITEMS

Footings

Many conventional footing designs are in use on the power lines of Alaska, such as grillages, concrete, and tamped earth where the soil conditions allow. In permafrost regions there are often stable soils where these designs are acceptable. Examples of designs used where the freeze-thaw cycle operates in unstable soil are illustrated by a photo or drawing and/or text in this section.

(1) Gravel backfill for wood-pole installations has been used successfully. Using local soil instead has produced severe frost - jacking in many installations. Figure 1 shows a typical gravel backfill construction unit, which has been used on Lines 4, 12, and 14.

(2) A bogshoe composed of a crosslog and outriggers to stabilize pole structures in muskeg has been used with limited success. The shoe works, but if frost jacking takes place on the pole, the whole assembly will tip with the prevailing winds. The design, shown in Figure 2, is used on line 3.

(3) Use of piling to reinforce poles or provide suitable pole stubs has a successful history. Figure 4 shows a typical installation for an H-frame wood pole line. These units have been used on lines 3, 4, and 5. Single and multiple steel H-pile have been used successfully for metal tower footings in muskeg and permafrost soils on lines 8, 11, 13 and for anchors on line 16 (Proctor and Meyers 1978).

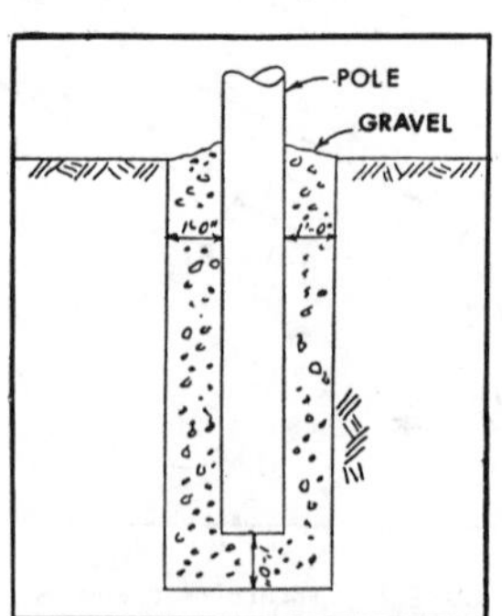

FIGURE 1 Gravel backfill for power poles

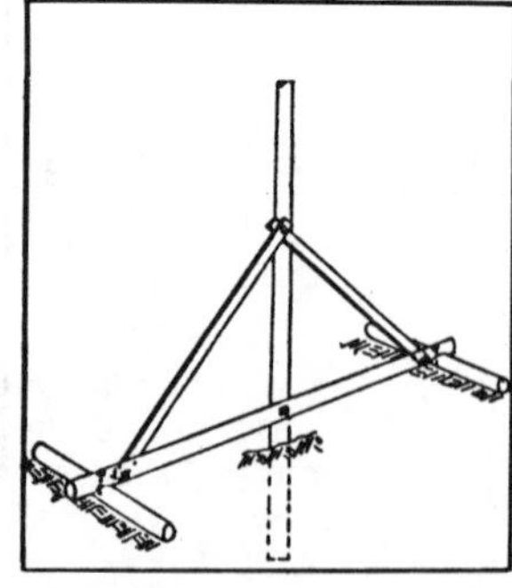

FIGURE 2 Bogshoe for power poles

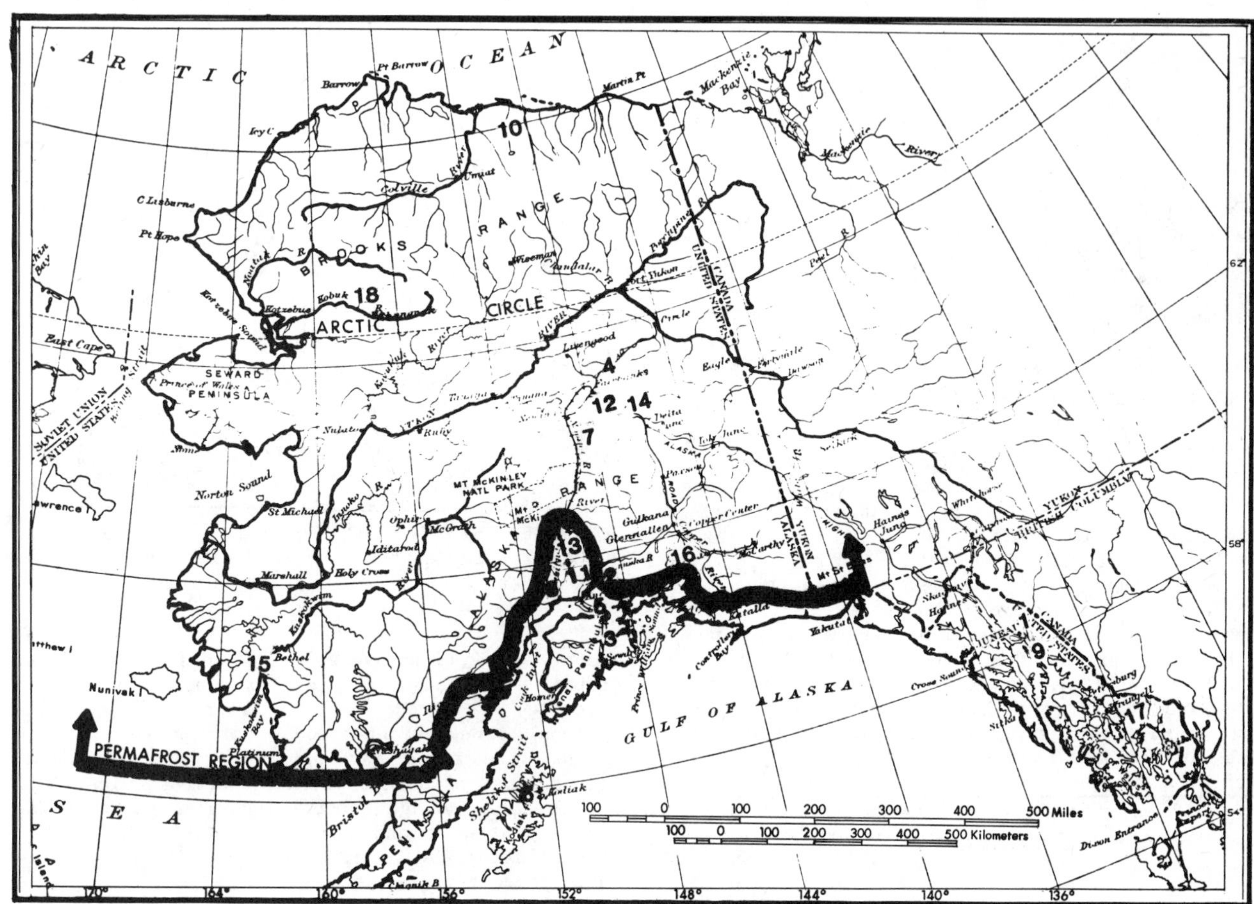

FIGURE 3 Key map

TABLE 1 Selected Alaska power lines

Lines	Year Energized	kV Level	Elev.-m Hi/Lo	Weather Criteria	Specific* Experience
1. Annex Creek - Juneau	1915	22	1060/10	1/	15, 16, 23
2. Eklutna-Anchorage-Palmer	1954	115	280/12	1/	1
3. Cooper Lake-Anchorage	1959	115	1060/5	1/ 4/	1, 3, 6, 12
4. FBKS to Hiway Park	1960	69	185/146	2/	1
5. Intn'l Sta-BuRec	1965	115	60/50	2/	4
6. Kodiak-Navy Tie	1966	69	300/10	2/ 4/	18
7. FBKS-Healy	1967	138	500/103	2/	8, 10, 11, 19
8. Beluga-MacKenzie Pt.	1968	138	60/5	2/	3, 20
9. Snettisham-Juneau	1973	138	300/10	2/ 4/	
10. Prudhoe Bay	1975	69	13/3	1/ 4/	5, 27
11. MacKenzie Pt.-Teeland	1975	230	110/12	2/	20
12. North Pole to FBKS	1976	138	135/145	2/	1, 6, 14
13. Teeland-Willow	1978	115	115/49	2/	3, 13, 20
14. FBKS-Big Delta	1978	138	630/146	2/	6, 9, 13
15. Bethel-Napakiak	1980	40	30/6	2/	22, 27
16. Glennallen-Valdez	1982	138	810/10	3/ 4/	3,20
17. Tyee-Wrangell-Petersburg	1983	138	530/-310	2/ 4/	20, 21
18. AVEC Systems	1968	12.5	120/10	2/	24

Footnotes of Table 1:

1/ 1.27 cm. radial ice + 0.383 KPa of wind at -18°C
2/ 1.27 cm. radial ice + 0.192 KPa of wind + NESC constant at -18°C
3/ 2.54 cm. radial ice + 0.192 KPa of wind at -18°C
4/ Additional criteria also used for design in specific locations or for extreme conditions.
* Refer to numbered paragraphs in text.

FIGURE 4 Reinforcing piling on wood poles ice encrusted tide flat

(4) A three-legged bogshoe (tripod) with a sliding necklace around the pole is in use on lines where severe frost action causes differential movement of pole and bogshoe (Figure 5).

(5) A sand-water slurry has been successful as backfill in regions of cold (-12°C) permafrost. The slurry is prepared and delivered unfrozen to each pole site where it is poured into the pole holes of the temporarily supported structure and allowed to freeze. Line 10 used this backfill.

(6) Extra depth of setting for poles installed in permafrost has worked well in many installations. A rule, referred to as the "... old Tsytovich rule..." (Sanger 1969), says that embedment in permafrost should be twice the thickness of the active (freeze-thaw) zone. With this guideline, extra depth has been used successfully with and without gravel backfill (see (1) above). Lines 10, 12, and 14 use extra-depth.

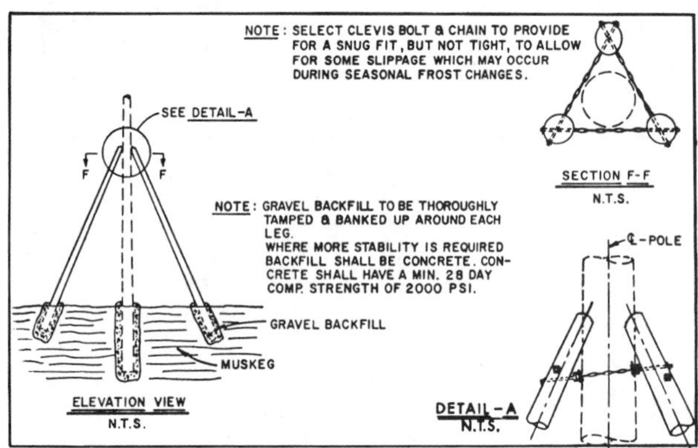

FIGURE 5 Tripod bogshoe with sliding necklace

(7) Pole bearing plates attached to poles or pole stubs (Figure 6) are being utilized in some installations in permafrost. Experience has been mixed with such devices (Sanger 1969). New efforts are reported, where the strength of the installation is increased by attaching plates to poles using spiked grid fasteners.

(8) A plastic boot for poles or pole stubs has been used in a few installations with mixed results. The purpose is to reduce the friction at the interface between the active layer and the pole or stub. This boot is installed on line 7 (Breckenfelder and Osborn 1965). See Figure 7.

(9) Soil stabilization using "self-powered" heat transfer devices to remove heat from soil is believed to have been started in Alaska in the 1950s by two individuals on separate tracks, a Salcha River homesteader named J. C. Balch, and an engineer named E. L. Long. The Balch tube (a liquid convection system), the Long pile (a liquid/

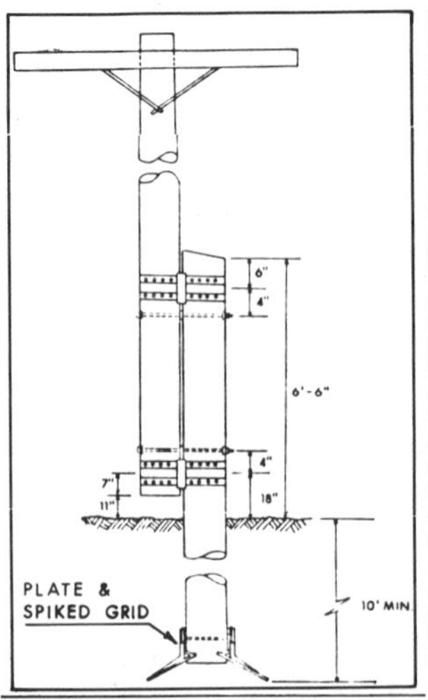

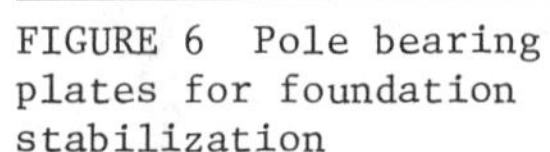

FIGURE 6 Pole bearing plates for foundation stabilization

FIGURE 7 Pile boot for wood pole stub

vapor convection system) and subsequent "frost tubes" have been applied to Alaska power line footings with success since that time. Three commercial products used today with power lines in Alaska are those manufactured by Arctic Foundations, Inc. of Anchorage, AK, Thermodynamics, Inc. of Seattle, WA and McDonnell-Douglas Astronautics Company in Richland, WA. A sketch of a typical installation is included as Figure 8. Line 14 uses some of these devices.

(10) A screw anchor footing has been used with varied results for aluminum towers on line 7 (Breckenfelder and Osborn 1965). Figure 9 is a sketch of the footing, and Figure 10 shows a footing which failed when one anchor jacked more than another. Other footings in this design continue to be successful.

(11) A concrete pad resting on a gravel pad on unstable ground has been successful in providing a stable footing for guyed metal towers. Figure 11 shows such a unit replacing a former failed screw anchor footing on line 7.

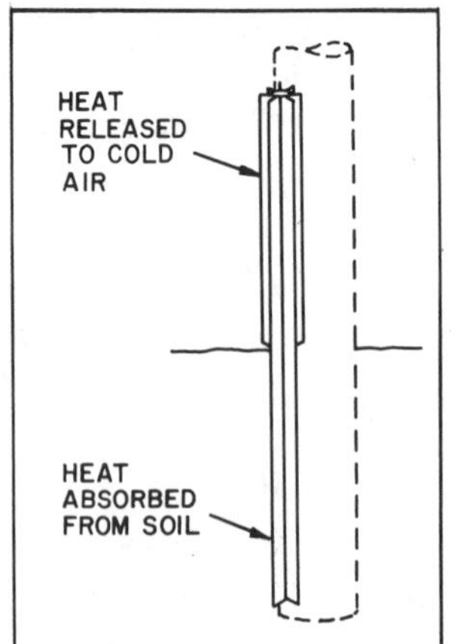

FIGURE 8 Cryo anchor for power pole stabilization

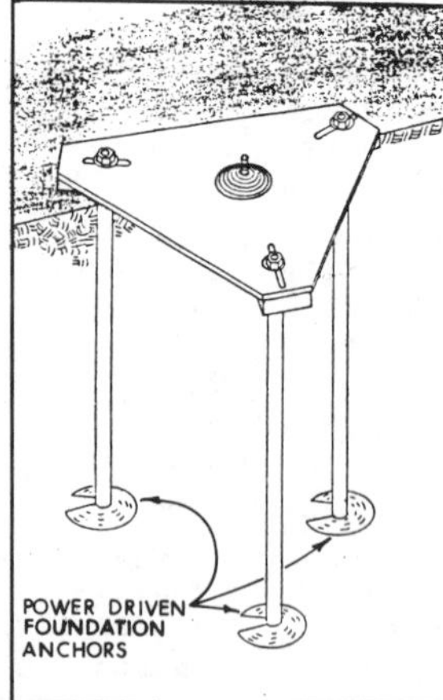

FIGURE 9 Tripod screw anchor tower footing

FIGURE 10 Screw anchor tower footing

FIGURE 11 Concrete pad on gravel

Anchors

Many conventional soil anchors used as standard practice in other areas are installed on power lines in Alaska. Examples include expanding anchors, cone anchors, plate anchors, and log anchors. As with footings, stable soil allows these practices even in permafrost regions. With unstable soils and permafrost some examples of successful anchoring techniques are described as follows:

(12) Prior to the advent of the power installed screw anchors a low grade concrete (or soil cement) backfill used to encase a conventional anchor (log, plate, etc.) placed in a watery, muskeg bog would give excellent results. The weight of approximately 1.2 to 2.4 m^3 of concrete and its ability to flow into the weave of fibre, muck and water and then set up has developed 5.5 to 11 tons of working strength. A number of these anchors on line 3 have survived the great Alaska earthquake of 1964.

(13) The power installed screw anchor has been used with success in muskeg and a variety of permafrost. It is believed that the screw anchor has first been tested and used in dense till on line 7 in 1965 (King and Crory 1965). Other installations followed as more powerful drilling motors became available (Retherford and King 1970). Anchors of this type are in use on lines 7, 8, 11, 12, 13, 14, and 17 (see Figure 12).

(14) Some permafrost soils are so hard that techniques for placing anchors include predrilling of a reduced size hole into which a screw anchor is installed. Line 12 has such anchors.

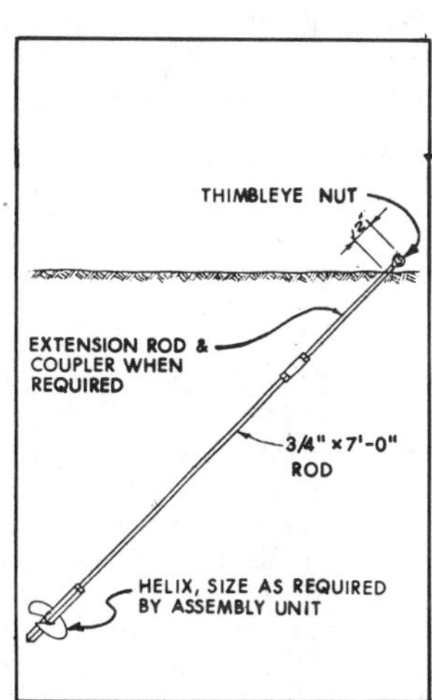

FIGURE 12 Screw anchor reinforced helix for glacial till

Structures

Conventional structures as known from standard practice in other areas such as single wood poles (lines 2, 3, 4, 5, 6, 12, and 14), guyed-Y (line 9), and four-legged self-supporting galvanized steel (lines 2, 3, 7, and 9) have been used for Alaska power lines. Rugged terrain, weather related phenomena, and effects of operating in permafrost soils have spawned innovative designs. One of the oldest power lines in Alaska (line 1) has experienced a broad range of these conditions. The perseverance of the operators of the line has generated numerous innovations (see Conductor) in their continuing efforts to maintain the line and adapt it to the environment (Wallenburg 1915, and Baum 1916). Severe wind, snow, and ice on high, steep slopes and ridges resulted in relocation of sections of line 1 and 9 to better sites. Some other unusual designs are described as follows:

(15) A few of the original steel towers of line 1 were subjected to creeping snow fields and severe unbalanced loading from broken conductors under heavy ice loads. Successful repair and preventive maintenance was accomplished by using timber splints to protect and reinforce the steel legs.

FIGURE 13 X-braced 4 pole wood structure

(16) A four-legged, X-braced wood pole structure was used on line 1 to resist the snow creep and high winds at other locations (Figure 13).

(17) Following the 1964 earthquake, emergency replacement of failed power line structures was made using a gravity-stabilized tripod arrangement (line 3). This structure was later used on a new line in a muskeg swamp (line 5). See Figure 14.

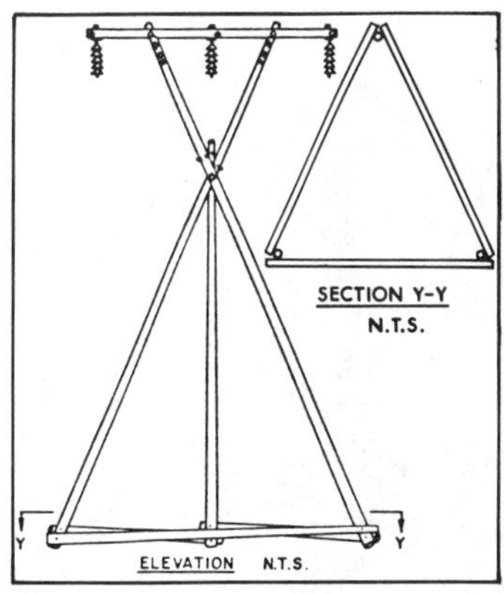

FIGURE 14 115 kV tripod gravity stabilized structure

(18) Strong winds moving up slope across a power line near Kodiak (line 6) lifted the conductor to make contact with its supporting cross-arm. A strut insulator added prevents excessive swing.

(19) The four-way guyed-Y tower of line 7 was installed in many locations

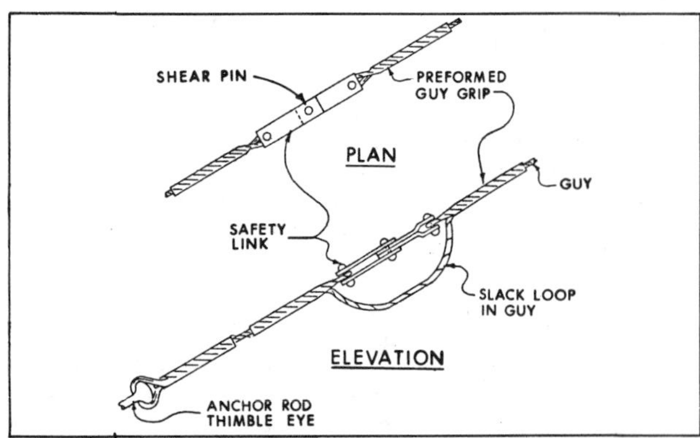

FIGURE 15 Guy safety link

where the frost conditions could produce different movements of tower footings and guys (Figures 9 and 10). A guy safety link was provided to allow a measured amount of slack to be added to a guy upon failure of a shear pin (Breckenfelder and Osborn 1965). See Figure 15.

(20) A guyed, hinged X-shaped metal tower design (Figure 16) was developed to provide a flexible structure tolerant of differential movements of tower footings and guy anchors. The guy yoke system allows unbalanced forces to distribute smoothly through the tower structural system with yoke failure prior to tower damage. Aluminum lattice and steel tube configurations have been used on lines 8, 11, 13, 16, and 17 (Rankin and Retherford 1967, LaRue and Retherford 1977).

(21) A guyed, hinged pi-shaped metal tower design using a guy yoke system similar to the X-tower described in (20) above is being used on lines with extremely long spans in the rugged fjords of southeastern Alaska. (Steeby et al 1979).

(22) A gravity stabilized A-frame structure was developed for use on roadless terrain and permafrost lands. It is part of a single wire electric power delivery system to minimize costs of electric lines in remote areas (Bettine and Retherford 1981, Retherford et al 1982). See Figure 17.

FIGURE 16 Guyed and hinged X-tower tubular steel

Conductor

A few unique experiences with power conductors in Alaska are described.

(23) Very heavy icing of the power conductors at high elevations (1000 m) on line 1 resulted in the installation of copper clad steel conductors with relatively high resistance so that electric current could heat the conductor and prevent the formation of ice (Baum 1916). This successful solution is produced by controlling the reactive current exchange between ends of the line. See Figure 18.

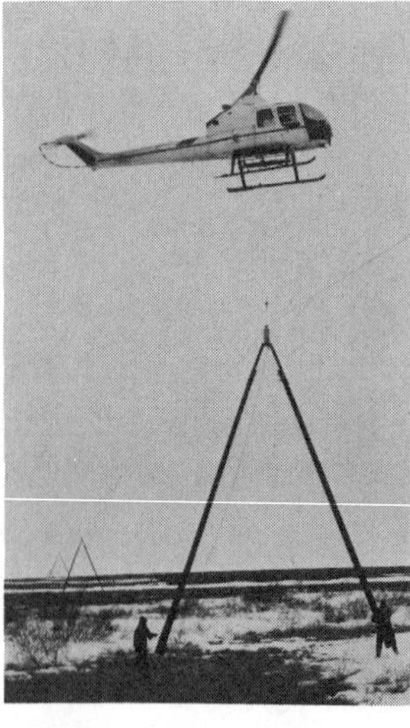

FIGURE 17 Gravity stabilized A-frame

(24) A wooden ground-lay utilidor enclosing primary, secondary, and grounding conductors has been in use since 1971 on line 18 with unsatisfactory performance. This duct originally replaced cables plowed into the ground, which were destroyed the action of the seasonal frost. Such a duct lying on the earth is difficult to protect from damage. A steel ground-lay utilidor replacing the wooden one has been under test since 1978 with improved results.

FIGURE 18 Heated conductor required

Grounding

Grounding of electric systems in Alaska is accomplished - as in other areas - through earth electrodes which are designed for the soil resistivity found. Large regions of Alaska have permanently frozen soils with a very high resistivity. One such area is Prudhoe Bay (line 10), and it has been given special attention by the operator - Sohio Alaska Petroleum Co. Another special situation is that of the single wire ground return electric transmission system now being demonstrated (Retherford et al 1982, Eaton 1974 and 1975). The following paragraphs describe some specific experiences regarding grounding:

(26) The 69 kV transmission system at Prudhoe Bay carries a neutral (ground wire) under the phase conductors and a phase wire "catcher" on

each structure to provide support and grounding should phase conductors come loose from insulators (line 10 Figure 25). The underrunning ground conductor provides for faster, surer relaying and added safeguards against accidental contact of phase conductors.

(27) The single wire ground return transmission line, in operation between Bethel and Napakiak, Alaska since 1980 (line 15) has primary earth connections at each end. At Bethel an abandoned steel well casing about 60 m deep extending through permafrost into other layers of soil provides a resistance of about 3.6 ohms. The electrode system at Napakiak consists of eight ground rods about 10 m long and 10 m apart connected by a 107 mm^2 bare copper wire in a trench about 0.6 m deep and provides a resistance of about 2.9 ohms. This ground grid is installed in a thaw bulb near the Kuskokwim River. Eaton (1975) suggests that if a hemispheric contact zone between a thaw bulb and permafrost is assumed, resistance (R) of the thaw bulb/permafrost contact can be estimated as follows:

R = p/2 pi r

where
p = resistivity of the permafrost
r = radius of the hemisphere
Identification of thawed zones within permafrost by electro-magnetic techniques was suggested and demonstrated on line 15 by D. Gropp (Gropp 1977). This method makes a rapid location of suitable thaw bulbs for further investigation as potential grounding electrodes in permafrost zones possible.

ACKNOWLEDGMENTS

This paper has been sponsored by the Arctic Division of International Engineering Company, Inc., a Morrison-Knudsen Company.

The author wishes to express his appreciation to the following persons: William A. Corbus and David G. Stone, Alaska Electric Light & Power Co.; Robert J. Cross, U.S. Department of Energy Alaska Power Administration; James A. Fillingame, Copper Valley Electric Association, Inc.; Robert L. Huffman, John Huber, Jr., Joe Killion, and Terry Terrell, Golden Valley Electric Association, Inc.; R. C. Farkas and Robert H. Shipley, Sohio Alaska Petroleum Company; Lloyd Hodson, Alaska Village Electric Cooperative, Inc.

REFERENCES

Baum, F.G., 1916, Suggested use of line conductor heating to prevent ice, letter to Alaska Gastineau Mining Co., in February 1916. Alaska Electric Light and Power Co. archives, Juneau, Alaska.

Bettine, F.J. and Retherford, R.W., 1981, Single wire ground return transmission for rural Alaska, Proceedings of the Specialty Conference on the Northern Community, ASCE, New York.

Breckenfelder, E.H. and Osborn, D.D., 1965, Report of selection of structures for 138 kV transmission lines of Golden Valley Electric Association, Inc., Fairbanks, Alaska. Stanley Engineering Company, Muscatine, Iowa.

Eaton, J.R., 1974, Single-wire ground-return transmission line electrical performance, report prepared for Robert W. Retherford Associates, Anchorage, Alaska.

Eaton, J.R., 1975, personal communication.

Eaton, J.R., Merritt, R.P., Rice, E.F., 1975, Electric power engineering in an arctic environment, Transactions of IEEE F75 523-1. New York.

Gropp, D.L., 1977, 1976/77 Geophysical survey of rivers of the Alaskan Arctic coastal plain between Prudhoe Bay and the Canadian border. Prepared for Alaskan Arctic Gas Pipeline Company, Anchorage, Alaska.

King, W.J. and F. Crory, 1965, Preliminary report on installation and testing of power installed screw anchors in permafrost and thawed silt, and 1966 supplemental letter report, prepared for Golden Valley Electric Association, Inc.

LaRue, D.S. and Retherford, R.W., 1977, Basic design data Willow project - 115 kV transmission line Teeland to Willow. Prepared for Matanuska Electric Association, Inc., Palmer, Alaska.

Proctor, D.O. and Meyers, C.W., 1978, Summary of basic design data, Solomon Gulch hydroplant to Glennallen 138 kV transmission line, prepared for Copper Valley Electric Association, Inc., Glennallen, Alaska. Miner and Miner, Inc., Greeley, Colorado.

Rankin, R.H., and Retherford, R.W., 1967, Basic design data 138 kV transmission line, Beluga Project. Prepared for Chugach Electric Association, Inc., Anchorage, Alaska.

Retherford, R.W., Bettine, F.J. and Thompson, C.J., 1982, Lower Kuskokwim single wire ground return transmission system, Phase II report, State of Alaska, Division of Energy and Power Development, Anchorage, Alaska.

Retherford, R.W. and King, W.J., 1970, A report on power-driven screw anchor applications in Interior and South Central Alaska. Prepared for Woodward-Clyde Associates (now Woodward-Lundgren and Associates), Oakland, California.

Sanger, F.J., 1969, Foundations of structures in cold regions. U.S. Army Cold Regions Research and Engineering Laboratory, Hanover, N.H.

Steeby, C.H., Burg, R., Retherford, R.W., 1979, Application for license to construct the Tyee Lake Hydroelectric project. Prepared for the Alaska Power Authority, Anchorage, Alaska.

Wallenburg, H.L., 1915, Outline of design of transmission line from Annex Creek to Sheep Creek (letter dated June 22, 1915). Alaska Gastineau Mining Co., Juneau, Alaska. Alaska Electric Light and Power Co. archives, Juneau, Alaska.

THE ROLE OF SPECIFIC SURFACE AREA AND RELATED INDEX PROPERTIES IN THE FROST HEAVE SUSCEPTIBILITY OF SOILS

Ross D. Rieke[1], Ted S. Vinson[2], and Daniel W. Mageau[3]

[1]Hart-Crowser & Associates, Inc., Seattle, WA 98102 USA
[2]Dept. of Civil Engineering, Oregon State University, Corvallis, OR 97331 USA
[3]R and M Consultants, Inc., Fairbanks, AK 99701 USA

Soil properties important to the frost heave phenomenon must be identified to establish which index property tests can aid in evaluating frost heave susceptibility. Following a consideration of existing frost heave theories, the frost heave mechanism was identified as being fundamentally related to the specific surface area of the soil. Specific surface area may be related to index property test results, specifically, the liquid limit and the clay mineralogy of the fine fraction. Laboratory frost heave tests were performed to develop relationships between index properties and frost heave susceptibility. Tests were performed on soil mixtures consisting of uniform sand with 5%, 10%, and 20% fines of different mineralogic composition, representing a range of specific surface areas. Over 30 frost heave tests were conducted on 11 distinct soil mixtures. The frost heave test consisted of freezing a soil column and noting the intake of water into the specimen as well as the frost heave. Using the concept of segregation potential, the frost heave susceptibility of a soil was found to increase with increasing percentages of fines, decreasing activity of the fine fraction, and, for a specific fine fraction mineralogy, increasing liquid limit of the fine fraction.

INTRODUCTION

Throughout the history of the development of frost heave theories, the specific surface area has been identified, either directly or indirectly, as a soil property important to the frost heave mechanism. Taber (1930) noted the importance of the size and shape of the soil particle to frost heave. Beskow (1935) noted Taber's theories and further considered the importance of the adsorbed layer of water on the surface of the soil particle. Using relationships between grain size, adsorbed water, and frost heave, Beskow developed one of the first frost heave susceptibility criteria.

Modern researchers have continued to note the specific surface area as a fundamental soil property in the frost heave mechanism. The models presented by Miller (1972, 1978) and Takagi (1980) indirectly identify the surface area characteristics of a soil as playing an important role in frost heave behavior. They suggested that water flowing to a forming ice lens must pass through a "frozen fringe" or "zone of diffused freezing," where free pore water is frozen. These concepts, along with present theories regarding flow through frozen soil (Hoekstra and Miller 1967), suggest that specific surface area is indeed important in the frost heave mechanism. Konrad and Morgenstern (1980, 1981) further suggest the importance of the frozen fringe. They found that the permeability of the frozen fringe was a controlling factor in the frost heave response of soils. Since Hoekstra and Miller (1967) suggest that the ease with which water migrates through frozen soil is related to the continuity of adsorbed water layers on the soil particle surfaces, Konrad and Morgenstern's theory indirectly notes the influence of specific surface area on the frost heave mechanism.

The relationships between specific surface area and index property tests are well known. By considering the fundamental nature of the Atterberg Limits (ASTM 423-66 and D424-59) and soil grain size distributions (ASTM D4422-63), the significance of the specific surface area to these index properties may be demonstrated. Specifically, Graboska-Olszewska (1970) concluded that the liquid limit of a soil (ASTM 423-66) is intimately related to its specific surface area. The relationship of grain size distributions to specific surface area may be inferred by noting the relationship of surface area to soil particle size and assumed particle shape.

In an attempt to better define the role of specific surface area and related index properties to the frost heave susceptibility of soils, a laboratory testing program was undertaken. The frost heave response of several soil mixtures representing a range of specific surface areas was determined in a frost heave test cell. Relationships between frost heave susceptibiliy, specific surface area, and Atterberg Limits were derived for the soils considered in the study. The results of the study are presented herein.

DESCRIPTION OF TEST SYSTEM AND PROCEDURES

The frost heave test consisted of freezing a soil specimen and noting the intake of water into the specimen as well as the change in vertical height. No attempt was made to duplicate field conditions in the laboratory test program. Although it is desirable to apply the results of this and other laboratory investigations to actual field situations, only relative frost heave susceptibilities, as measured in the laboratory, were determined. The frost heave tests were performed under "worst case" conditions to minimize the influence of variables other than specific surface area. To achieve conditions that were most conducive to frost heaving, the specimen was saturated and the free water surface was maintained at approximately the same level as the frost front.

The soils used in the laboratory study consisted of an extremely uniform fine sand (Astoria Sand), Hanover Silt with the plus No. 200 sieve (0.74 mm) sizes removed, and relatively pure kaolinite and montmorillonite. The D_{10} of the Astoria sand was .14 mm with a C_u of 1.4. The D_{10} of the silt was .006 mm with a C_u of 3.8. Kaolinite and montmorillonite were obtained from the Clay Mineral Society of America (CMSA), Columbia, Missouri. The specific surface areas for the kaolinite and montmorillonite were provided by the CMSA. The specific surface area of Hanover Silt was determined by considering its grain size distribution and employing specific surface area calculation procedures given by the Asphalt Institute (1979).

The frost heave cell used in this investigation is an adaptation of the cell described by Mageau (1978). The frost heave cell consists of a 30.5 cm (12.0 in.) long by 10.2 cm (4.0 in.) I.D., by 15.2 cm (6.0 in.) O.D. nylon barrel. The specimen is placed inside the cell and an aluminum top cap and bottom plate serve as constant temperature boundaries. The temperatures of the top cap and bottom plate are maintained at constant values by circulating constant temperature fluids through heat exchange mazes within the cap and plate. To aid in the boundary temperature control and to prevent radial heat flow, the cell was placed in a refrigerator whose temperature was maintained at approximately 2°C (36°F). A 50 ml (7.75 in.^{3}) buret was connected to the water intake line and the flow of water into or out of the specimen during a test was noted by the change in the water level in the buret. The change in the height (reflecting frost heave) of the specimen during the test was measured with a linear variable differential transformer (LVDT). The temperature along the length of the specimen was monitored with thermistors adjacent to the soil embedded in the wall of the frost heave cell barrel. Prior to freezing, the specimen was consolidated one-dimensionally under 50 kPa (7 psi) pressure and allowed to reach thermal equilibrium with the ambient refrigerator temperature.

INTERPRETATION OF FROST HEAVE TEST RESULTS

In conjunction with the development of a mechanistic frost heave theory, Konrad and Morgenstern (1980, 1981) presented a parameter referred to as the segregation potential (SP_o), which was defined as:

$$SP_o = \frac{V_o}{\text{grad } T} \qquad (1)$$

in which,

SP_o = segregation potential

V_o = water intake flux

grad T = temperature gradient across the frozen fringe.

At near steady-state conditions (i.e., very small frost penetration rate) this parameter, as determined in a frost heave test, was found to be constant for a given soil when the suction at the frost front is small (i.e., a small depth of unfrozen soil).

Further, they developed a relationship between segregation potential and the permeability of the frozen fringe such that the frost heave characteristics of a soil can be related to its segregation potential as determined under controlled laboratory test conditions. Specifically, a high segregation potential implies a high frost susceptibility. To compare the frost heave susceptibilities of various soils one needs only to compare segregation potential values as determined in the laboratory. A detailed explanation of the determination of SP_o from laboratory results is given by Mageau and Sherman in this volume.

TEST PROGRAM AND RESULTS

A total of 31 frost heave tests were performed on various combinations of sand, silt, and different types of clay. Astoria sand was used in each test together with fine fractions consisting of various combinations of silt, montmorillonite, poorly crystallized kaolinite, and well-crystallized kaolinite. Mixtures containing 5%, 10%, and 20% fine material were considered. In mixtures with a specific percentage of fines, the composition of the fine fraction was varied by considering different combinations of silt, montmorillonite, and the two types of kaolinite. The specific surface areas for the soil mixtures were determined as the weighted summation of specific surface areas for the various soil types comprising the mixture.

The segregation potential was calculated for each frost heave test using the measured intake of water and the temperature gradient in the frozen fringe. Thermistors close to the estimated location of the fozen fringe were used to establish the temperature gradient (Rieke 1982). The results from Test 21 were chosen to illustrate typical results. Figures 1, 2, and 3 show the frost front penetration versus time, water intake and heave versus time, and temperature gradients, respectively, during Test 21.

Table 1 presents the results of the 31 frost heave tests and liquid limit tests performed on the various mixtures of fines. Figures 4 and 5 show a well-defined relationship between segregation potential (SP_o) and specific surface area. Liquid limit versus specific surface area relationships for the three types of clay-silt combinations are shown in Figure 6.

Frost heave and index property tests were also performed on a specimen of silty sand (SM) obtained from a site on the North Slope of Alaska to allow a preliminary comparison to be made between the artificially prepared laboratory materials used in this investigation and remolded natural soils. The results of the frost heave and index

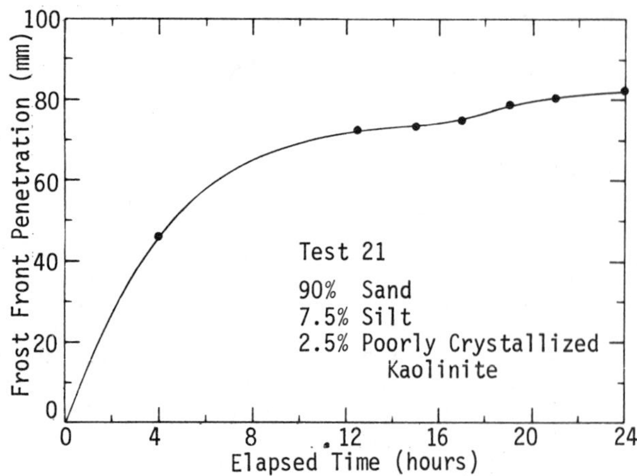

FIGURE 1 Frost front penetration versus time for test 21.

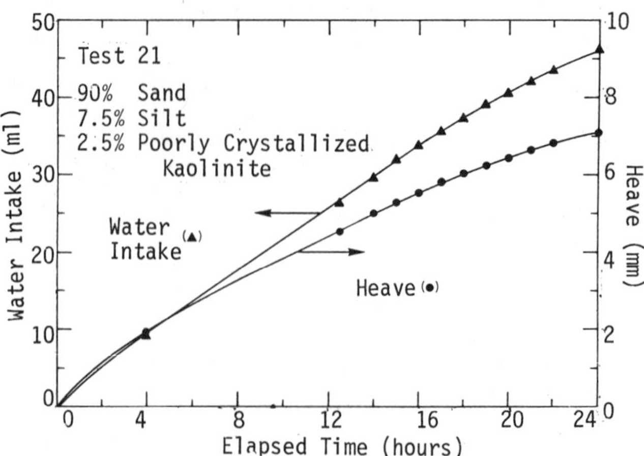

FIGURE 2 Heave and water intake versus time for test 21.

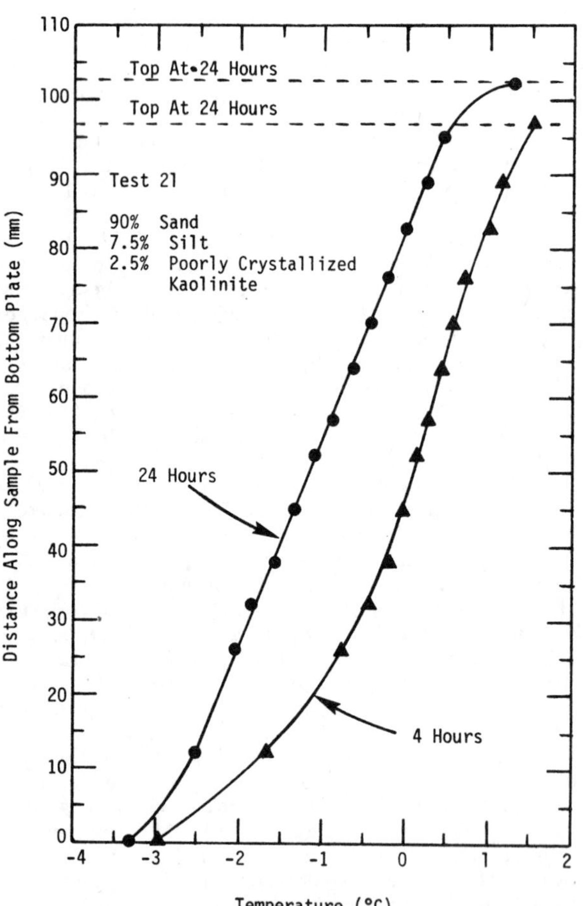

FIGURE 3 Temperature profile along specimen in test 21.

tests on this specimen are as follows: SP_o = 2682 mm^2/°C-sec; LL of fine fraction = 49.2%; % minus No. 200 sieve = 27%; and % minus .002 mm in fine fraction = 30%.

TABLE 1 Frost Heave and Index Property Test Results for Soil Mixtures.

Test	Percent Astoria Sand	Percent Hanover Silt	Percent Clay	Specific Surface Area m^2/g	Liquid Limit of Fine Fraction (LL_{ff}) %	Fines Factor[2] (R_f) %	Segregation Potential (SP_o) 10^{-6} $\frac{mm^2}{°C\text{-sec}}$
1	80	0	20 (M)[1]	19.5	135	14.8	2561
2	95	5	0	0.25	-	-	-
3	95	0	5 (M)	4.87	135	3.70	227
4	90	0	10 (M)	9.74	135	7.41	1565
5	90	10	0	0.50	-	-	293
6	80	20	0	1.0	-	-	1063
7	80	14	6 (M)	6.54	56	10.7	1846
9	90	7	3 (M)	3.27	56	5.36	841
10	90	6	4 (M)	4.20	66	6.06	1112
11	90	3	7 (M)	6.97	99	7.07	1338
13	95	1.5	3.5 (M)	3.48	99	3.54	302
14	95	3	2 (M)	2.10	66	3.03	212
15	95	3.5	1.5 (M)	1.64	56	2.68	295
16	95	1	4 (PK)[1]	0.99	58	6.90	1601
17	80	4	16 (PK)	3.96	58	27.6	4229
18	80	10	10 (PK)	2.85	40	25.0	3165
19	80	15	5 (PK)	1.93	29	17.1	2395
20	90	2	8 (PK)	1.98	58	13.8	2223
21	90	7.5	2.5 (PK)	0.96	29	8.53	1361
22	95	3.75	1.25 (PK)	0.48	29	4.27	541
23	95	3.75	1.25 (WK)[1]	0.31	26	2.88	478
24	80	15	5 (WK)	1.25	26	11.5	1916
25	80	4	16 (WK)	1.80	37	25.9	3870
26	95	1	4 (WK)	0.45	37	6.5	1176
27	80	12	8 (M)	8.39	66	12.1	2156
28	80	6	14 (M)	13.9	99	14.1	2159
29	90	5	5 (PK)	1.43	40	12.5	2189
30	95	2.5	2.5 (PK)	0.71	40	6.25	1036
31	95	0	5 (PK)	1.18	75	6.67	2319
33	90	0	10 (PK)	2.35	75	13.3	3353
34	80	0	20 (PK)	4.70	75	26.7	4737

[1] M = montmorillonite; WK = well-crystallized kaolinite; PK = poorly crystallized kaolinite.

[2] $R_f = \frac{(\%\text{fines})\ (\%\text{ clay sizes in fine fraction})}{LL_{ff}}$

DISCUSSION OF TEST RESULTS

Figures 4 and 5 show a strong correlation between segregation potential and specific surface area for soil mixtures with fine fractions consisting of montmorillonite-silt, poorly crystallized kaolinite-silt (PK), and well-crystallized kaolinite-silt (WK). A strong correlation between

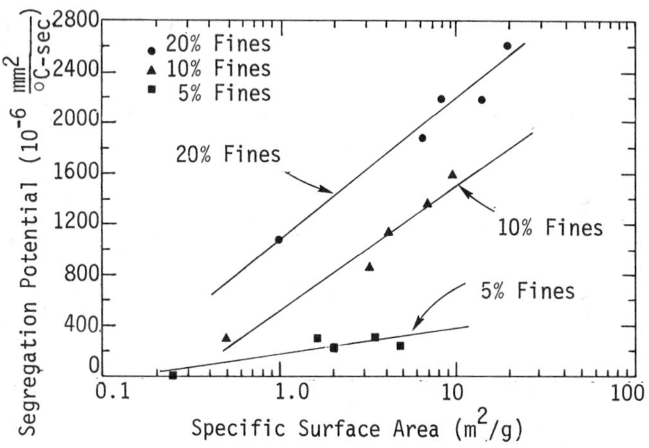

FIGURE 4 Segregation potential versus specific surface area for montmorillonite mixtures.

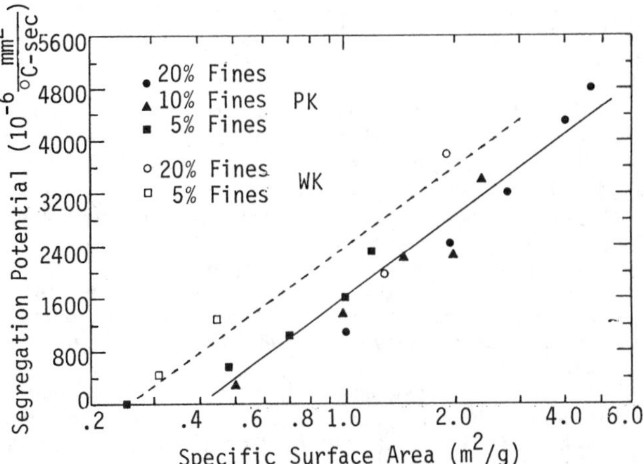

FIGURE 5 Segregation potential versus specific surface area for kaolinite mixtures.

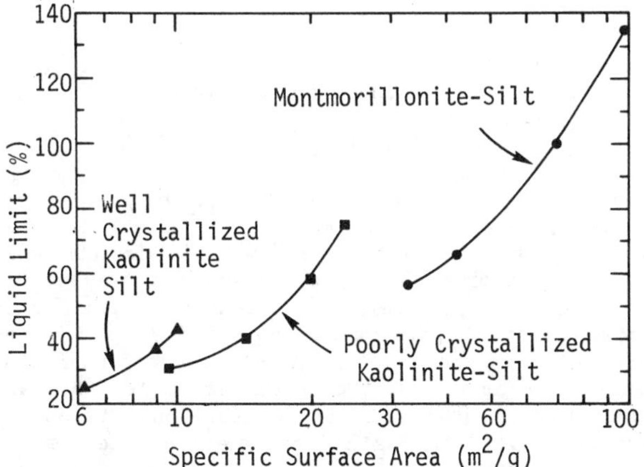

FIGURE 6 Liquid limit versus specific surface area.

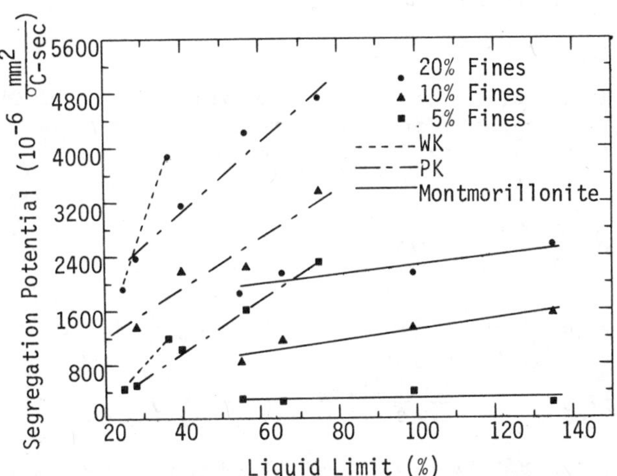

FIGURE 7 Segregation potential versus liquid limit of fine fraction for all tests.

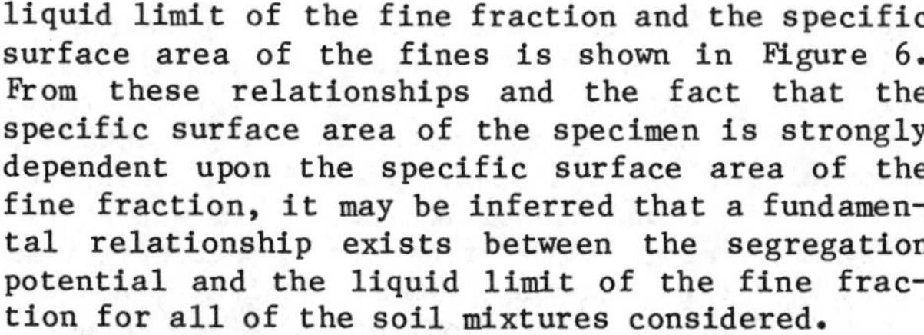

liquid limit of the fine fraction and the specific surface area of the fines is shown in Figure 6. From these relationships and the fact that the specific surface area of the specimen is strongly dependent upon the specific surface area of the fine fraction, it may be inferred that a fundamental relationship exists between the segregation potential and the liquid limit of the fine fraction for all of the soil mixtures considered.

Figure 7 shows the segregation potential versus the liquid limit of the fine fraction for montmorillonite-silt fines, poorly crystallized kaolinite-silt, and well-crystallized kaolinite-silt fines, respectively. The results presented indicate the percent fines has a much larger influence on the segregation potential than the liquid limit.

The results shown in Figure 7 also indicate that segregation potential is a function of the clay mineralogy of the fine fraction. It appears that mixtures with well-crystallized kaolinite exhibit greater segregation potentials than mixtures with montmorillonite. The segregation potential of mixtures with poorly crystallized kaolinite were intermediate to the other clay mineral mixtures.

These observations suggest that the frost heave susceptibility is dependent on the clay mineralogy of the soil. A convenient index property to identify the predominant clay mineral in a soil mass is the activity. The activity is classically defined as follows:

$$\text{Activity, } A = \frac{\text{Plasticity Index}}{\text{\% by weight finer than } 2\mu} \qquad (2)$$

(i.e. % clay sizes)

The definition of activity used in this investigation is different from that given in Eq. (2) in that the liquid limit is employed in the numerator in place of the plasticity index. Also, the activity is associated with the fine fraction only, rather than the entire sample. Thus,

$$\text{Liquid Limit Activity of the fine fraction, } A_{LL} = \frac{\text{LL of fine fraction, } L_{ff}}{\text{\% clay sizes in fine fraction}} \quad (3)$$

Considering the relative values of A_{LL} for calcium montmorillonite, poorly crystallized kaolinite and well-crystallized kaolinite, it is suggested that as the activity of the clay mineral in the soil decreases, the segregation potential and, hence, frost heave suceptibility increases. This conclusion is consistent with observations made by Lambe et al. (1969), who found that clays with high activities, specifically Na-montmorillonite, decreased frost heave when added to frost susceptible materials.

Because the segregation potential is strongly influenced by the percentage of fines in the soil and the liquid limit activity of the clay minerals in the fine fraction it is possible to develop a general parameter that relates index test results to segregation potential values. It was noted that the segregation potential increased as the percentage of fines (% minus No. 200 sieve (0.74 mm) increased, that is:

$$SP_o \propto \text{\% fines} \quad (4)$$

It was also noted that as the liquid limit activity of the fine fraction increased, the segregation potential decreased, that is:

$$SP_o \propto \frac{1}{A_{LL}} \, . \quad (5)$$

Combining Eqs. (4) and (5):

$$SP_o \propto \frac{\text{\%fines}}{A_{LL}} \quad (6a)$$

$$SP_o \propto \frac{\text{\%fines}}{LL_{ff}/\text{\% clay sizes in fine fraction}} \quad (6b)$$

$$SP_o \propto \frac{(\text{\%fines})(\text{\% clay sizes in fine fraction})}{LL_{ff}} \quad (6c)$$

Let us define the right side of Eq. (6c) as the fines factor, R_f, that is:

$$R_f = \frac{(\text{\%fines})(\text{\% clay sizes in fine fraction})}{LL_{ff}} \, . \quad (7)$$

The fines factor was calculated for each test and the results are presented in Table 1. Figure 8 shows a strong correlation between R_f and segregation potential. Also included in Figure 8 are the results of a frost heave test and index property tests performed on the field sample. It is encouraging to find a good correlation between the segregation potential and R_f for the natural soil. Thus, for the soils considered in this investigation, the results from index property tests (sieve analysis, liquid limits, and hydrometer analysis) may be related to the segregation potential with a reasonable degree of confidence.

As it is presently defined, R_f is the result of observations of test data and the attempt to relate the results of the observations to a common parameter. Thus R_f is an empirical parameter.

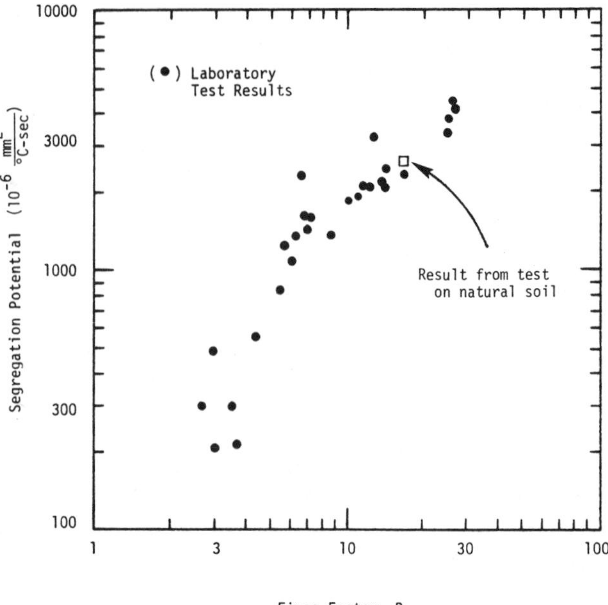

FIGURE 8 Segregation potential versus fines factor, R_f.

Empirical relations are useful in that they can serve to give an estimate of a desired parameter. In this regard R_f may aid in the more complicated problem of identifying frost susceptible soils under field conditions. Another role that empirical parameters can fulfill is in contributing to the development of a more fundamental understanding of the phenomena with which they are associated. This role is possible only if one is familiar with the basic development of the empirical parameter. Following this philosophy, the fundamental soil properties that affect R_f may be analyzed with respect to improving our understanding of existing frost heave models.

In the model presented by Konrad and Morgenstern (1980), the amount of the unfrozen film water between the surface of a clay particle and the frozen pore water is directly related to the ease with which water flows through the frozen fringe. Since the permeability of the frozen fringe was identified as the critical factor influencing the segregation potential of saturated soils, the abundance of unfrozen film water has a direct bearing on the frost heave susceptibility of the soil. The test results indicate, however, that although montmorillonite has a higher unfrozen water content than kaolinite, kaolinite has consistently higher segregation potentials. This fact suggests that the permeability of the frozen fringe is not solely a function of the abundance of unfrozen film water. Specifically, the ease with which water migrates through the frozen fringe is related to the rigidity of the unfrozen water films. Noting that the rigidity of the adsorbed water layers in kaolinite is less than that of adsorbed layers in montmorillonite (Grim 1962), the permeability of the frozen fringe should be greater for kaolinite than for montmorillonite.

R_f considers the effect of the unfrozen film water rigidity upon the segregation potential by noting that clays with low activities, such as kaolinite, have less rigid unfrozen water films and, hence, higher frozen fringe permeabilities and segregation potentials. Thus, assuming that the liquid limit activity of the fine fraction as defined in this investigation is an acceptable parameter to distinguish the rigidity of the unfrozen films, it can be seen that:

$$SP_o \alpha \frac{1}{A_{LL}} \tag{8}$$

This is identical to Eq. (5) that was initially used to develop R_f.

The specific surface area of a soil may be considered to be a function of the amount and nature of the fines in the soil. This is reasonable since the fine fraction, owing to the small grain sizes, controls the specific surface area of the soil. More specifically, the extreme values of specific surface area for clay greatly influence the specific surface area of a soil mass. Therefore, it is reasonable to expect a relation between the amount of fines in the soil and the continuity of the unfrozen water films, hence, the permeability of the frozen fringe. Noting the influence of the frozen fringe permeability on the segregation potential, the relationship between the amount of fines in a soil and the segregation potential of the soil may be expressed approximately as:

$$SP_o \alpha \text{ \%fines} \tag{9}$$

This relation is identical to that presented as Eq. (4) in the development of R_f.

SUMMARY AND CONCLUSIONS

Over 30 frost heave tests were conducted on 11 distinct soil mixtures. Using the concept of segregation potential, the frost heave susceptibility of a soil was found to increase with increasing percentages of fines, decreasing activity of the fine fraction, and for a specific fine fraction mineralogy, increasing liquid limit of the fine fraction. These relationships were summarized by introducing a term referred to as the fines factor, R_f, (re Eq. (7)). The fines factor allows index property test results to be related directly to the frost susceptibility of a soil, owing to the strong correlation found between segregation potential and the fines factor. Based on an analysis of the fines factor it may be demonstrated that the permeability of the frozen fringe and, hence, a soil's tendency to heave, are a function of not only specific surface area but also the mineralogy of the fine fraction. It is hypothesized that fine fractions with highly active clay minerals have less mobile unfrozen water films in the frozen fringe resulting in a reduced frozen fringe permeability.

ACKNOWLEDGMENT

The studies described herein were supported in part by the Department of Civil Engineering, Oregon State University. The support of the department is gratefully acknowledged. The efforts of Laurie Campbell and Nancy Platz, who assisted in the preparation of this paper, are greatly appreciated.

REFERENCES

Asphalt Institute, 1979, Mix design methods for asphalt concrete, Manual Series No. 2 (MS-2).

Beskow, G., 1935, Soil freezing and frost heaving with special application to roads and railroads: Swedish Geol. Soc., Series C, No. 375.

Graboska-Olszewska, B., 1970, Physical properties of clay soils as a function of their specific surface, in Proceedings of the First International Congress, International Association Engineering Geologists, Vol. 1, pp. 405-410.

Grim, R.E., 1962, Applied clay mineralogy, 2nd Edition, McGraw-Hill Book Company. pp. 166-173.

Hoekstra, P. and Miller, R.D., 1967, On the mobility of water molecules in the transition layer between ice and a solid surface: Journal of Colloid and Interface Science, Vol. 25, pp. 166-173.

Konrad, J.M. and Morgenstern, N.R., 1980, A mechanistic theory of ice lens formation in fine-grained soils: Canadian Geotechnical Journal, Vol. 17, No. 4, pp. 473-486.

Konrad, M.R. and Morgenstern, N.R., 1981, The segregation potential of a freezing soil: Canadian Geotechnical Journal, Vol. 18, No. 4, pp. 482-491.

Lambe, T.W., Kapler, C.W., and Lambie, T.J., 1969, Effect of mineralogical composition of fines on frost susceptibility of soils: Technical Report 207, U.S. Army Cold Regions Research & Engineering Laboratory, Hanover, NH.

Mageau, D., 1978, Moisture migration in frozen soil, M.S. Thesis, Department of Civil Engineering, University of Alberta.

Miller, R.D., 1972, Freezing and heaving of saturated and unsaturated soils: Highway Research Record No. 393, National Academy of Sciences, Washington, D.C., pp. 1-11.

Miller, R.D., 1978, Frost heaving in noncolloidal soils, in Proceedings of the Third International Conference on Permafrost, National Research Council of Canada, Vol. 1, pp. 707-709.

Miller, R.D., Loch, J.P.G., and Brelser, E., 1975, Transport of water and heat in a frozen permeameter, in Proceedings, Soil Science of America, Vol. 39, pp. 1029-1036.

Rieke, R.D., 1982, The role of specific surface area and related index properties in the frost susceptibility of soils, M.S. Thesis, Department of Civil Engineering, Oregon State University.

Taber, S., 1930, The mechanics of frost heaving: Journal of Geology, Vol. 38, pp. 303-317.

Takagi, S., 1980, The adsorption force theory of frost heaving: Cold Regions Science and Technology, Vol. 3, No. 1, pp. 57-81.

DETERMINATION OF THE THERMAL PROPERTIES OF FROZEN SOILS

D. W. Riseborough, M. W. Smith and D. H. Halliwell
Geotechnical Science Laboratories, Geography Department
Carleton University, Ottawa Canada K1S 5B6

This paper describes the development of a method for determining soil thermal properties (conductivity and diffusivity) at temperatures close to 0°C, based upon the transient heat flow from a cylindrical copper probe. A very small heat input and a very sensitive temperature measurement system allow the maximum temperature change in the soil sample to be kept to less than 0.1°C. The soil sample is first cooled to about -10°C, to ensure ice nucleation, and then determinations are done at closely spaced intervals on a warming curve. At each temperature, a least squares regression on the time-temperature data is used to determine the soil properties. The analytical solution is sensitive to both the soil conductivity and the apparent heat capacity; therefore the latter is also determined independently using the unfrozen water content curve obtained via time domain reflectometry. Results for two saturated silts are presented.

INTRODUCTION

In the temperature range just below 0°C, the thermal properties of soils containing water can change significantly with small changes in temperature. As the volume fractions of ice and water are temperature dependent, methods requiring prolonged and substantial heating of a soil sample cause a change in the properties being measured. Because of such limitations it has not been possible to reliably determine the thermal properties of soils in the temperature range 0°C to -2°C (e.g., Penner 1970). Unfortunately, for many practical studies of soil thermal regimes it is precisely this range of temperatures which is of concern. Kay et al. (1981) do present results for thermal conductivity, for one soil, at a temperature of -0.7°C.

This paper describes two stages of development of a technique, based on transient heat flow from a cylindrical probe, for the determination of the thermal properties of frozen soils (conductivity and diffusivity) in the temperature range 0° to -2°C, as well as beyond this.

Typically, conductivity probes are designed on the basis of a theoretical heat flow model which assumes a line heat source, with heat capacity of zero and infinite thermal conductivity (see Farouki 1981). This approach requires that the sample be heated continuously, until a rate of temperature rise which is linear with the logarithm of time is attained. The slope of T versus log(t) is then used to determine the conductivity. In frozen soils close to 0°C, this procedure is not feasible, since the unfrozen water content of fine-grained soils changes rapidly with small changes in temperature. As water changes phase, latent heat of fusion is absorbed or released. The apparent heat capacity of the soil is a function of the rate at which the change of state occurs, and this rate is a function of temperature. Also, ice and water differ in thermal conductivity. Therefore, as the volume fractions of ice and water are temperature dependent, the thermal properties of the soil mixture must change accordingly. Thus, any method requiring prolonged heating of a soil sample will cause a change in the properties being measured.

THE "CONDUCTIVITY" PROBE

The design of the probe is based on the analytical model of transient heat flow from a cylinder of infinite conductivity given by Jaeger (1956):

$$\Delta T = (Q/\lambda) \cdot G(\alpha,\tau) \quad (1)$$

where ΔT is the temperature change at the probe-soil interface and

$$G(\alpha,\tau) = (2\alpha^2/\pi^3) \cdot \int_0^\infty \frac{1 - \exp(-\tau u^2)}{u^3 \nabla(u)} du \quad (2)$$

where

$$\nabla(u) = (u \cdot J_o(u) + \alpha \cdot J_1(u))^2 + (u \cdot Y_o(u) + \alpha \cdot Y_1(u))^2 \quad (3)$$

and

$$\alpha = 2\ C/C_o, \quad \tau = \kappa t/a^2 \quad (4)$$

All symbols are defined in the notation list at the end of the paper. This formulation assumes perfect thermal contact between the probe and the soil. For a cooling period following a short pulse of heat, temperature change can be analyzed by the following:

$$G(\alpha,\tau) = (2\alpha^2/\pi^3) \cdot \int_0^\infty \frac{\exp(-(\tau-\tau_o)u^2) - \exp(-\tau u^2)}{u^3 \nabla(u)} du \quad (5)$$

for $\tau > \tau_o$, and where $\tau_o = \kappa t_o/a^2$ is the length of the heating period. Both the pulse method (equation 5) and the continuous heating method (equation 2) have been tested in our experiments.

In the theory for the standard needle probe, $G(\alpha,\tau)$ is simplified by assuming that the probe heat capacity and radius are zero, and by the use for sufficiently large times (e.g., see Farouki 1981). To use the method for smaller times - i.e. to limit ΔT and/or to determine the diffusivity - the complete form of $G(\alpha,\tau)$ must be used. A sensitivity analysis of $G(\alpha,\tau)$, using the parameters for our probe, indicated that the pulse method is primarily sensitive to soil conductivity, whereas the continuous heating method is strongly sensitive to both conductivity and diffusivity.

The probe radius and heat capacity enter into the solution of $G(\alpha,\tau)$, and therefore must be accurately known. The model also assumes that the thermal conductivity of the probe is infinite; in practice, this means the true conductivity must be very large relative to the conductivity of the material under test.

The probe, made of copper (λ = 380 W m^{-1} C^{-1}), is 220 mm long and 6.0 mm in diameter. Heat flows out radially from a coil of constantan resistance wire wound on a small copper rod which is inserted in a hole bored along the length of the probe; silicon heat sink compound was used to ensure good thermal contact. The temperature at the probe-soil interface is measured with a thin foil thermocouple (Omega type C01-T) referenced to the temperature bath. Temperature is recorded via a nanovoltmeter (Keithley Model 181) with a sensitivity of $\pm$ 50 nV. A small input of heat and the very sensitive temperature measurement system ($\pm$ 0.001°C) allow the maximum temperature rise in the soil sample to be kept to 0.1°C. This requires that the experiment be protected from thermal and electrical disturbances. The measurement system is guarded and grounded, and connections are protected from thermocouple effects. The system is shown in Figure 1.

A least squares regression is performed on the time-temperature record to determine the best fit of the soil thermal properties to these data. The method proved reliable in test materials (water in a weak gelatin solution, and ice).

In addition to the probe measurements, the unfrozen water content (θ_u) versus temperature (T) relationship, is determined for the soil, using the technique of time domain reflectometry, (TDR), (see Patterson and Smith 1981). These data are used to calculate the soil apparent heat capacity, which is then used to constrain the solution for conductivity from the heat probe data. So far, the technique has been applied to saturated samples, in order to reduce the possibility of moisture migration, and to ensure effectively zero contact resistance.

EXPERIMENTAL METHODS

Thermal property determinations have been undertaken for remolded samples of Caen silt and Castor sandy loam. (Pertinent physical properties for these soils are summarized in Figure 2). The soil sample was first frozen to about -10°C (to ensure ice nucleation), and thermal property determinations were made at various temperatures on a warming cycle.

A TDR probe was also inserted in the soil sample for determination of θ_u at each temperature (see Figure 2). Soil apparent heat capacity, (C_a) was calculated from:

$$C_a(T) = C_m(T) + \frac{L_f}{V_w} \cdot \left(\frac{d\theta_u}{dT}\right)_T \quad (6)$$

where $C_m(T)$ is the volumetric heat capacity of the soil-ice-water mixture, L_f is the latent heat of fusion, V_w is the specific volume of water, and ($d\theta_u/dT$) is the slope of the unfrozen water content curve at temperature T.

Determinations were first performed for Caen silt, using the pulse method. The procedure followed was to heat the probe for a short period (typically 15 s) and then allow it to cool to its original temperature. The temperatures during the cooling period were used to solve for the soil thermal properties via the regression procedure. However, at temperatures above -1°C, an unexpected problem was encountered. The time-temperature trace was higher for successive runs at the same temperature (Figure 3), thus yielding non repeatable results (cf. Kay et al. 1981).

This can be explained by hysteresis in the unfrozen water content temperature relationship (see Patterson and Smith 1981). When a sample is heated for the first time at any temperature, the warming curve of the unfrozen water content relationship is followed. However, once the sample begins to cool it will not return along this curve, but will follow some intermediate curve (Figure 4). When it is heated again it will follow a new warming curve, to return on another scanning curve, and so on. This interpretation was examined by placing a sample of Caen silt in a temperature controlled bath ($\pm$ 0.01°C) and subjecting it to alternating temperatures between -0.6°C and -0.25°C. Values of θ_u, determined using TDR (Figure 4), indicated hysteresis between successive runs at the same temperature. While the effect of hysteresis on the volume fractions of ice and water is small, its effect on the apparent heat capacity (which depends on $d\theta_u/dT$) will be large because of differences in $d\theta_u/dT$ on heating and cooling, and between successive runs.

Figure 5 shows a plot of conductivity versus T as determined for the Caen silt sample using the pulse method. At each temperature, the highest value of conductivity was determined on the first run, with values generally declining for each successive run. Successive values of conductivity were calculated using a value for C_a which became increasingly erroneous. Whilst the trend in conductivity values determined for the first run at each temperature seems consistent, the results above -3.9°C are unreliable.

It is concluded, therefore, that the heat pulse method cannot be used to determine soil thermal properties where hysteresis in the unfrozen water content relationship is large. The alternative method chosen was to heat the probe continuously for about 3 min, but at a very low rate (less than 1 W m^{-1}), in order, again, to limit the temperature rise to less than 0.1°C. Whilst it would be preferable to carry out a number of determinations at any one temperature, the hysteresis effect described above does not allow this. Instead, the procedure followed was to perform conductivity determinations at closely spaced temperatures on a warming cycle.

A sensitivity analysis of equation (2) for conductivity, diffusivity and heat capacity showed that markedly different combinations of these properties can produce probe heating curves which are indistinguishable, within the limits of measurement accuracy. Attempts to determine all properties simultaneously, therefore, are prone to error, since an incorrect estimate of one property will be offset by incorrect estimates of the other properties. The only property that can be determined independently is C_a. However, as $d\theta_u/dT$ approaches infinity (i.e. at very warm freezing temperatures), it becomes impossible to accurately estimate C_a. Therefore, the analytical procedure followed was to evaluate all three thermal properties, and satisfy the following criteria.

1. The thermal properties combine to produce the observed heating curve (to within $\pm$.001°C).
2. Each property plots as a smooth function of temperature (which seems physically reasonable.)

The data were first analysed without constraining the solution to $G(\alpha,\tau)$ for any of the thermal properties. The results of this were plotted. The calculated heat capacities were compared to those estimated using the TDR freezing characteristic data. A smoothed function was fitted to the combined heat capacity data. The smoothed values were then used to constrain $G(\alpha,\tau)$ in the determination of conductivity.

The results of this procedure are shown in Figures 6, 7, 8. Some scatter still remains in the values for conductivity, although the plot of diffusivity determinations is most satisfactory. For conductivity, the scatter is greatest at temperatures above -1°C; in this range, $d\theta_u/dT$ changes rapidly (see Figure 2), affecting estimates of C_a, hence λ. It appears that the accuracy of the method is limited by the accuracy with which $d\theta_u/dT$ (hence C_a) can be determined. Determination of more points on the θ_u versus T relationship, or direct (calorimetric) measurement of C_a, might improve the estimates. Kay et al. (1981) also present a method for determining soil thermal properties which requires knowledge of C_a. However, the only results they present above -2°C, are at -0.7°C. They indicate a wide scatter in their values.

It appears that the conductivity value at -1.9°C is simply anomalous, since estimates of C_a in this range should be reliable (where $d\theta_u/dT$ is small).

Finally, perhaps the best estimates of the thermal properties are achieved when all three plot as smooth functions (such as the line indicated in Figure 6). Therefore, additional smoothing might be justifiable, with a third criterion that the trend line passes through the values at 0° and -5°C, say. At these temperatures, the results are probably most accurate.

CONCLUSIONS

The continuous heating method described in this paper extends the temperature range for which the thermal properties of frozen soils can be determined. However, at temperatures where C_a changes rapidly, or becomes very large, further refinement in analysing the results may be desirable. The heat pulse method cannot be used in the range 0° to -2°C, because of hysteresis in the unfrozen water content versus temperature relationship.

NOTATION

T	= temperature.
ΔT	= temperature rise.
Q	= heat input per unit length.
λ	= soil thermal conductivity.
κ	= soil thermal diffusivity.
C	= soil volumetric heat capacity.
C_a	= apparent heat capacity.
C_o	= volumetric heat capacity of the probe.
a	= radius of the probe.
t	= time.
t_o	= length of heating period (pulse method).
τ	= $\kappa t/a^2$.
α	= $2.C/C_o$.
θ_u	= unfrozen water content (by volume).
$\exp(x)$	= e^x.
$J_n(x)$, $Y_n(x)$	= Bessel functions of order n, of the first and second kinds, respectively.

ACKNOWLEDGMENTS

The authors wish to thank L. Boyle for his technical assistance and D. Patterson and M. Hare for their help in the laboratory. The work was made possible by a research grant from the Natural Sciences and Engineering Research Council, Canada.

REFERENCES

Farouki, O.T. 1981. Thermal properties of soils. CRREL Monograph 81-1 United States Army Corps of Engineers Cold Regions Research and Engineering Laboratory, Hanover, New Hampshire. 136 p.

Jaeger, J.C. 1956. Conduction of heat in an infinite region bounded internally by a circular cylinder of a perfect conductor. Aust. J. Phys. 9, 167-179.

Kay, B.D., M. Fukuda, H. Izuta and M.I. Sheppard. 1981. The importance of water migration in the measurement of the conductivity of unsaturated frozen soils. Cold Regions Science and Technology, Vol. 5, p. 95-106.

Patterson, D.E. and M.W. Smith, 1981. The measurement of unfrozen water content by time domain reflectometry: results from laboratory tests. Canadian Geotechnical Journal, Vol. 18, pp. 131-144.

Penner, E. 1970. Thermal conductivity of frozen soils. Canadian Journal of Earth Sciences, Vol. 7, pp. 982-987.

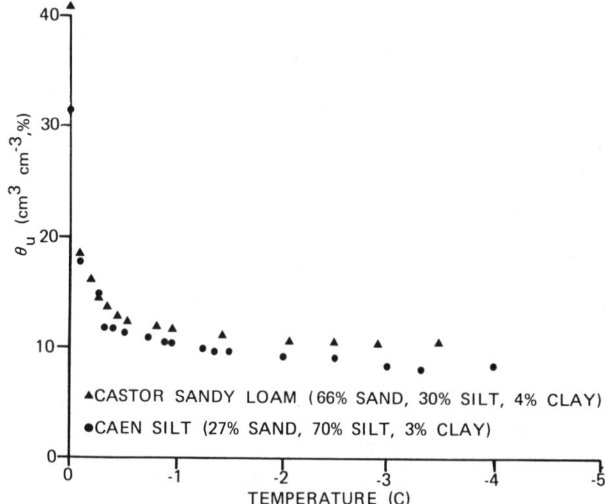

FIGURE 2 Unfrozen water content curves for test soils (determined by TDR).

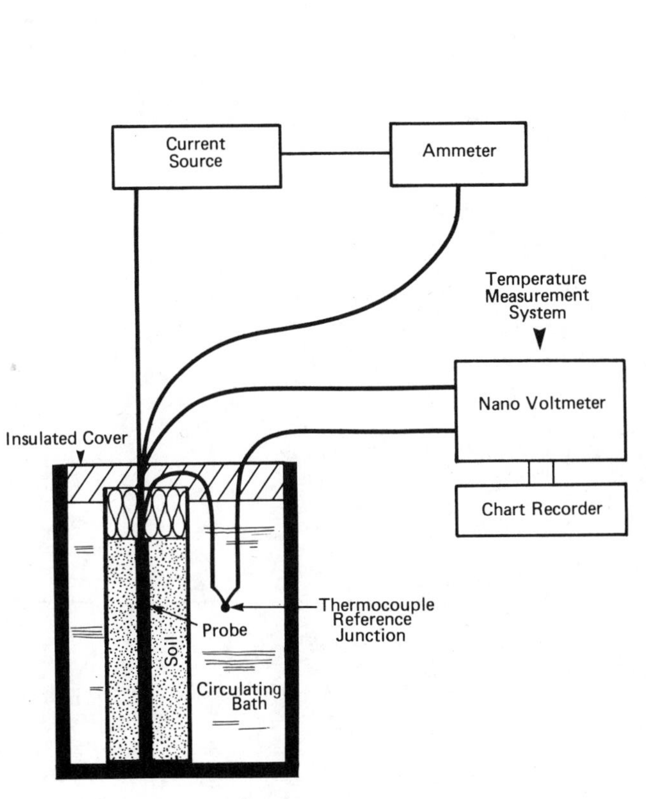

FIGURE 1 Schematic experimental setup.

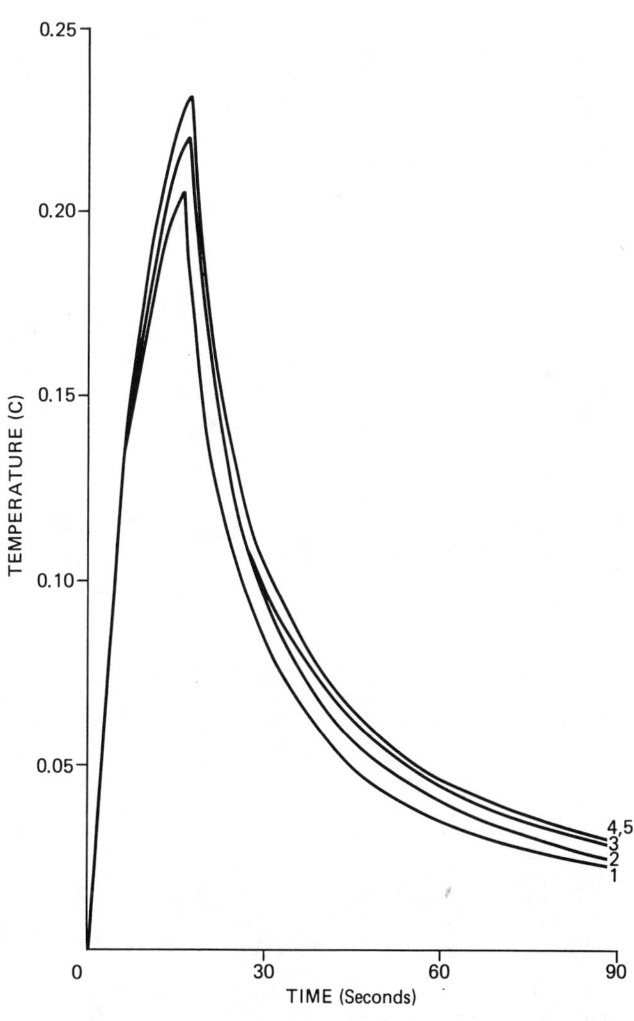

FIGURE 3 Time-temperature curves for Caen silt at -0.66°C (numbers refer to sequence).

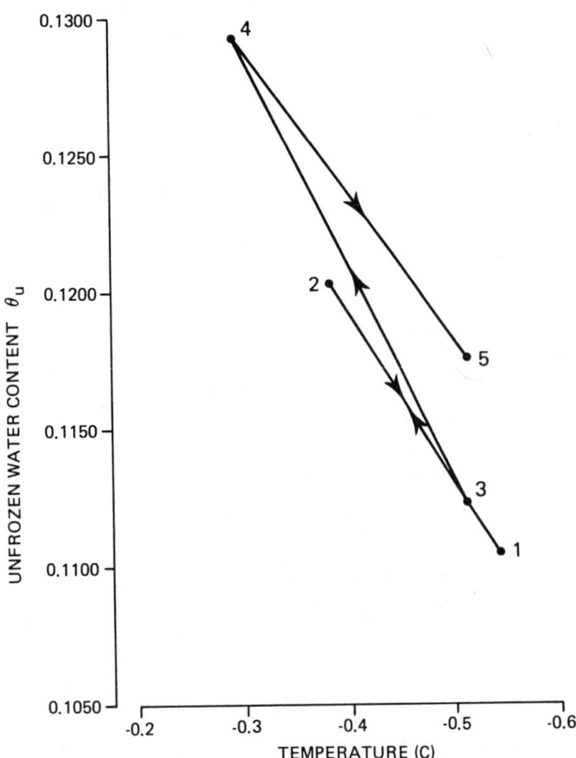

FIGURE 4 Freeze-thaw hysterisis in Caen silt (numbers refer to sequence - see text for explanation).

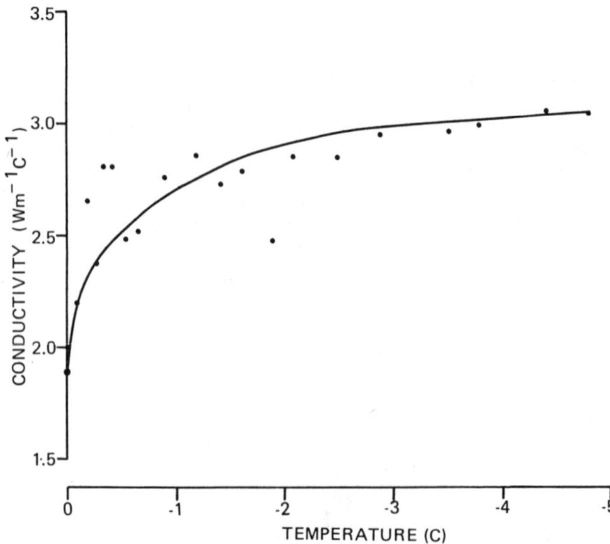

FIGURE 6 Thermal conductivity of Castor sandy loam, determined using continuous heating (line is a best fit by eye).

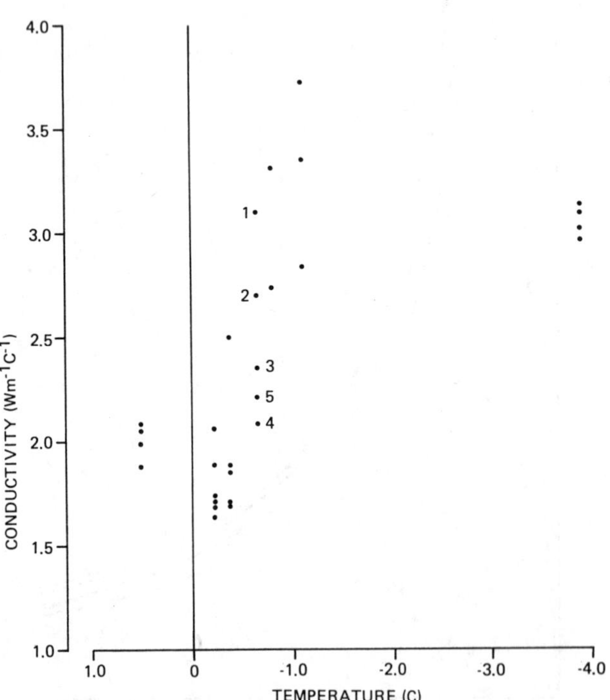

FIGURE 5 Thermal conductivity of Caen silt, determined using the pulse method (numbers refer to sequence).

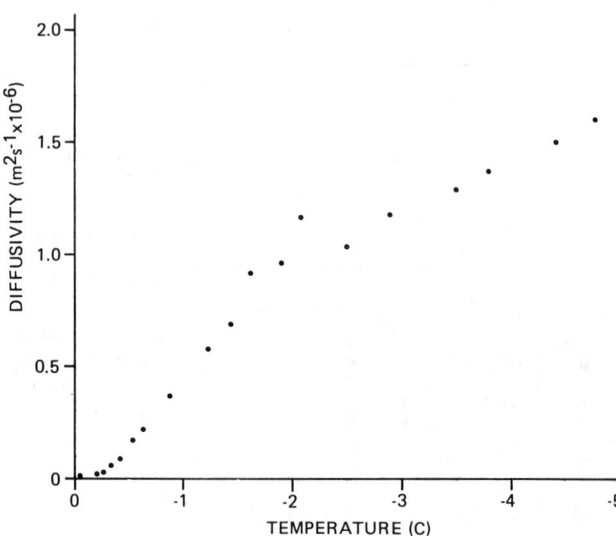

FIGURE 7 Thermal diffusivity of Castor sandy loam.

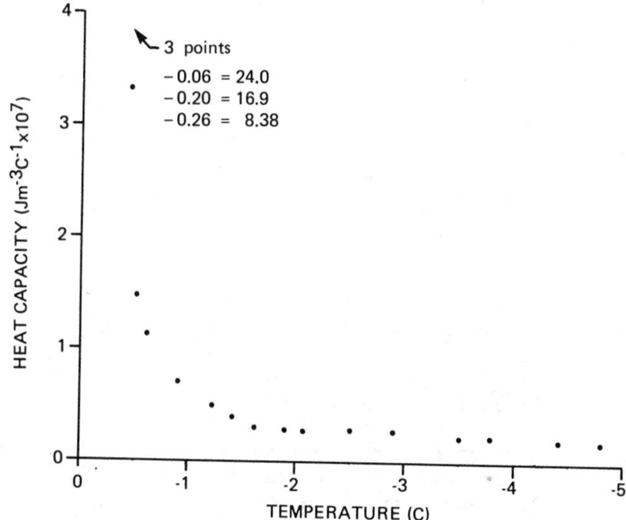

FIGURE 8 Heat capacity of Castor sandy loam.

ROCK STREAM DESERPTION*

N. N. Romanovskii and A. I. Tyurin

Geology Department, Moscow State University
Moscow, USSR

The authors have investigated rock streams by analyzing the frozen materials of which they are composed. On the basis of abundant data it was shown that rock streams are not chaotic accumulations of coarse material as was earlier considered but orderly geodynamic systems evolving according to certain rules. Three hypsometric belts have been distinguished, merging one into the next in orderly fashion. Each of these belts consists of a strictly defined combination of facies and subfacies differing both in structure and in the range of factors affecting its movement. The range of sediments occurring in a rock stream depends on its genesis, stage of development, and geological and geomorphological peculiarities. The data obtained by the authors have enabled them to elaborate scientifically substantiated recommendations as to the building of engineering structures on rock streams or in their zones of impact and to approach environmental protection in a rational fashion.

In recent years rock-stream (felsenmeer) research has acquired a great practical importance. This is connected with the intensive economic development of the regions which are subject to rock stream formation. Rock streams often have great areal extent. For instance, the areas occupied by rock streams in parts of Southern Yakutia and Northern Transbaikalia are as follows: (1) 1 to 5% in depressions and in other regions with weak bedrock, (2) 25 to 60% on Jurassic quartzite-like sandstones, and (3) 60 to 80% in the golets/alpine belt on Proterozoic and Archean metamorphic rocks.

The structure of rock streams is not uniform, and the processes leading to the formation of the coarse-fragmented material, its movement on slopes, and accumulation at the foot are also diverse. Various types of deserption (cryogenic, thermogenic, and sometimes hydrogenic) are among constantly active processes of movement of coarse-fragmented material of rock streams; the accompanying processes are frost heaving of fragments, viscoplastic flow of interstitial ice, and creep of the fragmented material in the seasonally thawed layer along the icy base. Each of these processes can act in certain parts of the rock stream and be absent in others, and have a direct interrelation with the structure of their profile. Different parts of rock streams are characterized by definite relationships between the thicknesses of the coarse-fragmented material and seasonally thawed layer by specific cryogenic structure, and by morphology.

It is possible to approach the study of rock streams using the method of frozen ground-facies analysis. The facies of rock streams were distinguished on the basis of the uniformity of the landforms and on morphology. General direction of the processes forming and transforming the coarse-fragmented material served as a principal character to distinguish the facies. Within the same facies it was necessary to distinguish subfacies on the basis of structural characteristics of the rock stream body in the cross-section (according to the presence or absence of silt horizons, character of sorting of fragments, cryogenic structure, relationships of its thickness with depths of the seasonally thawed and seasonally frozen layers and a set of movement factors).

In rock streams, which are mobile slope formations, three main "belts" are distinguished downslope. First, there is the belt of mobilization of coarse-fragmented material, its preparation for movement. There are still no rock streams as such, but an accumulation of coarse-fragmented material is formed that consequently is entrained into motion. The belt of mobilization is followed by the belt of rock streams sensu stricto, i.e. by mobile and rocky slopes and streams. Active movement (transit) of the rock stream mantle takes place here. The third belt is that of accumulation of coarse-fragmented material where movement downslope diminishes or stops completely. Internment or burial of rock streams by deposits of various origin, and their reworking by exogenic processes and transformation into other formations takes place here. In each belt a number of facies are distinguished which are grouped into three categories on the basis of their role in rock stream formation and for convenience of analysis: (1) facies of external supply belt, (2) facies of transit belt and (3) those of the accumulation belt.

The facies of external supply belt of rock streams may vary greatly with the geological structure of the area, the history of its development, and the relationships of the depths of the seasonally thawed and seasonally frozen layer and the mantle of loose deposits.

The facies of the external supply belt of rock streams pass downslope into those of the belt of movement (transit) of coarse-fragmented material. (1) The facies of the rock stream slope, (2) a group of facies of rock stream flows and (3) a group of facies of structural rock stream slopes are distinguished within the latter belt.

Facies of the rock stream slope is widely distributed within the study area. It is confined to

*See Editor's note on last page.

slopes between 3-10° and 30-35° in angle. The width of the rock stream is usually greater than or commensurable with its length, and the morphology of its surface is uniform.

The suffosion-destructive subfacies is characterized by coarse-fragmented material, represented by blocks, rubble and grus. It has weak sorting and passes into broken rock at the base. In the lower layers of coarse fragmented material occurs a small amount of filler which consists of sand of varied particle size and silty sandy loam. Because the thickness of the rock stream is less than the depth of the seasonally thawed or seasonally frozen layer, the bedrock is subjected to intense cryogenic weathering. The movement of coarse-fragmented material occurs mainly due to thermogenic and cryogenic deserption. Of great significance are disintegration of coarse fragmented material, its movement resulting from frost sorting, and suffosional removal of fines.

According to the observations in the Chulman Depression, the rate of movement of sandstone fragments of this subfacies on the surface of a slope with steepness of 25° is up to 2 cm/yr. Rapid movements of material do not occur. Creeping semi-disintegrated bedrock blocks are encountered within the suffosion-destructive subfacies which develops below the slump block facies.

The golets ice subfacies (in the rock stream slope facies) is usually located hypsometrically below the suffosion-destructive subfacies. The boundary between them has a complex configuration and is ill-defined in the microrelief of rock streams. The structure of rock streams of the golets ice subfacies is characterized by frost sorting of coarse-fragmented material without matrie in the seasonally thawed layer and by an ice-soil layer consisting of bedrock fragments cemented by golets ice below the seasonally thawed layer base. The ice content within this layer is different. In some cases ice fills only the voids in the coarse-fragmental material; in others, the rock mantle is "inflated", i.e. the blocks, rubble and grus are surrounded by the ice and the volume ice content reaches 70-90%. The golets ice includes small lenses of fines, layers and lenses of ice which differ in color, structure, content of mineral admixtures, and air bubbles of various shape. Small lenses of sand often duplicate the contours of bedrock blocks located in the ice. Flattened bedrock blocks change their orientation within the cross section of the ice-soil layer. In its lower part the planes of the fragments duplicate the position of the rocks in an undisturbed state. In the upper part of the section the position of the block planes approximates the direction of the general inclination of the slope surface. This is apparently a consequence of movement of the blocks in the ice-soil layer due to plastic deformations which are indicated by the structures of the enveloping ice.

The ice-soil layer includes planes marked by sprinkling of sand material and increased content of air bubbles that have a different dip with respect to the slope: subvertical, into the depth of the slope and along its dip. The genesis of these disturbances is apparently related to ruptures of the ice-soil layer as a result of volume-gradient deformations and to stress release occurring during movement of the coarse-fragmented material. These ruptures apparently play an important role in movement of rock streams.

The golets ice subfacies is characterized by processes of plastic flow of the ice-soil layer, and slow movement and slides of the coarse-fragmented material along the icy base. Such slides usually occur locally in places where the seasonally thawed layer base is sufficiently even, without a high content of large blocks frozen into the ice layer, which would hold the rock stream material even on steep slopes.

The accumulation of the golets ice in the perennially frozen base of a rock stream is possible while the mantle of coarse-fragmented material remains relatively constant or increases in thickness. Weathering and the disintegration and removal of fines downward is compensated by the more rapid transport of the fragmented material from above from the external or internal supply belts of the rock stream.

In the study area, the rate of movement of the fragmented surface material ranges from 1 to 3 cm/yr (the average being 2 cm/yr). The described golets ice subfacies is characterized by greater rates of creep of the fragmental mantle, in years of an increase of its thickness, and by rapid debris slides primarily on steep parts of the slope and on slopes undercut by rivers. The nonuniform movements of the rocks of the fragmented mantle of the golets ice subfacies leads to thawing, slumping and occurrence of depressions on some parts of the rock stream slope and, conversely, to an increase of the thickness of the fragmented mantle or to golets ice formation at the seasonally thawed layer base on other parts of the rock stream slope.

The siltation subfacies is confined to the lower parts of rock stream slopes and to the less steep parts of rock stream slopes with insignificant elongation or complex relief.

The cross sections of the rock streams which belong to this facies are composed of poorly sorted coarse fragmented material without filler to the depth of 0.2-0.5 to 1.0 m. Lower in the section occurs a layer of the coarse-fragmented material with dispersed filler whose content may reach 60-70%. Usually in the summer, the fines in the seasonally thawed layer are saturated and in a quick ground condition. In a frozen state the horizon has a massive and irregular lense-like bedded cryostructure. The thickness of the horizon with dispersed fines increases downslope. The seasonally thawed layer thickness does not exceed 1.5 m in this subfacies. Also within this subfacies, thermogenic deserption acts in the layer without filler and cryogenic deserption occurs in the entire seasonally thawed layer; heaving of the fragmented material, viscoplastic flow of the layer with fines, deluvial subsurface washouts, and accumulation of fines occur. Rapid movements and slides after heavy rains are also associated with this subfacies; these occur both on a wide front along with formation of depressions parallel to the slopes, and locally, forming stripes of the coarse-fragmented material. Local movement results in the occurrence of grooves of different length and width, with the cirques of separation in the upper part and heaved cones of indistinct shape in the lower one. Heaving of separated fragments also occurs.

The measured rate of heaving ranges from 2 to 20 mm/yr.

Facial variations occur either independently in short rock streams, or in combination, regularly replacing one another in long rock streams. Replacement of facies and subfacies in time occurs too. Replacement of subfacies is an objectively existing process determined by various conditions, such as resistance of rocks to weathering, steepness of slopes, undercutting of slopes by rivers, etc., as well as by climatic and frozen ground factors. Within the study area, the following combination of subfacies is most often encountered in the mature rock stream slope facies. The suffosion-destructive facies is located in the upper part of the rock stream slope. Hypsometrically lower there is the golets ice subfacies turning into the siltation one at the foot.

In the initial stage the facies of the rock stream slope is represented only by the suffosion-destructive subfacies which takes up the entire rock stream. The rock stream increases in length and thickness. As the thickness of the rock stream enlarges through both weathering of rocks and suffosion of fines, the rock porosity becomes higher and the quantity of golets ice being formed annually in the rock stream increases. The ice content is highest in the lower part of the section which is most washed out and characterized by maximum porosity. The increase in ice content leads to a reduction of the thickness of the seasonally thawed layer and to the transition of its lower part into perennially frozen ground. Consequently, the golets ice subfacies starts to form in the lower part of the rock stream slope. The growth of the golets ice subfacies proceeds upslope similarly to the "backward erosion". In cases where weathering of hard bedrock results in the formation of a small quantity of fines which are easily removed by surface runoff, the golets ice subfacies may replace the suffosion-destructive one completely. In this case, the belt of rock stream movement (transit) is again represented by only one subfacies.

Finally, the siltation subfacies is formed, irrespective of rock strength in the lower part of the rock streams developing on slopes not undercut by rivers. On rocks that are more resistant to erosion the siltation subfacies appears later, and develops more slowly than on less resistant rocks. The area of the siltation subfacies increases upslope at the expense of the other subfacies. In this case the rate of its formation decreases gradually, practically ceasing at a certain level. The remaining part of the rock stream, which is represented by the suffosion-destructive or the golets ice subfacies, is usually small and weathering of rocks within its limits cannot support the development of the siltation subfacies. If the amount of fines removed by the surface runoff is large, the siltation subfacies may replace all subfacies located upslope. Finally, the rock stream is degraded. However, hard rocks and high moisture content of the deposits may cause heaving of the coarse-fragmented material and the revival of the rock stream.

There is a group of rock stream flow facies. The rock stream flow facies include narrow, linearly extended rock streams the width of which is considerably smaller than their length. Within the study area, the rock stream-flows are represented by three facies: (1) rock stream rills, (2) rock stream rampart flow and (3) rock stream stripes. The distinctive features of the rock stream flow facies are: (1) the concentrated, directed movement of the coarse-fragmented material, (2) differences in structure not only downslope but also in the transverse section; these differences are related to orienting effect of the banks of the flow and other factors. Formation of rock stream-flows is determined by the occurrence of fractures extended downslope in the rock, or by systems of frost polygons (usually following tectonic cracks); the cracks following the surface gradient are most reworked (weathered, washed out, etc.). Because of this, the rock stream flow facies has slump blocks and the subfacies of differential weathering of rocks as an external supply belt.

The rock stream rill is represented by extended rock streams the width of which varies from 3 to 4 to several tens of meters; the side parts are raised compared to the axial part.

Within the boundaries of Southern Yakutia and Northern Transbaikalia, rock stream rill takes place along the lines of frost cracks where intense processes of suffosion of fines occurs. In this case the size of the cells in the network of polygons exceeds 15m; the large distance between the cracks prevents formation of the rock stream slope facies. The width of the rock stream rills varies from 3 to 6 m. The rock stream rill facies often occurs in depressions. This is most typical of Transbaikalia and Stanovoe Nagore Mountains.

The rock stream stripe facies is represented by linear, elongated, flat rock streams, which are not distinguishable in the relief of the slopes, are without a pronounced concentration of the flow of supra-permafrost water and without intensive suffosion. The cross section of the rock stream stripe subfacies is determined by its genesis and the stage of development. Depending on this it may or may not have a channel. If a channel is not formed, the cross section of the facies is filled almost completely with fines. For example, such facies are formed as a result of destruction of rock remnants, the banded coarse-fragmented material of which is shifted downslope. In the relief of the slope, the subfacies has the shape of the rock stream. The fragments are placed on their sides, their long axes oriented along the dip of the slope. Frost sorting is apparent in the cross section. If the rock stream stripe facies is formed on loose deposits the thickness of which exceeds the depth of the seasonally thawed or seasonally frozen layer, the rock stream flow appears to immerse into the slope according to its denudation. Coarse-fragments of the loose deposits composing the slope are included in the subfacies.

The rock stream rampart-flow facies represents elongated rock streams with an elevated central part and a depressed peripheral part. Characteristically, on steep areas and in the upper parts of slopes the immobile banks of the rock stream are raised as much as 1 m. On relatively flat areas and in the lower parts of the slopes the ramparts-flows form well defined positive microrelief. The sides of blocks are oriented along the dip of the slope; in the transverse section the blocks diverge fan-like from the center to the periphery. Distinct frost sorting of fragmented material is encountered in the vertical section. At the base of the seasonally thawed layer, the space between fragments is filled with saturated fines. The thickness of the horizon with fines increases downslope. An icy layer is usually found below the seasonally thawed layer without fines. In the upper parts of the rampart flows the seasonally thawed layer depth often exceeds the thickness of the rock stream mantle and the seasonally thawed layer includes broken rock. This is a typical suffosion-destructive subfacies. The occurrence of the greater part of the rock stream rampart-flow on a viscoplastic and heavily iced base leads to constant formation of debris slides and, consequently, to a change in configuration of the seasonally thawed layer base.

In many respects the structure and dynamics of the rock stream rampart facies are determined by the genesis of the rock stream. Within Southern Yakutia and Northern Transbaikalia, the formation of the rock-stream rampart facies (as well as of those of the rock stream rill and stripe) is closely associated both with weathering of bedrocks along tectonic lines and frost cracks and with the heaving of fragments to the surface. Surface runoff is concentrated in frost cracks penetrating considerably deep into the bedrock and follows its initial jointing. Periodical temperature fluctuations which extend deep along the cracks and high degree of water accumulation intensify weathering processes. Following the growth of veins of the coarse-fragmented material along the lines of cracks, the fragments start slowly move downslope and to be squeezed out toward the sides of polygons by heaving of ice formed in the seasonally thawed (seasonally frozen) layer, through temperature deformations, suffosional cones of finely-fragmented debris, gravitation and other factors. The rate of displacement of fragments is up to 1.5 cm/yr along the dip of a 6° steep slope and up to 1.2 cm/yr along the trend. Therefore, the growth of veins of the coarse-fragmented material brings about a specific qualitative change - transformation of veins into the rock stream.

There is a group of facies of structural rock stream slopes. These include a series of rock streams distributed in regions where presently or in the recent geological past, were active processes of frost riving and formation of polygonal-veined structures both in rocks and semirocks. Facies variations of this group of rock streams are formed by complex effects of rock stream formation processes on the rocky substrate composed of polygonal blocks. The intensity of rock stream formation is different on blocks and in depressions between them. The character of the forming rock streams varies with lithological characteristics of the rock, the intensity of surface runoff, and the depth of the seasonally thawed or seasonally frozen layer. This allows the recognition of a series of facies.

The facies of concealed-polygonal rock stream slopes is formed on surface of hard rock outcrops denuded of loose mantle and destroyed by a system of cracks. The most intense processes of frost weathering are encountered along cracks. The surface runoff is concentrated within the cracks and "pockets" of the deeper weathering zones are formed in the bedrock. In this way, the mantle of the rock stream fragmented material forms on the slope; its thickness is different in different parts. The thickness of the coarse-fragmented mantle both in the cracks and above the block-polygons changes as follows:

(1) 1.0-1.5 m in the upper part of the rock stream slope;

(2) 2.5-3.5 and 1.0-1.5 m in the middle part of the slope;

(3) in the lower part of the slope the thicknesses are similar and range from 2.5-3.0 m.

The relief of the facies is represented by a rock stream slope of medium steepness with the solid mantle of the coarse-fragmented material and with sink holes linearly extending downward. The fragments in the sink holes are of maximum size. In cross section the facies represents a series of furrows filled with coarse-fragmented material. In the downslope there are not only changes in relationships of different thicknesses of the coarse-fragmented material along the cracks and on the blocks, but also changes in the cross section structure and in the relationships of thicknesses of the fragmental mantle and the seasonally thawed (seasonally frozen) layer. This allows a recognition of a series of subfacies within the given facies. These subfacies are of complex nature. On the one hand, there are certain areas of which the structure is similar to that of the suffosion-destructive subfacies; grinding of rocks composing the seasonally thawed layer continues within the limits of blocks-polygons. At the same time, parts of subfacies with golets ice form along the lines of furrows.

The facies of banded concealed-polygonal rock stream slopes is represented by coarse-fragmented mantle with an uneven surface showing alternation of the fragmented material ramparts (the arrangement of which coincides with the trend of the slope) and relatively smoothed inter-polygonal areas. Elevation of ramparts above the surface ranges from 20 to 30 cm, reaching a meter in some cases. The width of the ramparts varies from 1 to 3 m. The surface of some ramparts is only slightly convex. Very coarse structure of jointing at the surface is an important character of the ramparts. The ramparts are often characterized by high porosity and look open. Downslope they become less distinct. In certain cases they merge.

Two cases of formation of this facies should be distinguished. (1) Along the polygonal network of ancient frost cracks in weathered bedrock forms a large amount of dispersed material (40-70%; sandy loam, loam) from which the fragments are heaved to the surface. (2) Bedrock is overlain by the mantle of loose deposits the thickness of which is less than that of the seasonally thawed or seasonally frozen layer. These loose deposits are subjected to frost riving. As a result, polygons with a medium size of 4x4 to 6x6 m have been formed. The surface runoff is concentrated along the polygonal depressions and zones of weathering are formed in underlying rocks. Higher moisture content in these depressions promotes intense heaving of fragmented material trending along the surface and causes formation of veins of the coarse-fragmented material. If the depth of seasonally thawing or freezing exceeds the thickness of the mantle of loose deposits, the growth of veins leads to formation of ramparts of coarse-fragmented material on the surface; the width of these ramparts is 4 to 5 m, and the height varies from 0.5 to 1.5 m. The coarse-fragmented mantle closes over the polygons. The structure of the facies of banded concealed-polygonal rock stream slopes is specific. A distinct asymmetric system representing alteration of veins of fragmented material and blocks-polygons appears under the coarse-fragmented material mantle. The thickness of veins coinciding with the trend of the slope reaches 2.5 to 3.0 m, and their width 2 to 3 m. The size of the veins extending downslope is considerably smaller. A series of subfacies can be recognized within this facies. Certain structure of the fragmented material composition and its processing is characteristic of the subfacies in which the destruction of rocks, included in the seasonally thawed layer is completed. The fragments of maximum size are located along the cracks and from the surface, as though enveloping the polygons.

The rate of and factors causing the movement of the fragmented mantle and coarse-fragmented mantle material is different in the different parts of the facies. The rate of the mantle movement on slopes with a steepness of 15° is about 0.4 cm/yr above polygonal remnants, and reaches about 1.5 cm/yr in the sinks extending downslope.

The present work is regarded as a first attempt at the facies analysis of rock streams. The authors have concluded that a systematical study of the mechanisms of movement of rock streams and evaluation of their engineering geologic characteristics should be carried out only when their specific facies variations are taken into account.

Rock streams are mobile slope formations having certain genesis, ordered structure and specific mechanisms of movement characteristic only to them. Our frozen ground-facies analysis has confirmed the necessity of distinguishing them as a specific genetic type of slope formations - that of the rock stream deserption.

Editor's note: Deserption is defined as a surficial creep deposit which results from a volume change such as is caused by frost heave and or swelling. It is a zonal type of deposit, not identical to either solifluction or gelifluction. Reference: Poppe, V. and J.E. Brown, 1976. Russian - English Glossary of Permafrost Terms: Canada, NRC, Research Technical Memorandum 117, 25 p.

For additional details on rock streams see the recently published book:

Tyurin, A.I., Romanovskii, N.N. and Poltev, H.F., 1982. Frost-facies analysis of rock streams. Moscow: Izdatel'stvo "Nauka." (Akademiya Nauk SSSR, Nauchnyi Sovet po Kriologii Zemli, Sibirskoe Otdelenie, Institut Merzlotovedeniya.)

FAIRBANKS WASTEWATER TREATMENT FACILITY: GEOTECHNICAL CONSIDERATIONS

J.W. Rooney[1] and J. H. Wellman[2]

[1]R&M Consultants, Inc., Anchorage, Alaska 99503 USA
[2]R&M Consultants, Inc., Fairbanks, Alaska 99701 USA

Subsurface investigation and geotechnical evaluation of the site of the Fairbanks wastewater treatment facility identified the soil profile, the extent of frozen ground, and other pertinent geotechnical parameters necessary for design and construction considerations. Primary geotechnical-related concerns included thaw settlement, extent of overexcavation, and minimizing groundwater infiltration into the excavation. The overlying ice-rich silt stratum was considered to be unsuitable for proposed site development because of its high thaw strain potential. The underlying gravel-sand stratum was considered to be suitable for foundation bearing, although the potential for limited thaw strain and seismically induced dynamic densification were of concern. From all indications, the structure has not exhibited any significant settlement during the period of operation. An example of differential movement attributed to a possible combination of soil thawing and heaving occurred when subsequent placement of the septic disposal building partially on silt outside the original excavation and backfill limits resulted in significant tilting of the structure.

INTRODUCTION

Prior to statehood, development of facilities within the Fairbanks floodplain region was typically accomplished on locations of unfrozen floodplain granular deposits. Notable exceptions are those documented in the literature by Terzaghi (1952) and Waterhouse and Sills (1952) on the Ladd Field (Fort Wainwright) power plant, by Terzaghi (1952) on the Eielson AFB hanger, and by Philleo (1966) on the University Park School addition. Site or performance data on other and more recent construction accomplished on frozen portions of these floodplain gravels have not been readily accessible or published. This paper presents, in case history format, information on site conditions at the Fairbanks wastewater treatment facility.

Based on overall project requirements, consultants for the City of Fairbanks selected the proposed sewage treatment plant site on the east side of Peger Road, approximately 7.2 km (4.5 mi.) from downtown Fairbanks (see Figures 1 and 2). Specific geotechnical concerns with proposed development at the site focused on defining: (1) the extent and limits of frozen ground, (2) surface water and subsurface groundwater conditions, (3) physical behavior of the soils (particularly frozen soil consolidation upon thawing and densification under dynamic loading), and (4) a sufficient area of frozen ground delineated so that the building could be positioned to allow all excavation "in the dry."

The wastewater treatment building, having approximate plant dimensions of 115.8 m (380 ft.) by 100.6 m (330 ft.), is a one-story concrete structure with below-ground areas for clarifiers, chlorinators, and sludge handling equipment. The building has multiple floor levels ranging from approximately 7.6 m (25 ft.) below to about 1.5 m (5 ft.) above the original ground surface. The approximate top of gravel and bottom of floor slab elevations are shown in Figure 3.

Thermal degradation of the frozen soils at the site, both during construction and throughout facility operational life, required that design should accommodate time-dependent changes associated with anticipated ground deformation. Preconstruction information indicated that development of the site was feasible; construction of the main facility was begun in late 1973 and was completed in 1976. Site instrumentation to monitor ground deformation was not installed. Performance of the building to date appears to be satisfactory, at least in terms of static thaw settlement. The potential effect of any dynamically induced settlement associated with any densification resulting from significant ground shaking remains unknown.

SITE CONDITIONS

General Topography and Geologic Setting

The sewage treatment plant site lies on an abandoned portion of the Tanana River floodplain, approximately 548.6 m (1,800 ft.) north of the active braided floodplain. The natural ground surface is relatively flat except for a few depressed thaw ponds and partially filled sloughs. The building site occupies an area between a thaw pond to the south and a marshy area to the north. Vegetation in the area is generally grass tussocks, moss, low brush, scattered groves of willows, and a few scattered birch, black spruce, and alder. Site construction grading has resulted in much of the area now being covered with granular fill with a relatively uniform surface.

Subsurface Soils

Subsurface soils at the site were identified in terms of two major soil strata; these being

(1) surficial ice-rich organic silts and (2) underlying clean (lower silt content), relatively ice-free gravels and sands. All borings drilled in the immediate vicinity of the defined building site indicated soils were frozen below the 2-3-foot active layer to total depth drilled. Boring locations are shown in Figure 2 and a log of Test Hole F-12 is presented in Figure 5.

Ice-rich organic silt extended from original ground surface to depths varying from 1.07 m (3.5 ft.) to 3.81 m (12.5 ft.). The natural moisture content of samples from these silts was high, (ranging from 10% to 61% and dry densities of the frozen samples were relatively low, ranging from 913 kg/cu.m (57 pcf) to 1,410 kg/cu.m (88 pcf).

The underlying gravel-sand stratum was found to alternate between sandy zones and gravelly zones. Moisture contents of the gravel-sand samples ranged from 7% to 31%; the sand density range was from 1,426 kg/cu.m (89 pcf) to 1,570 kg/cu.m (98 pcf), and the gravel density range was from 1,922 kg/cu.m (120 pcf) to 2,243 kg/cu.m (140 pcf). Based on boring results, the lateral extent of the sand substrata was considered to be relatively smaller than that of the gravel. Some occasional minor silt seams were also encountered in the gravel-sand strata; however, their extent and influence was felt to be small.

Permafrost

Permafrost in the immediate area of the site was classified as discontinuous. The permafrost was considered to be warm (i.e. between -2.0 and 0°C) and therefore sensitive to surface disturbance of the natural vegetation. Thermokarst depressions, resulting from thaw of ice-rich organic silt soil, were evident in the surrounding area. Unfrozen soils were encountered at the edge of a thaw pond and along the southern boundary of the initially proposed building location. The unfrozen-frozen soil contact zone was delineated with subsequent drilling and the building location was shifted slightly more than 15.2 m (50 ft.) north in an attempt to position the building excavation limits entirely within an area of frozen soil.

The near-surface organic silt stratum generally was found to contain large amounts of visible ice within individual 6-inch sample increments, with estimates ranging from 0 to 100%. The ice existed as individual crystals, crystal aggregates, and lenses and veins up to 0.15 m (6 in.) thick. Little visible ice was observed in the sand and gravel strata. Much of the coarse granular material appeared to be clean (very low silt content) and dry frozen, with only minor amounts of ice in voids between sand and gravel particles.

Water Table and Surface Drainage

Excavation and below-ground construction activities had the potential of being more time-consuming and costly if the proposed work had to be accomplished with a high water table condition. Emphasis was placed on defining groundwater conditions and, as much as possible, verifying that placement of the building completely within a frozen core area would likely minimize water-related problems that could result from infiltration through pervious zones or taliks during construction.

Surface drainage conditions in the immediate area were relatively poor due largely to the dense organic mat cover, the impervious nature of the frozen organic silt, and the flat terrain. While ponds did not exist directly within the building location, ponded water did exist, at least during the summer, in topographically low areas north and south of the site.

A high water level was found in the unfrozen soils at the edge of a thaw pond near the south boundary of the proposed building excavation. It was considered possible that the level of the water reflected a hydrostatic water table that could provide an almost infinite supply of water that might flood the excavation.

A number of test holes were drilled to further delineate and verify the extent of the frozen zone within which it was proposed to locate the building. Test Holes F-5 through F-11 were drilled to locate the northern boundary of the previously mentioned pond. Only Test Hole F-9 encountered unfrozen ground and high groundwater conditions.

DESIGN CONSIDERATIONS

General Concerns

The effect of construction of major facilities on ice-rich fine-grained frozen soils and the effect of consequent thaw-induced subsidence have been reported in numerous documents. Although a number of projects have been constructed on frozen granular deposits, only limited information has been published on the thaw settlement performance of frozen coarse-grained material within the Fairbanks area of the Tanana River floodplain deposits. The general perception that these granular deposits can be assumed to be thaw stable, such as presented by Pewe (1982) and in the Trans Alaska Pipeline System stipulations (1974), is relative; it is cautioned that thaw performance must be related directly to structure tolerances and site-specific conditions.

Some engineering problems that might result from thaw of these coarse granular materials and require assessment are described below.

(1) The extent of consolidation resulting from thawing of frozen gravel and sand is primarily dependent upon soil density, ice content and silt content derived from soil deposition processes and past depositional history. The influence of soil arching in reducing thaw settlement within the soil mass through stress and strain redistribution is also important; however, any subsequent earthquake-induced dynamic densification, may cause further settlement.

(2) Thawed granular material, with low-to-moderate densities, may liquefy when subjected to earthquake-induced shear stresses.

(3) Substantial lateral pressures and buoyancy effects may develop against below-ground structures and walls as the soil thaws and densifies around a building and as probable

perched ground or trapped surface water surrounds the structure.

Construction problems related to frozen coarse granular deposits involve several concerns.

(1) Excavation procedures and economic removal of frozen silt and bonded vs. unbonded frozen gravel. The influence of frozen soil temperature (warm vs. cold) and time of year can critically influence the extent of excavation effort and related earthwork economics.
(2) Earthwork operations "in the wet" are typically more costly than when performed "in the dry." When operations are planned to be performed in the dry, water infiltration into the excavation may adversely alter or impede planned excavation efforts or schedules, particularly when attempted during the winter construction period.

Thaw Consolidation

Field tests performed with the consultant supervision of Karl Terzaghi (1952) in the early 1950s on the power plant at Ladd Field provide important practical in situ data on the thaw consolidation behavior of the Tanana floodplain frozen granular deposits. Laboratory thaw consolidation tests made on the frozen granular cores obtained from the Fairbanks wastewater treatment facility site reported by Wellman and Nelson (1972) and laboratory and in situ field tests accomplished during studies on thaw performance of related frozen granular material for the Trans Alaska Pipeline System route, reported by Luscher and Afifi (1973) and Nelson et al. (in this volume) have provided further confirmation of potential thaw consolidation behavior.

Results of laboratory uniaxial thaw consolidation tests performed on frozen coarse-grained samples obtained from the site are summarized in Table 1. All frozen core samples were recovered using diamond core barrels with refrigerated diesel oil as a drilling fluid. The fluid was maintained at temperatures slightly below freezing (approximately equal to in situ temperatures). The recovered cores were 75 mm (3 in.) in diameter and, depending on sample condition, core segments had a length-to-diameter ratio ranging from 1 to 2.

The uniaxial thaw consolidation tests were performed according to the following procedures. Sample preparation was accomplished in a cold room; frozen core samples up to 4 in. long were fitted into rigid cylindrical rings after being trimmed to planar surfaces and to a 63.5 mm (2.5 in.) diameter. The samples were then brought into the -6.6°C testing room, loaded uniaxially in a constant air pressure loading frame under a pressure approximately equal to the sample overburden weight and then thawed. One-end drainage of any excess meltwater was permitted through a porous stone placed on top of the sample.

In general, the gravels were found to have frozen dry densities greater than 1,922 kg/cu.m (120 pcf) and the assumption was made that higher density, clean gravels (approximately less than 5% silt content) of alluvial origin exhibited little if any consolidation upon thawing. The sandy layers within the granular stratum gave some cause for concern because the recovered samples were found to have lower frozen dry densities, sometimes as low as 1,442 kg/cu.m (90 pcf). The thaw consolidation tests reported in Table 1 were performed on the sand samples obtained from the sand substrata intermixed with the gravel-sand stratum.

The results of the thaw consolidation tests indicated that limited static compression of the sand samples occurred during thaw under equivalent overburden load. The minimum recorded strain was 0.2% while the maximum strain value was 3.5%. These results compare reasonably well with those found for similar materials tested in the laboratory during the Trans Alaska Pipeline System routes studies. Potential additional strain effects resulting from dynamic loading were not assessed during either of these laboratory test programs.

Earthquake Concerns

The effect of earthquake ground motions was assessed with regard to potential for dynamically induced densification of the granular material and to the susceptibility of the material to liquefaction. The gravel substrata within the sand-gravel stratum were not considered to be susceptible to liquefaction since in their frozen state they were found to be at approximately 70-90% of their maximum relative thaw state density.

Because of limited sample material availability, evaluation of the sand substrata was based on two relative density tests (ASTM Specification D2049-69) performed on a composite sand sample with a unified soil classification of SP. The results of the tests and a list of individual samples that formed the composite sample are included in Table 2. Also included in this tabulation are the computed relative density values of all test hole samples having gradations similar to the composite sample. Engineering analysis of the sand samples' relative densities (average relative density approximately 44%) and the in situ stress and water table conditions based on Seed et al. (1971) (1972) indicated that the sands could withstand an earthquake with the equivalent of 10 stress cyles of approximately 0.1 g peak acceleration before liquefaction might occur.

Since occurrence of moderate-scale earthquakes (Richter magnitude 5.5 to 7.0) were anticipated in the immediate area and during the 30-year project design life and because of the number of variables involved, it was assumed that dynamic settlement could cause the sands to densify to approximately 70% of their maximum relative density. This increase in density implied a consolidation strain of about 4% in the sand substrata. Assessment of the approximate upper limit extent of equivalent vertical sand substrata indicated that about 3 m (10 ft.) of looser sand might be involved within the upper 12.1 m (40 ft.) of influence area for expected long-term thaw degradation below the building. Thus, it was felt that this equivalent thickness of sand could result in an upper limit of about 0.12 m (0.4 ft.) of total settlement.

Experience obtained from the Ladd Field power plant site also provides some additional in situ ground performance data for the same type of deposit. Although detailed site and sample data were not reviewed, information presented by Waterhouse and Sills (1952) indicated the frozen ground area of the Ladd Field power plant "to be composed of river-deposited stratified sandy gravel, the porosity and grain size of which varied erratically." The porosity, ranging from 15% to 40%, averaged 25% and the silt content had an average of 6%. The in situ thaw test site was reported by them to have produced approximately 0.4% static thaw strain for 6.1 m (20 ft.) of thaw, and dynamic ground-shaking effects generated by controlled blasting produced an average additional strain of 1.3% and thus a total thaw consolidation strain of approximately 1.7%. Terzaghi (1952) indicated "the measured value decreased from 1.8% at a depth of 3.9 m (13 ft.) below original ground surface to very small values at a depth of 9.1 m (30 ft.). He also reported relatively similar results from one test performed at the site of a large hangar at Eielson AFB. Here, maximum thaw strain near ground surface was found to be 2.5% and became negligible at depth.

CONCLUSIONS

It was recommended that the frozen silt be removed from within the entire plant area and excavation extend down into the gravel-sand stratum at least 1.52 m (5 ft.) below footing or floor slab grade, whichever resulted in a lower elevation. The proposed building bottom-of-slab elevation together with the approximate elevation of the ice-silt/gravel-sand strata interface, indicated in Figure 3. These soil strata are shown on the selected profile section in Figure 4. Given consideration of anticipated ground thawing during the construction and operational period and the influence of soil arching at greater depth, it was estimated that under static loading conditions the maximum total and the differential settlement would be on the order of approximately 25.4 mm (1 in.) and 19.1 mm (0.75 in.), respectively. Under earthquake loading conditions, it was felt that total and differential settlement would be increased by about 50.8 mm (2 in.).

Construction excavation was completed during the 1974-1975 winter season. Earthwork excavation was accomplished in the dry using dozers with ripper teeth (to break up the frozen soil) without encountering any significant difficulty or groundwater problems. Granular backfill was subsequently placed and compacted as specified to the approximate slab base elevations shown in Figure 3.

The Fairbanks wastewater treatment facility was completed in 1976 and the structure has not been reported to exhibit any evident deformation because of ground settlement; however, the potential compaction influence resulting from earthquake-induced ground shaking remains unknown.

In late 1975, construction of the septic disposal building was begun separate from the main treatment plant. This building, with footing excavations barely visible in the May 1, 1976, air photo of Figure 1, was placed at the edge but within the originally identified frozen ground segment; however, total excavation of the organic silt stratum apparently had not extended to this area.

Differential movement was noted in late 1976 and subsequent subsurface investigation Fowler (1977) revealed that some of the organic silt stratum had not been removed below footing depth within most of the building area. Differential movement of 0.12 m (0.4 ft.) was measured at that time. This building deformation was attributed to a possible combination of thawing of any frozen portions and frost heaving of unfrozen portions of the organic silt stratum found in three of the four test holes drilled in 1977 adjacent to the footings and at each corner of the 5.5 m (18 ft.) by 9.8 m (32 ft.) building. As reported in Pewe (1982), subsequent corrective measures, extension of concrete piers to bearing on the gravel stratum and placement of a horizontal blanket of insulation at the base and around the building perimeter have apparently stabilized the building movement.

REFERENCES

Fowler, J.D., 1977, Wastewater Treatment Facility, Phase II, Septic Disposal Building, Fairbanks, Alaska; Letter from R&M Consultants, Inc. to Philleo Engineering and Architectural Service, dated February 24, 1977.

Luscher, U. and Afifi, S.S., 1973, Thaw consolidation of Alaskan silts and granular soils, in Permafrost--North American Contribution to the Second International Conference, Yakutsk: Washington, D.C., National Academy of Sicences, pp. 325-334.

Pewe, T.L., 1982, Geologic hazards of the Fairbanks Area, Alaska, Alaska Geological and Geophysical Surveys Special Report 15, College, Alaska.

Philleo, E.S., 1966, Guides for engineering projects on permafrost, in Proceedings--Permafrost International Conference, NRC Publ. 1287: Washington, D.C., National Academy of Sciences.

Right of Way Lease for the Trans Alaskan Pipeline, State of Alaska, Amerada-Hess et al. March 7, 1974, Exhibit A - Stipulations for the R.O.W. Lease for the Trans Alaska Pipeline System, A-29, Para. 3.3.1.

Seed, H.B., and Idriss, I.M., 1971, Simplified procedure for ealuating soil liquefaction potential, vol. 97, No. SM9, Sept., p. 1249-1273.

Seed, H.B., and Silver, M.L., 1972, SEttlement of dry sands during earthquakes, Vol. 98. No. SM4, April, p. 361-397.

Terzaghi, K., 1952, Permafrost: Journal of the Boston Society of Civil Engineers, vol. 39, p. 36-50.

Waterhouse, R.W., and Sills, A.N., 1952, Thaw blast method prepares permafrost foundation for Alaska power plant: Civil Engineering, vol. 22, p. 126-130.

Wellman, J.W., and Nelson R.A., 1972, Soil and Foundation Investigation; Proposed Fairbanks Sewage Treatment Plant, R&M Engineering and Geological Consultants, R&M Project No. 21114.

Figure 1. Aerial Photograph of Site and Building, May 1, 1976

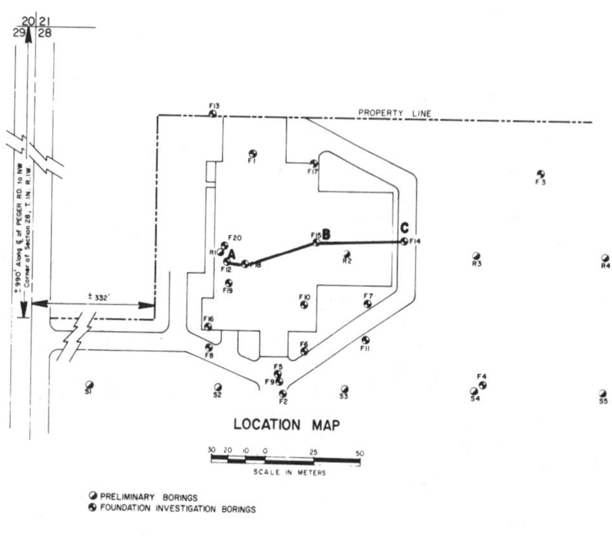

Figure 2. Site and Test Hole Location Plan

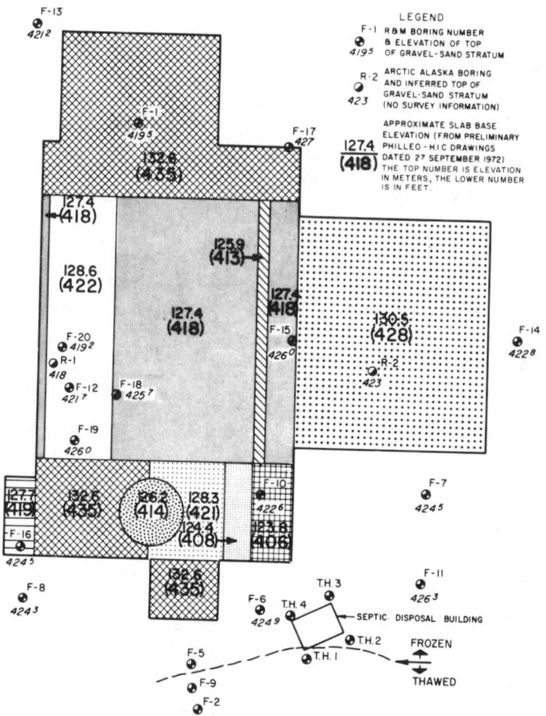

Figure 3. Approximate Building Slab Base/Top of Gravel Elevations

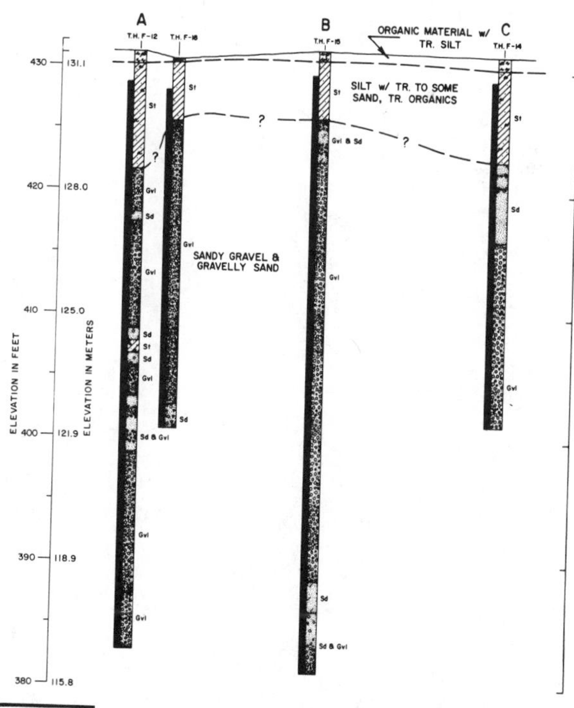

Figure 4. Soil Profile Section

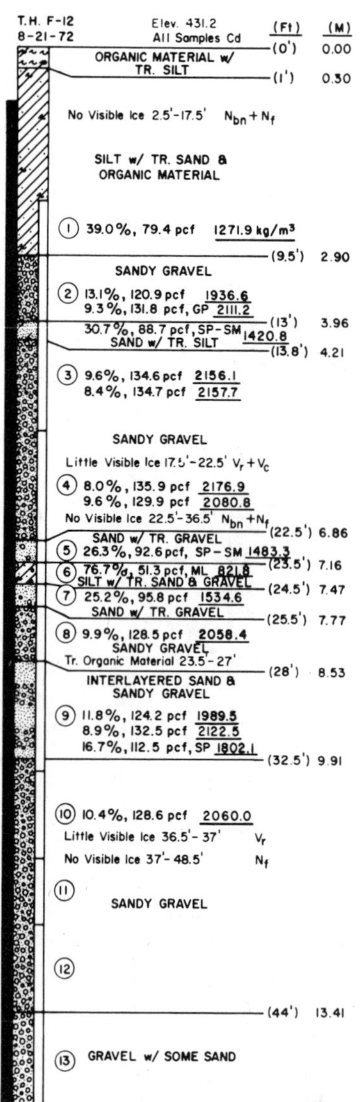

Figure 5. Test Hole F-12

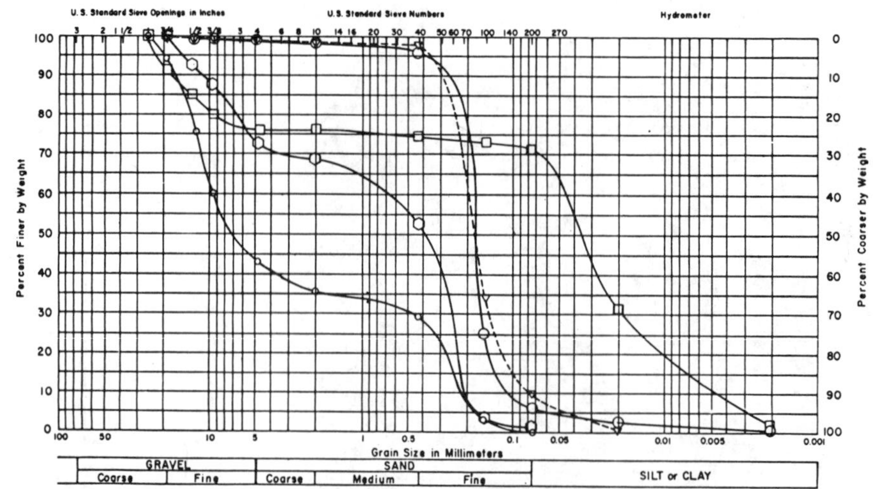

SAMPLE NO.	MOISTURE CONTENT	DRY DENSITY	LL	PI	CLASSIFICATION & DESCRIPTION	
F12-2B	9.3				Grey Sandy Gravel;	GP
F12-3A1	30.7		NV	NP	Grey Sand with Trace Gravel and SiH;	SP-SM
F12-5	26.3		NV	NP	Grey Sand with Trace Gravel and SiH;	SP-SM
F12-6	76.7		37	NP	Grey Silt with Some Sand and Gravel;	ML
F12-9B	16.7				Grey Gravelly Sand;	SP

Explanation of symbols: ∘ F12-2B, O F12-9B, □ F12-6, O F12-13A1, △ F12-5.

Figure 6. Laboratory Test Data - Test Hole F-12

TABLE II

RELATIVE DENSITY TEST RESULTS

Sample Composition

Samples F-12-4A, F-12-5, F-12-7, F-12-9A1, F-12-10, F-13-4A & F-17-13 were used for the composite sample.

Gradation of Composite Sample

100% passing #40 sieve
45% passing #60 sieve
15% passing #80 sieve
6% passing #100 sieve
0% passing #200 sieve

Minimum Density Results
Trial 1: 1393.6 kg/m3
Trial 2: 1392.0 kg/m3
(Use 1393.6 kg/m3)

Maximum Density Results
Trial 1 (dry): 1656.3 kg/m3
Trial 2 (dry): 1664.3 kg/m3
Trial 2 (wet): 1672.3 kg/m3
(Use 1665.9 kg/m3)

Relative Density of In-Situ Soil Samples:

Sample No.	Dry Density (kg/m3)	Relative Density (%)
F-1-7	1507.3	46
F-1-13	1565.0	67
F-12-3	1420.8	12
F-12-5	1483.3	37
F-14-4	1545.8	60
F-15-1	1508.9	47
F-15-11	1468.9	27
F-16-8	1525.0	53
F-17-3	1446.5	22
F-17-6	1552.2	63
F-17-10	1480.1	34
Average	1550.9	44

TABLE I

LABORATORY TEST DATA-THAW CONSOLIDATION

Boring and Sample Nos.	Depth m	Frozen Phase: Dry Dens kg/m3	Frozen Phase: Moist Cont. %	Frozen Phase: Vol. Ice %	Frozen Phase: Void Ratio	Frozen Phase: Sat %	Thawed Phase: Dry Dens kg/m3	Thawed Phase: Moist Cont. %	Thawed Phase: Void Ratio	Thawed Phase: Sat %	Exc Ice %	Sett %	SP GR	Uni Class	Ice Mode
1-07	4.27-5.03	1414.4	29.0	41.0	0.907	86.3	1436.9	28.6	0.878	88.0	0.5	1.5	2.70	SP	VC NBN
12-05	6.86-7.16	1380.8	29.2	40.4	0.939	83.3	1392.0	27.9	0.925	81.0	1.8	0.7	2.68	SP SM	NBN
12-06	7.16-7.47	1065.2	41.9	44.7	1.466	75.2	1105.3	41.1	1.379	78.3	0.9	3.5	2.63	ML	NBN VC
14-04	3.20-4.27	1539.4	23.3	36.0	0.759	83.3	1544.2	21.9	0.755	78.7	2.2	0.2	2.71	SP	NBN
15-11	13.26-14.02	1403.2	28.4	39.9	0.945	82.1	1416.0	26.3	0.927	77.4	3.0	0.9	2.73	SP	NBN
16-08	7.77-8.08	1286.3	25.0	32.2	1.075	62.2	1326.3	23.6	1.013	62.1	1.9	3.0	2.67	SP	VC NBN
17-06	5.49-5.79	1440.1	24.8	35.7	0.832	78.7	1449.7	24.0	0.819	77.2	1.2	0.7	2.64	SP SM	NBN
17-10	8.08-8.53	1457.7	26.9	39.3	0.817	87.3	1460.9	25.5	0.813	83.0	2.1	0.2	2.65	SP	NBN

ACTIVE LAYER ENERGY EXCHANGE IN WET AND DRY TUNDRA OF THE HUDSON BAY LOWLANDS

Wayne R. Rouse

Department of Geography, McMaster University, Hamilton, Ontario, Canada L8S 4K1

Results of weekly calculations of ground heat flux for Churchill, Manitoba, made from September 1981 to December 1982 are presented. Measurement sites included a variety of well-drained upland, semiupland and wet lowland locations, a forested site, and an ice-cored palsa.

Net radiation, Q^*, was measured at representative sites in summer and at one site in winter with the latter measurement being applied with corrections to all sites. The ground heat flux, Q_G, was divided into sensible and latent heat components and was determined from soil moisture and temperature measurements.

Snow depths were shallow on open tundra sites and deep at the forested and a forest-influenced site. As a result, final snowmelt at the latter sites occurred 7 weeks later than at the former ones. Q^* was similar for all sites with a maximum variation of 14% from the all-site average. Active layer depths on upland sites were double those in the wet lowland. During the thaw season, Q_G was almost the same for all sites and averaged 12% of Q^*. Q_G during freezeback fully compensated for the net radiational heat loss in winter.

The similarity in Q_G for such a heterogeneous groups of sites is explained in terms of thermal conductivities and vertical temperature gradients. The results are seen as having some general applicability and show that Q_G is an important component of the surface energy balance.

INTRODUCTION

The energy expended in melting the active layer in permafrost terrain is important in determining the depth of the active layer, in determining how much ice in the active layer will be released to take part in hydrologic processes, and in influencing differences in frost table depths, which will determine the direction of subsurface water flow. The degree of thawing and the release of water is important in thermokarst activity. The magnitude of ground heat storage is important in calculating how much energy remains for other processes, such as evaporation and sensible heating of the air. At least some of the energy absorbed during the thaw season will be released during freezeback and will ameliorate surface heat loss and be a thermal input to the base of a snowpack.

The ground heat flux, Q_G, might be expected to be variable according to terrain type, groundwater content, and the depth and longevity of snow cover. For example, in the Churchill area, an early study by Brown-Beckel (1957) indicated that active layers in different terrain types could vary from 55 to 335 cm in nonrocky soils. The depth of winter snow cover has been shown to strongly affect active layer depths by a number of researchers as summarized by Outcalt and Brown (1977). At snow depths and snowpack longevity are highly variable depending on vegetation type and topography is well known and is documented for the high arctic at Resolute by Woo and Marsh (1978) and in the subarctic at Schefferville by Adams and Roulet (1982).

There are a number of ways that Q_G can be measured or estimated. In addition to direct measurement with heat flux plates, which are not reliable in permafrost terrain, there is the vertical gradient approach in which

$$Q_G = k \frac{\Delta T}{\Delta z} \quad (1)$$

where k is the thermal conductivity and $\Delta T/\Delta z$ is a time-averaged vertical temperature gradient in the soil. Inherent problems in the practical application of this approach for long-term estimates involve experimental difficulties in determining k and the fact that it varies over time with changing soil ice and water content.

An alternate and preferred approach is to calculate the heat storage over time as

$$Q_G = C(\Delta T/\Delta t)\Delta z + L_f , \quad (2)$$

in which L_f is the latent heat of fusion involved in the thawing and freezing of soil water, C is soil heat capacity, and ΔT is the temperature change over the time period Δt averaged for a depth increment Δz, which ideally operates between the soil surface and the depth to the base of the annual thermocline. In equation (2), L_f can be determined from the volume and temperature of soil water (ice), and $\Delta T/\Delta t$ can be measured with few technical problems. C, however, must be divided into its component parts in which

$$C = C_s X_s + C_w X_w + C_i X_i , \quad (3)$$

where C is heat capacity, X is volume fraction of the soil, and s, w, and i refer to soil solids, water, and ice, respectively. C_s for sands, clay, and peat of varying density will be constant for a given site and horizon and can be obtained from the determinations of Kersten (1949). X_s can be determined by mechanical analysis or from $[1 - X_w(SAT)]$, where $X_w(SAT)$ is the saturated volumetric water content. It too is constant for a given site and horizon. X_w and X_i are the only important variables in equation (3) and must be measured. It is also important that one is able to differentiate between water and ice because C_w is more than twice the magnitude of C_i. It should be noted that because soil air has such a small heat capacity it can be ignored in equation (3).

This study employs equations (2) and (3) to determine the ground heat storage at eight sites of very different characteristics in the Hudson Bay lowlands near Churchill, Manitoba and relates this storage to measured and estimated net radiation at these sites. The study period extended from September 1981 to December 1982 and calculations are presented for weekly periods.

METHODS

Site

Churchill (lat. 58.75°N, long. 94.08°W) is located on the southwest coast of Hudson Bay in an area that possesses a tundra landscape and a climate similar in many respects to low arctic areas across the continent. Hudson Bay is ice-locked along this coast from mid-November to mid-June, and drifting ice frequently lingers offshore until the end of July. Thus in spite of its relatively low latitude the terrestrial snow cover persists on average until late May and cold air temperatures until July. The vegetation consists of short grass, shrubs, and ground lichen on upland raised beach systems, thick sedge grass growing on surface peat in wet lowland areas, which constitute the bulk of the landscape, and scattered open forest development, especially on upland ridges. This forest development, although similar morphologically to the tree line, represents an outlier since further inland and to the south the trees disappear until the main tree line is encountered. Soils range from deep sands underlain by clay in the beach ridges to clays overlain by peat, which can be up to 0.5 m thick in the low-lying areas.

The research reported in this paper was undertaken at a location 2 km inland from Hudson Bay at eight select sites along two separate transects as shown in Figure 1. Table 1 gives their characteristics. Transect 1 was located exclusively in open tundra. Site 1A, located on the upland beach system, consisted of deep sands down to 2 m, which represents the terminal depth of measurement for all sites. It has a 5 cm organic surface layer overlain by a dominant short grass vegetation. This terrain is well-drained and there is no surface runoff even during the heaviest rains. Site 1B is on the side slope of the beach ridge and consists of a 10 cm organic horizon overlying sand to 90 cm, which in turn gives way to clay down to the terminal depth. Vegetation consists of grasses, lichens, shrubs and mosses. Site 1B never has surface water but as thawing progresses, significant ponding of water above the frost table occurs. Sites 1C and 1D are similar except the latter has water at the surface throughout the summer, whereas the former undergoes some surface drying between rainstorms. Both sites have a 25 cm peat layer directly overlying clay. Vegetation is almost exclusively a thick sedge grass.

Transect 2 has more variety in its landscape features. 2A is located in an open forest on the upland ridge. The soils are similar to 1A except the surface organic layer is a little thicker. The trees are stunted black spruce with an average height of 6 m and an average trunk spacing of 2 m. Understory vegetation consists of shrubs and assorted lichen species. Site 2A acts as a snow trap accumulating a deep snowpack by late winter, which persists past the middle of June. Because soils

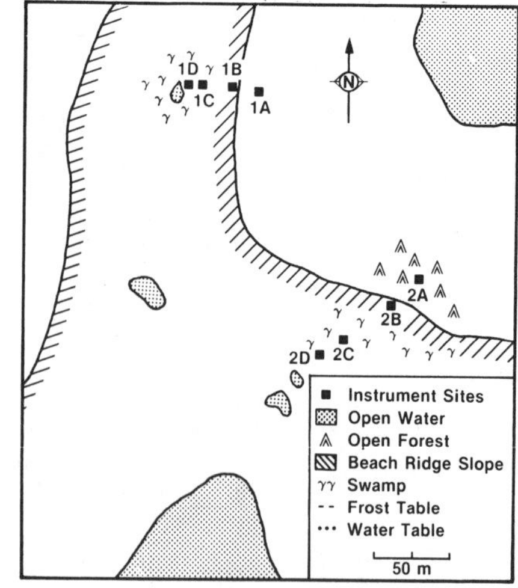

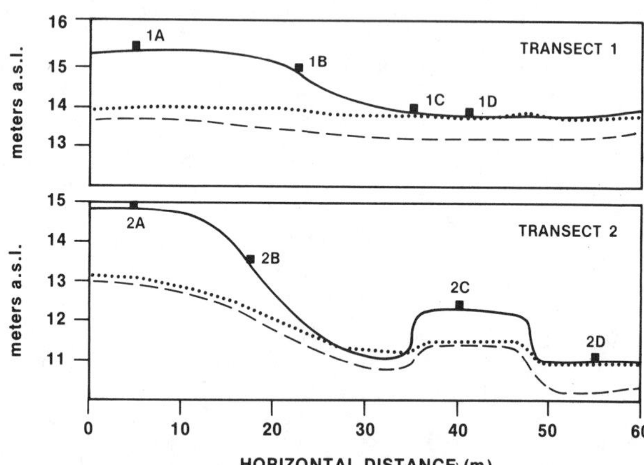

FIGURE 1 Research sites with a plan view of the basin and cross-sections of transects 1 and 2 showing the maximum active layer depth and the depth of the suprapermafrost groundwater.

thaw as the snowpack ripens, there is little surface runoff and the soils are saturated at final snowmelt. Site 2B on the south-facing side slope is similar to 2A except that the layer of sandy soil is thinner and there are no trees. The effect of the forest is, however, important. By virtue of braking the speed of the winds, snow is deposited early and deep at this site. It is, in fact, only after a deep snowpack has developed that buildup proceeds at site 2A within the forest. The deep and longlasting snow cover has important effects in keeping soil warm and in reducing the yearly total of net radiation. Site 2C is a palsa mound and was chosen for its uniqueness. It rises about 1 m above the surrounding wetlands, averages 10 m in diameter, has similar surface vegetation to site 1A, has a deep insulating peat layer, and boasts a solid ice core throughout the summer. In winter the snow cover was late developing in contrast to other sites.

For purposes of presentation the soils are grouped into pairs having similar soil topographic characteristics: 1A-2A, 1B-2B, 1C-2C, 1D-2D. It is noted that while 1C and 2C have similar soil profiles, their topographies are very different. The sands and clays both have bulk densities averaging about 1.6 x 10^3 kg/m^3 and porosities of .35 and

.40, respectively. The peats range in density from 0.2×10^3 at the surface to 0.4×10^3 at the peat-clay interface. Their porosities likewise range from 0.9 to 0.6.

The eight sites studied represent a very heterogeneous group of landscape types that are typical of those found elsewhere in the Hudson Bay lowlands, in central Keewatin, and in the Mackenzie Valley.

Measurement

Net radiation was measured with net pyrradiometers at sites 1A, 2A, 1C and 1D at 1 m above the ground and at site 2A at 1 m above the forest. The instruments were recently calibrated and should be accurate to within ±10%. The measurement period at sites was from late April to early September. In the winter period, measurements were made at one site similar to 1A and applied to all other sites except 2A. At 2A estimates were made according to a regression of net radiation on solar radiation derived for the April early May period in 1982 and 1979 and applied to the months of February, March and April, when there is substantial tree canopy absorption of solar radiation leading to larger Q^*. As a result, yearly totals of Q^* are larger at site 2A than at other sites. The summer measurements at site 1A were applied to 1B and 2B, those at 1C to 2C, and those at 1D to 2D.

The most important variable in equations (2) and (3) is the soil water and soil ice term as it is involved in both the sensible heat exchange in the soil Q_{SH} and the latent heat exchange Q_{LF}. Soil moisture was measured with a neutron probe at increments of 20 cm replicated at each site. The probe was calibrated for sands, clays and peats using gravimetric comparison. Since the neutron probe measures the H ion concentration in the soil, it cannot differentiate between liquid water and ice, which is a drawback to this measurement method. Soil temperature readings made it apparent when the main zero curtain effect ended, as rapid cooling or warming ensued. However, an assumption was made that at -5°C all soil water was frozen and that at 1°C it was all thawed. Since at most sites soil temperatures at maximum active layer depth were at least -10°C, total freezing is probably a safe assumption. However, at 2B, the warm soil hovered around -5°C depending on depth and this could introduce error into the calculations.

Soil temperatures were measured with thermisters at depths of 5, 10, 25, 50, 100, 125, 150 and 175 cm at the sites with maximum active layer depths and from 100 to 125 cm depths at the other sites. Measurements of soil temperature were made daily from April through August and every second week during the freezeback period and the winter. Soil moisture was measured every three days during thaw and every two weeks in the winter.

Air temperature was measured continuously with an aspirated thermistor in summer and taken from Environment Canada measurements at the Churchill airport in winter. Net radiation and temperature data were integrated for half-hour periods in a digital recorder and stored on magnetic tape for computer analysis at McMaster University.

Snow depths were measured at the individual sites with a meter stick and during the thaw period a grid along transects 1 and 2 was sampled regularly. The maximum active layer depths, groundwater depths, and topographic profiles shown in Figure 1, were obtained during an intensive field survey during a field course for senior students.

RESULTS

The mean monthly temperatures shown in Figure 2 indicate 5 months of thaw temperatures peaking in July-August around 11°C and 6 months of freezing temperatures reaching a minimum of -24°C in January. For the period of measurement the annual air temperature was -8.5°C. October is clearly a transitional season, with temperatures hovering around freezing; the winter period from November to April is clearly defined, as is the summer period May through September. Final snowmelt on the tundra was 3 weeks earlier than the long-term average, occurring at the end of April rather than near the end of May. Final melt in the forest in the third week in June is normal.

Rainfall, shown in Figure 3, occurred in October 1981 but was limited to the period May through September 1982. The yearly total was within 10% of the long-term average.

Snow depths are highly variable at different sites in normal fashion for the tundra. Table 1 indicates that the open tundra sites along transect 1 along with the palsa at site 2C accumulated the least snow as a result of wind erosion. This snow was deposited at the forested sites 2A and 2B, where depths in excess of 1 m were common. Site 2B was particularly notable for substantial snowpack development in early winter, and the effects on the ground temperature regime are evident. The snow at the forested sites persisted five weeks longer than at the tundra sites. Site 2D served as a minor accumulation area, with snow depths double those found at adjacent site 2C.

Seasonal net radiation totals are given in Table 1. The maximum yearly departure from the mean-measured Q^* for all sites was 14%. The forested site 2A has the largest net radiation, the result of a smaller albedo in winter, particularly during the high sun period in late winter. The minimum Q^* at site 2B resulted from the late-lying snowpack with its high albedo. Q^* measurements at the other sites are within 7% of each other and differences are largely accountable to different surface albedos. Positive net radiation at all sites except 2A begins in mid-March and persists into mid-October. For site 2A it begins about 5 weeks earlier due to strong canopy absorption of solar radiation, which is the topic of a separate report. At all sites except 2A and 2B there is an abrupt increase in Q^* that accompanies final snowmelt, and maximum values are reached in late June/early July. The negative Q^* in winter does not exceed 11% of the positive summer value and is substantially less for the forest.

As indicated in Figure 2 and Table 1, soil temperatures and active layer depths are highly variable between sites. The deepest active layers are achieved in the upland sandy soils at sites 1A and 2A and these are at least twice the depth of the wetter sites 1C, 1D, 2D, and the palsa at site 2C. The side slopes at 1B and 2B have intermediate depths, the latter being the deeper presumably due to its south slope exposure. The soils at site 2B have a longer thaw season due to early snow accumulation, which favours warm winter soil temperatures

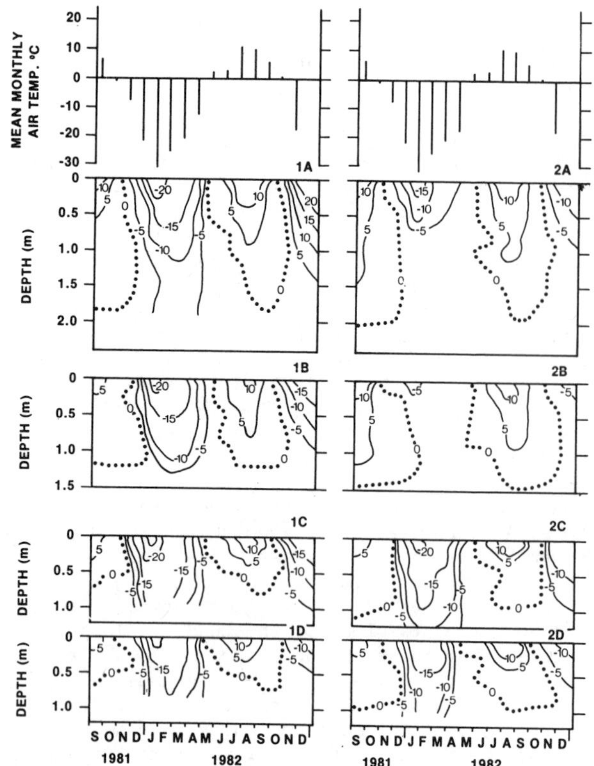

FIGURE 2 Monthly mean air temperature and the soil temperature at each site. Isotherms are at 5°C intervals. The dotted line denotes the position of the frost table.

and a prolonged zero curtain effect. At a depth of 0.5 m the coldest soils are those of the palsa at site 2C, and the warmest on an annual basis are at site 2B. The upland tundra site 1A has the largest annual temperature range.

Soil moisture is strongly influenced by soil type and topography and these are closely interrelated. The upland sites at 1A and 2A have gravel soils, and the lowland sites at 1D and 2D clay soils overlain by a thick layer of peat. Other sites are transitional between these extremes. Site 1A and 2A are the driest soils and also undergo the largest soil water loss from early to later summer. Sites 1D and 2D are the wettest sites with up to 90% soil moisture by volume in the surface organic layer, and they also show the least seasonal variability. The palsa site 2C is distinctive because the thick surface peat layer undergoes substantial summer drying.

The ground heat storage Q_G shown in Table 1 is subdivided into its components of the latent heat flux Q_{LF} and sensible heat component Q_{SH}. The following generalizations can be made. The wettest sites have the smallest latent heat exchange, which arises because they have the shallowest active layers and the largest temperature variation in the ice-rich near-surface permafrost. During the full thaw season Q_G makes up 10 to 13% the magnitude of Q^*. Early in the thaw season when both fluxes are large, the magnitude of Q_G is up to one-half that of Q^*. Total Q_G for all sites during the thaw season is very similar, with a maximum variability of 15% about the mean. The heat release during

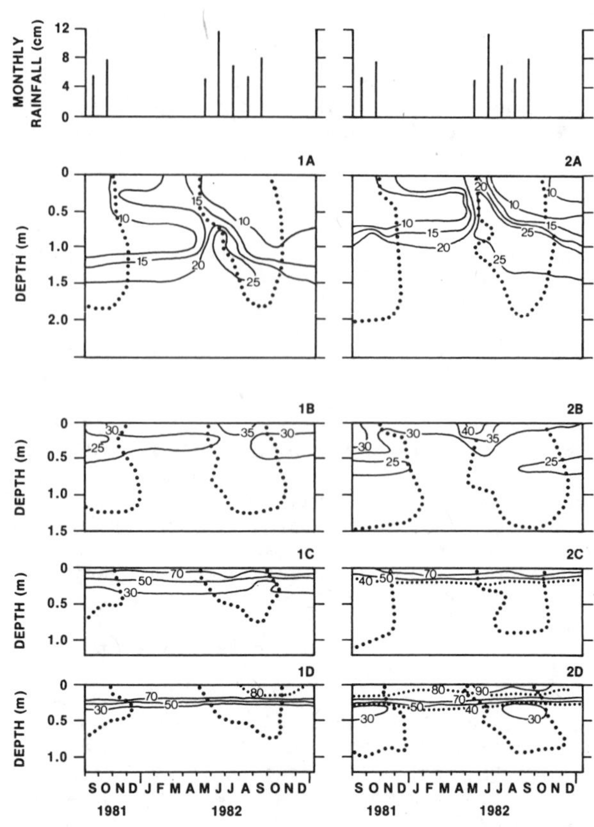

FIGURE 3 Total monthly rainfall and the volumetric soil moisture at each site. The isopleth interval is .05 in 1A, 2A, 1B, and 2B and .20 in 1C, 2C, 1D, and 2D, with fine dotted isopleths at nonregular intervals of 0.10 in the latter group.

freezeback is generally somewhat less than during thaw because, at least for this particular measurement period, the soils were drier in the fall than in the spring and the latent heat release was smaller. In terms of negative Q^* in winter, the heat released in the form of Q_G is equal to the radiational heat loss.

DISCUSSION

The most important results of this experiment are the large magnitude of Q_G in terms of Q^* and the uniformity of Q_G between sites. These sites are extremely varied in terms of vegetation type, soil type, soil moisture content, depth of active layer development, and depth and longevity of the snowpack.

In terms of the vertical heat transfer processes presented in equation (1) the two variables are the soil thermal conductivity k and the vertical temperature gradient $\Delta T/\Delta z$. The variable k is large for sands when they are even moderately wet and large for ice. It is moderate for wet clays and small for organic soils and for water and hence for saturated organic soils.

In early spring, k is large because the ground is frozen and $\Delta T/\Delta z$ across the atmospheric soil interface is large, so Q_G proceeds very rapidly,

TABLE 1 Characteristics and Energy Exchanges in the Eight Sites

	1A	2A	1B	2B	1C	2C	1D	2D
Soil type & depth(cm) P-Peat; S-Sand; C-Clay	P 0-5 S 6-200	P 0-10 S 11-120 C121-180	P 0-10 S 11-90 C 91-200	P 0-20 S 21-60 C 61-200	P 0-25 C 26-200	P 0-40 C 41-200	P 0-25 C 26-200	P 0-30 C 31-200
X_w(max)	S = .27	S = .28 C = .26	P = .37 S = .32 C = .34	P = .41 S = .30 C = .28	P = .81 C = .35	P = .79 C = .43	P = .88 C = .39	P = .91 C = .34
x_w(min)	S = .08	S = .06 C = .21	P = .28 S = .32 C = .34	P = .17 S = .23 C = .24	P = .66 C = .25	P = .47 C = .34	P = .64 C = .25	P = .79 C = .27
T_s(max) - 50 cm	10.0	9.2	6.2	7.0	3.1	1.0	1.2	4.1
T_s(min) - 50 cm	-16.4	-10.7	-15.3	-2.2	-16.4	-16.7	-15.0	-12.9
T_s(max-min) - 50 cm	26.4	19.9	21.5	9.2	19.5	17.7	16.2	17.0
z(AL)(cm)	171	182	125	144	61	92	58	80
z_s(max)(cm)	12	140	22	130	12	15	17	25
Q_G(TH)	148.8	157.8	184.8	146.0	134.2	184.1	151.7	182.7
Q_{LF}(TH)	113.5	137.9	155.2	127.0	80.4	131.7	100.6	122.4
Q_{LF}/QG(TH)	.76	.87	.84	.86	.59	.71	.66	.67
Q*(TH)	1414.3	1461.2	1414.3	1255.8	1329.7	1329.7	1379.7	1379.7
Q_G/Q*(TH)	.10	.10	.13	.11	.10	.13	.10	.13
Q_G(FB)	-117.9	-69.4	-176.0	-149.3	-128.2	-158.6	-124.4	-142.8
Q_{LF}(FB)	-77.6	-46.1	-152.3	-120.7	-71.5	-105.7	-68.9	-82.6
Q_{LF}/QG(FB)	.66	.67	.71	.81	.56	.67	.55	.58
Q*(TH)	-145.3	-39.1	-145.3	-133.0	-147.0	-147.0	-144.6	-144.6
Q_G/Q*(FB)	.81	1.77	1.21	1.12	.87	1.08	.86	.99
Q*(TH+FB)	1269.0	1422.1	1269.0	1122.8	1182.7	1182.7	1235.1	1235.1
Q*(FB)/Q*(TH)	-.10	-.03	-.10	-.11	-.11	-.11	-.11	-.11
Q_G(TH+FB)	30.9	88.5	8.8	-3.3	6.0	25.5	27.3	39.9

Note: X_w is volumetric soil moisture; T_s is soil temperature; z(AL) is active layer depth; z_s is snow depth; Q_G is total soil heat flux; Q_{LF} is latent heat flux; Q* is net radiation; TH is the thaw period; FB the freeze period. Fluxes are in MJ m^{-2} for the period.

particularly in the ice-rich soils such as 1C, 1D, and 2D. Even beneath the snowpack (such as at sites 2A and 2B) thawing proceeds, since the snow is at 0°C and the frozen soil is < 0°C and has a large thermal conductivity, particularly when snowmelt refreezes in the top soil layers.

In terms of uniformity in the total heat flux between dry upland sites with a deep active layer and wet lowland sites with a shallow active layer it is useful to compare 1A and 1D. Table 2 shows that in May, Q_G at 1D is substantially larger than at 1A, in June they are about the same, and in July and August the heat flux at 1D is considerably less than at 1A. This can be interpreted in terms of the thermal conductivities that are largest at 1D early in the season when there is ice in the surface soil layers but become smaller than at 1A later in the season because of the smaller k of the saturated surface peat layer.

It is instructive to compare sites 1A and 2A. Although a thick snow layer dominates 2A until mid-June, the soil heat flux is greater than at 1A in both May and June and the forest soils are thawed to 1.3 m by final snowmelt. This rapid thawing is a response to large thermal conductivity in the moist sand and to the fact that the soils at depth stay relatively warm throughout the winter. In July and August, Q_G at 2A is less than at 1A in spite of larger thermal conductivity because the vertical temperature gradients are less.

It is clear from Table 1 that neither the actual soil temperature nor the range of temperature change in the soil bears a relationship to the magnitude of the heat flux. Thus 2B, which is much warmer in winter than 1A, has the same magnitude of heat flux during the thaw season.

The soil stores summertime heat, which constitutes at least 10% of the net radiation flux. The release of this heat maintains relatively warm soils until the end of December and, in the event of a deep snow cover (as at 2B), throughout the winter season.

The results of this study should have more general applicability, since the varied terrain types studied are common in the Hudson Bay lowlands and in many areas of central Keewatin and the Mackenzie District. They are important to assessing evaporation using energy budget approaches, which often ignore Q_G in temperate latitude studies, but which cannot ignore it in permafrost areas. It is evident that warm soils can be achieved through effective thermal insulation of the surface in

TABLE 2 A Comparison of Sites 1A, 1D, and 2A During the Main Thaw Period

	1A				
Month	Q^*	Q_G	Q_{LF}	$\frac{Q_G}{Q^*}$	k
MAY	281.9	40.9	27.3	.15	1.00
JUN	423.7	45.9	37.5	.11	.8
JUL	383.5	32.9	29.1	.09	.7
AUG	323.3	16.4	19.6	.05	.6
	1D				
Month	Q^*	Q_G	Q_{LF}	$\frac{Q_G}{Q^*}$	k
MAY	277.4	64.0	49.6	.23	1.7
JUN	410.1	40.7	34.5	.10	.6
JUL	364.7	15.2	7.7	.06	.5
AUG	325.5	9.6	8.8	.03	.5
	2A				
Month	Q^*	Q_G	Q_{LF}	$\frac{Q_G}{Q^*}$	k
MAY	213.8	48.6	43.3	.23	1.8
JUN	399.1	63.9	54.6	.16	1.2
JUL	397.0	28.0	24.5	.07	1.0
AUG	331.8	14.4	15.6	.04	.7

Note: Q^* is net radiation; Q_G is soil heat flux; Q_{LF} is soil latent heat flux. flux values in MJ m^{-2} $month^{-1}$; k is thermal conductivity through the top 25 cm of soil in $Wm^{-1}{}^\circ C^{-1}$.

early fall, which would be at the end of August in the case of Churchill. This results in an augmentation of the thermal energy in the soil of at least 10% of the summer net radiation, as at site 2B.

CONCLUSIONS

As a result of much lesser snow depths final snowmelt on the open tundra preceded that in the forest by 7 weeks. However, snowmelt on the tundra was 3 weeks earlier than average. All other conditions were close to normal.

Net radiation was largest in the forest due to small albedos, especially in winter, and smallest at site 2B, where the longer-persisting snowpack maintained a large surface albedo into the summer solstice period. However, the maximum variation at any site was within 14% of the average. The total negative net radiation in winter did not exceed 11% of the positive summer total at any site.

Active layer depths in the sandy upland sites were double those of the lowland sites and, where snow accumulated early and stayed late, the soil presented the longest thaw season and warmest overall temperatures. Soil moisture was least on the upland sandy sites and also showed substantial summer drying, whereas in the wet sites it varied little over time and measured as large as 90% by volume in the surface low-density peat soils. The palsa was anomalous in many respects, having the coldest soils and also undergoing considerable drying in the surface peat layer.

During the thaw season, the ground heat storage was almost the same for all sites and averaged 12% of net radiation with site variations from 10 to 13%. Thus it is an important component in the energy balance at the surface. During freezeback, the heat energy released from the soil almost exactly compensates for winter season radiational heat loss associated with negative net radiation. The soil thus acts as a significant thermal regulator.

The similarity in total heat flux for such a variety of terrain types can be explained in terms of thermal conductivities and vertical temperature gradients and the large thermal conductivity of ice. The ice-rich wet sites thaw rapidly in spring due to large vertical temperature gradients and the large thermal conductivity of ice. However, with surface thawing the conductivity becomes that of saturated peat, which is not large, and the melting process is slowed, leading to shallow active layers. In contrast, the well-drained upland sandy soils remain moderately moist with less variable thermal conductivities and a more even thaw rate over the season. Because there is less ice to melt per unit volume, the active layers are deeper. Where the snow cover lies late in the forest-affected areas, thaw starts when the snow is still deep, because a temperature gradient between warm snowpack and colder soils exists and the frozen soils have large thermal conductivities.

The results of this study are seen as having more general applicability for wet or moderately wet low arctic terrain, since the processes that favour a large and relatively uniform ground heat flux are universal.

ACKNOWLEDGMENTS

This research has been continuously supported by the National Science and Engineering Research Council of Canada.

REFERENCES

Adams, W.P. and N.T. Roulet, 1982. Areal differentiation of land and lake snowcover in a small subarctic drainage basin. Nordic Hydrology 13, 139-156.

Brown-Beckel, D.K., 1957. Studies on seasonal changes in the temperature gradient of the active layer of soils at Fort Churchill, Manitoba. Arctic 10, 151-183.

Kersten, M.S., 1949. Laboratory research for the determination of the thermal properties of soils. U.S. Dept. Commerce, National Technical Information Serivce AD 712 516, 226 pages.

Outcalt, S.I. and J. Brown, 1977. Computer modeling of terrain modifications in the Arctic and Subarctic. In Symposium: Geography of Polar Countries, J. Brown (ed.) USACRREL, Special Report 77-6, 24-31.

Woo, M.K. and P. March, 1978. Analysis of error in the determination of snow storage for small high Arctic basins. J. App. Meteorol. 17, 1537-1543.

DESIGN CONSIDERATIONS FOR LARGE STORAGE TANKS IN PERMAFROST AREAS

W. L. Ryan

Ott Water Engineers, Anchorage, Alaska 99503 USA

Large water and/or oil storage tanks (50,000 gal. or more) are usually necessary for camps, sites, or communities in the Arctic. The design, construction, and operational problems for over 40 large water storage tanks in arctic and subarctic Alaska are reviewed and summarized. The majority of the difficulties encountered with the tanks fall into three areas: foundations, interior painting, and outside insulation systems. Foundation settlement or heaving has proven to be the most widespread problem with large storage tanks. Unfrozen lenses, "warm" permafrost, and incorrect adfreeze strengths have all contributed to the several instances of settlement. Piling and gravel pads have both received widespread use, depending on local soil conditions and the availability of equipment and materials. Another very common problem is the failure of the interior painting systems. These failures have occurred with both the vinyl and epoxy paint systems and tend to be caused by the lack of adequate curing temperatures, inadequate ventilation during curing, and the individual layers being applied too thick.

INTRODUCTION

Large water and oil storage tanks are necessary for camps, sites, or communities in the Arctic. Water storage tanks have several characteristics which make them more difficult to design and maintain than oil storage tanks and, thus, will be emphasized in this paper. The majority of the difficulties encountered with the large storage tanks fall into three areas: foundations, interior painting, and outside insulation systems.

The majority of the tanks constructed over the past 10 years have been welded steel tanks with sprayed polyurethane foam insulation. Occasionally, "bladder" tanks and wood stave tanks with strapped-on boardstock insulation have also been used. Table 1 presents a summary of many of the major water storage tanks that have been constructed in the cold regions of Alaska. The problems (if any) that have been experienced with them are included, along with the construction costs of tanks constructed prior to 1980 adjusted to 1980 costs considering 7% per year inflation.

FOUNDATIONS

Foundation design is heavily dependent on the soil conditions at the site. Piling and gravel pads comprise most of the foundations utilized. The foundations have proven to be the most widespread problem encountered with large storage tanks in Alaska. The type of foundation used depends on several factors: permafrost presence (and it's temperature) and soil properties; availability of construction materials such as gravel or piling; availability of equipment for construction; and the size and shape of the tank.

The presence of permafrost can simplify or complicate a foundation design. Due to the excessive weight that must be supported, hard-frozen, "cold" permafrost (less than -2°C) can best provide the support needed when using a pile foundation for tanks. The load carrying capacity of a pile placed in cold permafrost is considerably greater than that of one in unfrozen ground, which must essentially rely on friction and/or end-bearing. In fact, in warmer soils (nonpermafrost), it is often not feasible to place sufficient piling under the tank to support the weight considering friction and/or end-bearing alone. In the absence of permafrost, a gravel pad or post and pad foundation is usually designed. In cold permafrost locations, piles are usually used unless large quantities of gravel are readily available. If sufficient gravel is available (a rare occurrence in much of Alaska's Arctic), a pad can be designed with sufficient thickness to prevent thawing of the underlying frozen ground due to heat lost from the tank. Insulation can be added to the pad to reduce the amount of gravel required. Only extruded polystyrene insulation should be used in tank foundations because it is the most resistant to deterioration by moisture.

The most difficult design situation is when "warm" permafrost (above -2°C) or permafrost with thawed "pockets" (talik) is encountered. Warm permafrost usually occurs near the southern limits of the continuous permafrost areas. Talik areas can occur in any location containing frozen ground but are most abundant in coastal areas, where they are often caused by a concentration of salts. This accentuates the importance of a detailed test drilling program. Test drilling must be carried out before foundations are designed to determine if thawed pockets do, in fact, exist. Piles having a portion of their length in the thawed area may not provide the support needed and settlement will result. The water storage tank location in Wainwright, Alaska, encountered this condition. It may have also been a contributing factor to the settling problems experienced with tanks in Barrow and Kaktovik, Alaska. In warm permafrost areas, the active layer is also deeper and longer piles will be required to obtain the same adfreeze strength for support.

Any construction activity at the ground surface can have a lasting effect on the depth of the active layer. This must be considered during pile design. The construction should be carefully carried out using mats or pads to protect the surface moss and to retain its insulating value. Any degradation of the permafrost can cause severe settling problems in high moisture content situations. Subsequent degradation can also be caused by heat lost through the tank bottom. In warm permafrost areas, it is even more important to insulate the permafrost from this heat or to provide air circulation under the tank to carry it away to the atmosphere. This is usually best accomplished by placing the tank on piling 1 or 2 m above the ground, but it can also be accomplished by a thick, insulated gravel pad containing pipes or culverts for the circulation of air. These pipes can be fitted with blowers which can force circulation through them in the winter, but they must be deactivated and the culverts plugged to prevent circulation during the summer.

There are other solutions which are not in wide use because of the high construction and/or operation and maintenance costs. Two examples of other methods which have been used to remove this heat before it melts the underlying frozen ground are: refrigeration lines laid directly in the pad, and forced air or self-acting thermo-piles.

The method of pile placement is critical. The primary objective is to place the pile in the ground with the least amount of thermal disturbance to the permafrost. The preferred method is to slightly underdrill a hole and drive the pile. The disadvantage of this method is that it requires pile driving equipment. Thus, most piles are placed using a drill-slurry back method. However, if not done carefully, this method can thaw the permafrost immediately around the hole and significantly extend the period of time needed for freezeback. The piling can support only a small portion of the design load until the surrounding ground is frozen back and has dropped to the design temperature. In warm permafrost situations, this could take an entire winter or, more than one winter, depending on the thermal disturbance. With the drill-slurry back construction method, it is often necessary to mechanically freeze back the pile. It is essential that the temperature at various depths be determined prior to design and, if possible, at the time the ground is warmest. In calculating adfreeze bond strengths, differences in temperature from the top of the hole to the bottom should be integrated and not based only on an average temperature. The relationship between adfreeze strength and temperature is not linear.

Frost jacking of the piling in a tank foundation is not usually a problem because of the high loads. However, the design of a pile foundation must consider frost jacking prevention because it could be a problem if a tank should remain empty for a period of time. Frost heave can also create problems with the outer pile of a tank foundation which are usually only supporting a walkway that is not heavy enough to prevent the jacking.

Any coating, such as creosote, can significantly reduce the adfreeze strength and thus the load supporting capacity of a pile. This was a major contributing factor to the settlement problems encountered with a 2,160,000 liter water storage tank constructed on piles in Barrow, Alaska. The use of a bond breaking material on the pile in the region of the active layer should be encouraged to prevent frost jacking. However, it must not extend into the permafrost, or the load supporting area of the pile.

The active method of tank foundation construction will not be discussed at great length here, as it is not often used. This method consists of prethawing with excavation and replacement of the thaw unstable frozen ground and designing the tank for a small amount of subsequent settling.

Creep of frozen ground can be a significant factor in a piling tank foundation because of the high loading. The magnitude of creep is very temperature dependent and, thus, is a greater problem in warm permafrost than cold permafrost.

Foundation costs are extremely variable depending on the availability of equipment and materials, the size of tank, and the permafrost conditions. Typical 1980 construction costs (U.S. dollars) of a foundation for a 1,900,000 liter tank would be around $100,000 for a gravel pad (depending heavily on the availability of gravel) and $200,000 for a pile foundation and floor system.

TANK DESIGN AND MATERIALS

The most widely used tank in remote arctic areas is welded steel. Wood stave tanks have been successfully used in the subarctic areas but they usually have more drawbacks than advantages in the Arctic. The disadvantages of wood stave tanks follow: (1) By design, they "weep" to maintain their overall water tightness. This can create problems in the winter and it makes them difficult to insulate effectively. (2) They will leak if they are not properly constructed and there seems to be a lack of people trained to assemble them correctly. (3) They are usually not cost effective in sizes larger than 375,000 liters. (4) Wood stave tanks are designed to be kept full at all times. This is not possible in many of the remote locations, especially where the tanks are operated on a fill and draw cycle. Impermeable liners have been used inside tanks. However, they are usually not very successful as human and ice damage can cause punctures.

Wood stave tanks do have the advantage of being easily transported into remote areas and specialized equipment is not usually required for assembly. The wood does provide a certain amount of insulation, also.

Concrete tanks are seldom used in the Alaskan Arctic. They require large quantities of high-quality aggregate and a solid, nonsettling foundation (rock), both of which are usually not available.

Steel tanks can be either bolted or welded. Bolted tanks are not as popular in larger sizes. Any damage during shipping can make it impossible to align the bolt holes during erection and increase the tendency of the seams to "seep". They do have the advantage, however, that the interior coatings can be baked on in the factory and then touched up in the field.

Welded steel tanks in smaller sizes (less than 75,000 liters) can be shipped to the site prefabricated, with no field welding required. However, the larger tanks must be shipped in pieces and erected on site. Welders, qualified operators, and vacuum and radiographic equipment must be available to check for faulty welds.

At any location where the tank may remain empty or could reach the ambient temperature during the winter, special low-temperature welding procedures and steels with high impact strength at low temperatures should be used (ASTM A-516).

Steel tanks must also be painted, inside and out. Interior coatings have proven to be a major problem. Both vinyl and epoxy coatings have been used with varying success. Any interior treatment used in a tank storing potable water must be approved by the American Water Works Association (AWWA). The main problems that have occurred can be traced to improper application temperatures, solvent entrapment during curing, and the thickness of the individual coats. The keys to proper application appear to be assuring the surface is clean (sandblasted), dry, and warm (above 16°C), and that the total coating is applied in several thin coats instead of a few thicker ones. The AWWA, ASTM, and paint suppliers specifications should be consulted and followed. Exterior coatings are not as critical if sprayed-on urethane insulation is to cover the tank. In these cases, the outside is primed only, before applying the insulation.

Rubber "bladder" tanks have been used in some locations for oil and gasoline storage, usually on a temporary basis. Fiber glass tanks in the larger sizes are relatively new on the market and are not used extensively at the present time, but may be a consideration in the future.

Maintaining the water temperature above freezing in water storage tanks is a critical design consideration. Any ice forming in a tank can cause damage to coatings, attachments, controls, and even the tank itself. Heating systems and insulation should be designed to prohibit the formation of any ice and interior ladders and appurtenances should not be attached rigidly to the tank walls. Serious damage can result to the tank if ice should form on these fixtures. For example, the heat supply to a welded steel storage tank in Kotzebue, Alaska, failed, causing ice to form around the wall of the tank which included a stationary, interior ladder. Water continued to be drawn from the tank and the level dropped, leaving ice exposed. The hanging ice then broke loose and fell, tearing holes in the side of the tank where it pulled the ladder from the tank walls. The falling ice also damaged the bottom of the tank.

Surface ice can be prevented by operating the tank at a temperature above 4°C (point at which water is most dense), and wall icing can be prevented by keeping the contents mixed and above freezing. Outlet and inlet lines should be separated to help provide this circulation. Pumps should be utilized to provide circulation, if necessary.

INSULATION

All water storage tanks in arctic areas must be insulated or placed inside a heated building. Steel tank walls provide essentially no insulation value with the only "insulation" coming from the surrounding air film. Steel appurtenances should not penetrate the insulation, as ice will form on the inside of the tank at the point of the thermal break.

Urethane and polystyrene are the most common insulations used for large water storage tanks. In any installation where the insulation is likely to be in contact with water, extruded polystyrene is recommended. A typical location would be in the gravel pad under the tank. Polystyrene must be applied as boardstock and cannot be sprayed on in the field as can urethane. Urethane has a somewhat better insulating value providing it has an airtight covering to prevent the loss of the freon gas filling the voids. However, if it is not tightly covered, the freon escapes and is replaced by air and urethane's insulating value becomes nearly identical to that of polystyrene. Urethane will break down faster than extruded polystyrene if it is exposed to moisture and freeze-thaw cycles or ultraviolet light (sunlight).

The preferred method of insulating is to spray urethane on in the field and cover it with a low molecular weight, butyl membrane for vapor seal and ultraviolet light protection and then with aluminum siding for physical protection. A critical consideration with this procedure in cold regions is the ambient temperatures. The air temperature must be above 10°C for proper application. The weather must be calm (no wind) when spraying. Even the slightest wind can waste large amounts of insulation and cause considerable damage to the surrounding property such as buildings and automobiles.

Another often-used method of insulating tanks is to strap or bond boardstock, either urethane or polystyrene, to the outside walls of the tank. This method can be used where the weather conditions are not suited for a spraying operation. Problems have been encountered in high wind areas where pieces of insulation have been blown from the tank. The insulation, and its covering, must be well strapped to the tank using clips welded to the outside tank walls. Also, spray or pour foam must be used to fill the cracks between the insulation boards before the outside covering is added.

Boardstock, preferably extruded polystyrene, should be used to insulate under the tank also. This is especially important for tanks on gravel pads in permafrost areas to prevent the lost heat from degrading the permafrost. The amount and type used will be determined by the specific conditions at the site.

The thickness of insulation needed will vary with the climate and heating costs at the location and can be determined through an economic analysis.

CONSTRUCTION COST

Typical tank costs are presented in the attached table, adjusted to 1980 construction costs using a 7% per year inflation factor. Tank costs since 1979 are actual costs in the year constructed. The variances in the unit costs can be attributed to changes in the cost of steel, the location of the tank, and the size of the tank. Foundation costs are not included. Typical specifications for interior painting systems and exterior insulation can be obtained from the author or John Delapp with the Alaska Area Native Health Service. Mr. Delapp also supplied cost and experience information for tanks constructed in 1982.

TABLE 1 Alaska Water Storage Tanks.

Location	Year Const.	Tank Dim. (meters) Dia.	Ht.	Tank Size (liters [x10^6])	Approx. Weight of Steel (kg.)	Cost (U.S. $)* Total	$/ℓ.	Type of Use	Insulation Type, thickness, covering	Interior Paint	Ground Conditions	Foundation Type	Problems
Kotzebue	1967	25	10.7	5.67	N/A	N/A	N/A	Continuous Use	Sprayed urethane 7.6 cm	Vinyl	Frozen gravel	Gravel pad	Need interior repainted. No settlement problems.
Wainwright	1971	25.6	7.3	3.79	N/A	N/A	N/A	Annual Fill	Banded urethane boards, 8.5 cm; cracks filled with sprayed urethane, covered with elastomeric material.	Vinyl	Frozen silts and organics	Wood piling	Experiencing gradual settlement. Insulation on roof popping off from settlement. Cracks opening on wall insulation.
Barrow	1975	21.3	6.1	2.16	57,143	401,800	0.185	Continuous Use	Spray urethane 8.9 cm walls, 12.7 cm top, butyl covering.	Epoxy (orig.) Repainted United "Vabar"	Frozen silts, organics, with some free ice.	Treated piling	100% inside repaint done 1981. Massive foundation settlement upon initial filling in 1978. Posts and pads installed to stop settlement in 1979.
Kivalina	1975	18.3	7.3	1.92	53,968	374,500	0.196	Annual Fill	Beadboard in gravel/ foundation, 8.9 cm urethane boardstock walls, 10.2 cm urethane spray top, aluminum covering on walls.	Vinyl	Frozen gravel	Gravel pads	Originally boardstock roof blew off. Replaced with spray urethane.
Shungnak	1976	11.6	7.3	0.76	26,304	217,800	0.288	Monthly Fill	Sprayed urethane 8.9 cm, covered with butyl and paint.	Epoxy (orig.) now vinyl	Frozen sand overlain by 2 meters moist peat.	Gravel pad (removed peat)	Original epoxy rust "spotted" particularly at weld lanes. Tank repainted with vinyl in 1982.
Gambell	1976	9.1	6.1	0.38	16,780	122,200	0.322	Continuous Use	Sprayed urethane 8.9 cm, elastomeric coating. Aluminum sheathing installed 1981.	Epoxy	Pea gravel	Gravel	Interior paint failures recoated in 1981. Coating was "chalking" off and cracking. New paint is also blishered.
Savoonga	1976	9.1	6.1	0.38	16,780	138,000	0.365	Continuous Use	Sprayed urethane 8.9 cm, covered with elastomeric coating, aluminum sheathing installed in 1981.	Epoxy	Saturated frozen organics, ice lenses lava boulders.	Drill/drive steel pipe piles	Original interior paint failed. Recoated in 1981. Numerous small blisters developed after one year service of new paint.
Shishmaref	1976	12.8	9.1	1.14	33,107	231,000	0.203	Annual Fill	Sprayed urethane 8.9 cm covered with butyl.	Epoxy	Fine sand; probably frozen.	Sand pad	Interior inspected 1980. No major problems.
Grayling	1977	7.3	5.5	0.21	11,338	97,400	0.460	Continuous Use	Sprayed urethane 8.9 cm butyl.	Epoxy	Unfrozen silts and gravels.	Gravel pad	None known.
Twin Hills	1977	6.7	6.1	0.21	11,338	83,700	0.394	Continuous Use	Sprayed urethane 8.9 cm, butyl.	Epoxy	Unfrozen tundra mat underlain by thawed silts with some sand and gravel.	Gravel pad	None known.
Unalakleet	1977	24.4	6.1	3.78	99,320	346,700	0.092	Continuous Use	Sprayed urethane 8.9 cm, butyl.	Epoxy	Gravel	Gravel pad	Original paint failure interior. Repainted 1981. Recoating has blisters on floor.
Stebbins	1978	18.3	7.3	1.92	53,968	232,400	0.122	Annual Fill	Sprayed urethane 8.9 cm, butyl.	Epoxy	Sand & gravel gravel	Sand & gravel pad	None known.

TABLE 1 (Continued)

Location	Year Const.	Tank Dim. (meters) Dia.	Ht.	Tank Size (liters [x10^6])	Approx. Weight of Steel (kg.)	Cost (U.S. $)* Total	$/ℓ.	Type of Use	Insulation: Type, thickness, covering	Interior Paint	Ground Conditions	Foundation Type	Problems
Shaktoolik	1978	19.8	9.8	3.00	76,644	400,000	0.132	Fill & Draw (annual)	Poly. 8.9 cm spray foam. Sheet aluminum siding spray butyl, roof.	2 Coat epoxy	Beach gravel (frozen)	Select fill gravel pad with 5.1 cm polystyrene boardstock insulation.	Apparent settlement at utilidor 7/81.
Point Hope	1978	36.6	9.8	10.22	249,433	1,373,000	0.135	Fill & Draw (annual)	Poly. 12.7 cm spray foam. Sheet aluminum siding spray butyl, roof.	2 Coat epoxy	Beach gravel (frozen)	Select fill gravel pad with 5.1 cm poly. boardstock insulation.	Leaks at faulty welds in floor plate repaired in 1982.
Venetie	1978	14.6	7.3	1.23	37,188	339,900	0.277	Continuous Use (annual)	Poly. 8.9 cm spray foam. Sheet aluminum siding spray butyl, roof.	2 Coat epoxy	Silty sand underlain by gravel (frozen)	Select fill gravel pad with 5.1 cm poly. board stock insulation.	None known.
Kaktovik	1978	17.1	9.8	2.27	57,596	426,400	0.188	Annual Fill	Poly. 12.7 cm spray foam. Sheet aluminum siding spray butyl, roof.	2 Coat epoxy	Ice rich silts and gravel, frozen	Wood piling spacing 1.2 x 1.8 m, 4.9 - 5.5 m, deep	Interior piling failure -- slurry not installed. Corrected by adding interior post and pad supports, 1979.
Wales	1978	18.3	7.3	1.89	53,968	N/A	N/A	Annual Fill	Poly 8.9 cm spray foam.	2 Coat epoxy	N/A	Select fill gravel pad with 5.1 cm insulation	None known.
Tununak	1978	7.9	4.3	0.19	10,884	101,900	0.539	Fill & Draw (3/month)	Poly 8.9 cm spray foam.	2 Coat epoxy	Frozen silts and organics	1.5 m gravel pad, 0.61 m above grade, with 10.2 cm poly. board stock insulation.	None known.
Quinhagak	1978	9.8	4.9	0.36	16,326	132,000	0.365	Fill & Draw (monthly)	Poly 8.9 cm spray foam.	2 Coat epoxy	Frozen silts and organics	15.2 cm dia. steel piling 5.5 m deep spacing 1.2 x 1.8 m	None known.
Birch Creek	1979	7.3	7.3	0.30	14,059	154,600	0.510	Fill & Draw (annual)	Poly 8.9 cm spray foam. Sheet alum. siding spray butyl, roof.	2 Coat epoxy	Frozen sandy silt	Mud sill	None known, inspected in 1981.
Eek	1979	9.1	3.7	0.24	11,791	127,000	0.534	Fill & Draw (annual)	Poly 8.9 cm spray foam. Sheet alum. siding spray butyl, roof.	2 Coat epoxy	Frozen silts and organics	15.2 cm dia. steel piling 5.5 m deep spacing 1.2 x 1.8 m.	None known.
Nenana	1979	9.1	9.8	0.57	20,408	152,400	0.269	Fill on Demand	Poly. 8.9 cm spray foam. Sheet aluminum siding spray butyl, roof.	2 Coat epoxy	Gravel	Select fill gravel pad with 5.1 cm insulation.	Numerous small blisters after one year of service. Paint consultant studying problem.

TABLE 1 (Continued)

Location	Year Const.	Tank Dim. (meters) Dia.	Tank Dim. (meters) Ht.	Tank Size (liters [x10^6])	Approx. Weight of Steel (kg.)	Cost (U.S. $)* Total	Cost (U.S. $)* $/ℓ.	Type of Use	Insulation Type, thickness, covering	Interior Paint	Ground Conditions	Foundation Type	Problems
Karluk	1979	7.3	4.9	0.19	10,884	111,600	0.589	Continuous Use	No insulation.	2 Coat epoxy	Silty gravel	Select fill gravel pad.	None known. (Not inspected since initial filling.)
McGrath	1980	18.3	9.8	1.92	53,968	297,300	0.156	Fill & Draw (semi-annual)	Poly. 12.7 cm spray foam. Sheet alum. siding spray butyl, roof.	2 Coat epoxy	Silts & sands	Select fill gravel pad 1 m.	None known.
Nuiqsut	1980	33.5	9.8	8.59	204,082	950,000	0.111	Fill & Draw (annual)	Preformed sheets 12.7 cm (boardstock). Preformed sheets of aluminum.	5 Coat vinyl	Frozen silts and organics	Select fill gravel pad with insulaation mudsills & wood deck.	Interior partially inspected 12/81 while full. No defects noted.
Russian Mission	1980	7.6	4.9	0.22	11,338	108,000	0.483	Fill on Demand	Poly. 12.7 cm spray foam. Sheet alum. siding spray butyl, roof.	2 Coat epoxy	Silts	Select fill gravel pad.	None known.
Dillingham	1980	18.2	7.3	1.92	53,968	264,464	0.138	Continuous Use	Urethane spray 8.9 cm, with aluminum.	2 Coat epoxy	Silt & peat	Gravel pad with ring. ($108,500)	None known.
Kaltag	1981	7.3	2.5	0.10	11,400	104,000	1.040	Continuous Use	Urethane spray 8.9 cm, with aluminum.	2 Coat epoxy	Frozen silt	Mudsills on insulation gravel pad.	None known.
Kipnuk	1981	9.6	4.9	0.36	16,326	319,432	0.887	Fill & Draw	Urethane spray 8.9 cm, with aluminum.	2 Coat epoxy	Semi-froze high moisture silts, organics	Drive piling (steel-H piles)	None known.
Pilot Station	1982	6.1	5.5	0.15	8,843	138,830	0.926	Continuous Use	Urethane spray 8.9 cm, with aluminum.	2 Coat epoxy	Silt/sand/ gravel	Gravel pad with ring.	None known.
Mountain Village	1982	9.6	4.9	0.36	16,326	174,000	0.483	Continuous Use	Urethane spray 8.9 cm, with aluminum.	4 Coat epoxy mil. spec. (24441)	Silt	Gravel pad with ring.	None known.
Saint Michael	1982	16.6	7.3	1.52	46,485	348,900	0.230	Fill & Draw	Urethane spray 12 cm, with aluminum.	4 Coat epoxy (24441)	Frozen silt, organics	Mudsills on insulated gravel pad.	None known.
Deering	1982	14.0	9.6	1.52	46,485	298,260	0.200	Fill & Draw	Urethane spray 12 cm, with aluminum.	4 Coat epoxy (24441)	Beach gravels (frozen)	Gravel pad with ring.	None known.

* Costs of tanks constructed before 1980 were adjusted to 1980 U.S. dollars considering 7% per year inflation. Tanks constructed after 1980 include the actual construction cost in that year. Costs do not include foundations.

CREEP OF FROZEN SOILS ON ROCK SLOPES

A. V. Sadovsky and G. I. Bondarenko

Research Institute of Foundations and Underground Structures,
Moscow, USSR

This article presents the result of experimental investigations into the creep of frozen materials on rock slopes. Analysis of field data revealed that creep in frozen waste heaps in northern regions is a typical rheological process. Laboratory investigations included two types of tests: simple shear tests and tests of shear-with-skewing. More than 400 samples of soil frozen to the rock were tested. The mechanics of shear were investigated and the formation of two zones was established: a zone of contact, and an overlying zone. By using the data from the experimental studies and from field observations it was possible to develop a method of calculating the rate of creep of frozen waste heaps on rock slopes and to predict the failure of these waste heaps.

The creep of frozen soils along the inclined rocky bed has something in common with flow of glaciers. But there are some differences due to peculiarities of deformabilities of ice and frozen soil. Ice flows practically under any load, but frozen soil flows when the load exceeds the threshold load corresponding to the onset of visco-plastic flow.

Several studies were made to investigate the processes of frozen soils deformability when they slide along the inclined rocky bed. The sliding of ten weathered rock masses was investigated and laboratory tests of frozen soils adfreezed to the rock material were conducted.

While mining is proceeding in the open manner in hilly regions weathered rock masses are thrown down on the rocky slopes. Low temperature causes quick cooling of blasted rock and an accumulation of a large reservoir of cold within it. The frozen state is the main peculiarity of weathered rock masses in the northern regions. The mean temperature of observed frozen masses on the rocky slopes ranges from -0.6°C to -1.0°C, the height of slopes ranges from 170 m to 400 m, and the angle of inclining ranges from 37° to 51°. After the mass volume increases to a certain value, the sliding begins. As a rule a period of deformation proceeded the failure and creeping with constant speed lasted from 3 to 8 months. Analysis of field data showed the sliding of frozen masses to be the rheological process. Figure 1 shows the sliding to be a complicated process: the movement of "a" to the position of "a^1" shows the movement of the whole mass along the rocky slope, and shape of the mass surface during the sliding (curve $a^1b^1c^1$) reflects the creeping inside the mass.

More than 400 tests were conducted to investigate a long-term shear resistance of frozen soils adfreezed to the rock. Soil samples were prepared from coarse grained rock (maximal size of particles—5 mm). That soil was similar to sand, containing fine particles more than 8%; the water content was 0.13. Soil was adfreezed to the rock in the cold chamber with temperature -25°C. Then samples were kept during 15-18 hours at the testing temperature. The roughness of the rock surface was formed by ac. HCl. The roughness degree was estimated as a sum of cavities (in mm) per 1 m of length. Shear devices PRS-1 developed by Sadovsky were used for tests.

First, it was necessary to determine if the rupture always occurred along the contacting surface. For that purpose shear tests were conducted on rocky samples at five values of roughness and the influence of roughness on the adfreeze resistance was investigated. It was shown that the roughness increased both parameters of the shear resistance: namely friction and cohesion. Empirical dependence of both parameters from the value of roughness was obtained.

The investigations have shown the process of deformability along the contact surface proceeds like the process of deformability within the frozen soil (2). Creep curves and long-term resistance of frozen soil decrease with time because the cohesion is the main source of long-term resistance decrease. The change in the friction angle with time and temperature is negligible. But there are some differences: the rupture along the contact surface occurs over a shorter period of time than the rupture within the soil under the same loads, i.e., each stage of the creep takes place over a shorter period along the contact surface and the shear resistance along the contact surface at any given time is less than that within the soil mass.

Two types of shear tests were conducted: shear (along the contact surface) and shear-to-skewing tests (general shear) (Figure 2).

In the shear-to-skewing test a certain volume of frozen soil was deformed. Total and by-contact shear linear deformations (strains) were measured. Angular deformations of soil were determined as a ratio of strains to the proper height, h, of the skewed layer. Constant strain rates were determined from creep curves (Figure 3). Experimental results indicate that rupture always occurred along the contact surface since the contact surface creep begins under less loads than the creep within the soil mass. As a result, tertiary creep (or progressive stage of creep) begins earlier than that within the slide mass. For example, shear stress = 0.4 MPa led to creep with a contact strain rate equal to $1.4 \cdot 10^{-4} cm^{-1}$, but at the same time strains inside the soil were gradually damping.

The relationship between the angular strain rate and shear stress within the loaded boundary (thickness h_c in Figure 2) layer may be described by Bingham-Schwedoff's relation:

$$\gamma = \frac{1}{\eta} (\tau - \tau_\infty) \quad (1)$$

where τ = acting stress;
τ_∞ = long-term resistance; and
η = viscosity coefficient.

The viscosity coefficient for the boundary layer is practically equal to the viscosity coefficient for soil. Differences between strain rates of the boundary layer and of the soil are caused by differences of long-term resistances of abovementioned layers.

Linear slide speed (Figure 4), V, of any soil layer within the mass increases proportionally to distance from rocky bed and can be expressed by:

(a) for the boundary layer: as $V^c = \dot{\gamma}_c \cdot h_c$;
(b) for soil: as $V^s = \dot{\gamma}_s \cdot h_s$;

where h_c = thickness of the boundary layer, and
h_s = distance from rocky bed to the soil layer.

Linear slide speed at the top surface of the sliding mass is equal to:

$$V = V^c + V^s \quad (2)$$

Experimental results confirmed that the rupture occurs after deformation reaches a definite value. It was observed that the rupture occurred only along the contact surface and the time until rupture is proportional to the strain rate along this contact surface. This fact brought forth the possibility of forecasting time until rupture by using experimental or field observed strain rates.

Experiments showed that two zones are formed in the sliding mass: namely, the boundary layer and the soil located above the boundary layer; viscosities of both zones are similar, but shear resistances are different. As a result, tertiary creep and subsequent rupture occur along the contact surface.

To determine strain rates of frozen masses of rock rubble, we analyzed the creep of a single homogeneous layer resting on bedrock which extends downward indefinitely. It was assumed that the slope deformation process proceeds as simple shear in the direction parallel to the bedrock surface and the mass of rock rubble is a homogeneous frozen soil. Shear stress produced by weight of mass can be written as:

$$\tau = \gamma_o (H - y) \sin \alpha \quad (3)$$

where γ_o = bulk density;
H = thickness of the sliding mass;
y = distance from bedrock to the layer being analyzed; and
α = angle of inclination of bedrock surface.

It is known that the creep of frozen soil is caused by excess stress of: $\tau - \tau_\infty$, and for slopes:

$$\tau_\infty = C_\infty + \gamma_o (H - y) \cos \alpha \cdot tg\, \phi_\infty .$$

Strain rate at the surface of the sliding mass (where Y=H) was determined from Stroganov's (4) and Maslov's solutions:

$$V^s = \frac{H-D}{\eta} [\gamma_o \frac{H+D}{2} \cos \alpha\, (tg\alpha - tg\phi^s_\infty) - C^s_\infty] \quad (4)$$

$D = C^s_\infty \gamma_o \cos \alpha\, (tg\alpha - tg\phi^s_\infty)$ —thickness of upper passive moving layer in which $\tau < \tau_\infty$.

Similar to equation 4 strain rate in the boundary layer is equal to:

$$V^c = \frac{h_c}{\eta} [\gamma_o (H - \frac{h_c}{2}) \cos \alpha\, (tg\alpha - tg\phi^c_\infty) - C^s_\infty] \quad (5)$$

Strain rate on the surface of the sliding mass can be determined as a sum of a contact strain rate and a strain rate of the mass upper layer using formulas 2-5.

Calculations of strain rates for the secondary stage of creep were made using stress and deformation parameters determined from test results. Calculations showed that the creep speed of frozen heaps is very dependent on temperature. For example, an increase in temperature from -2°C to -1°C (H = 50 m, α = 50°) results in increasing the boundary layer strain rate from 0.35 to 32 mm per day and for the upper layer of the sliding mass, from 8 to 85 mm per day.

Comparison of both strain rates shows that strain rate inside the mass is to a great extent higher than contact strain rate. But if the temperature of the upper layers is below the temperature of the boundary layer, the creep inside the mass cannot begin. Since contact friction is less than friction inside the soil mass (ϕ^c_∞ = 18° and ϕ^s_∞ = 32°), creep of the boundary layer can begin at a lesser angle of inclination of bedrock. For example, if H = 30 m and the temperature is equal to 0.6°C, the contact creep begins at α = 26°, but creep inside of the soil begins only at α = 40°. This example shows how important it is to know the contact strain rate. Although very often the absolute value of the contact strain rate is small but the tertiary stage of creep develops along the contact surface.

It was mentioned that sliding of frozen masses of rubble begins after their volume increases to a certain value. The thickness of the mass to a greater degree influences the rate of strain. So, by increasing the thickness from 29 m to 40 m there will be an increase in the strain rate by more than ten times. This change in rate is in conformity with field observations of sliding masses of frozen rock rubble.

Figure 5 shows the graph describing the increasing sliding speed, V, of frozen mass in the process of increasing its thickness as the slide proceeds down the slope. Initial thickness of the sliding mass, H, at the beginning of the observation was 12 m, after 4 months, H = 20 m, after 6 months H = 28 m and after 8 months H = 34 m. The intermittent line on the graph shows the observed speed and the solid line shows the calculated speed. Comparison of those two lines confirms the satisfactory results of this study.

CONCLUSION

(1) The sliding of frozen soils over bedrock is a typical rheological process including three states of creep.

(2) Two layers are formed in the sliding mass: namely, the boundary layer and the soil mass above this layer. Long-term resistance of the boundary layer, as a rule, is less than that inside of the soil mass. So the rupture occurs along the contact surface.

(3) The slide speed at the surface of frozen mass is the sum of a contact strain rate and the strain rate of the upper thickness of the sliding mass.

(4) The results of this study can be recommended for forecasting the strain rates of frozen soils on the inclined bedrock using creep characteristics of frozen soils (C^S_∞, $tg\phi^S_\infty$) and contact characteristics (C^c_∞, tg^c_∞).

REFERENCES

Bondarenko, G. I., 1979, Prognozirovaniye skorosti opolzaniya merzlykh otvalov na skal'nykh sklonakh [The forecasting of slide rates of frozen dead rock heaps on rocky slopes], in Inzhenernoye merzlotovedeniye: Novosibirsk, Izdatel'stvo "Nauka," p. 197-204.

Bondarenko, G. I., and Sadovskiy, A. V., 1975, Prochnost' i deformiruyemost' merzlogo grunta na kontakte so skaloy [Stress and deformability of frozen soil at rock contacts], Osnovaniya, fundamenty i mekhanika gruntov, no. 3, p. 22-25.

Maslov, N. N., 1955, Usloviya ustoychivosti sklonov v gidrotekhnicheskom stroitel'stve [Stability conditions of slopes in hydraulic engineering]: Moscow-Leningrad, Gosenergoizdat, p. 467.

Rukovodstvo po opredeleniya fizicheskikh, teplofizicheskikh i mekhanicheskikh kharakteristik merzlykh gruntov [Manual for investigation of physical, thermophysical and mechanical characteristics of frozen soils]: Moscow, Stroyizdat, p. 37-42.

Stroganov, A. S., 1961, Vyazko-plasticheskoye techeniye gruntovogo sloya po naklonnoy poverkhnosti [Viscoplastic creep of the soil layer along an inclined surface], Inzhenernyy sbornik Instituta mekhaniki AN SSSR, v. 21, p. 132-143.

Vyalov, S. S., 1978, Reologicheskiye osnovy mekhaniki gruntov [Rheological principles of soil mechanics]: Moscow, Izdatel'stvo "Vysshaya shkola," p. 447.

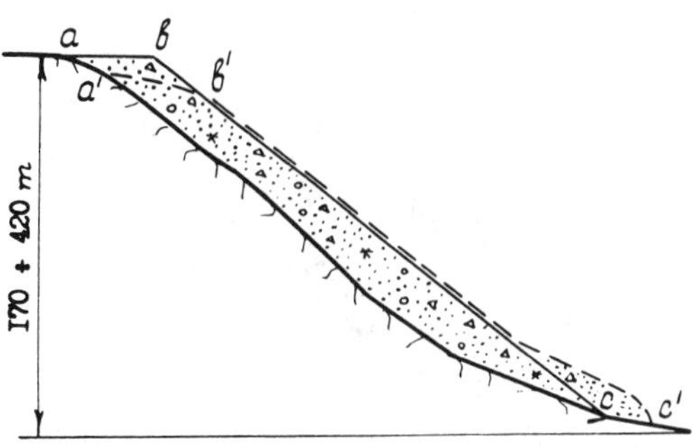

FIGURE 1 Positions of the frozen weathered rock mass on the slope. a,b,c - before sliding; a^1,b^1,c^1 - during sliding.

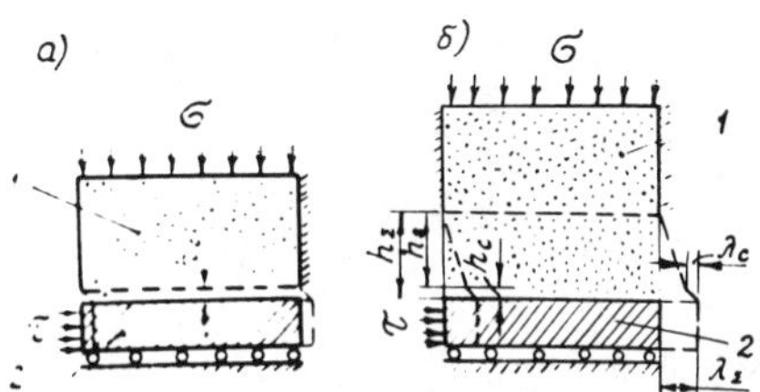

FIGURE 2 Schemes of shear tests. a) simple shear tests; b) shear-to-skewing test (general shear); 1 - frozen soil; 2 - rock.

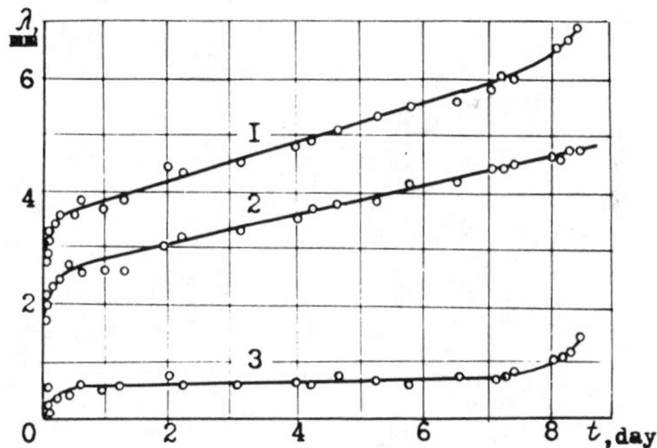

FIGURE 3 Creep curves for shear-to-skewing test. 1 - total deformation of frozen soil; 2 - deformations inside the frozen soil; 3 - by-contact layer's deformation.

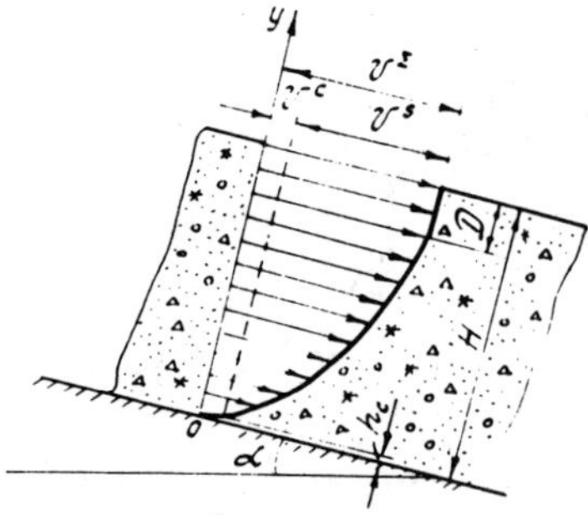

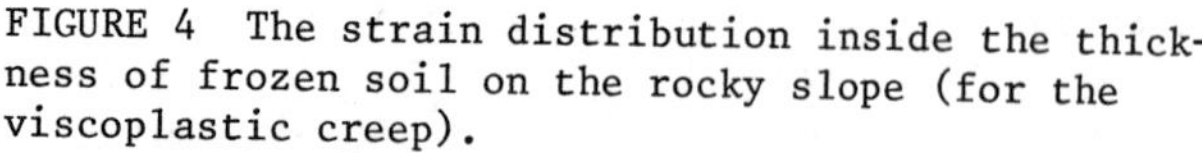

FIGURE 4 The strain distribution inside the thickness of frozen soil on the rocky slope (for the viscoplastic creep).

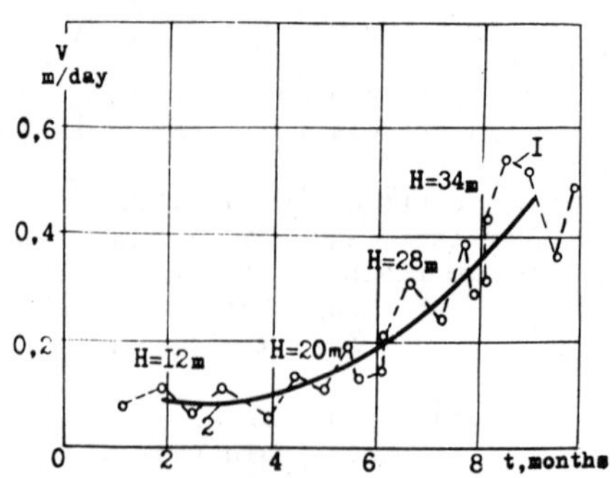

FIGURE 5 The influence of the heap thickness and of time on the sliding speed

ACCUMULATION OF PEAT CARBON IN THE ALASKA ARCTIC COASTAL PLAIN AND ITS ROLE IN BIOLOGICAL PRODUCTIVITY

Donald M. Schell and Paul J. Ziemann

Institute of Water Resources, University of Alaska, Fairbanks, Alaska 99701

Terrestrial peat on the coastal plain of arctic Alaska is estimated to be accumulating at a rate equivalent to about 7-18 g $C \cdot m^{-2} \cdot yr^{-1}$, or 10-20% of the annual net carbon fixation. The accretion of root matter and surficial vegetation as peat is inversely proportional to microbial activity and grazing pressure by herbivores since the depth of the permafrost horizon responds to the insulative properties of the vegetative mat. Basal peats along the coastline in the vicinity of Prudhoe Bay range from 8,000-12,000 years B.P. and their stable carbon isotope ratios indicate that vascular plants and mosses contribute the bulk of the organic matter. The aquatic habitats of the tundra represent active sites for peat oxidation and conversion to faunal biomass. Inputs of peat occur via streambank erosion, thaw lake expansion and coalescence, and coastline inundation and erosion. Stable carbon isotope ratios and carbon dating of surface lake sediments indicate that peat, rather than algal production or recent terrestrial vegetation, constitutes the bulk of the organic matter.

INTRODUCTION

The revegetation of the arctic Alaska coastal plain (Figure 1) following the Pleistocene has resulted in a mantling of virtually the entire landform in a layer of organic matter ranging up to 3 m thick. Although the short growing season and low ambient temperatures have restrained plant life to a thin microrelief, the productivity supports abundant fauna. Actively growing vascular plants, bryophytes, and lichens, or standing dead plant material from the immediately preceding seasonal growth are the obvious sources of food for herbivores. The close coupling of primary production and the population dynamics of consumers such as caribou, lemmings, and insects have received intensive study by groups such as the International Biological Program (IBP) and by researchers assessing the impacts of oil industry development in the Prudhoe Bay region. These studies have produced complex models and budgets describing energy flow and compartmentalizing the carbon pools of the tundra biome. Excellent collections of papers describing these phenomena are available in volumes by Tieszen (1978), Hobbie (1980), Brown et al. (1980), and Sonesson (1980). In general, however, the models describing carbon movement in arctic tundra ecosystems use as end products (and loss to the ecosystem energetics) the evolution of carbon dioxide or peat in the permafrost layer. Superficially these are obvious losses--carbon dioxide is a complete recycling to a starting component, and peat is evident as an end product accumulating beneath the vegetative mat and in the permafrost. Although the formation of ice wedges and landforms related to polygonal tundra can alter or erode the peat mat, the ubiquitous layer of peat throughout the coastal plain attests to the accumulation of peat as a growing pool of stored and unavailable energy in tundra carbon flow.

We question the validity of this premise. Through the use of stable and radioactive carbon isotope abundances in fauna from the coastal plain, we have demonstrated that the terrestrial and aquatic ecosystems are coupled through the production of peat and subsequent biological utilization (Schell 1983). The coastal tundra biome should not be viewed as separate terrestrial and aquatic ecosystems but instead considered as a whole with net energy flow from the terrestrial to the aquatic environment. Although Chapin et al. (1980) recognized the occurrence of this energy flow, they felt that the transfer represented only a minor fraction in the annual budget of the system and did not contribute to higher trophic level nutrition. In this paper we present estimates of the rates of carbon accumulation in coastal plain tundra and the subsequent movement from terrestrial to freshwater ecosystems. The role of peat carbon in the coastal plain ecosystem has been described elsewhere (Schell 1983). Details of the methodology are in Schell et al. (1982).

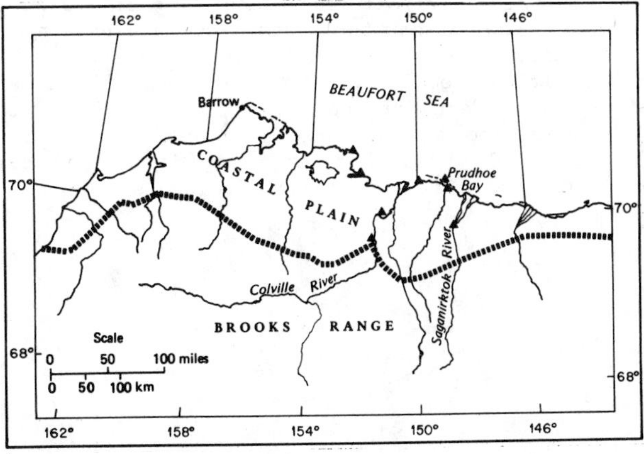

FIGURE 1 Coastal plain of the Alaska North Slope. Triangles mark soil section locations in Table 1.

Table 1. Peat Depths, Ages, and Carbon Standing Stocks at Arctic Coastal Plain Sampling Sites. Radiocarbon ages listed are ±200-300 years B.P.

Location		Peat Depth (cm)	Average Percent Carbon (dry weight)	Carbon Standing Stock ($kg{\cdot}m^{-2}$)	Basal Peat Age (yrs BP)	Carbon Accumulation Rate ($g{\cdot}m^{-2}{\cdot}yr^{-1}$)
Milne Point	70°31"N, 149°26'W	170	25.1	154	8,435	18.3
Peat Island	70°33'N, 149°24'W	160	18.6	98	9,056	10.8
Kavearak Point	70°31'N, 149°19'W	292	15.6	94	12,610	7.4
Oliktok Point #2	70°31'N, 149°52'W	198	33.8	143	8,550	16.7
Oliktok Point #1	70°31'N, 149°52'W	91	14.1	64		
Ocean Point	70°07'N, 151°20'W	130	7.6	52		
Atigaru	70°34'N, 151°45'W	50	5.6	21		
Sentinel Hill	69°53'N, 151°40'W	180	28.9	199		
Radiocarbon age sites:	X =	159 ± 73	18.7 ± 10.0	103 ± 59	9,663 ± 1,983	13.3 ± 5.1

METHODS

Terrestrial Primary Production

The data obtained from the IBP site at Barrow (Webber 1978) and Devon Island (Muc 1973) for above and below ground primary production led us to estimate an annual net production of 88 g $C{\cdot}m^{-2}{\cdot}yr^{-1}$. Webber cautions that most studies have been in productive sites and that dry impoverished sites may be more significant in the arctic landscape. Therefore, we have selected the above value as a midrange estimate on a scale of tundra productivities.

Grazing, decay, and transport remove most of the fixed carbon from the tundra surface before incorporation into the peat layer, and quantitative estimation of the net accumulation is very difficult. Chapin et al. (1980) compare three techniques to approximate net carbon inputs to the Barrow tundra and found that during the study period a net increase occurred of between 40 and 120 g $C{\cdot}m^{-2}$. Since that rate would have accounted for the observed organic matter in the soil within as little as 160 years, production/utilization ratios probably vary widely from year to year and in response to microenvironments.

The difficulties inherent in areawide projection of carbon fluxes from data obtained at specific locations and the measurement noise resulting from localized process studies led us to abandon that approach in favor of a more generalized estimation of net carbon flux based upon soil organic matter, peat depth, and basal peat age. The calculated inputs were then extrapolated areawide and compared to removal fluxes. Isotopic data were used independently to estimate the ecological significance of the physical processes.

Composition of Peat Soils

Detailed soil sections were taken at three sites near Simpson Lagoon, and several less detailed samplings were obtained from near Cape Halkett and from cut banks along the lower reaches of the Colville River. Additional data, previously collected from Simpson Lagoon (Lewellen 1973), are included here to illustrate the range of carbon contents involved.

Sample sites were chosen where erosion had exposed a fresh break in the permafrost. Pick and shovel were used to dig to undisturbed permafrost, and a vertical section was made at 10 cm intervals from beneath the vegetation surface mat to basal mineral soils. Samples were kept frozen until returned to the laboratory where moisture content was determined. Subsamples were analyzed for carbon, nitrogen, loss-on-ignition, radiocarbon content, and stable carbon isotope ratios.

Lake Sediments

Samples were collected either by Ekman dredge or by cores pushed by hand into the bottom, and sectioned horizontally in sufficient size to yield at least 5 g of carbon. After suspension in dilute hydrochloric acid to remove carbonates, samples were settled, decanted, rinsed with distilled water, and dried at 70°C in a vacuum oven.

Vegetation Samples for Isotopic Analysis

Fresh vegetation from tundra meadows, heaths, pond margins, and ponds was collected to establish current concentrations of radiocarbon and $^{13}C/^{12}C$ ratios in living primary producers. Samples were dried in vacuo at 80°C.

Fluvial Transport and Coastline Erosion

Riverbank, lakeshore, and coastline erosion constitute the major fluxes mobilizing peat incorporated into permafrost and dispersing it into a relatively warm oxygenated environment. Total quantities of particulate carbon in Colville River flow were calculated from U.S. Geological Survey (1979) data and from samples which we collected during the breakup period. These concentrations were extrapolated to include the entire Alaska Beaufort Sea drainages by multiplying the total slope runoff/Colville River runoff by factor of 1.7 (AEIDC 1974).

Samples of riverborne detritus for radiocarbon analyses were collected by suspending a 120 µm mesh plankton net in the Colville and Kuparuk rivers until sufficient material had been collected to assure 5 g of carbon. Samples were dried in vacuo prior tc analysis.

Total organic carbon concentrations were determined by summing the dissolved organic carbon and the particulate fraction which was collected on precombusted glass fiber filters. Temporal coverage for the hydrologic year was limited to the breakup season and single samplings in July and August. However, the bulk of the transported matter is carried during the breakup flooding (Arnborg et al. 1967), and our sampling was concentrated in that period.

Coastline erosion arises from thermoerosional niching during storm surges, and subsequent collapse and dispersion of the tundra blocks. Input rates of organic matter were obtained by multiplying the total volume of tundra eroded by the average organic carbon content of soil profiles. Lake shorelines are also subject to erosion, but the rates are dependent upon the lake size (wind fetch) and shoreline type. Comparisons of airphotos revealed that lakeshores were eroding much more slowly than the nearby seacoast in response to the much greater variations in water levels and longer fetches in the marine environment. Small ponds appear to expand by subsidence and thermokarst formation at rates independent of wind stress until large enough for appreciable wave action. Changes in small lakes and ponds were not discernible in the airphoto sequence, and erosion is not believed to be a significant means of carbon input to the water column in ponds.

Lake Primary Production

The shallow thaw lakes of the Arctic are ultra-oligotrophic and much less productive than the surrounding tundra. The Barrow ponds, for example, average 10 g $C{\cdot}m^{-2}{\cdot}yr^{-1}$ (Alexander et al. 1980). Ikroavik Lake, a typical mid-sized thaw lake, was estimated to fix 4.5 g $C{\cdot}m^{-2}{\cdot}yr^{-1}$. Other arctic lakes, tabulated by Alexander et al. (1980) ranged from <1 to 12.5 g $C{\cdot}m^{-2}{\cdot}yr^{-1}$ of phytoplankton production. Emergent macrophytes in the shallow margins of the lake are more productive per square meter than the lakes themselves in the cases noted above (Hobbie 1980). Benthic microalgae represent a larger carbon input than phytoplankton in the Barrow ponds, but little is known of their role in larger arctic lakes. Based upon the estimates of Alexander et al. (1980), we assigned an average production rate of 3 g $C{\cdot}m^{-2}{\cdot}yr^{-1}$ for benthic algae. The average primary production rate of coastal plain lakes (benthic and phytoplankton) was assumed at 8 g $C{\cdot}m^{-2}{\cdot}yr^{-1}$, a weighted mean for the ratio of large lakes (>1 km^2) to ponds wherein the higher production of benthic algae in the ponds is offset by a lower phytoplankton production. For calculation of coastal plain inputs, the area covered by lakes and ponds is taken as 50% of the total area (AEIDC 1975).

RESULTS

Major Annual Inputs of Carbon

Using the production estimates above, we calculated the magnitudes of these carbon inputs relative to each other and the major physically driven fluxes. At a mean annual primary production rate of 88 g $C{\cdot}m^{-2}{\cdot}yr^{-1}$, the terrestrial fixation far outstrips the aquatic. Given an equal distribution of aquatic and terrestrial area in the 75,000 km^2 coastal plain, 3.3 x 10^9 kg C is fixed by land plants compared to 0.3 x 10^9 kg C in the lakes and ponds. Pond margins, with the highly productive emergent macrophytes, have been included with the terrestrial producers. Of this production, grazers consume approximately 10-20% of the carbon fixed during the first year, and decomposers (fungi, bacteria) remove much of the rest (Flanagan and Bunnell 1980) over the course of several following seasons. The organic matter remaining after grazing and decomposition constitutes the peat column accumulation for that year.

Table 1 lists the basal peat ages and carbon standing stock at four sites where the soil section accumulated in place (Lewellen 1973). Unlike unfrozen bogs wherein compaction tends to increase bulk densities with depth (Clymo 1978), the formation of permafrost and the high ice content of the soils maintain the density except at polygon margins where elastic deformation occurs as the ice wedge grows (Billings and Peterson 1980). Peat accumulated at mean rates of 0.18-0.23 $mm{\cdot}yr^{-1}$ or 7.4-18.3 g $C{\cdot}m^{-2}{\cdot}yr^{-1}$ in these sites. If these rates are representative, they imply that 8-21% of the terrestrially fixed carbon is annually incorporated into the peat layer; this corresponds to a net loss to the biosphere of 0.3-0.7 x 10^9 kg $C{\cdot}yr^{-1}$ in the coastal plain alone if no remobilization occurs.

In contrast, the surface sediments from Gooseneck Lake (70°40'N, 152°47'W) and Lake C-2 (68°28'N, 156°44.5'W) contained terrestrially derived peat and yielded radiocarbon ages of 2,068 and 3,840 years B.P., indicating that the accumulation of recent carbon ceased as the thaw lake formed. Although the aquatic fixation rates add 0.3 x 10^9 kg C•yr^{-1} to the coastal lake ecosystem, utilization is nearly complete in the water column and surface sediments, leaving no excess for accumulation (Alexander et al. 1980).

The quantitative significance of the thaw lakes in the cycling of peat carbon remains uncertain. Shoreline erosion on large lakes from wave action disperses the peat across the bottom and into the water column. Airphotos clearly show plumes of eroding peat being spread by wave and current action across lakes. However, our coverage is too limited to allow an accurate estimate of the extent of lakes having sharply eroded banks. Almost all of the smaller lakes and many of the large lakes encroaching on previously drained thaw lake beds have very gently sloped margins protected by thick emergent vegetation. Here, apparently, thermokarst settling expands the lake size (Billings and Peterson 1980), and the tundra mat and underlying peat become the lake floor without being appreciably disturbed. The core sample from Lake C-2 (Table 2) indicates that the sediments had not been mixed to much extent. If the sediment ages represent tundra submerged about 4,000 years ago, the accumulation rate would have been 0.16 mm•yr^{-1}, very close to the rates obtained for the samples in Table 1. Once these thaw lakes drain, however, revegetation would cause resumption of peat accumulation with an age discontinuity occurring at the previous lake bottom level. All of the variables inherent in these processes prevent us from making a comfortable appraisal of the net flux of peat through the lake systems. Tentatively, however, we can approximate the magnitude. A shoreline erosion rate of 0.3 m•yr^{-1} on the ~500 lakes of >2 km diameter would thaw and suspend 1.5 x 10^5 m^3•yr^{-1} in the water column using an eroding bank height of 0.25 m. This material would contain some 26 x 10^6 kg C, a small amount relative to other potential fluxes. We assume that erosion in the small lakes is negligible. Even allowing a severalfold error in this estimate does not change the flux size relative to inputs by carbon fixation or fluvial transport of peat.

TABLE 2 Summary of carbon isotope data on Alaska North Slope primary producers, and detritus.

Sample Identification	$\delta^{13}C$	Years B.P.	^{14}C activity (% Modern)
DETRITUS			
Terrestrial and Freshwater			
MIL 78-3; Sagavanirktok River basal peat	-28.5	3,400	65.5
MIL 78-2; Pingok Is. basal peat	-28.3	8,432	35.0
MIL 78-1; Milne basal peat	-28.7	9,052	32.4
I-6838 Oliktok basal peat	--	8,550	34.5
I-6839 Kavearak Pt. basal peat	--	12,610	20.8
81-15; West Long Lake basal peat, USFWS camp	-27.24	9,805	29.5
79-12; particulate matter, Colv. R., 2 Jun 79	-26.39	900	89.4
79-14; particulate matter, Kuparuk R., 31 May 79	-26.89	2,683	71.6
79-11; particulate matter, Colv. R., 12-14 Jun 79	-27.01	2,375	74.4
81-6; sediments, 0-7.5 cm, Lake C-2, 27 Apr 81	-30.71	3,840	62.0
81-7; sediments, 7.5-18 cm, Lake C-2, 27 Apr 81	-29.61	4,400	57.8
81-16; sediments, Gooseneck Lake, 29 Aug 81	-27.09	2,068	77.3
PRIMARY PRODUCERS			
Terrestrial			
81-19; *Salix* spp.	-28.14		138.1
MIL 78-4; Live and standing dead tundra plants	-28.2		141.1
Freshwater			
81-8; *Nostoc* mats, pond near Prudhoe Bay, 27 Jun 81	-20.24		122.2
81-12; *Arctophila fulva*, Helmerick's Lake, 29 Aug 81	-26.85		121.1
81-13; *Carex aquatilis*, Helmerick's Lake, 29 Aug 81	-30.49		127.1

NOTE: Carbon-13/12 ratios are expressed in per mil relative to Pee Dee Belemnite standard. Carbon-14 is expressed as the radiocarbon age or percent modern relative to 95% NBS oxalic acid standard, normalized to $\delta^{13}C = -25^{o}/oo$.

Erosional and Fluvial Transport of Organic Carbon

The inputs of terrestrial carbon to the marine environment and their impact on nearshore marine food webs have been addressed elsewhere (Schell 1983). In order to estimate the major fluxes in the tundra ecosystem, the loss rates through erosion are included here. Several authors have measured coastline erosion rates (Lewellen 1970, 1973, Cannon and Rawlinson 1978, Hopkins and Hartz 1978), and S. Rawlinson (personal communication) has provided stratigraphic sections from approximately 100 sites along the Beaufort Sea coastline. Peat thicknesses are highly variable due to ice wedges and surficial relief, and usually range between 10 and 220 cm thick. The average peat thickness of 1.6 m found in areas of undisturbed accumulation contrasts with the overall average peat depth along the coast of 0.5 m. It was this latter value which we used to calculate the erosional losses. Based on eroded volumes, bulk (wet) densities and carbon content, we conclude that approximately 110 x 10^6 kg C•yr is washed into the coastal marine waters. This flux results in mineralization process as microbial activity oxidizes the peat in the relatively warm, oxygenated waters (Schell et al. 1982).

Cut bank erosion and fluvial transport of peat are a major flux of terrestrial carbon to freshwater and marine food webs. The Colville River discharge (12 x 10^9 $m^3 \cdot yr^{-1}$), represents about 30% of the drainage to the Beaufort Sea. The total organic carbon concentrations ranged from an average high during the breakup season of 19.3 mg C•ℓ^{-1} to 1.9 mg C•ℓ^{-1} during the low water season in August. We used a weighted flow corrected concentration of 14 mg C•ℓ^{-1} in calculating the average annual transport of 170 x 10^6 kg C•yr^{-1}. Stable and radiocarbon analyses of the particulate matter collected in May and June from the Kuparuk and Colville Rivers confirmed that peat, not surficial vegetation, constitutes most of the suspended organics (Table 2). Significantly, of the two Colville River particulate samples, the first (collected during maximum tundra runoff) contained more recent matter than the latter sample. These samples are very close to the mean radiocarbon content of the peat soil sections (63% percent modern) (Schell 1983).

DISCUSSION

Peat accumulation on the coastal plain annually incorporates 280-690 x 10^6 kg C•yr^{-1} into the permafrost soils. If these rates are also similar to accumulation rates in more upland sites, the north slope of Alaska accumulates 1.6-3.9 x 10^9 kg C•yr^{-1} less the loss from fluvial and erosional transport. Figure 2 depicts these fluxes projected for the entire North Slope (200,000 km^2). Although the errors inherent are compounded and this model is probably correct only in magnitudes, it serves to illustrate that the loss of carbon to the aquatic environment is a major factor governing the net accumulation of peat in tundra. Since erosional processes are widespread along the lakeshores and riverbanks, the aquatic biota have a readily available peat supply. Biological utilization of this refractory material by fungi and bacteria forms a trophic base upon which many of the apical organisms are ultimately dependent, and its availability on a year-round basis may increase its importance. Many shallow lakes and river channels where benthic algae would be more abundant are frozen to the bottom during winter months. Peat detritus collects in deeper quiet pools and lake basins, providing a perennially available source of food for insect larvae, which in turn are preyed upon by fish congregating in the ice-free deep waters. Butler et al. (1980), in assessing the energy requirements of chironomids, insect larvae heavily preyed upon by fishes and birds, concluded that they must be consuming detritus but could not conclusively demonstrate it in their situation. It appears, therefore, that even in situations where erosional inputs are small, the insect larvae take advantage of peat carbon along with the other sources available. The birds and fishes using the tundra lakes are thus assured of higher prey densities than would be normally available given the low and variable aquatic primary productivity of arctic lakes and ponds.

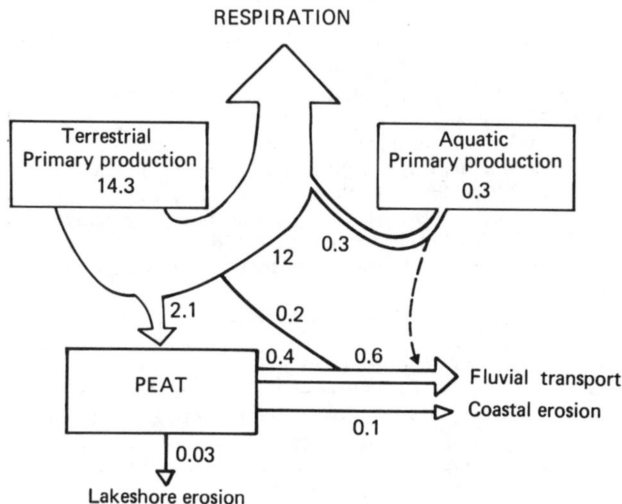

FIGURE 2 Box model of carbon flow for the North Slope (200 x 10^3 km^2). Fluxes are in 10^9 kg C•yr^{-1}.

REFERENCES

AEIDC, 1975. Alaska regional profiles, v. 2, Arctic region, Anchorage, Alaska: Alaska Environmental Information and Data Center, University of Alaska.

Alexander, V., Stanley, D.W., Daley, R.J., and McRoy, C.P., 1980. Primary producers: <u>in</u>

US/IBP synthesis series no. 13: Stroudsburg, Penn. Dowden, Hutchinson and Ross, Inc.

Arnborg, L., Walker, H.J., and Peippo, J., 1967. Water discharge in the Colville River, 1962: Geografiska Annaler, v. 28A, p. 195-210.

Billings, W.D., and Peterson, K.M., 1980. Vegetational change and ice-wedge polygons through the thaw-lake cycle in arctic Alaska: Arctic and alpine research, v. 12, p. 413-432.

Brown, J., Miller, P.C., Tieszen, L.L., and Bunnell, F.L., 1980. An arctic ecosystem: the coastal tundra at Barrow, Alaska: US/IBP Synthesis Series No. 12: Stroudsburg, Penn. Dowden, Hutchinson and Ross, Inc.

Butler, M., Miller, M.C., and Mozley, S., 1980. Macrobenthos, in US/IBP synthesis series no. 13: Stroudsburg, Penn. Dowden, Hutchinson and Ross, Inc.

Cannon, P.J., and Rawlinson, S.E., 1981. The environmental geology and geomorphology of the barrier island lagoon system along the Beaufort Sea coastal plain from Prudhoe Bay to the Colville River: Final Reports of Principal Investigators, NOAA/BLM, Boulder, CO. National Oceanic and Atmospheric Administration.

Chapin, F.S. III, Miller, P.C., Billings, W.D., and Coyne, P.I., 1980. Caribou and nutrient budgets and their control in coastal tundra, in US/IBP Synthesis Series No. 12: Stroudsburg, Penn. Dowden, Hutchinson and Ross, Inc.

Clymo, R.C., 1978. A model of peat bog growth, in Heal, O.W., and Perkins, D.F., eds., Production ecology of British moors and montane grasslands. Ecological Studies no. 27, New York, Springer-Verlag, Inc.

Flanagan, P.W., and Bunnell, F.L., 1980. Microflora activities and decomposition, in US/IBP Synthesis Series No. 12: Stroudsburg, Penn. Dowden, Hutchinson and Ross, Inc.

Hobbie, J.E., ed., 1980. Limnology of tundra ponds. US/IBP synthesis series no. 13: Stroudsburg, Penn. Dowden, Hutchinson and Ross, Inc.

Hopkins, D.M., and Hartz, R.W., 1978. Coastal morphology, coastal erosion and barrier islands of the Beaufort Sea, Alaska: U.S. Geological Survey open file report 78-1063.

Lewellen, R.I., 1970. Permafrost erosion along the Beaufort Sea coast. (Available from the author, P.O. Box 2089, Palmer, AK 99645.)

Lewellen, R.I., 1973. Special report (untitled) to the University of Alaska, Institute of Marine Science. (Available from the author, P.O. Box 2089, Palmer, AK 99645.)

Muc, M., 1973. Primary production of plant communities of the Truelove Lowland, Devon Island, Canada - sedge meadows: in Bliss, L.C., and Wielgolaski, F.E., eds., Primary production and production processes, tundra biome, Edmonton-Oslo, Tundra Biome Steering Committee.

Schell, D.M., 1983. Carbon-13 and carbon-14 abundances in Alaskan aquatic organisms: delayed production from peat in arctic food webs: Science, v. 219 (4 March, in press).

Schell, D.M., Ziemann, P.J., Parrish, D.M., and Brown, E.J., 1982. Food web and nutrient dynamics in nearshore Alaska Beaufort Sea waters: in U.S. Department of Interior/BLM/ NOAA, Environmental Assessment of the Alaskan Continental Shelf. Boulder, Colorado.

Sonesson, M., ed., 1980. Ecology of a subarctic mire, Ecological Bulletins No. 30: Stockholm, Sweden, Swedish Natural Science Research Council.

Tieszen, L.L., ed., 1978. Vegetation and production ecology of an Alaskan Arctic tundra: New York: Springer-Verlag, Inc.
U.S. Geological Survey, 1979. USGS water resources data for Alaska, Data report AK 79-1; Anchorage, Alaska.

Webber, P.J., 1978. Spatial and temporal variation of the vegetation and its productivity: in Tieszen, L.L., ed., Vegetation and production ecology of an Alaskan Arctic tundra: New York: Springer-Verlag, Inc.

SOLID WASTE DISPOSAL
NATIONAL PETROLEUM RESERVE IN ALASKA

John F. Schindler
Environmental Affairs
Husky Oil-NPR Operations, Inc.
5333 Fairbanks Street Suite 10
Anchorage, Alaska 99502

The disposition of solid waste has always been a problem for twentieth century man and in the permafrost areas of the Arctic the problem is compounded. The most utilized option has been to backhaul the debris to urban areas for disposal or recycling. In the mid-1940's the Navy's exploration program brought tons of supplies to the north slope of Alaska. This logistics deluge continued through 1953 when the construction of the Air Force's Distant Early Warning (DEW) line began, and since, there have been many years of continued government activities and their concurrent supplies. Under the Navy and later the U.S. Department of the Interior (USGS) an attempt was made to clean up and consolidate this debris. In the winter of 1979-1980 a pilot project was undertaken to bury the debris at two sites. A Cat train would travel over approved trails to the designated burials during the winter months when such travel was environmentally safe. A D-8 Caterpillar tractor with blade and ripper tooth was used. The overburden was stripped to one side and then the hole was dug using tooth and blade. When the hole was deep enough to take the debris and still be covered by a minimum of 2 feet of fill, the debris was pushed in, compacted, and everything covered. Once covered, the overburden was spread again, the scar seeded and fertilized, and the seed and fertilizer walked in with the Cat tracks. A seed mixture high in tundra bluegrass seed and standard 10-20-10 fertilizer was employed. The high rate of survival of the original plants in the overburden was surprising and gratifying. Some burials were difficult to locate from the air after only two growing seasons.

INTRODUCTION

In the Arctic, as elsewhere, man's concern for the environment evolved over many years from an almost total lack of concern, to the present scientific and technical approach. The early Eskimos of the Arctic Slope were quite nomadic, following the good hunting of the season, and discarded their refuse a few feet from their dwellings. However, the wastes were almost totally organic in nature and therefore biodegradable; although this was the practice for centuries it left no lasting damage to the environment.

It was not until the intrusion of the twentieth century and the introduction of metal that solid waste in the Arctic became a real problem. The sanitary landfill practices of disposing of solid waste that is economically feasibile in southern latitudes is extremely difficult in the north because of the permanently frozen ground. In addition, metals rust very slowly because the long season of cold severly limits the time available for the oxidation process. These factors, combined with the penchant of the Eskimo people to save scrap metals because they were exotic to their culture and therefore valuable, meant that it became the practice of all northern inhabitants to simply discard their solid wastes within sight of the front door. This practice of abandonment in place became the standard operating procedure for government agencies and private companies as well as individuals. The 55-gallon metal drum was so ubiquitous in the north it was referred to, only half in jest, as the offical flower of the tundra.

The Early Exploration Program 1944-1953

On February 27, 1923, President Harding signed Executive Order No. 3797-A creating Naval Petroleum Reserve No. 4 (now the National Petroleum Reserve in Alaska or the NPRA). Actual exploration of NPR-4 did not begin until World War II. On February 5, 1944, the Director of the Naval Petroleum Reserves sent a proposal to the Secretary of the Navy for exploration and test well drilling. On June 2 the President formally approved the project. The exploration program was well under way by the end of the year and continued until 1953, when it was suspended.

During the exploration and during later military construction, environmental repercussions of operations on the North Slope were not a major concern of the military or the American public at large. Debris associated with the programs was simply abandoned in place. Although most of the major concentrations were left behind at several exploratory drilling sites, DEW line sites, and at support bases like Barrow and Umiat, additional debris, such as wastes from seismic and geodetic survey camps, piles of empty barrels at fuel cache sites, and discarded broken or replacement equipment parts at Cat train maintenance stops, were abandoned at numerous locations. These locations were not noted in any historical record, or, if they were, the record was lost.

Numerous other programs, although of less significant magnitude, generated accumulations of debris within the reserve. These include several federal agencies: the U.S. Bureau of Land Management, the U.S. Geological Survey, and the Fish and

Wildlife Service. State agencies, auch as Fish and Game, have contributed a share, and private industry is not without blame.

The Exploration Program 1975-1982

The second exploration program was designed to be accomplished as much as possible during the winter months in order to minimize environmental impacts. Stipulations governed all operations and were based on the fact that even fragile tundra, when frozen under a foot of protective snow, is an exceedingly durable working platform.

Recent operations on the National Petroleum Reserve in Alaska were based from the 100-man camp at Lonely (POW-1). This camp was equipped with a chlorination/filtration plant to treat drinking water taken from nearby lakes and an extended aeration waste water treatment plant. The treated effluent was discharged to a small depression in the tundra. Sludge from the sewer plant, waste oils from camp operations, kitchen wastes, and combustible solid wastes were burned in the camp incinerator. The ashes, along with noncombustible materials, were disposed of in a State of Alaska approved sanitary landfill.

Drilling camps were, in essence, small versions of Camp Lonely without the sanitary landfill. All other utilities systems were about the same. Metal and ashes and any wastes beyond the capacities of the operating utilities were backhauled from the drilling camp to Camp Lonely for disposal in the appropriate manner.

Seismic and construction camps were mobile facilities, and the handling of utilities was accordingly different. Drinking water in the seismic camps came from melting snow or deep lakes when available. It was filtered and chlorinated. Some sewage facilities utilized electric incineration and others a physical chemical treatment system. Wastes were incinerated to the capacity of the units and otherwise backhauled to Lonely or a nearer facility.

Burn baskets were extensively utilized to open burn all wood and paper products. Putrescible wastes were disposed of in incinerators. Portable camps were allowed to dispose of small amounts of waste oils and other hydrcarbons by open buring under permits from the Alaska Department of Environmental Conservation.

Changes under Public Law 94-258

Although there existed environmental awareness and responsible management on the part of the Navy in the early stages of the exploration program, it wasn't until the Department of Interior assumed operational responsibility that the environmental program became more formally instituted. Public Law 94-258, the Naval Petroleum Reserves Production Act of 1976, transferred the exploration program from the Navy to the Department of Interior and charged the Secretary of the Interior to continue the exploration "in such a manner which will assure the maximum protection of . . . surface values to the extent consistent with the requirements of this Act for the exploration of the reserve" (Section 104.b). In response to this charge, the Department of the Interior instituted a Memorandum of Understanding (MOU) between the Bureau of Land Management, responsible for overall management of the Reserve, and the U.S. Geological Survey, which was responsible for the exploration. The MOU outlines the responsibilities and the steps taken annually in preparation for the proposed exploration operations.

The first step each year was to write a preliminary annual plan of operations. This plan identified and located the wells to be drilled and any alternate locations proposed. This proposal also discussed the general overall operational plan, including such specifics as which construction groups would build which sites and when, how they would travel to the sites, be resupplied, etc. In addition, the drill rig to be used was identified and the target depth and time schedule delineated. This broad plan was circulated for comment to interested entities of federal, state, and local governments as well as private organizations.

Once the general plan of operation was established, a site-specific environmental assessment (EA) was prepared for each drill site. The information for the EAs was gathered concurrently with the engineering survey during the summer months. This meant that as the construction and drilling specifications were developed for each site the environmental concerns could be considered early in the engineering design process. Early recognition of environmental factors meant better engineering, more economical operations and excellent environmental protection.

Summer surveys included the gathering of topographical and soils information, the identification of water sources and borrow areas, and the preliminary layout of the construction sites. These areas were then checked for archeological resources, fisheries resources, rare or endangered species, wildlife disturbance, and socio-economic perturbations. If necessary, water sources were changed or schedules of withdrawal modified to protect fisheries resources and borrow sites were moved to avoid disturbing archeological values. If this was not feasible, the site was excavated by professional archeologists to remove the resources for museum study. In the cases of the special areas of the Utukok caribou calving grounds or the raptors nesting in the Colville River area, operational schedules could be adjusted to minimize or eliminate any possible discommodations. All of these environmental concerns were addressed and resolved in the environmental assessments specific to each site.

History of the Cleanup Program

During the late 1960s, the deteriorating quality of the environment became a concern to the American public. Environmental organizations pointed out problems and made increasing demands for action. Against this background, the remote Arctic Slope of Alaska did not escape notice. When interest began to focus on the area for its important potential new energy sources, especially the Prudhoe Bay discovery, the area's wildlife and unique environment became of increasing concern.

The Navy included funds in its 1971 budget to initiate a cleanup of debris from the earlier exploration and military construction programs. The first Navy efforts were accomplished by the temporary assignment (2 weeks) of a company of SeaBees (CB - Construction Battalion) to the Barrow Camp in the fall of 1971. The manpower was used to police the immediate area of the camp and to manually pick up and stockpile as much debris as possible. Stockpiles were located on gravel beaches or roads where trucks and other equipment could be utilized.

Also in 1971, the oil industry operators of the Prudhoe Bay area initiated a summer cleanup program using rolligons on the tundra to pick up debris on the state lands that were under lease. This program was continued in the summers of 1972 and 1973.

The cleanup and rehabilitation program, initiated by the U.S. Navy in 1971, has continued in subsequent years. The Interior Department joined the effort during 1972 and conducted a cleanup program in the Barrow area. This program was unique in that a good deal of the work was conducted during the winter using small sleds pulled by snowmobiles.

In 1973, the Navy continued the effort with a program to clean up the Umiat Camp area. The low area east of the camp that had been used as a gravel source during exploration was used to receive the drums and debris, which were then covered with gravel. In addition, an effort was made to clean up the Point McIntyre area near Prudhoe Bay.

In the supplemental budget of the Congress for FY1974 the Navy was directed to begin a new exploration program in the reserve; the cleanup program was included as a line item in that budget. Efforts that spring centered on the Simpson Peninsula, and the debris was accumulated at the abandoned DEW line site, POW-A.

Early records of the cleanup program (1971-1975) are scattered and, although much cleanup was done, comprehensive annual reports were not prepared. Since Husky Oil-NPR Operations, Inc. assumed the responsibility (first under the Navy and later under the USGS) in December 1975, descriptions of the work accomplished have been somewhat more formalized.

The program continued under the Navy's direction until June 1, 1977, when jurisdiction of NPR-4 was transferred to the Secretary of the Interior, and the area was renamed the National Petroleum Reserve in Alaska (NPRA). The Secretary of the Interior assumed all programs and responsibilities formerly under the Navy, including the ongoing program for the cleanup and rehabilitation. Within the Interior Department, the U.S. Geological Survey was designated as the lead agency and the cleanup program was conducted as an adjunct to the exploration drilling program on the NPRA.

Cleanup-Summer 1976

During the summer of 1976, the cleanup program consisted of a great deal of hand labor. The major effort was picking up drums and debris and consolidating these materials at points accessible to vehicles, either immediately or later during winter when tundra traffic was possible. The cleanup program ran from June 28 to September 8 and a total of 331,300 pounds of debris were retrieved, 141,500 pounds of which were burned. In addition, 9,019 barrels were retrieved and stockpiled, 6,709 of these at POW-A. The area of cleanup stretched from Barrow to the Nechelik Channel of the Colville River. The following sites were cleaned as well as smaller collections of drums and debris found between these locations:

Barrow Gas Fields	POW-1 (Camp Lonely)
Iko Bay Drill Site	E. Teshekpuk Drill Site
Simpson Test Wells	Pitt Point
POW-A	Halkett Drill Site
Alaktak	POW-B
West Topagoruk	Fish Creek Drill Site

Cleanup Program-FY 1977

A trial program was conducted during the fall of 1976 to test the economic feasibility of winter cleanup (winter was defined for these purposes as November 1 to January 31). To accomplish the winter work it was necessary to provide a self-sustaining camp for an 18-man crew. Skull Cliff, an abandoned LORAN navigational site about 24 mi. south-southwest of Barrow was selected for the test effort. The site consisted of abandoned buildings, supplies, debris, and a collapsed 625-ft. tower scattered over an area of approximately 6 square mi. Work was conducted from October 26 to December 15, 1976, and during this period 2,280 barrels and 2,000 pounds of debris were picked up. The subzero temperatures were not as much a problem as the lack of daylight. Hard-packed snow and reduced light made much of the debris difficult to locate. The frozen contents of the barrels made them impossible to empty and heavy to handle. The slow progress due to these factors coupled with a lengthy mobilization period increased the cost per unit work accomplished, until it was decided to abandon the winter effort.

In the spring of 1977, cleanup efforts were concentrated at three sites: POW-A, a former DEW line site, Simpson test well, drilled during the 1947 program, and the Topagoruk test wells from the 1951 program. It was decided to pick up the debris from the Simpson and Topagoruk sites and retrograde everything to POW-A. The final plans for the disposition of the debris had still not been settled, but locating the debris at POW-A allowed the additional option of backhaul via barge to the south 48. A portable barrel crusher was used to consolidate the bulk.

Operations were begunon February 10 and were halted at the end of the first week of May. A total of 175 tons of debris were retrieved, 9,817 barrels were collected, and a total of 12,233 barrels were crushed (some of these having been retrieved previously). Although barrels were often filled with snow, covered, and frozen as in the fall work, the spring phase of the cleanup was more productive because of the increased daylight and the greater dependability and speed of the CATCO rolligons. In addition, many of the barrels at these sites were stacked above the frozen tundra and more accessible for retrieval.

Cleanup that summer was limited to the crushing of the barrels at POW-A and POW-B that had been stockpiled there that spring and during the previous summer's program. In addition, LIZ-C was cleaned up to prepare the site for use as a logistics base for the coming season of exploration.

In the FY 1977 program a total of 14,875 barrels were retrieved, 16,743 were crushed, and 485 tons of debris collected and burned or stockpiled.

Cleanup Program-FY 1978

The 1978 spring efforts at cleanup focused on the Skull Cliff and Topagoruk and East Topagoruk sites. Working out of Cat trains from April 1 through May 15, the cleanup crews stockpiled 2,120,000 pounds of debris on the beach at Skull Cliff. This included 236,000 pounds of tower, 200,000 pounds of crushed barrels, 1,380,000 pounds of cement and mud canisters from the exploratory program, and about 300,000 pounds of miscellaneous debris. Favorable weather contributed significantly to a highly productive season.

Summer cleanup of the old Navy sites commenced on June 8th and was accomplished at:

Skull Cliff tower site
Topagoruk and East Topagoruk
Wolf Creek Test Wels 1, 2, and 3
Weasel Creek Test Well
Square Lake Test Well
Titaluk Test Well
Knifeblade Test Wells 1, 2, and 2A
Mona Lisa Seismic Camp
East Oumalik Test Well
and an explosive cache discovered in the Ikpikpuk area

The summer's accomplishments were summarized as follows:

Barrels retrieved (uncrushed)	187,100 pounds
Barrels crushed	712,725 pounds
Combustible debris stockpiled	377,350 pounds
Combustible debris burned	924,310 pounds
Other debris	270,310 pounds

In excess of 1,235 tons of material and debris were burned or stockpiled.

Cleanup Program-FY 1979

Cleanup in FY 1979 was accomplished in the summer months only and the crews operated out of tent camps located near their work sites. This new method of operation proved to be very cost-effective because of the reduced helicopter and labor costs commuting to and from the work sites, and practically no lost time due to poor flying conditions. Work began at the old Brady site on the Kiligwa River with concurrent cleanup at Driftwood Creek and Liberator Lake. Efforts were then concentrated in the southeast area of the reserve. Combustibles that were stockpiled in 1978 were burned at Wolf Creek, Square Lake, Titaluk, and East Oumalik. Cleanup was accomplished at Fish Creek, Oumalik, Grandstand, and Gubic. The area of the Gubic test wells proved to have a great deal more debris than was originally estimated, including a burned-out rig, and the crews worked in the area for 32 days, better than one-third of the total summer's effort. During the summer of 1979 over 20 million pounds of debris were handled and a final 4,200,000 pounds of this material stockpiled for later burial.

Debris was stockpiled on dunnage on the tundra adjacent to possible burial sites. These sites were selected so as not to intercept any surface drainage, even the most ephemeral of water courses,

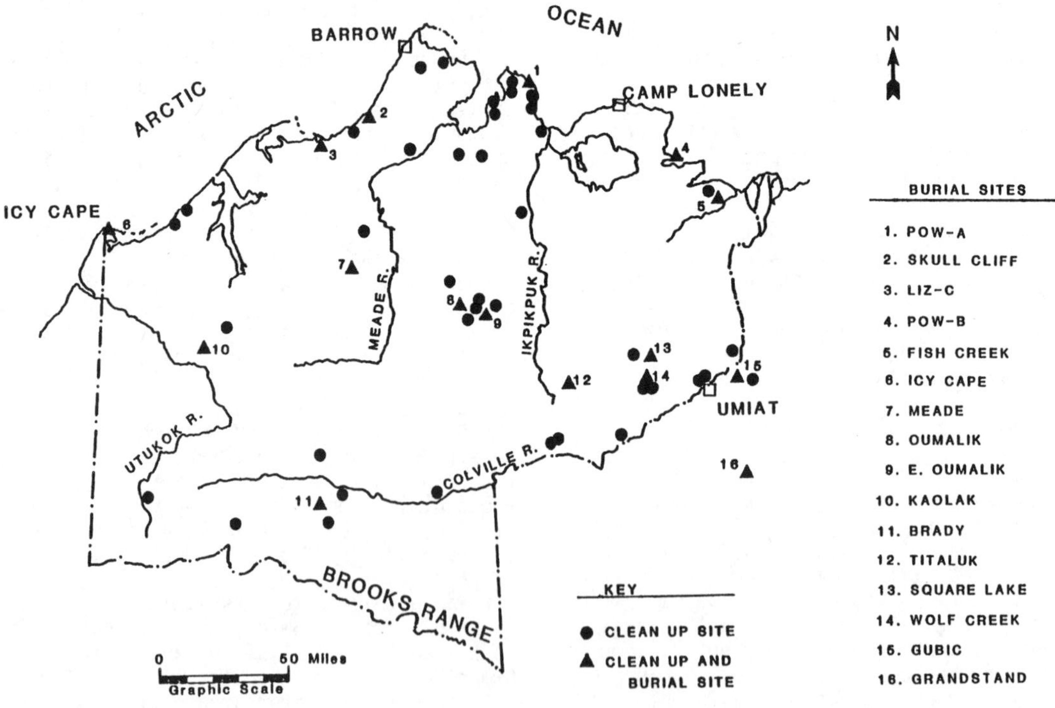

FIGURE 1 Cleanup and burial sites, National Petroleum Reserve in Alaska.

if possible. Any water erosion could uncover the buried debris and negate the entire program. The site was selected in relation to the surrounding terrain, on lower but not wet topography. The original intent was to avoid areas with soils of high ice content but subsurface ice is nearly ubiquitous on the North Slope, and once the program got under way the amount of ice did not seem to appreciably effect the cover of the debris. In actual practice, the presence of ice was desirable to ensure that ripping and excavation of the site was possible with the D-8 Cat. The lower sites were preferred as they were less visible and with more moisture, also easier to revegetate.

Cleanup Program-FY 1980

In the fall of 1979, the disposition of the consolidated debris from the cleanup of the old sites was still very much a question. It was decided to try burial at the two sites that were farthest from the ocean's edge as these would be expensive for marine retrograde. These sites were Grandstand and Gubic, both just off the reserve to the east and southeast of Umiat. The sites were on selected lands, so the prospect was discussed with the Bureau of Land Mangement and the prospective owner, the Arctic Slope Regional Corporation. When they agreed it was a feasible method of disposal, the permission of the Alaska Department of Environmental Conservation was obtained. A contract was let in early March 1980 for a Cat train with at least one D-8 Cat equipped with a ripper tooth and blade to do the burial effort. The idea was to strip away the organic overburden and stockpile it to one side; then excavate a pit of sufficient size to contain the debris; place the debris in the pit and compact it with the heavy equipment; and backfill with the original material so that a minimum if 2 ft. covered the debris. The overburden would then be spread over the site and seed and fertilizer distributed and walked in with the Cat. The effort went very smoothly.

The 1980 summer cleanup of old Navy sites and the cleanup and revegetation of drill sites of the current exploration program were combined under one contract. The season's operations got under way during the last week of May. An 18-man tent camp was set up at Oumalik;utilizing a Bell 205 helicopter for "lift" power to move the heavy timbers and other large debris, they began the cleanup of Oumalik, East Oumalik, Brady, Mona Lisa, and a number of new finds and old caches of explosives. The term "new finds" was used to designate any considerable amount of debris encountered locally within approximately 5 mi. of the site that was not scheduled in the work plan. It was decided to burn the explosives and their disposal was coordinated with the ADEC (Alaska Department of Environmental Conservation), the USGS, and other government authorities. By July 24, the tent camp was relocated to the Old Meade site and the crews helped with the cleanup at some current drill sites during the camp move.

Demobilization of the cleanup camp at Old Meade began on August 20 and the operation was out of the field by August 25. Although this was early according to schedules maintained in previous years, during the last days of operation, the camp was plagued by freezing water pipes and excess condensation in the tents. This season a total of 2,647,620 pounds of debris was stockpiled for burial; 1,817,300 pounds consolidated at Oumalik and 830,320 pounds consolidated at Old Meade. New finds at Oumalik totaled an estimated 389,000 pounds, and at Old Meade they totaled approximately 51,000 pounds.

In late September 1980, an inspection was made of the burial sites that were finsihed the previous spring to check for any erosion or settlement problems due to thaw. None was found. One unanticipated advantage to the burial program that was discovered at this time was the high survival rate of the original site vegetation. When the overburden was stripped prior to the excavation it was stockpiled to one side and then used to cover the area after filling. Because the stockpile was exposed to the winds and cold for only a short time, usually less than 48 hrs., and then protected by a cover of blown snow soon after replacement, quite a few individual plants survived. The cover was a jumble of organic clods, but almost all showed signs of recovery of the original vegetation, including willows and birches. These survivors meant that the recovery of the natural vegetative cover would occur far more quickly than at sites that were seeded with grass species only.

Cleanup Program-FY 1981

The 1980 spring burial efforts were considered very successful, and it was decided to bury all the stockpiled debris. During the last few days of February, the burial contractor mobilized his Cat train and equipment to Koluktak via rolligon following previously cleared trails. Koluktak was chosen for a jumping-off point because there was an ongoing drilling operation there, complete with camp and airstrip, and the location was convenient to the first burial site. The program got under way on March 9, 1981.

The contractor's task was to travel overland to the sites where the debris from previous cleanup programs had been stockpiled. Locations for burial were designated and archeological and environmental clearances had been obtained earlier. The contractor operated as in the pilot program and burial was accomplished at:

East Oumalik	Square Lake
Oumalik	Old Fish Creek
Old Meade	POW-B (Kogru)
Titaluk	POW-A (Cape Simpson)
Wolf Creek	

Because deep snow conditions prevented the train from reaching Brady, the POW-A site was substituted at the end of the season. The program was finished by April 19, and the contractor demobilized the train overland to Deadhorse.

Since the summer of 1981 was possibly the last opportunity to accomplish environmental work under the exploration program, the work scope was expanded to complete as much as time and weather would allow. Cleanup began at miscellaneous minor sites in June. High winds and fog in the western part of the reserve slowed down the establishment of the cleanup tent camp at Icy Cape but the operation was finally in place by July 2. A second siege of winds up to 50 knots hampered work for another week. Icy Cape was finished by July 28 and the camp was moved to the Kaolak site. The old sites cleaned that season were Icy Cape (LIZ-B), Peard Bay (LIZ-C), Kaolak, and numerous small finds.

Cleanup Program-FY 1982

The burial of the debris collected during the summer of 1981 was accomplished in the spring of 1982 at Skull Cliff, Brady, Kaolak, Icy Cape, and Peard Bay. The Skull Cliff operation was the major effort because an estimated 4,350,000 pounds of debris, which had been stockpiled on the beach for possible marine retrograde, had to be moved 1 mi. inland to the burial site. Howver, even with this rather large task, the program proceeded very smoothly. The contractor was working in the field on March 15, and began demobilization on April 17. Seed and fertilizer were applied to all sites as the crews departed.

REVEGETATION METHODS

The program for the revegetation of the burial sites was totally based on the program developed for the revegetation of the material (borrow) sites and drilling sites of the ongoing exploration effort. A review of the methods, materials, and reasons developed for that program is pertinent here.

Once a drilling site is cleaned up and and any planned civil work accomplished, the site is ready for seeding. Active seeding is done prior to late June or the first of July and dormant seeding after the first freezing temperatures of the fall, usually late August. At the beginning of the exploration program, the first attempts at revegetation employed the grass seed mixtures utilized by Alyeska for the northern areas of the Trans-Alaska pipeline. This was a mix by weight of: 30% annual rye grass, 10% boreal fescue, 10% redtop, 20% nugget blue grass, and 30% arctared fescue. This was applied at the rate of 50 lbs. per acre and fertilized with a commercial chemical fertilizer 10-20-10 (nitrogen-phosphorous-potassium) applied at the rate of 600-650 lbs. per acre. Seed was walked in with Caterpillar tractor tracks, or dragged in with a piece of chain-link fence. This seed mix, fertilizer, and application rates were found to be moderately successful at inland sites and unsuccessful at coastal sites.

A review of the available literature revealed there were as many theories and methods of revegetation as there were reports. Advice ranged from planting willows and birch seedlings by hand, to lush seeding of grasses, to sparse seeding of grasses, to no planting or seeding at all, to very heavy fertilization, to no fertilization, or almost any combination of these.

The goal of the revegetation program was that the seed and fertilizer selected and the methods of application would produce growth that would prevent erosion, both wind and water, establish a vegetative cover, both for aesthetics and biohabitat, yet not establish such a think vegetative cover as to deter the invasion of the original flora.

It was decided, in consultation with Drs. J. McKendrick and Wm. Mitchell of the Univeristy of Alaska Palmer Experiment Station, to use tundra bluegrass (Poa glauca), although the availability of this seed was quite limited. The bluegrass was used primarily in coastal areas, and a mix of tundra bluegrass, arctared fescue, nugget bluegrasss, and even annual rye was used in inland areas.

This native bluegrass species was grown from cultivars collected in the Sagwon Bluffs area of the Alyeska pipeline. Being native to the North Slope and the seed raised in Alaska, cultivation was very successful on the north slope. Using this seed also had the added advantage of not introducing an exotic into the ecosystem. The seed shows good growth success in both active and dormant seeding. It is an upright grass that grows in clumps of 4-8 or more stalks per base, which allows open space around the base of the plant for the invasion of native mosses, lichens, and grasses, yet prevents erosion. The height of the grass also enhances natural succession by providing a means of entrampment of local seeds.

The bluegrass seeds, or the seed mixture in inland areas, were dispersed at an application rate of 28-35 lb. per acre. This provided adequate coverage yet was not too dense as to prevent the return of other native species. All seed was applied with hand broadcasters, usually the type strapped to a man's chest or back.

For nutrients, the standard commercial shelf product, chemical fertilizer 10-20-10 or 10-20-20, proved to be completely adequate. Phosphorus was found to be severely limiting almost everywhere, so any fertilizer that was inexpensive, available, and provided phosphorus would do well. Not much advantage was found in greater amounts and lesser amounts did not appear to support adequate growth, so an application rate of 550-600 lb. per acre was decided upon. It was found that a second season's fertilizer application was very helpful in promoting increased survival. These were all qualitative observations. Randon sampling of the pads and surrounding areas did not detect any significant increase in soil acidity that could be traced to the fertilizer.

SUMMARY AND CONCLUSIONS

The objective of the cleanup program was environmental enhancement through the removal of as much as possible of the evidences of man's activities. To promote this objective, all operational methods were designed to be protective of the environment.

Actual cleanup was best accomplished in summer when debris was not obscured by snow or frozen in the tundra. Retrieval was done manually with the assistance of a hoist-equipped helicopter. All overland traffic involving heavy equipmeny was restricted to the winter when the tundra was frozen and covered with a protective layer of snow.

Selection of burial sites was made to avoid any surface drainage channels and be located on low but not wet terrain. Subsurface ice content does not appreciably affect the burial results and generally makes easier digging. Revegetation should use native grass seed if at all possible. Tundra bluegrass (Poa glauca) proved to be the most adaptable to the greatest variety of North Slope sites. Any commercial fertilizer with phosphorus will promote growth.

The excavation/burial process should be done as quickly as possible to limit the time the tundra overburden is exposed to winds and freezing temperatures. If possible, replace the snow cover or promote the collection of blowing snow over the disturbed surface to promote an increased survival rate of individual plants.

SOME NONTRADITIONAL METHODS FOR SEISMIC INVESTIGATIONS IN PERMAFROST

B. M. Sedov

North-East Interdisciplinary Science Research Institute,
Academy of Sciences, Magadan, USSR

Small taliks and ice masses, both occurring at shallow depth, as well as other features make refraction and reflection techniques rather ineffective for permafrost studies. Through the use of data on the physics of ultrasonic oscillations and on seismic wave velocities and dynamics observed in permafrost areas, some nontraditional seismic exploration methods are suggested. They are based on converted mono- or polytype Rayleigh waves generated at the permafrost-talik boundaries, reverberations over ice layers, and other parameters. These methods enable us to enlarge the scope of permafrost research. Furthermore, the utilization of high-amplitude Rayleigh waves ensures environmental protection.

INTRODUCTION

In permafrost studies, seismic prospecting was used, for the first time, exactly 50 years ago (see Bonchkovsky et al. 1937). Refracted waves were utilized to determine the position of the upper permafrost and first P wave velocities were measured. Trial pits revealed that at depths of 1.5 m the seismic method exaggeration was equal to 4-7 cm. That was remarkable for the beginning. Since then seismic techniques have made considerable progress in permafrost research. A review of the tasks facing seismic exploration techniques was presented by Scott et al. (1979). Despite evident achievements, certain problems remain to be solved by permafrost scientists which are due to the following: First, taliks occurring in frozen terrain commonly have sizes comparable with seismic wave lengths; their boundaries are vertical or subvertical and crop out at the surface while the nearby thaw zones are overlain by a frozen active layer. Seismic velocities are found to be similar or identical in ice and frozen material. Detection of massive ice in such cases is not possible with ordinary refraction or reflection methods. To find more adequate solutions to the geocryologic problems confronting seismic exploration, certain nontraditional approaches are needed.

PREREQUISITES FOR NONTRADITIONAL METHODS

In permafrost areas, the seismic properties of loose rocks depend on both their lithology and their physical (thawed or frozen) condition. In the latter case, seismic waves are characterized by high velocities and frequencies, while Rayleigh waves have high amplitudes (Sedov 1976a). At the thaw-frozen rock boundary, the P wave acoustic impedance changes in a stepwise manner due to velocity variations (up to a factor of 2.9 or more for R 12 in sands). The contrast is even higher in frozen and dry rocks. At the ice-frozen rock boundary, a density-induced change of acoustic impedance is possible for P waves up to a factor of 1.7 in clay and up to 3.1 times in sand. Except for the water-bottom boundary in marine seismic surveys, these conditions are not observed where permafrost is absent. The bottom effects on seismic waves are known widely (Burg et al. 1951). In ultrasonic physics, it has been shown that the energy of R waves is converted on rigidity-contrastic (metal-air) surficial inhomogeneities (De Bremaeker 1958, Viktorov 1966). Seismologists have also suggested that R waves can be dispersed by both positive and negative relief features. However, in all these cases, solid matter contacted air, i.e. a medium that actually does not transfer energy but returns all the energy back to the solid body.

EXPERIMENTAL DATA

With due consideration of geoseismic peculiarities of permafrost as well as ultrasonic physics and seismological data, seismic wave fields have been studied over variously shaped and sized and genetically different taliks and ground ice masses as well as lithologically different permafrost sections. *In situ* investigations were undertaken for seismic wave velocities in low-temperature rocks and for their dependences on temperature, ice content, age, origin, peculiar changes at the talik-permafrost boundaries, and other features (Sedov 1976a). Seismogeologic type columns and effective two-dimension models have been elaborated for permafrost areas which took account of the rock physical condition and densities in the profile (Sedov 1976c).

In view of space limitations, only some nontraditional methods for seismic permafrost research are dealt with in this paper.

Figure 1 demonstrates the influence exerted by the talik-permafrost contact on the seismic wave field (Dyuryagin et al 1975, Sedov 1978). Criteria for its recognition are inferred experimentally. P and R waves excited in frozen rocks propagate into talik as longitudinal PP_0 and RP_0 waves without other type waves generated at the contact. From thaw rocks, P_0 wave transfers the energy through the contact as a single-type P_oP wave and converted P_oR wave. Sound P_{air} wave forms converted P_{air} wave at the contact which propagates within the frozen rocks. As the waves impinge on the contact, it turns into a source of oscillations for the medium into which the energy is transferred. This is attended by velocity and frequency changes and by redistribution of the oscillation energy. For example, just as in the case of an explosion on frozen rocks, the P_oR wave amplitude is greater and its frequency smaller than those of the P_oP wave.

Whenever the horizontal extent of a talik (for instance, sub-channel talik) is comparable with the wave length, a different wave pattern can be observed. Figure 2 presents seismograms characteristic of the field over narrow (A) and broad (B) subchannel taliks overlain by a layer of seasonally frozen rocks measuring some 2.5 m in thickness (Sedov et al 1976). As the waves reach the taliks, they form converted PR and RP waves together with single-type passing waves, while the R wave forms, in addition, a RR wave. With other factors being equal, amplitudes of the PR, RP, and RR waves depend on the lateral extent of taliks.

Different-type high-amplitude converted waves are generated at the boundaries of sublacustrine taliks and from water cores of frost mounds, in steam-and-water thawing zones, pits, crack zones around blast holes, etc. Seismic wave fields over some of these inhomogeneous structures have been discussed previously (Roshchin et al 1975, Dombrovsky et al 1977, Sedov et al 1977). Despite the presence of similar features, each of the mentioned cases has its peculiar properties enabling recognition of an individual object.

The dependence of R-wave velocities on the lithology of perennially frozen rocks provides for determination of layer composition while pseudo-Rayleigh waves enable one to find the depths of occurrence and thicknesses, including the profiles where the velocities diminish downward (Sedov et al. 1978).

By changing the R wave length (changing the mass of charge or falling weight and vibrator effects) one can study the structure of perennially frozen rocks downward in the section. R waves have the greatest amplitudes with explosions on the surface. This enables one to reduce the amount of explosive used or, occasionally, to employ shocks or aerial explosions which spare the surficial layer of the ground.

Nontraditional seismic exploration methods have also been utilized in investigations of subpermafrost taliks occurring upon bedrock basements and covered by frozen rocks that are tens or even hundreds of meters thick (Bykov et al. 1978). Just as in the case of marine seismic surveys, water reverberations are observed above ground ice layers due to reflections of seismic waves from upper and lower boundaries (Sedov 1978). The "soft-bottom" formula and seismic velocities in the overlying frozen sediments enable one to determine the depth of the ice upper boundary with the aid of frequency value for the first reverberation wave. Ice thickness is found by the aid of the second reverberation wave and ice-related seismic velocities. Recognition of the second reverberation wave is facilitated by the oscillation attenuations, which are more rapid in frozen materials than in ice.

CONCLUSIONS

The described nontraditional seismic exploration techniques provide for broader geocryologic research works.

Employment of higher-amplitude R waves results in less violent excitation sources, which may be replaced by above-ground blasts. This means a lesser environmental hazard and more permissive seismic exploration techniques.

Nontraditional methods are useful tools for investigation of subsurface taliks characterized by dynamic size changes.

ACKNOWLEDGMENTS

The author wishes to express his hearty thanks to E. M. Anisimov, G. I. Roshchin, G. F. Chulkov, and other coworkers for assisting in the author's field studies. Special thanks go to Yuri Stoma without whose assistance this paper would never appear in English.

REFERENCES

Bonchkovskiy, V. F., and Bonchkovskiy, Yu. V., 1937, Studies of the applicability of a seismic method for determination of the depth of the upper border of permafrost (in Russian), Transactions of the Commission for Permafrost Studies, Vol V, Izd-vo, USSR Academy of Sciences, p. 131-164.

Burg, K. E., Eming, M., Press, F., and Stulken, E. I., 1951, A seismic wave guide phenomenon: Geophysics, v. 16, p. 594-612.

Bykov, I. A., Roschin, G. I., and Sedov, B. M., 1977, Measurements of Location of the lower border of permafrost by means of the KMPV seismic survey method, in Methodology of engineering - geologic investigations and mapping of permafrost (in Russian), Proceedings of the seminar, v. 2, Geophysical and mathematical engineering - geological studies, Yakutsk, p. 14-15.

De Bremaecker, G. C., 1958, Transmission and reflection of Rayleigh waves at corners: Geophysics, v. 23, p. 253-266.

Dombrovskiy, G. A., Tolpegin, Yu. G., Musin, L. M., and Sedov, B. M., 1977, Application of Rayleigh waves for selection of sites for autumn-winter openings in peat soils (in Russian): Kolyma, v. 8, p. 3-4.

Dyuryagin, B. S., and Sedov, B. M., 1975, On application of seismic surveys for the selection of thawed ground in loose deposits (in Russian): Kolyma, v. 12, p. 8-9.

Scott, W. J., Sellmann, P. V., and Hunter, J. A., 1979, Geophysics in the study of permafrost, in Proceedings of the Third International Conference on Permafrost, v. 2, Ottawa: National Research Council of Canada, p. 93-115.

Sedov, B. M., 1975, Application of seismic surveys for location of ground ice, in Materials on geology and minerals of the North-East of USSR (in Russian), v. 22, Magadan, p. 225-226.

Sedov, B. M., 1976a, Seismic wave rates in permafrost rock formations, in Geophysical methods for investigation of frozen ground (in Russian): Yakutsk, p. 107-117.

Sedov, B. M., 1976b, Seismological profiles and models for permafrost regions, in Geophysical methods of investigation of frozen ground (in Russian): Yakutsk, p. 139-143.

Sedov, B. M., 1978, Application of Rayleigh waves for location of surface taliks, in Reference Information, Ser. XV, Engineering survey in construction, Geophysical investigations (in Russian): Moscow, v 8 (73), p. 12-16.

Sedov, B. M., Chulkov, G. F., and Mishin, S. V., 1976, Experimental data on mutual exchange of longitudinal and surface waves, Izvestia of USSR Academy of Sciences, Physics of Earth (in Russian), v. 8, p. 95-99.

Sedov, B. M., Chulkov, G. F., and Musin, L. M., 1977, Application of Rayleigh waves for location and monitoring of taliks in permafrost, in Materials on geology and minerals of the North-East of USSR (in Russian): Magadan, v. 23 (1), p. 131-136.

Viktorov, I. A., 1966, Fundamentals of physics of engineering application of Rayleigh and Lamb supersonic waves (in Russian), Izd-vo, Nauka: Moscow, 176 p.

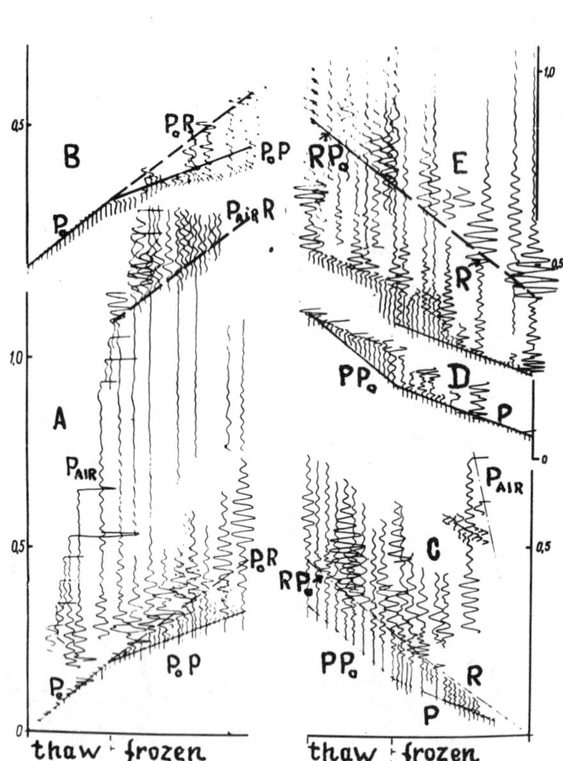

FIGURE 1 Seismic record demonstrating the influence exerted by talik-permafrost vertical contact in sandy gravels on the seismic wave field. Double-ended spread is 940 m long with 48 traces consisting of 1 geophone. In-line offset is 320 m (for B), 940 m (for D), and 500 m (for C). Records were obtained without AGC. Wave parameters: velocities: $V_P = 4550$ m/sec; $V_R = V_{P_oR} = V_{P_{air}R} = 2200$ m/sec; $V_{P_o} = 1850\text{-}1900$ m/sec; $V_{P_{air}} = 330$ m/sec; frequencies: $f_P = 100$ Hz; $f_R = 30$ Hz; $f_{PP_o} = 50$ Hz; $f_P = 65$ Hz; $f_{P_oP} = 100$ Hz; $f_{PP_o} = 37\text{-}40$ Hz; $f_{P_{air}} = 30$ Hz; $f_{P_{air}R} = 30$ Hz.

a.

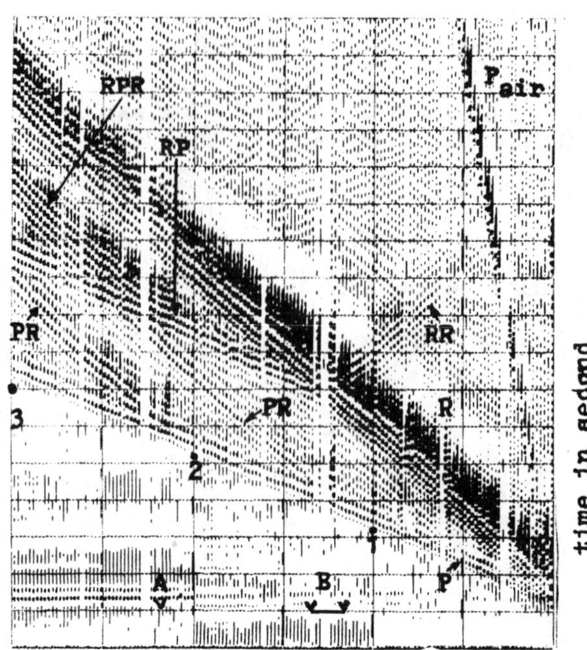

b.

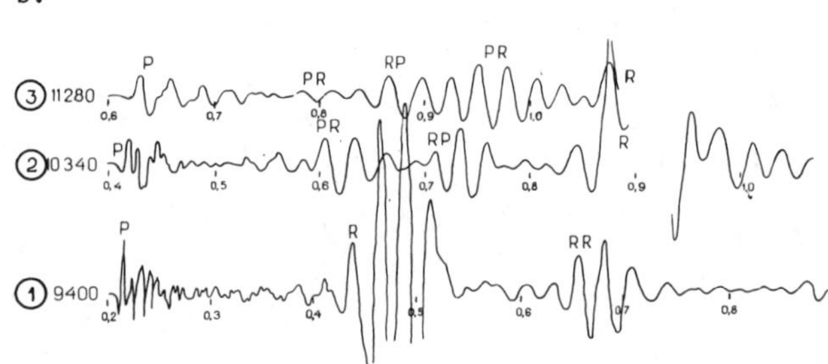

FIGURE 2 (a) Field record demonstrating the influence exerted by two subchannel taliks of different extent on the seismic wave field; the length of A talik is 20 m and that of B talik is 80-90 m. Single-ended spread is 2820 m long with 142 traces consisting of 1 geophone.
(b) Large-scale wiggle-trace presentation of the same data as in Figure 2a. The AGC is OFF. Additional wave parameters: velocities: $V_P = V_{RP} = 4550$ m/sec; $V_R = V_{PR} = V_{RR} = 2150\text{-}2200$ m/sec; frequencies: $f_P = 100\text{-}200$ Hz; $f_R = F_{RR}$ 30 Hz; lengths: $\lambda_P = 20\text{-}45$ m; $\lambda_R = 40\text{-}50$ m; $\lambda_{RP} = 100\text{-}115$ m.

ECOLOGICAL RELATIONSHIPS WITHIN THE DISCONTINUOUS PERMAFROST ZONE OF SOUTHERN YUKON TERRITORY

J.P. Senyk[1], E.T. Oswald[2]

[1]Lands Directorate, Environment Canada, Pacific Forest Research Centre, Victoria, British Columbia, V8Z 1M5

[2]Canadian Forestry Service, Environment Canada, Pacific Forest Research Centre, Victoria, British Columbia, V8Z 1M5

Recognizing the immediate need for baseline resource information to cope with present and anticipated exploitation and development, an exploratory level integrated resource survey was undertaken in the Yukon Territory. This survey, which identified 22 broad scale ecosystems (Ecoregions), was succeeded by more detailed ecological land classification studies (Ecodistrict, Ecosection) in a number of locations throughout the southern part of the Territory. During these surveys a number of ecosystems containing perennially frozen soils and/or that evolved as a result of permafrost degradation were examined and characterized. Several of these are described and relationships between the various ecosystem components discussed.

INTRODUCTION

Ecological Land Classification is one approach to partitioning the earths' natural resources into geographically bounded "parcels" (ecosystems) that have certain biological and physical ties, continuity and homogeneity in terms of climate, terrain, bedrock, soil, vegetation, water and fauna. The system involves an interdisciplinary team of scientists, relies heavily on the interpretation of various forms of remotely sensed imagery and is heirarchical in nature, allowing for ecosystem definition at six levels of generalization (Figure 1) each of which provides information of appropriate scope and detail for a particular level of management.

ECOPROVINCE . . . . 1:3 000 000 to 1:10 000 000

ECOREGION . . . . 1:1 000 000 to 1:3 000 000

ECODISTRICT . . . . 1:125 000 to 1:500 000

ECOSECTION . . . . 1:50 000 to 1:250 000

ECOSITE . . . . 1:10 000 to 1:50 000

ECOELEMENT . . . . 1:2 500 to 1:10 000

FIGURE 1 Ecological land classification heirarchy and commonly related map scales.

Initial survey work in the Yukon Territory recognized 22 ecoregions (Oswald and Senyk 1977). Subsequently, more detailed surveys were conducted in some of the southern ecoregions (Figure 2) (Oswald, compiler. 1979; Senyk, et al. 1981). These studies adopted the more or less formalized, hierarchical approach to ecological land classification of the Environmental Conservation Service Task Force (1981). Because the surveys conducted to date have been of an exploratory and reconnaissance nature, ecosystem recognition has been based largely on surface form, expression and composition of terrain, vegetation and water. Presence and surface expression of permafrost is used as an indicator of environmental condition,

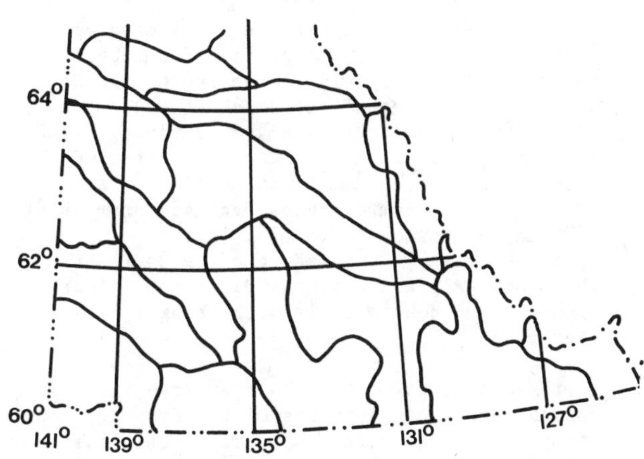

FIGURE 2 Ecoregion boundaries of southern Yukon Territory.

and coupled with observable changes in other components of the ecosystem, supplies additional evidence for ecosystem differentiation. Patterned ground, such as circles, polygons and stripes and other features such as peat plateaux, palsa, surface collapse or thermokarst scars can often be related to distinct differences in local and to some extent regional climates.

Fire disturbance has been nearly universal in southern Yukon. Its effects have been to create a mosaic of biophysical patterns in various stages of succession. Forecasting stable conditions after disturbance is difficult, particularly when consideration is given to the development of perennially frozen soils concomitant with other components of the ecosystem.

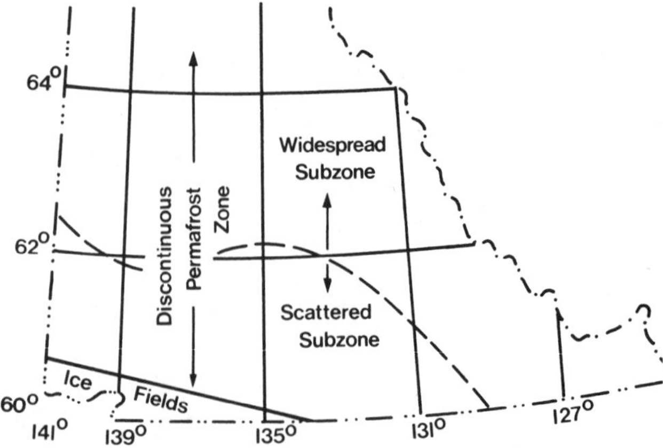

FIGURE 3 Permafrost Subzones of Southern Yukon Territory

DESCRIPTION OF AREA

Southern Yukon Territory is an area of rolling to hilly plateaux lands rimmed by high, rugged mountains on the southwest, north and east sides (Bostock 1965). The climate, described as subarctic continental (WAHL pers. comm.) shows the influence of the Pacific Ocean, specifically the Gulf of Alaska and the relatively frequent intrusion of mild air. Air flow is predominantly from the south and west, and as a result areas located to the south and west of mountain ranges receive considerably more cloud and precipitation than areas on the north and east sides. Annual precipitation increases in a northerly and easterly direction from about 220 mm per year immediately in the lee of the St. Elias and Coast Mountains to about 400 mm in the north and 750 mm in the Mackenzie Mountains to the east. Nearly half of the precipitation occurs during the summer months (June - August). Average annual temperatures decrease from about -1°C in the southwest to -7°C in the north and east. Temperature extremes increase in a northerly and easterly direction. These figures vary considerably with changes in elevation.

The area lies within the scattered and widespread permafrost subzones of the discontinuous permafrost zone, with the nearly continuous snow and ice covered St. Elias and Coast Mountains falling within a separate subzone (Brown in press) (Figure 3).

Except for the northwest corner, the area was glaciated during the Pleistocene (Hughes et al. 1969). Subsequently, nearly the entire area was covered with volcanic ash, which decreases in depth and particle size from the southwest to the east and north (Hughes et al. 1972).

Vegetation is typical of the northern sections of the Boreal Region (Rowe 1972), with Alpine Tundra occuring on all the higher elevation terrain. White spruce, lodgepole pine, alpine fir and black spruce are the common softwood species with tamarack having limited distribution; trembling aspen, balsam poplar and white birch are common hardwood tree species. White spruce occurs nearly universally throughout this part of the Yukon; black spruce is scarce to absent in the south centre. Tamarack is specific to the southeast, while alpine fir is present at the higher elevations throughout the eastern half. Lodgepole pine is common to abundant in east and central parts but generally absent in the west. Trembling aspen is common after fires throughout much of the area; balsam poplar on upland sites is most common in the west central part and white birch is scattered throughout, being mostly on cooler sites.

Soils in the southwest and south central parts are largely Eutric Brunisols, with Static and Turbic Cryosols occurring sporadically throughout but most commonly at higher elevations and in depressions (C.S.S.C. 1978). These grade to Dystric Brunisols and minor Luvisols and Podzols towards the east. Cryosols are common here in organic and imperfectly drained fine textured deposits at lower elevations. Cryosols become more prominent across the north occurring in a variety of materials and landscape positions. These are associated with Dystric and Eutric Brunisols and occasionally Podzols.

ECOLOGICAL RELATIONSHIPS

Mineral Soils

Permafrost occurs only sporadically at lower elevations, in the dry, warm climates in the lee of the Coast and St. Elias Mountains in the southwest and is confined largely to organic deposits, lower north aspect slope positions and fine textured soils. Of special interest in this area, are the thick glaciolacustrine deposits laid down during late Pleistocene time in the Takhini Valley (Lat. 60° 45'N, Long. 137° 35'W). In various locations throughout the central part of this valley the deposits consist of silts and silty clay materials to a depth of 60 m and greater (Kindle 1953; Hughes et al. 1972). Thaw lakes or thermokarst depressions are abundant in this area and are clearly visible on conventional airborne imagery and, with careful visual observation, on satellite scenes. Part of the basin was burned approximately 25 years ago and present vegetation after disturbance consists largely of willow, aspen, grass and minor white spruce and lodgepole pine regeneration. The undisturbed portion supports open to closed stands of white spruce and minor aspen with a nearly continuous ground cover of moss. Soils are Static Cryosols and minor Eutric Brunisols (Senyk et al. 1981).

Though the thaw processes are far more active in parts of the disturbed area, they are also acting on the undisturbed ecosystem and appear to have been operative for a considerable period of time. The area where activity is most noticeable covers approximately 40 to 45 km^2. On the eastern fringes of the basin, relic or stabilized thaw ponds are evident, though not abundant (Figure 4). They are generally small, 10 to 15 m in diameter and mostly shallow, 1 to 2 m, in depth. Some of these were likely initiated during the building of the Alaska Highway with the resultant disturbance of the thermal equilibrium of the underlying permafrost, others are "natural", still others due to fire disturbance. The terrain surface is nearly flat to slightly mounded.

FIGURE 4 Stabilized thaw pond initiated during construction of Alaska Highway

Permafrost occurrence, ice content and form appears to vary considerably from east to west. In the central part of the basin, depressional alkali flats are common (Figure 5). These bear strong resemblance to the "alases" described in Washburn (1973). The terrain surrounding these depressions is only slightly mounded and contains abundant surface cracks with permafrost at

FIGURE 5 Alkali encrusted depressions similar to Alases.

depths slightly greater than 1 m (Figure 8). Vegetation surrounding these depressions consists largely of aspen, white spruce and grass. The depressions are heavily crusted in carbonate salts with sedge and grass around their borders. Depth to permafrost was greater than 1.2 m beneath these depressions.

In this same vicinity and slightly to the west, thaw or thermokarst ponds are frequent under open to closed white spruce/feathermoss cover. The ponds are relatively uniform in size, ranging from about 35 to 45 m in diameter, with the collapsing banks averaging 2.5 to 3 m above water line (Figure 6). Occasionally thaw ponds have coalesced, partially infilled and presently support willow, sedge and grass. The uniformity in size of the thaw ponds might

FIGURE 6 Thermokarst depressions in "undisturbed" ecosystem.

suggest a similar time frame for initiation. Depth to permafrost is relatively uniform under the undisturbed vegetation, being in the vicinity of 0.7 m. Examination of the exposed banks indicated a relatively uniform ice form and content to depths of 1.7 to 2 m. No major lens or massive ice forms were encountered in the thaw ponds examined.

The collapse process proceeds relatively slowly, with sloughing material ravelling rather than collapsing. This material forms debris covered lower surfaces which insulate the underlying permafrost and slow the thaw process. The shading and insulating potential of the surrounding vegetation cannot be underestimated.

In the northwestern part of the basin, in a 25 year old burn, thaw collapse processes are extremely active. The terrain surface is slightly more mounded than the area previously discussed. Vegetation consists largely of willow, aspen, grass and sedge with minor white spruce regeneration. Massive ice, of a planoconvex form containing scattered air bubbles, underlies 1-1.2 m of windworked lacustrine sediments (Figure 7). The thaw layer extends to the upper contact with the massive ice. Though the exact extent of this condition was not determined, nearly identical collapse profiles extend over 3-4 km^2.

FIGURE 7 Wind-worked glaciolacustrine sediments overlying planoconvex ice.

There appears to be a very close resemblance to "pingo" ice form, however the pronounced mounded nature attributed to pingos is absent. What is probably being viewed here is a pingo cluster, some of which is in the latter stages of collapse. Stabilized collapse depressions surrounding these actively collapsing features are likely scars of former pingos. The collapse feature depicted, extends over 2-3 ha and has coalesced with adjacent smaller actively collapsing features. Bank collapse is abrupt, the banks being nearly vertical and 2-3 m in height. Surface debris occasionally collapses in large blocks leaving the vegetation relatively intact.

Probing outward from the slumping bank indicated a dished or concave nature to the debris covered ice surface extending towards the centre of collapse. Moist mineral soil at the ice contact indicates that a general downwasting of the ice is occurring. Water is being removed from the system through drainage developed by the coalescing of collapse features. The collapsed debris is water saturated and appears to flow towards the initial collapse centre. Ponds are present in all collapse depressions.

The ice contains entrapped air bubbles though they are not so abundant as to lend a milky appearance to the ice. The convex bands of ice are clearly visible though there does not appear to be a discontinuity or cleavage plane along these bands when pieces of ice are broken loose.

Organic Soils

In the southeast, wetlands are considerably more abundant than in the preceding region. In this part of the Yukon Territory, permafrost occurs in most organic deposits and imperfectly drained mineral soils.

The following example is taken from an area approximately 45 km north of Watson Lake (Lat. 60° 07'N, Long. 128° 49'W). First identified on 1:70,000 scale air photographs, largely through the abundance of collapse features and leaning trees, the wetland was later examined on the ground. The uplands surrounding the depression are comprised of morainal deposits, sandy loam to loam in texture, with a nearly closed stand of white spruce, lodgepole pine and feathermoss ground cover. The wetland located in an abandoned river channel is an extensive peat plateau (palsa) with its upper surface elevated 2.5 to 3.5 m above the collapsed surface. The collapse (thaw) processes appear to have been initiated at the upland-wetland interface and have been proceeding inward towards the centre of the depression. Scalloped or circular collapse features occur within and around the perimeter of the plateau (Figure 8). Most were the result of thawing of massive ice lenses during the progressive thaw of the perimeter. The presence of these "bulges" of massive ice would indicate that considerable differences in ice conditions exist within the plateau.

Probable cause of thaw acceleration is fire that swept through the area 60-70 years ago. Though the burn may not have been intense, it was sufficient to remove or kill part of the insulating moss layer and remove much of the shade-producing tree overstory. Vegetation on the plateau surface consisted of two distinct

FIGURE 8 Collapsing edge of peat plateau.

FIGURE 9 Collapse scar of peat plateau with with dead black spruce protruding from the surface.

age classes of black spruce with a uniform understory of Labrador tea, sphagnum, cloudberry and lichens.

The collapsing edges of this particular plateau revealed no evidence of mineral soils to depths of 2.5 m. A massive ice lens (pure ice) was examined along a collapse wall in one of the circular collapse areas. Its upper surface was within 1.3 m of the plateau surface. The ice was clear, containing only a few air bubbles. No other lenses of even comparable size were located elsewhere. The thaw layer was generally dry in the upper 7-10 cm of the surface with increasing moisture down to the contact with the frost table. The peat consisted of sphagnacious and woody material in the upper tier 0.4-0.6 m, poorly decomposed or fibric in nature. Underlying this deposit the clearly visible remains of sedge (fibric) continued to depths of 1.5 m, where the remains become less easily identifiable (mesic).

Open water generally occurs in the immediate vicinity of active collapse. At the outer fringes of these small thaw ponds, Sphagnum spp is usually the primary invader, while still further removed, sedges and woody plants are common. Often tree stems from the plateau surface stick above the area of collapse (Figure 9). Collapse and regrowth of these ecosystems appear to be cyclic, though it may take centuries to complete a single cycle, barring any disturbance. In a number of instances, permanently frozen, slightly elevated nodules have been encountered within the area of collapse some distance from the actively collapsing face.

Patterned Ground

Perhaps the most interesting forms in this area are those that result from intense frost action resulting in a great variety of relatively symmetrical surface ground features that are clearly visible, unhampered by high vegetation, in alpine environments. Slope orientation, soil types, precipitation patterns, etc. all strongly influence the occurrence and distribution of perennially frozen soils at higher elevations and hence the development of a variety of surface forms.

Attempts at relating specific forms and/or degree of development of patterned ground to broad environmental parameters, have met with little success. Increasing latitude results in a lowering of the altitudinal limits where these forms occur, but specific patterns appear to be tied more closely to local site conditions (i.e. soils, moisture regime, bedrock type, etc.), than to distinct changes in climate. Most of these patterned ground features have been well documented (i.e. Washburn 1973).

Of particular interest are the symmetrical, sorted features formed under water in ponds and lakes, most clearly visible and best developed at high elevations, (Figure 10) though not specific to these environments.

The narrow, stony beaches on the northwest side of Kusawa Lake (elevation 671 m, Lat. 60°30'N Long. 136°09'W) have regular "saw tooth" indentations, 2-2.5 m in length (Figure 11). Though the above water beach displays little or no easily identifiable frost-induced features (sorting, shattering), the underwater beach extension to a maximum depth of about 1.5 m has poorly developed stone bordered net features indicative of underlying permafrost and fairly intense frost activity. The regularity of the "saw tooth" features may indicate a direct relationship to these underlying permafrost conditions. Their appearance suggests that their formation is due to more than wave action and/or wind driven ice, though these may influence their genesis to some extent.

FIGURE 10 Sorted, net-like features on bottom of intermittent pond. Elevation 1540 m

FIGURE 11 "Saw-tooth" beaches, associated with underwater, poorly sorted nets Elevation 671 m

SUMMARY

Patterned ground and stable and collapse features resulting from the growth and/or degradation of permafrost can often be easily seen on various forms of remotely sensed imagery. Identification and examination of these features, during exploratory or reconnaissance level Ecological Land Surveys, has led to characterization of surface forms, their distribution, abundance and extent within a broad geographic framework. When coupled with other biological and physical components, the presence of these various forms aids substantially in ecosystem differentiation and characterization.

REFERENCES

Brown, R.J.E., (In Press), Permafrost in Canada, Map compiled by R.J.E. Brown for Hydrological Atlas of Canada.

Bostock, H.S., 1965, Physiography of the Canadian Cordillera, with Special Reference to the Area North of the Fifth-fifth Parallel, Dept. Energy Mines and Resources, Geol. Sur. Can. Mem. 237.

CSSC, 1978, The Canadian System of Soil Classification, Research Branch, Canada Department of Agriculture, Publ. No. 1646.

Environmental Conservation Service Task Force, 1981, Ecological Land Survey Guidelines for Environmental Impact Analysis, 42p. Ecol. Land. Class. Series., No. 13, Federal Environmental Assessment Review Office, Environment Canada, Ottawa.

Hughes, O.L., Campbell, R.B., Muller, J.E., and Wheeler, J.O., 1969, Glacial limits and flow patterns, Yukon Territory, south of 65 degrees north latitude, Dept. Energy, Mines and Resources, Geol. Sur. Can. Paper 68-34.

Hughes, O.L., Rampton, V.N., and Rutter, N.W., 1972, Quaternary Geology and Geomorphology, Southern and Central Yukon (Northern Canada), Guidebook, Field Excursion All, Internat. Geol. Cong. 24th Session.

Kindle, E.D., 1953, Dezadeash Map Area, Yukon Territory, G.S.C. Memoir 268.

Oswald, E.T. (compl.) 1979, Forest Resource Assessment of the Nisutlin Test Area, Yukon, P.F.R.C. Misc. Pub. CFS, Victoria, B.C.

Oswald, E.T. and Senyk, J.P., 1977, Ecoregions of Yukon Territory, 115 p. Rpt. BC-X-164, Canadian Forestry Service, Victoria.

Rowe, J.S., 1972, Forest Regions of Canada. Dept. Environ., Can. For. Serv. Publ. No. 1300.

Senyk, J.P., Oswald, E.T., Brown, B.N., and King, R.K., 1981, Ecological Land Classification and Evaluation of the Kusawa Lake Area, Yukon Territory, 178 p. Misc. Rpt. Canadian Forestry Service, Victoria, B.C.

Wahl, H., (Pers. Comm.) Climatologist, D.O.E., Whitehorse, Yukon Territory.

Washburn, A.L., 1973, Periglacial Processes and Environments. St. Martin's Press. New York.

SEASONAL THAWING OF PALSAS IN FINNISH LAPLAND

Matti Seppälä

Department of Geography, University of Helsinki, Hallituskatu 11-13
SF-00100 Helsinki 10, Finland

This study investigates the thickness of the active layer and its temporal, local, and regional differences in palsas - permafrost-cored peat mounds - in the bogs of northernmost Finland. Regional differences in the rate and amount of thawing are not found. The normal thawing depth per year is 55-70 cm. The minimum thickness of the insulating peat layer on palsas in Finland appears to be about 50 cm.

INTRODUCTION

It is characteristic of regions of discontinuous permafrost that only in places can the cold penetrate into the earth deep enough for it not to thaw completely during the summer. Not only low air temperature but also the thickness of the snow cover is a critical factor (Seppälä 1982). Snow is a good insulator. The quality of the soil and especially the thermal conductivity of the surface soil have a decisive effect on thawing. Gravel and sand have a high thermal conductivity. They freeze to considerable depth during the winter, but they also thaw fast during the summer. For example, in Finnish Lapland at the Kevo Subarctic Research Station more than 3 m of frost have been recorded, but in the summer it thaws completely (Seppälä 1976a). Frost develops much more slowly in peat (Seppälä 1982) than in mineral soil, but because the thermal conductivity of unfrozen dry peat is also much lower, the frost is much better preserved in peat than in mineral soil. This is the reason for the preservation of permafrost in palsas in northern peat bogs.

The first measurements of the summer position of the frost table were made on the Kola Peninsula by Kihlman in 1887 (from Kihlman's diary as cited by Rikkinen 1980). The thickness of the active layer was found to be 30-50 cm in August. In the Abisko region, Swedish Lapland, it was observed that the frost table on the July 10, 1911, was at the depth of 15 cm (Helaakoski 1912). In the Karesuando region, Sweden, it was observed that the frost table in July 1965 was some 40 cm deep and at the end of October about 60 cm deep (Forsgren 1966). According to Vorren (1967) the frost-free layer in palsas in northern Norway is 50-100 cm thick, while according to Åhman (1977) it varies from 65 to 100 cm. The present author has earlier described the thawing season 1974 of a palsa at Nunnanen (Seppälä 1976b); the results obtained are included in this study.

Interesting issues from the point of view of palsa development are, for example, (1) how deep the frost needs to penetrate into the peat bog for the formation of a palsa to start and (2) how thick the insulating layer of peat needs to be to preserve the frost over the summer.

This study investigates the thickness of the active layer of palsas in northernmost Finland and temporal, local, and regional differences in the thawing of palsas there.

METHODS OF INVESTIGATION

Observations were made during 9 years, 1972-1982 (Figure 4), by measuring the position of the frost table in palsas. An iron pole was forced into the ground through the thawed peat layer. Altogether 185 measurements were made on the top of palsas in 17 palsa bogs (Figure 1).

The pattern of thawing in different parts of the same palsa was studied at Nunnanen (1974) and at Utsjoki, Skallovarri (1975-1977). Several measurements were made at different points at intervals during the thawing seasons.

Thermal gradients in different parts of a palsa were also measured with a Wallac thermoprobe meter (see Seppälä 1976b).

If only one or a few measurements were taken in the same palsa bog, they were carried out on a horizontal surface on the top of palsas to eliminate the effect of solar exposition.

In this study no comparison was made of weather conditions with degree of thawing. We made our own weather observations on the Nunnanen (Seppälä 1976b) and Skallovarri palsa bogs, but the results of these will be published later and compared with the official weather recordings made at the closest observation points.

THAWING OF THE FROST IN DIFFERENT PARTS OF PALSAS

I do not think it is necessary to repeat the observations made on the Nunnanen palsa bog in 1974 (Seppälä 1976b), but we shall look in detail at a similar thawing development on Skallovarri palsa bog in 1975 (Figure 2)(Seppälä, in press).

Thawing starts in May on the top of the

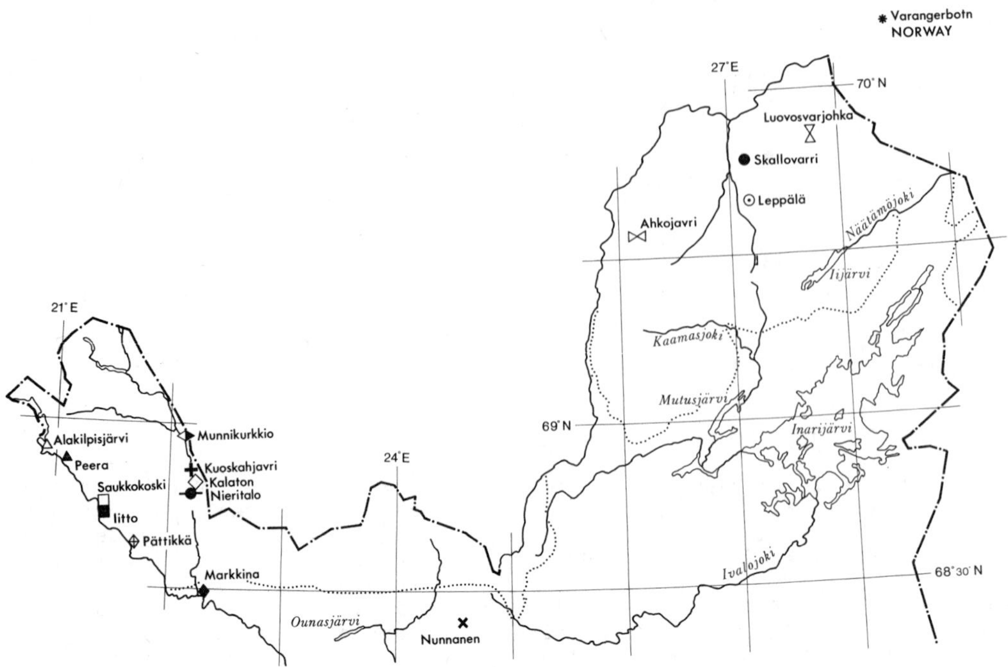

FIGURE 1 Location of the studied palsa bogs in Finnish Lapland. The dotted line indicates the northern limit of pine forest according to the Motoring road map of Finland 1978.

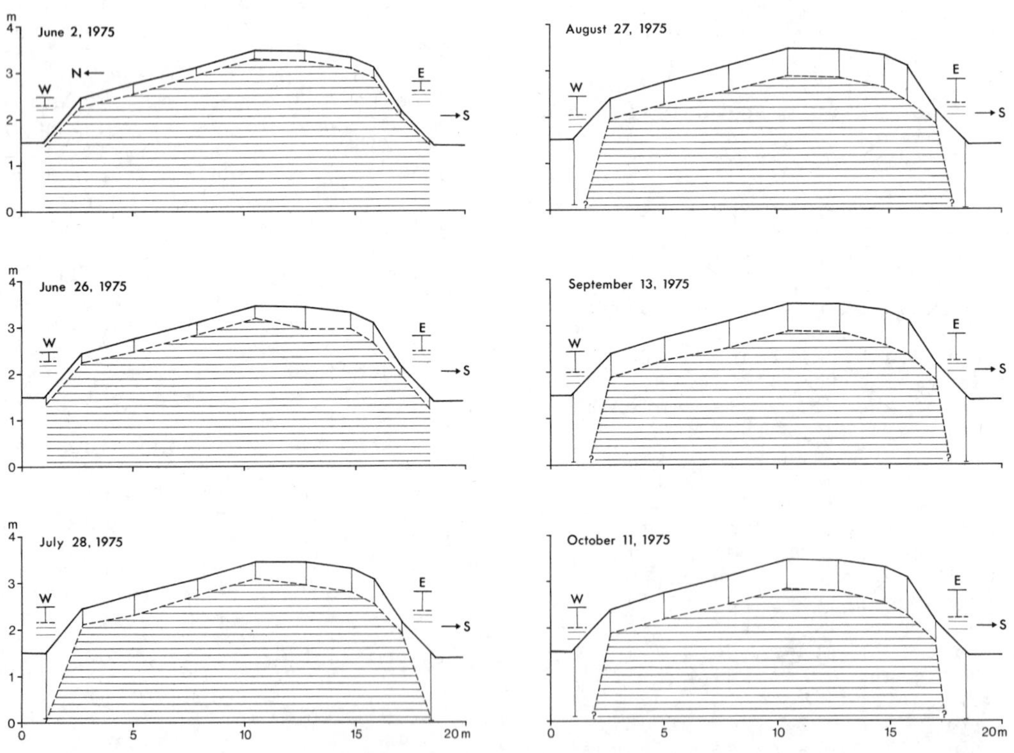

FIGURE 2 The position of frost table during the thawing season 1975 in Skallovarri. The hatched area indicates frozen parts. Vertical lines show the measurement points. W and E are the measurement points on the western and eastern parts of the palsa.

palsa when it becomes free of snow. In June and July, thawing is rapid, but in 1975 the rate of thawing decreased in August. Thawing penetrated deeper on the southern side of the palsa than on the northern side, although the differences are not great every year. Between east and west slopes the differences were slight. The east side thawed approximately as deeply as the top, while the western slope thawed least. Those slopes which are wet because of standing water at the foot of the palsa melt more quickly than drier slopes.

Local differences in the rate of thawing can be surprisingly great. At Skallovarri in 1975, for example, at two observation points, on two neighboring palsas separated by only 20 m, the frost table dropped by 28 cm at the first point and by only 7 cm at the second point during a period of 24 days in June (Figure 3). During the following month the table fell another 3 and 18 cm, respectively, so that by the end of July the difference in level between the two points was only 6 cm (Figure 3). A similar large difference in level between the two points was also noted at the beginning of the thawing season in 1976 and in 1977.

In some cases, during very hot weather, it was observed that the frost table moved upward somewhat. This may be explained by strong evaporation, which cools the lower wet layers.

Thermal gradients in different parts of a 4-m-high palsa at Pättikkä (68°37'30" N latitude, 21°45'E longitude) were measured on a warm summer day (June 23, 1973) (Figure 4). The frost table was much higher on the north side than on the south side. Differences between the top and the north side were not marked.

The lowest temperature at a depth of 10 cm (7.3°C) was found on the north side. On the south side, where there was vertical cracking of the surface and no vegetation,

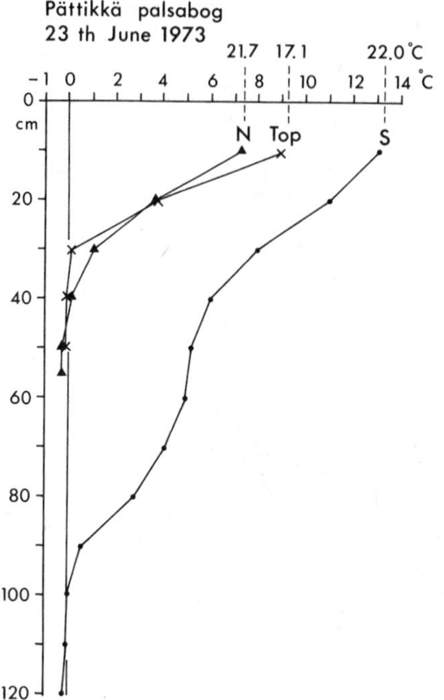

FIGURE 4 Temperature gradients of surface peat in a palsa in Pättikkä. Measurement points on the top, north edge (N), and south edge (S) of the palsa. The air temperature were recorded 20 cm above the surface at each point.

the temperature reached 13.0°C. At a depth of 20 cm on the top and on the north side the temperature were approximately the same. When measuring deeper layers from the top, the temperature dropped more steeply than in the other parts. The frost table on the top of the palsa lay at a depth of 50 cm, on the north side at 55 cm and on the south side at 120 cm. On the southern slope the

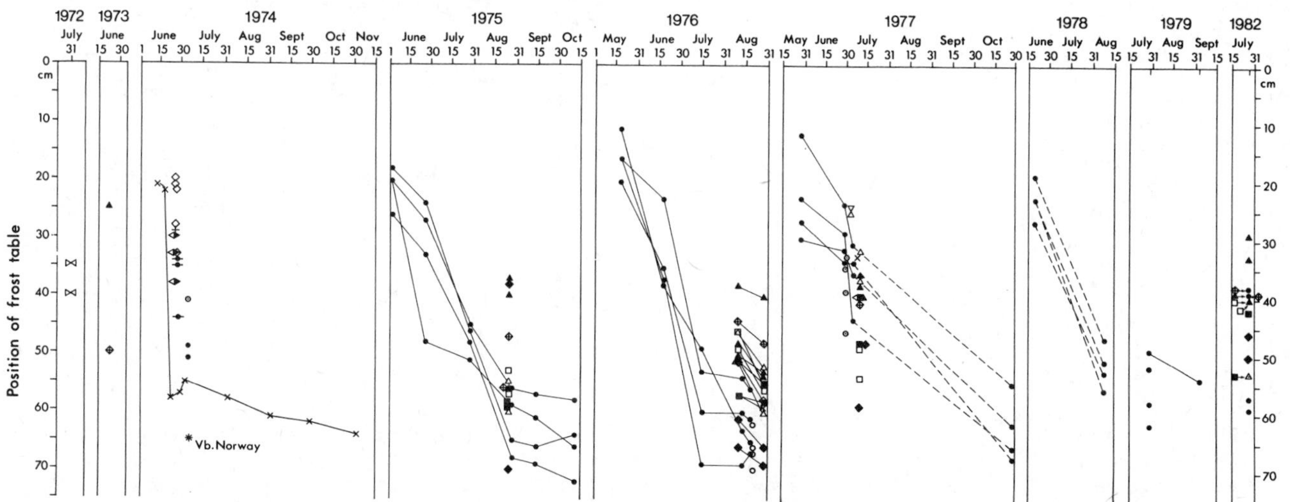

FIGURE 3 Observations made of the position of frost table in palsas. Lines join observations of same points. Dashed lines join those observations made at longer intervals and do not imply a linear thawing. Symbols indicate the observation points (see Figure 1).

temperature gradient from 40 to 60 cm depth was only $1^{o}C$.

Air temperature during the measurements ranges from 17.1^{o} to $22.0^{o}C$ 20 cm above the palsa surface (Figure 4).

On the same day (June 23, 1973) in this same palsa bog at Pättikkä the frost table in one palsa, 0.5 m high, lay at a depth of only 10 cm.

High day temperatures cause the asymmetrical shape of the core surface to become substantially more regular. Rain increases melting because the thermal conductivity of wet peat is much higher than that of dry peat.

Cracks on the palsa surface do not appear to have any great effect on the depth of thawing. Instead, they increase the possibility of peat blocks slumping; the palsa then loses its insulating cover of peat and a thermokarst depression is activated (Seppälä 1976b).

SEASONAL DIFFERENCES IN THAWING

In some years, thawing starts very fast, reaching great depths even by the end of June, as took place at Nunnanen in 1974 and at Skallovarri in 1976 (Figure 5). In 1977, thawing started especially slowly, but it reached almost the same level as in 1975 and 1976.

At the end of August 1975 and 1976 (Figure 6) the differences in thawing depths were very small at most measurement points. Only at Peera was the difference 14 cm at one point and at Markkina 31 cm, neither of which could be explained.

The greatest range between frost table

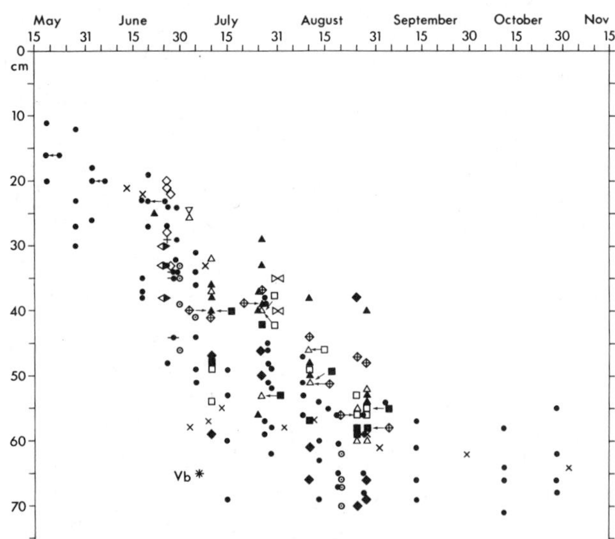

FIGURE 5 Summary diagram of the observations of the position of frost table in palsas in different years (Figure 4). Symbols indicate the recording points (Figure 1). Arrows show observations at the same site in diagram. Vb Varangerbotn.

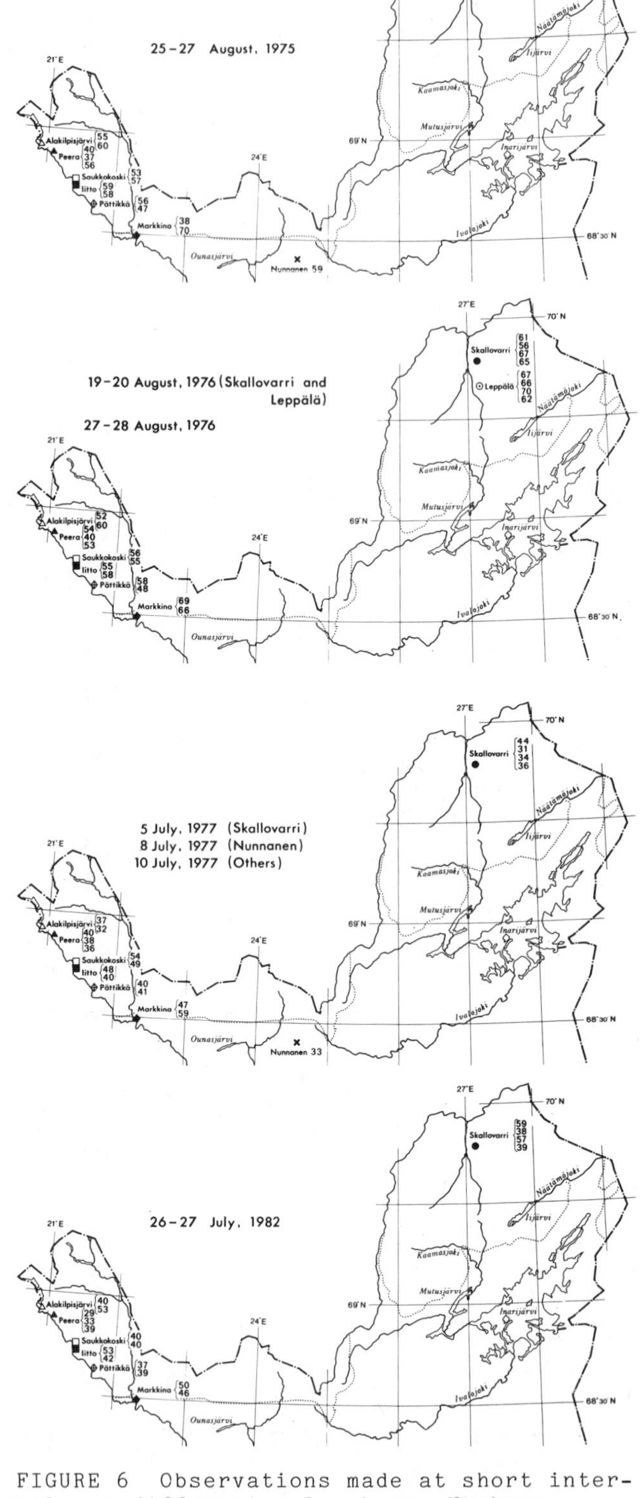

FIGURE 6 Observations made at short intervals on different palsa bogs. Numbers indicate the position of frost table in centimeters measured from the peat surface.

levels at the same points occurs in July. These differences in level were reduced by August, when in most years the differences were only a few centimeters.

REGIONAL DIFFERENCES IN ACTIVE LAYER

Regional differences in the rate and degree of thawing are not found. The differences between the various palsas were greater than the differences between palsa bogs (Figure 6). In general, low palsas melt less than higher ones in the same palsa bog. The normal thawing depth per year is 55-70 cm. This is the maximum thickness of the active layer on the top of most palsas in Finnish Lapland and is usually reached by the beginning of October (Figure 5).

In July only surprisingly small differences could be found in the thawing depth at different heights above the sea level. At Alakilpisjärvi and Skallovarri (Figure 1) the palsas had melted to very much the same depth (even less at Skallovarri in 1982)(Figure 6), although the altitude difference is almost 200 m (Table 1). Morphological differences in palsas, such as height, may explain the ranges in thawing rates better than, for example, altitude above sea level, which naturally affects the air temperature.

Vegetation on the palsa surface reduces thawing. Eroded old palsas thaw deeper than younger palsas with a covering vegetation.

The greatest difference in thawing depth was found between the Finnish and Norwegian (Varangerbotn, Vb, Figures 1, 4, and 5) palsas. In Varangerbotn one 3-m-high palsa had already thawed to a depth of 65 cm on July 6. This might be caused by its location only 50 m above sea level and close to the Arctic Ocean.

DISCUSSION

In northern Finland the frost in a flat-topped peat bog should penetrate to a depth of at least 50 cm before the frost is preserved through the summer to the next winter (Seppälä 1982). This happens only in places where the snow cover is thinner than normal. When a palsa rises above the peat bog surface, thawing starts earlier in spring and reaches deeper levels. The maximum thickness of the active layer on the top of most palsas in Finnish Lapland is 55-70 cm. If thawing goes deeper than 70 cm, as happens in some wind-eroded palsas, then the frost may fail to penetrate to the permafrost table. An unfrozen suprapermafrost layer then exists which may eventually destroy the frozen core of the palsa (cf. Salmi 1970, fig. 4).

The depths to which palsas thawed changed surprisingly little in different parts of Lapland and also remained very constant year after year. The summits of palsas covered with vegetation and located in different regions provide the best localities for comparing thawing measurements.

TABLE 1 Palsa bogs studied. See also Figure 1.

Palsa Bog Name	Approximate Location		Altitude a.s.l.(m)	Height (m)
Alakilpisjärvi	68°56'N	20°55'E	490	2-3
Peera	68°53'N	21°03'E	460	3-7
Saukkokoski	68°45'N	21°25'E	410	3
Iitto	68°44'N	21°25'E	405	5-6
Pättikkäkoski	68°38'N	21°43'E	395	1-2
Markkina	68°29'N	22°18'E	335	1-2
Munnikurkkio	68°55'N	22°10'E	440	6
Kuoskahjavri	68°51'N	22°14'E	440	1
Kalaton	68°50'N	22°15'E	440	1-3
Nieritalo	68°48'N	22°12'E	440	5-6
Nunnanen	68°25'N	24°35'E	320	2-4
Skallovarri	69°49'N	27°10'E	295	1-3
Leppälä	69°40'N	27°08'E	225	1-3
Luovosvarjohka	69°51'N	27°46'E	300	3
Ahkojavri	69°35'N	26°11'E	365	3
Varangerbotn (N)	70°15'N	28°30'E	50	3

The maximum depth of the frost table should be measured as late as possible in October-November, because heat penetrates very slowly into the peat.

The minimum thickness of the insulating peat layer on palsas in Finland appears to be about 50 cm. If the peat layer is thinner, palsas do not exist. Instead, there are small peat hummocks called pounu(s), characterized by a seasonally frozen core (cf. Salmi 1972, Seppälä 1979). Wind and rain erode the surface of the palsa, and as the peat layer gets thinner and vegetation disappears, then the insulation provided is no longer sufficient to preserve the frozen core of the palsa. This results in the "death" of the palsa and it forms a thermokarst hollow or pond in the peat bog.

ACKNOWLEDGEMENTS

In the field work I was assisted by J. Vainio, K. Lundén, V. Säntti, J. Rastas, A. Sujala, H. Nurminen, M. Sulkinoja, J. Heino, L. Koutaniemi, R. Åhman, and M. J. Clark. The figures were drawn in their final form by K. Lehto and copied by P. Mölsä, and the English of the text was checked by Ch. Grapes. The study was supported with travel grants from the National Research Council for Natural Sciences in Finland, Seth Sohlberg's Delegation (Societas Scientiarum Fennica), and the Rector of the University of Helsinki. To all these individuals and institutions I would like to express my sincere gratitude.

REFERENCES

Åhman, R., 1977, Palsar i Nordnorge: Meddelanden från Lunds Universitets Geografiska Institution, Avhandlingar, v. 78, 165 pp.

Forsgren, B., 1966, Tritium determinations in the study of palsa formation: Geografiska Annaler 48A, p. 102-110.

Helaakoski, A. R., 1912, Havaintoja jäätymisilmiöiden geomorfologisista vaikutuksista (Beobachtungen über die geomorfologischen Einflüsse der Gefriererscheinungen): Meddelanden af Geografiska Föreningen i Finland 9, p. 1-108.

Rikkinen, K., 1980, Suuri Kuolan retki 1887: Otava, Keuruu.

Salmi, M., 1970, Investigations on palsas in Finnish Lapland: Ecology of the Subarctic regions, Proceedings of the Helsinki Symposium 1966: Paris, UNESCO, p. 143-153.

Salmi, M., 1972, Present developmental stages of palsas in Finland: Proceedings of 4th International Peat Congress, Helsinki, v. 1, p. 121-141.

Seppälä, M., 1976a, Periglacial character of the climate of the Kevo region (Finnish Lapland) on the basis of meteorological observations 1962-71: Reports from the Kevo Subarctic Research Station, v. 13, p. 1-11.

Seppälä, M., 1976b, Seasonal thawing of a palsa at Enontekiö, Finnish Lapland, in 1974: Biuletyn Peryglacjalny v. 26, p. 17-24.

Seppälä, M., 1979, Recent palsa studies in Finland: Acta Universitatis Ouluensis, v. 82, p. 81-87.

Seppälä, M., 1982, An experimental study of the formation of palsas, <u>in</u> French, H. M., ed., Proceedings of the Fourth Canadian Permafrost Conference: Ottawa, National Research Council of Canada, p. 36-42.

Seppälä, M., in press, Palsarnas periodiska avsmältning i Finska Lappland: Geografisk Tidsskrift, v. 82.

Vorren, K-D., 1967, Evig tele i Norge: Ottar, v. 51, 26 pp.

REVEGETATION OF ARCTIC DISTURBED SITES BY NATIVE TUNDRA PLANTS

G. R. Shaver,[1] B. L. Gartner,[2] F. S. Chapin III,[2] and A. E. Linkins[3]

[1] The Ecosystems Center, Marine Biological Laboratory, Woods Hole, Massachusetts 02543 USA
[2] Institute of Arctic Biology, University of Alaska, Fairbanks, Alaska 99701 USA
[3] Department of Biology, Virginia Polytechnic Institute and State University, Blacksburg, Virginia 24061 USA

The aim of this research was to develop new methods by which the recovery of native vegetation might be promoted on development-related disturbances in northern Alaska. Current methods depend on an expensive and heavily supported introduction of nonnative grasses, often with detrimental effects on native plant recovery. The research included studies of both native plant population dynamics and the regulation of tundra nutrient cycles. The results showed that the single most important factor in native vegetation recovery is preservation or replacement of the upper organic layer of soil. The organic layer is important because (1) it contains a large, readily germinable native seed pool (2) native seed germination rates are higher in organic than in mineral soils (3) it reduces soil thaw and thermokarst erosion and (4) it reduces nutrient losses and nutrient movement in the soil, resulting in more normal patterns of nutrient availability. Future management practices should include conservation of soil organic matter as a top priority, with heavy fertilization and seeding only where erosion potential is great or the organic mat is lost.

INTRODUCTION

In the Arctic, large areas of native tundra vegetation are inevitably disturbed or eliminated by development. Proper management requires that those areas be revegetated and ultimately restored to a state as close to the original as is economically and logistically feasible. Current revegetation methods depend strongly on the use of plants that are not native to the disturbed regions, and often require expensive and labor intensive fertilization and planting procedures (Johnson 1981). The use of these non-native plants is usually effective in ameliorating problems such as rapid soil erosion, but it may actually reduce the restoration of native plant populations either directly through competition or secondarily through fertilization and other environmental modifications (Hernandez 1973, Chapin and Chapin 1980). Thus community restoration should incorporate native plants whenever possible, particularly if it can be done at a lower cost.

Current revegetation practices were developed when little was known about the reproductive biology and growth of native tundra plants. Methods are changing rapidly as new knowledge becomes available. The aim of this paper is to summarize recent research on ecological aspects of native tundra plants that are particularly relevant to revegetation management using native plants.

Two approaches were used in this research. The first involved a series of studies of native plant population dynamics, especially seed sources and seedling establishment on both natural and artificial disturbances. These studies established the basic patterns of seed production, seed germination, and growth in native tundra plants, and some of the major controlling factors. The second line of research dealt with controls on total primary production, nutrient cycling, and the regulation of species and growth form composition in whole tundra vegetation. Underlying this research was the assumption that natural revegetation and recovery from disturbance are largely mediated by changes in nutrient cycling. Detailed descriptions of the methods have been published in the original papers.

POPULATION DYNAMICS AND GROWTH

The establishment of native plants on disturbed sites depends on survival through all stages of the plant life cycle. For the purpose of this research these stages were divided into separate studies of seed production, seed storage, seed germination, seedling demography, and seedling growth. The species of particular interest was arctic cottongrass, *Eriophorum vaginatum* L., a tussock-forming sedge that is the dominant plant over much of the north slope of Alaska. Eriophorum species in general are also conspicuously successful as invaders of unmanaged disturbances.

An initial hypothesis was that revegetation by native plants might be promoted by fertilizing undisturbed areas adjacent to disturbances, to increase seed production and presumably also seed input to the disturbance. Primary production of tussock tundra is strongly nutrient limited (Shaver and Chapin 1980), and flowering of E. vaginatum is highly responsive to fertilization. For example, at Eagle Creek, Alaska, fertilization with NPK (Shaver and Chapin 1980) increased flowering from 11.3 ± 4.4 to 59.2 ± 12.6 inflorescences m^{-2} in 1978, 2 years after the fertilizer was applied. The number of seeds $inflorescence^{-1}$ increased from 15.3 ± 2.3 to 33.0 ± 2.4, for a total seed production of 173 seeds m^{-2} yr^{-1} in control plots and 1,954 seeds m^{-2} yr^{-1} in NPK fertilized plots.

The high seed production in fertilized plots in 1978 should be sufficient to increase significantly the seed rain onto adjacent disturbances. Eriophorum vaginatum seed are easily wind dispersed. However, flowering varies greatly from year to year; in control plots at Eagle Creek it ranged over 3 orders of magnitude, from 0.04 ± 0.02 to 19.0 ± 3.9 inflorescences m^{-2} between 1976 and 1982. Although fertilization increased flowering, the effect was greatest in high flowering years such as 1978. In years of low flowering fertilization had little or no significant effect. Because of this annual variation the effects of fertilization are unpredictable, and one cannot depend on fertilization of undisturbed tundra to supply seed to adjacent disturbances. Fertilization of adjacent areas may be useful as a supplementary technique.

A second potential seed source for use in revegetation is the pool of ungerminated seeds that may accumulate in the soil of undisturbed tundra ecosystems. Large buried seed pools have been found in many other ecosystems (Grubb 1977, Grime 1979), but their size and significance in tundra were unknown before this work began. At both Eagle Creek (McGraw 1980) and at Kuparuk Ridge, Alaska (Gartner 1982, Gartner et al. 1983), significant numbers of Carex spp. and Eriophorum spp. seed were found in the upper 10-15 cm of organic soil of undisturbed tussock tundra. These seed germinate readily when the soil is exposed and warmed by removal of vegetation.

In experimental disturbances at Eagle Creek and Kuparuk Ridge, most of the seedlings that appeared on the plots came from buried seed (Chester and Shaver 1982, Gartner et al. 1983). The seedlings could not have come from external seed rain because most of them appeared on the plots in June or early July, before any seed were produced or shed by plants in the surrounding undisturbed tundra. The seedlings appeared without any manipulation other than vegetation removal, suggesting that buried seed in the upper layers of soil are potentially an important source of seed in the restoration of disturbed sites using native plants.

The germination characteristics of native tundra seeds are typical of many disturbance-adapted species (Bliss 1971). In studies of Eriophorum vaginatum germination, in particular (Wein and MacLean 1973, Gartner 1982), the seeds generally have no strong constitutive dormancy mechanisms. Instead, Eriophorum seed germinate most readily under the conditions likely to occur only on disturbed sites, i.e., at soil temperatures of 20°C or higher. Germination is increased in the light, in accord with the higher light intensities at the soil surface that prevail on disturbances. These studies support the conclusion that periodic disturbance represents an important opportunity for seed reproduction in natural tundra, that native plants have been selected to respond to disturbance by maintenance of a large pool of seeds that germinate readily, and that this adaptation is of potential use in developing management schemes (Gartner 1982).

The native seedlings that appear in experimental disturbances are not evenly distributed. Eriophorum and Carex seedlings are much more abundant on organic substrates, whereas native grass species are more evenly divided between mineral and organic areas (Gartner 1982, Gartner et al. 1983). This pattern has two principal causes. Initially the most important is the distribution of seeds in the seed bank. Few or no seeds are stored in the deeper mineral layers of soil. Thus the deeper the removal of vegetation and soil, the fewer the remaining seedlings. Grasses are rare or absent from the buried seed pool (McGraw 1980, Gartner 1982), and their low seedling densities and even distribution may result from a dependence on external seed input. A second reason for the greater total abundance of seedlings on organic substrates is that seedling germination rates are higher and mortality rates lower there. When seeds of E. vaginatum and Calamagrostis canadensis were sown artificially on organic and mineral soils, the germination percentages were 3-5 times higher on the organic areas (Gartner 1982). Mortality rates were higher on mineral soils in part due to much greater needle ice action, which heaves seedlings out of the soil. Using toothpicks as model seedlings, Gartner (1982) found over 80% heaved out of the mineral soil after 2 years, but less than 20% heaved in the organic soil. Actual seedling mortality rates were lower but followed the same pattern.

Growth rates of the native grass and sedge seedlings are not significantly different on organic vs. mineral soils in the field. Thus the major differences in native plant cover after 4 years' growth on organic vs. mineral soils were due to seedling density differences rather than differences in growth rates (Gartner 1982).

The general conclusion from this portion of the research is that maintenance or replacement of the original organic mat is vital to the recovery of native graminoid species following disturbance, both because the organic mat already contains a major potential source of native plant seed and because it is the most favorable substrate for germination of seeds derived from external sources. This is in contrast to the normal practice of removing the organic mat and preparing a mineral seed bed, which is more favorable for both germination and growth of introduced nonnative grasses (Mitchell 1979).

NUTRIENTS, GROWTH, AND COMMUNITY REGULATION

Productivity of tundra vegetation is in general nutrient limited (Haag 1974, McKendrick et al. 1978, Shaver and Chapin 1980), and nutrient availability has major consequences for the distribution and abundance of native tundra plants. As in virtually all terrestrial ecosystems, physical disturbance to tundra disrupts the normal patterns of nutrient availability (Gersper and Challinor 1975). To study the effects of disturbance-related changes in nutrient availability in tundra, we combined descriptive comparisons of disturbed and undisturbed sites with controlled manipulation of nutrient availability in fertilization experiments. The research was organized around two basic questions: First, how might an increase in nutrient availability by fertilization improve the recovery of native plant populations following disturbance? Second, in the later stages of recovery how might manipulation of nutrient availability affect the composition of the native plant community?

Fertilization has various effects on native plant growth, depending on species, time since disturbance, and type of vegetation. On recently disturbed sites (less than 5 years old), fertilization increased growth of native _Carex_ and grass seedlings but not of _Eriophorum_ seedlings (Gartner et al. 1983). There was no difference in responsiveness between mineral and organic substrates. However, after 10 years' recovery on a bulldozed organic site at Eagle Creek, Chapin and Chapin (1980) found no long-term fertilization effect. Cover by native species, primarily _E. vaginatum_, was nearly 100% in both fertilized and unfertilized plots, whether or not seed of various nonnative grass species had been sown following the disturbance.

The lack of a long-term fertilization effect on native plant recovery on disturbed sites is consistent with the long-term vegetation responses on unmanaged disturbances such as vehicle tracks. Production and biomass of native plants on 7-10 year old vehicle tracks is as much as 2-5 times that of undisturbed tundra (Chapin and Shaver 1981). The high productivity of old disturbed sites is accompanied by an even greater increase in annual uptake of N and P (Wein and Bliss 1974, Chapin and Shaver 1981). The native vegetation of these sites is dominated by rapidly growing graminoids and occasionally some deciduous shrubs, all characteristic indicators of high nutrient availability.

Both the fertilization experiments and the descriptive studies of older disturbances suggest that nutrient availability is relatively non-limiting after about 5 years on a disturbance if the site is relatively moist and covered with an organic soil. Fertilization thus may improve native plant recovery in the first year or two but there is no need to continue the application except perhaps on dry mineral soils.

In fact, fertilization may cause a shift away from normal patterns of plant nutrient uptake. In _Eriophorum vaginatum_, acid phosphatase activity at the root surface is positively correlated with soil organic matter (Table 1), suggesting a well-developed ability to use organic phosphorus as a source of inorganic phosphates. In potted plants, _E. vaginatum_ growth rates were twice as high on organic than on mineral soils, despite nonsignificant differences in the field (Gartner 1982). The acid phosphate enzyme system is suppressed by the presence of inorganic phosphate. Addition of inorganic fertilizers or removal of organic soil horizons may reduce success of native plants by supressing an efficient nutrient uptake mechanism and altering competitive interactions in phosphorus uptake.

TABLE 1 Surface Acid Phosphatase Activity (Measured at Substrate Saturation Levels) on Roots of _Eriophorum vaginatum_ From Several Sites at Toolik Lake, Alaska.

% Soil Organic Matter	Phosphate Activity
6	6.2
17	16.2
60	46

Activity is expressed as nM PNP released hr^{-1} mm^{-2} root surface (n = 10). Methods are described in Antibus et al. (1981).

Once an adequate cover of native species is established, restoration of a natural community must involve manipulation of the relative abundances of species and growth forms. Nutrient availability is a key variable in this process, the major problem being that nutrient cycling rates are very rapid on old disturbances. Fertilization experiments have shown that when N and P are added to undisturbed tundra, the vegetation becomes more like that on old disturbances (Figure 1) (Shaver and Chapin 1980 and unpublished). Thus management of older disturbances should be focused on slowing down nutrient movement and decreasing nutrient availability.

Nutrient cycling on disturbed sites differs from undisturbed tundra in several aspects. Potentially one of the most important of these is soil temperature, which averaged about 2°C higher at 10 cm depth in old vehicle tracks than in undisturbed tundra (Chapin and Shaver 1981).

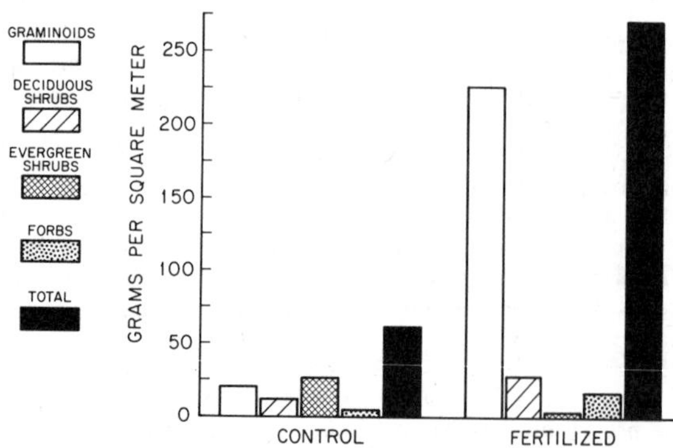

Figure 1 Primary production (g m^{-2}) of the major plant growth forms in control and fertilized areas of tussock tundra at Sagwon, Alaska, 4 years after the fertilizer was applied. Methods and rates of application followed Shaver and Chapin (1980).

Chapin and Shaver (1981) estimated the effects of a 2°C temperature change on individual nutrient cycling processes and concluded that although the soil temperature difference was significant it could not alone explain the higher productivity in vehicle tracks. Other factors such as differences in the rate of soil water and nutrient movement are now under investigation as explanations for the differences between disturbed and undisturbed sites.

Evidence for greater nutrient movement in vehicle tracks comes from a litter bag experiment (G. R. Shaver, F. S. Chapin, A. E. Linkins, and J. M. Melillo, unpublished) in which both N and P immobilization by Eriophorum litter was greater by a factor of 4-8 in wet vehicle tracks than in mesic undisturbed tundra (Figure 2). This greater immobilization is accompanied by much lower phosphatase and cellulase enzyme activity and higher soil respiration rates (Table 2; Herbein 1981). Furthermore, there is a more complete decomposition of cellulose in the vehicle track, as indicated by the lower ratios of exocellulase:endocellulase activity and the higher proportion of cellulose-derived CO_2 evolved from track soils (Table 2).

These results suggest that nutrient availability to plants might be reduced by management attempts to (1) lower soil temperature and depth of soil thaw, (2) increase nutrient immobilization in litter, (3) decrease soil oxygen, and (4) decrease both surface and subsurface soil water movement. All of these objectives might be achieved at least in part by increasing the thickness of the upper soil organic mat, and providing carbon-rich and nutrient-poor organic litter to the soil. This could be accomplished first by preserving as much as possible of the original organic soil and second by adding sawdust or some other heavy mulch to old disturbances. Experimental attempts to reduce nutrient availability by adding corn starch and sugar to undisturbed tundra have suggested but not confirmed that this approach should work (Shaver and Chapin 1980, Marion et al. 1982). The long term goal is the creation of

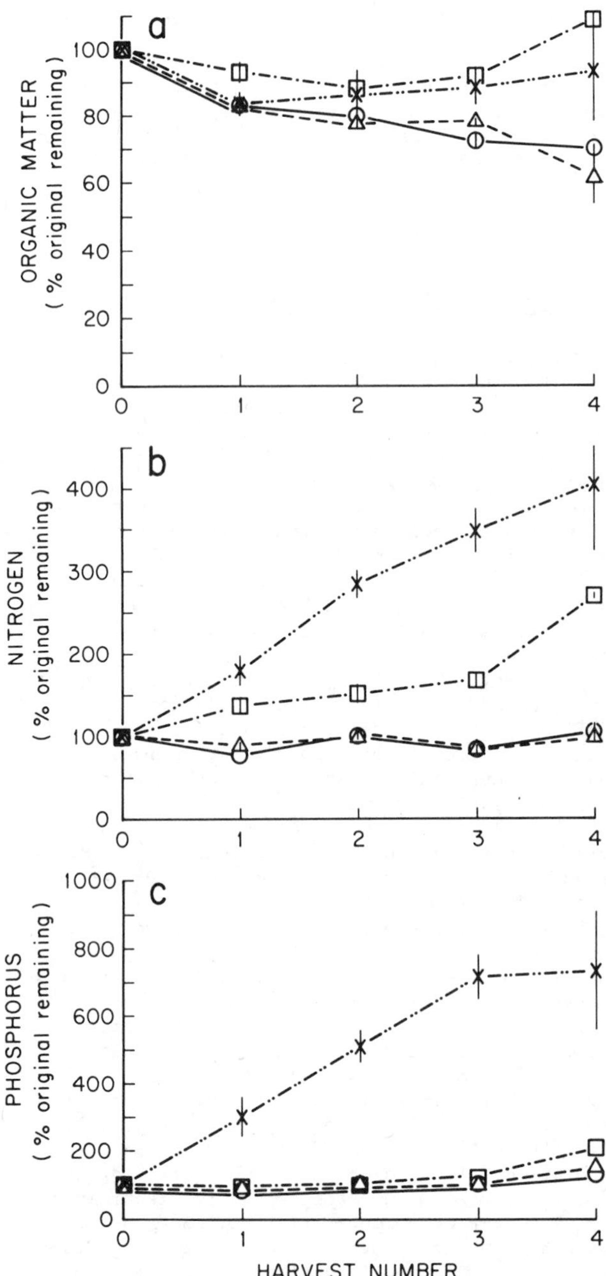

Figure 2 Changes in (a) total organic matter, (b) total N mass, and (c) total P mass in litter bags placed in and out of vehicle tracks at Slope Mountain, Alaska. Data are means ± SE, expressed as percentages of the original mass. Circles = mesic control, triangles = mesic track, squares = wet control, crosses = wet track.

TABLE 2 Substrate-Saturated Reaction Rates for Endocellulases, Exocellulases, and Phosphomonoesterases in the O_e-O_i Horizons of Tussock Tundra Soils at Slope Mountain, Alaska, and Associated Soil Respiration.

	Endocell-ulase	Exocell-ulase	Phospha-tase	Respir-ation	$\%^{14}CO_2$
Wet Control	90a	145a	21a	115a	19a
Wet Track	65b	137b	14b	182b	43b
Mesic Control	89a	142ab	16ab	108a	15a
Mesic Track	70b	135b	15b	175b	38b

Units for endocellulase activity are hr^{-1} gm dry wt^{-1}; for exocellulase, ug glucose equivalents hr^{-1} gm dry wt^{-1}; for phosphatase uM PNP hr^{-1} gm dry wt^{-1}; for respiration, ul CO_2 hr^{-1} gm dry wt^{-1}. The last column is $^{14}CO_2$ evolution from incorporated ^{14}C-cellulose, expressed as a percentage of total CO_2. N = 8 in all cases. Within each column, values followed by different superscript are significantly different at the 0.05 level. Methods followed Herbein (1981), Linkins and Antibus (1978), and Linkins and Neal (1982).

conditions that are more favorable for native species and more typical of the normal nutrient-limited tundra vegetation.

MANAGEMENT RECOMMENDATION

In the revegetation of arctic disturbed sites by native plants, the single most important factor is the preservation or replacement of the upper organic layer of soil. The organic layer is important because (1) It contains a large, readily germinable seed pool; (2) germination rates of native seeds are higher in organic than mineral soils; (3) it reduces soil thaw and thermokarst erosion; (4) in later stages of recovery, it may reduce nutrient movement and nutrient losses from the soil. Fertilization improves native plant recovery immediately after the disturbance but after 5-10 years it may prevent reestablishment of a normal species composition. Seeding with nonnative plants either has no effect on long-term recovery of native species, or the recovery of native species may be reduced. Nonnative species are particularly effective at reducing recovery of natives if the site is also fertilized.

Given these results, we recommend the following methods for management of development-related disturbances: (1) Replace the natural organic soil, particularly the upper layers which contain the most seed; (2) fertilize lightly only in the first year or two; (3) do not broadcast seed except where erosion potential is great or the original organic mat is lost; (4) do not fertilize old disturbances; (5) mulching or addition of peat to the soil of old disturbances may help to create a more natural soil, with a more natural species composition in the vegetation.

ACKNOWLEDGMENTS

This research was supported by the U.S. Army Research Office, with additional support from the U.S. Department of Energy and the U.S. Army Cold Regions Research and Engineering Laboratory.

REFERENCES

Antibus, R. K., J. G. Croixdale, O. K. Miller, and A. E. Linkins. 1981. Ectomycorrhizal fungi of Salix rotundifolia. III. Resynthesized mycorrhizal complexes and their surface acid phosphatase. Canadian Journal of Botany 59:2458-2465.

Bliss, L. C. 1971. Arctic and alpine plant life cycles. Annual Review of Ecology and Systematics 2:405-438.

Chapin, F. S., and M. C. Chapin. 1980. Revegetation of an arctic disturbed site by native tundra species. Journal of Applied Ecology 17:449-456.

Chapin, F. S., and G. R. Shaver. 1981. Changes in soil properties and vegetation following disturbance of Alaskan arctic tundra. Journal of Applied Ecology 18:605-617.

Chester, A. L., and G. R. Shaver. 1982. Seedling dynamics of cottongrass tussock tundra species during the natural revegetation of small disturbed areas. Holarctic Ecology 5:207-211.

Gartner, B. L. 1982. Controls over regeneration of tundra graminoids in a natural and a man-disturbed site in arctic Alaska. M.S. thesis, University of Alaska. 123 pp.

Gartner, B. L., F. S. Chapin, and G. R. Shaver. In press, 1983. Demographic patterns of seedling establishment and growth of native graminoids in an Alaskan tundra disturbance. Journal of Applied Ecology.

Gersper, P. L., and L. Challinor. 1975. Vehicle perturbation effects upon a tundra soil-plant system: I. Effects on morphological and physical environmental properties of the soils. Soil Science Society of America Proceedings 39:737-744.

Grime, J. P. 1979. Plant Strategies and Vegetation Processes. John Wiley and Sons, New York. 222 pp.

Grubb, P. J. 1977. The maintenance of species richness in plant communities: the importance of the regeneration nich. Biological Reviews 52:107-145.

Haag, R. W. 1974. Nutrient limitations to plant production in two tundra communities. Canadian Journal of Botany 52:103-116.

Herbein, S. B. 1981. Soil phosphatases: Factors affecting enzyme activity in arctic tussock tundra and Virginia mineral soils. M.S. thesis, Virginia Polytechnic Institute and State University, Blacksburg, Virginia. 104 pp.

Hernandez, H. 1973. Natural plant recolonization of surficial disturbances, Tuktoyaktuk Peninsula Region, Northwest territories. Can. J. Bot. 51:2177-2196.

Johnson, L. 1981. Revegetation and selected terrain disturbances along the Trans-Alaska pipeline. U.S. Army CRREL Report 81-12, 115 pp.

Linkins, A. E., and R. K. Antibus. 1978. Ectomycorrhizal fungi of _Salix rotundifolia_. II. Impact of surface applied crude oil on mycorrhizal respiration and cold hardiness. Arctic 31:381-393.

Linkins, A. E., and J. L. Neal. 1982. Soil cellulase, chitinase, and protease activity in _Eriophorum vaginatum_ tussock tundra at Eagle Summit, Alaska. Holarctic Ecology 5:135-138.

Marion, G. M., P. C. Miller, J. Kummerow, and W. C. Oechel. 1982. Competition for nitrogen in a tussock tundra ecosystem. Plant and Soil 66:317-327.

McGraw, J. B. 1980. Seed bank size and distribution of seeds in cottongrass tussock tundra, Eagle Creek, Alaska. Canadian Journal of Botany 58:607-611.

McKendrick, J. D., V. J. Ott, and G. A. Mitchell. 1978. Effects of nitrogen and phosphorus fertilization on the carbohydrate and nutrient levels in _Dupontia fischeri_ and _Arctagrostis latifolia_. In: _Vegetation and Production Ecology of An Alaskan Tundra_ (Ed. by L. L. Tieszen), pp. 509-537. Springer-Verlag, New York.

Mitchell, W. W. 1979. Three varieties of Alaskan grasses for revegetation purposes. Agricultural Experiment Station, University of Alaska, Fairbanks, Circular 32.

Shaver, G. R. and F. S. Chapin. 1980. Response to fertilization by various plant growth forms in an Alaskan tundra: Nutrient accumulation and growth. Ecology 61:662-675.

Wein, R. W., and L. C. Bliss. 1974. Primary production in Arctic cottongrass tussock tundra communities. Arctic and Alpine Research 6:261-274.

Wein, R. W., and D. A. MacLean. 1973. Cottongrass (_Eriphorum vaginatum_) germination requirements and colonizing potential in the Arctic. Canadian Journal of Botany 51:2509-2513.

A CLASSIFICATION OF GROUND WATER IN THE CRYOLITHOZONE

V. V. Shepelev

Permafrost Institute, Siberian Branch,
Academy of Sciences, Yakutsk, USSR

A subdivision is presented of the major types of groundwater in the cryolithozone, based on conditions of deposition and on the dynamics of development of the cryogenic aquicludes confining them. Among suprapermafrost waters (type 1) one may distinguish suprapermafrost vadose waters (subtype 1a) and suprapermafrost groundwater (subtype 1b). Intrapermafrost water (type 2) may also be subdivided into two subtypes. The first involves intrapermafrost water confined by horizontal cryogenic aquicludes (2a) and the second intrapermafrost water capable of vertical movements between frozen materials (2b). Subpermafrost waters (type 3) are subdivided into those in contact with the bottom of the permafrost (subtype 3a) and those not in contact with it (3b). The groundwater subtypes thus identified may be further subdivided into different categories and subcategories.

Much attention has generally been given to classification of gravitational ground waters in the permafrost region. This indicates the importance and significance of ground water in a variety of scientific and practical problems concerning the study of the conditions of distribution, formation, and discharge of ground waters of the permafrost zone, clarification of the aspects of their utilization and protection, as well as peculiarities of permafrost-hydrogeological mapping, zoning and accomplishment of water search and prospecting operations.

In 1939 N. I. Tolstikhin was the first to develop and formulate in his fundamental monograph a scientifically substantiated and, in its way, classical hydrogeological classification of ground waters in the permafrost region (Tolstikhin 1941). That classification is based on the principle of a spatial ratio of ground waters to the permafrost zone of rocks, according to which the ground waters are categorized into the three main classes: suprapermafrost, intrapermafrost and subpermafrost waters. The advances of the USSR in hydrogeological research on permafrost are quite justifiably being attributed to the emergence of this classification. Despite some criticism raised later (Meister 1955, Zelenkevich 1960, et al.) and new schemes of permafrost ground water classification (Baranov 1940, Ovchinnikov 1954) N. I. Tolstikhin's classification has been generally recognized and employed in hydrogeological exploration for nearly three decades.

However, the accumulation of new observational data acquired from hydrogeological and geocryological exploration efforts in the intensively developing northern regions of the country has contributed to a substantial improvement in our understanding of the conditions of formation, occurrence, and propagation of permafrost ground waters, as well as of the nature of their interaction with permafrost. This all demanded refinement and recomplement of the available classification. A number of new classification schemes have appeared, proposed by various researchers, most of which use as a basis N. I. Tolstikhin's classification (Obidin 1959, Kalabin 1960, Ponomarev 1960, Sukhodolsky 1964, Shvartsev 1964, Romanovsky 1966, Velmina 1970, Tolstikhin and Tolstikhin 1974, et al.).

Of all subsequent classifications general recognition received the scheme of permafrost ground waters, advanced by N. N. Romanovsky in 1966, somewhat refined by him later (Romanovsky 1967, 1978). Following N. I. Tolstikhin, N. N. Romanovsky retains the same basic principle of ground water classification according to the spatial position with respect to permafrost, by distinguishing supra-, sub-, and intrapermafrost waters and adding to these main types open talik and inpermafrost waters. This classification has found application in hydrogeological exploration and has been included in textbooks, manuals and methodical literature. Nonetheless, this classification fails to fully reflect the present understanding of permafrost development and the conditions of its interrelationship with ground waters. The main disadvantage is that it almost completely lacks consideration of the dynamic behavior of the permafrost zone.

The main hydrogeological features of the permafrost zone are associated with the appearance in rocks of a solid phase of ground waters and with its exceedingly dynamic occurrence in the annual and perennial cycles. Every year in the active layer ground ice changes into a liquid phase and vice versa in huge amounts which has crucial influence upon behavior not only of ground waters but also of surface waters of the permafrost zone, thereby determining the distinctive features of the conditions of their interrelation and regime.

In the lower parts of the permafrost layer the magnitude of such abrupt changes in the state of freezing of ground waters is considerably smaller. A steadier perennial change of the solid phase of ground waters into a liquid phase takes place in some regions or vice versa in others which, basically, depends on general climatic, paleoclimatic or hydrogeological conditions of particular regions of the permafrost zone. In this case also, however, seasonal changes of phase composition of ground waters are possible, being due to convective transfer of the free air through cracks and pores of frozen rock materials as well as to diffusion processes, various chemical reactions, etc.

Thus, the dynamic state of the permafrost zone is a key factor, determining hydrodynamic, hydrochemical and gas properties of ground waters, their regime and resources, contributing thus to the exceptional diversity of hydrogeological situations within even individual, not very large structures.

This, all taken together, indicates that the factor of dynamic behavior of the permafrost zone has to be taken into account in a hydrogeological classification of gravitational ground waters because otherwise it will not completely reflect the actual natural conditions and will not fulfill the requirements imposed on it.

I have elaborated a hydrogeological classification of permafrost zone ground waters, taking into account the factor of dynamic development of permafrost (Shepelev 1982). The subdivision of ground waters in this classification into the main types rests also on the principle of their position in a geocryological layering, according to which supra-, intra-, and subpermafrost waters are distinguished. As is apparent, the terminology for the main types of ground waters, except for the "open talik waters" type, remain the same as in N. N. Romanovsky's classification scheme. However, we attach to them a somewhat different meaning.

The classification we offer adopts the following definitions of the main types of ground waters:

Suprapermafrost waters have positive or negative temperatures and occur either above the top of a seasonally frozen rock layer (suprapermafrost vadose water) or above the top of permafrost (suprapermafrost ground water).

Intrapermafrost waters have positive or negative temperatures and occur within permafrost in lenses, layers, and unfrozen zones; they move either vertically or horizontally within the frozen ground or exist in a stagnant regime.

Subpermafrost waters have positive or negative temperatures and occur below the base of permafrost, either in contact with it or separated from it by water-resisting or permeable rocks (without contact).

Gravitational ground waters of the permafrost zone that have a subzero temperature are commonly termed "cryopegs" (bodies of liquid saline water below 0°C associated with permafrost, may also imply ice-free permafrost with saline pore water) (Tolstikhin 1971). Because subzero temperature can occur with all the main types of ground waters selected, one should distinguish among supra-, intra-, and subpermafrost cryopegs.

Each of the main types of ground waters are divided into subtypes, categories and subcategories, with respect to permafrost water-confining strata, their dynamic behavior, and conditions of occurrence.

Suprapermafrost waters (type 1) are divided into two subtypes. The first (1a) includes suprapermafrost waters occuring within the upper part of a seasonally frozen layer that is for them a temporal permafrost water-confining stratum. These waters exhibit the following typical features, they:

(1) occur only during a warm period of the year;
(2) are of small thickness and of irregular area;
(3) are fed through infiltration of atmospheric precipitation as well as thawing of ice occurring within a seasonally frozen layer and pore air condensation processes;
(4) are notable for an abruptly changing time boundary of the lower permafrost water-confining stratum, due to the summertime thawing of the seasonally active layer;
(5) are characterized by a complete lack of head.

These features indicate that suprapermafrost waters of this subtype are a peculiar variety of vadose water which, admittedly, includes waters that form in the zone of aeration on relative water-confining strata (Principles of Hydrology, 1980). The relativity and temporary occurrence of the water-confining stratum, represented by earth materials of the seasonally frozen layer, thawing during the summer, are obvious and, therefore, we have good cause to term the waters of the subtype under consideration the suprapermafrost vadose water, thereby emphasizing the high changeability of the uppermost part of the permafrost zone layering.

Subdividing these waters into categories is needed because their different lifetime depends on the interrelation of permafrost with the layer of seasonal thawing. Suprapermafrost vadose water in continuous permafrost ($1a_1$) can persist during the greater part of the warm period. In cases where the seasonally frozen layer has a substratum of unfrozen water-bearing earth materials, there takes place its rapid failure in view of the convective and productive thermal impact of the underlying suprapermafrost water. In this connection, the suprapermafrost vadose water in discontinuous permafrost ($1a_2$) lasts for a shorter period of time (from several days to one or two months) and because of a rapid break in continuity of the seasonally frozen layer it frequently occurs on frozen ground lenses.

Suprapermafrost ground waters generally have above-zero temperatures and, therefore, they exert a warming influence upon the underlying seasonally frozen ground, thereby contributing to its rapid thawing.

The second subtype (1b) includes suprapermafrost water, occurring within the upper part of permafrost which represents for the former a permafrost water-confining stratum, being uniform as to time and area. Ground water occurring on the uppermost steady-state water-confining stratum is commonly referred to ground waters. However, the suprapermafrost waters of the subtype in question are defined by some distinctive features and primarily by the possibility that they are able to acquire a temporary permafrost water-confining stratum (head) as well as to partially or completely freeze up in the winter. Consequently, they can be referred only to the suprapermafrost variety of ground waters.

According to the degree of interrelation with seasonally frozen ground, the suprapermafrost ground waters in the classification we offer are divided into three kinds: seasonally freezing ($1b_1$), partially freezing ($1b_2$), and seasonally nonfreezing ($1b_3$).

Depending on conditions of occurrence of taliks

which accumulate seasonally non-freezing suprapermafrost waters, these are subdivided into two subtypes: subaerial ($1b_3^1$) and subaquatic ($1b_3^2$). Seasonally non-freezing suprapermafrost ground waters of the subaerial saturation zone generally occur in the regions where water well confining ground is developed from the surface. Seasonally non-freezing suprapermafrost waters of the subaquatic saturation zone are widely spread within permafrost. They occur within taliks below a lake or below a river bed as well as in the shelf zone of Arctic seas where they exhibit a subzero temperature and high mineralization. In areas with man-made pollution and salinization a subzero temperature and high mineralization can be acquired by seasonally non-freezing suprapermafrost waters of the subaerial saturation zone and by partially freezing suprapermafrost ground waters.

All the above identified categories and subcategories of suprapermafrost ground waters seldom occur in isolation but most frequently they persist in conjunction in a close hydraulic interplay, producing complex permafrost-hydraulic structured suprapermafrost ground water basins or flows.

Intrapermafrost waters (type 2) with respect to frozen water-confining strata and occurrence properties are divided into two subtypes. The first (2a) includes intrapermafrost waters with horizontal boundaries of permafrost water-confining strata. The formation of these waters is associated with a variation of climatic, geomorphological, soil-vegetation, and other natural conditions, which initiate a partial perennial freezing of water-bearing taliks. The intrapermafrost waters of this subtype exhibit the following features:

(1) presence of cryogenic head, the magnitude of which is defined by the degree of freezing of water-bearing taliks;

(2) horizontal movement of ground waters or stagnant regime of filtration;

(3) predominant occurrence in between the strata.

Intrapermafrost waters of the subtype under consideration are subdivided into three kinds, depending on direction and rate of variation in the boundaries of permafrost water-confining strata. The first kind ($2a_1$) includes pressure intrapermafrost waters in the stage of stable perennial freezing; the second kind ($2a_2$) includes pressure-pressureless waters with departing boundaries of permafrost water-confining strata, and pressure waters with a relatively stable position of permafrost confining them are of the third kind ($2a_3$). Each of the categories selected is in turn divided into two subcategories: hydraulically non-isolated and isolated. Hydraulically non-isolated intrapermafrost waters generally exhibit horizontal filtration and a predominant above-zero temperature. Hydraulically isolated intrapermafrost waters of this subcategory have a stagnant regime and may exhibit both an above-zero and subzero temperature. In this regard they are similar to the concept of intrapermafrost waters in N. N. Romanovsky's (1966) classification scheme.

The second subtype (2b) of the suggested classification includes intrapermafrost waters with predominantly vertical boundaries of frozen water-confining strata. These waters in N. N. Romanovsky's classification scheme are referred to an independent type, "open talik waters" which is not fully justified. Indeed, like intrapermafrost waters of the first subtype, these waters also occur in between permafrost ground strata. The only distinctive feature for them is the fact that the permafrost confines water-bearing zones on sides rather than from above and below, thus contributing to vertical direction of filtration of these waters. However, this property cannot justify their being separated into an independent single type of permafrost ground water because, of the adopted principle of their typification.

The second subtype of intrapermafrost waters is subdivided into three categories, depending on the character of their vertical filtration. The first category ($2b_1$) includes pressureless intrapermafrost waters with a downward filtration through open taliks; the second category ($2b_2$) comprises pressure waters with an upward filtration; and the third category ($2b_3$) incorporates pressure-pressureless waters with a mixed regime of vertical filtration. The last category of intrapermafrost waters of the subtype under consideration is the most widely spread and occurs both within taliks below river beds and in water shed areas.

Depending on the conditions of occurrence of water-bearing open taliks all the selected categories of intrapermafrost waters of this second subtype are divided into two subcategories: subaerial and subaquatic. The latter are widespread because below river beds and below lakes and reservoirs are the most favorable geothermal conditions for the development of intrapermafrost taliks and for preventing them from freezing. Intrapermafrost waters of the subtype in question predominantly have above-zero temperatures.

Subpermafrost waters (type 3) in the classification we offer, as in N. N. Romanovsky's classification scheme, are divided into those in contact with the base of permafrost (subtype 3a) and those without contact (subtype 3b).

A typical feature of subpermafrost waters of the first subtype is that the permafrost for them acts as an upper permafrost water-confining layer, existence and dynamic position of which determine the peculiarity of their properties, composition and regime.

According to the dynamic state of the upper permafrost water-confining layer these waters are divided into three kinds: those with a degrading ($3a_1$) and an aggrading ($3a_2$) lower boundary of permafrost and those with a comparatively stable lower boundary of the upper frozen water-confining layer ($3a_3$).

As to the temperature, each selected category of subpermafrost waters of this subtype are divided into those with an above-zero and a below-zero temperature. Subpermafrost "cryopegs" are widespread on the Arctic coast of the USSR as well as in mainland regions where the fresh water zone has been completely frozen.

Subpermafrost waters of the second subtype make no direct contact with permafrost and therefore they normally have an above-zero temperature. According to their interrelation with permafrost these waters are divided into three categories. The first category ($3b_1$) includes pressure subpermafrost waters, separated from permafrost by a stable unfrozen lithological water-confining layer.

The piezometric level of these waters generally occurs above the base of permafrost and therefore there certainly is an interplay between them, if the relativity of water-confining strata is taken into account.

The second category ($3b_2$) comprises pressureless subpermafrost waters, separated from the permafrost base by water-bearing ground. Subpermafrost contact-free waters of this kind have substantial influence on the position of the lower boundary of permafrost, giving rise to its periodical oscillation due to a continual thermal convection of moisture-laden pore air and evaporation and condensation processes. The periodic variation of the lower boundary of permafrost can induce formation of a subpermafrost zone of highly jointed rocks.

Some cases of the occurrence of contact-free subpermafrost waters of this kind are reported, where the permafrost zone is represented by frozen rocks, in which ice only partially fills the porous space (Romanovsky, 1978). Such rocks are a good conductor of air and air moisture; therefore under such conditions subpermafrost waters exhibit high changeability of the water level and chemical composition.

The third category ($3b_3$) includes pressure subpermafrost waters, piezometric level of which is established below the base of permafrost. These normally are ground waters of deep water-bearing levels, and permafrost does not exert substantial impact on their regime, composition and properties.

Each of the selected main types of permafrost ground waters in a specific area under investigation may further be divided according to the character of water-holding rocks and their age. The need for such a subdivision is brought about by the specificity of hydrogeological work, its purpose and details.

In conclusion it must be emphasized that the ground water classification we have proposed in this study has a particular character because of only gravitational waters taken into account. To provide answers to some scientific and practical problems requires development of a general classification of permafrost ground waters that would comprise all phases of aggregation of water in rocks, occurring in different stages of cryogenesis.

REFERENCES

Baranov, I. Ya., 1940, The southern edge of the permafrost region, in Gidrologiya SSSR, v. XVII, book 2, Moscow-Leningrad: Gosgeolizdat.

Kalebin, A. I., 1960, Permafrost and hydrogeology of the north-east of the USSR, in Trudy VNII-I, v. 18, Magadan, 469 p.

Meister, L. A., 1955, On shortcomings of the classification of ground waters of the permafrost region, in Materials for the basic theories of the earth's crust frozen zones, Moscow: Izd. AN SSSR, p. 59-64.

Obidin, N. I., 1959, Classification of ground waters of the West-Siberian lowland and Siberian platform north of the Polar Circle, in Trudy of the Institute for Arctic Geology, v. 107, v. 12, Leningrad, p. 150-154.

Ovchinnikov, A. M., 1954, General hydrogeology, Moscow: Gosgeoltekhizdat, 384 p.

Ponomarev, V. M., 1960, Ground waters of a territory with a powerful stratum of permafrost rocks, Moscow: Izd. AN SSSR, 199 p.

Principles of hydrogeology, 1980, General hydrogeology, Novosibirsk: Nauka, 230 p.

Romanovsky, N. N., 1966, Permafrost ground water classification scheme, in Methods of hydrogeol. investigations and ground water resources of Siberia and the Far East, Moscow: Nauka, p. 28-42.

Romanovsky, N. N., 1967, Ground waters within permafrost and their interaction with permafrost, in General cryology, Moscow: Izd. Moscow University, p. 310-330.

Romanovsky, N. N., 1978, Ground waters within permafrost and their interaction with permafrost, Moscow: Izd. Moscow University, p. 352-376.

Shepelev, V. V., 1982, On hydrogeological classification of gravitationsl ground waters of the permafrost zone, in Methods of hydrogeol. exploration of the permafrost zone, Novosibirsk: Nauka, p. 5-27.

Shvartsev, S. L., 1964, Classification scheme of ground waters in areas with evolving permafrost, in Proceedings of the IV Meeting on Ground Waters of Siberia and the Far East, Irkutsk, p. 23.

Sukhodolsky, S. E., 1964, Formative properties of ground waters and their influence upon permafrost occurrence, in Geocryological conditions of the Pechora coal basin, Moscow: Nauka, p. 65-88.

Tolstikhin, N. I., 1941, Ground waters of the lithosphere frozen zone, Moscow-Leningrad: Godgeolizdat, 201 p.

Tolstikhin, N. I., 1971, Ground water classification, in Zap. Leningrad. gornogo in-ta, v. 62, 2, Leningrad, p. 3-15.

Tolstikhin, N. I., and Tolstikhin, O. N., 1974, Ground and surface waters in the permafrost zone, in General cryology, Novosibirsk: Nauka, p. 192-230.

Velmina, N. A., 1970, Properties of the hydrogeology of the lithosphere frozen zone, Moscow: Nedra, 348 p.

Zelenkevich, A. A., 1960, On the term "intrapermafrost waters," no. 2, Kolyma, p. 42-43.

A BRIEF INTRODUCTION TO PERMAFROST RESEARCH IN CHINA

Shi Yafeng and Cheng Guodong

Lanzhou Institute of Glaciology and Cryopedology, Academia Sinica
Lanzhou, People's Republic of China

The study of permafrost in China has progressed since the initial work in 1949, and is now at the stage of all-out development. The distribution and characteristics of permafrost in China have been analyzed, and taliks and other periglacial phenomena have been classified. Several kinds of permafrost maps have been compiled, including a comprehensive map showing the distribution of glaciers and frozen ground in China. A repeated segregation theory has been suggested to explain the formation of the massive ice layer beneath the permafrost table. Fossil periglacial phenomena are used to reconstruct the distribution of ancient permafrost. A series of applied parameters, including the thermal, physical, and mechanical parameters of typical soils, have been provided. An engineering classification system for frozen soils has been suggested. Some theories on frost-heaving, thaw settlement, heat transport, moisture migration, and attenuation creep of frozen soils and on the similarity criteria in model tests are recommended. Studies in road engineering, water conservation engineering, industrial and civil construction, and artificial freezing techniques for mining have been conducted. More than 2000 km of railways have been built in permafrost regions, and an oil pipeline over 1000 km long has been built from Golmud, Qinghai, to Lhasa, Xizang (Tibet).

The permafrost regions in China cover an area of about 2,150,000 km^2, second only to the USSR and Canada. If one includes ground that is seasonally frozen to depths greater than 0.5 m, the total frozen area occupies 68.6% of the entire territory of China. Such a vast area of frozen soil exerts a great influence on China's economy, but it will also expedite the study of permafrost in this country.

HISTORY

As early as 1633, the outstanding Chinese geographer Xu Xiake clearly described the lifting action of frost heaving; he called the block field on Wutai Mountain in northern China the "dragon-turned rock." Before 1949, however, only a few Japanese had made fragmentary observations of frozen ground in northeast China. The substantive and conscientious study of permafrost in China began only after the founding of the People's Republic in 1949.

The study of permafrost in China can be divided into four periods.

The Period of Initial Work (1949-1959)

After the founding of new China, various capital constructions in cold regions contributed much to our study of permafrost. In particular, the exploitation of forest resources in the Da Hinggan Ling in the Northeast and the contruction of water conservation projects and highways on the Qinghai-Xizang Plateau and between Urumqi and Ku'erle helped amass a large amount of information and experience with permafrost.

In 1954, a permafrost research group was established under the Department of Water Conservancy and Power Construction to develop techniques for measuring and testing permafrost, to study the freezing and thawing properties of soil, and to develop techniques for constructing compact embankments in winter. In 1956, Xin Kuide and Ren Qijia systematically reported the distribution of permafrost in Northeast China. In 1958, the Third Institute of Railway Design published "Engineering Geology and Railway Construction in Permafrost Regions," and in 1959, the Department of Water Conservancy and Power Construction published "Construction of Compact Embankments in Winter."

The Period of Growth (1960-1972)

The practical experience of the previous 10 years led to a deeper understanding of the importance of the study of permafrost. In 1960, a Research Department of Glaciology and Cryopedology was established under the Research Institute of Geography of the Chinese Academy (Academia Sinica). Special units were established for permafrost research in several governmental departments, including railways, building construction, coal mining, communication, and oil exploitation. Similar research units were also instituted in several colleges and universities.

During this period, the permafrost areas in the Qinghai-Xizang Plateau, Da Hingan Ling, and Qilian Shan and the seasonally frozen ground in the Northeast were systematically and scientifically investigated. Fixed observation stations were established and some substantial engineering experiments were made in such places as Tumen in Xizang; Muli, Reshui, and Jiangcang in Qinghai; and Da Hinggan Ling, Yakeshi, Genhe, and Daqing in the Northwest. The first low-temperature laboratories were built and various indoor tests were conducted. Several engineering problems that occur in frozen soil areas were initially solved by the voluntary application of the theory of cryopedology, including the utilization of frozen soil in laying seepage-prevention covering for earth embankments during the construction of several reservoirs, railway construction in Da Hinggan Ling and the Tian Shan, the construction of mining shafts in Qilian Shan, and the fundamental solution of the problem of foundation stability for industrial and civil construction in northern districts.

Artificial freezing was extensively applied in drilling engineering, and frost boils on highways and frost heave were successfully prevented by using lime earth.

"Investigation of Permafrost Along the Qinghai-Xizang Highway," published by the Research Department of Glaciology and Cryopedology of the Research Institute of Geography, Academia Sinica, is a representative work of this period.

The Period of Maturity (1973-1978)

During this period, a section of railway was successfully built in Qilian Shan near the lower permafrost limit with thick ground ice by applying the principle of protecting the permafrost; 8 years later, the road bed of this section of the railway is still stable. An oil pipeline from Golmud to Lhasa was also built in this period. The reconstruction of the Qinghai-Xizang Highway by covering the road surface with asphalt began.

A group of cryopedologists started to work all along the Qinghai-Xizang Highway, undertaking geological engineering prospecting and conducting extensive physical and thermological tests to obtain enough data to allow the definition of effective principles for the design and construction of the Qinghai-Xizang Railway. At the same time, a scientific and technological cooperative group was organized for the study of frost prevention in hydraulic architecture, with more than 30 units from water conservation and scientific research departments participating as well as colleges and universities in the north.

In 1975, a group researching road construction in the permafrost regions in China visited Canada, and in 1977, a Canadian permafrost investigation group headed by the late Dr. R.J.E. Brown paid a return visit to China. China's isolated study of permafrost had come to an end.

In 1975, the Lanzhou Institute of Glaciology, Cryopedology, and Desert Research of the Chinese Academy published "Permafrost," a compilation of the special knowledge of cryopedology. "Annals of the Lanzhou Institute" first published in 1976, and "Annals of Lanzhou Institute of Glaciology and Cryopedology, Academia Sinica (Permafrost)," published in 1981, made available to the world a part of our research results before 1978.

During this period, the first specialized curriculum in glaciology and cryopedology was offered by the Department of Geography at Lanzhou University. This was the starting point for training the younger generation in the special knowledge of glaciology and cryopedology.

In this period, a complete and systematic study of cryopedology took shape, including general cryopedology, permafrost mechanics and thermology, and engineering cryopedology.

The Period of All-Out Development (1978-Present)

After the 10 years of the Great Cultural Revolution, permafrost research in China entered a period of all-out development. 1978 is a year worth remembering in the history of permafrost study in China, for the Lanzhou Institute of Glaciology and Cryopedology was separated from the Lanzhou Institute of Glaciology, Cryopedology, and Desert Research of the Chinese Academy, and reorganized. The First National Academic Conference on Glaciology and Cryopedology was held in Lanzhou, and 68 articles and abstracts on cryopedology were presented. It was also in 1978 that China for the first time sent a delegation to attend the Third International Conference on Permafrost in Canada; China presented four articles.

In 1979, the first issue of the academic journal "Glaciology and Cryopedology" was published, greatly accelerating the study of cryopedology in China. In 1980, 62 papers and abstracts on cryopedology were presented at a seminar on testing and measuring techniques for glaciology and cryopedology, and the Society of Glaciology and Cryopedology was established under the Chinese Geographical Society. In 1981, when the Second National Conference on Cryopedology was held in Lanzhou, 185 papers and abstracts were presented, reflecting the progress in the study in all fields of cryopedology in China. Drs. J. Brown and Yin-Chao Yen from United States and Dr. Daisuke Kuroiwa from Japan attended this conference.

Since 1978, cryopedological investigations have been conducted in the Altai Shan, Tian Shan, Qilian Shan, and Hengduan Shan. Such activities as the paving of the Qinghai-Xizang Highway, drilling of shafts by artificial freezing, and prevention of frost damage in hydraulic constructions have led to greater achievements in both theoretical study and engineering practice. At the same time, studies of agricultural cryopedology and the ecology of frozen regions began. The movement of the study of cryopedology in China from qualitative to quantitative, from investigative to experimental, and from macrocosmic to microcosmic marked the beginning of China's all-out development of her permafrost regions.

PRESENT CONDITION OF STUDY

The study of cryopedology in China enters the 1980's with a record of success. I will now discuss our major achievements in permafrost research with emphasis on those attained since 1978.

General Cryopedology

Through investigation and study, the distribution and characteristics of permafrost in China have been defined, patterns of temperature dispersion and permafrost thickness have been analyzed, the genesis of taliks and frost-heaving topography has been described, permafrost maps have been compiled according to regional indexes, a classification system for frozen soil has been suggested, and the concept of the regularity of alpine permafrost zonation has been presented.

On the basis of these achievements, several kinds of permafrost maps have been compiled, among them a map showing the glacier and permafrost distribution in China (1:4,000,000), a map of permafrost regionalism in northeast China (1:3,000,000), and a map of permafrost along the Qinghai-Xizang Highway (1:600,000), as well as specialized maps of particular regions. The study of the history of permafrost formation on the Qinghai-Xizang Plateau and in northeast China has been initiated. Comprehensive hydrological surveys have been carried out in some permafrost regions, such as Qilian Shan, along the Qinghai-Xizang Highway, and in Da-Xiao Hinggan Ling, and several regional maps on comprehensive hydrogeology have been compiled. Progress has also been made in compiling geological maps for engineering use.

At present, the emphasis is on the study of periglacial phenomena. The theory of repeated segregation is offered to explain the formation of thick layer ground ice near the permafrost table. The idea of a "subperiglacial stage" has been suggested on the basis of the relation between glacial and periglacial phenomena. The viewpoint that the periglacial landscape on the plateau could be divided into maritime, subcontinental, and continental types has already been publicly recognized. Large amounts of ancient periglacial remains have been discovered in north China and have been widely used to reconstruct the paleoclimate and the boundaries of paleopermafrost in the Quaternary Age in China.

Physicomechanical and Thermological Properties and Processes of Frozen Ground

The physicomechanical properties and processes of frozen ground have been studied systematically. Through a wide range of laboratory and field tests, parameters have been compiled for the bearing capacity of frozen soil, coefficient of thawing settlement, coefficient of compression, frost-heaving amount, and frost-heaving force, as well as the classification of frozen ground for engineering purposes. Most of these have been included in the detailed design regulations. Relative equations have been suggested for the relationship between the normal frost-heave force, area, and the burial depth of foundations, as well as the distribution pattern and calculating diagrams for the horizontal frost-heave thrusting force along vertical space. Equations for the process of attenuation creep and for calculating creep deformation and the time-to-stable-state have been ascertained. A semi-empirical formula for the variation of the relaxation of frost-heave stress and of the modulus with the change of time has been obtained. Several properties of sandy gravel saturated with water during freezing, such as pore water pressure and frost heave amount, have been made clear. Model experiments on a single pile under vertical and horizontal loading have been carried out. The construction technology of driven, inserted, and filled piles and the bearing capacity of a single pile have been tested on the Qinghai-Xizang Plateau; and we have started to study the properties of moisture migration during freezing from the viewpoint of energy.

In the field of thermology, the specific heat, thermal conductivity, and diffusivity values of typical freezing and thawing soils were determined for several regions, problems of similarity criteria in model tests of frozen soil were solved, methods for calculating temperature fields and moisture migration under natural and artificial conditions were developed, the application of mathematical methods was expanded, and the application of thermological methods to calculate frost heave have begun.

Engineering Construction in Frozen Districts

Road engineering. In permafrost regions, 2000 km of railway have already been built. Railroad lines through permafrost regions have been successfully selected, and improved solutions have been found for such technical problems as the design and construction of the road bed in sectors with thick ground ice. Engineering experiments in the application of heat-preserving materials, both by model and actual construction, have been carried out. In the Hinggan Ling of northeast China and on the Plateau, methods have been developed to solve bridge and culvert foundation problems. Drilled and filled concrete piles have been used extensively and successfully. In tunnels in cold regions, where water seepage poses a serious problem, methods such as forced ventilation are used to melt ice, preserve heat, and drain water so that traffic is not interrupted.

Water conservation engineering. Throughout 13 provinces, municipalities, and autonomous regions in north China, damage to and destruction of canal systems through frost-heave deformation and thawing settlement pose a serious problem. Laboratory and field studies of pretective measures for engineering construction have yielded positive results. In lining canal beds, the technique of laying sand layers under the concrete slab effectively prevents frost damage. Using soft materials and thin membranes as seepage linings improves protection against the destructive action of frost heaving. The construction of simple beam structures and the technique of sinking slabs by steam-melting permafrost are effective ways to protect medium- and small-sized dams from frost damage. Using the frost-heave reaction to anchor foundations effectively prevents the uplifting of pilings. It is also effective to use non-frost-heaving sand-gravel as fill material in foundations while ensuring appropriate water drainage during freezing.

Industrial and civil construction. It has been recognized that, based on an understanding of the regularity of the main frost-heaving zone and the characteristics of restriction of frost-heaving under overload, a residual frozen layer could be preserved under foundations. The formula for determining the minimum burial depth for foundations was developed and has been included in the relevant specifications. In addition, filling sand-gravel at the lateral sides of foundations, preserving heat on the ground surface, and explosive extension of short-pile foundations are also successful in preventing frost heaving. In permafrost regions, the use of high-filled foundations, raised-ground foundations with ventilating pipes in the bedding layer, and foundations with suspended ventilation pipes are all useful techniques. Suspended ventilated foundations are generally used for heated buildings on ice-rich frozen ground. There are several effective methods to prevent the freezing of water pipelines, such as insulating suspended pipes or pipes on the ground surface, heating the well water, or putting the pipeline close to a hot pipeline. The problems of ensuring concrete strength and of construction technology under negative temperatures have already been solved.

Artificial freezing technique for mining. The maximum sinking depth by artificial freezing has already exceeded 400 m, reaching 358 m in loose layers. Comparatively systematic data have been obtained on thermal regime, freezing pressure, and so forth, in the shaft wall. Experiences have been gained in drilling techniques and artificial freezing. Methods to prevent the destruction of shaft walls and rupture of freezing pipes in medium-depth and deep shafts have been summarized. At present, a model is being prepared to test the creep of frozen soil under three-dimensional conditions.

Oil pipeline. The oil pipeline from Golmud to Lhasa, which transports refined oil across permafrost areas of the Qinghai-Xizang Plateau, has been operating safely for 9 years, proving that the construction designs of shallowly buried pipe, pile foundations, and hanging ventilated foundations at pumping stations are feasible and successful.

Prospecting and Testing Techniques

Geophysical prospecting techniques have been used extensively in the investigation of permafrost. The most commonly used is direct current prospecting. Direct current profiling, shallow seismic methods, and excited polarization are used to determine the permafrost table. The measurement of gas emissions for locating the talik is also being tested, and the use of radar equipment in investigation is planned. Photographs are widely used to study periglacial phenomena, and the trial use of satellite images for the interpretation of certain permafrost information is also in progress.

Measurements and testing techniques have been standardized for 14 fundamental physicomechanical and thermological parameters for the study of permafrost. A multipoint quartz crystal thermometer with an accuracy of 0.01°C has been developed. Neutron equipment is used to determine water content of permafrost. Ultrasonic techniques have been used to measure the modulus of elasticity, Parson's ratio, and unfrozen water content of frozen soil. The application of an electronic cryogenic cooling apparatus facilitates the experimentation.

Future Studies

A great amount of work has been done in the field of permafrost study in China, yet many problems remain to be solved. To raise the study of permafrost in China to a higher level, the following work has to be done:

1. Intensify the study of permafrost engineering. Solve the following problems in the near future:

- calculation of the critical height and the design height of the subgrade of an asphalted highway and of railways in permafrost regions,
- the design axiom of subgrade in zero-filling and cutting sections,
- the calculating theory of the temperature field and the strength of frozen walls in drilling with artificial freezing,
- principles and method for preventing frost heave in canals.

2. Concentrate on fundamental and theoretical study, to guarantee the steady development of research into the formation and distribution of high-altitude permafrost, the theory of ice formation, moisture migration, and rheologic principles.

3. Enlarge the scope of the research by the systematic development of engineering geological forecasts, environmental protection, and agricultural cryopedology in cold regions.

4. Improve prospecting methods, measurement, and testing techniques in permafrost study. Further enhance accuracy in measuring temperature, heat, water, force, and deformation; establish self-recording and telemetry; improve telemetry and remote sensing techniques; improve low-temperature laboratory equipment; and continue to establish fixed-point observation stations.

5. Increase national and international academic exchanges and accelerate the training of young cadres who specialize in permafrost.

REFERENCES

Chen Xiaobo, Jiang Ping, and Wang Yaqing, 1980, Pore water pressure of saturated gravel during freezing: Journal of Glaciology and Cryopedology, v. 2, no. 4, p. 33-37.

Cheng Guodong, 1982, The forming process of thick-layered ground ice: Scientia Sinica (Series B), v. XXV, no. 7, p. 777-788.

Chinese Society of Glaciology and Cryopedology and Lanzhou Institute of Glaciology and Cryopedology, Academia Sinica, ed., 1980, Selected proceedings from the Symposium of Measuring and Testing Techniques in Glaciology and Cryopedology in China: Journal of Glaciology and Cryopedology, v. 2, Special Issue.

Cui Chenghan and Zhou Kaijiong, 1982, Analaysis of frost damage at the medium bridge on the Yindaluchi: Journal of Glaciology and Cryopedology, v. 4, no. 1, p. 60-63.

Cui Zhijiu, 1980, On the periglacial mark in permafrost area and the relation between glaciation and periglaciation: Journal of Glaciology and Cryopedology, v. 2, no. 2, p. 1-6.

Division of Glaciology and Cryopedology, Institute of Geography, Academia Sinica, ed., 1965, Investigation of permafrost along Qinghai-Xizang Highway: Beijing, Science Press.

Dai Jingbo and Li Enying, 1981, Influence of snow cover to the ground temperature in the permafrost region in the northern part of the Great Xinan Mountain: Journal of Glaciology and Cryopedology, v. 3, no. 1, p. 10-18.

Ding Dewen, 1978, Calculation of talik under warm building and stable time: Kexue Tongbao (a monthly journal of science), no. 11.

Fan Ronghe and Yao Shangsen, 1982, Discussion on the formation and the trend of development of the perennial frost on Southern Qinghai-Northern Xizang Plateau: Journal of Glaciology and Cryopedology, v. 4, no. 1, p. 45-54.

Fan Xipeng, 1963, Permafrost in the eastern part of the Qinghai-Xizang Plateau: Geography, no. 4, p. 154-156.

Guo Dongxin, Wang Shaoling, Lu Guowei, Dai Jingbo, and Li Enying, 1981, Division of permafrost regions in Da-Xiao Hinggan Ling of Northeast China: Journal of Glaciology and Cryopedology, v. 3, no. 3, p. 1-9.

Huang Xiaoming, Yang Hairong, and Liu Tieliang, 1981, Some behaviours of seasonal freezing and thawing of natural ground in the interior of Qinghai-Xizang Plateau, in Proceedings of Symposium on Qinghai-Xizang (Tibet) Plateau, v. II, p. 1803-1816.

Lanzhou Institute of Glaciology, Cryopedology and Desert Research, Academia Sinica, ed., 1975, Permafrost: Beijing, Science Press. Available in English translation from National Research Council of Canada, NRC TT-2006.

Lanzhou Institute of Glaciology, Cryopedology and Desert Research, Academia Sinica, ed., 1975, Memoirs of Lanzhou Institute of Glaciology, Cryopedology and Desert Research, Academia Sinica: Beijing, Science Press, v. 1.

Lanzhou Institute of Glaciology and Cryopedology, Academia Sinica, ed., 1981, Memoirs of Lanzhou Institute of Glaciology and Cryopedology, Academia Sinica: Beijing, Science Press, v. 2.

Lanzhou Institute of Glaciology and Cryopedology, Academia Sinica, ed., 1982, Memoirs of Lanzhou Institute of Glaciology and Cryopedology, Academia Sinica: Beijing, Science Press, v. 3.

Lanzhou Institute of Glaciology and Cryopedology Academia Sinica, ed., 1982, Proceedings of the Symposium on Glaciology and Cryopedology held by Geographical Society of China (Cryopedology): Beijing, Science Press.

Lanzhou Institute of Glaciology and Cryopedology, Academia Sinica, ed., 1982, A Symposium of the Research of Permafrost on Qinghai-Xizang Plateau: Beijing, Science Press.

Lanzhou Institute of Glaciology and Cryopedology, Academia Sinica, ed., 1983, Proceedings of Second National Conference on Permafrost: Lanzhou, Gansu People's Press.

Lian Huisheng, Zhao Lin, and Wang Jiacheng, 1981, A preliminary experimental study on the instantaneous strength of frozen sand: Journal of Glaciology and Cryopedology, v. 2, no. 1, p. 37-40.

Liu Hongzu, 1981, On calcuation of normal frost heaving force: Journal of Glaciology and Cryopedology, v. 3, no. 2, p. 13-17.

Pei Wenzhong, 1956, The first find of solifluction process in China: Kexue Tongbao (a monthly journal of science), no. 11, p. 51-53.

Qiu Guoqing, 1980, On engineering geology survey in alpine permafrost regions of West China: Journal of Glaciology and Cryopedology, v. 2, no. 3, p. 17-23.

Qiu Shanwen, Li Fenghua, and Sui Xinlan, 1981, Periglacial landforms of the Changbai Shan: Journal of Glaciology and Cryopedology, v. 3, no. 1, p. 26-31.

Research Group of Qinghai-Xizang Highway, Ministry of Communications, 1979, On the stability of subgrade with asphalt pavement constructing in permafrost region on Qinghai-Xizang Plateau: Journal of Glaciology and Cryopedology, no. 2, p. 43-51.

Shi Yafeng, 1964, Study of glaciology, cryopedology and hydrology of dry air in China for five years: Kexue Tongbao (a monthly journal of science), no. 3.

Shi Yafeng, 1982, Some achievements of the Lanzhou Institute of Glaciology and Cryopedology (1978-1980): Journal of Glaciology and Cryopedology, v. 4, no. 3, p. 81-86.

Third Design Institute of Railway Ministry, 1958, Permafrost Engineering Geology and Railway Construction: Beijing, People's Railway Press.

Tong Boliang, 1982, Some features of permafrost on the Qinghai-Xizang Plateau and the factors influencing them, in Proceedings of Symposium on Qinghai-Xizang (Tibet) Plateau, v. II, p. 1795-1802.

Tong Changjiang and Shen Zongyen, 1981, Horizontal frost heave thrust acting on buttress constructions: Engineering Geology, v. 18, p. 259-268.

Wang Chunghe, 1982, Discussion on the formation of thick underground ice near the upper limit of permafrost: Journal of Glaciology and Cryopedology, v. 4, no. 2, p. 47-54.

Wang Jiacheng, Wang Shaoling, and Qiu Guoqing, 1979, Permafrost along the Qinghai-Xizang Highway: Acta Geographica Sinica, v. 34, no. 1, p. 18-32.

Wang Zhengqiu, 1980, Influence of grain-size constituent on frost heave in fine sand: Journal of Glaciology and Cryopedology, v. 2, no. 3, p. 24-28.

Wu Ziwang, 1982, Classification of frozen soils in engineering construction: Journal of Glaciology and Cryopedology, v. 4, no. 4, p. 43-48.

Xie Youyu, 1981, Periglacial landforms and regionalization in Northeast China: Journal of Glaciology and Cryopedology, v. 3, no. 4, p. 17-24.

Xin Kuide and Ren Qizia, 1956, Distribution of permafrost in Northeastern China: Geological Knowledge, no. 10.

Xu Xiaozu and Guo Dongxin, 1982, Compilation of the distribution map of frozen ground (1;4,000,00) in China: Journal of Glaciology and Cryopedology, v. 4, no. 2, p. 18-25.

Zhang Changqing, 1980, The discussion about "Permafrost classification for engineering:" Journal of Glaciology and Cryopedology, v. 2, no. 2, p. 58-60.

Zhang Weixin and Zhang Tingjun, 1982, Polygonal soil along the highway in Qinghai and Xizang (Tibet): Journal of Glaciology and Cryopedology, v. 4, no. 3, p. 77-80.

Zhang Yong and Cai Shiquan, 1981, Discussion on the time of formation of permafrost on Qingzang Plateau: Journal of Glaciology and Cryopedology, v. 3, no. 1, p. 32-37.

Zhou Chanqing, 1980, An introduction of achievements about the study of bases and foundations on the frozen ground in China: Journal of Glaciology and Cryopedology, v. 2, no. 1, p. 1-5.

Zhou Youwu and Guo Dongxin, 1982, Principal characteristics of permafrost in China: Journal of Glaciology and Cryopedology, v. 4, no. 1, p. 1-19.

Zhu Linnan, 1981, Analysis of temperature field of artificial frozen wall of a deep shaft: Journal of Glaciology and Cryopedology, v. 3, no. 4, p. 44-51.

A COMPREHENSIVE MAP OF SNOW, ICE, AND FROZEN GROUND IN CHINA (1:4,000,000)

Shi Yafeng and Mi Desheng

Lanzhou Institute of Glaciology and Cryopedology, Academia Sinica
People's Republic of China

On the basis of recent progress on the study of glaciers, frozen ground, and seasonal snow and ice, a preliminary and comprehensive map of snow, ice, and frozen ground in China has been compiled by the Lanzhou Institute of Glaciology and Cryopedology, Academia Sinica, to express the general distribution of these phenomena in our country. The map shows the following items: (1) mountain glaciers, including the continental and monsoon maritime types of glaciers as well as the current isopleth of the snow line and the hazardous places of glacial debris flows and flood from burst of glacial lakes, (2) seasonal snow, containing the stable and unstable snow cover regions, (3) seasonal ice, involving sea ice and river-lake ice, (4) permafrost, including continuous, discontinuous, and insular permafrost, (5) seasonally and instantaneously frozen ground, and (6) periglacial phenomena and Quaternary glaciation, such as the isopleth of the past snow line, blockfields, stone stripes, polygons, frost-heave hummocks, and involutions.

Since the 1950's, studies on glaciology and cryopedology have been carried out continuously by Chinese scientists. In the past 20 years or more, progress at different degrees has been achieved on mountain glaciers, permafrost, seasonal snow, seasonal ice, and seasonal frozen ground, as well as other related hazards including glacial debris flow, avalanche and snow drift, and also of periglacial phenomena and the action of Quaternary glaciation. In the past 2 years, a Comprehensive Map of Snow, Ice, and Frozen Ground in China has been preliminarily compiled by the Lanzhou Institute of Glaciology and Cryopedology to show the general distribution of these phenomena. Nearly 20 scientists participated in its compilation, examination, and verification. Mi Desheng is responsible for its design. The preliminary draft of the map was completed in 1983.

MOUNTAIN GLACIERS

The mountain glaciers in China are distributed in the high mountains west of 104°E, covering an area of about 56,500 km^2 with a water storage of about 5000 km^3 and a mean annual amount of melt water of 55X10^9 m^3 (Shi Yafeng and Yang Zhenniang, 1982), an invaluable water resource for the arid regions in Central Asia. The existing glaciers in China are distributed in the 12 mountain systems: i.e. Altay Shan, Tian Shan, Qilian Shan, Kunlun Shan, Pamirs, Karakoram, Qiangtang Plateau (northwestern part of Xizang), Tanggula Shan, Nianqing Tanggula Shan, Gangdisi Shan (i.e. Transhimalaya Mts), Himalayas, and Hengduan Mountains. Main areas of glaciers appear in the middle part of Tian Shan, west Kunlun Shan, Karakoram, the middle part of Great Himalaya, and Eastern Nianqing Tanggula, in which some long valley glaciers exceed 20, 30, or even 40 km in length. Glaciological investigations have been concentrated in the areas of Qilian Shan, Tian Shan, Hengduan Shan, and Altay Shan, as well as on the northern slope of middle Himalayas and the southeastern part of Xizang. According to their physical properties and geographical conditions, glaciers are divided into two main categories, namely the continental type and the monsoon maritime type.

From the Altay Shan to the northern slope of the Himalayas, nearly 80% of the glaciers in China belong to the continental type. Under the continental climatic conditions of Central Asia, this type of glacier has only a level of mass budget at 300-1000 mm/yr, and its formation of ice is mainly through the process of cold infiltration and congelation. The temperature in its active layer at the ice tongue has the characteristic of a cold glacier.

Monsoon maritime glaciers are mainly distributed in the southeastern part of Xizang (Eastern Nianqing Tanggula Shan and the eastern extremity of Himalaya) and Hengduan Mountains. Some of them have a level of mass budget at 2000-2500 mm/yr, and the formation of ice of this type is mainly due to the action of temperate infiltration (Shi Yafeng et al. 1980). Based on the study of Li Jijun (1982), the general demarcation line of these two types of glaciers is shown on the map from 29°N, 92°E extending northeastward to 31°N, 90°E and still eastward along the latitude through the Hengduan Mountains.

Data of glacier distribution and present snow line are taken mainly from the aerial topographic maps of 1960s and the Landsat images of 1970s, as well as from glaciological literatures. As to the regional variations of the present snow line shown on the map, except for some specific glaciers having observed data for some specific glaciers having observed data for ELA, generally the height of snow line is plotted on the aerial topographic map by using the method of H. Hess, simplifying the 2500 points totally selected for the height of snow line. Then the isopleth of snow line is drawn, the lowest is in the Altay Shan (2700-3200 m) while the highest on the northern slope of Mt. Qomolangma, the highest peak in the world (6000-6200 m) to the east of which on the same latitude in monsoon maritime climatic area the snow line descends again to 4600 m.

Correlated with some present glacial areas are the hazardous places of glacial debris flow and flood from burst of glacial lakes. Glacial debris flow concentrates densely in the southeastern part of Xizang and in the Congge'er Shan of the Pamirs, greatly damaging highway traffic, while flooding from burst of glacial lakes appears mainly in the southern part of Xizang and the Karakoram Mountains, catastrophically influencing villages and cities in the piedmont area.

SEASONAL SNOW

The study of seasonal snow in China has only developed in the past several years. Based on the snow records of 1600 meteorologic stations throughout the country, Li Peiji (1983) plotted the isopleth of annual average snow-covered days. Taking 60 days as the demarcation between a stable snow area and unstable snow area in winter, he found that the former accounts for 43% of the total area of China with an annual variation of C_v below 0.4, covering all parts of Northeast China and Inner Mongolia, northern part of Xingjiang, most part of Xizang Plateau, as well as some high mountains in the middle and southern parts of China. For example, the period of snow deposit in the Central Range in Taiwan also exceeds 2 months, that in the Altay of Xingjiang and in Da Hinggan Ling and the Changbai Shan of Northeast China all exceeds 150 days, having an annual variation within 0.1. In all these regions, the melt water of seasonal snow is an important source of surface runoff, greatly beneficial to crops and animal husbandry.

The unstable snow area with an average annual snow cover from 10-60 days is distributed in the extensive region between the Liaohe River in Northeast China and the latitudinal range of Qingling to Dabie Shan in Central China, where the winter snow keeps a certain period of days with an annual variation between 0.4-1.0. The very unstable snow cover area with an average period from 1-10 days has its southern boundary at the latitude of 25° to 27° (from Kunming, Guilin to Wenzhou) with an average annual variation above 1, where the snow deposit of certain years is particularly influential to the life of the people. As expressed by the old Chinese saying, "Fortune-giving show foretells a bumper harvest of a fortunate year". In a country like China with low precipitation in the winter and spring, snow deposits are favourable to agricultural production.

However, in the Tian Shan and the Altay Shan of Xingjiang, in the southeastern part of Xizang, and in several other high mountains, avalanches and snow drifts caused by seasonal snow seriously interrupt highway traffic, while snow storms and blizzards on the steppes of Inner Mongolia are very detrimental to animal husbandry.

Snowfall in China is often parallel with the advent of low temperatures, i.e. snow is thickest in January. But on the Qinghai-Xizang Plateau, snow falls more in the spring and autumn, and the snow deposit is less in severe winters.

SEASONAL ICE

Seasonal ice includes sea ice and river-lake ice. Based on the observed data from National Bureau of Oceanography, Li Peiji plotted on the map the limits of fixed and floating ice, the thickness of fixed ice, and the average annual frozen days. From the map we can see that sea ice is limited to the northern part of the East China Sea, i.e. along the coast of the Bohai and the Huanghai, and the duration of frozen period is generally 3-4 months. The extent of fixed ice is very small, limited to a few harbours with a maximum average thickness of 80 cm and a width of from several hundred metres to 3 km. Floating ice is the main form of sea ice in China, generally 10-40 km offshore, the farthest reaching 100 km with a thickness within 30 cm and a maximum floating speed at 0.6-1.8 m/s. In the years of 1936, 1947, and 1969, Bohai Bay was frozen along its coast, navigation was interrupted, and some constructions on the sea were destroyed.

River and lake ice appears in the vast area north of 30^oN. Based on data from 56 hydrologic stations, the average frozen period of rivers in North China and the maximum thickness of ice are also indicated on the map. In Altay Shan of Xingjiang and the Heilongjiang (Amour) River of Northeast China, the frozen period is over 150 days every year with maximum thickness of ice over 1.0 m, in the Haihe River of North China 40-80 days, maximum thickness of 0.6 m, for the Huanghe (Yellow River) most of the years 20-100 days, maximum thickness of 0.4-1.0 m, not frozen in particular years, and in the Huaihe River (lat. 32°-34°N) generally 10-40 days. Because the strength of cold air current in winter changes greatly year by year, the yearly variations of frozen time, frozen period, and thickness of ice in these rivers are also very great.

The frozen period of lakes in the interior of Qinghai-Xizang Plateau generally reaches about 6 months, while in Northeast China about 5 months. For example, the annual frozen period of the famous Qinghai Lake (Kokonor) at the Northeast corner of Qinghai-Xizang Plateau reaches 125 days and maximum depth of ice 0.44 cm.

PERMAFROST

Chinese permafrost researchers have investigated the permafrost in Da Hinggan Ling in Northeast China, along the Qinghai-Xizang Highway, in Qilian Shan, Tian Shan, and part of Altay Shan and Hengduan Mountains, and have compiled medium and small-scale maps of permafrost distribution, on the basis of which the important characteristics of permafrost have been primarily summarized (Zhou Youwu et al. 1982).

The present map, taking the data from actual investigations in the Altay Shan, Tian Shan, Qilian Shan, and along the Qinghai-Xizang Highway as basis, shows the result of analysis of the relation between permafrost and factors of the natural environment, and gives out the distributive boundary of permafrost. According to the study of Tong Boliang (1983 manuscript), the lower limit of insular permafrost in the Altai Shan, corresponds to the 2200 m contour, while that of the large stretch of continuous permafrost coincides with the 2800 m contour, where the snow cover on the ground surface is quite thick, elevating the lower limit of permafrost.

The snow deposit in the Tian Shan is comparatively less. According to the analysis of Qiu Guoqing and Huang Yizhi (1983 manuscript), the lower limit of the large stretch of continuous permafrost in the Tian Shan is about 2700 m on shady slope and about 3100 m on sunny slope. However, the lower limit of insular permafrost is still lowered 200-400 m.

In the Qilian Shan, the lower limits of the large stretch of continuous permafrost and insular permafrost in Reshui district at 37°25' are 3780 m and 3480 m respectively (Reshui Permafrost Study Group, 1976). According to the study of Luo Xiangrui (1983 manuscript) on the relationship between permafrost and the vegetation zone, the lower limit of permafrost is located in the zone of alpine meadow, and by using the Landsat images the localities of insular permafrost and continuous permafrost are plotted on the map.

Qinghai-Xizang Plateau is a major district of high-altitude permafrost in China. Along the Qinghai-Xizang Highway actual surveyed data were used, but for other areas the relation between permafrost and mean annual temperature were taken as the basis for mapping. According to the analysis of Xu and Guo (1982), in Xizang the mean annual ground temperature is 2.0-5.0°C higher than the mean annual air temperature. Thus, generally the isopleths of mean annual air temperature, -4 to -5°C and -2.0°C, could be taken as boundary exponents between the large stretch of continuous permafrost and insular permafrost and between insular permafrost and seasonal permafrost. In this way, the distribution of permafrost on the Qinghai-Xizang Plateau was plotted on the map. The height of the lower limit of permafrost in the Himalayas near the 28°N already reaches about 4900 m.

The climatic conditions of permafrost regions in Northeast China are wetter than in the plateaus of West China. Under the frequent attack of Siberian cold air currents, the lower limit of permafrost in the mountain regions of Northeast China is about 800-1000 m lower than that at the same latitude in west Tian Shan and Altai Shan. According to the map of permafrost distribution in Hinggan Ling in Northeast China (1:2,000,000) compiled by Guo Dongxin (1983), large stretch of continuous permafrost (north of -5°C isopleth, mean annual air temperature), insular thawing permafrost (between -5°C and -3°C isopleth, mean annual air temperature) and insular permafrost (between -3°C and 0°C isopleth, mean annual air temperature) are discernible on the map.

SEASONAL FROZEN GROUND

Actual investigative data are lacking on seasonal frozen ground, but inference is made from ground temperature records or from difference between ground temperature and air temperature records of the meteorologic stations. Xu and Guo (1982) took the lowest mean monthly ground temperature lower than 0°C with a duration over 1 month as an index for the existence of seasonal frozen ground. This value of ground surface temperature corresponds to the mean annual temperature of 8°C in Xizang with dry climate and small annual range in temperature, but may reach as high as 10° to 14°C in the southeastern part of China with its humid climate. This southern boundary of seasonal frozen ground generally runs from southwest to northeast, situated between 25°N-35°N, mainly covering large areas of Northwest, North, and Northeast China; over 3/4 of which has a seasonal frozen layer more than 0.5 m thick, greatly influencing highway, railway, channels, tunnels, agriculture, and engineering constructions in winter. To the south of the seasonal frozen ground zone, Xu and Guo (1982) sketched out another transient frozen ground zone with the extreme lowest ground temperature lower than 0°C and less than 1 month in year as its index generally corresponding to the mean annual air temperature 18.5° to 22.0°C.

PERIGLACIAL PHENOMENA AND QUATERNARY GLACIATION

Based on actually investigated data at more than 50 localities, Shi Yafeng and Ren Binghui (1980) carved out on the map the isopleth of the ancient snow line in the last glaciation of Late Pleistocene, which was only 2400-2600 m in the Altay Shan and rose to 5600-5800 m on the northern slope of Middle Himalaya (Mt. Qomolangma). Several famous cirques and terminal moraines used for calculation of the height of ancient snow line were also indicated on the map. According to the distributive regularity of snow line of last glaciation the possibility of the existence of Quaternary glaciers in the eastern part of China with an elevation lower than 3000 m has been expelled.

Periglacial phenomena from frost action, heaving and thawing, etc. such as blockfields, rock streams, polygons, frost heaving hummocks, involutions etc. spread extensively in China, of which a few places with typical patterns are indicated on the map by Cui Zhijiu (1980).

REFERENCES

Cui Zhijiu, 1980, Periglacial geomorphology, in Geomorology (one volume of Physical Geography of China), Chapt. 9, p. 269-300 (in Chinese), Science Press.

Guo Dongxin et al., 1981, Division of permafrost regions in Daxiao Hinggan Ling, Northeast China (in Chinese with English abstract), Journal of Glaciology and Cryopedology, v. 3, No. 3, p. 1-9.

Li Jijun and Zheng Benxing, 1982, Basic features of existing glaciers in Qinghai-Xizang Plateau (in Chinese), in Proceedings of the Symposium on Glaciology and Cryopedology held by Geographical Society of China (Glaciology), p. 14-17.

Li Peiji and Mi Desheng, 1983, Distribution of snow cover in China (in Chinese with English abstract), Journal of Glaciology and Cryopedology, v. 5, No. 4, (in press).

Reshui Permafrost Study Group, 1976, Features of the permafrost in Chaidaer area, Reshui, Qinghai, Memoirs of Lanzhou Institute of Glaciology, Cryopedology, and Desert research, No. 1, p. 60-75.

Shi Yafeng and Ren Binghui, 1980, Glacial geomorphology, in Geomorphology (one volume of Physical Geography of China), Chapt. 8, p. 202-232, (in Chinese), Science Press.

Shi Yafeng, Xie Zichu, Zheng Benxing, and Li Jijun, 1980, Distribution features and variations of glaciers in China, World Glacier Inventory (Proceedings of the Riederalp Workshop), Sept. 1978 IAHS Publ. No. 126, p. 111-116.

Shi Yafeng and Yang Zhenniang, 1982, Glacial resources and its effect on stream supply in China (in Chinese), Journal of Hydrology Special Issue (Water Resources), p. 6-12.

Xu Xiaozu and Guo Dongxin, 1982, Compilation of the distribution map of frozen ground 1:4,000,000 in China (in Chinese with English abstract), Journal of Glaciology and Cryopedology, v. 4, No. 2, p. 18-26.

Zhou Youwu and Guo Dongxin, 1982, Principal characteristics of permafrost in China (in Chinese with English abstract), Journal of Glaciology and Cryopedology, v. 4, No. 1, p. 1-19.

DESIGN AND CONSTRUCTION OF CUTTING AT SECTIONS OF THICK-LAYER GROUND ICE

Shu Daode and Huang Xiaoming

Northwest Institute of the Chinese Academy of Railway Sciences
Lanzhou, People's Republic of China

Cutting at sections of thick layer ground ice upsets the original ground layer's natural state of thermal equilibrium. To keep a new system stable, its thermal state, within a certain scope, must be restored to the original state or adapated to the conditions caused by the change of thermal state. Using this as the starting point for determining protective measures, the authors summarize the practical experience and experimental and research findings accumulated in China since 1960, especially recent ones made at sections with thick layers of ground ice on the Qinghai-Xizang Plateau, and present two major proposals for protecting permafrost in situ--the refilling and heat preservation method, based on the principle of similarity of heat power, and the method of supporting (sustaining) a structure that is based on the principle of local thawing self-burial stability. The design essentials of these techniques are described, and representative patterns of cross-sections are given. The influence of the harsh weather conditions on the Plateau on construction quality, technology, and machinery is discussed, and the main points of links in construction are emphasized.

Thick layers of ground ice in the region of the Qinghai-Xizang Plateau are one of the most serious barriers to constructing cuts in the area. The ice layer exposed in cut slopes and foundations after excavating will cause engineering hazards. Land-slides and foundation subsidence, and even slide after thawing, may force the abandonment of the project. To provide a stable and durable foundation, it is necessary to solve a series of problems that exist in design and construction. For cuts designed in the light of the first principle, that of keeping the permafrost in its frozen state, the problems to be addressed are the type of section, constructing season, as well as constructing technology.

THE EVIDENCE FOR THE MEASURES OF CONDUCTING A PROJECT AND THE PRINCIPAL DESIGNING PLAN

An appropriate approach to conducting an engineering project is bound to relate to the comprehensive consideration of the weather conditions and permafrost nature and take the deformation characteristics in foundation for road cause by natural force as the major evidence.

Regardless of the cause of formation of ground ice, the ice layer with the seasonally thawing layer over it has, under the effect of the natural environment over a long period, attained a relatively stable equilibrium. The temperature field, physical and thermal physical properties, storing, transmitting, and cycling of heat, and force system are in equilibrium. Excavating of cuts, however, will inevitably interrupt the equilibrium. To make a new system stable, i.e. to have engineering deformations controlled within permitted ranges of stability and durability, it is imperative that the thermal and mechanical state of the new system be restored, to some extent, to the original state or be adaptable to the changes brought about. This is the essential starting point for determining the measures to be taken in a new project, and the two principles of designing, either protecting the permafrost from thawing by refilling and preserving the thermal regime based on the principle of thermal similarity or the protecting the construction by using retaining structures, which is based on the principle of stabilizing through self-burial and drainage.

DESIGN OF REFILLING AND HEAT PRESERVATION METHOD

Types of Section and Processing Measures

Logical types of section and processing measures are necessary to meet the following requirements: reducing to the fullest disturbance and damage to the permafrost, reducing soaking and permeation of rainfall and harm caused by suprapermafrost water as far as possible, and reducing as far as possible the amount of work to facilitate construction and maintenance. Therefore, several matters are needed in excavating.

When an ordinary section is adopted, it should be excavated to the refilling depth and leave certain amount of extra excavation so that the maximum seasonally thawing depth can be kept above the boundary surface of refill (Figure 1).

A drainage system connecting the retaining

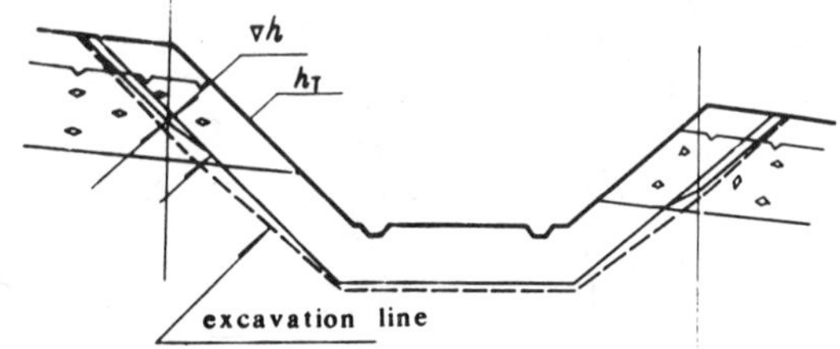

FIGURE 1 Sketch of typical cut section.
h_T is the thickness of refilling lower part of slope of cut;
∇h is the thickness of refilling in upper part of slope of cut.

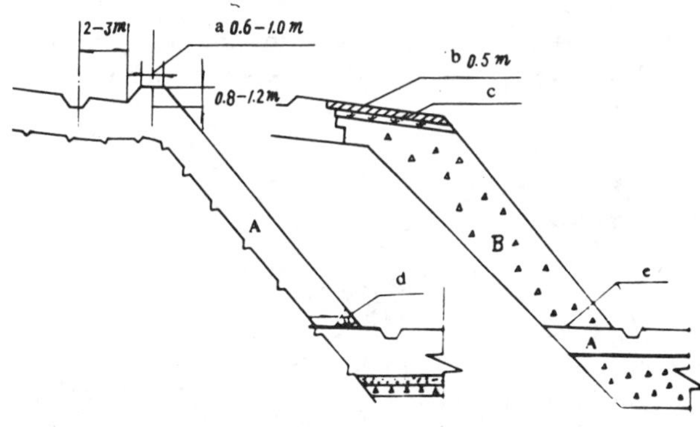

FIGURE 2 Sketch of retaining drainage, and water insulation facilities.
a is the retaining bank,
b is the layer of clayey soil,
c is the turf laid in an inverse order,
d is the stone pile laid without using mortar, e is the water insulation layer,
A is the clayey soil refilled,
B is the coarse granular soil refilled.

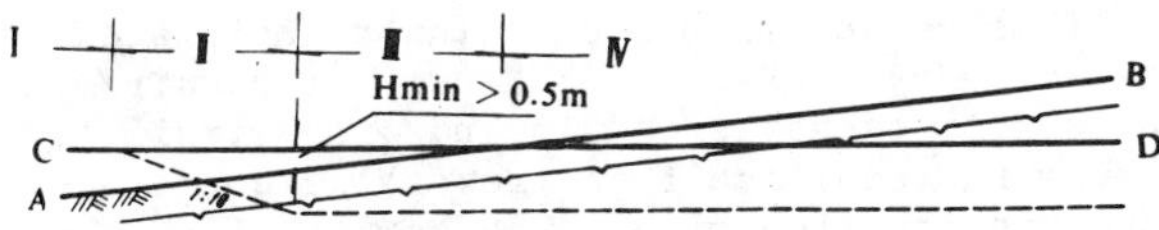

FIGURE 3 Transition from cut to embankment section.
I is the embankment sector,
II is the transition sector,
III is the zero-section,
AB is the original ground surface,
CD is the designed elevation.

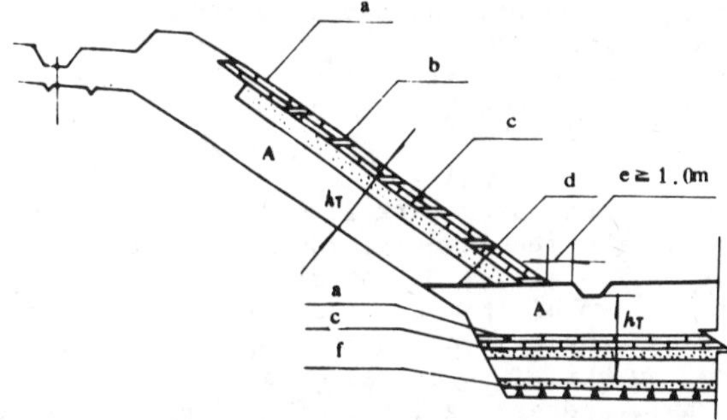

FIGURE 4 Cross-section showing heat insulation material and other construction details.
a is the heat insulation material,
b is the water-discharge openings,
c is the padding layer of sand,
d is the water insulation layer,
e is the platform of side ditch,
f is the broken stones, padding layer of coarse sand,
h_T is the thickness of heat preservation layer
A is the clayey soil refilled.

bank at the top of cut and the drainage ditch beyond the bank is built to prevent the slope from damage by suprapermafrost water flowing down from the upper part of slope (Figure 2).

A shallow and wide cross sectional side ditch is used with its bottom paved with pliable and tough water insulation material or clayey soil for convenient maintenance and unobstructed flow.

The slope of cut can be made stable by two feasible means---building a platform of side ditch or building a platform with a stone retaining pile, laid without using mortar, at the bottom of slope (Figure 2).

Pay special attention to the vertical transition and joint between refilled sector of cut, zero-section, and low embankment (Figure 3).

The following factors are considered about the measure of processing a project.

1. The refilling material is mainly taken from local materials, such as broken stones, pebble gravel, clayey soil, or turf. The protective measure of refilling with clayey soil, with a layer of turf paved over it, is advantageous to prevent the surface from sliding and reducing thawing depth.

2. To lessen the amount of refill for excavation, consideration should be given to pave industrial heat-preserving material alternately and in the course of construction lay a layer of coarse sand of certain thickness as isolating layer beneath the bottom of heat insulation plate. If heat insulation layer is placed over the surface of slope, water-discharge openings have to be provided (Figure 4).

3. To prevent the foundation from being damaged by water soaking in, it is necessary to construct a water-tight layer over proper parts of slope and foundation. The water content and density of filling material must be controlled.

4. If retaining bank is not set at the top of cut, the thawing depth there will be greater than in the middle of the slope by 6% to 10%, for the top of the cut is under the influence of heat sources from two sides. It is, therefore, essential to increase the thickness of the heat-preserving layer on the upper parts of the slope.

Determination of the Thickness of Heat-Preserving Layer to Be Refilled

The technique of heat preservation by means of refilling is based on the theory of thermal similarity. The various calculations and testing methods to determine the thickness of the refilling layer have to be connected to some extent with the equilibrium of local natural stratum in many respects, as boundary conditions, medium characteristics, etc. and should reflect the specific features of the cut section. This is the essential base for formulating the required calculations.

Equations are available which based on local natural permafrost table and revised in terms of orientation of slope, surface conditions, and the nature of filling material, etc. are used to calculate the refilling thickness of slope and foundation[1]; this also can be done by a statistical formula based on local average monthly air temperature in August[2]. In case industrial heat insulation material is used to pave over slope or foundation, a conversion equation that takes the theory of attenuation of amplitude as its basis is used instead[3].

Computation of Slope and Foundation Stability

The strength of foundation of cut in permafrost area can be computed with respect to the requirements in ordinary areas. As long as the thermal calculations are reliable, the maximum seasonally thawing depth will not exceed re filling thickness, and the foundation itself is effective enough to prevent the permafrost beneath it from being soaked by rain as well as suprapermafrost water of slope.

The stability of slope of cut constructed in warm seasons should be calculated by taking filling or excavating boundary surface as sliding surface. To avoid local instability, as surface slide, etc. caused by sudden increase of water content of slope which is brought about by rapid thawing of accumulated snow in warm season or rainstorm, it is necessary to pay attention to the choice of filling material and method of protection as well. The gradient of slope of cut located at region of thick layer of ground ice should be less than 1:1.5.

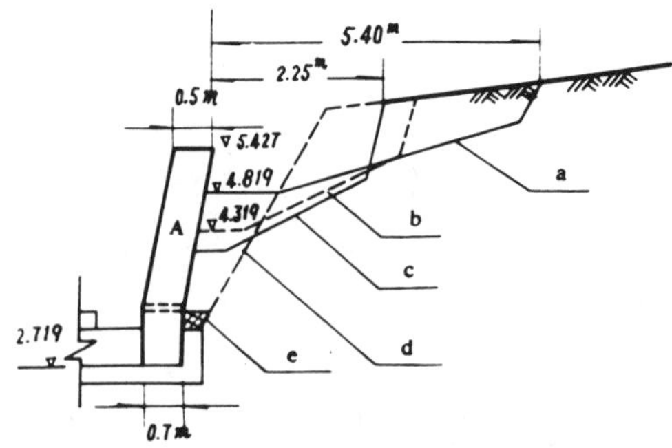

FIGURE 5 Sketch of retaining wall for cut allowing collapse and self-burial behind the wall.
a is the collapse contour measured in September, 1971,
b is the collapse contour measured in September, 1970,
c is the collapse contour while the retaining wall was completed in September, 1969,
d is the contour of slope excavated in August, 1968,
e is the concrete layer, 0.1-0.2 m thick,
A is the stone retaining wall laid with mortar.

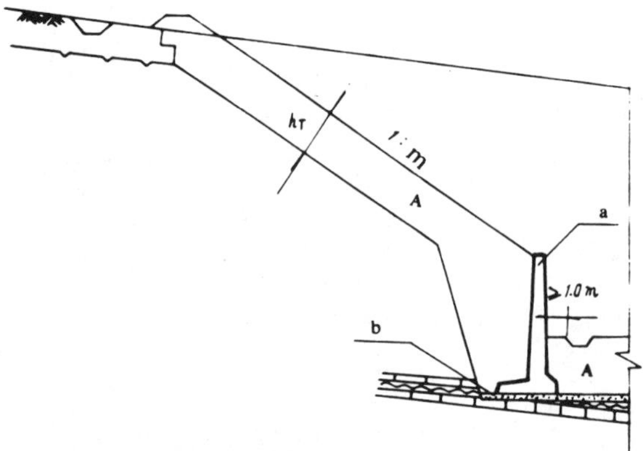

FIGURE 6 Sketch of section with retaining wall at the lower part and thermal protection of the upper part.
A is the clayey soil refilled,
a is the prefabricated erection reinforced concrete L-shaped retaining wall,
b is the sand padding, 0.1 m thick.

Design of Protection with Retaining Structure

This method is based on the principle of stabilizing by self-burial, hence the requirements needed to be satisfied include the stability of the structure itself and favorable conditions of the construction site. The merits of this method are reduction in cut section and reduction of the earthwork of cut and refilling. It is, in general, only adequate for shallow cut or deep cut with heat preservation (Figure 5 and 6). The main points considered in designing are: first, the method should normally be used in places where the horizontal slope of ground surface is gentle; second, the foundation of the retaining wall should be buried deeper than the permafrost table by 0.3-0.5 m or founded on bedrock; third, estimate the area of collapse of slope and the accumulated height of soil mass behind the wall, and design the section of retaining wall taking the pressure yielded by the soil accumulated behind the wall as the main evidence; fourth, discharge openings should be arranged at different heights of the wall and the verticle drainage capability of cut is required to be enhanced.

For deep cuts, the adoption of such sections that are capable of retaining at the lower part and protecting the upper part can lessen the amount of excavation and refill, only the material refilled behind the wall is required to have sufficient thickness. In designing, the factors to be considered include the stability under the action of horizontal frost heaving pressure and the thickness of heat-preserving layer.

CONSTRUCTION OF CUT

The most serious threat encountered in building cuts in regions of thick ground ice has been the thawing-freezing mud flow, which seriously delays the planned speed of construction and even makes the section can not shaped. There had been such an example on the Qinghai-Xizang Plateau where a test cut had to be abandoned because of damage to over 60% of its length by thawing. This was due to neglect of heat preservation in the course of construction (Figure 7). It is, therefore, an important principle to get rid of or reduce to the fullest extent the disturbance of heat-thawing while constructing in the region of thick ground ice.

Construction Season

In permafrost areas it is favorable for excavating and shaping in cold seasons, for refilling in warm season. Thus construction seasons

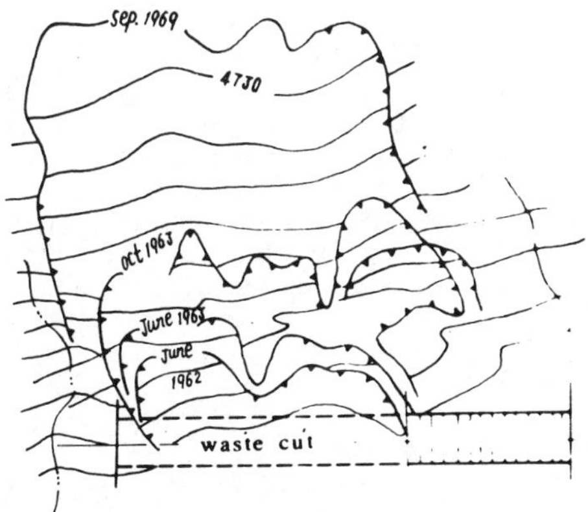

FIGURE 7 Deformation of cut caused by heat-thawing landslide in 1960.

are relevant to rationally solving the relation between excavating and shaping and the source of filling material, which are contradictory with each other.

If coarse granular soil is taken as refilling material and concentrate borrow is adopted, it is suggested to carry on construction in cold seasons so as to improve construction quality and avoid possible harm.

If construction has to be carried out in warm seasons, it is imperative not to do it in August or September, when rainfall is most frequent and heat-thawing effect is most active. The proper time under such conditions is likely summer and late autumn, and heat preservation is still worth to be noticed even then.

Construction going beyond the year, i.e. excavating and shaping in Autumn and refilling in warm seasons of next year, is beneficial to the stability of cut, which can meet the different requirements of both excavating.

Construction Technology

Construction technology is the key problem connected with the success or failure of building cuts. The working procedure should be planned and linked up as a whole, and carried out successively. It consists of the following four steps: planning---excavating---refilling---leveling.

Planning before construction includes construction organization design, setting construction schedules, building construction roads, delineating borrow areas and transport routes, and preparing construction facilities and materials, and establishing construction team, etc.

Prior to completion of planning, it is imperative to protect construction site and the surrounding natural vegetation. Rashly breaking ground is not allowed.

Pay attention to the natural drainage of the construction site. Built permanent drainage device is permitted before construction.

It is necessary to adopt high-speed mechanized construction in excavation to achieve quick completion.

Loosening soil with machinery or boring demolition are the efficient ways of excavating cuts. Deep hole demolition is feasible. Several points are to be noticed:

Borings are arranged according to the principle of excavating section to achieve shaping by demolition in once so as to shorten the working time to the fullest, thus lowering the effect of heat-thawing.

Parallel operation of drilling and cleaning up earth work is recommended. Sector-by-sector construction is effective for long cuts (50-100 m for each sector). Cleaning up after demolition and drilling the last sector are carried out simultaneously.

Great attention must be paid to waterproof and freezing-protecting of explosive. Explosive of good waterproof and freezing-protecting quality, e.g. J-2 pulp-like explosive, is satisfactory.

Damage of vegetation before drilling is strictly prohibited.

Procedure of excavating: excavate foundation first then slope for shallow cuts, and vice versa for deep ones. A drainage trough of certain width must be left over foundation surface for both cases to prevent mud flow from being accumulated in cuts.

The earthwork loosened by demolition or the loosener is cleaned up by bulldozer or loaded or transported by other means.

The loosened earthwork lying at the top of cuts, which can be reused, should be transported 30 m beyond the excavating limit; the ice layer containing soil and frozen soil of rich ice beneath permafrost table should be, according to the length of cut, bulldozed vertically at once or through horizontal passages sector-by-sector and piled up at certain place beyond the cut. While discarding waste soil, notice not to leave unfavorable effect on draining up silt during refilling and hidden trouble.

The building of horizontal passage should be done at the same time with loosening soil, the adequate interval between them being round 100 m. For cuts shorter than 200 m, excavation is carried out on both ends and horizontal passages are set up at the lower part of them; for long cuts over 200 m excavation is carried out sector by sector and an additional horizontal passage is equipped in the mid of the cut. Bulldozing is done from top to bottom.

When excavating comes to the stage of refilling, protection measure, covering in the day time and removing the cover in the night, should be adopted to exposed ice layer to reduce the effect of heat-thawing.

After the completion and cleaning up of cuts excavated in warm seasons, entire refilling must be done immediately to prevent the cut surface from being exposed too long. The points for attention are:

The necessity of sufficient refilling material supply. Refilling material should first be paved over both side slopes along the cut and then get downward up to the bottom step by step to provide good heat preservation. Compaction is needed immediately after paving of refilling material, especially at the time when it is likely to rain.

The sequence of refilling is from inner to outer and from top to bottom, thereby allowing the clean up of the thawed mud of the sector followed.

Attention should be paid to refilling and compaction side slopes. When compaction equipment is not available, by hand compaction must be resorted to layer by layer. The gradient of slope is adjusted by means of paving the slope.

Levelling includes the following steps: cleaning up the soil left after paving slope, cleaning up the side ditches, levelling off the platforms of side ditches and foundation surface, and the shaping of cuts, etc.

SUMMARY

In regions of ground ice, which is considered as the specially harmful geological area, the design as well as construction of cuts are most complicated and arduous and need a comprehensive consideration of local weather and permafrost conditions, topography, vegetation, the run of the cuts, and the nature of stratum, etc., and it is, therefore, impossible to have a unique method for designing section and construction. The article only presents an essential discussion on several main points, and many other problems, such as measures of water-insulation and drainage, construction technology, choice of machine, and commentary on work efficiency, etc. are left for further discussion so as to obtain a plan more applicable.

REFERENCE

1) Huang Xiaoming, 1979, "An experience equation for determining the thickness of heat-preserving layer of cut in permafrost area of the Qinghai-Xizang Plateau".

2) Liu Teliang, 1981, "Method of determining the depth of natural permafrost table and its use in cut engineering".

3) Huang Xiaoming and Yang Hairong, 1981, "Use of heat-insulating material in heat-preserving layer of foundation for railroad".

AN EXPERIMENTAL STUDY OF CRYOGENIC FACTORS AFFECTING GEOLOGICAL PROCESSES IN PLACER FORMATION

Yu. V. Shumilov

Permafrost Institute, Siberian Branch,
Academy of Sciences, Yakutsk, USSR

It was determined that rocks subjected to stress for long periods at negative temperatures experience a reduction in stability. This process accelerates the disintegration of the rocks in the hypergenesis zone due to pressure release unloading, weathering, and erosion. This effect should be taken into account in mining, construction, and geological investigations. It was demonstrated that weathering of rocks in the cryolithozone is accompanied by the simultaneous formation of unconsolidated materials of varying particle size. Cryogenesis results in intense removal of ions of some elements in water and thus increases the mineral content of the ground water. Migration of matter in the cryolithozone is clearly demonstrated both in ice and in water. Elements differ greatly with regard to their mobility, a factor which distinguishes the cryolithozone from other climatic zones.

INTRODUCTION

Advances of our country in developing the theory of permafrost rocks, including their structure, morphology, and evolution in space and time, make it possible to employ observed geocryological relationships in a broad range of scientific, technological, engineering, and agrobiological activities. One such application is the prospecting for mineral deposits in the permafrost zone. Presently available detailed investigations of cryogenic mechanisms and processes, and of properties of syngenetic and epigenetic residue, refer only to the formation of unproductive deposits. The formation, structure, and occurrence in the permafrost zone of mineral deposits such as placers represent a virtually unexplored field from the point of view of the cryology and geology of mineral deposits. Progress in this direction (Shilo, 1981) so far includes only general statements about the interrelation of the placer formation with cryolithogenesis. To render these insights effective in search and prospecting operations requires more detailed knowledge of cryogenic factors and their role in placer formation and other geological processes.

During the last decade we have been conducting an experimental study of the primary factors of placer formation including the cryogenic (Shumilov, 1981). A study was made of three groups of cryogenic phenomena and mechanisms: physical breakdown of monolithic homogeneous frozen rock and minerals; frost weathering of rock materials associated with gold, platinum, tin, and diamond deposits; and the formation of secondary haloes of gold and attendant element dispersion in conditions of the permafrost zone.

PHYSICAL BREAKDOWN OF FROZEN ROCKS

Experiments were aimed at modeling physicochemical weathering in the Arctic, using temperature controlled chambers of "Synthesis" type and "Nema" type. The determining factor was chosen to be a cyclic alteration of phase states of water in its transition from 25°C to -25°C and vice versa. Techniques were also employed to break samples through uniaxial compression, hydromechanical fragmentation, and impact. In these experiments variations of rock strength, porosity, mass loss, granulometric composition (once collapse was initiated), escape of material in the liquid phase, and related data were taken. Experimental results were compared with geological field observations.

The experiments have provided a detailed definition of the mechanism for cryogenic failure of rock, a process that originates within a rock long before it is subjected to temperature fluctuations, phase transitions of water, and the influence of other weathering factors (Shilo and Shumilov, 1981). It is known that when rock and ores crystallize and solidify, a large number of crystal dislocations, inhomogeneities, and other defects of the structure form within them. This primary condition is defined by a certain initial strength which is less than that of an ideal crystalline structure with no defects. The reason is that a portion of the atoms on the lattice sites are in a non-equilibrium (overstressed or understressed) state and tend to occupy a position of stability having the lowest energy level of thermal oscillation. The decrease of rock temperature in the case of permafrost formation leads to a greater degree of instability of non-equilibrium bonds in the crystalline lattices of minerals. Under mechanical influences present in upper levels of the Earth's crust (such as tectonic, seismic, tidal, etc.) frozen rock develops sites of micro-failure within mineral crystals in the form of incipient micro-cracks and the growth of existing cracks.

Thus, the process of rock breakdown should be considered as proceeding continuously in the region of subzero temperatures. The frozen state of rock under conditions of long stress with changing magnitude and sign enhances the fatigue of the material and leads to its destruction without any mass exchange. Micro-destruction in mineral crystals is particularly intensive with temperature fluctuations in the subzero temperature domain and when freezing and thawing occurs cyclically. These processes accompanied by phase transitions of water, if any, introduce additional energy which produces a kinetic fluctuation of unstable atoms and ions (Regal and Slutsker, 1973), destabilizing the atom-ion links within the material.

Experiments have confirmed that cryogenic destruction of rock is analogous to the brittle rupture of solid bodies as documented in technical physics. Destruction under the action of cryogenic factors proceeds discontinuously. During a time interval an accumulation of critical stresses occurs accompanied by the growth of micro-cracks which bring about a rupture along the most overstressed directions. The process then repeats itself. Sample mass loss and an increase of sample porosity also proceed discontinuously and simultaneously with the destruction.

It is important to note that cryogenic destruction evolves in two stages: precritical and post-critical. In the first stage the rock, despite previous cryogenic destruction, retains its external homogeneity and some residual strength as a consequence of purely mechanical engagement of grains, crystals, and aggregates. However, when such rock is subjected to weathering conditions on the ground it immediately enters the stage of post-critical destruction and rapidly changes to a disaggregated state under increases of stress, especially if there is a slight impact stress.

A similar situation arises when various structures (tunnels or various chambers) are built within ledge rock. If the rock is cryogenically altered, i.e. had been subject to a long stressed state at low temperatures or to periodical freezing and thawing, then the rock exhibits abruptly weakened mechanical and atom-molecular links within its structure. The rock structure approaches a critical point of internal stress where a minor increase of stress is sufficient for the rock to disintegrate, in the manner of brittle solids subjected to postcritical stress.

The influence of the degree of rock weathering upon rock strength is well known (Yarg, 1974). A typical feature of the phenomenon of cryogenic predestruction is the possibility that cryogenic effects may be present "behind the scenes" in rock that lacks external signatures of weathering decay, and that may appear monolithic and even possess considerable mechanical strength when tested in the domain of above zero temperatures. However, when such rock begins to bear a stress, there occurs an unforeseeable destabilization of its elements; its load-carrying capacity is abruptly decreased, with grave implications for the facilities under construction. There have been cases also of a considerable reduction in longevity of concrete structures built in areas with severe cold climate. One cause of these effects is the use of crushed stone for concrete aggregate which has been prepared from cryogenically modified rock. A further cause is an experimentally demonstrated increase of water corrosivity as it freezes, leading to a rapid physical and chemical weathering of unstable carbonate inclusions in rock.

In principle, the results of the experimental study of cryogenic destruction corroborate previous results. Similar data had been obtained from omnilateral compression of quartz samples beyond the stress at which the sample would be pulverized without deformation (Bridgeman, 1955). Latent processes of destruction of monolithic rock are suggested by the flaking of mining working walls exposed to mining pressure, and by other observations. The results show that in order to identify the field of application of the observed phenomena for technological and geological purposes, further investigations are pertinent. In particular, even in dedicated papers (Chistotinov and Shur, 1980; Tsytovich, 1979), cryogenic physical and geological processes in ledge rock have not previously been considered.

CRYOGENIC WEATHERING OF ORE MINERALS

The aim of another set of experiments was to study physical and chemical weathering of gold, tin, platinum, and diamond ores in conditions of the permafrost zone (Shumilov, 1981; Shilo, Shumilov, and Sporykhina, 1981; Prokopchuk, Shumilov, and Vazhenin, 1981). It has become apparent that dispersion of ore material in the permafrost zone proceeds from the earliest stages of weathering simultaneously at three levels: crude dispersion into particulates of crushed stone of gravel size and larger fragments, fine dispersion up to the size of pellet fractions, and chemical extraction of elements and their compounds from ore material. Within the layer of secular and seasonal temperature fluctuations, a weathering crust forms of the cryogenic type with inhibited development of chemical transformations but with a pronounced physical dispersion of ore material. Dispersion proceeds most strongly within the layer of seasonal thawing and initiates the liberation of valuable (placer-forming) minerals from the primary minerals of particulates. The latter are entrained into the movement of fragmented material downslope and produce haloes of secondary dispersion.

In contrast to lithogenetic situations where liquid water is available, the permafrost zone exhibits a sharp increase of exchange processes between eluvium and original substrate, between slope deposits and underlying rock, and between the river bed and its solid bed. The high velocity of material on the slope in the permafrost zone creates favorable conditions for feeding river beds with productive material. Cryogenic weathering contributes crucially to hydromechanical fragmentation of ore debris within the bed, resulting in the conversion of minerals to a free state and their subsequent concentration into a crudely inhomogeneous water-debris medium (a placer deposit). After the formation of a placer, syngenetic freezing of the productive layer contributes to subsidence of minerals into cracks of bedrock while epigenic freezing causes deformations and discontinuities of the productive layer.

The experimental results were used to evaluate the stability of different formative types of ore against cryogenic weathering (according to an index of formation of a fraction smaller than 1 mm in the weathering products). Comparison was made of seven formations: gold-quartz with low sulphide content, gold-quartz with moderate sulphide content, cassiterite-quartz, cassiterite-silicate, cassiterite-sulphide, dunite, and kimberlite. Assuming the stability of the first (most stable) formation to be 100, we obtain the following series of cryogenic stability: 100:9:62:29:3:14:2. This series indicates that within the permafrost zone the formative conditions of placers from different sources are sharply different and the search criteria for each mineral type of placer will differ substantially. The permafrost zone, as a whole, is found to be a site of intense production of disperse material that contributes to the formation of placers of a comparatively narrow mineral series.

CRYOGENIC CONCENTRATION OF FINE PARTICULATES OF MINERALS

The third group of laboratory experiments was aimed at a study of the chemical component of cryogenic weathering and of the behavior of finely divided mineral particulates. This work was carried out together with V. M. Pitulko and V. L. Sukhoroslov.

The mineralization of solutions produced during cryogenic weathering varied between 10 and 3000 mg/ℓ. These solutions contained the following elements (in decreasing order of concentration): chlorine, sulphur, magnesium, calcium, zinc, manganese, barium, nickel, copper, and sodium. Phase transitions of water led to an increase of mobility of arsenic, silver, manganese, nickel, titanium, antimony, molybdenum, and chromium.

The properties of cryogenic weathering were most pronounced in processes that form secondary haloes of dispersion within ore fields of gold and silver deposits (Pitulko, Shumilov, and Sukhoroslov, 1979). Detailed surveys (1:5000 and larger) reveal a zonal structure of secondary haloes that can be used as the criterion for evaluating the productivity of an ore outcrop. In this case we distinguish between secondary residual and secondary translated haloes, whose relative positions can be used to reveal the presence of a bedrock and its dimensions. In a single zone of anomalies the combination of ore complex elements (Au, Ag, As, Pb, Zn, Sb) and of antagonist-elements (Ti, Ni, Co, Ba, Mn, B, Sn and partly Mo) is indicative of the productivity of the source outcrop.

The experimental results have confirmed that up to 30-50 percent of ore components in the form of free primary minerals are concentrated in a small fraction of eluvium and deluvium of the cryogenic type. The predominance of a mineral form is particularly well-defined for gold. Fractions of 50-100 mkm size and of less than 50 mkm include up to 60 percent of extracted bullion gold. Up to 30 percent of gold particulates occur in aggregations with quartz and other minerals. Residual haloes of mineral forms in stripping are concentrated near the base of a seasonally thawing layer and also occur at the base of a suffosion level. Secondary accumulations of translated haloes, on the other hand, more frequently occur near the upper part of stripping.

The comparison of experimental laboratory and in situ measurements leads us to conclude that migration of material in the permafrost zone is actively manifested in the chemical composition of solid and liquid phases of cryogenic weathering. In this case we distinguish: salt migrators of rather mobile components (Zn, Mo, Ag, Ni, As, Mg, Sb, Ca); mechanical migrators of weakly mobile elements (Au, Sn, Ti, Al, and partially Pb); and mixed migrators (Fe, Cu, Bi, Ba) involved in migration both in a salt and mineral form. On the whole, analysis of the data shows that the behavior of material in the permafrost zone is defined by the following factors: permafrost conditions within watershed and slope deposits, the character of the surface and the microrelief; individual signatures of the micro-landscape situation (degree of turfing, the character of vegetation, exposure, inclination, etc.); composition of the ore source and its configuration and orientation with respect to relief elements.

The distribution of sorption-salt and mineral forms in ore haloes of the permafrost zone permits a diagnosis of residual and migrated haloes and anomalies. Cryogenic dispersion attributes to a high input rate of components into the deposit and to an effective development of lithochemical and hydrochemical anomalies.

Thus, there is a wide range of geological processes whose adequate interpretation in handling specific geological problems requires proper allowance for cryogenic factors and their further laboratory and in situ investigation.

REFERENCES

Bridgeman, P., 1955. The investigation of large plastic deformations and fault: Moscow, IL, 450 p.

Chistotinov, L. V. and Shur, Yu. L., 1980. Cryogenic physico-geological processes and their prognosis: Moscow, Nedra, 383 p.

Pitulko, V. M., Shumilov, Yu. V., and Sukhoroslov, V. L., 1979. Investigation of secondary dispersion haloes in gold-silver deposits in areas of cryolithozone, in Methods of geochemical prospecting: Prague, p. 170-175.

Prokopchuk, B. I., Shumilov, Yu. V., and Vazhenin, B. P.. Experimental data on kimberlite weathering and behavior in a water flow: DAN SSSR, v. 261, no. 2, p. 475-478.

Regel, V. R., and Slutsker, A. I., 1973. The kinetic nature of strength, in Physics today and tomorrow: Moscow, Nauka, p. 115-160.

Shilo, N. A., 1981. The principles of placer theory: Moscow, Nauka, 384 p.

Shilo, N. A., and Shumilov, Yu. V. On the effect of geological pre-destruction of rock and its practical aspects: DAN SSSR, v. 259, no. 5, p. 1188-1191.

Shilo, N. A., Shumilov, Yu. V., and Sporykhina, L. V. Physico-chemical weathering of the ores in laboratory conditions: DAN SSSR, v. 258, no. 6, p. 1432-1435.

Shumilov, Yu. V., 1981. Physico-chemical and lithogenetic factors of placer formation: Moscow, Nauka, 270 p.

Tsytovich, N. A., 1979. Soil mechanics: Moscow, Vysshaya shkola, 272 p.

Yarg, L. A., 1974. Weathering-induced changes of physico-mechanical properties of rock: Moscow, Nedra, 126 p.

EXPLOSIVE EXCAVATION OF FROZEN SOILS

James K. Simpson[1] and Peter M. Jarrett[2]

[1]Mobile Command Headquarters, St. Hubert, Quebec, J3Y 5T5, Canada
[2]Department of Civil Engineering, Royal Military College, Kingston, Ontario, K7L 2W3, Canada

Explosive cratering trials were carried out at four distinct sites in the vicinity of Inuvik, NWT. The primary aim was to correlate measured properties of the frozen soils with the cratering results. Two of the sites were underlain by coarse grained, sandy soils and two by fine grained silty soils. At each site ground temperatures and seismic properties were measured and core samples taken to allow laboratory measurement of moisture content, density, strength, and seismic properties. At each site seven craters were formed using 5 kg charges of ANFO explosive detonated at various depths. From crater measurements the optimum depths of charge burial and the maximum dimensions of the apparent and true craters were found. It was concluded that no single property is sufficient to characterize the soil with respect to its cratering behavior. Parameters derived from the seismic velocities appear to offer the best hope for possible characterization.

INTRODUCTION

Many previous studies concerning the explosive excavation or disintegration of frozen soils have been limited in scope to specific individual sites or mines. A search of the available literature reveals that in most instances the soil conditions have not been described in sufficient detail to allow behavioral correlations between the cratering results and the soil properties. One attempt at such a general correlation by Morgenstern et al. (1978) concluded that whether a soil was frozen or not appeared to make little difference to the relation between charge weight and optimum crater dimensions. Only the broadest differentiation was possible between soils, and this showed that in general larger craters are produced in fine grained soils than in coarse grained soils for a given weight of charge. A recent study by Smith (1980) points to the complex nature of the problem. Significant variations were observed in crater dimensions produced in a frozen silt compared to an unfrozen silt (moisture content about 30%). Only minor variations were observed between a frozen and a thawed gravel (moisture content 2-8%) and between the frozen gravel and the frozen silt. The problem therefore is that simple differentiation between soils as coarse or fine grained is insufficient to reflect the changes in probable cratering behavior.

With the uncertainty surrounding the descriptions of the soils in previous studies, it was decided that it would be of value to conduct basic cratering trials on different types of frozen soils and to attempt to correlate the cratering results with more extensive geotechnical properties. To this end cratering trials and geotechnical evaluations were carried out at four sites near Inuvik, NWT in February 1981. Limited results are presented in this paper but a detailed report was given by Simpson (1981).

CRATERING VARIABLES

There are three basic variables that determine the dimensions of the surficial crater produced by an explosive event. These are the energy and form of the explosive charge, the depth of burial of the charge beneath the surface, and the type of soil.

The energy released by a specific explosive may be represented as a function of the weight used (W). To maximize a particular crater property, such as its depth or volume, there is for a specific explosive charge an optimum depth of burial (d_c). This optimum depth of burial has been shown within reasonable limits of practical accuracy to vary linearly with the cube root of the weight of explosive used:

$$d_c = \text{Constant} \cdot W^{\frac{1}{3}} = C \cdot W^{\frac{1}{3}}$$

The constant C is known as the scaled depth of burial. The use of this constant in "cube root scaling" of explosive events is reasonably valid over a wide range of explosive weights. However, the optimum depth of burial may well be different for such individual factors as the diameter, the depth, or the volume of both the apparent and the true craters. Thus a series of constants, C, need to be defined for any specific situation.

In addition the dimensions of the crater produced by a specific weight of charge also respond to "cube root scaling," such that:

$$\text{Crater depth or diameter} = \text{Constant} \times W^{\frac{1}{3}}$$

and

$$\text{Crater volume} = \text{Constant} \times W$$

Thus another series of constants is required and these represent the scaled depth, scaled diameter, or scaled volume of the apparent or the true craters. It is the aim of most cratering trials to find experimentally the scaling constants for the crater dimensions and for the optimum depths of burial for particular soil conditions and explosives.

When considering the effects of the soil conditions on cratering, it is desirable to define some characteristic properties or behavior to which the scaling constants can be referenced. On a simple intuitive level one would expect cratering to be a combined function of the dynamic strength of the soil, the constitutive relationship between stress and strain and the soil density. Defining these properties, except for the density, is virtually impossible when consideration is given to the dynamic nature of the event. For a frozen soil the problem is more difficult than in an unfrozen soil.

Recognizing the difficulty of a "proper" solution, one may start at the other end of the spectrum and seek empirical groupings of soils that may have similar strength properties. This leads initially to simple differentiation or classification based on perhaps grain size and moisture content. For frozen soil the temperature should also be considered as an integral factor. Beyond this first empirical stage, "static" strength measurements may be made either on samples or using penetrometers in the hope that dynamic strengths will be relatively similar. These may be combined with seismic velocity and density measurements to attempt a more rational albeit still empirical approach. Support for some aspects of such an approach was given by Westine (1970) and Garg (1982), who both considered the seismic velocity to be a useful indicator of cratering behavior.

In this study attempts have been made to characterize the four sites studied using both the simple empirical approach and the somewhat more rational approach.

SITES STUDIED

Four sites south of Inuvik were selected for the trials. Two were composed of primarily fine grained soils (Sites FG1 and FG2) and two of coarse grained soils (Sites CG1 and CG2). Site FG1 was a natural undisturbed site, whereas the other three had all served as borrow pits. Basic soil classification information for the four sites is presented in Table 1.

Site FG1 was underlain by a fine grained, silty till with a thin surficial layer of peat. The moisture content for this soil was rather difficult to characterize with a single average value. The main reason for this was that limited zones existed with very noticeable ice stratification and yet the bulk of the soil was dense with a relatively low moisture content. In addition, where the surface peat layer was saturated it possessed a very high moisture content. The moisture content presented in Table 1 is an average for the basic soil without the ice stratification. The range of values measured including the stratified material was 18-55%.

TABLE 1 Site Classification Data.

Site		FG1	FG2	CG1	CG2
Soil Description	-	Fine grained glacial till with peat overcover and horizontal ice lenses	Clay shale with a reticulated ice pattern	Ice rich medium sand	Ice poor medium sand considerable organic content
Classification	-	ML-V_s	ML-V_r	SP-N_{be}	SP-N_f
Sand Fraction Silt Fraction Clay Fraction	% % %	25 53 22	10 70 20	90 10 00	95 05 00
D_{50}	mm	0.019	0.010	0.42	0.49
Liquid Limit	%	31.1	33.3	non-plastic	non-plastic
Plastic Limit	%	24.5	22.0	non-plastic	non-plastic
Average Moisture Content 0 to 1 m Depth	%	25.3	21.4	31.1	14.5
Average Bulk Density 0 to 1 m Depth	kg/m^3	1980	2040	1830	2090

TABLE 2 Average Strength Properties.

Site		FG1	FG2	CG1	CG2
Unconfined Compressive Strength	MPa	3.27	4.20	9.72	5.64
Failure Strain	%	13.4	4.3	3.0	4.9
Static Young's Modulus	GPa	0.19	0.35	0.85	0.41
Typical Mode of Failure	-	Plastic	Elasto Plastic	Elasto Plastic	Elasto Plastic

Site FG2 was formed by the base of a deeply excavated quarry. The "soil" appeared to be a brittle stratified shale or siltstone. Good fragmentation was observed at this site, which is believed to be a function of the brittleness and stratification.

Site CG1 was a sand pit used in the construction of the Dempster Highway. The water table was very close to the surface, giving a well bonded, ice-rich, medium sand.

Site CG2, also a sand pit, was better drained and the moisture content was considerably lower, giving a poorly bonded, unsaturated, medium sand. Sampling at this site proved difficult because of the poor bonding and the presence of a significant network of roots through the upper soil layers. The roots also appeared to affect the cratering, as a number of large slabs of soil were observed to be held together by the roots after the explosion.

In addition to the classification of the soils the following geotechnical program was carried out at each site:

1. Core samples of the soils at each site were obtained using a CRREL pattern sampler. These samples were shipped in the frozen condition to the University of Alberta in Edmonton, where unconfined compression tests were made on them in a cold room maintained at -8°C. The average results from these tests are presented in Table 2. The soil from site FG1 was notably more plastic than that from the other three sites as may be seen from the strain at failure of 13.4%

2. Seismic velocities were measured both in the laboratory and in the field. In the laboratory each unconfined compression sample was tested using Terrametrics ultrasonic equipment prior to the actual compression test. In the field, standard seismic equipment using a small primer charge as a source of disturbance was employed. The basic results together with certain derived parameters are presented in Table 3.

3. Ground temperatures were measured using strings of six thermistors placed at depths between 0 and 3 m. Typical results obtained midway through the trials, on a day when the air

TABLE 3 Seismic Velocities and Dynamic Moduli.

Site		FG1	FG2	CG1	CG2
Field Results					
Dilation Velocity	km/sec	2.682	2.857	3.871	3.810
Shear Velocity	km/sec	1.718	1.714	2.143	1.905
Young's Modulus	GPa	11.34	14.65	21.51	20.19
Poisson's Ratio	-	0.169	0.221	0.278	0.333
Laboratory Results					
Dilation Velocity	km/sec	2.231	2.846	4.194	3.842
Shear Velocity	km/sec	1.240	1.413	2.014	1.898
Young's Modulus	GPa	6.53	10.91	20.03	10.11
Poisson's Ratio	-	0.276	0.336	0.350	0.339

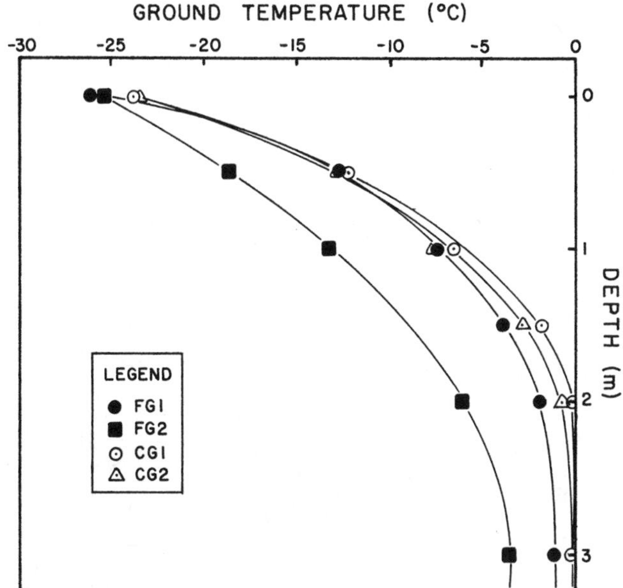

FIGURE 1 Ground temperature profiles.

temperature was -32°C, are presented in Figure 1. Site FG2 is significantly colder than the other three. This is representative of the conditions as measured.

CRATERING TRIALS

Snow was cleared from each site in December and again at the end of January immediately prior to the cratering trials. Seven craters were formed at each site with a sufficient variation in depth of charge burial to enable the optimum depths of charge burial for both apparent and true crater dimensions to be found. Because of the limited extent of certain of the sites, charges were formed using only 5 kg of AMEX II (Ammonium Nitrate Fuel Oil) explosive detonated with 0.45 kg Procore III primers and EB electric blasting caps. The explosives were charged in 16.5 cm holes at a free pour density of 890 kg/m^3. Such charges may be considered spherical in action.

Prior to and after cratering, surveys were made of the ground surface along set cross sections to allow determination of the apparent crater dimensions. Unfortunately during the operation the only two back-hoes within a few hundred miles broke down. Measurements of the true crater dimensions were still obtained but by hand excavation of the debris. It is felt that this led to lower estimates of the true crater volumes than would have been obtained by machine excavation.

The results for the apparent crater volumes are presented in scaled form in Figure 2. From these graphs the optimal conditions indicated in Table 4 are obtained.

The results for the true crater volumes are presented in scaled form in Figure 3. These results produce the optimal values given in Table 5.

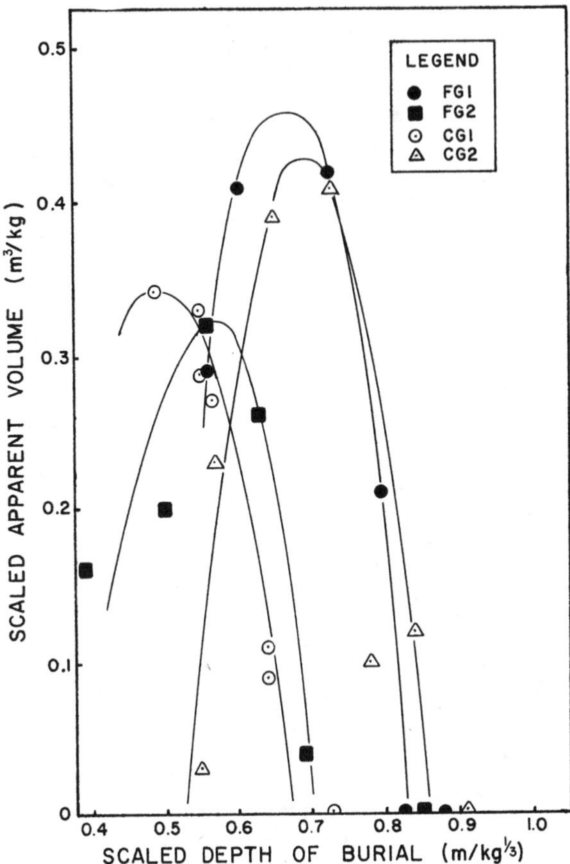

FIGURE 2 Scaled results from apparent craters.

TABLE 4 Optimum Results for Apparent Craters.

Site	Scaled Apparent Volume m^3/kg	Optimum Scaled Depth of Burial m/$kg^{\frac{1}{3}}$
FG1	0.46	0.68
FG2	0.33	0.58
CG1	0.34	0.49
CG2	0.43	0.69

CORRELATIONS AND DISCUSSION

The relative behavior of the scaled diameters and depths followed very similar trends to those of the scaled crater volumes indicated in Tables 4 and 5. In general, sites CG2 and FG1 produced similar results with CG2 tending to be slightly higher. Except for the apparent depth and apparent volume, site CG1 always produced the smallest craters, although the results were reasonably similar to site FG2.

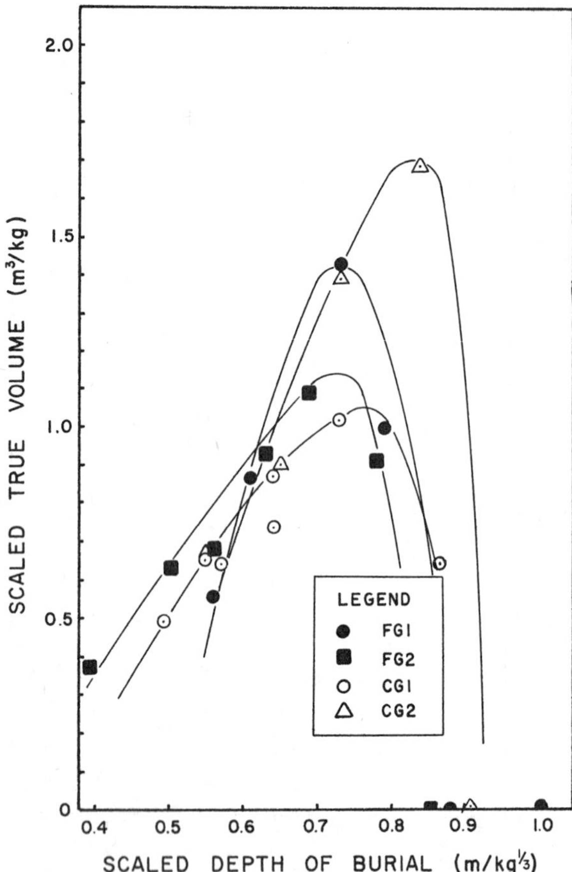

FIGURE 3 Scaled results from true craters.

TABLE 5 Optimum Results for True Craters.

Site	Scaled True Volume m^3/kg	Optimum Scaled Depth of Burial $m/kg^{\frac{1}{3}}$
FG1	1.42	0.73
FG2	1.14	0.73
CG1	1.06	0.78
CG2	1.70	0.83

No single soil property sufficiently characterizes the soils' behavior to give a usable correlation with basic cratering results. Looking initially at simple factors:

1. At CG1, the site with the lowest density, the smallest craters were produced. This is the opposite of what might be expected.
2. For the coarse grained soils the higher moisture content site, CG1, produced the smallest crater, whereas for the fine grained sites the higher moisture contents at FG1 produced larger craters. Qualitatively the anomaly may be explained for the coarse grained soils in terms of the better bonding and strength of the saturated soils at CG1 compared to the poor bonding in the unsaturated soil at CG2. This leads one to expect that a correlation may be developed with the soil strength!
3. This, however, did not show up, as the soil at FG2 was apparently weaker than that at CG2 even though the craters at FG2 were much smaller. Two factors may be affecting this correlation. At site CG2, sampling was very difficult, with many attempts producing little more than broken pieces. The samples obtained must by "natural selection" have been from areas of better bonded, stronger material and may therefore be unrepresentative of the in situ average strength. The second factor was the colder ground temperature measured at site FG2. All laboratory strength tests were carried out at a constant temperature of -8°C. Relatively, therefore, the in situ strength at FG2 may be expected to be greater. However, even accepting such rationalization, one cannot explain the closeness of the cratering results between FG2 and CG1 in view of the considerable strength difference between the soils.
4. The seismic velocities are also confusing. The two coarse grained sites had very similar values, which were larger than those at the fine grained sites. The fine grained sites also produced similar results.

Thus if an all-encompassing explanation is to be obtained, it will have to be in terms of some combination of properties. Westine (1970) suggested the term (ρc^2) as an indicator of the soil's "compressibility" or "constitutive effect," where ρ was the soil density and c the seismic dilation wave velocity. Once again CG2 appears the most serious anomaly, as its value of ρc^2 is the highest. At least the other three sites show a logical progression for this combined value. Another similarly based term would be the dynamic Young's modulus derived from the soil density and seismic velocities. Values are given in Table 3, and again, if CG2 could be ignored, then some limited consistency of pattern emerges between the moduli and the cratering volumes in Tables 4 and 5. One further possibility, which seems logical and arises as a grouping from a dimensional analysis, is to divide a strength term for the soil by the dynamic modulus value. This produces what might be thought of as an "elastic failure strain."

King and Garg (1982) suggested that correlation existed between the compressive strength and either the dynamic Young's modulus or dilation wave velocity. The results for the unconfined compressive strengths and dilation wave velocities obtained in the laboratory tests for this study are presented in Figure 4. A measure of agreement is observed. Certainly this type of agreement gives further credence to the concept that seismic velocities may be a possible general indicator of constitutive behavior as suggested by Westine (1970). From a practical point of view this certainly would be a desirable outcome.

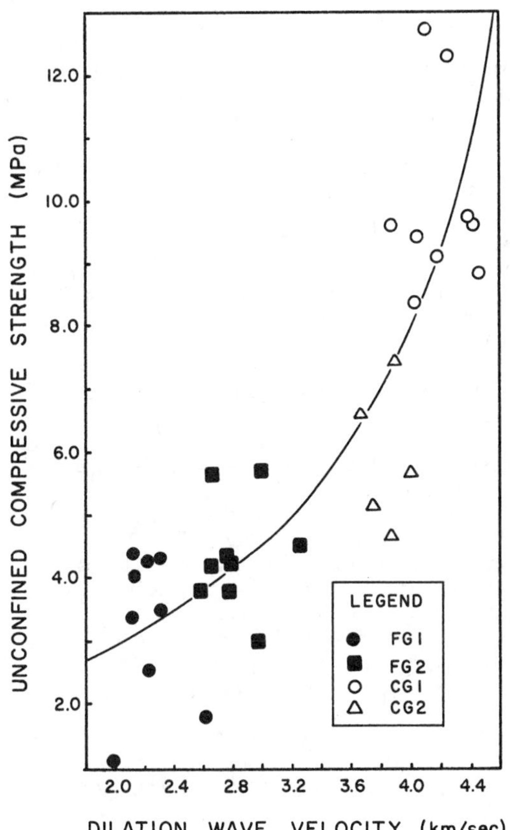

FIGURE 4 Strength vs dilation wave velocity from laboratory samples.

As a summation one can only consider the situation qualitatively. The four sites tested were certainly very different. In terms of many of the factors the fine grained soils were consistent when viewed alone and similarly the coarse grained soils could also be explained. Therefore, as in most other soil problems, it appears useful to divide the soils by grain size and treat coarse and fine grained separately. The next factor seems to be the degree of bonding or perhaps saturation. If the soils are well bonded, then they will be more resistant to cratering. Having grouped the soils in terms of grain size and bonding, it may then be more reasonable to expect factors such as strength or seismic properties to provide a correlation.

CONCLUSIONS

1. No single geotechnical property of the frozen soils correlated well enough with cratering behavior to allow for explosive designs to be made based on variation in that property.
2. Variation in the frozen soil properties does make a significant difference in cratering behavior. Maximum variations of 55% in true crater volume and 22% in its optimum scaled depth of burial and 31% in apparent crater volume and 42% in its optimum scaled depth of burial were observed in this series of trials.
3. Derived parameters based on the seismic velocites of the frozen soil are considered to provide the best possibilities for correlation, especially if soils are dealt with in more limited groupings.

ACKNOWLEDGMENTS

Many people and agencies assisted this study with logistical support and the loan of equipment. The authors particularly wish to thank those persons who assisted on site, Messrs. Hogg, Withers, Mast, and Nolan from DRE Suffield and especially Messrs. J. Dipietrantonio and L. Harvey from the Royal Military College, who bore the brunt of the arduous work involved. The University of Alberta through Dr. D. Sego provided facilities to perform the laboratory testing, and the Inuvik Research Laboratories through Dr. P. Lewis provided everything from a soil laboratory to a shovel.

Financial support for the study was provided through a grant from the Chief of Research and Development of the Department of National Defence.

REFERENCES

Garg, O.P., 1982, Recently developed blasting techniques in frozen iron ore at Schefferville, Quebec, in French, H.M., ed., Proceedings of the Fourth Canadian Permafrost Conference: Ottawa, National Research Council of Canada.

King, M.S., and Garg, O.P., 1982, Compressive strengths and dynamic elastic properties of frozen and unfrozen iron ore from northern Quebec, in French, H.M., ed., Proceedings of the Fourth Canadian Permafrost Conference: Ottawa, National Research Council of Canada.

Morgenstern, N.R., Thomson, S., and Mageau, D., 1978, Explosive cratering in permafrost, in Report to Defence Research Establishment Suffield, Alberta (unpublished).

Simpson, J.K., 1981, Explosive cratering in frozen media, in M.Eng Thesis, Royal Military College of Canada, Kingston, Ontario.

Smith, N., 1980, High explosive cratering in frozen and unfrozen soils in Alaska, in CRREL Report 80-9, Hanover, New Hampshire, USA.

Westine, P.S., 1970, Explosive cratering, in Journal of Terramechanics, v 7, No 2.

DEEP ELECTROMAGNETIC SOUNDING OVER THE PERMAFROST TERRAIN IN THE MACKENZIE DELTA, N.W.T., CANADA

Ajit K. Sinha and L. E. Stephens
Geological Survey of Canada, 601 Booth Street, Ottawa, Ontario, Canada K1A 0E8

A number of deep electromagnetic (EM) soundings using a multifrequency and a transient system were carried out over several locations near Tuktoyaktuk, N.W.T., during the spring of 1982. The locations were near drill holes that had already been surveyed with various geophysical logs. The multifrequency EM system used a vertical magnetic dipole as the transmitter with 128 frequencies from 1 Hz to 60 kHz for the detection of conductive horizons to a depth of 700 m from the surface. The transmitter for the transient soundings was a large Turam-type loop that can transmit square wave pulses with two base frequencies of 3 and 30 Hz.

The motivation for the experiments was to test a relatively inexpensive and environmentally acceptable geophysical technique for detecting horizontal layers at large depths, especially the contact between ice-bonded permafrost and the underlying unfrozen sediments, which varies in depth from 250 m to 650 m in the Mackenzie Delta area. The multifrequency EM sounding interpretations generally agreed well with well-log information at most sites. The time-domain results, on the other hand, detected major interfaces but could not resolve minor discontinuities. This could be attributed to the fact that the resolving capability of the time-domain system was somewhat less than that for the multifrequency system. The experiments demonstrated the viability and usefulness of conducting routine electromagnetic soundings for detecting and delineating horizons within areas of thick continuous permafrost.

INTRODUCTION

In the past few years, several electrical and electromagnetic (EM) methods have become available for routine mapping and delineation of continuous and discontinuous permafrost zones in the arctic regions of Canada and Alaska (Annan and Davis 1976, Arcone et al. 1978, Rossiter et al. 1978). These methods generally yield good estimates of the permafrost thickness and conductivity when the permafrost is relatively thin, i.e., less than 50 m thick. However, in areas of thick and continuous permafrost, other methods with greater depth penetration must be used. This study deals with the use of two deep electromagnetic sounding methods, one frequency-domain and the other time-domain, for mapping permafrost to depths of 700 m in the Mackenzie Delta, N.W.T., Canada.

The motivation for the study was to test and evaluate a relatively inexpensive and environmentally acceptable geophysical technique for detecting interfaces at large depths, especially the contact between the ice-bonded permafrost and the underlying unfrozen sediments. This interface is at depths varying from 250 m to 650 m in the Mackenzie Delta area. All test areas were chosen to be within 75 km from Tuktoyaktuk, N.W.T., where the facilities of the Polar Continental Shelf Project are available. The tests were carried out near deep oil and gas exploration wells that had already been surveyed by thermal and other geophysical logs for detecting various geological horizons, including the permafrost/thawed-zone contact, and were carried out during the spring of 1982 when the ground was frozen. Hence the active layer did not pose a problem in our study.

DESCRIPTION OF SURVEY TECHNIQUES

Multifrequency EM System

The multifrequency deep EM sounding system (Maxi-Probe) used for the test surveys consists of a vertical magnetic dipole transmitter of large moment and a receiver to record simultaneously the vertical and the radial components of the total magnetic fields at a distance. The transmitter loop placed on the ground behaves like a vertical dipole when the receiver is placed at a distance of at least 5 times its diameter. The transmitter is powered by a 2.5 kW gasoline-driven generator. The transmitter operator may select up to 128 discrete frequencies from 1 Hz to 60 kHz from the transmitter console. The receiver system consists of a horizontal and a vertical coil enclosed in a fiberglass, spherical shell to reduce wind noise. The shell is normally placed on a fiberglass tripod for stability. By using a built-in bubble-level and a directional arrow on the antenna, it is possible to ensure that the measured fields are true vertical and radial components of the magnetic field.

The transmitter and the receiver consoles are usually synchronized with the help of two crystal clocks at the beginning of the survey. This eliminates the need for a connecting cable between them for phase reference. The receiver operator normally records the ratios of the vertical field component H_z to the horizontal field component H_x and their phase differences at each frequency. Since absolute values of field components are not required, it is not necessary to keep the current constant or the geometry of the transmitter loop in any particular configuration. This simplifies field operation over rugged terrain.

The depth of investigation with this system is a function of frequency, electrical parameters of the ground, transmitter-receiver separation, and the strength of the transmitter dipole. The lower limit of usable frequency is normally dictated by the extent of atmospheric and cultural noise. Since the frozen ground was quite resistive, it was possible to sound to depths about 1-1.5 times the transmitter-receiver separation. The field results were initially interpreted by graphical means (Sinha 1979) and finally by comparing the field plots of $|H_z/H_x|$ versus frequency with computer generated responses of multilayered earth sections to obtain a good match. The plotting point is taken to be the middle point between the transmitter and the receiver. The system has been described by Sinha (1979, 1980, 1983).

Transient EM System

The deep sounding transient EM system (EM-37) consisted of a large square loop transmitter 450 m by 450 m lying flat on the ground. A steady current, most commonly 15 to 20 amperes is passed through the loop long enough to allow turn-on transients to dissipate. The steady current is then abruptly terminated in a controlled fashion. The rapid reduction of the transmitter current in a very short time causes the magnetic flux to change rapidly, thereby inducing an electromotive force (e.m.f.) in the ground conductors. This causes eddy currents to flow in the ground, and these currents decay in characteristic fashion depending on the conductivity, size, and shape of the conductor. The time rate of change of the decay field may be measured by a mobile receiver loop that may be oriented in three orthogonal directions to record the decay in three directions.

The decay of the field is measured at 20 preset time intervals or channels after the current is turned off. In practice, the fields are measured over many cycles and stacked to increase the signal/noise ratio. The system operates with two base frequencies of 3 and 30 Hz. Thus, if both frequencies are used, data over a total of 30 channels (from 0.089 msec to 72 msec) are recorded, since the last 10 channels of 30 Hz are identical with the first 10 channels of 3 Hz.

A 5-HP gasoline generator powers the transmitter loop. There are four possible ways of synchronizing the primary and secondary pulses, the most stable of which involves the use of an oven-warmed crystal oscillator. McNeill (1980) has described the system in detail. In late times, i.e., when the decay current pattern has stabilized, the decay voltage components are related to the apparent resistivity of the ground by a simple relation. Hence the decay values over 30 channels yield 30 apparent resistivity values of the ground at different time intervals. The field results are interpreted by comparing field plots of apparent resistivity versus $(\text{time})^{\frac{1}{2}}$ with a set of precomputed master curves for two- and three-layer media or by direct comparison with sets of computer generated responses over multilayered earth sections. Kaufman (1977) has adequately described the procedure.

GEOLOGY OF THE AREA

Field tests with the two EM systems were carried out at seven sites in the Mackenzie Delta, as shown in Figure 1. The Tertiary and Quaternary sediments of the Mackenzie Delta consist of fluviodeltaic clastic sediments eroded and transported from bordering, uplifted highlands to the west and south of Mackenzie basin (Young et al. 1976). The reworking and sorting of sediments by basinward progradation of delta formations and the meandering channels of the Mackenzie river and episodic marine transgressions create a very diversified sequence of sediments in the area. The sediments are more uniform east of Tuktoyaktuk than those west of it, possibly because of their offshore prodeltaic origin. In this paper results will be presented from one site west of Tuktoyaktuk (#1) and two sites east of it (#6 and #7).

FIELD RESULTS

Multifrequency EM Survey

Figure 2 illustrates a plot of the raw field data from site #1 (YA-YA P-53) obtained with the multifrequency system using a transmitter-receiver

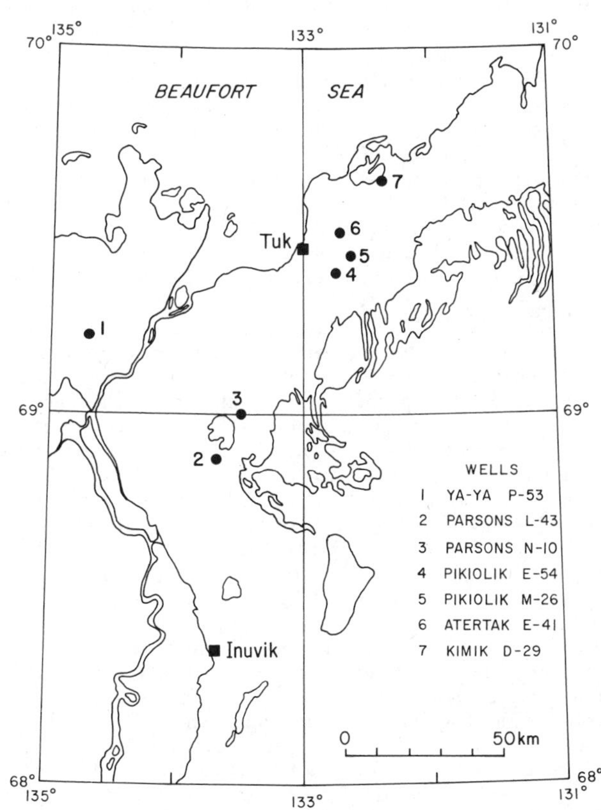

FIGURE 1 Locations of the test sites in the Mackenzie Delta.

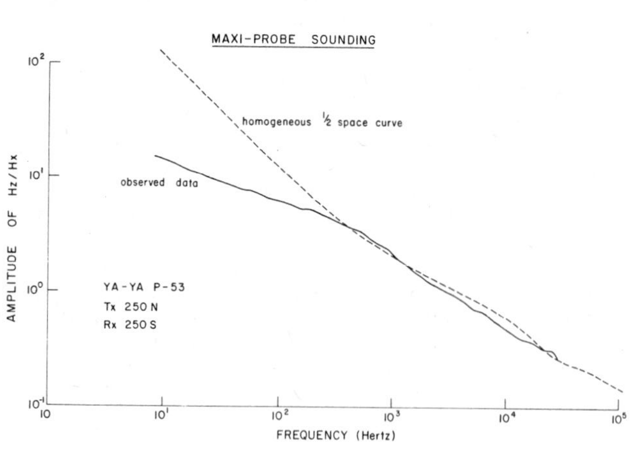

FIGURE 2 Raw field data with the multifrequency EM system at YA-YA P-53.

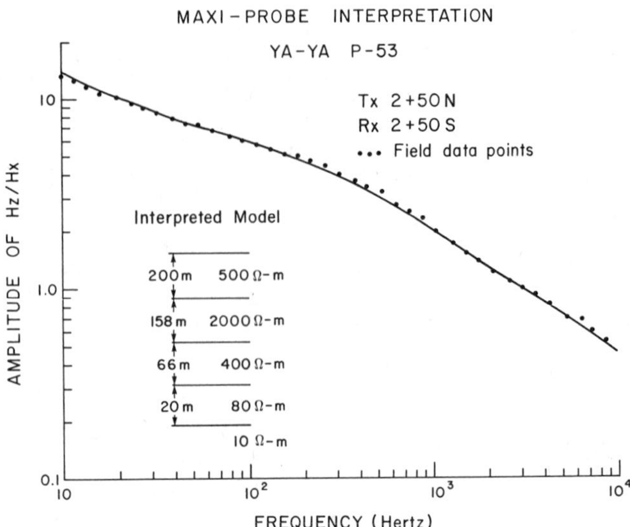

FIGURE 3 Interpretation of multifrequency EM data over YA-YA P-53 for T_x and R_x in a N-S line.

separation of 500 m. The amplitudes of H_z/H_x are plotted on the ordinate against frequency in the abscissa. The dashed line represents the response of a homogeneous half-space. As the frequency is decreased, greater depths are probed by the system. Any divergence of the field plot from the half-space curve represents a departure of the medium from this half-space model. By comparing such field plots with precomputed half-space and multilayered curves plotted with a nondimensional parameter proportional to frequency along the abscissa, it is possible to obtain the depth and resistivity parameters of a multilayered medium (Sinha 1979).

Figure 3 illustrates a plot of the same data (N-S direction) after correcting for the effects of altitude differences between the transmitter T_x and the receiver R_x (Sinha 1980). The data were inverted in terms of resistivity and thickness parameters of a multilayer earth after comparison with responses from several computer generated models. The results indicate the presence of a conductive layer at a depth of 444 m from the surface. Borehole geothermal surveys (Taylor and Judge 1977) have indicated the depth of the permafrost in the hole to be 433 m. Figure 4 illustrates the corrected field response over the same location with the transmitter and the receiver in a E-W direction, keeping their separation constant. The conductive horizon is now interpreted to be at a depth of 435 m. The good agreement between the two depth values when the transmitter-receiver system is rotated by 90° indicates that the layers are more or less horizontal and attests to the repeatability of the survey technique. The upward hump in the field values in Figure 3 could not be exactly modelled on the computer, thereby indicating the presence of a two- or three-dimensional resistive body in the area.

Figure 5 shows a section of a deep induction log at the same drill hole. The log shows a decrease in resistivity from a range of 10-30 ohm m at depths of 325-400 m to about 7 ohm m at a depth of about 405 m presumably where thawing begins. The resistivity of the permafrost however exhibits wide fluctuations possibly due to the presence of conducting materials like clay lenses at various depths. Thus the multifrequency interpretations differ from those from thermal and induction logs by less than 10 per cent.

Figures 6 and 7 show the field ratio versus frequency plots at two other sites, Atertak E-41 and Kimik D-29 (see Figure 1), respectively, with transmitter-receiver spreads of 600 m. In Figure 6, the depth sounding indicates a conductive layer at 565 m with another possible conductor at 750 m. Thermal measurements have indicated the permafrost/thawed-zone interface to be at a depth of 535 m at this site. At Kimik D-29 site (Figure 7) conductive features have been mapped at depths of 265 m, 330 m, and 635 m. Apparently there are several conductive horizons other than the unfrozen layer at this site, consisting possibly of clay and mudstone horizons. Geothermal data indicate the permafrost to be 663 m deep at this site, which differs from the interpreted value from multi-frequency interpretation by about 4 per cent.

Transient EM Survey

Figures 8 and 9 show the corrected field plots for two transient EM soundings in the near zone at Atertak E-41 and YA-YA P-53 respectively. The transmitter consisted of square loops with sides of 450 m, and the receiver was located near the centre of the loop. Using both frequencies of 3 and 30 Hz,

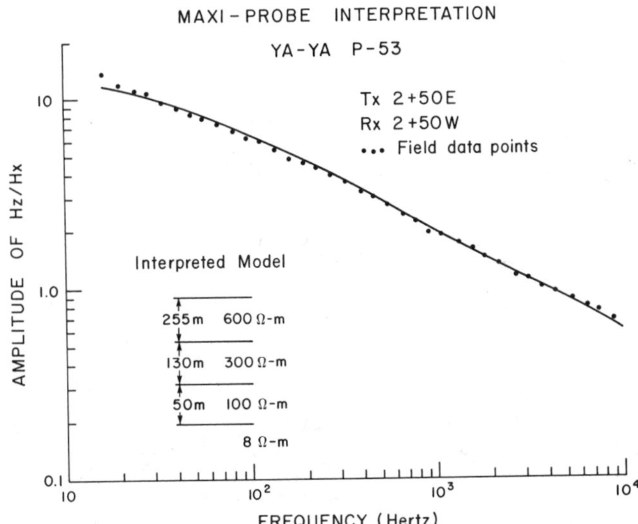

FIGURE 4 Interpretation of multifrequency EM data over YA-YA P-53 for T_X and R_X in a E-W line.

it was possible to obtain the decay voltages and hence apparent resistivities over 30 channels. For plotting purposes, apparent resistivities were plotted along the ordinate and $(\text{time})^{\frac{1}{2}}$ along the abscissa. The field data were interpreted in terms of resistivity and thickness of a layered earth by comparing them with a set of computer generated response curves. The models with the best match are indicated in Figures 8 and 9.

At Atertak E-41 (Figure 8) the depths to the conductive layers are at 370 m and 530 m, respectively. The bottom layer with an interpreted resistivity of 2 ohm m is close to the permafrost depth of 535 m from thermal logs and 565 m from the multifrequency sounding results. Figure 9, showing the depth sounding curve at YA-YA P-53 yields a three-layer medium with depths at 315 m and 440 m from the surface. This agrees well with thermal logs and multifrequency interpretation at this site. Table 1 indicates a comparison of interpreted results from the three sites with multifrequency and transient soundings with the thermal logs.

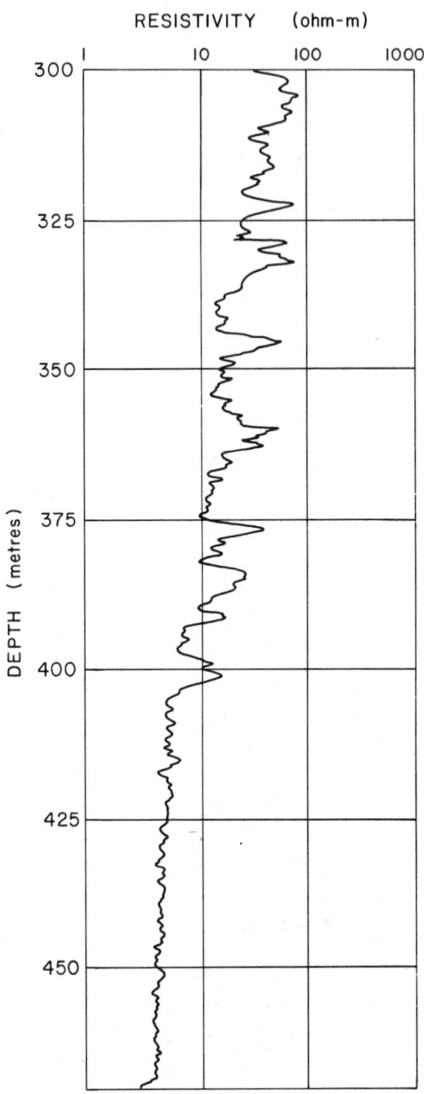

FIGURE 5 Deep induction log inside drill hole YA-YA P-53.

TABLE 1 Comparison of interpreted permafrost depths with thermal logs, Mackenzie Delta, N.W.T., Canada.

Drill Hole Location	Permafrost Thickness (m)		
	Multifreq. Survey	Transient Survey	Thermal Logs
YA-YA P-53	435-444	440	433
Atertak E-41	565	530	535
Kimik D-29	635	625	663

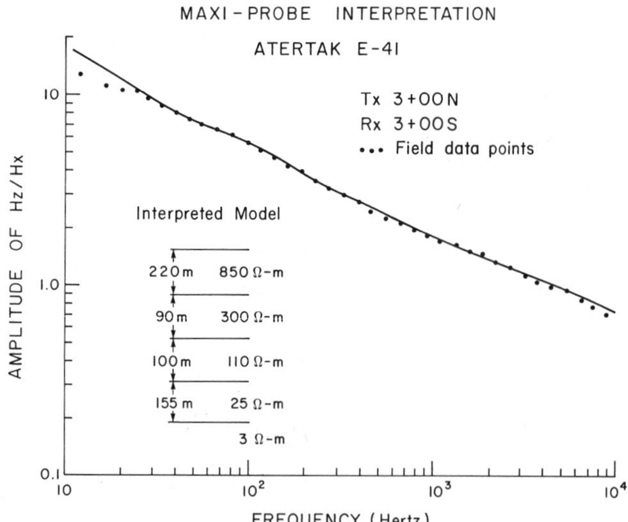

FIGURE 6 Interpretation of multifrequency EM data over Atertak E-41.

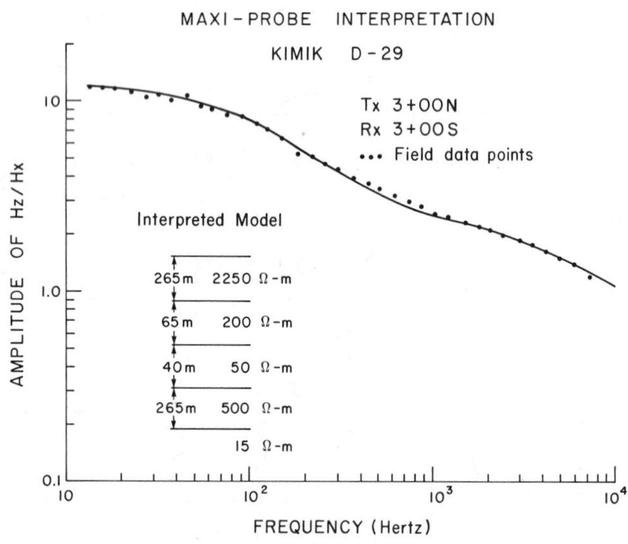

FIGURE 7 Interpretation of multifrequency EM data over Kimik D-29.

EM-37 SOUNDING CURVE
ATERTAK E-41
Interpreted Model
370m 500 Ω-m
160m 62.5 Ω-m
2 Ω-m
••• Field data points
APPARENT RESISTIVITY (ohm-m)
(time)$^{1/2}$ in sec$^{1/2}$

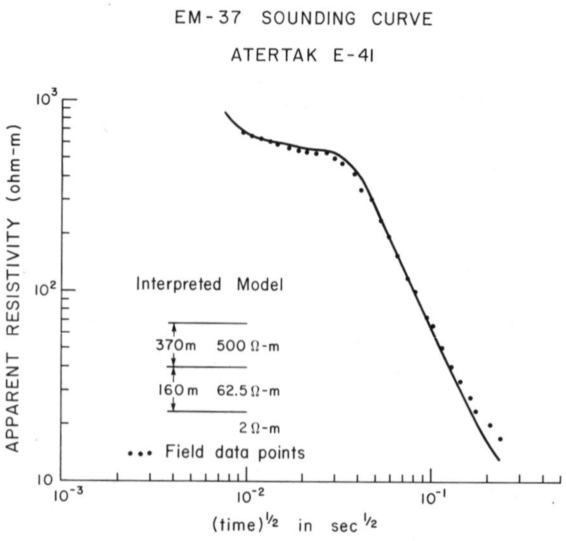

FIGURE 8 Interpretation of transient EM data over Atertak E-41.

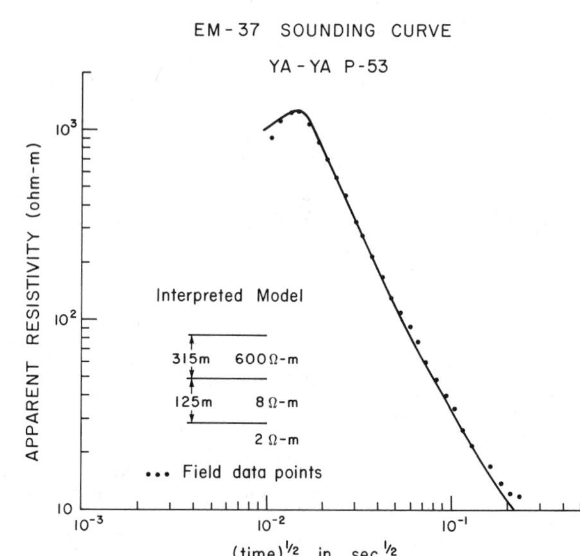

FIGURE 9 Interpretation of transient EM data over YA-YA P-53.

It was observed at a few sites that interpreted permafrost depths obtained by EM depth soundings tended to be somewhat less than those obtained from thermal logs. This may normally be explained by the relationship between the temperature and resistivity of various geological materials (Hoekstra and McNeill 1973). Another possible reason may be that ice-bonded permafrost thaws at a shallower depth than that predicted from thermal logs because of the presence of electrolytes but their temperatures still remain below 0°C. Therefore, the material below that depth may still be called permafrost but electrically speaking, that will be a more conductive medium.

SUMMARY AND CONCLUSIONS

Experimental results with deep EM sounding indicate that both frequency and time-domain systems may be used over the permafrost terrain of the Mackenzie Delta with a fair degree of reliability. The multifrequency system has somewhat higher vertical resolving power to identify thin beds of contrasting resistivity. The use of large Turam-type loops for the transient EM systems lowers the resolving power for that system. On the other hand, the use of large loops as transmitters makes the transient system less susceptible to terrain effects, which may be substantial in the multifrequency system.

The agreement between the interpreted depths of the permafrost/thaw-zone interface from EM sounding results and thermal logs were usually within 5% in almost all sites. It is obvious, however, the interpreted model parameters are one of several possible models that would satisfy the field data because of the phenomenon of equivalence. Multifrequency EM systems seem to be more susceptible to lateral inhomogeneities since T_X-R_X separation has to be larger than the loop size for transient systems for similar depth penetration. This is obvious at YA-YA P-53 where the interpreted highly resistive layer in Figure 3 was missing in Figure 4 when the T_X-R_X set up was rotated by 90° presumably because the hump in the field plot in Figure 3 was caused by a resistive body of limited extent. In terms of cost and productivity, the two systems are comparable. The field equipment is heavy for both systems requiring helicopter transport. On a good day, it was possible to carry out 3-4 soundings with either system with a three-man crew.

The success of the two relatively inexpensive and environmentally acceptable geophysical techniques to predict the depth of permafrost to depths up to 700 m may lead to their use in connection with exploration for oil and gas in the Mackenzie Delta. Such surveys may be carried out on a routine basis for siting of drill holes, and the results may be used by seismic exploration crews in making corrections to seismic time sections and for interpolating permafrost thicknesses between drill holes.

ACKNOWLEDGMENTS

We wish to acknowledge the logistical help rendered by the Polar Continental Shelf Project. We also thank Canada Oil and Gas Lands Administration for allowing us to use information regarding the logs for one of the drill holes. We are also grateful to Geoprobe Ltd., Toronto for lending us one of their Mark III receivers during the field tests.

REFERENCES

Annan, A. P. and Davis, J. L., 1976, Impulse radar sounding in permafrost: Radio Science, v. 11, no. 4, pp. 383-394.

Arcone, S. A., Sellmann, P. V. and Delaney, A. J., 1978, Shallow electromagnetic geophysical investigations of permafrost, in Proceedings of the Third International Conference on Permafrost, v. 1: Ottawa, National Research Council of Canada, pp. 502-507.

Hoekstra, P. and McNeill, J. D., 1973, Electromagnetic probing of permafrost, in Permafrost -- The North American Contribution to the Second International Conference, Yakutsk: Washington, D.C., p. 517-526.

Kaufman, A. A., 1977, The theoretical basis of transient sounding in the near zone: Report on DSS Contract OSU76-172, 111 pp. Available from GSC.

McNeill, J. D., 1980, Applications of transient electromagnetic techniques: Technical Note TN-7, Geonics Ltd., Toronto, 17 pp.

Rossiter, J. R., Strangway, D. W., Koziar, A., Wong, J. and Olhoeft, G. R., 1978, Electromagnetic sounding of permafrost, in Proceedings of the Third International Conference on Permafrost, v. 1: Ottawa, National Research Council of Canada, pp. 567-572.

Sinha, A. K., 1979, Maxi-Probe EMR-16: A new wide-band multifrequency ground EM system, in Current Research, Part B, Geological Survey of Canada, Paper 79-1B, pp. 23-26.

Sinha, A. K., 1980, A study of topographic and misorientation effects in multifrequency electromagnetic sounding: Geoexploration, v. 18, pp. 111-133.

Sinha, A. K., 1983, Deep multifrequency EM sounding at a site near Bowmanville, Ontario, in Current Research, Part A, Geological Survey of Canada, Paper 83-1A, pp. 133-137.

Taylor, A. E. and Judge, A. S., 1977, Canadian geothermal data collection -- Northern wells 1976-77, Geothermal series No. 10, Energy, Mines and Resources Canada, Ottawa, 193 pp.

Young, F. G., Myhr, D. W. and Yorath, C. J., 1976, Geology of the Beaufort-Mackenzie basin, Geological Survey of Canada, Paper 76-11, 65 pp.

SUMMER STREAMFLOW AND SEDIMENT YIELD FROM DISCONTINUOUS-PERMAFROST HEADWATERS CATCHMENTS

C. W. Slaughter, J. W. Hilgert, and E. H. Culp

Institute of Northern Forestry, USDA Forest Service,
Fairbanks, Alaska 99701

The presence of permafrost in a catchment system affects the hydrology and water quality of streamflow. Summer streamflow and concentration of suspended sediment in streamflow from first-order catchments in the discontinuous-permafrost taiga of central Alaska (latitude 65°10'N) have been analyzed. In terms of resultant streamflow, permafrost-underlain terrain is much more responsive to precipitation inputs than is permafrost-free terrain, and proportion of permafrost, with concomitant cold, thick organic layers overlying mineral soil, is a primary determinant of differing streamflow characteristics in headwaters catchments. A permafrost-free first-order stream consistently has higher (summer) baseflow than does an adjacent permafrost-dominated first-order stream. A small permafrost-dominated catchment has consistently lower suspended sediment concentrations than does a similar-size virtually permafrost-free catchment. During storm-free periods, suspended and dissolved loads are consistently less than 5 $mg \cdot l^{-1}$ in both basins. Concentrations are commonly an order of magnitude higher during storm events. Sediment and streamflow relationships are highly variable in this stream system.

INTRODUCTION

Less than two decades ago, virtually no published information existed concerning hydrology of headwaters, first-order streams in Alaska's discontinuous-permafrost regions. Covert and Ellsworth (1909) published hydrologic reconnaissance measurements of some small streams, but most historical (and present day) continuous streamflow measurements are of larger rivers.

Dingman (1966, 1971) provided the first modern, quantitative analysis of hydrologic conditions in a North American discontinuous-permafrost catchment. Dingman (1971) showed that a small (1.8 km^2 basin) stream had quite prolonged stormflow recessions when compared to similar-sized basins in more temperate settings, and suggested that the concept of variable-source-area runoff production (Hewlett and Hibbert 1966, Dunne 1978) is applicable to subarctic watersheds. Kane et al. (1978b) stated that "...permafrost areas are responsible for contributing a large percentage of the peak flow for both snowmelt and rainfall runoff events" in subarctic Alaska. Streamflow measurements in upland basins with differing proportions of permafrost provided further evidence that permafrost presence strongly affects headwaters runoff patterns (Slaughter and Hilgert 1978).

Hydrology of small catchments in the subarctic and Arctic has received increased attention in recent years, particularly in Canada. Anderson (1974) showed that streamflow from a 31-km^2 basin in the Mackenzie Delta, N.W.T., had hydrograph recessions and precipitation/runoff patterns similar to those reported by Dingman (1971), and suggested that stream response was strongly influenced by permafrost and "widespread occurrences of peaty topsoil and mosses." Ambler (1974), Jasper (1974), Newbury et al. (1979), Vincent and Russell (1979), and Marsh and Woo (1979), among others, have reported on high-latitude catchment studies in recent years.

Possible effects of frozen ground are considered in detail by Dingman (1975). Permafrost (perennially frozen ground) can act effectively as an aquiclude, confining groundwater and restricting aquifer recharge; in the discontinuous-permafrost zone this effect is localized and water is yielded to the surface through unfrozen zones (Kane and Slaughter 1973, Nelson 1978). Infiltration is severely restricted by fine-grained frozen soils (Dingman 1975, Kane et al. 1978a). Permafrost-underlain slopes would be expected to yield water more rapidly to streamflow than would non-permafrost slopes, other factors held equal, because of restricted water infiltration. Taiga permafrost-underlain sites are commonly mantled with a thick (0.1-0.5+ m), virtually continuous living and dead organic layer which has strong influences on hydrologic response (Dingman 1971). This permafrost/surficial organic mat condition must contribute to the "flashy, responsive storm hydrographs" observed in Glenn Creek (Dingman 1971) and in catchment C-3 in this study. Dingman (1971) also attributed prolonged storm-flow recessions (compared with more temperate basins) to storage and subsequent slow drainage of water in the surficial organic layer.

Stream ecology and quality are affected by the land-water interface (Cummins 1974, Hynes 1975), and by sediment introduction to streams. Analysis of undisturbed streams draining catchments of varying area, permafrost, aspect, slope, and vegetative cover should provide better understanding of hydrologic-sediment relationships within subarctic stream/catchment systems.

This study was undertaken to determine hydrologic and stream system quality characteristics of first-order subarctic catchments encompassing permafrost-dominated and permafrost-free landscapes. Results are being utilized in research concerning the consequences of land and resource management practices for taiga watershed systems.

Table 1. Catchment characteristics (adapted from Bredthauer and Hoch (1979) and Lotspeich and Slaughter (1981))

Catchment	Drainage area, km^2	Elevation range, m	Dominant aspect	Area distribution (%) by elevation range (m)				Dominant vegetation	Drainage density, km/km^2	Proportion of permafrost, %
				<305 m	305-487 m	488 m-639 m	>640			
C-2	5.18	329-738	S	0	29.0	38.0	33.0	Deciduous	0.70	3.5
C-3	4.55	305-770	NE	0.1	39.5	51.4	9.1	Coniferous	0.73	53.2
C-4	8.34	240-773	SSE*	5.9	27.3	50.9	15.9	Mixed**/	0.70	18.8

*/Extreme headwaters of catchment C-4 are northeast-facing.

**/Mixed: Deciduous forest on south, southwest slopes, coniferous forest on north, northeast slopes.

METHODS

Study Area

Field work was conducted in the Caribou-Poker Creeks Research Watershed (CPCRW), a 110-km^2 Experimental Ecological Reserve located at 65°10'N latitude, 35 km north of Fairbanks, Alaska. The study area is in the "Interior" climatic zone of Alaska, characterized by large diurnal and annual temperature extremes and low precipitation. The long-term mean annual precipitation at Fairbanks (elevation 132 m) is 285 mm; mean January and July temperatures are -24.4°C and +17.1°C, respectively. Mean annual precipitation at CPCRW is estimated at 500 mm; the average ratio of summer (May-September) precipitation at CPCRW to precipitation at Fairbanks is 1.47:1 (Haugen et al. 1982), consistent with precipitation/elevation gradients observed elsewhere in central Alaska (Santeford 1976). The estimated mean annual temperature at CPCRW varies from -4.9°C in the Caribou Creek valley (elevation 240 m), to as warm as -1.2°C at mid-elevation (480 m) on south-facing slopes (Haugen et al. 1982).

CPCRW lies in the Yukon-Tanana Uplands (Wahrhaftig 1965). Bedrock metamorphic Precambrian schists are mantled with Quaternary aeolian silts of varying thickness, commonly less than 1 m, and rock outcrops are encountered on ridges at higher elevations. Periglacial activity is evidenced by solifluction lobes and slumps on north-facing permafrost-underlain slopes. Soils of south-facing (permafrost-free) slopes are silt loams or gravelly silt loams, free from permafrost; north-facing slopes and valleys include extensive areas of permafrost-underlain silt loams (Rieger et al. 1972). The vegetation of the basin is heterogenous, reflecting soils, aspect, elevation and fire history. Permafrost-underlain, north-facing and valley sites are generally occupied by black spruce (*Picea mariana* (Mill.) B.S.P.) forest stands, with riparian (valley) settings dominated by willow (*Salix* sp.), dwarf birch (*Betula nana* L.), and stunted stands of black spruce and tamarack (*Larix laricina* (DuRoi) K. Koch). Warmer, better-drained south slopes support extensive stands of aspen (*Populus tremuloides* Michx.), birch (*Betula papyrifera* Marsh), lesser areas of white spruce (*Picea glauca* (Moench) Voss), and commonly a major understory component of alder (*Alnus crispa* (Ait.) Pursh.) (Troth et al. 1975).

Hydrologic data

The catchments studied are designated C-2, C-3 and C-4; pertinent characteristics are summarized in Table 1. Summer precipitation was recorded in the lower Caribou Creek valley (240 m elevation). Summer streamflow data were obtained from fiberglass Parshall flumes (23-cm throat in C-2 and C-3, installed in summer 1977; 45.7-cm throat in C-4, installed in fall 1979) equipped with waterstage recorders (Slaughter 1981a). Data analyzed in this report were obtained during the summer (ice-free) streamflow season, after breakup and snowmelt runoff. Acquisition of streamflow data during breakup has been hindered in these headwaters streams by extensive aufeis accumulation at the gaging sites (Slaughter 1982), and flows during breakup are often estimated using time-lapse photography (Slaughter 1981b).

Summer streamflow data continuity was good during the three years reported here (1978, 1980, 1981). The 1979 hydrographs had poor data continuity, and were not included in this analysis.

Non-filterable Residue (Suspended Sediment) Data

Non-filterable residue (suspended sediment) was sampled using pre-washed 500-ml widemouth Nalgene bottles. Weekly grab samples were taken at the center of each Parshall flume throat, and were not depth and width-integrated because of extremely shallow water depths at most sampling times. Storm-flow samples were obtained using automated water samplers, with the sampler intake nozzle suspended in the flume throat. A liquid-level actuator was placed in each flume, with the timer set to sample at 6-hour intervals following exceedance of a pre-determined flow level. The samplers were also manually activated weekly, at the time of grab sampling, to obtain a simultaneous collection for technique comparison.

The non-filterable residue (NFR) samples were returned to the lab and analyzed for turbidity and specific conductance, then filtered through tared, pre-ignited 0.45 micron Gelman micro-quartz fiber filters. The filter with residue (>0.45 micron) was dried to a constant weight at 105°C, and weighed again to determine NFR. If residue concentration exceeded 25 $mg \cdot l^{-1}$, the filter was ashed at 550°C, wetted, dried, and weighed to determine the amount of volatile material (American Public Health Association 1975).

Estimates of sediment production were calculated using mean daily flow and the mean of the residue samples taken in that day, because the NFR collections represented such a small sample both in terms of volume and time. Total estimated sediment production in each stream was then calculated as residue per unit area and residue per unit volume of stream discharge.

RESULTS AND DISCUSSION

Haugen et al. (1982) showed that 1978 and 1980 were summers of below-average precipitation, based on long-term Fairbanks climatic records, and thus reflect relatively dry conditions. June and July 1981 were quite wet and cumulative summer 1981 precipitation was 70% higher than the previous years.

Streamflow records of 1978, 1980 and 1981 were examined to determine whether previously reported (Slaughter and Hilgert 1978) hydrologic patterns are consistent over the more recent study period. Qualitative inspection of seasonal hydrographs and storm unit-area hydrographs ($l \cdot sec^{-1}km^{-2}$) (Figure 1) immediately shows the disparity in behavior between catchments C-2 and C-3. The storm of late June 1978 (Figure 1a), observed in both Caribou and Poker Creeks (Haugen et al. 1982, Figs. 9 and 10), was prominent in the C-3 hydrograph. Peak mean daily flow (24 June) was 61.5 $l \cdot sec^{-1}km^{-2}$; the instantaneous C-3 peak for that storm was 69.3 $l \cdot sec^{-1}km^{-2}$. Response to the same storm event in the south-facing C-2 catchment was later and relatively subdued, with the peak mean daily flow (27 June) being 8.0 $l \cdot sec^{-1}km^{-2}$; instantaneous C-2 peak flow was 11.9 $l \cdot sec^{-1}km^{-2}$. Basin-averaged total storm yields (22 to 30 June 1978) were 4.6 mm and 22.9 mm for C-2 and C-3, respectively.

Similar patterns are seen in storm hydrographs from 1980 and 1981. The modest storm of August 1980 (Figure 1b) produced virtually no streamflow increase in C-2, slight increase in C-4, and a prominent hydrograph rise in C-3. Storm period (120 hours) yields were 4.3 mm, 12.5 mm and 5.9 mm for C-2, C-3, and C-4, respectively. The larger July 1981 storm (Figure 1c) resulted in pronounced and synchronous flow peaks in the three streams; unit-area C-2 flows exceeded those of C-4. As in previous years and other storms, catchment C-3 yielded much higher unit-area peak flows than C-2 or C-4: storm period (312 hours) yields were 22.6 mm, 29.5 mm, and 16.2 mm for C-2, C-3, and C-4, respectively.

Flow-duration curves provide another method for comparing streamflow patterns (Dunne and Leopold 1978). Figure 2 presents flow-duration curves based on mean daily unit-area streamflow, for the years 1978, 1980 and 1981 (C-4 in 1980 and 1981 only). Any point on a curve indicates, for the time period used in its construction, the percent of time (X-axis) during which that flow (Y-axis value) was equalled or exceeded. The 1978 plot (Figure 2a) shows that at flows in the range of 3 to 5 $l \cdot sec^{-1}km^{-2}$, the C-2 and C-3 catchments are in fair agreement; the curves are approximately coincident in the 40-50% time equalled or exceeded range (in the following discussion, "time equalled or exceeded" is simply termed "duration"). At lower flows, the C-3 curve dropped relative to the C-2 curve, indicating that C-3 experienced relatively more low-flow days in 1978 than did C-2. For flows above 5 $l \cdot sec^{-1}km^{-2}$ the C-2 curve shows moderate responses, while the C-3 curve indicates a high proportion of flows equalling or exceeding the

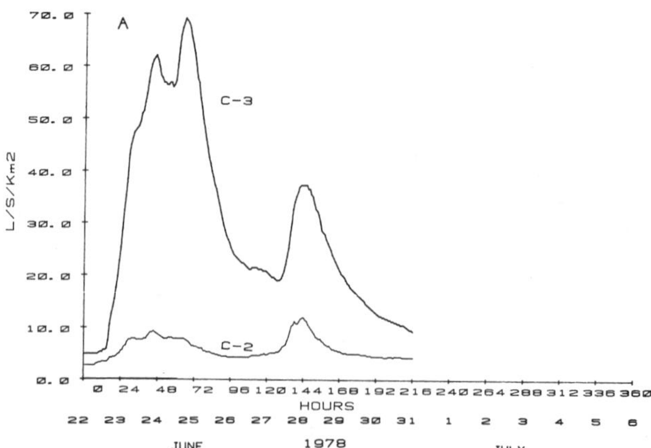

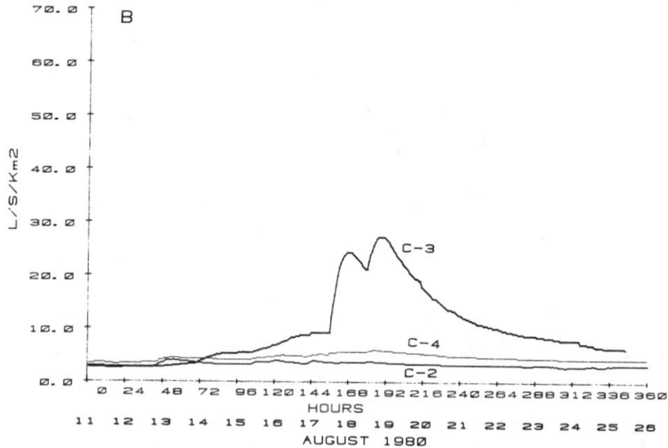

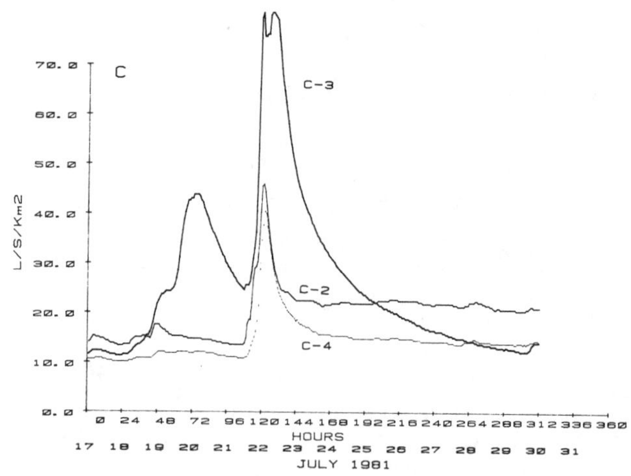

Figure 1 Storm hydrographs for (a) 1978, (b) 1980, and (c) 1981.

highest levels of C-2. The flow at the 20% duration was about 5 $l \cdot sec^{-1}km^{-2}$ for C-2, compared to 10 $l \cdot sec^{-1}km^{-2}$ for C-3; flow at the 10% duration was about 6 $l \cdot sec^{-1}km^{-2}$ and 17 $l \cdot sec^{-1}km^{-2}$ for C-2 and C-3, respectively.

1980 was a summer of generally lower flows than 1978, as reflected in Figure 2b. The flow-duration curves for C-2 and C-3 again differ markedly at high flows (low duration). At 20% duration, the C-2 flow was less than half that of C-3; at 10% duration, the C-2 flow was approximately 25% that of C-3. At low flows (high durations) the flow-duration curves for C-2 and C-3 were quite similar in 1980.

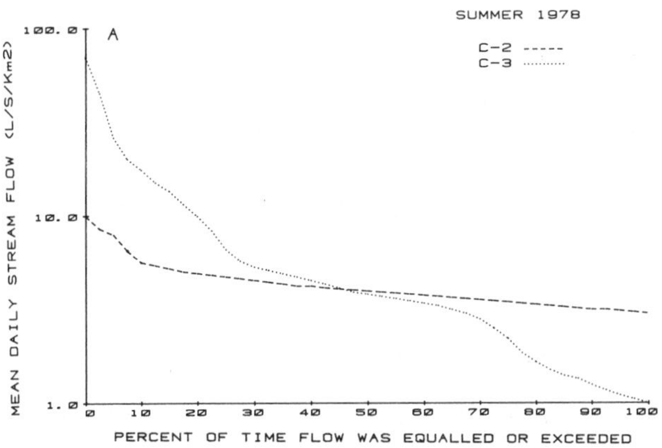

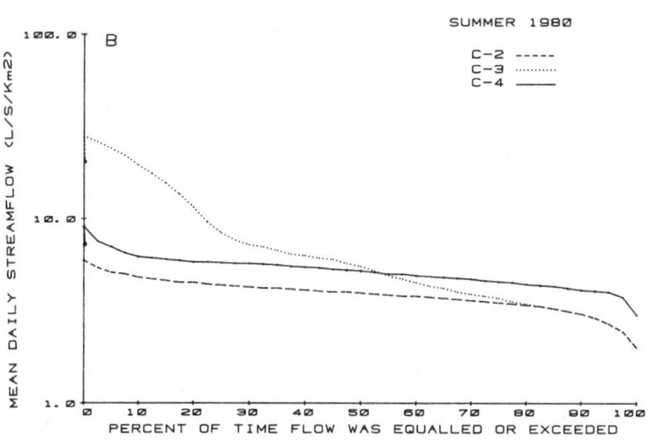

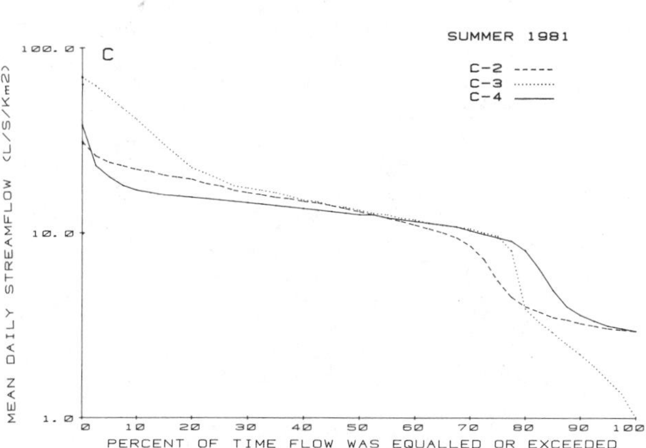

Figure 2 Flow-duration curves for (a) June 1978, (b) August 1980, and (c) July 1981.

The 1981 summer was markedly wetter (Table 2), a condition reflected in the curves of Figure 2c. The 50% duration point, nearly coincident for C-2 and C-3, is approximately 13 $l \cdot sec^{-1}km^{-2}$. Little difference between C-2 and C-3 is evident at lower flows, except that C-3 shows appreciably lower flows above 80% duration. At higher flows, the relative position of the curves is similar to that in 1978 and 1980, although C-2 evidenced higher unit-area flows than in the earlier years. At 20% duration there was little difference between C-2 and C-3, but at 10% duration the C-2 value was only 50% that of C-3.

It is evident that the streamflow pattern of contrasting catchments initially suggested, that the north-facing, permafrost-dominated C-3 basin has higher peak flows and lower low flows than the south-facing C-2 basin, is consistent during both dry (1978, 1980) and relatively wet (1981) summers.

Despite its larger size, the C-4 catchment is intermediate between C-2 and C-3 in terms of permafrost and aspect (Table 1). The northeast-flowing headwaters of C-4 presumably are strongly influenced by permafrost terrain, while major downstream slope areas are south-facing and permafrost-free. The hydrographs and the flow-duration curves suggest that streamflow patterns of C-4 may be considered "intermediate" between those of C-2 and C-3. In both 1980 and 1981, the flow of C-4 closely paralleled that of the permafrost-free C-2 basin. In the drier 1980 summer, the C-2 and C-4 unit-area flows were virtually coincident through the entire measurement period. In 1981 the patterns were again similar, although C-2 maintained a higher unit-area base flow than did C-4 (which in turn had higher unit-area base flow than C-3). The 1980 C-4 flow-duration curve closely follows that for C-2, but is displaced upward on the Y-axis. This indicates that streamflow response to precipitation was similar to that of C-2, but for any given duration the C-4 stream had a higher unit-area flow. During the wetter 1981 summer, the pattern was less consistent (Figure 2c). At durations above 70% the C-4 flows exceeded those of C-2, but at low durations the unit-area flows of C-4 were less than those of C-2. In both 1980 and 1981, the flow-duration curves indicated that the behavior of C-4 is more like that of C-2 than that of permafrost-dominated C-3.

The years 1980 and 1981 saw increased emphasis on NFR storm response monitoring. On a unit area/unit streamflow basis, NFR yields (Table 2) in the relatively dry 1980 season were highest in C-4 and lowest in C-3 (1.00 and 0.25 $g \cdot m^{-3}km^{-2}day^{-1}$, respectively) with C-2 having 0.34 $g \cdot m^{-3}km^{-2}day^{-1}$. The wetter 1981 season showed increased residue yields, with C-3 having the highest mean yield and C-2 the lowest (12.04 and 2.80 $g \cdot m^{-3}km^{-2}day^{-1}$, respectively), and C-4 having an intermediate yield of 7.71 $g \cdot m^{-3}km^{-2}day^{-1}$. The much higher NFR yields in the C-3 stream are an apparent result of more pronounced hydrologic response to storm events (Jasper (1974) reported that

Table 2. Non-filterable residue (suspended sediment) determinations.

Year:	1978		1980			1981		
	C2	C3	C2	C3	C4	C2	C3	C4
Measurement period:	June22 through October 1 (102 days)		June 7 through August 31, (76 days)			June 9 through September 23 (107 days)		
Total seasonal discharge: ($m^3 \times 10^5$)	1.56	2.40	1.01	1.88	2.69	5.71	6.59	8.08
Mean daily discharge: (m^3 day^{-1} $\times 10^3$)	1.53	2.35	1.33	2.48	3.54	5.34	6.16	7.55
Mean daily unit-area discharge: (m^3 km^{-2} day^{-1})	295.37	517.13	256.16	545.09	424.64	1030.07	1353.55	904.97
Seasonal NFR* (Kg):	50.24	48.13	179.39	215.71	2252.32	1553.18	5862.02	6881.63
Mean NFR concentration: ($g \cdot m^{-3}$)	0.32	0.20	1.78	1.14	8.37	2.72	8.90	8.52
Daily NFR production: ($kg \cdot day^{-1}$)	0.49	0.47	2.36	2.84	29.64	14.52	54.79	64.31
Daily area-weighted NFR production rate: ($g \cdot m^{-3}$ km^{-2} day^{-1})	0.06	0.04	0.34	0.25	1.00	2.80	12.04	7.71

*/NFR=non-filterable residue.

about 80% of total summer sediment yield occurred during storm runoff events in a catchment at 65°23'N in the Mackenzie Mountains, N.W.T.). When total residue yields for the 1981 season are compared (Table 2), C-3 and C-4 have high values (5862 and 6882 kg, respectively), while C-2 had only 1553 kg. The drastic dissimilarity of residue yields from the similar-sized C-2 and C-3 basins emphasizes the difference in catchment response.

Analysis of data from the 1978, 1980 and 1981 ice-free seasons suggests that in high-discharge water years NFR (suspended sediment) is yielded at a much greater rate in watersheds with relatively higher proportions of permafrost. Assuming that this observation is consistent with other watersheds within the discontinuous-permafrost zone, the effects of land management practices on sediment yield must be evaluated in the context of differing "natural" sediment production rates of watersheds. This "natural" sediment yield is affected by both permafrost and by precipitation, with major storm events and higher baseflows producing markedly higher yields of sediment.

CONCLUSIONS

Summer streamflow and sediment yield in subarctic Alaska are affected by presence of permafrost in headwaters catchments. A permafrost-dominated first-order stream has higher peak streamflow, higher storm-flow suspended sediment concentrations, lower base (non-storm) streamflow, and lower base-flow suspended sediment concentrations than does a nearby, virtually permafrost-free first-order stream. Such consistent differences in streamflow and sediment yield in undisturbed, close-proximity catchments suggest that (1) detailed understanding of streamflow and sediment production from upland watersheds throughout the discontinuous-permafrost taiga is needed; (2) regulation of land and resource management practices on the basis of possible consequences for streamflow and water quality must be accompanied by knowledge of the "natural", pre-management behavior of taiga watersheds, paying specific attention to effects of permafrost-underlain landscapes on catchment/stream systems; (3) extrapolation of hydrologic and water quality data from existing records, which are largely for major rivers, to lower-order (upstream) basins must include awareness of natural variation in hydrologic regimen of subarctic headwaters catchments as demonstrated in this study.

REFERENCES

Ambler, D. C., 1974, Runoff from a small Arctic watershed, p. 45-49, in Permafrost Hydrology, Proceedings of Workshop Seminar, 1974: Canadian National Committee, International Hydrological Decade, Environment Canada, Ottawa.

American Public Health Association, 1975, Standard methods for the examination of water and wastewater: American Public Health Association, Washington D.C. 1193 p.

Anderson, J. C., 1974, Permafrost-hydrology studies at Boot Creek and Peter Lake Watersheds, N.W.T., p. 39-49, in Permafrost Hydrology, Proceedings of Workshop Seminar, 1974: Canadian National Committee, International Hydrological Decade, Environment Canada, Ottawa.

Bredthauer, S. R., and D. Hoch, 1979, Drainage network analysis of a subarctic watershed, Special Report 79-19: USA CRREL, Hanover, N.H. 9 p.

Covert, C. C., and C. E. Ellsworth, 1909, Water-supply investigations in the Yukon-Tanana Region, Alaska, 1907 and 1908, Fairbanks, Circle, and Rampart Districts, Water-Supply Paper 228: Geological Survey, Washington, D.C. 108 p.

Cummins, K. W., 1974, Structure and function of stream ecosystems, Bioscience 24: 631-641.

Dingman, S. L., 1966, Hydrologic studies of the Glenn Creek watershed near Fairbanks, Alaska, Special Report 86: USA CRREL, Hanover, N.H. 30 p.

Dingman, S. L., 1971, Hydrology of the Glenn Creek Watershed, Tanana River Basin, central Alaska, Research Report 297: USA CRREL, Hanover, N.H. 111 p.

Dingman, S. L., 1975, Hydrologic effects of frozen ground, Special Report 218: USA CRREL, Hanover, N.H. 55 p.

Dunne, T., 1978, Field studies of hillslope flow processes, in M. J. Kirkby (ed), Hillslope hydrology: John Wiley and Sons, London.

Dunne, T., and L. B. Leopold, 1978, Water in environmental planning: W. H. Freeman and Co., San Francisco. 818 p.

Haugen, R. K., C. W. Slaughter, K. E. Howe, and S. L. Dingman, 1982, Hydrology and climatology of the Caribou-Poker Creeks Research Watershed, Alaska, CRREL Report 82-26: USA CRREL, Hanover, N.H. 35 p.

Hewlett, J. D., and A. R. Hibbert, 1966, Factors affecting the response of small watersheds to permafrost in humid areas, p. 275-290, in International symposium on forest hydrology: Pergamon Press, N.Y.

Hynes, H. B. N., 1975, The stream and its valley, Verhandlungen Internationalen Verein Limnologie 19: 1-15.

Jasper, J. N., 1974, Hydrologic studies at "Twisty Creek" in the Mackenzie Mountains, N.W.T., p. 263-281, in Hydrologic aspects of northern pipeline development, Report No. 74-12: Task Force on Northern Oil Development, Environment Canada, Ottawa.

Kane, D. L., and C. W. Slaughter, 1973, Recharge of a central Alaska lake by subpermafrost groundwater, p. 458-462, in Permafrost: The North American Contribution to the Second International Conference: National Academy of Sciences, Washington, D.C.

Kane, D. L., J. D. Fox, R. D. Seifert, and G. S. Taylor, 1978a, Snowmelt infiltration and movement in frozen soils, p. 201-206, in Proceedings of the Third International Conference on Permafrost, v. 1: National Research Council of Canada, Ottawa.

Kane, D. L., R. D. Seifert, and G. S. Taylor, 1978b, Hydrologic properties of subarctic forest soils, IWR-88: Institute of Water Resources, University of Alaska, Fairbanks, Alaska. 49 p.

Lotspeich, F. B., and C. W. Slaughter, 1981, Preliminary results on the structure and functioning of a taiga watershed, IWR-101: Institute of Water Resources, University of Alaska, Fairbanks, Alaska. 94 p.

Marsh, P., and M. Woo, 1979, Annual water balance of small high Arctic basins, p. 536-546, in Canadian Hydrology Symposium: 79-Proceedings: National Research Council of Canada, Ottawa.

Nelson, G., 1978, Hydrologic information for land-use planning, Fairbanks vicinity, Alaska, Open-file Report 78-959: U.S. Geological Survey, Anchorage, Alaska.

Newbury, R. W., K. G. Beaty, J. A. Dalton, and G. K. McCullough, 1979, A preliminary comparison of runoff relationships and water budgets in three experimental lake basins in the continental, sub-Arctic and Arctic climatic regions of the Precambrian Shield, p. 516-535, in Canadian Hydrology Symposium: 79-Proceedings: National Research Council of Canada, Ottawa.

Rieger, S., C. E. Furbish, D. B. Schoephorster, H. Summerfield, and L. C. Geiger, 1972, Soils of the Caribou-Poker Creeks Research Watershed, interior Alaska, Technical Report 236: USA CRREL, Hanover, N.H. 10 p.

Santeford, H. S., 1976, A preliminary analysis of precipitation in the Chena Basin, Alaska, Technical Memorandum NWS AR-15: National Oceanic and Atmospheric Administration, Anchorage, Alaska.

Slaughter, C. W., 1981a, Use of pre-fabricated Parshall flumes to measure streamflow in permafrost-dominated watersheds, Research Note PNW-382: Pacific Northwest Forest and Range Experiment Station, USDA Forest Service, Portland, Oregon. 8 p.

Slaughter, C. W., 1981b, Snowmelt runoff estimation with time lapse photography, Pacific Northwest Section, American Geophysical Union: Ellensburg, Washington.

Slaughter, C. W., 1982, Occurrence and recurrence of aufeis in an upland taiga catchment, p. 182-188, in Proceedings of the Fourth Canadian Permafrost Conference (The Roger J. E. Brown Memorial Volume): National Research Council of Canada, Ottawa.

Slaughter, C. W., and J. W. Hilgert, 1978, Upland stream system quality in the taiga, Pacific Northwest Regional Meeting, American Geophysical Union: Tacoma, Washington.

Slaughter, C. W., and D. L. Kane, 1979, Hydrologic role of shallow organic soils in cold climates, p. 380-389 in Canadian Hydrology Symposium: 79-Cold Climate Hydrology. Proceedings: National Research Council of Canada, Ottawa.

Soulis, E. D., and D. E. Reid, 1978, Impact of interrupting subsurface flow in the northern boreal forest, p. 225-231, in Proceedings of the Third International Conference on Permafrost, v. 1: National Research Council of Canada, Ottawa.

Troth, J. L., F. J. Deneke, and L. M. Brown, 1975, Subarctic plant communities and associated litter and soil profiles in the Caribou-Poker Creeks Research Watershed, interior Alaska, Research Report 330: USA CRREL, Hanover, N.H. 25 p.

Vincent, D. G., and S. O. Russell, 1979, Estimation of small basin flood flows in the discontinuous-permafrost zone, p. 230-238, in Canadian Hydrology Symposium: 79-Proceedings: National Research Council of Canada, Ottawa.

Wahrhaftig, C., 1965, Physiographic divisions of Alaska, Professional Paper 482: U.S. Geological Survey, Washington, D.C. 52 p.

PERMAFROST SENSITIVITY TO CLIMATIC CHANGE

M. W. Smith and D. W. Riseborough

Geotechnical Science Laboratories, Geography Department,
Carleton University, Ottawa, Ontario, Canada K1S 5B6

Possible large-scale climatic warming, due to the so-called greenhouse effect, has obvious implications for the stability of permafrost conditions. This paper presents a preliminary examination of the possible effects of climate change on the ground thermal regime in the zone of discontinuous permafrost. Caution must be exercised in extrapolating a warming trend in the atmosphere to the ground. The significance of atmospheric warming for permafrost depends on the accompanying increase in the ground surface temperature and the proximity of ground temperatures to $0^{o}C$. The characteristics of natural surfaces determine the interaction between climate, microclimate, and ground thermal conditions. A numerical microclimatic model based on the surface energy balance, has been used to investigate the range of ground temperature response under a uniform climate, due to variation in site conditions. Site factors included slope, albedo, wetness, roughness, snow cover, and soil thermal properties. Site wetness and snow cover were the most sensitive factors. The potential for permafrost degradation during 25 years of climatic warming was then simulated for various sites using climatic data for Whitehorse, Yukon Territory. In one part of the analysis, three different patterns of warming were analysed at a single site. The distribution of warming during the year had relatively little effect on long term permafrost degradation. Much greater differences in degradation resulted from variation in site conditions, for the case of uniform warming.

INTRODUCTION

This paper discusses the possible effects of climate change on the ground thermal regime in the zone of discontinuous permafrost. The possibility of large-scale climatic warming, due to the so-called greenhouse effect, has obvious implications for the stability of permafrost conditions, although this has been generally ignored in northern geotechnical considerations. A warming trend could be significant in areas of discontinuous permafrost, for, as the mean annual air temperature rises, some permafrost at the southern limit could thaw and ground ice would melt (e.g., see Thie 1974, Mackay 1975, Chatwin 1981). Judge (1973) suggests that permafrost up to 30 m thick in the Mackenzie Valley thawed completely during the warming period from approximately 1850 to 1950. Recent results from Alaska show that permafrost is beginning to warm up, in response to higher winter temperatures, and that active layer thicknesses are increasing (University of Alaska 1982, p. 16). It is important, therefore, to attempt some determination of the nature, rate, and magnitude of ground temperature changes that might occur as a result of climate change over the next 30-50 years.

As a result of human activity, carbon dioxide concentration is increasing in the atmosphere, and could double by the middle of the next century. There is, as yet, no definitive answer to what climatic structure would exist in a 2 X CO_2 world. However, the basic effects of a global warming, concentrated in the Arctic, and increased precipitation seem to be generally accepted (e.g., see Harvey 1982). Such changes will have significant and profound effects on the arctic environment. However, the nature and extent of these effects have yet to be determined; in fact, some may be indeterminable until they actually occur. For example, whilst thermal boundaries would move northward, increased precipitation could mean deeper snow cover, which would act to complicate the effects on ground thermal conditions. In addition, whilst winter temperatures could rise by as much as $15^{o}K$, summer increases could be much less.

Caution must be exercised, then, in extrapolating a warming trend in the atmosphere to the ground. The specific effects on permafrost conditions of any large-scale climate change will be modulated, perhaps strongly, by local microclimatic and lithologic conditions. The significance of any climate change depends on the magnitude of any accompanying increase in the ground surface temperature, and on the proximity of ground temperatures to $0^{o}C$. The change in surface temperature will depend upon the way in which energy is dissipated at a site. For example, at a wet site, additional energy would result largely in increased evaporation and have little effect on sensible heating. At a very dry site, however, additional energy input would be mostly utilized in sensible heating. The characteristics of natural surfaces determine the interaction between climate, microclimate, and ground thermal conditions, and between climate change and ground thermal response.

Because of the complex relationship between permafrost and climate, prediction of the rate and magnitude of ground temperature change is not simple. This paper discusses the use of a numerical microclimatic/ground temperature model based on the surface energy balance (Outcalt 1972, Outcalt et al. 1975, Smith 1977) to (1) illustrate the range of ground temperature response under a homogeneous climate, due to variation in site conditions, (2) analyse the susceptibility of permafrost to degradation due to various climate changes, and (3) analyse the role of site conditions in modulating permafrost degradation under climatic warming. A flow diagram of the model is shown in Figure 1; it identifies that changes in the ground thermal regime, and hence in the thickness and extent of permafrost, can result from changes in climate and/ or local conditions. Before proceeding with the results of an analysis, various field data are first presented to demonstrate these interrelationships.

MICROCLIMATIC INFLUENCES ON GROUND TEMPERATURE CONDITIONS

Wide variations in ground thermal conditions are known to occur within a small area of uniform climate. In areas where mean annual ground temperatures are close to 0°C, local factors can determine whether permafrost is present or not. Ground temperature data from Smith (1973, 1975) show that substantial differences exist between various sites in the Mackenzie Delta. Whilst the mean annual air temperature there is -9° to -10°C, mean annual surface temperatures range from about -4.2° to above 0°C over a small area. In fact, permafrost is over 60 m thick at some sites, but absent at others, due to local factors. Figure 2 shows the annual regimes at 50 cm depth for five sites, the details of which can be found in Smith (1975). The following points are noteworthy.

1. Site 1 (no vegetation) is warmest in summer but cools the most in winter (virtually no snow cover).
2. Site 4 (open vegetation canopy) is warmer than site 5 (dense canopy) all year round.
3. The regimes at sites 2 and 3 are distinctly different from the others, there being a virtual absence of the winter cooling wave because of deep snow cover.
4. Ground temperatures during the second winter were everywhere higher than in the previous year, even though monthly air temperatures were 2° - 9°C lower. This is because at all sites snow depths were greater and, moreover, snow was on the ground by the end of September.

Such data show that, although certain temperature differences arise because of vegetation effects, the most significant factor to the ground thermal regime in permafrost regions is snow cover (see Goodrich 1982), which can determine whether permafrost is present or not.

Table 1 presents 2-m ground temperatures at 10 closely spaced sites in the Takhini Valley, near Whitehorse, Yukon Territory, in the zone of discontinuous permafrost. The data (taken for approximately the warmest time of year) show a range of 3.5°C; permafrost is present at five sites, but at most of these it is very close to 0°C. The warmest sites include those where snow accumulates in winter. Permafrost in this area would presumably be very susceptible to a climatic warming, and therefore ground temperature observations here will be continued in the future.

TABLE 1 2-m Temperatures (September 1982), Takhini Valley, Yukon Territory.

Site	Temp. (°C)	Site description
1	-0.1	High ground, well drained (Snow free?)
2	1.3	Low, well drained, small poplar
3	2.3	Low, with bushes up to 1 m (Snow trap?)
4	-0.1	Poplar forest
5	2.6	Similar to 3
6	-0.1	Valley side (east-facing)
7	-0.1	Similar to 6
8	1.5	Willow, 1.5 m high
9	-0.9	Spruce forest
10	2.2	Open grassland

The data in Figure 2 show the variation in ground temperatures over less than a 2-year period. In Figure 3 are shown 1.06 m (3.5 foot) temperatures at four closely spaced sites at Churchill, Manitoba, for almost a 4-year period (data adapted from Brown 1978). It is interesting to note that

1. Permafrost is present at sites 2 and 3 but absent at 1 and 4.
2. Intra-annual variation is large at site 1 (high thermal diffusivity), moderate at 2 and 3, and small at 4 (high heat capacity).
3. Interannual variation is greater for winter than for summer, except at site 4. Variation is large at site 1, small at site 4, and intermediate at 2 and 3 (see note 2 above).
4. The summer of 1975 was the coolest at site 1, the warmest at site 2, and intermediate at 3 and 4 (little variation).
5. For sites 1 and 3, the summer of 1976 was warmer than 1975; for sites 2 and 4, 1975 was warmer than 1976.
6. Site 2 was cooler than 3 in summers 1973, 1974, and 1976, but warmer in 1975.
7. At site 2, the winter of 1975 was colder than 1974; at 4, 1975 was warmer; at 1 and 3 they were about the same.
8. At sites 1, 2, and 3 the winter of 1976 was colder than 1975; at 4 they were about the same.

These data illustrate the variability of ground thermal response to climatic conditions, and to climate variation. They demonstrate the difficulty of assessing the sensitivity of permafrost to climate change, and the need to unravel, first, the interactions of climate, microclimate, and ground thermal conditions.

SIMULATION RESULTS

This part of the paper discusses the application of a numerical climate/microclimate/ground temperature model to analyze the variability in ground thermal conditions and the magnitude and rate of response to climate change. The strategy for this is outlined in Figure 1. The local surface energy regime (microclimate) is modeled as the interaction between atmospheric (climate), surface, and subsurface conditions. This, in turn, provides the boundary condition governing the ground thermal regime. Hence, ground temperatures are explicitly linked to the microclimate, which is itself linked to the climate.

Soil moisture is modeled as a simple reservoir, the level of which serves to limit evaporation, via the equilibrium model. Soil thermal properties are fully temperature dependent, and moisture dependent in the surface layer. Except for snowfall, monthly normal climate data for Whitehorse, Yukon Territory, were used in the analysis. Snow accumulation was specified on a daily basis.

In part one a sensitivity analysis was conducted, the results of which serve as an indication of the range of possible ground thermal conditions created by variations in local (site-specific) factors under a uniform climate. These factors included albedo, slope and aspect, surface wetness, aerodynamic roughness, snow cover, and lithologic conditions, which were varied in turn in the model. The effect on the mean annual surface temperature is summarized in Table 2.

TABLE 2 Simulation of Changes to Mean Annual Surface Temperature by Changing Various Site Factors.

Site Factors		Mean Annual Surface Temp. Normal	ΔT
SFC LAYER	PEAT	+1.4	1.6
	SILT	0.0	1.4
SURFACE ROUGHNESS (Cm)	.01	+0.5	1.5
	0.1	+0.2	1.5
	1.0	0.0	1.5
	10.0	-0.2	1.3
ALBEDO	.1	+0.2	1.4
	.2	0.0	1.4
	.3	-0.2	1.4
	.4	-0.5	1.4
	.5	-0.7	1.4
WETNESS	0.0	+7.2	2.8
	0.2	+3.3	2.4
	0.4	+1.0	2.1
	0.6	0.0	1.4
	0.8	-1.6	1.1
	1.0	-3.0	1.8
SLOPE, ASPECT	30°N	-1.0	1.3
	10°N	-0.4	1.4
	0°	0.0	1.4
	10°S	0.0	1.9
	30°S	+0.9	2.0
SNOW COVER	½ NORMAL	-2.2	1.8
	NORMAL	0.0	1.1
	2XNORMAL	+0.5	0.2

Note: Values in column 1 are referred to 0°C at the standard site (underlined) for the normal climate

In addition, the surface temperature change for the same site conditions, but for a uniformly 2°K warmer climate, is also shown (ΔT). The authors recognize the limitations of varying site-specific factors independently, since they are actually cross-correlated in nature. The same holds true for varying air temperature alone, since climatic variables are also cross-correlated. To some extent, therefore, the results discussed in this section are a sensitivity test of the model, rather than of nature per se.

The main results of the sensitivity analysis are as follows.

1. Variations in every microclimatic parameter result in some variation in surface temperature. The variation is smallest for aerodynamic roughness (0.7°K for three orders of magnitude) and albedo (0.9°K for a five-fold change). However, such variations could be significant enough in the discontinuous permafrost zone to determine whether permafrost was present or not. The most sensitive factor is the surface wetness, since evaporation is such an important term in the energy balance. Snow cover is also important.

2. The difference in surface temperature due to the variation within a single microclimatic factor is, in some cases, greater than the change caused by air temperature warming. For example, variations in surface wetness, snow cover, wind speed, and slope angle result in differences in surface temperature greater than that due to air temperature warming.

3. The range in response to the 2°K warming among all sites simulated was 0.2° to 2.8°K (column ΔT, Table 2). Thus a simple correlation between air temperature changes and ground temperature changes is not likely to occur in nature.

In the second part of the analysis, three climatic warming patterns were investigated (based on Harvey 1982): (1) all months increased by 0.2°K/year, (2) an annual warming concentrated in the winter months (0.6°K/year), and (3) same as (2) but with double winter precipitation. The effect of each of these on mean annual surface temperature and active layer development at the standard site was simulated for 25 years. The results of these simulations are summarized in Figures 4 and 5.

1. Permafrost degradation proceeds more rapidly under a climatic change which is uniform in all months. Where the warming is concentrated in the winter months, the snow cover insulates the ground from the warming trend. The maximum difference at any time between active layer depths for the uniform trend and the winter trend is about 50 cm.

2. After 25 years, permafrost degradation associated with a winter warming trend alone does not appear to be significantly different from that associated with a winter warming trend accompanied by increased winter precipitation. Comparison of surface temperature trends shows that mean annual surface temperature of the site with increased snow cover is initially higher, but as the simulation proceeds, increasing amounts of winter precipitation fall as rain, thereby both reducing

the insulation effect and increasing evaporation, and thus lowering surface temperatures. Longer term simulations should produce greater degradation at sites without increased precipitation, where the effect of evaporation will not dominate.

In the last part of the analysis, four different model sites were subjected to a 25-year warming of 0.2°K per year in each month. Naturally, it is important to establish an equilibrium profile as the initial condition. Because of the thermal offset effect in ground temperature profiles (see Goodrich 1978), which is a function of soil thermal properties, the relationship between surface and subsurface temperatures is somewhat complicated. Equilibrium profiles for the four model sites (Figure 6) were established by repetitively running an annual simulation using normal Whitehorse climatic data. Whilst the profiles may not be precise replications of actual conditions, the marginal nature of permafrost they portray would seem to be quite representative for the region.

Figure 7 shows the lowering of the permafrost table at the four sites over a 25-year uniform climatic warming.

1. Variations in site characteristics produce larger changes to the rate of permafrost degradation than variations in the details of the climatic trend (compare Figures 4 and 7).
2. Where the subsurface material is changed to pure ice, active layer depths and the rate of degradation are greatly reduced, although the mean annual surface temperature is higher than for the standard site.
3. At the sites without ice rich lithology, degradation proceeds rapidly due to the lower heat required for melting in the ground materials. Degradation accelerates in the middle years, presumably because the whole profile has been warmed.

The authors have alluded to the limitation of using arbitrary changes in individual variables to represent climate change. In subsequent analyses, it is planned to synthesise future climatic data via stochastic simulation.

CONCLUSIONS

The implications of climatic warming to permafrost conditions must be viewed in the context of the complex interactions of climate, microclimate, and ground temperature conditions. This paper has presented a strategy for analyzing these interactions and for incorporating them into an assessment of the impact of climate change on permafrost. Climatic warming would be undeniably serious for permafrost conditions; however, the results presented here indicate that site effects can be as important as climatic factors in determining ground temperature response to environmental change.

ACKNOWLEDGMENTS

Some of the work reported here has been supported by a research contract with the Department of Energy, Mines and Resources, and a research grant from the Natural Sciences and Engineering Research Council, Ottawa. The authors also wish to thank G.H. Johnston, National Research Council, for supplying some of the field data presented. The assistance of Margo Burgess in various refinements to the computer model is gratefully acknowledged.

REFERENCES

Brown, R. J. E., 1978, Influence of climate and terrain on ground temperatures in the continuous permafrost zone of northern Manitoba and Keewatin District, Canada, in Proceedings of the Third International Conference on Permafrost, v. 1, p. 15-21: National Research Council of Canada.

Chatwin, S. C., 1981, Permafrost aggradation and degradation in a sub-arctic peatland: Unpub. M.Sc. Thesis, University of Alberta, Edmonton, 163 pp.

Goodrich, L. E., 1978, Some results of a numerical study of ground thermal regimes, in Proceedings of the Third International Conference on Permafrost, v. 1: p. 29-34, Ottawa, National Research Council of Canada.

Goodrich, L. E., 1982, The influence of snow cover on the ground thermal regime: Canadian Geotechnical Journal, v. 19, p. 421-432.

Harvey, R. C., 1982, The climate of arctic Canada in a 2 X CO_2 world, Atmospheric Environment Service, Canada, Report No. 82-5, 21 pp. & figs.

Judge, A. S., 1973, The thermal regime of the Mackenzie Valley: observations of the natural state, Environmental-Social Program, Task Force on Northern Oil Development, Report No. 73-38, 177 pp.

Mackay, J. R., 1975, The stability of permafrost and recent climatic change in the Mackenzie Valley, N.W.T.: Geological Survey of Canada. Paper No. 75-1, Part B, 173-176.

Outcalt, S. I., 1972, The development and application of a simple digital surface climate simulator: Journal of Applied Meteorology, v. 11, p. 629-656.

Outcalt, S. I., C. Goodwin, G. Weller and J. Brown, 1975. Computer simulation of the snowmelt and soil thermal regime at Barrow, Alaska. Water Resources Research, v. 11, p. 709-715.

Smith, M. W., 1973, Factors affecting the distribution of permafrost, Mackenzie Delta, N.W.T.: Ph.D. Thesis, University of British Columbia, 186 pp.

Smith, M. W., 1975, Microclimatic influences on ground temperatures and permafrost distribution, Mackenzie Delta, N.W.T.: Canadian Journal of Earth Sciences, v. 12, p. 1421-1438.

Smith, M. W., 1977, Computer simulation of microclimatic and ground thermal regimes: Department of Indian Affairs and Northern Development, ALUR Report No. 75-76-72, 74 pp.

Thie, J., 1974, Distribution and thawing of permafrost in the southern part of the discontinuous permafrost zone in Manitoba: Arctic v. 27, p. 189-200.

University of Alaska, 1982. Annual Report of Research, 1981-1982, 122 p.

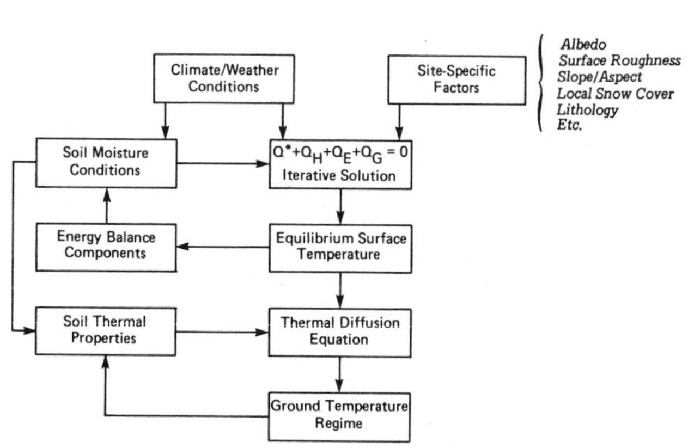

FIGURE 1 Flow diagram for the simulation model (Q^* = net radiation, Q_H = convection, Q_E = evaporation, Q_G = ground heat flux).

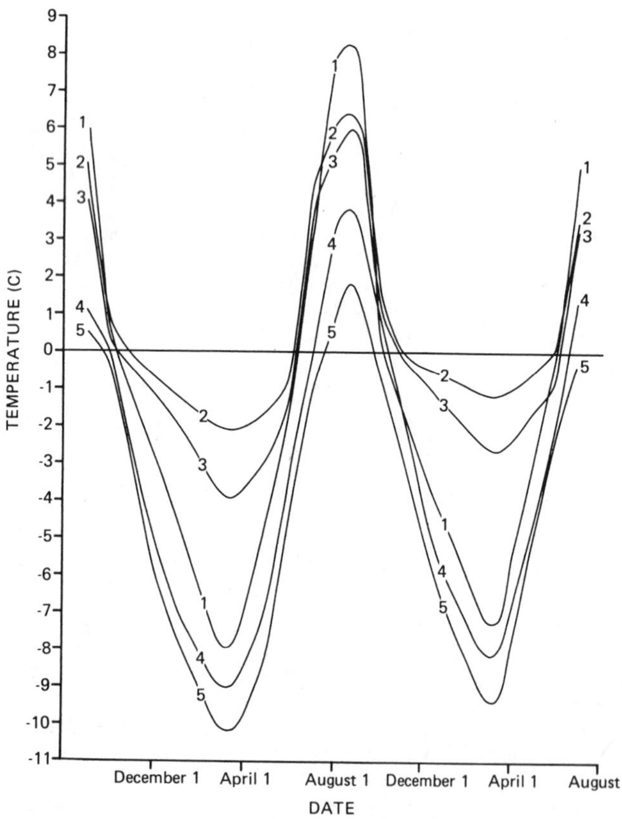

FIGURE 2 Ground temperatures at 50 cm depth for five sites in the Mackenzie Delta, N.W.T. (after Smith 1975).

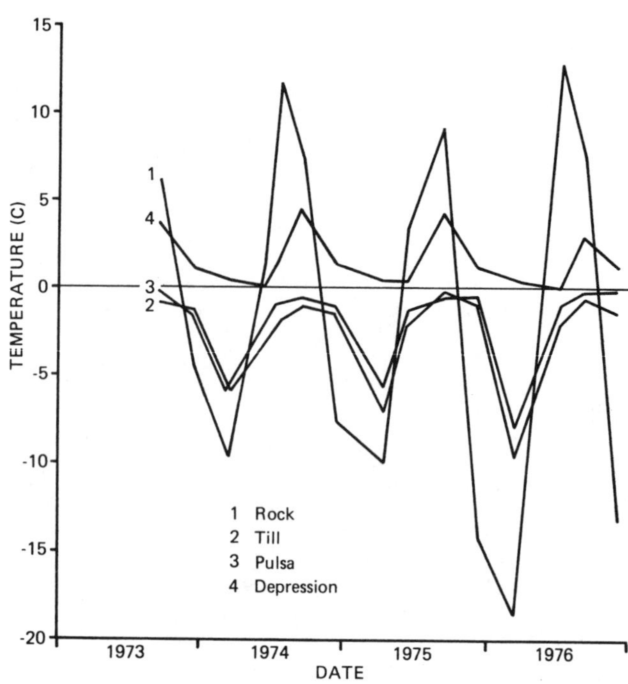

FIGURE 3 Ground temperatures at 1.06 m depth for four sites at Churchill, Manitoba (data adapted from Brown 1973).

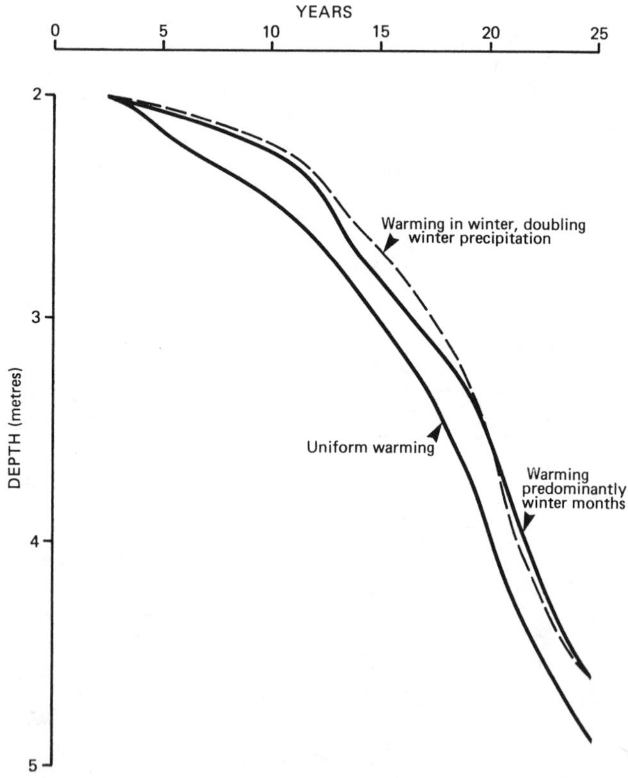

FIGURE 4 Annual maximum active layer development under various climatic warming trends.

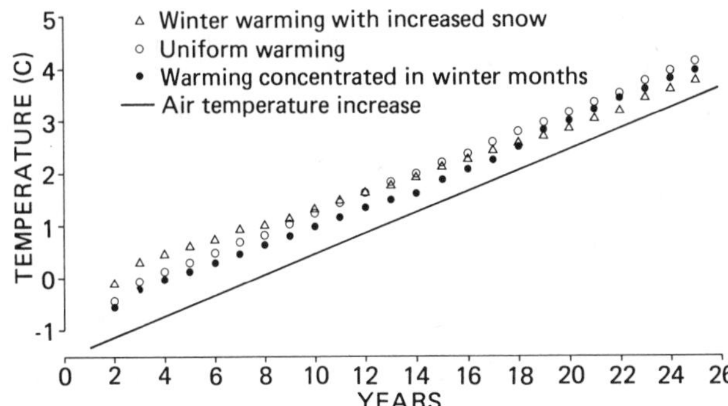

FIGURE 5 Change in mean annual ground surface temperature for various climatic warming trends.

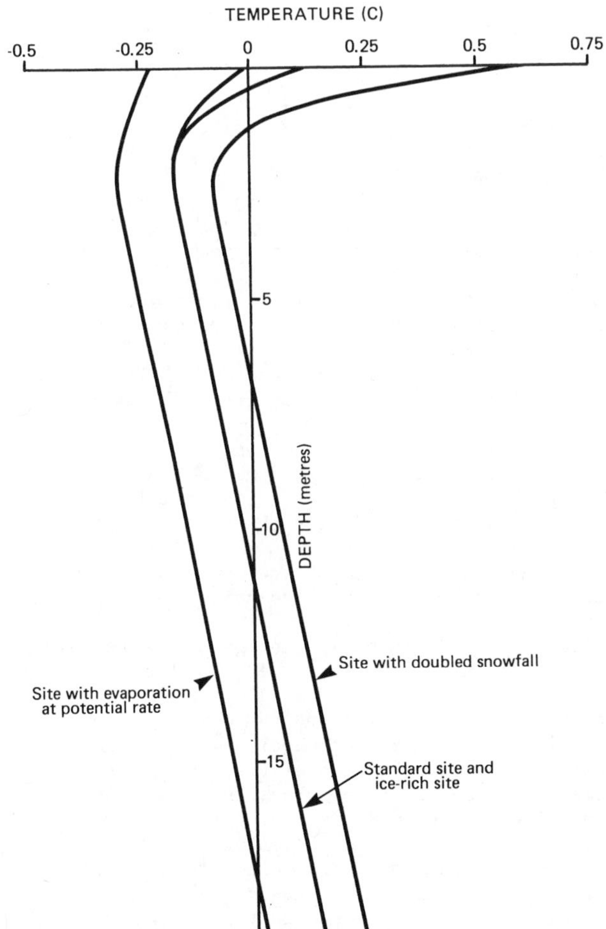

FIGURE 6 "Equilibrium" mean annual ground temperature profiles for four model sites.

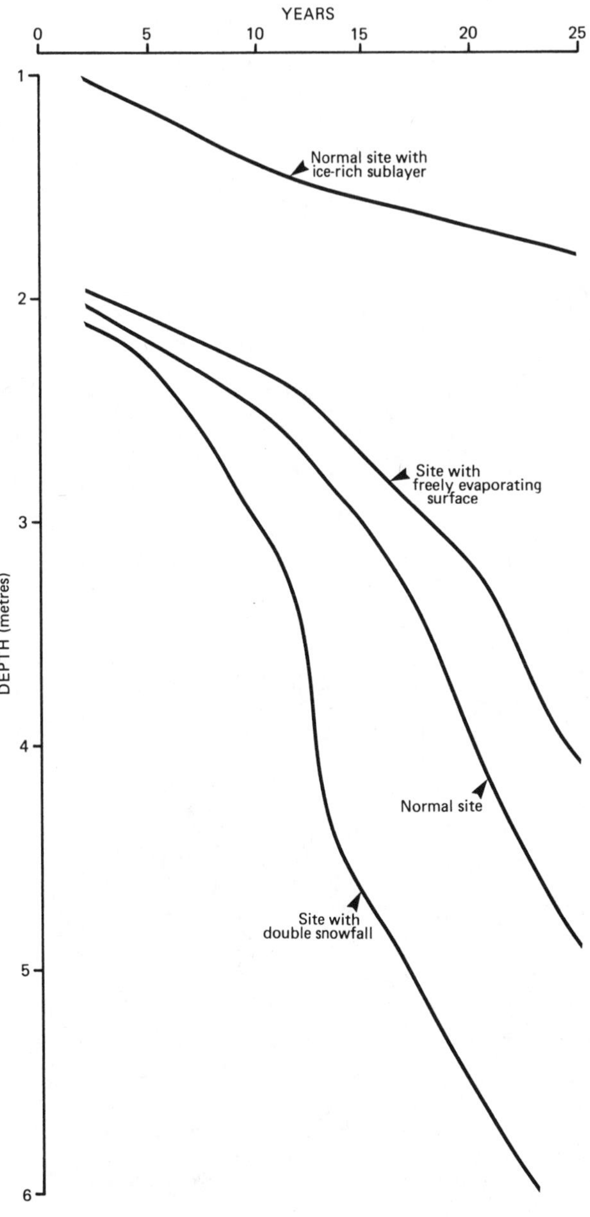

FIGURE 7 Annual maximum active layer development at four model sites under a uniform climatic warming trend.

ENGINEERING DESIGNS FOR LAYING PIPELINES IN PERMAFROST AREAS AND BOGGY TERRAIN IN THE NORTH

V. V. Spiridonov

All-Union Scientific Research Institute on Construction of Pipeline Mains (VNIIST), Moscow, USSR

On the basis of extensive research and prototype construction the USSR has accumulated considerable experience in the area of northern pipeline construction; engineering designs have been developed and standard documents produced to solve the problems of building large-diameter pipelines in northern regions. Systems for laying pipelines that are new in principle have been developed for various conditions. These include an above-ground system with slightly curved sections; an above-ground system with partial compensation for longitudinal strains for anchoring the pipeline on heaving or waterlogged sections; an underground system on supports for floodplain sections with permafrost conditions; a system where the exposed pipeline is laid on the surface with compensating sections; and finally, various modifications to all of these. Methods of analyzing the various systems and recommendations as to engineering design have been developed.

The urgent need for more and more energy resources for the successful development of all economic areas makes the development of the gas and oil industry of great priority. Major gas and oil deposits have been found in the northern regions, with most of the deposits far beyond the Arctic Circle. The main consumers of gas and oil, however, are in the temperate and southern climates. The construction of high-capacity pipeline systems has become a major challenge to the USSR, the USA, Canada, and other countries. The range of constructed pipelines that are 1220-1420 mm in diameter increases steadily with each passing year. At the same time the ratio of the lines laid in the northern regions also increases.

CONSTRUCTION IN NORTHERN REGIONS

Pipeline construction in the northern regions depends on climatic, hydrogeological, geocryological, technological, and economic factors. The northern regions of the USSR are characterized by long winters with low temperatures (down to -65°C), frequent snowstorms and winds up to 40 m/s, prevalence of permafrost, an appreciable number of bogs and water bodies, territories, the long-lasting polar night, a poorly developed infrastructure, and the lack of a permanent road network.

Pipeline mains as engineering constructions are unique in the nature and intensity of their interaction with the environment due to their length, surface area, the mass of the transported product, and its heat content, as well as the volume of the ground with disturbed natural structure. During their lifetime, in all their parts, pipelines are subjected to very high tensions approaching the standard characteristics of metal strength. Even moderate deviation of the operating conditions from the initial design parameters bring the system into a state of extreme tension.

Pipeline construction significantly changes the environment, and there appears to be a specific problem of long-term prognosis and control of the interaction between the pipeline and the environment. The existing dynamic balance of the biosphere of the northern regions is unsteady, so it is easily upset by any intervention.

Prognosis for the northern regions is highly complicated due to the many factors involved and the many interconnections of the processes and phenomena of natural development.

Therefore, the following considerations must be taken into account when constructing pipelines in all northern regions, including areas with permafrost:

(1) the thermal and mechanical interaction of the pipeline with freezing-thawing ground, with water (in passing), and with the surrounding atmosphere;

(2) the change in the physicomechanical properties of the ground during processing, freezing, and thawing;

(3) the change in the ground water content along the route of the pipeline;

(4) the impact of territorial development on geocryological processes and the change in the perennial frozen ground condition due to changes in vegetation, redistribution of the snow cover, formation of new foci for the waters over the permafrost, and so forth.

Permafrost presents a number of phenomena that make the construction and maintenance of pipelines difficult, such as ground subsidence, heaving, landslides, thermoerosion, the formation and development of ice coating, and frost fissures in the ground.

Therefore, a specific feature of such projects in permafrost regions is the formulation of a quantitative long-term prognosis of geocryological changes due to the pipeline construction and maintenance that affirms technological design, construction techniques, and transport regulations for gas and oil. This prognosis should not be passive, but must provide for future changes to the system.

Engineering and geological studies for pipeline construction in permafrost regions are the basis for design, and require, apart from routine operations, intensive geocryological research along the entire pipeline route, using the full range of photography and geophysical methods.

PIPELINE CONSTRUCTION IN NORTHERN REGIONS

The length of large-diameter (1220-1420 mm) gas and oil pipelines with compressors and pumping stations is considerable. They are laid in permafrost regions, wet and boggy territories with a large number of rivers and streams and their floodplains, without roads, and under unfavorable climatic conditions. Their construction is a complex scientific and engineering problem that must be solved with due regard for the most recent experience in preliminary studies, design, implementation, and maintenance of similar constructions both in the Soviet Union and abroad.

Most gas pipelines from northern deposits to the central regions have to cross permafrost (a complex, multicomponent ecosystem) and boggy and wet areas. In many cases, the permafrost may be classified as unsteady, and there are sites where violation of its temperature and humidity modes may turn it into a liquified mass. There may be an increase in or the appearance of landslides or thermokarst phenomena at these sites, as well as washouts, heavings, and the formation of permafrost belts, depending on the changes introduced into the 'natural life' of the region in the course of pipeline construction and maintenance. Those phenomena must be foreseen in the prognostic studies and must be taken into account in engineering design, construction management, and maintenance requirements.

The route may also have sites with relatively satisfactory characteristics from the viewpoint of the requirements for laying pipelines. Along the route, considerable diversity of geocryological conditions, landscape relief, water content, and other factors necessitate the use of different systems of pipeline construction at different sites, based on the requirement for high reliability at minimal cost and labor. Any solution should be justified, above all, from the point of view of equal reliability along the entire pipeline system, and both cost-effectiveness and technological reliability.

There are two principles to be applied in the construction of pipelines on permafrost. The first principle is maintaining the frozen state of foundations during construction and maintenance. The second allows thawing of the permafrost during construction and maintenance. Accordingly, underground, surface, and aboveground systems for laying pipelines are depending on the conditions along the route and the product transport method. Laying pipelines in complex geocryological conditions should take into account the possibility that the restraining capability of the ground may change, destabilization, changes in ground temperature during construction and maintenance, and the processes and phenomena in the ground surrounding the pipeline, as well as the dynamic wind load on aboveground pipelines. These and other features required principally new engineering designs of pipelines for various construction conditions.

The Soviet Union has constructed and is maintaining pipelines in the extreme North and in permafrost regions. We have accumulated considerable experience based on a large volume of scientific research and prototype construction of a number of complexes, which allows for the reliable design and construction of large-diameter (1220-1420 mm) pipelines.

At present, about 600 km of gas pipelines 529 mm in diameter have been built and are maintained in the permafrost region near Yakutsk, out of which 250 km is aboveground. More than 600 km of aboveground 720 mm gas pipeline and 300 km of lesser diameter, and 40 km of underground 429-529 mm diameter gas pipelines have been built in the floodplains of the rivers near the city of Norilsk. The fourth leg of this gas pipeline system is currently being designed and built. Gas pipelines 1220-1420 mm in diameter were built and are maintained on the Medvezhie-Nadim-Punga, Urengoi-Chelyabinsk, and Urengoi-Nadim routes. There is a total of more than 12,000 km of large-diameter pipelines in the complex engineering and geocryological conditions of northwest Siberia.

The eleventh five-year plan (1981-1985) includes construction of six pipeline mains from the Urengoi gas deposit, a total of more than 22,000 km. A 1220-mm-diameter pipeline 1285 km long was built in Alaska. Of this length, 640 km are aboveground on steel supports and 630 km are underground. US and Canadian experts have designed an underground large-diameter gas pipeline with refrigerated gas especially for laying in permafrost.

To substantiate engineering designs for the northern parts of Canada and for Alaska, experimental stations and testing grounds were built with pipeline legs of 1220 mm pipes, and a large complex of studies has been conducted concerning pipeline/environment interaction, pipe resistance to cold, techniques for pipeline construction and maintenance, as well as the development of construction technology.

Research and analysis of the results of these studies, as well as the design, construction, and maintenance of pipeline systems in the northern parts of Canada and in Alaska show that although these regions are climatically similar to the northern regions of west Siberia, many differences have been observed in geographical, geological, geocryological, and hydrogeological conditions. The US and Canadian engineering designs are highly labor-intensive and costly.

Traditional methods cannot be used to lay pipelines in boggy regions and in perennially frozen ground. Perennially frozen ground may change its characteristics in the area of construction due to disturbance of the thermal and hydrogeological regimes. When the frozen ground thaws, there may be considerable subsidence, formation of thermokarst sinks, solifluction phenomena, landslides, cave-ins, mud flows, and avalanches. Most such regions are characterized by ground heaving when the seasonally thawed layer is freezing.

The engineering methods, technology, and maintenance techniques have an essential bearing on the stability and durability of the pipeline and other constructions located in its vicinity. In this connection, the prognosis of the interaction of the pipeline with the permafrost and problems of geocryological control are of great significance in solving problems of pipeline design, selection of pipeline structures, construction technology, and maintenance techniques.

ENGINEERING DESIGNS APPLIED IN PIPELINE CONSTRUCTION IN THE SOVIET UNION

There are no precedents for constructing a long (3000 km) pipeline 1420 mm in diameter during a single construction season under the complex engineering and geological conditions found in western Siberia. Unlike Alaska and Canada, western Siberia lacks the large-aggregate and gravel ground required for road construction, ground replacement in pipeline trenches, and banking under surface constructions, as well as for producing construction structures. It also lacks a network of permanent roads for freight hauling and temporary construction roads.

In the USSR, the study of the problems of pipeline construction in northern regions involved a large number of scientific research and industrial organizations. Many years of research have produced technical solutions and standardized documents allowing construction of large-capacity pipeline systems in those regions.

In the Scientific Research Institute on Construction of Pipeline Mains (VNIIST), studies on pipeline construction in regions with perennially frozen ground were started in 1959. Experimental studies on the construction and operational parts of the Messoyakha-Norilsk gas pipeline have been carried out since 1968.

Research is currently underway at the 8-km looping experimental site in the Ukhta area, where 1220 mm pipes are buried underground, banked up on the surface, and aboveground on roller and sliding supports. There are also special testing grounds.

Concurrent with the studies on operational and experimental parts of gas pipelines, research is being carried out at test rigs, with special devices, on models using analog and digital computers, and there is a wide spectrum of theoretical studies.

The primary directions of research are engineering designs and calculation methods, prognosis of thermal and mechanical interaction of pipelines with the environment, the technology of gas transport, construction technology, and machinery design.

Experimental and theoretical studies have investigated the aboveground laying of 529 and 720 mm gas pipelines on suspended and sliding supports and of 1220 mm pipes on roller and sliding supports, underground and surface laying of pipelines of various diameters, and surface pipelines without embankments. Also designed and studied are diverse support structures (wooden, reinforced concrete, metal, slab, pile, thermopiles, with prestressed tension members, suspending, sliding, roller, etc.), as well as the loads on the supports, coefficients of pipeline displacement resistance, stress resistance of supports, and ground foundations.

The dynamic stability of aboveground pipelines under various effects including wind, stress resistance, and resilience of curved elements and pipelines of different contours with bent and welded laterals that have small and great curvature radii were studied, as well as stability and strength for all types of pipeline-laying, restraining and ballasting capacity of grounds, and reinforced concrete weights.

The design and construction of various devices have been developed and studied, including devices for preventing pipeline escillations for anchoring and ballasting underground pipelines in areas of frost heave and subsidence, and new types of anticorrosive insulation for pipelines, improved methods for construction of supports, anchoring stabilizers, and other constructions for permafrost regions.

The thermal and mechanical interaction of pipelines with the surrounding ground has been studied, including temperature fields in the ground, aureoles of freezing and thawing, gas temperature changes along pipelines of various engineering designs, the dynamics of thermal and mechanical interaction processes under various conditions of long-term pipeline maintenance, the processes of foundation heaving and settling, as well as the stresses to the pipelines brought about by these processes.

Geocryological processes along the routes of the pipelines in the North, such as thermoerosion, thermokarst, solifluction, ice crust formation, have also been studied.

Many years of complex studies resulted in the development of new systems for laying pipelines in the regions of the extreme North and permafrost grounds, including an aboveground system with slightly curved sites, surface and underground systems with partial compensation of longitudinal strains and special anchoring devices for heaving and wet sites, an underground system on supports for floodplain sites with permafrost, an open ground system without banking-up with compensating portions, laying on float supports, and their various modifications applied to concrete conditions.

It should be noted that aboveground and surface systems of laying with slightly curved aboveground sites have been comprehensively studied and checked in the course of industrial operation. The open aboveground system of laying is the most economical of all systems applied to date.

The aboveground laying without embankments may be applied widely in permafrost regions, as well as over bogs where placing the pipeline below the surface leads to appreciable construction and technological difficulties.

Methods of analysis were worked out for various systems of pipeline-laying, as well as recommendations for engineering design. In particular, the construction of roller supports that convey minimal loads to the foundations was studied, making it possible to use light foundations for the supports, such as slabs or piles, and avoiding transfer of dynamic loads to the pipeline. Constructions were designed for thermopiles and high supports with tension members to pass over narrow gullies and ravines.

To prevent oscillations in aboveground pipelines, special devices have been designed and studied that allow greater distance between supports, thereby lowering the cost of the whole construction.

In a number of cases it has been necessary to lay pipelines underground in unstable heaving ground (e.g., in wide river floodplains where ice drift is possible. In these cases the pipeline is fixed by means of patented anchoring devices (USA p.3903704 and Canadian p.988927). Also developed and patented was a construction of reinforced concrete ballasts that increase the restraint on buried pipelines that has been quite cost-effective and allows the development of new systems of pipeline laying with partial compensation of strains (USA p. 4166710 and Canadian p. 1076373).

Methods have been developed to analyze the thermal and mechanical interaction of pipelines with the surrounding ground, to predict and control the geocryological processes brought about by pipeline construction and maintenance, as well as methods of technological design for large-diameter gas pipelines of various constructions and in diverse climatic conditions.

These studies have led to the solution of problems of pipeline construction in the extreme North. However, they must be continued at the research stations with experimental sites of 1220-1420 mm pipes and on operational large-diameter gas and oil pipelines under various natural and climatic conditions, engineering designs, and technology of gas, oil, and oil products transport. These studies are necessary since advances in other fields of science and technology make it possible to steadily improve designs in the field of pipeline construction, increasing their reliability and lowering costs.

All the developed methods of analysis and design were thoroughly studied under laboratory conditions, as well as on operating pipelines and are now incorporated in practical pipeline design and construction in the USSR.

INVESTIGATIONS AND IMPLICATIONS OF SUBSURFACE CONDITIONS BENEATH THE TRANS ALASKA PIPELINE IN ATIGUN PASS

J. Michael Stanley[1] and John E. Cronin[2]

[1]Alyeska Pipeline Service Co., Anchorage, Alaska USA
[2]Shannon & Wilson, Inc., Fairbanks, Alaska USA

In June 1979, a leak was detected in the Trans Alaska Pipeline in Atigun Pass, located in the east central Brooks Range. A major subsurface investigation was performed for the purpose of providing detailed subsurface information as input to the design of remedial measures to repair the leak and to detect any other problem areas. Another area of pipe settlement was found on the south side of the pass in an insulated section of the pipeline. Subsurface exploration revealed thawing of the foundation materials, apparently as a result of subsurface water flow. Large-diameter diamond coring with refrigerated fluids was found to be the only exploration method capable of providing samples of adequate quality for thaw strain testing. Thaw strain test data, in conjunction with information on subsurface conditions obtained by refrigerated coring, allowed estimation of the settlement that might occur if the pipeline foundation materials were to continue to thaw. Data from test holes and other investigations during 1980 and 1981 have allowed determination of the need for additional repair work and a rational prediction of future pipeline performance.

INTRODUCTION

The completion of the Trans Alaska Pipeline in 1977 was a major achievement in arctic engineering and construction. Almost 1,290 km in length, the pipeline delivers crude oil from the oil fields of Prudhoe Bay to a tanker loading facility at the Port of Valdez (see Figure 1). Luscher (1977) and Liguori et al. (1978) provide a summary of the design and construction of the pipeline.

Engineering problems associated with permafrost were a major concern during pipeline design and construction. The 122 cm diameter pipeline was generally buried only in areas where permafrost was absent or was judged to be thaw-stable. In areas of non-thaw-stable permafrost, the pipeline was elevated on vertical support members. Because of the greater difficulty and expense involved in aboveground construction, initial route selection attempted to optimize the amount of belowground pipe.

At a few locations, subsurface conditions required special construction techniques. Atigun Pass was one of these locations.

MONITORING THE PERFORMANCE OF THE PIPELINE

Where burial of the pipeline prevents direct observation, data collected using a number of techniques are available to Alyeska engineers. Level surveys of monitoring rods attached to the top of the pipe, when compared with as-built surveys, allow determination of pipe settlement. Surveys of the pipe interior with a corrosion/deformation pig can detect pipe shape changes, which may indicate unsatisfactory pipe movement. (A pig is a device that fits inside the pipeline and is moved through the line by the force of the fluid flowing through the line.) Aerial reconnaissance of the pipeline right-of-way can reveal ground surface disturbances related to pipe settlement. In addition, where thermistor strings have been installed, future thermal performance may be predicted using computer techniques.

In the latter situation, where the thermal regime beneath the pipe is changing, accurate information on subsurface soil and rock conditions is critical to determination of possible future performance of the pipeline. Qualitative information from test hole logs is essential, and quantitative information from laboratory testing of samples is also an aid.

HISTORY OF THE PIPELINE IN ATIGUN PASS

Background

Atigun Pass, with a crest elevation of nearly 1,450 m, is the highest point on the Trans Alaska Pipeline. Located in the east central Brooks Range 265 km south of Prudhoe Bay and 160 km north of the Arctic Circle (see Figure 1), the pass is underlain by permafrost to great depth.

Construction of the pipeline in the pass took place in late 1976. Preconstruction subsurface investigations revealed large amounts of segregated ice, both in unconsolidated colluvial and glacial deposits and in weathered and fractured zones in the underlying bedrock.

Were it not for the hazards of avalanches, slush flows, and rock slides, the pipeline would have been constructed aboveground through these two ice-rich areas on the north and south sides of the pass. Instead the pipe was buried in an insulated box designed to thermally isolate the warm pipeline from the underlying ice-rich soil and rock. This box surrounded the pipe with over 50 cm of rigid foam insulation, designed to limit thawing to 0.3 m beneath the pipeline. About 0.5 km of pipe on the north side of Atigun Pass and 1.3 km on the south side were buried using this insulated box design (Figure 2).

Geology and Subsurface Conditions

Atigun Pass is typified by unconsolidated deposits of various origins overlying a bedrock sequence of quartzites and argillites. The unconsolidated deposits consist of varying thicknesses of talus, rock glacier, avalanche, and landslide deposits in addition to glacial till. The colluvial material is typically quite angular and contains material of all fractions from clay size to boulders, the coarser fractions predominating. These deposits are generally perennially frozen in their undisturbed state and inclusions of clear ice are common.

Pipeline Performance

In June 1979, a leak developed in the pipeline on the north side of Atigun Pass. The leak was found to be due to buckling of the pipe at a location about 40 m outside the southern end of the north insulated box (Figure 3). An extensive subsurface investigation carried out at that time to determine remedial measures revealed that the buckling had been caused by thawing of a limited zone of ice in the colluvial soils and fractured bedrock.

Since 1979, detailed monitoring of pipeline performance, in conjunction with gathering of additional subsurface information, has allowed Alyeska engineers to assess the condition of the pipe and predict its future performance. While conditions have necessitated extensive remedial measures in some instances, no further leaks have occurred. The largest remedial effort was required at the northern end of the south insulated box; it is described in detail in a later section.

SUBSURFACE INVESTIGATIONS IN ATIGUN PASS

Preconstruction

During the design and construction phases of the pipeline project, a total of about 200 exploratory test holes were drilled in Atigun Pass for purposes of route selection, design, and design verification. Almost all of these were drilled using some version of air rotary techniques, in which the test hole is logged primarily on the basis of drilling action and classification of cuttings ejected with the air. None of these exploratory holes indicated that potential problems existed in the areas later experiencing settlement nor did any of them indicate a need for further exploration.

South Repair Area-1979

Following the investigation of the leak area on the north side of Atigun Pass in June 1979, additional test holes were drilled in other areas, including 13 test holes near the northern end of the south insulated box. These 13 test holes were all drilled by logging cuttings from an air track. After drilling was completed, the holes were equipped with thermistor strings to measure ground temperatures.

This investigation revealed subsurface conditions as depicted in cross-section in Figure 4, with thaw extending much deeper than design criteria allowed. In addition, abundant groundwater was encountered in the test holes, which penetrated thawed materials beneath the south insulated box. Groundwater was also encountered in holes drilled adjacent to the uninsulated pipe above the box.

South Repair Area-1980

To provide detailed information for analysis and remedial action as necessary, Alyeska authorized a subsurface exploration program in the spring of 1980. Conventional auger drilling and drive sampling techniques were not capable of penetrating the bouldery overburden and indurated bedrock in the pass. Differentiation between bouldery overburden and the abundant soil-filled fractures in the bedrock was difficult with air rotary drilling techniques. For these reasons, diamond core drilling was used in the 1980 drilling program.

Almost 310 m of exploratory test holes were drilled in the repair area in 1980. Half of the holes were drilled using diamond core drilling techniques. Half of these utilized refrigerated drilling fluids to maintain the original thermal state of the core samples.

The diamond coring employed a triple-tube, PQ-size, wireline core barrel. The relatively large diameter PQ core (82.6 mm core diameter) was chosen to minimize the effect of thermal disturbance and fluid erosion on the core. The wireline system allowed relatively trouble-free advancing of the test hole through fractured rock or other caving materials, while the triple-tube core barrel design facilitated recovery of highly fractured or colluvial material in a relatively undisturbed state.

For refrigerated coring, a solution of propylene glycol and water was cooled to temperatures in the range of -2° to -6°C by an electrically-powered refrigeration compressor coupled to a heat exchanger system. In general, frozen core recovery was excellent, including recovery of ice lenses in the colluvial soils (see Figure 5). Some erosion of ice lenses in hard bedrock was found to be unavoidable.

Since loss of freezing-point-depressing drilling fluids in the soils adjacent to the pipeline might prevent later refreezing, coring in these areas used fresh water at about 15°C. Ice and some fines were washed away in this process, but core recovery was sufficient to differentiate between colluvial soils and bedrock.

When the primary purpose of a test hole was the installation of temperature or groundwater measuring instrumentation, air rotary drilling techniques were used. Both conventional and casing-advancing, down-hole-hammer drilling allowed rapid penetration of both frozen colluvium and bedrock. Limited geotechnical information was obtained by logging drill penetration and cuttings from these test holes.

The use of refrigerated and fresh water diamond coring techniques greatly improved the state of knowledge concerning subsurface conditions in the repair area. The thickness of colluvial soils was found to be somewhat greater than had been revealed by the 1979 air drilling program. In addition, refrigerated coring provided the first accurate assessment of the extent of ice in both the colluvium and the bedrock.

Ground temperature and groundwater monitoring instrumentation was installed in most test holes, depending on whether the ground was frozen or thawed. Thermistor strings were installed in 2.5 cm (nominal) plastic pipe filled with a propylene glycol/water mix. Slotted plastic pipe of the same size served as standpipe piezometers for observation of groundwater levels.

Laboratory testing of samples from the pass consisted primarily of standard index property tests. Because of the wide variation of grain sizes and the abundance of coarse gravels and cobbles, it was difficult to predict thaw settlement using parameters such as moisture content, density, and grain size. For this reason, specially fabricated cells that matched the core diameter were used to test representative samples for thaw strain (see Figure 6). After trimming with a diamond saw, the samples were subjected to a uniaxial load provided by dead weight, and strain upon thawing was measured with a dial indicator. Thaw strains as high as 55% were measured.

DATA ANALYSIS AND DETERMINATION OF REMEDIAL MEASURES

Abundant data were available from the test hole program, including test hole logs, thermal data from thermistor strings placed in the test holes, and water levels measured in piezometers. On the basis of these data, it was clear that thawing had exceeded design criteria for a significant distance along the insulated box. The presence of still-frozen, potentially ice-rich colluvial deposits below the existing thaw depth caused concern for the future stability of the pipeline.

Investigation of the area indicated that runoff water from snowmelt and rainfall was entering the backfilled pipe ditch near the top of the pass. Flowing through the ditch along the uninsulated pipe, it was being heated by the warm (about 55°C) oil. This heated groundwater then apparently flowed under and around the insulated box and appeared responsible for the observed thawing.

Alyeska engineers developed a repair program that was implemented in the summer of 1980. The program consisted of surface drainage to minimize water influx to the area and grouting to cut off the water flow beneath the insulated box followed by mechanical refrigeration and installation of self-refrigerating heat pipes to refreeze the ground. This program is described by Thomas et al. (1982). Figures 3 and 4 show the area of the pass where this work was undertaken, and Figure 7 shows this work being implemented in the summer of 1980.

For this repair program it was necessary to analyze a variety of data including pipe and soil movement monitoring, ground temperature measurements, groundwater levels, and groundwater movement information. By analyzing the data from this array of instrumentation, it was possible to (1) develop a repair program, (2) monitor the effectiveness of the program as the various components were implemented and judge their interaction, and (3) monitor the effectiveness of the repairs in subsequent years.

Pipe and Soil Movement

Pipe and soil movement data were obtained from several sources: (1) the elevation of the top of pipe exposed for repairs or monitoring rod installation, (2) the elevation of the top of a monitoring rod of known length attached to the buried pipe, and (3) inclinometer/extensometers founded in bedrock, some of which were also attached to the side of the pipe. Pipe movement data (the difference in elevation between the as-built top of pipe elevation and the present top of pipe elevation) determined from monitoring rods were used in the initial determination that there was a problem that required remedial action. Later, data collected from the rods showed that there was continuing movement and allowed estimation of the magnitude of the movement and the rate. After installation of the inclinometer/extensometer instrumentation, it was possible to very precisely measure pipe and soil movement in three dimensions. This allowed determination of movement vectors and enabled the engineers to analyze the stability of the pipe and soil independently.

Surface Water

The influence of surface water, as noted above, was evident from observations that water disappeared into the ground near the top of the pass except in times of very high runoff. In addition, rainfall events in the pass caused increases in the groundwater elevations observed in the standpipe piezometers in the repair area within a few hours of the onset of the rain. The amount of surface water that could enter the pipe ditch was reduced by channeling the surface water into control ditches.

Groundwater

Analysis of the data from the standpipe piezometers installed in the test holes and the pumping tests performed on test wells installed in the project area led to the conclusions that (1) the groundwater flow along the pipe uphill (north) of the repair area was causing the ground thawing and subsequent pipe settlement, (2) groundwater velocities through the area were too high for passive ground freezing to be effective, and (3) active ground freezing efforts would be severely effected by the flow through the site. On the basis of these conclusions, it was determined that it would be necessary to grout the area prior to any refreezing efforts.

A grouting program was designed with three components. An upper grout curtain would cut off groundwater flow above the insulated box. Grouting of the soils around the box would reduce porosity and allow artificial freezeback. Finally, a lower grout curtain would confine slurry grout within the limits of the repair area. The grout curtain locations are shown in Figures 3 and 4.

During the grouting and ground freezing programs, extensive monitoring of the piezometers and test wells was performed. The data were used to determine when sufficient grout had been injected to immobilize the groundwater or to reduce the velocities sufficiently to allow ground freezing to be effective. Later, groundwater data were used in the determination of the long-term effectiveness of the grouting and ground freezing programs.

Ground Temperature

Initially, data from the thermistor strings installed in the test holes was used in the determination of the cause of the pipe settlement in this area. Temperature measurements north of the insulated box section indicated that ground water temperatures were as high as 21°C near the pipeline. This led to the conclusion that flowing groundwater was the agent of thaw and subsequent settlement of the pipeline immediately to the south. Secondly, thermal data were used in determining the limits of thawed ground that could be grouted. Thermal data revealed that it was not feasible to install the upper grout curtain because of the extreme size of the thaw bulb at that location (greater than 15 m wide by 20 m deep). Third, the thermal data were used during the ground freezing program to determine the effects of the freezing efforts and to determine when the shutdown criteria for the freezing operation had been met. Finally, long-term thermal data are being gathered to provide information on the performance of the repairs.

EFFECTIVENESS OF REMEDIAL WORK

General

Analysis of data collected during 1981 and 1982 indicates that the repairs and remedial measures effected during 1980 and enhanced by additional work during 1981 (primarily surface drainage control) stabilized the northern end of the south insulated box. The maximum depth of thaw in the repair area was above the bottom of the insulated box during 1981 and 1982. In 1980 it had extended to depths greater than 10 m below the box. No significant degradation of the artificially refrozen zone occurred during the summer months, although it had been anticipated that the northernmost portion of the site would experience some thawing. Contingency plans were prepared for installation of a series of dewatering wells across the valley north of the insulated portion of the pipeline and for installation of a grout curtain at the northern end of the insulated box. The wells were installed during 1981 but were only test-pumped. Because of the excellent performance of the repairs accomplished during 1980 and 1981, there was no need to operate the wells after the test pumping was completed. The wells were not operated at all during 1982 and the upper grout curtain was not installed.

Verification by Subsurface Investigation

Subsurface conditions adjacent to the south insulated box south of the 1980 repair area were investigated by a drilling program in the summer of 1981 consisting of 270 m of diamond coring. Over half this amount was cored using refrigerated fluids. One purpose of this program was to assess the effectiveness of the remedial work in reducing the flow of heated groundwater (and consequent thaw settlement) beneath the insulated box. The investigation showed that groundwater flow was reduced, improving conditions downstream of the repair area.

CONCLUSION

Experience with the Trans Alaska Pipeline in Atigun Pass has proven the necessity for reliable subsurface geotechnical information in predicting the performance of structures in permafrost terrain. In coarse-grained colluvial deposits and frozen bedrock such as are found in the pass, diamond core drilling with refrigerated fluids was found to be the only practical method to obtain both this information and samples of suitable quality for laboratory thaw strain testing. Subsurface soils information in conjunction with data from instrumentation has allowed assessment of conditions, development of a repair program, and prediction of future performance of the pipeline.

ACKNOWLEDGMENTS

The permission of the Alyeska Pipeline Service Company to publish this information is gratefully acknowledged. Data have been drawn from reports on the 1980 and 1981 investigations by Shannon & Wilson, Inc., on the 1979 investigations by Quadra Engineering, and on the 1980 remedial work by Woodward-Clyde Consultants.

REFERENCES

Liguori, A., Maple, J. A., and Heuer, C. E., 1978, The construction of the Alyeska pipeline, in Proceedings of the Third International Conference on Permafrost, v. 2: Ottawa, National Research Council of Canada.

Luscher, U., 1977, Geotechnical aspects of the Trans Alaska Pipeline System, in Proceedings of the Ninth International Conference on Soil Mechanics and Foundation Engineering, Case History Volume: Japan, Japanese Society of Soil Mechanics and Foundation Engineering.

Thomas, H. P., Johnson, E. R., Stanley, J. M., Shuster, J. A., and Pearson, S. W., 1982, Pipeline stabilization project at Atigun Pass, in Proceedings of the Third International Symposium on Ground Freezing: Hanover, New Hampshire, U. S. Army Cold Regions Research and Engineering Laboratory.

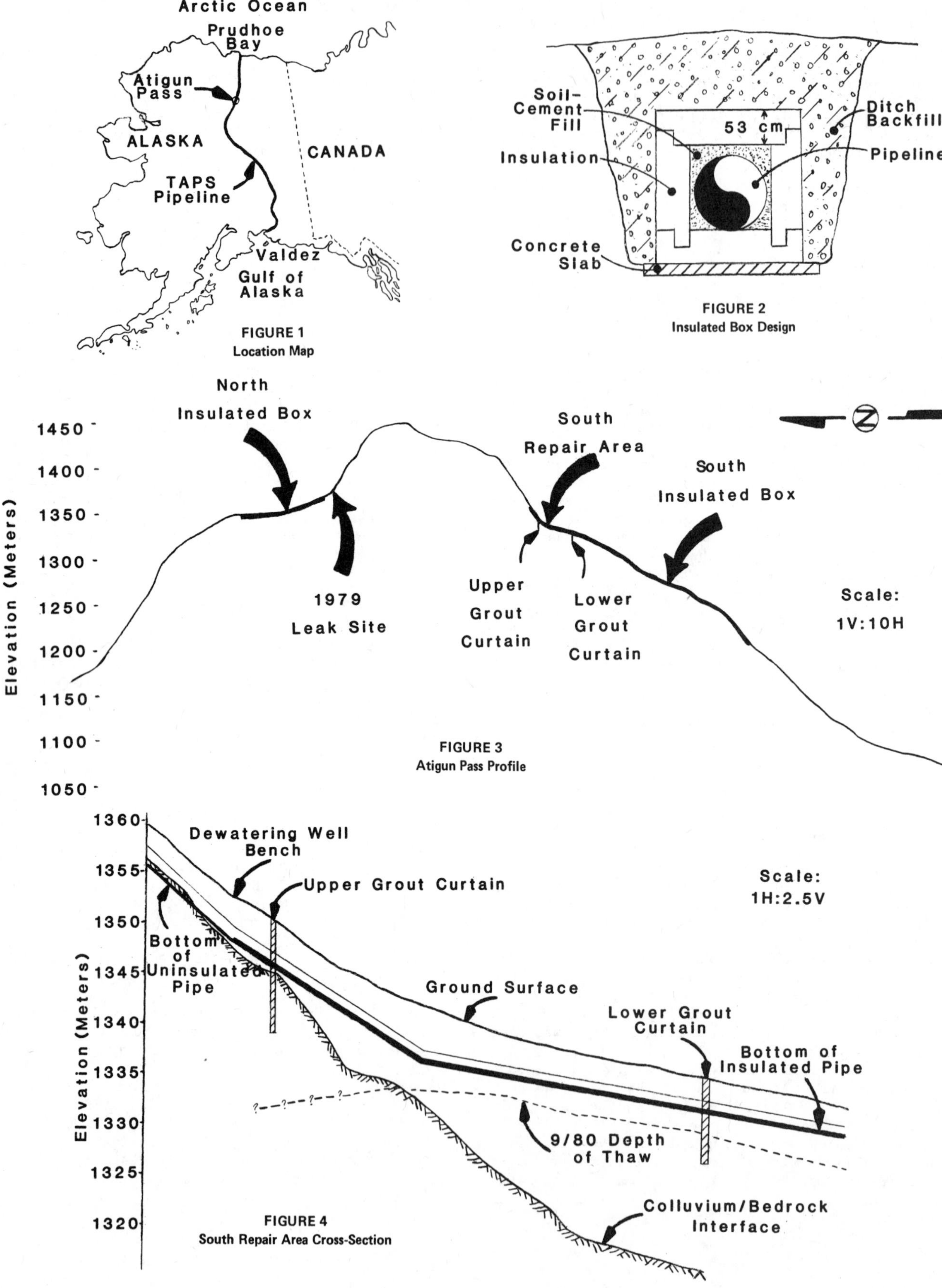

FIGURE 1
Location Map

FIGURE 2
Insulated Box Design

FIGURE 3
Atigun Pass Profile

FIGURE 4
South Repair Area Cross-Section

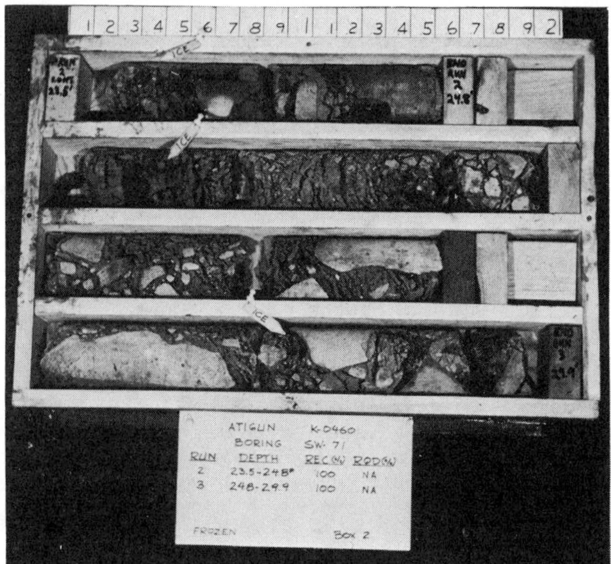

FIGURE 5 Frozen Core

FIGURE 6 Thaw Strain Test Cell

FIGURE 7 South Repair Area (Looking South)

MOISTURE AND TEMPERATURE CHANGES IN THE ACTIVE LAYER OF ARCTIC ALASKA

M. G. Stoner, F. C. Ugolini and D. J. Marrett
College of Forest Resources, University of Washington
Seattle, Washington 98195 USA

Occasional deep percolation of meteoric water in the foothills of the Brooks Range is critical to the genesis and physical behavior of well-drained Spodosols. In situ measurements of soil water tension and temperature were made during three consecutive summers at two sites in northern Alaska, one in the boreal forest and one in the arctic tundra. Soil climate responses to changes in weather conditions are explained with additional data on soil physical properties measured in the laboratory. Important factors controlling depth of wetting front penetration include: precipitation, vegetation cover, soil texture and antecedent soil moisture content. Coarse textured soils in the arctic tundra show leaching of the profile with only 5-7 mm of precipitation. Medium textured soils in the boreal forest require in excess of 25 mm of rain to overcome high evapotranspirative demands. Stable soil temperature profiles develop by July, exhibiting a sharp decrease in temperature with depth. These conditions are abruptly altered by a significant pulse of heat transmitted through the active layer during summer storms. This study shows that deep percolation is an important mechanism of transport for soil water, solutes, and heat at well-drained sites in permafrost affected regions.

INTRODUCTION

Accurate interpretation of observed changes in the active layer is especially important in permafrost affected regions. Many workers have studied the environmental factors controlling active layer depth and behavior in relation to permafrost stability and have found that sparse vegetation, convex landscape topography, and coarse soil texture favor deep thaw penetration (Drew et al. 1958, Brown 1973, Gold and Lachenbruch 1973, Linell and Tedrow 1981). Soil development at these deeply thawed northern sites is also affected by these site-specific conditions. In terms of latitudinal distribution, Spodosols are found in well-drained areas of the boreal forest, and Inceptisols and Entisols tend to dominate upland positions in the arctic tundra (Tedrow et al. 1958). However, repeated reports of Spodosols well beyond present treeline in Alaska and other arctic regions (Brown and Tedrow 1964, Ugolini 1966, James 1970, Moore 1974, Tedrow 1977, Everett 1979) have raised questions about the relations between soil processes and environmental conditions at high latitudes. A leaching water regimen, needed for podzolization, seems unlikely in the excessively harsh and limiting arctic climate. Brown and Tedrow (1964) suggested that Podzols (Spodosols) in northern Alaska may be relict features of a warmer and wetter paleoenvironment. More recently, it has been postulated that podzolization has been a continuous process in the boreal forest, but has occurred in the past and/or sporadically in limited areas of the arctic tundra (Ugolini et al. 1981a). This hypothesis is based in part on the climatic transition across the Brooks Range from the relatively warm and wet boreal forest to the relatively cold and dry arctic tundra (Berg et al. 1978). However, the effect of this climatic and biotic transition on active layer processes and soil development has not been adequately addressed.

Evaluation of the physical behavior of these apparently anomalous arctic Spodosols in relation to current environmental conditions is reported in this paper. The study required instrumentation of two sites, one in the boreal forest, where Spodosols normally occur, and one in the arctic tundra. Data collection included daily monitoring of soil moisture and temperature fluctuations in response to changing weather conditions during the summer. Additional data on soil physical properties determined in the laboratory are used to explain these observed soil climate variations.

SITE DESCRIPTION

Field investigations were conducted at two locations in the Brooks Range of northern Alaska which have been previously studied and extensively characterized (Brown 1966, Goldstein 1981, Ugolini et al. 1981a, Ugolini et al. 1981b). The boreal forest site is at 600 m elevation near Walker Lake (67°09'N, 154°22'W). It is near latitudinal treeline (*Picea glauca*, *P. mariana*) with a ground cover of ericaceous species and lichens. The arctic tundra site is at 610 m elevation in the Okpilak River valley (69°26'N, 144°02'W) and is 100 km north of present treeline with a lichen mixed-heath vegetation assemblage. Parent material at both sites is glacial drift approximately 13,000 years old (Hamilton and Porter 1975, Sable 1977, Hamilton 1979). The drift consists of medium textured mica schist in the boreal forest and extremely coarse textured granitic clasts in the arctic tundra. Remarkable soil morphologic similarities between sites are evident despite major lithologic, biotic and climatic differences (Ugolini et al. 1981a).

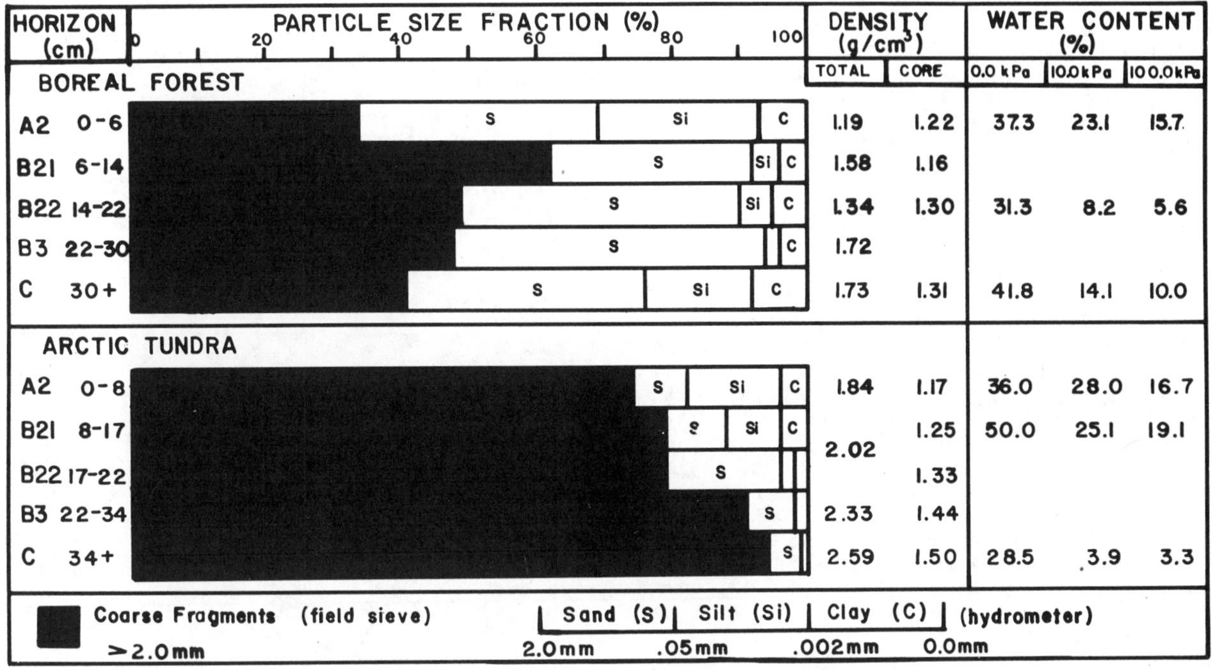

FIGURE 1 Particle size fraction, bulk density by large irregular hole (total) and small undisturbed sample (core), and moisture desorption at boreal forest and arctic tundra sites in Alaska.

METHODS

Soil climate responses to changes in weather conditions were measured daily at 1000 hrs for several weeks at each site during the summers of 1980-1982. Active layer depth was in excess of 1 m at both sites during the month of July for each of the study years. Weather data included air max/min temperatures, precipitation, relative humidity, and cloud cover. Soil temperatures were measured to 0.1°C at seven depths to a maximum of 1 m with thermocouples. Soil moisture tension was measured to 0.1 cm Hg (1.5 cm H_2O) at three depths corresponding to the A2, B2, and B3 horizons by mercury manometer tensiometers. Three soil moisture and temperature profiles were instrumented at each site.

Soil physical properties were measured by genetic horizon in the field and on samples in the laboratory. Measurements taken on site included bulk density by large irregular hole and percent particles larger than 2.0 mm. Small undisturbed cores (5.0 cm dia. x 3.0 cm) were collected for field moisture content, ovendry bulk density, and moisture desorption. Large soil columns (11.5 cm dia. x 50.0 cm) were used for saturated hydraulic conductivity and ponded-head infiltration. The column from the boreal forest was collected intact in the field; the arctic tundra column was packed by genetic horizon to bulk density approximating that measured between large cobbles and stones. Particle size analysis was done by the hydrometer method. All laboratory experiments were run at room temperature.

RESULTS AND DISCUSSION

Soil Texture

This study has shown that climate, vegetation and soil texture are critical in controlling active layer processes. Texture, bulk density, and water holding capacity values through the profiles are summarized in Figure 1. Coarse fragments (greater than 2.0 mm) are much more prevalent in the arctic tundra soil, especially in the B3 and C horizons, but textural multi-layering is evident in both soils. Both soils have significant components of silt and clay in the A2 horizons which provide much of the similarity in their morphologies. The texture of the <2.0 mm fraction of each horizon is important in determining water holding capacity and flow rates. Although the A2 horizons retain substantial moisture at 10 kPa (23.1% in the forest and 28.0% in the tundra), large differences occur in the lower horizons. At the boreal forest the C horizon is finer than the overlying B horizons, but at the tundra the B3 and C are much coarser than the upper horizons. These textural discontinuities impede downward infiltration; however, at the tundra site the dramatic increase in pore size with depth is more significant under conditions of low soil tension (Baver et al. 1972).

Moisture Flow

The importance of texture is further illustrated by moisture flow studies of these layered soils.

The saturated conductivity and infiltration capacity of porous media are functions of several factors, but mainly pore size. In multilayered soils the conductivity is limited by the layer with the smallest pore size; in our soils this feature is found at the surface. It was observed that ponded-head infiltration into dry (<5% moisture) soil columns was slowed at the lower boundary of this surface layer. Measured hydraulic conductivities were almost identical for each column: 1.53×10^{-5} m-s^{-1} for the forest and 1.55×10^{-5} m-s^{-1} for the tundra. The curves for cumulative infiltration and wetting front penetration versus time are shown in Figure 2. Water moves rapidly into the surface layer after saturation of the organic mats, due to a combination of capillary tension and gravity. Flow characteristics are very similar for the upper 22 cm of soil for both sites. Initial wetting of the columns is rapid, then slows as the higher matric potential in the A2 must be overcome. Once sufficient pore pressure is attained, wetting proceeds into lower horizons. Deeper wetting in the tundra column is rapid as gravitational water cascades through the large pores in the B3 and C horizons. Only 2.55 cm of water is needed to wet the column to 47 cm. The boreal forest soil responds more slowly at depth, requiring 5.86 cm of water to wet the column, more than twice as much as in the tundra soil. Thus soil texture at these two sites is a critical factor controlling deep percolation of meteoric water through the profile.

Observed variation in soil climate in response to weather conditions can be explained more completely with this information on soil texture, water holding capacity, and infiltration characteristics. Representative data on air and soil conditions for the 1981 field season are graphed for each site in Figure 3. During dry periods at the arctic tundra, moisture losses are due to evapotranspiration from the A2 horizon and downward unsaturated flow from the C. Under these conditions the B2 horizon (16 cm) remains wet (50 cm H_2O) despite soil matric tension of 145 cm H_2O in the A2 (7cm) and as high as 330 cm H_2O in the B3 (30cm). Soil tension in the B2 is sufficient to prevent downward flow into the coarse B3. This textural arrest of the wetting front may be important to pedogenesis as heat and solutes associated with the moving water are conserved in the upper body of the soil. Percolation through these textural barriers does occur, however, as is evidenced after 3 days of light orographic precipitation (July 4-6) totaling only 0.9 cm of water. A subsequent rain of 1.0 cm on July 11 was then sufficient to saturate the soil to at least 30 cm and cause intense downward leaching. Thus a sequence of light rains and fogs, common during July, can lead to deep saturated flow through the profile. This moisture regimen was documented several times at the arctic tundra site during the study period.

At the boreal forest, soil tension can be as high as 200 cm H_2O at 8 cm depth while only 80 cm H_2O at 35 cm depth, indicating that hydraulic gradients may at times be upward in the active layer. This is due to the relative humidity, often as low as 40%, the high transpirative demands of the forest and the capillary continuity of the medium textured soil. Wet periods in the boreal forest are initiated in the summer by intense convectional precipitation events. For example, on July 27 the profile was saturated after 4.0 cm of rain, but drying of the surface is shown over the next several days despite interception of an additional 2.3 cm of rain. Frequent showers of smaller magnitude often result in partial wetting of the profile. Limited penetration of the wetting front may also have pedogenic consequences in the boreal forest as solutes are repeatedly leached through just the upper portion of the soil.

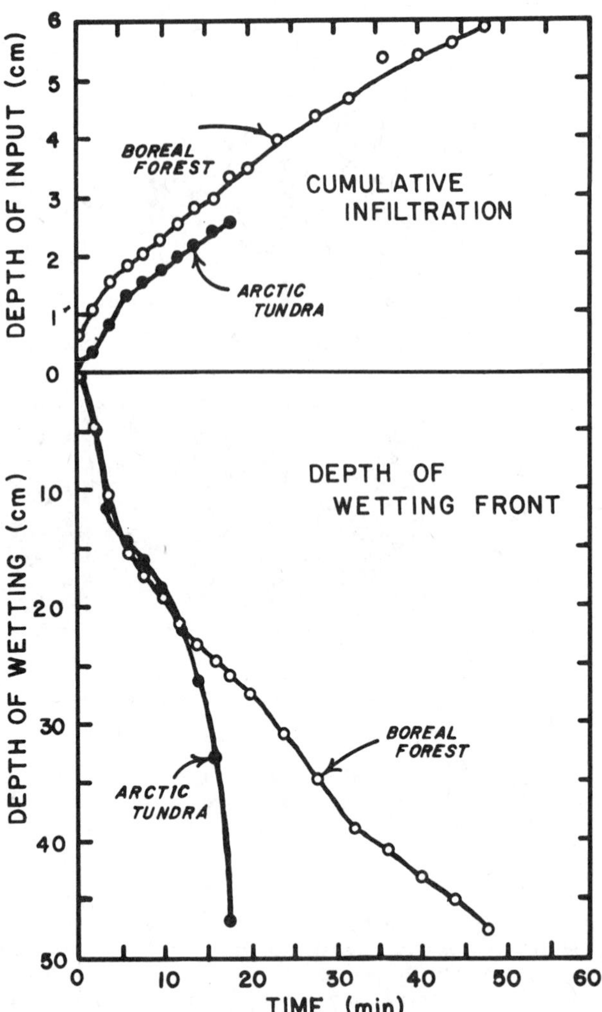

FIGURE 2 Cumulative infiltration and wetting front penetration on dry soil columns from the arctic tundra and boreal forest, Alaska.

Deep percolation through the active layer is clearly shown to be a normal hydrologic event in both of these northern Spodosols. Antecedent soil moisture content in the upper horizons is more critical in the arctic tundra where precipitation during the summer is much less than in the boreal forest. Coarse textured soils in barren convex positions of the tundra site allow repeated leaching of the profile. In the boreal

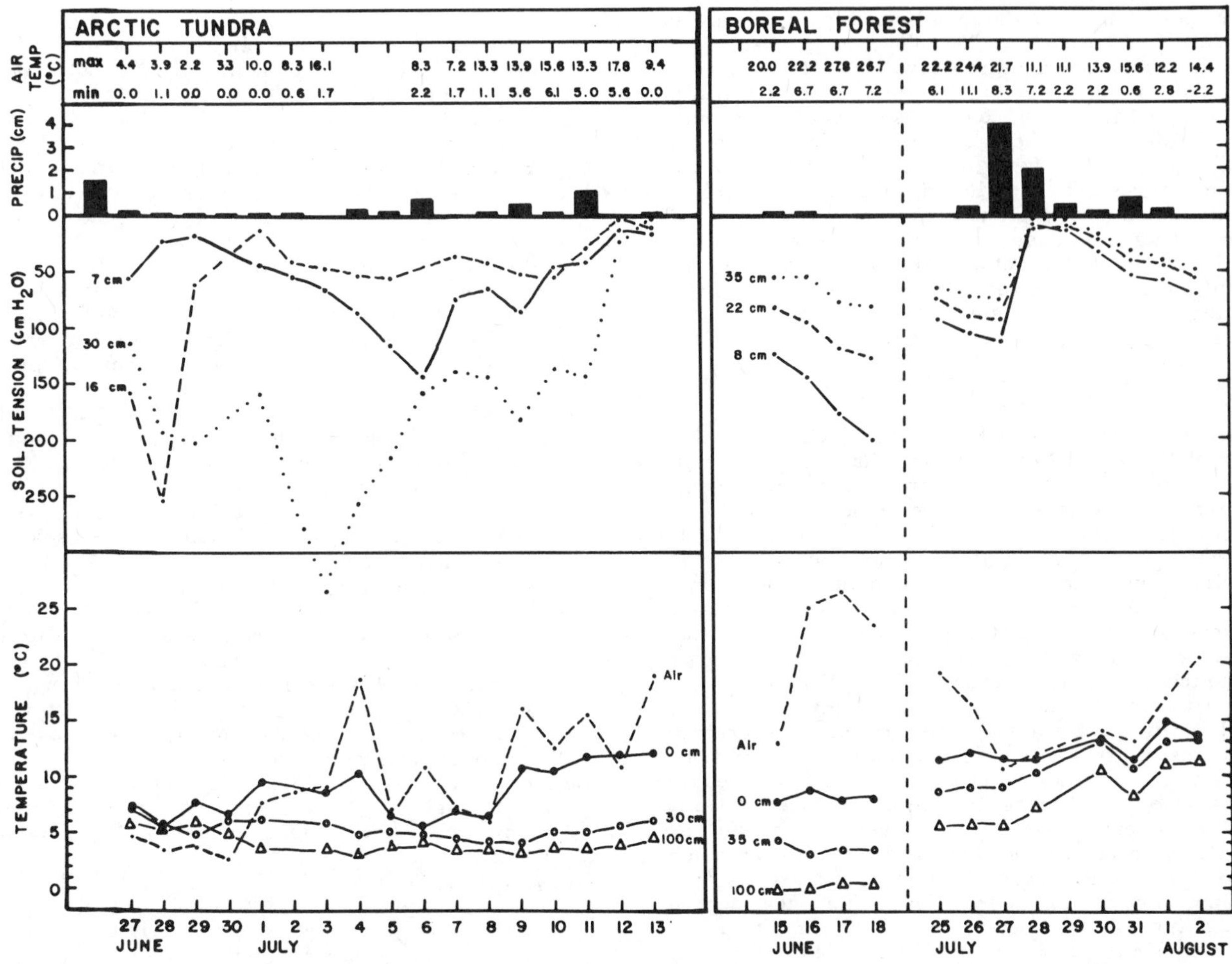

FIGURE 3 Soil moisture and temperature changes in response to weather conditions in northern Alaska. Measurements were taken daily at 1000 hrs during the summer of 1981.

forest, intense thunderstorms can regularly overcome vegetational demands and cause deep percolation even in these medium textured soils.

Soil Temperature

Soil temperatures are controlled by various factors including climate, organic layer insulation and thermal conductivity of the soil. Organic layer insulation is most effective at our sites during dry periods, particularly at the boreal forest where the surficial debris is thicker and continuous. Daily soil temperature fluctuations in the active layer are relatively small below the soil surface and a short-term steady state condition is attained by midsummer (Figure 3). These intervals of relatively uniform temperatures are abruptly broken by percolative events when temperature sensors indicate the rapid transmisson of heat downward through the soil. Measured soil temperature increases during these episodes were as high as 5°C in the boreal forest and 1°C in the arctic tundra even at 100 cm depth. Heat inputs from 6 cm of precipitation in the boreal forest and 2 cm in the arctic tundra with air-soil temperature differences of 3°C total 0.75 MJ-m $^{-2}$ and 0.25 MJ-m $^{-2}$, respectively. These caloric inputs are not enough to cause temperature changes of this magnitude (Linell and Tedrow 1981). Nor is this occassional small heat input from precipitation significant in relation to total heat absorbed by the soil on a daily basis from global radiation in July, which can be conservatively estimated at 2.3 MJ-m $^{-2}$ at the boreal forest and 3.5 MJ-m^{-2} at the arctic tundra (Carlson and Kane 1973, Ng and Miller 1977, Ryden 1978, Goldstein 1981). Since there is no decrease in surface temperature associated with the increase at depth, redistribution of heat in the soil mass must also be discounted (Gavril'yev et al. 1973). Consequently, these measured temperature increases must be localized to a portion of the soil which is dramatically affected by percolating water, as are the thermocouple sensors. During wet episodes rapid flow prevents temperature equilibrium between relatively warm meteroric water and the bulk soil mass, however, surfaces of particles respond quickly to this pulse of heat. As moisture flow slows, pore water and particle surfaces equilibrate with the thermal mass of the bulk soil, and measured temperatures return to normal steady state values. Although precipitation does not add appreciable heat to the soil body per se, the temporarily elevated temperatures of the liquid phase and particle surfaces can be significant not only for soil weathering reactions, but also for transpiration and net photosynthetic activity of treeline and tundra plant species (Billings et al. 1977, Tranquilini 1979, Goldstein 1981).

CONCLUSIONS

This study gives direct evidence that percolation of meteoric water through the solum and active layer is a contemporary hydrologic event, both in the northernmost boreal forest and arctic tundra of Alaska. There is no reason to question podzolization as a present phenomenon in arctic environments due to moisture limitations. In well-drained northern locations downward pulses of heat associated with summer storms may also create temporary conditions in the soil more conducive to biotic and mineral reactions.

Data also confirm that climate, vegetation, and soil texture control soil moisture and temperature changes in the active layer. Climatic conditions in the southern foothills of the Brooks Range are capable of overcoming the high moisture demands of the boreal forest in midsummer. Although the measured precipitation during the study was much less in the arctic tundra, losses from evapotranspiration were low due to sparse vegetation and persistent low clouds and fog. In coarse textured soil with a moist surface, precipitation of 5-7 mm can induce a percolative event through the profile. It is not, therefore, a coincidence that Spodosols in the tundra are limited to convex landscape positions, coarse textured parent material, and thin vegetation where repeated leaching can occur.

Further research of this type along more detailed transects and with more complete data collection is required for mathematical modeling and prediction of active layer responses to climate at well-drained locations in the arctic.

ACKNOWLEDGMENTS

The authors wish to thank Ron Sletten for his assistance in the field. This research was supported by National Science Foundation grant (DPP) 80-05795.

REFERENCES

Baver, L. D., Gardner, W. H., and Gardner, W. R., 1972, Soil Physics, New York, John Wiley and Sons, Inc.

Berg, R., Brown, J., and Haugen, R. K., 1978, Thaw penetration and permafrost conditions associated with the Livengood to Prudhoe Bay Road, Alaska, *in* Proceedings of the Third International Conference on Permafrost, v. 1: Edmonton, National Research Council of Canada.

Billings, W. D., Peterson, K. M., Shaver, G. R., and Trent, A. W., 1977, Root growth respiration and carbon dioxide evolution in an arctic tundra soil, Arctic and Alpine Research, vol. 9, p. 129-137.

Brown, J., 1966, Soils of the Okpilak River region, Alaska, Cold Regions Res. Eng. Lab. Research Report 188, Hanover, New Hampshire.

Brown, J., and Tedrow, J. C. F., 1964, Soils of the Northern Brooks Range, Alaska:4, well-drained soils of the glaciated valleys, Soil Science v. 95, p. 187-195.

Brown, R. J., 1973, Influence of Climatic Terrain Factors on Ground Temperatures at Three Locations in the Permafrost Region of Canada, *in* Permafrost--The North American contribution to the second International Conference, Yakutsk: Washington, D.C., National Academy of Sciences.

Carlson, R. F., and Kane, D., 1973, Hydrology of Alaska's Arctic, *in* Weller, G. and Bowling, S. A., eds., Climate of the Arctic, Fairbanks.

Drew, J. V., Tedrow, J. C. F., Shanks, R. E., and Karanda, J. J., 1958, Rate and Depth of Thaw in Arctic Soils, Transactions, American Geophysical Union, v. 39, p. 697-701.

Everett, K. R., 1979, Evolution of the soil landscape in the sand region of the Arctic Coastal Plain as exemplified at Atkasook, Alaska, Arctic, v. 32, p. 207-223.

Gavril'yev, P. P., Ugarov, I. S. and Rysakov, Z. H., 1973, Processes of heat and moisture exchange in flooding of meadows in Central Yakutia, *in* Permafrost--The USSR contribution to the Second International Conference, Yakutsk, Washington, D.C., National Academy of Sciences.

Gold, L. W., and Lachenbruch, A. H., 1973, Thermal conditions in permafrost--A review of North American literature, *in* Permafrost--The North American contribution to the Second International Conference, Yakutsk: Washington, D.C., National Academy of Sciences.

Goldstein, G. H., 1981, Ecophysiological and demographic studies in white spruce (Picea glauca) at treeline in the Central Brooks Range of Alaska, doctoral thesis, University of Washington, Seattle.

Hamilton, T. D., 1979, Late Cenozoic glaciations and erosion intervals, north-central Brooks Range, *in* Johnson, K. M., and Williams, J. R., eds., The United States Geological Survey Circular 804-B, p. B27-B29.

Hamilton, T. D. and Porter, S. C., 1975, Itkillik glaciation in the Brooks Range, Northern Alaska, Quaternary Research, v. 5, p. 471-497.

James, P. A., 1970, The soils of the Rankin Inlet area, Keewatin, N.W.T., Canada, Arctic and Alpine Research, v. 2, p. 293-302.

Linell, K. A., and Tedrow, J. C. F., 1981, Soil and Permafrost Surveys in the Arctic, Oxford, Clarendon Press.

Moore, T. R., 1974, Pedogenesis in a subarctic environment: Cambrian Lake, Quebec, Arctic and Alpine Research, v. 6, p. 281-291.

Ng, E., and Miller, P. C., 1977, Validation of a model of the effect of tundra vegetation on soil temperatures, Arctic and Alpine Research, v. 9, p. 89-104.

Ryden, B. E., 1978, Energy fluxes related to the yearly phase changes of water in the tundra, *in* Proceedings of the Third International Conference on Permafrost, v. 1: Edmonton, National Research Council of Canada.

Sable, E. G., 1977, Geology of the western Romanzof Mountains, Brooks Range, northeastern Alaska, U.S. Geological Survey Professional Paper 897, Washington, D.C., U.S. Government Printing Office.

Tedrow, J. C. F., 1977, Soils of the polar landscape, New Brunswick, Rutgers University Press.

Tedrow, J. C. F., Drew, J. V., Hill, D. E., and Douglas, L. A., 1958, Major genetic soils of the Arctic Slope of Alaska, Journal of Soil Science, v. 9, p. 35-45.

Tranquilini, W., 1979, Physiological ecology of the alpine timberline, Springer Verlag, N.Y.

Ugolini, F. C., 1966, Soils of the Mesters Vig District, northeast Greenland, I. The Arctic Brown and related soils, Medd. Gronland, Bd. 196 Nr. 1, p. 1-22.

Ugolini, F. C., Zachara, J. M., and Reanier, R. E., 1981a. Dynamics of soil-forming processes in the arctic, *in* Proceedings Fourth Canadian Permafrost Conference: Ottawa, National Research Council of Canada.

Ugolini, F. C., Reanier, R. E., Rau, G. H., and Hedges, J. I. 1981b. Pedological, isotopic and geochemical investigations of the soils at the boreal forest and alpine tundra transition in northern Alaska: Soil Science v. 131 p. 359-374.

PLEISTOCENE DIAPIRIC UPTURNINGS OF LIGNITES AND CLAYEY SEDIMENTS AS PERIGLACIAL PHENOMENA IN CENTRAL EUROPE

Horst Strunk

University of Saarbruecken, Department of Geography
6600 Saarbruecken, Federal Republic of Germany

Diapirlike intrusions of lignite into gritty or sandy topsoil, several meters thick, are widespread in the central European lignite fields and wherever clayey sediments lying near the surface. These structures occur outside the limits of the Scandinavian ice sheet. This eliminates the assumption that their formation was primarily caused by the weight of the ice. Rather, the upturnings were caused by density inversions during the degradation of permafrost. The light lignite, water-saturated by the freezing process, rose up--much the same as saline domes--into the overlying layers having greater specific weight. Previously-frozen lignite has an agglomerate structure, resembling "coffee grounds". Disturbed layers thus provide information about the minimum depth of permafrost penetration. If it did not penetrate as far as the lignite, then the latter did not become upturned. The boundary between agglomerate and undisturbed lignite appears to mark the limit of maximum frost penetration and thus the lower limit of the permafrost. The maximum thickness of Pleistocene permafrost in central Europe decreases rapidly from several dozen meters in the northeast to less than 10 m in the west and southwest.

INTRODUCTION

In central and western Europe, permafrost aggradation and degradation took place several times during the Pleistocene. A number of phenomena are known to occur only in the presence of permafrost (cf. Washburn 1979) and are found today as they leave definite traces behind even after the ice has melted. In many other cases unfortunately no signs can be found in the subsoil to indicate whether or not permafrost had formed. It is harder still to obtain reliable evidence on the thickness of the permafrost.

PROBLEMS

Because the Scandinavian ice sheet advanced to Europe several times during the Pleistocene (cf. Liedtke 1981), a southerly movement of the peripheral periglacial area must be assumed synchronous to the advance of the ice. There was a reverse direction, when the ice sheet retreated. It remains unclear, however, whether degradation of the permafrost came about beneath the ice. In any case, a different southerly and southwesterly limit to the extent of the permafrost must be assumed for each ice age. Only a small number of paleoclimatic or paleogeographic studies have been made so far for the Pleistocene of central and western Europe. These exist only for the "Weichselian" (i.e., late Devensian) ice age, approximately 30.000-13.000 B.P. for Britain (Williams 1975, Watson 1977). For the western parts of central Europe there is a synthesis by Karte (1981), who points out that diverging opinions still exist about the "Weichselian" permafrost extent in this area. Intrusions of lignite into the topsoil provide evidence to substantiate the occurrence and maximum depth of permafrost in each region, and, depending on the soil profile formation, infer the age of the intrusion phases.

DISTRIBUTION AND APPEARANCE

Intrusions of lignite into overlying sediments occur in all the lignite fields in central Europe, provided the upper limit of the lignite is not more than a few dozen meters below the recent surface. A small number of similar clay intrusions also exist. Because their occurrence has an extremely detrimental effect on the quality of the lignite and its exploitation, and also makes draining the pits a problem, attempts to describe and explain the intrusions go back nearly 150 years. The literature covers more than 120 titles: more recent summaries include Viete (1960a, b), Ruchholz (1977), and Eissmann (1978, 1981). The intrusions may be wide dome shaped, pointed cones, pillar shaped, and moundlike or wallike (Eissmann 1978). Enclosed depressions set within a moundlike surrounding frame with either round or oval outlines are also widespread (Strunk 1982). The base of the upturnings in central Germany lies at a depth of up to 150 m beneath Quaternary topsoil (Fries 1933). Intrusion heights are more than 50 m (Viete 1960a), but not greater than 10 m in the southeast (Heim 1971) and extreme west of Germany (Kaiser 1958, Quitzow 1958). They occur a long way outside the maximum extent of the Scandinavian ice sheet.

DEVELOPMENT

The deformations are partly "load casts" (Kuenen 1953), brought about by gravitational compensatory movement caused by the instability of lighter hydroplastic layers being affected by overlying strata with a greater specific weight (Dzulynski 1966, Anketell et al. 1969). Load casts up to a few decimeters height are known from all eras of earth history, starting with the Precambrian up to the present day (e.g., Allen 1982, Pettijohn and Potter 1964). Frequently, the lignites have penetrated the topsoil and become diapirs. These differ from the small forms described here only in their dimensions, and not in their origin of formation, i.e., density inversion (Heim 1958). They constitute a worldwide phenomenon (Ramberg 1972). The most spectacular examples of compensatory movement due to density inversion must surely be the saline domes (Lotze 1938, Balk 1949, Trusheim 1957).

Lignite from mining has a water content of 55-65%, as was established by Potonié (1930), Haase and Pflug (1958), and Spelter (1968). It should be kept in mind however, that the mining areas are given deep-drainage treatment before mining is begun and the water content of the lignite will thus be considerably higher below the groundwater level. Despite the fact that the Tertiary lignite deposits with high water content are covered by more than 300 m of gravelly sediments which may be old as Tertiary in age, no indication are found in the literature that the lignite intrusions in central Europe are older than the Pleistocene. Eissmann (1978) came in the same conclusion.

Bajor (1958) and Berger (1958) are exceptions since they describe intrusions of fine sand into lignite in the soft coal region of the German Rhineland. However, these were caused presumably by the transgression of the North Sea in the Miocene period and are not genetically linked to the lignite intrusions.

The considerable difference in density between lignite at 1.2-1.4 kg dm^{-3} and gravel at 2.0-2.4 kg dm^{-3} (Ruchholz 1977) was not enough to trigger density inversions between the lighter lignite and the heavier topsoil. However, these density inversions occurred several times in the Quaternary period. This is why it was conjectured that autoplastic lignite movement (Lehmann 1922, Weigelt 1928, Wölk 1937) and the permafrost of the Pleistocene period caused the upturnings (Grahmann 1934, Pruskowski 1954, Kaiser 1958, Schenk 1964). Ruchholz (1977), Eissmann (1978, 1981), and the author hold that the autoplastic intrusions of lignite are density inversions occurring in watersaturated, thawing permafrost (e.g., French 1979, Vandenberghe and Van den Broek 1982). Much the same conditions occur in peat deposits (Czudek and Demek 1973), where extensive lenses of segregational ice form during the freezing process. If the lower frost limit does not reach the base of the lignite deposit, already containing a good deal of water at the outset, (this being the general rule in the western parts of central Europe), the freezing of lignite takes place in an "open system" (Beskow 1935, Dücker 1939, Peltier 1953). Unlimited amounts of water keep reaching the freezing front. Horizontal crevices filled with sand from the topsoil, 8 cm thick and up to 17 m long (Keilhack 1921, Magalowski 1963), together with the fine flaky structure of the lignite (Keilhack 1921, Fliegel 1937, Kaiser 1958) indicate the former presence of lenticular segregational ice in the permafrost. Due to the formation of ice in the freezing lignite, the latter takes on an agglomerate structure. In extreme cases, this produces a "coffee grounds" type disintegration (Hübner 1979). Corresponding observations can be found in almost all descriptions of lignite upturnings. The degradation of the permafrost produces diapirlike compensatory movements of the water-saturated hydroplastic lignite into the heavier overlying sediments, which sink at the same time. They must also have been hydroplastic, because, as the author has observed, rocks accumulate in the gravel-filled depressions between the diapirs in the bottom of the hollows. Hunger and Seichter (1955) observe the same. This compensation of density inversions is referred to as "mollisol diapirism" by Eissmann (1978).

The idea of pure gravitational density inversions did not win wide acclaim until recently, and conflicted with the concept of upturnings occurring due to the load pressure of the ice sheets (e.g., Eissmann 1978). But, even for areas within the maximum extent of glaciation, Eissmann (1981) provides evidence of the autoplastic ascent of the lignite in several places, without being affected in any way by the ice sheet. Here, diapirism occurs--and this is clearly substantiated by stratigraphic observations--temporally independent of glaciation, with diapirs still forming in the last (i.e., Weichsel) ice age, when ice stagnated 100 km to the north. Lehmann (1922) made similar observations.

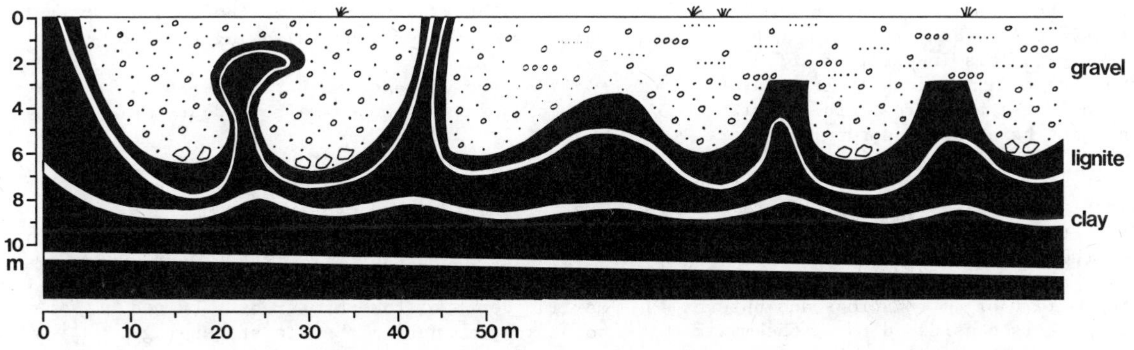

Figure 1 Deformation of lignite and thin, light layers of clay, caused by density inversion, Zievel claypit, German Rhineland. Scale is given in meters.

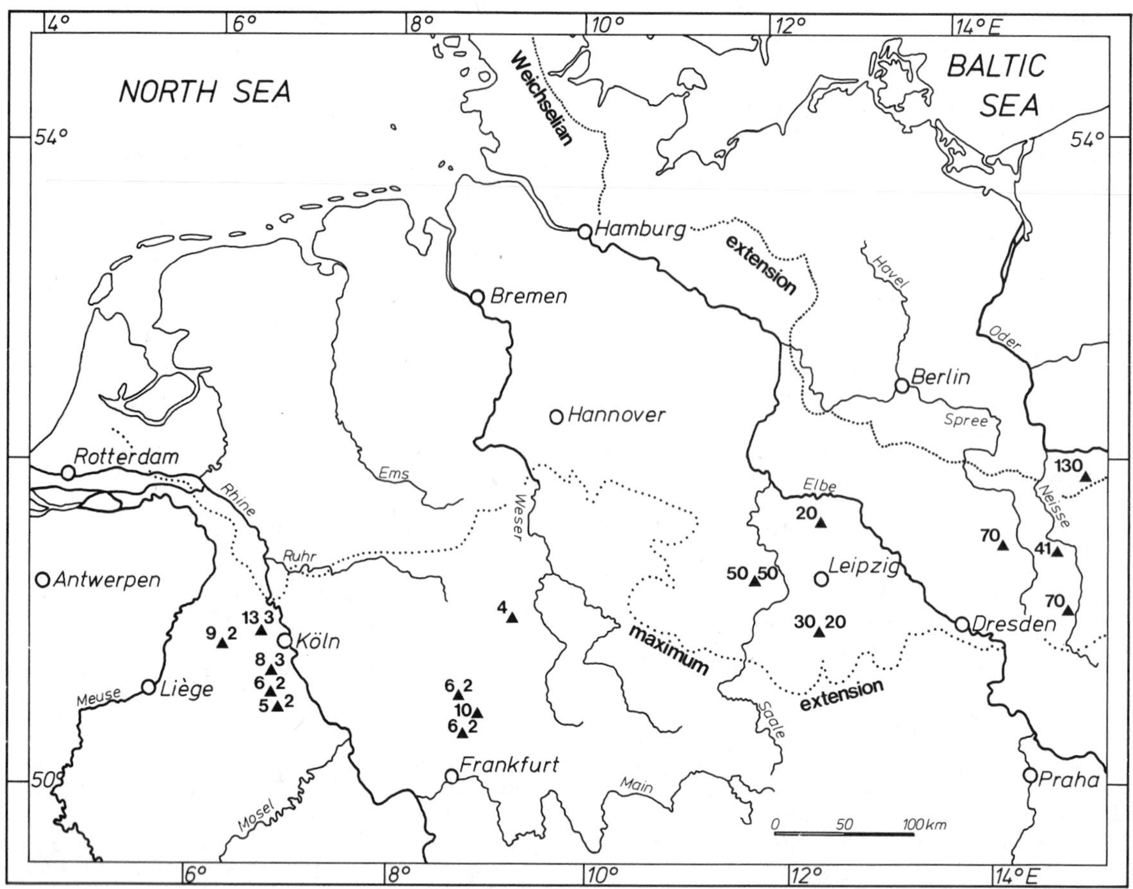

Figure 2 Depth in meters of permafrost in central Europe, with the left-hand value indicating the maximum Pleistocene thickness and the right-hand value the thickness during the last (i.e., Weichsel) glaciation. Pointed lines mark maximum extension of the ice sheets and the extent of the Weichselian ice sheet.

PERMAFROST DEPTH

As described above, previously frozen lignite is, to a greater or lesser extent, agglomerate. In areas of diapiric structures, it has a flow structure. The thin, light layers of clay, normally deposited undisturbed in lignite layers, are also affected by these upturnings, as can be observed in any mine (see Figure 1). The lowest limit of these upturnings runs through the deposits evenly or in slightly discordant waves. According to Eissmann (1981) and Maarleveld (1976) this lowest limit of deformations is the lowest limit of the active layer. With degradation of the permafrost, this limit moves down and must have caused upturnings there. It is striking that the lowest limit in the northern part of central Europe is 2-4 times deeper than in the south and west. This signifies, according to Eissmann (1981), that the ground thawed to a greater depth close to the ice sheet than further to the south and west. That is thought unlikely. If, however, the lowest limit of deformations and upturnings in the lignite is considered to be the maximum depth of the permafrost, then this interpretation better fits the observations made thus far. The thickness of the permafrost would have decreased as distance from the ice sheet increased, and with it, of course, the depth of the upturnings caused by permafrost-degradation.

PALEOGEOGRAPHIC ASPECTS

If the observations made of the lowest limit of the upturnings caused by permafrost and the depth of the lignite deposits below the surface are put into spatial order, the following results:

1. As the distance from the ice sheet increases, the thickness of the permafrost decreases.

2. If lignite does not posses upturnings, the maximum permafrost depth must have been less than the thickness of the overlying sediments.

3. Without further stratigraphic evidence, the lowest limit of disturbed lignite provides only the maximum depth of the permafrost attained at some point in time during the Pleistocene age.

4. By considering the overlying layers that can be dated (e.g., gravel or interglacial soils)--pos-

sibly in a discordant manner--the minimum age of the permafrost degradation can be determined. If the intrusions are syngenetic, the age of the permafrost degradation corresponds to the age of the covering sediment.

The permafrost thicknesses for central Europe, obtained in the way described above, are shown in Figure 2 (in part according to Eissmann (1981)), with the left-hand value indicating the maximum Pleistocene thickness and the right-hand value the thickness during the last (i.e., Weichsel) glaciation. The data are given in meters.

ACKNOWLEDGMENTS

The author thanks E. Bibus, K. Heine, H. D. Lang, W. Schirmer and G. Seidenschwann for helpful discussions and unpublished observations. He also thanks H. M. French for revising the English text.

REFERENCES

Allen, J. R. L., 1982, Sedimentary structures, their character and physical basis, v. II: Amsterdam, Oxford, New York, Elsevier Scientific Publication Co., p. 1-669.

Anketell, J. M., Cegla, J., and Dzulynski, St., 1969, Unconformable surfaces formed in the absence of current erosion: Geologica Romana, v. 8, p. 41-46.

Bajor, M., 1958, Beobachtungen über Fazies, synsedimentäre Tektonik und Schwimmsandintrusionen in der Grube Neurath (Niederrhein): Fortschritte der Geologie Rheinland und Westfalen, v. 1, p. 119-125.

Balk, R., 1949, Structure of Grand Saline salt dome, Van Zandt County, Texas: Bulletin of the American Association of Petroleum Geologists, v. 33, p. 1791-1829.

Berger, F., 1958, Flözsandauswaschungen und Schwimmsand-Intrusionen im Tagebau Frimmersdorf-Süd: Fortschritte der Geologie Rheinland und Westfalen, v. 1, p. 113-118.

Beskow, G., 1935, Soil freezing: Sveriges Geologiska Undersöknisg, Ser. C, v. 375, p. 1-145.

Czudek, T., and Demek, J., 1973, Die Reliefentwicklung während der Dauerfrostbodendegradation: Rozpravy Ceskoslovenské Akademie Ved, v. 83, p. 1-69.

Dücker, A., 1939, Untersuchungen über die frostgefährlichen Eigenschaften nichtbindiger Böden: Forschungsarbeiten aus dem Straßenwesen, v. 17, p. 1-79.

Dzulynski, S., 1966, Sedimentary structures resulting from convection-like pattern of motion: Rocznik Polskiego Towarzyatwa Geologicznego, p. 3-21.

Eissmann, L., 1978, Mollisoldiapirismus: Zeitschrift für angewandte Geologie, v. 24, p. 130-138.

Eissmann, L., 1981, Periglaziäre Prozesse und Permafroststrukturen aus sechs Kaltzeiten des Quartärs: Altenburger Naturwissenschaftliche Forschungen, v. 1, p. 1-171.

Fliegel, G., 1937, Erläuterungen zur Geologischen Karte von Preußen und benachbarten deutschen Ländern, Lieferung 283, Blätter Frechen, Köln, Kerpen, Brühl: Berlin, Preussische Geologische Landesanstalt, p. 5-132.

French, H. M., 1979, Periglacial Geomorphology: Progress in Physical Geography, v. 3, p. 264-273.

Fries, W., 1933, Tertiär und Diluvium im Grünberger Höhenrücken: Jahrbuch des Halleschen Verbandes für die Erforschung der mitteldeutschen Bodenschätze und ihre Verwertung, N.F., v. 12, p. 167-198.

Grahmann, R., 1934, Spät-und postglaziale Süßwasserbildungen in Regis-Breitingen und die Entwicklung der Urlandschaft in Westsachsen: Mitteilungen aus dem Osterlande, N.F., v. 22, p. 14-44.

Haase, F., and Pflug, H. D., 1958, Fazies und Brikettierbarkeit der niederrheinischen Braunkohlen: Fortschritte der Geologie Rheinland und Westfalen, v. 2, p. 613-632.

Heim, A., 1958, Beobachtungen über Diapirismus: Eclogae Geologicae Helvetiae, v. 51, p. 1-33.

Heim, D., 1971, Lateritische Basaltverwitterung und Bauxit im Bereiche der Wetterau-Schwelle: Abhandlungen des Hessischen Landesamtes für Bodenforschung, v. 60, p. 251-257.

Hübner, J., 1979, Frostdynamische Erscheinungen in der Braunkohlenlagerstätte Wallendorf: Zeitschrift für angewandte Geologie, v. 25, p. 365-367.

Hunger, R., and Seichter, A., 1955, Glazigene Deformationen in der Braunkohle von Gräfenhainichen: Freiberger Forschungshefte, v. C 21, p. 24-39.

Kaiser, K. H., 1958, Wirkungen des pleistozänen Bodenfrostes in den Sedimenten der Niederrheinischen Bucht: Eiszeitalter und Gegenwart, v. 9, p. 110-129.

Karte, J., 1981, Zur Rekonstruktion des weichselhochglazialen Dauerfrostbodens im westlichen Mitteleuropa: Bochumer Geographische Arbeiten, v. 40, p. 59-71.

Keilhack, K., 1921, Die abbaustörenden Einlagerungen und Verunreinigungen in den Braunkohlenflözen der Lausitz: Braunkohle, v. 20, p. 481-489.

Kuenen, Ph., 1953, Graded bedding in general and some observations in West-Central Wales: Verhandelingen Konikl. Nederlandse Akademie Wetenschapten, Afdel. Natuurk., Sect. I, v. 20, p. 1-47.

Lehmann, R., 1922, Das Diluvium des unteren Unstruttales von Sömmerda bis zur Mündung: Jahrbuch des Halleschen Verbandes für die Erforschung der mitteldeutschen Bodenschätze und ihre Verwertung, N.F., v. 3, p. 89-124.

Liedtke, H., 1981, Die nordischen Vereisungen in Mitteleuropa: Forschungen zur deutschen Landeskunde, v. 204, p. 1-307.

Lotze, F., 1938, Steinsalze und Kalisalze, Geologie, Die wichtigsten Lagerstätten der "Nichterze", v. III: Berlin, Borntraeger, p. 1-634.

Maarleveld, G. C., 1976, Periglacial phenomena and the mean annual temperature during the last Glacial time in the Netherlands: Biuletyn Peryglacjalny, v. 26, p. 57-78.

Magalowski, G., 1963, Glazigene Flözdeformationen im Tagebau Spreetal und ihr Einfluß auf die Braunkohlenqualität: Bergakademie, v. 15, p. 185-188.

Peltier, R., 1953, Contribution à l'élaboration

d'une théorie capillaire du gel des sols routiers, in Proceedings of the Third International Conference on Soil Mechanics and foundation Engineering Switzerland 53, v. II: Zürich, Schweizerische Gesellschaft für Bodenmechanik und Fundationstechnik, p. 128-132.

Pettijohn, F. J., and Potter, P. E., 1964, Atlas and glossary of primary sedimentary structures: Berlin, Göttingen, Heidelberg, New York, Springer-Verlag, p. 1-370.

Potonié, R., 1930, Über den Muskauer Faltenbogen, seine Oberflächenformen und deren Abhängigkeit von der Beschaffenheit und Tektonik der Braunkohle: Jahrbuch der Preussischen Geologischen Landesanstalt, v. 51, p. 392-416.

Pruskowski, P., 1954, Eiszeitliche Bodenfrostbildungen im Deckgebirge der rheinischen Braunkohle: Beiträge zur Rheinkunde, v. 2, p. 41-54.

Quitzow, H. W., 1958, Verwerfungen und pseudotektonische Faltungen im Hauptflöz der Ville zwischen Liblar und Brühl: Fortschritte der Geologie Rheinland und Westfalen, v. 2, p. 645-649.

Ramberg, H., 1972, Inverted density stratification and diapirism in the earth: Journal of Geophysical Research, v. 77, p. 877-889.

Ruchholz, K., 1977, Zur Genese gravitativer Schicht-und Sedimentkörper-Deformationen in Vereisungsgebieten: Wissenschaftliche Zeitschrift der Ernst-Moritz-Arndt-Universität Greifswald, v. 26, Mathematisch-naturwissenschaftliche Reihe No. 1/2, p. 49-57.

Schenk, E., 1964, Das Quartärprofil in den Braunkohlentagebauen bei Berstadt und Weckesheim (Wetterau): Notizblatt des hessischen Landesamtes für Bodenforschung, v. 92, p. 270-274.

Spelter, M., 1968, Tektonisch bedingte oder durch unterirdische Oxidation der Kohle hervorgerufene Lagerungsstörungen der Jüngeren Hauptterrasse auf der Ville im rheinischen Braunkohlenrevier: Zeitschrift der deutschen Geologischen Gesellschaft, v. 118, p. 162-173.

Strunk, H., 1982, Zur pleistozänen Reliefentwicklung talferner Areale der Eifel-Nordabdachung: Arbeiten aus dem Geographischen Institut der Universität des Saarlandes, v. 32, p. 1-109.

Trusheim, F., 1957, Über Halokinese und ihre Bedeutung für die strukturelle Entwicklung Norddeutschlands: Zeitschrift der deutschen Geologischen Gesellschaft, v. 109, p. 111-151.

Vandenberghe, J., and Van den Broek, P., 1982, Weichselian convolution phenomena and processes in fine sediments: Boreas, v. 11, p. 299-315.

Viete, G., 1960a, Zur Entstehung der glazigenen Lagerungsstörungen unter besonderer Berücksichtigung der Flözdeformationen im mitteldeutschen Raum: Freiberger Forschungshefte, v. C 78, p. 1-257.

Viete, G., 1960b, Über die Genese der glazigenen Deformationen der mitteldeutschen Braunkohlen und die Möglichkeit ihrer Vorhersage in neuen Grubenfeldern: Freiberger Forschungshefte, v. C 80, p. 13-24.

Washburn, A. L., 1979, Geocryology--A survey of periglacial processes and environments: London, Edward Arnold Ltd., p. 1-406.

Watson, E., 1977, The periglacial environment of Great Britain during the Devensian: Philosophical Transactions of the Royal Society of London, v. B 280, p. 183-198.

Weigelt, H., 1928, Die Kohlenaufpressungen in den Geiseltal-Gruben "Leonhardt", "Pfänderhall" und "Rheinland": Jahrbuch des Halleschen Verbandes für die Erforschung der mitteldeutschen Bodenschätze und ihre Verwertung, N.F., v. 7, p. 68-97.

Williams, R. B. G., 1975, The British climate during the last Glaciation. An interpretation based on periglacial phenomena, in Wright, A. E., and Mosely, F., eds., Ice ages, ancient and modern: Geological Journal, Special Issue, v. 6, p. 95-120.

Wölk, E., 1937, Faltungserscheinungen in der niederrheinischen Braunkohle: Braunkohle, v. 36, p. 405-410, 423-431.

LARGE-SCALE DIRECT SHEAR TEST SYSTEM FOR TESTING PARTIALLY FROZEN SOILS

Brian J. A. Stuckert and Larry J. Mahar

Ertec Western, Inc.
Long Beach, California 90807 USA

The design and construction of offshore earth-fill structures in the arctic environment requires the determination of mechanical properties of the construction material under appropriate thermal conditions. Salt-water saturated materials exhibit a zone of partially frozen soil at the frozen/unfrozen soil interface. To define the critical failure surface and governing shear strength in this zone, the effects of partial freezing on the mechanical behavior, as a freezing front advances, must be well understood.

A large-scale direct shear apparatus was developed to test the shear strength of a soil sample through an undisturbed partially frozen inter-facial zone. The apparatus has the capability of shearing a 30 cm square, frozen specimen using a wide range of shear displacement rates. The specimen is instrumented with thermistors to provide the capability to monitor the thermal regime as the partially frozen zone progresses toward a predetermined shear plane in a 30 cm deep sample. The shear box is constructed from transparent plexiglas material which minimizes thermal disturbance and permits visual inspection of the sample during shearing. A sample preparation system was designed to provide a specimen having a reproducible thermal condition while thermal and mechanical disturbance during handling is minimized.

The paper presents details of the design of the large-scale direct shear apparatus and testing procedures. Typical results from tests conducted on freshwater and seawater-saturated gravels are shown, demonstrating the usefulness of system and procedures for determining the strength of partially frozen soils.

INTRODUCTION

The design of offshore earth-fill structures constructed in arctic waters requires knowledge of the mechanical response of the construction material to anticipated loading conditions. Lateral ice loading can result in three potential failure modes: (1) localized edge failure, (2) foundation failure at or below the mudline, (3) shear failure in the fill material. The latter mode will likely occur as a gross horizontal shear failure at or near the frozen/unfrozen interface (Prodanovich 1979). Due to the presence of saline pore fluid, a significant zone of poorly bonded ice exists at this interface in typical granular fill material (Mahar et al. 1982). It is of interest to understand the mechanical behavior within this zone when subjected to vertical overburden and subsequent horizontal shear. Development of appropriate design strength parameters requires that anticipated loading and thermal conditions be modeled.

An ideally suited laboratory test to characterize horizontal shear strength in a soil sample is the direct shear test. Roggensack and Morgenstern (1978) performed direct shear tests on natural fine-grained permafrost, and Thomson and Lobacz (1973) performed direct shear tests at the thaw interface in freshwater saturated silt; however, the results are not applicable to the design of artificial arctic islands because of the coarse-grained nature of the construction material and the effect of a saline pore fluid on strength behavior. The only data available on the strength of saline-saturated granular materials is in the form of triaxial test results (Ogata et al. 1982, Sego et al. 1982); however, these studies do not investigate the strengths under the relatively warm conditions that exist within the freeze-front zone. The present study represents an effort aimed at investigating the strength within this zone.

The direct shear test simulates anticipated field loading conditions; however, conventional direct shear equipment is not suitable for testing the coarse-grained material of interest because of the small shear box size. Large samples must be tested to minimize particle size effects on test results. A large-scale direct shear machine and sample preparation system has been developed, allowing a large sample to be subjected to a wide range of repeatable thermal and loading conditions.

The equipment and procedures developed to test frozen and partially frozen coarse-grained material within and near the frozen/unfrozen interface zone are described below. Typical test results from a preliminary testing program are provided.

TEST APPARATUS

The large-scale direct shear system described

here was designed to permit inducing shear failure at a controlled plane within the partially frozen freeze-front zone. The system was designed to incorporate the following capabilities:

1. ability to induce shear failure at a prescribed plane in a sample at a controlled rate of displacement

2. sample size large enough to eliminate particle size effects associated with testing gravel material

3. adjustable shear gap to accommodate maximum particle size of material being tested

4. ability to maintain desired thermal conditions during direct shear setup and testing

5. ability to simultaneously freeze multiple samples uniaxially under well-controlled environmental conditions

6. remote monitoring of freeze-front progression and shear testing to minimize operator time inside cold room

7. ability to apply vertical load during shear testing to simulate in situ stress conditions

The large-scale direct shear test apparatus comprises a sample preparation system and a direct shear test system in which large samples (30 cm x 30 cm x 30 cm) are frozen uniaxially in a controlled-environment chamber and tested to failure under direct shear loading. Temperature monitoring of the sample allows direct shear testing when the desired thermal condition exists at the shear plane. The sample is then placed in the shear apparatus, subjected to an overburden pressure and sheared at a constant rate of displacement. Horizontal load, vertical load, horizontal deformation, vertical deformation, and temperatures are monitored during the shear test. All data from the freezing and shearing portions of the test are gathered and stored on tape, and selected data are plotted using a computer-based data acquisition system.

Sample Preparation System

The sample preparation system is shown schematically in Figure 1. The purpose of this system is to induce uniaxial freezing by application of a surface air step temperature. A wide range of step air temperatures (0 to -25°C) can be controlled to provide sample temperature distributions and freezing rates so that tests can be run for specific, repeatable thermal conditions at the shear plane.

The system is situated in a cold room in which air temperatures can be controlled to within ±1°C. Up to three samples can be prepared simultaneously inside a second environmental chamber in which temperatures are controlled to within ±.2°C. The cold-room and preparation-chamber temperature can be varied to provide air temperatures from -25°C to + 20°C.

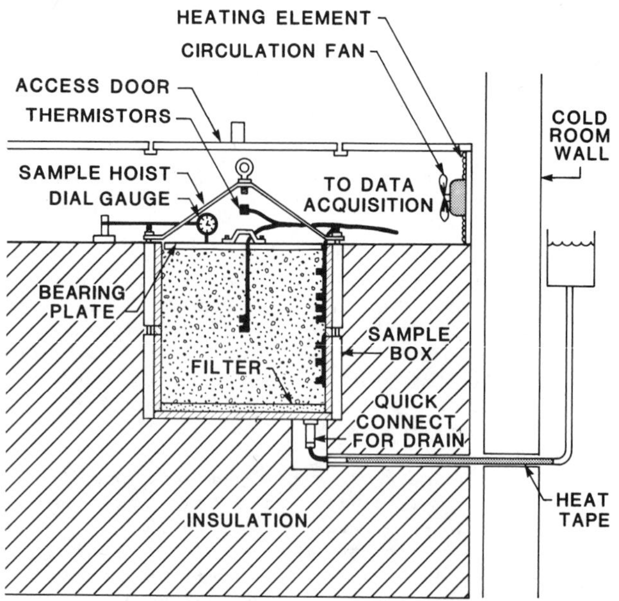

Figure 1 Sample Preparation System

The preparation chamber consists of a polystyrene foam insulation base, with cubical receptacles to house specially designed sample preparation and testing containers, covered by a plexiglas environmental chamber. The air temperature is controlled using a proportional temperature controller that supplies a current to a heating wire running the perimeter of the chamber. Four fans provide adequate circulation to maintain constant thermal conditions throughout the chamber. Samples are surrounded by a minimum of 30 cm of styrofoam, which ensures predominantly vertical heat flow and hence uniaxial freezing. Heated drain lines from the sample box allow expulsion of pore water to eliminate sample expansion during freezing.

The sample preparation and testing containers, shown in Figure 2, are made of two halves of a cubical plexiglas box joined by a membrane that prevents water leakage through the shear gap and allows relative displacement of the box halves during shear. Spacers are inserted in the shear gap and the upper and lower halves of the box are clamped against the spacers to minimize physical disturbance during transport prior to testing.

Direct Shear Test Machine

The shear testing system was designed to induce a horizontal shear failure using a controlled rate of displacement in a frozen or partially frozen soil sample while minimizing thermal disturbance during setup and testing. The apparatus, shown schematically in Figure 3 and pictured in Figure 4, is composed of four components: (1) horizontal reaction frame, (2) vertical load system, (3) shear box assembly, and (4) power train. Together, this system is capable of shearing a 30 cm square, completely frozen gravel sample over a wide range of displacement rates under vertical loads up to 2,270 kg, while maintaining desired thermal conditions established during sample preparation. The components of the apparatus are discussed below.

Figure 2 Sample Container

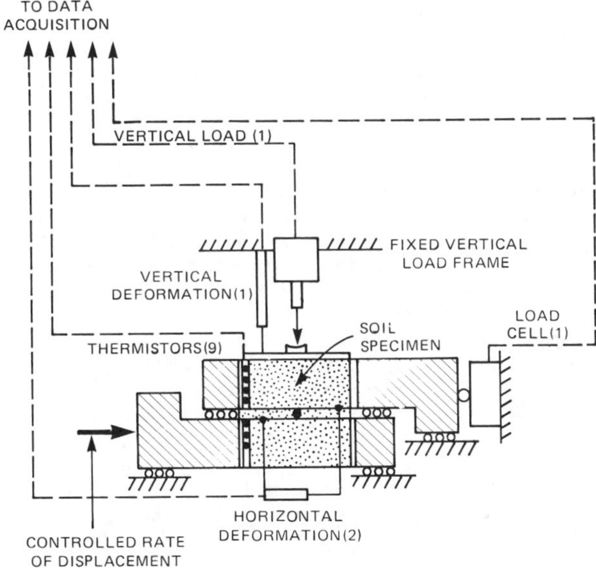

Figure 3 Schematic of Large Scale Direct Shear Apparatus

1. Horizontal reaction frame--The reaction frame is made of two structural-steel end sections tied together by four 3.8-cm diameter, high-strength steel tension rods 300 cm in length. It is capable of providing a 90-tonne horizontal reaction. The reaction frame is mounted on a base frame that distributes concentrated vertical loads to the cold room floor.

Figure 4 Large Scale Direct Shear Test Machine

2. Shear box assembly--The shear box assembly (Figure 5) is designed to withstand a horizontal compressive loading of 90 tonnes and allow shear displacements up to 5 cm. The shear box is constructed of transparent plexiglas, providing the required structural strength and insulating the sample to minimize thermal disturbance during setup and testing. The material also allows visual observation of the sample during shear. Tolerance between the sample box and shear box was minimized to eliminate sample rotation during shear. A slide bearing system provides axial alignment of both halves of the shear box, while maintaining static stability of the system. The shear box halves are separated by an adjustable bearing mount to allow a shear plane gap from 0 to 2.5 cm to accommodate testing of variety of maximum soil particle sizes. The ends of the shear box are fitted with steel plates to distribute

Figure 5 Shear Box Assembly

forces from the load cell and loading ram over the plexiglas face. A spherical interface between the plates and the ram and load cell ensures transmittal of a horizontal load.

3. Vertical load system--Vertical loads up to 2,270 kg are provided by a pneumatic ram cylinder fixed to a reaction beam. The system is mounted on the horizontal tension rods, and vertical tension rods tied to the base frame provide the necessary reaction. The vertical frame is mounted on linear slide bearings to maintain a vertical load centered on the sample throughout the test. Although the bottom half of the shear box is tied to the drive system to provide the relative shear displacements, small horizontal displacements of the upper half of the sample do occur due to system compliance.

4. Power train-- The power train is composed of a variable-speed electric motor, speed reducer, shaft assembly, and high capacity-screw jack. The motor and speed reducer are located outside the cold room, being connected by a drive shaft assembly to the screw jack mounted on the horizontal load frame inside the cold room. Once set at the desired speed (range: 30-1,800 rpm), the motor transmits torque as required directly to the speed reducer, which reduces at a constant ratio of 2,537:1. The shaft assembly drives the worm gear (6.4 turns/cm) of the screw jack. The entire system is designed to provide a 90-tonne horizontal thrust at a displacement rate in the range of .112 cm/min to .0019 cm/min.

Instrumentation and Data Acquisition System

A microcomputer-based data acquisition system is capable of monitoring temperatures of three samples during freezing and load-displacement during shearing. All raw data are stored on tape, reduced data are printed out, and selected data are plotted for quick reference.

All samples are instrumented with 10 thermistors, 9 at set 1.2- to 2.5-cm spacing providing internal soil temperatures and 1 monitoring the chamber air temperature 2 cm above the sample. The rapid-response thermistors are waterproofed and have certified accuracies to within ±0.1°C. During freezing, thermistor data are monitored at selected time intervals (typically 1 hr), stored on tape, printed out, and temperature versus depth plots are generated to observe progression of the freezing front or zone. Vertical deformation of the sample during freezing is monitored using a dial gauge (± 0.0002 cm).

During shear testing the shear test sample is monitored at selected time intervals (typically a 10-s scan rate). Sample temperatures are recorded to quantify thermal disturbance, if any, during testing. In addition, vertical load, vertical deformation, horizontal load, and horizontal shear deformation are recorded and printed. Relative shear displacements are measured between the top and bottom halves of the sample box at the shear plane, on both sides of the sample. Average shear displacement is plotted against horizontal load throughout the test.

TEST PROCEDURE

The test procedure consists of three separate stages: (1) sample preparation, (2) freezing phase, (3) shear test phase. The procedures are summarized as follows:

1. A membrane is installed in the 1.3-cm shear gap. Sample box halves are separated by 1.3-cm spacers that are removed prior to testing. Hoisting rods are inserted on the outside of the sample, and are used to clamp against the spacers, thereby reducing disturbance during subsequent transport. A string of thermistors is placed at the desired spacing on the inside of the sample box wall. A pea gravel drain, overlain by filter paper, is placed in the bottom of the box, and the air-dried test soil is placed in 5-cm lifts at the desired density. Two interior thermistors are placed down the middle of the sample, one an inch above and one directly at the shear plane. Saline fluid, prepared at the desired salinity (ASTM Standardized Sea Salt) and checked with a conductivity meter (±1 ppt), is used to saturate the sample slowly from the base under approximately 30 cm of head. Saturation levels are typically 95% ±5%. A top bearing plate is then placed, and the sample is put into a cooler and brought to an isothermal condition of approximately -5°C. The sample is then transported to the cold room testing facility, while great care is taken to ensure minimal physical disturbance during transport.

2. The sample is then placed in the freezing preparation chamber. A heated drain line is hooked up to the base of the sample and the desired air temperature is established and maintained to within ±.2°C throughout the freezing phase. The top perimeter of the sample is sealed against the surrounding insulation using insulation tape to minimize horizontal heat loss and thus achieve uniaxial freezing.

The test time is then selected based on the sample temperature profile. For the testing of distilled-water-saturated samples a distinct freeze front exists and is clearly identified from the temperature profile. Tests can be performed when the freeze front is at the desired location relative to the shear plane. Because a distinct freeze front does not exist in a saline-saturated soil, the progression of a desired test temperature, rather than depth of freeze front, is monitored to define testing conditions.

Typical thermal data during preparation are shown in Figure 6 for a saline-saturated sample. For the particular saline test shown the desired test temperature was -2.3°C. Also shown is a typical temperature profile generated during freezing of a distilled-water-saturated sample.

The distilled-water sample was tested when the freeze front coincided with the shear plane. The temperature profiles generated allow testing precisely when the temperature in the middle of the sample is at the desired test temperature. The outside thermistor string provides the progression of the temperature profile, while the interior thermistor 5 cm above the middle of the shear plane is used to assess uniaxial freezing close to the shear plane, as depicted in Figure 7. The interior thermistor at the shear plane is used to provide an exact testing time. This thermistor is removed at the end of the freezing to avoid thermistor failure during shear testing, and thermal disturbance during sample transport and testing is assessed with the remaining eight thermistors. At the end of the freezing phase, the sample is hoisted to the direct shear machine.

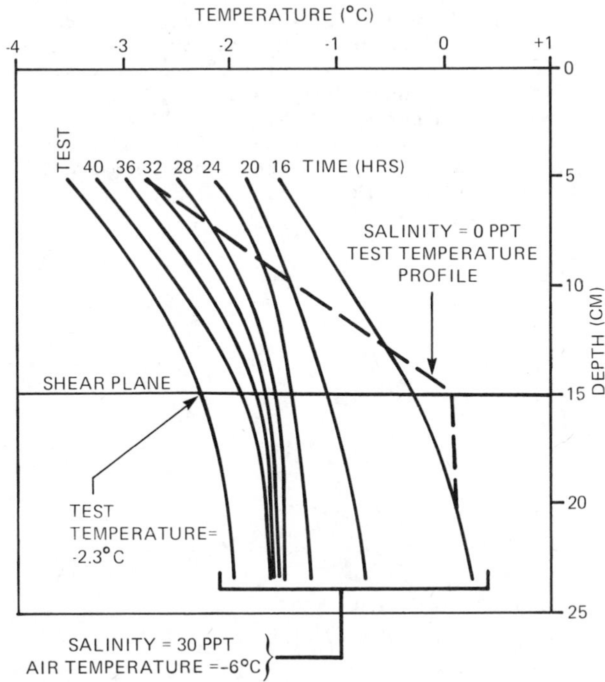

Figure 6 Temperature Profiles During Freezing

3. When the sample is ready for testing, it is removed from the preparation chamber and placed in the shear box. The hoisting rods and spacers are removed, and the LVDT mounts are attached to the sample box. The vertical load frame is centered over the top of the sample, and the desired vertical load is applied. The sample is then subjected to a constant shear displacement rate to obtain peak and residual strengths. Upon completion of the test, a portion of the frozen sample is removed for salinity and moisture-content determinations.

TYPICAL TEST RESULTS

Typical test results are presented for three well-graded sandy-gravel samples (maximum grain size diameter of 1.3 cm), similar to arctic gravel island construction material. The first sample was completely unfrozen and was performed as a reference test. A second test was performed on a distilled-water-saturated sample, with the freezing front coincident with the shear plane. The results for these two tests were essentially identical, implying that the unfrozen strength is appropriate for the strength at the freeze front for distilled-water-saturated material. A third test was performed on seawater (30 ppt) saturated material, with a sample temperature at the shear plane of -2.3°C. This is assumed to correspond to the cold portion of the poorly bonded ice zone.

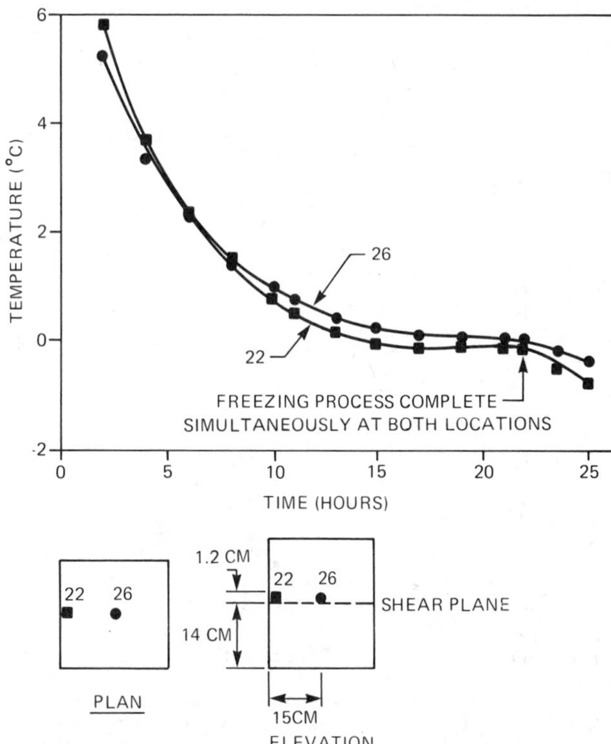

Figure 7 Assessment of Uniaxial Freezing

The results of distilled-water versus-seawater-saturated material are shown in Figure 8 for comparison. Note that a 15% gain in peak strength over the unfrozen state is obtained for the partially frozen saline sample, and the residual strength is essentially that for unfrozen material.

The test system described herein is currently being used in a major, joint industry/proprietary research study to investigate the shear strength of saline-saturated sands and gravels for use in arctic exploration and production islands. Although the detailed results and conclusions of this study are not yet available for publication, the large-scale direct shear test has proven to be a very effective method for determining the shear strength of soils in the region close to the frozen/unfrozen interface.

SUMMARY

It has been shown that the need exists to characterize the mechanical behavior, near and within the freeze-front zone that is present in saline-saturated granular material, for design of earth-fill structures constructed in arctic marine environments. To do so, in situ stress, thermal, and loading conditions must be modeled using typical construction soil.

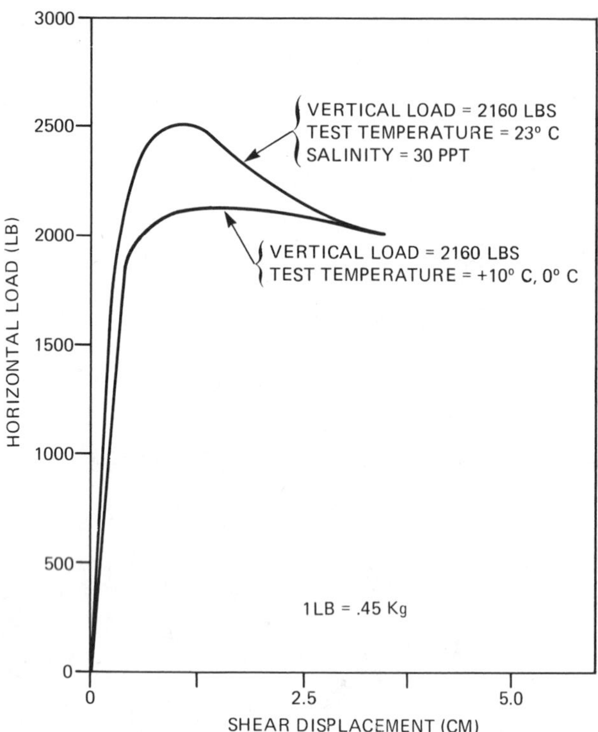

Figure 8 Typical Load-Displacement Results

The large-scale direct shear testing system described allows uniaxial freezing of large soil samples and subsequent direct shear testing when desired thermal conditions have been obtained at the shear plane. It has the capability to induce horizontal shear failure, a principal design failure mode, allowing simultaneous application of a vertical load to simulate overburden pressures present in situ. This testing system can be used to characterize the mechanical behavior within the freeze-front zone of saline-saturated granular material.

Typical test results are presented for three well-graded sandy-gravel samples. Two distilled-water-saturated samples, one completely unfrozen and the other with the freeze front coincident with the shear-plane, were found to exhibit very similar mechanical behavior. A saline-saturated (30 ppt) sample tested with a temperature of -2.3°C at the shear plane exhibited at 15% increase in peak strength and no increase in residual strength over the unfrozen material.

REFERENCES

Mahar, L. J., Vinson, J. S. and Wilson R., 1982, "Effects of Salinity on Freezing of Granular Soils," Proceeding of the Third International Symposium on Ground Freezing (ISGF), Hanover, New Hampshire, USA.

Ogata, N. Yasuda, M. and Kataoka T., 1982," Salt Concentration Effects on Strength of Frozen Soils," Proceedings, Third ISGF, Hanover, New Hampshire, USA.

Prodanovich, A., 1979, "Geotechnical Criteria," Technical Seminar on Alaskan Beaufort Sea Gravel Island Design, Exxon Corporation, Anchorage, Alaska, USA.

Roggensack, W. D. and Morgenstern, N. R., 1978, "Direct Shear Tests on Natural Fine-grained Permafrost Soils, "Proceedings, Third International Conference on Permafrost, Edmonton, Canada.

Sego, D. C., Schultz, T., and Banasch, R., 1982, "Strength and Deformation Behavior of Frozen Sand With a Saline Pore Fluid," Proceedings, Third ISGF, Hanover, New Hampshire, USA.

Thomson, S. and Lobacz, E. F., 1973, "Shear Strength at a Thaw Interface," North American Contribution, Second International Permafrost Conference, Yakutsk, USSR.

DEFORMABILITY OF CANALS DUE TO FREEZING AND THAWING

Sun Yuliang

Water Power Research Institute, Northeast Design Institute,
Ministry of Water Conservancy and Electric Power
Changchun, People's Republic of China

The destruction of canals by frost heave and thaw settlement is a serious problem in the northern regions of China. To lay down effective measures for preventing frost damage, the author has investigated the deformation characteristics of canals during freezing and thawing periods in situ. The results show that frost heave differs along the cross section of a canal: it is greatest at the bottom, smaller on the slope, and least on the top. The length of the slope is shortened during freezing and extended during thawing, and depends mainly on the amount of frost heave at the foot of the slope. Freezing and thawing will cause the collapse of canals and destroy the lined pavement, due to the distortion of concrete slabs, soil erosion, and thawing collapse. Based on the data, the authors have worked out formulas for estimating both the amount of slope shortening and the displacement of concrete slabs, and propose measures to protect canals from frost damage.

To study the reasons of the canal's frost damage for the lay down methods for preventing it, we went into field observations at 1979-1982 in some irrigation districts of Jilin province. Observed canals included soil canals, cement lining canals, plastic film lining canals, etc. This paper is the summation of these observations.

DEFORMATION IN CANALS DURING PERIODS OF THAWING AND FREEZING

Distribution of Frost Heave Along the Cross-Section on a Canal

Frost heave along the cross-section of a canal is unequal due to the effects of the shape of the canal, differing temperature conditions due to differences in height, and, most importantly, the diversity of moisture conditions. The drainage at the top of a canal is better than at the bottom, the moisture content at the top is less than at the bottom, and the distance to the level of ground water is greater at the top of the canal than at the bottom. Thus the conditions of moisture replenishment during freezing periods is different. The result is that the largest frost heave is at the bottom, less heave occurs in the middle part and the least heave occurs at the top of canal.

Investigations and observed data show that such deformation characteristics are abundant and universal and are the principal form of cross-sectional deformation in canals.

In some deeply excavated canals the distribution of frost heave along the cross-section is greater in the middle or at the upper part and smaller at the lower part. This behavior has been observed when the level of ground water is higher than or approaching the canal bottom, and at the same time the soil beneath the canal has a greater density and smaller permeability than soil beneath the sides. For example, at a certain section on the Xiangyang reservoir east main canal in Yushu county of province Jilin, the moisture content in the soil is over 30%, the maximum frost heave actually measured is 43 cm, the corresponding frozen depth is 149 cm, frost heave ratio is as large as 40.5%, and the distribution of frost heave takes the form of larger at the upper part and smaller at the lower. Figures 1, 2, and 3 show the distribution of maximum frost heave along the cross-sections of Songqian No. 1 main canal, Xiangyang west main canal and Xiangyang east main canal, respectively. TABLE 1 gives characteristics of the three observed canals.

In engineering practice, canals with cross-sectional frost heave amount larger at the lower

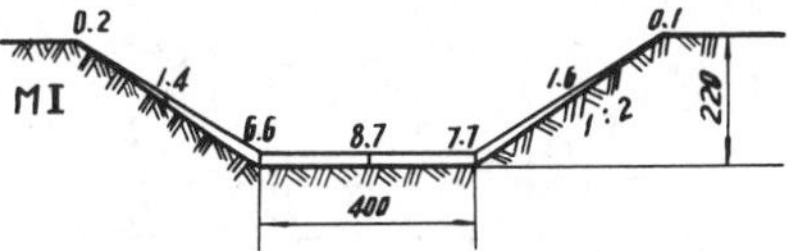

FIGURE 1 Distribution of frost heave amount on the cross-section of a section of earth canal on the Songqian No. 1 main canal, in cm.

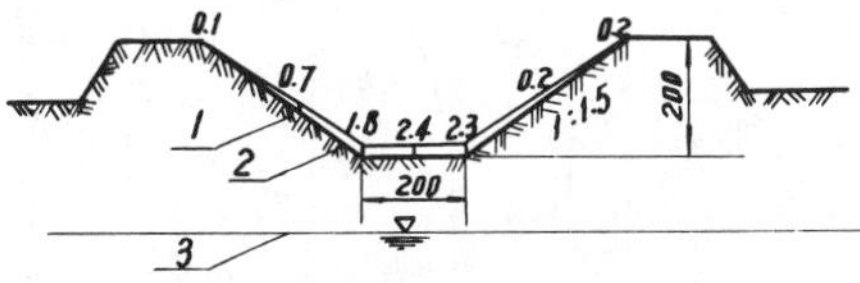

FIGURE 2 Distribution of frost heave amount on the cross-section of the Xiangyang west main canal, in cm. 1 Line of cross-section of the canal after freezing.2. Line of cross-section of the canal in original state.3.Ground water level.

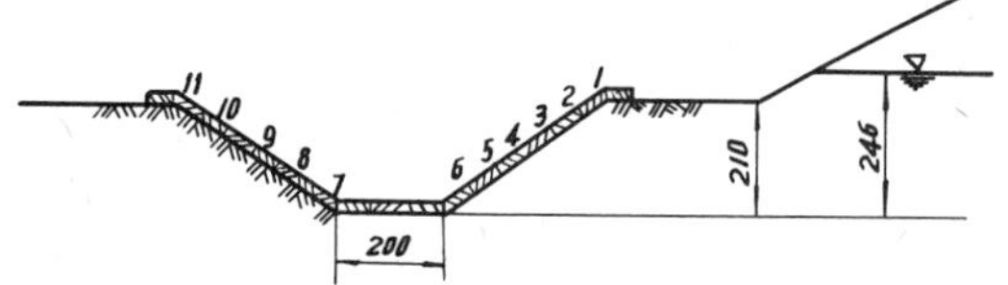

No.	1	2	3	4	5	6	7	8	9	10	11
Frost heave amount	366	368	364	331	312	298	283	287	302	306	262

FIGURE 3 Distribution of frost heave on the cross-section of the Xiangyang east main canal, in mm.

TABLE 1 Characteristics of three canals.

Name of canal	Songqian No1.main canal	Xiangyang west main canal	Xiangyang east main canal
Flow at canal head (m^3/sec.)	5.0	0.6	0.4
Width of bottom (m)	4.0	1.2	1.2
Depth of canal (m)	2.5	2.0	2.1
Ratio of canal (m)	2.0	1.5	1.5
Class of soil	MI	CI	CI
Depth of ground water (m)	14-15 Below ground surface	1.3 Below bottom of canal	2.4 Above bottom of canal
Method of construction	Excavating	Excavating & filling	Excavating

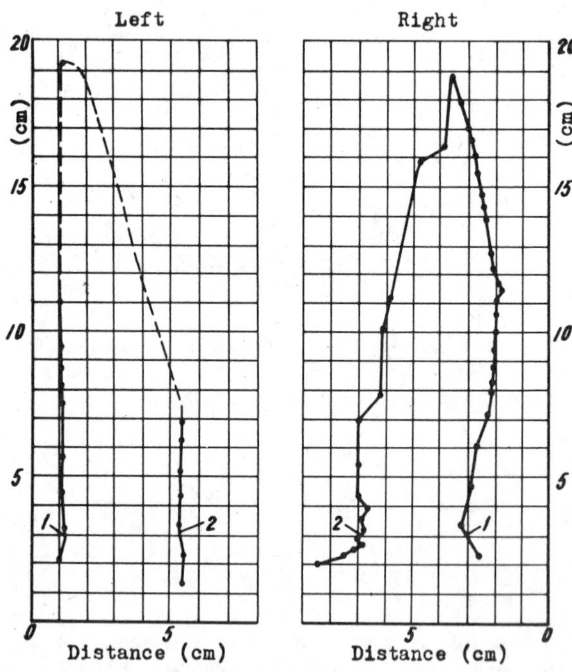

FIGURE 4 Process curve of freezing and thawing at the base of slope, Xiangyang east main canal. 1 Frost heave, 2 Thawing process.

part and smaller at the upper are more common, therefore this article will lay emphasis on discussing the regularity of this kind of deformation.

Regularity of Deformation Along Canal Cross-section During Freezing and Thawing.

Frost Heave Deformation at the Bottom of the Canal: During freezing, the locus line of frost heave deformation at the toe of the slope is a straight line perpendicular to the canal bottom, as shown in Figure 4. Such frost heave deformation at the toe of the slope indicates that during the period of freezing the width of the canal bottom basically remains unchanged or changes only little. These observations also illustrate that the connecting point of the canal bottom and the slope on the whole has no lateral displacement. As seen from the distribution of frost heave amount, it appears that there is a little more frost heave in the middle of the bottom than at both ends. Frost heaving at the toe of the slope is restrained by the slope, while the restraint action in the middle part of the bottom of the canal is relatively small.

Frost Heave Deformation and Displacement Direction on the Slope Surface: As mentioned above, when the distribution of frost heave amount on the slope surface is larger at the lower part and smaller at the upper, ie. the frost heave amount at the base of the slope is larger than on the slope and on the top of slope, the frost heave ratio on the lower part is all along larger than that on the upper part. As the frozen layer thickens and its strength increases, frost heaving at the canal bottom affects the deformation of each point on the slope. As seen from Figure 5, for point 5 at the toe of the slope, because of its nearness to the canal bottom, at only 18.3 cm, its frost heave condition is actually just the same as that at the bottom, therefore its orientation is vertical to the bottom of the

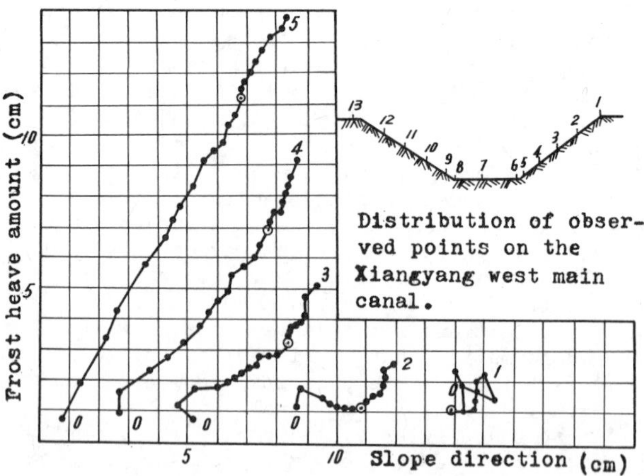

FIGURE 5 Locus of the displacement of observed points on the face of right side slope on the Xiangyang west main canal. 1,2,3,4,5....number of observing points, ⊙.....observed points on 6 Jan. 1981.

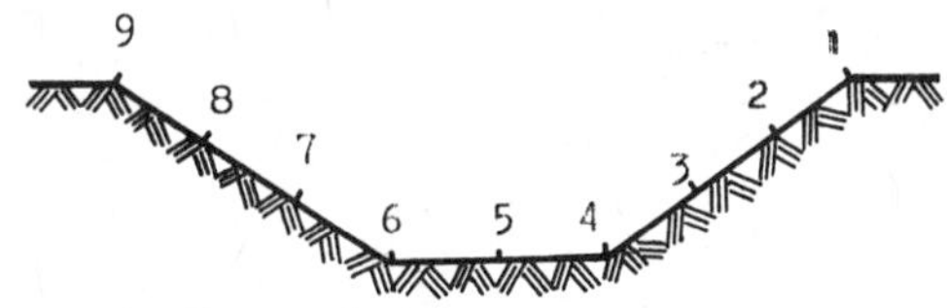

FIGURE 6 Observed points.

canal. The orientation of displacement of other points on the slope surface retains an including angle with the slope surface smaller than 90°, and the value of θ smaller while nearing the top of the slope. At point 2 near the top before Jan. 6, $\theta=0$ ie the displacement orientation is mainly along the direction of slope. The deformation value of each point on the slope surface, in addition to its own frost heave, also includes part of the deformation from the upward thrust caused by frost heaving at the canal bottom.

Shortening of Slope During Freezing: Although there is a difference between the frost heaving at the upper and lower parts of the canal slope, there may be no lateral displacement at the toe of the slope during the period of freezing. Therefore, as the cross-sectional line at the canal bottom is lifted during freezing, the length of the canal slope surface line is shortened. The amount of the shortening is related to the difference between the frost heave amount at the toe of the slope and at the top. The larger the difference, the greater the amount of shortening, and vice versa. TABLE 2 gives a practical example showing the amount of displacement and the shortening amount along the slope direction of all the measured points on the right slope throughout the freezing period at the Xiangyang west main canal.

TABLE 3 shows the shortening amount of distances between measured points on the same cross-section from 1 Dec. 1981 to 11 March 1982. This could further prove that in the course of freezing, the canal slope shortens while the width of the canal bottom basically stays unchanged.

Though the line of the slope deformation during freezing is slightly curved, it could be taken as linear with only little effect on the length of the slope surface. In this way, the shortened amount could be expressed approximately by the

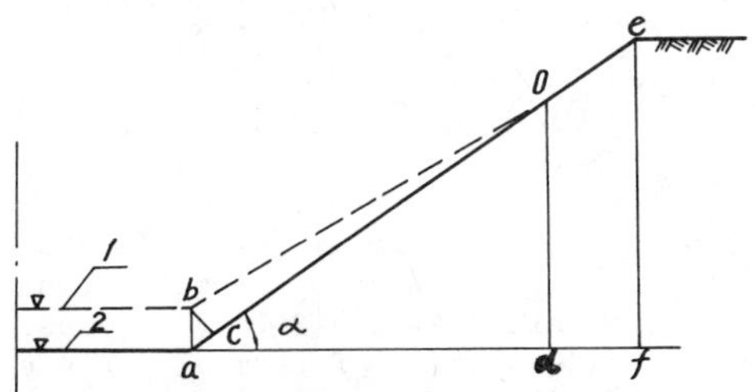

FIGURE 7 Diagram of contracted amount on the side slope. 1 Line of cross-section of the canal after freezing, and 2 Line of cross-section of the canal in original state.

TABLE 2 Displacement and Contraction Between the Observed Points on the Side Slope of Xiangyang West Main Canal.

Point	Displacement (cm)		Difference of displacement (cm)
	Plumbing	Slope facing	
1	0.5	-0.5	2.6
2	2.6	3.1	1.0
3	5.6	4.1	1.8
4	10.6	5.9	1.6
5	14.1	7.5	
Total displacement (cm) (Contraction of side slope)			7.0 (up to Mar. 20)

TABLE 3 Measured Distance between Observed Points on the Slope Face of Xiangyang West Main Canal. (1981 — 1982)

Position	Observed points	Contraction of side slope (cm)	
		Sectional	Total
Right side canal slope	1 - 2	2.4	3.0
	2 - 3	1.1	
	3 - 4	+0.5	
Bottom of canal	4 - 5	0.2	-0.3
	5 - 6	-0.5	
Left side canal slope	6 - 7	1.6	4.4
	7 - 8	0.9	
	8 - 9	1.9	

following formula:

$$\Delta B = B - \sqrt{B^2 - 2Bh.\ \mathrm{Sin}\alpha + h^2} \qquad (1)$$

Where ΔB is the shortened amount of slope, in cm

B is the distance between two calculated points, in cm,

α is the original slope angle of the canal, in degrees. (See Figure 7),

when $\frac{h}{B} \leq 0.1$, eq 1 could be approximately expressed by:

$$\Delta B = h.\ \mathrm{Sin}\alpha \qquad (2)$$

For example, on 20 March 1981, the frost heave amount measured at the foot of the right slope of the canal is 14.8 cm, and the frost heave amount at the top of the canal is 0.5 cm, giving a value of difference of frost heave amount h=14.3 cm. The calculation gives $\Delta B = h.\ \mathrm{Sin}\alpha = 14.3 \cdot \frac{1}{1.8} = 7.9$ cm while the value from actual measurement is 7.0 cm (see Table 2).

Deformation of Canal Cross-Section During Thawing: The side slope of a canal shortens during the period of freezing, but in the process of thawing it generally resumes its shape. Following the thawing and settlement of the toe of the slope, the canal bottom with the down-slide of the thawed layer and the thawed layer at the canal bottom is compressed and narrowed. TABLE 4 shows the actual measured data at the Xiangyang west main canal. Under the condition of a frost heave amount difference value at 16.3 cm, the right slope surface with a length of 314 cm and

TABLE 4 Measured Distance between Observed Points on the Cross-section of Xiangyang West Main Canal.

Position	Observed points*	Distance between observed points (cm)			Difference (cm)	Total difference (cm)
		Mar.16	Apr.13	Apr.29		
Right side canal slope	1 - 2	80.5	81.8	83.5	3.0	6.7
	2 - 3	70.0	73.0	71.2	1.2	
	3 - 4	76.0	73.0	80.0	4.0	
	4 - 5	58.0	58.8	59.0	1.0	
	5 - 6	30.0	29.3	27.5	-2.5	
Bottom of canal	6 - 7	102.5	100.0	100.0	-2.5	-6.8
	7 - 8	106.5	104.5	102.2	-4.3	
Left side canal slope	8 - 9	40.5	41.3	41.1	0.6	7.1
	9 - 10	46.5	42.0	47.5	1.0	
	10 - 11	50.0	52.0	52.0	2.0	
	11 - 12	71.0	74.0	75.5	4.0	
	12 - 13	30.0	29.5	29.5	-0.5	

* See Figure 5 for numbers of observed points.

the left slope surface with a length of 238 cm have a downward slide of 6.7 cm and 7.1 cm, respectively. During the same period, the width of the canal bottom contracts 6.8 cm.

Under the condition of a very large frost heave amount and a very high moisture content in frozen soil, as in Figure 8, downward slide or slope slipping of the embankment of the canal often occurs.

During thawing, the moisture in the upper part of the soil drains away through the pores in the surface layer of the thawing soil, and the upper part of the soil settles under its own weight and is compacted, reducing its permeability. The drainage condition of the thawing soil in the deep layer is due to heaving of a frozen layer below it and a thawing layer with small permeability over it. Consequently, this thawing soil layer with high moisture content slides down along the surface of its underlying frozen layer, simultaneously creating slope slipping with the thawed soil above it. Such a hazardous freezing-thawing phenomenon frequently occurs at the sections of deep excavation at pumping stations and canals, giving great trouble to the management of engineering works. Every year large amounts of silt deposits must be removed before irrigation to facilitate channelling water into the fields.

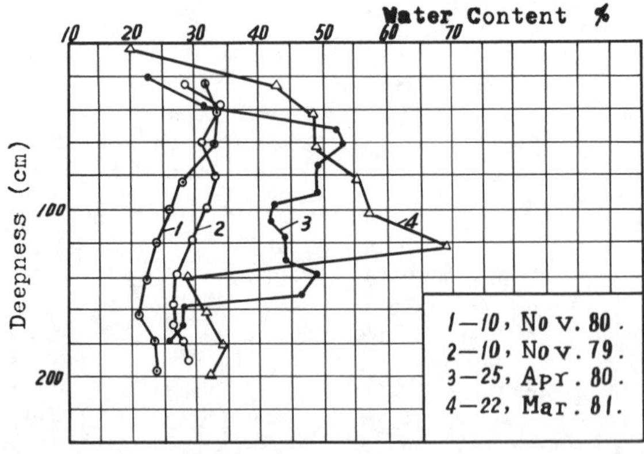

FIGURE 8 Variation of water content before and after freezing, Xiangyang east main canal.

FROST HEAVE DEFORMATION OF CANAL AND LINING PAVEMENT

To reduce the canal seepage loss, resist erosion, and ensure the stability of the canal, various types of lining pavement should be provided for canal. However, in cold regions, since the lining usually is not adaptable to frost heave deformation, frost damage problem is generally present in engineering works. There are a lot of different kinds of materials and different types of lining for canal, here we will lay emphasis on discussing concrete plate lining for trapezoid canal, which is a representative canal type widely used in engineering practice and has more frost damage problems.

The Main Cause of Destruction

The difference of frost heave amount between two certain points at the foot of and on the top of slope or on the slope surface arouses the contraction of the slope and causes the bottom face of the lining plate to receive a contracting pressure resulting from the freezing force between the lining plate and soil of the slope surface. These are the main reasons for the destruction of lining plate. Under such condition of reception of force, the lining plate would be subjected to different forms of destruction of different types.

When the lining plate bends and force is received at a deviated centre, the lining plate will be broken at point 0 as Figure 9, furthermore, it will heave up and hollow out in the process of slope shortening as Figure 10. When there is longitudinal expansion joint between

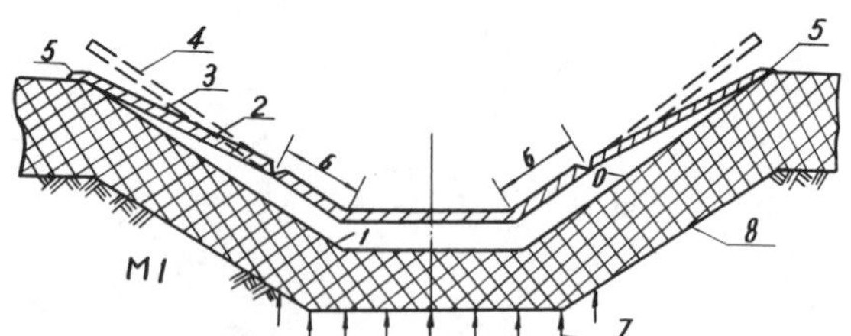

FIGURE 9 Deformation of lining plate in the process of freezing. 1 Line of cross-section of the canal in original state, 2 Line of cross-section of the canal after freezing, 3 Lining plate of slope after bending, 4 Lining plate hollowed out down below, 5 Sealed top with cement sand grout, 6 Lining plate and soil bed frozen together, 7 Direction of frost heave at the base of freezing soil bed, 8 Frozen soil bed.

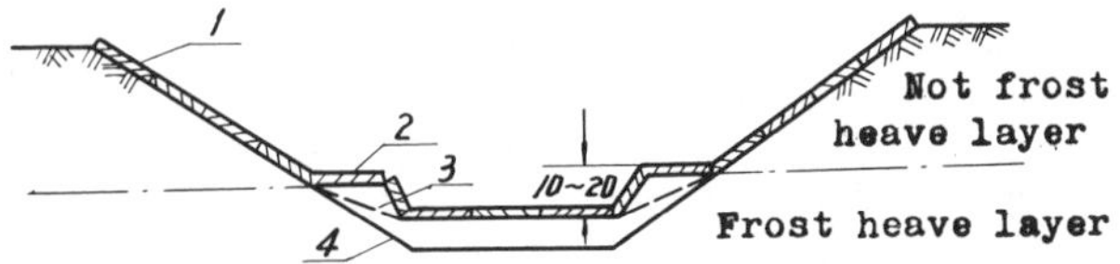

FIGURE 10 Lining plate bulging and hollowed out down below, in cm. 1 Lining plate in 50X50X8, 2 Lining plate swelling and caving, 3 Line of cross-section of the canal after freezing, 4 Line of cross-section of the canal in original state.

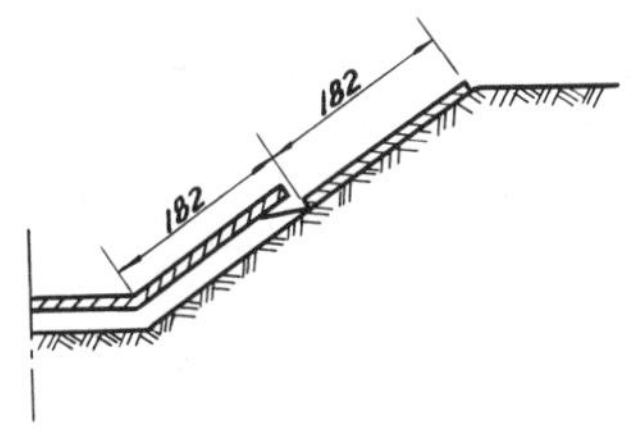

FIGURE 11 Displacement of concrete plate at middle seam, by action of frost heaving, in cm.

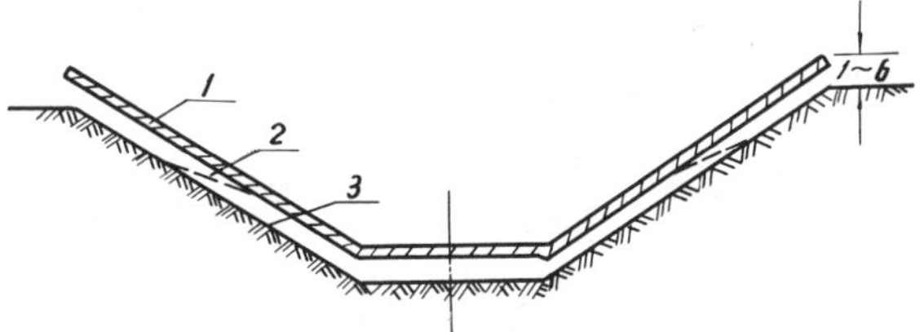

FIGURE 12 Bulging of lining plate at top of the embankment, in cm. 1 Lining plate, 2 Line of cross-section of the canal after freezing, 3 Line of cross-section of canal in original state.

lining plates on the slope surface, and if the plate above the joint remains unmoved while that below it heaves up gradually following the frost heaving of the slope surface, then the plates on the opposite sides of the joint will be staggered as (FIGURE 11). When the plate on the upper part does not stick very strong with the base soil by freezing or plastic thin film or sand gravel underlying layer is provided, then comparatively thick surface plate will move upward along the slope surface in a whole piece and protrude (Figure 12). When the unequality of frost heave amout along the slope surface is comparatively great, the middle and upper part of the lining plate on the slope surface would often hollow out as shown in Figure 9. If the plate is rather thin and the section hollowed out is rather long, then the plate will fail and break under its own weight. If prefabricated plates are used for protecting the surface, several terraces may take place.

The destruction location of the lining plate from bending is related to the deformation state of the slope surface during freezing and the restrictive condition of the frozen layer to the plate body, and generally it is at a place about 40-60 cm above the bottom of the canal. After the appearance of crack, the heaving up and hollowing out of the lining plate will occur in the process of continuous frost heave displacement at the bottom. The frost heave hollow-out amount can be calculated by the following formula:

$$H=\frac{1}{2}\sqrt{2B.\ \Delta B-B^{2}} \qquad (3)$$

Where H is the frost heave hollow-out amount in cm, B and ΔB share the same symbols in eq (1).

The frost heave hollow-out amount H is related to B and ΔB. When B=100, 200, and 400 cm, and if ΔB=5.5 cm, the amount of frost heave deformation aroused would be 16.3, 23.0, and 33.0 cm respectively. It could be seen that a small frost heave amount might bring about a big heave deformation. In the investigation of frost damage, it could often be seen that the concrete plate near the foot of the slope heaves up outwards and vertically to the slope surface and the height of the upheaving is about 10-20 cm, a result of such frost heaving action. Therefore, in the design of canal lining, measures should be taken to prevent the heaving-up and hollowing out of the plate as far as possible.

Types of Lining

To alleviate the damage of frost heaving to the lining, it is necessary to raise the seepage prevention property of the lining so as to lower the moisture content and the frost heave amount in the base soil on the one hand and to select the lining materials that are adaptable to the frost heave deformation in the base soil on the other hand. In cold regions, soft lining is preferable. Engineering practice shows that using thin membrane, plastic thin film, asphalt felt, etc. to prevent permeation and prefabricated concrete plate as protective layer is a better way for lining. Its adaptability to deformation depends on the length of the prefabricated plate and the material used to fill the seams. Previous calculation indicates that under the same conditions of frost heaving, the amount of upheaving and hollowing out of prefabricated plate is directly proportional to the length of plate (along the direction of slope). With the smaller length of plate, the up-heaving and hollowing out amount is smaller and subsequently its adaptability to the frost heave deformation is higher. We are of the opinion: the length of the prefabricated plate running in the same direction of the slope should be 50 cm for the canal with greater frost heaving amount, while 100 cm for the canal with smaller frost heaving amount. The seam between the prefabricated plates should be at about 1.5 cm which is to be filled with stiff asphalt sand mortar, a mortar won't be squeezed out at high temperature, serves as a buffer between plates during frost heaving and restores the prefabricated plates to their original position under their own weight during thawing in the base soil. If the plates are directly connected and cement mortar is used for filling the seams, they are easily breakable, and once up-heaved and hollowed out, for the great friction between plates, restoration to their original position following thawing is impossible, thus reducing their protective ability to the thin film.

Lining of canal, besides as a means to prevent seepage, is also used to guard against erosion and wash out from freezing and thawing. If a layer for antifiltration and drainage is added under the antiseepage thin film, under certain condition, the action of slope slipping during freezing and thawing could also be resisted. If a layer for antifiltration and drainage at 15-20 cm thick is added to the canal lining for small size side slope, the consolidation by drainage in the surface soil layer could be accelerated during thawing. It would together with the lining resist slopeslipping during thawing in the deep layer. For the stability of side slope, the degree of slanting of the slope may be reduced. For the canal with deep excavation, especially when the level of ground water is much higher than the bottom of the canal, for the sliding force is too great and lining is ineffective to prevent sliding of the slope, then ground pipe should be considered for use.

REFERENCES

Orlov, V.O., 1962, Frost heave of fine-grained soils. Moscow: U.S.S.R. Academy of Sciences Publishing House, p. 170-172.

Orlov, V.O., Dubnov, Yu., and Merenkov, N.D., 1974, Heave of freezing soil and its influence on foundations of structures. Leningrad: Stroyizdat, p. 138-150.

VENTIFACTS AS PALEOWIND INDICATORS IN A FORMER PERIGLACIAL AREA OF SOUTHERN SCANDINAVIA

Harald Svensson

Geographical Institute, University of Copenhagen, Copenhagen, Denmark

Studies of ancient wind activity have been performed in a coastal area of southern Sweden, situated below the highest shoreline of Late Weichselian time. The area is composed of fluvioglacial and eolian deposits. Cultivated fields show very distinct patterns (crop marks) of cast ice-wedge polygons. Pavements of intensely blasted stones are found below an eolian sand cover. Many ventifacts are sculptured with two facets. Larger, well-anchored ventifacts have been studied in detail (fluting and grooving) to determine the active wind direction. The main directions of wind-blasting are from the ENE-ESE and W-SW sectors. The east-facing facets are the most numerous and generally also represent the greatest reshaping of the stones. According to shoreline chronology, the area was deglaciated about 13,000 B.P. and constituted a periglacial environment. During this period the area was constantly influenced by steady, easterly winds from the nearby ice cap, shaping the east-facing facets. The west-facing facets are considered to have been formed during the Allerød period (11,800 to 10,900 B.P.).

LOCATION AND EMPIRICAL FINDINGS OF THE PERIGLACIAL ENVIRONMENT INVESTIGATED

The research area (Figure 1) constitutes a plain at the Laholm Bay, southern Sweden (N56^{0}30', 13^{0}E), situated below the highest shoreline (55-60 m a.s.l.) of Late Weichselian time and composed of fluvioglacial sand and gravel, which was reworked by the sea during the isostatic upheaval of the land. Formerly having been a heath, the area is now intensely cultivated.

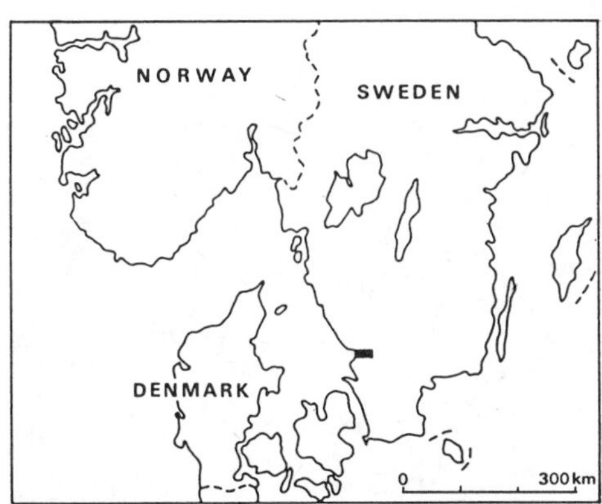

FIGURE 1 Index map. The location of the area investigated is shown by the black rectangle.

Indications of large-scale polygon patterns were first observed as faint lines on aerial photographs (Svensson 1962). Later the presence of polygons was demonstrated as crop marks (Figure 2) outlining patterns of higher-growing cereals (barley and oats). In vertical sections cut through the lines, funnel-formed casts show up (Figure 3), so the observations clearly confirm the existence of a fossil pattern og ice-wedge polygons. In some parts of the area, dry valleys occur whose existence can clearly be associated with the polygon pattern that indicates a thermokarst origin of the valley

FIGURE 2 Oblique aerial view of fossil ice-wedge polygons in cultivated fields.

FIGURE 3 Pseudomorph of ice wedge. Due to higher moisture content the infilling contrasts clearly with the coarser side soil.

trains. In total, the geomorphologic evidence proves the area to have been underlain with permafrost during a period after the withdrawal of the Weichselian ice sheet.

PRESENT-DAY WIND ACTIVITY

The top soil of the cultivated area is mostly made up of sand, and sections indicate the origin to be wind accumulated. At the Laholm Bay, coastal dunes are active, and by means of aerial photographs, contours of fossil dunes can be observed also in some fields located at a greater distance from the sea. Stones are not frequent in the cultivated fields, but some stones found show wind-blasted surfaces.

During dry seasons, especially in early spring, sand drift normally occurs for some days. The blasting may thus be a present-day phenomenon, but it may also be of an older date. Anyhow, these blasted stones cannot be used for a determination of active wind directions, as they have been turned again and again during cultivation.

PAVEMENTS OF OLD WIND-BLASTING

During the research of the fossil periglacial features of the Laholm area, gravel pits have regularly been observed for studies of sections since 1962. During these inspections a pavement of wind-blown stones was found at the village of Våxtorp, 14 km from the present coast, and later also at the village of Veinge, 12 km farther to the north and at a distance of 8 km from the coast. The altitudes are 54 and 55 m a.s.l. respectively, which means a few meters below the highest shore level of Late Weichselian time.

At both localities the pavements are composed of well-rounded, 5-20 cm blocks of granite and gneiss. Geomorphologically they belong to a fluvioglacial delta complex accumulated at the highest shoreline during the deglaciation. The blocks are well anchored in the underlying gravelly material and have for a period of time constituted a sea-washed zone during the isostatic upheaval of the area. Later the stone pavement was exposed to deflation and thereafter preserved under a cover of wind-blown sand. In the sometimes more than 1 m thick sand cover, fossil frost structures occur (Svensson 1973a), a fact that proves the wind-blasting not to be recent or subrecent but belonging to an early period of an arctic climate.

The wind-blasting of the pavement is evidenced by a smooth polish with furrows, or elongated hollows starting from rock structures or mineral aggregates in the stone surface. The intensity of the blasting is, however, most conspicuous in the resulting facets which, because of the well-rounded form of the stones, constitute a distinctly outlined new element of the surface (Figure 4).

PREPARATION AND SELECTION OF STONES FOR ANALYSIS OF ACTIVE WIND DIRECTION

From the first inspection of the pavements it was obvious that many of the wind-worn stones show-

FIGURE 4 Ventifacts from the Våxtorp gravel pit.

ed facets. In order to determine whether any significant directions of wind erosion could be established, 325 blocks were collected at the Våxtorp locality and 400 at the Veinge locality. At Våxtorp the blocks were dug out in a vertical section, the strike and dip of the block surface were marked and noted, and the blocks brought for analysis in a laboratory with adequate light conditions. Out of the 325 blocks, 100 were chosen for detailed analysis. Criteria of the choice were (besides marked wind-blasting) that the blocks were larger than 10 x 10 x 5 cm and had an almost plane bottom surface. Taking into consideration that the pavement is anchored in a gravel surface, it is supposed that the two characteristics are a fair guarantee that the blocks were hardly moved, turned, or dipped during deflation. At the Veinge locality the pavement was uncovered and the study thus made in a horizontal section (Figure 5). The selection and analysis of 100 stones were made in situ in the pavement, as adequate light conditions were provided on the sunlit ground surface. Otherwise the same criteria for selection of specimens were used as at the Våxtorp locality.

FIGURE 5 Part of uncovered pavement at Veinge. Most of the larger stones are modeled with facets.

It is worth noting that among the large number of blocks observed, no "Dreikanter" were found. The formation of "Dreikanter" is usually explained by the assumption that a stone was turned more times, facing new sides toward the eroding wind. Ventifacts with two facets have been found, one facing eastward and one westward, the east-facet always being the larger one.

DIRECTIONS OF ERODING WIND

The wind-worn microforms, wind furrows and elongated hollows (Figure 6), make possible a determination of active wind direction(s) on each stone. The directions are measured with a compass within 2^0, and the values placed in sectors of 22.5^0. From Table 1 it is evident that there exist two maxima for each locality, one main and one secondary maximum. At Våxtorp the main maximum is centered in the E-sector and at Veinge in the ENE-sector. Because of fewer observations the secondary maxima have not the same degree og significance. It is, however, worth noting that they are found in westerly sectors: at Veinge in the WSW-sector and at Våxtorp in the SW-sector.

FIGURE 6 Close-up view of block in situ abraded by easterly wind (from the left).

TABLE 1 Initial Directions for Wind-Blasting

	N	NNE	NE	ENE	E	ESE	SE	SSE	S	SSW	SW	WSW	W	WNW	NW	NNW
Veinge	1	3	18	33	27	12	3	0	1	0	4	7	4	5	1	0
Våxtorp	0	3	8	10	34	20	17	3	1	1	13	8	7	4	4	0

Number of investigated blocks is 100 at each site. Several blocks show wind-blasting from more than one direction.

Based on the frequent occurrences and the intense reshaping of the blocks, it can be stated that the blasting from the east has been far more active than the abrasion by westerly winds. A few ventifacts have been found on which an east-facing facet meets a westward facet in an intervening edge. On such stones the westward facet was developed over the eastward facet, which thus gives their relative age, namely the westward facet was last formed.

CHECK OF THE DETERMINATIONS OF ACTIVE WIND DIRECTIONS

After detailed studies of ventifacts at Våxtorp and Veinge had been performed, two new observations were made that can be used to test the reliability of the measurements of active wind direction on the ventifacts.

The first observation was made in the gravel pit at Våxtorp, where a huge ventifact (side lengths, 4.4, 3.2, 3.1, and 2.1 m; height, 2.0 m) is exposed. The upper part of the block was embedded in wind-blown sand, whereas the basal part lay enclosed in the original beds of fluvioglacial gravel. Because of its dimensions and stable foundation this block must be considered stationary under the deflation phase and its records of wind direction consequently of greater reliability than those of the smaller ventifacts. The block is intensely blasted and contains a lot of facets, the largest being 1.7 m wide (Figure 7). On the eastern, upper side of the block a very distinct direction of wind erosion is found all over the surface, namely from S76E. This is in good agreement with the result of the measured directions on ventifacts in the same gravel pit, where the most frequent observations were from E-ESE (Table 1). Furthermore, the good agreement between the wind direction observed on the block and in the sample of wind-blown stones justifies the methods used of measuring the direction of eolian abrasion in a great number of ventifacts, well anchored in the ground, to get a fairly good reconstruction of a main wind direction.

On the western side of the huge block the main direction of wind-blasting is found to be from N57W. This direction does not coincide with the secondary maximum of Table 1. The disagreement is probably due to the fact that the total number of observations of westerly wind-worn stones is too small.

Also, for the Veinge locality a chance of testing the measured wind direction appeared. Two

FIGURE 7 Part of large wind-eroded facet on the huge ventifact at Våxtorp.

kilometers from the pavement investigated an outcropping gneiss surface was found facing southeast and showing intense wind-blasting with strictly parallel wind furrows from N86E. As the main maximum of the measurements in the wind-worn pavement at Veinge is centered in the sectors E-ENE (Table 1), it again confirms the possibility of reconstructing a main wind direction from measurements on a large number of wind-blasted specimens. No check of the western wind-blasting found in the Veinge pavement could be performed in the outcropping gneiss surface, as it had no slope exposed to the west. The existence of blasting from the west in rocks of the Veinge area is, however, documented in a paper by Mattsson (1957), but without giving a more precise direction (number of degrees) of the active wind. Mattsson also observed an easterly, possibly periglacial wind activity, and from the coastal area further to the north, Hillefors (1969), besides frequent observations of easterly wind-blasting, mentions wind erosion from the west at lower levels.

CONCLUDING REMARKS CONCERNING PALEO-ENVIRONMENT AND AGE OF WIND EROSION

To decide the character of the environment in which the wind-blasting acted, it is at first important to state that the ventifacts are situated just below the highest shoreline of the area developed by wave erosion in Late Weichselian time. It means that the pavements once constituted a beach area. The fluvioglacial material below the pavement contains casts of ice wedges, which demonstrates that the area was underlain with permafrost. Furthermore, the cover of wind-blown sand shows thin infillings of the character of sand wedges, which indicates that the area was affected by frost fissuring also after the deposition of the eolian sand. From these facts it is quite clear that the wind-blasting took place in an arctic environment shortly after the isostatic upheaval of the innermost parts of the coastal plain.

From marginal deposits accumulated during the deglaciation of the area, it can be stated that the front of the receding ice cap had a NNW-SSE direction in the coastal area (Mohrén and Larsson 1968, Mörner 1969), indicating that the Våxtorp and Veinge localities had been subject to periglacial conditions outside the ice cap during its retreat eastward.

The steady easterly winds responsible for the intense wind-blasting of the pavements were caused by the ice cap, either as an element in the general circulation from a high-pressure cell, or simply as a catabatic effect just outside the ice margin. During this first period of wind-blasting the abrasive material may largely have been snow (ice crystals) at very low temperatures.

The younger westerly wind-blasting corresponds to a later period of deglaciation, when the glacial high pressure had moved northward, and west-wind systems could penetrate into the area. At that time, vast areas of sand deposits were set free from the sea due to the isostatic upheaval of the land and could supply material for the blasting as well as for the burying of the pavements.

Chronologically the eastern wind-blasting belongs to an early period of Late Weichselian time. According to a shoreline diagram of the Swedish west coast constructed by Mörner (1969), the Våxtorp and Veinge areas were raised above sea level 12,900 and 12,700 years B.P., respectively, which may thus be considered the earliest starting time of the intense eastern wind activity.

A chronological determination of the western wind-blasting is more difficult. As the covering sand shows frost fissures, the blasting must have been followed by a period of arctic climate. This fact points to the Allerød chronozone (11,800 to 10,900 B.P.), a period in which a west-wind regime is considered to have started. Furthermore, the Allerød period was followed by a period of cold climate, the Younger Dryas chronozone. That the area was really affected by frost-fissuring during this period is also documented by the existence of fossil ice-wedge polygons in the western, lower parts of the coastal plain (Svensson 1973b). To reach a more than hypothetical dating of the western wind-blasting, however, more observations of wind-worn blocks or outcrops, especially in lower levels of the area, are needed.

ACKNOWLEDGMENTS

The author is indebted to Mrs. Kirsten Winther and Mrs. Inge Hansen at the Geographical Institute, Copenhagen, for improving the English text and preparing the camera-ready copy, respectively, as well as to Mr. R. Laszlo at the Geographical Institute, Lund, for the photographic reproductions.

REFERENCES

Hillefors, Å., 1969, Västsveriges glaciala historia och morfologi: Meddelanden från Lunds Universitets Geografiska Institution. Avhandlingar 60, p. 1-319.

Mattsson, Å., 1957, Windgeschliffenes Gestein im südlichsten Schweden und auf Bornholm: Svensk Geografisk Årsbok 33, p. 49-68.

Mohrén, E. och Larsson, W., 1968, Beskrivning till kartbladet Laholm: Sveriges Geologiska Undersökning. Ser. Aa 197, p. 1-123.

Mörner, N. A., 1969, The Late Quaternary history of the Kattegatt sea and the Swedish coast: Sveriges Geologiska Undersökning. Ser. C 640, p. 1-487.

Svensson, H., 1962, Ett mönster i marken: Svensk Geografisk Årsbok 38, p. 95-104.

Svensson, H., 1973a, Fossil sand-wedges on the Laholm plain on the western coast of Sweden: Geologiska Föreningen i Stockholm förhandlingar 95, p. 144-145.

Svensson, H., 1973b, Distribution and chronology of relict polygon patterns on the Laholm plain, the Swedish west coast: Geografiska Annaler 55A, p. 159-175.

Åkerman, J., 1980, Studies on periglacial geomorphology in West Spitsbergen: Meddelanden från Lunds Universitets Geografiska Institution, Avhandling 89, p. 1-297.

HEAT AND SALT TRANSPORT PROCESSES IN THAWING SUBSEA PERMAFROST AT PRUDHOE BAY, ALASKA

D. W. Swift, W. D. Harrison and T. E. Osterkamp
Geophysical Institute, University of Alaska, Fairbanks, Alaska 99701

The thickness of the subsea thawed layer near the West Dock, Prudhoe Bay, Alaska, and its temperature, pore water salinity and pore water pressure are described. The pressure data are the first of their kind and provide good evidence for non-hydrostatic behavior that is probably due to pore water motion. Present knowledge of heat and mass transport mechanisms controlling the thawed layer development is summarized. Heat transport is conductive, but salt transport (necessary because temperatures are negative) is by pore water motion, probably on the order of a few tenths of a meter per year. What drives the motion is still uncertain. Driving by the release of relatively fresh, buoyant water by thawing must occur, but it may not provide enough energy to explain the observations. Driving by surface wave action seems unimportant, at least at West Dock.

INTRODUCTION

During the past decade or so of high interest in subsea permafrost, it has become increasingly obvious that subsea permafrost is a different material from its subaerial analogue. The most obvious reasons are the transient nature of subsea permafrost, and the important and complex role of salt in its evolution. The importance of salt is particularly obvious when ice-bearing subaerial permafrost is inundated by shoreline retreat, and a thawed layer develops beneath the sea bed even though mean annual sea bed temperatures are negative, over most of the Beaufort Sea shelf of Alaska. Much of the salt necessary for thawing in this negative temperature regime probably enters the sediments after inundation, although the transport of salt obviously would not be necessary in sediments that were sufficiently salty initially. Since lithology can influence the thaw rate by several orders of magnitude, it is clear that physical and chemical controls on the salt transport process must exist. In near shore regions where new subsea permafrost formation is rapid because of rapid shoreline retreat, its evolution in response to the salty, relatively warm sea water is correspondingly rapid. Prediction of the response of subsea permafrost to natural or man-made changes in sea bed boundary conditions given the lithology, is a task that cannot yet be accomplished because of an inadequate understanding of salt transport mechanisms.

This paper focuses on the observations and interpretations from the West Dock area, Prudhoe Bay, Alaska, that have a more or less direct implication for the heat and salt transport mechanisms operating in the thawed layer there. The data are described in greater detail in a series of reports, the most relevant of which are Osterkamp and Harrison 1976, 1980, 1982a; Harrison and Osterkamp 1977, 1981a. Additional data reduction has led to some small differences between the reports and this paper. Additional research has been carried out by several other investigators in this area (Sellman, 1980; Harrison and Osterkamp, 1982; Smith et al. and Morack and Rogers, in press). Data were obtained along a single line bearing N 31.5° E from North Prudhoe Bay State Number One well (Figure 1). The sediments are thick in this area, exceeding the permafrost thickness on shore of 560 m at the well. They consist mainly of alluvial material overlain by several meters of recent silty marine sediments. The shoreline retreat rate is roughly 1 m per year. The water is shallow, reaching 3 m depth about 3 km from shore. At the sites from which the data in this paper were obtained, the spring sea ice thickness usually equals or exceeds the water depth, although several hundred meters or more from shore it may occasionally be lifted by tide.

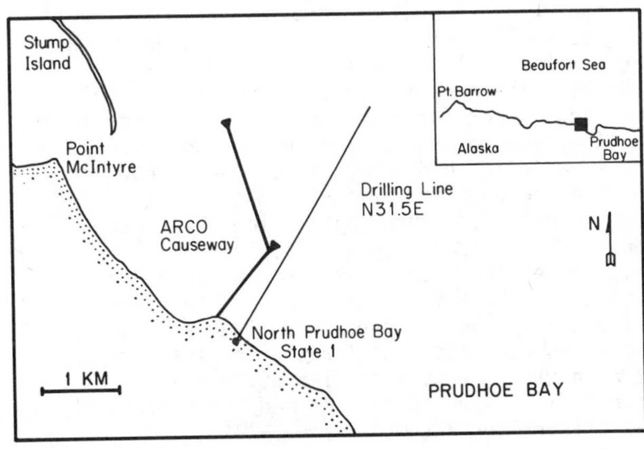

FIGURE 1 Location map of study area.

RESULTS

We begin with a summary of the results which give the most obviously direct information about heat and salt transport processes in the thawed layer at the West Dock. Old results are briefly summarized; new ones are described in more detail.

Theory

It has been shown (Harrison and Osterkamp 1978) that convection of the pore water in the thawed layer should occur at the West Dock, basically because relatively fresh and therefore buoyant water is generated by thawing of ice at the bottom of the thawed layer. This will be referred to here as gravity driven convection. To be effective the hydraulic conductivity of the sediments must not be too small.

Phase Boundary Definition

Present drilling and probing data indicate that the transition between ice-free and ice-bearing sediments at the bottom of the thawed layer, which we refer to as the phase boundary, is sharp at the West Dock. This implies that thawing is occurring in response to salt transport through the thawed layer, and not predominantly in response to salt that was present before the permafrost was submerged by changing shoreline position (Harrison and Osterkamp 1982).

Phase Boundary Shape, Shoreline Retreat Rate, and Mean Annual Temperature

Although there is an uncertainty of a factor of 2 or more, the shoreline retreat rate at the West Dock is on the order of 1 m yr^{-1}. To the extent that conditions have not changed in the last few thousand years or so, the offshore distribution of the thawed layer thickness, the phase boundry shape, can be interpreted to give the thaw rate over past time. The shape is shown in Figure 2. Little thawing seems to have taken place within 400 m of shore, but in the next 40 meters there is a "ramp" area where the phase boundary drops rapidly and more or less linearly. Beyond this the phase boundary is almost parabolic out to several kilometers from shore (Harrison and Osterkamp 1982, Osterkamp and Harrison, 1982b). The behavior of the ramp area is not yet understood, because its steep slope implies a thaw rate too rapid to be understood with any reasonable sea bed temperature model. However, beyond the ramp the parabolic shape must be due to the expected approximate $\sqrt{\text{time}}$ thickness dependence, and it seems fairly safe in this region to estimate thaw rate from phase boundary shape and shoreline retreat rate. The result (typically several hundredths of a meter per year) is much too fast for the necessary salt transport to be by molecular diffusion, given the typical mean annual sea bed temperature of -1°C (Harrison and Osterkamp 1978).

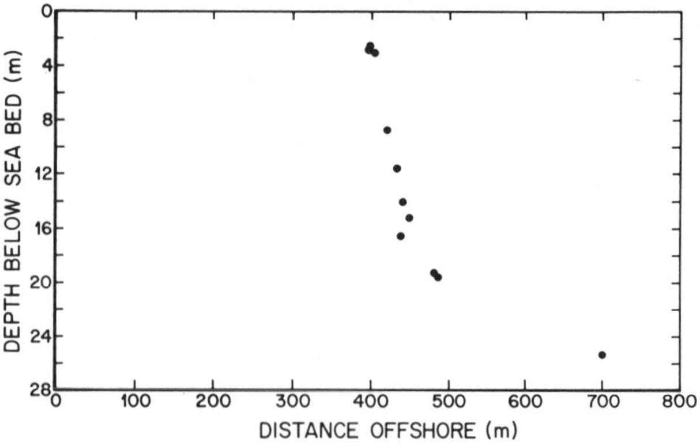

FIGURE 2 Depth to the base of the thawed layer.

Hydraulic Conductivity and Porosity

Hydraulic conductivity probably controls the transport regime through the thawed layer. Several profiles of it have been measured through the thawed layer in the course of pore water sampling, by determining the rate at which water enters the pore water sampling probe (Harrison and Osterkamp 1981b). These measurements tend to indicate a fairly constant distribution of hydraulic conductivity with depth, typically from 1 to 10 m yr^{-1}, which implies a scatter of perhaps a factor of 3 about some mean value. In the formulation of the simplest theory for heat and mass transport through the thawed layer based on Darcy's law (Harrison 1982), the porosity enters as well, but when it can be considered constant, only the combination, hydraulic conductivity to porosity ratio, enters. Unfortunately, probing does not determine porosity, and a value of 0.4 characteristic of deep samples on shore (Lachenbruch et al. 1982) is taken as representative. A typical hydraulic conductivity to porosity ratio is then 10 m yr^{-1}.

The uncertainty in porosity, the possible importance of anisotropy in hydraulic conductivity, and the method of measurement of hydraulic conductivity all leave something to be desired. The measurement problem is due to the possibility of clogging of the probe filter and filter guard through which the pore water is collected, and the fact that the incoming flow rarely obeys Darcy's law. The possibility of significant systematic errors in hydraulic conductivity to porosity ratio therefore still seems open.

Temperature

The linearity of temperature profiles below the depth of seasonal fluctuations suggests that heat transport, unlike salt transport, is essentially diffusive, or conductive; that is, it is not influenced appreciably by pore water motion. This means that the Peclet number for heat, a rough measure of the ratio of convective to diffusive transport, is significantly less than one, which translates into a typical vertical component of pore water velocity significantly less than 1 meter per year (Harrison and Osterkamp 1982, Osterkamp and Harrison 1982). Although not completely convincing, because of complications due to seasonal temperature variations and other factors, there is some evidence for curvature in some temperature profiles that could possibly be due to pore water motion on the order of several tenths of a meter per year (Figure 3).

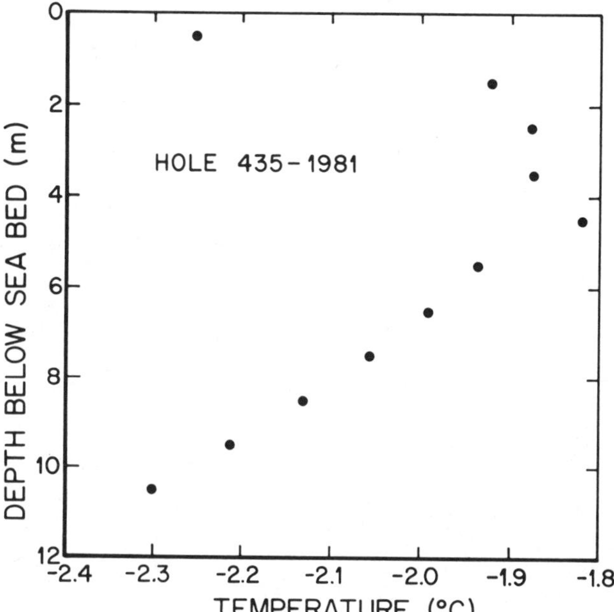

FIGURE 3 Temperature profile through the entire thawed layer thickness in hole 435-1981, June 2, 1981. The leading digits in the hole number represent the distance in meters offshore from a fixed mark near the tundra edge; it is the same reference mark used in the 1980 holes. Except for the point at about 4.5 m depth the temperatures are probably within a few hundredths of a degree of equilibrium.

Pore Water Chemistry

Although there are detectable differences, the pore water chemical composition resembles that of sea water (Iskandar et al. 1978, Page and Iskandar 1978). This may be evidence for a fairly efficient pore water flushing mechanism.

Pore Water Salinity

Pore water samples were collected through the thawed layer using lightweight driving equipment and specially designed sampling probes (Harrison and Osterkamp 1981b), and the electrical conductivity was measured in the lab to precision and accuracy of 0.5% or better. Since, as noted previously, the chemical composition is similar to that of sea water, salinity can be estimated from conductivity by standard methods. Recently measured pore water conductivity profiles exhibit most of the characteristics of those determined previously, notably the small gradients expected in a strongly convecting system. The large gradient observed in a very thin layer above the phase boundary, at the bottom of the thawed layer, is interpreted as a boundary layer over which the transport regime changes from convective to diffusive. The lower parts of two profiles from holes about 421 and 441 m from shore are shown in Figures 4a and 4b, along with probe driving information.

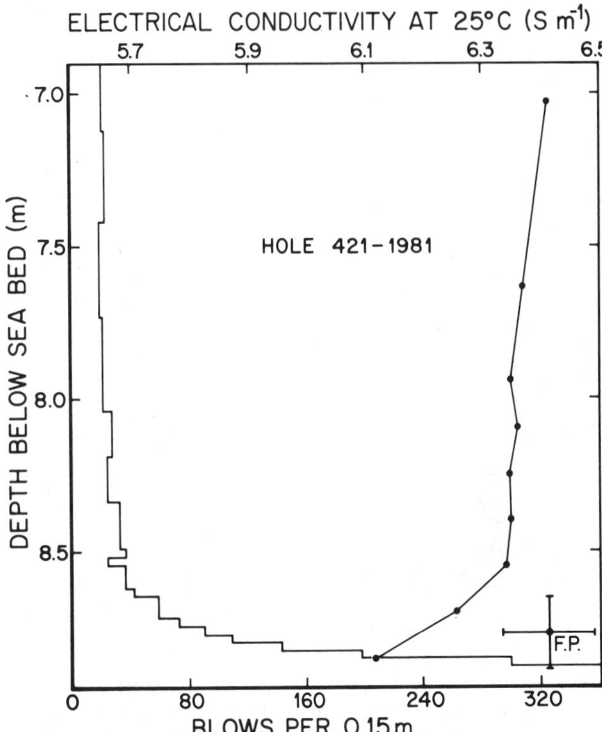

FIGURE 4a Pore water electric conductivity (right profile) and blows required to drive the sampling probe 0.15 m (left profile) at hole 421-1981, May 31June 1, 1981. Length of sampler was 0.1 m. The point labelled F.P. is the conductivity at the phase boundary as estimated from temperature data in hole 420-1981, 1 m closer to shore

The boundary layer feature is not so obvious as Figure 4 would indicate, because once the probe tip encounters ice, there is enough probe driving energy available to melt some of it, possibly diluting the samples and giving an erroneously low conductivity at the deepest point. Nevertheless, the conductivity has already begun to decrease at the second greatest depth, which is a consistent feature. However, the magnitude of the conductivity jump across the boundary layer cannot be estimated very accurately from these data. The result of an attempt to improve the estimate by determining the conductivity exactly at the boundary from measured temperature and freezing point data is labelled F.P. in Figure 4. The error bars are for a temperature uncertainty of 0.03 K, which may be optimistic, because while the agreement in Figure 4b is satisfactory, that in Figure 4a is not. It is concluded, however, that the conductivity (or salinity) jump across the boundary layer is typically very small, not much more than a few percent, which is similar to previous estimates at different sites (Harrison and Osterkamp 1982).

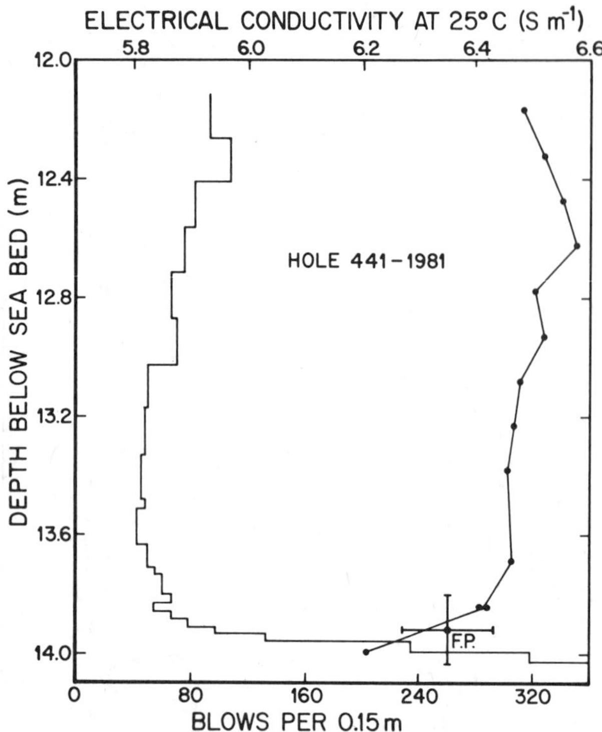

FIGURE 4b Pore water electrical conductivity (right profile) and blows (left profile) at hole 441-1981. May 28-30, 1981. The point labelled F.P. was estimated from temperature data in the same hole.

If convection of the pore water is the dominant salt transport mechanism, it is likely that the typical vertical component of velocity considerably exceeds the thaw rate (≈ 0.03 m yr^{-1} 700 m from shore). Otherwise a convecting parcel of water would have to give up most of its salt as it approaches the phase boundary, a condition that seems unlikely because of the small salinity variation in the thawed layer.

Pore Water Pressure

The most direct evidence so far for pore water motion comes from a pore water pressure profile measured through the thawed layer at one site. It was obtained with a commercial pneumatic piezometer connected to the filter assembly of a pore water sampling probe. The result from a hole about 441 m from shore is presented in Figure 5 as the difference between pressure and hydrostatic pressure. The latter was calculated from the density profile implied by the known electrical conductivity (salinity) of the pore water; density is almost independent of temperature under the prevailing conditions. The result shows a small deviation from hydrostatic pressure, which appears to be a real effect, but with two qualifications. First, the pressure gauge used was calibrated before and after use to standards traceable to National Bureau of Standards, but the transducer was not. However, the nonhydrostatic effect exceeds by an order of magnitude any effect of system nonlinearity as specified by the manufacturer (Sinco, Seattle, WA). Second, the effect could be caused by a deviation of the probe from the vertical of 10°, which seems extremely unlikely but not impossible. The nonhydrostatic effect, amounting to about 0.23 m head over a depth of about 14 m, appears to be real.

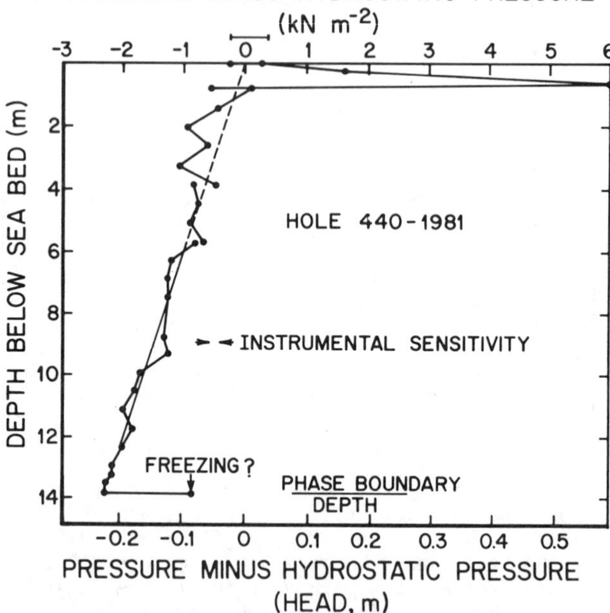

FIGURE 5 Pore water pressure minus hydrostatic pressure in hole 440-1981, May 27-30, 1981.

Certain other observations are also important. First, tide was measured by observing the level of water in a nearby hole through the sea ice, referenced to a probe driven securely into the sea bed; typical peak-to-peak amplitude was about 0.15 m. It is interesting that pore water pressure variations of approximately the same amplitude and phase occurred at least below the 5 m depth, although the ice did not float at the hole, and there was probably some seasonal ice formation in the sea bed above 1.45 m depth. The tide data used to correct the piezometer pressure data, which were obtained at different times and therefore at different tides. Second, failure to obtain expected system resolution (7 mm head) at depths less than 4 m may have been instrumental or operator related, or may have been due to the time varying in situ conditions. Third, an exponentially decaying pressure transient after driving the probe was observed at depths up to 3.3 m. Fourth, the point marked "freezing" in Figure 5 was obtained after the piezometer had been at the phase boundary for 34 minutes and could be due to freezing in or around the piezometer. Fifth, the high pressure at 0.54 m depth is probably related to the seasonal freezing in the shallow sediments indicated by temperature and conductivity data.

The solid line is a least squares fit to the data below 5.72 m. Its slope indicates an average head gradient of -0.016, which would indicate a downward vertical component of pore water velocity, the opposite direction to what would be caused by thaw consolidation, for example. Given a representative value of 10 m yr^{-1} for the hydraulic conductivity to porosity ratio discussed earlier, the implied velocity component would be 0.16 m yr^{-1}, which is of the expected order of magnitude in the light of previous discussion, and therefore suggests that the error in hydraulic conductivity to porosity ratio is not large.

Summary

The general conclusion is the following: Salt transport is rapid relative to molecular diffusion. The mechanism is probably convection of the pore water, with a typical vertical component of velocity on the order of a few tenths of a meter per year.

GRAVITY DRIVEN CONVECTION

Pore water motion can provide the necessary rapid salt transport, and good evidence for its occurrence now exists. However, to predict anything from first principles such as lithology and sea bed boundary conditions, we must understand what drives that motion. For example, is the energy source "external", driven by motion of the overlying sea water, or is it "internal", with the energy source below the sea bed? In the latter case the likely energy source would be the gravitational energy of the relatively fresh, and therefore buoyant, water released by thawing at the bottom of the thawed layer. This process will be called gravity driving, and we can investigate whether the energy it can supply is sufficient to explain the observations. A two-dimensional model is used.

If the pore water motion is governed by Darcy's law, it is straightforeward to show that the frictional power (F) dissipated in the pore volume of the thawed layer is given by

$$F = \int\int \frac{\rho_r g}{k} v^2 \, dxdy \qquad (1)$$

where g is the gravitational acceleration, k is the ratio of hydraulic conductivity and porosity, ρ_r is a reference density used in defining the head unit, v is the pore water velocity, and x and y are the horizontal and vertical coordinates. The power (E) available from gravitational potential energy release is given by

$$E = \int \beta \, \Delta\rho \, g \, Y \frac{dY}{dt} dx \qquad (2)$$

where $\beta < 1$ is the fraction of H_2O in the solid state just below the phase boundary ($\beta < 1$ if the salt content is unequal to zero there), $\Delta\rho$ is the density difference between the water at the sea bed and that at the phase boundary, Y is the thawed layer thickness, and dY/dt is the thaw rate. The maximum energy would be available if $\beta = 1$, and if $\Delta\rho$ had its maximum value ($\Delta\rho_{max}$) corresponding to fresh water at the phase boundary. In fact, the available mechanical energy will usually be considerably less, because of its transformation to chemical energy by diffusion. When spatial variations of Y and dY/dt are ignored, Equations (1) and (2) give the following constraint on the rms velocity ($\sqrt{v^2}$):

$$\sqrt{v^2} < \left[\frac{\Delta\rho_{max}}{\rho_r} \frac{dY}{dt} k \right]^{1/2}$$

For an apparently typical sea bed salinity of 43 $^0/_{00}$ dY/dt ≈ 0.03 m yr^{-1} (700 m from shore), and k ≈ 10 m yr^{-1} (Harrison and Osterkamp 1982) this gives

$$\sqrt{v^2} < 0.1 \text{ m yr}^{-1}$$

Since any convecting pore water velocity field is probably highly non-uniform, v could be considerably less than $\sqrt{v^2}$,

The net result of all of these factors is that conservation of energy applied to gravity driving seems to imply mean velocities considerably less than 0.1 m yr^{-1}. This seems too low in the light of the conclusion at the end of the previous section.

DISCUSSION

Although this calculation raises serious doubt that the gravity driving mechanism by itself will be able to explain the observations, we do not feel that it is entirely ruled out at the present stage of field observation and interpretation. A more complete interpretation, which includes the possible effect of convective driving from the sea bed by salinity variations there, awaits our completion of a numerical calculation of the pore water velocity field. In the meantime, it is clear that other driving mechanisms should be considered.

This has been done already for the most obvious of the possible "external" mechanisms, the driving of pore water motion by sea surface wave action, and its effect on salt transport into the sea bed. Harrison et al. (1983) have shown that this process, via a mechanism known in ground water hydrology as mechanical dispersion, is important where the hydraulic conductivity is characteristic of sand, but not at the West Dock where it is much smaller. This is another illustration of how the physics of the transport mechanisms through the thawed layer depends critically on hydraulic conductivity.

It should be emphasized that the brief summary in the second section does not describe all that is known about permafrost in the West Dock area, so there are probably other clues to the processes operating there if one knew how to interpret them. Possible examples are the uniformity of the phase boundary salinity, higher salt concentration in the pore water than in normal sea water, and the presence of significant amounts of brine below the phase boundary. None of these features is understandable within the framework of present ideas. We are forced to the conclusion that despite some progress, the evolution of subsea permafrost under natural conditions is not well understood, even in the intensely studied West Dock area. It seems to follow that new ideas are needed to predict the effects on subsea permafrost of future man-made disturbances, such as hot pipelines and well bores.

ACKNOWLEDGEMENTS

We are grateful to all who have helped with the field observations, particularly to Robert Fisk, Victor Gruol and Masayuki Inone. Primary financial support was from NSF grant DPP 77-28451. Support from NSF grant DPP 81-20471, and from the NOAA-BLM Outer Continental Environmental Assessment Program, is also acknowledged.

REFERENCES

Harrison, W. D., 1982, Formulation of a model for pore water convection in thawing subsea permafrost: Mitteilungen der Versuchsanstalt fuer Wasserbau, Hydrologie und glaziologie an der Eidgenoessischen Technischen Hochschule Zuerich, Nr. 57.

Harrison, W. D. and Osterkamp, T. E., 1977, Subsea Permafrost: Probing, thermal regime and data analysis; in Environmental Assessment of the Alaskan Continental Shelf, Annual Reports, v. 17, p. 424-466.

Harrison, W. D. and Osterkamp, T. E., 1978, Heat and mass transport processes in subsea permafrost 1. An analysis of molecular diffusion and its consequences. Journal of Geophysical Research, v. 83, p. 4707-4712.

Harrison, W. D., Osterkamp, T. E., 1981a, Subsea permafrost: Probing, thermal regime and data analysis; in Environmental Assessment of the Alaskan Continental Shelf, Annual Reports, v. 7, p. 291-401.

Harrison, W. D. and Osterkamp, T E., 1981b, A probe method for soil water sampling and subsurface measurements: Water Resources Research, v. 17, p. 1731-1736.

Harrison, W. D. and Osterkamp, T. E., 1982, Measurements of the electrical conductivity of interstitial water in subsea permafrost. in Proceedings of the Fourth Canadian Permafrost Conference: Ottawa, National Research Council of Canada, p. 229-237.

Harrison, W. D., Musgrave, D. and Reeburgh, W. S., 1983, A wave induced transport process in marine sediments: Journal of Geophysical Research, in press.

Iskandar, I.K., Osterkamp, T. E., and Harrison, W. D., 1978, Chemistry of interstitial water from the subsea permafrost, Prudhoe Bay, Alaska, in Proceedings of the Third International Conference on Permafrost, v. 1, p. 93-81: Ottawa National Research Council of Canada, p. 92-98.

Lachenbruch, A. H., Sass, J. H., Marshall, B. V. and Moses, T. H., Jr., 1982, Permafrost, heat flow, and the geothermal regime at Prudhoe Bay, Alaska: Journal of Geophysical Research, v. 87, p. 9301-9316.

Morack, J. L. and Rogers, J. C., 1983, Acoustic velocities of nearshore materials in the Alaskan Beaufort Sea, P. Barnes, E. Reimnitz, D. Schell Eds. in The Beaufort Sea (in press).

Osterkamp, T. E. and Harrison, W. D., 1976, Offshore permafrost: Drilling, boundary conditions, properties, processes and models; in Environmental Assessment of the Alaskan Continental Shelf, Annual Reports, v. 13, p. 137-256.

Osterkamp, T. E. and Harrison, W. D., 1980, Subsea permafrost: Probing, thermal regime and data anlysis; in Environmental Assessment of the Alaskan Continental Shelf, Annual Reports, v. 4, p. 497-677.

Osterkamp, T. E. and Harrison, W. D., 1982a, Subsea permafrost: Probing, thermal regime and data analysis; in Environmental Assessment of the Alaskan Continental Shelf, Annual Reports, in press.

Osterkamp, T. E. and Harrison, W. D., 1982b, Temperature measurements in subsea permafrost off the coast of Alaska, in French, H.M., ed., Proceedings of the Fourth Canadian Permafrost Conference: National Research Council of Canada, Ottawa, p. 238-248.

Page, F. W. and Iskandar, I. K., 1978, Geochemistry of subsea permafrost at Prudhoe Bay, Alaska: U.S. Army Cold Regions Research and Engineering Lab, Special Report 78-14.

Sellmann, P. V., 1980, Regional distribution and characteristics of bottom sediments of Arctic coastal waters of Alaska. Review of current literature: U.S. Army Cold Regions Research and Engineering Laboratory Special Report 80-19.

Smith, P., Hartz, R. and Hopkins, D., 1980, Offshore permafrost and shoreline history as an aid to predicting offshore permafrost conditions, in Environmental Assessment of the Alaska Continental Shelf, Annual Reports, v. 4, p. 159-255.

EXCAVATION RESISTANCE OF FROZEN SOILS UNDER VIBRATING CUTTINGS

Shingi Takasugi[1] and Arvind Phukan[2]

[1]Research Engineer, Technical Research Center, Komatsu Ltd. Kanagawa, Japan
[2]Professor of Civil Engineering, University of Alaska, Anchorage, Alaska, U.S.A.

This paper presents the dynamic laboratory experimental studies of frozen sands and silts at temperatures varying from -2°C to -11°C; using various cutting velocities from 2mm per second to 18mm per second. The 6Hz vibration mode used was varied from 1.3mm amplitude to 4mm amplitude. Theoretical analyses of the cutting resistance of frozen soils under vibration cutting are formulated to explain the failure characteristics obtained during the experimental work.

Energy ratio versus the number of contacts per inch of cutting (Nc) plots show a similar tendency for both silty and sandy frozen soil samples; that is, as the energy ratio increases, Nc increases. The contact length between the soil and the tool in a vibration cycle is inversely proportional to Nc. At high Nc, the contact length is too small to cause the soil failure and thereby increases the vibration energy. Relationships between the energy ratio and Nc at various amplitudes for both soil types, show a greater energy requirement at 4mm amplitude of vibration than at 1.3mm amplitude, in spite of the identical draft energy reduction at both vibration amplitudes. It is possible that by inducing vibration of the cutting tool, a reduction in draft energy can be obtained and thereby excavation process can be efficiently executed in frozen ground.

INTRODUCTION

Construction of foundations and installation of underground structures in the arctic and subarctic regions often require excavation in frozen ground. Various investigators (Mellor, 1977; Nishimatsu, 1972; Harrison, 1972 and Wismer et al., 1972) have reported the results of theoretical as well as experimental studies on the mechanics of cutting soils and rocks under different parameters such as width of tool, rake angle, cutting depth and speed. Smith et al. (1971), Barkan (1974), Burton et al. (1977) and Gohlieb et al. (1981) reported the results of experimental studies on coal, rock and soil cutting by vibrating tools under different test conditions of vibration amplitude, frequency, tool size and shape. Most of these reports indicated that the cutting resistance was reduced significantly with the use of the vibrating cutting mode. Since this cutting mode requires additional energy to vibrate the cutting tool, the total energy consumption per unit length of soil cut has to be compared with the energy required for static cutting to evaluate the efficiency of vibration cutting. Very little data is available regarding static as well as dynamic cutting characteristics of frozen soils. Excavation resistance of frozen soils under static loadings was studied by the authors (Phukan and Takasugi, 1982) and this work reported here is an extension of this earlier research work.

The paper summarizes the laboratory studies of dynamic cuttings of frozen sand and silt samples at temperatures varying from -2°C to -11°C. A vibration cutting mechanism was designed and used to operate at 6Hz vibration mode. The amplitude of vibration was varied from 1.3mm to 4mm to obtain the relationship between the energy ratio and the number of contacts per inch of cutting for both frozen soil types. Theoretical analyses are formulated to evaluate the failure characteristics of frozen soils under vibration cutting.

LABORATORY STUDY

Equipment

The vibrating cutting device consisting of a 3/4 H.P., DC motor with a gear reduction box, fly wheel and toggle linkage was designed and prepared as shown in Fig. 1. This device is able to produce up to 10 Hz frequency with 12.7mm amplitude vibration to the cutting tool. The vibration device was attached to the cutting tool, the sample feed device was the same one used in the static cutting test reported by Phukan and Takasugi (1982). Due to the characteristics of the toggle mechanism, the shape of the vibration was not a pure sinusoidal wave. A typical wave shape generated by this device is shown in Fig. 2.

Measurement

The horizontal force acting on the cutting tool was measured by strain gages. Two sets of four strain gages were glued on link CB (Fig. 1). Prior to the experiment, the calibration was obtained between strain and force.

A Gilmore acceleration pick-up was installed on the cutting tool and the output signal was integrated and amplified by a Gilmore Model NS04 integral amplifier. This output signal was used to activate the strain gage circuit. The output power (force x velocity) was monitored by a Hewlett-Packard plotter with respect to time. A second set of strain gages was used to monitor the force acting on the cutting tool, and the force was recorded on a Houston Instrument X-Y recorder with respect to the tranverse length.

Since the output power of Model NS04 is too small to activate the strain gage circuit, a Condyna analog computer GP6 was connected between the integral amplifier and the strain gage circuit. Details of the measurement system are given by Takasugi (1982).

Test Procedure

Initially, the samples were placed in the walk-in cold room for 24 hours to attain the required test temperature. The temperature of the soil samples were monitored until a constant desired temperature ($\pm$ 0.5°C) was obtained. After they remained for 8 hours at constant temperature, the prepared samples were tested.

Special attention was given to obtaining a flat surface on the sample to ensure a constant cutting depth for the entire 0.25m sample length. Each sample was cut by 6 test runs at various cutting velocities from 2.5mm to 20.3mm at a particular vibration amplitude. One of the six test runs did not employ the vibration mode.

The power and the cutting resistance were monitored as the sample was moved against the vibrating cutting tool.

Test Results and Discussion

The vibration energy was measured by means of the vibration velocity (V) and cutting resistance (F). The vibration energy (E_v) may be written as:

$$E_v = \int F \cdot V \, dt \qquad (1)$$

similarly, the draft energy without vibration (E_d) is given by:

$$E_d = \int F ds \qquad (2)$$

The energy ratio (R_e) is defined as:

$$R_e = \frac{E_v + E_d^v}{E_d} \qquad (3)$$

where E_d^v is the draft energy vibration.

The number of contact per inch of cutting is calculated as:

$$N_c = \frac{f}{V'} \qquad (4)$$

where

f : vibration frequency (cps)

V' : cutting velocity (cm/s)

The draft energy ratio is defined as:

$$R_d = \frac{E_d^v}{E_d} \qquad (5)$$

The energy ratio and the draft ratio against N_c for frozen silt and Ottawa sand are determined at different amplitudes and freezing temperatures and the results are shown in Figs. 3 to 5, inclusive.

The reason for the reduction in cutting resistance because of vibration of the cutting tool is not well understood. There appears to be no general agreement about the fundamental cause of this resistance reduction. Some of the more common hypotheses (Barkan, 1974) are as follows:

(1) In vibration cutting, the actual contact time between the soil and cutting tool is much smaller than the overall cutting time. This reduces the friction between the soil and the cutting tool and results in reduction of cutting resistance.

(2) An interaction between the frequency of the vibrating cutting tool and the natural frequency of the soil shear plane formation results in a decrease of the cutting resistance.

(3) The pressure wave caused by the vibrating tool breaks the ice bond matrices of the frozen soil near the tool, thus reducing the soil shearing strength, and hence, reducing the resistance of the cutting tool.

No one hypothesis explains more than 60% of the reduction in resistance. A combination of the above hypotheses is required for a full explanation of the cutting resistance reduction.

The energy ratio versus N_c relationship (Fig. 3) show a similar tendency for both samples; that is, the increase of the energy ratio as the N_c increases. The contact length between the soil and the tool in a vibration cycle is inversely proportional to the N_c. At higher N_c, the contact length is too small to cause the soil failure and that increases the vibration energy.

The relationship at various amplitudes between the draft ratio and N_c is shown in Fig. 4 and Fig. 5 for frozen sand and frozen silt, respectively. These plots show a greater energy requirement at 3.81mm amplitude of vibration than at 1.27mm amplitude, in spite of the identical draft energy reduction at both vibration amplitudes.

Results at different test temperatures are also shown in Fig. 3 for sand and silt. These plots are almost identical and indicate that the temperature, within the range tested, does not have a significant effect on the results.

THEORETICAL ANALYSIS

In this analysis, the vibration is assumed to be a pure sinusoidal wave. Every time the cutting tool contacts the soil, the failure of the soil occurs.

The acceleration, velocity and displacement of the cutting tool are expressed as:

$$\text{Acceleration,} \quad X = A \cdot W^2 \cdot \cos(W \cdot t) \tag{6}$$

$$\text{Velocity,} \quad X = A \cdot W \cdot \sin(W \cdot t) = V \tag{7}$$

$$\text{Displacement,} \quad X = A \cdot (1 - \cos(W \cdot t)) = V \cdot t \tag{8}$$

where

V = cutting velocity, cm/s

W = angular frequency, rad/s

A = vibration amplitude, cm

The relationship among displacement, velocity and acceleration are shown in Fig. 6.

Consider the first cycle after t = 0; the tool velocity is zero at both t_0 and t_r:

$$X(t_0) = A \cdot W \cdot \sin(W \cdot t_0) = V = 0 \tag{9}$$

$$X(t_r) = A \cdot W \cdot \sin(W \cdot t_r) = V = 0 \tag{10}$$

The times at which these occur are given by:

$$t = \frac{1}{W} \sin\left(\frac{V}{A \cdot W}\right)$$

or

$$t_0 = \pi/W + \frac{1}{W} \sin\left(\frac{V}{A \cdot W}\right) \tag{11}$$

$$t_r = 2\pi/W - \frac{1}{W} \sin\left(\frac{V}{A \cdot W}\right) \tag{12}$$

$$t_2 = t + T \tag{13}$$

where

$$T = 2 \cdot \pi/W$$

The time t_1 when the cutting tool begins to contact is calculated from the following equation by trial and error.

$$V \cdot t_0 = A \cdot (1 - \cos(W \cdot t_0)) = V \cdot t_1 + A \cdot (1 - \cos(W.t_1)) \tag{14}$$

It is assumed that the resistance acting on the cutting tool is the same as in the static condition.

Then

$$\text{Draft Energy Ratio} = \frac{t_2 - t_1}{T} \tag{15}$$

The vibration energy in a cycle may be calculated as:

$$E_v = \int_{t_1}^{t_2} F \cdot A \cdot W \cdot \sin(W \cdot t)\, dt \tag{16}$$

The vibration energy per inch cutting is:

$$E = E_v \cdot f/V \tag{17}$$

where f = frequency of the vibration

Then the energy ratio is:

$$E_r = \frac{E_v \cdot f}{V \cdot F} = \frac{t_2 - t_1}{T} \tag{18}$$

The equation (18) is plotted in Fig. 6 against the number of contacts per inch cutting, which may be expressed as:

$$N_c = f/V$$

Fig. 6 shows that except for small values of E_v, the energy ratio can not be less than unity. The calculated draft ratio shows a similar tendency of the test results.

CONCLUSIONS

1. In every test condition, the cutting resistance was reduced with vibration by more than 60% in both soils. The reduction of cutting increased as the number of contacts per inch cutting increased.

2. The 1.27mm amplitude vibration and 3.81 mm amplitude vibration showed the same percentage reduction of cutting resistance, but the 3.81mm amplitude vibration required more energy than the 1.27mm amplitude vibration.

3. It was found that the temperature of the samples did not affect the test results at the range tested.

4. At optimum conditions, which are 1.27mm amplitude and about 12.7mm/s cutting velocity, 50% reduction in the energy required for cutting were measured in both soils.

5. Once the vibration system fails to operate properly due to overloading, the static cutting resistance, which is about three times more than dynamic resistance, acts on the vibration system. This may produce inefficient cutting.

6. Most tractors have more than enough Horsepower to counteract the cutting resistance. In most cases, the limitation of the drawbar force of a tractor depends on the traction that its treads can develop, which is controlled by the tractor weight and soil condition. By inducing vibration of the cutting tool, a reduction in the draft can result, thus allowing smaller-sized tractors to be employed to do the same job.

7. It is recommended that further static and dynamic cutting tests should be carried out on frozen soils at additional temperatures. The geometry of the cutting tool should be investigated to arrive at a rake angle for minimum cutting resistance. The effect of ice bonding on the behavior of frozen soils under different frequencies and amplitudes of vibration needs to be investigated. Such research can lead to a major improvement in excavation techniques for frozen soils.

ACKNOWLEDGMENT

This report was made possible by financial support received from Komatsu Ltd. Japan.

REFERENCES

Barkan, D. D., 1974, Vibration methods in Construction, CRREL Draft Translation TL 446.

Burton, R. T., and Ikrainetz, P.R., 1977, Experimental Investigation of Vibrating Blades in Frozen Soils, S.A.E. Transactions, V. 86.

Gohlieb, L. and Moore, P. J., 1981, Vibratory Cutting of Brown Coal, Journal of Terramechanics, V. 18, p. 335-339.

Harrision, W. L., 1972, Soil Failure Under Inclined Loads. CRREL Research Report 303.

Mellor, M., 1977, Mechanics of Cutting and Boring, Part IV: Dynamics and Energetics of Parallel Motion Tools, CRREL Report 77-7.

Nishimatsu, Y., 1972, The mechanics of Rock Cuttings, Int. Journal of Rock Mechanics and Mining Science, V. 3, No. 3.

Phukan, A. and Takasugi, S., 1981, Excavation Resistance of Artificially Frozen Soils, Proc. Third International Conference on Ground Freezing, Hanover, New Hampshire, p. 35-40.

Smith, J. L., Hillman, K., and Flikke, A. M., 1971, Experimental Analysis of Vibratory Tillage, ASAE paper, p. 71-639.

Takasugi, S., 1982, Failure Characteristics of Frozen Soils Under a Simulated Excavation Technique, M.Sc. Thesis. University of Alaska, Fairbanks, Alaska.

Wismer, R. D. and Luth, H. J., 1972, Rate Effect in Soil Cutting, Journal of Terramechanics, V. 3, No. 3.

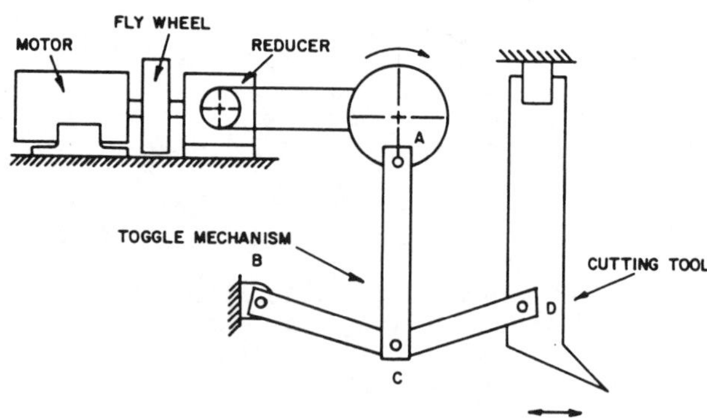

FIG. I VIBRATION CUTTING MECHANISM

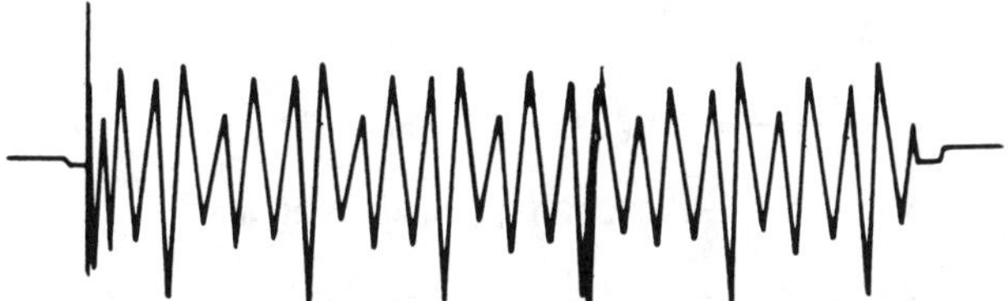

FIG. 2 TYPICAL VIBRATION SHAPE

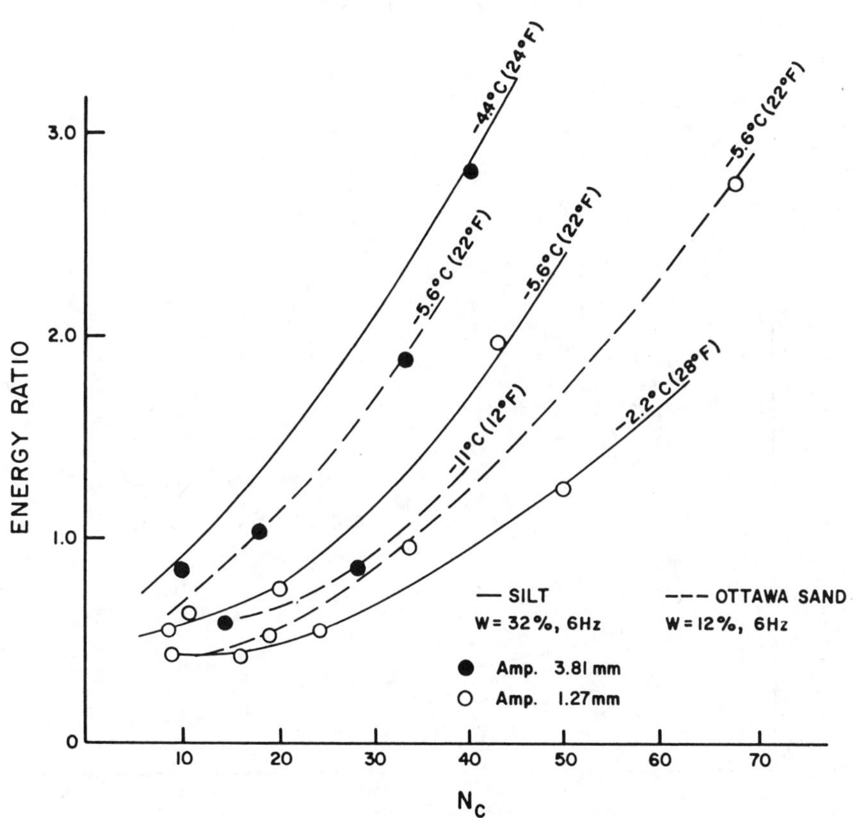

FIG. 3 ENERGY RATIO Vs N_C

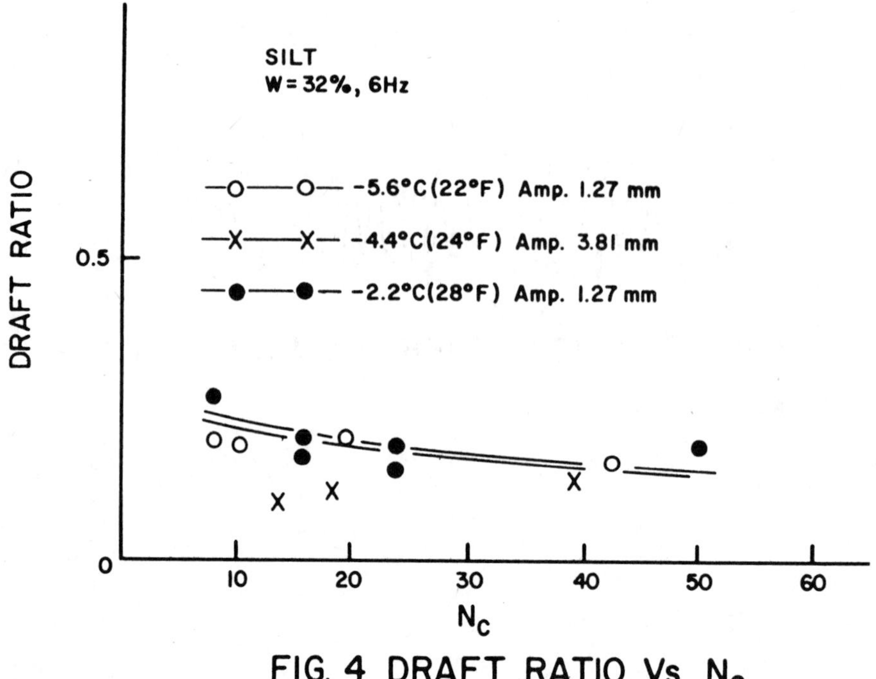

FIG. 4 DRAFT RATIO Vs N_c

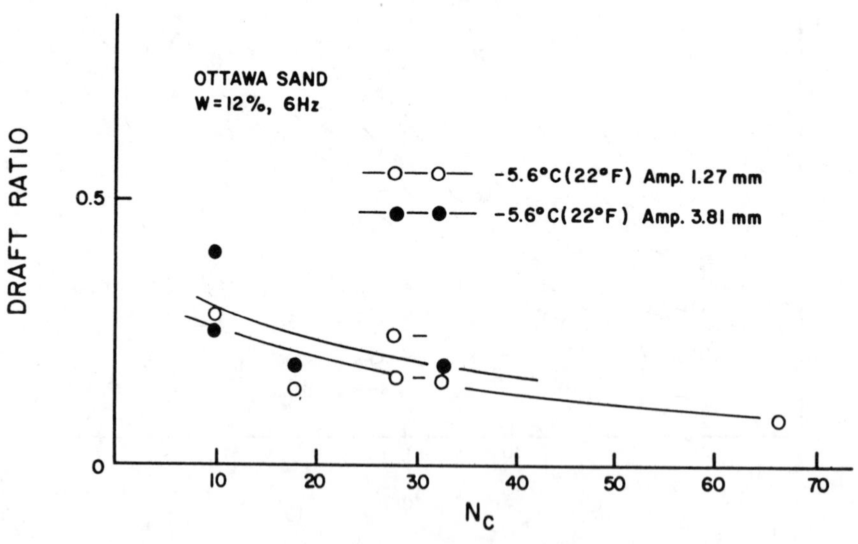

FIG. 5 DRAFT RATIO Vs N_c

WINTER CONSTRUCTED GRAVEL ISLANDS

Rupert G. Tart, Jr.

Woodward-Clyde Consultants, Anchorage, Alaska 99503 USA

This paper addresses a theoretical assessment of the performance of winter constructed gravel islands with respect to settlement. The basis of the material presented is our work during the recent winters in which we have inspected several gravel islands as they were constructed in the Alaskan Beaufort Sea as well as other winter constructed gravel pads. Construction procedures used to build the islands are summarized, and the effect of technique and construction materials on the subsequent performance of the fill is addressed. Our observations indicate that total island settlement potential is primarily dependent on the in-place dry density of fill. The actual settlement and the rate of settlement were found to depend on the thermal environment. We also found that in at least one case the island fill was stratified into three distinct layers shortly after construction. This paper presents opinions developed regarding the formation of these strata and how the performance of the island relates to the layers formed. The methods used to make these observations and the methods used to estimate thaw strain are also covered.

INTRODUCTION

Designing for settlements within foundation fills has always been an important concern. This is especially true in the arctic regions when frozen borrow sources are used in fills without providing time for the materials to thaw between excavation and placement. In frozen soils, the pore water (ice) is a solid and the dry density of a fill constructed from reworked frozen deposits will decrease as the natural water content of the material increases. When these fills thaw, substantial settlements can occur. This paper addresses this phenomenon as it pertains to winter constructed gravel islands in the Beaufort Sea.

Exploration drilling offshore in the Alaskan Beaufort Sea is typically performed from man-made gravel islands built with frozen materials. Because thaw strains of close to 25% have been recorded within portions of existing man-made islands (Scher 1982), it is critical to be able to estimate settlement during design. Fairly accurate estimates of settlement within a frozen fill can be made if the as-placed densities of the material are known. All settlements in granular, fills including thaw, creep, and consolidation, are dependent on the dry density of the in-place material. However, there are many variables that control in-place density such as soil gradation, moisture content of the undisturbed soil, environmental conditions at the time of construction (temperatures, precipitation, etc.), and the medium in which the fill material is placed, air or water.

Winter constructed gravel islands are typically built between January and April. Construction generally begins as soon as the sea ice is sufficiently thick to build ice roads capable of supporting gravel hauling equipment. The procedure to build the islands is basically to: (1) construct an ice road to the site, (2) cut and remove the sea ice at the site, (3) haul gravel and dump it through the ice, and (4) once the gravel has surfaced above the water line, spread and compact the material in thin lifts until the design freeboard is reached.

The remainder of this paper will focus on the performance, in terms of settlements, of man-made islands built during the winter with frozen fill in the Alaskan Beaufort Sea. In addition, techniques are discussed for predicting settlement within man-made islands built with frozen gravels.

ABOVE WATER FILL

The above water fill is typically placed in lifts and compacted; the final dry density being a function of the materials, ice content, and the amount of compactive effort during placement.

To estimate the dry density, the following rationale has been used. The in-place fill consists of solid borrow and newly created voids which form during placement. This solid borrow would contain both mineral and ice. The fill voids would contain air in the above water zone of the fill and seawater in the below water zone of the fill. Since the gravel sources used were almost always found to be saturated, we have assumed there were no air voids in the frozen chunks that were excavated. Thus, knowing the moisture content and specific gravity of mineral solid, we can determine the density and void ratio of the borrow material. Details of these calculations are given by Tart (1982).

Estimates of e_f, the fill void ratio considering both ice and mineral as solids, can be made by assuming the dumped frozen chunks of borrow material act as individual grains and form a mass with a void ratio similar to void ratios of dumped sands and gravels. In loose conditions, a realistic value of e_f probably falls between 0.3 and 0.6.

Using an assumed specific gravity of solids, G_s, = 2.7, and the rationale described above, Figure 1 was developed. From Figure 1 estimates of dry density of a fill constructed of saturated frozen chunks can be made, if e_f can be estimated and the borrow moisture content are known.

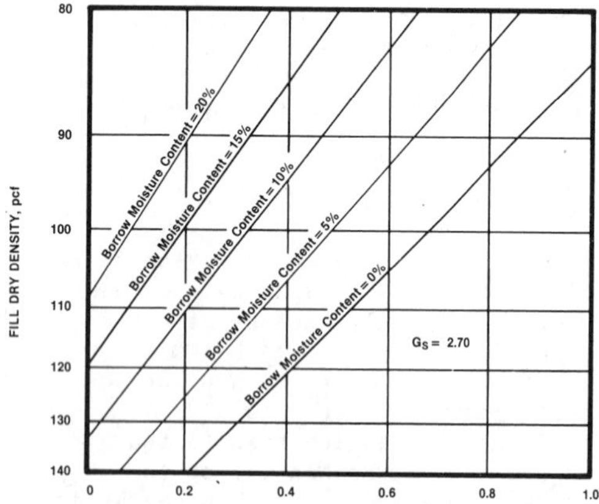

FIGURE 1 Chart for estimating as-placed density

Laboratory and field data by Clark (1970) and others have reported that the dry density of a frozen fill is inversely proportional to the moisture (ice) content as indicated in Figure 1.

Further, our experience has shown that the response of frozen materials at a given moisture content to compactive effort in the field can be simulated in the laboratory using procedures similar to those used for unfrozen materials. The frozen compaction tests differ from standard laboratory compaction tests in that the samples remain frozen and the moisture content is not changed. Thus, the maximum and minimum densities determined by laboratory tests can be used to predict in situ field densities by using an approach very similar to the approach used in thawed fills.

As the above water fill is placed conventional field density tests can be made. Sand cone density tests can be relatively accurate. However, since the chunks can be large, a foot or more in some cases where the borrow moisture content is high, caution must be exercised in using the sand cone to assure the volume filled with sand is representative of the volume of material removed and weighted. Nuclear density tests can be made as long as the instrument is not left in the cold for extended periods and care is used to prevent densification as the probe is driven into the ground.

A third method for obtaining in-place densities is the chunk density test which was described previously by Tart and Luscher (1981). In this method the zone to be tested is sprinkled with water. After the water freezes, a chunk of material is removed and a bulk density determination can be made of the frozen chunk. Moisture contents are determined from samples adjacent to but outside the flooded area.

For above water fills where borrow source moisture contents were around 10%, average as-dumped densities have been observed to be around 90 pcf. (Dry densities as low as around 70 pcf have been observed when borrow moisture contents were near 30%.) The 90 pcf density fill reached could be compacted to densities of 105-110 pcf.

Our field observations indicate that the effect of compactive forces applied decreased significantly with depth in a given lift. The density in the upper 2-5 in. was usually close to the maximum dry density as determined in the laboratory. The next 2-4 in. was usually a transition zone, and the density of the remaining bottom portion of the lift appeared to have not been affected by compaction equipment.

Further, we could not observe any significant increase in density that could be related to a specific type of compaction equipment. Equipment used included a grid roller, a vibratory roller, and a sheepsfoot roller. Most of the increase in density took place when the dumped gravel was spread and leveled. Compaction tests taken immediately after spreading gave densities that were generally as high as those where compaction equipment had been used, indicating that the use of compaction equipment was not effective. From this, we believe that careful spreading of the material in thin lifts followed by rolling with the gravel hauling equipment will result in as much productive compactive effort as the use of any of the above-mentioned standard compaction equipment.

BELOW WATER FILL

Winter construction of the below water portion of the island is accomplished by simply dumping the frozen gravel through a hole in the ice. In the past there has been no effort to mechanically compact the below water fill. By the time the gravel is mined and hauled to the construction site, it has cooled to roughly the ambient temperature (-15°F in late January at Prudhoe Bay). When this gravel is dumped in the sea, which has a temperature near 29°F, it begins to remove heat from the water with which it has contact, and some additional ice is formed.

This new ice may be only a surface coating or may fill a significant portion of the available void space. Standard thermal calculations can be used to estimate the amount of ice that would be formed by estimating the temperature of borrow (gravel/ice) mass when it stops dropping and estimating the temperature of the seawater trapped in the voids. Since convection would cause warming of the borrow material as it drops further through the water, one would expect more ice in the upper portions of the fill.

Similar to above water fill, the material underwater can be assumed to fall into a loose state with a void ratio somewhere between 0.4 and 0.6. By assuming a degree of saturation, usually 100%, an unfrozen moisture content can be estimated.

The formation of additional ice in the fill as it is dumped in the water is suggested by the steep island slopes formed below water reported by Galloway et al. (1982). Observed above water, the angle of repose of the frozen material was about 35-45° (32° thawed). However, underwater the island slopes were found to be between 60° and 80° from horizontal.

The existence of the steep underwater slopes seems to indicate the presence of additional ice and to imply that slumping will occur when the material thaws during the summer months. In addition, thaw or degrading of sea ice formed in the frozen fill could occur in the winter months at the base of the fill in deep water. The average underwater slope angle also appears to be not only a function of temperature and initial ice (freshwater) content, but also water depth. While we have not investigated this in detail, it appears as the water depth increases, the average slope angles will decrease, possibly due to warming of the gravel.

The design material volume of the island should be sufficient to allow the slumping of slopes from the initial average slope of around 0.6 horizontal to 1 vertical (0.6H:1V) to the summer average slope of around 3H:1V. This can be accomplished by making sacrificial beaches around the island perimeter which contain enough material to allow the slumping to 3H to 1V without losing required working space on the island slough. Another method of accomplishing this same goal is to groom the slopes to 3H:1V during construction.

Actual measurements of the underwater fill densities could not be obtained. For saturated fills the most definitive method of establishing the in situ density would be to obtain undisturbed samples of the fill. Until recently, the techniques used for this have been unsuccessful. Drive sampling densifies and/or remolds the material. Rotary drilling with mud melts the ice and disturbs the particle structure in the samples. Even with refrigeration, there is some melting and physical alteration to the sample perimeter. In addition, the core water is frequently contaminated by drilling fluid.

Recognizing the problems with existing sampling techniques, a series of borings were made through the island fills to mudline and split spoon samples were obtained. The densities measured on the limited samples retrieved from the below water averaged about 80-85 pcf, with moisture contents about 25-30%. This suggests saturations of only about 75%. This saturation seems low, since water was observed to flow into the boreholes, which would indicate saturations closer to 100%. Assuming 100% saturation, the dry density should be expected to average about 90-95 pcf.

Additional below water sampling should be done as well as laboratory modeling to better determine the actual condition of frozen materials dumped in cold seawater. During the winter of 1982 an experimental technique was used to core the frozen and partially frozen gravel fills. Very cold air and a thin kerfed bit were used which allowed the bit to cut with minimal friction and grinding. This technique produced a frozen core which looks very much like a concrete core, with ice replacing the cement. The experiment proved the mechanics of the technique to be sound. However, because there was not enough moisture in the gravel samples to securely bond the particles when cooled, no undisturbed samples were recovered.

ISLAND ZONES

Various borings made shortly after completion of construction indicated that the fill material existed in up to three zones of individual conditions of moisture content, phase, and structure depending on the water at the island. The material above the waterline had approximately the same moisture content it had before it was excavated. Below the waterline, the moisture contents increase to near saturation. However, the percent frozen moisture appeared to decrease with depth. The two underwater zones are distinguished by the change in unfrozen moisture content. Figure 2 illustrates these three zones. The figure also presents a theoretical gravel-ice- water matrix of the material in each zone at that time.

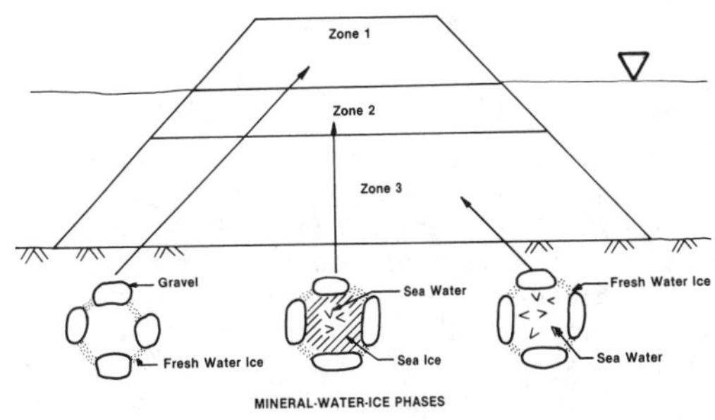

FIGURE 2 Three zones of gravel island

One explanation of the formation of the two underwater zones is that when the gravel particle is dumped in the water, it quickly cools the surrounding water, forming additional ice. However, the longer it is moving (dropping) in water, the more it warms, approaching the seawater temperature. As this happens, the additional ice formed, plus some of the initial ice content, may melt. Thus, below a certain depth, the fill has more thawed moisture content than frozen. Near the water surface the amount of unfrozen pore water can be very low. The physical separation of these two zones is apparent when a large-diameter hole is drilled. Several feet below the water, surface water flows into the hole. We estimate that near this elevation the unfrozen moisture content approaches 50%. Below this elevation the unfrozen moisture content increases.

An advancing "freeze front" or "freeze plane" can also contribute significantly to this development of two distinct underwater zones. The construction technique used on several islands has been to develop a full diameter surface of the island to an elevation near sea level, then build to a design freeboard elevation. During this period that the island surface is near sea level, the "freeze front" has the opportunity to advance into the upper portion of the below water fill. In one case we have observed water flowing into a shallow (3-5 ft) pit made during construction when the island elevation was at elevation +2 to +3 ft. After about 3 weeks and the addition of 10-13 ft more fill, this "apparent" freeze level was observed at an elevation near -8 ft.

This seems to be a somewhat rapid advance rate, and it seems possible that water flowing into the earlier shallow pit could have come from water entrapped as the gravel was pushed over the edge of the island surface during construction. More observations are necessary to be more conclusive about the zone formation process.

The existence of the three zones has only been observed in the period immediately after the fill material is placed (1-2 months). Borings made through one island a year after construction indicated that changes had occurred in the state of the pore waters. Analyses of the data are not yet available to explain the changes.

SETTLEMENT MECHANICS

Each of the three zones should be considered separately in characterizing the total settlement in the island. Scher (1982) has presented data indicating that the unit thaw strain in remolded frozen gravel fills is a function of the in-place dry density. However, this relationship has only been tested in gravels with no unfrozen moisture content. From Scher (1982), thaw strains in Zone 1 have ranged from roughly 10-25% depending upon the fill dry density and the maximum density the thawed fill can obtain. During the first summer, thaw may progress to depths of 6-8 ft, and maximum settlements of between 1.5 and 2.0 ft have been observed. The rate of settlement through Zone 1 would follow closely the advance of the thaw plane. After the first thaw period, additional settlements in Zone 1 would probably be minor.

Zone 2 is, for all practical purposes, insulated from summer warming by the air and the lack of a significant hydraulic gradient across the island, making it unlikely that there would be much warm water flowing through this zone. Thus, we believe it is relatively stable, thermally. However, as will be shown later, some settlement does occur in this zone. This settlement is most likely a result of creep between the ice and gravel particle contacts due to the overburden and possible warming from the above water fill.

Although the exact mechanics of the settlements in Zone 2 are not known, estimates of settlement potential can be made if the dry density can be determined. First the material around the well bore may thaw, in which case the thaw strain versus initial frozen dry density relationship (Scher 1982) can be used. For the remaining fill, it would be reasonable to assume that with time the in situ ice has filled the voids. If the ice content is known, a void ratio and dry density can be estimated. Here the settlement potential is tied to the creep rate of the pore ice. Long term settlements could approach settlements resulting from thaw, whereas, short term settlements are likely to be very small.

In Zone 3, the high unfrozen moisture content, over time, will probably react with and erode the bonding ice. As this bonding breaks down, the loose saturated fill should begin a slow settlement similar to that observed in the consolidation of fine-grained deposits. Similar to Zone 2, accurate settlement predictions in Zone 3 cannot be made unless the exact structure and mechanics of movements have been defined. Estimates can be made assuming that the fill density would approach that of a loose thawed gravel. However, the length of time for this to happen is unknown. Settlements in the fill around the well bore, if they thaw, should be predictable using the aforementioned thaw strain versus initial frozen dry density relationship.

FIELD SETTLEMENT OBSERVATIONS

Settlement information described above was based on observations we have made at several locations. The most effective method of measuring we have used is the Sondex system in which a flexible casing containing metal rings at designated spacings is placed in a large-diameter (8 in.) drill hole and then backfilled with clean pea gravel. As settlement occurs, the rings move in the zones which are moving, and the rings are detected using a sonde, which detects the relative position of each ring electronically.

To obtain the temperature profile for correlation with these movement data, thermistor strings can be installed in 1-in. PVC tubing filled with glycol.

To measure lateral movement of the islands, slope indicator casings can be installed. These casings are most easily installed in conjunction with the Sondex casings. The slope indicator casings will provide the stiffness for the flexible Sondex casing and prevent it from collapsing. It would also provide a one-location source of both lateral and vertical movement data.

In Figure 3, typical temperature curves that could be observed from a winter constructed gravel island during the first summer after construction are plotted. In Figure 4 the expected shape of the accumulated settlement versus depth curve is presented. This curve represents settlement that could be expected to occur during the summer thaw season. The rate of settlement in a stratum can be implied from the slope of the settlement curve. Our observations indicate that the seafloor sediments settle at a slower rate than any portion of the fill. In Zone 3 the rate of settlement is more than that of the natural sediments but less than that in Zone 2. In the lower portion of Zone 1, where active thaw settlement is occurring, the slope is the steepest.

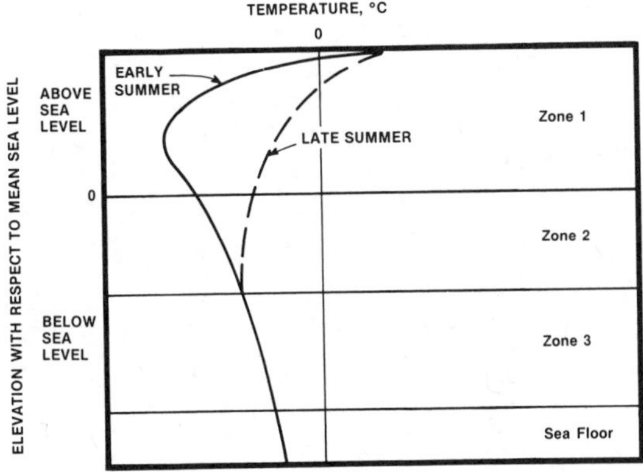

FIGURE 3 Typical temperature profiles

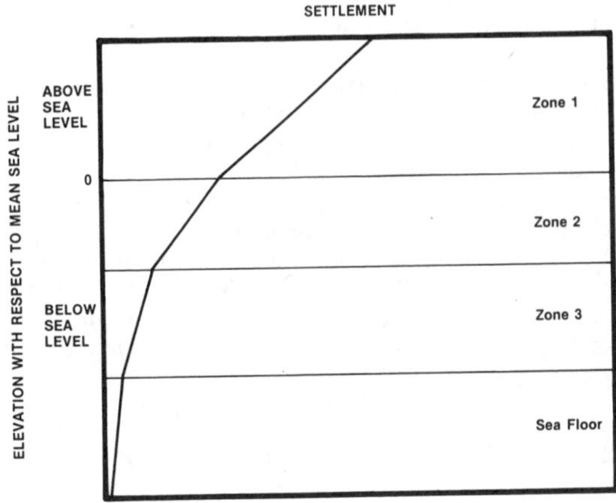

FIGURE 4 Typical settlement profile

From this, we believe that settlement in Zone 3 is occurring as a result of the slow degradation of the frozen material in this portion of the fill as a result of the high percentage of unfrozen moisture. Much of the Zone 3 settlement probably occurred during construction, before measurements could be made. In Zone 2, the additional unfrozen moisture probably resulted in a looser structure initially which also is degrading as a result of gradual infiltration of the seawater. This results in a slightly higher rate of settlement in this zone. Above sea level in Zone 1, the high rate of settlement is a result of thaw settlement resulting from the advance of the thaw front from the surface of the island. In this zone, even though there was some compactive effort placed on the fill which increased its initial density 5-20 pcf above that of the below water fill, the influence of the warming air through this nonsaturated portion of the fill results in an accelerated rate of settlement. Near the ground surface in Zone 1 the settlement rate decreases to near zero because thaw settlement occurred in this zone prior to instrument installation.

CONCLUSIONS

The experience reported herein demonstrates that short-term, above water settlement in winter constructed islands can be reasonably predicted once the in-place density is determined, the maximum dry density is defined, and the depth of thaw is established. Further methods for estimating these densities have been presented based on in situ moisture content measurements and in-place void ratio approximations. Long-term thaw strains in the below water portions of the fill could be large if in-place densities are found to be low. However, it is likely that some of the settlement in these lower zones occurs during construction and that the rate of subsequent settlement in these zones is relatively low.

Winter constructed gravel islands should be studied further to confirm settlement estimates and long-term stability predictions. These studies may include long-term monitoring and detailed sampling of the fills at various time intervals to relate the amount of settlement to the changes in the in-place density. In addition, because of the relatively undefined structure and condition of the below water fill, laboratory tests could be used to model the response of frozen soil fills placed in cold seawater.

REFERENCES

Clark, J. I., 1970, Cold weather compaction of soils, in Proceedings of the 23rd Canadian Geotechnical Conference in Banff: Ottawa, Canadian Geotechnical Society.

Galloway, D. E., Scher, R. L., and Pradanovic, A., 1982, The Construction of man-made drilling islands and sheetpile enclosed drillsites in the Alaskan Beaufort Sea, in Proceedings of the 14th Annual OTC in Houston, Texas, OTC 4335: New York, American Society of Civil Engineers.

Scher, R. L., 1982, Thaw settlement prediction in drill sites built with frozen gravels in Proceedings of the 14th Annual OTC in Houston, Texas, OTC4337: New York, American Society of Civil Engineers.

Tart, R.G., and Luscher U., 1981, Construction and performance of frozen gravel fills in Proceedings of the Specialty Conference on the Northern Community: A Search for a Quality Environment, ASCE, Seattle, Washington: New York, American Society of Civil Engineers.

Tart, R.G., 1982, Winter constructed islands in the Beaufort Sea in Proceedings of Canadian Society for Civil Engineering, 1982 Annual Conference, Edmonton, Alberta: Ottawa, Canadian Society for Civil Engineering.

SHORELINE REGRESSION: ITS EFFECT ON PERMAFROST AND THE GEOTHERMAL REGIME, CANADIAN ARCTIC ARCHIPELAGO

Alan Taylor[1], Alan Judge[1] and Dan Desrochers[2]

[1]Department of Energy, Mines and Resources, Earth Physics Branch, Ottawa, Canada, K1A OY3

[2]Department of Geography, University of Ottawa, Canada, K1A 6N5

In the Queen Elizabeth Islands of the Canadian Arctic Archipelago, the late Quaternary event with the most profound influence on ground temperatures is shoreline regression accompanying post-glacial isostatic uplift. The coastal margins of many islands have experienced hundreds to several thousand years of submergence since 8,000 years ago. The effect on the geothermal regime is far from subtle because of the large contrast between arctic air temperatures and sea temperatures.

Permafrost thicknesses measured today reflect this surface temperature history. Two deep wells 1 km apart on Cameron Island have measured permafrost thicknesses of 726 and 660 m; the geothermal analysis attributes the difference entirely to the first, and higher, site having emerged from the sea around 7,000 years B.P., about 2,000 years before the second. Inland sites on the Sabine Peninsula of Melville Island may have, in a similar lithology, permafrost twice as thick as at coastal sites. The geothermal analysis explains this variation in terms of a simple sea regression model derived from emergence curves published for the region.

INTRODUCTION

In a pioneer permafrost study, Misener (1955) examined a deep temperature profile at Resolute in the central part of the Canadian Arctic Archipelago and calculated a high outward flow of heat. Terrestrial heat flow originates largely from radioactive decay deeper within the crust and upper mantle, and this value for Resolute was considered anomalous for the geologic setting. This prompted Lachenbruch (1957) to undertake an extensive study of the thermal effects of the ocean on permafrost and on the geothermal regime in arctic regions. The Resolute temperature cable (site "O" in area E, Figure 1) is about 350 m inland from the modern shoreline to which the sea has regressed since the last glacial epoch. Lachenbruch showed that the enhanced heat flow could be attributed partly to the proximity of the ocean and partly to a geothermal regime still reflecting the much higher surface temperature conditions when the site was submerged following deglaciation.

For over a decade following the Resolute measurements, few other deep ground temperature data were available for studying the influence of a past surface temperature history on present subsurface temperatures. Concurrent with the accelerated petroleum exploration in the Canadian Arctic since the late 1960's, an extensive data set of subsurface temperatures has become available (Judge 1973a, Taylor et al. 1982, and preceding volumes referenced in the latter). In this paper, the data set will be surveyed for wells that reflect the shoreline regression history, and the effect on permafrost and present ground temperatures will be calculated at a number of sites. A full palaeoclimatic reconstruction at all the sites will be left to a later paper.

THE LATE QUATERNARY HISTORY

Much interest has developed over the past decade on the extent of glacial ice cover in the Queen Elizabeth Islands during the late Wisconsin (e.g., Blake 1970, Prest 1970, Paterson 1977, Sugden 1977). Geological evidence favours a lack of continuous ice cover over the central portion of the region for the past few tens of thousands of years (England 1976, Hodgson 1982). The break-up of this ice may have begun by 11,000 BP but most of the inter-island channels were freed of glacial ice by 9,000 BP, probably all by 8,000 BP (Blake 1972). Many islands exhibit marine terraces to 100 m above present sea level and the dating of materials found in these beaches suggests emergence of such magnitudes from the sea has occurred since that time. Curves drawn for various areas using these dates demonstrate the shoreline regression history for the Archipelago (Henoch 1964, Walcott 1972, McLaren et al. 1978; Figure 3).

The gentle topographic relief of some of the islands has resulted in land several kilometres from the present ocean having experienced this recent bimodal history (e.g. Sabine Peninsula, Figure 2). Because of the large contrast in sea temperatures, this history of inundation followed by the present subaerial conditions represents a major thermal event for the coastal margins of many of the islands.

FIGURE 1 The Canadian Arctic Archipelago showing the distribution of sites discussed in this paper. The major geologic provinces are labeled by open lettering. A and B represent the Sverdrup Basin. The stipled area is the Sabine Peninsula (Figure 2). C is the Franklinian Geosyncline, D the Arctic Platform, E the Cornwallis foldbelt, and F the Coastal Plain. The measured permafrost thicknesses are in metres and the Earth Physics Branch file numbers are in brackets (Taylor et al. 1982).

A further treatment of Holocene history and its geothermal implications is given in Desrochers (1983).

THE GEOTHERMAL ANALYSIS

Temperature changes impressed upon the earth's surface are preserved for considerable periods of time because of the low thermal diffusivity of rock and are propagated downwards with an amplitude that diminishes exponentially with depth and with a phase that lags the causative change at the surface. The geothermal temperatures measured today may be considered to have imbedded in them the steady state thermal regime typical of the geologic environment and a superposition of transient effects arising from a time series of surface temperature variations. In traditional heat flow studies, these transient effects are stripped away to leave a residual thermal regime that reflects the underlying geology (e.g. Birch 1948); here we find particular value in the magnitude of these transient contributions themselves, as they relate to our understanding of Holocene history.

The classical theory of heat conduction is used to calculate the effects of these surface temperature variations on subsurface

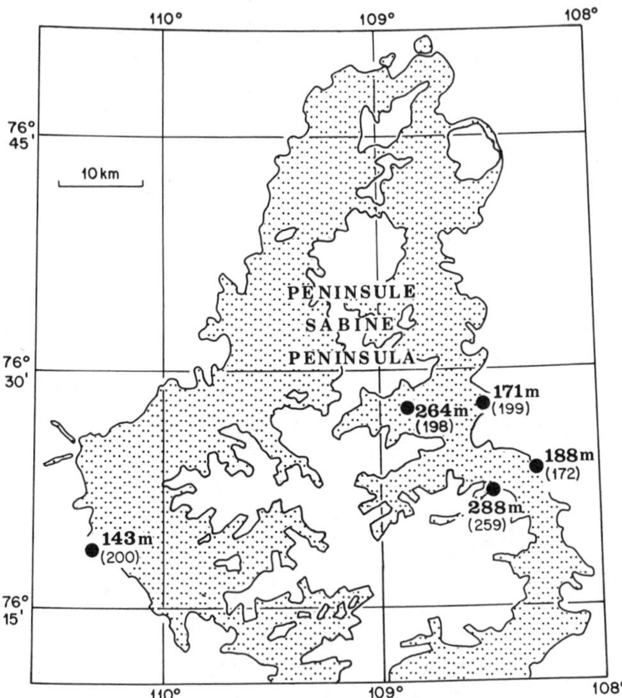

FIGURE 2 The Sabine Peninsula area, Melville Island. The stipled area is presently below 60 m ASL, and was subject to late Wisconsin marine submergence (McLaren and Barnett 1978). Five sites with deep temperatures are shown.

temperatures. The method is basically one-dimensional and the linear equations may be solved analytically for many problems. Gold and Lachenbruch (1973) suggest that other heat transfer mechanisms can be disregarded in thick permafrost because groundwater is largely immobilized either as ice or by surface forces. The latent heat of a phase change may be neglected in considering the overall thermal regime if there is not appreciable movement of an ice-rich permafrost boundary.

The theory will not be outlined here; concise summaries appear in Jaeger (1965) as it applies to traditional heat flow corrections, and in Gold and Lachenbruch (1973) for permafrost environments in particular.

THE DATA

Precise temperature profiles to depths of hundreds of metres are available for more than 40 sites in the Canadian Arctic Archipelago (Table 1 and Figures 1 and 2). Present mean surface temperatures are largely independent of elevation or geographic position throughout the region. Consequently, the wide variation in permafrost thicknesses may be attributed to the lithology (through the thermal conductivity) and the past surface temperature history at each site (Judge 1973b). For our calculations, estimates of thermal conductivity have been made from measurements in our laboratory on rock chips recovered during drilling and from porosities calculated from well logs; the general method has been described by Sass et al. (1971).

RESULTS

The two Bent Horn wells on Cameron Island (sites 196 and 286, area C of Figure 1) have measured permafrost thicknesses of 726 m and 660 m respectively, although they are only 1 km apart. We hypothesize that the difference in permafrost thickness may be attributed to the shoreline regression in the area. Both sites would have been sub-sea from the time when the inter-island channels were freed of glacial ice until the time of emergence. We assume 8,000 years BP for the first date as a conservative estimate (Blake 1972). Emergence curves (Walcott 1972) suggest that the site at higher elevation, number 196, emerged about 7,000 years ago, and the lower site about 2,000 years later (Table 2). To test the hypothesis, we assumed surface temperatures while inundated by the cold sea were 15 K higher than when subaerial, and applied the climatic correction (Jaeger, 1965) to data from the Bent Horn wells. The analysis predicts temperature profiles that would be expected today, had the marine episode not occurred. The permafrost thicknesses extracted from these calculated temperatures are

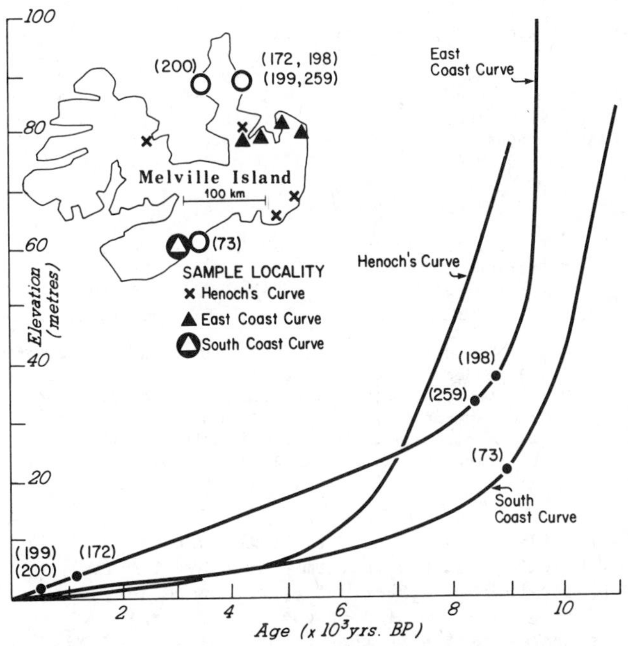

FIGURE 3 The three principal emergence curves for Melville Island (Henoch 1964, McLaren and Barnett 1978). The inset depicts the localities where samples were collected to establish the curves and the locations of the six geothermal sites on Melville Island.

TABLE 1 Geothermal Sites in the Canadian Arctic Archipelago

EPB No.	Site Name	Latitude N	Longitude W	Temperature Logs #	t_2/t_1	z max (m)	Permafrost Depth to 0°C (m)	Elev (m)	Inferred marine limit (m ASL)	Time since emergence (yrs BP)
Area A: Eureka Upland										
197	Neil 0-15	80° 44.6'	83° 4.8'	6	57	777	E549	497	25(1)	
175	Gemini E-10	79° 59.4'	84° 4.2'	8	23	876	E501	126	25(1)	
97	Fosheim N-27	79° 36.9'	84° 43.3'	1	.02	367	300+	562		
166	Mokka A-02	79° 31.2'	87° 1.2'	7	20	442	EX500	253	25(1)	
Area B: Central Sverdrup Basin										
169	Louise Bay 0-25	78° 44.9'	102° 42.0'	3	19	672	256	69	100(1)	
171	Dome Bay P-36	78° 25.9'	103° 15.8'	1	12	570	X660	154		
155	Kristoffer Bay B-06	78° 15.3'	102° 32.0'	7	25	866	E445	15	100(1)	6350 (1,5)
170	Thor P-38	78° 7.8'	103° 15.2'	6	105	568	E336	5	100(1)	3300 (1,5)
86	Hoodoo Dome H-37	78° 6.5'	99° 45.6'	5	18	655	E303	156	100(1)	
87	Wilkins E-60	77° 59.3'	111° 21.7'	1	1	580	271+	64		
195	Linckens Island P-46	77° 45.8'	97° 45.4'	3	27	518	E253	1		
256	Sutherland 0-23	77° 42.9'	102° 8.5'	5	7	477	E315	21	100(1)	7250 (1,5)
291	Cornwall 0-30	77° 29.8'	94° 39.0'	2	7	496	E362	20	70-100(3)	7100 (1,5)
258	Pat Bay A-72	77° 21.0'	105° 27.0'	5	41	488	300+	17	90(2)	8000 (2,5)
91	Jameson Bay C-31	76° 40.2'	116° 43.7'	3	13	732	E483	58		
	Sabine Peninsula									
199	Drake E-78	76° 27.3'	108° 29.4'	6	166	277	E171	2	60(4)	700 (4)
198	Drake D-68	76° 27.1'	108° 55.7'	2	.5	781	E264	37	60(4)	8650 (4)
172	Drake B-44	76° 23.1'	108° 16.1'	6	111	352	E188	4	60(4)	1350 (4)
259	Drake D-73	76° 22.1'	108° 29.5'	5	135	410	E288	33	60(4)	8325 (4)
200	Hecla I-69	76° 18.7'	110° 23.3'	5	39	738	E143	2	60(4)	700 (4)
Area C: Franklinian Geosyncline										
196	Bent Horn N-72	76° 21.8'	103° 58.2'	6	14	869	E726	63		7000 (6)
286	Bent Horn F-72A	76° 21.5'	103° 58.2'	4	37	837	E660	43		5000 (6)
257	Pedder Point D-49	75° 38.2'	118° 48.3'	4	91	551	E341	101		
Area D: Arctic Platform/Coastal Plain										
158	Brock I-20	77° 59.7'	114° 33.9'	5	49	840	E422	16		
99	Devon E-45	75° 4.3'	91° 48.3'	5	40	113	X600+	244		
73	Winter Harbour	74° 48.1'	110° 30.6'	6	19	605	E535	22	60(4)	
168	Dundas C-80	74° 39.0'	113° 23.0'	6	28	667	E577	240	60(4)	
92	Garnier 0-21	73° 40.9'	90° 36.8'	1	.02	610	500+	369		
Area E: Cornwallis Foldbelt										
157	Polaris (5 sites)	75° 24.0'	96° 56.0'	3	37+	197+	X335+	to 38		
0	Resolute 1	74° 41.0'	94° 53.8'	1	?	198	X380	10	25(1)	1800 (6)
55	Lobitos Resolute L-41	75° 40.7'	94° 44.6'	3	34	172	EX600	61	25(1)	7250 (6)

Notes:
(a) EPB No. is the Earth Physics Branch file number used throughout this paper.
(b) Under "Temperature Logs" is listed the number of logs available on the well, the figure t2/t1, (the ratio of the time between the latest log and well completion to the drilling time) and the maximum depth logged.
(c) Permafrost thickness determined from a series of logs ("E") or from extrapolated temperatures ("X").
(d) Marine limit and emergence data from (1) Blake (1970), (2) Hodgson (1981), (3) Hodgson (1982), (4) McLaren and Barnett (1978), (5) St-Onge (1965), and (6) Walcott (1972).

essentially identical for the two sites (Table 2), thereby confirming the hypothesis.

The thermal effects of recent shoreline regression are most pronounced at several geothermal wells in the Sabine Peninsula of Melville Island (Figure 2). The position of the present surface elevations of the geothermal sites on the East Coast curve in Figure 3 suggests a wide range of emergence dates (Table 2). The temperature profiles measured today (Figure 4) reflect this recent thermal event through enhanced temperature gradients in the upper few hundred metres, a pronounced curvature and thinner permafrost. The effects are most pronounced in comparison with the Bent Horn wells (numbers 196 and 286) and with the Winter Harbour site (number 73); the latter may have had no marine episode in the Holocene.

The large changes in temperature, gradient and heat flow attributed to the simple shoreline regression history (Table 2) is calculated for site 172 (Figure 5). The high gradients are

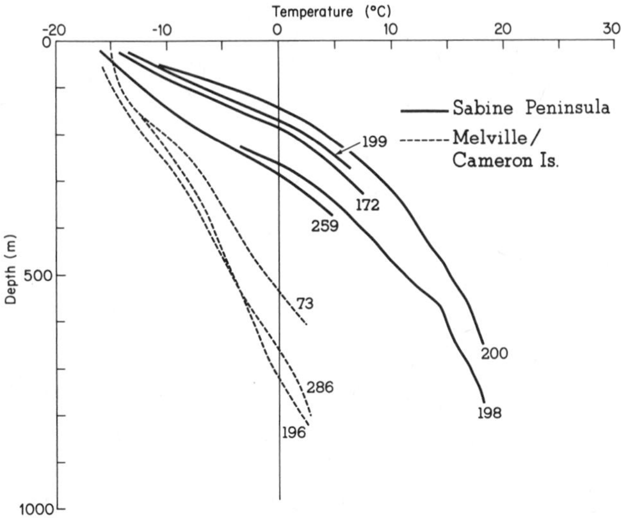

FIGURE 4 Temperature profiles for sites in the Sabine Peninsula and for Cameron Island, 100 km to the east (Figures 1, 2).

reflected in an unusually high apparent heat flow, an effect similarly noted in the Resolute data (Misener, 1955). The calculated temperature profile suggests that permafrost would be considerably thicker if the site had not experienced the lengthy marine inundation; without this event, permafrost would be expected to be up to double the present thickness at the three coastal wells (numbers 172, 199, and 200; Table 2). At the site further inland close to the marine limit (number 259), the thermal regime may be considered to be in quasi-equilibrium with present surface temperatures.

The pronounced peak in temperature gradient

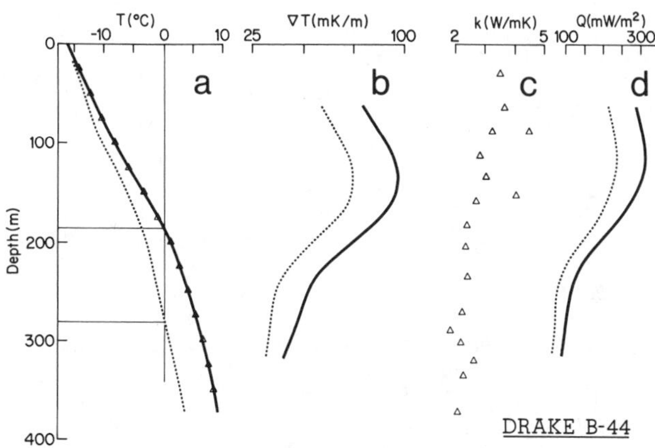

FIGURE 5 Temperature (a), temperature gradient (b), thermal conductivity (c) and heat flow (d) for the Drake B-44 well (site 172, Figure 2). Solid curves show measured values; the dashed curves show values corrected for approximately 6,650 years of submergence.

TABLE 2 Coastal Sites with an Appreciable Period of Marine Submergence

EPB No.	Site Name	Duration of recent history(1) sub-sea (years)	subaerial (years)	Permafrost thickness meas.(2) (m)	corr.(3) (m)
73	Winter Harbour	short	8000	535	540
155	Kris-toffer B-06	1650	6350	445	465
170	Thor P-38	4700	3300	336	N/A
196	Bent Horn N-72	1000	7000	726	750
286	Bent Horn F-72A	3000	5000	660	742
172	Drake B-44	6650	1350	188	280
199	Drake E-78	7300	700	171	300
200	Hecla I-69	7300	700	143	340
259	Drake D-73	short	8000	288	288

Notes: (1) see text; references, Table 1.
(2) measured permafrost thickness
(3) permafrost thickness corrected for effects of shoreline regression.

and heat flow is common at similar depth intervals in all the Sabine wells. It is consistent with an unusually cold climatic episode in the past century at the termination of the Little Ice Age, but has not been adequately resolved at present.

CONCLUSION

A simple geothermal analysis has demonstrated that substantial thermal effects of Arctic shoreline regression are reflected in today's deep ground temperatures and permafrost thicknesses. Arctic geothermal sites must be analyzed carefully for these and similar historical surface temperature variations if calculated values of terrestrial heat flow are representative of the geological environment. Alternatively, the geothermal analysis provides a method to confirm glacial and palaeoclimatic models.

Our limited data suggest that coastal areas of much of the central Arctic Islands will show similar thermal effects of shoreline regression accompanying glacial rebound. A two-dimensional numerical model, constrained by measured temperature profiles, would provide insight into the thermal signature of shoreline regression.

ACKNOWLEDGMENTS

This short analysis is based on data collected over the past 15 years and published in Taylor et al. (1982) and preceding volumes referenced therein. We thank the various oil exploration companies, principally Panarctic Oils Ltd., for their cooperation in that data collection, and for the logistic support to the program offered by the Polar Continental Shelf Project, EMR Canada. Assistance in fieldwork and thermal conductivity measurements was provided by V. Allen, M. Styles, D. Williams, L. Wilson, and A. Wilkinson. Assistance in manuscript and diagram preparation was given by R. Decosse, R. DeLaunais, E. Gélinas, and M. Whissell. We thank the reviewers for their helpful suggestions, especially D. Hodgson for insight into the Holocene history.

This paper is a contribution of the Earth Physics Branch, no. 1062.

REFERENCES

Birch, Francis, 1948, Effects of Pleistocene climatic variations upon geothermal gradients, American Journal of Science 246, 729-760.

Blake, W., Jr., 1970, Studies of glacial history in Arctic Canada, I. Pumice, Radiocarbon Dates, and Differential Postglacial Uplift in the Eastern Queen Elizabeth Islands, Canadian Journal of Earth Science, 7, 634-664.

Blake, W., Jr., 1972, Climatic implications of radiocarbon-dated driftwood in the Queen Elizabeth Islands, Arctic Canada. in, Vasari, Y., Hyvarinen, H. and Hicks, S., Ed., Climatic changes in Arctic areas during the last ten thousand years. A symposium held at Oulanka and Kevo, Finland, Oct. 1971, Acta Univ. Oulu, A3, Geologica 1, 77-104.

Desrochers, D.T., 1983, Permafrost and its relationship to late Quaternary history, High Arctic Islands, University of Ottawa, Ottawa, Canada, M.A. Thesis, in preparation.

England, J., 1976, Late Quaternary glaciation of the Eastern Queen Elizabeth Islands, N.W.T., Canada: Alternative Models, Quaternary Research 6, 185-202.

Gold, L. W. and Lachenbruch, A. H., 1973, Thermal conditions in permafrost--a review of North American literature, in, Permafrost--The North American contribution to the Second International Conference, Yakutsk: Washington, D.C., National Academy of Science, 3-25.

Henoch, W. E. S., 1964, Postglacial marine submergence and emergence of Melville Island, N.W.T., Geographical Bulletin 22, 105-126.

Hodgson, D. A., 1981, Surficial geology, Lougheed Island, Northwest Arctic Archipelago, in, Current Research, Geological Survey of Canada paper 81-1C, p. 27-34.

Hodgson, D. A., 1982, Surficial materials and geomorphological processes, western Sverdrup and adjacent islands, District of Franklin. Geological Survey of Canada paper 81-9, 44 p.

Jaeger, J. C., 1965, Application of the theory of heat conduction to geothermal measurements, in, W. H. K. Lee, Ed., Terrestrial heat flow, American Geophysical Union, Geophysical Monograph, Series No. 8, Washington.

Judge, A. S., 1973a, Deep temperature observations in the Canadian North, in, Permafrost--The North American contribution to the Second International Conference, Yakutsk: Washington, D.C., National Academy of Science, 35-40.

Judge, A. S., 1973b, The prediction of permafrost thickness, Canadian Geotechnical Journal, 10, 1-11.

Lachenbruch, A. H., 1957, Thermal effects of the ocean on permafrost, Bulletin of the Geological Society of America 68, 1515-1530.

McLaren, P. and Barnett, D. M., 1978, Holocene emergence of the south and east coasts of Melville Island, Queen Elizabeth Islands, Northwest Territories, Canada, Arctic 31, 415-427.

Misener, A. D., 1955, Heat flow and depth of permafrost at Resolute Bay, Cornwallis Island, N.W.T., Transactions, American Geophysical Union 36, 1055-1060.

Paterson, W. S. B., 1977, Extent of the late-Wisconsin glaciation in northwest Greenland and northern Ellesmere Island: a review of the glaciological and geological evidence. Quaternary Research, 8, 180-190.

Prest, V. K., 1970, Quaternary geology of Canada, in, R. J. W. Douglas, ed., Geology and economic minerals of Canada, Energy, Mines and Resources, Ottawa, 676-764.

St-Onge, D. A., 1965, La géomorphologie de l'Ile Ellef Ringnes, Territoires du Nord-Ouest, Canadian Department Mines & Technical Surveys, Geographical Branch, Paper 38, 58 p.

Sass, J. H., Lachenbruch, A. H., and Monroe, J. R., 1971, Thermal conductivity of rocks from measurements on fragments and its application to heat flow determinations. Journal of Geophysical Research, 76, 3391-3401.

Sugden, D. E., 1977, Reconstruction of the morphology, dynamics, and thermal characteristics of the Laurentide ice sheet at its maximum. Arctic and Alpine Research, 9, No. 1, 21-47.

Taylor, A. E., Burgess, M., Judge, A. S. and Allen, V. S., 1982, Canadian geothermal data collection--northern wells 1981. Geothermal Series Number 13, Earth Physics Branch, Energy, Mines and Resources, 153 pp.

Walcott, R. I., 1972, Late Quaternary vertical movements in eastern North America--quantitative evidence of glacio-isostatic rebound, Reviews of Geophysics and Space Physics 10, 849-884.

THERMOKARST FEATURES ASSOCIATED WITH BURIED SECTIONS OF THE TRANS-ALASKA PIPELINE

Howard P. Thomas[1] and John E. Ferrell[2]

[1]Woodward-Clyde Consultants, Anchorage, Alaska 99503 USA
[2]Alyeska Pipeline Service Company, Anchorage, Alaska 99512 USA

Following startup in 1977, thermokarst ground surface features started to appear along buried segments of the Trans-Alaska Pipeline. Features observed included longitudinal cracks and various depressions. Proper interpretation of these features became an important element of pipe settlement investigations which were initiated in 1979. Most of the features were relatively shallow and were not related to pipe thaw settlement. However, a few features provided an advance indication of developing critical pipe settlement conditions. These advance indications enabled the planning and undertaking of timely remedial actions.

INTRODUCTION

Natural and manmade thermokarst features are relatively common in Arctic and sub-Arctic regions (see Mackay 1970, French 1979, Sweet and Connor 1980). Examples of natural causes are fires, river bank undercutting and coastal retreat. Well-known manmade causes involve stripping or compression of the tundra moss mat as a result of plowing, bulldozing or vehicular movements. The potential effects of burying the Trans-Alaska Pipeline in permafrost terrain were recognized well before its actual construction (Lachenbruch 1970).

The pipeline, which carries warm crude oil, was allowed to be buried in permafrost soils only where the pipe itself could be supported on thaw-stable soils or bedrock. Following startup in 1977, thermokarst ground surface features started to appear in several segments of the buried pipeline. These features were formed by thawing of massive ice or ice-rich soils overlying the thaw-stable soils. Contributory factors included the additional disturbance to the thermal regime caused by the presence of the gravel workpad and, in some cases, thermal erosion caused by surface water.

This paper describes thermokarst features observed during pipeline surveillances in 1979 and 1980, and comments on their engineering significance. Selected photographs are presented to illustrate typical features.

FEDERAL STIPULATIONS

The operating temperature of the Trans-Alaska Oil Pipeline varies between 38 and 63°C. This elevated temperature is maintained throughout the length of the line by the effects of friction heating. To ensure adequate support for the 48-in.-diameter pipeline, the federal stipulations for its construction (U.S. Government 1974) allowed it to be buried only in initially-thawed or thaw-stable materials: "Below the level of the pipe axis, the ground shall consist of competent bedrock, soil naturally devoid of permafrost, or, if frozen, of thaw-stable sand and gravel." In a footnote, thaw-stable sand and gravel was defined as "Material within classes GW, GP, SW and SP but with less than 6 percent by weight passing the No. 200 U.S. Standard sieve." The stipulations did not prohibit ice-rich permafrost above the level of the pipe axis.

The federal technical stipulations provided the basis for justification of the buried pipeline construction mode. Mode designations were assigned during mile-by-mile design on the basis of borings and landform analysis. These designations were continually updated as alignment adjustments and new borings were made. The buried mode was further verified during construction by continuously logging the pipe ditch by qualified geologists as soon as it was excavated (see Luscher and Thomas 1979). Application of this process included drilling over 4,000 borings resulting in about 600 km of the pipeline (half of the total alignment) being buried.

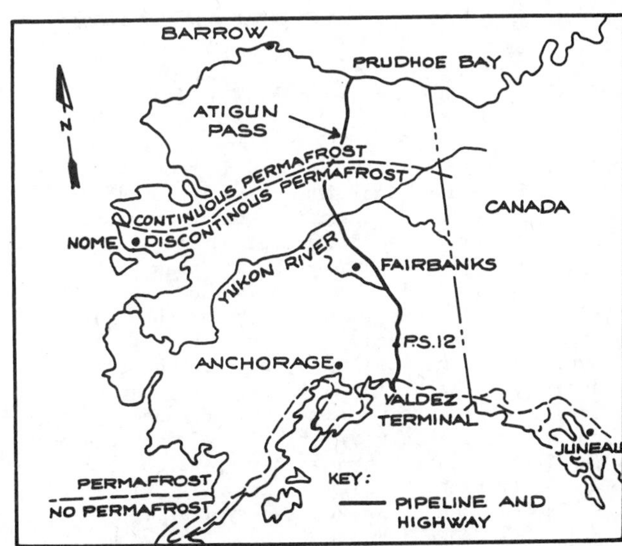

FIGURE 1 Route of the trans-Alaska pipeline

Buried sections varied in length from 150 m up to a maximum of 100 km. As shown in Figure 1, north of the Brooks Range, the line traverses continuous permafrost; from the Brooks Range south to about Pump Station 12, the line traverses discontinuous permafrost; and, south of Pump Station 12, the line traverses about 100 km of non-permafrost terrain.

THAW BULB GROWTH AND PIPE SETTLEMENT

In those segments where the pipe was buried in frozen ground, a thaw bulb started to grow around the pipe at the time of startup in June 1977. Early thermal simulations based on conductive heat flow estimated thaw bulb growth rates shown on Figure 2.

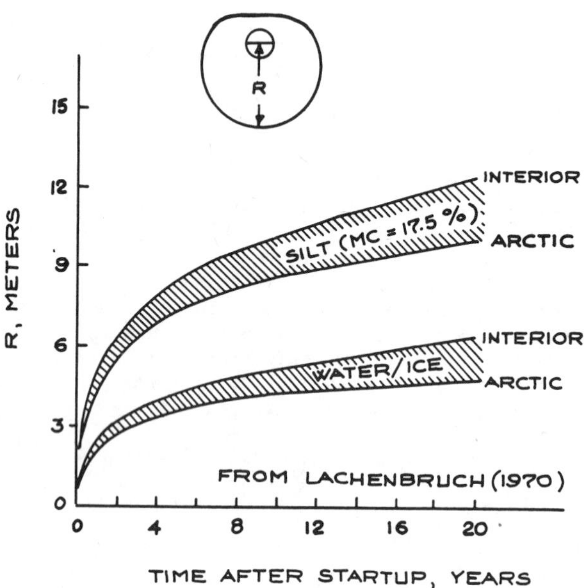

FIGURE 2 Growth of thaw bulb with time

Actual thaw bulb growth depended strongly on local conditions including initial ground temperature, soil stratification, groundwater flow, and presence of a gravel workpad. A North Slope example of the shape of the thaw bulb six years after pipeline startup is shown on Figure 3. At this location, the presence of a high water table and about 6 m of highly-pervious gravel immediately beneath the pipe has accelerated lateral growth of the thaw bulb as a result of convective heat flow. Presence of the gravel workpad on the right side of the pipe has resulted in slightly more thawing (also more winter freezeback) on that side.

By 1979, the thaw bulb had typically progressed to a depth of 3 to 6 m below the pipe, and to a similar distance laterally. In buried pipeline sections, pipeline settlements accompanied the thawing, but these were less than about 30 cm (5 to 10 percent thaw strain) over at least 95 percent of the buried pipeline.

At two widely-separated locations (Mileposts 166 and 734), large settlements (approaching 1.5 m) occurred which caused the pipe to buckle and leak in June of 1979 (see Johnson 1981). The pipe was quickly repaired and stabilized at these locations. Subsequent investigations revealed that the cause of the leaks was thaw settlement of unstable permafrost which had not been identified during construction. At Milepost 734 just north of Pump Station 12, the pipe was laid across an "island" of ice-rich silt, a near-circular remnant feature about 90 m in diameter. The gradual thaw around the warm pipeline caused the frozen silt to thaw and consolidate, resulting in loss of support for the pipe. The Milepost 166 leak at Atigun Pass was caused by melting of thick ice lenses in frozen weathered bedrock beneath the pipe and consequent loss of pipe support.

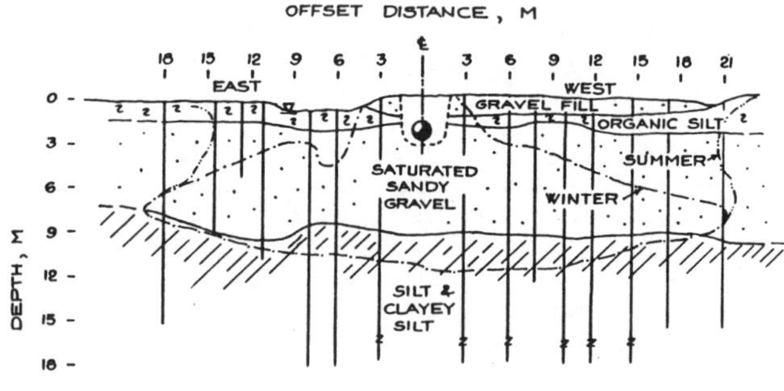

FIGURE 3 Milepost 34 thaw bulb cross-section - 1983

SPECIAL SURVEILLANCES

Concurrent with the 1979 leak investigations, intensive mile-by-mile studies were undertaken of all buried pipeline sections north of Milepost 740. These studies comprised both office reviews of design and construction documents and special field inspections of the buried pipeline sections.

Two special geotechnical surveillances of the pipeline were conducted in 1979. The first took place from June 27 to July 4 and concentrated on disturbance areas identified by Alyeska field personnel. Some of the areas were visited by helicopter, while others were reached using a four-wheel drive vehicle. Numerous photographs were taken, ground features were measured, detailed field notes were recorded, and tentative causes for the disturbances were advanced.

The Phase I studies demonstrated the usefulness of field observations and office records for evaluating buried pipeline foundation conditions. The objective of Phase II was to extend the general process developed in Phase I. Following the first field surveillance, a detailed office review was made of information available on the belowground pipeline sections.

Particular attention was paid to the continuous soil logs which had been prepared during construction of the ditch bottom and sidewalls.

Utilizing a detailed itinerary, the second special surveillance was conducted by helicopter from September 19 to 21. Each suspect area was circled in the air and on-the-ground inspections were made at most of the sites. Several unscheduled stops were also made to inspect suspicious ground disturbances. Once again, detailed field notes were recorded and numerous photographs were taken.

OBSERVATIONS

Typical ground features observed in the field inspections were cracks and depressions. The cracks were usually longitudinal (parallel to the pipe), were often 10 cm or more wide and 1 m or more deep, and sometimes ran for 100 m or more along the buried pipe (see Plates 1 and 2). The cracks occasionally ran right along pipe centerline, but were more often offset 3 to 6 m from pipe centerline. Sometimes there were parallel cracks on either side of the pipe, other times on one side only (as in Plates 1 and 2). Depressions were of the thermokarst type, sometimes abrupt (see sinkholes on Plates 3 and 4) sometimes gentle (see undulation on Plate 5), sometimes water filled (see ponding on Plates 6 through 8). Concentric cracking was sometimes evident around depressions.

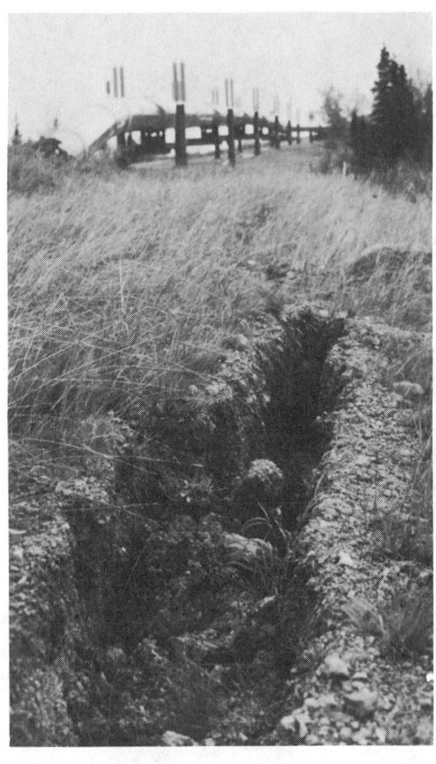

PLATE 2 Ground crack at MP 393

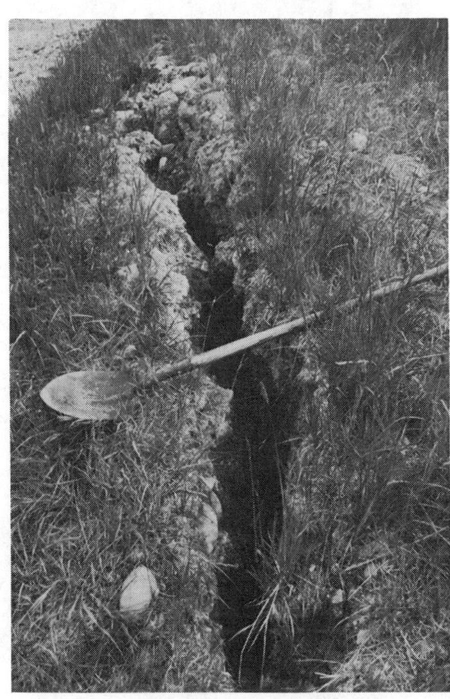

PLATE 1 Ground crack at MP 734 leak site

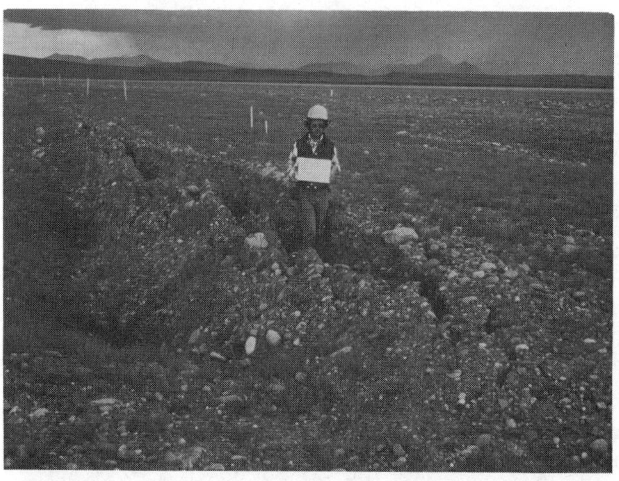

PLATE 3 Sinkhole at MP 109

PLATE 4 Sinkhole at MP 65

PLATE 5 Sinkhole at MP 45

PLATE 6 Sinkhole at MP 19

PLATE 7 Sagpond at MP 574

PLATE 8 Sagpond at MP 735

MECHANISMS

Several possible mechanisms were advanced for development of the observed ground features:

(1) Settlement of ditch backfill;
(2) Settlement of in-situ soil outside ditch walls;
(3) Compression of natural soils below ditch bottom.

Of these, only (3) could cause pipe settlement. (Compression of the relatively-thin pipe bedding layer was generally minor.)

Longitudinal cracking along the buried pipe is schematically depicted in Figure 4A. Cracking 1.5 to 3 m from pipe centerline usually reflected the edge of the pipeline ditch; likewise, cracking 4.5 to 6 m from pipe centerline usually reflected the edge of the thaw bulb. The asymmetric geometry associated with the presence of the workpad explains why cracking sometimes appeared on one side of the pipe but not on the other.

Wide and/or deep thaw bulbs were observed at several locations. As was the case at Milepost 34, these anomalies were generally associated with groundwater flow/convective heat flow.

Causes (1) and (2) are depicted schematically in Figures 4B and 4C, respectively. As suggested by Metz et al (1982), ditch backfill settlement typically resulted from poor compaction of the backfill or thawing of backfill which was placed in a frozen state. The most frequent cause of ground disturbance was thawing of ice-rich overburden overlying either thaw-stable gravels or competent bedrock. Abrupt features such as cavities and sinkholes were generally explained by melting of ice wedges and other massive ice inclusions in or beyond the ditch walls. Other causes of disturbance features included poor backfilling of bellholes made for remedial welds and hydrotesting.

The following points emerge from the foregoing discussion:

- Pipe settlement could only be caused by thawing of ice-rich permafrost beneath the pipe.
- Severe ground disturbance, even close to pipe centerline, did not necessarily imply pipe distress.
- It was difficult to tell from surface features alone if pipe settlement had occurred.
- The deeper the pipe was buried, the more difficult it was to interpret ground surface features in terms of pipe distress.
- Regrading sometimes obliterated past ground disturbance features, but complete absence of disturbance features in a buried pipeline segment was generally taken as a favorable indication.
- Presence of thermokarst features in a segment designed for thawed soil conditions was taken as an a priori cause for concern.
- Presence of transverse cracking and/or severe undulations along pipe centerline (see Figure 5) were cause for special concern as these had a high likelihood of being related to pipe settlement.

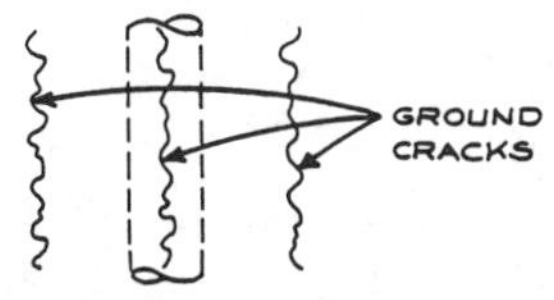

A) PLAN

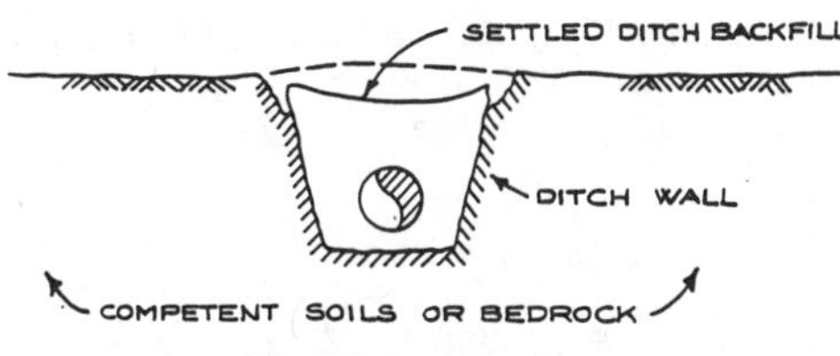

B) PROFILE 1

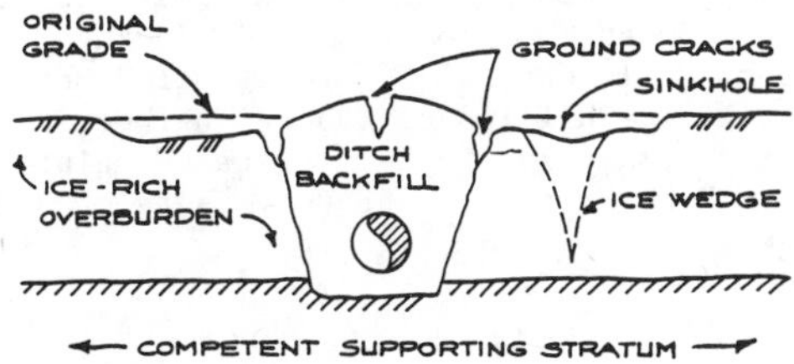

C) PROFILE 2

FIGURE 4 Causes of ground surface disturbance features

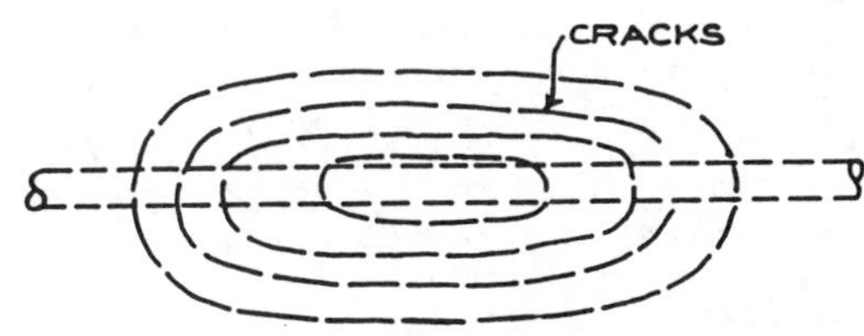

PLAN

SURFACE DEPRESSION

PROFILE

FIGURE 5 Surface feature associated with buried pipe settlement

Ground surface features at the leak sites were quickly obliterated by priority construction activities associated with containment of spilled oil, excavation of the buried pipe, and completion of temporary repairs. However, at the MP 734 site, a prominent longitudinal crack about 4.5 m west of pipe centerline was observed and photographed (see Plate 1). Probably because of snow cover, but perhaps also because the pipe was buried under 4 m of soil cover, no ground disturbance features were reported at the MP 166 site. It was concluded that it would have been quite difficult to anticipate pipe failure at the two leak locations based solely on visual surface disturbance observations.

Subsequent special surveillances were performed along the pipeline alignment in 1980 and 1981. An important feature of these surveillances was the obtaining of photographs which could be compared from year to year to see whether site-specific conditions were worsening or stabilizing. Staking of pipe centerline in critical areas prior to these surveillances aided in interpretation of ground disturbance features.

SUMMARY AND CONCLUSIONS

Numerous thermokarst features appeared along buried sections of the Trans-Alaska Warm Oil Pipeline following startup in 1977. Typical features observed were longitudinal cracks and various depressions. The most dramatic features appeared in the first two or three years after startup, when the thaw bulb was growing rapidly and was intersecting massive ground ice inclusions. These features have since been filled with gravel as part of Alyeska's ongoing right-of-way maintenance program.

Primary causes for the ground surface disturbance features were identified as:

- Thaw settlement of ice-rich permafrost overlying thaw-stable materials on which the pipe was founded;
- Settlement of ditch backfill above the pipe (frozen, wet, or uncompacted backfill);
- Inadequate backfilling of post-construction excavations such as bellholes for welds and excavations made during hydro-testing.

Some instances of ground disturbance appeared to be, and were later confirmed to be, due to local occurrences of unstable permafrost beneath the pipe. In a parallel investigation, both the MP 166 and the MP 734 leaks were attributed to settlement of the pipeline due to thawing of unstable permafrost. An ongoing mile-by-mile review focused on identifying such occurrences by evaluating a combination of design and construction records and field disturbance observations. Questionable pipeline segments so identified were excavated so that monitoring rods could be attached to the top of the pipe. This program succeeded in locating and stabilizing eight areas where the pipe was approaching buckling curvature.

By 1983, the thaw bulb had generally progressed to a depth below the pipe where soil ice contents are reduced, and the rate of thaw penetration has slowed to a point where the pipe settlement rate is generally quite small (less than 15 mm/yr).

Inspection and knowledgeable interpretation of thermokarst features proved to be a valuable tool in identifying potential pipe settlement areas along buried sections of the Trans-Alaska Pipeline. These observations provided important input into ongoing engineering evaluations which enabled early identification of pipeline sections requiring remedial actions (e.g., stabilization).

REFERENCES

French, H. M., 1979, Permafrost and ground ice in Man and Environmental Processes, Dawson Westview Press, Ch. 9.

Johnson, E. R., 1981, Buried oil pipeline design and operation in the Arctic - Lessons learned on the Alyeska Pipeline: 37th Petroleum and Mechanical Conference, Dallas, Texas.

Lachenbruch, A. H., 1970, Some estimates of the thermal effects of a heated pipeline in permafrost: Geological Survey Circular 632.

Luscher, U., and Thomas, H. P., 1979, Geotechnical issues and answers during construction of the trans-Alaska pipeline: Journal of Energy Resources Technology, Transactions of the ASME, v. 101, p. 128-133.

Mackay, J. R., 1970, Disturbances to the tundra and forest tundra environment of the western Arctic: Canadian Geotechnical Journal, 7., p. 420-432.

Metz, M. C., Krzewinski, T. G., and Clarke, E. S., 1982, The Trans-Alaska Pipeline System workpad - An evaluation of present conditions, in Proceedings of the Fourth Canadian Permafrost Conference, p. 523-534.

Sweet, L. and Connor, B., 1980, The Farmers Loop sinkhole: The Northern Engineer, v. 12, no. 4, p. 6-10.

U.S. Government, 1974, Technical Stipulations for Agreement and Grant of Right-of-Way for Trans-Alaska Pipeline.

ACKNOWLEDGEMENTS

Permission of Alyeska Pipeline Service Company to publish this paper is gratefully acknowledged. Also gratefully acknowledged are helpful comments of colleagues including E. R. Johnson of Alyeska, J. M. Stanley of Dames and Moore, W. T. Phillips of Quadra Engineering, Inc., P. Velasco of Gulf Interstate Engineering and Dr. Hugh French of the University of Ottawa, Canada.

THE ECOLOGY OF PERMAFROST AREAS IN CENTRAL ICELAND AND THE POTENTIAL EFFECTS OF IMPOUNDMENT

Thora Ellen Thorhallsdottir

Institute of Biology, University of Iceland, Grensasvegur 12, Reykjavik, Iceland

In the central highland of Iceland, extensive hydroelectric developments are now planned. Small areas were flooded in 1981, but there are concerns over the ecological effects of a much larger reservoir, planned at a later stage. This would involve inundation of parts of the Thjorsarver Reserve, comprising the most extensive and richest tundra vegetation in Iceland. In 1981, a long term study of the ecology of the Thjorsarver tundra was initiated, with emphasis on the vegetation and soils and the potential effects of impoundment. Permanent transects were set up in disturbed and undisturbed areas for detailed recording of vegetation and soil parameters. This paper reports the preliminary results on the distribution and thickness of permafrost, active layer development, ground water and soil characteristics, but only brief descriptions of the vegetation.

INTRODUCTION

Central Iceland is a highland plateau with glaciers and isolated mountains rising above a generally flat or gently rolling terrain. It is characterized by black basaltic sand and lava with extremely sparse vegetation. Continuous vegetation cover is limited to areas with a dependable water supply. Permafrost in Iceland is more or less confined to this central region, where, according to Priesnitz and Schunke (1978), it is found at altitudes between 460 and 720 m. The highland vegetation of Iceland has been described by Steindorsson (1945, 1964, 1966, 1967). A survey and classification of Icelandic palsas is given by Schunke (1973). Priesnitz and Schunke (1978) discuss aggradation and degradation of palsas with reference to climatic fluctuations. Other reports of palsas and periglacial phenomena in Iceland include Thorarinsson (1951), Friedman et al. (1971) and Bergmann (1972). These investigations, however, are largely limited to general descriptions or qualitative observations. There is no detailed information on the extent and thickness of permafrost, active layer development, groundwater status and soil characteristics. With plans for extensive hydroelectric developments in central Iceland, the need for a functional understanding of the ecology of these areas becomes all the more pressing.

Several major hydroelectric projects involve the Thjorsa River, the largest glacial river in Iceland. In the first instance, all its easterly tributaries will be diverted south, through a series of man-made lakes and finally into the Thorisvatn Reservoir (Figure 1). The first of these lakes, Dratthalavatn, was flooded in 1981, and partly submerged some vegetated land. At a later stage, there are plans to dam the uppermost part of the Thjorsa proper. The resulting reservoir would inundate parts of the Thjorsarver tundra. Thjorsarver is a collective name for the largely continuous vegetation immediately south of Hofsjökull Glacier and the more discrete and isolated vegetation further south and east (Figure 1). They are now a nature reserve, comprising the richest tundra vegetation in Iceland, as well as being the world's largest breeding ground of the pink-footed goose (e.g. Kerbes et al. 1971). Recent experience from Alaska and Canada has shown how susceptible tundra ecosystems may be to physical disturbances and impoundment (e.g. Kerfoot 1974, Rickard and Brown 1974, Newbury et al. 1978, Baxter and Glaude 1980, Newbury and McCullough 1982). There is concern

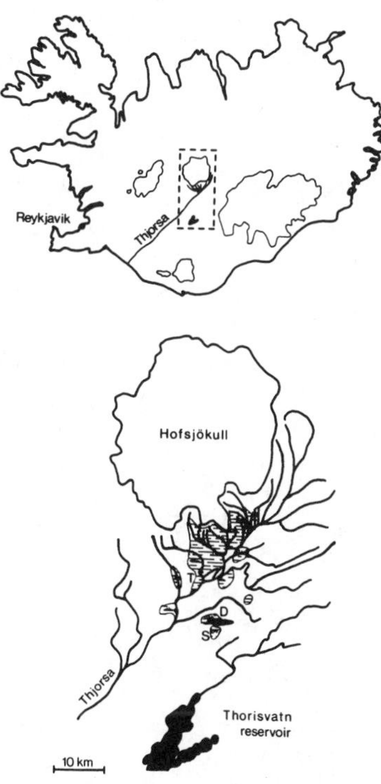

FIGURE 1 (Top) Location of Thjorsarver and the Thjorsa River in central Iceland. (Bottom) The area in detail. Vegetated land is dashed.
T = Tjarnaver; D = Dratthalavatn; S = Storaver

over the potential ecological effects of a large reservoir in the Thjorsarver Reserve.

The purpose of the research described here is twofold. It is a long term investigation of the general ecology of the Thjorsarver tundra with particular emphasis on the dynamics of the palsa areas. Second, the project is intended as a pilot study to assess the ecological impact of a large reservoir in the Thjorsa River, partly through monitoring the effects of the much smaller Dratthalavatn. Preliminary results are reported here with emphasis on the soil features.

CLIMATE AND GENERAL DESCRIPTION OF THE AREA

Thjorsarver lie at an altitude of about 600 m a.s.l. The generally flat and undulating plateau is broken by gently rolling hills. These are partly moraines and drumlins, and partly eroded bedrock, mainly basalt and sediment. Small mountains, bordering the area to the east and northeast usually have a palagonite core. Extensive esker complexes are found. Moraine covers most of the area, overlain by fine sand and in places by glacio/fluvial and lacustrine sediments (Tryggvason and Einarsson 1965, Tomasson and Thorgrimsson 1972, Kaldal 1981).

Mean monthly and annual temperatures, annual precipitation and general climatic information is given by Jonsson (1978) (Table 1).

TABLE 1 General Climate of the Thjorsarver Area for the Years 1966-1975, modified from the table of Jonsson (1978). Values are partly derived from the nearest permanent weather station at Hveravellir (approximately 50 km north west of Thjorsarver, altitude 642 m). For further information, see Jonsson (1978)

Mean monthly air temperatures (°C)		
January	-5.9	**Precipitation** 800 mm/yr
February	-6.3	**First total snow cover** September/October
March	-5.6	
April	-2.6	
May	1.1	**Last total snow cover** late May
June	5.0	
July	7.3	
August	6.7	**Earliest total ablation** early June
September	3.2	
October	-1.0	
November	-0.5	**Number of days of freezing** about 220
December	-6.2	
Mean annual air temperature	-0.8	

The largest part of the Thjorsarver tundra forms a rough triangle south of the Hofsjökull glacier (Figure 1). It is flanked by the main Thjorsa River to the south and two of its contributaries to the east and west. Numerous other glacial tributaries traverse the vegetation, giving rise to tall sedge (*Carex rostrata* and *C. lyngbyei*) fens in the wettest parts. Palsas are mainly found in slightly drier areas, and are more numerous and diverse in form than elsewhere in Iceland. Mosses, (especially *Racomitrium canescens* and *Drepanocladus uncinatus*) and lichens are conspicuous on the palsas, which also carry a diverse flora of higher plants, notably willows (*Salix glauca* and *S. herbacea*), forbs such as *Armeria maritima*, *Polygonum viviparum*, *Saxifraga caespitosa*, *S. hirculus* and *Silene acaulis*, and grasses such as *Festuca* spp and *Poa* spp. *Carex rariflora*, *Eriophorum angustifolium* and *E. scheuchzeri*, as well as *Calamagrostis neglecta*, often fill the depressions between palsas. *Salix glauca*, *S. lanata* and *Racomitrium canescens* dominate the driest parts. For a complete floral list and species distribution maps, see Johannsson et al. (1974).

East of the Thjorsa River, areas of similar vegetation are found, but less extensive and occurring more as isolated patches in depressions in the Sprengisandur Desert. They are generally drier and less diverse vegetationally.

The initial stages of hydroelectric development only affect the land east of the Thjorsa, whereas the proposed Thjorsa reservoir inundates vegetation both east and west of the river.

MATERIALS AND METHODS

The project started in 1981. Seven permanent transects were constructed by Dratthalavatn Lake and another seven in various vegetation types east of Thjorsa. In 1982, a single transect was set up in Thjorsarver proper, west of the river. The transects were usually 105 m long, and marked at either end by 1.5 m long aluminium poles.

On level terrain, each transect was further divided at 15 m intervals by 70 cm long wooden pegs. At each of these, the soil profile was examined, and the following soil variables recorded: 1) thickness of the organic layer; 2) mean and maximum rooting depth; 3) qualitative assessment of rooting density; 4) active layer thickness in early July (8 to 12), and late August (18 to 19) by five recordings with a metal auger; 5) the thickness of permafrost. This could only be done with the auger where the ice was 40 cm or thinner. Measurements were made on permafrost thickness of six palsas only, using a COBRA super drill (BORROS AB Sweden).

Local topography was mapped by theodolite readings every 15 m. Vegetation was sampled as percentage shoot frequency in 0.5 x 0.5 m quadrats divided into 25 grid quadrats with 10 cm sides. Recordings were made along a 5 m belt transect at every 15 m interval. Mosses were collectively grouped because of difficulties in field identification. Mosses on the drier heath were probably mostly *Racomitrium canescens*.

In palsa bogs, the topography called for a different sampling procedure. Soil measurements were taken at the top and edges of all the palsas and in the depressions between them, where possible. The distribution of permafrost was mapped, and depth of active layer measured, as described above, on the top and sides of all palsas and in the depressions between them. Local topography was mapped, as above, to show the outlines and maximum height of the palsas. Vegetation was recorded, as above, except that a belt transect running the entire length of the palsa was used. Recordings were made in the depressions, where possible.

Groundwater was measured in five transects: three by Dratthalavatn Lake, numbered 3 (gravel slope), 4 and 5 (moss/willow earth hummocks); one in a cottongrass/sedge plain (transect 7); and one in a willow/moss heath west of Thjorsa River (transect 9). Water level was measured in 60 or 100 cm long perforated plastic tubes by lowering a sealed (varnished) oak cube, attached to a string, into the tube. Water level was measured every 4 days from July 12 to August 18 by the lake, on July 12 and August 19 on transect 7, and on August 13 on transect 9.

RESULTS AND DISCUSSION

Vegetation

The most common species in six of the transects are shown in Table 2. The gravel slope with its sparse cover, had the greatest diversity of species. Mosses had very high frequency values (0.75-1.0) in all the communities. *Calamagrostis neglecta* was fairly ubiquitous. *Salix glauca* and *Festuca* spp, probably mostly *F. rubra*, but including some *F. vivipara*, were conspicuous in drier vegetation. Cottongrass and sedges dominated the wetter parts.

In a qualitative survey of palsas in Tjarnaver, that on photographic evidence were at least seven years old, no live plants, except *Carex rariflora* and *Eriophorum* spp., were found. They must be remnants from a wetter stage, indicating the slow successional processes. Since the majority of palsas in the area carry a diverse flora, very distinct from that of the bogs, it is likely that they are at least several decades old, but not less than about 10 years old, as claimed by Priesnitz and Schunke (1978). Further, typical palsa vegetation in Thjorsarver is not xerophilic (e.g. Priesnitz and Schunke 1978) but includes species normally associated with moist conditions, such as *Salix herbacea*, *Calamagrostis neglecta* (Table 2) and *Saxifraga hirculus* (see also Johannsson et al. 1974).

Soil Characteristics

The organic layer. There was no organic layer on the gravel slope. Under all continuous vegetation, decomposing organic matter was found, and this was usually sharply delimited from underlying sand. Mosses were invariably the chief constituents. The mean thickness of the organic layer on the moist heath was almost twice that of the palsas,

TABLE 2 The Vegetation on Six of the Permanent Transects in July (transects 1-8) and August (transect 9) 1982. Frequency values for species with 50% frequency in 0.25 m^2 quadrats are listed. N = sample size

Species	Dratthalavatn			Storaver		Tjarnaver
	Transect 1 Dry willow/ moss heath. Few herbaceous dicots. N 63	Transect 3 Gravel slope with very sparse vegetation cover N 70	Transect 5 Moss/willow earth hummocks with *Salix herbacea/Anthelia* depressions N 70	Transect 7 Relatively dry sedge/ cottongrass plain N 50	Transect 8 Palsas with ponds N 103	Transect 9 Willow/moss heath with earth hummocks and numerous dicots N 70
Salix glauca	0.94	-	0.91	-	0.88	0.88
S. herbacea	-	-	-	-	0.68	0.68
Festuca spp.	1.0	1.0	0.83	-	-	0.67
Poa pratensis	0.55	-	-	-	-	-
Calamagrostis neglecta	-	0.74	-	0.92	0.98	0.90
Carex rariflora	-	-	-	0.72	-	-
C. lachenalii	-	-	-	0.62	-	-
Carex bigelowii	-	-	-	-	-	0.88
Eriophorum angustifolium	-	-	-	0.86	-	-
E. scheuchzeri	-	-	-	0.86	-	-
Luzula spicata	-	-	0.50	-	-	-
Koenigia islandica	-	-	0.53	-	-	-
Polygonum viviparum	0.98	-	-	-	0.61	-
Thalictrum alpinum	-	-	-	-	-	0.50
Silene acaulis	-	0.61	0.51	-	-	-
Cerastium alpinum	-	0.58	-	-	-	-
Armeria maritima	-	0.60	0.50	-	-	-
Saxifraga caespitosa	-	0.51	-	-	-	-
Equisetum arvense	-	-	0.77	-	-	-
mosses	1.0	0.75	0.97	0.9	0.91	1.0

Nomenclature follows the Flora Europea.

with the moist heath intermediate (Table 3). Very wet vegetation posed methodological problems, but in five sites examined in the cottongrass/sedge community in transect 7, the organic layer was always about 17 cm thick. These values are somewhat

TABLE 3 Mean Organic Layer Thickness in three Types of Plant Communities in Thjorsarver in July 1981. N = sample size; S.D. = standard deviation

Community	N	Mean Thickness (cm)	S.D.
Moist moss/ willow heath	12	12.2	4.8
Dry heath	15	9.8	1.6
Palsas (tops)	31	6.3	2.5

low compared with the figures of 10 to 30 cm, given by Bliss (1981) as typical for the Low Arctic. For palsas in the Northern and Southern part of the Canadian discontinuous permafrost zone, Zoltai and Tarnocai (1975) recorded average peat thicknesses of 267 and 355 cm resepctively. These values are not directly comparable to Thjorsarver, since the Icelandic palsas are not forested, but they are nevertheless low. This may reflect a generally young vegetation, or possibly repeated cycles of erosion and revegetation, with the palsas being gradually eroded by sand and finer aeolian deposits, rather than collapsing and leaving a thermokarst pool.

Soil and rooting patterns. The materials underlying the organic layer consisted of alternating layers of fine sand and pumice with an occasional peat layer. On two sites on the gravel slope, however, coarse gravel and pebbles were found at depths of 17 and 25 cm. This layer extended down to permafrost at about 68 cm. No gleying or horizon differentiation was evident in any of the profiles. Poor horizon development is characteristic of Icelandic soils in general (Johannesson 1960). Depth to moraine or more consolidated material underneath has not been fully assessed, but in one site examined at Dratthalavatn Lake and another in Tjarnaver west of Thjorsa River, this was at least 2 and 3 m respectively. Rooting was sparse. Most roots lay in, or just below, the organic layer. Scattered roots extended to just above the permafrost.

The distribution of permafrost. The permafrost is thin and discontinuous. It was more or less confined to densely vegetated areas, although a 10 to 20 cm thick ice layer was still present on the gravel slope (transect 3) on August 18, 1982. In early July, there was ice under all vegetation. Very wet bogs, where surface water remained through all of the growing season, and very sandy areas, especially where Salix lanata was conspicuous, were ice free towards the end of summer. Elsewhere, the ice measured from about 10 to 30 cm in late August, except in palsas where the ice was much thicker. The ice lenses recorded varied in thickness from about 1.1 m in the smallest palsa to about 3.1 m in the largest. It is not possible to generalize from the small number of palsas drilled (6), but the permafrost may never be more than a few metres thick. The relatively small size of the palsas may also reflect this. Most rise 0.6-2.0 m above their surroundings, although occasionally they reach heights of about 2.5 m.

Groundwater. It is highly likely that groundwater status is the most critical single factor affecting both the occurrence and type of vegetation in Thjorsarver. True groundwater is difficult to measure in the presence of permafrost. For large scale measurements of groundwater table in the region, see Hjartarson (1981). Of interest here is simply depth to a water table accessible to plants, and the term is used in that limited context only. In the transects by Dratthalavatn Lake, monitored every three days, the water levels gradually subsided from July 12 to August 5, but changed little after that (illustrated in Figure 2 for two of the transects). The pattern was similar in all

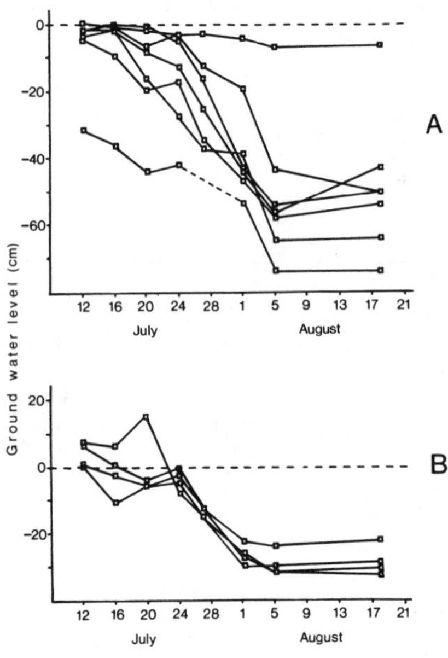

FIGURE 2 Changes in groundwater level from July 12 to August 18 in transect 3 on the gravel slope (A), and between the earth hummocks on transect 5 (B).

three transects, although the underlying materials were appreciably coarser on the gravel slope. The decrease varied from 23-39 cm between the earth hummocks on transect 5 (which had surface water at most sites in early July), from 43-61 cm on the gravel slope, and from 17-31 cm in the cottongrass/sedge plain on transect 7. Groundwater levels at the end of the growing season are shown in Table 4. The sites on transect 5 may have been influenced by their low elevation above Dratthalavatn Lake, and the values may be higher than typical for Salix herbacea/Anthelia depressions. Mean groundwater depth was greatest on the gravel slope, followed by

TABLE 4 Depth to Groundwater on August 13 (transect 9), August 18 (transects 3 and 5) and August 19 (transect 7), 1982. All the sites included for transect 3 were elevated more than 1 m above Dratthalavatn Lake. The sites on transect 5 were 0.6-0.8 m above the lake level. N = sample size; S.D. = standard deviation.

Transect	Community	N	Mean depth (cm)	S.D.
3	Gravel slope	6	57.3	9.7
5	Moss/willow earth hummocks	5	33.2	8.7
7	Cottongrass/ sedge plain	6	34.5	2.0
9	Willow/moss heath	6	45.1	8.1

the willow/moss heath, and finally by the cottongrass/sedge plain.

Active layer development. The depth of the active layer on August 18 and 19 is shown in Table 5. There was no significant difference between active layer depth on the palsas and other sites with continuous vegetation cover. On the gravel slope, the active layer was on average about 10 cm thicker. If, on the other hand, increases in active layer depth between early July (9-12) and late August (18-19) are compared, then this is slightly lower for the palsas than for the other closed vegetation (Table 6). Both are significantly lower than the gravel slope. The figures in Table 4 agree well with those given for active layer depth at Stordalen, Abisko, Sweden (40-80 cm), and for Truelove Lowland, Devon Island (30-100 cm) (see Ryden 1981).

TABLE 5 Active Layer Depth at the End of the Growing Season (August 18-19) 1982. Data are presented separately for palsas and for the gravel slope, which had very sparse vegetation cover. N = sample size; S.D. = standard deviation.

	N	Mean depth (cm)	S.D.
Sites with continuous vegetation cover	48	60.3	9.4
Palsas	13	58.8	8.0
Gravel slope	7	69.5	9.4

The depth of seasonal thaw was always greater than the organic layer thickness, hence the permafrost was confined to the underlying layers of fine sand. The presence of Dratthalavatn Lake (flooded in 1981) did not influence active layer depth, but on the gravel slope, there were indications that backwater might affect melting of permafrost from below.

TABLE 6 Changes in Active Layer Depth from Early July (9-12) to Late August (18-19) 1982. Date are presented separately for palsas and for the gravel slope, which had very sparse vegetation cover. N = sample size; S.D. = standard deviation.

	N	Mean change (cm)	S.D.
Sites with continuous vegetation cover	48	18.5	7.4
Palsas	13	15.4	3.9
Gravel slope	7	28.6	13.5

The Potential Effects of Impoundment

The presence of permafrost may indirectly cause catastrophic shoreline erosion in inundated areas, since a stable slope is not established with ground ice being continually exposed by wave action (Newbury et al. 1978, Newbury and McCullough 1982). Here, the outlines of ground ice generally followed the shoreline of Dratthalavatn Lake. Slumping and overhanging, as described by Newbury et al. (1978) and Newbury and McCullough (1982) has only been observed on a small scale where a cluster of palsas formed the shoreline of Dratthalavatn Lake. Since the lake was only flooded in 1981, it is premature to draw conclusions about potential erosion. It seems unlikely, though, that the thin and discontinuous permafrost will significantly contribute to erosion in the manner described above. The soil properties give more cause for concern. The spongy organic layer and the sparse roots will hardly provide much resistance to erosion. The underlying fine sand layers are similarly susceptible to wind and wave erosion alike.

ACKNOWLEDGMENTS

This work is part of a larger project on various aspects of the biology of the Thjorsarver Reserve, and is carried out under contract to the National Power Company of Iceland.

The help of the following in collecting field data is acknowledged: Th. Eysteinsson, B. Gunnlaugsson, H. Sigurjonsdottir, S. Snorrason and A. G. Thorhallsdottir. The National Energy Authority kindly lent staff and equipment for COBRA drilling of palsas.

A. Gardarsson and Th. Hafstad offered advice and constructive criticism on the preparation of the manuscript.

REFERENCES

Baxter, R. M. and Glaude, P., 1980, Environmental effects of dams and impoundments in Canada. Experience and prospects: Ottawa, Bulletin 205, Department of Fisheries and Oceans, pp. 34.

Bergmann, B., 1972, Um rustir a hunvetnskum heidum: Reykjavik, Natturufrædingurinn, v. 42, p. 190-198.

Bliss, L. C., 1981, North American and Scandinavian tundras and polar deserts, in Bliss, L. C., Heal, O. W. and Moore, J. J., eds, Tundra ecosystems: a comparative analysis, Cambridge, Cambridge University Press, p. 8-24.

Friedman, J. D., Johansson, C. E., Oskarsson, N., Svensson, H., Thorarinsson, S., and Williams, R. S., 1971, Observations on Icelandic polygon surfaces and palsa areas. Photointerpretation and field studies: Geografiske Annaler, v. 53, p. 115-145.

Hjartarson, A., 1981, Kvislaveita 5. Vatnafars-athuganir: Reykjavik, Orkustofnun OS-ROD 7624, pp. 85.

Johannesson, B., 1960, Islenskur jardvegur: Reykjavik, Bokautgafa Menningarsjods, pp. 120.

Johannsson, B., Kristinsson, H., and Palsson, J., 1974, Skyrsla um grasafrædirannsoknir i Thjorsarverum 1972: Reykjavik, Orkustofnun OS-ROD 7415, pp. 153 (with English summary).

Jonsson, M., 1978, Vedurfar i Thjorsarverum: Reykjavik, Orkustofnun OS-ROD 7804, pp. 5 (with English summary).

Kaldal, I., 1981, Kvislaveita 4. Jardgrunnskort: Reykjavik, Orkustofnun IK-81/02, pp. 3.

Kerbes, R. H., Ogilvie, M. A. and Boyd, H., 1971, Pink-footed geese of Iceland and Greenland: a population review based on aerial nestins survey of Thjorsarver in June 1970: Wildfowl, v. 22, p. 5-17.

Kerfoot, D. E., 1974, Thermokarst features produced by man made disturbances to the tundra terrain, in Fahey and Tompson, eds, Research in polar and alpine geomorphology, Third Guelph Symposium on Geomorphology, p. 66-72.

Newbury, R. W., Beaty, K. G. and McCullough, G. K., 1978, Initial shoreline erosion in a permafrost affected reservoir, Southern Indian Lake, Canada, in Proceedings of the Third International Conference on Permafrost, v. 1: Ottawa, National Research Council of Canada, p. 833-839.

Newbury, R. W. and McCullough, G. K., 1982, Shoreline erosion and restabilization in the Southern Indian Lake reservoir: Canadian Journal of Fisheries and Aquatic Sciences (in press).

Priesnitz, K, and Schunke, E., 1978, An approach to the ecology of permafrost in central Iceland, in Proceedings of the Third International Conference on Permafrost, v. 1: Ottawa, National Research Council of Canada, p. 473-479.

Rickard, W. E. and Brown, J., 1974, Effects of vehicles on arctic tundra: Environment and Conservation, v. 1, p. 55-62.

Ryden, B. E., 1981, Hydrology of northern tundra, in Bliss, L. C., Heal, O. W. and Moore, J. J. eds, Tundra Ecosystems, a comparative analysis, Cambridge, Cambridge University Press, p. 115-137.

Schunke, E., 1973, Palsen und Kryokarst in Zentral-Island: Nachrichten der Akademie der Wissenschaften Göttingen, Mathematische-Physische Klasse, 4, pp. 38.

Steindorsson, S., 1945, Studies on the vegetation of the central highland of Iceland, in The Botany of Iceland, v. 3., 14, Copenhagen, Munksgaard.

Steindorsson, S., 1964, Um halendisgrodur Islands. Fyrsti hluti: Flora, v. 2, p. 5-49 (with English summary).

Steindorsson, S., 1965, Um halendisgrodur Islands. Annar hluti: Flora, v. 4, p. 49-93 (with English summary).

Steindorsson, S., 1966, Um halendisgrodur Islands. Thridja grein: Flora, v. 5, p. 53-92 (with English summary).

Steindorsson, S., 1967, Um halendisgrodur Islands. Fjordi hluti: Flora, v. 5, p. 53-92 (with English summary).

Tomasson, H. and Thorgrimsson, S., 1972, Nordlinga-alda. Geological report: Reykjavik, Orkustofnun, ROD, pp. 19.

Tryggvason, T. and Einarsson, Th., 1965, Greinargerd um jardfrædi Thjorsarvera; Nordlingaalda-Soleyjarhöfdi: Reykjavik, Atvinnudeild Haskolans, pp. 17 (with English summary: On the geology of the Thjorsarver-area upper Thjorsa, Southern Iceland).

Zoltai, S. C. and Tarnocai, C., 1975, Perennially frozen peatlands in the western arctic and subarctic of Canada: Canadian Journal of Earth Sciences, v. 12, p. 28-43.

STRATIGRAPHIC EVIDENCE FOR VARIABLE PAST PERMAFROST CONDITIONS AT CANYON VILLAGE BLUFF, NORTHEAST ALASKA

Robert M. Thorson

University of Alaska Museum and Geology/Geophysics Program
907 Yukon Drive, Fairbanks, AK 99701 USA

The Canyon Village Bluff, a 30 m high gravelly exposure located along the Porcupine River (northeast Alaska), contains stratigraphic evidence of four fossil ice-wedge-cast horizons within alluvial sediments. Radiocarbon dating and geomorphic features in the area indicate that all ice-wedge-cast horizons are more than 35,000 years old and may be late Quaternary in age. Each ice-wedge horizon is interpreted to represent one cold-warm oscillation of soil temperature which could have been caused by any number of factors. These features indicate that the bluff locality has been very sensitive to conditions causing soil-temperature changes for a prolonged period of time.

INTRODUCTION

Only a few localities in Alaska are known where definitive evidence for fossil permafrost has been found in controlled stratigraphic contexts older than the Wisconsinan Stage (Péwé 1975, 1966). In many cases, ancient periglacial features are well documented, but unambiguous evidence for ancient permafrost is lacking. This paper presents evidence for the development of multiple relict permafrost horizons of late Quaternary age at one locality along the Porcupine River valley in northeastern Alaska.

Between 1979 and 1981 an interdisciplinary research team from the University of Alaska Museum conducted a reconnaissance archeological survey along the Porcupine River valley between the Canadian border and Fort Yukon (Figure 1). Numerous excellent exposures of Quaternary sediments in river bluffs were mapped and described during the survey. On the higher older bluffs, which consisted largely of gravelly alluvium, several silty cryoturbated zones and horizons exhibiting ice wedge casts (IWC) were observed.

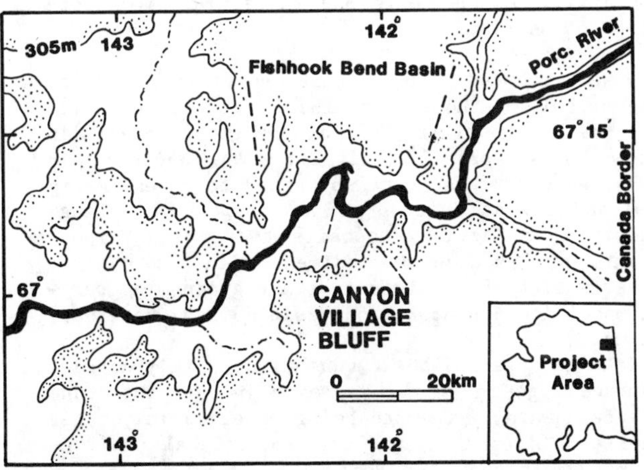

FIGURE 1 Map of the Porcupine River area showing localities mentioned in the text.

The most complete sequence of these periglacial horizons is at Canyon Village Bluff (hereafter referred to as the bluff) located approximately midway between the Canadian border and the Yukon Flats (Figure 1). In this bluff, evidence for six stratigraphically distinct intervals characterized by periglacial conditions are present, four of which contain unambiguous evidence for ancient permafrost.

The Porcupine River, one of the largest tributaries of the Yukon River, originates in central Yukon Territory and flows westward into Alaska 220 km to its mouth at Fort Yukon. Its westward course through Alaska crosses four distinct physiographic zones: (1) a narrow straight deep canyon (Upper Ramparts), (2) a broadly circular alluvial basin (Fishhook Bend), (3) a straight, narrow, shallow canyon (Lower Ramparts), and a broad alluvial plain (Yukon Flats). The bluff occurs as a high cutbank on the outside of a prominent meander near the center of Fishhook Bend basin.

The present course of the Porcupine River was established in late Quaternary time as the result of overflow from large proglacial lakes in Yukon Territory (Hughes 1969, Thorson and Dixon, in press). Prior to its capture of drainage from the Yukon Territory, the Porcupine River in the area of the bluff was only a modest-sized stream that was carving a broad erosional valley along its upper course and carrying gravels to Fishhook Bend (Thorson and Dixon, in press). During the initial overflow event, which occurred sometime prior to 31,000 years B.P., these older alluvial sediments were first buried by a thin sheet of gravel deposited by braided streams and later deeply incised, preserving the older gravels well above the level of subsequent fluvial activity.

The bluff area has a rigorous continental climate. Mean annual temperature averages about 7°C (Péwé 1975) with a mean January temperature of -30°F or less and a mean maximum July temperature of more than 70°F (Selkregg, n.d.). Vegetation on the bluff top is a dense spruce forest that occurs as part of the lowland spruce-hardwood forest ecosystem (Joint Federal-State

Land Use Planning Commission 1973). Permafrost is presently widespread throughout the area, which lies near the northern edge of the zone of discontinuous permafrost (Hopkins et al. 1955). At Fort Yukon, permafrost is at least 98 m thick (Péwé 1975). Modern ice wedges are not exposed at the bluff but are widespread in lowland alluvial plains bordering the river.

CANYON VILLAGE BLUFF

The bluff is a 1.9 km long, 30-35 m high east-facing bluff that offers nearly continuous exposure. The lower two thirds of the bluff consist of weakly indurated fluvial and lacustrine sediments of Tertiary age (Brosgé and Reiser 1969) which are well exposed and slope steeply to the river. The upper 13-15 m of the bluff exposes unconsolidated alluvial and eolian sediments of Quaternary age. The lower part of the Quaternary section is very sandy fine gravel that ravels rapidly and produces poor exposures. Prominent silty zones, which are especially common near the top, stand as nearly vertical faces. Nowhere along the bluff in 1980 was there a continuous vertical exposure. The best exposure is about 900 m from the eastern end of the bluff, where all units were exposed within a horizontal distance of 50 m. Shifting east and west of the median line of section did not introduce much error, because all prominent unit contacts within this area are horizontal. A measured stratigraphic section at the bluff, and a description of the units at this point, are presented in Figure 2 and Table 1, respectively. Particularly resistant congeliturbated zones and ice-wedge-cast (IWC) horizons occur at the same height over a horizontal distance of several hundred meters. Significant erosional disconformities would also be horizontal as well, but none were detected.

Interpretation of Units

Unit 1 is interpreted to represent lacustrine and fluviodeltaic sediments which were built into a basin now occupied by Fishhook Bend. Following deposition the beds were weakly indurated and tilted as much as 9°. The contact between Unit 1 and the overlying gravels is horizontal and only locally scoured, suggesting that it represents a broad erosional surface formed beneath an ancient stream.

Units 2, 3, and 5 are interpreted as lateral accretion deposits laid down by a relatively low-energy meandering stream during basin aggradation. Modest discharge is suggested because (1) the stream was competent to transport sediment no coarser than fine gravel and sand, (2) large-scale trough cross-bedding and deep channeling are absent, and (3) the fine-grained interbeds are highly discontinuous and lenticular. These silty organic-stained interbeds probably represent channel margin and overbank deposits that accumulated locally during general aggradation. Some apparently represent incipient paleosol development.

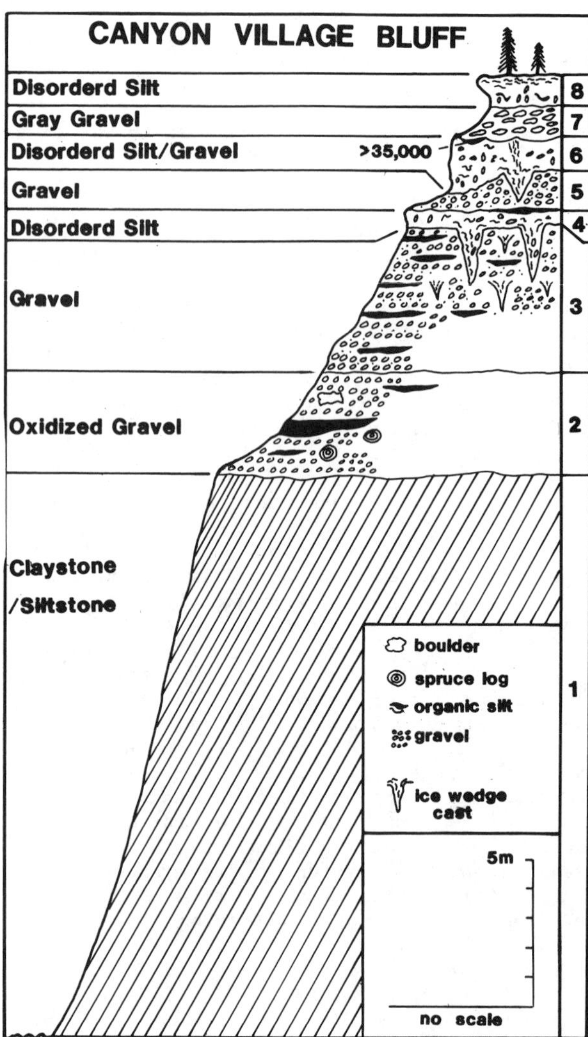

FIGURE 2 Measured stratigraphic section at Canyon Village Bluff. Refer to Table 1 for description of the units and to Figure 1 for location of section.

Well-rounded chert and quartzite pebbles identical to those from local Tertiary deposits comprise an estimated 60% of the gravel clasts in the lowest gravel (Unit 2). The decrease in abundance of these pebbles upward to an estimated 40 and 25% in Units 3 and 5, respectively, suggests that Tertiary deposits became less important as a source of sediment with time. This observation is consistent with a gradual aggradation of the Fishhook Bend Basin during Quaternary time.

Large, stained but unweathered, weakly abraded spruce logs in Unit 2 decrease in apparent abundance upward, with none being observed in younger units. Given the similar depositional environments inferred for Units 2-5, the absence of spruce in higher units may indicate a local extinction of spruce from this vicinity. The

TABLE 1 Description of Units at Measured Stratigraphic Section at Canyon Village Bluff.

Unit	Thickness (m)	Description
1	19	Claystone and siltstone: Laminated to massive gray siltstone and claystone interbedded with sand and well-rounded fine gravel. Weakly indurated. Unit dips 9° to northwest.
2	3.5	Oxidized gravel: Rounded to subrounded pebble- to small cobble-sized polymictic clasts in a subangular-rounded mixed sand matrix. Mean maximum clast diameter 2-4 cm. Commonly weakly bedded as subhorizontal slightly imbricate zones, with occasional cross-bedded lenses where sandy. Smaller pebbles are clasts of exceptionally well rounded chert and quartzite. Several large basalt boulders to 35 cm diameter occur within fine gravel matrix and exhibit percussion scars and occasional scratches. Unit contains common large spruce (Picea) logs to 15 cm diameter which have roots intact, slightly abraded appearance, and are stripped of bark. Logs diminish in abundance upward. Black organic silty lenses with flat tops and rounded lenses are present in channeled zones. Entire unit stained strong brown, with matrix less oxidized than clasts. Sharply defined horizontal slightly channeled lower contact.
3	4.5	Lower brown gravel: Rounded to subrounded pebble- to small cobble-sized polymictic clasts in a loose subangular to rounded sand matrix. Mean maximum diameter is 4-5 cm. Contains slightly less rounded and coarser clasts than Unit 1. Subhorizontally bedded and weakly imbricate, with rare trough cross bedding in sandy zones. Common flat-topped lenses of organic-stained silty sand, one of which appears to have incipient soil development. Unidentified mammal bone fossils are rare and retransported wood fragments common. Multiple wedge-shaped zones about 30 cm wide occur with a spacing of about 7 m along a definite horizon about 2 m below top of unit. Wedge forms defined by orientation of tabular clasts and organic zones that penetrate downward into subhorizontal bedding. Minor slumping within wedge forms. Several less well defined isolated wedge forms occur above the prominent horizon. Unit slightly stained to a brown color.
4	1	Lower disordered silt: Unit highly variable in thickness and appearance. Dominantly grayish brown massive unbedded silt and sand mixed with gravel near sharp disturbed lower contact. The orientation of organic lenses and tabular clasts is chaotic and contorted. Unit occasionally overlain by discontinuous 20 cm thick reddish brown organic zone which overlies a thin discontinuous gray horizon. Projecting from the base of unit are multiple vertical wedge-shaped structures that are typically 30-40 cm wide and spaced at about 10 m intervals along the bluff face. Wedge forms penetrate and crosscut underlying gravel to a maximum depth of 3 m and are defined by tabular and elongate clasts, organic lenses, wood fragments, and slump zones oriented parallel to wedge walls within a mixed silty sand matrix. At top of several wedge forms host sediments are deformed upward near margins. Lower contact abrupt and deformed.
5	1	Upper brown gravel: Rounded to subrounded pebble- to small cobble-sized polymictic gravel in a subangular to rounded coarse sand matrix. Mean maximum clast diameter 3-5 cm. Bedding commonly expressed as sub-horizontal orientation of stones. Contains small abraded wood fragments and no more than 25% well-rounded chert and quartzite fragments. Unit slightly stained to a light brown color. Subhorizontal slightly channeled lower contact.
6	2	Disordered silt and gravel: Unit variable in thickness and appearance. Dominantly grayish brown mixed silt, sand, granules, and fine gravels. Lower and upper parts are finer textured, with a central zone containing more gravel. Granule zones, elongate cobbles, and organic lenses show complex chaotic orientations near base, with many clasts having vertical orientation. Projecting from the base of unit are multiple wedge-shaped structures that are typically 50-100 cm wide and spaced at wide intervals along the bluff. Wedge forms penetrate and crosscut underlying strata to a depth of several meters. Wedge forms are similar to those in Unit 4. Several 10-30 cm wide zones of nonvertical clast orientations and silty segregated layers widen upward from the tops of large wedge forms to the channeled upward contact. At one locality along the upper contact a 0-5 cm thick bed of organic silt with sparse wood fragments yielded a 14C date of greater than 35,000 years B.P. (BETA-1823). Abrupt disturbed lower contact.
7	1.5	Gray gravel: Subangular to subrounded pebble- to cobble-sized clasts largely of sedimentary rocks in a very sparse, loose coarse sand-granule matrix. Mean maximum diameter 5-7 cm. Clasts are dominantly tabular to discoidal in shape with a strong southwest imbrication. Bedding generally absent, but a subhorizontal fabric is present. Upper 30 cm is weakly oxidized, increasing in intensity upward. Carbonate encrustations occur at base of nearly all clasts. Lower contact very sharp and moderately channeled.

TABLE 1 (Continued)

Unit	Thick-ness (m)	Description
8	1	Upper disordered silt: Unit dominantly mixed silt and fine sand, but is silty near base and sandy near top. Brown organic silt lenses to 3 cm thick with subparallel stained zones common, especially near base. Organic lenses greatly convoluted, especially near base, where gravel clasts, mixed from the lower contact, show chaotic orientations. Upper 20 cm is a peaty organic zone below spruce forest.
Total	33.5	

presence of anomalously large, subangular, scratched and battered basalt boulders in the fine sediment matrix of Unit 2 suggests boulder transport and modification by river ice under periglacial conditions. Similar phenomena occur along the Porcupine River today, during times of violent spring breakup.

The lowest Quaternary gravel (Unit 2) is oxidized to a strong brown color. Although the color contrast between Units 2 and 3 is sharp, there is no direct sedimentologic evidence for a unit contact, no apparent increase in the amount of weathering upward toward the Unit 2-3 contact, and no zones of clay alteration. Instead, oxidation seems equally pervasive throughout the entire unit. The horizontality of the Unit 2-3 contact, the absence of a weathering gradient, the presence of unweathered spruce logs, and the pervasive oxidation of these permeable sandy gravels suggest that groundwater circulation above the flat contact with impermeable Tertiary sediments may be responsible for the staining. This favored hypothesis requires unfrozen conditions at some time following deposition of Unit 3. Permafrost at the bluff must have been absent, very thin, or restricted to a great depth at this time.

Units 4, 6, and 8 are significantly finer textured than the gravel units (Table 1). All are disordered to some extent by congeliturbation, which makes their origin difficult to determine. Woody fragments and organic silt layers occur in each unit. The surface unit (Unit 8) is clearly eolian in origin because no opportunity exists for fluvial deposition and because deposition is occurring now. Unit 4 is similar in texture to Unit 8 and may also be eolian, at least in its upper parts. Unit 6 has disordered interbeds of fine gravel, suggesting a probable fluvial origin. Silty zones within Unit 6 may be either eolian or fluvial. Units 4 and 6 may represent fluvial overbank sediments completing two cycles of fluvial aggradation associated with units 3 and 5, respectively.

The oldest evidence for permafrost conditions at the bluff is a horizon of small IWC within the gravel of Unit 3 about 7 m below the bluff top (Figure 2). Small wedge-shaped features higher in Unit 3 are also interpreted as small IWC, but their lack of regularity makes them suspicious. Two pronounced horizons of IWC that penetrate the underlying gravels are exposed at the bases of Units 4 and 6. These wedge-shaped structures are interpreted as IWC because they exhibit downwarping of host sediments at their upper margins, irregular slump structures within the casts, and a prominent regularity in spacing along the bluff of several meters. These features are most characteristic of IWC (Black 1976, Péwé et al. 1969).

Above the base of both IWC horizons, organic lenses and elongate clasts are either highly contorted or randomly oriented, suggesting that considerable cryoturbation has occurred in these fine-grained deposits. Because they overlie the IWC, these cryoturbated zones are interpreted as former active layers. The thickness of each cryoturbated zone represents a minimum thickness of the former active layers. In the case of Unit 4 the incipient paleosol at the top of the cryoturbated zone suggests that the active layer may have been only 50-70 cm thick.

The youngest episode of wedge growth is indicated by narrow zones of vertically oriented clasts that begin at the top of the massive IWC in Unit 6 and that widen upward. These features are interpreted as either IWC or soil wedges (Black 1976). If they are IWC, they represent either incipient growth of ice wedges within Unit 6 or are the roots of larger former ice wedges that have been removed by erosion. An incipient ice-wedge origin is preferred because there is no evidence for significant erosion and because they lie directly above the former wedges.

Unit 7 originated in a manner very different from the other gravel units. The scoured nature of its lower contact, its relatively low textural and compositional maturity, its sheetlike uniformly imbricate character, and the braided appearance of its surface elsewhere in the valley (Thorson and Dixon, in press) suggest that it was deposited by a high discharge stream during glacial lake overflow from the Yukon Territory. Rapid incision of the Porcupine River after deposition of Unit 7 resulted in isolation of the present bluff area as a terrace at least 10 m above the active floodplain.

Age of Units

Only one radiocarbon date of greater than 35,000 14C yr B.P. (BETA-1823) is available for Canyon Village Bluff. This date, which was obtained from "fresh-looking" small wood fragments from a dense peaty silt at the very top of Unit 6, indicates that all four IWC horizons are older than 35,000 years. There is no evidence to suggest that significant erosional disconformities exist below the dated horizon or that prolonged intervals of weathering occurred on exposed surfaces. Rather, deposition of Units 2-6 appears to represent intermittent aggradation. Deposition of sediments by a migrating meandering stream apparently alternated with intervals when the ground surface was exposed to cryoturbation, eolian and/or overbank deposition, and contraction cracking with attendant growth of ice wedges at depth. The unweathered character of all units indicates that significant soil development did not occur, as for example during a long warm interglaciation. The presence of spruce logs near the base of the exposure suggests either interstadial or interglacial conditions because forest was probably absent from interior Alaska during full glacial conditions (Hopkins et al. 1981, Westgate et al. 1983). Based on these considerations the Unit 2-6 sequence is here interpreted to span all or part of the interval from the last interglaciation (Sangamon; marine isotope stage 5e) to the Middle Wisconsin Interstadial (marine isotope stage 3). This interval is dated globally from about 125,000 to 32,000 years B.P. (Shackelton and Opdyke 1973).

It is possible that the wood fragments dated at greater than 35,000 years B.P. were redeposited from older wood-rich units in the valley. This possibility is largely ruled out by many other 14C dates in more controlled contexts which indicate that Unit 7 must be older than about 31,000 years B.P. (Thorson and Dixon, in press).

The surface eolian cap (Unit 8) is clearly of recent origin. It resulted from eolian deposition on the terrace by sediments derived largely from the exposed bluff face and from more distant areas. The upward coarsening within this unit can be explained by the progressive approach of the source area (bluff face) to its present position. If Unit 8 is Holocene in origin and the underlying gravel is older than 31,000 years B.P., an interval of at least 20,000 years is not represented by sediments. Weathering on the gravel surface and calcification of the clasts must represent at least part of this interval. Nondeposition of eolian sediments and/or deflation may have occurred during this interval, but no evidence of wind abrasion was found. Apparently, ice wedge growth, significant weathering, or cryoturbation did not occur at Canyon Village Bluff during the maximum of the last glaciation (late Wisconsin).

CLIMATIC IMPLICATIONS

Units 2-6 at the bluff are interpreted as probably late Pleistocene in age and as having formed during episodes of alluvial deposition that alternated with episodes of cryoturbation. Intervals of congeliturbation and paleosol development on exposed surfaces probably spanned the greater share of time during basin aggradation. Development of ice wedges following deposition of Units 3, 4, and 6 must have been followed by melting and collapse; otherwise the wedges would still be present today. Furthermore each IWC horizon must have developed by thawing prior to the formation of the next higher one; otherwise the deposits overlying the IWC horizons would have been deformed as well. For example, ice wedges formed during the interval represented by Unit 6 cut through the underlying undeformed gravels which overlie the IWC of Unit 4. Thus, each IWC horizon can be interpreted as evidence for one cold-warm oscillation of temperature within the near-surface soil, the cause(s) of which are unknown.

Changes in soil temperature may result from a number of interacting factors, all of which may be operative in this case (Washburn 1980, Black 1976). Fluctuations in ground ice stability may accompany sedimentary cycles associated with meandering streams. Changes in air temperature (either mean annual or during a critical period), changes in the degree of thermal insulation from winter snowpack or ground vegetation, changes in vegetation caused by local wildfires, and changes in the thermal conductivity of the soil may all influence the stability of permafrost. Although the causes of the four cycles of permafrost formation and degradation at the bluff are unknown, their oscillatory nature indicates that the Fishhook Bend area has been in a sensitive regime with respect to the formation and preservation of ground ice since the last interglaciation. This interpretation is consistent with observations by Matthews (1980) and Morlan (1980) suggesting oscillatory climates in the northern Yukon Territory during the same time interval.

The complete absence of IWC horizons or significant cryoturbation of the youngest gravel at the bluff is puzzling when compared with the multiple IWC horizons beneath it and because of the widespread presence of large modern ice wedges in the area. The most plausible explanation for this phenomenon is that ice wedges in gravel can only develop locally under at least occasionally saturated conditions on active floodplains or low terraces. The rapid incision that followed deposition of Unit 7, the presence of an oxidized zone on its surface, and the carbonate encrustations within Unit 8 indicate that this gravel has not been poorly drained enough for ice wedge development to occur.

ACKNOWLEDGMENTS

Field studies at Canyon Village Bluff were supported by the National Geographic Society through a grant to the University of Alaska Museum. E. J. Dixon, project leader, was helpful in numerous ways. Field assistance by D. Hunter, D. C. Plaskett, and K. H. Thorson is greatly appreciated. Field consultations with D. M. Hopkins, O. L. Hughes, J. V. Matthews, N. Rutter, C. Schweger and A. Dyke greatly helped in interpretation of the bluff exposure. T. D. Hamilton and R. D. Reger, who reviewed this manuscript, provided many helpful suggestions.

REFERENCES

Black, R. F., 1976, Periglacial features indicative of permafrost: Ice and soil wedges: Quaternary Research, v. 3, p. 3-26.

Brosgé, W. P., and Reiser, H. N., 1969, Preliminary geologic map of the Coleen Quadrangle, Alaska: U.S. Geological Survey Open File Report 370, scale 1:250,000.

Hopkins, D. M., et al., 1955, Permafrost and ground water in Alaska: U.S. Geol. Survey Professional Paper 264-F, 146 p. with plates.

Hopkins, D. M., Smith, P. A., and Matthews, J. V., Jr., 1981, Dated wood from Alaska and the Yukon: Implications for forest refugia in Beringia: Quaternary Research, v. 15, p. 217-249.

Hughes, O. L., 1969, Pleistocene stratigraphy, Porcupine and Old Crow Rivers, Yukon Territory: Geol. Survey of Canada Paper 69-1, p. 209-212.

Joint Federal-State Land Use Planning Commission for Alaska, 1973, Major Ecosystems of Alaska (Map at scale 1:2,500,000).

Matthews, J. V., Jr., 1980, Possible evidence of an early Wisconsinan warm interstadial in East Beringia: Abstracts of the 6th Biennial Meeting of the American Quaternary Association, Orono, Me., p. 130-131.

Morlan, R. E., 1980, Taphonomy and archaeology in the upper Pleistocene of the northern Yukon Territory: A glimpse of the peopling of the New World: Archaeological Survey of Canada Paper 94, 398 p.

Péwé, T. L., 1966, Ice wedges in Alaska--classification, distribution, and climatic significance, in Natl. Acad. Sci. and Natl. Research Council, Proceedings of the First International Permafrost Conference, Lafayette, Ind, p. 76-81.

Péwé, T. L., 1975, Quaternary geology of Alaska: U.S. Geol. Survey Professional Paper 835, 145 p.

Péwé, T. L., Church, R. E., and Andresen, J. J., 1969, Origin and paleoclimatic significance of large scale polygons in the Donnelly Dome area, Alaska: Geol. Soc. America Special Paper 109, 87 p.

Selkregg, L. L., n.d., Alaska Regional Profiles, Volume VI, Yukon Region: University of Alaska Arctic and Environmental Data Center, 346 p.

Shackleton, N. J., and Opdyke, N. D., 1973, Oxygen isotope and paleomagnetic stratigraphy of Equatorial Pacific cores v28-238: Oxygen isotope temperatures and ice volumes on a 10^5 and 10^6 yr scale: Quaternary Research, v. 3, p. 39-55.

Thorson, R. M., and Dixon, E. J., in press, Alluvial history of the Porcupine River valley: role of overflow of glacially impounded lakes in Canada: Geological Society of America Bulletin.

Washburn, A. L., 1980, Permafrost features as evidence of climatic change: Earth Science Reviews, v. 15, p. 327-402.

Westgate, J. A., Hamilton, T. D., and Gorton, M. P., 1983, Old Crow Tephra: A new late Pleistocene Stratigraphic marker across north-central Alaska and western Yukon Territory: Quaternary Research, v. 19, p. 38-54.

LOESS-LIKE YEDOM-COMPLEX DEPOSITS IN NORTHEASTERN USSR; STAGES AND INTERRUPTIONS IN THEIR ACCUMULATION; AND THEIR CRYOTEXTURES

S. V. Tomirdiaro

North-East Interdisciplinary Scientific Research Institute,
Academy of Sciences, Magadan, USSR

Loess layers of a yedom complex with a special cryotexture accumulated during the glacial periods in the cryoarid landscapes of Northeastern USSR, while during the interglacials buried soil profiles and thermokarst facies were formed in cryohumid landscapes. This pattern corresponds in principle to that of the classic loess deposits of the periglacial regions of Europe.

Covers or remnants of covers of unique loess-like deposits containing syngenetic polygonal ice wedges of Late Pleistocene age are widely distributed in the Northeastern USSR and Alaska. Soviet authors define this association of rocks as "yedoma complex" or "loess-ice formation" (Tomirdiaro 1980) while the Alaskan (Fairbanks) counterparts describe it as eolian loess (Goldstream Formation; Péwé 1975).

Special investigations made by the author for many years in Yakutia and Chukotka indicate that in the more complete yedoma sections two ubiquitous chronostratigraphic horizons are recognizable; the latter comprise proper loess-like unredeposited beds with syngenetic ice wedges and is referred to as the Zyryanka and Sartanian cryochrones. The former comprise chronostratigraphic horizons of peat, soil, and thermokarst facies and belong to Kazantsevsky, Karghinsky or Hologenic thermochrones (Figure 1).

Presently, this stratigraphic scheme was approved as a regional rank division by the All-Union Interdepartmental Conference on Development of Stratigraphic Schemes for the Quaternary Deposits in the Eastern USSR held in Magadan in March 1982. The above-mentioned chronostratigraphic horizons were given local names corresponding to the stratotypes identified for them. Thus, a layer of loess-like deposits containing syngenetic ice wedges of the Ice time ($Q^2{}_{III}$) was named Oyagossky Horizon (type section: the Yagossky Yar yedoma on the bank of Dmitry Laptev Strait); a layer of buried soil, peat, and lake-thermokarst facies of the Karghinsky warm interstadial ($Q^3{}_{III}$) was named Molotkovsky Horizon (type section: lake-thermokarst deposits near Molotkovsky Kamen' rock on the Anyui River); and a layer of loess-like deposits containing syngenetic ice wedges of Sartanian glaciation ($Q^4{}_{III}$) was named Mus-Khaya Horizon (type section: the upper part of the Mus-Khaya yedoma section on the Yana River). Correspondingly, the previously common term "yedoma suite" has been replaced by "yedoma series."

During the Interdepartmental Conference of 1982, it was pointed out, in particular, that since the Oyagossky and Mus-Khaya horizons of proper loess-like deposits of the yedoma series should be considered as having accumulated during the maximal worldwide climatic deterioration and aridization in the Pleistocene ($Q^2{}_{III}$) and ($Q^4{}_{III}$), i.e. during the epochs of the commonly known global eolian loess accumulation in landscapes of the permafrost/periglacial ("mammoth") tundra-steppes of Northern Eurasia and America, that these deposits must be of a universal and uniform cryogenic-eolian nature. In the case of the yedomas from Chukotka and Yakutia this concept has been developed by the author (Tomirdiaro 1972, 1978, 1980) and Academician Shilo (Shilo and Tomirdiaro 1981) as well as by Péwé, Journaux, and Stuckenrath (1977). Investigations by Giterman and Kaplina as well as by the author (Tomirdiaro 1980) showed that, as evident from numerous representative spore-pollen analyses in sections of the yedoma series, both horizons of proper loess-like deposits (the Oyagossky and Mus-Khaya horizons in Figure 1) accumulated in areas of very dry steppe conditions, which were even devoid of shrubs. Layers of fossil soil, peat, and thermokarst facies of the Molotkovsky and Holocenic horizons accumulated in fundamentally different landscapes that were wetter, boggy, and forested (Figure 1). This conclusion results from palynologic analyses of fossil remains of theriofauna and fossil insect fauna.

All these drastic and cardinal changes of paleolandscape environments during the time of accumulation of different chronostratigraphic horizons of the yedoma complex (the yedoma series) were reflected, as shown by the author's investigations, in their cryologic structure (Figure 1).

The general scheme is as follows:

(1) Deposits occurring at the base of the yedoma complex (the yedoma series) may belong to any time prior to the Late Pleistocene. However, ancient soil, peat, and lake-alas thermokarst deposits of the Kazantsevsky interglacial ($Q^1{}_{III}$) are often found. Ice veins are epigenetic and wedge-like.

(2) The Zyryanka age ($Q^2{}_{III}$) of the Oyagossky horizon in the deposits of the yedoma complex is established by the combination of the following features:

(a) steppe or tundra-steppe spore-pollen spectra typical of cold periods of ice epochs (Tomirdiaro 1980);

(b) universal distribution of syngenetic ice wedges and eolian loess-like sandy loams also related to the ice epochs;

(c) abundant bone remains of Late Paleolithic mammoth fauna with ages that are beyond the limits of radiocarbon dating.

This unit is characterized by broad ice veins up to 6 m wide; there are common thick schlieren or ribbonlike parallel-concave continuous cryotextures (Figure 1).

(3) The Karghinsky ($Q^3{}_{III}$) age of the next, Molotkovsky, horizon is supported by many radiocarbon dates of buried peat and soil containing wooden macroremnants. These dates are within the period of 45-40 to 25-24 thousand years B.P. (Tomirdiaro 1980, Shilo and Tomirdiaro 1981). A bed of loess-like deposits occurring between the peat horizons (Figure 1) is characterized by steppe spore-pollen spectra and is referred to drier and colder Mid-Karghinian time. This unit exhibits thick schlieren, flat-layered cryotextures. The lower- and upper-lying horizons of peat and fossil soil are characterized by forest spectra and belong to Early and Late Karghinsky warm stages. Here loess-accumulation and ice vein development were interrupted although permafrost persisted. Its cryotexture is massive and ice veins are epigenetic and wedgelike throughout.

(4) The Sartanian age ($Q^4{}_{III}$) of the third layer from the base, i.e. Mus-Khaya horizon, is evident from radiocarbon dates of roots of herbaceous steppe vegetation. For example, 3 dates were obtained at the Mus-Khaya type section: the sample taken from 2 m depth below the top of the exposure (MAG-137), was dated as 11500±210 yr B.P., and the sample taken from 15.5 m depth was dated as 23360±720 yr B.P. (Tomirdiaro 1980). In the Duvanny Yar exposure, that is the key section for all yedomas of the continental subarctic type identified by us, a sample of rootlets of herbaceous vegetation (MAG-592) taken from 3 m depth, below the top of a 35 m high bluff, was dated at 15850±150 yr B.P. Even in Central Yakutia, near the city of Yakutsk at an elevation of 50 m above the Lena River, samples of analogous rootlets (IM-432, IM-431, IM-430) were taken from depths of 3, 8, and 14 m below the surface and gave dates of 13600±900 yr, 17800±850 yr, and 18700±990 yr B.P.

In all sites of the Mus-Khaya horizon, the spore-pollen complex is characteristically that of paleolandscapes of extremely dry Artemisian steppes and Selaginellan semideserts (Tomirdiaro 1980). A drastic increase in dryness and higher rates of loess-accumulation are also registered in the Mus-Khaya horizon of the subarctic type yedomas from both Mamontova Gora on the Aldan River in the south to Duvanny Yar on the Kolyma River in the north. There is a domination of exclusively microschlieren syngenetic flat-layered (microlenticular) cryotextures (filament lenticules of ice in the rock) and a sharp decrease in the width of ice veins (down to 1-2 m) as compared to that of the veins from the Oyagossky horizon (Figure 1).

(5) The Holocene age (Q_{IV}) of the yedoma upper horizon was determined by numerous absolute datings (Tomirdiaro 1980). A subhorizon identified here underwent thawing during the Holocene and was subsequently incorporated by the permafrost; its thick schlieren cryotextured rocks are peculiar to humidified and epigenetically frozen terrain. In addition, a modern soil layer is recognizable here. The whole horizon is characterized by forest and tundra-forest spore-pollen spectra.

The development of microschlieren (microlenticular) and discontinuous cryotextures in the Mus-Khaya horizon is definite indication that this type of rock was transformed into permafrost without having been humidified. A slow-rate freezing of fine-grained soils that have been saturated leads to the production of thick schlieren of segregated ice as well as ice-rock ribbons in them, i.e. to creation of sediments characteristic of the Oyagossky horizon of Zyryanka time. The Zyryanka time was less humid and, as demonstrated by the European data, the eolian loess-accumulation during it was 10 times slower than during the Late Würm ($Q^4{}_{III}$). This decrease in sedimentation promoted a drastic widening of ice veins within the Oyagossky horizon (Figure 1). Simultaneously, relatively higher humidity than that of Sartanian time (though still very continental and climatic) led to periodic manifestations of embryonic thermokarst and to the corresponding wetting of eolian deposits, i.e. to the eventual formation (subsequent to second freezing) of deposits with thick schlieren cryotextures (Tomirdiaro 1980). However, the wetting process sometimes did not involve the whole sequence of accumulated eolian matter. Hence there are frequent cases of interbedding of thick schlieren and microschlieren within the Oyagossky horizon ($Q^2{}_{III}$) in the polygonal rock blocks (Figure 1). As far as the genesis of the ice veins is concerned, high porosity values as well as other features indicate sublimation origin (Tomirdiaro 1980). The fact that large ice veins are created under very cryoarid modern climates in eolian deposits has been established by recent investigations of permafrost sand dunes occurring in the islands of the Canadian Arctic Archipelago (Pissart et al. 1977).

In the Arctic, eolian matter is deposited at a slower rate than within the Oyagossky horizon of the subarctic type yedomas. Besides, the slow-rate deposition took place both during the Zyryanka and Sartanian time. In our opinion, that was precisely the reason for the ice veins of the Arctic type yedomas to be unusually broad (as wide as 8-9.5 m). As compared to the yedomas from subarctic areas, they exhibit a reversed ratio between ground ice and enclosing rock. It is during descriptions of such yedomas that a previous hypothesis was put forward implying the distribution of buried glaciers in northern Yakutia. Since these entities are confined exclusively to the Arctic zone, we recognized them as yedomas of specific Arctic or shelf type (Tomirdiaro 1980).

The region where these 90% ice yedomas are distributed is restricted to the northernmost part of the Yana-Indigirka lowland as well as to the islands of New Siberian Archipelago; they are also preserved on the islands lying in the Lena delta area and in the Anabar-Olenyok Maritime Lowland. The southern limit of this region lies, as demonstrated by our reconnaissance surveys, at about 72° of northern latitude (Tomirdiaro 1982).

We undertook careful examinations of the following key sections for ice yedomas: Oyagossky Yar on the shore of Dmitry Laptev Strait, Khaptashinsky Yar on the shore of the East Siberian Sea, and Bykovskaya yedoma in the Lena River delta (Tomirdiaro et al. 1982).

Huge ice veins measuring 8-9.5 m in width and up

to 40 m in height are usually overlain here by a very thin soil veneer measuring 0.5-0.6 m. This layer often thaws during summer and the thawing front penetrates down into the vein ice. This results in its gradual latent thawing. Subsequently, intervein polygonal rock blocks get involved and hence elevated above the yedoma surface; this, in turn, leads to the development of the block-baidzherakhi microrelief on the surface that is typical of this yedoma type (Tomirdiaro et al. 1982). This microrelief enables a rough mapping of the area where they are distributed (Tomirdiaro et al. 1982). Where the veneer is especially thin (e.g. the Oyagossky and Khaptashinsky units), the Sartanian part of the yedoma (the Mus-Khaya horizon) has already thawed away. However, in the Bykovsky ice yedoma (Bykovsky Peninsula, the Lena River delta) we managed to discover a Sartanian horizon of the Arctic type yedomas preserved below a thicker soil-rock layer. Absolute datings of 22070±410 yr B.P. (LU-1263) were obtained here by the analyses of rootlets of fossil herbaceous vegetation recovered at 8 m depth from the top of a 32 m bluff (Tomirdiaro 1982). It was found that, as distinct from the yedomas of subarctic type mentioned earlier, the Mus-Khaya horizon ($Q^4{}_{III}$) here also contains huge ice veins as wide (up to 9.5 m) as those from the Oyagossky unit ($Q^2{}_{III}$). However, despite the mammoth size of the ice veins, the cryogenic texture here also generally changed from the thick schlieren and ribbonlike parallel-concave member in the Oyagossky horizon to the microlenticular horizontal-layered texture in the Mus-Khaya horizon (Tomirdiaro et al. 1982). This is definite indication that the cessation of embryonic thermokarst during the Sartanian time was related primarily to the general climatic dryness of that cryochrone and not to the ground ice formed in any one region.

All in all, the extensive belt of shelf type ice yedomas found by us in the Eastern Arctic testifies to a pronounced "gulf" in the process of loess accumulation which is so intensive in the cryochrones of Late Pleistocenic Eurasia. Actually, whenever the Arctic yedomas of loess-ice type measuring 20-25 m in height undergo complete thawing, a layer of mineral matter is left that is only 2-2.5 m thick.

In terms of eolian genesis, this can be explained by the domination of northeastern and eastern winds produced by the Arctic anticyclone during the ice epochs. With such a circulation pattern, the Arctic winds could not carry much dust since they did not meet, on their way, any mountainous edifices within the Arctic ocean which were covered by ice and fine dust at that time. In the meantime, the winds that blew over the coastal areas of the mainland picked up large volumes of dried dust from the vegetation-free slopes of mountainous slopes in Chukotka and Yakutia. This concept is substantiated by comparative grain-size analyses of rocks from yedomas belonging to the subarctic and Arctic types (Tomirdiaro 1980, Tomirdiaro et al. 1982). It is no accident that the former are made up of rather coarse sandy loams and the latter of fine-grained loams or even silty clays, i.e. perhaps, eolian silts/sediments of the ancient atmosphere of the cold "loess" epochs of the Late Pleistocene. Even the yedoma from Bykovsky Peninsula occurring at the foot of the Kharaulakh mountain range consists of 85% ice and the rocks from polygonal blocks or "pillars," also made up by loams (Tomirdiaro et al. 1982). Therefore, in this area, the mineral matter was also introduced from the north and not from the south; the sediment was not transported from the adjacent southern mountains; on the contrary, it was blown to the piedmont from the north. Besides, the Arctic winds carried the pollen rain. That is why the spore-pollen diagrams (Kaplina 1979, Tomirdiaro 1980) showed that steppe grass pollen prevails (Figure 1) and tree pollen is almost totally absent from the yedomas of both types in the Yagossky and Mus-Khaya horizons. This is a very peculiar feature since these vegetation species, especially such Northeast Asian endemic species as Populus suaveolens, Chosenia macrolepis and others, managed to persist during the same ice epochs in areas immediately to the south, i.e. in the mountainous valleys of Northeastern Asia.

REFERENCES

Kaplina, T. N., 1979, Spore-pollen spectra of deposits from the ice complex of maritime lowlands of Yakutia, Izv. AN SSSR, Ser. geogr., no. 2, p. 85-93.

Péwé, T. L., Journaux, A., and Stuckenrath, R., 1977, Radiocarbon dates and Late Quaternary stratigraphy from Mamontova Gora, unglaciated Central Yakutia, Siberia, USSR, Quaternary Research, no. 7.

Pissart, A., Vincent, J., and Edlund, S., 1977, Dépôts et phènomènes èoliens sur lile de Banks, Territorien du Nord-Ouest, Canada, Canadian Journal of Earth Sciences, National Research Council, Canada, v. 14.

Shilo, N. A., and Tomirdiaro, S. V., 1981, Paleogeography and absolute geochronology of Late Pleistocene in Northeastern Siberia, in Izv. AN SSSR, Ser. Geogr., no. 3.

Tomirdiaro, S. V., 1972, Permafrost in mountainous lands and lowlands, Magadan Publishing House, 172 p.

Tomirdiaro, S. V., 1978, Natural processes and development of permafrost terrain, Moscow: Nedra Publishers, 145 p.

Tomirdiaro, S. V., 1980, The loess-ice formation of Eastern Siberia during Late Pleistocene and Holocene, Moscow: Nauka Publishers, 184 p.

Tomirdiaro, S. V., 1982, The Arctic and subarctic types of permafrost loess and recognition of yedoma formations of the shelf and continental types, in Natural processes in the USSR territory during Late Pleistocene and Holocene, Moscow, p. 134-141.

Tomirdiaro, S. V., Chyornen'ki, B. I., and Bashlavin, D. K., 1982, The loess-ice formation of the shelf type and the Oyagossky Yar exposure, in The permafrost and geological evolution and paleogeography of lowlands in Northeastern Asia. Transactions of the N-E Interdiscipl. Sci. Res. Inst., FE Sci. Cntr., USSR Acad. Sci., Magadan, p. 30-53.

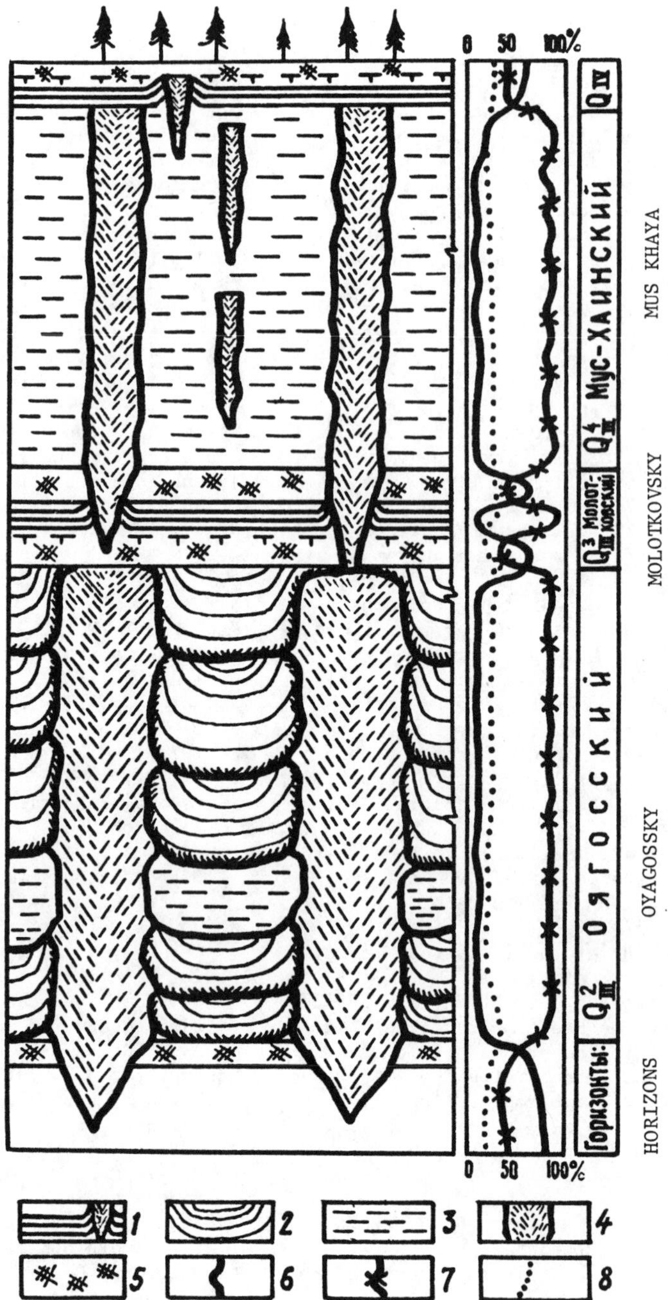

FIGURE 1 A scheme for the stratigraphic and cryolithic division of deposits belonging to yedoma (ice) complex.
1 - loess-like sandy loam with horizontal-layered thick schlieren; 2 - loess-like high-ice sandy loam with thick schlieren (ribbon) parallel-concave continuous cryotexture; 3 - loess-like low-ice sandy loam with microschlieren discontinuous lenticular cryotexture (ice lenticular filaments); 4 - syngenetic bodies of Pleistocenic ice veins; 5 - autochthonous peat and soils; 6 - relative percentage curve for tree and shrub pollen; 7 - same as (6) for non-tree vegetation; 8 - same as (6) for spores.

FROZEN GROUND IN THE ALTAI MOUNTAINS OF CHINA

Tong Boliang, Li Shude, Zhang Tingjun, and He Yixian

Lanzhou Institute of Glaciology and Cryopedology, Academia Sinica
People's Republic of China

Frozen ground in the Altai Mountains of China can be divided into three zones: seasonally frozen ground, sporadic permafrost, and widespread permafrost. In seasonally frozen ground, the depth of the thawed layer is 0.8-1.0 m in the peat layer, 0.9-1.6 m in the loam layer, and over 1.9 m in sandy gravel sediments; the mean annual ground temperature is no lower than -1.0°C. The sporadic permafrost zone is situated between 2200 and 2800 m a.s.l. In general, permafrost islands are distributed in peat layers in marshy depressions and in caves on northfacing slopes. Some palsas with 1- to 2-m ice layers were found. The widespread permafrost zone is situated in the high mountain zone above 2800 m a.s.l. The permafrost thickness is estimated to be less than 400 m, and the mean annual ground temperature is higher than -5°C. The lower limits of permafrost in the last glacial period and Neoglaciation were 750-900 m and 400-600 m, respectively, lower than at present. Permafrost in the sporadic zone was formed during the Neoglaciation, while the lower part of the permafrost layer in the widespread permafrost zone was probably formed during the last glacial period.

The glacial and permafrost expedition organized by Lanzhou Institute of Glaciology and Cryopedology, the Chinese Academy of Sciences, made the first investigation of permafrost in the Altay Mountains within the borders of China in 1980. The region investigated is situated in the southern side of the middle part of the Altay Mountains, from 49°8' N in the north, southeastward to about 46° N at a distance about 450 km from NW to SE within the border of China. It is 150 km wide in the northwestern part, and approximately 80 km in the southeastern part. It gradually decreases in height from north to south and from northwest to southeast. In the vicinity of watershed in the western part of the investigated region lie the boundaries of three countries: the People's Republic of Mongolia, the Soviet Union, and the People's Republic of China. It is also the highest area in the whole Altay Mountains, with several mountain peaks over 4,000 m high, including Youyi Feng, the highest (4,374 m). Around Youyi Feng is the most widespread area of glaciation in the Altay Mountains, including the largest glacier in the Altay Mountains in China, the Kalasi Glacier. In the southeastern part of the mountain range (about 3,200-3,500 m), no modern glaciers occur.

CHARACTERISTICS OF FROZEN GROUND

Field data indicate that permafrost occurs above 2,200 m in the Altay Mountains within the border of China, while seasonally frozen ground zone lies below 2,200 m (Fig. 1). Seasonally frozen ground is found in the lower part of middle mountain zone (2,000-2,200 m), the low mountain zone (1,200-2,000 m), hill zone (800-1,200 m) and front-range plain (below 800 m) (Xinjiang Scientific Expedition 1978). Only under surface water bodies or adjacent to heated buildings, seasonally frozen ground is absent. The mean annual ground temperature varies from 4°C to 7°C in this zone. The mean annual air temperature varies from 3-4°C in the front-range plain, 2-3°C in the hill zone, and 0 to -4°C in the low mountain zone (Fig. 2). The maximum frost penetration in sandy clay is about 1-1.5 m in the front-range plain, 1-2 m in the hill zone, and 1.5-2.5 m in the low mountain zone. Seasonally frozen ground occurs for 5 months, from the beginning of November to the end of March. The time of initial freezing in the plain areas is about two weeks later than in the mountain areas. The maximum depth of freezing occurs at the end of March and it thaws during the last ten days of May. There is little difference in freezing rate in different geomorphological units within the seasonally frozen zone (only 25-35% variation, see Fig. 3). It is remarkable that the frozen depth at Qinghe (1,218 m) in the low mountain-hill zone (Fig. 4) is much greater than that at Sentasi (1,900 m) in the middle mountain zone (Fig. 5). This demonstrates the effect of an air temperature inversion layer in winter, producing lower air temperatures in the zone below 1,200 m. The precipitation increase with elevation is also important. The annual precipitation increases from 100-200 mm in the plain area and 200-300 mm in the low

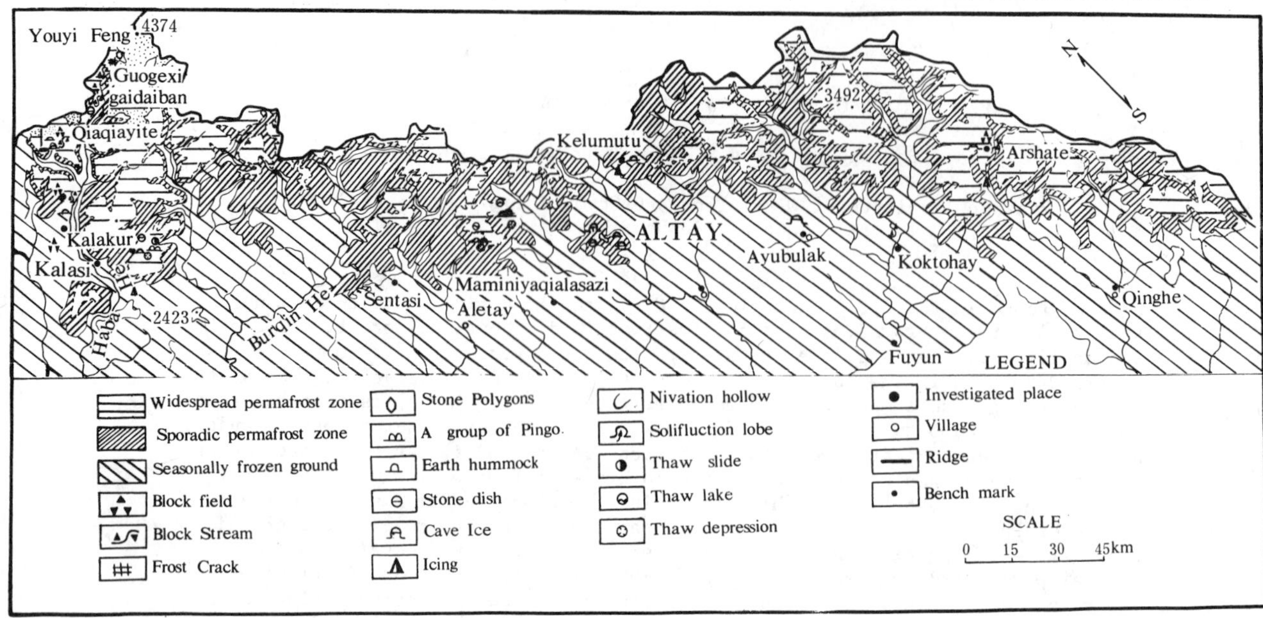

FIGURE 1 Permafrost map of the Chinese Altay.

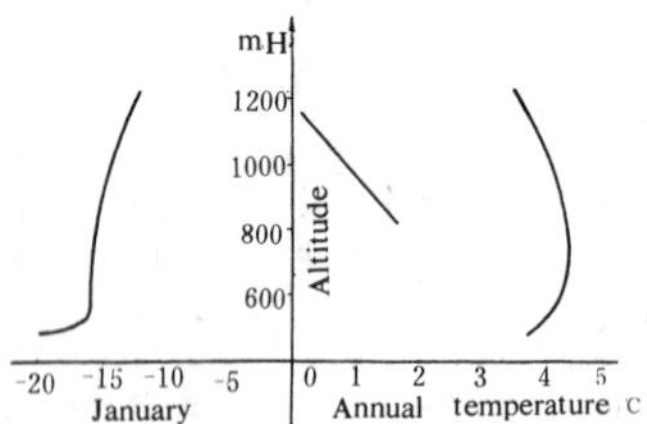

FIGURE 2 Relation between the mean air temperature and altitude in the Altay Mountains.

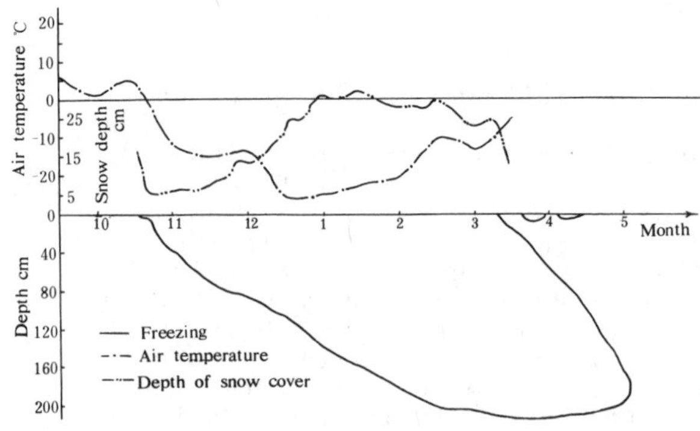

FIGURE 4 Process of frost penetration during 1979-1980 at Qinghe in the Altay Mountains.

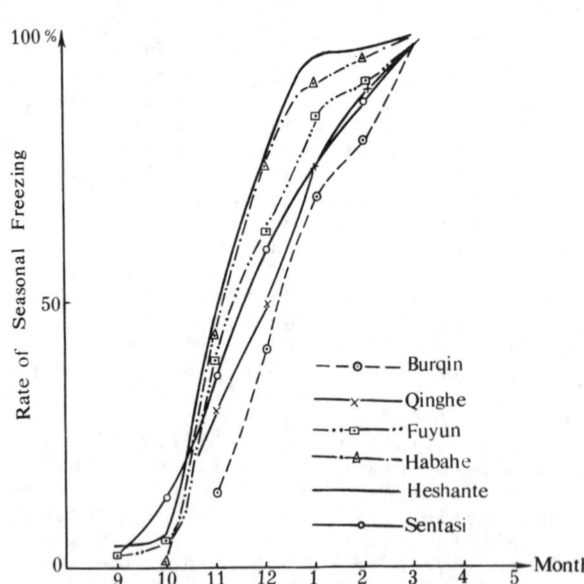

FIGURE 3 Rate of frost penetration in the Altay Mountains.

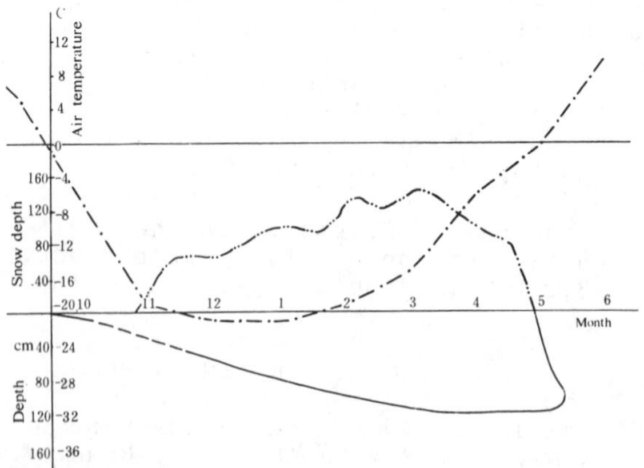

FIGURE 5 Process of frost penetration during 1960-1961 at Sentasi in the Altay Mountains.

mountain zone, to 400-500 mm in the middle mountain zone. More importantly, the depth of snow accumulation varies with the increase of elevation. Snow cover has a thickness of only 10-30 cm in the plain area, 100-200 cm in the low mountain zone (i.e. Sentasi), and up to 150-250 cm in the middle mountain zone. On the slopes, snow cover usually has a thickness of 60-170 cm, but in the snow accumulation area adjacent to a wind gap, it increases to 400-500 cm, covering the tops of electric poles, i.e. at Qiong Kur (2,600-2,700 m in Fuhai county). At a small town in mountain area of Fuyun county with 170 cm mean annual snow cover, the seasonally frozen layer is only 50 cm in depth, while without snow cover, the depth of seasonal freezing increases up to 150 cm. The mean the mean annual air temperature is 2.0-4.5°C in both the front-range plain and the hill zone, but the difference between freezing and thawing index is 1,200-1,600°C-day at the former and -190 to -1253°C-day at the latter, the former being twice as high as the latter. These are the results of the facts that where the snow cover is thick, the period of snow accumulation is long Hence, the heat-preserving and heat-insulating effect of snow during the period of abrupt drop of temperature in winter in the investigated region is obvious (Fig. 6). Snow cover increases the mean annual ground temperature and decreases the annual amplitude of ground temperature variation, as a result, decreasing the depth of the seasonally frozen layer (Lanzhou Institute of Glaciology and Cryopedology and Desert Research 1975). In the Altay Mountains within the border of the Soviet Union, snow cover 30-35 cm thick makes the ground temperature at a depth of 30 cm 25°C higher than that without snow cover, and a thickness of 50-60 cm 37°C higher than the air temperature (Shulekin 1954).

Periglacial phenomena within the seasonally frozen zone are: eolian dune, spring icing, and river icing in low mountain zone, seasonal cave ice in the lower part of middle mountain zone, which has a thickness of 60-70 cm, appearing in the middle of October and disappearing at the beginning of May

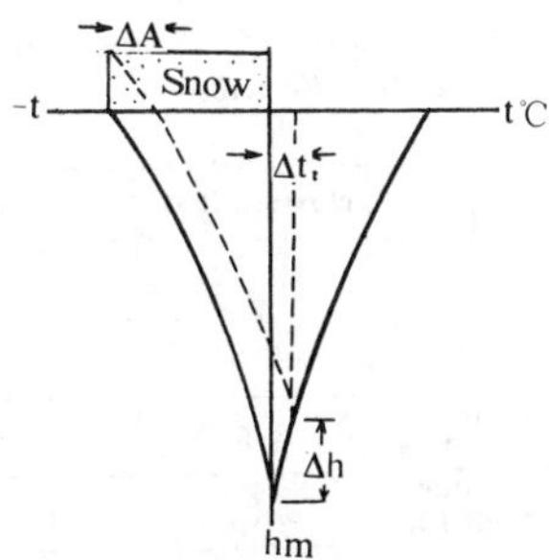

FIGURE 6 Influence of snow cover on freezing depth (h), mean annual ground temperature (t), and annual amplitude (A).

next year. Occasionally block slopes can be found on the north-facing slope in the western bank of Kalasi lake.

The sporadic permafrost zone is situated from 2,200 m to 2,800 m in the middle and upper part of middle mountain zone. It has an area of approximately 6,300 km^2, covering 25% of the area in the entire investigated region. Permafrost islands are present in peat layers in marshy depression and in caves on the north-facing slopes. The area of peat distribution is generally no greater than 15 km^2. For example, Kalakur (48°44'N, 88°11'E, 2,200 m), 12 km to the east of Kalasi town, has a peat layer 2 m thick in a marshy lake 4 km^2 in area. More than 200 palsas are found around its southern end, some being circular or elliptical and some U-shaped. They are 2-6 m high, with an area of 1,200 m^2. Luxuriant herbaceous plants grow on their top. Under the ground surface, there is 30-70 cm of peat overlying alternative underlying gravel loam and coarse sand. On August 27, 1980, the thawed depth was 85-110 cm. On a 1780 m^2 palsa, D.C. resistivity sounding indicated that the permafrost has a thickness of 4.7 m (He Yixian 1983). The adjoining palsa contains ground ice 1-2 m thick and permafrost 5-6 m thick (Li Peiji 1979). According to weather data from the adjacent Sentasi (1,900 m) and Aletay (735 m) weather stations, the calculated mean annual air temperature at Kalakur is -5.4°C, compared with about -6.8°C in the eastern part of the Altay Mountains. So far as we know, this is the lowest temperature at the lower limit of alpine permafrost in China, 2-3°C lower than that on the Qinghai-Xizang Plateau and in Qilian Shan. Altitude of the lower limit in the sporadic zone is at least 400-652 m higher than that ever calculated (Lanzhou Institute of Glaciology and Cryopedology 1975, Xu Xiaozu 1982). This is because the annual precipitation is over 500 mm and the depth of snow cover is 150-250 cm in the middle mountain zone, greatly decreasing the heat exchange between air ground surface during the seven months of cold season. According to Maksimova (1977), snow cover 100 cm thick can raise the mean annual ground temperature by 12.8°C (mean annual air temperature -8.9°C, annual amplitude of air temperature variation 52°C, cold season 215 days). It promotes permafrost retreat and makes the lower limit of permafrost rise up to the present elevation of 2,200 m.

In the natural cave (Qiaqiayite 2,660m) formed by glacial sediments on the north-facing slope and in adit or vertical shaft of the mining area (Qiongkur 47°45'N, 88°50'E, 2560 m) in the middle mountain zone, cave permafrost has a thickness of 19 m and the ground temperature varies between -0.2 to -0.5°C. Cave ice may be 1.4-2.0 m thick, and occurs in various forms such as ice stalagmites and stalactites as well as forming underground icing.

Cave permafrost is greatly influenced by faults and groundwater. In the adit of Qiongkur, there is an inverted fault 2 m wide. No permafrost exists in the upfaulted block, but permafrost develops in the downfaulted block.

The depth of the seasonally thawed layer in this zone is 80-100 cm in peat, 90-160 cm in loams and over 190 cm in sands and gravels. The mean annual ground temperature is no lower than -1.0°C.

More periglacial phenomena are found in this zone than in seasonally frozen ground zone. They include palsas, earth hummocks, icings, cave ice, stone dish, block stream, block slopes and gelifluction. This is not only due to cold climatic conditions, but also due to richer precipitation in this zone than at lower elevations. Large palsas develop in peatland located in the marshy depressions. The steep slopes provide conditions suitable for the development of stone dish, block stream, block slopes and gelifluction. Small-scale polygons only appear in glaciofluvial loam deposits on the sides of a glacier flow. For example, there are some small polygons at a height of 2460 m in front of Kalasi Glacier with a diameter of about 70 cm, side length of 35-45 cm and cracks of about 1 cm.

The widespread permafrost zone is situated above 2,800 m in the high mountain zone with an area of approximately 4800 km^2, covering 19% of the area of the entire investigated region. The indicators of permafrost exist on the south-facing slopes. Permafrost is found in every geomorphological unit except a few with structural and geothermal taliks. According to D.C. resistivity sounding data, permafrost does not exist below 2,800 m on the south-facing slope in Kalasi Valley (49°10'N, 87°49'E). On August 10, 1980, at 2,780m, ice temperature at the depth of 6 m was -2.2°C in the adjacent Kalasi Glacier, but at 3,130 m elevation, the ice temperature at the depth of 6 m was -3.7°C (Fig. 7, Wang Lilun et al. 1983). During the middle ten days of July, firn 1 m thick was still seen on the north-facing slope at 2,820 m in Qiaqiayite (48°59'N, 87°8' E). Snow temperature at a depth of 26 cm was -0.6°C, although the air temperature was 12.5°C. Therefore, there is no doubt that permafrost exists under firn. At the end of September, 1980, cave ice was 180 cm thick on the south-facing slope at 2,995 m in the Arshate mining area in the north of Qinghe county in the eastern part of the investigated region. The ice temperature was -2.0°C. Hence, in the whole investigated region, permafrost occurs extensively above an altitude of 2,800 m. The mean annual temperature in this zone is lower than -9.4°C. According to the data in a hole drilled in snow, the annual precipitation is 700-800 mm, decreasing to 600-700 mm in the east. Although there is much wind, the thickness of snow cover is generally greater than 1 m. The depth of seasonally thawed layer is still greater than 180 cm in sandy gravel soil. Using the data in the north-side of the Altay Mountains within the border of the Soviet Union, it can be estimated that the thickness of permafrost is less than 400 m and the mean annual ground temperature is higher than -5°C (Shache 1978).

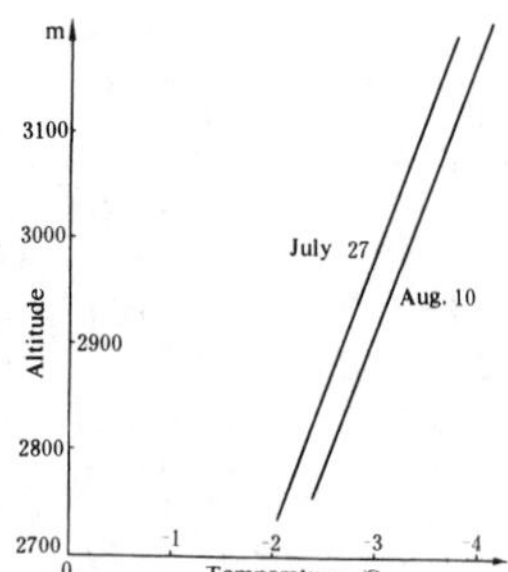

FIGURE 7 Relation between altitude and temperature at the depth of 6 m in Kalasi Glacier (1980).

PERMAFROST LOWER LIMIT IN ALTAY MOUNTAINS

The permafrost region in the Altay Mountains within the border of China is contiguous with the permafrost region in the Altay Mountains within the border of the Soviet Union and the People's Republic of Mongolia. It is situated in the southern fringe of the Eurasian permafrost region (Shache 1978, Mel'nikov 1974).

Within the border of the Soviet Union, the lower limit of permafrost on the north-facing slopes is at the height of 1,400-1,500 m at 49-50°N on the northside of the Altay Mountains, on the south-facing slopes, it is 2,300 m (Shache 1978). It basically corresponds with data within the border of China. The altitude of lower limit of widespread permafrost has a tendency to undulate, e.g. at 49°10'N, 87°40'E in the west, the altitude of the lower limit is at 2,800 m, but at 47°7'N, 90°18'E in the east, the lower limit is still at about 2,900 m, in spite of being

TABLE 1 Comparison of climatic data between the western and eastern Altay Mountains, 1961-1970.

	Content	Western Part	Eastern Part
Direction of Wind	Winter	SN-SSE	NE-N
	Summer	WS-NW	WNW-W
Air Temperature	Mean Annual Temperature	3.5	−0.2
	Annual Amplitude of Temperature	31	41.7
Precipitation	Annual Precipitation	200.8	159.6
	Period of Snow Cover	10−4	10−4

He shante 47° 06' N 86° 37' E 1225.2m a.s.l.
Qinghe 46° 40' N 90° 23' E 1218.2m a.s.l.

2° further south. This results from difference of climatic conditions between the eastern and western regions (Table 1). Because the investigated region is situated in the hinterland of the Eurasian Continent, only does the Arctic air move along Ebi river, eastward to the Ertix River then extends to the Altay Mountains within the border of China. Therefore, the annual precipitation, depth of snow cover and its effects in the west are greater than those in the east. The eastern region is controlled by the Mongolian and Siberian cold high pressure, and northeast and north winds prevail in winter. In January, the mean air temperature is -23.2°C in Qinghe county, but it is -12.1°C in Heishantou which is situated to the northwest of Qinghe. The difference of mean annual air temperature between these two areas is 3.7°C. The west and northwest winds prevail in summer in the entire region. The difference of temperature between the eastern and western parts is very small. In the east climate is colder and the climate continentality is stronger than that in the west. Therefore, the lower limit of widespread permafrost in the eastern part is at a lower elevation than that in the western part. It is estimated that the lower limit of sporadic permafrost between the eastern and western parts has also similar condition.

The kinds and distribution of vegetation are determined by the ecological environment; particularly climate and the hydrothermal condition of soil are very important factors. Thus a certain relationship exists between permafrost and vegetation. It is helpful to use the distribution of vegetation type to explore the extent of permafrost. In compiling the small scale permafrost map, it is worth trying to use this method. The alpine meadow vegetation with "five-flower" meadow small but very bright-coloured flower mainly grows in the western part. Its lower limit is at 2,300 m and it goes up to 2,600 m to the eastern part. In comparison with the elevation of the lower limit of alpine permafrost, the difference in the western part is only 100 m. It is estimated that the difference in the eastern part is less than the allowable error for compiling the map. Therefore, it is possible to use alpine meadow vegetation to indicate the distribution of permafrost in the Altay Mountains.

HISTORY OF PERMAFROST DEVELOPMENT IN ALTAY MOUNTAINS

The development history of permafrost in the Altay Mountains within the border of China is closely related to Quaternary geological history. According to C^{14} data, peat layers in the middle mountain were formed at 5,900± 150 B.P., a post-glacial hypsithermal product. Thus the palsas must be Neoglacial, but the lower part of the widespread permafrost layer probably formed before post-glacial climatic optimum period. Permafrost at that time developed more extensively than at present. Very few newborn palsas could be found in the present palsa group. They now mainly occur around the south of the lake basin, i.e. on the north-facing slope. However, there are a lot of U-shaped and annular thaw lakes. Permafrost in the marshy basin is only found underneath the palsa; no permafrost exists 10 m away from the palsa mounds. According to the weather data of recent more than 20 years, comparing data from 1971 to 1978 with the data from 1957 to 1970, the mean annual air temperature has risen 0.1-1.0°C, especially in the eastern part. It is thus very obvious that since the beginning of the nineteenth century, the climate has been basically turning to a warm one (Zhu Kezhen 1972). Permafrost has been correspondingly degenerating. Because of such natural geographical environment, permafrost possesses unstable properties of high ground temperature, decreased thickness, etc. in the investigated region.

Basing on close relationship of paragenesis between glacier and permafrost and the exposure of many paleoglacial remains in the Altay Mountains, the extent of ancient permafrost is estimated from the correlation between the firn line and the lower limit of permafrost in the same district. At present, the firn line of Kalasi Glacier is at the height of 3,250 m, a difference of 1,050 m in comparison with the lower limit of sporadic permafrost. There are three layers of paleo-cirques at the heights of 2,500 m, 2,630 m, and 2,750 m respectively at Qiaqiayite in the vicinity of Kalasi Lake, and at the heights of 2,600-2,700 m, 2,900-3,100 m and 3,100-3,400 m within the border of Qinghe county in the eastern part, and the present firn line is at 3,500 m. Thus, by calculation, the lower limits of paleo-permafrost were at the heights of 1,450 m, 1,580 m, and 1,700 m in the west, and at the heights of 1,300-1,450 m, 1,600-1,800 m and 1,800-2,100 m in the east respectively. From the heights of paleo-cirques, we know the lower limit of paleopermafrost was at the height of 1,800-2,100 m during the Little Ice Age, not far from the present lower limit of permafrost. The lower limit of permafrost in the Neoglacial was at 1,600-1,800 m, i.e. 400-600 m lower than the present. However, during the last glacial period (Q_3), it would have descended to 1,300-1,450 m, some 750-900 m lower than the present lower limit. According to these results, it can be estimated that the mean annual air temperature during Neoglacial and the last glacial period was 2-3°C and 4-5°C respectively lower than that at present.

CONCLUSIONS

Basing on the above discussion, the distribution of permafrost in the Altay Shan within the border of China can be divided into three zones: seasonally frozen ground (below 2,200 m a.s.l.), sporadic permafrost (2,200-2,800 m a.s.l.), and widespread permafrost (above 2,800 m a.s.l.). Block fields, block streams, block slopes and icing are dominant in the entire region, with some palsas, earth hummocks and cave ice in some places.

Sporadic permafrost must be Neoglacial. Permafrost possesses unstable properties of high ground temperature and decreased thickness. The upper part of the widespread permafrost layer must be Neoglacial too, but the lower part probably formed before the post-glacial Hypsithermal times.

The extent of ancient permafrost can be estimated from the correlation between the firn line and the lower limit of permafrost in the same district. The lower limit of paleopermafrost in the late Pleistocene was at 1,300-1,450 m, in the Neoglacial period at 1,600-1,800 m, and in the Little Ice Age at 1,800-2,100m. The mean annual air temperature was 4-5°C, 2-3°C and 0.5°C respectively lower than that at present.

Snow cover is thick in Altay Shan. It greatly decreases the heat exchange between air and ground surface in the cold season, raises the mean annual ground temperature, decreases the thickness of seasonally frozen layer and promotes degeneration of permafrost.

Climatic continentality in the eastern part of studied area is stronger than that in the western part. It is favourable to the existence of permafrost. Therefore the lower limit of sporadic and widespread permafrost zone in the eastern part is at a lower elevation than that in the western part.

There is a certain relationship between permafrost and vegetation. It is helpful to use the lower limit of alpine meadow distribution to indicate the lower limit of sporadic permafrost in the Altay Shan.

REFERENCES

Li Peiji, 1981, Palsa in the Altay, J. of Glaciology and Cryopedology, Vol.3, No.3. pp.69-56.

Lanzhou Institute of Glaciology, Cryopedology and Desert research, 1975. Permafrost. Science Press, in Chinese.

Maksimova, L.N., Chizhov, A.B., Malamed, V.G. Medvegev, A.V. Dubrovin, V.A. and Dunaeva, E.N., 1977, On influence of snow cover on ground temperature and seasonally freezing (With computer), Frozen ground and snow cover, Science Press, Moscow, pp. 127-135.

Mel'nikov, P.I.ed. 1974. Geocryological conditions in the People's Republic of Mongolia. Science Press, Moscow. pp. 30-49.

He Yixian, 1983, Determination of Permafrost distribution in the Altay Mountains with electric prospecting data, J. of Glaciology and Cryopedology Vol.5, No.4.

Shache, M.M., 1978, Geocryological conditions in alpine region of Altay Sanyang, Science Press, Moscow.

Shulekin, A.M., 1954, Soil climate and snow cover, Acad. Sci. U.S.S.R., Moscow.

Wang Lilun, Liu Chahai, Kang Xincheng, You Genxiang, 1983, Basic Characteristics of Modern glacier in the Altay Mountains of China, J. of Glaciology and Cryopedology, Vol.5, No.4.

Xinjiang Scientific expedition, 1978, Xinjiang geomorphology, Science Press, pp.43-53.

Xu Xiaozu and Guo Dongxin, 1982, Compilation of the Distribution Map of Frozen Ground 1:4,000,000, Journal of Glaciology and Cryopedology, Vol.4, No.2. pp.18-26.

Zhu Kezhen, 1972, Preliminary Research of Climate change in recent 5000 years in China, Acta Archaeologia Sinica No.1. pp.1-25.

RESEARCH ON THE FROST-HEAVING FORCE OF SOILS

Tong Changjiang and Yu Chongyun

Lanzhou Institute of Glaciology and Cryopedology, Academia Sinica
People's Republic of China

To investigate the normal frost-heaving forces that act upon foundations, frost-heave tests were conducted under various conditions in both field and laboratory. Test results showed that the frost-heaving force substantially depends upon the frost susceptibility and water content of the soil, temperature, frost penetration depth, and the area of the bearing plates. A number of empirical equations are presented to evaluate the influence of these factors on frost-heaving forces. Based on model tests, the authors propose that the normal frost-heaving force be considered to be composed of two parts--the normal frost-heaving force produced by the restrained ground body beneath the bearing plates and that produced by the unrestrained ground body surrounding the restrained soil column. Therefore, the authors conclude that the value of the normal frost-heaving forces is not very great, but approaches a certain "stable" limit.

In the past 20 or 30 years, many researchers have extensively studied the normal frost heave pressure acting upon foundations during the freezing of their subsoils, and many behavioral relationships and influencing factors have been discussed. In recent years, some researchers have reported methods for roughly estimating the normal frost heave pressure and the distribution of vertical stresses at the freezing front, and published their maximum experimental values. Since 1965, extensive research has been performed on the frost heave pressure both in our laboratory and in the field near by (in the Fenghuo Mountains on the Qinghai-Xizang (Tibet) Plateau, Muli in the Qilian mountains, Mangui in the Da Hinggan Ling of Northeast China, etc.). The main factors influencing frost heave pressures, the effect of the adjacent soil and the correction coefficients for the depth and width of foundations in engineering designs, etc. have been studied in addition to their laws of annual variation and their distribution. Consequently, the laws governing the change of normal frost heave pressures and their formation concept can be further understood. However, their formation and calculation method still need further study.

EXPERIMENTAL RESULTS AND DISCUSSION

Influence of Soil Characteristics on the Normal Frost Heave Pressures

The experimental data show that the influence of soil particle size on the normal frost heave pressure is quite prominent. The normal frost heave pressure decreases as the sand grain size increases (Table 1) under almost the same temperature and sufficient supply of water to the lower part of soil. In sands with content of sandgrain smaller than 0.1 mm, the effect on the normal frost heave pressures is more evident.

In the same way, the normal frost heave pressure increases as the specific surface area of soil increases. When the specific surface area of soil increases to 62.5 m^2/g, the normal frost heave pressure decreases correspondingly.

According to experimental results under similar conditions of temperature, moisture, and bearing plate area, the values of the normal heave pressure of different types of soil are arranged in the following order: sand clay and silt > clayey > sand > clay > fine sand > coarse sand.

Influence of Soil Moisture on Normal Frost Heave Pressure

In the freezing process, frost heave and frost heave pressures are dominated by the initial water content before frost and the migration of water during the freezing process. The experiments show that, under the same soil and temperature conditions (Figure 1), normal frost heave pressure increases with the increase of soil moisture content until the normal heave pressure of ice is reached. In a closed system, the relationship between the normal frost heave pressure and initial content of soil moisture before frost can be well described by the empirical formula:

$$\sigma_{no} = ae^{-\frac{b}{w}}$$

where σ_{no} is the normal heave pressure acting upon the foundation bearing plate with the dimension of km/cm^2; w is the initial moisture content in %; a and b are

TABLE 1 Normal Frost Heave Pressure of Different Types of Soil

	Fine sand (grain size) (mm)				Loess	Red clay
	2-1	1-0.5	0.5-0.25	0.25-0.1		
Content of particle size smaller than 0.1 mm (%)	4.36	7.77	6.94	16.04		
Specific surface area (m^2/g)	5.2		17.0	19.5	31.8	62.5
Normal frost heave pressure (kg/cm^2)	0.054	0.076	0.096	1.345	8.02 (average)	6.66 (average)

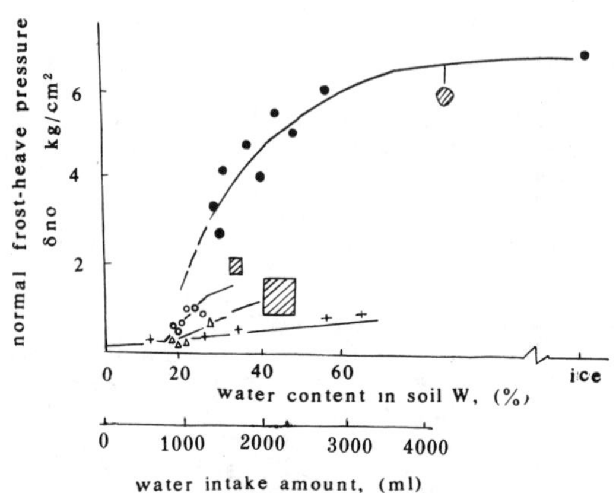

FIGURE 1 Relationship of normal frost heave pressure to water content and water intake amount.

experimentally determined coefficients related to soil, temperature, and area of foundation plate.

In an open system, though the initial water content before frost affects the normal frost heave pressure to a certain extent, the amount of water supplied during the freezing is more important. Experimental results show that the greater the amount of water supplied, the larger the normal frost heave pressure will be. Thus, the relationship can be approximately expressed by the following linear function:

$$\sigma_{no} = \sigma^{o}_{no} + KQ$$

where σ^{o}_{no} is the normal frost heave pressure without migrating water in kg/cm^2; Q is the amount of migrating water in the freezing process in ml; K is an experimentally determined coefficient related to the soil and freezing conditions.

Hence, the amount of water intake or discharge during freezing is the key factor to determine the increase or decrease of the value of the normal frost heave pressure.

Relationship Between Normal Heave Pressure and Soil Temperature

The gradient of ground temperature is closely related to the speed of the movement of the freezing front in the upper layer of soil. Under the same condition, the higher the freezing rate the less the frost heave ratio and the normal frost heave pressure. It's the same the other way around. Their relationship can be expressed as follows:

$$\xi_\sigma = \frac{d\sigma_{no}}{dt} = A + \frac{B}{\sqrt{U_f}}$$

where ξ_σ is the increasing ratio of the normal frost heave pressure in kg/cm^2; U_f is the frost penetrating ratio in mm/hr.; A and B are experimentally determined factors related to soil, moisture, and area of bearing plate; and t is the time in hours.

It must be pointed out that this equation is only suitable to the normal frost penetrating ratio. When the frost penetrating ratio is smaller than a certain value ($U_f < U_{f1}$), the changes of the rheology of ice body and freezing point of water are caused by the normal frost heave pressure, lowering the increasing ratio of the normal frost heave pressure. Inversely, when the frost penetration ratio is quite large ($U_f > U_{f2}$), the possibility of moisture movement is quite small and the normal frost heave pressure is only caused by the freezing of pore water.

In a closed system, normal frost heave pressures change with temperature. This law can be expressed as follows:

$$\sigma_{no} = \alpha e^{-\frac{\beta}{t}}$$

where t is the soil temperature in negative °C and α and β are experimentally determined factors related to soil type, moisture, and area of bearing plate.

That is to say, normal frost heave pressure increases exponentially with the decrease of soil temperature, which coincides with that the unfrozen water content of frozen soil decreases exponentially with the decrease of soil temperature. From the relationship between the unfrozen water content and soil temperature, the relationship between the normal frost heave pressure and the unfrozen water content in frozen soil can be deduced:

$$\sigma_{no} = \alpha \left(\frac{m}{w_u}\right)^{\frac{\beta}{n}}$$

where m and n are experimentally determined coefficients.

It is evident that the normal frost heave pressure is inversely proportional to the unfrozen water content in frozen soil.

Relationship Between Normal Frost Heave Pressure and Frost Susceptibility

Experimental data show that the normal frost heave pressure of soil with different frost susceptibility is closely related to the frost heave of soil under foundation bearing plate. Figure 2 shows that it increases proportionally to the frost heave ratio of the soil. Hence, it can be expressed in a parabolic formula:

$$\sigma_{no} = x\eta^{y}$$

where η is the frost heave ratio of soil in % and x and y are factors related to soil moisture and area of bearing plate.

On logarithmic curve, three turning points can be found, i.e. η=0.9, 3.2 and 5.8%. In China, these three points exactly reflect the three stages of frost heave of soil: low heave, heave, and high heave.

According to the result of measurements obtained both indoor and in the field, the normal frost heave pressure is directly related to the restrictive condition of frost heave deformation of soil layer. The higher the degree of restraint, the larger the normal frost heave pressure (Figure 3). This relationship can be described as follows:

$$\sigma_{no} = \frac{M - \lg\eta}{N}$$

where M and N are experimental factors.

Although the restrictive condition of frost heave deformation of soil layer is different, the deformation of ground body around foundation continues to take place. The farther the distance from the observing place to the foundation, the closer it is to the natural deformation of soil by frost heaving.

Based on the deformation curve of ground body around foundation plate, a formula can be deduced from the above relationship:

$$\sigma^{o}_{no} = \frac{1}{R} \ln \frac{\eta_o}{\eta_i}$$

$$\text{or } \eta_i = \eta_o \, e^{-R\sigma^{i}_{no}}$$

where η_o is the natural frost heave ratio in %; η_i is the frost heave ratio when the normal heave pressure being σ^{i}_{no} in %; and R is an experimental factor related to soil, moisture etc.

Normal Frost Heave Pressure Changes with Depth of Foundation

Generally speaking, with the increase of depth of foundation buried, the thickness of freezing soil layer under foundation plate decreases, as does the normal frost heave pressure. Based on the observational data on the Qinghai-Xizang Plateau and in Northeast China, the law of variation of normal frost heave

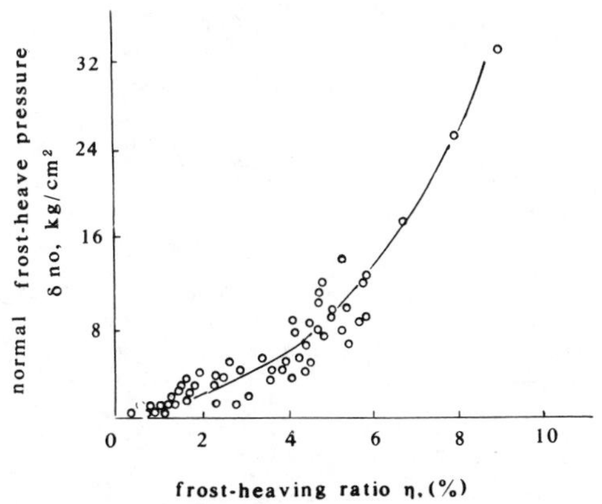

FIGURE 2 Relationship between normal frost heave pressure and frost-heave ratio of soil.

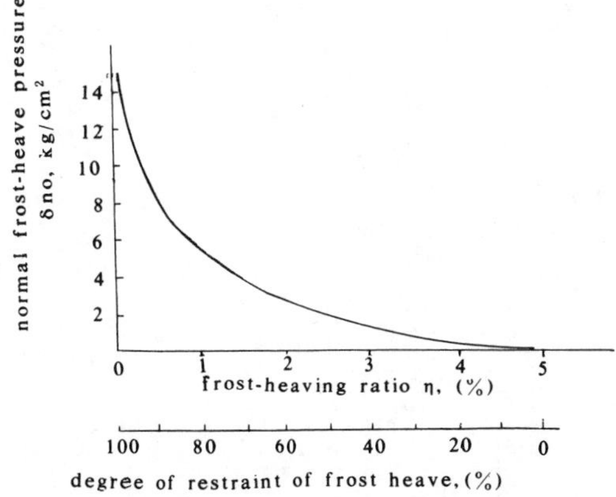

FIGURE 3 The normal frost heave pressure under degree of restraint of frost heave.

pressure can be approximately expressed in the following formula:

$$\sigma^z_{no} = \alpha_o H_z^{\beta_o}$$

where σ^z_{no} is the normal frost heave pressure when the depth of foundation buried being z in kg/cm^2; H_z is the relative depth, i.e. the percentage ratio of the depth z of the foundation buried and the maximal freezing depth:

$$H_z = \frac{z}{H_f} \times 100\%$$

H_f is the maximal freezing depth in m; it should be pointed out that this maximal freezing depth must include the incremental thickness of freezing soil layer due to the effect of foundation; z is the depth of foundation buried in m; α_o and β_o are experimentally determined factors related to the soil, moisture, and area of bearing plate.

Normal frost heave pressure reduces with the decrease or reduction of the thickness of the freezing soil layer under the foundation plate. This law is affected by two factors: the thickness of freezing layer and the change of soil temperature. The former may regard the normal frost heave pressure as a superimposed form of frozen layers at a definite thickness and the latter increases with the increase of the depth and thus the normal frost heave pressure presents a corresponding decrease.

Influence of Bearing Plate Area on Normal Frost Heave Pressure

Under almost the same boundary condition, the value of the normal frost heave pressure is different with the different area of foundation plate. The experimental results both in the laboratory and in the field show that the smaller the area of bearing plate the larger the normal frost heave pressure; and when area of bearing plate is quite enormous, it finally tends to a stable value related to certain conditions (such as soil type, moisture, temperature, depth of frost penetration, and area of bearing plate). When the value of the normal frost heave pressure tends to a sharp segment, it's corresponding plate area is 400-1600 cm^2 (Figure 4).

Therefore, the relationship between normal frost heave pressure and the area of the foundation plate can be well described by the expression:

$$\sigma_{no} = a_F \, e^{\frac{b_F}{F}}$$

where F is the area of foundation plate in cm^2 and a_F and b_F are experimentally determined factors related to soil type, moisture, temperature etc.

In a common case, the foundation plate above the earth is inevitably acted upon by the frost heave of ground body restrained by the bearing plate and of ground body unrestrained outside the bearing plate. According to the experimental results in the laboratory, this "stable value" is near the value of normal frost heave pressure caused by the ground body restrained by the bearing plate under similar conditions of soil, moisture, and frost penetration. The experimental results also show that the frost heaving of the ground body outside the bearing plate affects significantly the normal frost heave pressure under the bearing plate and that the normal frost heave pressure increases with the increase of ratio of sample diameter to plate diameter, until the ratio is more than 6 to 8 times.

So, it is considered that normal frost heave pressure measured at the test site

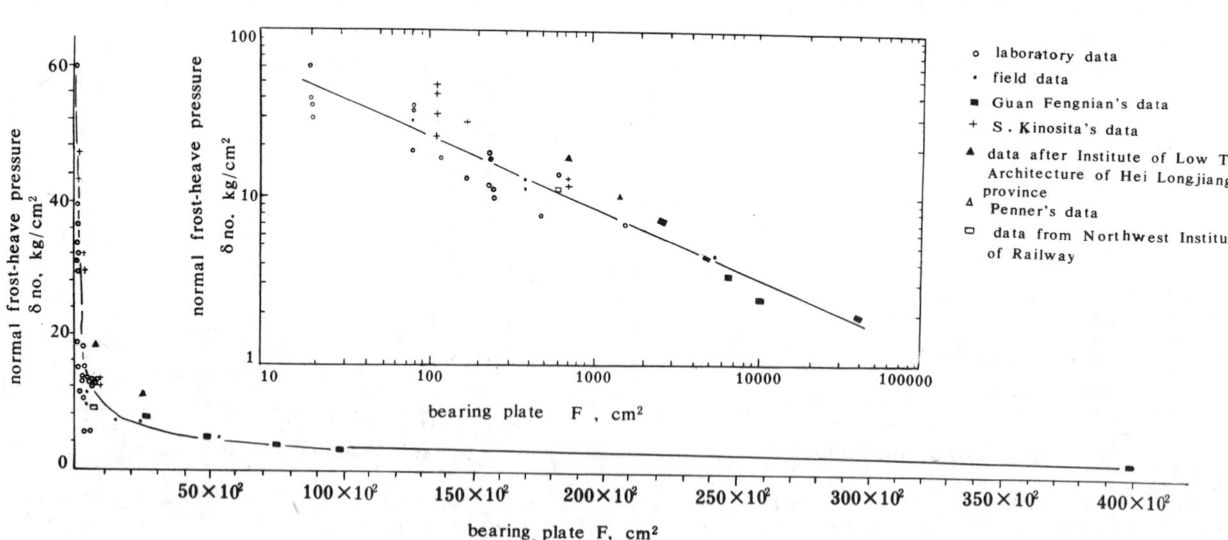

FIGURE 4 Relationship between normal frost heave pressure and bearing plate.

consist of two parts — i.e., the normal frost heave pressure produced by the restrained ground body directly lying beneath the bearing plate and that produced by the unrestrained ground body. It is also considered that the former must be a constant value under certain boundary conditions.

To sum up, the normal frost heave pressures are the results of various factors. They aren't a huge value uncomparable.

Through the research of correlation factors it is known that the value can be comprehended. Hence, we can study the "standard value" of the normal frost heave pressure of different soil and moisture with unified test method and under approximate boundary conditions, such as the restrained condition of soil, area of bearing plate, and frost penetration retio. The standard value after being corrected can be used in engineering design.

ACKNOWLEDGMENT

The authors would like to thank their colleagues — Zhang Jiayi, Wu Ziwang, Zhu Yuanlin, Liu Jiengshou, Ma Zhixue, Chi Jianmei and Yin Yaming — for their assistance.

REFERENCES

Tong Changjiang, 1982, Frost heaving force of foundation, in Proceedings of the Symposium on Glaciology and Cryopedology Held by Geographical Society of China (Cryopedology), p. 113-119. Beijing, China, Science Press.

Zhou Changqing, 1980, An brief introduction of research on frozen subsoils and foundation in China, Journal of Chinese Glaciology and Cryopedology, v. II, No. 1, p. 1-5.

Tong Changjiang and Yu Chongyun, 1982, On the relationship between normal frost heaving forces and areas of bearing plates, Journal of Chinese Glaciology and Cryopedology, v. IV, No. 4, p. 49-54.

Orlov, V. O., 1977, Frost heave of freezing soil and its influence on the foundation of construction, Leningrad, Stroyizdat.

Kinosita, S., 1967, Heaving force of frozen soils, in Oura, E.H., ed., Physics of Snow and Ice, Hokkaido University, p. 1345-1360.

Penner, E., 1970, Frost heaving forces in Leda clay, Can. Geotech. J., 7(1), p. 8-16.

IN-SITU DIRECT SHEAR TESTS AT THE FREEZE/THAW INTERFACE AND IN THAWED SOILS

Tong Zhiquan

Northwest Institute of the Chinese Academy of Railway Sciences
Lanzhou, People's Republic of China

To build railroads in permafrost regions requires an understanding of the shear strength of soils at the freeze/thaw interface while evaluating the stability of embankment slopes, foundations, and cut slopes. Considering the relatively high water content in the freeze-thaw transition zone in soil, many researchers have suspected that the freeze/thaw interface could be the weaker surface of shearing. To clarify this, the authors conducted indoor and outdoor direct shear tests at the interface and in thawed soils. Results both in situ and in the laboratory showed that the shearing strength at the freeze/thaw interface is greater than in thawed soils, as are both the interfrictional angle and the cohesion force. The weaker surface of shearing was not at the freeze/thaw interface, but in thawed soil with high water content. Therefore, to evaluate the stability of roadbeds in permafrost regions, shear-strength values should be taken from thawed soils with high water content.

When building a railroad in a permafrost region we must know shearing strength at the frozen-thawing boundary interface in order to compute the stability of embankments, foundations and cuts. Because of the comparatively rich water content in the frozen-thawing transition zone of thawing soil, we were concerned that the frozen-thawing boundary could be the weak surface of shearing. Certain experiments were conducted to check this expectation. Because of the different experimental conditions between laboratory and field sites, we conducted large in-situ direct shearing tests on clayey soil at the frozen-thawing interface and on completely thawed ground and compared the result with those of laboratory tests.

IN-SITU LARGE DIRECT SHEARING TEST

Description of Test Site

The region of Mt. Fenghuo is a permafrost region through which the planned Qinghai-Xizang Railroad Line will pass. This region was chosen as the location to carry out in-situ direct shear tests on a frozen-thawing interface and on fully thawed ground. The ground surface there was covered by turf and was underlain by a clayey soil containing crushed stone 3-5 mm in diameter. The depth of the natural permafrost table was within 1.1-1.4 m of the surface. This region is one where the permafrost has been relatively stable for a long time, the annual average temperature being within -4 to -7°C. The water content distribution from the ground surface to the frozen-thawing boundary appears as a K-type (Figure 1) with the larger values occuring at 0.2 m depth and near the frozen-thawing boundary. Tests were carried out under conditions of natural water content. The different dry bulk densities γ_d of the soil were: 1.55 g/cm^3 for the first test, 1.50 g/cm^3 for the second and 1.47 g/cm^3 for the third. Although the test site was composed of a clayey soil, the size distribution and the physical properties of the soil were a little different in accordance with different test-pit. The ranges of the size distribution of soil and its physical properties are given below. The dimension of soil particles vary within: 0.5-0.05 mm (35-49%), 0.05-0.005 mm (32.6-39%), and below 0.005 mm (18.4-26%); the physical indexes vary within the range: Liquid limit 24.4-35.7%; plastic limit, 14.4-22.5% and index of plasticity, 9.5-13.2%.

Testing Facilities

The equipment used for the test was a large direct shearing device capable of conducting experiments in situ on samples with areas as large as 1,000 cm^2, being 31 times that of normal laboratory samples.

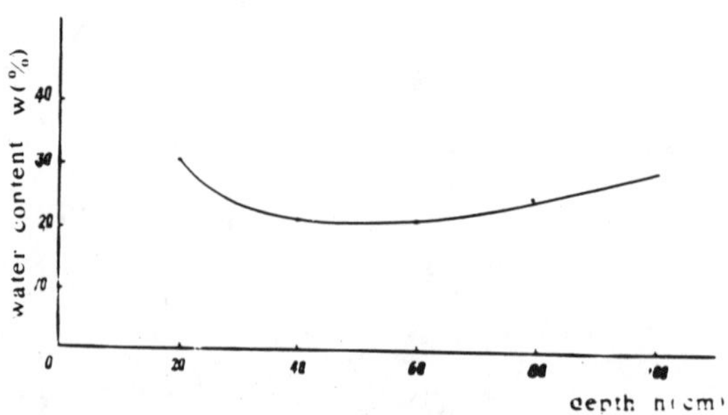

FIGURE 1 Distribution of Water Content from Ground Surface to Frozen-thawing Boundary.

Test Procedures

First a test-pit was excavated having a working surface of 2 x 1.4 m^2. This area is appropriate for both the fully thawed ground and the frozen-thawing interface cases. The test was conducted according to Chinese standards for in-situ large undrained quick shearing except that certain precautions were made for the case of the frozen-thawing interface. The tests were conducted as follows:

a. After excavation to a depth of 0.2 m above the frozen-thawing interface, a soil cylinder 40 cm in diameter was left in middle of the pit for preparing the sample.

b. A soil cylinder of 35.7 cm diameter was trimmed with the shearing ring of the device and left perpendicular to the frozen-thawing interface. The shearing ring was then pressed gradually downward until it reached the frozen-thawing surface. The direct shear apparatus was then installed and adjusted as quickly as possible.

c. Protection from thermal disturbance must be especially stressed during the installation of the testing device and during the conduct of the shear test to ensure that the shearing plane is coincident with frozen-thawing interface throughout the entire test.

d. The interface was measured and sampled immediately after shearing and its water content was determined.

Test Results

The results of the in-situ large shear tests on the clayey soil frozen-thawing interfaces and the fully thawed ground are listed in Table 1.

For the conditions of natural water content and natural density, a comparison of the shearing strengths at the frozen-thawing interface with the shear strengths of completely thawed ground is shown in Figure 2. The information in Figure 2 is from Table 2.

It is seen from the test that the shearing strength of a clayey soil frozen-thawing interface is greater than that for completely thawed ground.

TABLE 1 In-situ Large Direct Shear Test Data for the Frozen-thawing Interface and Fully Thawed Ground

Test No.	Test condition	Water content before shearing* w (%)	Void ratio before shearing e	Shearing strength under different vertical pressures τ (kg/cm^2)				Internal friction angle* ϕ	Cohesion* C (kg/cm^2)
				0.5 kg	1.0 kg	1.25 kg	1.5 kg		
I-1	Frozen-thawing interface	21.3	0.74	0.301	0.473		0.698	20°48'	0.110
I-2	Fully thawed ground	20.9		0.206	0.332		0.458	14°08'	0.080
II-1	Frozen-thawing interface	27.5	0.80	0.288	0.425	0.498		16°40'	0.140
II-2	Fully thawed ground	27.3		0.244	0.362	0.432	0.474	12°50'	0.135
III-1	Frozen-thawing interface	31.1	0.82	0.277	0.383		0.537	14°55'	0.135
III-2	Fully thawed ground	30.0		0.226	0.321	0.362		10°33'	0.130

* w = water content by weight, C = cohesion in kg/cm^2 and ϕ = angle of internal friction in degrees and minutes.

TABLE 2 Laboratory Small-scale Direct Shear Tests Results for Clayey Soil Frozen-thawing Interface and Completely Thawed Soil

Test No.	Test condition	Water content before shearing w (%)	Compacted dry bulk density γ_d (g/cm^3)	Shearing strengths under different vertical pressures τ (kg/cm^2)			Internal friction angle ø (degree)	Cohesion C (kg/cm^2)
				1 kg	2 kg	3 kg		
I-1	Frozen-thawing interface	17.14	1.55	0.850	1.480	2.110	32°20'	0.210
I-2	Completely thawed soil			0.635	1.120	1.600	25°53'	0.150
II-1	Frozen-thawing interface	20.74	1.55	0.650	1.180	1.722	28°22'	0.110
II-2	Completely thawed soil			0.450	0.820	1.195	20°33'	0.070
III-1	Frozen-thawing interface	22.50	1.55	0.520	1.000	1.470	25°10'	0.050
III-2	Completely thawed soil			0.320	0.630	0.940	17°13'	0.10

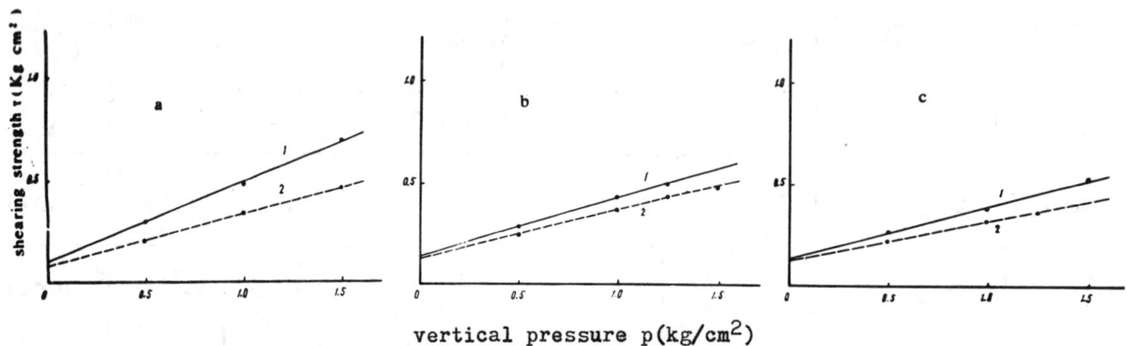

FIGURE 2 A Comparison of Shearing Strength at a Frozen-Thawing Interface and Completely Thawed Soil Obtained by Large In-Situ Direct Shearing Tests.

The frozen-thawed boundary interface forms in the course of the thawing of frozen ground. Frozen ground is composed of four phases——solid (mineral particles), plastic (ice), liquid (water unfrozen) and gas (air) with ice as the basic link. The ice is significantly anisotropic, such that the mechanical strength is maximum in the direction of the crystal axis. Because the four phases in frozen have different thermal conductivities and their thawing rates are different, the frozen-thawed interface formed during thawing is not an ideal smooth surface but an uneven and rough one. The shear place intersects ice crystals on frozen-thawing surface at different angles. This case does not exist in completely thawed ground, therefore, the internal friction angle of the frozen-thawing interface is greater than that of the completely thawed ground. Furthermore, the cohesion of the former is also a bit larger than that of the latter. It is now clear from the above results that the main reason that the shearing strength of the frozen-thawing interface is larger than that of the completely thawed ground is because of the larger internal friction angle the former possesses.

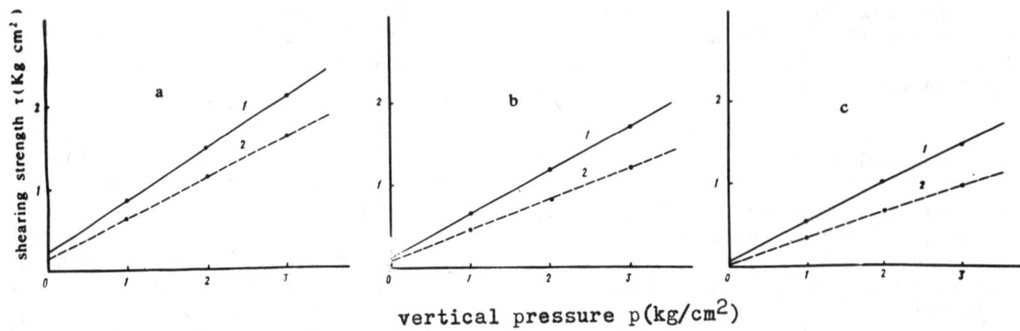

FIGURE 3 Comparison between Shearing Strength of Frozen-Thawing Interface of Clayey Soil and Completely Thawed Clayey Soil.

SMALL-SCALE LABORATORY DIRECT SHEAR TESTS

Soil Properties and Sample Preparation

The soil used for the laboratory direct shear test was the same clayey soil taken from the area of Mt. Fenghuo in the permafrost region of the Qinghai-Xizang Plateau. The granulometric composition and physical properties of which are the same as those of the soil taken on the spot where the in-situ tests were conducted.

The samples used were remoulded, and their densities were so controlled as normally done in controlling density of fill base materials for roads, i.e. the samples were compacted with a small tamper under the conditions of identical density but at different water contents.

Method of Testing

The common laboratory undrained rapid shear test was adopted as the method of testing. Samples were frozen and then thawed from the top down with an infrared bulb until the frozen-thawing interface coincided with the preferred shear place. The samples were then sheared.

Test Results

The results of the laboratory small-scale shear tests on the clayey soil frozen-thawing interface and the completely thawed soil are presented in Table 2.

A comparison between the shearing strength of the frozen-thawing interface of the clayey soil and the completely thawed clayey soil under fixed water contents and identical densities is shown in Figure 3. The information in Figure 3 is from Table 2.

The test indicates once again that the shearing strength of the frozen-thawing interface is greater than that of the completely thawed soil and the internal friction angle and cohesion of the former are also larger than those of the latter respectively.

COMPARISON BETWEEN IN-SITU LARGE AND LABORATORY SMALL-SCALE DIRECT SHEARING TESTS

A common result of both tests is that the shearing strength of the frozen-thawing interface is greater than that of the completely thawed ground. The slight difference between the results is that the indexes of shearing strength obtained in the laboratory tests are the greater, which is probably because the samples tested in laboratory were composed of remoulded soil, had a smaller area of shear and had undergone only a single freeze-thaw cycle. Whereas the in-situ tests were conducted under natural conditions, the samples tested were much larger and had undergone many freeze-thaw cycles. Thus the different test conditions along with the natural variation of soil structure probably caused the differences in the test result.

Moreover, as we found in laboratory tests, the internal friction angle and cohesion of samples decreased with increase in water content. However, an abnormal phenomenon occurred in the course of in-situ tests---the cohesion obtained in tests II and III was greater than in test I despite the larger water content. This is probably because there were a few grass roots near the shear surface of tests II and III, which acted as a reinforcement.

CONCLUSIONS

1. Under the same conditions the shearing strength of a frozen-thawing interface is greater than that of the completely thawed ground. This implies that the former is not the weak surface of shearing strength, while, in fact, the weak surface lies in fully thawed ground with high water contents.

2. The weak zone of a slope in a permafrost region lies in the completely thawed soil with high water contents. Therefore, it is imperative to choose the shearing strength indexes of

completely thawed soil of high water content in computing the stability of embankment slopes, foundations, and cut slopes for constructing railroad beds in permafrost areas.

REFERENCE

Lanzhou Institute of Glaciology and Cryopedology Academia Sinica, 1975, Cryopedology

N.A. Tsytovich, 1973, The Mechanics of Frozen Ground, p. 446.

Stanley Thomson and Edward F. Lobacz, 1973, Shear strength at a thaw interface, in Permafrost---The North American contribution to the Second International Conference, p. 419-426.

The Department of Water Conservancy and Water Power of the People's Republic of China, 1980, Rules for the Performance of the Soil Test.

The Northwest Institute of China Academy of Railway Sciences, 1979, The researches of the construction of the railway through the permafrost region of the Qinghai-Xizang Plateau

CLASSIFICATION OF ROCK STREAMS

A. I. Tyurin

Geology Department
Moscow State University, USSR

Rock streams (felsenmeer) are classified according to their origin. Rock stream formation occurs under a certain combination of a number of climatic, lithological and geomorphological conditions. A number of processes participate in the formation of coarse-fragmented material of rock streams and its movement, namely: rock weathering, heaving of fragments from the weathered substrate, suffosion of fine sediments, interstitial ice formation, and deserption. Each rock stream is composed of a certain set of facies units, which are predetermined by its origin. Therefore, the suggested classification of rock streams is based on the principal mechanisms governing the formation of the coarse-fragmental mantle which constitutes rock streams. During the rock stream formation often one of the abovementioned rock-stream forming factors predominates. Four types of rock streams can then be distinguished: 1) "rock weathering" type, 2) "coarse-fragmented heaving material" type, 3) "fine sediment-suffosion" type, and 4) "gravity" type. If factors dominate during the rock-stream formation, then the following additional types of rock streams are formed: 5) "rock weathering and coarse-fragmented heaving material" type and 6) "rock weathering and fine sediment-suffosion" type. According to the nature of source areas feeding the rock streams, 14 subtypes are identified. The sources feeding the rock streams can occur uniformly throughout the area, along the system of cracks and also locally, at rock escarpments and remnants.

The classification of rock streams should take into account all the major aspects of their formation; schemes of rock stream classification in the literature are, chiefly, of a morphological character.

The rock stream formation occurs under a certain combination of a number of climatic, lithological and geomorphological factors. Among the conditions required for the rock stream (felsenmeer) formation are the following:

(1) Availability of slopes with steepness ranging from 2-3° up to the value that does not exceed the angle of repose of coarse-fragmented material.

(2) Occurrence of fissured but weathering-resistant moraine rocks close to the current surface. The thickness of seasonally thawed or seasonally frozen layers should exceed the thickness of the loose surficial deposit mantle. The rock stream formation can also occur in deep bedrock, provided lithological characteristics of seasonally thawed and seasonally frozen layers are favorable.

(3) Considerable fluctuations of temperatures on the substrate rock surface, causing a thermogenic deserption.

(4) Congelation and segregated ice formation within the seasonally thawed zone resulting in a cryogenic deserption. Frost heaving of stones and suffosion due to subsurface flow contribute to intensive development of rock streams.

Rock stream formation is widespread in mountain areas, characterized by relatively mild, moderate amplitude upward movement (frost heaving), a severe (nival) climate and the presence of frozen substrate. Rock streams are rarely encountered in flatlands and under conditions of alpine relief, where purely gravity processes (rockfalls, talus) prevail.

Frost-facies analysis made possible the classification of rock streams suggested here. There are 3 stages in the process of rock stream formation: (1) origin, (2) progressive development and (3) attenuation. The first stage ends when a coarse-fragmented material layer with the simplest deserption mechanism of movement is formed on the slope surface. This can take place as a result of various processes; for example, weathering of rocks outcropping at the current surface; weathering of rocks underlying a shallow layer of loose deposits with subsequent heaving of fragments to the surface; heaving of coarse-fragmented material from dispersed substrate. The initial and subsequent structures of rock streams are different. During the initial stage of formation, the rock stream is represented only by the suffosion-destructive subfacies; the interstitial ice subfacies appears afterwards and is gradually replaced by the siltation one (Romanovskii and Tyurin, 1979). The composition of the rock-stream subfacies in the course of its later development stage directed downslope is as follows: (1) the subfacies of rock weathering and heaving of fragments to the surface, (2) heaving and (3) siltation subfacies. Engineering-geologic characteristics of each subfacies and of their combinations are different. The abovementioned rock stream subfacies do not cover the diversity of rock streams. The formation of a subfacies depends on (1) its nature, (2) relationship between the depths of the seasonally thawed layers and loose deposits, (3) water availability on the slope and subsurface flow intensity, (4) connection with external sources of material (talus, rockfalls, avalanches, etc.), (5) the processes in the seasonally thawed layers, etc. Each rock stream is composed of a strictly defined subfacial combination, but all of the subfacies are predetermined by a primary subfacies, i.e. by the genesis of the rock stream coarse-fragmented material. Because rock-stream facial composition is determined by their genesis, it is necessary to classify them according to this attribute. The point is to understand the basic mechanism governing the formation of coarse-fragmented mantle of rock streams. On this basis six types and 14 subtypes of rock streams are recognized. The first type includes rock streams formed as a result of hard rock weathering followed by the production of small quantities of dispersion material—"rock streams of weathering." The second type is represented by rock streams formed during the rock weathering with the subsequent heaving of fragments to the surface—"rock streams of weathering and heaving." The third type consists of rock streams being the product of the fragmented material heaving to the surface—"rock streams of heaving." The fourth type includes rock streams formed due to rock weathering and suffosion of dispersed filler material—"rock streams of weathering and suffosion." The fifth type is made up of rock streams formed as a result of suffosion reworking of fragmented material—"suffosion rock streams." The sixth type—"gravity rock streams"—are formed from gravity deposits. Rock-stream subtypes are defined on the basis of their relationship to the material sources feeding the rock streams. Fragmented material can be supplied uniformly throughout the area, along the system of fissures and locally by rock outcrops.

The first type of rock streams is divided into 3 subtypes: (1) rock streams of uniform weathering are formed during the weathering of rocks outcropping on the current surface slopes. Rock outcropping on gentle slopes, however, is rare. The same subtype includes also rock streams formed as a result of weathering of rocks underlying shallow unconsolidated deposits. When the first layer of the coarse-fragmented material is formed, loose deposits vanish; they collapse into the forming layers and are washed out with subsurface flow. They do not affect the rock stream formation process. Rock streams of weathering combine two subfacies: suffosion-destructive and interstitial ice. The siltation subfacies is limited in area and frequently missing.

(2) Rock streams resulting from differential weathering of rocks along systems of cracks are represented by the structural rock stream slope facies. Within the upper parts of slopes rock streams can sometimes be represented by the facies of rock-stream flows, insignificant in extension and width. Downslope the suffosion-destructive subfacies of a rock stream is transferred into the interstitial ice subfacies, which is encountered locally, and marks the initial crack systems.

(3) Rock streams produced due to local frost shattering of rocks (escarpments and remnants), are characterized by lithological heterogeneity. In cases where escarpments are capped by hard rocks, frost processes proceed most efficiently at the escarpment base. It is there that maximum wetting occurs along with the greatest development of processes associated with the suffosion of the dispersed material. Parallel retreat of the escarpment takes place. At the same time the coarse-fragmented mantle is preserved on sloping surfaces (3-5°); this mantle later on develops independently as a rock stream. The following set of subfacies is usually observed in the direction away from the escarpment. The suffosion-destructive subfacies, which has the form of a "frost face," is followed by the siltation one and, further on by the heaving subfacies. Rock streams formed due to the erosion of bedrock outcrops are more frequently represented by the facies of rock-stream slopes, than by the facies of rock stream flows.

There are two subtypes constituting the second type of rock streams: these are the result of (1) frost weathering (uniform throughout the area) and (2) heaving of fragments to the surface. The given subtype of rock streams is the combination of three subfacies. The upper part of the slope is occupied by a subfacies, the name of which corresponds to that of the rock stream; the rock stream was represented by this subfacies at the beginning of its formation. Downslope it is replaced by the coarse-fragmented material heaving subfacies. The thickness of the seasonally thawed or seasonally frozen layers here corresponds to or is slightly less than the thickness of the loose deposit mantle. The third siltation subfacies appears at the later stage of formation, destroying the heaving subfacies. Within a given subtype other combinations of subfacies seem also likely. The siltation subfacies can be transformed into the heaving subfacies; this means transformation of "weathering and heaving" type of a rock stream into the "heaving" one. The second subtype includes rock streams made up as a result of differential weathering of rocks along joints with the subsequent heaving of fragments to the surface. Rock streams of this subtype are composed of two groups of facies of (1) concealingly polygonal rock stream slopes and (2) flows. Facies of concealed polygonal rock-stream slopes are marked by (1) rock weathering along frost cracks and heaving of fragments to the surface (concealed polygonal remnants) and (2) complex combination—with interstitial ice, weathering and heaving, as well as siltation; sometimes the heaving subfacies is formed.

The third type of rock streams includes two subtypes: (1) rock streams formed due to uniform heaving of coarse-fragmented material from unconsolidated deposits. (These rock streams make up the rock stream slope facies, represented by subfacies of uniform heaving.) This subfacies always turns into a siltation one downslope; (2) rock streams generated by selective heaving of coarse-fragmented material along fissures. (This rock-stream subtype, unlike the preceding one, includes rock streams represented by a group of facies of (1) structural rock stream slopes and (2) flows.) The structural rock stream slope is usually characterized by the concealed polygonal outcrops subfacies with predominant heaving along frost cracks. Downslope this subfacies changes for the siltation one. Rock-stream flow facies usually include the rock stream stripe facies, sometimes replaced by the rock stream rampart-flow facies downslope.

The fourth type of rock streams comprises two subtypes: (1) uniform rock weathering and suffosion-reworking of unconsolidated deposits. These are formed by weathering of rocks which yields small quantities of fines and is followed by intense surface and subsurface flows. Rock streams are represented by the rock stream slope facies, including a complete set of subfacies from the suffosion-destructive to the siltation one; (2) rock streams formed as a result of selective rock weathering along cracks, as well as due to suffosion reworking of unconsolidated deposits. Each given subtype of rock streams includes the facies of rock stream slopes and flows. The rock stream slope can be represented downslope by the subfacies of concealed-polygonal remnants with prevailing weathering and crack suffosion, which are replaced by the siltation subfacies. Most often rock stream flows are represented by the rock stream rill facies, replaced downslope by the rock stream stripe facies, which may change into the rock stream rampart-flow facies.

The fifth type of rock streams has also two subtypes: (1) rock streams formed as a result of widespread suffosion reworking of loose fragmented deposits. (This rock stream subtype differs mainly from the abovementioned type in that it lacks additional supply of fragmented material on account of rock weathering.) The suffosion of dispersed material in seasonally thawed or seasonally frozen layers is accomplished both in its upper and lower part. Gradually, the whole section gets washed out, producing the rock stream mantle. The siltation subfacies formed downslope may be transformed into the suffosion one; (2) rock streams formed due to suffosion working of loose deposits along fissures. These are typical of the slopes with angles of up to 15° and where the mantle of alluvial-deluvial deposits exceeds 4 m in thickness. The thickness of the seasonally thawed or seasonally frozen layers here does not usually exceed 2.5-3.0 m. Each subtype is represented by the facies of both rock stream slope and flow. The rock stream slope facies is formed on the subfacies of concealed-polygonal remnants, under the dominating influence of suffosion along frost cracks. Downslope the rock stream slope facies is replaced by the suffosion subfacies of insignificant extension (up to 5-10 m). This is then replaced by the siltation subfacies.

The sixth type is the gravity rock stream, among which three subtypes are distinguished: (1) rockfall-talus (colluvial) rock streams formed at the base of rock walls or on slopes with the steepness less than the angle of repose of coarse-fragmented material. The upper part of the section of colluvial deposits is actually an originating rock stream. Further on, processes of suffosion and frost sorting will form a complete rock stream section to the full thickness of the seasonally thawed or seasonally frozen layers; (2) the formation of "sliding" rock streams is stimulated by the sliding of rock blocks. Heaving or detached from the massif, blocks can slide down the dispersed substrate in the form of "wandering stones." Accumulation of the blocks on the slopes causes the formation of the "film" coarse-fragmented material mantle. The upslope is divided into separate blocks—"point rock stream"; (3) avalanche rock streams. These rock streams are the result of avalanche transport of the coarse-fragmented material. Their formation takes place locally, at the periphery of the avalanche alluvial fan. The major part of the fragmented material gets saturated with fines, which is the reason why rock streams are formed after the preliminary reworking of drifting products during the processes of suffosion and frost heaving.

It is not easy to determine the basic rock stream forming process. For example, the heaving of the coarse-fragmented material along frost cracks and the suffosion of dispersed fillers are always combined. The morphology of the frost crack surface helps to determine which processes prevail. Where the coarse-fragmented material along the crack rises above the polygonal surface or is at the same level, the prevalence of the heaving over the suffosion is suggested. The negative surface indicates an inverse correlation. The processes of weathering, heaving and suffosion have equal contributions in the formation of some rock streams. This is reflected in the combinations of subfacies.

The above classification takes into consideration the genetic aspect of rock stream formation. This classification can be supplemented with a number of auxiliary classifications and schemes, for example, allowing for rock stream morphology, its topographic position, stage of development, intensity of movement, and connection with material supply regions. All the auxiliary schemes should be considered as derivatives of the genetic classification.

Determining the position of rock streams in the genetic classification of each region will help develop scientifically-grounded recommendations for their utilization.

REFERENCES

Romanovskii, N. N., and Tyurin, A. I., 1974. Facies characteristics of rock streams of Southern Yakutia and Northern Transbaikalia. Geologia, no. 4, Vestnik Moskovskogo Universiteta.

Editor's note: For additional details on rock streams see the recently published book:

Tyurin, A.I., Romanovskii, N.N. and Poltev, H.F., 1982. Frost-facies analysis of rock streams. Moscow: Izdatel'stvo "Nauka." (Akademiya Nauk SSSR, Nauchnyi Sovet po Kriologii Zemli, Sibirskoe Otdelenie, Institut Merzlotovedeniya.)

A COMPARISON OF SUCCESSIONAL SEQUENCES FOLLOWING FIRE ON PERMAFROST-DOMINATED AND PERMAFROST-FREE SITES IN INTERIOR ALASKA

Keith Van Cleve, Forest Soils Laboratory,
University of Alaska,
Fairbanks, AK 99701

Leslie A. Viereck, Institute of Northern Forestry,
Fairbanks, AK 99701

The structure and function of upland interior Alaskan forest ecosystems has been examined across two secondary successional sequences. One, the most common in interior Alaska, follows fire in black spruce stands on permafrost sites. The other, less common sequence, follows fire on warmer, generally south aspect sites and passes through a shrub and hardwood stage to white spruce.

On black spruce sites, the thickness of the forest floor is a key factor responsible for maintaining moist soil, cold soil temperatures and permafrost. Soil degree days range from 1250 to 1000 degree days 1 to 5 years following fire. In mature black spruce temperature ranges from 500 to 800 degree days. The active layer thickness in later stages of this succession is usually from 40 to 60 cm. Following fire it increases for several years and may reach 1 meter in thickness. In the south aspect white spruce sequence, soil temperature approaches 1400 degree days in heavily burned forests in contrast with 900 to 1000 degree days in mature white spruce.

On the black spruce successional sequences, cold soil temperature results in lowest rates of soil biological activity, nutrient cycling, and, in turn, lowest tree productivity. Net annual above-ground tree production ranges from about 7 metric tons per hectare on the warmer, south aspect hardwood-white spruce sequence to less than 1 metric ton per hectare on the black spruce sequence.

INTRODUCTION

There are many patterns of succession in the boreal forest related to such factors as type of disturbance and the state factors of ecosystem development (Jenny 1980). In this paper we compare secondary succession in upland forests developing on south compared with north aspects. Upland forest communities largely reflect secondary successional processes dominated by fire. Topography is an important state factor controlling nutrient cycling and succession. Warm, south aspect sites support productive, more rapid nutrient cycling communities in contrast to cold, low-productivity, slow nutrient cycling in north slope communities. Along with fire, permafrost is an additional physical control of nutrient cycling and productivity on cold aspects. Fire acts as a rapid decomposer and, depending on intensity of burn, replenishes or depletes nutrient pools. Early successional stages reflect this geochemical control with regard to nutrient cycles. Permafrost acts as an effective seal to inhibit drainage and nutrient loss. Community development may proceed through stages of rapid growing willow, alder, aspen and birch with subsequent dominance by nutrient conservative white spruce or black spruce. Depending on burn severity and seed source, black spruce may replace itself on cooler level or north aspect sites. Wildfire frequency of less than 100 years makes determination of the climax vegetation difficult and hypothetical, but there is a definite tendency in the older black spruce stands toward increased development of permafrost and moss and lichen layers with a corresponding drop in density and productivity of the black spruce.

Regardless of successional track, nutrient cycling proceeds from a state of available soil nutrient pool enrichment for P and cations following fire to nutrient depletion as these elements are accumulated in biomass. During this time period nitrogen increases, as reserves lost in fire are replenished, mineralized, then gradually decline as N is accumulated in biomass.

The buildup of a thick organic mat which results in colder soil temperatures and a tying up of the available nutrients is important in determining the productivity and the vegetation that ultimately occupies the site. The presence of permafrost, the slow decomposition rates, the increasing importance of moss layer productivity compared with tree productivity, the palludification of some sites, the development of a bog and forest cycle, and the high frequency of fires are all factors that have led to a questioning of the classical concepts of succession and climax by many northern ecologists. These concepts will be discussed in this paper in relationship to the nutrient cycle and ecosystem processes. Methods employed in soil and vegetation sampling and in chemical analysis of these materials are found in Van Cleve et al. (1981), Van Cleve et al. (in press) and Viereck et al. (in press).

VEGETATION

The sequence of secondary succession on contrasting upland sites has been described in some; detail in Van Cleve and Viereck (1981). On dry upland sites, especially south-facing slopes, the mature forest cover type is white spruce, paper birch, aspen, or some combination of these species. On contrasting cold, north-facing slopes the dominant forest type is black spruce although extensive birch and willow stands may, at times, follow fire on these sites. Succession following fire usually follows distinctly different sequences on the two contrasting sites as is illustrated in Figures 1-4.

UPLAND WHITE SPRUCE SUCCESSION

In the typical fire sequence on dry upland sites the following stages have been recognized and described (Figure 1). Following the fire and an intervening few days to weeks when no live vegetation is present (Stage I), there develops an herb and seedling stage (Stage II). The usual first invasion is by light seeded species such as fireweed (Epilobium angustifolium) and willow shrubs, especially Salix bebbiana Sarg. and S. scouleriana Barratt. A number of species resprout from stumps and roots, especially shrubs such as Viburnum edule (Michx.) Raf., Rosa acicularis and the willows, and aspen and birch. The next stage in the developing vegetation is dominated by shrubs (Stage III), primarily willows, and the deciduous tree saplings (Figure 1). Density of birch and aspen may be as great as 30,000 stems/ha and shrub cover as much as 50%. Spruce seedlings may be abundant at this time but their slow growth makes them relatively inconspicuous in the stand.

In the dense hardwood stage (Stage IV), young aspen and birch trees form a dense canopy that tends to shade out much of the understory that has developed since the fire. Heavy leaf litterfall prevents the development of a continuous moss layer but moss species of the mature spruce stages such as Hylocomium splendens and Pleurozium schreberi become scattered on mounded areas. This stage usually occurs during the period of 25-50 years following the fire (Figure 1). For the next 50 years deciduous trees dominate the site (Stage V), although white spruce may become conspicuous in the understory. Both the aspen and birch stands become more open and the density drops to about 700 trees/ha for aspen and 300 for birch.

At about 100 years after the fire, white spruce becomes dominant (Stages VI-VII), often with a components of birch trees. Tree densities are approximately 500/ha. As the stands become older, the hardwoods are less abundant. The greatest change is the development of a dense and thick layer of feathermoss. The mature spruce stage, with only scattered remnant birch and with a continuous moss mat, is reached at about 200 years following the fire (Figure 1).

UPLAND BLACK SPRUCE SUCCESSION

Revegetation following fire in the black spruce type is usually rapid (Figure 2). Stage I is brief: within weeks of the fire, sprouts from underground plant parts of shrubs and herbs are abundant. Even where burning is severe and all of the underground plant parts are killed, invasion by mosses and liverworts and the establishment of seedlings of light-seeded species is usually accomplished by the spring of the following year. The first stage of revegetation, the seedling-herb stage (Stage II), usually lasts from 2-5 years, depending on the site and fire conditions. Areas of bare mineral soil are covered with Marchantia polymorpha, Ceratodon purpurus, Polytrichum commune Hedw., Epilobium angustifolium and other light-seeded herbaceous species.
Spruce seedlings usually become established during this stage, often in the Marchantia and Ceratodon mats. Plant cover during this stage increases quickly from 0 to as much as 40 or 50%. It is during this stage that most of the vascular species that will continue on through the vegetation sequence are established (Figure 2). In the shrub stage (Stage III), the shrubs that have originated from sprouts and shoots continue rapid growth and dominate the vegetation, but the herb layer also continues to increase.

The tree canopy begins to dominate at from 25-30 years following the fire (Figure 2). The young stands of 40-60 years are dense (Stage IV), with as many as 4,000 to 7,000 trees and saplings/ha; and seedlings may be as dense as 12,000/ha. The most significant event that occurs during this stage is the invasion and rapid development of the feathermosses, Hylocomium splendens and Pleurozium schreberi, which along with some Sphagnum species, may develop as much as 50% cover. With the establishment of these mosses, there begins the development of a thick organic layer. Once the tree canopy is well established, the changes in the vegetation sequence are slower and more subtle. During the older stages, the tree canopy is mostly closed although the density is less, averaging 2,200 stems/ha. The moss layer remains about the same as in the younger tree stage except for an increase in feathermosses and a decline in both fruticose foliose lichens.

In Alaska, the development of the black spruce type into a mature stand of over 100 years follows without any major changes. Tree densities are about the same, 1,700/ha for black spruce, and a few paper birch may persist into the mature stage. The spruce tend to be in clumps produced by layered branches, and there are more openings in the canopy. Because of this, the shrub layers, especially the low shrubs, are better developed than they are during the 60- to 90-year period when the canopy is more closed. Moss cover in mature stands is dense and covers nearly 75% of the forest floor, but the lichen cover continues to decline on these mesic sites so that total lichen cover of both foliose and fruticose forms average only 2%.

SOIL TEMPERATURE AND FOREST FLOOR THICKNESS

In general, compared with the upland white spruce, the stages comprising the upland black spruce successional sequence display the thickest forest floors. As a result, black spruce ecosystems are characterized by the coolest, wettest forest floors and mineral soils (Van Cleve et al. 1983). These conditions dictate that fire removes a portion of the accumulated forest floor in black spruce but seldom burns all of this surface organic layer. In contrast, fire on south-facing slopes consumes all or most of the organic layer.

Following fire in black spruce, the organic layer depth may be only 4-8 cm. The change in surface albedo results in soil warming and increased active layer depth. This phenomenon may last well into succession or until spruce is reestablished and feathermoss covers nearly 100% of the ground surface. The organic layer thickness in mature white spruce may approach 8-12 cm in contrast to 16-20 cm in mature black spruce (Figure 1a and 2a).

In the first year following fire, soil heat sums may reach 1,250 in black spruce and exceed 2,000 in white spruce. Temperature summation above 0°C for period May 20 to September 10 for the 10 cm soil depth). These values decline to between 500 and 800 and in excess of 1,000 degree days, respectively, in mature stages of the two forest types (Figures 1a and 2a).

No permafrost is encountered in the south aspect successional sequence. Active layer thickness in mature black spruce ranges from 40 to 60 cm. Following fire, this increases for several years and may exceed a meter, returning to original depths 25-50 years following disturbance when spruce and feathermoss are established.

PRODUCTIVITY

Our data base for standing crop biomass immediately following fire in the two successional sequences is not complete. Observations at one white spruce site 5 years following fire indicate a standing crop biomass of dead white spruce or nearly 29 kg·m^{-2} (Figure 1b). Estimates from black spruce ecosystems indicate that aboveground biomass may be reduced by up to one-half or more in tree and forest floor components.

Recovery of the south aspect ecosystems is rapid following fire and reflects sprouting from tree and shrub root systems (Stages I and II). Maximum biomass accumulation, about 25-30 kg·m^{-2}, occurs between 100 and 200 years (Stages V and VI, Figure 1b). Beyond 200 years this total will decline in the over mature forest because of tree death.

Recovery of black spruce ecosystems is slower following fire. Fifteen to sixteen years following fire, live standing crop of aboveground tree biomass may equal 340 g·m^{-2} (Figure 2b). This is approximately 6% of the tree biomass accumulated in black spruce at 100 years, and 50% of the biomass accumulated in postfire aspen (688 g·m^{-2}) at about 20 years (Van Cleve and Oliver 1982). Maximum standing crop of tree biomass, aboveground, is also attained between 100 and 200 years but may reach only 10 kg·m^{-2}, 1/3-1/2 the maximum accumulated in the white spruce sequence.

A large portion of the organic matter in mature black spruce (Stages V and VI) is encountered in the forest floor (Figure 2b). The forest floor may contain 3-4 times the biomass encountered in tree tops plus roots (13 vs. 3.4 kg·m^{-2}; Van Cleve et al. 1981). In white spruce ecosystems the forest floor does not appear to comprise more than 1/4-1/3 the biomass (6-8 kg·m^{-2}) accumulated in aboveground tree parts (Figure 1b).

Mineral soil is an additional important organic matter reservoir. In black spruce ecosystems, mineral soil carbon content ranges from 4 to nearly 10 kg·m^{-2} compared with 3-6 kg·m^{-2} in mineral soil across the white spruce successional sequence. A significant correlation exists between soil carbon content and heat sum ($r = 0.72$, $\alpha = 0.05$) for stands sampled across both successional sequences (Viereck and Van Cleve, in press). Carbon content declines at the rate of 5.4 g·m^{-2} per 1 degree day increase in heat sum.

Probably the greatest contrast between these two successions lies in annual production. Maximum annual aboveground tree production in white spruce succession, 700 g·m^{-2} in Stage IV aspen, is nearly 5 times greater than maximum production (150 g·m^{-2}) encountered in Stage VI black spruce (Figures 1c and 2c). Mature white spruce (Stage VI) annually produce over twice as much biomass aboveground (390 g·m^{-2}) compared with black spruce. An additional contrast between the two successions lies in the fact that in mature black spruce successional stages (VI), bryophytes may annually produce nearly the same, or slightly more, organic matter than that produced by the trees aboveground. For the mosses this may amount to 120 g·m^{-2} compared with 100 g·m^{-2} for the trees. In mature stages of the white spruce succession, bryophytes may annually produce 1/3 the organic matter produced aboveground by trees. In earlier stages of this succession, the nonvascular plants are an insignificant component of the forest ecosystem.

Across the combined sequences, annual aboveground tree production is closely related to soil heat sum ($R^2 = 0.93$). Low productivity (< 200 g·m^{-2}), black spruce sites correspond with 700 degree days or less. Higher productivity aspen, birch, and white spruce (> 300 g·m^{-2}) are found on sites where the heat sum ranges from 800 to 1,300 degree days.

Vascular plant litterfall does not exceed 40-50 g·m^{-2} per year in black spruce compared with maxima exceeding 200 g·m^{-2} in the white spruce succession (Figures 1c and 2c).

NUTRIENT CYCLING

In nearly every category of ecosystem structure and function, with the exception of nonvascular plants and the forest floor, the various stages of the white spruce succession show a greater accumulation and a more dynamic flow of chemical elements through system compartments compared with the black spruce succession. Responding to greater annual production, more nutrient elements are accumulated at a faster rate in trees of the various stages of the white spruce succession (Figures 3 and 4). For example, in advanced stages of succession, approximately twice as much N (25 $g \cdot m^{-2}$) and P (3 $g \cdot m^{-2}$) and 4 times as much K (23 $g \cdot m^{-2}$) is accumulated in white as in black spruce trees. On the average, between 7 and 9 times more N (4.3 $g \cdot m^{-2}$), P (0.5 $g \cdot m^{-2}$) and K (2.4 $g \cdot m^{-2}$) are found in annual increment of trees across the white spruce succession. Furthermore, between 3 and 4 times more N, P, and K is required to support annual aboveground white spruce compared with black spruce tree production (Figures 3a-3c and 4a-4c).

The more dynamic nature of south aspect ecosystems is further emphasized when nutrient flux in litterfall is considered. From 3 to 10 times more N, 5 to nearly 30 times more P, and 5 to 20 times more K is returned to the forest floor in vascular plant litter in the white spruce succession (Figures 3d and 4d).

The generally greater accumulation of forest floor biomass results in total pools of N, P, and K in this ecosystem compartment being approximately equal (63, 7, and 20 $g \cdot m^{-2}$, respectively) in the black spruce and white spruce successions. However, the average concentration of these elements in black spruce (0.8%, 0.09% and 0.2%, respectively) is significantly less than that encountered in white spruce (1.2%, 0.1% and 0.4%, respectively). The genetic horizons of the forest floor also tend to be most acidic in the black spruce succession with pH ranging from 5.5 in the 01 layer to 4.1 in the 022 layer. The 022 horizon in black spruce generally displays highest lignin concentration (15 to >20%) compared with forest types encountered in the white spruce succession (8-15%).

Mineralization of forest floor organic matter and release of chemical elements for recirculation within the ecosystem also displays marked contrast between the two successions. The residence time (pool size/input to pool) ranges between 2 and 3 times longer for biomass (50 years), N (61 years), P (99 years), and K (74 years) in the forest floor of the black spruce succession.

DISCUSSION AND CONCLUSIONS

From previous discussion it is obvious that marked similarities and differences exist between the two successional sequences. One overriding similarity is the cold-dominated nature of the regional taiga climate. This condition determines that temperatures favorable for biogeochemical activity may occur only during the 140-day period from mid-April to mid-September. Within this interval the 90-day period during which tree growth may occur may be restricted by insufficient water. This is especially the case from mid-June to mid-July, the period of most favorable temperature and light regimes.

The state factor topography (Jenny, 1980) modifies microclimate with the result that the north aspect black spruce sequence is the coolest and wettest (least solar radiation) while the white spruce sequence occurs on warmer, drier south aspects. The drainage restriction imposed by permafrost helps to maintain higher soil and forest floor moisture regimes in black spruce. Substantial nutrient reserves may be "locked up" in frozen soil, while leaching losses may occur as soil solution flows across the permafrost interface on north aspects. In well-drained south aspect soils, little or no moisture loss through mineral soil may occur because evapotranspiration leaves no moisture for leaching.

Observations of the seasonal course of soil moisture using the neutron probe and measurements of water and nutrient flux using tension plate lysimeters indicate little or no moisture or nutrient flux out of south aspect soil profiles which are up to a meter deep over bedrock. With the possible exception of spring runoff of snow meltwater over the frozen soil, and during an undetermined but probably short period (up to 5 years) following fire, the south aspect successional sequence may be essentially closed to nutrient loss.

It is clear that soil temperature, as modified by thickness of forest floor, exerts a dominant control over productivity and nutrient cycling regardless of successional sequence. The most productive forests and the greatest flow of nutrients which support tree growth are encountered in the warm, south aspect white spruce succession. In the black spruce succession, temperature interacts with forest floor chemistry (concentration of N, P, K, lignin, acidity) to slow organic matter decomposition with the result that substantially reduced supplies of important nutrient elements are available for tree growth. Furthermore, bryophytes are effective competitors for the limited nutrient supply. However, black spruce make efficient use of limited resources through long-term foliage retention (> 20 years) and internal recycling of nutrients within the tree.

REFERENCES

Jenny, H. 1980. The soil resource, origin and behavior. Springer-Verlag, New York. 377 pp.

Van Cleve, K., L. Oliver, R. Schlentner, L. A. Viereck and C. T. Dyrness. In press. Productivity and nutrient cycling in taiga forest ecosystems. Canadian Journal of Forest Research.

Van Cleve, K., C. T. Dyrness, L.A. Viereck, J. Fox, F. S. Chapin III, and W. Oechel. 1983. Taiga ecosystems in interior Alaska. Bio-Science 83:39-44.

Van Cleve, K., and L. K. Oliver. 1982. Growth response of postfire quaking aspen (*Populus tremuloides* Michx.) to N, P, and K fertilization. Canadian Journal of Forest Research 12:160-165.

Van Cleve, K., and L. A. Viereck. 1981. Forest succession in relation to nutrient cycling in the boreal forest of Alaska. Page 185-210 in West, D. C., H. H. Shugart and D. B. Botkin, eds., *Forest Succession, Concepts and Application*. Springer-Verlag, New York.

Van Cleve, K., R. Barney, and R. Schlentner. 1981. Evidence of temperature control of production and nutrient cycling in two interior Alaska black spruce ecosystems. Canadian Journal of Forest Research 11:258-273.

Viereck, L. A., and K. Van Cleve. In press. Some aspects of vegetation and temperature interaction in Alaska Forests. The Potential Effects of Carbon Dioxide-Induced Climatic Changes in Alaska. The Proceedings of a Conference. Fairbanks, Alaska, April 7, 8, 1982. School of Agriculture and Land Resources Management, University of Alaska, Fairbanks. Miscellaneous Publication 83-1.

Viereck, L. A., C. T. Dyrness, K. Van Cleve, and M. J. Foote. In press. Vegetation, soils and forest productivity in selected forest types in interior Alaska. Canadian Journal of Forest Research.

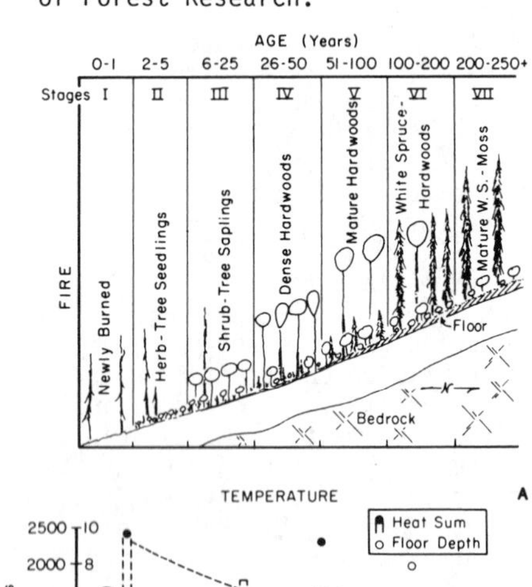

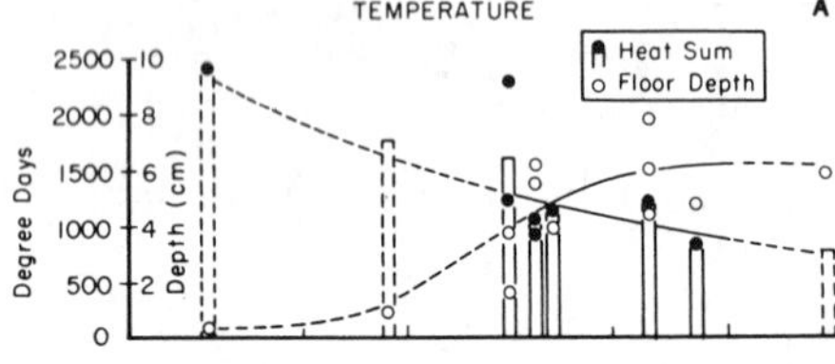

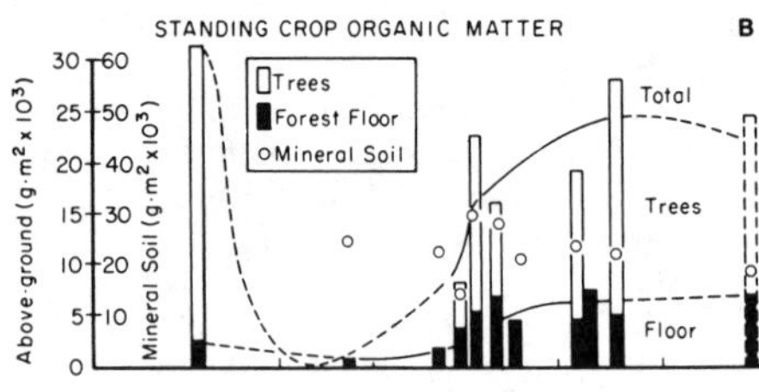

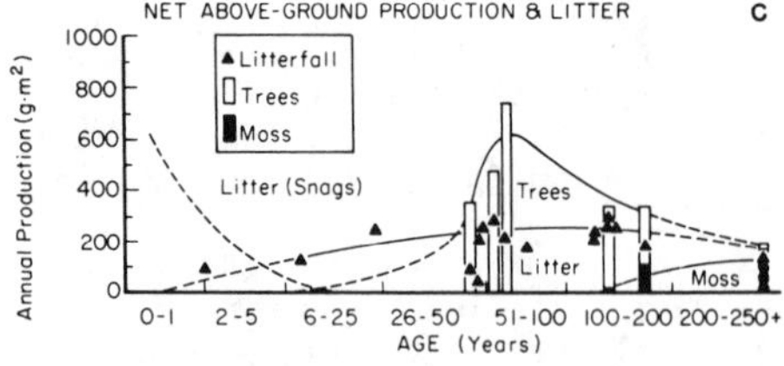

Figure 1. Upland white spruce succession.
A. Heat sum, forest floor depth.
B. Aboveground and mineral soil organic matter.
C. Vascular plant litterfall, net annual aboveground production for trees, mosses.

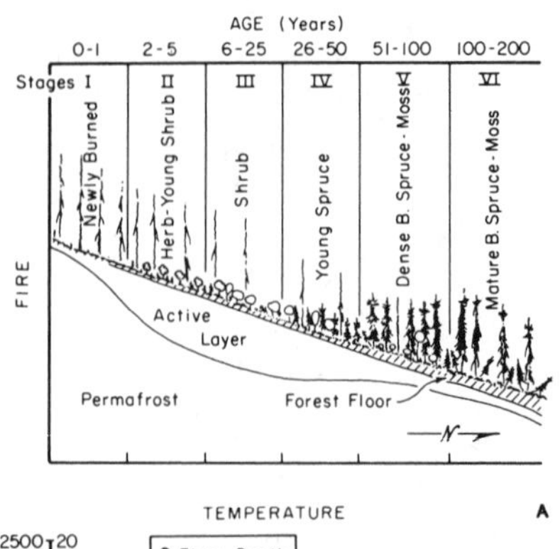

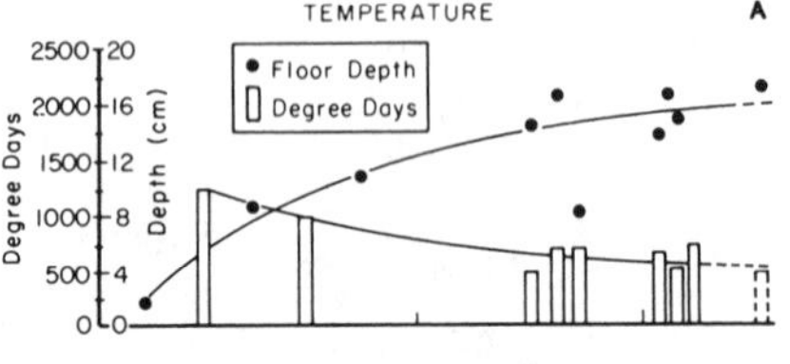

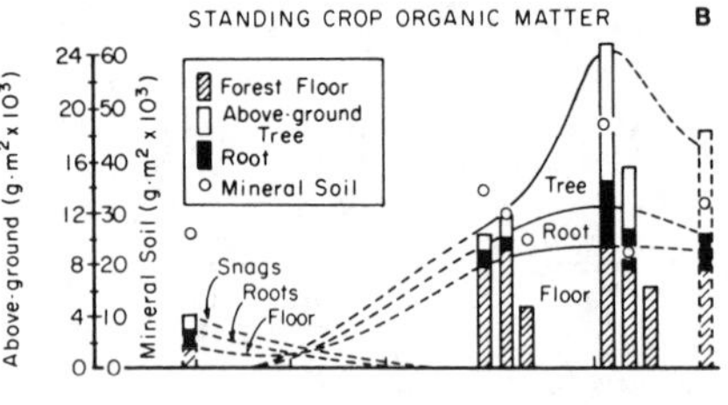

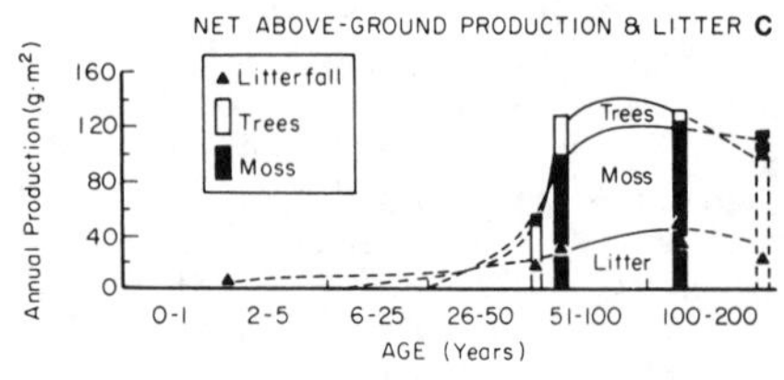

Figure 2. Upland black spruce succession.
A. Heat sum, forest floor depth.
B. Aboveground and mineral soil organic matter.
C. Vascular plant litterfall, net annual aboveground production for trees, mosses.

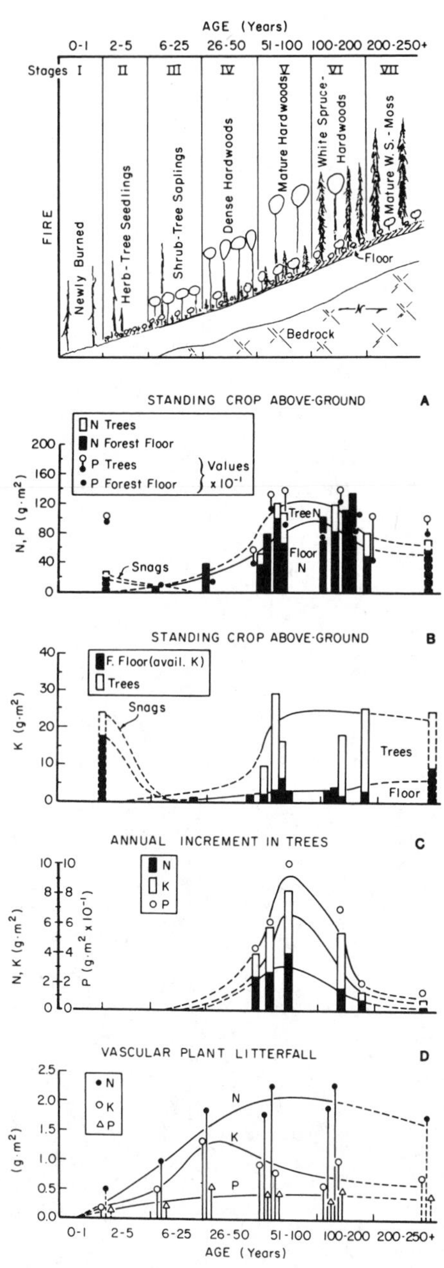

Figure 3. Upland white spruce succession.
A. Standing crop N, P.
B. Standing crop K.
C. Annual aboveground N, P, K increment in trees.
D. N, P, K in vascular plant litterfall.

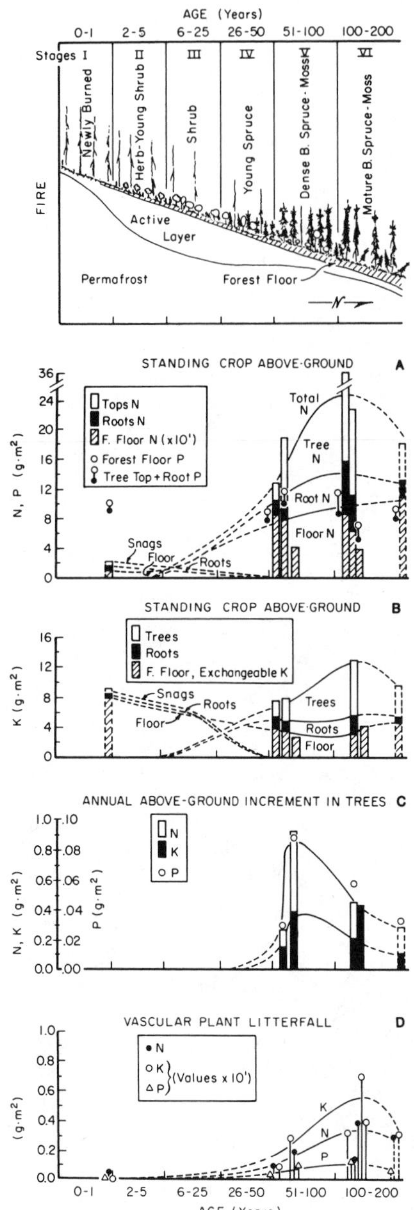

Figure 4. Upland black spruce succession.
A. Standing crop N, P.
B. Standing crop K.
C. Annual aboveground N, P, K increment in trees.
D. N, P, K in vascular plant litterfall.

GROUND MOVEMENTS AND DENDROGEOMORPHOLOGY IN A SMALL ICING AREA ON THE ALASKA HIGHWAY, YUKON, CANADA

Robert O. van Everdingen[1] and Harriet D. Allen[2]

[1]National Hydrology Research Institute, Environment Canada
Calgary, Alberta, Canada T3A 0X9
[2]Department of Geography, The University of Calgary
Calgary, Alberta, Canada T2N 1N4
Present address: Geography Department, Cambridge University, England CB2 3EN

Perennial groundwater discharge in a tributary of Donjek River causes severe icing problems at km 1817.5 (mile post 1130) on the Alaska Highway. Surveys of a section across the icing area revealed vertical ground movements of up to 0.92 m between winter and summer. Subsurface ice bodies up to 0.75 m thick appear to form in some winters, degrading in the following summer(s). The ground movements and the ice bodies are indicative of the formation and degradation of frost mounds. Airphotos of the study area indicate that icing activity was either induced or enhanced by construction of the highway. The distribution of reaction wood in white spruce growing on the valley bottom indicates several episodes of tilting, in various directions. The reaction-wood chronology shows that ground movements occurred in this area long before construction of the highway. Construction of the highway may, however, have affected the rate of groundwater discharge and the magnitude of icing and frost-mound activity.

INTRODUCTION

Problems caused by formation of icings along the Alaska Highway in Yukon and Alaska were first described by Eager and Pryor (1945) and later by Thomson (1966). It appeared that highway contruction in a number of instances either induced the icing formation or increased the extent of icing activity. Icing problems were found to be most prevalent and troublesome where the highway crossed small streams fed by perennial discharge of groundwater. One such crossing, where control of icing problems became very costly, has been studied since 1979 (van Everdingen 1982a).

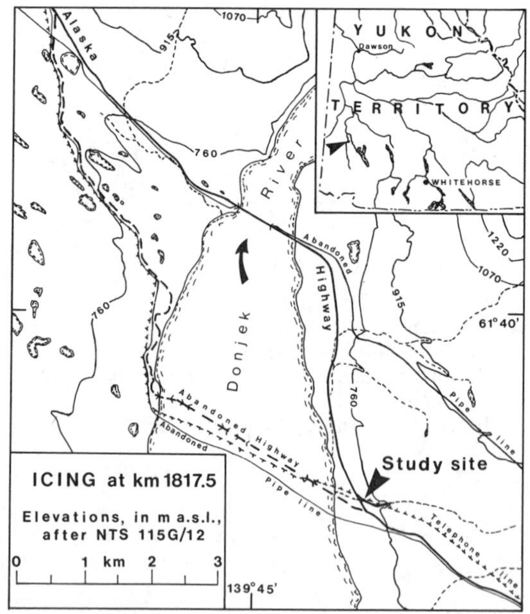

FIGURE 1 Location of study site.

The study site is located on the northeast edge of Shakwak Valley, at km 1817.5 (near old mile post 1130), about 4.5 km southeast of the crossing of Donjek River (Figure 1). The highway crosses a small tributary of Donjek River near the lower end of a stream-cut ravine about 740 m above sea level. The stream originates 2.5 km northeast of the icing site at an elevation of about 870 m; it discharges into Donjek River some 350 m west of the highway crossing at an approximate elevation of 710 m. The stream was called "Burlap Creek" by van Everdingen (1982a), because of the presence of a number of burlap fences put up for icing-control.

The original 1943 alignment of the highway did not cross Burlap Creek (see Figure 1). The existing alignment was constructed during the period 1949-1951 when the present Donjek River bridge was built. Observations in 1956 did not indicate any problem at the study site (S. Thomson, personal communication 1980), and serious icing problems, including plugging of the 1.4 x 1.6 m wooden culvert by ice, and icing extending across the highway, did not occur until the early sixties. In late 1981, the highway embankment was raised and widened, and new culverts were installed (see Figure 2).

The water supply to the icing area, the buildup of icing during the 1979-1980 winter (recorded by timelapse photography), and its maximum extent and thickness were described by van Everdingen (1982a). The presence of ground ice upstream of the highway, and vertical ground movements detected during 1980, led to the tentative conclusion that frost mounds develop locally beneath the icing. The present paper describes additional observations on the type and magnitude of ground movements in the icing area, followed by the results of airphoto and dendrogeomorphological studies aimed at establishing whether the occurrence of these ground move-

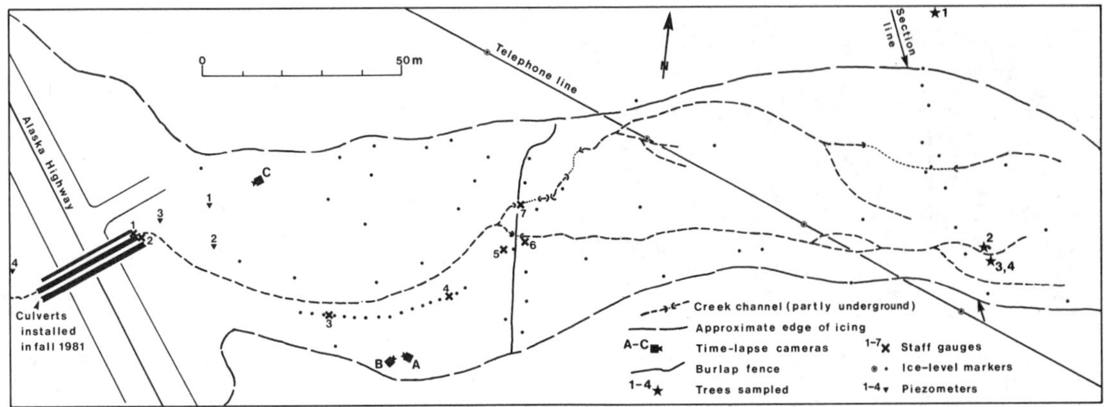

FIGURE 2 Sketch map of icing area at km 1817.5, Alaska Highway, Yukon.

ments is related to the construction of the highway.

GROUND MOVEMENTS AND EROSION

Surficial materials in the study area consist mostly of morainic deposits containing a high proportion of sand, gravel, and boulders. In the stream valley proper, gravelly till is overlain by tephra (White River ash) and peat. No bedrock exposures exist in the vicinity of the site.

A number of low mounds and low, wide ridges have formed in the icing area each winter since 1979. During spring and summer, fractures penetrating the peat and ash layers developed along the crests of the ridges (Figure 3). Massive ice was exposed in several places at the bottom of such fractures. A test pit dug in the fall of 1980 revealed 75 cm of clean ice, overlain by 40 cm of mixed peat and volcanic ash and 30 cm of peat; the ice rested on saturated sandy gravel (van Everdingen 1982a). Similar observations of ground movements and ground ice in icing areas were reported by Kane (1981) from Alaska.

FIGURE 3 Fracture along crest of low frost ridge, exposing volcanic ash and ground ice (small divisions on rod in cm).

Repeated level surveys have since been made of a section across the icing area some 200 m upstream from the highway crossing (Figure 2), using markers installed above the maximum icing level on a number of trees, to determine maximum thickness of the icing and the extent of vertical ground movements. The results of surveys in September 1980 and in March and July 1981, presented in Figure 4, show a maximum icing thickness in the section of 2.43 m, maximum vertical ground movement of 0.92 m, and maximum differential movement of 0.14 m/m. These observations confirm that frost mounds similar to frost blisters develop locally in the icing area.

Figure 2 indicates the approximate positions of gullies that are being eroded, through the peat and ash deposits to the water-bearing gravelly till or sandy gravel, by surface runoff (meltwater and groundwater discharge). Repeated observations indicate that the position of the gullies is controlled by the distribution of frost mounds and by the position of crest fractures in frost ridges.

The water discharged farther up the valley disappears underground again in a number of places (Figure 2), flowing through underground channels formed by subsurface erosion, or "piping", of volcanic ash. Small secondary deposits of the ash are found below the outlets of some of the underground channels (Figure 5).

During the early stages of this project, it was assumed that construction of the highway embankment across the stream valley caused gradual compaction that reduced transmissivity of the underlying unconsolidated water-bearing materials. This would

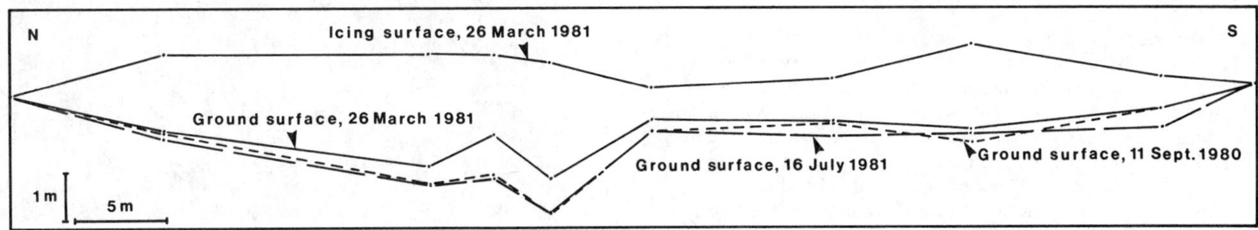

FIGURE 4 Cross section showing results of level surveys on 19 September 1980, 26 March and 16 July 1981.

FIGURE 5 Volcanic ash redeposited after erosion by water flowing below the organic mat.

have led to increasing water-table heights upstream of the highway and eventually to groundwater discharge, gullying, and formation of icings, and possibly, frost blisters. The latter phenomena have been observed earlier in similar situations along the Alaska Highway (Eager and Pryor 1945, p. 64) and Dempster Highway (van Everdingen 1982b, p.263). It is possible that deeper penetration of seasonal frost in the embankment contributes to the restriction of groundwater flow (Carey 1973, p. 31).

AIRPHOTO STUDY

As a first check on the assumed relationship between embankment construction and the observed groundwater-discharge phenomena, all airphotos available for the area were checked, starting with the most recent ones.

The only color photography of the area, taken on 4 April 1978 (Kenting NW 5478-4/5), shows minor icing activity over the first 200 m above the crossing. Photos taken on 29 May 1964 (A18395-97/99), show the icing extending as far as 340 m above the crossing. Photos of 30 July 1957 (A15728-112/114) were taken too late in the year to show any remaining ice, but the icing area is clearly recognizable on these photos.

Photos taken on 19 July 1949, before realignment of the highway for the new bridge (A11000-265/267), do not give any indication of icing activity. The earliest airphotos of the area (A1089-104-29/31), taken in 1942, before the start of highway construction, could not be obtained.

The results of the airphoto study suggest that icing activity (and possibly also ground movements) in the area started after realignment of the highway during the 1949-1951 period.

DENDROGEOMORPHOLOGY

The occurrence of frost mounds in the icing area affects the vegetation, as evidenced by numerous "drunken" white spruce (*Picea glauca* [Moench] Voss, Figure 6). Development of gullies can have similar effects on trees that start leaning towards the gradually widening channels. In addition, exposure of parts of their root systems may lead to suppression of growth. The unusual proportion of standing dead or dying trees in the icing area is likely the result of the annual submersion of root systems. Raup (1951) observed mortality of up to 25% of standing trees in similar situations elsewhere. Willows and alders, with higher inundation tolerance, are not noticeably affected.

The tilting of trees as a result of ground movements and gully development generally induces formation of *reaction* wood, in response to the unidirectional gravitational stress caused by the tilt. In gymnosperms (most coniferous species), the reaction wood develops around the downward side of the tilt as compression wood (Fritts 1971, Kramer and Kozlowski 1979).

Reaction wood appears in cross sections as dense, darker-colored, asymmetric or eccentric crescents that are usually widest in the first year after tilting. Both the annual width and the length of the horns of the reaction-wood crescents diminish in subsequent years as the tree recovers, until only normal wood is produced when the tree top is growing straight up again. This process will be interrupted if tilting recurs in either the same or a different direction. Multiple tilting in different directions will produce complex reaction-wood patterns, as well as the compound trunk curvatures displayed by the tree in Figure 6.

The presence of reaction wood in trees thus generally indicates that tilting has taken place, and should therefore make it possible to determine the timing and direction of tilt (e.g., Shroder 1980, Zoltai 1975). Samples from several white spruce in the icing area on Burlap Creek were collected in 1981 for a study of the distribution of reaction wood and relative ring widths, to determine whether such dendrogeomorphological evidence would reveal any correlation between highway construction and the occurrence of ground movements in the icing area.

Ring counts and width measurements were done in two or more directions on each cross-section sample (Figure 7), using 25 times magnification. The results are presented in Figure 8.

Tree No. 1 (Figure 7a and 8a), on the lower north slope of the valley, was sampled to provide background information on climatic effects that

FIGURE 6 Tilted tree No. 4 near the surveyed section, showing compound trunk curvature.

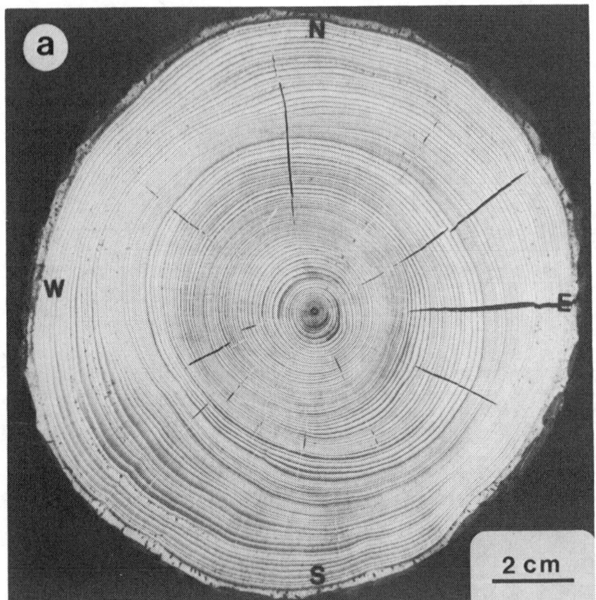

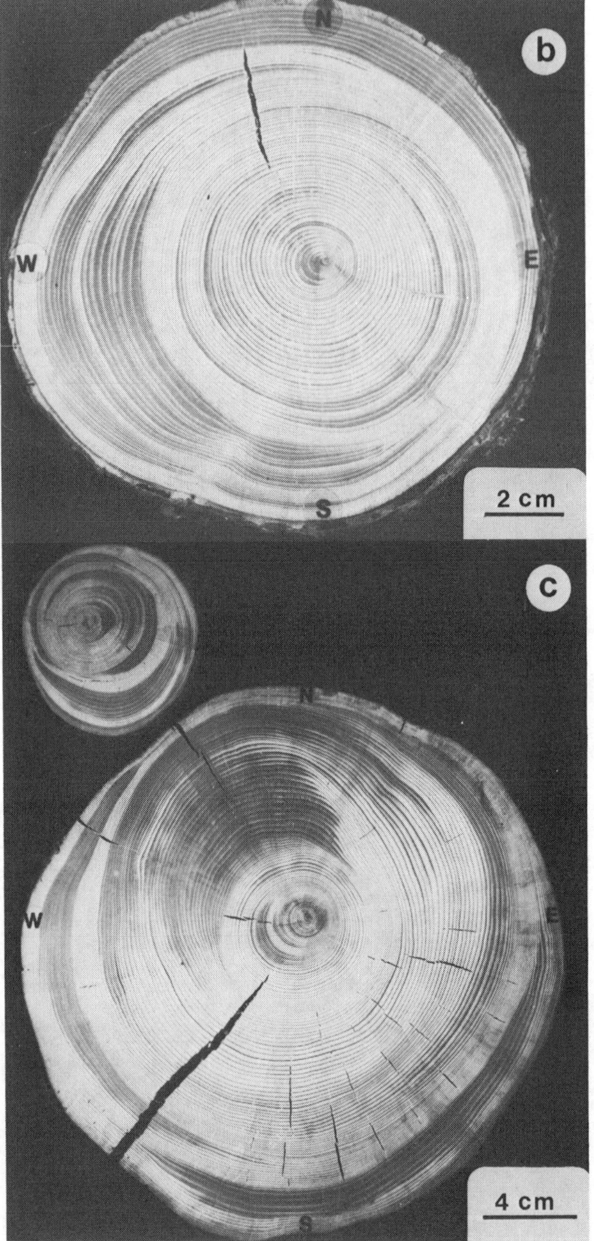

FIGURE 7 Cross sections of trees cut in the study area in July 1981 (top views). Locations indicated in Figure 2. (a) Tree No. 1. (b) Tree No. 2. (c) Twin trees No. 3 and 4.

might be reflected in variations of ring width in the trees in the icing area. Minor reaction wood in the southeast quadrant in Figure 7a suggests that this tree may have experienced slight tilting.

Tree No. 2 (Figures 7b and 8b), located on the north bank of one of the gullies, and trees No. 3 and 4 (Figures 7c and 8c), growing from a common bole on the south bank of the same gully, are examples of the influence of ground movements. The ring counts for the cross-sections of trees No. 3 and 4 differ by only 10 rings, and their tilting histories are similar; ring-width graphs are therefore shown for the larger stem only (Figure 8c).

The variations in ring width in tree No. 1 may reflect climatic influences on growth, but the curve does not correlate well with other chronologies derived from the Slims River area, about 95 km southeast of Donjek River, and used for dendroclimatology (Drew 1975, Allen 1982). Inspection of the minimum-width curves for trees Nos. 2, 3 and 4 shows that the variations in ring width in both tree No. 1 and the Slims River area are poorly represented by the trees in the icing area.

The distribution of reaction wood in trees No. 2 and 4 indicates several episodes of tilting, in widely varying directions. The two trees were most recently tilted in approximately opposite directions, away from the intervening gully, rather than towards it as might have been expected. This is interpreted as indicating the recent presence of a frost mound or ridge at the site of the gully; this interpretation is supported by the presence of ground ice below the peat on the banks of the gully.

The results further indicate that tilting of trees in the icing area occurred long before construction of the Alaska Highway. There is some indication in the tree-ring record, nevertheless, that highway construction may have had an effect on ground movements after 1949. Table 1 lists ratios between maximum and minimum ring widths for individual reaction-wood peaks indicated by letters in Figures 8b and 8c. The ranges of these ratios before and since 1949 show no overlap in either tree. In tree No. 2 the average ratio increased from 2.13 before 1949 to 5.64 after 1949; in tree No. 4 the average increased from 3.68 to 21.64.

The tree-ring records thus suggest that the magnitude of ground movements in the icing area increased following highway construction between 1949 and 1951. This likely reflects increased groundwater potentials in the icing area, which would contribute to the development of higher (and possibly steeper) frost mounds.

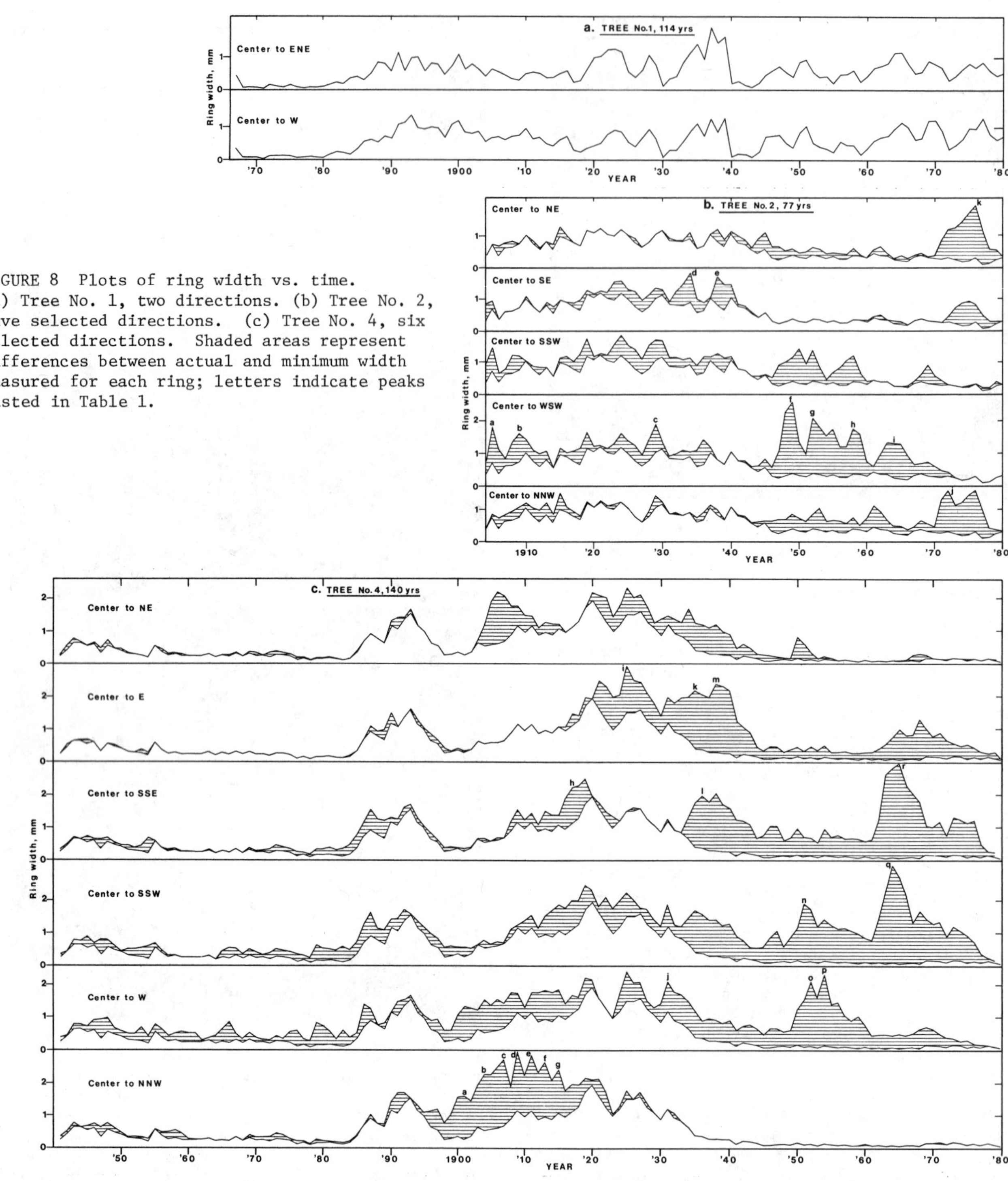

FIGURE 8 Plots of ring width vs. time. (a) Tree No. 1, two directions. (b) Tree No. 2, five selected directions. (c) Tree No. 4, six selected directions. Shaded areas represent differences between actual and minimum width measured for each ring; letters indicate peaks listed in Table 1.

RECENT CONSTRUCTION

In late 1981, the highway crossing of Burlap Creek was reconstructed. The roadway was widened from 8.5 to 13.7 m approximately, and the grade level was raised from about 2.9 to about 4.0 m above the level of the streambed at the upstream side of the highway; the basal width of the road embankment was increased from 20 m to 30 m. These changes can be expected to cause further compaction and reduced transmissivity in the underlying materials. Consequently, one can expect a further increase in the rate of groundwater discharge upstream of the highway, as well as increased icing activity and frost-mound development.

As part of the reconstruction, the wooden box culvert was replaced by two main culverts (1.4 m x 30 m, rolled section, steel) and an overflow culvert (0.60 m x 26 m). One of the large culverts has contained ice throughout 1982; the other contained some ice as late as July 22 in 1982. The effectiveness of the overflow culvert is limited,

TABLE 1 Ratios of maximum and minimum ring widths for selected reaction-wood peaks for trees in the Burlap Creek icing area.

Peak	First reaction ring	Peak year	Ratio
Tree No. 2 (Figure 8b)			
a	1905	1905	1.83
b	1908	1909	2.50
c	1928	1929	2.17
d	1932	1934	2.05
e	1938	1938	2.11
f	1947	1949	6.67
g	1951	1952	5.44
h	1958	1958	4.56
i	1963	1964	5.55
j	1971	1972	5.21
k	1974	1976	6.43
Tree No. 4 (Figure 8c)			
a	1900	1901	5.69
b	1900	1904	4.24
c	1900	1907	3.97
d	1900	1909	2.56
e	1900	1911	2.44
f	1900	1913	2.80
g	1900	1915	2.16
h	1916	1917	2.00
i	1925	1925	1.96
j	1931	1931	1.72
k	1934	1935	5.37
l	1934	1936	5.11
m	1934	1938	7.86
n	1950	1951	14.83
o	1950	1952	14.14
p	1950	1954	15.43
q	1962	1964	28.80
r	1962	1965	35.00

because its top is only 14 cm above the top of the active main culvert. It is therefore likely that changes in the magnitude of icing formation in the creek upstream of the highway will primarily reflect the influence of the increase in loading by the highway embankment, rather than any improvement in the drainage characteristics of the crossing.

CONCLUSIONS

Available airphotos suggest that icing activity in Burlap Creek above the Alaska Highway became noticeable only after rerouting of the highway across the creek in the period 1949-1951.

Dendrogeomorphological evidence revealed that tilting of trees in the icing area (resulting from growth and decay of frost mounds and ridges) was a recurring phenomenon long before the highway was built. Rerouting of the highway across Burlap Creek appears to have caused a large increase in the magnitude of the phenomenon.

The increased icing activity and frost-mound development resulted from an increase in groundwater potentials upstream of the highway, caused primarily by compaction of water-bearing materials underneath the highway embankment.

The recent modifications in the embankment at the crossing can be expected to enhance both the icing activity and the development of frost mounds.

In order to avoid creation of serious icing or frost-mound problems, northern highway (and pipeline) crossings of small streams fed by perennial groundwater discharge should be designed in such a way that their effect on upstream groundwater potentials is minimal. Installation of subdrains may be necessary in a number of cases.

ACKNOWLEDGEMENTS

J. A. Banner, National Hydrology Research Institute, assisted in the field work for the project. W. P. Hidinger and J. W. Flummerfelt, Yukon Territorial Government, provided information on highway construction and maintenance in the study area. The Kluane Lake Research Station, Arctic Institute of North America, served as a base for part of the field studies.

REFERENCES

Allen, H. D., 1982, Dendrochronological studies in the Slims River Valley, Yukon Territory: MSc. Thesis, The University of Calgary.

Carey, K., 1973, Icings developed from surface and ground water:U.S. Army Cold Regions Research and Engineering Laboratory, Monograph III-D3, 67p.

Drew, L. G. (editor), 1975, Tree-ring chronologies of western North America, vol VI, Western Canada and Mexico: Laboratory of Tree-Ring Research, University of Arizona, Tucson, Chronology Series 1, 30p.

Eager, W. L., and Pryor, W. T., 1945, Ice formation on the Alaska Highway: Public Roads, vol. 24 (3), p. 55-74, 82.

Fritts, H. C., 1971, Dendroclimatology and dendroecology: Quaternary Research, vol. 1, p. 419-449.

Kane, D., 1981, Physical mechanics of aufeis growth: Canadian Journal of Civil Engineering, vol. (8), p. 186-195.

Kramer, P. J. and Kozlowski, T. T., 1979, Physiology of woody plants: Academic Press, New York, 811p.

Raup, H. M., 1951, Vegetation and cryoplanation: Ohio Journal of Science, vol. 51, p. 105-106.

Shroder, J. F., Jr., 1980, Dendrogeomorphology: a review and new techniques of tree-ring dating: Progress in Physical Geography, vol. 4, p. 161-188.

Thomson, S., 1966, Icings on the Alaska Highway: Proceedings-Permafrost International Conference, National Academy of Sciences, Washington, D.C., NRC Publication 1287, p. 526-529.

van Everdingen, R. O., 1982a, Management of groundwater discharge for the solution of icing problems in the Yukon: The Roger J. E. Brown Memorial Volume, Proceedings-Fourth Canadian Permafrost Conference (H. M. French, Editor), National Research Council of Canada, p. 212-226.

van Everdingen, R. O., 1982b, Frost blisters of the Bear Rock spring area near Fort Norman, N.W.T.: Arctic, vol. 35(2), p. 243-265.

Zoltai, S. C., 1975, Tree ring record of soil movements on permafrost: Arctic and Alpine Research, vol. 7(4), p. 331-340.

ICE-WEDGE CASTS AND INVOLUTIONS AS PERMAFROST INDICATORS AND THEIR STRATIGRAPHIC POSITION IN THE WEICHSELIAN

Jef Vandenberghe

Institute for Earth Sciences, Free University, P.B. 7161, 1007 MC Amsterdam, Neth.

Ice-wedge casts and large size involutions are typical periglacial features in the Weichselian deposits of northwestern Europe. The ice-wedge casts consist of a central part with vertical lamination and a faulted marginal part. The casts are often arranged in a subparallel pattern, which represents an initial stage in the development of a true polygonal network. They are always closely associated with overlying intense involutions with amplitude of ca. 1.5 m. These involutions are often flat-bottomed and developed in a symmetrical way. They are explained as "periglacial load structures". Degradation of the ice-rich permafrost top caused a state of oversaturation in poorly drained areas. Consequently all cohesion was lost and a reversed density gradient originated. Other types of smaller sized deformations do not require the presence of a permafrost.

In the Weichselian sediment series two levels of ice-wedge casts connected with intense involutions are recorded. They point to permafrost conditions at the start of the Middle Weichselian and at about 20-25.000 years BP. In between these two periods climates were relatively milder.

INTRODUCTION

Relics of periglacial processes have been studied many times in the Late Pleistocene eolian deposits of Western Europe (references in Vandenberghe 1983). In this contribution only wedge structures and involutions that occur in the Weichselian coversand plains in northern Belgium and the southern part of the Netherlands will be considered (Figure 1). All these structures are thus formed in loosely packed eolian fine sands and loams which may be locally reworked.

Together with the pollen analyses and the sedimentary structures the periglacial phenomena may give definite indications on the paleo-environment and the paleoclimate. However, many periglacial phenomena reported from various publications are not well dated. Therefore the position of the periglacial processes must be fixed in a stratigraphic framework.

WEDGE CASTS

The conspicuous wedge casts in the study area consist of two genetically different parts: a lower and an upper one (Figure 2). The lower part extends over 1 to 1.5 m and can be subdivided in an external and an internal section. The internal section 15 to 25 cm wide shows a characteristic, wedge-like lamination with parallel arrangement. It is obviously filled with sediment coming from the adjacent layers and not brought in from the surface (e.g. by wind) at the time of the wedge formation. This may be derived from the sediment layering, while additional proof comes from sediment-petrological analyses. The external section consists of blocks which have been displaced downward stepwise along extension faults. It does not show any flow structures.

Vertical lamination and block faulting are limited to the lower wedge form of the cast. The upper part of the cast (ca. 1.5 m deep) shows no clear boundaries but reflects the flow pattern of the sinking part of an involution. Originally sedimentary structures are strongly disturbed, but generally a concave bending may be recognized (Figure 2). At the transition to the lower zone sediments which have sunk down penetrate further in the central section of the lower part of the wedge for a short distance (10 to 25 cm). Thus the two parts form a unique feature with a total depth of 2.5 to 3 m created in the same time span.

These wedge casts are rather common in Western-Europe and are generally interpreted as ice-wedge casts. However, the described individual features of the casts have to be explained in terms of processes, while the supposed former existence of ice wedges has to be demonstrated. Although Black (1976) states that "criteria to prove the former existence of ice wedges outside present permafrost regions are not available" the same author and Dylik & Maarleveld (1965) give a number of "indicators suggesting the presence of former ice wedges".

The existence of permafrost is Black's (1976) first criterion. The main group of wedge casts could be dated in Belgium and The Netherlands rather accurately as formed between 20 and 25.000 years BP (Vandenberghe 1983). During approximately the same period large pingo's existed in The Netherlands (De Gans 1981). They prove the presence of permafrost at that time in the concerned region. The existence of a permafrost table has also been proven by the large involutions penetrating to a common depth of ca. 1.5 m over hundreds or thousands of metres (Vandenberghe & Van de Broek 1982; next section).

One of the most remarkable features of the described casts is the blockwise movement to the centre. This needed a considerable cohesion of the de-

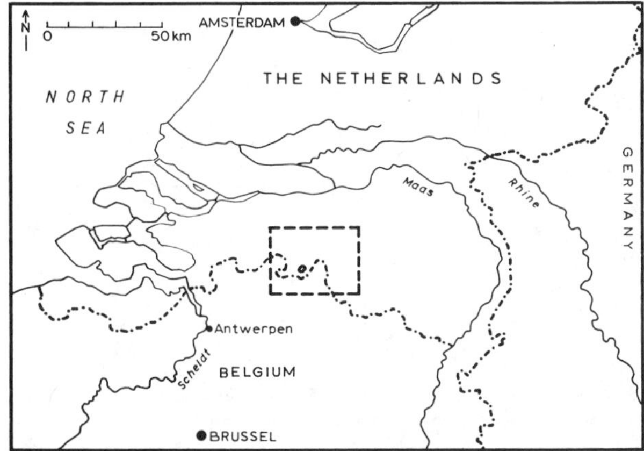

FIGURE 1 Situation map.

posits which often consist of loamless, loosely packed, water-containing sands. Apparently the movement as blocks was only possible in a solid state, which only could be a frozen state. As has been shown the displacements took place 1.5 to 3 m below the surface, this means in the permafrost because active layers did not reach this depth. On the other hand, the lowering in a graben-like structure points to a release of lateral pressure. This might be caused by tensional forces due to local collapsing or tectonic faulting nearby. However, such phenomena have never been observed in the vicinity of the wedge casts in the numerous exposures that have been investigated. A much better explanation for the pressure release may be found in the disappearance of the original wedge material. The start of the melting of ice in a wedge will induce tensile faulting in the still frozen ground. At the same time the ground surrounding the wedge will become liquefied and consequently fill the space left by the melting ice. As melting of the ice is an episodic feature, temporarily narrow fissures originate between the ice core and the adjacent layers giving rise to the characteristic vertical lamination of the lower central part of the cast. It does not seem to be disturbed noticeably by the sinking movement of the blocks. Therefore it is thought that the internal structures as well as the external ones develop progressively and rather simultaneously during the ice-wedge melting.

In accordance with Black (1976) flow and slumping of the surrounding saturated fine sediments must occur when the ice wedges are melting (lower central part and upper part). However, it has already been said that according to the field observations further infilling took place by block fall (external section). Moreover, one of the characteristics of ice-wedge casts reported by Dylik & Maarleveld (1967) are the "traces of collapsing from the wall, often combined with small faults". In contrast to the observed infilling structure, the degradation of supposed tensional cracks should show fluidization features in water-saturated sediment or cause homogeneization in dry, poorly cohesive sands. However, in both cases and in the described deposits block faulting is difficult to explain.

Another important criterion used by Black (1976) for recognizing ice-wedge casts is the local preservation of the characteristic upward bending of the

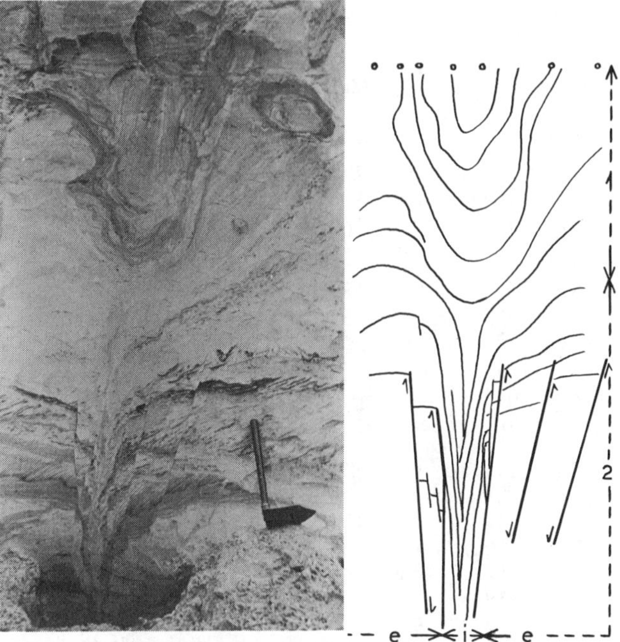

FIGURE 2 Ice-wedge cast with clear internal (i) and external section (e) in the lower part (2); concave bending in the upper (1) involuted part (from Vandenberghe & Krook (1981) with permission of 'Geologie en Mijnbouw'.

layers adjacent to the original ice wedge. However, Dylik & Maarleveld (1967) argue that downturned contact layers are characteristic of "fossil ice fissures". In the study area no fossil upturned structures have been found. In our opinion it is impossible that the upturned pressure structures should not be destroyed by flow and slumping in the space left by the disappearing ice.

Finally, both Black (1976) and Dylik & Maarleveld (1967) state that ice wedges are occurring in polygons with large diameter. Up to now three cases have been investigated. A first one, described by Gullentops & Vandenberghe (1981), reveals a clear polygonal pattern with distances of 7 to 10 m and rather orthogonal intersection. In two other cases, however, a subparallel arrangement of ice wedge casts is observed. These casts are completely of the same type as those in the first case. Analogous to the linear patterns of initial cracks on a progressively drying clay ("incomplete mud cracks"), the subparallel pattern of ice-wedge casts may represent the start of non-sorted polygon propagation. According to Lachenbruch (1962) "an orthogonal intersection suggests that one of the cracks predates the other". Where horizontal stresses are anisotropic initial cracks develop in a systematic array and only afterwards secondary cracks induce the orthogonal pattern (Lachenbruch 1962). From this reasoning it can be concluded that at some places the development of the polygonal pattern of Weichselian ice-wedge casts was only in its initial stage while elsewhere a real polygonal network already had been formed.

INVOLUTIONS

We have distinguished three categories of involutions which have a genetic significance and which are not recognized in other descriptive classifications (e.g. Maarleveld 1976).

a) folds of wide wavelength and small amplitude involving rather weak disturbances (a_1); irregular solitary structures (a_2);

b) large (generally 1.1 to 1.8 m high), strongly folded convolutions; they are often flat-bottomed, occur over larger areas and are developed in a symmetrical way;

c) small undulations and foldings (generally less than 0.6 m) which may occur isolated (c_1) or in groups (c_2); mostly regularly developed; single "tear-structures" are examples of the former group, while the well-known "druipstaarten" or "drop structures" are examples of the latter group.

a) The first category occurs often in situations where originally sediments were not laterally homogeneous (a_2) or the original lateral homogeneity has been disturbed previously e.g. by involutions of the second and third categories (a_1). This means that lithology and structure, and consequently the physical parameters, are changing in the horizontal section. In such a situation unequal cryostatic pressures may develop due to different freezing rates and volume changes. In this manner differential motions may be initiated. Examples of a_1-structures have been illustrated by Vandenberghe (1983) and a_2-structures by the experiments of Pissart (1970). Furthermore, this process requires a rather intensive freeze-thaw alternation and sufficient soil moisture. The significance of these structures for paleoclimatic reconstruction is rather small.

b) The repetitive and symmetrical development of the second category points to events with more than local significance. However, the original homogeneous horizontal layering precludes explanations where lateral heterogeneity is a necessary condition as in (a). Thus there is no reason to suppose lateral cryostatic pressures nor differential volume changes during the freezing to have produced the observed involutions. Furthermore, the deformations of this category show distinct flow characteristics which are difficult to explain if the intrusions should penetrate into frozen material (French 1976).

Load structures seem to be the best explanation (so-called "load casts" of Kuenen (1958)). Moreover the similarity of the observed field structures with the experimentally obtained load structures is striking (e.g. Anketell et al. 1970). However, two conditions have to be fulfilled for the process. Besides a reversed density gradient, it is necessary that the carrying capacity of the lower sediments is so small that the overlying material can sink down into it. Such a reversed density gradient was not demonstrated by density measurements on the actual sediment series, even when water saturated (Vandenberghe & Van den Broek 1982). Also the high shear strengths of the sediments in the present conditions, observed by the same authors, would oppose

FIGURE 3 Flat-bottomed involution; forms part of a regular series on top of a former permafrost.

any movement. Thus load structures require another setting than the present one. In this respect the flattened bottom (Figure 3) and the constant depth of many involutions are important indications. They point to the existence of an impervious layer. Moreover, in all studied cases water-saturated conditions - necessary to obtain reversed density gradients and low shear strengths - are difficult to imagine in the absence of an impervious layer. Remarkably, there are no lithological contrasts at the base of the involutions. Such a flat impervious horizon in otherwise permeable sediments points to the existence of former permafrost. This is confirmed by ice-wedge casts starting downward from the base of the convoluted zone.

As the relatively ice-rich top of the permafrost (Pollard & French 1980) is melting, large amounts of water are linerated. Then, in a flat landscape with fine sediments poorly drained oversaturated conditions result because the permafrost is acting as an impervious layer. When the meltwater volume exceeds porosity, the lower sediment is made less dense and a reversed density gradient may originate. Moreover, the excess pore water pressures cause the loss of intergranular contacts leading to liquefaction so that cohesion disappears. Thus the two conditions for load structures are present ("periglacial load structures"). Recently Washburn et al. (1978) obtained pocket structures due to vertical forces arising from natural density differences in thaw-frost experiments. Original density differences, however, are not a prerequisite as real involutions also occur in completely homogeneous sediments.

Undoubtedly close connection exists between ice-wedge casts and involutions (see also previous section and figure 2). At the top of a melting ice wedge a downward movement of the overlying soil will be initiated preferentially (Figure 2). The number of up- and downward moving cells is, however, not dependent on the presence of ice wedges provided enough water was produced at the top of the permafrost. Moreover, ice-wedge casts may be deformed, especially at their top, or completely disturbed. Consequently the number of ice-wedge casts is less than the number of original ice wedges and stands in no relation with the number of individual involutions. The origination of the involutions of this category in the yearly developing active layer is contradicted firstly by their amplitude which is larger than the supposed thickness of the active layer during the cold periods in the concerned regions (Maarleveld, 1981). Secondly, a recurrent involution process in the active layer leads to a completely homogenized, structureless upper layer contrary to the regular and symmetrical structures which are observed.

c) Small scaled but rather strongly folded involutions are also formed as load structures (c_1 and 2). However, there is no indication of the presence of permafrost (only small amplitude, no flattened bottom, no associated ice-wedge casts). Although the development of load structures requires conditions of oversaturation (see b), this situation may be seasonal and thus due to local and shallow frost and melt action.

Conditions of oversaturation due to melting (categories b and c) can only be realized in sediments of low permeability. This explains the intensity of involutions in loamy sediments with impeded drainage. On the other hand well-drained coarser sediments, even in the presence of permafrost, show only weaker and/or scarcer deformations.

STRATIGRAPHIC POSITION

As periglacial features are excellent paleoclimatic indicators, it is important to place them in a stratigraphic sequence (Figure 4). The qualitative estimate of the climatic environment of the Weichselian Pleniglacial is based on criteria derived from the periglacial phenomena (Maarleveld 1976) and on vegetation characteristics as revealed in pollen diagrams. Most striking is the occurrence of two stratigraphic levels of ice wedge casts which are associated with overlying large and intense involutions. They represent two periods with permafrost conditions (Vandenberghe & Krook 1981). On top of each level of involutions an extensive desert pavement occurs. The older cold phase is to be situated between 50.000 years BP and the Brørup-interstadial (Vandenberghe & Van den Broek 1982). This corresponds with the age of 62.000 to 70.000 years BP attributed to the cold phase at the beginning of the Middle Weichselian by Mook & Woillard (1982). The maximum age of the younger cold phase is 26.000 years BP, while it is older than the final development of the overlying gravel bed and the Upper Pleniglacial coversands on top of it. Probably this cold period corresponds to the maximum of the last big extension of ice sheets and glaciers dated at about 18 to 22.000 years BP by many authors. Between the two ice wedge levels only scattered periglacial phenomena are found: e.g. small involutions, narrow frost cracks, etc. The long interval between the two very cold periods was thus relatively milder with some minor oscillations. This has also been found at several occasions in North-America where the beginning of the Middle Wisconsin interstadial complex is dated at 65.000 BP. The period between ± 20.000 BP and the end of the Pleniglacial was in any case a very dry one, characterized by the total absence of fluvial activity and the presence of eolian sands and desert pavements. Although indications of permafrost -

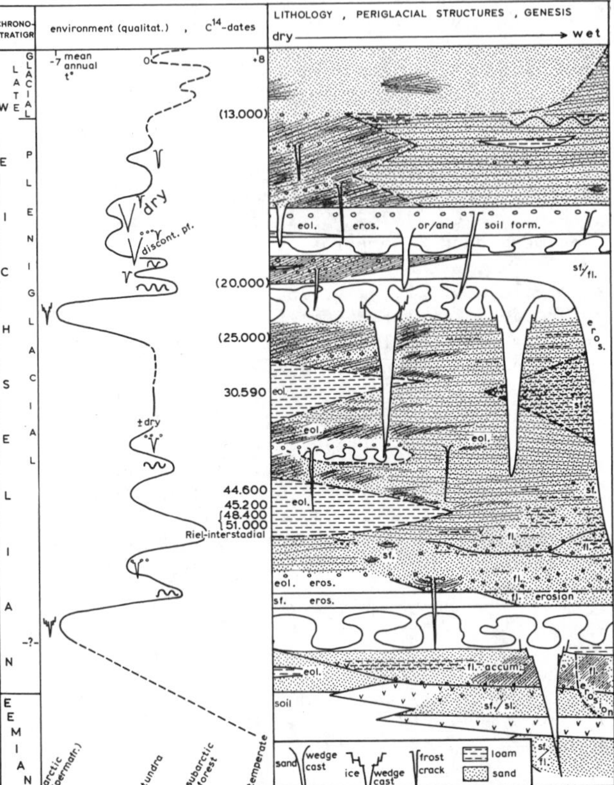

FIGURE 4 Stratigraphy of Weichselian deposits and environmental conditions in the southern Netherlands and northern Belgium.

apart from a few sand wedge casts - are missing, it was a very cold period. Probably for reasons of dryness and not for reasons of temperature all permafrost phenomena which need ice are absent. In accordance with Mook & Woillard (1982) the first cold maximum in the Weichselian corresponds with the deep-sea stage 4 (between 61 and 73.000 years BP), the succeeding long milder period with oygen-isotope stage 3 (between 29 and 61.000 years BP) and the second cold maximum with the middle and final part of stage 2 (29 to 11.000 years BP).

REFERENCES

Anketell, J.M., Cegla, J., and Dzulinski, S., 1970, On the deformational structures in systems with reversed density gradients: Annales de la Société Géologique de Pologne, v. 40, p. 3-30.

Black, R.F., 1976, Periglacial features indicative of permafrost: ice and soil wedges: Quaternary Research, v. 6, p. 3-26.

De Gans, W., 1981, The Drentsche Aa Valley system: Doctorsthesis, Free University, Amsterdam.

Dylik, J., and Maarleveld, G.C., 1967, Frost cracks, frost fissures and related polygons: Mededelingen van de Geologische Stichting, N.S. no. 18, p. 7-21.

French, H.M., 1976, The periglacial environment: London, Longman.

Gullentops, F., and Vandenberghe, J., 1981, Ramsel claypits, *in* Gullentops, F., Paulissen, E., and

Vandenberghe, J., Fossil periglacial phenomena in NE-Belgium: Biuletyn Peryglacjalny, v. 28, p. 350-352.

Kuenen, P.H., 1958, Experiments in geology: Transactions of the Geological Society of Glasgow, v. 23, p. 1-28.

Lachenbruch, A.H., 1962, Mechanics of thermal contraction cracks and ice-wedge polygons in permafrost: GSA Special Paper, number 70.

Maarleveld, G.C., 1976, Periglacial phenomena and the mean annual temperature during the last glacial time in the Netherlands: Biuletyn Peryglacjalny, v. 26, p. 57-78.

Maarleveld, G.C., 1981, Summer thaw depth in cold regions and fossil cryoturbation: Geologie en Mijnbouw, v. 60, p. 347-352.

Mook, W.G., and Woillard, G., 1982, Carbon-14 dates at Grande Pile: Correlation of land and sea chronologies: Science, v. 215, p. 159-161.

Pissart, A., 1970, Les phénomènes physiques essentiels au gel, les structures périglaciaires qui en résultent et leur signification climatique: Annales de la Société Géologique de Belgique, v. 93, p. 7-49.

Pollard, W.H., and French, H.M., 1980, A first approximation of the volume of ground ice, Richards Island, Pleistocene Mackenzie delta, NWT, Canada: Canadian Geotechnical Journal, v. 17, p. 509-516.

Vandenberghe, J., 1983, Some periglacial phenomena and their stratigraphical position in Weichselian deposits in The Netherlands: Polarforschung, v. 53, in press.

Vandenberghe, J., and Krook, L., 1981, Stratigraphy and genesis of Pleistocene deposits at Alphen (southern Netherlands): Geologie en Mijnbouw, v. 60, p. 417-426.

Vandenberghe, J., and Van den Broek, P., 1982, Weichselian convolution phenomena and processes in fine sediments: Boreas, v. 11, p. 299-315.

Washburn, A.L., Burrous, C., and Rein, R. Jr., 1978, Soil deformation resulting from some laboratory freeze-thaw experiments, in Proceedings of the Third International Conference on Permafrost, v. 1: Ottawa, National Research Council of Canada, p. 756-762.

CRYOHYDROCHEMICAL PECULIARITIES OF ICE WEDGE POLYGON COMPLEXES IN THE NORTH OF WESTERN SIBERIA

Yu. K. Vasilchuk and V. T. Trofimov

Moscow State University, Moscow, USSR

General conclusions are drawn from new data derived from 280 samples of wedge ice and 210 samples from the deposits bounding the wedges which were analyzed for the chemical composition of their dissolved salts. Slightly mineralized and strongly mineralized ice of Pleistocene and Holocene ages was identified. It is emphasized that saline sea water is involved in the formation of highly mineralized syngenetic wedge ice in the northern areas of Western Siberia.

INTRODUCTION

Most investigators believe that chemical composition of syngenetic ground ice is closely related to the hydrochemical conditions existing at the time of formation and serves as a reliable criterion for paleogeocryological constructions. The data on chemical composition of ice wedges of some regions of the USSR are, however, rather scarce (e.g. Volkova and Romanovskyi 1974, Kondratjeva et al. 1976, Lakhtina 1978, Gasanov 1981, Anisimova 1981). The data on hydrochemical composition of ice wedges of Western Siberia are also incomplete. They are discussed in the works of Danilov, Solomatin and Shmideberg (1980) and Trofimov et al. (1975). Because of this scarcity of chemical data, the authors analyzed the composition of water-soluble salts in ice wedges and in the deposits enclosing them from the Yamal and Gydan Peninsulas.

The primary objective of this study, undertaken in a vast and almost inaccessible area with few exposures, was to analyze the chemical composition of ice wedges from different thermal zones which were enclosed in deposits of different age and origin. To date, 280 samples collected from ice wedges and 210 samples from the deposits enclosing the ice wedges have been analyzed. In order to determine if variations in chemical composition, both in cross-section and along the length of an ice wedge occur, the samples were collected along vertical and horizontal transects and over a "net pattern." It was established that the ice is often divided into mineralogically distinct zones within the wedge form that vary along the length and cross section (Figures 1 and 2). This phenomenon is most pronounced within cross-sections characterized by a higher salinity. Configuration of the zones of different salinity is not always regular; these are often essentially asymmetric. Their occurrence is apparently due to successive changes in depositional and hydrochemical conditions at the time of ice wedge formation.

It should also be stated that localization of the area of ice wedge cracking over a period of time long enough to create a zone of a certain salinity is needed to form such zones in the body of a wedge. Without such a localization, the zones with distinct differentiation of salinity are not formed because ice of small, infilled cracks either entirely "desalinate" the ice of a wedge (if these are fresher) or increase its general salinity.

RESULTS

Low salinity of ice wedges has been confirmed by most data from previous studies. This has led to the general conclusion that the salinity of ice wedges does not exceed 0.1 g/ℓ (Anisimova 1981). Although most ice wedges analyzed in this study are characterized by a rather low salinity, some sections exhibit higher mineralization in syngenetic ice wedges (Figure 3).

Ice wedges with an overall mineralization of more than 0.2 g/ℓ accounts for more than 10% of those in the north of Western Siberia (Figure 3B). Further, in 28% of the samples, the content of the chlorine ions exceeds 0.02%, while in 8% of the samples it is higher than 0.1% (see Figure 3C). The two characteristics, namely, high mineralization and a high chloride content, are probably the most convincing indicators of paleo-conditions. According to other components, whose distribution is also heterogeneous, the indication of paleo-conditions of ice wedge formation is complicated and less distinct. The content of sodium and potassium ions is particularly significant—it is less than 0.02% in more than 80% of the samples (see Figure 3D).

In spite of the rather rare occurrence of salinity variations in ice wedges, such variations should not be neglected, because they are direct indicators of marine or lagoonal marine regimes of sedimentation during their formation. It is also interesting that the relationship of more and less saline ice in Pleistocene and Holocene syngenetic wedges is somewhat different.

Thick, Late Pleistocene syngenetic ice wedges occurring in II-IV sections of marine and lagoonal marine (bay) terraces in the north of Western Siberia are characterized by low salinity. According to Pitjeva's classification (1978), all are fresh, mainly of hydrocarbonate- or chloride-sodium composition. Mineralization of ice increases from 0.02 to 0.1 g/ℓ in wedges occurring in sediments of marine-lagoon deposits, to 0.1-0.5 g/ℓ in those of marine deposits. Zones of different mineralization were isolated within Late Pleistocene syngenetic wedges, otherwise characterized by a rather low salinity. For example, in a thick ice wedge located in Late Pleistocene strata of organic and mineral deposits ranging in age from 30,000 to 22,000 radiocarbon years BP near the Seyakha settlement in the center of the East Yamal, mineralization increases upwards from 0.04-0.08 to 11-0.14 g/ℓ. This increase may indicate an increase in the amount of saline bay water forming the syngenetic wedge. A comparable increase occurs in analysis of aqueous extracts from organic and mineral deposits enclosing the wedges. The upwards increase of mineralization from 1.02 to 1.18-1.60 g/kg is mainly the result of higher contents of sodium and potassium ions, which are typical salts in sea water.

Among the Late Pleistocene wedges, those with low mineralization are rather common. Even mineralization of syngenetic ice wedges is often low in the strata of marine terraces; the amount of dry residue does not exceed 0.05 g/ℓ. The lowest mineralization is found in the upper parts of wedges, including relict Late Pleistocene wedges and Holocene epigenetic ice that was formed more recently (Trofimov and Vasilchuk 1982, Vasilchuk 1982). These facts are important when analyzing ice wedges in the field. As is well known, the upper "desalinated" part of wedges is most frequently sampled in exposures, wells and some prospect holes.

On the whole, the salinity of ice wedges enclosed in Holocene strata is somewhat higher than those in Late Pleistocene strata (Figures 3 and 4). On the other hand, it varies greatly in the ice of wedges formed in the alluvial deposits and, on the other, in marine and lagoon deposits (Figure 4).

Study of Holocene syngenetic wedges in floodplains and the lowest level river terraces revealed that these are fresh in the upper reaches of rivers, their mineralization being lower than 0.1 g/ℓ (Figure 5A). Downstream, salinity tends to increase to a considerable extent. In the mouths of rivers in the strata of the lowest level terraces of laidas (lagoon-marine) and floodplains, the mineralization of ice in wedges is often more than 0.2 g/ℓ, sometimes reaching values of about 0.8-1.2 g/ℓ (Figure 2). Wedges fall within the class of slightly mineralized ones, according to Pitjeva's classification (1978).

Cryohydrochemical analysis proved to be most informative in comparing the mineralization of ice wedges of floodplains and laidas (lagoon-marine) in the north of Western Siberia (Figure 4). The number of samples analyzed by the authors, from the wedges of these two types of Holocene sediments, is approximately the same (Figure 4A). The occurrence of ice wedges with a value of dry residue exceeding 0.2 g/ℓ in the Holocene alluvial strata of floodplains proved to be close to zero (Figure 4B); they are, however, encountered in more than 22% of the samples taken from Holocene lagoonal-marine and marine strata of laidas (Figure 4C). Mineralization exceeding 0.4 g/ℓ was found in 16% of the samples collected from the strata of laidas. There is no doubt that a value of mineralization exceeding 0.2 g/ℓ should be regarded as a direct indication of marine or bay water participation in the formation of wedge ice. But, it is also rather common that rather fresh syngenetic ice wedges occur even in the relatively saline Holocene marine deposits (Figure 5B).

To explain this phenomenon, it is necessary to study the mechanism of formation of syngenetic wedges in the subaqueous regime. The solution of this problem may be found by comparing mineralization of syngenetic ice wedges presently forming on floodplains and laidas and that of the water sources which are likely to flow into the frost cracks. The analysis reveals that:

(1) Mineralization of ice filling the present frost cracks (those of the current year) on floodplains and laidas is often close to zero and in most cases does not exceed 0.02 g/ℓ. Ice fillings with higher mineralization are probably formed also, but none was encountered in this study.

(2) Mineralization of water from rivers and lakes ranges from 0.05 to 0.15 g/ℓ, increasing slightly in the mouths of valleys where salty lakes are likely to occur and river water may be strongly salinated during high tide. The salinity of the Kara Sea water if 7-16 g/ℓ, even near the coast.

(3) Mineralization of an ice cover, both within the Kara Sea coastal parts and the bays, is always less than that of the original water source, practically never reaching the value of 1 g/ℓ. The study of Saveljev (1980) explains this phenomenon. It should be noted that the ice samples were collected in summer during the period of ice destruction, i.e. when "freshness" of ice is the highest (Saveljev 1980). In autumn during the initial stage of ice formation, its mineralization may be much higher.

(4) Chemical composition of supragelisol water of the seasonally thawed layer has rather similar values of mineralization, varying from 0.07 to 0.14 g/ℓ. Sometimes, but not very often, however, lenses of water with mineralization of up to 3.5 g/ℓ are encountered at the base of the seasonally thawed layer (e.g. at the divides of the Pemakodayakha River headwaters, North Yamal (Trofimov et al. 1975).

(5) The amount of soluble salts in the deposits enclosing the wedges is generally much higher than that in the wedge ice. It is seldom lower than 1-2 g/kg in alluvial and marine lagoon deposits, but 2.5-5 g/kg (sometimes up to 10-20 g/kg) in marine (coastal-marine) deposits.

Proceeding from the above, the origin of differences in salt content of ice wedges in different depositional conditions are as follows. Within high terraces and divides, it is exclusively atmospheric water that freezes within the cracks in ice wedges (those of the epigenic type). On high floodplains and laidas, small ice wedges are also formed from atmospheric water flowing into the frost cracks (if the crack is open to the surface) or from the water of the seasonally thawed layer (in the case of intrastratal frost cracks). Water of higher mineralization may occasionally penetrate into the cracks when there is a salty lake nearby or as the result of an extremely active tide or, more often, a surge. Such tides and storm surges may occur only in summer when the sea or bay surface is free from ice. By this time, however, the majority of the frost cracks are already closed and thus, in only a few instances will this water freeze within the ice wedge.

The data obtained on cryohydrochemical

variability testify to the possibility of formation of syngenetic ice wedges under subaqueous conditions. In northern regions of the Yamal and Gydan peninsulas, polygonal ground has been repeatedly observed under the water surface on the bottom of shallow lakes. It is seen at a few localities beneath bays and the shoreline zone of the sea. Less often, polygonal ground occurs in the shoals of river channels. Undoubtedly, a similar situation was present in the Pleistocene and Holocene periods. But subaqueous polygonal ground may be inherited, being formed after flooding of sediments containing ice wedge polygons that had formed previously under subaerial conditions. This does not rule out the possibility of the subaquatic growth of wedges. A complex structure of syngenetic wedges in saline marine (lagoon-marine) sediments with their differentiation into separate wedges reflected in their general salinity (Figures 1 and 2) and the high content of water soluble salts (including chlorides), indicate that these syngenetic wedges may have been formed under subaquatic conditions in the nearshore zone of bays and seas.

Certainly, the question arises: why do syngenetic wedges that are so poorly salinated occur in marine, saline deposits? To discuss this question, one should consider that the frost cracking of wedges under a shallow sea takes place in winter when the overlying water layer is frozen. The ice cover may fracture above the wedges, forming open cracks that extend to the surface. As a result, the fractures may become filled with ultra-fresh snow that, on the one hand, prevents the penetration of sea water into the frost cracks and, on the other, greatly desalinates it. But it does not exclude the possibility that saline sea water may penetrate into the body of wedges along the fractures. It is known that the ice covering the sea contains a significant amount of brine. Of importance is the fact that the lowest temperature of freezing for $CaCl_2$ is -55°C (Saveljev 1980) and for NaCl it is -22.6°C. During a year, significant amounts of brine percolate into the sea ice during the process of metamorphism. As shown by the data collected by Saveljev (1980), a higher temperature in the ice cover during spring (March-April) leads both to an increase in the liquid phase and formation of capillary veins inside the ice through which intensive percolation of brine moves downwards. Chlorine ions are the first to migrate. Water of the brines often mixes with snow in the frost cracks, causing desalination. Passing lower into the wedge, it freezes as temperatures lower, forming a small elementary wedge or zone of ice whose salinity is one tenth or one fifteenth of that of the sea water (original) and one third or one fourth of the salinity of the ice cover. Its mineralization, however, is often one or two orders higher than that of atmospheric moisture that fills frost cracks under subaerial conditions. This occurs even after the water has flowed down the walls of frost cracks in rather saline rocks.

CONCLUSION

Although there are various factors preventing the formation of highly saline, syngenetic ice wedges, the results of this study indicate that rather saline (with mineralization exceeding 0.2 g/ℓ) syngenetic ice wedges occur in a series of outcrops of Late Pleistocene and Holocene sediments. The presence of saline ice wedges makes it possible to identify the genesis of the sedimentary strata as marine or lagoonal-marine and testifies to extremely severe climatic conditions at the time of deposition, otherwise subaqueous growth of ice wedges would have been impossible.

REFERENCES

Anisimova, N. P., 1981, Cryohydrochemical peculiarities of the permafrost zone [Kryogidrokhimicheskye osobennosti merzloi zony], Novosibirsk: Nauka, 153 p.

Danilov, I. D., Solomatin, V. I., and Shmideberg, N. A., 1980, Chemical composition of ground ice as an index of conditions of their formation and origin of enclosing rocks [Khimicheskyi sostav podzemnykh ldov kak pokazatel uslovyi ikh formirovaniya i genezisa vmeschayuschykh porod], Prirodnye usloviya Zapadnoi Sibiri, v. 7, Moscow: Izd-vo Mosk. universiteta, p. 119-126.

Gasanov, Sh. Sh., 1981, Cryolithological analysis [Kryolitologicheskyi analis], Moscow: Nauka, 196 p.

Kondratjeva, K. A., Trush, N. I., Chizhova, N. Ya., and Rybakova, N. O., 1976, Description of the Pleistocene deposits in the exposure of Mus-Khai of the Yana River [Kharacteristike pleisticenovykh otlozenyi v obnazhenii Mus-Khaya na r. Yane], Merzlotnye issledovaniya, v. XV, Moscow: Izd-vo Mosk. universiteta, p. 60-63.

Lakhtina, O. V., 1978, Physical and chemical properties of rocks of the Kolymskaya Lowland [Phyzicheskye i khimicheskye svoistva gruntov Kolymskoyi nizmennosti], Trudy PNIIIS, v. 54, Moscow: Stroiizdat, p. 13-54.

Pitjeva, K. E., 1978, Hydrogeochemistry (formation of groundwater chemical composition) [Gidrogeokhimiya (formirovaniye khimicheskogo sostava podzemnykh vod], Moscow: Izd-vo Mosk. universiteta, 328 p.

Saveljev, B. A., 1980, Structure and composition of natural ice [Stroeniye i sostav prirodnykh ldov], Moscow: Izd-vo Mosk. universiteta, 280 p.

Trofimov, V. T., Badu, Yu. B., Kudryashov, V. G., and Firsov, N. G., 1975, The Yamal Peninsula [Poluostrov Yamal], Moscow: Izd-vo Mosk. universiteta, 278 p.

Trofimov, V. T., and Vasilchuk, Yu. K., 1982, Cryolithogenesis of alluvial-marine deposits in the Yamal-Gydan province in Holocene. Abstracts of XI Congress of INKVA, v. 3, Moscow.

Vasilchuk, Yu. K., 1982, Regularities of development of engineering geology conditions in the north of Western Siberia in the Holocene period [Zokonomernosti razvitiya inzhenerno-geologicheskikh uslovyi severa Zapadnoi Sibiry v golocene], Avtoref. kand. giss, Moscow: Nauka, 27 p.

Volkova, V. P., and Romanovskyi, N. N., 1974, Chemical composition of ground ice in the Quaternary deposits in the southern part of the Yano-Indigirskaya Lowland [O khimicheskom sostave podzemnykh ldov v chetvertichnykh otlozheniyakh yuzhnoi chasti Yano-Indigirskoi nizmennosti], Problemy kryolitologyi, v. IV, Moscow: Izd-vo Mosk. universiteta, p. 199-202.

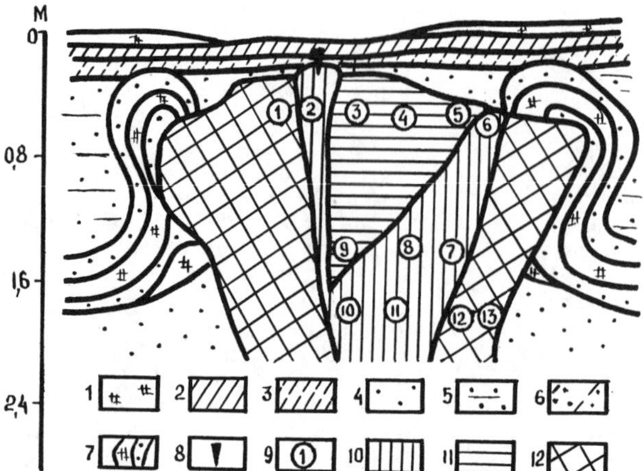

№ обр	Сухой остаток	HCO_3^-	Cl^-	$Na^+ + K^+$
1	0,14	0,02	0,03	0,01
2	0,07	0,01	0,02	0,02
3	0,09	0,02	0,01	0,01
4	0,11	0,01	0,02	0,02
5	0,10	0,01	0,02	0,02
6	0,07	0,01	0,02	0,01
7	0,07	0,02	0,03	0,02
8	0,07	0,01	0,03	0,02
9	0,08	0,02	0,03	0,02
10	0,06	0,01	0,02	0,01
11	0,07	0,02	0,02	0,02
12	0,60	0,01	0,32	0,16
13	0,64	0,02	0,35	0,17

FIGURE 1 Cross-section showing zones of ice of different salinity within a syngenetic ice wedge located in the Holocene deposits of the high laida of Gydan Bay, Eastern Yawaji, 0.2 km north of the Mongatalyangjakha River mouth. 1 - peat; 2 - loam; 3 - sandy loam; 4 - silty non-stratified sand; 5 - horizontally stratified sand; 6 - fine non-stratified sand; 7 - deformed member of interstratified sand and peat at the contact with the ice; 8 - ice of the presently growing small wedge; 9 - the place of sampling and the number of sample; 10-12 - ice of different mineralization: the value of dry residue is lower than 0.08 g/ℓ (10), ranges from 0.08 to 0.12 g/ℓ (11) and exceeds 0.12 g/ℓ (12).

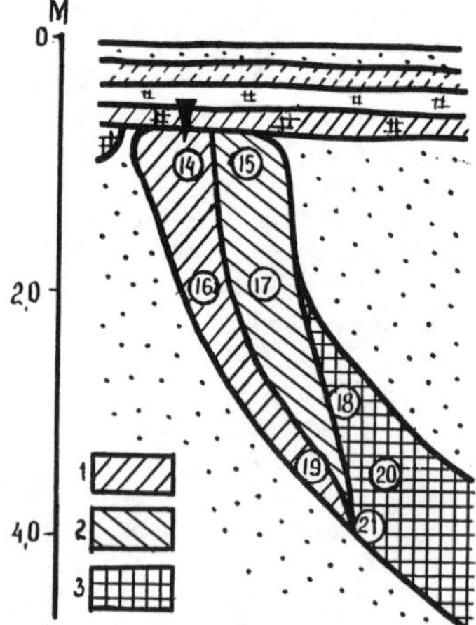

№ обр	Сухой остаток	HCO_3^-	Cl^-	$Na^+ + K^+$
14	0,46	0,02	0,22	0,13
15	0,71	0,02	0,34	0,20
16	0,48	0,01	0,23	0,14
17	0,58	0,02	0,27	0,16
18	1,24	0,02	0,64	0,37
19	0,43	0,02	0,19	0,11
20	0,82	0,05	0,40	0,23
21	1,08	0,04	0,51	0,30

FIGURE 2 Cross-section showing zones of ice of different salinity within a syngenetic ice wedge in the Holocene deposits of the lowermost marine terrace on Belyi Island (northwestern coast in the vicinity of the polar station). 1-3 - ice of different salinity: the value of dry residue ranges from 0.4 to 0.5 g/ℓ (1), from 0.5 to 0.8 g/ℓ (2) and exceeds 0.8 g/ℓ (3). The other symbols correspond to those of 1-9 in Figure 1.

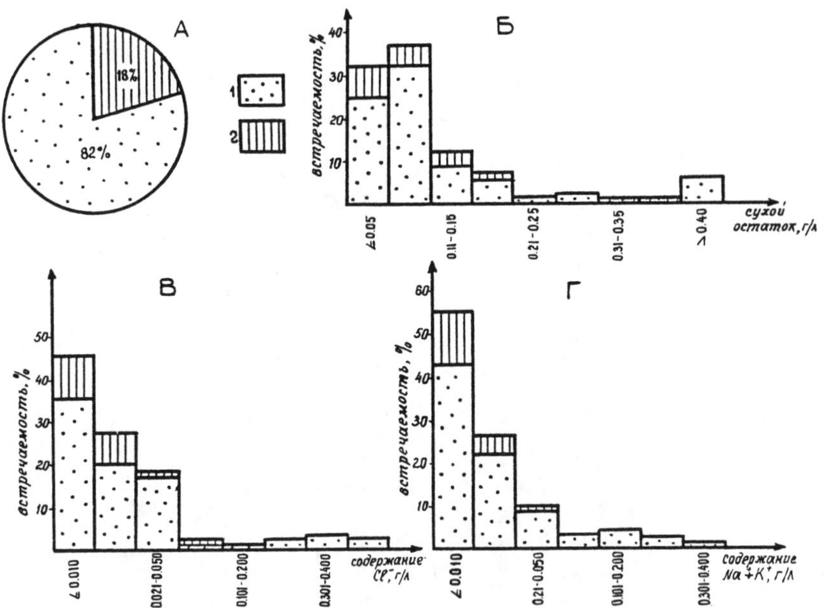

FIGURE 3 Relationship of the number of samples analyzed from the Pleistocene and Holocene ice wedges (A, total number of samples is 281) and histograms presenting distribution of dry residue values (B), chloride cations content (C), and total content of sodium and potassium anions (D). 1 - in ice wedges of the Holocene period; 2 - in ice wedges of the Late Pleistocene period.

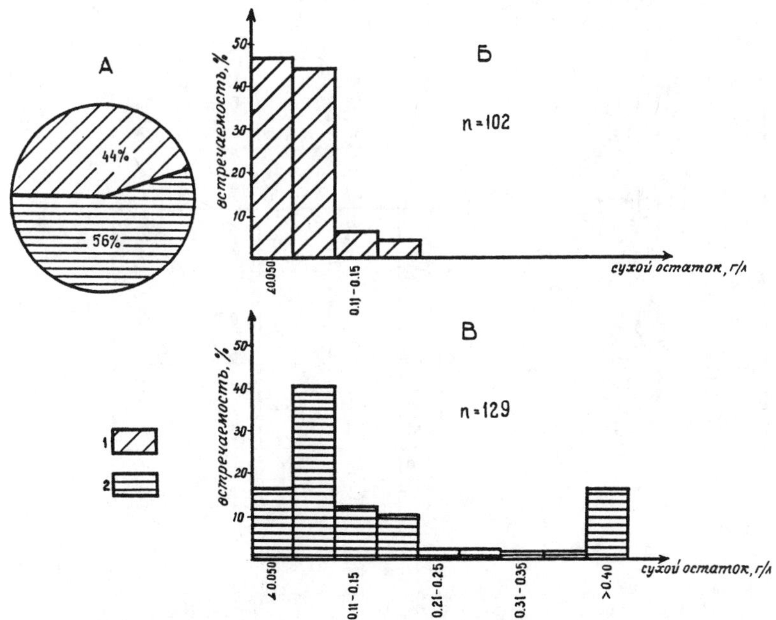

FIGURE 4 Relationship of the number of samples analyzed from ice wedges (A) in marine and lagoon-marine (1) and alluvial (2) Holocene deposits and histograms of distribution of dry residue values of ice wedges in alluvial (floodplain) strata (B) and in marine and lagoon-marine (laida) strata (C).

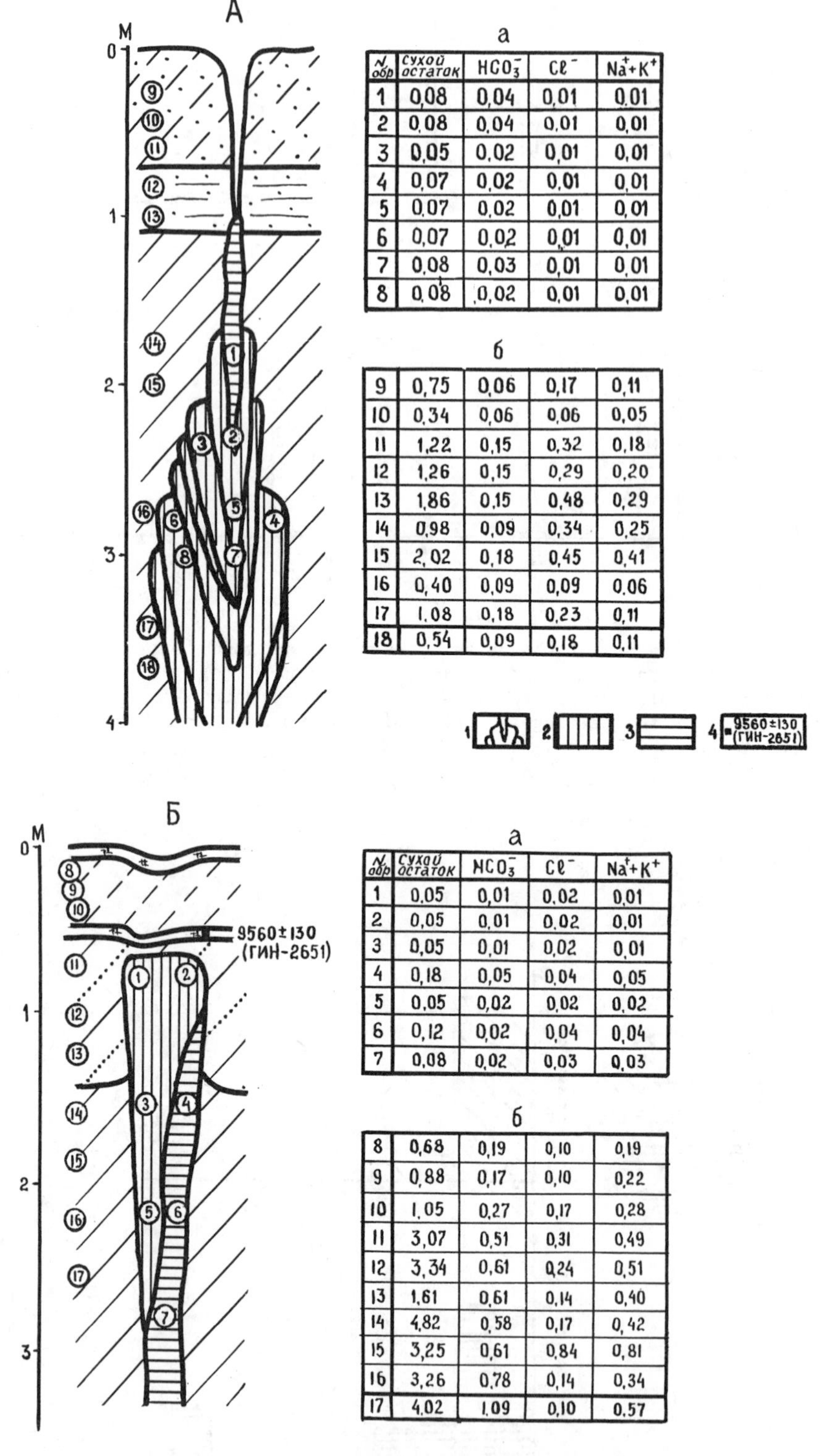

А

а

№ обр	Сухой остаток	HCO_3^-	Cl^-	$Na^+ + K^+$
1	0,08	0,04	0,01	0,01
2	0,08	0,04	0,01	0,01
3	0,05	0,02	0,01	0,01
4	0,07	0,02	0,01	0,01
5	0,07	0,02	0,01	0,01
6	0,07	0,02	0,01	0,01
7	0,08	0,03	0,01	0,01
8	0,08	0,02	0,01	0,01

б

№ обр	Сухой остаток	HCO_3^-	Cl^-	$Na^+ + K^+$
9	0,75	0,06	0,17	0,11
10	0,34	0,06	0,06	0,05
11	1,22	0,15	0,32	0,18
12	1,26	0,15	0,29	0,20
13	1,86	0,15	0,48	0,29
14	0,98	0,09	0,34	0,25
15	2,02	0,18	0,45	0,41
16	0,40	0,09	0,09	0,06
17	1,08	0,18	0,23	0,11
18	0,54	0,09	0,18	0,11

Б

а

№ обр	Сухой остаток	HCO_3^-	Cl^-	$Na^+ + K^+$
1	0,05	0,01	0,02	0,01
2	0,05	0,01	0,02	0,01
3	0,05	0,01	0,02	0,01
4	0,18	0,05	0,04	0,05
5	0,05	0,02	0,02	0,02
6	0,12	0,02	0,04	0,04
7	0,08	0,02	0,03	0,03

б

№ обр	Сухой остаток	HCO_3^-	Cl^-	$Na^+ + K^+$
8	0,68	0,19	0,10	0,19
9	0,88	0,17	0,10	0,22
10	1,05	0,27	0,17	0,28
11	3,07	0,51	0,31	0,49
12	3,34	0,61	0,24	0,51
13	1,61	0,61	0,14	0,40
14	4,82	0,58	0,17	0,42
15	3,25	0,61	0,84	0,81
16	3,26	0,78	0,14	0,34
17	4,02	1,09	0,10	0,57

FIGURE 5 Cryohydrogeochemical peculiarities of ice wedges within alluvial deposits in a floodplain in the middle reaches of the Tanama River, Central Gydan (A) and within marine deposits of the lowermost terrace on the western coast of Yamal Peninsula, 10 km north of the Kharasawaya River (B). a - chemical composition of wedge ice; b - chemical composition of deposits enclosing wedges; 1 - orientation of stratification in a wedge (enclosure of ice wedges of different texture); 2-3 - dry residue values of less than 0.08 g/ℓ (2) and more than 0.08 g/ℓ (3); 4 - location of the sample dated by the ^{14}C technique. Note: Dry residue content in aqueous extract from sediments is given in g/kg for comparison to mineralization (g/ℓ) in ice.

LONG-TERM USE OF FROST TUBES TO MONITOR THE ANNUAL FREEZE-THAW CYCLE IN THE ACTIVE LAYER

Leslie A. Viereck and Deborah J. Lev

Institute of Northern Forestry, USDA Forest Service,
Pacific Northwest Forest and Range Experiment Station
Fairbanks, Alaska 99701

ABSTRACT

In 1968, frost tubes were installed into the permafrost in a white spruce-black spruce/*Ledum groenlandicum*/*Hylocomium* stand near Fairbanks, Alaska, and the annual cycle of freezing and thawing was monitored from then through 1982. The annual freezeback of the active layer began in mid-October. The date the active layer was completely frozen varied with snow depth and air temperature: complete freezing was as early as December 13 and as late as January 28. Snow cover was usually permanent by October 8 and the last snow was usually melted by May 6. Thawing from the surface downward in the spring was directly related to the thaw index. Thawing began as early as April 20 and as late as May 7. Maximum thaw depth ranged from 54 to 61 cm, with an average of 58 cm. Footsteps of the observers compacted some vegetation along their route resulting in a slight depression in the lower surface of the active layer under their trail, but there was no significant difference in the thaw depths adjacent to the frost tubes after 14 years. Use of frost tubes is a reliable way to determine the annual freezing and thawing cycles in the active layer for periods of at least 14 years.

INTRODUCTION

In interior Alaska, permafrost underlies most of the forested stands in low-lying areas and on north-facing slopes. Seasonal variation in the rate and depth of both annual thawing and freezing of soil may be important to ecosystem processes, but little information is available on such soil activity. Information is accumulating on the effects of fire and other disturbance on the thickness of the active layer (Mackay 1977a; Racine 1980; Viereck 1973, 1982; Viereck and Schandelmeier 1980; Wein and Bliss 1973). Changes in mean annual temperature are predicted for northern regions, a result of increasing atmospheric CO_2. It is therefore important that baseline information be obtained on thickness of the active layer in natural stands in many areas. Baseline data are especially important in areas such as Fairbanks, where an increase of 2°-3°C mean annual temperature might result in considerable melting of permafrost. Thie (1974) found that in a region near the southern limits of permafrost, the area of permafrost had been reduced from 60 to 15% by climate amelioration in the past 120 years. Permanent or long-term monitoring of the active layer will be important in showing the effects of climatic change on permafrost.

This paper reports on the annual freezing and thawing cycle of the active layer over a 14-year period (1969-82). In 1968, three frost tubes were installed in a 70-year-old mixed stand of black spruce and white spruce located about 5 km north of the Fairbanks campus of the University of Alaska, at an elevation of 215 m. Their installation was part of a study to determine how effective frost tubes were in measuring the depths of soil freezing and thawing. Design of the frost tubes and a report of their performance is presented in Rickard and Brown (1972).

SITE DESCRIPTION

The frost tubes were installed in a stand classified as *Picea glauca*-*P. mariana*/*Salix glauca*/*Ledum groenlandicum*/*Pleurozium schreberi* (Viereck and Dyrness 1980). The stand developed after a fire in about 1914. White spruce trees were dominant, with a density of 2,000/ha, an average diameter of 6.0 cm, and a height of 10 m. A few black spruce trees, with a density of 400/ha and an average diameter of 6.4 cm, were interspersed. Ages of the trees ranged from 50 to about 70 years. The tree canopy was about 30 percent.

A tall shrub layer (>2 m), consisting primarily of decadent willows (*Salix glauca* L.) has a cover value of 20% with a stem density of approximately 15,000 stems/ha. A low shrub layer (<1 m), primarily *Ledum groenlandicum* Oeder., *Vaccinium uliginosum* L., *Vaccinium vitis-idaea* L., and *Spiraea beauverdiana* Schneid., forms a nearly continuous cover on the forest floor. A thick carpet of feathermosses, both *Hylocomium splendens* (Hedw.) B.S.G. and *Pleurozium schreberi* (Brid.) Mitt. covers most of the area, except in depressions where *Sphagnum warnstorfianum* DuRietz is common. *Peltigera aphthosa* (L.) Willd. is the only abundant lichen, locally forming mats on the feathermosses.

The soil in the area is a Minto silt loam (Rieger et al. 1963). These soils are developed in micaceous silty material and are often overlain by ice-rich permafrost. A typical Minto silt loam has approximately 10% sand, 80% silt, and 10% clay. At the study site, permafrost underlies the soil, but thermokarst is conspicuous adjacent to the site. The overlying organic layer varies from 24-40 cm in thickness and comprises an 01 (F) layer 15-20 cm thick over an 02 (H) humus layer, 5-10 cm thick. The living moss layer averages about 5 cm thick, ranging from 2 cm under the spruce trees to 10 cm in the most dense hummocks.

METHODS

Each frost tube consisted of an outer 3.4-cm diameter polyethylene tube, sealed at the bottom and permanently frozen into the permafrost, and extending 1 m above the moss mat surface. An inner, removable polyethylene tube, 2 cm in diameter, contained a fluorescein-saturated sand mixture that changed from green to yellow-pink when freezing. Extreme care was taken to keep disturbance of the natural vegetation to a minimum when installing the tubes. Three snow stakes were also installed adjacent to the tubes.

Readings were made at irregular intervals during the 14-year study. During especially important periods, such as the first thawing in May or June and freezing in October and November, the site was visited weekly. Another important period was in mid-December to late January when the soil profile became completely frozen. The site was also visited each week during that period. After the tubes became frozen the site was visited once a month or after major snow storms, to determine snow depth. In middle to late summer, when changes in thaw depth were only 1-2 cm per week, the site was visited at 2-week intervals to minimize disturbing the vegetation and surface organic layers.

Because visits to the sites were irregular, data were organized by pentads, or 5-day periods. There were thus 73 pentads for each year. Because visits to the site were infrequent during late winter, data on snow depth were summarized by groups of 3 pentads; in other words, 15-day or half-month periods. This proved especially useful in summarizing the 14 years of data.

Two indexes, one for freezing and one for thawing, were calculated using 0°C as a base and daily mean temperatures from the National Weather Bureau station at the Fairbanks airport. A day with an average temperature of +5°C would contribute 5° to the thaw index, whereas a day with an average temperature of -5°C would contribute 5° to the freeze index. These are indexes; actual temperatures at the site could have varied by several degrees from those at the Fairbanks airport, especially during periods of extreme cold and strong temperature inversions.

In 1977, mosses, lichens, herbs, and low shrubs on the site were described using three circular plots centered on the frost tubes. Density and diameter of trees and tall shrubs were determined on one 250-m^2 circular plot centered on the middle frost tube (no. 2).

In September 1981, 18 m^2 of the study area was divided into a 25-cm grid and the vegetation mapped. At each corner of the grid, distance was measured to the moss surface and to the base of the active layer. From this profile of the vegetation surface and the lower surface of the active layer, the configuration of the upper boundary of the permafrost layer could be determined. In addition, the thicknesses of the living moss, 01 (F) and 02 (H) layers, and mineral soil to permafrost were recorded at 18 points in the grid. During the summer of 1981, the depth to the frozen layer was probed with a steel rod at 10 points in a line adjacent to the frost tubes and at 10 points in the trail created by the observers. These measurements were made 12 times during the summer of 1981.

RESULTS

Thawing

Figure 1 shows the seasonal progression of freezing and thawing at the study site. Thawing from the surface downward in the spring is directly related to air temperature (thaw index) and the persistence of snow from the previous winter. Usually thawing of the upper 10 cm of soil occurs before all of the snow has melted. Thawing began as early as April 20 in 1970 and as late as May 7 in 1982, with an average beginning thaw date of May 1. Thawing is rapid, with 50% of the thaw depth reached by June 30, and 75-80% by the end of July. The maximum thaw depth, which ranged from 54 to 61 cm and averaged 58 cm, occurred from mid-August to mid-September.

The thaw depth closely follows the accumulated thaw index. The first days with an average temperature above 0°C usually occur in mid-April, although it may be late April in some years. In 1972, thaw did not begin until May 5.

There is a close linear relationship between thaw depth and the time since thaw began for the period from mid-April to the end of August. There is also a linear relationship between thaw depth and the thaw index. A regression of the thaw depth against the thaw index gives a coefficient of determination (R^2) value of 0.97. The regression equation for the thaw depth is

$$Y = 0.0675T + 6.9591, \qquad (1)$$

where Y = the depth of thaw, and T = the thaw index in °C.

Maximum thaw depth for the year varied by 7 cm during the 14-year period. When maximum thaw depth is compared with the total accumulated thaw index for the year, there is no correlation. This may indicate that other factors, such as the freezing index from the previous winter or precipitation from the summer, may influence the maximum thaw depth for any given year.

It is possible to draw only general relations between late-lying snow and the thawing of the

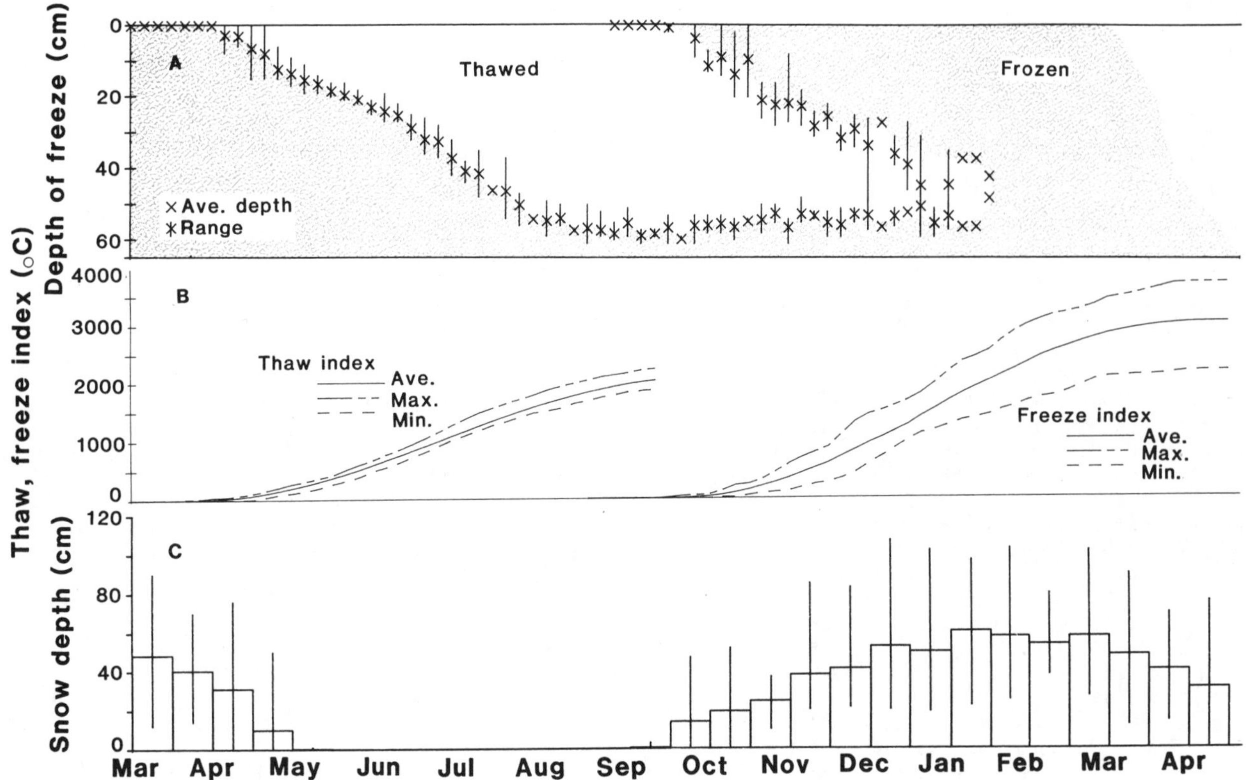

FIGURE 1. Annual freeze-thaw profile of the study site (A), thaw and freeze indexes (B), and snow depth (C). Points are average and range for the 14-year period, 1969-1982. Time is shown in pentads (73 five-day periods per year) and by months.

soil. For tundra regions, Goodwin and Outcalt (1975) used a computer model to show that winter snow depth affected both the time of active layer thaw and the maximum depth to which it thawed. In our study in the Fairbanks area, where spring thawing is rapid and the summer thawing season long, snow depth the previous winter did not seem to affect timing of thaw. For example, in 1971, year of the heaviest and latest snow during the study, soil began to thaw on May 5, even though snow remained on the ground for an additional 17 days. In a light snow year such as 1980 or 1981, soil began to thaw on May 5, but the snow had completely melted by that date.

Freezing

The annual freezeback of the active layer from the surface downward begins during October. Freezing is usually rapid and the whole active layer can be frozen as early as December 13. The average date, however, is January 7. The latest date of total freezeback was January 28 (in 1979). The fall and winter of 1978-1979 were relatively warm and there was only a moderate (50-cm) snow accumulation by late January.

Annual freezeback shows more variability than does the thawing. This is probably related to two factors. First, the annual freeze index has a much wider range and standard deviation than does the thaw index (freezing index averaged 2998°C, ranging from 2162° to 3673°C, with a standard deviation of 513; thaw index averaged 2019°C, ranging from 1858° to 2226°C, with a standard deviation of 109). The second factor is variation in timing of snowfall, especially early and heavy snow before the ground is deeply frozen. Thick and early snowfall insulates the ground from cold air temperatures and slows the descent of freezing into the ground.

As with thawing, the depth of the freezing line showed a close linear relationship to the elapsed time since freezing began for the period from early October through December. It also showed a linear relationship to the freezing index. Using only the freezing index as a variable against the depth of freeze line, the following equation was obtained:

$$Y = 0.02167F + 10.977 \qquad (2)$$

where Y = the depth of freeze in cm and F = the freezing index in °C. The coefficient of determination (R^2) value for this regression was 0.645.

There is also some annual freezing of the active layer upward from the top of the permafrost. This 5-10 cm of freezing upward begins in mid-September and continues until it meets the freezing line descending from the surface in mid-December to late January.

The effect of snow depth on the freezing rate of soil is difficult to express mathematically because time of snowfall, density of snowpack, and timing of the cold spells (i.e., before or after a heavy snow) all influence the rate of freezing. Figure 1 shows the average snowfall and range averaged for periods of three pentads for the 14 years of the study. Permanent snow cover usually occurred by October 8 (range: September 23 to October 25) and the last snow was usually gone from the site by May 6 (range: April 28 to May 20). Maximum snow depths at the site varied from 27 cm during the winter of 1969-1970 to 107 cm in the winter of 1970-1971. The maximum snow depth averaged 60 cm for the 14-year period. Adding snow as a second variable in the regression equation for freeze depth increases the R^2 value by only a small amount.

The effect of snowfall can best be shown in Figure 2, which compares depth of snow, freezing index, and depth of freeze line for two contrasting winters--1970-1971, when snow accumulation reached 100 cm by mid-December, and 1980-1981, when snow depth reached only 28 cm in early December and settled to 22 cm by mid-January. The freezing index was about the same through the end of December (1315°C for 1970 and 1345°C for 1980). January 1971 was much colder than January 1981, however, with an accumulated freezing index of 2377°C in 1971 compared with 1579°C in 1981. In spite of the similarity in freezing indexes in the early part of the two winters, the active layer froze much more slowly in 1970 than in 1980. Large amounts of early-season snow provided insulation to the soil and slowed freezing in 1970.

Effects of human disturbance

Despite the care taken in installing the frost tubes and in visiting the site, a trail developed about 70-100 cm from the tubes. When measurements were taken in 1981 (described in "Methods"), it was obvious that the trail had caused a linear depression in the vegetation; in the depression, sphagnum moss was replacing the feathermosses and shrubs.

Probes taken during summer 1981 indicated the depth of thawing from the surface of the vegetation downward was not significantly greater under the observers' trail than it was under the undisturbed vegetation. A statistical comparison of the differences in thaw depth was not attempted because the data did not meet the criteria for an appropriate test. Examining thaw depths determined both by the probing and frost tubes indicates the values are similar except for May 16, when the difference between the average of measurements of the three frost tubes and the 10 probe measurements is 4.8 cm. The probe measurements taken in the trail are similar to those taken in the undisturbed vegetation, but because the surface is lower, they indicate a slight increase in the thaw depth under the trail.

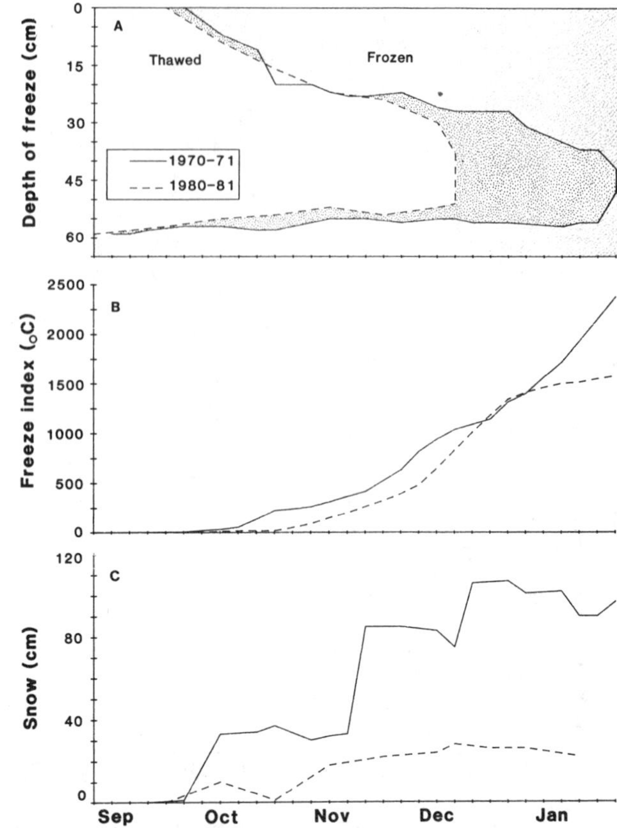

FIGURE 2. Freeze-thaw profile (A), freeze index (B) and snow depth (C) for September 3 to January 30, 1970-1971 and 1980-1981.

Profiles taken across the trail at each frost tube showed that there has been some deeper thaw into the original permafrost as a result of the trail (Figure 3). This thawing of permafrost under the trail does not seem to have affected the thaw depth of the active layer at frost tubes 1 and 2 but may have resulted in a slight increase in the thaw depth at frost tube 3. Human disturbance is always a factor in studies of permafrost under natural conditions. Such disturbance makes it extremely difficult to monitor natural changes in the active layer over long periods of time.

DISCUSSION

Brown and Grave (1979) summarize information on measurements of the active layer in the taiga and tundra of North America and the Soviet Union. They give maximum thaw depths for vegetation types and geographic areas but do not provide information on the variation in the active layer thickness during the thawing season or year-to-year variations. Most studies on which they report measured the thickness of the active layer in 1 year only, usually from mid-August to mid-September, the time previously determined (Brown et al. 1969, Dingman and Koutz 1974) as that of maximum seasonal thawing. Rouse (1982)

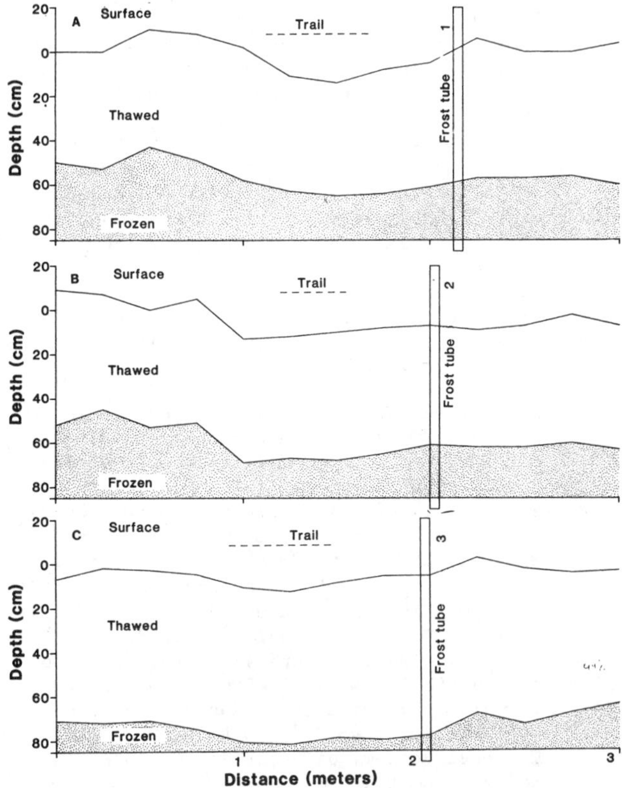

FIGURE 3. Profiles of the forest floor and active layer adjacent to the three frost tubes, showing the effect of the trail on thaw depth.

did show both the seasonal thaw and freeze line for forested and tundra areas near Churchill, Manitoba. He also showed that there was some variation in the active layer thickness during the 2 years of his study, which was related to the length-of-thaw period as well as to net radiation received at the site.

Dyrness (1982) reported a gradual increase over a 4-year period of about 8 cm in the active layer thickness (actually reported as the thickness of the thawed mineral soil) of control plots in a black spruce stand. Dyrness suggested that the annual increase in active layer thickness could probably be attributed to removal of the overstory trees even though the forest floor vegetation was undisturbed. Viereck (1973) reported a similar change in active layer thickness of a black spruce stand in which the overstory had been removed for a study on biomass. In a study of the effects of an oil spill on permafrost in a black spruce stand in the Fairbanks area, Johnson et al. (1980) showed an increase in thawing over a 2-year period of about 20 cm following an oil spill, but only a 0.4-cm difference between years in an undisturbed control.

A number of complex models have been developed for predicting thaw and freeze depths using thermal conductivity and diffusivity of the soil and heat fluxes at the surface and in the soil (for example McGaw et al. 1978, Outcalt et al. 1975, Scott 1966). Most of these models are for tundra. These models could not be used in our study because no measurements of temperature or solar radiation were made at the site.

Some attempts have been made to correlate the depth of thawing with more gross aspects of climate, solar inputs, and topography. Dingman and Koutz (1974) showed a weak correlation in one watershed in interior Alaska between thickness of the active layer and an equivalent factor of latitude, and a slightly higher correlation between the thickness of the active layer and a potential insolation index. Brown and Grave (1979) report an increase in active layer thickness at Prudhoe Bay, Alaska, as a function of distance along a strong summer temperature gradient from the coast to warmer inland localities. Dyrness and Grigal (1979) found changes in the active layer thickness along a slope transect that were related to slope position, vegetation, and organic layer thickness. More detailed observations are needed on variations in active layer thickness with time, with disturbed and undisturbed vegetation, and with environmental factors to aid in the development of predictive models.

Comparison of Measurements

Comparison of thaw depths measured by steel probe and frost tube show close agreement between the two methods. Mackay (1977b) has shown that probing may overestimate active layer thickness in some soil types because the probe enters the frozen zone with little or no resistance. He found, however, that in fine soils with high ice content, as in our study, the probe did measure thaw depth accurately. The frost tube, however, can measure freezeback in the fall, which is impossible with the steel probe.

SUMMARY AND CONCLUSIONS

Frost tubes are useful devices for measuring annual freezing and thawing in undisturbed vegetation and were successfully used in this study for 14 years. There is good correlation between depth of thaw, and a thaw index determined as accumulated degree days above 0°C. Freezeback of the active layer in winter is more variable than thaw because of a wider range in the freeze index and the timing and depth of snowfall. Human disturbance to the site was minimal and had an insignificant effect on the thaw depths after 14 years.

ACKNOWLEDGMENTS

We thank Mr. Gordon Herreid who gave permission to conduct the study on his property, and Mrs. M. J. Foote who took field measurements from August 1970 to August 1971.

REFERENCES

Brown, J., and Grave, N. A., 1979, Physical and thermal disturbance and protection of permafrost, in Proceedings of the Third International Conference on Permafrost, v. 2, p. 51-91: Ottawa, National Research Council of Canada.

Brown, J., Rickard, W., and Victor, D., 1969, The effect of disturbance on permafrost terrain. USACRREL Special Report 138. 13 p: Hanover, NH, U.S. Army Cold Regions Research and Engineering Laboratory.

Brown, R. J. E., 1971, Characteristics of the active layer in the permafrost region of Canada, in Proceedings of a Seminar on the Permafrost Active Layer: Technical Memorandum 103, p. 1-7: National Research Council of Canada.

Dingman, S. L., and Koutz, F. R., 1974, Relations among vegetation, permafrost, and potential insolation in central Alaska: Arctic and Alpine Research, 6(1): 37-42.

Dyrness, C. T., 1982, Control of depth to permafrost and soil temperatures by the forest floor in black spruce/feathermoss communities, Research Note PNW-396, 19 p.: Portland, OR, USDA Forest Service, Pacific Northwest Forest and Range Experiment Station.

Dyrness, C. T., and Grigal, D. F., 1979, Vegetation-soil relationships along a spruce forest transect in interior Alaska: Canadian Journal of Botany, 57(23): 2644-2656.

Goodwin, C. W., and Outcalt, S. I., 1975, The development of a computer model of the annual snow-soil thermal regime in arctic tundra terrain, in Weller, G., and Bowling, S. A., ed. Climate of the Arctic--Proceedings, AAAS-AMS Symposium, 15-17 August, 1973, Fairbanks, Alaska, p. 227-229.

Johnson, L. A., Sparrow, E. B., Jenkins, T. F., Collins, C. M., Davenport, C. V., and McFadden, T. T., 1980, The fate and effects of crude oil spilled in subarctic permafrost terrain in interior Alaska, USACRREL Report 80-29, 67 p.: Hanover, NH, U.S. Army Cold Regions Research and Engineering Laboratory.

McGaw, R. W., Outcalt, S. I., and Ng, E., 1978, Thermal properties and regime of wet tundra soils at Barrow Alaska, in Proceedings of the Third International Conference on Permafrost, v. 1, p. 48-53: Ottawa, National Research Council of Canada.

Mackay, J. R., 1977a, Changes in the active layer from 1968 to 1976 as a result of the Inuvik fire, in Report of the Activities, Part B, Paper 77-1B, p. 273-275: Geological Survey of Canada.

Mackay, J. R., 1977b, Probing for the bottom of the active layer, in Report of Activities, Part B; Paper 77-1A, p. 327-328: Geological Survey of Canada.

Outcalt, S. I., Goodwin, C. W., Weller, G., and Brown, J., 1975, Computer simulation of the snow melt and soil thermal regime at Barrow, Alaska: Water Resources Research, 11(5): 709-715.

Racine, C., 1980, Effects of a tundra fire on soils and plant communities along a hillslope in the Seward Peninsula, Alaska, USACRREL Report 80-37, 21 p.: Hanover, NH, U.S. Army Cold Regions Research and Engineering Laboratory.

Rickard, W., and Brown, J., 1972, The performance of a frost-tube for the determination of soil freezing and thawing depths: Soil Science, 113(2): 149-154.

Rieger, S., Dement, J. A., and Sanders, D., 1963, Soil Survey Fairbanks Area Alaska, Series 1959 No. 25, 41 p. and 24 maps: Washington, DC, U.S. Department of Agriculture. Soil Conservation Service.

Rouse, W. R., 1982, Microclimate of low Arctic tundra and forest at Churchill, Manitoba, in French, H. M., ed., Proceedings of the Fourth Canadian Permafrost Conference (The Roger J. E. Brown Memorial Volume), p. 68-80: Ottawa, National Research Council of Canada.

Scott, R. F., 1966, Factors affecting freeze or thaw depth in soils, in Proceedings of the Permafrost International Conference, NRC Publ. 1287, p. 63-267: Washington, DC, National Academy of Sciences, National Research Council.

Thie, J., 1974, Distribution and thawing of permafrost in the southern part of the discontinuous permafrost zone in Manitoba: Arctic, 27(3): 189-200.

Viereck, L. A., 1973, Ecological effects of river flooding and forest fires on permafrost in the taiga of Alaska, in Permafrost--The North American contribution to the Second International Conference, Yakutsk, p. 60-67: Washington, DC, National Academy of Sciences.

Viereck, L. A., 1982, Effects of fire and firelines on active layer thickness and soil temperatures in interior Alaska, in French, H. M., ed., Proceedings of the Fourth Canadian Permafrost Conference (The Roger J. E. Brown Memorial Volume), p. 123-135: Ottawa, National Research Council of Canada.

Viereck, L. A., and Dyrness, C. T., 1980, A preliminary classification system for vegetation of Alaska, General Technical Report PNW-106, 38 p.: Portland, OR, USDA Forest Service, Pacific Northwest Forest and Range Experiment Station.

Viereck, L. A., and Schandelemeier, L. A., 1980, Effects of fire in Alaska and adjacent Canada--a literature review. BLM Alaska Technical Report 6, 124 p.: U.S. Department of the Interior, Bureau of Land Management.

Wein, Ross W., and Bliss, L. C., 1973, Changes in the Arctic Eriophorum tussock communities following fire: Ecology, 54(4): 845-852.

DYNAMIC PROPERTIES OF NATURALLY FROZEN SILT

Ted S. Vinson[1], Charles R. Wilson[2], and Peter Bolander[3]

[1]Dept. of Civil Engineering, Oregon State University, Corvallis, OR 97331 USA
[2]Alaska Test Labs, Anchorage, Alaska 99502 USA
[3]U.S. Forest Service, Eugene, Oregon 97401 USA

Resonant frequency and cyclic triaxial tests were conducted on naturally frozen silt samples. Dynamic elastic properties (expressed in terms of dynamic Young's and shear moduli) and energy absorbing properties (expressed as damping ratio) were determined. The influence of various dynamic loading parameters (confining pressure, temperature, frequency, and strain amplitude), water content, and anisotropy on dynamic properties was evaluated. Confining pressure and water content were generally found to have little effect on either dynamic moduli or damping ratio. The test results indicate dynamic moduli decrease with increasing strain amplitude and increase with increasing frequency and ascending temperature. The damping ratio generally increases with increasing axial strain amplitude and decreases with increasing frequency and descending temperature. The soil/ice structure (i.e., lens thickness, orientation, and spacing) does not appear to influence the dynamic properties of naturally frozen silt.

INTRODUCTION

To overcome the problems associated with obtaining and transporting frozen soil samples to the laboratory for testing, many researchers and practicing engineers have used reconstituted frozen samples to evaluate the behavior of naturally frozen soils. While these studies have contributed to a phenomenological understanding of the behavior of frozen soils, the specific applicability of the test results to solve field problems is questionable. In the words of Anderson and Morgenstern (1973), "This is particularly so because the structure of frozen soil, in situ, differs markedly from that of material prepared in the laboratory." In recognition of this fact, emphasis is now being placed on in situ tests, field performance studies, and laboratory studies that employ samples of frozen soils taken in situ. A modest amount of data from this work has been published to evaluate the thaw-consolidation behavior, creep behavior and strength, and thermal conductivity of naturally frozen soils. Only a meager amount of data, however, has been obtained on the dynamic properties of naturally frozen soils, and much of this information is associated with test conditions that are not applicable to all classes of wave propagation problems.

As part of a long-term study to evaluate the dynamic properties of naturally frozen soils over a range of test conditions associated with wave propagation problems of frozen soil deposits (e.g., foundation vibrations, geophysical exploration, blasting, ground response analyses during strong motion earthquakes), resonant column and cyclic triaxial tests were performed on specimens of naturally frozen silt. Parameters that could influence the dynamic properties of naturally frozen silt, such as confining pressure, temperature, frequency, strain amplitude, water content, and soil/ice structure, were considered. The results from the study are reported herein.

NATURALLY FROZEN SILT SAMPLES

The silt samples associated with the research program were obtained in the U.S. Army Cold Regions Research and Engineering Laboratory (USA CRREL) Permafrost Tunnel located at Fox, Alaska, 16 km north of Fairbanks. The walls of the tunnel expose perennially frozen silts of Pleistocene age. Sands and gravels bonded with interstitial ice are also visible, but to a lesser extent. It is generally accepted that the silts are aeolian in origin. The silts were transported from a broad treeless flood plain of the Tanana River (Sellmann 1967).

The samples were collected at several locations in the USA CRREL permafrost tunnel. To evaluate anisotropy of dynamic properties, 7.6 cm diameter samples were obtained by core drilling with a Haines auger on axes parallel, perpendicular, and at 45° to the existing horizontal strata. Based on the grain size distribution, Atterberg limits and a visual inspection of ice features, the samples were classified as ML-OL-Vs, MH-OH-Vs, ML-OL-Vr (Linell and Kaplar 1966). The average water contents for the 20 samples employed in the study ranged from 56 to 109%.

DESCRIPTION OF TEST SYSTEMS AND PROCEDURES

The current research program is unique in that it employs both the resonant frequency and cyclic triaxial testing methods on each specimen. This permits a direct comparison of the results obtained from the two methods. Further, use of both test systems permits an evaluation of dynamic properties over a range of test conditions associated with all classes of wave propagation problems of frozen soil deposits (Vinson 1978).

The resonant column test system applies a nondestructive, steady state, vibratory load to a right cylindrical frozen soil specimen in either the longitudinal or torsional mode. The test system employed in the present study, shown in Figure 1, is similar to the system developed by Stevens (1975). The theoretical considerations associated with resonant column testing are given by Stevens (1975) and Vinson (1978). Briefly, complex Young's or shear moduli are determined from the frequency at resonance. Complex Young's moduli are determined for a specimen excited in the longitudinal mode, while shear moduli are obtained from the torsional mode of vibration.

To conduct a test, specimens were freeze-bonded to the top and bottom platens of the test system. A thin film of cool water was sprayed on each specimen and allowed to freeze to prevent sublimation during experimental setup and testing. Three thermistors were freeze-bonded to the specimen to monitor temperatures during the test. A triaxial cell was placed over the specimen and, if required, a confining pressure was applied in the cell using commercial grade nitrogen gas. Temperature control during testing was maintained with an environmental chamber placed around the test system. The test system and environmental chamber were located inside a walk-in cold room.

To establish a resonant condition the operator vibrated the specimen at the base with an electromagnetic motor in either the longitudinal or torsional mode, with the frequency at a low level. As frequency was increased, the signals from accelerometers attached to the top and bottom platen were monitored for resonance. At resonance the ratio of the top to bottom platen accelerometer outputs is a maximum, the first maximum being associated with the fundamental resonant frequency. Successive maximums indicate harmonics.

The theoretical considerations associated with cyclic triaxial testing have been presented by Vinson (1978). Specimens were subjected to a cyclic axial load, and specimen load and deformation were recorded. From the load-deformation response dynamic Young's modulus and damping ratio were determined.

The cyclic triaxial system employed in the present study is shown in Figure 2. Cyclic triaxial tests were conducted on specimens after the resonant column testing was complete because of the potentially destructive nature of the cyclic triaxial test. Following removal from the resonant column test system, the specimen was placed in the cyclic triaxial grips. Cool water was injected between the grip and the specimen to insure bonding. Following a 7-day set period, the specimen and grip assemblage were attached to a load frame located inside a walk-in cold room. An environmental chamber was used to control temperature; a triaxial cell was employed to apply a confining pressure.

The resonant column and cyclic triaxial test systems were designed to accommodate a right cylindrical specimen with a diameter of 7.2 cm and a length of 30 cm. A swing metal lathe inside a walk-in cold room was used to trim the smaller diameter test specimens from the 7.6 cm diameter field samples. The ends of the test specimen were faced off perpendicular to the longitudinal axis using a steady rest.

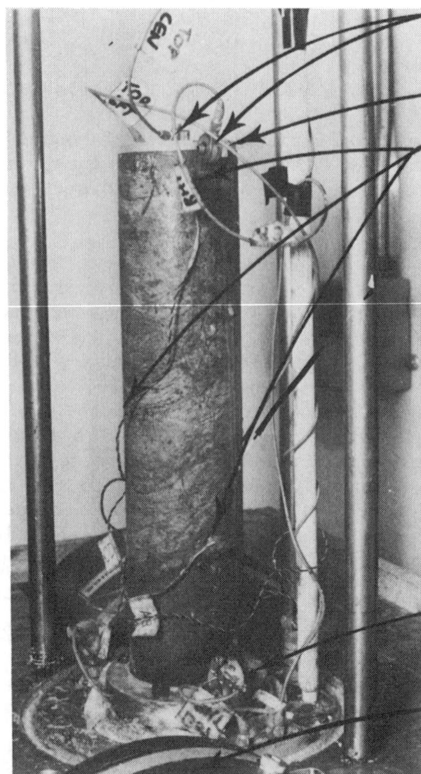

FIGURE 1 Resonant column test system.

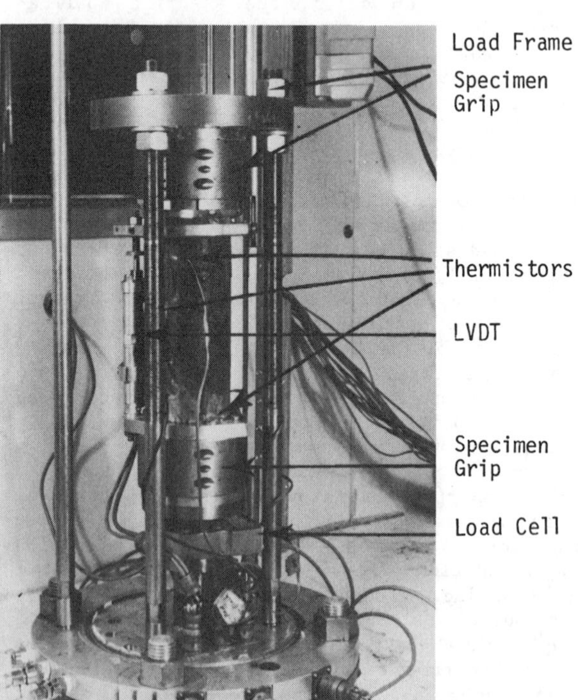

FIGURE 2 Cyclic triaxial test system.

TEST PROGRAM AND RESULTS

Resonant column and cyclic triaxial tests were conducted according to test histories which allowed an evaluation of the influence of confining pressure, temperature, strain amplitude, and frequency to be made for each test specimen (Wilson, 1982). For the various combinations of test parameters, the dynamic Young's modulus ranges from 0.7 to 16.5 GN/m^2, the dynamic shear modulus from 0.7 to 5.5 GN/m^2, and the damping ratio from 0.01 to 0.26. Following dynamic testing the ice lens thickness, orientation, and spacing were noted for each specimen, and the average water content was determined.

Dynamic Moduli Test Results

Representative relationships between dynamic moduli and confining pressure are shown in Figure 3. In general, it may be noted that there is a slight decrease in moduli with confining pressure. The decrease is generally less than 5%. This result is unexpected based on previous research to evaluate dynamic moduli of frozen cohesionless soils, which indicate no change in dynamic moduli with increasing confining pressure (Vinson et al. 1977). Overall, the decrease in modulus is considered to be negligible, and data for all confining pressures were combined to develop relationships with other dynamic loading parameters.

Representative relationships between dynamic moduli and temperature are shown in Figure 4. Kaplar (1969) and Stevens (1975) show that the decrease in dynamic moduli is relatively small for frozen silts for the temperature range -17 to -4°C. Therefore, the temperature range of -10 to -1°C is considered sufficient to evaluate the effect of temperature on the dynamic moduli. As

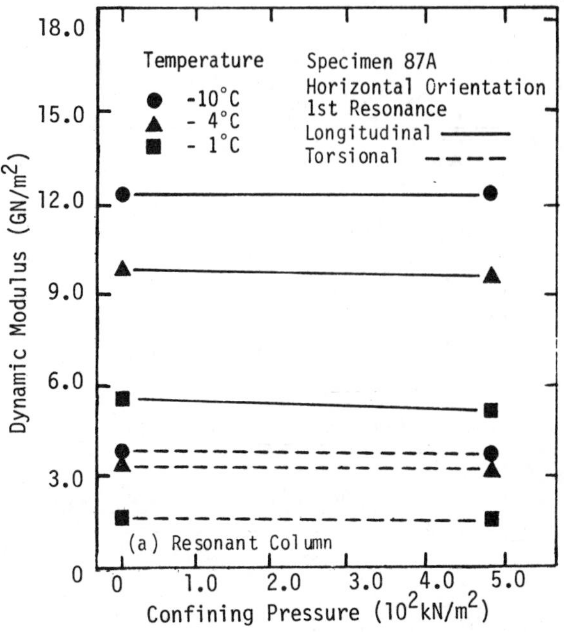

FIGURE 3 Dynamic modulus versus confining pressure.

expected, dynamic Young's and shear moduli decrease with ascending temperature. The rate of decrease increases with ascending temperature. These results are in good agreement with previous results obtained on reconstituted artificially frozen silt specimens (Vinson et al. 1977). The relationship between dynamic moduli and temperature for fine-grained soils has been shown to be directly related to the unfrozen water content (Nakano and Froula 1973). Also, the increase in

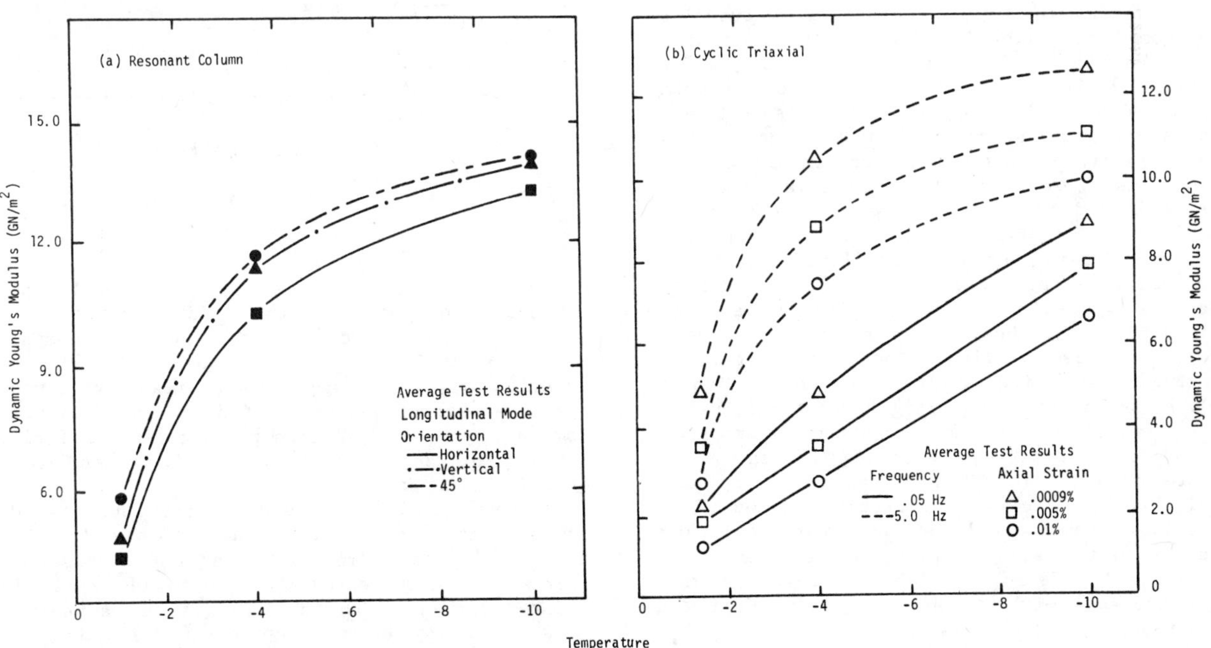

FIGURE 4 Dynamic modulus versus temperature.

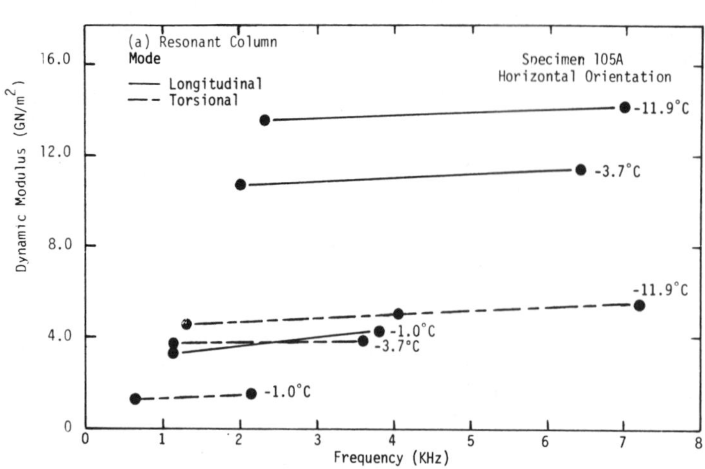

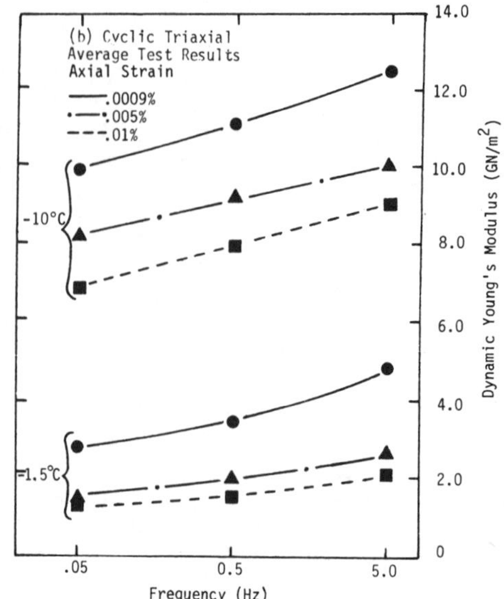

FIGURE 5 Dynamic modulus versus frequency.

dynamic moduli may be due to the increase in the stiffness of ice with decreasing temperature. Haynes (1973) attributes the strength increase in frozen silt with descending temperature in part to the strength increase in the ice matrix with descending temperature.

Representative relationships between dynamic moduli and frequency are shown in Figure 5. In the high frequency range associated with the resonant column test results there appears to be a slight increase in dynamic moduli with increasing frequency. In the low frequency range associated with the cyclic triaxial test results, dynamic moduli increase with increasing frequency. The increase in modulus in this range of frequencies is in good agreement with the results from previous investigations (Vinson et al. 1977). Further, the influence of frequency on dynamic modulus is an indication of time-dependent effects, such as stress relaxation or creep, on the dynamic elastic properties of viscoelastic materials such as frozen soils. Since relaxation effects will tend to cause increasing stress reduction with time under a constant strain, it follows that for a strain-controlled cyclic triaxial test a decrease in strain rate will result in a corresponding increase in stress reduction over a cycle of loading and, thus, a decrease in dynamic Young's modulus. Also, the ice matrix is strengthened with increasing strain rate (Haynes 1973). The increase in strength and stiffness of the ice matrix undoubtedly contributed to the increase in modulus with increasing frequency.

Representative relationships between dynamic moduli and axial strain amplitude are shown in Figure 6. Dynamic moduli are not appreciably affected by axial strain amplitudes in the range associated with resonant column testing. Dynamic moduli decrease with increasing strain amplitude in the range 10^{-3} to 10^{-2}%.

The results shown in Figure 6 are unique in that moduli of identical test specimens are com-

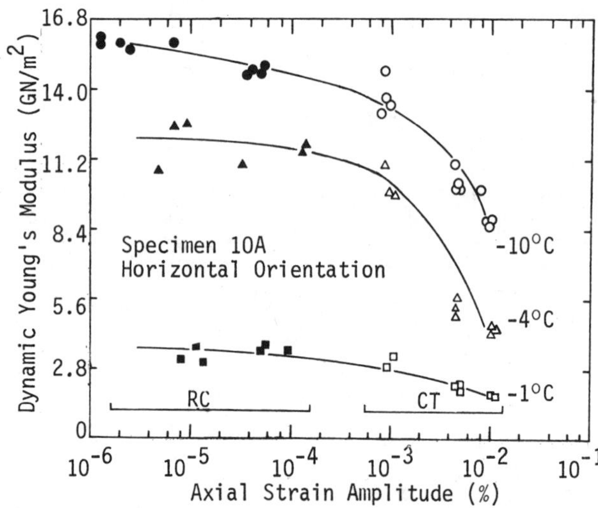

FIGURE 6 Dynamic modulus versus axial strain amplitude.

pared. Hence, differences in dynamic moduli associated with water content (or density) and soil/ice structure do not influence the comparison. Recognition of the strain dependence of dynamic properties may be important to ground response predictions associated with several categories of wave propagation problems, specifically, earthquake and blast loadings of frozen ground deposits (Vinson 1978).

The relationship between dynamic Young's modulus and water content is shown in Figure 7. At -4 and -10°C dynamic moduli appear to decrease with increasing water content. Dynamic moduli may increase slightly with increasing water content at -1°C. It may be more appropriate to conclude, however, that the influence of water content at warm temperatures is negligible. The thickness,

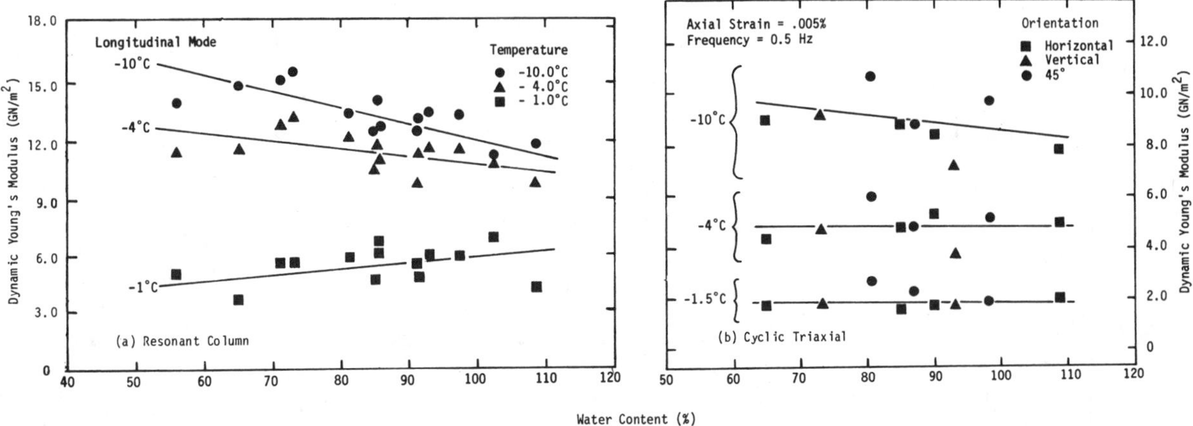

FIGURE 7 Dynamic modulus versus water content.

orientation, and spacing of the ice lenses does not affect dynamic moduli.

The test results shown in Figure 7 are associated with test specimens with visible ice lenses from 0.3 to 5 mm in thickness. For many of the specimens the ice lenses were randomly or irregularly oriented; for other specimens the ice lenses were stratified or distinctly oriented. The soil/ice structure could not be correlated to dynamic modulus. In the authors' opinion the soil/ice structure did not appear to influence the values of the dynamic Young's modulus for the range of structures represented by the test specimens employed in the study.

The effect of specimen orientation on dynamic moduli has been shown in Figures 4a, and 7b. Dynamic moduli appear to be slightly greater for samples cored at the 45° orientation compared to samples cored horizontally or vertically.

Damping Ratio Test Results

Emphasis in this paper has been given to dynamic moduli of frozen silt. Damping ratios were also determined in the test program. A summary of the test results associated with this work follows.

Representative relationships that illustrate the influence of dynamic loading parameters on damping ratio are shown in Figure 8. The damping ratio is not appreciably affected by strain amplitude in the range of strains associated with resonant column testing. Damping ratio increases with increasing axial strain amplitude in the range of strains associated with cyclic triaxial testing. The damping ratio decreases with ascending temperature and increasing frequency. The decrease in damping ratio with descending temperature may be a function of both the increase in the strength of ice and the decrease in unfrozen water content with decreasing temperature. The influence of frequency on damping ratio may be a function of creep effects due to sliding between ice or soil particles, or both, during dynamic loading. In the same way that stress relaxation effects influence the relationship between dynamic moduli and frequency, creep would tend to cause greater displacements between grains and thus, greater values of damping ratio at lower frequencies than at higher frequencies. No definitive relationship between sample orientation and damping ratio could be established. In general, however, the differences in damping ratio associated with sample orientation were small and may be considered to be negligible. While there is scatter in the test results it appears damping ratio increases slightly with increasing water content and confining pressure. The increase is negligible, however.

CONCLUSIONS AND PRACTICAL SIGNIFICANCE OF TEST RESULTS

The dynamic properties obtained in the present study were found by Wilson (1982) to be in good agreement with the results obtained by previous investigators who evaluated dynamic properties of reconstituted artificially frozen silt. Any differences between the dynamic properties obtained in the present study compared to dynamic properties obtained in previous studies could be explained by differences in testing techniques or silt material characteristics. The major factors that influence dynamic properties of frozen silt identified in the present study (i.e. strain amplitude, frequency, and temperature), were reinforced through the comparison with the test results from previous studies.

Further, the test results from the present study and the comparison with the results from previous investigations suggests that the dynamic properties of frozen silt are influenced primarily by soil material characteristics (e.g., grain size distribution, soil mineralogy) and, to a lesser extent, by water content. The soil/ice structure (i.e., lens thickness, orientation, and spacing) does not appear to influence dynamic properties of frozen silt. Based on this conclusion, it would appear that it is not necessary to test specimens obtained from naturally frozen in situ samples if artificially frozen test specimens are prepared in the laboratory at the same water content as the in situ material.

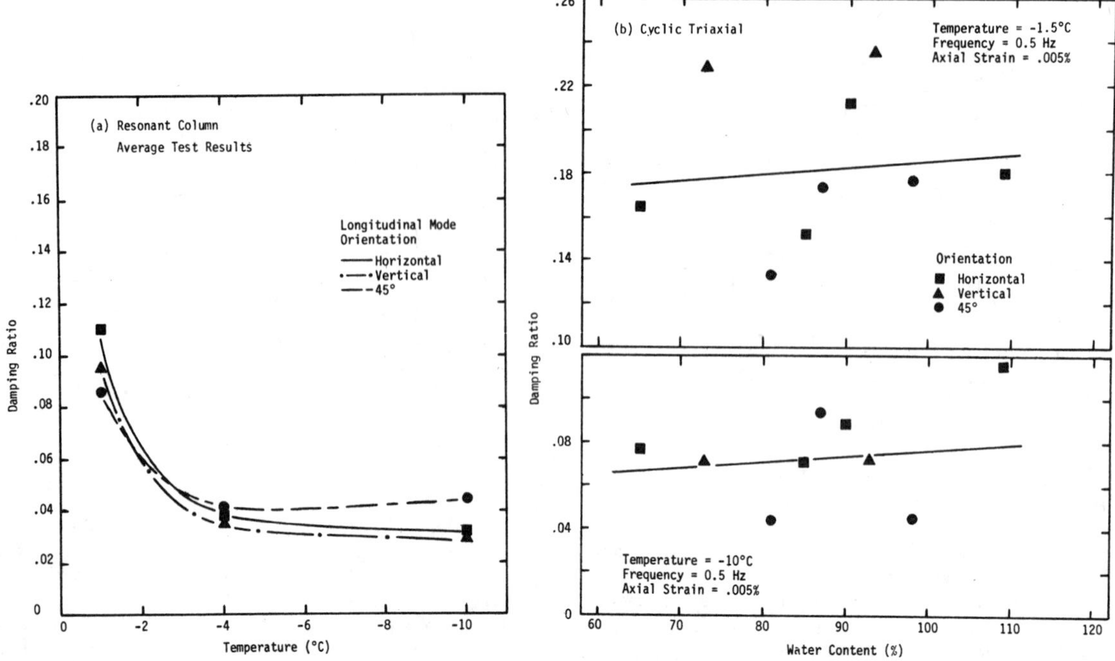

FIGURE 8 Damping ratio versus temperature and water content.

ACKNOWLEDGMENT

The studies described herein were supported by grant no. CME-7919697 from the National Science Foundation for the project entitled, "Dynamic Properties of Naturally Frozen Soils." The support of the Foundation is gratefully acknowledged. The efforts of Laurie Campbell and Nancy Platz, who assisted in the preparation of this paper, are greatly appreciated.

REFERENCES

Anderson, D.M. and Morgenstern, N.R., 1973, Physics, chemistry, and mechanics of frozen ground, in the North American Contribution to the Second International Conference on Permafrost, National Academy of Science, pp. 257-288.

Haynes, F.D., 1973, Strength and deformation of frozen silt, in Proceedings of the Third International Conference on Permafrost, V. 1: Ottawa, National Research Council of Canada, pp. 656-661.

Kaplar, C.W., 1969, Laboratory determination of dynamic moduli of frozen soils and ice, Research Report 163, U.S. Army Cold Regions Research and Engineering Laboratory, Hanover, NH.

Linell, K.A. and Kaplar, C.W., 1966, Description and classification of frozen soils, Technical Report 150, U.S. Army Cold Regions Research and Engineering Laboratory, Hanover, NH.

Nakano, T., and Froula, N.H., 1973, Sound and shock transmission in frozen soils, in the North American Contribution to the Second International Conference on Permafrost, National Academy of Science, pp. 359-369.

Sellmann, P.V., 1967, Geology of the USA CRREL permafrost tunnel, Fairbanks, Alaska: Technical Report 199, U.S. Army Cold Regions Research and Engineering Laboratory, Hanover, NH.

Stevens, H.W., 1975, The response of frozen soils to vibratory loads: Technical Report 265, U.S. Army Cold Regions Research and Engineering Laboratory, Hanover, NH.

Vinson, T.S., Czajkowsi, R., and Li, J., 1977, Dynamic properties of frozen cohesionless soils under cyclic triaxial loading conditions: Report No. MSU-CE-77-1, Divison of Engineering Research, Michigan State University.

Vinson, T.S., 1978, Response of frozen ground to dynamic loadings: Chapter 9 in Geotechnical Engineering in Cold Regions, McGraw-Hill Book Company, Inc., pp. 405-458.

Wilson, C.R., 1982, Dynamic properties of naturally frozen Fairbanks silt, M.S. Thesis, Department of Civil Engineering, Oregon State University.

THAW PLUG STABILITY AND THAW SETTLEMENT EVALUATION FOR ARCTIC TRANSPORTATION ROUTES: A PROBABILISTIC APPROACH

C. L. Vita
R&M Consultants, Inc., Seattle, Washington
c/o P.O. Box 6087, Anchorage, Alaska 99503 USA

A unified, probabilistic approach to the evaluation of thaw plug stability and thaw settlement is outlined for practical application to arctic or subarctic transportation routes. The approach can explicitly account for uncertainties, geotechnical variability, and limitations in available geotechnical data/information. Using Bayesian probability concepts, landform identification, and statistical characterization of landform soil property parameters, it can provide site-specific estimates of thaw plug stability and thaw settlement. The methodology is suited to situations where regional or local landform soil property parameters can be statistically well known, but site-specific data may generally be either unavailable or sparse. Estimates can be based only on landform information or systematically updated and improved with site-specific data to the extent they are available. Estimates include the mean and the variance of both thaw plug stability limit equilibrium factor of safety (FS) and average thaw settlement magnitudes (TS); using these, estimates can be made of the probability of instability (Pf) and of reliability/probability levels on the settlement estimates, including the probability of any given estimate being above or below a given critical value. Differential settlement is not treated.

INTRODUCTION

Segment-by-segment evaluation of potential thaw plug stability and thaw settlement along arctic or subarctic transportation routes must address uncertainties associated with sparse site-specific subsurface soil data, loading conditions, subsurface thermal response, and the mechanics of stability. The purpose of this paper is to outline a practical yet rigorous approach for (1) systematically quantifying these uncertainties, and (2) rationally augmenting sparse site-specific subsurface soil data with data available from other sites having the same geologically characteristic landform; these are integrated as illustrated with results of a case study. Procedures presented here provide geotechnical tools for project planning, analysis, and design of an arctic or subarctic transportation route.

THAW PLUG STABILITY--FACTOR OF SAFETY

The resistance of a thawing slope to mass movement is commonly characterized by a limit equilibrium factor of safety (FS) for the expected critical failure surface A with FS defined as the ratio C/D, where C (capacity) is the sum of all forces resisting movement along A, and D (demand) is the sum of all forces contributing to movement along A. In theory, if FS could be evaluated with absolute precision, it would give a true measure of the future stability of a slope: $FS \geq 1.0$ would imply certain stability and $FS < 1.0$ would imply certain movement. In practice, FS cannot be evaluated with such precision due to uncertainties in the failure mechanism, soil conditions, loading conditions and thermal response; these uncertainties make FS a random variable.

FS can be modeled as a site-specific random variable and evaluated probabilistically using estimates of the mean or expected value of FS, E[FS], and the variance of FS, V[FS], using a second-order approximation of the mean, a first-order approximation of the variance (Benjamin and Cornell 1970), and assuming a reasonable probability density function (PDF) model, PDF[FS], e.g., lognormal:LN, normal:N, or beta:B. In the case study PDF[FS] is assumed to be N for $1.0 \leq FS \leq E[FS]$. Given E[FS], V[FS], and PDF[FS], the estimated probability of instability or "failure," Pf, a random variable defined as P[FS<1.0], can be calculated as Pf=P[FS<1.0/E[FS], V[FS],PDF[FS]]. Whenever appropriate, multiple estimates of Pf based on alternative sets of reasonable assumptions or interpretations can be weighted and averaged, and/or used to form a PDF on Pf using a beta distribution (Harr 1977), and then assessed for sensitivity and technical/economic implications. In general, uncertainty in Pf can be made as small as desired by obtaining sufficient information. Also, a (PDF-free) probabilistic stability margin, U, can be defined as U=(E[FS]-1)/SD[FS]; SD[FS] is the standard deviation of FS, equal to the square root of V[FS] (the standard deviation of any variable equals the square root of its variance).

The above procedure can be followed for any mathematical form of FS generally defined by

$$FS = fs(X_i) + e \qquad i=1,2...k \tag{1}$$

The function $fs(X_i)$ represents the stability model C/D based on the X_i random variables used to model geotechnical, geometric, and loading conditions; X_i may include effects of slope geometry, soil shear strength, depth of failure surface, groundwater conditions, excess pore pressures, massive ice, dead loads (embankment and soil weight) and live loads, stress distribution effects, thaw plug geometry, and so forth. Variable e accounts for bias and random error (determined, if adequate data is available, by calibration of predicted performance with observed field performance), reflecting potential inaccuracy and uncertainty in the stability model $fs(X_i)$ itself. E[FS] and V[FS] are then estimated via

$$E[FS] = fs(\bar{x}_1, \bar{x}_2, \ldots \bar{x}_k) + \bar{e} + \frac{1}{2}\sum_{i=1}^{k}\sum_{j=1}^{k} \left.\frac{\partial^2 fs}{\partial x_i \partial x_j}\right|_m C[X_i,X_j] \quad (2a)$$

$$V[FS] = \sum_{i=1}^{k}\sum_{j=1}^{k} \left.\frac{\partial fs}{\partial x_i}\right|_m \left.\frac{\partial fs}{\partial x_j}\right|_m C[X_i,X_j] + V[e] \quad (2b)$$

$C[X_i,X_j]$ are the covariances of the parameters of Equation 1. All derivatives are evaluated at the means, m, of the X_i parameters, $\bar{x}_i$ (a bar above a variable denotes its average or expected value). $\bar{e}$ accounts for bias and V[e] accounts for random error in $fs(X_i)$.

Using an effective stress Mohr-Coulomb failure model with attention focused on the uncertain soil strength parameters, $fs(X_i)$ can be written as the sum of two components: $(G \cdot C' + F \cdot \tan \phi')$, where C' and $\tan \phi'$ are the average effective soil strength parameters along the critical failure surface; G is a function of all X_i related to cohesion (C'), and F is a function of all X_i related to friction ($\tan \phi'$). Note that G and F could include reduction factors to account for the effect of massive ice on FS (e.g., as presented by Vallejo 1980). Application of Equations 2a and 2b yields

$$E[FS] = E[G] \cdot E[C'] + E[F] \cdot E[\tan \phi'] + \bar{e} \quad (3a)$$

$$V[FS] = E[G]^2 \cdot V[C'] + E[F]^2 \cdot V[\tan \phi'] + Z \quad (3b)$$

Here, Z represents in summary form all uncertainty from Equation 2b, including random error in the model, other than due to V[C'] and V[$\tan \phi'$].

Cost-effective procedures for estimating the expected values and variances of C' and $\tan \phi'$ may be based on regression analysis results: the soil strength parameters are predicted from the standard soil index properties of representative samples recovered (site-by-site) during subsurface boring and sampling investigations. The necessary predictive stochastic relationships between soil strength properties and soil index properties and the associated statistical parameters can be developed from a geotechnically and statistically designed strength testing program using a limited, but acceptable, number of generally representative soil samples. This approach can form the entire basis for estimating C' and $\tan \phi'$ for a given slope or it can supplement any available site-specific strength data. The model used here is based on average frozen dry density, $\bar{\gamma}_{df}$, as the predictive soil index property, whereby

$$\tan \phi' = \tan[\sin^{-1} \beta \cdot \bar{\gamma}_{df}] \quad (4a)$$

$$C' = \rho \cdot \bar{\gamma}_{df} / \cos \phi' \quad (4b)$$

in which ρ and β are data-based regression constants.

AVERAGE THAW SETTLEMENT

The estimated average thaw settlement potential of a stratum or strata of average thickness h is TS, where $TS = h(1-\bar{\gamma}_{df}/\bar{\gamma}_{dt})_h$; $(1-\bar{\gamma}_{df}/\bar{\gamma}_{dt})_h$ is the average thaw strain (Crory 1973) over the stratum and $\bar{\gamma}_{df}$ and $\bar{\gamma}_{dt}$ are stratum frozen and thawed dry densities averaged over the thickness h. An estimate of the mean and variance of TS where $X_1=h$, $X_2=\bar{\gamma}_{df}$, and $X_3=\bar{\gamma}_{dt}$ is given by

$$E[TS] = h \cdot (1-\bar{\gamma}_{df}/\bar{\gamma}_{dt})_h + \frac{1}{2}\sum_{i=1}^{3}\sum_{j=1}^{3} \left.\frac{\partial^2 TS}{\partial x_i \partial x_j}\right|_m C[X_i,X_j] \quad (5a)$$

$$V[TS] = \sum_{i=1}^{3}\sum_{j=1}^{3} \left.\frac{\partial TS}{\partial x_i}\right|_m \left.\frac{\partial TS}{\partial x_j}\right|_m C[X_i,X_j] \quad (5b)$$

Confidence intervals on TS for given probability/reliability levels including estimates of the probability of TS being above (or below) any given value can be estimated using E[TS], V[TS] and an appropriate PDF[TS] model (N, LN, or B). Also, the coefficient of variation of TS, given by CV[TS]=SD[TS]/E[TS], gives a simple and useful measure of uncertainty in TS.

LANDFORM-BASED ESTIMATION OF SITE-SPECIFIC SOIL PARAMETERS

As developed above, the estimation of FS and TS requires prediction of average frozen dry densities for each site. Landform-based procedures (Kreig and Reger 1976, Vita 1982, 1983a, 1983b) provide a rational approach for making meaningful predictions where site-specific subsurface data are sparse or nonexistent: (1) the landforms occurring at a site are identified by airphoto interpretation and field exploration, (2) site-specific subsurface data are collected, (3) these data are statistically augmented by subsurface data obtained from the same landform at all other sites in the region to create an expanded data base, from which (4) landform/site data estimates of site-specific soil property parameters can be made.

A model for making estimates by this procedure of any site-specific soil property parameter, $\theta(S)$, is as follows. Without site-specific data available, estimates of $\theta(S)$ can be based only on landform data available from other sites. As results of site-specific data become available from field/laboratory programs, they are summarized by a sample likelihood function (giving the relative likelihoods of $\theta(S)$ given the site-specific data), and Baye's theorem is used to make an updated landform/site data estimate of $\theta(S)$, $\theta(S)''$, based on the landform (prior data) and the available site-specific data (sample data).

Here $\theta(S)$ is $\bar{\gamma}_{df}(S)$ in a stratum. In general the observed PDF of γ_{df} will closely resemble an inverse lognormal distribution, which may be transformed to a normal distribution; for most practical purposes directly assuming a normal distribution is adequate. For γ_{df} following a bell shaped PDF, using results modified from Vita (1982), the updated landform/site data estimates of the mean and variance of $\bar{\gamma}_{df}(S)$ are given by

$$E[\bar{\gamma}_{df}(S)]'' = \frac{\bar{x}(S)\cdot n + \bar{x}(Lf)\cdot\alpha}{n+\alpha} \qquad (6a)$$

$$V[\bar{\gamma}_{df}(S)]'' = \frac{\{(n-1)\cdot s(S)^2 + (w-1)\cdot s(Lf)^2 + n\cdot w\cdot[\bar{x}(Lf)-\bar{x}(S)]^2/(n+2)\}}{(w+n)\cdot(w+n-3)} \qquad (6b)$$

where: n is the number of available samples from the site; $\bar{x}(S)$, $s(S)^2$ and $\bar{x}(Lf)$, $s(Lf)^2$ are the sample mean and variance of the site and landform γ_{df} data respectively; α is the estimated ratio of $V[\gamma_{df}(Lf)]$ to the variance of the observed distribution of $E[\bar{\gamma}_{df}(S)]$ for all other sites in the subject landform; and w is calculated from Equation 7. Equations 6a and 6b are based on Bayesian probability theory (Raiffa and Schlaifer 1961) assuming (as observed) that γ_{df} (or its transform) is normally distributed with unknown mean and variance. Prior sample weights α and w are conservatively chosen assuming (1) for n=0 (no sample data): $E[\bar{\gamma}_{df}(Lf)]$ is the best estimate of $E[\bar{\gamma}_{df}(S)]''$; the best estimate of $V[\bar{\gamma}_{df}(S)]''$ is the variance of $E[\bar{\gamma}_{df}(S)]$ for all sites comprising the landform; and $V[\gamma_{df}(S)]$ is equal to the landform data-based estimate of $V[\gamma_{df}(Lf)]$ for N available samples from the landform; and (2) α is conservatively estimated from available data (α=1 being most conservative) and is less than the ratio of $V[\gamma_{df}(S)]$ to $V[\bar{\gamma}_{df}(S)]$. It then follows that w can be calculated from

$$\left\{\frac{N-3}{N-1}\right\}\cdot\frac{(w+n-1)}{(w+n)\cdot(w+n-3)} = \frac{1}{\alpha+n} \qquad (7)$$

Note that α is dependent on site size (ranging from 1.0 for sample-sized "sites" to $\sqrt{N}$ for a landform-sized site); on landform geological/statistical characteristics (α=1 reflects greatest site-to-site variability, α being relatively larger for statistically homogeneous landforms); and on the quality, quantity, and statistical uncertainty of available multiple site data (α=1 reflects greatest uncertainty).

Site data-based estimates are made (utilizing site-specific data only) by neglecting landform prior data and letting α=w=0. Note that Equation 6b is undefined for site data-based estimates where $n \le 3$; the assumption that $\bar{\gamma}_{df}(S)$ is uniformly distributed over a maximum possible range R introduces the least possible bias and gives $V[\bar{\gamma}_{df}(S)] = R^2/12$, centered at $\bar{x}(S)$, which is at least as conservative as is the estimate of R ($R \le 2{,}400 kg/m^3$). In the following case study R=1,100, 800, and 640 kg/m^3 for n=1,2, and 3, respectively. Site data-based estimates of E[FS], V[FS], and Pf are then compared with updated estimates using landform prior data. If they are significantly different (Vita 1982, 1983a, 1983b), the landform prior data can be correspondingly de-emphasized in estimates of Pf by decreasing α and/or weighting and averaging estimates or by forming a PDF on Pf using a beta distribution.

The foregoing tacitly assumes that sampling and testing biases and limitations are controlled and that any skewing affects on parameters are adequately rectified. Also, for landforms characteristically composed of significantly different soil types (e.g., tills), landform parameters may be made soil type-dependent and updated site-specific estimates conditioned on observed soil types. Further, updated site-specific estimates for situations where landforms are uncertain and/or occur in complex associations can be made by conditioning estimates on the expected probability of occurrence of the various potential conditions (the sensitivity of such estimates to the uncertainty, particularly the potential extreme conditions, should be properly evaluated).

REPRESENTATIVE RESULTS AND CASE STUDY

Figure 1 displays generally representative characteristics of the above equations, and includes specific results for a several-acre permafrost site located near Fairbanks in a landform composed of colluvial silt. The site was selected for illustration because it had sufficient soil data to compare predictions based on typically limited data (small n) with predictions based on substantial data (large n--representing much more soil data than would be available for a typical segment site along a route).

A total of n=40 samples (each from one borehole) were obtained from the potential critical failure zone at 3-4m depth. Sample-by-sample results for the first 10 (n=1,2...10) samples are shown; results based on all 40 samples are shown for comparison. Actual site and landform γ_{df} data have been used; other parameters are considered representative but are assumed. Two cases are illustrated: Case 1 (Figure 1c) actual site data with α=3, representing a relatively statistically homogeneous landform; and Case 2 (Figure 1d) actual sample γ_{df} data reduced by 320 kg/m^2 (20 pcf) with α=1, representing a statistically variable landform or one where so little information on other sites is available that a minimum α is considered prudent. Case 2 simulates a site which is in fact substantially more ice rich than the landform as a whole. Both updated landform/site data estimates ($\alpha \ge 1$) and site data-based estimates (α=w=0) are shown.

Case 1 (actual γ_{df} data) shows the efficiency of the updated landform/site data method when the site is in fact similar to the landform: uncertainty in FS and TS is significantly decreased at small n relative to the site data-based estimate. Therefore, the site as is can be qualified as acceptable or unacceptable based on a given Pf criterion with less site-specific data.

Case 2 (actual γ_{df} data minus 320 kg/m^2) shows the effect of a "negative surprise" site--one that is in fact much more ice-rich (and therefore less stable) than the landform as a whole. The benefits of the updated landform/site data method for Case 2 sites are derived through comparison with the site data-based predictions: at small n, potentially significant differences become obvious--suggesting the need for further exploration, analysis, and/or changed design to achieve the desired Pf design criterion.

Figure 1 suggests important characteristics and implications of FS and TS as site-specific random variables. For example, to achieve any given Pf criterion, E[FS] cannot be constant for every slope but must be related to the uncertainty in site-specific conditions, e.g., as reflected by SD[$\bar{\gamma}_{df}$]; similarly for settlement, site-specific effects of uncertainty must be considered. Also, from Figure 1c and Figure 1d it is clear that (as is common to all probabilistic assessments of complex, variable and uncertain states of nature) Pf is itself a random variable, and that uncertainty in Pf will tend to decrease with increasing data (and can be made as small as desired by obtaining sufficient information).

In all cases the potential effect of unanticipated negative conditions should be considered since, regardless of specific technique, subsurface conditions are never completely defined in exploration or analysis--such that critical geotechnical details may go undetected or be inadequately interpreted and eventually result in unsatisfactory performance. Clearly, sound interpretive and integrative judgment is fundamental to successfully developing and utilizing geotechnical evaluations in the design process. Nevertheless, any technically responsible method for evaluating stability must be clearly defined and rational, must objectively account for uncertainty, and must be capable of objective calibration or validation based on observed field performance. Ultimately, any requirements to reduce impacts on facilities from unanticipated negative conditions must rely on appropriate, resilient design.

REFERENCES

Benjamin, Jack R. and Cornell, Allen C., 1970, Probability, statistics, and decision for civil engineers, McGraw-Hill, New York.

Crory, Frederick E., 1973, Settlement associated with thawing of permafrost, in Permafrost-The North American contribution to the Second International Conference, Yakutsk: Washington, D.C., National Academy of Sciences.

Harr, M.E., 1977, Mechanics of particulate media - A probabilistic approach, McGraw-Hill, New York.

Kreig, Raymond A. and Reger, Richard D., 1976, Preconstruction terrain evaluation for the Trans-Alaska pipeline project, in Geomorphology and engineering, Donald R. Coates, Ed., Dowden, Hutchinson, and Ross.

Raiffa, H. and Schlaifer, R., 1961, Applied statistical decision theory, Harvard University Press, Cambridge.

Vallejo, Luis E., 1980, A new approach to the stability of thawing slopes: Canadian Geotechnical Journal, V. 17, pp 607-612.

Vita, Charles L., 1982, A landform-based probabilistic methodology for site characterization, in Proceedings Nineteenth Annual Symposium on Engineering Geology and Soils Engineering, Idaho State University.

___________, 1983a, The systematic use of geotechnical data for characterization of transportation routes," in Proceedings Twentieth Annual Symposium on Engineering Geology and Soils Engineering, Idaho Department of Transportation, Boise, Idaho.

___________, 1983b, A landform-based probabilistic methodology for geotechnical characterization and analysis of transportation routes, unpublished manuscript available from the author.

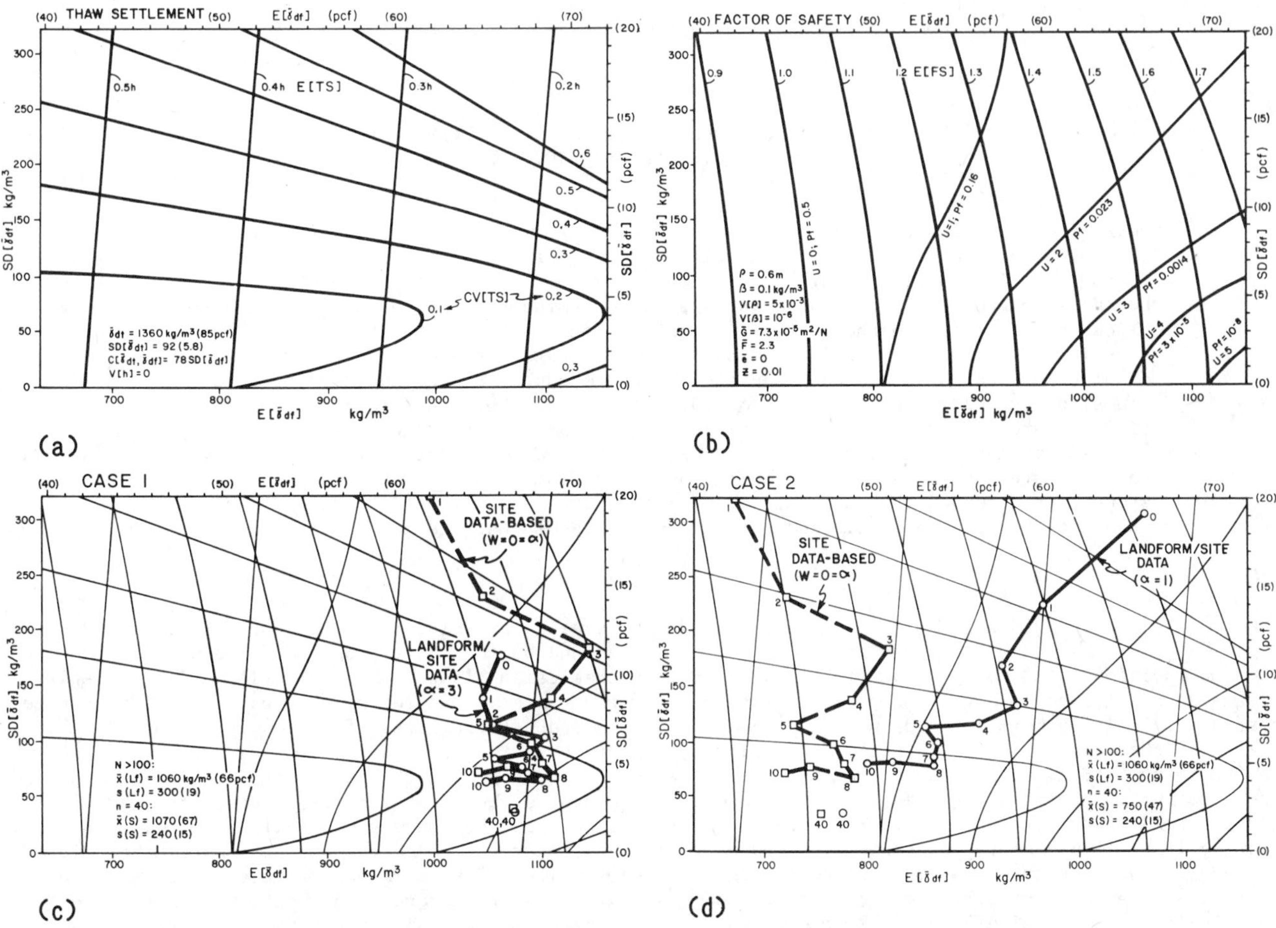

FIGURE 1 Representative results for a stratum of frozen silt, plotted as a function of the estimated expected value, E[$\bar{\gamma}_{df}$], and estimated standard deviation, SD[$\bar{\gamma}_{df}$], of the (unknown) average frozen dry density of the stratum, $\bar{\gamma}_{df}$(S). (a) Average Thaw Settlement, TS, (b) Factor of Safety, FS, Stability Margin, U, and Probability of Failure, Pf, (c) Case Study 1: sample-by-sample results for a site-specific stratum that is similar to the landform as a whole, i.e., $\bar{\gamma}_{df}(S) \approx \bar{\gamma}_{df}(Lf)$, (d) Case Study 2: sample-by-sample results for a site-specific stratum that is substantially more ice-rich than the landform as a whole, i.e., $\bar{\gamma}_{df}(S) << \bar{\gamma}_{df}(Lf)$. Note numbers (1,2,3,...10,40) next to box and circle symbols on case study results indicate number of samples from the site, n.

STONE POLYGONS: OBSERVATIONS OF SURFICIAL ACTIVITY

John D. Vitek

Department of Geography, Oklahoma State University,
Stillwater, Oklahoma 74078 USA

Small stone polygons, approximately one meter between centers, have developed in a 675 m^2 depression at 3,865 m in the Blanca Massif of the Sangre de Cristo Mountains in southern Colorado, USA (37°35' N., 105°30' W.). The absence of vegetation and human disturbance permitted observations to determine the rate of horizontal displacement of stones in the centers of the polygons. Photographs taken of the centers in August 1975, 1978, 1979, 1980, and 1982 provide the evidence from which displacement can be measured. Gaps in the photographic record occurred because the site was underwater in August of 1976, 1977, and 1981. A sample of 173 identifiable stones, greater than 1.3 cm in length, moved an average of 4.76 cm during a 7 year period; one stone moved 18.3 cm during these observations. The average rate of movement per stone was 0.68 cm per year from 1975 to 1982. In general the stones moved toward the gutters. Clast motion can be attributed to heaving and thrusting by frost action, needle ice, and perhaps lacustrine ice.

INTRODUCTION

The locations of stone polygons in alpine areas can be divided into two categories: sites that are relatively well drained and sites in depressions that are periodically covered by water. In the first situation, the stone polygons are large, often exceeding several meters in diameter, and regarded as relict landforms. The vegetation growing in the fines of the center of a polygon hinders, if not prevents, movement of rock fragments. Furthermore, stability of this micro-landform is indicated by lichens on rocks in the gutter. In the second location, the absence of vegetation can frequently be explained by the presence of water. Stone polygons are common micro-landforms in depressions subjected to periodic flooding and in the shallow areas of alpine lakes (Dionne 1974 and Shilts and Dean 1975). Because the water level in the lakes or depressions can fluctuate greatly, opportunities exist to study the surficial pattern and movement of clasts in the centers of stone polygons. The absence of vegetation plus the availability of water ensures the action of frost processes on these landforms.

Previous researchers have used several techniques to assess change in stone polygons over time. Cailleux and Taylor (1954) studied the superposition of photographs of stone polygons exposed over 13 years and concluded that no change had occurred. The same authors also observed that painted stones moved more than 10 cm in 2 years in the center of one stone polygon. Tallis and Kershaw (1959) noted that small stone polygons, 45-60 cm in diameter, formed in one winter and were obliterated by wet weather in the following summer. Pissart (1964) measured the movement of 6 stones over 17 years in the center of a disrupted stone polygon. One stone moved 21 cm in this study. French (1976) summarized the methods that have been employed to monitor the amount of movement in stone polygons. These methods include the use of painted objects, the deformation of layers, leveling and direct measurement, vertical and ground photography, frost heave recorders, geochemical analyses, and carbon-14 dating. Despite the variety of efforts, few generalizations can be made about the rates of motion for clasts in the centers of stone polygons.

The purpose of this research is to document the amount of horizontal displacement of surficial stones in the centers of small stone polygons that are located in a depression on a moraine at 3,865 m. With a 7 year record, this research can provide answers to questions such as: What is the rate of stone displacement in the centers of stone polygons, and what is the predominant direction of displacement? Moreover, the rates and direction of stone displacement may provide additional evidence for evaluating the hypotheses of stone polygon formation. Finally, accurate estimates of the impact of frost action are becoming more significant as human activities continue to encroach upon alpine areas.

STUDY AREA

Within the Sangre de Cristo Mountains of southern Colorado, USA (37°35' N., 105°30' W.), a small depression, 675 m^2, at 3,865 m (Figure 1) contains stone polygons with approximately one meter between centers (Figure 2). The depression is located in the upper portion of the northeast-facing Dutch Creek cirque, a compound cirque that was probably ice-free since the end of the Pleistocene. No evidence of Neoglacial ice exists in the vicinity. Glacial till on the floor of the elongated cirque is composed of primarily felsic and hornblende gneisses and some metabasalt and metagraywacke, all of Precambrian age.

The site, first observed in August 1975, is devoid of lichens and shrubs, although a few clumps of grass have occasionally sprouted in the centers of the polygons. The general absence of vegetation can be explained, in part, by the presence of

FIGURE 1 An oblique aerial view of the study area in the Sangre de Cristo Mountains of Colorado. The arrow points to the area with stone polygons.

water. Standing water was observed in the gutters in 1975, although the level of the water was 25 cm below the surface. Maximum water level in the depression is defined by the vegetation around the depression. In the summer of 1976, for example, the water level coincided with the limit of vegetation. In 1981 during a rainstorm, the water level rose more than 25 cm in less than 48 hours. Observations on May 1, 1977, and May 6, 1981, at the beginning of spring melt, revealed lacustrine ice over the depression. Sufficient melting had already taken place and I was not able to determine if the ice was frozen to the surface of the stone polygons.

The fines in the center of the stone polygons are on an average 30% sand, 39% silt, and 31% clay. Additional textural analyses are planned when the site is excavated. Detailed studies of mineralogy have not been completed but will be useful in explaining the source of silts and clays. Extensive eolian activity in the San Luis Valley, west

FIGURE 2 A view of the stone polygons in a depression at 3,865 m. An ice axe, height of 80 cm, provides a measure of scale.

of the study area, is one hypothesis for the source of this fine material. Within the drainage area of the depression, only one small channel has developed, thereby suggesting that the influx of fines has had an effect on the development of stone polygons. This relationship will be investigated at a later date.

The study site is located in a tundra environment approximately 200 m above the tree line. No evidence has been observed to suggest that the study site was once forested. The nearest climatic station, the Great Sand Dunes National Monument, is 16 km northwest but is too low (2,499 m) to be of significant value in assessing the nature of the weather and climate at the depression. Although the climate station on Niwot Ridge in the Front Range is 270 km north, a similar elevation (3,750 m) and aspect on the landscape permit interpolation of general conclusions. Andrews (1973) reported precipitation averages 102 cm yr^{-1} and temperature averages $-3.8^{o}C$ on Niwot Ridge. Frost action is a very effective agent of surficial change when the freezing point, $0^{o}C$, is crossed frequently during the periods when water is available. The presence of talus, stone stripes, rock glaciers, and a patterned fen (Vitek and Rose 1980) attest to the effectiveness of frost action in the study area.

METHODOLOGY

The rates of many changes in a periglacial environment are perceptible only with long-term observations. The micro-landforms in the study area have had approximately 9,000 years in which to evolve, hence the actual rates of development and change may be extremely slow. Stone polygons represent one type of response to the interaction of atmospheric conditions and processes with lithologic and topographic variables. In effect, stone polygons have developed at this location rather than some other form of patterned ground. Knowledge of the types and rates of change in the stone polygons may provide information about past and future development.

The lack of sites in the Sangre de Cristo Mountains in which stone polygons may be active contributed to the conservative approach for the analysis of these forms. With repetitive photography, the stone polygons can remain undisturbed until such time that details of internal characteristics are essential to the interpretation of surficial activity. Painted stones were not added to the centers of the polygons to detect the rates of displacement in order not to disrupt the system. Recently, Caine (1981) reported that the addition of painted objects biases the measurements of change. Without paint, more effort is necessary to trace the movement of stones, but the results should provide more accurate estimates of the natural process.

The small size of these stone polygons allowed the centers to be photographed on one frame of a 35 mm camera with a 50 mm lens, thus minimizing the distortion introduced by photography (Figures 3 and 4). A 150 mm reference scale was placed within each stone polygon to aid in the alignment of photographs on a yearly basis. An assumption was made that the position of larger stones in the gutters would not change significantly over the

FIGURE 3 A 1982 (t_7) photograph of the stone polygon center referred to in the text as center A. Scale is 150 mm (6 in).

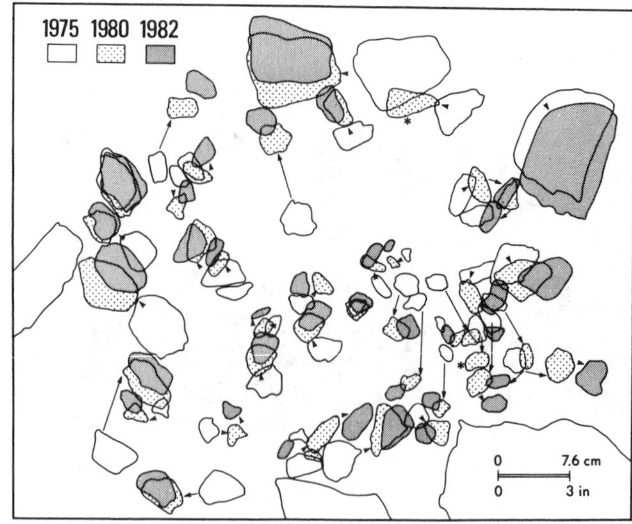

FIGURE 5 Horizontal displacement of stones monitored in center A from 1975 to 1982. The arrows indicate the direction of clast movement and the asterisks indicate clasts that could not be located in 1982.

length of the study. Although the pattern of gutters and centers appears fixed, stones in the gutters may be disturbed by frost action, rodents, or people. Alignment of photographs confirms, however, that the disruption of stones in the gutters is minimal.

Laboratory experiments have proven that stones move to the surface and toward the edge of experimental systems subjected to alternate freeze-thaw (Corte 1962, 1966; Kaplar 1965; Mackay and Burrous 1979). Nicholson (1976) recently used field observations to conclude that a combination of the circulatory movement model and the radial movement model can explain many forms of patterned ground. Although the photographic method used in this analysis was designed to evaluate only the horizontal component of motion in a three dimensional system, the appearance of new stones on the surface through time also confirms the presence of vertical motion.

FIGURE 4 A 1982 (t_7) photograph of the stone polygon center referred to in the text as center D. Scale is 150 mm (6 in).

Maps of stone position in the centers of polygons were created from photographic enlargements (28 x 36 cm). Maps from different time periods were aligned using a Kail reflecting projector to correct scale distortions introduced in the production of the photographs. For simplicity and accuracy, the stones that were mapped exceeded 1.3 cm in length because they could be easily identified from year to year. By comparing color slides taken during the 7 years, only 10 stones from the initial sample have been lost. The linear distance a stone moved was calculated as the distance the center of the stone moved relative to its initial position. Although stones are three dimensional objects, motions such as rotation or tilt have not been evaluated in this study.

ANALYSIS

Positions of the stones within four stone polygons (labeled A, B, C, and D) were recorded on August 6, 1975 (t_0). Although shadows were present when these photographs were exposed in midafternoon, they have not posed a major problem in interpreting change. In subsequent years, however, the photographs were taken during overcast conditions to eliminate any effect that shadows might have in masking true position. Because the depression was covered with water in August 1976 (t_1) and 1977 (t_2), the centers were not photographed for the second time until August 1978 (t_3). At this time initial photographs were taken of 12 additional centers. All 16 centers were photographed in August 1979 (t_4), August 1980 (t_5), and August 1982 (t_7).

Summary statements for this analysis include changes in the horizontal position of stones in four centers from 1975 (t_0) to 1980 (t_5) and from

1980 (t_5) to 1982 (t_7). Yearly summaries cannot be produced because of the gaps in the photographic record. In retrospect, comparison of two photographs for total change introduces less error than the summation of four changes between five photographs or six comparisons over seven years. The resultant maps of change (see Figure 5 for an example of the map of center A) were summarized in tabular form (Table 1). Total horizontal displacements on centers A (Figure 3) and B were greater than on centers C and D (Figure 4). In essence, centers A and B have similar physical characteristics in that smaller stones comprise the centers compared to scattered larger stones on centers C and D. The larger stones should offer greater resistance to the forces of motion, hence the average total displacement would be less. The range of distances moved by individual stones on sites A and B also supports the conclusion of greater resistance to movement in sites C and D.

For comparative purposes, horizontal displacement can also be expressed as a rate of movement on a yearly basis. Table 2 provides values based upon a 5 year record, a 2 year record, and the average per year based upon 7 years. Centers A (Figure 3 and Figure 5) and B exhibit virtually the same rates of change for the 5 year and 2 year periods. Although stones moved across centers C and D (Figure 4) at the same rate for 5 years, during the last 2 years, the rate of movement in center C tripled while that in center D only doubled. The mean values derived for 7 years must be evaluated carefully because one value, high or low, can skew the results. Similar rates at centers A and B over seven years may be accurate measures of change because the sites have similar physical characteristics. Greater resistance to motion in the form of larger stones on centers C and D would yield lower rates of change. Increases in the rates of change between 1980 and 1982 for centers C and D, however, suggest that additional observations are necessary before mean values are accepted as accurate.

Movement of stones can be attributed to three processes: frost action (heaving and thrusting), the growth of needle ice, and lacustrine ice. On Niwot Ridge, Colorado, Fahey (1974) measured 25-30 cm of vertical displacement in frost boils, i.e., areas without vegetation but where sufficient water was available. Because these stone polygons are devoid of vegetation and have an ample water supply, frost action should generate the forces necessary to move stones. The centers of the stone polygons are a matrix of fine material surrounded by rock gutters, so the freezing plane may penetrate from the sides in addition to the top, thereby affecting the forces on the surface stones. The presence of other periglacial landforms in the immediate vicinity confirms that frost action is an effective agent of change in this area.

The impact of needle ice on surface phenomena has been documented by many researchers, including Mackay and Matthews (1974), Hastenrath (1977), Heine (1977), and Meentemeyer and Zippin (1981). Needle ice was observed in the centers of several stone polygons in May 1981. Unfortunately, the depression contained sufficient water and ice to cover the centers studied in this analysis. Several other centers, however, that were above the level of the water exhibited needle ice. Small stones were uplifted by the ice, whereas larger stones appeared undisturbed. Moreover, needle ice has been observed in June, July, and August over the past 7 years in conjunction with other investi-

TABLE 1 Summary of Measurements for Seven Years: 1975-1982.

Site	N(1975-1982)	Avg. total displacement per sample	Standard deviation	Range of values (1975-1982)
A	41 - 39	5.85 cm	2.79 cm	1.3 - 13.0 cm
B	43 - 40	6.24 cm	3.79 cm	0.9 - 18.3 cm
C	46 - 42	3.80 cm	2.12 cm	0.6 - 8.2 cm
D	43 - 42	3.14 cm	1.85 cm	1.0 - 7.5 cm
	173 -163	4.76 cm sample^{-1} years^{-1}		

TABLE 2 Rate of Displacement Per Sample Per Year.

Site	N	1975-1980	N	1980-1982	1975-1982
A	41	0.82 cm	39	0.87 cm	0.84 cm
B	43	0.89 cm	40	0.90 cm	0.89 cm
C	46	0.36 cm	42	1.00 cm	0.54 cm
D	43	0.37 cm	42	0.65 cm	0.45 cm
		0.60 cm		0.85 cm	0.68 cm sample^{-1} year^{-1}

gations in the vicinity of the stone polygons. Based upon these observations, needle ice is considered as one of the processes responsible for the movement of stones in the centers of the polygons.

The final mechanism that may contribute to stone movement is lacustrine ice. Ice shove ridges are common in areas subjected to the forces of expanding ice (Washburn 1980). Although no definitive ridge exists around this depression, lacustrine ice was observed in 1977 and 1981. At the time of these observations, however, spring melting had already begun and the ice cover was incomplete. Given the rigors of the climate at this elevation, lacustrine ice may freeze to the surface of the stone polygons and contribute to stone movement upon expansion and cracking or upon breakup in the spring. Wind-driven ice could also dislodge stones. In summary, all three processes probably contribute to the horizontal movement of stones in the centers of stone polygons.

CONCLUSION

Monitoring 173 stones in the centers of four stone polygons for 7 years revealed an average total horizontal displacement of 4.76 cm, or an average yearly rate of displacement of 0.68 cm. The observed maximum horizontal displacement was 18.3 cm, although some of the lost stones could have been moved greater distances. Based upon the variability observed in these four centers, additional information is necessary to determine the relationship between stone size and distance moved. This analysis traced stones greater than 1.3 cm in length. A future project will calculate displacement based upon stone size. Moreover, additional studies with stones smaller than 1.3 cm are possible, such as the changes in percent of surface covered by stones in response to the vertical migration of stones.

Through August 1982, four centers of stone polygons had been photographed for 7 years and 12 others for 4 years. Data collection via photographs will continue for several more years before additional conclusions are drawn. Calculating mean values from short records could result in significant error. While the values given for average total displacement and yearly rate of change are accurate based upon the procedures employed, only additional observations can confirm if the sample period was representative of the rates of change.

ACKNOWLEDGMENTS

I would like to express my appreciation to David R. Butler, Oklahoma State University, and John R. (Rick) Giardino, Texas Tech University, for their helpful comments on the draft of this manuscript. I wish to thank the Department of Geography at Oklahoma State University for a portion of the financial support for this project. Finally, to Peg Vitek, Andy Deutch, Sue Berta, Mark Gregory, Susan Panciera, and Neil Salisbury, a sincere thank you for your support in the field.

REFERENCES

Andrews, J.T., Webber, P.J., Thorn, C., and LeDrew, E., 1973, Tundra environment on Niwot Ridge, Colorado Front Range: Field Guide No. 7, Rocky Mountain Section, Geological Society of America.

Cailleux, A., and Taylor, G., 1954, Cryopédologie, études des sols gelés: Expeditions Polaries Francaises, Missions Paul-Emile Victor IV: Paris, Hermann and Cie, Actualités Scientifiques et Industrielles 1203.

Caine, N., 1981, A source of bias in rates of surface soil movement as estimated from marked particles: Earth Surface Processes and Landforms, v. 6, p. 69-75.

Corte, A.E., 1962, Horizontal sorting. The frost behavior of soils: Laboratory and field data for a new concept: U.S. Army Cold Regions Research and Engineering Laboratory, Res. Rep. 85.

Corte, A.E., 1966, Particle sorting by repeated freezing and thawing: Biuletyn Peryglacjalny, v. 15, p. 175-240.

Dionne, J.C., 1974, Crysols avec triage sur rivage et fond de lacs, Quebéc central subarctique: Revue de Géographie de Montréal, v. 28, p. 323-342.

Fahey, B.D., 1974, Seasonal frost heave and frost penetration measurements in the Indian Peaks region of the Colorado Front Range: Arctic and Alpine Research, v. 6, p. 63-70.

French, H.M., 1976, Current field measurements concerning the nature and rate of periglacial processes: Biuletyn Peryglacjalny, v. 25, p. 79-91.

Hastenrath, S., 1977, Observations of soil frost phenomena in the Peruvian Andes: Zeitschrift für Geomorphologie N.F., v. 21, p. 357-362.

Heine, K., 1977, Zur morphologischen Bedeutung des Kammeises in der subnivalen Zone randtropischer semihumider Hochgebirge: Zeitschrift für Geomorphologie N.F., v. 21, p. 57-78.

Kaplar, C.W., 1965, Stone migration by freezing of soil: Science, v. 149, p. 1520-1521.

Mackay, J.R. and Burrous, C., 1979, Uplift of objects by an upfreezing ice surface: Canadian Geotechnical Journal, v. 16, p. 609-613.

Mackay, J.R. and Matthews, W.H., 1974, Needle ice striped ground: Arctic and Alpine Research, v. 6, p. 79-84.

Meentemeyer, V. and Zippin, J., 1981, Soil moisture and texture controls of selected parameters of needle ice growth: Earth Surface Processes and Landforms: v. 6, p. 113-125.

Nicholson, F.H., 1976, Patterned ground formation and description as suggested by low Arctic and subarctic examples: Arctic and Alpine Research, v. 8, p. 329-342.

Pissart, A., 1964, Vitesse des mouvements du sol au Chambeyron (Basses Alpes): Biuletyn Peryglacjalny, v. 14, p. 303-310.

Shilts, W.W. and Dean, W.E., 1975, Permafrost features under Arctic Lakes, District of Keewatin, Northwest Territories: Canadian Journal of Earth Science, v. 12, p. 649-662.

Tallis, J.H. and Kershaw, K.A., 1959, Stability of stone polygons in North Wales: Nature, v. 183 (no. 4659), p. 485-486.

Vitek, J.D. and Rose, E.M., 1980, Preliminary observations on a patterned fen in the Sangre de Cristo Mountains, Colorado, U.S.A.: Zeitschrift für Geomorphologie N.F., v. 24, p. 393-404.

Washburn, A.L., 1980, Geocryology: New York, Halsted Press.

A HIERARCHICAL TUNDRA VEGETATION CLASSIFICATION ESPECIALLY DESIGNED FOR MAPPING IN NORTHERN ALASKA

Donald A. Walker

Institute of Arctic and Alpine Research
University of Colorado, Boulder, Colorado 80309 USA

This paper presents a tundra vegetation classification scheme that is designed for describing vegetation at four levels: (1) very-small-scale maps, (2) LANDSAT-derived maps, (3) photo-interpreted maps, and (4) plant community descriptions. A system of nomenclature is described that links the four levels.

INTRODUCTION

Land-use planning in tundra regions utilizes knowledge of vegetation more than any other terrain factor. The vegetation gives insight to a host of environmental variables, many of which are related to permafrost, including soil properties, depth of the active layer, temperature regime and snow regime. There are three primary methods of interpreting vegetation: (1) plant community descriptions at ground level, (2) aerial photographs, and (3) multi-spectral satellite data. Currently there is no classification system that relates the map units from one method to those of the other two. Viereck and Dyrness (1980) developed a hierarchical method of vegetation classification for Alaska, but it is not specifically designed for mapping and is particularly difficult to apply to LANDSAT-dervived classifications. The classification scheme presented here (Table 1) meets three basic criteria:

- At the LANDSAT level, the land cover units are based on those characteristics of the vegetation that can be classified consistently from LANDSAT data.
- At lower levels, the classification system is consistent yet flexible enough to describe the great variety of tundra communities. At the community level, the system is open-ended so that units that do not accurately describe the vegetation of a given area need not be used.
- The lower level units can be grouped within the higher level units with a minimum of overlap so that there is clear compatibility between levels.

The highest classification level, Level A, is very general and useful for very-small-scale vegetation maps of Alaska. Level B consists of LANDSAT-level land cover units that can be interpreted using digital multi-spectral satellite data. Level C consists of vegetation subunits that can be interpreted from aerial photographs if supplemented with adequate ground truth. Level D consists of individual plant communities, determined by ground surveys. The following discussion presents the classification system "from the ground up," starting with level D.

LEVEL D--PLANT COMMUNITY NAMES AND UNITS FOR VERY-LARGE-SCALE MAPS

Level D units describe specific vegetation classes that correspond approximately to the stand types of Marr (1967) the associations of Daubenmire (1952) or Braun-Blanquet (1932), and the plant community or community type of Whittacker (1967). At this level there are many units and the system is open, such that any newly described vegetation community can be easily added. The nomenclature used for describing vegetation at this level always follows fixed guidelines. The following discussion explains the nomenclature system for plant communities and noncomplex map units and then for complex map units. Complex map units contain two or more distinct vegetation communities, and each community covers at least 30% of the map unit. Level D is appropriate for very-large-scale maps of small areas (e.g. a 1:1,000-scale map of a 5 acre ecology study site).

Noncomplex Units

Plant community names have four parts that are always arranged in the following sequence: (1) a site moisture term, (2) the dominant plant taxa, (3) the dominant plant growth forms, and (4) an overall physiognomic descriptor. The site moisture term can be dry, moist, wet or aquatic. These are subjective terms based on the soil moisture at the end of the growing season. The site moisture term is followed by the names of the dominant plant taxa, one or more from each of the representative shrub, herb, and cryptogam layers of the canopy. The number of taxa is kept to the minimum required to adequately distinguish the community from others on the map; the total normally does not exceed six.

The dominant growth forms follow next and can be any of the following: (1) tall shrub (>1.5 m), (2) low shrub (0.2 to 1.5 m), (3) dwarf shrub (<0.2 m), (4) sedge, (5) grass, (6) rush, (7) tussock sedge, (8) forb, (9) moss, (10) crustose lichen, and (11) fruticose lichen. The term graminoid is used when two or more of the dominant

TABLE 1 Hierarchical Classification Scheme for Tundra on the Arctic Coastal Plain and Foothills of Northern Alaska.

Level A VERY SMALL SCALE UNITS	Level B LANDSAT LAND COVER UNITS (suggested map colors)	Level C PHOTO-INTERPRETED MAP UNITS	Level D TYPICAL PLANT COMMUNITIES
A. Water	I. Water (light blue)	Ia. Water	No vegetation
B. Wet Tundra	II. Very Wet Tundra (dark blue)	IIa. Shallow Water (pond margins)	No vegetation
		Noncomplex subunits: IIb. Aquatic Graminoid Tundra	Aquatic Arctophila fulva Grass Tundra Aquatic Carex aquatilis Sedge Tundra
		IIc. Aquatic Forb Tundra	Aquatic Hippuris vulgaris, Caltha palustris, Menyanthes trifoliata Forb Tundra (aquatic tundra, inland areas)
		Common complex subunits: IId. Water/ Tundra Complex: (pond complex)	Typical communities listed under Ia, IIa, IIIa, IIIb, and Va
	III. Wet Tundra (dark green)	Noncomplex subunits: IIIa. Wet Sedge Tundra	Wet Carex aquatilis, Scorpidium scorpioides Sedge Tundra (wettest facies of wet alkaline tundra) Wet Carex chordorrhiza, Eriophorum scheuchzeri, Potentilla palustris Sedge Tundra (wet acidic tundra - inland areas) Wet Carex aquatilis, Eriophorum angustifolium, Pedicularis sudetica ssp. albolabiata, Drepanocladus brevifolius Sedge Tundra (wet alkaline tundra) Wet Eriophorum angustifolium, Dupontia fisheri, Campylium stellatum Graminoid Tundra (wet acidic tundra, coastal areas)
		IIIb. Wet Graminoid Tundra (wet saline Tundra)	Wet Carex subspathacea, Puccinellina phryganodes, Stellaria humifusa, Cochlearia officinalis Sedge Tundra
		Common Complex Subunits: IIIc. Wet Sedge Tundra/ Water Complex (pond complex)	Typical communities listed under Ia, IIa and IIIa
		IIId. Wet Sedge/ Moist Sedge, Dwarf Shrub Tundra Complex (wet patterned-ground complex)	Typical communities listed under IIIa and Va
C. Moist Tundra	IV. Moist/ Wet Tundra Complex (light green)	Common Complex Subunits: IVa. Moist Sedge, Dwarf Shrub/ Wet Graminoid Tundra Complex (moist patterned-ground complex)	Typical communities listed under Va and IIIa
	V. Moist or Dry Tundra (tan)	Noncomplex Subunits: Va. Moist Sedge, Dwarf Shrub Tundra	Moist Carex bigelowii, Eriophorum angustifolium ssp. triste, Dryas integrifolia, Salix reticulata, Tomenthypnum nitens, Thamnolia subuliformis Sedge, Dwarf Shrub Tundra (moist alkaline tundra) Moist Luzula arctica, Poa arctica, Saxifraga cernua, Salix planifolia ssp. pulchra, Dicranum elongatum, Ochrolechia frigida Graminoid, Dwarf Shrub, Crustose Lichen Tundra (moist coastal acidic tundra) Moist Carex aquatilis, Eriophorum angustifolium ssp. triste, Salix planifolia ssp. pulchra, Campylium stellatum Sedge, Dwarf Shrub Tundra (moist acidic tundra, wetter facies) Moist Carex bigelowii, Dryas integrifolia, Lupinus arcticus, Salix lanata ssp. richardsonii, Arctagrostis latifolia, Equistetum arvense, Tomenthypnum nitens, Sedge, Dwarf shrub, Forb Tundra (moist non-tussock alkaline tundra)
		Vb. Moist Tussock Sedge, Dwarf Shrub Tundra	Moist Eriophorum vaginatum, Dryas integrifolia, Salix reticulata, S. arctica, Tomenthypnum nitens, Thamnolia subuliformis, Tussock Sedge, Dwarf Shrub Tundra (alkaline tussock tundra) Moist Eriophorum vaginatum, Dryas integrifolium, Salix planifolia ssp. pulchra, Salix reticulata, Hylocomium splendens, Ptilidium ciliare, Cetraria cucullata Tussock Sedge, Dwarf Shrub Tundra (neutral to slightly acidic tussock tundra)
		Vc. Dry Dwarf Shrub, Crustose Lichen Tundra (Dryas tundra)	Dry Dryas integrifolia, Carex rupestris, Oxytropis nigrescens, Salix reticulata, Ditrichum flexicaule, Lecanora epibryon Dwarf Shrub, Forb, Crustose Lichen Tundra (Dryas river terraces) Dry Dryas integrifolia, Astragalus alpinus, Oxytropis borealis, Salix reticulata, Distichium capillaceum, Lecanora epibryon Dwarf Shrub, Forb, Crustose Lichen Tundra (Dryas river terraces)
		Vd. Dry Dwarf Shrub, Fruticose Lichen Tundra (Dry acidic tundra)	Dry Dryas octopetala, Arctostaphylos alpina, Empetrum nigrum Salix phlebophylla, Rhytidium rugosum, Alectoria nigricans Dwarf Shrub, Fruticose Lichen Tundra (dry acidic tundra on kames and moraines in foothills) Dry Salix rotundifolia, Pedicularis kanei, Luzula arctica, Polytrichum sp. Alectoria nigricans, Cetraria islandica Dwarf Shrub, Fruticose Lichen Tundra (dry acidic tundra near coast)

TABLE 1 (Continued)

Level A VERY SMALL SCALE UNITS	Level B LANDSAT LAND COVER UNITS (suggested map colors)	Level C PHOTO-INTERPRETED MAP UNITS	Level D TYPICAL PLANT COMMUNITIES
		Common Complex Subunit: Ve. Moist Graminoid, Dwarf Shrub Tundra/ Barren Complex (frost-scar complex)	Typical communities listed under Va and Vb plus either completely barren, frost-scars or communities such as: Dry Saxifraga oppositifolia, Dryas integrifolia, Chrysanthemum integrifolium, Juncus biglumis, Arctagrostis latifolia, Ochrolechia frigida Barren (alkaline frost scars)
	VI. Moist Tussock Sedge, Low Shrub Tundra (brown)	Noncomplex Subunit: VIa. Moist Tussock Sedge, Low Shrub Tundra (acidic tussock tundra)	Moist Eriophorum vaginatum, Salix planifolia ssp. pulchra, Betula nana ssp. exilis, Ledum palustre ssp. decumbens, Vaccinium spp., Sphagnum spp., Cladonia spp. Tussock Sedge, Low Shrub Tundra
		Complex Subunits: VIb. Moist Tussock Sedge, Low Shrub Tundra/ Tall Shrub Complex (alder tundra savanna)	Typical communities listed under Vb and VIa plus widely spaced Alnus crispa
		VIc. Moist Tussock Sedge, Low Shrub/ Wet Low Shrub Tundra Complex (water track complex) Note: This complex may appear as sub-unit of VI or VII depending on the density of water tracks.	Typical communities listed under VIa and VIIIa
	VII. Moist Shrub-rich Tundra (dark brown)	Noncomlex Subunits: VIIa. Moist Low Shrub, Tussock Sedge Tundra (shrubby tussock tundra)	Moist Salix planifolia ssp. pulchra, Betula nana ssp. exilis, Eriophorum vaginatum, Ledum palustre ssp. decumbens, Vaccinium spp., Sphagnum spp. Low Shrub, Tussock Sedge Tundra
		VIIb. Moist Dwarf Shrub, Moss Tundra (Sphagnum-rich dwarf shrub tundra)	Moist Rubus chamaemorus, Ledum palustre spp. decumbens, Betula nana, spp. exilis, Vaccinium spp., Sphagnum spp., Cladonia spp. Dwarf Shrub, Moss Tundra
		Complex Subunit: VIIc. Moist Tussock Sedge, Low Shrub/ Wet Low Shrub Tundra Complex (water track complex -- see note under VIc.)	Typical communities listed under VIa and VIIa
D. Shrubland	VIII. Shrubland or Shrub Tundra (red)	VIIIa. Wet Low Shrub Tundra	Wet Salix planifolia pulchra, Betula nana ssp. exilis, Sphagnum spp. Low Shrub Tundra (wet willow tundra) Wet Betula nana ssp. exilis, Sphagnum spp. Low Shrub Tundra (wet birch tundra)
		VIIIb. Moist Low Shrub Tundra	Moist Betula nana spp. exilis, Ledum palustre ssp. decumbens, Salix planifolia ssp. pulchra, Vaccinium spp., Cladonia spp. Low Shrub Tundra (moist birch tundra) Moist Betula nana ssp. exilis, Vaccinium uliginosum, Potentilla fruticosa, Shepherdia canadensis, Salix spp., Festuca altaica Low Shrub Tundra (south facing shrub tundra in foothills) Moist Alnus crispa, Betula nana ssp. exilis, Salix spp. Low Shrub Tundra (alder shrub tundra)
		VIIIc. Moist Shrubland (closed riparian shrubland)	Moist Salix alaxensis, Salix spp. Tall Shrubland (willow riparian shrubland) Moist Betula nana ssp. exilis, Betula glandulosa Low Shrubland (birch riparian shrubland)
E. Partially Vegetated and Barren	IX. Partially Vegetated (violet)	Riparian areas: IXa. Dry, Barren/ Low Shrub Complex (open riparian shrubland)	Typical communities and ground cover listed under VIIIc and Xa.
		IXb. Dry Barren/ Dwarf Shrub, Forb Grass Complex (forb-rich river bars)	Typical communities listed under Vc, Xa, also mixed forb grass and dwarf shrub communities such as: Dry Bromus pumpellianus, Festuca rubra, Astragalus alpinus, Androsace chamaejasme, Salix ovalifolia Grass, Forb, Dwarf Shrub Tundra (forb-rich river bars) Dry Dryas integrifolia, Artemisia borealis, A. glomerata, Salix ovalifolia, Androsace chamaejasme Dwarf Shrub, Forb Tundra (Dryas river bars near arctic coast)
		IXc. Dry Barren/ Forb Complex	Dry Epilobium latifolium, Artemisia arctica, Wilhelmsia physodes Forb Barren (active river channels)
		IXd. Dry Barren/ Low Shrub Forb Complex (open riparian shrubland)	Dry Salix alaxensis, Salix spp. Hedysarum spp. Astragalus alpinus, Equisetum arvense, Oxytropis campestris, O. borealis, Anemone parviflora Low Shrub, Forb Tundra (river bars inland)
		Sand Dunes: IXe. Dry Barren/ Grass Complex (sand dune grassland)	Dry Elymus arenarius Grass Tundra (sand dune grassland)
		IXf. Dry Barren/ Dwarf Shrub, Grass Complex (sand dune steppe)	Dry Artemesia borealis, A. glomerata, Deschampsia caespitosa Trisetum spicatum Dwarf Shrub, Grass Tundra (sand dune steppe)
		IXg. Dry Barren/ Low Shrub Complex (sand dune scrub)	Dry Salix alaxensis, S. glauca, Elymus arenarius, Carex obtusata, Dryas integrifolia Low Shrub, Tundra (sand dune scrub)

TABLE 1 (Continued)

Level A VERY SMALL SCALE UNITS	Level B LANDSAT LAND COVER UNITS (suggested map colors)	Level C PHOTO-INTERPRETED MAP UNITS	Level D TYPICAL PLANT COMMUNITIES
		Beaches, river deltas, and estuaries: IXh. Wet Barren/ Wet Sedge Tundra Complex (barren/ saline tundra complex)	Typical ground cover listed under IIIb
		IXi. Dry Barren/ Forb, Graminoid Complex (coastal barrens)	Dry Cochlearia officinalis, Stellaria humifusa, Puccinellia phryganodes, P. andersonii, Salix ovalifolia, Potentilla pulchella Forb, Graminoid Tundra (coastal saline barrens)
		Mountainous areas: IXj. Dry Barren/ Dwarf Shrub, Graminoid Tundra Complex (dry alpine tundra)	Typical ground cover listed under Xd, Vc, or the following, among many others: Dry Dryas octopetala, Salix phlebophylla, Carex microchaeta, Kobresia myosuroides, Saxifraga bronchialis, Hierochloe alpina, Potentilla hyparctica, Minuartia arctica Dwarf Shrub, Graminoid Tundra (dry alpine tundra)
		IXk. Moist Barren/ Moss, Forb, Dwarf Shrub Tundra (moist alpine tundra)	Moist Hylocomium splendens, Saxifraga bronchialis, Saxifraga tricuspidata, Salix phlebophylla, S. chamissonis, Cladonia spp., Moss, Forb, Dwarf Shrub Tundra
	X. Light-colored Barrens (Note: Most areas classed as barrens are likely to have some vegetation but ground cover is less than 30%) (black)	Xa. River gravels	Completely barren or with typical communities listed under IXb, IXc, IXd
		Xb. Sand dunes	Typical communities listed under IXe, IXf, IXg
		Xc. Barren gravel outcrops	Typical communities listed under Vd or the following, among many others: Dry Dryas octopetala, Lupinus arcticus, Potentilla biflora, Smelowski calycina, Saxifraga tricuspidata, Salix phlebophylla, Silene acaulis Dwarf Shrub, Forb Barren (gravel outcrops)
		Xd. Talus slopes and blockfields	Dry Rhizocarpon spp., Lecidea spp., Umbilicaria spp., Cetraria spp. Crustose Lichen Barren (blockfields and talus)
		Xe. Gravel roads and pads	Completely barren
	XI. Dark Colored Barrens (gray)	XIa. Wet mud	Completely barren or with communities listed under IIIb
		XIb. Wet or dark-colored gravels	Completely barren
		XIc. Bare peat	Mostly barren areas along the coast caused by storm surges or man-made disturbances, communities listed under IIIb
		XId. Talus slopes and block fields	Same as Xd
F. Ice	XII. Ice (white)	XII. Ice	Completely barren

grass-like plants are in different families. Only the growth forms contributing at least 30% of the readily visible ground cover are included in the community name.

The last portion of the community name is the physiognomic descriptor, which is a term that applies to the appearance of the general vegetation landscape. The term tundra is used for most arctic and alpine nonforested areas with generally continuous ground cover. The term barren is used in areas where there is less than 30% ground cover. The term shrubland applies only to shrub-covered areas that are traditionally not considered tundra, such as dense riparian shrubs along large rivers. Shrub dominated vegetation in water tracks that are common in the foothills are generally considered shrub tundra, as are shrub-dominated units on mountain slopes and on open flat terrain. Examples of community names can be found in the right hand column of Table 1.

Complex Units

Complexes of vegetation are particularly common in the Arctic, where patterned ground is prevalent. Areas where complexes are mapped include ice-wedge polygons, sorted block fields, strangmoor, water tracks, frost-scar areas, and solifluction stripes and lobes. Often one community is consistently associated with a particular element of the surface form, such as polygon rims, while another community is consistently found on another element, such as the polygon basins and troughs. A consistent method of describing complexes utilizes the basic community nomenclature described above. For example, the following description is for a map unit in a foothill area with water tracks.

Water-track complex:

a) Interfluves and upland areas: Moist Eriphorum vaginatum, Salix planifolia ssp. pulchra, Ledum palustre ssp. decumbens, Sphagnum sp., Cladina arbuscula Tussock Sedge, Low Shrub Tundra.
b) Water tracks: Wet Salix planifolia ssp. pulchra, Betula nana ssp. exilis, Carex aquatilis, Sphagnum sp. Low Shrub Tundra.

Note that the community names follow descriptions of the microsites on which they occur, and the complex is named according to the dominant patterned-ground feature or landform. The unit description includes only those plant communities that are associated with distinctive patterned-ground elements (e.g. polygon rims, water tracks, polygon troughs etc.) and that cover more than 30% of a map unit.

LEVEL C--PHOTO-INTERPRETED MAP UNITS

Level C can be used for photo-interpreted maps at scales from 1:6,000 to 1:63,360. On aerial photographs there are two main characters that are useful for identifying tundra vegetation. The first is color or a gray tone. The darkness of tone is often indicative of the moisture status of the site. Darker areas are normally wet, and lighter areas tend to be moist or dry due to an abundance of erect dead graminoid vegetation and/or crustose lichens. There are, of course, exceptions to this. Sometimes dry areas will also be dark due to barren peat or an abundance of dark-colored fruticose lichens, such as _Alectoria nigricans_ and _Cornicularia divergens_, or wet areas may be light-toned due to marl on pond bottoms. On color-infrared photographs, color is important. For example, red tones are indicative of deciduous shrubs and are important in interpreting categories of tussock tundra vegetation with varying amounts of shrub cover.

The second useful character is texture. Many textures are indicative of surface forms and thus are useful for recognizing vegetation complexes. The presence of ice-wedge polygons, frost boils, solifluction lobes, strangmoor, blockfields, talus, and rugged rocky terrain can be recognized on the basis of texture. On very-large-scale photographs, texture can also be helpful in identifying shrub vegetation and cottongrass tussocks.

Photo interpretation of tundra vegetation is difficult because nearly all the communities are low growing and the clues for distinguishing units are frequently quite subtle. It should be stressed that the critical element for accurate vegetation maps is _extensive ground reference data_. With adequate ground experience, site moisture regime and dominant plant growth forms can normally be interpreted.

Noncomplex Units

The species composition of tundra vegetation can very rarely be reliably interpreted from aerial photographs. Thus at Level C, the nomenclature drops the plant taxa names and consistently uses the remaining parts of the nomenclature outlined for Level D, i.e., the site moisture term, the dominant plant growth forms, and the physiognomic descriptor. An example of a Level C unit is Moist Tussock Sedge, Low Shrub Tundra. Other examples may be found in Table 1.

Complex Units

Complex units are treated in a similar fashion with the term _complex_ attached to the end of the unit name and the components of the complex separated by a slash (/). An example for a low-centered ice-wedge polygon complex is Wet Sedge/Moist Sedge, Dwarf Shrub Tundra Complex. The physiognomic term _tundra_ is included only for the last portion of the complex. The physiognomic term for the first portion of the complex is included only if it is different from the last. The first part of the complex name is the dominant portion. The Level C equivalent of the water track complex mentioned in the previous section is Moist Tussock Sedge, Low Shrub/Wet Low Shrub Tundra Complex. The term _water track complex_ could be used as a shorter synonym in general discussion. For the formal map unit titles, however, every attempt should be made to use the complete names since this increases the amount of information available on the map and makes all the units comparable.

LEVEL B--LANDSAT-INTERPRETED MAP UNITS

LANDSAT methods have certain advantages over photo interpretation. These include the digital format of the data, and the speed with which maps of large areas can be made. The minimum LANDSAT mapping area is one pixel or picture element that corresponds to a ground area of 0.44 ha (1.1 acre). This is considerably smaller than minimum map unit size at all but the very largest photo-interpreted map scales.

The big disadvantage of LANDSAT methods is that the final map units are based solely on surface reflectance. Promising methods that may aid in interpretation of tundra vegetation from LANDSAT data include: 1) using multiple LANDSAT scenes from several seasons, and 2) use of digitized landform and terrain data from geographic information systems. The combination of spectral reflectance and terrain information can be used to produce computer models that are capable of interpreting more vegetation units than can be mapped with spectral data alone. There have been some attempts to use digital elevation data from topographic maps to help model problem categories on the basis of slope aspect and elevation (for example, Justice et al. 1981). These methods have not, however, been used extensively on the Arctic Slope due largely to the very flat landscape where interpolation of elevation values from widely spaced topographic contours can produce inaccurate interpretations.

A classification for LANDSAT-derived maps should recognize the limitations of the data. There are two primary characters of the northern Alaskan vegetation that affect its spectral reflectance and are most important with regards to LANDSAT-derived vegetation classifications. These are the amount of water on the surface and the percentage of deciduous shrubs in the vegetation canopy. Numerous other factors, such as the total percentage of plant cover, the amount of erect dead graminoid vegetation, the color of the substrate, the amount of lichen cover, and the nutrient status of the site, also affect the reflectance. Figure 1 is a cluster diagram for a typical LANDSAT scene from northern Alaska illustrating the spectral signatures in two bands for the major Level B classes. The 12 Level B units are based primarily on moisture status, the amount of shrubs in the canopy, and, in the case of the partially vegetated and barren units, the total percentage of plant cover. A full discussion of the units can be found in Walker et al. (in press).

LEVEL A--VERY-SMALL-SCALE MAP UNITS

Level A consists of only six units that are useful for very general vegetation maps of Alaska. The units are Water, Wet Tundra, Moist Tundra, Shrubland, Partially Vegetated and Barren, and Ice. These units are comparable to the classes used for the major ecosystem map of Alaska (Joint Federal-State Land Use Planning Commission 1973) and the USGS land cover classification

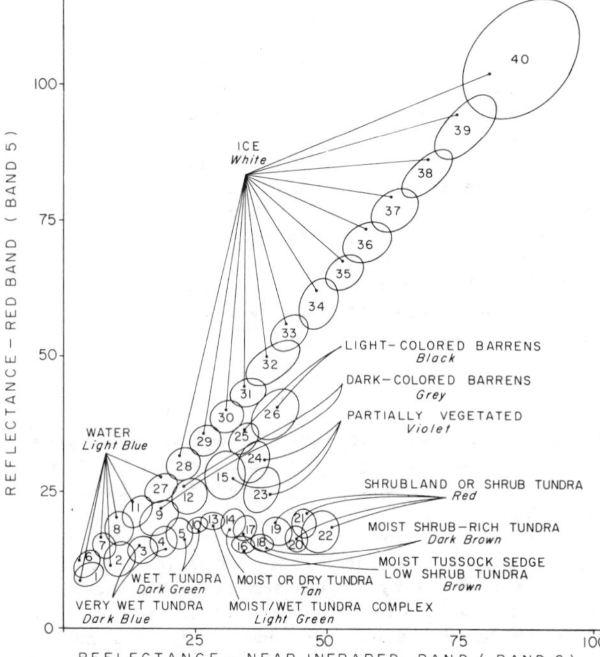

FIGURE 1 Cluster diagram for a LANDSAT scene of the Prudhoe Bay region, Alaska (scene no. 21635-21044), bands 5 and 6. The land cover designations and map colors indicate how the clustered were grouped in the final classification. Each ellipse encloses 80% of the pixels assigned to the respective cluster. The clustering algorithm is part of the EDITOR LANDSAT analysis software system used on the TENEX-DEC System PDP 10 computer available from Bolt Beranex and Newman Inc., Boston, Mass. (Courtesy of USGS Geography Branch, Moffet Field, California.)

system for remote sensor data (Anderson et al. 1976).

CONCLUSION

The hierarchical classification scheme presented here offers a first approximation at a link between two methods of vegetation mapping that are being widely used in northern Alaska--one based on LANDSAT technology and the other based on photo interpretation. It ties both of these methods to a comprehensive means of describing tundra vegetation on the ground. It is presently a flexible system that will undoubtedly continue to evolve as more experience is gained in mapping tundra vegetation.

ACKNOWLEDGMENTS

This work was sponsored by the U.S. Army Cold Regions Research and Engineering Laboratory (CRREL). The classification system was developed during contracts with CRREL, the Sohio Alaska Petroleum Co., and ARCO Alaska, Inc. The U.S. Fish and Wildlife Service (USFWS), the U.S. Geological Survey (USGS), and the North Slope Borough have provided extensive logistical support and funds. The USGS Geography Branch at Moffet Field, Calif., and Reston, Va., was responsible for the technical expertise required for the LANDSAT classifications. Patrick J. Webber, Director of INSTAAR and others too numerous to mention here have helped me in developing this classification.

REFERENCES

Anderson, J. R., Hardy, E. E., Roach, R. E. and Witmer, R. E., 1976, A land cover classification system for use with remote sensor data: USGS Professional Paper 969, 28 pp.

Braun-Blanquet, J., 1932, Plant Sociology: the study of plant communities (English translation): McGraw-Hill, New York, 439 p.

Daubenmire, R., 1951, Forest vegetation of northern Idaho and adjacent Washington and its bearing on concepts of vegetation classification: Ecological Monographs vol. 22, p. 301-330.

Joint Federal-State Land Use Commission for Alaska, 1973, Major ecosystems of Alaska. Fold-out map. (Scale 1:2,500,000).

Justice, C. O., Wharton, S. W., and Holdben, B. N., 1981, Application of digital terrain data to quantify and reduce the topographic effect of Landsat data: International Journal of Remote Sensing, vol. 2, p. 213-230.

Marr, J. W., 1967, Ecosystems of the east slope of the Front Range in Colorado: University of Colorado Studies, Series in Biology, No. 8, 134 p.

Viereck, L. A. and Dyrness, C. T., 1980, A preliminary classification system for vegetation of Alaska. U.S. Department of Agriculture, Forest Service, Pacific Northwest Range and Experimental Station, General Technical Report PNW 206. 38 p.

Walker, D. A., Acevedo, W., Everett, K. R., Gaydos, L., Brown, J. and Webber, P. J., 1982, Landsat-assisted environmental mapping in the Arctic National Wildlife Refuge, Alaska: U.S. Army Cold Regions Research and Engineering Laboratory, Hanover, N.H. 03755, CRREL Report 82-37, 59 p.

Walker, D. A., Acevedo, W., Everett, K. R., Gaydos, L., and Webber, P. J., (in prep.), Landsat-derived vegetation map of the Beechey Point Quadrangle, Arctic Coastal Plain, Alaska.

Whittacker, R. H., 1967, Gradient analysis of vegetation: Biological review, vol. 42, p. 207-264.

THE INFLUENCE OF SUBSEA PERMAFROST ON OFFSHORE PIPELINE DESIGN

David B. L. Walker[1], Don W. Hayley[2], and Andrew C. Palmer[3]

[1]Sohio Petroleum Company, 100 Pine Street, San Francisco, CA 94111
[2]EBA Engineering Consultants Ltd., Edmonton, Alberta, Canada
[3]RJBA Holland BV, The Netherlands

The way in which permafrost affects the design of land-based pipelines carrying hot fluids is well documented and has been the subject of a considerable amount of engineering analysis in the last decade. This paper considers the related problem of designing offshore pipelines in areas where subsea permafrost is encountered. A thermal analysis using a finite element model is described, and results based on measured soil data are presented for the cases where a pipeline is supported in a causeway and where it is trenched into the seabed. This is followed by a discussion of settlement analysis, considering such questions as the determination of thaw strain and the influence of arching in the soil. The selection of a criterion to determine "allowable" pipeline curvatures and the question of treating the statistical variability of soil and permafrost conditions along the pipeline route are also covered.

INTRODUCTION

The thawing of subsea permafrost below a warm submarine pipeline creates a potential hazard if it leads to excessive pipeline settlement. Several solutions have been advanced; these include the elevation of the pipeline above the seabed on piles or in a berm and the provision of an active refrigeration system. While these techniques have extensive applications on land, they are unlikely to be economic for all but short lengths of offshore pipelines in shallow water.

The preferred solution is to trench the pipeline into the seabed (to protect against ice scour) and, by providing sufficient insulation around the pipe, to restrict the thawing of permafrost and the resulting settlement to an acceptable degree. The way in which this can be calculated is the subject of this paper. Results of a thermal analysis are presented, leading into a discussion of the important considerations governing the mechanical response of the pipeline and the surrounding thawed material.

Since it is not intended that a site-specific design problem should be addressed, hypothetical pipeline sizes and operating temperatures have been selected in the analysis. However every attempt has been made to use measured soils and climatic data to make the study representative of inshore Alaskan Beaufort Sea conditions.

THERMAL ANALYSES

Thermal Modeling

Transient heat conduction through the seabed was predicted using a two-dimensional finite element model (Hwang et al. 1980, Hwang 1976). A number of one-dimensional analyses were also carried out to determine initial conditions and to match these to measured data. The model geometry remained fixed during each simulation and no adjustments were made to accommodate pipeline settlement.

Properties of soils required for geothermal analyses include latent heat, specific heat and thermal conductivity. Geotechnical engineering practice is to evaluate soil thermal properties indirectly by reference to known correlations to soil index properties such as texture, moisture content and bulk density. With seabed soils, salinity must also be considered, since this affects not only the soil freezing temperature, but also its latent heat due to the presence of unfrozen water.

The salinity of ice-bonded permafrost is generally on the order of 10 ppt or about a third of that for normal seawater; however the salinity of unfrozen seabed soils can exceed that of seawater by up to 25% (Harrison and Osterkamp 1982). The relationship used in this study between soil freezing point and salinity is derived from the sea ice model presented by Ono (1966).

The latent heat of soils is computed by multiplying the latent heat of water by the volume of water that must change phase in order to thaw the soil. This volume equals the total moisture content minus the moisture content that remains unfrozen at the particular soil temperature. The unfrozen moisture is considered as the sum of the moisture absorbed by soil minerals (McGaw and Tice 1976) and the unfrozen moisture within the brine pockets of pore ice (Ono 1966).

The relationship between unfrozen water content and temperature for varying salinities is shown in Figure 1. The soil is typical of the fine grained seabed permafrost within the study region and can be classified as a sandy silt with a total water content of 32%. The latent heat required to thaw this soil will typically be 50-75% of that required for a similar fresh water soil.

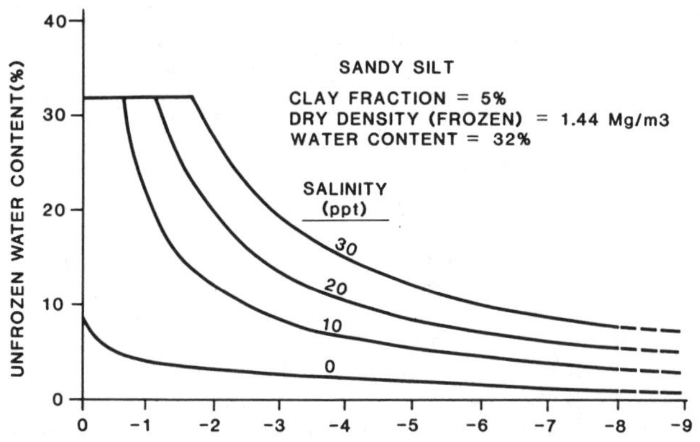

FIGURE 1 Effect of salinity on unfrozen water content.

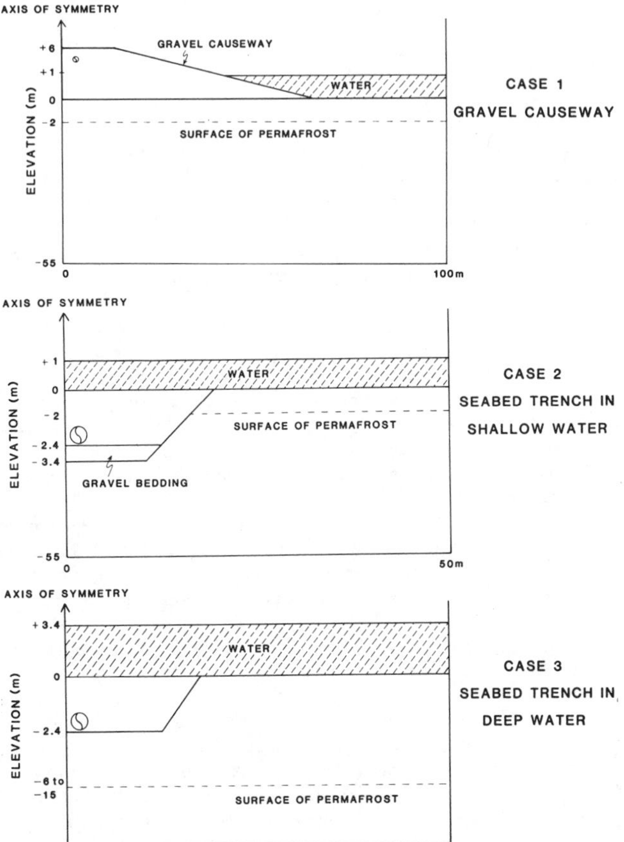

FIGURE 2 Model geometries (not to scale).

Salt rejection is observed in frozen soils at temperatures below -8°C. This process was not modeled since temperatures are primarily in the range of -1.5 to -3.5°C.

Thermal conductivity of cohesionless soils was estimated using Johansen's method as described by Farouki (1981). The estimated conductivity provided a reasonable comparison with the limited data available from proprietary field studies.

The specific heat of ice-bonded permafrost was computed as a weighted average of the soil grains, unfrozen moisture and frozen moisture.

Model Geometries and Boundary Conditions

Figure 2 shows the geometries of the three cases under consideration. Water depth is an important variable. Cases 1 and 2 consider shallow conditions, i.e. water depths less than 2m. Due to the presence of bottom-fast ice during the winter, the mean annual seabed temperature is substantially lowered and hence relatively cold permafrost (-3°C) is encountered within 2m of the surface. Typical soils data for shallow conditions are summarized in Table 1.

Case 3 is typical of deep water (greater than 2m) where the seabed temperature is warmer and remains approximately constant throughout the year. In this case, the depth to the relatively warm permafrost (-0.9°C) is treated as a variable.

In each case, twin 16 in (406mm) diameter pipelines are modeled carrying hot oil at a temperature of 66°C. Insulation thickness is treated as a variable. A vertical axis of symmetry midway between the pipeline allows modeling of half of the actual section.

The causeway was assumed to have a crest width of 15.2m with side slopes of 1 in 10. The freeboard was set at 4.6m in a water depth of 1m. The pipelines were buried 1.4m below the surface, 1.05m on either side of the axis of symmetry.

The boundary conditions for the exposed surface of the causeway were derived from meteorological data on measured air temperatures, wind speed, snow depths and solar radiation. Figure 3 illustrates the boundary condition applied to the submerged surface of the causeway and to the adjacent seabed. This was obtained from a separate one-dimensional analysis, where, by adjusting conductivities, seabed temperatures were derived that matched measured data.

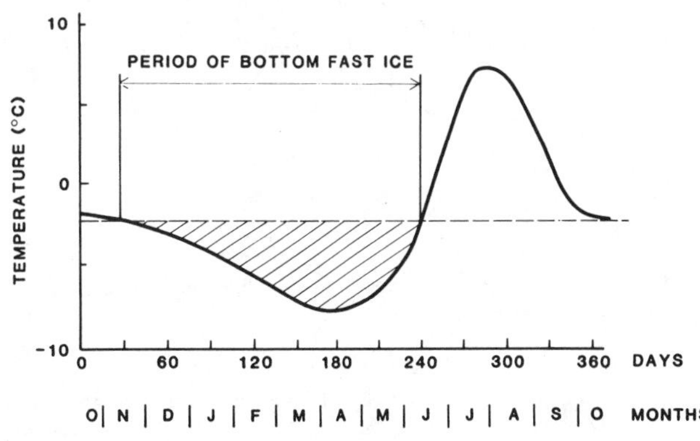

FIGURE 3 Boundary condition used on submerged surfaces for shallow water cases.

Constant temperature and adiabatic boundary conditions were prescribed for the base and vertical surfaces of the grid respectively. Initial seabed temperatures were established from the one-dimensional analysis to compare closely with available measured data.

An alternative configuration to the causeway consists of the pipelines buried directly in the seabed. A water depth of 1m was selected for Case 2, where the pipelines were assumed to be operating in a backfilled ditch with 1m of granular bedding provided below the pipes. The bottom of the pipe was placed at a depth of 2.4m in a trench with a base width of 6m and side slopes of 1 (V) in 2 (H) (see Figure 2).

Case 3 considered a location in 3.4m of water, where the depth to ice-bonded permafrost was treated as a variable, between 6 and 15m below the seabed. This was achieved by adjusting soil salinities and hence freezing points. The granular bedding was eliminated below the pipe as shown in Figure 2. The boundary conditions were similar in both analyses with the exception of the exposed surface below sea level. For the deeper water case where seasonal freezing does not extend to the bottom, the temperature was varied over a 1°C range (cf. 15°C in shallow water).

Causeway Results

Thermal simulations were run for pipes having 0, 50 and 100mm of insulation. Figure 4 indicates the extent of thaw after 5 years of operation. (Abrupt changes in the thawing isotherm position reflect changes in soil properties.)

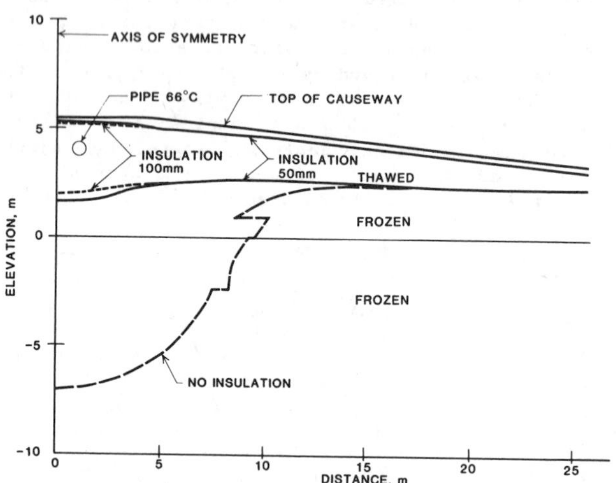

FIGURE 4 Causeway condition during September of year 5.

Permafrost aggradation under the causeway is observed along the full width of the causeway. With insulated pipe, the entire causeway freezes each winter, while the bare pipe maintains a thaw bulb all year. During the summer months, approximately 2m of gravel thaws out on the flanks of the causeway, and under the insulated pipes thawing extends down to about 4m from the surface, but is still contained within the causeway itself.

From the above, it is apparent that to prevent thaw settlement, the provision of 50mm of insulated coating around the pipelines is adequate. The possibility of frost heave however, due to the presence of the causeway itself promoting the freezing of seabed soils, cannot be overlooked.

Seabed Trench Results

A full parametric study has not been conducted at this stage; only water depth, depth to permafrost surface and insulation thickness are treated as variables.

Figure 5 compares the worst case, that is where the permafrost table lies 6m below the seabed (Case 3a) with the next most severe condition, where the pipeline trench intercepts the permafrost 2m below the seabed (Case 2). Dropping the permafrost table to 9 and 14 m beneath the seabed (Cases 3b and 3c) produced progressively less rapid thawing.

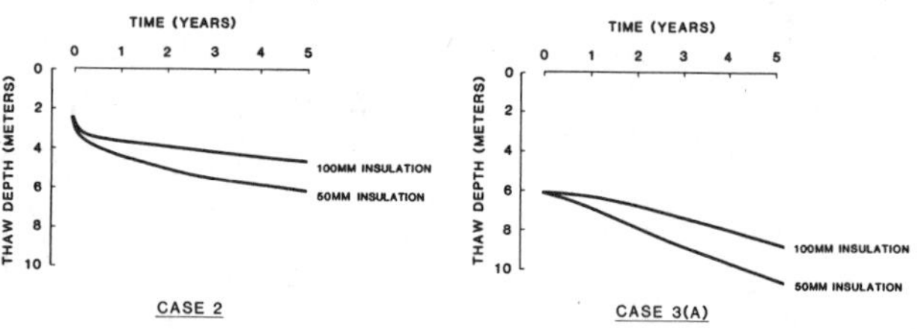

FIGURE 5 Depth of thaw below seabed (midway between pipes) versus time.

Two phenomena are observed here: dropping the permafrost table reduces thawing by introducing more unfrozen material acting like insulation around the pipe. In the first of the Case 3 simulations however, this is more than offset by the warmer initial ground temperatures when compared with Case 2.

Case 3b (permafrost table initially 9m below seabed, 100mm insulation) showed that only 1.2m of thaw had taken place after 5 years, and none at all was observed in Case 3c when the table was dropped to 14m.

A typical cross-section of the thaw bulb is given in Figure 6, showing the Case 3a results after 5 years.

Unlike the causeway analysis, it is not possible to reach definite conclusions at this stage about the acceptability of the degree of thawing observed. In the worst case, even with 100mm of pipeline insulation, up to 4.5m of permafrost was thawed beneath the pipe before steady state was reached. Nevertheless it should be recognised that from our existing borehole data this represents unusual conditions (that is, encountering "warm" permafrost only 6m below the seabed). For more typical conditions where the permafrost table is 10-15m into the seabed, steady state was reached after 1-2m had been thawed. Throughout, it should be recognised that conservative assumptions about heat loss to the ocean have been made, in that complete backfilling of the trench with uniform seabed material is assumed.

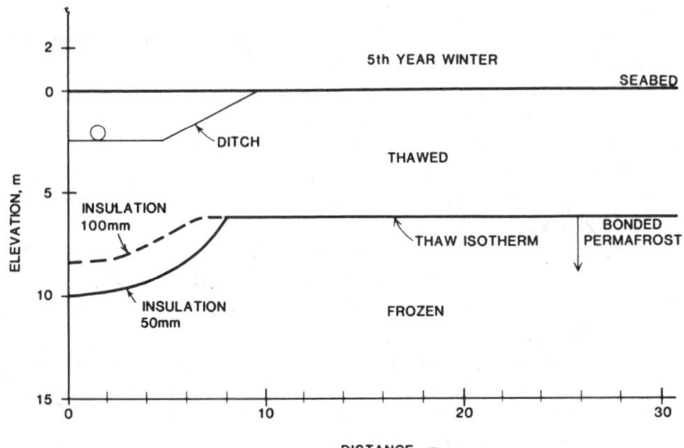

FIGURE 6 Case 3 thaw bulb profile.

PIPELINE SETTLEMENT

Introduction

The prediction of the size of the thaw bulb under a warm pipeline overlying permafrost is a relatively straightforward computation, provided accurate soils data are available. The settlement calculation on the other hand is more complex and there are reasons to believe that simplified models may be overly conservative. Three aspects of the problem will be discussed here:

- the computation of the deflected shape of the trench profile.
- the interaction of the pipe with the trench and the determination of an acceptable pipeline curvature.
- the statistical treatment of soil variables.

Settlement

There are two aspects of this problem which any model must address, namely the computation of thaw strain and the influence of soil arching.

In a continuum model of a frozen soil, the total stress is shared between an effective

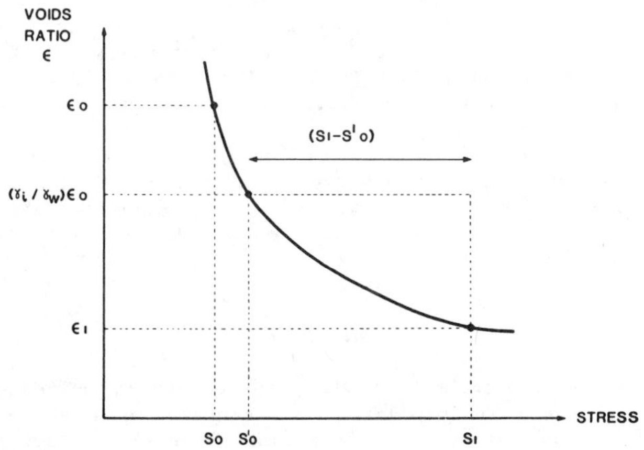

FIGURE 7 Voids Ratio versus Stress.

stress in the soil skeleton and an ice stress. A simple one-dimensional model of thaw consolidation is described by Figure 7, where the axes are stress and voids ratio, and the continuous curve expresses the relation between stress and voids ratio observed in a conventional consolidation test on unfrozen soil, in which the effective stress is increased continuously, and free and complete drainage of pore water is allowed. If the soil is frozen at voids ratio e_o, and then subjected to a total stress s_1, the effective stress is s_o and the ice stress s_1-s_o. If now the soil thaws, the voids ratio decreases to $(\gamma_i/\gamma_w)e_o$, where γ_i and γ_w are the densities of ice and water, and the effective stress increases to s'_o. Since the total stress is still s_1, a pore pressure $s_1-s'_o$ is induced. If drainage is not permitted, this portion of the total stress can continue to be carried by the pore water, and no settlement will occur. If drainage can occur, the pore pressure falls, and the total stress is progressively transferred away from the pore water, so that the effective stress increases to s_1 and the voids ratio falls to e_1; the corresponding thaw strain is $(e_o-e_1)/(1+e_o)$. In some soils with segregated ice, s_o is almost zero, while in coarse-grained soils, the consolidation curve is almost parallel to the stress axis, so that thaw strain is almost independent of s_1. Only then does it make sense to speak of thaw strain as a material property, and usually it will depend on the final stress level.

Morgenstern's theory of one-dimensional consolidation introduces the concept of thaw-consolidation ratio, a measure of the extent to which pore pressures in the thawed region can dissipate in relation to the speed of advance of the thaw front. Like all idealizations, it leaves out factors that may sometimes be important; for example, it takes no account of unfrozen water in the frozen soil, or of partial saturation, or, like our own thermal model, no account of thermal diffusion or convective interaction between the diffusion of heat and water.

The stress field under a pipeline is obviously more complicated than in one-dimensional consolidation. In some areas, deformation may be dominated by shear strains, which do not imply volume change, do not require pore water to diffuse, and therefore can occur as an immediate response to stress change. A complex progressive interaction develops during thaw consolidation, as pore pressures diffuse and the relative magnitudes of shear and volumetric strain alter, both with time and with position within the stress field.

Thawing under a submarine pipeline induces a complex three-dimensional situation, but many of the features of the simple model persist. Firstly, settlement induced by large volumetric strains implies drainage, and therefore will not occur at once, but only after times of order L^2/c_{vc}, where L is the drainage path length and c_{vc} the coefficient of consolidation; this would still be the case even if thawing were instantaneous, and drainage rather than thaw rate will often determine the rate of development of thaw settlements. Secondly, the extent of thaw strain depends on the stress field; if stress redistribution during thawing reduces the increases in effective stress that would occur otherwise, the thaw strains and settlements will be correspondingly reduced.

Figure 8 is a schematic transverse and longitudinal section of a submarine pipeline in a backfilled trench in seabed soil containing a mass of ice-rich material subject to large thaw strains. When the pipeline goes into operation, the ice-rich material begins to thaw, and consolidation begins immediately beneath the pipe, at a. Arching will reduce the effective stress at a (and therefore reduce the thaw strain) by redistribution of the vertical compressive stress laterally, away from the "softer" area of thaw strain, so that the stress is increased on either side at b and c . To a lesser extent, arching also redistributes load longitudinally. A second effect results from the flexural stiffness of the pipeline, which resists bending and tries to bridge the area of thaw settlement, so that instead of being carried by soil underneath, the weight of the pipe and its contents (and some of the overburden) is transmitted longitudinally by shear forces in the pipe to areas d and e. Finally, arching also occurs in the backfill above the pipe, so that its weight is partially redistributed.

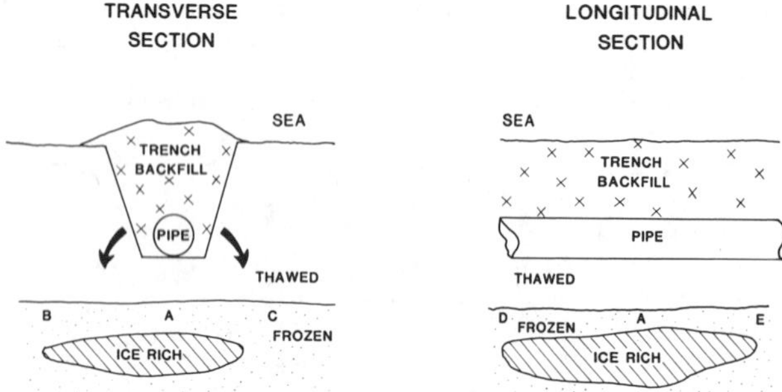

FIGURE 8 Pipeline schematic.

All of the above effects serve to reduce soil deflections under the pipe and if ignored, there is a serious risk that an analysis may be excessively conservative.

Behaviour of the Pipeline

It can be seen from Figure 8 that the effect of thawing permafrost under a pipe depends on the horizontal extent of the permafrost feature. This will determine the importance of the arching effects described above and will directly influence the behaviour of the pipeline, since the allowable deflection is clearly a function of the span length over which it is observed.

It is not hard to show that a critical span length (and hence permafrost feature length) exists for any given vertical soil deflection, such that the pipeline curvature is maximised. This geometric consideration offers the potential for a reduction in the amount of analysis required, since only critical spans need be addressed.

A realistic model for determining allowable pipeline curvatures (and hence settlement) takes as its limiting condition the onset of a condition which threatens environmental damage or the continuing operation of the pipeline, by the initiation of buckling of the pipe wall and the possibility of excessive local strains leading to crack initiation and rupture.

The pipeline is a robust structure and can be bent a long way, far beyond yield, before any damage that significantly affects its operation will occur. A land pipeline is relatively thin; the Alaska pipeline has a diameter/thickness ratio of 96, but even so has been found to withstand very large curvatures before rupturing. A submarine pipeline is comparatively thick; typically the pipelines considered in the thermal analysis have a diameter of 406mm and a wall thickness of 13mm, giving a diameter thickness ratio of 32 and a buckling curvature of $0.082m^{-1}$, corresponding to a radius of curvature of 12m. It follows that a realistic assessment of the risk of pipeline damage associated with thaw settlement need not be based on an allowable stress concept, but can allow much larger bending strains based on an allowable percentage of the strain to cause buckling. In this it follows the recent development of code requirements for submarine pipelines.

Statistical Analysis

There are two alternative approaches to the problem of interpreting site information about the horizontal variability of ice content and thaw strain. In one approach, previously described by Palmer (1972), the variability of settlement from point to point is represented by a probability density distribution, and the horizontal variability is represented by an autocorrelation function. The autocorrelation and the distribution are inputs to a statistical simulation of a trench bottom settlement profile, and the response of the pipeline to the simulated profile is determined by structural analysis. The alternative approach is to use statistical methods to estimate an extreme (probable worst case) ice feature, and then to determine the response of the pipe to that feature. For the reasons explained above, the length of the feature is important, and calculation of the extreme feature must take account of pipe response. This method is analogous to the "design wave" approach to hydrodynamic analysis of forces on offshore structures.

It is important to understand that both approaches are statistical, and that the difference between them lies in logical sequence rather than in using statistics in one instance and not in the other.

CONCLUSIONS

- To predict accurately the effect of bringing a warm pipeline into contact with subsea permafrost, it is essential that good soils data be obtained from the field. Parameters such as soil salinity, moisture content, soil type, initial temperature distribution and occurrence of ice-bonded material will substantially affect the results.

- In the particular cases analysed, it was apparent that a causeway provides an adequate solution to the problem of thaw settlement.
- When seabed trenches were considered, it was clear that the occurrence of warm permafrost features 5-6m from the surface in "deep" water conditions presents a greater potential for thawing than the existence of cold permafrost at the level of the seabed itself in "shallow" water conditions.
- From considerations of thaw strain and the influence of soil arching, it is concluded that simplified models that fail to adequately account for these affects may seriously overestimate the magnitude of the thaw settlement.
- Compared with most land pipelines, typical submarine pipelines have substantial buckling resistance and can tolerate considerable settlements. Criteria to determine "allowable" settlement should not be limited to elastic behaviour.
- A statistical analysis of variability of soil and permafrost conditions along a pipeline route is required to determine an acceptable design.

REFERENCES

Farouki, O. T., 1981, Thermal properties of soils: CRREL Monograph 81-1, December.

Harrison, W. D., and Osterkamp, T. E.. 1982, Measurement of the electrical conductivity of interstitial water in subsea permafrost, in Proceedings of the Fourth Canadian Permafrost Conference (The Roger J. E. Brown Memorial Volume): Ottawa, National Research Council of Canada.

Hwang, C. T., 1976, Predictions and observations on the behaviour of a warm gas pipeline on permafrost, in Canadian Geotechnical Journal: Vol. 13, No. 4, pp. 452-480.

Hwang, C. T., Seshadri, R., and Krishnayya, A. V. G., 1980, Thermal design for insulated pipes, in Canadian Geotechnical Journal: Volume 17, No. 4, pages 613-622.

McGaw, R. W. and Tice, A. R., 1976, A simple procedure to calculate the volume of water remaining unfrozen in a freezing soil, in Proceedings of Soil Water Problems in Cold Regions: Edmonton, September, pp. 114-122.

Ono, N., 1966, Specific heat and heat of fusion of sea ice, in Proceedings of International Conference on Low Temperature Science, Physics of Snow and Ice, Vol. I, part I: the Institute of Low Temperature Science Publication, pp. 599-610.

Palmer, A. C., 1972, Settlement of pipelines on thawing permafrost, in Transportation Engineering Journal: ASCE, Vol. 98, No. TE 3, Proceedings Paper 9090, August 1972, pp. 477-491.

TABLE 1 Soils data modeled in cases 1 and 2.

DEPTH BELOW SEABED (m)	MATERIAL OR SOIL DESCRIPTION	WATER CONTENT %	DRY DENSITY Mg/m³		FREEZING TEMPERATURE °C	QUARTZ CONTENT %	CLAY FRACTION %	PORE WATER SALINITY ppt	THERMAL CONDUCTIVITY W/m°C At -5°C		SPECIFIC HEAT kJ/kg°C At -5°C		LATENT HEAT OF SOIL AT TEMPERATURE -8°C MJ/m³
			FROZEN	UNFROZEN					FROZEN	UNFROZEN	FROZEN	UNFROZEN	
0.0-2.1	Silty Sand (SM)	33	1.43	1.51	-1.7	80	5	30	2.92	2.01	1.26	1.59	119
2.1-7.6	Silty Sand (SM)	31	1.47	1.55	-0.4	80	5	8	3.43	2.09	1.13	1.55	135
7.6-10.7	Sand (SP)	25	1.60	1.68	-0.8	90	0	15	3.79	2.63	1.09	1.42	117
10.7-13.7	Silty Sand (SM)	22	1.68	1.76	-0.3	80	5	6	3.68	2.53	1.05	1.34	111
>13.7	Gravely Sand (SP)	12	2.05	2.15	-0.4	90	0	7	4.88	4.05	0.92	1.09	74
	Gravel fill above water	3	2.10	2.10	0.0	90	0	0	1.83	2.58	0.75	0.84	21
	Gravel fill below water	12	2.05	2.05	-1.9	90	0	35	4.50	3.63	0.96	1.09	60
	Pipe insulation above water	--	0.06	0.06	--	--	-	--	0.02	0.02	0.92	0.92	--

EROSION IN A PERMAFROST-DOMINATED DELTA

H. Jesse Walker

Louisiana State University, Baton Rouge, Louisiana 70803

Field trips to the Colville River delta have provided the opportunity to repeatedly examine and monitor the types and rates of erosion occurring at a number of locations. Field measurements have been made in several distinct environmental situations, including high banks composed of Gubik materials, peat banks in locations of lake tapping, a pingo and adjacent lake fill, and the head of a mid-channel bar. In all cases permafrost is involved, and in some, ice wedges are important. Rates of retreat are variable. During the past 30 years a number of banks have been eroding at average rates of between 1 and 3 m per year; rates that have a high annual variability. At times of block collapse, retreats of up to 12 m may be almost instantaneous. However, in areas of block collapse the average annual rates of erosion are usually similar to those elsewhere because collapsed blocks serve as buffers to further retreat for 1 to 4 years.

INTRODUCTION

In 1961 a long-term research project on the hydrology, geomorphology, climatology, and nearshore oceanography of the Colville River delta in Arctic Alaska was initiated. Between 1961 and 1982, 10 field seasons made possible the collection of data on the types and rates of erosion at a number of deltaic locations. Several reports have been issued dealing with different aspects of the river channels and banks (Ritchie and Walker 1974, Walker 1969, Walker 1978, Walker 1983, Walker and Arnborg 1963, and Walker and Morgan 1964). None has concentrated on rates of erosion, which is the emphasis of this paper.

THE DELTA: RIVERBANKS AND EROSION FACTORS

The Colville River delta, the largest delta in Arctic Alaska, is situated in the zone of continuous permafrost. Two of its many distributaries, the Main Channel (about 45 km long) and the Nechelic or West Channel (about 40 km long) carry nearly all (90% during breakup and 99% during normal flow) of the water. Their banks are composed of various materials, including gravels, sands, silts, peat, and ice. Genetically and morphologically they are equally as varied. Part of the Nechelic Channel borders the Gubik Formation, which is more than 50,000 years old at this site (Hopkins, personal communication, 1980). The other banks are much younger (probably none more than 4,000 years old) and include such forms as sand dunes, interdune swales, old lake beds, mid-channel and point bars, mud flats, ice wedges, and pingos. A characteristic they all have in common is permafrost.

Field data collected in 1971 show that 59% of the banks are erosional, 35% depositional, and 6% neutral (Ritchie and Walker 1974). Erosion dominates along right banks (72%), whereas deposition is most common along left banks (54%).

The possibility of bank erosion in the Colville River delta is limited to about 4 months per year. During the other 8 months most riverbanks are protected by ice, snow, and frozen ground. In some situations snow drifts become thick enough to last well into the flood season.

Some of the water in the channels is sufficiently deep, so that it does not freeze to the bottom. Such subaqueous banks, even though permafrost free, are not subjected to erosion during winter because water flow is minimal. In the main channels, river water is replaced by seawater and water temperatures beneath the ice drop below 0°C.

Although wind does have some erosive effect on river bars and banks, most modifications are caused by the mechanical and thermal energy of moving water and the thermal energy from the air and solar radiation. These two processes have been referred to as "thermal abrasion" and "thermal erosion" by Aré (1978).

In the case of the Colville River delta, these processes are usually initiated in May at the time the melt season begins. Snow melting, river-ice breakup, and active layer thaw are all important and once initiated, proceed rapidly. Extreme flooding, which ranges from less than 1 m at the front of the delta to over 5 m at its head, normally lasts from 2 to 3 weeks. During this short period of time much of the annual erosion occurs.

Although the dominant dynamic processes affecting erosion are thermal abrasion and thermal erosion, there are a number of other factors that affect its character and rate. Type, texture, ice content, height, and orientation of the bank; direction, frequency, and intensity of the wind, especially as they affect wave action; and duration of water contact with the riverbank are all important variables.

RIVERBANKS AND EROSIONAL RATES

Six sites representing different bank types, orientations, and rates of erosion have been selected for analysis. These include banks representing (1) the Gubik Formation; (2) thick, dense peat; (3) an old lake bed and pingo; (4 and 5) tapped lake entrances; and (6) the head of a mid-channel island with duneswales (see Figure 1).

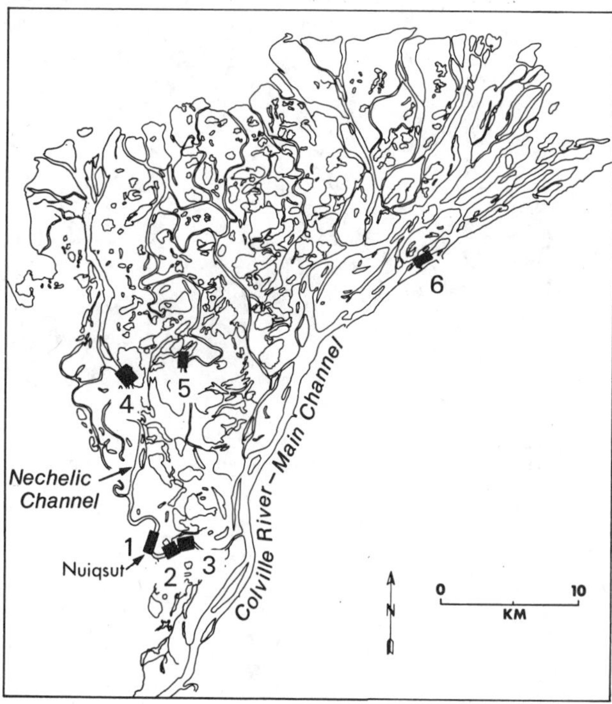

FIGURE 1 Map of the Colville River delta showing the location of erosion-monitoring sites.

The Gubik Formation (Site 1)

The Gubik Formation, which borders part of the left bank of Nechelic Channel, is being rapidly eroded at two cutbank locations. At these locations the bank rises some 8 m above normal river stage and is never overtopped by floodwater. The Gubik is composed of gravels; sands; silts; organic matter, including some large forms, such as mammoth tusks and tree trunks; and ice of various types, especially ice wedges.

Floodwater, which begins to flow before river-ice breakup, first melts the snow at and below the water surface and then begins to develop a thermoerosional niche (Walker and Arnborg 1963) that deepens and widens as flooding continues. In the case of the Gubik this niche may be cut into the bank by as much as 10 m, leaving a massive cornice hanging above (Figure 2). In some cases, depending mainly on fluctuations in stage, multiple niches may develop (Figure 3). Overdeepening of these niches leads to block collapse (Figures 4 and 5). Normally such collapse occurs along ice wedges. Thus, individual blocks represent portions, and in some cases nearly all, of particular ice-wedge polygons (Figure 4). If collapse does not occur, sloughing from a rapidly thawing niche roof and oversteepened bank seals off the niche (Figure 2). During summer, thaw and sloughing gradually convert the collapsed blocks into pyramidal piles and reestablish the angle of repose of the riverbank that was destroyed during flooding. This process establishes a new bank position that will be preserved until the next flood season. The pyramidal piles serve as buffers and help reduce the rate of retreat of the bank behind them. The two types of bank retreat, more rapid block collapse and slower thaw and sloughing, tend to offset each other over a period of years, so that in plan view long-term retreat forms a bank with a relatively smooth arc (Figure 4).

FIGURE 2 Thermoerosional niche, cornice, and sloughing material in the Gubik Formation.

FIGURE 3 Two niches and eroding ice wedge in the Gubik Formation.

From 1949, when the first aerial photographs were made of this part of the delta, to 1982, the maximum retreat along this bank has been about 45 m for an average of nearly 1.4 m per year (Figures 6, 7, and 8). However, during any one year at any particular location, the actual retreat has been as much as 10 m. At the present rate of erosion along the northern half of this bank, the river will shift about one river width in 75 to 100 years

(Figure 8). At such rates the total volume of material eroded from this kilometer-long section is about 20,000 m^3 per year.

FIGURE 4 Collapsed block in the Gubik Formation.

FIGURE 5 Collapsed blocks in the Gubik Formation.

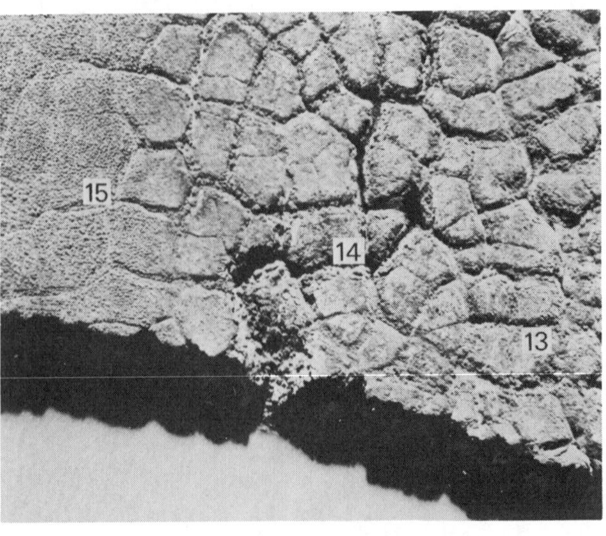

FIGURE 7 Air photo (1970) of the tundra surface formed on the Gubik Formation at section locations 13, 14, and 15.

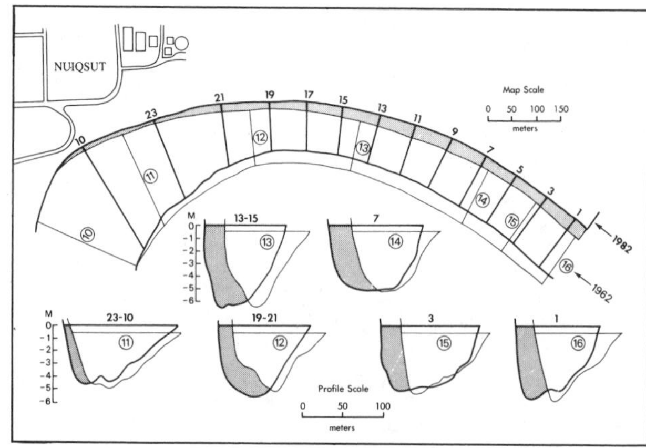

FIGURE 8 Map of and sections across the Nechelic Channel near Nuiqsut showing channel migration between 1962 and 1982.

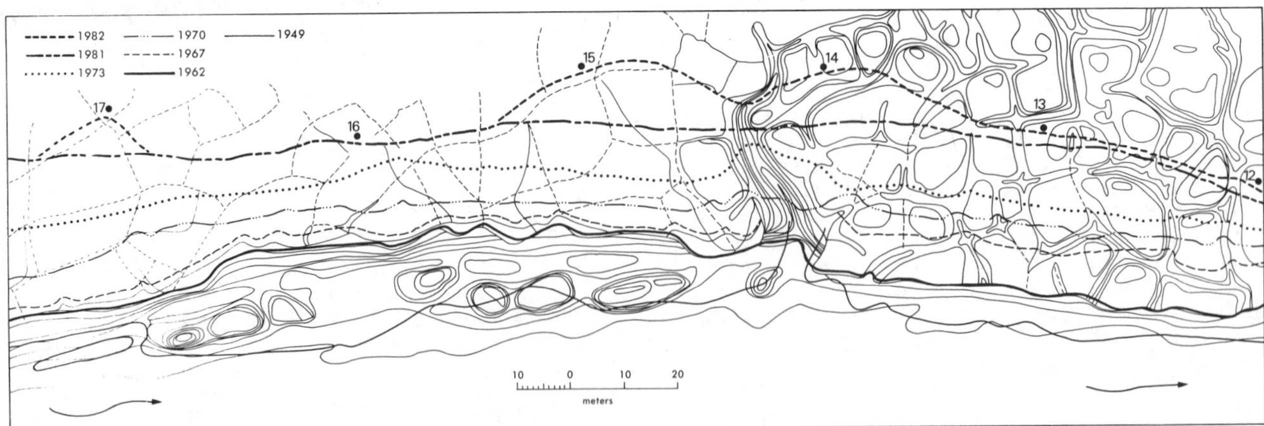

FIGURE 6 Map of site 2 showing bank positions during a 33-year period for section locations 12 through 17.

Thick, Dense Peat (Site 2)

Most of the right bank of the Main Channel and portions of the right bank of the Nechelic Channel are composed of thick layers of dense peat that support extensive ice-wedge polygon systems (see Figures 9 and 10).

Normally, erosion along such banks is slower than along the perdominantly mineral banks, such as those in the Gubik Formation. Thermerosional niches do develop in peat banks, although they generally do not develop to a depth that causes block collapse (Figure 10). Exceptions do occur, especially when silt or sand layers lie within or at the base of the peat blocks, as was the case of the example shown in Figure 11.

The rate of erosion of ice wedges, which may occupy as much as 20% of the bank face (Figure 10), is faster than that of the separating peat. Such differential retreat results in a serrated bank, a type that is typical of most peat banks in the delta. Ice gouging and shaving by blocks of ice transported during breakup are sufficiently common along peat banks that these banks normally have a relatively smooth surface.

FIGURE 11 Peat block just before collapse. The niche developed in a basal sand layer.

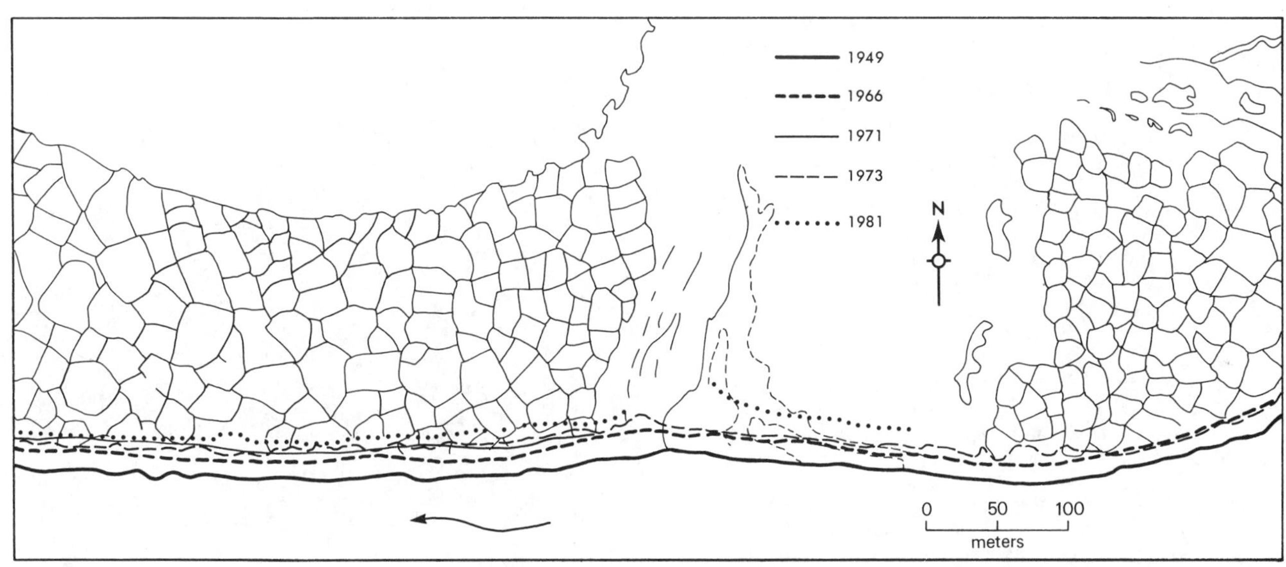

FIGURE 9 Map of site 2 along the right bank of the Upper Nechelic Channel. The serrated edge of an eroding lake shoreline is also depicted.

FIGURE 10 Peat bank typical of site 2. Note ice wedge beneath flag.

Although the niches that develop in this type peat bank are not deep, thawing of their roofs causes widening. The peat shreds and clumps that drop from the roof (Figure 10) are easily removed during high water.

The rate of erosion along the section of site 2 averaged between 0.8 and 0.9 m per year during a 32-year period. This particular bank is subjected to ice jams almost every year and, therefore, probably has a higher average erosion rate than most peat banks in the delta.

Old Lake Bed and Pingo (Site 3)

There are many locations within the delta where the river is now cutting into old lake beds. Most

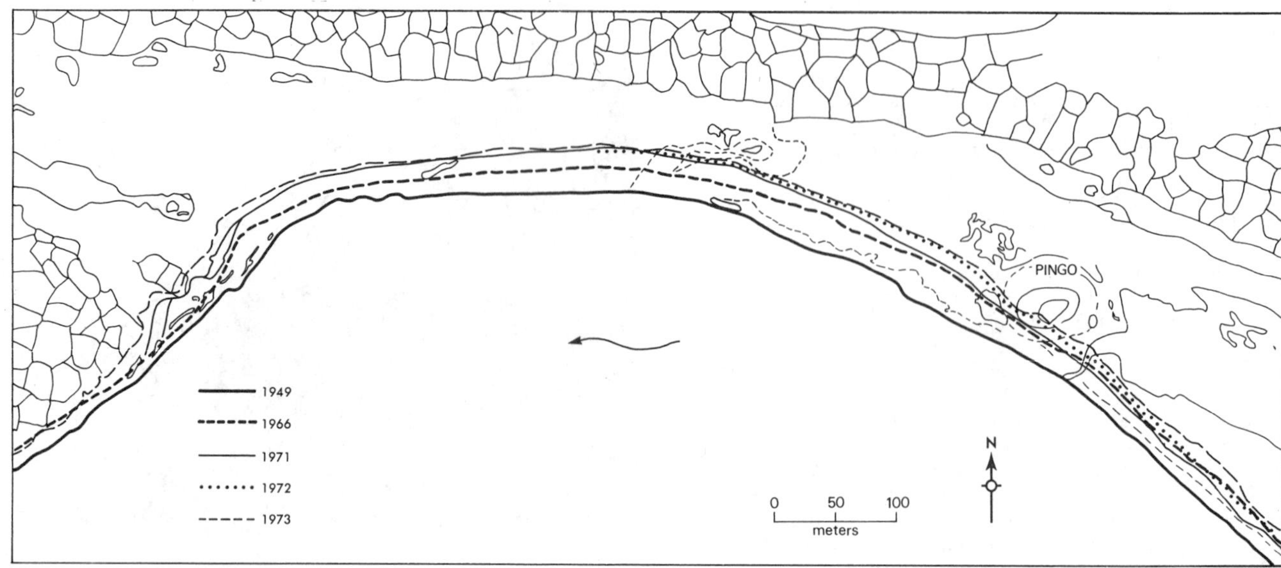

FIGURE 12 Map of site 3 showing drained lake bed and pingo.

of these banks are composed of fine sands, silts, clays, and organic matter. This material is easily thawed and eroded and in most instances has relatively small ice wedges. The lake bed represented at site 3 is a river meander that was abandoned, converted into a lake, filled with sediment, and eventually tapped (see Figure 12). Since tapping, the river continued to cut into the old lake basin and by 1980 had eroded more than halfway through. Accompanying erosion has been a continuing buildup of the old lake bed because each year's flood brings a new layer of sediment as the bed is overtopped.

The present rate of erosion of this former lake bed is 1.8 m per year; a rate that is uniform along its entire length. At the present rate of erosion the old lake bed will be completely breached by the year 2000.

At the time the long oxbow lake was tapped, a pingo formed near the middle (Figure 12). This pingo has been eroding at a rate slightly less than that of the surrounding lake bed. Sloughing from the pingo face during erosion at water level is sufficient to keep the ice core covered. At the present rate of cutting, the pingo will be destroyed within about 50 years.

Tapped Lake Entrances (Sites 4 and 5)

One of the many results of river migration is the frequent tapping of delta lakes. Within the delta, lakes at various stages of tapping, filling, and destruction are numerous (Walker 1978). Site 4 (Lake Nanuk) and site 5 (unnamed lake) represent two relatively recent examples. Lake Nanuk (Figure 13) was tapped before the first aerial photographs were taken of the Colville delta (1948), whereas the lake at site 5 was tapped in 1971.

The bank on the south side of the Lake Nanuk entrance channel is composed of dense peat, that on the north side, of peat overlying sand. The Nanuk entrance is migrating northward at about 1 m per year, although in some years the erosion is more (see Figure 14).

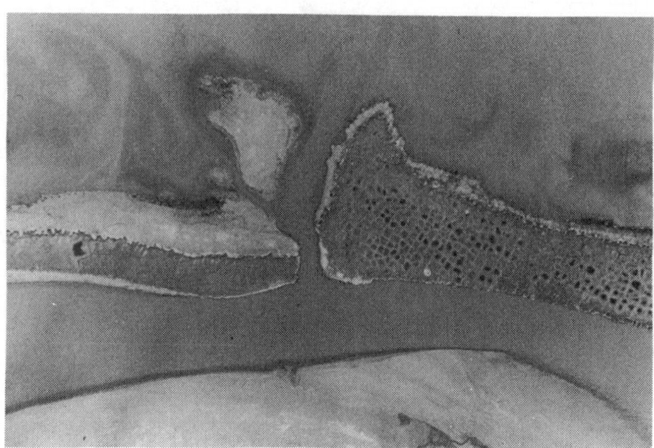

FIGURE 13 Entrance to Lake Nanuk.

FIGURE 14 Peat overhang on north side of entrance to Lake Nanuk.

At site 5 there has been little change in the entrance, both sides of which are thick peat, between the time of tapping (1971) and the present. One of the peat blocks that collapsed into the channel at the time of tapping had not been completely destroyed by 1980.

Mid-Channel Island (Site 6)

The last example is the head of a mid-river island located in the Main Channel. The island is composed of sand dunes and interdune swales. These swales are mainly sandy peats in which small ice-wedges have developed (see Figures 15 and 16). This location receives the brunt of the river flow during flood and is impacted heavily by ice during breakup. Although a thermoerosional niche develops, it is modified rapidly by ice shove. The combination of thermal abrasion, thermal erosion, and ice gouging are responsible for an erosion rate of about 2.2 m per year at the head of the island. Rapid retreat of the island is also bringing about a gradual change in shape. Thirty years ago the island was much more pointed than it is today.

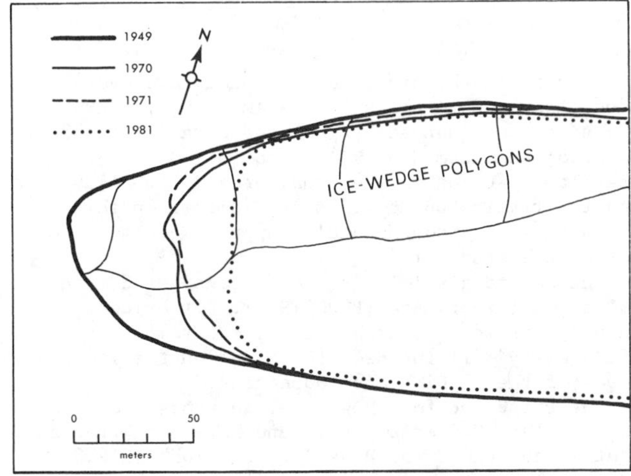

FIGURE 15 Map of island at site 7.

FIGURE 16 Photograph of erosion at the south end of the island at site 6.

SUMMARY

Monitoring bank retreat at a number of cut-bank locations within the Colville River delta has shown that erosion rates vary with bank composition, bank height, and degree of exposure to erosional and transportational processes. Long-term annual average rates range from much less than 1 m in thick, dense peat to over 2 m in highly mineralized banks such as those formed in the Gubik Formation.

Although this paper has concentrated on erosion of deltaic distributary banks, it should be noted that in general the processes discussed are also operative along the margins of other arctic water bodies. For example, high rates of erosion can occur along the shorelines of lakes, lagoons (Lewellen 1970), and the Arctic Ocean (Aré 1978 and McDonald and Lewis 1973).

ACKNOWLEDGMENTS

Data used in this report were collected during a number of field seasons made possible by grants and other support from the Office of Naval Research, the Naval Arctic Research Laboratory, the North Slope Borough, and Louisiana State University.

REFERENCES

Aré, F. E., 1978, The reworking of shorelines in the permafrost zone, in Permafrost--The USSR Contribution to the Second International Conference: Washington, D.C., National Academy of Sciences, p. 59-62.

Lewellen, R., 1970, Permafrost erosion along the Beaufort Sea coast: Denver, University of Denver, 25 p.

McDonald, B. and Lewis, C., 1973, Geomorphic and sedimentologic processes of rivers and coasts, Yukon coastal plain: Ottawa, Geological Survey of Canada, 245 p.

Ritchie, W. and Walker, H. J., 1974, Riverbank forms of the Colville River delta, in J. C. Reed and J. E. Sater, eds., the Coast and Shelf of the Beaufort Sea: Washington, D.C., The Arctic Institute of North America, p. 545-562.

Walker, H. J., 1969, Some aspects of erosion and sedimentation in an arctic delta during breakup, in Hydrologie des deltas: Association Internationale de'Hydrologie Scientifique, p. 209-219.

Walker, H. J., 1978, Lake tapping in the Colville River delta, in Proceedings, Third International Conference on Permafrost, v. 1: Ottawa, National Research Council of Canada, p. 233-238.

Walker, H. J., 1983, The Colville River delta, a field guide, prepared for the Fourth International Conference on Permafrost.

Walker, H. J. and Arnborg, L., 1963, Permafrost and ice-wedge effect on riverbank erosion, in Proceedings, First International Permafrost Conference: Washington, D.C., National Academy of Sciences, p. 164-171.

Walker, H. J. and Morgan, H. M., 1964, Unusual weather and riverbank erosion in the delta of the Colville River, Alaska: Arctic, v. 17, p. 41-47.

SORTED PATTERNED GROUND IN PONDS AND LAKES OF THE HIGH VALLEY/TANGLE LAKES REGION, CENTRAL ALASKA

James C. Walters

Department of Earth Science, University of Northern Iowa
Cedar Falls, Iowa 50614 USA

Well-developed sorted patterned ground is widespread on gently sloping shores and bottoms of shallow lakes and temporary ponds in the High Valley/Tangle Lakes region of the Amphitheater Mountains, on the south side of the central Alaska Range. This area is within the zone of discontinuous permafrost and has a mean annual air temperature of -4°C. The numerous lakes and ponds occupy depressions in middle to late Quaternary silty tills which blanket the valleys. Sorted patterned ground forms include stone circles, stone polygons, stone pits, and stone stripes, with circles and polygons being most common. All patterned ground sites, whether presently underwater or subaerially exposed, occur in situations of fluctuating water levels. It is suggested that the features formed by frost sorting, frost heaving, and mass displacement when lake shores and pond bottoms were subaerially exposed during drier periods. Although the patterned ground appears to have developed at some time in the past, many sites show evidence of being at least semi-active today.

INTRODUCTION

Patterned ground features sometimes occur subaqueously (Conrad 1933, Lundqvist 1962, Mackay 1967, Dionne 1974, 1978, Shilts and Dean 1975). These features are commonly assumed to have originated subaerially and to later have been submerged. However, observations by Mackay (1967), Shilts and Dean (1975), and Washburn (1980, p. 125) suggest that in some situations patterned ground may form underwater. Sorted patterned ground is widespread in shallow lakes and ponds of the High Valley/Tangle Lakes region of south-central Alaska. This paper examines the sorted patterned ground features in this area and discusses their possible origin.

Over 300 lakes and ponds containing sorted patterned ground along their gently sloping shores and bottoms were identified by aerial reconnaissance, air photo interpretation, and field observations. The numerous lakes and ponds occupy depressions in the silty till which blankets the area. Underwater patterned ground is found throughout the broad valleys, from the lowest elevations (along the Tangle Lakes) of 870 m up to the highest elevations (in High Valley) of 1,200 m.

DESCRIPTION OF THE STUDY AREA

Glacial Geology

Located on the south side of the central Alaska Range, the High Valley and Tangle Lakes areas exist as broad valleys within the Amphitheater Mountains (Figure 1). These valleys were glaciated four times during the Quaternary (Péwé 1961), the glaciers advancing into the area from the Alaska Range to the north. Péwé (1961) considers the earliest glacial advance to be early to middle Quaternary age. Although till from this advance is not exposed in the area, the ice must have overridden the 1,800 m peaks of the Amphitheater Mountains as evidenced by isolated erratics found on top of an unnamed peak just to the east of the upper Tangle Lakes in the study area. Péwé (1961) reports erratics up to 4.5 m in diameter on this peak and also Paxson Mountain, approximately 12 km east of the study area.

The second glacial advance is considered middle to late Quaternary age (Péwé 1961). It evidently did not override the Amphitheater Mountains but left a silty till in the valleys and on the lower slopes of the Amphitheater Mountains.

The third and fourth glacial advances took place in late Quaternary time and have been grouped together and named the Denali glaciation by Péwé (1961). These two advances are closely related in extent as well as age. They moved through gaps in the Amphitheater Mountains and were joined by local cirque ice. Deposits of these glaciers, especially the late Denali advance, cover much of the earlier glacial material. The ice of the late Denali advance thinned and stagnated during retreat, leaving ice-contact features such as kames, kettles, eskers, crevasse fillings, and pitted surfaces. The Tangle Lakes area is a region of well-developed ice-stagnation topography.

Climate

The High Valley/Tangle Lakes region is within the zone of discontinuous permafrost and has a mean annual air temperature of approximately -4°C. The area experiences a continental climate, which can be generally characterized as dry with extreme summer and winter temperatures, light precipitation, and light surface winds. Local weather records kept at Paxson (approximately 17 km east of the study area) from 1969 to 1973 give the following information: mean maximum temperature 1.1°C, mean minimum temperature -12.2°C, record high 29.4°C,

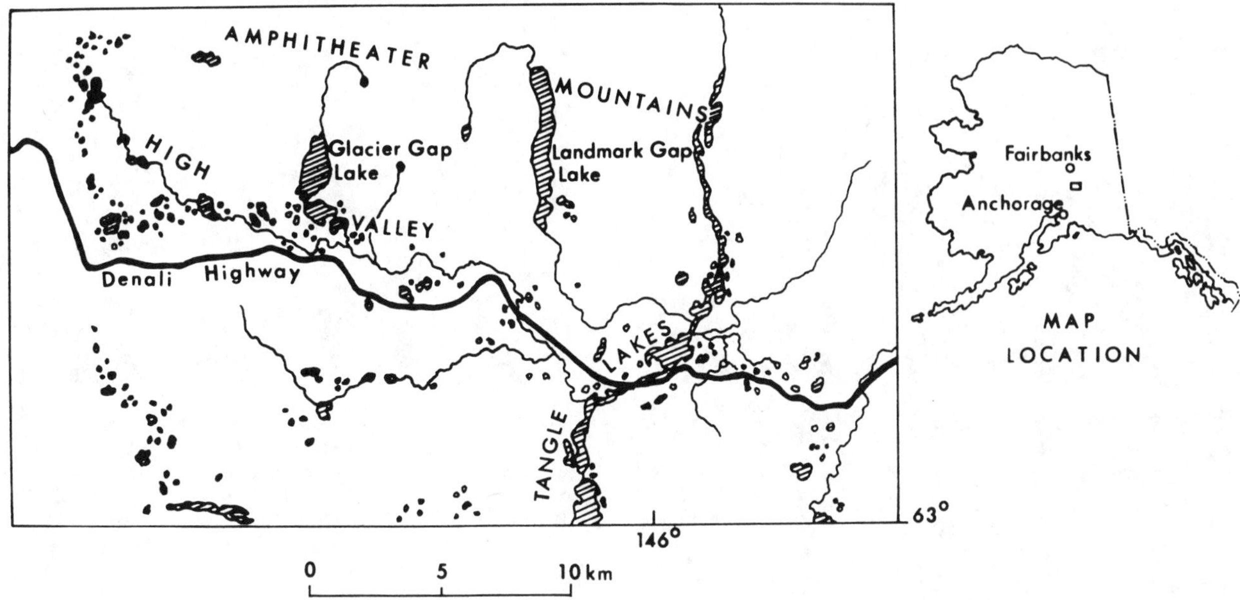

FIGURE 1 Map of the study area.

and record low -50.5°C. Mean annual rainfall during this period of time was 44.2 cm, and mean annual snowfall was 278 cm (Miller et al. 1976).

DESCRIPTION OF THE PATTERNS

Lakes and ponds are widespread in the study area, and sorted patterned ground features are associated with many of them. Most of the shallow ponds display sorted patterns along their entire bottoms, whereas in the larger lakes the patterns are typically restricted to the shallower shelf areas along the lake margins, The ponds and smaller lakes are simply kettles and other depressions in hummocky till; the larger lakes, specifically Glacier Gap Lake and Landmark Gap Lake (see Figure 1), occupy basins cut by ice as glaciers flowed southward out of the Alaska Range.

Sorted patterned ground forms include stone circles, stone polygons, stone pits, and stone stripes (Figure 2). Circles and polygons are the most common forms and vary in diameter from about 0.4 - 2 m. These types of patterns are found on level to very gently sloping (< 2°) surfaces. Stone pits are less common and occur in similar situations, but stone stripes are found perpendicular to shorelines on slopes of 3 - 10°. No patterns were observed on slopes greater than 10° or in water depths of more than about 2 m. Trenches excavated through sorted circles revealed that stones in the border areas tend to have their long axes vertical, and scattered pebbles and cobbles near the surface in the central areas show the same tendency. The surfaces of the central areas are usually convex, with as much as 20 cm of relief between the center and the tops of stones in a bordering trough.

The fine-grained material associated with the patterns is predominantly a loam or silt loam. It is characterized by a low liquid limit (20 - 26%) and a very low plasticity index (1 - 5%), conditions common in subaerial nonsorted circles (mudboils) and sorted circles. During the summer, this material is quite fluid and easily flows in response to agitation. This thixotropic effect makes trenching of patterns a very difficult chore, but it also favors the processes of mass displacement, frost sorting, and frost heaving, which can be initiated by slight increases in cryostatic pressure during fall freezeup.

The sorted patterned ground sites can be grouped into three manin categories: those which are unvegetated and frequently underwater, those partially vegetated and occasionally inundated, and those almost completely vegetated, with evidence of inundation only rarely. In sites with patterns which are frequently underwater, boulders and cobbles making up the borders of the patterns are covered with greenish-black stain, apparently biochemical in origin. Staining is not present on those stones or portions of stone surfaces which are buried in the adjacent fines (Figure 3). It is also absent on the surfaces of large boulders where they project above water level. Sites which are often subaerially exposed show a less intense staining of stone surfaces, and in addition the stones typically have some lichen growth on them. It is not uncommon to find lichens on rock surfaces now 10 - 25 cm underwater.

All patterns, whether presently underwater or subaerially exposed, occur in situations of fluctuating water levels. Field observations over a 2-year period and examination of air photos from past years show that smaller lakes and ponds in which the patterns exist are occasionally subaerially exposed. In the larger lakes, the patterns are usually restricted to the shallower shelf areas along the lake margins which are sometimes emerged.

Patterned ground is developed in each of the three tills found in the area. As mentioned

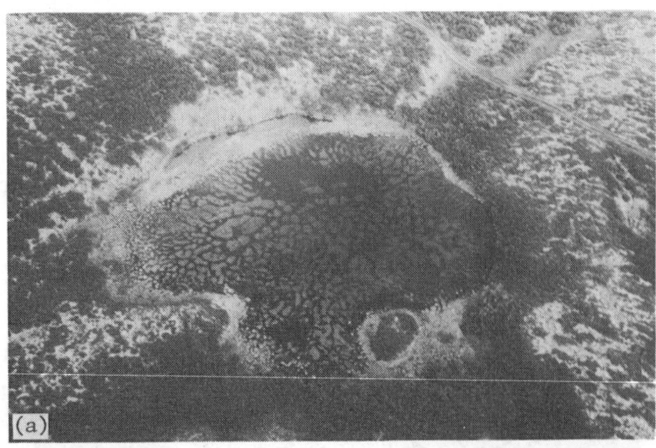

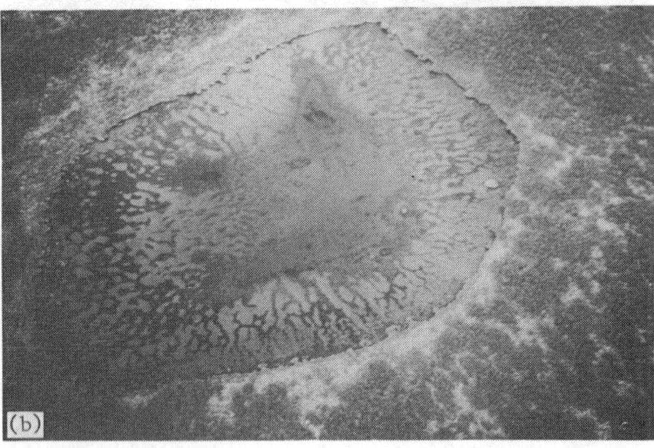

FIGURE 2 (a) Aerial photograph of sorted patterned ground in a pond of the Tangle Lakes area. Stone circles, stone polygons, stone pits, and stone stripes are present. An off-road vehicle trail in the upper right corner provides scale. Photo taken July 26, 1982. (b) Aerial photograph of sorted patterned ground in a pond of the High Valley area. A variety of patterns is present, but there is a lack of patterns in the deeper water of the pond center; only some stone pits and individual boulders are found here. The large boulder projecting above the water surface on the right side of the pond is about 1.5 m across. Photo taken July 26,1982.

earlier, till of the earliest glacial episode (early to middle Quaternary age) is not known in the area, the advance being recognized only by large erratics found on peaks on the Amphitheater Mountains. Although all sorted patterned ground types occur in each of the three tills, they do vary somewhat depending on the material present.

Patterns in Till of Middle to Late Quaternary Age

In most of the study area, deposits of this glacial advance have been covered by drift of later glacial activity. However, much of the High Valley region escaped later glaciation and thus contains a rather extensive area of this earlier till. Underwater sorted patterned ground is not as common in this till as it is in the later tills, and only 32 patterned ground sites were examined. In the shallow lakes and ponds, sorted patterns are

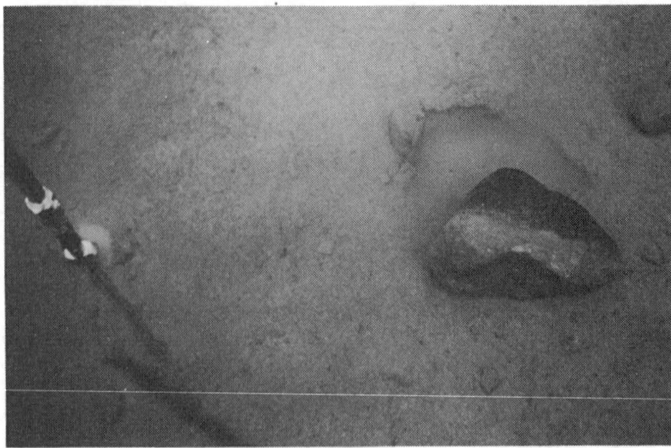

FIGURE 3 A cobble, which had been partially embedded in fines of the central area of a sorted circle, removed to show the extent of staining on its upper surface. Staining is not present where it was embedded in the fine-grained material. The cobble is 15 cm long on its longest axis. Water depth is approximately 25 cm.

found along their entire bottoms, but larger and deeper lakes display patterns only along their shallower shelf areas. The pebbles, cobbles, and boulders making up the borders of the patterns are mostly subangular to subrounded and 10 - 40 cm in diameter. The fine material in the centers of the patterns is basically a silt loam.

It is common to find sorted patterned ground on old emerged lake terraces bordering the present lakes. The terraces are no more than about 0.5 m above the present shorelines and are almost completely vegetated with *Salix* shrubs and grasses, making the patterns difficult to recognize. A trench dug across a vegetated stone circle showed a moderately well-developed soil profile (Table 1, Figure 4). These vegetated patterns on emerged terraces appear to be inactive.

Patterns in Till of Late Quaternary Age (Early Denali Advance)

Till of this glacial advance is found along the borders of the High Valley area and in the center of High Valley where glacier ice flowed through Glacier Gap. Moraines of this advance contain numerous lakes and ponds, and subaqueous patterned ground is associated with many of them. The patterned ground is quite well developed, and 165 sites were examined.

The pebbles, cobbles, and boulders along the borders of the patterns are mostly subangular to subrounded (85 - 90%), with only a few being angular or rounded (10 - 15%). These border stones show quite a range in diameter, but 15 - 40 cm is common (Figure 5). The larger stones typically occur in the middle of the borders with the smaller stones at the margins. Trenches excavated across patterns subaerially exposed on lakeshores and pond bottoms showed no development of soil profiles, but near-vertical mottle streaks are commonly found in the center of the patterns (Figure 6). The texture in the area of the mottles is a loam to a clay loam, just as it is for the surrounding material.

TABLE 1 Soil Profile in the Center of a Sorted Circle on an Emerged Lake Terrace.

Cover: Vegetation in the center of the pattern consists of grasses, forbs, mosses, and lichens with clumps of Salix to 0.5 m; some Ledum decumbens and Vaccinium uliginosum to 15 cm.

Horizon	Depth (cm)	Description
01	5-3	Undecomposed to partially decomposed organic matter; mainly Sphagnum, grasses, and lichens.
02	3-0	Dark reddish brown (5 YR 3/2 moist) partially to well decomposed organic matter; some silt; abrupt smooth boundary.
A1	0-2	Very dark grayish brown (10 YR 3/2 moist) silt loam; moderate fine granular structure; friable, slightly sticky, slightly plastic; abundant very fine to medium roots; some pebbles poking up from below; clear wavy boundary.
B2	2-7	Brown (10 YR 4/3 moist) gravelly loam; common distinct dark brown (7.5 YR 4/4 moist) mottles; massive structure; friable, slightly sticky, slightly plastic; common fine to medium roots; gradual boundary.
Cg	7-27	Grayish brown (2.5 Y 5/1 moist) very gravelly loam; some distinct dark brown (7.5 YR 4/3 moist) mottles; massive structure; friable, nonsticky, slightly plastic; encountered water table at 29 cm.

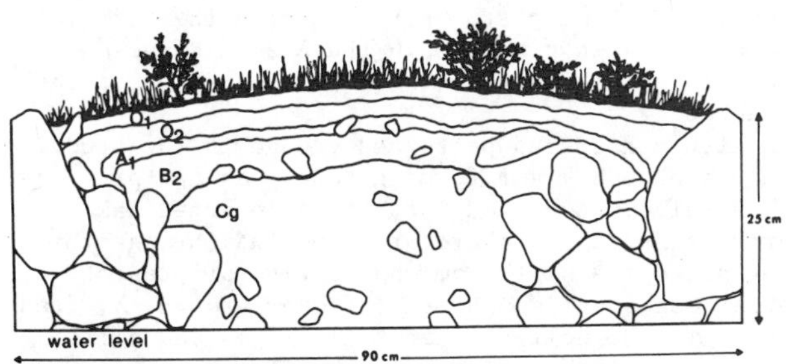

FIGURE 4 Cross-section of a sorted circle on an old emerged lake terrace. See Table 1 for a description of the soil profile. Smaller gravels are not included in the sketch. Drawn from field sketches and photographs.

FIGURE 5 Sorted stone circles on the floor of an emerged pond. The surface of the central area is convex and slopes toward the coarse borders. Note the larger stones in the middle of the bordering troughs and the smaller stones along the margins. Note also that the smaller stones along the inner margins and in the central areas of the patterns show unstained surfaces. These stones appear to have been recently exposed and have not yet developed staining on their surfaces.

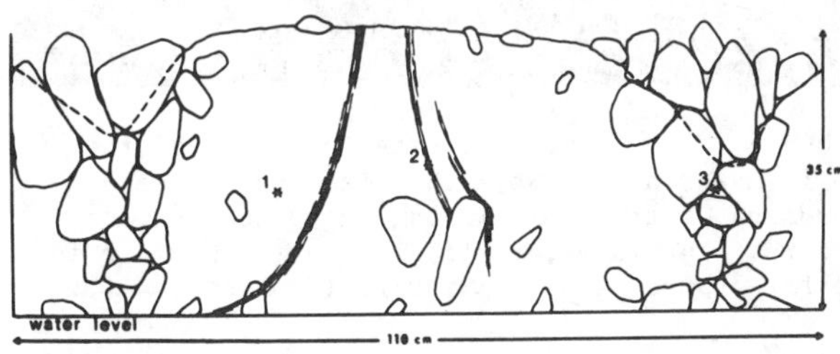

FIGURE 6 Cross-section of a sorted circle on the floor of an emerged pond. Sample 1 = dark grayish brown (10 YR 4/2 moist) to brown (10 YR 4/3 moist) gravelly loam; sample 2 = strong brown (7.5 YR 4/5 moist) to yellowish red (5 YR 4/6 moist) loam to clay loam, 1.5 - 2 cm wide mottle streaks; sample 3 = brown (10 YR 4/3 moist) sandy clay loam. Note the vertical orientation of stones in the troughs. Smaller gravels are not included in the sketch. Drawn from field sketches and photographs.

Samples of the fine material just below the coarser material in the troughs is a sandy clay loam. Color of the mottles is strong brown (7.5 YR 4/5 moist) to yellowish red (5 YR 4/6 moist). Color of the surrounding material is dark grayish brown (10 YR 4/2 moist) to dull brown (10 YR 4/3 moist).

Patterns in Till of Late Quaternary Age (Late Denali Advance)

This is the most extensive glacial deposit in the study area. All of the Tangle Lakes area and the Glacier Gap area of High Valley are covered with till of the late Denali advance. Well-developed patterned ground is abundant, and a total of 125 sites were investigated.

In the Tangle Lakes sites the borders of the patterns contain cobbles and boulders which are subrounded to rounded (80 - 90%), with some subangular (5 - 15%) and some well rounded (5 - 10%). Diameters of the border stones range from 15 to 50 cm, with some up to 1 m (Figure 7). Patterns in the High Valley area display cobbles and boulders along the borders which are mostly subangular to subrounded (80 - 90%), with some angular or rounded (10 - 20%). Diameters typically range from 10 to 35 cm (Figure 8).

FIGURE 7 Sorted stone circles in a pond of the Tangle Lakes area. Note the size and degree of roundness of the boulders making up the pattern borders. The large, partially emerged boulder in the foreground is approximately 1 m long.

FIGURE 8 Sorted stone circles in a small pond of the High Valley area. Note the smaller size and greater angularity of the stones making up the pattern borders in this area as compared to the previous figure. The pole is in 30.5 cm (1 foot) divisions. Water depth is about 10 cm at the pole.

Trenches were dug across some patterns subaerially exposed along lakeshores. They revealed conditions very similar to those shown in trenches excavated across patterns occurring in till of the early Denali advance. No soil development was apparent, and although small circular mottles occur in some exposures, no mottle streaks were observed. The texture of the fine-grained material in the center of the patterns varies from a clay to a silty clay, and in color it varies from brown (10 YR 4/3 moist), to dark yellowish brown (10 YR 4/4 moist), to weak red (2.5 YR 5/3 moist).

FORMATION OF THE PATTERNS

It is clear that the underwater sorted patterned ground features probably form subaerially rather than subaqueously, as evidenced by the fact that all patterns occur in situations of fluctuating water levels. The features are fundamentally no different from subaerial sorted patterned ground except for their occurring underwater. The silty till in which they are developed has a low liquid limit and a low plasticity index. As stated earlier, these characteristics are important factors in the processes of mass displacement, frost sorting, and frost heaving, processes responsible for the formation of sorted patterned ground (Washburn 1980, p. 166-170). Although these periglacial processes may vary in degree of importance for different types of sorted patterned ground features in the same area or for the same types of features in different areas, freeze-thaw and ice segregation are the underlying causes. Over a period of time, coarser material is forced to the outside of the pattern and the fine-grained material is concentrated in the center. During wet years or wet portions of certain years, the depressions in which the patterns occur are filled with water, but a cover of water is not essential to the formation of the patterned ground.

The High Valley/Tangle Lakes sorted patterned ground has been observed by T. L. Péwé over a period of several years, and he is inclined to suggest that they have not formed underwater (T. L. Péwé, personal communication, 1982). Dionne (1974, 1978) has described very similar features in subarctic Quebec and concludes that they formed subaerially through frost heaving and frost sorting in the recent past. The High Valley/Tangle Lakes patterned ground appears to also have formed at some time in the past, perhaps during the very late Quaternary when glaciers in the Alaska Range and small local glaciers in the Amphitheater Mountains advanced.

Although sorted patterned ground in the study area is probably not forming today, as the presence of inactive, vegetated patterns on emerged lake terraces suggests, there are some indications that many sites are still somewhat active at present. Permafrost studies show that frozen ground is often absent in the well-drained till ridges and hills, but fine-grained sediments in shallow depressions are often frozen below depths of 0.5 - 1 m. The thickness of the frozen layer is unknown, but it is probably on the order of several meters. Active nonsorted circles (mudboils), small sorted circles, and sorted steps commonly exist in low spots, on ridges, and ridge slopes, respectively, in the immediate vicinity of the underwater patterned ground sites. Depth to the frost table in these situations during the month of July 1982 was measured at 24 - 55 cm. The presence of active periglacial features and a fairly shallow active layer immediately adjacent to ponds and lakes with underwater patterned ground, suggests that frozen ground underlies the shallow water-filled depressions as well. It also suggests that periglacial processes might be presently active on the shores and bottoms of the lakes and ponds when they are subaerially exposed.

Although long-term studies are necessary to determine if the patterns are still active, observations over a 2-year period indicate that some frost action and movement is taking place. Boulders along the margins of patterns are sometimes found which have been split or fractured, apparently by frost action. Cobbles and pebbles with unstained surfaces or portions of their surfaces unstained are often found along the inner margins of the patterns as well as in the central areas along with the fines (Figure 5). These stones have been recently exposed and have not yet developed staining on their surfaces.

CONCLUSIONS

The depressions in which sorted patterned ground occurs in the High Valley/Tangle Lakes region are ideally suited for the development of such features. The silty till, with its low liquid limits and low plasticity indexes, promotes frost sorting, frost heaving, and mass displacement, processes responsible for the formation of sorted patterned ground. It is suggested that the patterns formed through such periglacial processes when lake shores and pond bottoms were subaerially exposed during drier periods. The fact that the patterns are often covered with water is apparently not essential to the origin of the features, except that extra moisture is available in these sites and the periodic cover of water inhibits vegetative growth. The lack of vegetative cover in frequently inundated patterned ground sites causes them to resemble active patterned ground. Most sites are probably inactive, but many areas of patterned ground appear to be experiencing some frost action and movement at present.

ACKNOWLEDGMENTS

I am grateful to T. L. Péwé, J.-C. Dionne R. D. Reger, and Lou Waller for providing helpful information. Personnel at the Geophysical Institute at the University of Alaska in Fairbanks kindly searched their files for pertinent aerial photographs and allowed me the use of their facilities to examine them. I was ably assisted in the field by B. J. Walters and M. J. Kuhlman. This study was partially supported by a University of Northern Iowa Summer Faculty Fellowship.

REFERENCES

Conrad, V., 1933, Ein Unterwasser-Strukturboden in den Ostalpen: Gerlands Beitrage zur Geophysik, v. 40, p. 353-360.

Dionne, J.-C., 1974, Cryosols avec triage sur rivage et fond de lacs, Québec central subarctique: Revue de Géographie de Montréal, v. 28, no. 4, p. 323-342.

Dionne, J.-C., 1978, Formes et phénomènes périglaciaires en Jamésie, Québec subarctique: Géographie Physique et Quaternaire, v. 32, no. 3, p. 187-247.

Lundqvist, J., 1962, Patterned ground and related frost phenomena in Sweden: Sveriges Geologiska Undersökning, Stockholm, ser. C., no. 583.

Mackay, J. R., 1967, Underwater patterned ground in artificially drained lakes, Garry Island, N.W.T.: Geographical Bulletin, v. 9, no. 1, p. 33-44.

Miller, W. D., Aukerman, R., and Fletcher, R. C., 1976, The Denali Highway information plan for the United States Bureau of Land Management: BLM Anchorage District Office, Anchorage, Alaska.

Péwé, T. L., 1961, Multiple glaciation in the headwaters area of the Delta River, central Alaska: U.S. Geological Survey Professional Paper 424-D, p. D200-D201.

Shilts, W. W., and Dean, W. E., 1975, Permafrost features under arctic lakes, District of Keewatin, Northwest Territories: Canadian Journal of Earth Science, v. 12, p. 649-662.

Washburn, A. L., 1980, Geocryology: New York, Halstead Press.

PERIGLACIAL ACTIVITY ON THE SUBANTARCTIC ISLAND OF SOUTH GEORGIA

D. W. H. Walton and T. D. Heilbronn

British Antarctic Survey, Natural Environment Research Council,
Madingley Road, Cambridge, United Kingdom, CB3 OET

ABSTRACT

Very little information is available on periglacial features and activity on the subantarctic islands. From measurements on South Georgia gelifluction is observable to a depth of 12 cm at very active sites but most sites show movement only to 8 cm. For small scale stripes sorting is generally 5-6 cm deep with the majority of sorting activity in autumn. Altitude appears to affect sorting through the incidence of freeze-thaw cycles, with east facing sites showing most activity. Stripe orientation is not related to wind direction. The rate of clast movement downslope was linearly related to clast length and site slope.

INTRODUCTION

Although a recent bibliography on antarctic soils and periglacial features listed 386 references (Walton 1980), relatively few of these relate to periglacial activity. Until recently, on the subantarctic islands only one investigation has been completed on the rates of periglacial processes (Smith 1960). An unpublished survey (Thom 1980) of the variety and extent of patterned ground on South Georgia (latitude 54-55°S, longitude 36-38°W) identified this lack of process data as an important unfilled link in describing the genesis of the periglacial features.

South Georgia (Figure 1) is a heavily glacierized island with a cold, moist subantarctic climate (Smith and Walton 1975). The bedrock is mainly fine-grained tuffs and greywackes (Stone 1980) and is highly fissile. Large areas of glacial till of varying ages are exposed in summer (Clapperton 1971). Studies were carried out on the north coast near Cumberland Bay to establish rates of downslope movement in sorted stone stripes with respect to aspect, altitude, slope, and season. A recent paper (Heilbronn and Walton 1983) describes the morphology of these and other periglacial features in detail.

METHODS

A series of sites with small scale stone stripes and a 14° slope, all in the vicinity of Cumberland Bay, were selected to cover a range of altitudes and the four cardinal aspects (Table 1). Twenty cm strings of 1 cm long wooden dowels and 10 cm strings of 2 cm long metal roller bearings were inserted vertically into the soil and excavated after 1 and 2 years to measure downslope movement of a vertical section of the profile. At several other sites of different slopes, regular photographs were taken of 11 fixed 50 x 50 cm quadrats. From these, measurements were made of downslope movement of 8-21 native clasts in each quadrat with a length axis

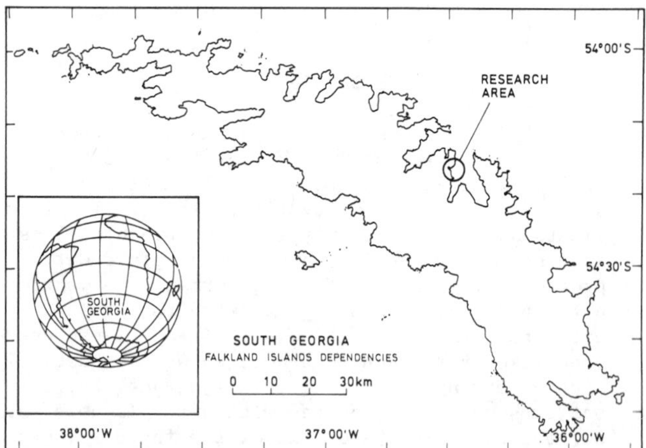

FIGURE 1 Map showing the position of South Georgia, and the research sites.

TABLE 1 Characteristics of Sites Used for Dowel and Roller Strings

Sites	Aspect	Altitude	Strings
MV	east	75	roller
RR	northwest	83	roller
TR	south	120	roller
DM	southwest	190	roller
ZB	southwest	225	roller
N	north	125	dowel
S	south	120	dowel
E	east	125	dowel
W	west	140	dowel

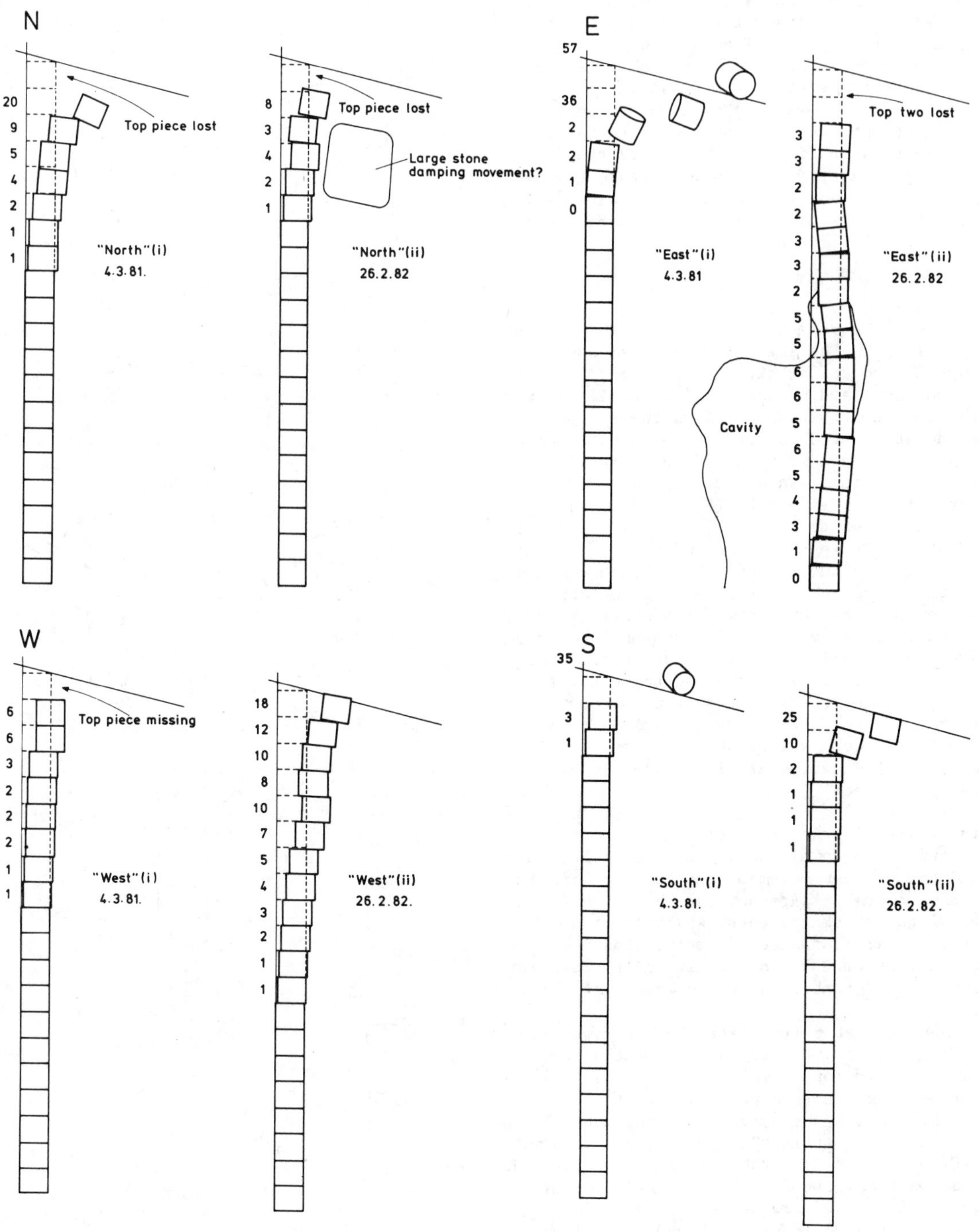

FIGURE 2 Downslope movement of the soil profile in sorted stripes at four sites of different aspects, using wooden dowel strings, after one (i) and two (ii) years. Site details in Table 1. Figures are mm movement.

from 7 to 85 mm. The data for downslope movement for up to 21 clasts in each fixed quadrat were regressed against clast size, giving a mean movement rate per unit clast length for each site. To investigate the relationship between movement and angle of site, the slope of each regression line (downslope movement factor) was plotted against the slope of each site and a linear regression fitted. These data complemented a long term study over 6 years (1975-1981) of the annual downslope movement of 34 introduced clasts on a 6-10° slope of SW aspect.

RESULTS

Activity with Depth

Downslope movement at all the sites was slow, except close to the surface. All the dowel (Figure 2) and roller strings (Figure 3) showed major movement and, occasionally, complete loss of the top one or two units from the string. Loss was attributed to needle ice formation and the subsequent gravitational displacement downhill when the ice crystals collapsed.

At all sites except ZB (Figure 3), year-by-year movement of the top roller was similar, with the distance in year 2 about twice that in year 1. Movement in the rest of the profile was less predictable over the 2 years with TR showing least movement and DM most. Even after 2 years, there was little evidence of movement below 8 cm at any of the roller sites. Evidence from the dowel sites (Figure 2) supported these observations, although at W site slight movement was detected down to 12 cm after 2 years. The lower part of the second string at E unfortunately entered a soil cavity, and its movements are therefore suspect.

Activity with Aspect and Altitude

Amongst the dowel sites the west facing one showed the deepest movement and the south facing the least overall movement. At the north facing site, slightly more movement was evident after the first year than after 2 years. For most dowel strings, needle ice formation in the first year was sufficient to heave out at least the top dowel.

Assuming that movement of the top dowel and top roller was principally due to needle ice formation, downslope displacement of the top roller bearing showed a positive relationship with altitude (Figure 4) at all except the MV site. Using the standard adiabatic lapse rate (0.6°C/100 m) the temperature gradient between RR and ZB would have been -0.85°C, likely to have caused an increased frequency of needle ice formation which would have enhanced downslope surface movement.

It is considered that the enhanced movement at the highest sites is not an aspect effect since both ZB and DM face southwest and RR faces northwest, a much more favourable aspect in terms of the radiation environment and yet producing only 12% of the movement measured at ZB. The similarity in results between RR and TR (south aspect) also supports the relationship between altitude and movement, rather than aspect and movement for surface material.

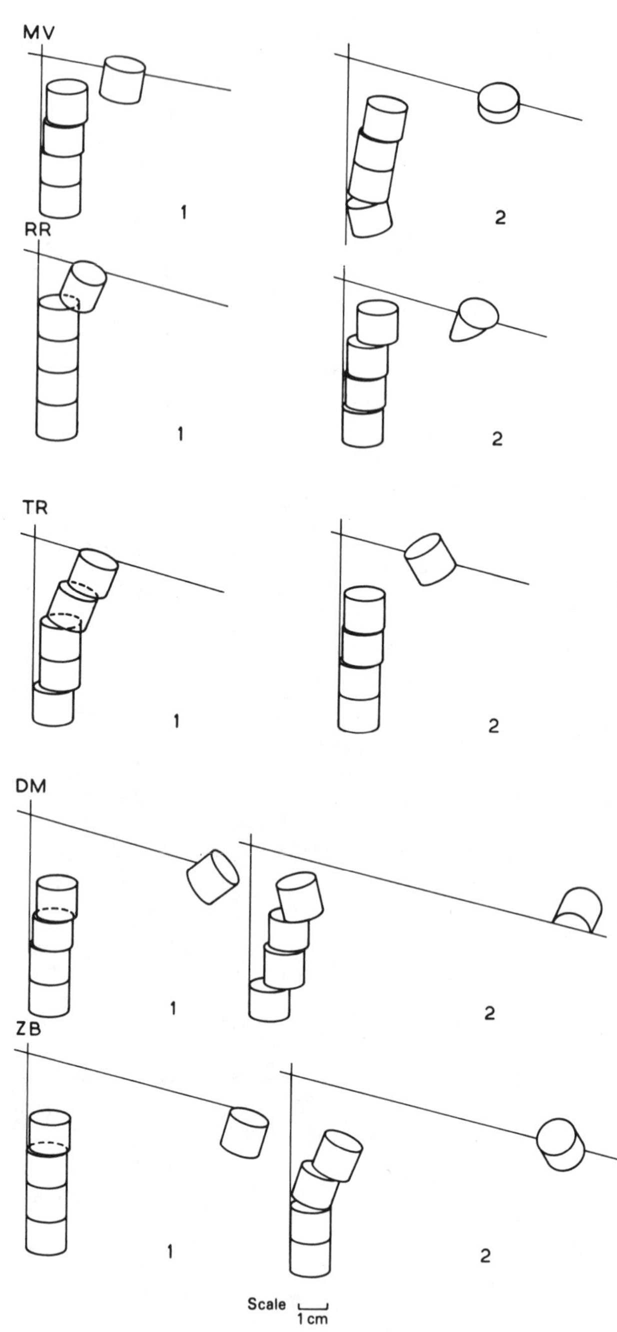

FIGURE 3 Downslope movement of the soil profile in sorted stone stripes at five sites of different altitudes, using metal roller bearings, after one (1) and two (2) years. Site details in Table 1. Drawings are half life size.

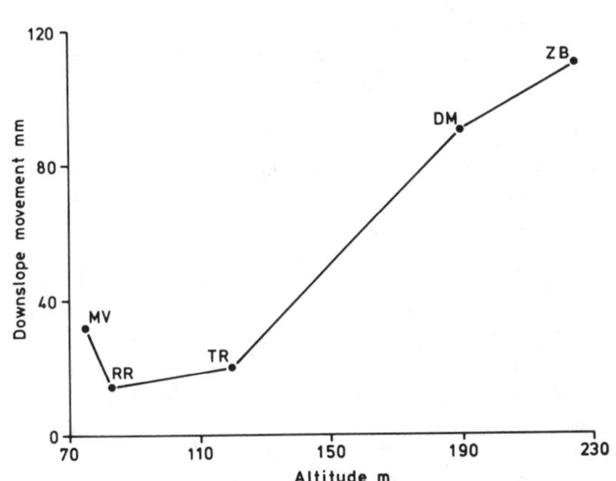

FIGURE 4 Annual downslope movement of the top roller bearings (0-2 cm) at five sites of different altitudes. Sites as in Table 1.

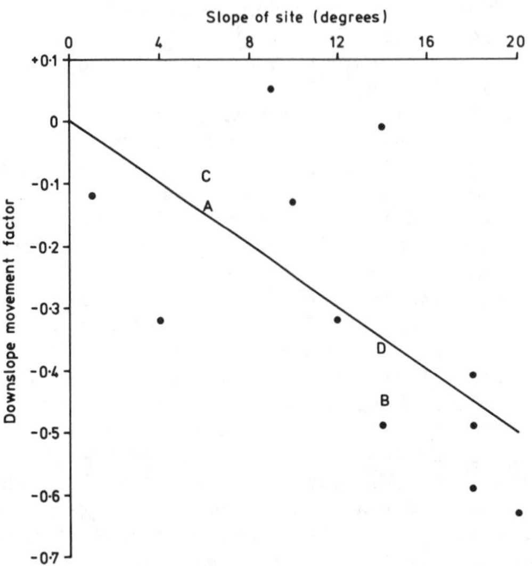

FIGURE 5 Regression of a mean annual downslope movement factor on slope at eleven sites. Sites A-D were not used in the regression - see text. (Regression y = - 0.025 x, r = - 0.64, P = 5%).

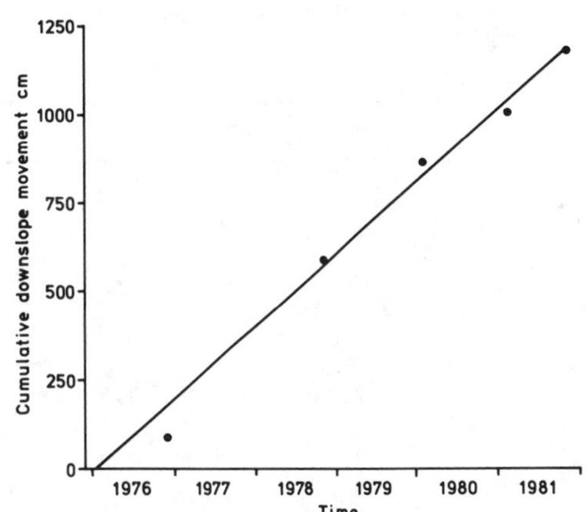

FIGURE 6 Cumulative downslope movement in sorted stone stripes of 34 introduced clasts over a period of 6 years. (Regression y = 17.21 x - 41.86, r = 0.994).

Activity with Slope

Downslope movement of native clasts was monitored at 11 sites ranging in slope from 1° to 20°. Regression lines for clast length against movement were fitted to each data set and the significance of the fitted lines varied from p=10% to p=0.1% at different sites. At all except two sites there was a steady decrease in annual movement with increasing clast size. The slope of these regression lines was taken as an indication of the mean rate of movement per unit clast length for each site and these regression slope values (downslope movement factor) are plotted against site slopes in Figure 5. The correlation coefficient for the regression line in Figure 5 is -0.64 but this includes the data for all 11 sites. If the two sites which did not show a decrease in annual movement with increasing clast size are excluded the correlation coefficient rises to 0.90.

The fitted linear regression in Figure 5 is thus a simple testable model for a preliminary investigation of the relationship between clast size and slope angle for sorted stripes on South Georgia. To test the model, values for regression slopes were calculated for clasts at three other sites not used in deriving the original relationships. For one site (in Bore Valley), movements during 1975-1976 (A) and 1978-1981 (C) were calculated separately. A second Bore Valley site was calculated for 1980-1981 (D), and a site at Gull Lake (B) for 1975-1976. When plotted in Figure 5, these all fall close to the regression line and within 90% confidence limits, suggesting the model is a reasonable one.

Activity with Time

The data from the longterm study site in Bore Valley (Figure 6) show total cumulative downslope movement for 34 clasts on the 7 m transect over a 6 year period from five unequally spaced field measurements. Most clasts moved sideways as well as down the slope, and no clear difference in final position occurred between those clasts originally in fines and those in coarse material. Movement downslope varied from zero to 19 cm within 1 year, and for individual clasts there were considerable between-year variations. After the first year the mean annual rate of clast movement generally fits the regression line quite well. This is interesting, because over this period considerable differences were observed yearly in both snow depth and duration and presumably therefore in the number of freeze-thaw cycles. Detailed examination of the annual rates shows that within the first year, movement (98 cm) was much less than in the immediately following years (264 cm), while for 1980, movement slowed to 135 cm, increasing again in the final year to 210 cm. Records from the meteorological station 2 km away at King Edward Point suggest that snow lie differed between years, with the winter snow in 1976 and 1980 persisting for longer than in the winters of 1977-1979. At this site the mean clast movement per year over 6 years was 58 mm, with a range of annual means from 29 to 78 mm. Measurement made on a steeper (14°) and wetter site in Sphagnum Valley from January 1981 to January 1982 gave an annual mean movement of 177 mm. Although initial movements of tracer clasts are sometimes much higher than those of native clasts (Caine 1981), it seems likely that much of the difference occurred because of a longer period of solifluction in Sphagnum Valley caused by continued drainage into the site throughout much of the summer period.

TABLE 2 Mean Monthly Temperatures at South Georgia

	1958	1980	1981
J	5.4	5.2	5.8
F	3.2	5.9	7.4
M	3.6	3.7	3.2
A	2.7	3.1	3.5
M	-0.8	0.9	0.5
J	-1.2	0.0	-0.7
J	-2.4	-3.6	-0.6
A	0.5	-3.3	0.4
S	1.6	-0.7	0.8
O	2.4	2.6	0.6
N	4.2	3.2	5.1
D	3.4	3.9	5.2
$\bar{x}$	1.9	1.7	2.6

values are in ^{o}C

DISCUSSION

In the only previous paper to examine periglacial activity on South Georgia, Smith (1960) reported on the movement of surface stones and stakes. On a 21° slope, Smith found, in 1956, mean surface downslope movement of 470 mm for twenty 10 cm marked stones. Although this was measured over only 7 months, this corresponds to an annual movement because during the remaining 5 months the ground was frozen and snow covered.

On the steepest site examined in the present study, a 20° slope, and with the largest clasts used (70 mm long), annual movement was only about 60 mm. With the exception of the Sphagnum Valley site, downslope movement of 80 mm clasts did not exceed 105 mm at any site. Recent detailed studies by Caine (1981) of the movement of introduced clasts has shown that an overestimation of up to 300% in the rates of surface movement is possible immediately following the introduction of the tracer clasts. Smith used large clasts, which exacerbate the problem, because they take longer to incorporate into the surface layer, and no settling period was allowed before starting his measurements. Smith's data for surface movement are likely to be considerable overestimates, even if his site was a very wet one. Interestingly, in the first year, movement at the Bore Valley site was lower than in any of the following years, in direct contradiction to Caine's findings.

Smith also assessed a component of mass wasting, gelifluction, by using stakes buried to various depths. He considered that mass flow took place down to 25 cm on a site with a 21° slope. At none of the present sites used to study activity with depth was mass flow observed below 12 cm even after 2 years. Two explanations seem possible here. First, it is possible that Smith's stake site was much wetter than any of those in present use, and this would greatly have assisted mass flow. Second, a comparison of snow duration between his site in 1957 and our sites in 1980-1982 shows that his site did not gain significant snow cover until early July, almost 2 months later than at our sites. His site would therefore almost certainly have been exposed to both lower temperatures and more autumnal freeze-thaw cycles than ours. A comparison of mean monthly temperatures for the 3 years (Table 2) shows that although the mean annual temperature was lowest in 1980, the critical period in April, May, and June was appreciably colder in 1958 than in 1980 and 1981.

Although the dowel string showed deepest activity at the west facing site, the most pronounced surface activity, i.e., that attributable to needle ice, was at the east facing site. This agrees with Mackay and Matthews (1974), who found that differential thaw of needle ice in striped ground was concentrated in the period when the sun was lying in the southeast (equivalent to northeast in the Southern Hemisphere).

This indirect evidence for the effect of insolation in the production of striped ground on South Georgia is in direct contrast to the conclusions reached by Hall (1979) for Marion Island and Iles Kerguelen (Hall, 1983). Hall found that stripes on Marion Island were almost always aligned northwest, parallel with the wind. Although on Kerguelen the relationship was not quite as clear, Hall did find evidence for a correlation between katabatic winds and stripe alignment, and he considered that stripe development was generally most marked on sites exposed to wind. Although he could not provide an explanation for this, he dismissed directional melting by insolation as being of minor significance due to high and persistent cloud cover over the subantarctic islands. Bellair (1969), on the other hand, suggested that freeze-thaw cycles were less significant than wet-dry cycles in producing sorted stone stripes parallel to the major wind direction on Iles Kerguelen and Iles Crozet. On South Georgia there is no evidence for stripe alignment parallel to either the general direction of prevailing winds or local katabatic winds. Insolation is however much higher than on Marion Island and it is possible that wind is only an important formative factor under continuously cloudy conditions.

The constant downslope movement of surface material creates an unstable substrate for plant growth. Heilbronn and Walton (in press) have shown that on South Georgia limited colonization of sorted stone stripes is possible. Although the binding of the roots slows plant movement downslope to less than 50% of that shown by native clasts of the same size, the plants appear to be of little significance in stabilizing the gelifluction activity until their roots reach down to at least 10 cm. In agreement with Washburn (1979), there is no clear evidence on South Georgia that plant-bound gelifluction material is the principal cause of the production of benches and lobes.

CONCLUSIONS

Measurements of downslope movement in small-scale sorted stone stripes on South Georgia suggest that gelifluction is the major process, observable down to 12 cm at very active sites but with most sites only showing movement down to 8 cm. Altitude exerts a significant effect on surface sorting through an increased frequency of needle ice formation, with east facing slopes showing most activity. Sorting in these stripes is generally only 5-6 cm deep. The rate of downslope movement was linearly related to clast length and to the slope of the site. Mean annual movement rate for 34 clasts over 6 years at one site was related to the duration of snow lie. Stripe orientation was not related to wind direction as on other subantarctic islands.

REFERENCES

Bellair, P., 1969, Soil stripes and polygonal ground in the sub-Antarctic islands of Crozet and Kerguelen, in Pewe, T.L., ed., The periglacial environment: Montreal, McGill-Queen's University Press, p.217-222.

Caine, N., 1981, A source of bias in rates of surface soil movement as estimated from marked particles: Earth Surfaces Processes and Landforms, v.6, p.69-75.

Clapperton, C. M., 1971, Geomorphology of the Stromness Bay-Cumberland Bay area, South Georgia: British Antarctic Survey Report, v.70, p.281-289.

Hall, K., 1979, Sorted stripes orientated by wind action: some observations from sub-Antarctic Marion Island: Earth Surface Processes and Landforms, v.4, p.281-289.

Hall, K., 1983, Sorted stripes on sub-Antarctic Kerguelen Island: Earth Surface Processes and Landforms, v.8. p.115-124

Heilbronn, T. D., and Walton, D. W. H., 1983, The morphology of some periglacial features on South Georgia and their relationship to temperature and snow lie, British Antarctic Survey Bulletin, v.61

Heilbronn, T. D., and Walton, D. W. H., in press, Plant colonization of actively sorted stone stripes in the sub-Antarctic: Arctic and Alpine Research.

Mackay, J. R., and Matthews, W. H., 1974, Needle ice striped ground: Arctic and Alpine Research, v.6, p.79-84.

Smith, J., 1960, Cryoturbation data from South Georgia: Biuletyn Periglacjalny, v.8, p.72-76.

Smith, R. I. L., and Walton, D. W. H., 1975, South Georgia, sub-Antarctic, in Rosswall, T. and Heal, O.W., eds., Structure and function of tundra ecosystems: Ecological Bulletin, v.20, p.399-423.

Stone, P., 1980, The geology of South Georgia: IV Barff Peninsula and Royal Bay areas: British Antarctic Survey, Scientific Report, v.96, p.1-45.

Thom, G., 1980, Patterned ground in South Georgia, Antarctica, unpublished Ph.D. thesis, University of Aberdeen.

Walton, D. W. H., 1980, An annotated bibliography of Antarctic and sub-Antarctic pedology and periglacial processes: British Antarctic Survey Data, v.5, p.1-75.

Washburn, A. L., 1979, Geocryology: London, Edward Arnold.

IMPACT OF FREEZE-THAW OF SWAMP ON AGRICULTURAL PRODUCTION IN THE SANJIANG PLAIN OF CHINA

Wang Chunhe

Changchun Institute of Geography, Academia Sinica
People's Republic of China

The Sanjiang Plain in northeast China is rich in water and soil resources, but as a result of the interaction of several natural factors, about 2 million hectares of cultivated and uncultivated land are subject to the development of swamps, which results from the hydrogeological and thermal properties of soils, the distribution of rainfall, and, especially, the characteristics of the freeze-thaw process. Swamps and frozen soil are interdependent and promote each other. The soils are silty with a permeability ranging from 0.0013 to 0.635 cm/day, so they are poorly drained; the thermal conductivity of humic and peaty layers covering the surface is only 0.0011-0.0023 cal/cm sec deg, and its heat capacity is only 0.4-0.5 cal/cm^3 deg. Rainfall is concentrated in the period before the freezing season, so seasonally frozen ground is usually saturated and the thawing process during spring is very slow. Spring meltwater does not drain well vertically because of the poor permeability of the ice-rich unthawed active layer. This reduces the harvest, influences transportation, and damages engineering construction. Through practice, valuable experience has been gained in using bulldozers to promote thawing, drilling through the frozen layers to allow water to drain, and utilizing the freeze-thaw action in drought years to preserve soil moisture.

GEOGRAPHIC CONDITIONS

The Sanjiang alluvial plain, a wide tract of the confluence plain of three rivers, the Heilongjiang, the Songhuajiang and the Wusulijiang in the northeastern part of China, covers an area of over 42,500 square kilometers, rich in fertile soil and water resouces, and is one of the important commodity grain bases of China. Because the plain, in tectonics, belonged to the Hejiang interior fault basin, it has experienced differential wide subsidence since the Mesozoic, with low and smooth relief, gradient ratio of 1/10,000 to 3/10,000, and generally 50-60 meters above sea level. The surface layer of the east plain is a 3-17 m layer of clayey soil, while in the west plain the soil is sandy loam and fine sand. Below the surface layer is a unified water-bearing layer with 5-10 m of fine sand and 50-200 m of gravel. Below that are lacustrine mud stone with gravel and sand, continental coal-series stratum, and neutral and acidic volcanic and sedimentary rock of the Mesozoic and the Cenozoic.

The area has a humid climate, with annual mean air temperature of 1.4-3.0°C. The temperature in the coldest months is -18 to -21 °C, the extreme lowest temperature is -42 °C and the freezing season lasts 7 to 8 months with a freezing depth of 1.5 to 2.5 m. Annual precipitation is 500 to 700 mm, but non-uniformly distributed. It often rains in autumn, and freezing occurs before discharge of flood water. As a result of the effects of these natural factors, swamps develop. There are currently more than 29,000,000 mu of swamp land and swampy wilderness, amounting to 45.74% of the total area of the whole plain.

RELATIONSHIP BETWEEN FREEZE-THAW AND SWAMP

The surface layer of the eastern part of the plain consists mostly of clayey soil, formed by glimmerton, smalite and askanite. The specific gravity is 2.67-2.72, wet unit weight is 1.7-2.6 gm/cm^3, and dry unit weight is 1.47-1.72 gm/cm^3. Because the soil is fine, the texture is heavy and more than 60% is powdery soil with partical size diameters 0.05-0.005 mm (Figure 1). The coefficient of permeability is generally 0.0013-0.635 cm/day. The thermal conductivity of 10-20 cm of grass root and peat layer on the surface

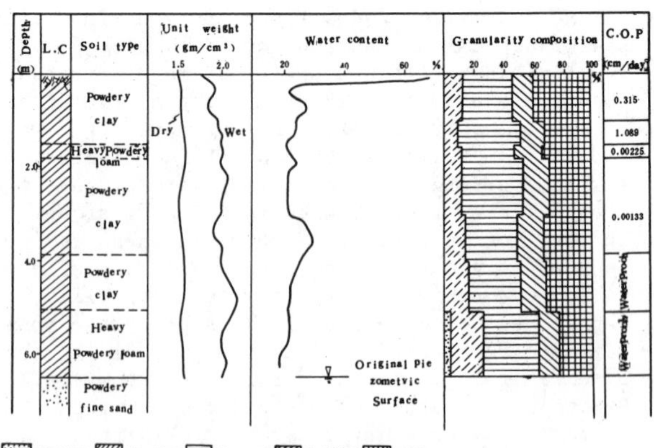

L.C. - Lithological Characters
C.O.P. - Coefficient of Permeability

Figure 1. Clay property of swamp land on Qiangin Farm.

layer is 0.0011-0.0023 cal/cm.sec.deg, 2-4 times smaller than normal soil and rocks. The heat capacity is 0.4-0.5 cal/cm.deg (swamp Dept. 1981) Because of dense swamp vegetation, thick snow accumulation, stagnant water in the swamp, and a great quantity of latent heat released while freezing, freezing rate is slower than farmland, soil and rocks, and frezing surface is rather stable. While freezing, influenced by temperature gradient and by the comprehensive action of various stresses such as adsorption capacity of ice crystals, differences of films, chemically potential differences, capillary force etc, water migrates to form ice-rich layer; the water content of soil increases by 10-19% over its value before freezing as water moves towards freezing front (Figure 2). This redistribution and enrichment of soil water contributes to the development of swamp.

Since the porosity of the grass roots and peat layer is 72-93%, being just like a sponge, the water retention is high, generally 149-552%, and saturation water retention is up to 830-1030% (Swamp Dept, 1976), 10-20 times higher than mineral soil and bedrock. The water capacity of saturated frozen grass roots and peat layer is very high. While thawing, much heat is consumed to slow thawing rate. Thawing depth in swamp land is 0.5-1.0 m less than on farmland. Generally the date of complete thawing of the frozen active layer is 30-50 days later than dry farmland, this is favourable, of course, to the preservation of frozen soil. Because of sunshade, reflection and transpiration of thick swamp vegetation in summer, and the consumption of heat in evaporating stagnant water, the ground temperature in swamp is 6-8°C lower than meadow and 8-13°C lower than bare farmland. Daily variation of temperature is small, vertical gradient is great, the frozen soil is protected. Since a large amount of heat energy is received and consumed in ground ice melting, the frozen active layer can not be thawed quickly. Precipitation and meltwater of ice and snow do not penetrate well through such a frozen layer, but remain on the surface and in the upper layers of soil, and freeze in the autum. As a result, excessively wet swamp forms. These distinctive thernodynamic and hydrogeologic properties of swamp slow thawing rate, protect the frozen layer in the soil, and promote the genesis and development of swamp.

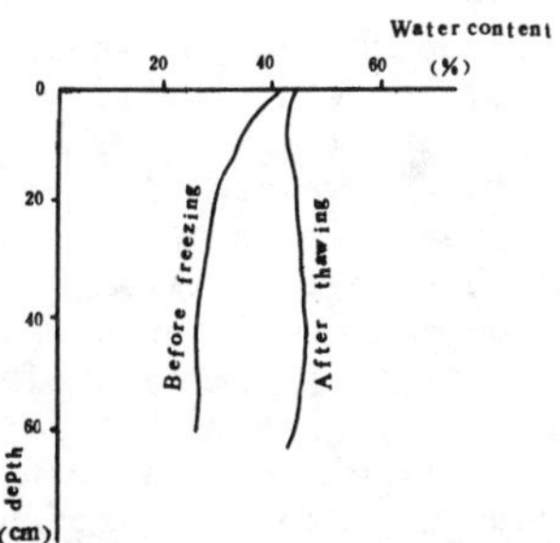

Figure 2. Curves of water contents before freezing and after thawing.

IMPACT OF FREEZE-THAW ON AGRICULTURAL PRODUCTION

Spring Waterlogging Caused by Autumn Flood Water

Precipitation in the area is only 500-700 mm, but it is unseasonably distributed during the year, being highly concentrated in autumn. Rainfall in September and October is generally about 20% of the total annual precipitation; In some years it is 36-40% of the total. There were 14 years of autumn waterlogging among the last 26 years of observation. Autumn flood water freezes on the surface and upper part of the soil layer, which not only interferes with autumn harvest, but also causes spring waterlogging the next year. Over-wet soil and thick snow cause the soil to freeze slowly and form an ice-rich layer, intensifying spring waterlogging. Next spring, the meltwater of ice, snow and the ice-rich layer itself can not permeate through frozen layer, but remain on the surface to cause serious spring waterlogging (Figure 3). Spring sowing is delayed, and crop output is reduced greatly. In the most serious cases, spring sowing can not be done at all. According to the Jiansanjiang State Farm Authority, serious spring waterlogging in 1957, 1960, 1963, 1973 and 1982 was caused by autumn flood water in the previous years. So a bitter lesson was learned, "Autumn flood water cause spring waterlogging and such a spring waterlogging will last two years."

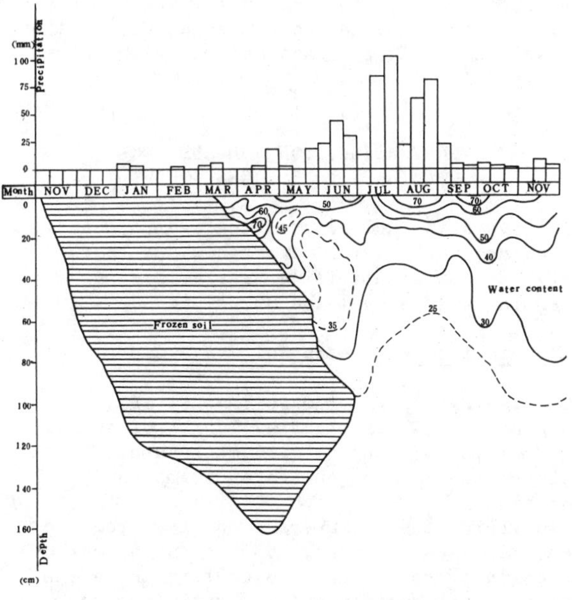

Figure 3. Water regime in soil of swampy wilderness. (Based on data of Test Station for Draining Waterlogging of Survey Designing Institute of Northeast General Agricultural Reclamation Bureau.)

Snow Calamity

The snowfall of the plain is generally 40-70 mm. There are 120-140 days annually with snow on the ground . Maximum observed snow thickness was 40-68 cm, and on some roads between forests the snow was more than one meter thick. On the roadsides near forests, wind blew snow to form snow drifts. This resulted in snow blocking roads and difficult transportation in winter. Snow meltwater in Tongjiang-Fuyuan district is 15-20% of runoff (The Office of Leading Group 1976). In some places snow meltwater caused over-wet soil to intensify spring waterlogging and influence spring sowing. In certain years the first snow covered crops. For example, it snowed 40 mm on October 11, 1972, covering a considerable amount of cut-down soybean, which was harvested with difficulty.

Freeze Hazards

Freeze-thaw greatly damaged buildings and other structures. When the ground around poles become over-wet and the post holes fill with water, poles have been frozen out of ground in winter, and also have subsided and tilted in spring; some have even fallen or been uprooted. Saturated soft muck and silt in swamp deposits swell on freezing, and subside or buckle on thawing. This has caused crevices in walls of houses, doors and windows which are twisted askew, gables inclined outwards and sunk inwards, heaved or sunken terraces, and even the collapse of some houses. Because of unreasonable expansion while freezing and subsidence while thawing, culvert pipes were apt to come apart. Freeze-thaw can also deform bridges, cause canal percolation, and seriously distort roads and railways, blocking traffic.

PREVENTION, CONTROL AND UTLLIZATION OF FREEZE-THAW

In developing and controlling swamps, people have gained much usefull experence and learned costly lessons on utilizing freeze-thaw, and preventing freeze and snow calamities.

Use of Machines on Frozen Soil of Swamp

In digging 113 km of main channel of the Bielahong River and constructing 226 km of railway from Fulitun to Qianjin, good methods have been developed for working on frozen swamp land soil by mechine.

Shovelling thawed layers properly. From the second ten days of March, some dozens of bulldozers began to remove snow to contribute to the thawing of the surface sod. At that time the highest ground temperature was 7.9 ℃, and the lowest was -21.6 ℃ (Survey Designing Institute 1980). The surface was thawed in daytime, but frozen at night. Bulldozers can push aside the layer being thawed. After removing all thawed mud and sod, one can construct on hard soil below 1.0 m.

Sawing ice into small pieces and taking them away by machines. When the frozen layer was less than 50 cm thick, the frozen soil can be cut into 2.0 × 1.2 m or 1.0 × 1.0 m pieces, and these frozen soil clods can be taken away, then construction work can proceed on dry soil directly.

Using ice as a wall to intercept flow. If a construction site is needed, the frozen layer around it can be uncovered and a ditch 30-50 cm wide can be dug. If the ditch is filled with ice, the ice can be used as a wall to intercept flow onto the site.

There are also some additional methods: for example, using a dam to divert meltwater, followed by the removal of ice away after freezing.

Drilling Through Frozen Layers to Allow Drainage

If spring meltwater of ice and snow and precipitation can not infiltrate through frozen layers nor drain away because of low and flat relief, this will cause spring waterlogging which seriously delays spring sowing. In the west part of the plain, there is 0.5-3.0 m of underground water section in water-bearing strata. One can blast or drill through the frozen layer in low catchment areas to make meltwater flow into the underground water-bearing strata. This removes the standing water, ensuring timely spring sowing.

Thinning Surface Snow to promote thawing

Snow cover in the area always thaws slowly. Meltwater often caused stagnant water locally, hampering spring sowing and transportation. These problems can be overcome by using different machines in the spring to spread and thin snow on the surface, or by rolling and mixing snow and soil. Because dirty snow can absorb more heat than white snow, the thawing rate is speeded and spring sowing can be carried out properly to ensure a bumper harvest.

Predicting Spring Drought

Weather forecasts and agricultural production records have been used to predict spring drought the following year. When drought is anticipated, the following proceedure can be used. After harvest in autumn, soybeen stubble is not harrowed and fields are not plowed. This enhances the capture of wind-blown snow in the fields. The freezing rate of the soil is thus slowed to promote water migration while freezing, causing an ice-rich layer to form. Next year, leveling the field and spring sowing properly, using water from the thawing of the ice-rich layer, and packing the ground firmly to preserve soil moisture will reduce the effect of drought and permit a good harvest.

REFERENCES

Swamp Department of Changchun Institute of Geography, Academia Sinica, 1981, A preliminary approach to the vicissitudes of natural environments of Sanjiang plain and its rationalized development: Acta Geographica Sinica, Vol.36, No.1, Science Press.

Swamp Department, 1976, Geography periodical, Geography Institute of Jilin Province No.1

The office of leading group for controlling the Sanjiang Plain of Heilongjiang provice,March, 1976. The comprehensive control program for the Sanjiang Plain of Helongjiang Province.

Survey Designing Institute of General State Farm Bureau of Heilongjiang Province, May,1980, Beidahuang Survey and Design.

PROPERTIES OF FROZEN AND THAWED SOIL AND EARTH DAM CONSTRUCTION IN WINTER

Wang Liang, Xu Bomeng, and Wu Zhijin

Water Power Research Institute, Northeast Design Institute
Ministry of Water Resources and Electric Power
People's Republic of China

To solve problems of winter construction of earth dams, we conducted an experimental study on paving sloping clay core, making frozen clay blanket, and laying sloping weathered sand core for cofferdam in winter, and made an extensive investigation on the construction of medium and small-sized earth dams. We found that the pace of winter construction can be speeded up while still maintaining quality on the basis of the following characteristics of frozen and thawed soil: (1) The compactibility of filled soil under negative temperature can be as good as that under positive temperature when soil is in the supercooled state. (2) The permeability of thawed soil is higher than that of unfrozen soil, the consolidation process is faster, and there are no significant differences between the quick shear strengths of the two soils. Therefore, the mechanical properties of the clay soil with water content close to the plastic limit shows no significant change after compacting and freezing. (3) When it is submerged in water, frozen clayey soil thaws and crumbles quickly on its surface, filling in voids between frozen masses and increasing its density. It can be used for construction. (4) Roughening has no notable influence on the compressibility and permeability of the binding layers and on their shearing strength; it can be omitted for binding layers.

The filling construction of earth dams is made difficult by the freezing of soil, especially when the compactibility of filled soil in winter and the impairment to engineering due to the change in the properties of soil during freezing and thawing are not exactly known. Filling construction in winter with clayey soil is generally suspended prolonging the period of construction, increasing costs, and subsequently delaying the efficient use of the engineering works.

Considering the requirements of practical engineering construction, we first carried out the experimental study on paving sloping clay core in winter, making frozen clay blanket, laying sloping weathered sand core for cofferdam in winter, etc. At the same time, we also made an extensive investigation on the construction of medium and small-size earth dams in winter. This paper introduces some results in experiments on the freezing and thawing properties of soil and their application to the construction of earth dams in winter.

COMPACTIBILITY OF SOIL IN LOW TEMPERATURE

For the cementation power of ice, frozen soil has high ability to counter outside load. Therefore, in winter construction, filling entirely with frozen soil is unadvisable. But if the frozen soil mass is surrounded by warm soil, the compactibility will not be greatly affected. Besides, the warm soil will melt part of the frozen soil and alleviate the bad effect of frozen soil. So a definite amount of frozen soil might be mixed in filling, and generally not exceeding 30% is recognized as suitable. But the allowable content of frozen soil is related to the properties of soil, its moisture content, the ambient temperature, the method of rolling and the process of construction, etc. Figure 1 shows a curve of compressive properties of soil (Table 1) at Baishan cofferdam.

It indicates that a higher frozen soil content (over 30%) is allowable.

In the process of winter construction, the freezing of filled soil is relevant to the duration of construction, the thickness of filled soil, the super cooling of soil, etc. As shown in Figure 2, the temperature variation after soil filling indicates that when air temperature is -11°C the temperature of clayey soil (moisture content being 17%, equivalent to plastic limit) at a depth of 5 cm under the effect of heat preservation of the soil layer merely drops to 0 °C after 4 hours.

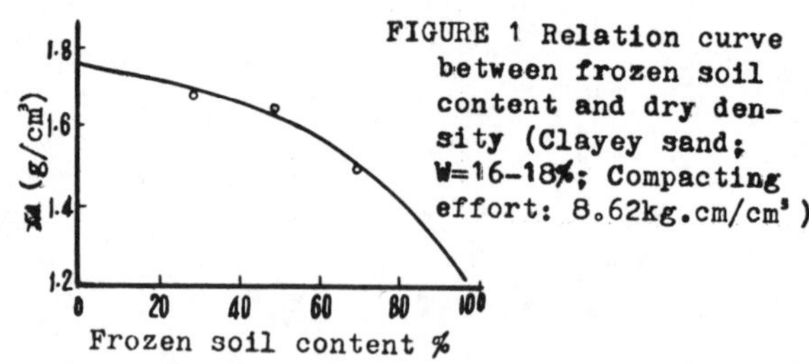

FIGURE 1 Relation curve between frozen soil content and dry density (Clayey sand; W=16-18%; Compacting effort: 8.62kg.cm/cm³)

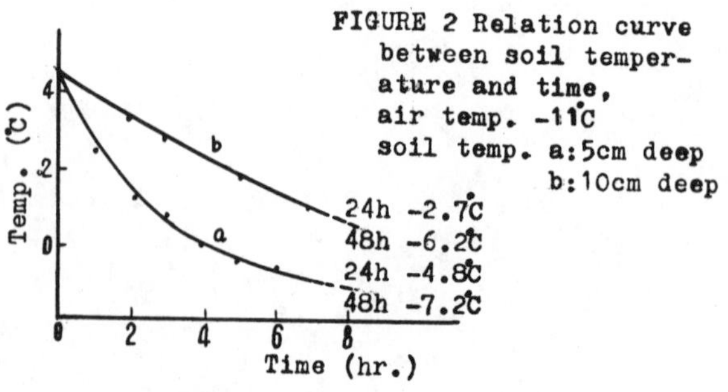

FIGURE 2 Relation curve between soil temperature and time, air temp. -11℃ soil temp. a:5cm deep b:10cm deep

It is known that soils of different types with different moisture contents have different freezing temperatures. In general, the stable freezing temperature of clayey soil is lower than 0°C. Moreover because of the supercooling nature of soil, before ice crystals are formed, the soil temperature will be even lower than its stable freezing temperature, and lasts a certain period of time. The super cooling soil still has a higher compressibility than warm soil. For example, the clayey soil at Qinghe reservoir (see Table 1 for its physical properties), when its moisture content is 21.5%, has a super cooling temperature at -2°C to -3°C under an ambient temperature of -5°C to -6°C, and the duration of super cooling comes up to several tens of minutes (Figure 3).

Though the time is quite short, yet in the fight for compactibility of soil, it is still very profitable and preferable. From the above experiment it is sufficient to understand that, in winter construction, as suitable heat preservation measures for the clayey soil of the earth embankment are taken, and proper arrangement for construction procedure to shorten the time for loading, transporting, laying and compressing soil is affected, sufficient utilization of the same quality of compactibility of filling soil as that in the construction under ordinary temperature could be attained.

TABLE 1 Physical properties of several kinds of soil

Location	Gradation			Plasticity		
	0.05	0.05 0.005	0.005	W_L	W_P	I_P
Qinghe	27	46	27	35	19	16
Longfeng-shan	14-29	41-54	25-35	31-45	21-25	12-23
Baishan	61	22	17			

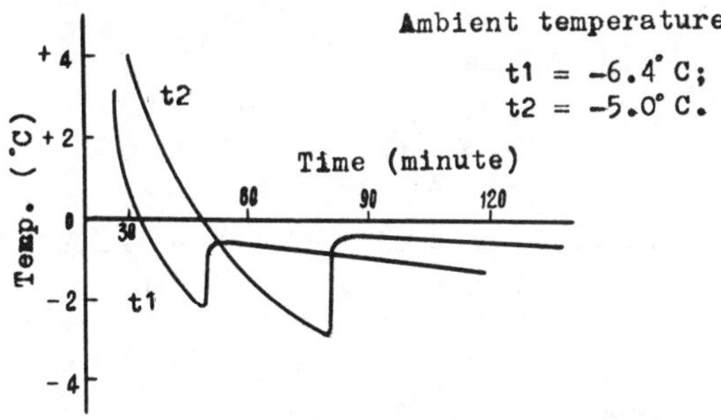

FIGURE 3 Freezing process of Qinghe soil sample

PERMEABILITY AND MECHANICAL PROPERTIES OF FREEZING-THAWING SOIL

In the process of freezing and thawing, soil undergoes a series of complicated changes. Under the action of negative temperature, the clayey soil mass with certain moisture content is expected to have moisture migration and phase equilibrium changes within it . Simultaneously, for the heterogeneity of temperature, freezing form and structure in it, the soil will form ice laminated layers in horizontal, vertical and inclining direction with different thickness making the soil structure layered and net-veined. The formation of ice laminated layers divides the soil mass into many small collective bodies, increases the volume of soil and diminishes its density, on the one hand, and the moisture content of these collective bodies decreases and the density increases on the other hand. The new structure of soil divided and cemented by many ice laminated layers brings about very complicated changes in the nature of soil after its thawing. Moreover, such changes vary with the state and process of the soil after thawing.

Permeability of Freezing-Thawing Soil

Clayey soil with different freezing structures at the beginning stage of thawing can have its moisture quickly drained through the crevices between between the mineral collective bodies, therefore its permeability is higher that that of unfrozen soil. In Table 2, from a group of experimental results at Qinghe reservoir, we can see that when the soil begins to thaw, its permeability coefficient is about 50 times that of the unfrozen soil in average.

TABLE 2 Permeability coefficient of freezing-thawing and non-frozen soil

Condition of sample	Non-frozen soil	Freezing and thawing soil
W%		17.9 21.0 21.0
r_d g/cm³	1.68 1.69 1.69	1.69 1.69 1.69
K cm/sec	7.56 3.07 1.57 $\times10^{-9}$ $\times10^{-9}$ $\times10^{-8}$	1.42 7.53 3.16 $\times10^{-7}$ $\times10^{-7}$ $\times10^{-7}$

Along with the increase in moisture content, the ice laminated layers or crevices will also increase and thicken, thus greatly raising the permeability coefficient of the freezing and thawing soil. For example, under the same dry density, while the moisture content increases from 18% to 21%, the permeability coefficient increases about 2 to 5 times.

After thawing of the soil, under the action of load, the crevices formed by the ice-laminated layers gradually close up and the filled soil will be gradually drained and consolidated, while the collective bodies take in moisture and swell. Eventually, the structure of soil gradually becomes homogeneous, the permeability coefficient of soil diminishes, and the ability of seepage prevention of the soil is improved. The result of experiment with frozen soil of Qinghe earth dam shows that at a dry density of 1.68g/cm³, the permeability coefficient is 1.64×10^{-7}cm/sec., while under a pressure of 0.5-1.0kg/cm², the permeability coefficient reduces to 3.9×10^{-8}cm/sec.

Compressibility of Freezing-Thawing Soil

When frozen soil thaws, though the mineral collective body encircled by ice laminated layers takes in water and swells, this process is comparatively slow, while the ice laminated layer between the collective bodies will first melt up and the melted water will drain quickly, making the collective bodies get closer together and the soil mass more compact. Of the two processes the latter is dominant. Moreover, such compacting of soil can be effected under the action of its own weight. Thus, freezing and thawing soil has a quicker process of consolidation and a greater compactibility. Figure 4 shows a group of results of compressive tests on Qinghe clay soil.

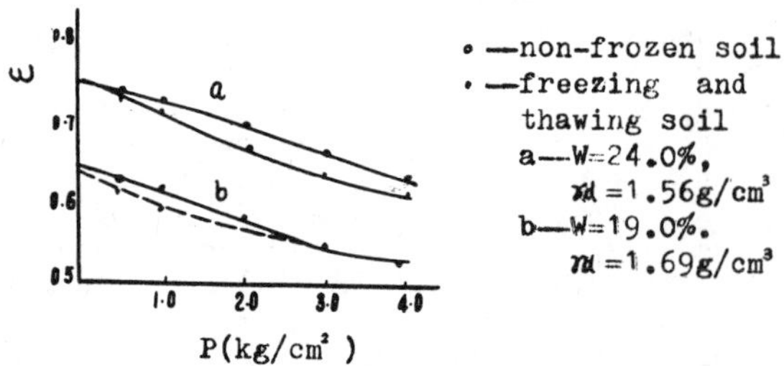

FIGURE 4 Compressibility of freezing and thawing soil and non-frozen soil

From curve(a) in the figure we can see that when moisture content is 24% and the load is 0 - 1kg/cm^2 its compression coefficient is two times that of unfrozen soil. This indicates that soil with a higher moisture content after freezing - thawing may give excessive great settlement, bringing detrimental effects to engineering works. When the moisture content in soil corresponds to plastic limit, its compressibility after freezing and thawing has nearly no difference as compared with that of unfrozen soil (Figure 4 b).This is because the soil under this moisture content in the process of freezing has no prominent phenomenon of moisture migration and ice segregation, its volume undergoes no frost heaving, its freezing structure is in a perfect entity, and after thawing its property has only a very small change compared with the original unfrozen soil.

Shear Strength of Freezing and Thawing Soil

The drainage condition of soil is an important factor in the shear strength of freezing-thawing soil. Though the initial moisture content of soil is relatively high, under certain load and drainage conditions the freezing-thawing soil has a greater compressibility, thus increasing its dry density, sometimes even to such an extent that it would be higher than that before freezing and consequently enlarging its shearing strength. As shown in the curve of Figure 5, at a moisture content of 24%, the consolidated quick shearing strength of freezing and thawing soil is even higher than that of unfrozen.

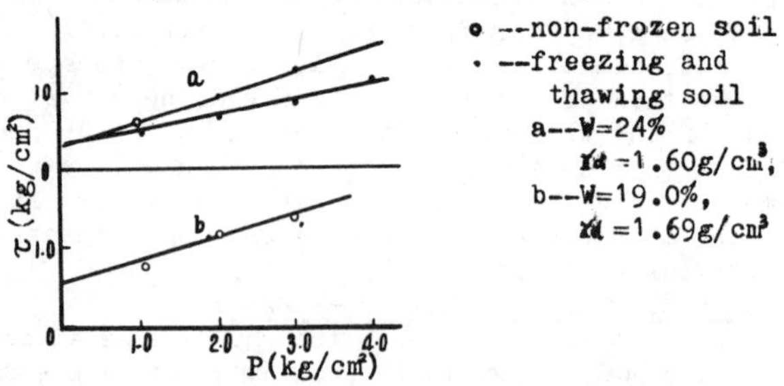

FIGURE 5 Consolidated quick shear strength of freezing-thawing soil and non-frozen soil (consolidate 3 hr.)

Under certain conditions, this property of improving strength at the later stage of filling could be properly utilized in practical engineering.

However, if drainage is poor and the water in soil could not be drained quickly, then under the action of load there would be greater pore water pressure, which greatly lowers the shearing strength in freezing and thawing soil. Many reports at home and abroad expound that the quick shearing strength of freezing and thawing soil is lower than that of unfrozen, about half of the unfrozen soil. At the initial stage of thawing, these loose collective bodies always peel off on the slope surface layer by layer with the melted water flowing. When there is a water flow passing freezing and thawing erosion follows. Sometimes, though there is no peeling off on the surface, with the deepening of melted layer, the soil at the upper layer may gradually consolidate due to better drainage condition and the permeability coefficient also lessens, making water in the just thawing layer difficult to drain and causing prominent lowering of strength on the thawing surface. As a result, the slope becomes unstable, and with the deepening of thawing layer, the range of sliding slope also enlarges. In north China regions, the destruction of side slope from above mentioned reason could often be seen. Therefore, to the side slope with exit of ground water or relative high moisture content, appropriate measures must be taken to prevent such slope slipping, which often happens in the time of thawing.

From Figure 5b it could be seen that, when the moisture content in soil approaches plastic limit, the consolidated quick shearing strength of the freezing and thawing soil, just as in the case of its compressibility, is not different from that of unfrozen soil.

From above mentioned properties of freezing and thawing soil, we can see that when the moisture content of soil is at plastic limit, no prominent change will occur in the structure and other properties in the freezing and thawing soil.

Using unfrozen soil with a plastic limit moisture content in quick construction at negative temperature one can get the same good quality of compactness as at positive temperature and the quality of the filled soil will not be affected by freezing and thawing action.

THE COMPACTIBILITY OF LAYERED FROZEN SOIL MASSES AFTER SUBMERGED IN WATER AND THAWING.

Crumbling Nature of Frozen Soil Mass in Water

Tests on the crumbling of frozen soil in water show that frozen clayey soil with a definite moisture content, when submerged in water, thaws and crumbles with a much higher speed and a much larger amount than that of warm soil. Figure 6 is the results of crumbling tests of a group of clayey soil (CI), the curves in which show that generally the amount of crumbling of frozen soil is over 80%, while that of warm soil not over 40%; and that for frozen soil put into water after thawing, its amount of crumbling between those of frozen soil and warm soil. With the lengthening of the time of thawing, the degree of cementation among those collective bodies increases, its amount of crumbling in water will go down close to that of warm soil.

This favourable property of crumbling of frozen soil comes mainly from the following fact: when the frozen soil cut into many small collective bodies by the ice laminated layers is submerged in warm water, the ice laminated layers quickly thaw and the collective bodies subject to peeling off layer by layer in small pieces. The higher the water temperature, the greater the speed of crumbling. Moreover, the crumbled soil bodies of the frozen clayey soil is different from that of the warm soil, for the migration of moisture during freezing, the moisture content decreases and its density increases in the collective bodies. (Table 3)

It is clear from above, freezing improves the crumbling property of clayey soil, making the method of pouring the earth into water adaptable for construction.

TABLE 3 Properties of soil mass collapsed in water

Condition of sample	Frozen soil	Collapsed soil mass
γ_d (g/cm³)	1.50	1.60-1.66
W %	27.9	23.8-25.3

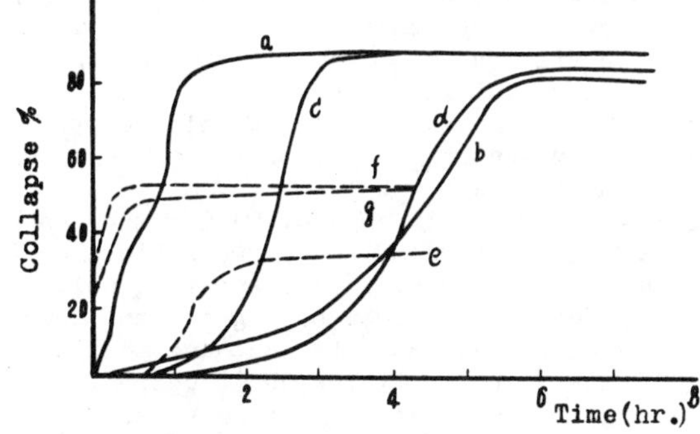

Frozen soil:(soil temperature -18 ℃)
a: W=18.5%, water temperature t =5-6 ℃;
b: W=18.2%, " " " t =0-0.5 ℃;
c: W=25.0%, " " " t =4.0 ℃;
d: W=26.5%, " " " t =1.0 ℃;
Warm soil: e--W=24.0%, γ_d =1.60 g/cm³;
Freezing and thawing soil:f,g--w=25.5-25.7%

FIGURE 6 Relation of collapse percentage of soil in water and time

Thaw-Subsidence and Consolidation Process of Layered Frozen Soil Masses

After the submergence of layered frozen soil masses in water, first, quick thawing happens on the surface of the frozen masses and at the points of contact, and the collective layers of the soil crumbled down and peeled off fill up the pores and crevices. At the same time relative displacement and squeezing among the frozen masses occurs.These make the big pores and crevices as the layered frozen soil masses (about 30%) shrink.The process of thawing and peeling off develops continuously until the shrinking pores and crevices among the frozen masses are entirely filled up. The soil bodies filled in the pores and crevices under the action of overlying load quickly increase their density, however still smaller than the natural density.Then, the frozen soil masses continuously thaw, but the collective bodies already stop crumbling,melt water quickly drains, making the collective bodies get even closer together. The density of this part of soil will be greater than the natural density. Since those capable to keep equal to or larger than the natural density often occupy an important part, the average density of the layered frozen soil masses after settlement and compacting approaches the natural density.With the increase of the size of the frozen soil masses and the improvement of grading distribution, their density will still increase.The above process could be finished in a quite short time and was proved by a large amount of experimental results both in laboratory and on field. Figure 7 shows an experiment of settlement and compacting of stacked-up body of frozen soil in thawing with a thickness of 1.28m.

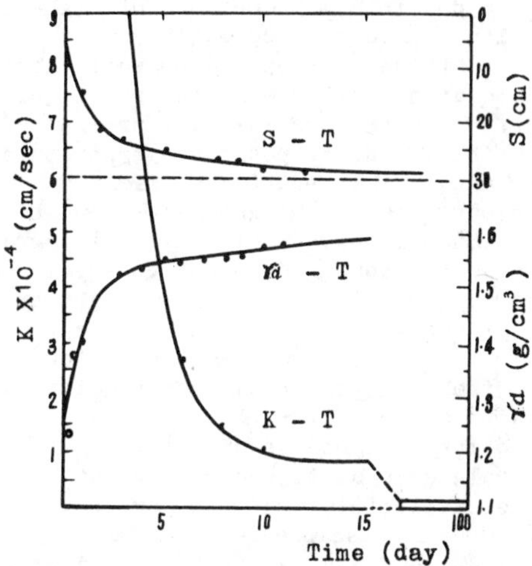

FIGURE 7 Thawing process of piling frozen soil mass during submersion in water

The testing sample is taken from the Longfengshan reservoir (Table 1), the moisture content of the frozen mass is 24%, porosity of large pores is 28.3%, the water temperature for testing is 3 - 7 ℃,the initial density after stacking up is 1.24 g/cm³,settlement and compacting is fundamently finished in 5 - 6 days,and the density attained is 1.55 g/cm³.Later, under its own weight and the pressure of permeability, it continues the process of settlement and compacting on one hand, and the collective bodies with lower water content take in water from the crevices and swell on the other. At this time, the increase in density of the filling body is not great, but the structure tends to be more homogeneous.The permeability of water changes from passage along the concentrated crevices to passage through the homogeneous pores.The figure shows that on the tenth day the permeability coefficient is 1×10^{-4} cm/sec, while after 100 days it reduces to 1.1×10^{-5} cm/sec. During this period, the density of soil has little change.

If frozen soil mass is submerged in water after its thawing, then the mass body is dispersed into a loosely stacked collective body which will take in water and lower its density, thus making the settlement compacting body have a smaller density and larger permeability.

The above mentioned property of frozen soil makes it possible to use the soil of higher viscosity in the construction of horizontal seepage prevention blanket by the method of dumping earth into water or submergence in water.

CONSTRUCTION OF ROLLED EARTH DAM IN WINTER

With the knowledge of above mentioned freezing and thawing properties of frozen soil, if proper measures are taken, the successful construction of earth dam in winter is perfectly possible. For large size construction since the considerably strict requirements of engineering quality, essentially warm soil should be used for construction, the moisture content of soil materials should near the plastic limit so as to keep the soil temperature as high as possible,successive working procedures should be connected to each other as close as possible, the methods of high-speed construction are to be adopted,the super cooling temperature of the soil should be sufficiently utilized, and the soil compacting should be performed before it freezes.Thus the similar effect of compactibility under positive temperature is also obtainable under negative temperature.

At the earth dam of Qinghe reservoir, Liaoning Province, which is 39.4 m high with sand gravel shell, a clayey sloping core was built. The main soil material used is medium plastic clayey soil (C I). It was at the beginning of the winter season, the air temperature was not below - 10°C, and the method of high-speed construction was adopted.With sheep-foot roller, it could be compressed to the designed density (1.66 g/cm³). If the temperature is still lower, the guaranteed rate of compactibility would be lowered, so a simple warm shed house provision was used. A warm shed is wind-proof and can increase soil temperature, but it is difficult to raise the temperature. When the outside temperature is -20°C, the temperature on the soil surface in the warm shed is -12°C. If unfrozen soil (a little freezing in the process of transportation and compressing) is used in construction, the quality requirements can generally be satisfied.

In order to solve the problem of construction under low temperature in open air, tests with heavy tamping plate (2.2 ton) have been carried out. It is a round iron tamper with 7 sheep-feet. In experiments, the height of drop is 2.5 -3.0 m, and different beating conditions are applied. The result of tests shows that with a filling thickness of 70 - 80 cm, tamping 5 - 8 strokes, air temperature at -12 °C, the temperature of soil higher -2.5 °C and frozen soil mass content at 20%, the density of minimum compactness of all satisfies the requirements of engineering design. Even at air temperature -28.4°C, with some freezing in the surface layer, the compactness of designed density is also reached. But if after filling, tamping to required compactness has not been executed in time, soil materials are frozen or frozen masses are concentrated in the bottom part, the filling soil could not be tamped to required compactness even when the thickness of filling is reduced to 50 cm and the tamping strokes increased to 24 times.

Generally if rolling is applied in construction, before filling, the compressed layer should be roughened first to facilitate the binding between the two layers. But roughening in winter construction is not only difficult but also takes time, thus further lowers the soil temperature and is not favourable for filling. Large amount of experimental study show whether roughening or not has no notable influence both on the compressibility and permeability of the binding layers, and on their shearing strength.The only effect is the increase of shearing deformation at maximum shearing strength, which would not give damage to the construction. Therefore,the process of roughening for the binding layer could be omitted, so that the labour for roughening could be saved, the construction be speeded up, and the quality of winter filling be improved. In the construction of earth dams at Qinghe, etc.In ordinary seasons, this process of roughening had been cancelled,and in the period of their long-term running,no question has revealed.

In the construction of cofferdam at the Baishan water-power station, the height of the dam being 25 m, the weathered sand was used to make seepage prevention body. Before winter, the method of filling soil into water was applied, utilizing the property of soil material in water, i.e. its quick dispersing, settling and compacting, the dry density reached 1.6 g/cm³and permeability coefficient $K = AX10^{-4} - 10^{-6}$ cm/sec. While in winter, for the construction of the part above water, the content of frozen soil before rolling and compressing was controlled lower than 50%, compressed density is 1.7 g/cm³, at this monent, $K = AX10^{-4} - 10^{-5}$ cm/sec. During the construction of the cofferdam, heavy truck (total weight about 40 ton) used in transporting soil did the work of rolling and compressing and the construction advanced successfully. But due to the very low temperature in winter construction, it was difficult for some very particular part in the filling to perfectly meet the requirement. So attention was paid to the quality of the filter. In the course of running, though particular part of filling showed insufficiency in compactness and presented the phenomenon of concentrated seepage flow, yet the soil particles dispersed while getting in touch with water had not been carried away by seepaging water flow, and were still settled and compressed gradually, exhibiting the ability to resist permeation. In the course of operation for several years, the cofferdam is in perfect condition.

To the small size earth dam generally below 10 m high, if the construction is to be made in the open air, usually a mixture of frozen soil and warm soil is used in the building of dam body and a certain process of compressing and compacting is to be done. So far as there is no perforating through hole in the dam body giving place to concentrated seepage, generally speaking safe running is ensured. But the moisture content in the soil material should not be too high, otherwise, in the process of thawing, slope slipping may occur at some part of the dam body.

In winter construction, it is perfectly possible to excavate unfrozen soil from some comparatively deep layer for concentrated usage as a part of soil material in building the dam body so that the dam may have better ability to resist seepage and the safety of the dam be warranted. In the course of impoundment, some part of the foreslope of the dam with poor compactness of filling may also get settling and compacting in the submerging water and gradually exert its ability to prevent seepage. But for its low strength, the slope on the upper reaches side should be made gentler.

CONSTRUCTION OF FROZEN SOIL BLANKET IN WINTER

Based on the characteristic of settling and compacting of layered frozen soil masses in water while thawing, successful experience had been obtained first in utilizing frozen soil for construction of blanket at Longfengshan reservoir in Heilongjiang. The height of the dam is 27 m, its length 716 m, the length of the blanket is 70 - 90 m, the dam foundation is a Quaternary alluvial sand gravel layer, and its thickness is 10 - 27m.

The frozen soil masses used in building blanket were taken from the field of explosion, with various sizes, generally 30 - 50 cm, larger ones reaching 70 cm. In the course of construction, frozen masses were directly thrown into water, and in ceratin sections the frozen soil masses were first stacked up and then water was filled, and frozen soil masses underwent thawing, settling and compressing. The dry density of frozen soil was about 1.55 g/cm^3, and for the blanket after thawing, settling and compressing, even if the samples taken between the frozen masses had their dry density (except in the 50 cm surface layer) reaching over 1.50 g/cm^3, and their permeability coefficient reaching AX10^{-5} - AX10^{-6}cm/sec, having certain ability to prevent seepage.Further higher density in the blanket could be obtained by using frozen soil of better grading so as to reduce the amount of large size pores in the stacked up body.

If the method of construction by first stacking up the frozen soil masses and then filling water to melt them is used, the filling of water should be done as soon as possible so as to make the frozen soil masses thaw and settle in water (not to let the frozen soil masses melt before the water is filled). While filling water, its level should be kept under the surface of the stacked up body at a certain height so that the stack can sufficient exert the action of its own weight in the thawing, settling and compressing of frozen soil.

Besides, the characteristic of easy collapsing of frozen soil in water is also used in some reservoirs, to construct the blanket by breaking ice on the surface of the reservoir and pouring frozen soil into the water.

In the construction of blanket from frozen soil, the construction procedure is simple, the operation is easy, a good way to overcome the difficulties in attaining required compactness in filling in winter. It is significant in early exhibition of engineering effectiveness, and lowering the engineering cost. The frozen soil blanket at the Longfengshan reservoir had been built more than 20 years ago, and is running normally. This method is also applied in the construction of other reservoirs, such as East Is Red in Heilongjiang, Yinhe, etc. and good effects are obtained, indicating that using frozen soil in the construction of blanket is a successful method.

CONCLUSION

1. Under the action of negative temperature the water migration will occur in clayey soil with a certain moisture content and the soil will form ice-laminated layers, dividing the soil mass into many small collective bodies. The collective bodies have lower moisture content and higher density than the original soil mass before freezing. When thawing the moisture can be quickly drained through the crevices between the mineral collective bodies, therefore compared with the unfrozen soil, the frozen soil has higher permeability, greater consolidation coefficient and settlement, lower quick-shear strength and higher consolidation quick-shear strength, better crumbling property in water.

2. High-speed filling construction of the clayey soil with moisture content close to plastic limit makes it possible to utilize upper-cooling temperature of the soil. In this way the same compacted quality would be obtained as at positive temperature. The mechanical properties of the soil could remain unvaried in the main before freezing and after thawing and engineering works will not be impaired.

3. The blanket constructed by dumping frozen soil masses or submergence of the layered frozen soil masses in water would have a density close to the natural soil and certain impermeability. These methods can also be used to strengthen blankets.

4. The seepage prevention body of cofferdam made by using weathered sand with loam with certain content of frozen soils in winter can meet the requirements of engineering when the heavy trucks are used for rolling and compressing. But it is necessary to pay attention to the quality of the filter back the seepage prevention body to prevent from failure at particular part by seepaging water flow.

In conclusion provided the characteristics of frozen soil and the freezing and thawing soil are correctly used and proper measures are taken for different engineering conditions, filling construction with clayey soil in winter can guarantee the quality of engineering works. The long-term operation of the engineering works indicates that filling construction of earth dams in winter is applicable.

REFERENCES

Ministry of water resources and electric power, P.R.C., 1958, The rules for construction of rolled earth dam (in Chinese), p. 22-24.

Shenyang Hydraulic Research Institute, 1959, Construction of rolled earth dam in winter (in Chinese), Press of Water Conservancy, p. 8-14.

Tetsuaki Nagasawa, Yusuharu Umeda, Influences of freezing and thawing processes on soil strength no. 60, Proceedings of agricultural engineering society, Japan, p. 19-23.

Tsytovich, N. A., 1975, Thaw settling of frozen ground, <u>in</u> the Mechanics of Frozen Ground, Washington, D.C., p. 215-218.

PALSAS AND CONTINUOUS PERMAFROST

A. L. Washburn

Quaternary Research Center, University of Washington, Seattle, Washington 98195 USA

This report describes peaty mounds in continuous permafrost, especially near Resolute, Cornwallis Island, Arctic Canada. Dome-shaped permafrost mounds here are up to about 60 cm high and 8 m in diameter. They are covered with tundra vegetation, have an active layer some 20 cm thick, and are characterized by a permafrost core of icy peat and/or silt. Some mounds also occur as similarly low but more irregular forms, others as parts of low plateaulike rises up to 1,600 m^2 in area. Most of the observed mounds are in or near present or former drainage lines that remain moist much of the summer. Field relations suggest that the ice of some mounds is primarily segregation ice, that the ice of others may include injection ice, and that some mounds are related to disintegration of plateaulike rises. It is concluded that most of the peaty Resolute mounds that are some 50 cm or more high and 2 m or more in diameter should be regarded as palsas.

INTRODUCTION

The term palsa originally designated "...a hummock rising out of a bog with a core of ice" (Seppälä 1972). Although there are now a variety of usages of the term, all pertain to perennial mounds, mostly in discontinuous permafrost. However, palsas or palsalike forms are also becoming known from the zone of continuous permafrost, as reviewed below, including occurrences in the vicinity of Resolute, Cornwallis Island, Arctic Canada.

EARLIER OBSERVATIONS

Palsas or palsalike features within the zone of continuous permafrost have been reported from Svalbard, the Soviet Union, Arctic Canada, and Alaska.

In Nordaustlandet, Salvigsen (1977) observed peat mounds in two areas, the largest mounds having a height of ∿1 m and diameters of 3-4 m. One mound was characterized by ice up to 70 cm thick beneath a peaty cover some 30 cm thick, of which 11-12 cm consisted of a sand layer. Salvigsen questioned whether the term palsa should be applied to these forms.

In Spitsbergen, Federoff (1966 p. 172-173) reported "palses" at Sassendal. They were elliptical, had a median height of 50 cm and short diameters of 2-3 m, and were associated with organic soil. At Kapp Linné, Åkerman (1973) and Jahn observed palsalike mounds that were "...up to 1.5 m high, built of segregation ice and occur in the area of polar moors" (Jahn 1975, footnote, p. 99); Åkerman (1980 p. 79, 1982 p. 47-48) described the same mounds as palsas.

In the Soviet Union, palsas in continuous permafrost occur in the southern part of the Yamal and Taymyr peninsulas and in Siberia, judging from Grigoriew's (1925 p. 346-348) discussion of peat-earth mounds (Torf-Erdhügel) in Siberia (some of which, however, were seasonal frost mounds), and comparison of Frenzel's (1959, Abb. 15, p. 1028) map of palsa bogs and flat hilly bogs (Palsmoore and Flachhügelmoore) with Soviet maps of permafrost distribution (cf. Washburn 1979, Figure 3.3, p. 25; Figures 3.9-3.10, p. 31-32). More specifically, continuous permafrost palsas, 0.5-3.5 m high and rarely exceeding 15 m in diameter (exceptionally up to 100 m), occur in the vicinity of the Viljui hydroelectric power station in western Yakutia (Åkerman 1982 p. 45-46).

In Arctic Canada, Jankovska and Bliss (1972, 1977) described peat mounds 1-2 m high and 5-15 m in diameter, with peat 1.5-2.0 m thick, occurring as high-center polygons on the north coast of Devon Island. W. Barr's report of "Peat mounds, 2-3 m high containing 30-cm thick layers of ice...." on Devon Island (in R. J. E. Brown 1973 p. 29) refers to the same area.

On Bathurst Island, Blake (1974 p. 235, 239) investigated two peat mounds. The largest was ∿1.4 m high and 3.7 m in diameter and was frozen below a depth of 40 cm on August 11, 1963. The mound consisted of peat and ice lenses to a depth of ∿3.3 m. Below this, ice extended to the bottom of the core hole at a depth of 5.0 m. The other mound was frozen below a depth of 21 cm on July 2, 1963. Peat extended to a depth of 2.6 m, below which was sand.

On northern Ellesmere Island, Flügel and Mäusbacher (1981) described two peat mounds. The largest was 40 cm high, ∿5.5 m in diameter, and had a frozen core in early August. King (1981) described three types of peat mounds in the same area, some up to 2 m high and as much as 20 m in diameter.

On Amund Ringnes Island, Hodgson (1982 p. 28) reported palsas, which he described as

> Small ice-cored mounds, 1 to 5 m in diameter and up to 1 m high...in some level and poorly drained fine grained deposits insulated by a complete moss/lichen cover.... A 75 cm-high mound that was cored...comprised a 20 cm organic-rich active layer, 60 cm of icy fines and sand, and 35 cm of pure ice, underlain by fines having low ice-content.

The mounds were commonly 50 cm high, and beneath some there was as much as 1 m of ice (Hodgson 1982 p. 21).

Bird (1967 p. 203) described ice-cored peat mounds, oval to elongate and 1-2 m high and 15 m or more long, from several places in continuous permafrost in Arctic Canada. Although perennial, the mounds were apparently short-lived forms whose core consisted entirely of massive clear ice. Somewhat similar but much smaller forms less than 50 cm high were described from Banks Island by French (1971).

In northern Alaska, Nelson and Outcalt (1982 p. 39-43) observed palsas or palsalike mounds adjacent to the trans-Alaska pipeline where road construction had interfered with the natural drainage. They believed that "Palsas can no longer be characterized as geographically restricted to the equatorward permafrost boundary, or even to the discontinuous zone" (Nelson and Outcalt 1982 p. 43).

PEATY MOUNDS IN THE RESOLUTE AREA, CORNWALLIS ISLAND

Peaty mounds in a variety of forms occur in the immediate vicinity of Resolute (latitude 74°43'N, longitude 94°57'W) (Figures 1 and 2). The bedrock is exclusively dolomite and limestone (Thorsteinsson and Kerr 1968). The region has been glaciated, and much of it was formerly submerged. Deposits due to glacial, marine, slope, and weathering processes (Cruickshank 1971) provide a general debris cover. Frost-wedged bedrock is everywhere near the surface and is locally exposed over considerable distances. Much of the Quaternary history is still obscure, but areas below ∿100 m emerged from the sea less than 10,000 years ago.[2]

The climate is cold and arid. The mean annual air temperature and precipitation at Resolute (1951-1980) are -16.6°C and 131 mm. Permafrost is ∿400 m (1,300 ft) thick (R. J. E. Brown 1970, Table 1, p. 10) and, depending on the nature of the surface cover and soil, the active layer ranges from some 20 to 80 cm thick, and possibly up to 100 cm in places. Because of the climate and carbonate bedrock, vegetation is sparse, some areas being almost completely barren except in depressions and downslope from long-lasting snowdrifts. In areas remaining wet for considerable periods, sedge-moss meadow communities prevail (Edlund 1982). The soils involved have been mapped as tundra and bog soils (Cruickshank 1971, Figure 2, p. 199). The mounds are restricted to these soils but are much less widely distributed.

Several low peaty mounds are well developed on the south side of Five-Mile Lake (Figure 2). The surface in their vicinity is flat and essentially horizontal. Except for occasional patches of bare ground, there is a general but commonly thin vegetation mat, some 5-10 cm thick, overlying stony silty sand. Turf hummocks 10-15 cm high and 10-30 cm across are present, along with comparably low but irregular turf ridges up to 2 m long. A 60-cm-high peat mound with marginal cracks (Figure 3) was cored to a depth of ∿125 cm with a modified CRREL 3-in. (7.2-cm) core barrel. The core sections, which were various shades of brown, revealed the stratigraphy and plant remains indicated in Tables 1 and 2.

An area of low mounds occurs at the north base of "Airport Ridge" (Figure 2). The mounds are so low they could easily escape notice, being only some 20 cm high (Figure 4). They lie amid vegetated stony silt and are characterized by a mossy cover, whose surface vegetation was dying in places as shown by a grayish tinge. On August 2, 1981, the frost table lay at a depth of ∿20 cm except in bare spots, whereas beyond the mounds it was at a depth of ∿30 cm in mossy vegetated areas and ∿40 cm in bare gravelly areas. The mounds were present throughout the 1981 and 1982 thaw seasons. Coring of a mound to a depth of 47 cm on July 15, 1982, revealed the stratigraphy indicated in Table 3, as recorded in the field by B. Hallet, C. Gregory, and R. Anderson.

Slightly farther east an east-draining shallow depression is bordered by an irregular, low longitudinal peaty mound some 50 cm or more high (Figure 5), lying parallel to the depression but divided into sections, in part by vehicle tracks. The mound is perennially frozen beneath a thin active layer.

A few, more prominent mounds occur between McMaster River and the low hills to the north (Figure 2). Some of these mounds comprise a plateaulike rise in the drainage depression of a small stream; the rise is 1-2 m high, ∿1,600 m^2 in area, and may be, in part, an erosion remnant left by the stream. The highest mound stands prominently above the rise (Figure 6). Other mounds here are lower, are less prominently dome shaped, and are separated by polygonal depressions. Some of these lower mounds show shell-bearing silt patches at their crest.

ORIGIN

It is generally agreed that palsas in discontinuous permafrost are due to frost heaving and that in many places they originate by differential heaving where snow cover is especially thin or lacking (Seppälä 1982). The heaving of palsas is usually ascribed to segregation ice but has also been reported as due to injection ice. Additionally, palsas have been ascribed to the disintegration of a peaty plateau.[3] A summary of these various concepts is presented elsewhere (Washburn 1983).

With respect to segregation ice, the areas where the Resolute mounds occur were formerly submerged, and near-surface permafrost would not have developed until emergence. Thus growth of segregation ice from upward-drawn water would have been possible as the surface sediments froze downward. However, this source for the mound ice seems improbable, considering the time needed for peat development and for mound inception and growth. Possibly general thinning of a thicker active layer may have permitted gradual growth of segregation ice. In any event, movement of unfrozen water within permafrost, a process that is becoming increasingly widely recognized (Cheng 1982, Mackay 1983), would have permitted this as peat accumulated and the active layer thinned locally because of increased insulation provided by the peat. As the permafrost table rose locally, such ice could build up by water moving downward as well as upward. The mounds on the south side of Five-Mile Lake appear to conform to this model, because their topographic position does not favor the availability of water under hydraulic pressure and development of injection ice.

On the other hand, water under hydraulic pressure between a downward-freezing active layer and the permafrost table would be favored at the base of a slope. This situation could lead to injection ice as well as segregation ice beneath and within a peaty layer. To the extent that the ice delayed thawing, aided by lower temperatures because of a thinner snow cover accompanying mound growth, the ice content and mounding might become perennial. Because of their topographic position and clear ice layers with vertical bubble trains, the low mounds at the north base of "Airport Ridge" may be such embryonic forms involving both injection ice and segregation ice.

The low mounds on the small plateaulike rise between McMaster River and the hills to the north are probably due to permafrost cracking and separation of peaty centers, and are more appropriately regarded as high-center, nonsorted polygons than incipient palsas. However, the highest mound here and another nearby are more isolated, much more palsalike, and may be aggradation forms rather than owing their origin to permafrost cracking.

Are palsalike mounds in continuous permafrost really significantly different from palsas in discontinuous permafrost? The term palsa has been used in a variety of ways, and in an attempt to clarify the question of "What is a palsa?" the following broad definition, which encompasses most of the usages, has been suggested (Washburn, 1983):

> Palsas are peaty permafrost mounds, ranging from c. 0.5 to c. 10 m in height and exceeding c. 2 m in average diameter, comprising (1) aggradation forms due to permafrost aggradation at an active-layer/permafrost contact zone, and (2) similar-appearing degradation forms due to disintegration of an extensive peaty deposit.[4]

Thus defined, palsa is a useful general term. It can be made specific by modifiers, as needed, describing (1) morphology (dome-shaped, etc.), (2) occurrence (discontinuous or continuous permafrost), (3) constitution (mainly organic or mainly mineral soil), (4) ice type (segregation-, injection-, or mixed-ice), and (5) origin (aggradation or degradation forms).

Accordingly, the Resolute and other perennial peaty mounds described above as exceeding ∿50 cm in height and 2 m in diameter are here accepted as continuous-permafrost palsas.[5] Also, more specifically, the Five-Mile Lake mound would be a segregation-ice (?) organic palsa. This work supports Åkerman's (1982) view and earlier suggestions that palsas occur in continuous as well as in discontinuous permafrost.[6]

NOTES

[1]I am most grateful to the Director, G. Hobson, and the staff of the Polar Continental Shelf Project of the Canadian Department of Energy, Mines and Resources for excellent support. I am also indebted to J. Åkerman, J. Brown, M. Eronen, H. French, B. Hallet, R. Mackay, F. Nelson, S. Outcalt, A. Pissart, M. Seppälä, C. Tarnocai, and S. Zoltai for helpful discussions--views on palsas differ, and responsibility for conclusions is mine alone; also to E. B. Leopold, B. M. Tucker, D. Vitt, and D. Clemens for determining plant remains in the palsa core from Five-Mile Lake, and to M. Stuiver for radiocarbon dating of the palsa core and the marine shells.

[2]Marine shells from an altitude of 102 m on "Airport Ridge" show a radiocarbon age of 9610 ± 40 years B.P. (University of Washington Quaternary Isotope Laboratory No. QL-1612). "Airport Ridge," so named for the purposes of this report, is the north-south trending ridge on the east side of the Resolute Airport.

[3]Not necessarily a peat plateau in the sense of a peatland requiring more than 40 cm peat accumulation (National Wetland Working Group 1981).

[4]Where high-center peaty polygon forms can be identified with certainty, they should be so designated and not termed palsas.

[5]High-center polygons excluded (see Note 4).

[6]Åkerman's (1982) article was received after submission of the present paper. I am grateful to the editor for permission to make some changes and insert reference to Åkerman's article and to the National Wetland Working Group's (1981) report.

REFERENCES

Åkerman, H. J., 1973, Palsstudier vid Kapp Linné, Spetsbergen, in Svensson, H., et al., Studier i periglacial geomorphologi på Spetsbergen: Lunds Universitets Naturgeografiska Institution, Rapporter och Notiser Nr. 15, p. 54-69.

Åkerman, H. J., 1980, Studies on periglacial geomorphology in West Spitsbergen: Meddelanden från Lunds Universitets Geografiska Institution, Avhandlingar 89.

Åkerman, H. J., 1982, Observations of palsas within the continuous permafrost zone in eastern Siberia and in Svalbard: Geografisk Tidsskrift, v. 82, p. 45-51.

Bird, J. B., 1967, The physiography of Arctic Canada: Baltimore, The Johns Hopkins Press.

Blake, W., 1974, Periglacial features and landscape evolution, central Bathurst Island, District of Franklin: Geological Survey of Canada Paper No. 74-1B, p. 235-244.

Brown, R. J. E., 1970, Permafrost in Canada: Toronto, University of Toronto Press.

Brown, R. J. E., 1973, Ground ice as an initiator of landforms in permafrost regions, in Fahey, B. D., and Thompson, R. D., eds., Research in polar and alpine geomorphology--Proceedings: 3rd Guelph Symposium on Geomorphology, 1973: Norwich, England, University of East Anglia.

Cheng, G., 1982, The forming process of thick layered ground ice: Scientia Sinica, Series B, v. 25, p. 777-788.

Cruickshank, J. G., 1971, Soils and terrain units around Resolute, Cornwallis Island: Arctic, v. 43, p. 195-209.

Edlund, S. A., 1982, Plant communities of Cornwallis, Little Cornwallis and associated islands: Geological Survey of Canada Open File map 857 (1:250,000).

Federoff, N., 1966, Les sols du Spitsberg Occidental, in Audin, ed., Spitsberg 1964 et premières observations 1965: Lyon, Centre National de Recherches Scientifiques (CNRS).

Flügel, W.-A., and Mäusbacher, R., 1981, Beobachtungen zu organischen Kleinformen im Oobloyah Tal, N-Ellesmere Island, N.W.T., Kanada, in

Barsch, D., and King, L., eds., Ergebnisse der Heidelberg-Ellesmere Island-Expedition: Heidelberger Geographisches Arbeiten, Heft 69, p. 559-563.

French, H. M., 1971, Ice cored mounds and patterned ground, southern Banks Island, western Canadian Arctic: Geografiska Annaler, v. 53A(1), p. 32-38.

Frenzel, B., 1959, Die Vegetations- und Landschaftszonen Nord-Eurasiens während der letzten Eiszeit und während der postglazialen Wärmezeit--I. Teil: Allgemeine Grundlagen: Akademie der Wissenschaften und der Literatur (Mainz), Abhandlungen der Mathematisch-Naturwissenschaftlichen Klasse, Jahrgang 1959, Nr. 13.

Grigoriew, A. A., 1925, Die Typen des Tundra-Mikroreliefs von Polar-Eurasien, ihre geographische Verbreitung und Genesis: Geographische Zeitschrift, v. 31, p. 345-349.

Hodgson, D. A., 1982, Surficial materials and geomorphological processes, western Sverdrup and adjacent islands, District of Franklin: Geological Survey of Canada Paper No. 81-9.

Jahn, A., 1975, Problems of the periglacial zone (Zagadnienia strefy peryglacjalnej): Warsaw, Państwowe Wydawnictwo Naukowe.

Jankovska, V., and Bliss, L. C., 1972, Pollen and macroscopic analysis of a peat mound profile, Truelove Lowland, in Bliss, L. C., ed., Devon Island I.B.P. Project High Arctic Ecosystem, Project Report 1970 and 1971: Department of Botany, University of Alberta, p. 105-112.

Jankovska, V., and Bliss, L. C., 1977, Palynological analysis of a peat from Truelove Lowland, in Bliss, L. C., ed., Truelove Lowland, Devon Island, Canada: A high arctic ecosystem: Edmonton, Alta., The University of Alberta Press.

King, L., 1981, Typen von Torfhügeln im Gebiet der Oobloyah Bay, N-Ellesmere Island, N.W.T., Kanada: Polarforschung, v. 51, p. 201-211.

Mackay, J. R., 1983, Downward water movement into frozen ground, Western Arctic coast, Canada: Canadian Journal of Earth Sciences, v. 20, p. 120-134.

National Wetland Working Group, 1981, Wetlands of Canada--Terres humides du Canada: Environment Canada, Ecological Land Classification Series No. 14.

Nelson, F., and Outcalt, S. I., 1982, Anthropogenic geomorphology in northern Alaska: Physical Geography, v. 3, p. 17-48.

Salvigsen, O., 1977, An observation of palsa-like forms in Nordaustlandet, Svalbard: Norsk Polarinstitutt Årbok 1976, p. 364-367.

Seppälä, M., 1972, The term 'palsa': Zeitschrift für Geomorphologie, Neue Folge, Band 16, p. 463.

Seppälä, M., 1982, An experimental study of the formation of palsas, in French, H. M., ed., Proceedings of the Fourth Canadian Permafrost Conference (The Roger J. E. Brown memorial volume): Ottawa, National Research Council of Canada.

Thorsteinsson, R., and Kerr, J. W., 1968, Cornwallis Island and adjacent smaller islands, Canadian Arctic Archipelago: Geological Survey of Canada Paper No. 67-64.

Washburn, A. L., 1979, Geocryology--A survey of periglacial processes and environments: London, Edward Arnold.

Washburn, A. L., 1983, What is a palsa?, in Poser, H., ed., Mesoformen des heutigen Periglazialraumes (Geomorphologisches Symposium, Kehr bei Göttingen, 1982), Proceedings: Göttingen, Akademie der Wissenschaften, in press.

TABLE 1 Core Stratigraphy of Peat Mound on South Side of Five-Mile Lake.[+]

Depth (cm)		
0-5	1.	Mossy surface vegetation
5-20	2.	Fibrous peat--grading down to
20-35	3.	Icy fibrous peat, fibers have dominantly horizontal orientation, ice disseminated, ice content* (20-25 cm) = 566%--grading down to
35-86	4.	Icy fibrous peat like 3 but fibers have dominantly wavy, in places flamelike, orientation, ice disseminated, ice content (61-71.5 cm) = 707%--grading down to
86-99.5	5.	Icy fibrous peat, more ice than in 4, ice both disseminated and in irregular clear ice masses†--grading down to
99.5-111.5	6.	Icy pebbly silt, minor content of peat fibers but many nearly vertical; irregular clear ice masses dominant, some measuring up to ∿2 x 4 x 7 cm--grading down to
111.5-124.5	7.	Increasingly pebbly silt similar to 6 but with fewer peat fibers, contains shell fragments, ice content# (111.5-118.5 cm) = 228%, (118.5-124.5 cm) = 42%.

+Radiocarbon dating at the University of Washington Quaternary Isotope Laboratory provided peat ages as follows: Depth 20-25 cm, 1680 ± 60 years B.P. (QL-1739); depth 61-71.5 cm, 2430 ± 60 years B.P. (QL-1740); depth 111.5-124.5 cm, 5410 ± 50 years B.P. (QL-1741).

*Ice content given as $(W_w/W_s) \cdot 100$.

†The clear ice masses from here to depth 124.5 cm contained isolated spherical bubbles irregularly distributed, also (in the larger ice masses) dominantly vertical bubbles up to 2 cm long and a fraction of a millimeter in diameter. No bubble trains noted.

#Because of the greater dry density of gravel than peat, comparison of ice contents on a percent weight basis can be misleading and even opposite on a volumetric basis.

TABLE 2 Plant Remains in Core from Peat Mound on South Side of Five-Mile Lake. Spot checks by E. B. Leopold and D. Clemens (pollen and spores) and B. M. Tucker (mosses)--Department of Botany, University of Washington; surface sample determined by D. Vitt--Department of Botany, University of Alberta.

Depth (cm)	
Surface	Brachythecium turgidum
18-20	Rare moss spores; 1 Caryophyllaceae pollen grain
44-46	Orthothecium chryseum (Schwaegr.) B.S.G.; abundant moss spores; abundant Rhizopods
59-61	Drepanocladus sp.; abundant moss spores; Rhizopods very abundant; also present: resting cysts of dinoflagellate algae, 1 Selaginella spore
71.5-73.5	Cratoneuron sp.; moss spores very abundant; also present: Rhizopods, resting cysts of dinoflagellate algae, Selaginella sp. megaspores
86-88	Cratoneuron sp.; moss spores common; also present: Rhizopods, rare fern spores
99.5-101.5	Abundant moss spores; also present: Pediastrum-type green colonial algae, 1 fern spore
101.5-103.5	Small moss spores; also present: Rhizopods, 1 Pinus pollen grain, 1 trilete spore

Note: The moss genera are characteristic of wet environments; Drepanocladus grows in water. From 59-61 cm downward there is a decrease in Rhizopods and algae, both of which require moist environments. The Selaginella and fern spores suggest the presence of vascular plants.

TABLE 3 Core Stratigraphy of Low Mound at North Base of "Airport Ridge."

Depth (cm)		
0-1.5	1.	Mossy cover
1.5-4	2.	Single ice lens (vertical crystals 1-10 mm long, 0.5-mm diameter) with vertical bubble cylinders (3-4 mm long, 2-4 mm apart) becoming spherical downward
4-8	3.	Icy silt in subhorizontal, approximately equal layers (up to 5 mm thick) of ice and silt, some ice lenses clear, some bubbly
8-11	4.	Ice, largely sediment free. Clear ice contained cylindrical bubbles
11-16.5	5.	Icy silt, 3-4 subhorizontal and 5-6 subvertical, interconnected ice bodies, lowest ice lens (1 cm thick) had cylindrical bubbles in bottom 2-3 cm--grading down to
16.5-21	6.	Icy silt, 3 irregular subhorizontal ice lenses (3-4 mm thick) with a few near-vertical ice veins, 4 soil layers (6-10 mm thick)--grading down to
21-28	7.	Icy silt, thin ice lenses (1-6 mm thick except near 27 to 28-cm depth where ice bodies were nearly equidimensional), soil layers quite variable (2-10 mm thick)
28-29.5	8.	Icy silt with finely dispersed ice and a few vertical ice veins
29.5-31.5	9.	Icy silt with bubbly ice lenses
31.5-32.5	10.	Icy silt, soil with ice lenses
32.5-35	11.	Icy peat with finely dispersed ice
35-38	12.	Icy peat with thin ice lenses--grading down to
38-47	13.	Icy silt, ice lenses (0.5-1.4 cm thick) with a few bubble zones

a

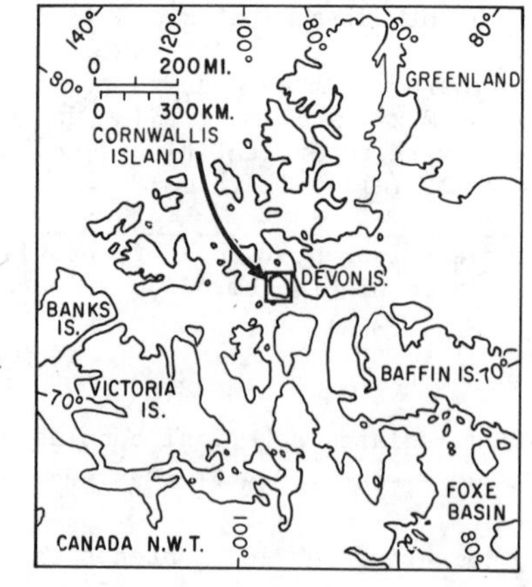

b

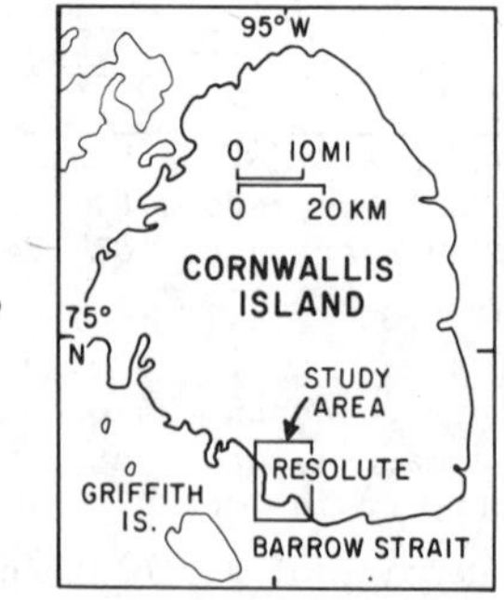

FIGURE 1 (a) Location of Cornwallis Island. (b) Cornwallis Island showing location of Resolute area.

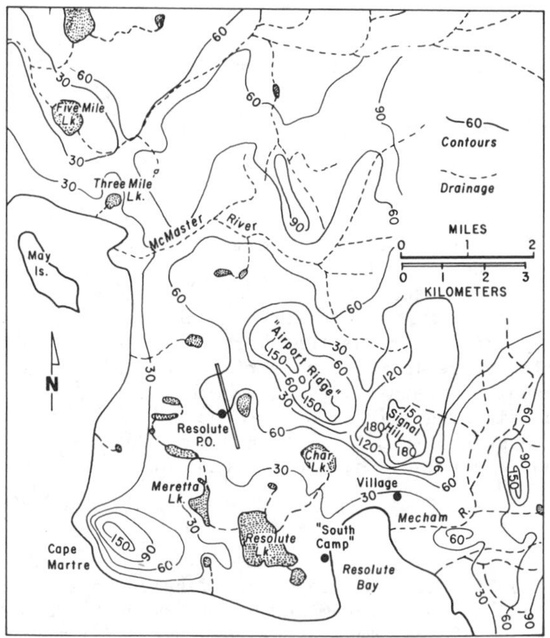

FIGURE 2 Resolute area.

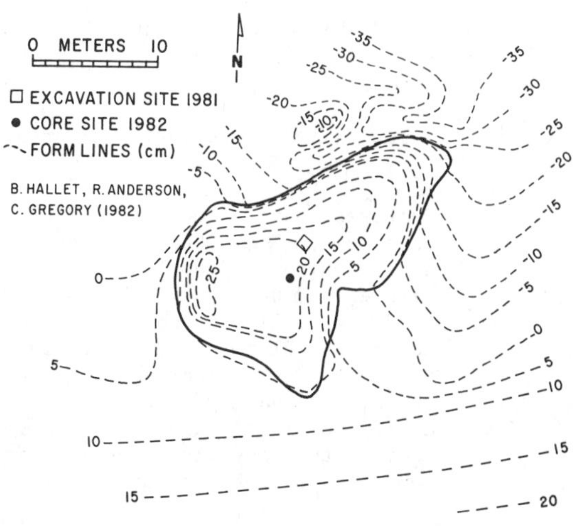

FIGURE 4 Sketch map of low mound at north base of "Airport Ridge."

FIGURE 3a Peat mound on south side of Five-Mile Lake.

FIGURE 5 Eroded longitudinal peaty mound bordering drainage depression east of low mound in Figure 4.

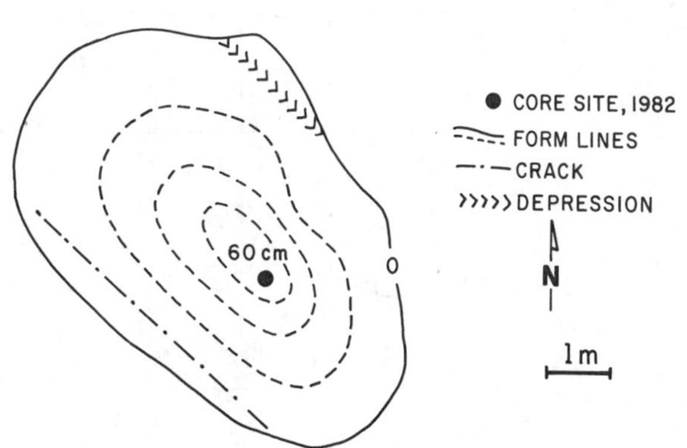

FIGURE 3b Sketch map of peat mound on south side of Five-Mile Lake.

FIGURE 6 Peaty mound on small plateaulike rise in drainage depression between McMaster River and low hills to north.

PALEOCLIMATIC INFERENCES FROM RELICT CRYOGENIC FEATURES IN ALPINE REGIONS

William J. Wayne

Department of Geology, University of Nebraska, Lincoln, Nebraska 68588 0340 USA

Features indicative of intensive frost action and of permafrost observed in the Ruby, Schell Creek, and Snake ranges of Nevada and the Cordón del Plata of the Mendozan Andes, which include lobate and small linguoid (ice-cemented) rock glaciers, sorted circles and stripes, garlands, and stone-banked lobes, permit a reconstruction of some of the paleoclimatic conditions adjacent to the ice in alpine areas that were glaciated during the Pleistocene. The distribution of paleocryogenic forms on surfaces where snow cover is minimal suggests that the mean annual air temperature above and adjacent to the valley glaciers in these desert mountain ranges in both hemispheres during the last Pleistocene glaciation would have been 5°-6°C lower than the present ones at the same altitudes. During the Holocene neoglaciations, the lowering of temperature probably was 1.5°-2°C.

INTRODUCTION

Temperature anomalies during the Pleistocene glacial maxima varied widely. Many investigators (e.g. Flint 1971, p. 725) indicate that the mean annual air temperature overall was about 5°C lower than the present. The overall average temperature of the oceans 18,000 B.P. was only 2.3°C below that today (CLIMAP 1976), but temperature as much as 18°C lower than the present have been interpreted from paleocryogenic features in central Europe, and anomalies nearly that great have been reported for the areas marginal to the North American ice sheet (Washburn 1980). In alpine regions, the ridges that stood above valley glaciers undoubtedly were colder than they are today, but few studies have provided data with which to estimate conditions that existed there when ice tongues were at their maximum extent.

Permafrost should have been widely distributed above glacial ice in many mountainous regions, and preserved landforms characteristic of former permafrost should provide a guide to some of the paleoclimatic conditions in those regions. Cryogenic landforms that might be used for such reconstructions would have formed primarily in sites where, because of wind or orientation, snow cover was minimal; sites that received greater amounts of moisture would have been protected so that cryogenic features would be uncommon and those observed probably would not reflect the maximum changes in temperature (Brown 1969, Nicholson and Granberg 1973, Harris and Brown 1978). Features reported in this study were observed in the Ruby, East Humboldt, Schell Creek, and Snake Ranges of Nevada, U.S.A. (Figure 1); a few observations made in the Cordón del Plata, Mendoza Province, Argentina are included for comparison. The types of active, inactive, and relict cryogenic features observed in these mountains include ice-cored and ice-cemented rock glaciers; stone-banked and turf-banked lobes; sorted circles, nets, and stripes; and string bogs.

Washburn (1980, p. 123) and Harris (1982) have summarized the temperature ranges that have been suggested under which some cryogenic landforms develop (Table 1). Ice wedges develop in fine-grained sediments at mean annual temperatures of -3°C to -5°C and lower, but some small (diameter <2m) circles and nets seem to form with temperatures between +5°C and -2° (Goldthwait 1976). Stone stripes and stone-banked gelifluction lobes are comparable forms on steep slopes. Turf-banked gelifluction lobes are common in regions of permafrost, but also may form where mean annual temperatures are slightly above 0°C. String bogs do not require perennially frozen ground for their formation.

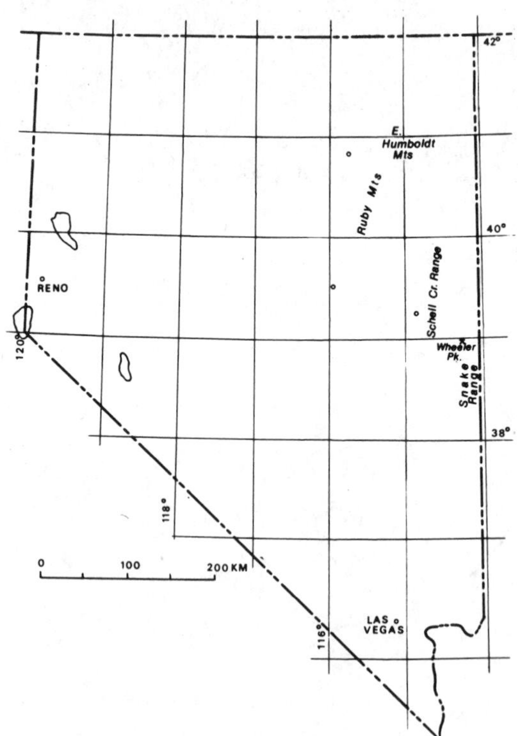

FIGURE 1 Index map of Nevada

TABLE 1 Environmental Conditions Under Which Cryogenic Features Commonly Observed In Dry Alpine Regions Form (1)

	Temperature (°C)	Precipitation and/or Moisture Conditions
Ice wedges	-6 to <-15	<600 mm
Cryoplanation	-12 to ?	- - - -
Large sorted circles and stripes (2 m dia,)	-3 to <-15	Saturation
Small sorted circles and stripes (2 m dia.)	0 to -3	Saturation
Rock glaciers, ice-cemented	-4 to <-10	>500mm, <1,500mm
Rock glaciers, ice-cored	+1 to <-10	- - - -
Gelifluction lobes	+2 to <-10	Saturation
String bogs	+3 to -1 (?)	Saturation

(1) Modified from Washburn (1980, p. 123) Reger and Péwé (1976), Harris (1981b, 1982), and based in part on discussions with A. E. Corte.

The Ruby-East Humboldt Mountains are in east-central Nevada, between 40° and 41°N latitude (Figure 1); to the southeast, between 38°30' and 40°N latitude, are the Schell Creek and Snake Ranges. The crest of the Ruby-East Humboldt Range exceeds 3,400 m in some places, that of the Schell Creeks is 3,500 m, and Wheeler Peak, the highest point of the Snake Range, reaches 4,000 m in altitude. Precipitation in east central Nevada, most of which falls in the winter, is about 250 mm on the piedmont but rises to 1,200 mm above 3,000 m. The altitude of the mean annual 0°C isotherm in the Ruby Mountains has been calculated to be 3,500 m. If the 150 m/degree latitude southward rise in the 0°C isotherm noted for Siberian mountains by Nekrasov (1978) should be close to the figure for Nevada, the 0°C isotherm at Wheeler Peak, 1°45' of latitude south of the Lamoille station in the Ruby Mountains, would be about 3,760 m, or 240 m below the peak. Records are inadequate to determine whether inversions, such as those reported by Harris and Brown (1982) for the Rocky Mountains of Alberta, exist on these mountains. Shaded slopes are at least 2°-3° colder than the mean air temperature at the same altitude. Across exposed surfaces, wind velocities are great throughout the year, so the winter snow cover is thin in many places.

ACTIVE CRYOGENIC FORMS

Active cryogenic features in the Nevada Mountains studied are few. A rock glacier occupies a north facing (shaded) cirque floor on Wheeler Peak in the Snake Range, from 3,430-3,275 m, and turf-banked lobes are common on the slopes above 3,600 m; most of the surfaces above 3,600 m would best be described as block slopes (White 1976). A few peaks in the Schell Creek Mountains approach a zone of potential permafrost, although no perennially frozen ground has been recognized in the range.

The crest of the Schell Creek Mountains, like the northern part of the Snake Range, is dominantly quartzite that has been shattered by frost action. Fossil patterned ground extends as low as 2,800 m, but the only active cryogenic features seem to be block slopes and block streams; neither requires permafrost to form. The highest peaks of the Ruby Mountains fail to reach the altitude of 0°C mean annual air temperature, and no permafrost has been found in the range. Active cryogenic forms include block streams at 3,000 m and string bogs between 2,750 and 3,000 m (Wayne 1982).

In the dry Central Andes of Mendoza Province, Argentina, small active glaciers are present above 4,000 m in many of the cirques. Sporadic permafrost in the form of active ice-cemented rock glaciers is present in shaded cirques and valley walls above about 3,500 m (Wayne 1981b). Frozen ground was encountered below an active layer one m thick in late January, and is associated with stone-banked gelifluction lobes, sorted and nonsorted stripes, sorted nets, and gutters that may have formed over ice wedges from 3,900-4,200 m.

The calculated 0°C mean annual isotherm is 3,300 m, the altitude of the lowest of the active rock glaciers. These rock glaciers are the termini of small debris-covered glaciers and probably contain some glacial ice. Active ice-cemented rock glaciers, which are part of the body of active discontinuous permafrost (Barsch 1978), are present only above 3,500 m, where the calculated mean annual air temperature is about -1.5°C. Observed summer temperatures along the shaded slopes and cirques where these landforms develop are approximately 3°C lower than are the air temperatures in more open areas, so it is likely that the local temperatures needed for development of ice-cemented rock glaciers are in the range of -3° to -5°C (Table 1).

PLEISTOCENE-HOLOCENE GLACIAL ACTIVITY

Each of these ranges has been glaciated at least two times during the later Pleistocene and from one to three phases of Neoglacial activity have been recognized (Corte 1957; Sharp 1938; Wayne 1981a; Wayne 1982; Wayne and Corte 1983). In the Ruby-East Humboldt Mountains, the oldest glacial deposits represent the Lamoille Glaciation, which is being correlated with Bull Lake/Illinoian deposits and Oxygen Isotope Stage 6 (Wayne in press). The younger Angel Lake moraines are Pinedale/Wisconsinan equivalents, and are correlated with Oxygen Isotope Stage 2. No Neoglacial ice tongues formed, but rock glaciers have been active at least two times during the Holocene, the most recent probably during the Little Ice Age.

The glacial chronology of the Snake and Schell Creek Ranges has been examined in less detail; moraines equivalent to the Lamoille and Angel Lake deposits of the Ruby Mountains are present in both, however, as are inactive and fossil rock glaciers of both Wisconsinan and Holocene age. Two and possibly three Holocene rock glacier or glacier advances are recorded in the post-Wisconsinan deposits at Wheeler Peak. Unfortunately, no buried organic matter has been found, so radiocarbon dating of these deposits has not been possible.

DISTRIBUTION OF CRYOGENIC FEATURES INDICATIVE OF FORMER PERMAFROST

During the Pleistocene glaciations, both lower temperatures and higher precipitation than the present combined to form or greatly expand glaciers in these dry mountain ranges. Where snow cover was sufficiently thin to allow the lowered air temperature to affect the ground, several kinds of landforms characteristic of, although not necessarily restricted to permafrost regions developed during the glaciations. Relict forms recognized in the field include rock glaciers, both ice-cored and ice-cemented, sorted circles and nets, debris islands, sorted stripes, stone-banked and turf-banked lobes, and surfaces that may have formed by cryoplanation.

Collapsed rock glacier debris is present in three areas of the Ruby-East Humboldt Range where the crest exceeds 3,000 m. One group of two is south of Pearl Peak at 40°13'N latitude. A second group, numbering 13, is in the northern part of the Ruby Mountains between 40°25' and 40°40', in the region where glacier tongues were longest. The third group of seven is in the East-Humboldt Mountains at the north end of the range, between 40°53' and 41° latitude. Of the 22, I have examined 11 in the field.

Most (17) of the rock glaciers are lobate forms that undoubtedly developed as a result of mobilization of frozen scree and hence were part of the permafrost body when they formed (Barsch 1978; Haeberli 1978). Fifteen of the 17 are oriented between N 30° W and N 45° E. None are oriented toward the south.

Of the five linguoid rock glaciers, one, the most southerly of the range, has the form of a collapsed debris-covered glacier that developed the structure of a rock glacier, as have many in the Mendozan Andes (Wayne 1981a). The other four probably formed as debris falls on the waning Angel Lake ice tongues and became rock-glacierized.

Most of the 22 features identified as fossil rock glaciers were active during the Angel Lake glaciation and show only a single phase of development, but 4 of the lobate ones and 3 linguoid rock glaciers have been active at least twice, most recently during one or more of the Neoglacial cold periods (Figure 2). Those of Angel Lake age range from 2,680 m near the north end of the range (latitude 41°N) through 2,780 m in the central part (latitude 40°35'N) and 2,870 for the most southerly in the range (latitude 40°13'N). Rock glaciers identified as early Holocene reached 2,750 m in the central East Humboldt Range and 3,000 m in the main section of the Ruby Mountains. One late Holocene rock glacier in the Ruby Mountains expanded to 3,050 m, and another that may have formed as a rock fall over ice during the earlier Lamoille glaciation is lower than the rest at 2,700 m.

Patterned ground is well developed in the East Humboldt Range, but only small areas are present in the Ruby Mountains to the south. With two exceptions, all the patterned ground is regarded to have formed during Angel Lake Glaciation. No troughs or gutters suggestive of relict ice wedges were recognized in any of the Nevada mountains. Most of the patterned ground observed consists of sorted circles, sorted stripes, and garlands, all

FIGURE 2 Relict lobate rock glacier 150 m wide and 340 m long in northeast-facing cirque, Right Fork of Lamoille Creek, Ruby Mountains; late Angel Lake with Holocene reactivation phase.

FIGURE 3 Stone stripes and garlands on southwest-facing slope, Third Boulder Creek, East Humboldt Range

FIGURE 4 Cryoplanation (?) surface above cirques, north of Wines Peak, Ruby Mountains, altitude 3,125-3,175m.

large, but on some slopes nonsorted stripe patterns were seen. A few large sorted circles or debris islands have formed on gently sloping surfaces at 3,025 m above Angel Lake. Stone stripes are striking features on south- and west-facing slopes of 25°-33° (Figure 3) but are absent from north-facing slopes in the same valleys. They are best developed on medium grained granitoid rocks, which yield large blocks, and the soil stripes are dominantly silt. Gutter to gutter distances average 15 m. Their lower margins, in many places marked by garlands, seem to represent the trim line of the glacier along the slope. In some valleys, 3 or 4 lines of boulder accumulations show up above the lowest set, marking successively lower trim lines. The farthest south that stone stripes were observed is on the slopes of Wines Peak, at about 40°31'N latitude, where strongly developed stripes of Angel Lake age formed on a north-facing slope. Indistinct stripes, possibly of Lamoille age, are on a smooth surface above the cirques from 3,125-3,175 m in altitude (Figure 4). Stripes composed of boulders that are deeply weathered and may be Lamoille in age are also present at 2,700 m in the lower part of Lamoille Canyon.

In the Schell Creek Mountains, rock glacier debris of the Holocene cold periods is limited to altitudes above 3,100 m (Figure 5), but those of Wisconsinan age are lower. A sorted net of late Wisconsinan age was observed at one place on a northwesterly-facing slope at 3,000 m. Sorted stripes on a south-facing slope terminate in garlands at 2,750 m in altitude.

On Wheeler Peak, Holocene inactive and relict rock-glacier debris extends beyond the active rock-glacier front to about 3,230 m (Figure 6), and some of the late Wisconsinan topography as low as 3,150 m probably formed as an ice-cored rock glacier rather than a moraine. Much of the peak above about 3,300 m is covered with frost-riven debris and would be best referred to as block slopes (Figure 7). Even though much of the frost-shattering undoubtedly developed during the Wisconsinan and, perhaps, earlier glaciations, the part of the peak higher than 3,760 m is above the calculated mean annual 0°C isotherm and the process surely is active now.

FIGURE 5 Lobate rock glacier, Holocene, Schell Creek Mountains.

FIGURE 6 Active, inactive, and relict rock glaciers on Wheeler Peak, Snake Range.

FIGURE 7 Wheeler Peak, Snake Range, block slopes and block streams; active rock glacier shown in Figure 6 is below snowbank in cirque.

DISCUSSION

In the Ruby Mountains, Nevada, the absence of ice-wedge phenomena and the paucity of large sorted circles suggests that either the snow cover exceeded 0.50 m over much of the range (Harris 1981a) or that the mean annual air temperature above 3,000 m probably did not drop much lower than -5°C. Smooth upland surfaces with sorted circles and stripes at 3,200-3,300 m in the middle of the range, which surely were as wind-swept during glaciations as they are now, tend to corroborate this estimate. Presence of most stone stripes on south- or west-facing slopes and their absence on north- and east-facing slopes probably reflects the thickness of winter snow cover as much as temperature. They do suggest, though, that temperatures on the slopes above the glaciers probably were no higher than about -3°C at 2,800 m in the north part of the range. Relict rock glaciers of the ice-cemented type are almost wholly in cirques that face between east and northwest. If these locations average 4°C colder than the mean annual air temperature at the same altitudes as observed in the Cordón del Plata, Argentina (Wayne 1981b), the 0° mean annual isotherm at 41°N latitude at the time they formed, probably 12,000-13,000 years ago, would have been about 2,600 m, approximately 5°C colder than the present mean annual temperature at that altitude.

Only relict rock glaciers, sorted stripes, and a few sorted circles in the Schell Creek Range provide reasonably reliable guides to the Pleistocene temperatures there. On the slopes of South Schell and Taft Peaks (latitude 39°20'N), rock glaciers reached 2,930 m, which suggests a mean annual air temperature of -1°C to -2°C, again about 5°C lower than the present calculated temperature of +4.3°C. A similar difference is suggested by the altitudes of relict cryogenic features on Wheeler Peak (latitude 39°N).

For comparison in the Cordón del Plata (latitude 33°S), patterned ground on wind-swept surfaces that protruded above the Vallecitos (=Wisconsinan) glaciers at 3,500 to more than 4,000 m are covered with large sorted circles, sorted and nonsorted stripes, boulder lobes, and trenches that suggest ice wedge development. Such features at and above 4,010 m in bouldery debris derived from quartzites suggest a mean annual air temperature over these surfaces of about -10°C or lower. This would be a reduction of at least 5°-6°C below the present calculated mean annual temperature of -4°C, a figure not significantly different from the 4.5°-5° lowering calculated for 2,500 m in the same region from the distribution of relict rock glaciers (Wayne 1981a).

The existence of Holocene rock glacier debris in the Ruby and Schell Creek Mountains and its presence well below the present active rock glacier on Wheeler Peak indicates that a significant reduction in temperature did take place in these mountains during Neoglaciation, probably with little change in precipitation. In the northern part of the Ruby Mountains older Neoglacial rock glaciers reached 3,000 m; in the same area, two rock glaciers were active during the later Neoglaciation at 3,050 m. These altitudes suggest a 2°-3°C reduction in mean air temperature during the two Neoglaciations.

On Wheeler Peak an earlier Neoglacial advance brought an active rock glacier to 3,260 m. During a somewhat later Holocene glaciation, a glacier left a small end moraine at 3,300 m. The Little Ice Age of the past few centuries brought the rock-glacier front to 3,360 m, just slightly beyond its present position. The temperature reduction on Wheeler Peak during the earliest Neoglaciation would have been about 2°C, during the Little Ice Age about 1.5°C, based on the position of the rock glacier fronts.

In the Cordón del Plata all three Neoglaciations seem to have accompanied about the same temperature reduction. Data from rock-glacier altitudes and the character and distribution of patterned ground suggest a mean annual air temperature on the order of -1.5° to -2.0°C at 3,300-3,500 m.

The use of cryogenic features provides a maximum figure for paleotemperatures. In this evaluation, temperatures above alpine glaciers were at least 5°-7°C lower than the present during the Wisconsinan glaciations. These temperatures are relatively close to the 5°C figure that has been suggested by some authors.

ACKNOWLEDGEMENTS

The field work in Argentina that provided some of the data in this report was done in 1980 and 1983 while I was attached to the Instituto Argentino de Nivología y Glaciología in Mendoza, Argentina and was supported by NSF grants. I would like to thank Arturo E. Corte of IANIGLA for the opportunity to discuss some of the ideas, and Troy Péwé and R. F. Black for their comments on the manuscript. Naomi Wayne was my field assistant in both the Nevada and Mendoza investigations.

REFERENCES

Barsch, D., 1978. Active rock glaciers as indicators for discontinuous alpine permafrost, an example from the Swiss Alps, in Proceedings of the Third International Conference on Permafrost, National Research Council of Canada, v. 1, p. 348-352.

Brown, R.J.E., 1969. Factors influencing discontinuous permafrost in Canada, in Péwé, T.L., ed., The periglacial environment--past and present. Montréal, McGill-Queens University Press and Artic Institute of North America, p. 11-15.

CLIMAP Project Members, 1976. The surface of the Ice Age Earth. Science, v. 191, p. 1131-1137.

Corte, A. E., 1957. Sobre geología glacial Pleistocénica de Mendoza. Anales del Departamento de Investigaciones Científicas, Universidad Nacional de Cuyo, toma 2, fasc. 2, p. 1-27.

Flint, R. F., 1971. Glacial and Quaternary Geology. New York, John Wiley and Sons, Inc., 892 pp.

Goldthwait, R. P., 1976. Frost sorted patterned ground: a review: Quaternary Research, v. 6, p. 27-35.

Haeberli, W., 1978. Special aspects of high mountain permafrost methodology and zonation in the Alps, in Proceedings of the Third International Conference on Permafrost, Ottawa, National Research Council of Canada, p. 379-384.

Harris, S. A., 1981a. Climatic Relationships of permafrost zones in areas of low winter snow-cover. Arctic, v. 34, p. 64-70.

Harris, S. A., 1981b. Distribution of active glaciers and rock glaciers compared to the distribution of permafrost landforms, based on freezing and thawing indicies. Canadian Journal of Earth Sciences, v. 18, p. 376-381.

Harris, S. A., 1982. Identification of permafrost zones using selected permafrost landforms, in Proceedings of the Fourth Canadian Rermafrost Conference. Ottawa, National Research Council of Canada, p. 49-58.

Harris, S. A. and Brown, R. J. E., 1978. Plateau Mountain: A case study of alpine permafrost in the Canadian Rocky Mountains, in Proceedings of the Third International Conference on Permafrost, Ottawa, National Research Council of Canada, p. 385-391.

Harris, S. A., and Brown, R. J. E., 1982. Permafrost distribution along the Rocky Mountains, in Proceedings of the Fourth Canadian Permafrost Conference. Ottawa, National Research Council of Canada, p. 59-67.

Nekrasov, I. A., 1978. On the formation of the cryolithozone in the mountains of eastern Siberia, in Permafrost; Second International Conference, USSR Contributions, Washington, D.C., National Academy of Sciences, p. 21-24.

Nicholson, Frank, and Granberg, H. B., 1973. Permafrost and snowcover relationships near Schefferville, in Permafrost, Second International Conference, North American Contribution. Washington, D.C., National Academy of Sciences, p. 151-158.

Reger, R. D., and Péwé, T. L., 1976. Cryoplanation terraces: indicators of a permafrost environment. Quaternary Research, v. 6, p. 99-109.

Sharp, R. P., 1938. Pleistocene glaciation in the Ruby-East Humboldt Range, northeastern Nevada. Journal of Geomorphology, v. 1, p. 296-324.

Washburn, A. L., 1980. Geocryology, a survey of periglacial processes and environments. London, Edward Arnold Ltd., 406 pp.

Wayne, W. J., 1981a. La evolución de glaciares de escombros y morrenas en la cuenca del Río Blanco, Mendoza. VIII Congreso Geológico Argentino, Actas, tomo 4, p. 153-166.

Wayne, W. J., 1981b. Ice segregation as an origin for lenses of non-glacial ice in "ice-cemented" rock glaciers. Journal of Glaciology, v. 27, p. 506-510.

Wayne, W. J., 1982. Fossil patterned ground and rock glaciers in the Ruby Mountains, Nevada. Geological Society of America, Abstracts with Programs, v. 14, p. 353.

Wayne, W. J., in press. Glacial Chronology of the Ruby-East Humboldt Mountains, Nevada. Quaternary Research.

Wayne, W. J., and Corte, A. E., in press, 1983. Multiple glaciations of the Cordón del Plata, Mendoza, Argentina. Palaeogeography, Palaeoclimatology, Palaeoecology.

UNDERGROUND CAVITIES IN ICE-RICH FROZEN GROUND

Paul C. Weerdenburg[1] and Norbert R. Morgenstern[2]

[1] EBA Engineering Consultants Ltd., Edmonton, Alberta, Canada T5L 2M7
[2] Department of Civil Engineering, University of Alberta, Edmonton, Alberta, Canada T6G 2G7

An analytical study of closure behavior and stress changes surrounding unlined and lined circular openings in warm, ice-rich permafrost has been undertaken. The creep law of polycrystalline ice has been used to describe the constitutive behavior of fine-grained, ice-rich permafrost soils. The results show that shallow openings in warm permafrost experience very large strains above the crown of opening. Placement of a tunnel liner will reduce the tunnel closure to an insignificant level. A new analysis of the in situ deformation behavior of the Fox Tunnel near Fairbanks, Alaska, has shown that the flow law for polycrystalline ice does not yield an upper bound solution to the observed room closure measurements. However, it is argued that the in situ deformations resulted from creep and plastic yielding.

INTRODUCTION

Underground chambers have been used extensively for civilian and military purposes since very early in history. Military applications of underground space date back to Roman times. Traditionally, mining operations have made extensive use of underground space and will continue to do so in the future. The current world energy situation has brought about an increased awareness amongst the civilian population of the tremendous advantages of utilizing underground space as a means of conserving energy.

The U.S. Army Cold Regions Research and Engineering Laboratory (USA CRREL) has investigated the use of underground space in frozen ground as early as 1955. During the summers of 1955-1957, the Lower Tuto Tunnel was excavated into the Greenland Ice Cap at Camp Tuto, Greenland (Abel 1961). The objective of this tunnel was primarily to provide a research facility to study the deformation behavior of glacial ice and assess the feasibility of excavating tunnels and rooms in ice to utilize the space for storage of perishable food, military equipment, etc. With the experience gained from the Lower Tuto Tunnel, the Upper Tuto Tunnel was excavated the following two summers near the original tunnel as a prototype shelter to house a 25 man camp and related support facilities under ice. This facility was closed down during the summer of 1962 due to excessive deformations of the openings, attributed primarily to warming of the surrounding ice medium (Russel 1961).

During the summers of 1959 and 1960, a third tunnel was driven into cold frozen till near Camp Tuto, Greenland. Personnel of USA CRREL report that this tunnel is virtually the same size as it was when it was excavated.

The Fox Tunnel, near Fairbanks, Alaska, was excavated to evaluate the use of underground space in warm permafrost. During the winters of 1963-1965 USA CRREL excavated a near horizontal tunnel 100 m long into frozen, ice-rich Fairbanks silt. During the winter of 1968 to 1969, in a cooperative study with the U.S. Bureau of Mines (USBM), an inclined winze was excavated off of the main adit to gain access to the gold-bearing gravel deposit underlying the frozen silt. USA CRREL and the USBM excavated one room each at the end of the inclined winze. Each room had frozen gravel walls and the ceiling was ice-rich Fairbanks silt. In situ deformation studies carried out in conjunction with both phases of this project revealed that tunnel closure becomes excessive when the ambient air temperature in the tunnel is near the soil temperature of -1° to -2°C (Swinzow 1970).

CONSTITUTIVE BEHAVIOR OF ICE-RICH FROZEN SOIL

Ice is the most important component controlling the rate, time, and temperature dependent properties which characterize frozen soil. Ice can be present both as discrete segregated structures and more uniformly in the pore spaces of the soil mass.

The load-deformation behavior of ice-rich frozen soil is dominated by the presence of segregated ice. Laboratory evidence of Savigny (1980) shows deformations are localized along segregated ice in natural permafrost soils. Thus, it would seem appropriate to use the creep behavior of ice as an upper bound to the time dependent deformations in ice-rich frozen soil.

The constitutive equation for steady state creep of polycrystalline ice is most commonly represented empirically by a simple power law of the form

$$\dot{\epsilon} = A\,\sigma^{n} \qquad (1)$$

where $\dot{\epsilon}$ denotes the strain rate, σ denotes the stress and A and n are the experimentally derived creep modulus and creep stress exponent, respectively. The creep modulus is constant for a given ice type and temperature. Considerable experimental evidence is available to show that the power creep law represents the steady state creep data in the low to intermediate stress range (Sego 1980).

An extensive review of the secondary creep data was carried out by Morgenstern et al. (1980). Based on this review, the authors presented the values shown in Table 1 for the parameters A and n in the power law.

TABLE 1 Creep Parameters for Polycrystalline Ice

Temperature (°C)	A (kPa^{-3} yr^{-1})	n
-1	4.5×10^{-8}	3.0
-2	2.0×10^{-8}	3.0
-5	1.0×10^{-8}	3.0
-10	5.6×10^{-9}	3.0

Creep data for undisturbed ice-rich permafrost tested at the stress and temperature range of interest in geotechnical practice are scarce. Roggensack (1977), McRoberts et al. (1978), and Savigny (1980) tested fine-grained ice-rich permafrost from the Mackenzie River valley. These data are shown in Figure 1. The flow law shown in Figure 1 is that for polycrystalline ice as proposed by Morgenstern et al. (1980). With a few exceptions, the laboratory data fall within a narrow band at stresses exceeding 75 kPa. Thus, it can be concluded that the flow law for ice represents a reasonable upper bound to the constitutive behavior of fine-grained ice-rich permafrost.

FINITE ELEMENT CREEP ANALYSIS

The finite element method is now widely used in geomechanics to determine the stress and deformation states in a nonlinear viscous medium. An incremental, initial strain finite element procedure was adopted to solve steady state creep problems in frozen soil. Details of development of this procedure are given by Weerdenburg (1982).

The incremental procedure is based on Hoff's (1954) analog, which simply states that the stresses in a nonlinear viscous medium can be obtained directly by solving for the stresses in a nonlinear elastic body. The following assumptions are made in order to compute the increments of creep strain:

1. There is no volume change associated with creep strains.
2. For an isotropic material, the principal directions of strain rate and stress tensors coincide.
3. The creep rate is independent of any superimposed hydrostatic state of stress.
4. The generalization of the uniaxial creep laws to the multiaxial state of stress should recover the uniaxial relationship for the case of a uniaxial stress.

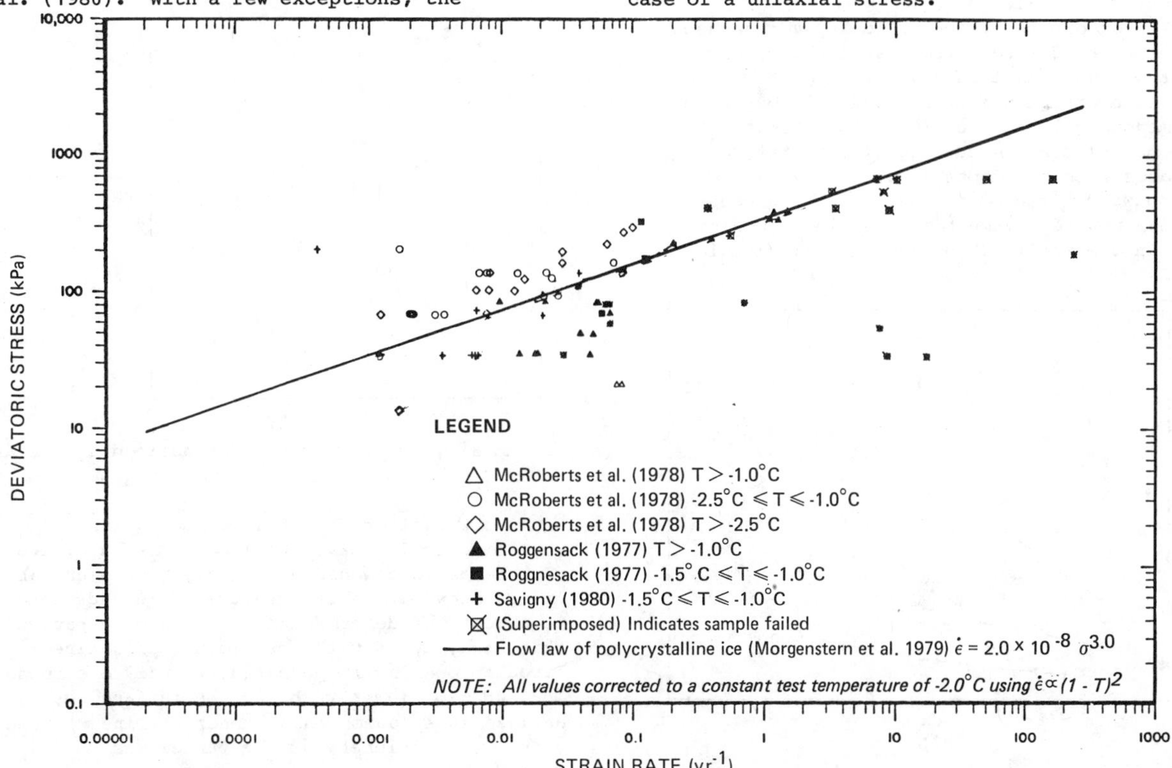

FIGURE 1 Steady-state creep data from ice-rich glaciolacustrine clay from various locations in the Mackenzie River Valley, N.W.T. (Savigny, 1980).

UNDERGROUND CIRCULAR OPENINGS IN PERMAFROST

Major design considerations of tunnels in frozen ground are stand-up time, change in diameter of the unlined opening with time, and change in pressure on a tunnel liner with time. The first two factors will influence the decision on whether or not to support the underground opening with a tunnel liner. The third factor will provide guidelines for dimensioning the tunnel liner. A limited parametric study of shallow circular tunnels in permafrost was carried out to study closure phenomena and the stress changes in the frozen soil surrounding the opening.

The first part of this study deals with the closure behavior of an unlined circular opening excavated at various depths in warm ice-rich permafrost. The height to diameter (H/D) ratios of the circular openings varied from 1.5 to 4.0. Warm ice-rich permafrost was chosen as the host soil medium as it would present an upper bound solution for the closure rates.

A homogeneous soil profile was assumed for all analyses. Material properties were assigned on the basis of data reported in the literature. The flow law chosen for the frozen soil used in this study was

$$\dot{\epsilon} = 2 \times 10^{-8}\ \sigma^3 \qquad (2)$$

where the units for stress and strain rate are kPa and yr^{-1}, respectively. This flow law is identical to the flow law for ice at -2°C (Morgenstern et al. 1980).

A plot of the closure velocity versus depth of overburden for the crown, springline, and invert of the tunnel is shown in Figure 2. The depth of overburden in this figure is expressed nondimensionally as the H/D ratio. For all three points considered, the closure rate increases with the depth of overburden. Also, the downward movement of the crown is greater than the upward movement of the floor, creating a sag in the roof. Thus, the initial circular opening undergoes a transition to an elliptical shape as the frozen soil creeps inward. At greater depths, this transition will be less pronounced because the relative change in overburden stress from the crown to invert will be smaller. Based on these preliminary results, it would appear that a circular opening located 9-10 diameters below the surface would deform axisymmetrically. This is somewhat deeper than a circular opening in an elastic soil or rock.

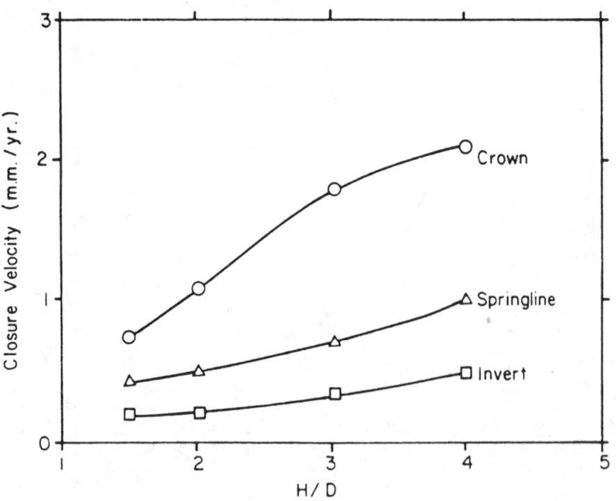

FIGURE 2 Tunnel closure velocity versus depth of overburden.

A vector plot depicting the magnitude and direction of velocity throughout the region analyzed for a tunnel at H/D = 2.0 is shown in Figure 3. The relative magnitude is given by the length of the vector. As expected, the frozen soil flows in toward the opening. The largest velocity components occur along the periphery of the tunnel wall. The transformed cross section is shown on a grossly exaggerated scale by the dashed line connecting the velocity vectors around the tunnel wall.

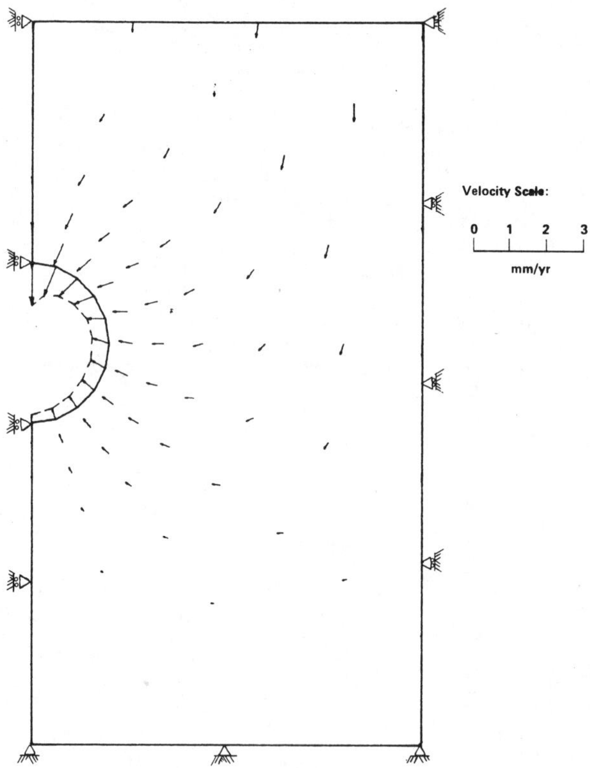

FIGURE 3 Velocity vectors for unlined tunnel at H/D = 2.0.

It was clear from the foregoing that shallow circular cavities located in warm, ice-rich permafrost would ultimately become unstable unless some mitigative measures were undertaken to arrest the deformations. The second part of this study dealt with the closure of a lined circular opening in permafrost. Only one tunnel located two diameters below the surface was analyzed. A 100 mm thick concrete liner placed around the periphery of the tunnel was considered.

The velocity vector diagram for the lined tunnel is shown in Figure 4. As shown in the

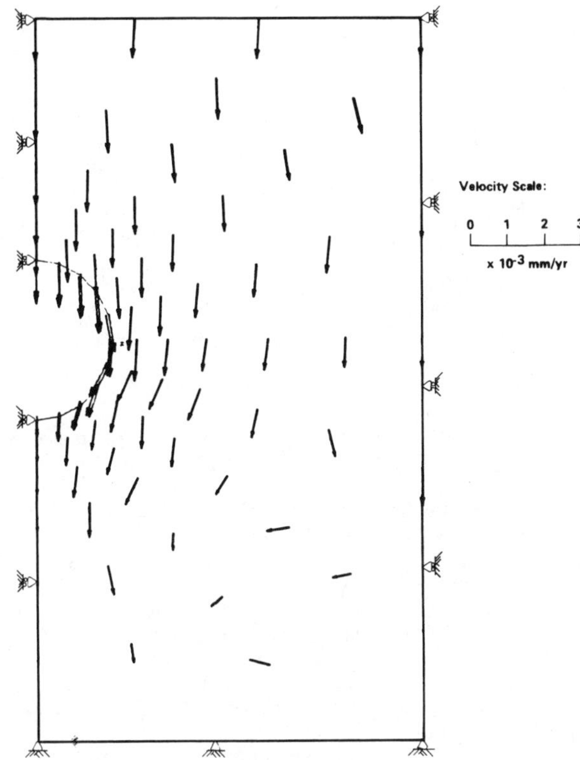

FIGURE 4 Velocity vectors for lined tunnel at H/D = 2.0.

figure, the liner plays a significant role in altering the velocity distribution in the frozen soil surrounding the opening. In the vicinity of the tunnel, the creeping frozen soil is deflected around the tunnel liner since the stiffness of the concrete is much greater than that of the frozen soil. The tunnel liner also serves to drastically reduce the downward velocity of the crown of the tunnel. In this case, the crown velocity is reduced by three orders of magnitude for the same tunnel with no liner present.

IN SITU CREEP AT THE FOX TUNNEL

In situ creep movements of the ice-rich Fairbanks silt were evaluated independently by Thompson and Sayles (1972) and Pettibone (1973).

Pettibone (1973) studied the in situ creep behavior of the USBM room at the Fox Tunnel. After 37 months, a total of 300 mm of floor to roof deformation was recorded. Vertical separation and expansion of the ice-rich silt between the ceiling and the 1.8 m depth into the roof accounted for 20% of the total room closure.

The vertical closure measurements recorded in the USA CRREL room are shown in Figure 5 (Thompson and Sayles 1972). The vertical closure rate increased rapidly during the first year of observations due to the general warming trend caused by installation of the insulated

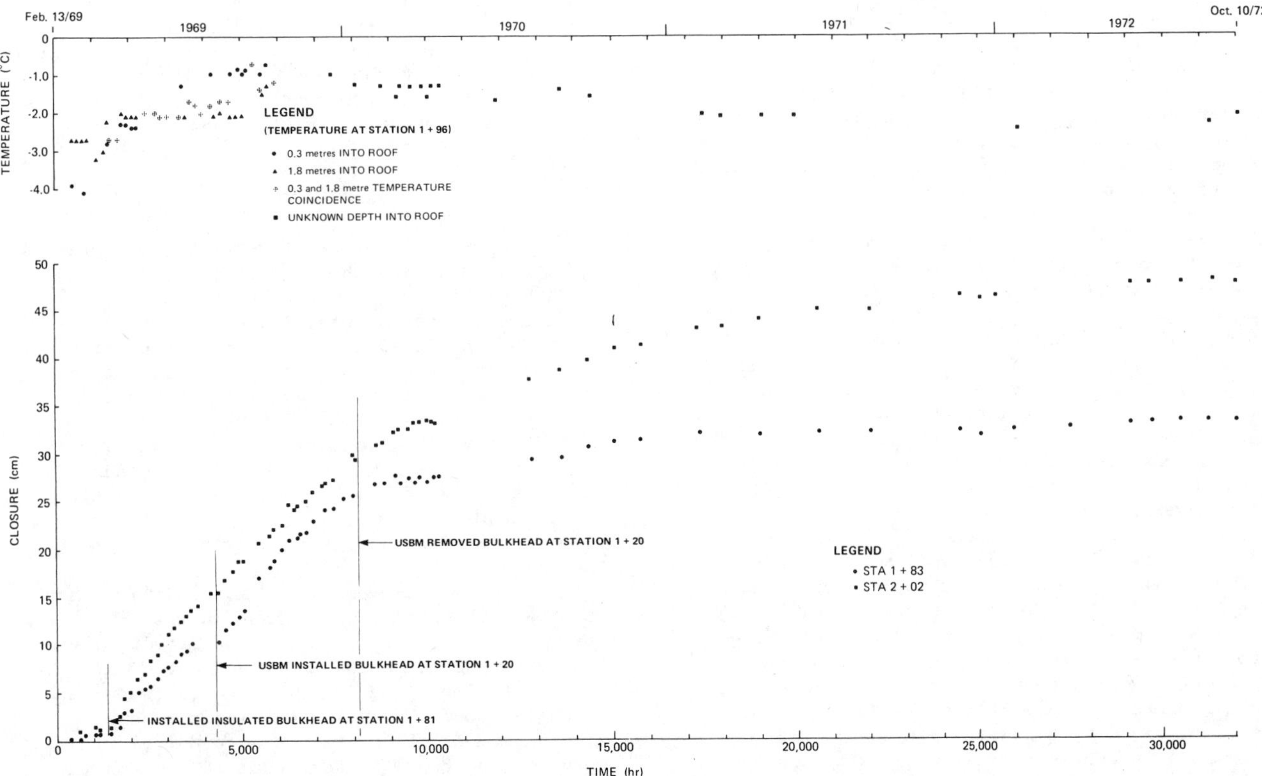

FIGURE 5 Vertical closure of USA CRREL room in Fox Tunnel.

bulkheads. Following removal of the USBM bulkhead, the closure rate decreased as a consequence of cooler permafrost temperatures. The same general behavior was noted in the USBM room. The vertical closure rate continued to attenuate for the entire period that closure measurements were recorded. After 3 1/2 years of observations, 480 mm of vertical closure had taken place.

Thompson and Sayles (1972) reported the results of a finite element simulation of the closure of the USA CRREL room. Laboratory samples were obtained for uniaxial creep testing. The authors used a simple power law to predict the in situ behavior of the room for the first year of observations. They concluded that the form of the flow law predicted the in situ creep behavior quite well. However, the back-calculated flow law predicted a creep rate 3.3 times faster than had been established in the laboratory on the same frozen silt. A few comments can be made regarding the analysis carried out by Thompson and Sayles (1972).

A new analysis of the USA CRREL room has been carried out. The purpose of this simulation was to assess the validity of the simple power law for ice as an upper bound solution for the in situ deformations in the frozen Fairbanks silt.

The creep properties of the silt were assigned the values of pure ice at -2°C as given by Equation 2 (Morgenstern et al. 1980). The frozen gravel walls were assumed to behave elastically based on the fact that they were very dense in situ and retained grain-to-grain contact with no massive ice forms present (Sellman 1972).

The predicted vertical steady state velocity using the simple power law given by Equation 2 to model the creep behavior of the frozen silt was 1 mm/yr. In this case, the simple power law for ice does not represent a valid upper bound for the in situ creep behavior of the frozen silt. However, an examination of the vertical closure data presented in Figure 5 clearly shows that steady state conditions did not exist during the first year of observations as assumed by Thompson and Sayles (1972). It is entirely possible that the overall closure measurements that were recorded consisted of both plastic and creep flow. Plastic flow in this sense refers to time dependent failure of the frozen soil.

Excavation of the USA CRREL room has decreased the vertical stress at the ceiling from overburden to zero. This stress release is accompanied by a corresponding increase in the stresses in the adjoining soil mass above the gravel walls. Consider the two elements shown in Figure 6. The vertical stress in element A will be a small fraction of its initial value before excavation. The vertical stress in element B will have to increase to maintain equilibrium. Assume that the strength of frozen soil is given by a linear Mohr Coulomb failure envelope and the vertical stress decrease in element A is equal to the stress increase in element B. Plotting these stress increments on a simple vertical-horizontal principal stress

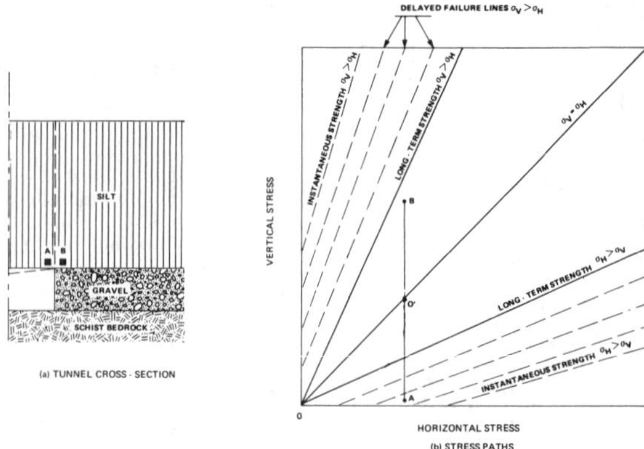

FIGURE 6 Stress paths for soil elements above an underground opening.

plot, as shown in Figure 6, shows that the stress release in element A could exceed the long term $\sigma_H > \sigma_V$ failure line. However, an equal stress increase in element B lies below the long term $\sigma_V > \sigma_H$ failure line. Although the actual stress paths followed by soil elements surrounding a rectangular opening are quite complex, this simplified approach illustrates the interaction between strength and deformation in frozen soil.

An additional factor contributing to the discrepancy between the in situ and the laboratory flow law is that closure instrumentation installed in the roof and floor could not discriminate between the vertical deformations in the gravel walls and the downward creep of the ice-rich silt. The strain rate measured in the laboratory on the frozen silt samples was independent of the vertical deformations occurring in the gravel walls.

CONCLUSIONS

The case histories of cavities in frozen ground reported in the literature, while limited in number, indicate that these underground cavities can be successfully completed. However, in situ deformation studies carried out to date indicate that deformations can become severe when a tunnel is excavated in warm, ice-rich permafrost.

A comprehensive review of the constitutive behavior of ice and ice-rich soil suggests that the creep law of polycrystalline ice constitutes a reasonable upper bound to the constitutive behavior of fine-grained, ice-rich permafrost soils.

The results of preliminary analytical study indicate that shallow openings in warm permafrost experience very large strains above the crown of the opening. Placement of a concrete liner after excavation will reduce the tunnel closure to an insignificant amount.

A reanalysis of the in situ deformation behavior of the USA CRREL room at the Fox Tunnel has shown that the flow law for polycrystalline ice did not yield an upper bound solution to the observed room closure measurements. More comprehensive instrumentation is needed in the future to discriminate between the various deformation processes that affect closure of highly stressed underground cavities in frozen ground.

REFERENCES

Abel, J.F., 1961. Ice tunnel closure phenomena. U.S. Army Snow Ice and Permafrost Research Establishment. Technical Report 74, 37p.

Hoff, N.J., 1954. Approximate analysis of structures in the presence of moderately large creep deformations. Quarterly of Applied Mathematics, Vol. 12(1), pp. 49-55.

McRoberts, E.C., Law, T.C. and Mirray, T.K., 1978. Creep tests on undisturbed ice-rich silt. Proceedings 3rd International Permafrost Conference, Edmonton, Vol. 1, pp. 539-545.

Morgenstern, N.R., Roggensack, W.D., and Weaver, J.S., 1980. The behavior of friction piles in ice and ice-rich soils. Canadian Geotechnical Journal, Vol. 17(3), pp. 405-145.

Pettibone, H.C., 1973. Stability of an underground room in frozen gravel. Proceedings, North American Contribution, 2nd International Permafrost Conference, Yakutsk, USSR, pp. 669-706.

Roggensack, W.D., 1977. Geotechnical properties of permafrost soils. Unpublished Ph.D. Thesis, University of Alberta, 346p.

Russel, F., 1961. An under-ice camp in the arctic. U.S. Army Cold Regions Research and Engineering Laboratory, Special Report 44, 14p.

Savigny, K.W., 1980. In-situ analysis of naturally occurring creep in ice-rich permafrost soil. Unpublished Ph.D. Thesis, University of Alberta, 286p.

Sego, D.C.C., 1980. Deformation of ice under low stresses. Unpublished Ph.D. Thesis, University of Alberta, 368p.

Sellman, P.V., 1972. Geology and properties of materials exposed in the USA CRREL permafrost tunnel. U.S. Army Cold Regions Research and Engineering Laboratory, Special Report 177.

Swinzow, G.K., 1970. Permafrost tunnelling by a continuous mechanical method. U.S. Army Cold Regions Research and Engineering Laboratory, Technical Report 221, 36p.

Thompson, E.G., and Sayles, F.H., 1972. In-situ creep analysis of room in frozen soil. Journal of the Soil Mechanics and Foundation Division, A.S.C.E., Vol. 98, pp. 899-915.

Weerdenburg, P.C. 1982. Analytical study of time-dependent deformation in permafrost. Unpublished M.Sc. Thesis, University of Alberta, 146p.

DIURNAL FREEZE-THAW FREQUENCIES IN THE HIGH LATITUDES: A CLIMATOLOGICAL GUIDE

Ruth L. Wexler

U.S. Army Corps of Engineers
U. S. Army Engineer Topographic Laboratories
Fort Belvoir, Virginia 22060 USA

This study provides relatively simple climatological models for determining the incidence of frost days (minimum daily temperature ≤ 0° C), ice days (maximum daily temperature ≤ 0° C) and freeze-thaw days (minimum daily temperature ≤ 0° C, maximum daily temperature > 0° C) throughout Alaska, E. Siberia, Iceland, and Greenland. Both area and station models yield estimates of the frequency of diurnal freeze-thaw cycles per month or year. The various models demonstrate the relationships between daily freezing conditions and the different temperature regimes. The results should improve understanding of periglacial activity and provide a means of predicting possible climatic effects on the construction of buildings, roads and airport runways.

INTRODUCTION

The impact of diurnal freeze-thaw cycles on soils, rocks, roadbeds, and construction materials is of much concern to the agriculturist, hydrologist, geologist, and transportation or construction engineer. As noted by Troll (1958), such alternation of freezing and thawing not only affects the particular size structure of the soil but also causes erosion, runoff, or flooding and reduces the bearing strength of the surface layer. The vulnerability of the ground depends, among other variables, on the amount of available moisture and whether the ground is bare or is covered with vegetation, ice, or snow.

The surface covering affects the albedo which in turn is largely responsible for local differences in extreme temperatures, that is the daily maximum and minimum, the determining factors of daily freeze-thaw. Daily freezing conditions may be defined as consisting of frost days (min ≤ 0° C), ice days (max ≤ 0° C), and freeze-thaw days (min ≤ 0° C, max > 0° C). The interrelationship among these three variables forms the basis of this study. The principal question is: What incidence of diurnal freeze-thaw cycles may be expected at a given site per given interval of time?

This paper presents information on daily freezing conditions in the permafrost and contiguous regions of Alaska, E. Siberia, Iceland, and Greenland. Also included are several German stations representing a range of elevations for the comparison of the effects of altitude with those of latitude. As in an earlier study by Wexler (1982), a number of guides are offered for estimating the respective frequencies of frost days, ice days, and freeze-thaw days per month or year for any given site from routine climatological parameters.

BACKGROUND

The geographical distribution of the annual number of diurnal freeze-thaw cycles has been determined for various countries or sections: United States, Canada, Poland, Japan, the Arctic, Europe, and the U.S.S.R (Fraser 1959, Hastings 1961, Hershfield 1972, Pelko 1970, Russell 1943, Shitara 1970, Visher 1945, Wexler 1982, Williams 1964). Annual or monthly frequencies of frost days, ice days, and freeze-thaw days have been correlated respectively with mean daily minimum temperatures, mean daily maximum temperatures, and a combination of both (Fraser 1959, Hershfield 1972, Shitara 1970, Wexler 1982).

DATA

Frequencies of diurnal freeze-thaw cycles are not readily available. Many climatic summaries list frost days, but few list ice days. In the past information on freeze-thaw cycles has sometimes been simulated (Hastings 1961, Visher 1945). In this paper, all the analyses are based on actual observations of frost days and ice days, the data for which were obtained from a variety of sources, mainly: the U.S. Department of Commerce (1980), the Danske Meteorologiske Institute (1947-1965), and U.S. Air Force Environmental Technical Applications Center (1978). All the temperatures referred to in this paper were from standard weather shelters at 1.5 to 1.8 m above ground.

FROST DAYS AND ICE DAYS

For a network of stations in a given region observations were obtained of: The mean daily minimum temperature, N, the mean daily maximum temperature, M, the number of frost days, F, and the number of ice days, I, per month and year. From these observations, simple linear regression models were determined such that F may be derived from N and I may be derived from M for any site within the specified region per given interval of time. Figure 1 gives examples of these regression plots for annual data for Alaska, E. Siberia, Iceland, and Greenland. Figure 2 contains similar plots for monthly data for May, Greenland and April for the other areas. All of the days of the

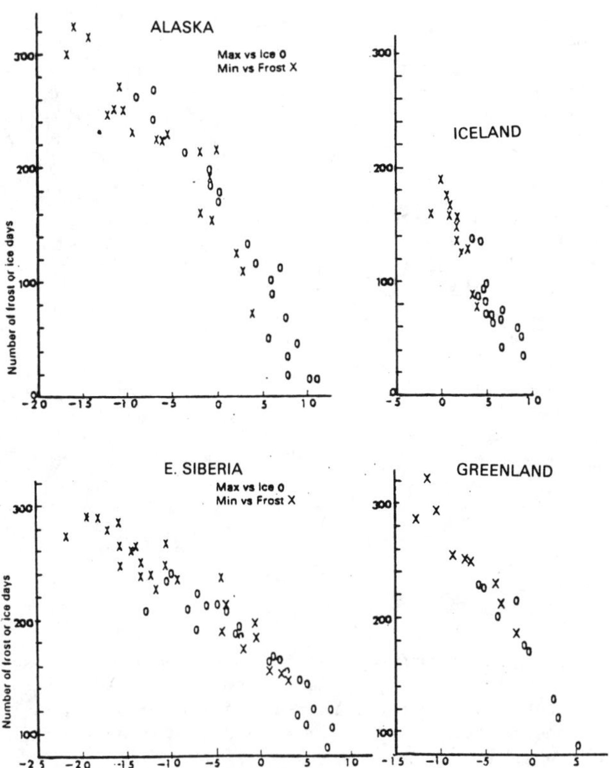

FIGURE 1 Annual data for selected regions.

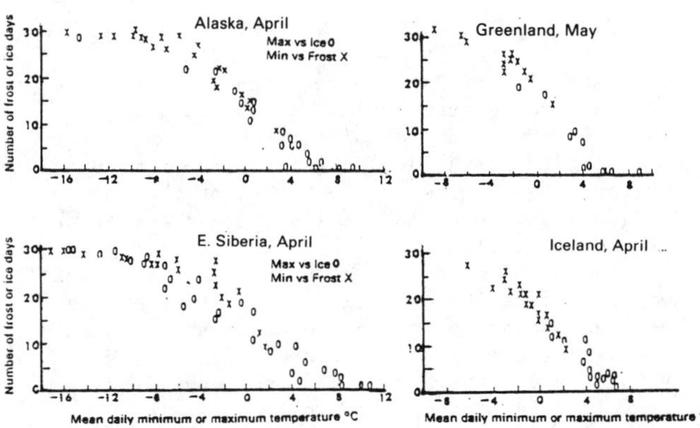

FIGURE 2 Selected monthly data.

TABLE 1 Constants a and b for Monthly and Annual Regression Equations

	$F = a_1 + b_1N$		$I = a_2 + b_2M$	
	a_1	b_1	a_2	b_2
		ALASKA		
JAN	21.1	-0.6	13.8	-1.4
FEB	19.3	-0.7	11.9	-1.3
MAR	20.2	-0.9	12.4	-1.9
APR	14.9	-1.8	13.3	-1.4
MAY	14.4	-3.1		
SEP	15.3	-2.6		
OCT	14.0	-1.8	15.8	-2.1
NOV	15.9	-1.3	12.5	-1.5
DEC	20.7	-0.7	13.0	-1.1
ANN	159	-9.7	166	-13.8
		E. SIBERIA		
MAR			16.8	-1.5
APR			14.8	-1.7
MAY	15.1	-1.9	14.6	-1.7
OCT	14.7	-1.9	15.2	-1.5
NOV			16.6	-2.1
ANN	183	-5.6	167	-6.1
		ICELAND		
JAN	19.4	-1.6	15.6	-1.4
FEB	16.7	-1.3	14.2	-2.2
MAR	17.4	-1.4	16.0	-2.1
APR	15.8	-2.1	15.9	-2.3
MAY	15.6	-2.6	9.7	-1.0
JUN	11.5	-1.7		
SEP	12.0	-1.8		
OCT	16.1	-3.0	13.9	-2.1
NOV	14.9	-1.8	14.1	-2.2
DEC	20.1	-1.2	15.9	-2.1
ANN	176	-20.2	180	-17.9
		GREENLAND		
JAN			23.5	-0.4
FEB			17.0	-0.6
MAR			18.4	-0.8
APR			14.6	-1.5
MAY	21.1	-1.4	14.6	-2.1
JUN	16.3	-2.7		
JUL	7.5	-1.4		
SEP	14.3	-2.5		
OCT			16.2	-2.8
NOV			18.7	-1.3
DEC			21.2	-0.7
ANN	227	-5.3	168	-12.4

month were usually below freezing if the mean maximum temperature was $< -6^{\circ}$ C and above freezing if the mean minimum temperature was $> 6^{\circ}$ C. The corresponding regression equations are of the form:

$$F = a_1 + b_1 N \qquad (1a)$$

and

$$I = a_2 + b_2 M \qquad (1b)$$

The parameters a and b depend on the data: They serve as constants for any data set. Table 1 lists values of these constants for monthly and annual data for the above regions. The coefficient of determination, r^2, for the various equations (implied in Table 1) ranged from .50 to .99 with few exceptions: .24 for February, Greenland and .36 for March, E. Siberia. (The value of r^2 indicates the quality of fit between F and N or I and M: 1.0 = excellent fit, 0 = no fit).

Stations were chosen in a given area so as to represent as large a range of temperature as possible. Each pair of equations as (1a) and (1b) is therefore applicable throughout the entire area, with only the parameters M and N changing from station to station.

The parameters (constants) a and b may vary a little with the period of record. If the computations are carried out for M and N in degrees Celsius (as for Table 1), then the value of a equals 160-185 days for the annual data or 12-18 days for the monthly data. In other words, F(I) equals half the days per interval of time if the minimum (maximum) temperature is 0°C. An exception is for the annual number of frost days for Greenland, with a = 227. For the annual data, the absolute value of parameter b appears to decrease with continentality, as from 20 for Iceland, a maritime climate, to 6 for E. Siberia, a highly continental climate.

With respect to the monthly data, the value of a tends to be closest to 15 for the months of the transitional seasons, the spring or the fall, when diurnal freeze-thaw cycles are most frequent in this latitude (See Figure 3). Usually the colder the month the higher the value of a, the warmer the month the lower the value of a. Those months for which no constants are given in Table 1 are either too warm (no ice days or frost days) or too cold (all days are frost or ice), therefore no linear relationship. In the computations of the monthly data, the number of frost (ice) days were limited to $1 < F(I) < (n-1)$ where n = total number of days per month.

A comparison between the observed and the estimated annual numbers of frost days or of ice days for test stations is given in Table 2. Values of r^2 for the quality of fit between the observed and estimated frequencies were >.8 for each set of stations.

FREEZE-THAW DAYS

The number of freeze-thaw days, Z, per given interval of time, a parameter not available in climatic summaries, may be found by a variety of methods, as by direct counting (Williams 1964) or by correlation with ΔT (Fraser 1959, Vischer 1945). In the case of just the crossover of the freezing level, as in this study, Z is simply the difference between the number of frost days and the number of ice days, or

$$Z = F - I \qquad (2)$$

For each area under consideration, the individual equations for annual F and I were first obtained by means of the appropriate constants in Table 1. The annual Z may be expressed as follows:

Alaska

$$Z = 14M - 10N - 7 \qquad (2a)$$

E. Siberia

$$Z = 6(M-N) + 16 \qquad (2b)$$

Iceland

$$Z = 18M - 20N - 4 \qquad (2c)$$

Greenland

$$Z = 12M - 5N + 59 \qquad (2d)$$

A comparison between the observed and the estimated annual Z (Table 2) shows somewhat greater discrepancies than in the cases of F and I, the factors on which Z depends. Values of r^2, which indicate the quality of fit between estimated Z and observed Z range from .60 to .76 compared to > .90 for F and I (except for Iceland for which r^2 was .84).

Equation (2b) implies that Z may be correlated directly with (M-N) or ΔT for E. Siberia. Fraser (1959) found a similar relationship for Canada, although the cycles he investigated were for a larger temperature span, -2° to 1° C, rather than just across the freezing level.

STATION MODELS OF DIURNAL FREEZE-THAW CYCLES

Station models of percent days per month with diurnal freeze-thaw cycles from January to December are given for selected stations in Greenland, Alaska, Iceland, the U.S.S.R, and Germany (See Figure 3). For most of the high-latitude stations, especially those close to the Arctic Circle, the peak incidence of diurnal freeze-thaw cycles occurs during the transitional seasons of spring and fall. The stations are arranged so as to show a gradual change in pattern from an extremely cold climate as Barrow, Nord, or Polar Station where diurnal freeze-thaw cycles prevail only in summer to the relatively warm stations of Annette or Vladivostok, which have relatively long summers with no freezing and the peak frequencies of freeze-thaw cycles in the winter. The German Stations from Zugspitze (2962 m) to Munich (532 m) reflect the effect of altitude on the incidence of diurnal freeze-thaw cycles. The station model for Zugspitze is somewhat similar to that for Barrow (or Polar Station). The plots for the Icelandic stations resemble that of Fichtelberg (or Brocken). At Vladivostok, despite the low latitude, January and December are too cold for freeze-thaw, whereas in Iceland, freeze-thaw occurs all winter.

TABLE 2 Estimated and Observed Annual Numbers of Frost Days, Ice Days, and Freeze-Thaw Days.

Station	Air Temp °C	Diurn Temp Range ΔT°C	Estimated Number: Frost Days	Ice Days	Freeze-thaw Days	Observed Number: Frost Days	Ice Days	Freeze-thaw Days
Alaska								
Barter Is.	-12.2	5.9	310	292	18	318	260	52
Summit WSO	-3.9	8.2	239	163	76	251	177	74
Fairbanks	-3.6	11.4	252	137	115	237	157	80
Gambell Is.	-3.4	5.6	221	174	47	256	197	59
Aniak	-2.2	9.5	228	130	98	226	142	84
Naknek	1.5	8.6	187	86	101	204	99	105
Anchorage	1.6	9.2	189	80	109	208	110	98
Cordova	3.1	8.6	171	64	107	184	53	131
Seward	4.0	6.6	152	65	87	160	62	98
Sitka	5.7	7.3	139	38	101	140	21	119
E. Siberia								
O. Chetyrekhstolbovoy	-13.6	5.0	269	240	29	288	242	46
Dzhardzhan	-12.5	7.2	269	226	43	261	220	41
Zyryanka	-11.4	8.3	266	215	51	253	213	40
Ust Yvdoma	-9.2	11.7	265	189	76	252	181	71
Guga	-3.6	17.2	247	134	113	239	145	94
Nogliki	-2.0	10.5	219	145	74	220	154	66
Sukhanovka	-0.6	10.0	210	138	72	211	149	62
Poronaysk	0.0	8.8	204	138	66	196	136	60
Grosserichi	1.2	7.7	194	134	60	188	128	60
Dolinsk	2.3	9.9	194	119	75	192	119	73
Sarychevo	2.5	5.0	179	134	45	189	110	79
Bukhta Preobrazheniya	4.2	10.5	185	105	80	161	79	82
Iceland								
Modhrudalur	0.0	6.6	243	121	122	247	139	108
Raufarhofn	2.5	3.8	164	98	66	177	88	89
Blonduos	3.4	4.5	155	76	79	166	71	95
Egilsstadhir	3.4	6.7	176	56	120	178	72	106
Saudharkrokur	3.6	5.0	154	67	87	143	63	80
Akureyri	3.9	5.6	154	56	98	158	73	85
Haell	4.2	6.1	154	47	107	170	58	112
Loftsalir	5.9	3.9	97	40	57	80	33	47
Greenland								
Thule	-11.1	8.4	342	253	89	314	242	72
Upernavik	-7.3	6.7	292	210	82	284	210	74
Jakobshavn	-4.5	6.7	264	173	91	250	183	67
Umanak	-2.8	5.6	242	159	83	253	185	67
Godthaab	-0.9	6.1	225	130	95	254	141	113
Ivgtut	0.8	7.2	214	102	112	221	105	116
Narssag	1.7	6.7	203	94	109	222	96	126

Another set of station models is shown in Figure 4. This time the abscissa is the mean monthly temperature, however, the ordinate is the same as that of Figure 3, namely the average percent days per month with freeze-thaw. In general, for a given station the daily freeze-thaw cycles per month tend to increase as the mean monthly temperature approaches 0° C. The frequencies per given temperature vary from one station to another. Nevertheless, certain of the models may sometimes serve several stations or groups of stations.

In particular each model shows the temperature limits for its freeze-thaw regime. These limits, as well as the peak amplitudes, are climate dependent.

Station models of this type have also been obtained for numerous other stations in the high and mid-latitudes. A few of these models, as well as an equation which approximates the manual plots in Figure 4, has been given previously (Wexler 1982). See Appendix for the equation. For a highly continental station, as Verkhoyansk (not shown), the freeze-thaw temperature regime extends

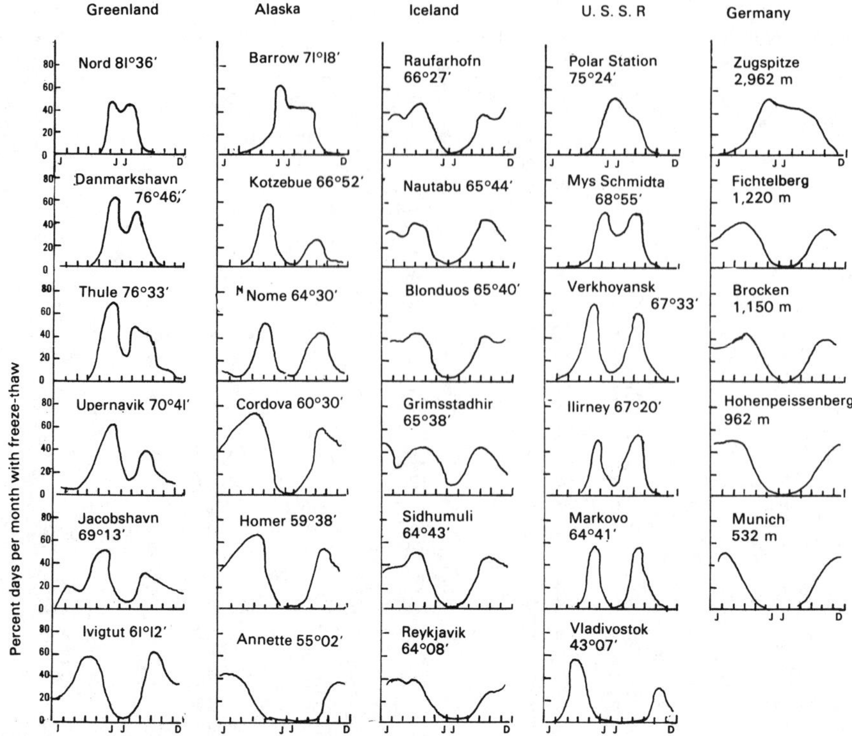

FIGURE 3 Percent days per month with freeze-thaw

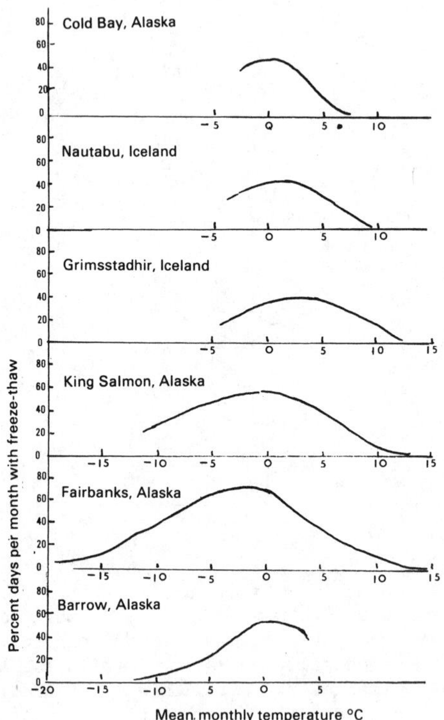

FIGURE 4 Percent days per month with freeze-thaw vs. mean monthly temperature.

from about +16 to -16° C with an amplitude of about 68% at about 0° C. Most stations in Iceland have an amplitude of about 40% to 50% with a positive temperature range of about 10° C. Cold Bay, Kodiak, and Annette in Alaska have freeze-thaw regimes similar to stations in Iceland. On the other hand, interior stations in Alaska have much higher amplitudes, about 70% to 80% as Fairbanks, McGrath, and Gulkana, with temperature ranges from 12° C to -16° C. For a number of stations in Greenland the plots (not given) show much greater irregularities of pattern than for Iceland or Alaska, possibly because of the relatively short periods of records.

SUMMARY

Several mathematical and graphical models were presented for estimating the frequencies of frost days, ice days, and freeze-thaw days per month or year for stations in Alaska, E. Siberia, Iceland, and Greenland. Once linear regression equations are determined for the derivation of frost days and ice days, respectively, per given situation, these equations are then applicable to any site within the area, given only M and N for the site. The frequencies of diurnal freeze-thaw cycles then may be readily obtained from the difference between the frequencies of the frost days and the ice days.

Two types of station models were provided, the first the conventional annual cycle of monthly freeze-thaw and the second, the same data plotted per mean monthly temperature. The percent freeze-thaw for a given temperature varies from one station to another. Although the models tend to be distinctive, each depending on latitude (solar elevation), altitude, continentality, and local conditions nevertheless certain of the models may be used to represent groups of stations.

The study shows that diurnal freeze-thaw cycles may not be derived from mean temperatures alone: Essential parameters are the mean daily minimum and mean daily maximum temperatures. The latter yield definitive information concerning frost days, ice days, and freeze-thaw days. The results indicate that periglacial activity might be better correlated with the mean minimum and/or mean maximum daily temperatures per year or month rather than mean temperatures.

REFERENCES

Danske Meteorologiske Insititute, 1947-1965, Meteorologisk Arbog, Part 2, Greenland.

Fraser, J. K., 1959, Freeze-thaw frequencies and mechanical weathering in Canada. Arctic, v. 12, p. 40-52.

Hastings, A. D., 1961, Atlas of Arctic Environment: Headquarters, Quartermaster Research and Engineering Command, Natick, Mass.

Hershfield, D. M., 1972, An investigation into the frequency of freeze-thaw cycles: University of Maryland, thesis.

Pelko, I., 1970, Przymrozki W. Polsce W. Dziesieciolecíu: Reports and studies of the Geographic Institute, University of Warsaw, Chair of Climatology, p. 95-104.

Russell, R. J., 1943, Freeze and thaw frequencies in the United States: Transactions American Geophysical Union, Part I p. 125-133.

Shitara, H., 1970, On winter days and ice days in Japan: Japanese Progress in Climatology, p. 85-99.

Troll, C., 1958, Structure soils, solifluction, and frost climates of the earth, SIPRE Translation.

U.S. Air Force Environmental Technical Applications Center (ETAC), 1978, Summarized data (unpublished) of frequencies of occurrence of maximum and minimum daily temperatures for several hundred stations.

U.S. Department of Commerce, 1980, Local climatological data annual summary with comparative data (stations in Alaska).

Visher, S. S., 1945, Climatic maps of geologic interest: Bulletin of the Geologic Society of America, v. 56, p. 713-736.

Wexler, R. L., 1982, A general climatological guide to daily freezing conditions: frost days, ice days, and freeze-thaw days: Ft. Belvoir, Va., U.S. Army Corps of Engineers.

Williams, L., 1964, Regionalization of freeze-thaw activity: Annals of the American Association of Geographers, v. 14, p. 597-611.

Wilson, C., 1969, Climatology of the Cold Regions, Northern Hemisphere II. Cold Regions Science and Engineering Monograph, 1-A3b.

APPENDIX

An equation which roughly approximates the manual plots in Figure 4 is given by:

$$y = a \sin Q \qquad (3)$$

where

y = % days per month with freeze-thaw

a = amplitude of curve (peak % monthly freeze-thaw)

Q = 1.57 $(T + b)/b$

b = $T_{y=a} - T_{y=o}$

$T_{y=a}$ = temperature at peak % (usually 0°C)

$T_{y=0}$ = temperature at which freeze-thaw is nil = cut-off temperature

T = mean monthly temperature (°C)

NOTE: If the peak % frequency is not at 0° C, let d = departure from 0° C and substitute T-d for T.

ROCK GLACIERS—PERMAFROST FEATURES OR GLACIAL RELICS?

W.Brian Whalley

Department of Geography, The Queen's University of Belfast,
Belfast, BT7 1NN, United Kingdom

The two main models of rock glacier composition, ice-rock mixture and glacier ice cores, are briefly examined. It is argued that there seems to be no evidence which suggests that talus can normally become a rock glacier just by the addition of ice. Talus exists at high slope angles but shows no signs of glacier-like movement. All that appears necessary for a rock glacier to form is relict glacier ice and sufficient debris cover to preserve the core for long periods. As they are such thin bodies of ice, they can flow only slowly; this is sufficient to explain their behavior and distribution. Although permafrost can be present in an area of rock glaciers and may prolong the existence of the ice core, it is not necessary for their formation.

INTRODUCTION

Rock glaciers have been phenomena of interest and controversy since they were described in the early part of this century. Despite much recent work (Washburn 1979), many aspects of rock glacier behavior, formation, and flow mechanisms are still unknown and the paper by Wahrhaftig and Cox (1959) still holds considerable sway, even though some of its conclusions have been questioned. There are problems because of the the difficulties of examining rock glaciers in detail and much must be inferred from visual inspection. Tautologically, if it looks like a rock glacier then we may call it so, but qualification may be necessary if something specific is known, e.g., a glacier ice core. This paper is concerned mainly with valley or cirque floor rock glaciers (Outcalt and Benedict 1965) rather than the valley wall type, and the simple term 'rock glacier' alone will be used to identify them. This paper is not primarily concerned with flow mechanisms but discusses problems relating to the occurrence and preservation of ice in rock glaciers.

DISTRIBUTION OF ROCK GLACIERS

Rock glaciers are predominantly alpine phenomena; in Europe, many exist in the Alps (e.g. Barsch 1977, Barsch et al 1979, Haeberli 1975, 1978) although they appear to be very infrequent in Norway (Whalley 1976, Griffey and Whalley 1979). In Scandinavia difficulties in terminology have arisen with respect to "rock glaciers" and "ice-cored moraines" (Barsch 1971, Østrem 1971). In Sweden, transitional forms between the two exist (Whalley 1974) and in Iceland, several examples of rock glaciers have been found (Whalley 1974, Eyles 1978). The range of environments in which rock glaciers are found in North America is reviewed in Washburn (1979) and in White (1976) and of examples in South America by Corte (1976, 1978).

It is usually considered that rock glaciers indicate the presence of permafrost (Barsch et al 1979, Haeberli 1978). Thus,

> The climatic snow line and the lower limit of active rock glaciers do not run parallel but diverge from oceanic to continental environments. The general conditions for the formation of rock glaciers are an adequate supply with coarse debris, a favourable topography, and last but not least a climate cold enough to produce alpine permafrost.(Höllermann 1980)

The work of Wahrhaftig and Cox (1959) is taken by some authors to be the definitive framework within which to study rock glaciers (Barsch 1971, 1977, Barsch 1978, White 1976). Thus in this interpretation the presence of an active rock glacier is a definite indicator of permafrost. To use the presence of rock glaciers in this way depends upon their having been formed from ice-rock mixtures (Barsch 1978). Some authors (e.g. Washburn 1979) acknowledge that there may be several types of rock glacier mechanics and basic structures to be found: an interstitial ice-rock mixture and a debris-covered remnant glacier (sandwich) are the two main models used.

W.B. Whalley (1974) has suggested that there are difficulties in accepting the mixture model on mechanical grounds. In essence, he argued that an ice-rock mixture has too high a 'yield strength' to flow plastically under the stresses which would be expected from their topography. Conversely, a very thin glacier ice core, protected from melting by overlying debris, would show flow velocities commensurate with those found for rock glaciers.

It is difficult to envisage valley floor rock glaciers being built up by filling the interstices of talus slopes with ice as some have suggested (Barsch 1969) Talus slopes, often bearing snow patches are usually quite distinct from active rock glaciers, which flow at right angles to them. There is a substantial difference in angle of the surface of these features; talus slopes are generally at 35° - 40° whilst the general rock glacier surface is rarely greater than 10°. Two notions arise from this. Why, if melting snow percolates into the talus slopes and then refreezes, does talus not form valley side rock glaciers? With an angle of twice (at least) that of the valley floor rock glacier, the shear stresses in talus might be expected to allow the ice in the interstices of the mixture to flow. This rarely seems to be the case. Conversely, it is the rather low-angled rock glaciers which move. Thus the latter must develop a much lower basal shear stress than the talus slopes.

The assumption must be that rock glaciers can move because in fact they are still glaciers and possess thin ice cores which are protected from melting by accumulated debris. The ice core deforms according to the flow law for ice and actual rock glaciers show close agreement with predictions from the flow law (Whalley 1974). Although drilling (e.g. Barsch et al 1979) has found ice-rock mixtures this does not show that the whole thickness of the rock glacier is an ice-rock mixture. In the sandwich model, rock glaciers are viewed as being different from ordinary glaciers only in the amount of surface debris carried and, consequently, in their response to climatic changes. This last factor includes the mass balance of the original glacier and debris-induced modifications. There is a gradual transition between active glaciers, composed entirely of ice and with no debris cover, through glaciers with a heavy debris load to rock glaciers. The final stage is a stagnant rock glacier from which the once-present ice core has melted. This viewpoint is much simpler than the possible origins proposed by J.P.Johnson (1973) or P.G.Johnson (1980) as it supposes that substantially all the ice which is responsible for rock glacier flow is of true glacial origin. It is no longer a problem to explain the origin of the ice although the way in which the ice is preserved does need explanation.

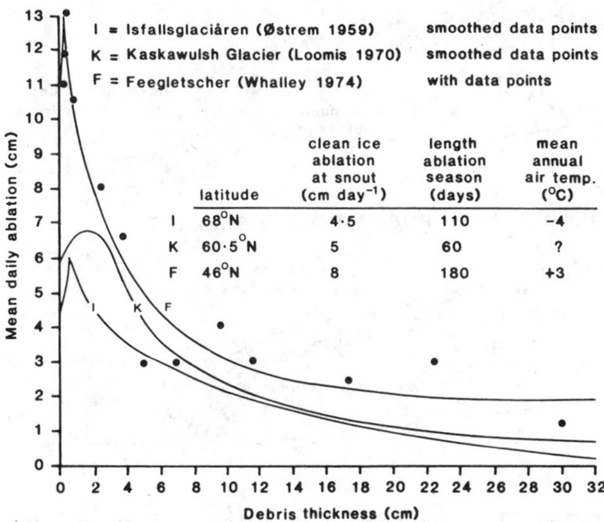

FIGURE 1 Curves showing relations between ablation and debris cover for three glaciers at various latitudes.

PRESERVATION OF ICE CORES

Debris cover over an ice surface protects the ice below; the amount of protection depends mainly upon climatic factors. Østrem (1959, 1964) and Loomis (1970) give what few data there are on this problem. Figure 1 shows their curves as well as one for Feegletscher (Switzerland), which at 2,100 m lies well below any permafrost. Local climate and debris thickness are primarily responsible for the preservation of ice cores. If sufficient debris can accumulate over an ice core and ablation cut to a minimum then the core can be preserved for great lengths of time (Østrem 1959, 1964). Geothermal heat flow will, in the absence of permafrost, allow some ice to melt at the base of the core but this will be rather small, of the order of a few mm per year (Paterson 1981) and would probably not approach the amount of surface melting. Melting due to the frictional flow of ice is likely to be negligible because rock glacier flow rates are very low. There is a general lack of data at rock glacier sites to indicate the amount of melting although several authors have reported streams emerging from the snouts of rock glaciers (Corte 1978, Johnson 1981).

EFFECT OF DEBRIS ON GLACIERS

A debris cover suddenly imposed on the ablation area of a glacier reduces the ablation rate and, for a given glacier in a steady state, this means that the ablation area will no longer be effective in balancing the accumulation. The extension of a glacier to compensate for such an apparent loss of ablation area is seen in some glaciers in the Alps (Renaud 1964) as well as in more northerly latitudes (Bull and Marangunic 1968).

Debris falling in the accumulation area of a glacier ultimately emerges in the ablation area and may reduce ice ablation as just described. Snow input must greatly exceed any debris input so that an ice-rock mixture could not be formed in this way.

Debris accumulates at a glacier snout, producing a slow response of the glacier with respect to modified ablation rates. Any glacier extension will depend upon the climatic control of ablation rates as well as upon changes in accumulation and the debris thickness. If we consider a glacier which is losing mass (i.e., a negative ice mass balance) but whose snout is advancing due to an increase in debris cover, then a steady state may be achieved very rapidly. For an uncovered glacier there would be a continued retreat, dependent on the length of time during which the mass balance is negative. For two identical glaciers, both with the same negative net mass balance but one free from debris and the other heavily charged at the snout, the latter will keep its snout position very much longer than the clean glacier—entirely due to the reduction of snout ablation. In practice, there will be many variables operating over time; however, no comparisons of this type have been done. Generally, the more debris the lower downvalley the snout position will be. If ice discharge is restricted or debris supply very great then the behavior is modified and a rock glacier may form (Whalley 1976, Griffey and Whalley 1979).

There will come a time when the ice supply is totally cut off from the glacier ice below the debris due to the disparity of ablation rates.

Before this happens there will be a thinning of the exposed ice; this is the reason for the spoon-shaped hollow noted by many authors and is sometimes acknowledged as such (Potter 1972). In an extreme case, a sufficient cover of debris over a considerable glacier area could result in a glacier advance even though the glaciers in the area are generally retreating. Such a case is described by Michaud and Cailleux (1950) and is commonly seen in the rock glaciers in the Hindu Kush mountains (Whalley, unpublished data).

In many cases of glacier retreat the debris cover is too thin and the ice discharge too great to provide any reduction of the ablation rate in the ablation area; hence the thinning is rapid. However, if debris accumulation at the snout is substantial or the ice discharge is reduced then the effect of the debris on the glacier dynamics will become increasingly important. As long as there is contiguity between accumulation area ice and the debris-covered ice of the core, then the system will behave in a predictable way, very similar to a glacier with little or no debris (Griffey and Whalley 1979). If a glacier does get cut off from its accumulation area, then it may exist for a long time, dependent upon the thermal conditions, but will eventually decay (Whalley 1979, Griffey and Whalley 1979).

EFFECTS AT THE ROCK GLACIER SNOUT

Active rock glaciers are usually so-called because of the fresh appearance of the debris at the snout, i.e., occasional rockfalls and an unstable surface inclined approximately at the resting angle of the material. If the snout is not fresh, then the rock glacier is usually considered to be inactive.

Yearly movements of the debris at the front of an active rock glacier represent, according to the interstitial ice model, the amount of debris discharging through the system. In a steady state ice-cored rock glacier this is also true but frontal movement is a function of both debris discharge and the amount of ablation of the ice core. If extra ice arrives at the snout (which then advances, due to a positive mass balance or reduction in ablation), then an exaggerated debris discharge would be calculated. Such an advance is a normal consequence of glacier movement. The exact relationship between ablation, the forward movement of the ice, and the debris quantity is uncertain without detailed measurements of ice core melting characteristics. It has been assumed (Barsch 1977) that the volume of a rock glacier represents the quantity of debris removed from cliffs above. However, this is only true if the debris-ice mixture model applies and it is not so if there is a glacier ice core. Similarly, calculation of annual debris discharge at an advancing snout would also seem to be in error if the interstitial-ice model is used. Reactivated rock glaciers (Wahrhaftig and Cox 1959) show fresh, steep, snouts though there may be no apparent movement of the rock glacier. Clearly, the appearance of the rock glacier front is not a good guide to what is happening to the system as a whole.

Figure 2 illustrates three possibilities of frontal change with the glacier ice core model. In Figure 2a the ablation balances the ice income and the net position of the snout is static. At the very front however, surface debris does move from stable to unstable positions due to slight seasonal variation of position. If the ice flow is weak and the ice thin, then movement about the mean ice front position is small and the face ultimately may become relatively stable with very little movement of the surface debris.

In Figure 2b there is a decreasing ice input for a given climate and this gives a general surface lowering. However, debris is still added to the snout if the ice remains active. Decreasing surface velocity can still give a fresh snout appearance. Increasing debris thickness at the snout gives increased protection to the ice so that there may be a return to equilibrium and, ultimately, a low-angled and stable snout.

A

GLACIER IN STEADY STATE

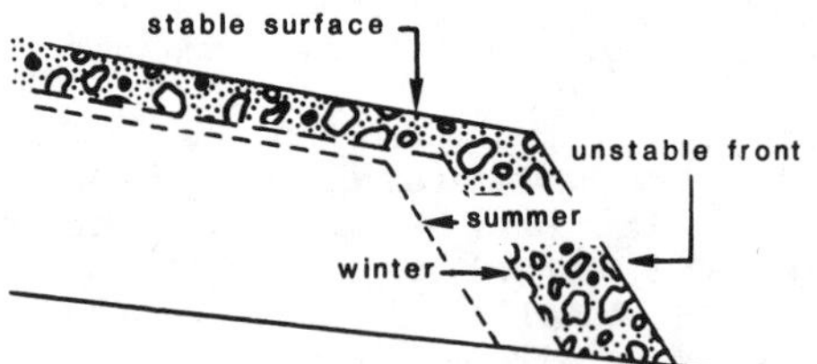

B

GLACIER WITH NEGATIVE MASS BALANCE

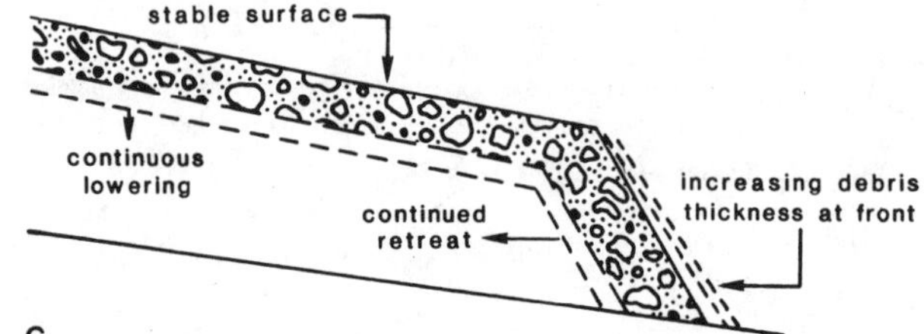

C

GLACIER WITH POSITIVE MASS BALANCE

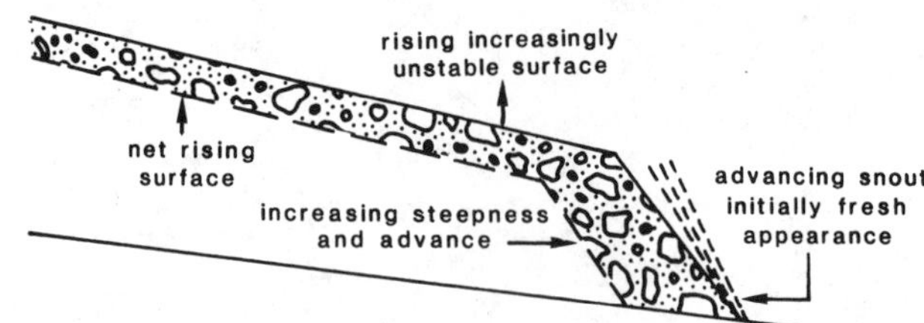

FIGURE 2 Changes at the fronts of rock glaciers with changing mass balances. See text for details.

In Figure 2c any lifting of the ice surface through a positive ice mass balance is most pronounced at the snout and the ice velocity increases. Debris dumping is accelerated but thins the protective debris layer relative to its volume; thus ice ablation increases. This ultimately gives increasing stability as debris covering the ice toe thickens again.

What these three basic regimes do not show is the complex way in which the nature of any individual rock glacier may alter due to changes in the ice mass balance or debris input. Which balance dominates depends upon factors such as debris supply and glacier length. For instance, a kinematic wave bringing down 'extra' ice will not carry a corresponding debris supply; conversely, a rockfall will produce a substantial debris increment.

A variety of situations exist relating the protective nature of a debris cover to the ice mass balance. There is considerable complexity of dynamics as well as topography for rock glaciers but it would appear that the sandwich model only needs changes in two basic parameters—debris and ice supplies—to accommodate them. We do not need to imply gross changes of climate for the formation of a rock glacier but rather adjustments between the two supplies.

THE ROLE OF PERMAFROST

The presence of permafrost could almost permanently protect an ice core. In such cases it is possible that the glacier ice core may move very slowly downhill under its own weight plus that of the overburden debris. It is possible that a variety of forms such as protalus lobes and the lobate rock glacier of Wahrhaftig and Cox (1959) could be formed in this way. The ice mass could be glacial (Brown 1925) or snowbank (Outcalt and Benedict 1965); it would merely have to be sufficiently thick to flow and be protected by a debris cover.

In W.F.Thompson's (1962) discussion of Balch ventilation, the idea of addition of ice to a glacier surface through covering debris is not proposed but there is no reason why melting snow on the surface of the debris should not melt, percolate through the rocks, freeze on top of the glacier ice and be preserved by the Balch mechanism. In this way it would be possible to build up debris encased in ice near the equilibrium line. Such superimposed ice would be very similar in origin to that found on polar glaciers (Müller 1962) and would act in the same way, lowering the equilibrium line to below the visible snow line. In the ablation area, melting of the true glacier ice below the debris/ice mantle would be further reduced as this "superimposed" ice is melted instead. This is one link with the interstitial ice hypothesis; various authors have reported an ice-rock mixture above glacier ice (Brown 1925, Outcalt and Benedict 1965). In the main hypothesis of the sandwich model this is merely an addition, possible where there are permafrost or near permafrost conditions, but it is not the main driving force of the rock glacier. If interstitial ice and rock debris are found on a rock glacier; the likelihood is that glacier ice exists below. In discussing the rock glaciers at Grubengletscher, Whalley (1979) has found and interprets a glacier ice core whilst Barsch et al (1979) have found interstitial ice. It is probable that this is an example of interstitial (superimposed) ice overlying a glacier ice core.

If the ablation area of a rock glacier-ice glacier system is well-covered with debris, then even a very small debris-free accumulation area (and net accumulation) will be sufficient for ice replenishment, both conventionally and by superimposed ice within the rock debris.

CORRIE GLACIERS AND ROCK GLACIERS

Proponents of the interstitial ice model for rock glacier flow do not consider that the proximity of a true corrie glacier indicates a connection between the two , either at present or in the past. This idea seems untenable, not only because the mechanics demand a solid, though thin, ice layer which can remain there (due to its protective cover) but also because of the relationship of ice and debris mass balances.

For a given climatic regime, a rock glacier is most likely to form where the corrie glacier is small rather than large (Whalley 1974). The formation is due to the large amount of rock debris falling onto the ablation area in relation to the activity of the glacier in removing it. This effect can be seen in a comparison of the rock glaciers and corrie glaciers of the Colorado Front Range (Whalley 1974).

Rock glaciers in northern Iceland show close relations of corrie glaciers to the mass balances of debris and ice (Whalley 1974). The debris input from the disintegrating cliffs is sufficient to keep the glaciers covered, even though the ice discharge is high. In various examples examined by the author no superimposed ice was seen, although in one case glacier ice could be traced all the way from the edge of the debris covered zone to the snout. Although there is plentiful evidence of sporadic permafrost in Iceland (Priesnitz and Schunke 1978) there is no evidence related in any way to the rock glaciers found in the Akureyri region or to the example described by Eyles (1978). The rock glaciers descend to around 800 m, the only explanation being that they are intimately related to corrie glaciers and not to the presence of permafrost.

CONCLUSIONS

There is evidence for the existence of rock glaciers in many parts of the world; they may occur in regions of sporadic, discontinuous or continuous permafrost. This does not necessarily mean that there is any implicit connection between them. It has been argued that mean annual temperatures near zero may favor refreezing of snowmelt. This adds to the ice supply of ice-cored rock glaciers in a way analogous to superimposed ice found on many northern glaciers.

There is opportunity for rock glaciers to form directly from talus but this is rarely, if ever, the case. They can form (to give lobate or valley-side forms) where the debris supply is great and where it spills onto a small glacier, glacierette or large snowpatch (Whalley 1979). There is evidence that rock glaciers are closely related to corrie glaciers, even if the former have now ceased to exist. It is usually stated that the evidence for lack of glaciers (but presence of periglacial conditions) in the Pleistocene is lack of moraines. Conversely, it can be argued that rock glaciers actually are the moraines and that this is what would be expected because of the close relationship between rock debris and ice mass balances in their formation. Terminal moraines, rock glaciers, and protalus ramparts are related forms, only differing in topography.

The complex and varied forms of rock glaciers are often assumed (e.g., Johnson 1974) to indicate the operation of a number of mechanisms. A much simpler explanation is achieved if the sandwich model, based upon glacier ice preservation, is accepted. Much still needs to be explained about rock glaciers and their relation to climate. It would appear that they are not good indicators of climatic conditions and may exist in their present form substantially unchanged for long periods. Just how long depends upon such factors as local climate (mean air temperature and radiation), debris cover thickness and any contiguity with an ice accumulation area. Thus they are also insensitive in their relationship to permafrost of whatever type. The presence of permafrost may help to preserve a rock glacier but does not cause it.

ACKNOWLEDGMENTS

I thank various people who have assisted with field work or discussion at various times: Peter Birkeland, Jack Ives, Gunnar Østrem, Helgi Björnsson, Malcolm Clark, Tony Escritt, and Nigel Griffey. Jim McGreevy and a reviewer made valuable suggestions about the paper.

REFERENCES

Barsch, D., 1969, Studien und Messungen an Blockgletschern in Macun, Unterengadin: Zeitschrift für Geomorphologie, Supplementband, v. 8., p. 11-30.

Barsch, D., 1971, Rock glaciers and ice-cored moraines: Geografiska Annaler, v. 53A, p. 203-206.

Barsch, D., 1977, Nature and importance of mass wasting by rock glaciers in alpine permafrost environments: Earth Surface Processes, v. 2, p. 231-245.

Barsch, D., 1978, Active rock glaciers as indicators for discontinuous alpine permafrost. An example from the Swiss Alps, _in_ Proceedings of the Third International Conference on Permafrost, v. 1: Ottawa, National Research Council of Canada, p. 348-353.

Barsch, D., Fierz, H., and Haeberli, W., 1979, Shallow core drilling and borehole measurements in the permafrost of an active rock glacier near the Grubengletscher, Wallis, Swiss Alps: Arctic and Alpine Research, v. 11, p. 215-228.

Brown, W. H., 1925, A probable fossil rock glacier: Journal of Geology, v. 33, p. 464-466.

Bull, C., and Marangunic, C., 1968, Glaciological effects of debris slide on Sherman Glacier, _in_ The great Alaskan earthquake of 1964, Hydrology, National Academy of Science, Washington, publication 1603, p. 309-317.

Corte, A.E. 1978, Rock glaciers as permafrost bodies with debris cover as an active layer. A hydrological approach. Andes of Mendoza, Argentine, _in_ Proceedings of the Third International Conference on Permafrost, v.1: Ottawa, National Research Council of Canada, p. 262-269.

Eyles, N., 1978, Rock glaciers in Esjüfjoll nunatak area, South East Iceland: Jökull, v. 28, p. 53-56.

Griffey, N. J., and Whalley, W. B., 1979, A rock glacier and moraine complex, Lyngen Peninsula, North Norway: Norsk Geografisk Tidskrift, v. 33, p. 117-124.

Haeberli, W., 1975, Untersuchungen zur Verbreitung von Permafrost zwischen Füelapass und Piz Grialetsch (Graubünden): Mitteilung der Versuchsanstalt fur Wasserbau, Hydrologie und Glaziologie, 17, pp. 221.

Haeberli, W., 1978, Special aspects of high-mountain permafrost methodology and zonation in the Alps, _in_ Proceedings of the Third International Conference on Permafrost, v.1: Ottawa, National Research Council of Canada, p. 378-384.

Höllermann, P., 1980, Altitudinal limits in high Mountains, Summary of a symposium, _in_ Jentsch, C. and Liedtke, H., eds., Hohengrenzen in Hochgebirgen, Arbeiten aus dem Geographische Institüt der Universität des Saarlandes, v. 29, p. 389-392.

Johnson, J. P., 1973, Some problems in the study of rock glaciers: _in_ Fahey, B. D., and Thompson, R. D., eds., Research in polar and alpine geomorphology, Third Guelph symposium on geomorphology, Norwich, Geo Abstracts, p. 84-94.

Johnson, P. G., 1974, Mass movement of ablation complexes and their relationship to rock glaciers: Geografiska Annaler, v. 56A, p.93-101.

Johnson, P. G., 1980, Rock glaciers: glacial and non-glacial origins: International Association of Scientific Hydrology, Publication 126, p. 285-293.

Loomis, S. R., 1970, Morphology and ablation processes on glacier ice: Arctic Institute of North America Research Report 57, p. 1-65.

Michaud, J., and Cailleux, A., 1950, Vitesses des mouvements du sol au Chambeyron (Basses Alpes): Comptes Rendus de l'Académie des Sciences, v. 230, p. 314-315.

Müller, F., 1962, Zonation in the accumulation areas of the glaciers of Axel Heiberg Island, N.W.T., Canada: Journal of Glaciology, v. 4, p. 302-311.

Østrem, G., 1959, Ice melting under a thin layer of moraine and the existence of ice in moraine ridges: Geografiska Annaler, v. 41, p. 228-230.

Østrem, G., 1964, Ice-cored moraines in Scandinavia: Geografiska Annaler, v. 46, p. 282-337.

Østrem, G., 1971, Rock glaciers and ice-cored moraines, a reply to D. Barsch: Geografiska Annaler, v. 54A, p. 76-84.

Outcalt, S. I., and Benedict, J. B., 1965, Photo interpretation of two types of rock glacier in the Colorado Front Range,USA: Journal of Glaciology, v. 5, p. 849-856.

Paterson, W. S. B., 1981, The physics of glaciers:

Oxford, Pergamon Press.
Potter, N., 1972, Ice-cored rock glacier, Galena Creek, Northern Absaroka Mountains, Wyoming: Geological Society of America Bulletin, v. 83, p. 3025-3058.
Priesnitz, K., and Schunke, E., 1978, An approach to the ecology of permafrost in central Iceland, in Proceedings of the Third International Conference on Permafrost, v. 1: Ottawa, National Research Council of Canada, p. 473-479.
Renaud, A., 1964, Les variations des glaciers suisses 1963 a 1964: Les Alpes, v. 40, p. 284-295.
Thompson, W. F., 1962, Preliminary notes on the nature and distribution of rock glaciers relative to true glaciers and other effects of the climate on the ground in North America: International Association of Scientific Hydrology, Publication 58, p. 212-219.
Wahrhaftig, C., and Cox, A., 1959, Rock glaciers in the Alaska Range: Geological Society of America Bulletin, v. 70, p. 383-436.
Washburn, A. L., 1979, Geocryology, London, Arnold.
Whalley, W. B., 1974, Rock glaciers and their formation as part of a glacier debris-transport system: Geographical Papers, Reading, v. 24.
Whalley, W. B., 1976, A rock glacier and its relation to the mass balance of corrie glaciers, Strupbreen, Troms: Norsk Geografisk Tidsskrift, v. 30, p. 51-55.
Whalley, W. B., 1979, The relationship of glacier ice and rock glacier at Grubengletscher, Kanton Wallis, Switzerland: Geografiska Annaler, v. 61A, p. 39-61.
White, S. E., 1976, Rock glaciers and blockfields. Review and new data: Quaternary Research, v.6, p. 77-97.

COLD-MIX ASPHALT STABILIZATION IN COLD REGIONS

Antoni S.E. Wojcik[1], Peter M. Jarrett[2] and Anthony N.S. Beaty[2]

[1]Construction Engineering Unit, CFB Winnipeg, Westwin, Manitoba, R2R 0T0, Canada
[2]Department of Civil Engineering, Royal Military College,
Kingston, Ontario, K7L 2W3, Canada

Problems arising with unbound aggregate runways in the northern environment are identified and arctic paving experience briefly reviewed. Following consideration of a range of possible stabilizing agents, cut-back bitumen was identified as being the most promising for conditions in the high arctic. A modified Marshall method of mix design for cold-mix stabilization of northern aggregates near to 0°C is described. Laboratory and field test results are presented and the effects, on ground temperature, of unbound, stabilized and white painted stabilized pavements are reported. It is concluded that cold-mix stabilization, at temperatures little above freezing, is feasible using the method described.

INTRODUCTION

All of Canada's major population centres are found close to its southern border. These are served by transportation links which run essentially along an east-west axis. However, the great land mass of Canada lies to the north of this populated region and extends almost to the North Pole. This vast area has a very low population but contains increasingly important mineral and energy reserves. The provision of amenities to the population, the development of the resources and the exercise of sovereignty can only be assured if a reliable transportation system exists throughout the region.

The principal means of transportation in the region is the aircraft. Unbound aggregate runways form by far the most common landing sites. These runways may, depending on the quality of local aggregate, suffer from the following problems during the warm season:

- rutting of the surface
- formation of dust clouds causing poor visibility and loss of fines
- flying stones with the potential to damage aircraft
- imperfect drainage due to deterioration of the profile

In view of the importance of such runways to the communities they serve, it would be desirable to develop a cost effective form of surface stabilization to alleviate these common problems. The stabilization methods must be applicable under the local climatic conditions and with the local aggregates.

ARCTIC PAVING EXPERIENCE

A number of runways in the North have successful, conventional, hot-mix asphalt pavements. These include Inuvik, Frobisher Bay, Whitehorse and also Thule in Greenland. One problem considered in design has to be the effect on the ground thermal regime of imposing a heat absorbing black surface. In Thule this problem was avoided by painting the asphalt surface white to prevent thawing of unstable soils lying beneath the runway. However, for most small communities a full hot-mix asphalt pavement is not economically justifiable. One therefore turns to less demanding methods. Transport Canada is believed to have laid trial sections of cold-mix asphalt at Rankine Inlet to observe its effect on the thermal behaviour. No results have been published, however. The main street of Inuvik was successfully stabilized using a mix-in-place asphalt stabilization method. An MC250 cut back bitumen was used at between 5 and 6 percent by mass of aggregate. In this installation, insulation was placed beneath the pavement system where thaw-unstable soils were present. This case is the only major northern use of cold-mix stabilization known to the authors. One probable reason for this is the Asphalt Institute's (1977) guideline that cold-mix not be used below about 10°C.

In the work reported in this paper, various possibilities for stabilization in the Arctic region were considered and the specific case of a gravel runway at Canadian Forces Station Alert on Ellesmere Island was studied in detail. Full details of the work have been given by Wojcik (1982).

CHOICE OF STABILIZER

The stabilizers and dust palliatives considered in general were:

Fuel or Road Oils which can be used as short term dust palliatives but require frequent renewal as they have low residual bitumen after evaporation and may be washed out by rainfall.

Lime stabilization requires more fines than were present in the aggregate at the specific site considered, but also in general, the full development of the cementing action intrinsic to lime stabilization requires higher temperatures than prevail in the north.

The use of Portland cement to form soil-cement is feasible. This was shown by construction of a small trial section at Alert. However, its long term durability is not known and fears have been expressed concerning maintenance difficulties should the hard, brittle, cement-stabilized material begin to break up under the thermal regime of the high arctic.

Lignosulphonates are chemical by-products of the pulp and paper industry that have been used as dust palliatives. They are water soluble and so are prone to leaching from the stabilized material. However, under the relatively dry conditions common in many arctic areas, this may not be a severe problem. Certainly laboratory and small scale field trials conducted during this project show this method to be of potential interest and worthy of further study.

Bituminous emulsions will work at low temperatures. This again was shown in a small scale field trial at Alert. However, bituminous emulsions break down if subject to freezing. Logistically, it is difficult to guarantee that the material would not be subjected to freezing during shipment to the north. This form of bitumen was therefore discounted.

Cut-back bitumen was considered to offer the most promise especially for the high arctic location under consideration. At the site the mean daily temperature is above freezing only during July and August and only in July is the mean minimum daily temperature above freezing. The construction season therefore lasts only about six weeks with the mean temperature between 0°C and 4°C. Further details of the climate and ground thermal regime have been given by Taylor et al. (1982) for that period. Because of these low temperatures, only low viscosity cutbacks were considered. The results presented from laboratory and field trials deal with a Gulf Oil RC-30 cutback and a Primer. The products are similar but have the following slight differences:

Property	Primer	RC-30
Kinematic Viscosity at 60°C, cs	26.2	33.0
Minimum Residue by volume, %	60.0	65.0
Solvent type	Gasoline	Heavy Reformate

MODIFIED MARSHALL MIX DESIGN

No widely accepted method is available for the design of cold-mix asphalt pavements. Lefebvre (1966) and Head (1974) used modifications of the Marshall method commonly used for hot-mix design. A similar approach was adopted for this study. Marshall samples were compacted using 75 hammer blows per face, then cured and tested in a cold room maintained at 4°C, Wojcik (1982). This temperature was selected to represent the mean daily temperature at Alert in July. The grading of the crushed rock aggregate used at Alert and in these mixes is given in Figure 1.

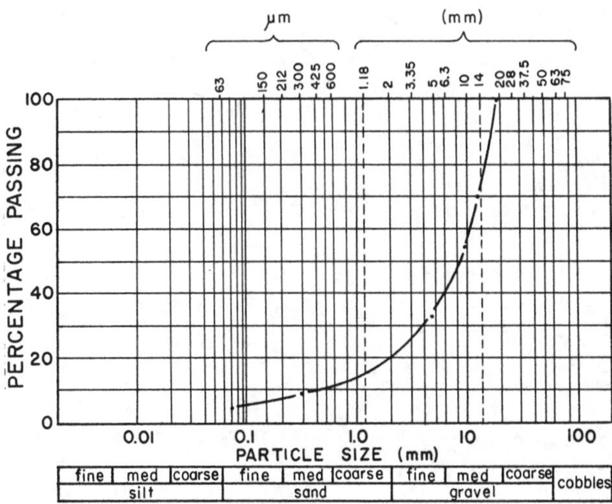

FIGURE 1 Typical grain size distribution for aggregate from CFS Alert crusher.

The design aircraft is the Lockhead Hercules transport, operating at a tire pressure in the range 0.55 to 0.69 MPa and having a maximum take-off mass of 70,400 kg. For design purposes, this aircraft falls within load classification group IV. Consideration of these characteristics and of values typically cited for hot-mix asphalt tested at 60°C led to the following target values for the cold-mix design:

Minimum stability	4000 N
Maximum flow	4 mm
Voids in the mineral aggregate	14%
Air voids	3-5%

From the results obtained with binder contents varying between 3 and 6 percent by weight of aggregate it appears that 5 percent binder content is optimal in meeting these design criteria. The following are the results at that percentage:

Property	RC-30	Primer
Bulk density (kg/m^3)	2330	2360
Air voids (%)	4.1	4.2
VMA (%)	12.2	12.3
Marshall stability (N) (3 day curing)	5900	5800
Marshall flow (mm)	5.3	4.0

Samples prepared over the same range of bitumen contents were also subjected to freeze-thaw cycling. Whilst no criteria are available with which to compare results certainly those samples with 4 percent or less were inadequately bound and showed signs of wear after 12 cycles.

LABORATORY TESTING PROGRAM

Having decided on an optimum mix, a laboratory testing program was instituted to study the effects of variation of the mixing and compaction temperatures and to study the rate of curing. Sets of identical Marshall samples were prepared and compacted at temperatures of 0°C, 4°C and 10°C. They were weighed and then cured at 4°C for periods of up to sixty days. Periodically samples were reweighed and the Marshall test performed. Results of stability and solvent loss against curing time are shown in Figure 2 for the case of the Primer mixed at 4°C. These results

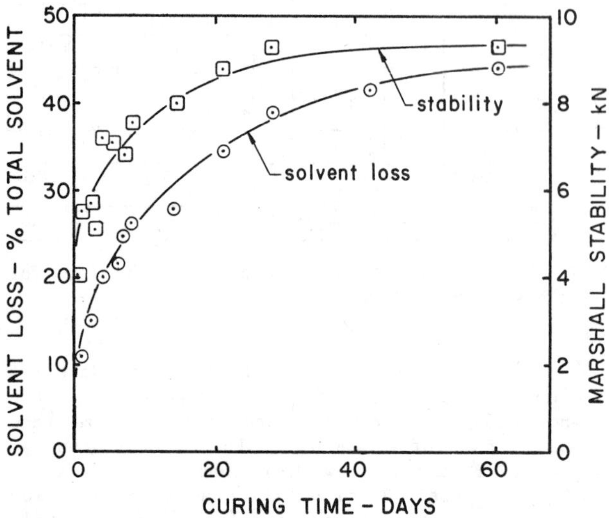

FIGURE 2 Solvent loss and Marshall stability versus curing time for Primer-aggregate specimens prepared at 4°C.

are typical of all series for both the binders used and show the expected parallel development of strength with solvent loss. In general the rate and extent of curing is slower than that experienced at higher temperatures but is still thought to be adequate to allow construction of a useable pavement. Mixing temperature did not markedly affect the results. Stabilities were perhaps 10 percent higher and mixing easier at 10°C compared to 0°C. Indeed at 0°C with the more viscous RC-30, hand mixing was difficult and some cases of incomplete coating were noted. No consistent variation in compacted density was obtained over the range of temperature investigated.

FIELD TRIALS

In the summer of 1980, a series of preliminary small trial sections 10 m long by 5 m were constructed on a road at Alert. These included Lignosol, soil cement, bituminous emulsion, fuel oil, chip seal and control sections. Mixing was achieved using a grader to turn windrows of aggregate on which the stabilizer had been sprayed and placed. From these trials and the results of the laboratory testing program, it was decided that cold mixing at low temperature was both possible and potentially a useful method of stabilization. Therefore in July 1981, a series of larger test sections was constructed on the turning circle at the end of the runway in Alert.

In brief, an 8 cm thick pad of compacted crushed rock aggregate was laid over an area of approximately 20 m by 60 m. The aggregate was compacted at its optimum moisture content of 6 percent. Four lanes, each 3.8 m wide, were laid out on this pad using 5 cm by 10 cm timbers as dividers. Stabilized aggregate was then prepared using a Cedarapids pug mill with a 700 tonne per hour capacity. For the small trial sections constructed, this machine was operated for only brief periods at it slowest production rate. This fact, combined with problems in calibrating the bitumen flow rate and the limited quantity of bitumen available on site, led to mixes that were not proportioned exactly as originally intended. However, mixes were prepared over a wide range of bitumen contents. These mixes were then laid in the lanes on the compacted pad, screeded and raked flat using the timbers as a guide and compacted using a 9 tonne vibrating roller. The as-built sections with their various bitumen contents, as determined by bitumen extraction tests, are shown in Figure 3. The construction was carried out over a period of 3 days during which time the mean temperature was approximately 3.4°C. The aggregate temperature was between 0°C and 1°C.

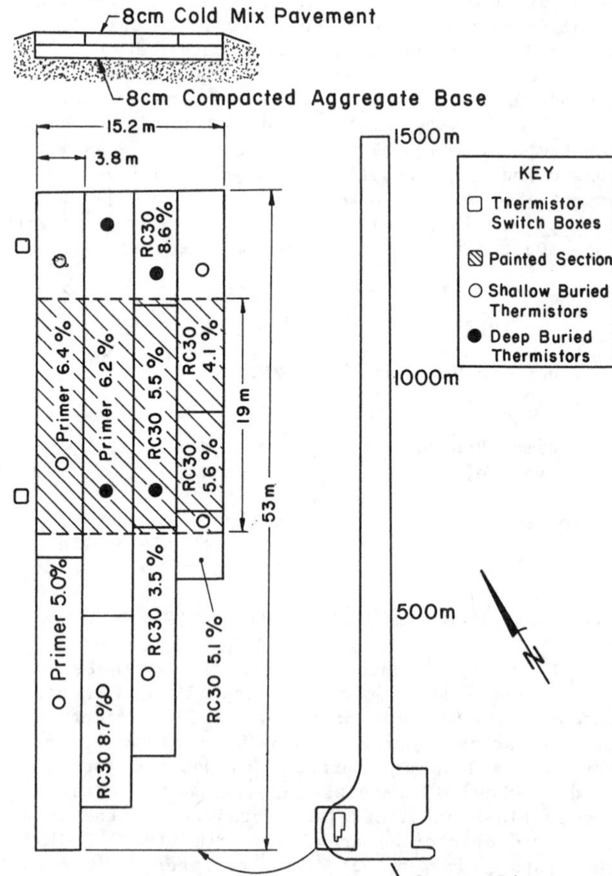

FIGURE 3 Layout of field trial sections as built.

The following construction details are of significance: The mixes looked good with an even distribution of the bitumen over the particles. Even the more viscous RC-30 mixed well. This was attributed to the fact that the pug mill imparts more energy to the mixing than could be achieved by hand in the laboratory. Compaction of all sections was monitored using a nuclear-density gauge and confirmatory sand-cone tests. Density was measured after each double pass of the roller. A comparison of the field densities after six roller passes and the laboratory densities from the Marshall test is given in Figure 4. This

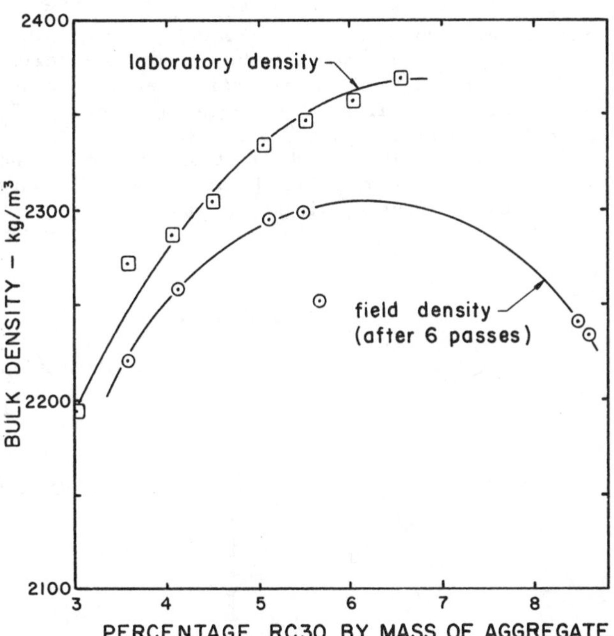

FIGURE 4 Laboratory and field bulk densities for RC30 cutback-aggregate mixtures.

indicates that approximately 97 percent of laboratory compaction was achieved in the field. Vibration was observed to be of definite benefit. Bleeding occurred in the sections with high bitumen contents (8.7% and 8.6% RC-30, 6.4% Primer).

After compaction, field CBR tests were made on each section as a measure of the strength. It was realized that CBR is not a normal test for asphaltic materials, but under the circumstances it was the only test practically possible. The rate of strength gain with time was studied by periodically repeating the field CBR tests. Typical results for three of the RC-30 sections are shown in Figure 5. These results are felt to be representative. The scatter is attributed to variations in temperature of the surface with time and the size of aggregate compared to the CBR plunger as well as normal variation attributable to non-homogeneity. The trends shown were, however, consistent. The sections with a high bitumen content (8.6% and 8.7% RC-30) were very soft and clearly contained excessive bitumen. Also, they were not curing very rapidly. The low

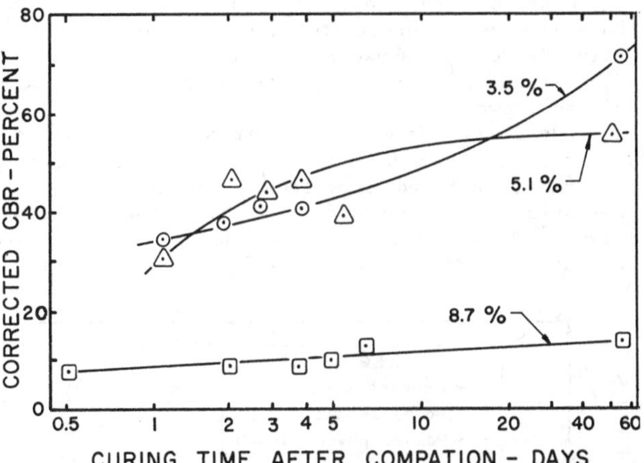

FIGURE 5 Corrected field CBR values versus curing time for RC30 cutback-aggregate trial sections.

bitumen content specimens (3.5% and 4.1% RC-30) were the strongest, as they were in the laboratory. However, there was insufficient bitumen present in these sections to coat the particles fully. This affects the permeability of the stabilized material and in the long term its durability also. In September when the fifty day CBR measurements were made, ice crystals were observed within the fabric of these sections. The ice may have been a factor in the relatively high strength measured but would certainly also represent the first stage in a break down process. The sections with bitumen contents close to 5 percent represented the "design" condition. The degree of compaction achieved in the field was good and yet the strength values were rather lower than expected, with some early softness noticed in subjective judgment of the sections. From Figure 5, it is seen that the CBR was 30 percent after one day and about 45 percent after 4 days. The curing rate then slowed considerably. A correlation was attempted in the laboratory between Marshall Stability in Newtons and CBR. For stabilities between 4000 and 6500, the following relationship held with a correlation coefficient of 0.94:

$$\text{Marshall Stability} = 36.5\ \text{CBR} + 1925$$

For a CBR of 45 percent, by extrapolation this gives a stability of 3568 N which is rather less than the desired minimum and considerably less than the equivalent stability obtained in the laboratory. This may be due to a combination of a slower rate of curing in the field and variability between the field and laboratory methods of measuring CBR. This latter point is considered the most probable.

It was also decided to investigate the effect of the paving on the ground thermal regime. To provide a wider range of information, part of the paved area was painted white. "Deep" strings of thermistors were installed every 30 cm to depths of approximately 1.5 m. In addition, thermistors were placed within the stabilized layer and also in the compacted base course. "Shallow" thermistors were placed only in the asphalt and base courses. The thermistor locations and the painted area are shown in Figure 3. An additional deep string of thermistors was placed at the edge of the runway.

Readings were taken from the time of construction until freeze-up in September. Figure 6, compiled from these results, shows the variation with time of the depth to the freezing plane. This indicates a significant increase in the depth of thaw beneath the black paved section

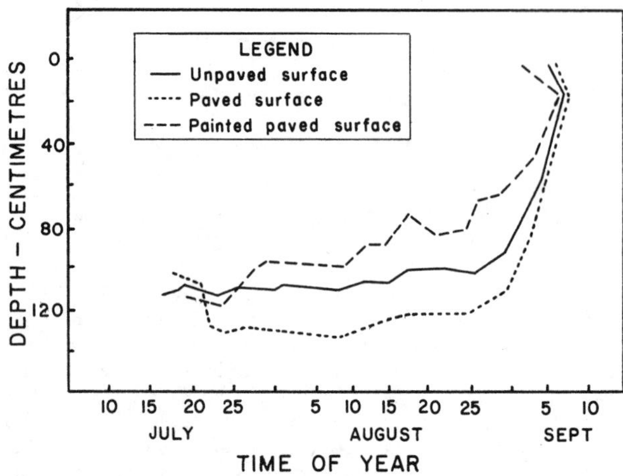

FIGURE 6 Depth of thaw under the unpaved runway, the unpainted pavement and the white painted pavement.

and a decreased depth of thaw beneath the white painted section as as compared to that beneath the gravel runway which is grey-brown in colour. Comparison of the air temperature with that of the pavement surface showed that the temperature of the white painted section was close to or less than the air temperature whereas the black section was considerably warmer than air temperature for most of the time. It should be noted that these results represent only a portion of the thawing season and so should not be considered indicative of a full season's thaw depth.

CONCLUSIONS

1. It proved possible to mix, lay and compact cold-mix asphalt under arctic conditions with operational temperatures close to 0°C.

2. The modified Marshall method of mix design yielded a bitumen content which proved satifactory in field construction and reliably predicted achieved field densities.

3. The overall results verified the viability of cut-back bitumen for construction of a stabilized pavement under arctic conditions. Further observations will be required to determine the long-term performance of the stabilized pavement.

REFERENCES

Asphalt Institute, 1977, Asphalt Cold-Mix Manual, in Manual Series No. 14 (MS-14): Asphalt Institute, College Park, Maryland.

Head, R.W., 1974, An informal report on cold mix research using emulsified asphalt as a binder, in Proceedings Association of Asphalt Paving Technologists, v. 43, p 101-131, Williamsburg, Virginia.

Lefebvre, J.A., 1966, A suggested Marshall method of design for cutback asphalt-aggregate paving mixtures, in Proceedings of the Eleventh Annual Conference of Canadian Technical Asphalt Association, v. XI, p 135-181.

Taylor, A., Brown, R.J.E., Pilon, J., Judge, A.S., 1982, Permafrost and the shallow thermal regime at Alert, N.W.T., in Proceedings of the Fourth Canadian Permafrost Conference (The Roger J.E. Brown memorial volume), p 12-22.

Wojcik, A.S.E., 1982, Asphalt stabilization of the CFS Alert runway, M.Eng Thesis, Royal Military College of Canada, Kingston, Ontario.

BASIN WATER BALANCE IN A CONTINUOUS PERMAFROST ENVIRONMENT

Ming-ko Woo, Philip Marsh, and Peter Steer

Department of Geography, McMaster University, Hamilton, Ontario, Canada L8S 4K1

The hydrology of a basin in the continuous permafrost region of Canada was studied for 6 years to establish seasonal and annual water balances. Emphases were placed upon the spatial and temporal variations of snowmelt, rainfall, evaporation, streamflow and active layer storage capacity. Spring melt released considerable meltwater which could not be accommodated by a thinly thawed active layer. High runoff resulted. Evaporation was active concurrently but from the snow-free portion of the basin. Summer rainfall was often of low intensity. Storage capacity increased as the frost table receded and much of the rain could be held in the active layer to maintain evaporation and baseflow. Over the years, snowfall constituted about three-quarters of annual precipitation, about 80% of which was consistently removed by runoff. Annual evaporation was small and net change in stroage for the 6 year period was also of low magnitude.

INTRODUCTION

The water balance for an arctic drainage basin is

$$M + R - E - Q = ds/dt, \qquad (1)$$

where M is snow and ice melt, R is rainfall, E is evaporation, Q is runoff, and ds/dt is change in storage. Dingman (1973) regarded the water balance to be a fundamental environmental characterization of a geographic region. Yet, for the arctic, the paucity and the inaccuracy of the hydrometeorologic data render water balance computations difficult. One consequence is that most attempts were restricted to individual components of the water balance.

Another problem is the lack of spatially distributed information. Most studies made use of single point measurements to represent basin gains or losses. This is inadequate because topographic variations often cause uneven distribution of snow or rainfall. Similarly, evaporation determined for a station cannot be substituted for basin evaporation without accounting for the daily changing basin snow cover (Woo et al. 1981). Further, all water balance studies have ignored the presence of permafrost at shallow depths. A thinly thawed active layer provides limited storage capacity and encourages runoff (Woo and Steer 1982), and a deepening of the frost table consumes energy for thawing which otherwise is used to warm the ground or to increase evaporation.

This study recognizes the need to consider all the components of equation (1) in attempting a basin water balance; to deal with the areal variations of hydrometeorologic variables; and to continue the research over several years to ensure representativeness of results. Our objectives are therefore to establish the seasonal and annual water balance for a small basin in the High Arctic, and to relate the water balance to the hydrologic conditions of the continuous permafrost region.

STUDY AREA AND METHOD

The McMaster River basin (area 33 km^2) was selected for this study because it is centrally located in the Canadian Arctic archipelago and is close to a weather station at Resolute (74°45'N, 94°50'W). The basin has a rolling topography, with elevation ranging between 85 and 200 m a.s.l. (Figure 1). The dominant surface materials are gravel and loam with only patches of tundra vegetation scattered over an otherwise barren landscape. The weather station reports a January mean minimum of -36°C and a mean July maximum of 7°C. Continuous permafrost is maintained, usually at 0.2-0.7 m below the ground surface. The snow cover lasts for about 9 months each year, and such hydrologic activities as evaporation and runoff are confined between late June and early September.

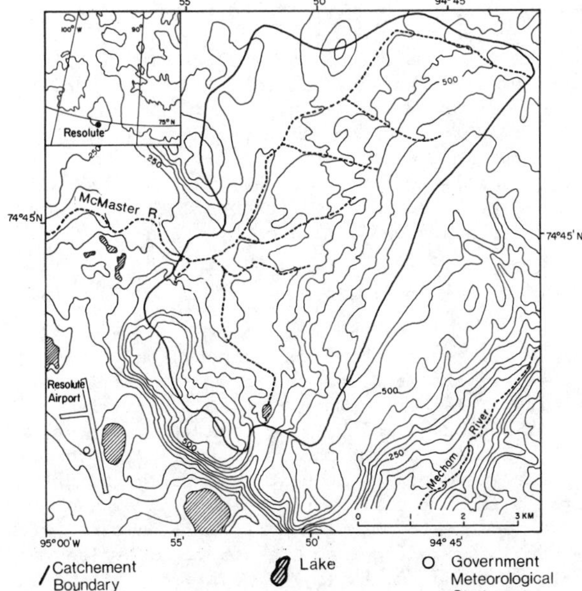

FIGURE 1 Topography of McMaster basin with contours at 50' intervals.

This study was carried out between 1976 and 1981. Emphases were placed upon the spatial variation of hydrometeorologic quantities in the basin. Extensive snow surveys were conducted in late winter (May or June) using a method described by Woo and Marsh (1978). In summer, a network of 5-15 gages was deployed to establish the spatial variation of rainfall. Results of these measurements were used to adjust the precipitation data reported by the Resolute weather station.

Snowmelt was determined at a central site using the energy balance approach (Woo et al. 1981). In extending the computation to other parts of the basin, corrections were made for temperature and vapor pressure differences (Brutsaert 1975). To obtain mean melt rate for the basin, n terrain units were first recognized, and the mean premelt snow storage held in each unit was determined by the late winter snow survey. When melt began, basin snowmelt (M) on day t is

$$M(t) = \sum_{i=1}^{k} m_i(t)\, a_i / \sum_{j=1}^{n} a_j, \qquad k \le n, \tag{2}$$

where k of the n units remains snow covered on day t, m_i is melt rate for terrain unit i calculated by the energy balance approach, a_i is the area occupied by unit i, and there were n = 24 units for McMaster basin. The amount of snow remaining at unit i was updated daily by

$$ps_i(t) = ps_i(t-1) + p_i(t) - m_i(t), \qquad ps_i(t) \ge 0, \tag{3}$$

where $ps_i(t)$ is the snow stored at unit i on day t, and $p_i(t)$ is new daily snowfall. When $ps_i = 0$, terrain unit i becomes snow free and the bare ground would be incremented by a_i.

Evaporation was computed using the Priestley and Taylor (1972) approach,

$$e = \alpha[\Delta/(\Delta + \gamma)](Q^* - Q_G)/L_v\rho_w, \tag{4}$$

where e is evaporation, Δ is the slope of the temperature-saturated vapor pressure curve, calculated as function of air temperature (Dilley 1968), γ is the psychrometric constant, Q^* is net radiation, Q_G is ground heat flux, L_v is the latent heat of vaporization, ρ_w is water density, and α is an empirical coefficient which varies with surface soil moisture (Marsh et al. 1981). Net radiation was measured with a Swissteco net radiometer. Ground heat flux was estimated by (Rouse 1982)

$$Q_G = \int_{z=0}^{z_0} c\,\frac{\partial T}{\partial t}\,dz + L_f\rho_w I\,\frac{\partial z_f}{\partial t}, \tag{5}$$

where z_0 is the depth at which the annual soil temperature amplitude equals zero, c is soil heat capacity, T is soil temperature, L_f is the latent heat of fusion, I is the fractional ice content, and z_f is the depth of the frost table. Field evidence showed that the heat that warms the ground (first term on the right-hand side) can be neglected because its magnitude is small compared with the heat required to thaw the active layer (second term). The thaw depth (z_f) τ days after ground thaw began, can be obtained as (Woo 1976)

$$z_f(\tau) = \beta\sqrt{\tau}, \tag{6}$$

where β is a coefficient, empirically found to be 0.1 for gravel soils in the Resolute area. Field observation showed that the soil pores immediately below the frost table were usually filled with infiltrated meltwater which subsequently refroze (Woo and Heron 1981). The ice content can therefore be approximated by the soil porosity (ε) which averages 0.37 for the study basin. Then,

$$Q_G \approx [z_f(\tau) - z_f(\tau - 1)]\, L_f\rho_w\varepsilon. \tag{7}$$

Evaporation occurs only in the snow-free areas, and daily evaporation for the basin (E) is

$$E(t) = \sum_{i=1}^{m} e_i(t)\, a_i / \sum_{j=1}^{n} a_j, \tag{8}$$

where m terrain units are snow free on day t.

Discharge was measured twice daily using the velocity-area method during spring when the channel was not yet stabilized (Woo and Sauriol 1980). When a stable channel was established, river stage was recorded by a Leupold-Stevens Type F recorder and the stage was converted into discharge using empirically derived rating curves.

RESULTS

Water Balance During Melt Period

Snow is unevenly distributed at the end of winter. Figure 2 gives an example of snow storage in McMaster basin at the end of May 1978. In a typical year, areas with little snow included hilltops and exposed flat areas, but topographic depressions such as valleys and gullies had considerable snow. Snowmelt often began in June and progressed quickly, fragmenting the snow cover where the pack was thin. Several processes then occurred simultaneously. The snow-covered areas yielded meltwater, while the bare ground experienced evaporation. The shallow thawed zone of the bare areas had little storage capacity, and the excess water moved rapidly downslope as surface flow (Woo and Steer 1982). Upon reaching the deep valley snowpacks, the bulk of the flow was absorbed as liquid water storage. It was when these deep packs were partly saturated that channels were initiated within, above, or below the snow barrier to convey the first streamflow (Woo and Sauriol 1980).

Being controlled by the above processes, the water balance provides a quantitative expression of the hydrologic conditions. Figure 3 is an example drawn from 1979. When the snow cover was extensive, meltwater was released from almost the entire basin and the weighted basin melt contribution was high. Gradually, a shrinking snow cover diminished the water supply, accompanied by increasing basin evaporation due to the enlarged bare area. The combined magnitude of snowmelt and the occasional rainfall, however, overwhelmed the evaporation loss.

Meltwater can be held in deep snowpacks or in the thawed zone of the active layer. On any day t after the initiation of melt, the amount of water storage (s_t) is

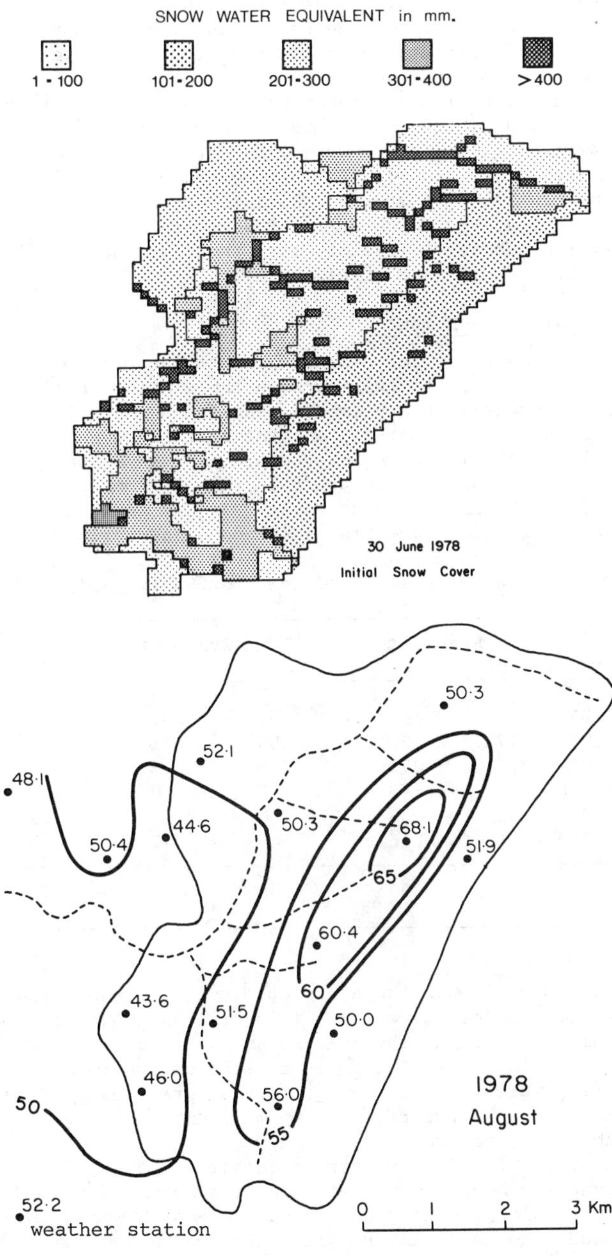

FIGURE 2 (Top) Snow distribution by terrain units immediately before snowmelt in 1978. (Bottom) Total rainfall in mm for August 1978.

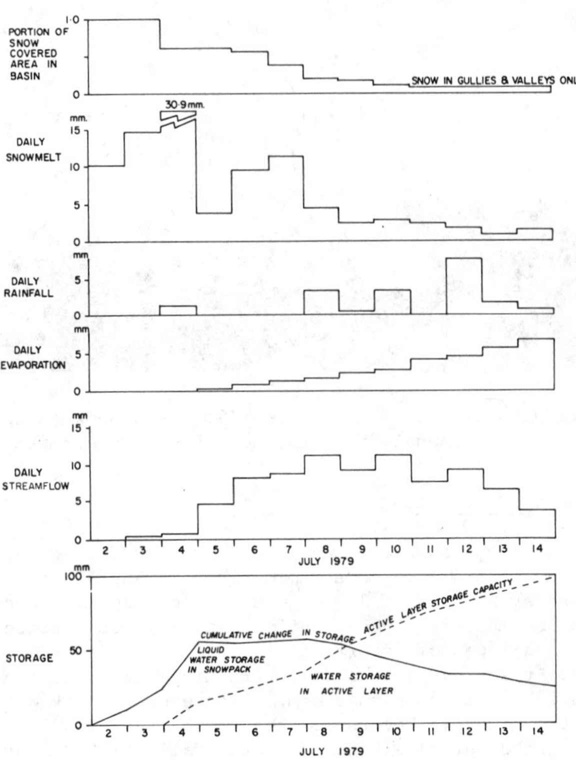

FIGURE 3 Water balance of the snowmelt period, 1979.

$$S_t = \sum_{i=1}^{t} (M_i + R_i - Q_i - E_i), \tag{9}$$

where M, R, Q, E are daily melt, rainfall, runoff, and evaporation from the basin. The storage capacity of the active layer at any point in the basin on day t (SC_t) is

$$SC_t = z_f(t)\ \varepsilon, \tag{10}$$

where z_f is depth of the frost table, obtained by equation (6), and ε is soil porosity. The mean basin storage capacity can be obtained by weighting against the fractional snow-free area in the basin. Figure 3 shows that in 1979, S > SC until July 9, indicating that the excess water was probably stored in the snowpack. Field observation confirmed a gradual hydrostatic water level rise in the valley snowpack until flow began in channels carved through the snow (Figure 4). Once begun, streamflow removed the bulk of the storage. Simultaneously, the active layer storage capacity increased as the frost table receded. Surface flow ceased gradually as SC > S. The melt season ended with a depletion of the snow cover and a steady drying of the ground.

FIGURE 4 Initiation of streamflow in a snow-filled channel after the deep pack was saturated.

Summer Water Balance

Although snowfall can occur any time during the year, rainfall is more common in July and August. Most events are of low magnitude and sometimes cannot be measured by conventional means. Such trace rainfall was of importance in certain years (Woo and Steer 1979). Spatially, rainfall distribution varied from storm to storm. Orographic effect was not noticeable because of the low basin relief. Over the entire summer, mean rainfall for McMaster basin was often similar to the value recorded at the government weather station (Figure 2).

Continued thawing of the active layer increased its storage capacity which was able to absorb most of the low rainfall without yielding immediate runoff. Only occasional high rainfall produced significant increase in flow. An example of the summer water balance is given as Figure 5. After the spring of 1978, the basin became snow free except along steep valley walls. Streamflow was low on most nonrainy days. Despite the extensive bare surfaces, evaporation was limited by low radiation energy for high latitudes and by the drying ground. The rainstorm of August 2-3 deposited 17 mm of rain, 14 mm of which was taken up by active layer storage. Most of the remaining rainwater went to streamflow. Unliked the melt season, such high flow was short lived and its magnitude was lower than the spring floods.

At the beginning of September, most of the storage was depleted by evaporation and by continuous low flow. Streambeds were often dry when winter arrived, offering little chance for ice formation. The low level of storage also prevented aufeis formation. In this regard, rivers of the High Arctic behave differently from those of the sub-Arctic (e.g., Kane et al. 1973).

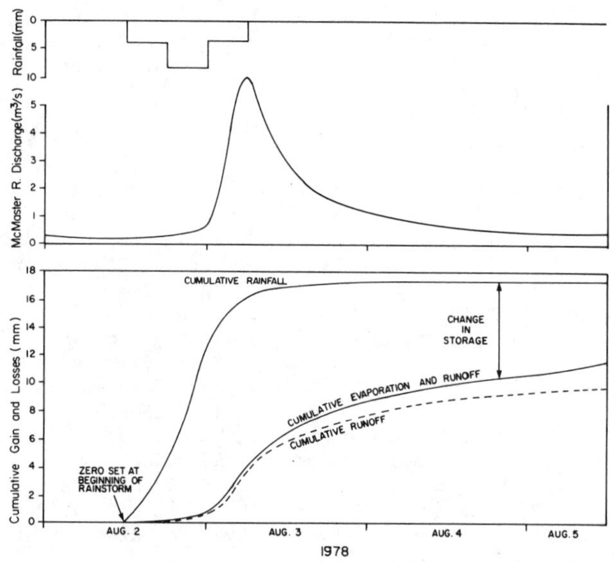

FIGURE 5 Water balance before, during, and after a rainstorm, 1978.

Annual Water Balance

The annual water balance computation considers the water year to span between the first significant snowfall in September (when the snow would remain for the winter), and the arrival of snowfall in the following September. In the 6 years of study, annual precipitation ranged from 120 - 273 mm (Table 1), about three-quarters of which was snowfall, suggesting a preponderant influence of snowmelt upon the hydrologic cycle. Streamflow accounted for 80% of annual water loss, most of which was during the melt season. A high ratio of runoff to precipitation is typical of impermeable areas which, in our cases, is the permafrost. With considerable runoff, only a small amount is available for evaporation.

TABLE 1 Water Balance of McMaster Basin (mm/yr.)

	Precipitation	Runoff	Evaporation	Storage Change
1975-1976	191	161	31	-1
1976-1977	120	155	31	-66
1977-1978	273	213	38	23
1978-1979	178	143	30	5
1979-1980	165	130	51	-16
1980-1981	215	148	47	21
Mean	190	158	38	Total-34

The active layer storage capacity was highly dynamic, being very limited during spring but increasing steadily as the bare areas thawed. From middle to late summer, active layer storage sustained evaporation and baseflow, and by September the residual storage was low. . Water balance confirms that the year to year change in basin storage was usually less than 10% of precipitation (Table 1). An exception was 1977 which was a dry year with a large storage deficit. One possible reason for such a large deficit is an underestimation of precipitation due to many trace rainfall events (Woo and Steer 1979).

CONCLUSION

The major conclusions of this paper are

1. Spatial variation of hydrologic quantities plays a major role in the water balance of polar drainage basins. Of particular importance is the melt season when snowmelt and evaporation occur simultaneously but at different parts of the basin.

2. Snowfall constitutes a large portion of annual precipitation and most of the annual runoff is generated by snowmelt. Such hydrologic prominence of snow quantitatively verifies Church's (1974) conclusion that rivers such as the McMaster manifest a distinct arctic nival regime.

3. Hydrologic effect of permafrost is transmitted through the dynamic active layer storage capacity. A shallow active layer in spring enhances rapid runoff. As the frost table recedes, storage capacity increases and much of the summer rain can be stored to maintain evaporation and to sustain low discharge.

ACKNOWLEDGMENTS

Financial support was provided by the Natural Sciences and Engineering Research Council of Canada. The generous logistical support of the Polar Continental Shelf Project, Canadian Department of Energy, Mines and Resources, is gratefully acknowledged.

REFERENCES

Brutsaert, W.,1975, On a derivable formula for long-wave radiation from clear skies. Water Resources Research, 11, 742-744.

Church, M., 1974, Hydrology and permafrost with reference to northern North America. Proceedings Workshop Seminar on Permafrost Hydrology, Canadian National Committee, IHD, Ottawa, 7-20.

Dilley, A.C., 1968, On the computer calculation of vapor pressure and specific humidity gradients from psychrometric data. Journal of Applied Meteorology, 7, 717-719.

Dingman, S.L., 1973, The water balance in arctic and subarctic regions - annotated bibliography and preliminary assessment. U.S. Army, Cold Regions Research and Engineering Laboratory Special Report 187, 131 pp.

Kane, D.L., Carlson, R.F. and Bowers, C.E., 1973, Groundwater pore pressures adjacent to subarctic streams. in: Permafrost: The North American Contribution to the Second International Conference, Yakutsk, USSR. Washington: National Academy of Sciences, 453-458.

Marsh, P., Rouse, W.R. and Woo, M.K., 1981, Evaporation at a High Arctic site. Journal of Applied Meteorology, 20, 714-716.

Priestley, C.H.B. and Taylor, R.J.,1972, On the assessment of surface heat flux and evaporation using large-scale parameters. Monthly Weather Review, 100, 81-92.

Rouse, W.R.,1982, Microclimate of low arctic tundra and forest at Churchill, Manitoba. in: French, H.M. (ed.) Proceedings of Fourth Canadian Permafrost Conference. Ottawa: National Research Council of Canada, 68-80.

Woo, M.K., 1976, Hydrology of a small Canadian High Arctic basin during the snowmelt period. Catena, 3, 55-168.

Woo, M.K. and Heron, R.,1981, Occurrence of ice layers at the base of High Arctic snowpacks. Arctic and Alpine Research,13, 225-230.

Woo, M.K., Heron, R., and Steer, P., 1981, Catchment hydrology of a High Arctic lake. Cold Regions Science and Technology, 5, 29-41.

Woo, M.K. and Marsh, P., 1978, Analysis of error in the determination of snow storage for small High Arctic basins. Journal of Applied Meteorology, 17, 1537-1541.

Woo, M.K. and Sauriol, J., 1980, Channel development in snow-filled valleys, Resolute, N.W.T., Canada. Geografiska Annaler, 62A, 37-56.

Woo, M.K. and Steer, P., 1979, Measurement of trace rainfall at a High Arctic site. Arctic, 32, 80-84.

Woo, M.K. and Steer, P., 1982, Occurrence of surface flow on arctic slopes, southwestern Cornwallis Island. Canadian Journal of Earth Sciences, 19, 2368-2377.

RELATIONSHIPS BETWEEN RUNOFF GENERATION AND ACTIVE LAYER DEVELOPMENT NEAR SCHEFFERVILLE, QUEBEC

Richard K. Wright
Department of Geography, McGill University
805 Shebrooke St. West, Montreal, Quebec
H3A 2K6 CANADA

The generation of runoff in areas underlain by permafrost is similar to that of temperate regions, but is distinguished by the seasonal changes in the extent, shape, and thickness of the areas generating runoff as well as the storage and transfer of small amounts of water by frost-related mechanisms. Studies conducted near Schefferville, Quebec have shown that these mechanisms can significantly alter the development of the active layer. The development of the active layer partially controls the distribution of saturated surfaces in a given catchment, and thus its hydrologic response to runoff-generating events. This phenomenon is similar to the variable-source area runoff generating mechanism of temperate regions, except that the ratio of stormflow to baseflow is also partially controlled by the depth and extent of the active layer. A detailed study of a small catchment at Schefferville has demonstrated that consideration of these effects can increase the precision of models of runoff-generation from such catchments.

INTRODUCTION

The movement of moisture within the active layer is an important component of both the moisture and thermal balances in northern regions. The rate of moisture movement through the active layer largely controls the distribution of saturated surfaces in a given catchment, which inturn, controls the rate at which runoff leaves the catchment. At the same time, the vertical and horizontal fluxes of moisture within the active layer transfer both sensible and latent heat, thereby influencing the rate and depth of thaw. There is therefore a degree of feedback between the rate of runoff generation and the rate of active layer development. This paper will present some observations of the processes of active layer development at Schefferville, and discuss the implications of those processes for modelling subarctic water balances.

The results presented in this paper are only a part of the information obtained in a study carried out near Schefferville, Quebec during the period 1976-1979. The primary goal of the project was the accurate representation of the water balance of a lichen tundra underlain by permafrost. In line with that project, some data were collected about the ground thermal regime. The results of the water balance investigations have been reported elsewhere (Wright 1981,1982). This paper will present, in more detail, some observations of the development of the active layer and the generation of runoff.

ACTIVE LAYER DEVELOPMENT

The study's primary objective was accurate characterization of the water balance, but data about the ground thermal regime were required for the accurate estimation of the amount of suprapermafrost groundwater. It was initially thought that point determinations of the depth of the active layer could be made from interpolation of the discrete temperature measurements along a thermocable while the areal estimates of the depth of the active layer could be achieved by applying a known relationship between maximum active layer depth and the distribution of vegetation at the surface (Nicholson 1978). Significant problems were encountered, however, due to the importance of latent energy transfer, which had been implicitly ignored in the estimation procedures.

The transfer of sensible and latent energy can result from moisture movement in either the liquid or gaseous phases. The transfer of sensible heat by water vapor is probably of little importance in cold regions, but the transfer of sensible heat by liquid water movement appears to be of considerable importance. Although it can be shown that the vertical transfer of sensible heat is of little moment, Lewis' (1977) study at Timmins 4 indicates that the lateral concentration into the very permeable wetlines (areas of slight to moderate topographic depression and concentrated subsurface flow) of the latent and sensible heat associated with suprapermafrost groundwater can lead to active layers up to 10 m deep. Nicholson (1978) improved the accuracy of his model of permafrost distribution in the Schefferville area by adding a groundwater flow component.

The problem of adequately explaining the physics of latent heat transfer within a partially-frozen soil is one that has attracted considerable attention in recent years. Much of the attention has been focussed on accurately modeling the phenomenon in the laboratory and, while considerable success has been achieved (e.g., Jame and Norum 1980), it is not clear whether such models are applicable under the range of conditions normally found in the field.

One of the effects associated with latent energy transfer is the well-known "zero-curtain", where a significant portion of the freezing or thawing active layer assumes temperatures at the freezing point due to the large amount of latent energy involved in the fusion of water. The near-zero temperatures produced during the phase change of the soil water reflects conditions under which water exists in two phases. Under those conditions, it is difficult to accurately determine the thermal and hydraulic properties of the ground.

The relative changes in moisture content and phase over the depth of the active layer also influence the rate and depth of thaw. The magnitudes of the thermal conductivity and apparent specific heat will vary over the depth of the soil column according to changes in the composition and structure of the inorganic matrix, the amount and distribution of organic matter, and the proportions and distribution of frozen and unfrozen water within the soil column. Since the variation of soil moisture content is a function of both depth and time, the rate and depth of thaw is a function not only of the heat flux at the surface, but is also a function of the distribution and state of water within the active layer.

In general, models of active layer development (e.g. Nakano and Brown 1972, Smith 1975, McGaw et al 1978) have assumed that soil thermal properties are double-valued; i.e. one value would hold in the 'frozen' state and another value would hold in the 'thawed' state. The transition between the two states was presumed to be instantaneous, occurring when the soil temperature passed through the freezing point. The data collected in this study clearly demonstrates that heat transfer is not solely by conduction and that the active layer is not sharply demarcated into frozen and unfrozen zones.

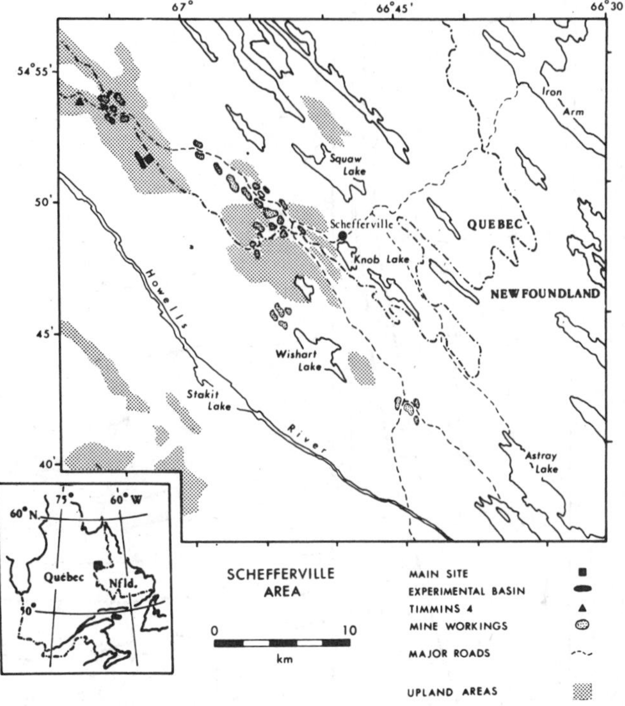

FIGURE 1 Location of Schefferville and the study sites.

THE FIELD AREA

The bulk of the results presented in this paper were collected near Schefferville, Quebec (54°51'N, 67°01'W,; 520 m a.s.l.). Located within the Labrador Trough in central Ungava, the area is composed largely of Poterozoic metamorphosed rocks (Figure 1). The region is characterized by broad, permafrost-free valleys and long, narrow ridges underlain by extensive permafrost. The valley bottoms are largely covered by spruce woodlands and lakes, while the ridgetops are dominantly lichen tundra. Mean annual air temperature at Timmins 4 is -6.2°C. Mean annual precipitation at the Schefferville townsite (15 km SE of the main research site) is 785 mm, of which 407 mm falls as rain and the rest falls as snow (Barr and Wright 1981).

The main research site (Hematite, Figure 2) is at an altitude of 685 m a.s.l. The area is typical of lichen-heath tundra in central Ungava. The surface over 80% of the site is a 5-10 cm thick lichen mat (predominantly Cladonia and Cladina spp.) with scattered dwarf birch (Betula glandulosa). The remaining 20% of the surface is bare, frost-scarred ground, partially paved with frost-shattered debris. The bedrock, which is typically overlain by 1-2 m of very stony till, is weakly metamorphosed sediments. Permafrost underlies the entire site to a depth of at least 30 m. The annual depth of thaw ranges from 2.0 m under thick lichen mat to 3.0 m under bare surfaces. Two areas of special interest at Hematite are locations One and Two. Both are located in an area of lichen-heath tundra, but the surface of One is slightly convex and is stripped of vegetation. Location Two is relatively flat and has an undisturbed lichen mat.

Data were collected at the main site from December 1976 through August 1979, though the frequency of observation varied from daily during the summers of 1977 and 1978 to monthly or bimonthly during the winter. Intensive data collection was carried out from May through September in 1977 and 1978 as well as frequent observations during the freeze-up of 1977. Daily observations (during the thaw season) included air temperature and humidity at screen height, ground temperatures, incoming and net radiation, evapotranspiration, and rainfall (Wright 1981). In addition, soil moisture contents were monitored with a neutron moisture probe through the freeze-up of 1977 and the summer of 1978.

Observations at the main site in 1978 were paralleled by observations of the water budget of a small basin adjacent to the main site (Bazilchuk 1979). The basin is, like most of the Schefferville region, strongly controlled by the underlying geology. Permafrost is estimated to underlie approximately 80% of the basin, being absent beneath the lake and possibly the lower reaches of the creek. The distribution of vegetation in the basin is controlled by aspect, exposure, and drainage. Low-lying, often-saturated areas are dominated by birch-willow scrub while better sites are dominated by lichen-heath tundra.

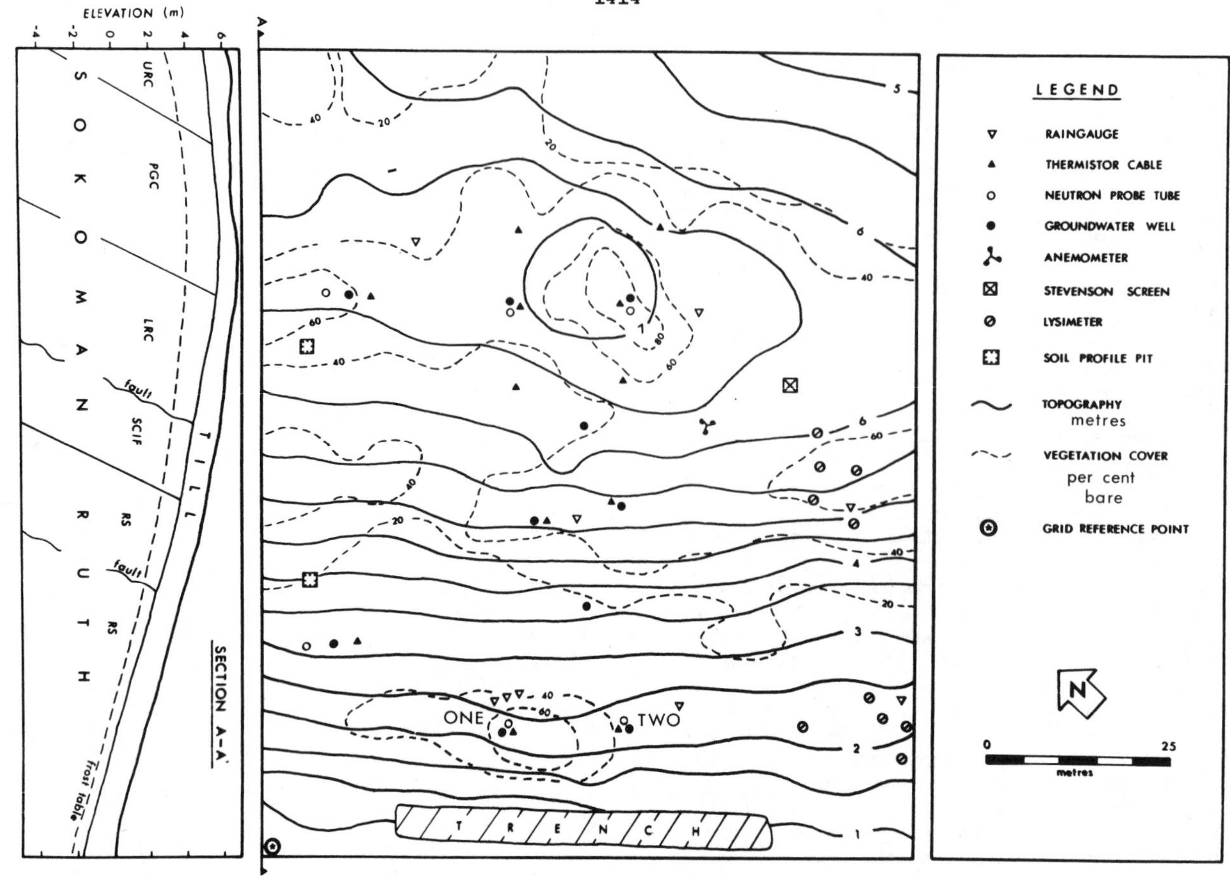

Figure 2 Map and cross-section of Hematite, the main research site. Geological data courtesy IOCC.

RESULTS

Freeze-up

In the Schefferville region, when the surface of the mineral soil freezes, an essentially impermeable barrier is formed between the atmosphere and the active layer. The development of a well-defined zero-curtain was observed at all cables at Hematite in 1977. Zero-curtain temperatures observed during freeze-up ranged from -0.02°to -0.08° C with the most frequently observed value being -0.06° C.

The magnitude of the freezing point depression increases as the solute content of the soil water increases and as the grain size of the matrix decreases. It can be shown that (in the active layer) the effect of solutes in the pore water is trivial, but it is difficult to accurately characterize the effect of a given grain size distribution. The values observed at Hematite are slightly lower than the values observed at Timmins 4 (Nicholson 1978), but the slight difference is probably due to instrumental error, though it may also be partially attributable to the deeper, finer-grained till cover at Hematite.

The unsaturated part of the active layer is usually at or near field capacity at the beginning of freeze-up because that period tends to be cool and moist in central Ungava. In addition, late season rainfall is usually augmented by the melting of an early snowpack and that may lead to a large soil moisture content as freeze-up commences. It seems unlikely, however, that freeze-up could be completed with soil moistures in excess of field capacity unless the drainage of the soil were poor or nonexistent. For example, the complete melting of a 15 cm snowpack and heavy rainfall (approximately 20 mm) in late October 1977 produced a saturated zone up to 2 m thick at Hematite, but, despite a relatively rapid freeze-up, the saturated zone quickly disappeared.

The reduction of the saturated zone during freeze-up is largely due to suprapermafrost groundwater flow, but there is also a minor amount of moisture transport from the lower part of the active layer towards the surface. The upwards transport result from the water potential gradient set up by the freezing process itself. The gradient is caused by the attenuation of the interfacial liquid films on the soil particles as the ice crystals forming within the pores abstract the less tightly-held soil water (Dirksen and Miller 1966). It follows that a finite amount of moisture transport should take place within a freezing soil; such transport has been detected in the laboratory (e.g. Hoekstra 1967, Jame and Norum 1980), but only limited field observations are available (e.g. Woo and Heron 1981).

Field data demonstrating these phenomena are shown in Figure 3, which depicts the moisture and temperature data collected at Hematite from November 1977 to February 1978. The precision of the temperature measurements is quite high (± 0.03°C), but the interpolations between the discrete

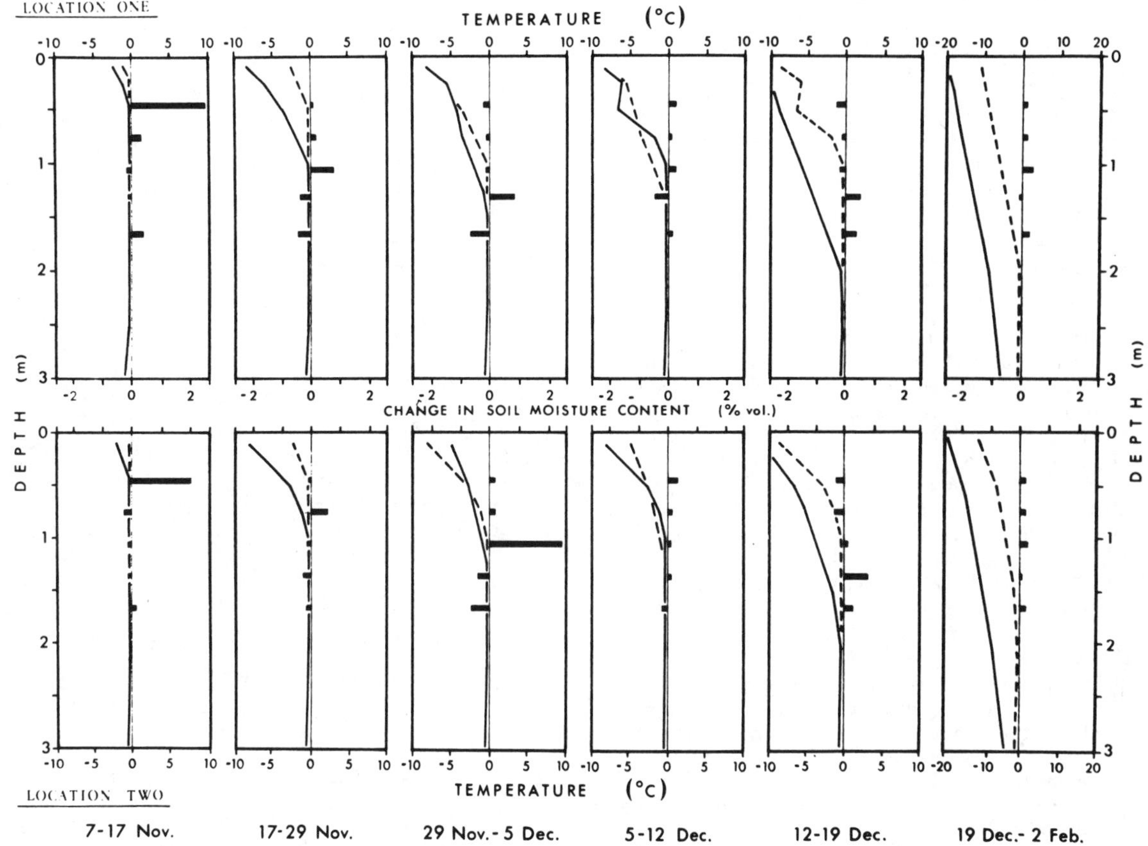

FIGURE 3 Temperature and soil moisture profiles observed at Locations One and Two during freeze-up 1977. Dashed lines represent the temperature profile on the first date indicated while the solid line is for the second date indicated. The bars represent the change in soil moisture content between the two indicated dates.

temperature measurement points are approximate, particularly in the vicinity of the freezing front. The changes in moisture contents, though relatively small, show quite clearly that the moisture transport is correlated with changes in the ground thermal regime. The large transfers took place between 12 November and 5 December and between 12 and 29 December. Both of those periods were characterized by rapidly decreasing air and surface temperatures, leading to strong temperature gradients toward the surface.

The data clearly demonstrate that there is detectable moisture transport during freeze-up and that those changes are closely associated with the ground temperature field within the active layer. The changes in moisture content are presumably the result of the unsaturated flow induced by the increased soil matrix suction developed during the freezing process. However, attempts by the author to numerically model the presumed relationships between soil moisture contents and the temperature field have been unsatisfactory, primarily because of an inability to adequately characterize the properties of the active layer during the freezing process.

Thaw Season

It was initially thought that soon after the ground began to thaw there were three zones within the active layer that possessed significantly different thermal and hydraulic properties: unsaturated unfrozen, saturated unfrozen, and unsaturated frozen. The difference in thermal properties between the saturated and unsaturated unfrozen materials is probably not very large because of the genrally low porosity of the till and bedrock in the Schefferville area, but the higher thermal conductivity and lower specific heat of the frozen ground should give it a significantly higher thermal diffusivity with respect to the unfrozen materials. It follows that the temperature wave should be strongly attenuated with depth, especially in the frozen material. That conclusion is based, however, on the assumption that the transfer of heat is solely by conduction and that there is no significant transfer of heat by moisture transport. The data collected at Hematite during June and July of 1978 clearly demonstrate that those assumtions are not warranted.

Although the phenomena indicating latent heat transfer were observed at all main thermistor cables during 1977, 1978, and 1979, the following discussion is based on the data from Locations One and Two during 1978 since those sites and periods have the most complete data records.

Following the completion of freeze-up in 1977, soil moisture contents remained essentially constant until the following spring. Ground temperature patterns observed during the spring of 1978 were characterized by decreasing amplitude of the temperature wave with depth, indicating that heat transfer was primarily conductive.

In early June, however, when active layer depths were approximately 35 cm and completely saturated, abrupt and substantial increases in temperature were observed at Location Two. Those temperature increases coincided with increases in moisture content of the still-frozen part of the active layer. A week later, a very similar sequence of events was observed at Location One, when percolation from a thick saturated zone led to large moisture increases in the still-frozen ground and those moisture changes were coincident with sharp temperature increases in the still-frozen ground. The amount of percolated water was quite large, equivalent up to 30 mm of water at Location One. The magnitude of the temperature increase was not attenuated with depth, but was instead a function of the initial temperature with respect 0°C (Figure 4). The closer the initial temperature to the freezing point, the smaller was the temperature increase. This is attributed to the sharp increase in the apparent specific heat of frozen materials as the freezing point is approached (Williams 1967).

A related phenomenon was the existence of a zero-curtain throughout the thaw season. Its presence was indicated by a pause (at -0.02°to -0.08°C) in the passage of each temperature measurement point through the freezing point. The transition was frequently observed to coincide with the occurrence of a thick saturated zone in the thawed portion of the active layer. In periods when the saturated zone was thin or absent, the zero-curtain appeared to descend slowly. The zero-curtains were observed at every thermocable, although the duration of a particular temperature measurement point at the freezing point varied.

These phenomena indicate that there is significant penetration of suprapermafrost groundwater into the still-frozen parts of the active layer. When the amounts of water are large, the zero-curtain is well developed, but when the pore pressures at the thawing front are small or negative, only a very small amount of water penetrates the still-frozen material. Presumably, such percolation is self-limiting, since the penetration and refreezing of the water would tend to reduce the effective permeability of the soil.

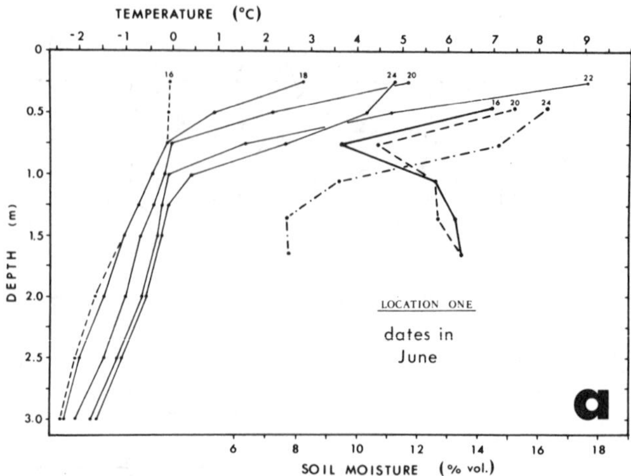

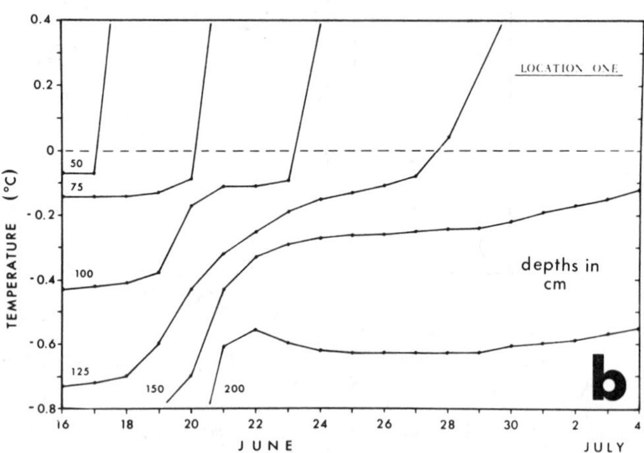

Figure 4 (a) Temperature and soil moisture profiles observed at Location One during June 1978. (b) Temperature patterns through time at Location One during June 1978.

DISCUSSION

It is clear that heat transfer in at least some frozen soils is not solely by conduction and that the traditional view of the active layer is not always adequate; the transition from frozen ground to unfrozen is shown to be gradational in nature both thermally and hydraulically. It follows, then, that simple models of active layer development will not give accurate results unless latent heat transfer is somehow inhibited (e.g., the frozen material is effectively impermeable). Heat budget models based upon time-dependent heat storage will give adequate results only if the zero-curtain zone and the variation in moisture storage can be accurately approximated or if the time interval is sufficiently long that those effects can be neglected.

For example, in non-wetline areas (i.e. those areas that are little affected by suprapermafrost groundwater flow), there is a good correlation between the percentage of bare ground in a given area and the maximum active layer depth (Nicholson 1978), which indicates that conduction of surface heat is the dominant mode of heat transport over an entire thaw season. On the other hand, in the low-lying wetline areas the saturated zone tends to be much thicker, so the pore pressures at the thawing front are much higher than on adjacent slopes. The higher pore pressures, in conjunction with the higher permeabilities of the wetlines, leads to significant amounts of suprapremafrost groundwater percolating to relatively great depths within the still-frozen ground. The influence of latent heat transfer therefore extends well below the zone in which conductive transfer is dominant and the magnitude of the latent heat transfer probably exceeds that of conduction. It is for these reasons that the active layer depths in the wetlines do not correspond to the vegetation distribution at the surface. Accurate prediction of active layer depths along the wetlines requires some measure of the magnitude of the latent heat transfer, either through a budget type of procedure (e.g., Lewis 1977, Wright 1981) or through actual knowledge of the pore pressure distribution within the wetline.

There are two main areas in which the development of the active layer can strongly influence the generation of runoff in subarctic areas. First and most important is in controlling the effective depth to which percolating water can penetrate. The depth of the effectively permeable portion of the active layer strongly controls the distribution of saturated surfaces within a given catchment and that, in turn, controls the rate at which runoff reaches the outlet of the catchment. Thus, the rate and depth of thaw in a given area can influence the response time of that area to a given runoff-generating event.

The second area in which development of the active layer affects the water balance is through the percolation of suprapermafrost groundwater into the still-frozen part of the active layer. As noted above, the amounts percolated at Location One equalled up to 30 mm of water. If the same processes take place along the wetlines, the amount of abstracted water could be quite significant. That such percolation does take place is indicated by the very deep wetlines themselves. Water stored in this fashion would presumably be released only when the depth of thaw had reached the depth of the percolation. Attempts to incorporate such an effect in subarctic water balance models have been partially successful (Wright 1982). Good correlation is achieved between observed and predicted runoff volumes, but few data are available on subsurface moisture storage over the course of a thaw season, so it is difficult to assess the realism of such a model.

CONCLUSION

The results presented here constitute strong qualitative evidence of the importance of latent heat transfer to the development of the active layer. It is clear that such effects can significantly affect the rate and depth of thaw and, indirectly, the rate of runoff generation. The rate and depth of thaw control the rate at which runoff is generated and, in turn, patterns of moisture flow and storage influence the thermal behavior of the ground.

These results represent only the first step in quantifying the thermal and hydrologic behavior of the active layer. The prime needs at this time are for a more detailed picture of of the distribution of heat and moisture within a wetline over an entire thaw season and the development of a model which can adequately describe the three-dimensional flow of heat and moisture within a partially thawed active layer. It may well be that the present numerical models of these processes can be adapted to the far less homogeneous conditions of the field. That will certainly require a more complete description of the temperature and moisture distributions within a wetline, a project which the author has recently begun in central Ungava.

ACKNOWLEDGMENTS

The author would like to thank the Iron Ore Company of Canada, McGill University, and the governments of Quebec and Canada, all of whom contributed to the success of this research. Thanks also to N. Bazilchuk, D. Barr and many others at the McGill Lab at Schefferville.

REFERENCES

Barr, D.R. and Wright, R.K., 1981. Selected climatological data, 1955-1980, for the Schefferville 'A' station. McGill Subarctic Research Paper, 32, 117-134.

Bazilchuk, N.R., 1979. The water balance of a small high subarctic basin near Schefferville, Quebec. unpub. B.Sc. thesis, McGill University, 139 pp.

Dirksen, C. and Miller, R.D., 1966. Closed system freezing of unsaturated soils. Soil Sci. Soc. Amer. Proc., 30, 168-173.

Hoekstra, P., 1967. Moisture movement towards a freezing front. Int. Ass. Sci. Hydr., Publ. No. 78, 411-417.

Jame, Y.W. and Norum, D.I., 1980. Heat and mass transfer in a freezing unsaturated porous medium. Water Resources Res., 16, 811-819.

Lewis, J.S., 1977. Active layer depths and suprapermafrost groundwater in a small subarctic catchment, Schefferville, Quebec., unpub. M.Sc. thesis, McGill University, 160 pp.

McGaw, R.W., Outcalt, S.I., and Ng, E., 1978. Thermal properties and regime of wet tundra soils at Barrow, Alaska. Proc. 3rd Int. Conf. on Permafrost, Nat Res. Council, Ottawa, 47-53.

Nakano, Y. and Brown, J., 1972. Mathematical modeling and validation of the thermal regimes of two tundra soils, Alaska. Arctic and Alpine Res., 4, 19-38.

Nicholson, F.H., 1978. Permafrost distribution and characteristics near Schefferville. Proc. 3rd Int. Conf. on Permafrost, Nat. Res. Council, Ottawa, 427-433.

Smith, M.W., 1975. Numerical simulation of microclimatic and active layer regimes in a high arctic environment. Arctic Land Use Res. Program, Dept. Indian and Northern Affairs, Publ. No. 74-75-72, 29 pp.

Williams, P.J., 1967. Properties and behaviour of freezing soils. Norwegian Geotech Inst., Publ. No. 72, 119 pp.

Woo, M.K. and Heron, R., 1981. Occurrence of ice layers at the base of High Arctic snowpacks. Arctic and Alpine Res., 13, 225-230.

Wright, R.K., 1981. The water balance of a lichen tundra underlain by permafrost. McGill Subarctic Res. Paper, 33, 109 pp.

Wright, R.K., 1982. Modeling the thaw season runoff in Nouveau-Quebec. Naturaliste Canadienne, 109, 469-479.

A STUDY OF THERMAL CRACKS IN FROZEN GROUND

Xia Zhaojun

Department of Hydraulic Engineering, Northeast Agricultural College
Harbin, Heilongjiang, People's Republic of China

The formation of thermal cracks damages roads and hydraulic construction and causes seepage and other problems. Thermal cracks usually occur in early winter, when the snow cover is still thin and the surface temperature falls rapidly to -20°C or lower, and they may reach 1.5 m in depth and 10 to 16 cm in width. The particle composition, water content, temperature gradient, freezing depth, and freezing penetration rate of soil as well as the snow cover and vegetation all affect crack formation. Thermal cracks occur when the maximum contractional stress σ_{max} reaches the ultimate tensile strength σ_0^ℓ of the soil. The distance between cracks is directly proportional to σ_0^ℓ and inversely proportional to the temperature gradient and linear contractional coefficient of frozen soil. The width of the crack is proportional to the distance between cracks and to the temperature gradient. The depth of the crack during one freeze-thaw cycle is proportional to the freezing depth, and is related to the minimum temperature and the temperature when the crack occurs. The main stresses can be determined and used to predict crack formation.

For small-scale hydraulic engineering facilities, such as pools and reservoirs, elevated canals and ditches, or embankments, frost cracks can be the cause of problems such as spring seepage, scouring, soil flows, and piping. This is especially obvious in the case of irrigation and drainage systems, at the level of ditch crossing or of culverts under elevated ditches, where the increase in contact plane reduces the cross-section area of the upper structure, the possibility of frost crack forming at the top of the lower structure increases and subsequent risks of leakage or even collapse also increase. Therefore, the study and analysis of frost cracks is necessary to help us acquire the theoretical and practical knowledge for predetermination, detection, and prevention. Only on this basis can concrete measures be developed, including selection of less-susceptible, environment-compatible building materials, control of pre-frost water content, and local thermal insulation to reduce thermal and mechanical frost-period factors. The objective of this research is therefore to lessen frost damage and to ensure the safe operation of civil structures under severe climatic conditions.

CHARACTERISTICS OF CRACK FORMATION

In the winter, under the influence of subzero temperatures, a portion of the water in the ground turns into ice, which causes an increase in volume and frost heave phenomena. After the completion of this phase-change, i.e. when the unfrozen ground has turned into frozen ground, if the air temperature continues to fall, the ground will be affected by unhomogeneous contractional deformations and contractional stresses, which will be largest at the soil surface.

When surficial stresses are higher than the tensile strength of the frozen ground, the ground surface is pulled apart and cracks appear. We call such cracks "frost cracks", or "cracks", as they will be referred to in this paper.

The curve in Figure 1 shows the relationship between the contractional stresses in frozen ground and the depth of freezing. For ease of analysis and calculation, we can in a first step reduce this relationship to a primary straight-line function. The curve also shows that the distribution of the contractional stress as

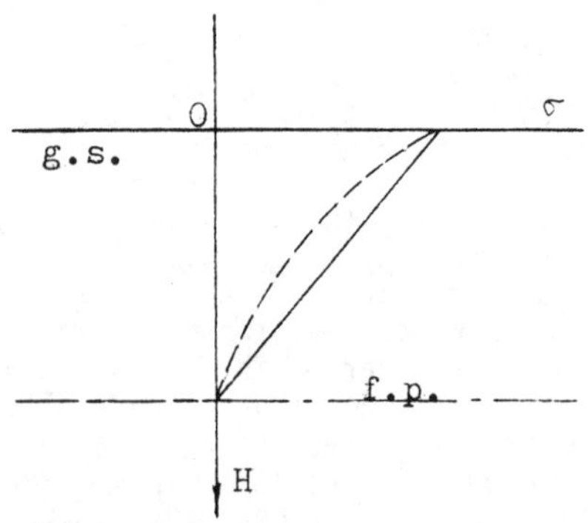

FIGURE 1 Contractional stress distribution along depth.

a function of depth of freezing takes the form of a triangle with its apex down, the contractional stress being zero on the freezing plane in the ground and reaching its highest value at the ground surface.

In areas of seasonally frozen ground, frost cracks occur mostly in the beginning of winter, when the snow cover is thin and ground temperature falls rapidly to -20°C to -25°C. They can reach 1.5 m in depth, 10 to 16 cm in width, and they are wedge-shaped, with the opening wider than the bottom. They occur more frequently on roads and runways and around canals, dams, and embankments, whereas they are rarely found under thick snow covers.

In homogeneous ground, when the boundary conditions make frost penetration unidirectional in half space, intersecting frost cracks are usually at right or close-to-right angles. As the subzero temperatures keep falling, and the number, depth, and length of the cracks increases, and the soil surface might be hacked into blocks with irregular sides.

Cracks in seasonally frozen ground are not as frequent and their development, in terms of both size and continuity, is not as complete as in permanently frozen ground where, due to longer freezing cycles and greater temperature gradients, cracks appear in larger numbers, are deeper and wider, and form more developed networks. This is especially evident in mountainous areas where, in addition to vertical cracks, one can also observe horizontal, or slanting cracks; the intersection of these cracks often creates favourable conditions for avalanches and landslides, especially during melt-water runs in the spring. Even in flat terrain, however, frozen ground cracks can damage roads, highways, and airport runways, or to the points of connection with the ground of hydraulic engineering buildings, as well as to canals, dams, embankments, and buried pipelines.

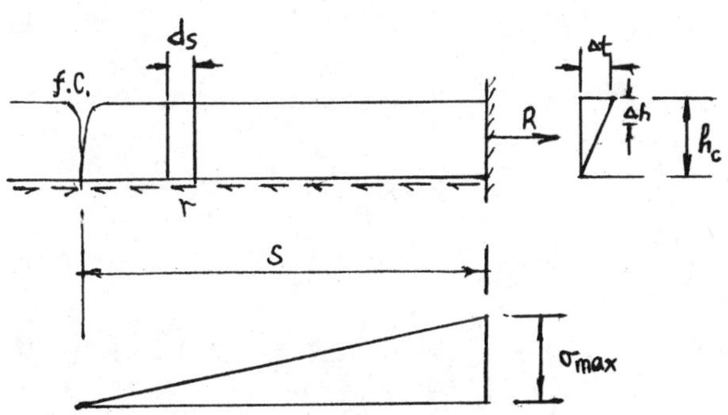

FIGURE 2 Crack spacing and width calculations

CRACK SPACING, WIDTH, AND DEPTH CALCULATION

With the continuous falling of subzero temperatures, contractional stress increases in the superficial ground layer. When the main contractional stress reaches the unidirectional tensile strength limit $[\sigma_0^l]$ of the frozen ground, there is a possibility of cracking. When the former becomes higher then the latter, cracks must occur. After the cracks have been formed, due to the stress concentration as well as to the increase of the frost contact plane and its thermic effects, there is an acceleration of the directional development of the cracks both in depth and in width.

Crack Spacing Calculation

Let us assume an ideal strip of homogeneous ground elongated in the direction of the main stress, subjected to normal motion limitations by the friction between frozen and unfrozen soils, and with a cross-section equal to 1 unit. We sample a segment of minute length ds. When the temperature gradient reaches grad(t), the contractional stress is σ and, in the whole strip, the accumulated contractional stress is:

$$\sigma_{max} = \int_0^S \sigma \cdot ds = \sigma \cdot S \qquad (1)$$

$$\sigma_{max} = S \cdot E \cdot \alpha \cdot (h_c - \Delta h) \cdot grad(t) \qquad (2)$$

where

$$grad(t) = \frac{\Delta t}{h_c} = \frac{t_2 - t_1}{h_c} \qquad (3)$$

When $\sigma_{max} > [\sigma_0^l]$ superficial ground breaks and cracks occur. Let us now take the critical state $\sigma_{max} = [\sigma_0^l]$

From the above, we can deduce the basic formula to calculate the distance between two cracks;

$$[\sigma_0^l] = n \cdot S \cdot E \cdot \alpha \cdot (h_c - \Delta h) \cdot grad(t) \qquad (4)$$

or

$$S = [\sigma_0^l] / n \cdot E \cdot \alpha \cdot (h_c - \Delta h) \cdot grad(t) \qquad (5)$$

where s is the distance between two cracks;

n is the modifying factor for soil type, water content , vegetation and snow cover. Though its value is

$n \leq 1$, assume a value $n = 1$ for rough calculations;

α is the linear contraction factor of frozen soil;

$[\sigma_o^L]$ is the unidirectional tensile strength limit of the frozen ground;

S is the crack spacing

grad(t) is the temperature gradient.

We can also obtain;

$$S = \tfrac{1}{2} \cdot n \cdot E \cdot \alpha \cdot h_c^2 \cdot \text{grad}(t) \,/\, [\tau] \qquad (6)$$

where h_c is the depth of the crack of frozen ground;

$[\tau]$ is the shearing strength limit of the frozen ground.

Crack Width Calculation

By width, or width of opening, we designate the longest distance between the two sides of one frost crack measured at the surface. To any one crack, it is a function of the temperature gradient; if the temperature gradient is a constant, the crack width is a constant as well. When the distance between two cracks is known, the width of the opening can be obtained from the following formula:

$$\Delta L_{max} = n \cdot L \cdot \text{grad}(t) \cdot (h_c - \Delta h) \cdot \alpha \qquad (7)$$

where ΔL_{max} is the maximum width at the ground surface;

L is the distance between two cracks.

Crack Depth Calculation

In the winter, with the continuous decrease of subzero temperatures, there is also a continuous deepening of frost cracks

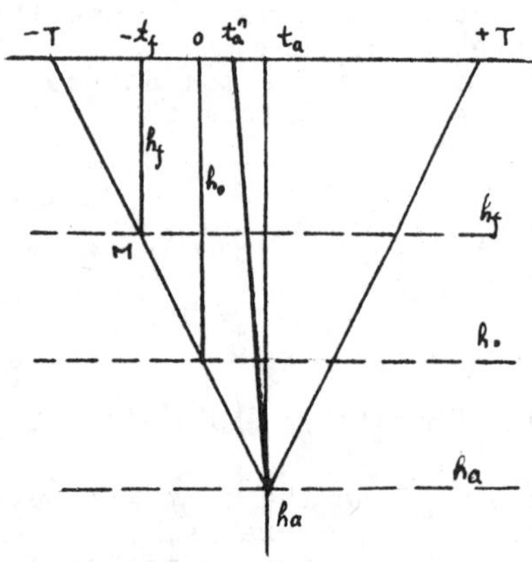

FIGURE 3 Crack depth calculation

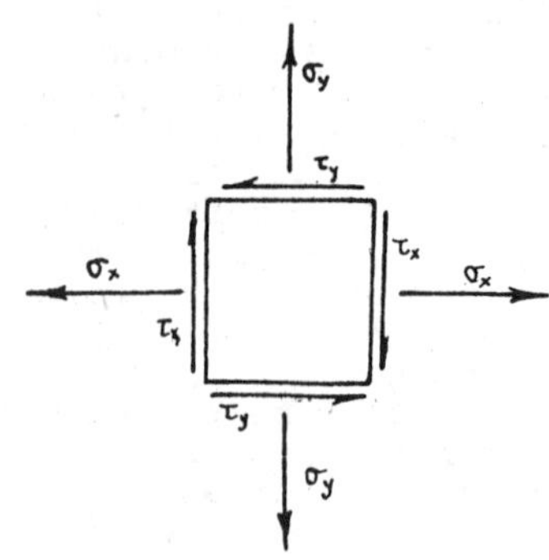

FIGURE 4 Stresses acting on element.

and at a certain depth from ground surface, the temperature of the frozen ground layer keeps decreasing, which causes new tensile stresses. When this layer reaches the breaking temperature of the soil surface, it is subjected to the same cracking and thus the cracks keep developing in depth.

In order to study further the question of crack depth, we have first drawn a graph to show, for analytical purposes, the correlation between the average depth of the active layer calculated over years and temperature (Fig. 3).

In the case when such a graph is drawn for an area of seasonally frozen ground, we have first indicated on the graph the temperature of cracking $-t_c$, we have drawn a vertical line which meets at M the oblique linking point -T to h_a. Therefore, the length of the line from $-t_c$ to M is the maximum depth of the crack. From Figure 3, we can obtain:

$$h_c = \frac{-T - (-t_c)}{-T} \times h_o \qquad (8)$$

where h_c is the depth of the crack;

$-T$ is the lowest temperature;

$-t_c$ is the temperature at which cracks occur;

h_o is the depth of the winter frozen layer.

Since
$$h_o = \sqrt{\frac{\lambda^+ \cdot t_F \cdot \tau}{Q_L}} \qquad (9)$$

where λ^+ is the heat conduction coefficient of freezing ground;

t_F is the temperature at which the ground surface freezes;

τ is the longest frost

period;

Q_L is the phase-change temperature.

The above relationship can also be written

$$h_c = \frac{-T-(-t_c)}{-T} \cdot \sqrt{\frac{\lambda^{+} \cdot t_F \cdot \tau}{Q_L}} \quad (10)$$

To simplify, we can use $\xi = \frac{-t_c}{-T}$ and write the above formula as follows:

$$h_c = B \cdot (1 - \xi) \cdot h_o \quad (11)$$

where $\xi = \frac{-t_c}{-T}$ (12)

B is the modifying coefficient for which we can assume a value B = 1 for rough calculations.

MECHANICAL ANALYSIS OF FROST CRACKS

The width, depth, and spacing of frost cracks are related to the intensity of thermal propagation, i.e. thermal currents intersecting isotherms at right angles. We take a small cubic sample unit at the ground surface, with its sides parallel to the thermal currents and to the isotherms. The stresses to which the sample unit is submitted during the process of ground surface freezing are shown in Figure 4.

These stresses vary according to the distance to the surface and to temperature as well as to soil conditions. In figure 4, we can also see that the stresses are oriented along two perpendicular axes. There are 2 groups of pulling stresses (σ_x and σ_y) and 2 groups of shearing stresses (τ_x and τ_y), with $\tau_x = \tau_y$.

In such conditions, stress intensity can be obtained by the method illustrated in Figure 5 (stress round calculation method). From Figure 5, we can obtain

$$\sigma_1 = \frac{\sigma_x + \sigma_y}{2} + \sqrt{\left(\frac{\sigma_x - \sigma_y}{2}\right)^2 + \tau_x^2} \quad (13)$$

$$\sigma_2 = \frac{\sigma_x + \sigma_y}{2} - \sqrt{\left(\frac{\sigma_x - \sigma_y}{2}\right)^2 + \tau_x^2} \quad (14)$$

The angle between the main stresses σ_1 and σ_x is

$$\alpha = \frac{1}{2} \operatorname{arctg} \frac{2\tau}{\sigma_y - \sigma_x} \quad (15)$$

where σ_1 is the first main stress;

σ_2 is the second main stress.

As to the value of σ_x and σ_y, it suffices to measure arbitrarily two perpendicular ground surface contractional stresses, calling one σ_x and the other σ_y (Usually the greater of the two is called σ_x and the lesser σ_y); then, by using the above method and formulas, calculate the position, orientation, and intensity of the maximum and minimum stresses. Then, by comparing the maximum stress thus obtained and the maximum tensile strength under the same temperature conditions, it is possible to determine whether or not a frost crack will occur and, further, to predict its direction and size.

Using the above stress-determination method, the following cases are usually encountered:

a. The measured values for the perpendicular stresses σ_x and σ_y are not equal. This is the most common case.

b. The measured values for σ_x and σ_y are equal; in that case, the value of the main stresses σ_1 and σ_2 is as follows:

$$\sigma_1 = \sigma_x + \tau_x \quad (16)$$

$$\sigma_2 = \sigma_x - \tau_x \quad (17)$$

c. In the two measured stresses, one is equal to zero, that is $\sigma_x = 0$ or $\sigma_y = 0$, but not the other; in that case,

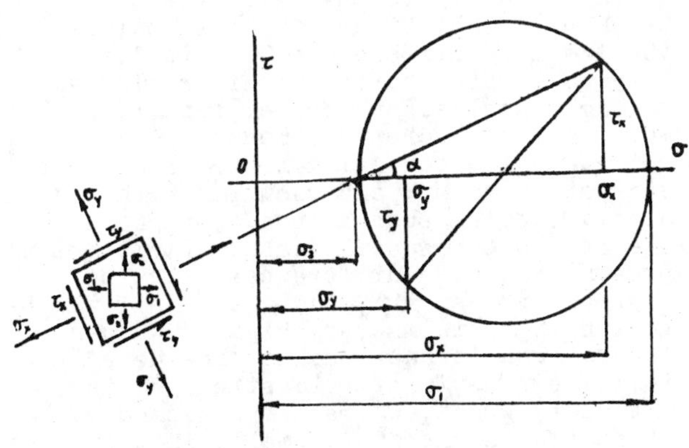

FIGURE 5 Graphic analysis of the main stresses in frozen ground in homogeneous half-space conditions

σ_1 and σ_2 are determined as follows:

$$\sigma_1 = \frac{\sigma_x}{2} + \sqrt{\left(\frac{\sigma_x}{2}\right)^2 + \tau_x^2}$$

$$\sigma_2 = \frac{\sigma_x}{2} - \sqrt{\left(\frac{\sigma_x}{2}\right)^2 + \tau_x^2}$$

The angle between the main stresses σ_1 and σ_x is

$$\alpha = \frac{1}{2} \operatorname{arctg} \frac{-2\tau}{\sigma_x}$$

As to the values σ_x and σ_y are equal, the angle between the main stresses σ_1 and σ_x is

$$\alpha = \frac{1}{2} \operatorname{arctg} \frac{-2\tau}{\sigma_x - \sigma_x} = 45^\circ$$

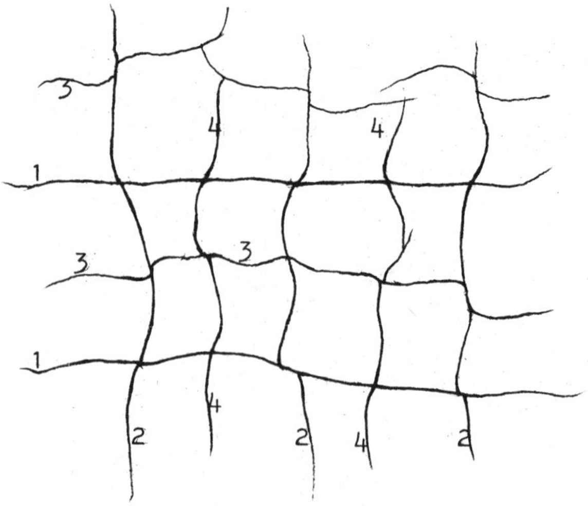

It is also worth noticing that in the process of frost-induced contraction, when the main stresses reach or exceed the maximum limit of tensile strength of the frozen ground, the ground breaks and cracks appear. The cracks thus formed are perpendicular to the first main stress and cause a sudden release of the first main stress. At that time, the second main stress, perpendicular to the first, is still building: a substantial change then occurs and this second main stress becomes the first main stress and keeps growing until a crack perpendicular to it is formed. As the process is repeated, it leads gradually to the development of crack networks. Since the angle between the two main stresses is 90°, the cracks they cause also cut at right or close-to-right angles, the former conforming more to the theoretical model.

When the composition, water content, vegetation cover, and snow layer thickness of civil works such as roads, runways, elevated ditches and canals, or embankments present tangible differences with the surrounding terrain, and when their width-to-length ratio is very high or, under certain conditions, can be considered infinite, the first main stress σ_1 is extremely great, whereas the second main stress σ_2 is extremely small. In that case, one may consider the tensile strength as unidirectional. At the same time, the first main stress σ_1 is basically oriented along the length axis of these civil works and the frost cracks that occur are nearly all perpendicular to it.

FIGURE 6 The formation of frost networks. The numbers represent the orders of frost-crack formation.

CONCLUSION

For some small-scale projects of hydraulic engineering such as pools and reservoirs, elevated canals and ditches, or embankments, the thermal cracks of frozen ground are the major sources resultiong soil current and piping in embankment in the spring, especially for the crossing projects of hydraulic engineering of agriculture.

The formation of frost cracks can be detrimental to construction projects in cold regions. In China, cases where construction workers neglected to take adequate protection measures against frost have been observed. The subsequent formation of cracks has resulted in fissures up to 3 cm wide in the ground floor and basement floors, necessitating major repairs. This is not only waste of time and capital, but has a detrimental effect on the building's life span; it can be avoided by a careful study of the frost crack susceptibility of the ground and materials and by taking appropriate preventive measures at the planning stage.

REFERENCES

Xia Zhaojun, 1980, A Study on Determining Elastic Modules of Frozen Ground in Field Conditions, in Proceedings of National Conference on Hydraulic Engineering: Harbin, China.

Geographical Society of China, 1981, Proceedings of 2nd National Conference on Permafrost: Lanzhou, China,

DEVELOPMENT OF PERIGLACIAL LANDFORMS IN THE NORTHERN MARGINAL REGION OF THE QINGHAI-XIZANG PLATEAU SINCE THE LATE PLEISTOCENE

Xu Shuying, Zhang Weixin, Xu Defei, Xu Qizhi, and Shi Shengren

Department of Geology and Geography, Lanzhou University
People's Republic of China

Based on field investigations and C^{14} dating, there have been three cold periods since the Late Pleistocene: (1) the main Würm Glaciation, when the air temperature was about 7° to 8°C lower than at present, and the lower limit of the periglacial belt was at 2200-2300 m a.s.l.; (2) the second stage of the Middle Holocene, when the air temperature was about 2.5°C lower than at present, and the lower limit of the periglacial belt was at about 3300 m a.s.l.; and (3) the Neoglaciation at the end of the Middle Holocene, when the air temperature was about 2.7°C lower than at present, and the lower limit of the periglacial belt was at about 3200 m a.s.l. Between these three cold periods, there were warm stages with air temperatures 1°-2°C higher than at present, and the periglacial lower limits were at about 3500 to 3900 m. During the cold periods, many periglacial phenomena were formed, namely frost-heave mounds, sand wedges, involutions, blockfields, spline cryoplanation terraces, and sorted and nonsorted polygons. The modern periglacial belt can be divided vertically into three zones: the zone of frozen weathering and of freeze-thaw denudation in the upper part, the zone of freeze-thaw creep and frost heaving in the middle, and the zone of frost heaving and heat thawing in the lower part.

The northeast marginal region of the Qinghai-Xizang plateau refers to the peripheral area bordering the northeast of the plateau. It covers the eastern part of Qinghai plateau, the eastern section of Qilian mountains and the western marginal region of Gansu Centre Basin, etc. (Fig. 1)

The northeast boundary of permafrost on the plateau crosses the northern hillside of Ela mountain at the eastern end of Kunlun mountain at elevation of 3,800-4,000 m. In this region, only Qilian mountains, western Qinling mountain and other areas of high altitude where permafrost develops belong to the contemporary periglacial belt. This region was the area of both altitude and latitude periglacial zones during the Quaternary Ice Age. Many former periglacial phenomena are left in the present-day non-periglacial belt. Some fossil periglacial features were formed under colder climatic conditions than currently experienced in the modern periglacial belt.

PERIGLACIAL LANDFORMS OF MAIN WÜRM GLACIATION OF LATE PLEISTOCENE

The Würm glaciation of the late Pleistocene can be divided into two substages: early Würm glaciation and main Würm glaciation. The records of seafloor sediments and levels of the Yellow and East China Seas both prove that the climate of the main Würm glaciation was colder. During this glacial age, the snowline of the mountain areas dropped to about 1,500 to 1,700 m in the mountain areas around the Qinghai-Xizang plateau. The precipitation today is abundant, with 900-1,000 m in the general continental glacier areas, (Zheng Benxing et al., 1981). Field observations indicate that the snowline drop in Lenglongling mountain amounted to 800 m. Glaciation was extensive (Shi Yafeng et al., 1981). In the main Würm glaciation,

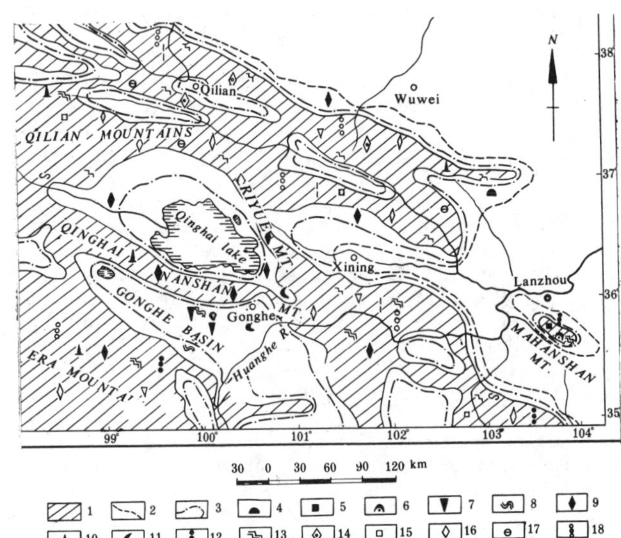

Fig. 1. The distribution of periglacial landforms northeast marginal region of the Qinghai-Xizang plateau. 1. modern periglacial belt; 2. periglacial belt low limit of main Würm-glaciation; 3. periglacial belt low limit of neo-glaciation; 4. fossil pingos; 5. fossil blockfields; 6. fossil nivation hollows; 7. fossil ice wedges; 8. fossil involutions; 9. fossil patterned ground; 10. fossil tors; 11. fossil periglacial sand dunes; 12. fossil stone streams; 13. fossil freezethaw mud-flow terraces; 14. stone circles; 15. blockfield; 16. patterned ground; 17. heat-thawing hollows; 18. stone streams; 19. stone stripes; 20. nivation hollow and freeze-thaw debris cone; 21. involutions; 22. freeze-thaw mud-flow terraces.

the altitudes of fossil cirques on the northern slope of Nanshan mountain, Qinghai, was 3,800 to 3,900 m. This is 900 to 1,000 m lower than the calculated modern climatic snow line. Such lowering of the snow line is bound to bring about the lowering of the periglacial belt. As a result, wide areas remain of the main Würm glaciation in this region. Descriptions of the periglacial land forms of the Gonghe Basin and the foot hills of the Maomao and Mahan mountain since the main Würm glaciation are as follows.

The Gonghe Basin

The area below about 3,000 m in the basin is non-periglacial. Wang Jiyu et al (1979) reported involutions of late Pleistocene age at Taggemu in the centre of the basin. Recently we have found sand wedges on the surface of Tara 1, the high terrace of the Yellow River in the basin. The altitude there is 2,950 m. The sand wedges formed in layers of alluvial gravel with loess. The wedges are filled mainly with gravel. The enclosing material is 17,300±250 years B.P. proving that the wedges were the products of the main Würm glaciation.

In the Gonghe Basin, there are two different kinds of sand dunes of different ages. The older kind developed extensively on the Tamai terrace which is on the left bank of Qiabuqia river. This terrace was formed at approximately the same time as Tara 1. The base of the terrace consists of layer of river-lake facies, the Gonghe group, which was formed during the early-middle Pleistocene. It is covered with layers of alluvial gravel and loess. On top of the loess layer are older sand dunes whose tops have developed a fossil soil layer. C-14 dating of this layer is 6,180± 80 years B.P. It is, therefore, the product of the high temperature period of the mid-Holocene. Based on this, the older sand dunes are inferred to be the result of periglacial wind action of the Würm glaciation. At that time, fossil sand dunes were formed in many other parts of the world, (Goudie 1981; Xu Shuying et al. 1982).

The Maomao Mountain Foot Hills

Maomao mountain, the eastern extension of Lenglongling mountain, is 3,949 m above sea level. The lower boundary of the periglacial belt is about 3,700 m above sea level. At the foot of the Maomao mountain is Songshantan, a faulted basin of Quaternary age. In the south is the Shihuigou brook, flowing south into the Zhuanglang river. Near Hongshanzui, two terraces of the brook are well developed. The second terrace is 10 to 15 m above the river level. Below this terrace is a gravel layer with horizontal bedding. The overlying layer of alluvial loess contains several thin humic, dark-gray layers. A thicker layer in between arches upward to form a buried fossil pingo whose horizontal length is 25 to 30 m. The height is about 5 m. The altitude there is 2,540 m, 1,160 m lower than the lower boundary of the contemporary periglacial belt. The layer over this deformation is 31,100±1,500 years B.P. This proves that the fossil pingo may have been formed during the main Würm glaciation.

Mahan Mountain

Mahan mountain, located on the southern part of the Lanzhou Basin, is 3,670 m above sea level. The mean annual air temperature at the summit is about -2°C. The extent of the present-day periglacial belt is small and periglacial processes are not well developed. According to temperature measurements there may be no permafrost present. The existing periglacial processes are mainly freeze-thaw weathering, thaw-freezing flow and frost heaving. Tors and blockfields found on the flat cryoplanation level at the summit, together with cryoplanation terraces and nivation hollows, are all fossil periglacial features. The blockfields in the summit area consist of large blocks 1 to 2 m in size, resulted from the frost shattering of gneissic granite. The necessary condition are extreme cold and a long freeze-thaw period (Embleton 1975). If cracking caused by freezing is at its maximum at -22°C, this proves that the blockfields on top of Mahan mountain may not be the products of the present day climatic conditions. Also, they did not form during the Neoglacial age when the temperature was only -2°C to -3°C lower than today. The blocks are stable with their surface covered with mosses. Some blocks bear frost clefts and experience "frost-heaving action", both of which are probably the results of the modern periglacial climate. Cui Zhijiu (1981a) believes that the altitudes of the blockfields in various places of the Qinghai-Xizang Plateau are generally about 200 to 250 m lower than their contemporaneous snowlines. Thus it can be deduced that when the blockfields on Mahan mountain were formed the snowline should have been 3,800 to 3,850 m. During the main Würm glaciation, Lenglongling mountain, which is 2°C higher in latitude than Mahan mountain had a snowline of 3,600 m and the Qinghai Nanshan mountain, which is 0.5°C higher in latitude than Mahan mountain had a snowline of 3,800 to 3,900 m. Both snowlines agree with that suggested for the Mahan mountain when the blockfields were formed. Therefore it can be presumed that the blockfields on top of Mahan mountain also formed during the main Würm glaciation.

PERIGLACIAL LANDFORMS OF HOLOCENE PERIOD

The Holocene period underwent cold stages. During these cold stages the snowlines and the periglacial lower boundaries in China's western mountain areas lowered between 100 to 300 m. For example, the snowline of Lenglongling mountain was about 200 m lower than today's during the Neoglacial age. Though the periglacial zone in the region has reduced greatly, it was still larger than today's. Therefore many periglacial phenomena formed during the Holocene cold stages are still found in existing non-periglacial areas. Radiocarbon dating indicates two periods of cold stages during the Holocene. The periglacial phenomena formed during the cold stages can be found in the following areas.

The Gonghe Basin

The Shazhuyu River flows into the Dalianhai lake and forms a closed river system. Since the lake level has lowered since the Holocene, and river -lake facies have emerged at the river mouth, groups of small sized sand wedges and involutions are found in the deposits at depths of 0.4 to 0.5 m beneath the ground surface. The enclosing deposits are light yellow and dark gray silt and clay. The wedges are filled with reddish clay and small gravels which are 7,750± 90 years old B.P. The involutions are 0.5 m to 0.8 m under the ground surface. Both the overlying and the underlying layers have very fine horizontal bedding. The involutions assume the form of gently symmetrical folded curves. According to the C-14 dating, the overlying and underlying layers are 7,090±185 and 8,350±100 B.P. years old each. This proves that the involutions were formed in the same period as the sand wedges.

Riyueshan Mountain and Qinghai Nanshan Mountain

Both mountain areas are over 4,000 m above sea level, with the lower boundary of the existing periglacial belt lying between 3,600 to 3,700 m above sea level. On the flat and gentle slope at the pass of Riyueshan mountain (3,450 m) there are extensive areas of non-sorted polygons, 1 to 2 m or 5 to 8 m in diameter. The clefts around the polygons have, however, been filled with post-glacial fine-grained mineral soil. Between 1.5 to 2.0 m under the ground surface is a paleo-soil layer, with rich humus. It is dated at 4,920±80 years B.P. As the non-sorted polygons are over this soil layer, they were developed earlier. Therefore, they are the products of the Neoglacial age. On the gentle, flat summit level of 3,500 m a.s.l. in Qinghai Nanshan mountain where the Qinghai-Xizang Highway passes, there are frost heaving grassy mounds. Some are decaying. No waterlogged areas have been left but instead, grass meadows. 1.8 to 2.3 m under the ground surface is a well developed palaeo-soil layer of subalpine meadow type which is 3,590±90 years old B.P. This proves that the frost heaving grassy mounds were formed during the Neoglacial age.

Qinghai Lake Basin

The level of the Qinghai Lake is 3,196 m above sea level, and the lake basin is generally no more than 3,400 m above sea level. The mean annual air temperature there is 0.9°C to 2.7°C. Though the basin belongs to the modern non-periglacial region, many fossil periglacial remains are found in the lakeshore areas. For example, the valleys near the lake-like Daotanghe, Buhahe, and Heimahe rivers have many non-sorted polygons. The polygons are usually 3 to 4 m in diameter. Besides the polygons there are also freeze-thaw mud-flow terraces and frost heaving grassy mounds in the valleys. Various signs indicate that they are all non-active.

Concentrated pyramid-shaped dunes are distributed on the east bank of the lake from mouth of Ganzihe river in the north to the foot of Manlong mountain in the south. On the watershed between the Datang and Langmahe rivers active dunes have formed over the old dunes which have been already covered by a layer of paleo-soil which is 3,960± 100 years old B.P. At present plants have begun to grow sporadically on these new dunes, which shows that the stage of strong wind action has passed. Therefore, we think that new dunes formed during the Neoglacial age. The history of the Qinghai Lake indicates that during the middle and late Pleistocene the lake level progressively rose. It was not until the Holocene that the lake level began to drop, associated with uplift of the adjacent mountain areas and as the climate became arid. The wind action then began to erode the Holocene lake deposits. Hence the old dunes in Qinghai Lake basin should have been formed during the early cold stage of the Holocene and should be more than 3,960±100 years old.

Mahanshan Mountain

In this mountain the blockfields, tors, and altiplanation platforms were formed during the main Würm glaciation. Other features such as snow-eroded hollows and terraces of freeze-thaw mud flows are the products of the Holocene cold stages. C-14 dating of a paleo-soil from 0.8 m under the ground surface shows it is 1,860±95 years old. It is inferred that snow-eroded hollows may have formed during the Neoglacial age.

THE MODERN PERIGLACIAL BELT AND ITS VERTICAL ZONATION

Since the permafrost on the Qinghai-Xizang plateau is situated in the area of lower-middle altitude, the lower boundary of the permafrost is approximately the same as the mean annual air temperature, which is -2°C to -3°C. Accordingly, based on the -2° mean annual air temperature isotherm and using field data to amend, the distribution of modern periglacial belt is shown in Fig.1. The zonation can be devided into three periglacial belts.

1. The upper belt. The mean annual temperature of part of Datong mountain within this belt is -4.5°C to -5°C and of Tuolai mountain -4.5°C to -8.2°C. The periglacial phenomena are mainly tors, altiplanation platforms, nivation hollows and freeze-thaw debris cones, blockfields, and polygons.
2. The middle belt. The mean annual temperature of the parts of Datong mountain and Tuolai mountain within this belt is -3.5°C to -4.5°C. The periglacial forms are chiefly terraces of freeze-thaw mud flows, tongue of mud flows and frost heaving mottled grounds.
3. The lower belt. The mean annual temperature here is -2°C to -3.5°C. There is comparatively weak frost heaving and freeze-thaw mud flow activity. There is also widespread solar thawing which produces sliding collapse and the formation of small ponds and marshy areas.

PERIGLACIAL HISTORY SINCE MAIN WÜRM GLACIATION

The great number of relic periglacial phenomena found in the northern region of the Qinghai-Xizang Plateau enable discussion of periglacial history of this region.

During the main Würm glaciation the most significant periglacial phenomena are relic frost heaving mounds, relic sand wedges, and relic blockfields.

Washburn (1973) believes that pingos or frost heaved mounds, which indicate the existence of permafrost, develop in polar or subpolar areas where the mean annual air temperature is -2.2°C to -5.6°C. The mean annual air temperature of the areas in the permafrost belt on the Qinghai-Xizang Plateau where pingos extensively develop is -3°C to -5°C. (Wang Shaoling et al, 1981; An Zhongyuan et al, 1980) If we take the mean annual air temperature of -4°C as the necessary temperature under which the frost heaved mounds are able to form, then the difference between the main Würm and the present temperature is 7.5°C.

Another indicator showing the existence of formed permafrost are relic sand wedges. Pewe (1963) believes they exist where the mean annual air temperature is below -6°C to -8°C, while Washburn believes no less than -5°C. In view of the smaller size of the sand wedges in Gonghe Basin during the main Würm glaciation, and taking -5°C as the temperature indicator for their growth, we can see that the air temperature of the Gonghe Basin during the main Würm glaciation was 7°C lower than today's, and the lower periglacial boundary was about 2,400 m above sea level.

As said above, very cold climatic condition must have existed for the formation of the blockfields as found in Mahan mountain. On top of the present Tuolaishan and Zoulangnashan mountains which are 4,500 m above sea level and where the mean annual air temperature is about -5°C, no blockfields are formed with such great blocks as are found in Mahan mountain. In west Datan area of eastern Kunlun mountain, contemparory blockfields composed of blocks of 2 to 3 m in dimensions develop on mountain summit level at 4,950 to 5,000 m above sea level. In the Qomolangma area, they only form on summit levels over 5,800 to 6,000 m. Based on calculations, the mean annual air temperature in these areas is about -10° C. If we take -10°C as the air temperature condition under which the blocks were formed on Mahan mountain, then the amplitude between the main Würm and the present temperature is 8°C. Thus it is estimated that the lower boundary of the periglacial belt at that time was 2,220 m.

The periglacial development and environmental evolution of the Holocene can be devided into the following stages.

The Early Holocene Stage

This stage, 12,000 to 8,000 years ago, was the warm climatic stage at the beginning of the post-glacial age. In Gonghe Basin there are paleo-soil layers widely distributed 2 m under the ground surface of the footslope in front of Qinghai Nanshan mountain. Among them the soil buried near Qanbulu temple is 11,540±150 years B.P. At the place where the Shazhuyu River flows into Dalianhai lake, the top of the thick layer of lake facies which is 5 m above the lake surface is 8,350±100 years B.P. This indicates that during the early Holocene stage, Dalianhai lake experienced a period of high lake levels. The analysis of Holocene spore pollen in lake facies of the Qinghai Lake basin demontrates that the woody plant pollen accounts for 48% of the total. In view of the large quantity of poplus sp. and salix sp., each of which accounts for 16% of the total pollen, and comparing their ecological environment with that of the present day poplus and salix, it is calculated that the mean annual air temperature at that time was about 1°C higher than today. Therefore, the altitude of the lower boundary of the periglacial belt was 160 m higher than present day.

The Middle Holocene Stage

This stage which is 8,000 to 2,500 years ago can be further divided into the following stages:

1. The first cold stage. This stage, which is 8,000 to 7,000 years ago, is a cold fluctuating period. In Daxingan mountain (Xie Youyu et al) and Daihai lake, Inner Mongolia (Zhou Tingru et al) there are involutions which formed between 7 and 8 thousand years ago. Cui Zhijiu (1981b) reported relic pingos in the Kunlun mountain. In the Qinghai-Xizang region, the involutions and sand wedges near the mouth of the Shazhuyu River in the Gonghe Basin and the paleo-periglacial dunes in the Qinghai Lake basin were all produced during this stage. Since the involutions and sand-wedges produced in the Gonghe basin are small, which suggests the climatic conditions were not severe, and since it is also possible that they were formed during a short period of time, their altitude of 2,860 m above sea level may not represent the lower boundary of the periglacial zone of the whole cold stage. Generally speaking, the periglacial dunes are in the external areas of the periglacial belt. Therefore, the altitude of 3,300 m above sea level is tolerance as the lower limit of the periglacial belt. During this cold stage, the temperature should have been about 2.5°C lower than today's, and altitude of the lower boundary 400 to 500 m lower than today's.

2. The high temperature stage. This stage which is 7,000 to 3,500 years ago is period of the middle Holocene when the climate was warm and humid. Peat formed during this stage. The water levels of lakes increased. Thick deposits of lake facies sediments were formed. For instance, the water level of Dalianhai lake in Gonghe Basin was over 10 m higher than today, and that of Qinghai Lake was at least 20 m higher than today. Spore-pollen shows that salix sp. still accounted for 5% of the total amount. The mean annual air temperature during this stage ought to have been about 1.8°C higher than today, and the lower boundary of the periglacial belt about 300 m higher than present day.

3. The second cold stage. This cold stage, which is 3,500 to 2,500 years ago, is also called the Neoglacial stage. The periglacial phenomena formed during this stage are widely distributed and of various types. The snow line value of Lenglongling mountain, which is the eastern part of Qilian mountain, drops about 200 m during this stage and temperature was accordingly about 1.2°C lower than today's. The lowest altitude of periglacial phenomena in the Qinghai Lake basin is 3,200 m a.s.l., the mean annual air temperature was 2.7°C lower and the lower boundary of the periglacial belt ought to have been 500 to 600 m lower than today's.

Judging from this, the Neoglacial age was the coldest stage in the Holocene.

The Late Holocene Stage

About 2,500 years ago the world came into the late Holocene stage where the climate was comparatively warm. With the rise of the lower boundary of the periglacial belt, paleo-soils again developed widely in the non-periglacial belt. The C-14 dating of the paleo-soils in the nivation hollow on Diaoling, Mahan mountain shows that the soil was formed about 1,860±90 years ago. In order to make comparisons, Table 1 sums up the periglacial development and the changes in the altitudes of the lower boundary of the periglacial belt.

TABLE 1: The development of periglacial landforms and the altitude changes of periglacial belt in the northeastern region of Qinghai-Xizang Plateau since main Würm glaciation.

Age			Events	Age of Formation (Years B.P)	Altitudes of periglacial boundary (m a.s.l.)	Temperature change	Elevation change of lower boundary of periglacial belt
Holocene period	Late Holocene		Paleosoil on Diaoling of Mahanshan	1860±95	3,700-3,800		
Holocene period	Middle Holocene	Neoglaciation	Frost heaving grassy mounds on Qinghai-Nanshan	3,590±90 after		Dropped temperature about 2.7°C	Fell about 500-600 m
			Periglacial sand dunes on the east bank of Qinghai Lake	3,960±100 after			
			Nonsorted polygons, solifluctions and frost heaving grassy mounds on the banks of Qinghai Lake	?			
			Nivation depressions and solifluctions on Diaoling of Mahan mountain	1,860±95 ago	3,200±		
			Nonsorted polygons on Riyue mountain	4,920±80 after			
			Periglacial sand dunes in Gonghe Basin	4,955±105 after			
		High temperature	Paleosoil of the watershed of Daotanghe River	3,960±100		Increased temperature about 1.8°C	Rose about 300± m
			Paleosoil on the Qinghai-Nanshan	3,590±90			
			Paleosoil of the pass of Riyue mountain	4,920±80	3,500-3,600		
			Paleosoil of the platform of Gonghe Tamai	6,180±80			
		Cold	Sand wedges in Gonghe Shazhuyu River	7,750±90		Dropped temperature about 2.5°C	Fell about 400-500 m
			Sand dunes in the watershed of Daotang River	3,960±100 ago			
			Involutions of Gonghe Shazhuyu River	8,350±100 after	3,300±		
	Early Holocene	Warm	Peat of Gonghe Qangbulu area	11,540±150		Increased temperature about 1.0°C	Rose about 160-180 m
			Layer of lake facies of Gonghe Dalianhai Lake	8,350±100	3,900±		
Late Pleistocene period	Main Würm glaciation		Frost heaved mounds (pingos) of the foot of Maomao mountain	31,100±1,500		Dropped temperature about 7.0°C-8.0°C	Fell about 1,400-1,500 m
			Sand dunes of the Gonghe Basin	17,300±250 after			
			Involutions of Gonghe Tanggem area	?			
			Sand wedges of Gonghe Tala 1	17,300±250	2,200-2,300		
			Blockfields, tors, and altiplanation platform	?			

ACKNOWLEDGEMENTS

The writers are indebted to the C-14 labs of Lanzhou University and of Guiyang Geochemistry Research Institute of the Academy of Science of China for the C-14 dating. The text was edited by H. M. French, University of Ottawa, Canada.

REFERENCES

An Zhongyuan, 1980, Formation and evolution of permanent ice mound on Qinghai-Xizang Plateau Journal of Glaciology and Cryopedology, v. 2, no. 2, p. 25-30 (in Chinese).

Cui Zhijiu, 1981a, Some progress in the research of periglacial landform in China; Journal of Glaciology and Cryopedology, v. 3, p. 70-77 (in Chineses).

Cui Zhijiu, 1981b, Basin characteristics of periglacial landforms in the Qinghai-Xizang Plateau: Scientia Sinica, no. 6.

Goudie, A. S., 1981, Environmental change: London, Oxford Univ. Press.

Lanzhou Institute of Geology, Academia Sinica et al., 1979, Report on Comprehensive investigations in the Qinghai Lake; Beijing, Science Press. (in Chinese).

Pewe, T. L., 1963, Ice wedges in Alaska classification, distribution and climatic significance: Geol. Soc. Am. spec, pap. 76.

Shi Yafeng et al., 1981, Physical geography of China (geomorphology): Beijing, Science Press, p. 202-254 (in Chinese).

Wang Jiyu et al., 1979, Quaternary stratigraphy of the Gonghe Basin in the Qinghai, Geological Review, v. 25, no. 2, p. 15-20 (in Chinese).

Wang Shailing et al., 1981, On the pingos along both banks of the Qingshui River on Qinghai-Xizang Plateau, Journal of Glaciology and Cryopedology, v. 3, no. 3, p. 55-62 (in Chinese).

Washburn, A. L., 1973, Periglacial processes and environments: London, Edward Arnold Ltd.

Xu Shuying et al., 1982, Aeolian sand deposits in the Gonghe Basin, Qinghai Province, Journal of

Zheng Benxing et al., The evolution of the Quaternary glacier in the Qinghai-Xizang Plateau and its relationship with the uplift of the plateau in studies on the period, amplitude and type of the uplift of the Qinghai-Xizang Plateau: Beijing, Science Press. (in Chinese).

A PRELIMINARY STUDY OF THE DISTRIBUTION OF FROZEN GROUND IN CHINA

Xu Xiaozu and Wang Jiacheng

Lanzhou Institute of Glaciology and Cryopedology, Academia Sinica
People's Republic of China

Based on field data and statistical calculations, a distribution map of frozen ground in China has been compiled. It was found that the different types of frozen ground are located in different climatic zones. Permafrost is located mainly in the northern temperate zone, seasonally frozen ground in the middle and southern temperate zones, and intermittently frozen ground in subtropic areas. The values of mean annual air temperature at the lower or southern limit of permafrost depend on the differences between the annual mean ground temperature and the air temperature; these differences vary with districts because of the variation in solar radiation. The altitude of the lower limit of permafrost depends on four factors: latitude, the change rates of annual mean air temperature to latitude and altitude, and the difference between the annual mean ground temperature and the air temperature at that point. Differences of altitude between the snow line and the lower limit of permafrost vary with climatic districts: 1000-1500 m for arid districts, 500-1300 m for semi-arid districts, 250-1000 m for subhumid districts, and 0-400 m for humid districts.

Several papers have discussed the distribution of permafrost in China, but except for one (Zhou and Guo 1982), they deal with particular regions of China.

Based on data from field investigations and using statistical methods, a map of the distribution of frozen ground in China has been compiled at a scale of 1:4,000,000. From this work, it was found that the distribution of frozen ground in China is characterized by specific environmental factors.

HORIZONTAL DISTRIBUTION OF FROZEN GROUND

Frozen ground in China is classified into three types according to its duration, i.e. permafrost, seasonally frozen ground, and intermittently frozen ground. Permafrost can in turn be divided into two zones, a continuous and a discontinuous zone, according to the soil types on the ground surface.

In the continuous zone, permafrost occurs at the ground surface in all soil types, from fine-grained to coarse-grained soil. The area distribution of continuous permafrost is more than 50% in the Da-Xiao Hinggan Ling, northeast China, and more than 70-80% in the high mountains and on the Qinghai-Xizang Plateau.

In the discontinuous zone, permafrost occurs only in peat or in wet fine-grained soil. The area percentage of permafrost in this zone is less extensive than in the continuous permafrost zone.

The criteria for classification of frozen ground are shown in Table 1.

Permafrost is mainly distributed on the Qinghai-Xizang Plateau, in the Pamirs, in the high mountains of western China, and in the Da-Xiao Hinggan Ling and on the tops of high mountains in northeast China, such as Dashi Shan, Changbai Shan, and the Zhangguangcai Ling.

The continuous permafrost zone of the Qinghai-Xizang Plateau can be divided into three subzones according to the continuity of the permafrost in horizontal distribution: a nearly continuous subzone from the northern slopes of the Kunlun Mountains to the southern slopes of the Tanggula Mountains; a wide-spread subzone on both sides of the valley of the Za'gya Zangbo River in Tibet; and a sporadic subzone from the south bank of the valley of the Yarlung Zangbo River to the Himalayas. The discontinuous permafrost zone is narrow in the north

TABLE 1 Classification of frozen ground.

Type	Limiting ground-surface temperatures (°C)	Minimum duration	Mean annual air temperature (°C)
Permafrost	Mean annual temperature <0	≥2 yrs	Continuous -2.4 - -5 Discontinuous -0.8 - -2
Seasonally frozen ground	Mean monthly minimum temperature <0	≥1 month	8-14
Intermittently frozen ground	Extreme minimum temperature <0	<1 month	18.5-22

and wide in both the eastern and southern Qinghai-Xizang Plateau.

The discontinuous permafrost zone in the high mountains of western China is also very narrow, but it is widely distributed in the Da-Xiao Hinggan Ling mountains of northeast China, especially in Xiao Hinggan Ling.

The southern limit of seasonally frozen ground runs from Wakuhe (25°14'N, 97°52'E) northeast to Lianyun Harbour (34°30'N, 119°20'E). Seasonally frozen ground is also distributed on the tops of mountains, such as Dabie Shan, Laiyang Shan, and Yu Shan.

The southern limit of intermittently frozen ground is nearly identical to the Tropic of Cancer. There is no frozen ground south of this line except on high mountains.

The area of each type of frozen ground in China is shown in Table 2. Frozen ground covers 98% of the total area of China, and permafrost covers approximately 20%.

The general distribution of frozen ground shows definite zonal patterns in latitude and altitude (Figure 1). From north to south, as latitude decreases, the type of frozen ground gradually changes from continuous permafrost to intermittently frozen ground. The pattern of this horizontal zonation is extremely clear in eastern China.

From north to south, as altitude decreases, the zonal structure of different types of frozen ground changes. First, there is a two-laminate structure with permafrost and seasonally frozen ground, as in the Tian Shan and the Altai Shan. Then, a three-laminate structure with permafrost and seasonally and intermittently frozen ground occurs, as in the Himalayas, followed by another three-laminate structure with seasonally and intermittently frozen ground and unfrozen ground, such as in the mountains of Taiwan. Finally, a two-laminate structure with intermittently frozen and unfrozen ground appears, as on Mt. Wuzhi on Hainan Island in the Gulf of Tonkin. This pattern of vertical zonation is readily apparent in the western region of China.

TABLE 2 Areal distribution of frozen ground.

Type	Area (10^4 km)	Total area of China (%)
Permafrost	198.8	20.7
Seasonally frozen ground	521.7	54.3
Intermittently frozen ground	229.1	23.9

DISTRIBUTION OF FROZEN GROUND VS. ENVIRONMENTAL FACTORS

Climate

By comparing the distribution of frozen ground to the climate zones, we find that the different types of frozen ground are located in the different climatic zones, and the limits of the frozen ground classifications are nearly the same as the lines of climatic districts (Central Bureau of Climate 1979).

Permafrost is located mainly in the northern temperate zone and in the mountain regions of the middle temperate zone; seasonally frozen ground is found in the middle and southern temperate zones and in the mountain regions of the northern subtropics; intermittently frozen ground occurs in the subtropics and in the mountain regions of the northern tropics.

In Da-Xiao Hinggan Ling, continuous permafrost is located mainly in the humid region; discontinuous permafrost extends through the humid, subhumid, and subarid regions. Permafrost essentially appears in the subarid region of the Altai Shan, extends into the subarid and the arid regions of the Tian Shan, and is located in the arid regions of the Qilian Shan and the southern Qinghai-Xizang Plateau and the subhumid regions of the eastern Qinghai-Xizang Plateau.

Mean Annual Air Temperature

Information published both in China and abroad shows that the mean annual air temperature at the lower (southern) limit of permafrost differs from place to place. This is mainly due to differences in ground-surface temperature, which in turn are the result of differences in solar radiation (Figure 2).

The mean annual air temperature at the lower limit of permafrost in China and in similar areas of the world is shown in Table 3; we can see that the values for discontinuous permafrost in China are basically the same as those in other places, but are higher than in Gongbu Himalaya, in Nepal. This phenomenon can be explained by differences in total annual solar radiation. The mean annual air temperature at the lower limit of continuous permafrost in China is much higher than that of the continuous permafrost in arctic and subarctic regions, because in our classification the distribution of continuous permafrost does not take into account the influence of such factors as slope orientation, snow cover, and local heat resources.

From Figure 2, we can see that the mean annual values of total solar radiation around the world, including America, are within the range of the mean annual values of the western and eastern regions of

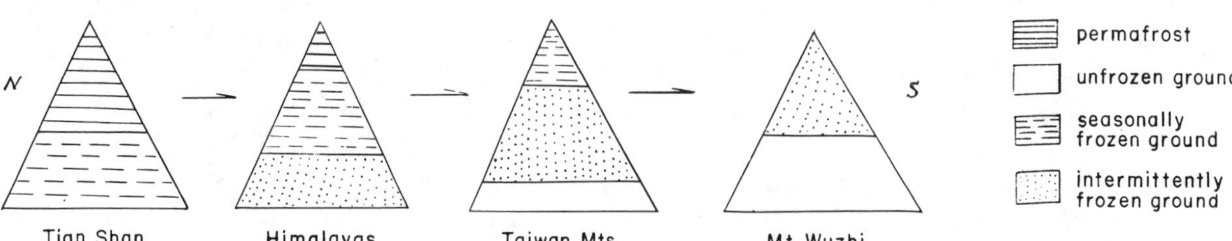

FIGURE 1 The laminate structure of types of frozen ground (after Gorbunov 1978).

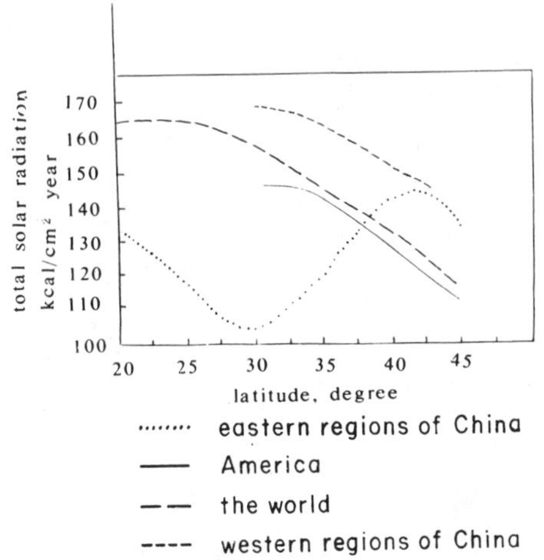

FIGURE 2 Total solar radiation vs. latitude.

TABLE 3 Mean annual air temperature at the lower limit of permafrost, °C.

Region	Types of permafrost	
	Discontinuous	Continuous
Arctic or subarctic	-1.1	-8.3
Canada	0 - -1.0	
Rocky Mountains	-1.0	
Mt. Fujiyama (Japan)	-1.4 - -1.8	
Gongbu Himalaya (Nepal)	-2.8 - -3.4	
China	-0.8 - -2.0	-2.4 - -5.0

(Source: Brown and Péwé 1973).

TABLE 4 Total solar radiation vs. ground-surface temperature.

Station	Location			Soil type	T_a^1	Q^2	T_s^3
	Latitude	Longitude	Altitude				
Mangui	51°59'N	122°06'E	900 m	Peaty clay	-5.3	105	-4.3
Muli	38°15'N	99°12'E	4091.2 m	Peaty clay	-5.6	155	-3.7
Tumen	32°51'N	91°34'E	4930.5 m	Peaty clay	-5.2	195	-3.6

[1] T_a is the mean annual air temperature, °C.
[2] Q is the total annual solar radiation, kcal/cm^2.
[3] T_s is the mean annual ground-surface temperature, °C.

China. Total solar radiation in western China increases with the decrease in latitude, so it is natural that the values in Gongbu Himalaya are higher than in western China.

The differences in total solar radiation result in differences in ground-surface temperature (Table 4). In general, where the mean annual value of total solar radiation is large, the ground-surface temperature is high. Simultaneously, where the mean annual ground-surface temperature is high, and given that all other conditions are the same, the mean annual air temperature must be low if frozen ground is to occur.

To sum up, the altitude of the lower limit of frozen ground depends on four factors: latitude, the ratios of mean annual air temperature to latitude and to altitude, and the difference between mean annual air temperature and ground-surface temperature (or the mean annual air temperature at the lower limit of frozen gound). This can be expressed by the empirical formula

$$H = (A-T-B_1L)/B_2$$

where: H = the altitude of the lower limit of frozen ground, in units of 100 m
A = a constant, controlled by regional climatic conditions, in °C
L = the latitude, in °N
B_1 = the ratio of mean air temperature to latitude, in °C/°lat.
B_2 = the ratio of mean annual air temperature to altitude, in °C/100 m

Using this formula, we obtained the values for the lower altitude limit. The difference in the lower altitude limit between continuous and discontinuous permafrost ranges from 265 m to 600 m. The differences are approximately 300 m in arid mountain regions and 500 m in humid regions.

Snow Line Altitude

The difference between the lower altitude limit of discontinuous permafrost and the altitude of the snow line varies with climatic districts (Table 5). From arid regions to the subhumid, the altitude of the lower limit of discontinuous permafrost is lower than the altitude of the snow line, but in the humid regions this may be reversed.

The differences between the altitude of the snow line and the altitude of the lower limit of the discontinuous permafrost are as follows: 1000-1500 m for arid regions, 500-1300 m for subarid, 250-1000 m for subhumid, and 230-400 m for humid.

TABLE 5 Lower altitude limit of permafrost vs. the altitude of snow line.

Location	Latitude	Longitude	H_s^1	H_p^2	ΔH^3	District
Lhunze	29°30'N	94°30'E	4800-5200	4850	-50-350	Humid
Nu Shan	28°20'	98°35'	5300-5400	5000	300-400	
Yulong Shan	27°03'	100°08'	4500-5000	4730	-230-270	
Kunlun Shan	36°30'	90°-91°	5200-5400	4400	800-100	Subhumid
Queer	32°	98°50'	5100	4850	250	
Himalayas						
(N. slope)	28°30'	85°40'	6000	4900	1100	Subarid
(S. slope)	28°30'	85°40'	5500	4900	600	
Mt. Everest	28°20'	87°	5800-6200	4900	900-1300	
Mt. Palipaohanli	28°	88°54'	5500	5000	500	
Mt. Nyainqentanglha	30°06'	90°20'	5800	4800	1000	
Mt. Kangrinboqe	31°	81°30'	6000	4750	1250	
Qilian Shan	38°	95°-96°	4400	3400	1000	
	38°	101°	4000	3400	600	
Tian Shan	42°30'	81°-83°	3600	3000	600	
Altai Shan	49°	87°35'	2700	1500	1200	
Tian Shan	44°	83°30'	4000	2700	1300	Arid
	43°	94°30'	4200	2900	1300	
Kunlun Shan	37°	85°	5800-5900	4400	1400-1500	
Tanggula Shan	33°06'	91°	5400-5500	4400	1000-1100	

[1] H_s is the altitude of the snow line, m.
[2] H_p is the lower altitude limit of the permafrost, m.
[3] ΔH is the difference between H_s and H_p, m.

CONCLUSIONS

There are three types of frozen ground in China: permafrost, seasonally frozen ground, and intermittently frozen ground. Of these, seasonally frozen ground is the most common, as indicated by the percentage of area it covers.

The distribution of frozen ground in China follows horizontal and vertical patterns. From north to south, the types of frozen ground vary from permafrost to intermittently frozen ground, and the 2-3-3-2 laminate structures are observable.

The permafrost in the Da-Xiao Hinggan Ling has the characteristics of the high-latitude type, and latitude is the main factor controlling its distribution. In western China, permafrost has the characteristics of the alpine type, and altitude is the main factor controlling its distribution.

The distribution of frozen ground is closely related to environmental factors. The lower altitude limit of frozen ground depends on four factors: latitude, mean annual air temperature, and the ratios of mean annual air temperature to latitude and to altitude.

ACKNOWLEDGMENTS

We wish to express our gratitude to Professor Shi Yafeng, director of our institute, who directed this work, and to Prof. Zhou Youwu and Wu Zhiwan for their many helpful suggestions in the preparation of this paper.

REFERENCES

Brown, R.J.E., and Péwé, T.L., 1973, Distribution of permafrost in North America and its relationship to the environment: A Review, 1963-1973, in Permafrost, The North American Contribution to the Second International Conference, Yakutsk: Washington, D.C., National Academy of Sciences.

Central Bureau of Climate, 1979, Climatic maps of the People's Republic of China: Atlas Publishing House.

Grobunov, A.P., 1978, A survey of alpine permafrost: Arctic and Alpine Research, v. 10, no. 2.

Qiu Guoqing and Zhang Changqing, 1981, Distributive features of permafrost near the Kuyxian Daban in the Tianshan Mountains, Memoirs of Lanzhou Institute of Glaciology and Cryopedology, Academia Sinica: no. 2, Publishing House of Science.

Wang Jiacheng, Wang Shaoling, and Qiu Guoqing, 1979, Permafrost along the Qinghai-Xizang Highway: Acta Geographica Sinica, v. 34, no. 1.

Zhou Youwu and Guo Dongxin, 1982, Principal characteristics of permafrost in China: Journal of Glaciology and Cryopedology, v. 4, no. 2.

Xu Xiaozu and Guo Dongxin, 1982, Compilation of the distribution map of frozen ground at 1:4,000,000 in China: Journal of Glaciology and Cryopedology, v. 4, no. 2.

DETERMINATION OF ARTIFICIAL UPPER LIMIT OF CULVERT FOUNDATION IN PERMAFROST AREAS OF THE QINGHAI-XIZANG PLATEAU

Ye Bayou and Yang Hairong

Northwest Institute of the Chinese Academy of Railway Sciences
Lanzhou, People's Republic of China

In permafrost regions, the bearing capacity of foundation soil changes with periodic thawing and freezing. Many shallowly buried culvert foundations have been destroyed by uneven subsidence from thawing and frost heaving. The authors present a method for calculating the safe burial depth of culvert foundations and the approximate maximum thaw depth in accordance with specific features of the permafrost areas of the Qinghai-Xizhang Plateau. This paper presents the calculation of maximum thawing depth, provided thawing of culvert foundation is permitted, and the design principle for keeping the ground beneath a culvert foundation in the unfrozen state. To simplify design, empirical formulas are recommended as reference as well. Comparing the calculated results with data from an experimental culvert in the Fenghou Mountains, the authors give the relation curve of the water-passing surface temperature of an experimental culvert vs. the ground surface and atmospheric temperatures of the local weather station.

The brooks and river valleys on the Qinghai-Xizang Plateau are generally 4,000-5,000 m above the sea level. The distribution of permafrost at small-sized river valleys is continuous horizontally and vertically, and the supraperma-frost table is about 2-3 m. The rainfall in this area reduces gradually from south to north. It is solid precipitation mainly. The precipitation concentrates from 6 to 9 months, among which precipitation in 7-8 months is about 50% of annual precipitation. So time for precipitation is short and strength is great. Thus, the flow in most of the culverts in this area is intermittent stream except for few culverts with constant stream during thawing.

Large numbers of the culverts built in this area before belong to the shallow-buried foundations. Many of them are subjected to failure of different degree. In order to determine correctly the burying depth of culvert foundation and the alternative depth of foundation soil, the law of artificial upper limit of foundations after the culverts built is mainly discussed in this paper.

CALCULATION OF THE MAXIMUM THAWING DEPTH OF SOIL BENEATH FOUNDATION

Derivation of Equations

Under the effect of daily and seasonal variation of atmospheric temperature, ground surface temperature will fluctuate regularly. The simplest pattern of temperature fluctuation appears in the form of sine or cosine, i.e.

$$t_F = \bar{t}_F + A_F \cos\left(\frac{2\pi\tau}{T}\right) \qquad (1)$$

where t_F and $\bar{t}_F$ are the ground surface tempera- and its average (in $^{\circ}$C), respectively; A_F is the amplitude of fluctuation of the ground surface temperature ($^{\circ}$C); τ is the time, in hours; T is the annual fluctuating cycle (T = 8760 hr).

As ground surface temperature varies sinusoidally or according to the law of cosine and without heat source in underground, the temperature at any given depth also varies in the same way. t_F in eq 1 can be substituted by t_x--- temperature at depth x and the average temperature $\bar{t}$ depth x can be assumed as a constant. Let $A_F = A_0$ and draw into temperature wave $\theta = t_x - \bar{t}$, we have, for the depth of perennial ground temperature, i.e. the depth not affected by temperature wave, the lower boundary condition $\theta(x = \infty, \tau) = t_x - \bar{t} = 0$. Then, by eq 1, the upper boundary condition will be:

$$\theta(x = 0, \tau) = A_0 \cos\left(\frac{2\pi}{T}\tau\right).$$

For ground surface under natural conditions, the distribution of initial temperature in ground layer factually does not affect on the temperature field of it, because radiation equilibrium and heat equilibrium act on it cyclically in an unlimited long period of time. Thus, it is clear, there isn't initial condition but boundary condition then. If the ground layer is uniform on nature of soil and thermal physical indexes do not vary with temperature, the differential equation for heat conducting is

$$\begin{cases} \dfrac{\partial\theta}{\partial\tau} = a\,\dfrac{\partial^2\theta}{\partial x^2} \\ \theta(x = 0, \tau) = A_0 \cos\left(\dfrac{2\pi\tau}{T}\right) \\ \theta(x = \infty, \tau) = 0 \end{cases} \qquad (2)$$

The specific solution of eq. 2 is:

$$\theta(x,\tau) = A_0 e^{-x\sqrt{\frac{\pi}{Ta}}} \cos\left(\frac{2\pi}{T}\tau - x\sqrt{\frac{\pi}{Ta}}\right)$$

while, if ground surface temperature varies sinusoidally, it becomes

$$\theta(x,\tau) = A_0 e^{-x\sqrt{\frac{\pi}{Ta}}} \sin\left(\frac{2\pi}{T}\tau - x\sqrt{\frac{\pi}{Ta}}\right)$$

Using the above equations to solve the necessary time τ_1, taken for temperature of culvert foundation to get to 0°C and assuming the thickness of foundation is δ, we have:

$$\sin\left(\frac{2\pi}{T}\tau_1 - \delta\sqrt{\frac{\pi}{T a_2}}\right) = 0$$

Through simple calculation,

$$\tau_1 = \sqrt{\frac{T}{4\pi a_2}}\,\delta \qquad (3)$$

where a_2 is the thermal coefficient of diffusivity of foundation material (m^2/hr); δ is the thickness of the foundation, including culvert pipe, in meters. If culvert pipe and foundation material differ, the thickness of the pipe shell must be converted into the thickness of the foundation material $\delta_1 \cdot (\lambda_2/\lambda'_1)$. Then the total thickness of foundation material will be

$$\delta = \delta_2 + \delta_1\frac{\lambda_2}{\lambda'_1} \qquad (4)$$

where λ'_1 and λ_2 are the thermal conductivities of pipe and foundation material respectively (see Figure 1).

Foundation soil begins to thaw after time τ_1, which is within the thawing period, has elapsed. The thawed depth is calculated by Rukiyanov's (1963) equation:

$$h = \sqrt{\frac{2\lambda_1\theta_0}{Q_0 + \frac{1}{2}C_1\theta_0}(\tau - \tau_1) + S^2} - S$$

where S is the equivalent thickness of surface layer. If the thermal conductivity of foundation material is λ_2, according to the viewpoint of equivalent heat resistance the thickness of the first layer, foundation, can be converted into the equivalent thickness of the second layer, foundation soil, then $\delta(\lambda_1/\lambda_2)$. Substitute S = $\delta(\lambda_1/\lambda_2)$ along with eq. 3 into the above equation, we have the maximum thawed depth of foundation soil within the section of culvert body as:

$$h = \sqrt{\frac{2\lambda_1\theta_0}{Q_0 + \frac{1}{2}C_1\theta_0}\left(\tau - \delta\sqrt{\frac{T}{4\pi a_2}}\right) + \left(\delta\frac{\lambda_1}{\lambda_2}\right)^2} - \delta\frac{\lambda_1}{\lambda_2} \qquad (5)$$

where τ is the cumulative time in thawing period, hr; δ is the thickness of culvert foundation in meters, (calculated by eq. 4); Q_0 is the latent heat fusion of permafrost (kcal/m^3); λ_1, C_1 is the thermal conductivities of foundation soil beneath culvert foundation

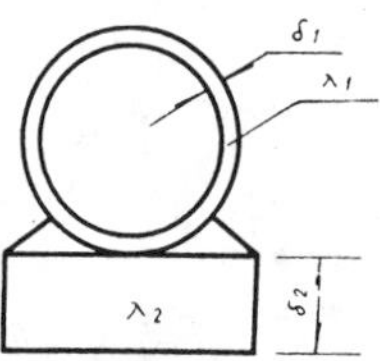

FIGURE 1 Sketch of culvert foundation.

(kcal/m.hr. °C) and volume heat capacity (kcal/m^3) respectively; λ_2, a_2 are the thermal conductivity of foundation material of culvert (kcal/m.hr. °C) and thermal coefficient of diffusivity respectively; θ_0 is the average positive ground surface temperature of water-passing surface of culvert (°C).

In order to study the effects of the large-diameter culvert through the high filling of the soils on the foundation temperature field and the change of artificial upper limit, we built a concrete rectangle culvert of 2 x 2.5 m in the Mount Fenghuo region in Qinghai-Xizang Plateau. The topics developed with this testing project are blasting excavation of foundation pit and the construction technology of concrete foundation prefabricated unit of large size, etc. The culvert is 26.32 m long. The highest height of the filling soils above it is 6.0 m. The observation net of ground temperature, water temperature, atmospheric temperature inside it and ground surface temperature of water-passing surface is set up inside. The relative curves in Figures 2, 3, 6 are obtained through the information of continuous observation these years. The values adopted are the observed data from 5-9 months of annual thawing time. The value of every point in these figures is the average value of a month.

Since flowing surface water in Qinghai-Xizang Plateau mainly originates from solid-state falling water and thawed snow, the water temperatures are lower. Therefore, it is worth while to use the results of long-term observation data on the testing culvert at Mt. Fenghuo in this area, which is located in this region, for reference.

1. Temperature θ_0 of Culvert with Intermittent Stream

It is seen from Figure 2 that the temperature of the side wall of the body section of the test culvert at Mt. Fenghuo is slightly lower than local atmospheric temperature, on account of that there occasionally exists the influence by small amount of flow, it is, therefore, suggested that the mean value of average monthly positive temperature reported by local weather station be taken as θ_0 for calculating, supplemented by necessary safety coefficient according to particular conditions.

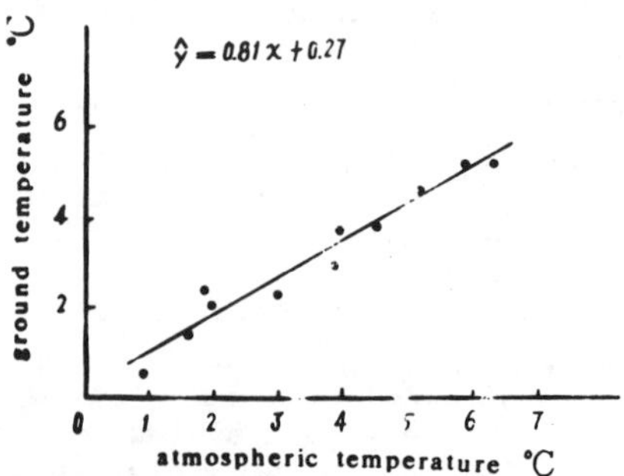

FIGURE 2 Relation between side wall temperature of testing culvert at Mt. Fenghuo and atmospheric temperature by local weather station.

TABLE 1 Various Calculating Indexes of Test Culvert at Mt. Fenghuo.

Locality	τ (hr)	Q_o (kcal/m^3)	λ_1 (kcal/m.hr.^{o}C)	C_1 (kcal/m .^{o}C)	λ_2 (kcal/m.hr.^{o}C)	a_2 (m^2/hr)
Opening No.4 (crushed stone concrete foundation)	3240	18816	1.38	630	1.1	2.75×10^{-3}
Opening No.5 (ceramsite concrete foundation)	3240	18816	1.38	630	0.5 (0.75)	1.5×10^{-3} (2.25×10^{-3})

2. Temperature Θ_0 of Culvert with Constant Stream

The ordinary flow area of the test culvert at Mt. Fenghuo is 0.25 m^2. According to the long-term observation data, we sorted out the mutual relation between temperature Θ_0 of water-passing surface of culvert body, atmospheric temperature reported by Mt. Fenghuo weather station, and ground temperature (Figure 3). In this article, the values of Θ_0 taken in calculating were all calculated by regression equation given by Figure 3 based on ground surface temperature provided by local weather station.

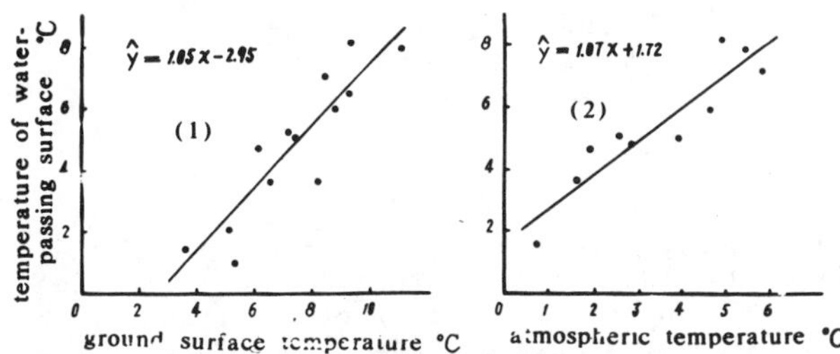

FIGURE 3 Relation between temperature of water-passing surface of culvert body, atmospheric temperature by local weather station and ground temperature.

Correlation between Calculated Results and Field Conditions

Given: A reinforced concrete rectangular culvert of 2 x 2.5 m^2 area in Mt. Fenghuo region of Qinghai-Xizang Plateau, the thickness of foundation of its body being $\delta_2 = 0.5$ m; Prefabricated crushed-stone concrete was used as the foundation of section from inlet to center line of the culvert; Prefabricated ceramsite concrete was used as the foundation of section from center line of the culvert to outlet. All of the culvert pipes were made of reinforced concrete, the thickness of its shell $\delta_1 = 0.23$ m and the thermal conductivity $\lambda'_1 = 1.33$ kcal/m. hr. ^{o}C.

The contrast between results calculated and those tested is shown in Table 2. One thing to be mentioned here is that water movement after setting up the culvert and variation of thermal conductivity of ceramsite concrete after its being buried were not taken into account. On condition that the thermal conductivity of ceramsite concrete is increased by 50%, the corresponding results calculated can be found in Tables 1 to 3 in parentheses.

TABLE 2 Comparison between Calculated Thawed Depth and Measured Depth*.

Year	Temperature (^{o}C)		Opening No. 4		Opening No. 5	
	Surface temperature by local weather station	Temperature of water-passing surface of culvert	Thaw depth claculated (m)	Thaw depth measured (m)	Thaw depth calculated (m)	Thaw depth measured (m)
1978	7.44	4.86	1.49	1.40	1.23,(1.37)	1.40
1979	7.34	4.76	1.48	1.25	1.21,(1.36)	1.25
1980	6.79	4.18	1.41	1.20	1.17,(1.30)	1.30
1981	7.38	4.80	1.49	1.60	1.23,(1.37)	1.50

* Measured values from spots of measuring hole by means of insertion (measuring distance =0.5 m). The value of the thawing depth includes the thickness of the culvert foundation and pipe.

TABLE 3 Comparison of Calculated Depth of Foundation and That of Living Example, Keeping Foundation Ground Frozen.

Foundation	year	Temperature of water-passing surface of inlet and outlet, θ_0 (°C)	Necessary burial depth of foundation calculated (m)	0°c Position in foundation measured (m)	Volume heat capacity of foundation material, rC_2 (kcal/m³)
Culvert inlet	1980	4.93	1.87	1.57	400
	1981	5.38	1.96	1.88	
Culvert outlet	1980	4.52	1.06, (1.49)	1.25	334
	1981	4.91	1.12, (1.58)	1.55	

CALCULATION OF CULVERT FOUNDATION THICKNESS NEEDED TO KEEP FOUNDATION SOIL FROZEN

While building a culvert in a permafrost area, the subsoil beneath the foundation of culvert body may thaw to a certain extent, provided it is refilled with coarse granular soil or other effective measures are adopted. The soil at the foundation of inlets, outlets, the wing wall, and the foundation beneath the culvert itself must be kept in a frozen state. Under such conditions, theory of energy equilibrium is used to derive equations for calculating the necessary thickness of foundation.

Derivation of Equation

a. According to the observed data of ground temperatures and strata beneath the culvert, it is assumed that the temperature within the seasonally thawing layer is distributed in a straight line, while that within permafrost layer is roughly expressed by eq. 6.

$$\theta = \theta_d \cdot \operatorname{erf}\left(\frac{x-\delta}{2\sqrt{a_1\tau}}\right) \qquad (6)$$

where θ_d is the perennial ground temperature, which is unaffected by temperature fluctuation; a_1 is the thermal coefficient of diffusivity of foundation soil. The other symbols are the aame as above. The ground temperature curve during the thawing period is shown in Figure 4.

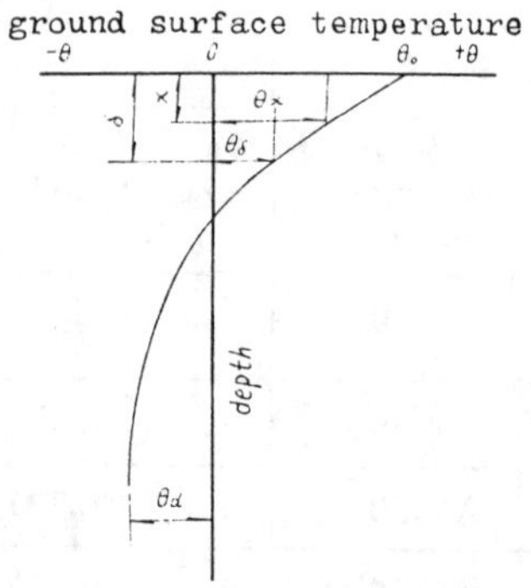

FIGURE 4 Sketch of ground temperature curve in thawing period.

b. The temperature within the thawing layer can be considered to be distributed in a straight line. From the similar triangles in Figure 4, we see that

$$\frac{\theta_0-\theta_x}{\theta_x-\theta_\delta} = \frac{x}{\delta-x}$$

By this the temperature at a given depth of foundation is solved as:

$$\theta_x = \theta_0 - \frac{\theta_0-\theta_\delta}{\delta}x \qquad (7)$$

Equation 7 suits the first group of boundary conditions, when $x = 0$, $\theta = \theta_0$; $x = \delta$, $\theta = \theta_\delta$; where θ_δ is the temperature of the foundation bottom. In the light of Fourier's theorem, the heat gained by culvert foundation of unit length under the condition of temperature difference $= \theta - \theta_\delta$ can be obtained by eq. 7 as:

$$Q_1 = q\tau = -\lambda_2 \frac{\partial\theta}{\partial x}\Big|_{x=\delta}\cdot\tau = \frac{\lambda_2}{\delta}(\theta_0-\theta_\delta)\tau$$

To prevent the foundation soil beneath the culvert foundation from thawing, the foundation bottom temperature θ_δ must be 0°C, thus the above statement becomes:

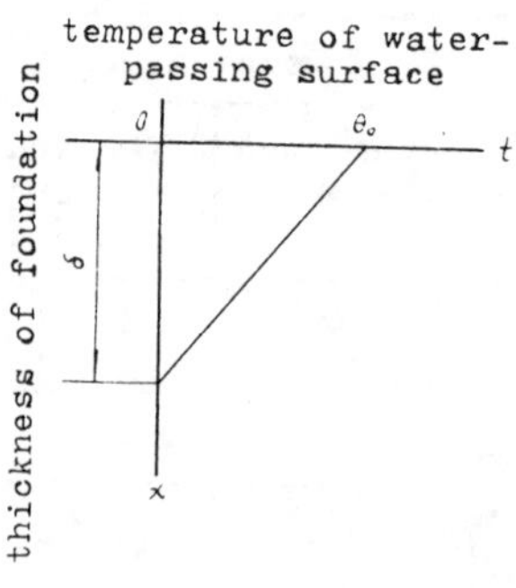

FIGURE 5 Heat consumed in temperature rise of culvert foundation.

$$Q_1 = \frac{\lambda_2}{\delta}\theta_0 \tau \quad (8)$$

c. The heat consumed in raising the temperature of the foundation material is equivalent to the product of the area of the triangle in Figure 5 by the volume heat capacity of the material, rC_2:

$$Q_2 = \frac{1}{2}\theta_0 \delta \gamma C_2 \quad (9)$$

d. According to Fourier's theorem, the quantity of heat flow leaving from foundation bottom and getting into the surface of the frozen area is calculated by eq. 6 as:

$$q = -\lambda_1 \frac{\partial \theta}{\partial x}\bigg|_{x=\delta} = \frac{-\lambda_1 \theta_d}{\sqrt{\pi a_1 \tau}}$$

It is seen from the above equation that heat flow q is a function of time τ, and if we substitute the negative value of perennial ground temperature of permafrost layer into the equation and change its sign, the total heat from thawing area to frozen area in the whole course of thawing, i.e. in a successive period of τ hours will be:

$$Q_3 = \int_0^\tau q\,d\tau = \int_0^\tau \frac{\lambda_1 \theta_d}{\sqrt{\pi a_1 \tau}}\,d\tau = \frac{2\lambda_1 \theta_d}{\sqrt{\pi a_1}}\sqrt{\tau} \quad (10)$$

e. In the light of Law of Conservation of energy and on account of lateral dissipation of foundation, we have $KQ_1 = Q_2 + Q_3$, where K is a revision coefficient. Based on the field investigation, it is suggested that when λ_1 (of foundation material as crushed stone concrete) = 1.1, K be 0.7; or $\lambda_1 \doteq 0.5$ (as ceramsite concrete), K be 0.8. Thus we have

$$K\frac{\lambda_2}{\delta}\theta_0 \tau = \frac{1}{2}\theta_0 C_2 \delta \gamma + \frac{2\lambda_1 \theta_d}{\sqrt{\pi a_1}}\sqrt{\tau}$$

Noticing $a_2 = \lambda_2/rC$ and rearranging the above equation, we get the formula to calculate depth of culvert foundation:

$$\delta = \frac{\sqrt{\frac{2\theta_0^2 \lambda_2^2}{a_2}K + \frac{4\lambda_1^2 \theta_d^2}{\pi a_1}} - \frac{2\lambda_1 \theta_d}{\sqrt{\pi a_1}}}{\gamma C_2 \theta_0}\sqrt{\tau} \quad (11)$$

where the symbols have the same physical meanings as stated above except for λ_1 and a_1, which have to be the thermal conductivity and thermal coefficient of diffusivity of foundation soil within the range from foundation bottom downward up to depth of annual change zone of ground temperature. For layered foundation soil composed of different media, the method of weighted average is used to calculate the average of λ_1 and a_1.

Determination of Temperature θ_0 of Water-passing Surface of Culvert Inlet and Outlet

Little information has been accumulated so far for making a quantitative evaluation of a culvert with intermittent water flow. However, the temperature of the inlet and outlet foundation may be only a bit higher than the atmospheric temperature, since they are only slightly affected by sunshine. As for culverts of constant flow, regression equation of mutual relation between temperature of water-passing surface of inlet and outlet and that of ground surface reported by local weather station has been obtained, referring to the data observed at Mt. Fenghuo test culvert (Figure 6). This equation is used as reference for designing projects in the Qinghai-Xizang Plateau permafrost area.

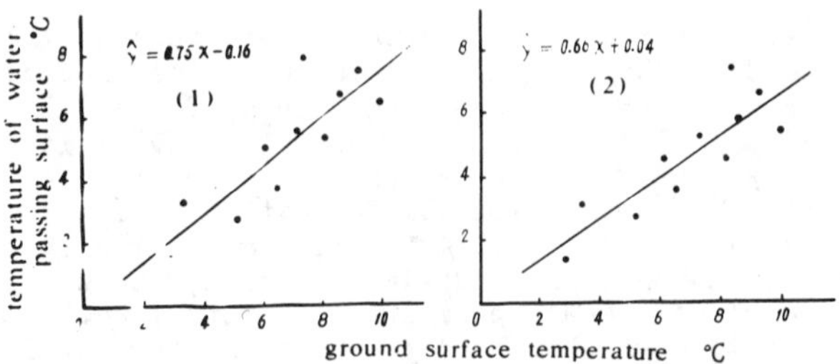

FIGURE 6 Relation between temperature of water-passing surface of inlet(1) and outlet(2) of Mt. Fenghuo testing culvert and ground surface temperature provided by local weather station.

Comparison between calculated results and data measured given in the above example of Mt. Fenghuo testing culvert, the foundation of its inlet and outlet are composed of crushed stone concrete and ceramsite concrete respectively both 3 m deep. The perennial ground temperature θ_d at the spot where the culvert is located is -3.2°C (Note: Its positive value is taken while substituted into the equation); the thermal conductivity and thermal coefficient of diffusivity of the foundation soil (the layer downward up to depth of annual change zone of ground temperature are λ_1(= 1.19) and a_1 (= 2.66 x 10^{-3}) respectively; the other calculating indexes are listed in Table 1. The results calculated are presented in Table 3, in which the temperature (θ_0) of water-passing surface of culvert inlet and outlet were calculated by regression formula on the basis of the ground surface temperature by local weather station. The successive thawing periods were all the same, i.e. τ = 3240 hr.

EMPIRICAL FORMULA

Determination of the Formula

Based on the analysis of weather data in permafrost area along with consulting information abroad, it is found that thawing depth of foundation soil is proportional to the square root of the positive cumulative temperature of local atmosphere or ground surface temperature, and the thawing depth curves of ground of different thermal conductivities in the same region almost tend to be a group of parallel lines; again, thawing depth of foundation soil is dependent on the specific yield as well as the dimension of opening of culvert. Thus the experience formula is written in the form of

$$h = K\lambda\sqrt{\Sigma T \tau} \quad (12)$$

where h is the thawed depth beneath culvert (m), including the thickness of culvert pipe, foundation depth as well as thawed depth of foundation soil underlying; λ is the thermal conductivity (kcal/m. hr.°C). The weighted average of culvert pipe, foundation and foundation soil is taken in calculating; $\sum T\tau$ is the thawing index of air temperature (°C day), namely, the sum of product of average positive temperature each month times number of days of the month. It is best to take the 10-year average thawing index of air temperature by local or neighboring weather station; K is an experience coefficient concerning the diameter of culvert inversely calculated in accordance with the measured data of culverts of different apertures (Table 4).

TABLE 4 K Values of Different Culvert Apertures.

Aperture of Culvert (m)	2.0	1.0	0.75
Experience Coefficient K	0.075	0.050	0.040

A test was made for comparing the calculated maximum thawing depth of foundation soil of Mt. Fenghuo test culvert by using the 5-year (1976-1980) average thawing index of air temperature provided by Mt. Fenghuo weather station with the measured data, which resulted in a relative error of 15%. As for the existing culverts, only the data of thawing depth of foundation sides were present, however, the calculated results basically tallied with the variation of thawing depth of culvert sides measured. For instance, the thawed depths measured in 1977 on the side of Mt. Fenghuo Pass Highway Culvert and Wudaoliang Highway Culvert were 0.2 m and 0.16 m deeper than that measured in 1976 respectively, the error being no more than 4% compared with the variation amplitude of thawing depth calculated (Table 5). In Table 5 the numerator values are the thawed depth calculated in accordance with the temperatures measured in testing boring at culvert side while the denominators are those calculated based on the annual thawing indexes of air temperature in the same years.

Use of equations if foundation soil is kept frozen. To design a culvert keeping foundation soil beneath its body foundation in frozen state,

TABLE 5 Measured and Calculated Thawing Depth.

Locality		No. of opening (m)	Annual thaw depth					Calculated value by 5-year average thawing index of air temperature
			1976	1977	1978**	1979	1980	
Mt. Fenghuo Test culverts	2.0 diam.	4 5	1.52 1.22	1.63 1.68	1.40 1.40	1.25 1.25	1.20 1.30	1.78* 1.48
Mt. Fenghuo pass highway culvert	1.0 diam.		0.98/ 1.00	1.18/ 1.24				
Wudaoliang highway culvert	0.75 diam.		0.85/ 0.85	1.01/ 1.00				
Qingshuihe highway culvert	1.0 diam.		0.92*/ 1.00					

* Obtained by means of experience coefficient of culvert having same apertures.
** A new temperature-testing boring set up in 1978.

TABLE 6 Results Calculated by eqs. 12, 13, and 14.

Year	Concrete foundation		Potter granule concrete foundation		Concrete foundation		Potter granule concrete foundation	
	Conversion formula	Empirical formula (12)	Conversion formula	Empirical formula(12)	Conversion formula	Empirical formula (12)	Conversion formula	Empirical formula(12)
1976	1.38	1.30	0.91	0.70				
1977	1.82	1.60	1.23	0.78				
1978	1.89	1.64	1.29	0.80	1.51	1.52	1.29	0.70
1979	1.63	1.47	1.08	0.70				
1980	1.70	1.52	1.14	0.74				

we still use eq. 12 to calculate the necessary foundation thickness, at the moment the value of λ is the thermal conductivity of foundation material substituted into eq. 12. Since the foundation thicknesses of the culverts along Qinghai-Xizang Highway are generally not greater than 0.5 m, there is not sufficient practical information for verification. Here the Sahunianch (1958) converting Formula is used to present a collateral evidence, i.e.

$$h_3=h_2\sqrt{\frac{a_3}{a_2}}=h_2\sqrt{\frac{\lambda_3 C_2 \gamma_2}{\lambda_2 C_3 \gamma_3}} \qquad (13)$$

where h_2 is the thawing depth of foundation soil, in meter; h_3 is the thickness of foundation material converted from thawing depth of foundation soil, in meter; a_2, C_2, r_2 are the thermal coefficient of diffusivity (m^2/hr), specific heat (kcal/kg.°C) and unit weight (kg/m^3) of foundation soil respectively; a_3, C_3, r_3 are the thermal coefficient of diffusivity (m^2/hr), specific heat (kcal/kg, °C) and unit weight (kg/m^3) of foundation material respectively.

This total thickness of foundation after being converted by Sahuniach formula (1958) is

$$h=h_1+h_3 \qquad (14)$$

where h_1 is the predetermined foundation thickness.

The results calculated by eqs. 12, 13, and 14 are listed in Table 6 for comparison. The value from the empirical formula in Table 6 is derived from eq. 12, while the value from the conversion formula is derived from eqs. 13 and 14. The left hand numbers in the table are calculated results based on the thawing index of the air temperature each year, while those on the right are based on 5-year average thawing index of air temperature and the maximum measured thawing depth.

CONCLUSION

The permafrost area in Qinghai-Xizang Plateau is the source of the Yangzi River basin, where the river system is very developed. The culverts of road engineering built in this area are more and the engineering quantities is large. So it is important to study the change law of thawing depth after the culverts built and to determine correctly the burying depth of culvert foundation and alternative thickness of foundation soil.

In this paper, the thawing depth and the burying thickness of foundation calculated by equations are zero (°C) places in the center line of transverse section of culvert.

Equation 5 is considered by one dimension, not corrected. Time that the temperature of foundation bottom needs to reach °C is determined by attenuation equation of temperature. Thus, the calculating error increases with the increase of foundation thickness. The equation is only used to calculate the maximum thawing depth beneath the culvert body. It can't determine the shape of thawing circle.

In addition to that, the effect during temperature fluctuating is not considered in the equations given in this paper. So, these equations should be used according to the actual condition.

A REVIEW OF CRYOGENIC STRUCTURE AND TEXTURE IN FINE-GRAINED ROCKS AND SOILS

E. D. Yershov, Yu. P. Lebedenko, O. M. Yazynin, Ye. M. Chuvilin,
V. N. Sokolov, V. V. Rogov, and V. V. Kondakov

Geology Department
Moscow State University, USSR

The paper presents the results of investigations into the mechanics and laws governing the formation of cryogenic structure and texture in soils during freezing or thawing. A number of thermophysical, physico-chemical and physico-mechanical processes were studied theoretically and experimentally using a range of modern techniques. In the light of the results obtained, conditions for the formation and development of segregated ice layers and for the creation of various cryogenic textures were examined. A system of classifying cryogenic textures is proposed, which is unique in its potential to explain the formation of various types of cryogenic textures within a standard framework based on the major classification parameters and conditions of segregated ice formation. The paper considers peculiarities in the formation of cryogenic structure as observed both in laboratory experiments and in the field. A close relationship is demonstrated between the structural parameters, their degree of variability, and the intensity of the processes accompanying the freezing and thawing of unconsolidated rocks. The peculiar microstructure of naturally frozen unconsolidated rocks of various origins is described. The proposed classification system of cryogenic structures takes into account the type of ice cement, the pattern of bonds between the structural elements, and the nature of the areas of contact.

MECHANISM OF CRYOGENIC TEXTURE FORMATION

Cryogenic structure and texture formation in fine-grained rocks and soils is due to complex thermophysical, physico-chemical and physico-mechanical processes which essentially transform the composition, structure, and properties of freezing, thawing and frozen soils. As shown earlier (Yershov, 1977; Yershov et al., 1978), to determine the mechanism and kinetics of segregated ice formation (in other words, the conditions for cryogenic texture formation) one must know the relationship between heat exchange and water migration in freezing and thawing ground (i.e., the thermal conditions for cryogenic texture formation).

It is also necessary to consider the physico-mechanical environment in which segregated ice layers appear and grow. Cryogenic texture formation is impossible if the essential thermal conditions of coupled flows of heat and water within the frozen and thawed parts of the freezing or thawing soil are not present. In particular, segregated ice layers are formed in freezing fine-grained soils at the cost of migrating water whenever the amount of heat withdrawn exceeds that of heat emitted at the phase transition of water segregated during freezing. The thermal conditions vital to cryogenic texture formation, nevertheless, do not include all the physico-chemical and physico-mechanical processes occurring in freezing and thawing fine-grained soils (namely, shrinkage, swelling - heaving, structurization, appearance of zones with high structural strength, etc.). Thermal conditions alone fail to show the location and nucleation point of ice layers with different orientation (for instance, parallel or perpendicular to the freezing front). The conditions sufficient for revealing the generation and growth of segregated ice layers are the physico-mechanical conditions under which cryogenic textures are formed, since they determine when the local strength of the soil is exceeded and microlayers of ice come into being.

The mechanism of forming ice microlayers can be generally described as follows. When a fine-grained soil is freezing, the thawed part is dehydrated owing to water migration to the frozen part and undergoes a process of structural formation and shrinkage resulting in new structural units of soil. Zones of concentrated strain evolve along the boundaries of these structural units. Within such zones the soil water is under tension (less than atmospheric pressure) compared to other sites of the soil. Owing to the pressure gradient, water migrates to the zones of concentrated strain. When these zones appear in the freezing portion of the soil, the shrinkage strain sharply increases (Yershov, 1979), and the swelling-heaving strains caused by water migration and by ice crystallization in the freezing area become intensive. Local cohesion of the soil is broken along the zone of concentrated strain, and the water contained there passes into an unstressed state (i.e. its total thermodynamic potential increases stepwise) and the water rapidly transforms into ice. Microlayers of ice emerge in a definite region of the freezing soil, becoming most highly developed near the border at which the shrinkage and swelling-heaving strains change their directions. This region is the negative temperature range (-0.2°C to -0.4°C).

Conditions for specific features of the emergence of segregated ice layers parallel and perpendicular to the freezing front and the kinetics of their evolution have been described in detail (Yershov, 1977, 1979; Yershov et al., 1978). It was shown that ice layers parallel to the freezing front result from cleavage (or displacement) stresses (P_{cl}) which are due to the variously directed strains affecting the dehydrating and the swelling (heaving) parts of the rock. Taking into account the disjoining pressure of fine water films (P_d^f), one can determine the region of soil where ice layers parallel to the freezing front may appear. To this end, the relationship ($P_{cl} + P_d^f$) and the local cohesion of the soil (shear strength P_{coh}^{sh}) are used together with the value of actual pressure (P_{act}) of the overlying soil sequence:

$$P_{cl} + P_d^f > P_{coh}^{sh} + P_{act} \quad (1)$$

The region of further growing ice layers may be identified by means of a similar relationship:

$$P_{cl} + P_{coh}^{sh} > P_{coh}^{s-i} + P_{act} \quad (2)$$

where P_{coh}^{s-i} is the cohesion of ice layers and frozen soil ($P_{coh}^{s-i} < P_{coh}^{sh}$).

Vertical segregated ice layers are related to the developing tensile stresses (normal stresses, P_n) specified by the difference between the skrinkage stresses and stresses of swelling-heaving ($P_{sw-heav}$). The area of P_n is the one below the border at which the shrinkage and swelling-heaving stresses change their directions. This area embraces some part of the freezing horizon and the unfrozen part of the soil. The regions where vertical microlayers of ice can originate and grow further can be found by the following relationships:

$$P_{shr} - P_{sw-heav} > P_{coh}^{tens} + P_{act} \quad (3)$$

$$P_{shr} - P_{sw-heav} > P_{coh}^{s-i} + P_{act} \quad (4)$$

where P_{coh}^{tens} is the local tensile strength of the rock ($P_{coh}^{s-i} < P_{coh}^{tens}$). It should be noted that regarding the appearance and growth of vertical ice layers (unlike horizontal ones) the disjoining effect of fine water films is neglected, since water migration to vertical zones of concentrated strains is chiefly due to the gradient of P_n inasmuch as the horizontal isothermal plane has no temperature gradient providing a gradient of P_d^f.

CLASSIFICATION SCHEME FOR CRYOGENIC TEXTURES

The formation of cryogenic textures is determined by a great number of interrelated processes which can be divided into thermophysical, mass exchange, physico-chemical, and physico-mechanical processes. However, cryogenic textures of certain types can form only under specific conditions. These include the following conditions: the lithological features of rocks; conditions providing for water migration to the frozen region of freezing and thawing fine-grained soils (thermophysical and mass exchange conditions); conditions determining the appearance of migratory-segregated ice interbedding parallel or perpendicular to the front of freezing or thawing (physico-mechanical conditions); and finally, conditions specifying the incidence and thickness of layers. The thermophysical and physico-mechanical conditions for formation of cryogenic textures have been described above. As to the incidence of ice layers in fine-grained soils of homogeneous composition, structure, and properties, the case for ice interbedding parallel and perpendicular to the freezing or thawing front is determined by gradients of cleavage and normal strains, respectively, according to the following relationships:

$$l_{par} = f\left(\frac{I}{grad.\ P_{cl}}\right) \quad (5)$$

$$l_{perp} = f\left(\frac{I}{grad.\ P_n}\right) \quad (6)$$

where l is the distance between the interbeds of segregated ice. The thickness of segregated ice layers (h) is found from the relationship

$$h = f\left(\frac{I \cdot \Delta X}{V}\right), \quad (7)$$

where I is the density of water flow migrating to ice interbeds, V is the freezing rate, and ΔX is the vertical thickness of the region where horizontal or vertical segregated ice layers can develop.

The above notions on cryogenic texturization have been used to base a classification scheme for cryogenic textures in terms of their emergence and development conditions (Yershov, 1977, 1979).

Among streaky textures, classes are determined according to specific geological and genetic features of friable deposits and their lithology. Fine-grained soils of heterogenic composition, structure, constitution, and properties give rise to a class of inherited cryogenic textures, whereas homogeneous soils develop a class of superimposed cryogenic textures. In the first case, the pattern of cryogenic textures depends on the lithological-facial features of the freezing or thawing friable deposits. The genetic systematization of the second class of soils involves the thermophysical, physico-chemical, and physico-mechanical processes occurring in the specific geological and geographical environment.

A class of bulky cryogenic textures typical of soils with a broad gradation (from coarse to fine-grained) is formed when the physico-mechanical conditions are not satisfied (I,3). This class includes interstitial and basal-solid types of cryogenic textures which can be formed both with and without water migration.

Finally, cryogenic textures of contact and film-solid types develop when the thermophysical and physico-mechanical conditions for segregated ice layer formation are not satisfied.

The types of streaky superimposed cryogenic textures (for homogeneous materials) are determined by physico-mechanical conditions under which ice layers are formed parallel or perpendicular to the front of freezing (or thawing). Versions of such conditions specify the variety of existing types of cryotextures within a particular class (porphyry-like, incompletely and completely developed layered and reticulate cryotextures).

Heterogeneous fine-grained soils may reveal combined physico-mechanical conditions similar to those which are found in the case of superimposed segregated textures. However, such conditions are usually also satisfied for the various inhomogeneities, providing for the origination and evolution of inherited ice layers. Considering the effect of inhomogeneities on the thermophysical, mass exchange, physico-chemical, and physico-mechanical processes, three types of inherited cryogenic textures can be identified. Textures of defective strength evolve in friable deposits of homogeneous composition which have zones of displacement, bearing, overmoistening, thinning. etc. Such zones determine the configuration of ice layers confined within soils of defective strength. Contact-stressed cryotextures reveal ice layers confined to contacts of fine-grained soils with different composition, structure, constitution, and properties. When ice layers are formed, the soil ruptures most easily along the contacts of heterogeneous layers, because the thermophysical, mass exchange, physico-chemical, and physico-mechanical processes in each of them differ. And, finally, there are cryogenic textures where the configuration of ice layers depends on the disposition, geometry, and material composition of foreign inclusions such as pebble, boulders, peat lenses, etc., within the heterogeneous fine-grained soil.

The classification scheme specified for each type of cryogenic texture the incidence and thickness of ice layers. To this end, relationships 5, 6, and 7 are used. The scheme is unique because it provides a uniform basis for explaining the formation of various types of cryotextures in freezing and thawing fine-grained soils by using the principal classification parameters of the segregated ice separation. It therefore becomes possible to assess quantitatively the great variety of geological and geographical factors of the environment.

CRYOGENIC STRUCTURIZATION

In laboratory experiments, intense water exchange and ice layer formation in freezing soils were induced by maintaining temperature conditions close to the natural ones (the cooling surface temperature was -6°C to 12°C). Water exchange was excluded through rapid acceleration of freezing at very low freeze-through temperatures (-30°C to -60°C). Such conditions are known to fix water as ice in situ (without obvious migration) and to yield massive cryotextures. The technique helps to identify the effect of mass transfer on structurization and also to assess the changes in the structure in relation to different factors. In the investigation of cryogenic structurization, such parameters were considered as the size and quantitative ratio of structural elements (organic-mineral particles and aggregates, ice inclusions, and pores), their shape and interrelation in clay soils of different composition and properties under diverse conditions of freezing-thawing. These investigations yielded the following results.

When water-saturated soils are freezing such that water transfer from the unfrozen to the freezing region is intense, the mineral skeleton of the unfrozen part becomes dehydrated and shrunken irrespective of the rate of water inflow to the soil samples (with or without aquifers). The highest degree of dehydration (and of shrinkage) was observed in soils in which the water exchange was most intensive. Studies based on a microaggregate analysis have shown that the porosity of soils undergoes important changes largely in the region of highest dehydration and intensive shrinking strains, i.e. at the boundary of freezing. In unfrozen regions lying rather far from the freezing boundary the porosity does not change much. It is common for microaggregate compositions in the dehydrated unfrozen region of soils for the fraction of coarse dust and fine sand to increase while that of fine dust (and, more rarely, of clay particles) decreases.

The reduction in porosity owing to aggregation and coagulation of fine-grained materials results from the dehydration of the mineral skeleton. As the films of bound water around the particles become thinner and more dessicated, the particles draw closer and their interaction increases, i.e. their structural bonds become stronger. Within the freezing and frozen horizons, the growing aggregation levels off as the aggregates undergo destruction caused by crystallization and temperature stresses. In an electron-microscopic analysis the traces of ruptured primary particles and aggregates can be easily discerned by their distinct broken contours and the abundance of fine-grained material along the boundaries of the elements.

Low temperature freezing of fine-grained soils (without water exchange) mainly disperses structural elements. Such dispersion is particularly characteristic of grounds with high freezing rate and rapid water-ice phase transitions. In these cases we regard the temperature-crystallization destruction as the principal mechanism of structural disintegration. The porosity of low-temperature soils rises also during their thawing.

MICROSTRUCTURE OF NATURAL FROZEN FINE-GRAINED SOILS

Microstructure of freezing soils with different initial porosities is characterized by an increasing size of aggregate and a decreasing porosity of soil with ice-cement. This effect is most obvious in clays. Enlargement of aggregates in loams and sandy loams is weaker because water exchange is lower and shrinkage is less intense. In clay soils, the mineral composition determines the water exchange and ice formation during freezing, as well as the morphology of the cryogenic structure. The peat content of fine-grained soils proportionally contributes to the enlargement of ice inclusions. Organic-mineral material in freezing grounds yields large, shapeless aggregates (up to several mm in size). Freezing soils with organic material develop basal forms of ice-cement.

The effect of chemical composition on microstructurization is manifested primarily with salinity. Low soil salinization with monovalent cations ($10^1 - 10^2$%) practically gives no aggregation, providing for homogeneous microstructurization of soils and diminishing the size of ice-cement inclusions. Higher soil salinity enhances aggregation because the salts cryostallizing from the freezing solutions cement the soil particles. Multivalent cations (especially Fe^{3+}) augment the aggregation of particles in the mineral skeleton and enlarge the inclusions of ice-cement inasmuch as water migration and interactions of structural elements increase.

CLASSIFICATION SCHEME OF CRYOGENIC STRUCTURES

The specific microstructural features of freezing, thawing, and frozen soils discussed in this paper testify to the great range of variously combined parameters of cryogenic structures. In view of this, it is hardly possible today to offer a scheme to rank cryogenic structures with regard to all features indicative of frozen soils. Nevertheless, it is feasible to provide a classification reflecting the principal structural parameters typical for groups of cryostructures. Among such characteristics are the type of ice-cement, and the pattern of bonds between structural elements and their contacts (Figure 1). In terms of the ice-cement type, cryostructures are specified in accordance with the total water content of the rock. Higher water contents alter the type of ice-cement, which can be epitaxial-pellicular, cuff-like, pellicular, interstitial, or basal.

Epitaxial-pellicular ice-cement is formed when the water content is lower than maximum hygroscopicity ($W < W_{m.h.}$); cuff-like ice-cement evolves with $W > W_{m.h.}$, and pellicular ice-cement develops when the water content causes capillary breakage ($W'_{c.b.}$). Interstitial ice-cement appears when the pores become completely filled with ground moisture, while with W exceeding the water content of swelling (W_{sw}), basal ice-cement is formed. Systematization of cryostructures in terms of the bonds between their structural elements is based on the content of unfrozen water determining the coherence and, indirectly, the force interactions of structural elements separated by water films of diverse thickness. It is also possible that at very low temperatures the bound water completely freezes, forming ice contacts. If so, ice-aggregating contacts transform into ice-epitaxial, and ice-coagulating ones—into highly epitaxial contacts.

The classification parameters discussed above rank various groups of cryogenic structures, including microtextures of fine-grained soils.

In terms of their contacts, cryogenic structures are systematized according to the distance between interacting structural elements and the state of contacts of water layers. Groups of contacts in frozen soils can be identified as water-free, water, ice, and water-ice contacts (Figure 2). Water-free contacts exist in overdense soils where the particles of the mineral skeleton directly contact one another (the phase type of contact), and also in heavily salinized fine-grained soils in which the particles are cemented with various salts (the crystallizing type of contact). The group of water contacts includes aggregating and coagulating ones. In the case of aggregating contacts, the particles are separated by weakly bound water; in the case of coagulating contacts the water is strongly bound.

With bound water partially freezing out, the above contacts transform into water-ice ones, both of ice-aggregating and ice-coagulating type. Stabilization of the mineral skeleton during freezing of bound water of different categories (from weakly to strongly bound) depends on the character of the ice-cement bonds. In terms of soil temperature, which determines the unfrozen water content, the following types of cryogenic structures can be defined: (a) the weakly bound ones observed at temperatures from 0°C to the freezing point of weakly bound water (t_1) when the unfrozen water content exceeds the maximum molecular water capacity ($W_{m.m.w.c.}$); (b) the bound structures found within the range from t_1 to the freezing point of strongly bound water (t_2) when $W_{unfr.}$ is less than W_{mmwc} but greater than the water content of maximum hygroscopicity W_{mh}; and (c) the strongly bound structures developing at t_2 and lower temperatures, with $W_{unfr.}$ being less than or equal to W_{mh}.

REFERENCES

Yershov, E. D., 1977. Mechanism of migrational-segregational ice formation in freezing and thawing fine-grained rocks: Vestnik moskovskogo universiteta. Ser. geologicheskaya, no. 3, p. 97-108 (in Russian).

Yershov, E. D., 1977. Classification of cryogenic textures in frozen soils in terms of their emergence and evolution conditions: Vestnik moskovskogo universiteta, ser. geologicheskaya, no. 4, p. 78-84 (in Russian).

Yershov, E. D., 1979. Water transfer and cryogenic textures in fine-grained soils: Moscow, Izdatelstvo moskovskogo universiteta (in Russian).

Yershov, E. D., Cheverev, V. G., Lebedenko, Yu. P., and Shevchenko, L. V., 1978. Water migration, structurization and streaky ice formation in frozen and thawing soils. Proceedings of III International Conference on Geocryology: Edmonton, Canada, v. 1, p. 175-180.

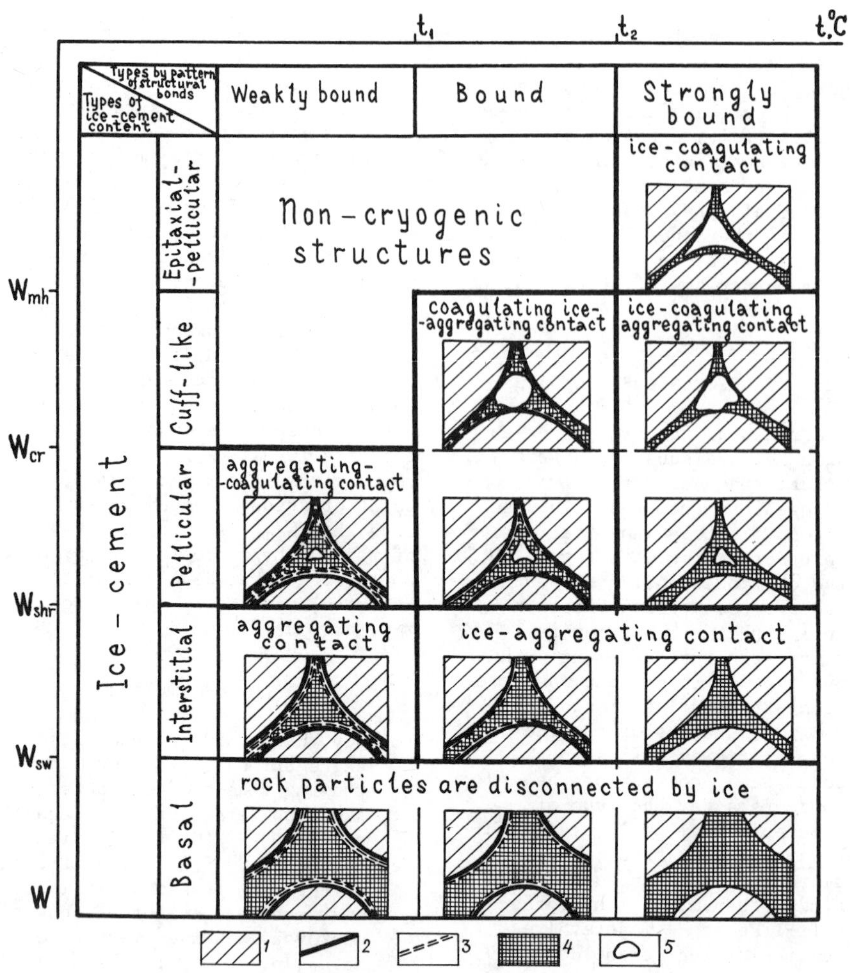

FIGURE 1 Principal scheme of groups of cryogenic structures in frozen soils (based on classification parameters of their cryogenic components). Conventional symbols: 1—mineral particles; 2—strongly bound water; 3—weakly bound water; 4—ice; 5—pores; t—rock temperature; t_1 and t_2—freezing temperature of weakly bound and of strongly bound water, respectively; W—water content of the rock; W_{mh}—maximum hygroscopicity; W_{cr}—critical water content at the transition of cuff-like ice into pellicular ice; W_{shr}—water content of shrinkage limit; W_{sw}—water content of swelling.

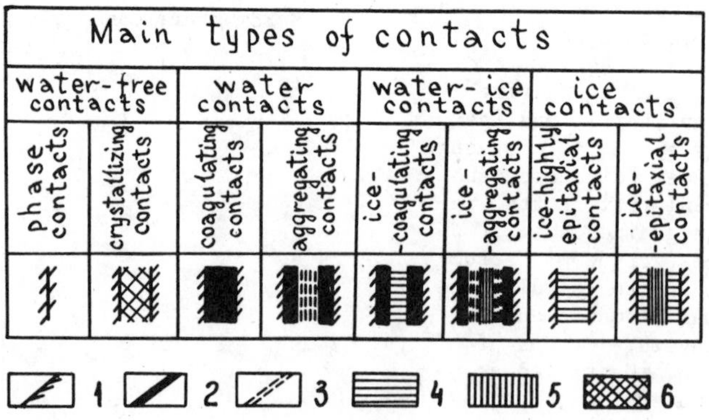

Main types of contacts							
water-free contacts		water contacts		water-ice contacts		ice contacts	
phase contacts	crystallizing contacts	coagulating contacts	aggregating contacts	ice-coagulating contacts	ice-aggregating contacts	ice-highly epitaxial contacts	ice-epitaxial contacts

1 2 3 4 5 6

FIGURE 2 Classification of main types of contacts in frozen fine-grained soils. Conventional symbols: 1—mineral surface; 2—strongly bound water; 3—weakly bound water; 4—highly epitaxial ice; 5—epitaxial ice; 6—cementing substances (salts, etc.).

WATER REDISTRIBUTION MEASUREMENTS IN PARTIALLY FROZEN SOIL BY X-RAY TECHNIQUE

Kiyoshi Yoneyama, Takeshi Ishizaki and Nobuaki Nishio

R & D Institute, Tokyo Gas Co., Ltd., Tokyo, Japan

ABSTRACT

This paper reports the results of water redistribution measurements in partially frozen soils under overburden pressure. The authors developed an experimental apparatus using X-ray technique to measure the water flow both in the unfrozen and frozen part of a freezing saturated soil under overburden pressure. The experimental method consists of taking X-ray photographs of the sample in which small lead shots are embedded, determining the location of small circular images of lead shots on the film and measuring the dilatations or the consolidations in each part of the sample by comparing them with those of subsequent films. During the experiment the temperatures of end plates were held constant. After the experiment was started, a large ice lens was formed around the center of the sample. Ice segregation was also observed in the region that was colder than the center of the sample. The amount of dilatations due to ice segregation was measured quantitatively with displacements of lead shots. The consolidation of the unfrozen part was observed. And the water flow both in the unfrozen and in the frozen part was calculated.

INTRODUCTION

Several models for frost heave have been proposed to describe water flow, heat flow, and ice segregation in freezing soils. A theory of coupled heat and water flow in freezing unsaturated soils without overburden pressure was proposed by Harlan (1973) and Taylor and Luthin (1976). Experiments along these lines of soil moisture movement were carried out by Dirksen and Miller (1976) using the gravimetric method and by Hoekstra (1966), Jame and Norm (1976), and Fukuda et al. (1980) using the gamma ray technique. These experimental results verified the validity of the coupled model for the frost heave phenomenon.

On the other hand, in designing Liquefied natural gas in-ground storage tanks and performing artificial ground freezing, saturated soils under overburden pressure must be taken into account. Theoretical studies of freezing soils under this condition have been carried out by Hopke (1980), Gilpin (1980), and Takashi (1982). Experimental research of water movement in freezing soils, including the frozen part, has been performed by Mageau and Morgenstern (1980) using the gravimetric method and by Loch and Kay (1978) using the gamma ray technique. However, only few such experiments have been made.

The present authors developed an experimental apparatus using X-ray technique to measure the water flow both in the unfrozen part and in the frozen part of a freezing saturated soil under overburden pressure. The X-ray technique permits us to observe the dilatation in the frozen part of the sample with a high accuracy. Consequently, experimental results obtained by using the X-ray technique might provide useful data to improve the theoretical researches for frost heaving of saturatated soils under overburden pressure.

EXPERIMENTAL METHOD

Experimental Apparatus

The complete test configuration is given in Figure 1. The soil sample is contained in a plexiglass cylinder with a wall thickness of 2 cm and an inside diameter of 12.5 cm. The temperature of the end plates is controlled by a thermoelectric cooling control system. The drainage tube from the upper end plate is connected to a volume change indicator which measures the water inflow and outflow from the soil sample. A dial gauge is set on the upper end plate to measure the total frost

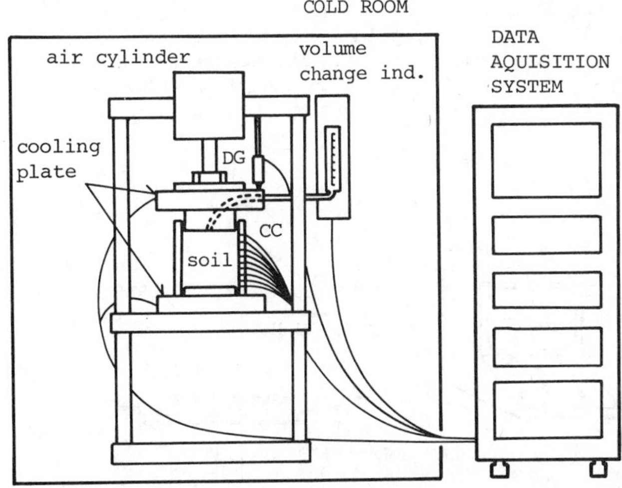

Figure 1 Experimental apparatus

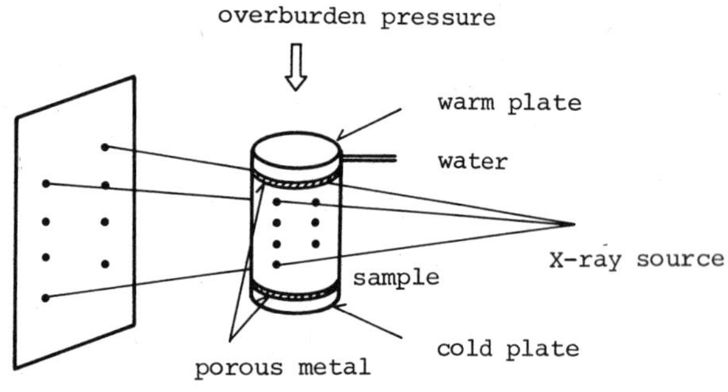

Figure 2 X-ray instrument

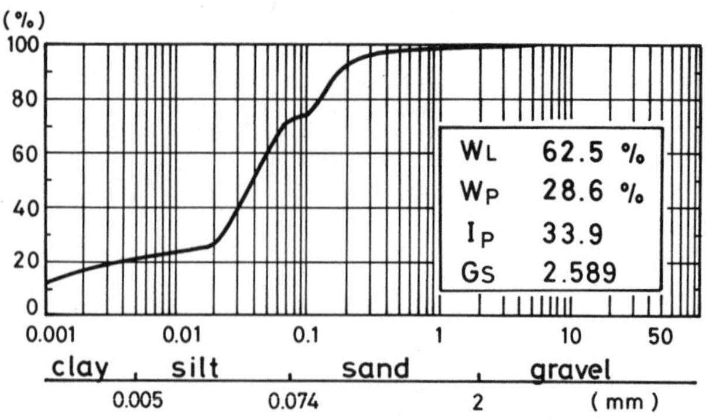

Figure 4 Properties of silty loam

heave amount. Copper-constantan thermocouples are located at 1-cm intervals along the column to determine the soil temperature profile. The load can be applied to the sample through the top piston to the maximum capacity of 4 t. The soil container is covered with thermal insulation, to prevent radial heat exchange between the sample and air, and is placed in a room with a constant temperature of +1 °C. A schematic diagram of the sample and X-ray apparatus is shown in Figure 2. Small lead shots (diameter D = 2 mm) were embedded in the sample at intervals of about 5 mm in two vertical rows on a single plane. Lead shots for position reference are located on the outside of the sample. The sample is located between the X-ray source and the X-ray film holder. The X-ray source is positioned at the elevation of the growing ice lens to avoid parallax errors.

X-ray Film Reader

An X-ray film reader is used to determine accurately the location of small circular images of lead shot. A schematic diagram of the apparatus is shown in Figure 3. The apparatus is composed of a 45 x 45 cm film carriage, a two-axis D.C. motor drive system, a projection system, and a C.C.D. (charge coupled device) light intensity measuring system. An X-ray film is placed between two optically flat glass plates. The camera has 1,500 pieces of C.C.D., which is a semiconductor to transfer the light intensity into an electric signal. The accurate center of the image is determined by its light intensity profiles. Thus the accurate coordinates of lead shot images are measured. The accuracy of reading is 30 microns. A detailed explanation of the apparatus is given by Takagi et al. (1983).

Test Procedure

A silty loam was used in this experiment. The various properties of the soil are summarized in Figure 4. The frost heave test was performed in the following sequence. First the test sample was saturated with pure water. Lead shots were embedded in a sample in the manner mentioned above. The sample was then consolidated to the desired applied load. After the sample preparation was finished, the lower side plate was cooled down, and the sample was frozen from the bottom upward. During the experiment, both end plates were held at a constant temperature. X-ray photographs of the sample were taken at initially set time intervals. These photos were obtained using an X-ray tube with a source current of 5 mA at 195 kV and an exposure time of 2 min. The expansion in the frozen part of the sample was measured by reading accurate coordinates of lead shot images and by comparing them with those of subsequent films. Data from the thermocouples, the volume change indicator, and the dial gauge were collected periodically by a digital data aquisition system.

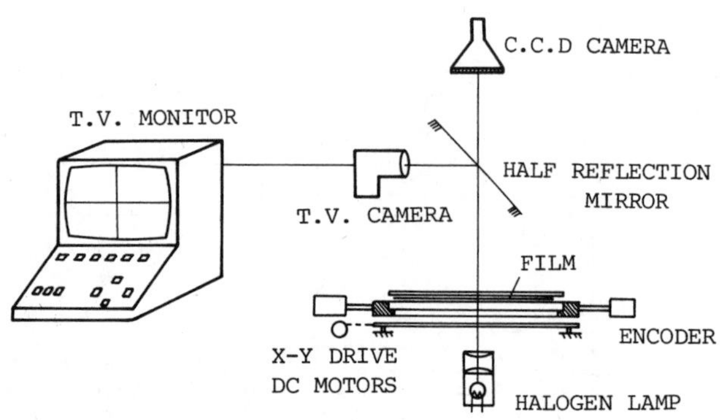

Figure 3 X-ray film reader

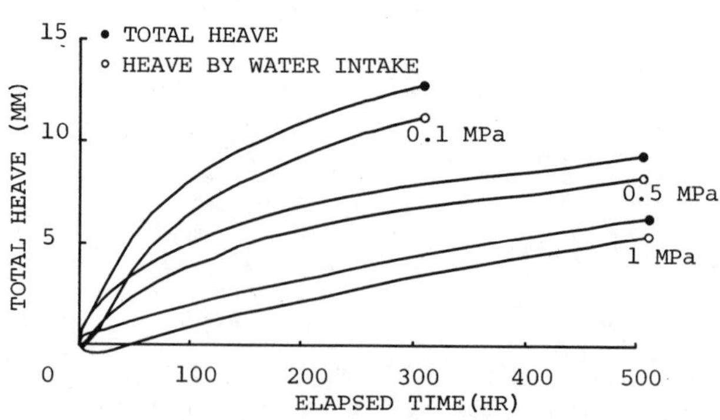

Figure 5 Total heave and heave by water intake

TEST RESULTS AND DISCUSSION

Total Heave and Heave by Water Intake

Three experiments (Test 1, Test 2, Test 3) were carried out with the sample in normally consolidated condition. The overburden pressures were 0.1, 0.5, 1 MPa; the initial sample heights were 94, 86, 104 mm; and the durations were 300, 550, 500 hours for Test 1, Test 2 and Test 3, respectively. The temperatures of the upper and the lower plate were maintained at +5°C and -5°C, respectively. Figure 5 shows both changes of the total heave amount and the heave amount by water intake as a function of time. As the applied load was increased, the rate of water migration into the sample descreased. The rates of the total heave coincide with the rates of the heave by water intake after about 24 hours from the start of the run. This coincidence is due to the completion of in situ water freezing by frost penetration in about 24 hours. This was confirmed by X-ray photos and the temperature profiles of the sample.

Temperature Change in Sample

The soil temperature changes for Test 2 are shown in Figure 6. As porous metal plates with a thickness of 5 mm are attached to both ends of the sample, the temperatures of both sample ends are a little different from those of the cooling plates. The measured temperatures of the upper and the lower end of the sample were +4.7°C and -4.7°C, respectively. The temperature profile established a semisteady state after about 24 hours from the start. However, the soil temperature decreased gradually and was shown to change after 500 hours.

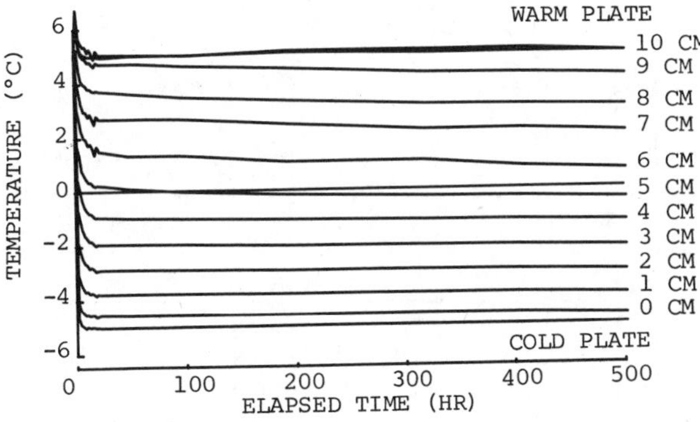

Figure 6 Temperature changes (P = 0.5 MPa)

Displacements in Sample

Displacements of lead shot in the sample for Test 2 are shown in Figure 7. The curves indicated as -1°C, -2°C, -3°C, and -4°C show the displacements of lead shot where the average temperatures are the indicated values. The temperatures at the locations of the lead shot decreased by about 0.2°C between 24 and 500 hours from the start. The U.P. curve shows the displacements of lead shot in the unfrozen part. The line with dots shows the total heave amount. This is consistent with the displacements in the unfrozen part. In the frozen part, greater displacements took place in the warmer layer than in the colder layer. The segregation speed of the large ice lens that formed in the center of the sample decreased with time. However, the segregation rates of the small ice lenses that formed in the region colder than that in which the large ice lens had formed did not decrease in the same way, and were almost constant throughout the duration of this experiment.

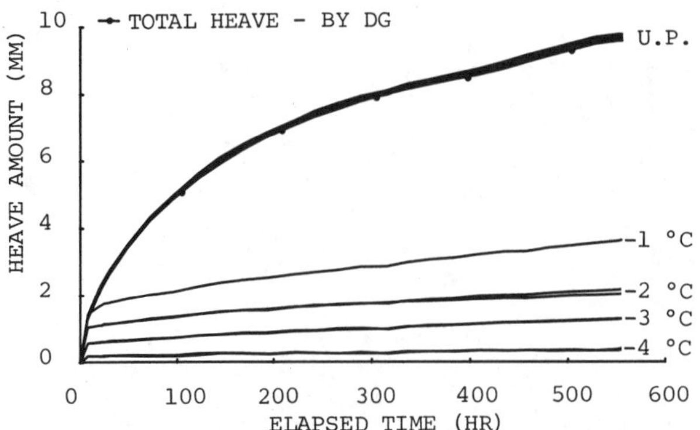

Figure 7 Displacements in sample versus elapsed time (P = 0.5 MPa)

X-ray Films and Other Data

X-ray photos, temperature profiles, displacements of lead shots, and dilatations between lead shots for Test 2 are shown in Figure 8. A, B, C, and D show the results after 24, 96, 315, and 502 hours from the start, respectively. A1 shows a print of the X-ray film of the sample after 24 hours from the start. The dots in the sample are the images of the embedded lead shot. The dots on the outside of the sample show the images of lead shot used as reference points. The horizontal white band in the middle of the sample indicates the image of the segregated ice lens. The thickness of the ice lens increased gradually. The thicknesses of the ice lens for A1, B1, C1, and D1 are 0.6, 3.0, 5.3, and 6.1 mm, respectively. In C1 (315 hours) the other ice lens images can be seen at the distances of 15 and 35 mm from the bottom of the sample. These ice lenses are seen to have increased their thicknesses to D1 (502 hours).

A2, B2, C2, and D2 show the temperature profiles at each elapsed time. The dashed line shows the temperature profiles after 24 hours from the start. The horizontal line shows the location of the upper side of the large ice lens. According to the modified Clausius-Clapeyron equation, ice lenses are considered to segregate at or below -0.45°C when the overburden pressure is 0.5 MPa (Hoekstra, 1969). The vertical line shows -0.45°C. The temperature of the upper side of the large ice lens determined from the temperature profile is close to the expected value.

A3, B3, C3 and D3 show the displacements of lead shot at each elapsed time. The dashed line shows the displacements of lead shot from the start to 24 hours. The dots show the displacements of the upper end plate measured by the dial gauge. And it is noticed that the upper part of the large ice

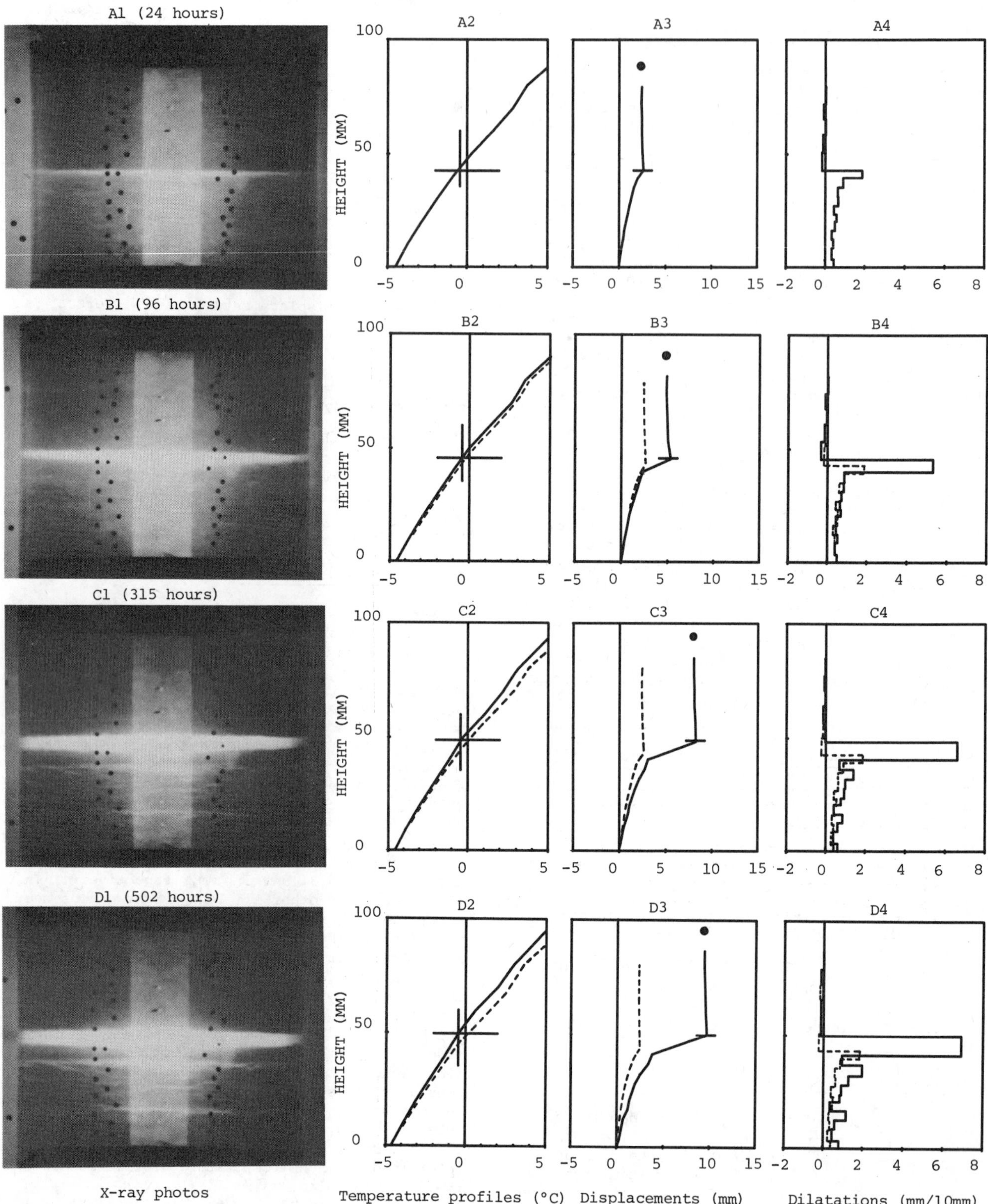

Figure 8 X-ray photos and other data (P = 0.5 MPa)

lens displaced more than the part near the warm end. This means that the unfrozen part has consolidated after the experiment was started.

A4, B4, C4, and D4 show the dilatations between lead shots at each elapsed time. As the distance between lead shots is different in each location, the amount of dilatation was normalized so that the distance between every lead shot is 10 mm. The dashed line shows the dilatations between lead shots from the start to 24 hours. The area between the solid line and the dashed line shows the dilatation of the sample after 24 hours. Large dilatation is shown at the distances of 15 and 35 mm from the bottom, as well as in the center of the sample (C4, D4). This corresponds to the ice segregation in each location which is shown in the X-ray films (C1, D1).

Consolidation due to Soil Freezing

Figure 9 shows the displacement change with various lapses of time for Test 1. The dots show the displacements at the warm end measured by the dial gauge. The horizontal line shows the location of the upper side of the large ice lens. The unfrozen part of the sample has consolidated greatly since the start of this experiment. Line TH shows the displacement of lead shot after the sample was thawed. Figure 10 shows the displacement change for Test 2. In the unfrozen part, the consolidation can also be seen. However, its amount is smaller than that for Test 1. The amount of the thaw consolidation is also small.

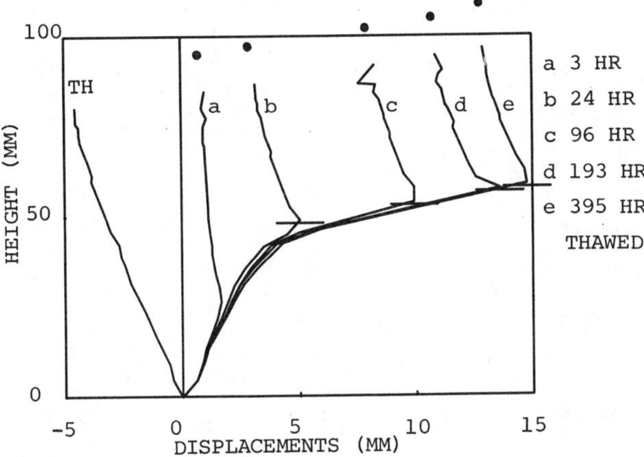

Figure 9 Displacements in sample (P = 0.1 MPa)

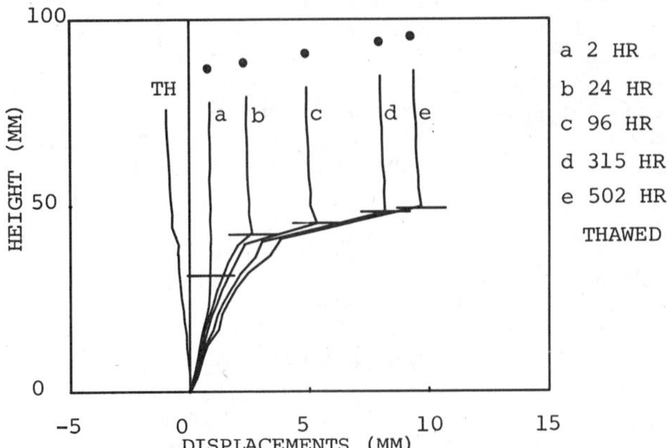

Figure 10 Displacements in sample (P = 0.5 MPa)

Water Flow Rate in Sample

Figure 11 shows the relationship between the water flow rate and the temperature in the sample. As the temperature profile both in the unfrozen part and in the frozen part is almost linear, the horizontal line (the temperature axis) can also be considered to represent its location. F1 shows the average flow rate between 96 and 192 hours, and F2 shows that between 407 and 502 hours. The flow rates in the unfrozen part for F1 and F2 are 1.8 x 10^{-3} and 7.6 x 10^{-4} cm/h, respectively. These values are consistent with the measured water intake rates. The flow rate changes abruptly at the location of the large ice lens. Although the flow rate in the unfrozen part for F2 reduced to 40% of that for F1, the flow rate in the frozen part for F2 did not reduce in the same way. This tendency can also be seen in Figure 7. The flow rate for F2 also changes greatly at the locations of -1.5°C and -3.5°C. This is caused by the segregated ice lens at each location as shown in Figure 8.

CONCLUSIONS

The water movement and the ice segregations were observed by using a recently devised X-ray technique. The experiments were performed for a saturated soil sample under overburden pressure. Ice lens segregation was observed in the region that was colder than the center of the sample where the large ice lens had formed. The dilatation in the frozen part due to ice segregation was measured quantitatively by the precise determination of the locations of lead shot images on X-ray films (Figure 7 and Figure 8). The consolidation of the unfrozen part due to soil freezing was observed

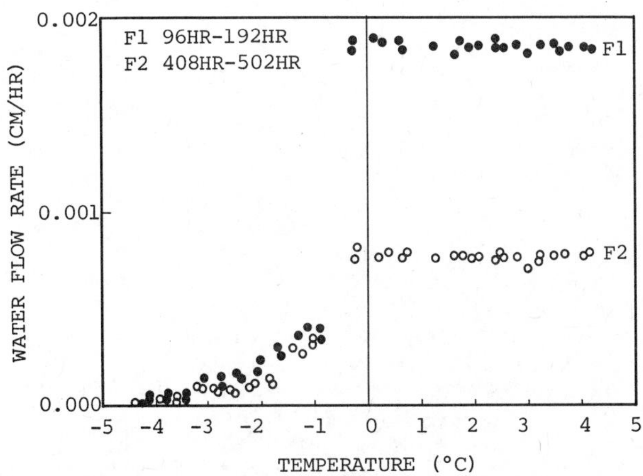

Figure 11 Water flow rate in sample (P = 0.5 MPa)

(Figure 9 and Figure 10). The water flow both in the unfrozen and in the frozen part was calculated (Figure 11). This frost heave apparatus using X-ray technique is found to provide various data on saturated soils under overburden pressure, and these data will be used to develop the theoretical studies of the frost heave phenomenon.

ACKNOWLEDGMENT

The authors wish to thank S. Kinoshita and M. Fukuda of the Institute of Low Temperature Science for their valuable suggestions.

REFERENCES

Dirksen, C., and Miller, R. D., 1966, Closed-system freezing of unsaturated soil: Soil Science Society of America, Proceedings, v. 30, p. 168-173.

Fukuda, M., Orhun, A., and Luthin, J. N., 1980, Experimental studies of coupled heat and moisture transfer in soils during freezing: Cold Regions Science and Technology, v. 3, p. 223-232.

Gilpin, R. D., 1980, A model for the prediction of ice lensing and frost heave in soils: Water Resources Research, v. 16, No. 5, p. 918-930.

Harlan, R. R., 1973, Analysis of coupled heat-fluid transport in partially frozen soil: Water Resources Research, v. 9, No. 5, p. 1314-1323.

Hoekstra, P., 1966, Moisture movement in soils under temperature gradients with the cold side temperature below freezing: Water Resources Research, v. 2, No. 2, p. 241-250.

Hoekstra, P., 1969, Water movement and freezing pressures: Soil Science Society of America, Proceedings, v. 33, p. 512-518.

Hopke, S. W., 1980, A model for frost heave including overburden: Cold Regions Science and Technology, v. 3, p. 111-127

Jame, Y. W., and Norm, D. I., 1976, Heat and mass transfer in freezing unsaturated soil in a closed system, in Proceedings of the Second Conference on Soil-Water Problems in Cold Regions: Edmonton, Alberta, Canada, p. 400-406.

Loch, J. P. G., and Kay, B. D., 1978, Water redistribution in partially frozen, saturated silt under several temperature gradients and overburden loads: Soil Science Society of America, Journal, v. 42, p. 400-406.

Mageau, D. W., and Morgenstern, N. R., 1980 Observations on moisture migration in frozen soils: Canadian Geotechnical Journal, v. 17, p. 54-60.

Takagi, N., Nishio, N., Yoneyama, K., and Shimamura, K., 1983, Development of a strain measurement system for soils and its application to sand around a buried pipe: Soils and Foundation (to appear).

Takashi, T., 1982: Analysis of frost heave mechanism.

Taylor, G. S., and Luthin, J. N., 1976, Numeric results of coupled heat-mass flow during freezing and thawing, in Proceeding of the Second Conference on Soil-Water Problems in Cold Regions: Edomonton, Alberta, Canada, p. 155-172.

CONDITIONS OF PERMAFROST FORMATION IN THE ZONE OF THE BAIKAL-AMUR RAILWAY

S. I. Zabolotnik

Permafrost Institute, Siberian Branch,
Academy of Sciences, Yakutsk, USSR

On the basis of statistical analysis of data derived from many years of observations conducted at 170 weather stations in the region, quantitative relations between the basic environmental parameters were obtained. They enable us to explain the formation of taliks in the western and eastern fringes of the zone and also the extreme low temperatures (down to -7 to -12°C) and the extreme thickness (700-1300 m) of the permafrost in the alpine areas of the Udokan and Kodar ranges. It was established that continuous permafrost occurs in the regions where the mean annual air temperature is lower than -7°C. In those parts of the zone where the mean annual temperature falls between -3 and -7°C, either seasonally frozen ground or permafrost occurs. As a rule, no permafrost forms in regions where the mean annual air temperature is above -3°C.

The Baikal-Amur Railway stretches more than 3.5 thousand kilometers, from the Taishet station of the main Trans-Siberian Railway in the west to the Pacific coast in the east. It crosses vast regions of southern Siberia and the Far East—regions of complex relief including flat country as well as ranges of mountains of different altitude and degree of distribution.

Numerous investigations carried out along the railway itself and in territories adjacent to it (Bibliografiya ... 1978, Bibliografiya ... 1982) have shown that the permafrost regime of the region is extremely complex. The majority of the region is dominated by continuous permafrost, whose temperature varies between values close to 0°C to -10°C and below while its thickness ranges, respectively, from several meters to 500 m and more. At the same time, in the extreme west and in the eastern part of the Baikal-Amur Railway the permafrost exhibits a discontinuous or island-shaped nature while in the Amur River valley to the south of Komsomolsk, it occurs only in the form of individual islands or as short-term permafrost (The Baikal-Amur Railway, 1979).

Such a wide range of variation of the permafrost zone characteristics is largely due to the climatic peculiarities both of the entire region and of its parts in isolation. In order to reveal these peculiarities an analysis was made of many years of observations from all weather stations located within the railway zone—a belt up to 200-250 km across, extending eastward from 101° EL along the Baikal-Amur Railway. At the beginning of construction work on the railway, nearly 170 weather stations were operative within this zone, including about 30 stations in the immediate vicinity of the railway (Directories on Climate in the USSR, v. 22-25, 1966).

This number of weather stations is obviously insufficient to provide a detailed understanding of all parts of such a diverse region. Statistical methods do, however, help to reveal both the general trends in particular climatic parameters and their correlation with one another.

In analyzing correlations between individual quantities, it was assumed that the relationship was linear and obeyed the equation:

$$y = ax + b \qquad (1)$$

where the regression coefficients a and b were determined with a least-squares technique (Smirnov and Dunin-Barkovsky 1959 and Fisher 1958).

One of the principal factors that determines the degree of severity of climate and indicates the possibility that permafrost is present is the annual average air temperature. In assessing this factor in different parts of the region, we must recall that the Baikal-Amur Railway passes mainly in the latitude direction so that the zone of the Baikal-Amur Railway, stretching from west to east by more than 40° L, lies completely within 48 and 58° NL.

The analysis of the dependence of the annual average air temperature (t_a) on longitude of the site (ϕ) shows that it is substantially different for the western and eastern parts of the railway (Figure 1), and the natural demarcation line between these parts occurs at the eastern spurs of the Stanovoi upland.

The railway, moving eastward from Taishet, crosses first the northern part of the Lena-Angara plateau and the near-Baikal depression where the terrain varies between 200 and 400 m above sea level. Then it enters the area of mountain formations of the Stanovoi upland, gradually rising up to 800-1300 m altitude. Naturally, in such conditions the mean annual air temperature steadily decreases in this direction to 120° EL (see Figure 1, line 1). The dependence of the air temperature on longitude of the site in this part of the zone is described by the relation:

$$t_a = 0.34\phi + 31.79. \qquad (2)$$

The lowest air temperatures are observed in the

eastern part of the Stanovoi upland within high-altitude (from 1000 to 1500 m ASL) valleys and hollows of the Yankan, Kalarsky, Udokan and Kodar ridges (Directory ..., v. 23, 1966), whose several tops reach 2000-3000 m above sea level.

The eastern part of the railway leaves the Stanovoi upland and crosses the southern spurs of the Stanovoi Ridge, the Upper-Zeya and Amur-Zeya plains, and the Turan, Bureys and Badzhalsky Ridges before arriving in Komsomolsk-on-Amur. Finally, it surmounts the northern spurs of the Sikhote-Alin to approach the Pacific. It is easy to see that in this direction, except for some local parts, a steady decrease of elevation is experienced, down to sea level. As one moves eastward, the annual average air temperature continuously increases (see Figure 1, line 2) with the longitude-dependence of temperature being described by the relation

$$t_a = 0.28\phi - 40.61. \qquad (3)$$

The considerable range of air temperature fluctuations (±-4°C) is a function of latitude and true altitude (Zabolotnik 1978).

On considering the fluctuations of the annual air temperature within the entire zone of the Baikal-Amur Railway, it is apparent that owing to the specific features of the relief, the air temperature on the average decreases by 3.4°C from west to east with every 10 degrees of longitude along the Bratsk-Udokan stretch while on the contrary that along the Udokan-Sovetskaya Gavan stretch increases by 2.8°C for each 10 degrees of longitude.

The relationships of annual average air temperature fluctuations within the region are mainly due to the different conditions of the formation of wintertime (subzero) temperatures.

The summertime average (above-zero) air temperatures (t_s) vary between 9 and 13°C throughout the Baikal-Amur Railway, with rare exceptions. On the average, the summertime air temperature ranges from 10.5°C in the region of Bratsk to 11.6°C in the region of Komsomolsk-on-Amur (Figure 2, line 3) and its relationship to the longitude of the site is described by the relation:

$$t_s = 0.03\phi + 7.47. \qquad (4)$$

Such a minor variation of the summertime air temperature (0.3°C by every 10 degrees of longitude) cannot, of course, have substantial influence upon the formation of an annual air temperature and indicates the relative homogeneity of the summertime conditions of the entire zone of the Baikal-Amur Railway.

A completely different picture presents itself in the distribution within the railway zone of average wintertime air temperatures (t_w). Their dependence on longitude of the site is practically similar to the one for the air temperature and is expressed through the following relationships: for the western part of the zone (see Figure 2, line 4):

$$t_w = -0.34\phi + 19.09 \qquad (5)$$

while for the eastern part (see Figure 2, line 5):

$$t_w = 0.27\phi - 53.32 \qquad (6)$$

The calculation of dependences (4-6) neglected the data from weather stations situated in the immediate vicinity (up to 50 km) of the Tatar Strait coast, where the influence of the Pacific is far larger than elsewhere. That influence is expressed in a substantial decrease (to 8.5 to 9°C) of average summertime temperatures and in an abrupt increase (up to -9 to -10°C) of average wintertime temperatures (see Figure 2).

The fluctuation of annual average air temperatures within the zone of the Baikal-Amur Railway, predominantly in the subzero range (from -2 to -10°C), indicates the severity of the climatic conditions of the region. However, the subzero air temperature, in itself, is not yet a necessary condition for the formation of permafrost. It is known, for example, that there are taliks even near the Chulman settlement where the annual average air temperature is -9.4°C (Directory ..., v. 24, 1966). The permafrost forms only in areas with a subzero annual average temperature of the ground surface, that is below snow and plant covers.

The warming effect of a snow cover on underlying earth materials is fairly well known (The Foundations of Geocryology, 1959, Dostovalov and Kudryavtsev, 1967, General Geocryology, 1974, 1978, Geocryology, 1981), but determining the magnitude of this effect for a specific area presents a problem. The difficulty arises due to substantial snow cover variation (by 10 to 13 cm), even within one area, depending on a variety of local factors (open or wind-protected site, field, forest, etc.) as well as yearly variations. Therefore, it can be determined only quite roughly.

Within the zone of the Baikal-Amur Railway the winter-average thickness of the snow cover, inferred from 10-day data obtained at weather stations (Directory ..., v. 22-25, 1968) varies between 7 and 45 cm, with maximum thicknesses observed at the western and eastern ends of the zone (Figure 3).

The dependence of snow cover thickness (Z, cm) on longitude of the site is described by the following relationships:
in the western part (see Figure 3, line 6)

$$Z = -0.96\phi + 127.56, \qquad (7)$$

in the eastern part (see Figure 3, line 7)

$$Z = 0.89\phi - 91.61. \qquad (8)$$

Consequently, the snow thickness, on the average, decreases by one cm in the western part of the Baikal-Amur Railway zone for every degree of longitude while that in the eastern part, on the contrary, increases by about the same value.

Of interest are the simultaneous decreases of both annual average air temperature and snow cover thickness in the middle part of the region. Such a combination of natural factors is highly favorable for the development of the lowest temperatures of the ground surface and therefore for the formation of permafrost of large thickness. It is, therefore, not unusual that we have observed within the high-altitude part of the Udokan and Kodar Ridges the lowest temperatures (as low as -7 to -12°C) and the largest thickness (700 to 1300 m) of permafrost in the entire zone of the Baikal-Amur Railway (Nekrosov, Zabolotnik, et al., 1967).

It is possible to determine a particular effect of the snow cover thickness on the underlying ground thickness, using stationary measurements of ground and air temperatures and snow cover thickness. In the Baikal-Amur Railway zone, of the nearly 170 weather stations discussed herein, such measurements are being carried on only at 26 stations. The measurements from these stations indicate that while protecting grounds from the wintertime cooling-down, the snow cover increases the annual average temperature of ground surface by 3 to 7°C. Analysis of the relationship between the difference of annual average temperatures of ground surface and air and the snow cover thickness shows that depending on the amplitude of fluctuations of annual temperatures and snow density the snow cover's warming effect for average wintertime thicknesses of 10, 20 and 30 cm is, respectively, 2.5 to 4.5°C, 4.5 to 8°C and 6.5 to 11.5°C.

Such a large effect of the snow cover on the formation of ground surface temperature can lead, in some areas where snow accumulates constantly or when very snowy years are experienced, to the formation of talik zones at very low air temperatures.

The determination of specific values of the warming effect of the snow cover makes it possible, using air temperatures nearly always obtainable on the basis of the data from the nearest station, to infer the range of ground surface temperature fluctuation. Therefore, prior to conducting field investigations, it is possible to establish the likelihood of the existence of talik zones or masses and islands of permafrost as well as to evaluate their thickness.

Similar information can be obtained using the data from the weather stations which take systematic measurements of ground temperature together with air temperature. Although the number of such weather stations is currently small (31 stations within the zone under consideration), the correlation between annual average air temperatures (t_a) and ground surface temperatures ($t_{g.s.}$), expressible through the relation:

$$t_{g.s.} = 0.77t_a + 4.05, \qquad (9)$$

reveals some features of the formation of the ground temperature condition.

According to relation 9, the ground surface temperature, with some time delay, increases with increasing air temperature. On the average, it reaches 4°C at an air temperature equal to 0°C and becomes subzero only when the air temperature is below -5.2°C (Figure 4, line 8). However, due to the great inhomogeneity of the natural conditions in different areas of such a vast region, the same surface temperature is able to form at air temperatures varying within ±2°C (Figure 4, lines 9 and 10). Therefore, at the present stage of studies it appears possible to determine only the air temperature ranges at which the conditions are favorable for the formation of either seasonally frozen soils or permafrost.

Analysis of the relationship between the temperatures of air and ground surface shows that at an annual average air temperature below -7°C, the ground surface temperature within the Baikal-Amur Railway zone is practically always below zero. Consequently, in the areas of the zone having such an air temperature, there must be continuous permafrost.

As mentioned previously, the annual average air temperatures in the region are able to reach -10°C and lower values. In such conditions, the annual average temperature of the ground surface will vary between -3 and -8°C and the thickness of permafrost will be 150-400 m at the mean gradient of 20 grad/km. In extreme conditions, there are much lower temperatures and permafrost thicknesses in excess of 2 to 3 times.

When the annual average air temperature varies from -3 to -7°C and the region is dominated by areas with such temperatures, either seasonally frozen grounds or permafrost form, depending on the relationship of some regional factors, because the ground surface temperature is able to vary from +3 to -3°C. The permafrost thickness (at the same gradient) can be quite different, from 1 or 2 to 150 m.

In regions where the annual average air temperature is above -3°C, generally no permafrost is present because the ground surface temperature here is above zero (see Figure 4). Exceptions to this are areas, occupied by "mari" (swampy sparse larch forests interrupted by hummocky bogs), in which owing to the cooling effect of moss-peat cover, reaching 2°C and more (Balobaev, Zabolotnik, et al., 1979), mainly thin (of up to 50 m) permafrost zones with annual average ground temperatures of about -2°C occur.

REFERENCES

The Baikal-Amur Railway, 1979, Geocryological map/ V.V. Gogichaishvili, L. P. Golubykh, S. I. Zabolotnik et al., Moscow: GUGK SSSR, 2 s.

Balobaev, V. T., Zabolotnik, S. I., Nekrasov, I. A., Shastkevich, Yu. G., and Shender, N. I., 1979, Dynamics of ground temperature of the North Amur region during development of its territory, in Technogenic landscapes of the North and their recultivation, Novosibirsk: Nauka, p. 74-88.

Bibliografiya po geokriologii zony BAMa, 1978, Nekrasov, I. A., and Chuvashova, I. I., in Permafrost of the BAM zone, Novosibirsk: Nauka, p. 81-119.

Bibliografiya po geokriologii zony BAMa (second communication), 1982, Chuvashova, I. I., Yakutsk: Permafrost Institute SO AN SSSR, 36 p.

Directory on climate of the USSR, 1966, v. 22, p. II, Leningrad: Gidrometeoizdat, 360 p.

Directory on climate of the USSR, 1966, v. 23, p. II, Leningrad: Gidrometeoizdat, 318 p.

Directory on climate of the USSR, 1966, v. 24, p. II, Leningrad: Gidrometeoizdat, 398 p.

Directory on climate of the USSR, 1966, v. 25, p. II, Leningrad: Gidrometeoizdat, 312 p.

Directory on climate of the USSR, 1968, v. 22, p. IV, Air moisture, atmospheric precipitation and snow cover, Leningrad: Gidrometeoizdat, 280 p.

Directory on climate of the USSR, 1968, v. 23, p. IV, Leningrad: Gidrometeoizdat, 328 p.

Directory on climate of the USSR, 1968, v. 24, p. IV, Leningrad: Gidrometeoizdat, 352 p.

Directory on climate of the USSR, 1968, v. 25,

p. IV, Leningrad: Gidrometeoizdat, 276 p.

Dostovalov, B. N., and Kudryavtsev, V. A., 1967, General geocryology, Moscow: Izd. Moskovskogo universiteta, 404 p.

Fisher, R. A., 1958, Statistical methods for research workers, Moscow: Gosstatizdat, 268 p.

The Foundations of geocryology (permafrost science), Part I, 1959, General geocryology, Moscow: Izd. AN SSSR, 460 p.

General geocryology, 1974, Ed. by Melnikov, P. I., and Tolstikh, N. I., Novosibirsk: Nauka, 292 p.

General geocryology (permafrost studies), second edition revised and supplemented, 1978, Ed. by Kudryavtsev, V. A., Moscow: Izd. Moskovskogo universiteta, 464 p.

Geocryology (abridged course), 1981, Ed. by Kudryavtsev, V. A., Moscow: Izd. Moskovskogo universiteta, 240 p.

Nekrasov, I. A., Zabolotnik, S. I., Klimovsky, I. V., and Shastkevich, Yu. G., 1967, Permafrost of the Stanovoi upland and Vitim tableland, Moscow: Nauka, 168 p.

Smirnov, N. V., and Dunin-Barkovsky, I. V., 1959, Abridged course on mathematical statistics for technical applications, Moscow: Fizmatgiz, 436 p.

Zabolotnik, S. I., 1978, Application of the linear correlation method in analysis of conditions of formation of seasonally frozen grounds and permafrost, in Heat exchange in permafrost landscapes, Yakutsk: Permafrost Institute SO AN SSSR, p. 104-118.

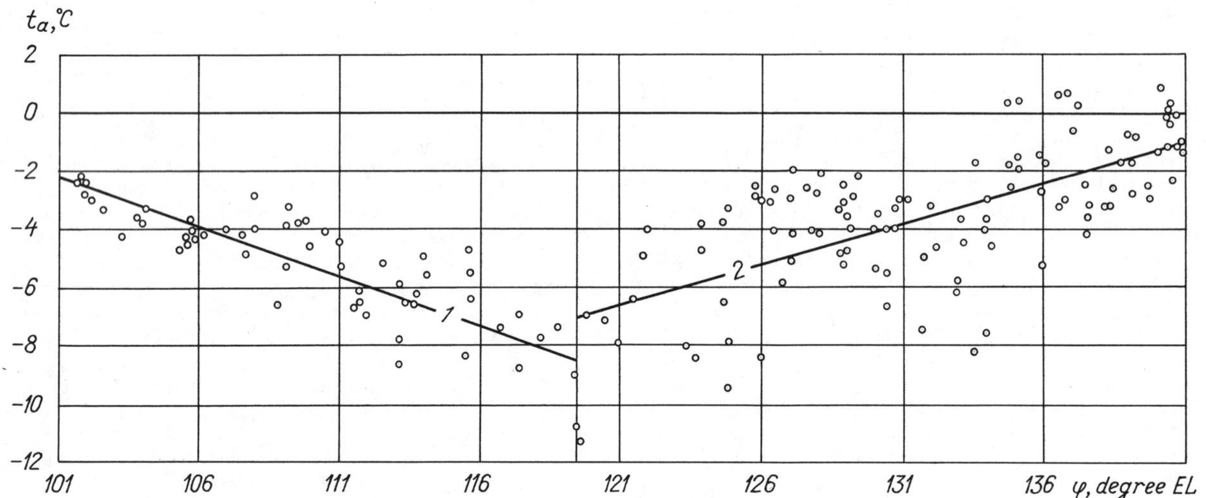

FIGURE 1 Annual average air temperature dependence on longitude of the site in the western (1) and eastern (2) parts of the Baikal-Amur Railway zone. 1 and 2 are the regression lines.

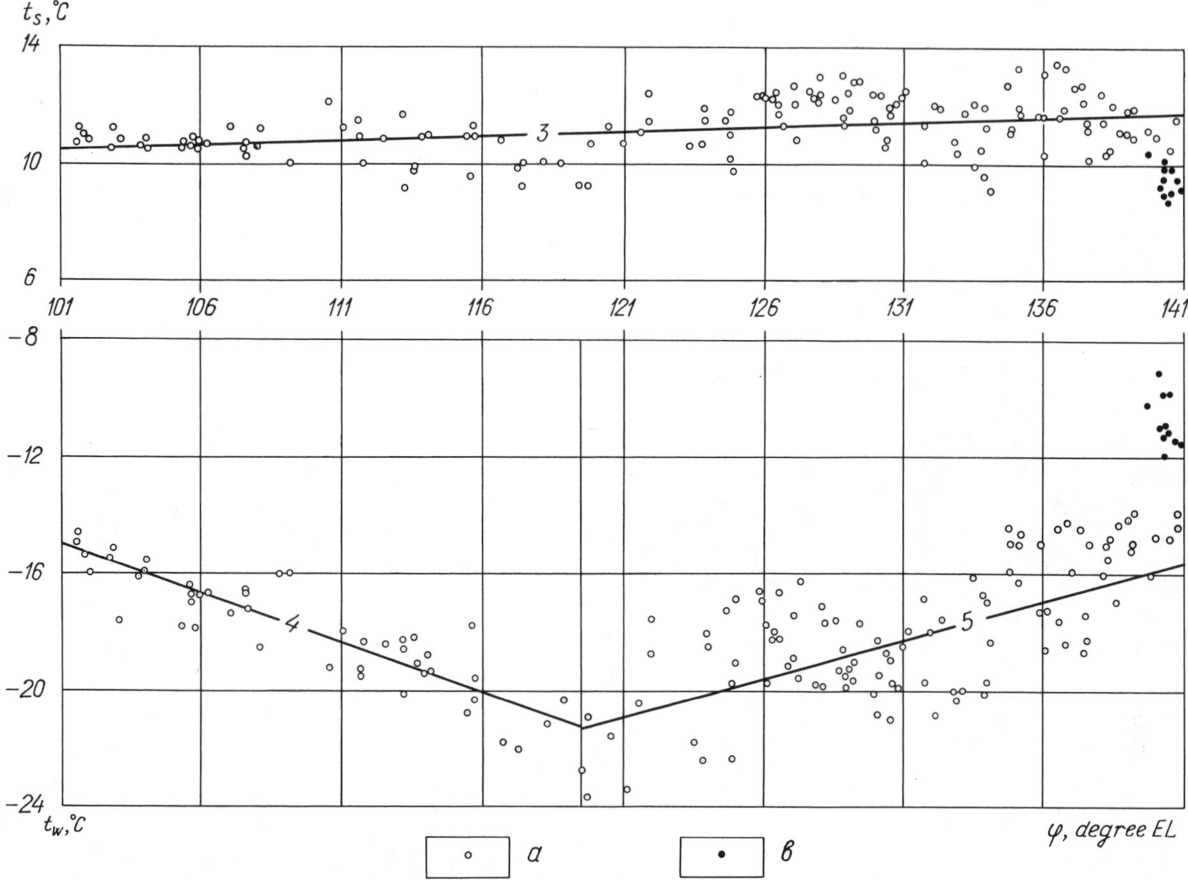

FIGURE 2 The dependence of average summertime (3) and wintertime temperature in the west (4) and east (5) of the Baikal-Amur Railway zone on longitude of the site. (a) - weather stations, used to calculate correlational relationships; (b) - weather stations, neglected in calculation.

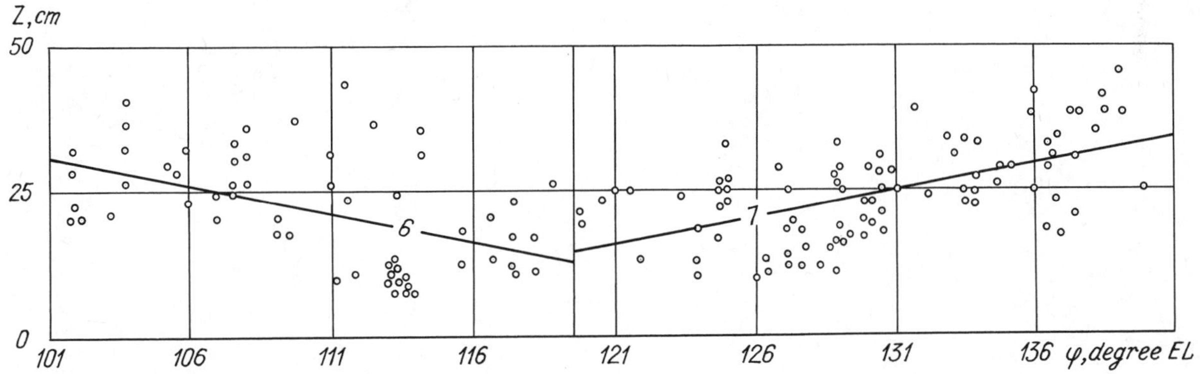

FIGURE 3 The dependence of wintertime-average thickness of snow cover in the west (6) and east (7) of the Baikal-Amur Railway zone on longitude of the site.

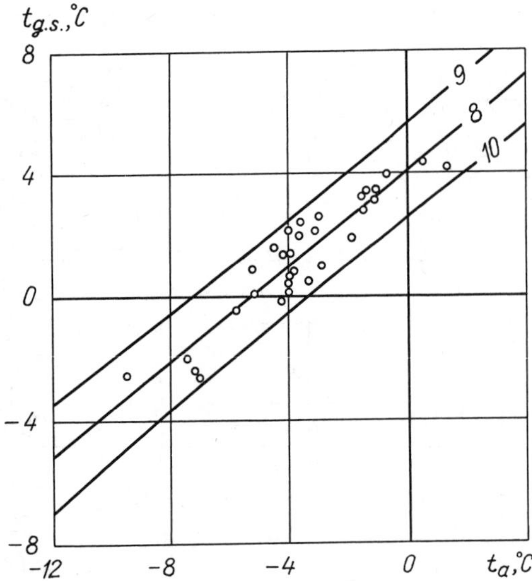

FIGURE 4 Relationship between the annual average air temperature (t_a) and the ground surface temperature ($t_{g.s.}$). (8) - regression line, (9) and (10) are the 95% confidence intervals.

A MATHEMATICAL MODEL FOR THE VISCOPLASTIC DEFORMATION OF FROZEN SOILS

Yu. K. Zaretskiy[1] and A. G. Shchobolev[2]

[1]Scientific Research Center of Hydroproject Institute, Moscow, USSR
[2]Civil Engineering Institute, Moscow, USSR

A rheological model for frozen soil and ice, based on a theory of viscoplastic hardening, is suggested. The critical equations take the form of non-holonomic incremental relationships and describe the behavior of frozen ground and ice with time both in pre-limiting and limiting states. The model is formulated in such a way that no special experimental work is required to derive its parameters. Design parameters may be determined from triaxial compression tests. The results of numerical calculations of the behavior of frozen soils under triaxial compression are presented for various loading trajectories and for cases involving both decaying and progressive creep. The predicted soil behavior agrees well with experimental data. Results of applying the proposed model for solving the boundary problem using the finite elements method are also presented.

Frozen soils show pronounced rheological properties resulting from the presence of ice layers and massive frozen water in their composition. Deformability of frozen soils may be described in terms of viscoplastic hardening theory based on the following assumptions.

For an elastic viscoplastic medium the strain tensor may be represented as

$$\varepsilon_{ij} = \varepsilon_{ij}^{e} + \varepsilon_{ij}^{vp}$$

where ε_{ij}^{e} and ε_{ij}^{vp} are the components of elastic and viscoplastic strain tensors, respectively.

Following Naghdi and Murch (1963) we assume the existence of a piecewise smooth instantaneous loading surface which may be described by the following relationship,

$$f_o^{(\tau)} = f_o^{(\tau)}(\hat{\sigma}_{ij}, \varepsilon_{ij}^{vp}, \chi_i, k_i, T) = 0, \quad \tau = 1,2 \ldots \quad (1)$$

where $\hat{\sigma}_{ij}$ are the components of active stress tensor, $\hat{\sigma}_{ij} = \sigma_{ij} - \tau_{ij}$, σ_{ij} are the components of the applied stresses, τ_{ij} are the components of internal stresses, $\tau_{ij} = f(\varepsilon_{ij}^{vp}, \dot{\varepsilon}_{ij}^{vp})$ at $\dot{\varepsilon}_{ij}^{vp} = 0$, $\tau_{ij} = 0$, ε_{ij}^{vp} are the components of viscoplastic strains, χ_i are the hardening parameters, T is the frozen soil temperature, k_i are the material constants, $\dot{\varepsilon}_{ij}^{vp}$ is the velocity of viscoplastic strains,

$$\dot{\varepsilon}^{vp} = \frac{d\varepsilon_{ij}^{vp}}{\alpha t}$$

Region of elastic behavior of the medium corresponds to negative values of instantaneous loading surface

$$f_o^{(\tau)} < 0 \quad (2)$$

For the stabilized state (with zero viscoplastic strain velocity) the instantaneous loading surface coincides with the stabilized surface $f_s^{(\tau)} = 0$ which corresponds to the loading surface introduced in plastic hardening theory. The equation of the stabilized surface has the form

$$f_s^{(\tau)} = f_s^{(\tau)}(\sigma_{ij}, \varepsilon_{ij}^{vp}, \chi_i, k_i, T) = 0 \quad (3)$$

Some features of accumulation mechanism of viscoplastic strains accounted for by introduction of the instantaneous loading surface may be illustrated by a simple mechanical model (Figure 1). It is obvious that the strain of the mechanical model is equal to a sum of elastic, e^{e}, and viscoplastic, e^{vp}, components. The applied stress σ is distributed between viscous and plastic elements, respectively. The stress in the plastic element will be called "active," $\hat{\sigma}$, while the stress in the viscoplastic element will be called "internal," r. It is clear that the plastic element's deformation is due to the magnitude of active stress $\hat{\sigma}$, rather than total stress σ. In the general case, the process of plastic strains' accumulation is described by the instantaneous loading surface, which depends on the active stress tensor. The internal stress tensor takes account of a delay in accumulation of plastic strains with respect to the moment in time of applying the stress. In general, this tensor depends on the value of viscoplastic strains and on their velocity.

The instantaneous loading surfaces have the following geometric interpretation (Figure 2). The stabilized surface I-I corresponds to the stressed state M_I on the plane of invariants $\sigma_i \div \sigma$. Under repeated loading by a vector M_IM_2 the surface I-I finally occupies the position II-II, provided the deformation process is stabilized. If the equation of instantaneous loading surface is written in terms of active stresses, the loading point position in different time moments is determined by vectors M_IM_2, M_IM_2'',... The components of these vectors depend on the velocity of viscoplastic strains and they are equal to $\hat{\sigma}_{ij} = \sigma_{ij} - \tau_{ij}$. Instantaneous loading surfaces II-II, II"-II",... pass through the loading points M_2M_2",... In the stabilized state, when $\tau_{ij} = 0$, the loading point coincides with the point M_2, while the instantaneous surface coincides with the stabilized surface II-II. If the instantaneous surface is written in terms of applied stresses, the point M_2 is passed through by instantaneous loading surfaces I-I, I"-II" (at different moments in time), the position of which depends on viscoplastic strain velocity. It can be easily seen that the surfaces I-I, I"-I",... are parallel to the surfaces II-II, II"-II",...

It should be noted that an experimenter who would like to check the associative law, would come to the conclusion that the vector of plastic strain increment at the loading point (M_2) changes its direction, so the point may be classified as singular.

Let us introduce criteria of unloading, loading and neutral loading for the piecewise smooth instantaneous loading surface $f_o = 0$ (Ivlev and Bykovtsev 1971, Zaretskiy and Lombardo 1982). In accordance with equation 3 the material is assumed to be elastic, $f_o^{(\tau)} < 0$ inside the instantaneous surface.

The unloading process occurs if the active stress velocity is such that both viscoplastic strain velocity and increments of parameters $d\chi_i$ are equal to zero*: $\dot{\varepsilon}_{ij}^{vp} = 0$, $d\chi_i = 0$

$$f_o^{(m)} = 0,\ \dot{f}_o^{(m)} \equiv \frac{\partial f_o^{(m)}}{\partial \hat{\sigma}_{ij}} \dot{\hat{\sigma}}_{ij} = \frac{\partial f_o^{(m)}}{\partial \sigma_{ij}} (\dot{\sigma}_{ij} - \dot{\tau}_{ij}) < 0,\quad f_o^{(j)} < 0, \tag{4}$$

where m and j are different and enumerate all the values of indices r.

Under neutral loading the active stress velocity is such that the end of the vector $\hat{\sigma}_{ij}$ remains at the instantaneous loading surface. In this case $\dot{\varepsilon}_{ij}^{vp} = 0$, $d\chi_i = 0$, $f_o^{(m)} = 0$,

$$\dot{f}_o^{(m)} \equiv \frac{\partial f_o^{(m)}}{\partial \hat{\sigma}_{ij}} \dot{\hat{\sigma}}_{ij} = \frac{\partial f_o^{(m)}}{\partial \sigma_{ij}} (\dot{\sigma}_{ij} - \dot{\tau}_{ij}) = 0 \tag{5}$$

where m does not enumerate all the values of indices r, since the unloading process may take place for some regular parts of the surface.

Under the loading process the active stress velocity is such that the following relations hold: $\dot{\varepsilon}_{ij}^{vp} \neq 0$, $d\chi_i \neq 0$, $f_o^{(m)} = 0$,

$$\dot{f}_o^{(m)} = \frac{\partial f_o^{(m)}}{\partial \hat{\sigma}_{ij}} \dot{\hat{\sigma}}_{ij} + \frac{\partial f_o^{(m)}}{\partial \varepsilon_{ij}^{vp}} \dot{\varepsilon}_{ij}^{vp} + \frac{\partial f_o^{(m)}}{\partial \dot{\varepsilon}_{ij}^{vp}} + \frac{\partial f_o^{(m)}}{\partial \chi_i} \dot{\chi}_i = 0,$$

$$\dot{f}_o^{(m)} = \frac{\partial f_o^{(m)}}{\partial \hat{\sigma}_{ij}} \cdot \dot{\hat{\sigma}}_{ij} = \frac{\partial f_o^{(m)}}{\partial \sigma_{ij}} (\dot{\sigma}_{ij} - \dot{\tau}_{ij}) = 0 \tag{6}$$

---

* Since $\frac{\partial \hat{\sigma}_{ij}}{\partial \sigma_{ij}} = 1$ and τ_{ij} is independent of σ_{ij} we may write:

$$\frac{\partial f_o^{(m)}}{\partial \hat{\sigma}_{ij}} \dot{\hat{\sigma}}_{ij} = \frac{\partial f_o^{(m)}}{\partial \sigma_{ij}} (\dot{\sigma}_{ij} - \dot{\tau}_{ij}) .$$

where m does not enumerate all the values of indices r, since unloading or neutral loading processes may take place for some parts of the surface.

It may be shown that Mizes's maximal principle is valid within the framework of viscoplastic hardening theory (Zaretskiy and Lombardo 1982). The principle states that any given parameters ε_{ij}^{vp}, χ_i and any given value of strain velocity components $\dot{\varepsilon}_{ij}^{vp}$, the inequality

$$\hat{\sigma}_{ij}\dot{\varepsilon}_{ij}^{vp} > \hat{\sigma}_{ij}^{*}\varepsilon_{ij}^{vp} \tag{7}$$

takes place, where $\hat{\sigma}_{ij}$ are the active stresses, corresponding to the given value ε_{ij}^{vp}, $\hat{\sigma}_{ij}^{*}$ are the components of any possible stressed state which satisfies the condition $f_o(\hat{\sigma}_{ij}^{*},\ldots) < 0$.

Equation 7 may be written in another form:

$$(\sigma_{ij} - \tau_{ij})\dot{\varepsilon}_{ij}^{vp} \geq (\sigma_{ij}^{*} - \tau_{ij})\dot{\varepsilon}_{ij}^{vp} \tag{8}$$

or

$$\sigma_{ij}\dot{\varepsilon}_{ij}^{vp} \geq \sigma_{ij}\dot{\varepsilon}_{ij}^{vp} \tag{9}$$

It immediately follows the validity of associated law of flow in spaces of both active and applied stresses. Thus we may write:

$$\dot{\varepsilon}_{ij}^{vp} = \dot{h}\frac{\partial f_o}{\partial \hat{\sigma}_{ij}} = \dot{h}\frac{\partial f_o}{\partial \sigma_{ij}} \qquad \left(\frac{\partial \hat{\sigma}_{ij}}{\partial \sigma_{ij}} = 1\right) \tag{10}$$

For piecewise smooth instantaneous loading surface it follows from generalized Koiter's law that

$$\dot{\varepsilon}_{ij}^{vp} = \sum_i \dot{h}^{(i)} \frac{\partial f_o^{(i)}}{\partial \hat{\sigma}_{ij}} = \sum_i \dot{h}^{(i)} \frac{\partial f_o^{(i)}}{\partial \sigma_{ij}}\ ; \quad \left(\frac{\partial \hat{\sigma}_{ij}}{\partial \sigma_{ij}} = 1\right) \tag{11}$$

A procedure of experimental determining the time domain of recoverable strains, i.e. of determining the position of instantaneous loading surface, was presented in detail by Zaretskiy and Lombardo (1982).

Within the framework of viscoplastic hardening Zaretskiy (1980) formulated master equations describing transient creep of unfrozen soils. On the basis of these equations, he solved the stability problem of a dam made of soil materials.

While describing the rheological behavior of ice and frozen soil, it is sometimes necessary to consider both transient and progressive characters of strains' evolution in time.

Based on a stochastic approach Zaretskiy and Lombardo (1982) suggested an equation of instantaneous loading surface describing the process of ice deformation and destruction. This equation may be written in a modified way as follows:

$$f^o = \sigma_i - \sigma_{i(t)}^{*}\,\phi(\varepsilon_i^{vp}) - \frac{m}{\psi(t)}\,\ell n\left(1 + \frac{\dot{\varepsilon}_i^{vp}}{\dot{\varepsilon}_i^{*}}\right) = 0 \tag{12}$$

where σ_i is the applied stress intensity and $\sigma^*_{i(t)}$ is the long-term strength of soil. In the case where the long-term strength curves are similar $\sigma^*_{i(t)}$ may be approximated by the function

$\sigma^*_{i(t)} = \sigma^*_{i(\infty)}(1 + \frac{æ^*}{\psi(t)})$ which depends on the sum of main stresses $\sigma_{i(\infty)} = c_{(\infty)} + \sigma tg\phi_{(\infty)}$, $\psi(t) = t^{\delta}$, $\delta < 1$. If there is no similarity, the long-term strength values are to be introduced into computational program in table form and found for each value of t and σ by linear interpolation. In equation 12 $\varphi(\varepsilon_i^{vp})$ is the hardening function,

$\varphi(\varepsilon_i^{vp}) = \frac{\varepsilon_i^{vp}}{A + B \cdot \varepsilon_i^{vp}}$, A and B are parameters which depend on average stress; $\varphi(\varepsilon_i^{vp}) < 1$ after reaching the long-term strength limit the hardening function becomes $\varphi(\varepsilon_i^{vp}) \equiv 1$; $\dot{\varepsilon}_i^*$ is the minimal velocity of viscoplastic strains, which characterizes the transition into the limiting state

$\dot{\varepsilon}_i^* = \frac{1}{\eta}\dot{\psi}(t)$; t is the internal time, which corresponds to real time in case of active loading process, and to the time of viscoplastic deformation of the material in case of unloading process.

The equation of instantaneous surface includes time-dependence explicitly. This is due to the chosen form of the function, describing the long-term strength of ice, and not to a disadvantage of viscoplastic hardening theory. Zaretskiy and Chumichev (1980) suggested a condition of the long-term strength which does not depend on time explicitly. Although mathematical difficulties have not permitted realization of the condition in practical calculations until recently, the solving of this problem has already begun.

The model of ice proposed describes its deformability in both pre-limiting and limiting states. The volume creep flows due to the hydrostatic compression not being taken into account, the behavior of ice inside the instantaneous surface is assumed to be elastic.

A comparison of theoretical prediction of the ice creep with experimental data was presented by Zaretskiy and Chumichev (1980) for the case of one-dimensional compression of a sample. The model of ice formulated makes it possible to take into account the influence of loading regime on the ice deformability. Figure 3a shows loading diagrams of a sample under one-dimensional compression, while Figure 3b shows creep curves calculated on the basis of the model proposed. It is important to note that decrease of stresses does not always lead to unloading (curve 4). This is due to the fact that equation 4 of the unloading process is written for active stresses rather than applied stresses. Therefore, the unloading process will occur provided the applied stress vector finds itself inside instantaneous loading surface, which in this case coincides with the stabilized surface. The instantaneous surface changes its position in time, therefore equal decrease of stresses σ_1 occurring in various time moments may lead to unloading (curve 3) and may not (curve 4).

Based on the model of ice formulated, Groshev and Chumichev (1982) solved the problem of the ice field pressure on stationary supports.

To describe the frozen soil deformability under isothermal conditions the piecewise smooth instantaneous loading surface is used (Zaretskiy and Lombardo 1982, Shchobolev 1982). An analytic expression for regular parts has the form:

$$f_o^{(1)} = \sigma - \sigma^*,\quad f_o^{(2)} = \sigma_i - \sigma_{i(t)}\varphi(\varepsilon_i^{vp}) - \frac{1}{\psi(t)}\ln\left(1 + \frac{\dot{\varepsilon}_i^{vp}}{\dot{\varepsilon}_i^*}\right);\quad f_o^{(3)} = \sigma - p \tag{13}$$

where σ^* is the strength limit with respect to tension; the other notation is the same as in equation 12.

Volumetric strain of the frozen soil may be represented as

$$\varepsilon_v^{vp} = \varepsilon_{v(o)}^{vp} + \varepsilon_{v(i)}^{vp} \tag{14}$$

where $\varepsilon_{v(o)}^{vp}$ is the volumetric strain, caused by hydrostatic pressure, and $\varepsilon_{v(i)}^{vp}$ is the volumetric strain, caused by tangential stress intensity. Volumetric strain of frozen soil, which is determined by hydrostatic pressure and by tangential stress intensity in pre-limiting state is usually small, so for the sake of simplicity it may be neglected.

The function P depends on the ratio $\varepsilon_i^{vp}/\varepsilon_{v(i)}^{vp}$ and on the tangential stress intensity σ_i. Zaretskiy and Gorodetskii (1975) found that the dilatant part of volumetric strain and shear strain intensity develop in synchronism, their ratio being independent of time. This condition may be written mathematically as

$$\frac{\varepsilon_i^{vp}}{\varepsilon_{v(i)}^{vp}} = \beta(\sigma)\cdot\sigma_i;\quad \sigma_i \le \sigma^*_{i(\infty)}$$

$$\frac{\varepsilon_i^{vp}}{\varepsilon_{v(i)}^{vp}} = \beta(\sigma)\cdot\sigma^*_{i(\infty)};\quad \sigma_i \ge \sigma^*_{i(\infty)} \tag{15}$$

Neglecting volumetric strains caused by tangential stress intensity at $\sigma_i < \sigma^*_{i(\infty)}$, we may write the function P as follows:

$$p = \frac{\varepsilon_i^{v}/\varepsilon_{v(i)}^{vp} - C_1}{C_2} \tag{16}$$

where C_1 and C_2 are the material parameters, which depend on σ. The model parameters can be determined by means of soil tests under triaxial. For the model to be used it is necessary to determine:

- the long-term strength curve $\sigma^*_{i(t)} = \sigma_{i(\infty)}(1 + \frac{æ^*}{\psi(t)})$ with the account of the dependence $\sigma_{i(\infty)} = f(\sigma)$;

- the hardening function $\phi(\varepsilon_i^{vp})$ based on stabilized values of viscoplastic strains at $\sigma_i < \sigma^*_{i(t)}$;

- the function $p(\varepsilon_i^{vp}, \varepsilon_{v(i)}^{vp}$ and the minimal velocity of viscoplastic strains $\dot{\varepsilon}_i^{vp}$;

- the value σ and moduli of elasticity G^e and K^e.

A picture of the instantaneous loading surface for a frozen soil is shown in Figure 4. Figure 5 shows a comparison of computational results obtained using the model presented, with experimental data (Gorodetskii 1975) of frozen soil tests at the deviatoric stress path.

Thus, the master equations proposed for a frozen soil have non-holonomic incremental character. The model enables one to describe the soil behavior in both prelimiting and limiting states with account of the loading path. The model does not require special investigations for determining its parameters. The design parameters may be found from triaxial compression tests by means of triaxial apparatus.

The model formulated served as a basis for solving the evolution problem of strength and deformability of a frozen retaining structure by means of the finite elements method. The solving algorithm was constructed on the basis of the initial strains' method (Zienkiewicz 1975), while separation of strains into elastic and viscoplastic components was done by a procedure suggested by Zaretskiy and Lombardo (1982).

The computational diagram is shown in Figure 6 for a case of shaft sinking by 2 m stopes. Fixation of the top end section by means of tubing was assumed to be rigid. The calculations were made for Callovian silty sandy loam, the pressure on the frozen retaining structure was assumed to be distributed uniformly and to be equal to 5.0 MPa which corresponds to the rock pressure at the depth about 500 m. The bottom was assumed to be completely frozen and locked in position. The soil temperature beyond the cross-section of the frozen retaining structure was taken constant and equal to 50°F.

Figure 7 shows the profile of the frozen soil pressure on the tubing for time moments t = 15 minutes and t = 6 hours. Figure 8 shows bending of the shaft walls for the same time moments.

The maximal displacement of walls of the frozen retaining structure may be found in terms of an analytic solution (Vyalov, Zaretskiy and Gorodetskii 1981) by a formula

$$\frac{B}{a} = [1 + \frac{(1-m)}{\bar{A}} (1-\xi)(\frac{h}{a})^{1+m}(\frac{a}{u_a})^m]^{\frac{1}{1-m}} \tag{17}$$

where the following values of parameters for Callovian silty sandy loam are to be used at temperature 50°F: $m = 0.27$; $t_p = 6$ hours,

$\bar{A}(t_p) = 3^{-\frac{1+m}{2}} A(t_p)$, $A(t_p) = 6.4$ MPa, $\xi = 0$ or 0.1 depending on fixation degree of the computational cylinder's tops. The results of calculations by the equation 17 mean that the frozen retaining structure's displacement at the time moment t = 6 hours varies from 12 cm to 8 cm, while the corresponding observed result is 9 cm. Thus the results obtained are in good agreement with analytic solutions.

Further improvements of the approach, which is directed to a description of viscoplastic behavior of soils, require detailed theoretical developments and more experimental investigations of rheological behavior of soil under conditions of complex stressed states for various loading paths and regimes.

REFERENCES

Gorodetskii, S. E., 1975, Frozen soil creep and strength in complex stressed state: Bases, Foundations and Soil Mechanics (in Russian), no. 5, p. 39-43.

Groshev, M. I., and Chumichev, B. D., 1982, The ice creep and analysis of the ice field pressure on stationary supports. Theses of reports at IV All-Union Symposium on Soil Rheology: Samarkand, p. 95-96.

Ivlev, D. D., and Bykovtsev, G. I., 1971, A theory of hardenable plastic bodies: Moscow, Nauka.

Naghdi, P. M., and Murch, S. A., 1963, On the mechanical behavior of viscoelastic plastic solids: J. Appl. Mech., v. 30, p. 321-328.

Shchobolev, A. G., 1982, A model of viscoplastic deformation of frozen soils. Theses of reports at IV All-Union Symposium on Soil Rheology: Samarkand, p. 183-184.

Vyalov, S. S., Zaretskiy, Yu, K., and Gorodetskii, S. E., 1981, Strength and creep analysis under artificial freezing: Leningrad, Stroyizdat.

Zaretskiy, Yu. K., 1980, A new concept of viscoplastic soil flow, in Proceedings of the III All-Union Symposium on Soil Rheology: Leningrad, Publ. Erevan University, p. 58-73.

Zaretskiy, Yu. K., and Gorodetskii, S. E., 1975, The frozen soil dilatance and development of the deformation theory of creep: Hydrotechnical Building, no. 2, p. 15-18.

Zaretskiy, Yu. K., and Chumichev, B. D., 1982, The short-term creep of ice: Novosibirsk, Nauka, Siberian Department.

Zaretskiy, Yu. K., and Lombardo, V. n., 1982, Mathematical models and method of analysis of viscoplastic soil deformation. The report at IV All-Union Symposium on Soil Rheology: Samarkand (in press).

Zaretskiy, Yu. K., and Lombardo, V. N., 1982, Soil dam statics and dynamics: Energoizdat (in press).

Zaretskiy, Yu. K., Lombardo, V. N., and Scherbina, V. E., 1982, An analysis of viscoplastic strains of soil dams. The report at the IV All-Union Symposium on Soil Rheology: Samarkand (in press).

Zienkiewicz, O., 1975, Finite elements method in technology: Moscow, Mir.

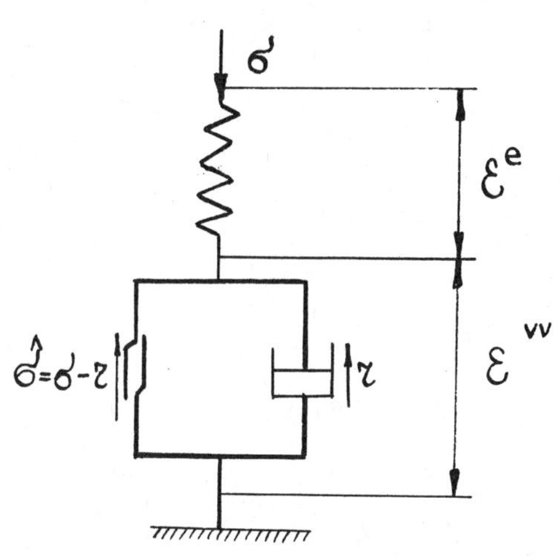

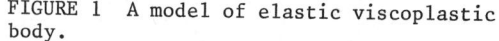

FIGURE 1 A model of elastic viscoplastic body.

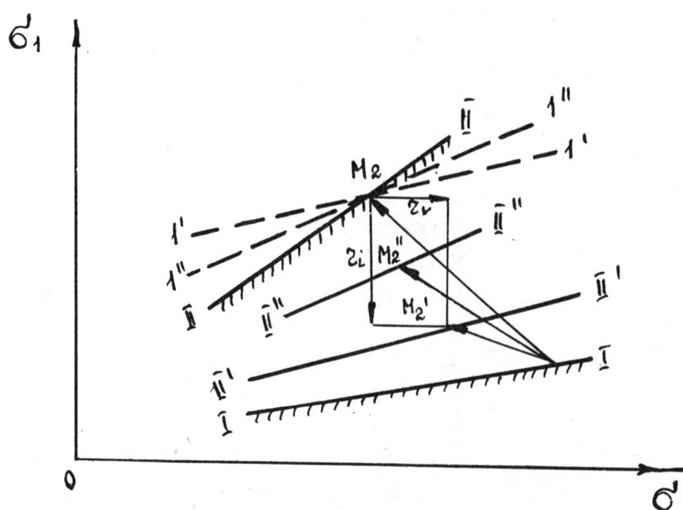

FIGURE 2 A geometric interpretation of instantaneous loading surfaces.

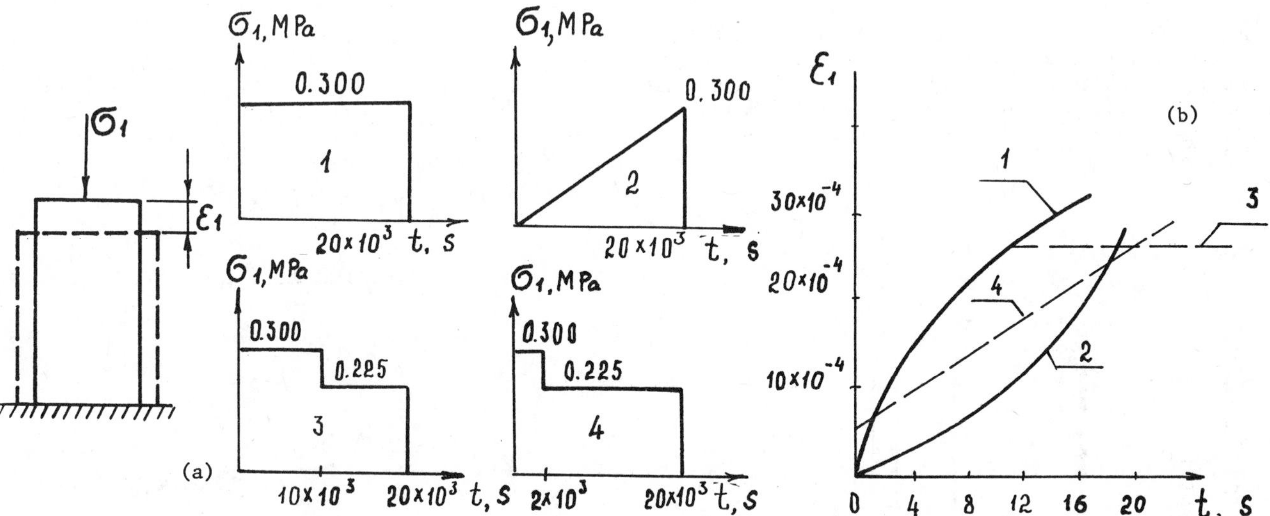

FIGURE 3 Creep curves under one-dimensional compression:
(a) the loading diagram and regimes;
(b) the time-evolution of a deformation ε_1.

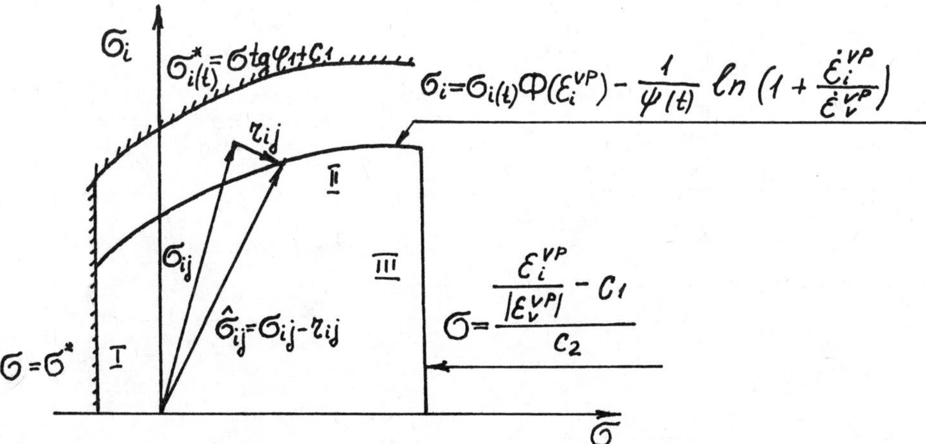

FIGURE 4 Instantaneous loading surface.

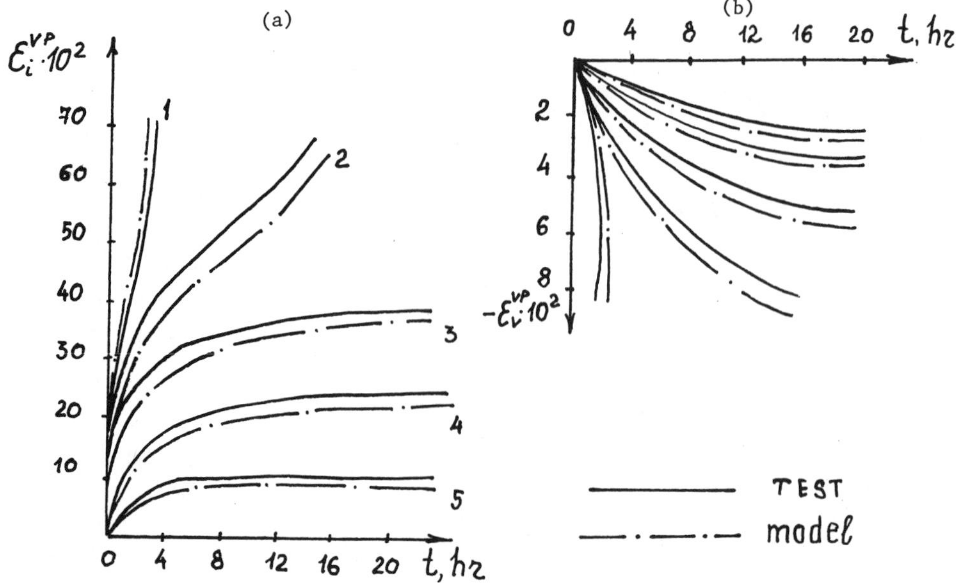

FIGURE 5 Creep curves for shear (a) and volumetric (b) strains (in case (b) volumetric compaction strains are not taken into account) of a sample at hydrostatic pressure σ = 3.0 MPa for various values of shear intensity σ_i: 1 - 3.9; 2 - 3.47; 3 - 3.03; 4 - 2.6; 5 - 2.16 MPa.

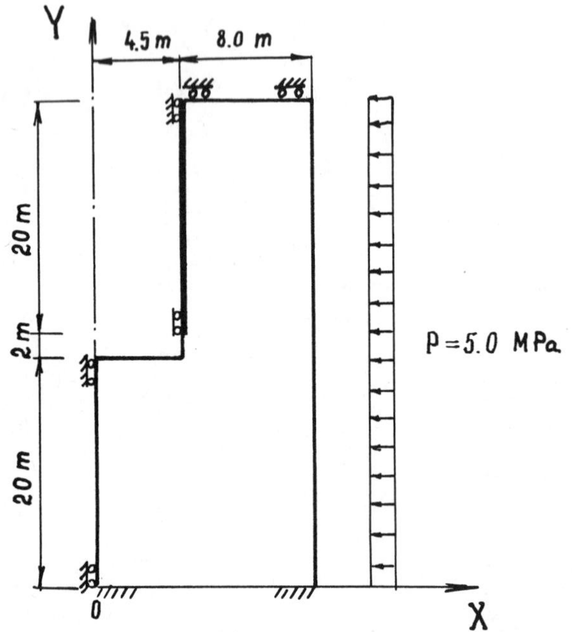

FIGURE 6. Computational diagram of frozen retaining structure.

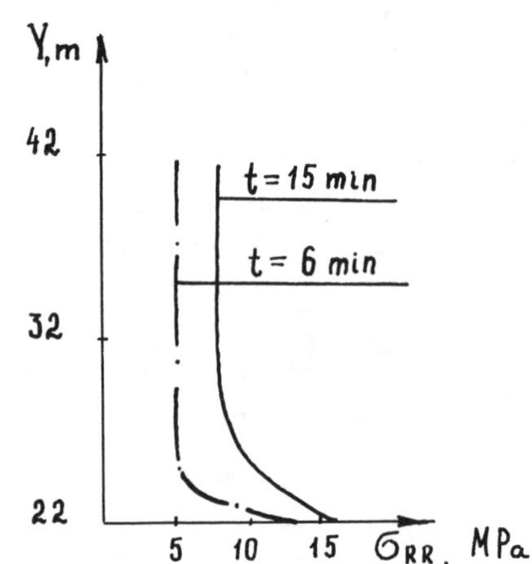

FIGURE 7 The profile of the pressure on the support.

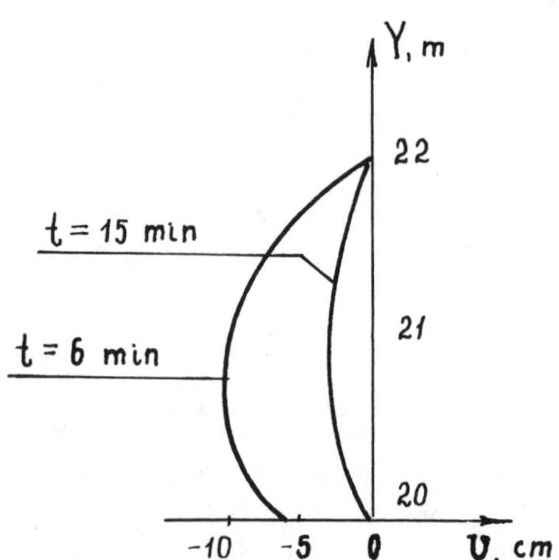

FIGURE 8 Radial displacements of horizontal wall of locking portion of the shaft.

AIR DUCT SYSTEMS FOR ROADWAY STABILIZATION OVER PERMAFROST AREAS

John P. Zarling, P.E.[1], Billy Connor, P.E.[2], and Douglas J. Goering[3]

[1]Department of Mechanical Engineering, Duckering Building
University of Alaska, Fairbanks, Alaska 99701
[2]State of Alaska, Alaska Department of Transportation and Public Facilities
Division of Planning and Programming, Research Section
2301 Peger Road, Fairbanks, Alaska 99701-6394
[3]Department of Mechanical Engineering, Duckering Building
University of Alaska, Fairbanks, Alaska 99701

This paper presents the results of both an experimental and analytical research program undertaken to develop design criteria for air duct systems. These systems have been used in Alaska as a method of preventing degredation of permafrost beneath highway embankments. An experimental duct was assembled and instrumented to determine the relationship between air flow rates and temperature difference, heat transfer rates, air duct length, stack heights, etc. A finite element computer model has also been used to investigate the placement of the air duct under the roadway. Optimum placement of the air ducts would allow sufficient winter cooling of the ground so that degradation of the underlying permafrost would not occur during the summer thawing season. Thermal profiles resulting from the finite element simulations showing the effects of air duct placement are presented. From the data presented in this paper, the design engineer should find it easier to design air duct systems for roadway stabilization in permafrost areas.

INTRODUCTION

The existence of permafrost in northern Canada and Alaska has required roadway designers to pay special attention to the thermal regime of the ground. In the past, many highway projects have employed varying thicknesses of gravel fill, or in some cases foam board type insulation with gravel fill to thermally protect the underlying frozen ground.

The purpose of using gravel fill or using insulation with gravel fill is to maintain the active layer within the non-frost-susceptible fill material. Generally, the thickness of the fill required to thermally protect the underlying permafrost increases as the mean annual soil surface temperature approaches 0°C. The thickness of the fill can be reduced by the use of foam plastic board insulation. Another source for structural degradation of the roadway is the embankment rotation due to thaw settlement. In many cases, even though the active layer is maintained in the fill material below the central portion of the roadway, the sloping of the fill material to smaller thicknesses out to the toe of the roadway section allows this region to experience thermal degradation of the original permafrost. This degradation can be further accentuated by the insulating effect of drifted or plowed snow over the roadway sideslopes which prevents the complete freeze-back of the active layer leaving a talic. The thaw settlement which results due to these effects usually shows up as lateral cracking on the wearing surface of the roadway.

A potential solution to this problem is the use of naturally ventilated air ducts placed in berm at the toe of the road, Figure 1. Air duct systems have also been used to stabilize foundations under buildings and tanks, Sanger (1969) and Nixon (1978). These ducts are usually .25 m to .5m diameter corrugated metal pipe (culvert) with a short sloping or vertical inlet section leading to a long horizontal section placed in the problem zone connected to a vertical outlet section or stack. Cold air flows in the inlet, is heated by the ground surrounding the buried horizontal section and then flows out the vertical stack. Flow in the duct system is maintained by the "chimney" or "stack" effect. The warm air in the vertical outlet stack being less dense is bouyed upward by the cold ambient air establishing flow. Flow is maintained as long as the air is heated by the ground, i.e. the ground temperature exceeds the ambient air temperature.

During the summer when the ambient air temperature is higher than the ground temperature, flow ceases as the cold air now inside the culvert remains trapped there because of its greater density. So, theoretically, the air duct ventilation system is a totally passive device, extracting thermal energy from the ground during the winter accelerating the freeze-back time and then becoming dormant or inoperative during the summer months when thawing from the surface occurs. Wintertime winds will increase the cooling effect (by increased air flow) especially if the inlet and outlet sections have been designed to take advantage of the prevailing wind direction. However, summertime winds can have a detrimental effect if they are allowed to cause flow through the buried ducts.

This paper discusses an analytical and experimental study which has investigated the parameters that affect the performance of air duct systems.

ANALYSIS

A schematic diagram of an air duct system is shown in Figure 1. The static pressure difference, ΔP_s, created by the stack effect is:

$$\Delta P_s = g(\rho_i - \rho_o)h \cong (ghP_a/R)\left[(1/T_i)-(1/T_o)\right] \quad [1]$$

For the stack shown, if no heating occurs in the inlet section and no cooling occurs in the outlet section, then h is the stack height and ΔP_S can be calculated using equation [1]. A larger temperature difference between the cold and warm air or a longer (taller) stack will increase the stack effect pressure difference resulting in a greater air flow. This stack effect pressure difference is balanced by the velocity pressure $P_V = \rho V^2/2$ and the frictional pressure drop, ΔP_f

$$\Delta P_f = \frac{\rho f L_e V^2}{2D} \quad [2]$$

The Darcy-Weisbach friction factor, f, is determined based on the Reynolds number of the flow within the duct and the relative roughness of the duct. The sum of the velocity pressure and frictional pressure loss is equal to the stack effect pressure difference. Substituting the equations for ΔP_f, P_v and ΔP_s into the above expression and rearranging yields an expression for the average velocity

$$V = \left\{[4gh/(1+fL_e/D)][(T_o-T_i)/(T_o+T_i)]\right\}^{1/2} \quad [3]$$

The friction factor must be known in order to use equation [3] to calculate the velocity. Therefore, an iterative method of solution is necessary. It is recommended that a trial value of the friction factor of .07 be used to arrive at a first approximation to the velocity. An improved value of the friction factor can be determined from Figure 2 once the Reynolds number has been calculated. This iterative process is continued until the desired accuracy is achieved.

If an annular ring of frozen ground of radius R exists around the buried air duct, then the quasi-steady state heat transfer rate is

$$Q_c = \frac{(T_f - T_a)}{\dfrac{1}{h_c\pi Dl} + \dfrac{\ln(2R/D)}{2\pi kl}} \quad [4]$$

The heat transfer rate given above is set equal to the thermal energy liberated in the freezing process

$$Q_c = 2\pi RL\, l\, \frac{dR}{dt} \quad [5]$$

Equating equations [5] and [6], rearranging and integrating, yields the following relationship:

Figure 1. Field Installation of Air Duct

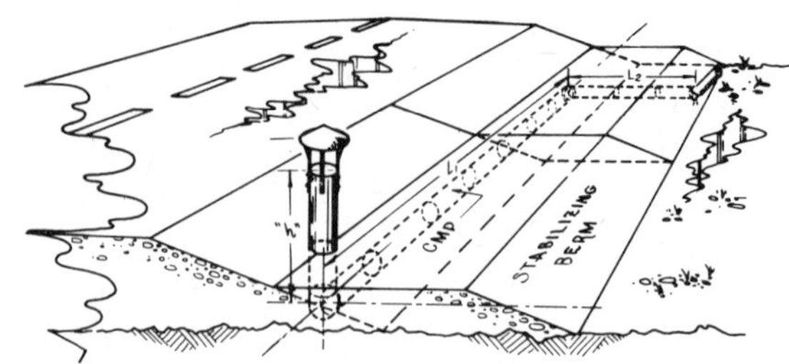

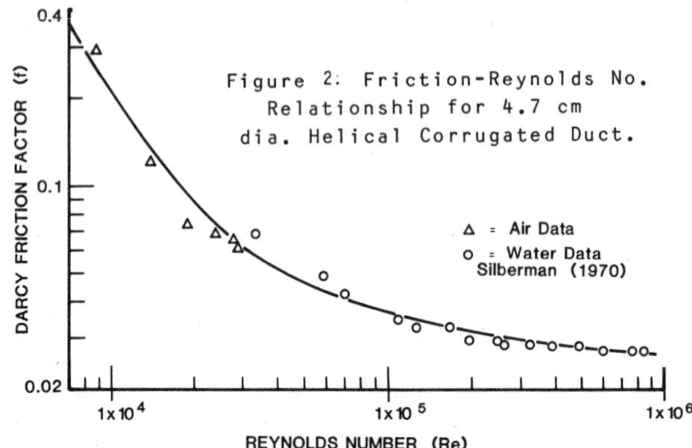

Figure 2: Friction-Reynolds No. Relationship for 4.7 cm dia. Helical Corrugated Duct.

$$F.I. = L\left[\frac{[R^2-(D/2)^2]}{2}\left(\frac{2}{h_cD} - \frac{1}{2k}\right) + \frac{R^2\ln(2R/D)}{2k}\right] \quad [6]$$

This solution neglects the sensible energy changes in the ground and as a result overpredicts the size of the frozen annulus for a given F.I. An improved approximation for the freeze radius to account for the non-steady temperature distribution and specific heat effects is given by Harlan and Nixon (1978) as

$$R' = R(1-.12Ste)^{\frac{1}{2}} \quad [7]$$

where R is the initial approximation for freeze radius given by equation [6] and R' is the solution of increased accuracy. The Stefan number defined as the ratio of sensible to latent heat or

$$Ste = C_tT_a/L \quad [8]$$

FIELD INSTALLATION

In 1973 a research project was initiated under the Highway Planning and Research Program of the Federal Highway Administration to study the benefits of several alternate embankment designs in controlling permafrost thaw under embankment slopes. Insulation, toe berms and air duct sections were constructed in 1974 at Bonanza Creek, approximately 40 kilometers west of Fairbanks on the Parks Highway.

The roadway embankment was constructed over undisturbed muskeg, underlain by ice-rich silt containing segregated ice. The permafrost temperature was measured to be between -3°C and -1°C.

Twenty centimeter diameter galvanized corrugated metal pipe ducts were installed. The inlet ends were placed above the maximum snow cover. The 15 to 30 meter long buried sections were sloped slightly upward toward the 3 meter stack. The conclusions of this study were reported by Esch (1978).

Air ducts in combination with insulation proved to provide maximum protection against thaw. Air ducts without insulation provided significantly less protection. However, after studying the temperatures along the duct, Esch concluded that the ducts were too long to provide efficient cooling along their entire length. This claim can be substantiated by Figure 3. As the air flows toward the outlet, its temperature approaches the ground temperature. Near the outlet of the duct, the temperature differential is quite small resulting in negligible cooling of the surrounding soil. It appears that this duct should have been about 10 meters shorter.

Heat flow can be estimated using the equation for cylinderical heat flow

$$Q = \frac{T_o - T_d}{\ln (2R/D)/2\pi k} \qquad [9]$$

The soil density at the site is approximately 1600 kg/m³ with a 15% moisture content by weight and a thermal conductivity of 1.5 W/m °C. Using the pipe radius as 0.10 m and the outside soil radius of 0.40 m, the heat flow per unit length at any point along the duct can then be calculated. The heat transfer rate along the pipe for January 1980 is plotted in Figure 4. The total energy transferred can be calculated by integrating the heat transfer rate per unit length over the entire length of air duct or,

$$Q_t = \int Q \, dx = 381.1 \text{ W} \qquad [10]$$

Combining the First Law of Thermodynamics and the mass flow rate relationship yields

$$V = Q_t / A \rho c_p \Delta T \qquad [11]$$

Using the thermal properties of air at 0°C and the Q_t value above, gives the air speed in the duct as

$$V = 0.33 \text{ m/s}$$

Field measurements have indicated approximately 0.3 m/s air speed at ΔT of 30°C.

COMPUTER MODELING

An air duct system was modeled on the University of Alaska Computer Network (UACN) Honeywell Computer using the DOW Chemical Model, "Finite Element Heat Conduction Program", Wang (1979). The model is designed to solve a two dimensional heat conduction problem, assuming there is no temperature gradient in the third direction. The boundary temperature may be either fixed or time-varying. Each nodel point can be either a heat source or a heat sink.

The program further assumes that the phase change occurs between 0°C and -1°C. Latent and

Figure 3. Ground Temperature at Bonanza Creek, January 1980.

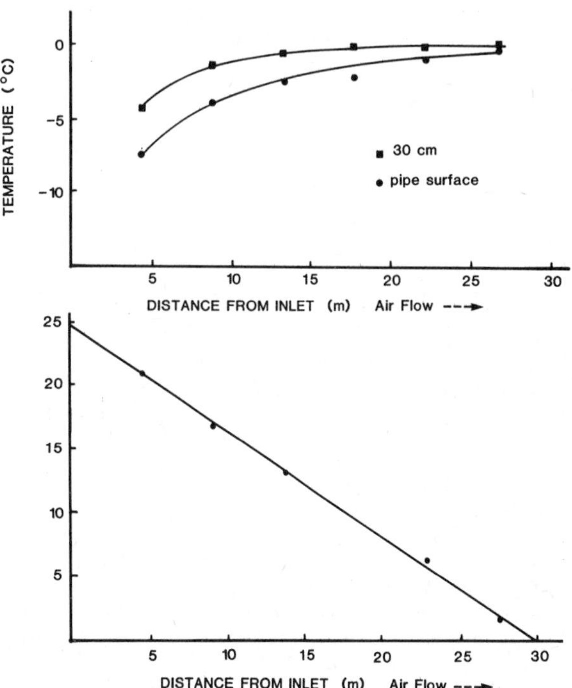

Figure 4. Heat Flow from Duct, January 1980.

sensible heats have been combined for this region to yield an apparent specific heat. (Most fine grained soils exhibit some subcooling in the freeze-thaw process as well as latent heat liberated over a temperature range due to unfrozen water content). The thermal conductivity of the soil can be varied in the x and y directions. Here it was assumed uniform for the same soil type but varied from layer to layer.

The air duct system in the embankment is shown in Figure 5. The embankment is comprised of three layers with the duct placed at the surface of the original ground. The properties of each layer are indicated on Figure 5. A 0.6 meter snow cover was placed on the embankment slopes and the berm between October 1 and April 15 of each year. The pavement surface was left bare throughout the year. The surface temperature was modeled by the cosine function:

$$T_s = \overline{T}_s - T_v \cos (wt - w\phi) \qquad [12]$$

The following values were used;

$\overline{T}_s = -3°C$, $T_v = 20°C$, and $\phi = 10$ days.

An initial run was made without the ducts to determine the effect of the embankment on the underlying permafrost. As might be expected, degradation of the premafrost was noted beneath the shoulder. This would be evidenced on the pavement surface as a longitudinal cracking. Additional runs were made with the air duct using

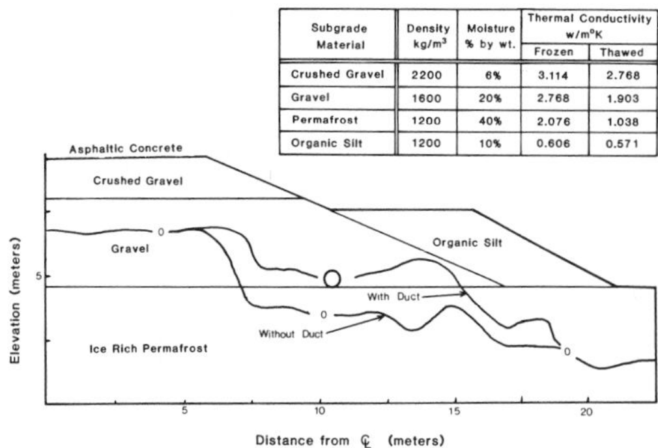

Subgrade Material	Density kg/m³	Moisture % by wt.	Thermal Conductivity w/m°K	
			Frozen	Thawed
Crushed Gravel	2200	6%	3.114	2.768
Gravel	1600	20%	2.768	1.903
Permafrost	1200	40%	2.076	1.038
Organic Silt	1200	10%	0.606	0.571

Figure 5. Finite Element Simulation in August.

a convective heat transfer coefficient of 9.7W/m²-°C for the pipe and an air temperature equal to outdoor ambient. As seen in Figure 5, the duct raised the 0°C isotherm into the embankment, thereby preserving the permafrost.

It should be noted that the computer model provides an upper bound to the field installation. The duct in the model was larger and placed more effectively under the embankment. Ideally, the duct should be placed far enough under the embankment to prevent the degradation of permafrost beneath the shoulder material.

EXPERIMENTAL SYSTEM

An experimental facility was constructed during 1982 in order to augment field data obtained at Bonanza Creek. The experimental facility, Figure 6, consisted of an above ground convection duct, an adjustable electric heating circuit, and data acquisition system. Helically corrugated metal culvert 30 cm in diameter (60 cm plate and 7 x 1.5 cm corrugations) was used for construction of the duct. The horizontal section was 12.5 meters in length with a one meter vertical inlet section and a vertical outlet stack of variable length. Heat input was accomplished via copper clad heating tape wrapped in the corrugations on the outside of the duct. Three separate heating circuits spanning the horizontal section of the duct were employed (see Figure 6). In order to reduce heat transfer directly to the outside air and allow a stable duct temperature to be maintained, 5 cm of fiberglass insulation were wrapped around the outside of the horizontal section covering both the duct and heating tapes. Each of the tapes was powered by a "Variac" variable transformer and monitored by an individual power transducer. This allowed for flexibility not only in overall heat input but also in the heating profile along the duct.

Thermocouples were used to measure both air temperatures within the duct, and temperatures of the duct surface. Temperature and heat input data was collected with a Hewlett-Packard 3054A data acquisition system interfaced with the power transducers and temperature probes.

Headloss-flowrate tests were run on a 12.5 m horizontal section of the 30 cm diameter duct. Pressure drop data at flowrates covering Reynolds numbers from 7,000 to 30,000 were recorded. Darcy-Weisbach friction factors were calculated from the

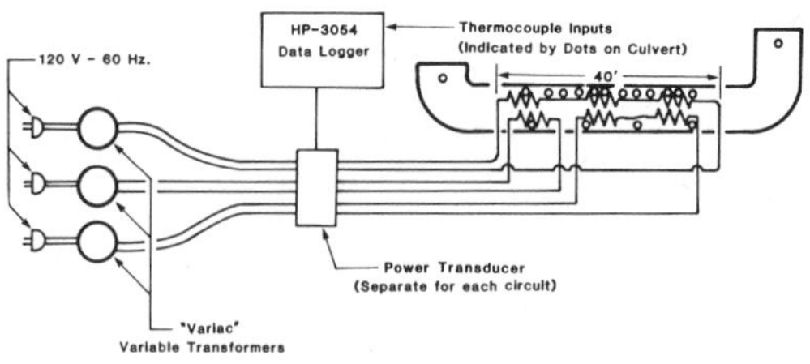

Figure 6. Schematic of Experimental Duct.

data and the results have been plotted in Figure 2. Data from tests conducted with water in 30 cm diameter helical culverts by Silberman (1970) is also shown in this figure.

A series of tests were conducted as outlined in Table 1.

Table 1 - Experimental Tests--Duct Configurations

Test	Stack Height (m)	Description
1	2.10	no insulation on stack, no weather caps.
2	3.35	no insulation on stack, no weather caps.
3	4.60	no insulation on stack, no weather caps.
4	3.35	no insulation on stack, weather caps installed.
5	3.35	insulated stack, no weather caps.

Each of the experiments was run for six twenty-four hour periods at different power input levels. Temperatures and power inputs were scanned and recorded hourly during each period.

Reduction of the recorded data was accomplished by inspecting it for periods of stable operation. Usually a stable outside air temperature resulted in stable duct operation although this was not always the case, particularly during windy weather. In general, one twenty-four hour period resulted in three or four usable scans.

The amount of thermal energy absorbed by the air inside the duct was calculated by subtracting the heat lost through the insulation directly to the outside air from the total heat input recorded by the data logger. The value for the thermal resistance of the insulating layer covering the heat tapes and duct was obtained by heating the duct with airtight, insulating covers at the inlet and outlet. In this situation all power input by the heat tapes must escape through the insulating layer. Consequently, a thermal resistance value can be calculated simply by noting the heat input and the temperature drop across the layer. Once this resistance is known it is possible to calculate the amount of heat being absorbed by air flowing through the duct as described above. This value can then be used in conjunction with various temperatures throughout the duct to calculate the mass flow rate of duct air, Reynolds number, and the convection heat transfer coefficient as shown below:

$$h_c = Q_{air}/A_s\ (T_d - T_a) \quad [13]$$

$$m = Q_{air}/(c_p \Delta T)$$

$$Re = mD/A\mu$$

Finally the experimental results were presented as plots of m versus Q_{air} and h_c versus Reynolds number.

Experimental results presented here were obtained during winter of 1981/1982 on the University of Alaska Fairbanks campus. Outside air temperatures ranged from -5°C to -43°C. The heating circuits were adjusted to deliver a uniform heat flux along the entire horizontal section of the duct. Initial experimentation was hampered by difficulty in establishing proper air flow in the duct. By introducing a 1.5% slope to the horizontal section with the outlet at the high end this problem was eliminated.

As expected, a taller stack results in a larger draft head which in turn produces larger mass flow rates, Figure 7. Figure 8 shows that insulating the outlet stack with 5 cm of paper faced fiberglasss has a detrimental effect on duct operation and is backed-up, at least in part, by observations conducted at Bonanza Creek. Figure 9 is a plot of h_c versus Re with data from all five experiments included. Theoretical h_c values calculated using the Reynolds analogy relating fluid friction and heat transfer are in good agreement with Figure 9. Results obtained for duct operation with weather caps on the inlet and outlet openings reveal reductions in mass flow rates of up to 20%. The weather caps used in this experiment were of the inverted cone type with a 15 cm gap between the cone and duct opening. This effect will vary depending on the geometry of the cap installation; therefore, results have not been presented graphically. Finally, stack efficiencies have been calculated by dividing the theoretical velocity given in [3] by the observed velocities. An average efficiency of 86% was obtained with a standard deviation of 3.5%.

Contrary to what was expected, insulating the outlet stack did not improve duct performance by maintaining a higher air temperature in the stack and thereby increasing the draft head. A possible explanation is that when no insulation is present on the outlet stack the stack itself will have a relatively low temperature, consequently the air film on the inside surface will be at a lower temperature and therefore lower viscosity. On the other hand, with insulation on the outside of the stack, a higher surface temperature results in a more viscous air film on the inside surface of the stack. Therefore, insulating the stack results in higher frictional losses in the stack which may be substantial enough to overshadow the increases in the draft head.

DESIGN PROCEDURE

A design procedure can be developed based on the information presented in this paper. If a talik of radius R' needs to be frozen back, then equations [6] and [7] can be used to determine the required freezing index. Figure 9 would be used to choose a convective heat transfer coefficient.

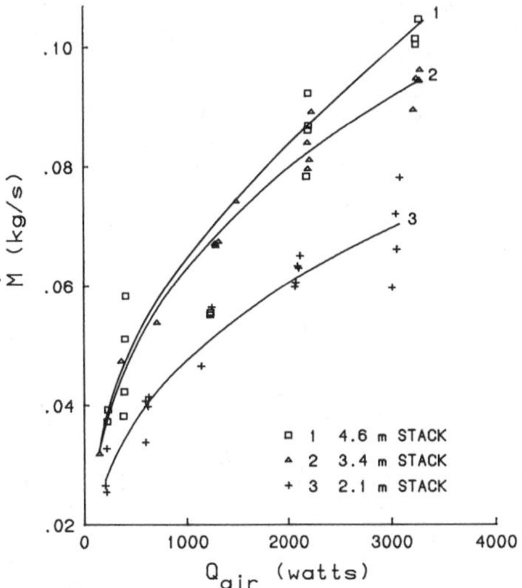

Figure 7. Effect of Stack Height.

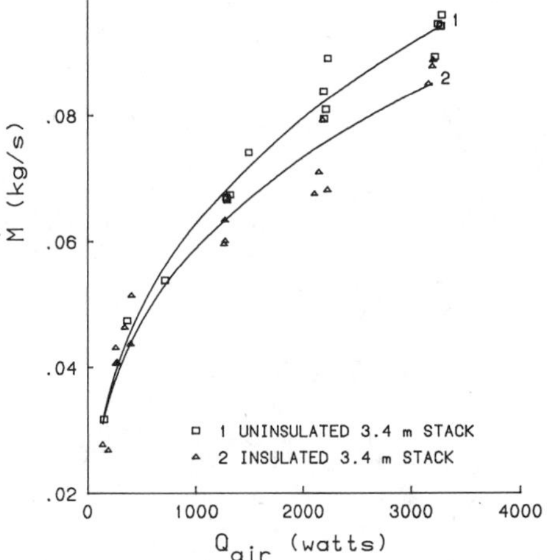

Figure 8. Effect of Stack Insulation.

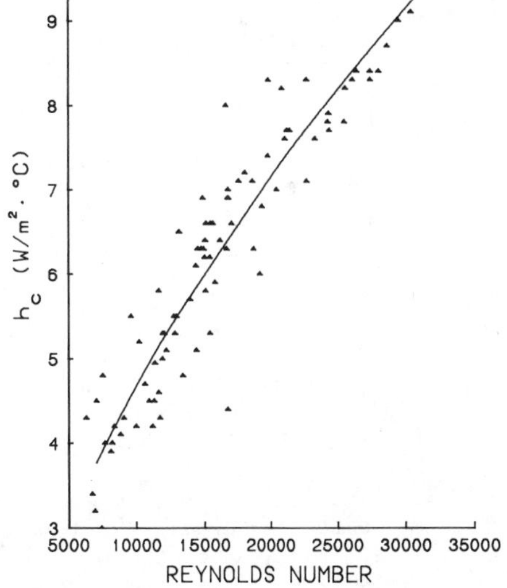

Figure 9. Convective Heat Transfer Coefficient.

If this freezing index was less than the air freezing index for the site then a dual air duct system should be considered. Next, the maximum permissible temperature leaving the stack can be calculated by dividing the freezing index from equation [6] by the length of the freezing season. The total energy removed from the ground can be estimated by determining the heat transfered in freezing and sensibly cooling a talik of radius R'. The average heat removal rate is found by dividing the total energy removed by the length of the freezing season. Then the average air speed in the duct can be calculated by equation [11]. The equivalent length of air duct is then calculated using equation [3] with the appropriate friction factor taken from Figure 2 and the stack height assumed. Finally, the length of the buried section is calculated by subtracting the inlet length, outlet length, and fitting losses from the equivalent length.

CONCLUSIONS

An analysis of an air duct system has been presented. The weak points in this analysis were lack of data on friction factors and heat transfer coefficients for air flow in the helical ducts and the stack efficiency. This data has been determined from the experimental program and presented. The finite element analysis was used to determine the optimum location for air duct systems which is centered in the expected talik. Finally, a design procedure has been outlined which can be used to estimate system size.

ACKNOWLEDGEMENTS

The authors would like to acknowledge the Federal Highway Administration for their support on this project, under the Highway Planning and Research Program.

REFERENCES

Esch, D.C., 1978, Road embankment design alternatives over permafrost, Applied Techniques for Cold Environments, Vol. 1, ASCE, New York, N.Y.

Wang, F.S., 1979, Mathematical modeling and computer simulation of insulation systems in below grade applications, ASHRAE--DOE Conference--Thermal Performance of Exterior Envelopes of Buildings, Orlando, Florida.

Johnson, P.R. and C.W. Hartman, 1969, Environmental Atlas of Alaska, Institute of Water Resources, University of Alaska, College, Alaska.

Harlan, R.L. and J.F. Nixon, 1968, Ground thermal regime, in Geotechnical Engineering for Cold Regions, McGraw Hill, New York, N.Y.

Silberman, E., 1970, Effect of helix angle on flow in corrugated pipes, Journal of Hydraulics Division, Vol. 96, HY11, ASCE, New York, N.Y.

Sanger, F.J., 1969, Foundations of structures in cold regions, U.S. Army CRREL, Monogr. III-C4, Hanover, N.H.

Nixon, J.F., 1978, Geothermal aspects of ventilated pad design, Proceedings Third International Permafrost Conference, Vol. 1, National Research Council of Canada, Ottawa, Canada.

LIST OF SYMBOLS

e = stack efficiency, 0 - 1
ϕ = lag time, days
ρ = density of air, kg/m^3
μ = dynamic viscosity, kg/m-s
A = cross-sectioned area of duct, m^2
A_s = area of heated duct surface, m^2
c_p = specific heat of air, kJ/kg-k
C_t = volumetric specific heat, kJ/m^3-k
D = inside duct diameter, m
f = Darcy-Weisback friction factor
FI = freezing index, °C-day
g = gravitational acceleration, m/s^2
h = effective height of stack, m
h_c = convective heat transfer, W/m^2-k coefficient, W/m^2-°C
i = cold air
k = soil thermal conductivity, W/m-°C
l = length, m
L = volumetric latent heat kJ/m^3
L_e = equivalent length of duct, m
m = mass flow rate, kg/s
o = warm air
P = pressure, N/m^2
P_a = average pressure between warm and cold air
ΔP_f = frictional pressure drop, N/m^2
ΔP_s = stack effect pressure difference, N/m^2
P_v = velocity pressure, N/m^2
Q = heat flow/unit length, W/m
Q_c = total heat flow, W
R = radius of frozen annulus, m
$\overline{R}$ = gas constant, J/kg-k
Re = Reynolds number
Ste = Stefan number
t = time, s
T_a = temperature of air in duct, °C
$\overline{T}_a$ = average temperature of air in heated section of the duct, °C
T_d = average temperature of heated duct surface, °C
T_f = temperature at the freezing point, °C
T_o = temperature 0.3 m from duct surface
T_s = ground surface temperature, °C
$\overline{T}_s$ = mean annual ground surface temperature
T_v = seasonal temperature variation
ΔT = change in temperature of air flowing through duct
V = average air velocity, m/s
w = $2\pi/365$

PRELIMINARY EXPERIMENTAL STUDY OF WATER MIGRATION AT THE ICE/SOIL INTERFACE

Zhang Jinsheng and Fu Rong

Lanzhou Institute of Glaciology and Cryopedology, Academia Sinica
People's Republic of China

On the basis of the theory of adsorption and Darcy's law, and by using the contact method, samples of Qinghai-Xizhang red clay and Lanzhou yellow silt were tested to determine their individual amounts of unfrozen water and the hydraulic conductivity and diffusivity in unsaturated soil (without ice in soil at negative temperature). The results show the following regularities: The amount of unfrozen water increases with time and approaches a certain constant value; different soils have different adsorption rates and different processing curves. At high temperatures, the adsorption rate is high, and the unfrozen water content is also large, while the rate of curvature approaching the final value is smoother. At low temperatures, the final value of unfrozen water is also smaller. During the entire process, diffusivity is not constant, but is a function of water content. The accuracy of the results depends on the interval of measurement used in the experiment.

Owing to the existence of a certain surface energy on the surface of soil mineral constituents, a certain amount of water molecules can be adsorbed on it and a water film with considerable thickness may be formed. Under negative temperature, such water film coexists with ice crystals in frozen soil, and is called unfrozen water.

The surface energy of ice is directly affected by temperature, and it has a constant value under a given temperature. Thus when ice crystals and soil particles coexist, no matter how much the total water content may be, there should be a corresponding amount of existing unfrozen water, which depends upon the surface energy of ice. Hence, the quantity of unfrozen water relates to temperature directly.

According to the adsorption theory, when different substances come into contact, the instability of thermodynamics induced by surface energy kept on the soils surfaces will cause molecular interchange at its boundary faces (Watanabe 1973). The larger the surface area, the greater such instability. So under the same temperature conditions, the sum of unfrozen water in soil of fine particles is generally a little larger. Therefore, when perfectly dry soil comes into contact with ice, a part of the water molecules will be adsorbed on the surfaces of soil particles in the forms of sublimation and direct removal. The increase of adsorbed water amount is a function of soil surface energy and temperature. To a specific soil, the amount of adsorbed water is just the amount of unfrozen water, and the final amount of adsorbed water is only a function of temperature.

From this principle, the adsorptive nature of soil may be utilized to determine the quantity of unfrozen water in soil and the parameters of water movement in unsaturated soil.

METHOD

The tested soil is in the form of a slab, thoroughly dried under constant temperature and fixed pressure and put into mutual contact with ice or ice-rich frozen soil. To guarantee good contact, the contact surface should be smoothed beforehand. It would be all right to measure the increments of water content in the soil at regular intervals. This method is called the Contact Method. The sample weight lightened on the test process will cause error for the result, the error will prior correct before data processing.

According to the adsorption principle, the amount of adsorbed water with time obeys the following law (de Boer 1953):

$$G_t = G_\infty (1 - e^{-k_d \cdot t}) \qquad (1)$$

where t is time; G_t is the adsorbed amount (in weight) at time t; G_∞ is the adsorbed amount theoretically only reached after an indefinitely long time; k_d is a constant which, related to ambient conditions (temperature, pressure, etc.) and contact circumstances, soil surface area and permeability, and viscosity of water.

We have used the red clay of Qinghai-

TABLE 1 Basic Indexes for Two Kinds of Soils.

Kind of soil	Specific weight	Liquid limit	Plastic limit	Specific surface area (%)	Salt content (%)
Red clay of Qinghai-Xizang	2.76	22.0	15.0	31.8	0.08
Yellow silt of Lanzhou	2.70	19.9	13.6	41.9	0.194

TABLE 2 Particle Diameter Analyses for Two Kinds of Soils.

Kind of soil	Percentage of various particle sizes (diam in mm) 1-0.5	0.5-0.25	0.25-0.1	0.1-0.05	0.05-0.01	0.01-0.005	0.005-0.002	0.002
Red clay of Qinghai-Xizang	2.70	2.02	5.00	1.68	26.30	11.59	17.41	33.30
Yellow silt of Lanzhou		0.15	1.92	17.84	51.89	6.10	5.29	16.81

Xizang and yellow silt of Lanzhou as the examples to determine separately their individual amounts of unfrozen water and water movement parameters in unsaturated soil. The basic indexes of various kinds of soils are shown in Table 1 and Table 2.

RESULTS

Experimental results show that the amount of unfrozen water increases with time and approaches a certain constant value. Figure 1 shows the change in weights of unfrozen water at -3.0°C of the samples. To the same kind of soil, under different temperatures, the curve showing the process of change in amount of unfrozen water is different. Figure 2 shows the process of change in amount of Qinghai-Xizang red clay at three different temperatures.

From these two figures, it is not difficult to see that different soils have different adsorption rates and different processing curves. The shapes of various curves are different too. For the same kind of soil, under high temperature, the adsorption rate is high, and the final value of unfrozen water content is also big. Meanwhile the rate of increase approaching the final value is smaller. Under low temperatures, rapidly approach-

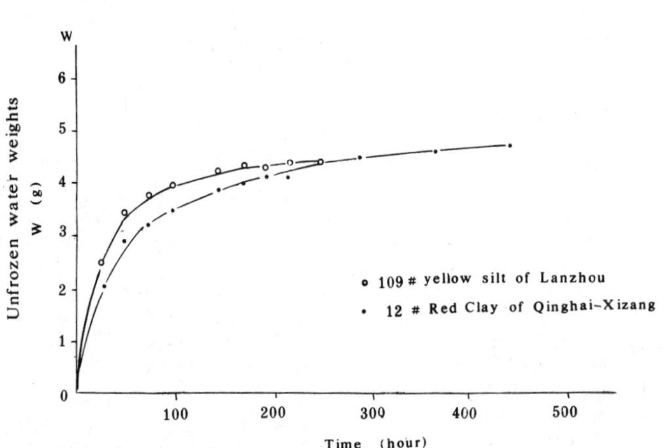

FIGURE 1 Changes of unfrozen water weights at -3 degree centigrade of two kinds of soil.

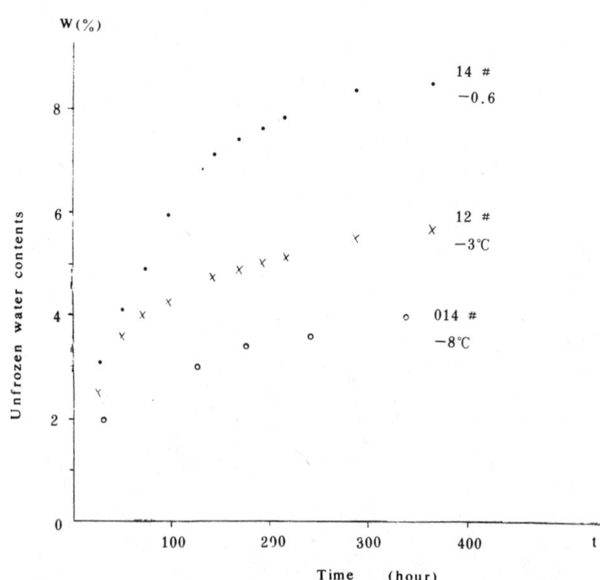

FIGURE 2 Changes of unfrozen water contents of the Red Clay of Qinghai-Xizang under three kinds of low temperatures.

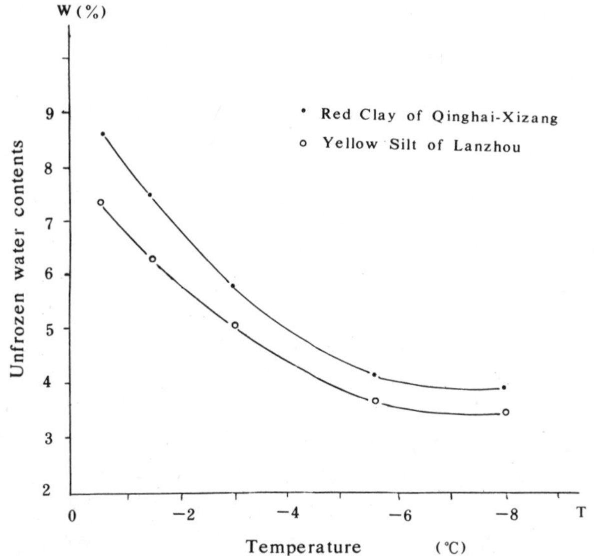

FIGURE 3 Unfrozen water contents and temperatures of two kinds of soil.

ing its terminal, the final value of unfrozen water is smaller too.

Figure 3 shows the relation between unfrozen water contents and temperatures of two kinds of soils from tested results obtained by the contact method (Zhang Jinsheng and Fu Rong, 1983).

EQUATION OF WATER MIGRATION

Moisture moves and its flux can be expressed in the same form of Darcy's law.

$$q_x = -K(\theta)\frac{\partial \psi}{\partial x} \tag{2}$$

where K(θ) is unsaturated water conductivity and varies with the water content; θ is volumetric water content and ψ is the soil water potential. If the unsaturated condition is described by Darcy's law of diffusive equation, we can obtain

$$q_x = -D(\theta)\frac{\partial \theta}{\partial x} \tag{3}$$

and

$$\frac{\partial \theta}{\partial t} = \frac{\partial}{\partial x}\left(-D(\theta)\frac{\partial \theta}{\partial x}\right) \tag{4}$$

where D(θ) is diffusivity of unsaturated water soil; t is time; and D(θ)=K(θ)/C, C is specific water capacity, and

$$C = \frac{\partial \theta}{\partial \psi} \ .$$

The diffusivity D(θ) changes with time. If ignoring the change of diffusivity between the two measuring times of water weight in sample, we can obtain

$$\frac{\partial \theta}{\partial t} = D(\theta)\frac{\partial^2 \theta}{\partial x^2}$$

$$t = 0, \quad \theta = 0$$
$$t = \infty, \quad x = 0, \quad \theta = \theta_\infty \tag{5}$$
$$t = \infty, \quad x = H, \quad \frac{\partial \theta}{\partial x} = 0$$

where θ_∞ is the volumetric unfrozen water content of the sample under certain temperature; H is half of the sample length. If $\bar{t}=D.t/H^2$, $\bar{x}=x/H$, then eq 6 can be rewritten as follows:

$$\frac{\partial \theta}{\partial \bar{t}} = \frac{\partial^2 \theta}{\partial \bar{x}^2}$$

$$\theta(\bar{x},0) = 0$$
$$\theta(0,\bar{t}) = \theta_\infty \tag{6}$$
$$\left.\frac{\partial \theta}{\partial \bar{x}}\right|_{\bar{x}=1} = 0$$

Using Laplace transform, eq 6 converted as

$$\frac{\partial^2 \bar{\theta}}{\partial \bar{x}^2} = P\bar{\theta} \tag{7}$$

$$\bar{\theta}(0,P) = \frac{\theta_\infty}{p} \tag{8}$$

$$\left.\frac{d\bar{\theta}}{d\bar{x}}\right|_{\bar{x}=1} = 0 \tag{9}$$

and the solution can be obtained

$$\theta = \frac{ch(1-\bar{x})\,p}{ch\ p}\ \frac{\theta_\infty}{p}$$

After the Laplace inversion, we obtain

$$\theta = \theta_\infty \quad 1 - \frac{4}{\pi}\sum_{k=1.3.5\ldots}^{\infty}\frac{\sin\frac{k\pi}{2}}{k}\ e^{-(\frac{k\pi}{2})^2\cdot\bar{t}}$$

so

$$\frac{\partial \theta}{\partial \bar{x}} = -2\theta_\infty \sum_{k=1.3.5\ldots}^{\infty} \cos\left(\frac{k\pi}{2}\bar{x}\right) e^{-(\frac{k\pi}{2})^2\cdot\bar{t}}$$

so

$$\left.\frac{\partial \theta}{\partial \bar{x}}\right|_{\bar{x}=0} = -2\theta_\infty \sum_{k=1.3.5\ldots}^{\infty} e^{-(\frac{k\pi}{2})^2\cdot\bar{t}} \tag{10}$$

Illustrated installation of sample in Figure 4. If A is the contact area of ice-soil, then eq 3 can be rewritten as

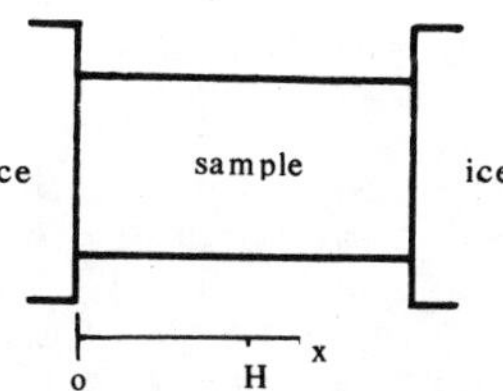

FIGURE 4 X axis and sample.

follows:

$$Q = -A \frac{D(\theta)}{H} \left(\frac{\partial \theta}{\partial \bar{x}}\right) \Bigg|_{\bar{x} = 0} \qquad (11)$$

where Q is the quantity of water migration in the sample. Hence, the water weight in the sample at time t is

$$G_t = \int_0^t Q.dt \qquad (12)$$

Substitute eq 10 and 11 into eq 12, we have

$$G_t = \int_0^t \frac{A\ D\ \theta_\infty}{H} \sum_{k=1 \cdot 3 \cdot 5 \cdots\cdots}^{\infty} 2e^{-(\frac{k\pi}{2})^2 . \bar{t}}\ dt$$

so

$$G_t = \int_0^{\bar{t}} \frac{A\ D\ \theta_\infty}{H} \frac{H^2}{D} \sum_{k=1 \cdot 3 \cdot 5 \cdots\cdots}^{\infty} 2e^{-(\frac{k\pi}{2})^2 . \bar{t}}\ dt$$

The resulting equation can be written as

$$G_t = G_\infty\left(1 - \sum_{k=1 \cdot 3 \cdot 5 \cdots}^{\infty} \frac{8}{\pi^2} \frac{e^{-(\frac{k\pi}{2})^2 . \frac{t \cdot D}{H^2}}}{K^2}\right) \qquad (13)$$

As converges rapidly, we take k=1, and

$$Ln(G_\infty - G_t) = Ln\left(\frac{8G_\infty}{\pi^2}\right) - \frac{\pi^2}{4} \frac{t \cdot D}{H^2} \qquad (14)$$

after transformed the unit of time and became as

$$D(\theta) = \frac{96H^2}{\pi^2 t} Ln \frac{\frac{8}{\pi^2} G_\infty}{G_\infty - G_t} \quad (cm^2/day) \qquad (15)$$

Using a pressure membrane extract instrumentation, we can determine the relation between soil water potential and volumetric water content. As a result, the soil-water characteristic curves can be determined. Consequently, the conductivity K is reckoned (Xie Shenchuan 1982).

Some outcomes of experiments for diffusivities are shown in Figure 5, and part of the numerical value of D and K are shown in Table 3.

In fact, the diffusivity is not a constant during the entire process, but a function of water content. Eq 15 is an approximate equation, the accuracy of results depends on the interval of measurement in experiments. In addition, the length of the sample cannot be excessively large.

CONCLUSION

The contact method used to determine unfrozen water content in frozen soil is a simple technique, but the period of experiment is comparatively long.

With the help of moisture-time relationship, we can obtain the conductivity and diffusivity of unsaturated water (without ice in soil at negative temperature).

The method cannot reflect the influence of salt on unsaturated water.

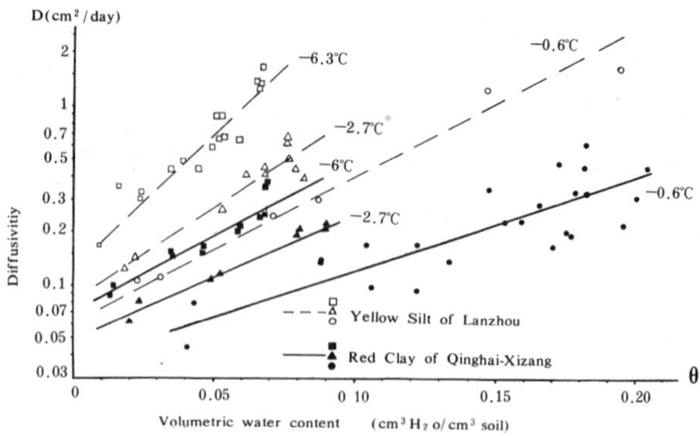

FIGURE 5 Diffusivities of two kinds unsaturated soils and volimentric water contents.

TABLE 3 Diffusivity D (cm^2/day) and Conductivity K ($X10^4$ mm/day) of Unsaturated Water Soil.

Soil type	Temperature (^oC)		Volumetric water content ($cm^3\ H_2O/cm^3$ soil) 0.01	0.07	0.10	0.15	0.20
Red clay of Qinghai-Xizang	-0.6	K				1.83	0.40
	-0.6	D	0.04	0.08	0.12	0.22	0.40
	-2	D	0.05	0.13	0.18		
	-4	D	0.06	0.19			
	-6	D	0.08	0.26			
Yellow silt of Lanzhou	-0.6	K				8.58	31
	-0.6	D	0.07	0.23	0.40	1.03	2.70
	-2	D	0.09	0.34	0.54		
	-4	D	0.13	0.61			
	-6	D	0.16	0.71			

REFERENCES

de Boer, J. H., 1953, The dynamical character of adsorption, Oxford University Press, p. 225.

Watanabe Nobuatsu, Watanabe Akira, and Tamai Yasukatsu, 1973, Surface and boundary, Japan, Kyoritsu Press, p. 26.

Xie Shenchuan, 1982, Determination of water movement parameters in unsaturated soil: Chi. Journal of Hydrogeology Engineering Geology. 63, p. 8.

Zhang Jinsheng and Fu Rong, 1983, Analysis of several method for determining the unfrozen water content, in Proceedings of the Second National Conference on Permafrost (Selection).

LONG-TERM RESISTANCE OF ANCHORS IN PERMAFROST

Zhang Luxing and Ding Jingkang

Northwest Institute of the Chinese Academy of Railway Sciences
Lanzhou, People's Republic of China

Pull-out tests on reinforced concrete anchors with a nominal diameter of 10 cm were conducted at field research stations on the Qinghai-Xizang Plateau to investigate the long-term resistance of anchors in permafrost. Based on analysis of the load-displacement behavior of the anchors tested, the authors present the concept of gradually advancing failure of the adfreezing strength between the anchors and their surrounding soils. The authors consider that there exists an "effective length" for anchors in permafrost, which may be of great significance in the study of piles and anchors in permafrost. Test results show that the ultimate resistance of permafrost anchors depends primarily upon the length and diameter of the anchors; soil temperature, composition, and moisture; and the loading rate. From tests it was found that the displacement rate of an anchor in permafrost can be expressed by $\dot{\gamma} = K(\tau - \tau_\infty)^n$, where τ_∞ is the long-term resistance of anchors in permafrost.

In order to build the light anchor retaining wall in permafrost areas, we made the long-term resistant test of anchors in Qinghai-Xizang Plateau in 1979. The reinforced concrete anchors prefabricated from 5 to 22 cm diameter and from 50 to 250 cm long were used in the test. The 72 anchors were tested in all.

The anchors were set up by following:

Inserting the anchors in the holes drilled of 15-32 cm diameter, then backfilling the cracks with the saturated middle-coarse sand.

We were testing with method of constant loading rate and constant load. The loads were loaded by screw block. The displacement was measured by dial indicator.

ANALYSIS OF THE TEST RESULTS

The Stress Distribution of Anchors Under External Load

Buried in permafrost, the anchors were backfilled with saturated sand and frozen into place by the surrounding permafrost . The load acting on the anchor is transferred to the sand and permafrost. Figure 1 shows the radial distribution of stress in the anchor system when under axial load and Figure 2 gives the distribution of stress on shearing surface of anchor system along depth. It is seen from Figure 1 that stress decreases rapidly in radial direction with increase in distance from anchor center and becomes zero at a distance of 25-45 cm from the center. From Figure 2 (test on No. 42 anchor of 10 cm diameter), we see that, when the load is comparatively small, the shearing stress as acting on shearing surface are a maximum in the upper part, attenuating rapidly with increasing depth, and becomes zero at certain depth. It appears that, under the maximum load, the shape is approximately trapezoidal. Such a redistribution of stress indicates that the adfreeze strength in the upper part of the shearing surface has already been destroyed, attaining the maximum value before the load on the anchor reached its maximum. As the maximum load is attained, the adfreeze strength of the lower part of shearing surface reaches its maximum while that of the upper part remains at some residual value.

Figure 3 gives a load-time-deflection curve of No. 42 anchor obtained under gradually applied loads. It is seen from that the creep curves under 11,000 kg and 12,000 kg loads appear in the shape of stair, which is different from the others. This kind of creep curve is a reflection of gradually advancing failure of the anchor.

The "Effective Length" and "Conversion Effective Length" of Anchor

It is shown from gradually advancing failure that there exists an "effective length" of anchorarm in permafrost, which refers to the transmitting depth of the shearing strength at certain spot of shearing surface of anchor system when it attains the ultimate long-term adfreeze strength (see Figure 4). The effective length of anchor remains constant provided its diameter, ground temperature, and medium remain unchanged, i.e. the resistant forces of the anchor are irrelevant to its overall length in cases where the latter is longer than the "effective length".

However, it is shown by experiment that adfreeze strength can be divided into two parts, peak and residual. After failure, there still exists the action of the latter between the anchor and the backfill sand, as indicated by the

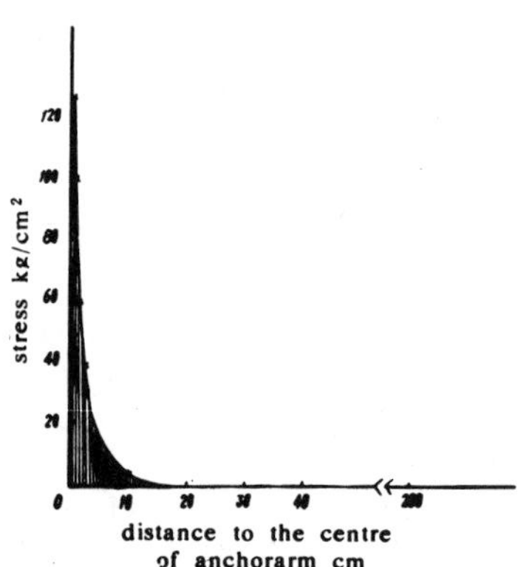

Figure 1. The radial distribution of stress in anchor system in permafrost. (Calculating by finite element. P = 2,000 kg, anchor of 10 cm diameter, backfill of 7 cm thickness.)

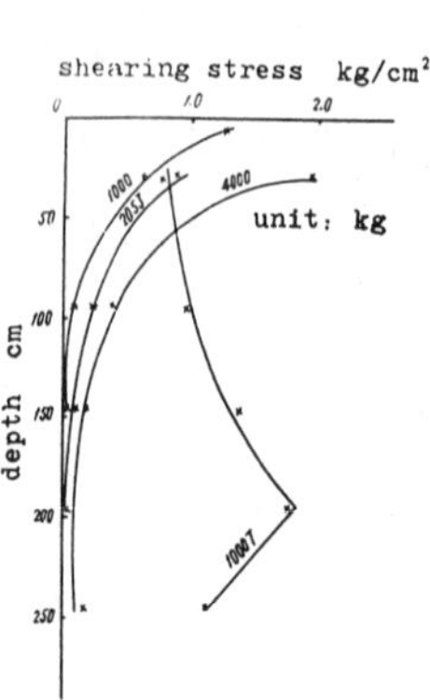

Figure 2. The distribution of stress on the boundary surface between anchor and filled sand along depth.

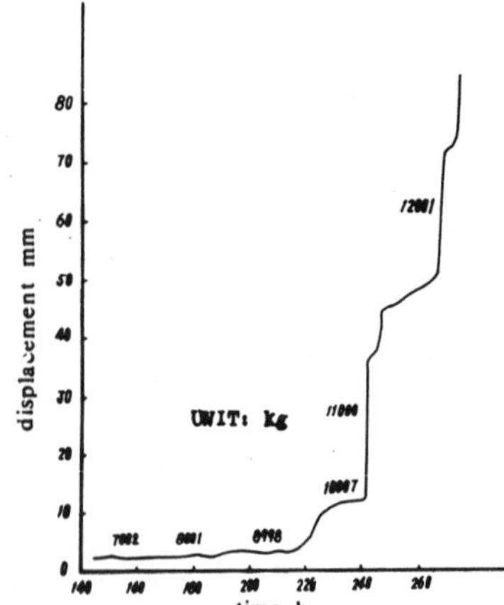

Figure 3. Stair-shaped creep curve of anchorarm system.

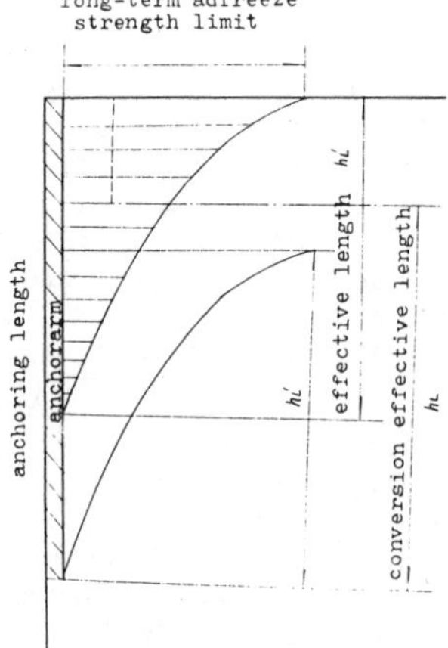

FIGURE 4 The sketch of effective length and conversion effective length of anchor.

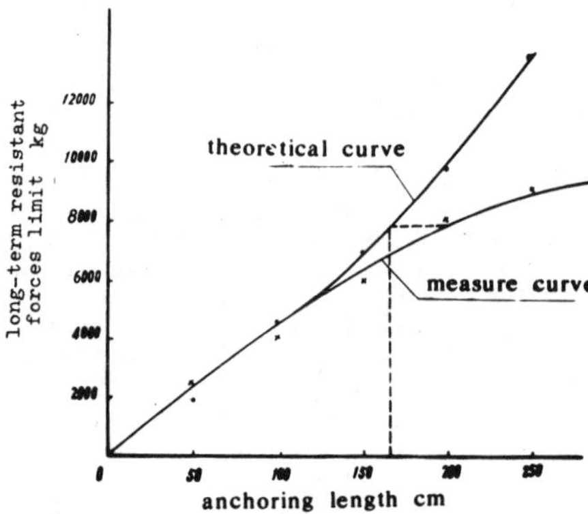

FIGURE 5 The conversion curve of the conversion effective length of anchorarm

dotted line in Figure 4. Therefore, the total resistant forces rise correspondingly as the anchor length increases and the "conversion effective length" can be used on account of the action of residual adfreeze strength. The action of the residual adfreeze strength increases with increase of anchor length. As a result, the "conversion effective length" increases as anchor length increase.

The "conversion effective length" can be calculated by the following:

1. Set the value of conversion ultimate long-term adfreeze strength on the basis of the average ground temperature around the anchoring section of anchor;

2. Calculate the theoretical ultimate long-term resistant force under such conditions by multiplying the above value by the frozen surface area of the anchor;

3. Plot the above obtained values on a diagram of the ultimate long-term resistant forces of anchor versus the anchoring lengths to achieve the conversion curve of anchor length (Figure 5);

4. On Figure 5 draw a straight line parallel to the horizontal axis from the tested curve to the theoretical curve, which intersects the latter at a point and the projection of the point on the abscissa is the value of the required conversion effective length of anchor.

The values of conversion effective length of anchor h_L obtained by the above method are listed in Table 1.

The ratio of conversion effective length of anchor to the actual anchor length is called the conversion coefficient of anchor length. It decreases with increasing anchor length (Table 1). Within the range of anchor lengths tested from

TABLE 1 Conversion Coefficient of Anchorarm Length

Anchor	Length of Anchoring part* (cm)	Average ground temperature (°C)	Tested ultimate long-term resistant force (kg)	Calculated ultimate long-term resistant force (kg)	Conversion effective length h_L (cm)	Conversion coefficient of anchorarm length $K_L = h_L/h$
33	50	-2.30	2504	2295	50	1.00
34	100	-2.50	5000	4762	100	1.00
35	150	-2.55	6974	7416	140	0.93
36	200	-2.74	8485	10684	172	0.86
37	250	-2.25	9995	13716	190	0.76

* Diameter of all anchors is 10 cm

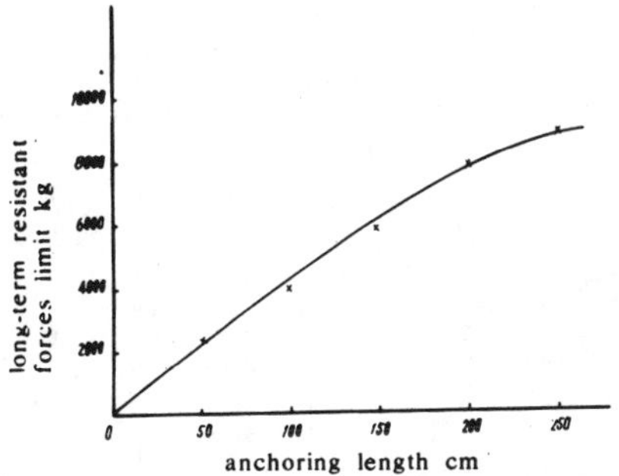

FIGURE 6 Relation between the ultimate long-term resistant force of anchorarm and anchoring length.

100 to 250 cm, the relation between the two is considered to be linear and is expressed by the following equation.

$$K_L = 1.168 - 0.001665\ h \qquad (1)$$

where K_L is the conversion coefficient of anchor length; h is the actual anchoring length of anchor in permafrost.

The conversion effective length of the anchor, h_L is calculated by the following equation:

$$h_L = K_L h = 1.168h - 0.001665\ h^2 \qquad (2)$$

where K and h are as defined in equation 1.

Ultimate Resistant Force of Anchor vs Its Actual Anchoring Length

When the diameter of anchor, the constituents of medium around anchor, water content, and temperature are kept constant and the loading rate is uniform, the ultimate resistant force of the anchor is a function of anchor length, i.e. it increases with the increase of anchoring length (Figure 6) and gradually approaches a maximum value---the maximum loading value determined by the yield strength of the reinforcing rod of anchor. The relation is expressed by the following equation:

$$P = \sigma A(1 - e^{-Kh}) \qquad (3)$$

where P is the ultimate resistant force of anchor, kg; σ is the yield stress of reinforcing rod, cm^2; A is the cross-section of reinforcing rod, cm^2; h is the anchoring length of anchor, cm; and K is a testing parameter. K=0.00698 when A=3.8 cm^2 and the diameter of the concreate anchor equals 10 cm.

It is seen from Figure 6 and equation 3 that to increase resistance force by increasing the anchor length is rational only within a limited range, i.e. when the anchor length exceeds a certain length, there is only a small increase in resistant force.

The nonlinear relation between resistant force and anchor length is directly related to the gradually advancing failure to the adfreeze strength in the upper reaches of the anchor. Since the effective length of anchor is constant provided ground temperatures, medium conditions, and the diameter of the anchor remains unchanged. An increase in anchor length is only beneficial to utilize a greater section of the anchor having residual strength.

Ultimate Resistant Force vs Diameter of anchor

The ultimate resistant force of anchor vs. the increase in anchor diameter, appearing in hyperbolic curve, is shown in Figure 7.

However, since the increase of anchor diameter is accompanied by the increase of rigidity, the decrease of effective length of anchor, under the same ground temperature and medium conditions, will bring about a decrease in the adfreeze stress distribution pattern with depth

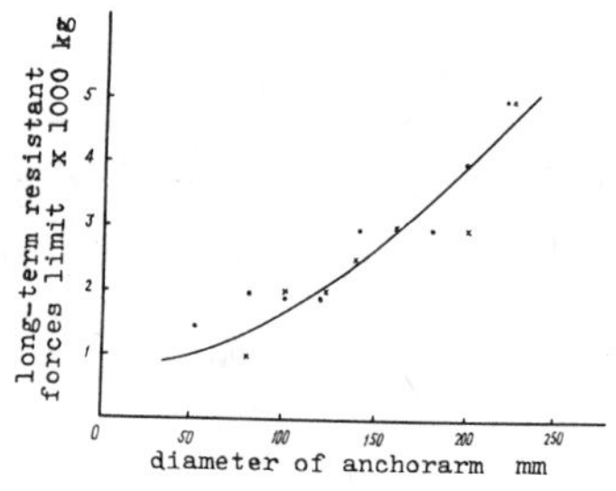

FIGURE 7 Ultimate long-term resistant force vs. diameter of anchors.
. Group A (No. 10-18 anchors),
* Group B (No. 19-27 anchors).
The length of all anchors is 100 cm.
The temperature is from -0.9°C to -1.3 °C.

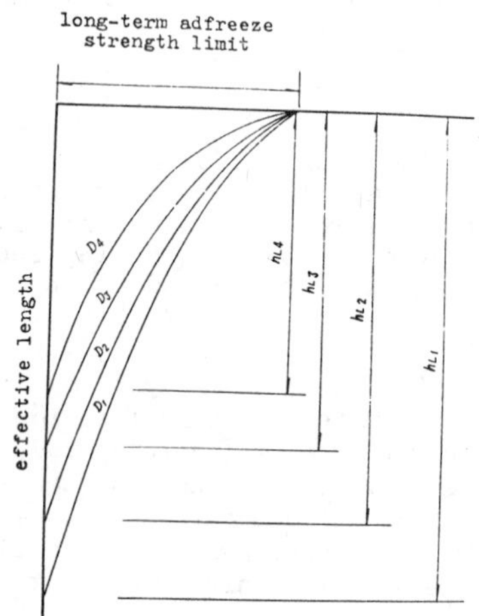

FIGURE 8 The sketch of effective length vs diameter of anchor.

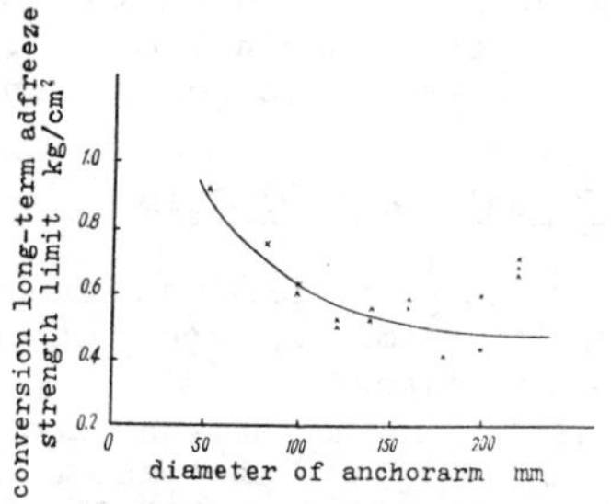

FIGURE 9 Conversion ultimate long-term adfreeze strength vs diameter of anchor

TABLE 2 Coefficient of Effective Diameter

Anchor diameter	Conversion ultimate long-term adfreeze strength	Coefficient of effective diameter K_d
5	0.92	1.44
8	0.71	1.11
10	0.64	1.00
12	0.59	0.92
14	0.55	0.86
16	0.52	0.82
18	0.51	0.80
20	0.50	0.78

(Figure 8). Therefore, the increase in anchor diameter will result in the weakening of conversion ultimate long-term adfreeze strength (Figure 9)*.

The effect of anchor diameter on the conversion ultimate long-term adfreeze strength can be expressed in terms of the coefficient of effective diameter. The so-called coefficient of effective diameter is the ratio of the conversion ultimate long-term adfreeze strength obtained in accordance with different diameters of anchor to that based on a fixed diameter of 10 cm, represented by K_d.

$$K_d = \frac{\tau_d}{\tau_{10}} \tag{4}$$

where τ_d is the conversion ultimate long-term adfreeze strength of anchor of diameter, d, in cm; τ_{10} is the conversion ultimate long-term adfreeze strength of anchor of 10 cm diameter.

The coefficient of effective diameter decreases with increase of the diameter (Table 2), which is expressed by equation 5.

$$K_d = K_{d_o} + \frac{a}{D} \quad \frac{b}{D^2} \tag{5}$$

where D is the diameter of anchor, in cm; K_{d_o}, a, b is the testing parameters. $K_{d_o} = 0.5457$, a = 4.6125, b = -0.7023 for reinforced concrete anchor buried in frozen sand.

Ultimate Resistant Force of Anchor vs Temperature

When the diameter of the anchor, backfill, and water content remain constant, the ultimate resistant force increases linearly with a de-

* Conversion ultimate long-term adfreeze strength is equal to the ultimate resistant force of anchor divided by the frozen area of anchor.

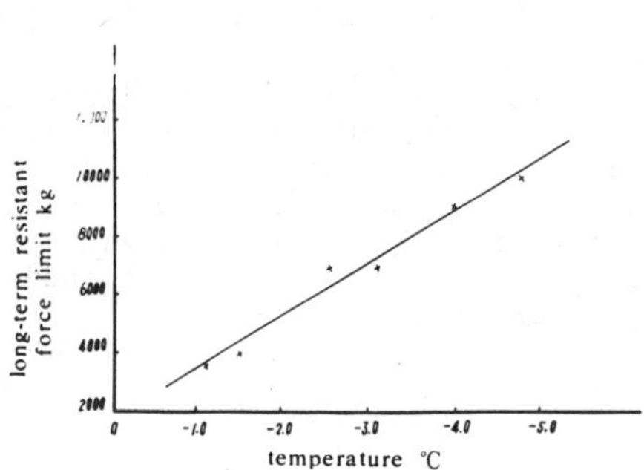

FIGURE 10 Relation between ultimate resistant force of anchorarm versus temperature

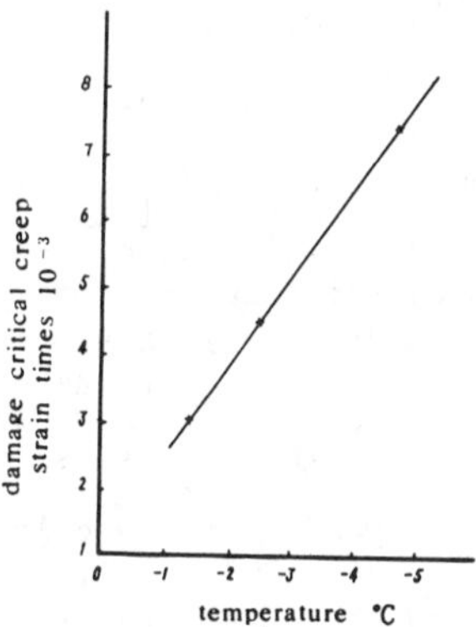

FIGURE 11 Damage creep critical strain versus temperature

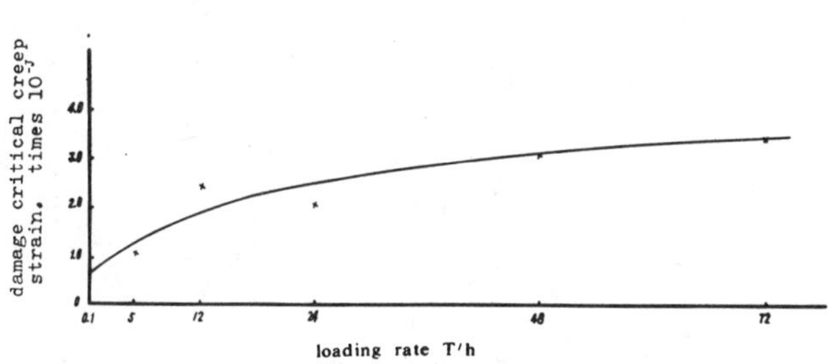

FIGURE 12 Damage creep critical strain versus loading rate

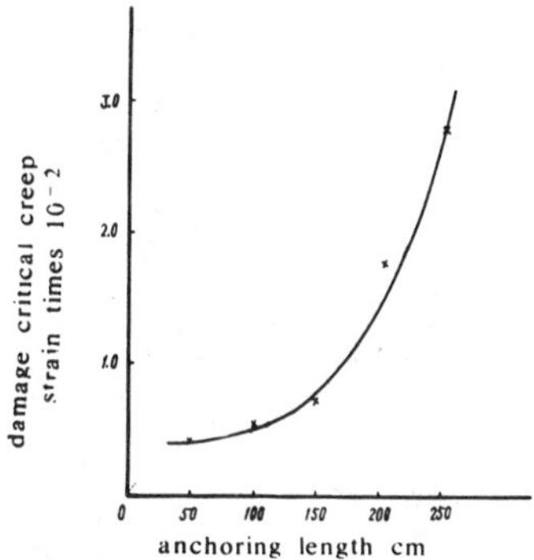

FIGURE 13 Damage creep critical strain versus anchoring length of anchor

crease in medium temperature (Figure 10). The relation between the conversion ultimate long-term adfreeze strength and temperature can be expressed by a linear equation, as:

$$\tau = A + m\,t \qquad (6)$$

where τ is the conversion ultimate long-term adfreeze strength, kg/cm^2; A is the testing parameter, kg/cm^2; m is the testing parameter, kg/cm^2, Deg.; t is the absolute value of negative temperature, Deg.. For mediums of saturation sand the following values of testing parameters can be used:

$$A = 0.06 \text{ kg/cm}^2; \quad m = 0.56 \text{ kg/cm}^2\text{, Deg.}.$$

Equation 6 can be used to evaluate the conversion ultimate long-term adfreeze strength of saturation sand with reinforced concrete anchor 10 cm in diameter and 100 cm in length at different temperatures.

The Attenuation Creep Critical Displacement and Damage Creep Critical Displacement of Anchor System in Permafrost

Field tests show that when the creep displacement of anchor in permafrost under the action of an external load reaches a certain value, the displacement of the anchor becomes uniform with time. This value is called the attenuation creep critical displacement of anchor system. When the anchor system comes to uniform flow and creep displacement reaches another maximum value, a transit of anchor from uniform flow to advancing flow begins. The latter maximum value is called the damage creep critical displacement of anchor system.

According to the results of experiments, the attenuation creep critical displacement of reinforced concrete anchor buried in saturation sand at temperature of -1.85 to -2.8°C, is about 2 to 4 mm.

The damage creep critical strain versus temperature is shown in Figure 11, which increases with a decrease in temperature and is approximately linear.

Damage creep critical strain limits increase with increasing loading rates, as shown in Figure 12.

In addition to the above mentioned factors, damage creep critical strain also is relevant to the anchoring length of anchor, i.e. increases with an increase of anchor length, as shown in Figure 13.

To sum up, the damage creep critical displacement of anchors is a specific feature relative to the anchor system, but its variation is very complicated, since it is influenced by a number of factors. It tends to be a constant, however, provided temperature, testing method, anchor length, and medium conditions remain unchanged.

CONCLUSION

1. The failure of the adfreeze strength of anchors in permafrost takes place gradually along the frozen boundary surface between the anchor and backfill. After failure of the adfreeze strength, there still exists the residual adfreeze strength at the boundary surface.

2. The resistant force of anchor is the function of the length, diameter of anchor, temperature, water content of medium and acting time of load. It increases with increase of anchor length and diameter. When the anchor length exceeds a certain length, there is only a small increase in resistant force.

3. There exists an "effective length" of the anchors in permafrost, which is influenced by the anchor diameter, temperature and constituents of medium and so on. In the calculations of resistant force of anchor, we can only consider it.

4. The attenuation creep critical displacement and damage creep critical displacement of anchor system are a specific feature relative to it. When the displacement of anchor is less than damage creep critical displacement, the failure of anchor will not take place under the action of an external load.

ACKNOWLEDGMENT

Lou Anjing, Zhou Aihua, Jiang Jiazheng, Cui Fengling, and Huang Geodian participated in the work.

A STUDY OF THE CALCULATION OF FROST HEAVING

Zhang Shixiang and Zhu Qiang

Water Conservancy Department of Gansu Province,
Lanzhou, People's Republic of China

In recent years, numerous irrigation canals lined with concrete and masonry slabs have been built in regions of seasonally frozen ground in Gansu, China. Many of them, however, have been damaged by the heaving action of frozen soil. So it is of great importance to find a practical way to determine frozen depth and frost heave under various conditions. Frost heaving is known to be a comprehensive process of heat conduction and moisture migration. The authors describe the moisture migration process by the theory of unsaturated soil water movement, developed since the 1950's, and an improvement over the taxonomic method prevalent in the past study of soil water. A stepwise calculation method for one-dimensional frost penetration and heaving is derived by simplifying the boundary conditions of heat and moisture transport and temperature distribution. This method provides not only the maximum frozen depth and frost heave, but also describes the freezing process and the heaving ratio along the frozen depth. Analytical formulas for deeply and shallowly seated ground water levels are presented with illustrating examples.

INTRODUCTION

In recent years, numerous irrigation canals lined with concrete and masonry slabs have been built in seasonal frost regions of Gansu, China. However, many of them were damaged or partially damaged by the heaving action of the frozen soil. So it is of great importance to find a practical way to determine the frozen depth and frost heave under various conditions. Since the 1970's the numerical method for simulating the heat and water transfer and heaving processes have been developed by many investigators (Harlan, 1973; Guymon and Luthin, 1974; Taylor and Luthin, 1978; Jame and Norum, 1980). From the practical viewpoint, however, because of the complexity of this method, it has not been applied extensively to the design of large-scale canal constructions. In addition, many of the current analytical methods for predicting frost heave only deal with heat and mass transfer processes separately and cannot meet the various soil and ground water conditions.

This paper describes the moisture migration process and sets up its equation by the theory of water movement in unsaturated soil. To obtain a calculation method that will be simple enough for practical engineering use and to be able to reflect the various temperature and moisture conditions of soil during freezing, we will first make some simplifications to the boundary conditions of the heat and moisture transport equations as well as temperature distribution in the frozen soil. An analytical formula using a stepwise calculation method can be derived by solving these equations simultaneously.

WATER MIGRATION

According to the theory of unsaturated soil water movement (Hiller, 1971), the soil water potential is composed of the matric potential ψ_m, pressure potential ψ_p, gravitational potential ψ_g, temperature potential ψ_T, osmotic potential ψ_{os}, i.e.

$$\psi = \psi_m + \psi_p + \psi_g + \psi_T + \psi_{os}$$

Hereinafter, the study is limited only to the matric and gravitational potential. The moisture flux can be expressed as the same form of Darcy's law:

$$q = - K(\psi_m) \cdot \nabla\psi \tag{1}$$

where $K(\psi_m)$ is unsaturated water conductivity and varies with the water content.

During freezing, the water migration occurs mainly in the unfrozen zone. According to the experimental and simulated results presented by Taylor and Luthin (1978), Jame and Norum (1980), water movement in the frozen zone is insignificant. In the unfrozen zone, the occurrence of both water migration and frost heave depend on whether the potential in the unfrozen soil below the freezing front is greater than ψ_o, the potential at the freezing front. Because the initial water content for frost heaving, w_o, is a content below which neither frost heaving nor water migration will take place, ψ_o may be regarded as the potential corresponding to w_o. In turn, we can determine ψ_o by w_o from the characteristic curve of soil water. On the basis of the experimental work of Wu et al. (1981), w_o is about 0.84 w_p, where w_p is the plastic limit of water content.

In this paper, only a one-dimensional case will be studied. If the soil water potential is taken as that of a unit weight, we have:

$$q = - K(\psi_m)(\frac{\partial m}{\partial x} + 1) \tag{2}$$

$$\frac{\partial w}{\partial t} = \frac{\partial}{\partial x} (D(w) \frac{\partial w}{\partial x}) + \frac{\partial K(w)}{\partial x} \quad (3)$$

where x is the vertical coordinate, positive downward, D(w) is the diffusivity of unsaturated soil water, a function of water content.

1. In the case of deeply seated ground water, the soil is relatively dry and gravitational potential can be neglected as compared with matric potential, then equation 3 can be rewritten as follows:

$$\frac{\partial w}{\partial t} = \frac{\partial}{\partial x} (D(w) \frac{\partial w}{\partial x}) \quad (4)$$

Assuming that soil is semi-infinity, the initial water content and diffusivity are uniform along the depth and are equal to w_B and D, respectively. Where D is the weighted mean value of D(w). According to Crank (1956), for a drying process, D can be expressed as:

$$D = \frac{1.85}{(w_B-w_o)^{1.85}} \int_{w_o}^{w_B} D(w)(w-w_o)^{0.85} dw \quad (5)$$

In addition, as shown before, the water content at freezing front is w_o, thus the boundary and initial conditions can be written as:

$$w(t, h) = w_o, \; w(t,\infty) = w_B, \quad t > 0$$

$$w(0, x) = w_B, \quad x > h$$

where h is the frozen depth. Then the analytical solution of equation 4 is:

$$w(t, x) = w_o + (w_B - w_o) \operatorname{erf} (\frac{x-h}{2\sqrt{Dt}}) \quad (6)$$

The water migration flux toward the freezing front is:

$$q = (w_B - w_o)\sqrt{\frac{D}{\pi t}} \quad (7)$$

2. In the case of shallowly seated ground water, the freezing process is divided into a number of time steps. Water movement can be approximately considered as a steady flow for each step. In equation 2, substitute soil water tension S for $(-\psi_m)$, after rearranging, this gives:

$$x = \int_0^S \frac{dS}{1 + q/K(S)} \quad (8)$$

At the present time, the theoretical relation between K(S) and S has not yet been found. In general, they are related by an empirical formula derived from experimental data taking the following form:

$$K = \frac{K_s}{CS^m + 1}$$

where K_s is the saturated water conductivity and C and m are constants. When m = 1, 1.5, 2, 3, 4, the analytical solution of equation 8 can be derived. For example, if m = 2, we have:

$$\frac{K_s}{\sqrt{Cq(K_s + q)}} t_g^{-1} (S \sqrt{\frac{Cq}{K_s + q}}) = x \quad (9)$$

At the freezing front, x = H-h, S = S_o, where H is the depth of the ground water table, and we can obtain an implicit equation for the water flux:

$$h = H - \frac{K_s}{\sqrt{Cq(K_s + q)}} t_g^{-1} (S_o \sqrt{\frac{Cq}{K_s + q}}) \quad (10)$$

HEAT CONDUCTION

The heat conduction equations during freezing are as follows:

$$\frac{\partial u_f}{\partial t} = a_f \frac{\partial^2 u_f}{\partial x^2}, \quad h > x > 0, \quad t > 0 \quad (11)$$

$$\frac{\partial u_T}{\partial t} = a_T \frac{\partial u_T}{\partial x^2}, \quad x > h, \quad t > 0 \quad (12)$$

Assuming the freezing temperature of soil is zero, then at the phase changing point we have:

$$u_f(h, t) = u_T(h, t) = 0 \quad (13)$$

$$\lambda_f \frac{\partial u_f}{\partial x}\Big|_{x=h} - \lambda_T \frac{\partial u_T}{\partial x}\Big|_{x=h} = 333\gamma(q+(w-w_o)\frac{dh}{dt}) \quad (14)$$

where u is the temperature, λ is the heat conductivity, a is the thermodiffusivity, γ is the specific weight of water, and the subscripts f and T are for the frozen and thawed zones, respectively. The unit of latent heat is taken as J/g.

For practical purposes, it would be better to find an approximate solution for equations 11 to 14. First, the temperature increment in the period Δt is assumed to be linearly distributed along the depth. During the steady frost penetration period, the error caused by this assumption is an infinitesimal of higher order. Then we can obtain:

$$\Delta u_f = u_o - \frac{\Delta u_o - \Delta u_f'}{h} x \quad (15)$$

where Δu_o, $\Delta u_f'$ is the temperature increment in the period Δt at the surface (x = 0) and the frozen depth (x = h), respectively (see Figure 1).

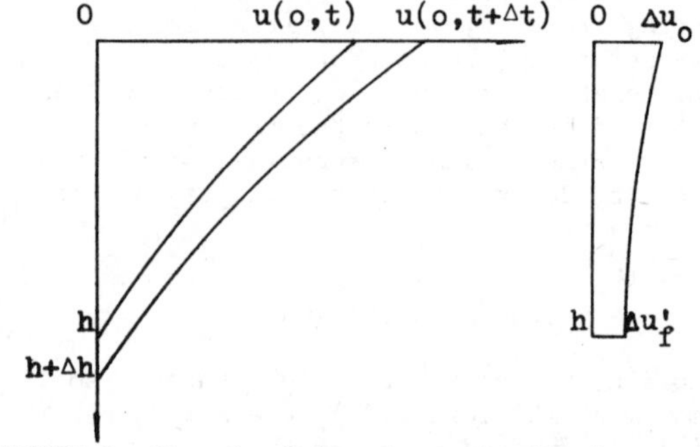

FIGURE 1 Sketch of Δu along the depth.

If we rewrite the left side of equation 11 in a differential form and substitute equation 15 into it, we have:

$$a_f \frac{\partial^2 u_f}{\partial x^2} = \frac{\partial u_f}{\partial t} \doteq \frac{\Delta u_o}{\Delta t} - \frac{\Delta u_o - \Delta u_f'}{h\Delta t} x$$

Integrate the above equation over x with the boundary conditions: $x = 0$, $u_f = u_o$; $x = h$, $u_f = 0$. After rearranging, we have:

$$u_f = \frac{-(\Delta u_o - \Delta u_f')}{6a_f h\Delta t} x^3 - \frac{\Delta u_o}{2a_f\Delta t} x^2 - \left(\frac{u_o}{h} + \frac{2\Delta u_o + \Delta u_f'}{6a_f\Delta t} h\right) x + u_o \tag{16}$$

$$\left.\frac{\partial u_f}{\partial x}\right|_{x=h} = \frac{h\Delta u_o + 2u_o\Delta h}{6a_f\Delta t + 2h\Delta h} - \frac{u_o}{h} \tag{17}$$

In the unfrozen zone, the initial condition: $u_T(x, 0) = u_a$, $x > h$, is assumed and it should be noted that the temperature at the upper boundary, i.e. the freezing point, drops to zero. We can obtain the solution of equation 12 as:

$$u_T(x, t) = u_a \mathrm{erf}\left(\frac{x-h}{2\sqrt{a_T t}}\right) \tag{18}$$

$$\left.\frac{\partial u_T}{\partial x}\right|_{x=h} = \frac{u_a}{\sqrt{\pi a_T t}} \tag{19}$$

where u_a is the mean temperature of soil before freezing.

Substituting equations 7 and 17 which are for the deeply seated ground water table, or 10 and 19, which are for the shallowly seated ground water table, into equation 14, we then obtain the expression of dh/dt. Furthermore, the unfrozen water still exits in the negative temperature, mainly in the range of 0°C to -5°C. To consider the effect on frost penetration precisely is difficult for an analytical solution. According to the result of numerical method (Nixon et al., 1973), we can regard the latent heat is to be suddenly released at temperature zero without a significant error to the freezing process. So we can modify the second term $(w - w_o)$ in the right parentheses of equation 14 with $(w - w_{-5})$, where w_{-5} is the unfrozen water content at -5°C. Thus the basic formula can be expressed as follows:

$$\lambda_T\left(\frac{h\Delta u_o + 2u_o\Delta h}{6a_f\Delta t + 2h\Delta h} - \frac{u_o}{h}\right)\Delta t = 333\gamma((w - w_{-5})\Delta h + q\Delta t) + \lambda_T \frac{u_a\Delta t}{\sqrt{\pi a_T t}} \tag{20}$$

DEEPLY SEATED GROUND WATER TABLE

Substituting q determined by equation 7 into equation 20 gives:

$$\lambda_f\left(\frac{h\Delta u_o + 2u_o\Delta h}{6a_f\Delta t + 2h\Delta h} - \frac{u_o}{h}\right)\Delta t = 333\gamma(w_B - w_{-5})\Delta h + (w_B - w_o)\sqrt{\frac{D}{\pi t}}\Delta t) + \lambda_T \frac{u_a}{\sqrt{\pi a_T t}}$$

This is a quadratic equation. After rearranging, we get the solution of Δh:

$$\Delta h = \left(\tfrac{1}{2}\lambda_f h\Delta u_o - 3a_f\left(333\gamma(w_B - w_o)\sqrt{\frac{D}{\pi t}} + \frac{\lambda_T u_a}{\sqrt{\pi a_T t}} + \frac{\lambda_f u_o}{h}\right)\right)/1000\gamma a_f(w_B - w_{-5}) \tag{21}$$

When u_o is constant, equation 21 can be integrated to yield h as follows:

$$h = \left(\sqrt{\frac{-2\lambda_f u_o}{333\gamma(w_B - w_{-5})}} - \frac{333\gamma(w_B - w_o)\sqrt{D} + \lambda_T u_a/\sqrt{a_T}}{594(w_B - w_{-5})}\right)\sqrt{t} \tag{22}$$

If the latent heat of migrating water and the heat conducting toward the freezing front from the unfrozen zone are neglected (the second term in the parentheses), and the unfrozen water is assumed to be non-existent, then the well-known Stefan's formula can be derived from the above equation:

$$h = \sqrt{\frac{-2\lambda_f u_o t}{333\gamma w_B}}$$

We can see that Stefan's formula is a particular form of equation 22 under special condition.

When u_o changes with time, we must divide the freezing period into a number of intervals. The increment of frozen depth in a certain period can be calculated from the frozen depth at the end of the last period. Water migration flux in the various period can be defined by equation 7, and the frost heave is:

$$\Delta H_h = 0.09(w_B - w_{-5})\Delta h + 1.09\sqrt{\frac{D}{\pi t}}(w_B - w_o)\Delta t \tag{23}$$

where ΔH_h is the heave value in the various period. In general, the first term may be neglected.

Example 1. At a given locality, the ground water is deeply seated and the soil is silty loam:

$\lambda_T = 0.0105$J/cm·s°C, $\lambda_f = 0.0151$J/cm·s°C,

$a_T = 3.82 \times 10^{-3}$cm^2/s, $a_f = 7.41 \times 10^{-3}$cm^2/s,

$w_B - w_o = 0.13$, $w_B - w_{-5} = 0.26$,

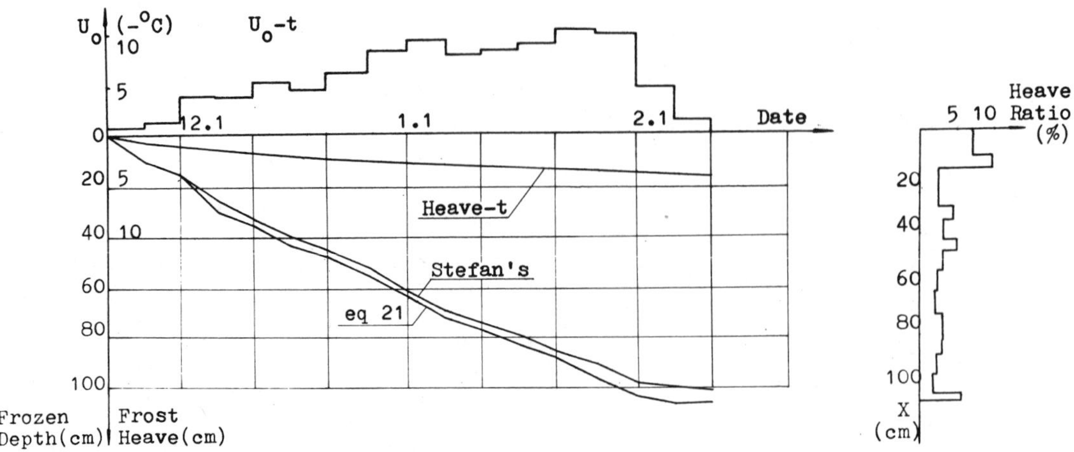

FIGURE 2 Frozen depth and frost heave versus time and heave ratio distribution for the deeply seated ground water.

$D = 1.16 \times 10^{-4} cm^2/s$, $u_a = 8.2°C$.

The surface temperature changing with time is shown in Figure 2.

The time step is taken as 5 days, and the results are shown in Figure 2.

We can see that in the case of deeply seated ground water, the variation between results from equation 21 and Stefan's formula is insignificant. Moreover, the frost heave ratio in the upper zone is greater than in the lower zone; this conclusion can fit many observed data in situ (Zhu, 1980).

SHALLOWLY SEATED GROUND WATER

Under this condition, equation 20 should be modified. As shown in Figure 3, when frost penetration increases from h to $h + \Delta h$, the original water content that has changed phase is the shaded area between curve 1-1 and 2-2, so the first term of the right side of equation 20 should be rewritten as:

$$333\gamma\left(\left(\frac{2}{3}w_s + \frac{1}{3}w_o - w_{-5}\right)\Delta h + q\Delta t\right)$$

where w_s is the saturated water content. Then we can get the solution of Δh under this condition:

$$\Delta h = \frac{\frac{1}{2}\lambda_f h \Delta u_o - 3a_f \Delta t\left(333\gamma q + \frac{\lambda_T u_a}{\sqrt{\pi a_T t}} + \frac{\lambda_f u_o}{h}\right)}{1000\gamma a_f\left(\left(\frac{2}{3}w_s + \frac{1}{3}w_o\right) - w_{-5}\right)} \quad (24)$$

The water flux q can be defined by a relation curve between q and h drawn from the implicit equation 10.

Example 2. The ground water table is 120 cm below the surface.

$w_s = 0.51$, $w_o = 0.24$, $w_{-5} = 0.11$,

$K_s = 1.16 \times 10^{-5} cm/s$, $c = 2.5 \times 10^{-3} cm^{-2}$,

$m = 2$, $S_o = 370$ cm.

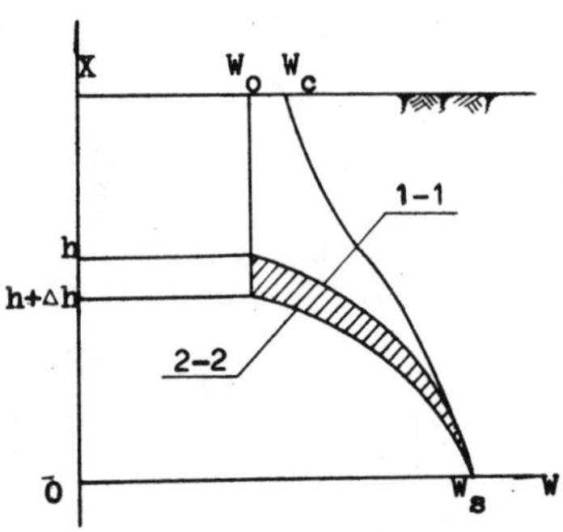

FIGURE 3 Sketch illustrating the phase change of the original water.

The other conditions are the same as in example 1. The results are shown in Figure 4. We can see that the frozen depth is much less and the frost heave is much greater than that under the deeply seated ground water condition. The result indicates that adopting Stefan's formula in this case may cause a great error. The heave ratio in the upper zone is much less than in the lower. This conclusion fits well with the field experiment and has an important effect on the perfect selection of the anti-heave measure in engineering (Zhu, 1980).

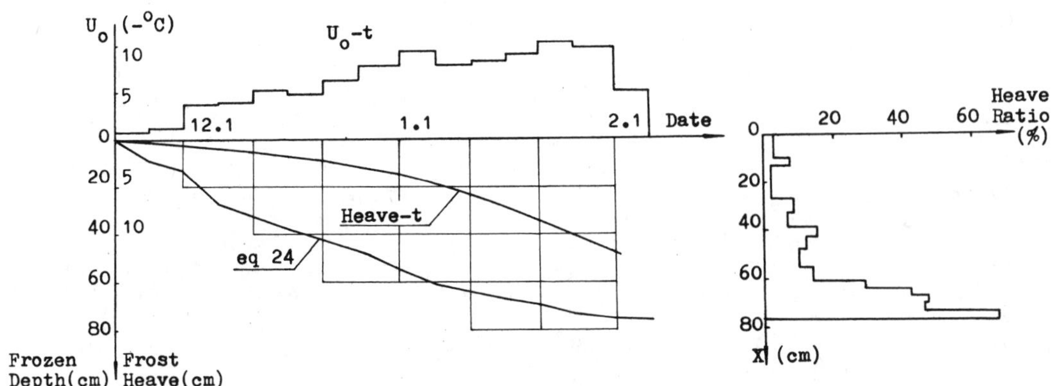

FIGURE 4 Frozen depth and frost heave versus time and heave ratio distribution for the shallowly seated ground water.

REFERENCES

Crank, J., 1956, The mathematics of diffusion: London, Oxford Press.

Guymon, G. L. and Luthin, J. N., 1974, A coupled heat and moisture transport model for arctic soils: Water Resources Research, v. 10, p. 995-1001.

Harlan, R. L., 1973, Analysis of coupled heat-fluid transport in partially frozen soil: Water Resources Research, v. 9, p. 1314-1323.

Hiller, D., 1971 (Chinese translation from English, 1981), Soil and water: physical principles and processes, Beijing, Agro. Press.

Jame, Y. W. and Norum, D. I., 1980, Heat and mass transfer in a freezing unsaturated porous medium: Water Resources Research, v. 16, p. 811-819.

Nixon, J. F. and McRoberts, E. C., 1973, A study of some factors affecting the thawing of frozen soils: Canadian Geotechnical Journal, v. 10, p. 439-452.

Taylor, G. S. and Luthin, J. N., 1978, A model for coupled heat and moisture transfer during soil freezing: Canadian Geotechnical Journal, v. 15, p. 548-555.

Wu, Z. W., Zhang, J. Y., Wang, Y. Q. and Sheng, Z. Y., 1981, An experimental research of the frost heave of soils, _in_ Memoirs of Lanzhou Institute of Glaciology and Cryopedology, Academic Sinica, No. 2, p. 82-96.

Zhu, Q., 1980, Characteristics of frost heaving in concrete lined canals and the measure of replacing subsoils with sand-gravel in Gansu, _in_ Proceedings of the 2nd National Conference on Permafrost, Lanzhou, to be published.

THE PERIGLACIAL ENVIRONMENT OF THE LATE PLEISTOCENE ALONG THE QINGHAI-XIZANG HIGHWAY

Zhang Weixin, Xu Shuying, Xu Qizhi, and Shi Shengren

Department of Geology and Geography, Lanzhou University
People's Republic of China

There are several kinds of periglacial phenomena on the Qinghai-Xizang Plateau. Polygon and wedge-shaped casts 1 to 50 m in diameter were formed during the last glacial period, and the involutions buried at depths of 3 to 11 m may have been formed during the Neoglaciation. Based on C^{14} dating, pollen analysis, and periglacial phenomena, the paleogeographic environment during the Late Pleistocene can be reconstructed. About 35,000 years ago, the Qinghui He Basin and other high plains in southern Qinghai were an interperiglacial environment, and carbonate nodules were formed in lake deposits; the upper parts of the Kunlun and Tanggula Shan were periglacial environments. From 26,000 to 14,000 year ago, the mean annual air temperature was about -7° to -8°C. No carbonate nodules developed in the lake basins, but large polygons and sand wedges were formed on terraces along some rivers, and glaciers continued to grow in the high mountains. The lower limit of the periglacial belt in the northern part of the Qinghai-Xizang Plateau descended to 3200-3400 m a.s.l. From 14,000 to 12,000 years ago, the climate was still rather cold, as indicated by undecayed buried plants.

As is well-known, the Qinghai-Xizhang plateau is an area of strong uplift of altitude since the Quaternary period (Li Jijun et al, 1979). According to the data which have obtained, from geologic (Wen Shixuan 1981), geomorphic (Zheng Benxing et al, 1981), and paleontologic (Kong Zhaochen et al, 1981), and many other evidences demonstrate that the altitude of the plateau during the late Pleistocene was generally more than 4,000 m a.s.l., and the mountain areas were higher, usually more than 5,000 m. This high-altitude topography of the plateau first brought about a drop of the temperature with the rise of the plateau surface. Along with the coming of the cold climate of the late Pleistocene on a global scale, a superimposed cold happened on the plateau. The combination of the two kinds of cold put the plateau surface into a periglacial environment of long duration. This has been proved by a series of periglacial phenomena, ancient and existing, biologic and non-biologic, recently found on the south Qinghai plateau.

During the 4 summers from 1974-1977, we all worked along the Qinghai-Xizhang Highway, and paid, to a certain degree, attention to a series of periglacial phenomena there. Later we made reports on some of them (Zhang Weixin 1979a; 1979b; 1982). It is based on these reports, on the date of the assemblages of spore pollen in the deposit of 20 m deep drilling hole, and absolute C-14 dating that makes this special discussion on several changes in the environment of south Qinghai plateau during the late Pleistocene. According to these data the paleogeographical environment in this area has been reconstructed.

Qingshui River area is located near the 68 rd station of Qinghai-Xizhang Highway. To the north are Kunlun Shan and to its south the Tanggula Range; both mountains are more than 5,500 m a.s.l. and about 60 and 400 km away from the area respectivery. To the east and west is the hinterland of southern Qinghai plateau. During the Quaternary period, this area was a structure fault basin. The deposit of the Quaternary is over 200 m thick. At present there remains a small lake to the southeast of this area, and the former river bed from the southern slope of Kunlun Shan down to the lake basin is indistinct.

This area, has mean air temperature in January is -17.3°C, in July of 5.4°C; the mean annual temperature is -5.5°C, and for more than seven months in the year it is under 0°C. The annual precipitation there is 267 mm.

PERIGLACIAL REMAINS OF THE LATE PLEISTOCENE

Up to now, the periglacial phenomena of the late Pleistocene we found in the area on the plateau south of Qinghai and north of Xizhang, and the periglacial remains are great in variety, of type, scale, and content.

Polygon and Wedge-Shaped Casts

During Würm glaciation period large polygons were widely formed on the plateau (Zhang Weixin; Xu Shuying et al, 1981; Zhang Linyung 1981). The biggest of slopes of these developed on the southern and northern slopes of the Tanggula Range and on the banks of the valleys of some big rivers. They were usually more than 50 m in diameter. There are some of sand wedges in various shapes extensively distributed under the terraces surface, the biggest being over 3 m broad at the upper and 1.8 m deep. C-14 dating of the silt in the wedges is 24,000 to 15,000 years old (Gao Dongxin 1979; Zhang Weixin 1983). Therefore, the sand wedge is the product of the main Würm-glaciation.

Involution

This is the best preserved form of former and modern periglacial on the plateau surface. In 1976 we found in 20 m by drilling hole under the ground surface 6-11 m many well-developed ground ice layers and marlmud interbeddings, which are in form of involution. In the southern part of the Tanggula Range, there is also well-developed involution 2 m under the ground surface of the place east side of the highway near Liangdaohe. The brown humic soil under the ground surface is 545±60 years old by its C-14 dating, which shows that the involution was formed during the Neoglaciation period.

Stair Surface on Slopes

During our research on the southern part of the Tanggula Range, we found a kind of well-developed bedrock stair surface on the mountain slopes between Ando Xian (county) and Nagqu Xian (Heihe county). It is different from the river-eroded bedrock terraces. The landform is distributed on the convex slopes at 4,500 to 4,800 m a.s.l. The rock step surface slant slightly forward. They are mostly covered with a thin layer of slope debris, and with angle of block rocks are accumulated before the tread surface, but at another place the bedrock are exposed, at the back border of the tread the bedrock becomes steepen only a little higher. Upwards from there the slope becomes gentle and its surface is covered with a thin layer of soil, and the ground surface is covered with a high cold of meadow. This kind of landform is nonactive under the present cold climatic condition. This landform is quite similar to the cryoplanation terrace developed in the periglacial environment of high latitude areas reported by Péwé et al (1981) and Reger et al (1976). It might have developed during the cold stage of the late Pleistocene. Landforms with such features have also been found in Qinlian Shan areas and on the southern slopes of Mt. Datong and Mt. Longlongling, which are distributed at piedmont, 4,000 to 4,200 m a.s.l.

FEATURES OF THE PALEOGEOGRAPHIC ENVIRONMENT

From that 20 m hole in the Qingshui River area, the following assemblages of spore pollen are found in the layer of 18 to 20.25 m deep: Leptogramma, Pinus, Picea, Abies, Cedrus, Tsuga, Larix. Ephedra, Gramineae, Allium, Alnus, Quercus, Ulmus, Chenopodium, Nitraria, and Compositae. According to the analysis of their biological character and their required ecological geographic conditions, these plants are not likely to grow in the same natural geographic region. because of it is different of biological feature of the different plants and of ecologic geographical condition, which their required. So the proper explanation for this is that they were distributed in the areas from plateau surface of warm and dry environment to mountains of

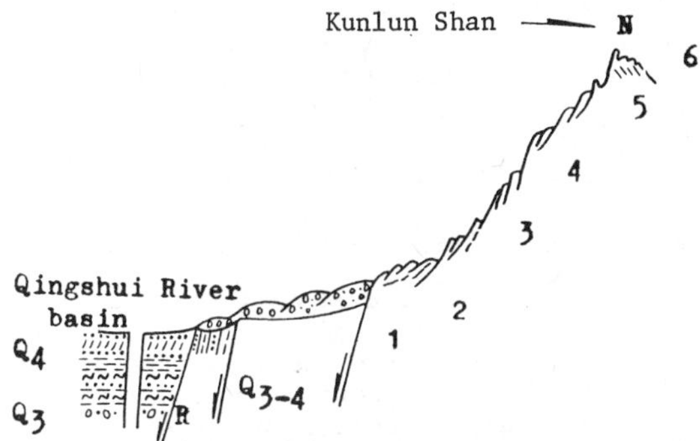

FIGURE 1. Paleolandscape belts. 1) Lake basin, 2) Gravel-Gobi steppe, 3) Broadleaf trees, 4) Broadleaf-coniferous, 5) Alpine meadow and periglacial belt, 6) Alpine snowline.

warm-cool and humid climatic conditions, similar to the natural landscape from Gansu corridor to Qilianshan area today. The spore-pollen of some of the plants probably have carried down by flowing water from Kunlunshan and accumulated in basin. According to their biologic features, these landscape belt may be divided as follows: the lake basin belt of Qingshui River; the belt of aquatic and wet plants on the banks; the belt of gravel-gobi steppe before mountains; the belt of broad leaf trees on low mountains and in valleys; the belt of broad leaf trees mingled with coniferous trees in mountain areas; the alpine meadow and periglacial belt; the alpine snow-ice belt (Figure 1).

The Lake Basin Belt

The lake basin of the Qingshui River is a fault basin of Cenozoic Era, the Quaternary deposit being over 200 m thick. The unearthed stratum from the 20 m hole is a set of deposits of river-lake facies formed during the late Pleistocene. The deposit features and the spore pollen in the deposits show that the environmental conditions were different from the interperiglacial climates on the plateau surface to the periglacial climates on the mountains.

The Belt of Gravel-Gobi Steppe before the Mountains

As a result of the new tectonic movement, the altitude of Kunlun Shan uplift very high, the great alluvial fans were distributed before the southern foot of Kunlun Shan. Because of the severity of the ecological environmental conditions, there are some drought-enduring plants, such as Ephedra, Nitraria, and Compositae growing there. This shows that the climatic condition then was warm-dry, similar to the natural landscape at the southern border of Qaidam basin at the northern foot of the Kunlun Shan.

The Broadleaf Trees Belt of Low Mountains and Valleys

This belt is distributed mainly in the areas of low mountains and valleys at the southern foot of Kunlun Shan. As the environment of plants vary with an increase in altitude and landform scattered broadleaf trees of warmth-resisting and drought-enduring types such as Ulmus, Alnus, Betula, and a few Cupressus etc. grow on the sunny slopes of drier growing condition of low mountains and valleys.

Broadleaf-Coniferous Belt in Mountain Areas

The belt is distributed on the slopes at higher altitudes, the lower of which the climate becomes temperate cool and the atmospheric humidity increases. The plants which can grow in such environment are Betula, Quercus, Larix, Tsuga, and Cedrus. At the upper part of this belt the climate becomes cooler and the humidity greater; therefore, the number of trees of coniferous genus increases. At the place where the altitude still higher instead of Picea and Abies gradually.

Alpine Meadow and Periglacial Belt

This belt is above the forest belt, with a mean annual air temperature below 0°C. Alpine meadow takes the place of the needle-leaved forests. Up to the upper part of the mountain areas is periglacial environment where various periglacial processes and phenomena have developed.

Alpine Snow and Ice Belt

This belt is distributed on top of mountain areas of still higher altitude where all kinds of glaciers and glacial landforms have developed.

Quaternary Age				Depth m	C-14 yrs BP.	Stratum
Quaternary	Holocene	Post-glacial age	Recent	2	2,580±120	
				4	12,950±270 14,250±160	
	Late Pleistocene	Main Würm	Periglacial	6		
				8		
				10	26,200±890	
		Interglacial	Sub-periglacial	12		
				14		
				16	34,700±280	
				18		
				20		
		Early Würm	Periglacial	22		
	Middle Pleistocene					

Relics of leaves & roots of plants

Ground ice & involution

Carbonate nodules

Sand & peaty soil

● Sample points

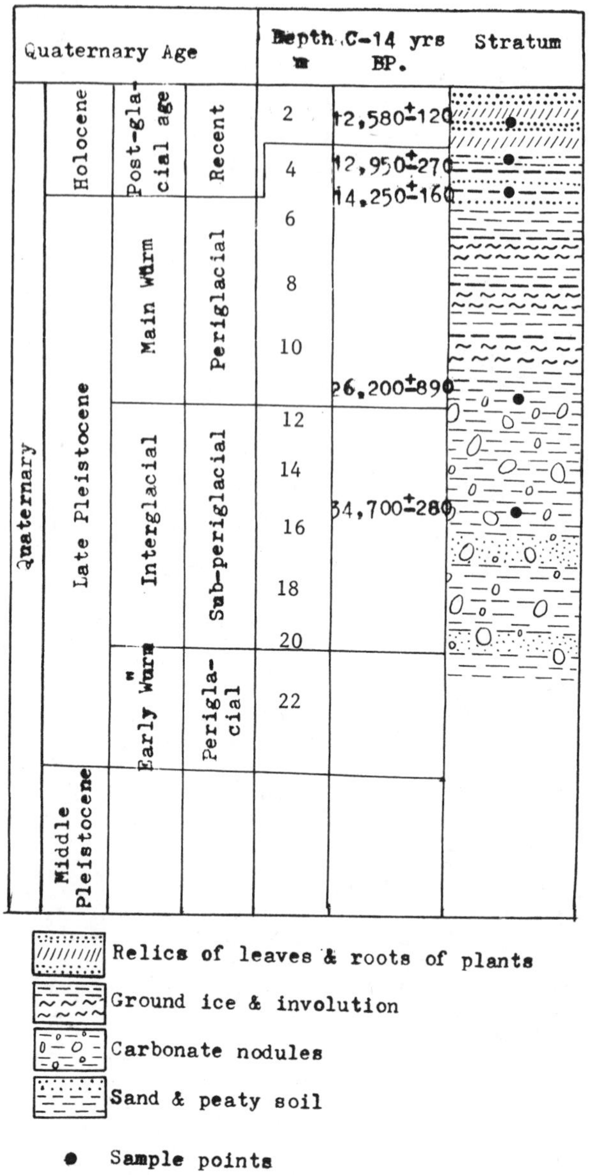

FIGURE 2. Geologic profile from drill record of Qingshui River Basin.

DISCUSSION

Characteristics and Evolutionary History of the Deposit of River and Lake Facies

According to the features of the deposit from that 20 m deep hole in Qingshui River, the assemblages of spore pollen and the absolute age by C-14 dating in it, this deposit can be divided into three layers.

The lower layer is between 11.78 to 20.25 m under the ground surface. The upper section of it, which is between 11.78 to 14 m, is of marlmud with a few ice crystals in it. The colour at the top of this part is dark-gray, which changes gradually into black at the lower part. Probably because of the influences of ground water, the record of this part is in melted form with a comparatively great content of carbonate nodules. The lower section of it, which is between 14 to 20.25 m, is of blue gray and light gray marlmud with coarse sand and limestone pebbles in it. This shows that the lake at the time when this part was formed still had rivers flowing into it. C-14 dating of the blue gray marlmud layer from 14 to 14.7 m under the ground surface shows that it is 34,700 ± 2,800 years old (Figure 2).

The middle layer, which lies 4.32 to 11.78 m under the ground surface, is composed of interlayers of gray-white and blue-gray marlmud and a yellow green one with a few pebbles and coarse sand layers in it, and the underground ice which is 5 to 7 cm thick being well-developed in it. Its frost deformation structure (involution) is very clear. Down from there, it gradually turns into a marlmud and sand layer which contains carbonate nodules. In the part from 9.37 to 11.78 m in this layer the carbonate nodule is 26,200 ± 890 years old by its C-14 dating.

The upper layer, 1 to 4.32 m under the ground surface is, at the upper part, composed of sheets of brown roots and leaves of plants mixed with a little amount of sand and clay, and, at the lower part, a layer of blue-gray ooze mixed with sheets of medium-size and fine sands as well as with humic sheets of half-rotten leaves and roots of plants. Still lower it turns into an interlayer of sand and ice. The blue-gray ooze layer, whice is 2.73 to 2.90 m under ground surface is 14,250 ± 160 years ago, and the ooze layer, which is 1.57 to 2.52 m under the ground surface, is 12,950 ± 290 years old by C-14 dating.

The Quaternary deposit in the lake basin discussed above and the features of spore pollen in it reflect that the lake basin of the Qingshui River area of the late Pleistocene underwent three stages.

The early stage was the period when all the deposit of the lower layer were formed and the spore pollen of plants in it was accumulated. According to C-14 dating of the blue-gray marlmud between 14 to 14.7 m, this period is 35,000 years ago. This sedimentary material of Kunlun Shan col-basin from 40 to 20 m of drilling hole (upwards) under the ground surface is a layer of fine and coarse sand at the lower part, and a layer of subclay at the upper part. The spore pollen of plants in this part contains Ephedra, Artemisia, and Chenopodiaceae, whose composition is similar to the landscape of the southern part of Gansu Corridor. This part also shows that the environmental condition is semiarid steppe (Tang Lingyu et al, 1979). Thus it can be seen that the deposit in the lake basin was formed from 35,000 years ago when the environment was semiarid down to the climate at that time became very dry, the lake getting less water, and the amount of evaporation being great. The lake water therefore became concentrated and became rich in carbonate. As a result, deposit of marl containing large amount of carbonate was formed. The spore pollen of coniferous and broadleaf trees in this layer was probably carried down by flood from warm-cool mountain areas nearby. On the whole, this stage demonstrates the environmental features of the interglacial period.

The middle layer of the deposit in the lake basin was formed during this period. The lower section of it is composed of marlmud and sand with a few carbonate nodules. Upwards the layer turns gradually into the upper part of marlmud of grayish white, bluish gray, and yellowish green, in which sheets of coarse sand are mingled. This upper section is empty of carbonate nodules. The environmental condition reflected is warm and dry climatic condition evolving into cold climatic condition, the evaporation being lessened, and the chemical process weakened. From then on, it came into one cold periglacial environment, i.e. the Würm glaciation period.

The part formed during the late stage was the upper layer of the deposit. Its lower part is of ooze layer, which turns upwards into ooze of bluish gray which is mixed up with medium-sized sand sheets and gray ooze sheets containing half-rotten roots and leaves of plants. The upper part is a layer containing well-preserved brown roots and leaves of plants, with a little sand in the roots. The ooze layer of bluish gray, 2.73 to 2.9 m under the ground surface, is 14,250 ± 160 years old by its C-14 dating, and the sandy ooze layer, 1.57 to 2.52 m under the ground surface, is 12,950 ± 290 years old by its C-14 dating. These show that the water supply to the lake was lessened and the lake became shallow while the air temperature was still cold. The roots and leaves of aquatic plants growing on the lake bank accumulated without decomposition. Therefore this must have been at the stage of the Würm glaciation.

The analysis of the deposit from that hole in Qingshui River lake basin and the biological features of the spore pollen in it shows that during the early stage, about 35,000 years ago, the environmental condition in which the deposit in Qingshui River was formed is a landscape of warm and dry steppe in areas of plateau; and a landscape of cool, humid forest steppe in low mountain areas changing gradually towards cold periglacial and glacial landscapes in top of the high mountains.

About 26,000 to 14,000 years ago, the climate grew very cold with the coming by degrees of the Würm glaciation. In this period, the lake deposit changed from marlmud containing carbonate nodules of the early stage into marlmud without any content of carbonate nodules. Glaciers developed throughout the Kunlun Shan and Tanggula Range, and polygons and river-bank sand wedges developed widely on the terraces of valleys and the plains.

About 14,000 to 12,000 years ago, owing to the plateau surface uplift by new tectonic movement, the altitude is increased, the cold of the Würm glaciation in the plateau area continued. As a result, the roots and leaves of aquatic plants accumulating 1.5 m under the ground surface of the lake basin and were not able to decompose. The cold periglacial environment lasted until the early Holocene. During the mid-Holocene period of high temperature, the temperature on the plateau increased slightly; it dropped again during the late Holocene. Therefore, the periglacial environment has been preserved until now.

Northern Lower Boundary of Periglacial Belt on Qinghai Plateau during Late Pleistocene

The highest peak in the Tanggula Range is 6,000 to 6,500 m a. s. l. The average height of the range is about 5,800 m. The modern snowline at 5,000 to 5,400 m, while the fossil cirques are at 5,000 to 5,200 m. The average height of fossil cirque is about 5,100 m. But the amount of uplift of plateau surface in Holocene is about 800 to 900 m, deduct this number from 5,100 m, then the height of fossil forms is about 4,300 to 4,200 m. The Fenghuo Shan area in the northern part of Tanggula Range is 4,600 m high, with a mean annual air temperature of -6.6°C today. Calculated according to the rate of the progressive decrease of 0.5°C per hundred metre, the mean annual air temperature near the existing snowline of the Tanggula Range is about -10°C to -11°C. The value is the temperature of fossil cirques formed. Accordingly it is deduced that the northern lower boundary of the paleo-periglacial environment in the plateau south of Qinghai was about 2,100 to 2,200 m during Würm glaciation.

Kunlun Shan is approximately 6,000 m a.s.l. at the central part and 5,600 to 5,800 m on an average. The height of the modern snowline at 5,200 to 5,400 m and 5,300 m on an average. The fossil cirques were at 4,800 to 5,200 m during the Würm glaciation, the average height at about 5,000 m. But the amount of uplift of plateau surface in Holocene is about 800 to 900 m, the number is deduct from 5,000 m, then the fossil cirques at 4,300 to 4,100 m.

Xidatan, the fault valley, in the northern

part of Kunlun Shan is 4,200 m a.s.l., and the modern mean annual air temperature is about -3°C. Calculated according to the rate of the progressive decrease of temperature mentioned above the mean annual air temperature near the modern snowline of Kunlun Shan is -9°C to -10°C, the temperature value is similar to that near the fossil snowline to some extent. Hence, it is inferred that the lower boundary of the periglacial belt on the northern slope of Kunlun Shan was 2,100 to 2,300 m during the Würm glaciation. This height is in accordance with that of the lower boundary of the northern part of the periglacial belt of Tanggula Range, which is deduced from the lower boundary of the fossil cirques on it. This height is lower than that of Nachitai.

Nachitai, which is at the central reaches of Kunlun River, is 3,500 m a.s.l. On the southern bank of Kunlun River in this area, a layer of involution is found on the first terrace. The plant roots and the stems underlying the layer are 4,900 ± 100 years old by its C-14 dating (Fan Ronghe 1982), which demonstrates that the layer was formed during the Neoglacial Age. Similarly, such well developed structure was also found on the second terrace, which was possibly formed earlier, and is perhaps the product of Würm glaciation. The altitude at which the periglacial remains in this area lie fully proves that the northern lower boundary of the periglacial belt of Würm glaciation once reached the height of 2,100 to 2,300 m.

As is stated above, the involution layer in Liangdaohe area took shape in the Neoglacial Age. The mean annual air temperature today is about -3°C there. According to this value, it is proved that the mean annual air temperature under which the involution layer on the first and second terraces were formed was no higher than -3°C. However, the [illegible] mean annual air temperature is about +0.5°C in Nachitai. Thus we can see that the mean annual air temperature of Nachitai during Würm glaciation was 3°C to 3.5°C lower than today, meanwhile, the mean annual air temperature was between -8°C to -9°C on the plateau surface south of Qinghai and north of Xizang of 4,500 m during Würm glaciation. Therefore large polygons of over 50 m in diameter and the sand wedges developed under them were formed on the southern slope of Tanggula Range as well as on the western bank of the Tongtian River, on the northern bank of the Tuotuo River, on the river terraces of Fenghuo Shan area, all of which are to the north of Tanggula Range.

CONCLUSIONS

1. The fossil periglacial remains and the modern periglacial phenomena formed under the periglacial environment of the Würm glaciation and the Holocene are widely distributed on the plateau surface and mountains.

2. The hinterland area on the plateau of southern Qinghai was a warm and interglacial environment about 35,000 years ago when glaciers and amphitheater terrain was left in the areas of Tanggula Range and Kunlun Shan, and about 26,000 to 14,000 years ago, great polygons with sand wedges under the periglacial environment widely developed on the terraces of all the big rivers, and about 14,000 to 12,000 years ago, the periglacial environment on the plateau surface is the lasted of Würm glaciation.

3. The northern lower boundary of the periglacial belt on the plateau was 2,100 to 2,300 m during Würm glaciation.

ACKNOWLEDGMENTS

The writers express their appreciation to T.L. Péwé, professor of the Geology Department of Arizona University, Chairman of Fourth International Conference on Permafrost, to Li Jijun of Department of Geology and Geography of Lanzhou University, and to Cheng Guodong of Lanzhou Glaciology and Cryopedology Research Institute of the Chinese Academy of Sciences for their precious suggestions. Gratitude is also expressed to Yang Huiqou in Lanzhou Geology Research Institute of the Chinese Academy of Science for the analysis of the spore pollen, to the Northwest Railway Research Institute of the Ministry of Railway of China for drilling holes, and to Laboratory C-14 of the Geology and Geography Department of Lanzhou University for analysis of sediment age by C-14 dating, as well as to Han Pinlian for the diagrams.

REFERENCES

Fan Ronghe and Yao Shangsen, 1982, Discussion on the formation and the trend of development of the perennial frost on southern Qinghai-northern Xizhang Plateau (Tibet), Journal of Glaciology and Cryopedology. v.4, No. 1.

Gao Dongxin, 1979, Sand wedges on Qinghai-Xizhang plateau, Journal of Glaciology and Cryopedology, V. 1, No. 1.

Kong Zhaochen, Liu Lansuo and Du Naiqiu, 1981, Neocene Quaternary from the Kunlunshan to the Tanggula Range and the uplift of the Qinghai Xizhang Plateau, Studies on the Period, Amplitude and Type of the Qinghai-Xizhang Plateau. Science Press.

Li Jijun, Wen Shixuan, Zhang Qingsong, Wan Fubao, Zheng Benxing and Li Binyuan, 1979, Approaches on the Period, Amplitude and Type of the Uplift of the Qinghai-Xizhang Plateau. Sinica,6

Péwé, T.L., Reger, R.D., 1981, Cryoplanation Terraces. Geological and Ecological Studies of Qinghai-Xizhang Plateau. V. 2.

Reger, R.D., Péwé, T. L., 1976, Cryoplanation Terraces. Indicators of a Permafrost Environment. Quaternary Research. V. 6. No. 1.

Tang Linyu and Wan Yu, 1979, Quaternary Lake Sedimentary Spore-pollen Assemblages and its Significance in Kunlunshan Col-basin of Qinghai. Spacial Journal of Lanzhou Institute of Glaciology, Cryopedology,Desert, Academia Sinica. Science Press.

Wen Shixuan, 1981, Some Remarks on the Uplift of Xizang Based on Stratigraphical and Paleontological Evidences. Studies on the plateau Amplitude and Type of the Uplift of the Qinghai Xizang Plateau. Science Press.

Xu Shuying and Zhang Linyuan, 1981, A Geomorphological Analysis of Ages and Amplitudes of the Tanggula Range Upheaval. Studies on the Period Amplitudes and Type of the Uplift of the Qinghai-Xizang Plateau. Science Press.

Zhang Linyuan, 1981, Glaciers at the Source Region of Tuotuo River in the Upper Reaches of the Changjiang and Their Evolution. Journal of Glaciology and Cryopedology. V.3,No.1.

Zhang Weixin, 1979a, Some Indicative Features of Permafrost Along the Highway in South Qinghai and North Xizang. Journal of Lanzhou University.

Zhang Weixin, 1979b, The Characters of Periglacial Landforms in Fenheshan area of the Qinghai Xizang Highway-line. Journal of Glaciology and Cryopedology. V. 1. No. 1.

Zhang Weixin and Zhang Tiengjun, 1982, Polygonal Soil Along the Highway in Qinghai and Xizang (Tibet). Journal of Glaciology and Cryopedology. V. 4. No. 3.

Zheng Benxing, Mou Yunzhi, Li Jijun, 1981, The Evolution of the Quaternary Glacier in the Qinghai Xizang Plateau and its Relationship with the Uplift of the Plateau. Studies on the Period, Amplitude and Type of the uplift of the Qinghai Xizang Plateau. Science Press

CALCULATING THE THAWED DEPTH BENEATH HEATED BUILDINGS IN PERMAFROST REGIONS

Zhao Yunlong and Wang Jianfu

Scientific and Technical Institute of Qiqihar Railway Bureau
Qiqihar, Heilongjiang Province, People's Republic of China

Determining the maximum thawed depth beneath the foundation of a heated building in the permafrost region is an important subject in engineering cryopedology. It is directly related to the materials, depth, and type of foundation. Based on the formula suggested by Ding Dewen, the authors have developed a new formula for calculating the thawed depth in the midpoint of the building span, which represents the maximum thawed depth. In this formula, the following factors have been taken into account: the effect of the thermal resistance of the floor, the thermal current under the ground, and modifications corresponding to the transition from two-dimensional calculation, as well as empirical coefficients for calculating the thawed depth beneath wall foundations. At the end of the paper, there is a monograph for determining the thawed depth. The difference between the calculated and observed values of thaw depth is less than 8%.

INTRODUCTION

In the permafrost region, the foundation of a heated building is built on the perennially frozen soil. Because of the heating inside the building, the ground beneath it thaws gradually and the intensity of this thawing process becomes weaker as time goes on until equilibrium is reached. The limiting surface designating this equilibrium expands in summer, Figure 1, and contracts in winter Figure 2, within a certain range.

If the foundation of a heated building is designed improperly, the thawing of the frozen ground may cause sinking and deformation of the building due to differential settlement between various parts of the building having exceeded a certain limit. These excessive nonuniform settlements are caused by the decrease of bearing strength of the foundation base. The bearing capacity of the soil varies due to the thawing of the perennially frozen ground in which the foundation is built. This is one of the important subjects in engineering cryopedology.

The conditions for thermal equilibrium at the boundary surface of phase transformation are used and the generalized Green's formula is often employed in establishing integral differential equations, using potential equations instead of thermal conduction equations. Almost all the equations for calculating the thawed depth include transcendental functions and it is difficult to consider the effect of the thermal resistance of floor and that of the heat flow under ground. Therefore, it is a major problem to find a numerical solution directly from analysis.

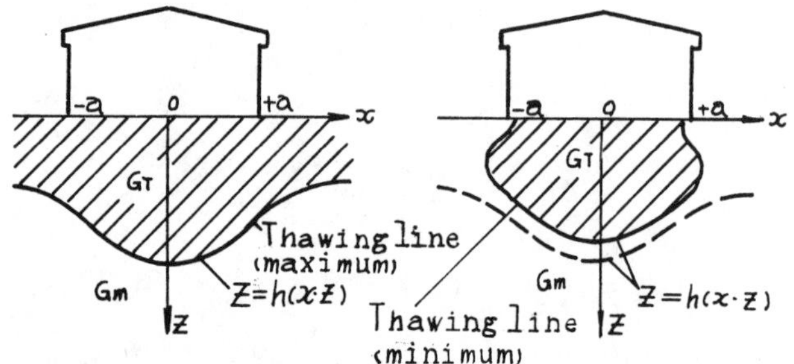

FIGURE 1 In summer. FIGURE 2 In winter.

Taking into account the effect of the thermal resistance of floor and the heat flow under ground, a new approximate formula for calculating the thawed depth at the mid-point of the building has been derived, using the method of iteration and the formula suggested by Ding Dewin (1978).

In this formula, modifications corresponding to the transition from two-dimensional to three-dimensional calculation have been taken into account and empirical coefficients have been used in calculating the thawed depth beneath the wall foundations. The difference between calculated and observed values is less than 8%.

In this article, nomographs for practical use are constructed simplifying the laborious process of manual calculation, so the thawed depth can be found directly from the nomograph. Thereby the basis for designing the foundation of heated building in permafrost region is given.

THAWING OF THE GROUND BENEATH HEATED BUILDINGS

If a heated building is constructed on permafrost, heat is transmitted downward which thaws the frozen ground. The intensity of thawing decreases gradually with time and depth, until a state of relative equilibrium is reached. After that, the thawed depth of the ground at this

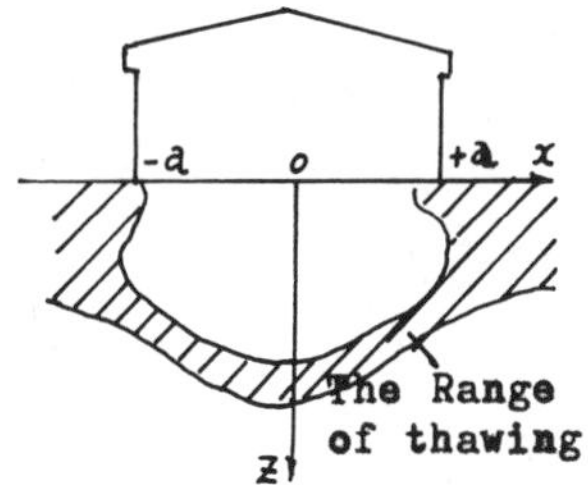

FIGURE 3 The range of thawing

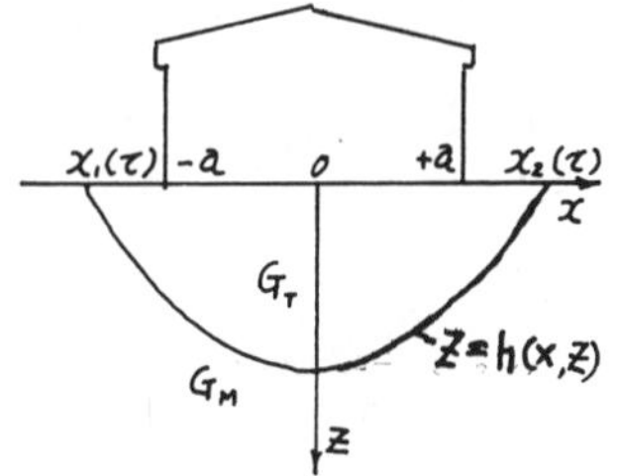

FIGURE 4 The geometry for the problem

site will vary with the season within a certain range as shown in Figure 3, if there are no unusual changes in such conditions as room temperature, thermal conductivity of the ground, thickness of the floor, etc.

In the summer, the frozen soil below the earth's surface outside the building (active layer) has already thawed and linked up with the thaw bowl inside the building, thus the boundary surface of equilibrium takes an open shape as shown in Figure 1, and the thawed depth of the ground has reached its maximum value.

In winter, the thawed soil below the earth's surface outside the building freezes again, and the thawed depth inside the building draws back due to the horizontal loss of energy. Thus the boundary of equilibrium takes a closed shape as shown in Figure 2, when the thawed depth has reached its minimum value. Although the thawed depth in winter is different from that in summer, under ordinary conditions the difference is not great during a year. In very rare instance this difference has reached several tens of centimeters. The amount of this variation fluctuates within a certain range as shown in Figure 3, and this fluctuation is basically steady.

BASIC THEORY

The geometry for the problem is shown in Figure 4. The basic assumptions are as follows:

1. The soil is homogeneous. If the soil is heterogeneous, it should be made homogeneous by a weighted average.
2. The temperature field is quasi-steady. At any given time τ the temperature distribution as a function of the space variables will satisfy the potential equation. Thus, we would substitute potential equation for the thermal conduction equation.

According to the above-mentioned assumption. The functions $t_T(X,Z,\tau)$, $t_M(X,Z,\tau)$ for the temperature distributions of the thawed and frozen regions satisfy the potential equations.

$$\Delta t_T=0 \qquad (X,Z)\in G_T \qquad (1)$$

$$\Delta t_M=0 \qquad (X,Z)\in G_M \qquad (2)$$

The general cases are shown in Figure 4. The upper-boundary condition are:

$$t_T|_{Z=0}=f_T(X,\tau),\quad X\in\Gamma_T,\quad \Gamma_T: Z=0, X_1(\tau)<X<X_2(\tau) \qquad (3)$$

$$t_M|_{Z=0}=f_M(X,\tau),\quad X\in\Gamma_M,\quad \Gamma_M: Z=0, X>X_2(\tau),\quad X<X_1(\tau). \qquad (4)$$

When X_1 and X_2 approach infinity, the upper-boundary function is

$$t_T|_{Z=0}=f_T(X,\tau),\quad X\in\Gamma_T,\quad \Gamma_T: Z=0 \quad -\infty<X<+\infty .$$

On the interface of the phase, namely, on the $Z=h(X,\tau)$ temperature and heat flux equality exist.

$$t_T|_{Z=h(X,\tau)}=t_M|_{Z=h(X,\tau)}=t_0 \qquad (5)$$

$$\therefore \quad \lambda_T\frac{\partial t_T}{\partial n}=\lambda_M\frac{\partial t_M}{\partial n}.$$

Where $\frac{\partial t}{\partial n}$ is the derivative along the normal. In the layer of frozen soil in the natural condition, the heat flow is $q=\lambda_M g$. (6)

Where g is the gradient of the temperature in the permafrost layer.

Now, define the converted temperature t_x

$$t_x=\frac{\lambda_M}{\lambda_T}t \qquad (7)$$

With the bounded solutions $t_T(X,Z,\tau)$ and $t_M(X,Z,\tau)$ of equations (1)~(7) solve the unknown interface $Z=h(X,\tau)$ by Green's function and the generalized Green's formula.

Take Green's function

$$g'(x,x_0,Z,Z_0)=\frac{1}{4\pi}\ln\frac{(x-x_0)^2+(Z+Z_0)^2}{(x-x_0)^2+(Z-Z_0)^2} \qquad (8)$$

It is easy to know that the g' is to satisfy

$$\Delta g'=-\delta(x-x_0)(Z-Z_0) \qquad (9)$$

$t_T(X,Z,\tau)$ and $g'(X,X_0,Z,Z_0)$ are used in the generalized Green's formula within the G_T and can be written as follows

$$\oint_{\Gamma_T}\left(g'\frac{\partial t_T}{\partial n_T}-t_T\frac{\partial g'}{\partial n_T}\right)d\Gamma_T=\iint_{G_T}\left(g'\Delta t_T-t_T\Delta g'\right)dG_T. \qquad (10)$$

Thereupon, when the surface temperature f(x) and $Z_0=h(X_0)$ are known, the solution of equation (1) can be written as:

$$(t_T)=\frac{1}{\pi}\int_{-\infty}^{+\infty}f(X)\frac{h(X_0)\,dX}{(X-X_0)^2+h^2(X_0)} \qquad (11)$$

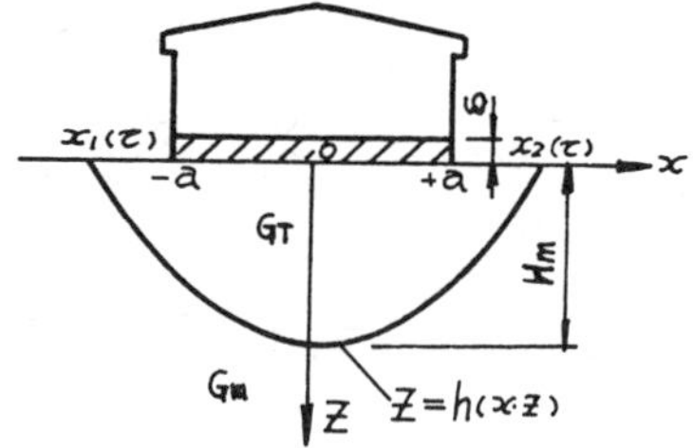

FIGURE 5 Equivalent layer

Considering the boundary condition we split the integral to obtain:

$$(t_\tau) = \frac{1}{\pi}\int_{X_1}^{X_2} f_\tau(x)\frac{h(X_0)dX}{(X-X_0)^2+h^2(X_0)} + \frac{1}{\pi}\int_{-\infty}^{X_1}\frac{\lambda_M}{\lambda_T}f_{M_1}(x)\frac{h(X_0)dX}{(X-X_0)^2+h^2(X_0)}$$
$$+\frac{1}{\pi}\int_{X_2}^{+\infty}\frac{\lambda_M}{\lambda_T}f_{M_2}(x)\frac{h(X_0)dX}{(X-X_0)^2+h^2(X_0)} = 0 \; . \tag{12}$$

Solving the equation of implicit function of $Z_0=h(X_0)$, we can get the shape of the stable thawed bowl.

In equation (12), considering the mean annual temperature value f(X) at the different locality X on the ground surface, we take $X_1=-a$; $X_2=+a$ (as a is the span of the building), f_τ= constant (the indoor mean annual temperature), f_{M1} and f_{M2} are also constants (the outdoor surface mean annual temperature), when both sides are the same, namely, $f_{M1}=f_{M2}=f_M$.

Substituting them into equation (12), we get

$$\frac{\lambda_M}{\lambda_T}\cdot\frac{f_{M_1}+f_{M_2}}{2}+\frac{1}{\pi}\Big[(f_\tau-\frac{\lambda_M}{\lambda_T}f_{M_1})\,tg^{-1}\frac{a+X_0}{h}+$$
$$+(f_\tau-\frac{\lambda_M}{\lambda_T}f_{M_2})\,tg^{-1}\frac{a-X_0}{h}\Big]=0 \tag{13}$$

When the maximum thawed depth is at the centre (as $X_0=0$) of the span of the building, equation (13) can be written as:

$$H = a.ctg\left(\frac{\pi}{2}\,\frac{F}{F-f_\tau}\right) \tag{14}$$

where

$$F=\frac{\lambda_M}{\lambda_T}\cdot\frac{f_{M_1}+f_{M_2}}{2}$$

The equation (14) had been obtained by Ding Dewin and Lunardini.

When considering the indoor floor, a layer of insulation exists as shown in Figure 5.

δ is thickness of layer of the floor of the building, λ_δ is thermal conductivity and the coordinates are also shown in Figure 5.

In this problem, when Z=0, the mean annual temperature values ranging from $-T<Z<+T$ have been chosen, then, it can be calculated by equation (14).

Under stable condition, one-level approximate is made. Suppose that, the distribution of temperature T in the center line by sectional linearity is expressed by

$$t_\tau\Big|_{Z=0} = \begin{cases} A_\delta\cdot Z + T_\delta & (\delta \geq Z \geq 0); \quad (15)\\ A_T\cdot Z + T_T & (0 \geq Z \geq -H_S). \quad (16)\end{cases}$$

Determination of A_δ, A_T, T_δ, T_T from the following group of the linear equation

$$A_\delta\cdot\delta + T_\delta = f_\tau \;; \tag{17}$$
$$A_T\cdot(-H_S) + T_T = 0. \tag{18}$$

$T_\delta=T_T$ (the temperature is equal on the contact surface Z=0);

$\lambda_S A_S=\lambda_T A_T$ (the heat flow is equal on the contact surface).

Thus, to solve equations (15)~(18) we obtain:

$$T_\delta=T_T=\frac{H_s\cdot f_\tau}{H_s+\frac{\lambda_T}{\lambda_\delta}\delta}=f_\tau-\frac{f_\tau\cdot S}{H_s+S} \tag{19}$$

$S=\frac{\lambda_T}{\lambda_\delta}\delta$ (Thickness of equivalent layer).

Therefore, the formula of the maximum thawed depth at the centre of clear span of the building can be rewritten as a transcendental equation

$$H_S=a\cdot ctg\left[\frac{\pi}{2}\;\frac{F}{F-f_\tau+\frac{f_\tau\cdot S}{H_s+S}}\right] \tag{20}$$

Thus, the numerical solution is solved by the method of successive approximation. This method of approximate solution of algebraic equation is given that, as follows

$$\text{take}\quad \frac{f_\tau}{F-f_\tau}\cdot\frac{S}{H_S+S}=D \tag{21}$$

D is a small parameter of dimensionless equation (20) can be written as

$$H_S=a\cdot ctg\left[\frac{\pi}{2}\cdot\frac{F}{F-f_\tau}\cdot\frac{1}{1+D}\right]\doteqdot a\,ctg\left[\frac{\pi}{2}\right.$$
$$\left.\frac{F}{F-f_\tau}\cdot(1-S)\right] \tag{22}$$

Solving the above equation we obtain

$$H_S=\frac{1}{2}\left[H-S+\sqrt{(H+S)^2+\frac{2\pi F f_\tau}{(F-f_\tau)^2}\cdot\frac{S}{a}(H^2+a^2)}\right. \tag{23}$$

On this basis, considering simultaneously the heat flow q under the ground, namely, adding the converted value $G\,H_S=\frac{\lambda_M}{\lambda_T}g\,H_S$ to the equation (20), it can be rewritten as

$$H_G=a\cdot ctg\left[\frac{\pi}{2}\;\frac{(F+G\cdot H_S)(H_S+S)}{(F-f_T)(H_S+S)+f_T\cdot H_S}\right] \tag{24}$$

when it is transmitted from two-dimension to three-dimension, the coefficient K of rectification should be multiplied to it, then, we can get the value of H_M

$$H_M=k\cdot H_G \tag{25}$$

CALCULATION OF THE MAXIMUM THAWED DEPTH OF THE GROUND

The point of maximum thawed depth of ground beneath the building in the permafrost region is mostly located near the mid-point of the span of building or slightly deviated from this position to that of the heat source. So that the thawed depth at the mid-point of the span of building can be used in calculation instead of the maximum thawed depth, and the maximum difference between these two values does not exceed 2%.

The formula for calculating the maximum thawed depth of ground is given as follows:

$$H_M = k \cdot H_\phi$$

In which H_M is the maximum thawed depth of ground, in m; H_ϕ is the thawed depth of ground when considering the effect of the thermal resistance of floor and the heat flow under ground; (as the values of F,G,S,a, f_T etc. are known, the value of H_ϕ can be found directly from the nomograph), in m.

When consulting the nomograph to find the value of H_ϕ the meanings of the symbols shown in the nomograph are:

F is the converted value of the annual mean surface of ground temperature outdoors, °C, is equal to

$$F = \frac{\lambda_M}{\lambda_T} f_M \quad \text{or} \quad \frac{\lambda_M}{\lambda_T} \cdot \frac{f_{M_1} + f_{M_2}}{2};$$

f_M, f_{M_1}, f_{M_2} is annual mean surface of ground temperature outdoors, °C; f_T is annual mean atmospheric temperature indoors, °C;

λ_M, λ_T is thermal conductivity of frozen soil and of the thawed soil (kcal/m, °C,h); G is the converted value of temperature gradient of the frozen layer $G = \frac{\lambda_M}{\lambda_T} g$ (g is the value of temperature gradient of the frozen layer and it is a negative value, ordinarily taken as 0.01~0.02); S is the equivalent thickness in meters (m) of the heat insulation layer;

$$S = \lambda_T R_n = \lambda_T \left(\frac{1}{\alpha} + \frac{\delta_i}{\lambda_i} + \cdots \quad \cdots + \frac{\delta_n}{\lambda_n} \right) \qquad (26)$$

where α is coefficient of heat emission, 7.5 is adopted; δ_i is thickness (m) of the material of the i-th layer; λ_i is thermal conductivity (kcal/m, °C,h) of the material of the i-th layer; δ_n is thickness (m) of the material of the n-th layer; λ_n is thermal conductivity (kcal/m, °C,h) of the material of the n-th layer; a is the half of clear span of the building (m); K is the coefficient considering the effect of the ratio of the length and width of building on the thawed depth in the mid-point, see TABLE 1.

TABLE 1 Value of K.

L /a		1	2	3	4	5	6	7	8	9
F	0.40	0.49	0.66	0.76	0.83	0.87	0.90	0.93	0.96	1.00
F-f	0.30	0.56	0.75	0.84	0.89	0.93	0.94	0.96	1.00	1.00
	0.20	0.61	0.80	0.89	0.93	0.95	0.97	1.00	1.00	1.00

CALCULATION OF THAWED DEPTH BENEATH WALL FOUNDATION

The formula for calculating the thawed depth H_a beneath wall foundation is given as follows:

$$H_a = K_a \cdot H_M \qquad (27)$$

where H_M is the maximum thawed depth of the ground (m); K_a is the coefficient considering the effect of orientation of the building.

TABLE 2 Determining the value of Ka

The location of Ka	The value of Ka under condition listed below		
	R=0.30	R=0.60	R=1.00
K_{a_1}	0.80	0.70	0.65
K_{a_2}	0.70	0.65	0.60
K_{a_3}	0.55	0.50	0.50
K_{a_4}	0.65	0.65	0.60
K_{a_5}	0.60	0.60	0.60
K_{a_6}	0.70	0.70	0.70

where

$$R = \frac{\lambda_M f_M}{\lambda_T f_T}$$

In which the meanings of the symbols a are the same as that in formula (25).

EXAMPLE

Now, a building of region of Mangui is made by way of example .

1. Clear length and clear width of the building are 4.6 m and 27.6 m.

2. The geological conditions of engineering are as follows (See TABLE 3)

3. Other values

f_T is 17 °C; f_M is -5.5 °C; λ_T = 1.2 kcal/m·°C·h; λ_M =1.29 kcal/m °C h; g is 0.02 °C/m; a=2.3 m L=13.8 m.

$$\frac{L}{a} = \frac{13.8}{2.3} = 6.00, \; K = 0.955$$

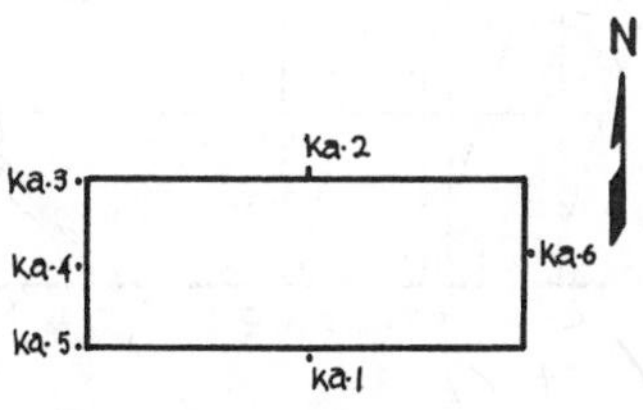

FIGURE 6 The location of Ka.

TABLE 3 Geological conditions of engineering

Thickness of layer of soil	Kind of soil	Moisture content (%)	Volume weight (T/M)	Thermal conductivity (kcal/m·°C·h)		The weighted average of thermal conductivity (kcal/m·°C·h)	
				λ_T	λ_M	λ_T	λ_M
0.4	Manual filling	12	1.60	1.06	1.12		
1.3	Loam and gravel 20%	18	1.70	1.10	1.10	1.20	1.29
2.1	gravel	13.9	1.90	1.46	1.45		
1.2	Loam and gravel 10%	39.5	1.30	0.89	1.27		

4. Conversion values

$$F = \frac{\lambda_M}{\lambda_T} f_M = 1.075 \times (-5.5) = -5.913 \ ^\circ C \quad ;$$

$$G = \frac{\lambda_M}{\lambda_T} g = 1.075 \times (-0.02) = -0.022 \ ^\circ C/m \quad ;$$

$$\frac{F}{F-f_T} = \frac{-5.913}{-5.913-17} = 0.258$$

5. Thermal resistance of the floor is with brick laid on flat, the thickness is 0.06 m, the $\lambda_\delta = 0.70$ kcal/m·°C·h.

$$R_n = \frac{1}{\alpha} + \frac{\delta_i}{\lambda_i} = \frac{1}{7.5} + \frac{0.06}{0.70} = 0.219$$

$$S = \lambda_T R_n = 1.2 \times 0.219 = 0.263 \text{ m}$$

The values of $\frac{F}{F-f_T}$, a, S, G are known, the value of H_G can be found directly from the nomograph. $H_G = 5.11$ m.

$H_M = K \cdot H_G = 0.955 \times 5.11 = 4.87$ m.

H_M is 4.87 m and observed value is 5.0 m, the difference between the calculated and observed value is 2.6%.

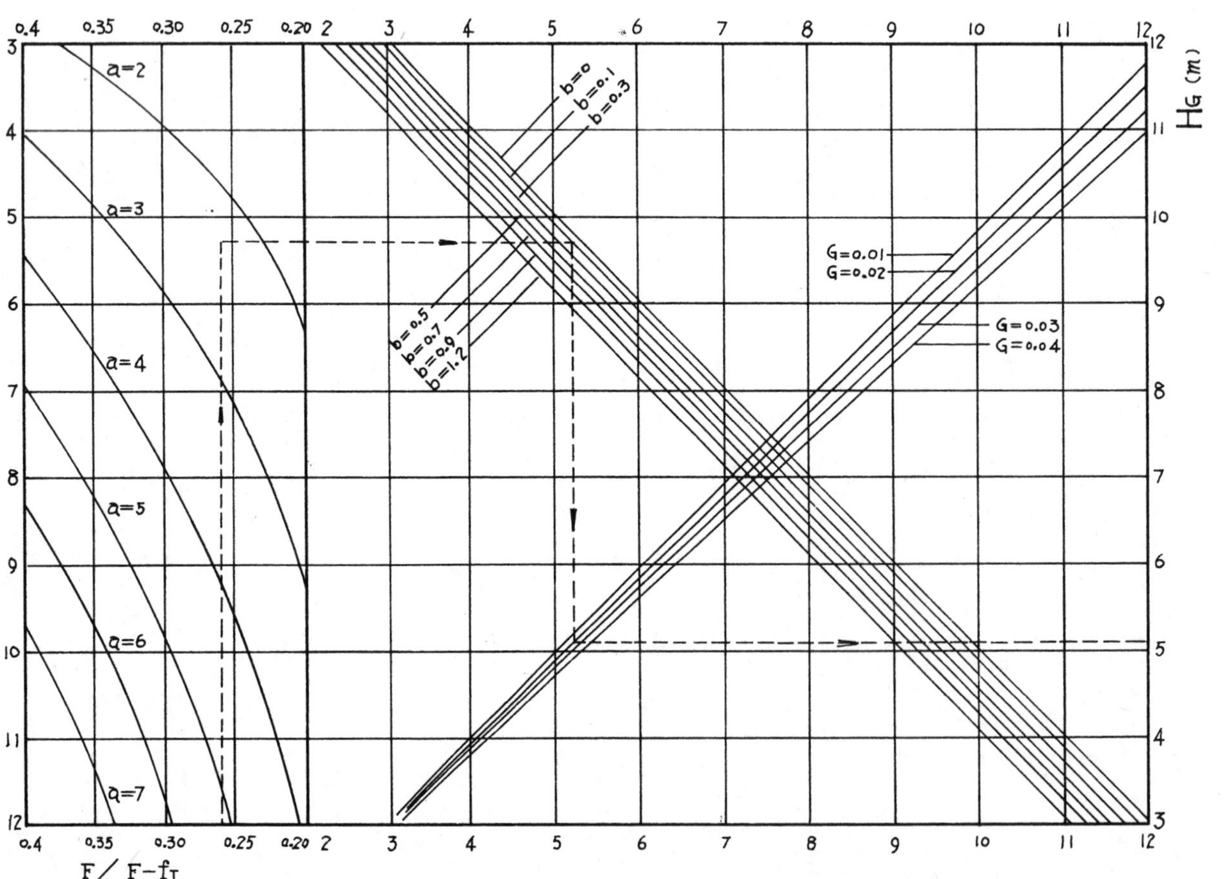

FIGURE 7 Nomograph of calculatious

CALCULATION OF THE THICKNESS OF INDOOR FLOOR (THE THERMAL INSULATION LAYER)

To design the indoor floor for heat insulation is used to control unthaw or a little thaw of perennially frozen soil beneath heated building, so that, the heat insulating layer must be calculated according to thermodynamics. In general, according to H converted to the heat insulating layer by various formulas. In this paper the following formula is submitted for reference of calculation.

Under the stable condition, one-level approximation is made, according to equations (15)~(18) and the condition at Z = 0, $T_s = T_T$, $\lambda_s A_s = \lambda_T A_T$ solve the group of equations when the frozen foundation is unthawed, namely, $H_M = 0$

we obtain $$S = \frac{\lambda_s f_T}{\lambda_T A_T} \qquad (28)$$

When the frozen foundation is thawed admittedly for H namely, $H_M = H$,

we obtain $$S = \frac{\lambda_s f_T}{\lambda_T A_T} - \frac{\lambda_s}{\lambda_T} H. \qquad (29)$$

where A_s and A_T is the gradient of temperature of equivalent layer of indoor floor and the foundation of thawed soil (°C/m);

λ_s and λ_T is the thermal conductivity of the equivalent layer of indoor floor and the foundation of thawed soil (kcal/m, °C,h);

S is thickness of equivalent layer (m);

f_T is the mean annual indoor temperature (°C).

REFERENCES

Ding Dewen. and Luo Xuebo., 1978, Discussion of calculation of thawed ground beneath building in permafrost region, in China, The Journal of Lanzhou University: 1978.6, P. 7-40.

Jumikis, A.R. 1978, Graphs for disturbance-temperature distribution permafrost and heated rectangular struction, in Proceedings of the Third International Conference on permafrost, V. 1: Ottawa, National Research Council of Canada, P. 589-596.

Lunardini, V.J., 1982, Conduction phase change beneath insulated heated or cooled structures, CRREL Rept. 82-22.

SOME FEATURES OF PERMAFROST IN CHINA

Zhou Youwu and Guo Dongxin

Lanzhou Institute of Glaciology and Cryopedology, Academia Sinica
People's Republic of China

The permafrost that underlies 2,150,000 km^2 of China can be divided into two broad categories: high-latitude permafrost and high-altitude (alpine) permafrost. High-latitude permafrost is found in the northeast, and its distribution and characteristics depend chiefly upon latitudinal zonation. The mean annual ground temperature increases by 0.5°-1°C/1°N, and permafrost varies in thickness from 5-20 m in the south to 50-100 m at the northernmost limit. High-altitude (alpine) permafrost occurs in the high mountains and plateaus of western China and is vertically zoned. The mean annual ground temperature varies from -0.1° to -2.5°C at the lower limit to -4° or -5°C at the higher limit. Correspondingly, the permafrost increases in thickness with elevation. Alpine permafrost also shows latitudinal variation, however. As the latitude increases, the lower limit of the alpine permafrost descends with a variable of about 150-200 m/1°N. The latitudinal and vertical zonation of permafrost distribution may be modified by a zonal factor. On the Qinghai-Xizang Plateau, some taliks are the result of geothermal anomalies and surface and ground water; and in northeast China, temperature inversions in winter, together with the swampy process, makes for well-developed permafrost in the valleys rather than on the higher crests. Horizontal ground-ice layers are widely developed on north-facing slopes and in well-vegetated intermontane swampy depressions, especially in fine-grained soils. In coarse-grained sediments, there is cemented and cemented-segregated ground ice. In perennially frozen bedrock, there is some vein-shaped ground ice in cracks.

Regional investigation on permafrost in China began in the early 1950's. With the national economy developing and with the specific institutions being set up, systematic scientific explorations have been carried out since 1960 in the vast permafrost regions, including the Da-Xiao Hinggan Ling of Northeast China, along the Qinghai-Xizang Highway, Qilian Shan, and Tian Shan and Altay Shan of West China.

DISTRIBUTION, TEMPERATURE, AND THICKNESS OF FROZEN GROUND

Permafrost underlies 2,150,000 km^2 or 22.3% of the territory of China. The total area of perennially and seasonally frozen ground (the depth is over 0.5 m) occupies 68.6%. Generally, permafrost in China can be divided into two broad categories: permafrost in high latitudes and permafrost in high altitudes (Fig.1).

The permafrost region in northeastern China is part of the southmost zone of the Eurasian continental high latitude permafrost area. The distribution of permafrost depends on latitudinal zonation. The mean annual ground temperature rises southward. Generally, with a 1° decrease in latitude, the mean annual air temperature and the mean annual ground temperature will increase $1^{o}C$ and $0.5^{o}C$, respectively. To the south, the continuity of permafrost goes from predominantly continuous to island permafrost. Permafrost thickness varies from 50 to 100 m in the northmost part, to 5 to 20 m at the south limit (Table 1, Guo 1981).

The southern limit of permafrost is between 46.6 and 49.4^{o} N; the mean annual air temperature in the middle section is $0^{o}C$; at the western section, between $0^{o}C$ and $-1^{o}C$; and at the eastern section the air temperature is between 0°C and $+1^{o}C$. In mountains to the south of this limit, such as the Huangganglian Shan, Changbai Shan, Wutai Shan and Taibai Shan, alpine permafrost occurs.

In West China, permafrost occurs in high mountains, such as the Qilian Shan, Tian Shan, and Altay Shan. With rising elevation, the permafrost temperature at the depth of zero amplitude varies from $-0.1^{o}C$ at the lower limit to -2.5^{o} or -4--$5^{o}C$ at the higher; correspondingly, permafrost varies in thickness from several meters to 100 to 200 m or more (Table 2). For instance, in the Reshui Region of the Qilian Shan, with an elevation increase of 100 m, the ground temperature decreases $0.6^{o}C$ and the permafrost thickness increases 14 to 21 m. The temperature gradient at an elevation of 2,700 to 3,000 m in the Tian Shan is the same as in the Reshui region; however, the thickness increases as much as 31 m/100 m. The lower limit of alpine permafrost ranges from 2,200 m in the north (Altay Shan) to 3,700 m in the south (Qilian Shan).

The Qinghai-Xizang Plateau has the highest and largest permafrost region in

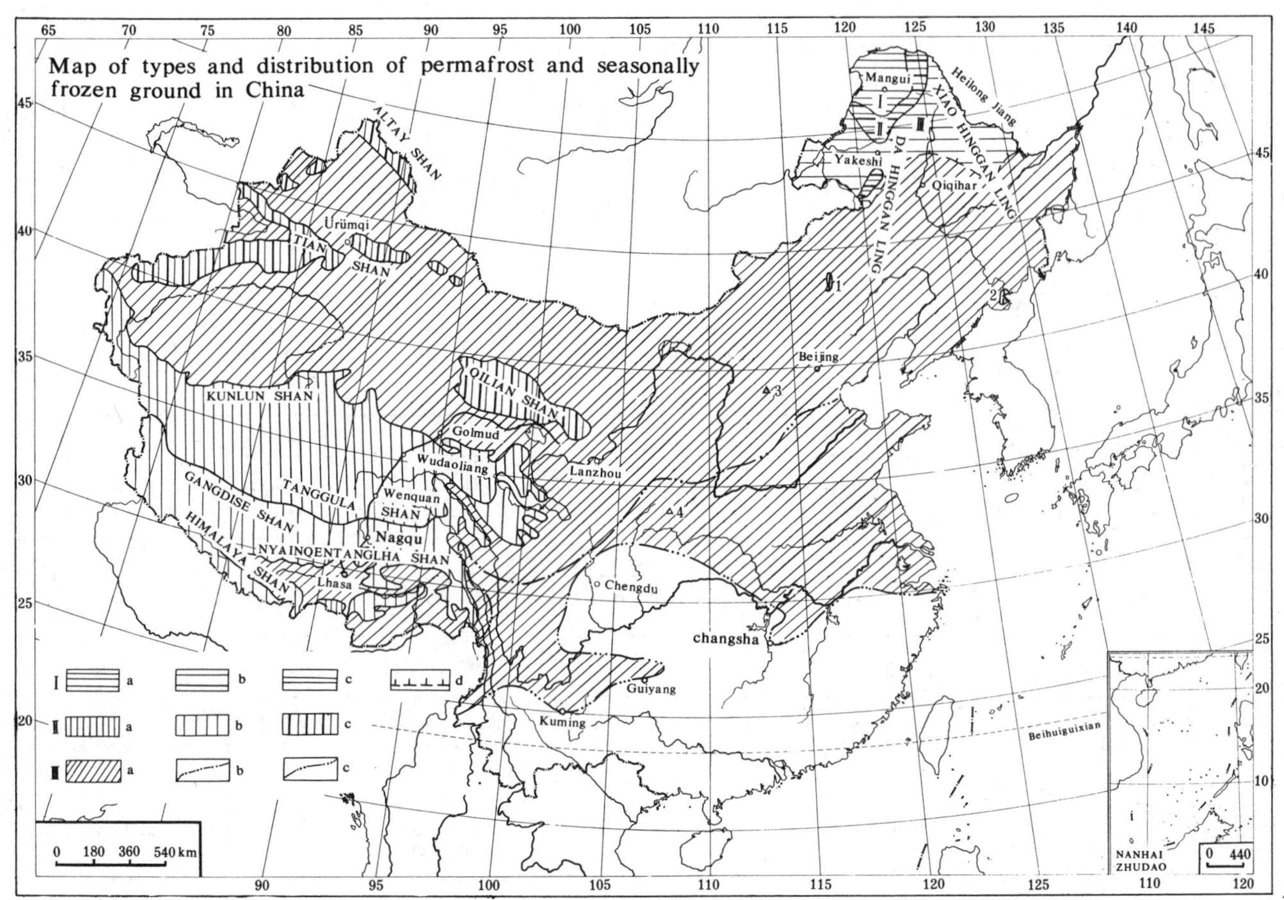

FIGURE 1 Map of types and distribution of permafrost and seasonally frozen ground in China
I. High latitude permafrost: a—Predominantly continuous zone; b-Discontinuous zone; c—Island permafrost zone; d—Southern limit of permafrost.
II. High altitude permafrost: a-Continuous zone of permafrost on the Plateau; b-Isolated permafrost on the Plateau; c—Alpine permafrost.
III. Seasonally frozen ground: a—Seasonal zone of frozen ground; b—Southern boundary of frozen ground (0.5 m thick); c—Southern limit of frozen ground.
1—Huanggangliang Shan; 2—Changbai Shan; 3—Wutai Shan; 4—Taibai Shan.

TABLE 1 Main Characteristics of Permafrost in Northeast China

Divisions	Mean annual air temperature (°C)	Mean annual ground temperature (°C)	Percentage of area occupied by permafrost (%)	Thickness of permafrost (m)	Depth of zero amplitude of annual ground temperature (m)
I	-5	-3.5--1 (lowest-4.4)	Predominantly continuous permafrost 70-80	100-50	
II	-5--3	-1.5--0.5	Discontinuous permafrost 50-60	50-20	12-16
III	-3-0	-1.0-0	Island permafrost 10-30, southern edge <5	20-5	

TABLE 2 Temperature and Thickness of Permafrost in West China

Regions		Elevation (m)	Mean annual ground temperature (°C)	Depth zero amplitude of ground temperature (m)	Thickness of permafrost (m)	Source
Altay Shan		over 2,800	lowest -4.0—-5.0		thickest 100-200	derived from Gorbunov, 1978 a, 1978 b; Shats, 1978; Mel'nicov, 1974.
Tian Shan	Kuixiandaban	2700 3000-3200	-0.1—-0.2 -2.0—-2.5		16 110-150	Qiu Guoqing, 1981.
	Motuoshala	3200-3400	-2.6		120-170	*Tan Qingfeng, 1973.
Qilian Shan	Muri	4000-4300	-0.6—-2.3	13-17 sometimes 9	30-95	*Muri Team, Academia Sinica, 1966.
	Hongshuiba	3829,4032			79.3,139.3	Guo Pengfei, 1983.
	Reshui	3480-4050	-0.1—-1.5 and lower	10-14 sometimes 6-7	0.6-11.0 2.0-90	Reshui Team, 1976; Guo Pengfei, 1983.
	Haiyan, Menyuan	3500-3600			5-35 generally 10-20	Fan Xipeng, 1963.
Qinghai-Xizang Plateau	Qinghai-Xizang Highway: Continuous permafrost zone					Map of permafrost along Qinghai-Xizang Highway (1:600000), 1983.
	High plain and river valley	4500-4650	0—-0.5 -0.5—-1.5	9-10	0—25 25—60	
	High mountains and hills	4700-4900	-1.5—-3.5 lower -3.5	12-16	60-120 120	
		over 4900	-4—-5		140—175	Cui Zhijiu,Zhou Youwu, 1980.
	Isolated permafrost zone					Wang Jiacheng, 1979. Above mentioned Map of permafrost (1:600000)
	Xidatan	4200	0—-0.5		<20	
	Amdo-Nagqu	4500-4700	0—-0.5		<25	

*from unpublished reports.

the low-middle latitudes of the world. Generally, this plateau inclines southeastward topographically, and the precipitation and air temperature become lower northwestward.

Along the Qinghai-Xizang Highway, which traverses the plateau from north to south, the lower limit of isolated permafrost in the Kunlun Shan in the north is at 4150 to 4200 m. As the elevation increases per 100 m, the ground temperature decreases 0.6 to 1.0°C. Above 4350 m continuous permafrost occurs. The lower limit of isolated permafrost rises southward and coincides approximately with the -2 to -3°C isotherm of mean annual air temperature. In the continuous permafrost regions, at an elevation of 4500 to 4900 m, the mean annual ground temperature ranges from zero to -3.5°C. The maximum thickness measured is 128.1 m. (Zhou 1965; Cui and Zhou 1980; Zhou and Guo 1982; Tong 1965, 1981; Wang et al. 1979).

In most of the alpine permafrost regions the lower limit of permafrost on south-facing slopes is higher than on north-facing slopes. However, the Himalaya Shan are an exception. According to calculations based on meteorological data and glacial ice temperature (Shen 1975; Xie and Wang 1975), it could be estimated that the lower limit of permafrost on north-

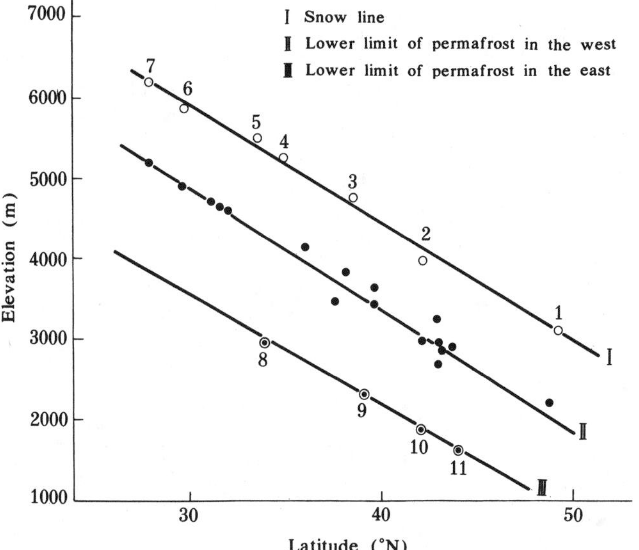

Fig.2. Relationship between elevation of the lower limit of permafrost distribution and latitude

1. Altay Shan 2. Tian Shan 3. Qilian Shan 4. Kulun Shan 5. Tanggula Shan 6. Nyainqentanglha Shan 7. Himalaya Shan 8. Taibai Shan 9. Wutai Shan 10. Changbai Shan 11. Huanggangliang Shan

facing slopes is at about 5100 to 5300 m, and is 300 to 400 m higher than on south slopes (Fujii 1980). This can be explained by the fact that the north slope is much drier than the south.

As mentioned above, the altitude in the mountains and plateau of West China is a decisive factor for permafrost development and conservation there; however, latitude also has an important effect on alpine permafrost (Fig. 2).

1. The snow line and the lower limit of alpine permafrost have a close relationship to latitude. With the increase in latitude, the elevation of the lower limit of the snowline and permafrost becomes lower. From the north to the south, the lower limit rises 3000 m.

2. The lower limit of permafrost in the west at a given latitude is 1150 to 1300 m higher than in the east.

3. The variation in lower limit of permafrost is 150 m/1° of latitude in the high mountains both in the east and west.

4. Snowline is usually higher than the lower limit of permafrost; the difference between them is 1000 to 1100 m.

In the Qinghai-Xizang Plateau, the permafrost temperature rises 0.5 to 1.0^{o}C per 100 to 200 km southward; correspondingly, the thickness decreases 10 to 20 m for 100 to 200 km.

Because of the combination of the vertical zonation and latitudinal variation, the permafrost distribution in this plateau is very complex.

INFLUENCE OF REGIONAL FACTORS ON PERMAFROST

The latitudinal and vertical zonation mentioned above is modified by intrazonal factors, including geologic structure, neotectonic movement, thermal regime, drainage, lithological characteristics, relief, vegetation and local climatic conditions.

The main geologic structure line stretches W-E in the Qinghai-Xizang Plateau; the secondary structure line is N-S, stretching with several fold-fault zones or rotating layers that are connected with the pressure-twisted faults. Such a tectonic pattern has important effects on permafrost development.

First, there are many geothermal anomalies at the intersection of two structure groups, forming cool springs, hot springs, gas fountain springs, and the structure-geothermal taliks. The cool springs, though their temperature is only 1 to 4° or 7°C, can make the permafrost warmer and thinner, and even form thawed channels, such as those seen at the Xidatan, Budongquan (Unfrozen Spring) of the Kunlun Shan. From the Kaixin Ling to the Amdo, a 260 km section along the Qinghai-Xizang Highway, there are many hot springs with temperatures of +20 to +30°C, or even +70 to +80°C, and a series of structure-geothermal taliks.

Second, the tectonic pattern determines the skeleton of the landform and controls the development of the hydrological system. The geothermal background, together with the thermal action of surface water, makes a 1.5 to 2.0 km wide talik in Buqu River Valley. There are complete taliks found under the Tuotuo River and Tongtain River, 1000 and 4000 m wide, respectively. Under and surrounding the lakes on faults, such as the Yaxing-cuo Lake and Basicuo Lake, there are structure-lacustrine taliks with diameters from several hundred meters up to 2-3 km.

Third, tectonic movement controls the landform differentiation and the depositional condition, which affect permafrost development. For example, the neotectonic movement in the Kunlun Shan and Tanggula Shan is rather strong, so the Xidatan and Wenquan Basins are filled with coarse-grained sediments; but in the middle of this plateau, the rising and sagging movement during the Neocene and Quaternary periods is rather weak; therefore, a series of fault-basins were filled with fine-grained sediments. Correspondingly, the latter was a more favorable condition for the development of permafrost and ground ice.

Lithological characteristics, marsh formation, and slope orientation all have a close relationship to permafrost development. When the other conditions are similar, the mean annual ground temperature of perennially frozen clay and sandy clay is usually 1 to 3°C lower than that of bedrock or gravels. The existence of a coal layer raises the ground temperature and geothermal gradient. According to the temperature measurement in wells, the geothermal gradient in permafrost of West China is 1.5-2.0°C/100 m, in general; however, under the influence of a coal layer, the gradient increases 4 to 5°C/100 m. The marsh and moss cover can promote the development of permafrost, especially in those places where the temperature is near 0°C. In the Da Hinggan Ling, the temperature in a marshland is 1 to 2°C lower than in other kinds of landscapes. Therefore, near the southern limit of the permafrost zone, the marshland could be considered as the indicator of isolated permafrost. On the other hand, in those places with a mean annual ground temperature near 0°C and with a zero geothermal gradient, permafrost is very unstable; therefore, clearing of land results in the increase of ground temperature and lowering of the permafrost table. Moreover, even in Region I of Da Hinggan Ling, which is far from the southern limit, permafrost is regenerating due to the interference of mankind; this is indicated by the fact that the temperature-depth curves with a negative gradient have been found there.

The lithological characteristics and water content have an important effect on the seasonal freezing/thawing process, making them vary in large range within a short distance. Generally, the maximum freezing depth ranges from 2 to 8 m, and the thawing depth is 1-4 m in unsolid Quaternary sediments, 8-10 m in bedrock. The seasonal layer is able to link up with the permafrost, however, near the lower/south limit of permafrost, or at those places with the supra-permafrost water unfrozen in winter, the permafrost will not connect with the seasonally thawing layer.

Difference of slope orientation and steepness could make the alpine permafrost distribution unsymmetrical. For instance, in Tian Shan, the lower limit of permafrost on south-facing slopes is 400 m higher than on north-facing slope (Qiu and Zhang 1981). In the Qilian Shan, there is a 210 m difference between north and south (Guo Pengfei 1983), and in the Qinghai-Xizang Plateau, the temperature difference between north-facing and south-facing slopes is 1.7 to 2.4°C, and the difference in permafrost thickness is 50 to 70 m. Also, ground ice development conditions vary on different slopes. For example, in the Fenghuo Shan Region, it has been found that at the lower part of north-facing slopes the ice layer is as thick as 2 to 5 m, while on the south-facing slopes with the same elevation, only a few discontinuous thin ice lenses occur (Wang Jiacheng et al. 1979).

The winter temperature inversion covers a considerable area of Northeast China, from Heilong Jiang in the north to Nenjiang and Hailar in the south; this has an obvious effect on the development of permafrost. Moreover, the marshland at the bottom of the valley is usually covered by moss, peat, or other well-developed vegetation; the combination of inversion and marshland makes the permafrost develop well at the lowland of a valley but not at the top of hills. For example, in the Amur region (Dai Jingbo, 1982), at the bottom of a valley, the mean annual ground temperature is as low as -4.2°C; at the lower part of a gentle slope, -1.7°C; on the north-facing steep slope, -0.1°C, and at the foot of a steep south-facing slope, +1.0°C. Usually, the ground temperature in lowlands is 1 to 2°C (max. 4 to 5°C) lower than that at a high divide.

GROUND ICE

Ground ice is widespread in the permafrost and has a close relation to relief, lithological characteristics and moisture supply.

1. On gentle north-facing slopes and in intermontane marsh depressions with well-developed vegetation, horizontal ice layers are usually well-developed, especially in clayey soil or peat with high water content. It is known that the upper surface of an ice layer is always parallel to the ground surface and approximately coincides with the permafrost table. The depth of the ice layers ranges from 0.9 m to 3.0 m in the Qinghai-Xizang Plateau, and from 0.3 to 1.5 m in the Da Hinggan Ling.

The horizontal ice layers can be divided into three types, according to their thickness: thin, from several millimeters to several centimetres; thick, several tens of cm; and the thickest, up to 6 to 7 m thick, distributed within the depth of zero amplitude of annual ground temperature, belonging to cemented-segregated ice.

There are two kinds of pingo ice: intrusive and segregated ice. In the famous huge pingo in the Pass Basin of the Kunlun Shan, there are four ice layers at the following depths: 1.3 to 14.6 m; 26.3 to 29.1 m; 30.0 to 34.9 m; and 42.1 - 43.0 m. The first one contains a little amount of silt, but the silt content becomes smaller with increasing depth. Between the thick ice layers, there is vein-structure permafrost composed of sandy clay, clayey sand, and fine sand, of which the ice content by volume is 50 to 70%.

2. In coarse-grained sediments, such as moraine, alluvial-pluvial gravels, and eluvial debris, the ground ice usually be-

longs to cemented and cemented-segregated types, forming conglomerated cryogenetic permafrost.

3. In perennially frozen bedrock, there is some vein-shaped ground ice filling in the cracks.

It has been pointed out that in Galaya of Da Hinggan Ling, the ice veins reach the depth of 54 m; the largest one is 20 cm wide. Besides, the ice layer/veins develop along the original stratification or weathered cracks, forming the reticulated structure of permafrost.

Up to now, no ice wedges have been found in China. However, on the Qinghai-Xizang Plateau, on the second-order terrace of the Beilu, Tuotuo, Tongtian and Buqu Rivers, there are many sand wedges, from several tens of centimeters up to 2 to 4 m deep. According to the C^{14} dating, these may be the traces of ice wedges formed in late Pleistocene time (Guo 1979).

SUMMARY

1. Permafrost in China can be divided into two broad categories: permafrost in high latitudes and in high altitudes. The former is in northeast China and depends chiefly on latitudinal zonation; the latter occurs in the high mountains and plateaus in west China, and is vertically zoned.

2. A series of azonal factors play important roles in the development of frozen soils; among them, temperature inversion in winter and peat promote permafrost development in lowlands of the Northeast; geologic and hydrogeologic factors result in many kinds of taliks in the Qinghai-Xizang Plateau.

3. Most of the alpine permafrost in China belongs to the continental type, because the lower limit of permafrost is generally 800 to 1000 m lower than the local snowline; however, permafrost occurring on the south-facing slopes of the eastern section of Himalaya Shan, in southeastern Xizang, and in the Hengduan Shan belongs to maritime, with the lower limit near or higher than the local snowline.

4. Permafrost in the Northeast has a southern limit coinciding with the -1 to +1°C isotherm and extends to latitude 46.6°N. The lower limit of permafrost on Qinghai-Xizang Plateau coincides with the -2 to -3°C isotherm.

5. By comparing the thickness and ground temperature of permafrost in the Northeast with that on the Qinghai-Xizang Plateau, it is known that the permafrost on the latter is more stable.

REFERENCES

Cui Zhijiu, Zhou Youwu, 1980, Permafrost and periglacial phenomena, c.9, in Physical geography in China (Geomorphology). Science press.

Dai Jingbo, 1982, Characteristics of ground temperature in permafrost areas in the northern part of Great Xinan Mountain, Journal of Glaciology and Cryopedology, 4(3), pp.53-63.

Fujii, Y. 1980, Distribution of alpine permafrost in the northern hemisphere and environmental conditions, Snow and Ice, 42(1). pp.41-52.

Guo Dongxin, 1979, Sandy Wedges, Journal of Glaciology and Cryopedology, 1(1).

Guo Dongxin, Wang Shaoling, Lu Guowei, Dai Jingbo and Li Enying, 1981, Division of permafrost regions in Daxiao Hinggan Ling Northeast China, Journal of Glaciology and Cryopedology, 3(3), pp.1-9.

Guo Pengfei, 1983, Permafrost in Qilian Shan area, in Proceedings of the Second National conference on Permafrost, Gansu People's press.

Qiu Guoqing, Zhang Changqing, 1981, Distributive Features of Permafrost near the Kuixian Daban in the Tian Shan , No.2 Memoirs of Lanzhou Institute of Glaciology and Cryopedology, Academia Sinica, pp. 1-16, Science press.

Shen Zhibao, 1975, Air temperature and its variation with elevation on the northern slope of Qomolangma Feng, in Reports of Science Expedition in District of Qomolangma Feng, 1966-1968 (Meteorology and Solar Radition), Science press.

Tong Boliang, 1965, Permafrost regionalization and characteristics from Kunlun Mt. Pass to Xidatan Valley, in Permafrost Expedition Along Qinghai-Xizang Highway, Science press.

Tong Boliang, 1981, Some features of permafrost on the Qinghai-Xizang Plateau and the factors influencing them, in Proceedings of Symposium on Qinghai-Xizang (Tibet) Plateau (Beijing, China), Vol.II, Science press, Beijing;Gordon and Breach, Science Publishers, INC. New York.

Wang Jiacheng, Wang Shaoling and Qiu Guoqing, 1979, Permafrost Along the Qinghai-Xizang Highway, Acta Geographica Sinica, 34(1),pp.18-32.

Xie Zichu, Wang Zongtai, 1975, Temperature of ice layers in the Rongbu Glacier, in Reports of Science Expedition in District of Qomolangma Feng, 1966-1968 (Existing Glaciers and Geomorphology), Science press.

Zhou Youwu, 1965, Permafrost along Qinghai-Xizang Highway, in Permafrost Expedition Along Qinghai-Xizang Highway, Science press.

Zhou Youwu, Guo Dongxin, 1982, Principal characteristics of permafrost in China, Journal of Glaciology and Cryopedology, 4(1),pp.1-19.

STABILIZATION OF HIGHWAY SUBGRADE IN PLATEAU PERMAFROST REGION OF CHINA

Zhu Xuewen, Chen Guojing, and Zhange Bintao

Highway and Transportation Research Institute, Ministry of Communication
People's Republic of China

According to the permafrost, engineering, and geological conditions, the permafrost regions on the Qinghai-Xizang Plateau can be divided into four categories: (I) poor in ice, with thaw-settlement coefficient of frozen soil A less than 5%; (II) with a moderate ice content, and A is 5-10%; (III) rich in ice, and A is 10-40%; (IV) with a massive ice layer beneath the permafrost table, and A > 40%. In applying these classifications to highway construction, the authors find that in Category I, no special techniques are needed to deal with frost damage. In Category II, cutting is generally avoided, but when it is unavoidable, the side slope should be gentle and the strength of the subgrade should be ensured. In categories III and IV, no cutting is allowed, since the natural ground surface should not be disturbed within 10 m of the roadside; the height of the embankment should not be lower than the critical value. Based on observation and calculation, a minimum thickness of embankment is recommended: for category I, the height of the road has no relation to permafrost and frost action. For category II, the height of embankment is 0.8-1.2 m; for category III, 0.9-1.5 m; and for category IV, 1.1-1.7 m, depending on what kind of fill material is to be used.

Permafrost regions in China are scattered in the northern part of the Major and Minor Xinggan Ling, the Qinghai-Xizang Plateau, and the alpine areas of western China. The permafrost grounds discussed in this article are those located in the Kunlun Shan and the plateau to its south with an altitude higher than 4200 m above sea level.

The climate of this region is cruel. In the thin air at high altitude, the oxygen content is only about one half of that at sea level. The annual mean air temperature is -2°— -7°c, and the frozen season is as long as 7 - 8 months a year. Even in the warm season (July and August), negative temperatures often occur at night.

The presence of permafrost and the severe climate cause a series of problems to the construction of asphalt pavement. It is necessary to find out the proper depths of embankment of the preservation of permafrost and the measures for prevention against frost action. According to the investigations taken in 1973-1980, several recommendations were made for the design of subgrade under an asphalt pavement.

GEOLOGICAL REGIONALIZATION OF HIGHWAY CONSTRUCTION AND PERMAFROST PROTECTION

The Qinghai - Xizang Plateau Permafrost Region belongs to one of the general highway natural regions of China. It may be divided into two subregions: continuous permafrost and discontinuous (sporadic) permafrost.

The continuous permafrost subregion that lies in the north of the plateau can be classified into three conditions: high mountains, low mountains and rolling hills, and level land in the valley (valley plain and terrace), according to their topographic and geomorphologic features.

With regard to the regional engineering geological conditions, those of the valley plain zones are the best, and there are thawing zones along big rivers. In the zones of low mountains and rolling hills, top soils are finer, moisture contents are higher, thawing and freezing actions are much more intensive, underground ice is more developed, and marsh lands are often found at the flat ridge or in depressions among mountains. All these phenomena make the engineering geological conditions most unfavorable. In the high mountain zones, the situation lies between the other two. But when the route is passing over the mountain ridge where the undulation of the terrain is flatter, the geological conditions are similar to that of the zones of low mountains and rolling hills.

The discontinuous (sporadic) permafrost subregion is on the south of Tangula Shan. The main unfavorable geologic phenomenon is marshy lands, and the route through it should be avoided whenever possible.

Underground ice is abundant in the plateau permafrost region. The engineering properties of frozen soils after thawing are substantially different according to their moisture content. Thus, the classification of frozen soil in highway engineering in the permafrost area depends upon the total water content of the frozen soil and its coefficient of settlement in thawing. The depth of the active layer, types and ice content (or the total water content) of the frozen soils must be defined by boring, and for design and construction purposes, roads are generally divided into sections according to the total water content of the frozen soils within the depth of 1 meter from the upper boundary of the permafrost horizon. The classification of the road sections is shown in Table I, and the appropriate requirements for the preservation of the frozen state of the permafrost are also proposed. (1) For the cut sections and/or the sections where the hydrological and geological conditions are unfavorable, detailed geologic survey and special design must be taken.

MINIMUM DEPTHS OF EMBANKMENTS FOR ASPHALT PAVEMENTS

Solar radiation in the plateau permafrost region is much stronger than in the high latitude region. In the case of asphalt pavement, because of the higher adsorbing capacity of road surface from sunlight, the surface temperature of pavement and the accumulated heat in the road bed beneath it will be higher and more than that of the gravel pavement.

Thus, the thaw penetration will be deeper, and thaw settlement will also occur, if no appropriate preservation measures to the frozen state of permafrost were employed.

Practice shows that the settlement of the road bed caused by paving with black top can be prevented by raising the height of the embankment. However, under difficult conditions due to severe climate, raising the height of embankment from borrow on plateau would be very expensive.

Different thicknesses of embankment had been built at the altitudes of 4500-4600 m above sea level near 35°N in 1973-1976 to find out what is the proper thickness required. The data obtain from observation indicate that the thaw penetration of the active layer sank into the shape of a "disk" after the asphalt surface is placed(Fig1). For fillings with silty and clayey soil, an actual sinking of 39-43 cm has resulted (h, in Fig. 1) for loam or sandy soils; it is about 1.2 times of the mentioned value. The pavement on top of the embankment having 1.2 m thick at the section of category IV (see Table 1) still kept in good condition for nine years, although there was some light but uniform settlement. For the section of category I and II, the embankment with 0.8 m thick will be enough.

Two observation sites were also established near the experimental road sections in 1973 and 1977. The difference between the gravel and the asphalt pavements were observed. Surface temperatures were measured by glass surface temperature gauges used by the meteorological station 3 times a day for 2 years (1979-1980). The data at the depth of 5 cm (just beneath the asphalt pavement) were used.

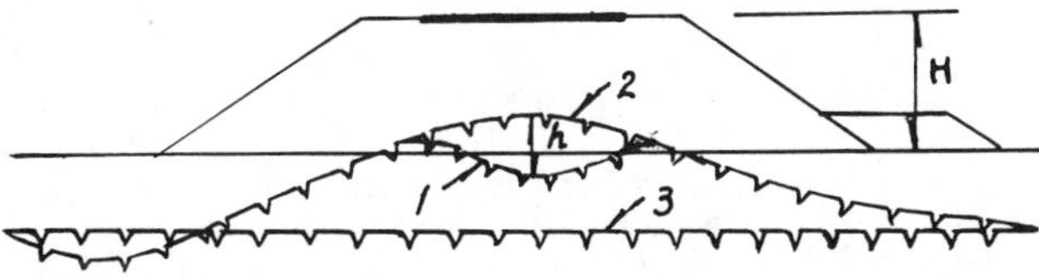

FIGURE 1 Variation of upper boundary of permafrost horizon after the construction of embankment and asphalt pavement. H, height of embankment; h, deepened thawing penetration after paving asphalt pavement.
—v—v—v— upper boundary of permafrost horizon. 1, natural; 2, after the construction of embankment; 3, after paving asphalt pavement.

The data of our observation and the meteorological stations show that there is a good correlation between the air thawing index and surface thawing index of the sandy gravel ground. This correlation is shown in Fig. 2, and can be expressed as equation 1.

$$N_g = \frac{I_{s-a}}{I_a} = 0.850 + 984 I_a^{-1} \quad (1)$$

Where N_g is the N-factor of gravel ground; I_a is the air thawing index, in C°- days; I_s - g is the surface thawing index of sandy gravel ground, in C°- days.

From 1, the surface thawing index of sandy gravel for various areas of this region can be calculated according to the long term air temperature data, and the N-factor is much greater than in high latitude areas.

The ratio of the surface thawing index I_s of gravel pavement against sandy gravel ground is 0.98 obtained from observation, and is considered to be the same, while the I_s of asphalt pavement against gravel pavement is 1.26 - 1.38. There will be a little difference between the years of observation and different filling heights. But these values are very close to those obtained from Alaska and the USSR. (3,5)

By these correlations and taking into account the effect of temperature variation during the service life of the pavement, the design surface thawing index of asphalt pavement can be defined. The results are shown in Table 2.

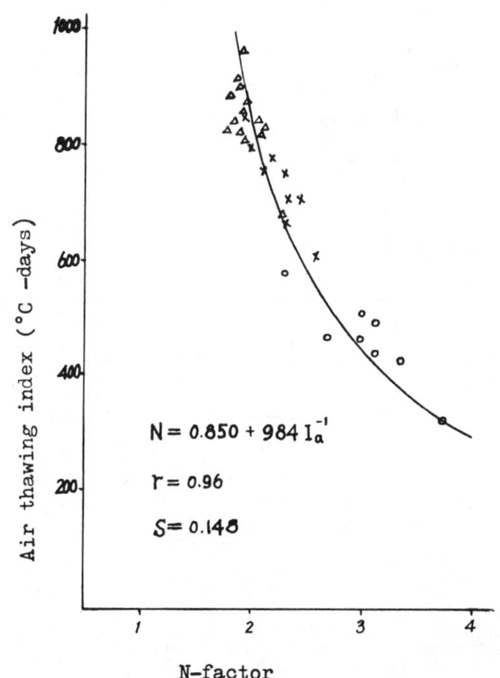

FIGURE 2 N - factor of natural sandy gravel ground verse the air thawing index in plateau permafrost region.

TABLE 1 Classification of road sections according to the preservation requirements

Category of road section	Conditions of frozen soil at the upper boundary: Ice content	Thaw-settlement coef. A%	Preservation requirements
I	Low	5	Design and construction according to common seasonally frozen region
II	Moderate	5-10	Cutting is generally avoided. When unavoidable, side slope should be gentle and measures be taken to ensure the strength of the subgrade. Distance between borrow pit and the foot of the side slope should be not less than 5 m.
III	Rich	10-40	Height of embankment should meet the preservation requirement. Generally no cutting is allowed, measures should be taken to ensure the stability of the subbase and the side slope if cutting is unavoidable. Principally, natural ground surface within 10 m at the road side should not be disturbed. Distance of borrow pit should be at least 10m apart from the foot of the side slope, depth of excavation should not be greater than 80% of thaw penetration of local permafrost soil.
IV	Ice strata with soil inclusion	>40	Principally, cutting is not allowed. Height of embankment should meet the preservation requirements. Natural ground surface within 10m at the road side should not be disturbed. Earth borrowing from the road side should be avoided when possible, otherwise the borrow pit should be located as given above. For intercepting surface water retaining dikes are preferable to intercepting ditches. Side ditches along road side should be avoided.

TABLE 2 The design surface thawing index of asphalt pavement

Site No.	I	II
Latitude	35°17'N	33°57'N
Altitude(m)	4612	4533
Annual mean air temperature C°	-5.7	-4.3
Mean air thawing index (C°-days)	465	744
Surface thawing index of asphalt Pavement (C°- days)	2240	2484

Thus, the effect of an asphalt pavement on the top of gravel pavement can be estimated by equation 2.

$$h = \sqrt{\frac{48K}{L}}\left(\sqrt{I_{s-a}} - \sqrt{I_{s-g}}\right) \qquad (2)$$

where h is the depth of thaw penetration deepened by paving asphalt pavement, in m; I_{s-a} and I_{s-g} are the surface thawing index of asphalt and gravel pavements respectively, in C°-days; K is the coefficient of thermal conductivity, in kcal/m.hr. C°; L is the latent heat of the soil, in kcal/m^3.

It is found that (for homogeneous soil system):

1. For fine grained soils, the variation of the value of h is not significant for a same soil type

TABLE 3. Critical thickness of embankment in the plateau permafrost region (m)

Material for filling \ Subgrade soils	Silty, clayey, and sandy soils	Coarse sand, sandy gravel, gravel, and stony soils
Local soils	0.8–0.9	1.0–1.2
Sandy gravel	1.1–1.2	1.0–1.2

having the moisture content and density changed within a conventional range. They range from 43 to 52 cm for silty and clayey soils, and from 56 to 66 cm for sandy soils, but somewhat deeper for coarse grained soils that range from 66 to 102 cm.

2. Although there exists a difference in the mean air thawing index between site I and site II, the surface thawing indexes of asphalt pavement are essentially the same.

Because the temperature conditions of site I and site II can be approximately considered as the margin cases of the plateau between the Tanggula Shan and Kunlun Shan, the critical thickness of embankment for preservation of permafrost in this area may be evaluated.

For the case of layered systems, the problems can be solved with the method used by Sanger (1973).

Figures shown in Table 3 had been developed using the data from the road investigation and observation sites(2). Results had been justified that it is applicable for conventional local cases.

Different factors of safety are adopted for different categories of road section. The minimum depths of embankment suggested to prevent thermo-thawing damages are shown in Table 4.

The pavement may be considered as a part of the preservation layer, and the proper thickness will be determined by the soil engineer in the field. For sections where the hydrological and geological conditions are very unfavourable, special design should be taken.

The annual rainfall of the southeast part of this region is 400-600 mm, but decreases to 200 mm from south to north, and decreases to less than 100 mm from east to west. The warm season is short and the temperature is low. Therefore, the soil forming process is very slow. These factors make the soil of the active layer usually coarse in size and moderate in moisture (less than liquid limit, and a large part of it ranged from 0.7 -0.8 of the liquid limit). It is possible to build the embankment with local soils and drop the cost of construction. As the heat conductivity of local soils is less than that of gravel, the thickness of embankment may be thinner. But in this case, proper gradation must be kept in mind.

FROST HEAVE, FROST BOILS AND THEIR PREVENTIVE AND REMEDIAL MEASURES

In this region, sand, gravel, and sandy soil predominate. Investigation revealed that only very small amounts of accumulated water occured in sandy subgrade soil with medium water content during freezing. Therefore, the stability of the subgrade will generally not be affected. While in subgrade

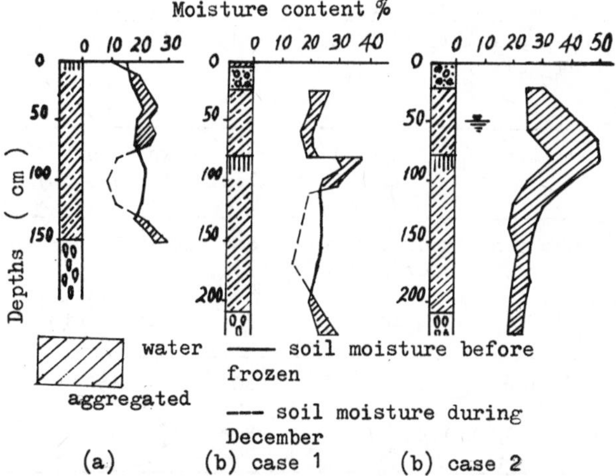

FIGURE 3 Soil moisture aggregation during the freezing period in plateau permafrost region. (a) natural ground, without external water supply to the system during freezing period; (b) embankment filling with silty soil on the grassland; case 1, without external water supply to the road system, case 2. external water came from the accumulating water in the borrowing pit.

with silty soil, the accumulated water during freezing is apparent.

It was also revealed in the investigation that without external water supply to the system, the moisture content of the entire active layer of soils remains essentially unchanged during the freezing period, and the upper and lower parts of the layer collected the water, while the moisture of the central part decreases. (Fig. 3a) The water aggregating phenomenon varies as the structural types of subgrade and pavement change, but their basic pattern is similar to that of the natural ground (Fig.3 b case 1).

In the presence of external water supply, it can be seen that the moisture content of the whole subgrade soil increases (Fig.3 b, case 2). The amount of frost heave is evidently dependent on the ground water level or the level of the accumulated water along the road side before or during the freezing period. Maximum frost heave up to 131 mm has been observed on highways in this region. Hence preventive and remedial measures must be taken.

Measures for the prevention and remedy of frost heave and frost boils currently adopted include improvement of the drainage system, raising the height of the embankment, and provision of a sand gravel bedding course.

Sand gravel bedding course has an obvious drying affect on its underlying base soil. It can not only serve as a part of the pavement structure, but also improve the performance of the subgrade, so that it has been widely adopted. The thickness of such bedding course should not be less than 20 cm, preferably extending right through the whole width of the cross-section.

When the accumulated water along the road side is difficult to drain or ground water effects still exist in winter times, for the sake of pro-

TABLE 4 Minimum thickness of stabilizing course for prevention of freezing damages (for silt and silty clay)

Height of embankment (m)		Minimum thickness of stabilizing course (cm)	
Above road side water level	Above ground water level before freezing	For asphalt concrete	For other types of asphalt pavements
1.4-1.6	2.0-2.2	60	45
1.1-1.4	1.7-2.0	70	50
0.8-1.1	1.4-1.7	80	60

tection from frost heave the total thickness of the whole pavement structure, including the sand gravel bedding course, should not be less than the values listed in the Table 4.

CONCLUSION

The stability of highway subgrade is an essential problem in the construction of asphalt pavement in the plateau permafrost region. For the prevention of thawing settlement and frost action, proper depths of embankment is needed. Practice has been shown that it is possible to build the embankment with local soils of moderate moisture content. The minimum depths of embankment recommended for road sections according to the preservation requirements as well as the measures for preventing frost heave, may be a feasible and economical approach for our country. Because of the insufficient experience for the construction of asphalt pavement in this area, further study will be needed.

ACKNOWLEGEMENT

Information referred to in this article was presented by the Qing-Zang Plateau Highway Research Section. The authors wish to acknowlege all the researchers, engineers, and workers of the section. We also wish to acknowlege the organizations and individuals who gave much help in our investigation.

REFERENCES

Qing-Zang Plateau Highway Research Section, 1978, Construction of asphalt pavement in plateau permafrost region: Beijing, Report of Highway and Transportation Research Institute, the People's Republic of China.

Zhu Xuewen, 1981, Study of minimum depths of embankment under an asphalt pavement in plateau permafrost region: Beijing, Report of Highway and Transportation Research Institute, the People's Republic of China.

V.J. Lunardini, 1978, Theory of N-factors and correlation of data, in Proceedings of the Third International Conference on permafrost, v.1: Ottawa, National Research Council of Canada.

F.J. Sanger, 1973, Degree-days and heat conduction in soils, Permafrost - The North American Contribution to the Second International Conference, Yakutsk: Washington, D.C., National Academy of Sciences.

A.A. Malyshev, 1974, Highway subsoil under northern conditions. Moscow: Transport Publishers.

CREEP BEHAVIOR OF FROZEN SILT UNDER CONSTANT UNIAXIAL STRESS

Zhu Yuanlin[1] and D. L. Carbee[2]

[1]Lanzhou Institute of Glaciology and Cryopedology, Academia Sinica
People's Republic of China
[2]U.S. Army Cold Regions Research and Engineering Laboratory
Hanover, New Hampshire 03755 USA

A series of unconfined compression creep tests was conducted on remolded, saturated frozen Fairbanks silt at constant-stress (σ) and constant-temperature conditions. It was found that a sudden change occurred in the slope of log $\dot{\varepsilon}$ vs. $1/\sigma$ curves at almost the same minimum strain rate (about $10^{-6}\ s^{-1}$) for various test temperatures. Therefore, the authors suggest that the creep of frozen soil be classified into two types: short-term creep and long-term creep. Different constitutive and strength-loss equations are presented for each type of creep. The criterion of creep failure of frozen soil is considered to have the general form of $\dot{\varepsilon}_m \times t_m^n = \varepsilon_f$, where n is a material constant dependent only upon water content, ε_f is the failure strain, and t_m is the time to failure in minutes. On the basis of Assur's creep model (1980) and this criterion, a creep equation was derived that can describe the entire process of creep of frozen soil.

The design of stable structures on permafrost or in artificially frozen ground requires an understanding of the strength and deformation characteristics of frozen soils. The creep of frozen soil is a complicated physicomechanical process governed by various factors such as applied load, temperature, and material properties (such as composition, ice content, and texture).

The main external factors that influence the creep response of frozen soil are stress and temperature. To evaluate these influences quantitatively so as to predict the process of creep deformation and strength-loss of frozen soil, this investigation was performed under various stress levels and at various temperatures varying from -0.5 to -10°C. Three soil densities were used to investigate the influence of density on creep behavior, but this paper deals primarily with the data obtained from tests on the specimens with a dry density of from 1.16 to 1.24 g/cm^3 (correspondingly, with a water content of 40.0% to 44.8%). The complete results will be included in a CRREL report entitled "Creep and Strength Behavior of Frozen Silt in Uniaxial Compression," which is now in process.

EXPERIMENT

Test Material

The material used in this investigation was a remolded silt from the USACRREL experimental permafrost tunnel at Fox, near Fairbanks, Alaska. The physical property indicators are as follows: plastic limit of 34.16%, liquid limit of 38.4%, organic content of 5.49%, and a specific gravity of 2.68. The silt is classified as ML with approximately 94% passing the #200 mesh sieve and 17% finer than 0.01 mm.

Test Specimens

Each soil specimen was compacted in uniform-density layers, vacuum-saturated, and quick-frozen unidirectionally in an open system. The frozen specimens were 7 cm (2.75 in.) in diameter and end-machined to 15.24 cm (6 in.) long.

Test Method

The unconfined compression creep tests were performed using the constant-stress creep test apparatus designed by Sayles (1968). It allows the applied load to increase proportionally to the increase in the deformation of specimens so that the test (true) stresses can remain constant during test.

RESULTS AND DISCUSSION

$\dot{\varepsilon}_m$ - σ Relationship and Constitutive Equations

As reported by many researchers (Assur 1980, Mellor and Cole 1982, Martin et al. 1981, etc.), we consider that creep failure occurs when the creep rate reaches its minimum value $\dot{\varepsilon}_m$. The time to the minimum rate is defined as the time to failure t_m, and the strain at the minimum is the creep failure strain ε_f.

Experiment indicates that for a particular material the minimum strain rates are closely related to temperature and stress. In Figure 1 the minimum rates $\dot{\varepsilon}_m$ are plotted against stresses σ in logarithm for the material tested at various temperatures. They are obviously not straight lines, but a family of similar curves. It is especially interesting to note that they suddenly increase their slopes at a certain minimum rate as stresses

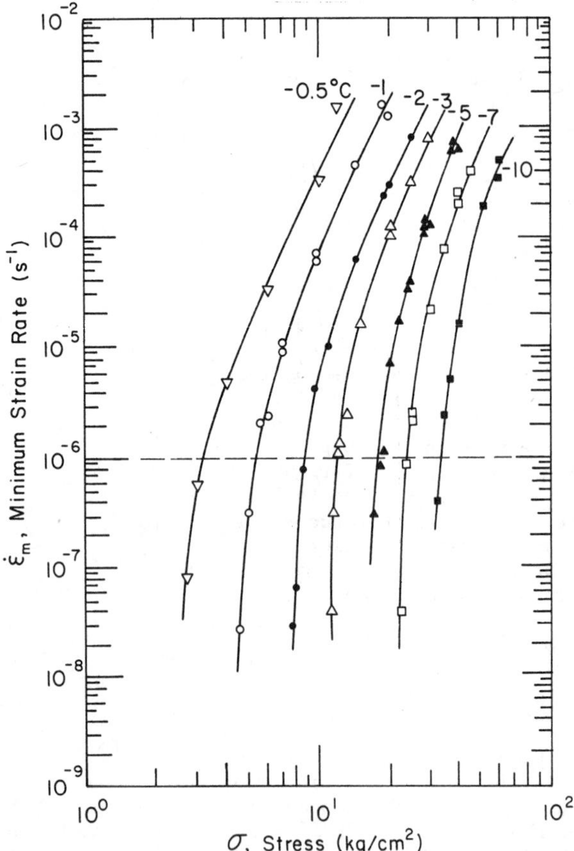

FIGURE 1 Plot of $\dot{\varepsilon}_m$ vs. σ in logarithm.

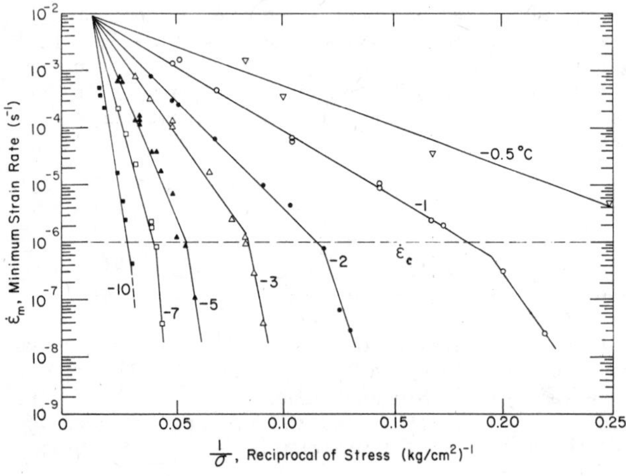

FIGURE 2 Plot of log $\dot{\varepsilon}_m$ vs. $1/\sigma$.

decrease. This feature can be seen more clearly by plotting log $\dot{\varepsilon}_m$ vs. $1/\sigma$ as shown in Figure 2. In this graph a family of bilinear curves is obtained, and they all deflect at almost the same minimum strain rate (about 10^{-6} s^{-1}). This implies that there is a critical creep rate, $\dot{\varepsilon}_c$, that may be considered the minimum rate for distinguishing between two fundamental types of creep of frozen soil. The authors therefore suggest that the creep of frozen soil be classified into short-term creep, which has minimum strain rates greater than $\dot{\varepsilon}_c$, and long-term creep, which has minimum strain rates less than $\dot{\varepsilon}_c$.

Based on Figure 2, the constitutive equation can be well described by the following exponential equations. For short-term creep,

$$\dot{\varepsilon}_m = \dot{\varepsilon}_* \exp[-k(\frac{1}{\sigma} - \frac{1}{\sigma_*})] \tag{1}$$

where $\dot{\varepsilon}_*$ and σ_* are material constants independent of temperatures. By linear regression and standard deviation analysis, we obtain $\dot{\varepsilon}_* = 8.84 \times 10^{-3}$ s^{-1} and $\sigma_* = 71.4$ kg/cm^2. The parameter k reflects the influence of temperature and can be determined by $k = 53.1\ \theta^{0.72}$ with $-0.5 \geq -\theta \geq -2$°C, and $k = 42.4\ \theta^{1.02}$ with $-2 \geq -\theta \geq -10$°C, where θ is the absolute value of temperature, in °C.

For long-term creep,

$$\dot{\varepsilon}_m = \dot{\varepsilon}_c \exp[-k'(\frac{1}{\sigma} - \frac{1}{\sigma_c})] \tag{2}$$

where $\dot{\varepsilon}_c$ is the critical creep rate, equal to 10^{-6} s^{-1} for the material tested, and σ_c is the critical creep strength in kg/cm^2, which corresponds to the stresses under which $\dot{\varepsilon}_m = \dot{\varepsilon}_c$. It varies with the temperature: $\sigma_c = 5.2\ \theta^{0.78}$. The parameter k' also varies with temperature: $k' = 125.7\ \theta^{1.10}$.

To explain this phenomenon, the authors propose that these two types of creep underlie different deformation mechanisms: short-term creep is controlled by dislocation creep, while long-term creep is controlled by glide creep (Zhu and Carbee, in press).

t_m-σ Relationship and Long-Term Strength

Figure 3 presents the plot of log t_m vs. $1/\sigma$ curves for the tested material at various test temperatures. These curves show a similar pattern to that shown in Figure 2, and can be well described by, for short-term creep,

$$t_m = t_* \exp[k_1(\frac{1}{\sigma} - \frac{1}{\sigma_*})] \tag{3}$$

and for long-term creep,

$$t_m = t_c \exp[k_1'(\frac{1}{\sigma} - \frac{1}{\sigma_c})] \tag{4}$$

where t_* and t_c are material constants independent of temperature. For the material tested, $t_* = 0.19$ min and $t_c = 912$ min. k_1 and k_1' are a function of temperature:

$k_1 = 49.7\ \theta^{0.75}$ with $-0.5 \geq -\theta \geq -3$°C

$k_1 = 30.7\ \theta^{1.18}$ with $-3 \geq -\theta \geq -10$°C, and

$k_1' = 117\ \theta^{1.1}$ with $-0.5 \geq -\theta \geq -10$°C

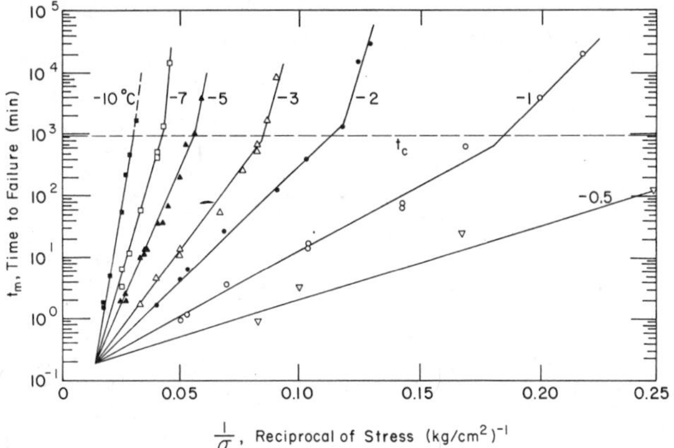

FIGURE 3 Plot of log t_m vs. $1/\sigma$.

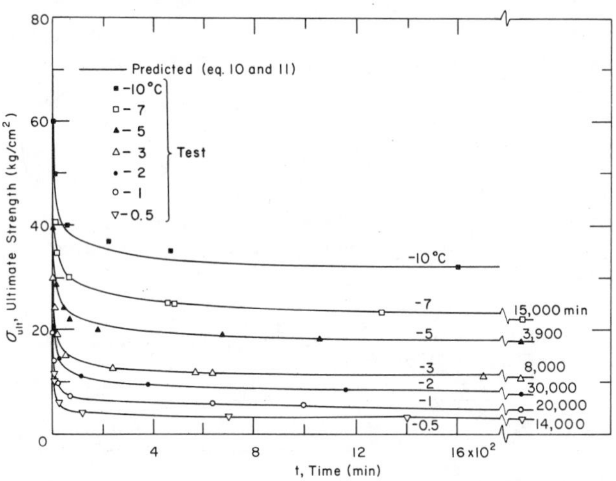

FIGURE 4 Comparison of the predicted strength-time curves with test data.

By the definition of creep failure, the creep stress σ in Equations 3 and 4 is the ultimate strength of the frozen soil (denoted as σ_{ult}) at the time of load action (t) equal to t_m. Then, from Equations 3 and 4 we derived the following strength-loss equations, for $t < t_c$:

$$\sigma_{ult} = \frac{k_1 \sigma_*}{\sigma_* \ln(t/t_*) + k_1} , \tag{5}$$

and for $t \geq t_c$:

$$\sigma_{ult} = \sigma_c \frac{k_1' \sigma_c}{\ln(t/t_c) + k_1'} , \tag{6}$$

which can be used to predict the ultimate strength of frozen soil at any given time.

The curves predicted by Equations 5 and 6 and the data points are compared in Figure 4, showing a very good agreement.

It is clear that one should adopt different equations to evaluate ultimate strength according to the duration of loading. For the prediction of long-term strength, Equation 6 should be used.

Following the usual criterion, if we take the 100-year strength as the ultimate long-term strength (σ_{1t}), then from Equation 6 we have the following relation for frozen Fairbanks silt:

$$\sigma_{1t} = 3.49\ \theta^{0.87} \tag{7}$$

where σ_{1t} is in kg/cm^2 and θ is in °C.

The values of σ_{1t} predicted in this investigation are very close to those predicted for remolded frozen Hanover silt by Sayles (1974), but higher than those for undisturbed frozen silt reported by McRoberts et al. (1978).

$\dot{\varepsilon}_m - t_m$ Relationship

Figure 5 shows a set of log $\dot{\varepsilon}$ vs. log t curves for the tested material under various stresses at -5°C. It is clear that all minimum points F in this figure distribute very close to a straight line. This means that there is a definite relationship between the minimum strain rate $\dot{\varepsilon}_m$ and the time to failure t_m. Figure 6, which shows all of the minimum points for all test temperatures, indicates that this relationship is not influenced by temperature, and can be described by a simple power-law equation:

$$\dot{\varepsilon}_m = c\ t_m^{-n} \tag{8}$$

where n is a constant dependent only upon water content, and c depends upon the water content and the unit of time (if $n \neq 1$). The values of n and c, with the unit of time in minutes for different water content groups, are given in Table 1. We note that the value of n for ice-rich frozen soil is very close to that of polycrystalline ice (Mellor and Cole 1982).

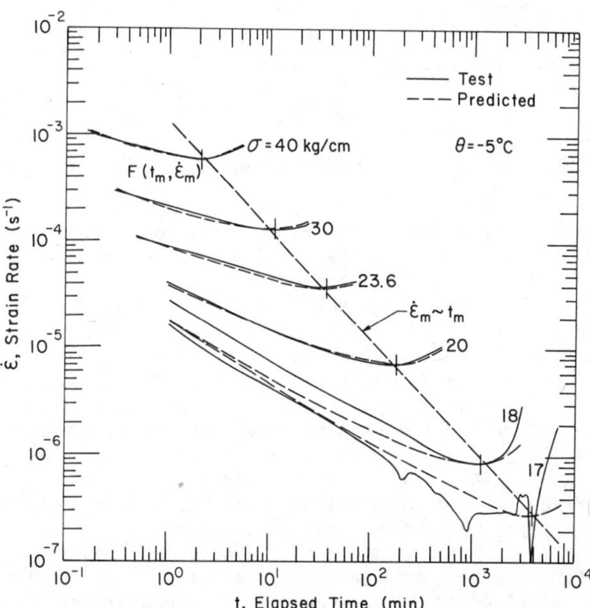

FIGURE 5 A set of log $\dot{\varepsilon}$ vs. log t curves under various stresses at -5°C.

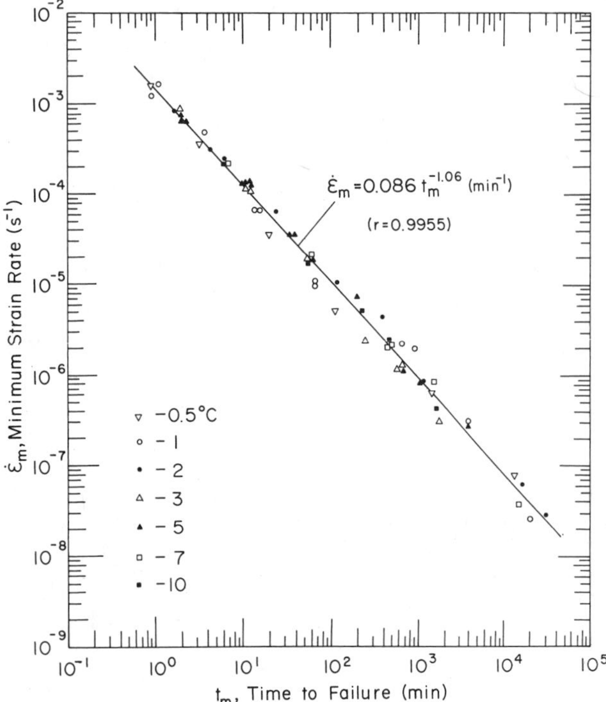

FIGURE 6 Relationship between $\dot{\varepsilon}_m$ and t_m.

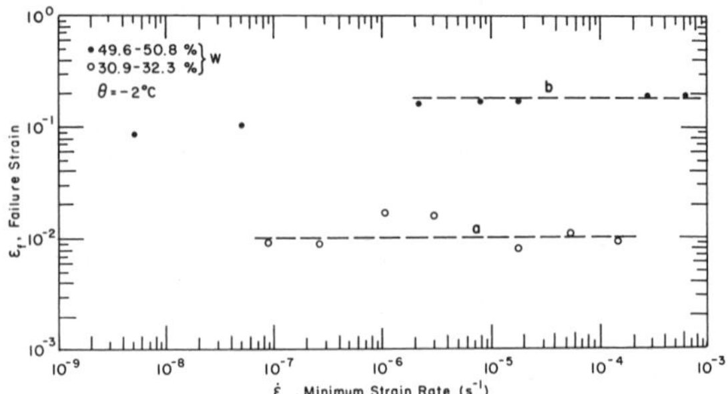

FIGURE 7 Failure strain as a function of minimum strain rate at -2°C for (a) w = 49.6-50.8% and (b) w = 30.9-32.3%.

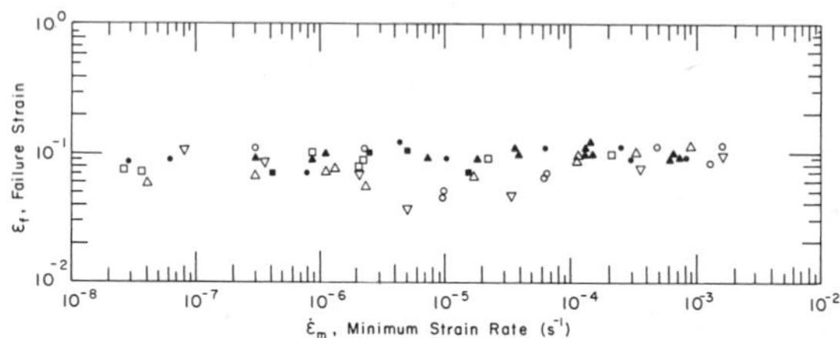

FIGURE 8 Failure strain as a function of minimum strain rate for w = 40-44.8% at various temperatures.

TABLE 1 Values of c and n in Equation 8.

W,%	n	c
49.6-50.8	1.01	0.0083
40.0-44.8	1.06	0.0860
30.9-32.3	1.16	0.1800

TABLE 2 Averaged values of ε_f for tested material at different water contents.

Water content (%)	ε_f
49.6-50.8	0.0117
40.0-44.8	0.0870
30.9-32.3	0.1760

Creep Failure Strain and Failure Criterion

Creep failure strain ε_f as a function of minimum strain rate $\dot{\varepsilon}_m$ for different water-content groups is plotted in Figures 7 and 8, respectively. Clearly, it is substantially governed by the water content of the frozen soil: the higher the water content, the less the failure strain. It is interesting to note that for high (Figure 7a) and medium (Figure 8) water content, the creep failure strain is nearly independent of the strain rate and temperature, while for low water content (Figure 7b), it depends upon the strain rate: it decreases slightly with decreasing strain rate within each $\dot{\varepsilon}_m$ range of the two types of creep discussed in this paper, and significantly decreases as $\dot{\varepsilon}_m < 10^{-6}$ s^{-1}. Again, this evidence supports the viewpoint that the two types of creep are controlled by different deformation mechanisms. The average values of ε_f for high and medium water-content groups, together with that of the low water-content group for short-term creep, are given in Table 2.

Comparing the values of failure strain with those of coefficient c in Equation 8, we can see that they are very close to each other. Thus, from Equation 8 we have the following creep failure criterion:

$$\dot{\varepsilon}_m \times t_m^n = \varepsilon_f \tag{9}$$

where t_m is in minutes, and $\dot{\varepsilon}_m$ is in min^{-1}.

Evidently, the criterion $\dot{\varepsilon}_m \times t_m = \varepsilon_f$, which was proposed by Ladanyi (1972) and Assur (1980), is a particular case (n=1) or an approximate form of criterion 9.

Creep Model and Prediction of Creep Strain

To predict the creep deformation of frozen soils, a number of models have been proposed by different investigators. It is worth noting that a similar creep model was recently proposed, from different perspectives, by Assur (1980), Fish (1980),

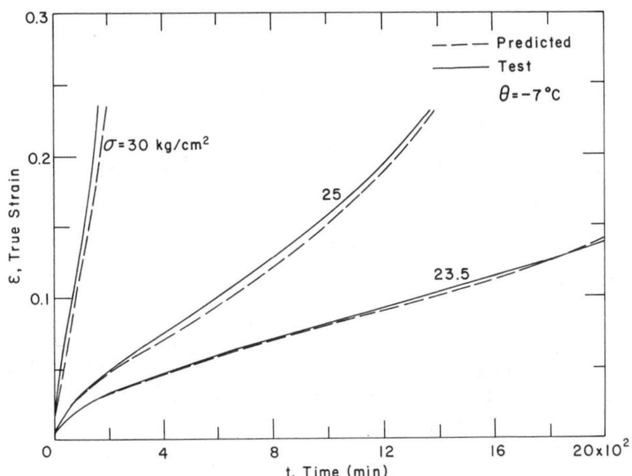

FIGURE 9 Comparison between the predicted and test creep curves for short-term creep.

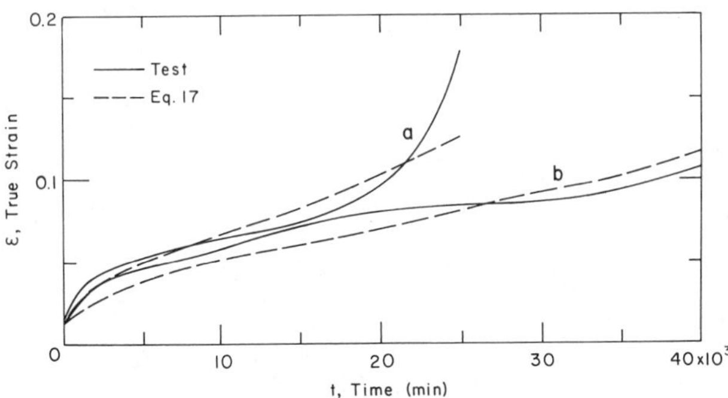

FIGURE 10 Comparison between the predicted and test creep curves for (a) σ = 22 kg/cm^2, θ = -7°C and (b) σ = 7.7 kg/cm^2, θ = -2°C.

and Ting (1981), that can describe the entire process of creep of frozen soils. Among them, Assur's model is the simplest one. It is based upon physical reasoning in a differential equation with the resulting integral:

$$\dot{\varepsilon}(t) = \frac{\dot{\varepsilon}_m}{e^{\beta}} \left(\frac{t_m}{t}\right)^{\beta} e^{\beta t/tm} \quad . \qquad (10)$$

We found that the values of β were not directly related to temperature, but were closely related to stress. They can be expressed in terms of t_m as follows for short-term creep:

$\beta = 0.33$ when $t_m < 30$ min

$\beta = 0.23\, t_m^{0.11}$ when $t_m \geq 30$ min

The curves predicted by Equation 10 and the test data are compared in Figure 5. It is clear that Equation 10 can well describe the creep process for short-term creep, but does not describe the process as well for long-term creep. It is difficult to determine β in Equation 10 for long-term creep. The authors suggest that, as a first approximation for long-term creep, one determines the value of β by fitting Equation 10 to the test data within the primary creep stage. From our test data, β has an average value of 0.6.

Integrating Equation 10 and taking into consideration Equation 9 and instantaneous strain ε_0, we derived the following creep equation:

$$\varepsilon(t) = \varepsilon_0 + \frac{\varepsilon_f}{e^{\beta}\, t_m^{n-1}} \left(\frac{t}{t_m}\right)^{1-\beta} e^{\beta t/t_m} \left[\frac{1}{1-\beta} - \frac{\beta t/t_m}{(1-\beta)(2-\beta)} + \frac{(\beta t/t_m)^2}{(1-\beta)(2-\beta)(3-\beta)} - \ldots\right] \qquad (11)$$

where t_m is determined by Equations 3 and 4.

Comparing the predicted curves with the test data, we can see that Equation 11 can well predict the entire process of short-term creep (Figure 9). It does not predict long-term creep as well (see Figure 10), but as a first approximation it is still applicable for predicting the long-term creep deformation of frozen soil.

CONCLUSIONS

It can be concluded from this investigation that:

1. According to applied stress levels, the creep of frozen soil can be classified into two types: long-term creep and short-term creep. Each obeys different constitutive equations and stress relaxation equations.
2. The 100-year strength of remolded, saturated, frozen Fairbanks silt with a water content of from 40% to 44.8% can be determined by $\sigma_{1t} = 3.49\, \theta^{0.87}$, where θ is the absolute temperature in °C and σ_{1t} is in kg/cm^2.
3. For a certain type of failure mode, the creep failure strain of frozen soil is almost stress- and temperature-independent, but it is closely related to ice content.
4. The creep failure criterion of frozen soil takes the general form of $\dot{\varepsilon}_m \times t_m^n = \varepsilon_f$, where n depends only upon water content and t_m is in minutes.
5. Equation 11 can be used to predict adequately the entire process of short-term creep, and can also provide a rough estimation of creep deformation for long-term creep.

ACKNOWLEDGMENTS

This study was sponsored by the U.S. Army Corps of Engineers at the Cold Regions Research and Engineering Laboratory. The senior author expresses his appreciation to CRREL for providing the facilities to conduct this study. The authors are grateful to the CRREL staff for their help, especially to F. H. Sayles, Dr. A. Assur, and W. F. Quinn for their instructive discussions.

REFERENCES

Assur, A., 1980, Some promising trends in ice mechanics, in Physics and Mechanics of Ice, International Symposium: Copenhagen, Springer-Verlag.

Fish, A.M., 1980, Kinetic nature of the long-term strength of frozen soils, in Proceedings of the Second International Symposium on Ground Freezing: Trondheim, Norwegian Institute of Technology, p. 95-108.

Ladanyi, B., 1972, An engineering theory of creep of frozen soils: Canadian Geotechnical Journal, v. 9, p. 63-80.

Martin, R.T., Ting, J.M., and Ladd, C.C., 1981, Creep behavior of frozen sand (final report part I): Cambridge, Massachusetts Institute of Technology, p. 122-129.

McRoberts, E.C., Law, T.C., and Murray, T.K., 1978, Creep tests on undisturbed ice-rich silt, in Proceedings of the Third International Conference on Permafrost, v. 1: Ottawa, National Research Council of Canada, p. 539-545.

Mellor, M., and Cole, D.M., 1982, Deformation and failure of ice under constant stress or constant strain-rate: Cold Regions Science and Technology, v. 5, p. 201-219.

Sayles, F.H., 1968, Creep of frozen sands, CRREL Technical Report 190: Hanover, N.H., U.S. Army Cold Regions Research and Engineering Laboratory.

Sayles, F.H., 1974, Creep of frozen silt and clay: CRREL Technical Report 252: Hanover, N.H., U.S. Army Cold Regions Research and Engineering Laboratory.

Ting, J.M., 1981, The creep of frozen sands -- qualitative and quantitative models (final report, part II): Cambridge, Massachusetts Institute of Technology, p. 241-259.

Zhu Yuanlin and Carbee, D., in press, Creep and strength behavior of frozen silt in uniaxial compression, CRREL report: Hanover, N.H., U.S. Army Cold Regions Research and Engineering Laboratory.

UNDERGROUND UTILIDORS AT BARROW, ALASKA: A TWO-YEAR HISTORY

Winston L. Zirjacks[1] and C.T. Hwang[2]
[1]Frank Moolin & Associates, Anchorage, Alaska, U.S.A.
[2]EBA Engineering Consultants Ltd., Edmonton, Alberta, Canada

The City of Barrow has no areawide piped water or sewerage facilities. Water is pumped from the upper portion of the Isatkoak Lagoon to the treatment facilities. Treated water is then hauled to residences via private carriers. Many residents haul ice from clean water sources south of town. Sewage wastes are collected at residences in holding tanks or plastic bags, then trucked to a landfill. The North Slope Borough is improving these conditions by installing piped sewerage and water facilities in an underground utilidor system. Due to the unprecedented size of this project and the compatibility of a heated utilidor buried in ice-rich permafrost, an instrumented test section was built and operated in ice-rich soil. The objectives are to verify thermal modeling used in the design and to investigate the interaction between a heated utilidor and the surrounding permafrost. This paper presents and evaluates the data thus far obtained.

Barrow, Alaska, located on the Arctic Coastal Plain, is the northernmost community in the United States, with a population of approximately 3,800. The town is a modern arctic Eskimo community serving as the governmental center for the large North Slope Borough (Figure 1). Whaling and game hunting still constitute a major portion of the subsistence of the town's inhabitants.

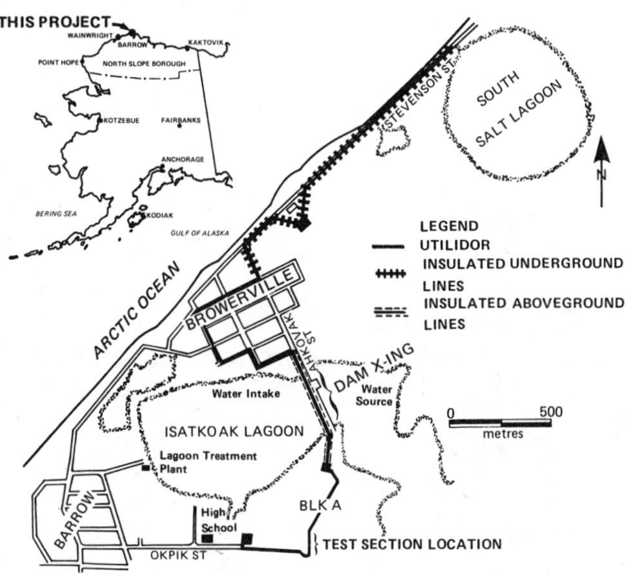

Figure 1 Location map.

Permafrost underlies the entire Arctic Coastal Plain. The bottom of the permafrost zone in the Barrow area lies 300 - 400 m below the surface. Polygonal or patterned ground is common in the area. The permafrost consists of ice-rich silt and contains numerous ice wedges and other ice inclusions (Black 1964). The area is characterized by typical arctic tundra growth of moss and sedges. The active layer of seasonal freezing and thawing is approximately 0.5 m in the undisturbed areas and varies in the disturbed areas in a manner proportional to any gravel fill that has been placed over the tundra.

The City of Barrow has not had areawide piped water or sewerage facilities in the past. Water is pumped from the upper portion of the Isatkoak Lagoon to treatment facilities in Barrow. The treated water is then hauled to residences via private carriers, and many residents haul ice from clean water sources south of the town for potable water uses. Sewage wastes are collected at residences in holding tanks or plastic bags then transferred to a "honey bucket" truck for deposition at the landfill at the South Salt Lagoon.

The North Slope Borough is improving these conditions by installing piped sewerage and water facilities in an underground utilidor system. Due to the unprecendented nature of this project, in terms of constructability and the compatibility of a heated utilidor buried in ice-rich permafrost, an instrumented test section was constructed and operated in ice-rich soil in order to verify thermal modeling used in the design and to investigate the interaction between a heated utilidor and the surrounding permafrost (Cerutti et al, 1982).

This paper presents and evaluates the data obtained from the test section during the last two years.

UTILIDOR DESIGN

The utilidor is constructed with 5 cm x 15 cm Douglas fir lumber placed with the wide side of the boards against one another. Wheel loads as well as soil overburden and structure dead loads were considered in the design. The utilidor sections are 3 m long and contain six steel tie rods. The utilidor is designed to contain water, sewerage, electrical, cable TV, and telephone transmission systems. The interior will be kept above freezing with heated, circulating water, and space is provided for human access for maintenance (Figure 2).

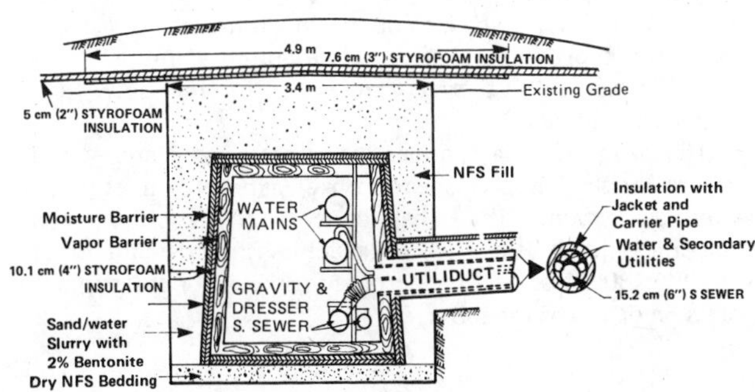

Figure 2 Typical utilidor section.

The exterior of the utilidor is covered with an impermeable vapor barrier, which is covered with 100 mm (4 in.) of insulation. The vapor barrier is designed to eliminate damage due to frost formation within the insulation layers and to also contain warm water within the utilidor from leaks or line breaks. An impermeable moisture barrier is then placed over the utilidor insulation to direct small amounts of free moisture away from the warming front created by the utilidor. Access to the utilidor is through prefabricated metal manholes.

Service to residences from the utilidor mains will be via utiliducts consisting of preinsulated polyethylene pipe. Water services will be circulated to residences with small circulating pumps located within the utilidor. All sewage wastes will ultimately be deposited in a facultative facility located in the South Salt Lagoon (Figure 1).

After the utilidor structure is installed, a slurry consisting of sand, water, and bentonite (2%) is placed to the midheight of the utilidor, and compacted dried gravel is placed above the slurry to the surface. A similar method is used for the utiliduct installation.

THERMAL DESIGN

The compatibility of a buried utilities system with the surrounding permafrost is of utmost importance in the design of the system. Sensitivity of the ice-rich permafrost to thermal degradation requires intensive engineering analyses as well as stringent construction procedures to ensure that the ground thermal regime around the utilities system remains stable. Design considerations are: (1) frost heave, thaw settlement, and buoyancy problems as a result of seasonal freezing and thawing of the surrounding soils and/or underground ice wedges; (2) structural integrity in response to the freeze/thaw mechanism; (3) possible groundwater seepage effects on the thermal regime; (4) possible freezing of the utility pipes within the utilidor or utiliduct during shutdown or maintenance periods; and (5) the influence on the ground thermal regime of neighboring structures, roadways, etc.

The EBA geothermal model (Hwang 1976) was used to perform extensive thermal analyses with respect to: (1) the insulation configuration of the utility structures; (2) the effect of burial depth; (3) soil stratigraphy, such as areas of ice-rich versus low-ice-content soils; (4) freezing point depression due to groundwater salinity; and (5) upset weather conditions, such as two successive "warm" years following construction and their long-term effect on the ground thermal regime. The warm-year simulation uses the maximum monthly air temperatures of the Barrow climatic data, while the normal year uses mean air temperature values.

Input parameters to the geothermal model are: (1) Barrow meteorological data, which include air temperature, wind velocity, snow depth, shortwave radiation, and evapotranspiration at the ground surface; (2) thermal properties of the soils; and (3) utilidor temperatures.

The design criteria considered are: (1) soil/fill around the structure should stay permanently frozen as to prevent seasonal freeze/thaw action and (2) half or more of the utilidor and utiliduct should lie below the maximum depth of penetration of the -1.1°C (30°F) isotherm to account for possible freezing point depression due to soil salinity.

TEST SECTION

To evaluate construction techniques and verify the thermal design parameters, part of the utilidor located in Block A (Figure 1) was isolated and instrumented as a test section in May 1981. This section was heated to the design interior operating temperature of 4°C to 9°C.

The instrumentation consisted of pressure cells, piezometers, and thermistor strings installed at various locations around the utilidor (Figures 3 and 4). Interior vertical and horizontal displacements were measured with extensometers at three locations.

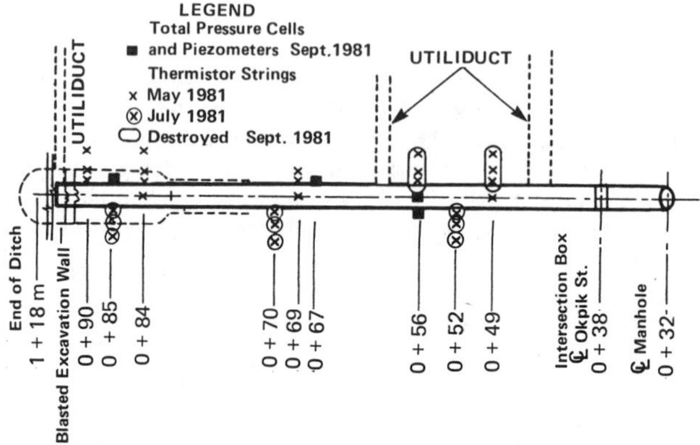

Figure 3 Test section plan.

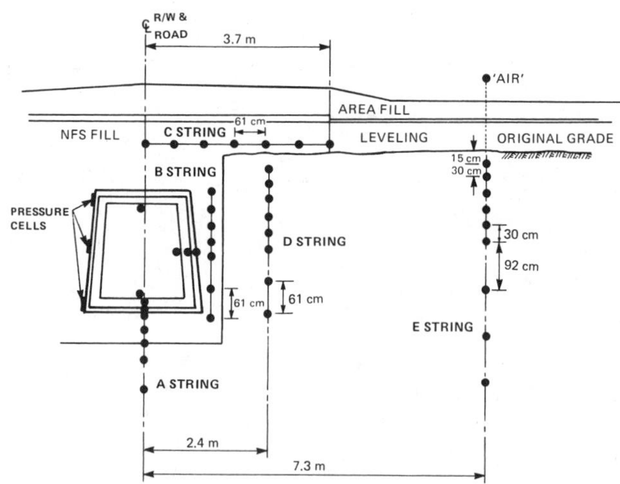

Figure 4 Thermistor string location.

The test section was constructed in an undisturbed tundra area, which contains 45 - 60 cm (1.5 - 2 ft.) of organic rich silt (tundra) overlying massive ice and ice-rich sandy silt to a depth of 4.6 m (15 ft.). Beneath this material, sandy silt and silty sand are present.

The test section trench was excavated by three methods that were used for construction evaluation. These methods were: (1) using a large trenching machine (ROCSAW) to cut the trench borders and blasting the interior (Figure 5); (2) using the ROCSAW to cut one trench border and blasting the remainder; and (3) blasting only. Method (1) produced a clean trench with vertical walls (Figure 6), but methods (2) and (3) produced wider trenches with irregular walls (Figure 7). The dashed line in Figure 3 indicates the limits of trench excavation that resulted from method (2) (Sta. 0 + 75M) and method (3) (Sta. 0 + 84M - Sta. 1 + 18M). Figure 8 shows the installation of the utilidor.

Figure 5 Rocsaw.

Figure 6 Typical section showing ice.

Figure 7 Rocsaw cut and blasted section.

Figure 8 Utilidor installation.

GROUND TEMPERATURES

Ground temperatures in the backfill adjacent to the utilidor during the summer after construction are shown in Figure 9. It can be seen that the closer the thermistor station is to the overexcavated area, the warmer the ground temperature observed. This is due to the warming effect on nearby ground, by water ponding in the overexcavated area. Water table measurements in the area by standpipe probing were found to be about the same elevation as the water bodies around the test site. The source of water is believed to be the melting of the active layer. Drainage is allowed through the active layer and the gravel area fill. As the water was pumped out of the utilidor, its volume was approximately related to the increase in air temperature and corresponding precipitation. However, after the winter of 1981, all thermistors indicated freezeback of the trench backfill throughout the test section, as expected.

Figures 10 to 14 show satisfactory comparisons between the predicted and observed temperatures in 1982. The initial temperature condition used in the model simulation was based on the ground temperatures on 13 May, 1981, as predicted by the model for undisturbed ground rather than those measured values shown in Figure 9. In spite of this and the ponding effect by groundwater immediately after construction, the trench backfill freezes back through the first winter and exhibits ground temperatures close to those predicted by the model, indicating domination by the conduction over the convection mode in the ground heat transfer process. The reason for colder ground temperatures than predicted for March 1982 is partly attributed to: (1) winter activity inside the utilidor for piping installations, as indicated by large fluctuations of the utilidor temperatures over the winter of 1981 - 1982 (Figure 15) and (2) the opening of the utiliduct trenches, which were excavated and left open over the same period (Figure 3). As the utilidor operating temperatures stabilized later, the correlation between the predicted and measured temperatures improves, as indicated by Figure 16 for 11 January, 1983.

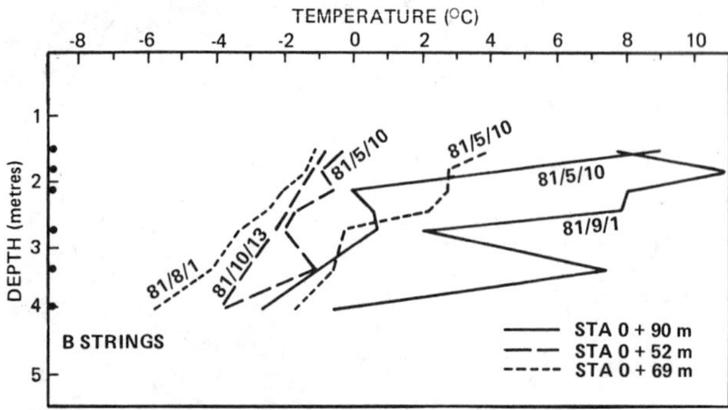

Figure 9 Temperature after construction.

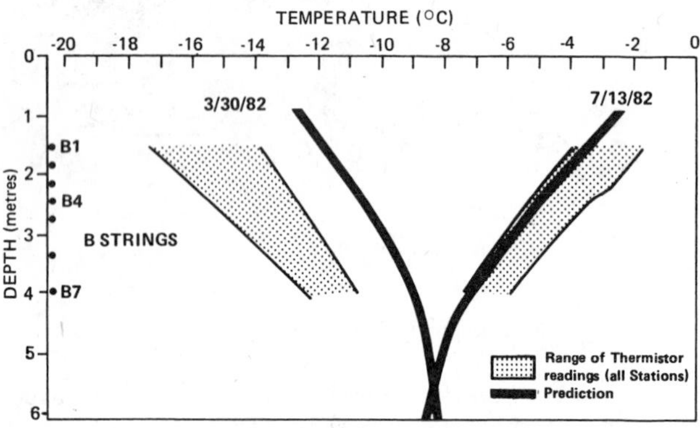

Figure 10 Predicted vs observed temperatures.

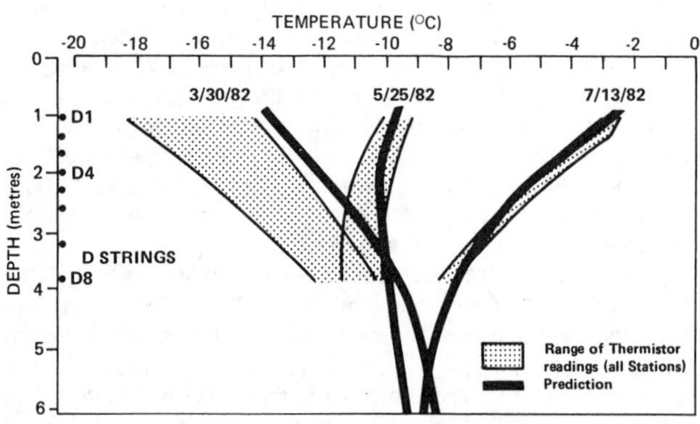

Figure 11 Predicted vs observed temperatures.

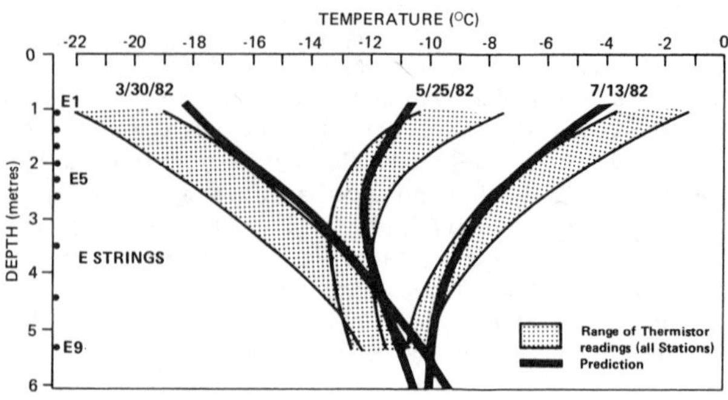

Figure 12 Predicted vs observed temperatures.

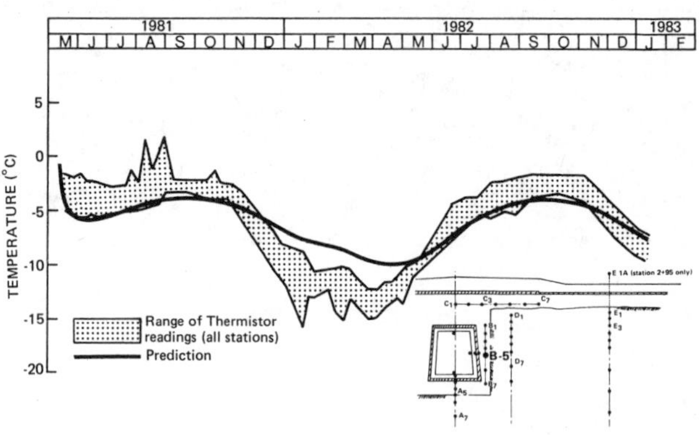

Figure 13 Predicted vs observed temperatures (thermistor B-5 for all stations).

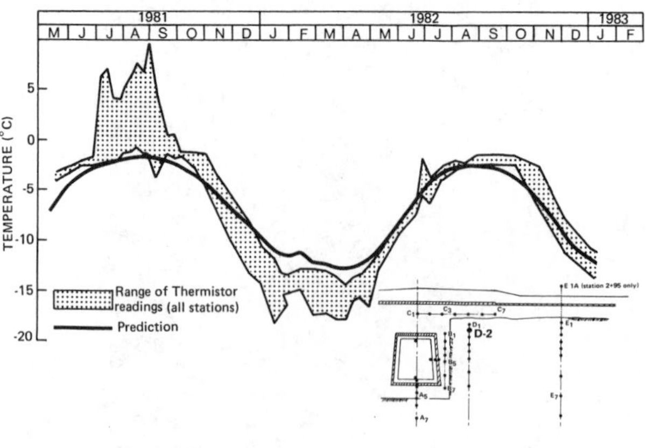

Figure 14 Predicted vs observed temperatures (thermistor D-2 for all stations).

DISPLACEMENT

The movements of the utilidor were measured with an extensometer at three locations. The measurements were made at the centerline and midheight points by connecting the extensometer to permanent hooks installed for this purpose. The data (Figure 17) seem to indicate that the utilidor is contracting and expanding with time. The maximum horizontal length occurs in March and April, when the permafrost temperature is at its minimum; and the minimum horizontal length occurs in September and October, when the permafrost temperature is at its maximum. It appears that the utilidor may be experiencing the effects of volumetric changes in the surrounding permafrost due to seasonal temperature changes. The vertical displacement in conjunction with the horizontal seems to support the interpretation that the utilidor expands sideways during spring and contracts during fall. It is emphasized that further data are required to confirm this cyclic action. However, the data indicate a cross-sectional displacement of 0.5 cm for the utilidor structure, which is well within the design limit of 2.5 cm. No meaningful data were obtained by the pressure cells and piezometers.

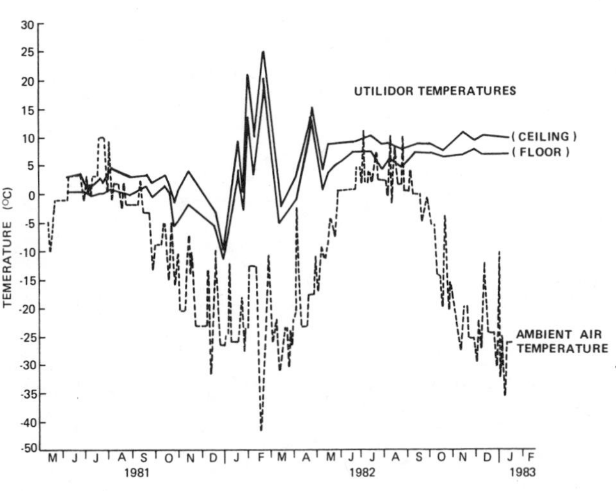

Figure 15 Utilidor and ambient temperatures.

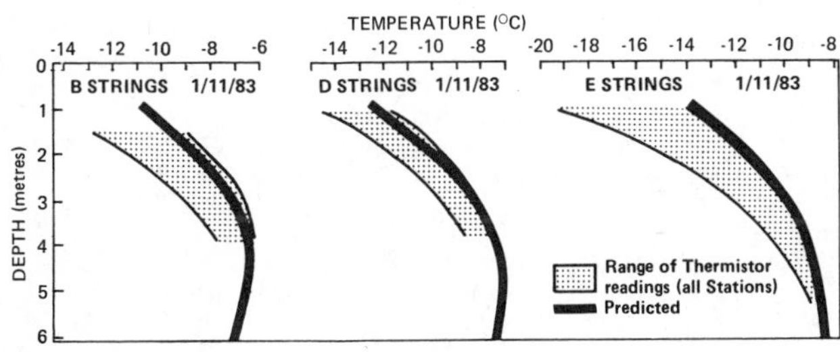

Figure 16 Predicted vs observed temperatures.

CONCLUSIONS

The performance of the test section to date has confirmed the adequacy of the design parameters used in both thermal and structural designs, as indicated by the ground temperatures and utilidor displacement data obtained. The construction experience obtained at the test site indicates that the ROCSAW technique proved to be the most effective excavation procedure, as it provides minimum disturbance and thus enhances rapid freeze-back of the trench backfill. Water ponding effects near the over-excavated area indicated that the groundwater movement should be kept to a minimum. As a result, trench plugs consisting of a 5:1 sand:bentonite mixture have been placed at 30 m intervals along the trench length in all utilidor installations in order to inhibit the migration of runoff water.

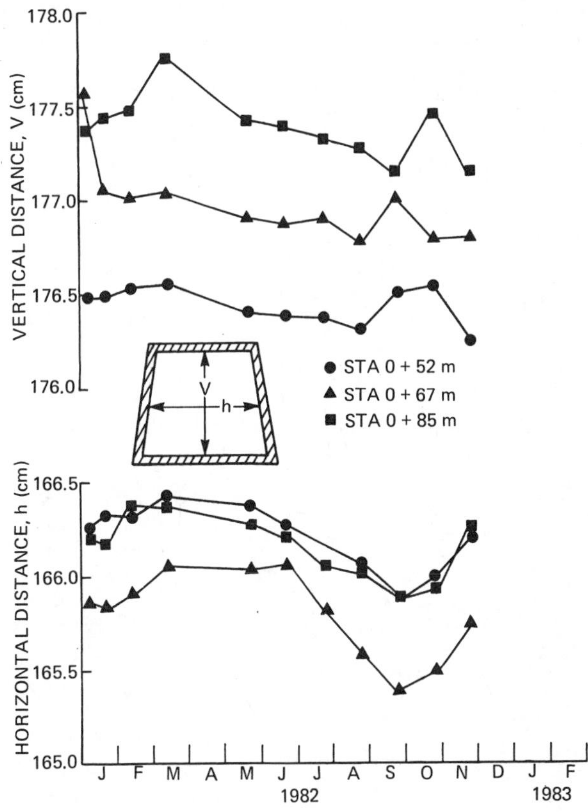

Figure 17 Extensometer readings.

Continuous monitoring of the test section is being maintained. It is felt that data obtained will be valuable in the preparation of maintenance guidelines for the utilidor system.

REFERENCES

Black, R.F., 1964, Guhik formation of quaternary age in northern Alaska: U.S. Geological Survey Professional Paper 302-C.

Cerutti, J.L., Zirjacks, W.L., Hwang, C.T., and Brugger, D.E., 1982, Underground utilities in Barrow, Alaska, in proceedings of the Third Symposium on Utilities Delivery in Cold Regions, Edmonton, Alberta, Canada, May 25 - 26.

Hwang, C.T., 1976, Predictions and observations on the behaviour of a warm gas pipeline on permafrost: Canadian Geotechnical Journal, v. 13, p. 452 - 480.

Author Index

General Subject–Senior Author Index

PATTERNED GROUND AND FROST FEATURES

PERMAFROST DISTRIBUTION AND PROPERTIES

PERMAFROST HISTORY

PILES AND CASINGS

PIPELINES

PLANETARY PERMAFROST

SOIL MECHANICS

SUBSEA PERMAFROST

SURFACE DISTURBANCE AND RECOVERY

THERMAL PROPERTIES AND PROCESSES

WATER TRANSPORT

WEATHERING

Available Publications of International Permafrost Conferences

National Research Council, 1966. *Permafrost: International Conference Proceedings*. NRC Pub. 1287. Washington, D.C.: National Academy of Sciences. Available on microfilm in The Cold Regions Science and Technology bibliography #25-3138.

National Academy of Sciences, 1973. *Permafrost: The North American Contribution to the Second International Conference, Yakutsk*. Washington, D.C.: National Academy of Sciences. $50.00

National Academy of Sciences, 1978. *Permafrost: USSR Contribution to the Second International Conference*. Washington, D.C.: National Academy of Sciences. $19.50

National Research Council of Canada, 1978. *Proceedings of the Third International Conference on Permafrost*, 2 volumes. Ottawa, Ontario: National Research Council of Canada. $35.00

National Research Council of Canada, 1980. *Third International Conference on Permafrost, English Translations of the Soviet Papers*, 2 parts. Ottawa, Ontario: National Research Council of Canada. $35.00

University of Alaska, 1983. *Permafrost: Fourth International Conference. Abstracts and Program*. Fairbanks: University of Alaska. $10.00

National Academy of Sciences, 1983. *Permafrost: Fourth International Conference, Proceedings*. Washington, D.C.: National Academy Press. $65.00